International Exhibition & Conference for Power Electronics, Intelligent Motion, Renewable Energy and Energy Management (PCIM Europe 2016)

Nuremberg, Germany
10 - 12 May 2016

Volume 1 of 3

ISBN: 978-1-5108-2530-7

Printed from e-media with permission by:

Curran Associates, Inc.
57 Morehouse Lane
Red Hook, NY 12571

Some format issues inherent in the e-media version may also appear in this print version.

Copyright© (2016) by Mesago PCIM GmbH
All rights reserved.

Printed by Curran Associates, Inc. (2016)

For permission requests, please contact Mesago PCIM GmbH
at the address below.

Mesago PCIM GmbH
Rotebuehlstrasse 83-85
70178 Stuttgart Germany

Phone: 49 711 619 460
Fax: 49 711 619 4690

info@mesago.com

Additional copies of this publication are available from:

Curran Associates, Inc.
57 Morehouse Lane
Red Hook, NY 12571 USA
Phone: 845-758-0400
Fax: 845-758-2633
Email: curran@proceedings.com
Web: www.proceedings.com

International Exhibition & Conference for Power Electronics, Intelligent Motion, Renewable Energy and Energy Management (PCIM Europe 2016)

Nuremberg, Germany
10 - 12 May 2016

Volume 1 of 3

pcim
EUROPE

International Exhibition and Conference
for Power Electronics, Intelligent Motion,
Renewable Energy and Energy Management

Nuremberg, 10 – 12 May 2016

Proceedings

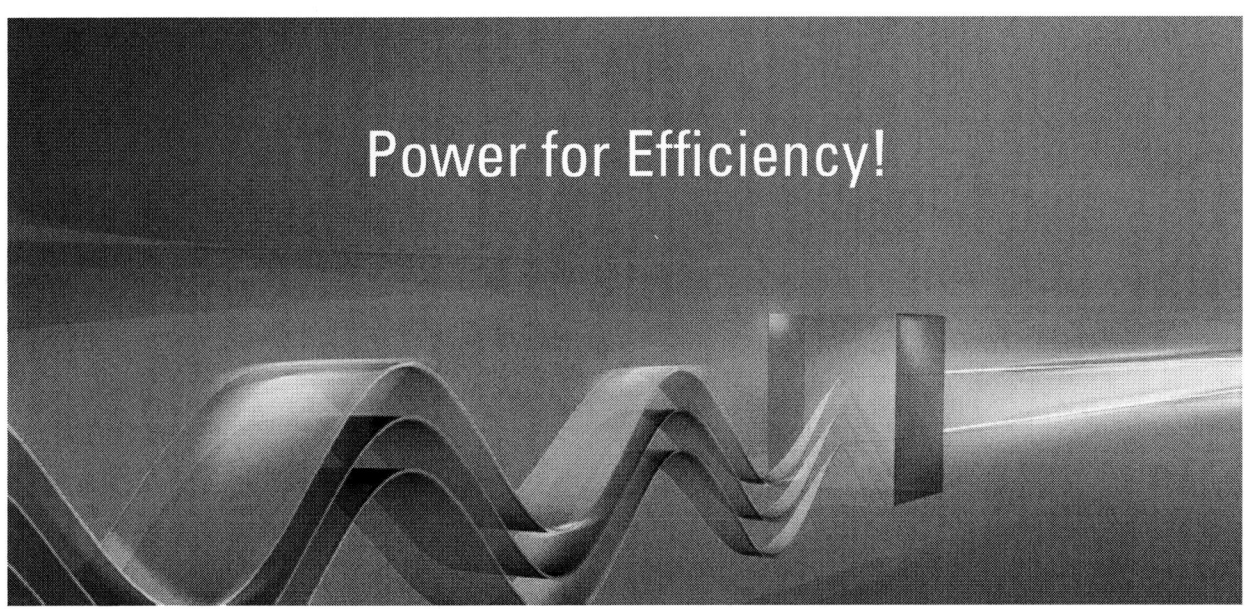

Power for Efficiency!

→ Contents　　　　　　　　→ Authors

Organizer:
Mesago PCIM GmbH, Stuttgart
www.pcim-europe.com

mesago
Messe Frankfurt Group

Board of Directors:
Prof. Leo Lorenz, ECPE, Germany

Jean-Paul Beaudet, Schneider Electric, France
Dr. Eric Favre, IMI Precision Engineering, Switzerland
Prof. Johann Walter Kolar, ETH Zürich, Switzerland
Prof. Philippe Ladoux, University of Toulouse, France
Prof. Jose Mario Pacas, University of Siegen, Germany
Prof. Uwe Scheuermann, Semikron Elektronik, Germany

PCIM Europe 2016, 10 – 12 May 2016, Nuremberg, Germany

Dear PCIM Europe participants,

It is my pleasure to welcome you to the PCIM Europe 2016 Conference in Nuremberg. This important event serves as a technical and scientific platform for engineers and researchers engaged in all fields related to power components, power converter technologies and future development of power electronics.

The technical program for this year's conference highlights advanced technologies for power semiconductor devices and passive components control. Additionally it highlights drive strategies for high efficient power converters, drive systems for e-vehicles and renewable energy technologies. New material for semiconductor devices, reliability issues on power modules and on system levels as well as designs to manage ultrafast switching devices in the circuit set up, form the backbone of the PCIM Europe Conference.

The keynote papers cover the development trend for power semiconductor devices, including packaging technologies, developments in solid state transformers as well as trends on solar systems. The highlights of the conference include the special sessions on challenges in smart lighting, passive components and traction drivers for e-cars as well as a panel discussion on "The Smart Future of Power Electronics".

With its high level technical program and discussion platform, this year's PCIM Europe Conference will provide you with an overview of the key technology development trends in power electronics and inspire you to pursue new business opportunities.

I wish you an enjoyable and successful conference, packed with new ideas for your future business.

Leo Lorenz

General Conference Director, Germany

PCIM Europe 2016, 10 – 12 May 2016, Nuremberg, Germany

Advisory Board PCIM Europe 2016

General Conference Director PCIM Europe

Prof. Leo Lorenz, ECPE, D

Board of Directors PCIM Europe

Jean-Paul Beaudet, Schneider Electric, F

Dr. Eric Favre, IMI Precision Engineering, CH

Prof. Johann Walter Kolar, ETH Zürich, CH

Prof. Philippe Ladoux, University of Toulouse, F

Prof. Jose Mario Pacas, University of Siegen, D

Prof. Uwe Scheuermann, Semikron Elektronik, D

Consultatory Board Members

Prof. Friedrich-Wilhelm Fuchs, Christian-Albrechts-University of Kiel, D

Prof. Josef Lutz, Chemnitz University of Technology, D

Members

Bodo Arlt, A Media, D

Prof. Francisco Javier Azcondo, University of Cantabria, E

Prof. Dr.-Ing. Mark M. Bakran, University of Bayreuth, D

Dr. Pavol Bauer, Delft University of Technology, NL

Dr. Reinhold Bayerer, Infineon Technologies, D

Werner Berns, Texas Instruments, D

Prof. Frede Blaabjerg, Aalborg University, DK

Serge Bontemps, Microsemi PMP Europe, F

Eric Carroll, EIC Consultancy, F

Bruce Carsten, Bruce Carsten Associates, USA

Daniel Chatroux, CEA-LITEN, F

Dr. Silvio Colombi, General Electric, CH

Hilmar Darrelmann, Darrelmann + Partner Ingenieure, D

Prof. Enrique J. Dede, University of Valencia, E

PCIM Europe 2016, 10 – 12 May 2016, Nuremberg, Germany

Prof. Drazen Dujic, Power Electronics Laboratory, EPFL, CH

Prof. Dr.-Ing. Hans-Günter Eckel, University of Rostock, D

Prof. Hans Ertl, Vienna University of Technology, A

Prof. J. A. Ferreira, Delft University of Technology, NL

Dr. Petar J. Grbovic, Huawei Technologies, D

Dr. Steffan Hansen, Danfoss Solar Inverters, DK

Prof. Klaus F. Hoffmann, Helmut-Schmidt-University, D

Edward Hopper, MACCON, D

Ionel Dan Jitaru, Rompower, USA

Prof. Dr.-Ing. Nando Kaminski, University of Bremen, D

Dr. Ulrich Kirchenberger, STMicroelectronics, D

Dr. Philip C. Kjaer, Vestas Wind Systems, DK

Christopher Kocon, Texas Instruments, USA

Dr. Jacques Laeuffer, Dtalents, F

Prof. Stéphane Lefebvre, SATIE, F

Romeo Letor, STMicroelectronics, IT

Prof. Andreas Lindemann, Otto-von-Guericke-University Magdeburg, D

Dr. Stefan Linder, Alpiq, CH

Prof. Marco Liserre, Christian-Albrechts-University of Kiel, D

Prof. Dr.-Ing. Martin März, Fraunhofer IISB, D

Dr. Gourab Majumdar, Mitsubishi Electric, J

Klaus Marahrens, SEW-EURODRIVE GmbH, D

Prof. Elison Matioli, POWERlab, EPFL, CH

Dr. Mike Meinhardt, SMA Solar Technology, D

Dr. Thomas Neyer, Fairchild Semiconductor, D

Prof. Yasuyuki Nishida, Chiba Institute of Technology, J

Geraldo Nojima, Eaton Corporation, USA

Yasuhiro Okuma, Fuji Electric, J

Prof. Dr. Nejila Parspour, University of Stuttgart, D

Dr. Robert J. Pasterczyk, Schneider Electric, F

Dr. Volker Pickert, University of Newcastle, UK

Prof. Bernhard Piepenbreier, University of Erlangen, D

Dr. Munaf Rahimo, ABB Switzerland, CH

Dr. Kaushik (Raja) Rajashekara, University of Dallas at Texas, USA

Chris Rexer, ON Semiconductor, USA

Katsuaki Saito, Hitachi Europe, GB

Franck Sarrus, Mersen France, F

Andrew Sawle, International Rectifier, GB

Achim Scharf, Techmedia International, D

Dr. Hubert Schierling, Siemens, D

Dr. Manfred Schlenk, Infineon Technologies D

Prof. Manfred Schrödl, Vienna University of Technology, A

Prof. Dr.-Ing. Walter Schumacher, Braunschweig University of Technology, D

Dr. Yasukazu Seki, Fuji Electric, J

Prof. Toshihisa Shimizu, Tokyo Metropolitan University, J

Christopher A. Soule, Thermshield, USA

Elmar Stachorra, KoCoS Power Grid Services, D

Dr. Peter Steimer, ABB Switzerland, CH

Dr. Bernhard Strzalkowski, Analog Devices, D

Dr. Wolfram Teppan, LEM SA, CH

Prof. Giuseppe Tomasso, University of Cassino and South Lazio, I

Joël Turchi, On Semiconductor, F

Dr. Yoshiyuki Uchida, Japan Fine Ceramics, J

Prof. Alfredo Vagati, Politecnico di Torino, I

Dr. Peter Wallmeier, Delta Energy Systems, D

Prof. Dehong Xu, Zhejiang University, CN

Prof. Peter Zacharias, University of Kassel, D

Honorary Board

Prof. Helmut Knöll, University for Applied Sciences Würzburg-Schweinfurt, D

Prof. Jean-Marie Peter, France

Prof. Gerhard Pfaff, University of Erlangen, D

Prof. Alfred Rufer, EPFL, CH

PCIM Europe 2016, 10 – 12 May 2016, Nuremberg, Germany

Table of Content PCIM Europe 2016

Keynotes

Keynote: Welcome to the Post-Silicon World: Wide Band Gap Powers Ahead...................................... 31
Dan Kinzer, Navitas Semiconductor, USA

Keynote: Smart Transformers – Concepts-Challenges-Applications.. 32
Johann Walter Kolar, ETH Zürich Power Electronic Systems Laboratory, CH

Keynote: Trends of Solar Systems and their Integration in Electricity Networks 33
Jens Merten, National Solar Energy Institute, F

SiC Devices

**Ultra-low (1.25 mΩ) On-Resistance 900V SiC 62 mm Half-Bridge Power Modules Using New
10 mΩ SiC MOSFETs**.. 34
Jeffrey Casady, Vipindas Pala, Edward Van Brunt, Brett Hull, Sei-Hyung Ryu, Gang-Yao Wang,
Jim Richmond, Scott T. Allen, Dave Grider, John W. Palmour, Peter Killeen, Brice McPherson,
Kraig Olejniczak, Brandon Passmore, David Simco, Wolfspeed, USA

Evolution of SiC Products for Industrial Application 42
Naoyuki Kizu, Satoru Nate, Mineo Miura, Noriaki Kawamoto, Kazuhide Ino, Rohm, J; Masaharu Nakanishi,
Rohm Semiconductor, D; Nobuhiro Hase, Rohm Semiconductor, USA

Advanced Protection for Large Current Full SiC-Modules... 48
Eugen Wiesner, Eckhard Thal, Mitsubishi Electric Europe, D; Andreas Volke, Karsten Fink, Power
Integrations, D

Switching Performance of a 1200 V SiC-Trench-MOSFET in a Low-Power Module............................. 53
Daniel Heer, Daniel Domes, Dethard Peters, Infineon Technologies, D

Module Materials

Beyond Thermal Grease, Enhancing Thermal Performance and Reliability 60
Sanjay Misra, Henkel, USA

High Thermal Conductivity Silicon Nitride substrate for Power Semiconductor Applications 64
Dai Kusano, Gen Tanabe, Yoshiyuki Uchida, Japan Fine Ceramics, J

Thermal Management of Future WBG Devices using Two-Phase Cooling 72
Shailesh Joshi, Ercan Dede, Toyota Research Institute North America, USA

**Highly Reliable and Lead-Free High Power IGBT Modules Using Novel Copper Sintering
Die Attachment** 78
Akitoyo Konno, Takaaki Miyazaki,Yuusuke Yasuda, Osamu Ikeda, Hiroshi Nakano, Toshiaki Morita,
Hiroshi Houzouji, Mutsuhiro Mori, Masato Nakamura, Yoshihiko Koike, Hitachi, J

7

PCIM Europe 2016, 10 – 12 May 2016, Nuremberg, Germany

Magnetics & Inductors

A New Generation of Nanocrystalline Magnetic Cores with Very Low Magnetic Losses 84
Bashar Gony, Stephane Camus, Julia Hill, Frederic Pottier, Aperam Alloys Amilly, F

The Fe-based Glassy Alloy Powder Core Inductor for the Boost Converter by GaN HEMT and SiC SBD 1 MHz Operation 91
Mitsunao Fujimoto, Yutaka Naitoh, Takao Mizushima, ALPS Green Devices, J

Accurate Calculation of AC Losses of Inductors in Power Electronic Applications 99
Ranjith Bramanpalli, Würth Elektronik Eisos, D

Thin-Film Based Microtransformer Suitable for High Switching Frequency Power Applications 107
Dragan Dinulovic, Mahmoud Shousha, Martin Haug, Würth Elektronik eiSos, D; Sebastian Beringer, Marc C. Wurz, Leibniz University Hannover, D

DC/AC and AC/DC Converters

Design of High Efficiency High Power Density 10.5 kw Three Phase On-Board-Charger for Electric/Hybrid Vehicles 113
Gang Yang, Valeo, F; Eirik Draugedalen, Torbjorn Sorsdahl, Hui Liu, Roar Lindseth, Valeo Powertrain Energy Conversion, NO

A Bridgeless, Quasi-Resonant ZVS-Switching, Buck-Boost Power Factor Correction Stage (PFC) 120
Markus Scherbaum, Manfred Reddig, University of Applied Sciences Augsburg, D; Ralph Kennel, Technical University of Munich, D; Manfred Schlenk, Infineon Technologies, D

A high efficiency 5.3 kW Current Source Inverter (CSI) Prototype using 1.2 kV Silicon Carbide (SiC) Bi-Directional Voltage Switches in hard Switching Mode 128
Jérémy Martin, Anthony Bier, Stéphane Catellani, Luis Gabriel Alves-Rodrigues, Franck Barruel, Commissariat à l'énergie atomique et aux énergies alternatives, F

Comparison of the EMC and Efficiency Characteristics of Hard and Soft Switching Three-Level Inverters 136
Manfred W. Gekeler, Stefan Schreitmüller, Gunter Voigt, HTWG Konstanz University of Applied Sciences, D

Special Session "Passive Components"

Drive System Loss Reduction by Allpole Sine Filters 144
Dennis Kampen, BLOCK Transformatoren-Elektronik, D; Michael Burger, SEW Eurodrive, D

Harmonic Filtering in Variable Speed Drives 148
Luca Dalessandro, Xiaoya Tan, Andrzej Pietkiewicz, Martin Wüthrich, Norbert Naeberle, Schaffner, CH

A just Comparison of Ferrite and Nanocristalline Common Mode Chokes 156
Jörn Schliewe, Christian Paulwitz, Stefan Weber, Epcos, D

Modelling an Anti-Ferroelectric Ceramic Capacitor for Time- and Frequency-Domain Simulations of Power Systems 164
Stefan Schefler, Markus Koini, Stefan Weber, Epcos, D; Markus Puff, Epcos, AT

PCIM Europe 2016, 10 – 12 May 2016, Nuremberg, Germany

SiC Reliability

Breakdown of Gate Oxide of 1.2 kV SiC-MOSFETs Under High Temperature and High Gate Voltage 172
Menia Beier-Möbius, Josef Lutz, Chemnitz University of Technology, D

Avalanche Robustness of SiC MPS Diodes 180
Thomas Basler, Roland Rupp, Rolf Gerlach, Bernd Zippelius, Infineon Technologies, D; Mihai Draghici, Infineon Technologies, AT

Compact, Low Loss and High Reliable 3.3 kV Hybrid Module 188
Satoshi Kaneko, Naoyuki Kanai, Motohiro Hori, Nakazawa Masayoshi, Hideaki Kakiki, Yasushi Abe, Yoshinari Ikeda, Eiji Mochizuki, Fuji Electric, J

DC/DC Converters I

Isolated Synchronous Forward Controller with Integrated Feedback Loop and Adjustable Dead Time for High Efficiency DC/C Converter 195
Bernhard Strzalkowski, Analog Devices, D; Subodh Madiwale, Gabriele Bernardinis, Michael Daly, Analog Devices, USA

Extreme High Efficiency Non-Inverting Buck-Boost Converter for Energy Storage Systems 202
Zhe Yu, Holger Kapels, Fraunhofer Institute ISIT, D; Klaus F. Hoffmann, Helmut Schmidt University, D

A High-Efficiency Bidirectional GaN-HEMT DC/DC Converter 210
Michael Ebli, Martin Wattenberg, Universtiy of Reutlingen, D; Martin Pfost, University of Innsbruck, AT

Control Converters

Direct Delta Sigma Signal Processing for Control of Power Electronics 216
Michael Homann, Axel Klein, Walter Schumacher, Technical University of Braunschweig, D

Finite Control Set Model Based Predictive Control of a PMSM with Variable Switching Frequency and Torque Ripple Optimization 224
Fernando David Ramirez Figueroa, Mario Pacas, University of Siegen, D

Frequency- and Mode-Adaptive Control of DC-DC Converter for Efficency Improvement 232
Lukas Keuck, Farshid Almai, Sven Bolte, Norbert Fröhleke, Joachim Böcker, University of Paderborn, D

Control Techniques in Intelligent Motion Systems I

Load Torque Estimation in Repetitive Mechanical Systems by Using Fourier Interpolation 240
Van Trang Phung, Mario Pacas, University of Siegen, D

Fast Current Waveform Calculation Algorithm for a Six Phase Switched Reluctance Machine 248
Jacek Borecki, Bernd Orlik, University of Bremen, D

Process Requirements-Based Adaptive PWM for Improved Efficiency of Machine Tool Feed-Drives 256
Matthias Braband, Florian Frick, Armin Lechler, Alexander Verl, ISW – University of Stuttgart, D

PCIM Europe 2016, 10 – 12 May 2016, Nuremberg, Germany

DC/AC Converters

Comparison of Bidirectional T-Source Inverter and Quasi-Z-Source Inverter for Extra Low Voltage Application 264
Thomas Baier, Bernhard Piepenbreier, Friedrich-Alexander-University Erlangen, D

Benchmarking of SiC JFET and SiC MOSFET Modules for the Application in Medium Power Traction Converters 272
Andreas März, Roman Horff, Teresa Bertelshofer, Mark-M. Bakran, University of Bayreuth, D;
Martin Helsper, Siemens, D

New Bus-bar Topology to Suppress the Current Imbalance of Parallel-connected IGBT Modules for High Power Railway 280
Naoki Sakurai, Masaoki Konishide, Yasuhiko Kohno, Hitachi, J

GaN Converters

EMI Investigation in a GaN HEMT Power Module 288
Xiaoshan Liu, Francois Costa, Bertrand Revol, Cyrille Gautier, ENS Cachan- SATIE, F

Ultra-High Frequent Switching with GaN- HEMTs using the Coss-Capacitances as non-dissipative Snubbers 296
Hubert Berger, Luis Alfonso Fernández-Serantes, Wolfgang Stocksreiter, Gerald Weis, University of Applied Sciences Joanneum, AT

eGaN® FET based 6.78 MHz Differential-Mode ZVS Class D AirFuel™ Class 4 Wireless Power Amplifier 304
Michael de Rooij, Yuanzhe Zhang, Efficient Power Conversion Corporation, USA

High Frequency, High Temperature designed DC/DC Coreless Converter for GaN Gate Drivers 312
Yohan Wanderoild, Dominique Bergogne, CEA Leti, F; Hubert Razik, University Claude Bernard Lyon, F

Monolithic GaN-on-Si Half-Bridge Circuit with Integrated Freewheeling Diodes 319
Richard Reiner, Patrick Waltereit, Beatrix Weiss, Matthias Wespel, Michael Mikulla, Rüdiger Quay, Oliver Ambacher, Fraunhofer-Institute IAF, D

Module Design

Resin Encapsulation Combined with Insulated Metal Baseplate for Improving Power Module Reliability 326
Shinsuke Asada, Satoshi Kondo, Yusuke Kaji, Hiroshi Yoshida, Mitsubishi Electric Corporation, J

An Experimental Study on the Thermal Performance of Double-Side Direct-Cooling Power Module Structure 331
Akira Matsushita, Ryuichi Saito, Takeshi Tokuyama, Takashi Kimura, Hitachi Automotive Systems, J;
Kinya Nakatsu, Hitachi, J

New Transfer Mold DIPIPM™ Utilizing Silicon Carbide (SiC) MOSFET 336
Yazhe Wang, Kiyoto Watabe, Shinji Sakai, Toshikazu Tanioka, Mitsubishi Electric Corporation, J

A 1700 V-IGBT module and IPM with new insulated metal baseplate (IMB) featuring enhanced Isolation Properties and Thermal Conductivity 342
Takuya Takahashi, Yoshitaka Kimura, Hiroshi Yoshida, Hidetoshi Ishibashi, Yoshitaka Otsubo, Mitsubishi Electric Corporation, J

PCIM Europe 2016, 10 – 12 May 2016, Nuremberg, Germany

NHPD² (Next High Power Density Dual) with Next Generation Chip Suitable for Low Internal Inductance Package 348
Daisuke Kawase, Kazuhito Nagashima, Tomohisa Hirayama, Katsunori Azuma, Seiichi Hayakawa, Hitachi Power Semiconductor, J; Katsuaki Saito, Hitachi Europe, GB

Power Electronics in Transmission Systems in Smart Grids

An Optimized Hybrid-MMC for HVDC 355
Viktor Hofmann, Mark-M. Bakran, University of Bayreuth, D

Selective HVDC Transmission Line Breaking for Bus Bar Applications under Reduced Expenses 363
Rene Sander, Michael Suriyah-Jaya, Thomas Leibfried, Karlsruhe Institute of Technology, D

Decentralized Controller for Modular Multilevel Converter 371
Seleme Isaac Seleme Jr., Federal University of Minas Gerais – UFMG, BR; Luc-André Grégoire, Marc Cousineau, Philippe Ladoux, University of Toulouse, F

Power Converters for PV Systems with Energy Storage: Optimal Power Flow Control for EV's Charging Infrastructures 379
Mauro Di Monaco, Umberto Abronzini, Ciro Attaianese, Matilde Arpino, Giuseppe Tomasso, University of Cassino and South Lazio, I

Special Session "E-Mobility"

Modular and Comfortable Electromobility 386
Mihai Dragan, Bernhard Budaker, Jonathan Brix, Fraunhofer Institute IPA, D

Power Hardware-in-the-Loop Emulation of Permanent Magnet Synchronous Machines with Nonlinear Magnetics – Concept & Verification 393
Alexander Schmitt, Jan Richter, Michael Braun, Martin Doppelbauer, Karlsruhe Institute of Technology, D

Multimode Charging of Electric Vehicles – A combined Concept with Multiple Use of Components and Strategies for Decreasing Power Losses, Weight and Volume 401
Marco Jung, René Marklein, Georgios Lempidis, Jonas Steffen, Axel Seibel, Jörg Kirchhof, Roland Gaber, Fraunhofer Institute IWES, D

Design of Contactless Energy Transfer System for an Electric Vehicle 410
Mike Böttigheimer, Nejila Parspour, Marco Zimmer, Anna Lusiewicz, University of Stuttgart, D

High Power Semiconductor

3300 V HiPak2 modules with Enhanced Trench (TSPT+) IGBTs and Field Charge Extraction Diodes rated up to 1800 A 417
Chiara Corvasce, Maxi Andenna, Liutauras Storasta, Sven Matthias, Arnost Kopta, Munaf Rahimo, Luca De Michielis, Silvan Geissmann, Raffael Schnell, ABB Switzerland, CH

Durable Design of the New HVIGBT Module 425
Nobuhiko Tanaka, Kenji Ota, Shuichi Kitamura, Shinichi Lura, Keiichi Nakamura, Mitsubishi Electric Corporation, J; Eugen Wiesner, Eckhard Thal, Mitsubishi Electric Europe, D

PCIM Europe 2016, 10 – 12 May 2016, Nuremberg, Germany

The 62Pak IGBT Module Range Employing the 3rd Generation 1700 V SPT++ Chip Set for 175 °C Operation... 432
Sven Matthias, Chiara Corvasce, Athanasios Mesemanolis, Emilia Gustafsson, Charalampos Papadopoulos, Arnost Kopta, Silvan Geissmann, Martin Bayer, Raffael Schnell, Munaf Rahimo, ABB Switzerland, CH

Extended Power Rating of 1200 V IGBT Module with 7G-RC-IGBT Chip Technologies 438
Misaki Takahashi, Souichi Yoshida, Akira Tamenori, Yasuyuki Kobayashi, Osamu Ikawa, Fuji Electric, J; Daniel Hofmann, Fuji Electric Europe, D

Multi Level Converters

DC-DC Converter based on the Asymmetric Multistage Stacked Boost Architecture with Feed-Forward Control for Photovoltaic Plants 445
Georgios Mademlis, Aristotle University of Thessaloniki, GR; Gina Steinke, Alfred Rufer, EPFL – Ecole polytechnique fédérale de Lausanne, CH

A Novel Submodule Concept for Modular Multilevel Converters 453
Benjamin Ruccius, Nicola Burani, Dirk Malipaard, Fraunhofer Institute IISB, D; Marek Galek, Siemens, D

Electro-Thermal Design of a Modular Multilevel Converter Prototype 461
Emilien Coulinge, Alexandre Christe, Drazen Dujic, EPFL – Ecole Polytechnique Fédérale de Lausanne, CH

DC/DC Converters II

Non-isolated Three-Port DC/DC-Converter for ±380VDC Microgrids 469
Yunchao Han, Julian Kaiser, Leopold Ott, Matthias Schulz, Fabian Fersterra, Bernd Wunder, Martin März, Fraunhofer Institute IISB, D

Wide Voltage Input Range Insulated Current Fed Buck Flyback-Forward for HV/LV Power Conversion in Electric/Hybrid Vehicle 477
Reda Chelghoum, Luis De Sousa, Larbi Bendani, Valeo, F; Daniel Sadarnac, CentraleSupelec, F

SiC JFET Cascode Enables Higher Voltage Operation in a Phase Shift Full Bridge DC-DC Converter 484
Jonathan Dodge, United Silicon Carbide, USA

Lamp Ballasts Lighting Systems

Detailed Comparison of One Stage Topologies for LED Lighting Applications 492
Alexander Pawellek, Thomas Dürbaum, Friedrich-Alexander-University Erlangen, D

Low Cost High Density AC-DC Converter for LED Lighting Applications 501
Simon Nigsch, Janosch Marquardt, Kurt Schenk, University of Applied Sciences NTB Buchs, CH

Design Optimization for a High Power-Density, Wide Output, High Frequency LLC Resonant Converter for Lighting Applications 509
Janosch Marquart, Simon Nigsch, Kurt Schenk, University of Applied Sciences NTB Buchs, CH

Sensorless Motor Control

Computationally efficient Anisotropy-Identification based on a Square-Shaped Injection Pattern 518
Peter Landsmann, Dirk Paulus, Sascha Kühl, Ralph Kennel, Technical University of Munich, D

PCIM Europe 2016, 10 – 12 May 2016, Nuremberg, Germany

Estimation of the Excitation Current and the Rotor Resistance of an Externally Excited Synchronous Machine with an Inductively Supplied Excitation Coil 526
Stefan Köhler, Bernhard Wagner, Technical University Nuremberg Georg Simon Ohm, D;
Stefan Endres, Fraunhofer Institute IISB, D

High Speed Sensorless Control of a Synchronous Motor with Kalman Filter 534
Philipp Niedermayr, Alpitronic, I; Silverio Bolognani, Luigi Alberti, University of Padova, I;
Reiner Abl, BMW, D

Software Tools and Applications

Physical Modeling and High-Fidelity Simulation of the Transient Behavior of Multiply-Contacted Power Busbars 543
Vanessa Basler, Wolfgang Hölzl, Gerhard Wachutka, Technical University of Munich, D

Small-signal Output Impedance Modeling of Intersil's R4TMTechnology 550
Yi Huang, Chun Cheung, Intersil Corporation, USA

An Approach of Reinforcement Learning Based Lighting Control for Demand Response 558
Xinxing Pan, Brian Lee, AIT, IE

Cosmic Ray & Ruggedness

Cosmic Ray Failure Mechanism and Critical Factors for 3.3 kV Hybrid SiC Modules 566
Tetsuya Nitta, Yoko Sakiyama, Raita Kotani, Tomoki Inoue, Ryoichi Ohara, Kenya Sano,
Masakazu Yamaguchi, Toshiba, J; Georges Tchouangue, Toshiba Electronics Europe, D

Benefits of Increased Cosmic Radiation Robustness of SiC Semiconductors in large Power-Converters 573
Christian Felgemacher, Samuel Araujo Vasconcelos, Christian Nöding, Peter Zacharias,
University of Kassel, D

Passive IGBT Turn-off during Short-circuit Type V 581
Jan Fuhrmann, Hans-Günter Eckel, University of Rostock, D; Sebastian Klauke, Infineon, D

High-Current Power Cycling Test-Bench for Short Load Pulse Duration and First Results 588
Guang Zeng, Christian Herold, Menia Beier-Möbius, Josef Lutz, Technical University of Chemnitz, D;
Sascha Kubera, Rodrigo Alvarez, Siemens, D

Issues in Testing Advanced Power Semiconductor Devices 596
Gabor Farkas, Mentor Graphics, HU; Zoltan Sarkany, Marta Rencz, BME, HU

Special Session "Smart Lighting"

Smart Lighting – Requirements for Modern Lighting Systems and Expected Trends 604
Diederik de Stoppelaar, LightingEurope, BE

Special requirements on a SMPS for LED-Lighting Purposes by an Example of an Individual High-End LED-Driver Solution 605
Florian Müller, Stefan Raithel, Vossloh-Schwabe Lighting Solutions, D

The seven challanges of LED lighting 612
Claudio Adragna, Francesco Ferrazza, Giovanni Gritti, STMicroelectronics, I

PCIM Europe 2016, 10 – 12 May 2016, Nuremberg, Germany

Highly Flexible Single Stage Flyback Quasi-Resonant Digital Controller for Advanced LED Applications 620
Marc Fahlenkamp, Infineon Technologies, D

New and Renewable Energy Systems

Carrier-Based Modulation Technique to Reduce Low Frequency Ripple at the Partial Dc-Link Voltages of a Three-Level/-Phase/-Wire NPC Converter Applied to Future Dc Bipolar Active Distribution Networks 621
Joabel Moia, Marcelo Heldwein, Federal Institute of Santa Catarina – IFSC, BR

FPGA Based Direct Model Predictive Power and Current Control of 3L NPC Active Front Ends 629
Zhenbin Zhang, Ralph Kennel, Technical University of Munich, D

Scalable Insulated DC/DC Converters for Safe and Efficient Coupling of Fuel Cells, Electrolyzers and DC Grids 637
Bernd Seliger, Stefan Matlok, Stefan Zeltner, Fraunhofer Institute IISB, D

SiC MW PV Inverter 645
Maja Harfman Todorovic, Ljubisa Stevanovic, Gary Mandrusiak, Brian Rowden, Fengfeng Tao, Philip Cioffi, Jeffrey Nasadoski, Rajib Datta, GE Research Center, USA; Fabio Carastro, Tobias Schuetz, Robert Roesner, GE Research Center, D

Power Electronics in Automotive

A Performance Comparison of a 650 V Si IGBT and SiC MOSFET Inverter under Automotive Conditions 653
Teresa Bertelshofer, Roman Horff, Andreas März, Mark-M. Bakran, University of Bayreuth, D

A Generic Topology for Electrical Energy Storage Systems 661
Christoph Marxgut, Helbling Technik, Dli

Pulse Width- and Frequency Modulated DC/DC Converter for Hybrid- and Electrical Vehicles 669
Magnus Böh, Andreas Lohner, Technical University of Cologne, D; Christoph Engelhard, HELLA KGaA Hueck

New High Power Density Modules for EV/HEV Applications 677
Seiichiro Inokuchi, Shoji Saito, Arata Izuka, Hata Yuki, Shinji Hatae, Mitsubishi Electric Corporation, J; Khalid Hussein, Mitsubishi Electric Europe, D

Module Technology

Fault-Tolerant B6-B4 Inverter Reconfiguration with Fuses and Ideal Short-On Failure IGBT Modules 683
Michael Gleißner, Mark-M. Bakran, University of Bayreuth, D

Statistical Evaluation of Current Imbalance in Parallel Devices 691
Uwe Scheuermann, Semikron Elektronik, D

Batch Purity in Semiconductor Power Modules 698
Christian Aggen, Henning Ströbel-Maier, Matthias Mau, Jürgen Laue, Marco Bäßler, Danfoss Silicon Power, D

PCIM Europe 2016, 10 – 12 May 2016, Nuremberg, Germany

Novel Technique to Reduce Substrate Tilt & Improve Bondline Control between AIN Substrate & AlSiC Baseplate in IGBT Modules 704
James Booth, Paul Mumby-Croft, Matthew Packwood, Kim Evans, Andy Dai, Dynex Semiconductor, GB; Karthik Vijay, Indium Corporation, GB

Drive Strategies in Power Converters

Communicating Gate Driver for SiC MOSFET 712
Christophe Bouguet, Nicolas Ginot, Christophe Batard, University of Nantes, F

New Ultra Fast Short Circuit Detection Method Without Using the Desaturation Process of the Power Semiconductor 720
Stefan Hain, Mark-M. Bakran, University of Bayreuth, D

High Power, High Frequency Gate Driver for SiC-MOSFET Modules 728
Gunter Königsmann, Reinhard Herzer, Sven Buetow, Matthias Rossberg, Semikron Elektronik, D

Integrating a real-time Tvj calculation into an IPM 735
Stefan Schmies, Peter Lahl, Wolfram Kruschel, Matthias Lassmann, Infineon, D

Energy Storage

12 V Lithium Ion Starter Batteries 742
Hans-Georg Schweiger, Enrique Machuca, Jonas Löchel, Technical University of Ingolstadt, D

Electric Vehicles Batteries Modeling Analysis Based on a Multiple Layered Perceptron Identification Approach 750
Sender Rocha dos Santos, Thais Tóssoli de Sousa, Alex Pereira França, CPqD, BR

An Efficient Implementation of a Reconfigurable Battery Stack with Optimum Cell Usage 757
Martin Wattenberg, Reutlingen University, D; Martin Pfost, University of Innsbruck, AT

Comparative Study and Evaluation of Passive Balancing Against Single Switch Active Balancing Systems for Energy Storage Systems 763
Iosu Aizpuru, Unai Iraola, Jose Mari Canales, Ander Goikoetxea, Mondragon University, ES

Control Techniques in Intelligent Motion Systems II

Voltage Levels Comparison and System Optimization for Electric Drives in Hybrid Vehicles 772
Quentin Werner, Daimler, D; Serge Pierfederici, Noureddine Takorabet, University de Lorraine, F

Real-Time Capable Model Predictive Control of Permanent Magnet Synchronous Motors Using Particle Swarm Optimisation 780
Oliver Wallscheid, Joachim Böcker, University of Paderborn, D; Ulrich Ammann, Esslingen University of Applied Sciences, D

A Modular Multilevel Matrix Converter for High Speed Drive Applications 788
Dennis Bräckle, Felix Kammerer, Mathias Schnarrenberger, Marc Hiller, Michael Braun, Karlsruhe Institute of Technology, D

Smart Supercapacitor based DC-link Extension for Drives offers UPS Capability and acts as an Energy Efficient Line Regeneration Replacement 796
Jens Onno Krah, Markus Höltgen, David Langhals, Technical University of Cologne, D; Nico Sieweke, Christoph Klarenbach, Beckhoff Automation, D

PCIM Europe 2016, 10 – 12 May 2016, Nuremberg, Germany

MOSFET, IGBTs, Freewheeling Diodes

New LV Wide SOA Power MOSFET Technology for Linear Mode Operation 804
Filippo Scrimizzi, Gaetano Bazzano, Daniela Cavallaro, Marco Comola, Giuseppe Consentino,
Stefania Fortuna, Giuseppe Longo, Gaetano Pignataro, STMicroelectronics, I

Field Stop Trench IGBT Process Parameter Calibration for Advanced Predictive Prototyping 813
Mehrdad Baghaie Yazdi, Hermann Fischer, James Victory, Fairchild Semiconductor, D;
Detlef Conrad, Synopsys, D

Best-in-class 1200 V IGBT for High Frequency Applications 819
Ramakrishna Tadikonda, Jorge Cerezo, Chiu Ng, Infineon Technologies Americas, USA

**Extra Electro-Thermal Performance of 1700V IGBT with the Latest 7th Generation Chipset/
Package Technologies** 824
Thomas Heinzel, Fuji Electric Europe, D; Mutsumi Sawada, Shinichi Yoshiwatari, Hiroaki Ichikawa,
Yuichi Onozawa, Osamu Ikawa, Fuji Electric, J

**Parameter Extraction for PSpice Models by Means of an Automated Optimization Tool –
An IGBT model Study Case** 831
Carlos Gomez Suarez, Francesco Iannuzzo, Paula Diaz Reigosa, Ionut Trintis, Frede Blaabjerg,
Aalborg University, DK

**800 V Super Junction MOSFET (HV-DTMOS IV) with Better Trade-Off Between Switching Loss
and dVDS/dt** 839
Hiroyuki Irifune, Hiroshi Ohta, Kaga Toshiba Electronics Corporation, J; Hiroaki Yamashita, Hideyuki Ura,
Kenji Mii, Masato Nashiki, Naotsugu Kako,Toshiba Corporation Semiconductor, J; Georges Tchouangue,
Toshiba Electronics Europe, D

Highly Robust 1700 V Diodes Fabricated on 8"Line Using Optimized Proton Implanted Buffer 845
Maolong Ke, Haihui Luo, Ian Deviny, Xiaoping Dai, Jianwei Huang, Guoyou Liu, Dynex Semconductor, GB

Loss and Softness Optimized IGBT-Diode System for Fast-Switching Applications 850
Christian Müller, Stefan Buschhorn, Infineon Technologies, D

Intelligent Power Modules

Protection Features of Intelligent Power Module against Transient State 858
Taehyun Kim, Minsub Lee, Junbae Lee, Daewoong Chung, Infineon Technologies Power Semitech, ROK

**New High Level Integrated Intelligent Power Module with Three Phase Inverter and Power Factor
Correction Topologies Optimized for Home Appliance** 863
Hyosang Jang, Byoungho Choo, Junbae Lee, Minsub Lee, Daewoong Chung, Infineon Technologies
Power Semitech, ROK

New DIPIPM+TM Series Module with All-in-one Integrated 871
Yuancheng Zhang, Xiankui Ma, Hongtao He, Gaosheng Song, Ming Shang, Mitsubishi Electric &
Electronics, CN

Improvement of System Level Power Density of 15 A / 600 V Intelligent Power Modules 877
Jonathan Harper, ON Semiconductor, D

Optimization of FREDFET-based µIPMTM for very Low Power Motor Drive Applications 882
Rajeev Krishna Vytla, Danish Khatri, Brian Sun, Infineon Technologies North America, USA

PCIM Europe 2016, 10 – 12 May 2016, Nuremberg, Germany

A novel Transfer Molding Intelligent Converter Inverter Brake IGBT Module (DIPIPM+) with Integrated Level Shifting Control ICs 889
Marco Honsberg, Mitsubishi Electric Europe, D; Teruaki Nagahara, Mitsubischi Electric Corporation, J; Eric R. Motto, Powerex, USA

An Automatic IGBT Collector Current Sensing Technique via the Gate Node 895
Jingxuan Chen, Andrew Shorten, Wai Tung Ng, University of Toronto, CA; Masahiro Sasaki, Tetsuya Kawashima, Haruhiko Nishio, Fuji Electric, J

High Voltage Devices

Design and Characterisation of Optimised Protective Thyristors for VSC Systems 903
Michael Spence, Ashley Plumpton, Colin Rout, Alan Millington, Richard Keyse, Dynex Semiconductor, GB

Cathode Emitter vs. Carrier Lifetime Engineering of Thyristors for Industrial Applications 911
Jan Vobecky, Marco Bellini, Karlheinz Stiegler, ABB Switzerland, CH

Experimental Results of a Large Area (91 mm) 4.5 kV "Bi-Mode Gate Commutated Thyristor" (BGCT) 917
Thomas Stiasny, Umamaheswara Reddy Vemulapati, Martin Arnold, Munaf Rahimo, Jan Vobecky, ABB Switzerland, CH; Christian Kähr, Norbert Hofmann, University of Applied Sciences Nordwestschweiz, CH

Effect of Self Turn-On during Turn-On of HV-IGBTs 924
Patrick Münster, Daniel Lexow, Hans-Günter Eckel, University of Rostock, D

An Innovative 6500 V HVIGBT with High Robustness 932
Bo Hu, Gaosheng Song, Mitsubishi Electric & Electronic, CN

New High Power 3.3 kV / 1500 A IGBT Module Packaging 938
Daohui Li, Wei Zhou, Fang Qi, Matthew Packwood, Yangang Wang, Steve Jones, Xiaoping Dai, Dynex Semiconductor, GB

The LinPak High Power Density Design and its Switching Behaviour at 1.7 kV and 3.3 kV 945
Samuel Hartmann, Fabian Fischer, Andreas Baschnagel, Harald Beyer, Raffael Schnell, Christian Treier, ABB Switzerland, CH

Power Converter GTO to IGBT Upgrade – a New Life for Traction Converters 953
Luis Sequeira, Augusto Franco, Adriano Carvalho, Nuno Freitas, Nomad Tech, PT

Packaging Technologies and Materials

Aspects of Reliability Improvement for Large Area Power Semiconductor Devices through Sintering 957
Dmitry Titushkin, Alexey Surma Proton-Electrotex JSC, RUS; Michiel De Monchy, Anna Lifton, Alpha, RUS

Analysis of Interface Structure and Composition of Cu/Al_2O_3 for the High Stability of DBC (Direct Bonded Copper) 961
Hyunwoo Kim, Jaehoon Jung, Hanna Choi, Kisoo Jun, KCC Corporation, ROK

Power Stack – Advantages and Reliability of an Aluminum Based Stacked Power Module 969
Chris Burns, AB Mikroelektronik, AT

Evaluation of Metal-Matrix composites Baseplates with anisotropic thermal Conductivity Inserts 976
Fabian Streb, Infineon Technologies, D; Henning Zeidler, Michael Penzel, Andreas Schubert, Thomas Lampke, Technical University of Chemnitz, D

PCIM Europe 2016, 10 – 12 May 2016, Nuremberg, Germany

Analysis of Packaging Impedance on Performance of SiC MOSFETs 984
Andrew Lemmon, Levi Gant, The University of Alabama, USA; Sujit Banerjee, Kevin Matocha, Monolith
Semiconductor, USA

Low-Stress Silicone Encapsulant for Reliable Power Conversion Devices 992
Guy Beaucarne, Eric Vanlathem, Dow Corning Europe, B; Lu Zhou, Dow Corning, CN; Kent Larson,
Dow Corning Corporation, USA

Pumping out Failure Free Package Structure 996
Junji Yamada, Yoshitaka Otsubo, Mitsubishi Electric Corporation, J; Satoshi Miyahara, Mitsubishi
Electric, D

High power IGBT Module with New AlN Insulated Substrate 1001
Hiroyuki Nogawa, Akira Hirao, Yoshitaka Nishimura, Takashi Saitou, Yuuta Tamai, Fumihiko Momose,
Eiji Mochizuki, Yoshikazu Takahashi, Fuji Electric, J

Nanosilver Paste for Low Pressure Die Attach: A Turn Key Process 1009
Francesc Masana, Barcelona Semiconductors, ES

New Silicone Gel Enabling High Temperature Stability for next Generation of Power Modules 1017
Thomas Seldrum, Eric Vanlathem, Vincent Delsuc, Dow Corning, BE; Hiroji Enami, Dow Corning Toray, J

**A New Ag Paste Composed by Nano and Micro-Ag Particles prepared Simultaneously and
Application as Die-attachment Materials** 1021
Katsuaki Suganuma, Jinting Jiu, Hao Zhang, Shunsuke Koga, Shijo Nagao, Osaka University, J

Packaging and Reliability

Reliability of Double Side Silver Sintered Devices with various Substrate Metallization 1027
Francois LeHenaff, Alpha Assembly Solutions, D; Gustavo Greca, Paul Salerno, Oscar Khaselev,
Monnir Boureghda, Jeffrey Durham, Anna Lifton, Alpha Assembly Solutions, USA; Olivier Mathieu,
Martin Reger, Rogers Germany, D; Zoltan Sarkany, Weikun He, Joe Proulx, John Parry, Mentor Graphics,
UK; Jean Claude Harel, Satyavrat Laud, Renesas Electronics America, USA

New Interconnect Materials: For Future High Reliable Power Module Assembly 1035
Stieven Josso, Henkel Electronics, BE

**Encapsulation of Smart Power Electronic Devices – Thermal Degradation and Dielectric
Behavior** 1040
Tina Thomas, Technical University of Berlin, D; Karl-Friedrich Becker, Klaus-Dieter Lang, Fraunhofer
Institute IZM, D

**Improvement of the Mechanical Properties of Sn-Ag-Sb Lead-Free Solders: Effects of Sb
Addition and Rapidly Solidified** 1046
Mohammed Gumaan, Rizk Shalaby, Mustafa Kamal, Mansoura University, AE; Esmail A. Ali, University
of Science and Technology Yemen, YE

Health-Monitoring of IGBT Power Modules using repetitive Half-sinusoidal Power Losses 1055
Marco Denk, Mark-M. Bakran, University of Bayreuth, D

Reliability Investigation on SiC BJT Power Modules 1063
Alexander Otto, Sven Rzepka, Fraunhofer-Institute ENAS, D; Eberhard Kaulfersch, Berliner Nanotest &
Design, D; Sophia Frankeser, Technical University of Chemnitz, D; Klas Brinkfeldt, Swerea IVF AB, SE;
Olaf Zschieschang, Fairchild Semiconductor, D

PCIM Europe 2016, 10 – 12 May 2016, Nuremberg, Germany

Investigation of the Influence of Ageing Processes on Thermal Characteristics of an IGBT Power Module by Means of Transient Thermal Analysis 1072
Tobias von Essen, Berliner Nanotest & Design, D; Stefan Stegmeier, Gerhard Mitic,
Siemens, D

Test Setup for Multistress Characterization of Insulation Degradation Mechanisms in Electric Drives 1073
Davide Barater, Alessandro Soldati, Giorgio Pietrini, Giovanni Franceschini, Università degli Studi di Parma, I; Chris Gerada, Michael Galea, University of Nottingham, GB; Fabio Immovilli, Raw Power, I

Integration of a Measurement Circuit to determine Junction Temperatures of IGBTs in a Three-phase Converter 1081
Bastian Strauss, Andreas Lindemann, Otto-von-Guericke-University, D

A Recuperation Topology for Power Device Testing 1089
Tomas Krecek, ON semiconductor, CZ

Electrolytic Capacitor Age Estimation using PRBS-Based Techniques 1095
David Hewitt, James Green, Jonathan Davidson, Martin Foster, David Stone, University of Sheffield, GB

On-line Monitoring for Diagnosis on Traction Transformer for Rolling Stock Application 1103
André-Philippe Chamaret, SNCF, F; Toufann Chaudhuri, ABB Sécheron, CH

Cooling

Direct-Water-Cooled Next High Power Density Dual (nHPD2) Considering Inverter Layout 1110
Keisuke Horiuchi, Yuichiro Konishi, Mutsuhiro Mori, Daisuke Kawase, Hitachi, J; Katsuaki Saito,
Hitachi Europe, GB

Heat Pipes used as Heat Flux Transformers and for Remote Heat Rejection 1118
Devin Pellicone, Jens Weyant, Advanced Cooling Technologies, USA

New Class of Graphite TIMs provide Performance and Reliability 1126
Prasanth Subramanian, Alex Augoustidis, GrafTech International, USA

Heat Pipe System Development for Railway Application working with speed Motion Convection 1132
Thomas Albertin, Atherm, F

High Performances Passive Two-Phase Loops for Power Electronics Cooling 1139
Vincent Dupont, Cyrille Billet, Thomas Nicolle, Calyos, BE

Thermal Modelling and Management for increasing the Power Density in High Current Power Electronic Systems 1147
Marco Schilling, Benjamin Köhnlechner, Ulf Schwalbe, Tobias Reimann, Technical University of Ilmenau, D

Packaging and Characterization of Silicon SiC-based Power Inverter Module with Double Sided Cooling 1155
Charles-Alix Manier, Hermann Oppermann, Lothar Dietrich, Christian Ehrhardt, Fraunhofer-Institute IZM, D; Zoltán Sárkány, Budapest University of Technology and Economics, HU; Bernhard Wunderle, Technical University of Chemnitz, D; Wilhelm Maurer, Infineon Technologies, D; Radoslava Mitova, Klaus-Dieter Lang, Technical University of Berlin, D

PCIM Europe 2016, 10 – 12 May 2016, Nuremberg, Germany

Sensors, Control and Protection

Digital Adaptive Control Approach to Dynamic Response Improvement for Compact PFC Rectifiers 1163
Trong Tue Vu, George Young, Eisergy, IE

Nonlinear Output Characteristic of DAB Converter caused by ZVS Transition 1171
Martin Jagau, Michael Patt, Technologienetzwerk Allgäu, D

Efficiency Maximization for Half-Bridge LC Converter through Automatic Dead Time Tuning 1178
Vittorio Crisafulli, ON Semiconductor, D; Gianluca Fazio, On Semiconductor Italy, I;
Diego Hernandez Gutiérrez, CH

Improved Finite Control Set Model Predictive Control with Fixed Switching Frequency for Three Phase NPC Converter 1186
Margarita Norambuena, Hang Yin, Sibylle Dieckerhoff, Technical University of Berlin, D;
Jose Rodriguez, University Andres Bello, CL

Current Sensorless Totem-pole Bridgeless Power Factor Corrector 1194
Felipe López, Francisco Azcondo, Paula Lamo, Alberto Pigazo, University of Cantabria, ES

State Space Model for n-Parallel Connected DC-DC Converters with Predictive Current Control Strategy 1201
Aditya Shekhar, Pavol Bauer, Laurens Mackay, Laura Ramirez-Elizondo, Delft University of Technology, NL

Parameter-Independent Battery Voltage Control Based on Virtual Capacitor Emulation 1209
Andoni Urtasun, Ernesto L. Barrios, Pablo Sanchis, Luis Marroyo, Public University of Navarre, ES

FPGA Digital Control for VSI Nonlinearity Effect Compensation 1217
Mauro Di Monaco, Umberto Abronzini, Ciro Attaianese, Matilde Arpino, Giuseppe Tomasso, University of Cassino and South Lazio, I

Offline Non Isolated Converter Protection 1224
Cathal Sheehan, Bourns Electronics, IE; Roberto Scibilia, Texas Instruments, D

Optimisation of Shunt Resistors for Fast Transients 1232
Melanie Adelmund, Christian Bödeker, Nando Kaminski, University of Bremen, D

High Bandwidth Current Sensors as an Enabler for Advanced Control Techniques 1240
Rolf Slatter, Sensitec, D

Rotational Speed Measurement Based on Avago ADNS-9800 Laser Mouse Sensor 1248
Cheng Liu, Yanan Xu, Ji-Gou Liu, Hui Sun, Chenyang Technologies, D; Ralph Kennel, Technical University of Munich, D

Low EMI high efficiency converters

Practical EMI Control in a Power Component Design Space 1253
David Bourner, Vicor Corporation, USA

Converter Switching Noise Reduction for Enhancing EMC Performance in HEV and EV 1262
Ho Tae Chun, SeungHyun Han, ChangHan Jun, JeongYun Lee, JaeWon Lee, JeeHye Jeong,
JeongHong Joo, JinHwan Jung, Hyundai Motors Company, ROK

Efficiency and Vibration Observations of a Symmetrical Six-Phase Drive applying Interleaved Space Vector Modulation 1270
Daniel Glose, Peng Qian, Ralph Kennel, Technical University of Munich, D

PCIM Europe 2016, 10 – 12 May 2016, Nuremberg, Germany

High Efficiency Three-Phase-Inverter with 650 V GaN HEMTs 1278
Jennifer Lautner, Bernhard Piepenbreier, Friedrich-Alexander-University of Erlangen, D

**A Large Input Voltage Range 1 MHz Full Converter with 95 % Peak Efficiency for
Aircraft Applications** 1286
Nicolas Quentin, SAGEM, F; Remi Perrin, INSA de Lyon, F; Christian Martin,
Charles Joubert, Ampere Laboratory, F; Louis Grimaud, Safran Group, F;
Rolando Burgos, Dushan Boroyevich, CPES/Virginia Tech, USA

Integrating Depletion-Mode SiC VJFETs into Production Motor Drives 1294
Michael Mazzola, James Gafford, Mississippi State University, USA; Gerald W. Godbold, Hyperion
Technology, USA

Higher Light Efficacy in LED-Lamps by Lower LED-Current 1300
Reinhard Jaschke, Klaus F. Hoffmann, Helmut Schmidt University Hamburg, D

Synchronized Switching and Active Clamping of IGBT Switches in a Simple Marx Generator 1305
Martin Sack, Martin Hochberg, Georg Müller, Karlsruhe Institute of Technology, D

High Efficient and Lightweight Auxiliary Power supply with new SiC Power Device 1311
Ryosuke Nakagawa, Mitsubishi Electric Corporation, Japan

A New Behavioral Model for Accurate Loss Calculations in Power Semiconductors 1315
Ajay Poonjal Pai, Tomas Reiter, Infineon Technologies, D; Martin März, Fraunhofer Institute IISB, D

High Speed Electronic Over Current Breaker for DC-Grids without Additional Sensing 1324
Alexander Würfel, Johannes Adler, Nando Kaminski, University of Bremen, D;
Anton Mauder, Infineon Technologies, D

Wireless Power Transmission with High Efficiency for Extensive Applications 1332
Markus Rehm, IBR Ingenieurbüro Rehm, D

Motors and Motor Drives

**Prevention of Traction Drives Stator Insulation Faults Based on Overvoltage Reduction
Utilizing Active Edge Shaping** 1339
Clemens Zöller, Thomas Hausberger, Mathias Blank, Tobias Glück, Hans Ertl, Andreas Kugi,
Technical University Vienna, AT; Markus Vogelsberger, Bombardier Transportation Austria, AT

Noise & Vibration Levels of modern Electric Motors 1345
Christoph Stuckmann, Maccon, D

**Development Platform and Techniques for the Rapid Implementation of High Performance
Drives** 1353
Christian Balke, Simon Wiedemann, Maccon, D

Functional Safety for Integrated Circuits used in Variable Speed Drives 1361
Tom Meany, Analog Devices ERDC, IE

**A Sytem Approach To Understanding The Impact of Non-ideal Effects In A Motor Drive
Current Loop** 1369
Jens Sorensen, Analog Devices, USA; Dara O'Sullivan, Analog Devices, IE

Gate Driver as Part of the Inverter Safety Concept: Optimizing Inverter's Design 1377
Laurent Beaurenaut, Infineon Technologies, D; Peter Sinn, Robert Bosch, D

PCIM Europe 2016, 10 – 12 May 2016, Nuremberg, Germany

Passive Components

Estimation of Ripple and Inductance Roll off when using Powdered Iron Core Inductors 1383
Gautham Ram Chandra Mouli, Pavol Bauer, Miro Zeman, Delft University of Technology, NL;
Jos Schijffelen, Power Research Electronics, NL

Optimized DC Link for Next Generation Power Modules 1391
Michael Brubaker, Terry Hosking, SBE, USA; Tomas Reiter, Infinoen Technologies, D;
Laura D. Marlino, Madhu S. Chinthavali, Oak Ridge National Laboratory, USA

**In-Circuit-Characterization of Ceramic Capacitor with Anti-Ferroelectric Material for Voltage
Source Inverters** 1400
Jürgen Kropp, Mark-M. Bakran, University of Bayreuth, D

Operability of Metallized Polypropylene Capacitors under High Pressure 1408
Magnar Hernes, Ole Christian Spro, SINTEF Energy Research, NO; Volker Gleitner, Electronicon
Kondensatoren, D

Application of High-Voltage 750 V Aluminum Electrolytic Capacitor in Inverter 1416
Kezhuang Yu, Mingkai Peng, Mianwei Qiu, Shenzhen Zeasset Electronic Technology, CN

Analytic Loss Calculation for E-Core Inductors including the End Windings 1421
Johannes Heseding, Axel Mertens, Leibniz University Hannover, D

The Applicability of Nanocrystalline Stacked and Block Cores for Power Electronics 1428
Cezary Swieboda, Marian Soinski, Marcin Kwiecien, Magneto, PL; Wojciech Pluta, Czestochowa
University of Technology, PL; Jacek Leszczynski, AGH University of Science and Technology, PL

The Benefit of Formed or Compacted Litz-Wire Coils 1434
Tobias Appel, STS, D; Hans Rossmanith, Friedrich-Alexander-University of Erlangen, D

**Development of a 100 kW, 20 kHz Nanocrystalline Core Transformer for DC / DC Converter
Applications** 1439
Kapila Warnakulasuriya, Carroll & Meynell Transformers, GB; Farhad Nabhani, Vahid Askari,
Teesside University, GB

Simulation of a 3-Phase Common- and Differential Mode Inductor on a Four-Limb Core 1447
Michael Owzareck, BLOCK Transformatoren-Elektronik, D; Nejila Parspour, University of Stuttgart, D

Design Procedure for Pot-Core Integrated Magnetic Component 1455
Martin Foster, University of Sheffield, GB; Andrew Fairweather, Grant Ashley, VxI Power, GB

Investigation of Core Losses under Different Conditions Applying the Cross Power Method 1463
Boris Hudoffsky, PMK Mess- und Kommunikationstechnik, D; Chihiro Okinori, IWATSU Test Instruments, J;
Jürgen Trüller, HF Instruments, D

A Finite Element Simulation of Nanocrystalline Tape Wound Cores 1471
Christian Scharwitz, Holger Schwenk, Johannes Beichler, Werner Loges, Vacuumschmelze, D

SiC and GaN

**An Insightful Evaluation of a 650 V High-Voltage GaN Technology in Cascode and Stand-Alone
Transistors** 1479
Jaume Roig, German Gomez, Frederick Declercq, Filip Bauwens, On Semiconductor, BE;
Manuel Fernandez, Diego Gonzalez, University of Oviedo, ES

PCIM Europe 2016, 10 – 12 May 2016, Nuremberg, Germany

Static Characterization of Discrete State-of-the-Art SiC Power Transistors 1487
Michael Meisser, Horst Demattio, Thomas Blank, Karlsruhe Institute of Technology, D

Analytical Losses Model for SiC Semiconductors dedicated to Optimization Operations 1494
Gnimdu Dadanema, Francois Costa, ENS Cachan - SATIE, F; Jean-Luc Schanen, Yvan Avenas,
G2ELAB, F; Christian Vollaire, Laboratoire Ampere,F

**Current Measurement and Gate-Resistance Mismatch in Paralleled Phases of High Power SiC
MOSFET Modules** 1503
Roman Horff, Teresa Bertelshofer, Andreas März, Mark-M. Bakran, University of Bayreuth, D

Gate Drive Strategies of SiC Cascodes 1511
Anup Bhalla, Xueqing Li, Shirley Zhang, United Silicon Carbide, USA

State-of-the-art of HF Soft Magnetics and HV/UHV Silicon Carbide Semiconductors 1518
Geraldo Nojima, Faete Filho, Eaton Corporation, USA; Paul Ohodnicki, DOE-National Energy
Technology Laboratory, USA; Alex Leary, Michael E. McHenry, Carnegie Mellon University, USA

**Comparison of Unipolar Silicon Carbide Power Transistors Used in High Switching Frequency
Inverter Topologies** 1528
Sebastian Fahlbusch, Nizar Sahli, Sebastian Klötzer, Ulf Müter, Björn Schäning, Klaus F. Hoffmann,
Helmut Schmidt University Hamburg, D

ST SiC MOSFETs in 1 MHZ DC-DC Converter 1536
Luigi Abbatelli, Giuseppe Catalisano, STMicroelectronics, I

Towards a One Nano-Henry Power Module for SiC and GaN 1541
Jacques Laeuffer, Dtalents, F

Scalable SiC Cascode Power Blocks 1549
Jonathan Dodge, Matt Grady, Ke Zhu, Anup Bhalla, United Silicon Carbide, USA

**High Power Density, High Efficiency 380v to 52v LLC Converter Utiliziing Emode GaN
Switches** 1555
Moshe Domb, Infineon, USA

Gate Drive Units

**A Low Impedance Drive Circuit to Suppress the Spurious Turn On in High Speed Wide
Band-Gap Semiconductor Halfbridges** 1562
Franz Stubenrauch, Norbert Seliger, Doris Schmitt-Landsiedel, University Rosenheim, D

Isolated Gate Driver for High Current/ High Speed FET-Converters 1570
Florian Kapaun, Rainer Marquardt, Universität der Bundeswehr Munich, D

Simple Gate-boosting Circuit for Reduced Switching Losses in Single IGBT Devices 1578
Martin Hochberg, Martin Sack, Georg Müller, Karlsruhe Institute of Technology, D

**Stability and Performance Analysis of a Voltage Controlled Resistor Circuit for Wide Band-gap
Device Gate Drivers** 1584
Alessandro Soldati, Giorgio Pietrini, Davide Barater, Carlo Concari, Università degli Studi di Parma, I

State of the Art of Gate-Drive Power Supplies for Medium and High Voltage Applications 1592
Layal Ghossein, Piotr Dworakowski, SuperGrid Institute, F; Hervé Morel, Florent Morel, Ampère, F

PCIM Europe 2016, 10 – 12 May 2016, Nuremberg, Germany

The Optimized Gate Driver Design Techniques for IGBT Properties and Downsizing in Eco-Friendly Vehicle 1600
KangHo Jeong, SangChul Shin, KiYoung Jang, JinHwan Jung, KiJong Lee, JiWoong Jang, Hyundai Motors, ROK

A Revisit to Resonant Gate Driver and a New Driver to Improve EMI vs. loss Tradeoff for SiC MOSFET 1608
Chi-Ming Wang, Toyota Motor Engineering & Manufacturing, USA

Application and Design Considerations of CoolMOS™ CFD2 and EiceDRIVER™ IC in Motor Drive Application 1615
Wolfgang Frank, Michael Wendt, Infineon Technologies, D; Sam Abdel-Rahman, Infineon Technologies Americas, USA

AC-DC Converters and Power Supplies

4D-Interleaving of Isolated ISOP Multi-Cell Converter Systems for Single Phase AC/DC Conversion 1622
Matthias Kasper, Michael Antivachis, Dominik Bortis, Johann Walter Kolar, ETH Zürich, CH; Gerald Deboy, Infineon Technologies, AT

Battery Charger Based on a Triple-LCp Resonant Converter 1631
Christian Branas, Francisco Azcondo, University of Cantabria, ES; Juan C. Viera, Manuela González, University of Oviedo, ES

High Efficient Flyback Converter with SiC-MOSFET 1639
Johann Austermann, Tim Stuckmann, Holger Borcherding, University of Applied Sciences Ostwesfalen-Lippe, D

PCB Integration of a Magnetic Component dedicated to a Power Factor Corrector Converter 1647
Herault Guillaume, Mercier Adrien, Stéphane Lefebvre, Denis Labrousse, ENS Cachan-SATIE, F

Evaluation of TCM and CrCM modulation for Totem Pole PFC 1655
Haihua Zhou, Wenduo Liu, Eric Persson, Infineon Technologies Americas, USA

System Concept and Model-Based Optimization of High-Current Variable-Voltage Chopper-Rectifiers 1662
Zhiyu Cao, Holger Fahnert, Jürgen Schiele, AEG Power Solutions, D; Jitendra Solanki, Norbert Fröhleke, Joachim Böcker, University of Paderborn, D

Evaluation of a Unidirectional Three-Phase Rectifier Based on the Third Harmonic Injection Concept in Comparison to a VIENNA Rectifier 1669
Markus Makoschitz, Hans Ertl, Technical University of Vienna, AT; Michael Hartmann, Schneider Electric Power Drives, AT

SiC Improves Switching Losses, Power Density and Volume in UPS 1677
Nikolai Epp, Christian Schulte-Overbeck, Zhiyu Cao, Michael Lemke, Lothar Heinemann, AEG Power Solutions, D

Optimization of 12 Pulse and 18 Pulse Rectifier Systems by the Selection of Optimum Parameters for Magnetics 1685
Kapila Warnakulasuriya, Carroll & Meynell Transformers, GB; Farhad Nabhani, Vahid Askari, Teesside University, GB

PCIM Europe 2016, 10 – 12 May 2016, Nuremberg, Germany

DC-DC Converters I

Analysis of the Flyback Converter Utilizing a Transformer with Stepped Air-Gap 1693
Panagiotis Mantzanas, Daniel Kübrich, Markus Barwig, Thomas Dürbaum, Friedrich-Alexander-
University of Erlangen, D

Novel Method for the Estimation of Switching Losses in Resonant Converters 1701
Christian Oeder, Markus Barwig, Thomas Dürbaum, Friedrich-Alexander-University of Erlangen, D

Active Dead-Time Optimization for wide Range Flyback Active-Clamp Converter 1707
Sebastien Larousse, Nacer Abouchi, Remy Cellier, Institut des nanotechnologies de Lyon, F;
Hubert Razik, Laboratoire Ampere, F; Philippe Volay, Centralp, F

**Energetic Macroscopic Representation (EMR) and Control Scheme for the Asymmetric 4-Stage
MSBA** 1713
Gina Steinke, Alfred Rufer, EPFL - Ecole Polytechnique Fédérale Dde Lausanne, CH

**Adjustable 20 kW Full-SiC Electronic Load with Energy Recovery for Medium-frequency
Inverter** 1721
Fabian Denk, Karsten Haehre, Julian Koerner, Rainer Kling, Wolfgang Heering, Karlsruhe Institute
of Technology, D

A New High Frequency Transformer for UPS 1728
Michael Schmidhuber, Manfred Wohlstreicher, Michael Baumann, Markus Schmeller, SUMIDA
Components & Modules, D

DC/DC-Converter for Modular Coupling of 48 V Battery Packs to a High Voltage DC Bus 1733
Michael Eberlin, Milad Mohammad Hossein Khani, Fraunhofer Institute ISE, D

High Dynamic Current Source for LED Light and Data Transmission Applications 1741
Karl Edelmoser, Technical University of Vienna, AT; Felix Himmelstoss, Technikum Vienna, AT

**Design Methods for LLC Converter considering Buck and Boost Mode with Limited Frequency
Range for Wide Input Voltage Range** 1749
Dustin Funk, Tobias Reimann, Technical University of Ilmenau; Ulf Schwalbe, ISLE Steuerungstechnik
und Leistungselektronik, D

The Behavior of Electro-Magnetic Radiation of Storage Inductor in DC-DC Converters 1757
Ranjith Bramanpalli, Würth Elektronik Eisos, D

DC-DC Converters II

A New High Frequency Ferrite Material for Gan Applications 1764
Herbert Jungwirth, Michael Schmidhuber, Michael Baumann, Markus Schmeller, SUMIDA Components
& Modules, D

Multi-Stage LLC Resonant Converters designed for Wide Output Voltage Ranges 1769
Chi Wa Tsang, Chris Bingham, University of Lincoln, GB; Martin Foster, Dave Stone, University of
Sheffield, GB; John Leach, Castle, GB

Application Advantages and Disadvantages of Modern Fast Switching MOSFETs in VRM 1777
Zhiyang Chen, Ann Starks, ON Semiconductor, USA

**Medium to Low Voltage DC/DC Resonant Converter with SiC SCRs and Nanocrystalyne
Magnetic Core Transformer** 1785
Iñigo Martinez de Alegria, Angel Luis Perez, Madaci Mansour, Jon Andreu Larrañaga, University of
the Basque Country, ES; Kerdoun Djallel, GLEC Constanine 1, DZ

PCIM Europe 2016, 10 – 12 May 2016, Nuremberg, Germany

GaN Active-Clamp Flyback Converter with Resonant Operation Over a Wide Input Voltage Range 1792
Nicolas Quentin, SAGEM, F; Remi Perrin, Cyril Buttay, INSA de Lyon, F; Christian Martin, Charles Joubert, Ampere Laboratory, F; Bertrand Lacombe, Safran Group, F

Demonstration of superior SiC MOSFET Module performance within a Buck-Boost Conversion System 1800
Maximilian Slawinski, Tim Villbusch, Daniel Heer, Marc Buschkühle, Infineon Technologies, D

High Efficiency and High Power Density Boost / Buck Converter with SiC JFET Modules for Advanced Auxiliary Power Supply in Trolleybuses 1808
Miroslav Hruska, Skoda Electric, CZ; Martin Jara, West Bohemian University, CZ

Development of a 12 kW isolated and bidirectional DC-DC Converter dedicated to the More Electrical Aircraft: The Buck Boost Converter Unit (BBCU) 1814
Pascal Asfaux, Jeremy Bourdon, Airbus Operation, F

Reverse Mode Application of Sine Amplitude Converters 1822
David Bourner, Vicor Corporation, USA

DC-AC Converters

Influence of the Configuration of the Load Cable on Switching Characteristics of IGBTs 1829
Lars Middelstaedt, Dennis Richter, Andreas Lindemann, Otto-von-Guericke-University, D; Arendt Wintrich, Semikron, D

Improved Power Decoupling Scheme for Single-Phase Grid-Connected Differential Inverter with Realistic Mismatch in Storage Capacitances 1837
Wenli Yao, Xiaobin Zhang, Northwestern Polytechnical University, CN; Xiongfei Wang, Poh Chiang Loh, Frede Blaabjerg, Aalborg University, DK

Technical Approach: Interleaved, Folding, Interpolating Dual-Path Adiabatic Autotransformer Based Power Converter 1845
John Wood, Ed Shelton, Silicon Contact, GB; Kevin Rathbone, Robotae, GB; Mehdi Baghadadi, Patrick Palmer, University of Cambridge, GB

Design and Performance Evaluation of a Three Phase AC Power Source with Virtual Impedance for Validation of Grid Connected Components 1846
Peter Jonke, Johannes Stöckl, Hans Ertl, AIT-Austrian Institute of Technology, AT

Design and Testing of a Modular SiC based Power Block 1853
Maja Harfman Todorovic, Rajib Datta, Ljubisa Stevanovic, Xu She, Philip Cioffi, Gary Mandrusiak, Brian Rowden, Paul Szczesny, Jian Dai, Tony Frangieh, GE Research Center, USA

A Novel Method to simulate the Control-to-output Transfer Function of Resonant Converters 1857
Julian Dobusch, Christian Oeder, Thomas Dürbaum, Friedrich-Alexander-University of Erlangen, D

A Study of the Thermal and Parasitic Optimization of a Large Current Density Highly Parallelized Three-Phase Reference Board for Motor Drive Applications 1864
Mehrdad Baghaie Yazdi, Xiaomin Wu, Peter Haaf, Klaus Neumaier, Fairchild Semiconductor, D

PCIM Europe 2016, 10 – 12 May 2016, Nuremberg, Germany

AC-AC and Multilevel Converters

Trends in Residential and Industrial Induction Cooking: Topologies and Power Devices for High Efficiency 1865
Vittorio Crisafulli, ON Semiconductor, D

Direct Power Control for a Grid Connection of a Three Phase Z-Source Inverter 1873
Manuel Steinbring, Mario Pacas, University of Siegen, D

Interleaved Series Input Parallel Output forward Converter with Simplified Voltage Balancing Control 1880
Kaspars Kroics, Alvis Sokolovs, Linards Grigans, Ugis Sirmelis, Riga Technical University, LV

Fault-Tolerant Behaviour of the Three Level Advanced-Active-Neutral-Point-Clamped Converter ... 1888
Sidney Gierschner, David Hammes, Jan Fuhrmann, Hans-Günter Eckel, University of Rostock, D;
Max Beuermann, Siemens, D

Cell Voltage Balancing Controller for the Modular Multilevel Converter Arm using Symmetrical Transformation 1896
Andrey Dudin, Aaron Fischer, Thomas Ellinger, Jürgen Petzold, Technical University of Ilmenau, D

Isolated low-power multi-output DC-DC Converters with Heterogeneous Loads for an Efficient Supply of Modular Power Electronics Systems 1904
Arthur Singer, Thomas Weyh, Florian Helling, Universität der Bundeswehr Munich, D; Arun Jeyaprakash,
Technical University of Munich, D; Stefan Götz, Duke University, USA

A wire based communication interface for Medium and High-Voltage Converters 1912
Marek Galek, Manuel Blum, Siemens, D

An Auxillary Power Supply with integrated Communication Capability for Medium and Highvoltage Applications 1919
Manuel Blum, Marek Galek, Siemens, D

FPGA Based Direct Model Predictive Current Control of PMSM Drives with 3L-NPC Power Converter 1926
Zhenbin Zhang, Christoph Hackl, Ralph Kennel, Technical University Munich, D

IGBT Power Module in Three-Level Neutral Point Clamped Type 2 (NPC2, T-NPC, Mixed Voltage) Topology in Short Circuit Modes 1934
Vladan Jerinic, Kevin Lenz, Reiner Hinken, Danfoss Silicon Power, D

Efficiency Verification Power Circulation Method of a High Power Low Voltage NPC Converter for Wind Turbines 1942
Berthold Benkendorff, Friedrich W. Fuchs, Christian-Albrechts-University, D; Toke Franke, Danfoss
Silicon Power, D

Control of the Actively Balanced Capacitive Voltage Divider for a Five-Level NPC Inverter-Estimation of the Intermediary Levels Currents 1949
Alfred Rufer, Nelson Koch, Nicolas Cherix, EPFL – Ecole Polytechnique Fédérale de Lausanne, CH

Automotive Applications

Automotive Power Module Technologies for High Speed Switching 1956
Shinichiro Adachi, Takuma Kouge, Souichi Yoshida, Hiroshi Miyata, Daisuke Inoue, Yoshikazu Takamiya,
Hideto Kobayashi, Akira Nishiura, Fumio Nagaune, Fuji Electric, J; Thomas Heinzel, Fuji Electric Europe, D

PCIM Europe 2016, 10 – 12 May 2016, Nuremberg, Germany

Power Semiconductors for the Automotive 48 V Board Net 1963
Felix Hüning, University of Applied Sciences Aachen, D

Status and Advances in Electric Vehicle's Power Modules Packaging Technologies 1970
Itxaso Aranzabal, Asier Matallana, Oier Oñederra, Iñigo Martinez de Alegría, David Cabezuelo,
University of the Basque Country, ES

A Highly Integrated Full SiC Six Pack Power Module for Automotive Applications 1979
Bao Ngoc An, Viktor Wegelin, Martin Bernd, Benjamin Leyrer, Michael Meisser, Horst Demattio,
Thomas Blank, Marc Weber, Karlsruhe Institute of Technology, D; Johannes Kolb, Schaeffler Technologies, D;
Jochen Altstadt, Kortec, D

**Automotive-grade P-channel Power MOSFETs for Static, Dynamic and Repetitive Reverse
Polarity Protection** 1987
Filippo Scrimizzi, Giuseppe Longo, Giusy Gambino, STMicroelectronics, I

Isolated On-Board DC-DC Converter for Power Distribution Systems in Electric Vehicles 1992
Sven Bolte, Joachim Böcker, Norbert Fröhleke, University of Paderborn, D

Innovative Solution of Static and Dynamic Contactless Charging Station for Electrical Vehicles 1999
Nikolay Madzharov, Valeri Petkov, Technical University of Gabrovo, BG

Combining an External Rotor Motor with Vernier Concept for Drives in Intralogistics 2007
Matthias Thesseling, Tao Liu, Lenze SE, D; Hans-Joachim Wendt, Lenze Drives, D

**Dynamic Modeling and Optimal control for Series-parallel Drivetrain based on Lithium-ion
battery** 2014
Tedjani Mesbahi, Ecole Centrale de Lille, F; Moudrik Meradji, Gaolin Wang, Dianguo Xu, Harbin
Institute of Technology, CN

**Smart Diode and 4-Switch Buck-Boost Provide Ultra High Efficiency, Compact Solution for
12-V Automotive Battery Rail** 2019
Vijay Choudhary, Mathew Jacob, Texas Instruments, USA

On-Chip Temperature Measurement: A new Approach for Optimizing Automotive Inverters 2027
Laurent Beaurenaut, Inpil Yoo, Infineon Technologies, D

Renewable Energy Systems

Resonant load Emulator for Distributed Energy Resources to test Anti-islanding Algorithms 2035
Daniel Heredero-Peris, Fernando Jorge-Ques, Daniel Montesinos-Miracle, Universitat Politécnica de
Catalunya, ES; Tomàs Lledó-Ponsati, TeknoCEA, ES

Low Voltage Ride Through (LVRT) Capability of an Enhanced DFIG System 2043
David Velasco, Jesús López, Public University of Navarre, ES

Control and Modulation for Loss Minimization for Dc/Dc Converter in Wind Farm 2050
Catalin Gabriel Dincan, Philip C. Kjaer, Aalborg University, DK

**A Variable Step Size Perturb and Observe Method Based MPPT for Partially Shaded
Photovoltaic Arrays** 2058
Jawad Ahmad, Filippo Spertino, Paolo Di Leo, Alessandro Ciocia, Politecnico di Torino, I

**Renewable Electricity Conversion and Storage: Focus on Power to Gas process, EMR
Modelling and Simulation** 2066
Ahmed Remaci, Octavian Curea, Christophe Merlo, Amélie Hacala, ESTIA, F; Vincent Guerre, Local
Energy Alternative & Fair, F

PCIM Europe 2016, 10 – 12 May 2016, Nuremberg, Germany

Balancing Current and Efficiency Modelling of Single Switch Active Balancing Systems for Energy Storage Systems 2074
Iosu Aizpuru, Unai Iraola, Jose Mari Canales, Ander Goikoetxea, Mondragon University, ES

A Control Strategy for Multiple Energy Storage Devices for Power Leveling of Renewable Energy Systems 2083
Koji Kato, Yoichi Ito, Sanken Electric, J

Examining Contrasting Excitation Modes within Battery Characterisation using Maximum Length Sequences 2090
Andrew Fairweather, VxI Power, GB; David Stone, Martin Foster, University of Sheffield, GB

Comparison Between Standard and Innovative Solutions to exchange Energy between High Energy Storage Systems 2098
Laurent Garnier, Daniel Chatroux, Sébastien Carcouet, Julien Dauchy, University Grenoble Alpes CEA LITEN, F

EMI, Harmonics, Filters

Simulation and Experimental Analysis of Non-Linear Loads from Residential and Educational Buildings 2106
Gabriel Nicolae Popa, Angela Iagar, Corina Maria Dinis, Politehnica University Timisoara, RO

A Digital Predictive Constant Frequency Controller for High Frequency 3-Phase Silicon Carbide PFC Rectifier 2114
Marcelo Schupbach, Cree, USA; Binod Agrawal, Navneet Mangal, CREE India Private Limited, IN

DC-link Harmonic Content in Double Two-Level Inverter for Permanent Magnet Synchronous Motor Drive Systems – Comparison and Analysis 2122
Toktam Khani, Michael Patt, Technologienetzwerk Allgäu, D

Hybrid Filter With an Optimized Switching Method of the Compensation Capacitors and Predictive Active Filter Control 2129
Swen Bosch, Heinrich Steinhart, HTW Aalen, D

Active Mains Filters with Combined Feed-Forward and Feed-Back Control 2137
Felix Himmelstoss, Technikum Vienna, AT; Karl Edelmoser, Technical University of Vienna, AT

Influence of the Zero Sequence Voltage on the Design of a Series Active Filter 2145
Andreas Reinhold, Rolf Grohmann, HTWK Leipzig, D; Uwe Rädel, Jürgen Petzoldt, Technical University of Ilmenau, D

Electromagnetic Emissions in High Density and Fast GaN Switched Half Bridges with Resonance Filter Structures 2151
Wolfgang Stocksreiter, Hans List, Hubert Berger, Gerald Weis, Markus Krenn, FH Joanneum, AT; Günter Engel, CeraCap, AT

Energy Transmission and Grid

AC or DC Grid for Railway Stations? 2159
Lilia Galai Dol, Efficacity, F; Alexandre De Bernardinis, French Institute of Science and Technology for Transport, Development and Networks, F

Solid-State Transformer Modeling for Analyzing its Application in Distribution Grids 2167
Christoph Hunziker, Nicola Schulz, University of Applied Sciences Northwestern Switzerland, CH

PCIM Europe 2016, 10 – 12 May 2016, Nuremberg, Germany

Hybrid Reactive Power Compensation System for Grid Code Compliance in Renewable Energy Power Plants 2175
Gianluca Postiglione, Antonio Raso, Giovanni Borghetti, Francois Pezet, Nidec ASI, I

A new shunt connected HVDC Tap Based on a Highly Efficient Resonant Cascade Converter 2181
Andre Birkel, Mark-M. Bakran, University of Bayreuth, D

Analysis of Voltage and Current Unbalance in a Multi-Converter Topology for a DC-Based Offshore Wind Farm 2189
Thomas Lagier, SuperGrid Institute, F; Philippe Ladoux, University of Toulouse, F

Experimental Demonstration of a Solid-StateDamping Resistor for HVDC Applications 2197
Konstantin Vershinin, Ikenna Efika, David Trainer, Colin Davidson, Alstom Grid, GB; Nick Wright, Amit Tiwari, Newcastle University, GB

High Precision Loss Measurement at HVDC Converter 2204
Helmut Weiß, Technical University of Leoben, AT; Bernhard Grasel, Dewesoft, AT

A Fast Methodology for Solving Power Flows in Hybrid AC/DC Networks: The European North Sea Supergrid Case Study 2211
Rodrigo Teixeira Pinto, Monica Aragues-Penalba, Andreas Sumper, CITCEA-UPC, ES; Christian Alejandro Leon-Ramirez, Elmer Sorrentino, University Simon Bolivar, VE

Panel Discussion "The smart future of power electronics"

Using Smart Converter to obtain Traction-Machine Insulation Health State Information 2219
Markus Vogelsberger, Bombardier Transportation Austria, AT; Clemens Zöller, Jörg Bellingen, Thomas Wolbank, Technical University of Vienna, AT

Manuscripts which were handed in late

Investigation of the Influence of Ageing Processes on Thermal Characteristics of an IGBT Power Module by Means of Transient Thermal Analysis 2226
Tobias von Essen, Berliner Nanotest & Design, D; Stefan Stegmeier, Gerhard Mitic, Siemens, D

A Study of the Thermal and Parasitic Optimization of a Large Current Density Highly Parallelized Three-Phase Reference Board for Motor Drive Applications 2231
Mehrdad Baghaie Yazdi, Xiaomin Wu, Peter Haaf, Klaus Neumaier, Fairchild Semiconductor, D

Welcome to the Post-Silicon World: Wide Band Gap Powers Ahead

Dan Kinzer
Navitas Semiconductor, USA
dan.kinzer@navitassemi.com

ABSTRACT

In the early 1980's, industry pioneers transformed the silicon bipolar transistor into the mass production MOSFET. The 1990's saw the adoption of IGBT, and the 2000's brought superjunction MOSFET into volume. Now, an even more fundamental transition is underway as Silicon is overtaken by the introduction of SiC and GaN high performance, wide band-gap power products. The breakthrough performance of qualified products, with the subsequent application benefits of size and cost, is now fully appreciated by power system designers. As the 'eco-system' continues to mature, with new enabling topologies, control ICs and magnetic solutions, the transition will accelerate and transform the power industry. Examples will be taken from applications such as electric vehicles, renewable energy, power supplies, and battery chargers.

INTRODUCTION

In high power (multi-kW) applications like electric vehicle drive and solar string inverters, IGBTs are being surpassed by 1,200V SiC FETs which utilize a cost-effective vertical structure to enable high efficiency, simplified architectures. In lower power (~300W) solar micro-inverters, high frequency 650V GaN systems are under evaluation. As manufacturing volumes increase and costs fall, low voltage (100V) Si used in solar optimizers may also be replaced by GaN devices. (Fig. 1)

Material properties of WBG device, such as high critical E-field and high carrier mobility, enable SiC and GaN to achieve huge improvements over Si in current-handling and high-frequency operation. Advances in low-inductance packaging and a transition from discrete power and drive implementation to co-packaging and monolithic integration simplify and optimize end applications, enabling new form-factors and lower system costs.

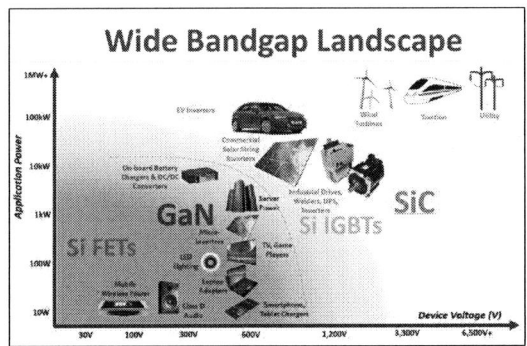

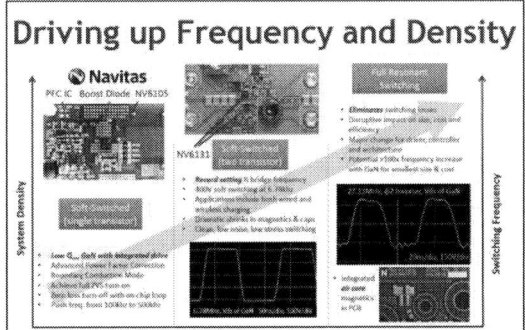

Figure 1: Power applications by technology   Fig.2: Switching frequency drives density

Off-line AC-DC converters and inverters are about to undergo massive transformation, as new GaN power integrated circuits are introduced to the market. Switching frequencies >5MHz in bridge topology and >25MHz in single ended resonant topologies are demonstrated. (Fig. 2) As the SiC and GaN 'eco-system' continues to develop and mature, with enabling topologies, packages, new control ICs and new magnetic solutions (materials and implementations), the transition to wide band-gap materials will continue to accelerate and transform the power industry.

© VDE VERLAG GMBH · Berlin · Offenbach

For information on the keynote

Smart Transformers - Concepts/Challenges/Applications
Smart and/or Solid-State Transformers (SSTs) are formed by power electronics interfaces at the MV input and LV output side, which are linked through a MF transformer. Accordingly, SSTs show a high power density and are offering high controllability. Therefore, SSTs are seen as key elements of future smart microgrids, and are well suited for replacing bulky LF transformers of locomotives. However, the connection to MV, the high overall complexity / realization costs, and the potentially lower efficiency are still major challenges for practical applications.

please contact the speaker directly:

Professor Johann Walter Kolar

ETH Zürich Power Electronic Systems Laboratory
Power Electronic Systems Laboratory
Physikstr. 3 ETL H23
8092 Zürich
Switzerland

Phone: +41 44 632 28 34
E-Mail: kolar@lem.ee.ethz.ch

For information on the keynote

Trends of Solar Systems and their Integration in Electricity Networks
With their constant cost reduction, solar systems have become already today an economically viable alternative in many regions of the world. While the cost for solar modules has been drastically reduced in the last years, the cost reduction potential for the balance of system components, like for example power conditioners, still needs to be exploited. On the other side, the massive penetration of variable solar energy systems challenges the management of the electricity networks. This speech outlines trends for solving these issues.

please contact the speaker directly:

Dr. Jens Merten

Commissariat à l´énergie atomique et aux énergies alternatives
Institut National d'Energie Solaire
Avenue du Lac Léman 50
73375 Le Bourget-du-Lac
France

Phone: +33479792205
E-Mail: jens.merten@cea.fr

Ultra-low (1.25mΩ) On-Resistance 900V SiC 62mm Half-Bridge Power Modules Using New 10mΩ SiC MOSFETs

Jeffrey, Casady, Wolfspeed, a Cree Company, 3028 E. Cornwallis Road, RTP, NC 27709, USA, jeff.casady@wolfspeed.com

Vipindas Pala, Wolfspeed, a Cree Company, USA, vipindas.pala@wolfspeed.com
Edward Van Brunt, Wolfspeed, a Cree Company, USA, edward.vanbrunt@wolfspeed.com
Brett Hull, Wolfspeed, a Cree Company, USA, brett.hull@wolfspeed.com
Sei-Hyung Ryu, Wolfspeed, a Cree Company, USA, sei-hyung.ryu@wolfspeed.com
Gang-Yao Wang, Wolfspeed, a Cree Company, USA, gangyao.wang@wolfspeed.com
Jim Richmond, Wolfspeed, a Cree Company, USA, jim.richmond@wolfspeed.com
Scott T. Allen, Wolfspeed, a Cree Company, USA, scott.allen@wolfspeed.com
Dave Grider, Wolfspeed, a Cree Company, USA, dave.grider@wolfspeed.com
John W. Palmour, Wolfspeed, a Cree Company, USA, john.palmour@wolfspeed.com
Peter Killeen, Wolfspeed, a Cree Company, USA, peter.killeen@wolfspeed.com
Brice McPherson, Wolfspeed, a Cree Company, USA, brice.mcpherson@wolfspeed.com
Kraig Olejniczak, Wolfspeed, a Cree Company, USA, kraig.olejniczak@wolfspeed.com
Brandon Passmore, Wolfspeed, a Cree Company, USA, brandon.passmore@wolfspeed.com
David Simco, Wolfspeed, a Cree Company, USA, david.simco@wolfspeed.com

Abstract

For the first time, a new 900V, 10mΩ SiC MOSFET chip is fabricated, tested, and assembled in a >400A, ½ bridge power module, with only 1.25-2.5mΩ on-resistance at 25°C, depending on the number of chips per switch position (i.e., eight or four, respectively). The SiC MOSFET chip had a measured breakdown > 1kV, and a specific $R_{DS(ON)}$ of 2.3mΩ•cm². The chips were then assembled in 16 power modules, and characterized up to 175°C. Only a 40-50% increase in $R_{DS(ON)}$ was measured with a temperature increase from 25°C to 150°C. With no knee voltage, conduction losses relative to comparably rated Si IGBT power modules can be reduced up to 70%.

1. Introduction

1.1. SiC potential impact on conduction losses in systems

SiC MOSFETs are currently being investigated for very low conduction losses at moderate frequencies, in addition to its advantages in high frequency applications. Lower-frequency applications such as Electric Vehicle (EV) drivetrain and motor-drives can exploit SiC for conduction efficiency significantly more than comparable Si IGBT devices [1-3]. With no knee voltage in the on-state, SiC MOSFETs can offer much lower conduction losses at light load.

These benefits in the EV drivetrain can extend vehicle range by up to 10%, and compress power control electronics up to 80% [3]. Relative to standard Si IGBT power modules, at light-load conditions where many drive applications run for the majority of the time, conduction losses can be up to 70% lower for similarly rated SiC modules. The impact to EV platforms can vary widely with system architecture andtype of vehicle, but an approximate calculation shown in Figure 1 illustrates the significance. Using US Environmental Protection Agency drive cycles, with an assumed 90kW IPM motor, synchronous rectification of the SiC MOSFETs with no external anti-parallel diode, the SiC MOSFETs can reduce EV drivetrain inverter losses by 65% to 73%. Graphically this result is shown in Figure 1.

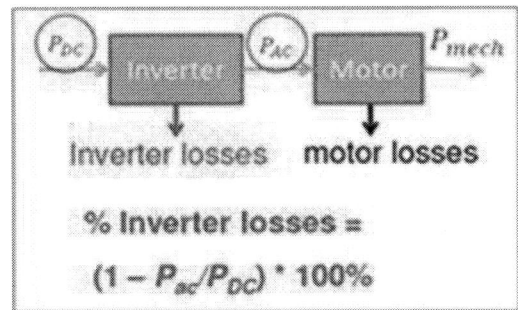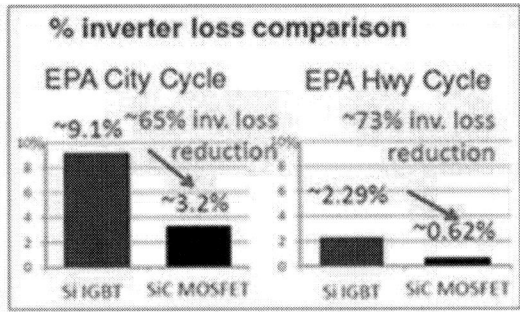

Fig. 1. Example impact of replacing 600V Si IGBT-based inverters with 900V SiC MOSFET based inverters in a 90kW IPM motor. Inverter losses can be reduced by 65-73% for EPA city cycle and highway cycle driving.

Further benefits to reducing inverter losses this significantly would be in the regeneration drive, reducing weight, reducing thermal management, and battery cost for a given range. To achieve these savings, high-performance SiC power MOSFET chips and modules need to be developed which enable very low conduction losses over vehicle drivetrain normal operation at light-load. What is presented here, to the best of our knowledge, is a record low 1.25mΩ on-resistance (low conduction losses) for a 900V 62mm ½ bridge power module. The switching energy and $R_{DS(ON)}$ is characterized up to 175°C, and initial High Temperature Reverse Bias (HTRB) testing is shown up to 175°C.

1.2.   New 900V SiC 10mΩ SiC MOSFET chip design for low conduction losses

The SiC MOSFET chip used here is a newly developed 900V, 10mΩ SiC MOSFET using a planar DMOS design recently described elsewhere [4]. While retaining the reliable DMOS structure (see Fig. 2), specific $R_{DS(ON)}$ is driven down to 2.3mΩ•cm^2, significantly less than competing Si and SiC MOSFET products in this voltage range. The total chip area of the 900V MOSFET is 4.38mm x 7.28mm, with top-level over-layer Al metal, 4µm thick, for standard Al wire-bonding on top-side gate and source.

1.3.   Static & dynamic characterization of SiC MOSFET chip over temperature

Initial die characterization is shown in Figure 3 using a simple TO-247-3L package, with seven stitched eight-mil (203.2µm) diameter wire bonds on the source pad, and a single five-mil (127µm) diameter wire bond on the gate pad. In this limited package, ~1.5mΩ of package resistance is added, there is ~15nH of inductance with no source Kelvin contact, and the wire-bond limits the maximum I_{DS} to about 75A DC.

The measured $R_{DS(ON)}$ over temperature is positive, but much reduced relative to Si MOSFETs. As a comparison, a 650V, 19mΩ Si superjunction Si MOSFET [5] is shown relative to the 900V, 10mΩ SiC MOSFET in Fig. 3a. At 150°C, the 650V Si MOSFET rise in $R_{DS(ON)}$ is

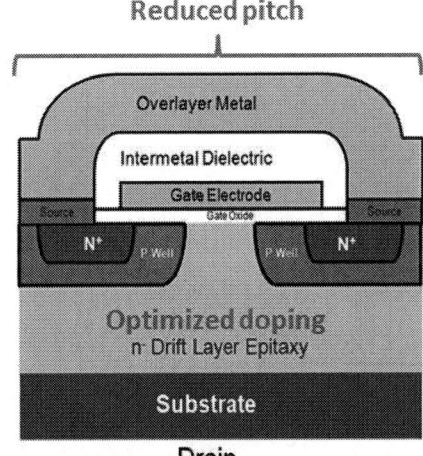

Fig. 2. Device cross-section of 900V, 10mΩ SiC MOSFET above. Top overlayer metal is 4µm thick Al, backside metal is Ni:Ag (0.6µm:0.8µm thick).

1.5x higher than the 900V SiC MOSFET. Over this same temperature range, the Si MOSFET increases by 2.4x, an approximately 60% steeper slope than the SiC MOSFET.

In Fig. 3b, the on-state of the 900V, 10mΩ SiC MOSFET is contrasted with a 650V Si IGBT [6] as a function of forward voltage drop (V_{DS}/V_{CE}). As illustrated in the graph, at low voltage (<2V forward voltage drop), in the so-called "light load" condition, the on-state conduction losses are significantly lower for the SiC MOSFET. The Si IGBT has a "knee voltage" which delays turn-on until approximately 0.8V, whereas the majority-carrier SiC MOSFET, despite being rated 250V higher blocking at 900V, can significantly lower the light load forward voltage drop up to 125A.

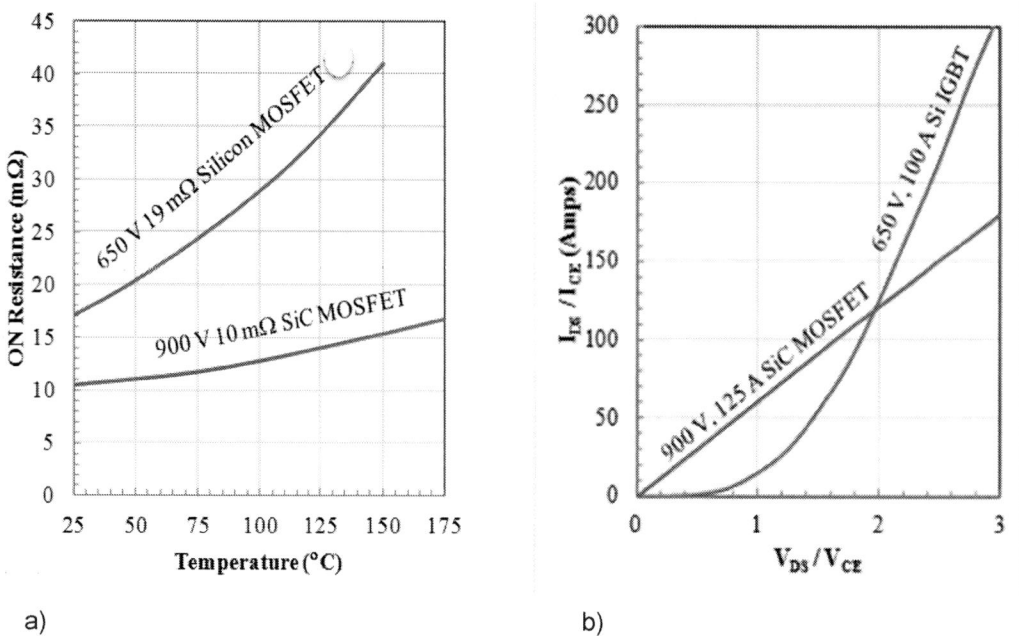

a)   b)

Fig. 3. A comparison of the 900V, 10mΩ SiC MOSFET on-resistance compared to a 650V, 19mΩ Si MOSFET from 25-175°C is shown in Fig. 3a. The SiC MOSFET, despite 250V higher blocking, has a much lower positive temperature coefficient than the Si MOSFET, offering lower conduction losses with increasing temperature. In Fig 3b, the same SiC MOSFET is contrasted with a lower-rated 650V Si IGBT. Up to 125A, the SiC MOSFET has significantly lower forward voltage drop (and thus conduction losses) [4].

For the chip shown in Fig. 2, switching energy measurements were performed, but using a TO-247-4L package with a Kelvin source contact to minimize the feedback of the source inductance to the gate. Adding the Kelvin source contact can lower the switching losses significantly, and is more representative of actual performance inside modules where Kelvin source contacts are common. As SiC MOSFETs are fast devices with high dv/dt, a Kelvin source contact is an effective method to reduce the impact of source inductance feedback on the gate signal.

The switching measurement data is shown in Fig. 4, switching the gate from -4V to +15V, and drain from 800V to 0V. At 60A, only 0.7mJ is measured with an external gate resistor of 2.5Ω. Fig. 4 also illustrates switching losses as a function of external gate resistance. Larger $R_{G(EXT)}$ can smooth current waveforms, and provide easier EMI filtering, but at the cost of higher losses.

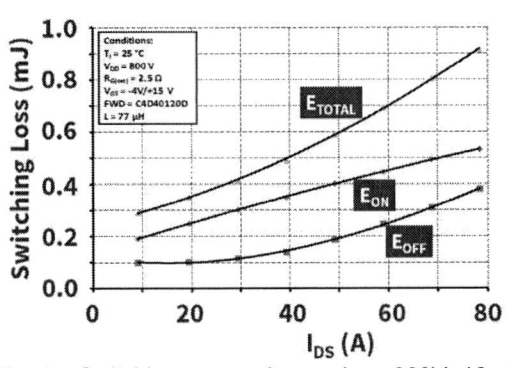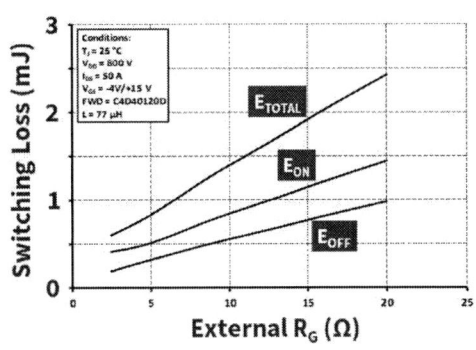

Fig. 4: Switching energy losses in a 900V, 10mΩ SiC MOSFET in TO-247-4L package with Kelvin source contact. Die size is 4.38mm x 7.26mm, with the same layout shown in Fig. 2. Measurements were performed at 25°C, 800V, SiC free-wheeling diode, V$_{GS}$ from -4V to +15V [4].

2. 900V SiC ½ Bridge Power Module Results

2.1. 900V, 2.5mΩ SiC MOSFET multichip power module

Using a 62mm module footprint, with 4nH total inductance, two ½ bridge versions of a 900V SiC power module were constructed, with 1.25 and 2.5mΩ. In this section, the 2.5mΩ module data is shown below. Each switch position contains four of the 900V, 10mΩ SiC MOSFETs per switch position for the 2.5mΩ module (Fig. 5a). 16 switch positions were measured and data (median 3.6mΩ R$_{DS(ON)}$ at 150°C) are shown in Fig. 5b at 150°C.

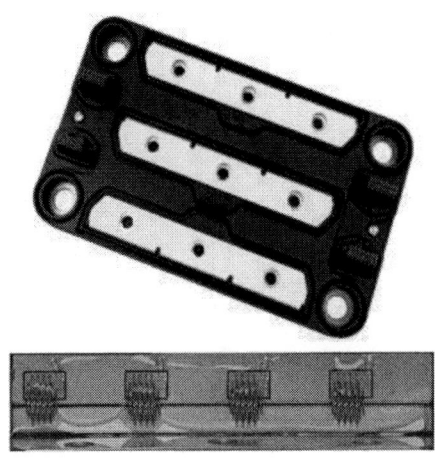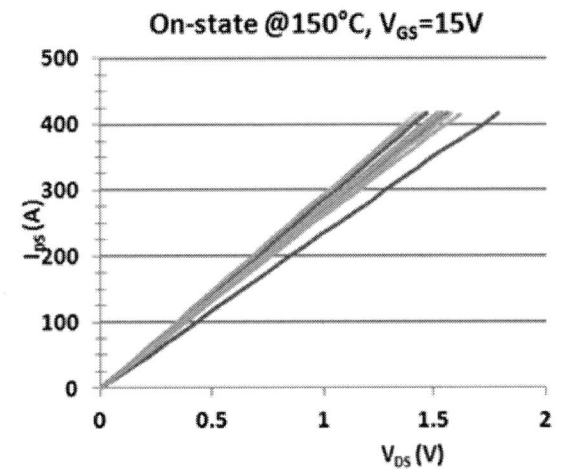

Figure 5a: 900V, 2.5mΩ, ½-bridge power module in 62 mm footprint. Each switch position contains four 10mΩ SiC MOSFETs.

Figure 5b: Measured on-state 150°C current-voltage characteristics of eight, 900V ½-bridge 62mm SiC MOSFET power modules. The number of 900V, 10mΩ SiC MOSFETs is four MOSFETs per switch position.

To compare with Si, a comparably rated 650V, 430A rated EconoDUAL™3 Si IGBT power module, (FF450R07ME4_B11) was examined. Using measured data for the SiC module and datasheet values of the Si module, a comparison is shown in the table below. Despite 250V higher blocking voltage, the SiC module still offers significant (10-20 times lower) advantages in body diode recovery, gate charge, and reverse transfer capacitance. It also offers symmetrical 3rd quadrant conduction. Additionally, per Fig. 5b, ultra-low on-state losses for the eight SiC power MOSFET modules with median 3.6mΩ at 150°C, were measured.

These on-state losses will allow motor drive and EV traction drive inverter markets to exploit a new way to reduce system losses.

Key parameters of the 900V SiC power MOSFET ½ bridge module compared to a 650V Si IGBT ½ bridge module are shown in the table below. Part number XAB350M09HM3 is a preliminary part number as this is a development product.

Parameter	Wolfspeed XAB350M09HM3	FF450R07ME4_B11
Package	HT-3000 (custom)	EconoDUAL3™
Blocking voltage (V)	900	650
T_{JMAX} (°C)	175	175
$R_{DS(ON)}$ (mΩ) (25°C/150°C)	2.5 / 3.6	N/A
I_{DS} at 150°C (A)	405	430
Q_G (nC)	648 (162x4)	4800
Q_R at 150°C (µC)	2.016 (0.504x4)	35.5
Input cap, C_{iss} / C_{ies} (nF)	15.72 (3.93x4)	27.5
Rev transfer cap, C_{rss} / C_{res} (pF)	72 (.018x4)	820

High Temperature Reverse Bias (HTRB) testing of 900V, 2.5mΩ SiC MOSFET ½ bridge module

For pre-qualification testing, a small number of modules were measured under HTRB testing. Six modules (eight transistors per module, or 48 transistors total), were measured at rated voltage (900V) for 1,000 hours, 150°C, with zero failures. After the testing (no failures were observed during or post testing), five of the same six modules (40 transistors) were subjected to an additional 850 hours of stress at 175°C. Again, no failures were observed either during or in post-stress testing.

In Fig. 6 below, the leakage currents of the power modules were measured as a function of V_{DS} up to 900V prior to testing start (Sept. 17, 2015), after the 1000 hours at 150°C (Dec. 21, 2015), and again after 850 hours at 175°C (Feb. 16, 2016).

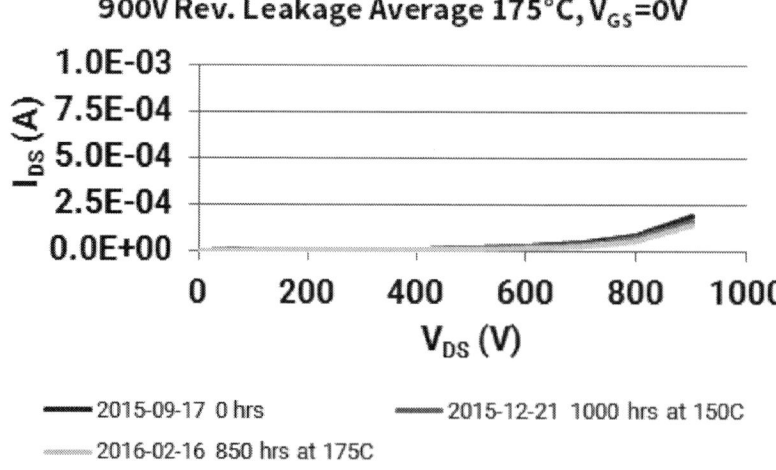

Fig. 6: 900V, 2.5mΩ, ½ bridge SiC MOSFET module leakage measurements prior to HTRB testing (purple line), post 1000 hours 150°C testing (orange line), and post additional 850 hours 175°C HTRB testing. Leakage currents decreased slightly after each HTRB test.

2.2. 900V, 1.25mΩ SiC MOSFET multichip power module

DC Measurements

Using the same module platform as in section 2.1, ½ bridge power modules were assembled with eight transistors (900V, 10mΩ SiC MOSFET chips) per switch position, or 16 total transistors per module. A total of eight modules were assembled. The average DC on-state I-V curves are shown below with gate-source voltages applied from 0V to +15V in +3V increments. In Fig. 7 below, the average on-state characteristics of the modules is shown at 25°C (left graph) and 175°C (right graph). In both cases, 800A current carrying capability is demonstrated.

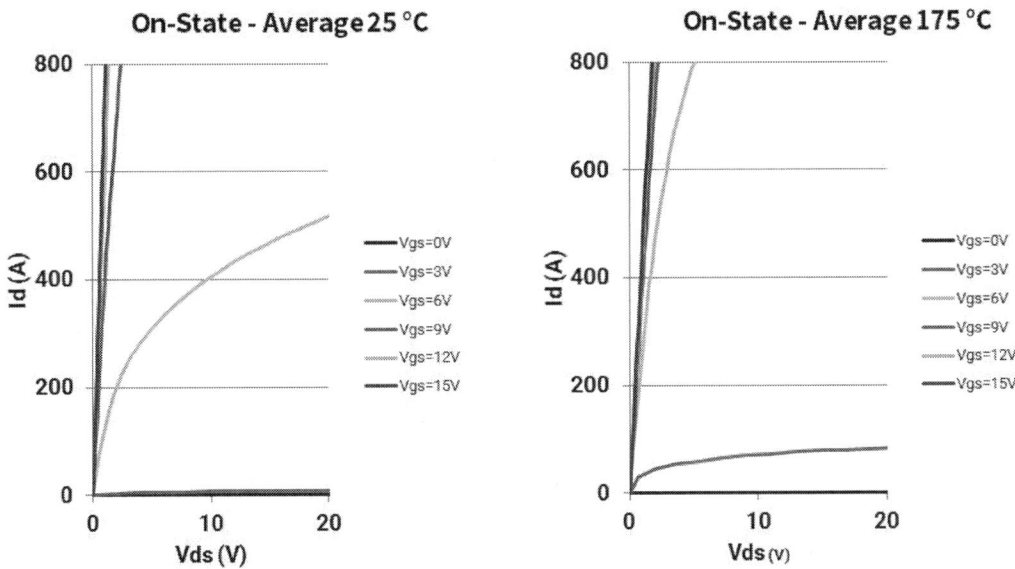

Fig. 7: DC on-state I-V curves for 900V, 1.25mΩ, SiC MOSFET ½ bridge module at 25°C (left) and 175°C (right). V_{GS} stepped from 0V to +15V in +3V increments.

Next the $R_{DS(ON)}$ of the 900V SiC MOSFET ½ bridge modules was measured as a function of drain current, up to 800A, at 25°C and 175°C. In Fig. 8 below, the measured $R_{DS(ON)}$ is shown for each of the eight modules at 25°C (left graph) and 175°C (right graph). For low I_{DS} (~50A), approximately 1.25mΩ median $R_{DS(ON)}$ was measured at 25°C, increasing to a median of 1.35mΩ at 800A, where undoubtedly self-heating was contributing to a temperature rise. A significant portion of the $R_{DS(ON)}$ at these very low values is from the module lead resistance.

At 175°C, the $R_{DS(ON)}$ was measured to be <2.25mΩ, even at 800A, for all modules assembled. These on-state values would allow for significant inverter loss savings in EV drivetrain applications as discussed earlier.

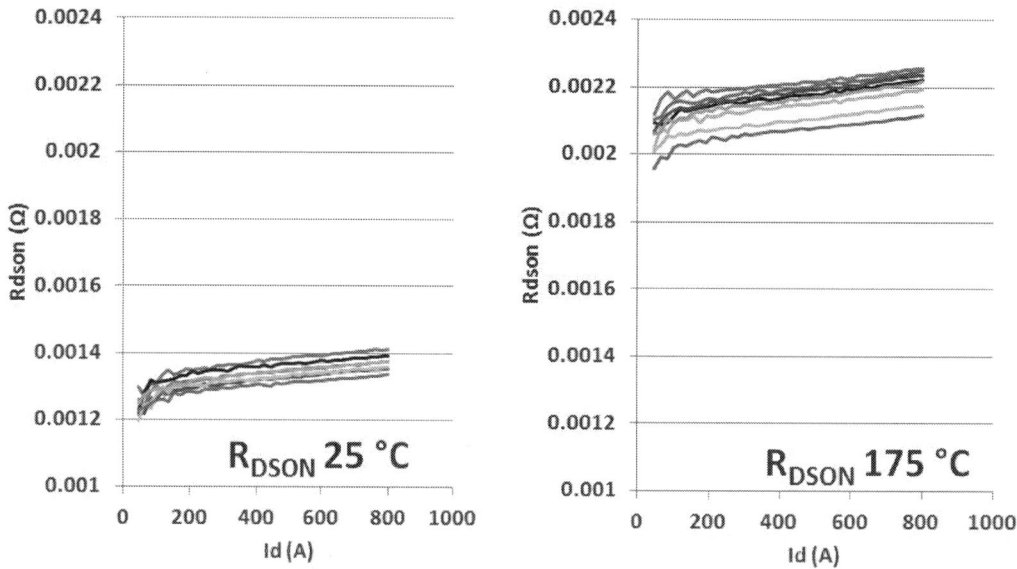

Fig. 8: $R_{DS(ON)}$ measurements as a function of current up to 800A for a 900V, 1.25mΩ, ½ bridge SiC power MOSFET module. 16 transistors per module or eight transistors per switch position. Measurements at 25°C (left) and 175°C (right). V_{GS} set to +15V.

Switching Measurements

Finally switching measurements were made on the 900V, 1.25mΩ SiC MOSFET ½ bridge power module. An internal gate resistor of 1Ω was placed in series with each MOSFET. The external resistance was varied at 1Ω, 2.5Ω, and 5Ω. Naturally the total switching energy increased with higher external gate resistance, but the slower dv/dt also allowed for less overshoot and ringing in the switching waveforms. Even with 5Ω of external gate resistance, a total switching energy of only 52mJ was measured up to 750A, 600V switching (see Fig. 9). The switching energy in the MOSFET based module is fairly temperature independent, although data is not shown for brevity in this publication. The switching waveforms (turn-on and turn-off) are shown in Fig. 10, with 5Ω of external gate resistance. It is important to note these modules do not have any external anti-parallel SiC diodes.

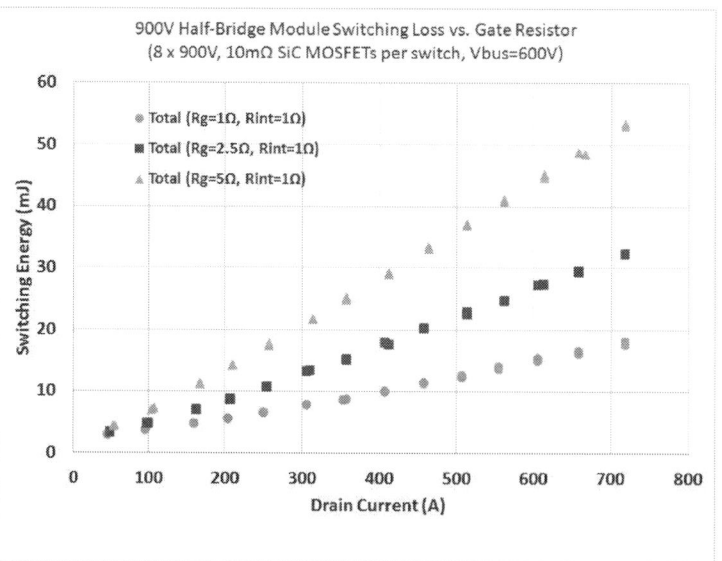

Fig. 9: Switching energy measurements as a function of current up to 800A for a 900V, 1.25mΩ, ½ bridge SiC power MOSFET module. 16 transistors per module or eight transistors per switch position. Measurements at 25°C with a 600V bus voltage and V_{GS} switched from -4V to +15V.

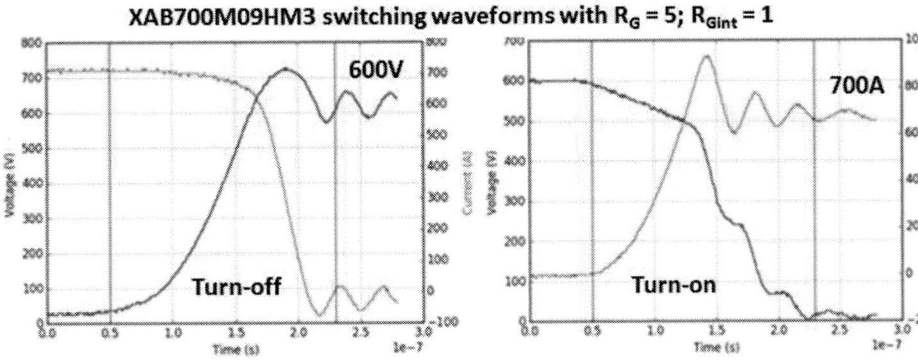

Fig. 10: Switching waveforms for 900V, 1.25mΩ, half-bridge SiC module (XAB700M09HM3) shown above with R_G of 5Ω, between 50A and 700A, and 0V and 600V.

3. Summary

The benefits of using SiC MOSFET based modules were briefly discussed in terms of inverter loss reduction in the EV drivetrain. For bus voltages of 400-700V, a 900V rated SiC MOSFET based module could prove to be ideally rated for these applications.

For the first time, a new 900V, 800A, 1.25mΩ SiC ½ bridge power module was designed, assembled and characterized to 175°C. Using a newly developed 900V, 10mΩ SiC MOSFET chip, only a 50% increase in $R_{DS(ON)}$ was measured from 25°C to 150°C. This performance contrasts favorably with competitive silicon offerings of MOSFETs or IGBTs. HTRB testing on lower current modules for 1000 hours at 150°C, and an additional 850 hours at 175°C on the same modules, revealed no degradation in leakage currents.

At 800A, 175°C, the 1.25mΩ power modules increased in $R_{DS(ON)}$ to only a maximum of 2.10-2.25mΩ for the modules measured, and a significant portion of that resistance was from the lead frame of the module. With these low on-state losses, these new 900V SiC power modules offer interesting options for system-level improvements in EV drive train applications.

References

[1] K Shirabe, M Swamy, J Kang, M Hisatsune, M Das, R Callanan and H Lin, "Design of 400V Class Inverter Drive Using SiC 6-in-1 Power Module," ECCE, 15-19 Sept. 2013.
[2] Kimimori Hamada, "Installation of all-SiC Inverter System to Hybrid Electric Vehicle," International Conference on SiC and Related Materials, 4-9 Oct. 2015.
[3] R. Colin Johnson, "SiC/GaN Poised for Power," *EE Times*, 1 Sept. 2015.
[4] V. Pala, *et al*, "Record-low 10mΩ SiC MOSFET in TO-247, rated at 900V," IEEE Applied Power Electronics Conference (APEC), accepted for publication, 20-24 Mar. 2016.
[5] Infineon 650V, 19mΩ MOSFET, part number IPZ65R019C7, http://www.infineon.com/dgdl/Infineon-IPZ65R019C7-DS-v02_00-en.pdf.
[6] Infineon 650V 100 A IGBT, Part No. IGZ100N65H5, http://www.infineon.com/dgdl/Infineon-IGZ100N65H5-DS-v02_01-EN.pdf.

Acknowledgment

The information, data, or work presented herein was funded in part by the Office of Energy Efficiency and Renewable Energy, U.S. Department of Energy, Award No. DE-EE0006920.

The authors acknowledge the assistance of Drs. Chingchi Chen and Ming Su from Ford Motor Company in assisting our understanding of automotive drivetrain application requirements, and EPA drive-cycle losses.

Evolution of SiC products for industrial application

Naoyuki Kizu, ROHM Co.,Ltd. , Japan, naoyuki.kizu@mnf.rohm.co.jp
Satoru Nate, ROHM Co.,Ltd. , Japan, satoru.nate@dsn.rohm.co.jp
Mineo Miura, ROHM Co.,Ltd. , Japan, mineo.miura@mnf.rohm.co.jp
Masaharu Nakanishi, ROHM semiconductor Gmbh, Germany, masaharu.nakanishi@de.rohm europe.com
Nobuhiro Hase, ROHM semiconductor U.S.A., LLC, U.S.A., nhase@rohmsemiconductor.com
Noriaki Kawamoto, ROHM Co.,Ltd. , Japan, noriaki.kawamoto@mnf.rohm.co.jp
Kazuhide Ino, ROHM Co.,Ltd. , Japan, kazuhide.ino@mnf.rohm.co.jp
…

Abstract

Since commercial manufacture of Silicon Carbide (SiC) started in 2010, we have been extensively expanding its product line-ups. This paper covers recently released, new SiC products and technical updates which reflect the demands of industrial applications. First, our newly released 1700V SiC-MOSFETs driven by SiC-optimized flyback control Integrated Circuits (ICs) for auxiliary power supplies show an outstanding performance compared to conventional Si-based solutions. Second, we have continuously added new products to our SiC-SBD and SiC-MOSFET product line-ups. As with the existing line-ups, our new products also exhibit excellent reliability in high temperature and high humidity environments—which are desirable characteristics for industrial applications. Finally, our new generation SiC-SBD is reported, which has both best-in-class low forward voltage and improved current surge robustness compared to its predecessor.

1. 1700V SiC-MOSFET solution for auxiliary power supply

1.1. Background

Commercial uses of SiC-MOSFETs used to be limited to some extreme applications which allow high device prices, but are now expanding toward high volume, commercial applications such as solar inverters and kW-range power supplies thanks to improved performance to cost balance. However, despite the growing acceptance of SiC-MOSFETs, the performance of SiC-MOSFETs may not be fully realized unless they are used with proper driver ICs. To cope with this bottleneck, we have been developing and commercializing SiC-optimized driver ICs.

We recently released 1700V SiC-MOSFET targeting auxiliary power supplies for industrial equipment, which typically have 3-phase AC input ranging from 400V to 690V. Likewise to extract the best performance, we also released a flyback control IC which is optimized to drive SiC-MOSFETs for auxiliary power supplies.

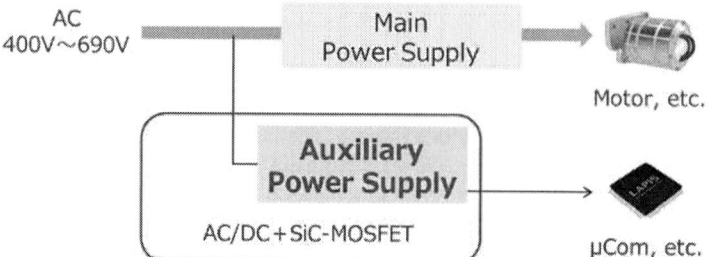

Fig. 1. Block diagram of typical industrial equipment with auxiliary power supply

1.2. 1700V SiC-MOSFET characteristics

Generally, IGBTs provides good balance between high voltage and low resistance ($V_{CE(sat)}$) but is not a choice for several topologies such as switching mode power supplies due to slow turn-off speed. In this type of application, Si based MOSFETs have been the device of choice despite the drawback of increasing on-resistance along with the increase of breakdown voltage.

Because of its high breakdown electrical field, SiC-MOSFET keeps low on-resistance per chip area even with high breakdown voltage. Figure 3 compares on-resistance between our newly developed SiC-MOSFET and market available Si-MOSFET, which is commonly used in auxiliary power supplies in industrial equipment. On-resistance ($R_{DS(on)}$) of the 1700V SiC-MOSFETs is much smaller than that of the 1500V Si-MOSFET – 0.75Ω / 1.15Ω for SiC-MOSFET vs 9Ω for Si-MOSFET, even though chip size of SiC-MOSFET is 17 times smaller than that of Si-MOSFET.

TO-268-2L   TO-3PFM

Figure 2: 1700V SiC-MOSFET products in two discrete packages

Figure 3: On-state V-I characteristics of 1700V 1.15Ω SiC-MOSFET, compared with 1500V 9Ω Si-MOSFET (dotted line)

1.3. 1700V SiC-MOSFET operated with SiC-optimized flyback control IC

Although SiC-MOSFETs have superior characteristics, they need optimized gate drive control. Gate-source voltage is the most significant differences between SiC-MOSFETs and Si-MOSFETs. Our SiC-MOSFETs are optimized with driving voltage of 18V to turn-on while Si-MOSFETs generally require 10V. Our newly released Quasi-resonant flyback control IC is optimized to drive our SiC-MOSFET with adjustable gate-source voltage and under voltage lockout (UVLO) protection, and consequently can draw out the performance benefits of SiC-MOSFETs. Figure 4 compares conversion efficiency between the SiC-MOSFET and the Si-MOSFET using a reference circuit. The SiC-MOSFET improved conversion efficiency by 6% at full load compared to the Si-MOSFET, which suggests potential for downsizing of the power block.

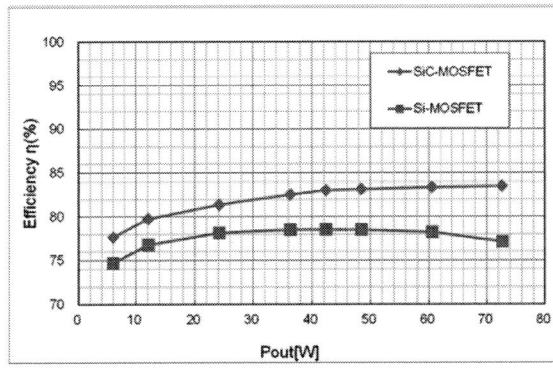

Figure 4: Comparison of conversion efficiency between Si-MOSFET and SiC-MOSFET using a reference circuit

2. Line extension of SiC products and product reliability

2.1. Recent line extension of SiC products

Besides 1700V SiC-MOSFET introduced in the previous section, we recently added several new SiC-SBDs and SiC-MOSFETs.

To meet market demands for higher current, we expanded the SiC-SBD product line-up (2nd gen SiC-SBD) to support higher rated current – 650V class SiC-SBD now ranges up to 100A current rating and 1200V and 1700V class SiC-SBD cover up to 50A current rating. As with our existing SiC products, these new SiC-SBD products with high current rating exhibit excellent reliability in high temperature and high humidity environments.

The most noteworthy product release in our SiC product line-up in 2015 was the world-first commercialization of the SiC "Trench" MOSFET, which is our 3rd gen SiC-MOSFET. As our group previously reported, we developed a double-trench, SiC MOSFET structure, which has both gate and source trench [1]. A conventional trench device structure shown in Figure 5 (a) suffered from poor gate oxide reliability due to high electric field concentration at the bottom corners of the trench structure. Our solution to this technical challenge was to form the source trench with the gate trench as shown in Figure 5 (b), thus named "double trench structure". The most prominent feature of this new product, our 3rd gen SiC "trench" MOSFET, is a 50% reduction in on-resistance per chip area (RonA) compared to the existing planner SiC-MOSFETs (2nd gen SiC-MOSFET) in all temperature range as Figure 6 shows. RonA of 1200V class 3rd gen SiC "trench" MOSFET is $4.1m\Omega cm^2$, and that of 650V class 3rd gen SiC "trench" MOSFET is $3.1m\Omega mm^2$. The initial product line-up covers $22m\Omega$ to $40m\Omega$ for 1200V class and $30m\Omega$ for 650V class.

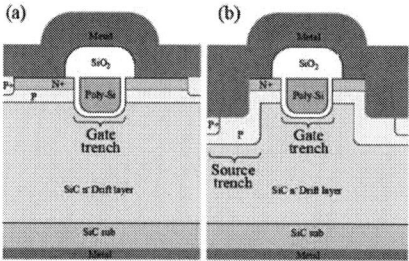

Figure 5: Schematic cross section of (a) conventional single-trench structure and (b) double-trench structure with source trench and gate trench

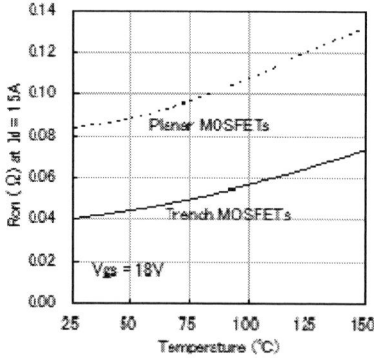

Figure 6: On-resistance of planner MOSFET (2nd gen) and trench MOSFET (3rd gen) with the same die size

2.2. High Temperature High Humidity Reverse Bias Test (H3TRB)

Industrial applications often require guaranteed reliability in high temperature and high humidity environments because of the very harsh conditions where they are sometimes installed – solar inverters installed in high altitude regions would be a good example. High temperature and high humidity reverse bias (H3TRB) test is commonly used as a good indicator of reliability against failures caused under this type of condition. H3TRB reliability at die level is critically more important for case type power modules than epoxy-molded TO packages because silicone gel-filled module cases do not protect dice from humidity penetration very well.

Figure 7 shows H3TRB test results for our newly released 1700V 50A rated 2nd gen SiC-SBD, which was assembled in silicone gel-filled power module, under the conditions of Ta=85°C 85%RH and reverse biased at 1360V. As shown in the graph, leakage current kept stable for the test duration of 3000hrs. The H3TRB test results for 1200V 50A 2nd gen SiC-SBD and 1200V 40mΩ 3rd gen SiC "trench" MOSFET are also shown in Figure 8 and Figure 9 respectively, which again indicated stability of device against H3TRB test.

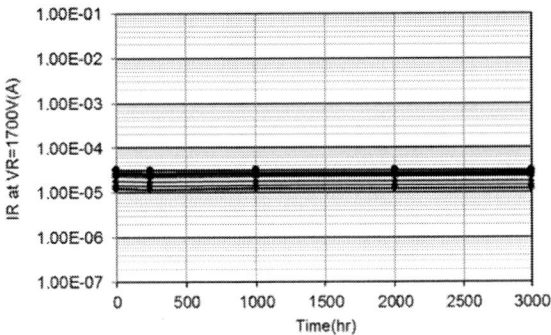

Figure 7: H3TRB test result of 1700V 50A rated 2nd gen SiC-SBD assembled in Si-gel filled power module, at Ta=85°C 85%RH, 1360V reverse biased (80% of max V_R)

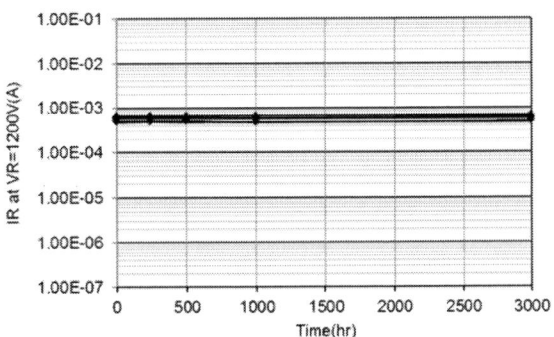

Figure 8: H3TRB test result of 1200V 50A rated 2nd gen SiC-SBD assembled in Si-gel filled power module, at Ta=85°C 85%RH, 960V reverse biased (80% of max V_R)

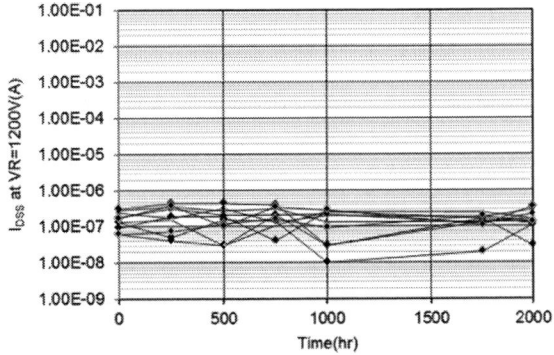

Figure 9: H3TRB test result of 40mΩ 1200V rated 3rd gen SiC "trench" MOSFET assembled in Si-gel filled power module, at Ta=85°C 85%RH, 960V reverse biased (80% of max V_{DSS})

3. New generation SiC SBD

3.1. Background of product development

SiC-SBDs have been widely used for switching mode power supplies (SMPS) as a better replacement for Si fast recovery diodes in power factor corrector (PFC) circuits. For this application, use of by-pass diodes is recommended to prevent SiC-SBDs from failures caused by inrush current as SiC-SBDs have lower current-surge robustness compared to Si based fast recovery diodes. However, some switching mode power supply designs do not allow the addition of by-pass diodes due to cost and/or space constraints. As a result, improvement of SiC-SBD's current surge robustness has been often requested. SiC-SBDs with Merged PiN Schottky (MPS) structure were developed to meet this demand for improved current surge robustness. Compared to MPS, pure Schottky type SiC-SBD (like our 2nd gen SiC-SBD) has an advantage of lower forward voltage but has a drawback of lower robustness against current surge than MPS.

So far SiC-SBD with both improved current surge robustness and low forward voltage has never existed, which is our target to achieve with our 3rd generation device. The 3rd gen SiC-SBD is to be released in April 2016 with package variations for both through-hole and surface-mount packages.

3.2. Characteristics

Comparisons between our 2nd and 3rd gen SiC-SBDs are summarized in Table1 below.

Our 3rd gen SiC-SBD has even lower forward voltage (V_F) than our 2nd gen SiC-SBD with the same device rating, especially at high temperature. Figure 10 compares V_F between 2nd gen and 3rd gen 650V, 10A rated SiC-SBDs at 25°C and 150°C. The 3rd gen SiC-SBD also realized lower leakage current than the 2nd gen SiC-SBD as shown in Figure 11 although there is a general trade-off relationship between forward voltage and leakage current.

Another key feature—current surge robustness—is significantly improved compared to the 2nd gen. Non-repetitive forward surge current peak (I_{FSM}) is doubled with the 3rd gen SiC-SBD.

		ROHM 3rd gen SiC-SBD	ROHM 2nd gen SiC-SBD
Absolute maximum ratings			
I_{FSM} 10msec, sin wave TO-220AC package		82A	40A
Electrical characteristics			
$V_F(V)$ Typ.	Tj=25°C	1.35	1.35
	Tj=150°C	1.44	1.55

Table.1. Characteristics comparison between 2nd gen. and 3rd gen. SiC-SBD

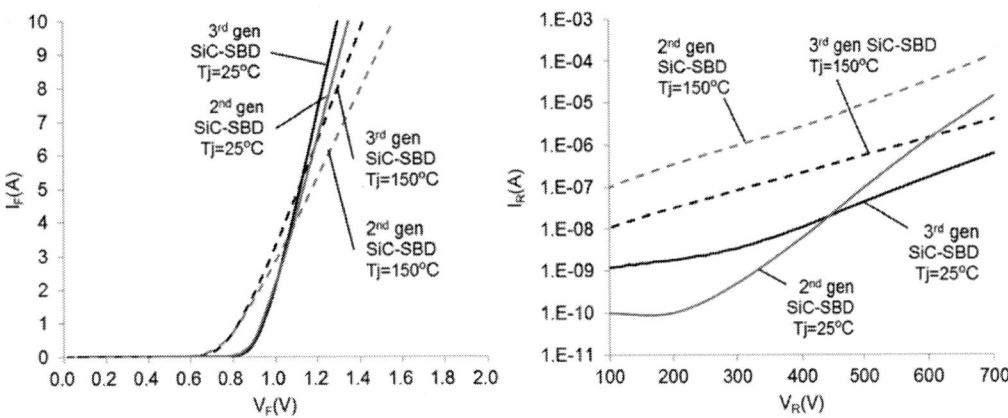

Figure 10: Forward voltage of 650V 10A rated 2nd gen and 3rd gen SiC-SBD at 25°C and 150°C

Figure 11: Leakage current of 650V 10A rated 2nd gen and 3rd gen SiC-SBD at 25°C and 150°C

Conclusion

SiC devices have been a focus of the power semiconductor market due to the realization of dramatically improved material characteristics over Si. For these reasons, we have continuously developed SiC-SBD and MOSFET devices over the past 10 years. Furthermore, many recent applications required us to propose suitable products for realistic scenarios by expanding on the current selection of SiC devices in use today. As a comprehensive semiconductor company, we are well situated to introduce the combination of controller IC and 1700V SiC-MOSFET for industrial power supplies and contribute to improved efficiency in these applications. Additionally, expanding on our line-up of robust devices for high temperature and high humidity environments has led to the evolution of a new generation of devices which is realized in the lower $R_{DS(ON)}$ trench-gate SiC-MOSFET and low V_F SiC-SBD with higher current surge robustness. As SiC market commercialization accelerates, we will continue to deliver a sense of presence through the advancement of our product offerings.

References

[1]   R. Nakamura, Y. Nakano, M. Aketa, N. Kawamoto and K. Ino, PCIM Europe 2014,   pp 441-447

Advanced protection for large current full SiC-modules

Eugen Wiesner, Mitsubishi Electric Europe B. V., Germany, Eugen.Wiesner@meg.mee.com
Dr. Eckhard Thal, Mitsubishi Electric Europe B. V., Germany, Eckhard.Thal@meg.mee.com
Andreas Volke; Power Integrations GmbH, Germany, Andreas.Volke@power.com
Dr. Karsten Fink, Power Integrations GmbH, Germany, Karsten.Fink@power.com

Abstract

This paper presents an advanced overcurrent detection and short circuit protection method for large current full-SiC modules using current sense source terminals. Test results obtained with a newly developed gate driver (for SiC modules) are presented and discussed herewith. This new driver incorporates overcurrent detection, soft-shut-down and active overvoltage clamping.

1. Introduction

The new 800A 1200V full SiC module - FMF800DX-24A was developed [1; 2] for high power applications allowing either high switching frequencies (in the range of 30 to 100 kHz) or high efficiency or high power densities.

Employing SiC technology facilitates a drastic reduction in the switching losses [3] compared to the Si IGBT. On the other hand, the static losses should be carefully adjusted without sacrificing the ability to handle short circuit conditions. The low on resistance of the SiC MOSFET is inversely proportional to the chip short circuit capability. The Fig. 1 shows the trade-off between the SiC MOSFET on resistance $R_{DS(on)}$ and the short circuit capability E_{SC} of the chip.

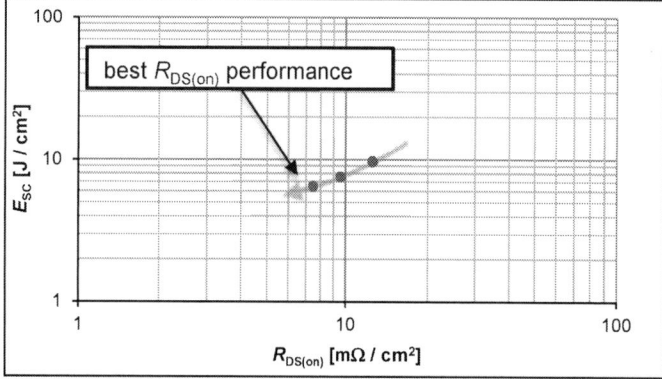

Fig. 1. Trade off between SiC MOSFET on resistance and short circuit capability

Taking into account the limited SC-endurance capability of today's SiC MOSFET-chips (in the range of a few microseconds) the availability of a separate current sense terminal is a promising option for reducing the response time and accordingly the energy dissipated during SC-turn-off in the MOSFET chip. Furthermore the MOSFET on resistance can be tuned for lower values. By using this option, overcurrent conditions (adjustable to any level) can be detected easily and appropriate countermeasures for SC-turn-off can be initiated in the gate driver. As a result the SC-current level and the SC-energy dissipated in the MOSFET can be remarkably reduced.

2. Full SiC-MOSFET short circuit capability

The FMF800DX-24A uses eight connected parallel 100 A SiC MOSFET-chips for one switch. The typical gate source threshold voltage $V_{GS(th)}$ is in the range of 1 V. For the short circuit capability of the module the total distribution of $V_{GS(th)}$ must be taken into consideration. Fig. 2 shows the short circuit limitation of an FMF800DX-24A module as a function of the gate source threshold voltage $V_{GS(th)}$. The measured maximum short circuit time (solid line) is between 3 µs and 4 µs. The measured maximum short circuit energy is about 18,4 J. Based on the device limiting constrains in the datasheet, a maximum short circuit time of $t_{SC(max)} = 2$ µs is specified which shall not be exceeded by the user. But by using the conventional SC-detection method which detects the desaturation of the MOSFET, it is not easy to realize a safe turn-off during such a short period. This was the motivation to propose an advanced SC-protection method employing current sense source terminals.

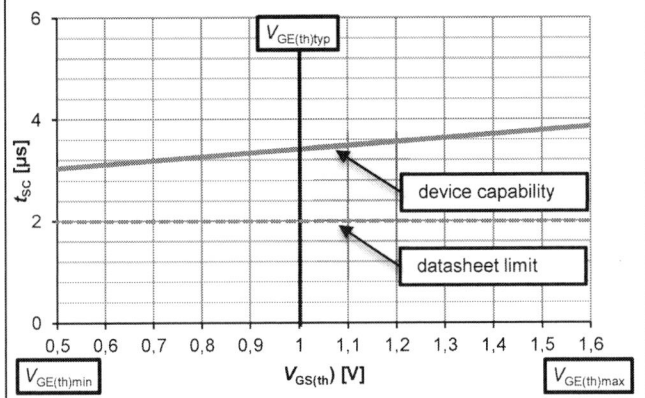

Fig. 2. Short circuit capability of FMF800DX-24A full SiC Module. V_{GS}=+15 V, T_J=150 °C, V_{DD}=850 V

3. Advanced method for detecting overcurrent or short circuit

The 100A 1200V SiC MOSFET chips used in the FMF800DX-24A have an isolated source area on top of the source metallization. This small source area is connected to the sense terminal. Thus an earmarked portion of the total source is provided at the sense terminal. A picture of the MOSFET chip and its equivalent simplified electrical circuit is shown in Fig. 3.

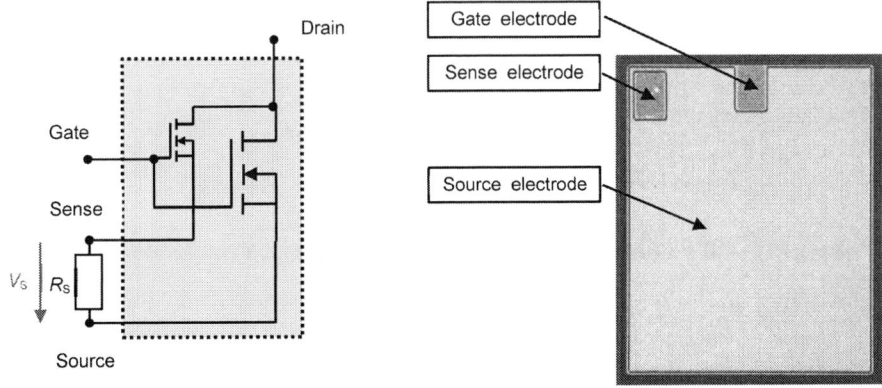

Fig. 3. SiC MOSFET chip with current sense terminal

The monitored source sense voltage across the sense and source terminals can be used for detecting overcurrents. The current through the sense resistance is proportional to the main source current. The ratio between the sense current and main source current is in the range of 1:61500. It has to be noted that the sense voltage (V_S) across the shunt resistance (R_S) depends on the junction temperature (T_J) as shown in Fig. 4. By considering this current dependency and T_J – dependency, an appropriate shunt resistor value (R_S) can be selected setting the needed overcurrent trip level.

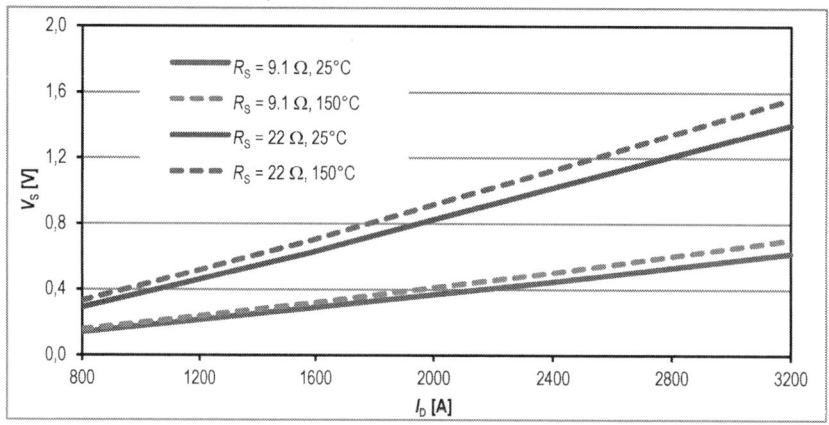

Fig. 4. Relationship between drain current and sense voltage

4. Gate driver realization for FMF800DX-24A protection

The efficiency of the advanced protection method of FMF800DX-24A by current sensing was practically realized by using the reference design RDHP-1417 with implemented core gate driver 2SC0435T [4; 5] providing overvoltage protection and overcurrent detection. For the overvoltage protection, active clamping is employed [6]. For overcurrent detection, the designated sense voltage triggers the soft shut down function of the gate driver. Principle equivalent circuit of the gate driver employed is shown in Fig. 5.

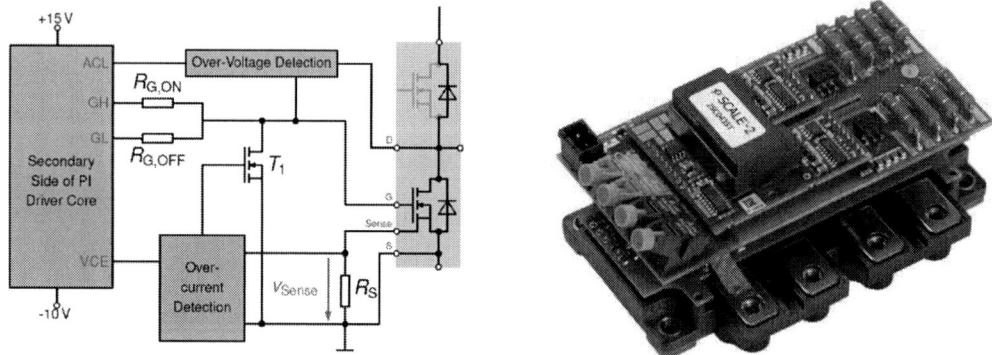

Fig. 5. A photograph of the gate driver employed (right) and its corresponding equivalent circuit diagram (left).

5. Overvoltage protection

Turning-off high load currents with a high di/dt will result in large over voltage spikes across the device being switched. If the power module is operated below the set overcurrent trip level, it is recommended to add additional measures to the gate driver in order to prevent the turn-off over voltages from exceeding the rated break-down voltage of 1200 V. The preferred method is the implementation of Active Clamping along with an additional dv/dt feedback (in this case the gate driver does not actively turn-off the power module and no fault signal is reported to the host controller).

Fig. 6 shows turn off waveforms at 800 A and 1600 A with FMF800DX-24A device. In both cases the overvoltage protection limits the overvoltage spike safely below the break-down voltage of 1200 V.

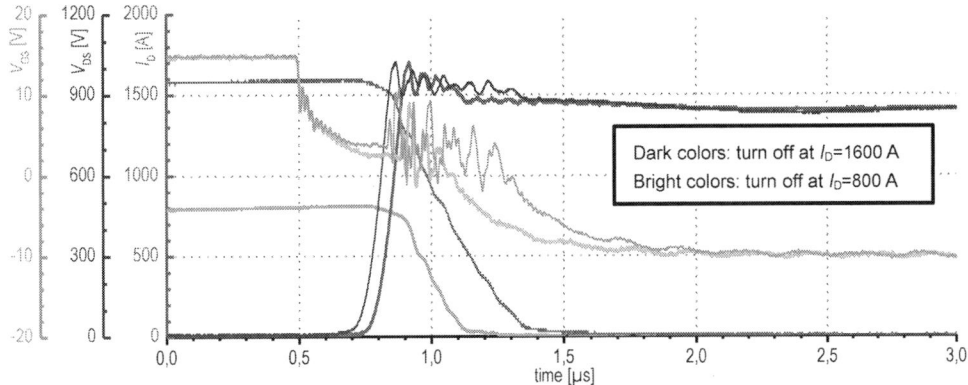

Fig. 6. MOSFET turn off wave forms at nominal and two times the nominal module current.

Conditions: V_{DD}=850 V, T_J=25 °C.

6. Overcurrent detection

At the evaluated gate driver the overcurrent trip level was set to about 2000 A (2,5 times the rated current). Once the trip level is reached, the gate driver actively turns-off using the soft shut down function and transmits a fault signal to the host controller. Fig. 7 shows an example of a turn-off event where the actual load current is slightly below the trip level of the overcurrent detection circuitry. Furthermore Fig. 7 shows the turn off event at the point of overcurrent detection and soft shut down for comparison. The power module has no significant overvoltage spike at such a soft shut down event.

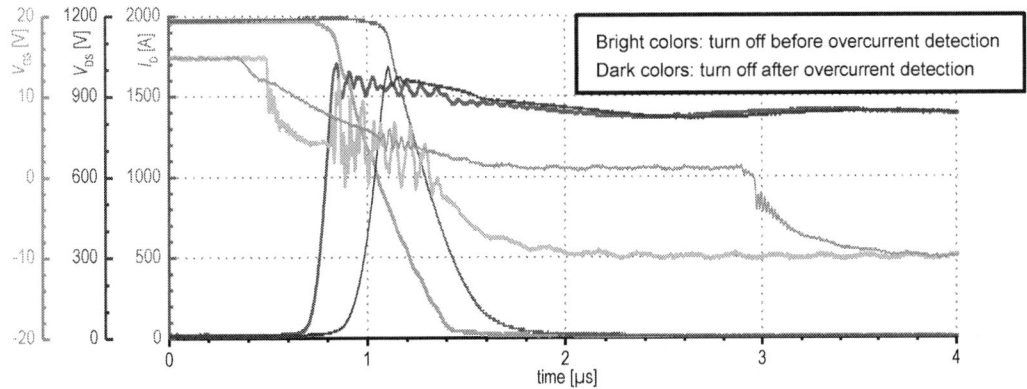

Fig. 7. Overcurrent detection and soft shut down at 2,5 times of rated current

Conditions: V_{DD}=850 V, T_J=25 °C.

© VDE VERLAG GMBH · Berlin · Offenbach

7. Short circuit protection

To verify the effectiveness of the overcurrent detection in case of a short circuit event, the short circuit test was performed at a DC-link voltage of V_{DD}=850 V and a short circuit inductance (L_{SC}) of about 170 nH. The result of this measurement is shown in Fig. 8. The overcurrent was detected at about 2000 A after which an immediate soft turn off was initiated. The short circuit time was about 1µs. And the calculated short circuit energy (E_{SC}) is 0,65 J. The two critical aspects: short circuit time and short circuit energy are significantly lower than the critical boundaries with respect to the module. By using the soft shut down function, overvoltage during a short circuit turn off event was maintained well below 1000 V.

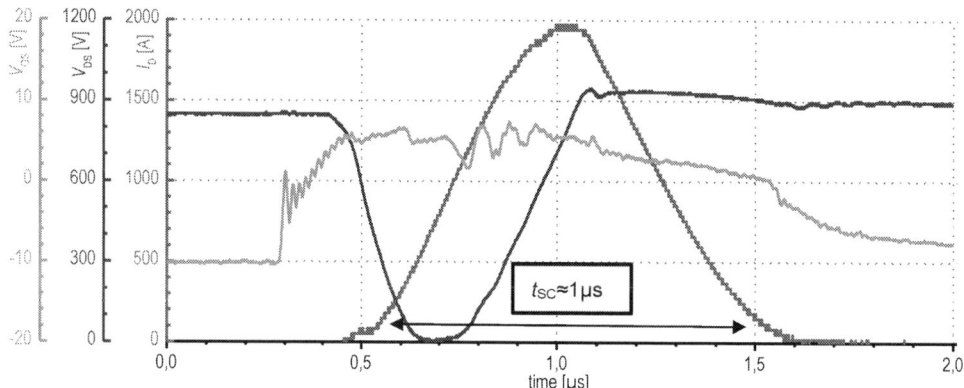

Fig. 8. Evaluation result during a short circuit event. Conditions: V_{DD}=850 V, T_J=25 °C.

8. Conclusion

The newly developed Mitsubishi Electric 800A 1200V full SiC Module FMF800DX-24A offers a unique ultra-compact solution for safely operating highly efficient power converters in the range of several 100kW. For this full SiC module, Power Integrations has developed a gate driver reference design (RDHP-1417). It uses the current mirror source terminal of FMF800DX-24A for providing overcurrent detection in short circuit conditions. Furthermore the implemented active clamping and soft shut down functions are able to limit overvoltage spikes when high currents are turned-off.

References

[1] Mitsubsihi Electric Corporation: Mitsubishi Electric to Begin Shipment of Silicon Carbide Power Module Samples. Press release No. 2687, Tokyo, July 9, 2012
[2] Thal, E.; Masuda, K.; Wiesner, E.: New 800A/1200V Full SiC Module. Bodo's Power Systems, April 2015, p.24-28
[3] Wiesner, E.; Joko,M.; Masuda, K.: New 1200V full SiC module with 800A rated current. EPE'15 Europe, ISBN: 9789075815238
[4] Power Integrations: 2SC0435T Datasheet. https://igbt-driver.power.com/products/scale-2-driver-cores/2sc0435t/
[5] Power Integrations: SCALE™-2 2SC0435T Description & Application Manual. https://igbt-driver.power.com/products/scale-2-driver-cores/2sc0435t/
[6] Rüedi H.; Thalheim J.; Garcia O.: Advantages of Advanced Active Clamping. Power Electronics Europe, Nov/Dec 2009, p.27-29

Switching performance of a 1200 V SiC-Trench-MOSFET in a low-power module

Daniel Heer, Infineon Technologies AG, Germany, Daniel.Heer@Infineon.com
Daniel Domes, Infineon Technologies AG, Germany, Daniel.Domes@Infineon.com
Dethard Peters, Infineon Technologies AG, Germany, Dethard.Peters@Infineon.com

Abstract

In modern power electronic systems, fast switching semiconductors like unipolar diodes and transistors are required in order to reduce losses or to save filter size, weight and cost by using higher switching frequencies. In this paper, results of a detailed characterization of a 1200 V SiC-Trench-MOSFET are presented. The device under test is a half-bridge module with an on-resistance of 45 mΩ per chip resulting in a rated current of 25 A per chip. In general, the SiC-Trench-MOSFET shows superior performance in terms of switching behavior and overall losses. The analysis confirms the full controllability of the voltage slopes in turn-on and turn-off by means of the gate resistor. Compared with the SiC-JFET [1], the SiC-Trench-MOSFET allows a significant reduction of the switching losses. The device concept of the SiC-Trench-MOSFET shows considerably suppressed parasitic turn-on under typical operating conditions. This results in drastically reduced recovery losses leading to very low total switching losses.

1. Introduction

A challenge of the development of SiC-Power-MOSFETs is to balance gate-oxide reliability and low on-resistance. Due to the MOS-interface defect structure the MOS-channel resistance contributes largely to the total on-resistance. For achieving an acceptable on-resistance, the inversion in the channel region is forced by a design using high electric fields, even in the gate oxide. Hence there is the risk of too high gate oxide stress which can lead to poor long term reliability [2]. The Infineon SiC-Trench-MOSFET-concept combines both, an attractive on-resistance and an optimization against too high gate oxide field stress.

2. Device design philosophy

Switching losses of SiC-MOSFETs are usually quite low and especially almost temperature independent. Today R&D activities focus on the area specific on-resistance as the major benchmark parameter for a given technology. For 4H-SiC based planar MOSFETs being manufactured on the so called silicon face of the SiC-wafer, one has still to deal with an extraordinary high interface trap density close to the conduction band. This ends up at very low channel mobilities and therefore high contributions of the channel to the total on-resistance. Even the progress made many years ago by the introduction of nitride oxides was not able to eliminate this drawback in an acceptable manner. The high defect density is reflected in various peculiarities of SiC-MOSFET based devices. One example are weak transconductance characteristics in comparison to silicon based power MOSFETs, often in combination with a low threshold voltage of around only 2 V instead of 4 V and above as usually this is the case in silicon based power devices. Another implication is an abnormal temperature behavior of the on-resistance. While physics usually indicate an increase of

R_{DSon} with temperature due to increased scattering effects, some available SiC-MOSFETs show no or even negative temperature coefficients. An observed way to overcome this dilemma is to increase the applied electric field across the oxide in on-state above values being usually used in silicon based MOSFET-devices. Such high fields in the oxide in the on-state can potentially accelerate the wear out. This can be seen as a long term reliability risk, in particular with respect to the high defect density of SiC-substrates.

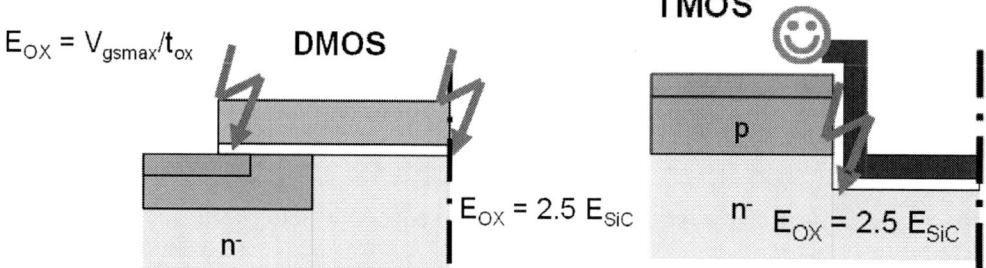

Fig. 1 (Left) Typical structure of planar MOSFET showing two sensitive areas with respect to oxide field stress. (Right) Typical structure of a Trench-MOSFET, critical issue is the oxide field stress at the trench corners.

Based on those considerations, it is obvious that planar MOSFET-devices in SiC have actually two sensitive areas with respect to oxide field stress, as sketched in Fig. 1 (Left): 1st the usually discussed stress in reverse mode in the area of highest electric field close to the interface between drift region and gate oxide and 2nd the overlap between gate and source which is stressed in on-state. The authors expect that the stress in blocking mode is the less critical one since it can be mitigated by a proper device design, e.g. shielding by p-layers. In addition, in the practical operation of a device nearly never the full field will be applied. Even in the most critical target application regarding DC reverse bias stress over a longer period of time like solar booster circuits maximum 1000 V will occur with 800 V being the normal average bias level. However, a high electric field in on-state is seen as more dangerous since no device design measures are in place which could reduce the field stress during on-state.Thus, the overall goal is to combine the low R_{DSon} potentially offered by SiC with an operation mode where the part remains in the well investigated safe oxide field conditions. In the on-state, this can be realized today by moving away from the planar surface with its high defect density towards other more favorable surface orientations. MOS-channels on the so called a-face of SiC which lies 90° inclined towards the planar surface offer a factor of minimum 10 lower defect densities. Thus, one widely investigated approach is to use a trench based structure, similar to nearly all modern silicon power devices. This approach does not only open up the door towards a low channel resistance, also the options for cell shrink are extended compared to a planar device design, resulting in a more economic material utilization and therefore again a lower area specific on resistance compared to DMOS based components. Attractive $R_{DSon}*A$ values can be achieved without the need to apply critical oxide fields in on-state.

However, in trench-based components usually the field stress on the oxide at trench corners is a very critical issue (see Fig. 1 (Right)), and especially in SiC this can be a killer argument since in the semiconductor it is required to utilize the about 10* higher critical electric field of SiC nearly completely in order to leverage the benefits of the structure. However, there are various possibilities in place to realize an effective shielding of the critical areas, e.g. by deep pn-elements. Thus, in contrast to the on-state dilemma in the DMOS, the off-state challenge can be addressed by a smart design.

A Trench-MOSFET design was developed which limits the electric field in the gate oxide in on-state as well as in off-state. An attractive specific on-resistance of 3.5 mΩcm² for the voltage class 1200 V is provided, achievable even in mass production in a stable and reproducible way. The on-resistance is already achieved at V_{GS} of 15 V. The gate-source-threshold voltage is close to 4 V. These boundary conditions are the baseline for transferring quality assurance methodologies established in the silicon power semiconductor world in order to guarantee FIT rates expected in industrial and even automotive applications.

3. Characterization

The device under test is a SiC-Trench-MOSFET with typical $R_{DS,on}$ of 45 mΩ at V_{GS}=15 V and T_{vj}=25°C. A pure ohmic character is observable in the 1st and 3rd quadrant at V_{GS}=15 V. The threshold voltage of tested devices amounts V_{GSth}=3.5 V (at 25°C, V_G=V_D, I_D=1 mA). The typical off-state-voltage is V_{GS}=-5 V. In off-state, there is an internal pn-diode for reverse conduction in case of half bridge interlock.

3.1. Switching behavior

The analysis of the switching behavior of the SiC-Trench-MOSFET is done testing a half-bridge module, equipped with a single chip per leg. This guarantees the analysis of switching behavior excluding influences arising from the parallel connection of chips. The characterization was performed by utilizing the double-pulse method switching an inductive load. The internal pn-diode of the SiC-Trench-MOSFET is used for reverse conduction, no additional diode is necessary. In Fig. 2, the turn-on waveforms of the SiC-Trench-MOSFET are shown. The comparison of the drain-source voltage slope dv_{DS}/dt from low- and highside-MOSFET results in a significant difference. There are -24 kV/µs for the lowside and 56 kV/µs for the highside-switch. This mismatch can be explained by the influence of the stray inductance of the power circuit [3].

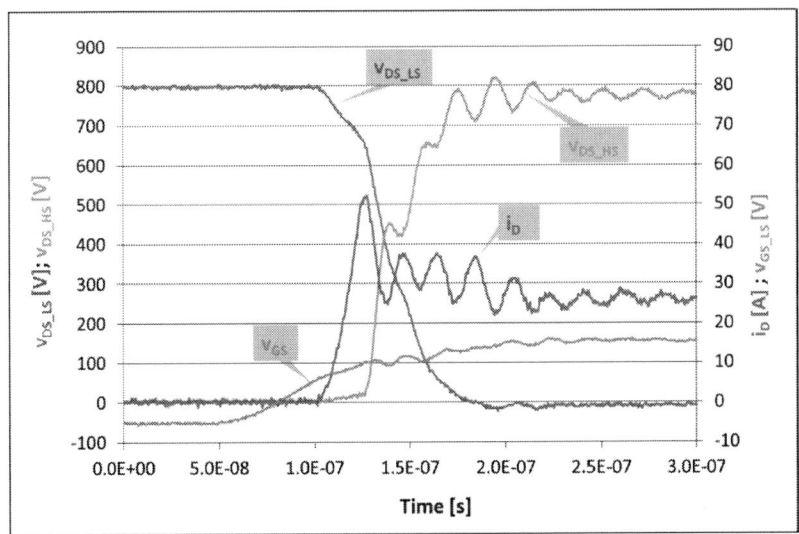

Fig. 2: Turn-on SiC-Trench-MOSFET, half-bridge module with single chip per leg, T_{vj}=25°C, V_{CC}=800 V, I_D=25 A, R_G=10 Ω, resulting dv_{DS}/dt=-24 kV/µs for the active switch, E_{on}=838 µJ

In Fig. 3 the turn-off waveforms of the SiC-Trench-MOSFET are presented. While the drain-source voltage increases, a dip in the drain current can be observed. This is caused by the effect of the discharge of the output-capacitance from the highside-MOSFET and the parasitic capacitance of the inductive load as long as dv_{DS}/dt is present. Therefore the active switch loses a big portion of the conducted current, although the current commutation to the diode is not possible yet. This has a direct impact on the turn-off losses. Further investigation confirms, in fast turn-off conditions, the di_D/dt seems not to be controllable by the gate resistor. In fact, the channel-current is controllable but the influence of the output-capacitance dominates the total current. At the end of the turn-off-process an undershoot of the drain current can be observed. The reason for this is also to be found in the effect of the output-capacitance. After reaching the maximum in v_{DS}, the output capacitance is discharged while drain-source voltage decreases.

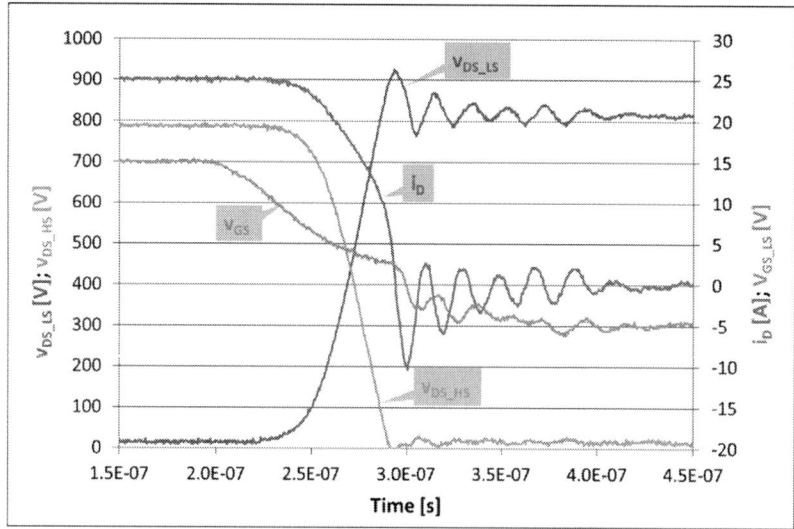

Fig. 3: Turn-off SiC-Trench-MOSFET, half-bridge module with single chip per leg, T_{vj}=25°C, V_{CC}=800 V, I_D=25 A, R_G=10 Ω, resulting dv_{DS}/dt=23 kV/µs for the active switch, E_{off}=312 µJ

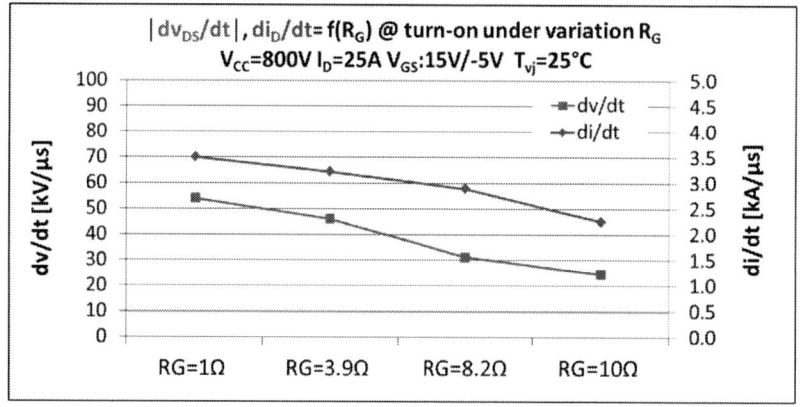

Fig. 4: dv_{DS}/dt and di_D/dt at turn-on of SiC-Trench-MOSFET under variation of R_G

In Fig. 4 and 5 the dv_{DS}/dt in turn-on und turn-off as a function of external gate resistor is presented. The analysis shows the full controllability of the device. With further increase of the external gate resistor, values of 5 kV/µs and below are possible e.g. used in drives applications.

The dynamic losses of the 1200 V SiC-Trench-MOSFET as a function of the gate resistance are presented in Fig. 6. The total losses are dominated by the turn-on losses. With increasing gate resistor from 1 Ω to 10 Ω, the $|dv_{DS}/dt|$ in turn-on und turn-off are decreasing to 50% compared with the value at R_G=1 Ω. With increasing R_G, the dynamic losses change from 800 µJ to 1200 µJ.

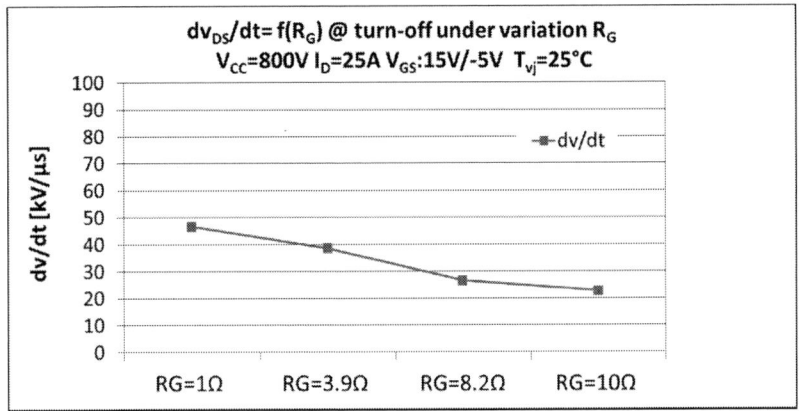

Fig. 5: dv_{DS}/dt at turn-off of SiC-Trench-MOSFET under variation of R_G

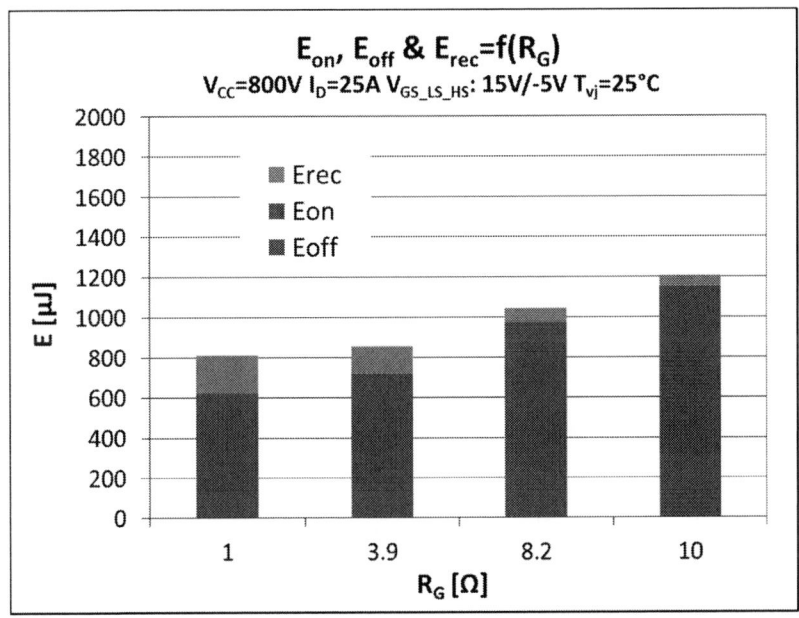

Fig. 6: Dynamic losses of SiC-Trench-MOSFET at T_{vj}=25°C, V_{CC}=800 V, I_D=25 A as a function of R_G

3.2. Comparing SiC-JFET and SiC-Trench-MOSFET

In Fig. 7 the dynamic losses of the SiC-JFET [1] are compared with the dynamic losses of a SiC-Trench-MOSFET. The characterization is done under the same conditions, laboratory setup and module-type is similar. During the turn-off, a $dv_{DS}/dt=-38$ kV/µs could be achieved for both switches with $R_G=3.9$ Ω.

At turn-on, the SiC-Trench-MOSFET shows $dv_{DS}/dt=-46$ kV/µs for $R_G=3.9$ Ω. Although the JFETs gate resistor was decreased down to 1 Ω, a $dv_{DS}/dt=29$ kV/µs was achievable only. Thus, the gate resistors for comparing the turn-on losses were chosen to be $R_G=1$ Ω for the JFET and $R_G=3.9$ Ω for the MOSFET, respectively. The analysis identified a significant difference in the recovery losses for both SiC-switches. The reason for this is that the recovery losses of the SiC-JFET are dominated by the parasitic turn-on of the passive switch [3], whereas the SiC-Trench-MOSFET E_{rec} is caused by displacement current only. In total, the SiC-Trench-MOSFET shows only 45% of the switching losses of the SiC JFET.

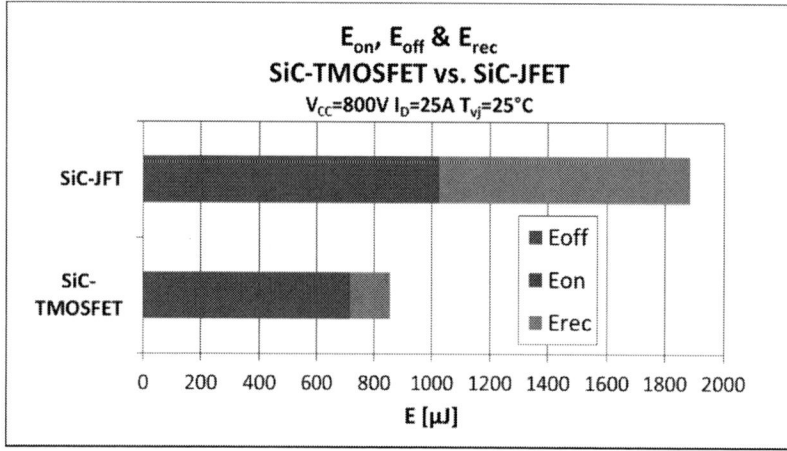

Fig. 7: Dynamic losses of normally-on SiC-JFET vs. SiC-Trench-MOSFET at $T_{vj}=25°C$, $V_{CC}=800$ V, $I_D=25$ A

3.3. Switching performance of four chips per leg in parallel

In Fig. 8 the dynamic losses of a half-bridge module with a considerably higher power handling capability are presented. The module is equipped with four SiC-TMOSFETs in parallel per leg, resulting in a nominal current of approx. 100 A. The characterization is done at $T_{vj}=150°C$, $V_{CC}=800$ V, $I_D=100$ A under variation of the gate resistance. The dynamic losses change from 3.9 mJ to 9.7 mJ.

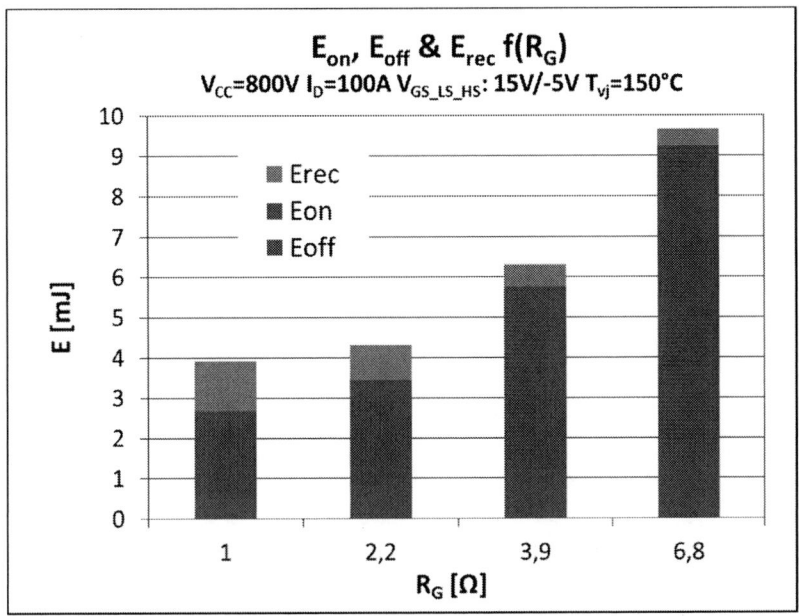

Fig. 8: Dynamic losses of a half-bridge module equipped with the SiC-Trench-MOSFET at T_{vj}=150°C, V_{CC}=800 V, I_D=100 A as a function of R_G

4. Conclusion

In this paper the results of a detailed characterization of the 1200 V SiC-Trench-MOSFET are presented. The concept of Infineon's SiC-Trench-MOSFET combines an attractive on-resistance with an optimized design against too high gate oxide field stress which provides IGBT like gate oxide reliability. The SiC-Trench-MOSFET shows superior performance in terms of switching behavior and losses. The analysis confirms a full controllability of the voltage slopes of turn-on and turn-off transients. The current slopes for turn-on can be controlled by the gate resistor. In turn-off, the di_D/dt is determined by parasitic capacitive effects. Compared with the SiC-JFET, the SiC-Trench-MOSFET shows a further improvement regarding switching losses. Total dynamic losses of the SiC-Trench-MOSFET are less than half the total switching losses of the SiC-JFET.

5. Acknowledgment

The authors would like to thank all Infineon colleagues for their contribution which made this paper possible, in particular the members of the concept-, technology- and module-development team at the Infineon sites: Erlangen, Villach and Warstein.

6. References

[1] D. Domes et. al.: 1st industrialized 1200 V SiC JFET module for high energy efficiency applications, PCIM 2011, Nuremberg, Germany

[2] J. Lutz et. al.: Reliability of SiC Devices, ECPE Workshop Power Semiconductor Robustness, ECPE SiC & GaN User Forum 2014, Warwick, Unit Kingdom

[3] D. Heer et. al.: SiC-JFET in half-bridge configuration – parasitic turn-on at current commutation, PCIM 2014, Nuremberg, Germany

PCIM Europe 2016, 10 – 12 May 2016, Nuremberg, Germany

Beyond Thermal Grease:

Enhancing Thermal Performance and Reliability

Dr. Sanjay Misra, Henkel, USA, sanjay.misra@henkel.com

Abstract

Power electronics based on silicon devices must operate below 125 C and IGBTs under 150 C. Future SiC devices could extend this to 200 C. Thermal management of power electronics requires interfacing the package to a heat sink using a thermal interface material (TIM). In general this interface is the crucial in terms of the impact on overall thermal impedance and long term reliability. While several greases provide good end of line performance, pump out and separation can degrade thermal performance. Phase change materials as well as gel like, cure-in-place, TIMs can not only match end of line performance of greases, but significantly enhance long term reliability. Finally, significant automation in the case of these TIMs can enhance productivity and manufacturing. In this presentation we will present options for phase change materials – both roll processed as sheets and die-cut parts as well as compounds that can be screened/stenciled by the end user. In addition, new developments in thin bond line, high thermal conductivity, cure-in-place gels for high reliability.

1. IGBT Thermal Challenge

The thermal stack up in an IGBT is a sum of several resistances in series. This is shown in the schematic below. A key compnent of the thermal stack up is the thermal interface material – typically a thermal grease – as shown in Fig.1

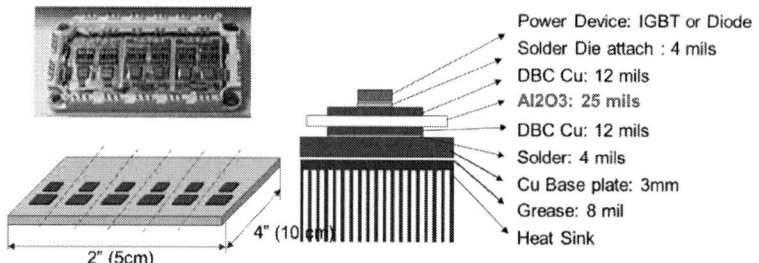

Fig. 1. Typical IGBT device and power assembly using multiple IGBTs

The effective thermal bond line can be quite significant given the size and warpage of the assembly. The thermal resisatnce of the interface material can therefore a significant portion of the total. This is ilustrated below in Fig.2 which shows that the end-of-line thermal resistance of the thermal interface can be 20-40% of the total.

© VDE VERLAG GMBH · Berlin · Offenbach

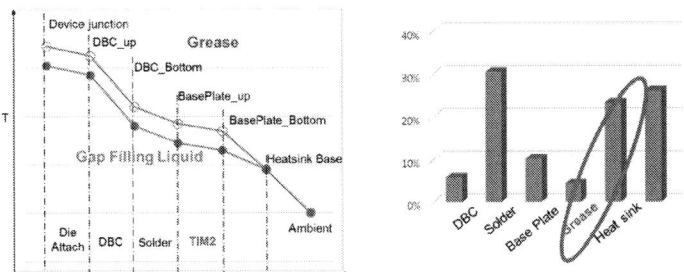

Fig. 2. Thermal simulation of a typical IGBT stack up

The thermal stack up includes materials of several different CTEs and, in addition, there are temperature gradients within the whole assembly. As the assembly goes through power and thermal cycling these can lead to cyclic mechanical load on the interface. The thermal material can pump out of the interface thereby degrading the thermal resistance and consequently the device performance. Therefore the thermal interface material should be chosen carefully to balance and optimize lifetime thermal performance.

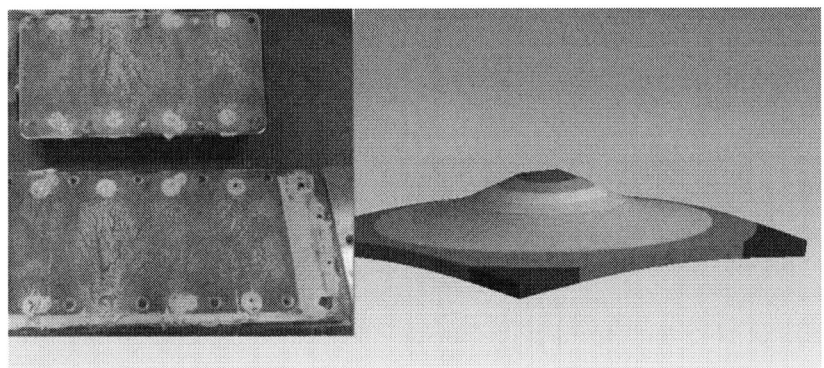

Fig. 3 Simulated warpage during thermal cycling and associated pump out

2. Thermal Interface Material Design

For thermal performance and long term reliability the TIM has to be designed to optimize several interdependent variables. The three main criteria are (1) Thermal conductivity of the TIM, (2) the rheology of the TIM, i.e. the deformation behavior of the material under stress and, (3) response to long term cycling. The interface between surfaces has gaps on two different length scales. The first is small-scale roughness typically O (1 μm) – from which air is eliminated by flow and wetting by the interface material. The second is related to larger gaps - O (100 -1000 μm) - due either to the non-planarity of surfaces and poor co-planarity. The thermal interface material needs to be able to conform to the surfaces, with a low external stress to produce deformation without straining the electronic components.

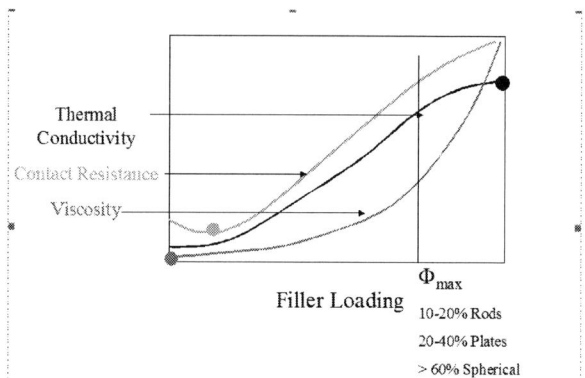

Fig. 4. Balancing thermal interface material properties

The thermal performance and rheology of thermal interface materials depend of both the polymer microstructure as well as the morphology and loading of the filler material. These properties are interdependent and cannot be manipulated in isolation. Since optimal performance depends on a combination of thermal conductivity and rheology one must match the application carefully to formulate the TIM.

3. Application Performance

Thermal interface materials are available in several general categories:

Thermal Greases: These are relatively lower viscosity polymer liquids that have been loaded with thermally conductive particulates. These are thermoplastic, i.e. they will deform continuously under external stresses without limit. They unfortunately are not only difficult and messy in manufacturing but also tend to migrate out of interface due to pump out – creating reliability issues.

Phase Change: These are materials generally solid at room temperature and melt at the phase change temperature. These are available as sheets or rolls and also as screenable or stencilable compounds. They are easier to work with, can be automated and are generally better in terms of pump out resistance. In manufacturing, they may need to be over-torqued or re-torques after phase change.

Curable/Reactive Liquids: These may be either adhesives or form-in-place gap fillers and may be one component (1-k) or two component (2-k). These are thermoset materials that crosslink into a network that does not deform. These perform best – performance like grease but strong pump out resistance. In addition there is little mess and great opportunity for tailoring properties for automated manufacturing.

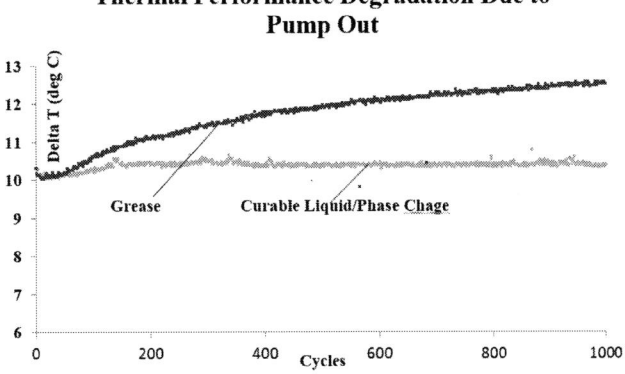

Fig. 5. Effect of pump out on various TIMs

4. Summary

For best thermal performance 1-k and 2-k curable gap filling liquids are the best alternative to greases. Phase change compounds offer an attractive alternative as well from reliability perspective.

5. References

1. S. Narumanchi, M. Mihalic, K. Kelley and G. Eesley. *Thermal Interface Materials for Power Electronics Applications.* NREL CP-540-42972, July 2008.
2. M. Schultz. *Thermal Interface – An Inconvenient Truth.* Bodo's Power System. June 2010.
3. Y. Nishimura, M. Oonota, F. Momose. *Thermal Management Technology for IGBT Modules.* Fuji Electric Review, FER-56-2-079, 2010.
4. J. E.Sergent, A. Krum. *Thermal Management Handbook: For Electronic Assemblies.* Mc Graw Hill 2000.
5. D. M. Biggs. *Thermally Conductive Polymer Compositions. Polymer Composites,* 1986, Vol 7, No 3, p.125.

High Thermal Conductivity Silicon Nitride substrates for Power Semiconductor Applications

Dai Kusano[1], Gen Tanabe[1], Yoshiyuki Uchida[1], Hideki Hyuga[2], You Zhou[2], and Kiyoshi Hirao[2]

[1]Japan Fine Ceramics Co. Ltd., Sendai 981-3203, Japan, kusano@japan-fc.co.jp
[2]National Institute of Advanced Industrial Science and Technology, Nagoya 463-8560, Japan

Abstract

The importance of IGBT power modules has been increasing in a variety of applications such as motor control, robotics, traction, power control systems for solar and wind power generations. Especially the rapid spread of hybrid and electric vehicles is supported by the development of power semiconductor devices such as the IGBT. In order to release the heat generated by these power semiconductor devices, insulating substrates with high thermal conductivity are of increasing importance. Recently silicon nitride has attracted much attention as a substrate material for power semiconductor devices because of its excellent mechanical properties and high intrinsic thermal conductivity. Though conventional sintered silicon nitride has excellent mechanical characteristics, its thermal conductivity is limited as high as around 90W/(m·K). In order to overcome the limitation, we worked on the development of a silicon nitride substrate with higher thermal conductivity.

1. Introduction

Packing density of power modules are rapidly increasing as their application field expands, in particular in the automobile industry. In order to guarantee the stable operation of the power module, heat release technology in the module becomes very important. Now, material of Al_2O_3, ZrO_2 toughened Al_2O_3, AlN and Si_3N_4 are being used for insulating substrates of IGBT power modules (Fig. 1).

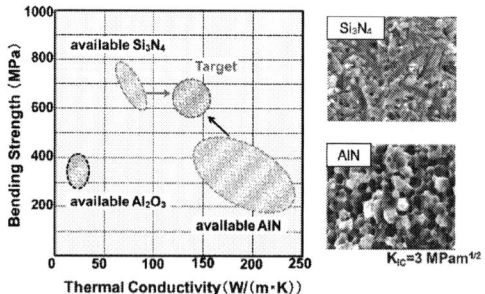

Fig. 1 Characteristics comparison of available ceramics substrates

Especially, AlN has been used as a substrate material for power modules used in harsh environments because of its high thermal conductivity at 180W/(m·K) or higher[1]. In general, a copper (or aluminum) foil as an electrode is bonded to a ceramic substrate via a high temperature process, which causes large residual thermal stress in the substrate. The mechanical properties of AlN are not sufficiently high, which may result in low reliability of power modules. The electronic industry, therefore, is eager to seek alternative substrate materials. Si_3N_4 is well known as a typical structural ceramic which exhibits many excellent properties such as high strength and high toughness[2]. However, thermal conductivity of Si_3N_4 is generally lower than AlN. If thermal conductivity of Si_3N_4 can be improved, Si_3N_4 substrate with high thermal conductivity as well as excellent mechanical properties can be provided as a more suitable substrate material for power modules required long lifetime. In this paper, Si_3N_4 substrates with enhanced thermal conductivity fabricated by newly developed sintering process of reaction-bonding are introduced.

2. Thermal conductivity of silicon nitride

2.1 External factors

The β-Si_3N_4 single crystal itself has been proven to possess a high intrinsic thermal conductivity over 200W/(m·K)[3]. However, the thermal conductivity of sintered silicon nitride is generally much lower than the predicted value of high purity β-Si_3N_4 single crystal. That is because imperfections in β-Si_3N_4 grains such as point defects as well as grain boundaries and secondary phases with low thermal conductivity negatively affect the thermal conductivity of sintered Si_3N_4(Fig 2).

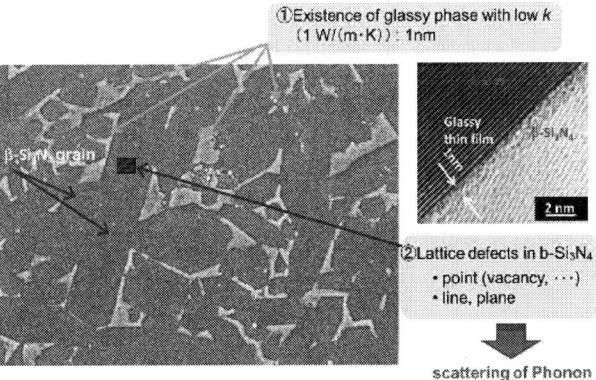

Fig. 2 External factors of decreasing thermal conductivity of Si_3N_4

Thermal conductivity of currently available Si_3N_4 substrates are 60 to 90 W/(m·K) as shown in Table 1, which is much lower than the intrinsic value as described above.

A research group at AIST payed attention to lattice oxygen as a harmful lattice defect scattering phonons for degrading the thermal conductivity of Si_3N_4. Lattice oxygen of about

1.2mass% is already contained in the α-Si_3N_4 powder generally used as the raw material for conventional Si_3N_4 sheets. Therefore, it is very difficult to control the oxygen content at a desirable level or lower.

Table 1 Thermal conductivity of available Si_3N_4 substrates

		A	B	C
Density	g/cm³	3.3	3.3	3.3
Thermal Conductivity	W/(m·K)	90	90	58
Bending Strength	MPa	600	650	850

* From manufacturers' leaflets

2.2 Sintered reaction-bonded silicon nitride (SRBSN)

Recently, a research group at AIST has succeeded in fabricating Si_3N_4 with both high thermal conductivity and high mechanical strength via a reaction bonding followed by post sintering process from Si powder. Zhou et al.[4] fabricated SRBSN materials by nitriding a high-purity Si powder, in place of Si_3N_4 powder used in conventional process, doped with Y_2O_3 and MgO additives, followed by post-sintering. The record-high thermal conductivity of 177W/(m·K) has been obtained in Si_3N_4 fabricated by this SRBSN method.[5]

3. Effect of metallic impurities on thermal conductivity of SRBSN

From a standpoint of industrial application of the SRBSN, it is of great importance to know what level of metallic impurities in the raw powder is tolerable. Our group has, therefore, implemented systematic investigations on the effect of metal impurities[6, 7]. Here we will report some results regarding Al and Fe impurities which are most popular impurities contained in industrial grade Si powder.

3.1 Influence of Al impurity

AlN powder (F1 grade, Tokuyama Co. Ltd., Japan) was added in high purity Si powder to study on influence of Al impurity on the thermal conductivity. The sintered bodies were fabricated by method show in Fig.3 and Table 2

Table 2 Sample number and additive amount of Al

Sample number	Y_2O_3[mol%]	$MgSiN_2$ [mol%]	Al [mass%]
Al-0	2	5	-
Al-0.01	2	5	0.01
Al-0.1	2	5	0.1
Al-0.2	2	5	0.2
Al-0.4	2	5	0.4

© VDE VERLAG GMBH · Berlin · Offenbach

PCIM Europe 2016, 10 – 12 May 2016, Nuremberg, Germany

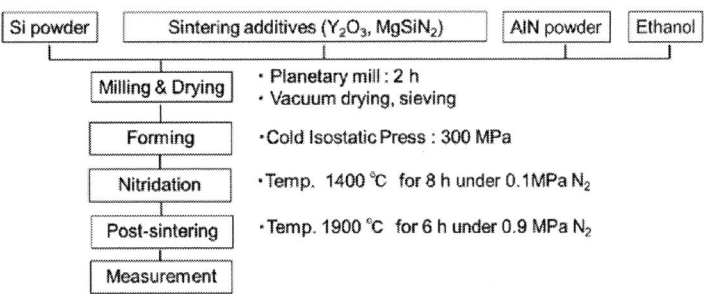

Fig. 3 Experimental procedure

The thermal conductivity decreased with increasing AlN additive content as shown in Fig. 4.

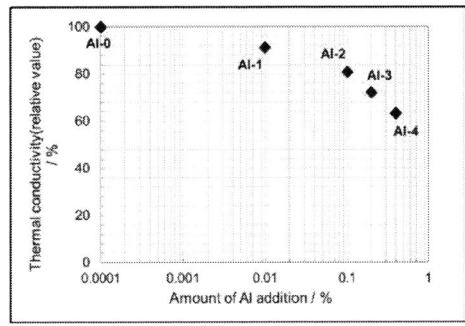

Fig. 4 Thermal conductivity of sintered bodies

This is conjectured to result from the solid solution of Al and O in crystalline β-Si_3N_4, whereby the reaction (1) occurred, and SiAlON formed and became a scattering factor of phonons [8].

$$(6-z)\,\beta\text{-}Si_3N_4 + zAlN + zAl_2O_3 \rightarrow 3Si_{6-z}AlzOzN_{8-z} \quad \cdots (1)$$

From above results, the Al impurity have to be less than 0.1% as the range that does not affect the thermal conductivity.

3.2 Influence of Fe impurity

Fe powder was added in high purity Si powder to study its influence on the thermal conductivity as shown in Table 4. High purity Fe powder (purity>99.9%, impurities: Co 1ppm, Cr 2ppm, Ni 1ppm, Kojundo Chemical Laboratory Co., Ltd., Japan) was added to the starting powder mixtures as shown in Table 3. The experimental procedure is similar to Fig. 3.

Table 3 Sample number and additive amount of Fe

Sample number	Y_2O_3 [mol%]	$MgSiN_2$ [mol%]	Fe [mass%]
F 0	2	5	-
F 0.1	2	5	0.1
F 1	2	5	1.0
F 5	2	5	5.0

© VDE VERLAG GMBH · Berlin · Offenbach

The XRD patterns of the sintered bodies are shown in Fig.5. In the specimens of F1 (Fig.5c) and F5 (Fig.5d) $SiFe_x$ (SiFe and $SiFe_2$) and Si_2N_2O could be observed. XRD spectra of the sintered specimens revealed the existence of iron silicide, and the strength of the peaks of iron silicide increased with the increasing amount of the added iron in the starting powder mixtures.

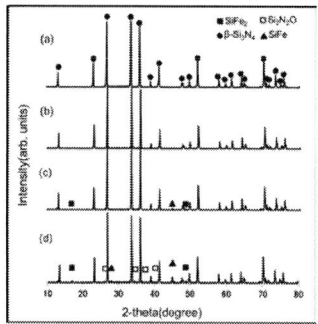

Fig.5 The XRD patterns of the SRBSNs: (a) F0, (b) F0.1, (c) F1, (d) F5

Thermal conductivity decreased with increasing amount of iron additions as shown in Fig.6.

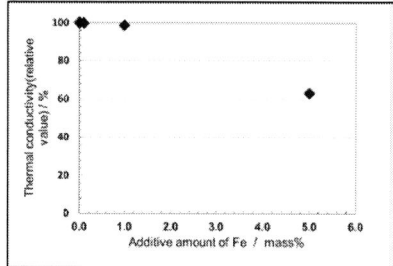

Fig.6 Relation between Fe additive amount and thermal conductivity

Especially, the F5 specimen exhibited very low thermal conductivity. This resolute, for the oxidation of Fe during milling, the increase in the oxygen content and liquid phase by addition of Fe powder was considered to have caused the decreased thermal conductivity.

From above results, the Fe impurity have to be less than 0.1% as the range that does not affect the thermal conductivity.

4. Development of the SRBSN process

We have developed the SRBSN processes expandable to the volume production as illustrated in Fig. 7. Also suitable raw Si powders have been carefully selected by considering our studies on the external factors on the thermal conductivity. The process is composed by (1) preparation of raw materials and ball milling, (2) sheet molding by the doctor blade method, (3) degreasing of green sheets, (4) nitriding of green sheets, (5) sintering of nitride sheets. One of the key factors of degrading thermal conductivity is the increasing amount of oxygen in the process. SRBSN substrates have to be fabricated by controlling the amount of oxygen in the process.

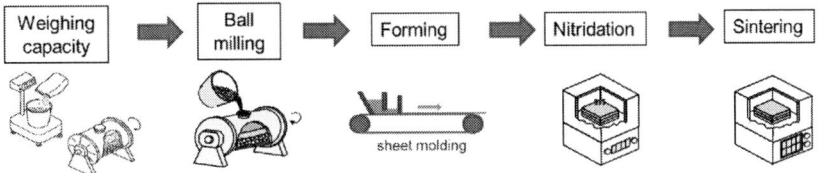

Fig.7 SRBSN process flow

4.1 Control of oxygen content

The silicon powder may be oxidized during the ball milling process. Because, it becomes finer particles and the surface area is increased. Therefore, we have measured the oxygen content as a function of ball milling time. The oxygen content increases with increasing milling time as shown in Table 4, which has to be considered in determining the SRBSN process conditions.

Table 4 Oxidation content (relative value) as a function of milling time

		Milling time			
		0h	6h	12h	24h
Si powder	Oxygen content	1.00	1.20	1.72	2.48

4.2 Green sheet process

The green sheet is formed with an automated doctor blade system. The slurry, the oxygen content of which is controlled, is adjusted its viscosity suitable to make coating. The green sheet is dried in the dry zone that temperature adjustment has been done beforehand. The fabricated green sheet was shown in Fig. 8. In optimizing of the grinding condition, viscosity and dry process condition, we succeeded in fabricating a green sheet without any defects.

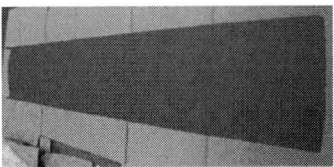

Fig.8 The fabricated green sheet

4.3 Nitride body and Sintered body

The nitriding reaction of the silicon powder is an exothermal reaction. Therefore it is necessary to control a heating profile so that silicon powder does not melt. If the thermal profile is not appropriately controlled, one can have some failures in the nitride sheet. The fabricated nitride with and without failures are shown in Fig.9.

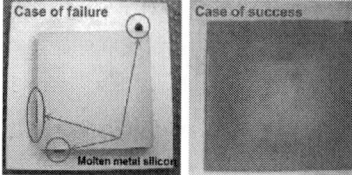

Fig.9 Case of failure and success of the nitriding reaction

After the nitridation with the optimized conditions, the sintering treatment is applied. Fig. 10 shows fabricated SNBSN sheets. The thermal conductivity and mechanical properties are shown in Table 5. We were successful to control the thermal conductivity of the SRBSN substrates at around 90W/(m·K), compatible to the existing material, or higher value at around 120W/(m·K), for example, by adjusting the fabrication conditions.

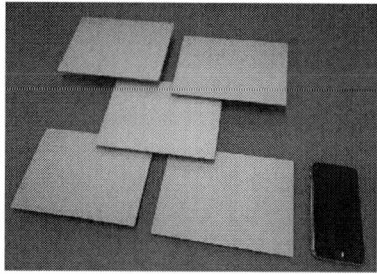

Fig.10 Fabricated SRBSN substrate

Table 5 Characteristics of developed the SRBSN substrate

| | Sintered body | | | | |
	Density [g/m³]	Thermal conductivity [W/(m·K)]	Bending strength [MPa]	Fracture toughness [MPa·m¹/²]	Dielectric Breakdown Strength [kV/mm]
*¹Available Si₃N₄ substrate	3.3	90	600	6	>15
*²Developed of SRBSN substrate 1	3.3	90	650	6	>15
*²Developed of SRBSN substrate 2	3.3	120	580	9	>15

*¹From manufacturers' leaflets
*²Representative value

5. Summary

Fabrication of silicon nitride substrates with thermal conductivity higher than conventional silicon nitride materials by using the SRBSN method is presented. The experimental results are summarized as follows.

· A silicon nitride with a record-high thermal conductivity of 177W/(m·K) was fabricated by using the SRBSN method[5].

· Influences of major impurity elements on the thermal conductivity of the silicon nitride ceramics fabricated by the SRBSN method were investigated.

- The SRBSN process expandable to volume production in near future has been developed
- Silicon nitride substrates with thermal conductivity of 120W/(m·K) were successfully fabricated by the SRBSN process using an industrial grade silicon powder as a raw material.
- We were successful to control the thermal conductivity at 90W/(m·K), compatible to the existing material, or higher value at 120W/(m·K), for example, by adjusting the fabrication conditions.

6. Acknowledgment

A part of this work was supported by Council for Science, Technology and Innovation(CSTI), Cross-ministerial Strategic Innovation Promotion Program (SIP), "Next-generation power electronics/Consistent R&D of next-generation SiC power electronics" (funding agency: NEDO)

7. References

(1) G. A. Slack, J. Phys .Chem. Solids., 34, 321 (1973)

(2) P.F. Becher, et al., J. Am. Ceram. Soc., 81, 2821 (1998).

(3) J.S. Haggerty, et al., Ceram. Eng. Sci. Proc., 16, 475-487 (1995).

(4) Y. Zhou, et al., Int. J. Appl. Ceram. Technol., 5,119-126 (2008).

(5) Y. Zhou, et al., Adv. Mater., 23, 4563 (2011).

(6) D. Kusano, et al., Int. J. Appl. Ceram. Technol., 11, 534 (2014).

(7) D. Kusano, et al., Int. J. Appl. Ceram. Technol., 10, 690 (2013).

(8) Y. Okamoto, et al., J. Mater. Res., 13, [12] 3473 (1998).

(9) N. Otogawa, et al.,Thin Solid Films, 461 223–226 (2004).

Thermal management of future WBG devices using two-phase cooling

Shailesh N. Joshi and Ercan M. Dede
Toyota Research Institute of North America, 1555 Woodridge Ave, Ann Arbor, MI 48105, USA

Abstract

The two-phase cooling performance of two different porous structures, namely, open tunnel (OPT) and closed tunnel (CLT), were tested and compared. The results show that the boiling performance of the OPT porous structure is higher when compared to that of the CLT structure. Specifically, the OPT structure shows higher heat transfer coefficient values. Furthermore, while the pressure drop trend is same for both structures, the OPT structure exhibits slightly lower pressure drop relative to the pressure drop for the CLT structure. From both two-phase cooling performance and pressure drop, it is concluded that the OPT structure has superior performance compared to the CLT structure and is better suited for thermal management of future power-dense wide band-gap devices.

1. Introduction

To meet future high density electronic packaging requirements of electric/hybrid vehicles, wide band-gap (WBG) power semiconductor devices, such as silicon carbide (SiC) and gallium nitride (GaN) are being widely researched. These devices have advantages, such as providing higher breakdown voltage, higher switching frequency, and lower switching loss. However, they also pose severe challenges in terms of cooling as the heat fluxes dissipated by these devices may exceed 250 W/cm^2. Such high heat fluxes cannot easily be accommodated using conventional cooling technologies such as air-cooling or single-phase liquid cooling, especially in a compact space. One potential cooling technology to dissipate high heat fluxes is phase-change heat transfer or two-phase cooling.

Two-phase cooling is under current investigation by the present authors [1-4] for thermal management of automotive power electronics. One of the challenges with two-phase cooling is the effective removal of vapor from the cooler. In an earlier study [1], vapor entrapment inside the cooler was identified and its effect on the overall two-phase heat transfer process was studied. A subsequent study was conducted [2] to mitigate this vapor entrapment issue by carefully designing the geometry of the cooler. To improve the two-phase performance further, porous coatings may be applied on a multi-scale heat spreading surface. Recently, enhancement in the two-phase heat transfer coefficient (HTC) was reported, where jet impingement technology was combined with porous coated pin-fin surfaces [3]. A 79% enhancement in the HTC at a heat flux of 50 W/cm^2 was reported. A further enhancement of 1.8X was achieved by reducing the sub-cooling from 10 K to 5 K. Another study by M. Rau et al. [8] reports enhancement in two-phase heat transfer using similar pin-fin porous coated surfaces in combination with jet impingement. The jet array, along with porous coating, enhanced the critical heat flux (CHF) 1.89-2.33X. Before these recent studies, limited research was conducted using such porous surfaces in combination with jet impingement technology, and an even smaller number of investigations existed on effective removal of vapor from the porous surface for CHF enhancement. In a somewhat related study [10], the researchers developed two-dimensional (2-D) and three-dimensional (3-D) modulated copper (Cu) powder coatings that were tested under pool boiling conditions. In [10], the authors found that the modulated coatings provided 2-3.3X higher performance relative to plain surfaces.

The present study reports the two-phase cooling performance of porous coated surfaces in combination with jet impingement technology. In Section 2, an explanation of the test set-up including the test section and porous coated surfaces is provided. A dielectric coolant R-245fa (Honeywell International, Inc.) is used for experimental evaluation of the two surface structures. The first surface structure has closed tunnels that confine the vapor flow within the tunnels, while the second surface has open tunnels to allow un-confined flow towards the outlet. An estimation of heat losses is then provided in Section 3 followed by the data reduction procedure in Section 4. Experimental heat transfer and pressure drop results are given in Section 5. Conclusions follow in Section 6.

2. Experimental Test Set-up

The experimental facility developed for the two-phase jet impingement experiments is shown schematically in Figure 1. The main components of the closed flow loop consist of the pump, sub-cooler, test section, condenser, and receiver (i.e. reservoir). A brief description of the loop follows. The fluid is pumped from the receiver by a positive displacement pump (Micropump Inc., GJ 21 series) to provide a fixed mass flow rate of 0.010 kg/sec. A sub-cooler is located downstream of the flowmeter to control the sub-cooling of the coolant entering the test section. For all tests, a constant sub-cooling of 5 K is maintained. The sub-cooler is a liquid-to-liquid shell and tube type heat exchanger (HX). The R-245fa coolant flows through the tubes while water from a first chiller (Polysciences, Inc.) enters the shell side. The shell side coolant temperature is controlled to maintain a fixed coolant temperature entering the test section during the experiments. Downstream of the test section is a dual-pass shell and tube HX to condense the two-phase mixture exiting the cooler. A 5 kW chiller (Thermo Fisher Scientific, Inc.) circulates water at a fixed inlet temperature on the shell side to condense R-245fa vapor flowing through the tubes. The receiver shown in Figure 1 stores a fixed amount of R-245fa coolant so that only liquid coolant enters the pump. The outer surface of the receiver is wrapped with a PID-controlled band heater that maintains a constant coolant temperature of 318 K (i.e. 45 °C) inside the receiver, and hence, constant pressure inside the closed loop.

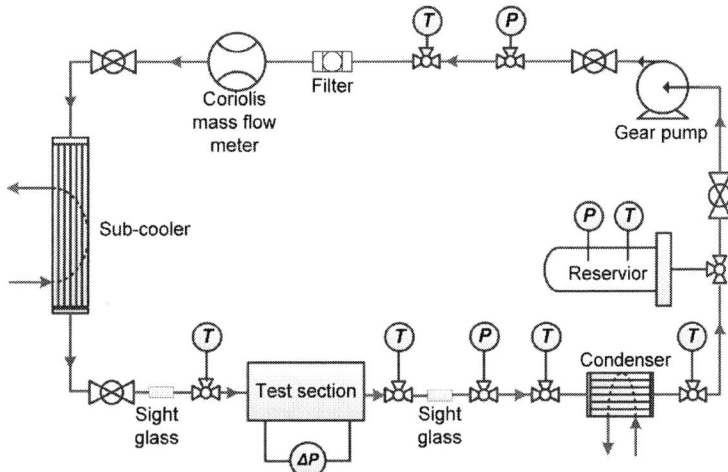

Figure 1: Closed two-phase test loop.

2.1 Test Section

A submerged jet impingement cold plate concept is used to evaluate two-phase performance of the porous surfaces. The details of the test section, jet plate, and heat spreader follow.
The test section consists of a single-device jet impingement cooler, as shown in Figure 2. The cooler consists of five main parts, namely the upper, middle, and lower outlet manifolds plus the jet orifice plate and Cu heat spreader. The upper manifold has both inlet and outlet ports for the cooler assembly. After passing through the inlet and jet orifices to form liquid jets, the coolant impinges on the Cu heat spreader. The outflow from the impinging jets (a liquid and vapor mixture in two-phase operation) flows horizontally outward in confined fashion between the jet orifice plate and heated Cu plate, where it is then removed through the three-piece outlet manifold. The sloped and vertical nature of the outlet flow paths in the cooler ensures efficient removal of vapor from the heated surface. The test section parts are constructed out of polycarbonate to reduce heat loss to the environment.

© VDE VERLAG GMBH · Berlin · Offenbach

A jet orifice plate consisting of an array of 5 x 5 jet orifices is used for all tests. Each orifice is round in shape with a diameter of 0.75 mm. The jet-to-spreader distance is fixed at 2.5 mm. A nichrome resistive element heater (shown in Figure 2) is scribed directly to the underside of the heat spreader using a 100 μm thick magnesium aluminate dielectric thermal interface material, $k = 14.7$ W/mK, positioned between the heating element and Cu surface. The heater dimensions are approximately 19 mm x 19 mm and the assembly includes a 0.8 mm diameter Type K thermocouple (calibrated to ± 0.5 K accuracy) to measure the temperature at the center of the device. A PEEK block is secured on the back side of the Cu heat spreader (i.e. the heated side) to reduce heat loss from the heater to the environment.

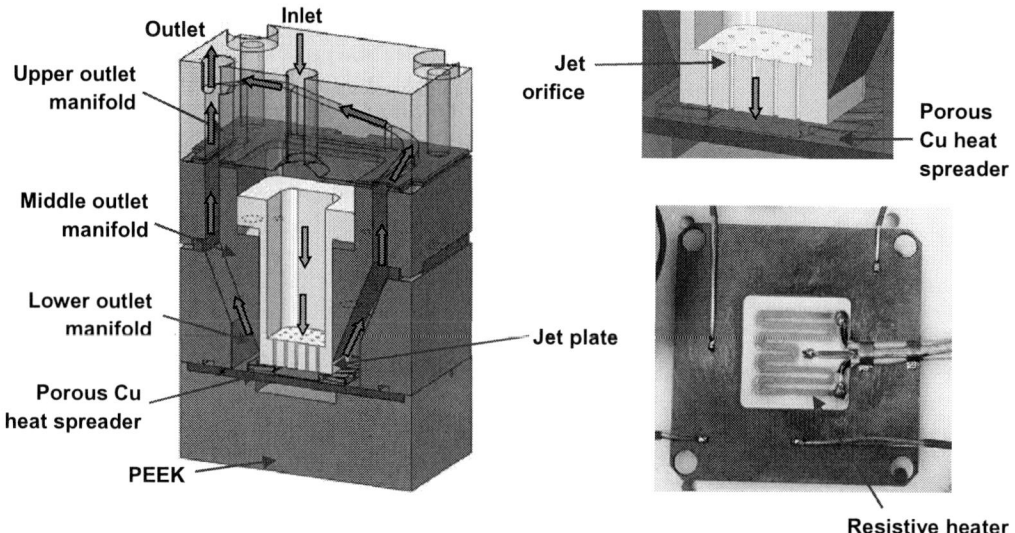

Figure 2: Two-phase jet impingement cooler (left) with zoomed view of orifice plate (top right) and heater detail (lower right).

2.2 Porous Coated Surfaces

Two different porous coated surface structures, namely the OPT and CLT surfaces shown in Figure 3, were fabricated and tested. In each case, the surface is designed such that 5 x 5 liquid jet array empties into the centre square cavity where boiling then occurs. To facilitate vapor outflow, the tunnels in each surface structure design are oriented radially outward. For reference, the over-

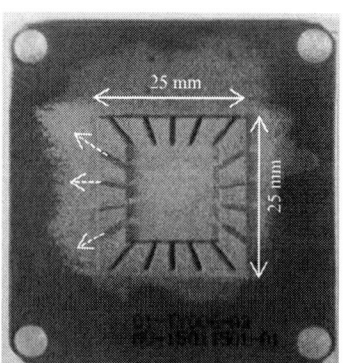 

Figure 3: Porous coated surfaces including OPT (left image) and CLT (right image). Note that the dashed arrows indicate flow outlet direction.

all footprint of the OPT and CLT porous surfaces are each 25 mm x 25 mm in size.

The porous surfaces were fabricated using a powder metallurgy technique wherein micron-sized particles with particle diameter ranging between 75 μm -100 μm were sintered inside a graphite mould. The resulting pore size is around 100 μm as determined using a scanning electron microscope. Details about the sintering method are found in [3].

3. Heat Loss Estimation

Before post-processing the heat transfer data, heat loss from the cooler needs to be estimated. A 3-D, steady-state heat conduction model of the cooler is built in COMSOL v.4.2a [11]. The model accounts for the heat input from the resistive device to the Cu heat spreader, conduction out of the bottom and sides of the heat spreader, and heat lost to the environment via free convection on the external boundaries of the cooler. Details regarding the numerical strategy used for the mesh dependency study may be found in [2]. A linear curve fit is then used to correlate the cooler heat loss, q_{loss}, to the heater temperature, T_{heater}, (in °C) and is given by Equation 1,

$$q_{loss} = 0.0919 \times T_{heater} - 3.281. \tag{1}$$

To determine the experimental heat loss at a given total input power, the numerical heat loss correlation, Equation 1, obtained from the finite element analysis is used. The power dissipated by convection via the two-phase jet impingement cooler is determined as,

$$q_{conv} = q_{total} - q_{loss}, \tag{2}$$

where q_{conv} is the heat dissipated by the cooler, q_{total} is the total electrical heat input to the heater, and q_{loss} is the heat lost from the cooler to the environment. Depending on the amount of total heat input applied, the calculated heat losses range from approximately 3% to 14%. Please note that the following heat flux values reported herein account for heat losses.

4. Data Reduction

The heater is electrically powered by a 40 A, 60 V DC B&K precision power supply. The heat flux, $q"$, is defined as the convective power, q_{conv}, calculated using Equation 2, divided by the heater area, A_{heater}, (19 mm x 19 mm) and is given by

$$q" = q_{conv}/A_{heater}. \tag{3}$$

An overall effective heat transfer coefficient, h, based on the fluid inlet temperature, T_{in}, is defined via Equation 4,

$$h = q"/(T_{heater} - T_{in}), \tag{4}$$

where T_{heater} is the temperature of the heater (measured by the thermocouple located at the center of the device).

The pressure drop, ΔP, across the cooler is measured using a differential pressure transducer (Omega) via the inlet and outlet ports identified in Figure 2. The pressure drop is given as

$$\Delta P = P_{in} - P_{out},$$

where P_{in} is the inlet pressure and P_{out} is the outlet pressure.

An experimental uncertainty analysis is carried out, per [12], to determine the overall uncertainties in the derived quantities. Considering the individual errors in voltage and temperature measurements, a propagation of error analysis is performed, which yields overall uncertainties of ±1.1% in heat flux and ±5 - 12% in heat transfer coefficient.

5. Heat Transfer and Pressure Drop Results

The two-phase tests were performed at a fixed flow rate over a heat flux range of 3-200 W/cm^2. The heat transfer results are compared using boiling and heat transfer coefficient curves. Figure 4 shows the comparison of boiling performance between OPT and CLT surfaces. The results show that the heat transfer performance of the two surfaces is fairly close to each other up to a heat flux of 25 W/cm². Beyond 25 W/cm² the performance of the OPT surface is higher than the CLT surface. Additionally, the CHF value reached by the OPT surface, 200 W/cm², is slightly higher than the CLT surface, 193 W/cm².

Figure 5 shows a comparison between the HTC for the OPT and CLT surfaces. Observe that for heat fluxes greater than 25 W/cm², the HTC value of the OPT surface is higher than that for the CLT surface. For example, at 137 W/cm² the heat transfer coefficient for the OPT surface is about 12% higher than the HTC for the CLT surface. Furthermore, the CLT surface shows a more abrupt transition into CHF compared to a gradual transition for the OPT surface.

Figure 6 shows a comparison of pressure drop between the OPT and CLT surfaces. Observe that the variation of the pressure drop is quite similar for both surfaces over the entire range of heat fluxes tested. For heat fluxes less than 25 W/cm², the pressure drop is fairly constant at ~1.1 kPa. At heat fluxes greater than 25 W/cm², the pressure drop begins to gradually increase due to increased vapor generation. A similar pressure drop trend is reported by authors in [2]. Furthermore, in the two-phase regime, the pressure drop of the OPT surface is lower than that for the CLT surface. For example, at a heat flux of 137 W/cm² the pressure drop of the OPT surface is about 6%

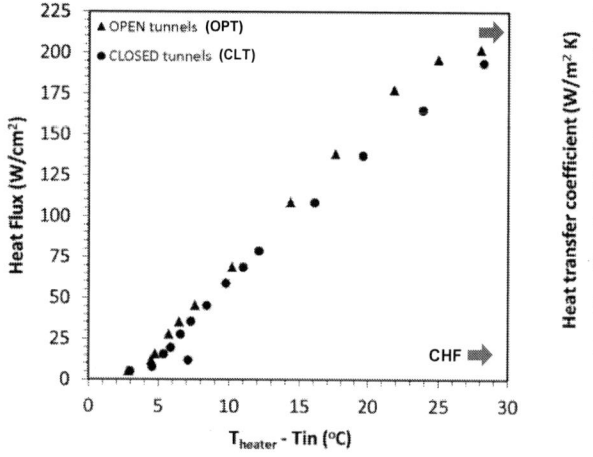

Figure 4: Comparison of boiling curves.

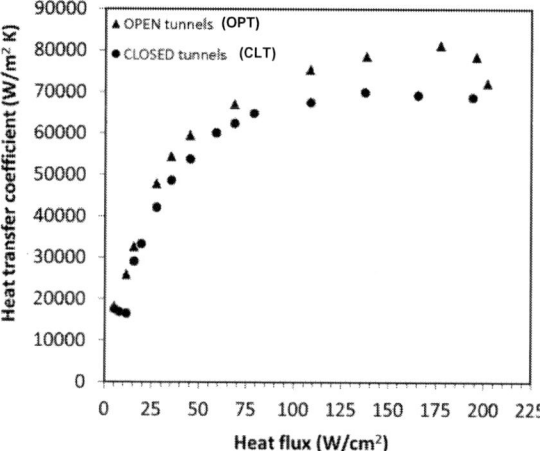

Figure 5: Comparison of heat transfer coefficient.

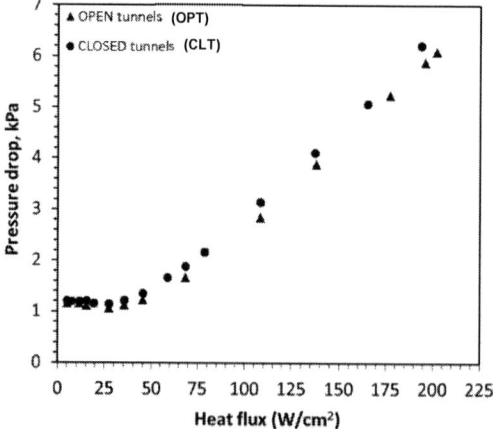

Figure 6: Comparison of pressure drop.

lower compared to the pressure drop for the CLT surface. Logically, the increase in the pressure drop may be attributed to flow confinement from the closed tunnels.

6. Conclusions

The focus of this article was to present experimental two-phase heat transfer and pressure drop results for OPT and CLT surfaces using a submerged jet impingement cooler in combination with a 5 x 5 jet array. A dielectric coolant R-245fa was used to evaluate the performance of the porous surfaces. The results indicate that, in general, the boiling performance of the OPT porous structure is higher when compared to that for the CLT structure. The boiling performance of two surfaces follow a similar trend up to a heat flux of 25 W/cm^2, however, beyond 25 W/cm^2 the performance of the OPT surface is higher than that for the CLT surface. Furthermore, the CHF value attained using the OPT surface is 200 W/cm^2 compared to 193 W/cm^2 for the CLT surface. Additionally, the heat transfer coefficient of the OPT surface is higher than CLT surface in the entire two-phase region. For example, at 137 W/cm^2 the heat transfer coefficient for the OPT surface is about 12% higher than the CLT surface. Finally, the pressure drop of the OPT surface is lower than that for the CLT surface. The increased pressure drop of the CLT surface is due to confinement inside the porous surface.

Since the coefficient of performance for a cooler is typically proportional to heat transfer and inversely proportional to pumping power (or pressure drop for a fixed mass flow rate), the OPT surface structure appears to be a more favourable design. As a consequence, the OPT structure may be preferred for the thermal management of high heat flux wide band-gap power electronics.

References

1. S. N. Joshi, M. J. Rau, E. M. Dede, and S. V. Garimella, A study of a multi-device jet impingement cooler with phase change using HFE-7100, ASME Summer Heat Transfer Conference, HT2013-17059, Minneapolis, July 2013.
2. S. N. Joshi, M. J. Rau, and E. M. Dede, An experimental study of a single-device jet impingement cooler with phase change using HFE-7100 and a vapor extraction manifold, ASME International Mechanical Engineering Congress & Exhibition, IMECE 2013-63249, San Diego, Nov. 2013.
3. S. N. Joshi and E. M. Dede, Effect of sub-cooling on performance of a multi-jet two phase cooler with multi-scale porous surfaces, Int. J. Therm. Sci., 87 (2015) 110-120.
4. F. Zhou, S. N. Joshi, and E. M. Dede, Visualization of bubble behavior for jet impingement cooling with phase change, IEEE Intersociety Conference on Thermal and Thermomechanical Phenomena in Electronic Systems (ITherm), Florida, May 2014.
5. M.S. Sarwar, Y.H. Jeong, and S.H. Chang, Subcooled flow boiling CHF enhancement with porous surface coatings, Int. J. Heat Mass Tran., 50 (2007) 3649-3657.
6. D. Schäfer, R. Tamme, and H. Müller-Steinhagen, The effect of novel plasma-coated compact tube bundles on pool boiling, Heat Transfer Eng., 28 (2007) 19-24.
7. S. Hsieh, G. Huang, and H. Tsai, Nucleate pool boiling characteristics from coated tube bundles in saturated R-134a, Int. J. Heat Mass Tran., 46 (2003) 1223-1239.
8. M. Rau, S. V. Garimella, E. M. Dede, and S. N. Joshi, Boiling heat transfer from an array of round jets with hybrid surface enhancements, J. Heat Transf., 137 (2015) 071501.
9. P. Bai, T. Tang, and B. Tang, Enhanced flow boiling in parallel microchannels with metallic porous coating, Appl. Therm. Eng., 58 (2013) 291-297.
10. D.H. Min, et al., 2-D and 3-D modulated porous coatings for enhanced pool boiling, Int. J. Heat Mass Tran., 52 (2009) 2607-2613.
11. COMSOL AB, COMSOL Multiphysics ver. 4.2a, Stockholm, 2011.
12. S. Kline and F. McClintock, Describing uncertainties in single-sample experiments, Mech. Eng., 75 (1953) pp. 38.

Highly Reliable and Lead-Free High Power IGBT Modules Using Novel Copper Sintering Die Attachment

Akitoyo Konno, Takaaki Miyazaki, Yuusuke Yasuda, Osamu Ikeda, Hiroshi Nakano, Toshiaki Morita, Hiroshi Houzouji, Mutsuhiro Mori, Hitachi, Ltd., Research & Development Group, Japan, akitoyo.konno.ea@hitachi.com
Masato Nakamura, Yoshihiko Koike, Hitachi Power Semiconductor Device, Ltd., Japan

Abstract

Novel Cu sintering die attachment material has been developed to obtain highly reliable and lead-free power module which can be operated at high temperature. Conventional Pb rich solder and Cu sintering bonding techniques were compared by thermal cycling test under the condition of -40 to 200°C, then the tested sample applied with Cu sintering bonding indicated no delamination after 1k cycles whereas that with Pb rich solder showed delamination. Moreover, at power cycling test, Cu sintering bonding did not break down up to 10 times higher cycle than that of Pb rich solder.

1. Introduction

Semiconductor power modules for hybrid vehicles, trains, and renewable energy conversions tend to be packed much densely and produce higher heat per unit volume than those of previous generations. Therefore, low loss devices and bonding materials with higher heat dissipation and thermal stability are strongly demanded to operate chips stably with long-term reliability. Regard with highly heat-resistant die attachment materials, Ag sintering materials have been widely developed [1-6], however, Ag is expensive to produce die attachment materials and precious metal electrode such as Au or Ag is necessary to achieve strong bonding by Ag sintering technique. To solve these problems and satisfy the demands, Cu sintering die bonding material, which can be bonded to non-precious metal such as Cu or Ni has been developed so as to lower the material cost [7]. In this paper, reliability of Cu sintering technique will be indicated by comparing with conventional Pb rich solder and Ag sintering.

2. Advantages of copper sintering die attachment material

2.1. Properties comparison of die attachment materials

Table 1 shows thermal and mechanical properties of Pb rich solder, as well as Ag and Cu bulks. For high bonding reliability, lower coefficient of thermal expansion and higher yield

Table 1. Properties of die bond materials

	Coefficient of thermal expansion (1E-6/K)	Young's modulus (GPa)	Yield stress (MPa)	Thermal conductivity (W/mK)	Melting point (°C)
Pb rich solder	17	12	59	24	280
Ag	19.7	76	262	427	960
Cu	16.6	117	310	398	1083

stress and melting point are essential as die attachment materials. As shown in table 1, it is recognized that Cu is preferable from the viewpoint of above mentioned, although the properties are different between sintered body and bulk body, it is assumed the relative relations are stored. Therefore, Cu sintering bonding technique is thought to be the best solution for highly reliable power module.

2.2. Bonding strength on non-precious metal

For cost reduction, it is important for die attachment materials to be capable of bonding with non-precious metals because the electric-conductive layer on ceramic substrates or lead frames generally consist of Cu, Ni-plated Cu or Ni-plated Al. If it is difficult to bond with non-precious metals, it is necessary that the electric-conductive layer and chip electrode is coated with a precious metal.

Figure 1 shows the relationship between the shear strength and the bonding temperature for Ni-plated Cu disk bonded by Cu sintering, Ni-plated Cu disk bonded by Ag sintering and Au-plated Cu disk bonded by Ag sintering. The shear strength of Ni-plated Cu disk bonded by Ag sintering was inadequate with the bonding temperature under 400°C. On the other hand, the shear strength of Ni-plated Cu disk bonded by Cu sintering and Au-plated Cu disk bonded by Ag sintering show higher shear strength around 20MPa with the bonding temperature over 250°C. This suggests that Cu sintering technique enables semiconductor chips to be bond on low cost substrate or lead frame which is not coated with precious metals such as Au or Ag.

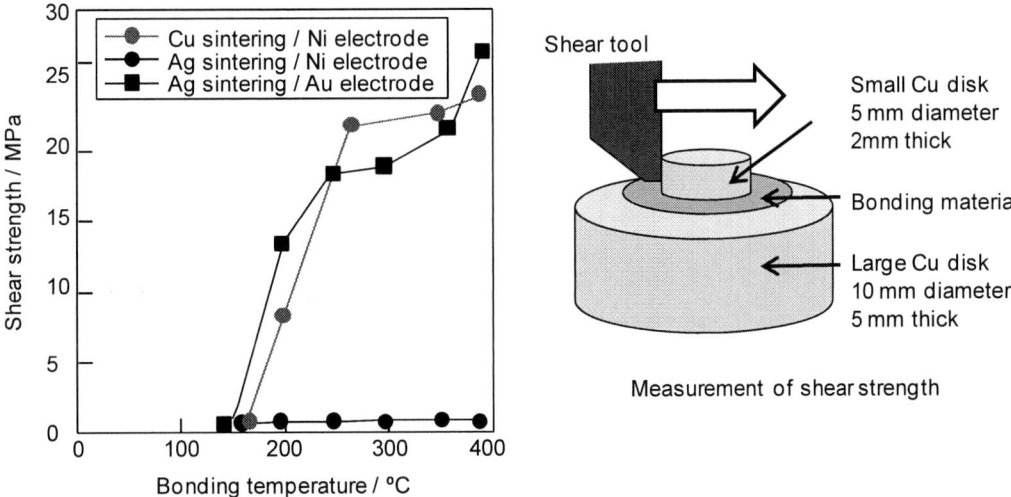

Fig 1. Relationship between bonding temperature and bonding strength.

2.3. Power semiconductor chip bonding with ceramic substrate by Cu sintering

Figure 2 shows the cross-sectional SEM image of sintered Cu layer between the electric-conductive layer on the AlN substrate and the metallization layer on the bottom electrode of the semiconductor chip. In analogy with Ag sintering [1-2], the sintered Cu layer whose shape is porous is generated from submicron Cu particles. As mentioned above, Cu sintering is capable of bonding with non-precious metals. So in this sample, the electric-conductive layer on the AlN substrate is bare Cu and the metallization layer of the semiconductor chip is Ni plated.

Figure 3(a) shows the cross-sectional TEM image of the interface between Ni plate layer and sintered Cu layer, and figure 3(b) shows the Cu mapping image of corresponding area by using TEM-EDX. As can be seen in Fig. 3(b), Cu diffuses through the grain boundary of Ni

© VDE VERLAG GMBH · Berlin · Offenbach

crystals into Ni plate layer. We think that this phenomenon is a cause of realizing a strong metal-bonding between Cu and Ni.

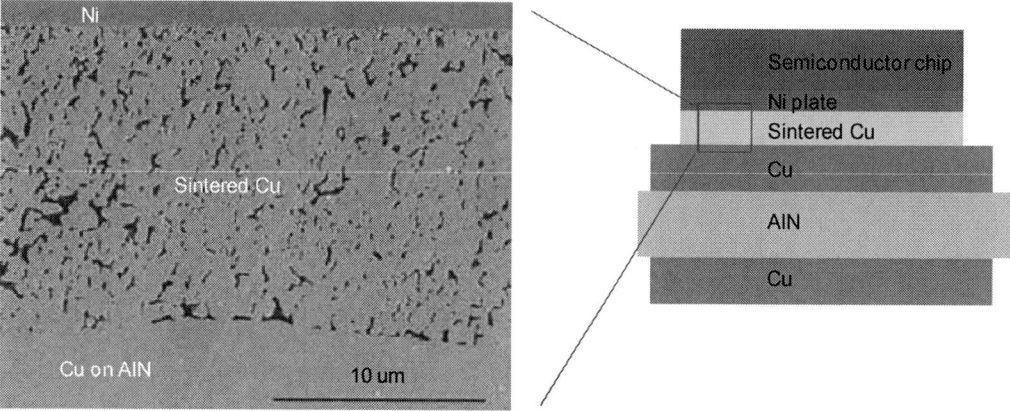

Fig2. Cross-sectional SEM image of sintered Cu layer between Cu layer on AlN substrate and Ni layer coated on bottom electrode of semiconductor chip

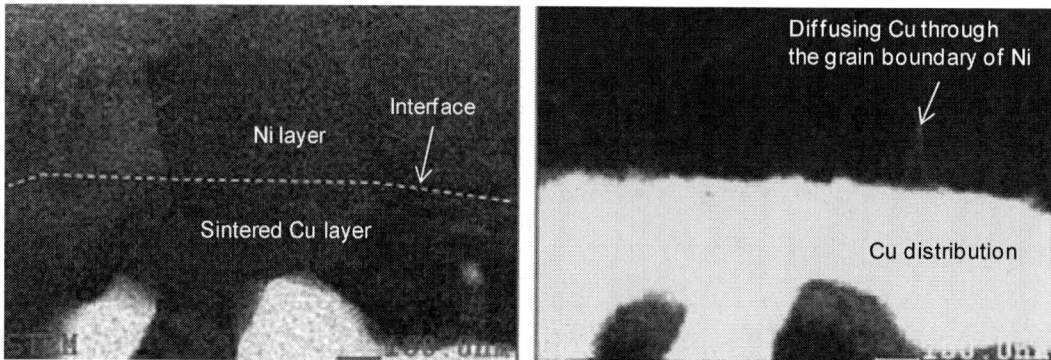

Fig3. Bonding state of sintered Cu and Ni ((a) cross-sectional TEM image of the interface between Ni plate layer and sintered Cu layer, (b) Cu mapping image by using TEM-EDX)

3. Reliability test

3.1. Thermal cycling test

Table 2 shows the ultrasonic testing images of the bonding layer under the semiconductor chip on Ni coated electrode after the thermal cycling test under the condition of -40°C to 200°C for Pb rich solder and Cu sintering bonding. After 1k cycles, the bonding layer of Cu sintered indicated no delamination whereas that of Pb rich solder indicated delamination in the peripheral areas of the semiconductor chip.

Table 2. Ultrasonic testing images of the bonding layer under semiconductor chip at thermal cycling test under the condition of -40°C to 200 °C

Die bonding	Pb rich soldered	Cu sintered
Electrode surface metalize	Ni	Ni
Initial		
After 1k cycles		

3.2. Power cycling test

Figure 4 shows the result of the power cycling tests of one chip module without base plate bonded with Cu sintering and Pb rich solder under the condition of delta Tj 125K and initial Tj 175°C and 150°C respectively. As can be seen in Fig. 4, the tested sample of Pb rich soldered broke down at around 50,000 cycles, on the other hand that of Cu sintering did not break down at over 550,000 cycles.

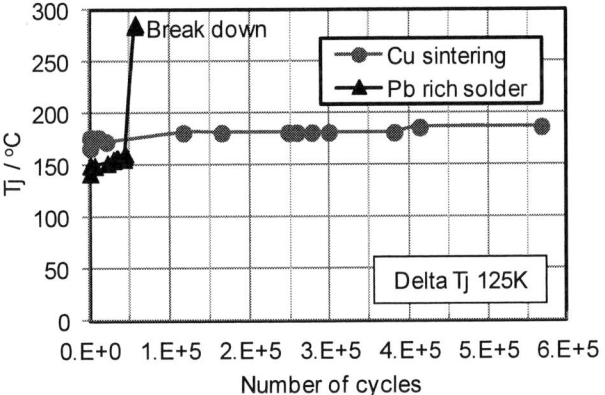

Fig4. Result of power cycling tests of one chip module without base plate bonded with Cu sintering and Pb rich solder under the condition of delta Tj 125K and initial Tj 175°C and 150°C respectively.

4. Prototype module evaluation

3.3kV 1800A high power IGBT prototype modules applied new devices which have being developed as Hitachi's next generation and Cu sintering bonding technique have been developed, which is shown in Fig. 5.

Figure 6 shows switching wave forms of developed 3.3kV 1800A prototype module under the condition of VCC=1,800V, Ic=1800A, Ls=80nH, RG(on/off)=4.7Ω/5.6Ω, VGE=±15V, Tj=150°C, which indicates that the prototype module can successfully switch large current at high temperature.

Fig5. Developed 3.3kV1800A prototype module applied Hitachi's next version devices and Cu sintering bonding technique

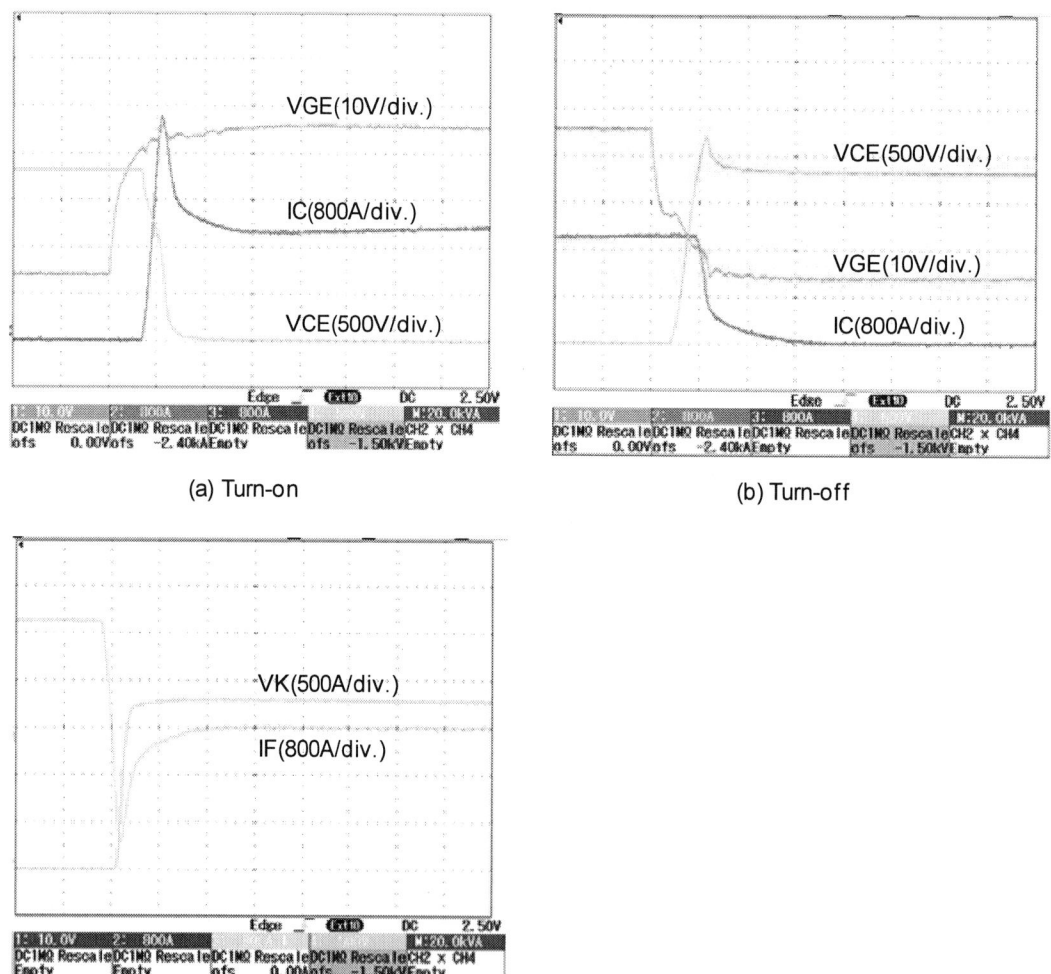

Fig6. Switching wave forms of developed 3.3kV1800A prototype module under the condition of VCC=1,800V, Ic=1800A, Ls=80nH, RG(on/off)=4.7Ω/5.6Ω, VGE=± 15V, Tj=150°C, applied new devices which have being developed as Hitachi's next generation and Cu sintering bonding technique

© VDE VERLAG GMBH · Berlin · Offenbach

5. Conclusion

As described above, Cu sintering bonding material has appropriate properties and potential of cost reduction for highly reliable and lead-free power module. At the thermal cycling test and power cycling test, Cu sintering bonding indicates further reliable performance in comparison with Pb rich solder. Using new devices which have being developed as Hitachi's next generation and Cu sintering bonding technique, we have developed 3.3kV 1800A prototype power module which can operate at high temperature and has low loss properties.

6. References

[1] Dmitry Titushkin et al., "Advantages of High Power Fast Thyristors and Diodes Produced by Means of Low Temperature Sintering of Silver Paste" PCIM Europe 2015, Proceedings, pp.899-905.

[2] Hao Zhang et al., "Reliability Improvement of high Temperature Sintered Ag Die-Attachment by Adding Sub-micron SiC Particles" PCIM Europe 2015, Proceedings, pp.80-87.

[3] Toshiaki Morita et al., "Study of Bonding Technology Using Silver Nanoparticles" Jpn. J. Appl.Phys. 47 (2008) 6615-6622.

[4] Yuusuke Yasuda et al., "Low-Temperature Bonding Using Silver Nanoparticles Stabilized by Short-Chain Alkylamines" Jpn. J. Appl. Phys. 48 (2009) 125004.

[5] Toshiaki Morita et al., "Bonding Technique Using Micro-Scaled Silver-Oxide Particles for In-Situ Formation of Silver Nanoparticles" Mater. Trans. 49 (2008) 2875-2880.

[6] Yuusuke Yasuda et al., "Low-Temperature Bonding of Silver Derived from Silver–Oxide Particles to Nickel" Mater. Trans. 54 (2013) 1063-1065.

[7] Toshiaki Morita et al., "New Bonding Technique Using Copper Oxide Materials" Mater Trans. 56 (2015) pp. 878-882.

PCIM Europe 2016, 10 – 12 May 2016, Nuremberg, Germany

A New Generation of Nanocrystalline Magnetic Cores with Very Low Magnetic Losses

Bashar, GONY, Aperam Alloys, France, bashar.gony@aperam.com
Stephane, CAMUS, Aperam Alloys, France, stephane.camus@aperam.com
Julia, HILL, Aperam Alloys, France, julia.hill@aperam.com
Frederic, POTTIER, Aperam Alloys, France, frederic.pottier@aperam.com

Abstract

In this paper, we present a new generation of Nanocrystalline magnetic cores (kµ) with low permeability and low losses. The kµ magnetic cores provide high frequency stability of permeability compared with other magnetic products. This material also exhibits high B-H linearity until saturation which is not the case of most magnetic cores. These properties make kµ magnetic cores the product of choice in fields such as embedded electronics for electric (EV) and hybrid (HEV) vehicles.

1. Introduction

For many applications including current sensors, electronic filters, differential interrupters, smart watt hour meters and energy storage, regular Nanocrystalline cores have been widely used as a promising magnetic material [1], [2].

Nanocrystalline material has intrinsic properties that effectively help the design of downsized passive components for low and high frequency applications. Their innate advantages such as medium saturation, low thickness – high resistivity and low magnetic losses, and the ability to design different shapes of hysteresis cycle have motivated various industries to further extend their field of application.

Important resources have been devoted in the last decade to decreasing the relative permeability μ_r range of nanocrystalline wound cores down to lower ones: regular $Fe_{74}Cu_1Nb_3Si_{15}B_7$ flat loop shaped nanocrystalline cores provide µr as low as 20 000 through magnetic field annealing however it remained until now difficult to get smaller µr without airgap.

The kµ magnetic cores are a new generation of patented low permeability Nanocrystalline cores [3], [4], based on stress annealing processing. They exhibit high B-H linearity until saturation (at 1.2 T), low magnetostriction (a few ppm) and low magnetic losses, competing against ferrites or powder cores products providing both low μ_r and no-airgap [5], [6].

In this paper, the properties of kµ magnetic cores (low magnetic losses, stability of permeability, and high linearity until saturation) are compared to another commercial magnetic core (MPP).

2. Magnetic losses

The magnetic losses of kµ cores, as shown in Fig.1 versus induction at 10 kHz and compared to MPP materials (permeability range: 14 - 550), are low, and moreover lower than those of MPP by a few tenth of %. As a difference the MPP core contains about 80% of Nickel and exhibits moderate saturation (<0,9T compared to nanocrystalline material at 1.25T) which limits its use.

© VDE VERLAG GMBH · Berlin · Offenbach

PCIM Europe 2016, 10 – 12 May 2016, Nuremberg, Germany

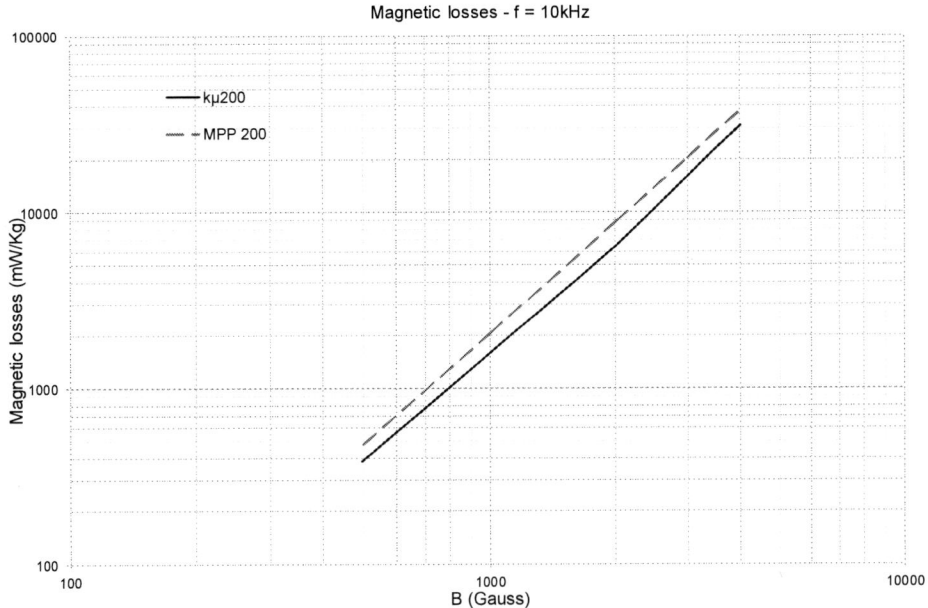

Fig. 1. Magnetic losses in kµ magnetic core compared to MPP at f = 10 KHz, with a given relative permeability of 200.

Increasing the frequency up to 10 kHz, the magnetic losses of the 2 materials become similar (Fig. 2). The electrical resistivity of MPP is about 75 µΩ.cm compared with 115 µΩ.cm in nanocrystalline core, and may explain such increasing losses difference with frequency.

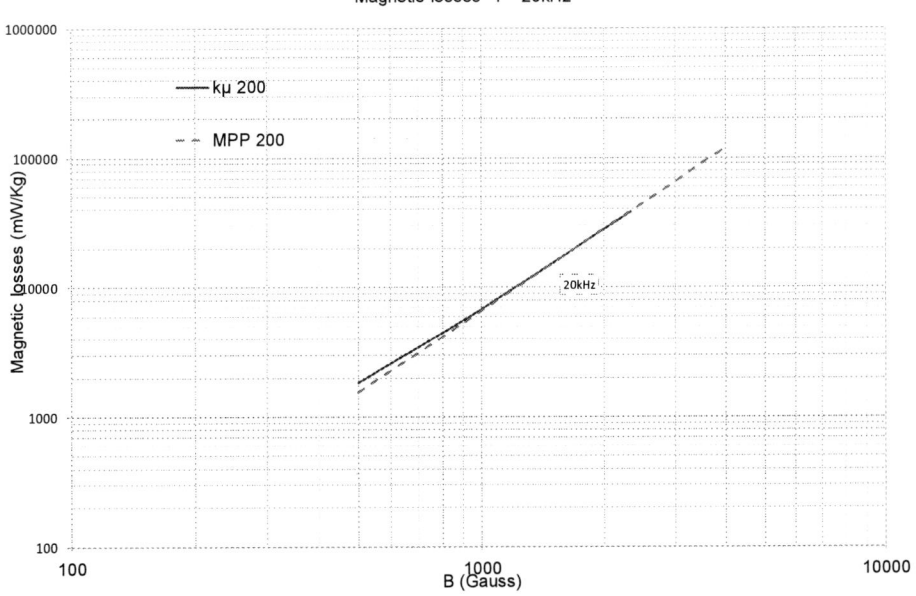

Fig. 2. Magnetic losses in kµ magnetic core compared to other products at f = 20 KHz

In other words, we can say that for the frequencies between 10 - 20 KHz, the kµ magnetic cores have magnetic losses lower or similar to MPP magnetic cores.

© VDE VERLAG GMBH · Berlin · Offenbach

85

3. Stability of permeability

The kμ magnetic cores keep constant permeability for a wide range of frequencies compared to MPP products. This means that dissipative phenomena are weaker in the case of nanocrystalline cores kμ, consistent with the lower magnetic losses shown in Fig.1 at medium frequencies. In example of Fig.3 the μr magnitude is 200, which corresponds to a cut off frequency higher than 12MHz for kμ cores, <5MHz for MPP.
For this reason kμ cores are a good solution in the domain of high frequency.

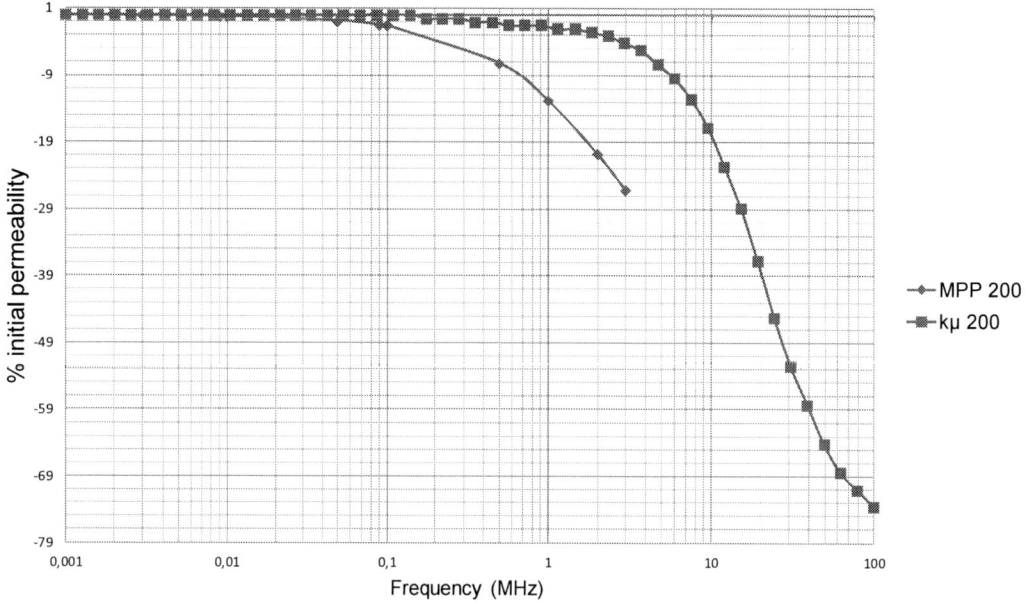

Fig. 3. Permeability versus frequency kμ, and MPP

Fig. 4 shows the permeability versus la frequency for two different kμ magnetic cores (kμ200, and kμ3000). The skin effect depth is the principal raison of this phenomenon:

$$\delta = \sqrt{\frac{\rho}{\pi . f . \mu_0 . \mu_r}}$$

Where:
ρ: Electrical resistivity of nanocrystalline (115 $\mu\Omega. cm$)
f: Frequency (Hz)
μ_0: Vacuum permeability ($4\pi \times 10^{-7} H/m$)
μ_r: Relative permeability

The cutoff frequency f_c obtained when $\delta = e/2$ (e = thickness of the nanocrystalline tape)
We can find that for kμ200, and kμ3000 the value of f_c is respectively 12.5 MHz, 0.8 MHz.
The skin effect has a more important influence for the kμ magnetic core with a relatively high permeability.

PCIM Europe 2016, 10 – 12 May 2016, Nuremberg, Germany

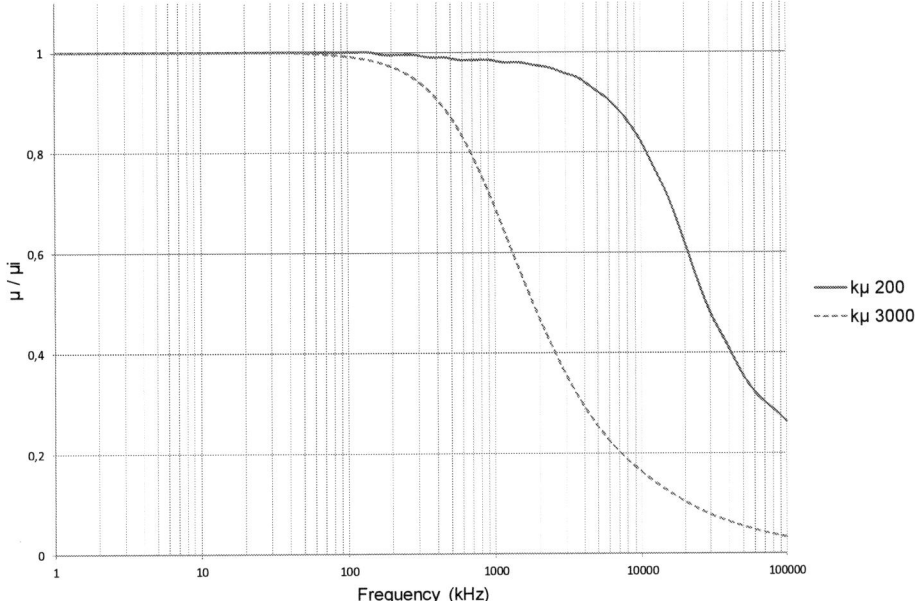

Fig. 4 . Permeability versus frequency kμ200, and kμ3000

Fig. 5 presents permeability versus magnetic field for kμ and MPP magnetic core. We can clearly see that the permeability stays stable until 60 Oe (3978 A/m) for kμ magnetic core and about 7 Oe (577 A/m) for MPP magnetic core.
This characteristic becomes very important for applications which need a high magnetic field.

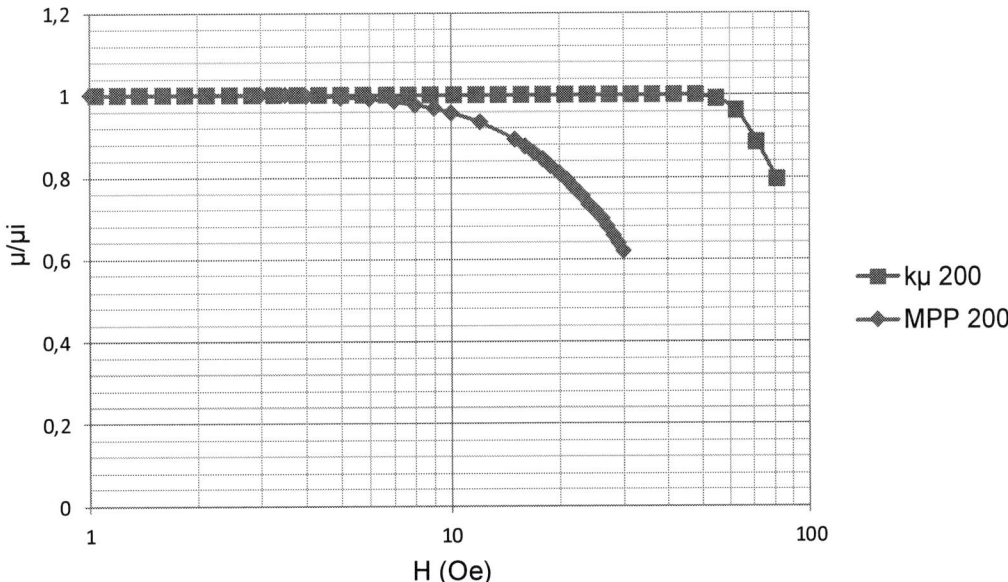

Fig. 5. Permeability versus magnetic field kμ, and MPP

Fig. 6 shows permeability versus magnetic field for two different permeability values (kμ 200, and kμ 3000).
The value of the magnetic field where the permeability decreases is about 60 Oe, and 4 Oe for kμ 200, and kμ 3000 respectively.

© VDE VERLAG GMBH · Berlin · Offenbach

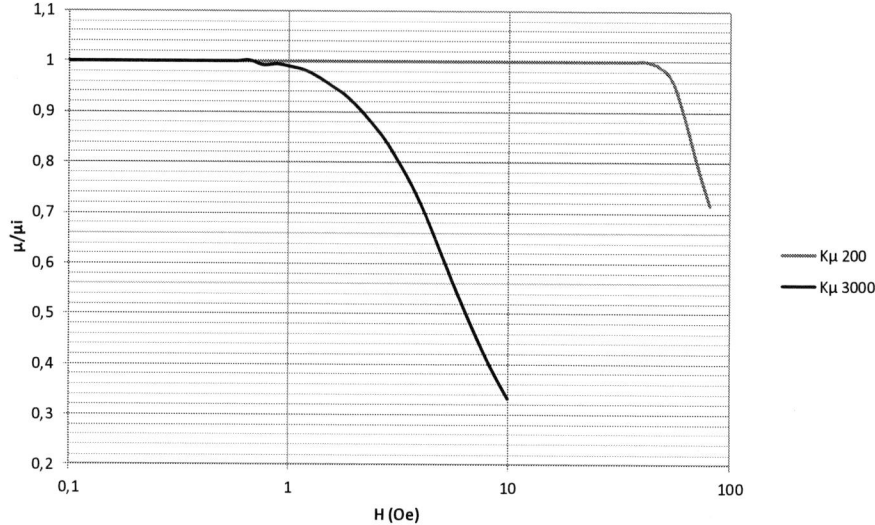

Fig. 6. Permeability versus magnetic field kµ200, and kµ3000

4. High saturation and linearity

Fig.7 shows B-H curves for kµ and MPP magnetic cores. We can see the high linearity of these materials before saturation at 1.2 T for kµ magnetic core and about 0.8 T for MPP. This high saturation value for kµ magnetic core when compared to MPP is one of the important parameters to take into account when designing applications. The high saturation value is one of the essential requirements of the magnetic core for some applications such as pulse power core, power transformer, and common mode choke.

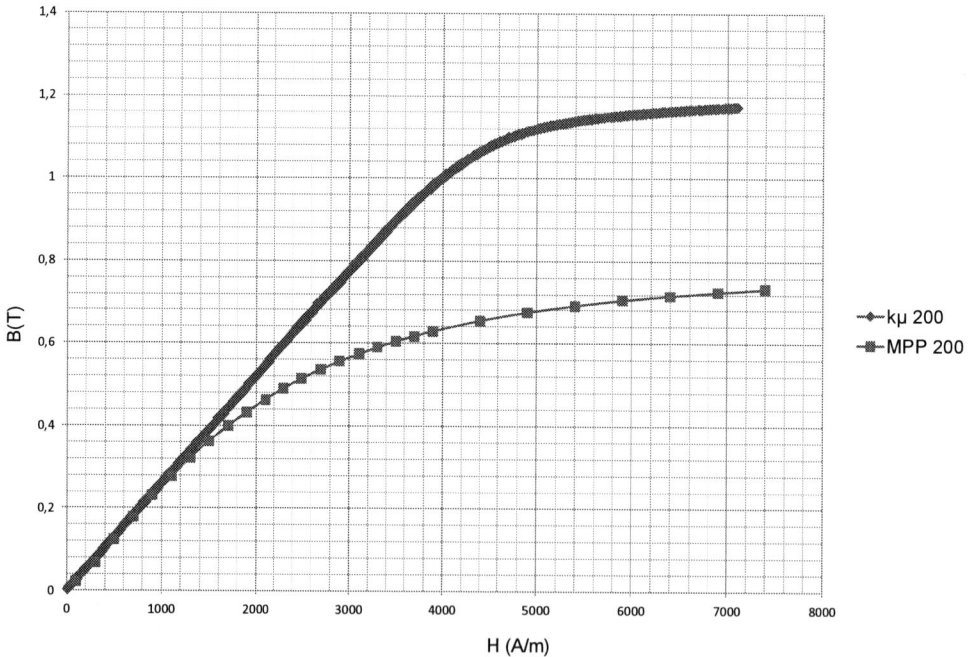

Fig.7 : B-H curve of kµ, and MPP magnetic cores

The high linearity is also an important parameter of the magnetic core for some applications such as current transformers.

For all power electronics applications, the lower the magnetic losses in the magnetic material, the higher the efficiency of the application.

For example, the kμ magnetic core can be used to filter the noise produced inside a power supply in order to protect itself and the device connected to it. For the DC-DC converter in electric vehicles, an end noise filter is needed to eliminate the different harmonics which can damage the device. In some applications, the kμ magnetic core can replace ferrites cores without needing an air gap. The absence of an airgap in the kμ magnetic core reduces the perturbation produced in the electric device compared to ferrite magnetic core [7], [8].

5. Conclusion

In this paper, we presented a new generation of Nanocrystalline magnetic cores produced by Aperam Alloys (kμ). The high saturation value, high stability against frequency, and very low losses of kμ cores allow the design of electronic components that respond to the requirements of ever more demanding markets and applications such as embedded electronics for EV and HEV.

References

[1] O. Ishii, Y. Miyaguchi and S. Kambe "Application of Fe-Based Nanocrystalline Ribbon as a Noise Filter", IEEE TRANSACTIONS ON MAGNETICS, *VOL 35*, NO. 5, 1999.

[2] Y. Liu, Y. Han, S. Liu, and F. Lin "Pulse Magnetic Properties Measurement and Characterization of Fe-Based Nanocrystalline Cores for High-Voltage Pulse Magnetics Applications", IEEE TRANSACTIONS ON POWER ELECTRONICS, *VOL 30*, NO. 12, 2015.

[3] ALVES FRANCISCO; DESMOULINS JEAN BAPTISTE; HERISSON DIDIER; BENCHABI ADEL; WAECKERLE THIERRY; FRAISSE HERVE; BOULOGNE BRUNO, N° de brevet: FR2823507.

[4] WAECKERLE THIERRY; SIMON FABIEN; ALVES FRANCISCO; SAVE THIERRY; DEMIER ALAIN, N° de brevet: FR2877486 (A1).

[5] T. WAECKERLE – T. SAVE – B. VACHEY – D. GAUTARD – H. FRAISSE « Compared nanocrystalline and ferrite optimized common mode choke for RFI filters suitable as compact components» conference EPE-PEMC 2004 RIGA, 11th International Power Electronics and Motion Control Conference, 2004.

[6] F. ALVES - C. RAMIARINJAONA - S. BERENGER - R. LEBOURGEOIS - T. WAECKERLE « High frequency behavior of magnetic composites based on FeSiBCuNb Particles for power electronics », IEEE Transactions on Magnetics, 2002, vol. 38, n° 5, part 1, pp 3135-3137.

[7] T. WAECKERLE - H. FRAISSE - D. GAUTARD « Nanocrystallized cores for ground fault circuit breakers », Magnetism and Magnetic Materials, 254-255 (2003).

[8] T. Waeckerlé – T. Save – B. Vachey – D. Gautard « Strong volume reduction of common mode choke for RFI Filters with the help of nanocrystalline cores : design and experiments», SMM17, Soft Magnetic Materials, 2005.

The Fe-based glassy alloy powder core inductor for the boost converter by GaN HEMT and SiC SBD 1MHz operation

Mitsunao Fujimoto, Alps Green Devices Co. Ltd., Japan, mitsunao.fujimoto@alpsgd.com
Yutaka Naito, Alps Green Devices Co. Ltd., Japan, yutaka.naito@alpsgd.com
Takao Mizushima, Alps Green Devices Co. Ltd., Japan, takao.mizushima@alpsgd.com

Abstract

The highest efficiency as 97.5% could be obtained on the 300W boost converter which was operated at 1MHz switching and discontinuous mode. It consisted of GaN HEMT, SiC SBT and the new Fe-based glassy alloy called "Liqualloy™" inductor. Furthermore, the temperature raise of inductor part was lower than that of the other conventional inductors which were evaluated in this work. These excellent results are based on both low core loss and high saturation flux density of the Liqualloy™ inductor. This is suitable for the power converting efficiency improvement even under higher frequency and large ripple current operation.

1   Introduction

It has been secured for a long time to realize both smaller size and higher efficiency for a power supply. In generally, however, we have had to give it up because properties of each electric device which constitute the power supply are not suitable for achieving both small size and high efficiency. Now a day, one of solutions for that is to use the GaN HEMT and SiC SBD [1]. It is well known that the GaN HEMT and SiC SBD can perform at higher frequency and large current without remarkable increase in switching losses. On the other hand, the magnetic passive components such as inductive devices have a big problem because pretty large core losses are generated in case of high frequency and large ripple current conditions.   In order to solve this problem, a serious effort has been done to reduce core loss at higher frequency. Until today, such ideal soft magnetic material has been strictly limited in terms of low flux density.

It has been known that magnetic cores consisted of Fe-based amorphous alloy is suitable for inductive applications because of its good soft magnetic properties [2]. However, the shapes of amorphous alloys for these applications have been usually limited to sheet, wire and films because of their low glass forming ability, hence the core shape has been restricted [3,4]. Furthermore, generally speaking, resistivity of these alloys is so lower than any other oxide magnetic materials, that these cores have had large core loss arisen from generation of large eddy current at higher frequency range. In order to obtain the amorphous alloy cores of any desired shape using at higher frequency range, great efforts have been devoted to prepare them by powder metallurgy [5-8]. It has been, however, pretty difficult to obtain such cores, because they must be fabricated by consolidating process of high pressure at lower temperature without crystallization [9-11]. In addition to that, it is hard to eliminate inner stress sufficiently by annealing at low temperature without crystallization. Accordingly, new amorphous alloys that have high thermal stability and high deformability are desired to obtain a magnetic core for a high frequency drive magnetic devices.

In the second half of the nineteen eighties, many bulk amorphous alloys have been formed in multicomponent Mg- [12,13], Ln- [14,15], Zr- [16,17] and Zr-Ti-based [18] (Ln=lanthanoide metal) alloy systems. These bulk amorphous alloys have a wide supercooled liquid region more than 60K below crystallization temperature. The appearance of the large supercooled liquid region ($\triangle T_x$) implies a high resistance against crystallization, leading to a high glass forming ability. Additionally, it has further been reported that deformation and working processes are easier in the $\triangle T_x$ region because of its low viscosity and ideal Newtonian flow [19]. These bulk amorphous alloys always satisfy the following empirical rules for achievement of large glass-forming ability: multicomponents with significantly different atomic size ratios higher than approximately 12 % and negative heats of mixing [20-23]. According to these rules, we have tried to find a new bulk Fe-based glassy alloy [24] and already reported that Fe-Al-Ga-P-C-B-(Si) amorphous alloy had a wide $\triangle T_x$ exceeding 60K below crystallization temperature and relatively high electrical resistivity (ρ) of about 160 μm [25]. Recently, we also reported that the metal composite type powder core which was made of Fe based glassy alloy powder called "Liqualloy™" was effective to improve the efficiency in the POL 500kHz DC/DC convertor for VRM [26-28]. These phenomena were based on the excellent soft magnetic properties and its high electrical resistivity for the Liqualloy™ inductors.

In this paper, we intend to present the comparison data between Liqualloy™ and any other commercial inductors which are tested on the 1MHz operating boost converter by GaN HEMT and SiC SBD operation.

2. Experimental procedure

Multicomponent FeCrPCBSi alloy was used as "Liqualloy™". The Liqualloy™ was prepared by water atomize method. After the binding material as an adhesive was mixed to the powder, it was consolidated to a $\varphi20 \times \varphi12 \times 7t$ troidal core under a uniaxial pressure at room temperature. The troidal cores were annealed at 573~623 K for 3.6 ks in N_2 atmosphere using electrical furnace. Micro-structure of the core body was observed by HRTEM and XRD. Structural thermal stability was measured by differential scanning calorie meter (DSC). A static electrical property like inductance (L) was measured by a precision impedance analyzer under a maximum applied field of 2.4kA/m. And the core losses under various conditions were measured by IWATSU SY-8218 B-H loop analyzer. Saturation magnetic flux density was evaluated by VSM at 800kA/m. The dynamic characteristics like the efficiency (η) of the boost converter using each inductor made of various kinds of magnetic materials were measured at 1MHz switching by GaN HEMT (Transphorm inc.) and commutating by SiC SBD. Test condition was 120~270V input voltage and 390V output voltage, 0.77A output current. Figure 1 shows the circuit diagram to evaluate the various inductors.

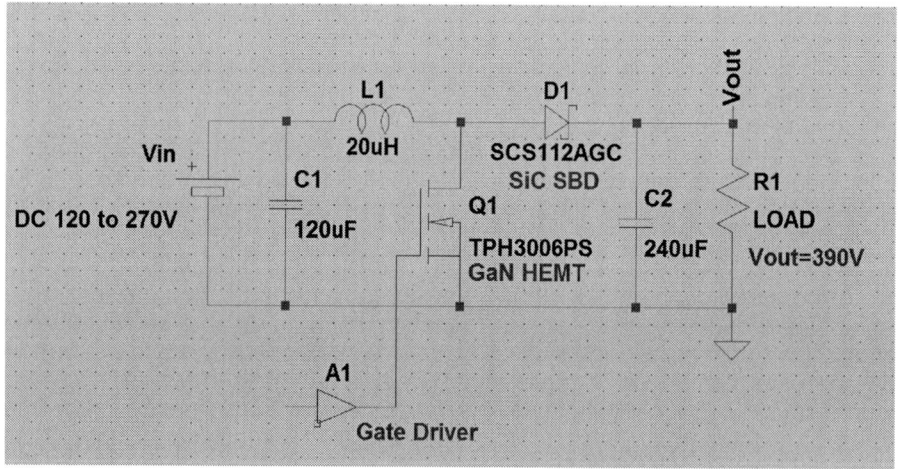

Fig. 1 Boost converter circuit diagram to evaluate various inductors.

3. Results

Figure 2 shows the outer appearance and X-ray diffraction pattern of the Liqualloy™ troidal core. No luster and smooth surface of the core can be observed. No contrast based on the precipitation of the crystalline phase is also observed after optimized annealing. This fact indicates that this core is consisted of only homogeneous amorphous phase.

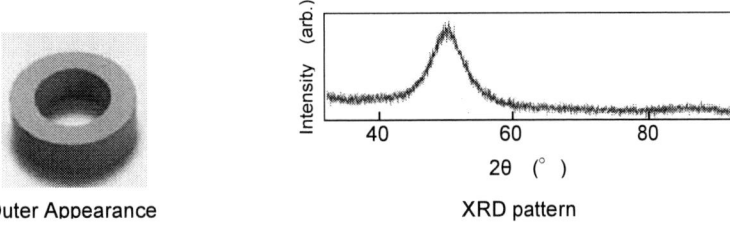

Outer Appearance   XRD pattern

Fig. 2   Outer appearance and XRD pattern of the Liqualloy™ φ20×φ12×7t toroidal core.

3-1. Static magnetic properties comparison between Liqualloy™ and conventional magnetic materials

Changes in the core loss of the Liqualloy™ powder core as a function of frequency at Bm=50mT and maximum magnetic flux density at 1MHz are shown in figure 3. The data of the conventional Fe based amorphous alloy, NiFe and FeAlSi powder cores are also shown for comparison.

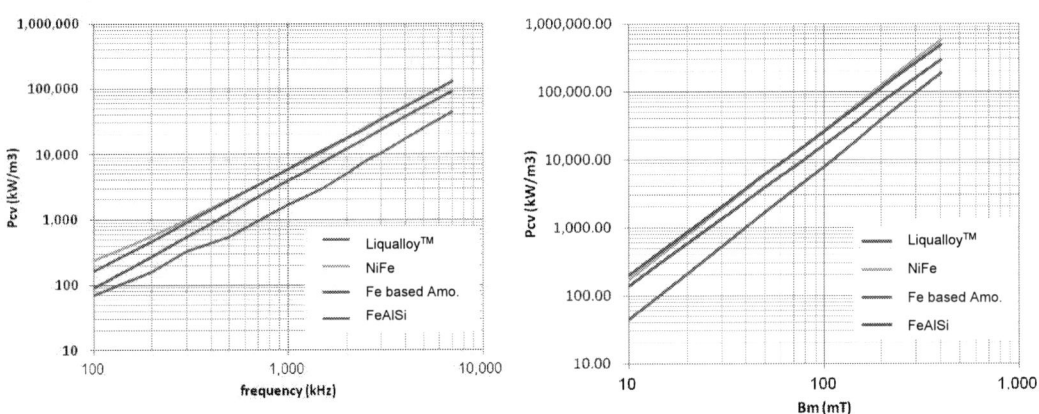

Fig. 3   Changes in the core loss of the Liqualloy™ powder core as a function of frequency at Bm=50mT and maximum magnetic flux density at 1MHz. The data of the other conventional powder cores are also shown for comparison.

The core loss of the Liqualloy™ is much lower than that of the other materials in the measurement range. The difference of the core loss between the Liqualloy™ and conventional powder core becomes larger in the higher frequency and larger Bm range. The low core loss of the Liqualloy™ is based on its good soft magnetic properties and much high electrical resistivity.

Table 1 shows the saturation magnetic flux density (B_S), core loss (P_{cv}) at 1MHz and the maximum flux density (B_m) of 50mT and the resistivity for various cores. The B_S of the Liqualloy™ core is larger than that of the FeAlSi alloy powder core, but is lower than that of the conventional Fe based amorphous alloy and NiFe cores. The Liqualloy™ core shows the lowest core loss in these evaluated cores. This is based on its excellent soft magnetic properties at a higher frequency range such as 1MHz and high resistivity. To reduce the core loss at high frequency range, it is effective to increase of the resistivity of the core because the eddy current loss is dominant factor of the core loss at higher frequency range. The

Liqualloy™ powder core has the highest resistivity about 3kΩm in the evaluated cores shown in table 1. Then the core loss of the Liqualloy™ is the minimum value because of its low hysteresis loss and eddy current loss.

Figure 4 shows the comparison data of the DC bias characteristics of the various inductors which are powder cores with winding. And table 2 shows the initial inductance (L), number of winding and DC resistance (Rdc) of each inductor. Since initial permeability(μ') of the various cores were not the same, initial inductance was adjusted to 20μH by means of number of windings. The μ' of Liqualloy™ is lower than that of the other cores, then more winding is needed. Accordingly, Rdc of the Liqualloy™ inductor is higher. L of the inductors which are evaluated in this study decreases with increase of dc bias current.

Table 1 Comparison of flux density, core loss and electrical resistivity for various powder

Core	flux density (B_S)	core loss (P_{cv}) @1MHz 50mT	electrical resistivity (ρ)
	(T)	(kW/m^3)	(Ωm)
Liqualloy™ powder core	1.3	1,460	3,000
Fe based amorphous powder core	1.5	3,844	120
NiFe alloy powder core	1.5	5,979	21
FeAlSi alloy powder core	0.9	6,062	17

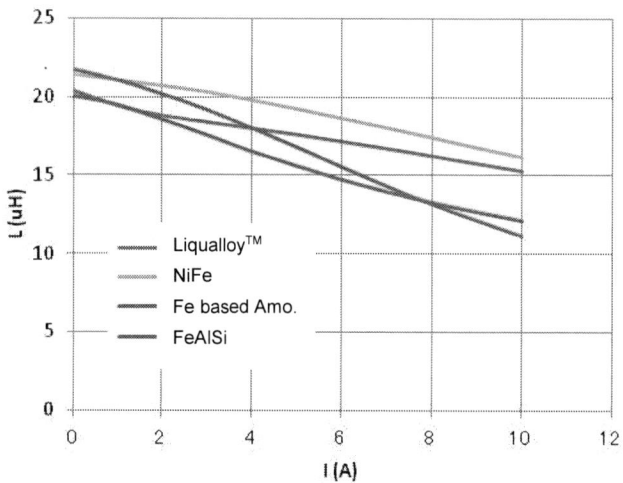

Fig. 4 D.C. bias characteristics of the various kinds of powder cores at 1MHz.

Table 2 Number of windings, Inductance (L) at 1MHz and D.C. resistance (Rdc) for the various inductors.

Inductor	Number of windings	L@1MHz (uH)	Rdc (mΩ)
Liqualloy™	28	20.43	31.6
Fe based Amo.	21	20.05	25.6
NiFe	19	21.07	23.1
FeAlSi	17	21.02	21.0

3-2. Dynamic characteristics of the test board operated by GaN HMET using various inductors

Figure 5 shows the relation between the efficiency of the boost converter and output power for using each inductor. In the case of using FeAlSi powder inductor, the efficiency cannot be evaluated because of rising temperature of inductor over output power of 150W. The efficiency of the Liqualloy™ is higher than that of the other inductors in the small output power region. The maximum efficiency as 97.5% can be obtained for using the Liqualloy™ inductor. It is considered that this phenomenon is based on its low core loss at 1MHz. In the case of using the conventional Fe based amorphous alloy inductor, the efficiency is 95.8%, but it was operated only for 220sec. because of its large heat generation over temperature of 80°C. Low efficiency and short time operation of 145sec. at output power of 300W are observed when the NiFe inductor is applied. Poor efficiency of the 84% and only short time operation of 100sec. at output power of 150W can be done for the FeAlSi alloy powder inductor originated from its high core loss at 1MHz.

Table 3 shows the data of the efficiency for the evaluation board at the output power of 100W and 300W when the each inductor is applied.

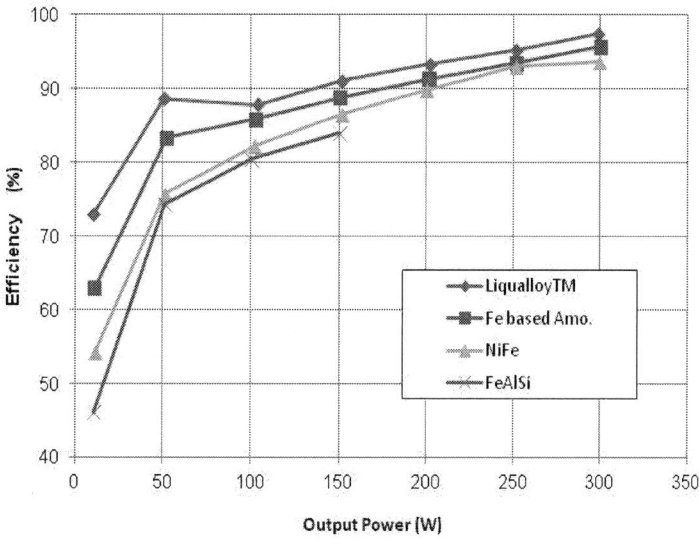

Fig. 5 Changes in efficiencies of the boost converter as a function of output power when the inductors made of the Liqualloy™, the Fe based amorphous and the FeAlSi are applied, respectively.

Table 3 Efficiency of the evaluation board at the output power of 100 W and 300 W when the each inductor is applied.

Inductor	Efficiency (%)	
	Po=100W	Po=300W
Liqualloy™	87.9	97.5
Fe based Amo.	85.9	95.8
NiFe	82.2	93.7
FeAlSi	80.5	-

The Liqualloy™ inductor shows the largest efficiency in any other inductors at the both output power. It has been known that the dominant factor to the efficiency is core loss in the small output power range and the dc resistance of the winding in the high output power range. Then the large efficiency difference between Liqualloy™ inductor and any other inductors in the low output power range is based on the extremely low core loss of the

Liquallot™ inductor at the 1MHz. On the other hand, the reason why a difference of the efficiency becomes small at a large output power range is that the resistance of the winding becomes large in the case of Liqualloy™ inductor because of low permeability. In this condition, the Liqualloy™ is the best inductor for achieve the highest efficiency.

4.   Discussion

As described above, it is considered that the highest efficiency of this evaluation board power supply using the Liqualloy™ power inductor is based on the low core loss of its inductor. Then the core losses of each inductor under the operating condition were simulated by means of a finite element analysis from the dependence of core loss on frequency and B_m. Figure 6 shows the typical wave form of the inductor L1 and GaN HEMT Q1 at the evaluation board shown in figure 1. Control IC cannot be obtained for commercial availability, and then input voltage is controlled to constant value of 390V to fix the duty factor as 0.28 when the load resistance is varied. The wave forms of an end voltage for an inductor and voltage between drain and source for a GaN HEMT are shown as ① and ②. The current wave of the inductor is also shown as ③. This evaluation board is operated under discontinuous mode to reduce switching loss. Sinusoidal and ringing wave form before each rectangular voltage wave of an inductor is based on a parasitic capacitance, inductance of GaN HEMT and a stray inductance, capacitance of the PWB. The wave form of the current shows nearly a triangle shape.

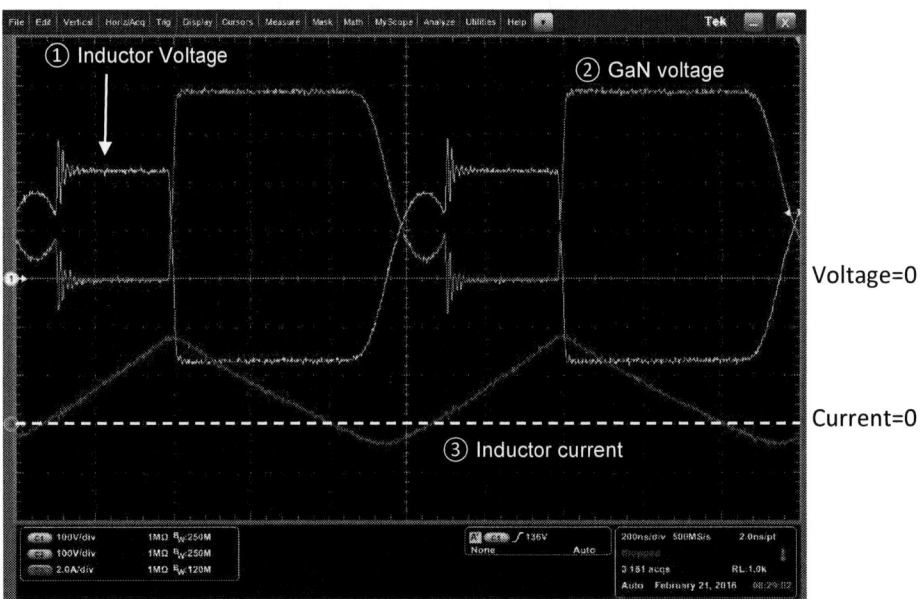

Fig. 6 Wave forms of the Liuqlloy™ inductor on the evaluation circuit operated GaN H*EMT* and SiC SBT at 1MHz.

From the wave forms using various power inductors, core losses in the working state ware simulated. Simulation results are shown in table 4. Though a circuit loss which attracted core loss from overall loss theoretically should indicate the same value in the case of using any inductors, various values can be seen in this table. It is considered that it is caused by increasing switching loss because zero current switching cannot be achieved by ringing of the inductor voltage.

© VDE VERLAG GMBH · Berlin · Offenbach

Table 4 Calculation results of the circuit loss, core loss and ratio of core loss against total loss.

Inductor	Efficiency (%)		Circuit Loss (W)		Core loss (W)		Circuit loss without core loss (W)		Rate of core loss against total loss (%)	
	Po=100W	Po=300W	Po=100W	Po=300W	Po=100W	Po=300W	Po=100W	Po=300W	Po=100W	Po=300W
Liqualloy™	87.9	97.5	14.7	7.6	1.6	2.1	13.1	5.5	10.8	28.0
Fe based Amo.	85.9	95.8	17.0	13.2	6.3	8.9	10.7	4.3	36.9	67.3
NiFe	83.2	93.7	22.1	20.0	12.6	13.5	9.5	6.5	56.9	67.4
FeAlSi	80.5	-	24.4	-	17.4	-	7.0	-	71.1	-

From this simulation results, core loss of the Liqualloy™ inductor is 1/4 ~ 1/10 times smaller than any other evaluated powder inductors. The rate of the core loss against the total loss also becomes smaller than that of the any other inductors. These facts indicate that the highest efficiency using the Liqualloy™ inductor is based on its low core loss

Accordingly, the Liqualloy™ inductor is the best to achieve the high efficiency for this application which is operated by GaN HEMT and SiC SBD at 1MHz.

5. Conclusion

The highest efficiency of the 97.5% at the output power of 300W can be obtained when the Liqualloy™ power inductor made of new glassy alloy powder is applied to the evaluation board operated by GaN HEMT and SiC SBD at 1MHz. It is based on its significantly low core loss at the working state. This is originated from its good soft magnetic properties and high resistivity. Therefore, it can be said that the Liqualloy™ inductor is one of important devices to achieve high efficiency for the power supply operated by new generation IC operated at higher frequency.

6. Literature

[1] U. K. Mishra, P. Parikh and Y. F. Wu., *PROCEEDINGS-IEEE* 90.6 (2002): 1022.

[2] C. H. Smith, "Applications of rapidly solidified soft magnetic alloys', in Rapidly solidified alloys; processes, structures, properties, applications' (ed. H.H. Liebermann), New York, Marcel Dekker (1993), 617.

[3] "Materials Science of Amorphous metals" (ed. T. Masumoto), Ohmu Yokyo (1982).

[4] M. Hagiwara, A. Inoue and T. Masumoto, Metall. Trans., A 13A(1982), 373.

[5] R. Hasegwa, R. H. Hathaway and C. F. Chang, J. Appl. Phys.,57(1985), 3566.

[6] S. Minakawa and T. Masumoto, IEEE Trans. Magn.,MAG-23(1987), 3245.

[7] I. Ohtsuka et all, J. Mag. Soc. Japan, 21(1997),617.

[8] I. Endo et all, IEEE Trans. Magn., 35(1999),3385.

[9] C. F. Cline and R. W. Hopper, Scr. Metall., 11(1977),1137.

[10] L. E. Murr et all, Scripta METALLURGICA, 17(1983), 1353.

[11] D. G. Morris, Rapidly Quenched Metals, 2(1984)1751.

[12] A. Inoue, K. Ohtera, K. Kita and T. Masumoto, Jpn. J. Appl. Phys., 27(1988), 2248.

[13] A. Inoue, M. Kohinta, A. P. Tsai and T. Msumoto, Mater. Trans. JIM,30(1989), 378.

[14] A. Inoue, T. Zhang and T. Masumoto, Mater. Trans. JIM, 30(1989), 965.

[15] A. Inoue, H.Yamaguchi and T. Masumoto, Mater. Trans, JIM, 31(1990), 104.

[16] A. Inoue, T. Zhang and T. Masumoto, Mater. Trans. JIM, 31(1990), 177.

[17] A. Inoue, T. Zhang and T. Masumoto, Mater. Trans. JIM, 32(1991), 1005.

[18] A. Peker and W. L. Johnson, Appl. Phys. Lett., 63(1993), 2342.

[19] A. Inoue, Y. Kawamura, T. Shibata and K. Sasamori, Mater. Trans., JIM,37(1996), 1337.

[20] A. Inoue, Mater. Sci. Forum, 179-181 (1995), 691.

[21] A. Inoue, Mater. Trans. JIM, 36(1995), 866.

[22] A. Inoue, Sci. Rep., RITU,A42(1996),1.

[23] A. Inoue, Mater. Sci. Eng., A226-228 (1997), 357.

[24] A. Makino, A. Inoue and T. Mizushima, Mater. Trans. JIM, 41(2000), 1471.

[25] S. Yoshida, T. Mizushima, A. Makino and A. Inuoe, J. Magn. Soc. Jpn.,23(1999), 871.

[26] T. Mizushima, H. Koshiba, Y. Naito and A. Inoue, J. Jpn. Soc. Powder Powder Metall., 54(2007), 768.

[27] H. Koshiba, Y. Naito, T. Mizushima and A. Inoue, Materia,47(2008),39.

[28 T. Mizushima, H. Koshiba, Y. Naitoh and A. Inoue, J. J. Soc powder and powder metallurgy, 55,(2008), 146.

Accurate Calculation of AC Losses of Inductors in Power Electronic Applications

Ranjith Bramanpalli
Würth Elektronik eiSos GmbH, Max-Eyth Str.1, 74638, Waldenburg, Germany
ranjith.bramanpalli@we-online.com

The Power Point Presentation will be available after the conference

Abstract

This paper discusses the new method to calculate AC losses of power Inductors (filter, storage & coupled Inductors) in power electronic applications with rectangular driven voltage waveforms. AC loss is the loss occur in the core and windings of a power Inductor caused by AC flux swing. The new model is proposed to overcome the disadvantages of Steinmetz equation and its extensions. The experiments conducted on various materials (MnZn, NiZn, iron powder & WE-superflux) have shown the model to be very accurate over wide range of frequency and duty cycle. Dependability on core manufacturers for core loss charts will be avoided.

1. Introduction

In Switch Mode Power Supply (SMPS), magnetic component is one of the crucial component. Hence predicting the losses, behavior and temperature rise accurately is critical. To predict temperature rise and behavior, core loss needed to be calculate accurately.

Historically, core losses are estimated by using Steinmetz power equation (1) and later its extensions Modified (2) or Generalized Steinmetz equations (3).

$$P_v = K \times f^\alpha \times B_{pk}^\beta \qquad (1)$$

$$P_v = \left(K \times f_{eq}^{\alpha-1} \times B_{pk}^\beta \right) \times f \quad (2) \quad \text{where} \quad f_{eq} = \frac{f}{2\pi \times (DC - DC^2)} \quad (2)$$

$$P_v = \left(K \times f_{eq}^\alpha \times B_{eq}^\beta \right) \quad (3) \quad \text{where} \quad B_{eq} \text{ is } \frac{1}{4} \int_0^T |\frac{dB}{dt}| dt \quad (3)$$

© VDE VERLAG GMBH · Berlin · Offenbach

Where P_v is the core loss for unit volume, f is the frequency, B_{pk} is peak flux density of a sinusoidal excitation. f_{eq} being the equivalent frequency with respect to change in the duty cycle (DC) of non-sinusoidal waveforms. K, α & β are the constants, derived from core loss graph shown in Figure 1 which is provided by core manufacturers. The data given in figure 1 for core losses usually includes the effects of both hysteresis and core eddy currents.

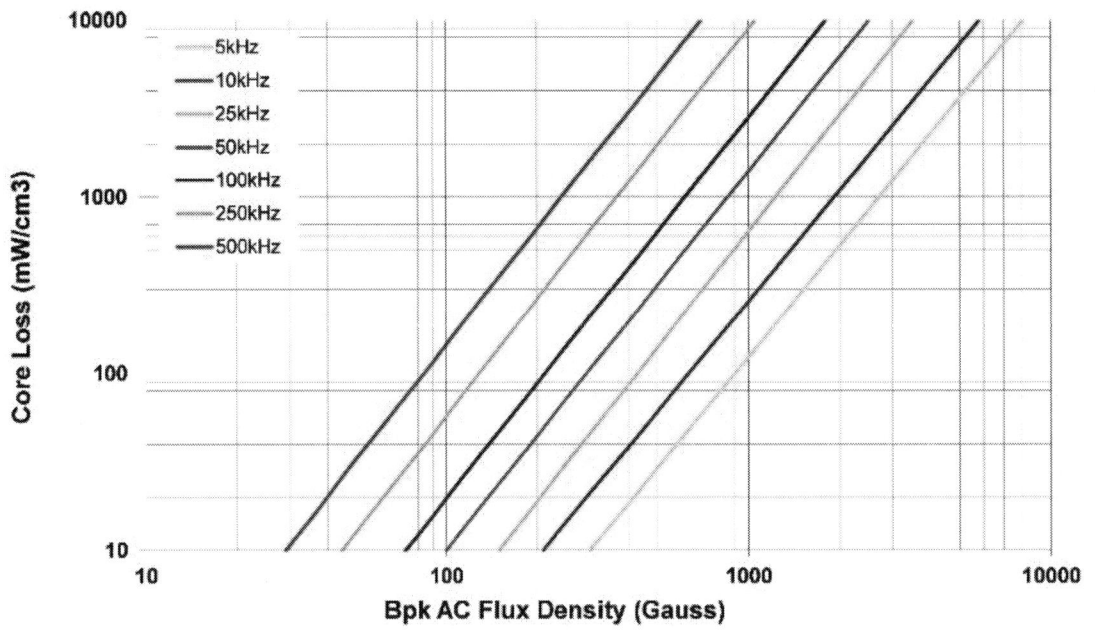

Figure 1: Core Loss graph plotted against peak flux density at different frequencies

The major drawback of the Steinmetz equation is that it is mostly valid for sinusoidal excitation. This is a huge drawback because in most power electronic applications, the Inductor is usually exposed to non-sinusoidal flux waveforms. Even though the extended Steinmetz models try to compensate for non-sinusoidal flux waveforms but as the empirical data (figure 1) is purely based on sinusoidal excitation, the accuracy of these extended models are limited to a narrow range of frequency, duty cycle, flux density and materials. Another disadvantage is that, to produce the empirical shown in figure 1 requires complex and expensive test set up. There also exist some other models, developed by core manufacturers which works best with the cores only manufactured by them.

2. The Würth Elektronik AC Loss Model

In order to utilize the Steinmetz equation for non-sinusoidal waveforms to estimate core losses, Würth Elektronik eiSos has developed state of the art model, allowing customers to effectively select the inductor and optimize the system. The model is based on the empirical data, derived from real time application set up.

Test Setup:

A DC-DC converter as shown in figure 2a is utilized to produce an empirical data. A pulsating voltage is applied over the Inductor and the power Input, P_{in} and Power output P_{out} are measured, then $P_{Loss} = P_{in} - P_{out}$ is estimated and then AC loss of the Inductor P_{AC} is separated. This process is repeated over wide range of parameters including variation of peak flux density swing, frequency, ripple current, etc., to produce the empirical data shown in figure 3.

In the Würth Elektronik model the total loss of the Inductor is divided into two separate losses as AC loss and DC loss. The loss in addition to DC power loss occurred due to AC flux swing in the coil and the core is termed as AC loss.

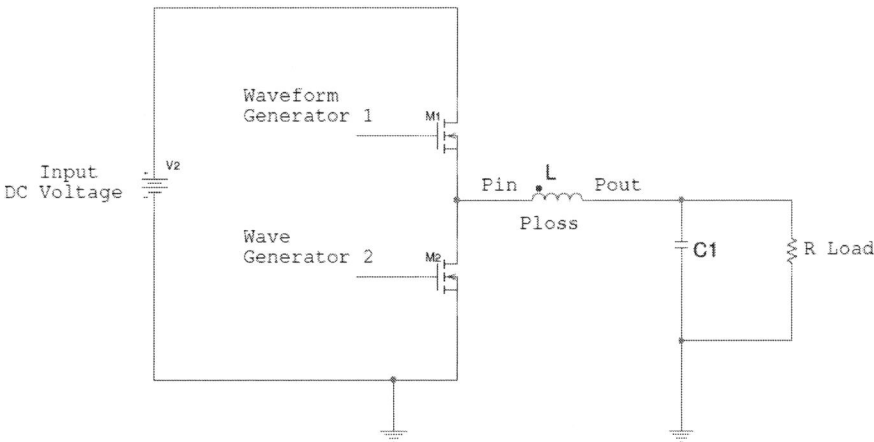

Fig: 2a

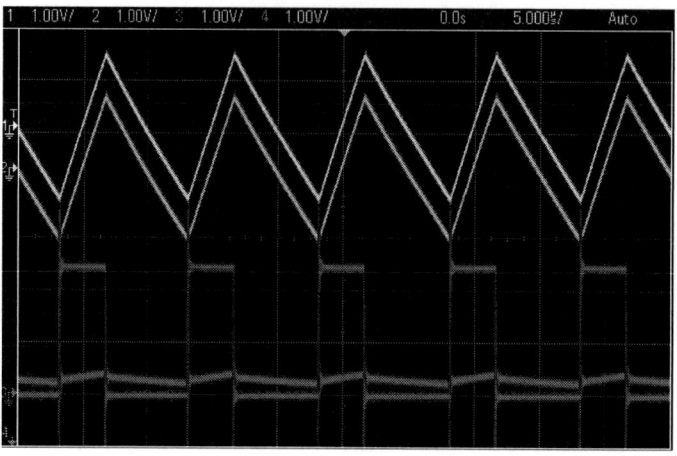

Fig: 2b

Fig 2a & 2b: Setup of DC-DC Converter for Loss Determination and resulting scope screenshots

Typically, the empirical data is generated by core manufactures as shown in figure 1 is by applying sinusoidal excitation to the ring core, overdriven from + to - saturation and the hysteresis loop area represents energy loss shown in figure 4a.

But in Würth Elektronik model, the inductor(instead of core itself) is driven by a much smaller rectangular waveform with peak flux density limited by core losses to a minor hysteresis loop just like in power electronic applications as shown in figure 4b.

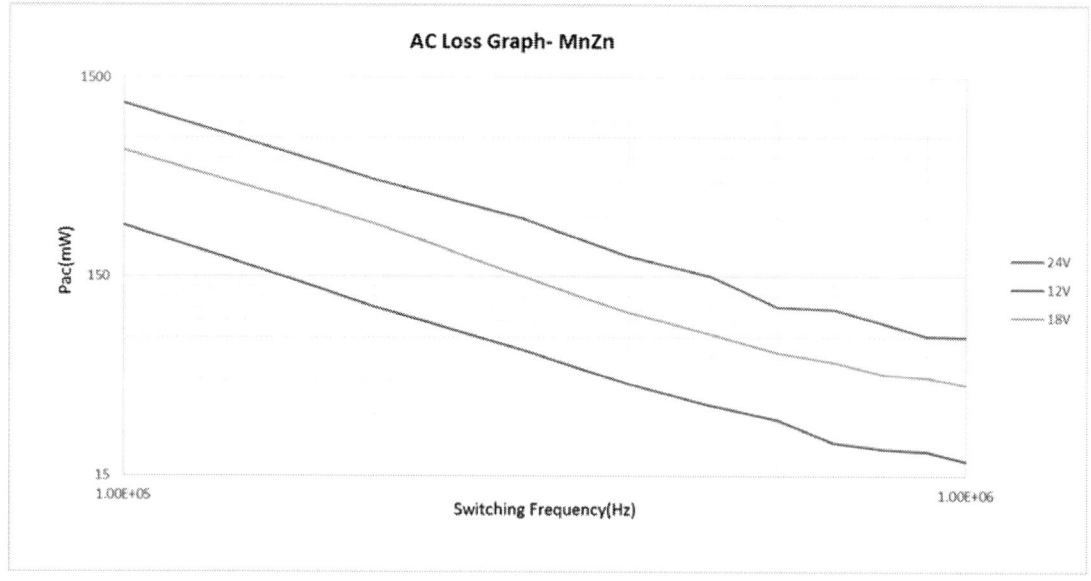

Figure 3: Core loss chart plotted using empirical data from DC-DC converter

Power loss depends on how many times per second the hysteresis loop is traversed. Thus, hysteresis loss varies directly with frequency. The hysteresis loop changes shape somewhat with waveform shape, current or voltage drive, and temperature. The minor loop area depends on the voltage above the Inductor. Due to this minor loop approach followed by Würth Elektronik AC loss model to produce empirical data, has proved to be robust, accurate over wide range of parameters like frequency, ripple current and duty cycle.

Finally, this empirical data is then used to create an equation to calculate AC loss which is the function of (4)

$$P_{AC} = f(V, Fsw, k1, k2, ...) \quad (4)$$

Where V_{ind} is the voltage measured over the inductor, T_{on} is the turn-on time of switch M1.
N & A_e is the number of turns of an inductor and surface area of the core respectively. K_0, α & β are the constants derived from core loss chart shown in figure 3.

Since, the empirical data is purely based on real time parameters with accurate estimation of losses for any given Duty Cycle being achieved. This model also includes AC winding losses, estimates loss for metal alloy and iron powder materials. The model is also applicable over wide range of frequency, as the constants of the power equation are derived over a wide range with respect to the flux swing.

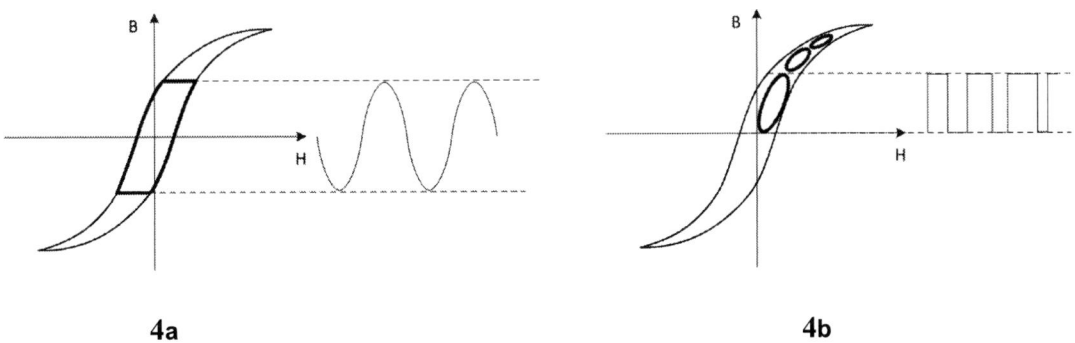

Fig 4a & 4b: Typical modelling used for Steinmetz equation and it's extensions on the left and Würth Elektronik's Minor loop approach on the right

3. Performance

The AC loss for various materials like WE-Super flux , Iron powder, NiZn, MnZn, etc., are measured at a wide range of duty cycles, frequencies and other parameters compared with theoretical models as shown in figures 5 to 8. In the below charts compares the real measurement data versus Core/AC losses determined using Steinmetz power equation(s), Würth model & other models like loss separation model.

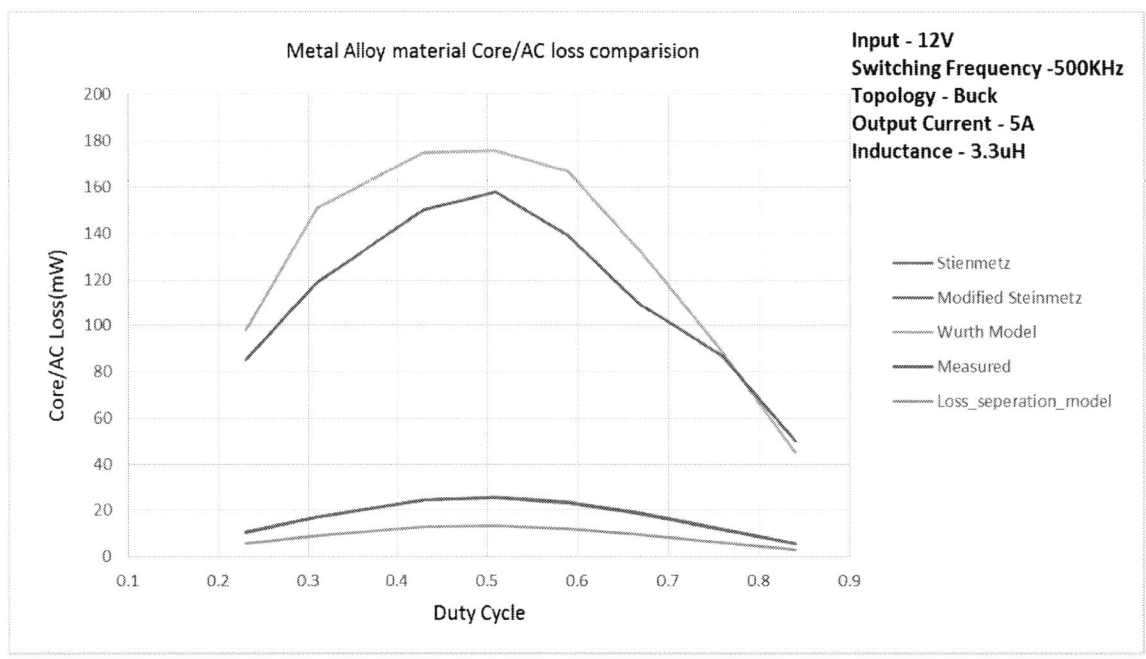

Fig 5: Loss determination using different approaches at various duty cycles for Metal Alloy Inductor

PCIM Europe 2016, 10 – 12 May 2016, Nuremberg, Germany

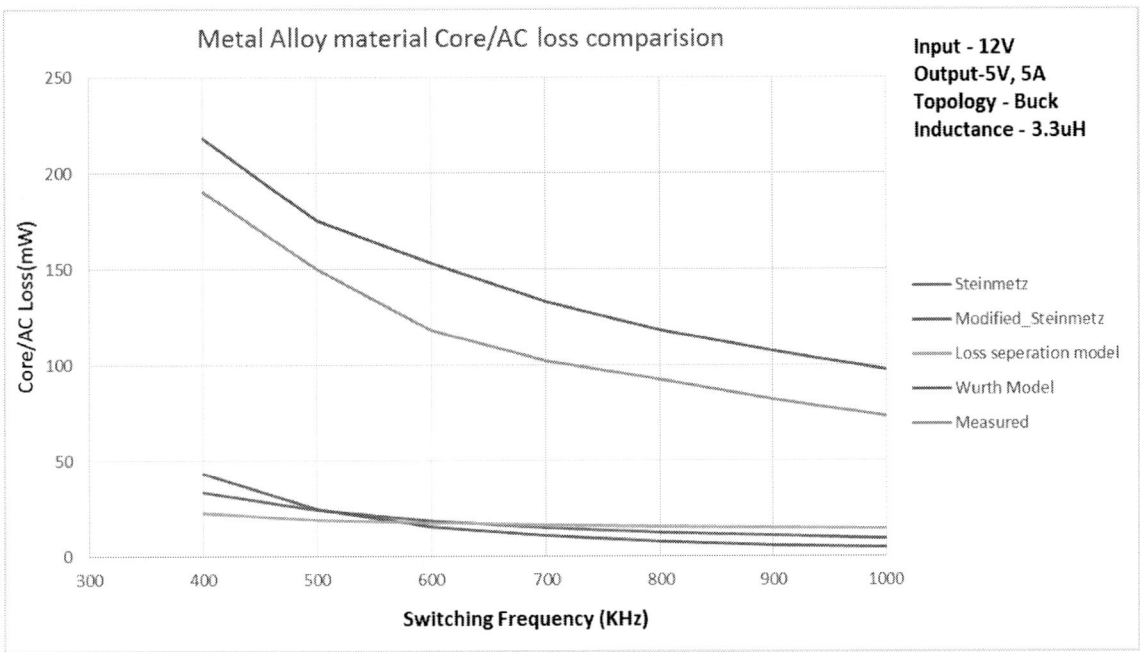

Fig 6: Loss determination using different approaches at various switching frequencies for Metal Alloy

Inductor

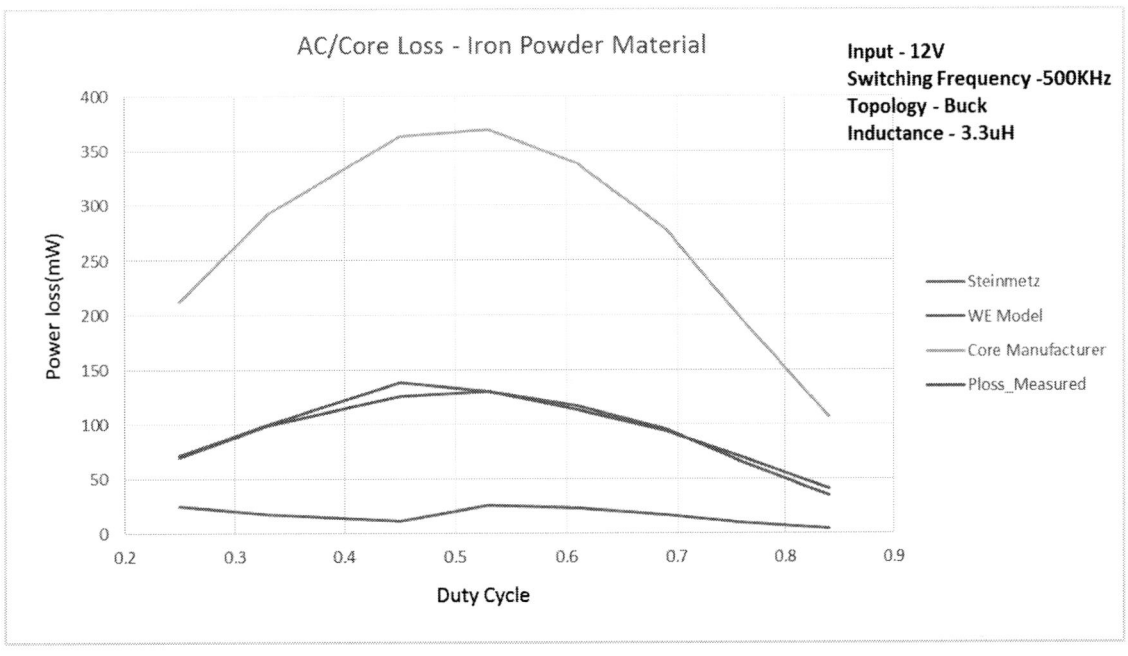

Fig 7: Loss determination using different approaches at various duty cycles for Iron Powder Inductor

© VDE VERLAG GMBH · Berlin · Offenbach

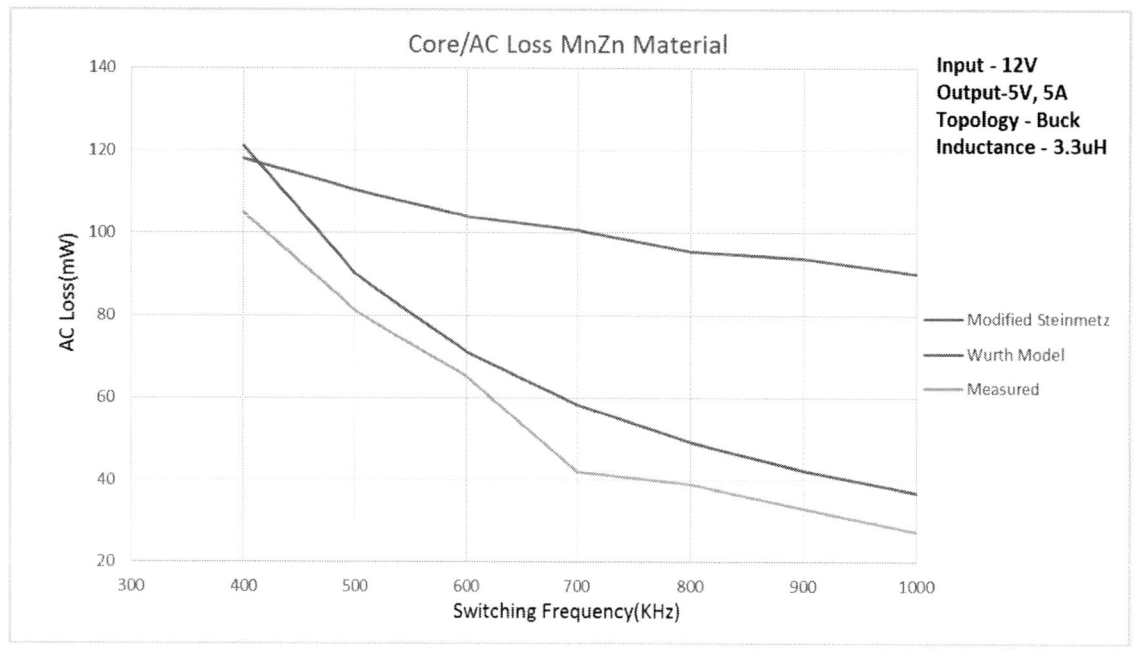

Fig 8: Loss determination using different approaches at various switching frequencies for MnZn Inductor

4. Summary

The Würth Elektronik's AC loss model is accurate and practical for determining AC losses. The model has been experimentally verified over wide range of frequency, ripple currents and duty cycles and proved to be robust.

References

- Magnetics Design for Switching Power Supplies by Lloyd H. Dixon
- On the law of hysteresis by C.P. Steinmetz
- "Calculation of losses in ferro- and ferrimagnetic materials based on the modifiedSteinmetz equation" by Reinert, J.; Brockmeyer, A.; De Doncker, R.W.
- "Improved calculation of core loss with nonsinusoidal waveforms by Jieli Li; Abdallah, T.; Sullivan, C.R.
- The Standard for measuring SMPS Inductors by Mike Wens & Jef Thone, Mindcet NV, Belgium

Thin-Film Based Microtransformer Suitable for High Switching Frequency Power Applications

Dragan Dinulovic, Würth Elektronik eiSos, Germany, dragan.dinulovic@we-online.de
Mahmoud Shousha, Würth Elektronik eiSos, Germany, mahmoud.shousha@we-online.de
Martin Haug, Würth Elektronik eiSos, Germany, martin.haug@we-online.de
Sebastian Beringer, Institute for Micro Production Technology, Leibniz Universität Hannover, Germany, beringer@impt.uni-hannover.de
Marc C. Wurz, Institute for Micro Production Technology, Leibniz Universität Hannover, Germany, wurz@impt.uni-hannover.de

Abstract

In this paper, the design, the fabrication, and the characterization of on silicon integrated microtransformers will be presented. The microtransformer is suited for power applications at high switching frequency towards to 100 MHz. This device has stable L vs. f characteristic up to 50 MHz. The design is improved regarding to the electrical resistance and current capability. The microtransformer shows an inductivity of about 50 nH and can be applied for current higher than 1 A.

1. Introduction

The recent trend in power electronics is the increasing of a switching frequency. This sets the new requirements for inductive components regarding to size, inductance, and current capability. Inductors and transformers should provide smaller inductivity and smaller size. Thin-film fabricated micro microinductors and microtransformers are ideal components for fulfill this requirements. Both, microinductors and microtransformers are recently in many research works main topic of development [1].

The main challenge besides the integration of the magnetics on silicon is achieving of good device properties like inductance, resistance, losses and finally the good efficiency at high frequencies. Different types of microtransformers devices were reported before. Two basic device types regarding to the coil design are solenoid and spiral transformer type.

The solenoid microtransformers devices were fabricated using multilayer coil surrounding the magnetic core [2]. The magnetic core is closed and is fabricated in one step, therefore the fabrication of such devices is not complex, but the axial anisotropy cannot by applied on these devices.

Many approach of transformer are spiral type transformer. These devices show easier coil de sign, but the magnetic core is complex [3, 4]. Some microtransformers are developed without magnetic core (air-core microtransformers) [5], which are suitable for very high frequencies a nd can be used for signal transfer.

Our device presented in this work is a solenoid type microtransformer. First design of this kin d of microtransformer is presented before [6].

The aim of this work is to improve design of microtransformer device especially to improve th e device parameter at higher switching frequencies up to 30 MHz. Therefore, the fabrication procedure of device was improved, with aim to reduce the coil losses.

2. Design

The layout of the microtransformer was simulated and optimized using Finite Element Method (FEM). The software tool Ansys Maxwell® was applied. To find an optimum between design and technology issues, the technological aspects of the thin-film fabrication also have to be taken into account during the simulations. The optimal design for the magnetic core is an oval core. The size of four coils was restricted by complete size (length of 2.5 mm) of a microtransformer device. As a complete size of a microtransformer an EIA standard size 1008 was chosen. Therefore, whole microtransformer device features a footprint of 2.5 mm x 2 mm. The simulated optimal design of the microtransformer is shown in the Fuigure 1.

The aim of the simulation was to develop a design with as small as possible electrical resistance and at the same time the design should have maximal electrical inductance. As an optimal design, a system with electrical inductance of about 50 nH and resistance of about 200 mΩ was chosen. Based on fabrication parameters (i.e aspect ratio and flank angle, defined by the applied photolithography processes) and requested device size, the coil design was defined. The coil consists of 5 turns. The cross-section of a coil turn is 60 µm (width) x 20 µm (height).

The microtransformer consists of a closed Co-Fe magnetic core and four coils. Two coils are on the primary and two coils are on the secondary transformer side. The microtransformer has only 6 contacting pads, where the middle pad is common pad for two coils.

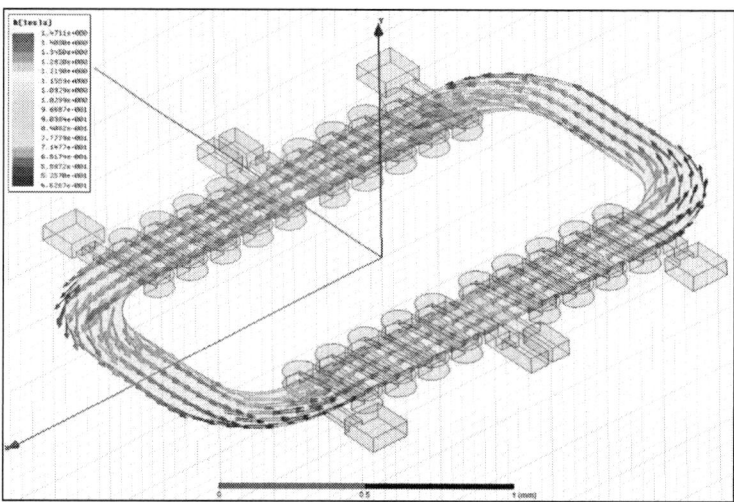

Fig. 1. FEM-simulation of microtransformer

3. Fabrication

The microtransformer device is fabricated using thin-film technology. As a magnetic material for the transformer core, Co-Fe alloy is chosen based on excellent magnetic properties. Fe-Co alloy features saturation flux density B_s up to 2.3 T. The thickness of a magnetic core should be very thin to avoid core losses. Co-Fe magnetic core is deposited by Co-Fe electroplating. Coils are fabricated using Cu electroplating. Coils and magnetic core are embedded in polyimide.

The micro transformer is fabricated atop of 4" 500 µm thin oxidized silicon wafer substrates. The first seed layer for electroplating 20 µm copper in a sulfuric acid electrolyte is realized by physical vapor deposition (PVD) of 50 nm Cr and subsequent 200 nm Au. Then the contact pads and the pilar-like sides of the coil windings are made with 10 µm Cu electroplating. The seed layer is removed with ion beam etching. The etching process is stopped at the silicon dioxide surface.

The copper structures are embedded into copper compatible, photosensitive spin-on polyimide. After spinning the polyimide on it is dried with double step hotplate softbake process: first at 70°C for 10 minutes and finally 10 minutes at 100°C. The polyimide is cured on a hotplate at 350°C under a nitrogen atmosphere. The first embedding layer is levelled with CMP (Chemical Mechanical Polishing) avoiding the photo development in order to level the polyimide coating layer with electroplated copper structure. Additional the planarization will be serves the retention of core magnetic properties [7]. The CMP is made on polymer pad with slurry and 240 nm aluminum dioxide abrasive particles.

The next step is the electroplating of closed soft magnetic core. The seed layer consists of 50 nm Chrome and 200 nm permalloy. The soft magnetic core is 5 µm thick electroplated Co-Fe with mass ratio of 80 to 20. The seed layer is removed with ion beam etching. Often the cured polyimide surfaces have a residual electrical conductance, which may be brought through carbonization of polyimide [8, 9] and / or diffusion of metal ions into the polyimide layer.

The semicircle pillars of the coil windings are formed with subsequent copper electroplating. Therefore the seed layer of 50 nm chrome and 200 nm gold is used. The final electroplated copper thickness is 4 µm over the magnetic core. The system is encapsulated with the polyimide in the same manner as the first embedding step with the same heat treatment. The polyimide is photo-structured in order to access the square copper pillars from the face surface. Then the next seed layer consisting of 50 nm chrome and 200 nm gold is sputtered. The coil windings are joined with 20 µm electroplated copper. The system is encapsulated with photo structured polyimide and the conduction pads are strengthened with electrodeposited copper. Finally the wafer is diced into chips with footprint of 2.4 mm x 2 mm.

Completed microtransforemer device is shown in the Figure 2.

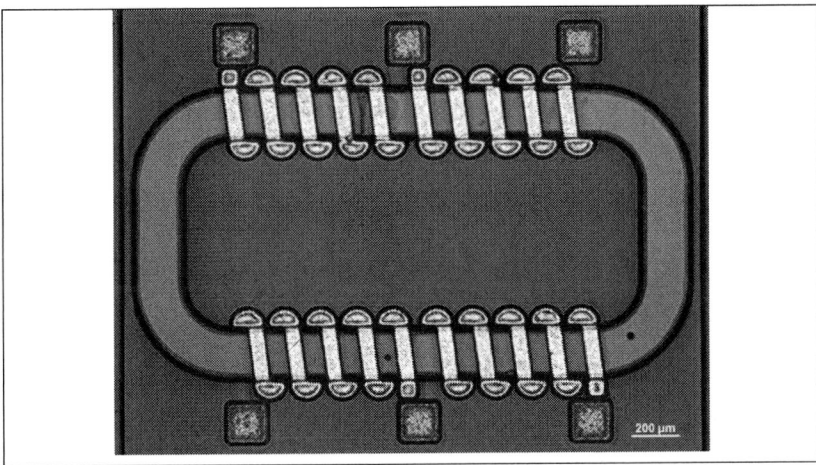

Fig. 2. Completed microtransformer device in thin-film technology

4. Test Results

Diced microtransformer chips were prepared for testing. Dies were mounted into open cavity QFN package and bonded. The QFN package 4 mm x 4 mm with 12 leads was used. Measurements were performed by applying Agilent Impedance Analyzer E4991A. The devices were measured with a signal oscillating level of 5 mA at 1 MHz. Figure 3 shows a characteristic of inductance and resistance of microtransformer depending on frequency.

If all four coils are connected in the series the measured inductance is about 50 nH. The inductance is very stable up to 50 MHz. Therefore, this microtransformer device is suitable for high switching frequency applications up to 50 MHz switching frequencies. The electrical resistance of one coil is about 60 mΩ. Four coils in series have then the resistace of about 250 mΩ. At frequency of 50 MHz. shows the microtransformer device the resistance of only 1.45 Ω.

PCIM Europe 2016, 10 – 12 May 2016, Nuremberg, Germany

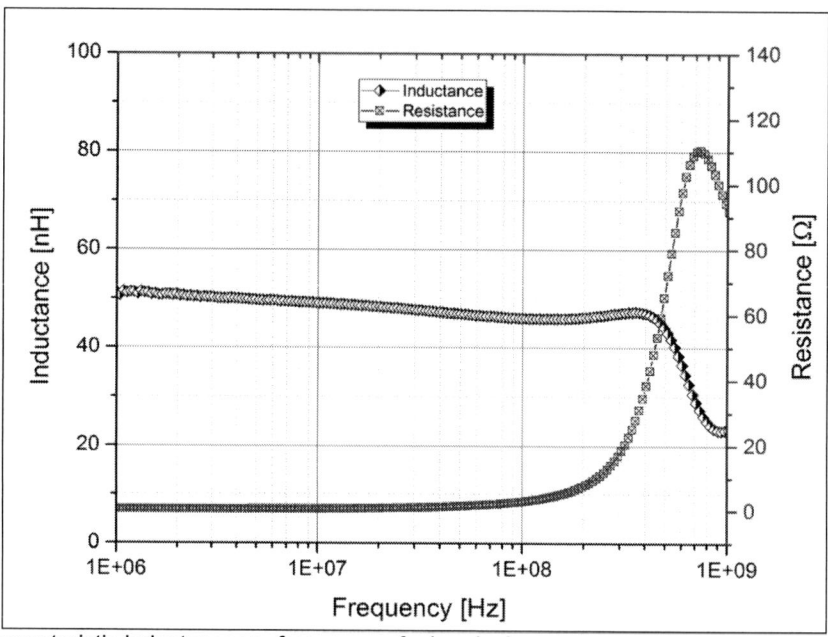

Fig. 3. Characteristic inductance vs. frequency of microdevice

Figure 4 shows characteristics of rated current I_R and saturation current I_{SAT} of microtransformer devices. Bothe measurements for defining the I_R and I_{SAT} are done for primary transformer side (two coils in series). The rated current for primary transformer side is 550 mA. At current of 550 mA the microtransformer heats up to 40°C degrees above the ambient temperature.

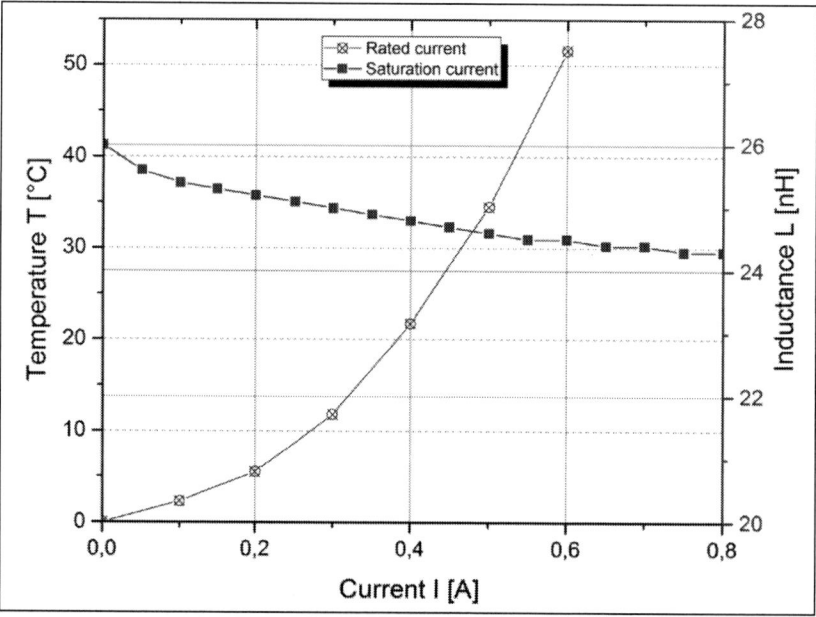

Fig. 4. Rated I_R and saturation current I_{SAT} characteristics of microtransformer

The saturation characteristic of microtransformer is very soft. The inductance decrease with increasing of the DC-Bias current is small. At DC-Bias current of 800 mA drops the inductance only 2 nH what is about 8 percent. The inductance drop of 30 percent (what is common value for power inductors) will be achieved at DC-Bias current of about 1.5 A

© VDE VERLAG GMBH · Berlin · Offenbach

The microtransformers parameters were also defined using S-Parameter measurement. Conventional techniques used to estimate transformer parameters, such as open circuit and short circuit tests, cannot be used to estimate micro-transformer parameters. The reason is these techniques are based on the assumption that the leakage inductance value is significantly small compared to the magnetizing inductance value which is not the case for micro-transformers due to its very small size.

Scattering parameters [10], also known as S-parameters, measurements is commonly used in microwave designs and can be used as an alternative to estimate micro-transformer parameters. In this approach, micro-transformer is considered as a two-port network and forward and reverse reflection coefficients, S11 and S22, and forward and reverse gains, S12 and S21, are measured with a vector network analyzer. Then impedance parameters, also known as Z-parameters, are calculated from S-parameters matrix using the transformation in [11]. Finally, Z-parameters can be directly related to the T-model of the transformer and hence transformer parameters can be estimated. The S-parameter measurement result is shown in the Figure 4.

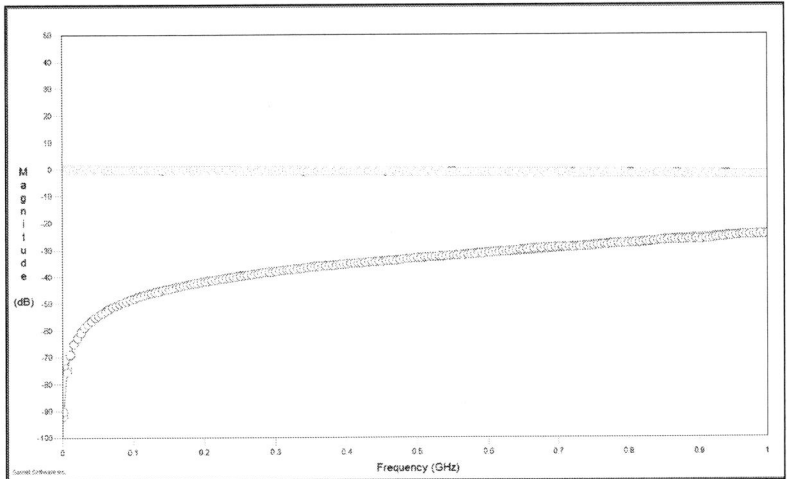

Fig. 5.   Amplitude of s-parameters [dB] vs. frequency

It can be seen that S11 matches S22 and S12 matches S21 since both sides of the transformer are almost identical. It should be noted that the same situation happens with Z-parameters. At 1 MHz was the inductance of 43 nH and resistance of about 180 mΩ measured.

5. Summary and Outlook

The microtransformer design (magnetics on silicon) was developed and fabricated. The microtransformer device was fabricated using thin-film technology. For the fabrication of magnetic core of microtransformer Co-Fe alloy by electroplating is applied. The microdevice shows very stable inductance characteristic up to frequencies higher than 50 MHz and device is suitable for applications at very high switching frequencies. The maximal inductance of device is about 50 nH (all four coils connected in series). Whole device with all four coils in series show the resistance of only 250 mΩ.

In the future works, the microtransformer device will be tested in DC-DC converter application at high switching frequencies up to 30 MHz. Next, the packaging of the device should be defined. First test using eWLB technology (Embedded Wafer Level Ball Grid Array) shows good results. Currently, a new packaging concept is in testing. For this approach the embedding technology in FR4 substrate is under development.

References

[1] C. O'Mathuna, N. Wang, S. Kulkarni, S. Roy, "Review of Integrated Magnetics for Power Supply on Chip (PwrSoC), in IEEE Transaction on Power Electronics, Vol. 27, No. 11, (2012), pp. 4799 – 4816.

[2] M. Xu, T. M. Liakopoulos, C. H. Ahn, "A microfabricated transformer for high-frequency power or signal conversion", IEEE Trans. Magn., Vol. 34, No. 4, (1998), pp. 1369–1371.

[3] C. Feeney, N. Wang, C. O'Mathuna, M. Duffy, "A 20-MHz 1.8-W DC–DC Converter With Parallel Microinductors and Improved Light-Load Efficiency", in IEEE Transaction on Power Electronics, Vol. 30, No. 2, (2015), pp. 771 – 779.

[4] N. Wang, R. Miftakhutdinov, S. Kulkarni, C. O'Mathuna, "High Efficiency on Si Integrated Micro-transformers for Isolated Power Conversion Applications", in IEEE Transaction on Power Electronics, Vol. 30, No. 10, (2015), pp. 5746 – 5754.

[5] F. Khan, Y. Zhu*, J. Lu*, J. Pal, D. V. Dao, "Micromachined Coreless Single-Layer Transformer Without Crossovers", in IEEE Magnetics Letters, Vol. 6 (2015), 6500404.

[6] D. Dinulovic, M. Kaiser, A. Gerfer, O. Opitz, M. C. Wurz, L. Rissing, "Microtransformer with closed Fe-Co magnetic core for high frequency power applications", in Journal of Applied Physics 115, (2014), 17A317.

[7] X. Xing, N. X. Sun, B. Chen, "High-Bandwidth Low-Insertion Loss Solenoid Transformers Using FeCoB Multilayers", in IEEE Transaction on Power Electronics, Vol. 28, No. 9, (2013), pp. 4395 – 4401.

[8] J. Davenas, G. Boiteux, X. L. Xu, "Role of the Modifications Induced by Ion Beam Irradiation in the Optical and Conducting Properties of Polyimide", in Nuclear Instruments and Methods in Physics Research B32 (1988), pp. 136 – 141.

[9] A. M. Ektessabi, S. Hakamata, "XPS Study of Ion Beam Modified Polyimide Films", in Thin Solid Films 377-378 (2000), pp. 621 – 625.

[10] Test & Measurement Application Note 95-1,"S-Parameters Techniques for Faster, More Accurate Network Design," Hewlett Packard, (1996).

[11] D. A. Frickey, "Conversions between S, 2, Y, h, ABCD, and T Parameters which are Valid for Complex Source and Load Impedances," in IEEE Transaction on Microwave Theory and Techniques, Vol.42, No.2, (1994), pp. 205 – 211.

Design of High Efficiency High Power Density 10.5kW Three Phase On-board-charger for Electric/hybrid Vehicles

Gang YANG, VALEO, France, gang.yang@valeo.com
Eirik DRAUGEDALEN, Torbjorn SORSDAHL, Hui LIU, Roar LINDSETH, VALEO, Norway, eirik.draugedalen@valeo.com

The Power Point Presentation will be available after the conference.

Abstract

The design of an electric/hybrid automobile oriented 10.5kW, AC-to-DC three phase on-board-charger is presented in this paper. In order to achieve full power ability when plugged into a three phase grid and single phase grid, three single phase AC/DC converters are paralleled together to offer full power delivery from AC grid to DC high voltage battery. Each phase AC/DC converter is composed by a PFC and a LLC resonant converter while the output current of each phase is regulated to an equal reference to get an equal power distribution. Benefiting from the ZVS characteristics of LLC resonant converter, the converter achieves high power efficiency and low volume. The total prototype performs 6L, and a high power density 1.75W/cm^3. Experimental results prove that a peak efficiency of 95.5% is obtained and efficiency is higher than 94% from 1.6kW to 10.5kW.

1. Introduction

On-board-charger is key equipment in electric/hybrid vehicles that transfers power from the AC power grid to the vehicle battery. Nowadays, available power chargers on the market are mostly single phase on-board-chargers with its AC charging cable plugged into single phase power grid, where the power of the charger is around 3kW-3.5kW. At recent times, in order to increase the charging speed and shorten the charging time, three phase charger is more and more demanding by car manufacturers to achieve high speed charging. Operating at high power at both single phase mode (when plugged to single phase grid) and three phase mode (when plugged to three phase grid) is frequently requested by automobile industries. However, traditional topology by a three phase PFC and a single DCDC converter cannot guarantee full power at single phase mode. In this paper, a new three phase charger topology based on a multi-phase parallel connection of different AC/DC power cells is proposed.

2. Converter Topology

As discussed in the introduction part, the frequently adopted three phase PFC topologies (three half-bridge or Vienna structure) show their inconveniences to achieve full power at single phase mode. In order to overcome this inconvenience, this paper proposed the concept of power cell duplicating to share equally the power no matter which kind of AC power grid. Topology of the proposed three-phase OBC is shown in the figure 1.

PCIM Europe 2016, 10 – 12 May 2016, Nuremberg, Germany

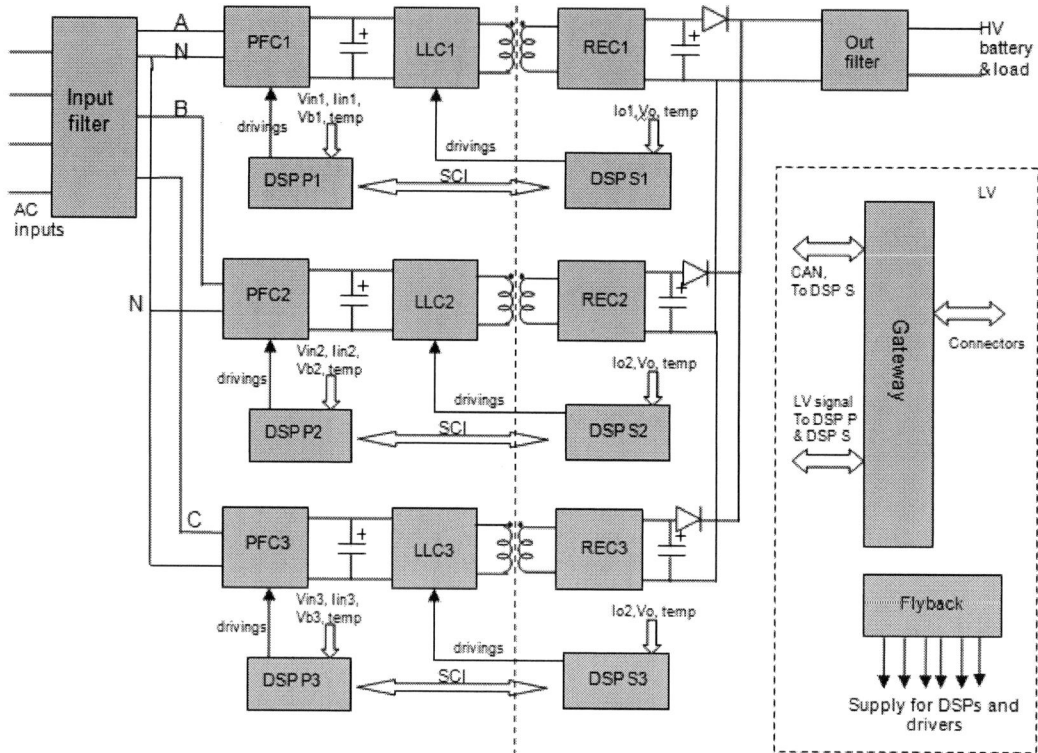

Fig. 1. Structure of the proposed three phase 10.5kW OBC

The three phase power charger contains three single phase power converter cells. Each converter cell contains itself a power factor corrector (PFC, operating at 90 kHz) and a DCDC converter (LLC resonant converter, 90-275 kHz), both are being controlled by its respective DSP. The two DSP controller cards are plugged into the power board and communicate with each other using SCI Bus. The primary DSP is responsible for controlling the input stage of the power cell, hence the input current of each phase and the charging of the Boost DC Link capacitors, while the secondary DSP regulates the DCDC converter to provide the desired output voltage and current according to the control parameters sent from Gateway controller. By regulating the output current of each cell to an identical current reference, the total output current is equally shared among three cells, thus guarantees an equal power distribution among all phases.

All the three modules together share the same input filter and the same output filter, this avoids duplicating the filter elements 3 times. A switch matrix is added in front of the input filter, which will decide the operating mode of the charger by automatically detecting the type of grid that the charger is plugged in. Inside the charger there is also a gateway card, which serves as an interface between the charger and the outside control system(s). The gateway communicates with the secondary controller card using a CAN bus interface. A Flyback auxiliary power supply delivers supply voltages for primary and secondary DSPs, and also for MOSFETs drivers.

By this power cell duplicating, each power cell is able to work quasi-independently without interfering the other power cells, thus the whole OBC is not only possible to work at three phase mode while connecting to (L1 L2 L3 N G) at 10.5kW, but also possible to work in single phase mode while connecting through (L1 N G), both with full power. It has to be noted that with this kind of configuration the OBC needs always to be connected to the Neutral. Loss of Neutral will disturb the input voltage and power sharing among different power cells

© VDE VERLAG GMBH · Berlin · Offenbach

and will trigger the AC over-voltage or input voltage shutdown which will stop the charger.

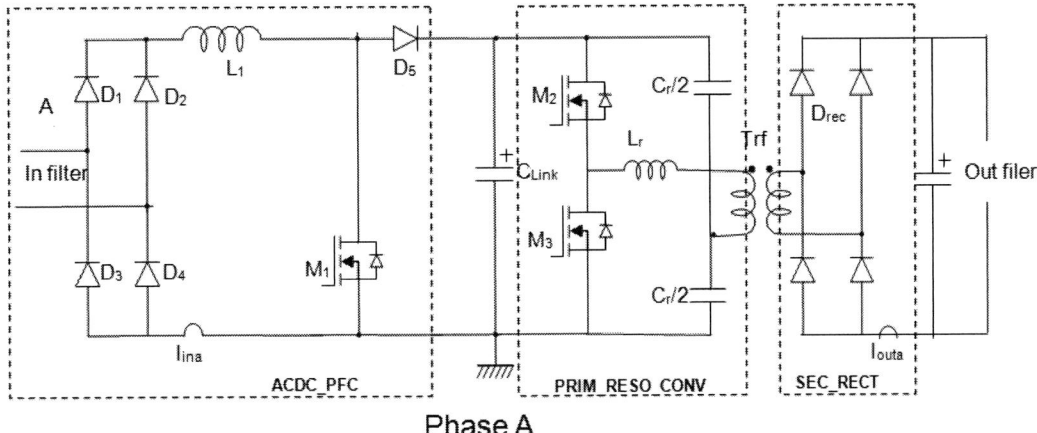

Phase A

Fig. 2. Power structure of the 3.5kW basic power cell

Figure 2 shows the power stage of AC/DC converter for one phase. Each phase is a double stage converter which contains an AC-DC stage and a DC-DC stage. The PFC topology is a traditional PFC with diode bridge (D_1 D_2 D_3 D_4) as input. PFC works at continuous current mode with hard switching. Inductor L1 is made of Amoflux material with distributed air-gaps which has a high saturation level. Diode D5 is selected to be a SiC 20A diode which has very low reverse recovery current thus very low switching losses. Link capacitors C_{LINK} are bulky aluminum capacitors at PFC output in order to reduce the line frequency voltage ripple. At LLC side, the topology used is a half bridge resonant topology, while the resonant tank is composed by a resonant inductor Lr, a transformer Trf with magnetizing inductance Lm integrated, and two resonant capacitors (Cr). The secondary of the transformer is connected to a full diode bridge. M2 and M3 are super-junction MOSFETs working in soft switching.

3. Converter Design

Transformer integration has a great impact for OBC's overall volume. A transformer with very high performance is designed in order to achieve low overall volume and high efficiency. The transformer's structure is presented at the following figure 3.

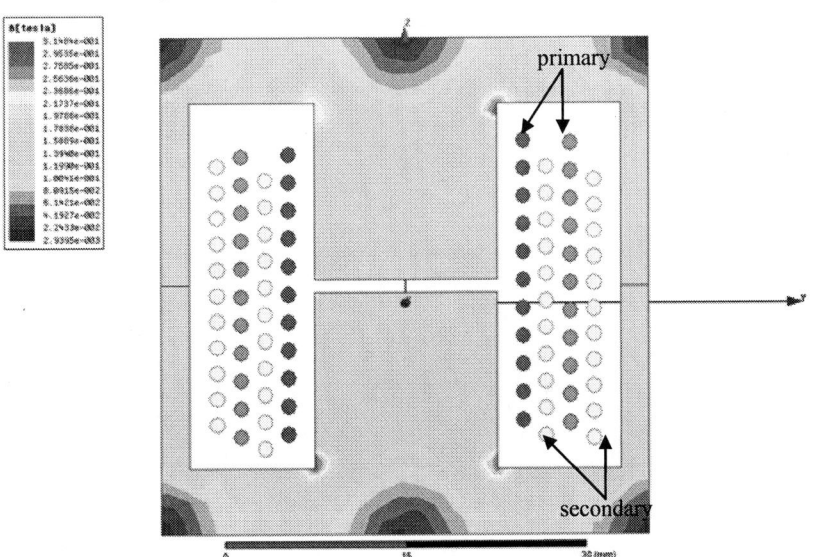

Fig. 3. Integrated transformer in 10.5kW OBC

Transformer is designed based on a PQ4040 core and 3C95 material with a central air-gap to generate the required magnetizing inductance Lm. Litz wire is used to reduce the proximity effect caused by fringing flux close to air-gap. Primary winding contains 11 turns with each turn having two wires in parallel and secondary winding contains 22 turns. Primary winding and secondary windings are interleaved layer by layer in order to reduce the leakage inductance and flux at the core window area. The performance of this transformer is summarized at the following Table.

Table 1. Transformer's loss breakdown simulated by Maxwell 3D simulation

		Point 1 highest conduction loss	Point 2 Highest core oss
Primary	Joule loss (W)	3.940	2.624
	Eddy current loss (W)	0.336	0.356
	Total (W)	4.276	2.980
Secondary	Joule loss (W)	4.477	2.605
	Eddy current loss (W)	0.302	0.226
	Total (W)	4.779	2.831
Core loss (W)		2.166	5.792
Total (W)		11.221	11.603

As reported from the table 1, transformer's total loss is kept around 12W. The core loss and copper losses are well equilibrated for different operating points.

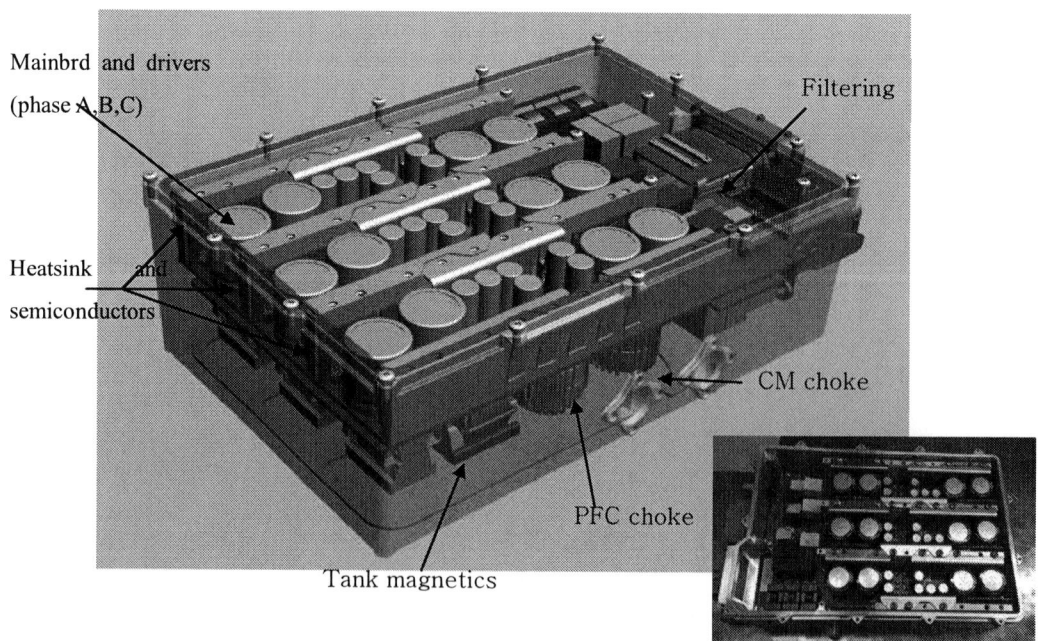

© VDE VERLAG GMBH · Berlin · Offenbach

Fig.4. Fig. 4. Prototype assembly of three-phase OBC, water cooled and power density 1.75W/cm3

The prototype integration is shown in the Fig.4. The charger is composed by a mainboard (top PCB), a filter board (bottom PCB) and a gateway card (not shown in this figure). Primary and secondary DSP boards are mounted on the mainboard. Semiconductor components are mounted vertically; fasten by clip belt to the heat-sink to get a good cooling effect. The PFC capacitors, output capacitors, drivers and microcontrollers are also mounted at the mainboard. On the filterboard (bottom PCB), all power magnetic components are placed into the cavity (close to the water cooling channel) to get a better water cooling effect. Power signals between these two boards are connected by board-to-board power connections. The total prototype design shown in the above figure demonstrates a very compact integration (6L, 300mmx200mmx100mm).

4. Experimental Results

Experiments are conducted in order to validate the correct operation of the proposed on board charger prototype. The PFC stage is a traditional Boost converter thus the waveform is rather well known. The high efficiency is mainly achieved by using the resonant DCDC converters. The following figure shows the key waveforms of LLC stage:

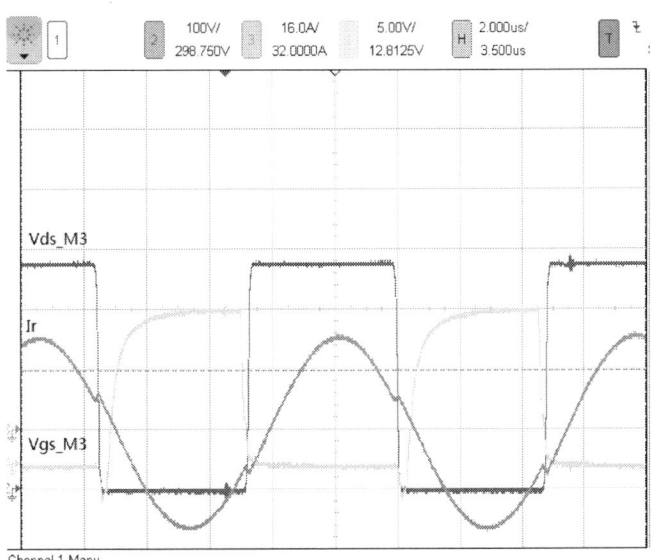

Fig 5. Key operational waveform for LLC signals at Vo=380V, Po=8kW

As reported by the waveform, before MOSFET Q3 starts to switch on, the remaining resonant current will discharge linearly the capacitor across M3. The drain-to-source voltage Vds of M3 falls done linearly to zero before M3 is switched on, thus the ZVS switching-on is achieved. At the switch-off, the parallel ZVS capacitor of MOSFET delays the voltage increase at MOSFET drain-source thus the quasi-ZVS switching off is also achieved. In all, the ZVS operation of the MOSFETs is assured during the created dead-time.

The advantages of using LLC converter into the DCDC stage are rather obvious: 1) current at the converter input and output is a sinusoidal waveform which contains mainly only the first harmonic and is easy to be filtered; 2) no voltage spike is found across the MOSFET drain-source thus avoids the risk of overvoltage breakdown; 3) converter is capable of operating at higher frequencies with very low losses on semiconductors. The efficiency of the whole charger is reported at the following figure.

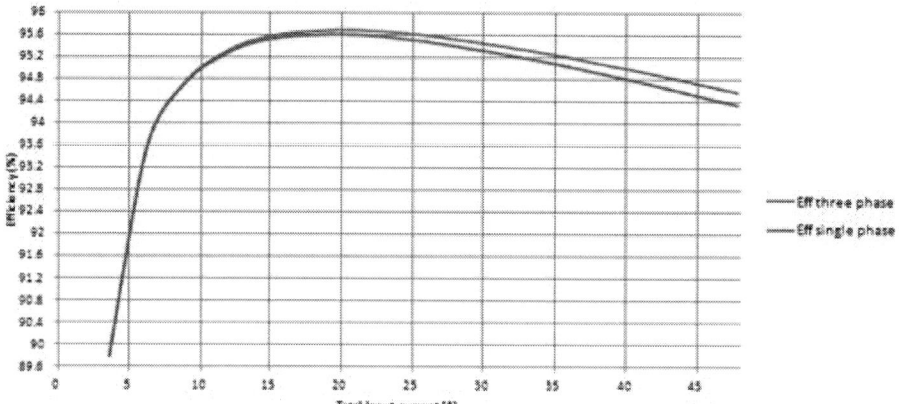

Fig. 6. OBC efficiency at single phase mode and three phase mode at Vo=350V.

As reported at figure 6, the converter performs very high efficiency at a large load range at both three phase mode and single phase mode. The conversion efficiency of the charger is maximal at 4.6kW, with a peak efficiency of 95.5%. Efficiency begins to decrease when load power exceeds 4.6kW. The charger's efficiency is higher than 94% from 1.6kW to 10.5kW. Efficiency at single phase mode is slightly lower than that at the three phase mode. That is because at single phase mode, charger will have a large neutral current 48A which will cause extra loss at the CM inductors and PCBs.

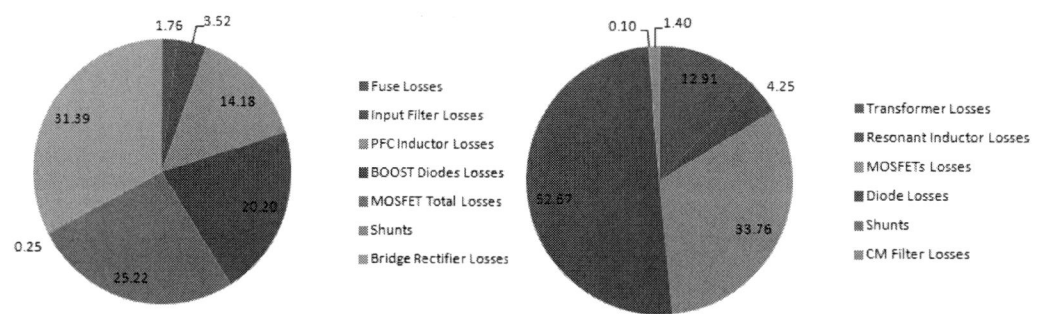

Fig.7. PFC loss breakdown (left) and LLC loss breakdown (right)

Figure 7 shows the loss repartition for PFC stage and LLC stage at the output voltage Vo=350V. The losses are nearly equally shared between PFC and LLC: at PFC stage is 96.5W and at LLC stage is 105.1W, which counts in total 5.4% efficiency loss. At the PFC, semiconductor losses take a great part: 31.4W for diode bridge, 25.2W for MOSFET and 20.2W for SiC power diode. The temperatures on MOSFETs and Diodes are the most critical parts which have the highest component loss and highest junction temperatures. At LLC, main losses are also at semiconductors: 33.7W on MOSFETs and 52.7W on diodes. But since these losses are shared by a couple of components thus the individual temperature on each component is not so critical.

Conclusions

In this paper, a three phase 10.5kW OBC prototype is described and presented in detail. Three independent power cells are putting in parallel in order to achieve full power capability on single phase mode and three phase mode. The designed integrated power transformer reduces the charger volume and keeps the charger at a high power density. Through a compact mechanical integration, the designed three phase OBC exhibits a very high power density and a good conversion efficiency at a wide load variation range. The efficiency of charger in single phase mode will still also have potentials to be optimized by switching-off one or two cells at light load.

References

[1] Bo Yang, "Topology Investigation for Front End DC/DC Power Conversion for Distributed Power System", PhD thesis of Virginia Polytechnic Institute and State University, 2003

[2] Gang Yang; Dubus, P.; Sadarnac, D., "Double-Phase High-Efficiency, Wide Load Range High- Voltage/Low-Voltage LLC DC/DC Converter for Electric/Hybrid Vehicles," in Power Electronics, IEEE Transactions on , vol.30, no.4, pp.1876-1886, April 2015

[3] J.-E. Yeon, W.-S. Kang, K.-M. Cho, T.-Y. Ahn and H.-J. Kim , "Multi-phase interleaved LLC-SRC and its digital control scheme" , Proc. Int. Symp. Power Electron. Elect. Drives Autom. Motion, pp.1189 -1193

[4] G. Yang, P. Dubus and D. Sadarnac , "Design of a high efficiency, wide input range 500kHz 25W LLC resonant converter" , J. Low Power Electron. , vol. 8 , no. 4 , pp.1 - 8 , 2012

A Bridgeless, Quasi-Resonant ZVS-Switching, Buck-Boost Power Factor Correction Stage (PFC)

Markus Scherbaum, Augsburg University of Applied Sciences, Germany,
   markus.scherbaum@hs-augsburg.de
Ralph Kennel, Technical University of Munich, Germany,
   ralph.kennel@tum.de
Manfred Reddig, Augsburg University of Applied Sciences, Germany,
   manfred.reddig@hs-augsburg.de
Manfred Schlenk, Infineon Technologies AG, Germany,
   manfred.schlenk@infineon.com

Abstract

This paper presents a new bridgeless, quasi-resonant zero voltage switching (ZVS), buck-boost power factor correction stage (PFC). The elimination of the input rectifier leads to a higher level of efficiency, especially at low input voltages. Several bridgeless PFC circuits already exist, but the proposed new converter also achieves ZVS in a wide operating range with significant reduced switching losses and some other benefits.

1. Bridgeless Power Factor Correction (PFC)

Power supplies should reach an increased power density in the future. In addition to improved operation efficiency, the reduced losses will also reduce the cooling effort. The costs must also be taken into account. One possible solution is the use of bridgeless power factor correction stages. Due to the elimination of the full-bridge input rectifier, the components in the current path are reduced and this results in fewer conduction losses. There are several bridgeless PFC topologies already published, for example the "Basic Bridgeless Boost PFC", "Dual Boost Bridgeless PFC", "Totem-Pole Bridgeless Boost PFC" or "True Bridgeless PFC Converter" [1][2]. Figure 1 shows the new approach.

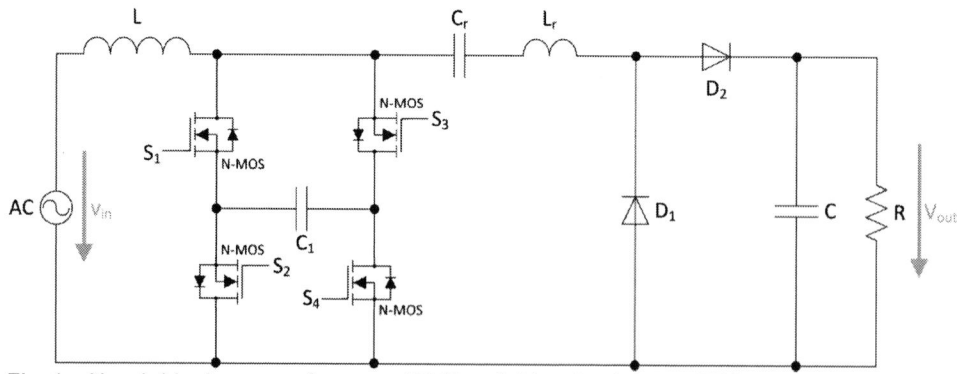

Fig. 1.   New bridgeless, quasi-resonant ZVS switching, buck-boost PFC

2. The existing "True Bridgeless PFC Converter"

The "Bridgeless Boost PFC" stage and its variants such as the "Dual Boost Bridgeless PFC" or "Totem-Pole Bridgeless Boost PFC", are well known and published in many papers. In a power supply, these PFC stages are often followed by an isolated DC-DC converter with a transformer to achieve an isolated output with a lower output voltage.

The "True Bridgeless PFC Converter" [2] is different. This converter could be extended to an isolated converter without adding an additional DC-DC stage. But there are also some disadvantages associated with the "True Bridgeless PFC Converter", which makes it difficult to easily operate this converter. This "True Bridgeless PFC Converter" is basically a boost converter, nevertheless a converter soft start with completely discharged output capacitor is technically possible. Unfortunately there will be some issues with a high resonant current amplitude or with inductor current freewheeling. Another problem exists during normal operation: The current commutation from the switch path to the resonant network during the switch turns off. The resonant inductor of the resonant network can't take over the input inductor current immediately. This results in an overvoltage across the switch during turn off. These critical issues are described in detail later.

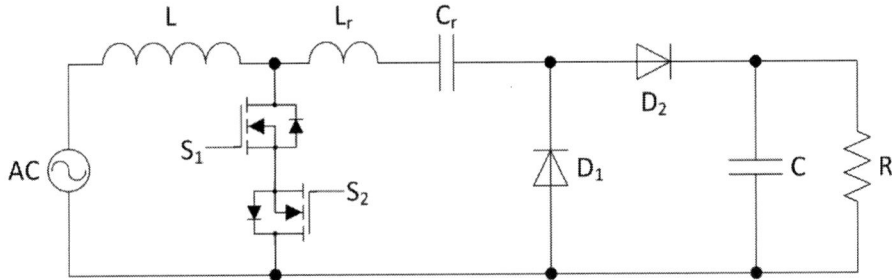

Fig. 2. "True Bridgeless PFC Converter" circuit diagram

The characteristic features of the circuit shown in Fig. 2 are a bi-directional switch, here modeled by two anti-serial MOSFETs S_1 and S_2 and a resonant $L_r - C_r$ network. The resonant frequency f_r of this $L_r - C_r$ network is chosen so that the resonating can completely occur during the ON-time t_{on} of the switch S_1 and S_2. The following equations will be applied, where f_{SW} is the converter switching frequency:

$$f_r = \frac{1}{2 \cdot \pi \cdot \sqrt{L_r \cdot C_r}} > \frac{1}{2} \cdot f_{SW} \tag{1}$$

$$t_{on} \geq \frac{1}{2} \cdot \frac{1}{f_r} = \frac{1}{2} \cdot T_r \tag{2}$$

2.1. Resonant current at converter start-up

At the beginning of the start-up phase, the output capacitor is completely discharged. The voltage across the resonant capacitor C_r is equal to the input voltage (Fig. 3). If the switch S is turned on and equation (2) is fulfilled, a high circular resonant current flows. This high current will stress the semiconductors and could saturate the resonant inductor L_r, which is typically designed to carry only the normal operation current. One solution for the unisolated converter could be a bypass path, which includes an additional diode and a resistor, to pre-charge the output capacitor C.

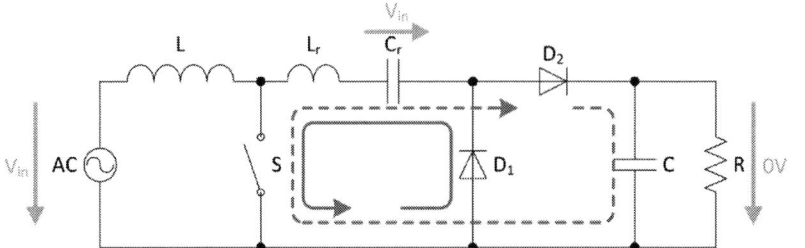

Fig. 3. "True Bridgeless PFC Converter" start-up phase with high resonant currents

2.2. Resonant current cut off

Another solution for a converter start-up phase is violating equation (2) and choosing a shorter ON-time t_{on}. The oscillating cannot occur completely during ON-time. The sinusoidal resonant current i_r is cut off before it reaches zero again. The switch is now open, but the resonant inductor still wants to force the current i_r in the same direction. There is no appropriate freewheeling path, which is shown in Fig. 4. This results in an overvoltage across the switch S if no additional measures are taken.

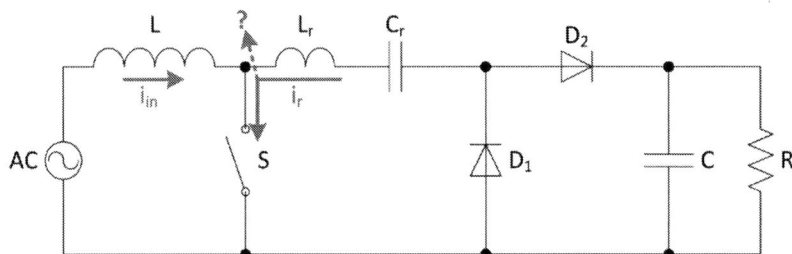

Fig. 4. No existing freewheeling path during resonant current cut off

2.3. Inductor current freewheeling during normal operation

The previously mentioned items only affect the start-up phase, however, there is also a critical point during normal operation. After the switch S is turned off, the input current i_{in} has to commutate form the switching path to the resonant path, as shown in Fig. 5. Unfortunately the resonant path includes the resonant inductor L_r. This inductor can't take over the current immediately. This results in a overvoltage across the switch S. Due to the small inductance value of L_r compared to L, a current commutation to the resonant inductor is possible. The commutation time depends on the allowed overvoltage. This overvoltage is added to the output voltage and defines the voltage stress of the switch S.

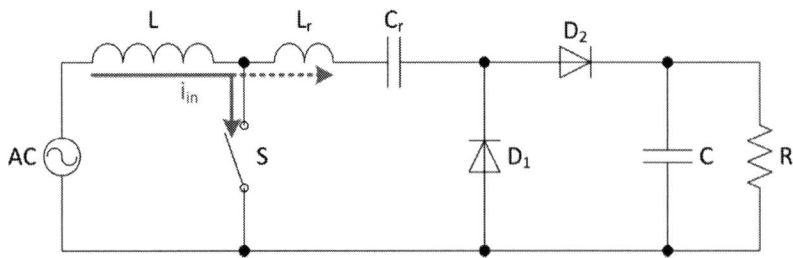

Fig. 5. Current commutation during normal operation

© VDE VERLAG GMBH · Berlin · Offenbach

One solution to protect the switch is the use of Z-Diodes to clamp the voltage to an acceptable value. However this method produces additional losses. To keep the efficiency high, non-dissipative clamp circuits may also be used [3]. Unfortunately these circuits often need more components and switches. The clamp switches must be controlled too. This could be quite complex.

3. The new bridgeless, quasi-resonant ZVS switching, buck-boost PFC

To solve these previously mentioned problems, a new converter topology is proposed and its operation principle is described. This circuit always offers a low impedance commutation path for the input inductor current, meaning there is no overvoltage across the switch during the turn-off transition. This PFC is a special buck-boost converter. A startup without additional component stress is possible. It should be noted that the converter voltage gain has no similarities with a conventional buck-boost converter. Compared to the known bridgeless topologies, the proposed topology has a negligible inrush current; an additional inrush current limitation is not necessary. Furthermore the circuit makes zero voltage switching possible and the switching losses are reduced significantly.

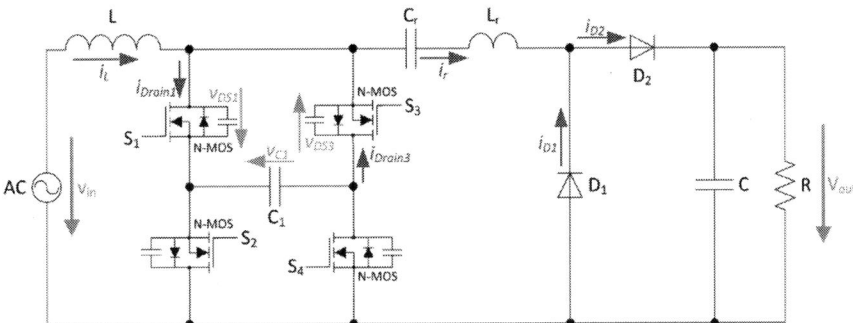

Fig. 6. Circuit diagram of the bridgeless, quasi-resonant ZVS switching, buck-boost power factor correction stage (PFC)

The converter shown in Fig. 6 operates directly at the AC line input. The AC line is connected via an input inductor L to a switching network (S_1, S_2, S_3, S_4) followed by a C_r-L_r resonant network. The switching network can conduct and block in both directions. It also provides a diode path to a capacitor C_1. Due to the alternating input voltage polarity, the switching network is built symmetrically. The MOSFETs are arranged in a way that follows them to build two half-bridges (S_1 and S_3, S_2 and S_4). Therefore two standard half-bridge drivers with a bootstrap high side supply can be used. Furthermore this converter can be extended to an isolated single stage converter if a transformer is inserted.

Compared to the "True Bridgeless PFC Converter" [2], the resonant frequency f_r of the C_1-C_r-L_r network is lower than the switching frequency f_{SW}. The following equation describes the relation between f_r and f_{SW}:

$$f_r = \frac{1}{2 \cdot \pi \cdot \sqrt{L_r \cdot \frac{C_1 \cdot C_r}{C_1 + C_r}}} < \frac{1}{2} \cdot f_{SW} = \frac{1}{2 \cdot T_{SW}} \tag{3}$$

This requirement can be extended to $f_r < f_{SW}$, if it is ensured that the resonant oscillation did not again pass the zero-crossing during a switching state.

3.1. Voltage gain

Figure 7 shows the voltage gain of a converter based on Fig. 6 with $C_1 = C_r = 470\,nF$, $L = 2\,mH$, $L_r = 65\,\mu H$, $f_{SW} = 100\,kHz$. The nominal output voltage is $V_{out} = 200\,V$.

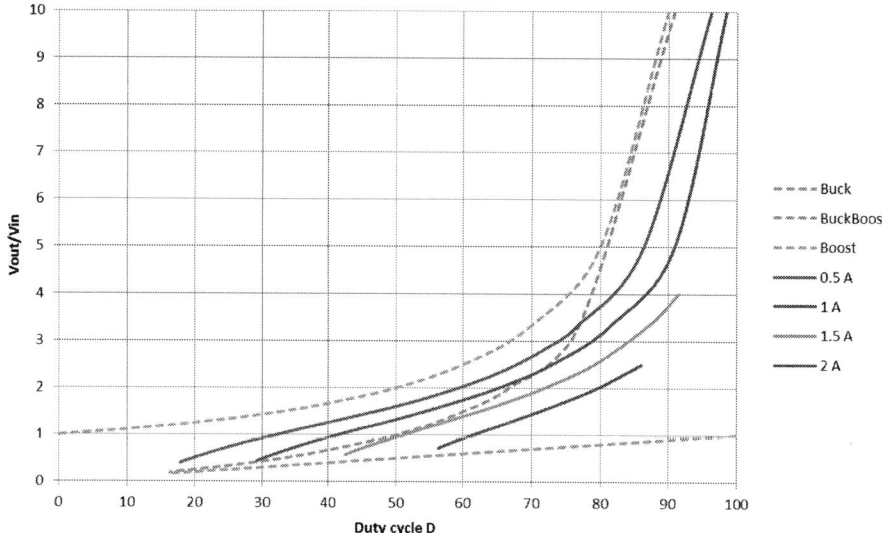

Fig. 7. Voltage gain dependent on duty cycle and input inductor current i_L

The converter has no similarities with a conventional boost, buck or buck-boost topologies (dashed lines). The voltage gain can be smaller or greater than unity gain and depends also on the input inductor current i_L. To get the same voltage gain with higher input current or higher load, the duty cycle has to be increased. The duty cycle is defined as the percentage of the switching period T_{SW} in which the ON-signal to the MOSFETs S_1 or S_2 is applied.

3.2. MOSFET control scheme

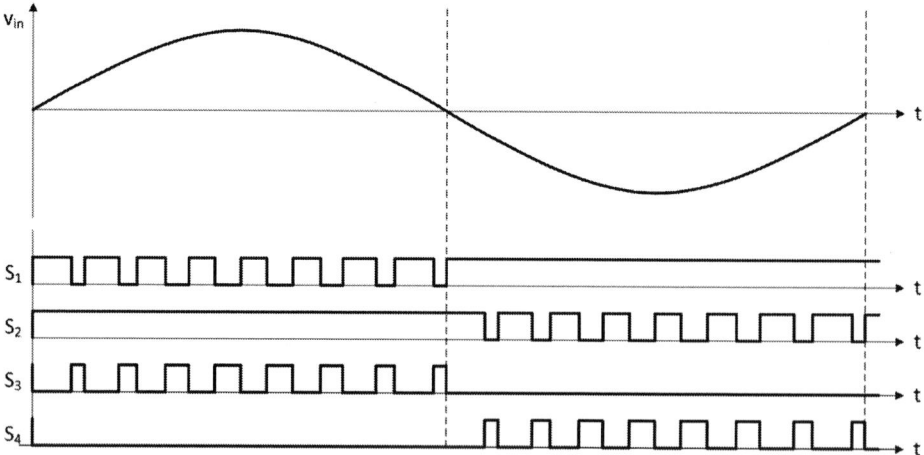

Fig. 8. MOSET ON-signals related to input voltage

The MOSETs S_1 and S_3 respectively S_2 and S_4 build a half-bridge, these MOSFETs are turned on complementary with a small interlock dead time. For positive input voltages v_{in}, S_2 is permanently switched on and S_4 is permanently switched off. S_1 and the complementary S_3 modulate with the higher switching frequency f_{SW} pulse with modulation (PWM) the shape of the input current. With negative input voltages v_{in}, the roles are reversed. This MOSFET control scheme is illustrated in Fig. 8.

3.3. Characteristic waveforms and zero voltage switching (ZVS)

The following description is based on voltages and currents marked in Fig. 6. The half-bridge interlock dead time is $400\,ns$, the component values are still $C_1 = C_r = 470\,nF$, $L = 2\,mH$, $L_r = 65\,\mu H$ and the switching frequency $f_{SW} = 100\,kHz$. Figure 9 illustrates a switching period at an operating point with an input voltage $V_{in} = +385\,V$, an output voltage $V_{out} = 200\,V$ and a load of $P_{out} = 200\,W$. Due to the positive input voltage polarity, S_2 is permanently switched on.

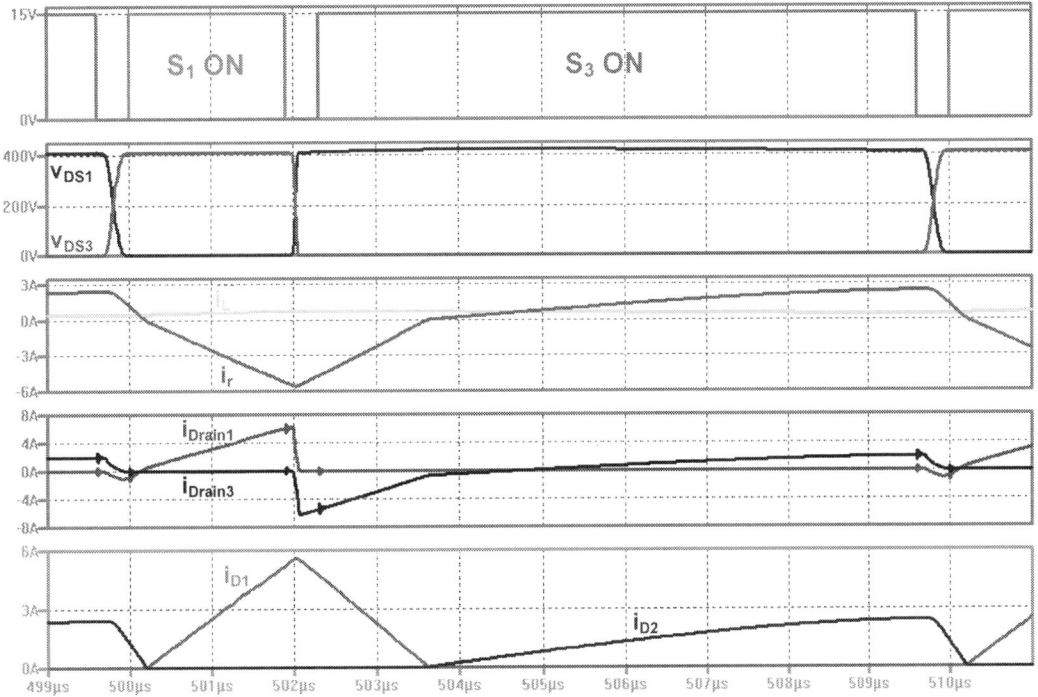

Fig. 9.   Characteristic Waveforms over one switching period

The duty cycle respectively the ON-time of S_1 is provided by a current controller, which ensures the average input inductor current i_L follows the input voltage v_{in} shape. At the turn off moment of S_1, its MOSFET output capacitance is charged from input inductor current i_L and the resonant current i_r. Simultaneously the MOSFET output capacitance of S_3 is discharged. If the output capacitance is completely discharged, the MOSFET Drain-Source voltage v_{DS3} is zero, the internal MOSFET diode starts to conduct. Now the MOSFET can be turned on without any turn-on losses and zero voltage switching (ZVS) is achieved.

The situation is similar during the turn off transition of S_3. There the resonant current i_r is larger than the input inductor current i_L, only the current difference between these two currents is responsible for the MOSFET output capacitance charge of S_3 respectively the discharge of S_1. The MOSFET can be ZVS switched on again, when the internal MOSET diode starts to conduct.

The ZVS process unfortunately depends on the amplitude of the charge/discharge current. At light load, the currents are too low to achieve a complete discharge of the MOSFET output capacitance. The MOSFETs have to be turned on with an only partially discharged output capacitance. In the worst case, the MOSFET has to be turned on completely in a hard switching transition. Unfortunately a PFC converter always has a region with low currents / transmitted power near the line voltage zero crossings. Here ZVS cannot be attained and the switching losses are not negligible.

But there is something that can help here. The MOSFET blocking voltage is equal to the voltage across the capacitor C_1. This voltage v_{C1} depends on the input and output voltage of the converter, the input current and the inductance of L_r. With the lower input current near the line voltage zero crossing, the MOSFET blocking voltage is reduced automatically. A hard MOSFET switching transition near the zero crossing is unavoidable, but this only happens if there is a reduced MOSFET Drain-Source voltage. This effect can be seen in the following measurement results in Fig. 11.

4. Hardware demonstrator measurements

A first hardware demonstrator setup was built for a wide input supply voltage range and an output DC voltage of $V_{out} = 200\,V$. Figure 11 shows the measurement results for an operation at an input line voltage $V_{in} = 120\,V_{rms}$ and $V_{in} = 230\,V_{rms}$. The input inductor current i_L is in phase with the input voltage v_{in} and a unity power factor is almost achieved. In addition the voltage v_{C1} across C_1 is shown over one line period.

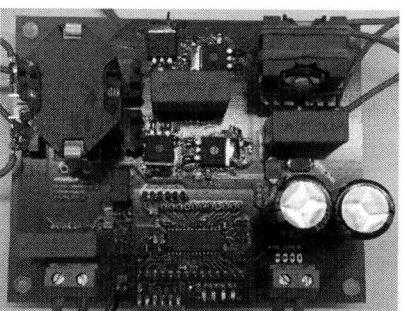

Input voltage range:	$V_{in} = 85\,V_{rms} - 264\,V_{rms}$
Output voltage:	$V_{out} = 200\,V$
Output power:	$P_{out} = 200\,W$
Switching frequency:	$f_{SW} = 100\,kHz$
Microcontroller:	$PIC24FJ32MC102$
Components:	$C_1 = C_r = 470\,nF,$
	$C = 200\,\mu F$
	$L = 2\,mH, L_r = 65\,\mu H$

Fig. 10. Experimental prototype (unisolated) and specifications

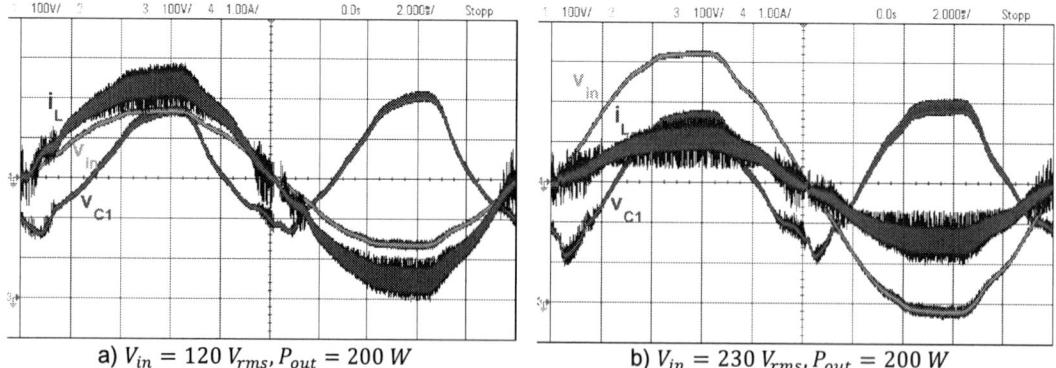

a) $V_{in} = 120\,V_{rms}, P_{out} = 200\,W$ b) $V_{in} = 230\,V_{rms}, P_{out} = 200\,W$

Fig. 11. Line voltage v_{in}, input inductor current i_L, and C_1 capacitor voltage v_{C1} measurements at
a) $V_{in} = 120\,V_{rms}$ and b) $V_{in} = 230\,V_{rms}$

The input inductor current i_L in Figure 11 consists of a sinusoidal average input current shape and a superimposed current ripple. The load dependent voltage gain (Fig. 7) affects the ripple current. At higher average input currents the duty cycle and thus the ON-time is increased. This leads to an approximately 20% greater current ripple at full load in this designed converter compared to a conventional boost PFC stage with the same input inductance value. But at a certain load point the current ripple getting smaller. The small current spikes are caused by the winding capacitance.

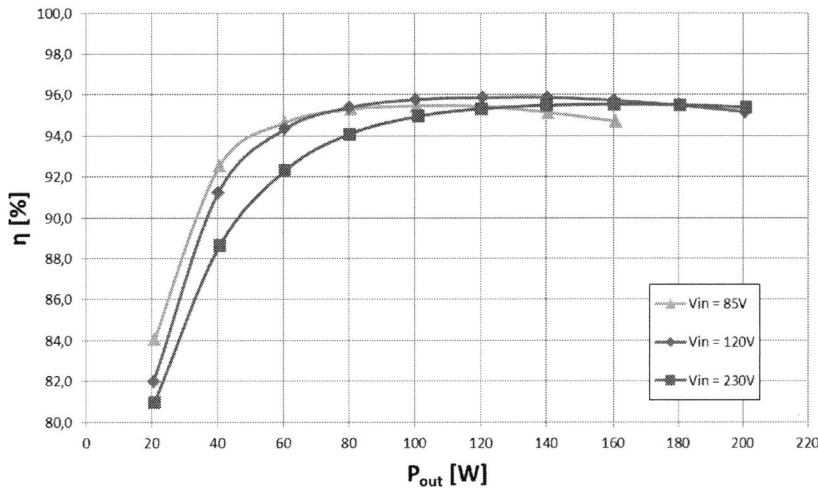

Fig. 12. Efficiency at different input line voltages

The efficiency is almost independent from the input line voltage. In contrast to conventional PFC stages, the efficiency is even higher at lower input line voltage in the light load range. The efficiency drop at light load is caused by the loss of ZVS in a wider range of the line period.

5. Conclusion

In this paper, a new bridgeless, quasi-resonant ZVS switching, buck-boost PFC topology was presented. The circuit may also be extended to an isolated converter if a transformer is inserted. The inrush-current is negligible and a fast converter soft-start is possible without additional component stress. One unique feature is the ability to work in buck mode or in boost mode, so the output voltage of this unisolated PFC converter is set to $V_{out} = 200\,V$. Furthermore the converter has no overvoltage across the semiconductor switches and offers zero voltages switching (ZVS) to reduce the switching losses.

6. References

[1]   D. Miller, R. Kennel, M. Reddig, "New methods for digitally controlled bridgeless PFC converters", *Proceedings of the Intellec 2013*, Hamburg, Germany, 13-17 Oct. 2013

[2]   S. Cuk, "True bridgeless PFC converter achieves over 98% efficiency, 0.999 power factor", *Power Electronics Technology Magazine*, July 2010

[3]   S. Nigsch, S. Cuk, K. Schenk, "Analysis, Modeling and Design of a True Bridgeless Single Stage PFC with Galvanic Isolation", *Proceedings of the APEC 2015*, Charlotte, NC, 15-19 March 2015

A high efficiency 5.3kW Current Source Inverter (CSI) prototype using 1.2kV Silicon Carbide (SiC) bi-directional voltage switches in hard switching mode

Jérémy, Martin, jeremy.martin@cea.fr
Anthony, Bier, anthony.bier@cea.fr
Stéphane, Catellani, stephane.catellani@cea.fr
Luis Gabriel, Alves-Rodrigues, luis.gabriel.alvesrodrigues@cea.fr
Franck, Barruel, franck.barruel@cea.fr

Commissariat à l'énergie atomique et aux énergies alternatives
Département des technologies solaires
Laboratoire des systèmes photovoltaïques
50 avenue du Lac Léman | F-73375 Le Bourget-du-Lac

This project has received support from the State Program "Investment for the Future" bearing the reference (ANR-10- IEED -0003)

Abstract

The market brings Silicon Carbide (SiC) transistors and diodes in the 1.2kV range [1-7], which can be implemented in solar switching power converters operating under AC-400V. In the scientific literature, SiC switches have been evaluated for a voltage source inverter operation [8], [9], but there is a lack of studies that address the implementation of SiC components running in a current source inverter (CSI). Characterizations of a voltage commutation cell against the commutated voltage have been discussed in the last paper presented at PCIM 2015 [10]. This article is based in the following of the previous one and treats mainly of the characterization of the serial association of a transistor and a diode required for CSI operation.

A test bench is developed at the Department of Solar Technologies (DTS) of the CEA. It allows to characterize SiC power semi-conductor TO-220/247 packaged operating in a current source inverter. Switching energies are measured and presented. Once the characterization results obtained, the loss coefficients are extracted and electro-thermal calculations of efficiencies against the switching frequency are finally given and discussed.

1. Introduction

A current source inverter topology is presented figure 1. A CSI operates at fixed current while the voltage is sinus modulated, so, the required switch is voltage-bidirectional and current-unidirectional [11]. Market brings 1.2kV SiC MOSFET and 1.2kV Schottky diode that can be serial associated to achieve a voltage-bidirectional switch.

Semi-conductors manufacturers provide only in theirs datasheets switching energies for voltage source inverter design and analysis, so a test bench has been built to characterize SiC components at fixed current and variable voltage.

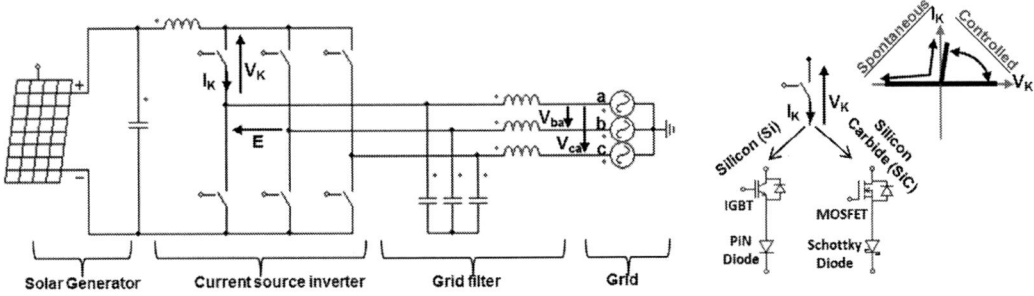

Fig. 1. Current-source inverter conversion chain

2. The test bench for switching energies characterization

The double pulse tester figure 2 and figure 4 is usually employed to characterize semiconductors switching losses [12-13]. A power supply is connected to a storage capacitor C_{bulk} followed by a converter leg and a pure inductor L_{load}. C_L is the load inductor equivalent paralleled capacitor which measure equals to 8pF on the 10MHz-50MHz range (considering a V_{D2} voltage variation last on the range 7ns to 35ns). This value is considered fairly negligible regarding the diode capacitor which value is evaluated between 2.5nF-150pF from 1V to 600V (Fig.3). Thus the impact of C_L is not considered in this paper.

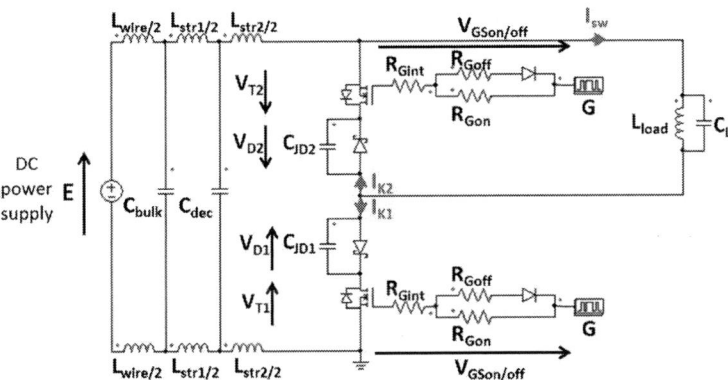

Fig. 2. Test bench for switching energies characterization

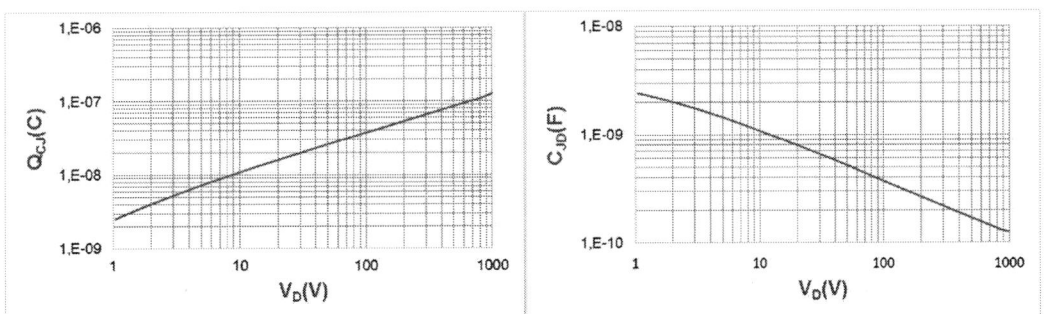

Fig. 3. Schottky diode capacitive charge measured with *Keysight B1506* at 1MHz

Two 1µF ceramic capacitors C_{dec} (Fig. 4) confers an instantaneous voltage source property as close as possible from the commutation cell reducing L_{str2}. Pearson 2877 are used for di/dt measurement. The manufacturer indicates a useable rise time of 2 nanoseconds and a 200MHz bandwidth [14]. Current transformers built of 10 turns around 3E27 ferrite toroids are connected to the anode sides; which limits as much as possible the stray inductor L_{str2}.

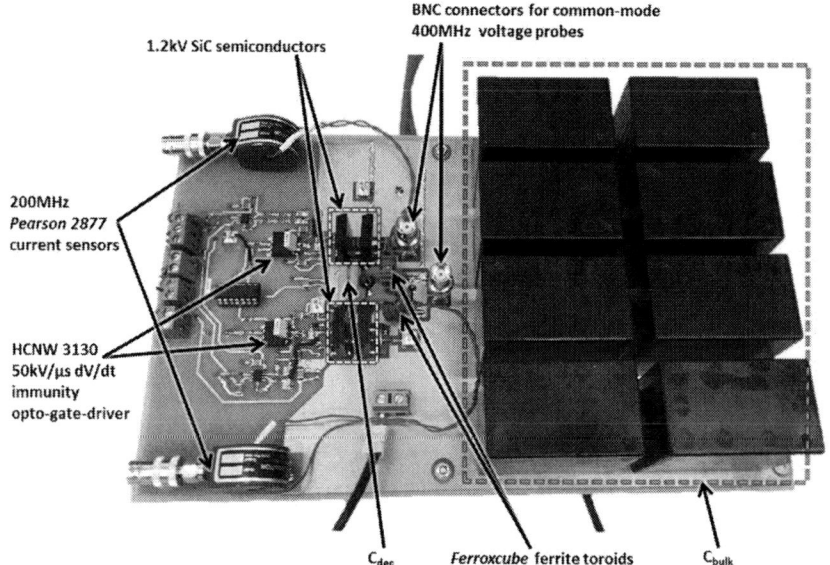

Fig. 4. Test bench details

3. Switching energies characterization results

3.1. The test bench setup

The bench uses a 1.2kV 25mΩ SiC MOSFET connected to a 54A SiC Schottky diode (Fig. 5). A 12Amp switched current corresponds to the ratio of 5.3kW by the maximum average input voltage thus for a modulation ratio equals to 0.9. The presented switching energies are measured by steps of 40V; and those equal to the average of the energies measured from 30 pulses. The di/dt and dv/dt switching speeds are given with respect to a change from 10% to 90% of the commutated magnitude excluding the oscillations.

1.2kV SiC MOSFET ratings $I_D@T_C$=25°C	$R_{Gon/off}(\Omega)$	$R_{Gint}(\Omega)$	$V_{GSon/off}(V)$	1.2kV SiC Schottky Diode ratings $I_F@T_C$=135°C	I_{sw}
90A	4,3/1,77	1.1	+20/-5	54A	12A

Fig. 5. Test bench tunings

3.2. Controlled turn-on

MOSFET turn-on waveforms are presented figure 6 with E=600V and I_{SW}=12A. On one hand, the dI_{T1}/dt of 1.74kA/µS cause a voltage deep of 200V through the stray inductor L_{str2} which is deduced equal to 115nH (1). On the other hand, the dV_{D2}/dt of 41kV/µs leads to a current overshoot equals to 20A.charging C_{JD2} which is calculated equal to 487pF (2) (3.2 times higher than expected Fig.3).

$$V_{T1} = E - L_{str2} \times (dI_{k1}/dt) \tag{1}$$

$$I_{k1} = I_{sw} + C_J \times (dV_{D2}/dt) \tag{2}$$

The switching energy E_{on_T} is calculated (3) and presented as a function of the commutated voltage. The energy when turned-on is equal to 140µJ for 12 Amp under 600V switched voltage and loss coefficients are extracted.

$$E_{on\_T} = \int_{t1}^{t2} V_{T1} \times I_{k1} \times dt \tag{3}$$

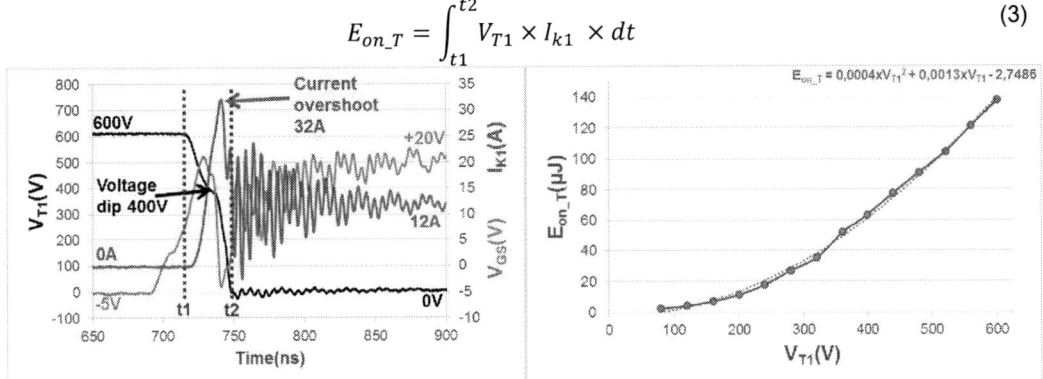

Fig. 6. T1 turn-on with external R_{Gon} = 4.3 Ω, 12 Amp switched current under 600V (left) Switched energies versus the switched voltage for I_{sw}=12Amp and loss coefficients (right)

3.3. Controlled turn-off

MOSFET turn-off is presented figure 7. The dV_{T1}/dt is reduced to 22.4kV/µs and equals the ratio of the transistor current minus the load current by the diode parasitic capacitor C_{JD2} which is discharging and evaluated equal to 535pF considering (4).

$$dV_{T1}/dt = d\,V_{D2}/dt = (I_{k1} - I_{sw})/C_{JD2} \tag{4}$$

The turn-off energy versus the switched voltage is calculated (5), given figure 7 and coefficients are extracted. The commutated energy is equal to 80 µJ for the point 12 Amp / 600V.

$$E_{off\_T} = \int_{t1}^{t2} V_{T1} \times I_{k1} \times dt \tag{5}$$

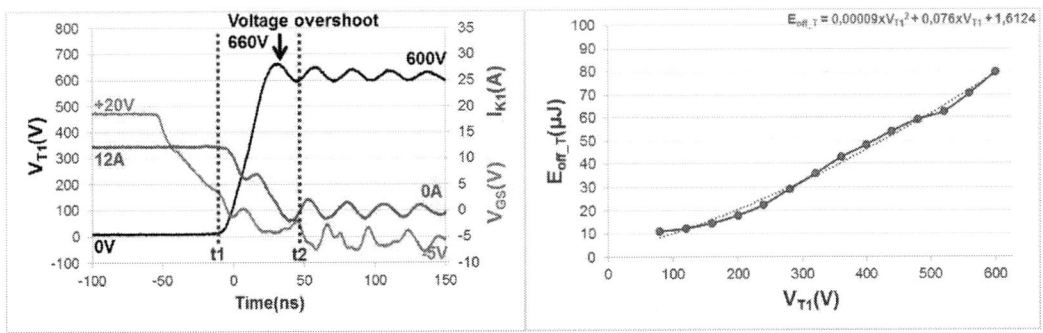

Fig. 7. T1 turn-off with external R_{Goff} = 1.77 Ω, 12 Amp switched current under 600V (left) Switched energies versus the switched voltage for I_{sw} = 12Amp and loss coefficients (right)

3.4. Spontaneous turn-on

The diode commutation waveform when turn-on is given figure 8. The switched energy is too low to be calculated (is measured equal to 1.9 µJ for the point at the highest magnitude).Thus, spontaneous turn-on losses will be not be taken into account for the loss calculations.

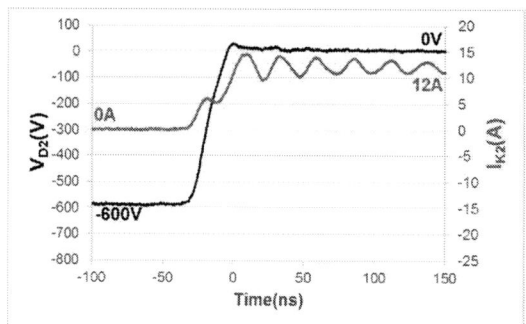

Fig. 8. Spontaneous turn-on of K2 for 12Amp under 600V

3.5. Spontaneous turn-off

The diode turn-off waveform is presented and calculated (6) figure 9. For a 12Amp switched current under 600V, the peak reverse current is equal to 16Amp and the commutated energy to 50µJ.

$$E_{off\_SD} = \int_{t1}^{t2} V_{D2} \times I_{k2} \times dt \tag{6}$$

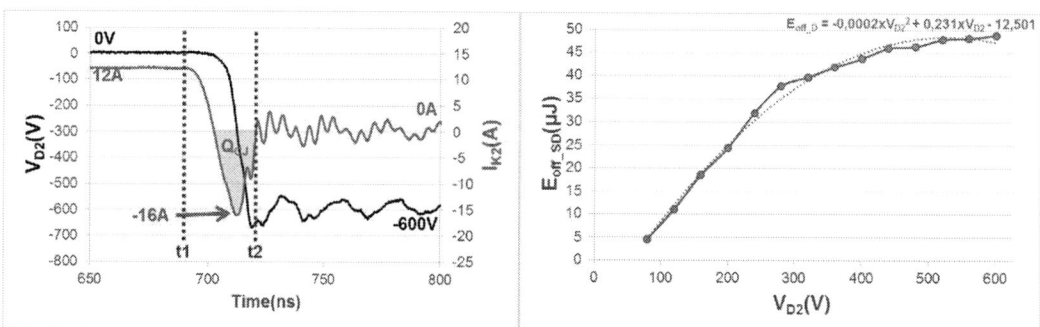

Fig. 9. Spontaneous turn-off of K2 for 12Amp / 600V (left) with external R_{Gon} = 4.3 Ω Switched energies as a function of the commutated voltage for I_{SW} = 12Amp and loss coefficients (right)

When the dI_{K2}/dt is modified (slowed-down from 1747A/µs to 893A/µs by increasing the MOSFET gate resistor from 4.3Ω to 13Ω), the peak reverse current is decreased from 16A to 10A, the capacitive charge Q_{CJ} waveform is modified and pass from 168nC to 133nC at 600V. The diode capacitive charge is integrated (7) and is presented as a function of the commutated voltage (Fig.10); this results are close to the ones obtained figure 3.

$$Q_{CJ} = \int_{tQ1}^{tQ2} -I_{k2} \times dt \tag{7}$$

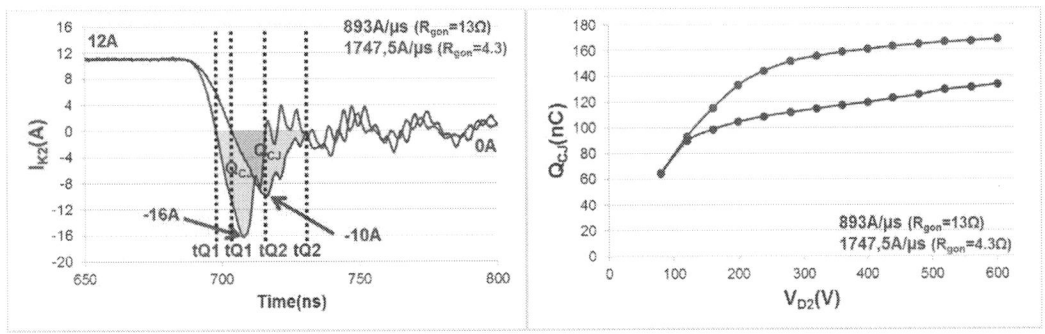

Fig. 10. Modified dI_{K2}/dt and peak reverse current (left) capacitive charge versus the switched voltage for I_{SW} = 12Amp (right)

4. Electro-thermal calculations [15-17]

4.1. Switched voltage and current for one grid period

Figure 11 illustrates the voltage V_{K1} and the current I_{K1} of a voltage-bidirectionnal switch operating in a three-phase 400V_{RMS} grid tied current-source inverter. MOSFET and diode conduction and switching power losses when used in a CSI can be then evaluated.

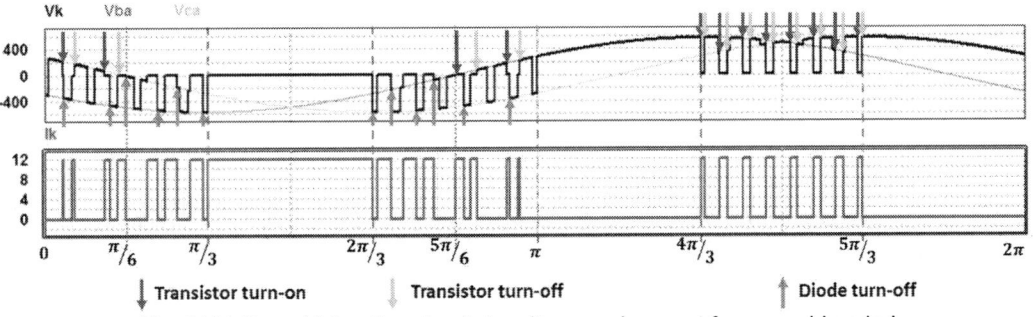

Fig. 11.Voltage-bidirectional switch voltage and current for one grid period

4.2. MOSFET and diode conduction losses

In a CSI, whatever the duty cycle, each voltage-bidirectionnal switch is on-state during two-sixth of the grid period. MOSFET conduction power losses are calculated using (8) with $R_{DS(on)}$ its on-state resistor.

$$P_{cond\_T} = \frac{2}{6} \times R_{DS(on)} \times I_K^2 \tag{8}$$

Schottky diode conduction power losses are calculated using (9) with V_T the voltage threshold and R_T the dynamic resistor.

$$P_{cond\_SD} = \frac{2}{6} \times (V_T \times I_K + R_T \times I_K^2) \tag{9}$$

$R_{DS(on)}$, V_T and R_T are given as function of the junction temperature in the devices datasheets.

© VDE VERLAG GMBH · Berlin · Offenbach

4.3. MOSFET and diode switching losses

MOSFET and diode turn-on and turn-off energies from characterization results points can be expressed as function of their respective voltages with order two polynomial regressions which coefficients are given in the table figure 12.

	Polynomial coefficients		
$E_{\_on\_T}(V_T)$	$a_{sw\_on\_T} = 395 \times 10^{-12}$	$b_{sw\_on\_T} = 1.33 \times 10^{-9}$	$c_{sw\_on\_T} = -2.75 \times 10^{-6}$
$E_{\_off\_T}(V_T)$	$a_{sw\_off\_T} = 89 \times 10^{-12}$	$b_{sw\_off\_T} = 76 \times 10^{-9}$	$c_{sw\_off\_T} = 1.61 \times 10^{-6}$
$E_{\_off\_SD}(V_{SD})$	$a_{sw\_off\_SD} = -219 \times 10^{-12}$	$b_{sw\_off\_SD} = 231 \times 10^{-9}$	$c_{sw\_off\_SD} = -12{,}5 \times 10^{-6}$

Fig. 12. Order 2 polynomial coefficients of switching energies as a function of the switched voltage

MOSFET and diode can be considered switching during a quarter of the sinusoidal grid line to line voltage wave at the switching frequency f_{SW}. With $v(\theta)$ the line to line instantaneous voltage as function of the angle θ in radian. MOSFET turn-on and turn-off power losses are calculated using equation 10 and 11; diode turn-off power losses are calculated using (12).

$$P_{sw\_on\_T} = \frac{f_{SW}}{2\pi} \times \int_0^{\pi/2} \left(a_{sw\_on\_T} \times v(\theta)^2 + b_{sw\_on\_T} \times v(\theta) + c_{sw\_on\_T} \right) d\theta \qquad (10)$$

$$P_{sw\_off\_T} = \frac{f_{SW}}{2\pi} \times \int_0^{\pi/2} \left(a_{sw\_off\_T} \times v(\theta)^2 + b_{sw\_off\_T} \times v(\theta) + c_{sw\_off\_T} \right) d\theta \qquad (11)$$

$$P_{sw\_off\_SD} = \frac{f_{SW}}{2\pi} \times \int_0^{\pi/2} \left(a_{sw\_off\_SD} \times v(\theta)^2 + b_{sw\_off\_SD} \times v(\theta) + c_{sw\_off\_SD} \right) d\theta \qquad (12)$$

4.4. Balance of power losses for a 5.3kW CSI

Using the previous equations, MOSFET and diode conduction/switching losses are calculated when operating in a 5.3kW three-phase 50Hz 400V$_{RMS}$ grid-tied CSI. The conduction/switching current is 12A and the switches junction temperatures are considered at 125°C. Balance of power losses and efficiencies, only considering the switches power losses, of a CSI based on 6 voltage-bidirectionnal switches is given in figure 13 for four switching frequencies: 25kHz, 50kHz, 100kHz and 200kHz. Below 200kHz power losses are mainly due to conduction losses and diode conduction losses are more than 2.5 times the MOSFET conduction losses.

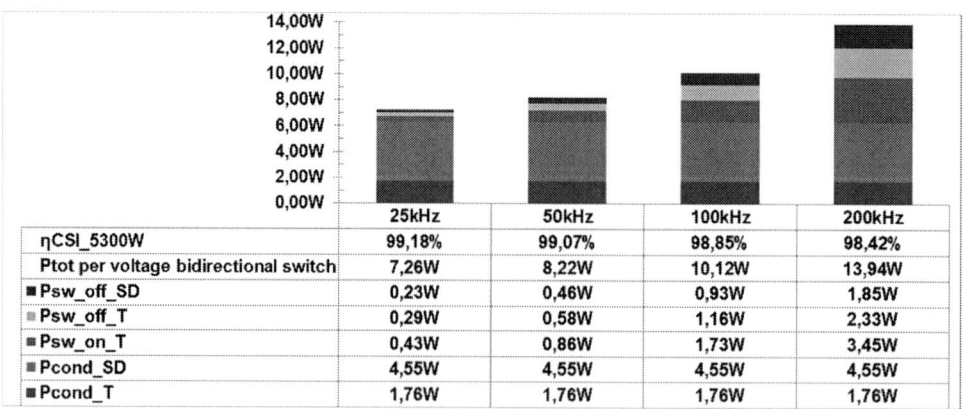

	25kHz	50kHz	100kHz	200kHz
ηCSI_5300W	99,18%	99,07%	98,85%	98,42%
Ptot per voltage bidirectional switch	7,26W	8,22W	10,12W	13,94W
■ Psw_off_SD	0,23W	0,46W	0,93W	1,85W
▨ Psw_off_T	0,29W	0,58W	1,16W	2,33W
■ Psw_on_T	0,43W	0,86W	1,73W	3,45W
■ Pcond_SD	4,55W	4,55W	4,55W	4,55W
■ Pcond_T	1,76W	1,76W	1,76W	1,76W

Fig. 13. Efficiencies and details of conduction and switching losses in transistors and diodes for a 12A DC current input 5.3kW three-phase 50Hz 400V$_{RMS}$ grid-tied CSI

5. Conclusions

This article shows firstly a methodology to study and evaluate the switching losses of SiC semiconductors implemented in a current source inverter, and secondly, models of loss to thermal design the current source inverter from these results. The characterizations results show that for a 1.74kA/μs switching speed (mainly limited by the physical introduction of parasitic inductors in the switching loop), the Schottky diode parasitic capacitor causes an overcurrent whose amplitude is equal to 2.7 times the load current, the SiC diodes could be problematic regarding the safe operating area if increasing switching speed. The transistor turn-off switching speed is highly influenced by the parasitic capacitor of the SiC diode again (for a load current of 12 Amp studied in this paper). Finally the main source of losses through the current source inverter is conduction losses in the diodes. If a CSI whose an efficiency superior as 99% is expected, important chip surfaces and so capacitive diode are required what is limiting for an increased switching speed. The good compromises must therefore be found between efficiency, current rating, and safe operating area.

6. References

[1]http://www.rohm.com/
[2]http://www.cree.com/
[3]http://www.infineon.com/
[4]http://www.genesicsemi.com/
[5]http://www.unitedsic.com/
[6]http://www.fujielectric.com/
[7]http://www.st.com/
[8]Joseph Fabre, Philippe Ladoux, Michel Piton "Opposition Method based test bench for characterization of SiC Dual MOSFET Modules" PCIM Europe 2013, 14–16 May 2013, Nuremberg
[9]R. Burkart, J. W. Kolar "Comparative Evaluation of SiC and Si PV Inverter Systems Based on Power Density and Efficiency as Indicators of Initial Cost and Operating Revenue" Proceedings of the 14th IEEE Workshop on Control and Modeling for Power Electronics (COMPEL 2013), Salt Lake City, USA, June 23-26, 2013
[10]Stéphane Catellani, Anthony Bier, Jérémy Martin, Luis Gabriel Alves-Rodrigues, Franck Barruel "Characterization of 1.2kV Silicon Carbide (SiC) semiconductors in hard switching mode for three-phase Current Source Inverter (CSI) prototyping in solar applications" PCIM Europe 2015, 19 – 21 May 2015, Nuremberg
[11]Abu-Khaizaran, M, Birzeit Univ, Bir Zeit, Palmer, P. "Commutation in a high power IGBT based current source inverter" Power Electronics Specialists Conference, 2007. PESC 2007. IEEE
[12]Callanan, Bob. Cree Inc "SiC MOSFET Double Pulse Fixture" February 2011
[13]G. Laimer, J.W Kolar "Accurate Measurement of the Switching Losses of Ultra High Switching Speed CoolMOS Power Transistor / SiC Diode Combination Employed in Unity Power Factor PWM Rectifier Systems"
[14]www.pearsonelectronics.com/pdf/2877.pdf
[15]Sahan, B.; Notholt-Vergara, A.; Engler, A.; Zacharias, P. "Development of a Single-Stage Three-Phase PV Module Integrated Converter" Power Electronics and Applications, 2007 European Conference on
[16]Mohr, M.; Fuchs, F.W. "Comparison of three phase current source inverters and voltage source inverters linked with DC to DC boost converters for fuel cell generation systems" Power Electronics and Applications, 2005 European Conference on
[17]Bierhoff, M.H.; Fuchs, F.W. "Semiconductor Losses in Voltage Source and Current Source IGBT Converters Based on Analytical Derivation" 2004 35th Annual IEEE Power Electronics Specialists Conference Aachen, Germany, 2004

Comparison of the EMC and efficiency characteristics of hard and soft switching three-level inverters

Prof. Dr. Manfred W. Gekeler, HTWG Konstanz, University of Applied Sciences,
Brauneggerstraße 55, D-78464 Konstanz, gekeler@htwg-konstanz.de
B. Eng. Stefan Schreitmüller, HTWG Konstanz, University of Applied Sciences,
Brauneggerstraße 55, D-78464 Konstanz, stefan.schreitmueller@gmx.de
Prof. Dr. Gunter Voigt, HTWG Konstanz, University of Applied Sciences,
Brauneggerstraße 55, D-78464 Konstanz, gvoigt@htwg-konstanz.de

Abstract

Three-level inverters are used in electrical drive systems, as grid infeed inverter in PV power plants or as active power line filters. Up to now so called hard switching topologies have been used. A new 'Soft Switching Three Level Inverter (S3L Inverter)' which is now available provides reduced switching losses and higher efficiency. In this paper the S3L inverter is compared with a hard switching T-type inverter topology (H3L inverter). S3L inverters provide higher efficiency and additionally advantages in electromagnetic compatibility due to the soft switching performance, especially when using the 'Super Soft Switching Three Level Inverter (SS3L Inverter)'.

1. Three- level inverters

Three-level inverters are well known. They are used in a power rage from approx. 50 kVA up to some MVA in electrical drive systems, as grid infeed inverters in PV power plants, in UPS installations or as active power line filters. A constant DC input voltage U_d is converted to a three-phase AC voltage system with adjustable frequency and amplitude. Fig. 1 shows a diagram of the working principle of such an inverter. The switches are power semiconductors, particularly IGBTs. The output terminal voltage may have one of the following 3 voltage states: $+U_d/2$, 0 or $-U_d/2$, which give the topology the name 'three-level inverter'. Typical pulse frequencies are set from some hundred Hz to approx. 18 kHz. Using pulse width modulation (PWM), pulse patterns are generated containing a fundamental of the required amplitude and frequency.

Different possible circuitry designs are described below.

1.1. Hard switching three-level PWM inverter in T-type design (NPC II inverter)

This topology has been presented already in 1980 [1]. Fig. 2 shows the circuitry topology of one single phase. The two 'outer' IGBTs V1 and V4 form together with their antiparallel diodes D1 and D4 a conventional 2 level PWM inverter. In combination with the IGBTs V2 and V3, and the respective diodes D2 and D3, a switchable path is set up to the centre tap of the input DC voltage U_d. With this, the terminal voltage u_0 may be set to zero as well.

This circuitry design is known as T-type inverter. It possesses particularly low conduction losses and therefore a high efficiency [2]. This is based on the fact that with the T-type inverter only one power semiconductor switch conducts when V1 / D1 or V4 / D4 are turned on. On the contrary, when using a NPC I (neutral point clamped) inverter, two switches conduct in the respective time, which doubles the value of the forward voltage and so the conducting losses [1]. Usually IGBTs are used as power semiconductor switches, recently IGBT modules.

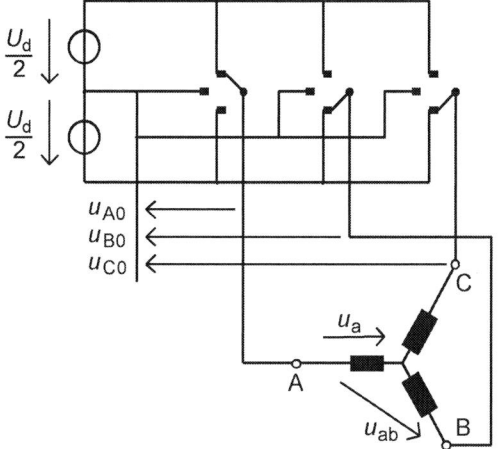

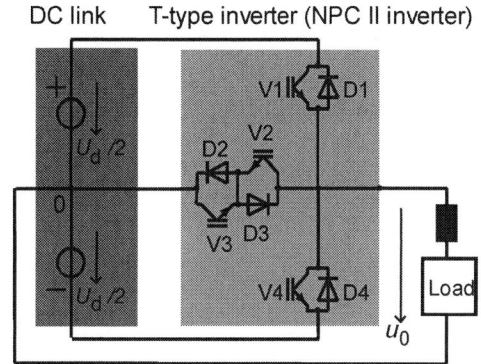

Fig. 2: Single phase hard switching three-level inverter (NPC II or T-type inverter)

Fig. 1: Schematic of a three-level inverter

In addition to the conducting losses, the switching losses have to be taken into account. Switching losses result during turn-on or turn-off of IGBTs when high values of current and voltage occur at the same time. This is called 'hard switching' [3]. Fig. 3a shows an example of the collector-emitter-voltage $u_{CE}(t)$ and the respective collector-current $i_C(t)$ when turning on the IGBT. It results in measured values of the turn-on energy E_{on} of 4,5 mJ.

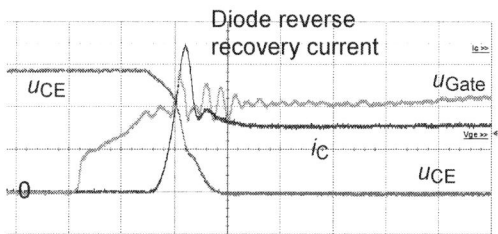

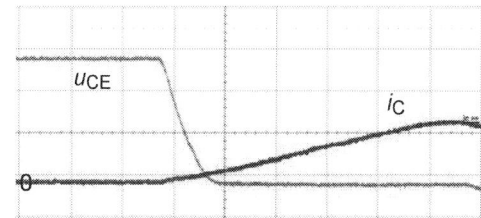

Fig. 3a: Hard switching turn-on
u_{CE} : 200 V/div; i_C : 50 A/div
u_{Gate} : 10 V/div; t : 200 ns/div
Turn-on energy E_{on} = 4,5 mJ

Fig. 3b: Soft switching turn-on
u_{CE} : 200 V/div; i_C : 50 A/div
t : 200 ns/div
Turn-on energy E_{on} = 0,1 mJ

These switching losses may not be neglected. Especially for higher switching frequencies, the average values of the switching losses may significantly exceed the conduction losses. This limits the switching frequency for this topology. To achieve high efficiency, the switching losses must be reduced. Up to now the standard solution for this is the reduction of the switching times. Especially when using transistors based on wide bandgap materials such as silicon carbide (SiC) or gallium nitride (GaN), significant progress has been achieved.

Short switching times reduce the switching losses as desired. The necessary high rate of rise of the current *di/dt* which induces voltage peaks at the parasitic inductances of the connecting lines is disadvantageous. These voltage peaks may reach high values, stressing the semiconductor switches and increasing the electromagnetic emission. The rate of rise of voltage *du/dt* using silicon-IGBT may reach up to 5 kV/µs, for SiC-IGBT it is significantly above 10 kV/µs.

© VDE VERLAG GMBH · Berlin · Offenbach

1.2. Soft switching three-level PWM inverter

The following section describes a soft switching PWM inverter which limits the switching losses as well as the rate of rise of voltages du/dt and currents di/dt. Therefore parasitic line inductances are less important, and electromagnetic compatibility improves significantly.

1.2.1. Soft switching three-level inverter (S3L inverter)

This new and patented [4] topology was presented first in 2011 [5], [6]. Fig. 4 shows the principle circuit diagram. Based on the T-type inverter (NPC II inverter) according to fig. 2, an additional sub-circuit consisting of 4 diodes, 2 capacitors and one inductor was implemented. This 'S3L snubber circuit', consisting only of passive elements, is shown in Fig. 4. The circuit achieves a reduction of switching losses by avoiding high values of collector-emitter voltage and collector current at the same time. It is remarkable that the snubber circuit does not need resistors to achieve the aim, as other snubber circuitries often do. Details of the working principle are described in [5] [6] [7] [8].

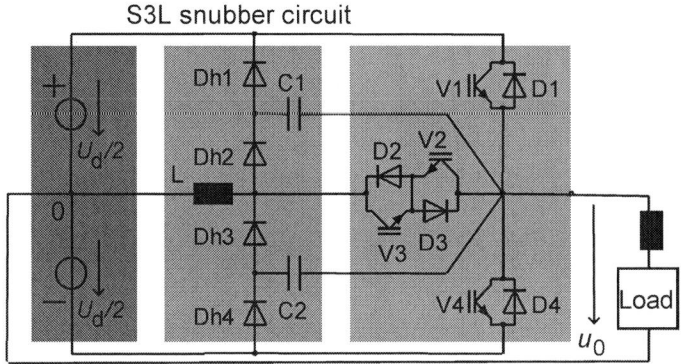

Fig. 4: Soft switching three-level (S3L) inverter

The example in fig. 3b shows the difference compared to the hard switching topology in fig. 3a: Collector voltage $u_{CE}(t)$ is reduced to the low value of the forward voltage in the conduction mode within 200 ns only, whereas the collector current $i_C(t)$ increases only slowly with limited di/dt. The simultaneous appearance of voltage and current at the IGBT is more or less prevented. So the losses, i.e. the product of voltage and current, can be limited to low values. The turn-on energy E_{on} has been reduced from 4,5 mJ in fig. 3a to 0,1 mJ in fig. 3b, i.e. by a factor of 45.

It has already been described [8] that all switching processes are soft switching, i.e. either the rate of rise of the current di/dt when turning on the IGBT is limited or the rate of rise of the voltage increase du/dt is limited when turning off the IGBT. So the switching losses are limited significantly. This is confirmed in measurements carried out with the prototype described in paragraph 2. With a rated power of 20 kVA the prototype shows values of di/dt < 27 A/µs and du/dt < 1000 V/µs. In some applications the values of du/dt < 500 V/µs are specified. This can be achieved by respective choice of capacitance and inductance of the snubber network. On the contrary the hard switching inverters with Si IGBTs result in values of up to du/dt = 5000 V/µs, with SiC IGBTs up to 50000 V/µs.

Especially the latter values du/dt result in high-power, high-frequency interference voltages. For the S3L inverter such high values du/dt only occur when turning on the 'outer' IGBTs. For all other switching processes du/dt is limited. This provides a significant reduction in interference voltages.

1.2.2. Super soft switching three-level inverter (SS3L inverter)

A simple additional modification of the S3L inverter reduces the described *du/dt* values when turning on the 'outer' IGBTs and with this the interference voltage. To achieve this, the time for turning on is lengthened by a reduction of the rate of rise of the gate-emitter-voltage. The simplest solution for this is the implementation of a gate-series resistor which is only active during turn on, as shown in fig. 5.

In hard switching inverters this solution causes a dramatic increase of the switching losses when turning on the IGBTs which is severely disadvantageous. In contrast for S3L inverters, the rate of rise of the current when turning on the IGBTs is limited by the inductance L (see fig. 3b and fig. 6a). Because of the slow increase of the collector current and the slow decrease of the collector-emitter-voltage, high values of current and voltage at the same time are comparatively limited (see fig. 6b). Compared to the original S3L inverter, the switching losses during turn on are increased slightly, but still significantly reduced compared to the hard switching inverter. This topology of the S3L inverter is called 'Super Soft Switching Three Level (SS3L) Inverter' (fig. 5).

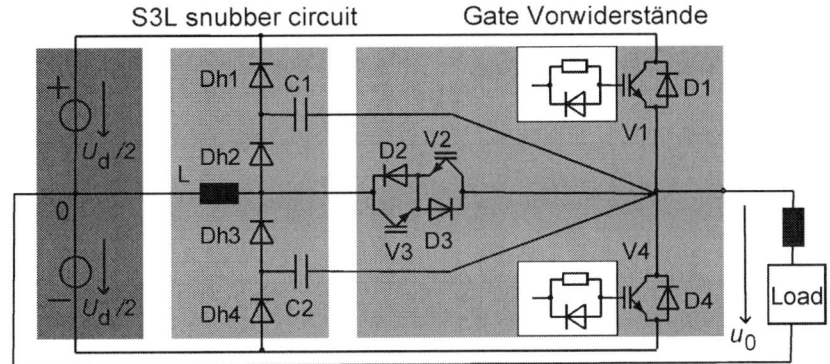

Fig. 5: Super soft switching three-level (SS3L) inverter
Gate resistors cause a slow turn-on of V1 and V4.

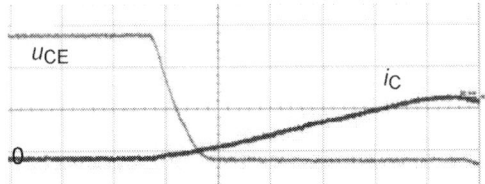

Fig. 6a: Turn-on of IGBT V1 in S3L Inverter
i_C : 50 A/div; u_{CE} : 200 V/div; t : 200 ns/div
Turn-on losses E_{on} = 0,1 mJ

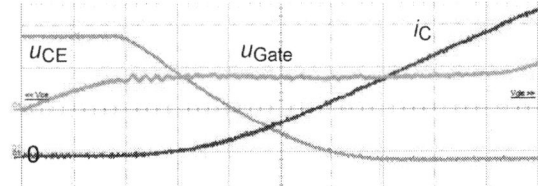

Fig. 6b: Turn-on of IGBT V1 in SS3L Inverter
i_C : 25 A/div; u_{CE} : 200 V/div; t : 200 ns/div
Turn-on losses E_{on} = 1,5 mJ

2. Comparison of EMC and efficiency

A 3-phase S3L prototype (20 kVA, 1000 VDC, 29 A_{rms}; fig. 7) was built to compare by measurements the different topologies H3L, S3L and SS3L including possible modifications of the control procedure. The semiconductor switches, the circuit board and later the load machines are always the same, so that the direct comparison is possible. Comparability of the results is given. The setup does not conform to the respective standards.

Fig. 8 shows the test setup [9]. The DC input voltage is provided by regulating transformer and an uncontrolled rectifier. The inverter feeds an induction motor, which is burdened mechanically by a DC machine.

The control procedure is realised as a sine-triangle-PWM being implemented in a DSP. Different operation modes are chosen for comparison: Constant pulse frequency of 8 kHz, 16 kHz or 32 kHz (i.e. switching frequencies 4 kHz, 8 kHz or 16 kHz for each single IGBT) and an additional procedure with variable pulse frequency (8 – 32 kHz). For the variable pulse frequency operation, the pulse frequency is varied within one period of the fundamental, so that the ripple of the output current was approximately constant [7], [8].

Fig. 7: S3L prototype
1000 VDC
20 kVA
29 A_{rms}

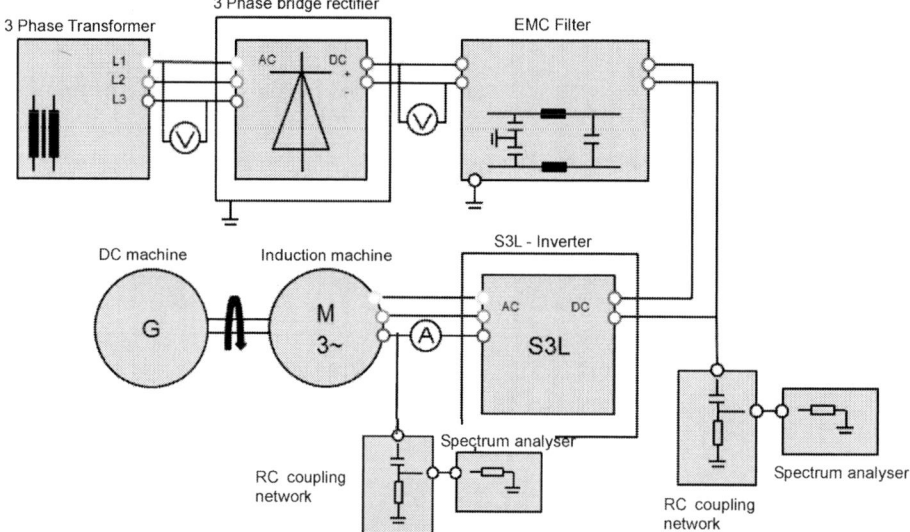

Fig. 8: Test setup including EMI measurement

2.1. EMC measurements

The interference voltage is measured in the frequency range of 150 kHz and 30 MHz at the DC supply and at the terminals of the induction load machine by use of a spectrum analyser Agilent E4403B / 9 kHz – 3 GHz. No EMI filters are implemented in the inverter. The setup does not conform to the standards, comparability of results is given.

In order to to find out especially the effect of the different control procedures, the spectrum of the inverter terminal voltage is measured additionally from one output terminal to the DC supply centre tap in the frequency range of 1 kHz and 400 kHz.

2.2. Measurements of efficiency

Input power and three phase output power is measured by a power analyser Yokogawa Digital Power Meter WT1600. Based on these results, the efficiency is calculated to compare the effect of topologies and control procedures.

3. Results of EMC comparison

3.1. Interference voltage at the DC input

Fig. 9 shows the comparison of the interference voltage for the three circuitry topologies H3L (Hard switching three level inverter), S3L (Soft switching three-level inverter) and SS3L (Super soft switching three level inverter). Pulse frequency is set to 8 kHz, the induction machine is operated with nominal torque and low speed. H3L and S3L show similar interference voltages, whereas the SS3L result is lower because of the limited *du/dt* .

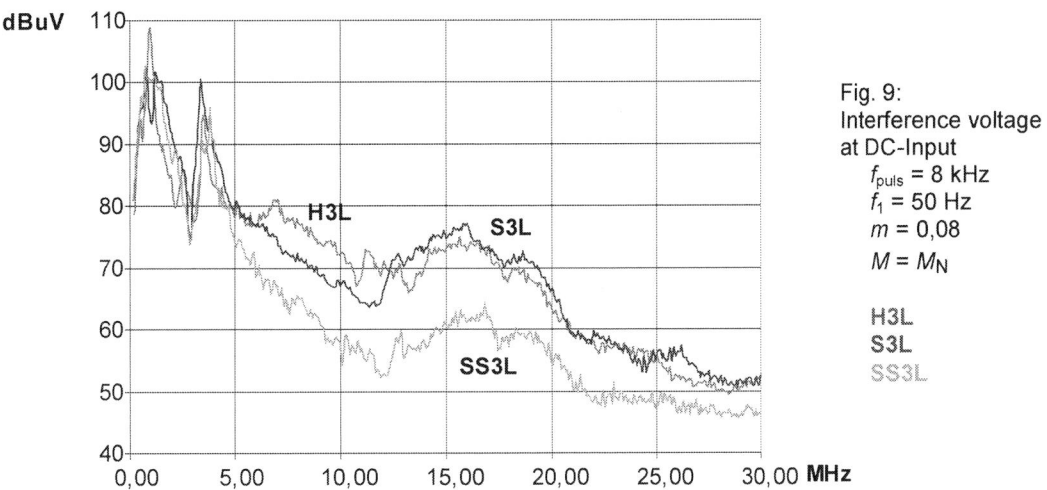

Fig. 9:
Interference voltage
at DC-Input
f_{puls} = 8 kHz
f_1 = 50 Hz
m = 0,08
$M = M_N$

H3L
S3L
SS3L

3.2. Interference voltage at the AC output

Fig. 10 is measured with the same operation condition as before. Again SS3L shows the lower distortion voltages particularly in the higher frequency range.

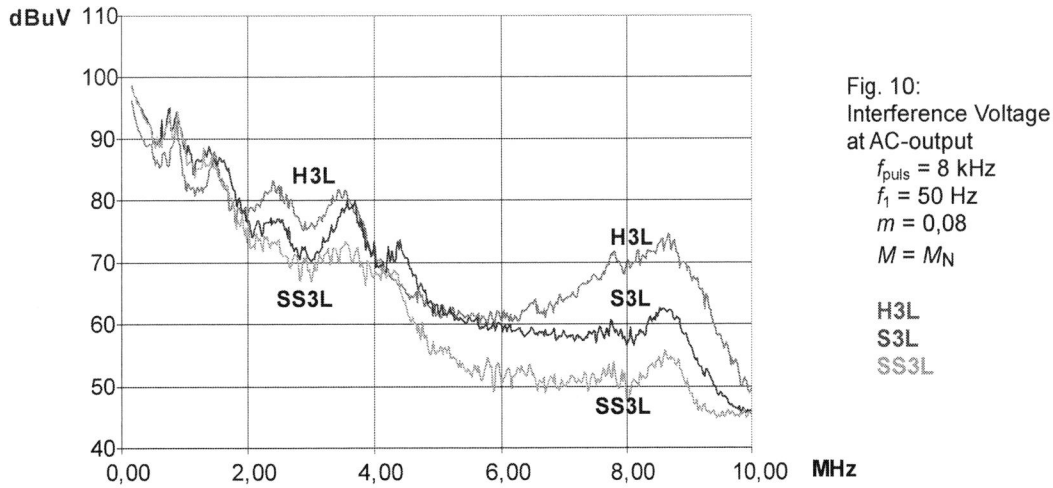

Fig. 10:
Interference Voltage
at AC-output
f_{puls} = 8 kHz
f_1 = 50 Hz
m = 0,08
$M = M_N$

H3L
S3L
SS3L

3.3. Effect of pulse frequency

Fig. 11 shows the effect of the different modes of control procedures on the interference voltage. Results are given for H3L inverter with 8 kHz pulse frequency, S3L inverter with 16 kHz and SS3L with 32 kHz. Additionally SS3L is operated with variable pulse frequency from 8 kHz to 32 kHz with average of 16 kHz. The latter shows similar interference as the S3L operated at 16 kHz. Interesting differences occur in the lower frequency range as described in paragraph 3.4.

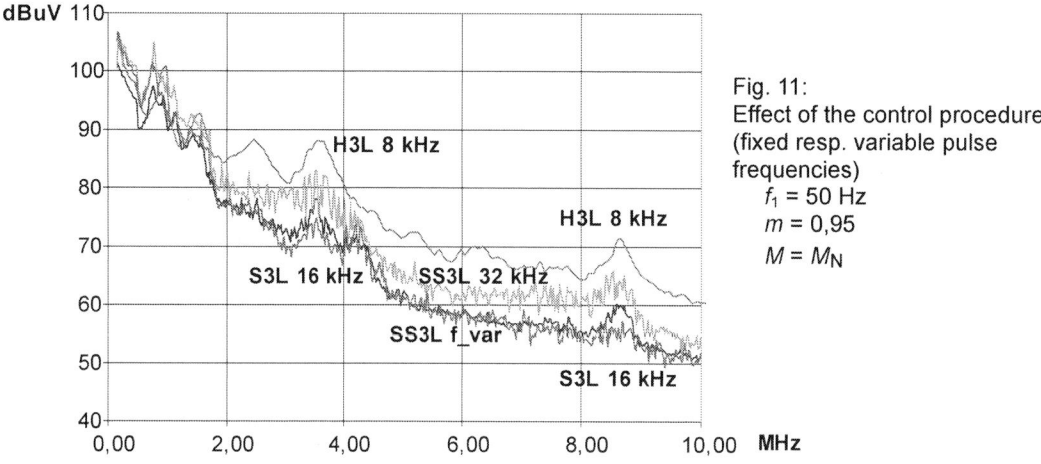

Fig. 11:
Effect of the control procedure (fixed resp. variable pulse frequencies)
f_1 = 50 Hz
m = 0,95
$M = M_N$

3.4. Effect of variable pulse frequency

Fig. 12 shows the spectrum for the interference voltage up to 1 MHz. The harmonics are obvious when operating the inverter with fixed pulse frequency of 22 kHz, whereas they almost disappear for variable switching frequencies (here: 12 kHz to 32 kHz).

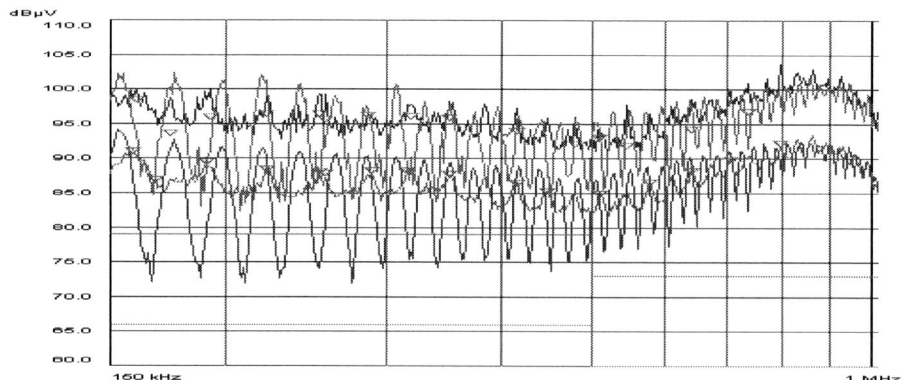

Fig. 12 Harmonics for constant and variable pulse frequency (peak/average)

3.5. Conclusion of EMC measurements

The S3L inverter shows only slightly lower interference voltage compared to the H3L inverter. This is based on the fact of the remaining existence of high *du/dt* values in S3L. These are advantageously suppressed in the SS3L topology.

4. Results of measurements of efficiency

As expected the soft switching topologies provide better efficiency compared to the H3L inverter. This has already been published e.g. in [7]. The advantages in EMI reduction of SS3L might cause a slight reduction in efficiency. Fig. 13 shows the efficiency for different output power.

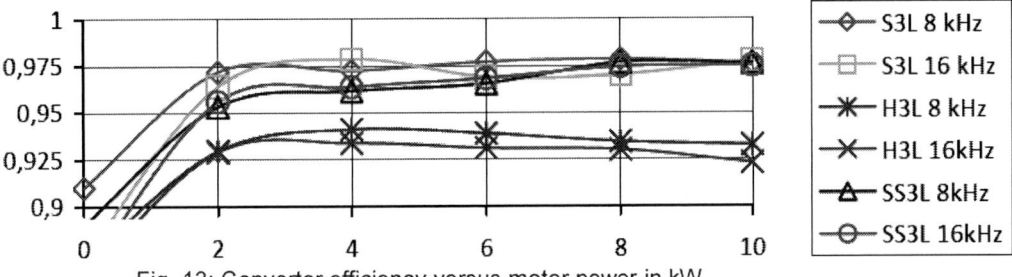

Fig. 13: Converter efficiency versus motor power in kW

5. Summary

The soft switching S3L and SS3L inverter provides significantly higher efficiency compared to the hard switching H3L inverter. Regarding the interference voltage, S3L inverters have only a small benefit compared to H3L inverters, whereas the SS3L is advantageous. Especially when operated with variable switching frequencies, the SS3L inverter is the most attractive solution.

6. Literature

[1] Nabae, H. Akagi and I. Takahashi: A New Neutral-Point-Clamped PWM Inverter, IEEE Transactions on Industry Applications, Vol. IA-17, NO. 5, September/October 1981, pp. 518-523

[2] Schweizer, M.; Kolar, J.W.: High efficiency drive system with 3-level T-Type inverter; EPE (European Conference on Power Electronics and Applications) 2011; August 2011, Birmingham, UK; ISBN 9789075815153

[3] Applikationshandbuch Leistungshalbleiter, Semikron International GmbH, Verlag ISLE, 2. Auflage 2015, ISBN 978-3-938843-85-7

[4] 3-Stufen-Pulswechselrichter mit Entlastungsnetzwerk; Deutsches Patent DE 10 2010 008 426 B4; Anmeldetag: 18.02.2010; Veröffentlichungstag der Patenterteilung: 01.09.2011; Patentinhaber: Hochschule Konstanz; Erfinder: Manfred W. Gekeler

[5] Manfred W. Gekeler: Soft switching three level inverter with passive snubber circuit (S3L inverter); Proceedings of the 2011 14th European Conference on Power Electronics and Applications (EPE 2011, Birmingham); ISBN 9789075815153, S. 1–10

[6] Manfred W. Gekeler: Weich schaltender 3-Stufen-Pulswechselrichter mit verlustfreiem Entlastungnetzwerk; Internationaler ETG-Kongress 2011 (ETG-Fachbericht 130 Teil B); 2011, ISBN 978-3-8007-3376-7, S. 264-270

[7] Manfred W. Gekeler: Soft Switching Three Level Inverter (S3L Inverter). Proceedings of the European Conference of Power Electronis and Applications (EPE 2013), 2013. ISBN 978-90-75815-17-7 and 978-1-4799-0114-2

[8] Manfred W. Gekeler: Optimierte PWM-Steuerung für Soft Switching Three Level Inverter (S3L Inverter). Internationaler VDE ETG-Kongress November 2013. ISBN 978-3-8007-3550-1

[9] Stefan Schreitmüller: EMV-Messung am S3L Inverter. Bachelorthesis HTWG Konstanz, Fakultät Elektrotechnik und Informationstechnik. February 2015

Drive system loss reduction by allpole sine filters

Dennis Kampen, BLOCK Transformatoren-Elektronik GmbH, Max-Planck-Str. 36-46, 27283 Verden, Germany, dennis.kampen@block.eu

Michael Burger, SEW-EURODRIVE GmbH & Co KG, Ernst-Blickle-Str. 42, 76646 Bruchsal, Germany, Michael.Burger@sew-eurodrive.de

Abstract

Drive system efficiency must be specified by PDS manufacturers according to new Ecodesign directive EN 50598. This evaluation Shows, how passive Motor filters can increase overall system efficiency of drive systems. Also the impact of the filters on the losses of each single component like inverter, cable and motor has been evaluated. The measurements were done with different cable lengths, switching frequencies, motor frequencies and loads.

1. Introduction

Allpole sine filters are installed between inverter and cable+motor. The switching frequency components of the inverter output voltage differential mode and common mode are nearly eliminated. In both, differential and common mode the resonance frequency of the filter is between the motor frequency and the switching frequency.

This investigation shows the effect of these filters on system efficiency and loss distribution between inverter, filter, cable and motor. Efficiency of drive systems becomes more and more important since directives like European eco design are forced into market [1].

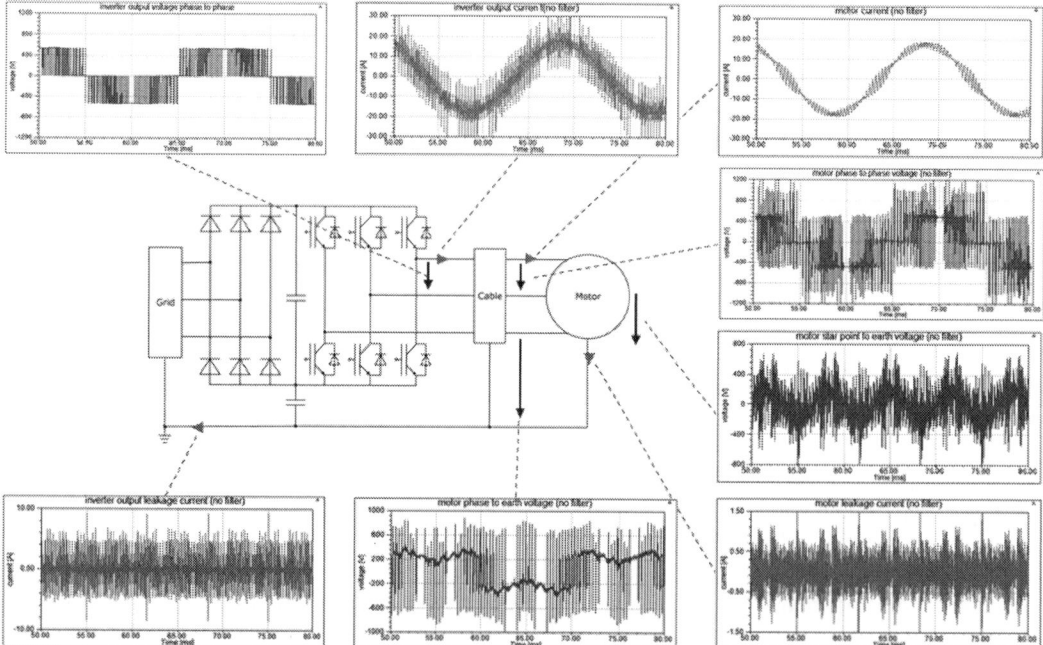

Fig. 1 Principle voltage and current waveforms without allpole sine filter

PCIM Europe 2016, 10 – 12 May 2016, Nuremberg, Germany

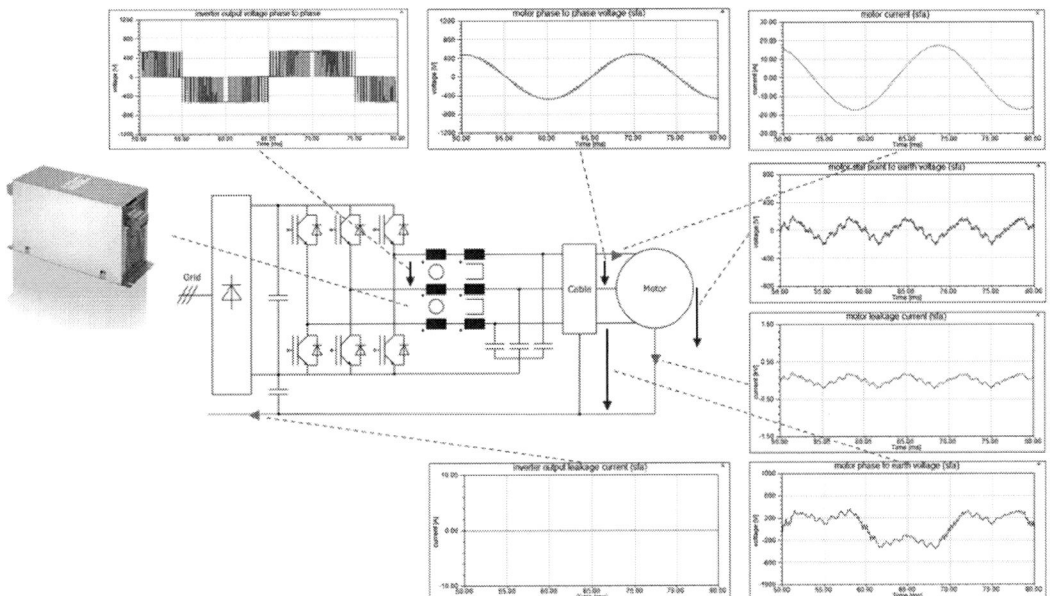

Fig. 2 Principle voltage and current waveforms with allpole sine filter

2. Measurement setup/Electric Circuit

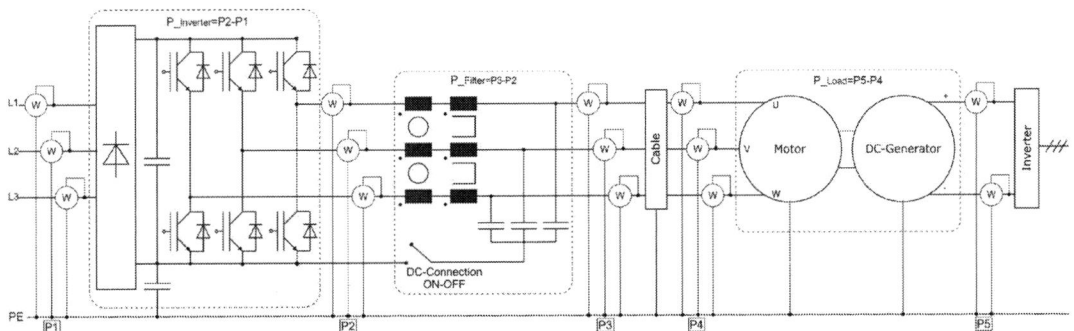

Fig. 3 electric circuit of measurement

3. Measurement results

3.1. Total drive system losses

Although additional losses are generated by inserting passive filters, the total drive system losses can be reduced, Table 1. Especially in case of partial load and long motor cables a significant increase of efficiency can be achieved.

© VDE VERLAG GMBH · Berlin · Offenbach

145

Table 1 effect of allpole sine filter on total system losses

Imot /A	fmot /Hz	fs /kHz	shielded cable/m	Without filter P_total /W	With filter P_total /W	loss savings by filter
7	5	8	10	246	237	3,66%
7	5	8	50	373	285	23,59%
22	50	4	10	2315	2396	-3,50%
22	50	4	50	2522	2573	-2,02%
22	50	8	10	2412	2436	-1,00%
22	50	8	50	2618	2623	-0,19%
22	50	8	100	2954	2849	3,55%
22	50	8	150	3276	3117	4,85%

3.2. Inverter losses

Inverter losses are reduced drastically independent of the operating point by inserting a filter into the system, Table 2. This leads to a lower derating of an inverter considering cable length, switching frequency or motor speed. Another option would be to build the inverter smaller and cheaper.

Table 2 effect of allpole sine filter on inverter losses

Imot /A	fmot /Hz	fs /kHz	shielded cable/m	Without filter P_inverter /W	With filter P_inverter /W	Loss Saving by filter
7	5	8	10	42	40	4,76%
7	5	8	50	81	51	37,04%
22	50	4	10	252	238	5,56%
22	50	8	10	335	289	13,73%
22	50	4	50	258	241	6,59%
22	50	8	50	338	295	12,72%
22	50	8	100	379	297	21,64%
22	50	8	150	388	298	23,20%

3.3. Cable losses

Cable losses are reduced by the filter especially at high cable lengths. i.e. at 150m cable length the reduction of losses by the filter is 11% at full load and even higher at partial load.

Table 3 effect of allpole sine filter on cable losses

Imot /A	fmot /Hz	fs /kHz	shielded cable/m	without filter P_cable /W	with filter P_cable /W	loss savings by filter
7	5	8	10	29	22	24,14%
7	5	8	50	108	51	52,78%
22	50	4	10	136	133	2,21%
22	50	4	50	329	319	3,04%
22	50	8	10	137	135	1,46%
22	50	8	50	330	318	3,64%
22	50	8	100	613	555	9,46%
22	50	8	150	913	811	11,17%

3.4. Motor losses

Nearly all additional motor losses because of PWM in comparison to direct grid operation can be eliminated with a filter. Table 2 shows the loss savings of motor + generator with a filter. It must be kept in mind, that the %-loss savings of the motor losses is much higher because the losses of the generator are much higher than the losses of the motor and independent of the filter. Considering other publications about motor losses, where the additional harmonic losses are identified to be 10%-20%. The results of this research confirm the literature percentages. With filter the motor lifetime will be increased drastically not only by reduced losses and heat, but also by less insulation material stress and the elimination of inverter caused bearing current damages. The motor could be built much smaller and cheaper by using less insulation and uninsulated bearings.

Table 4 Effect of allpole filter on losses of the load (losses motor+generator)

Imot /A	fmot /Hz	fs /kHz	shielded cable/m	without filter P_load /W	with filter P_load /W	loss savings by filter
7	5	8	10	175	105	40,00%
7	5	8	50	184	113	38,59%
22	50	4	10	1930	1897	1,71%
22	50	4	50	1935	1890	2,33%
22	50	8	10	1942	1902	2,06%
22	50	8	50	1950	1898	2,67%
22	50	8	100	1962	1903	3,01%
22	50	8	150	1975	1902	3,70%

4. Reference

[1] EN 50598 "Ecodesign for power drive systems, motor starters, power electronics & their driven applications", DRAFT 2014-01

Harmonic Filtering in Variable Speed Drives

Luca Dalessandro, Xiaoya Tan, Andrzej Pietkiewicz, Martin Wüthrich, Norbert Häberle

Schaffner EMV AG, Nordstrasse 11, 4542 Luterbach, Switzerland

luca.dalessandro@schaffner.com

Abstract

This paper discusses and presents results related to still open controversial issues in passive harmonic filters' design, such as the identification of filter resonant frequencies, the effects of interactions of harmonic filters with system impedances and the importance of damping elements. Benefits of passive rectifiers equipped with harmonic filters are compared to those of other rectifier front-ends typically used in VSD applications. An experimental comparative evaluation of harmonic performances of passive and active rectifiers is also provided.

1. Introduction

Passive three-phase rectifiers equipped with single tuned (ST) harmonic filters, as shown in Fig. 1, represent the most cost effective, technically simple and robust front end of variable speed drives with low effects on the mains.

Harmonic filter design procedures and solutions for complex harmonic filtering application problems have been documented in several publications in the past few decades [1-3]. Every new harmonic mitigation problem still deserves a careful system analysis and special attention needs to be paid on crucial design aspects such as:

- selection of tuning frequencies and damping factors,
- interactions between harmonic filter and system impedances,
- optimal number of harmonic trap branches.

Following sections provide further practical insight and considerations on these topics and, additionally, benefits of passive rectifiers equipped with harmonic filters are presented and compared to those of other rectifier front-ends typically used in VSD applications.

2. Single tuned and multi-tuned passive harmonic filters

Six pulse current converters are common rectifier interfaces for variable speed drive systems and they are harmonic producing loads [2]. Because grid harmonic current emissions are regulated at both equipment level, e.g. [3, 4] and PCC [5], a typical harmonic mitigation approach is to install single tuned (ST) harmonic filter close to these non-linear loads. An accurate harmonic filter design procedure requires taking into account several aspects, such as estimate of harmonic current injection for different system's loadings, short-circuit level, load demand and harmonic analysis study considering presence in the installation of components which can impact harmonics, as capacitor banks or reactors [2].

A ST filter featuring a sharp tuning close to the 5th harmonic is typically the first approach to reduce the harmonic distortion, since the 5th harmonic has higher amplitude for six-pulse converters. If the harmonic content is still above limit, the procedure is to implement further ST filters tuned at higher order harmonics [1, 2, 3].

PCIM Europe 2016, 10 – 12 May 2016, Nuremberg, Germany

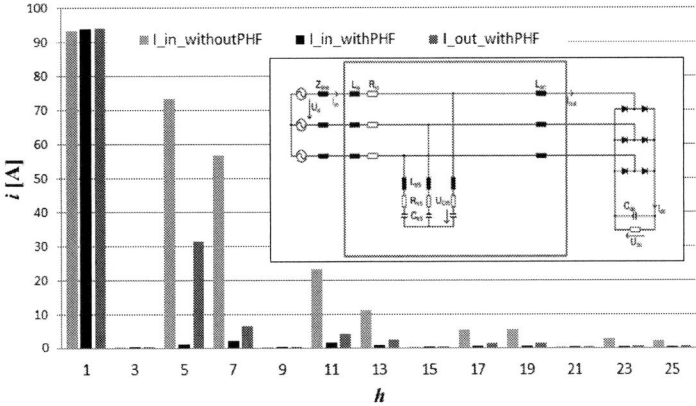

Fig. 1 Effect of a ST harmonic filter tuned at the 5th harmonic on the mains current i_{in} and on the passive rectifier input current or filter output current i_{out} for a 45 kW, 400V, 50 Hz system. THDi without and with ST harmonic filter is 103.5% and 3.5%, respectively.

The attenuation by multiple parallel harmonic traps is achieved at the cost of higher loading of the lower harmonic ST branch, tuned at the 5th harmonic, and results in higher load current respect to using a ST filter; this effect was also observed in the empirical design procedure presented in [2].

Fig. 2 shows the harmonic spectrum of load or input rectifier current and of the resulting current in the 5th harmonic trap, corresponding to two different passive harmonic filter (PHF) designs for a 55kW drive: one ST filter tuned at the 5th harmonic and a filter made out of the combination of four parallel branches each tuned at a different harmonic (5th, 7th, 11th, 13th). For both configurations, single trap and multi traps, the passive harmonic filters draw the same level of capacitive current, which is typically 20% of the line current.

Furthermore, there is no DC-link choke or rectifier's input choke installed in both cases. The 5th trap inductance in the multi-trap filter results to be three times larger than the one in the single trap model, 4.7mH vs.1.84mH. The 5th branch circuit of the multi-tuned PHF is much more stressed; at the same time it features a larger inductor than a single tuned PHF, which results into a more expensive passive filter solution. The THDi performance is sufficiently good with ST harmonic filter, 5.8% vs 3.3% of multi-tuned PHF.

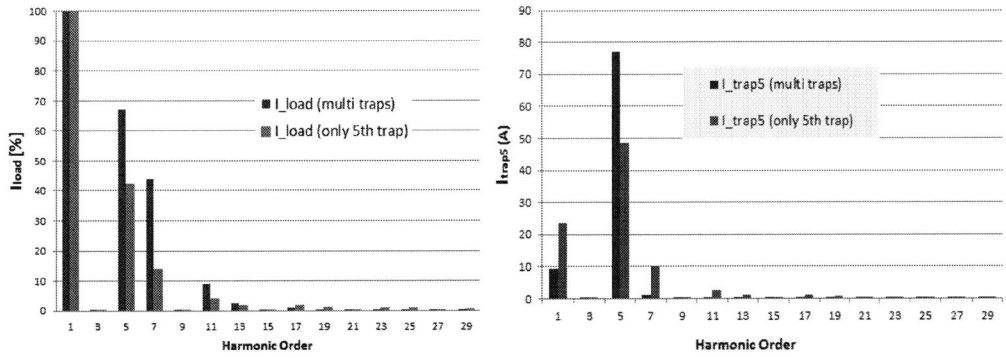

Fig. 2 Harmonic spectrum of load current and the resulting current in the 5th harmonic trap for the two different PHFs for a 55kW drive: one single filter tuned at the 5th harmonic and a multi-tuned PHF made out of four parallel branches each tuned to a different harmonic (5th, 7th, 11th, 13th).

© VDE VERLAG GMBH · Berlin · Offenbach

3. Interaction with system impedances

According to the power rating of the installation and the type of impedances present in the system, PHFs may introduce unwanted distortions and resonances and therefore their design requires a careful system analysis to avoid operational issues.

Fig. 3 shows the frequency characteristic of PHFs with different ratings, 1.1kW and 200 kW, installed in input of drives equipped with integrated EMI filters, in particular the X2 capacitor stage. Whilst for the larger rating PHF the interaction with the EMI filter does not cause variations of the frequency characteristic, the curve of the 1.1kW PHF presents a parallel resonance at about 600 Hz, which may result into an amplification of harmonic frequency sources in a range between the 11th and 13th order.

Such a resonance can hinder the correct functioning of the filter and it is a case of interaction of PHF with EMI filter capacitance which may result into a poorer performance.

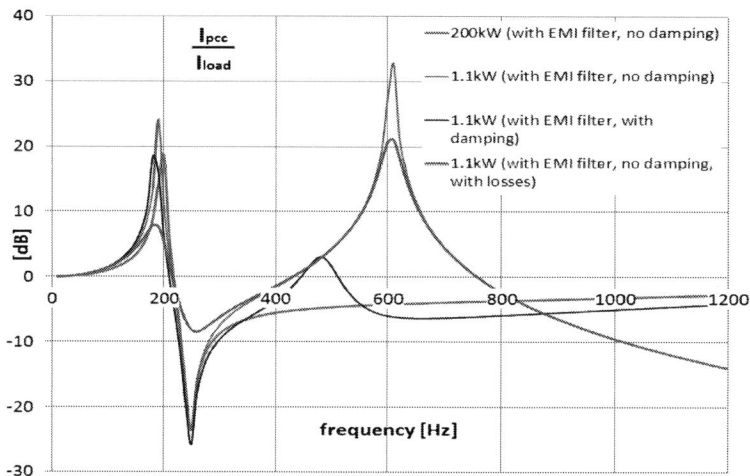

Fig. 3 Frequency characteristics of PHFs installed in input of drives equipped with EMI filter stage. The interaction of the 1.1kW PHF with the EMI filter capacitance causes an unwanted resonance. The use of adequate damping techniques or even a modification of the filter circuit may preserve correct operation.

The frequency and amplitudes values of the first parallel resonance (due to interaction with line impedance L_i) and of the series resonance are given respectively by [2]:

$$f = 1 / 2\pi \, (LC + L_iC)^{\frac{1}{2}}$$

$$Q = ((L + L_i)/C)^{\frac{1}{2}} / (R + R_i)$$

These two resonances are expected in the ST filter design; the series resonance results from the trap branch LC tuned at about the 5th harmonic, ($L_i = R_i = 0$ in the formula) and has the effect of attenuating the amplitude of the 5th harmonic source.

The parallel resonance, occurring at a lower frequency, is due to the interaction between trap circuit and line-side impedance L_i (see Fig. 7a) and it is selected outside the range of harmonic sources; if this condition is not verified, the magnitude of the parallel resonance can be reduced by modifying the filter circuit with additional resistive, damping elements or by varying the value of the line inductor [1].

The amplitude of parallel resonance due to interaction of the filter with EMI filter capacitors can be reduced by implementing a further damping branch [12], for instance in parallel to the line reactor. The attenuation effect of such a solution is visible in the plot of Fig. 3. The additional resistance and losses contribute to decrease the value of the Q factor corresponding to the parallel resonance, hence to flatten the shape of the filter frequency characteristic.

The accurate tuning of the harmonic filter is crucial to achieve the target attenuation. However, effects like tolerances, component aging and temperature dependent characteristics need to be considered during design and therefore a ST filter is typically tuned to a frequency lower than the harmonic to suppress. The most crucial tolerances are on components of trap branch as depicted in Fig. 4.

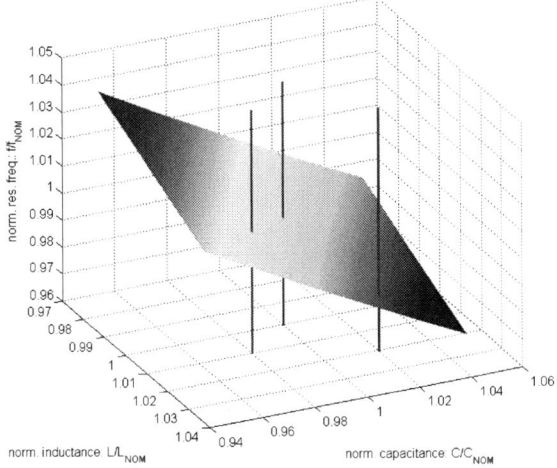

Fig 4 Plane of possible resonant frequencies for the 3-phase trap inductor in series with the capacitor bank, as function of the component tolerances. In particular, the lines parallel to the z-axis indicate the case of three limbs of the inductor having different values due to tolerance.

An overall system harmonic analysis is crucial to define the best filter solution [1-2] to fulfill the regulations [4-6]. Fig. 5 shows the outcome of an analysis realized with the PQS tool [13] for the case of multi-motor system.

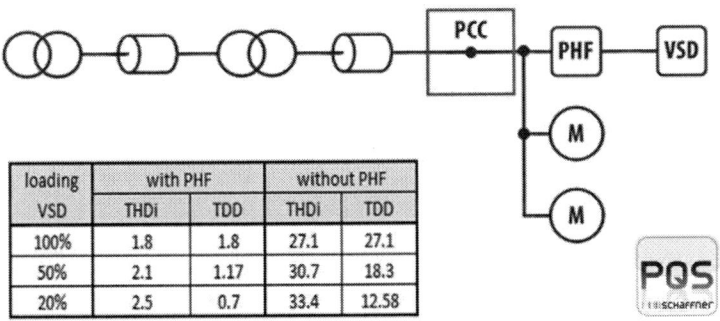

loading	with PHF		without PHF	
VSD	THDi	TDD	THDi	TDD
100%	1.8	1.8	27.1	27.1
50%	2.1	1.17	30.7	18.3
20%	2.5	0.7	33.4	12.58

Fig 5. Harmonic analysis of a multi motor system performed with PQS of Schaffner. In particular, only one motor is equipped with VSD, and the effect of VSD partial loading on the harmonic distortion is considered, with and without PHF. If the rating of the VSD is larger than the demand of both motors, then the resulting TDD at PCC and THDi for the VSD line have different trend.

4. Figures of merit of rectifiers for VSD applications

A state-of-the-art LV VSD is typically constituted by a diode rectifier front-end cascaded by a voltage source inverter which drives the electric motor. In order to comply with harmonic standards [4-6], the diode bridge is commonly equipped by a ST harmonic filter, able to reduce the THDi, typically below 5%. If the application requires regeneration back into the mains, active-front end (AFE) is used. An active rectifier is able to fulfill the harmonic requirements. The enhanced controllability has an immediate impact on cost, reliability and availability of the system. A compromise to improve the input current THDi is the use of multi-pulse rectifiers, typically 12 or 18 pulses, together with multi-winding input transformers [11]. This kind of rectifier stage increases considerably the complexity of the front-end and has the drawback of reduced performance in presence of imbalance. A further possibility to reduce the THDi of the drive is to actively compensate harmonics through an active harmonic filter (AHF), which is commonly a shunt device installed at the input terminals of a single or multi-drive system [3, 11]. Fig. 6 presents figures of merit, derived from [2,3,11] of rectifiers with low harmonic impact on the mains, typically used in VSD applications.

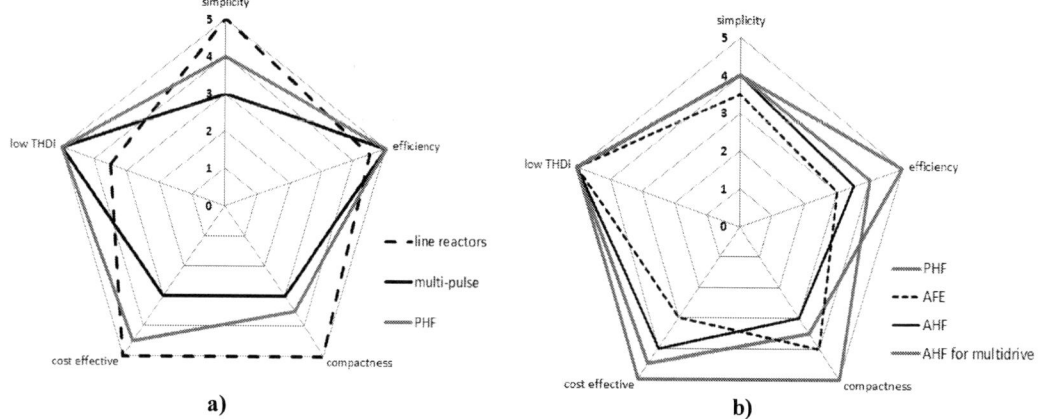

Fig. 6. Figures of merit of rectifiers typically used in VSD applications for nominal loading, a) passive interfaces and b) active interfaces as active harmonic filter (AHF) and active front end (AFE) compared to passive rectifier equipped with PHF. Line reactor in input of diode bridge is a further simple, cost effective solution.

5. Experimental verification

The harmonic performances of two rectifiers, typically used as input rectification stage for 2Q variable speed drive applications, were compared; in particular, the following rectifier topologies were considered:

- A passive rectifier, 3-phase diode bridge, without DC-link choke, equipped with a ST passive harmonic filter tuned at the 5th harmonic, designed for a target THDi < 5%.

- An active front-end (AFE), 3-phase, 2-level PWM rectifier equipped with LCL and EMI input filter stages and switching at a frequency of 1.5 kHz.

For both rectifiers, the input current and voltage are 95Arms and 400Vrms respectively, the mains frequency is 50 Hz, 75 kW the power rating. In addition, the current spectra were derived and current harmonics until the 40th order shown in Fig. 9.

PCIM Europe 2016, 10 – 12 May 2016, Nuremberg, Germany

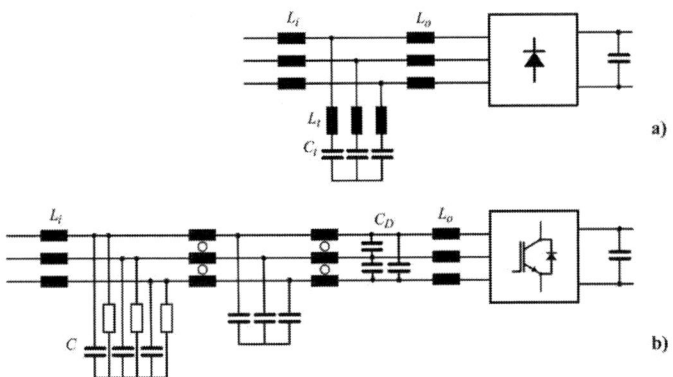

Fig. 7. Rectifiers and their respective input filter stages considered for the comparative evaluation of harmonic performance: a) passive rectifier, 3-phase diode-bridge featuring a single tuned (5th harmonic) passive harmonic filter and b) AFE equipped with LCL and EMI filters.

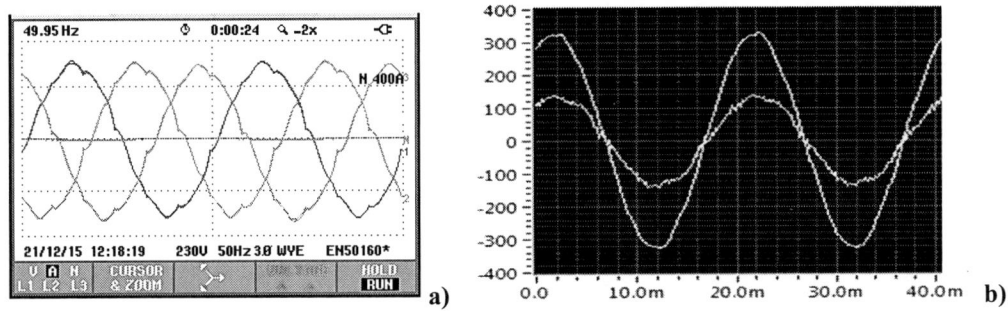

Fig. 8. Measured input rectifier current waveforms, a) passive rectifier with PHF and b) AFE equipped with LCL and EMI input filters. The amplitude of the measured current waveforms is the same, 95 Arms, the frequency is 50 Hz.

PHF	L_i	0.605 mH
	L_o	0.149 mH
	L_t	0.94 mH
	C_t	3x 440 uF wye
LCL	L_i	0.044 mH
	L_o	0.33 mH
	C_D	3x 40 uF delta
EMI	C	10 uF
	L_{CM1}	0.4 mH
	L_{CM2}	26 uH

Table 1. Parameters of the filters in input to the rectifiers, as shown in the circuits in Fig.7. Source impedance condition is same for both setups.

The total harmonic content (THC) and the reference current I_{ref} were assessed using the formulas as defined in [5] for both measured input rectifier currents shown in Fig.8. The THDi defined after [6] is calculated as well to estimate the harmonic performance; in the IEEE 519 the harmonic content is typically up to the 50th order.

© VDE VERLAG GMBH · Berlin · Offenbach

153

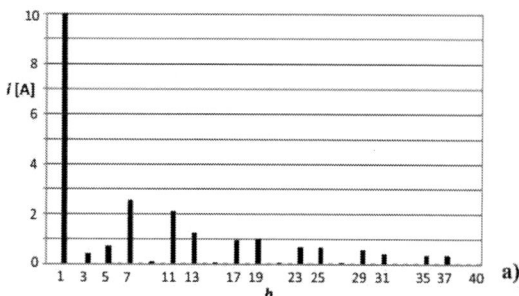 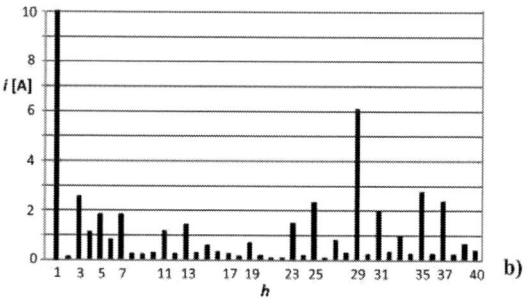

Fig. 9. Input rectifier current waveforms spectra. Harmonic frequencies are displayed until 40th, harmonic amplitudes are in a) passive rectifier with PHF and b) AFE equipped with LCL and EMI filters. The amplitude of the fundamental component (h=1) is 95Arms; the enlarged vertical axes cut the representation of these values.

	Passive rectifier + PHF	*AFE + LCL, EMI filters*
THC	4.34 Arms	6.38 Arms
I_{ref}	92.2 Arms	94.4 Arms
THC/I_{ref}	4.7 %	6.7%
THDi	4.7 %	7%

Table 2. Harmonic performance of the tested rectifier systems.

For this particular case study, the rectifiers and their respective input filters were commercially available equipment and were not modified for the tests. The measured harmonic performance at nominal loading is summarized in Table 2. The passive rectifier equipped with ST filter yields to a better THDi; nevertheless the PHF corresponding to the values in Table 1 is about 50% more expensive than the LCL combined with EMI filter.

However, for same target harmonic performance, e.g THDi = 4.7%, and same switching frequency, the cost of LCL and EMI filter stages would exceed the price of a PHF. Furthermore, a passive rectifier equipped with PHF results about 25% cheaper than a AFE featuring LCL and EMI filter, for a power rating ranging from few kW up to several hundreds of kW [7].

The higher contribution to harmonic distortion of an active rectifier is expected to occur at PWM switching frequency f_s in the kHz range. If f_s is lower than 2 kHz, corresponding to the 40th harmonic order, the switching ripple will contribute to increase the harmonic content as defined by the norms [5, 6]. It is good practice to verify for active rectifiers the amplitude of lower harmonic orders [9], which can increase, for instance, if low frequency signals injection techniques are applied to improve modulation [10].

On the other side, if f_s is higher than 2.5 kHz, the ripple components will be shifted in a range, 2-150 kHz, for which the compatibility limits are still not regulated. The IEC SC77A WG8 has the task of the standardization for 2-150 kHz, which will result in an updated standard, eg. IEC 61000-2-2 [8].

For an AFE the LC filter stage has to attenuate the ripple current. Further optimization of the filter components is possible and it consists in finding the trade-off between filter volume and overall system efficiency [9].

6. Conclusion

Passive harmonic filters in combination with three-phase passive rectifiers constitute the most cost effective, technically simple and robust front end of variable speed drives with low effects on the mains. To solve the harmonic mitigation problem and to properly design the harmonic filter, a careful system analysis is needed and dedicated simulation tools, like the PQS [13], help to take into account and verify all the most important design requirements.

The paper has shown that special attention needs to be paid on the selection of tuning frequencies and damping factors. Interactions between harmonic filter and system impedances can cause system resonance, which can be avoided by proper shaping of the filter frequency characteristic. Figures of merit for rectifier front-ends typically used in VSD applications were compared to show the superior advantages of passive rectifiers equipped with PHF. Finally, the experimental comparative evaluation of harmonic performances of passive and active rectifiers confirmed the economic and performance benefits of a passive filter solution.

7. References

[1] M. Allenbaugh, T. Dionise, T. Natali, Harmonic Analysis and Filter Bank Design for a New Rectifier for a Cold Roll Mill, IEEE Trans. Ind. Appl. Vol 49, No. 3, May 2013 pp 1161 - 1170.

[2] J.C. Das, Passive Filters- Potentialities and Limitations, IEEE Trans. Industry Appl. Vol 40, No. 1, Jan./Febr. 2004 pp 232- 241.

[3] H. Akagi, Modern active filters and traditional passive filters, Bulletin Polish Academy of Sciences, Vol. 54, No. 3, 2006, pp. 255-269.

[4] IEC 61000-3-2 EMC - Part 3-2: Limits for harmonic current emissions (equipment input current ≤ 16 A per phase)".

[5] IEC 61000-3-12. EMC - Part 3-12: Limits for harmonic currents produced by equipment connected to public low-voltage systems with input current >16 A and ≤ 75 A per phase.

[6] IEEE-519 2014, IEEE Recommended Practice and Requirements for Harmonic Control in Electric Power Systems.

[7] ABB low voltage drives, ACS800, Product Pricing List, *Online Available.*

[8] IEC 61000-2-2:2002, EMC - Part 2-2: Environment - Compatibility levels for low-frequency conducted disturbances and signalling in public low-voltage power supply systems

[9] J. Mühlethaler et al. Optimal design of LCL Harmonic Filters for Three-Phase PFC Rectifiers, IEEE Trans. Power Electr., vol. 28, No, 7, July 2013, pp. 3114-3125.

[10] L. Dalessandro et al. Center-Point Voltage Balancing of Hysteresis Current Controlled Three-Level PWM Rectifiers, IEEE Trans. Power Elect., Vol 23, No. 5, Sept. 2008, pp. 2477-2488

[11] B Bahrani, R Grinberg, Investigation of harmonic filtering for the state-of-the-art variable speed drives, Proc. 13th European Conf. on Power Electronics and Application EPE 2009, pp. 1-10.

[12] Harmonic Filter, Granted patent family US8115571 B2.

[13] Power Quality Simulator, http://pqs.schaffner.com

A just Comparison of Ferrite and Nanocristalline Common Mode Chokes

Schliewe, Jörn, EPCOS AG, Heidenheim, Germany, joern.schliewe@epcos.com
Paulwitz, Christian, EPCOS AG, Regensburg, Germany
Weber, Stefan, EPCOS AG, München, Germany

Abstract

Based on our EMC lab experience we look into the reason why ferrite common mode chokes outperform nanocristalline counterparts in many practical applications. For a fair comparison, designs with same impedance at 150 kHz crucial frequency have been investigated. At the expense of a 30% larger size the ferrite common mode chokes gain better EMI noise reduction for common mode and differential mode noise sources.

1. Introduction

Driven by megatrend applications like green energy, energy management, drive systems and electrical vehicles, EMI behavior of power electronic systems is of increasing importance. Common mode chokes (CMC) are crucial parts for filtering EMI from inverters in different applications. During the last 20 years nanocristalline materials emerged on the market and have been promoted as the magnetic material of choice for future CMCs [1-3]. Ferromagnetic nanocristalline core materials are known for outstanding permeability values at reasonable losses in the kHz range. They are mandatory, when high permeability values are needed at low frequencies. Because of their high resistive impedance up to high frequencies in the range of some MHz they are also promoted superior for EMI applications.

	Ferrite	Iron Powder	Nanocristalline Iron
Permeability	+	-	++
Saturation	-	++	+
Losses	+	+	+
Price	++	+	-

Table 1 is a rough summary of advantages and disadvantages of basic core materials and it implies that the only reason to use something else than nanocristalline iron cores would be the price. But is this true? Is it only the price why still almost every EMI filter is built with ferrite CMCs?

In contrast to the general statement to prefer nanocristalline materials instead of ferrite materials, we experience better performance of the ferrite based CMCs in our accredited EMC lab. Common mode currents significantly reduce the advantage of high permeability materials, which is already documented in literature [2]. Due to the multitude of available ferrite materials comparisons of a single parameter with "typical" ferrites are of limited use. It is possible to find good and bad ferrite materials for each application. In this paper we will deduce from practical examples the advantages and disadvantages of the specific materials.

2. Choice of samples

The choice of samples is a crucial factor in the discussion about pros and cons between ferrite and nanocrystalline CMCs. In literature, quite often the samples are changed for each measurement, hiding the true potential of the materials. To avoid such confusion we define the samples by focusing on the application in industrial EMI filtering. According to standard EN 61800-3 frequencies above 150 kHz are of interest. Since the impedance of the inductances increases with the frequency, the lowest frequency usually defines the necessary inductance of the CMC. For comparison we have therefore chosen CMC samples with the same impedance at 150 kHz. We accept a 30% larger size for the ferrite CMC to achieve a comparable DC-resistance.

Fig. 1: Nanocristalline M-046-02 2x7 turns (left) and ferrite R40x24x16 2x12 turns 1 mH (right)

For nanocristalline CMCs Magnetec Nanoperm cores M-046-02 and M-030-04 have been chosen. M-046-02 is from a high permeability nanocrystalline material with a permeability of 90 000 according to the data sheet. The permeability of M-030-04 in contrast, is given in the data sheet as 30 000. For ferrite CMC the toroidal core R40x24x16 of ferrite material T65 will be used. T65 has been chosen, to achieve good EMI performance for temperatures up to 100°C. Figure 1 shows the M-046-02 core with 2x7 turns and the T65 core with 2x12 turns (1 mH) both with 2.6 mΩ DC-resistance per winding. Compared to the 1 mH ferrite sample, the sample of M-046-02 is approximately 10% smaller in the largest dimension and approximately 20% smaller in volume.

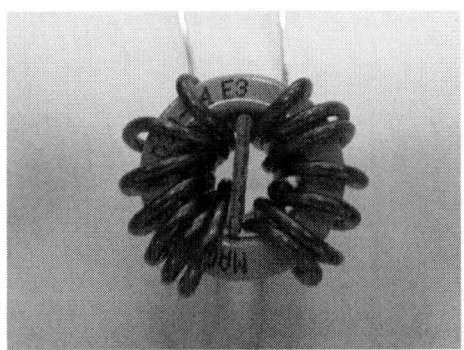 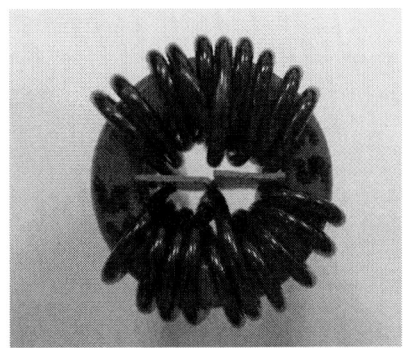

Fig. 2: Nanocristalline M-030-04 2x7 turns (left) and ferrite R40x24x16 2x10 turns 0.8 mH (right)

Figure 2 shows the M-030-04 core with 2x7 turns and the T65 core with 2x10 turns (0.8 mH) both with 1.8 mΩ DC-resistance per winding. Compared to the 0.8 mH ferrite sample, the sample of M-030-04 is approximately 15% smaller in the largest dimension and approximately 30% smaller in volume.

© VDE VERLAG GMBH · Berlin · Offenbach

PCIM Europe 2016, 10 – 12 May 2016, Nuremberg, Germany

3. Measured characteristics
3.1. High permeability nanocristalline CMC vs. ferrite 1 mH CMC

Figure 3 shows on the left side the CM inductance versus CM current for the M-046-02 nanocristalline and the T65 ferrite core with 2x12 turns. The inductance without DC-bias and at 100 kHz is in both CMCs approximately 1 mH. Due to the much lower permeability at low frequencies and the corresponding lower flux density, the ferrite core shows a significantly higher inductance for CM currents above 0.2 A.

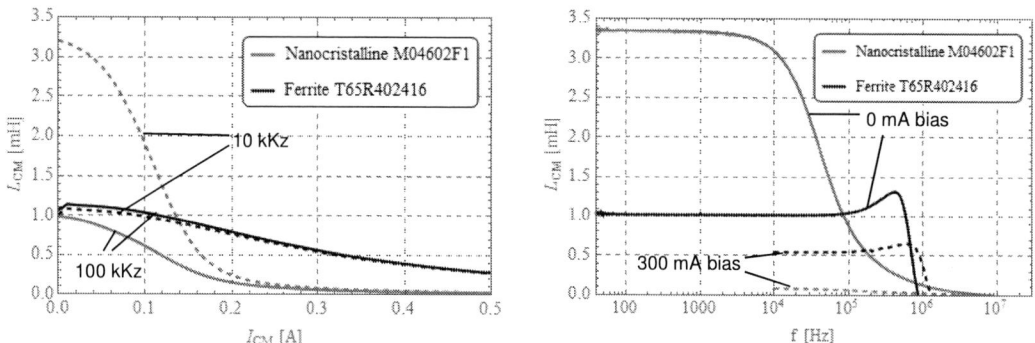

Fig. 3: CM inductance versus CM current (left, solid 100 kHz and dashed 10 kHz) and frequency (right, solid curves are without and dashed curves with 300 mA CM-dc-current).

On the right side, Figure 3 shows the CM inductance versus frequency. For the high permeability nanocristalline material the inductance at low frequencies is more than 3 times higher than that of the ferrite core. However, the inductance of this material starts to decrease approximately at 10 kHz. The remaining inductance at 100 kHz is approximately 1 mH. The reason is that eddy currents cause an opposing field which reduces the magnetic energy and therefore also the inductance. The ferrite choke has higher inductance values between 100 kHz and 900 kHz.

Figure 4 shows the impedance characteristic of the two CMCs for 25°C (left) and 100°C (right). Below 150 kHz the impedance of the nanocristalline core is much higher, due to its high permeability. At 10 kHz the eddy currents start to decrease the slope of the impedance, the losses start to dominate the impedance.

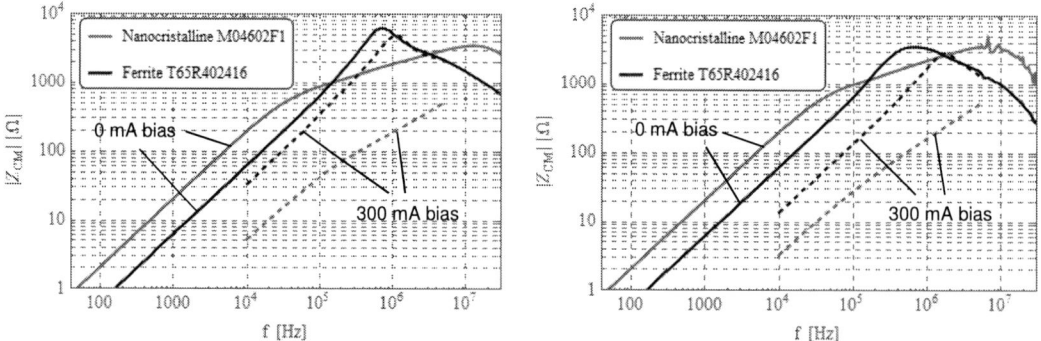

Fig. 4: Common mode impedance versus frequency at 25°C (left) and 100°C (right). Solid curves are without CM-DC-current and dashed curves with 300 mA CM-DC-current.

In the interesting frequency range above 150 kHz the ferrite shows a higher impedance up to a frequency of 3 MHz. In addition to the classical impedance curves we measured the

© VDE VERLAG GMBH · Berlin · Offenbach

impedance curves at a CM DC-bias of 300 mA. These curves are shown as dashed lines. Their frequency range is limited due to the limits of the decoupling network we used. The main result of these measurements is: with a low frequency CM current in the application the ferrite CMC shows a much higher noise attenuation compared to the high permeability nanocristalline CMC, because it is already partially saturated.

As nanocristalline material has a high Curie temperature and ferrite a much lower the saturation capability of ferrite decreases with increasing temperature. To verify the behavior in a more practical temperature range we measured also the 100°C curves on the right side of Figure 4. The advantage for ferrite is smaller at 100°C but still striking.

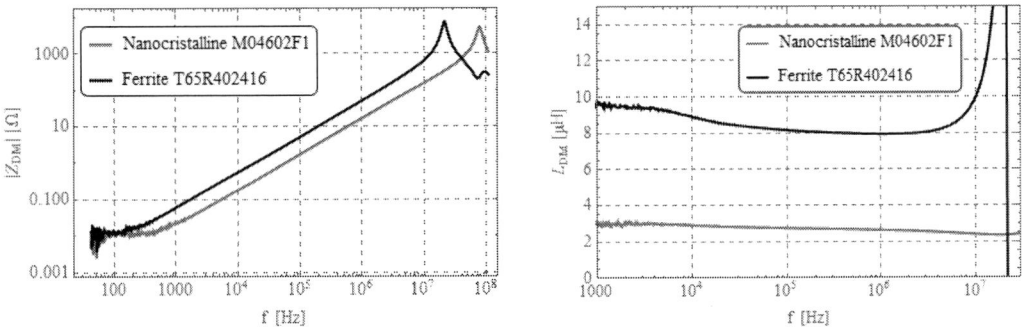

Fig. 5: Differential mode impedance (left) and DM inductance (right) versus frequency

Depending on the specific application in addition to common mode noise also differential mode noise can exceed the limits. Figure 5 shows the differential mode behavior of the two CMCs. Due to the larger number of turns, a 3 times higher differential mode inductance and a higher differential mode impedance make the ferrite CMC more advantageous.

3.2. Low permeability nanocristalline CMC vs. ferrite 0.8 mH CMC

We have seen in the previous Section that the high permeability of classical nanocristalline material in combination with eddy currents starting in the 10 kHz range can lead to reduced applicability especially if significant CM currents are present. In this Section we compare a so called low permeability nanocristalline CMC with a ferrite CMC.

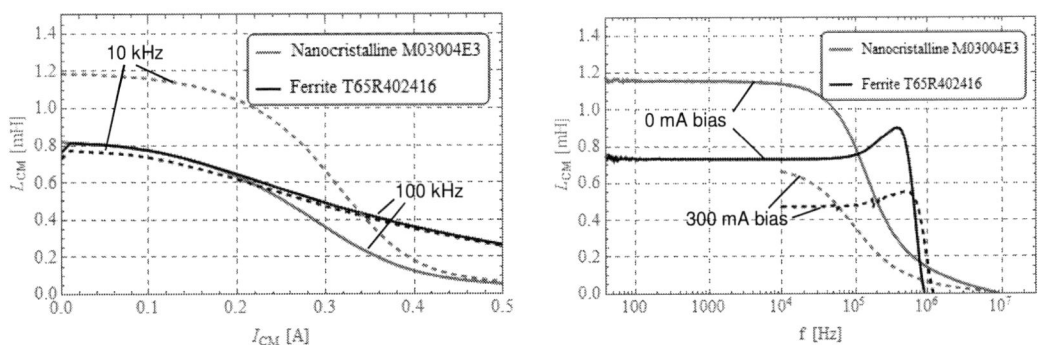

Fig. 6: CM inductance versus CM current (left, solid 100 kHz and dashed 10 kHz) and frequency (right, solid curves are without and dashed curves with 300 mA CM-DC-current).

On the left side of Figure 6 the CM inductance versus CM current is shown. The inductance of the low permeability nanocristalline CMC at 10 kHz is only 30% larger compared to the ferrite CMC. At 100 kHz both CMCs have an inductance of 0.8 mH. The low permeability nanocristalline CMC shows better saturation behavior than the high permeability counterpart. However, compared to the ferrite CMC the nanocristalline CMC inductance is significantly smaller for CM currents above 300 mA.

On the right side of Figure 6 the CM inductance versus frequency is shown. At around 50 kHz the CM inductance of the low permeability nanocristalline material starts to decline due to eddy currents. In spite of this higher limiting frequency the low permeability nanocristalline CMC has a lower CM inductance than the ferrite CMC in the frequency range between 100 kHz and 800 kHz. The dashed lines show the CM inductances at 300 mA CM DC-current. It can be seen that this current expands the beneficial range of ferrite CMC further.

Figure 7 shows the impedance characteristic of the two CMCs. In contrast to the high permeability comparison the impedances are now more comparable even though the general differences can still be seen. However, at room temperature and a CM current of 300 mA the ferrite CMC shows higher impedance between 100 kHz and 4 MHz.

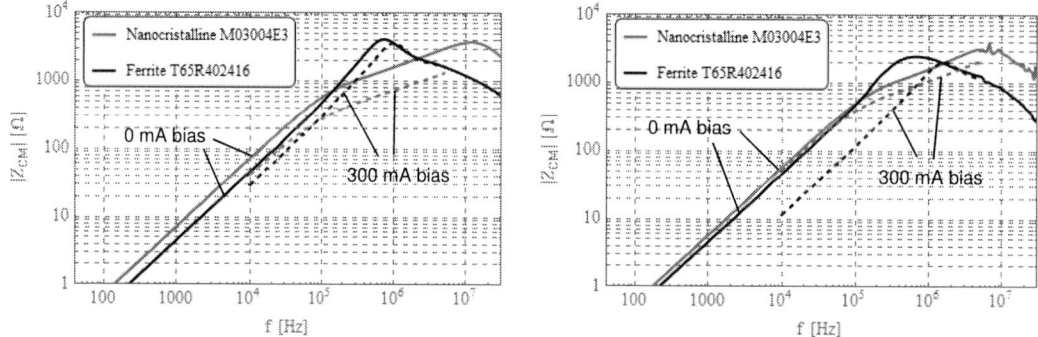

Fig. 7: Common mode impedance versus frequency at 25°C (left) and 100°C (right). Solid curves are without CM-DC-current and dashed curves with 300 mA CM-DC-current.

At temperatures around 100°C the ferrite CMC loses saturation capability leaving an advantageous range between 600 kHz and 3 MHz for the ferrite. After all we conclude that the CM performance of the low permeability nanocristalline CMC is comparable to the ferrite CMC performance.

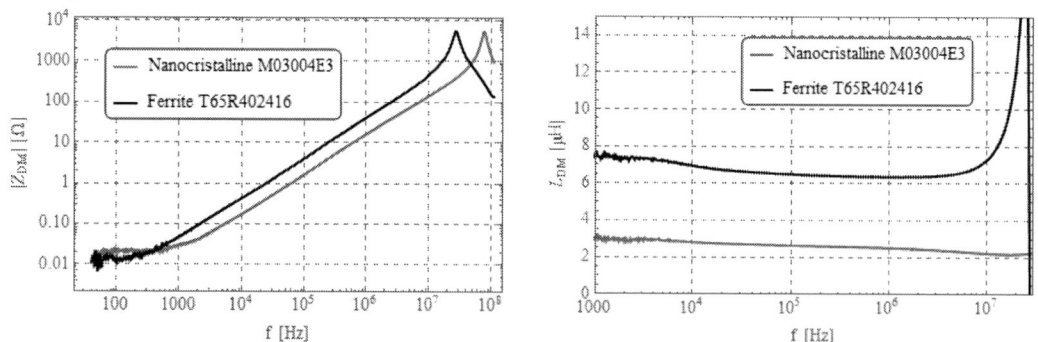

Fig. 8: Differential mode impedance (left) and DM inductance (right) versus frequency

Figure 8 shows the differential mode behavior of the two CMCs. The ferrite CMC shows more than twice of the differential mode inductance and therefore higher DM impedance than the low permeability nanocristalline CMC.

4. EMI behavior in a real application

Figure 9 shows the conducted emission measurement according to EN 61800-3 on a 750W inverter. In the frequency range above 150 kHz the inverter shows mainly CM noise which enables a fairly easy interpretation of the results and neglects the additional advantage of the ferrite CMC for DM noise suppression. The FFT receiver TDEMI 1G from Gauss Instruments has been used for the measurements, enabling full scans with QP detector and AV detector instead of the conventional pre-scan with PK detector and final measurement with QP detector only at selected points.

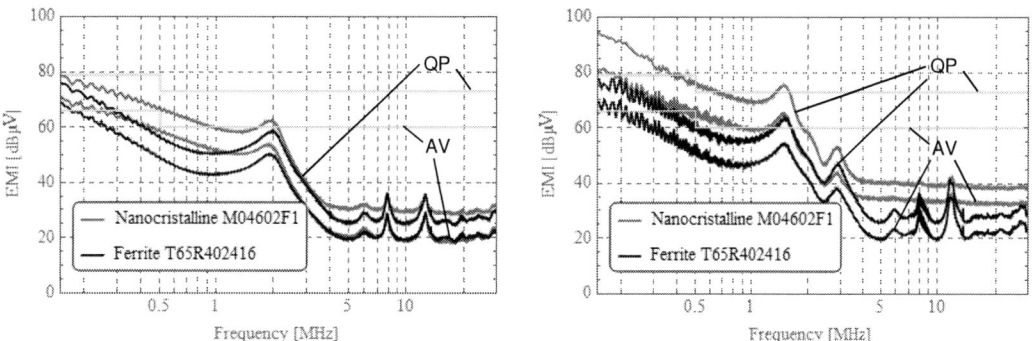

Fig. 9: Conducted emission measurements of high permeability nanocristalline CMC vs. ferrite 1 mH CMC at 27°C with CM-peak-currents smaller (left) and larger than 150mA (right).

For low CM-currents both CMCs show approximately the same behavior at 150 kHz. In accordance to the results above the ferrite CMC shows reduced EMI noise between 150 kHz and 3 MHz. For CM currents lager than 150 mA the ferrite CMC outperforms the nanocristalline CMC clearly. Both cores saturate partially but the ferrite core remains on a higher inductance level as depicted in Figure 3.

The measurements of Figure 9 prove nicely the direct correlation between the characterization from above and the practical relevance for the application. The following conducted emission measurements are performed with the same inverter conditions as the measurement on the right of Figure 9, leading to CM-peak currents larger than 150 mA, depending on the actual CMC.

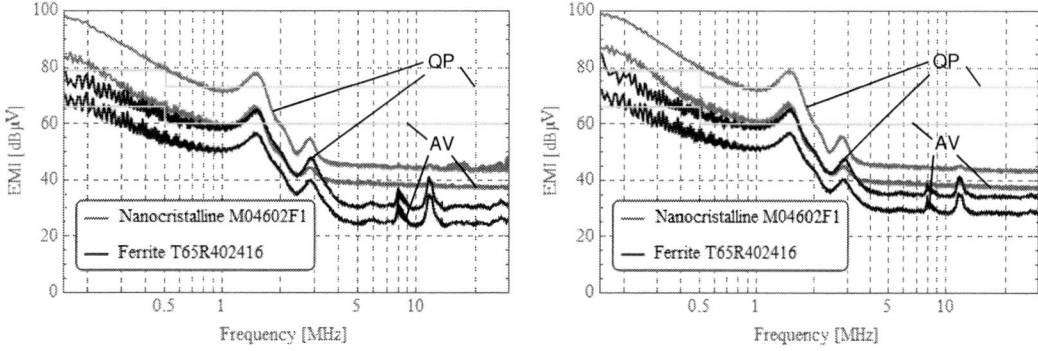

Fig. 10: Conducted emission measurements of high permeability nanocristalline CMC vs. ferrite 1 mH CMC at 80°C (left) and 100°C (right).

Figure 10 shows the comparison of the two 1 mH CMCs at 80°C and 100°C. For both CMCs the EMI noise level increases moderately with temperature. From the impedances in Figure 4 we expect a larger increase of the noise level for the ferrite CMC compared to the nanocristalline CMC. However, when using impedances for noise reduction, the achievable results seem more sensitive to changes in smaller impedances than to changes in larger ones. These results show that even if the nominal permeability of nanocristalline material remains constant in this temperature range, the impedance in partial saturation does not.

Furthermore we want to know, how the low permeability nanocristalline CMC behaves in comparison with the 0.8 mH ferrite CMC. On the left in Figure 11 one can see the conducted emission measurement at 27°C. In comparison with the 1 mH chokes in Figure 9 the 0.8 mH CMCs show an increased noise level, as expected for the lower inductance levels.

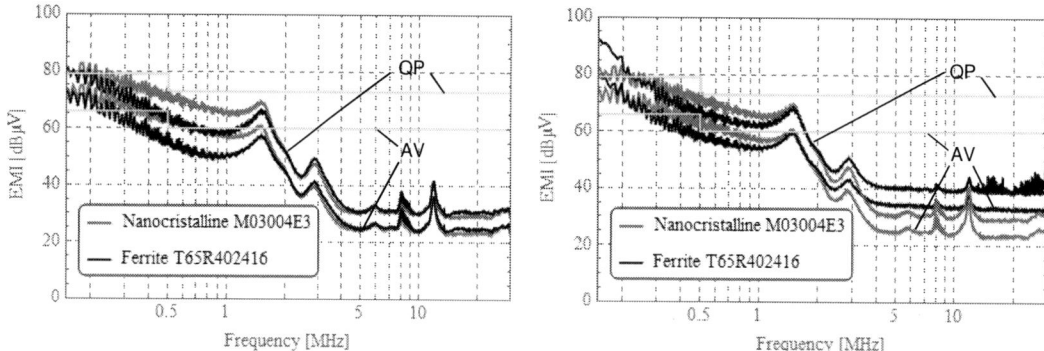

Fig. 11: Conducted emission measurements of low permeability nanocristalline CMC vs. ferrite 0.8 mH CMC at 27°C (left) and 100°C (right).

At 150 kHz both 0.8 mH chokes have the same performance, showing that the saturation level is comparable. In the frequency range between 150 kHz and 2 MHz the ferrite CMC shows a slightly better performance compared to the nanocristalline CMC.
At 100°C the ferrite CMC starts to saturate partially, leading to an increased noise level. However, only up to 250 kHz the ferrite CMC leads to higher noise levels compared to the low permeability nanocristalline CMC. We conclude that the low permeability nanocristalline CMC and the ferrite CMC yield fairly comparable results.

5. Conclusion

A comprehensive comparison of nanocristalline and ferrite common mode chokes has been performed. Both counterparts have been adjusted to the same impedance level at 150 kHz at room temperature and same DC-resistances. This is a fair approach because the necessary impedance at 150 kHz usually defines the common mode choke in practical applications. We found advantages and disadvantages for both materials. The nanocristalline CMCs are smaller in size (less than 30%) and show advantages for high temperature applications. Due to the multitude of ferrite materials the difference in both designs cannot be stated absolutely. The ferrite CMCs show clear advantages compared to high permeability nanocristalline material in the CM impedance between 150 kHz and approximately 3 MHz. Especially if low frequency CM currents are present, the ferrite shows at all frequencies higher impedances. CM currents at the switching frequency of practical systems and their harmonics are the usual case. Also at 100°C the ferrite CMC shows a better performance than the high permeability nanocristalline material.

In case of low permeability nanocristalline material the behavior of both CMCs is more or less comparable with no big advantages in the CM impedance.

Besides a lower price of the ferrite CMCs, one additional technical advantage of the ferrite CMC is the significantly higher differential mode impedance. This helps in applications with a strong differential mode noise source. A further advantage of the ferrite material is the flexibility in core shape. We have shown in [4] that by using an application optimized core shape design, further improvement of the DM noise reduction capability can be realized.

To summarize the results shortly we can state: beside the disadvantage of larger size many advantages for CM and DM noise reduction can be obtained by using ferrite CMCs.

References

[1] H. Schwenk, et. al., Actual and Future Developments of Nanocristalline Magnetic Materials for Common Mode Chokes and Transformers, PCIM Europe 2015, Nürnberg, Germany

[2] J. Beichler, Actual and Future Developments of Nanocrystalline Magnetic Materials for CMCs, ECPE Workshop: Innovations in Passive Components for Power Electronics Applications, Berlin, 2014

[3] M. Ferch, Nanokristalline Magnetwerkstoffe in der EMV, OTTI Seminar: EMV von Hochvolt-Antriebssystemen in Elektro- und Hybridfahrzeugen, Mai 2015, Regensburg, Germany

[4] J. Schliewe, M. Neudecker and S. Weber, Simulating Saturation Behavior of Common Mode Chokes, PCIM Europe 2015, Nürnberg, Germany

Modelling an anti-ferroelectric ceramic capacitor for time- and frequency-domain simulations of power systems

S. Schefler, M. Puff, M. Koini, S. Weber, EPCOS AG, stefan.schefler@epcos.com

Abstract

In this paper a level 2 model of a new ceramic DC-link capacitor is presented. The model identifies the most important characteristics like temperature, AC ripple and DC bias dependency of capacitance and losses. The model is implemented for time- and frequency-domain up to radio frequencies for EMC simulation. Non-linearities have been taken into account to be able to do system optimization for steady-state conditions of applications.

1. Introduction

In power applications the reduction in size and the increase of efficiency will be supported by new ceramic capacitor generations, as they have a huge capacitance density and high temperature capabilities. These parts are used in a wide range of areas including EMI reduction as well as DC link energy storage.

In the public funded project "InSeL" the BMBF supports the evaluation of the technology as EMI measures as close as possible to the power switches in e-mobility and industrial applications. As simulation is a powerful tool to optimize such systems it is the purpose of this work to create a simulation model which helps to understand the complex behavior of these types of ceramic capacitors in the system environment.

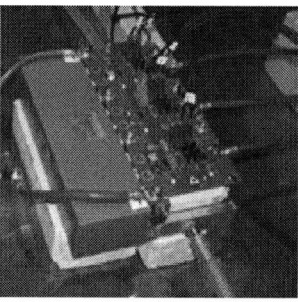 

Figure 1: Testing of power module integrated AFE ceramic capacitors in the project "Insel"

The key challenge is to model the strong dependence of the capacitance and its losses from the DC bias voltage. Also the dependence on ambient temperature and AC ripple needs to be taken into account. The research how to include the additional amplitude dependence which is defined by the hysteretic behavior of the material is still ongoing.

Acknowledgment This work is part of the project "InSeL", which is supported by the German Ministry for Education and Research (BMBF) 16EMO0034. Furthermore the authors want to thank all project partners for their support to evaluate the model.

2. Properties of the anti-ferroelectric ceramic capacitor

In contrast to small permittivity of conventional capacitors the ferroelectric and anti-ferroelectric (AFE) based ceramic capacitors employ a high permittivity in the range above 1000. These permittivities are highly nonlinear depending on the voltages applied to the capacitor.

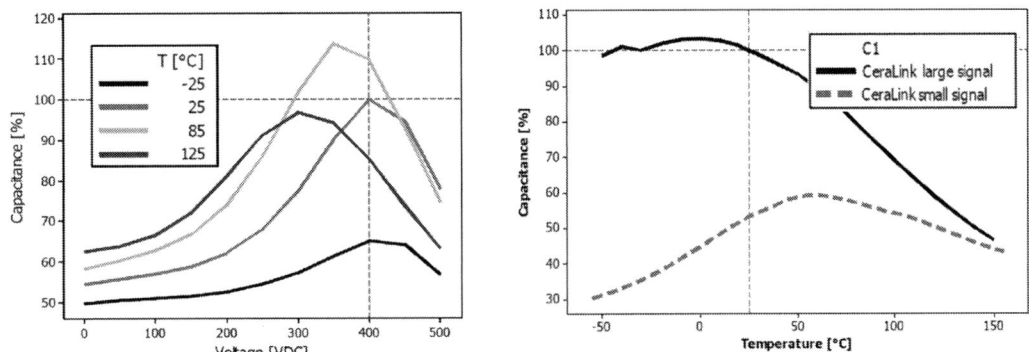

Figure 2: Effective capacitance depends on DC Voltage [1] and AC ripple [2]

While the capacitance of ferroelectric BTO capacitors decreases with the applied voltage [3], AFE capacitors reach their maximum storage capabilities at a certain DC voltage operating point (Figure 2, left) [2]. Beside the DC voltage the amplitude of the superimposed AC ripple voltage has an influence on the amount of polarization of the AFE ceramic like shown in Figure 2, right. With the CeraLink AFE ceramic capacitor series this influence is implemented in a component with outstanding pulse handling capabilities at DC-bus operating voltage. Above 70°C operating temperature, dependence of the effective capacitance on AC ripple decreases. From a system designer's view it is challenging to find the point of most efficient operation. There is a strong need to support them by providing extended simulation models.

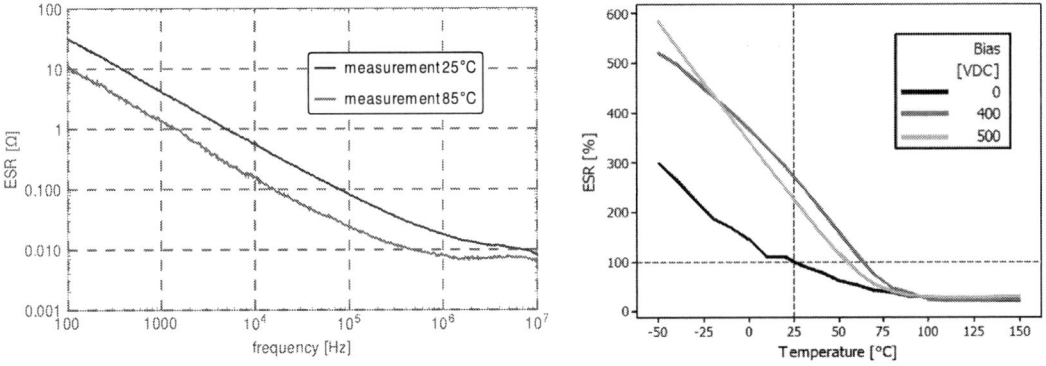

Figure 3: ESR(f) and ESR(theta) [1]

Figure 3 shows the resistance with respect to frequency and temperature. The inner dielectric and ohmic losses of the AFE ceramic decrease with frequency and are generally low due to copper based inner electrodes. With its good thermal conductivity the AFE ceramic capacitor provides a high current capability. [2, p. 2]

Losses decrease with rising operating temperature which makes it easy to stabilize the temperature of the component. Above 70°C – which is below the commonly expected temperature range of operation – ESR does not depend on DC bias anymore and no further temperature rise exists.

3. Equivalent circuit

A level 1 model of a capacitor is a capacitance with its effective C value. A level 2 model of an electronic component takes into account all parasitic behavior that is defined by the component itself. We define Level 3 modeling behavior of a system with inductive or capacitive coupling between relevant parts of the circuitry.

The level 2 model of an AFE ceramic capacitor comprises an effective capacitance C, an equivalent series resistance (ESR) modeling conduction and dielectric losses and an equivalent series inductance (ESL). Such a linear simulation model is valid or optimized for a single operation point. It is the goal of this work to enhance this model to show the major nonlinear effects of the AFE ceramic capacitor CeraLink at all possible points of operation discussed in the previous chapter.

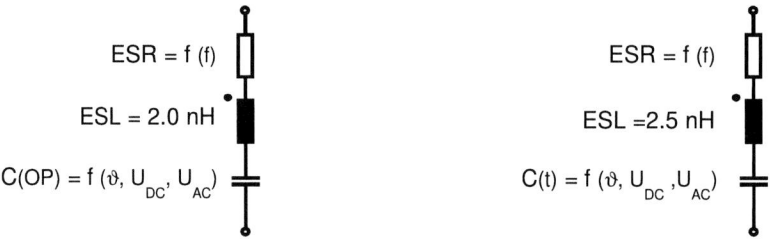

Figure 4: Equivalent Circuit for frequency-domain (left) and time-domain (right)

Figure 4 shows the models proposed for frequency- and time-domain. In both models the non-linearity of the component is taken into account full-scale. They differ in the selection of a fixed operating point dependent capacitance in frequency-domain and a time dependent adaption of the capacitance in time-domain.

The known relationship between effective capacitance and temperature ϑ, DC operating point and AC ripple voltage can be taken into account by a complex function.

3.1 Modeling ESL for time- and frequency-domain

Regarding the equivalent series inductance there is a frequency dependence due to the inner and outer inductive part. At higher frequencies the charge carriers tend to flow on the surface of the conductors. Thus, the inner inductive part disappears and the overall parasitic inductance decreases. This effect can be well shown by a FEM simulation like Figure 5.

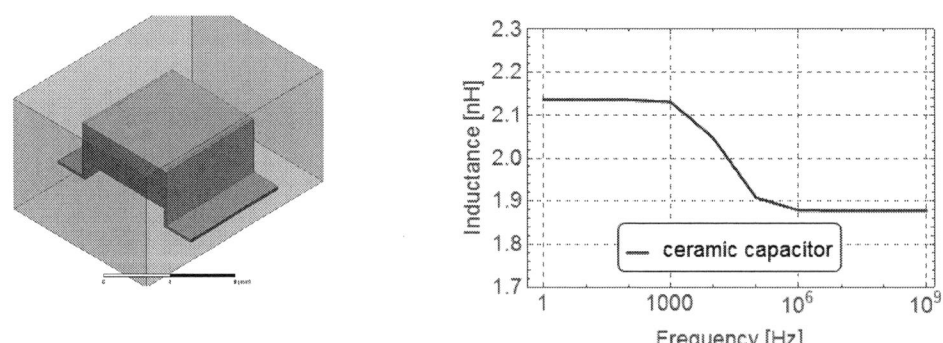

Figure 5: FEM calculation of the components inner and outer inductance

The components partial outer inductance is lower than 2.0nH. State-of-the-Art mounting the component in a system lets us expect ESL of 2.5nH at lower frequencies and 2.0nH at higher frequencies. A more exact modeling of ESL would be possible by a level 3 approach taking the system into account. Frequency-domain simulations are mostly used for EMC simulations above 100 kHz. Thus, we use 2.0nH for ESL in this case.

Differently to the frequency-domain model, the time-domain model will be mostly used for functional simulations with emphasis of the behavior at the lower frequency range. The equivalent series inductance (ESL) is based on the higher inductance value effective at lower frequencies up to 10 kHz in this case.

3.2 Modeling ESR for time- and frequency-domain

Looking at the temperature and DC operating point dependence of the resistive part, one can see in Figure 3 on the right that above 70°C a steady state is reached. The values of the ESR will thus be derived from the measurements at $0V_{DC}$ and 85°C. A further dependence of the resistive part needs not to be included into the model.

Looking at the equivalent series resistance in Figure 3 left there is a strong dependence on the frequency. The dielectric losses decrease with the frequency down to a minimum value defined by the material resistance of the conductive materials.

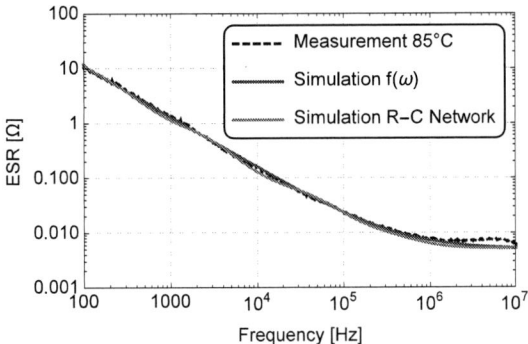

Figure 6: ESR Modelling

In Figure 6 it is shown, that the frequency dependent ESR can be simple approximated with a resistor proportional to $\frac{1}{\omega^k}$. This is not limited due to an infinitely rising resistor at lower frequencies. With the R||R-C||R-C||... ladder model the curve can be fitted in a certain range but rests at a maximum resistance at low frequencies.

3.3 Modeling effective capacitance in time- and frequency-domain

As described above there is a strong dependence of the capacitance value with respect to the DC operating voltage as well as to the temperature and superimposed AC ripple. So the capacitance of the simulation model has to follow these environmental conditions.

To analyze the variable capacitance value of this part an extensive measurement series over temperature and DC voltage at two different excitation levels have been performed. Out of this pool of data we extracted the capacitance values over DC voltage at all measured operating points.

To handle the data mathematically EPCOS developed a complex tangens hyperbolicus formula to describe the capacitor charge in dependence of the DC operating voltage. The deviation of this formula can get used to model the effective capacitance. For every measured operating point a set of parameters was generated.

In Figure 7 there is the calculated effective capacitance at a typical operating temperature vs. DC voltage for two different AC ripple voltages shown, compared to the measurement points.

PCIM Europe 2016, 10 – 12 May 2016, Nuremberg, Germany

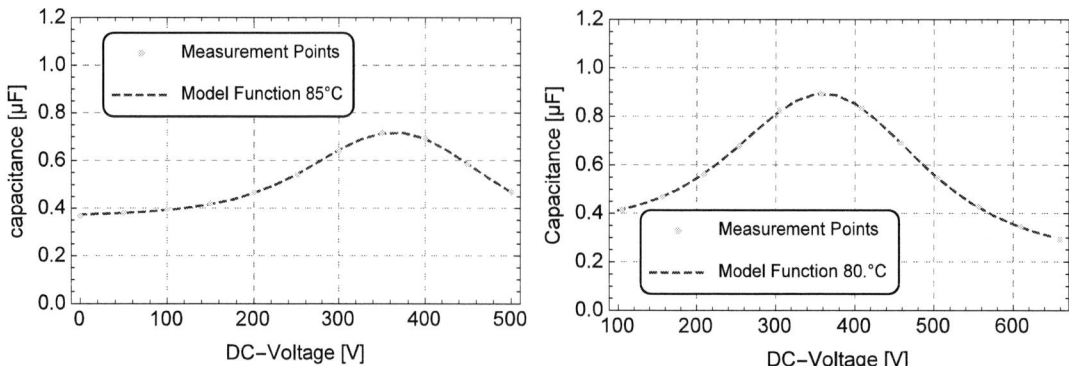

Figure 7: Fit-Curve for effective capacitance with $0.5V_{RMS}$ (left) and $7.0V_{RMS}$ (right)

The dependency regarding temperature and excitation was incorporated by fitting the table of parameters over temperature and then over excitation using mostly quadratic equations. The result is a complex formula giving capacitance values in dependence of only three parameters. The temperature, DC operation point voltage and AC ripple voltage.

4. Adaption as a simulation model for LTspice

The results given from the previous chapters have to get incorporated in direct useable simulation models. To have a common simulation platform this was done with LTspice [4] which was commonly decided in the BMBF project InSeL. Standard R, L and C Components were used as well as frequency and equation controlled current sources.

4.1 Frequency-domain model

The frequency dependent equivalent series resistance (ESR) of the capacitor decreases with the frequency up to its natural frequency. The minimal resistive part was modeled using a base resistor (Figure 8) and adding a frequency dependent current source to define the real part of the frequency behavior up to natural frequency.

In frequency-domain the capacitance can be modeled by using an ideal capacitor in the schematic. Its capacitance gets derived by the simulator directly using the developed formula.

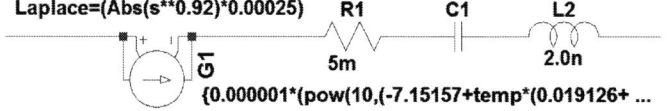

Figure 8: AC simulation model in LTspice

For the present this simulator does not give the possibility to take the result of a preceding internal operating point analysis for the AC analysis. So the DC operating point, the AC ripple voltage as well as the ambient temperature gets defined once at the beginning of the simulation and given as parameter values to the simulation model. With these values the capacitance value is calculated and stays fixed over the simulated frequency variation. As AC simulations are not capable to handle amplitude-nonlinearities there is no possibility to react on dynamic changes in AC ripple voltage. So this parameter has also to be defined before starting the simulation.

© VDE VERLAG GMBH · Berlin · Offenbach

4.2 Time-domain model

To create the same frequency dependent behavior of the equivalent series resistance (ESR) a frequency based function cannot be used here. To create this behavior analog in time-domain a R||R-C||R-C||... ladder model was used (Figure 9).

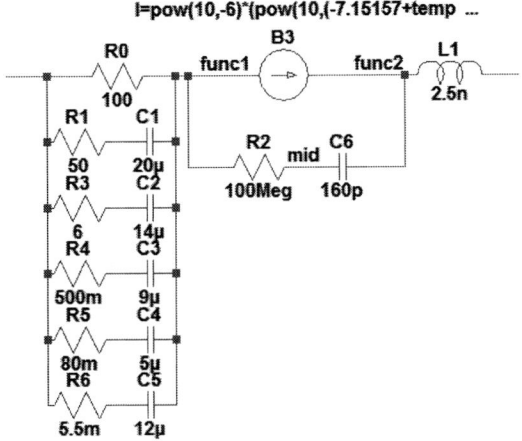

Figure 9: Time-domain simulation model in LTspice

The ambient temperature is given by using a parameter which is taken out of the environment variables or set manually in simulation.

To distinguish between the AC ripple and the DC operating point a highly resistive low pass filter was created. This gives a floating mean value over time for the DC operating point depending on the cutoff frequency of the low-pass.

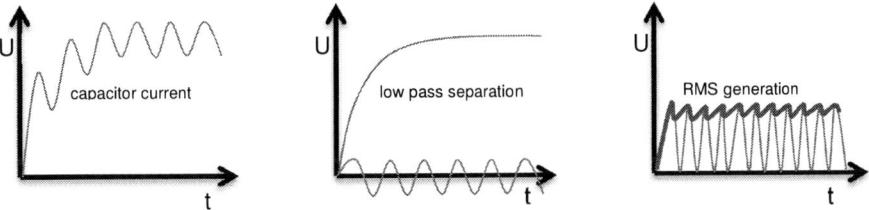

Figure 10: Time-domain simulation model in LTspice

To create a value for the AC ripple the voltage above the capacitor of the low-pass is getting converted to an effective value using analog integration in the simulation model. Depending on the time constant of the integrator the reaction velocity on AC ripple changes can be influenced. As the capacitor can be used for ripple voltages in between some Hz up to several MHz it is very advantageous that the time constant as well as the cutoff frequency get set automatically for a given external cutoff parameter.

$$i_C = C(u, \vartheta) \frac{\delta u_C}{\delta t} \tag{1}$$

To establish the variable capacitive behavior in time-domain simulation an arbitrary current source had to be used to model the capacitor. This source defines the current through the capacitor using the developed formula in dependence of the temperature and the actual AC and DC voltage above the capacitor. Using Equation (1) the simulator calculates the instantaneous capacitor current.

5. Comparison of measurement with simulation

5.1 Comparison in frequency-domain

In frequency-domain we can compare the behavior over frequency shown in Figure 11. The measurement equipment with 1m cable extensions and positioning in the oven at 85°C adds uncertainties in cable length and calibration. Thus a parasitic inductance of 9nH was substracted from the measurement results. This leads to a match of real and simulated behavior even at the natural frequency.

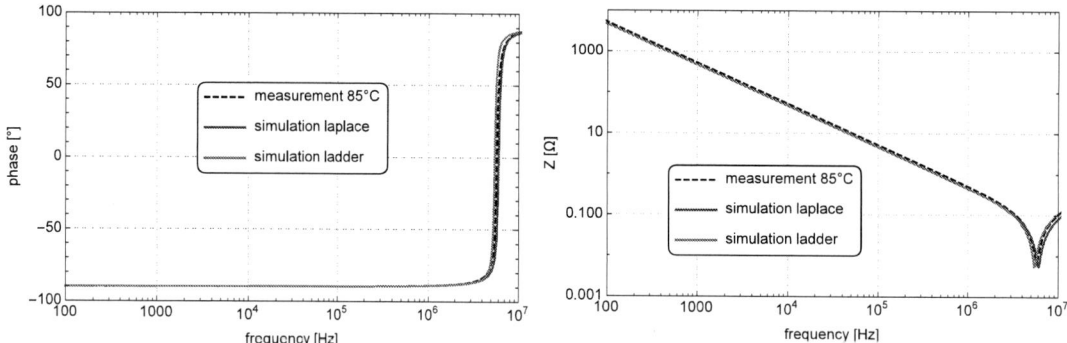

Figure 11: Comparison of small signal measurement and simulation with 0V DC at 85°C

Using the time-domain simulation model in frequency-domain one can see, that it also resembles the basic AC behavior.

Looking at the temperature, DC operating voltage and AC ripple dependence of the capacitor the AC simulation model directly uses the capacitance values given by the evaluated formula. These values are identical to the plot in Figure 12.

5.2 Comparison in time-domain

To test the simulation model in time-domain, a sine excitation with 10 kHz and 100 kHz was given to the capacitor sample at temperatures of 25°C and 85°C. The voltage amplitude was 3V with a 6V DC offset.

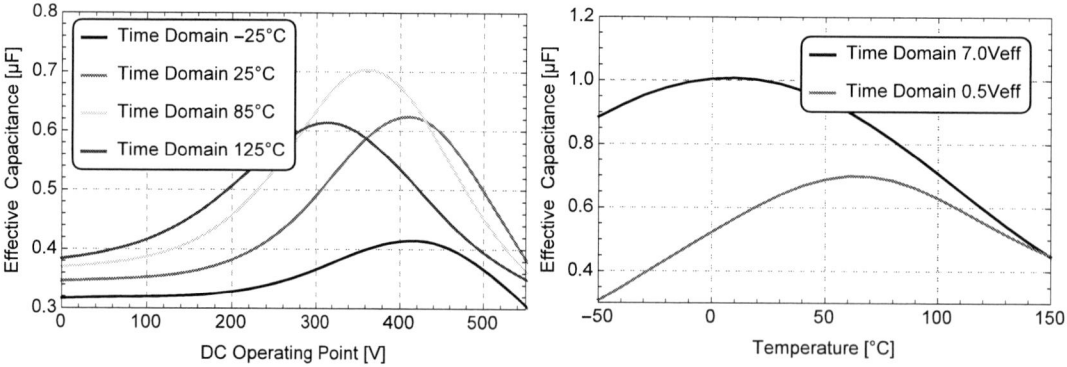

Figure 12: Time-domain measurement and simulation of sine excitation at 85°C

The measured capacitance and resistance values at 85°C and 10kHz of 0.14Ω and 0.31µH get very good resembled by time-domain simulation with 0.14Ω and 0.36µF.

Other major operating points like DC voltage up to 400V were not possible in this test setup. Evaluating the whole valid range of operation in steady state time-domain simulation leads to the effective capacitance values set shown in Figure 12. It clearly matches the measured behavior shown in Figure 2. We can see, that all comparisons between measurement and simulation confirm the proper performance of the simulation models.

5.3 Limitations

The simulation model is ready to use within the limitations of temperature between -50 and 150°C, between 0 and 600V_{DC} and 0.5 and 7.0V_{ACeff}.
As the time-domain simulation model dynamically distinguishes between DC operating voltage and AC ripple, the separating frequency has to get selected reasonable. Also there exists some kind of settling procedure to follow changes in AC ripple which might influence the simulation result in highly dynamic systems.

6. Conclusion

The simulation model presented in this work is the first step in the target to catch the numerous nonlinear effects of this type of ceramic capacitor. It enables simulations taking into account temperature, AC ripple and DC bias to analyze the system design in time- and frequency-domain. As a result of this research project it is possible to use this capacitor e.g. in DC-Link applications where the integration direct to the power modules looks promising. In these applications, simulations are a common development tool and the temperature and DC bias dependency of this model can help to find efficient and small-sized solutions.
Actually, the modeling approach presented is tested in innovative applications like e-mobility drive inverters and might be further improved soon [5]. The simulation model for this CeraLink capacitor is available at www.epcos.com.

References

[1] Datasheet:, „CeraLink capacitor for fast-switching semiconductors," EPCOS AG, München, 2014.

[2] J. Konrad, M. Koini, M. Schossmann und M. Puff, „New demands on DC link power capacitors," Paris, 2014.

[3] G. F. Engel, M. Koini, J. Konrad und M. Schossmann, New high current - high voltage ceramic power capacitors, 2013.

[4] „Linear Technology Homepage," Linear Technology, [Online]. Available: http://www.linear.com/designtools/software/#LTspice. [Zugriff am 14 September 2015].

[5] P. Hillenbrand, S. Tenbohlen, C. Keller und K. Spanos, Understanding Conducted Emissions from an Automotive Inverter Using Common-Mode Model, Dresden: IEEE EMC, 2015.

[6] J. Konrad, CeraLink - The next generation of power capacitors for high temperatures, Berlin: ECPE Workshop innovations in Passives, 2014.

[7] G. F. Engel, Introduction, Challanges and Trends in Power Electronics, Berlin: ECPE Workshop innovations in Passives, 2014.

[8] C. K. Campbell, J. D. van Wyk und R. Chen, „Experimental and Theoretical Characterization of an Antiferroelectric Ceramic Capacitor for Power Electronics," *IEEE Transactions on components and packaging technologies, Vol.25, No. 2 ,* 2002.

PCIM Europe 2016, 10 – 12 May 2016, Nuremberg, Germany

Breakdown of gate oxide of 1.2 kV SiC-MOSFETs under high temperature and high gate voltage

Menia Beier-Möbius, Professorship of Power Electronics and EMC, Technische Universität Chemnitz, Germany, menia.beier@etit.tu-chemnitz.de

Josef Lutz, Professorship of Power Electronics and EMC, Technische Universität Chemnitz, Germany, josef.lutz.@etit.tu-chemnitz.de

Abstract

This work analyzes the onset of gate oxide breakdown of different discrete 1.2 kV SiC MOSFETs for different gate voltage steps. The devices were tested at the same high temperature and the gate voltage was increased step by step. The research on this breakdown distribution is necessary to prove if SiC MOSFETs can fulfill the same reliability requirements for power electronic devices, which are also applicable for current IGBTs.

1. Introduction

SiC MOSFETs can be used in applications with high temperature, high switching frequency and high blocking voltage [1]. For these applications, a high reliability is required. Important for the reliability of SiC MOSFETs is the gate oxide quality. The commonly used material for gate oxide in Si and SiC devices is SiO_2. At a high electrical field especially, the Fowler-Nordheim tunneling contributes to a dielectric breakdown [2]. The Fowler-Nordheim tunneling current is higher in SiC than in Si [1]. Additionally, in an application at high switching frequency, gate overvoltage peaks were found [3, 4].

For SiC MOSFETs the gate oxide is thinner in comparison to Si MOSFETs due to the impact of oxide thickness on the channel resistance [5]. Therefore, the reliability of the gate oxide for high gate voltages must be well known.

2. Experimental

In this work, devices of three manufactures were tested, with manufacturer 1 set to M1, manufacturer 2 set to M2 and manufacturer 3 set to M3. Devices from M1 and M2 are commercial devices, whereas M3 devices are up to now not available in the market.

100 unstressed devices per manufacturer were tested at a constant test temperature of 150°C. The applied gate voltage V_G of the devices was increased during the test for every 168 h. After every 168 h, the test was interrupted for a re-measurement of the threshold voltage at room temperature. The gate voltage at the first step is the rated gate voltage V_{GUSE} according to the data sheet. At the second step, the voltage was set to the maximum use gate voltage V_{GMAX}. After this step, the gate voltage was increased in defined steps.

A graphic demonstration of the test strategy is shown in Figure 1. In this figure, the voltage step after every 168 h is shown. To be able to compare every manufacturer, the difference between applied gate voltage V_G and the use gate voltage V_{GUSE} of the test, V_G-V_{GUSE}, is

© VDE VERLAG GMBH · Berlin · Offenbach

used. Most of the publications use presentation with the electrical field. The electrical-field-determining requires a measurement of the gate oxide thickness. This measured thickness is the nominal gate oxide thickness without thinning. In comparison to the presentation with electrical field, the presentation with V_G-V_{GUSE} is better for the application. By using the V_G-V_{GUSE}-method, the test can be reproduced in every laboratory without extra microscopic preparation and analyses. V_G-V_{GUSE} is application-relevant due to the safety margin to V_{GUSE} with regard to gate overvoltage in applications [4].

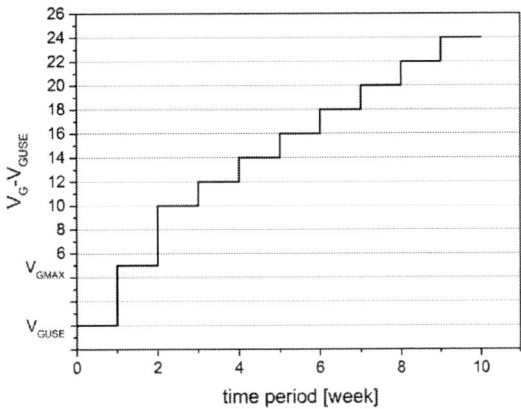

Fig. 1. Test process V_G-V_{GUSE}

Intrinsic and extrinsic failures can be detected by applying the V_G-V_{GUSE}-method due to the breakthrough distribution. "Extrinsic failures are attributed to defects or weak points in the oxide, whereas intrinsic failures represent the true capability of the oxide itself" [6]. Intrinsic failures are caused by a broken SiO_2-bond [7] or due to Fowler-Nordheim tunneling [8] and will occur at 10 MV/cm for SiO_2. Simplified, any extrinsic failure is considered as gate oxide thinning [9]. Figure 2shows gate oxide with d_{ox} as nominal thickness. d_{ox}'' and d_{ox}' are the reduced gate oxide thicknesses for the extrinsic failures. For the lowest thickness d_{ox}'' the breakdown of gate oxide thickness will occur first due to the highest electrical field at the same gate voltage.

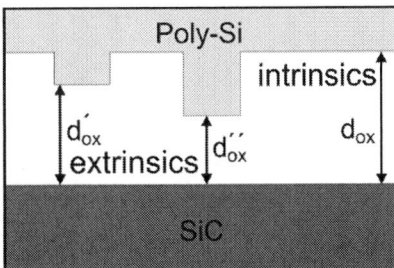

Fig. 2. Gate oxide thinning, d_{ox}'' and d_{ox}' demonstrates gate oxide thinning for extrinsic failures

A failure which occurs early in the failure distribution and at lower V_G-V_{GUSE} corresponds to extrinsic failures. If the most failures occur at a higher V_G-V_{GUSE}, an intrinsic failure limit is achieved.

© VDE VERLAG GMBH · Berlin · Offenbach

3. Results

3.1. Breakdown distribution

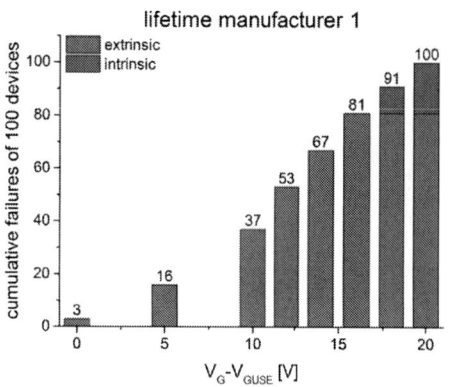

Fig. 3. Breakdown distribution of manufacturer 1

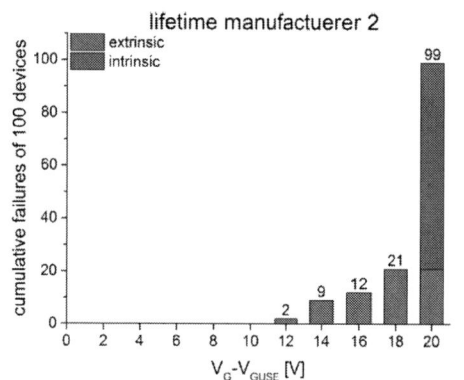

Fig. 4. Breakdown distribution of manufacturer 2

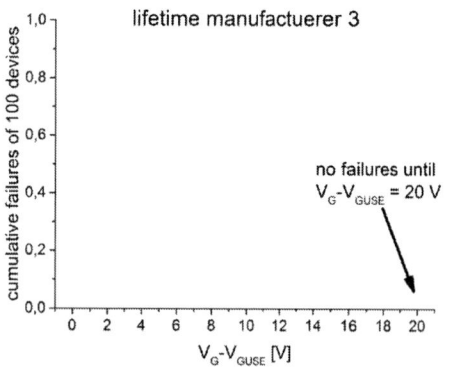

Fig. 5. Breakdown distribution of manufacturer 3

The breakdown distributions of three manufactures were found to be different. Figure 3 shows the failures of 100 devices versus the difference between applied gate voltage and use gate voltage for the first manufacturer. The red bars show the extrinsic failures and the green bars the intrinsic failures. At nominal gate voltage, three devices failed within the 168 h. Most of the failures of these devices occur at V_G-V_{GUSE} of 12V. Manufacturer 1 shows a high rate of extrinsic failures.

Figure 4 shows the failure distribution for manufacturer 2. In comparison to the M1-devices, the failure distribution of M2-devices is concentrated at higher voltages.

The gate oxide quality of the M2-devices is better than that of the M1-devices, as the extrinsic failure rate of the M2-devices is lower and the intrinsic failure rate at V_G-V_{GUSE} = 20 V is displayed.

The test of manufacturer 3 has not been completed, yet. The gate voltage was increased five times, so the difference between test gate voltage and use gate voltage is now 20 V. No extrinsic failures have occurred up to this time.

M1 and M2 show failures under V_G-V_{GUSE} = 12V. M3 does not show failures at these steps.

These results indicate a better gate oxide quality for M3 compared to M1 and M2.

Delivering devices with high gate oxide reliability requires a low rate of extrinsic failures.

3.2. Oxide breakdown estimation

For failure rate calculation, a Weibull analysis of the extrinsic failures was done. The cumulative distribution F is calculated by Bernard's approximation [10]

$$F = \frac{i-0,3}{N+0,4} = 1 - \exp\left(-\left(\frac{t}{T}\right)^\beta\right). \tag{1}$$

N is the total number of the tested devices and i is the cumulative number of defaulted device at the voltage step, T is the characteristic time, when 63.2% of devices fail, and β is the Weibull slope. The survival probability R is given by

$$R = 1 - F. \tag{2}$$

For calculation of the time to failure t_{VGUSE} at the used gate voltage V_{GUSE} the E-model [11] is used

$$t_{VGUSE} = t_{VG} \cdot \exp\left(-\frac{\gamma}{d_{ox}} \cdot (V_G - V_{GUSE})\right). \tag{3}$$

γ is the field acceleration factor, t_{VG} is the time to failure at the rated gate voltage and d_{ox} is the gate oxide thickness. Substituting (3) in (1) and using the natural logarithm results in

$$-\ln(1-F) = \left(\frac{t_{VG} \cdot \exp\left(-\frac{\gamma}{d_{ox}} \cdot (V_G - V_{GUSE})\right)}{T}\right)^\beta. \tag{4}$$

Figure 6 shows the plot of (4) for the different V_G-V_{GUSE} values with a logarithmic y-axis. For lifetime estimation an exponential fit function

$$-\ln(1-F) = a \cdot \exp\left(b \cdot (V_G - V_{GUSE})\right) \tag{5}$$

was used, this leads to the linear function for M1 and M2 . To forecast the failure rate for low V_G-V_{GUSE} values, (5) is solved for F

$$F = 1 - \exp\left\{-a \cdot \exp\left(b \cdot (V_G - V_{GUSE})\right)\right\} \tag{6}$$

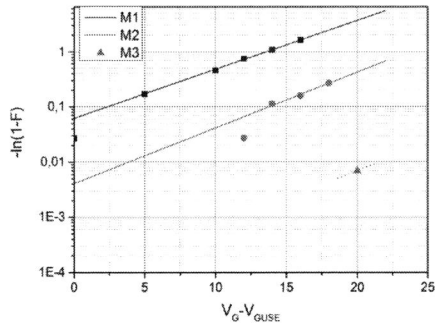

Fig. 6. Extrinsic failure rate – experimental results and fit for M1 and M2, worst case point for M3 if one failure would have occurred

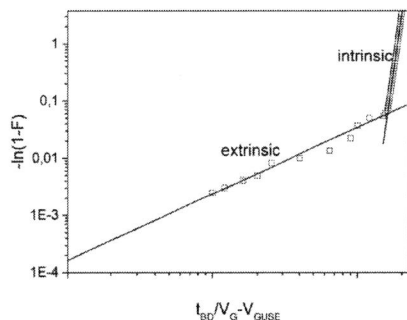

Fig. 7. Illustration of different failure rates for extrinsic and intrinsic failures

The failure rate can be determined by equation (5). For M1 and M2 the first failures were not used for lifetime estimation. In order to obtain the maximal parts per million (ppm) value for low V_G-V_{GUSE} by 150°C and 168h, multiply the –ln(1-F) fit value with 10^6. For V_{GUSE} by 150°C and 168h M1 has a maximal failure rate of 62190 ppm in comparison to M2 with 4100 ppm.

Both functions have similarities in their increase. The gate oxide failure rate of M1 is more than ten times higher than of M2. The $-\ln(1-F)$-value at V_{GUSE} is lower than the calculated fit. The reason is supposed be the burn in test of the manufacturers. For example, if M2 carries out burn in test at V_G-V_{GUSE} = 10V for some time, its failure rate at this voltage will be lower. M1 might perform burn in test too, but uses lower gate voltages than M2.

To be able to compare M3 with M1 and M2 the maximum possible $-\ln(1-F)$ for V_G-V_{GUSE} = 20V was calculated, assuming that one extrinsic failure occur. Figure 6 shows a more than ten time lower failure rate for M3 than M2. The dotted straight line demonstrates the potential distribution for extrinsic failures. In case of intrinsic failure the increase becomes steeper, see Figure 7. Up to now the onset of intrinsic failures is not evaluated, Figure 7 just shows an expectation.

3.3. Microscopic evaluation

After the test, a microscopic evaluation was carried out for some devices of manufacturer M1 and manufacturer M2. To be able to locate the failure region on the chip, a thermographic analysis was performed for three devices of each M1 and M2. Figure 8 shows a failed device of M1. The failure location is found at the gate runner (Figure 8) and at the edge of the gate pad for a further device. The failure of the device of manufacturer M2 is within the active area at the edge to one source bond wire foot, see Figure 9. An additional failure was found close to the bond foot and one random in the active area. Both figures show extrinsic failures.

Fig. 8. Failure spot manufacturer 1 (observed at $V_G=V_{GUSE}$ / 500 µA / 1Hz)

Fig. 9. Failure spot manufacturer 2 (observed at $V_G =V_{GUSE}$-5 V / 700 µA / 1 Hz)

3.4. Threshold-voltage-measurement

The test was interrupted for re-measurement of the Gate-Source threshold voltage at room temperature after every 168h. The measurement is performed due to a context between threshold voltage instability after bias stress and interface traps [12 to 17]. For threshold voltage-measurement, drain was connected to the gate, a current of 5 mA was impressed and the voltage was measured.

PCIM Europe 2016, 10 – 12 May 2016, Nuremberg, Germany

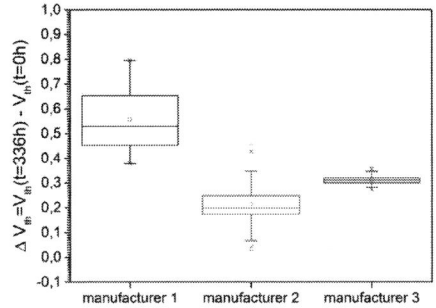

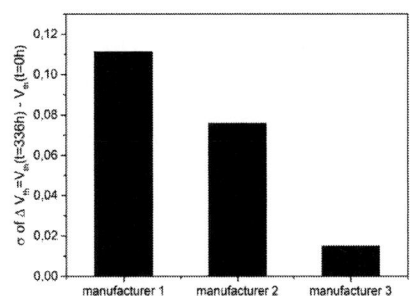

Fig. 10: Boxplot V_th-measurement after test with V_Gmax

Fig. 11. Sigma of the V_th-measurement after test with V_Gmax

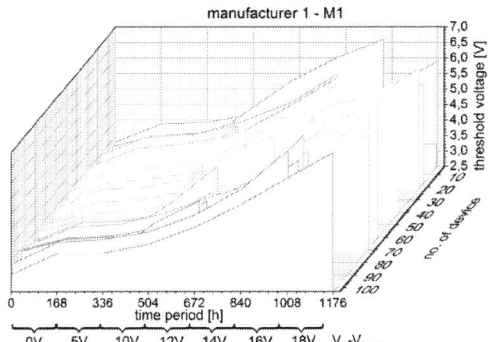

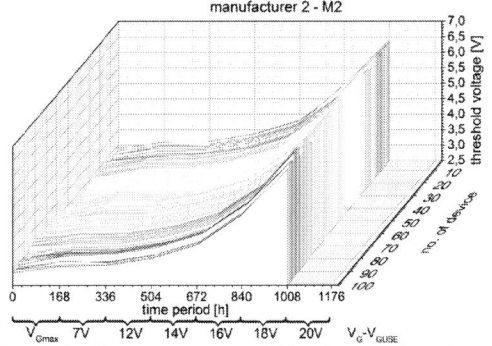

Fig. 12. Measurement of V_th during breaks of V_G stress test for manufacturer 1

Fig. 13. Measurement of V_th during breaks of V_G stress test for manufacturer 2

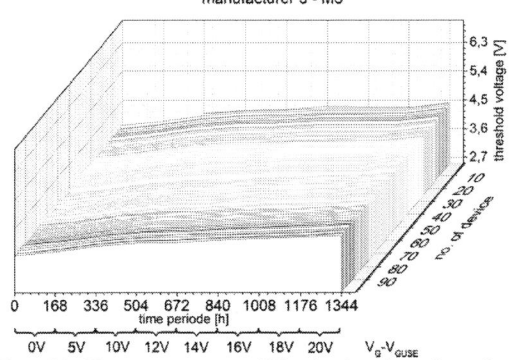

Fig. 14. Measurement of V_th during breaks of V_G stress test for manufacturer 3

Figure 10 displays a boxplot of the threshold voltage drift ΔV_{th} for the three different manufacturers. ΔV_{th} is defined as the difference between threshold voltage after stress $V_{th}(t=336h)$ and the threshold voltage before test start $V_{th}(t=0)$.

The box displays the values between 25% and 75% of all values. In comparison to M2 and M3, M1 shows the largest box, therefore the largest ΔV_{th} spread of 50% of values. M3 shows the smallest ΔV_{th} spread of 50% of values. A narrow band of all values is also given for M3. M2 shows the largest ΔV_{th} deviation between minimal and maximal found values, due to the outliers. Figure 11 shows the standard deviation σ of this ΔV_{th} for every manufacturer.

© VDE VERLAG GMBH · Berlin · Offenbach

177

M1 shows the highest σ, $\sigma = 0.11191$, followed by M2. The lowest σ is presented by M3 with a value of $\sigma = 0.01508$. Figure 12 until Figure 14 display the threshold voltage measurement for the different manufacturers. In comparison to M2 and M3, M1 shows a large spread of V_{th} increase for the tested devices. M1-devices present a high V_{th} increase up to factor of two before breakdown of gate oxide, see Figure 12.

4. Summary

A gate oxide stress test with the increase of the gate voltage was done for three different manufacturers. The purpose is to show the breakdown distribution for different gate over voltages.

Differences in breakdown distributions between discrete devices of three manufacturers for the same voltage class are shown. Two manufacturers present gate oxide breakdown for a gate overvoltage of 12V. The failure rate ppm for 168 h and the rated gate voltage for M1 and M2 was calculated. M1 shows a 62190 ppm compared to of 4100 ppm of M2. The ppm rate of M3 could not be calculated up to now due to the running test without any failure. In evaluation regarding application conditions, it must be considered that T_{vj} 150°C occurs only at condition of highest load, and additionally V_G is pulsed.

Differences of the threshold voltage drift ΔV_{th} were found during the threshold voltage drift investigation for the three manufacturers. M1 indicates the greatest standard deviation of ΔV_{th}. In summary, the devices demonstrate a high difference in reliability. One manufacturer shows a high extrinsic failure rate for low gate over-voltages of 10 V. Furthermore investigations for gate oxide reliability should be done and screening test with longer time and higher gate voltage are needed.

References

[1] M. Gurfinkel et al., "Time-Dependent Dielectric Breakdown of 4H-SiC/SiO2 MOS Capacitors", IEEE Transaction on Device and Materials Reliability, vol. 8, no. 4, 2008, pp. 635-641

[2] B. Schlund et al. „A New Physic-Based Model for Time-Dependent-Dielectric-Breakdown", Proc. of the International Reliability Physics Symposium, 1996, pp. 84-92

[3] M. März et. al., "Are modern power semiconductors too fast? Aspects how to deal with circuit parasitics", Proc. Bauelemente der Leistungselektronik und ihre Anwendung – 6. ETG Fachtagung, 2011, pp. 89-95

[4] P. Anthony et al., "High-Speed Resonant Gate Driver With Controlled Peak Gate Voltage for Silicon Carbide MOSFETs", IEEE Transactions on Industry Applications, vol. 50, no. 1, pp. 573-583.

[5] T. Nguyen et al., "Gate Oxide Reliability Issues of SiC MOSFETs Under Short-Circuit Operation", IEEE Transaction on Power Electronics, vol. 30, no. 5, 2015, pp. 2445-2455

[6] J.A. Cooper, Jr., "Oxides on SiC", IEEE/Cornell Conference on Advanced Concepts in High Speed Semiconductor Devices and Circuits, 1997, pp. 236-243

[7] J. W. McPherson et al., "Field-Enhanced Si-Si Bond-Breakage Mechanism for Time-Dependent Dielectric Breakdown in SiO_2 Dielectrics", Applied Physics Letters, vol. 71, no. 8, 1997, pp. 1101-1103

[8] T. Tomita et al., "Hot hole induced breakdown of thin silicon dioxide films", Applied Physics Letters, vol. 71, no. 25, 1997, pp. 3664-3666

[9] J. C. Lee et al., "Modeling and characterization of gate oxide reliability", IEEE Transactions on Electron Devices, vol. 35, no. 12, 1988, pp. 2268-2278

[10] D. Natarajan "Reliable Design of Electronic Equipment", Springer International Publishing, 2015

[11] T. Pompl et al., "Modeling of substrate related extrinsic oxide failure distributions", Proc. of the Reliability Physics Symposium, 2002, pp. 393-403

[12] M. Gurfinkel et al., Ultra-Fast Measurement of Vth Instability in SiC-MOSFETs due to Positive and Negative Constant Bias Stress", IEEE International Integrated Reliability Workshop Final Report, 2006, pp. 49-53.

[13] A.J. Lelis et. al, "Bias-Stress-Induced Threshold-Voltage Instability of SiC MOSFETs", Materials Science Forum. Vol.. 527-529, 2006, pp. 1317-1320.

[14] T. Okayama et. al, "Bias-stress induced threshold voltage and drain current instability in 4H-SiC DMOSFETs", Solid-State Electronics, vol 52, no. 1, 2008, pp. 164-170.

[15] A. J. Lelis et. al,"Time Dependence of Bias-Stress-Induced SiC-MOSFET Threshold-Voltage Instability Measurements", IEEE Transaction on Electron Devices, vol. 55, no. 8 ,2008, pp. 1835-1840.

[16] M. Treu et. al; "Reliability of SiC Power Devices and its Influence on Commercialization - Review, Status and Remaining Issues", IEEE International Reliability Physics Symposium, 2010, pp. 156 - 161 .

[17] R. Green et. al,"Applications of Reliability Test Standards to SiC Power MOSFETs", IEEE International Reliability Physics Symposium, 2011, pp. EX.2.1 - EX.2.9 .

Avalanche Robustness of SiC MPS Diodes

Thomas Basler, Infineon Technologies AG, Germany, thomas.basler@infineon.com
Roland Rupp, Infineon Technologies AG, Germany, roland.rupp@infineon.com
Rolf Gerlach, Infineon Technologies AG, Germany, rolf.gerlach@infineon.com
Bernd Zippelius, Infineon Technologies AG, Germany, bernd.zippelius@infineon.com
Mihai Draghici, Infineon Technologies Austria AG, Villach, mihai.draghici@infineon.com

Abstract

Silicon carbide merged pin-Schottky (MPS) diodes are predestined to withstand high avalanche energies due to deep implanted p^+-regions within the active area of the device. This paper shows unclamped-inductive-switching (UIS) measurements at state-of-the-art 650 V and 1200 V diodes of MPS type at different conditions, e.g. at Infineon's G5 SiC Schottky diodes. The dependency of the avalanche energy versus the applied inductance is studied in detail and the minimum of this characteristic is explained. Robust and weak diode designs are compared. Repetitive avalanche tests at challenging conditions show the robustness of the used MPS design. First studies at SiC MOSFETs show also a comparatively high avalanche capability.

1 Introduction

Driven by the target of lower V_F values, the base doping of a SiC Schottky diode has to be increased. This leads directly to higher electric field strengths at the Schottky barrier and thus to higher leakage currents. To overcome this drawback, the MPS design uses implanted p^+-regions to reduce the electric field at the Schottky barrier [1,2,3], see Fig. 1 a). For instance, the 5th generation of Infineon's SiC diodes use hexagonal p^+-regions to obtain an effective and homogenous shielding, as shown in Fig. 1 b).

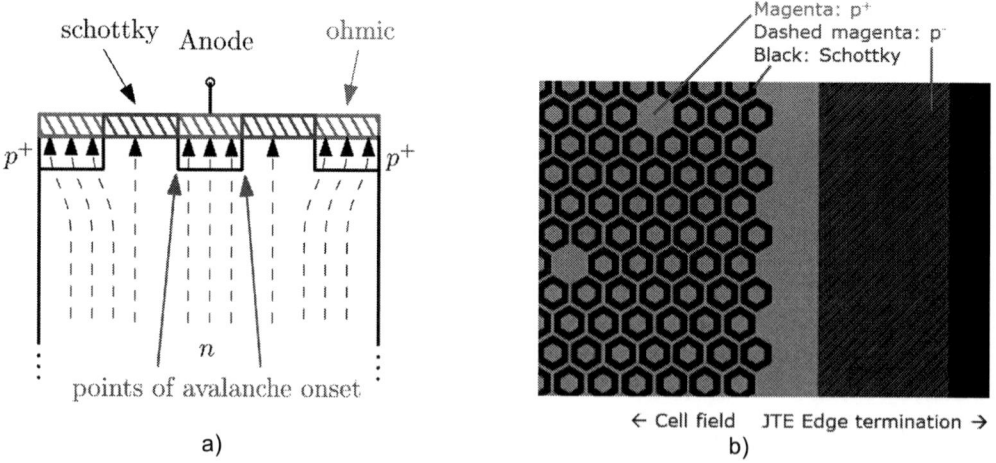

Fig. 1: a) Schematic cross section of an MPS diode structure, showing the onset points of impact ionization, dashed arrows show the current flow at high currents; b) Hexagonal cell structure of Infineon's 5th generation SiC diode with MPS design, black – Schottky area

Enabled by this structure, the avalanche breakdown occurs typically in the cell field of such devices below the p$^+$-regions. This is the base for achieving a good and stable avalanche behavior. Additional to that, the junction termination design must be chosen well to avoid early breakdown and a good ohmic contact to the p$^+$-regions is necessary for low on-state losses and a high surge-current ruggedness, see also [4,5,6]. The following investigations and measurements were performed on Infineon's latest MPS generation of 650 V and 1200 V SiC Schottky diodes. With the help of a robust avalanche capability, the diodes are self-protected against overvoltages or they may be used for protecting the typically non-avalanche rugged IGBTs. Also the reduction of voltage safety margin of power devices may be possible, e.g. to use a 900 V class device for a 600 V DC-link voltage in a two-level topology.

2 Unclamped-Inductive-Switching Behavior

To prove the good avalanche ruggedness of the MPS concept, UIS tests were performed. For that, the test diodes were connected in parallel to a 1700 V IGBT, see Fig. 2 a). When the IGBT turns off a pre-adjusted current, an overvoltage is induced across the switch and the diode runs into avalanche breakdown. A classical UIS test is typically performed at only 50 V DC voltage which is far away from practice. Therefore, UIS tests with DC voltages up to 800 V (for 1200 V diodes) were performed. Due to the low intrinsic carrier density of SiC, the tested diodes are predestined to block a high voltage after the UIS pulse.

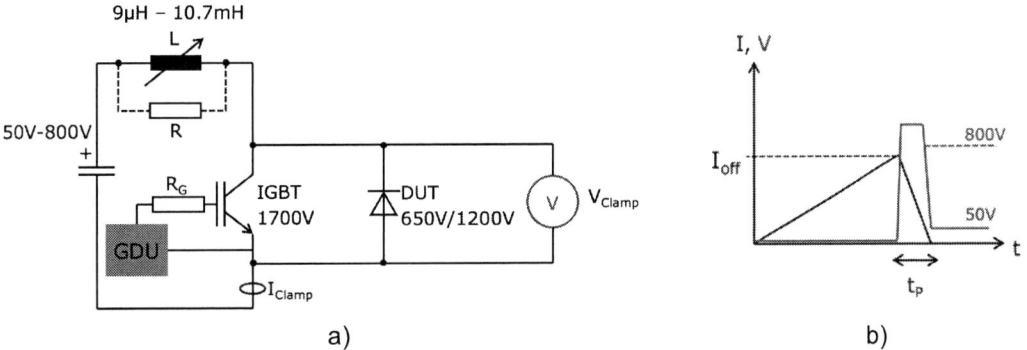

a) b)

Fig. 2: a) Test circuit with optional parallel resistor R to damp LC oscillations; b) Schematic current and voltage waveforms during UIS test, dissipated energy during UIS pulse: $E_{AV} = \frac{1}{2} \cdot L \cdot I_{off}^2$

For showing the energy variation versus the applied inductance the measurements where carried out between L=9 µH and L=10.7 mH. This changes the turn-off current and the clamping time t_P. Furthermore, repetitive avalanche tests with lower E_{AV} were performed, too.

Fig. 3 shows a first representative waveform of an 8 A rated 1200 V diode. During the last pulse before destruction $E_{AV,max}$=6.8 J/cm^2 was dissipated. With the help of the voltage waveform, the maximum junction temperature during clamping is estimated with $T_{j,max}$=230 °C in Fig. 3 b), assuming a homogenous temperature and current distribution. For this estimation, the temperature dependency of the breakdown voltage of this diode was measured in advance on a heat plate at I_r=30 mA with 0.33 V/K. However, the 230 °C cannot explain directly destruction. Current inhomogeneities and the anisotropy of impact ionization which is known for SiC devices (see e.g. [7]) are supposed to be the root cause for destruction.

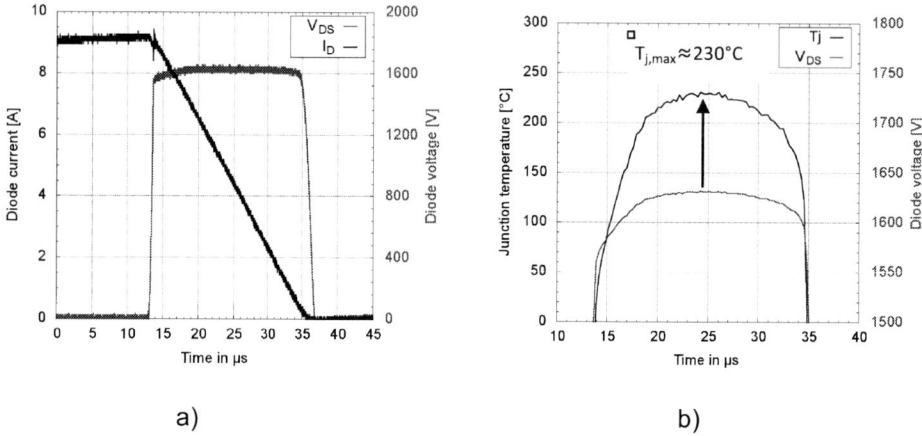

a)   b)

Fig. 3: a) Last UIS pulse of an 8A rated 1200V SiC MPS diode before destruction with $E_{AV,max}$=6.8 J/cm² at L=3.8 mH, V_{DC}=50V, T=25°C; b) Temperature estimation derived from the measured voltage waveform from Fig. 3 a), V_{BR}=f(T) was measured in advance on a heat plate with approx. 0.33 V/K.

3 Results from Diode Measurements

In a first step, different junction termination designs are compared. In Fig. 4 the failure patterns of a 20 A and 8 A rated 1200 V MPS diode are shown. The 20 A diode was destroyed directly at the beginning of an UIS clamping pulse and was not able to clamp any overvoltage. The early breakdown failure point could be found at the transition from active area to junction termination. In contrast to that, the 8 A diode was destroyed only after dissipating a larger amount of energy. The failure spot is randomly distributed within the active area which indicates a more robust design. The diodes have the same cell structure in the active area, only the junction termination has a different design.

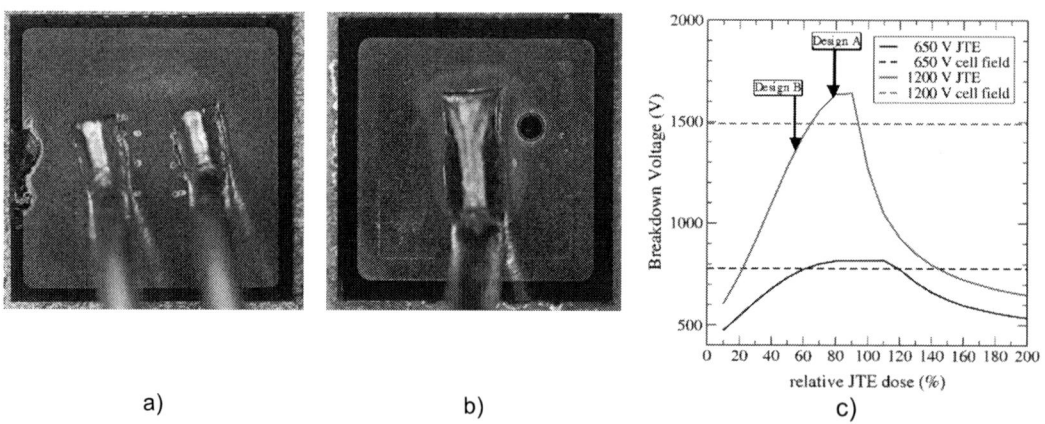

a)   b)   c)

Fig. 4: a) Early breakdown failure pattern of a 20A/1200V diode with an insufficient junction termination design; b) Failure pattern of a robust junction termination of an 8A/1200A diode after dissipating 6.9 J/cm² of energy; c) TCAD (Technology Computer Aided Design with Sentaurus Device) simulation results: breakdown voltage of cell field compared to breakdown voltage of p⁻-JTE, picture taken from [1]

© VDE VERLAG GMBH · Berlin · Offenbach

In [1], TCAD simulations showed for a p⁻-JTE (junction termination extension) that a specific dose windows must be selected to pin the breakdown point within the active area, see Fig. 4 c). If not, early breakdown occurs as happened for the diode in Fig. 4 a).

In a next step, diodes of a lower voltage class, namely 6 A/650 V devices with robust junction termination, were measured. Fig. 5 shows the last pulse before destruction. At room temperature an $E_{AV,max}$ of approx. 3.7 J/cm^2 could be achieved. At a junction temperature of 175 °C $E_{AV,max}$ is reduced by 40 % and the initial breakdown voltage value $V_{BR,start}$ is slightly higher, see Fig. 5 b). The failure spot was found within the active area.

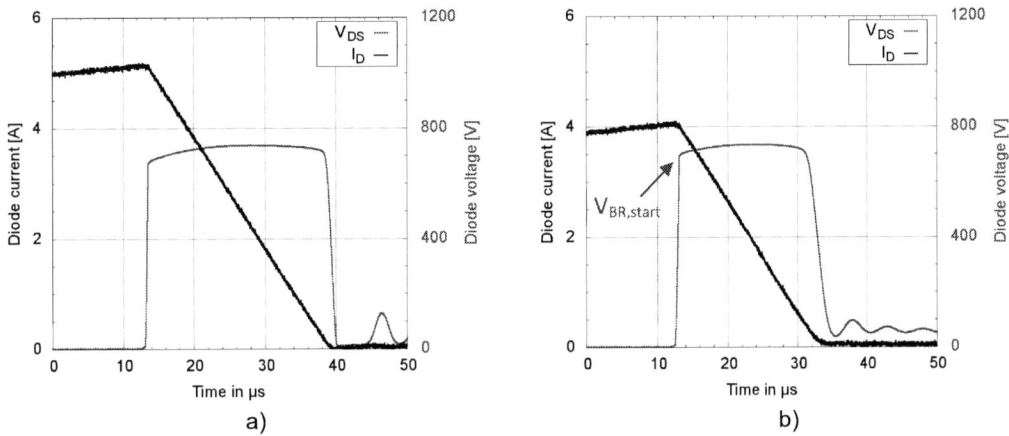

Fig. 5: UIS waveforms of 6 A/650 V MPS diode class, L=3540 µH, V_{DC}=50 V; a) At room temperature, $E_{AV,max}$=3.7 J/cm^2; b) At 175 °C, $E_{AV,max}$=2.2 J/cm^2

To make the clamping condition even more challenging, the 8 A/1200 V diode used in Fig. 3 was directly stressed with 800 V DC voltage after the clamping pulse, see Fig. 6. This would be a more practical condition.

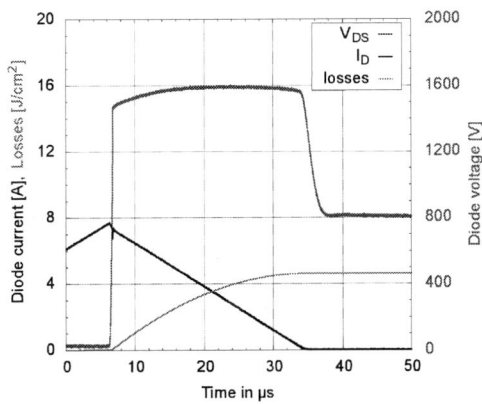

Fig. 6: UIS waveforms of 8A/1200V MPS diode class, L=3044 µH, V_{DC}=800 V, $T_{j,start}$=25°C, E_{AV}=4.6 J/cm^2

The diode showed no earlier destruction and could withstand approx. the same max. clamping energies as in Fig. 3. As stated above, the temperature rise during UIS is not high enough to raise the intrinsic-carrier density n_i to regions where the leakage current levels will lead to thermal runaway. And this even though the saturation-current term j_s is dependent on n_i^2 as given in the overall leakage current density j_r of a pn-junction (Schottky contact contribution neglected):

$$j_r = j_s + j_{SCR} = q \cdot \left(n_i^2 \cdot \frac{D_p}{L_p + N_D} + n_i \cdot \frac{w_{SCR}}{\tau_{SCR}} \right), \qquad (1)$$

with q – elementary charge, D_p – diffusion constant of holes, L_p – diffusion length of holes, N_D – base doping, w_{SCR} – space-charge region, τ_{SCR} – lifetime in the space charge region.

For a deeper investigation, the $E_{AV}=f(L)$ characteristic is used, see Fig. 7. In this Figure a minimum in $E_{AV,max}$ can be found for an 8 A/1200 V diode in a TO-220 housing at L=400 µH. Beside the burn mark in the active area, the failure picture shows strong molten anode metallization even the hexagonal cell structure has become visible, see Fig. 7 b). For higher inductances, $E_{AV,max}$ rises due to longer clamp pulse length, also shown in [3]. More and more energy can already be transferred to adjacent layers (solder layer, leadframe, etc.). For very low inductances $E_{AV,max}$ rises, too. The energy is only stored in the SiC semiconductor portion and cannot damage the anode metallization significantly which corresponds to the failure pattern in Fig. 7 b) at 9 µH. This principle behavior holds also for other voltage/current classes and is directly linked to the Z_{th} characteristic.

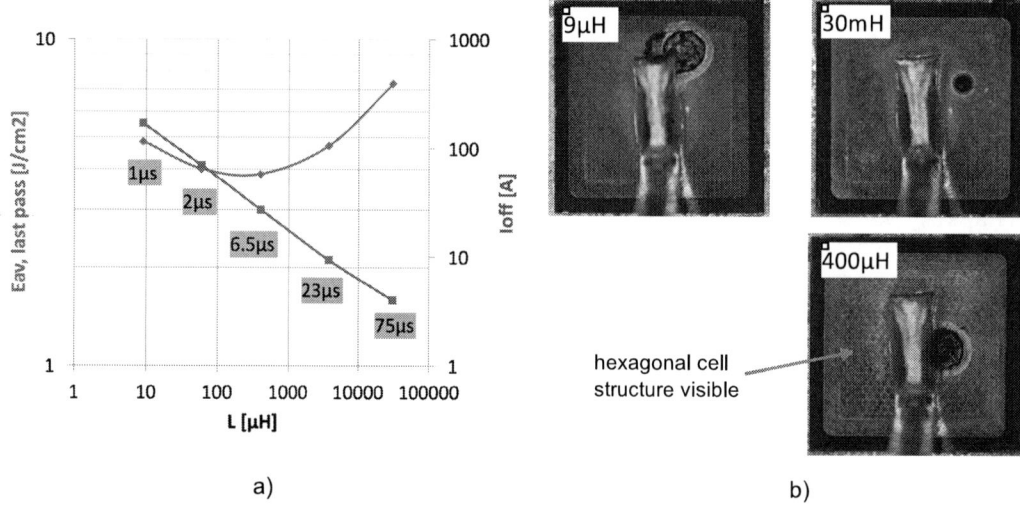

Fig. 7: a) Last-pass avalanche energy and turned-off current versus applied inductance (V_{DC}=50 V, 8 A/1200 V diode, T=25 °C); b) Chip failure pattern at different inductances, $E_{AV,max}$ and turned-off current

4 Stability Tests

To proof the overall stability during avalanche condition each 10.000 clamping pulses were performed at E_{AV}=2.85 J/cm^2 at different conditions. The leakage current, breakdown voltage and forward voltage drop were tested initially and after 10.000 pulses. Even at very challenging conditions of 144 A at the breakdown branch (=I_{off}) for an 8 A rated diode no degradation could be observed, see Table 1. The shielding by the p$^+$-regions is obviously sufficiently dimensioned.

Table 1: Static characterization of an 8 A/1200 V SiC MPS diode before and after 10.000 avalanche pulses at different conditions. No significant change in values was observed.

2.85 J/cm², 10k pulses	t_p=12.3 µs I_{off}=10.7 A		t_p=3.4 µs I_{off}=38 A		t_p=0.92 µs I_{off}=144 A	
	initial	end	initial	end	initial	end
I_R@1.2 kV [µA]	1.40	1.39	1.26	1.21	1.35	1.36
V_F@8 A [V]	1.46	1.46	1.49	1.48	1.47	1.47
V_F@80 A [V]	7.05	7.06	7.26	7.28	7.11	7.12

At the beginning it was mentioned that an avalanche-rugged MPS diode may be used to protect a non-avalanche rugged switch in circuits with higher parasitic inductance. In that case, the parallel diode has to clamp a certain amount of energy with switching frequency (e.g. in an inverter for high load-current cases). To test this, an 8 A/1200 V diode was tested under repetitive UIS configuration with a switching frequency of 4 kHz at 800 V V_{DC}. The test was continued until thermal equilibrium, see Fig. 8. During every single UIS pulse an energy of 28 mJ/cm² was dissipated. Also here, the MPS diode showed a very rugged behavior without destruction. However, the protection feature will lead to higher overall switching losses and should only be used at high- and over-load conditions.

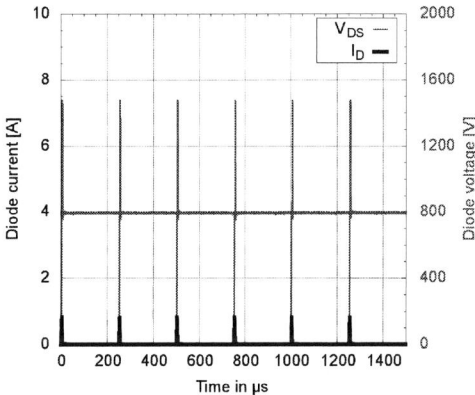

Fig. 8: Repetitive UIS test of an 8 A/1200 V MPS diode with a frequency of 4 kHz, E_{AV}=28 mJ/cm², V_{DC}=800 V, L=3 mH.

5 UIS Behavior of 1200V SiC MOSFET

Due to the deep implanted p⁺-regions the MPS design can be used as reference for other technologies like the SiC MOSFET (DMOS, Trench). Fig. 9 shows the same good avalanche capability of a 40 mΩ/1200 V Infineon MOSFET as for the actual SiC MPS diode. An even higher $E_{AV,max}$ of 9.7 J/cm² was found. Enabled by this, first 1200 V power switches with an assured avalanche capability may become available.

However, in future work the MOS-gate stability must be proven especially for repetitive avalanche condition since the critical field strength is 10 times higher for SiC compared to silicon. Gate-oxide degradation must be prevented.

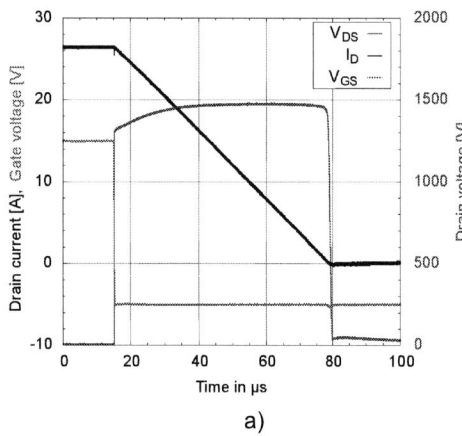 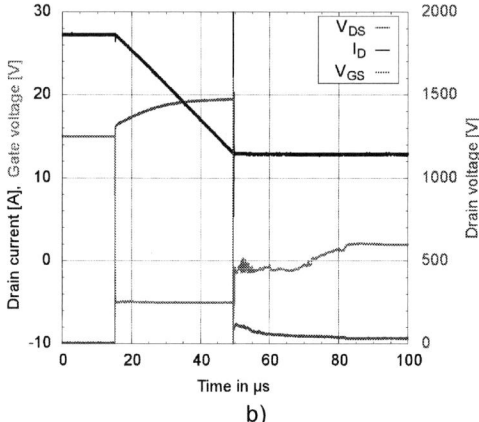

a)   b)

Fig. 9: UIS waveform of a 1200 V Infineon SiC MOSFET at V_{DC}=50 V, L=3.5 mH, T=25 °C; a) Last pulse before destruction with $E_{AV,max}$=9.7 J/cm²; b) Destruction pulse, fail during UIS pulse

6 Conclusion

The above measurements show that state-of-the-art SiC MPS diodes of 650 V and 1200 V voltage class can have a rugged and high avalanche capability. Table 2 summarizes the achievable avalanche energies at T=25 °C and L≈3 mH. A device with a thicker drift zone seems to be more robust. First investigations at SiC MOSFETs indicate a potential towards even higher avalanche capability.

Table 2: Comparison of avalanche energies between different technologies but same overall chip thickness.

	650 V SiC MPS diode	1200 V SiC MPS diode	1200 V SiC MOSFET
$E_{AV,max}$ @ ≈3mH	3.7 J/cm²	5 J/cm²	9.7 J/cm²

The SiC MOSFET shows the highest possible energy per area. Presumably, the point of the maximum field strength is shifted deeper into the device which may help to distribute the dissipated energy better within the semiconductor volume.

However, to achieve a high avalanche capability in general the junction termination design must be chosen well and a sufficient shielding mechanism has to be applied to protect e.g. Schottky contacts or MOS structures.

Enabled by a high avalanche capability, a higher level of power-electronic system protection will be reached. Internal and external overvoltages can be clamped and switched off.

© VDE VERLAG GMBH · Berlin · Offenbach

References

[1] Rupp et al. Avalanche Behaviour and its Temperature Dependence of Commercial SiC MPS Diodes: Influence of Design and Voltage Class. *26th International Symposium on Power Semiconductor Devices & IC's (ISPSD 2014)*, 2014

[2] M. Treu et al. A surge current stable and avalanche rugged SiC merged pn Schottky diode blocking 600V especially suited for PFC applications. *Materials Science Forum vol. 527-529: 1155-1158*, 2006

[3] Konstantinov, S. Jinman, S. Young, F. Allerstam and T. Neyer. Silicon Carbide Schottky-Barrier Diode Rectifiers   with high Avalanche Robustness. *International Exhibition and Conference for Power Electronics, Intelligent Motion, Renewable Energy and Energy Management (PCIM 2015)*, 2015

[4] S. Fichtner, S. Frankeser, R. Rupp, R. Gerlach, T. Basler and J. Lutz. Ruggedness of 1200V SiC MPS diodes. *Proceedings of the 26th European Symposium on Reliability of Electron Devices, Failure Physics and Analysis (ESREF 2015), Microelectronics Reliability. vol. 55. issues 9-10, pp. 1677-1681*, 2015

[5] S. Fichtner, J. Lutz, T. Basler, R. Rupp, R. Gerlach. Electro-Thermal Simulations and Experimental Results on the Surge Current Capability of 1200 V SiC MPS Diodes. *8th International Conference on Integrated Power Electronics Systems (CIPS 2014), Nuremberg, VDE Verlag Berlin, Offenbach, S. 438 – 443*, 2014

[6] O. Harmon, T. Basler and Fanny Björk. Advantages of the 1200V SiC Schottky Diode with MPS Design. *In Bodo's Power Systems, December issue, ISSN: 1863-5598*, 2015

[7] T. Hatakeyama. Measurements of impact ionization coefficients of electrons and holes in 4H-SiC and their application to device simulation. In *Phys. Status Solidi A, 206: 2284–2294. doi: 10.1002/pssa.200925213*, 2009

Compact, Low Loss and High Reliable

3.3kV Hybrid Power Module

Satoshi Kaneko, Naoyuki Kanai, Motohito Hori, Nakazawa Masayoshi,
Hideaki Kakiki, Yasushi Abe, Yoshinari Ikeda, and Eiji Mochizuki
Electronic Device Laboratory, Fuji Electric Co., Ltd.
Matsumoto, Nagano, Japan
e-mail:kaneko-satoshi@fujielectric.com

Abstract

Power electronics products are required to be small size, light weight, and high efficiency. To meet these trends, we developed 3.3kV hybrid power module with silicon carbide schottky barrier diode (SiC-SBD) and silicon insulated gate bipolar transistor (Si-IGBT). The total loss of the developed hybrid power modules are significantly reduced by applying SiC-SBD compared to conventional Si power modules. In addition, applying high-strength Sn-Sb solder improve high temperature reliability of the developed hybrid power modules and their operation temperature from 125 degree C to 150 degree C [1] [5]. With these technologies, 30% downsizing and light weight of power module are realized. In addition, the total loss is reduced by 21 % at a carrier frequency of 1 kHz.

1. Introduction

3.3kV power modules are key component of power conversion system for infrastructure especially in railways and wind power generations. In power conversion system of railways, required features are saving energy, small size, light weight and high reliability.   In order to satisfy these demands, it is necessary to improve the performance of 3.3kV power module. However, the Si device which is mounted 3.3kV power module approaches its physical limit. Therefore, it is hard to expect breakthrough of loss reduction so far and it is difficult to downsize the system. At this point, the Silicon Carbide (SiC) is more focused because the SiC has superior physical characteristics than the Silicon (Si).   The SiC can be expected to have lower on state voltage drop at the result of utilizing high breakdown electric field. In addition, it is possible to realize low switching loss by the SiC unipolar device. Considering these reasons, applying the SiC to power conversion system cause downsizing.

In railway application, better silence performance of power conversion system is required in order to achieve less noise and more ride quality. There is one solution that high carrier frequency operation of power module. However, device temperature is increased because total loss is proportionally increased with carrier frequency. High device temperature causes low reliability due to accelerating joint materials degradation. Therefore, in order to achieve high frequency operation, it is necessary to ensure high temperature reliability and high operation temperature simultaneously. Regarding these backgrounds, this paper describes the packaging technology which achieves small size and right weight.

2. Structure of Developed Hybrid Power Module
2.1 Effect of Hybrid Structure

3.3kV 1200A hybrid power module is developed with replacing Si free wheeling diode (Si-FWD) with SiC-SBD in order to reduce loss and to downsize power module. The external and cross sectional views of 3.3kV 1200A hybrid power module are shown in Fig. 1.

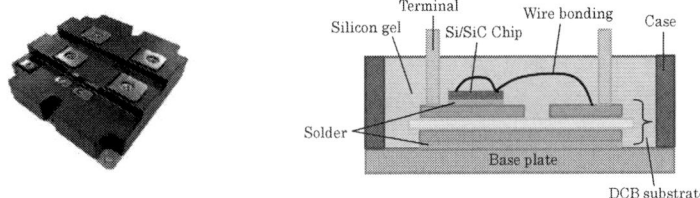

Fig. 1. The external and cross sectional views of 3.3kV 1200A hybrid power module

The significant loss reduction of power module with applying 3.3kV SiC-SBD instead of conventional Si-FWD is achieved. We have utilized high-strength Sn-Sb solder to ensure high temperature reliability and have been able to improve operation temperature. The reduction in generated loss and improvement of operation temperature have enabled to achieve a power density increase, and the footprint size of the module has been reduced by approximately 30% as compared with conventional Si power module as shown in Fig. 2.

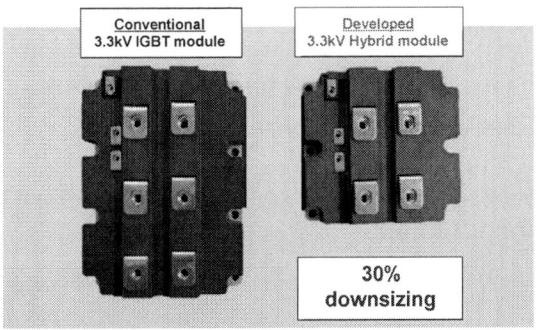

Fig. 2. Comparison between Conventional (left) and developed structure (right)

2.1.1 Forward characteristics

Fig. 3 shows the junction temperature (T_j) dependency of forward voltage (V_F) at the 1200 A rating. The V_F of the developed hybrid power module is about 30% smaller than that of conventional Si power modules at 25 degree C. On the other hands, for V_F at 150 degree C, that of the developed hybrid power module is about 30% larger than that of conventional Si

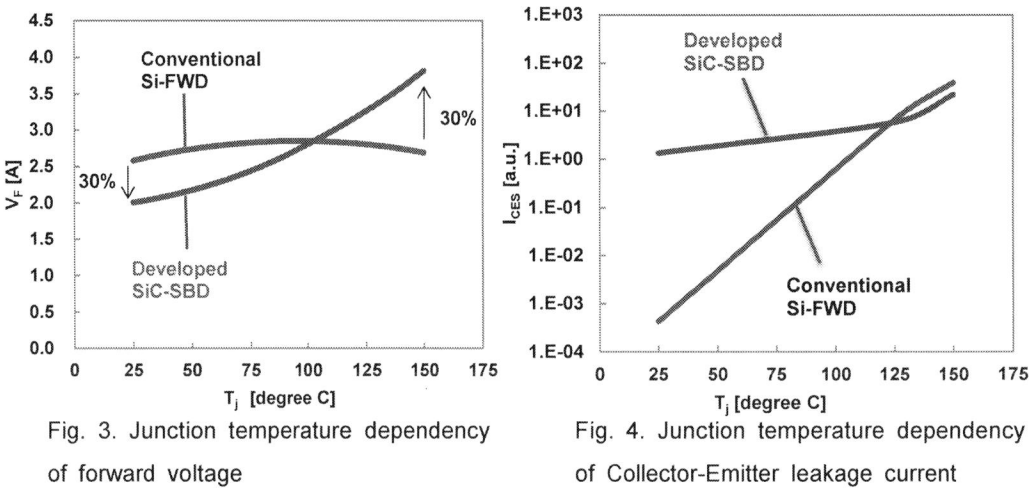

Fig. 3. Junction temperature dependency of forward voltage

Fig. 4. Junction temperature dependency of Collector-Emitter leakage current

power modules. However, while conventional Si power module exhibits negative temperature characteristics that V_F decreases as T_j increases, developed hybrid power module has positive temperature characteristics. The positive temperature characteristic causes increasing internal resistance as T_j increases, which results in equal sharing of the current through the chips connected in parallel. Accordingly, developed hybrid power modules are advantageous in multi-parallel connections. Fig. 4 shows the Collector-Emitter leakage current (I_{CES}) characteristics of developed hybrid power module and conventional Si power modules. The I_{CES} of the developed hybrid power module is about 11% smaller than that of conventional Si power modules at 125 degree C. In the case of the I_{CES} at 150 degree C, that of the developed hybrid power module decreases by about 44% that of conventional Si power modules. In addition, the I_{CES} of developed hybrid power module is nearly constant, in other words, its temperature dependency is small. From these results, the developed hybrid power module is capable of operating at high temperatures.

2.1.2 Switching characteristics

Switching waveforms of conventional Si power module and developed hybrid power module are compared in Fig. 5. The test was performed under V_{AK} (V_{CE}) =1800V, I_F (I_C) =1200A, and T_j =150 degree C. As compared with conventional Si power module, the developed hybrid power module exhibits a much lower reverse recovery peak current. This is because the SiC-

SBD is a unipolar device and there is no storage effect caused by minority carriers. In addition, the reverse recovery peak current of the SiC-SBD is reflected in the IGBT turn on current in the opposite arm and the turn on peak current can be significantly reduced as well.

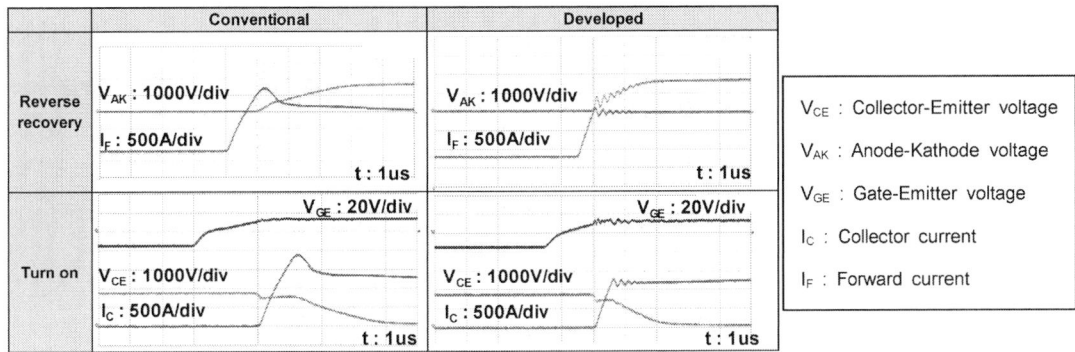

Fig. 5. Switching waveform of conventional Si power module and developed hybrid power module

Switching loss of conventional Si power module and developed hybrid power module are compared in Fig. 6. The junction temperature in this test was 150 degree C. The switching loss of the developed hybrid power module decrease by 30% compared to the conventional Si power module due to reduction of turn on loss (E_{on}) and reverse recovery loss (E_{rr}). Calculation results of the total loss including on-state voltage is shown in Fig. 7. The total loss of developed hybrid power module at a carrier frequency of 1 kHz and 10 kHz can be reduced by 21% and 37% as compared with conventional Si power module respectively. From these results, a significant loss reduction can be achieved in developed hybrid power modules in the region from a low frequency to a high frequency. Therefore, the developed hybrid power modules can be expected to be applied in a wide frequency range.

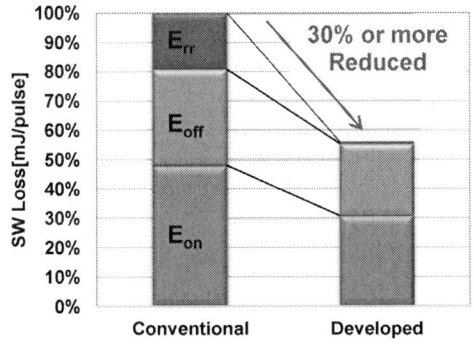

Fig. 6. Switching loss of conventional and developed structure

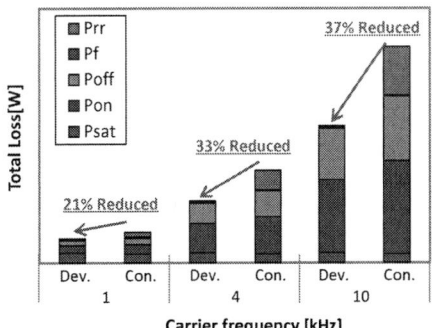

Fig. 7. Results of calculating inverter power loss

2.2 High Reliability Structure

Reducing the loss and size of the developed hybrid power modules are realized by replacing Si-FWD with SiC-SBD. Furthermore, to achieve high frequency operation by high temperature operation which is one of the advantage of SiC devices, the operation temperature of developed hybrid power modules are increased by 25 degree C compared to those of conventional Si power module and the packaging technology with high reliability at high temperature is developed. To achieve further high temperature reliability of developed hybrid power module, high-strength Sn-Sb solder is applied to the joint material under power chip instead of conventional Sn-Ag solder [4] [5]. Delta T_j power cycling test results of the hybrid power modules with high-strength Sn-Sb solder or conventional Sn-Ag solder is shown in Fig. 8.

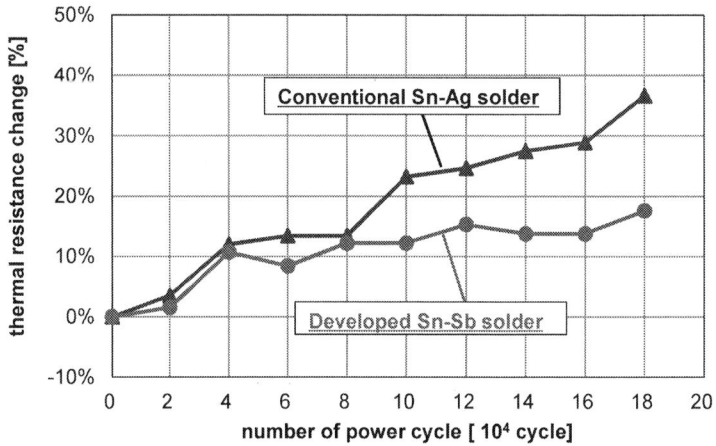

Fig. 8. Delta Tj power cycling test results

The horizontal and vertical lines indicate the number of cycle and thermal resistance change respectively. The test was performed under delta T_j: 25 degree C to 150degree C. It can be found that the crack is occurred at conventional Sn-Ag solder as shown in Fig. 9 and then thermal resistance of power module starts to increase at around 80,000 cycles. On the other hands, the thermal resistance of developed power module with high-strength Sn-Sb solder starts to increase at 160,000 cycles. The propagation of the solder cracks are assumed to be suppressed in the high-strength Sn-Sb solder. The high-strength Sn-Sb solders are strengthened by precipitation and solid solution strengthening [2] [3] [4] [5]. Sb Solid-dissolved in Sn matrix strengthen grain boundaries of Sn grains, and inter-metallic compound by adding third element precipitate in grain boundaries between Sn grains and strengthen Sn-Sb solders. Fig. 10 shows the planner morphology of high-strength Sn-Sb solder before and after delta T_j power cycling test. In Sn-Sb solder, concrete of inter-metallic compound and coarsening of Sn grains are inhibited, and the solder crack propagation are decelerated under the operation temperature of delta T_j: 25 degree C to 150 degree C.

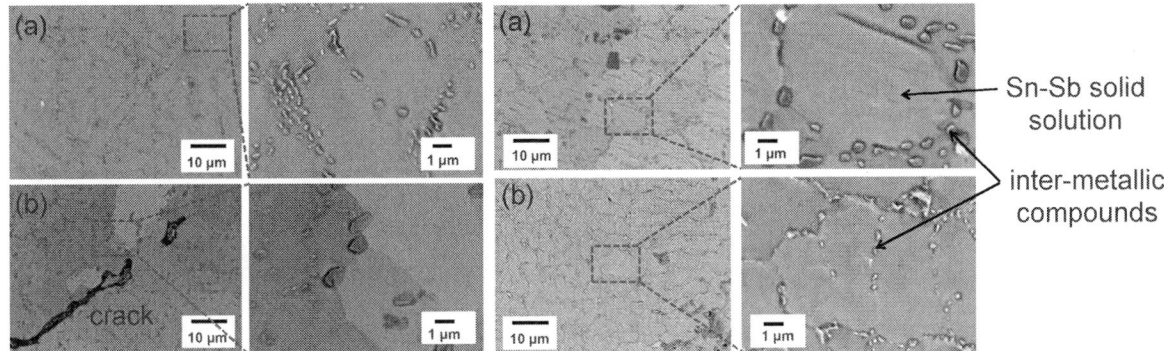

Fig. 9. Planar from of solder joint at Sn-Ag (a) before PC test (b) after PC test

Fig. 10. Planar from of solder joint at Sn-Sb (a) before PC test (b) after PC test

Capability of delta T_j power cycling test is shown in Fig. 11. The horizontal and vertical lines indicate the delta T_j and the number of cycle. In the case of the delta Tj: 25 degree C to 150 degree C, delta T_j power cycling capability of developed hybrid power module with applying the high-strength Sn-Sb solder increase five times higher than that of conventional Si power module.

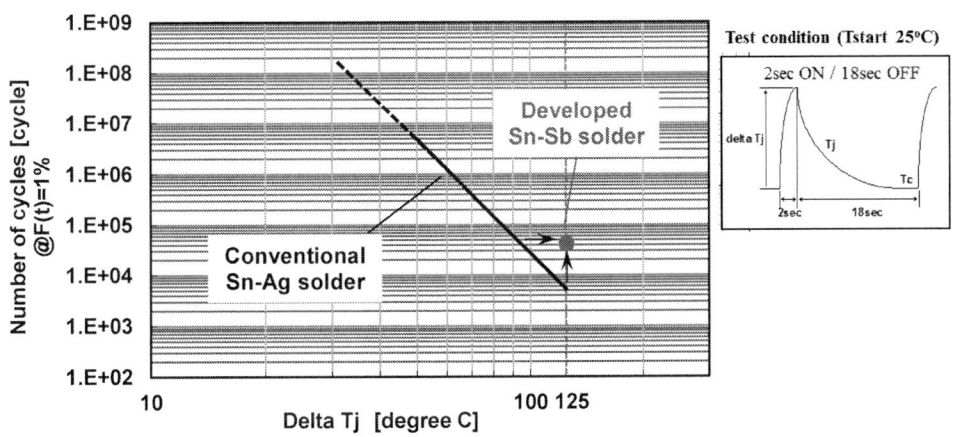

Fig. 11. Delta T_j power cycling capability

3. Conclusions

The significant loss reduction and the downsizing of power module with applying 3.3kV SiC-SBD instead of conventional Si-FWD are achieved. In addition, applying Sn-Sb solder with high strength improve the reliability of the developed hybrid power modules and their operation temperature from 125 degree C of the conventional Si power module to 150 degree C. With these technologies, 30% downsizing and light weight of power module are realized. In addition, the total loss is reduced by 21 % at a carrier frequency of 1 kHz.

Reference

[1] T. Saito et al "New assembly technologies for $T_j$max=175 degree C continuous operation guaranty of IGBT module", pp.455-461 Proceeding of PCIM Europe 2013.

[2] A. Morozumi, K. Yamada, T. Miyasaka, and Y. Seki, "Reliability of power cycling for IGBT power semiconductor modules," Proceedings IEEE, 36th Industry Applications Conference Vol.3 ,pp.1912-1918, 2001.

[3] Nishiura and A. Morozumi, Improved life of IGBT module suitable for electric propulsion system, Proceedings of the 24th EVS, Stavanger, May 13-16, 2009.

[4] T. Saito, Y. Nishimura, A. Morozumi, Y Tamai, F. Momose, E. Mochizuki, and Y. Takahashi, Investigation of New Joint Technology for High Temperature Operation and High Reliability of Power Module, Proceedings of the 20th Symposium on Microjoining and Assembly Technology in Electronics, Yokohama, Feb 4-5 2014.

[5] T. Saito, et al. "Novel IGBT Module Design, Material and ReliabilityTechnology for 175°C Continuous Operation," Proceedings of ECCE, 2014, pp.4367-4372.

Isolated synchronous forward controller with integrated feedback loop and adjustable dead time for high efficiency DC/C converter

Bernhard Strzalkowski, Bernhard.Strzalkowski@analog.com, Analog Devices, Wilhelm-Wagenfeld-Str.6 Munich / Germany

Subodh Madiwale, Subodh.Madiwale@analog.com, Bernardinis Gabriele Gabriele.Bernardinis@analog.com, Daly, Michael, Michael.Daly@analog.com, all Analog Devices, 3550 North First street San Jose, CA 95134. USA

Abstract

A new highly integrated controller topology for DC/DC converters with active clamp on the primary side and synchronous rectifiers on the secondary side of the converter is presented. The controller consists of two chips; one chip on the primary side, controlling the main power switch and active clamp switch as well as second chip on the secondary side controlling synchronous rectifiers. Both chips communicate with each other by means of integrated micro transformers ensuring galvanic isolation and bi-directional signal exchange. The micro transformers offer high speed communication between the both chips. Thus, fast closed feedback loop with minimum latency can be realized when transfer error signal from the secondary to the primary side. High speed data transfer ratio from primary-to-secondary side allows precise adjustment of dead time between main switch and synchronous rectifiers. Therefore, the high speed bi-directional communication helps to elevate the bandwidth of the DC/DC converter and increase the power efficiency.

1. Different method of closing feedback loop

1.1. Analog feedback loop realized by means optocouplers

Traditionally in a forward or flyback converter, a discrete opto-coupler is used in the feedback path to transmit analog error signal from the secondary side to the primary side to the DC/DC converter. In high power efficiency converter, a pulse transformer is used for transmitting the PWM signal from the primary side to control synchronous rectifiers on the secondary side. One serious weakness of opto-couplers is the current transfer ratio (CTR) degradation over time, over temperature and over forward photo diode current. The CTR degradation reduces converter's performance, because the CTR directly determine the feedback-loop gain. The stability and noise level of the output voltage and the dynamic load performance varies with the duty time and converter temperature. Hence, the opto-coupler needs be replaced every 5 to 10 years depending upon the manufacturing grade that determines the initial CTR. The optocouplers manufacture mostly specify CTR value at 25°C; the spread between minimum and maximum value is in range of 600%. To design power converter with specified transient response, the minimum CTR has to be taken into account. For calculating of the desired feedback-loop stability, the maximum CTR-value has to be used. The Fig.1 show the simulation results of Uout fast transient response for the minimum and maximum CTR-values for the state of the art DC/DC converter.

The voltage dip caused by step load is 1.2% of nominal voltage (5.0V) for CTR=240%, and is 2.2% for CTR=120%

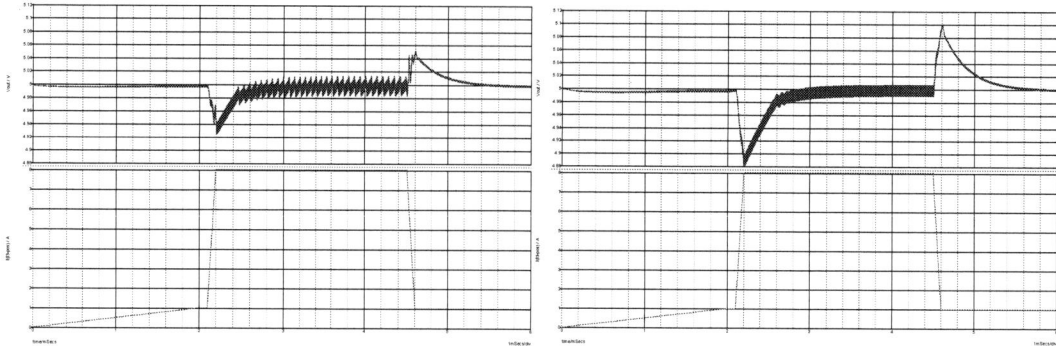

Fig. 1. Uout transient response simulation for CTR=240% (left waveform) and for CTR=120% (right waveform) Horizontal scale: 200us/div., Uout,: blue trace: 1A/div. Iout: black trace: 1A/div.

In low voltage application, such a difference in step load answer could lead to system breakdown. Thus new converter with high bandwidth feedback loop are demanded.

1.2. High speed, high accuracy digital feedback loop

High voltage micro transformer integrated on chip offers analog bandwidth in range of several GHz [1]. Therefore, very fast, analog or digital signal transmission can be realized. Not only accurate PWM or duty ratio transmission but also signals multiplexing and encoding can be realized. When considering the feedback loop approach, the use of pulse width modulation technique to encode the output voltage or error signal of DC/DC converter gives the benefit of an almost infinite resolution [2], [3]. There are no quantization effects. Due to finite encoding time and finite modulation/demodulation time some propagation delay occurs. The propagation delay causes appropriate phase loss that occurs in the DC/DC converter system. The total digital delay is the sum of the above mentioned quantities. In the investigated digital feedback loop circuit with considered bandwidth of 20kHz, the total propagation delay is in rang of 600ns which results in phase loss of only 600ns*20KHz*360= 4.32° [2].

Due to the constant sampling process of output voltage, the system is updated with the new level of current limit several times during the switching period. At a typical converter switching frequency of 300 kHz the output voltage is sampled 32 times per switching period. For a bandwidth of 20 kHz, the additional phase loss is only (1/300k)/32*20 kHz*360= 0.75°.

The approach of duty ratio modulation for error signal transmission over isolation barrier has been already realized in existing product ADUM3190 [4].The transmission bandwidth realized in the product is 400 kHz.

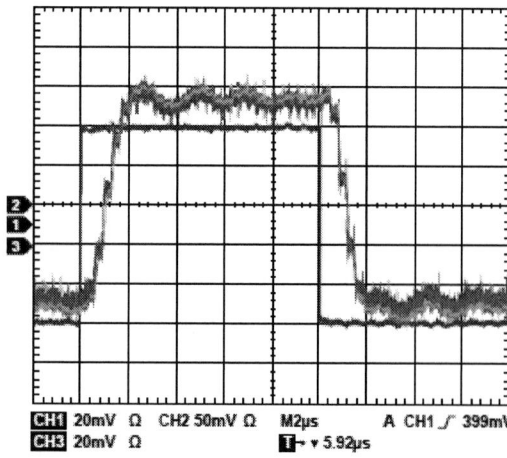

Fig. 2. Output square wave response of isolated error amplifier [4]

The high value of the signal bandwidth compared to optocouplers approach was the starting point for further development digital PWM controller with integrated, isolated feedback loop.

2. Realization of PWM controller with integrated feedback loop

2.1. Principal of chip arrangement and of operation

The primary side controller is a PWM current mode, fixed frequency active-clamp forward controller designed for isolated DC/DC power supply. The primary side pins provide functions for programming the switching frequency and maximum duty cycle, external frequency synchronization and slope compensation. The secondary side pins provide functions for differential output voltage sensing, compensation, OVP, power good, and programmable light load mode setting.

Two mico-transformers are integrated and enable bidirectional data transfer from either side of the two chips. The micro-transformers have a polyamide insulation material that provide upto 5kV isolation. Referring to Fig.3, the upper transformer (T1) transfers the feedback signal from the secondary to primary side. An OVP condition is also transmitted using the same transformer to trigger a fault response. The bottom transformer (T2) transfers signals from the primary to the secondary side for synchronous rectification and over current protection (to terminate the SR pulses). The micro-transformers eliminate external bulky signal transformers to control synchronous rectifiers and an opto-coupler to close the feedback loop. Therefore, the integrated micro transformer reduces system design complexity, cost, component count, and improving overall system reliability. An additional benefit includes a reduced component placement cost during production. The principal block diagram of the forward converter is presented in Fig.3.

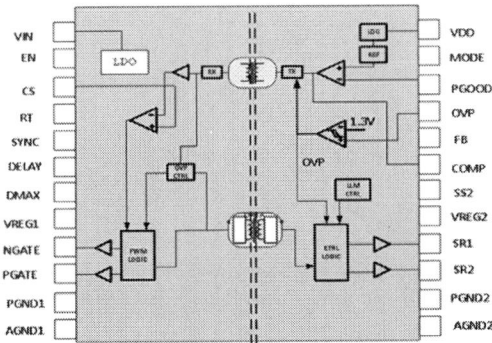

Fig. 3. Principal block diagram of a forward converter with integrated, isolated feedback loop

The output voltage on the secondary side is sensed and compared to a reference voltage for the closed loop regulation. A transconductance amplifier is used for compensating the loop and providing proportional and integral gain. As known in peak current mode control, the output of voltage error amplifier acts as the peak current limit of the switch. This signal is fed into the inverting input of the comparator while the sensed current limit is fed into the non-inverting input. The signal of the voltage error amplifier is encoded using combination of pulse width modulation techniques and on-off keying (OOK) at a frequency, which is much greater than the switching frequency. The system transmits the encoded error signal over the isolation barrier by transformer T1 to the primary side controller. The primary side controller generates PWM signals for the main switch, active clamp switch as well as for the synchronous rectifiers.

© VDE VERLAG GMBH · Berlin · Offenbach

The isolated PWM signals for the synchronous rectifiers SR1 and SR2 are transferred from primary-to secondary side by means of transformer T2.

The complete schematic of the new isolated synchronous active clamp forward controller with integrated feedback loop is presented in the Fig.4.

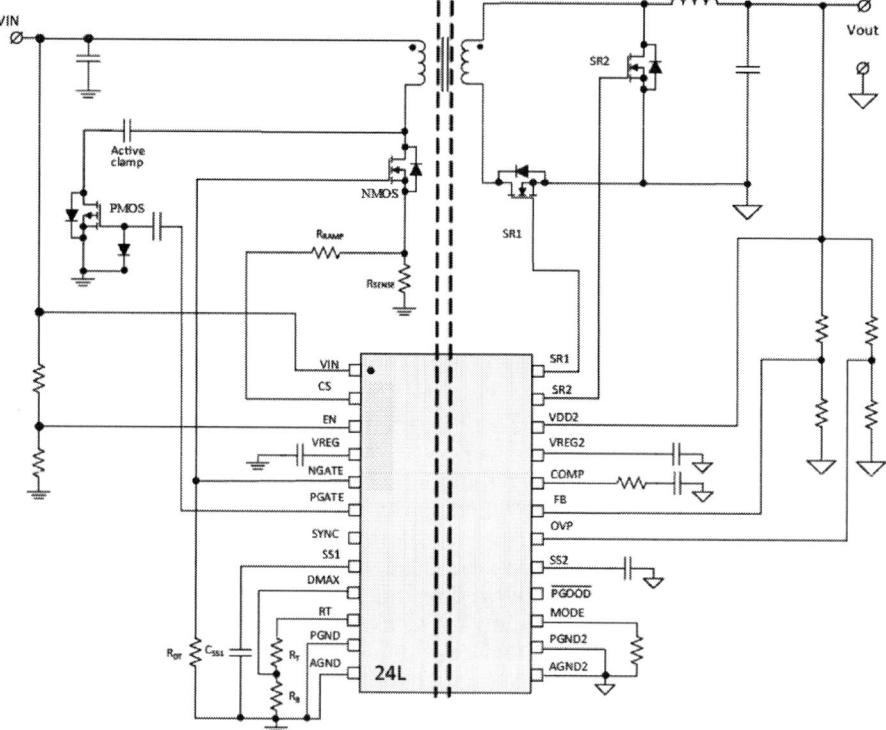

Fig. 4. Schematic of isolated forward converter with active clamp circuit and synchronous rectifier

Because of high band width of the micro transformer, additional signals can be combined with the PWM data stream. Signals such as input current protection, output over-voltage protection (OVP), under voltage lockout (UVLO), over temperature protection (OTP), light-load power saving mode (LLM), as well as gate delay and dead time are included to the inter-chip communication.

2.2. Test results

The high bandwidth communication between the primary side and secondary side controller allows fast exchange of information and precise signal timing, which are crucial to optimize the power efficiency and to keep to the output voltage within extremely tight limits. The precise signal timing and dead time adjustment among main switch (NMOS), active clamp switch (PMOS) and synchronous rectifiers SR1, SR2 ensure high power efficiency. In a forward converter, the magnetizing energy stored in the transformer core during the switch-on cycle must be demagnetized or reset to zero volt during the switch-off cycle. To reset the transformer core, an active clamp PMOS is turned on during the off cycle by returning the magnetizing current back to the input power stage. The new primary side controller allows programmable delay time between the PGATE rising and the NGATE rising to account for different input voltages, various MOSFET models and leakage inductance of the transformer. Also, sufficient gate delay between PGATE and NGATE could ensure ZVS (zero volt second) switching, which is important in reducing switching losses in the main MOSFET. To maximize power efficiency and to avoid cross conduction between the primary main switch NGATE and the secondary side synchronous rectifiers SR1 and SR2, appropriate delay times between SR1 and NGATE as well as between SR2 and NGATE are needed. These delay times can be programmed within several nanoseconds by means of external resistors.

© VDE VERLAG GMBH · Berlin · Offenbach

Test forward converter with active clamp circuit, nominal input voltage of 48V, nominal output voltage of 12V was built to measure the performance of the new controller concept (refer to Fig.3). The converter has been dimensioned for output power range of 100W at switching frequency of 300 kHz; all power switches are state of the art Silicon-MOSFETs. The efficiency plot with optimized timing among power switches PMOS, NMOS, SR1, SR2 is presented in Fig. 5.

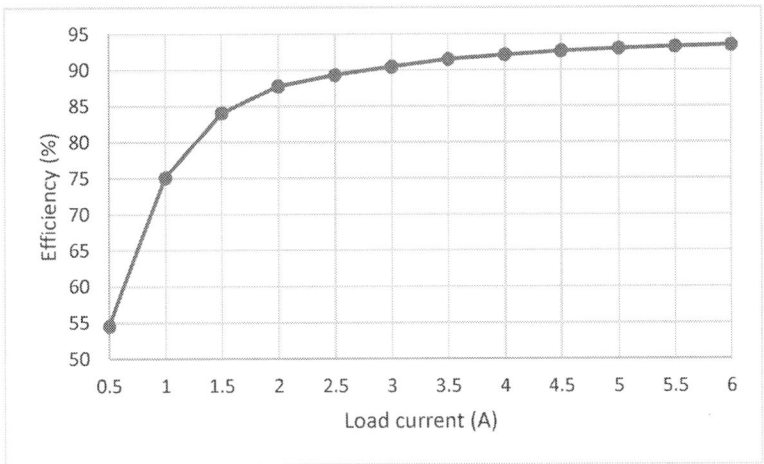

Fig. 5: Efficiency plot of the new isolated forward converter concept

The load regulation of the new converter exhibits very good stability. This is because of appropriate sampling procedure of output voltage and transfer processing to the primary side PWM controller.

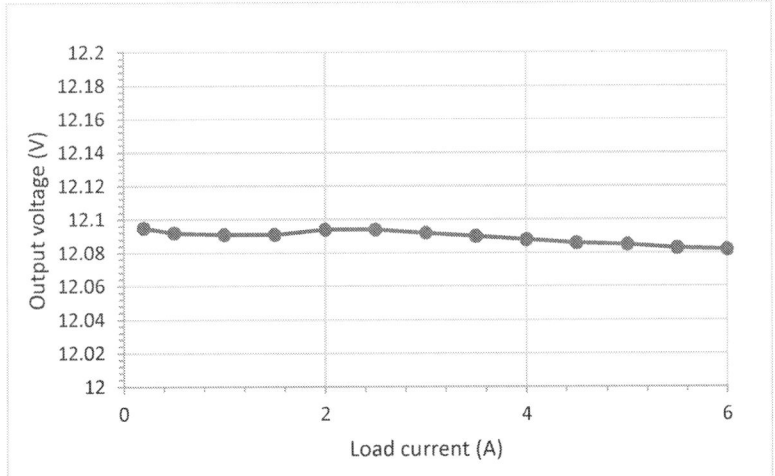

Fig. 6: Voltage regulation vs load of the new isolated forward converter concept

The test circuit shows very stable step load answer. The measured voltage under- and overshoots exhibit values 1.25% of nominal Uout at step load 25% of maximum Iout. The measured converter bandwidth for the same conditions was 15 kHz.

© VDE VERLAG GMBH · Berlin · Offenbach

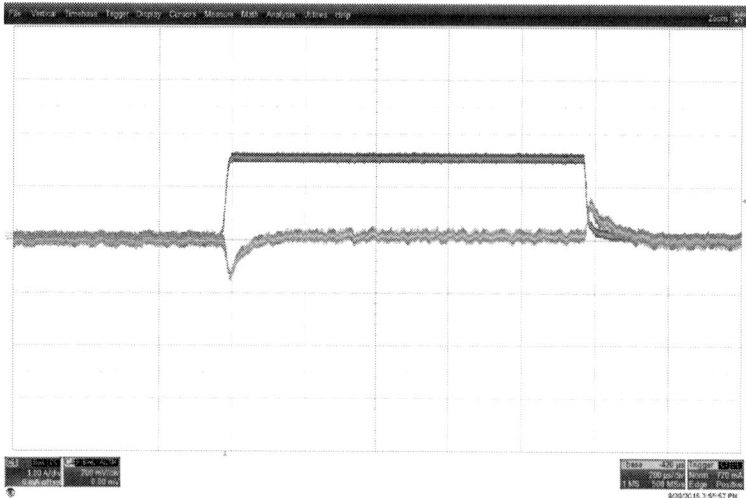

Fig. 7. Step-load (0%-25%Inominal-0%) response at 12 V output voltage of isolated forward converter exhibiting high bandwidth of the integrated feedback loop. Horizontal scale: 200us/div. Yellow trace: 1A/div. Green trace: 200mV/div.

The Body plot of the test converter is presented in Fig.8.

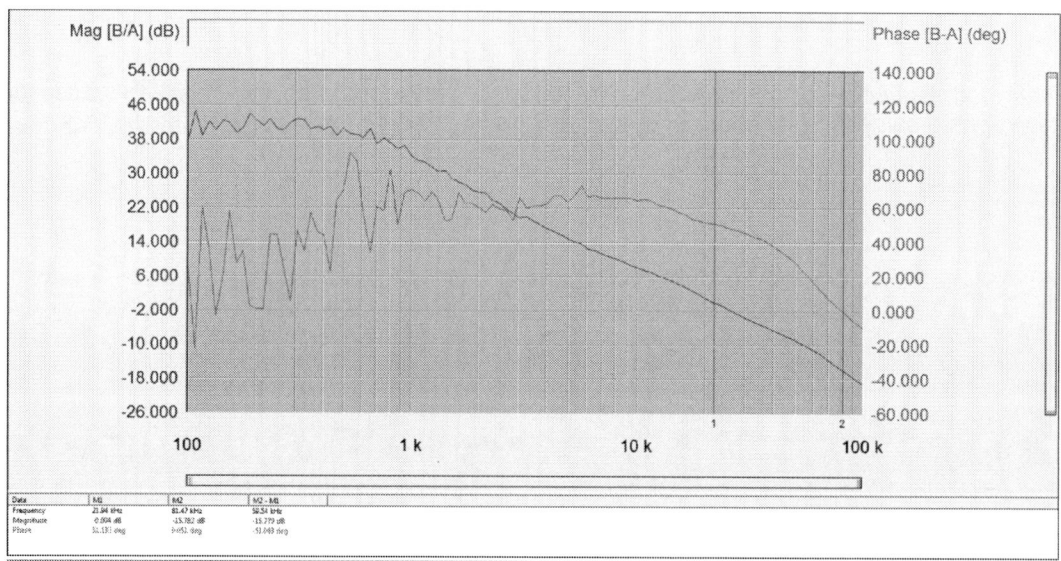

Fig. 8. Bode plot of the system with 21kHz crossover frequency, 51° phase margin and 15dB gain margin

Conclusions

High voltage micro transformers allows integration of high speed communication between the primary and secondary side of isolated converter. This allows to eliminate the bulky signal transformers for control of synchronous rectifiers and opto-couplers for closing the feedback loop. The micro-transformer integration allows reducing system design complexity, cost and component count, and improving overall system reliability. With the integrated isolators and drivers on both the primary and the secondary, the new chip-arrangement offers a compact system level design and yields a higher efficiency than a non-synchronous forward converter at heavy loads. We achieved outstanding load regulation- and load step answer performances.

© VDE VERLAG GMBH · Berlin · Offenbach

References

[1] Niknejad A., Meyer R. : *Analysis, Design, and Optimization of Spiral Inductors and Transformers for SI RF IC's*, IEEE Journal of Solid-State Circuits, Vol. 33, pp. 1470-1481, Oct. 1998

[2] Erickson, Robert W., and Dragan Maksimovic.: *Fundamentals of power electronics* Springer Science & Business Media, 2007.

[3] Corradini, Luca, Dragan Maksimović, Paolo Mattavelli, and Regan Zane.: *Digital Control of High-Frequency Switched-Mode Power Converters*. John Wiley & Sons, 2015.

[4] Data sheet ADuM3190: http://www.analog.com/media/en/technical-documentation/data-sheets/ADuM3190.pdf

PCIM Europe 2016, 10 – 12 May 2016, Nuremberg, Germany

Extreme High Efficiency Non-Inverting Buck-Boost Converter for Energy Storage Systems

Zhe Yu[1], Holger Kapels[1], Klaus F. Hoffmann[2]

[1] Fraunhofer Institute for Silicon Technology ISIT, Application Center Power Electronics for Renewable Energy Systems, Steindamm 94, 20099 Hamburg, Germany

[2] Helmut Schmidt University, Holstenhofweg 85, 22043 Hamburg, Germany

E-Mail: zhe.yu@isit.fraunhofer.de

Abstract

This paper presents a high efficiency bidirectional non-inverting buck-boost converter for energy storage systems. A new control concept for achieving high efficient power conversion within a wide power range is introduced. A 3 kW prototype is designed and tested with Si- and SiC-MOSFETs. Experimental results show that the prototype achieves a minimum efficiency of more than 98% at 5% load and a maximum efficiency of more than 99.4% at the operating points where the output voltage approaches the level of the input voltage.

1. Introduction

The bidirectional non-inverting buck-boost converter (NBuBoC), as shown in Fig. 1, is often employed to couple energy storages with power systems due to its wide voltage conversion ratio and simple topology [1], [2], [3]. However, the low power conversion efficiency is a significant drawback of this converter using conventional control methods.

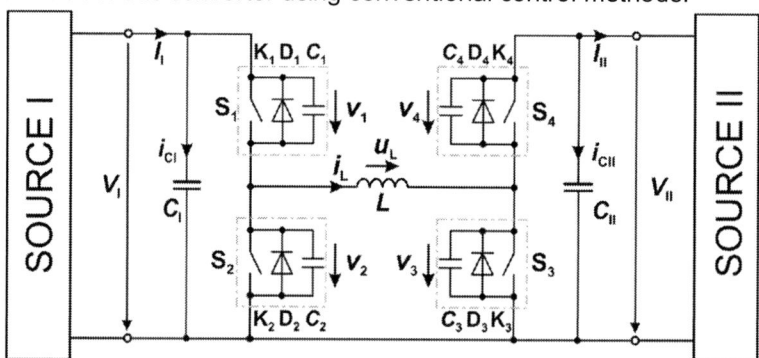

Fig. 1: Topology of the bidirectional non-inverting buck-boost converter

An improved control method with five different modes is introduced in [4]. The basic idea of the method is to halve the switching frequency when the output voltage approaches the input voltage. As a result, the switching losses in the critical transition phase between the step-up and step-down modes can be reduced by 50% theoretically. Nevertheless, the efficiency by using this control method is still strongly limited because of the hard switched semiconductor devices. Different control methods for soft-switching of the NBuBoC without the necessity to add any auxiliary components are presented in [5], [6] and [7]. It has been shown that the converter can be operated under zero-voltage-switching (ZVS) condition in steady state. However, it has not been disclosed, how the ZVS condition can be maintained if the operating point changes. In [8] a hybrid control concept for the NBuBoC is addressed. The converter operates in three modes and the switching frequency is adaptively changed, so that the semiconductor devices can always be switched under ZVS condition. As presented, the NBuBoC can achieve an efficiency of more than 99% in the step-up and step-down modes. However, the efficiency in the transition phase between the step-up and step-down modes still leaves room for improvement. In this paper a new control concept is presented.

© VDE VERLAG GMBH · Berlin · Offenbach

Using this concept the NBuBoC can achieve high efficiency overall, especially at the operating points where the output voltage approaches the level of the input voltage.

2. Control Concept

With conventional control methods the NBuBoC typically operates in several different modes depending on the voltage levels of SOURCE I and SOURCE II (Fig. 1). In contrast to this, using the proposed control concept no transition between different operating modes is needed any more. The switches S_1 and S_3, as shown in Fig. 1, are gated by the PWM signals with the same frequency and duty ratio. A time delay between the PWM signals is set. The voltage conversion ratio can be theoretically adjusted from 0 to infinity by changing the duty ratio of the PWM signals without the influence of the time delay. Additionally, by adaptively modulating the frequency and the time delay of the PWM signals depending on operating points the switches can always be switched under ZVS condition and at the same time the RMS value of the inductor current can be minimized. Consequently, the power conversion efficiency of the converter can be maximized for a wide power range.

2.1 Control signals and waveforms in steady state

The control signals and the waveforms of the inductor voltage and current in case that the power flows from SOURCE I to SOURCE II (case 1) are illustrated in Fig. 2. The switches S_1 and S_2 are respectively gated by the complementary signals PWM1H and PWM1L with the duty cycle T, while S_3 and S_4 are gated by PWM2H and PWM2L with the same duty cycle. The conduction durations of S_1 and S_3 are both t_{ON1}, whereas the ones of S_2 and S_4 are t_{ON2}. The dead times t_{D1} to t_{D4} are set between the complementary signals to prevent a short circuit. Furthermore, there is a time delay αT between PWM2H and PWM1H which is shorter than the conduction time t_{ON2} of S_2. The switch S_3 is turned on only when S_2 is conducting.

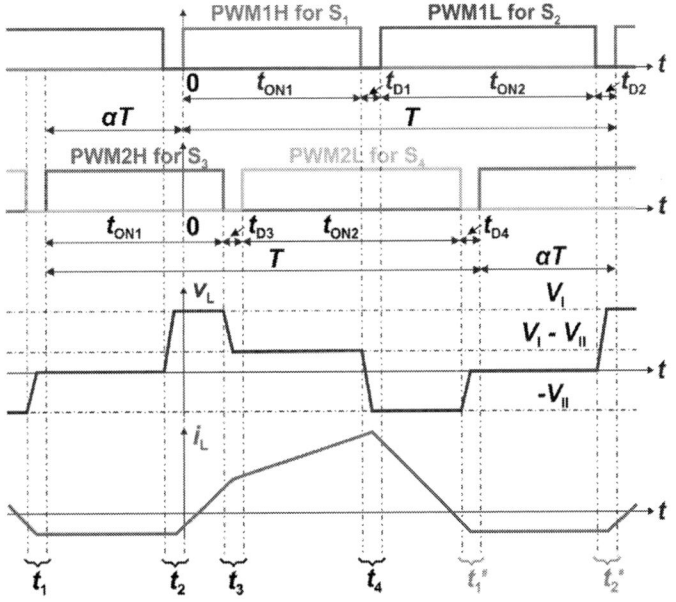

Fig. 2: Control signals and waveforms in steady state in case 1

As shown in Fig. 2, the time delay causes a waveform of the inductor voltage v_L with the four different levels V_I, V_I-V_{II}, 0 and $-V_{II}$. Accordingly, the inductor current varies with the slopes V_I/L, $(V_I$-$V_{II})/L$, 0 and $-V_{II}/L$ in the different time intervals and changes its flow direction twice within a duty cycle. The inductor current $i_L(t)$ is shaped by the four-level inductor voltage in the waveform with smaller RMS value in comparison with the sawtooth-shaped waveform.

© VDE VERLAG GMBH · Berlin · Offenbach

The control signals for the case in which the power flows from SOURCE II to SOURCE I (case 2) are shown in Fig. 3. In contrast to case 1, S_3 is turned on only when S_1 is conducting in case 2. Due to the symmetrical structure of the NBuBoC, the control principles in the two cases are very similar. Therefore, a further description for the case 2 will not be presented in this paper.

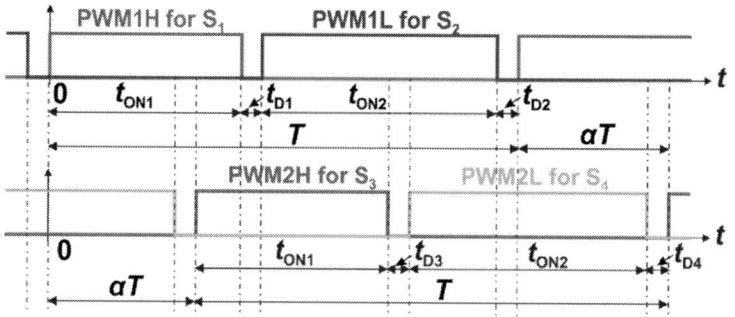

Fig. 3: Control signals for case 2

2.2 Zero-Voltage-Switching of the semiconductor devices

A switch used for the NBuBoC is normally a combination of an active semiconductor device with an external Schottky diode and a snubber capacitor. Schematically, each switch S_i in the converter consists of a controllable channel K_i, an antiparallel diode D_i and a capacitor C_i, as depicted in Fig. 1. The diode symbolizes the body diode of the used semiconductor device or the parallel connection of the body diode and an external Schottky diode. Similarly, the capacitor can be the parasitic output capacitance of the semiconductor device or the parallel connection of the parasitic capacitance with an additional snubber capacitor.

(a) during the dead time t_{D1} (b) during the dead time t_{D4}

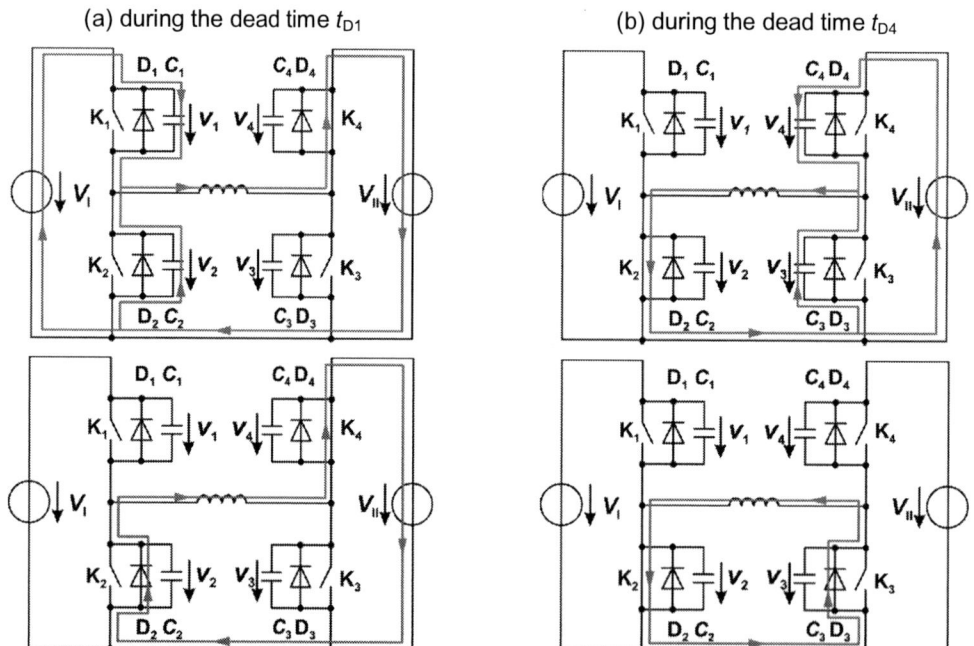

Fig. 4: Current flow in the NBuBoC during the dead times t_{D1} and t_{D4}

A basic principle for the ZVS of the switches is that the inductor current must not firstly commutate to the diodes after turning off the channels of the switches in an half bridge.

Instead, the inductor current must flow through the capacitors at first in order to charge and discharge them. For example, after turning off the channel K_1, the inductor current must commutate from K_1 to the capacitors C_1 and C_2. For that purpose, the inductor current must be positive during the dead time t_{D1}, as shown in Fig. 2. Otherwise, the inductor current would first flow through the diode D_1. Analogously, the inductor current must be positive during t_{D3} and be negative during the dead times t_{D2} and t_{D4} for the ZVS of the switches.

The current flow in the NBuBoC during the commutation phases within the dead times t_{D1} and t_{D4} is exemplarily shown in Fig. 4. After turning off S_1 the current flow commutes from the channel K_1 to the capacitors C_1 and C_2 due to the positive inductor current (Fig. 4 (a)). At this moment, the diode D_2 is still blocked by the voltage v_2 across C_2. Consequently, C_1 is charged and C_2 is discharged. After the voltage v_2 is reduced to zero volts, the current flow commutates to D_2. At the end of the dead time t_{D1} the channel K_2 is turned on at zero voltage. As shown in Fig. 4 (b), after turning off S_4 the current flow commutates from K_4 to C_3 and C_4 due to the negative inductor current, whereby C_4 is charged and C_3 is discharged. After the voltage v_3 is reduced to zero volts, the current flows through D_3. At the end of t_{D4} the channel K_3 is turned on at zero voltage.

In summary, for the ZVS of the switches the inductor current must be positive during the dead times t_{D1} and t_{D3} and be negative during t_{D2} and t_{D4}. Furthermore, the absolute value of the inductor current must be high enough while turning off the switches so that the capacitors can be completely charged and discharged within the corresponding dead times.

As is well known, the parasitic output capacitance of a semiconductor device is dependent on the voltage across the device. Therefore, the capacitance of a switch in the converter is voltage-dependent regardless of whether the switch is a single semiconductor device or a combination of a semiconductor device with a snubber capacitor. Assuming that the capacitance of the used switch is a function $f(v)$ depending on the voltage v across it and the inductor current is constant during the dead times, the minimum absolute values of the inductor current $i_L(t)$ during these dead times can be deduced, as shown in Tab. 1.

Tab. 1: The minimum absolute values of the inductor current during the dead times

	During t_{D1}	During t_{D2}	During t_{D3}	During t_{D4}
Minimum absolute value of $i_L(t)$	$\dfrac{2 \cdot \int_0^{V_I} f(v)\,dv}{t_{D1}}$	$\dfrac{2 \cdot \int_0^{V_I} f(v)\,dv}{t_{D2}}$	$\dfrac{2 \cdot \int_0^{V_{II}} f(v)\,dv}{t_{D3}}$	$\dfrac{2 \cdot \int_0^{V_{II}} f(v)\,dv}{t_{D4}}$

2.3 Voltage conversion ratio

In order not to effect the accuracy of the voltage and power conversion, the dead times between the complementary signals are normally very short and negligible in comparison to the conducting durations of the switches. Therefore, the dead times are not considered for the derivation of the voltage conversion ratio.

By equating the integral of the inductor voltage over one duty cycle to zero yields

$$0 \cdot \alpha T + V_I \cdot (t_{ON1} - \alpha T) + (V_I - V_{II}) \cdot \alpha T + (-V_{II}) \cdot (T - t_{ON1} - \alpha T) = 0$$

Defining the duty ratio $D = t_{ON1}/T$ and rearranging terms yields

$$V_{II}/V_I = D/(1 - D) \qquad \text{Eq. 1}$$

From Eq. 1 it can be derived that in steady state the voltage conversion ratio does not depend on the time delay αT between PWM1 and PWM2. Furthermore, the voltage conversion ratio can vary from 0 to infinity by changing the duty ratio of the PWM signals. No transition of different operating modes due to the change of the voltage levels will occur.

2.4 Optimal switching frequencies and time delays

Different combinations of switching frequencies and time delays can stabilize the NBuBoC at one single operating point (V_I, V_{II}, I_I, I_{II}). However, the power conversion efficiency at this operating point can only be maximized by an optimal one of these combinations. One way to find the optimal combination is to build the mathematical relationship between the total power losses in the converter and the combination of the switching frequency and the time delay. Then the optimal switching frequency and time delay can be obtained by minimizing the total power losses under ZVS condition of the switches. Nevertheless, the mathematical model will be very complicated in this way. Alternatively, a simpler strategy is to minimize the RMS value of the inductor current under ZVS condition. The basis for the feasibility of this strategy is that the power losses in the switches, in the inductor and in the capacitors are all determined by the RMS value of the inductor current when the switches are gated under ZVS condition.

As mentioned in the section 2.1, the inductor current i_L changes with four different slopes in the time intervals when the power flows from SOURCE I (input) to SOURCE II (output). According to Fig. 2 and neglecting the dead times (the dead times are equal to zero), the mathematical expression of the inductor current can be obtained for case 1

$$i_L(t) = \begin{cases} I_{L0} & t_1 \leq t < t_2 \\ \frac{V_I}{L}(t - t_2) + I_{L0} & t_2 \leq t < t_3 \\ \frac{V_I - V_{II}}{L}(t - t_3) + \frac{V_I}{L}(t_3 - t_2) + I_{L0} & t_3 \leq t < t_4' \\ \frac{-V_{II}}{L}(t - t_4) + \frac{V_I - V_{II}}{L}(t_4 - t_3) + \frac{V_I}{L}(t_3 - t_2) + I_{L0} & t_4 \leq t < t_1^{\#} \end{cases}$$

Eq. 2

where $t_1 = -\alpha T$, $t_2 = 0$, $t_3 = (D - \alpha)T$, $t_4 = DT$, $t_1^{\#} = (1 - \alpha)T$ and I_{L0} is the inductor current at the time t_1. The RMS value $I_{L,eff}$ of the inductor current can be derived from Eq. 2

$$I_{L,eff} = \sqrt{\frac{1}{T} \int_{t_1}^{t_1^{\#}} (i_L(t))^2 \cdot dt}$$

Eq. 3

$$= \sqrt{\frac{D}{3(1-D)L^2 f^2} \left(V_I^2 \alpha^3 - 3(1-D)(V_I^2 D + 2LV_I I_{L0} f)\alpha + (1-D)(V_I^2 D + 3LV_I I_{L0} f) \right) + I_{L0}^2},$$

where the parameter f defines the switching frequency. Assuming that the SiC-MOSFETs "C2M0080120D" (Cree Inc.) are used as switches for the NBuBoC and the converter operates in the power range "input voltage 400V, output voltage range from 150V to 500V, output current from 1A to 5A", the optimal switching frequency f and the time delay factor α are extracted by minimizing $I_{L,eff}$ under the ZVS condition of the switches, as shown in Fig. 5 and Fig. 6, respectively.

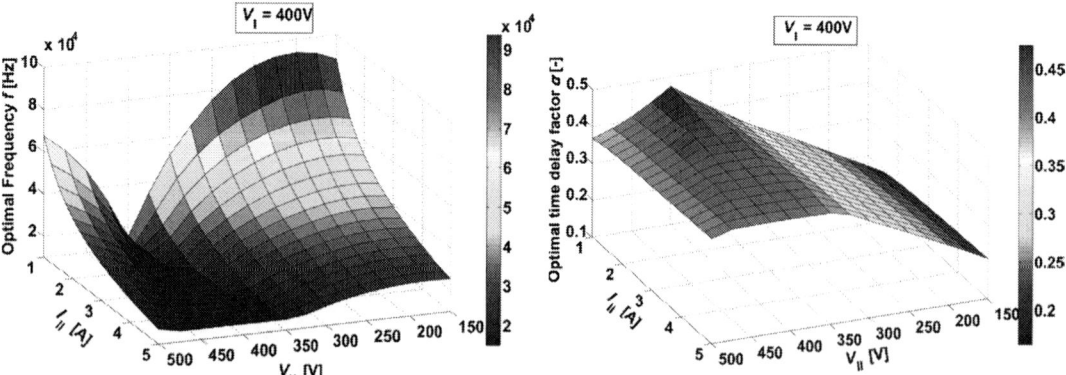

Fig. 5: Optimal switching frequency f Fig. 6: Optimal time delay factor α

According to Fig. 5 it can be extracted that the switching frequency is firstly increased with the increase of the differential between the input voltage V_I and the output voltage V_{II}. However, after reaching a maximum value the switching frequency is reduced at a small step. At the operating points where the output voltage is near to the level of the input voltage the switching frequency approaches its minimum value. Additionally, the switching frequency decreases with the increase of the output current I_{II}. Finally, the switching frequency is limited. The lower boundary is utilized to restrict the volume of the system and the ripple of the output voltage as well as to prevent acoustic harassments. The upper limit is determined by the maximum switching frequency of the used semiconductor devices. As shown in Fig. 6, the time delay factor α decreases with the increase of the differential between V_I and V_{II}.

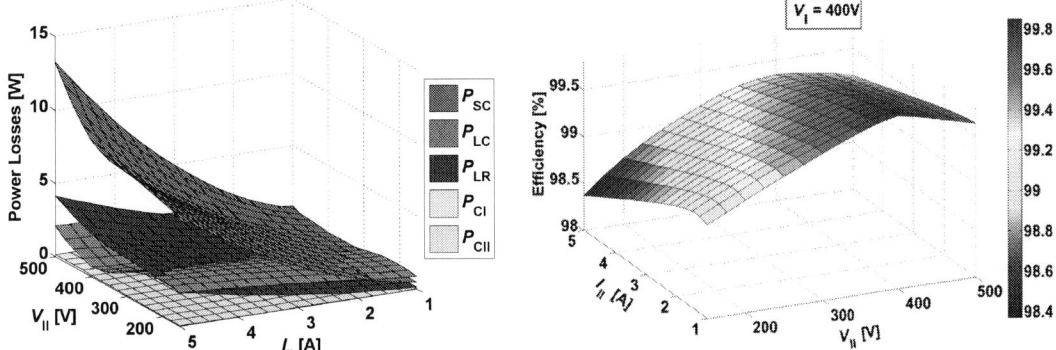

Fig. 7: Calculated power losses of the converter based on the SiC-MOSFETs "C2M0080120D"

Fig. 8: Calculated power conversion efficiency based on the SiC-MOSFETs "C2M0080120D"

The calculated power losses of the converter with the optimal switching frequencies and time delays are given by Fig. 7, where the conduction losses in the switches P_{SC}, the conduction loss in the inductor P_{LC}, the core loss in the inductor P_{LR}, and the conduction losses in the input capacitor P_{CI} and output capacitor P_{CII} are taken into account. In this calculation the switching losses are ignored due to ZVS of the switches. The conduction losses in the switches are dominant in the entire power losses, as shown in Fig. 7. Moreover, the core losses of the inductor are comparable to the conduction losses of the inductor due to the high ripple of the inductor current. The power losses in the input and output capacitors are very low in comparison with the losses in the inductor and the switches. The calculated power conversion efficiency is illustrated in Fig. 8. Theoretically, the NBuBoC can achieve an efficiency of more than 98.8 % at 5 % load by utilizing the proposed control concept. Efficiency of more than 99.7 % can be achieved at the operating points, where the output voltage is near to the level of the input voltage.

3. Experimental results

For the verification of the control concept, a prototype dimensioned for a maximum power of 3 kW and a maximum voltage of 600 V, as shown in Fig. 9, has been designed and tested in the power electronics laboratory of the Helmut Schmidt University in Hamburg. The gate driver circuits are isolated from the power part and their voltage levels for the turning-on and turning-off of the switches are adjustable for the testing with Si and SiC semiconductor devices. Fig. 10 illustrates the steady-state waveforms of the inductor current i_L and the gate-source voltages v_{GS1}, v_{GS2} and v_{GS3} within a duty cycle at three different operating points. The ripple of the inductor current is strongly depending on the differential between the input and output voltages. Fig. 11 shows the gate-source voltages v_{GS3}, v_{GS4} and the drain-source voltages v_{DS3} and v_{DS4} in the ZVS phase during the dead time t_{D4}. Before turning on S₃ the drain-source voltage v_{DS3} has been reduced to zero volts.

PCIM Europe 2016, 10 – 12 May 2016, Nuremberg, Germany

Fig. 9: A prototype for the verification
of the control concept

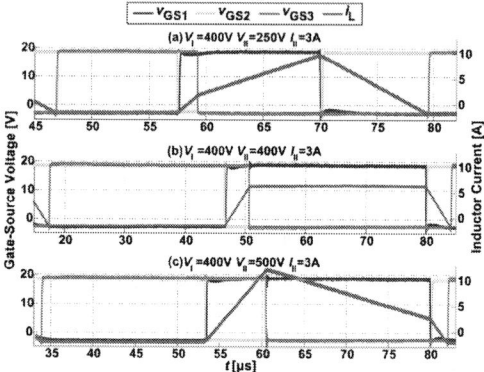

Fig. 10: Inductor current and gate-source voltages
at different operating points

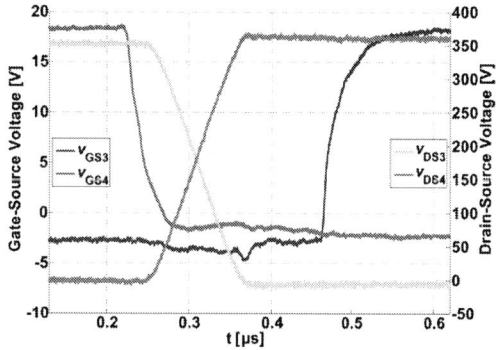

Fig. 11: ZVS phase during the dead time t_{D4} at
the operating point V_I=400V, V_{II}=360V, I_{II}=5A

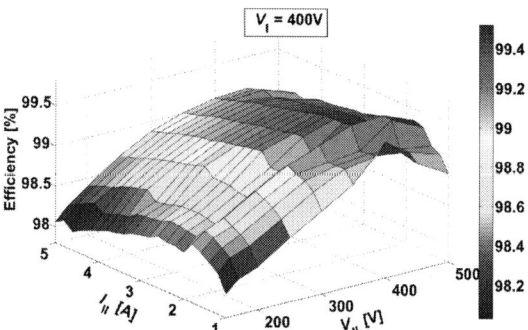

Fig. 12: Power conversion efficiency of the
prototype with SiC-MOSFETs "C2M0080120D"

The power conversion efficiency of the prototype with the SiC-MOSFETs "C2M0080120D" is shown in Fig. 12. By comparing Fig. 12 with Fig. 8, it can be observed that the measured efficiency is comparable to the calculated efficiency in the high power range. However, the calculated efficiency in the low power range is much higher.

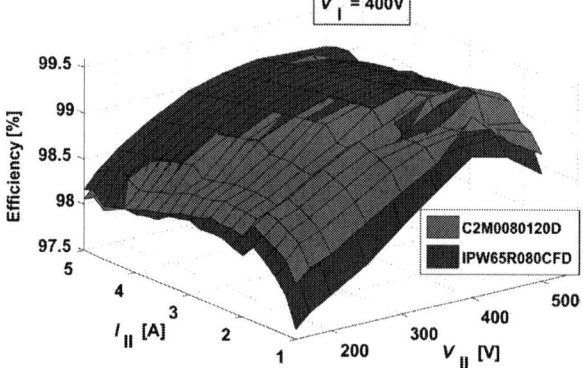

Fig. 13: Comparison of the efficiency of the prototype with Si- and SiC-MOSFETs

A main reason for the difference between the calculated and measured results is that the losses of the switches during the ZVS are not considered for the calculation of the efficiency. In fact, there are always some losses caused by charging and discharging the output capacitors of the switches, which is proportional to the output capacitances of the utilized switches and the switching frequency. However, the prototype with the SiC-MOSFETs can still achieve a minimum efficiency of more than 98 % at 5 % load.

© VDE VERLAG GMBH · Berlin · Offenbach

A comparison of the efficiency of the prototype equipped with SiC-MOSFETs "C2M0080120D" and Si-MOSFETs "IPW65R080CFD" (Infineon Technologies) respectively is illustrated in Fig. 13. Due to the much lower output capacitances of the SiC-MOSFETs and the high switching frequencies in the low power range (see Fig. 5) the losses caused by the ZVS of the SiC-MOSFETs are much lower than the ones caused by the ZVS of the Si-MOSFETs. Therefore, the efficiency of the prototype with the SiC-MOSFETs is higher in this power range. In contrast, the efficiency with the Si-MOSFETs is slightly higher in the high power range due to the lower $R_{DS(on)}$ of "IPW65R080CFD" in comparison with "C2M0080120D". In addition, the switching frequencies in the high power range are lower (see Fig. 5), which means that the losses caused by the ZVS of the switches have a smaller impact on the efficiency in the high power range.

4. Conclusion

This paper addresses a bidirectional non-inverting buck-boost converter using a new control concept for the coupling of energy storages with power systems. Due to the modulation of the switching frequency and the time delay between the PWM signals the proposed converter can maintain ZVS condition in a wide power range. Simultaneously, the conduction losses in the switches and the losses in the inductor can be significantly minimized. A 3 kW prototype has been designed and tested with Si- and SiC-MOSFETs, respectively. Experimental results show that by using SiC-MOSFETs the prototype achieves a minimum efficiency of more than 98 % at 5 % load and a maximum efficiency of more than 99.4 %. The efficiency with Si-MOSFETs is slightly higher in the high power range and lower in the low power range in comparison to the results with SiC-MOSFETs.

5. Acknowledgement

This work was funded by the Free and Hanseatic City of Hamburg (Hamburg City Parliament publication 20/11568).

6. References

[1] H. R. Karshenas, H. Daneshpajooh and A. Safaee, "Bidirectional DC-DC Converters for Energy Storage Systems," in *Energy Storage in the Emerging Era of Samrt Grids*, In Tech, 2011.

[2] D. Polenov, "DC/DC-Wandler zur Einbindung von Doppelschichtkondensatoren in das Fahrzeugenergiebordnetz," diss., Jun. 2009.

[3] Z. Yu, K. F. Hoffmann and H. Kapels, "Hocheffizienter DC/DC-Wandler zur leistungselektronischen Kopplung von Energiespeichern," in *NEIS-Konferenz Tagungsband*, Hamburg, 2013.

[4] H. Kobayashi, "Circuit of high efficient buck-boost switching regulator and control method thereof". US Patent 2012146594A1, 14 Jun. 2012.

[5] A. A. M. Esser, "Bidirectional Buck-Boost Converter". Patent US5734258, 31 Mar. 1998.

[6] P. Vinciarelli, "Buck-Boost DC-DC Switching Power Conversion". Patent US6788033B2, 7 Sep. 2004.

[7] S. Waffler and J. W. Kolar, "A Novel Low-Loss Modulation Strategy for High-Power Bidirectional Buck + Boost Converters," *IEEE Trans. Power Electron,* pp. vol. 24, no. 6, pp. 1589-1599, Jun. 2009.

[8] Z. Yu, H. Kapels and K. F. Hoffmann, "High Efficiency Bidirectional DC-DC Converter with Wide Input and Output Voltage Ranges for Battery Systems," in *PCIM*, Nuremberg, Germany, 2015.

A High-Efficiency Bidirectional GaN-HEMT DC/DC Converter

Michael Ebli,* Martin Wattenberg,* and Martin Pfost**
*Reutlingen University, Germany, **University of Innsbruck, Austria
michael.ebli@reutlingen-university.de

Abstract

This paper presents a compact 3 kW bidirectional GaN-HEMT DC/DC converter for 360 V to 400-500 V. A very high efficiency has been reached by applying a zero voltage turn-on in conjunction with a negative gate-source voltage, even though normally-off HEMTs are used. Further improvements were achieved by adapting the switching frequency to the load current and output voltage, as will be explained by means of the loss contribution of the specific elements for a constant and an adaptive switching frequency. Measurements have shown a high converter efficiency exceeding 99% over a wide output power range of up to 3 kW.

1. Introduction

Gallium Nitride (GaN) HEMTs are unipolar devices without minority charge storage and with very small parasitic capacitances. Hence, very fast switching transients are possible. These fast transients, however, can lead to high overvoltages due to parasitic inductances [1]. They also increase the risk of an undesired turn-on [2]. To prevent this, negative gate-source voltages can be applied even to transistors that are normally off, cf. [3].

A further advantage of a negative gate-source voltage is the faster turn-off of the GaN-HEMT, which allows a significant reduction of switching losses as has been shown for SiC-MOSFETs [4]. In combination with zero-voltage turn-on and an appropriate inductor, a high efficiency can be achieved. This supports the development of more compact power electronic systems, which is required to fully exploit the potential of modern GaN devices [5].

This paper presents an optimized bidirectional GaN-HEMT DC/DC converter with a primary voltage of 360 V and a variable secondary voltage between 400 V and 500 V. In Sec. 2, the converter topology and the gate driver design are discussed. Sec. 3 will stress two key aspects. First, the improvements gained by zero voltage turn-on in conjunction with a negative gate-source voltage will be presented in Sec. 3.1 and Sec. 3.2, respectively. Second, an adaptive switching frequency will be discussed, and the loss contribution to the specific elements will be highlighted in Sec. 3.3. Experimental results are presented in Sec. 4, showing a converter efficiency exceeding 99 % over a wide output power range.

2. Converter Topology and Gate Driver

The schematic of the bidirectional DC/DC converter is shown in Fig. 1(a). The converter can be operated in buck and boost mode. Fig. 1(b) shows the schematic of the gate driver. As already mentioned in Sec. 1, fast switching transients are possible with GaN-HEMTs. A negative gate-source voltage spike on $v_{GS,LS}$, however, is often observed during the turn-off of Q_{HS} in buck mode. This is because after the turn-off of Q_{HS}, in the deadtime (t_{DT}) before Q_{LS} is turned on, v_{SW} will decrease rapidly. This is caused by the negative inductor current i_L, which quickly discharges C_{DG} of Q_{LS}, resulting in a high dv/dt of v_{SW}. Since C_{GS} is comparatively small, the major part of the current flows through the gate driver path. The parasitic inductance of the gate driver path and the gate resistor R_g lead to the negative voltage spike. Measurements of v_{GS} with a reduced secondary voltage V_2 are shown in Fig. 2(a). It is important to note that the negative v_{GS} spike can lead to the destruction of the transistor Q_{LS}. Note that the Schottky diode D_g, which ensures a fast turn-off of the HEMT, is reverse-biased and does not impact the negative voltage spike.

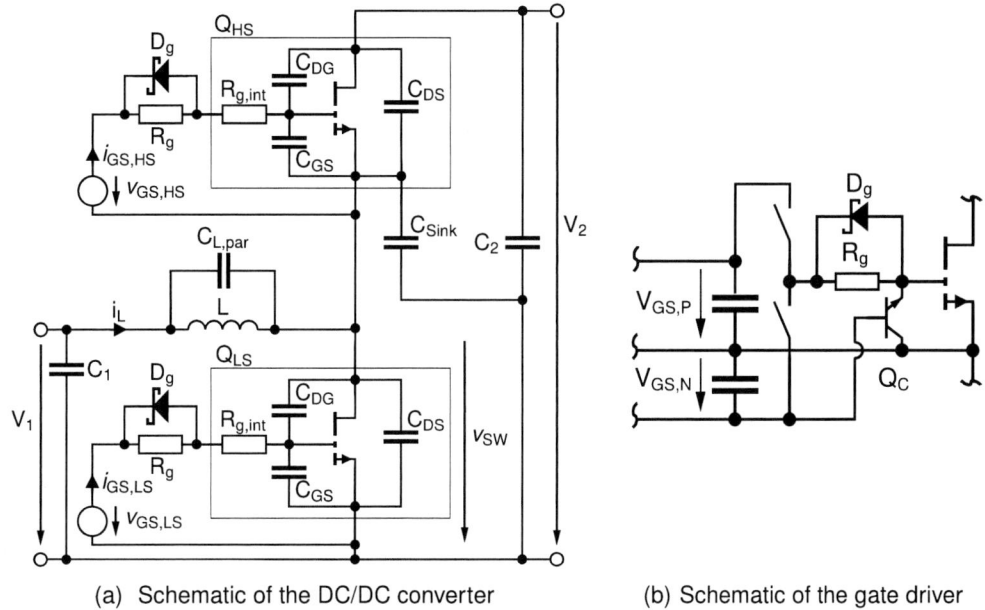

(a) Schematic of the DC/DC converter   (b) Schematic of the gate driver

Fig. 1: (a) shows the simplified schematic of the DC/DC converter. The gate drivers, allowing positive and negative gate voltages, are presented in more detail in (b).

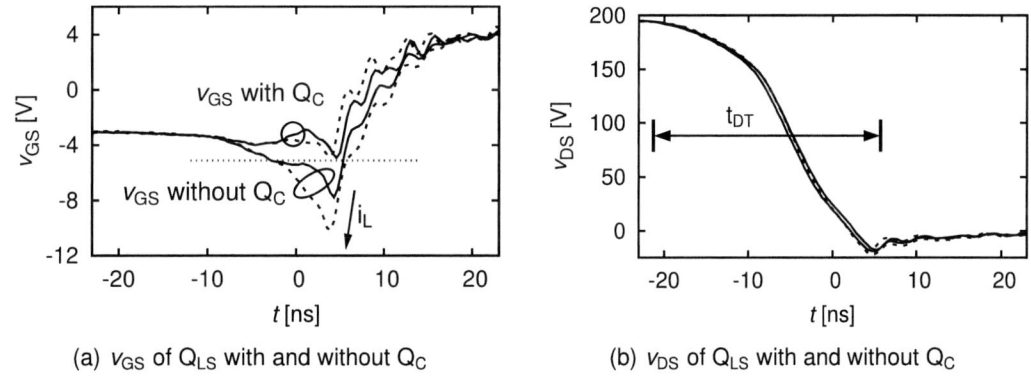

(a) v_{GS} of Q_{LS} with and without Q_C (b) v_{DS} of Q_{LS} with and without Q_C

Fig. 2: Measurements of v_{GS} (a) and v_{DS} (b) of Q_{LS} after the turn-off of Q_{HS} in buck mode for two different inductor currents i_L. During t_{DT}, a negative v_{GS} spike of Q_{LS} occurs, which can be reduced by adding a bipolar transistor Q_C to acceptable values of $v_{GS} > -5\,V$, denoted by the horizontal dotted line in (a).

To ensure that v_{GS} remains in its acceptable range, an NPN bipolar transistor Q_C was introduced. Q_C is only active if v_{GS} of Q_{LS} is below $V_{GS,N}$ minus the forward v_{BE} of the NPN transistor. The current through C_{DG} can now flow from gate to source via Q_C, hence the voltage spike is limited. Q_C has no impact to the switching speed of v_{DS} as shown in Fig. 2(b). It has to be noted that this v_{GS} spike can also be observed in boost mode. Then, v_{GS} of Q_{HS} shows a voltage spike similar to Fig. 2(a).

© VDE VERLAG GMBH · Berlin · Offenbach

3. Converter Optimization

3.1. Zero-Voltage Turn-On

A zero voltage turn-on of Q_{LS} in boost mode can be achieved by allowing a negative inductor current i_L. By turning off Q_{HS}, a negative i_L decreases the voltage at the switch node v_{SW} quickly by discharging the parasitic capacitance (C_{Par}). As soon as v_{SW} reaches zero, Q_{LS} can be turned on without significant losses. It is important to note that C_{Par} does not only comprise the output capacitance of Q_{LS} and Q_{HS}, cf. Fig. 1(a), but also the parasitic capacitance of the inductor L ($C_{L,par}$). Moreover, the capacitance between the switch node and the heat sink (C_{Sink}) has to be taken into account.

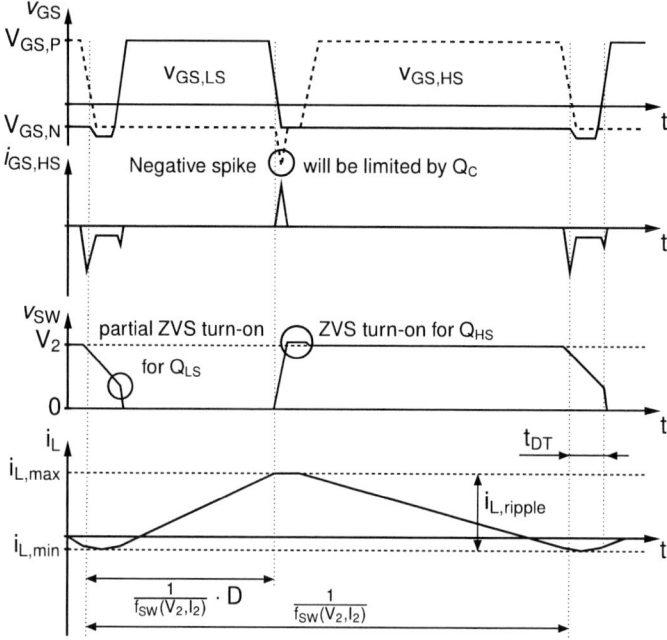

Fig. 3: The timing diagram of the converter in boost mode.

Zero-voltage turn-on of Q_{HS} can be obtained in a similar way. $i_{L,max}$ has to charge C_{Par}, so that v_{SW} will reach V_2. v_{DS} of Q_{HS} is now zero. With a higher value of $i_{L,max}$, C_{Par} will be charged much faster, respectively dv/dt is higher. This can result in a negative voltage spike of $v_{GS,HS}$ due to the discharging of C_{DG} of Q_{HS}, similar to the description in Sec. 2.

In terms of system efficiency, it can be attractive to allow a partial zero-voltage turn-on, so that a smaller value of $i_{L,min}$ can be chosen, if possible with an increased f_{SW}. Then, the losses in the inductor can be decreased as highlighted in Sec. 3.3.

3.2. Negative Gate-Source Voltage

Contrary to the turn-on, the turn-off of Q_{LS} and Q_{HS} has to be done in hard-switching condition. In boost mode, Q_{LS} has to be turned off with the maximum inductor current as drain current and $v_{DS} = V_2$, respectively. In buck mode, Q_{HS} has to operate with similar turn-off conditions. Since turn-on losses could be dramatically reduced by zero-voltage turn-on, the turn-off losses become a dominant fraction of the total switching losses.

As already mentioned in Sec. 1, [4] shows that SiC MOSFETs can be turned off with relative low losses. This can be explained by considering the channel current and the capacitance charging current separately. Only the turn-off of the ohmic channel leads to significant losses. Contrary to this, charging the output capacitance does not cause power dissipation.

A further decrease in turn-off switching losses was possible by using a negative gate supply. With this, the turn-off loss ($e_{\text{turn-off}}$) could be decreased from $11\,\mu J$ at $V_{GS,N} = 0\,V$ to $4\,\mu J$ at $V_{GS,N} = -1.5\,V$ (see Fig. 4). The decrease of the turn-off loss saturates at $V_{GS,N} = -1.5\,V$. Hence, it is reasonable to operate with a moderate $V_{GS,N}$ of $-3\,V$ to prevent high reverse conduction losses, which would occur for very large negative v_{GS}.

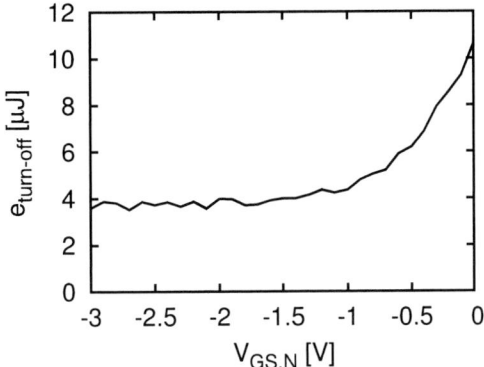

Fig. 4: Simulated turn-off energy as a function of $V_{GS,N}$. Measurements showing the corresponding increase of converter efficiency will be presented later in Fig. 6(a).

3.3. Loss Contribution

Fig. 5(a) and Fig. 5(b) compare the converter losses as determined from an accurate circuit simulation for an adaptive switching frequency, and a fixed switching frequency in boost mode with $V_2 = 500\,V$.

Adaptive Switching Frequency

To obtain the highest efficiency throughout the whole power range, the switching frequency f_{SW} has to be decreased continuously with rising power, ranging from $200\,kHz$ for smaller power levels to $120\,kHz$ at $3\,kW$. The contribution of the conduction losses of the inductor ($P_{L,cond}$) and of the core losses ($P_{L,core}$) is highest in the low power regime. It can be reduced by an increased f_{SW}, so that the current ripple of i_L is smaller. Q_{LS} is turned on with partial zero-voltage turn-on, so that the switching frequency can be set to a higher value to reduce the losses in the inductor. This reduces the total system losses, as already mentioned in Sec. 3.1.

Fixed Switching Frequency

With a constant switching frequency of $140\,kHz$ and a power below $1.5\,kW$, a zero-voltage turn-on for Q_{LS} was obtained, as described in Sec. 3.1. However, above $1.5\,kW$ $v_{DS,LS}$ cannot be discharged to zero anymore. Q_{LS} is now turned on with reduced voltage. A hard turn-on of Q_{LS} results for P_{out} above $3\,kW$. Furthermore, due to the higher DC bias of the inductor, the core losses increase.

© VDE VERLAG GMBH · Berlin · Offenbach

PCIM Europe 2016, 10 – 12 May 2016, Nuremberg, Germany

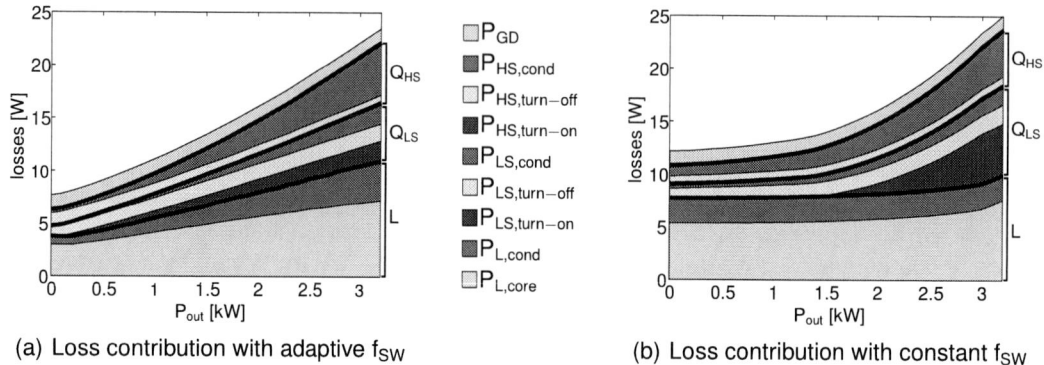

(a) Loss contribution with adaptive f_{SW}

(b) Loss contribution with constant f_{SW}

Fig. 5: Loss contribution in boost mode for (a) an adaptive switching frequency f_{SW} and (b) for a constant f_{SW} of 140 kHz.

The losses of Q_{HS} are mainly due to the conduction losses, see $P_{HS,cond}$ in Fig. 5. The switching losses $P_{HS,turn-on}$ and $P_{HS,turn-off}$ have only a minor impact. The small reverse conduction losses, due to $V_{GS,N}$ and t_{DT}, are included in the turn-on losses. The gate driver and microcontroller losses (P_{GD}) depend only slightly on f_{SW}.

4. Converter Efficiency Measurements

Fig. 6(a) shows the efficiency of the converter for different secondary voltages V_2. The negative gate-source voltage allows a reduction of the losses up to 10 %, which is shown in Fig. 6(a) with dotted lines. For a further increase of the efficiency, the switching frequency was adjusted depending on the secondary voltage and the load current. Fig. 6(b) shows the efficiency for two different (fixed) switching frequencies and for an adaptive switching frequency. The latter approach clearly exhibits better results because the loss contribution of the inductor and of the GaN-HEMTs are frequency-dependent, see Sec. 3.3.

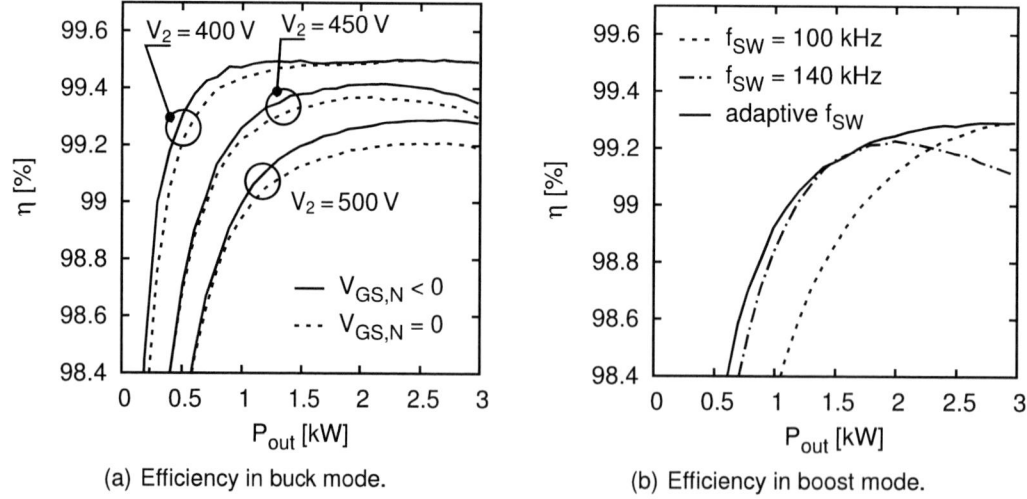

(a) Efficiency in buck mode.

(b) Efficiency in boost mode.

Fig. 6: (a) shows the efficiency of the converter in buck mode with and without negative gate-source voltage (solid and dotted lines, respectively). The converter efficiency in boost mode is shown in (b) for $V_1 = 360$ V and $V_2 = 500$ V for different fixed or adaptive switch frequencies.

© VDE VERLAG GMBH · Berlin · Offenbach

5. Conclusions

The presented bidirectional GaN-HEMT DC/DC converter shows an excellent performance, reaching a very high peak efficiency of 99.3% for an output power of 2.5 kW at a primary voltage of 360 V and a secondary voltage of 500 V. This was achieved by applying a negative gate-source voltage (even for HEMTs which are normally-off), in conjunction with an adaptive switching frequency. The loss contributions of the operation for constant and adaptive switching frequencies were compared, and the reasons for the increase in efficiency were highlighted. These results can be helpful for the development of highly-optimized systems.

6. References

[1] L. Hoffmann, C. Gautier, S. Lefebvre, and F. Costa. Optimization of the driver of GaN power transistors through measurement of their thermal behavior. *IEEE Transactions on Power Electronics*, 29(5):2359–2366, May 2014.

[2] Michael Meisser, Max Schmenger, and Thomas Blank. Parasitics in power electronic modules: How parasitic inductance influences switching and how it can be minimized. In *Proceedings of PCIM Europe 2015*, pages 1–8, May 2015.

[3] R. Mitova, R. Ghosh, U. Mhaskar, D. Klikic, Miao-Xin Wang, and A. Dentella. Investigations of 600-V GaN HEMT and GaN diode for power converter applications. *IEEE Transactions on Power Electronics*, 29(5):2441–2452, May 2014.

[4] Xuan Li, Liqi Zhang, Suxuan Guo, Yang Lei, A. Q. Huang, and Bo Zhang. Understanding switching losses in SiC MOSFET: toward lossless switching. In *IEEE 3rd Workshop on Wide Bandgap Power Devices and Applications (WiPDA)*, pages 257–262, Nov 2015.

[5] J. Popovic, J. A. Ferreira, J. D. van Wyk, and F. Pansier. System integration of GaN converters paradigm shift challenges and opportunities. In *8th International Conference on Integrated Power Systems (CIPS) 2014*, pages 1–8, Feb 2014.

PCIM Europe 2016, 10 – 12 May 2016, Nuremberg, Germany

Direct Delta Sigma Signal Processing for Control of Power Electronics

Michael Homann, Technische Universität Braunschweig, Germany, homann@ifr.ing.tu-bs.de
Axel Klein, Technische Universität Braunschweig, Germany, klein@ifr.ing.tu-bs.de
Walter Schumacher, Technische Universität Braunschweig, Germany, w.schumacher@tu-bs.de

The Power Point Presentation will be available after the conference.

Abstract

A new approach for the signal processing in control loops is the direct signal processing of a delta sigma ($\Delta\Sigma$) bit stream. The advantages are a high bandwidth in the control loop and lower timing constraints for measurement synchronization. In this paper, the main focus is on the performance improvement of delta sigma signal processing (DSSP) structures. Different architectures and modulator orders are analyzed regarding signal to noise and distortion ratio (SNDR) and dynamic performance of a single DSSP operation. Furthermore, the SNDR decline in a cascade of DSSP operation is examined and the step response of a phase current control loop of a permanent magnet synchronous machine using DSSP is shown.

1. Introduction

Mechatronic systems are usually controlled by discrete-time signal processing. The basic elements of a discrete control system are the analog digital converter (ADC), the control law processing the impulse series and the digital analog converter (DAC) or pulse-width modulation (PWM). Synchronous logic ensures the real-time execution of the control law and the correct timing of the ADC sampling with respect to the PWM frequency. Sampling and processing time of discrete-time control loops limit the small signal bandwidth [1]. Furthermore, a discrete-time control has to handle different clock domains. Set- and actual values are frequently transmitted by digital bus systems. The different clock rates of bus systems and digital control loops require a synchronization of the data words. The analog to digital conversion of values like phase current or position is increasingly carried out by oversampling ADCs in order to increase the effective number of bits. The low pass filter and decimation used in these ADCs also require a synchronization of the data words. The bandwidth and disturbance rejection of the phase current control of electrical drives with $\Delta\Sigma$ ADCs, an oversampling ADC, strongly depends on the correct timing of the control loop [2].

DSSP is a promising technique to overcome these problems. It combines a discrete-time digital control loop with the advantages of an analog control loop. This enables a hybrid signal processing with a high closed loop bandwidth since sampling effects can be neglected due to the extraordinarily high sampling rate. As a consequence, the control loop is compatible with different clock rates of bus systems or can even operate asynchronously with respect to bus cycle times. First DSSP methods emerged in the 80s [3], whereas DSSP in control loops [4, 5] followed roughly 20 years later. The fundamental concept is to use the high frequent bitstreams of $\Delta\Sigma$ modulators in control loops to mitigate the sampling effects to achieve a high bandwidth control system.

© VDE VERLAG GMBH · Berlin · Offenbach

The internal feedback of a $\Delta\Sigma$ modulator (DSM) in figure 1 creates a high frequent bitstream whose average value tracks the input signal. For simplicity, a first order $\Delta\Sigma$ modulator is shown. Usually, digital low pass filters are used following the modulator to attenuate the quantization noise and achieve a greater resolution in exchange for a lower sampling rate and bandwidth.

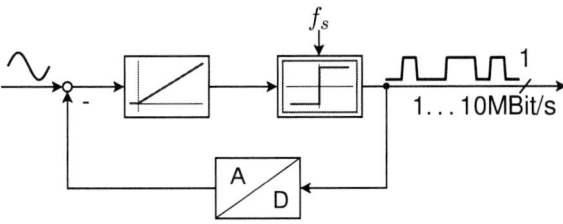

Fig. 1: Basic principle of a first order $\Delta\Sigma$ modulator

In DSSP control loops, the digital low pass filter is omitted since the bitstream already is a digital conversion of the analog signal, which features high information content. The DSSP operations directly modify the high frequent bitstream. The output of an ideal operation is the same as the output of a $\Delta\Sigma$ modulator with a modified analog input. All basic mathematical operations necessary for linear control loops like summation, scaling and integration can be implemented to operate directly on $\Delta\Sigma$ bitstreams. Linear systems can be designed using these single building blocks. $\Delta\Sigma$ modulators are used within this context in two ways. External $\Delta\Sigma$ modulators convert the analog or digital signals of set and actual values to bitstreams as depicted in figure 2. The DSSP operation is realized by an appropriate combination of multi-bit arithmetic followed by an internal $\Delta\Sigma$ modulator.

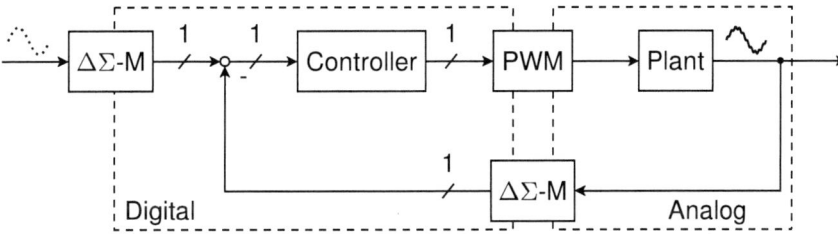

Fig. 2: Delta sigma signal processing control loop

All operations in the controller are performed in the bitstream domain resulting in a high frequent 1-bit input and output. The actuating variable of the controller can be directly applied to the plant in case of high frequent power electronic switches like gallium nitride (GaN) [6]. For power electronics switching in the 4 - 40 kHz range, hysteresis based PWM controllers [7, 8] are used to reduce the mean switching frequency.

2. Delta Sigma Signal Processing Operations

A comparison of different DSSP methods shows that the quanta decoding technique offers the best spectral performance [9]. It consists of a bitstream decoder using a local reference clock and a following internal $\Delta\Sigma$ modulator [5, 10]. In this paper, this method is called quanta decoding. Although the original concept [10] uses only first order internal $\Delta\Sigma$ modulator, it can be easily extended to higher order $\Delta\Sigma$ modulators. This enables the superior spectral performance compared to methods like delta adders [3], which are restricted to first order internal

$\Delta\Sigma$ modulators. The scaling element in figure 3 is used to demonstrate the principle of quanta decoding.

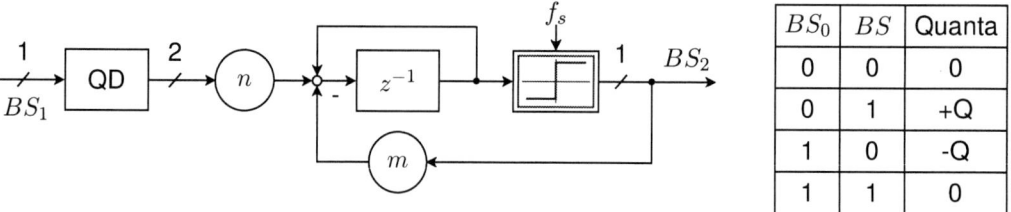

BS_0	BS	Quanta
0	0	0
0	1	+Q
1	0	-Q
1	1	0

Fig. 3: DSSP scaling element with quanta decoding

A $\Delta\Sigma$-modulator with zero input produces a bitstream that consists of alternating ones and zeros. This reference bitstream, also called zero bitstream BS_0, is used to decode the input bitstream BS into ternary signals according to the truth table in figure 3. The ratio of the quanta gain n and the value of the feedback gain m of the internal $\Delta\Sigma$ modulator determine the scaling gain V.

$$V = \frac{n}{2m} \tag{1}$$

The operation summation uses a second quanta decoder and adds the quanta before the internal $\Delta\Sigma$-modulator. A gain n for each quanta decoder allows a weighted summation. A discrete-time integrator after the quanta decoding enables a DSSP integration with the time constant T_i.

$$T_i = T_s \cdot \frac{n}{2m}, \quad T_s = \frac{1}{f_s} \tag{2}$$

3. Higher Order Internal Delta Sigma Modulator

First order $\Delta\Sigma$ modulators are used for DSSP since they are always stable as long as the input signal is in the valid input range defined by the feedback gain m [11]. The simple structure allows an efficient implementation with the feedback gain m as only parameter. Higher order $\Delta\Sigma$ modulators offer superior spectral performance and are less prone to idle tones [11]. Their implementation requires more resources due to additional integrators and parameters. The design of these parameters is a trade-off between the spectral performance, the stability and dynamic response of the $\Delta\Sigma$ modulator. For a linear analysis, the quantizer is replaced by a linear model, which consists of an additive white noise source and an effective quantizer gain [11]. An optimization of the maximum signal-to-noise ratio (SNR) for the $\Delta\Sigma$ modulator with a minimum number of independent parameters, abbreviated as (min. par.), leads to the parameters in [12]. This architecture is shown in figure 4. The Matlab $\Delta\Sigma$ toolbox [13] allows to design several $\Delta\Sigma$ modulator architectures based on the linear quantizer model. The cascade of integrators with distributed feedback and distributed coupling (CIFB) and its extension (dashed line) by resonators (CRFB) as shown in figure 5 are used in this paper. The CRFB architecture allows to optimize the zeros of the noise transfer function to achieve a better SNR. Other architectures like chain of integrators with weighted feedforward summation (CIFF) and its extension by resonators (CRFF) [11] are not considered as they are other implementations of the same signal and noise transfer function.

PCIM Europe 2016, 10 – 12 May 2016, Nuremberg, Germany

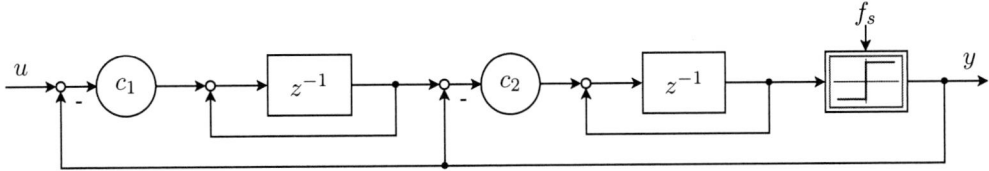

Fig. 4: Topology for second order $\Delta\Sigma$ modulator with minimum number of independent parameters

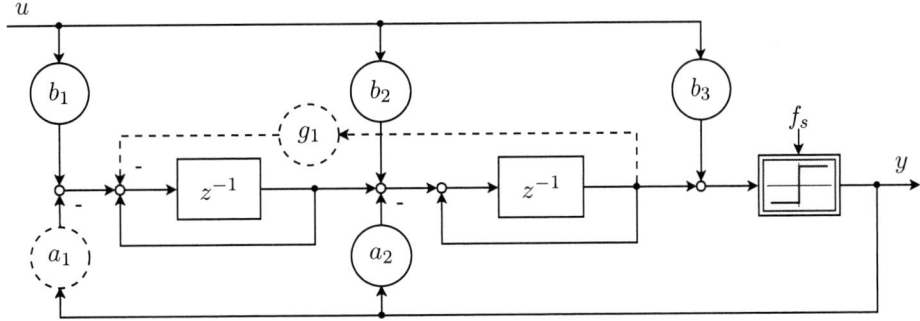

Fig. 5: Topology for second order $\Delta\Sigma$ modulator, toolbox, CIFB, CRFB - dashed line

3.1. Spectral and Dynamic Performance

Different higher order $\Delta\Sigma$ modulators are compared regarding spectral and dynamic performance including a first order modulator as a reference. For the spectral performance, several $\Delta\Sigma$ modulators are excited by a sine wave. A fast Fourier transform (FFT) with Hann windowing is applied to the bitstreams. The spectral analysis according to [14] results in the signal-to-noise and distortion ratio (SNDR) shown in figure 6.

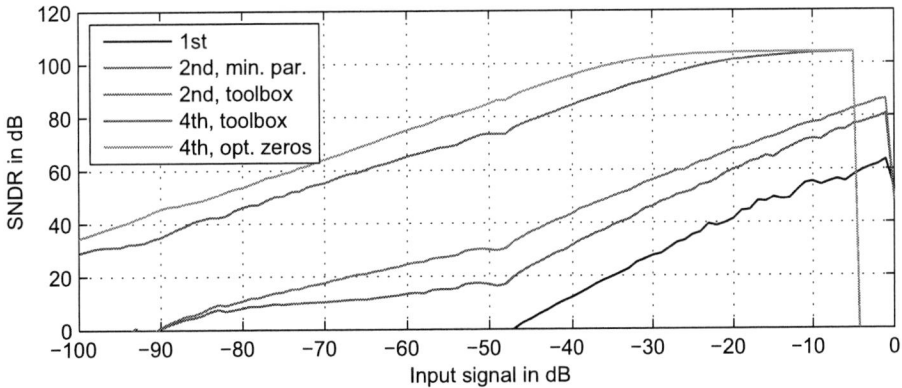

Fig. 6: SNDR for different $\Delta\Sigma$ modulators

For input signals close to the maximum input range, the SNDR of higher order $\Delta\Sigma$ modulators drops significantly. However, a first order $\Delta\Sigma$ modulator does not suffer from this SNDR drop, but has a worse performance regarding maximum SNDR and smallest input signal that leads to a positive SNDR. A second order $\Delta\Sigma$ modulator with a minimum number of independent parameters shows worse performance than the $\Delta\Sigma$ toolbox second order $\Delta\Sigma$ modulator. The

© VDE VERLAG GMBH · Berlin · Offenbach

219

CRFB architecture with optimized zeros features better SNDR performance than the design without optimization.

Step responses of the nonlinear $\Delta\Sigma$ modulators are used to characterize the dynamic performance. The same digital low pass filter is used for the bitstream of the $\Delta\Sigma$ modulator and the step response. The delay between a filtered ideal step response and the filtered $\Delta\Sigma$ modulator step response is evaluated at 90% of the final value. Table 1 shows the delay for first and second order $\Delta\Sigma$ modulators. A special case in the design of $\Delta\Sigma$ modulators with the $\Delta\Sigma$ toolbox is to use all b_i coefficients to achieve an ideal signal transfer function of 1 by generating zeros in the signal transfer function. Thus in this paper, this design is called additional zeros (AZ) in contrast to the design in which $b_1 = 1$ and the other b_i coefficients are zero resulting in a signal transfer function without zeros.

$\Delta\Sigma$ modulator	1st	2nd, min. par.	2nd, CIFB	2nd, CIFB-AZ
Delay in T_s	1	8	3	0

Tab. 1: Delay of different $\Delta\Sigma$ modulators

The second order $\Delta\Sigma$ modulator with a minimum number of independent parameters has the greatest delay, whereas the CIFB architecture with additional zeros has a delay smaller than one sampling time due to the ideal signal transfer function. With regard to the high sampling rate, the delay of a single $\Delta\Sigma$ modulator is negligible. In a cascade of operations, the overall delay has to be evaluated.

A basic concept in DSSP is to change the feedback gain m to achieve operations like scaling. The linear model of a CIFB/CRFB $\Delta\Sigma$ modulator predicts that the transient behavior of the step response depends on the feedback gain m since it influences the feedback coefficients a_i. However, the results of step responses of a second order nonlinear $\Delta\Sigma$ modulator show a different behavior. In simulation, the gain n is constant and the feedback gain m is varied so that a scaling factor of -20 dB to +20 dB is achieved. An analysis of the delay as described above shows that the delay is constant for all the scaling factors analyzed.

The difference between the linear and nonlinear model can be explained by a block diagram rearrangement. In figure 7, the feedback gain m of a scaling operations is shifted. As a consequence, the gain m scales the b_i coefficients, which leads to a different input range. The gain m also scales the last integrator before the quantizer. This only changes the number range of the integrators of the loop since the 1-bit quantizer detects the sign of the last integrator [11]. As the simulation of the step responses show, the variation of the feedback gain m has no influence on the delay of the step responses.

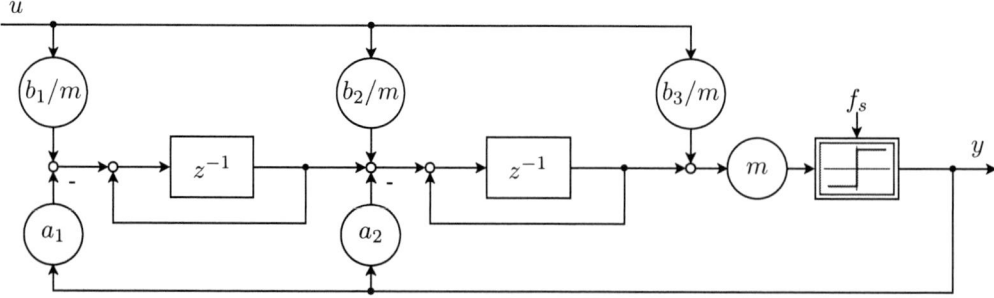

Fig. 7: Rearranged block diagram of a scaling operation

3.2. Quanta Decoding

The quanta decoding in DSSP [5, 10] requires a zero bitstream. For a first order $\Delta\Sigma$ modulator, this bitstream consists of alternating ones and zeros. A histogram of the distribution of groups of ones and zeros inside the bitstream is used to analyze the zero bitstream of higher order $\Delta\Sigma$ modulators up to an order of four. As figure 8 and further simulations show, the maximum number of ones and zeros in a group is two.

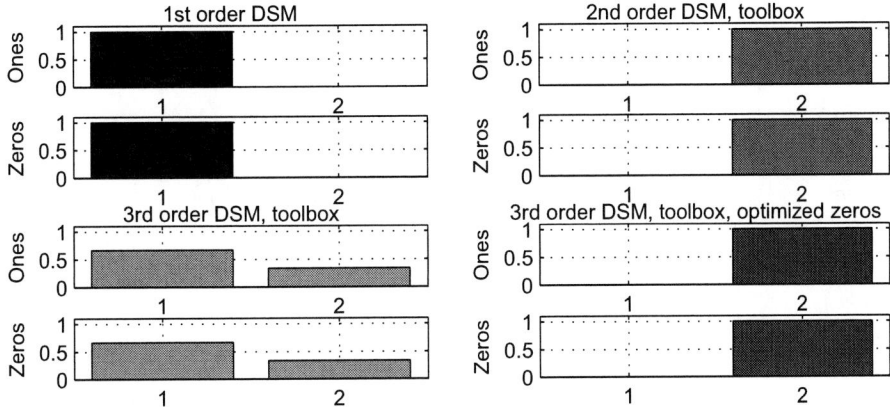

Fig. 8: Histogram of one and zero groups of zero bitstreams

The distribution changes significantly with different architectures. However, the average value of the zero bitstream is still zero. Therefore, a zero bitstream with alternating ones and zero is sufficient for quanta decoding from an average point of view.

3.3. Cascade of Operations

The implementation of a control law usually requires several of the basic operations scaling, summation and integration. Therefore, the SNDR decline in a cascade of operations is analyzed in simulation. A sine wave excites a cascade of DSSP operations with a scaling factor of one. Figure 9 shows the SNDR for various $\Delta\Sigma$ modulator architectures. Except for the toolbox design with additional zeros, the SNDR decreases for a increasing number of DSSP operations. After ten DSSP operations, the lowest SNDR still yields an effective number of bits of approximately 8.5. An increase of the order of the $\Delta\Sigma$ modulator significantly improves the SNDR. The toolbox design with additional zeros is a special case, since the SNDR is constant. Due to a signal transfer function of 1, the input signal is not altered and each $\Delta\Sigma$ modulator creates the same quantization noise. The bitstreams of the $\Delta\Sigma$ modulator with additional zeros are therefore identical. A simulation with a 1st order lag element as a DSSP operation leads to different bitstreams, but the SNDR is basically constant with a difference between maximum and minimum value of less than 1dB.

4. High Bandwidth Phase Current Control Loop of a PMSM

Experimental results show that a high bandwidth phase current control loop of a permanent magnet synchronous machine (PMSM) using DSSP and hysteresis based PWM controllers [8] can be achieved. For the DSSP operations, a second order $\Delta\Sigma$ modulator with minimum

PCIM Europe 2016, 10 – 12 May 2016, Nuremberg, Germany

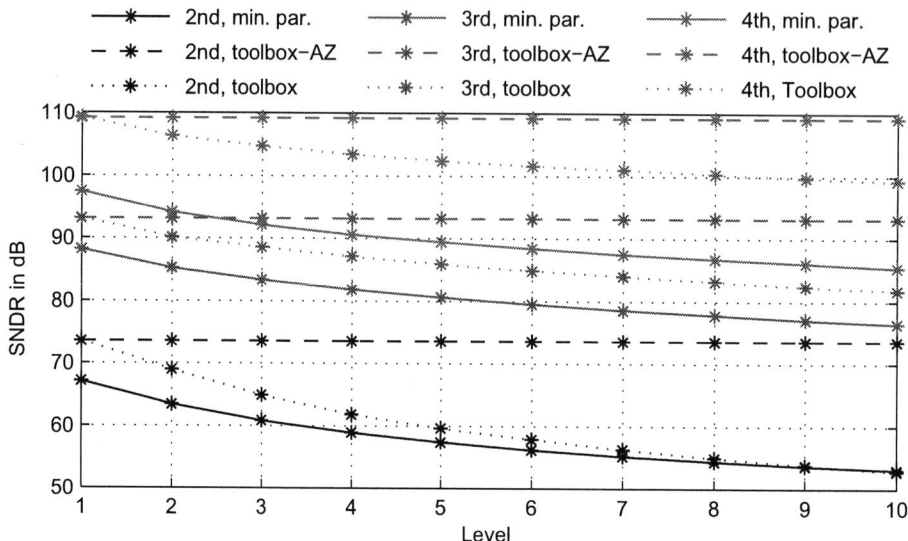

Fig. 9: Cascade of DSSP operations

independent parameters is used due to the efficient implementation of its binary coefficients. A low pass filter for both, set and actual values has to be introduced to evaluate the performance of the bitstream control. The PWM controller switches with an average frequency of 16 kHz. Therefore, a digital filter which averages over a $62.5\,\mu s$ period is used. The step response of the closed phase current loop is shown in figure 10. An averaging over 68 step responses suppresses noise. The slope of the step is finite due to the low pass filter. The step response shows no significant delay compared to the step which demonstrates a high bandwidth control with a settling time smaller than $62.5\,\mu s$.

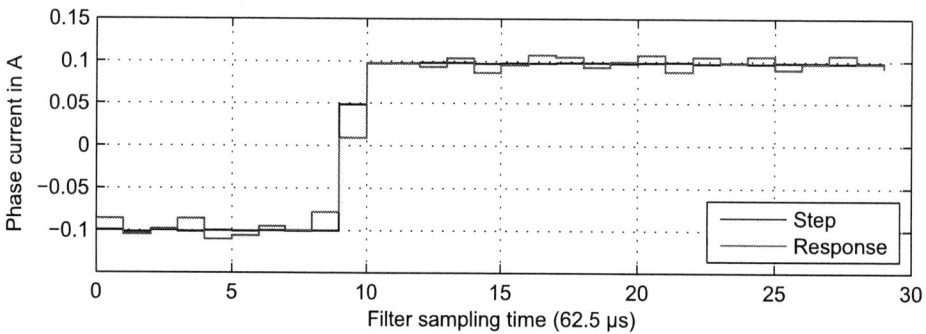

Fig. 10: Step resonse of DSSP phase current control loop (avagered over 68 step responses)

5. Conclusion

This paper analyzes the performance improvement of higher order $\Delta\Sigma$ modulators in DSSP operations. Several architectures of a single $\Delta\Sigma$ modulator have been studied regarding spectral and dynamic performance. An increase in modulator order leads to a better maximum SNDR. The dynamic performance, evaluated as the delay of the step response of the nonlinear model,

© VDE VERLAG GMBH · Berlin · Offenbach

strongly depends on the modulator architecture. For a scaling DSSP operation, it is shown that the dynamic performance is independent of the modulator feedback gain. The SNDR decline in a cascade of scaling DSSP operations depends on the order of the $\Delta\Sigma$ modulator. An example of the use of DSSP operations in the phase current control of a PMSM is given. The combination of the high frequent DSSP and a high bandwidth hysteresis based PWM controller lead to a highly dynamic phase current control loop.

6. References

[1] C. Gröling. Optimierungspotenzial bei Servoumrichtern für permanenterregte Synchronmaschinen (in German). Dissertation, TU Braunschweig, 2009.

[2] M. Homann, T. Noeßelt, and W. Schumacher. Aspekte der Strommessung in Drehfeldmaschinen mit Delta Sigma Umsetzern (in German). In *SPS/IPC/Drives*, pages 429–438, 2013.

[3] N. Kouvaras. Operations on delta-modulated signals and their application in the realization of digital filters. *Radio and Electronic Engineer*, 48(9):431–438, 1978.

[4] M. K. Kurosawa, M. Kawakami, K. Tojo, T. Katagiri, and T. Higuchi. Single-bit digital signal processing for current control of brushless dc motor. In *Industrial Electronics. ISIE. Proceedings of the IEEE International Symposium on*, volume 2, pages 589–594, 2002.

[5] N. Patel, G. Coghill, and Kiong Nguang Sing. Digital realization of analogue computing elements using bit streams. In *System-on-Chip for Real-Time Applications, Proceedings. The 3rd IEEE International Workshop on*, pages 76–80, 2003.

[6] Texas Instruments. Gan technology preview, lmg5200 80-v, gan half-bridge power stage, 2015. http://www.ti.com/lit/ds/symlink/lmg5200.pdf.

[7] J. B. Bradshaw. Bit-stream control of doubly fed induction generators. Dissertation, University of Auckland, 2012.

[8] M. Homann and W. Schumacher. Stromrichter und Computerprogramm. DE 102014108667 A1. Filing date: 8, Jun. 20, 2014., 2015.

[9] M. Homann, A. Klein, R. Kirchner, and W. Schumacher. Quasi-kontinuierliche Signalverarbeitung mit Delta Sigma Bitströmen in der Antriebstechnik - Ein Überblick (in German). In *Fortschritte in der Antriebs- und Automatisierungstechnik*, 2016.

[10] N. Patel. Bit-streams: Applications in control. Dissertation, University of Auckland, 2006.

[11] R. Schreier and G. C. Temes. *Understanding delta-sigma data converters*. IEEE Press, Piscataway, NJ, 2005.

[12] A. Marques, V. Peluso, M. S. Steyaert, and W. M. Sansen. Optimal parameters for delta digma modulator topologies. *Circuits and Systems II: Analog and Digital Signal Processing, IEEE Transactions on*, 45(9):1232–1241, 1998.

[13] R. Schreier. Delta sigma toolbox, 2000. http://www.mathworks.com/matlabcentral/fileexchange/19-delta-sigma-toolbox.

[14] M. Ortmanns and F. Gerfers. *Continuous-time sigma-delta AD conversion: Fundamentals, performance limits and robust implementations*, volume 21 of *Springer series in advanced microelectronics*. Springer, Berlin, 2006.

© VDE VERLAG GMBH · Berlin · Offenbach

Finite Control Set Model Based Predictive Control of a PMSM with Variable Switching Frequency and Torque Ripple Optimization

Fernando Ramirez, Universität Siegen, fernando.ramirez@uni-siegen.de
Prof. Dr.-Ing. Mario Pacas, Universität Siegen, pacas@uni-siegen.de

Abstract

In this paper a model based predictive control of a PMSM (Permanent Magnet Synchronous Machine) with variable frequency PWM is presented. Finite Control Set Model Based Predictive Control (FCS-MPC) with variable switching frequency without modulator has been reported in other research works [1], [2]. Yet the high performance of FCS-MPC can be combined with methods that use conventional modulation techniques with the purpose of optimizing some other variables, in the present case the torque ripple and the total harmonic distortion (THD) of the current. Furthermore, if the torque ripple can be predicted and controlled, a compromise between switching losses and maximum torque ripple can be found. This implementation can be realized by means of a FPGA, which combines very fast times of execution, parallelism and the pipelining of algorithms.

1. Introduction

Model Based Predictive Control (MPC) is today a well-established control technique and offers many advantages; particularly it can be easily applied to multivariable systems and non-linearities can be considered in a simple way [1]. The MPC techniques applied to power electronics have been classified in two main categories: Continuous Control Set (CCS) and Finite Control Set (FCS)-MPC. Because for power electronics converters, the number of switching states is in fact limited to a finite set, the FCS-MPC approach can be easily implemented. Usually considered as an advantage, FCS-MPC avoids using the modulation stage [3]. However, this approach needs a high sampling frequency to achieve a good performance, yet by using a modulator the switching frequency can be reduced while keeping the merits of MPC. Recently the advantages of working with variable switching frequency modulator-based control schemes had been presented in [4] and [5]. Nonetheless, the largest drawback of MPC is the large computational burden, especially for online implementations where calculations have to be done in real-time. Field Programmable Gate Arrays (FPGAs) contain millions of elementary cells and interconnections that are fully programmable by the final-user, allowing them to build tailor-made hardware architectures. As shown in the following, FPGAs have the advantage of parallelism and pipelining; therefore, in power electronics systems they allow the computation of complex or numerous algorithms that are needed in MPC schemes, while having very short times of execution.

2. Control Strategy

The drive to be examined and controlled consists of a three-phase Permanent Magnet Synchronous Machine (PMSM) fed by a three-phase two level inverter. The dynamic behavior of the PMSM is described by the well-known voltage equations, all referred to the d-q reference frame. More about the FPGA platform, were the control is implemented, can be found in [4] and [5]. The control structure shown in Fig 1 follows the same scheme of conventional Field Oriented Control (FOC). Current space phasors are obtained by measuring the phase currents and converted to the rotor reference frame. A super-imposed velocity controller provides the reference for i_q. The switching commands for the inverter are obtained by using an online adaptable SVM, capable of changing the sampling time on each period.

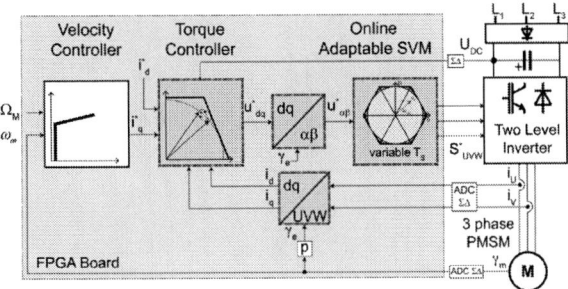

Fig 1.   Implemented Control Strategy

The proposed control scheme as shown in Fig 2 is based on a cost function that minimizes the electromagnetic torque error by minimizing the errors of i_d and i_q in the next sampling state. In addition, the cost function is extended in a way that the torque ripple is optimized by applying the most appropriate voltage to the machine. Moreover, the switching frequency is included in the calculation with a penalization. The objective is to maintain the quality of the control system while reducing the switching losses by decreasing the switching frequency.

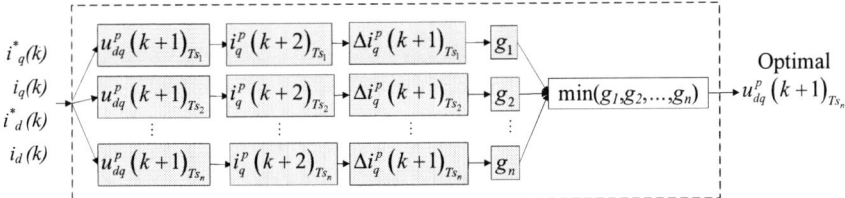

Fig 2.   Predictive Torque Control Scheme: Optimization of different switching periods T_{Sn}

The predicted voltage space phasors $u^p_{dq}(k+1)$ for different switching periods T_s are calculated by means of (1) and (2), then the current space phasors at the beginning of the next period $i^p_{dq}(k+2)$ are calculated by applying the forward Euler integration method to the voltage model as expressed in (3) and (4). In the next step, the Δi^p_q current ripples are calculated, thus the torque ripple can be included in the cost function. Finally, a cost function g that will be explained in the next sections is minimized. It considers the torque ripple, the frequency of operation and a penalization of the switching frequency.

$$u_d(k+1) = Ri_d(k) + \frac{L}{T_s}\left[i_d^* - i_d(k)\right] - \dot{\gamma}_e L i_q(k) \tag{1}$$

$$u_q(k+1) = Ri_q(k) + \frac{L}{T_s}\left[i_q^* - i_q(k)\right] + \dot{\gamma}_e \left[L i_d(k) + \psi_{d0}\right] \tag{2}$$

$$i_d^p(k+2) = i_d(k+1) + \frac{T_s}{L}\left[-Ri_d(k+1) + \dot{\gamma}_e L i_q(k+1) + u_d^p(k+1)\right] \tag{3}$$

$$i_q^p(k+2) = i_q(k+1) + \frac{T_s}{L}\left[-Ri_q(k+1) - \dot{\gamma}_e \left[L i_d(k+1) + \psi_{d0}\right] + u_q^p(k+1)\right] \tag{4}$$

2.1.   Current Ripple Prediction

For the proposed control scheme the well-known Space Vector Modulation (SVM) technique [7] was implemented. Conventionally each transistor on each leg will switch two times every switching period of the inverter. As it can be appreciated in Fig 3, the expected behavior of the currents can be derived from the modulator switching signals. Then as explained in further details in [5] and [8] the current deviation for each vector can be calculated by (5) and the current ripple prediction Δi^P_q is then calculated by (8).

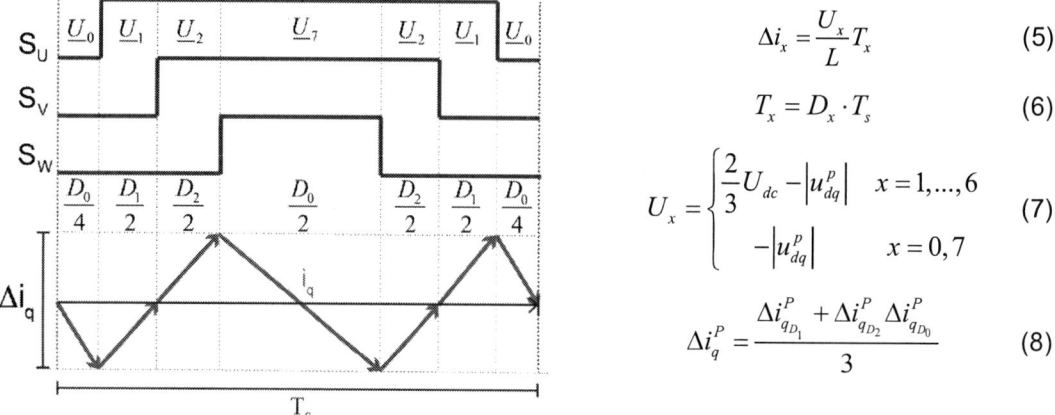

$$\Delta i_x = \frac{U_x}{L} T_x \tag{5}$$

$$T_x = D_x \cdot T_s \tag{6}$$

$$U_x = \begin{cases} \dfrac{2}{3} U_{dc} - \left| u^p_{dq} \right| & x = 1,...,6 \\[2mm] -\left| u^p_{dq} \right| & x = 0,7 \end{cases} \tag{7}$$

$$\Delta i^P_q = \frac{\Delta i^P_{q_{D_1}} + \Delta i^P_{q_{D_2}} \Delta i^P_{q_{D_0}}}{3} \tag{8}$$

Fig 3. Current ripple estimation: Current behavior derived from the modulator switching commands

2.2. Cost Function

As mentioned above, the optimization process is achieved by evaluating different alternatives i.e. the predicted values for different sampling times and respectively different switching frequencies. In each case, a cost function as defined in (10) is evaluated and minimized. The cost function contains the absolute errors of the currents of i_d and i_q, the weights w_q and w_d which are used for setting the dynamics of the control; the term Δi^P_q weighs the torque ripple, the term f_S is the value of the switching frequency and minimizes the switching losses by assuming that they are proportional to the inverter switching frequency.

The current terms are normalized to the i_N nominal current, as well as the f_S which is normalized to the maximum switching frequency f_{Smax}. The $f_{Lim}(i^P_q)$ and $f_{Lim}(i^P_d)$ terms defined in (11), penalize the absolute error of the i^P_q and i^P_d currents. For certain operation ranges, certain frequencies can be taken out of the minimization process using $f_{Pen}(e_{iq})$ defined in (12), similarly $f_{Pen}(\Delta i^P_q)$ defined in (13), allows some frequencies to be excluded from the minimization process if the torque ripple is larger than the penalization term Δi_{qLim}.

$$g_1 = \frac{w_q \left| i^*_q - i^P_q \right|}{i_N} + \frac{w_d \left| i^*_d - i^P_d \right|}{i_N} + \frac{\Delta i^P_{1_q}}{i_N} + \frac{f_S}{f_{S\max}} + f_{Lim}\left(i^P_q\right) + f_{Lim}\left(i^P_d\right) + f_{Pen}\left(e_{i_q}\right) + f_{Pen}\left(\Delta i^P_q\right) \tag{9}$$

$$f_{Lim}\left(i^P_{dq}\right) = \begin{cases} \infty, & \left| i^*_{dq} - i^P_{dq} \right| > i_{dqMax} \\ 0, & \left| i^*_{dq} - i^P_{dq} \right| \le i_{dqMax} \end{cases} \tag{10} \qquad f_{Pen}\left(e_{i_q}\right) = \begin{cases} \infty, & \left| i^*_q - i_q \right| > e_{i_{qLim}} \\ 0, & \left| i^*_q - i_q \right| \le e_{i_{qLim}} \end{cases} \tag{11} \qquad f_{Pen}\left(\Delta i^P_q\right) = \begin{cases} \infty, & \Delta i^P_q > \Delta i^P_{q_{Lim}} \\ 0, & \Delta i^P_q \le \Delta i^P_{q_{Lim}} \end{cases} \tag{12}$$

2.2.1. Integral Cost Function

One of the main drawbacks of model based control schemes is the sensitivity to inaccuracies in the parameters of the model, which can produce a steady state error. There are several techniques that can be used to eliminate this steady state error, given that the MPC lacks an integral part. As explained in [6], the use of a trajectory cost function that takes into account the history of the system by integrating the past errors can be used.

$$g_{i_{dq}}^{trajectory} = \frac{w_d}{i_N} \int_{t_{k-L}}^{t_{k+H}} \left| e_{i_d} \right| dt + \frac{w_q}{i_N} \int_{t_{k-L}}^{t_{k+H}} \left| e_{i_q} \right| dt = \frac{w_q \left| i_q^* - i_q^P \right|}{i_N} + \frac{w_d \left| i_d^* - i_d^P \right|}{i_N}$$

$$+ \frac{w_{\Sigma_{dq}}}{i_N} \left(\frac{e_{i_{dq}}(k-1) + e_{i_{dq}}(k)}{2} \cdot T_{s_{k-1}} + \frac{e_{i_{dq}}(k) + e_{i_{dq}}(k+1)}{2} \cdot T_{s_k} \right)$$

(13)

The final cost function is given by:

$$g_2 = g_{i_{dq}}^{trajectory} + \frac{\Delta i_{1q}^P}{i_N} + \frac{f_S}{f_{S\max}} + f_{Lim}\left(i_q^P \right) + f_{Lim}\left(i_d^P \right) + f_{Pen}\left(e_{i_q} \right) + f_{Pen}\left(\Delta i_q^P \right)$$

(14)

3. Experimental Results

A commercial DBC5CEFA7 Board [1] is used for the implementation of the digital control. For the measurement of the mechanical angular position of the shaft a sin-cos encoder with 2048 increments per revolution is used. The machine currents are measured using Hall-effect current transducers. $\Sigma\Delta$ ADCs were used for the measurements of currents, DC-link voltage and encoder signals. The inverter used is an INFINEON MIPAQ^{TM} Serve Series IFS150V12PT4, its maximum switching frequency f_{Smax} is 20 kHz. The DC-Link was 600 V_{DC}, the nominal current of the machine is i_N 4.1A, w_q = 1.0, w_d = 0.25, Σw_q = 1.0 and Σw_d = 0.25. Experimental results were carried out with inverter switching frequencies of {0.8, 0.9, 1.0, 1.111, 1.25, 1.333, 1.6, 1.777, 2.0, 2.5, 3.2, 4.0, 5.0, 6.666, 8.0, 8.888, 10.0, 13.333, 16.0, 20.0}, i_{dqMax} = 0.5A, e_{iqLim} = 0.5A, for this value the frequencies below 20kHz get out of the minimization process. Finally, the number of switching transitions that take place within one second was measured in order to find the average value of the switching frequency.

For the first experiment the speed reference was set to 1500 min^{-1} and Δi_{1qMax}^P was set to 0.25A (6.1% of i_N). As it can be seen in Fig 4, the dynamic response of the controller is as expected. The desired switching frequency response was also obtained; when a change in i_q^* happens, the highest switching frequency is selected, as can be seen in the right side of the Fig 4; for a few instants the switching period T_s will be the minimum (50µs), making the value of i_q almost equal to i_q^* in approximately 200µs or four T_s. Once the error between i_q^* and i_q is reduced, the whole set of solutions are considered in the minimization process. This will reduce the switching frequency until it settles down to an optimal value, which will depends on the maximum allowed torque ripple, reducing in this way the switching losses. A steady state error, typical of model based controllers lacking an integral component can also be seen. The average switching frequency was 15.888 kHz, even though this value is not in the set of possible solutions the controller alternates the possible values (in this case between 13.3 and 16 kHz) and create it. It can also be seen that the response on the current i_d is almost unaffected. With the aid of the FPGA and $\Sigma\Delta$ADCs, the currents were sampled every 1.6 µs, so the real form of the i_q could be measured and the real current ripple could be calculated; subsequently the current ripple prediction could be validated. As shown in Fig 5, it can be seen that the current ripple stays within the desired levels. Here the steady-state error can be more clearly seen, being around 1.5% of i_N. Both forms of the cost function g_1 and g_2 were tested, yet the integral cost function g_2 did not yield results different form g_1.

Next Δi_{1qMax}^P was set to 1.5 A (36.6% of i_N) as shown in 0, the steady state error increases, as i_d is mostly affected with a steady-state error of 0.2A (4.87% of i_N). A decrease in the switching frequency can also be noticed, even though the same high dynamic response when the change in i_q^* occurs can also be seen, the final average frequency was 4.0 kHz. The current ripple presented in Fig 7, shows that it is well contained within the desired limit.

PCIM Europe 2016, 10 – 12 May 2016, Nuremberg, Germany

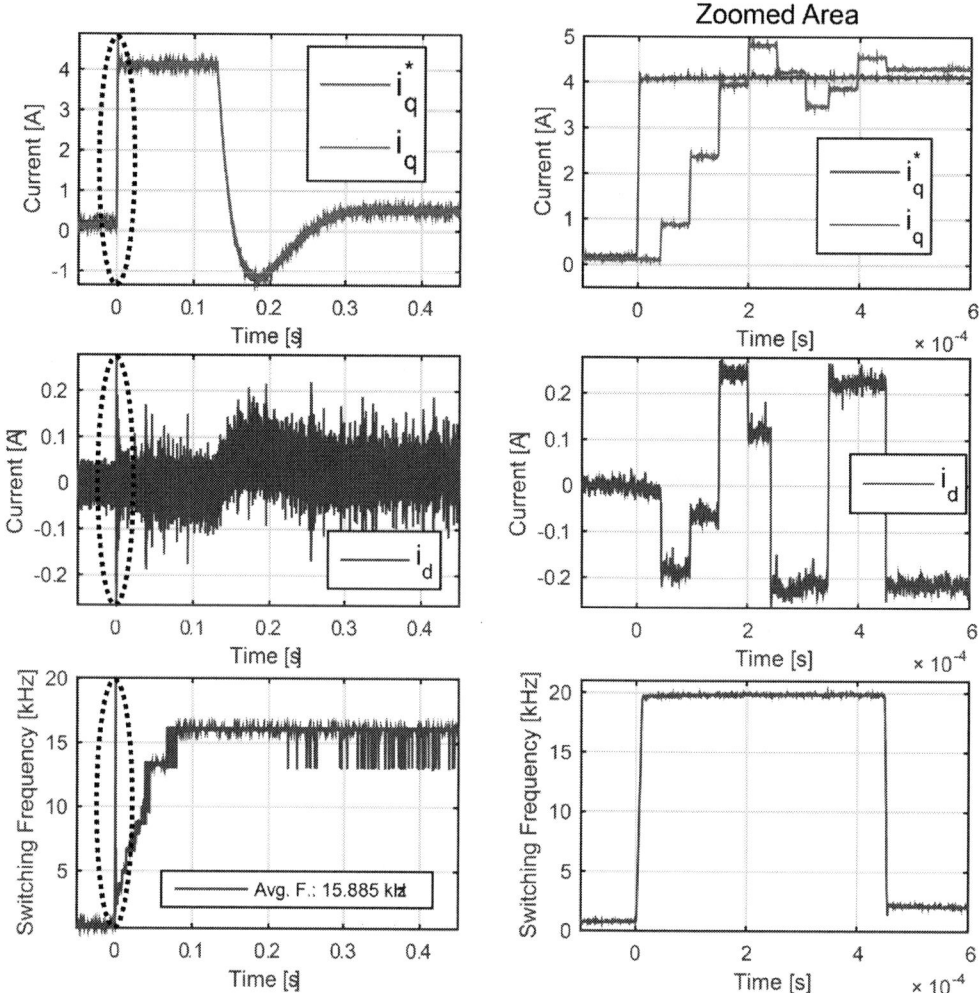

Fig 4.   Signals: i_q^* (blue), i_q (red), i_d (green) and f_s (magenta), n = 1500 min^{-1} and Δi^P_{qMax} = 0.25A

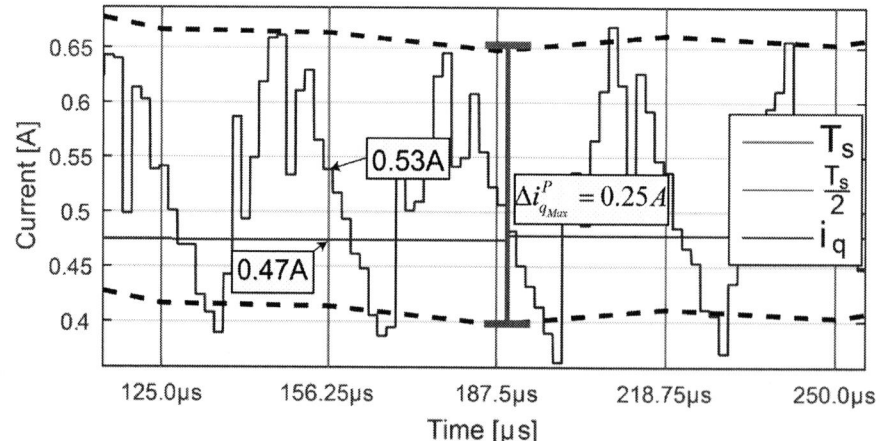

Fig 5.   Signals: i_q (red), T_s (blue) and $T_s/2$(red) n = 1500 min^{-1} and Δi^P_{1qMax} = 0.25A

© VDE VERLAG GMBH · Berlin · Offenbach

PCIM Europe 2016, 10 – 12 May 2016, Nuremberg, Germany

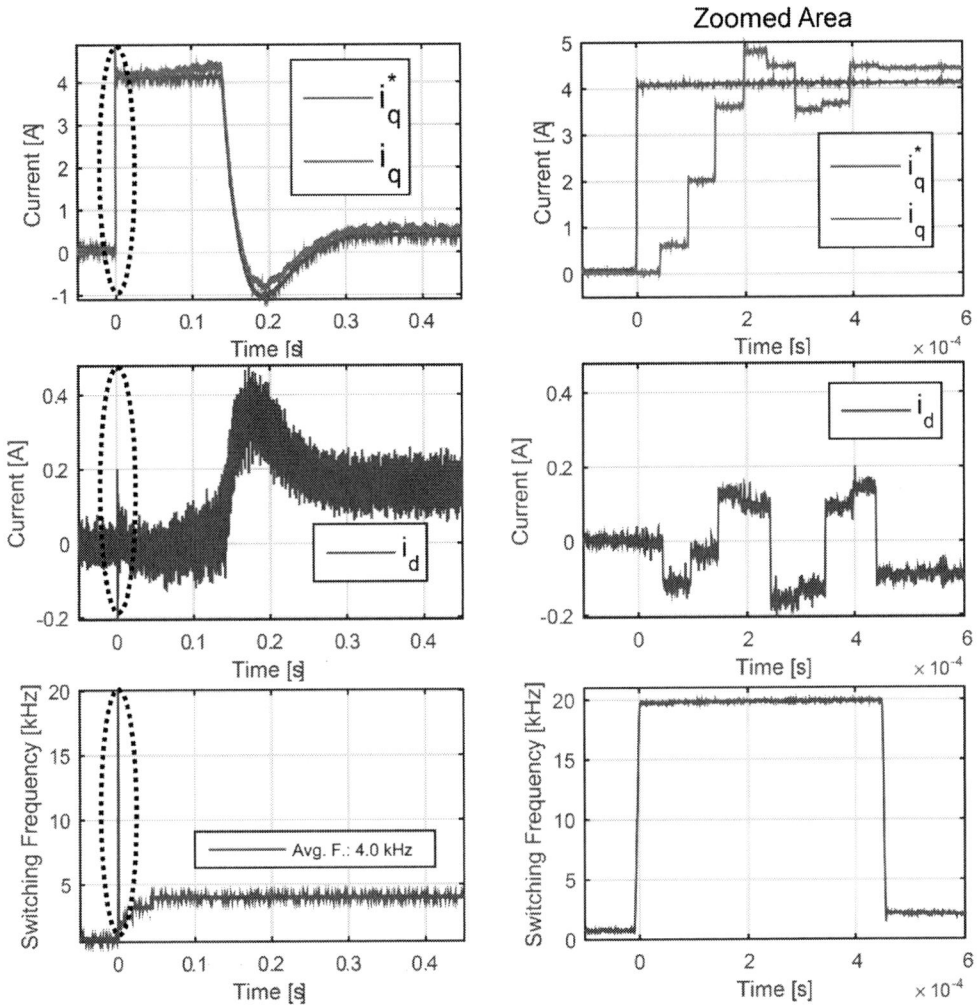

Fig 6.   Signals: i^*_q (blue), i_q (red), i_d (green) and f_s (magenta), n = 1500 min^{-1} and Δi^P_{qMax} = 0.41A

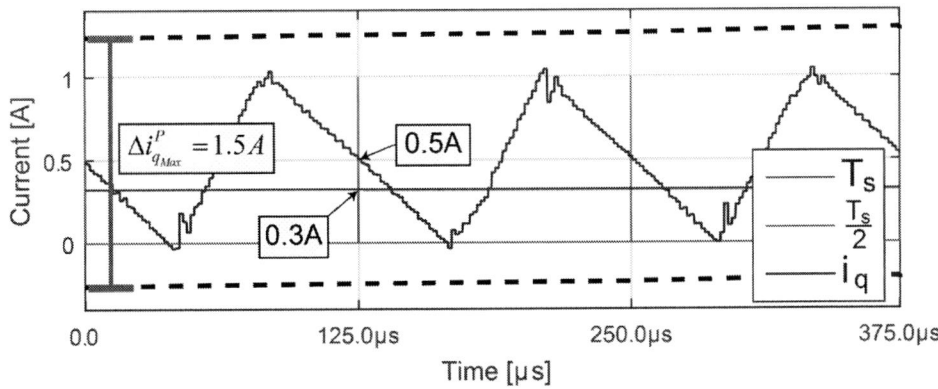

Fig 7.   Signals: i_q (red), T_s (blue) and $T_s/2$(red) n = 1500 min^{-1} and Δi^P_{1qMax} = 0.41A

For the next test the speed reference was set to 300 min^{-1} and Δi^P_{1qMax} to 2.05A (50% of i_N) as shown in Fig 8 the controller presents the same expected dynamic response, working at

© VDE VERLAG GMBH · Berlin · Offenbach

the maximum switching frequency for just a small period of time, the steady-state error was of (1.22% of i_N). The average switching frequency was 2.5 kHz and the current ripple (shown in Fig 9) stays within the predicted limit. Finally Δi^P_{1qMax} was set to 0.3A (7.32% of i_N), the average switching frequency was 4 kHz, the current ripple (shown in Fig 10) is also maintained within the desired limits.

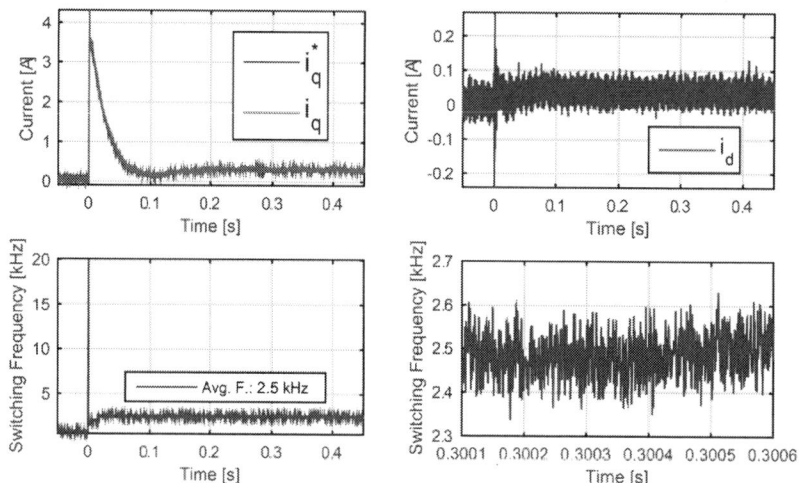

Fig 8.　Signals: i^*_q (blue), i_q (red), i_d (green) and f_s (magenta), n = 300 min^{-1} and Δi^P_{qMax} = 2.05A

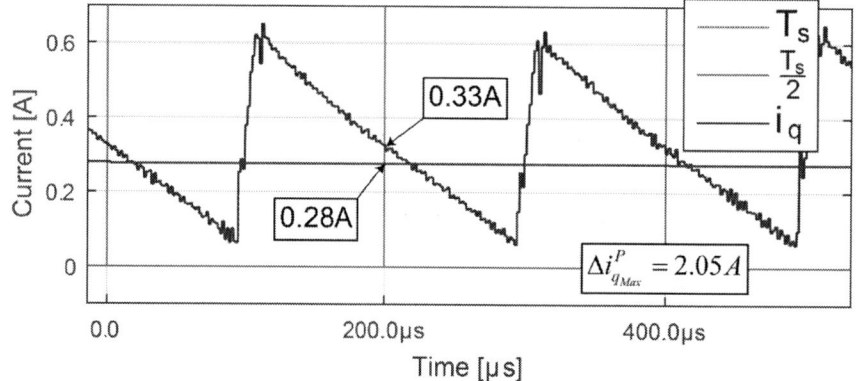

Fig 9.　Signals: i_q (red), T_s (blue) and $T_s/2$(red) n = 300 min^{-1} and Δi^P_{1qMax} = 2.05A

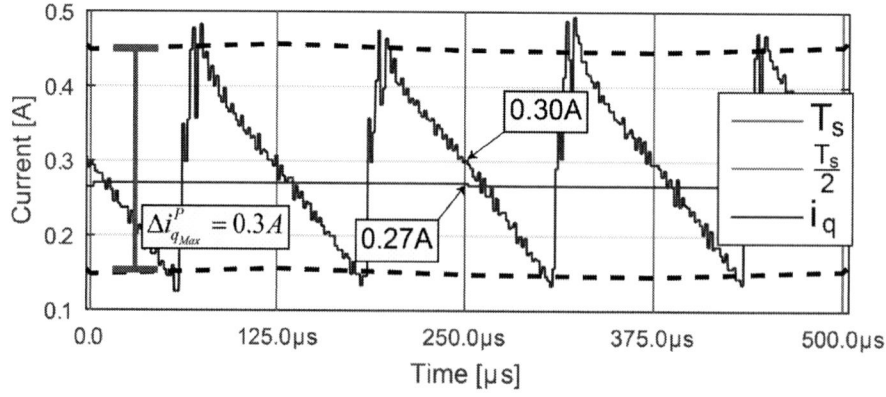

Fig 10.　Signals: i_q (red), T_s (blue) and $T_s/2$(red) n = 300 min^{-1} and Δi^P_{1qMax} = .3A

© VDE VERLAG GMBH · Berlin · Offenbach

Conclusions

A FCS-MPC with variable switching modulator-based frequency is presented; it is implemented in a FPGA platform suitable for the control of a PMSM. The main objective of using a FPGA was to exploit its parallel processing and pipelining capabilities as well as the option to customize the required peripherals. The benefit of the using the FPGA becomes evident given that it would have been impossible to execute the desired algorithm with a conventional microcontroller or digital signal processor solution.

The desired high dynamic of the control as well as the aimed switching frequency response were obtained. The controller will work at the maximum frequency just when a change in the torque reference is presented, after that it will reduce and settle to a value that will depend on the desired maximum torque ripple. This aims to reduce the switching losses on the inverter as it is well known that the switching losses are directly related to the switching frequency.

The steady-state error could be eliminated by using a method to identify and adapt the parameters of machine online, given that the integral cost function did not presented any differences in this work. Another possibility is the use of a PI current controller, as already presented in [4] and [5].

References

[1]   Rodriguez, J.; et al. "State of the Art of Finite Control Set Model Predictive Control in Power Electronics." IEEE Transactions on Industrial Informatics, 2013, Vol. 9-2.

[2]   Cortes, P. et al. "Predictive Control in Power Electronics and Drives." IEEE Transactions on Industrial Electronics, 2008, Vol. 55-12, pp. 4312 – 4324.

[3]   Vazquez, S. et al. "Model Predictive Control: A Review of Its Applications in Power Electronics." IEEE Industrial Electronics Magazine, 2014, Vol. 8-1, pp. 16 – 31.

[4]   Ramirez F., Pacas M. "FPGA Implementation of a Predictive Control for a PMSM with Variable Switching Frequency" ACEMP-OPTIM-ELECTROMOTION 2015.

[5]   Ramirez F., Pacas M "Model Based Control of a PMSM with Variable Switching Frequency and Torque Ripple Control" IECON2015 – Annual Conference of the IEEE Industrial Electronics Society, November 2015.

[6]   Pozo Marcelo "Finite Set Model Predictive Control of the PMSM with Sine-Wave Filter", Doctoral Thesis, Siegen, Germany, March 2015.

[7]   H. W. van der Broeck, H. C. Skudelny, and G. Stanke, "Analysis and realization of a pulse width modulator based on voltage space vectors," in Proc. IEEE IAS Annual Meeting, Denver, CO, 1986, pp. 244–251.

[8]   Krishnan, R. Permanent Magnet Synchronous and Brushless DC Motor Drives. CRC Press, USA, 2010.

[9]   devboards GmbH, http://www.devboards.de

PCIM Europe 2016, 10 – 12 May 2016, Nuremberg, Germany

Frequency- and Mode-Adaptive Control of DC-DC Converter for Efficency Improvement

Lukas Keuck, Farshid Almai, Sven Bolte, Norbert Fröhleke and Joachim Böcker
Paderborn University, Power Electronics and Electrical Drives, Paderborn, Germany
keuck@lea.upb.de, bolte@lea.upb.de, boecker@lea.upb.de

Abstract

It is common practice to operate Two-Quadrant-Converters with a constant switching frequency. However, the loss-optimal switching frequency strongly depend on the operation point thus for most operation points the selected frequency leads to higher losses than necessary. This paper presents a derived loss model which can be used to find the loss-optimal switching frequency. The presented model is easily computable on todays DSPs, thus allows frequency adaption on the run. A lab-prototype was built in order to verify the performance enhancement enabled by the proposed model. Experimental results are presented which show an efficiency improvement of approx. 1%.

1. Introduction

The Two-Quadrant Converter (2QC) remains the topology of choice for non-isolated power transfer between different voltage levels due to low number of passive components and low conduction losses in semiconductor devices (Fig. 1). Applications extend from microcontroller power supplies [1] to high-power automotive chargers [2]. In literature various names can be found for this topology, e.g. synchronous buck converter which may be misleading since the topology inherently allows also boost operation for negative average inductor current $\bar{i}_L$. In order to reduce the RMS current of the input and output capacitors, multiple 2QCs are often connected in parallel and are operated interleaved.

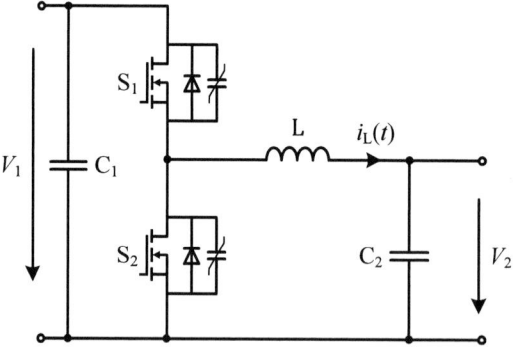

Fig. 1: Schematic of the Two-Quadrant Converter (2QC)

Once the 2QC hardware is designed, the key objective is to operate the converter for minimal losses. Today's state-of-the-art products are operated typically at a constant switching frequency which satisfies specified limits but does not necessarily lead to minimal losses [3].

© VDE VERLAG GMBH · Berlin · Offenbach

It can be beneficial, if switching frequency is varied with the operation point, with the objective of minimizing the losses. This paper presents a method in which loss-minimal switching frequency is found by a loss model which also distinguishes ZVS and hard-switching conditions, as explained later. The loss model includes the major loss-components: the conduction, the core and the switching losses which depend on the switching frequency and operation mode. As the load, input and output voltages must be taken as given, the switching frequency can be used as free optimization parameter. The switching losses depend directly on the switching frequency. However, due to the variation of the current ripple, the conduction and core losses also change. Further, switching losses are also influenced by the commutation current, which also changes as a result of change in switching frequency. This will lead to different operation modes that can be classified according to the shape of the inductor current.

As said earlier, the operation modes can be classified according to the inductor current waveform i_L [4]. When S_1 and S_2 are switched strictly complementary the converter is operated in continuous current mode (CCM). Depending on the mean load current, the input output voltages and the switching frequency we will observe two variants shown in Fig. 2, characterized by the fact whether the current will change its polarity or not, called CCM-ZVS and CCM-HS. In CCM-ZVS, any hard turn-on for both switches S_1 and S_2 is avoided since their body diodes conduct directly before the turn-on event, resulting in zero-voltage switch-on (ZVS) (Fig. 2(a)). Thus, CCM-ZVS allows to use Super-Juction MOSFETs (e.g. Infineon's CoolMOS) which comes with a significant parasitic output capacitance C_{oss} as also applied in [2]. In literature, the name triangular current mode (TCM) is also common for CCM-ZVS. CCM-ZVS leads to low switching losses, however, it brings a large current ripple and thus high RMS currents. The current ripple can be minimized by increasing the switching frequency so that the ZVS must be given up for S_2 and the converter enters CCM-HS (hard-switched) mode of operation (Fig. 2(b)). As the current ripple is smaller for CCM-HS than for CCM-ZVS it comes with smaller conduction losses.

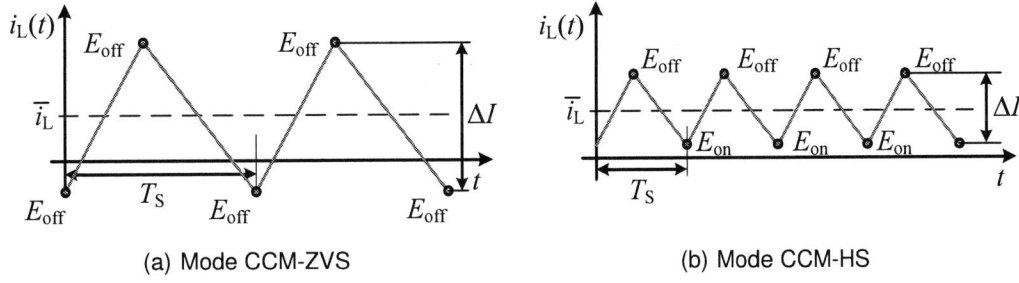

(a) Mode CCM-ZVS   (b) Mode CCM-HS

Fig. 2: Inductor current in mode CCM

Another way to decrease the high RMS current of CCM-ZVS is to operate the converter in discontinuous conduction mode (DCM) by switching S_2 off when the current reaches zero (Fig. 3) [4]. DCM is characterized by a time interval where neither S_1 nor S_2 conducts, thus the current i_L is zero for a certain time interval (Fig. 3). For light load, in particular, DCM leads to lower losses as with CCM [3]. When the converter is idling (no load) it is advantageous to change to pulse skipping so that the complete converter is turned off for several periods (burst-mode). In [3], DCM and pulse-skipping mode for light load conditions has been treated in detail. Though, a lot of good work on different operation modes of a 2QC has been published, work related to loss-optimal selection of the switching frequency has been not been widely reported. The objective of this paper is to exactly explore this zone.

PCIM Europe 2016, 10 – 12 May 2016, Nuremberg, Germany

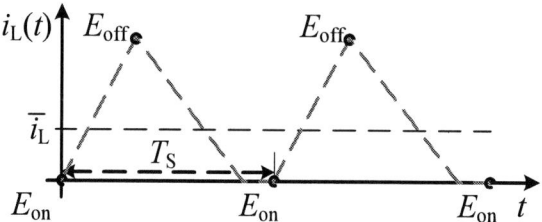

Fig. 3: Inductor current in mode DCM

2. Loss Model

In order to find the loss-optimal switching frequency, the total losses need to be estimated in dependence to the load condition, input/output voltage and temperature. The loss-optimal switching frequency could be found offline and stored in a look-up table which would require large DSP storage due to many model parameters. Alternatively, the loss-optimal switching frequency can also be computed on-the-run, using the power of todays microcontrollers and DSPs. This work considers an on-the-run computation based loss model which estimates the loss distribution across conduction, core and switching losses for a 2QC.

However, the loss model ignores the component specification limits during the computation process. The switching frequency cannot be selected freely but multiple constraints need to be considered. Often a maximal voltage ripple Δv_2 is specified which leads to a lower frequency limit. Also the inductor saturates at a certain current value i_L which defines another lower frequency limit for high power. Upper frequency limits exist due to the MOSFET properties.

2.1. Conduction Losses

Significant losses are due to ohmic resistances associated with various components, which are termed as conduction losses. These losses appear in all components but significantly in the MOSFETs and the copper windings of the inductor. If different MOSFETs are used for S1 and S2 their individual RMS currents need to be calculated. However, if S_1 and S_2 have equal on-resistances R_{DSon} the conduction losses of the MOSFETs and the inductor can be combined by the equivalent resistance $R_{\mathrm{cd}} = R_{\mathrm{DSon}} + R_L$ where R_L is inductor's winding resistance. In CCM (Fig. 2) the inductor RMS-current can be calculated by

$$I_{\mathrm{L}} = \sqrt{\bar{i}_{\mathrm{L}}^2 + \frac{\Delta I^2}{12}} \tag{1}$$

where ΔI is the current ripple given by $\Delta I = \frac{(V_1 - V_2)D}{Lf_s}$ and $\bar{i}_{\mathrm{L}}$ is the inductor average current. Neglecting the losses caused by skin and proximity effects, the conduction losses can be calculated by $P_{\mathrm{cd}} = I_{\mathrm{L}}^2 R_{\mathrm{cd}}$ which decreases for higher switching frequencies due to the lower current ripple.

© VDE VERLAG GMBH · Berlin · Offenbach

2.2. Core Losses

The current ripple causes losses within inductor's core since energy is required to change the magnetization. The core losses are estimated using the improved generalized Steinmetz equation (IGSE) [5]

$$P_{\text{fe}} = \frac{1}{T_{\text{s}}} \int_0^{T_s} k_{\text{i}} \left| \frac{db}{dt} \right|^{\alpha} \Delta B^{\beta - \alpha} dt \tag{2}$$

where k_i, α and β are the Steinmetz parameters, ΔB is the peak-to-peak flux density and $b(t)$ is the flux density. Assuming steady-state (2) can be simplified to

$$P_{fe} = k_{\text{i}} \Delta B^{\beta - \alpha} \left[\dot{b}_1^{\alpha} D + \dot{b}_2^{\alpha} (1 - D) \right] \tag{3}$$

where $\dot{b}_1 = \frac{V_1 - V_2}{N A_{\text{C}}}$, $\dot{b}_2 = \frac{V_2}{N A_{\text{C}}}$, $\Delta B = \frac{(V_1 - V_2) D}{N A_{\text{C}} f_s}$, D is the duty cycle, N denotes the no. of turns and A_{C} denotes the core cross section. The Steinmetz parameters k_{i}, α and β are determined by fitting core manufacturer's published data. Typical values for ferrite cores lie in the range $\alpha = 1...1.6$ and $\beta = 2...2.6$ so that the core losses typically decrease with increasing switching frequency, since $P_{\text{fe}} \propto \frac{1}{f_{\text{s}}^{\beta - \alpha}}$.

2.3. Switching Losses

Each commutation of the current causes losses which appear instantaneously while the current is changing between S_1 and S_2. These losses have to be considered either at MOSFETs' switch-on or switch-off, typically not both. This is because, when the body diode of MOSFET S_1 is conducting it can be switched on with practically no losses (ZVS). Also, when the MOSFET S_2 is switched off and the current remains in its body diode, the turn-off losses are negligible [4]. However, if S_2 is switched off and the current commutates to the body diode of S_1, significant turn-off losses appear. Also if the body diode of S_2 conducts and S_1 is switched on, the losses are significant. All of these cases are considered inherently in (4) by switching loss curves E_{on} and E_{off} of S_1 and S_2. These curves are identified by interpolation of measurements of the double-pulse test presented in [6] or [7] at different voltages V_1 and current levels i_L.

$$P_{\text{sw}} = f_{\text{s}} \left[E_{\text{on,S1}} \left(I_{\min}, V_1 \right) + E_{\text{off,S1}} \left(I_{\max}, V_1 \right) + E_{\text{on,S2}} \left(-I_{\max}, V_1 \right) + E_{\text{off,S2}} \left(-I_{\min}, V_1 \right) \right] \tag{4}$$

$$\text{with } I_{\max} = \bar{i}_{\text{L}} + \frac{\Delta I}{2} \text{ and } I_{\min} = \bar{i}_{\text{L}} - \frac{\Delta I}{2} \tag{5}$$

where identical curves can be used for S_1 and S_2 if equal MOSFETs are used. Exemplary curves for E_{on} and E_{off} are given in the appendix.

2.4. Gate-Driver Losses

Changing the switching frequency effects also the gate driver power consumption which should be considered to achieve minimal total losses. With acceptable accuracy the driver power consumption can be assumed to be proportional with the switching frequency ($P_{\text{gd}} = E_0 f_{\text{s}}$).

3. Experimental Verification

A 48V-14V DC-DC-converter with 600 W power rating for automotive application was designed to prove the proposed concept. Typical leaded MOSFETs with $R_{DS,on} = 12$ mΩ are used which are robust for hard commutation. For the inductor a compact ERU33-type choke qualified for automotive application was selected which delivers an inductance of $L \approx 12$ μH till $i_L = 50$ A.

Tab. 1: List of components of the lab prototype

PART	TYPE	TYPE NO.
Half Bridge $S_{1,2}$	Infineon OptiMOS 3	IPP120N20NFD
Input Capacitor C_1	3 x WIMA MKP10 0.33 μF	MKP1G033305F
Input Capacitor C_1	2 x TDK Electrolytic Capacitor 4700 μF	B41456
Inductor L	2 x TDK ERU33 6.0 μH Choke	P100761-A1-51
Output Capacitor C_2	3 x WIMA MKP10 0.33 μF	MKP1G033305F
Output Capacitor C_2	2 x TDK Electrolytic Capacitor 4700 μF	B41456
PWM-Generation	Piccolo controlSTICK 90 MHz	TMS320C28343

Next, the loss-model is parametrized according to the parameters described in Section 2. The list of parameters is included in the appendix. The accuracy of the loss model is evaluated by measuring the total losses using the precision power analyser Yokogawa WT3000. The measured total losses are compared with the estimation of the loss model at discrete operation points. Fig. 4(a) shows the verification for an output power of $P_2 = \bar{i}_L V_2 = 50$ W. Due to multiple simplifications, e.g. neglected skin and proximity effects, the total losses are estimated with moderate accuracy, however both measurement and loss model agree that minimal losses appear at $f_s \approx 95$ kHz in mode CCM-ZVS. Fig. 4(b) shows the waveforms corresponding to this loss-optimal case (in CCM-ZVS mode).

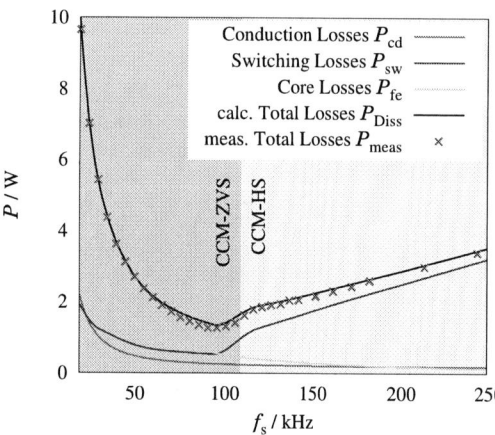

(a) Comparison of measured total losses with loss model for $P_2 = 50$W versus f_s

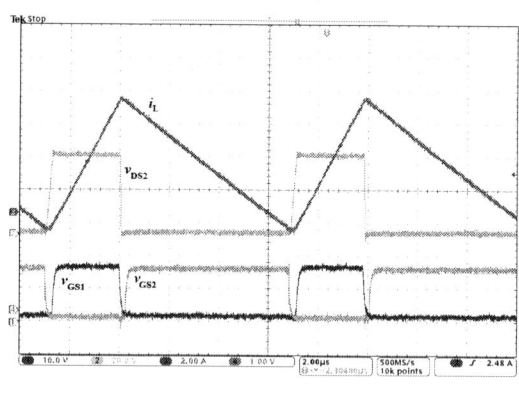

(b) Waveforms at the optimal operation point for $P_2 = 50$ W at $f_s = 95$ kHz in mode CCM-ZVS: red: i_L, light-blue: $v_{DS,2}$ and gate-source voltages

Fig. 4: Verification of the loss model at $P_2 = 50$ W, $V_1 = 48$ V and $V_2 = 14$ V

At elevated power levels CCM-ZVS is no longer the loss-optimal mode, since a huge current ripple is required to enter ZVS, resulting in high conduction losses. Fig. 5(b) shows the losses at

PCIM Europe 2016, 10 – 12 May 2016, Nuremberg, Germany

$P_2 = 400$ W where minimal losses occour at $f_s \approx 45$ kHz in mode CCM-HS which is estimated by the loss model with sufficient accuracy.

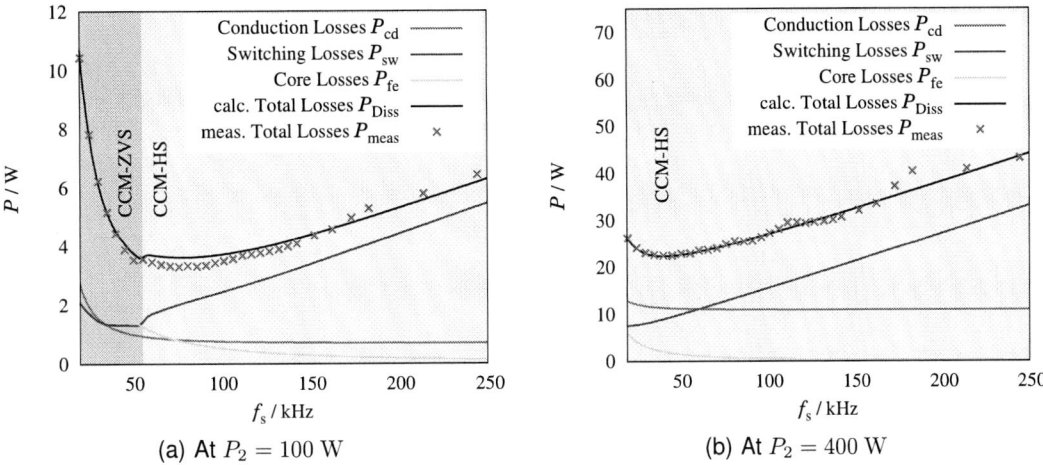

(a) At $P_2 = 100$ W

(b) At $P_2 = 400$ W

Fig. 5: Comparison of measured total losses with loss model at $V_1 = 48$ V and $V_2 = 14$ V

For the experimental design the transition between CCM-ZVS and CCM-HS is located at approximately $P_2 = 100$ W where both modes lead to similar losses (Fig. 5(a)). In this case changing the frequency f_s in the range of $50...110$ kHz practically does not change the total losses. Instead the losses are shifted from inductor's core to the MOSFETs.

Finally an optimization procedure needs to be applied in order to find the switching frequency which leads to minimal losses within the constraints. Various optimization algorithms can be employed which is not treated in this paper. In order to demonstrate the benefit of the adaptive frequency selection the loss-optimal switching frequency is found offline for this verification. Fig. 6 shows the measured efficiency improvement for the lab-prototype with the adaptive frequency selection compared with the standard 100 kHz frequency. At partial-load the converter is strictly operated in mode CCM-ZVS, at moderate load CCM-HS leads to lowest losses. For high power ($P_2 > 400$ W) the current limit of the inductor need to be considered by defining a lower frequency limit.

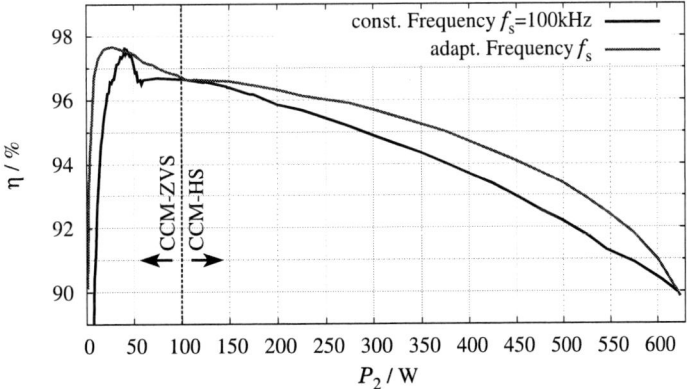

Fig. 6: Measured efficiency for constant switching frequency $f_s = 100$ kHz compared with measured efficiency when the loss-minimal frequency is employed within the constraints

© VDE VERLAG GMBH · Berlin · Offenbach

4. Conclusion

The proposed method of finding the loss-optimal switching frequency can be easily utilized to improve the efficiency performance of Two-Quadrant-Converters. Since only moderate computation efforts are needed to implement the proposed loss model, it can be easily realized on todays DSPs for online estimation of the total losses. The results obtained from the laboratory prototype give a proof of the concept with almost 1% increase in efficiency over a large power range. In this work, the dependency on temperature and component tolerances was neglected which should be a starting point for further investigations.

5. Appendix

The loss model of the verification is based on parameters listed below.

Tab. 2: Parameter list for the lab-prototype

$$R_{cd} = 13.3 \text{ m}\Omega \quad k_i = 5.8 \, \frac{\text{nWs}^\alpha}{\text{T}^\beta} \quad \alpha = 1.1 \quad \beta = 2.6 \quad E_0 = 3.4 \, \mu\text{J}$$

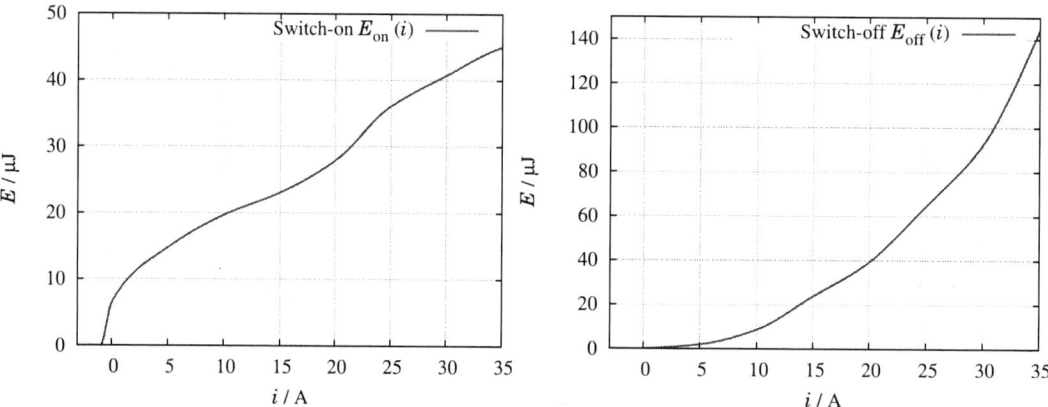

Fig. 7: Identified curves for $E_{on} = E_{on,S1} = E_{on,S2}$ and $E_{off} = E_{off,S1} = E_{off,S2}$ for $V_1 = 48$ V

6. Acknowledgments

This research and development project is funded by the German Federal Ministry of Education and Research (BMBF) within the Leading-Edge Cluster Intelligent Technical Systems OstWestfalenLippe (it's OWL) and managed by the Project Management Agency Karlsruhe (PTKA).

7. References

[1] M. Orabi; A. Abou-Alfotouh and A. Lotfi. Coss capacitance contribution to synchronous buck converter losses. *Power Electronics Specialists Conference*, pages 666–672, 2008.

[2] D. Christen; S. Tschannen and J. Biela. Highly efficient and compact dc-dc converter for ultra-fast charging of electric vehicles. *Power Electronics and Motion Control Conference (EPE/PEMC)*, 2012.

[3] A. V. Peterchev and S. R. Sanders. Digital multimode buck converter control with loss-minimizing synchronous rectifier adaptation. *IEEE Transactions on Power Electronics*, 21:1588–1599, 2006.

[4] Robert W. Erickson and Dragan Maksimovic. *Fundamentals of Power Electronics*. Springer, 2. Edition, 2001.

[5] J. Mühlethaler; J. W. Kolar and A. Ecklebe. Loss modeling of inductive components employed in power electronic systems. *Power Electronics and ECCE Asia (ICPE and ECCE)*, pages 945–952, 2011.

[6] S. Bolte; J. Baurichter; C. Henkenius; Norbert Fröhleke und Joachim Böcker. Verlustmodellierung und Effizienzoptimierung einer hart schaltenden, netzfreundlichen Pulsgleichrichterstufe (PFC). *Internationaler ETG-Kongress, Berlin, Germany*, 2013.

[7] Gerold Laimer and Johann W. Kolar. Accurate measurement of the switching losses of ultra high switching speed coolmos power transistor / sic diode combination employed in unity power factor pwm rectifier systems. *Swiss Federal Institute of Technologie (ETH) Zurich*, 2011.

Load Torque Estimation in Repetitive Mechanical Systems by Using Fourier Interpolation

M.Sc.E.E. Van Trang Phung, University of Siegen, Germany, vantrang.phung@uni-siegen.de

Univ.-Prof.Dr.-Ing. Mario Pacas, University of Siegen, Germany, pacas@uni-siegen.de

Abstract

This paper deals with the load torque estimation in repetitive mechanical systems by using Fourier interpolation. Two different solutions are presented, one is theory based and one utilizes a look-up table for the estimation of coefficients of a Fourier series while making interpolation of the load torque. A laboratory set-up with a horizontal slider-crank mechanism was built as an example of repetitive mechanical systems and was used to verify the effectiveness of the two methods. It has been determined that the look-up table based method yields better performance quality than the other method. The estimated load torque is also utilized as a compensation signal in a feed-forward control scheme to improve the quality of the speed response. It can be also used in the diagnostic for the rolling bearing faults in future works.

1. Introduction

The information of the load torque can be used for the mechanical analysis and performance improvement of mechatronics systems. In [1], it is considered as an input known a priori for an identification process, from which the mechanical parameters of a two-mass system are identified. Later, these parameters are used in an automatic tuning rule of a state-space controller [2]. The load torque is also utilized as a compensation signal to reduce the shaft torsional vibration [3] and to increase the robustness of a speed controller [4].

The load torque can be measured by using sensors, but these sensors lead to higher overall cost and less robustness of the system. In the literature several methods have been proposed for the load torque estimation. For this purpose a disturbance observer whose gain selection can be done by applying a pole placement method [5] [6] can be used. The Kalman filter technique has also been proposed [7] [8]. The pole placement method ensures good estimation accuracy if the noise associated with the measurement is low and the system's parameters are well known. Otherwise, the estimation accuracy can be unsatisfactory. In the case of the Kalman filter technique, the load torque is considered as a state variable and obtained as an output of the Kalman filter. However, this method has a disadvantage related to the selection of the covariance matrixes, as there is no general rule for choosing these matrixes.

Many mechanical systems like those found in production machines are characterized by repetitive cycles in which changes in the load torque and the moment of inertia are periodic. As a result, the demanded load torque changes periodically. A horizontal slider-crank mechanism is an example of such a system. The load torque in the mechanism is a periodic function of the crank angle, so it can be characterized by applying Fourier interpolation. The estimation of the load torque can be conducted by evaluating the coefficients of the Fourier series. Two methods including a theoretical method and a look-up table based method for the estimation of the Fourier coefficients are presented in this paper. The estimated load torque is used in a feed-forward control scheme to improve the quality of the speed control.

2. Horizontal slider-crank mechanism

2.1. Mechanical structure

The slider-crank mechanism consists of three main parts: a slider, a connecting rod and a crank as shown in Fig. 1(a). The reduced model depicted in Fig. 1(b) is used for the kinetic analysis of the mechanism. O, A, and B are the three articulated joints between the coupling, the crank, the connecting rod and the slider. G_c and G_{crod} correspond to the centers of gravity of the crank and the connecting rod, respectively. Since the mechanism has one degree of freedom, a two dimensional coordinate system (Oxy) can be used to describe the position of each single body. The angles of the crank and the connecting rod with respect to the horizontal axis are denoted by θ and ϕ. In addition, without generality, the surface on which the slider moves can be assumed to be the potential surface.

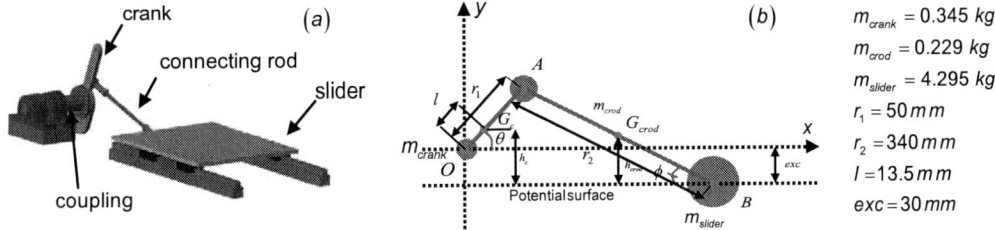

Fig. 1. Horizontal slider-crank mechanism: a) mechanical structure; b) reduced model

2.2. Moment of inertia

According to the kinetic analysis [9], the equivalent moment of inertia of the mechanism is given by:

$$J_e(\theta) = J_A + m_c r_1^2 \left[\sin\theta + \frac{\cos\theta(\lambda + k\sin\theta)}{\sqrt{1-(\lambda + k\sin\theta)^2}} \right]^2 \tag{1}$$

where: $J_A = J_{crank} + m_{crank}l^2 + 0.5 m_{crod} r_1^2$; $m_c = m_{slider} + 0.5 m_{crod}$; $\lambda = r_1/r_2$; $k = exc/r_2$; J_{crank} is the moment of inertial of the crank.

Normally, $\lambda \ll 1$ and $k \ll 1$. J_e in (1) can be considered as a multivariable function of λ and k. If only terms up to the first order of λ and k are regarded when representing J_e in form of Taylor series, (1) can be rewritten as:

$$J_e(\theta) = J_A + m_c r_1^2 \left(\sin^2\theta + \lambda\sin\theta\sin2\theta + 2k\sin\theta\cos\theta \right) \tag{2}$$

The total moment of inertia is defined by taking into account the moment of inertia of the motor (J_{motor}), the coupling ($J_{coupling}$) and the mechanism (J_e):

$$J_{total}(\theta) = J_{motor} + J_{coupling} + J_e(\theta) \tag{3}$$

2.3. Potential energy

The crank and the connecting rod are the two components storing potential energy. With respect to the potential surface, the total potential energy is given by:

$$W_p = W_{p,crank} + W_{p,crod} = m_{crank}gh_c + m_{crod}gh_{crod} = m_{crank}g(l\sin\theta + exc) + 0.5 m_{crod}g(r_1\sin\theta - exc) \tag{4}$$

where g is the acceleration of gravity.

2.4. Friction model

In the mechanism, friction mostly comes from the contact surface between the slider and the supporting frame. It can be divided into four components: Coulomb friction, stiction, viscous damping and Stribeck effect as shown in Fig. 2. As it can be seen, the Coulomb friction is constant and depends on the movement direction of the slider. The stiction, which is also known as break-away friction, has a non-zero value only when the slider is in standstill. The viscous damping is directly proportional to the slider velocity. The Stribeck effect is due to the use of lubrication and occurs at the translation boundary from partial to full fluid lubrication. The total friction is illustrated in Fig. 2(e).

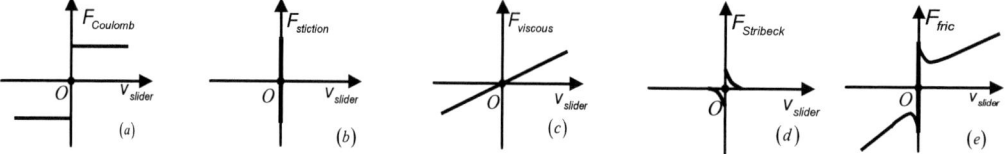

Fig.2. Friction model: (a) Coulomb friction; (b) stiction; (c) viscous damping; (d) Stribeck effect; (e) total friction

For mathematical description, the total friction can be defined by the sum of the Coulomb friction and the viscous damping:

$$F_{fric} = F_{Coulomb} + F_{viscous} = K_c \, \text{sgn}(v_{slider}) + K_v v_{slider} \tag{5}$$

where: $\omega = d\theta / dt$; $v_{slider} = -r_1\omega(\sin\theta + 0.5\lambda\sin 2\theta + k\cos\theta)$; K_c and K_v are the Coulomb and viscous constants; sgn(.) is the sign function.

3. Load torque estimation methods

3.1. Theoretical calculation of the load torque

The load torque demanded by the mechanical system is defined by the Euler-Lagrange equation [9]:

$$T_{load} = J_{total}\frac{d^2\theta}{dt^2} + \frac{1}{2}\frac{dJ_{total}}{d\theta}\left(\frac{d\theta}{dt}\right)^2 + \frac{dW_p}{d\theta} + F_{fric}\frac{dx_{slider}}{d\theta} \tag{6}$$

where: J_{total}, W_p, F_{fric} are given by (3), (4) and (5), and $\dfrac{dx_{slider}}{d\theta} = -r_1\left(\sin\theta + \dfrac{\lambda}{2}\sin 2\theta + k\cos\theta\right)$

Supposed that the crank is driven at a constant angular velocity: $\omega = d\theta / dt = const$, $d^2\theta / dt^2 = 0$ (6) can be rewritten as:

$$T_{load} = T_{inertia} + T_{Coulomb} + T_{viscous} + T_p \tag{7}$$

$T_{inertia} = 0.5 m_c r_1^2 \omega^2 (\sin 2\theta - 0.5\lambda\sin\theta + 1.5\lambda\sin 3\theta + 2k\cos 2\theta)$; $T_{Coulomb} = K_c r_1 |\sin\theta + 0.5\lambda\sin 2\theta + k\cos\theta|$

$T_{viscous} = K_v r_1^2 \omega (\sin\theta + 0.5\lambda\sin 2\theta + k\cos\theta)^2$; $T_p = (m_{crank}l + 0.5 m_{crod}r_1)g\cos\theta$

where $T_{inertia}, T_{Coulomb}, T_{viscous}, T_p$ are due to the change of moment of inertia, Coulomb friction, viscous damping and potential energy.

Eq. (7) shows that the load torque T_{load} is a periodic function of the crank angle, so it can be described in terms of trigonometric functions by utilizing Fourier interpolation. Furthermore, if the electric drive is designed properly, the demanded load torque by the mechanical system has the approximate value of the reference torque which is the output of the speed controller. Therefore, the reference torque is a good estimation of the load torque.

3.2. Fourier interpolation of the load torque

Since the control system works in discrete time-domain, the discrete Fourier series for the description of the load torque should be considered [10]:

$$T_{load} \approx \frac{a_0}{2} + \sum_{k=1}^{M-1} a_k \cos(k\theta) + \sum_{k=1}^{M-1} b_k \sin(k\theta) \tag{8}$$

where: $a_k = \sum_{i=0}^{N-1} \frac{2}{N} T_{load}(i\Delta\theta) \cos(k \times i\Delta\theta); b_k = \sum_{i=0}^{N-1} \frac{2}{N} T_{load}(i\Delta\theta) \sin(k \times i\Delta\theta)$; $\Delta\theta = 2\pi / N$; N is the number of samples per period.

The calculation of the Fourier coefficients can be implemented by using sliding windows. An increase of N means an increase of the windows resolution and yields more precise results. However, it demands more calculation efforts by the microprocessor or DSP. Therefore, N should be chosen properly before carrying out any calculation.

3.3. Theory based method

In this method, each Fourier coefficient is explicitly described by a function of the crank angle. For this purpose, the four components of the load torque in (7) are expressed in Fourier series. The two components $T_{inertia}$ and T_p are already trigonometric functions. When considering the real value of λ and k from the mechanical design, two other terms can be expressed in Fourier series whose coefficients are calculated and presented in Table I and Table II.

TABLE I

COEFFICIENTS OF $\left| \sin\theta + \frac{\lambda}{2}\sin 2\theta + k\cos\theta \right|$

Harmonic order	DC	1	2	3	4	5
Cos	0.64	0.06	-0.423	-0.04	-0.084	-0.01
Sin		0	0.074	0	0.031	0.0002

TABLE II

COEFFICIENTS OF $\left| \sin\theta + \frac{\lambda}{2}\sin 2\theta + k\cos\theta \right|^2$

Harmonic order	DC	1	2	3	4	5
Cos	0.51	0.0697	-0.492	-0.069	0	0
Sin		0.0061	0.0868	0.0061	0	0

Based on Table I and Table II, the load torque can be described in the following form:

$$T_{load} = \frac{a_0}{2} + \sum_{k=1}^{5} a_k \cos(k\theta) + \sum_{k=1}^{5} b_k \sin(k\theta) \tag{9}$$

where a_k and b_k are given in Table III.

In this process, M = 6 was chosen as a compromise between the estimation precision and the algorithm complexity.

To conduct the theoretical calculation, the Coulomb and viscous constants need to be identified. The identification is carried out by considering the relationship of these two parameters with the coefficient a_0 in (9). The procedure for this method is shown in Fig. 3.

3.4. Look-up table based method

In addition, the Fourier coefficients can be estimated by using look-up tables obtained by measurements on the mechanical system. The procedure to build such the tables includes two steps:

Step 1: The crank is driven at different angular velocities within its operational range $[0, \omega_{max}]$. In each case, a set of coefficients $[a_k, b_k]$ is obtained.

Step 2: For the calculation of each Fourier coefficient a two-dimensional look-up table is created from the data obtained in Step 1. The input of the tables is the crank angular velocity. Linear interpolation is used for reducing the size of the tables.

Fig. 4 shows the block diagram describing the idea of the method.

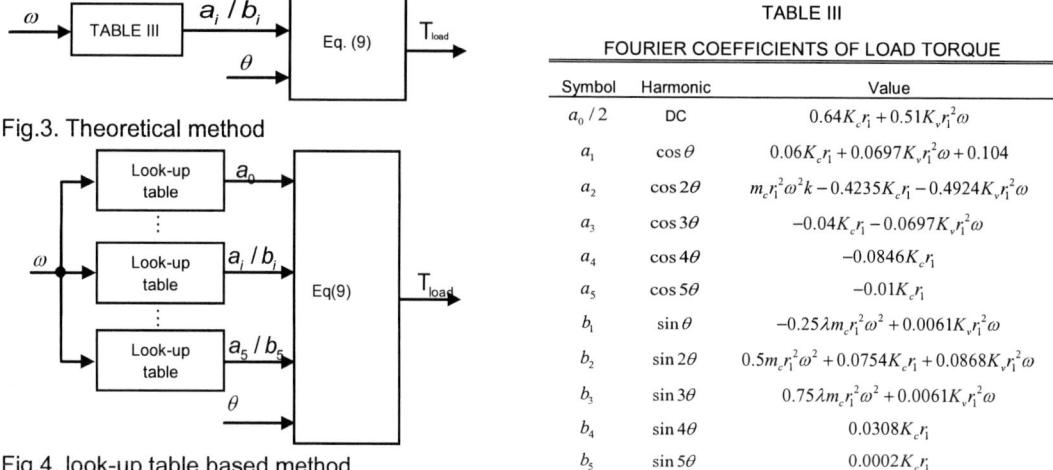

Fig.3. Theoretical method

Fig.4. look-up table based method

TABLE III

FOURIER COEFFICIENTS OF LOAD TORQUE

Symbol	Harmonic	Value
$a_0/2$	DC	$0.64K_cr_i + 0.51K_vr_i^2\omega$
a_1	$\cos\theta$	$0.06K_cr_i + 0.0697K_vr_i^2\omega + 0.104$
a_2	$\cos 2\theta$	$m_cr_i^2\omega^2k - 0.4235K_cr_i - 0.4924K_vr_i^2\omega$
a_3	$\cos 3\theta$	$-0.04K_cr_i - 0.0697K_vr_i^2\omega$
a_4	$\cos 4\theta$	$-0.0846K_cr_i$
a_5	$\cos 5\theta$	$-0.01K_cr_i$
b_1	$\sin\theta$	$-0.25\lambda m_cr_i^2\omega^2 + 0.0061K_vr_i^2\omega$
b_2	$\sin 2\theta$	$0.5m_cr_i^2\omega^2 + 0.0754K_cr_i + 0.0868K_vr_i^2\omega$
b_3	$\sin 3\theta$	$0.75\lambda m_cr_i^2\omega^2 + 0.0061K_vr_i^2\omega$
b_4	$\sin 4\theta$	$0.0308K_cr_i$
b_5	$\sin 5\theta$	$0.0002K_cr_i$

4. Results

4.1. Description of the experiment setup

A laboratory set-up was specially designed and built to verify the effectiveness of the two proposed methods. As it is shown in Fig. 5, the test bench is composed of two main parts: the mechanical part and the electrical part. The mechanical part is a horizontal slider-crank mechanism coupled to a motor drive equipped with a permanent magnet synchronous machine (PMSM). The PMSM is supplied by an inverter working at $f_s = 4kHz$. The control algorithm and data acquisition are carried out by a dSPACE 1104 board. The position and velocity of the crank are measured by using a 2048-pulse incremental encoder assembled in the driving motor.

4.2. Results

The mechanism was driven at 3 min[-1], 5 min[-1], 7 min[-1], 10 min[-1], 15 min[-1], 20 min[-1], 25 min[-1], 30 min[-1], 40 min[-1], 50 min[-1], 60 min[-1], 70 min[-1], 80 min[-1], 90 min[-1], 100 min[-1]. The 'Fourier' block in Fig. 5 follows the sliding window method with two inputs including the reference load torque and the crank angle. It is necessary to mention that the precision of the crank angle strongly affects the precision of the estimation algorithm. Therefore, the angle should be corrected during the commissioning mode of the driving motor. The coefficients obtained from the estimation process are depicted in Fig. 6.

PCIM Europe 2016, 10 – 12 May 2016, Nuremberg, Germany

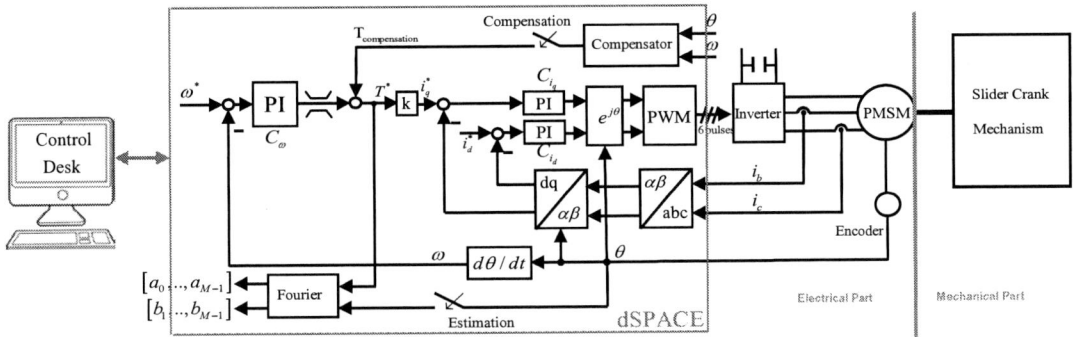

Fig.5. Control system of the slider-crank mechanism

Fig. 6 shows that the dependency of b_2 on the crank angular velocity resembles a parabola. This behavior agrees with the calculation of b_2 in Table III. Furthermore, it can be seen that the shape of the DC component a_0 is similar to that of the total friction force in Fig. 2(e). The linear region ranges from 25 min^{-1} to 100 min^{-1}. The straight blue line characterizing the linear behavior is obtained by connecting the two points located in the border of the linear region. The intersection of the straight blue line with the vertical axis helps to identify the Coulomb constant while its slope can be used to calculate the viscous constant.

$$K_v = 32.84\,Ns\,/\,m; K_c = 20N$$

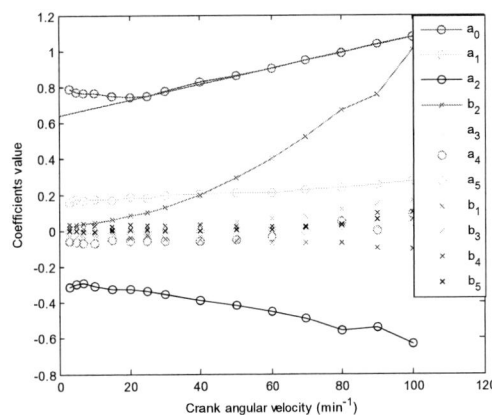

Fig. 6. Fourier coefficients of the load torque estimation

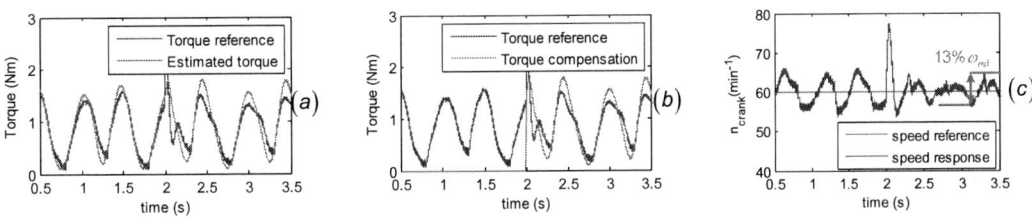

Fig. 7. Theoretical method: a) Estimated load torque; b) Load torque compensation; c) speed response

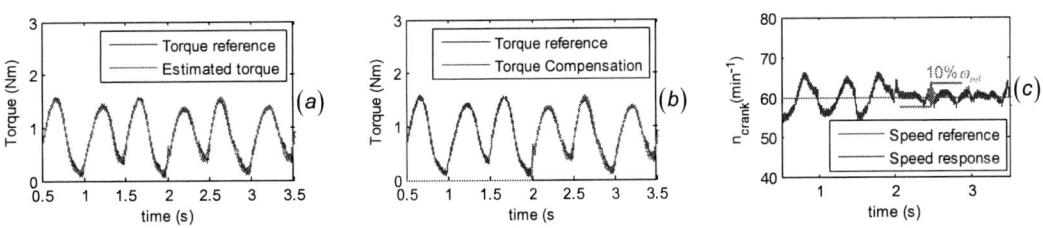

Fig. 8. Look-up table method: a) Estimated load torque; b) Load torque compensation; c) speed response

The theoretical method is carried out after the identification of the Coulomb and viscous constants.

© VDE VERLAG GMBH · Berlin · Offenbach

For the operation of the system with a feed-forward control scheme, the switch *"Compensation"* in Fig. 5 is closed. Fig.7 (a), (b), (c) show the speed response, the estimated load torque and the compensation load torque, respectively. Fig. 8 depicts the results corresponding to the look-up table based method. Fig. 7(c) and Fig. 8(c) illustrate that both methods improve the quality of the speed control in the steady state. However, the look-up table based method leads to smaller speed ripples as it yields more precise estimated results, see Fig. 7(a) and Fig. 8(a). One of the reasons for this behavior is, unlike the theoretical method, the look-up table method takes into account the nonlinear friction.

The effectiveness of the look-up table method has been verified by two other experiments associated with the changes of the mechanical parameters. For the calculation of the look-up tables, the reference speed was set to 55 min^{-1}. First r_1 was changed from 50 mm to 85 mm. The corresponding results are shown in Fig. 9. The increase of r_1 leads to an increase of the load torque as defined by (7). As it can be seen in Fig. 9(a), the look-up table based method still provides good estimation results. Fig. 9(c) shows that the speed ripple is significantly reduced in the feed-forward control scheme. For the second experiment, a mass of 1.7 kg was added to the slider. The corresponding results depicted in Fig. 10 demonstrate that the look-up table method still work well in this case.

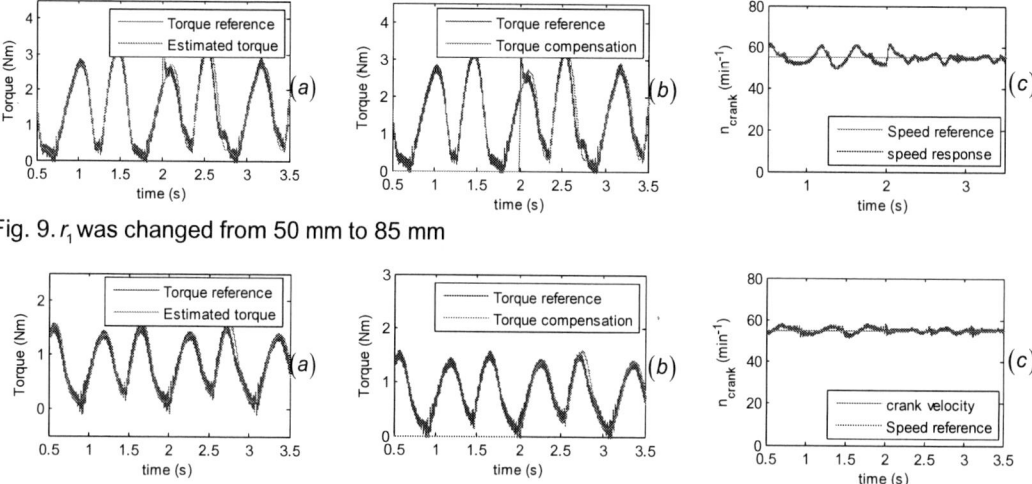

Fig. 9. r_1 was changed from 50 mm to 85 mm

Fig. 10. A mass of 1.7 kg was added to the slider

4.3. Dynamic Performance

The compensation signal has been tested in the dynamic state. Three types of compensators were tested:

i.   No torque compensation
ii. Estimated load torque from the look-up table method
iii. The compensator in ii) augmented with the acceleration component $J_{total} \dfrac{d\omega_{crank}}{dt}$.

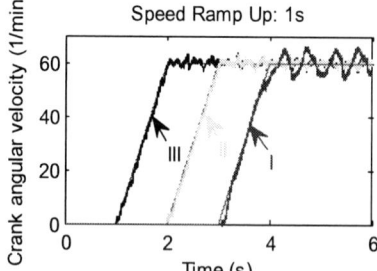

Fig.11. Performance of the three compensators

A ramp reference speed was used to verify the performance of the three compensators. Fig. 11 demonstrates that the speed responses have been improved by applying the load torque compensation. In addition, the compensator iii) works more effective than the compensator ii) only in the low speed region since the augmented term can be compared to other terms when the crank is driven at low speed.

5. Conclusion

Two methods for the estimation of the coefficients of the Fourier series used for the analytical description of the load torque in a mechanical system are proposed. It was determined that the look-up table based method works better than the theory based method as it takes into account the effects of the nonlinear friction. The estimated load torque was used in a feed-forward control scheme to improve the speed response. It was observed that the torque compensation signal helped to reduce the ripple of the crank angular velocity in both dynamic and steady states.

REFERENCE

[1] M. Hinkkanen S. E. Saarakkala, "Identification of two-mass mechanical systems in closed-loop speed control," in *IECON 2013 - 39th Annual Conference of the IEEE Industrial Electronics Society*, 2013, pp. 2905-2910.

[2] M. Hinkkanen S. E. Saarakkala, "State- Space Speed Control of Two-mass Mechanical Systems: Nanlytical Tuning and Experimental Evaluation," *IEEE Trans. Ind. Appl.*, vol. 50, no. 5, pp. 3428-3437, Sep. 2014.

[3] T. Matsuda, M. Kanno, K. Saito, T. Sukegawa T. Ohmae, "A Microprocessor-Based Motor Speed Regulator Using Fast-Response State Observer for Reduction of Torsional Vibration," *IEEE Trans. Ind. Appl.*, vol. IA-23, no. 5, pp. 863-871, Sep. 1987.

[4] N. Matusi M. Iwasaki, "Robust Speed Control of IM with Torque Feed-Forward Control," *IEEE Trans. Ind. Electron.*, vol. 40, no. 6, pp. 553-560, Dec. 1993.

[5] H. Iseki, K. Sugiura Y. Hori, "Basic consideration of vibration suppression and disturbance rejection control of muti-inertia system using SFLAC (state feedback and load acceleration control)," *IEEE. Trans. Ind. Appl.*, vol. 30, no. 4, pp. 889-896, Jul. 1994.

[6] T. Makino, H. Sato N. Matsui, "Auto-compensation of torque ripple of DD motor by torque observer," in *Conference Record of the 1991 IEEE Industry Applications Society Annual Meeting*, 1991, pp. 305-311.

[7] J. Deur, I. Kolmanovsky D. Pavkovic, "Adaptive Kalman Filter-Based Load Torque Compensator for Improved SI Engine Idle Speed Control," *IEEE Trans. Control Syst. Technol.*, vol. 17, no. 1, pp. 98-110, Jan. 2009.

[8] Z. Wang, C. Xia T. Shi, "Speed Measurement Error Suppression for PMSM Control System Using Self-Adaption Kalman Observer," *IEEE. Trans. Ind. Electron.*, vol. 62, no. 5, pp. 2753-2763, May 2015.

[9] F. Holzwelßig H. Dresig, *Dynamics of Machinery*. Berlin: Springer Berlin Heidelberg, 2010.

[10] R. Isermann, *Identifikation dynamischer systeme*. Berlin, Heidelberg: Springer, 1992.

© VDE VERLAG GMBH · Berlin · Offenbach

Fast current waveform calculation algorithm for a six phase switched reluctance machine

Jacek, Borecki, University of Bremen – Institute for Electrical Drives, Power Electronics, and Devices Otto-Hahn-Allee NW 1, 28359 Bremen, Germany, borecki@ialb.uni-bremen.de
Bernd, Orlik, University of Bremen – Institute for Electrical Drives, Power Electronics, and Devices Otto-Hahn-Allee NW 1, 28359 Bremen, Germany, borlik@ialb.uni-bremen.de

Abstract

A novel current waveform calculation method for six phase switched reluctance machines is presented. The approach described in this paper, allows fast offline computation of current waveforms that result in a ripple-free output torque. Moreover, the algorithm simplifies global optimization problems due to the low number of input parameters. Thanks to the sine base initial waveform, only two parameters are necessary to carry out local optimization for minimum current RMS value and/or other quantities.

1. Motivation

After significant price fluctuations of the rare earth permanent magnets (PM) in the recent years it became obvious that they are a vulnerable part of the supply chain in ever-growing need for high performance electric motors. Moreover PMs are sensitive to high temperature and complicated to handle in production environment. Alternatives to the PM based electric machines are, therefore, necessary. As a result of the above, the switching reluctance machines (SRM) started regaining the interest of the industry. The main reasons behind it are the construction specific advantages and the relatively low costs of the power electronics components nowadays. Lack of rare earth magnets and simple rotor construction based exclusively on laminated steel, ensure a cost effective production as well as extreme robustness. Additionally, if not energized, they do not affect other mechanically coupled machines which is highly desired in hybrid and redundant system applications. Furthermore, an extraordinary low mutual coupling effect in this type of machines allows the impact of phase short circuit failure on counter torque production to be neglected. The major disadvantage of the SRM on the other hand is the inherent torque ripple that influences the reliability of the mechanical components. However, advanced current calculation algorithms are proposed to address this problem [1]–[7]. Moreover, to take the advantage of SRMs fault tolerance, a six phase motor or two three phase motors with relative phase shift of 180 degrees of the electrical period can be used. Still, additional degrees of freedom in form of multiple, simultaneously interacting phases cause large performance degradation, extensive computing time or even inability to apply most of the algorithms. In order to overcome these limitations a new calculation method is proposed.

2. Functional principle of the algorithm

2.1. Necessary machine parameters

As for all of the advanced offline calculation methods the proposed algorithm requires detailed information about the machine as well. The following parameters are necessary:
- Torque dependency on current and rotor position Fig. 1
- Flux linkage dependency on current and rotor position
- Winding resistance
- Number of turns
- Number of pole pairs in stator and rotor

PCIM Europe 2016, 10 – 12 May 2016, Nuremberg, Germany

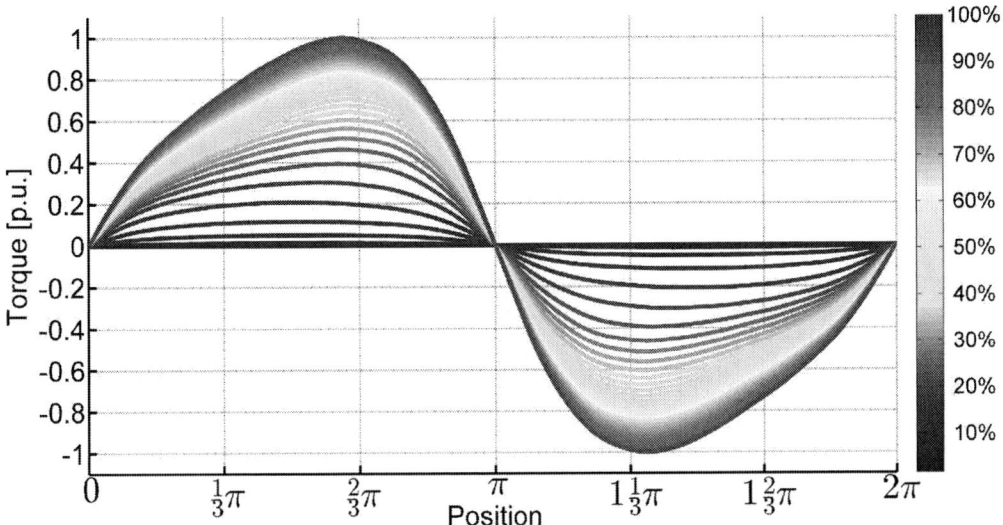

Fig. 1. Torque dependency on position and current (normalized to the nominal values)

This parameters can be obtained through analytical computations [8], [9], Finite Element Simulation (FEM) and a variety of measurement methods with [10], [11] and without rotor clamping [12], [13]. The most promising calculation algorithm proposed in [2] turned out to give relatively low energy efficient results (based solely on the current RMS value) in the low speed range when adapted to six phase SRM. However, the idea of choosing a well defined start point called critical angle was further explored.

2.2. Critical angle definition

Example torque waveforms for a constant current through the whole electrical period are presented in Fig. 2. Naturally, the development of negative torque should be avoided to ensure energy efficient operation, therefore, only the positive torque period is considered. Taking into account the characteristics in Fig. 1, an interesting particularity of SRM can be noticed, i.e. no matter how high the current supplied in the winding is, in the aligned and unaligned position the developed torque would be equal to zero. These particular positions are chosen in the proposed algorithm and are defined as critical positions, as only two phases are normally active in the torque production at this instance. Due to that observation

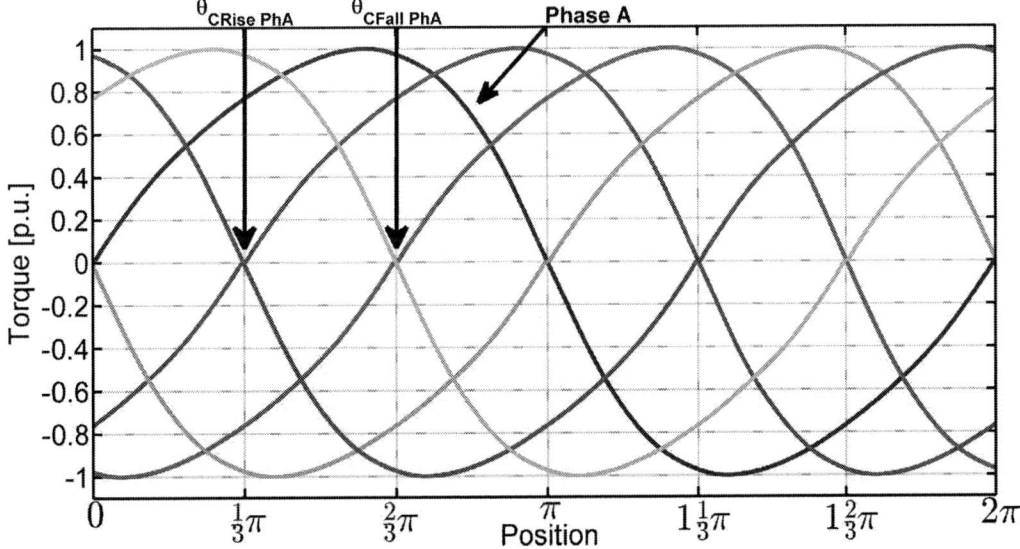

Fig. 2. Torque produced by all six phases at constant current (normalized to the maximum value)

© VDE VERLAG GMBH · Berlin · Offenbach

249

first optimization parameters can be defined, i.e. torque share factor between the considered phase and N-1 or N+1 phase, respectively. After choosing the torque sharing factor the torque values of each phase are known for all 6 critical angles. Next, based on the machine parameters mentioned in 2.1, current and flux linkage can be simply computed. In consequence six well defined points are found which are necessary for further calculations of the initial waveform.

2.3. Sine based initial waveform

Output torque ripple compensation can be achieved easily when the state of all less significant phases are known and the phase with highest torque/current ratio (most significant) is used for that purpose. In such a case an initial current waveform definition is necessary to start the compensation for ripple-free output torque. Following the example in Fig. 2, in order to compensate the torque generated by phase A between 1/3 π and 2/3 π, the torque waveform of phase F and B has to be defined. Because the two critical points are defined as optimization parameter, a waveform that could be described by minimum number of parameters and as close as possible to the final result must be found. Previous optimization methods for three phase machines [7] have shown that the results are relatively similar to a half period of a sinusoid. Such a sine waveform can be described with two points at a known angle and frequency. Based on the torque share parameter at the critical angles and the torque dependency of current and position, the current required to develop this torque can be calculated. That leads to only one additional parameter which is the frequency of the initial waveform. Examples of such an initial current waveform for variable frequency, at torque share values equal to 0,6 and 0,5 respectively are shown in Fig. 3 (normalized to the nominal value).

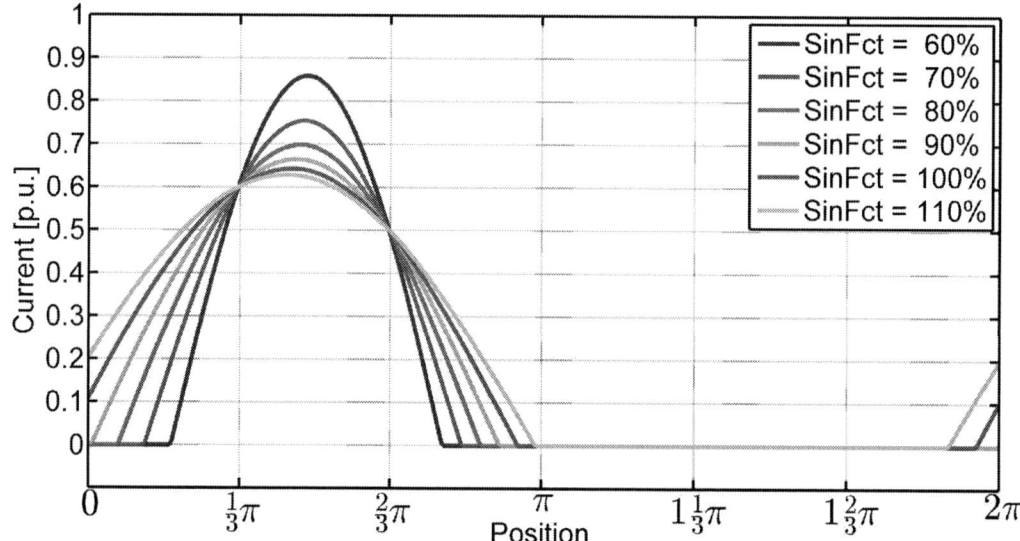

Fig. 3. Sine based initial current waveform examples

2.4. Torque ripple compensation

The torque ripple compensation procedure made in multiple stages is described as follows.
- Initial waveform calculation and its results (Stage 1)
- Voltage limiting of the initial current waveform (Stage 2a)
- Negative torque compensation at the critical angle (Stage 2b)
- Compensation with the most significant phase in the clockwise direction (Stage 3a)
- Compensation with the most significant phase in the counter-clockwise direction (Stage 3b)
- Compensation with a less significant phase (Stage 4)
- Dual phase compensation (Stage 5)

The third stage as well as the last two stages are mostly used either in the high speed operation where the voltage constrain is distinct or for the suboptimal parameters of the initial waveform.

Stage 1 - Initial results

After the initial waveform calculations, the resulting torque and flux linkage can be calculated. Computed flux linkage is used to calculate the voltage required to build up the initial current waveform. At this point there is still no constrain on the converter voltage i.e. DC bus voltage. These initial results are shown in Fig. 4 with a dashed line. Torque values are normalized to the torque reference value, voltage is normalized to the DC bus value and the remaining quantities i.e. current and flux linkage are scaled to their local maximum.

Stage 2a - Voltage limiting of the initial current waveform

In the following example shown in Fig. 4, lack of the voltage constrain can be noticed, as it rises largely above the value that can be supplied. In order to limit the current slope to the maximum voltage available the following procedure has to be carried out. In case of magnetization the maximum current step rise is calculated for each discrete position until the current reaches zero. The same procedure is done for the demagnetization part with the difference that the calculation direction as well as the voltage limitation is of the opposite sign.

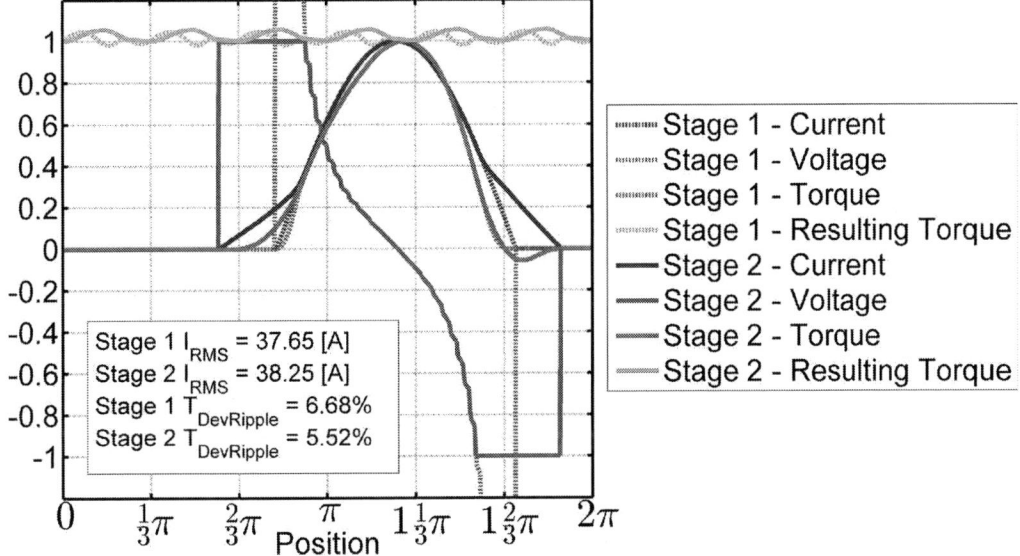

Fig. 4. Compensation comparison in "Stage 1" and "Stage 2"

Stage 2b - Negative torque compensation at critical angle

As already mentioned, at the position instance defined as critical angle only two phases are involved in the torque production namely phase N and N+1 if the angle on falling edge is considered. However, in the high speed range phase N+3 and N-2 can still be energized due to the insufficient DC bus voltage. This can be iteratively compensated by multiple run through the Stages 1 - 2b with adjustment of the local (at critical angle) torque reference value until the resulting torque will be in the set error limit range.

Stage 3 - Compensation with the most significant phase

At this stage the most significant phase is used to obtain the resulting torque equal the reference torque. According to the example shown in Fig. 2, for 1/6 of the electrical period only one phase has the best torque to current ratio. Due to the fact that the torque produced by the surrounding phases is calculated in the previous stages, the considered phase can be used to compensate for ripple-free output. The result of such compensation can be noticed with comparison to the result of stage 2 in Fig. 5. This compensation procedure can be done starting the computations on the rising edge (Stage 3a) as well as on the falling edge (Stage

3b). In both cases results will differentiate, if during the calculations voltage necessary to achieve the required current slope will exceed the DC bus voltage. In such a case, the resulting torque will not be fully compensated to get ripple-free output due to the voltage limitation.

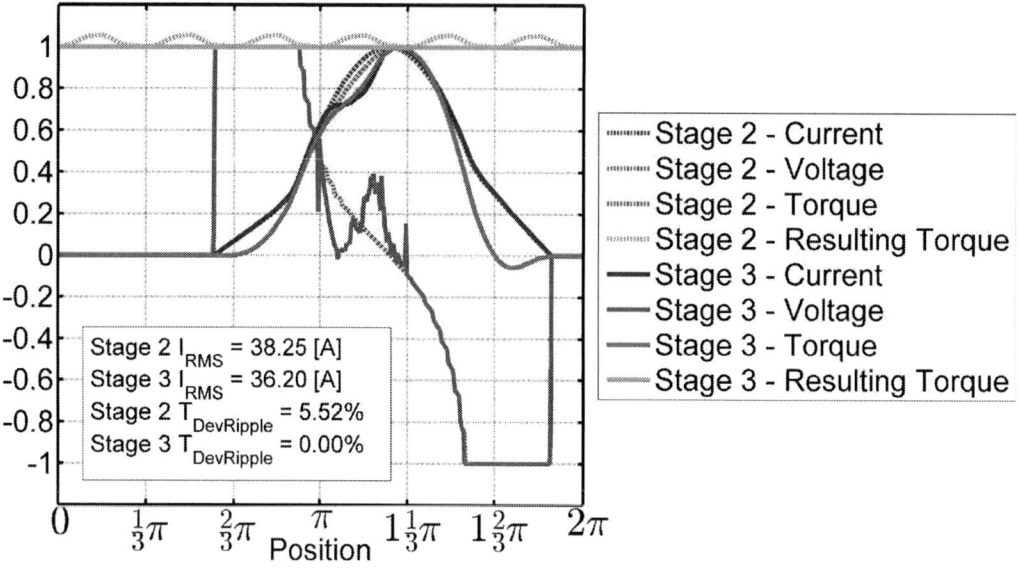

Fig. 5. Compensation comparison in "Stage 2" and "Stage 3"

Stage 4 - Compensation with less significant phase

When resulting output waveforms formed in the Stage 3a and/or 3b are not satisfactory, the next phase with highest torque to current ratio can be use to compensate the output torque. Naturally this can influence the torque before or after the voltage limiting respectively to the direction used in Stage 3 (a or b). As a consequence, the calculation done in Stage 3 has to be repeated on this waveforms in order to compensate for the newly added ripple. It is worth to note that depending on the actual position and voltage limitation some phases can only compensate for resulting torque under or above the reference torque. Based upon that, the adequate phase should be chosen for compensation at this stage.

Stage 5 - Dual phase compensation

Parallel to the Stage 4, an alternative method is used to obtain ripple-free output torque. Here, however, instead of relying on a single phase to compensate the particular torque ripple as it is done in Stage 4, both less significant phases namely N-1 and N+1 are used simultaneously with share coefficients equal to the actual torque ratio between those two phases.

2.5.   Calculation results

The following Fig. 6 and Fig. 7 are presenting snippets of the results, single phase and all six phases respectively. In Fig. 6 all of the waveforms are normalized to their maximum value, and only a single phase result is shown for the clarity. It can be noticed that the converter voltage limitation is considered as well. Fig. 7 on the other hand proves that the resulting torque is ripple-free, even if the converter voltage limitation causes a small negative torque at the demagnetization stage. All waveforms in this figure are normalized to the torque reference value. Moreover, the additional information such as required voltage waveform can be used in the controller structure as a fed-forward quantity to improve the performance of the current controller.

PCIM Europe 2016, 10 – 12 May 2016, Nuremberg, Germany

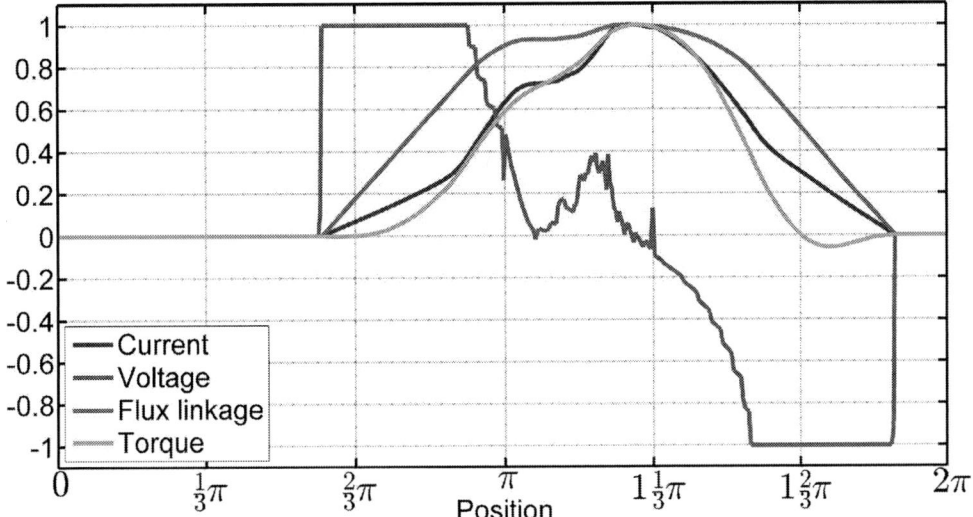

Fig. 6. Detailed result of a single phase (normalized to their maximum values)

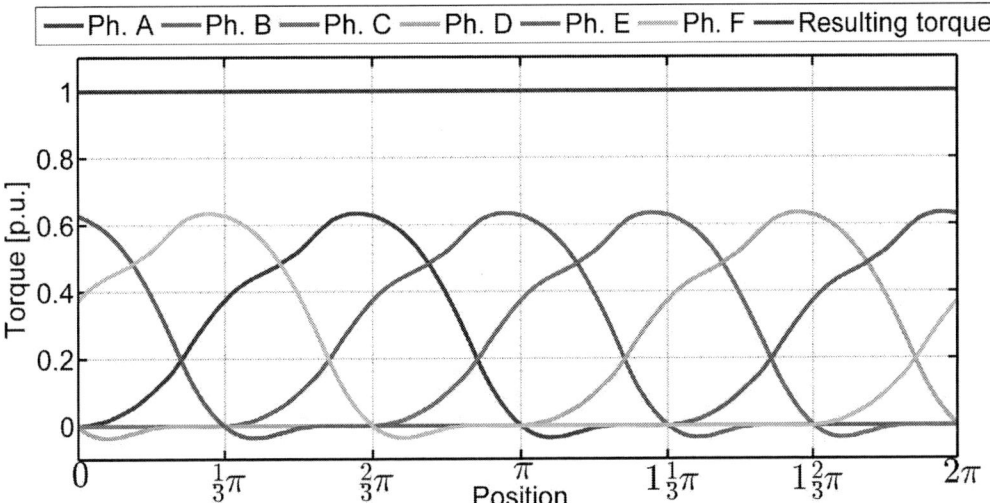

Fig. 7. Torque produced by all six phases and the resulting torque (normalized to the torque reference value)

2.6. Optimization

Number of optimization methods can be used based on this method in order to find the best current waveforms for each operating point. A finitely terminating method and genetic algorithm based method were carried out in this work. The optimization has been done in order to minimize the copper losses during the operation. An example of such a local optimization can be seen in Fig. 8 where the parameter influence is shown. However, this subject goes beyond the scope of this paper and will not be described in detail.

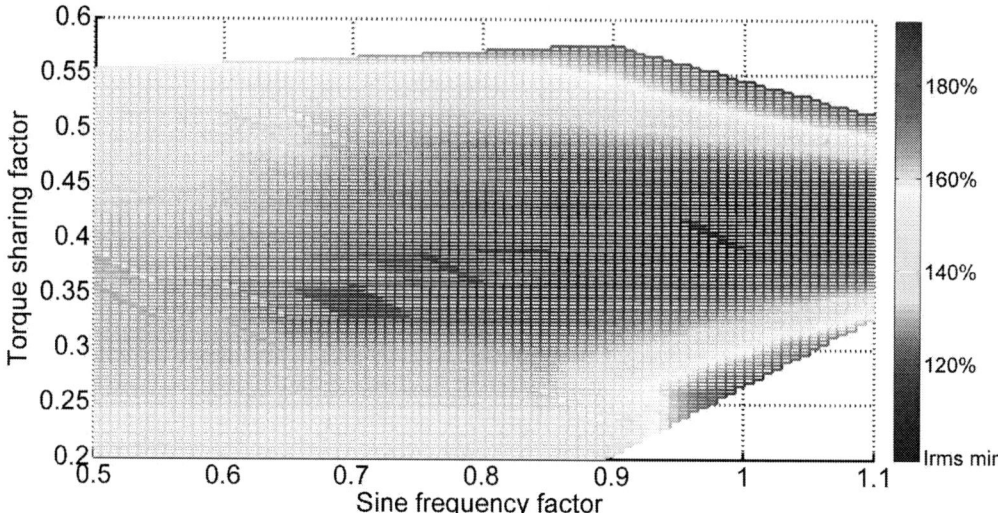

Fig. 8. Influence of the input parameters on the RMS current value for a single operating point

3. Conclusions

By variation of the torque share and sine wave frequency parameters, an optimal current waveform for ripple-free output and minimum current RMS value can be achieved. According to the simulation results, current waveform calculations for a six phase machine give a larger achievable range of speed than the same calculations made for two three phase machines where both are generating ripple-free torque at half of the required value. Furthermore, the computation time required for the method proposed in this publication is much shorter than the universal method based on dynamic programming proposed previously [7].

References

[1] P. C. Kjaer, J. J. Gribble, and T. J. E. Miller, "High-grade control of switched reluctance machines," *IEEE Transactions on Industry Applications*, vol. 33, no. 6, pp. 1585–1593, Nov. 1997.

[2] P. C. Kjaer and J. J. Gribble, "Instantaneous torque control," in *Electronic Control of Switched Reluctance Machines*, T. J. E. Miller, Ed. Elsevier, 2001, pp. 98–132.

[3] R. Mikail, I. Husain, Y. Sozer, M. S. Islam, and T. Sebastian, "Torque-Ripple Minimization of Switched Reluctance Machines Through Current Profiling," *IEEE Transactions on Industry Applications*, vol. 49, no. 3, pp. 1258–1267, May 2013.

[4] J. Ye, "Advanced Control Methods for Torque Ripple Reduction and Performance Improvement in Switched Reluctance Motor Drives," Thesis, 2014.

[5] J. Ye, B. Bilgin, and A. Emadi, "An Offline Torque Sharing Function for Torque Ripple Reduction in Switched Reluctance Motor Drives," *IEEE Transactions on Energy Conversion*, vol. 30, no. 2, pp. 726–735, Jun. 2015.

[6] X. Gao, X. Wang, Z. Li, and Y. Zhou, "A Review of Torque Ripple Control Strategies of Switched Reluctance Motor," *International Journal of Control and Automation*, vol. 8, no. 4, pp. 103–116, Apr. 2015.

[7] J. Borecki, M. Joost, H. Groke, and B. Orlik, "Current waveform optimization for ripple-free output torque of transverse flux reluctance machines," presented at the ACEMP - OPTIM - ELECTROMOTION Joint Conference, Side, Turkey, 2015.

[8] E. Pădurariu, L. Someşan, I. A. Viorel, C. S. Marţiş, and O. Cornea, "Switched reluctance motor analytical models, comparative analysis," in *2010 12th International Conference on Optimization of Electrical and Electronic Equipment (OPTIM)*, 2010, pp. 285–290.

[9] W. M. Arshad, T. Backstrom, and C. Sadarangani, "Analytical design and analysis procedure for a transverse flux machine," presented at the International Electric Machines and Drives Conference - IEMDC 2001, Cambridge, MA, USA, pp. 115–121.

[10] A. D. Cheok and Z. Wang, "DSP-Based Automated Error-Reducing Flux-Linkage-Measurement Method for Switched Reluctance Motors," *IEEE Transactions on Instrumentation and Measurement*, vol. 56, no. 6, pp. 2245–2253, Dec. 2007.

[11] R. Krishnan and P. Materu, "Measurement and instrumentation of a switched reluctance motor," in , *Conference Record of the 1989 IEEE Industry Applications Society Annual Meeting, 1989*, 1989, pp. 116–121 vol.1.

[12] S. A. Hashemi, A. Rashidi, and S. M. S. Nejad, "Analysis flux linkage measurement method without using rotor clamping device for switched reluctance motors," in *Power Electronics, Drives Systems Technologies Conference (PEDSTC), 2015 6th*, 2015, pp. 298–303.

[13] L. Shen, J. Wu, S. Yang, and X. Huang, "Fast Flux Linkage Measurement for Switched Reluctance Motors Excluding Rotor Clamping Devices and Position Sensors," *IEEE Transactions on Instrumentation and Measurement*, vol. 62, no. 1, pp. 185–191, Jan. 2013.

Process requirements-based adaptive PWM for improved efficiency of machine tool feed drives

Matthias Braband[1], Matthias.Braband@isw.uni-stuttgart.de
Florian Frick[1], Florian.Frick@isw.uni-stuttgart.de
Armin Lechler[1], Armin.Lechler@isw.uni-stuttgart.de
Alexander Verl[1], Alexander.Verl@isw.uni-stuttgart.de
[1] Institute for Control Engineering of Machine Tools and Manufacturing Units (ISW), Germany

Abstract

Feed drives of machine tools are usually controlled by cascaded PI controllers combined with Pulse Width Modulation (PWM). Based on the strictest performance and accuracy requirements, the drive control system is configured, often necessitating a high PWM-frequency. While improving the control bandwidth, the energy efficiency and lifetime of the power electronic decrease significantly due to switching losses. Analyzing the production process performed by the machine tools reveals that the requirements must only be met in critical process states, e.g. when the final surface is manufactured. Much laxer requirements are acceptable in other states. Therefore, a significant amount of energy could be saved if the minimal required PWM frequency would be used in each system state. To exploit this unused potential, an algorithm deciding on the frequency and a control system allowing to change the frequency during runtime are prerequisites. Both are addressed by the presented research. To decide on the frequency, an algorithm based on process knowledge is proposed. The key challenge of a control loop with a non-constant frequency are the parameters. An efficient way to derive these is presented. The validity of the parameter calculation as well as the energy savings are experimentally validated.

1. Introduction

Feed drives are crucial components of any machine tool since they directly affect their performance and the quality of the produced goods. For both, a high bandwidth of the control is necessary. The market is dominated by drive systems which are based on cascaded Proportional Integral (PI) controllers combined with Pulse Width Modulation (PWM), referred to as PWM-PI. In order to achieve the required bandwidth, a high PWM switching frequency is necessary [1]. To demonstrate this direct relation, the same feed drive was analyzed regarding its bandwidth using different PWM frequencies (5 kHz and 10 kHz). Figure 1 shows the resulting command response Bode plots, indicating a significant phase drop when reducing the PWM frequency. While increasing the PWM frequency is advantageous for the control performance, the efficiency of the used power electronic, usually IGBTs, decreases significantly. In an experimental test setup consisting of standard drive components, the energy efficiency was analyzed with different PWM frequencies. Measurements show that especially during low speed, the power consumption is up to 30% more when using the higher PWM frequency. The lifetime is affected in a similar way, as shown in [2].

In industrial applications, PWM frequencies are selected based on the strictest requirements, usually between 5 and 20 kHz and remain constant throughout. An analysis of the utilization

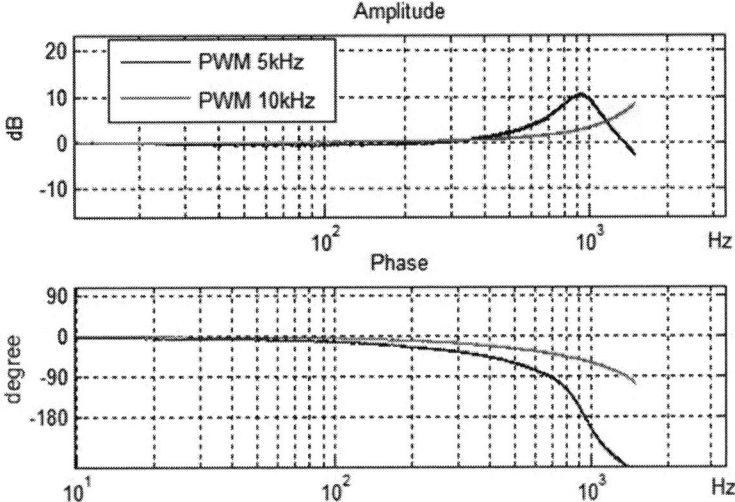

Fig. 1: Influence of different PWM frequencies on the current control bandwidth

of typical machine tools reveals that there are many states and phases of the production process in which a lower performance or accuracy is not disadvantageous for the final results. For example, this is the case during rough-machining, positioning or changing of the tool. These process states take up a significant amount of the production time as can be seen by analyzing a typical reference part [3].

The overall energy efficiency could be increased significantly by using the optimal PWM frequency at each moment in time. Various techniques are known from research utilizing non-constant PWM frequencies with different objectives. Two examples are the Flat-Top and Over-Modulation, which reduce the average frequency by switching only two instead of three phases per cycle. Other methods use a spectrum instead of a constant frequency [4] or manipulate the frequency based on current ripples [5], thermal load [6] or EMI [7]. Completely omitting a base frequency and using an entirely different control structure are methods like Direct Current Control [8].

While providing certain advantages, there are major drawbacks if these methods are applied to feed drives. The strict requirements regarding dynamic and robustness are not always fulfilled and the existing approaches do not exploit the full potential regarding the energy consumption. A basic reason is that if a method is not aware of the actual requirements of the process at a certain point in time, it is impossible to perform optimally. Approaches using different control structures are difficult to integrate into existing systems and have problems regarding their acceptance due to difficult parametrization.

The presented research focuses on a new process requirement-based approach to adapt the PWM frequency, keeping the standard cascaded PI controllers for easy parametrization. A key challenge when operating a controller with different PWM frequencies and therefore varying controller frequencies is the adaption of the controller parameters for each frequency. Since it is not acceptable to optimize the cascaded controller for each frequency, a method is required allowing an automated configuration of the control loops at different frequencies.

2. Online PWM Frequency Adaptable Control Structure

In order to operate the control loop at different frequencies, parameters are required for all desired frequencies. A method to derive these, based on the maximum PWM frequency, is presented. The basic idea is to get the same behavior in the time domain for every PWM frequency relative to the sample time. Therefore, the controller must have the same overshoot and damping properties. This is done by calculating the proportional and integral gain parameters based on the root locus for every PWM frequency.

The considered PWM frequencies for the adaptable control structure are

$$f_{PWM,i} = \frac{f_{max}}{i} \quad \text{with} \quad \{i \in \mathbb{N} | 1 \leq i \leq i_{fmin}\} \tag{1}$$

where f_{max} is the maximum frequency. i_{fmin} denotes the index for the minimum PWM frequency applied to the system. The current controller is synchronized to the PWM frequency using regular sampling, resulting in a sampling frequency for the controller of

$$f_{S,i} = 2 f_{PWM,i} \tag{2}$$

Changing the PWM frequency has an impact on the current control loop. Therefore, the parameters of the current, velocity and position loop must be adapted to the new switching frequency.

2.1. Current Control Loop

All new control values are derived from the parameters of the highest frequency. Thereby the natural frequency relatively to the sampling time is kept invariant. The transfer function of the current control loop in rotating d, q coordinates for a permanent magnet synchronous motor (PMSM) is given by

$$G_{cs}(s) = \frac{1}{R_s} \frac{1}{\tau s + 1} \quad \text{with} \quad \tau = \frac{L_s}{R_s} \tag{3}$$

where R_s and L_s are the stator resistance and inductance. By reducing the PWM switching frequency significantly, the assumption $f_{S,i} \gg \frac{1}{\tau}$ allowing to assume the controller as continuous becomes inapplicable. Consequently, the transfer function needs to be considered in the discrete z-space given by

$$G_{cs,i}(z) = \frac{1}{R_s} \frac{1 - e^{-\frac{T_{s,i}}{\tau}}}{z - e^{-\frac{T_{s,i}}{\tau}}} \quad \text{with} \quad T_{s,i} = \frac{2}{f_{PWM,i}} \tag{4}$$

The transfer function of the PI controller is transformed by applying the Euler-Backward substitution, resulting in

$$G_{cc,i}(z) = K_{pc,i} \left(1 + \frac{K_{ic,i} T_{s,i} z}{z - 1} \right) \tag{5}$$

As seen in Eq. 4 and 5, changing the PWM frequency will change the sample time $T_{s,i}$ of the discrete plant and also the PI current controller. Therefore, the controller parameters for the control loop and also for the outer loops must be adapted to the new switching frequency.

To determine the control parameters $K_{pc,i}$ and $K_{ic,i}$, it is assumed that the root of the current controller compensates the pole of the system. Based on this the parameters for the integral part of the controller are calculated by

$$K_{ic,i} = \frac{1 - e^{-\frac{T_{s,i}}{\tau}}}{T_{s,i}e^{-\frac{T_{s,i}}{\tau}}} \tag{6}$$

To get the same behavior relative to the PWM frequency in the time domain, the positions of the closed loop poles in the z-plane must be the same for every frequency. Based on the roots of the characteristic equation

$$1 + G_{cs,i}(z)G_{cc,i}(z) = 0 \tag{7}$$

for the highest PWM frequency, the values for the proportional part $K_{pc,i}$ are calculated by solving the characteristic equation for all other frequencies.

2.2. Velocity Control Loop

The parameters of the velocity loop must be adjusted in a similar way. The structure of the control loop is depicted in Fig. 2.

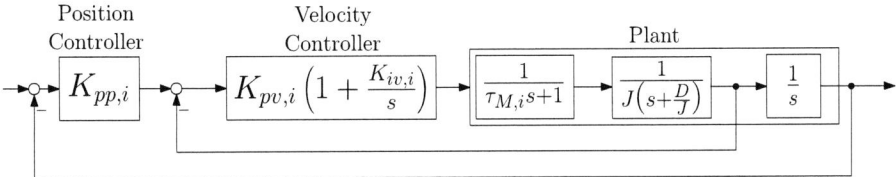

Fig. 2: Velocity and position control loop

The velocity and also later the position control loop can be assumed as continuous transfer functions since their sample frequency is much higher than the bandwidth of the velocity control loop. The continuous transfer function for the velocity PI controller is defined as

$$G_{vc,i}(s) = K_{pv,i}\left(1 + \frac{K_{iv,i}}{s}\right) \tag{8}$$

The plant, including the drive and the current controller, is given by

$$G_{vs,i} = \frac{1}{\tau_{M,i}s + 1}\frac{1}{J\left(s + \frac{D}{J}\right)} \tag{9}$$

where D denotes the damping and J the inertia of the motor including the axis. The current control loop is simplified to a PT1 with the equivalent time constant $\tau_{M,i}$. This time constant is defined by the step response of the current controller in the discrete time space which is calculated by

$$i_{d,q}(k) = -c \cdot i_{d,q}(k-1) - d \cdot i_{d,q}(k-2) + a \cdot u_{d,q}(k-1) + b \cdot u_{d,q}(k-2) \tag{10}$$

a, b, c and d are the coefficients given by the inverse z-transformation of the closed loop current controller and $i_{d,q}, u_{d,q}$ denotes the current and voltages of the controlled PMSM in the d, q

coordinates. The assumption that the root of the controller compensates the dominant plant pole is made for the velocity loop as well. For the compensated case the transfer function is given by

$$G_{vscomp,i} = K_{pv,i} \frac{1}{Js(\tau_{M,i}s + 1)} \tag{11}$$

where $\tau_{M,i}$ varies depending on the PWM frequency. According to the root locus plot, as seen in Fig. 3(a), only the cases with real poles and with a complex pole pair for the closed loop are relevant.

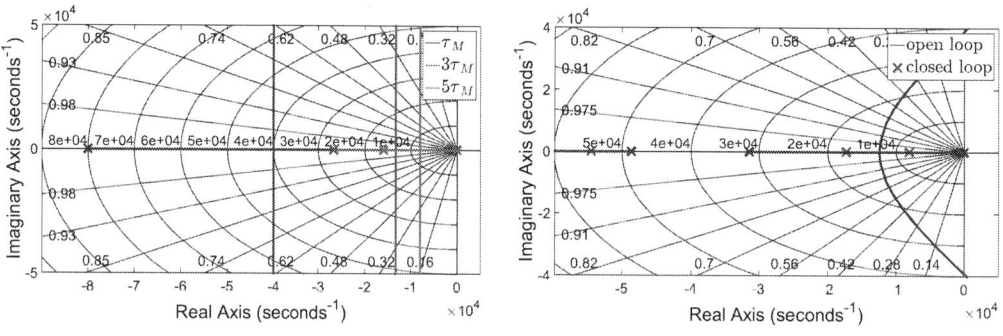

(a) Velocity loop for different time constants $\tau_{M,i}$ (b) Position Loop with $\tau_M = \frac{1}{80000}$

Fig. 3: Root locus plots of different velocity and position loops for the test setup with $D = 0.0005$Nms and $J = 0.65$kgcm^2.

To derive $K_{pv,i}$ for real roots, the ratio of the distance between the breakaway point of the root locus and the closed loop pole and between the breakaway point and the left pole $\frac{1}{\tau_{M,i}}$ must remain the same for all frequencies. The new closed loop poles are calculated by

$$s_{v,i} = \frac{s_{v,1} - s_{vba,1}}{\frac{1}{\tau_{M,1}} - s_{vba,1}} \left(\frac{1}{\tau_{M,i}} - s_{vba,i} \right) + s_{vba,i} \tag{12}$$

where the breakaway point $s_{ba,i}$ is given by

$$s_{vba,i} = \frac{1}{2\tau_{M,i}} \tag{13}$$

$K_{pv,i}$ is calculated by using the characteristic equation of the velocity loop and is given by

$$K_{pv,i} = -Js(\tau_{M,i}s + 1) \tag{14}$$

In the second case (complex poles) the design criteria is to keep the same damping for every PWM frequency. This implies that the relation between the real and the imaginary part must be equally maintained for every frequency. The real part is defined by the breakaway point and the imaginary part is derived by using

$$\pm \Im\{s_i\} = \pm \Im\{s_1\} \frac{s_{ba,i}}{s_{ba,1}} \tag{15}$$

The new closed loop poles are defined by the new breakaway point and the imaginary part calculated by Eq. 15.

2.3. Position Control Loop

For the position control loop it is assumed that the closed loop poles of the velocity loop are only real since no position or velocity overshoot should occur in machine tool feed drives. The resulting root locus plot is shown in Fig. 3(b). The corresponding transfer function is given by

$$G_{ps,i}(s) = K_{pp,i} \frac{K_{pv,i}}{J\tau_{M,i}s^3 + Js^2 + K_{pv,i}s} \tag{16}$$

To determine the new values of $K_{pp,i}$ the same rule as in Eq. 12 is applicable. Because there is one integrator in the open loop system Eq. 12 is reduced to

$$s_{p,i} = \frac{s_{p,1} - s_{pba,1}}{s_{pba,1}} s_{pba,i} + s_{pba,i} \tag{17}$$

2.4. Adaptive Control Law

The presented method to derive the controller parameters allows to operate a controller with different PWM frequencies and adjust the parameters accordingly. In order to utilize this capability with the objective to reduce the energy consumption, an additional controller is required deciding on the actual PWM frequency. Various approaches are currently being researched, however, a straight forward approach is to select the PWM frequency directly based on information about the process. For instance, in a milling process, there are a lot of machine states which have much laxer requirements regarding accuracy and dynamic than other states. Such states are, for example, driving to the tool changer, idle modes or jogging the axis without manufacturing. Also some manufacturing steps have no high accuracy requirements such as roughing a surface. This process information is known in the higher control layers and can therefore be provided. For example, a CNC control can inform the adaptive PWM controller about the required dynamic according to the actual process step.

3. Experimental Results

The adaptive PWM algorithm with automatic parameter calculation was implemented and tested on a FPGA-based platform [9] controlling an IGBT-based inverter system. The frequencies were selected by a high level control. A PMSM with incremental encoder feedback is used. The DC voltage is rated at 48V and the motor parameters are $L_s = 4.1\text{mH}$, $R_s = 0.25\Omega$, $D = 0.0005\text{Nms}$, $J = 0.65\text{kgcm}^2$ and a rated torque of $M_n = 1.15\text{Nm}$.

3.1. Validation of the Derived Parameters

To evaluate the correctness of the derived parameters, step responses for several frequencies and for each control loop are recorded. Fig. 4 shows the results. The assumption that no overshoot should occur is validated and the controller behaves as expected for every PWM frequency. The noise of the measured velocity is caused by the discrete derivation of the position signal. It is verified that the settling time of each controller, relative to the sample time of the current controller, is constant.

PCIM Europe 2016, 10 – 12 May 2016, Nuremberg, Germany

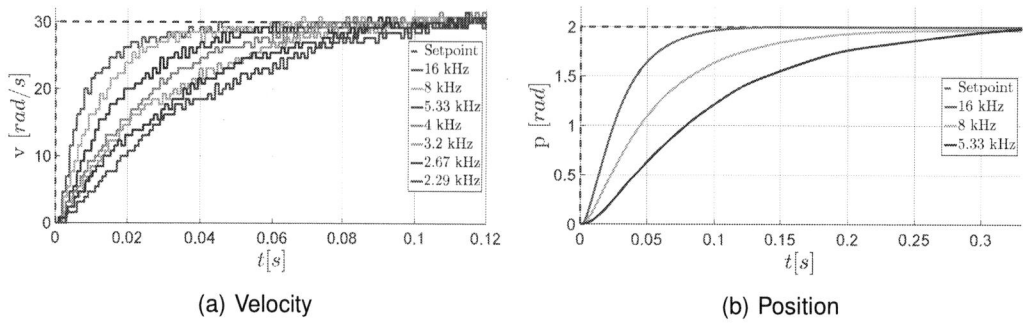

(a) Velocity

(b) Position

Fig. 4: Step responses using different PWM frequencies for a setpoint step with no additional load or disturbances

3.2. Evaluation of the Power Consumption

Energy measurements for different PWM frequencies under several working conditions were taken. Fig. 5(a) and 5(b) show the absolute and relative power saving for different motor currents using different PWM frequencies. The potential of power saving depends on the motor current. Relative to the load the switching losses are smaller at high currents. But the absolute values show, that for higher loads, more energy could be saved by reducing the PWM frequency since switching losses increase. Fig. 5(c) and 5(d) show the absolute energy saving for the velocity control loop.

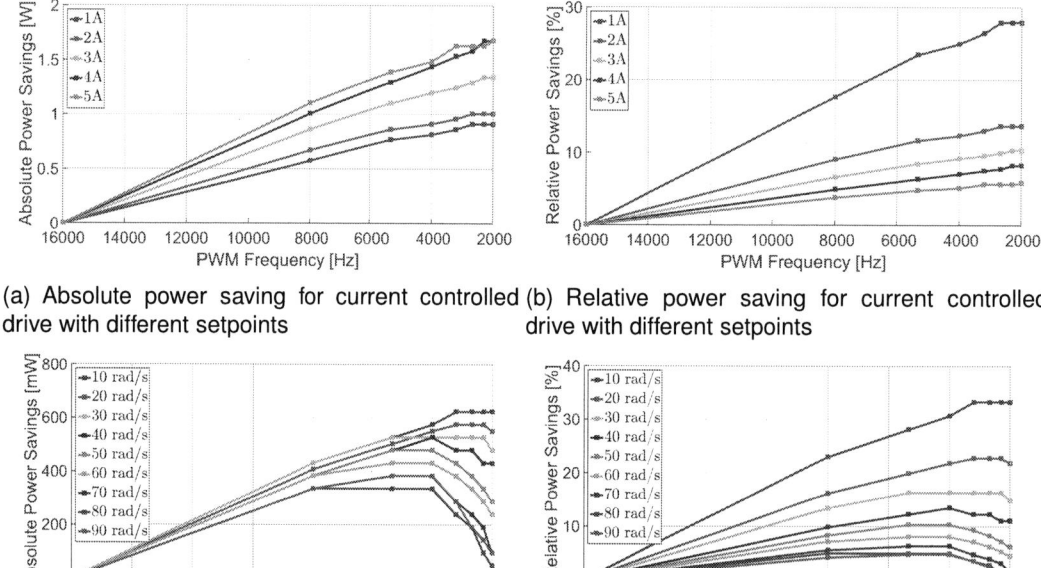

(a) Absolute power saving for current controlled drive with different setpoints

(b) Relative power saving for current controlled drive with different setpoints

(c) Absolute power saving for velocity controlled drive with different setpoints

(d) Relative power saving for velocity controlled drive with different setpoints

Fig. 5: Power saving for different operating points in current and velocity control mode

© VDE VERLAG GMBH · Berlin · Offenbach

These measurements were taken without any additional mechanical load at different motor speeds. The energy consumption is reduced in velocity control as well for certain PWM frequencies. Due to the harmonics caused by the PWM and larger inter-modulation products between the fundamental and the PWM frequency for lower PWM frequencies, there is a break even point where a lower switching frequency will not further reduce the energy consumption.

4. Conclusion and Outlook

In this paper, a method to derive the control parameters automatically for every control loop and PWM frequency based on the parameters for the highest frequency was presented. To prove the concept, experimental results for every control loop were presented. Additionally a first process requirement based adaptive PWM algorithm was introduced. It was shown that there is a significant potential for saving energy by reducing the switching losses of the IGBT module using lower PWM frequencies. In the future, the process based algorithm will be evolved and advanced system state based adaptive PWM algorithms will be developed. In addition, the influence of current harmonics needs to be reduced to improve the energy efficiency for applications with high velocities.

5. References

[1] J. Böcker. On the Control Bandwidth of Servo Drives. In *13th European Conference on Power Electronics and Applications, 2009*. IEEE, Piscataway, NJ, 2009.

[2] L. Wei, J. McGuire, and R. A. Lukaszewski. Analysis of PWM Frequency Control to Improve the Lifetime of PWM Inverter. *IEEE Transactions on Industry Applications*, 47(2):922–929, 2011.

[3] H.-H. Westermann, M. Kafara, and R. Steinhilper. Development of a Reference Part for the Evaluation of Energy Efficiency in Milling Operations. *Procedia CIRP*, 26:521–526, 2015.

[4] A. M. Trzynadlowski, K. Borisov, Y. Li, and L. Qin. A Novel Random PWM Technique With Low Computational Overhead and Constant Sampling Frequency for High-Volume, Low-Cost Applications. *IEEE Transactions on Power Electronics*, 20(1):116–122, 2005.

[5] D. Jiang and F. Wang. Variable Switching Frequency PWM for Three-Phase Converters Based on Current Ripple Prediction. *IEEE Transactions on Power Electronics*, 28(11):4951–4961, 2013.

[6] R. L. Kirlin, C. Lascu, and A. M. Trzynadlowski. Shaping the Noise Spectrum in Power Electronic Converters. *IEEE Transactions on Industrial Electronics*, 58(7):2780–2788, 2011.

[7] D. Jiang and F. Wang. Variable switching frequency PWM for three-phase converter for loss and EMI improvement. pages 1576–1583.

[8] M. Nemec, K. Drobnic, D. Nedeljkovic, and V. Ambrozic. Direct Current Control of a Synchronous Machine in Field Coordinates. *IEEE Transactions on Industrial Electronics*, 56(10):4052–4061, 2009.

[9] F. Frick, P. Zahn, A. Lechler, and A. Verl. Modular Design Approach for Model-Based Drive Control Systems on Reconfigurable Logic. *Applied Mechanics and Materials*, 704:380–384, 2014.

Comparison of Bidirectional T-Source Inverter and Quasi-Z-Source Inverter for Extra Low Voltage Application

Thomas Baier and Bernhard Piepenbreier
Friedrich-Alexander University Erlangen-Nuremberg
Chair of Electrical Drives and Machines
Cauerstr. 9, 91058 Erlangen, Germany
Email: thomas.baier@fau.de, bernhard.piepenbreier@fau.de

Abstract

Inverters with impedance source networks combine their one-stage energy conversion with the function as buck and boost converters. Due to this fact they can provide output voltages that are independent from the feeding source. This makes them very attractive to a wide range of applications. In order to evaluate the performance and find the inverter that matches the application best it is essential to compare different topologies of these inverters. This paper compares a bidirectional T-Source Inverter with a bidirectional Quasi-Z-Source Inverter for extra low voltage application. The comparison is based on the analytical analysis and the experimental results which lead to a quite contrary evaluation of the performances.

1.   Introduction

Traditional voltage source inverters (VSI) have certain limitations and problems. The output voltage is not independent from the input voltage and dead time is required to avoid short circuits between the upper and the lower switch of one phase leg. In 2002, a new inverter type that overcomes this limitations and problems was proposed by F.Z. Peng [1], [2]. The so called Z-Source inverter was the basis for a whole new family of inverters - the inverters with impedance-source networks. Up to the present day, a great variety of different topologies of these inverters has been proposed like the Z-Source inverter (ZSI) itself, the Quasi-Z-Source inverter (QZI) [3], [4] or the T-Source inverter (TSI) [5], [6], to name just a few. These inverters can be build as two-point or multilevel inverters. The great potential is enabled by their special topology consisting of inductors (magnetically coupled or uncoupled), capacitors, and diodes. In combination with a new operation state called 'shoot through' it is possible to boost the output voltage beyond the level of the input voltage. For regenerative operation an additional switch S_0 needs to be added. To find the inverter that fits best for a certain application it is necessary to compare different inverters with impedance source networks. First of all, this paper compares the QZI and TSI in an analytical analysis before the comparison is continued with the results of two experimental setups for an extra low voltage application.

2.   Analytical Analysis

The analytical circuit analysis of both inverters is based on their DC models that are illustrated in Fig. 1 and Fig. 2. In these figures, the three half bridges are reduced to only one phase leg and the load is represented by a constant current source conducting the current I_{Load}. The output voltage is called V_{out}. In steady state, the average inductor voltage and the average capacitor current respectively the capacitor charge are equal to zero. For the circuit analysis state space averaging is executed assuming ideal conditions, according to which voltage drops at the semiconductors are neglected and the transformer of the T-Source inverter assumed is to be perfectly coupled. One switching time period T_S is composed by the shoot through time T_{ST} and the non-shoot through time T_1 which is describes by (1). The non-shoot through time

PCIM Europe 2016, 10 – 12 May 2016, Nuremberg, Germany

itself is put together by the time of the active state T_{AS} and duration of the zero state T_{ZS} according to (2).

$$T_S = T_{ST} + T_1 \tag{1}$$

$$T_1 = T_{AZ} + T_{ZS} \tag{2}$$

t_{ST}, t_1, t_{AZ} and t_{ZS} are relative times referred to the switching period. If one cycle uses only active state or active state and zero state the inverter is operating in buck mode, hence the average output voltage is smaller than the input voltage according to (3). In order to achieve maximum output voltage it is reasonable to skip zero states in boost mode. In that case one switching period consists only of active state and shoot through state. The average output voltage in boost mode is higher than the input voltage (4).

$$\overline{V}_{out} \leq V_{in} \tag{3}$$

$$\overline{V}_{out} > V_{in} \tag{4}$$

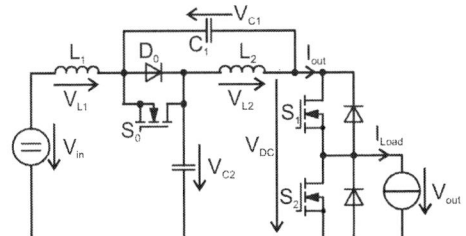

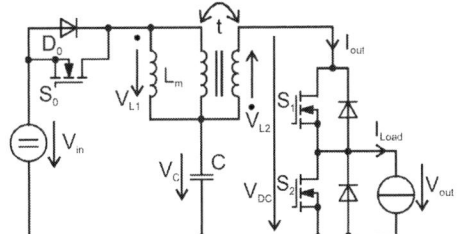

Fig. 1: DC model of the QZI   Fig. 2: DC model of the TSI

States	State of the Switches			QZI, V_{out}[V]	TSI, V_{out}[V]
	S_1	S_2	S_0		
Active State	On	Off	On	$V_{C1} + V_{C2}$	$(V_C(n+1) - V_{in})/n$
Zero State	Off	On	On	0	0
Shoot-Through	On	On	Off	0	0

Table 1: Switching States and Output voltages of QZI and TSI

2.1. Quasi-Z-Source Inverter

Using Kirchhof´s law and the condition for the steady state, the average capacitor voltages can be developed for the QZI. In non-shoot through mode, the input diode D_0 is conducting. In shoot through mode it is reverse biased. The switches S_1 and S_2 are gated on simultaneously and the phase leg is short circuited which leads to the equations (5) and (6). By rearranging these equations, the equations (7) and (8) can be developed. According to (9), the DC link voltage in non-shoot through state is the sum of the capacitor voltages. Using the developed relationships, equation (3) for the buck mode can be confirmed. As there is no shoot through in buck mode, the average capacitor voltage $\overline{V}_{C1}$ is equal to zero. $\overline{V}_{C2}$ corresponds to the input voltage V_{in}. The average output voltage in buck mode is expressed in equation (10). In boost mode with $t_{ST} > 0$, the average capacitor voltage $\overline{V}_{C1}$ rises above zero and $\overline{V}_{C2}$ exceeds the

© VDE VERLAG GMBH · Berlin · Offenbach

input voltage. In order to achieve maximum output voltage, the zero state is skipped in boost mode and the average output voltage equals $\overline{V}_{C2}$ according to equation (4).

$$\overline{V}_{L1} = (V_{in} - V_{C2})(1 - t_{ST}) + (V_{in} + V_{C1})t_{ST} = 0 \tag{5}$$

$$\overline{V}_{L2} = (-V_{C1})(1 - t_{ST}) + V_{C2}t_{ST} = 0 \tag{6}$$

$$\overline{V}_{C1} = \frac{t_{ST}}{1 - 2t_{ST}} V_{in} \tag{7}$$

$$\overline{V}_{C2} = \frac{1 - t_{ST}}{1 - 2t_{ST}} V_{in} \tag{8}$$

$$V_{DC} = \frac{1}{1 - 2t_{ST}} V_{in} \tag{9}$$

$$\overline{V}_{out,Buck} = t_{AS} V_{in} \tag{10}$$

Applying the remaining steady state condition that the capacitor current respectively the capacitor charge needs to be balanced, the average output current during shoot through state can be calculated by

$$\overline{I}_{out,ST} = 2\frac{1 - t_{ST}}{1 - 2t_{ST}} I_{Load}. \tag{11}$$

2.2. T-Source Inverter

Analog to chapter 2.1, the equations for the T-Source inverter are now developed. In non-shoot through states, the DC link voltage is calculated according to (14). For buck mode, the average capacitor voltage equals the input voltage and the mean output voltage can be described by (15). In boost mode, the average capacitor voltage exceeds the input voltage and the average output voltage is equal to the capacitor voltage. For a winding ratio of $n = N_1/N_2 = 1$, equation (5) equals (12) and equation (8) is equal to (13). In order to achieve a higher voltage gain, the winding ratio needs to fulfill $n > 1$.

$$\overline{V}_{L2} = ((V_{in} - V_C)/n)(1 - t_{ST}) + V_C t_{ST} = 0 \tag{12}$$

$$\overline{V}_C = \frac{1 - t_{ST}}{1 - (n + 1)t_{ST}} V_{in} \tag{13}$$

$$V_{DC} = V_C - V_{L2} = \frac{1}{(1 - (n + 1)t_{ST})} V_{in} \tag{14}$$

$$\overline{V}_{out,Buck} = t_{AS} V_{in} \tag{15}$$

Considering the steady state condition for the capacitor current, the average output current in shoot through state can be derived:

$$\overline{I}_{out,ST} = \frac{(1 - t_{ST})(n + 1)}{1 - (n + 1)t_{ST}} I_{Load} \tag{16}$$

2.3. Comparison of the Analytical Results

Referring to the equations derived in chapter 2.1 and 2.2, the transformer design is the main difference between these two inverters. For a turns ratio of $n = 1$, the two inverters are equal. With an increasing winding ratio, the TSI requires less shoot through time to reach a certain output voltage, in other words the boost factor rises. At the same time, the generated dc link voltage decreases, the higher the winding ratio the smaller the voltage stress of the semiconductors. Fig. 3 illustrates the described relation between shoot through time, winding ratio respectively inverter type and output voltage. Fig. 4 depictures the context of shoot through, winding ratio respectively inverter type and the dc link voltage. Another important requirement towards the semiconductors of the inverter is the ability to conduct the high shoot through currents. Again, the winding ratio of the TSI has great influence on that. Fig. 5 shows the correlation between shoot through time, winding ratio respectively inverter type and the average output current in shoot through state related to the load current. The semiconductors of the QZI need to conduct smaller currents for longer time periods. In contrast to that, the higher the turns ratio of the TSI the higher the output currents but the smaller the shoot through time. If the amplitude of the shoot through currents is weighted with their duration there are advantages for a TSI with a high turns ratio compared to the QZI. As a summary of the analytical analysis of the QZI and the TSI considering different winding ratios, it can be stated that the potential of the TSI with $n > 1$ exceeds the potential of the QZI. The properties of the TSI itself can be improved by using a higher turns ratio of the transformer.

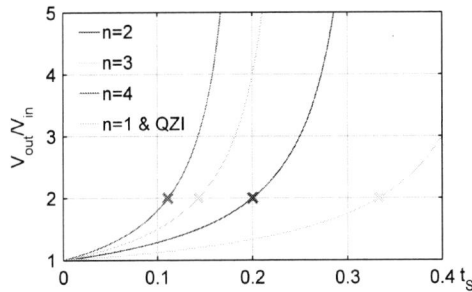
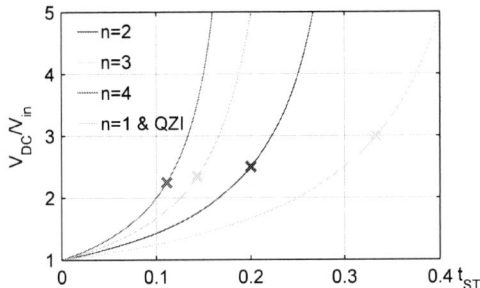

Fig. 3: V_{out}/V_{in} of the QZI and the TSI using different winding ratios, $V_{out}/V_{in} = 2$ is marked.

Fig. 4: V_{DC}/V_{in} of the QZI and the TSI using different winding ratios, $V_{out}/V_{in} = 2$ is marked.

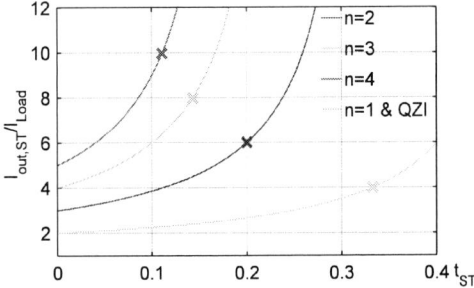

Fig. 5: $\bar{I}_{out,ST}/I_{Load}$ of the QZI and the TSI using differen winding ratios, $V_{out}/V_{in} = 2$ is marked.

3. Experimental Results

For the experimental investigation on the QZI and the TSI, two prototypes were built up. In order to assure a good comparability, the inverters were constructed in a modular way which is illustrated in Fig. 6. The permanent modules are the dc source and the inverter connected to the load. The exchangeable part is the impedance network of the QZI and the TSI. Detailed information about the experimental setup can be found in Table 2. The induction motor was used in delta configuration and the 3-phase inverter was controlled by a modified space vector modulation [7], [8], [9]. The realization of the modified space vector modulation can be achieved easily by shifting the switching actions from zero state to active state 1 and from active state 1 to zero state of the traditional space vector modulation about the shoot through time. In that way, additional switching actions can be avoided. The modified modulation sequence is pictured in Fig. 7.

In order to prevent failure of the semiconductors due to parasitic inductance in combination with hard switching action, a RCD snubber circuit was used [10], [11].

Fig. 8 and Fig. 9 are illustrating the output voltages of the inverters under light load conditions depending on the shoot through factor calculated according to the control equations (8) and (13). Referring to the TSI, it is clearly noticeable that the actual achieved output voltages are not reaching the requested voltage. In case of the QZI, the measured voltages are exceeding the requested voltage. The boost ability of the TSI is lower than expected. By contrast, the QZI exceeds it. Since the inductor current of the QZI is continuous and mosfets turn on very fast, it is possible to switch into shoot through state without any delay. Due to the fact that the turn-off time of mosfets is not as small as the turn-on time the shoot through ends slightly delayed. This delay causes the measured voltage to exceed the requested voltage. Assuming ideal conditions and no leakage inductance of the transformer, switching into shoot through state works without any

Input Voltage
$V_{in} = 60\,\text{V}$
Induction Motor
$P_N = 7{,}5\,\text{kW}$
$U_N = 110\,\text{V Y}$
$I_N = 51\,\text{A}$
$\cos\varphi_N = 0{,}87$
$n_N = 2950\,\text{Hz}$
Quasi-Z-Source Network
$C_1 = C_2 = 0.98\,\text{mF}$
$L_1 \approx L_2 \approx 27.05\,\mu\text{H}$
T-Source Network
$C = 1.96\,\text{mF}$
$L_1 = 61{,}22\,\mu\text{H}$
$k = 0{,}984$
$N_1/N_2 = 2$
Semiconductors
Mosfet SKM 180A020

Table 2: Configuration of the Experimental Setup

delay for the TSI as it works for the QZI. The condition to enter the shoot through state, an input current fallen down to zero and a reverse biased input diode, can be achieved immediately. Taking the leakage inductance of the transformer into account, the transition between the states is not possible without delay.

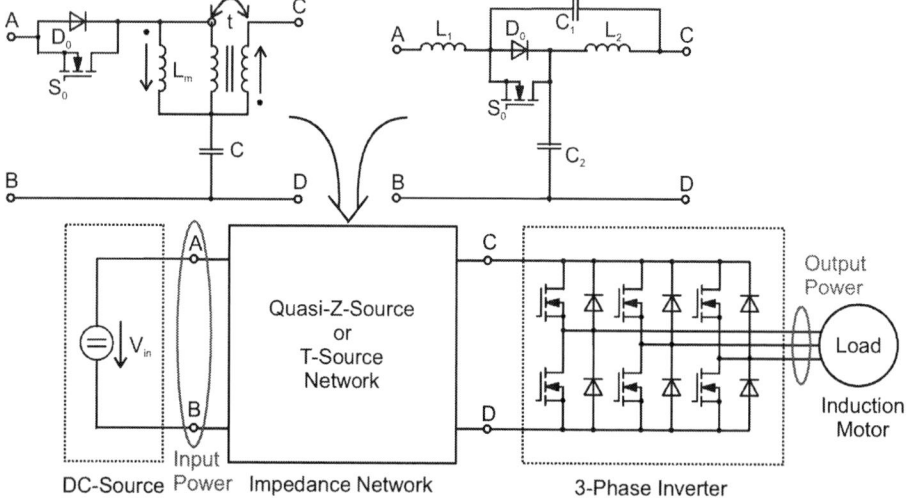

Fig. 6: Modular scheme of the experimental setup. Measuring points for the input and output power are marked.

The slew rate of the input current is determined by the leakage inductance. The magnitude it slews from is dependent on the load current, the shoot through factor, and the winding ratio. The time period that is needed to switch into shoot through state shortens the actual shoot through time and causes the measured voltages to fall below the requested voltage as shown in Fig. 8 [11], [12], [5].

Realizing the shoot through of the inverter is possible in different ways. The modified space vector modulation performs the shoot through by short circuiting one phase leg (index 1ph). By adding additional switching actions, all three phase legs (index 3ph) can be used to conduct the high shoot through currents simultaneously. Depending on the operation point and the case if the reduced conducting losses overweigh the additional switching losses, a three phase shoot through can improve the performance of the inverter. With the used dSpace system it was not possible to use the three phase shoot through with a shoot through factor smaller than 4%.

As pictured in Fig. 8 and Fig. 9 using a three phase shoot through leads to a higher output voltage compared with the voltage achieved with the one phase shoot through.

Starting with light load conditions illustrated in Fig. 8 and Fig. 9, the inverters are now investigated under high load conditions which is depictured in Fig. 10 and Fig. 11. Referring to the QZI, it can be observed that the output voltages are smaller compared to the voltages achieved under light load conditions. The output voltage reached with a one phase shoot through matches the requested voltage reasonably accurate. The three phase shoot through still exceeds the requested voltage but on a lower level. The lack of amplitude can be explained by higher currents and as a consequence, higher voltage drops at the semiconductors and parasitic resistances.

The performance of the TSI drastically suffers from the changed load conditions. As stated before, the transition into shoot through state is not possible without any delay. Since the delay is depending on the load, the changed load conditions are shortening the effective shoot through time even more compared with the light load condition. Operating with small shoot through times, there is even the possibility that the actual shoot through state cannot be reached depending on the load current. In that case, the shoot through causes only additional losses but no voltage gain. Fig. 10 shows that increasing the shoot through duty cycle has almost no effect on the output voltage under one phase shoot though operation. Implementing a three phase shoot through improves the boost ability and a slight voltage gain can be observed.

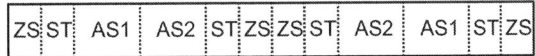

Fig. 7: Switching sequence of the modified space vector modulation

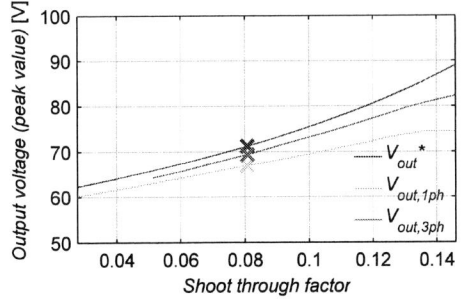

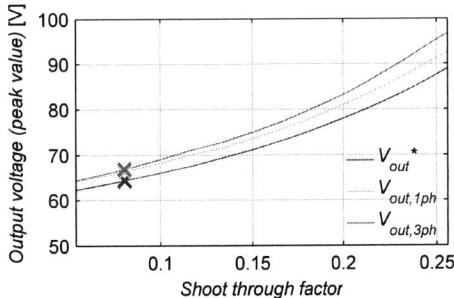

Fig. 8: Output voltage of the TSI controlled with the control equation according to the analytical analysis under light load conditions, $t_{ST} = 0.08$ is marked.

Fig. 9: Output voltage of the QZI controlled with the control equation according to the analytical analysis under light load conditions, $t_{ST} = 0.08$ is marked.

The performance of the TSI is way below the expectations. Due to the fact that the load has a great influence on the capability of the TSI, Fig. 12 illustrates the effect of an increasing shoot through factor on the output voltage under high load conditions more detailed. Working with a three phase shoot through, the output voltage increases continuously and reaches the desired voltage at $t_{ST} \approx 0.13$. The QZI reaches the same voltage level with approximately the same shoot through factor. Employing a one phase shoot through, the process of the output voltage is quite different. Up to a shoot through factor of $t_{ST} \approx 0.15$, the voltage rises slightly and reaches a maximum. After this maximum an increasing shoot through leads to an even lower output voltage and high losses. The requested voltage cannot be achieved. In contrast to that, the QZI reaches that voltage with a one phase shoot through of $t_{ST} \approx 0.15$.

As a last step, the efficiency factors of the two inverters are compared and depictured in Fig. 13. The efficiency of the QZI is significantly higher than the efficiency of the TSI. For small shoot through factors, the QZI operating with a three phase shoot through is slight advantageous against the one phase shoot through. With an increasing shoot through factor up to $t_{ST} \approx 0.25$ the one phase shoot through performs a bit better than the three phase shoot through. In general, there is no big difference between one phase and three phase shoot through. Referring to the TSI, the three phase shoot through is always better than the one phase shoot through. The higher the shoot through factor and the requested voltage, the higher the gap between the two efficiencies. Output voltages higher than 80V can only be reached with a TSI performing with a three phase shoot through.

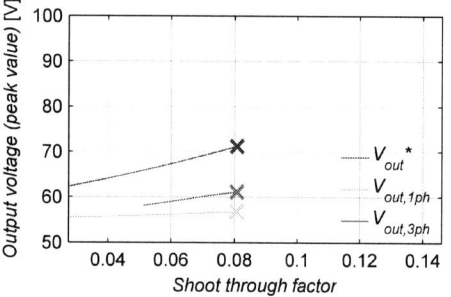

Fig. 10: Output voltage of the TSI controlled with the control equation according to the analytical analysis under high load conditions, $t_{ST} = 0.08$ is marked.

Fig. 11: Output voltage of the QZI controlled with the control equation according to the analytical analysis under high load conditions, $t_{ST} = 0.08$ is marked.

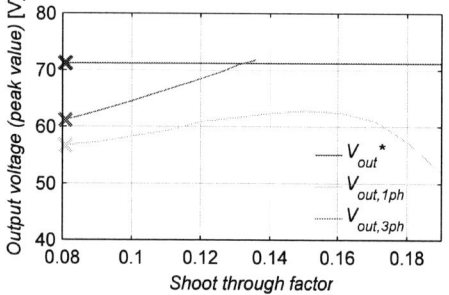

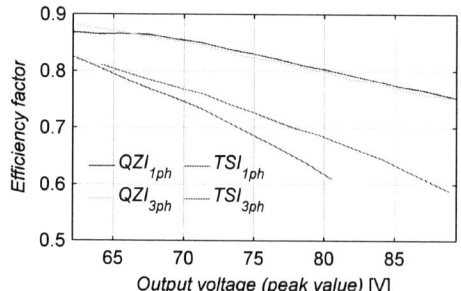

Fig. 12: Effect of an increasing shoot through on the output voltage of the TSI under high load conditionons. Starting with the shoot through calculated according to the control equation.

Fig. 13: Comparison of the efficiency factors of QZI and TSI for one phase and three phase shoot through under light load conditions.

Conclusion

This paper presents an analytical comparison of the QZI with the TSI. Theoretically, the potential of the TSI for $n > 1$ exceeds the potential of the QZI referred to the voltage stress of the semiconductors and the average shoot through current that needs to be conducted. The higher the turns ratio, the higher the analytical advantageous of the TSI in comparison with the QZI. Moreover, the winding ratio represents an additional degree of freedom with regard to the design of the TSI. When it comes to the practical application and the experimental setup, the results of the comparison are quite different. Due to the leakage inductance of the transformer of the TSI and the shoot through time lost while transiting into shoot through state, its performance is way below the theoretical expectations. The shown drawback would even be greater with a higher turns ratio of the TSI. In contrast to that, the QZI can withstand the expectations and easily outperforms the TSI. Under light load conditions, there are slight advantages performing with a one phase shoot through. Under high load conditions the three phase shoot through exceeds the results of the one phase shoot through.

References

[1] F. Z. Peng, "Z-Source Inverter," *Industry Applications Conference,* vol. 2, pp. 775-781, 2002.

[2] F. Z. Peng, "Z-Source Inverter," *IEEE Transactions on Industry Applications,* vol. 39, no. 2, pp. 504-510, 2003.

[3] J. Anderson und F. Z. Peng, „A Class of Quasi-Z-Source Inverters," *Industry Applications Society Annual Meeting,* pp. 1-7, 2008.

[4] K. Beer und B. Piepenbreier, „Properties and advantages of the quasi-Z-source inverter for DC-AC conversion for electric vehicle applications," *Emobility - Electrical Power Train,* pp. 1-6, 2010.

[5] R. Strzelecki, M. Adamowicz, N. Strzelecka und W. Bury, „New Type T-Source Inverter," *Conference on Compatibility and Power Electronics,* pp. 191-195, 2009.

[6] R. Strzelecki, W. Bury, M. Adamowicz und N. Strzelecka, „New Alternative Passive Networks to Improve the Range Output Voltage Regulation of the PWM Inverters," *Applied Power Electronics Conference and Exposition,* pp. 857-863, 2009.

[7] M. von Zimmermann, S. Labusch und B. Piepenbreier, „Bi-directional AC-AC Z-source inverter with active rectifier and feedforward control," *IEEE Energy Conversion Congress and Exposition (ECCE),* pp. 3180-3186, 2010.

[8] M. von Zimmermann, M. Lechler und B. Piepenbreier, „Z-source drive inverter using modified SVPWM for low Output Voltage and regenerating Operation," *European Conference on Power Electronics and Applications,* pp. 1-10, 2009.

[9] Y. Liu, B. Ge und H. Abu-Rub, „Theoretical and experimental evaluation of four spacevector modulations applied to quasi-Z-source inverters," *IET Power Electronics,* Bd. 6, Nr. 7, pp. 1257-1269, 2013.

[10] S. Dong, Q. Zhang, C. Zhou und S. Cheng, „Analysis and design of snubber circuit for Z-source inverter," *European Conference on Power Electronics and Applications (EPE'14-ECCE Europe),* pp. 1-10, 2014.

[11] Y. Siwakoti, P. Loh, F. Blaabjerg und G. Town, „Effects of leakage inductances on magnetically-coupled impedance-source networks," *European Conference on Power Electronics and Applications (EPE'14-ECCE Europe),* pp. 1-7, 2014.

[12] W. Qian, F. Z. Peng und H. Cha, „Trans-Z-Source Inverters," *IEEE Transactions on Power Electronics,* Bd. 26, Nr. 12, pp. 3453-3463, 2011.

PCIM Europe 2016, 10 – 12 May 2016, Nuremberg, Germany

Benchmarking of SiC JFET and SiC MOSFET modules for the application in medium power traction converters

Andreas März[1], Roman Horff[1], Teresa Bertelshofer[1], Martin Helsper[2], Mark-M. Bakran[1]
andreas.maerz@uni-bayreuth.de

[1] University of Bayreuth, Department of Mechatronics, Center of Energy Technology, Universitätsstr. 30, 95447 Bayreuth, Germany
[2] Siemens AG, Vogelweiherstr. 1-15, 90441 Nuremberg, Germany

Abstract

In this paper the behaviour of a SiC JFET power module is being analysed and the impact of its bodydiode behaviour on the performance of the SiC JFET is compared to a SiC MOSFET and a state of the art IGBT. Second it will be discussed how a low inductance DC-Link affect the dynamic losses of the JFET and whether the SiC JFET can benefit from a low inductance setup.

1. Introduction

One way of increasing the power density of modern traction converters is the use of SiC power semiconductors [1]. There have already been a few comparisons of different SiC power devices, like BJT, MOSFET or JFET on the basis of discrete devices [2]. During the last years SiC switches became available in conventional power modules [3, 4] opening the way for medium power application for SiC semiconductors. But the rather high internal stray inductance of conventional modules and the DC-link connection are often limiting the switching speed and thus the performance of SiC devices in such kind of housing.

In this paper an investigation is made on what impact a low inductance DC-links has on the performance of SiC JFET in a low inductance power module for medium power traction application.

2. Benefits of the SiC MOSFET under hard switching

For this comparison two SiC power devices in half-bridge configuration have been selected. For the SiC MOSFET a 300A/1700V Full-SiC MOSFET [3] with antiparallel schottky barrier diodes (SBD) was chosen.

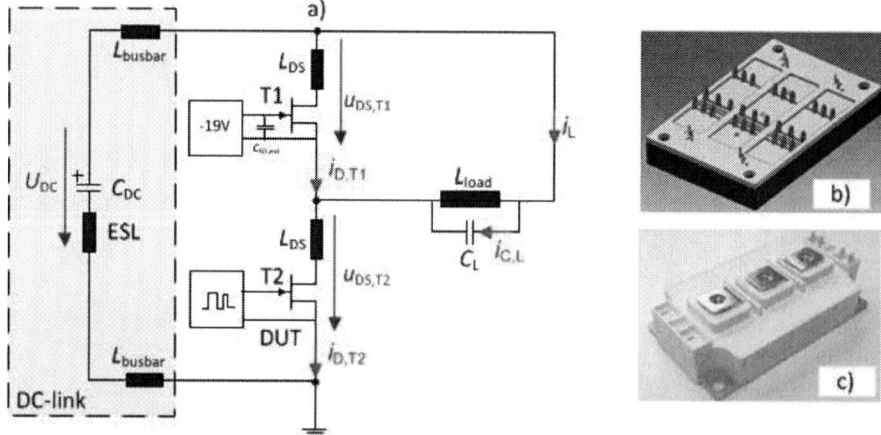

Fig. 1: a) Test circuit configuration b) low inductance half-bridge module with SiC JFET [6] and c) conventional 62mm half-bridge module with SiC MOSFETs [3]

© VDE VERLAG GMBH · Berlin · Offenbach

For the SiC JFET a low inductance module [5, 6], see Fig 1b) was used in this comparison, which is rated at 480A/1700V. It has a very low internal inductance of the module of $L_{\sigma,\text{module}}$ = 5nH. To compare the performance under hard switching conditions both devices have first been tested in the same test setup, which is a conventional DC-link. The product stray inductance times the rated current of the 62mm SiC MOSFET module together with test setup accounts to 34nH · 300A = 10.2µVs. For the SiC JFET in the same DC-link together with its low inductance module with pressfit connections, this product equals to 14nH · 480A = 6.7µVs, which is 33% less.

In order to benchmark the performance of both devices, they have been tested at their highest possible switching speeds. The SiC MOSFET was driven with $u_{\text{GS,MOSFET}}$ = -5V/+20V and no external gate-resistor. The SiC JFET being a direct driven, normally-on device was driven with $u_{\text{GS,JFET}}$ = -19V/0V and also no external gate-resistor. It is shown in [6] that the SiC JFET under test is very vulnerable to parasitic turn-on.

Due to an adverse ratio of $Q_{\text{GD}}/Q_{\text{GS}} > 1$ of the SiC JFET, the capacitive current through the miller-capacitance $i_{\text{c,GD}} = C_{\text{GD}} \cdot du_{\text{DS}}/dt$ at high switching speed (du_{DS}/dt) and gate inductance will lead to a temporary raise of the u_{GS} above the pinch-off voltage which causes the switch to turn-on unintentionally during the du/dt-phase of the switching transient. In the following this effect will be referred to as parasitic turn-on (PTO). Therefore, from the measures shown in [7, 8] that can be taken to suppress PTO, an increase of C_{GS} by adding an additional external capacitor $C_{\text{GS,ext}}$ to the gate was chosen. It was placed close to the gate in order to keep the impedance of gate circuit at a minimum. During the off-state of the switch an auxiliary MOSFET clamps the gate to the negative voltage of the capacitor bank of $C_{\text{GS,ext}}$. Even with this measure taken, the SiC JFET shows some parasitic turn-on, which can be seen in the difference of drain current overshoot at turn-on compared to the SiC MOSFET turn-on. There are also other methods, like integrating a current-source on the driver which provides a negative gate-current during the off-state in order to clamp the gate to its negative avalanche voltage. But these methods increase the effort and size and complexity of the gate drive unit, so that it is no longer proportional compared to the effort to drive a SiC MOSFET or an IGBT.

A comparison of the switching behaviour of both devices at equal voltage overshoot at turn-off is shown in Fig. 2a). It can be seen that the SiC JFET turn-off is much faster (higher du/dt), with sharper di_{D}/dt compared to the SiC MOSFET. For both devices the level of overvoltage due to stray inductance $di_{\text{D}}/dt \cdot L_{\text{S}}$ = 354.7V is the same, making this a fair comparison of both devices together with its package and DC-link.

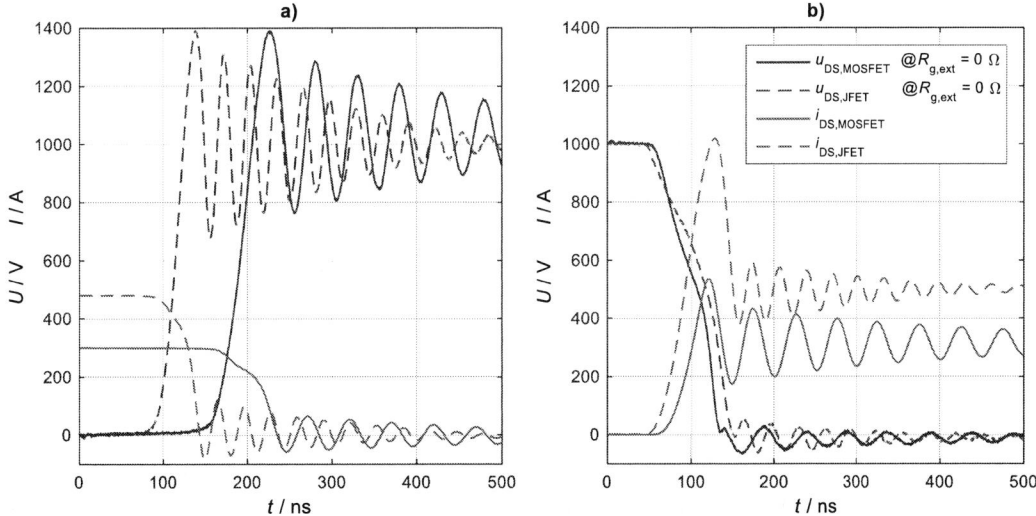

Fig. 2: Switching behaviour of SiC MOSFET and SiC JFET with the same overvoltage at turn-off a) and at turn-on b) @U_{DC} = 1000V, I_{DC} = I_{N}, T_{J} = 125°C

Comparing the switching losses at this point of operation one can see that at turn-off the SiC JFET relative losses E_{off}/I_N are about 40% less compared to SiC MOSFETs, see Fig. 3a). Looking at the absolute level of E_{off}, the loss of both devices are nearly the same, though the rated current of the SiC JFET is 60% higher. It has to be noted that the SiC MOSFET module is equipped with internal gate-resistors $R_{g,int}$ = 3.7Ω. Due to the layout and level of stray inductance of the 62mm module [3] internal gate resistors are necessary to prevent oscillation between paralleled chips. In essence with this housing the SiC MOSFET is slowed down through $R_{g,int}$ which limits the switching speed. Due to the symmetrical layout of the low inductance module [5] the SiC JFET isn´t equipped with any $R_{g,int}$ and therefore achieves a higher switching speed.

Regarding turn-on of T2, which is shown in Fig 2b), one can see that the switching speed of the SiC JFET is higher than that of the SiC MOSFET. The drain current i_D, however shows a big current overshoot at turn-on which is due to the parasitic turn-on behaviour of the SiC JFET T1 during du_{DS}/dt of the switch T2 and additionally the discharge of the parasitic load-capacitance C_L, which was identical for all switching tests. In spite of higher switching speed these additional losses, due to parasitic turn-on energy losses E_{PTO}, lead to high relative turn-on losses of the SiC JFET and bodydiode turn-off losses compared to the MOSFET, see Fig. 3a). As a consequence of a better ratio of both charges (Q_{GD}/Q_{GS}) stored in the input capacitance, the SiC MOSFET doesn´t show any parasitic turn-on tendency and therefore has the lower overall switching losses balance.

Regarding conduction losses in Fig. 3b) the JFET has the lower value of $R_{DS,on}$. This is due to the absence of the MOS-channel resistance which even at T_J = 125°C makes up about one third of the $R_{DS,on}$ of a planar 1.7kV SiC MOSFET.

At T_J = 125°C the IGBT [9] at rated current of the module I_N has an identical on-state voltage compared to the SiC JFET. At partial load the SiC JFET is much better than the IGBT. This breakeven with the IGBT is at $1/2 \cdot I_N$ for the SiC MOSFET

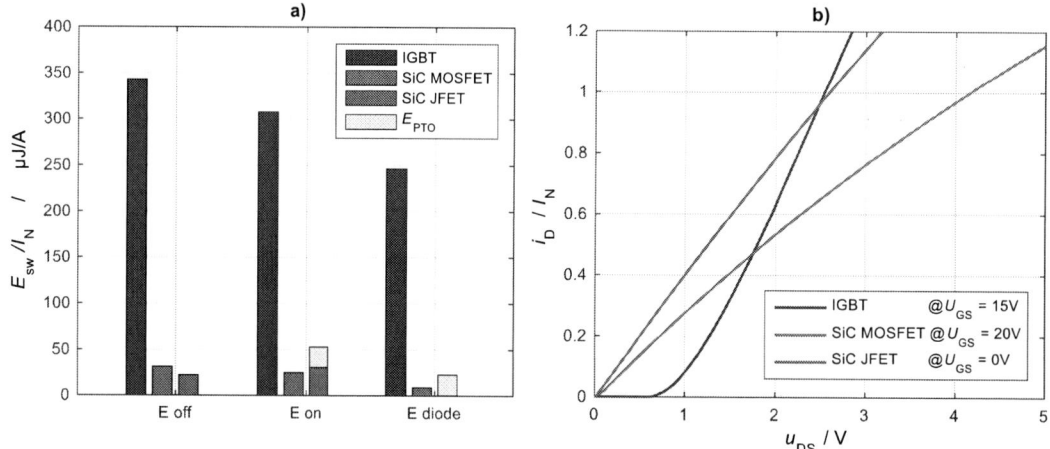

Fig. 3: Comparison of all devices relative to the rated current of the module a) switching losses E_{off}, E_{on} and E_{diode} respectively $E_{bodydiode}$ and in b) conduction losses @U_{DC} = 1000V, I_{DC} = I_N, T_J = 125°C

Switching and conduction losses affect the performance of each device in the application of a traction converter. In the continuing lines in Fig. 6 one can see a comparison of output power performance all three devices for a VSI.

When comparing the performance of these two SiC devices and the performance of the conventional IGBT for application in a two level VSI, one can see that for all switching frequencies the IGBT has a substantially lower output performance. This is due to the high $R_{th,JC}$ of its housing respectively its DCB which in this case is based on Al_2O_3 instead of AlN.

Comparing both SiC devices at equal $R_{th,JC} \cdot A_{chip}$ one can see that at switching frequencies above 20kHz the SiC MOSFET is the device of choice due to its low switching losses even though its switching speed in the conventional 62mm housing is a lot slower than that of the SiC JFET. Due to its superior conduction losses the SiC JFET enables higher output power below 20kHz.

Comparing the switching behaviour of both SiC devices as seen the commutation loop inductance has a strong effect on switching speed and respectively overvoltage and switching losses. As overvoltage is the main factor limiting the switching speed it is important to reduce stray inductance for fast switching devices inside the module but also outside. This includes the DC-link and its connection to the module.

Since the SiC JFET module has a very low internal module inductance investigations are been made on how beneficial a low inductance DC-link is for this unipolar device.

3. Impact of a low inductance DC-Link on the switching behaviour of the JFET

For this reason a low inductance DC-link was designed and fitted to the SiC JFET test module. With this the measured stray inductance across T2 was cut to L_S = 7nH. In Fig. 4 the switching behaviour of the SiC JFET module together with a low inductance DC-link and a standard DC-link is depicted.

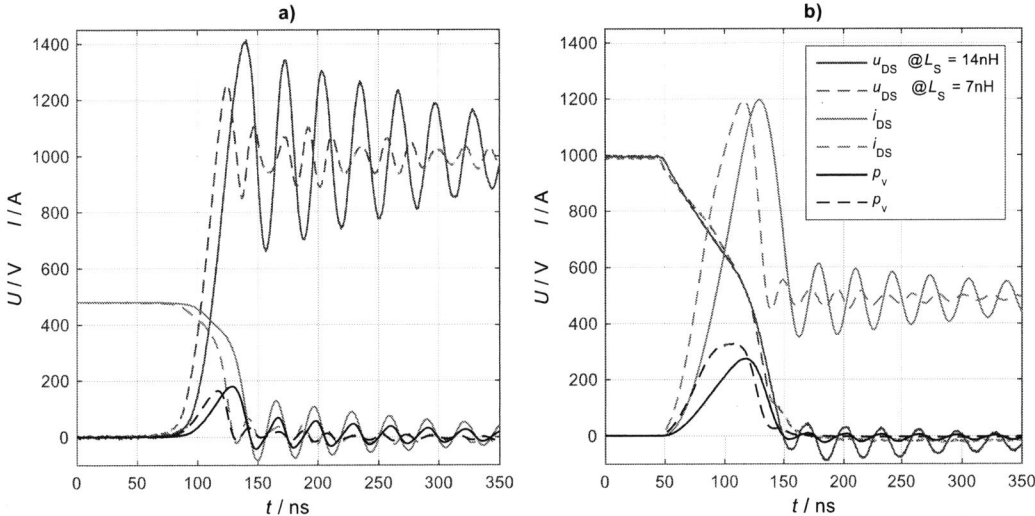

Fig. 4: Influence of a low inductive setup on the switching behaviour of a SiC JFET a) turn-off and b) turn-on @U_{DC} = 1000V, $I_{DC} = I_N$, with $C_{GS,ext}$, T_J = 125°C

This standard DC-link is made from electrolytic capacitors and a conventional busbar connection made from copper, and is therefore referred to as standard DC-link. Together with the JFET module this accounts to L_S = 14nH.

Regarding the turn-off in a low inductance DC-link, see Fig. 4a), a decrease in overvoltage of $\Delta U_{DS,peak}$ = -115V can be observed. Since the output capacitance C_{oss} of the SiC JFET module is unchanged the switching speed during the du/dt-phase of the switching transient is unchanged. During the di/dt-phase of the switching transient the device shows a

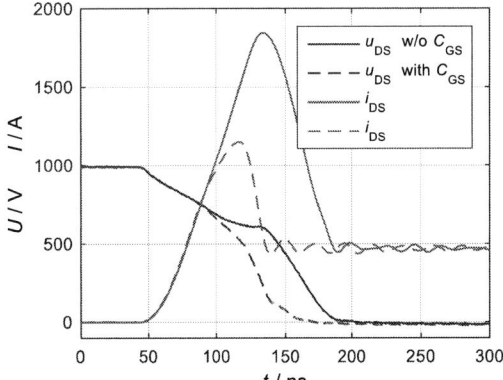

Fig. 5: influence of the negative gate clamp with $C_{GS,ext}$, in order to minimise PTO

10% higher switching speed with the low inductance DC-link. This reduces the measured switching losses by 24%. This can be seen in the difference of $P_V(t)$ between both waveforms. The current at turn-on is mostly limited by the commutation-loop inductance of the low inductance DC-link together with the module. As a result i_D rises according to its transfer characteristic 25% faster, regarding the di_D/dt_{10-90}. The maximum higher di_D/dt, which is well above 20A/ns, together with a lower level of commutation-loop inductance result in an inductive voltage drop which is of the same magnitude as with the standard DC-link, see Fig. 4b). With the du/dt as the cause for the top side switch to parasitically turn-on stays the same between both DC-links as shown. The value of the $I_{D,peak}$ as a result of PTO is identical. As a result the turn-on losses E_{on} increase by 28%. At the same time E_{diode} decrease by 20% due to the higher switching speed. With the turn-on losses being the dominant part, see also Fig. 3a) the total losses of the SiC JFET increase when changing to a low inductance DC-link due to PTO by 7%. This leads to a lower output power performance over switching frequency of the device, see Fig. 6.

In Fig. 5, a SiC JFET turn-on is demonstrated without a clamping capacitance. For this SiC JFET T2 was turned on at a low inductance DC-link without an additional capacitive clamping of the negative gate-voltage of T1.

In this case the maximum drain current reaches 1844A, which is 3.84 times the value of the switched current I_N. This leads to a massive increase in E_{diode} and E_{on} respectively would result in an output power performance which, with increasing switching frequency, drops almost as sharply as the performance of the IGBT, see Fig. 6.

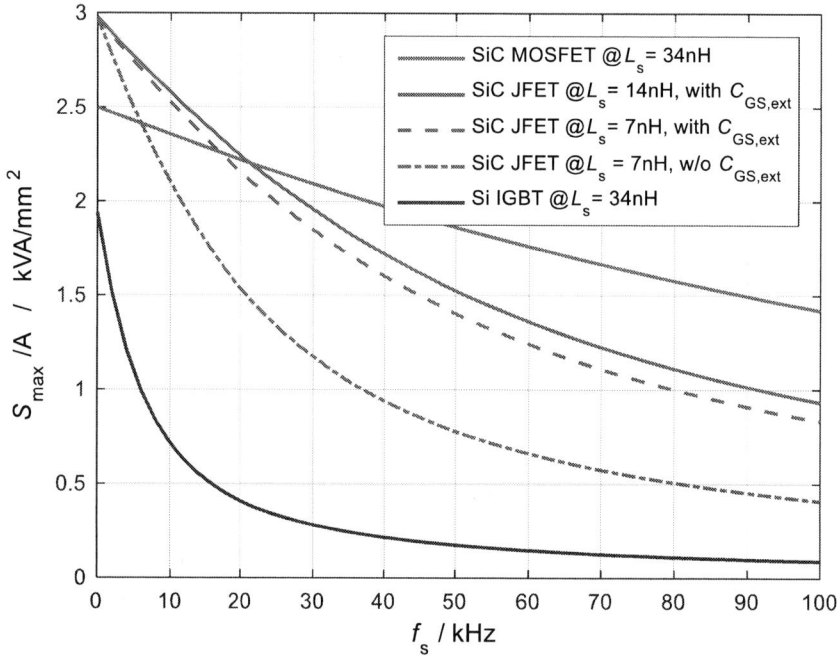

Fig. 6: Output power density of SiC JFET, SiC MOSFET and IGBT at different f_S at U_{DC} = 1000V, M = 1, I_{DC} = 480A, $cos\ \varphi$ = 0.8, T_J = 125°C

4. Can a low inductance DC-Link be exploited by the SiC JFET?

The output power comparison at maximum switching speed and equal DC-link voltage in Fig. 6, doesn´t consider the lower voltage exploitation when using a low inductance DC-link (lower $U_{DS,max}$). Consequently some investigation is made into whether a change in driving or circuit parameter can lead to a higher output power performance and therefore higher utilisation of low inductance DC-link by the SiC JFET.

Exploiting the obtained voltage margin can mean raising the DC-link voltage by 15% at I_N until $U_{DS,max}$ = 1150V is reached, see Fig. 7a). Of course higher DC-link voltages like U_{DC} = 1000V would also be possible with lower values of the load current, in this case for the same $U_{DS,max}$, the load current i_L cannot be higher than 240A. But since the model is regarded at maximum thermal exploitation this case of partial load is not considered.

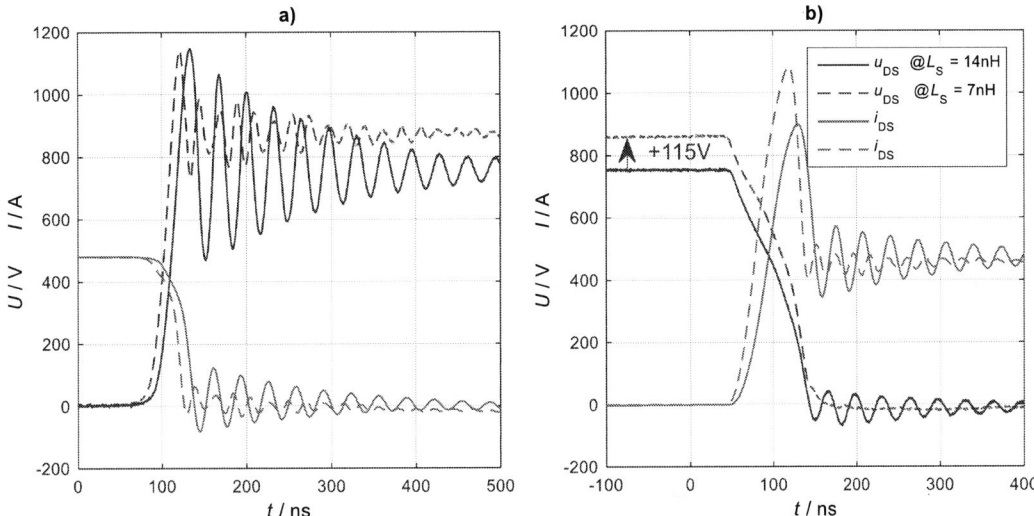

Fig. 7: With a low L_S setup the DC-Link voltage can be increased a) turn-off and b) turn-on @$U_{DS,max}$ = 1150V, I_{DC} = 480A, with $C_{GS,ext}$, T_J = 125°C

Comparing both switching waveforms at equal $U_{DS,max}$ at turn-off one can see that this increase of the DC-link voltage leads to a higher du/dt, which as a result increases the level of PTO compared to the comparison in Fig. 4b). As a result the amount of the dynamic losses of the device at turn-on increases. The same effect occurs when raising I_N and keeping U_{DC} constant as mentioned before. Therefore this 15% increase in output power gain at zero switching frequency is lost already at f_s = 35kHz because of the higher switching losses, referring to Fig. 9a) and Fig. 10a).

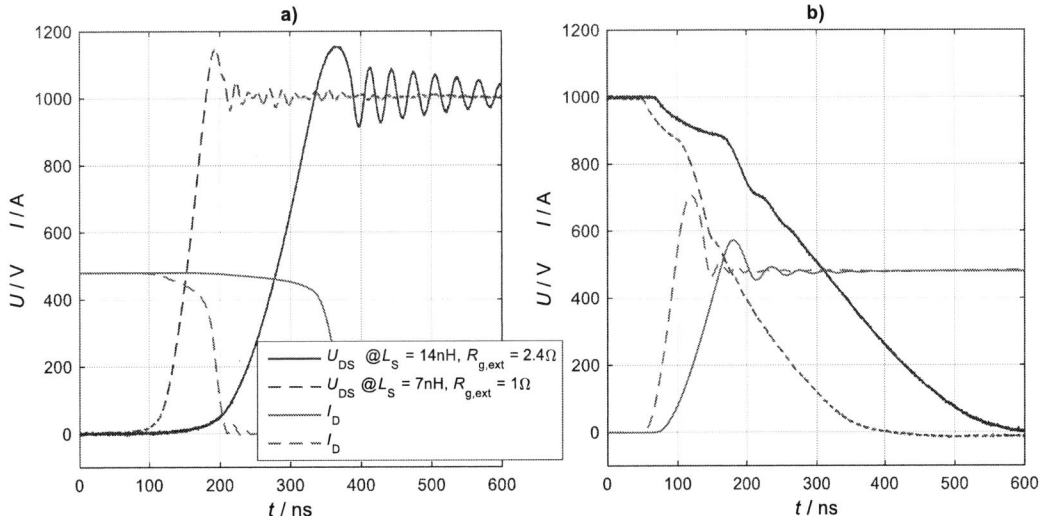

Fig. 8: Raising the DC-Link voltage by limiting the switching speed of the SiC JFET a) turn-off and b) turn-on @U_{DC} = 1000V, I_{DC} = 480A, with $C_{GS,ext}$, T_J = 125°C

If using the SiC MOSFET at maximum switching speed together with a standard DC-link the device can be operated at U_{DC} = 800V to get the same $U_{DS,max}$. If the performance of the MOSFET is calculated for this boundary conditions, it can be seen that already at frequencies >30kHz it has the best performance even with a level of stray inductance which is 5 times more than the SiC JFET together with a low inductance DC-link, referring to Fig. 10a).

In a second comparison of the benefit of a low inductance DC-link U_{DC} was set to 1000V and each device was slowed down by an external gate-resistor in order to meet $U_{DS,max}$ = 1150V. In Fig. 8, the behaviour of the SiC JFET with both DC-links can be seen. For the low inductance DC-link the device had to be slowed down only a little bit, which increases the turn-off losses but reduces the level of E_{PTO} compared to Fig 4b). If the module is applied to the standard DC-link the JFET must be slowed down even further, which reduces parasitic turn-on but for every value for $R_{G,ext}$ increases the total amount of switching losses of the SiC JFET, see Fig. 9b).

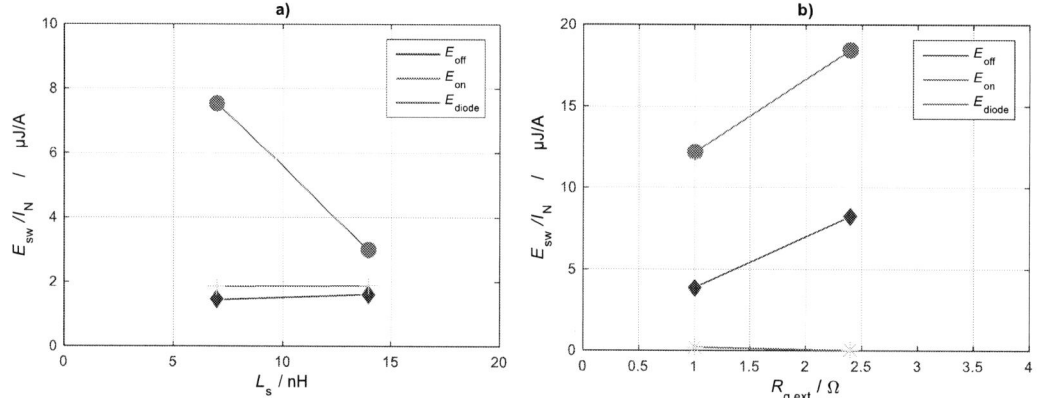

Fig. 9: SiC JFET switching losses at $U_{DS,peak}$ = 1150V a) maximum U_{DC} and maximum switching speed b) at U_{DC} = 1000V and limited switching speed with additional external gate resistor

As a result the SiC JFET with the highest switching speed is showing the better output power performance, see Fig. 10b). So in this case the performance of the SiC JFET can be exploited higher by making an effort to reduce the level of commutation loop inductance.

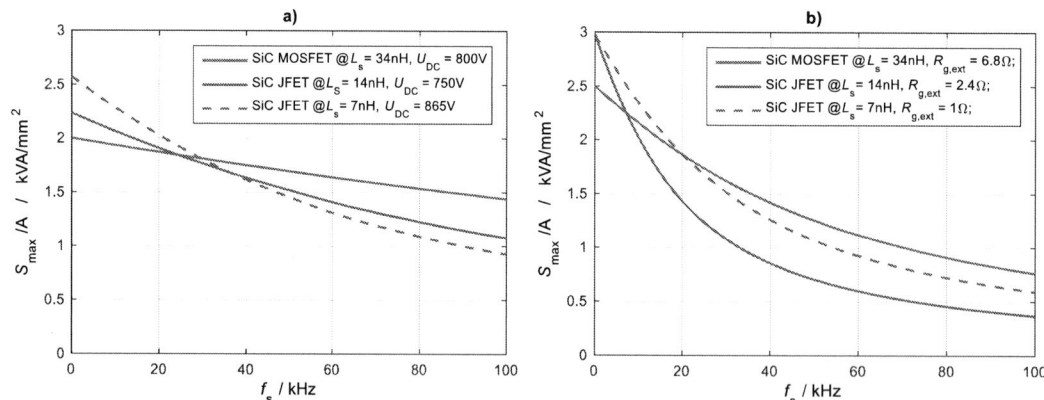

Fig. 10: Output power performance at $U_{DS,peak}$ = 1150V a) maximum U_{DC} and maximum switching speed b) at U_{DC} = 1000V and limited switching speed with additional external gate resistor

But even with an external gate resistor of 6.8Ω, at switching frequencies about 20kHz the SiC MOSFET in the conventional 62mm housing and no low inductance DC-link shows the best performance making it the device of choice for medium power, high frequency applications.

5. Conclusion

Improving the gate-drive unit of the SiC JFET in order to completely eliminate parasitic turn-on will decrease its switching losses and move the breakeven point between SiC JFET and SiC MOSFET further towards higher switching frequencies. The SiC MOSFET by constrast does not show a tendency of parasitically turning on and is easier to drive. If driving effort between both devices is targeted to be similar, PTO effect of the SiC JFET has to be solved by internal measures of the chip. Only in this case its switching losses would be comparable to the SiC MOSFET.

Regarding the low inductance power module which strictly follows the design rules in [4] it enables fast switching SiC devices to fully utilize their potential in terms of switching speed.
The general presumption that a lower stray inductance result in a higher switching speed of unipolar devices and by that also in lower switching losses due to a smaller overlap of current and voltage is not true for the SiC JFET. The current overshoot at turn-on due to parasitic turn-on of the SiC JFET under test leads to higher turn-on losses which exceed the decrease in E_{off}. Regarding the SiC JFET it can be seen that a very low inductance DC-link doesn´t improve the performance of the device in this configuration.
In view of the SiC MOSFET power modules this comparison shows what increase in switching speed might be possible if these chips where offered in such a low inductance module housing and by what extend a low inductance DC-link would enhance its switching speed. Therefore SiC MOSFETs in a low inductance housing promises the best performance and usability and for this reason are demanded from manufacturers.

References

[1]   M.-M. Bakran, A. Maerz, B. Laska, U.E. Krafft, O. Koerner and A. Nagel, „Latest development in increasing the power density of traction drives", IPEC, 2014.

[2]   S. Safari, A. Castellazzi and P. Wheeler, „Experimental and Analytical Performance of SiC Power Devices in the Matrix Converter", IEEE Transactions on Power Electronics, vol.29, no.5, 2014.

[3]   Cree inc: CAS300M17BM2 datasheet: www.cree.com, 2014.

[4]   Semikron: SKM500MB120SC datasheet: www.semikron.com, 2015.

[5]   R. Bayerer and D. Domes, „Power Circuit design for clean switching", CIPS, 2010.

[6]   D. Heer; R. Bayerer and D. Domes, „SiC-JFET in half-bridge configuration – parasitic turn-on at currrent commutation", PCIM, 2014.

[7]   E. Velander, A. Löfgren, K. Kretschmar and H.-P. Nee, "Novel Solutions for suppressing Parasitic Turn-on behaviour on Lateral Vertical JFETs", EPE, 2014.

[8]   D. Domes, R. Werner, R. Domes, W. Hoffmann and S. Krauß, "A New, Universal and Fast Switching Gate-Drive-Concept for SiC JFET based on Current Source Principle", 'PESC, 2006.

[9]   Infineon: FF300R17KE4 datasheet: www.infineon.com, 2014.

New Bus-bar Topology to Suppress the Current Imbalance of Parallel-connected IGBT Modules for High Power Railway

Naoki Sakurai, Hitachi, Ltd. Research & Development Group, Japan,
naoki.sakurai.bs@hitachi.com,
Masaomi Konishide Hitachi, Ltd, Rail Systems Company, Japan
masaomi.konishide.cy@hitachi.com
Yasuhiko Kohno, Hitachi, Ltd, Rail Systems Company, Japan
yasuhiko.kono.fn@hitachi.com

Abstract

Three types of current imbalance between IGBT modules of over a 1 MW inverter were found, which composed of the parallel-connected IGBT modules in double-pulse switching test. The mechanisms of these current imbalances were investigated by circuit analysis with newly developed circuit models considering the bus bar topology of the inverter. The analysis results indicated that the current imbalance was strongly affected by the coupling coefficient of the inductances between the converter or filter capacitor and the IGBTs. In order to suppress the current imbalance, a new bus bar topology with uniform coupling coefficients was developed by using circuit simulation considering a bus-bar equivalent circuit model calculated by electromagnetic-field analysis

1. Introduction

Because high-speed railway traction need to be faster, the electrical output power of these railway traction systems should be increased. In the high-speed railway systems mainly supplied with AC high voltage, the converter initially changes AC to DC, and then a motor is driven by the inverter, which converts DC to AC again. The converter and the inverter utilize insulated-gate bipolar transistors (IGBTs) as switching devices. An inverter composed of one mass-produced IGBT module in one arm is limited to 1 MW. Therefore, more than 1 MW inverter is composed of a parallel arrangement of IGBT modules in one phase [1]. Imbalance of chip currents in a IGBT module with multiple chips in parallel has been widely studied [2] [3] [4] [5]. And it was reported that the current in such parallel IGBT modules should not be concentrated on a specific IGBT module, even transiently [6] [7].

© VDE VERLAG GMBH · Berlin · Offenbach

In this study, the mechanism that generates a current imbalance in an inverter of the railway system with a parallel arrangement of IGBT modules is investigated, and a method to suppress this current imbalance is proposed.

2. Measurement results

2.1. Railway traction power system

A schematic view of a high-speed railway traction power system, which is supplied with AC voltage from an overhead wire, is shown in Figure 1. A circuit diagram of U phase unit of the inverter with a prototype bus bar is shown in Figure 2.

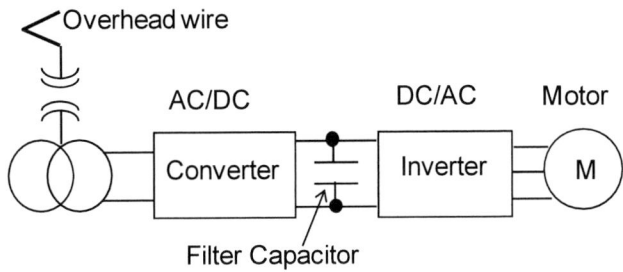

Figure 1: Schematic view of railway traction power system

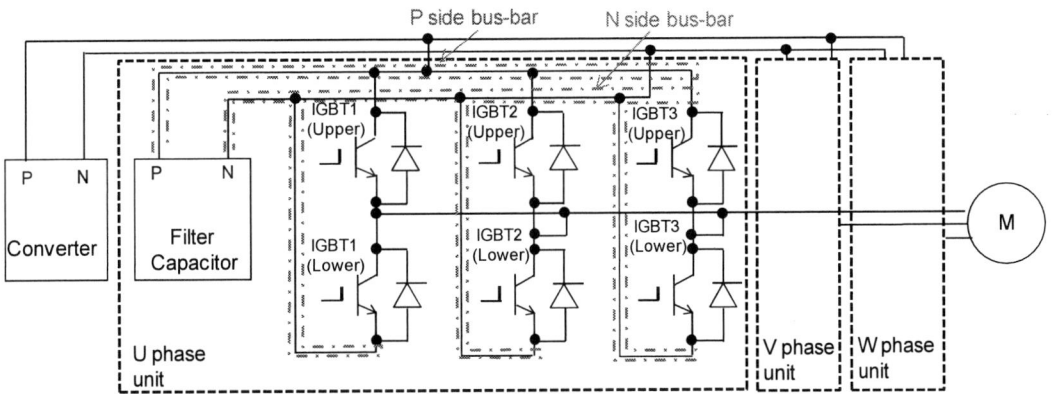

Figure 2: Circuit diagram of U phase unit of the inverter with a prototype bus bar

The U phase unit is composed of three parallel-connected 4.5-kV/1200-A IGBT modules. These IGBT modules are connected to a filter capacitor by P-side and N-side bus bars. The P-side and N-side bus bars have a laminated structure to reduce parasitic inductance by narrowing the gap between them. The IGBT modules are connected to the converter by the thick copper plate wirings. The inductances of the bus bars between the IGBT modules and

2.2. Measurements results

Measured waveforms obtained by a double-pulse switching test under an assumed overload are shown in Figure 3. The current imbalance between three IGBT modules is under 20% in the first turn-on period (period 1). In period 2, the gates of the IGBTs are turned off, and the currents of the IGBTs do not flow. In the second turn-off period (period 3), since IGBT1 is turned off faster than IGBT2 and IGBT3, the current of IGBT1 suddenly decreases faster than that of IGBT2, and the current oscillation results in creating a current imbalance of 50%. This current imbalance occurred in period1 and period3 is due to the fact that the current flows continually with the inductive load, but the current of IGBT2 and IGBT3 increase when the current of IGBT1 decreases when IGBT1 is turned off.

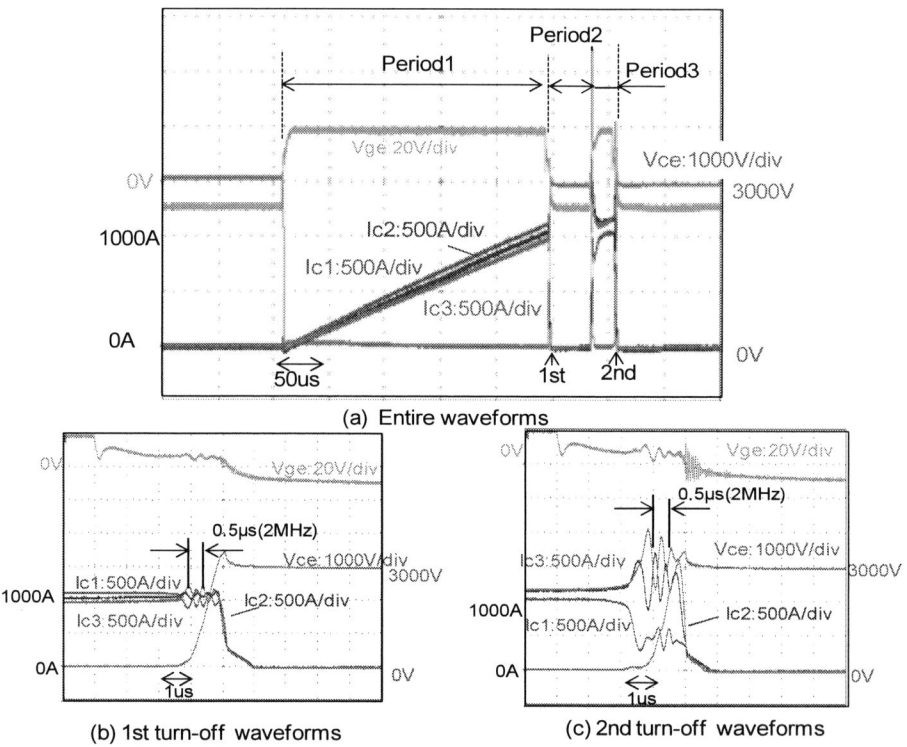

Figure 3: Waveforms measured by a double-pulse switching test using the prototype bus bar

3. Simulation

3.1. Simulation models and circuit

To examine the mechanism that produces the current imbalances described above, the transient current was analyzed with the newly developed circuit models considering the bus bar topology of the inverter. First, circuit models of parasitic inductance and resistance among three IGBTs, the bus bar connected to the filter capacitors, and the copper wirings connected to the converter were created by an in-house electromagnetic-field analysis tool, ECTAS(Eddy Current Analysis)[8][9], which calculated the stray inductances in consideration of the eddy current effect.

A schematic of the simulation is shown in Figure 4(a). The converter was modeled using a DC voltage source. A model of an IGBT which was provided by circuit simulator was used, And the parameters of the IGBT model were adjusted to be consistent with the measured changes in voltage (dV/dt), which occurred in turn-on and turn-off of the IGBT. The model of the N-side bus bar which is described with the netlist of the circuit simulator is shown in Figure 4(b). The netlist of the P-side bus bar is similar to that of the N-side bus bar. The netlist is composed of three inductances (L1c, L2c, and L3c) between the converter and the emitter terminals, three inductances (L1s, L2s, and L3s) between the filter capacitor and the emitter terminals, and 11 coupling coefficients (KL1c to KLc11) between these inductances. L1c, L2c, and L3c are much smaller than L1s, L2s, and L3s. The distance between the N-side bus bar and the P-side bus bar is significantly reduced by the laminated structure between the N-side and P-side bus bars, and the values of L1c, L2c, and L3c are reduced by the current-cancelling effect of the eddy current.

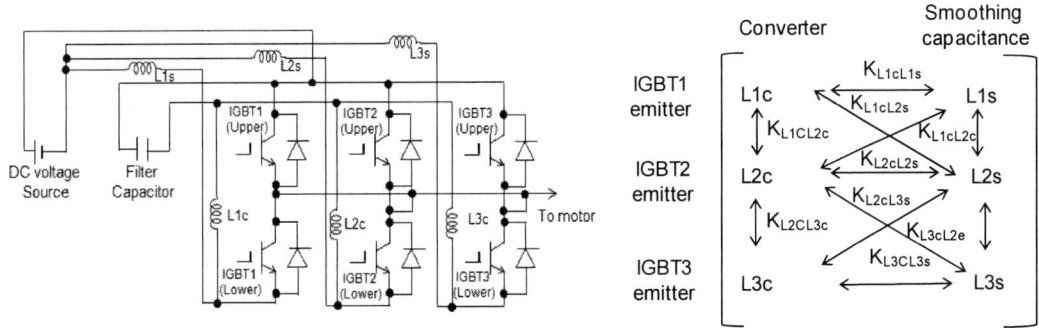

(a) Schematic of circuit simulation   (b) Conceptual netlist of N-side bus bar

Figure 4: Simulation circuit and model of N-side bus bar

3.2. Simulation results

Simulated waveforms obtained by a double-pulse switching test are shown in Figure 5. In periods 1 and 3, the simulated current is consistent with the measured current by the error of just 10%. Also, the simulation is reproduced the sudden decrease in the current of IGBT1 immediately after the second turn-on. In period1, the simulation results indicate that the collector currents of IGBT1 (Ic1), IGBT2 (Ic2), and IGBT3 (Ic3) have the relation as Ic3 > Ic2 > Ic1, and in period 3, as Ic1 $\doteqdot$ Ic2 > Ic3. These relations are reproduced by the measurements.

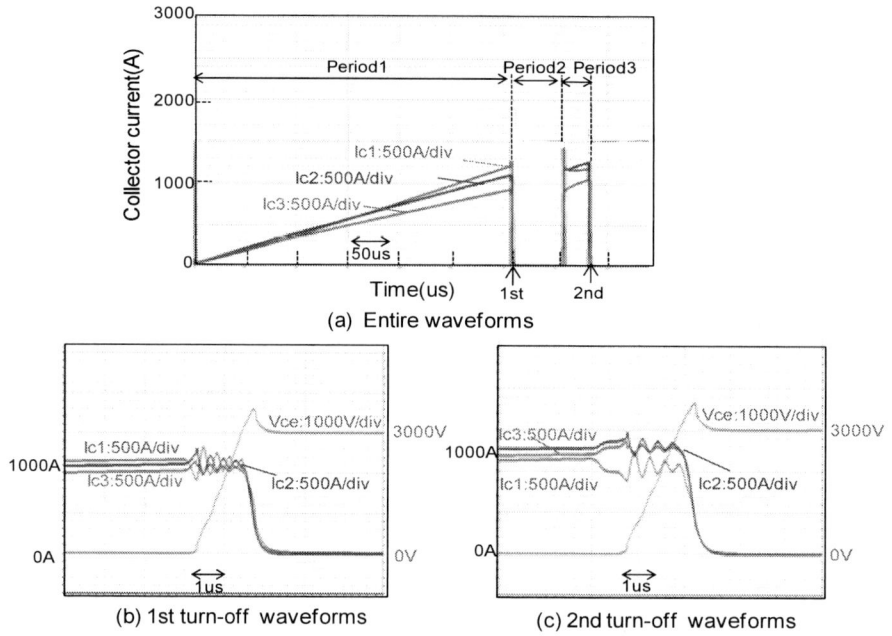

Figure 5: Waveforms simulated by a double-pulse switching test using the prototype bus bar

To investigate the mechanism that produces the current imbalance, the current flow through the filter capacitor and the converter was calculated. Collector currents Ic1, Ic2, and Ic3 of IGBT1, IGBT2, and IGBT3, respectively, the current of the filter capacitors, and the current to the inverter are shown in Figure 6.

The change in current is classified into the following three periods as also shown by the measurements. In period 1, the converter mainly supplies current to the IGBTs. In period 2, the capacitors are charged from the converter. In period 3, the currents from both the converter and the capacitors flow to the IGBTs. Soon after the IGBTs are turned off, the currents from the filter capacitors mainly flow. Gradually, the currents from the capacitors decrease, and the current from the converter increases. In period 1, the current imbalance is mainly determined by L1c, L2c, and L3c. Inductances L1c, L2c, L3c, and L1s, L2s, L3s are consistent by the error of less than 5%, and thus the current imbalance is small compared to that in period 1. In period 3, because the currents from both the converter and the capacitors flow to the IGBTs, coupling coefficients (KL1c, etc.) between the converter and the filter capacitors strongly affect the current flow. The results of the electromagnetic-field analysis show that the coupling coefficients are more than 200% each other. This variability of coupling coefficients is thought to be due to the difference between the distances of the N or P bus bars from the connection points of the IGBT emitter or collector terminals to those of the copper wirings connected to N or P terminals of the converter.

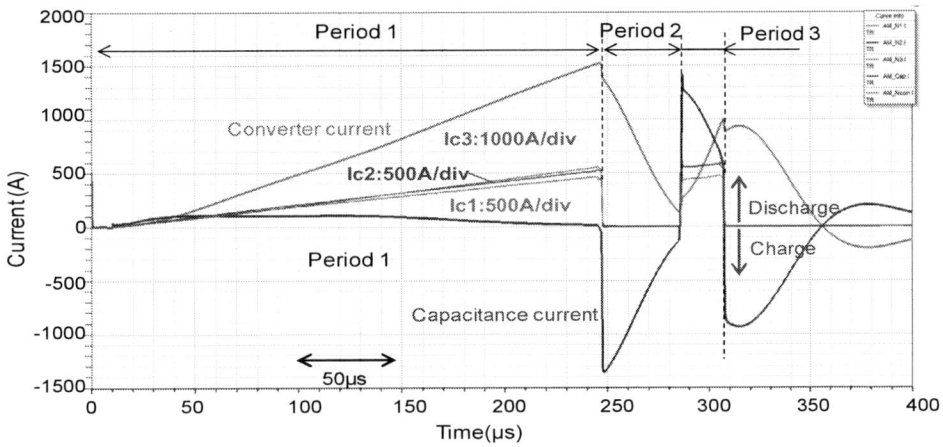

Figure 6: Collector currents of IGBT1, IGBT2, and IGBT3, the current from the filter capacitor, and the current to the inverter.

3.3. Improvement of current balance by new bus bar topology

To reduce the variability of the coupling coefficient, the distances between the connection points of the IGBT emitter or collector terminals to those of the thick copper plate wirings connected to the converter should be as equal as possible. Simulated waveforms obtained by a double-pulse switching test utilizing the newly developed bus bar topology are shown in Figure 7. In terms of both the simulation and measurement results, the figure shows that the current imbalance among the three IGBT modules is suppressed. This developed bus bar topology for suppressing current imbalance among the parallel-connected IGBT modules has been applied commercially.

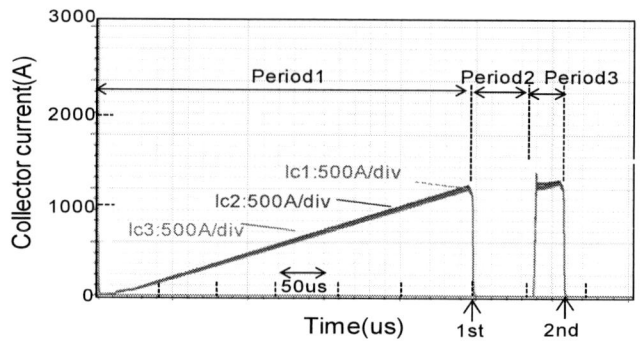

(a) Entire waveforms

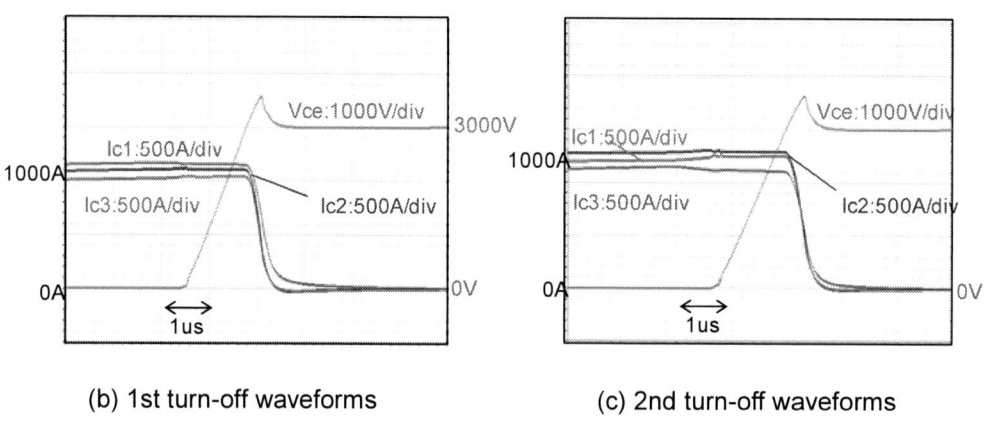

(b) 1st turn-off waveforms   (c) 2nd turn-off waveforms

Figure 7: Simulated waveforms of a double-pulse switching test
utilizing the newly developed bus bar

Conclusion

The mechanism that generated current imbalances among three IGBTs in double-pulse switching tests was studied. The results showed that the current imbalance was strongly affected by the coupling coefficients of the inductances between the converter or the filter capacitor and the IGBTs. On the basis of these results, a new bus bar topology with the uniform coupling coefficients was developed and evaluated.

References

[1] J. Ito; Y. Hagiwara; N. Yoshie, "Development of the IGBT applied traction system for the Series 700 Shinkansen high-speed train", Developments in Mass Transit Systems, 1998. International Conference on, pp. 25-30

[2] Xuesong Wang; Zhengming Zhao; Liqiang Yuan, "Current sharing of IGBT modules in parallel with thermal imbalance", Energy Conversion Congress and Exposition (ECCE), 2010 IEEE, pp. 2101 – 2108, pp. 850-856

[3] Rui Wu; Liudmila Smirnova; Huai Wang; Francesco Iannuzzo; Frede Blaabjerg, "Comprehensive investigation on current imbalance among parallel chips inside MW-scale IGBT power modules", Power Electronics and ECCE Asia (ICPE-ECCE Asia), 2015 9th International, pp. 1775-1779

[4] T. Ohi; T. Horiguchi; T. Okuda; T. Kikunaga; H. Matsumoto, "Analysis and measurement of chip current imbalances caused by the structure of bus bars in an IGBT module", Industry Applications Conference, 1999. Thirty-Fourth IAS Annual Meeting. Conference Record of the 1999 IEEE, pp. 1775-1779

[5] R. Saito; Y. Koike; A. Tanaka; T. Kushima; H. Shimizu; S. Nonoyama, "Advanced high current, high reliable IGBT module with improved multi-chip structure", Power Semiconductor Devices and ICs, 1999. ISPSD '99. Proceedings. The 11th International Symposium on, pp. 109-112

[6] Mikko Paakkinen; Didier Cottet, "Simulation of the non-idealities in current sharing in parallel IGBT subsystem", Applied Power Electronics Conference and Exposition, 2008. APEC 2008. Twenty-Third Annual IEEE, pp. 211-215

[7] Nan Chen, Muhammad Nawaz, Liwei Wang, "Dynamic Characterization of Parallel-Connected High-Power IGBT Modules", IEEE Transactions on INDUSTRY APPLICATIONS, Vol. 51, pp. 539–546 (2015)

[8] H. Fukumoto, Y. Kameoka; K. Yoshioka; T. Takizawa; T. Kobayashi, "Application of 3D eddy current analysis on magnetically levitated vehicles", IEEE Transactions on Magnetics, Vol. 29, Issue 2, pp. 1878–1881 (1993)

[9] S. Shirakawa; H. Fukumoto, "Analysis of electromagnetic wave propagation considering continuity of polarization current by using vector facet elements", IEEE Transactions on Magnetics, Year: 1997, Volume: 33, Issue: 2, pp. 1358-1361

EMI investigation in a GaN HEMT power module

Xiaoshan, LIU, SATIE, France, xiao-shan.liu@satie.ens-cachan.fr
François, COSTA, université Paris Est Créteil, SATIE, France, francois.costa@satie.ens-cachan.fr
Bertrand, Revol, ENS Cachan, SATIE, France, bertrand.revol@satie.ens-cachan.fr
Cyrille, GAUTIER, université Paris 10, SATIE, France, cyrille.gautier@satie.ens-cachan.fr

Abstract

A half bridge power module using 650V/30A GaN HEMTs on PCB substrate has been fabricated and investigated with different ceramic decoupling capacitors for in-module EMI filtering. The experiments have shown that, by using small value decoupling capacitors, both differential and common mode currents' spectra have been significantly reduced from conducted to radiate emission frequency range. Time domain modeling and simulation have been studied to revisit the experimental results. The placement of the common mode decoupling capacitor has been discussed. In-module filtering strategy has been suggested in the conclusion.

1. Introduction

1.1. Background

Thanks to the material's physical properties and the investment in the engineering and manufacturing infrastructure, power semiconductor devices based on Gallium Nitride (GaN), such as high-electron-mobility-transistor (HEMT), allows faster switching and less power losses comparing to Silicon (Si) and/or Silicon Carbide (SiC) devices. Therefore they are promising for high frequency, high current and high efficiency power conversion. With GaN HEMT's small die formed packages, high power density and modularized design can be achieved. Not only the active devices, but the gate drivers [1] and the inductive (L) and capacitive (C) elements [2] can be integrated in the power module to improve its electromagnetic performance.

However, GaN HEMT's fast switching results in high slew rate in voltage (dV/dt) and current (dI/t). These high dV/dt and dI/dt combined with parasitic LC elements in the power module will result in electromagnetic interference (EMI) noise in a wide frequency range, which will cause both conducted and radiated emission problems in differential mode (DM) and common mode (CM). The in-module EMI filtering strategy, by using low equivalent series inductance (ESL) multilayer ceramic capacitors (MLCC) to decouple the EMI noise sources where they occur, is an efficient way [3] [4] [5]. The impact on EMI of different values and placements of these MLCC will be studied in the following sections.

1.2. Investigated module

A half-bridge (HB) module using two 650V/30A e-mode GaN HEMTs (GS66508P) on two-layer PCB substrate is fabricated and investigated. The module's layout and equivalent circuit are shown respectively in **Fig. 1a** and **Fig. 1b**. The top copper is used as electrical connection and the bottom one is attached to a grounded heatsink. The four corners (GR) on the top copper are connected to the ground by screws. This configuration is similar to a real power module using an insulated metal substrate (IMS) or direct bonded copper (DBC). Therefore, the result of this study can be scaled to those modules. The parasitic capacitances between the top and bottom copper of the PCB are identified as Cp+, Cp- and Cout. MLCC Ccm+, Ccm- and Cdm are used for CM and DM decoupling. It should be noted that a relatively large parasitic inductance Lcm (~15nH each) has been formed in series with each Ccm, since the corresponding high frequency (HF) current loop is large (Ccm is far from the HB and the PCB is 1mm thick). All the parasitic elements' values are determined by measurements.

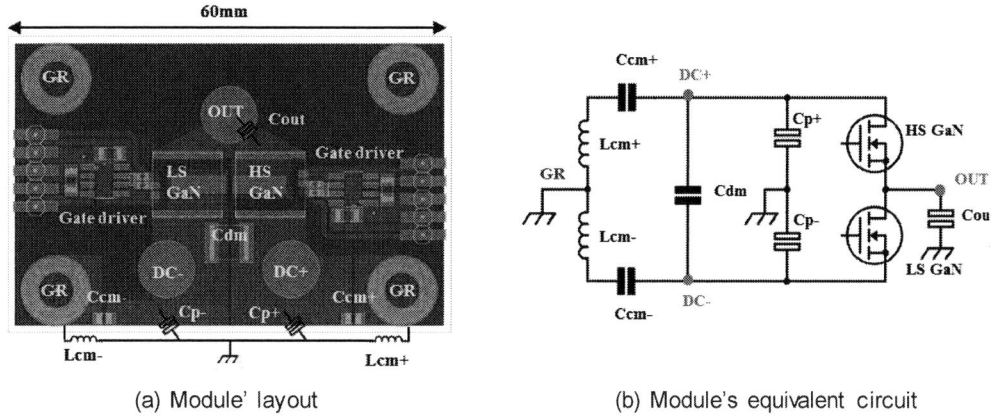

(a) Module' layout   (b) Module's equivalent circuit

Fig. 1. HB GaN module with in-module EMI filter MLCCs

2. Test setup and measurement procedure

2.1. Test setup

The HB GaN module is tested in synchronous buck mode. The equivalent circuit of the test setup is shown in **Fig. 2a** and the experimental setup is shown in **Fig. 2b**.

- Auxiliary board

The module is connected to an auxiliary board above (**Fig. 2b**). This board contains all the logic functions (dead-time generation and signal inversing) and the gate driver power supply (bootstrap for HS driver). It is important to filter the CM current passing through the lab power supply from the gate driver. Therefore, a 1mH CM choke is used between the lab power supply and the input of the gate driver power supply. In addition, a DC bank film capacitor (100µF) is connected to the module through this board. Obviously, this capacitor is not adapted to EMI filtering since it determines only the DC bus voltage ripple at switching frequency. All the voltage monitoring (Vds, Vgs for both high and low side) are settled by probe tip adapters. The turn-on and turn-off gate resistances are 15Ω and 2Ω respectively for an optimal switching performance. No current measurement is presented since it is beyond this paper's topic.

- LISN

Two LISN (9403-5-BP-10-BNC), which are adapted to standard DO160F, are used for 1) the noise isolation from the outside power network and 2) having fixed EMI propagation impedances in a relevant frequency range. In this paper, the DM and CM currents will be measured by two Pearson current sensors (model 6595). They are inserted between the output of the LISN and the input of the DC bank capacitor. Their configurations are shown in **Fig. 2c**; this enables separating the DM and CM currents. Both the LISN and the currents sensors are shielded in the same metal box.

- RL load

The HB GaN module is loaded with a resistance (47Ω) and an inductance (200µH). A non-inductive power resistance is chosen to eliminate the HF propagation effect; an air core inductance is made to reduce the load parasitic capacitance due to a ferrite core (**Fig. 2b**).

© VDE VERLAG GMBH · Berlin · Offenbach

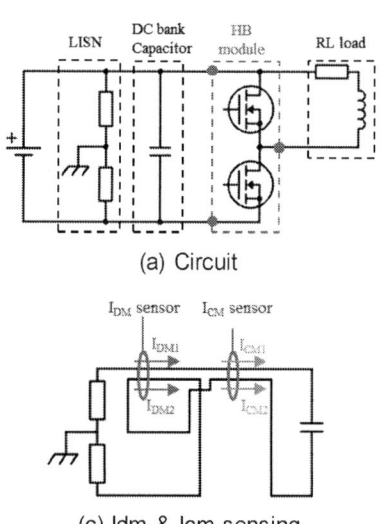

(a) Circuit

(c) Idm & Icm sensing

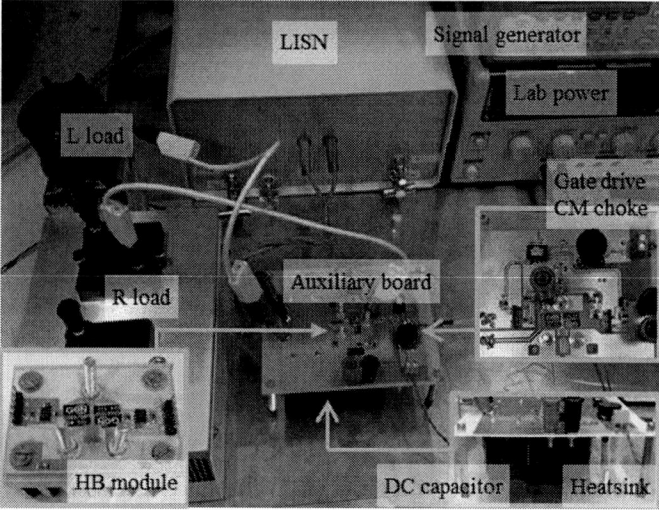

(b) Test setup

Fig. 2. Test setup

2.2. Measurement procedure

The in-module EMI filter capacitances include: 1) One MLCC Cdm for DM decoupling and 2) Two MLCC Ccm+, Ccm- for CM decoupling. These capacitors combine their effects with the parasitic capacitances Cp+ and Cp- as we will see later. However, in a strict point of view, one decoupling element will never only just reduce the EMI noise in one mode because noise transfer from one mode to the other cannot be avoided.

Since the placements of the MLCCs have been already fixed in the layout, only their values will be changed. The test procedure will be divided into the DM decoupling procedure (P1) and then the CM one (P2), which is shown in **Tab. 1**. These two test procedures allow investigating the impact of the in-module EMI filter capacitors of different values.

	Constant values	Varying values	Procedure name
DM decoupling procedure	Ccm+ = Ccm- = 0 nF Cp+ = 13 pF Cp- = 18 pF	Cdm = 0 nF	P1.0
		Cdm = 10 nF	P1.1
		Cdm = 100 nF	P1.2
		Cdm = 1 µF	P1.3
CM decoupling procedure	Cdm = 1µF Cp+ = 13 pF Cp- = 18 pF	Ccm+ = Ccm- = 0 nF	P2.0
		Ccm+ = Ccm- = 1 nF	P2.1
		Ccm+ = Ccm- = 10 nF	P2.2
Tab. 1. Test procedures			

During each procedure, both of the DM and CM currents are measured up to 100MHz for the reason that the EMI noise at HF, especially CM current, contributes to the radiated emissions even though conducted EMI is only considered to 30MHz. The frequency analyses are taken by a spectrum analyzer (Agilent 4195). The HB module is tested in 100V/1A, 200 kHz and duty circle of 0.5, due to its limiting heat sinking capability. Even though this working point is far from the nominal working condition, the variations of the quantities observed in this PCB module are relevant of what would be generated in a DBC or IMS module of the same layout.

© VDE VERLAG GMBH · Berlin · Offenbach

3. Experimental results and analyses

3.1. DM decoupling procedure

Fig. 3 to **Fig. 4** shows the measured spectra of DM and CM currents in different DM decoupling procedures (P1.0 to P1.2). It can be observed that:

- As stated in the previous study **[3] [6]**, the presence of HF decoupling capacitor Cdm creates a resonance peak. The increasing value of Cdm results in lower resonance amplitude and frequency (**Fig. 3a** and **Fig. 4a**). This resonance Fdm is due to the capacitor Cdm and all the parasitic inductances between the DC bank capacitor and Cdm. Beyond this resonance frequency, the DM noise amplitude is reduced. The reason is that Cdm close to the HB results in low impedance in the noise propagation path.

- The presence of Cdm reduces the CM noise spectra between 40MHz and 100MHz (**Fig. 3b** and **Fig. 4b**). This effect is explained in **Fig. 5**: only the DM EMI noise source Idm is considered in this equivalent circuit of the converter and the LISN. Assuming that the LISN impedances are perfectly balanced, the dissymmetry of the values of Cp+ and Cp- results in the mode transfer from HF DM to CM since the potential of the points A and B are not equal in HF. The presence of Cdm will reduce the HF DM noise and therefore the corresponding CM noise is reduced as well. As a conclusion, the presence of Cdm can significantly reduce the radiated emission caused by HF CM current.

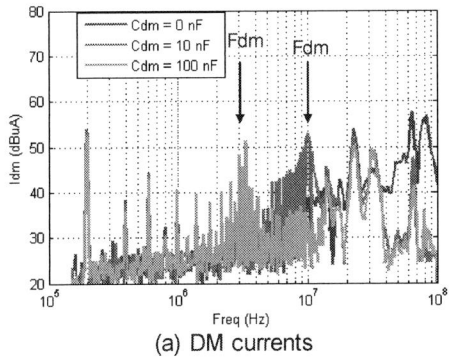

(a) DM currents

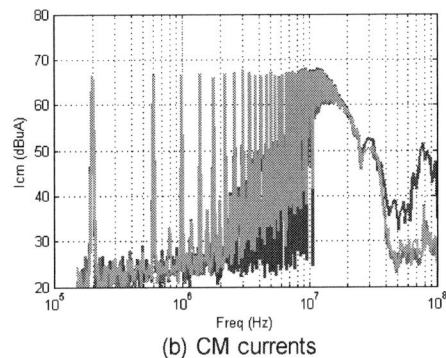

(b) CM currents

Fig. 3. Spectra of Idm and Icm in DM decoupling procedure (P1.0 to P1.2)

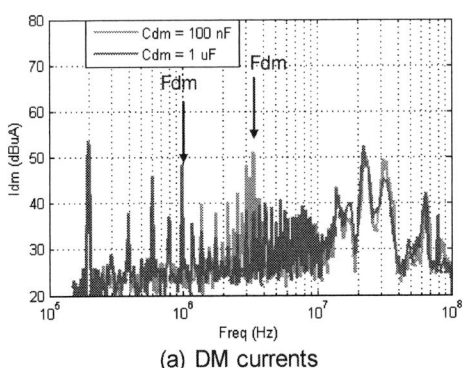

(a) DM currents

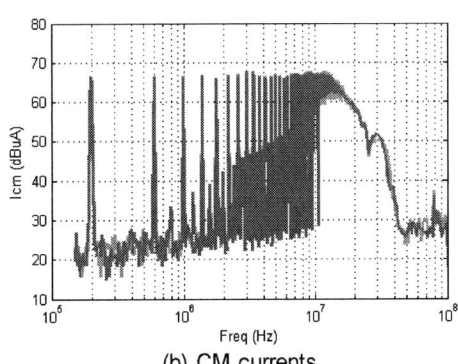

(b) CM currents

Fig. 4. Spectra of Idm and Icm in DM decoupling procedure (P1.2 to P1.3)

PCIM Europe 2016, 10 – 12 May 2016, Nuremberg, Germany

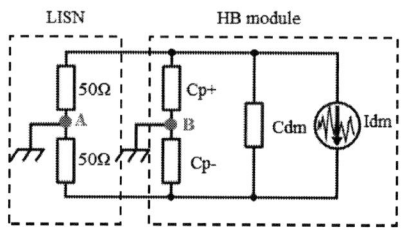

Fig. 5. HF mode transfer from DM to CM

3.2. CM decoupling procedure

Fig. 6 shows the measured spectra of DM and CM currents in different CM decoupling procedures (P2.0 to P2.2).

- It is observed that DM current spectrum between 3MHz up to 40MHz is reduced very significantly as Ccm capacitors are added in the module (**Fig. 6a**). This can be explained hereafter: when no CM decoupling capacitors (Ccm+ and Ccm-) are present, the parasitic capacitances Cp+ and Cp- will recycle only very partially the CM noise at high frequency. Moreover, since they are not equal, the currents I1 and I2 created by the CM noise source Vcm are not equal and therefore a HF DM current is formed by their difference in the LISN (**Fig. 7**). The added Ccm capacitors not only improve the symmetry of the CM impedance but also filter the CM noise at which frequency it is transferred to DM noise.

- With increasing Ccm capacitors' values pulls the noise valley towards the lower frequency (**Fig. 6b**). However, with a large Ccm value (10nF) there is a critical resonance between 10MHz and 40MHz. This is due to the relatively large parasitic inductance Lcm. It will be beneficial if its value could be reduced since the damping factor is proportional to $1/\sqrt{L}$ (see the section 4.3).

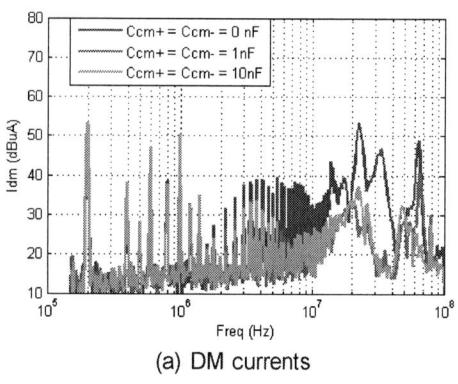

(a) DM currents

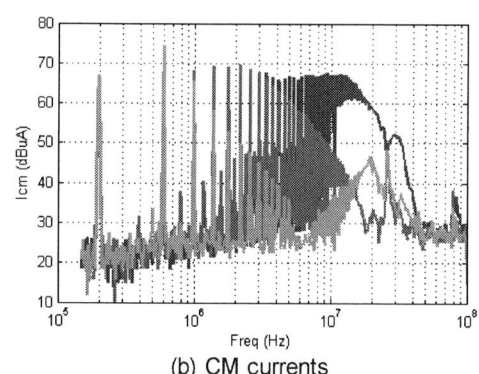

(b) CM currents

Fig. 6. Spectra of Idm and Icm in CM decoupling procedure (P2.0 to P2.2)

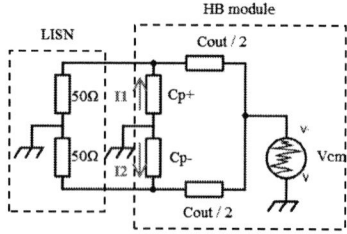

Fig. 7. HF mode transfer from CM to DM

© VDE VERLAG GMBH · Berlin · Offenbach

3.3. Final comparison

The final DM and CM EMI spectra comparison between the HB module without and with in-module decoupling capacitors is presented in **Fig. 8**. It can be observed that using 3 small values MLCCs (Cdm, Ccm+ and Ccm-) inside the module can reduce significantly the EMI noise from MHz up to 100MHz, which means a nice attenuation in both conducted and radiated EMI frequency range. In real applications, these in-module EMI filter MLCCs can be sized by consideration with the EMC standards and therefore down-sizing the total EMI filter at the input of the system.

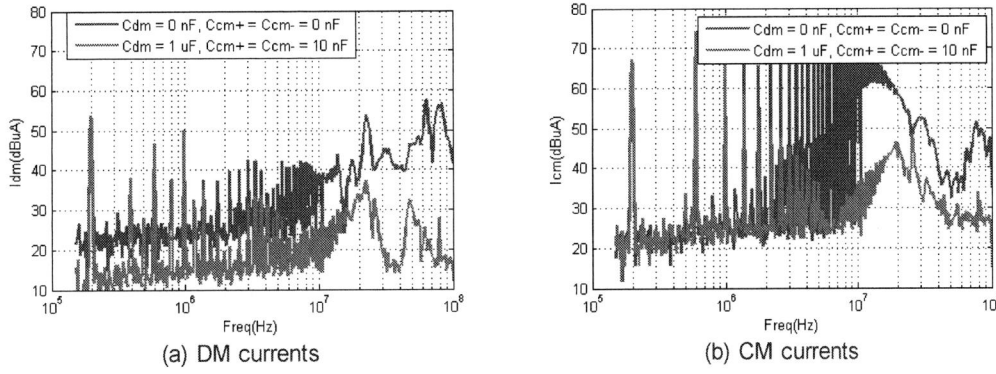

(a) DM currents

(b) CM currents

Fig. 8. Spectra of Idm and Icm before and after DM and CM decoupling procedure

4. Time domain modeling and simulation

4.1. Modeling

In order to verify the previous experimental analyses, time domain modeling and simulation followed by fast Fourier transfer (FFT) is executed to study the considered EMI phenomena. The modeling procedure involves 1) the EMI noise source and 2) its propagation path towards LISN. In time domain, the EMI noise sources are the switching devices (GaN HEMTs). In this paper, the commercial device model of GS66508P is used. The noise propagation paths are identified by lumped impedances in power loop and ground loop. An impedance analyzer (HP4194a) is used to measure these impedances. Calibration and compensation are carefully executed. The whole lumped circuit for simulation is shown in **Fig. 9**.

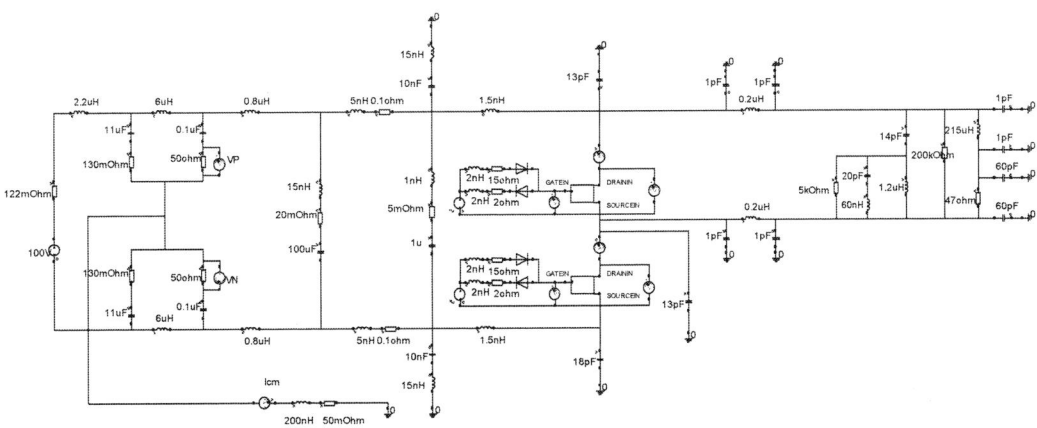

Fig. 9. Modeling circuit

© VDE VERLAG GMBH · Berlin · Offenbach

4.2. Simulation results

Fig. 10a shows the impacts of DM decoupling capacitors (Cdm) on DM current spectra and **Fig. 10b** shows the impacts of CM decoupling capacitors (Ccm) on CM current spectra. Comparing them respectively to the measurement results in **Fig. 3a** and **Fig. 5b**, the filtering effects of different decoupling capacitors are correctly simulated, particularly in CM mode. However, the DM noise resonance between 20MHz and 35MHz (**Fig. 3a**) are less perfectly simulated. This is probably due to the lack of the modeling of the gate driver CM propagation circuit not taken into account in the model of **Fig. 9**. However, Cdm resonance frequencies Fdm are correctly predicted (see arrows in fig. 10).

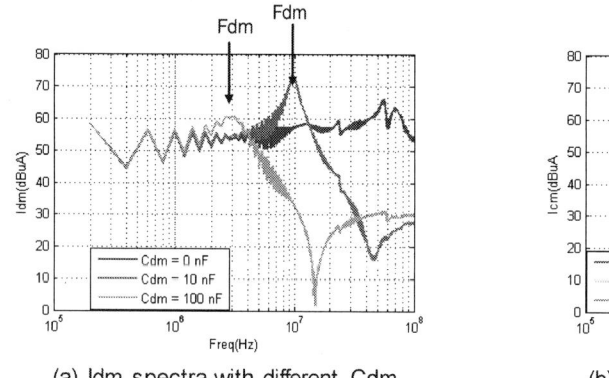
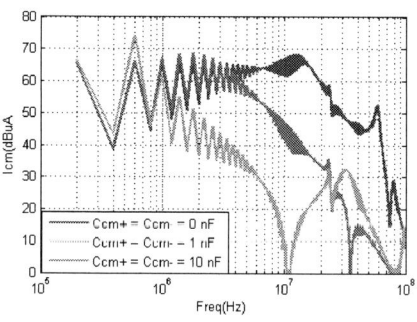

(a) Idm spectra with different Cdm   (b) Icm spectra with different Ccm

Fig. 10. Simulation results

4.3. Discussion on the placement of Ccm

As mentioned in section 1.2, since the CM decoupling capacitor Ccm is relatively far from the HB (the commutation cell), this results in a parasitic loop whose inductance Lcm is about 15nH for each capacitor Ccm. This parasitic inductance degrades the CM decoupling effect of Ccm in HF. The simulation is tested with different Lcm of 15nH and 5nH (configuration in procedure 2.2) and the CM currents' spectra are shown in **Fig. 11**. It is observed that the CM noise spectrum between15MHz and 100MHz is significantly reduced since the damping factor is higher by increasing$1/\sqrt{L}$. The improvement of Lcm can be realized by moving the Ccm capacitors as close as possible to the HB just as the capacitor Cdm.

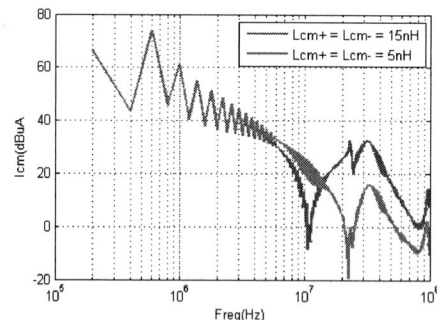

Fig. 11. Impact of Lcm on CM current spectra

5. Conclusion

By means of experiment and simulation, the significant benefit of using in-module EMI decoupling MLCCs has been proved. Conducted emissions have been reduced by using small value DM and CM MLCCs. This probably will improve the radiate emission as well. Therefore the in-module filtering strategy can be stated as follow:

- Minimizing the parasitic capacitance between the mid-point of HB and the ground (Cout);
- Balancing the impedance of DC+ and DC- (including the Cp+ and Cp-);
- Placing the low ESL DM decoupling capacitor (Cdm) as close as possible to the HB;
- Placing the low ESL CM decoupling capacitor (Ccm+ and Ccm-) as close as possible to the HB or in some way to limit their loop parasitic inductance (Lcm).

The values of these decoupling capacitors can be determined by experimental or simulated way, according the EMC standard and the system demand.

6. References

[1] D. Reusch, D. Gilham, Y. Su and F. C. Lee, "Gallium Nitride based 3D integrated non-isolated point of load module," Applied Power Electronics Conference and Exposition (APEC), 2012 Twenty-Seventh Annual IEEE, Orlando, FL, 2012, pp. 38-45.

[2] W. Zhang, Y. Su, M. Mu, D. J. Gilham, Q. Li and F. C. Lee, "High-Density Integration of High-Frequency High-Current Point-of-Load (POL) Modules With Planar Inductors," in IEEE Transactions on Power Electronics, vol. 30, no. 3, pp. 1421-1431, March 2015.

[3] Q. Liu, S. Wang, A. C. Baisden, F. Wang and D. Boroyevich, "EMI Suppression in Voltage Source Converters by Utilizing dc-link Decoupling Capacitors," in IEEE Transactions on Power Electronics, vol. 22, no. 4, pp. 1417-1428, July 2007.

[4] Robutel, R.; Martin, C.; Buttay, C.; Morel, H.; Mattavelli, P.; Boroyevich, D.; Meuret, R., "Design and Implementation of Integrated Common Mode Capacitors for SiC-JFET Inverters," in Power Electronics, IEEE Transactions on , vol.29, no.7, pp.3625-3636, July 2014

[5] Schanen, Jean-Luc; Roudet, James, "Built-in EMC for integrated power electronics systems," in Integrated Power Systems (CIPS), 2008 5th International Conference on , vol., no., pp.1-10, 11-13 March 2008

[6] Huang, Xudong, "Frequency Domain Conductive Electromagnetic Interference Modeling and Prediction with Parasitics Extraction for Inverters," PhD thesis of Virginia Polytechnic Institute and State University, 2004

[7] Hoene, Eckart; Ostmann, Andreas; Marczok, Christoph, "Packaging Very Fast Switching Semiconductors," in Integrated Power Systems (CIPS), 2014 8th International Conference on , vol., no., pp.1-7, 25-27 Feb. 2014

Ultra-High Frequent Switching with GaN-HEMTs using the C_{oss}-Capacitances as non-dissipative Snubbers

Luis Alfonso Fernández-Serantes, Hubert Berger, Wolfgang Stocksreiter, Gerald Weis
FH JOANNEUM, University of Applied Sciences, WerkVI-Strasse, 8605 Kapfenberg, Austria
LuisAlfonso.fernandezserantes@edu.fh-joanneum.at
hubert.berger@fh-joanneum.at
wolfgang.stocksreiter@fh-joanneum.at
gerald.weis@fh-joanneum.at

The Power Point presentation will be available after the conference

Abstract

The intent of building high efficient and very compact buck-converters with wide-band-gap transistors in a classic hard-switching operation is limited by the losses arising from dissipative reloading of the output capacitors C_{oss}. The switching cycle of a 650V GaN-HEMT half-bridge is analyzed in terms of the C_{oss}-related effects. A linearized model serves for deriving the relations and explaining how the C_{oss} can be deployed for minimizing the losses. This is achieved by using C_{oss} as an integrated non-dissipative snubber, providing soft switching for the internal transistor channel. EMI-measurements at 1MHz switching frequency are finally showing the reduction of radiated emissions when using soft-switching mode.

1   Introduction

Until now, silicon power devices are still the most used components in power electronic converters. Silicon transistors and diodes have been improved and optimized over the last decades. Their production costs have been reduced considerably [4]. Especially the super junction MOSFETs have become a very attractive solution for power converters by showing an almost optimal performance [5]. Further steps towards increased power density, however, require an improved switching behavior, which is mainly related to its inherent bipolar body diode, which is severely affecting any hard-switching operation.

The wide band-gap semiconductors (SiC- and GaN-based) have emerged as new promising devices for power electronics. Up to 600V the GaN high electron mobility transistor (GaN-HEMT) is showing best switching performance in comparison with SiC and Si-MOSFETs [2]. Due to their zero-reverse-recovery diode behavior, only the parasitic capacitances are setting limits for higher switching frequencies.

In order to clarify the influence of the parasitic capacitances on the switching performance of GaN-HEMTs the relations are derived at the base of a linearized model. The analysis is done for a half-bridge configuration in a synchronous buck converter. The turn-off as well as the turn-on process are described in detail in section 2. After this, a zero voltage switching scheme using the parasitic output capacitances as non-dissipative snubbers is analyzed in section 3, followed by the corresponding EMI measurements of a 650V GaN-HEMT half-bridge.

2 Switching Analysis

For analyzing the switching performance, a half-bridge in a DC-DC buck converter configuration is used (fig 1a). This half-bridge with inductive load is representing the base element of a great variety of converter circuits.

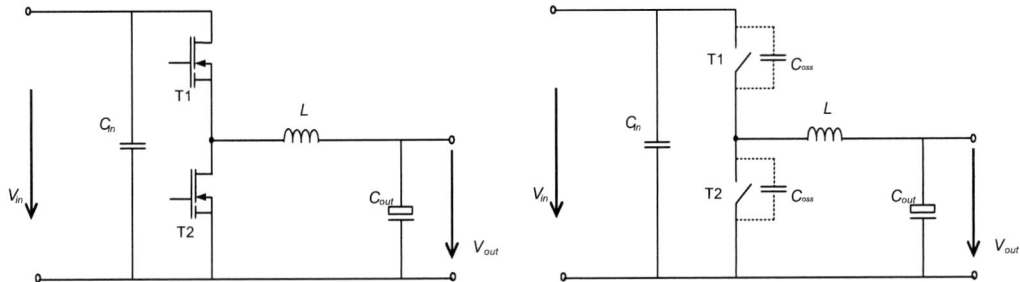

Fig 1: a) Synchronous buck converter with GaN-HEMTs (left) and b) simplified equivalent circuit

Fig. 1b) shows the simplified equivalent circuit where the switches shall have linear transitions of the (channel) resistance during turn-on and turn-off. The output capacitors (C_{oss}) are assumed to have voltage independent constant values, even if in real transistors they are showing a strongly non-linear behavior. The values can be interpreted as the equivalent time related C_{oss}.

The GaN-HEMTs represented in fig. 1a) include all the parasitic elements according to the model provided by the manufacturer. The parasitic capacitances are related to the different nodes of the transistors, so they are defined as gate-source capacitance C_{gs}, gate-drain capacitance C_{gd}, and drain-source capacitance C_{ds}. They are combined to two groups: input capacitance C_{iss}, and output capacitance C_{oss}. The input capacitance is defined as the sum of C_{gs} and C_{gd} and the output capacitance as $C_{ds} + C_{gd}$.

In a first step, the transitions of a pure switch without capacitances (fig. 1b) are shown. This is then complimented by adding the influence of an idealized output capacitance C_{oss}.

2.1 Turn-off Transition (of HS-Switch)

A linearized transistor model serves for deriving the basic relations of the transitions during turning-off the high-side switch.

Ideal Switches without Capacitances

In (2a) the parasitic capacitances are neglected and the channel resistance is assumed to change linearly causing also a linearly rising voltage, whereas the current still stays constant (*period t_{rv}*). Only when the drain-source voltage reaches the value of the dc-link voltage, the low side switch can take over the current, initiating the commutation. According to the simplified model, the current is now decreasing linearly, caused by an assumed hyperbolic increase of the drain-source resistance (*period t_{fi}*).

Ideal Switches with parallel Capacitances

Fig 2b shows these turn-off transients in presence of a constant C_{oss}, for both switches. In a half-bridge configuration, the capacitances of both HS- and LS-switch are reloaded (HS capacitance is charged and LS capacitance is discharged). The effective parasitic capacitance is therefore the sum of both. The voltage at the switched node increases proportionally to the drain-source resistance. The output current is now split into a component flowing through the ohmic channel and a component flowing via the effective C_{oss}. This capacitance is thereby acting as a turn-off snubber [3], storing a part of the energy, which has been previously fully dissipated in the channel. Due to the charging of the output capacitance, the turn-off process is taking longer than in the previous case.

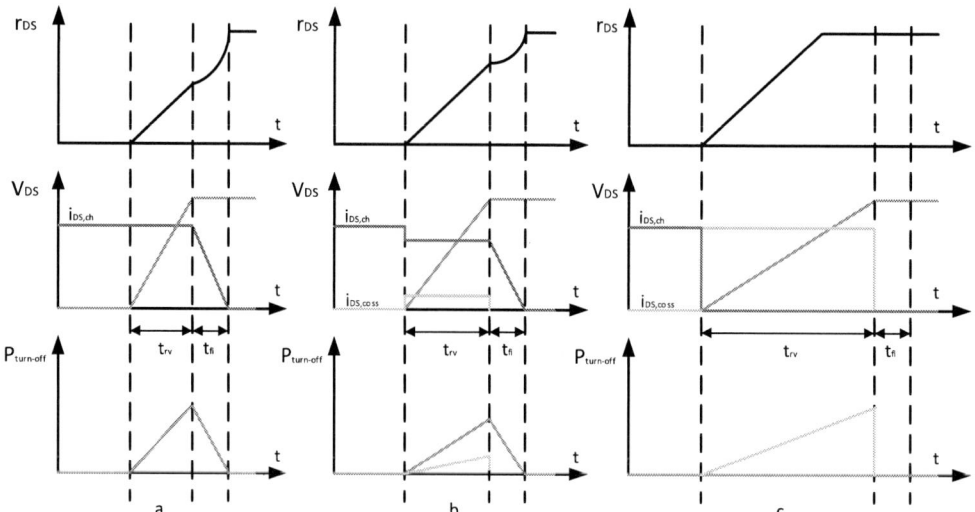

Fig 2: Turn-off switching transients: a) without C_{oss}; b) with C_{oss}; c) With optimally adjusted C_{oss}.

Ideal switches with optimally adjusted Capacitances

In a third case (fig 2c), we assume an ideal relation between the output current and the size of the C_{oss} in a way that the capacitors are taking over the whole current from the channel, resulting in a lossless turn-off transition. The voltage rise time is now only defined by the reload process of the capacitances and not by the change of the channel resistance.

2.2 Turn-on transition (of HS switch):

Ideal switches without capacitances (Fig.3a)

During turning-on the HS-switch the current commutes from the reverse conducting LS-switch to the HS-switch (period tri). The current rise is assumed to be linear related with a hyperbolic decrease of the drain-source resistance. During this period, the drain-source voltage is still constant. Only when the whole output current is taken over by the HS-switch, the voltage at the HS-switch starts decreasing. This voltage transition is again assumed to be linear related with a proportional decrease of the drain-source resistance (*period t_{fv}*).

Ideal switches with parallel capacitances (Fig 3b and 3c)

The additional capacitances are not involved in the current commutation, as the voltage is still constant in this transition phase (*period t_{ri}*). Once the output current is fully taken over by the HS-switch, the voltage starts decreasing linearly. The high side capacitance is now discharged and the low side capacitance is charged. Thereby both charging and discharging happens via the channel resistance of the HS-switch resulting in the dissipation of

$$E = \frac{1}{2} 2 C_{oss} \cdot V_{DS}^2 = C_{oss} \cdot V_{DS}^2$$

The amount of these losses is corresponding with the loss reduction during the previous turn-off transients. This is why the capacitances are neutral in terms of the overall calculation of the switching losses [1,2]. They are only shifting losses from the turn-off transition to the turn-on transition. Fig 3b and 3c are showing these turn-off transitions for different capacitance values.

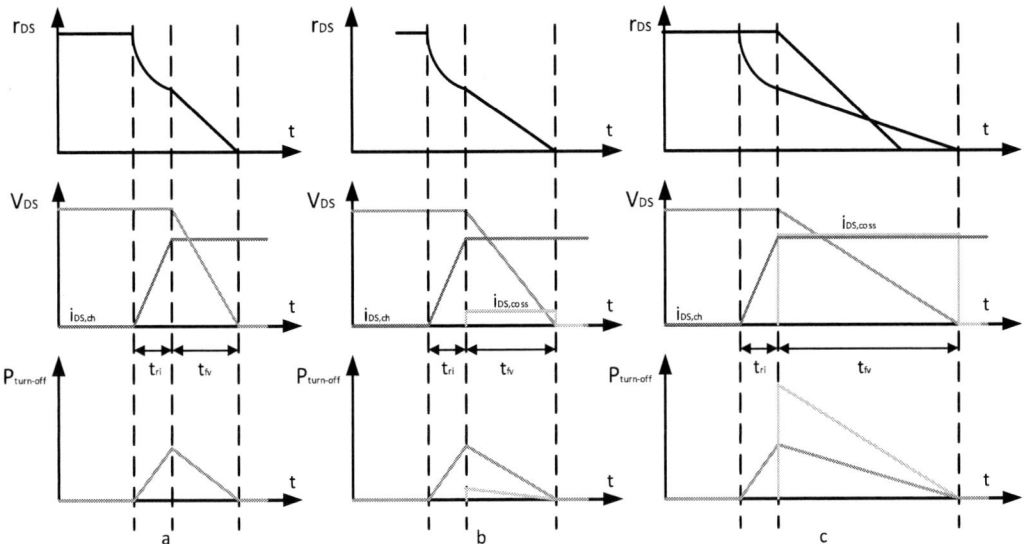

Fig 3: Turn-on switching transients of HS-switch: a) Without C_{oss}; b) With low C_{oss} value c) With high C_{oss} value

2.3 Switching-loss Mechanisms at different Loads

The loss mechanisms during the switching transients are dependent on the load conditions. Fig 4 is indicating three different load conditions of the analyzed buck converter. The previous analysis is only valid for relatively high average values of the inductor current (fig.4a). For lower inductor current values the C_{oss}-losses are not neutral any more in terms of the overall switching losses. The additional losses, which are arising during turn-on by reloading the capacitors via the channel of the HS-switch, are then no longer fully compensated by the reduction of turn-off losses (turn-off energy is lower than the energy stored in the capacitors).This means that at lower load the dissipative reloading of the output capacitors begin to dominate the switching losses. Any intent of building very high efficient converters by paralleling several transistors (which is then corresponding with low load operation of each single transistor) will be limited by the losses due to dissipative reloading of the output capacitors.

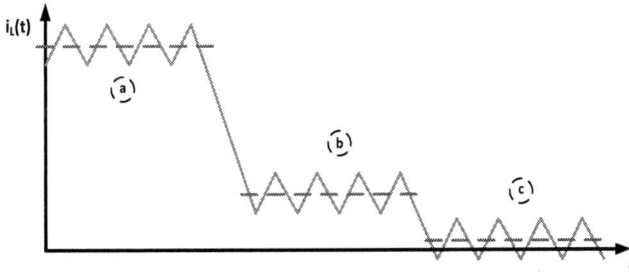

Fig 4: Inductor current waveforms at different loads

© VDE VERLAG GMBH · Berlin · Offenbach

The problem can be overcome by operating the buck-converter as shown in fig 4c, where the negative peaks of the ripple current are crossing the zero line. Because of the negative current during turning-off the LS switch, the current commutes immediately to the reverse conductive HS-switch. Thus, the parasitic capacitances will now act as non-dissipative snubbers also for the LS-switch like shown in fig 2b and 2c. This offers the possibility of turning-off the LS switch in a lossless way according to e.g. fig 2c representing an ideal soft switching.

3 Soft-Switching by using C_{oss} as a non-dissipative Snubber

The soft-switching method for buck-converters based on a bipolar current ripple has been described in textbooks already in the 1990s [3]. It did not really obtain relevance in combination with Silicon MOS-Fets. For GaN-HEMTs, however, it seems of particular interest, as it will allow very efficient and compact converter designs with switching frequencies above 1MHz even for designs with 650V devices. In real transistors, the output capacitance C_{oss} is highly non-linear. Fig.5 shows the voltage dependency of C_{oss} for GaN-HEMTs and Silicon MOSFETs (for both 650V devices as well as 100V devices). Silicon MOSFETs show a similar energy-related C_{oss} equivalent but a several times higher time-related C_{oss} equivalent as their GAN-HEMT counterparts. The time-related C_{oss} is defined as a "fixed capacitance that gives the same charging time as the real C_{oss} while V_{ds} is rising from 0% to 80% V_{dss}" [7]. As can be seen in fig. 5 silicon MOSFETs have a very high initial output capacitance and are therefore less attractive for this soft switching method, as they would need a several times higher negative current peak for achieving a similar short time for fully reloading their output capacitances.

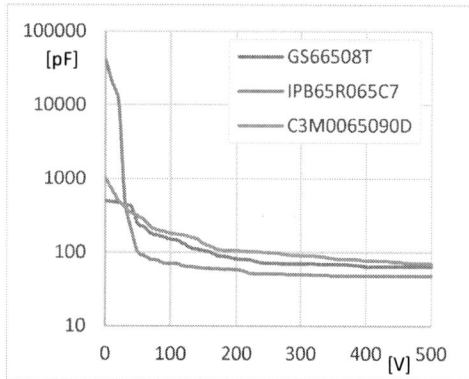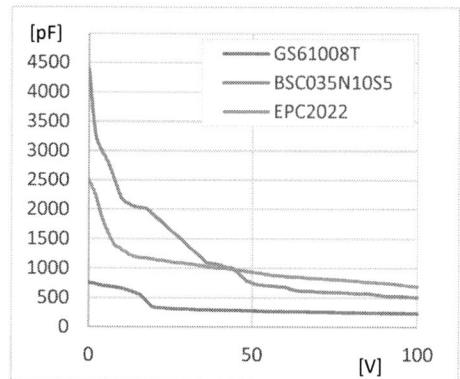

Fig 5: Voltage dependency of C_{oss} for different Transistor types
(650V types in the left diagram, 100V types in the right diagram)

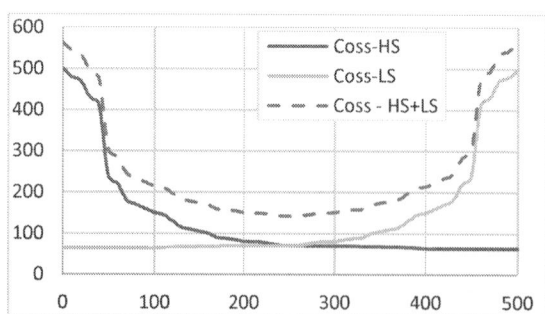

Fig 6: Effective output capacitance at the switched node of the half bridge as the sum of HS- and LS C_{oss} (for a half-bridge with two GaN-HEMTs GS66508T)

An increased initial capacitance (as long as it is not too high), however, is advantageous in terms of the resulting smooth voltage transition. The dashed line in fig 6. shows the corresponding symmetric characteristic of the effective capacitance in a half bridge, which is then responsible for the smooth voltage transition (fig. 7a). In order to guarantee a complete reloading of the capacitors, the negative peak of the current has to be high enough. The implementation of this switching scheme requires a more sophisticated control of the gate signals. In order to minimize losses as well as radiated emission, both the switching frequency as well as the interlock delay time shall be adapted to the actual duty cycle and the actual load (output current). For first implementations, powerful signal processing units have been used [6]. In order to achieve a more low-cost control solution, we expect that the principle functionalities of this switching scheme shall be implemented in a smart gate driver providing both an accurate detection of the negative current peak as well as the detection of the instant, when the reload of the capacitors is finished. This second feature corresponds with a so-called break-before make functionality, where the transistors are only allowed to turn on when the drain source voltage has become zero.

4 EMI measurements

The high frequent soft-switching of GaN-HEMT half bridges is not only reducing losses and filter size; the soft transition also minimizes the radiated emissions. In hard-switching the main source for radiated noise is the reload process of the C_{oss} capacitances via the transistor channel. This shorting of the parasitic capacitances results in extreme high *di/dt* values, which can only be handled by slower turn-on of the channel and by extremely compact design (minimized distance between HS- and LS switch).

Comparative EMI measurements have been performed with a DC-DC converter according to fig 1, using 650V GaN-HEMTs (GS66508P) from GaNSystems. The transistors were operated at frequencies of about 1MHz with a DC-link voltage of 200V both in soft switching as well as in hard-switching mode for comparison. The relatively low voltage was chosen because of the very high switching losses, which are arising during hard-switching at higher voltages. For the dc-link a 1,5µF cera-link capacitor from TDK-EPCOS was placed extremely close to the transistor half bridge. For the LC output filter the same cera-link capacitor is applied together with a single layer wound inductor (L=3,4µH / Litz wires with 0,08mm diameter for each Litz).

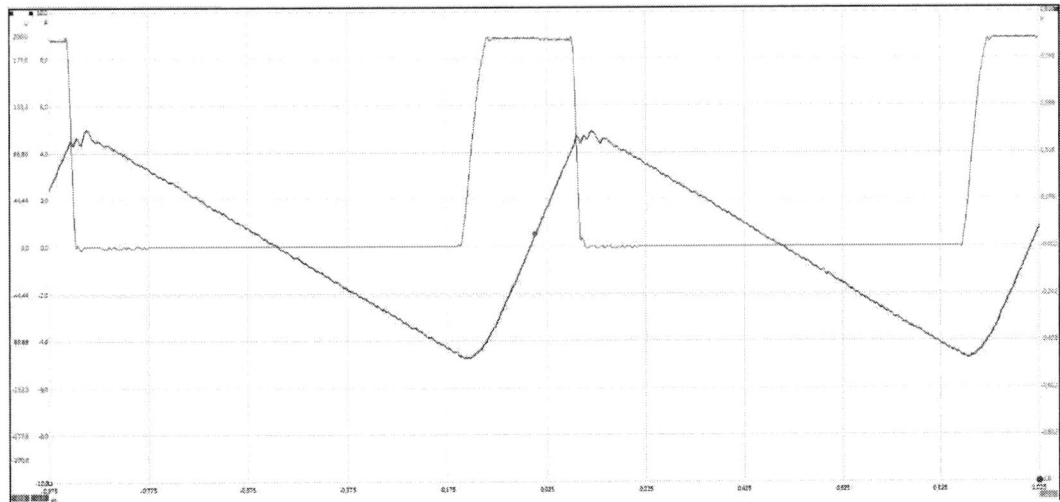

Fig 7a: Soft-switched DC-DC converter: f_{sw}= 984kHz, V_d=200V, duty cycle D= 0,22, I_{out} =0,5 A_{dc}
blue line: Voltage at the switched-node
red line: Current through the inductor *(pure AC value measured by a Rogowski coil)*

PCIM Europe 2016, 10 – 12 May 2016, Nuremberg, Germany

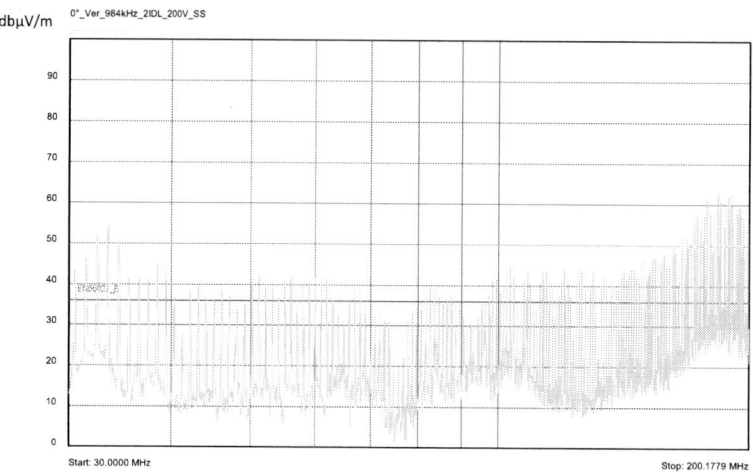

Fig 7b: EMI emission of soft-switched DC-DC converter according to 7a (without shielding or other EMI measures)

Fig 8a is showing the voltage and current waveforms for the DC-DC converter in hard-switching operation. During the relatively high interlock delay-time, the voltage starts to rise smoothly according to the soft-switching transition, but then the HS-switch is turned on resulting in a fast voltage rise even if the voltage rise time is relatively high for a typical hard switching operation, as the turn-on gate resistance is 20 Ohms. With a lower turn-on gate resistor the rise time can be decreased resulting in less switching losses. However, this would further increase the radiated emissions.

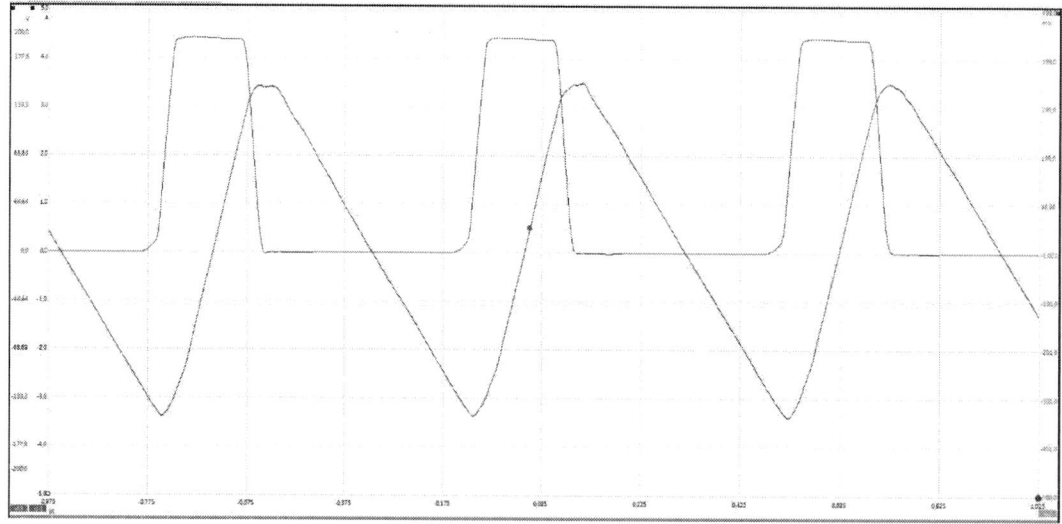

Fig 8a: Hard-switched DC-DC converter: f_{sw}= 1,567MHz, V_d=200V, duty cycle D= 0,22, I_{out} =1 A_{dc}
blue line: Voltage at the switched-node
red line: Current through the inductor *(pure AC value measured by a Rogowski coil)*

© VDE VERLAG GMBH · Berlin · Offenbach

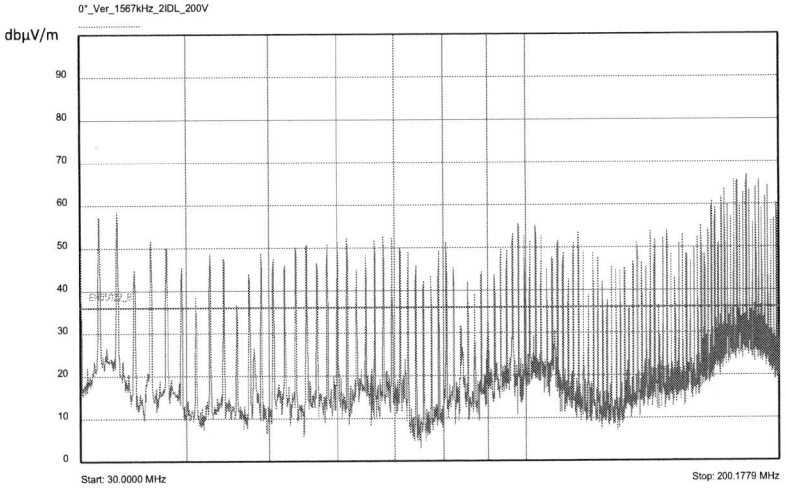

Fig 8b: EMI emission of hard-switched DC-DC converter according to 8a
(without shielding or other EMI measures)

Summary and Outlook

The influence of the parasitic output capacitances in a GaN-HEMT half-bridge configuration has been analyzed. The soft-switching scheme based on a bipolar ripple current will be a viable approach for ultra-high frequent switching of GaN-HEMTs. Comparative EMI measurements show that this switching scheme will reduce the levels of radiated emissions by about 12dBµV (fig 7b) in comparison with hard switching (fig 8b). FH JOANNEUM is therefore working on integrated smart gate drivers for the presented soft-switching scheme including a break-before make functionality for minimizing the reverse conducting phase.

References

[1] Lautner, J.; Piepenbreier, B., "Analysis of GaN HEMT switching behavior," in Power Electronics and ECCE Asia (ICPE-ECCE Asia), 2015 9th International Conference on , vol., no., pp.567-574, 1-5 June 2015. doi: 10.1109/ICPE.2015.7167840

[2] Wang, K.; Yang, X.; Li, H.; Ma, H.; Zeng, X.; Chen, W., "An Analytical Switching Process Model of Low-Voltage eGaN HEMTs for Loss Calculation," in Power Electronics, IEEE Transactions on , vol.31, no.1, pp.635-647, Jan. 2016. doi: 10.1109/TPEL.2015.2409977

[3] Power electronics: converters, applications, and design. Mohan, N. Undeland, T.M. Robbins, W.P. ISBN 9780471584087, 1995, Wiley

[4] Xiucheng Huang; Qiang Li; Zhengyang Liu; Lee, F.C., "Analytical Loss Model of High Voltage GaN HEMT in Cascode Configuration," Power Electronics, IEEE Transactions on , vol.29, no.5, pp.2208,2219, May 2014 doi: 10.1109/TPEL.2013.2267804

[5] Alatise, O.; Adotei, N.; Mawby, P., "Super-junction trench MOSFETs for improved energy conversion efficiency," Innovative Smart Grid Technologies (ISGT Europe), 2011 2nd IEEE PES International Conference and Exhibition on, vol., no., pp.1,5, 5-7 Dec. 2011 doi: 10.1109/ISGTEurope.2011.6162631

[6] Johann W. Kolar et al. "Approaches to overcome the Little box challenges" keynote in 37th International Telecommunications Energy conference, Osaka 2015.

[7] Costinett, D.; Maksimovic, D.; Zane, R., "Circuit-Oriented Treatment of Nonlinear Capacitances in Switched-Mode Power Supplies," in Power Electronics, IEEE Transactions on , vol.30, no.2, pp.985-995, Feb. 2015 doi: 10.1109/TPEL.2014.2313611

PCIM Europe 2016, 10 – 12 May 2016, Nuremberg, Germany

eGaN® FET based 6.78 MHz Differential-Mode ZVS Class D AirFuel™ Class 4 Wireless Power Amplifier

Michael de Rooij, Yuanzhe Zhang, Efficient Power Conversion, 909 N. Sepulveda Blvd. ste230, El Segundo CA, 90245, U.S.A., Michael.derooij@epc-co.com, Yuanzhe.zhang@epc-co.com

Abstract

The ongoing evolution of highly resonant wireless power solutions, enabled by eGaN FETs, continues in this paper where a 33 W capable AirFuel compatible Class 4 [1] power amplifier is presented. As the wireless power levels and charge surface areas increase, so do the design challenges. A 10 W eGaN FET zero voltage switching (ZVS) class D amplifier has been demonstrated as being capable of driving the entire AirFuel Class 2 [2] impedance range without additional circuitry [3]. Unfortunately, given the large increase in impedance range for AirFuel Class 4 systems, this may no longer be possible. This paper delves into the many challenges faced to realize a Class 4 wireless power amplifier solution that include, device thermals, device voltage limits, device selection, impact of timing, and design of support circuitry on the performance of the amplifier and devices. An experimental system is tested and the results show that despite the higher current and power levels, eGaN FETs continue to make inroads into realizing highly resonant loosely coupled wireless power solutions.

1. Introduction

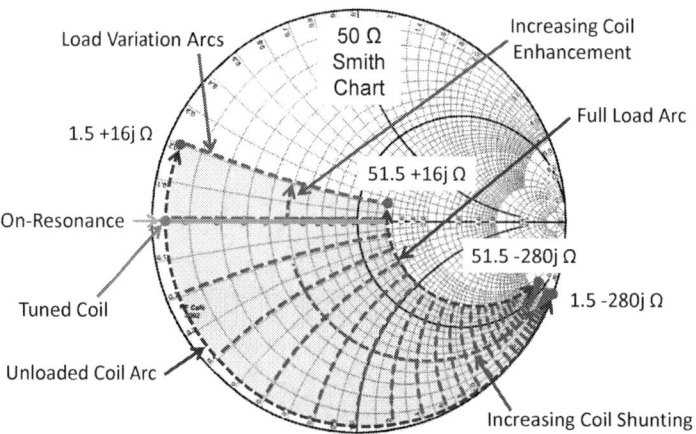

Fig. 1. Smith Chart (50 Ω) showing one of the AirFuel Class 4 coil impedance variation ranges [1].

AirFuel [4] Class 2 [2] and Class 3 [5] wireless power systems based on eGaN FETs have been demonstrated in the past using various amplifier topologies [3, 6, 7, 8, 9, 10, 11, 12, 13, 14, and 15]. Wireless power transfer is not exclusive to small handheld devices, but includes tablets and small laptop computers that require more power. The AirFuel Class 4 standard [1] supports up to 33 W total power and is suitable for powering up to three category 3 devices [4, 16], each rated to 6.5 W; one category 4 device rated at 13 W together with one category 3 device and one category 5 device rated at 20 W [4, 16]. The AirFuel Class 4 [1] standard's specifications are nearly double that of the Class 3 [5] and is differentiated by the wider impedance range shown in figure 1, higher RMS coil current of 1.375 A, and higher power at 33 W, that the amplifier needs to drive, significantly increasing the design challenges.

© VDE VERLAG GMBH · Berlin · Offenbach

1.1. Differential Mode ZVS Class D Amplifier

Since an eGaN FET based differential mode ZVS class D amplifier has previously demonstrated the ability to efficiently drive a wide impedance load range to the AirFuel Class 3 standard [3, 14], it is again chosen to drive a load to the Class 4 standard [1]. The power amplifier schematic for the differential mode ZVS class D amplifier is shown in figure 2 and comprises two single ended amplifier topologies connected back to back and operated at 180° phase with respect to each other.

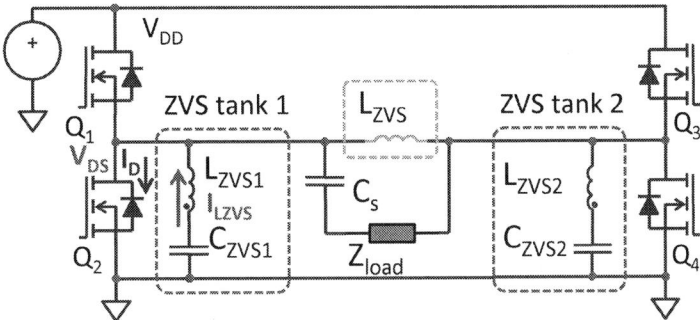

Fig. 2.   Differential-mode ZVS class D amplifier schematic.

Figure 3 shows waveforms for the various operating modes of the ZVS class D amplifier that can occur based on numerous load impedances and operating requirements. For capacitive loading the ZVS current is negated by the load and reverse conduction occurs prior to switching. The second condition is partial ZVS where there is insufficient current to complete self-commutation within the dead-time setting. The third condition is ZVS which yields the lowest losses and is the ideal operating condition. The last condition is post ZVS reverse conduction that occurs when self-commutation is completed within the dead-time setting due to the load contributing to the self-commutation current.

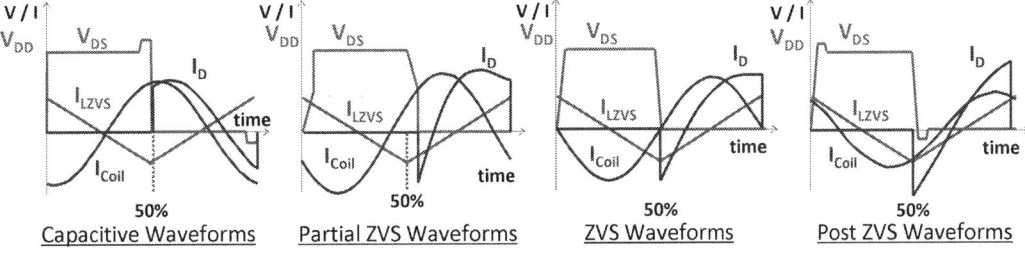

Fig. 3.   Differential-mode ZVS class D amplifier operating modes with waveforms.

1.2. Amplifier design

Before an amplifier can be designed the impact of the Class 4 standard on the amplifier needs to be understood. Driving a Class 4 tuned coil requires high voltage to deliver the required current or power under high reflected imaginary impedance conditions as shown in figure 4. This is the main reason a differential-mode amplifier is needed as it inherently has a higher supply voltage to RMS coil voltage gain. Wide impedance operating range capability is the key to keeping amplifier costs down by reducing the complexity and component count of adaptive tuning circuits to a minimum. Figure 3 shows that by using an 80 V limit for the FETs in a differential mode ZVS Class D amplifier, the expected impedance range capability can be ±50j Ω. However, other factors also need to be considered to verify the amplifier capability such as C_{OSS} losses, reverse conduction losses and more. The amplifier design procedure given in [3] can simply be followed by accounting for the differential mode of operation as supported by [17].

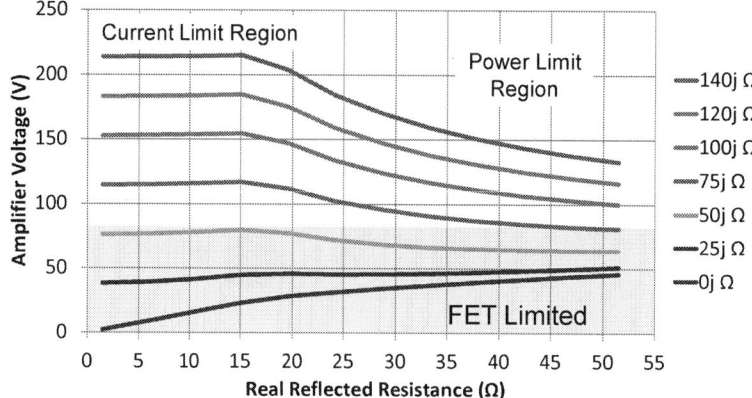

Fig. 4.   AirFuel Class 4 [1] tuned amplifier voltage as a function of real reflected impedance for various reflected imaginary impedances.

The higher operating current for the Class 4 amplifier requires lower $R_{DS(on)}$ FETs which results in higher C_{OSS} for the FETs and thus higher currents are needed to establish ZVS. In addition, any technique that can further enhance amplifier efficiency, such as the synchronous bootstrap FET power supply that eliminates Q_{RR} of the gate driver bootstrap diode [3, 14], must be used when possible.

For this evaluation, the EPC2007C eGaN FETs [18] are used for the amplifier. The ZVS tank inductance was calculated to be 390 nH with a dead-time setting of 8 ns and an amplifier supply voltage of 60 V.

2.   Amplifier performance analysis

An analysis of the impact of load impedance variation based on the Class 4 standard on the ZVS current was performed and the results are shown in figure 5. The graph shows 3 main regions of interest which are: (1) Partial ZVS, (2) Self-commutation followed by reverse conduction and, (3) Reverse conduction prior to full hard switching with maximum C_{OSS} losses.

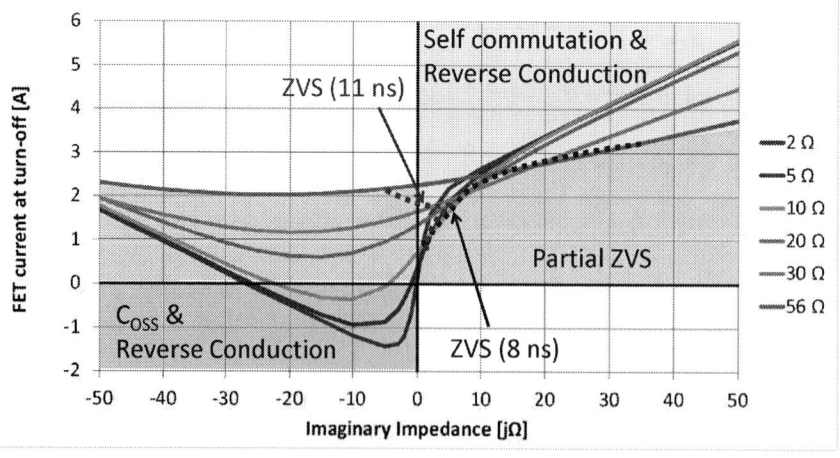

Fig. 5.   Impact of load impedance on FET ZVS current for the ZVS Class D amplifier.

As indicated by the red region, the capacitive load with low reflected load resistance negates the ZVS current completely, resulting in reverse current conduction in the FET being turned off and high C_{OSS} losses when the subsequent complimentary FET is turned on. The increase in reflected load resistance requires a higher supply voltage to the amplifier which in turn increases the current in the ZVS tank circuit, resulting in partial recovery of ZVS with increasing reflected load resistance.

Partial ZVS only occurs with positive current, i.e. current into the drain of the FET as indicated by the orange region, and occurs regardless of the sign of the reflected load reactance. However, a positive reactance adds to ZVS current thereby reducing the self-commutation time that can lead to reverse conduction prior to switching, indicated by the blue region.

Indicated by the dashed black trace for the 8 ns dead-time setting, is a small boundary region between the Partial ZVS and self-commutation regions where ZVS can be achieved. Further analysis using a 11 ns dead-time setting shows how ZVS shifts into the capacitive load region shown by the brown dashed trace. This is important as it can be used to optimize the efficiency of the amplifier over a wider load impedance range instead of at one fixed set point. In this case a dead-time of 8 ns yielded the longest ZVS trace on the graph and should result in the highest amplifier efficiency over the load range to be tested.

3. Experimental Verification

The EPC9065 differential-mode ZVS class D amplifier was used for experimental verification.

3.1. Experimental Setup

The experimental setup is the same as described in [3, 11] and comprises two main components, (1) an amplifier under test, and (2) a discrete programmable load with an adjustment range from 1.7 Ω though 57 Ω and from -60j Ω though +60j Ω.

Testing commences by setting the imaginary impedance to 0j Ω and adjusting the reflected load resistance over the full range. This test is operated to meet the current and power requirements of the AirFuel Class 4 standard. During testing the device and gate driver temperature were monitored to ensure that none exceeded 100°C. In addition, the supply voltage (V_{DD}) to the amplifier was limited to 80 V. If either condition were to be violated then that measurement was regarded as a fail. For each successive pass, the magnitude of the imaginary impedance would be increased until the compliance test failed.

3.2. Experimental Results

Total amplifier efficiency including gate drive power is shown in figure 6 for various reflected load impedances as a function of reflected load resistance variation. The amplifier is capable of exceeding 90% efficiency over a wide range of load resistance when operating on resonance (0j Ω). Over the imaginary impedance range of -15j Ω through +45j Ω the amplifier exceeded 85% efficiency for most of the full power load range.

The dead-time of 11 ns was originally selected to ensure complete ZVS transitions for no load operation with an amplifier supply of 40 V. Initial testing revealed difficulty operating beyond 40j Ω reactance. Further analysis of the converter revealed that operating with a dead-time of 8 ns would result in improved performance in the inductive reflected load impedance region. Figure 7 shows both the original measurements taken at 11 ns dead-time setting shown by the dotted traces and the revised measurements taken with a dead-time setting of 8 ns shown by the solid traces for two imaginary load impedance cases of 0j Ω shown in green, and +20j Ω, shown in red. The measurements confirm the analysis findings where a reduced dead-time setting results in decreased losses in the inductive region of the reflected load impedance range. Based on this finding, testing continued using only the 8 ns dead-time setting with the results over the full operating capability range as shown in figure 6. At 0j Ω with reduced dead times, partial ZVS occurs. As a result, the efficiency drops slightly, by less than 0.5%. At 20j Ω, about 1% to 3% efficiency improvement can be observed. This is desirable as the overall goal is to achieve high efficiency over a wide range.

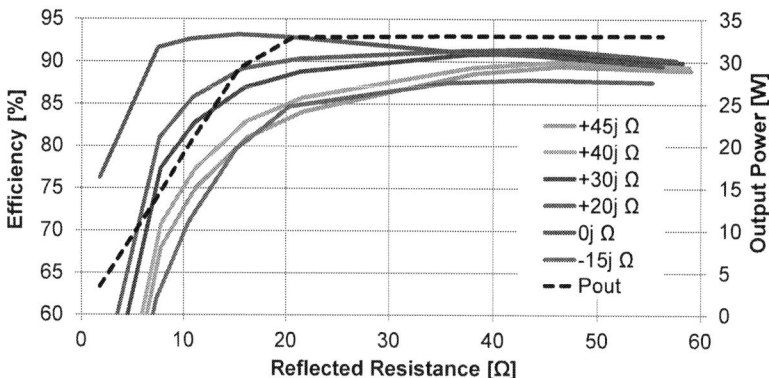

Fig. 6. Experimentally measured total amplifier efficiency driving a Class 4 load as a function of real reflected resistance load variation for various imaginary reflected load reactance's.

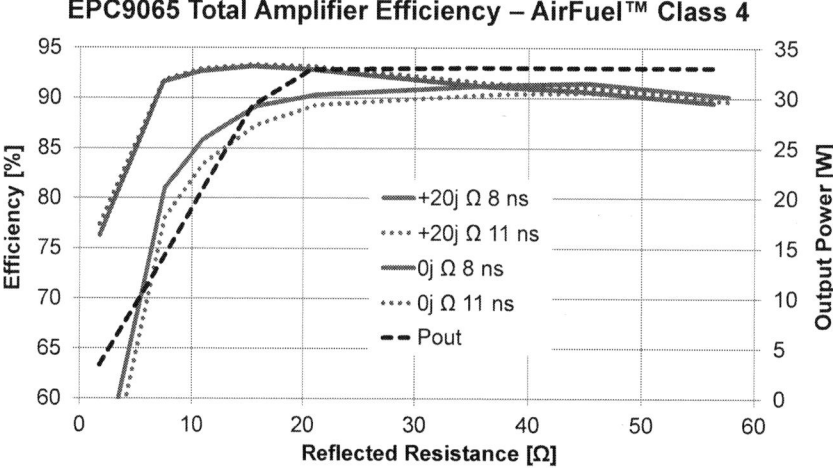

Fig. 7. Experimentally measured total amplifier efficiency driving a Class 4 load as a function of real reflected resistance load variation for 0j Ω and +20j Ω imaginary reflected load reactance and two different dead-times of 8 ns and 11 ns respectively.

4. Post Experimental Analysis

The inability of the amplifier to drive negative impedances beyond -15j Ω requires additional analysis. Using the FET characteristics and operating settings of the AirFuel Class 4 standard, an analysis of the amplifier was made to determine the total FET losses, and is shown in figure 8 for the EPC2007C [18] with L_{ZVS} = 390 nH, and the dead-time set at 8 ns. From figure 8 it is clear that the total FET power dissipation increases rapidly with increasing negative load impedance magnitude. Figure 8 also shows the thermal limit exceeded for heatsink free operation occuring beyond -15j Ω which is the same point determined experimentally. An increase in load reactance also shifts the point at which ZVS occurs for a given load resistance, and is also shown in figure 8.

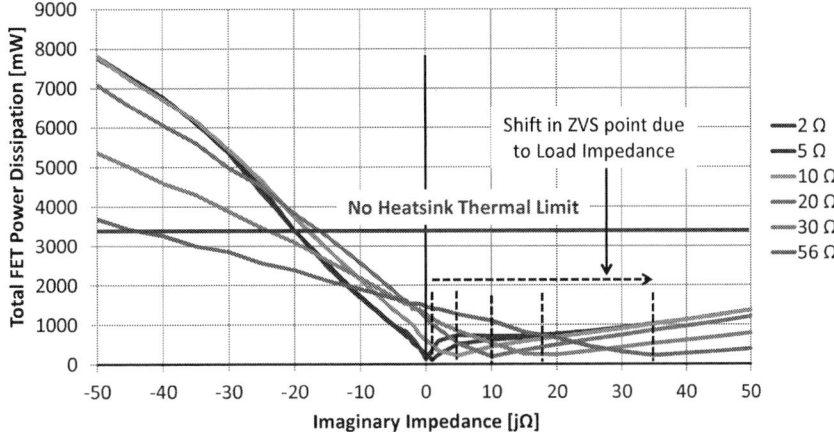

Fig. 8. Calculated total FET losses for a differential mode ZVS Class D amplifier using EPC2007C [18], L_{ZVS} = 390 nH and 8 ns dead-time.

Dead-time impacts the partial ZVS point as well as the duration of reverse current conduction, and for this reason a dead-time of 11 ns was also analyzed. An increase in dead-time increases the reverse conduction time and hence the losses. However, under partial ZVS conditions, a longer dead-time decreases the residual voltage at the time of switching due to the self-commutation current, resulting in a decrease in C_{OSS} losses. A balance between the two can thus be found that yields the lowest overall losses. Figure 9 shows the effect of both the increase in reverse conduction losses and decrease in partial ZVS losses as a result of an increase in dead-time from 8 ns, shown in grey tones, to 11 ns shown in color. This finding was validated by the experimental results of figure 7.

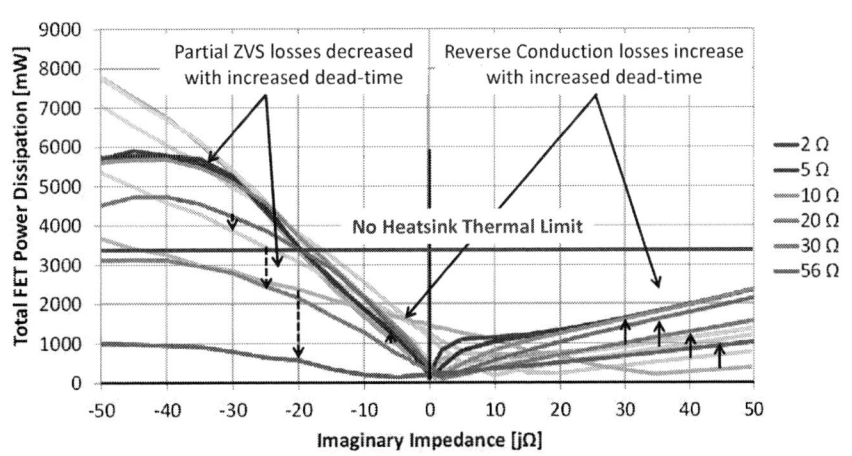

Fig. 9. Calculated total FET losses for a differential mode ZVS Class D amplifier using EPC2007C [18], L_{ZVS} = 390 nH and 11 ns dead-time with the 8 ns dead-time results in grey.

Finally, it is evident that the losses are dominated in the capacitive load region by C_{OSS} and thus a different FET, the EPC8010 [19], is selected for analysis. Since this FET has a significantly lower C_{OSS}, the ZVS current can also be reduced by increasing L_{ZVS} to 600 nH. Again the lower C_{OSS} results in a shorter self-commutation time so 2.5 ns was selected. The results of the analysis are shown in figure 10 in color together with the original EPC2007C results shown in grey scale. It is apparent that the EPC8010 has the potential to nearly meet the entire ±50j Ω load impedance range without the need for a heatsink and will become the subject of subsequent work.

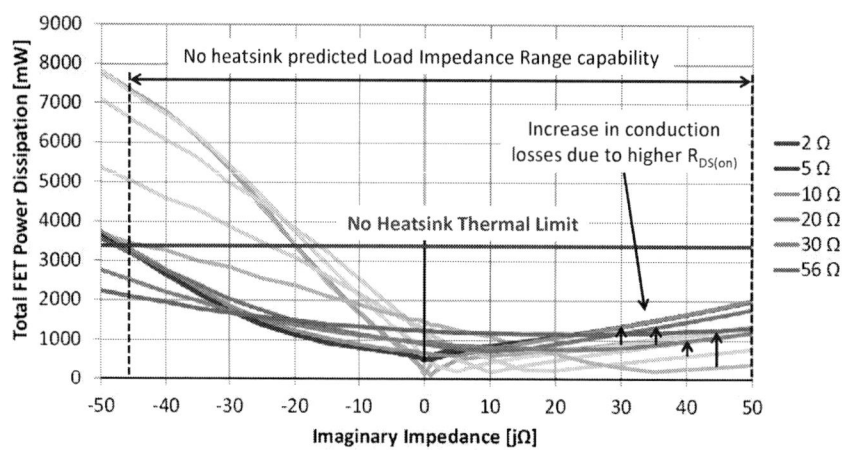

Fig. 10. Calculated total FET losses for a differential mode ZVS Class D amplifier using EPC8010 [19], L_{ZVS} = 600 nH and 2.5 ns dead-time compared with the EPC2007C [18] at 8 ns dead-time results shown in grey.

5. Conclusions

An experimental evaluation of an eGaN FET based differential ZVS Class D amplifier operating over a wide impedance range in compliance with the AirFuel Class 4 standard was presented in this paper. The amplifier was fitted with a synchronous FET bootstrap power supply capable of eliminating the reverse recovery losses associated with the internal diode of the gate driver bootstrap circuit. Using EPC2007C devices for the amplifier, an impedance drive range from -15j Ω through +45j Ω was achieved. Testing required that neither the FETs or gate driver temperature exceeded 100°C or that the voltage needed to drive the amplifier exceeded 80 V. Voltage capability predictions indicated that the amplifier should be capable of driving a reflected load impedance range from -50j Ω through +50j Ω and that the C_{OSS} of the chosen FETs was too high resulting in excessive losses in the capacitive load region.

Additional analysis of the amplifier circuit was conducted and shown that in the capacitive load region C_{OSS} is the most importance factor when selecting FETs and furthermore, that conduction losses decrease in this region allowing FETs with much higher $R_{DS(on)}$ to be used. This opens the possibility being able to drive the entire ±50j Ω reflected impedance range using EPC8010's as the FETs for the amplifier. While the conduction losses will increase in the inductive load region, they do not increase to the point that the thermal limit will be exceeded. Efficiency is further improved because gate driver power consumption will decrease substantially due to the lower C_{ISS} of the EPC8010. The largest benefits will be the elimination of the heatsink and the reduction in adaptive tuning circuit complexity yielding a substantial size and cost reduction.

6. References

[1] "A4WP PTU Resonator Class 4 Design - A2Tx@25mm 250x250 mm", A4WP standard document RES-15-0018 Ver. 0.5, June 9, 2015.

[2] A4WP PTU Resonator Class 2 Design - Spiral Type 140-90 A4WP standard document RES-14-0008 RES-14-0006 Ver. 1.2 June 26, 2014

[3] M. A. de Rooij, "Wireless Power Handbook," Second Edition, PCP, El Segundo, October 2015, ISBN 978-0-9966492-1-6

[4] R. Tseng, B. von Novak, S. Shevde and K. A. Grajski, "Introduction to the Alliance for Wireless Power Loosely-Coupled Wireless Power Transfer System Specification Version 1.0," IEEE Wireless Power Transfer Conference 2013, Technologies, Systems and Applications, May 15-16, 2013.

[5] A4WP PTU Resonator Class 3 Design – Spiral Type 210-140 Series (PTU 3-0001), A4WP standard document RES-14-0008 Ver. 1.2, June 30, 2014.

[6] A. Lidow, M. A. de Rooij, "Performance Evaluation of Enhancement-Mode GaN transistors in Class-D and Class-E Wireless Power Transfer Systems," Bodo Magazine, May 2014, pg. 56 - 60.

[7] M. A. de Rooij, J. T. Strydom, "eGaN® FET- Silicon Shoot-Out Vol. 9: Wireless Power Converters," Power Electronics Technology, pp. 22 – 27, July 2012.

[8] M. A. De Rooij and J. T. Strydom, "eGaN® FETs in Low Power Wireless Energy Converters," Electro-Chemical Society transactions on GaN Power Transistors and Converters, Vol. 50, No. 3, pg. 377 – 388, October 2012.

[9] M. A. de Rooij, "eGaN® FET based Wireless Energy Transfer Topology Performance Comparisons," International Exhibition and Conference for Power Electronics, Intelligent Motion, Renewable Energy and Energy Management (PCIM - Europe), May 2014, pg. 610 – 617.

[10] A. Lidow, "How to GaN: Stable and Efficient ZVS Class D Wireless Energy Transfer at 6.78 MHz," EEWeb: Pulse Magazine, Issue 126, pp. 24 – 31, July 2014.

[11] M. A. de Rooij, "The ZVS Voltage-Mode Class-D amplifier, an eGaN® FET-enabled Topology for Highly Resonant Wireless Energy Transfer," IEEE Applied Power Electronics Conference (APEC), March 2015.

[12] W. Chen, et al., "A 25.6 W 13.56 MHz Wireless Power Transfer System with a 94% Efficiency GaN Class-E Power Amplifier," IEEE MTT-S International Microwave Symposium Digest (MTT), pg. 1 – 3, June 2012.

[13] M. A. de Rooij, "Performance Evaluation of eGaN® FETs in Low Power High Frequency Class E Wireless Energy Converter," International Exhibition and Conference for Power Electronics, Intelligent Motion, Renewable Energy and Energy Management (PCIM - Asia), June 2014, pg 19-26.

[14] M. A. de Rooij, "Performance Comparison for A4WP Class-3 Wireless Power Compliance between eGaN® FET and MOSFET in a ZVS Class D Amplifier," International Exhibition and Conference for Power Electronics, Intelligent Motion, Renewable Energy and Energy Management (PCIM - Europe), May 2015.

[15] A. Lidow, J. Strydom, M. de Rooij, D. Reusch, "GaN Transistors for Efficient Power Conversion," Second Edition, Wiley, ISBN 978-1-118-84476-2.

[16] A4WP Wireless Power Transfer System Baseline System Specification (BSS), A4WP-S-0001 v1.3.1, February 25, 2015.

[17] S-A. El-Hamamsy, "Design of High-Efficiency RF Class-D Power Amplifier," IEEE Transactions on Power Electronics, Vol. 9, No. 3, pg. 297 – 308, May 1994.

[18] Efficient Power Conversion, "EPC2007C – Enhancement Mode Power Transistor," EPC2007C datasheet, December 2013 [Revised January 2015], [Online] Available: http://epc-co.com/epc/Products/eGaNFETs/EPC2007C.aspx

[19] Efficient Power Conversion, "EPC8010 – Enhancement Mode Power Transistor," EPC8010 datasheet, December 2013 [Revised January 2015], [Online] Available: http://epc-co.com/epc/Products/eGaNFETs/EPC8010.aspx

High Frequency, High Temperature designed DC/DC Coreless Converter for GaN Gate Drivers

Yohan Wanderoild[1], Dominique Bergogne[1], Hubert Razik[2]
[1]Univ, Grenoble Alpes, F-38000 Grenoble, France
CEA, LETI, Minatec Campus F-38054 Grenoble, France
http://www.cea.fr
[2] Universite Claude Bernard Lyon 1,
AMPERE UMR CNRS 5005, Villeurbanne, France
yohan.wanderoild@cea.fr

Abstract

This paper presents a functional prototype of a 200mW DC/DC converter to be used for a GaN transistor gate driver. The structure of the converter is presented. It is based on a resonant coreless transformer especially designed and tuned to operate at a frequency close to 6 MHz.

1. Introduction

The gate driver is the meeting point between the control circuit and the power component. In the recent years, GaN power transistors have made significant progress, increasing substantially the voltage stress imposed to the gate driver power supply. Signal coreless transformers have been used at high temperature [1], [2], environment required by several applications [3]. This papers studies the possibility of using coreless printed circuit board (PCB) transformers for the insulated power supply of GaN transistors gate drivers. Using coreless transformers requires a high working frequency, often out of the actual commercial integrated circuits specifications.

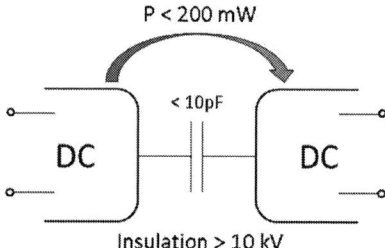

Fig. 1 Gate driver power supply basic specifications

2. Choosing a transformer

The high working temperature specification imposes the use of specific magnetic materials, some conventional transformers are limited to 100 °C, mainly because of the coating material chosen but also because of the Curie temperature of the magnetic material [4], nevertheless some specific materials can be used at temperature higher than 500 °C [5], [6].
There are several kind of materials which can be used [7].

Material	Saturation flux	Frequency behaviour	Temperature
Ferrite	--	>100MHz	-
Laminated cores	+	>10kHz	+-
Powder iron cores	+	>200kHz	+-
Amorphous alloys	+-	10 kHz	-
Nanocrystalline	++	Mhz	--
Air (coreless)	+++	+++	+++

Table 1; comparison of magnetic component used.

Some new non organic binding agents allow to work with powder cores at higher frequency and temperature but the use of magnetic material has a negative impact on the manufacturing process and induce losses. Working without any magnetic material afford to cut out losses mainly by hysteresis and eddy current, but the magnetic field is not contained, increasing the electromagnetic pollution and decreasing the coupling factor. The magnetic field can be contained thanks to a toroidal shape [8], [9], but a planar structure has been chosen for manufacturing purposes. Nevertheless based on SY. Hui (Ron) work [10], we can assume that at frequencies lowers than 10 Mhz, because of its size, the coreless is an extremely poor EMI source in front of the surrounding components.

The use of a square shape maximizes the inductance versus used space ratio, and produces a higher interwindings capacitance than the circle shape, lowering the resonant frequency [9].

3. Design of the coreless

The chosen manufacturing process for the PCB imposes a track width larger than 200µm and an insulation distance greater than 150µm. Several configuration have been simulated with Mohan's formulas [11].

The use of four layers allows to decrease the size of the transformer at a given inductance value but also decrease the coupling factor. Placing the coils in a staggered arrangement would result in a better coupling but at the expense of increasing the parasitic capacitance between the primary and the secondary coils.

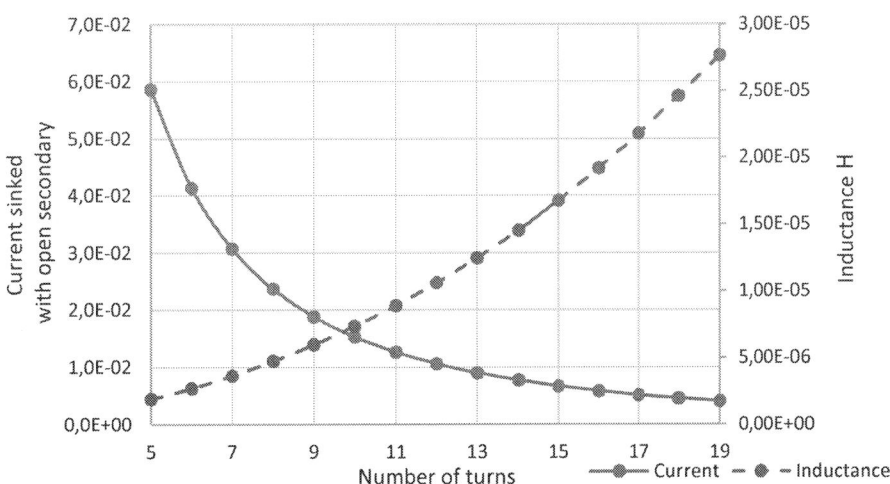

Fig. 2 Transformer inductance and current sinked by the transformer as a function of the number of turns.

The choice of 9 turns allows to have an inductance higher than 5uH equivalent to 50 Ohms at frequencies around 10 MHz, a lower value would have been too constraining for the integrated H bridge (see Fig 7). The middle layer defines the isolation between the primary and the secondary, they are separated by a 710µm insulation layer which provide a 14.2 KV insulation with FR4 material [5], this insulation can be improved by using a multilayer ceramic with a 65KV/mm break down voltage [12].

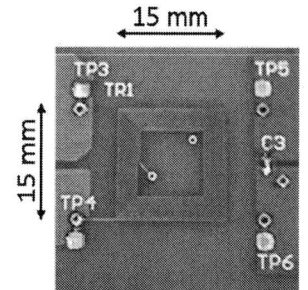

Fig. 3 Photograph of the PCB coreless transformer top view.

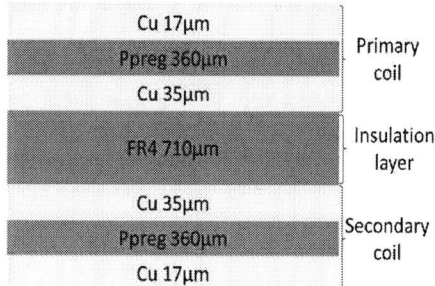

Fig. 4 PCB layer disposition

The transformer parameters have been measured with an impedance analyzer.
The procedure used to determine elements of the chosen model is fully detailed in [1].

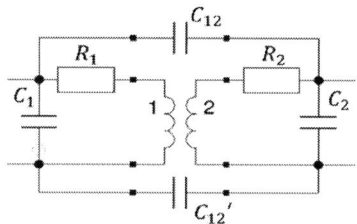

Fig. 5 Equivalent model of the transformer

Elements	Value
$C_{12}, C_{12}{}'$	4.1 pF
C_1, C_2	22 pF
R_1, R_2	3.41Ω
k	0.7
L_1, L_2	6.3 µH

Fig. 6 values measured on the experimental PCB transformer.

4. The converter structure

Many structures have been used to obtain insulated DC/DC converters [13], the structure chosen complies with the basic power electronics rule: "voltage source again current source only". As the structure exhibit a capacitor in parallel of the input and output inductive links must be used in order to protect the voltage source from peak currents. The parallel structure allow to use the parasitic intewindings capacitance as a part of the structure and not as a parasitic, but at the cost of being dependent of the load impedance [13]. This structure can handle a square waveform power supply, thanks to the low pass filter produced by the inductive links.

The converter structure is composed of a high frequency buffer, E_1, which produces a 10 Volts wave at 6 MHz, close to the resonance frequency of the transformer, in order to enable a soft switching behavior, at the resonant frequency the structure is equivalent to a resistor minimizing the losses in the conductors. Capacitors $C_1{}'$ and $C_2{}'$ have been added in order to decrease the transformer resonance frequency which is originally around 13MHz, the protection inductors L1 and L2 magnify the resonance, correctly chosen they improve the transmitted power.

The transformer's output is rectified by a Graetz bridge which can beneficially be replaced by an active one. Capacitor C_{out} is the output tank capacitor and R_{out} emulates the gate driver's consumption. L1 and L2 have been chosen the same and C1 C2 also in order to have a symmetrical structure.

PCIM Europe 2016, 10 – 12 May 2016, Nuremberg, Germany

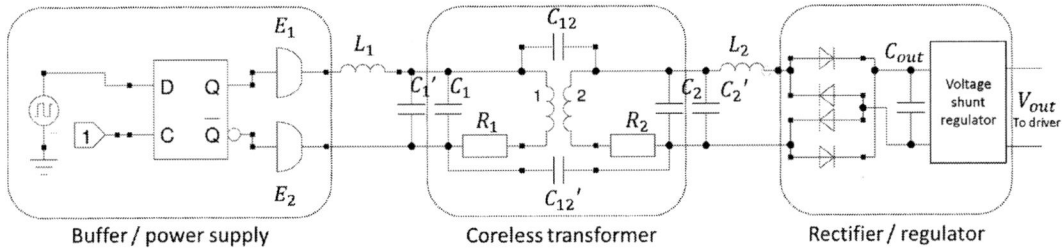

Fig. 7 Proposed coreless based gate driver power supply $F_{generator}$ = 6MHz, $L_1 = L_2$=2uH, $C_1 = C_2$=120 pF.

The use of a quasi-zero Ohms impedance source E_1 and E_2 Fig 7 is a major point, a 50 Ohms source would drastically reduce the transmitted power [14].

5. Optimization of the structure

The structure has been modeled with spice using the measured PCB transformer equivalent model Fig. 8.The optimization process has been performed in order to have the maximum out put power with a given frequency and power supply (10 V peak to peak). The values obtained does not maximize the efficiency. For practical reasons a simplified model has been used for the optimization.

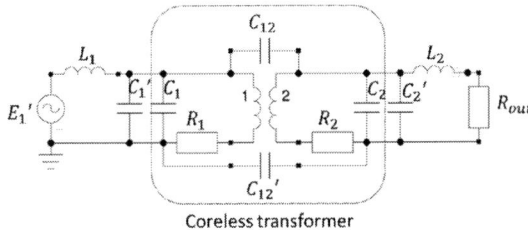

Fig. 8 Structure used for the optimisation

For each frequency there is an optimal value for L1, L2, C1 C2, and Rout.

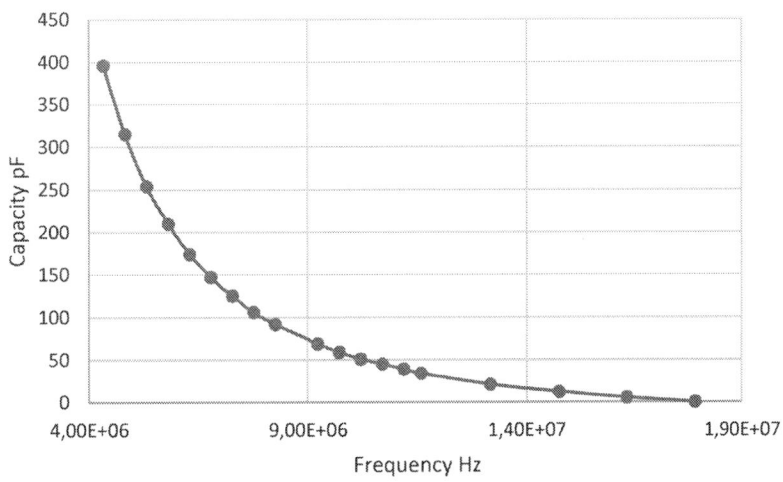

Fig. 9 Optimal simulated values for capacitorsC_1' and C_2' maximizing the output power versus frequency.

© VDE VERLAG GMBH · Berlin · Offenbach

6. Measurements

A prototype was build using the optimized values. Several differences between the simulation and the measurement have been noted. The resonant frequency measured on the PCB is 5.2 MHz which mean a 13% difference, however several explanations have been distinguished.

- the generator which produce less than 8,58V peak fully charged instead of 10V
- The tolerances of the components values.
- The parasitic brought by the cms capacitors, inductors, and their connections.

An output power of 220 mW has been experimentally measured, with an RMS output of 11 volts and a RMS current of 20mA

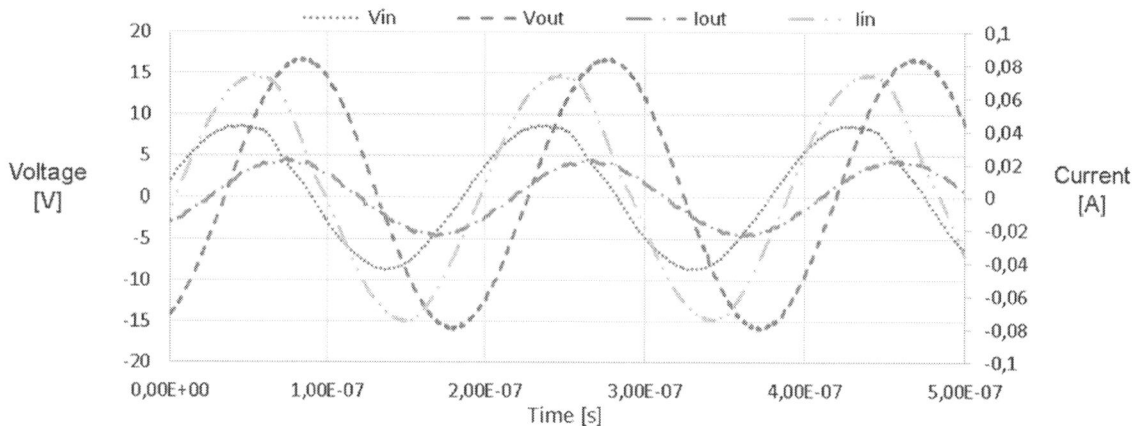

Fig. 11 Primary / Secondary measured on the prototype (Fig. 8) current and voltage waveform at maximum output power (200mW).

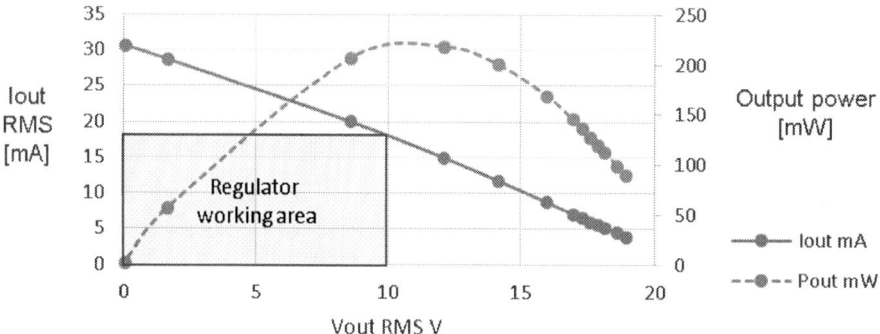

Fig. 10 Output characteristic of the model shown Fig. 8

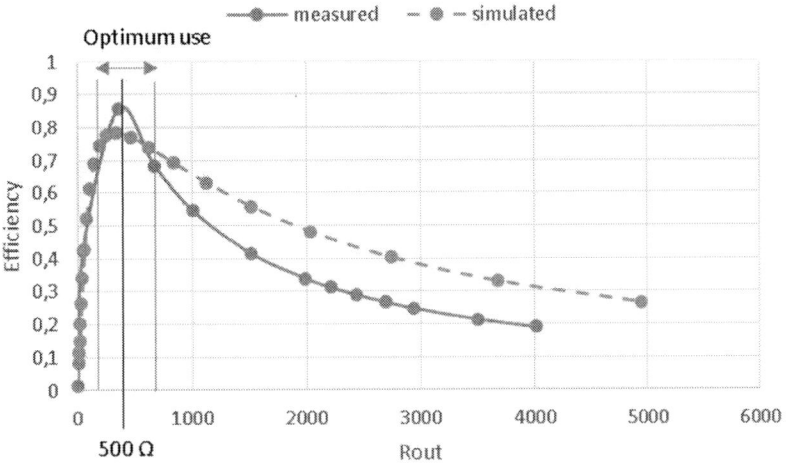

Fig. 12 Comparison between measurements and simulation based on Fig 8 model

7. Perspectives

On Fig. 9 it can be observed that increasing the frequency reduces the required values for C_1 and C_2, the interwinding capacitance of the coreless transformer. The same applies for required additional inductors L_1 and L_2. Further study of the compromise between the magnetizing current and the operating frequency should lead to a fully integrated solution concerning the transformer and the inductive links. Simulations have shown that taking different values between C1 and C2 the transmitted power can be doubled.

8. Conclusion

This preliminary work has shown the possibility of using a coreless transformer with an integrated H bridge in order to provide an insulated DC/DC converter for a gate driver. The transmitted power has been experimentally measured to 200 mW. The insulation is provided by the PCB substrate material. The parasitic insulation capacitance is lower than 10pF.

[1] D. Bergogne, C. Martin, B. Allard, K. El Falahi, G. Picun, H. Ezzeddine, and C. Pintout, "High Temperature Discrete Integrated Coreless Signal Insulator," in *2014 8th International Conference on Integrated Power Systems (CIPS)*, 2014, pp. 1–4.

[2] D. Bergogne, C. Martin, P. Bevilacqua, W. Zine, J.-C. Riou, H. Izzeddine, R. Meuret, and B. Allard, "Integrated coreless transformer for high temperatures design and evaluation," in *2013 15th European Conference on Power Electronics and Applications (EPE)*, 2013, pp. 1–8.

[3] C. Buttay, D. Planson, B. Allard, D. Bergogne, P. Bevilacqua, C. Joubert, M. Lazar, C. Martin, H. Morel, D. Tournier, and others, "State of the art of high temperature power electronics," *Mater. Sci. Eng. B*, vol. 176, no. 4, pp. 283–288, 2011.

[4] F. Qi, L. Xu, G. Zhao, and J. Wang, "Transformer isolated gate drive with protection for SiC MOSFET in high temperature application," in *2014 IEEE Energy Conversion Congress and Exposition (ECCE)*, 2014, pp. 5723–5728.

[5] M. Öztürk, E. Demirci, O. Gürbüz, S. Güner, V. Valeev, F. Vagizov, R. Khaibullin, and N. Akdoğan, "Formation of different magnetic phases and high Curie temperature ferromagnetism in Fe57-implanted ZnO film," *J. Magn. Magn. Mater.*, vol. 373, pp. 83–85, Jan. 2015.

[6] Z. Yang, D. Gao, J. Zhang, Q. Xu, S. Shi, K. Tao, and D. Xue, "Realization of high Curie temperature ferromagnetism in atomically thin MoS2 and WS2 nanosheets with uniform and flower-like morphology," *Nanoscale*, vol. 7, no. 2, pp. 650–658, Dec. 2014.

© VDE VERLAG GMBH · Berlin · Offenbach

[7] T. Sørsdahl, "Magnetic Design for High Temperature, High Frequency SiC Power Electronics," *157*, 2013.

[8] V. Ermolov, T. Lindstrom, H. Nieminen, M. Olsson, M. Read, T. Ryhanen, S. Silanto, and S. Uhrberg, "Microreplicated RF toroidal inductor," *IEEE Trans. Microw. Theory Tech.*, vol. 52, no. 1, pp. 29–37, Jan. 2004.

[9] P. N. Murgatroyd, "The optimal form for coreless inductors," *IEEE Trans. Magn.*, vol. 25, no. 3, pp. 2670–2677, May 1989.

[10] S. Y. R. Hui, S. C. Tang, and H. Chung, "Some electromagnetic aspects of coreless PCB transformers," in *30th Annual IEEE Power Electronics Specialists Conference, 1999. PESC 99*, 1999, vol. 2, pp. 868–873 vol.2.

[11] S. S. Mohan, M. del Mar Hershenson, S. P. Boyd, and T. H. Lee, "Simple accurate expressions for planar spiral inductances," *IEEE J. Solid-State Circuits*, vol. 34, no. 10, pp. 1419–1424, Oct. 1999.

[12] W. Huebner, S. C. Zhang, B. Gilmore, M. L. Krogh, B. C. Schultz, R. C. Pate, L. F. Rinehart, and J. M. Lundstrom, "High breakdown strength, multilayer ceramics for compact pulsed power applications," in *Pulsed Power Conference, 1999. Digest of Technical Papers. 12th IEEE International*, 1999, vol. 2, pp. 1242–1245 vol.2.

[13] C. Auvigne, P. Germano, Y. Perriard, and D. Ladas, "About tuning capacitors in inductive coupled power transfer systems," in *2013 15th European Conference on Power Electronics and Applications (EPE)*, 2013, pp. 1–10.

[14] I. Awai, "Basic characteristics of 'Magnetic resonance' wireless power transfer system excited by a 0 ohm power source," *IEICE Electron. Express*, vol. 10, no. 21, pp. 20132008–20132008, 2013.

Monolithic GaN-on-Si Half-Bridge Circuit with Integrated Freewheeling Diodes

Richard Reiner, Patrick Waltereit, Beatrix Weiss, Matthias Wespel, Michael Mikulla, Rüdiger Quay, and Oliver Ambacher

Fraunhofer-Institute for Applied Solid State Physics (IAF),
Tullastrasse 72, 79108 Freiburg, Germany, Phone: +49 761 5159-552,
Email: richard.reiner@iaf.fraunhofer.de

Abstract

This work presents the design, realization, and the characterization of a monolithic GaN-on-Si half-bridge circuit with integrated Schottky contacts as freewheeling diodes. The extrinsic- and intrinsic- layouts are realized, analyzed, and compared to other approaches. The high- and low-side switches feature an off-state voltage of 600 V, an on-state resistance of 120 mΩ, and a reverse resistance of below 150 mΩ at corresponding drain currents of 30 A. Furthermore, the switches achieve very low gate-charges of below 5 nC and reverse recovery charges of 12 nC. The on-state- and reverse-state-performances are benchmarked against other state-of-the-art power devices and compared to the theoretical limits.

Introduction

First commercially available Gallium Nitride (GaN) High Electron Mobility Transistors (HEMT) demonstrate high performance in terms of power performance and switching behavior [1]. In direct comparison to conventional silicon power devices GaN-HEMTs feature high breakdown voltages, low on-state resistances, as well as low gate charges [2]. However, GaN-HEMTs and its heterojunction technology should not only be used as a one-to-one silicon replacement. Beyond power performance this technology offers new additional opportunities of integration. In contrast to most conventional power technologies, as power MOSFETs or IGBTs, the GaN-on-Si heterojunction technology is of lateral nature. This property enables the integration of several power devices side-by-side on a single chip. Thus monolithically-integrated circuits can be realized on one power chip. With this technology reduced electric parasitics and improved reliability can be achieved, due to on-chip interconnections, as well as lower module cost due to a reduced effort of assembly technologies. The core topology in most converter applications is a half-bridge circuit, which consists of two transistors and two reverse diodes, as shown in the schematics in Fig. 1 a).

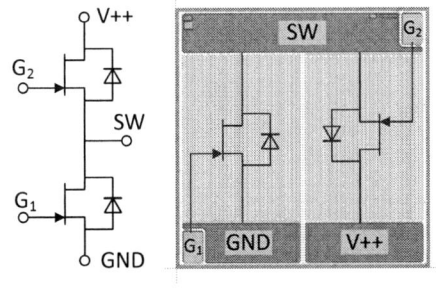
a)

b)

Fig. 1 a) Circuit and folded chip layout of a half-bridge with integrated reverse diodes. b) Assembled half-bridge chip (4×4 mm²) soldered on a direct copper bonded ceramic board.

In previous works first monolithic GaN-half-bridges have been shown. An integrated 600 V-3-phase inverter was published in [3]. Furthermore a commercially available half-bridge chip with voltages up to 100 V was presented in [4]. The high- and low-side switches of these chips are based on a conventional HEMT-structure without additional reverse diodes. Thus, significant losses are generated in the reverse-state, as analyzed in [5]. However, a hybrid connection of external diodes would introduce considerable parasitics. A monolithic integration of a conventional HEMT structure and additional Schottky diode fingers has been shown in [6]. However, this approach results in high chip area consumption.

This work presents a fully-monolithic GaN-half-bridge using a new chip layout with an improved intrinsic HEMT-structure with integrated reverse Schottky contacts.

Design and fabrication

The GaN-based half-bridge chip is designed to achieve fast switching rates due to a compact layout with low parasitics. The design is shown in Fig. 1. The half-bridge layout is folded in a way that a DC-link capacitor can be applied close to the chip. Furthermore, the reverse diodes are directly integrated into the HEMT-structure using an approach, which was introduced and explained in [7], and previously applied in [8]. A series of Schottky contacts are implemented between gate and drain and thereby enable a low-resistive freewheeling path for the reverse-state operation. In the off-state, the channel is depleted by the gate and the Schottky contacts are biased in reverse direction. In the on-state the electrons can flow around and under the separated Schottky contacts through the channel. In the reverse-state the electrons can flow through the drift region via the Schottky contacts across the field plate metallization to the source. This reverse current flow is independent of the gate-source voltage and the channel is conductive, even if the channel under the gate is depleted. Thus this structure (shown in Fig. 2 a) enables a low-resistive, high-current-capable and fast reverse recovery diode behavior without requiring a high amount of additional chip area. In Fig. 2 b) the output characteristic of the structure is directly compared to that of a conventional HEMT. The chip layout, which is shown in Fig. 1 using the intrinsic structure of Fig. 2 a), has been fabricated in a high voltage AlGaN/GaN-on-Si heterojunction technology [9]. Such a fabricated and assembled chip is shown in Fig. 1 b). The chip has a total area of $A = 4 \times 4$ mm² and a gate width of $W = 145$ mm per switch.

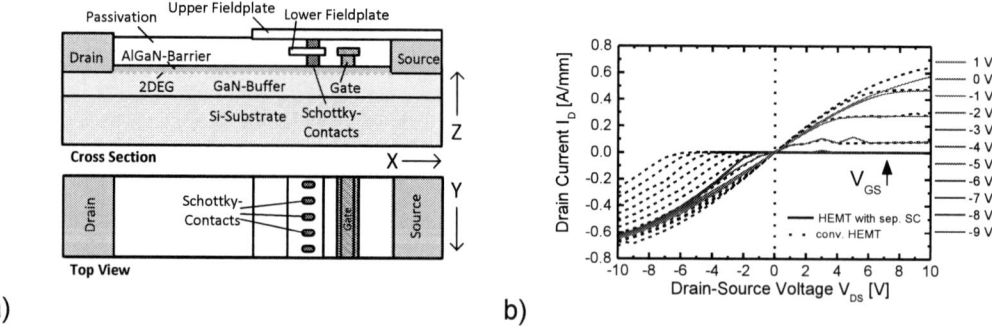

a)   b)

Fig. 2 a) Schematic cross section and top view of the intrinsic HEMT structure with integrated reverse Schottky contacts. b) Output characteristics of a HEMT structure with separated reverse Schottky contacts in comparison to the output characteristic of a conventional HEMT. The measurements have been made on small device test structures with a gate width of $W = 50$ µm.

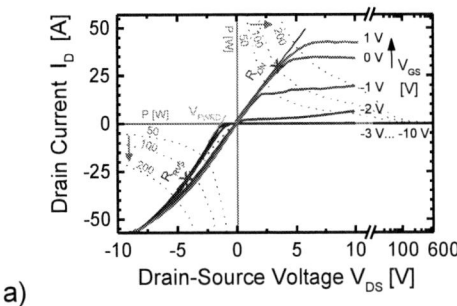 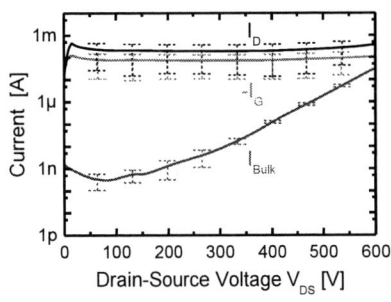

a) b)

Fig. 3 Static performance of a half-bridge HEMT with integrated reverse Schottky contacts. a) The output characteristics in the on -, off -, and reverse-state are measured in a pulsed on-wafer four-point measurements with t_{PLS} = 100 µs. b) Off-state leakage currents measured on-wafer up to 600 V.

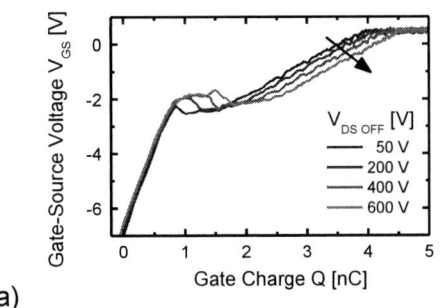 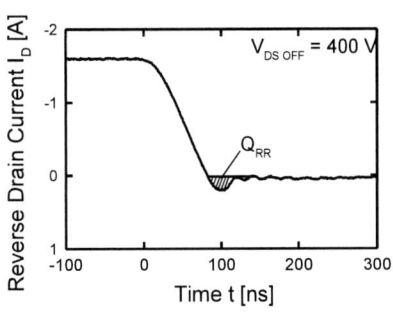

a) b)

Fig. 4 Dynamic performance of a half-bridge transistor with integrated reverse Schottky contacts. a) The gate-charge is measured during a turn-on event with off-state drain-source voltages up to $V_{DS\,OFF}$ = 600 V. b) The reverse recovery charge of the freewheeling diode was measured in a double pulse measurement setup at $V_{DS\,OFF}$ = 400 V.

Static measurement results

The performance of the fabricated chips has been characterized in on-wafer measurements. The results are illustrated in Fig. 3. High- and low-side switches show an identical behavior with an on-state resistance of R_{ON} = 120 mΩ at a corresponding drain current of I_D = 30 A and a dissipated power in the range of P = 110 W at this operation point. In the reverse-state the resistance is measured to be R_{RVS} = 150 mΩ at a corresponding drain current of I_D = -30 A, a gate-source voltage of V_{GS} = -10 V, and a corresponding power of P = 135 W. The reverse diode features a forward voltage of V_{FRWD} = -1.4 V with a gradient of R_{FRWD} = 93 mΩ.

Off-state measurements have been performed at an on-wafer mapping probe station. The results of the leakage currents are shown in Fig. 3 b) up to a voltage of V_{DS} = 600 V. All devices have been measured with grounded substrate. At an off-state voltage of 600 V the vertical buffer leakage is below I_{Bulk} = 50 µA. The gate leakage currents are in the range of I_G = -100 µA, and the drain leakage currents are below I_D = 500 µA. The drain leakage current is composed of the gate leakage currents, the buffer leakage currents, and the leakage currents of the reverse diode Schottky contacts. With a gate width of W = 145 mm the sum of all device leakage currents are below 5 µA/mm, which is appropriate for the use in power applications.

© VDE VERLAG GMBH · Berlin · Offenbach

Dynamic measurement results

Switching losses arise during the accumulation and depletion of carriers in the transistor channel. Due to the wide-bandgap properties GaN-switches feature low gate-charge quantities to switch the HEMT from the off-state into the on-state and back, as well as low reverse recovery charge quantities to turn-on and -off between off-state and reverse-state. These two switching charges have been characterized for these different switching types.

The gate-charge has been measured during a turn-on event in a pulse setup. The corresponding measurement results are shown in Fig. 4 a). The gate-charge is determined by calculating the time integral of the measured gate-current, which is calculated according to $Q_G = \int I_G(t)\, dt$. Afterwards the gate-charge curve is plotted as gate-source voltage V_{GS} as a function of the gate-charge Q_G. The depletion zone of the channel and therefore the charge which has to be displaced out of the channel depends on the applied off-state voltage. Therefore the curve was measured for different drain-source off-state voltages in ≈ 200 V steps up to a value of $V_{DS\,OFF} = 600$ V. A resistor with a value of $R_G = 120\ \Omega$ is connected between driver output and gate. Thus, the gate current $I_G(t)$ can be measured across this resistor using an oscilloscope. The maximal total gate-charge value for the half-bridge switches was measured to be $Q_G = 4.5$ nC at a corresponding off-state voltage of $V_{DS\,OFF} = 600$ V.

Furthermore, the reverse recovery charge has been characterized using a double pulse measurement setup. The turn-off behavior of the tested freewheeling diode is shown in Fig. 4 b). The pinched-off transistor at the high-side is turned-off from reverse-state into the off-state, with an voltage of $V_{DS\,OFF} = 400$ V. The calculation of the integral of the current below zero over time results in a reverse recovery charge of around $Q_{RR} = 12$ nC.

In-circuit operation

The complete functionality of the monolithic half-bridge chip is tested in a resonant demonstrator circuit. In the synchronous buck converter operation the two half-bridge transistors as well as the two reverse diodes carry currents of up to 5 A. At the switch node of the half-bridge module a storage inductor, a smoothing capacitor, and a resistive load are applied, as shown in Fig. 5 a). The inductor has a value of $L = 10$ µH, the capacitance of 3.5 µF, and the load resistor of $R_L = 170\ \Omega$. A DC-link capacitor is connected close to the chips between the pads of the supply voltage $V_{++} = 400$ V and the ground GND. The demonstrator operates at a high switching frequency of $f_{SW} = 1.23$ MHz and a duty cycle of 0.5 without any instabilities or overshoots observed. Fig. 5 b) shows the transient switch node voltage V_{SW} and the inductor current I_L. At the load R_{OUT}, an output voltage of $V_{OUT} = 196$ V and an output current of $I_R = 1.15$ A are measured. Thus the monolithic half-bridge is suitable as a power chip for high frequency converter applications.

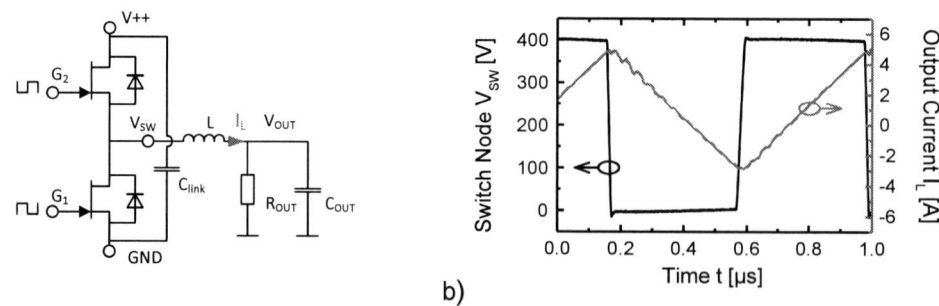

Fig. 5 a) Half-bridge circuit with output load elements. b) Transient voltage V_{SW} monitored at the switch node and current I_L measured at the storage inductance L.

Device benchmark

GaN power devices are predestined for efficient and fast switching power electronic systems. High switching frequencies enable the systems to use compact and low-value inductive and capacitive storage devices. Thus volume, weight and cost of power electronic systems are reduced. In this context the most important figure-of-merit (FOM) is the product of the on-state resistance and the switching charge $R_{ON} \cdot Q_{SW}$. The switching charge Q_{SW} is the most relevant device parameter that influences the switching losses in a switch-mode power-supply (SMPS). Whereas the on-state resistance R_{ON} is the crucial parameter of the static-losses. It was shown by A.Q. Huang in [10], that power devices with a low product of $R_{ON} \cdot Q_{SW}$ generate low static- and low dynamic- losses. Therewith, such devices are well-suited for compact and high-efficient power electronics applications.

The lowest product of $R_{ON} \cdot Q_{SW}$ can be approximated for high-voltage vertical channel devices, such as Schottky diodes or asymmetric doped pn-junction diodes, and it can be used as a theoretical limit for vertical semiconductor devices. It can be derived from the limit proposed by B.J. Baliga for the area-specific on-state resistance $R_{ON} \cdot A = 4 \cdot V_{BD}^2/(\varepsilon \cdot \mu \cdot E_C^3)$ [11], together with the parallel plate capacitance $C = \varepsilon \cdot A/d$. The parallel plate capacitance can be used to approximate the depletion capacitance of a diode, where the depletion length of such a vertical device is given by $d = 2 \cdot V_{BD}/E_C$. The breakdown voltage is denoted by V_{BD}, the critical electric field by E_C. The diffusion capacitance is neglected. Using these equations and $Q_{SW} = C \cdot V_{BD}$ the product of $R_{ON} \cdot Q_{SW}$ is given by equation:

$$R_{ON} \cdot Q_{SW} = 2 \cdot V_{BD}^2/(\mu \cdot E_C^2) \tag{1}$$

This expression can be used as the lowest achievable value for vertical channel devices, as shown in Fig. 6, and previously used in this manner in [7].

Half-bridge switches are affected by two different conduction losses and two different switching losses. The product of the on-state resistance and the gate-charge $R_{ON} \cdot Q_G$ is the relevant product for the switching operation between off-state and on-state, as shown in [10]. The product of the on-state resistance in reverse direction and the reverse recovery charge $R_{RVS} \cdot Q_{RR}$ is the relevant factor for the switching operation between off-state and reverse-state. In [12] $R_{RVS} \cdot Q_{RR}$ is introduced as a FOM for the reverse recovery losses. In many power devices the on-state resistance in forward direction is similar to the on-state resistance in reverse direction, and thus the values can be approximated as $R_{ON} \approx R_{RVS}$. This is possible since the electrons use the same drift path in forward and reverse direction. However, in conventional HEMT-structures this can only be assumed if the gate is open, with $V_{GS} > V_{TH}$. If the gate-source voltage is lower than the threshold voltage $V_{GS} < V_{TH}$, the on-state resistance in reverse direction dependents on the gate-source voltage $R_{RVS}(V_{GS})$. The current through the transistor in this state is illustrated with the dashed curve in Fig. 2 b). The intrinsic structure in this work solves this issue by implementing Schottky contacts, which provide a low-resistivity reverse conduction paths, thus R_{RVS} is largely independent of the gate-source voltage.

A performance comparison for the on-state as well as for the reverse-state operation is shown in Fig. 6 for Si-, SiC-, and GaN-semiconductor devices. Fig. 6 a) compares $R_{ON} \cdot Q_G$ for different power semiconductor technologies as a function of the breakdown voltage V_{BD}. It is shown that wide-bandgap semiconductor devices yield significantly lower values compared to their Si-counterparts. The half-bridge transistors in this works achieve an $R_{ON} \cdot Q_G = 0.54 \ \Omega$ nC, which is about one decade smaller than the best Si-devices achieved in this voltage class. The half-bridge transistors are comparable to the best state-of-the art GaN-HEMTs without integrated reverse diodes.

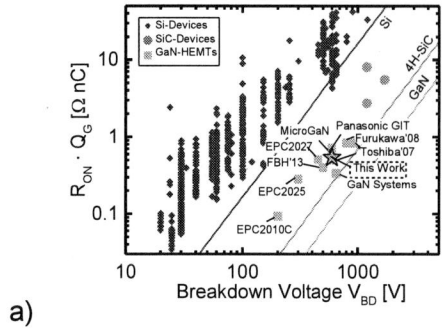 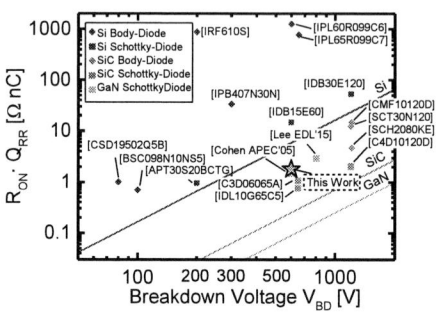

a)   b)

Fig. 6 Performance comparison of state-of-the-art power semiconductors. Values are taken from data sheets and publications. These values are compared to the performance of the transistors used in this work. Furthermore, the approximated $R_{ON} \cdot Q_{SW}$ - limit for vertical channel diodes, which is given in equation (1), is depicted in both diagrams for the compared semiconductor. a) Comparison of the FOM: $R_{ON} \cdot Q_G$ for efficient switching between on- and off- state. b) Comparison of the FOM: $R_{ON} \cdot Q_{RR}$ for efficient switching into reverse-state.

In Fig. 6 b) $R_{ON} \cdot Q_{RR}$ is compared for different freewheeling diode types as a function of the breakdown voltage V_{BD}. In many conventional power MOSFETs the body-diode is used as a freewheeling diode. The performance of the intrinsic body-diodes is relatively poor (Fig. 6 b)), which is why additional low-Q_{RR} Schottky-diodes have to be applied to operate the circuit efficiently. Schottky diodes show much lower reverse recovery charges. However, hybrid solutions are affected strongly by parasitics caused by interconnections, as bondwires. Therefore the major advantage of the approach in this work is the integrated Schottky contacts with low reverse recovery charge and low reverse resistance. The half-bridge transistors in this works achieve a value of $R_{RVS} \cdot Q_{RR} = 1.8\ \Omega$ nC, which is about one decade smaller than the best Si-Schottky diodes in this voltage class, and more than two decades smaller than the best Si-body diodes. GaN- and SiC-Schottky diodes achieve a similar reverse recovery performance.

Conclusions

The lateral GaN-on-Si HEMT-technology opens up entirely new possibilities for power electronics applications. The lateral technology allows a co-integration of several power components on one chip. Thus wide-bandgap performance is combined with high functionality. This work presents a new folded design of a half-bridge circuit on-chip, by using an advanced intrinsic HEMT-structure with integrated reverse Schottky contacts. In contrast to hybrid modules, a monolithic circuit features better switching performance, higher circuit reliability, and lower assembly costs. Compared to half-bridge circuits realized with a conventional HEMT-structure, this approach achieves a clearly improved reverse performance. Furthermore compared to a monolithic solution with separated Schottky-diode fingers, this design achieves higher chip area efficiency.

The half-bridge transistors with integrated freewheeling diodes achieve a low FOM: $R_{ON} \cdot Q_G$ value, which is about 10 times lower than state-of-the-art Si-devices, and a low FOM: $R_{RVS} \cdot Q_{RR}$ value, which is more than 100 times lower than state-of-the-art Si-devices using body-diodes as freewheeling diodes.

Thus the solution in this work demonstrates a high performance half-bridge device for fast switching, high density power electronics applications.

Acknowledgments

The authors would sincerely like to thank the staff at Fraunhofer IAF who was involved in the epitaxy, fabrication and characterization of the devices. This work was supported partly by the German Federal Ministry for Environment, Nature Conservation, Building and Nuclear Safety (BMU) through grant "GaNPV" FKZ: 0325529 and by the German Federal Ministry of Education and Research (BMBF) through grant "ZuGaNG" FKZ: 16ES0076K.

References

[1] E.A. Jones, F. Wang, B. Ozpineci, "Application-based review of GaN HFETs", IEEE Workshop on Wide Bandgap Power Devices and Applications (WiPDA), pp.24-29, 2014.

[2] R. Reiner, P. Waltereit, F. Benkhelifa, S. Müller, M. Wespel, R. Quay, M. Schlechtweg, M. Mikulla, O. Ambacher, „Benchmarking of Large-Area GaN-on-Si HFET Power Devices for Highly-Efficient, Fast-Switching Converter Applications," IEEE Compound Semiconductor Integrated Circuit Symposium (CSICS), pp.1-4, 2003.

[3] Y. Uemoto, T. Morita, A. Ikoshi, H. Umeda, H. Matsuo, J. Shimizu, M. Hikita, M. Yanagihara, T. Ueda, T. Tanaka, D. Ueda,"GaN Monolithic Inverter IC Using Normally-Off Gate Injection Transistors with Planar Isolation on Si Substrate", Intern. Electron Devices Meeting (IEDM), pp.1-4, 2009.

[4] A. Lidow, "GaN Transistors – Giving New Life to Moore's Law", IEEE International Symposium on Power Semiconductor Devices & IC's (ISPSD), pp.1-6, 2015.

[5] H. Zhang, R.S. Balog,"Loss Analysis During Dead Time and Thermal Study of Gallium Nitride Devices", IEEE Applied Power Electronics Conference and Exposition (APEC), pp.737-744, 2015.

[6] T. Kachi, M. Kanechika, T. Uesugi, "Automotive Applications of GaN Power Devices," IEEE Compound Semiconductor Integrated Circuit Symposium (CSICS), pp.1-3, 2011.

[7] R. Reiner, P. Waltereit, B. Weiss, M. Wespel, R. Quay, M. Schlechtweg, M. Mikulla, O. Ambacher, "Integrated Reverse-Diodes for GaN-HEMT Structures," IEEE International Symposium on Power Semiconductor Devices & IC's (ISPSD), pp.45-48, 2015.

[8] B. Weiss, R. Reiner, P. Waltereit, S. Muller, M. Wespel, R. Quay, O. Ambacher, "Monolithically-Integrated Mulitlevel Inverter on Lateral GaN-on-Si Technology for High-Voltage Applications," IEEE Compound Semiconductor Integrated Circuit Symposium (CSICS), pp.1-4, 2015

[9] P. Waltereit, R. Reiner, H. Czap, D. Peschel, S. Müller, R. Quay, M. Mikulla and O. Ambacher , "GaN-Based High Voltage Transistors for Efficient Power Switching", Physica Status Solidi, vol. 10.5, pp. 831-834, Feb.2013.

[10] A.Q. Huang, "New Unipolar Switching Power Device Figures of Merit," IEEE Electron Device Letters, vol. 25, no. 5, pp. 298-301, 2004.

[11] B.J. Baliga, "Power Semiconductor Device Figure of Merit for High-Frequency Applications", IEEE Electron Device Letters, Oct. 1989, vol.: 10, no.10, pp. 455 – 457.

[12] B.J. Baliga, "Advanced power MOSFET concepts", Springer, 2010.

Resin Encapsulation Combined with Insulated Metal Baseplate for Improving Power Module Reliability

Shinsuke Asada*, Satoshi Kondo*, Yusuke Kaji**, Hiroshi Yoshida*
*Power Device Works, Mitsubishi Electric Corporation, Japan,
E-mail: Asada.Shinsuke@db.mitsubishielectric.co.jp
**Advanced Technology R&D Center, Mitsubishi Electric Corporation, Japan

Abstract

A newly developed packaging technology in combination with direct potting (DP) resin and Insulated Metal Baseplate (IMB) is presented. DP resin is liquid epoxy resin so that epoxy resin encapsulation can be applied to power module with conventional case-type package. By combining with IMB, that is eliminating the solder layer between insulator and copper baseplate, DP resin encapsulation contributes to prevent the degradation of solder layer under the chips. As consequence, the new package is improving the capabilities of heat cycling and thermal cycling.

1. Introduction

As the market of power module is expanding, the current density of power module is increasing. In addition there is an increasing need for operations at higher temperature and larger temperature swings of power module, for example, at very slow rotation speed of a motor, nearly direct current or motor lock etc. Therefore the reliability of power module with larger temperature swing of package, such kind of heat cycling or thermal cycling are becoming more important in addition to power cycling. To meet the requirement we have proposed the new package structure that combines a developed Insulated Metal Baseplate (IMB) with new direct potting (DP) resin [1]. In this paper, the advantage of this package in reliability is reported.

2. Package structure

The conventional packaging structure shown in Fig.1(a) has ceramic substrates joined to a copper baseplate by soldering. The difference of coefficient of thermal expansion (CTE) between ceramic substrates and copper baseplate induces stress in the solder layer under the substrates by temperature swing. Those stresses are related to the heat cycling and thermal cycling capability thereby the maximum size of those ceramic substrates is limited.

Fig.1(b) shows the newly developed structure that consists of IMB and DP resin encapsulation. The IMB has an insulating resin sheet with high thermal conductivity, copper baseplate and thick copper foil. The CTE of resin insulator is designed close to copper. In this way the solder layer under ceramic substrates are eliminated by adopting IMB. And it is using a newly developed DP resin which is epoxy resin instead of silicone gel for encapsulation.

Since many years we are already manufacturing power modules adopting epoxy resin encapsulation with transfer mold package. And its effect of reducing strain of solder has been reported earlier [2]. However, this transfer mold technology has been applied to smaller size package from the standpoint of higher productivity.

At initial, the transfer mold resin is solid state, but DP resin is liquid state. This means that DP resin can be applied by dispensing method similar as silicone gel. Therefore, the epoxy resin encapsulation by using DP resin can be extended to larger conventional case-type package that has higher design flexibility (Fig. 2).

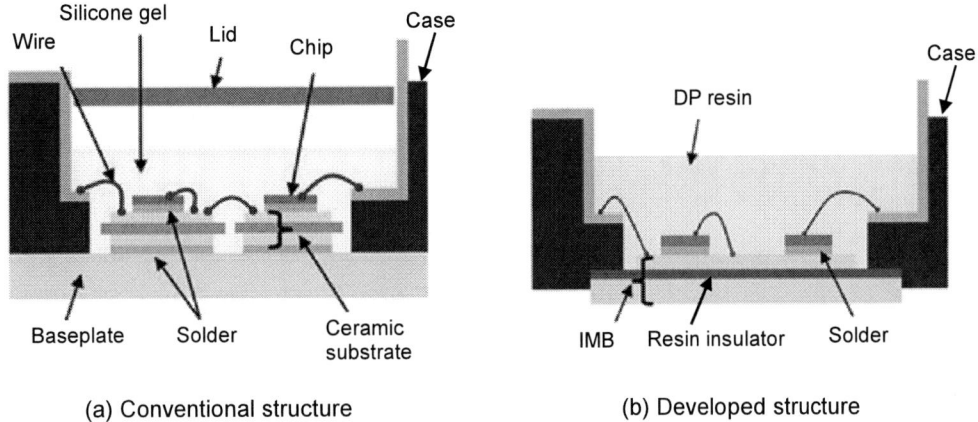

(a) Conventional structure   (b) Developed structure

Fig. 1. Cross section of package structure

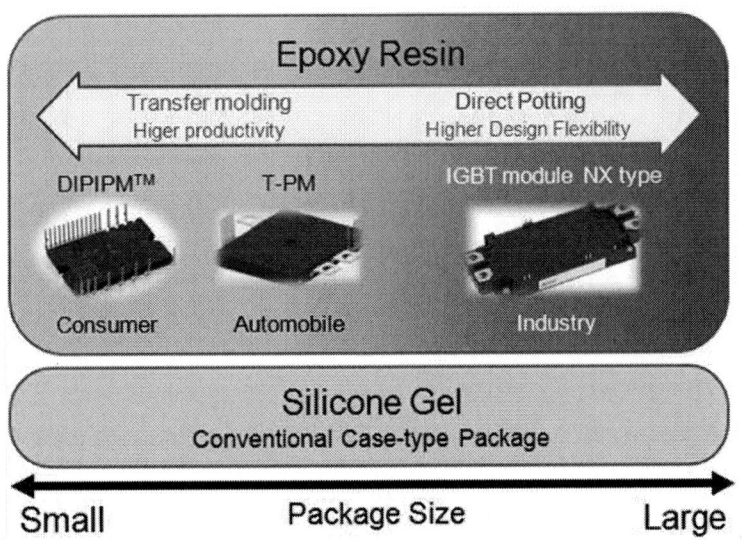

Fig. 2. Expand of epoxy resin encapsulation

3. Direct potting resin

The material of encapsulation for power module is required to have stability of its properties in the range of operating junction temperature (Tjop) of power module. Thermoset resin such kind of epoxy has a glass transition temperature (Tg). Its CTE, elastic modulus and strength of adhesive are usually changed at temperatures higher than Tg. Therefore it is necessary to design Tg above than Tjop. Fig. 3 shows a thermal mechanical analysis (TMA) chart of new developed DP resin. The resin is designed so that Tg is enough over the maximum of Tjop. This results in small deviation of CTE under the Tjop. The newly developed DP resin has also enough fluidity to encapsulate.

Because DP resin is enough rigid as a mechanical structure, it is important to consider balancing the CTE of DP resin, IMB and the other components. CTE of DP resin is basically designed to be matched to copper in IMB that is dominant structural member considering bimetal effect which is causing the warpage of baseplate. Simulated warpage variation in relation to CTE of DP resin is shown in Fig.4. CTE around 15ppm, that is close to copper, induces nearly no variation of warpage. The CTE is finalized by considering with variations of package size or internal layout.

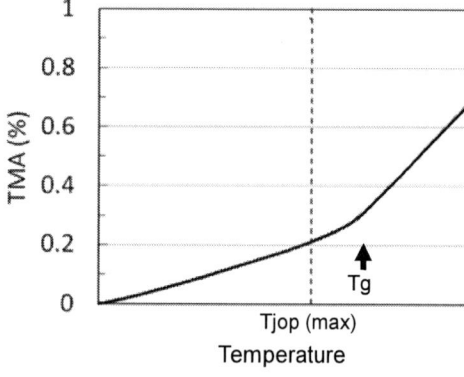

Fig. 3. TMA chart

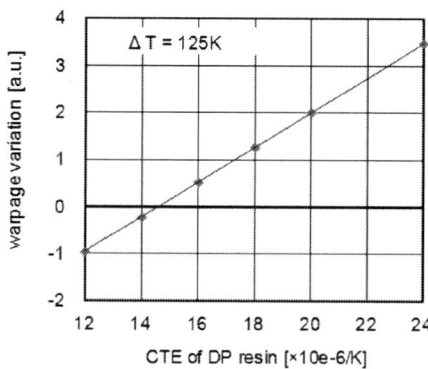

Fig. 4. Simulated warpage

4. Reliability of the new package

The new developed structure is designed to minimize the difference of CTE as described above. In addition to epoxy resin encapsulation can reduce the strain of solder layer under the chip [2]. This accomplished with higher reliability of heat cycling and thermal cycling. Fig.5 shows the Scanning Acoustic Tomography (SAT) images by heat cycling test (-40~+125deg.C) at initial, 300cycles and 600cycles. Conventional structure had degradation of the solder layer under the ceramic substrate after 300 cycles. On the other hand, new structure with DP encapsulation has only solder layer under the chip and no degradation after 600cycles was observed. This result is an effect of removing the substrate solder layer and of the epoxy resin preventing to progress solder cracks under the chip.

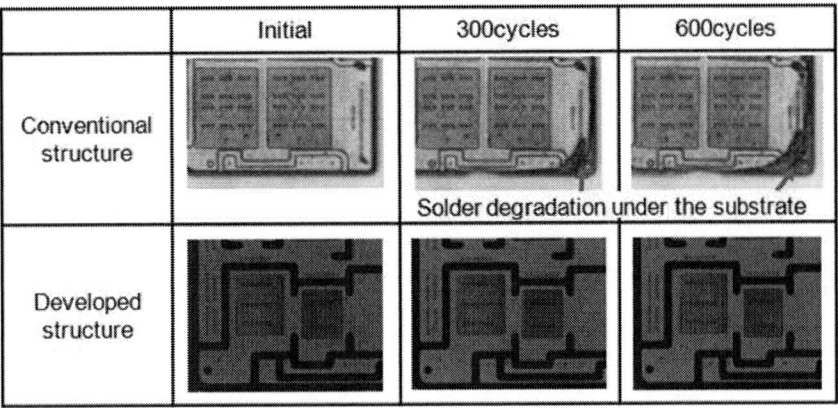

Fig. 5. Heat cycling test result (images by SAT)

Fig. 6 shows the result of thermal cycling test. The newly developed structure was tested until 40kcycles and has not failed yet. On the other hand, the conventional structure had degradation of solder layer under the substrate like the result of heat cycling. From this result, new package with DP resin and IMB increased the thermal cycling life dramatically comparing with conventional structure. That is contributed by removing a solder layer under the insulator and reducing the strain of solder layer under the chips.

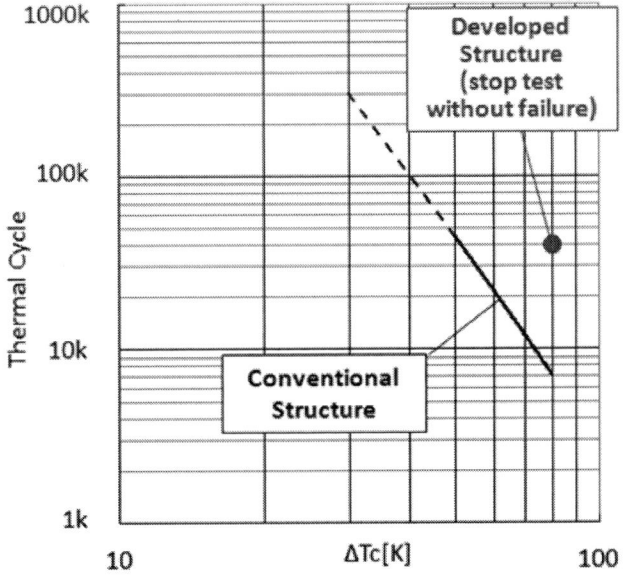

Fig. 6. Thermal cycling test result

5. Conclusion

New package structure combing DP resin and IMB has been developed. The DP resin has high fluidity to apply epoxy resin encapsulation to power module of conventional case-type package. The DP resin is optimized considering with Tg and CTE. New package structure has no solder layer under the insulator. In addition, degradation of solder layer under the chips is prevented by epoxy resin encapsulation. As a result, the capabilities of heat cycling and thermal cycling have been enhanced.

References

[1] Ohara, et. al., "A New IGBT Module with Insulated Metal Baseplate(IMB) and 7th Generation Chips", PCIM Nuremberg(2015).
[2] T.Ueda, et al., "Simple, Compact, Robust and High-performance Power module T-PM(Transfer-molded Power Module)", pp.37-40, ISPSD2010

An experimental study on the thermal performance of double-side direct-cooling power module structure

Akira Matsushita, Hitachi Automotive Systems, Ltd., Japan, akira.matsushita.uu@hitachi-automotive.co.jp

Ryuichi Saito, Hitachi Automotive Systems, Ltd., Japan, ryuichi.saito.pg@hitachi-automotive.co.jp

Takeshi Tokuyama, Hitachi Automotive Systems, Ltd., Japan, takeshi.tokuyama.xs@hitachi-automotive.co.jp

Kinya Nakatsu, Hitachi Research Laboratory, Hitachi Ltd., kinya.nakatsu.nq@hitachi.com

Takashi Kimura, Hitachi Automotive Systems, Ltd., takashi.kimura.th@hitachi-automotive.co.jp

Abstract

The structure of double-side direct water cooling power module and experimental study on thermal performance of the structure are presented in this paper. To meet the continuing demand for smaller and electrically efficient inverter, the authors have developed inverter using double-side direct water cooling power module structure. To evaluate the thermal performance of the new structure, thermal resistance, thermal cycling and power cycling tests were performed. Test results indicate that the power module has enough thermal performance to be used in automotive field. Finally, example of inverter using the double-side direct-cooling power module is presented.

Keywords: power module, double-side, direct water cooling, thermal performance

1. Introduction

In order for carbon dioxide reduction for global environment protection, automobile electrification such as HEV/PHEV/EV (electric drive vehicle) is more and more expanding. Inverter in the electric drive vehicle has an essential role, which convert the energy in battery from DC power to AC to drive well spread AC magnet motors, and convert also from AC to DC to charge excess energy of the vehicle to battery. Several efforts for inverter development to realize energy efficient performance, well controllable function, compactness to fit limited vehicle space, light weight, durable and low cost inverter was studied [1][2]. Novel power module and inverter technology to increase power density of inverter was also reported [3][4]. In this paper, the structural design of a double-side direct water cooling power module and experimental study result of thermal performance shall be presented.

2. Power module cooling structures

As a method to improve thermal resistance of a power module, direct cooling structure has been introduced and is already in use in power electronics technology [2]. Fig. 1 shows three different cooling structures of a power module, with relative thermal resistance of respective structure. Compared to conventional, single-side indirect cooling structure, single-side direct cooling structure can reduce thermal resistance to 70%, and applying double-side cooling structure reduces thermal resistance to 50%.

© VDE VERLAG GMBH · Berlin · Offenbach

PCIM Europe 2016, 10 – 12 May 2016, Nuremberg, Germany

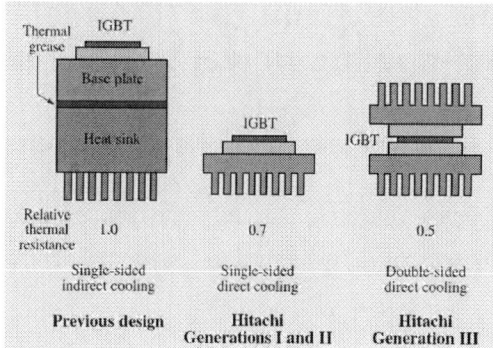

Fig. 1: Different cooling structures of a power module and relative thermal resistance.

3. Double-side direct cooling power module design

Outline photograph and cross section of the new double-side direct cooling power module is shown in Fig. 2 (a) and (b). The power module is composed of IGBT and diode chips, attached to lead frames on both sides of the chips, thus double-side cooling, and molded with resin. The chips are attached to the lead frame with standard, Pb-free solder. The molded assembly shall be placed inside a case with cooling pin fins on both sides of the case. The pin fins shall be exposed to cooling medium such as LLC (long life coolant). To reduce the thermal resistance and thus increase the power density of the power module, direct cooling structure is also applied to the double side cooling power module, eliminating the use of grease layer. Simplified stack-up structure, such as chip-lead frame-isolation-cooling fin structure was adapted, which does not apply thermal grease in order to achieve higher thermal conduction between different layers of stack-up structure.

Isolation material with relatively high thermal conductivity is used in between the mold assembly and the cooling fins. Conventional, single-side direct cooling power modules generally use ceramic-based material for the isolation material. This new power module uses epoxy-based material for this isolation material to achieve low thermal resistance and high reliability.

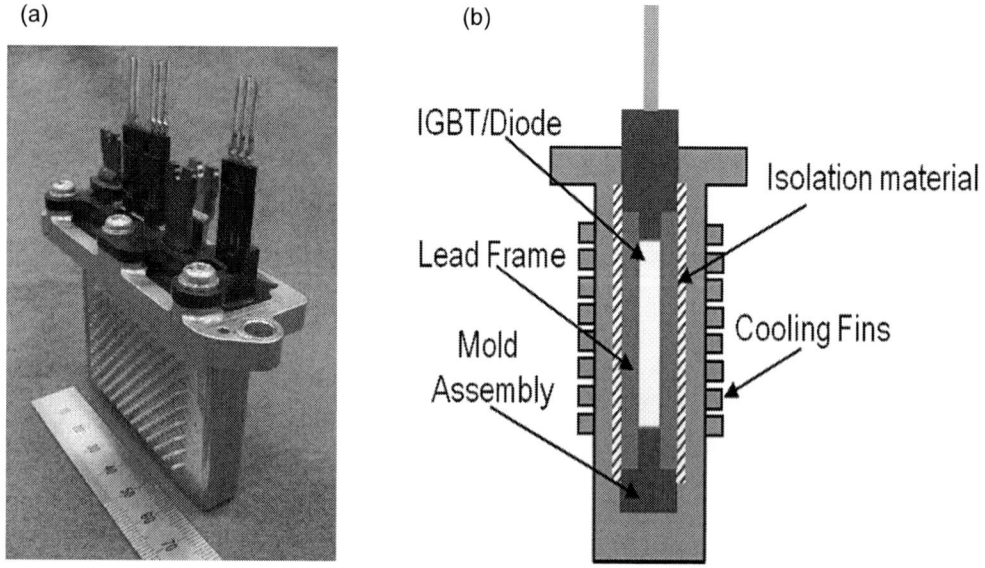

Fig. 2: (a) Outline photograph of a double-side direct cooling power module, (b) Schematic cross section of the power module.

© VDE VERLAG GMBH · Berlin · Offenbach

This power module also incorporates low loss, high performance IGBTs and diodes, with low-inductance bus-bar layout. The power module uses 2-in-1 configuration, improving flexibility in inverter layout. The power module also is excellent when assembling the inverter, with the bus-bars and control terminals located on one side of the power module [4].

By adapting this direct cooling structure, thermal resistance of thermal grease was excluded, which had been one of major portion of thermal resistance in case of conventional, indirect cooling power module structure. The direct cooling structure using epoxy-based material can also reduce the potential deterioration of the grease layer, causing thermal resistance increase. Improved thermal resistance enables higher current for better performance, lower LLC flow rate for less electricity consumption, or smaller chips for better price.

The heat generated by the power module is released to the cooling medium (LLC) through pin fins. Therefore, the pin fin design is an important aspect of the power module design. For this power module design, the pin fin parameters such as the diameter, the height, and the gap between pins were optimized using genetic-algorithm method to achieve both the heat dissipation and pressure drop targets [5].

4. Thermal performance evaluation

To achieve this low thermal resistance, the structure of the lead frame, isolation layer, and cooling fin are optimized. To determine the best structure, numerical analysis was first performed to estimate thermal resistance, and then experimental tests were conducted.

4.1 Thermal resistance study

Fig. 3 (a) shows analytical comparison of thermal resistance distribution within a single-side direct cooling structure and double-side direct cooling structure. The x-axis represents relative distance from the chip (at 0) to the cooling fins (at 1) of the power module. The distance is normalized to the maximum distance for each cooling structures. The points in graph represent structures or layers such as the chip, solder, lead-frame, isolation material, and the cooling fins. The y-axis represents relative thermal resistance, normalized to the thermal resistance of single-side direct cooling structure.

As shown in Fig. 3 (a) comparison, the overall thermal resistance for double-side cooling structure is nearly 60% of single-sided cooling structure. Smaller slope in Fig. 3 (a) graph means smaller thermal resistance, and the comparison shows that the double side cooling structure has smaller thermal resistance in each layer of stack-up structure.

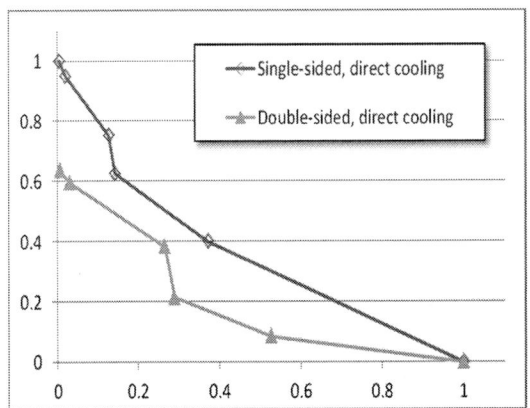
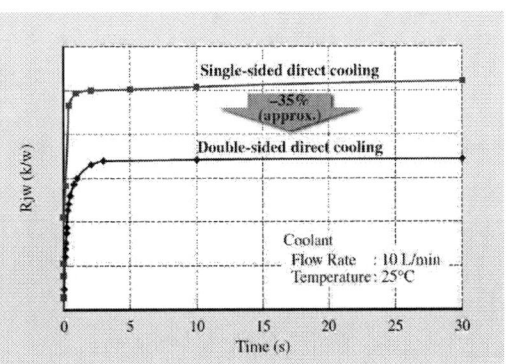

Fig. 3: (a) Analytical comparison of thermal resistance distribution between single-sided and double-sided direct cooling power modules. (b) Thermal resistance improvement from single-sided to double-sided direct cooling by measurement.

Fig. 3 (b) shows thermal resistance measurement example of single-side direct cooling to double-side direct-cooling power module at 10L/min, 25 deg C coolant condition. The result is a ~35% reduction in thermal resistance, and is in fair agreement to analysis.

4.2 Reliability study

The reliability of this new structure is evaluated through means of general test methods: thermal cycling and power cycling. Thermal cycling test was performed with a representative power module structure between temperature ranges of -40 to 125 deg C and -55 to 150 deg C. For power cycling experiment, the power modules were exposed to active temperature cycles. The maximum junction temperature Tjmax was set to 150 degree C, and lower side of the temperature was set to either 50 or 30 deg C. During the test, the power module was turned for duration long enough to reach junction temperature of 150 deg C, and then turned off long enough to cool down to 50 or 30 deg C.

From thermal cycling and power cycling test result, the robustness of the double-sided direct cooling structure is verified. Robustness of solder joint was also confirmed from post-testing examination of the solder joint. This good result is because of the molding structure, which can greatly reduce solder strain leading to solder fatigue failure. Electrical and thermal bias tests common to power modules are also performed, and passed successfully. These result indicate that the power modules are robust and reliable to be used for automotive applications.

5. Example of inverter using double-side direct water cooling power module

Fig. 4 and 5 are the examples of inverter which applied the double side direct cooling power module. Fig. 4 is standard inverter which volume is only 3.5L, current rating is possible at 325Arms or 400Arms in the same inverter outline by changing power module current rating. Fig. 5 is high performance inverter which is suitable for high power electrified drive applications. Each of the double-side cooling power module is a 2-in-1 configuration. Using three power modules forms a three-phase inverter, suitable for electric drive applications.

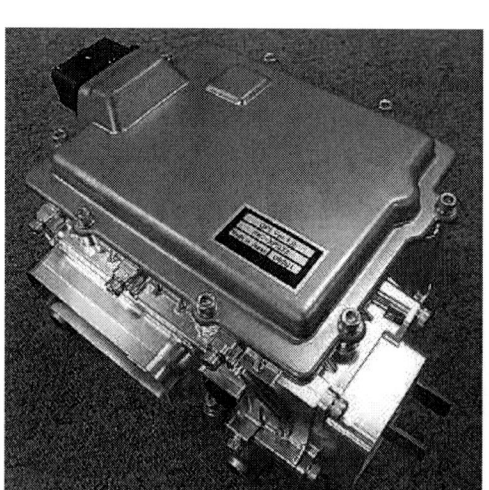

Fig. 4: Third generation standard inverter.

Fig. 5: Third generation inverter.

6. Conclusion

The structure of double-side direct water cooling power module is described. Simplified stacking structure is used to realize a direct cooling structure. Analytical and experimental study showed ~35% better thermal resistance compared to single-side direct cooling structure. Thermal cycling and power cycling tests are conducted to evaluate the robustness of the new structure. Test results indicate the power module has enough capability for automotive usage.

7. Acknowledgments

The authors would like to express their deep appreciation for the support and encouragement of this work to Mr. Ishikawa and Mr. H. Ishikawa, and would like to express their acknowledgement to the contribution of their colleagues to this work.

References

[1] S. Dieckerhoff, S. Guttowski and H. Reichl: "*Performance Comparison of Advanced Power Electronic Packages for Automotive Applications*", Automotive Power Electronics, 21-22 June 2006 Paris (APE 2006).

[2] K. Nakatsu et al: "*Next-generation Inverter Technology for Environmentally Conscious Vehicles*", Hitachi Review, Volume 61 Number 6 October, pp.254-258, 2012.

[3] R. Saito et al: "*High power density inverter technology for automotive applications*", Automotive Power Electronics, R-2013-01-13, April 2013 Paris (APE 2013).

[4] R. Saito et al: "*Enhanced flexibility of inverter design by applying 2 in 1 double side power module*", Automotive Power Electronics, April 2015 Paris (APE 2015).

[5] K. Horiuchi et al: "*Design of Cost-effective Water-Cooled Pinfin Heatsinks*", Proc. InterPACK2013, pp. IPACK2013-73194, July 2013 California (InterPACK2013).

PCIM Europe 2016, 10 – 12 May 2016, Nuremberg, Germany

New Transfer Mold DIPIPM™ utilizing silicon carbide (SiC) MOSFET

Yazhe Wang, Kiyoto Watabe, Shinji Sakai, Toshikazu Tanioka

Mitsubishi Electric Corporation, Power Device Works

1-1-1 Imajukuhigashi Nishi-Ku Fukuoka City 819-0192, Japan

Abstract

A new transfer mold type full Silicon Carbide (SiC) super-mini DIPIPM™ developed for small capacity motor drive system in both home appliances and industrial applications that higher energy-saving efficiency is needed is presented in this paper.

The new DIPIPM™ is designed to compatible with current super-mini series DIPIPM™ product package outline, while employing the Silicon Carbide (SiC) MOSFET power chip inside to help realize the power loss reduction and the miniaturization of the inverter drive system. SiC-MOSFET is expected to replace silicon IGBT because its on-state voltage at low current density and switching characteristics are superior to those of silicon IGBT. By applying the Silicon Carbide (SiC) power MOSFET chip technology, the power loss was reduced about 76% compared with conventional silicon type super-mini DIPIPM™ products.

1. Introduction

Silicon carbide (SiC) is an ideal material for power semiconductor application because it has three times the bandgap, thermal conductivity and ten times the dielectric breakdown field strength than silicon (Si). Si-IGBT (Insulated Gate Bipolar Transistor) is one of the popular power devices for high-voltage, high-current applications however, in low current operation the buit-in voltage drop due to p-n junction will generate more power loss. Furthermore, the conductivity modulation of IGBTs will generate tail current when devices are turned off, resulting in a significant switching loss. The normal conducting mode of SiC-MOSFETs do not need conductivity modulation to achieve low on-resistance and MOSFET generates no tail current in principle, thus SiC-MOSFET has much lower switching loss than IGBTs. Fig.1 shows the photograph of the new full SiC super-mini DIPIPM™.

Fig.1 Photograph of full Silicon Carbide (SiC) super-mini DIPIPM™

© VDE VERLAG GMBH · Berlin · Offenbach

Replacement of Si-IGBT with SiC-MOSFET without modification of gate drive circuits requires a high threshold voltage V_{th}. However the effective mobility in the MOSFET channel fabricated using conventional oxidation technique is much lower than the expected value and the low mobility is attributed to the high density electron traps at the SiO_2/SiC interface. A SiC-MOSFET has the trade-off between the threshold voltage and the mobility.

Mitsubishi Electric has successfully invented an unique method to realize effective V_{th} control of SiC-MOSFET in gate oxidation process and achieved high V_{th} (>4V) SiC-MOSFET device compatible with silicon IGBT.

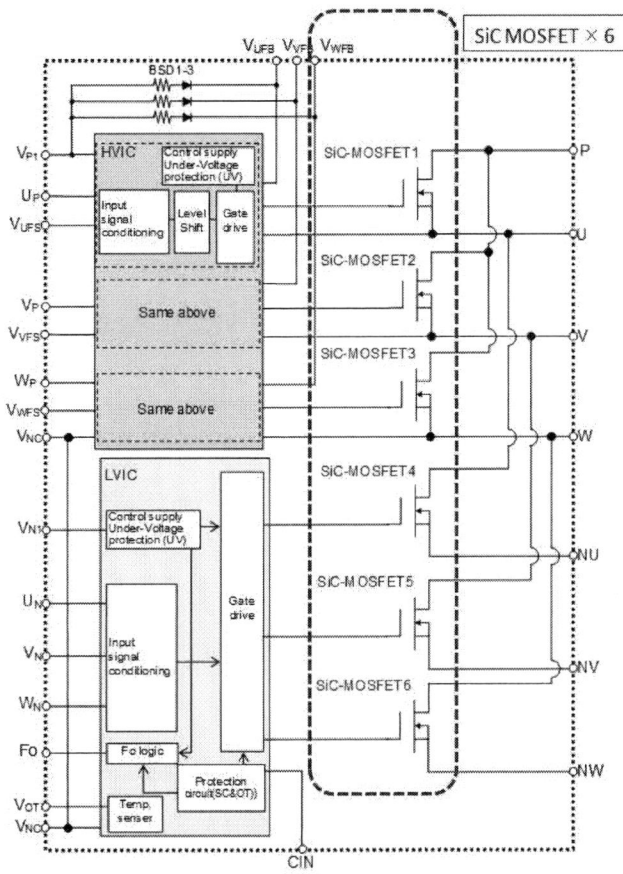

Fig.2 Internal block diagram of full Silicon Carbide (SiC) super-mini DIPIPM™

The new 600V class super-mini DIPIPM™ contains 6 silicon carbide (SiC) MOSFET power chips for typical 3-phase inverter drive system. The high voltage integrated circuit (HVIC) and low voltage integrated circuit (LVIC) are configured to drive and protect the power MOSFETs and the floating high side HVIC driver is supplied by embedded BSD (Bootstrap Diode) chips and outside bootstrap capacitors which make it possible to use only a single power supply in the inverter system. Fig.2 shows the internal block diagram of the new full SiC super-mini DIPIPM™.

© VDE VERLAG GMBH · Berlin · Offenbach

2. Performance improvement

The wide spread IGBT devices with high breakdown voltage usually have high conducting resistance. In order to reduce the conducting resistance, IGBT device needs a process to inject the minority carriers into the drift region however, these minority carriers generate tail current when IGBT turns off, resulting in a significant switching-off loss. SiC MOSFET device does not need such a process due to its much lower drift-layer resistance than silicon device and no tail current presents in switching. As a result, SiC MOSFET has much lower switching loss than IGBTs, which makes higher switching frequency and miniaturization of inverter drive system possible. The new full SiC super-mini DIPIPM™ product is capable to run at a carrier frequency as high as 50kHz and the switching loss is reduced 76% compared with its silicon counterparts, the super-mini Ver.6 DIPIPMs. Fig.3 shows the switching characteristic comparison.

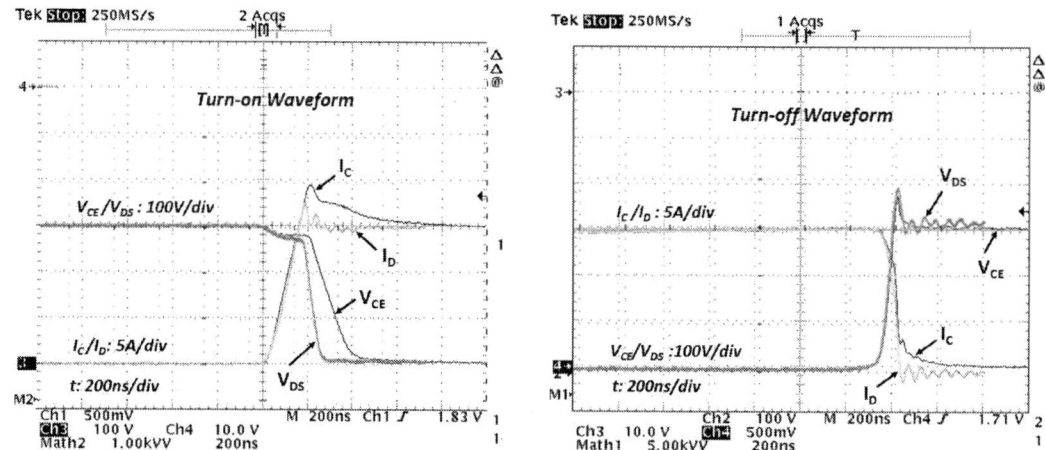

Fig.3 Switching wave form comparison

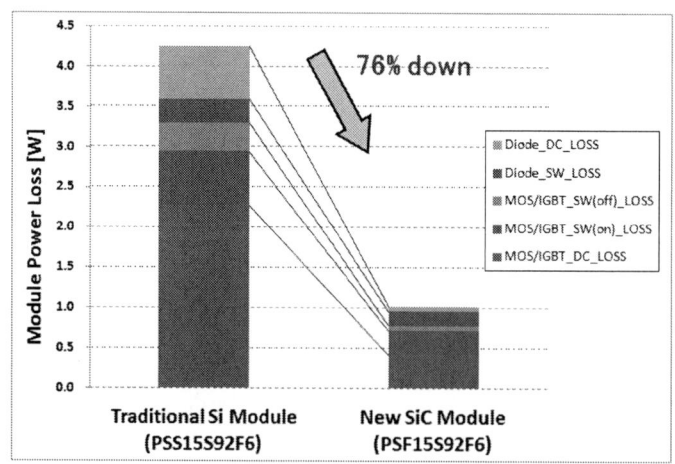

Fig.4 Power loss comparison

VCC=250V, VD=VDB=15V, fc=5kHz, Io=1.5Arms

P.F=0.95, MR=0.8, 2-phase PWM modulation

© VDE VERLAG GMBH · Berlin · Offenbach

By utilizing SiC-MOSFET chip technology, the new full SiC super-mini DIPIPM™ achieved a total of 76% power loss reduction compared with Mitsubishi Electric's latest silicon type super mini DIPIPM™ Ver.6 products. Fig.4 shows the module power loss comparison.

3. Integrated function

New SiC super-mini DIPIPM™ product has three BSD(Bootstrap Diode) chips and temperature measurement function(VOT) integrated in order to achieve lower total system cost by minimizing peripheral electronic parts and PCB board size.

The embedded BSD chip is designed with an integrated 60Ω current limiting resistor, lower resistance to ensure a more stable high-side power supply. Temperature measurement function (VOT) is realized by the laser-trimming technology powered high precision temperature converting circuit of the embedded control LVIC, and by following that analog output voltage signal, it is possible to use the DIPIPM™ at a junction temperature much closer to the tolerable maximum rating value. Fig.5 shows the IF-VF curve of BSD. Fig.6 shows the output characteristic of VOT.

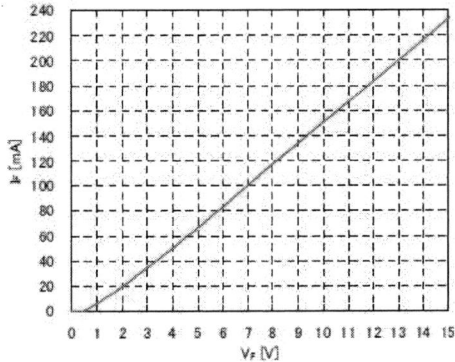

Fig.5 V_F-I_F curve for bootstrap diode

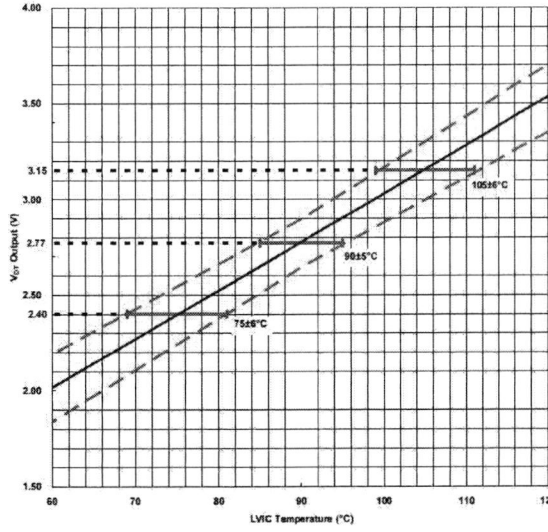

Fig.6 VOT-LVIC Temperature of new full SiC super-mini DIPIPM(15A/600V)

4. Electrical characteristics

The main electrical characteristics (inverter part and control part) of full SiC super-mini DIPIPM are shown in Table 1.

Table. 1. Main electrical characteristics of full SiC super-mini DIPIPM (Tj=25degC, unless otherwise noted)

Item	Symbol	Condition		Min.	Typ.	Max.	Unit
Drain-Source On-state voltage drop	$V_{DS(on)}$	$V_D=V_{DB}=18V$ $I_D=15A, V_{IN}=5V$	Tch=25degC	-	1.10	1.80	V
			Tch=125degC	-	1.10	1.80	
Source-Drain forward voltage	$V_{SD(off)}$	$V_D=V_{DB}=18V, -I_D=15A, V_{IN}=0V$		-	4.00	5.00	V
Switching times	t_{on}	$V_{DD}=300V, V_D=V_{DB}=18V$ $I_D=15A, Tch=125degC$ $V_{IN}=0 \leftrightarrow 5V$ Inductive load		0.70	1.30	1.85	µs
	t_{rr}			-	0.10	-	
	$t_{c(on)}$			-	0.10	0.36	
	t_{off}			-	1.50	2.10	
	$t_{c(off)}$			-	0.10	0.18	
Circuit current	I_D	Total of $V_{P1}-V_{NC}, V_{N1}-V_{NC}$	$V_D=18V, V_{IN}=0V$	-	-	3.5	mA
			$V_D=18V, V_{IN}=5V$	-	-	3.5	
	I_{DB}	$V_{UFB}-U, V_{VFB}-V,$ $V_{WFB}-W$	$V_D=V_{DB}=18V, V_{IN}=0V$	-	-	0.38	
			$V_D=V_{DB}=18V, V_{IN}=5V$	-	-	0.38	
Short circuit trip level	$V_{SC(ref)}$	$V_D=18V$		0.455	0.48	0.505	V
Temperature Output	V_{OT}	LVIC Temperature=90degC	Pull down R=5.1kΩ	2.63	2.77	2.91	V
		LVIC Temperature=25degC		0.88	1.13	1.39	
Control supply under-voltage protection	UV_{DBt}	Tch≦125degC	Trip level	10.0	-	12.0	V
	UV_{DBr}		Reset level	10.5	-	12.5	
	UV_{Dt}		Trip level	10.3	-	12.5	
	UV_{Dr}		Reset level	10.8	-	13.0	
Bootstrap Di forward voltage	V_F	IF=10mA, including voltage drop by limiting resistor		0.9	1.3	1.7	V
Built-in limiting resistance	R	for bootstrap circuit		48	60	72	Ω

5. Conclusion

A new super-mini DIPIPM™ has been developed by utilizing the silicon carbide (SiC) MOSFET power chip technology for applications that the highest energy saving effect is pursued. The new module has six SiC-MOSFET power chips and two control ICs integrated and achieved about 76% power loss reduction compared with conventional silicon type super-mini DIPIPM™ products. This will contribute for the building up of an energy saving and low carbon society.

References

[1] G. Majumdar, et al.,"A New Generation High Performance intelligent Power Module", 1992 PCIM Europe.

[2] S. Shirakawa, T. Iwagami, H.Kawafuji, M. Seo, K. Satoh "A New Version Transfer Mold-Type IPMs with Compact Package" 2005 PCIM China.

© VDE VERLAG GMBH · Berlin · Offenbach

[3]Toru Iwagami, Katsumi Satoh, Kou shomei, Hisashi Kawafuji, Shinya Shirakawa, Tomofumi Tanaka " A Development of 30A/600V Super mini DIPIPM™ " 2006 ipemc,

[4]Masayuki Furuhashi, Toshikazu Tanioka, Yuji Ebiike, Eisuke Suekawa, Yoichiro Tarui, Shinji Sakai, Naoki Yutani, Naruhisa Miura, Masayuki Imaizumi, Satoshi Yamakawa, T atsuo Oomori " Breakthrough in Trade-off between Threshold Voltage and Specific On-Resistance of SiC-MOSFETs" 2013, Proceedings of The 25th International Symposium on Power Semiconductor Devices & ICs, Kanazawa.

[5]Yazhe Wang, Kosuke Yamaguchi, Tomofumi.Takahashi, Kiyoto.Watabe " New Hybrid Large DIPIPM™ for PV Application with built-in SiC SBD and seventh-generation CS TBT™" 2015 PCIM Asia.

[6]Masahiro.Kato, Masataka.Shiramizzu, Tomofumi. Tanaka " An APF oriented Transfer Mold DIPIPM utilizing MOSFET with super junction structure" 2014 PCIM Europe.

A 1700V-IGBT module and IPM with new insulated metal baseplate (IMB) featuring enhanced isolation properties and thermal conductivity

Takuya Takahashi, Yoshitaka Kimura, Hidetoshi Ishibashi, Hiroshi Yoshida
and Yoshitaka Otsubo
Power Device Works, Mitsubishi Electric Corp., Japan,
e-mail(first author): Takahashi.Takuya@dp.MitsubishiElectric.co.jp

Abstract

An enhanced module insulating technology IMB (Insulated Metal Baseplate) that features a high thermal conductivity and insulation properties and its application to IGBT module and IPM are presented. New IMB improves the thermal conductivity of the insulating resin layer by approximately 50% compared to the conventional IMB by optimization of powder particle and resin material. Thickness optimization realizes the best balance of heat resistance and insulation capabilities. This is the first IMB solution for 1700V module which requires high heat dissipation and insulation. It is also fitted for IPM application. We have successfully confirmed new IMB's heat resistance and insulation capabilities suitable for these applications.

1. Introduction

In power electronics, power devices are responsible for power conversion and they are used in various applications such as motor control, wind power and UPS. In recent years, for system requirements of space and weight saving, the research of miniaturization of the power module has been progressing[1]. Since the chip current density of the power module is increased when downsizing, it is necessary to improve the module's heat dissipation. Furthermore, since the operation voltage of power devices is high, insulating structure responsible for heat dissipation and insulation plays an important role.

To meet the above requirements, IMB has been introduced[1]. However, if this conventional IMB is applied to high voltage 1700V module, thermal resistance becomes too large due to the necessity to increase the thickness of the insulating resin layer. Therefore we developed a new IMB that has high thermal conductivity and insulation capabilities at the same time and applied it to 7th generation 1700V IGBT module and IPM.

2. New IMB with high thermal conductivity and insulation properties

2.1 Advantages and disadvantages of the conventional IMB

Generally, an Al_2O_3 substrate is often used for the insulating structure of the power module. We adapted the Aluminium Nitride (AlN) substrate with high thermal conductivity, whose thermal resistance is 35% smaller than that of an Al_2O_3 substrate as shown in Fig. 1. However, it is difficult to improve the thermal conductivity of ceramic moreover. In addition, the stress of surface between ceramic and metal occurs by the mismatch of CTE (Coefficient of Thermal Expansion) between these materials as shown in table 1. Thus, thinner ceramic with better thermal resistance may suffer damage in the thermal profile.

On the other hand, the IMB structure has advantages that the ceramic substrate does not have. When CTE of IMB's insulating resin layer is designed to be equivalent to that of the metal, the stress caused by the mismatch of CTE is reduced. Therefore, the thickness of insulating resin layer can be thinner than that of the ceramic substrate and the thickness of the metal layer can be thicker. Since a thick metal layer can substitute the baseplate, it is possible to eliminate the solder layer under the substrate. As a result, the thermal resistance and thermal cycling capability can be improved[3]. Furthermore, the IMB size can be enlarged more than the ceramic substrate because there is flexibility in insulating resin layer. Thus, the high mounting density is achieved by removing the connecting wire between substrates and eliminating the wiring pattern.

However, IMB's heat conduction path which depends on the contact of ceramic grains in the insulating resin layer, the thermal conductivity is relatively low compared with the ceramic substrate. In order to apply the IMB to high insulation voltage needed for 1700V module, the thickness of the insulating resin layer becomes larger than that of 1200V module. Therefore, it is necessary to improve the thermal conductivity of the insulating resin layer and reduce the thermal resistance of modules.

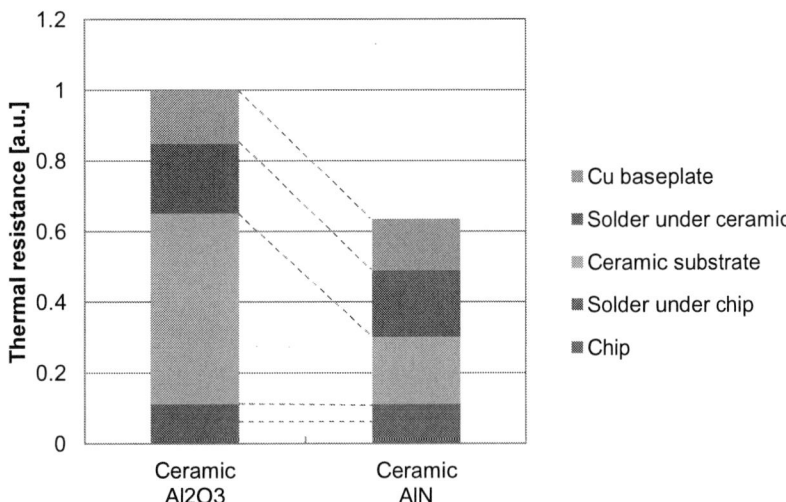

Fig.1 Simulation result of the thermal resistance of IGBT module

Material	CTE [ppm/K]
Ceramic	4.5 ~ 7.0
Cu	≈17
Insulating resin layer	≈17

Table 1 CTE (Coefficient of Thermal Expansion) of typical materials

2.2 Improvement of IMB properties

To achieve the improvement of thermal conductivity, it is necessary to increase the area of the heat conduction path, that is, the ratio of the ceramic particle, or to improve thermal conductivity of resin. Especially because the amount of resin shrinks when increasing the amount of particle, the press pressure required to remove voids increases due to lower fluidity. If the powder volume concentration exceeds the critical powder volume concentration, it causes the decrease of dielectric breakdown voltage and thermal conductivity because voids are left between the particles. Thus, materials of powder particle and resin and distribution of powder size are optimized to reduce voids and to keep high fluidity and press force at sheet molding. As a result, the voids are eliminated and the filling of particles in the insulating resin layer is increased, therefore it becomes possible to enhance both the thermal conductivity and withstand voltage characteristics.

Fig.2 shows the measurement result of thermal resistance of the insulating resin layer depends IMB on its thickness. The thermal resistance of this layer is improved by approximately 35% compared to the conventional IMB, which means the thermal conductivity of this layer is enhanced by 50%. New IMB's partial discharge inception voltage (PDIV) shows almost the same thickness dependence as the conventional IMB. Thus, the proposed IMB has been confirmed to have equivalent PDIV characteristics while improving the thermal conductivity.

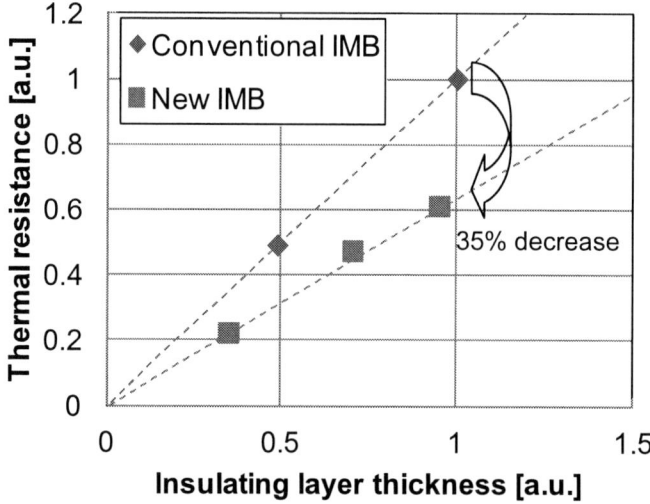

Fig.2 The measurement result of thermal resistance of the insulating resin layer in proposed and conventional IMB

3. 7th generation 1700V-IGBT module and IPM with the proposed IMB

As described above, the proposed IMB has excellent thermal resistance and insulation properties, so it can be applied to applications which can not be realized with the conventional IMB. 1700V IGBT module requires high isolation voltage at least 4000V. Thus, isolation layer thickness of new IMB for 7th generation 1700V IGBT module should be increased to satisfy required isolation voltage. By the new IMB's enhanced characteristics, it becomes possible to have the equivalent thermal resistance to the conventional IMB while satisfying isolation voltage requirements.

On the other hand, new IMB for IPM with 650-1200V voltage rating is optimized in another way. Since IPM is often used under hard operating conditions such as the lock mode operation in servo amplifier, it is strongly required to have high heat dissipation characteristics for IPM package. So, new IMB for 7th generation IPM is designed especially for low thermal resistance. Table 2 shows a summary of the proposed and the conventional IMB characteristics. It is confirmed by evaluation that the thermal resistance of the new IMB for the 1700V approximately 5% better than the conventional IMB, and the simulation result shows the new IMB for 650-1200V is 20% improved from the conventional IMB. Fig. 3 shows the outline structure of the proposed 1700V-IGBT module and IPM.

Moreover, by introducing 7th generation IGBT and Relaxed Field of Cathode (RFC) -planar anode diode[2,4] with optimized characteristics, the loss of the module itself is reduced. Snappy turn-off behavior is suppressed compared with the conventional 1700V module by the RFC-diode. Therefore, combination of 7th generation chips and the proposed IMB structure enables us to increase the current density of the module. It means that this combination can make the size of IGBT module and IPM smaller. Up to 49% size reduction of 1700V-IGBT and up to 55% size reduction of IPM can be realized compared with the conventional ceramic structure as shown in Fig. 3. By combining new IMB with resin encapsulation process named SoLid Cover technology (SLC) as described in Fig. 4, power cycle life is extended owing to the suppressed stress on wire bonding[3]. In addition, thermal cycling capability of the new IMB is better than ceramic substrate by virtue of IMB structure. It can be seen that the proposed IMBs are well fitted for the respective applications.

Structure	Insulation voltage	Thermal resistance with same chip size	
		Simulated result	Evaluated result [K/W]
Conventional IMB	2500V	1	0.130
Proposed IMB for 1700V IGBT module	4000V	0.95	0.124
Proposed IMB for 650-1200V IPM	2500V	0.82	-

Table 2 Summary of module characteristics with proposed and conventional IMB

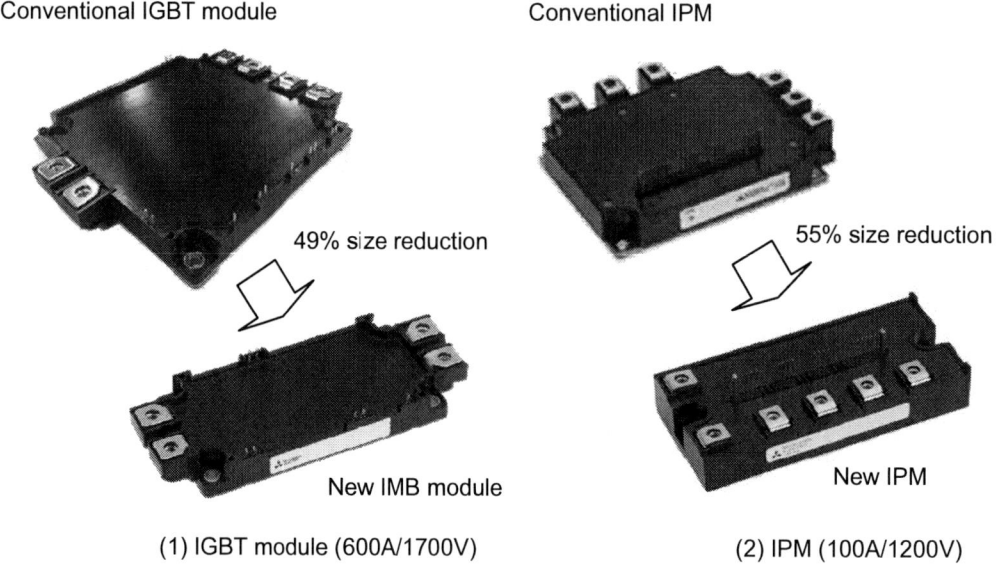

(1) IGBT module (600A/1700V)   (2) IPM (100A/1200V)

Fig.3 Outline structure of the proposed 1700V-IGBT module and IPM and comparison with conventional modules

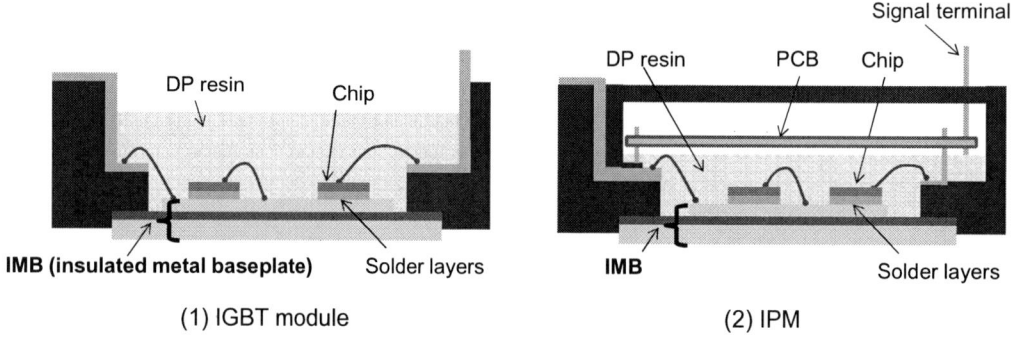

(1) IGBT module   (2) IPM

Fig.4 Cross-sectional view of packages of IGBT module and IPM

4. Conclusion

We have developed a new IMB that achieves a high thermal conductivity while maintaining the insulation properties. It is realized by the optimization of an insulating resin layer of IMB, this is, increasing the ratio of the ceramic particle and optimizing powder particle and resin material, its size distribution and pressing force at sheet molding. As a result, thermal conductivity of the insulating resin layer is improved by approximately 50% compared to the conventional IMB. Therefore, we are able to adapt it to 7th generation 1700V IGBT module that requires higher insulation voltage and to IPM with a strong requirement of lower thermal resistance. It not only contributes to satisfy such application's requirements, but also contributes to the reduction of module size by 55% with a combination with 7th generation

IGBT and diodes. Moreover, high reliability is achieved by combining the new IMB with resin encapsulation. The proposed IMB can be widely used for power modules and it can contribute to the evolution of power electronics.

References

[1] K.Ohara et. al., "A New IGBT Module with Insulated Metal Baseplate (IMB) and 7th Generation Chips", PCIM Nuremberg (2015)

[2] M.Miyazawa et. al., "7th Generation IGBT Module for Industrial Applications", PCIM Nuremberg (2014)

[3] Thomas Radke et. al., " More Power and Higher Reliability by 7th Gen. IGBT Module with New SLC-Technology", Bodo's Power Systems (2015)

[4] F.Masuoka et. al., "Great Impact of RFC Technology on Fast Recovery Diode towards 600V for Low Loss and High Dynamic Ruggedness", Proc. ISPSD 2012, Bruges, Belgium (2012)

nHPD^2 (next High Power Density Dual) with next generation chip suitable for low internal inductance package

Daisuke Kawase, Hitachi Power Semiconductor Device, Ltd., Japan,
daisuke.kawase.ey@hitachi.com
Kazuhito Nagashima, Hitachi Power Semiconductor Device, Ltd., Japan,
kazuhito.nagashima.zk@hitachi.com
Tomohisa Hirayama, Hitachi Power Semiconductor Device, Ltd., Japan,
tomohisa.hirayama.qn@hitachi.com
Katsunori Azuma, Hitachi Power Semiconductor Device, Ltd., Japan,
katsunori.azuma.em@hitachi.com
Seiichi Hayakawa, Hitachi Power Semiconductor Device, Ltd., Japan,
seiichi.hayakawa.kj@hitachi.com
Katsuaki Saito, Hitachi Europe Ltd. Power Device Division, United Kingdom,
katsuaki.saito@hitachi-eu.com

Abstract

In order to take full advantage of a Wide Band Gap (WBG) semiconductor device, it is essential to reduce the total stray inductance of a power circuit. Hitachi recently developed a low internal inductance package named "next High Power Density Dual" (nHPD^2). The package achieved a drastically lower internal inductance. A 75% reduction compared with an equivalent conventional high power package. On the other hand, the next generation package requires next generation IGBTs suitable for low internal inductance switching, realizing both low loss and an adequate control of dV/dt. In this paper the effects of next generation reduced feedback capacitance IGBTs in nHPD^2 are introduced. Furthermore, the resulting benefits of SiC FWD in nHPD^2 combined with the next generation small feedback capacitance IGBT are explored.

1. Introduction

WBG semiconductor is the overt device technology which demonstrates very low switching loss characteristics, not only for high frequency DC-DC convertor applications, but also for motor control in inverters [1]. However, when applied to motor control, switching oscillation is an inevitable problem to overcome. In order to mitigate this oscillation, slower turn-on and turn-off may be considered. Reducing switching speed deliberately will unfortunately ruin the benefit of a WBG device. Internal inductance reduction is effective to avoid the oscillation [2]. Reducing the inductance has been reported [3,4,5]. Figures 1(a) and 1(b) show the appearance and the equivalent circuit diagram of the nHPD^2 package respectively. Owing to the dual configuration and optimized terminal structure, the internal inductance had been reduced by 75% from conventional package [2]. Figure 1(c) shows the main terminal structure. 10nH is achieved by arranging the terminal area face-to-face with short distances between them. As for IGBT turn-off, low internal inductance realizes small peak voltage. Due to low internal inductance the turn-on loss becomes larger than conventional packages because the terminal voltage of IGBT is increased during turn-on period. Thus the next generation package requires next generation IGBT technology suitable for low internal inductance package. Miller period reduction using a small feedback capacitance is effective to reduce turn-on loss in a low inductance package. Furthermore, using a SiC FWD in a low inductance package with small feedback capacitance IGBT provides low switching loss without severe oscillation.

PCIM Europe 2016, 10 – 12 May 2016, Nuremberg, Germany

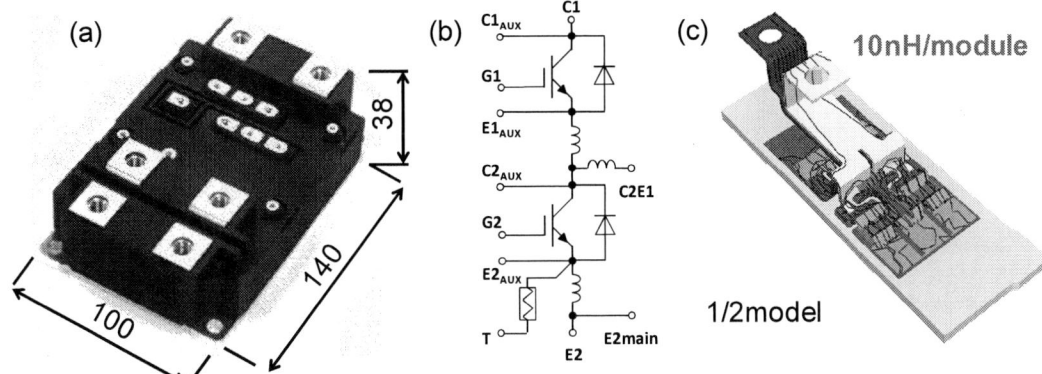

Fig. 1   (a) Appearance, (b) equivalent circuit diagram and (c) terminal structure

2. Small feedback capacitance IGBT

2.1.   Small feedback capacitance efficient

Figure 2 shows the schematic waveforms of IGBT turn-on and FWD recovery. In the example of nHPD[2], Vce during turn-on is larger than the conventional package because of low internal inductance (Ls) (comparing Vce1 and Vce2). Thus turn-on switching loss (Eon) becomes larger. In order to reduce Eon, a small gate resistance (Rg) should be selected. On the contrary small Rg makes FWD recovery dV/dt larger (comparing dV/dt1 and dV/dt2). Large dV/dt may damage motor insulation. To achieve small Eon and small FWD recovery dV/dt at the same time, smaller feedback capacitance (Cres) is suitable for next generation low Ls package technology. With a smaller Cres value, the Miller period becomes smaller (comparing Δt1 and Δt2). Thus a lower Eon is achieved without increasing dV/dt or using a smaller Rg. Reduced Cres IGBT is the basis for next generation IGBT adopting low Ls package.

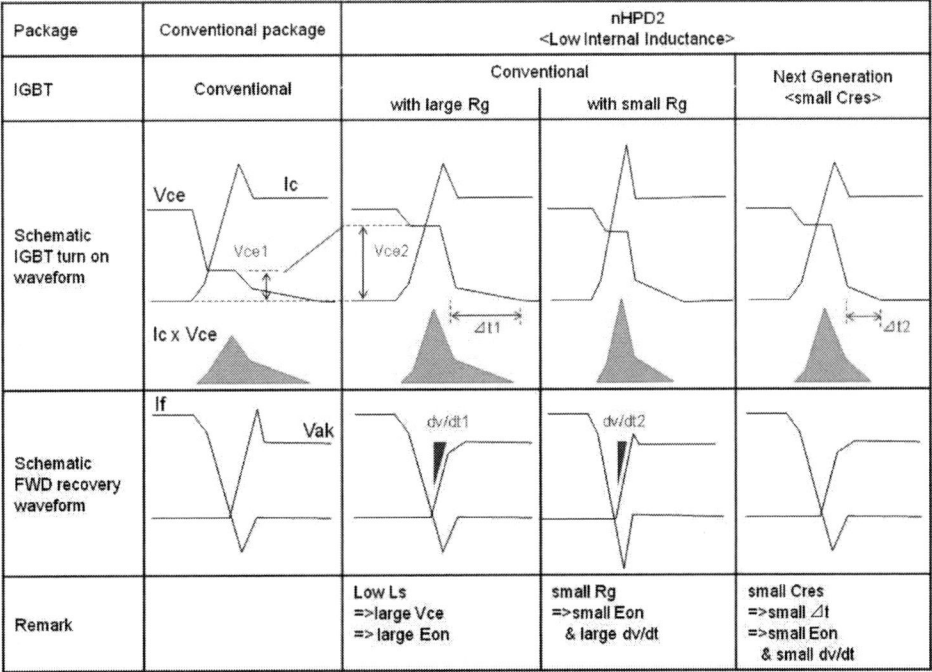

Fig. 2   Schematic waveforms of turn-on and recovery

© VDE VERLAG GMBH · Berlin · Offenbach

2.2. Simulations results

Figure 3 shows simulation waveforms for the succeeding generation adopting smaller Cres IGBT design. Comparing conventional IGBT and the next generation IGBT, it is evident that the Miller period of the next generation IGBT is shorter than conventional type in spite of smaller dI/dt.

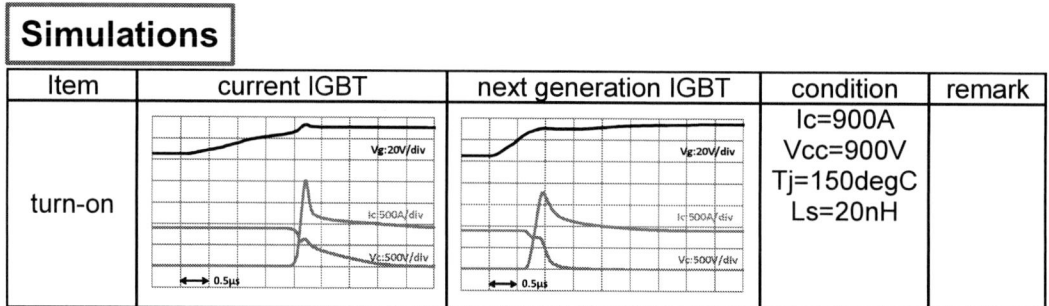

Item	current IGBT	next generation IGBT	condition	remark
turn-on	Vg:20V/div Ic:500A/div Vc:500V/div 0.5µs	Vg:20V/div Ic:500A/div Vc:500V/div 0.5µs	Ic=900A Vcc=900V Tj=150degC Ls=20nH	

Fig. 3   Comparison between conventional and next generation IGBT by simulation

3. Experimental results of small Cres IGBT

3.1   QG-VG characteristics

Figure 4 shows a comparison of the QG-VG characteristics of the conventional IGBT with the next generation small feedback capacitance IGBT. The charge of the plateau region corresponding to the Miller period has reduced 27% and the feedback capacitor reduction is confirmed.

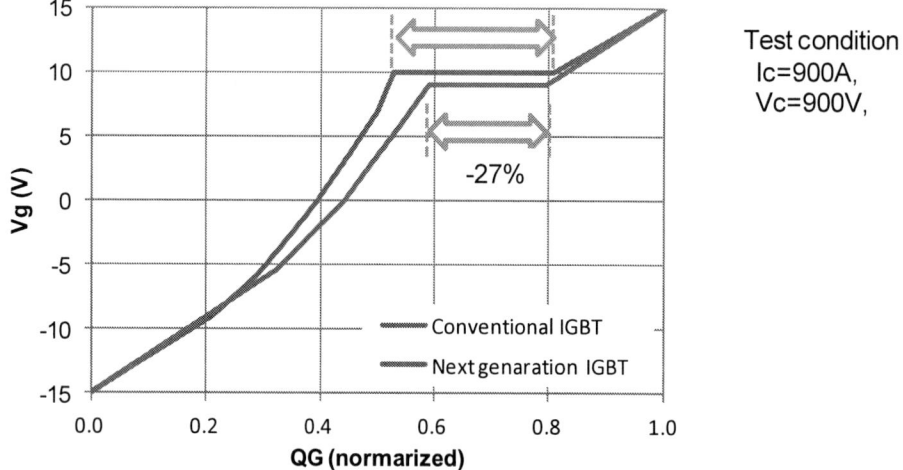

Fig. 4   Comparison QG-VG characteristics between conventional and next generation IGBT

3.2   Switching loss and FWD recovery dV/dt trade-off improvement

Figure 5 shows the relationship between IGBT Eon, FWD recovery loss (Err) and FWD recovery dV/dt. Conventional IGBTs and the next generation IGBTs were both mounted in separate 900A/1700V nHPD^2 modules. The next generation IGBT Cres was smaller than the conventional IGBT. The reduced Cres effect improved the trade-off relationship between Eon

+ Err and recovery dV/dt. Eon + Err was reduced by 28% at the same recovery dV/dt (12.5kV/us).

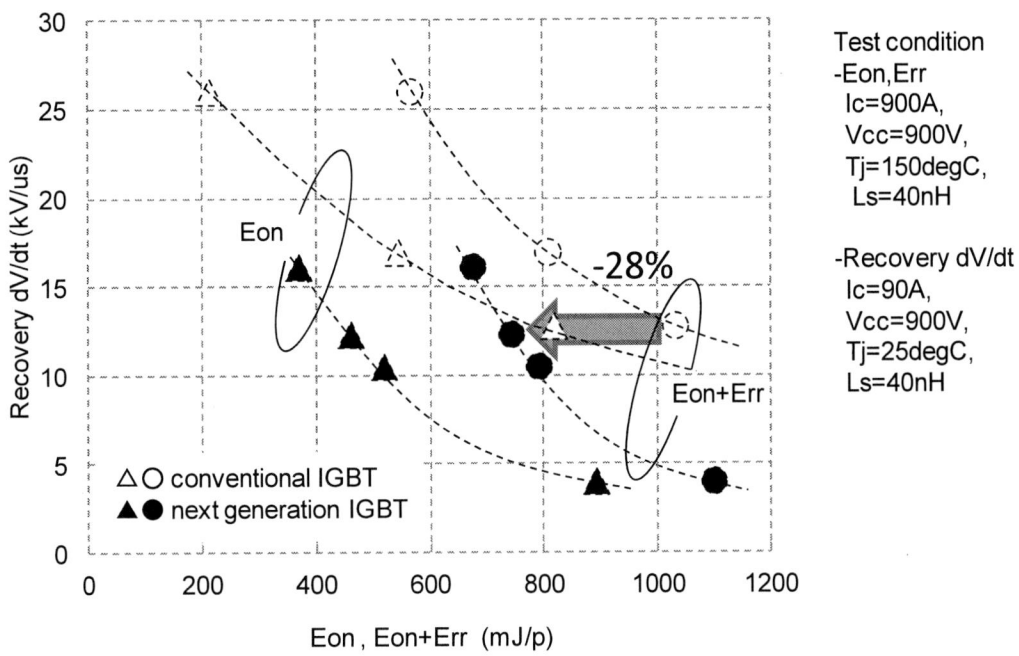

Fig. 5   Trade-off relationship between dV/dt and switching loss

Figure 6 shows the experimental results of IGBT turn-on and FWD recovery waveforms. Compared with the conventional IGBT, the next generation IGBT had a smaller Eon in spite of the same FWD recovery dV/dt.

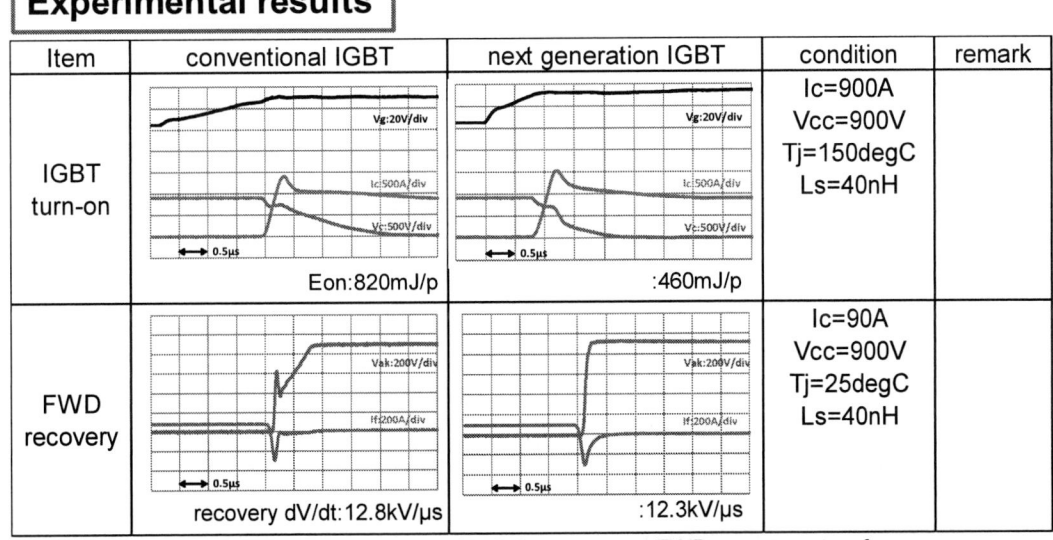

Fig. 6   Experimental results of IGBT turn-on and FWD recovery waveforms

3.3. Vce(sat) and Eoff trade-off improvement

Figure 7 shows the trade-off relationship between Eoff and Vce (sat). Next generation IGBT achieved a 17% reduction of Vce (sat) and 17% improvement in Eoff.

Figure 8 shows a high-current turn-off waveform of a 1700V 900A nHPD^2 module using next generation IGBT. Experimentally we confirmed that the RBSOA of the next generation IGBT was sufficient to turn-off twice the rated current.

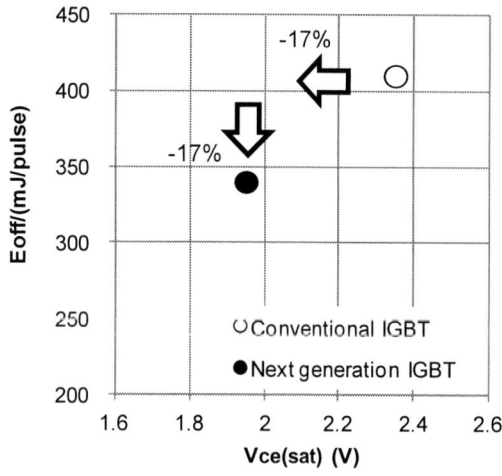

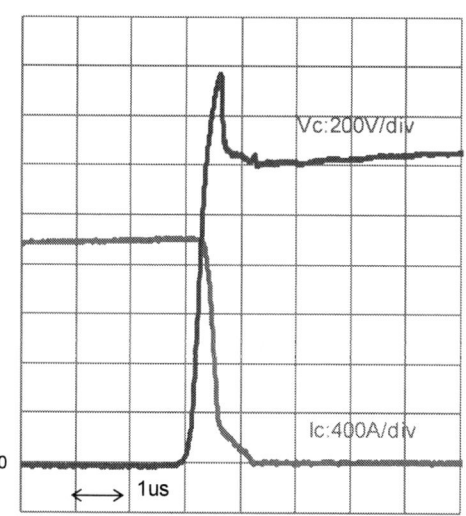

Test condition
-Vce(sat)   -Eoff
Ic=900A,Vge=15V,   Ic=900A,Vcc=900V,
Tj=150degC   Tj=150degC,Ls=40nH

Test condition
Ic=1800A,Vcc=1200V,
Tj=150degC,Ls=40nH

Fig. 7   Eoff - Vce(sat) trade-off relationship of next generation IGBT in nHPD^2

Fig.8   RBSOA waveform of next generation IGBT in nHPD^2

4. Wide Band Gap FWD

4.1. Switching waveforms

A Wide Band Gap FWD contributes to low switching loss due to small recovered charge in the case of high speed switching. With the low internal inductance of nHPD^2, the system designer is able to take the full advantage of Wide Band Gap technology. Due to low internal inductance the Wide Band Gap device can be switched at higher speeds, but importantly without oscillation.

Figure 9 shows the switching waveforms of Si IGBT + Si FWD and Si IGBT + SiC FWD, mounted in a 900A/1700V nHPD^2 module. The Si IGBTs were next generation type having low feedback capacitance in both cases. With reference to the FWD recovery waveforms, a reverse recovery current is not present for the SiC FWD and so FWD recovery loss becomes significantly lower. It should be noted that severe oscillation was not observed. Regarding IGBT turn-on waveforms, by improving SiC FWD recovery behavior, the switching loss of Si IGBT + SiC FWD becomes smaller than Si IGBT + Si FWD.

4.2. Electrical characteristics improvements

Table 1 shows the electrical characteristics of SiC hybrid nHPD^2. SiC FWD benefits from the point of view of switching losses are summarized in the table. Eon and Err are improved drastically due to small recovery capacitance of SiC FWD. We experimentally confirmed low switching losses suitable for high frequency operation.

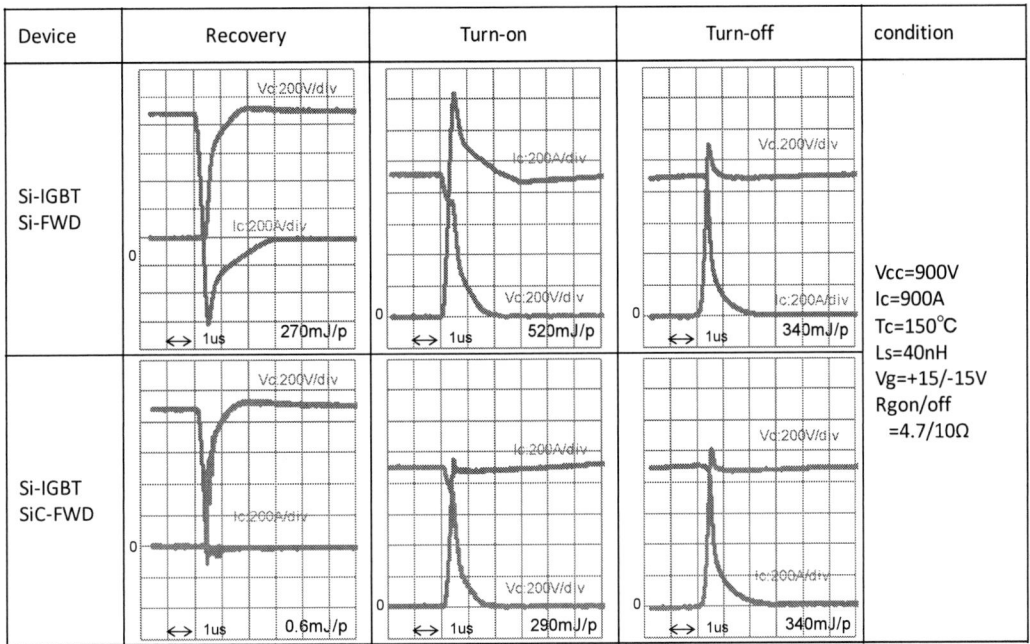

Fig.9   Comparison between waveforms of Si FWD and SiC FWD

Table1 Electrical characteristics of SiC hybrid nHPD[2]

Item	Si IGBT+ Si FWD	Si IGBT+ SiC FWD	Remark
Vce(sat)	1.95V	1.95V	
VF	1.75V	2.25V	+29% worse
Eoff	340mJ/p	340mJ/p	
Eon	520mJ/p	290mJ/p	-44% better
Err	270mJ/p	0.6mJ/p	-99.8% better
Eon + Err	790mJ/p	291mJ/p	-63% better

5. Conventional and next generation of SiC hybrid module

Figure 10 shows the SiC FWD hybrid module switching waveforms comparison between a conventional package with conventional IGBT and a low inductance next generation package with small feedback capacitance next generation IGBT. Recovery with a conventional combination shows oscillations which is not the case for the next generation combination. During turn-on, in spite of the smaller voltage drop (ΔVce), losses of next generation combination remain small due to the influence of shorter Miller period (Δt). Both low loss and low noise is achieved in the case of next generation. Next generation combination shows lower over-voltage peak (ΔVcep) and a small turn-off loss. Lower Eoff is due to lower over-voltage peak and IGBT characteristics. The low inductance package and small feedback capacitance IGBT demonstrates low loss and low noise can be achieved enabling the adoption of high-speed switching, this complimenting the performance of SiC FWD technology.

© VDE VERLAG GMBH · Berlin · Offenbach

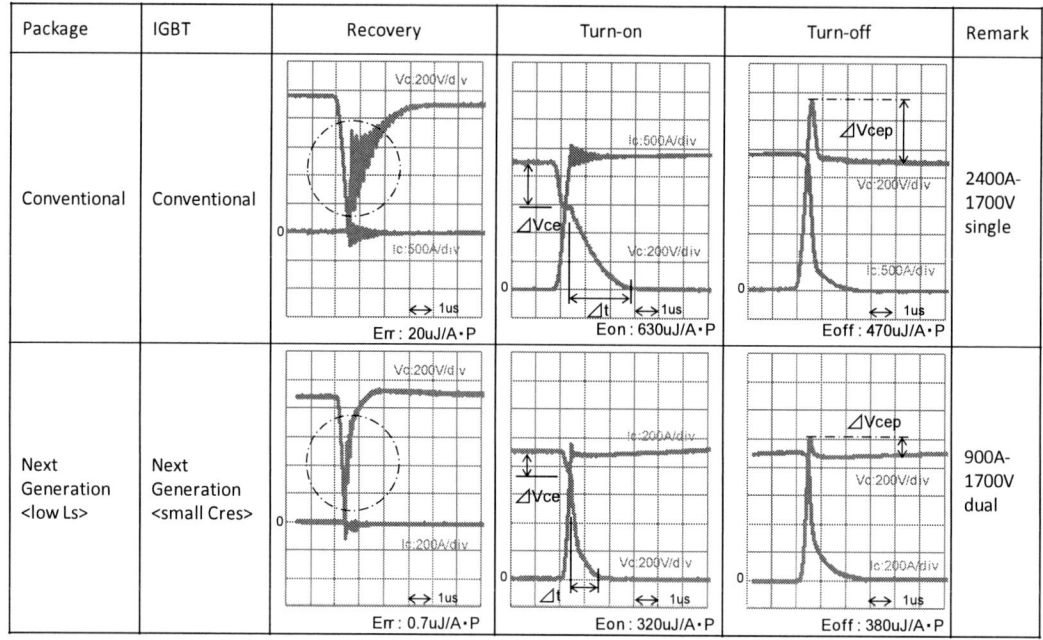

Test condition : Ic=Rated current, Vcc=900V, Tj=150degC
Ls=55nH [2400A-1700V single conventional package , Ls x Ic =132uAH]
40nH [900A-1700V dual next generation package , Ls x Ic =36uAH]

Fig.10 Switching waveform comparison of conventional and next generation SiC hybrid modules

6. Conclusion

The following was proved by both simulation and experimentation.

1) nHPD[2] with the next generation small Cres IGBT design improves the trade-off relationship between turn-on switching loss and FWD recovery dv/dt.

Small feedback capacitance IGBT is suitable for low internal inductance package.

2) nHPD[2] adopting a combination of the next generation IGBT and SiC FWD realizes further switching loss improvement based on the SiC FWD properties without severe oscillation.

3) The combination of nHPD[2] and the new IGBT chip is suitable to achieve a low internal inductance package solution, complimenting both package and switching device technology.

7. Reference

[1] K. Ogawa et al., "Traction inverter that applied SiC hybrid module"; PCIM2011

[2]D. Kawase et. al., High voltage Module with low internal Inductance for next Chip Generation – next High Power Density Dual (nHPD[2]), PCIM Europe 2015

[3]R. Schnell et.al., LinPak, a new low inductive Phase-Leg IGBT Module with easy paralleling for high Power density Converter Designs, PCIM Europe 2015

[4]G. Borghoff, Implementation of low inductive strip line concept for symmetric switching in a new high power module, PCIM Europe 2013

[5]A.Nagel, A New Standard IGBT Housing for High-Power Converters, EPE'15 ECCE Europe

An Optimized Hybrid-MMC for HVDC

Viktor Hofmann, Mark-M. Bakran, University of Bayreuth,
Center of Energy Technology, Germany, viktor.hofmann@uni-bayreuth.de

Abstract

The Modular Multilevel Converter (MMC) is a well suited converter topology for high-voltage direct current (HVDC) transmission. The converter performance is often discussed for a MMC consisting of only one cell type, such as full bridge or half bridge cells. This paper investigates the combined use of these basic cell types in order to achieve an improved MMC performance. Thereby the possible advantages and limiting factors of this hybrid MMC topology are demonstrated. Finally a possible modulation method for the different cell types is presented.

1. Introduction

The Modular Multilevel Converter (MMC) [1] (Fig. 1) provides a lot of features such as high output voltage quality, efficiency and easy scalability [2]-[4]. Due to its several advantages it is a well suited converter topology for high-voltage direct current (HVDC) applications [5]. The MMC consists of three phase legs, where each leg consists of an upper and a lower arm. Such an MMC arm is assembled with a large amount of series connected cells. These cells can be of different design and provide various advantages and disadvantages.

Basic and most common used cell types are half bridge (Fig. 1a) and full bridge cells (Fig. 1b). Half bridges offer low losses and a low amount of semiconductors to be installed for a given voltage rating. In contrast, full bridges have higher losses but provide e.g. a DC fault

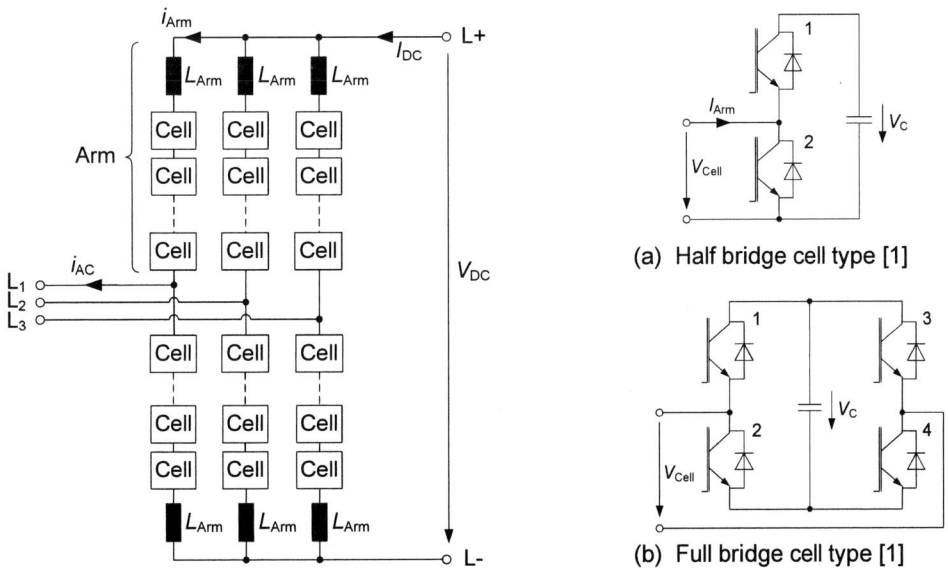

Fig. 1. Modular Multilevel Converter (MMC) with two different cell types [1].

ride-through capability. Furthermore, for the same DC voltage rating a higher power transfer can be realized with full bridges because of the higher thermal admissible current and the possibility of an over-modulation ($2\hat{V}_{AC}/V_{DC} > 1$).

In this paper a combined use of these basic cell types is investigated and the possible advantages as well as the limiting factors are demonstrated.

2. Functional Principle of the MMC

The voltage modulation of each MMC arm is realized by inserting or bypassing single cells. Half bridge cells allow two cell states. Its terminal voltage V_{Cell} can be either the positive capacitor voltage $+V_C$ or zero. With full bridge cells it is possible to modulate a positive voltage level $+V_C$, a negative voltage level $-V_C$ and the zero voltage level. Depending on the cell state and the direction of the arm current the cell capacitor is either charged or discharged. For a safe operation it is necessary that the net energy flow to the cell capacitors is zero.

The voltage of one phase leg is always equal to the DC link voltage V_{DC}. The phase-to-phase voltage of the AC side is given by the voltage difference of two neighbored arms. The modulated AC voltages of each MMC arm are phase shifted. The upper and the lower arm are phase shifted by π, and two neighbored arms are phase shifted by $2\pi/3$. The DC link current I_{DC} is equally distributed over the three phase legs and the AC side phase current is the difference of the upper arm current and the lower arm current. Since the load of each MMC arm is only phase shifted but otherwise symmetrical it suffices to consider one single MMC arm for the energetic analysis. The modulated arm voltage is given by

$$v_{Arm} = \frac{V_{DC}}{2} + \hat{V}_{AC} \cdot \sin(\omega t) \tag{1}$$

and the arm current can be expressed as

$$i_{Arm} = \frac{I_{DC}}{3} - \frac{1}{2}\hat{I}_{AC} \cdot \sin(\omega t - \varphi) \tag{2}$$

where φ is the angle between the AC arm voltage and the AC arm current. By introducing the modulation index m

$$m = \frac{2\hat{V}_{AC}}{V_{DC}} \tag{3}$$

and taking the energy balance into account

$$P_{DC} = V_{DC} \cdot I_{DC} \quad = \quad 3 \cdot \frac{\hat{V}_{AC} \cdot \hat{I}_{AC}}{2} \cos\varphi = P_{AC}, \tag{4}$$

the equations (1) and (2) can be written as

$$v_{Arm} = \frac{V_{DC}}{2}(1 + m \cdot \sin(\omega t)) \tag{5}$$

$$i_{Arm} = \frac{m}{4} \cdot \hat{I}_{AC} \cdot \cos\varphi - \frac{1}{2}\hat{I}_{AC} \cdot \sin(\omega t - \varphi). \tag{6}$$

3. Energetic Analysis

3.1. Basic analysis

In this chapter the energetic relationships are considered and the resulting restrictions are discussed. To describe the configuration of the Hybrid-MMC the following variables are introduced:

$$n_0 = \frac{\max(|v_{Arm}|)}{V_C}: \qquad \text{minimum required amount of cells in one single MMC arm,}$$

n_{HB}:   amount of installed half bridge cells in one single MMC arm,

n_{FB}:   amount of installed full bridge cells in one single MMC arm.

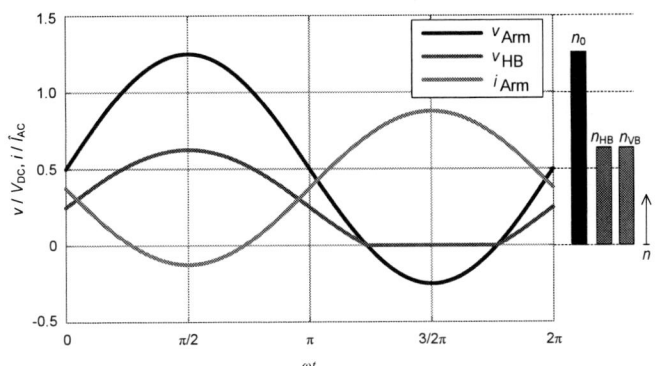

Fig. 2: Voltage and current profile of a single MMC arm.

At first an MMC is considered which consists of 50% full bridge cells and 50% half bridge cells ($n_{\mathrm{HB}}/n_0 = 0.5$, $n_{\mathrm{FB}}/n_0 = 0.5$). With a half bridge cell it is not possible to modulate a negative voltage level. Hence, the half bridge cells can only be discharged at a negative current level. For a stable operation of the MMC it is necessary to ensure that the net energy flow to the cell capacitors is zero. Therefore the power and energy balance is considered in the following.

Fig. 2 shows the voltage and current profile of a single MMC arm for a chosen modulation index $m = 1.5$ and a load angle $\cos\varphi = 1$. If a common modulation scheme is assumed the half bridge cells would modulate a voltage profile (v_{HB}) as it is shown in Fig. 2.

$$v_{\mathrm{HB}} = \begin{cases} 0 & \text{if } \omega t_1 \leq \omega t \leq \omega t_2 \\ \frac{n_{\mathrm{HB}}}{n_0} \cdot v_{\mathrm{Arm}} & \text{else} \end{cases} \tag{7}$$

with $\omega t_1 = \pi + \arcsin\left(\frac{1}{m}\right)$ (8)   and $\omega t_2 = 2\pi - \arcsin\left(\frac{1}{m}\right).$ (9)

ωt_1 and ωt_2 are the zero crossings of the arm voltage.

The positive arm voltage would be modulated equally distributed by half bridges and full bridges and the negative arm voltage would be modulated only by the full bridge cells.

However this distribution violates the energy balance criteria, as shown below. The instantaneous power p_{Arm} flowing through an MMC arm is the product of arm current and arm voltage

$$p_{\mathrm{Arm}} = v_{\mathrm{Arm}} \cdot i_{\mathrm{Arm}} = \left[\frac{V_{\mathrm{DC}}}{2}(1 + m \cdot \sin(\omega t))\right] \cdot \left[\frac{m}{4} \cdot \hat{I}_{\mathrm{AC}} \cdot \cos\varphi - \frac{1}{2}\hat{I}_{\mathrm{AC}} \cdot \sin(\omega t - \varphi)\right] =$$

$$\frac{1}{8} \cdot V_{\mathrm{DC}} \cdot \hat{I}_{\mathrm{AC}} \cdot (1 + m \cdot \sin(\omega t)) \cdot [m \cdot \cos\varphi - 2 \cdot \sin(\omega t - \varphi)]. \tag{10}$$

The load angle φ is zero in the considered scenario. Thus, the above-mentioned equation is simplified to the following equation

$$p_{\mathrm{Arm}}(\varphi = 0) = \frac{1}{8} \cdot V_{\mathrm{DC}} \cdot \hat{I}_{\mathrm{AC}} \cdot [\sin(\omega t) \cdot (m^2 - 2) + m \cdot \cos(2\omega t)]. \tag{11}$$

The instantaneous power p_{HB} flowing through the half bridge cells is calculated in the same way

$$p_{\mathrm{HB}}(\varphi = 0) = \begin{cases} 0 & \text{if } \omega t_1 \leq \omega t \leq \omega t_2 \\ \frac{n_{\mathrm{HB}}}{n_0} \cdot \frac{1}{8} V_{\mathrm{DC}} \hat{I}_{\mathrm{AC}}[\sin(\omega t) \cdot (m^2 - 2) + m \cdot \cos(2\omega t)] & \text{else} \end{cases} \tag{12}$$

Integrating (11) and (12) and substituting ωt_1 and ωt_2 with (8) and (9), the energy profile of the total MMC arm E_{Arm} and the half bridge cells E_{HB} can be expressed as

$$E_{\mathrm{Arm}} = \int_0^{\omega t'} p_{\mathrm{Arm}}(\varphi = 0)\mathrm{d}\omega t = \frac{1}{8}V_{\mathrm{DC}}\hat{I}_{\mathrm{AC}}\left[(m^2 - 2)(1 - \cos(\omega t')) + \frac{m}{2}\sin(2\omega t')\right] \tag{13}$$

$$E_{\mathrm{HB}} = \int_0^{\omega t'} p_{\mathrm{HB}}(\varphi = 0)\mathrm{d}\omega t =$$

$$= \frac{n_{\mathrm{HB}}}{n_0} \cdot \frac{1}{8} V_{\mathrm{DC}} \hat{I}_{\mathrm{AC}} \begin{cases} \left[(m^2 - 2)(1 - \cos(\omega t')) + \frac{m}{2}\sin(2\omega t') \right] & \omega t' < \omega t_1 \\[2ex] \left[(m^2 - 1)\sqrt{1 - \frac{1}{m^2}} + m^2 - 2 \right] & \text{if } \omega t_1 \le \omega t' \le \omega t_2 \\[2ex] \left[2(m^2 - 1)\sqrt{1 - \frac{1}{m^2}} + (m^2 - 2)(1 - \cos(\omega t')) + \frac{m}{2}\sin(2\omega t') \right] & \omega t' > \omega t_2 \end{cases}$$

$$(14)$$

The power and energy profiles of the half bridge cells and the whole MMC arm are shown in Fig. 3 for a fundamental cycle.

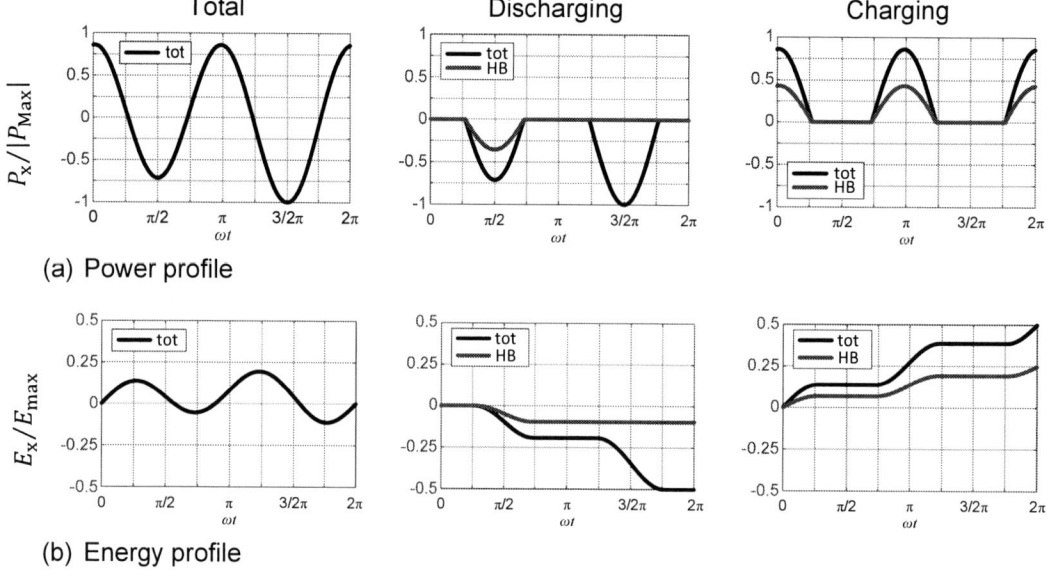

(a) Power profile

(b) Energy profile

P_{Max}: Maximum power value during a fundamental cycle   tot: total MMC Arm

E_{max}: Total amount of converted energy during a fundamental cycle   HB: All half bridge cells

Fig. 3: Power and energy profile of an MMC arm for a fundamental cycle.

The power profile of an MMC arm during a fundamental cycle is illustrated in the left column of Fig. 3(a). The middle column shows only the negative part, the discharging power. There are two periods in which the arm power is negative, but the half bridge cells (blue line) can only be discharged in one of them, since the arm voltage is negative only during the second period. The power profile of the positive arm power (charging power) is depicted in the right column.

The energy profile of an MMC arm during a fundamental cycle is illustrated in the left column of Fig. 3(b). The net energy flow through an MMC arm is zero in average. The discharging and charging energy are shown in the middle and the right column. The considered MMC consists of 50% full bridge cells and 50% half bridge cells. All half bridge cells are able to convert half of the charging energy (right column) but only about one-fifth of the discharging energy. Because the net energy flow has to be zero, the use of the half bridge cells is limited by the discharging energy for the considered point of operation and modulation scheme.

3.2. Dependency on the load angle

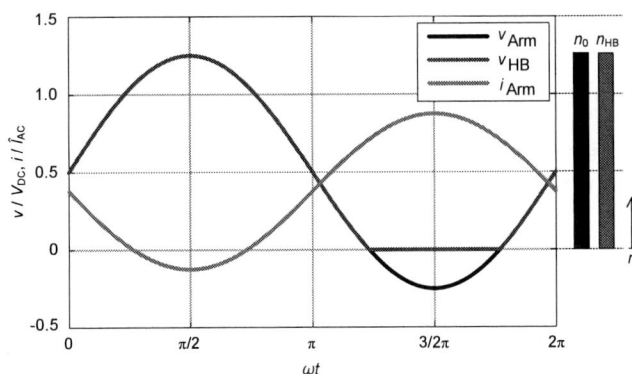

Fig. 4: Voltage and current profile of a single MMC arm.

It is investigated in this chapter how the maximum amount of energy, that can be converted by the half bridge cells depends on the load angle φ. Therefor an MMC configuration is considered with which the maximum arm voltage level could be modulated only by the half bridge cells ($n_{\mathrm{HB}}/n_0 = 1$). The arm voltage and arm current profiles for a total MMC arm are described by the equations (1) and (2) and the voltage profile which is modulated only by half bridge cells is described in (7). Fig. 4 illustrates these profiles for the considered MMC configuration and a modulation index $m = 1.5$ as well as a load angle $\varphi = 0$. The total amount of charging $E_{\mathrm{Arm,ch}}$ and discharging energy $E_{\mathrm{Arm,dch}}$ which are converted in a total MMC arm during a fundamental cycle are given by

$$E_{\mathrm{Arm,ch}} = \int_0^{2\pi} \beta_{\mathrm{ch,Arm}} \cdot v_{\mathrm{Arm}} \cdot i_{\mathrm{Arm}} d\omega t \tag{15}$$

$$E_{\mathrm{Arm,dch}} = \int_0^{2\pi} \beta_{\mathrm{dch,Arm}} \cdot v_{\mathrm{Arm}} \cdot i_{\mathrm{Arm}} d\omega t \tag{16}$$

with $\quad \beta_{\mathrm{ch},x} = \begin{cases} 1 & \text{if } v_x \cdot i_{\mathrm{Arm}} \geq 0 \\ 0 & \text{else} \end{cases}$ (17) $\quad$ and $\quad \beta_{\mathrm{dch},x} = \begin{cases} 1 & \text{if } v_x \cdot i_{\mathrm{Arm}} < 0 \\ 0 & \text{else} \end{cases}$. (18)

The total amount of charging $E_{\mathrm{HB,ch}}$ and discharging energy $E_{\mathrm{HB,dch}}$ which are converted only by the half bridge cells during a fundamental cycle are calculated by

$$E_{\mathrm{HB,ch}} = \int_0^{2\pi} \beta_{\mathrm{ch,HB}} \cdot v_{\mathrm{HB}} \cdot i_{\mathrm{Arm}} d\omega t \tag{19}$$

$$E_{\mathrm{HB,dch}} = \int_0^{2\pi} \beta_{\mathrm{dch,HB}} \cdot v_{\mathrm{HB}} \cdot i_{\mathrm{Arm}} d\omega t \tag{20}$$

The described equations can be calculated numerically. The maximum percentage of charging and discharging energy that can be converted by half bridge cells in dependence of the load angle φ is shown in Fig. 5. Only the minimum amount of the charging or the discharging energy can be used for a given load angle, due to the energetic requirements of

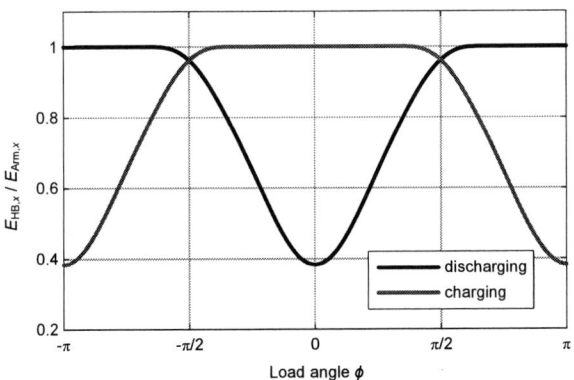

Fig. 5: Percentage of charging and discharging energy, that can be converted by half bridge cells ($m = 1.5$, $n_{\mathrm{HB}}/n_0 = 1$) in dependence of the load angle φ.

a stable operation. According to Fig. 5 these minimum values are located at $\varphi_1 = 0$ and $\varphi_{2,3} = \pm\pi$. Thus, these load angles characterize the energetic worst-case. Consequently, it suffices to consider a load angle $\varphi = 0$ for energetic reasons. At other load angles a higher percentage of energy can be converted by the same amount of half bridge cells. A different modulation index leads to different values of the energy percentage that can be converted by the half bridge cells but it does not change the location of the minimum at $\varphi_1 = 0$ and $\varphi_{2,3} = \pm\pi$.

3.3. Dependency on the modulation index

In this chapter it is analyzed how the maximum percentage of energy that can be converted by half bridge cells, depends on the modulation index m. The basis of the following investigation is an MMC configuration in which the amount of half bridge cells is $n_{HB}/n_0 = 1$. So, each full-bridge added allows an increase in m above 1. Furthermore the observations are performed for a load angle $\varphi = \varphi_1$. To determine the percentage of convertible energy, it is necessary to define the charging and discharging energy during a fundamental cycle of a total MMC arm and only of the half bridge cells. This is achieved by integrating (11) and (12) in sections. Thus, the charging and discharging energy of the half bridge cells can be calculated by

$$E_{HB,ch}(\varphi = \varphi_1) = \int_0^{\omega t_3} p_{HB}(\varphi = \varphi_1)d\omega t + \int_{\omega t_4}^{\omega t_1} p_{HB}(\varphi = \varphi_1)d\omega t + \int_{\omega t_2}^{2\pi} p_{HB}(\varphi = \varphi_1)d\omega t =$$

$$= \frac{n_{HB}}{n_0} \cdot \frac{V_{DC} \cdot \hat{I}_{AC}}{16} \cdot \left[\sqrt{(4-m^2)^3} + \frac{4}{m} \cdot \sqrt{(m^2-1)^3} \right], \tag{21}$$

$$E_{HB,dch}(\varphi = \varphi_1) = \int_{\omega t_3}^{\omega t_4} p_{HB}(\varphi = \varphi_1)d\omega t = -\frac{n_{HB}}{n_0} \cdot \frac{V_{DC} \cdot \hat{I}_{AC}}{16} \cdot \sqrt{(4-m^2)^3} \tag{22}$$

with $\quad \omega t_3 = \arcsin\left(\frac{m}{2}\cos\varphi\right) + \varphi \quad$ (23) $\quad$ and $\quad \omega t_4 = \pi - \arcsin\left(\frac{m}{2}\cos\varphi\right) + \varphi.$ (24)

ωt_3 and ωt_4 are the zero crossings of the arm current.

For the considered MMC configuration ($n_{HB}/n_0 = 1$), the maximum charging energy of the half bridge cells is equal to the charging energy of the total MMC arm:

$$E_{Arm,ch}(\varphi = \varphi_1) = E_{HB,ch}(\varphi = \varphi_1) \tag{25}$$

Due to energy-related requirements the discharging energy is equal to the negative charging energy of a total MMC arm. As shown in Fig. 5 the discharging energy is the limiting factor for the use of half bridge cells for the considered load angle $\varphi = \varphi_1$. Consequently the maximum percentage of energy that can be converted by half bridge cells is given by the convertible discharging energy

$$\frac{E_{HB,dch}(\varphi=\varphi_1)}{E_{Arm,dch}(\varphi=\varphi_1)} = \frac{n_{HB}}{n_0} \cdot \left(1 - \frac{\frac{4}{m} \cdot \sqrt{(m^2-1)^3}}{\sqrt{(4-m^2)^3} + \frac{4}{m}\sqrt{(m^2-1)^3}} \right) \tag{26}$$

This percentage is shown in Fig. 6 in dependency of the modulation index. For a modulation index $m \leq 1$ the total energy can be converted by half bridge cells. There is no negative arm voltage level and thus the half bridges can be used at any time. For higher modulation indices the use of half bridge cells is limited by the discharging energy and the period of time where the arm current level is negative, respectively. With higher modulation indices the negative arm current period decreases and the negative arm voltage period increases. Consequently the amount of energy which can be converted by half bridge cells decreases for higher modulation indices. For a modulation index $m \geq 2$, there is no negative arm current level (Eq. 6) and thus, half bridge cells cannot be used for the arm voltage modulation because it is not possible to discharge these cells. Fig. 6 shows that it is important to consider the energy relations and not only to install a certain amount of full bridge cells to handle the negative arm voltage level, when over-modulation is used.

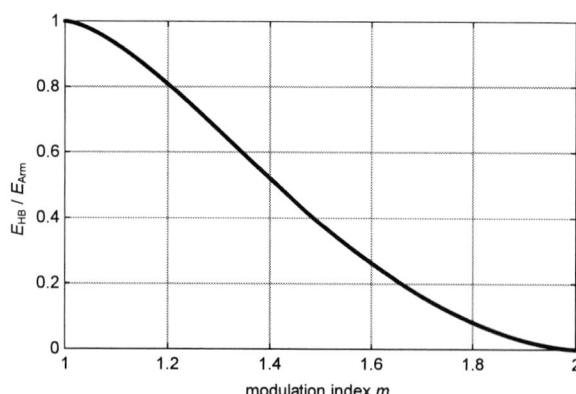

Fig. 6: Maximum percentage of energy that can be converted by half bridge cells in dependency of the modulation index m.

© VDE VERLAG GMBH · Berlin · Offenbach

4. Modulation Method

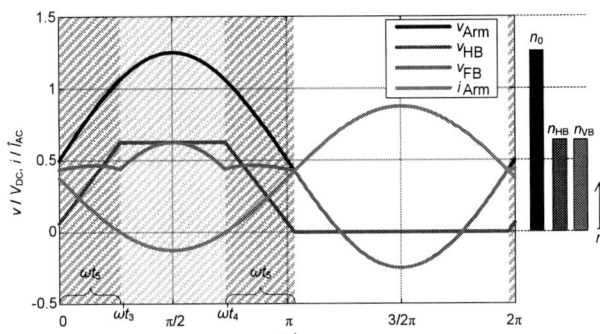

Fig. 7: Voltage profile of half bridge and full bridge cells for the presented modulation method.

The aim is to use as many half bridges as possible and as few full bridges as necessary for the operation. Half bridge cells offer lower losses and are cheaper than full bridge cells. In contrast, full bridges provide a blocking voltage during a DC fault and are able to modulate a negative voltage level. In the following, a possible modulation method is presented and will be discussed for an MMC which consists of 50% half bridge cells and 50% full bridge cells ($n_{HB}/n_0 = 0.5$) and a modulation index $m = 1.5$. The resulting voltage profiles of the different cell types are illustrated in Fig. 7. The proposed modulation method is subdivided into three sections. During the period of a negative current level (green shaded area) all half bridge cells are activated (modulation of a positive cell voltage) to ensure a maximum utilization of these cells. The energy $E_{HB,s1}$, which is converted by the half bridge cells in this section amounts

$$E_{HB,s1} = \int_{\omega t_3}^{\omega t_4} n_{HB} \cdot V_C \cdot i_{Arm} d\omega t = \frac{n_{HB} \cdot V_C \cdot \hat{I}_{AC}}{2} \cdot \left[m \cdot \left(\frac{\pi}{2} - \arcsin\left(\frac{m}{2} \right) \right) - 2\sqrt{1 - \frac{m^2}{4}} \right] \quad (27)$$

In the second section (red shaded area) the modulated voltage profile of the half bridge cells decreases to zero with a constant voltage gradient $\frac{dv_{HB}}{d\omega t} = \frac{-V_C \cdot n_{HB}}{\omega t_5}$. ωt_5 can be calculated numerically and is defined in such a way that the converted energy in the second section is equal to the converted energy in the first section. This second section is symmetrically attached to the first one and the formal condition is described by:

$$-\frac{1}{2} E_{HB,s1} = \int_{\omega t_3 - \omega t_5}^{\omega t_3} v_{HB} \cdot i_{Arm} d\omega t = \int_{\omega t_4}^{\omega t_4 + \omega t_5} v_{HB} \cdot i_{Arm} d\omega t \quad (28)$$

Finally the remaining arm voltage level v_{FB} is modulated only by the full bridge cells

$$v_{FB} = v_{Arm} - v_{HB} \quad (29)$$

In this way an energy balance can be ensured for both cell types and a minimum amount of full bridge cells has to be installed. Furthermore it should be noted that all half bridge cells are activated during the first section and in bypass mode after the second section. An active voltage modulation is only performed during the second section. This reduces the switching losses of the half bridge cells.

5. Conclusion

This paper investigates the combined use of half bridge and full bridge cells in an MMC in order to achieve an improved performance. The use of half bridge cells offer low losses and a low amount of semiconductors to be installed for a given voltage rating. Full bridge cells provide a DC fault blocking capability and allow an over-modulation to increase the power rating. A detailed energetic analysis was performed for the different cell types and limiting factors of this hybrid MMC topology are demonstrated. It is important to consider the energy relations and not only to install a certain amount of full bridge cells to handle the negative arm voltage level. Finally a possible modulation method was presented, which allows the combined use of the basic cell types for an active voltage modulation.

References

[1] R. Marquardt, "Modular Multilevel Converter: An universal concept for HVDC-Networks and DC-Bus-applications," *International Power Electronics Conference*, pp. 502-507, 2010.

[2] M. Perez, S. Bernet, J. Rodriguez, S. Lizana and R. Kouro, "Circuit topologies, modelling, control schemes and applications of modular multilevel converters," *IEEE Trans. Power Electron.*, vol. 30, no. 1, pp. 4-17, 2015.

[3] M. Malinowski, K. Gopakumar, J. Rodriguez and M. Pérez, "A survey on Cascaded Multilevel Inverters," *IEEE Trans. Ind. Electron.*, vol. 57. no. 7, pp. 2197-2206, 2010.

[4] M. Glinka and R. Marquardt, "A New AC/AC Multilevel Converter Family," *IEEE Trans. Ind. Electron.*, vol. 52, no. 3, pp. 662-669, 2005.

[5] S. Allebrod, R. Hamerski and R. Marquardt, "New transformerless scalable Modular Multilevel Converters for HVDC-transmission," *Power Electronics Specialists Conference*, pp. 174-179, 2008.

Selective HVDC Transmission Line Breaking for Bus Bar Applications under Reduced Expenses

Rene Sander; Karlsruhe Institute of Technology; Germany; rene.sander@kit.edu
Michael Suriyah; Karlsruhe Institute of Technology; Germany; michael.suriyah@kit.edu
Thomas Leibfried; Karlsruhe Institute of Technology; Germany; thomas.leibfried@kit.edu

Abstract

Comparable to AC systems, complex HVDC structures are likely to require more than one circuit breaker installation at particularly relevant nodes. A possible solution for bus bar applications at reduced expenses is presented here. Following the proposed principle by adding two diodes and one auxiliary switch per line, an unlimited number of connections can be protected with a single unipolar DC Circuit Breaker (CB) comparable to the plug-and-play principle. Thereby, a solution for cost-optimized protection without choice restrictions for HVDC CBs is made. The outlined benefits will be verified by simulations in MATLAB Simulink.

1. Introduction

Launching HVDC Multi-Terminal-System (MTS) integration as a step further towards low-loss power transmission, controllable load-flow and wide-ranging grid stabilizing services requires a functional protection system [1]. HVDC Circuit Breakers (CB) could offer an enhancement for reliable operation of MTS by isolating faulty parts. Conventional (AC) circuit breaker concepts are not suitable due to a lack of zero-current crossing and missed reaction time targets of less than five milliseconds. Therefore, a variety of those devices have been proposed recently [2]-[6], each solving the problem of DC current interruption on its own way. Those devices however, use high-priced components and in case of semiconductor-involved solutions, which apply for most proposals, require a high voltage DC hall. In general terms, HVDC CBs are likely to become a cost-intensive part within HVDC grids. However, the more complex DC transmission structures might grow, the more important cost-efficiency will become. This represents a contradiction since the necessity for protection grows with the system size as well.
However, most likely not every single line will require an individual HVDC CB. It is more probable that strategic nodes get protection during their stepwise interconnection with other MTS or simple extension by an additional converter. In that case, three or more lines are connected at one bus bar, which raises the question, whether unexploited optimization potential is left.

This work does not explicitly propose a new circuit breaker design, but offers an alternative to realize a bus bar protection system utilizing only one single HVDC CB with the aim of cost-reduction. This construction principle for transmission line interconnection offers the hybrid protection design as introduced in [2], which means a low loss auxiliary breaker and a high loss but even high power main breaker are integrated. This concept is taking advantage of the same basic principle with the difference of summarizing all individual currents, which are supposed to be interrupted. This bundling approach is realized again by the smaller breaking power of auxiliary breakers owned by each line and forcing the current into the common breaking path. The proposed bus bar protection method is presented and evaluated in the following.

© VDE VERLAG GMBH · Berlin · Offenbach

2. Design Proposal and (Short Circuit) Breaking Operation

As already indicated the proposal is bundling up currents for concentrated breaking. However, during normal operation all transmission lines are connected to bus bar N_0 as given in Fig. 1. Each line owns a current slope limiter, carried out by an inductive air-coil for example, and an Auxiliary Breaker (AB). The AB can be assumed as a series connection of a semiconductor-based load commutation switch and a fast disconnector similar to the design in [2]. From their nodes K_1 to K_N two connections to nodes N_1 and N_2 are created via diodes. It can be stated that in contrary to [7], a rectification design for unipolar switching is chosen. Thereby an arbitrary number of transmission lines can be connected to these nodes N_0, N_1 and N_2 allowing even subsequent installations via the bus connector shown in Fig. 1. The Main Breaker (MB) then again connects N_1 and N_2 with each other and necessary energy absorbers are set in parallel with an optional grounding. The MB can be of any design suitable for HVDC circuit breaking. However, due to fixed current direction a unipolar concept seems obviously preferable due to circuit simplification issues and costs.

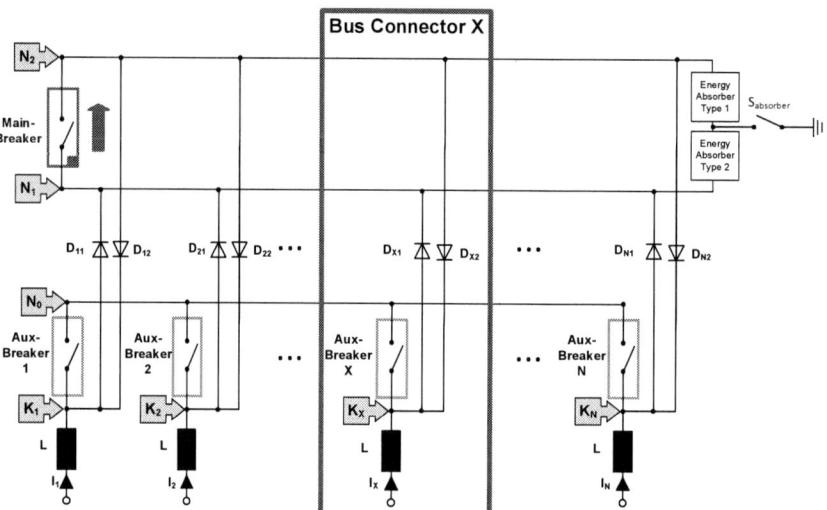

Fig. 1: Bus Bar Breaker Design Proposal with N lines protected by one single main breaker and optional grounding of related energy absorbers

During normal operation stage 1 is active, see Fig. 2. As soon as a breaking command is given to the bus bar protection system, its MB is closed and those AB associated with lines to be separated open. In stage 2 all relevant line currents have been fully commutated into the MB and transient voltage withstand capability of the AB is recovering. Afterwards final clearing is initiated by opening the MB. A post-breaking phase occurs and the corresponding currents are reduced to zero (Stage 3). In the end, only load currents are observable and the breaking operation is finished as shown in Stage 4.

The described process is suitable for any combination or number of lines to be interrupted as long as the ABs work properly and the MB is capable of interrupting the concentrated current. Furthermore, load currents of healthy lines and their related power transfer remain almost unaffected by the bus bar breaking operation since low-ohmic conduction can be ensured among them at all time, see blue marked load current at stages 1 - 4. Despite of that, comparable to single line protection transient switching voltage phenomena are generally observable and must be considered for system insulation coordination and converter control.

Since a separated line still remains part of the Bus Bar Breaker, this line has to be completely disconnected before a second breaking operation can be initiated. Otherwise it would be connected to the live system as soon as the MB recloses. Vice versa, if one or more separated lines shall be connected to the system, those outer disconnectors need to close and afterwards the MB and the AB work reversely until load current commutation is finished.

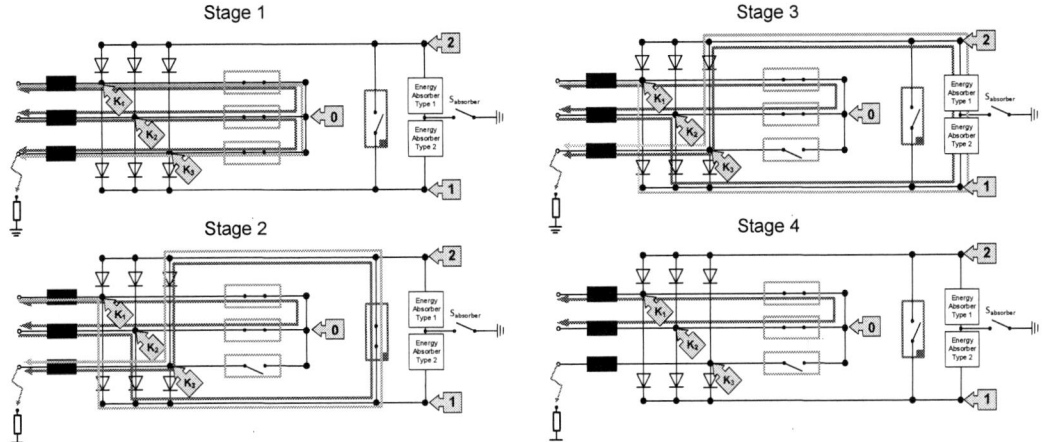

Fig. 2: Breaking Operation for a short circuited line connected via an inductor to node K₃

3. General Characteristics

All transmission lines are coupled to the Bus Bar Breaker by individual connectors. Its associated AB enables low conduction losses during normal operation, but also requires sufficient commutation voltage to force a short circuit current into the breaking path within an acceptable time period. Basic design evaluations for such a device have been developed in [8], whereas an awaiting increase of parasitic inductance due to expected spatial expansion may require further dimensioning and timing analysis for the load commutation switch.

The MB obviously has to cope with full transient switching voltages and superimposed currents during a breaking operation. Since a bus bar breaking operation is decoupled from individual line breaking in the first place, dimensioning of the MB and related turn-off ratings do not clearly indicate the Bus Bar Breaker's maximum turn-off capability. As shown in Figure 3, if the individual breaker is not feasible of interrupting the fault due to a lower turn-off capability, redundant breakers might still handle the situation since its fault current shares are obviously lower than the overall fault current. However, as long as technical and economic reasons do not guide into other directions, safety demands are assumingly required to offer a suitable performance for each line breaker to handle the most severe fault solely. This does not answer the question for a Bus Bar Breaker, which will be the most challenging situation to deal with. As long as only the faulty line requires an interruption operation, breaking power of the Bus Bar Breaker's MB remains at the degree of an individual line breaker. Otherwise, if all lines are supposed to be interrupted within a short circuit situation, the maximum current turn-off capability is given by summarizing all incoming or outgoing currents according to

$$I_{max} = \sum_{x=0}^{n} I_{x+} = \sum_{x=0}^{m} |I_{x-}| \quad \text{with } I_{x+} > 0, I_{x-} < 0 \text{ and n + m = N.} \quad (1)$$

Since short circuits in HVDC systems are characterized by steep current slopes, appropriate limitations are necessary. Considered inductors result in an impedance network and an equivalent overall inductance, whereas a multiple-line fault with half the total number of connections can be assessed to be the most severe one. The minimum current limiting inductance is then determined by

$$L_{min} = min\left(\frac{L}{n}+\frac{L}{m}\right) = L \cdot min\left(\frac{m+n}{m \cdot n}\right) = \begin{cases} L \cdot \dfrac{4}{N} & \text{for even N} \\[2mm] L \cdot \dfrac{4N}{N^2-1} & \text{for uneven N} \end{cases}. \tag{2}$$

Reliability is a technical benchmark for HVDC CB design deliberations, which seems to be lower for this bus bar concept than for individual line breakers. This is mainly indicated by the cascaded tripping order in Figure 3. A breaker failure is countervailed by the second level protection and entire line blocking. However, specific oversizing within a centralized component such as the MB has a lower impact on costs than spread improvements on every single device and thereby offers a direct opportunity for reliability enhancement by overall robustness increase and resulting (individual) error rate reduction. Therefore fundamental statements are rarely possible and application-oriented analysis then again require assured information about specific component lifetime and costs.

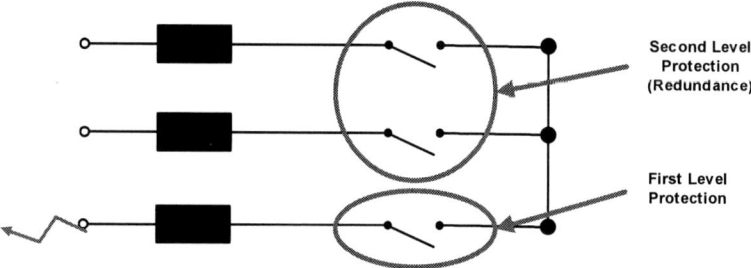

Figure 3: Redundancy for single line breakers

4. Functional Verification

Verification is carried out by simulations in MATLAB Simulink. Therefore a simplified model with four interconnected 500 kV converters in a back-to-back layout was built according to Fig. 4 and components from Fig. 5. Furthermore a parasitic inductance of L_{par} = 30 µH series-connected to the MB is considered.

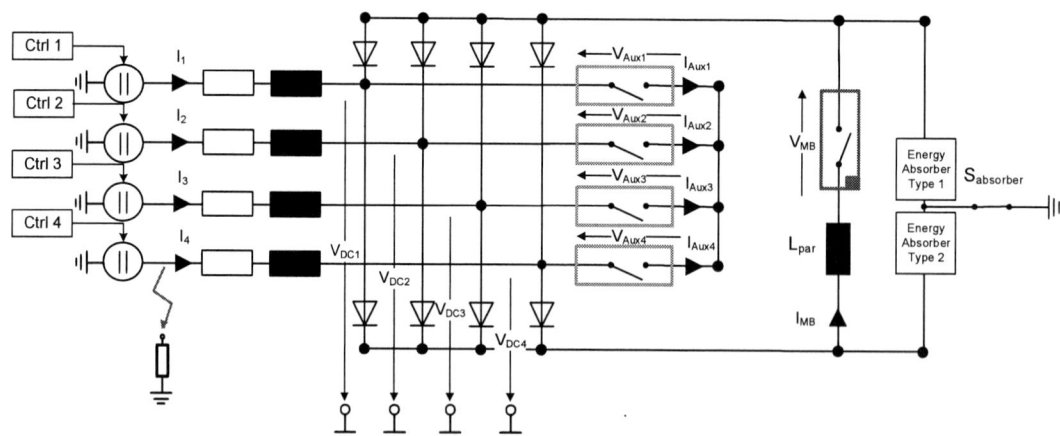

Fig. 4: Back-to-back layout with controlled voltage sources (scenario 1+2: all V-const / scenario 3: Line 2 V-const, Line 1+3+4 I-const)

PCIM Europe 2016, 10 – 12 May 2016, Nuremberg, Germany

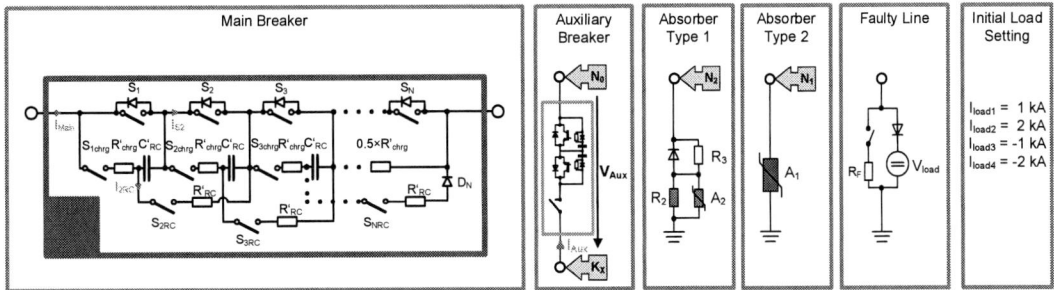

Fig. 5: Component definition for exemplary Bus Bar Breaker (from left to right: Main Breaker with three switching modules [6], Auxiliary Breaker [2], Absorber Type 1 [6] and Absorber Type 2 [6])

The proposed Bus Bar Breaker construction concept shall be investigated by three scenarios representing typical breaking operations:

1. short circuit interruption at line 4
2. load current interruption at lines 2 and 4
3. load connection at line 4

Fig. 6 shows scenario 1 with a low-resistive fault at line 4. The AB is successfully commutating the short circuit current into the MB, which is carrying the current until voltage recovery of the fast disconnector is completed. Thereby the parasitic inductance L_{par} is responsible for the commutation voltage peak in V_{aux4} given in Fig. 6d. After a certain safety delay breaking is initiated with a following current decrease period determined by integrated energy absorbers. Related voltage transients are over as soon as fault current shares have fully degraded and a new load current set point is reached. Due to grounding of energy absorbers, current decrease of the faulted system part is independently realized as given in Fig. 6a. Thereby the insulation system is exposed to minimum negative voltage stresses, which is proven in Fig. 6c.

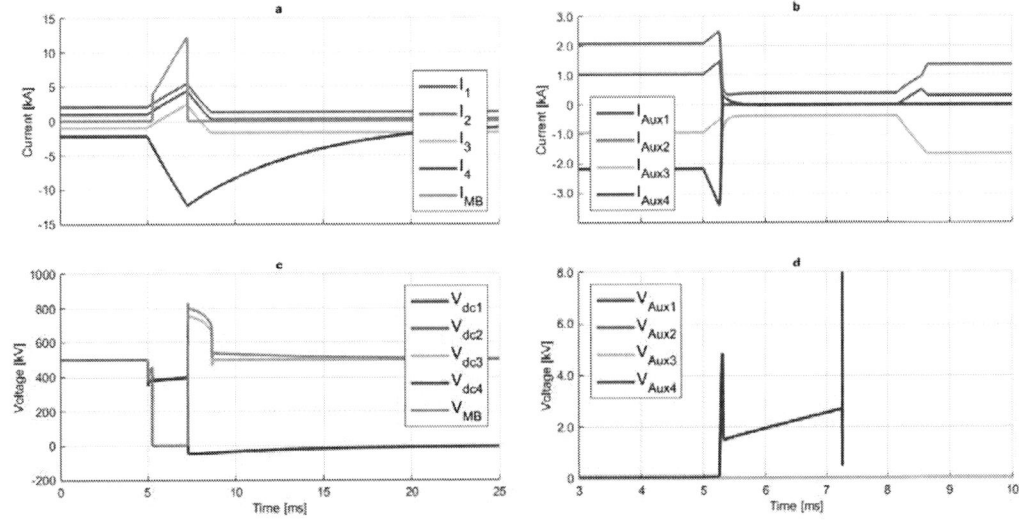

Fig. 6: Current and voltage curves during commutation into Main Breaker (b and d) as well as breaking (a and c) for interruption of faulted line 4 without superfluous disturbance on remaining lines (N=4)

In Fig. 7a a load shedding situation by successful interruption of lines 2 and 4 is given. While relevant ABs break their current shares, the MB is taking them as described in Eq. (1). Final breaking causes transient voltages, whereas in this particular situation with almost equivalent load situations for I_2 and I_4 other lines remain unaffected. This is due to the fact, that hardly

© VDE VERLAG GMBH · Berlin · Offenbach

367

any currents are shared between lines remaining connected and lines supposed to be interrupted. This is observable in Fig. 7d by voltage curves from AB 2 and AB 4 as well.

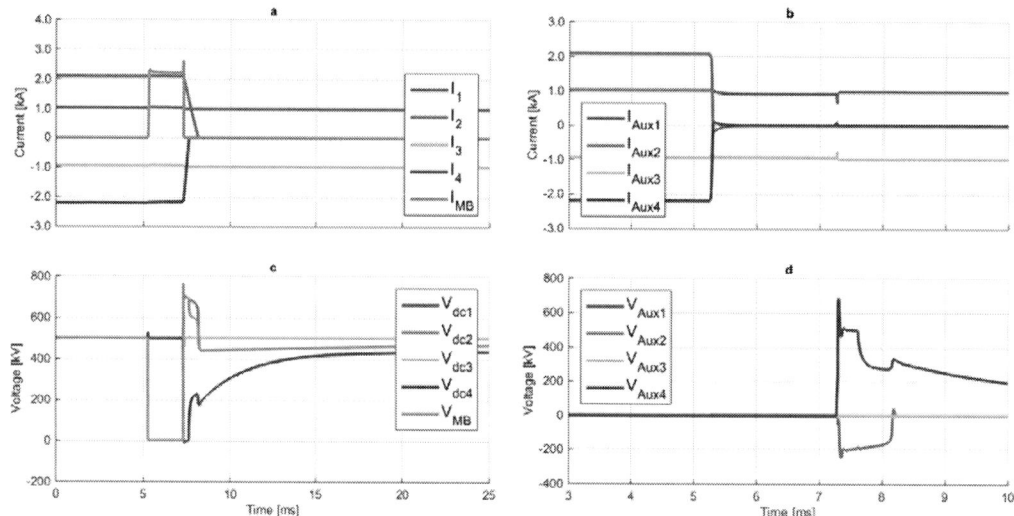

Fig. 7: Current and voltage curves during commutation into Main Breaker (b and d) as well as breaking (a and c) for separation of line 2 and line 4 (N=4)

The last scenario is modelled with current-controlled voltage sources at lines 1, 3 and 4 (line 2 can be assumed a dc voltage controlled converter), whereas for better visualization a voltage drop of 10 kV was initially added to line 4. Simulation starts in a preset load situation with line 4 in idle mode. In order to connect this line with the other three active system parts, the MB is closed after 5 ms. A negative current slope driven by the inherent current controller is observable in Fig. 8a. After the fast disconnector has contacted its metal plates a snubber capacitor discharge current in AB 4, as shown in Fig. 8b, occurs. After a safety delay the related load commutation switch closes and load current commutates into the AB. The MB can be opened again and thereby the connection process is terminated.

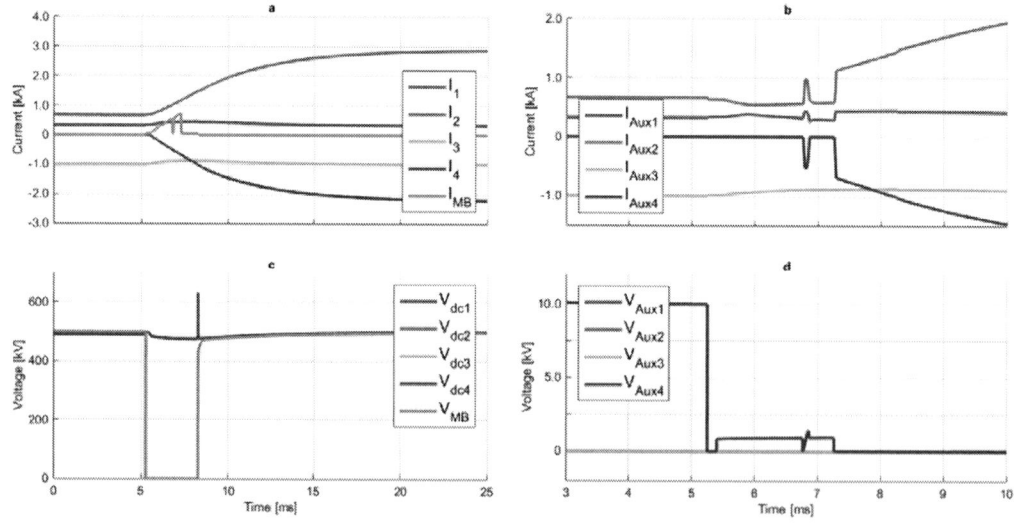

Fig. 8: Current and voltage curves during connection process of line 4

5. Economic Considerations

Overall lifetime expectancy, operational and investment costs directly influence economic impact. Regarding a power-loss-optimized operation, the Bus Bar Breaker with its load commutation switches and air-coils is not discriminated towards other individual line breaker concepts in hybrid design. Quite contrary to that, maintenance of centralized components such as MB and energy absorbers rather offers additional cost saving opportunities.

Furthermore, implemented diodes are characterized by typically low design complexity, high component lifetime and moderate costs per switching power. The implementation of two diodes, which is relevant for each node connection, hereby clearly presumes economic benefits against the alternative investment of a complete MB with equivalent switching power.

Thinking of HVDC grids in urban areas with rare transmission line routing opportunities, it cannot be ruled out that high power backbones occur. At such nodes superpositioning load currents could grow beyond state-of-the-art values if two, three or even more converters are feeding into one load area - or one transmission line respectively. This consequently demands for remarkably high breaking capability, which would apply for Bus Bar Breakers just as for individual line breakers protecting such a critical node. Bearing in mind that only unipolar stresses are applied to the device and the number of breakers is reduced to one single device, a comprehensive cost saving aspect can generally be imputed.

6. Conclusion

A novel concept for innovative circuit breaking at bus bars is presented. During normal operation, a low resistive conduction path enables cost-optimized current flow. For isolation of a selectable line or even a combination of more lines, its current is commutated into the Main Breaker and interrupted. The faulty line is disconnected by first opening the corresponding Auxiliary Breaker and thereby commutating the fault current shares of the remaining healthy lines into diodes and afterwards breaking with the unipolar Main Breaker. Finally, the fault current is decreased by jointly utilized energy absorbers. Load flow remains unaffected at all time. Furthermore only one Auxiliary Breaker and two additional diodes are necessary to protect each transmission line connected to the protective node, which most likely offers a considerable cost saving opportunity. Its general functionality is discussed and verified by simulations with MATLAB Simulink.

Centralized bus bar breaking calls for higher robustness of involved components than common solutions already do, but also lead to the very basic question how regulatory N-1 criterion and related error rate reduction in matters of redundant switching devices and redundancies within the device itself might be interpreted and solved for a prospective HVDC grid. This issue should be discussed in greater detail since the perspective of HVDC transmission systems as a single black box within an AC system will inevitably be insufficient as soon as it comes to MTS.

Regarding reliability an allegedly lower but more cost-effective safety level can be chosen by introducing a Bus Bar Breaker featuring full functional range. Last but not least, this solution is not limited to the field of HVDC. It can be implemented in low and medium voltage DC power distribution systems for realization of a sophisticated protection just as for HVDC. Thereby, DC protection in general is ready to move away from individual line breakers heading towards selective node interrupters with plug-and-play extensibility.

7. References

[1] C. M. Franck, "HVDC circuit breakers: A review identifying future research needs", IEEE Transactions on Power Delivery, vol. 26, no. 2, pp. 998–1007, 2011.

[2] J. Häfner & B. Jacobson, "Proactive hybrid HVDC breakers – a key innovation for reliable HVDC grids", Cigré Symposiums 2011, Bologna, 2011.

[3] Y. Wang & R. Marquardt, "Future HVDC-grids employing modular multilevel converters and hybrid DC-breakers", 15th European Conference on Power Electronics and Applications (EPE), 2013, pp. 1–8.

[4] W. Grieshaber; J.-P. Dupraz; D.-L. Penache & L. Violleau, "Development and test of a 120kV direct current circuit breaker", Cigré Paris, 2014

[5] T. Heinz; V. Hinrichsen; S. Kosse; J. Teichmann & E. Taylor, "Direct Current Interruption by a Current Zero Impulse of Constant Steepness", 19th International Symposium on High Voltage Engineering (ISH 2015), 2015

[6] R. Sander; M. Suriyah & T. Leibfried, "A Novel Current-Injection Based Design for HVDC Circuit Breakers", PCIM Europe 2015, 2015

[7] D. Ergin & H.-J. Knaak, "Apparatus for Switching Direct Currents in Branches of a DC Voltage Grid Node", Patent WO 2014/117813 A1, 2014

[8] A. Hassanpoor; J. Häfner & B. Jacobson, "Technical Assessment of Load Commutation Switch in Hybrid HVDC Breaker", IPEC 2014 Hiroshima, 2014

PCIM Europe 2016, 10 – 12 May 2016, Nuremberg, Germany

Decentralized Controller for Modular Multilevel Converter

Seleme Isaac Seleme Jr.[1], seleme@cpdee.ufmg.br
Luc-André Grégoire[2], luc.andre.gregoire@gmail.com
Marc Cousineau[2], marc.cousineau@laplace.univ-tlse.fr
Philippe Ladoux[2], philippe.ladoux@laplace.univ-tlse.fr

[1]Universidade Federal de Minas Gerais, Department of Electronical Engineering, Minas Gerais, Brasil

[2] Université de Toulouse – INPT, UPS, CNRS LAPLACE (Laboratoire Plasma et Conversion d'Energie, ENSEEIHT - 2, rue Charles Camichel – BP 7122, F-31071 Toulouse Cedex 7 – France

Abstract

In this paper a new decentralized control for modular multilevel converteris proposed. Capacitor voltagebalancing is achieved using only the voltage of the preceding and following submodules. Using the proposed method removes the need to sort capacitor voltages, and makes the complexity of the controller independent of the number of submodules. Furthermore, the decentralized control uses self-interleaving carriers, allowing to remove or to replace defective submodules without affecting the main control. Versatility of the proposed method is demonstrated using real-time simulation and experimental setup.

1. Introduction

Modular multilevel converter (MMC) topology has been growing in interest over the past decade not only for high voltage direct current (HVDC) application [1-3], but also for medium voltage or drive application [4]. Like in other multilevel topologies, levels are used to reduce both total voltage harmonic distortion (THD) [5] and stress on each semiconductors. Unlike Flying Capacitor or Neutral Point Clampedmultilevel converters, the total HVDC bus voltage is divided equally across the submodule(SM)capacitors.

Increasing the number of SM reduces the voltage stress but increases the

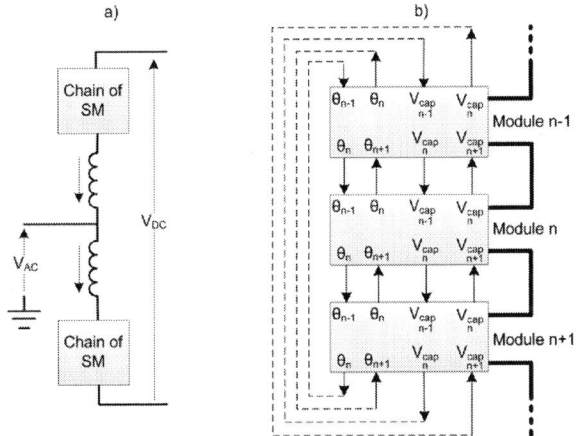

Fig.1. a) single phase MMC b) chain of submodules with decentralized control signals and power signals (where θ_i is the phase of the local carrier, $V_{cap_i}$is the voltage of the local capacitor).

number of capacitor voltagesto regulate. When dealing with a large number of SM, controller gets more complex, and large number of signals must be transferred to the main controller. In this paper, voltage regulation for each SM is decentralized[6, 7], which greatly reduces the complexity of the main control. Furthermore, in classical MMC controller, sorting algorithm must be used in order to balance capacitor voltages[8], which requires a lot of computation power. This is the main challenge of this topology as time required to compute proper switching partern increases as the number of SM increases. Fig. 1. a) shows a single phase MMC where a chain of N SM is represented by a single block. Fig. 1. b) shows the different signals that needs to be shared between SM in order to achieve decentralization of the controller. Using decentralized control makes the main control independent to the number of SM used, therefore SM can be dynamically added or removed upon fault or faillure.

© VDE VERLAG GMBH · Berlin · Offenbach

The paper is divided as follow. Section 2 present the proposed controllerand its implementation. Performance is demonstrated using real-time simulation and experimental results in section 3. Finally, conclusions are drawn and discussed in section 4.

2. Proposed controller

The controller is divided in two modules; the master control and the decentralized control. The master control generates modulation index for the SM to follow a current reference. The decentralized control regulates capacitor voltage of each SM by adding a correction to the modulation index from the master control.

2.1. Master control

The main objective of the master control is to regulate the current in the upper and lower arms of the converter. This is achieved using predictive control to generate a voltage reference for each arms, as shown in fig. 2.

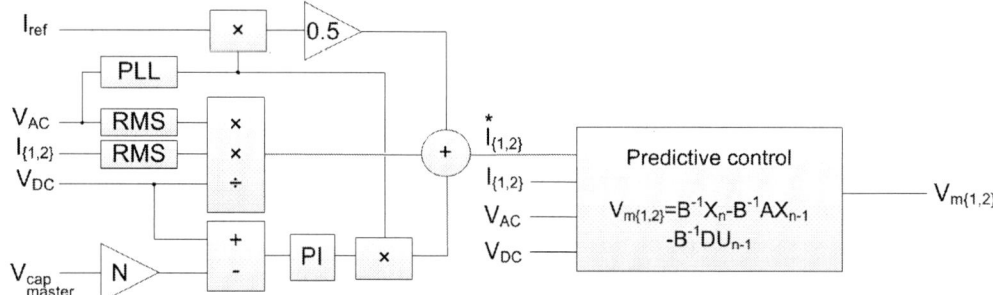

Fig. 2. Block diagram of main controller.

In fig. 2., voltage references (V_{m1} and V_{m2}) are obtained from a predictive control. The predictive control is concieved from the discrete time state-space model of the converter shown in (1), with a horizon length 1.

$$\overbrace{\begin{bmatrix} i_1(n) \\ i_2(n) \end{bmatrix}}^{I_{\{1,2\}}(n)} = A \begin{bmatrix} i_1(n-1) \\ i_2(n-1) \end{bmatrix} + B \overbrace{\begin{bmatrix} v_{m1}(n) \\ v_{m2}(n) \end{bmatrix}}^{V_{m\{1,2\}}(n)} + D \begin{bmatrix} V_{AC}(n) \\ V_{DC}(n) \end{bmatrix} \tag{1}$$

In (1), $I_{\{1,2\}}(n)$ are the future value of the arm currents, which can also be considered as the desired reference current. Modulation index required to follow the current reference is obtained by isolating $V_{m\{1,2\}}$in (1), giving the equation of the predictive control shown in fig. 2. Reference current is obtained from the different measurement of the circuit.I_{ref} is the amplitude of the desired current on the AC side. V_{AC} is the voltage measurement on the AC side. Unitary power factor is achieved by synchronizing I_{ref} with V_{AC} using a PLL. In steady-state, power transferred from the DC side to the AC side must be maintained, the required DC current reference is obtained from the RMS value of V_{AC}, $I_{\{1,2\}}$ and V_{DC}. Finally, as it will be later explained, one of the SM plays the role of master, and regulation of its capacitor voltage is obtained using a PI controller and considering V_{DC} and N, the number of active SM. The output of the PI controller is then also multiplied by the output of the PLL; to ensure an optimal energy transfer to the capacitor. Once $V_{m\{1,2\}}$ are obtained, they are used as modulation index. V_{m1} is used for the SM in the upper arm of the converter and V_{m2} for the SM of the lower arm of the converter. Decentralized control will modify the modulation index of each SM ensuring proper capacitor voltage regulation.

2.2. Decentralized control

The innovation of the proposed algorithm is the decentralization of the controller for the submodules capacitor voltage regulation. In classical method, the sum of every submodules capacitor voltage is required to determine the amount of energy to inject in the MMC. In the proposed method, capacitor voltage of only oneSM of each arm is used to determine the required energy; this SM is then refered to as the master. Every other SM, or slave, will regulate their capacitor voltages using the average of the capacitor voltages of their two immediate neighbors, as shown in fig. 3.

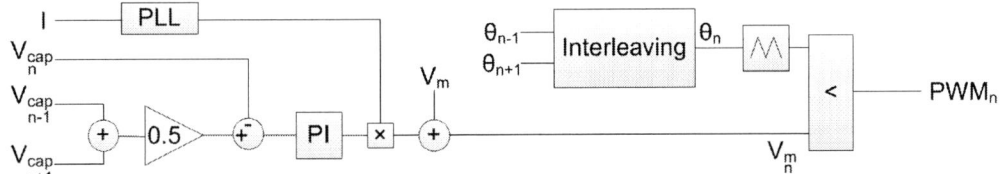

Fig. 3.Block diagram of decentralize controller for slave submodule.

The average of $V_{cap\ n-1}$ and $V_{cap\ n+1}$ is computed, and it is used as reference for $V_{cap\ n}$. The modulation index of each SM is corrected using the output of the PI, which is synchronized using the current in the arm and a PLL.The corrected modulation is then compared to a carrier. Phase of the carrier for each SM is obtained so it is interleaved between its two neighbors SM. The method to achieve self-alignment of the carrier is discussed in [9, 10].

Using this controller reduces the number of signals exchanged with the master control and therefore reduces communication delay. In the case where hundreds of modules areused, classical control requires acquisition by the master control of every SM. In the proposed method, regardless of the number of SM, the master control always needs the same number of signals. In the case of a three-phase converter, the master control only needs 16 measurements. Each SM would then receives capacitor voltage and phase of the carrier of its two immediate neighbors, and a common modulation index from the master control. They would also received a signal indicating whether they are active or not. This signal allows to dynamically remove SM while the converter is active, without modifying the controller. The correct operationof the control is demonstrated using real-time simulation and a 3kVA prototype.

3. Simulation and experimental results

The newly proposed method is tested under various conditions. Tests were made throughreal-time simulation first and then were validated using a prototype. During real-time simulation the simulated controller and the simulated converter were connected together using physical input and output. This allows a more realistic simulation where communication delays are considered. Once the real-time simulation is working, the simulated converter is replaced by the real one, keeping the exact same controller. This drastically reduces the required integration time in the laboratory[11]. For

Table 1 Simulation parameters

Power rating	3 kW
DC voltage	600 V
AC voltage	200 V_{rms}
Arm inductance	5 mH
SM capacitor	2 mF
Load resistance	33 ohms
Load inductance	5 mH
Number of SM per arm	9
Sampling time	12 μs

every case presented in this section, MMC is used as an inverter feeding an RL load. Model parameters are found in table 1.

3.1. Steady-State

This test shows the converter working in steady-state. The AC current reference is set to 8 A, and DC voltage is set to 600V. Fig. 4shows real-time simulation and experimental results for the different currentsin the converter.

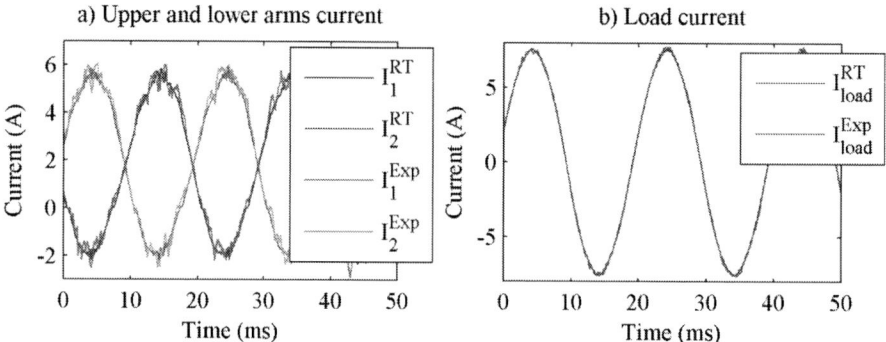

Fig. 4.Real-time simulation and experimental results a) for arms current and b) load current

Notice that subscript RT refers to real-time simulation results and Exp refers to experimental results. Load and arms current from the real-time simulation and the experimental results are superimposed, and relative error is less than 2%. The main difference observed is the noise seen on the experimental setup. Fig. 5shows the voltage reference for the master SM in real-time simulation and on experimental setup.

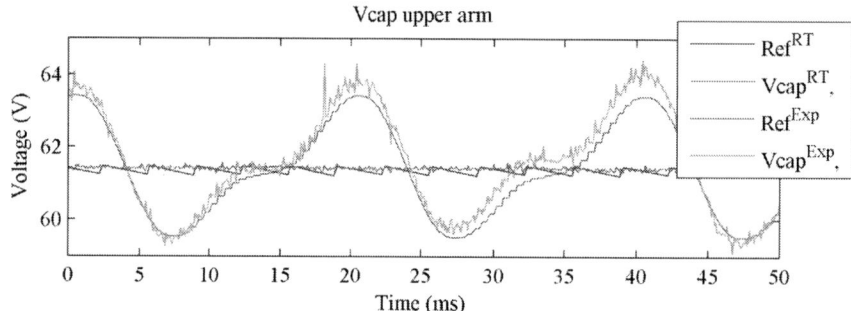

Fig. 5.SM capacitor voltage of the master for real-time simulation and experimental setup.

The capacitor voltage follows the reference with an ondulation of 1%. Relative error between the simulation and experimentation is again, at worst, 2%. Decentralized control keeps the remaining capacitor voltages grouped as shown in fig. 6. a) for the real-time simulation model and fig. 6. b) for the experimental setup.

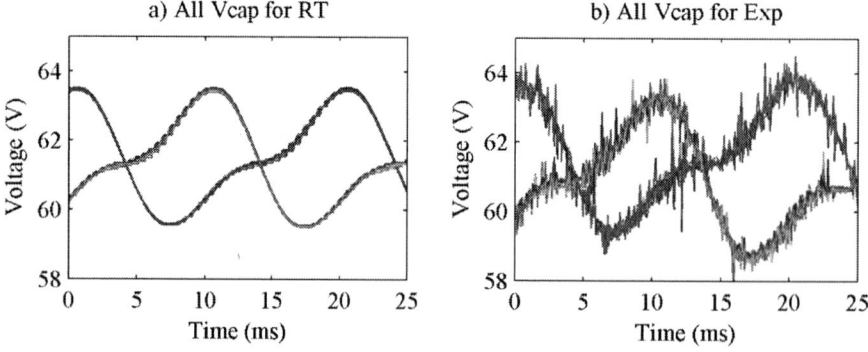

Fig. 6.Capacitor voltage for all SM a) real-time simulation b) experimental setup.

© VDE VERLAG GMBH · Berlin · Offenbach

PCIM Europe 2016, 10 – 12 May 2016, Nuremberg, Germany

Real-time simulation and experimental results are superimposed and relative error is less than 3% for every measurement. Current and voltage references are closely followed, and SM capacitor voltages are well balanced, demonstrating the proposed control effectiveness. Control behaviour during reconfiguration is demonstrated in the next section.

3.2. Converter reconfiguration

This test shows the converter behaviour when SM are removed and reinserted. SM can be added or removed if they are damaged. In classical control, it requires to modify part of the control, such as carrier or gain for PI controller.

3.2.1 Removing SM

After reaching steady-state, one SM is removed from the lower arm. There is one less SM which requires a new voltage reference. Fig. 7shows the reference voltage, capacitor voltage for the master and the deactivated SM.

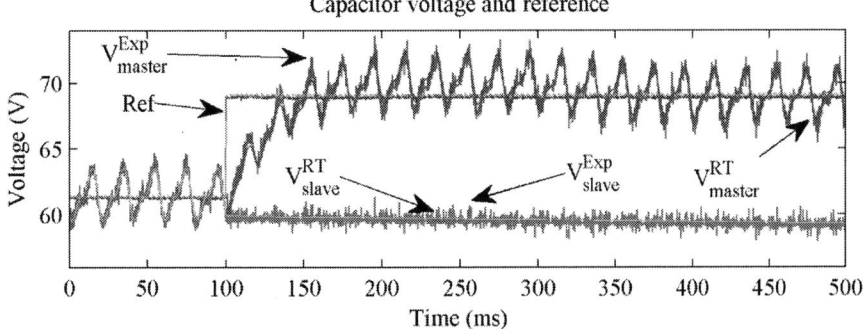

Fig. 7.SM capacitor voltage of the master for real-time simulation and experimental setup.

After being removed, the SM is slowly discharging, while the master jumps to follow the new reference. Voltage reference for the SM from the upper arm remains the same while the one from the lower arm increases. Capacitor voltage remained regulated as shown in fig. 8.

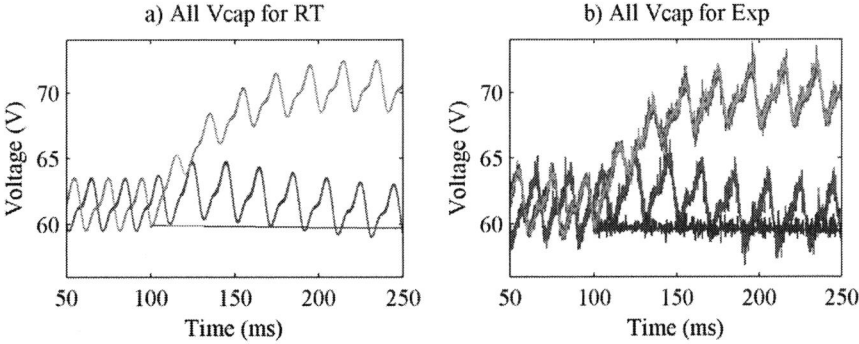

Fig. 8.Capacitor voltage for all SM a) real-time simulation b) experimental setup when SM is removed.

When a SM is removed, energy stored in the converter needs to be increased. This can be observed on the different currents in fig. 9.

© VDE VERLAG GMBH · Berlin · Offenbach

375

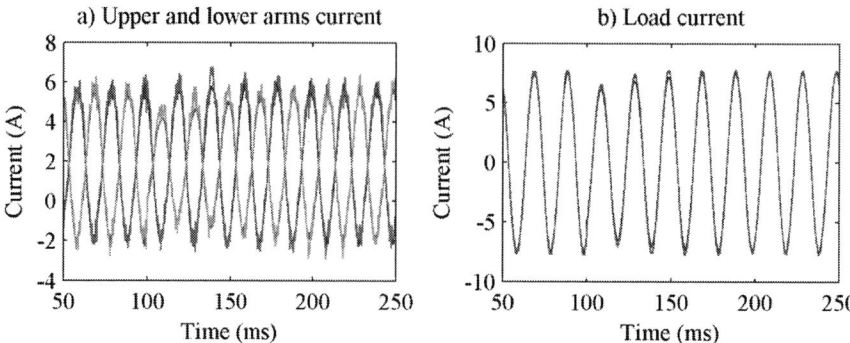

Fig. 9.Real-time simulation and experimental results a) for arms current and b) load current when SM is removed.

During transition, the DC component of the current reference remains the same while the AC component for the lower arm is reduced. The upper arm current is not affected by removing a SM in the lower arm.

3.2.2 Inserting SM

SM can also be inserted just as easily. Fig. 10shows the reference voltage, capacitor voltage for the master and the inserted SM.

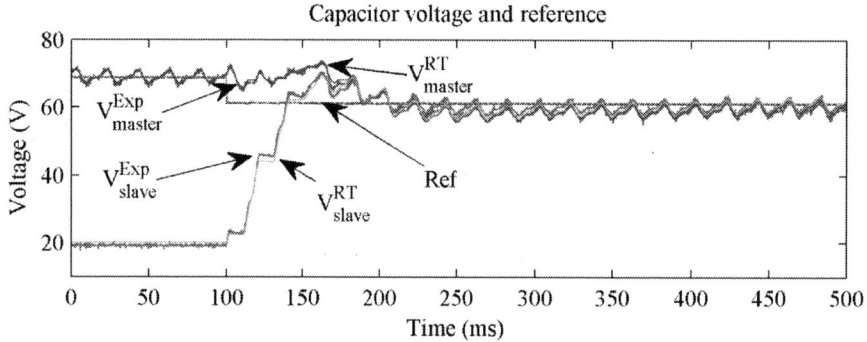

Fig. 10.SM capacitor voltage of the master for real-time simulation and experimental setup.

When a SM is inserted, a new reference is computedand the capacitor voltage is regulated in 5 cycles, since a slow regulation dynamic is expected. The remaining capacitor voltages from the lower arm reach their reference in the same amount of time. The SM from the upper arm remains unaffected. This is observed for the simulation and the experimental results in fig. 11.

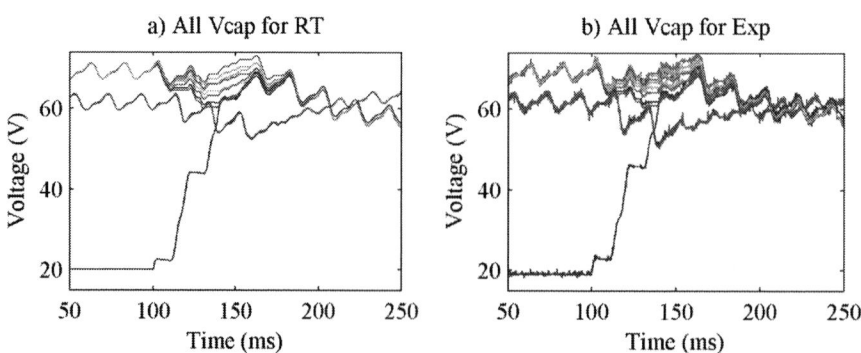

Fig. 11.Capacitor voltage for all SM a) real-time simulation b) experimental setup when SM is added.

Inserting a SM has also an impact on the current as that of removing one, as shown in fig.

© VDE VERLAG GMBH · Berlin · Offenbach

12.

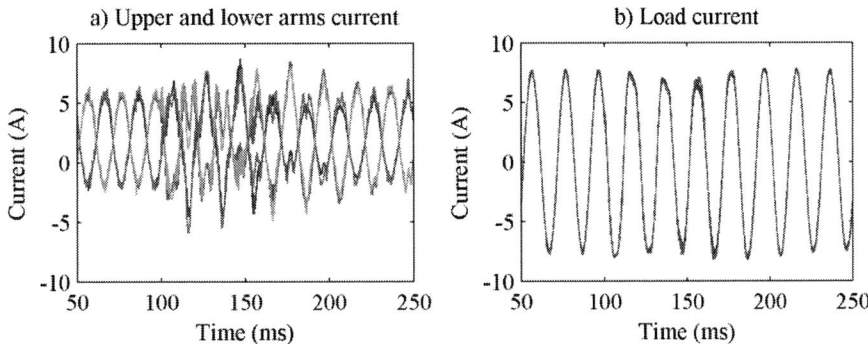

Fig. 12. Real-time simulation and experimental results a) for arms current and b) load current when SM is added.

When removing a SM, part of the energy stored in the converter is removed. To return to the same amount of stored energy, the current in the converter is increased. When a SM is inserted, the energy stored in the converter needs to be redistributed to the inserted SM, and therefore, an increase in the arm current is observed. But this hardly affects the load current; there is only a slight distortion for a few cycles.

4. Conclusion

In this paper, a new decentralize control method is proposed. The method allows to simplify the controller, since only one submodule in each arm needs to be regulated. The remaining submodules regulate their capacitor value to the one of their neighbors. Using the newly proposed algorithm, submodules can be removed an inserted without modifying any controller. Using real-time simulation results has allowed to reduce development time. Results obtained in real-time simulation are superimposed with the ones obtained with the experimental setup. Future work are focussed on expending the control to a three-phase converter.

Aknowledgements

This work has been supported by the Brazilian agency CAPES Foundation, granting a scholarship Proc. No.: 99999.002619/2015-06, to the first author.

References

[1] L.-A. Gregoire, W. Li, J. Belanger, and L. Snider, "Validation of a 60-Level Modular Multilevel Converter Model-Overview of Offline and Real-Time HIL Testing and Results," *Proc. of IPST,* pp. 14-17, 2011.

[2] W. Li, L.-A. Gregoire, and J. Bélanger, "Control and performance of a modular multilevel converter system," *CIGRÉ Canada, Conference on Power Systems, Halifax,* 2011.

[3] W. Li, L.-A. Grégoire, and J. Bélanger, "Modeling and control of a full-bridge modular multilevel STATCOM," in *Power and Energy Society General Meeting, 2012 IEEE,* 2012, pp. 1-7.

[4] M. Hagiwara, K. Nishimura, and H. Akagi, "A medium-voltage motor drive with a modular multilevel PWM inverter," *Power electronics, IEEE Transactions on,* vol. 25, pp. 1786-1799, 2010.

[5] J. Arrillaga, Y. H. Liu, and N. R. Watson, *Flexible power transmission: the HVDC options*: John Wiley & Sons, 2007.

© VDE VERLAG GMBH · Berlin · Offenbach

[6]   M. Cousineau and B. Cougo, "Interleaved converter with massive parallelization of high frequency GaN switching-cells using decentralized modular analog controller," in *Energy Conversion Congress and Exposition (ECCE), 2015 IEEE*, 2015, pp. 4343-4350.

[7]   M. Cousineau, P. Ladoux, and N. Serbia, "Circuit de conversion statique d'énergie électrique à architecture cascadée et à contrôle modulaire décentralisé," French Patent Pub. No. FR 15 60397 (INP Toulouse/CNRS), October2015.

[8]   A. Antonopoulos, L. Ängquist, and H.-P. Nee, "On dynamics and voltage control of the modular multilevel converter," in *Power Electronics and Applications, 2009. EPE'09. 13th European Conference on*, 2009, pp. 1-10.

[9]   M. Cousineau, M. Le Bolloch, N. Bouhalli, E. Sarraute, and T. Meynard, "Triangular carrier self-alignment using modular approach for interleaved converter control," in *Power Electronics and Applications (EPE 2011), Proceedings of the 2011-14th European Conference on*, 2011, pp. 1-10.

[10]   M. Cousineau and Z. Xiao, "Fully masterless control of parallel converter," in *Power Electronics and Applications (EPE), 2013 15th European Conference on*, 2013, pp. 1-10.

[11]   L. Gregoire, K. Al-Haddad, and G. Nanjundaiah, "Hardware-in-the-Loop (HIL) to reduce the development cost of power electronic converters," in *Power Electronics (IICPE), 2010 India International Conference on*, 2011, pp. 1-6.

Power Converters for PV Systems with Energy Storage: Optimal Power Flow Control for EV's Charging Infrastructures

Umberto Abronzini, University of Cassino and Southern Lazio, Italy, u.abronzini@unicas.it
Ciro Attaianese, University of Cassino and Southern Lazio, Italy, attaianese@unicas.it
Matilde D'Arpino, University of Cassino and Southern Lazio, Italy, m.darpino@unicas.it
Mauro Di Monaco, University of Cassino and Southern Lazio, Italy, m.dimonaco@unicas.it
Giuseppe Tomasso, University of Cassino and Southern Lazio, Italy, tomasso@unicas.it

Abstract

This paper presents an optimal power flow control for a micro-grid composed with EV's charging infrastructures, a PV system and a energy storage. The objective function of the power flow control aims to minimize the daily operating cost of micro-grid by means of a stochastic management of the power sources, which allows reducing the impact of EV's charging process on the main utility grid as well. Vehicle to Grid (V2G) is also considered in the power flow management.

1. Introduction

Large-scale deployment electro-mobility infrastructures can be guaranteed only if a proper integration in the main electric network system is achieved [1–3]. Also, the well known EV's *well to wheel* approach requires that renewable energy sources are used to supply directly or indirectly electric vehicles, in order to achieve a real "zero-impact mobility". A proper integration of renewable energy sources can guarantee high penetration of electric vehicles without strong reinforcements in the distribution network. The shift of the EV charging load to off-peak period and PV maximum generation phase appears a viable solution in the short term to address this issue. However, this could not meet the needs and behavior of EVs owners, thus, other solutions have been investigated. The integration of stationary Energy Storage Systems (ESSs) can considerably reduces the impact of EVs on the main grid. In fact, they can guarantee a decoupling between the charging infrastructure and main grid in terms of energy but, also, of power demand if ESS is used as power boost to achieve fast and ultra fast charging. Also, the V2G can guarantee a further distributed storage system for the smart-grid. Thus, micro-grid based on PV, ESS and V2G represents a valid configuration for guaranteeing the e-mobility penetration at zero-impact, but a proper power flow control technique has to be implemented to guarantee the optimal use fo the available energy sources [4–6]. Moreover, an optimal sizing of both PV system and ESS has to be performed in order to achieve a feasible return of investment (ROI), taking into account also the life cycle reduction of the ESSs. In [7], authors have already presented an optimal sizing and power flow control of multi-source power converter system for EV charging infrastructure. A deterministic optimization problem has been solved to obtain the system design with the minimum ROI. Moreover, the set point of the daily power flow control has been carried out with the aim to minimize the cost of the charging considering a definite scenario (electricity price, EVs charging profile, costs of system units, degradation cost of ESS and production of renewable source, unidirectional power flow for EV). In this paper, an improved power flow control based on a stochastic approach is proposed and the impact of the V2G operating mode on the micro-grid has been evaluated too. In detail, the V2G mode provisionally allows storing the vehicle energy during the V2G operation, in order to achieve power/energy time shifting to charge other vehicles connected to the infrastructures during peak energy cost periods. Of course, both technical and global cost constraints have to be taken into account, in order to guarantee the feasibility of the overall process. In particular, the balance between the economical revenue of the charging infrastructure and the provided service of the vehicle to the grid has been integrated in the power flow management, including also the life cycle reduction of the vehicle on-board storage unit. Therefore, the optimization function at the basis of the proposed power flow control allows feeding a group of EVSEs (Electric Vehicle Supply Equipments) connected to the micro-grid by optimizing the power flow coming from the main electric network, the PV system, the local ESS and other EVs connected to the grid in V2G mode (not necessarily at the same time) to reduce the impact on main grid and minimize the cost of charging process as well.

2. Smart micro-grids for charging electric vehicles

The proposed charging micro-grid, in its most general configuration, is based on a multi-source power conversion system, as shown in fig. 1. The power balance at the dc-link of the power conversion system is given by:

$$P_{DC,g} + P_{DC,PV} + P_{DC,BP} + P_{DC,SC} = P_{DC,L_{AC}} + P_{DC,L_{DC}} \tag{1}$$

where $P_{DC,g}$, $P_{DC,PV}$, $P_{DC,BP}$, $P_{DC,SC}$ are respectively the power of grid, PV field, battery pack (BP) and supercapacitors (SC) bank. On the vehicle side, $P_{DC,L_{AC}}$ and $P_{DC,L_{DC}}$ represents the EVSEs load power AC and DC, respectively. In case of the latter ones, a bidirectional power flow is considered to perform V2G. SC allows increasing the dynamics response of the power conversion system avoiding an overcharge either of the grid or of the electrochemical BP during load variations. BP can be used to store the energy when a surplus of PV production, lower price of grid electricity or V2G occur. AC/DC conversion unit can also be used to charge the BP directly from the grid. The stored energy can be successively used during high electricity price periods to minimize the cost of recharging process. An optimization of the power flows within a time frame has been developed with a stochastic approach, as shown in the following section. In particular, power reference of available power sources are chosen on the basis of an optimal power flow management.

3. Optimal power flow control for the charging micro-grid

The optimization problem is based on the evaluation of the EVSE's hourly operating cost c_e. With reference to fig. 1, the operating cost corresponding to the h-th hour is achieved by the sum of: cost of the energy drawn from the grid $E_g[h]$, BP degradation cost $c_{deg}[h]$ (cost linked to the charging-discharging cycle and Depth of Discharge) and revenue of EV owners in V2G mode ($c_{V2G}[h]$). It yields:

$$c_e[h] = c_g[h]E_g[h] + c_{deg}[h] + c_{V2G}[h] \tag{2}$$

where $c_g[h]$ is the electricity price per hour. The power to be supplied by each each source during the EV's charging process is carried out with the aim to minimize the daily operating cost C_e:

$$min \ C_e = \sum_{h=1}^{24} c_e[h] \tag{3}$$

© VDE VERLAG GMBH · Berlin · Offenbach

PCIM Europe 2016, 10 – 12 May 2016, Nuremberg, Germany

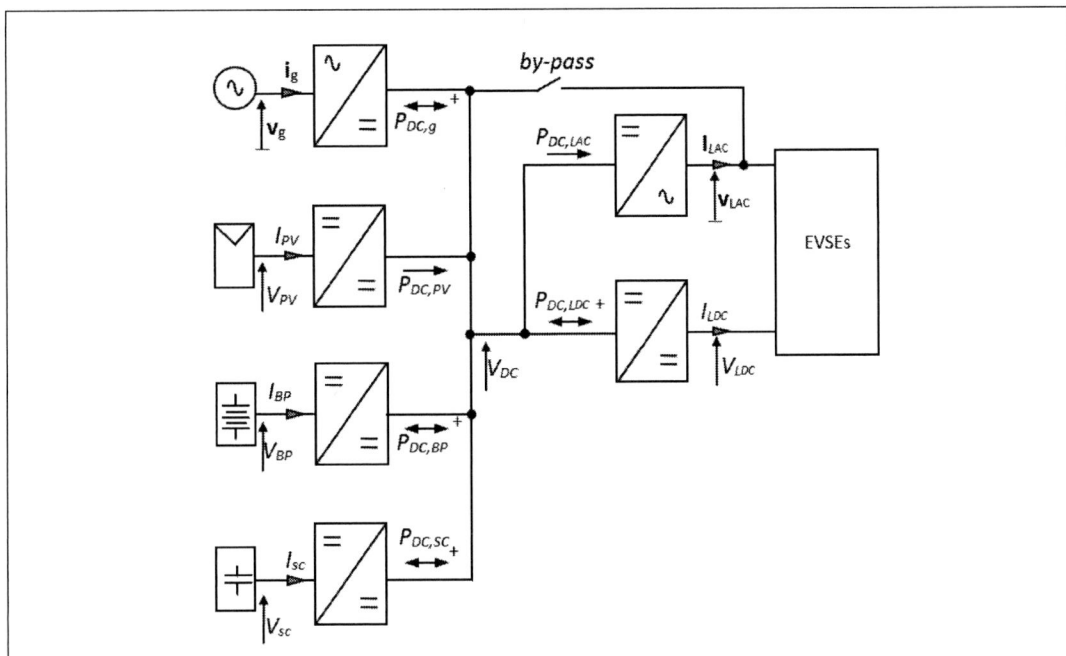

Fig. 1: Structure of the micro-grid for electric vehicles charging.

subject to (1) and:

$$P_s^m \leq P_s \leq P_s^M \qquad\qquad \text{with } s = g, PV \quad (4)$$

$$P_{BP,i}^m \leq P_{BP,i}[h] \leq P_{BP,i}^M \qquad\qquad \text{with } i = c, d \quad (5)$$

$$P_{V2G}^{m,k} \leq P_{V2G} \leq P_{V2G}^{M,k} \qquad\qquad \text{for } k\text{-th EV} \quad (6)$$

$$P_{PV}[h] \leq P_{mpp}[h] \qquad\qquad\qquad (7)$$

$$SOC_{[h]} = SOC_{[h-1]} + \left(\frac{-P_{BP,d}[h]}{\eta_d E_{BP}} + \frac{\eta_c P_{BP,c}[h]}{E_{BP}} \right) \Delta t \qquad\qquad (8)$$

$$SOC^m \leq SOC[h] \leq SOC^M \qquad\qquad\qquad (9)$$

$$SOC[1] = SOC[24] = SOC_{in} \qquad\qquad\qquad (10)$$

$$\Delta SOC_{V2G} \leq \Delta SOC_{V2G}^M \qquad\qquad\qquad (11)$$

The constraints in terms of the maximum power P^M and the minimum power P^m of the sources are reported from (4) to (5). Different minimum and maximum power are considered for the charging (c) and the discharging (d) process of BP. Furthermore, power constraints are set for the each k-th EV that works in V2G mode. Eq. (7) limits the produced power from the PV field

© VDE VERLAG GMBH · Berlin · Offenbach

up to the one of the maximum power point (P_{mpp}). Eq. (8) shows how the state of charge at h-th hour ($SOC[h]$) is function of the SOC in the previous hour, power and efficiency in charge or discharge mode and BP capacity. Moreover, the state of charge has to be kept within the range SOC^m-SOC^M, as shown in eq. (9), in order to ensure the life cycle performances of the BP. Eq. (10) guarantees the energy balance on the BP during each day, the SOC at end of each day (hour 24) must be equal at initial $SOC[1]$. The constraint of (11) ensures the energy coming from of EV in V2G is always below the maximum amount that is make available by EV. The optimization problem has been solved using four optimization loops. The procedure starts from the C_e achievable using just the main grid and PV field to supply EV charging process. In the first loop a possible reduction of C_e is tried by using the BP to store the (eventual) surplus of energy coming from the PV system and re-using this stored energy during the electricity high peak cost periods. The use of renewable source is considered with the highest priority since its energy cost is almost zero. In this calculation, the cost increment due to degradation of BP is also considered [7]. If after the first optimization loop a cost reduction is achieved, C_e is updated, a new power flow scheduling is set and new optimization loop is started. In the second loop, a further reduction of C_e is tried by evaluating a possible storing in the BP of the grid energy during low price periods. This energy can be used during the peak-period (grid time-shifting). In the third optimization loop, a reduction of the operating cost by means of V2G operation is considered. To identify the best working condition, two operating modes are analyzed: 1) vehicle energy flowing directly to other vehicles connected at that time to the charging infrastructure; 2) vehicle energy provisionally stored in the BP in order to perform the time-shifting and use that energy in peak-period. In the last optimization loop the energy injection to the grid is evaluated, with its relative costs and benefits. Several input variables to optimization problem are considered: price of electricity, energy production of PV plant, efficiency and cost of the BP, PV field and power conversion system and a penetration scenario of EVs. In [7] the unit commitment problem has been solved with a deterministic approach; here a stochastic solution of the problem is proposed. In fact, some input variables are intrinsically aleatory such as the PV generation, the load (EV's charging) profiles, numberer of EVs in V2G and their energy availability. The Monte Carlo method [8] has been adopted for the stochastic analysis. In detail, pseudo-random samples of the aleatory inputs are generated from their probability density function (PDF) and used successively to solve in deterministic way the optimization problem. The PDFs of the output variables are evaluated from the set of values achieved by means of numerical simulations. In particular, the solver identifies by means of the stochastic solution of the unit commitment problem when the stationary BP, with its initial SOC, and the battery packs of EVs in V2G can be use in order to minimize the operating cost of the charging infrastructure. Starting from the initial SOC, average value of the distribution of the PV generation, of the load profile and of the participation of EVs in V2G mode, the management system of the charging infrastructure works out the day-ahead scheduling of the power reference of the sources of the smart micro-grid. Moreover, the optimization algorithm is run in a quasi-real-time operation to adapt the power flow in case of perturbations during the real operating conditions. In this way it is possible a fine tuning of the management system on the basis of the gap between the forecast analysis and the working conditions for achieving the minimum operating cost and the respect of (10).

4. Experimental implementation and results

The flexibility of proposed numerical platform allows testing the performance for different sizes of the charging micro-grid that can be considered for a given scenario of EV penetration. To

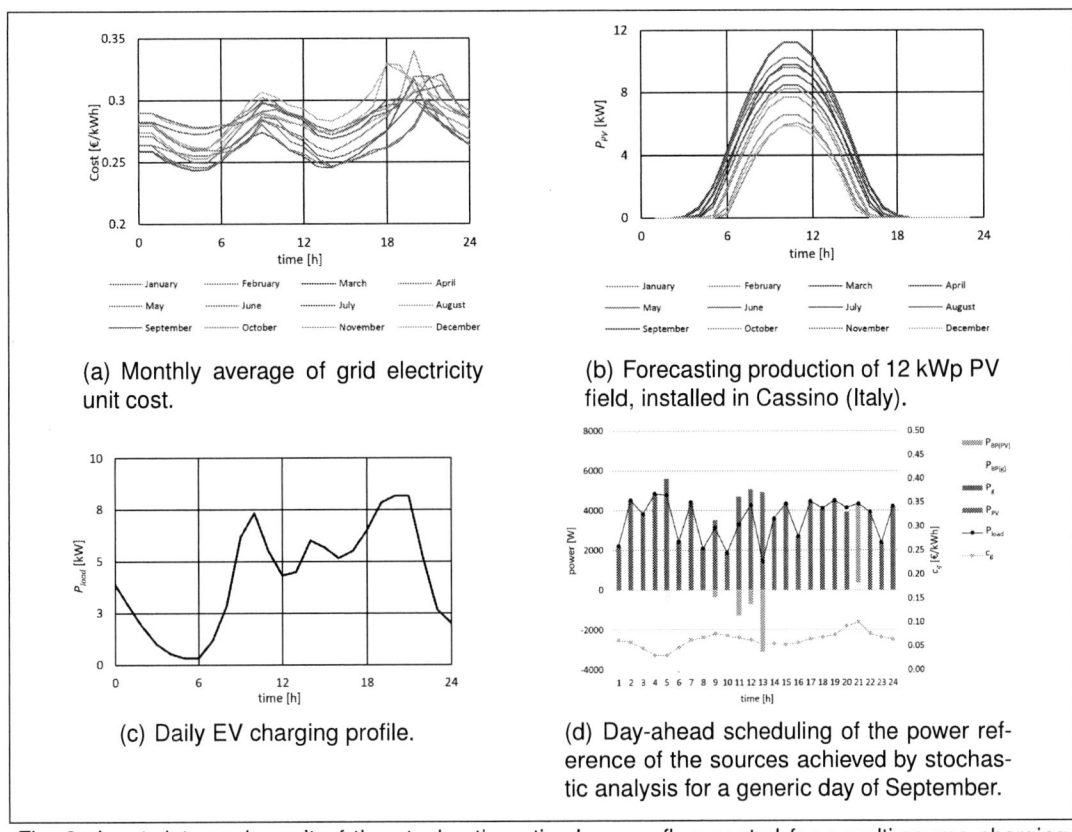

(a) Monthly average of grid electricity unit cost.

(b) Forecasting production of 12 kWp PV field, installed in Cassino (Italy).

(c) Daily EV charging profile.

(d) Day-ahead scheduling of the power reference of the sources achieved by stochastic analysis for a generic day of September.

Fig. 2: Input data and result of the stochastic optimal power flow control for a multi-source charging infrastructure.

validate the approach and performance of the optimal power flow management also from the experimental point of view, the small micro-grid of tab. 1 has been set-up. First, a stochastic analysis of the system has been performed by means of the simulation platform, which has been implemented by Matlab programming language. Figures 2(a) and 2(b) respectively show the electricity unit cost and the produced energy by 12 kWp PV field for a typical day of each month considered during the analysis. Fig. 2(c) shows a possible EV load curve in terms of average required power per hour. A lead-acid technology has been adopted with a capital cost equal to 0.10 €/Wh to implement the stationary BP. For EVs in V2G mode, a tariff for their electricity fed into the micro-grid has not been considered but just a degradation cost of their lithium-ion battery packs [7], considering a capital cost equal to 0.56 €/Wh. In the stochastic analysis of the optimization problem, Gaussian PDFs for each hour of the day have been defined for the PV production, for the load demand and for the presence of EVs in V2G mode. Starting from these PDFs, random samples of aleatory variables have been generated in order to obtain by Monte Carlo method the PDFs of output variables, which are needed to forecast the operating condition of the micro-grid. For the presented scenario, grid time-shifting and V2G modes have not been presented in the optimal solution of the unit commitment problem due to high degradation costs of battery packs. Thus, a reduction equal to 30% of their capital costs has been performed in order to try all operating modes as much as possible. However, also in this new scenario, not all the operating modes are presented in the stochastic solution calculated for the day-ahead. This is shown in fig. 2(d). In particular, the V2G modes are not convenient for the minimization of the daily operating cost. In the test condition, high

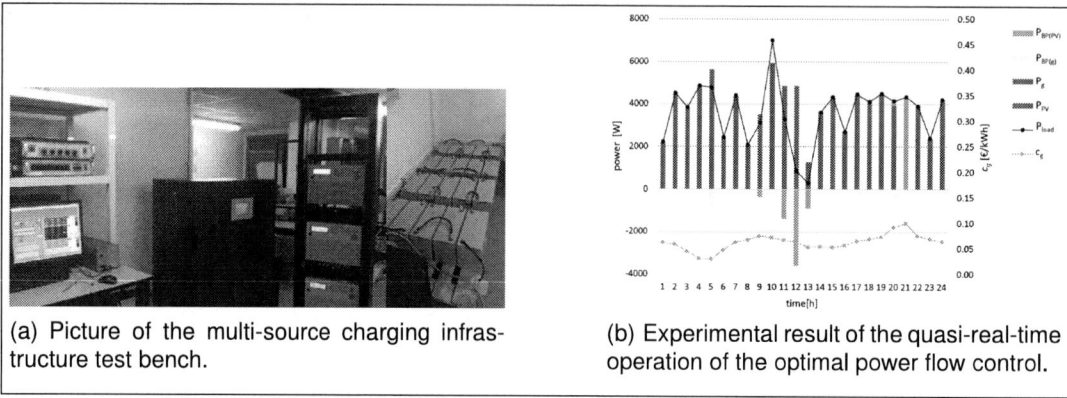

(a) Picture of the multi-source charging infrastructure test bench.

(b) Experimental result of the quasi-real-time operation of the optimal power flow control.

Fig. 3: Experimental implementation.

value of variance for the input variables has been considered. A prototype of the proposed system has been designed and implemented to validate quasi-real-time operation of the optimal power flow control as well. Fig. 3(a) shows a picture of multi source power converter, which is interfaced to the main grid (230 V, 50 Hz), PV field (12 kWp), stationary BP (20 lead-acid batteries FIAMM 12FGL120, 12 V 120 Ah) and three programmable electronic loads to emulate the EVs (CHROMA 63804 4.5 kW/45 A/350 V). The optimal power flow control has been implemented on industrial PC, which sends the power reference to control unit of multi-source power converter by means of Modbus-TCP/IP. The unit control of the converter includes the grid synchronization, the MPPT algorithm for PV source, the PQ control for the output power. The real time control of power flow during the current day (real time analysis) is reported in fig. 3(b). During the real time operation of the micro-grid a variations of PV production and load demand occur. Consequently, the optimal power flow control changes the power flow scheduling coming from previous stochastic analysis, to achieve the minimum operating cost. In particular, the purchased energy at 5 am with minimum price is used at 10 am (yellow bars), instead of 9 pm, in order to recover the PV energy production not directly used by loads. The light blue bars represent the PV energy stored into the BP. This energy is used to supply the charging process during high price period 8-9 pm.

5. Conclusion

A stochastic optimal power flow control for a multi-source charging infrastructure for electric vehicle with V2G operation is presented. The proposed control algorithm allows reducing the charging cost thanks to a suitable use of PV plant, ESSs and grid. Moreover, the impact of the charging power on the grid is minimized. A numerical simulation and an experimental test bench have been implemented to verify the performance of the optimal control in several scenarios.

6. References

[1] Alyona Zubaryeva and Christian Thiel. Paving the way to electrified road transport. *JRC Scientific and Policy Report*, 2013.

[2] Murat Yilmaz and Philip T Krein. Review of the impact of vehicle-to-grid technologies

P_g	$10kW$
P_{PV}	$12kW$
E_{BP}	$28.8kWh$
P_{load}	$10kW$

Tab. 1: Data of the charging micro-grid.

on distribution systems and utility interfaces. *Power Electronics, IEEE Transactions on*, 28(12):5673–5689, 2013.

[3] Miguel A Ortega-Vazquez. Optimal scheduling of electric vehicle charging and vehicle-to-grid services at household level including battery degradation and price uncertainty. *IET Generation, Transmission & Distribution*, 8(6):1007–1016, 2014.

[4] I Safak Bayram, George Michailidis, Michael Devetsikiotis, and Fabrizio Granelli. Electric power allocation in a network of fast charging stations. *Selected Areas in Communications, IEEE Journal on*, 31(7):1235–1246, 2013.

[5] Michail Vasiladiotis, Alfred Rufer, and Antoine Béguin. Modular converter architecture for medium voltage ultra fast ev charging stations: Global system considerations. In *Electric Vehicle Conference (IEVC), 2012 IEEE International*, pages 1–7. IEEE, 2012.

[6] Sanzhong Bai, Du Yu, and Srdjan Lukic. Optimum design of an ev/phev charging station with dc bus and storage system. In *IEEE 2010 Energy Conversion Congress and Exposition*, pages 1178–1184, 2010.

[7] U Abronzini, C Attaianese, M D'Arpino, M Di Monaco, A Genovese, G Pede, and G Tomasso. Multi-source power converter system for ev charging station with integrated ess. In *Research and Technologies for Society and Industry Leveraging a better tomorrow (RTSI), 2015 IEEE 1st International Forum on*, pages 427–432. IEEE, 2015.

[8] Athanasios Papoulis and S Unnikrishna Pillai. *Probability, random variables, and stochastic processes*. Tata McGraw-Hill Education, 2002.

Modular and Comfortable Electromobility

Dipl.-Ing. Mihai Dragan, M.Sc.; Dr.-Ing. Bernhard Budaker; Dr.-Ing. Jonathan Brix.
all: <name>.<surname>@ipa.fraunhofer.de
Fraunhofer Institute for Manufacturing Engineering and Automation IPA
Nobelstr. 12
70569 Stuttgart
Germany

The Power Point Presentation will be available after the conference.

Abstract

This paper gives a short overview about technologies that enable modular building blocks for electromobility applications. In the first section comfort is discussed as one of the key enablers for a broader distribution of electromobility. As a basic building block, a surface heating system based on carbon nanotubes (CNT) is presented. This surface heating system gives an example for a modular, easy to apply technology for electromobility. Modular building blocks for battery systems in the field of electromobility are presented in the second part of this paper. The current research state and the next working steps are presented in the final chapter.

Keywords: comfortable electromobility, mobility, surface heating, modular battery, powerline communication, intelligent cell.

1. Introduction

The last decade showed extensive research effort in developing new technologies and systems to reach the goal of high distribution of electromobility (e-mobility). The numbers of electric cars driving in German cities still seem to be very limited. In fact, when looking at the absolute figure like in [1], they are very limited in comparison to conventional cars [2]. As first step on the way to broaden the acceptance of electromobility, one has to analyze the reasons why potential customers still avoid using Battery Electric Vehicles (BEV) or even Hybrid Electrical Vehicles (HEV) during everyday life.

The most important topics when discussing e-mobility today are the affordability and the comfort when driving electric vehicles. The main challenges for enabling a comfortable way of using electric or general vehicles for individual mobility are:

- a good infrastructure that allows nearly unlimited individual mobility. In case of e-mobility this infrastructure includes the charging stations and the possibility of simple charging systems at home
- an easy and safe way of using the vehicle
- the possibility of using the vehicle under different climate conditions, in different surroundings, or generally under different boundary conditions
- an affordable price. This can be divided into cost when buying the vehicle and cost when using the vehicle, or generally the TOC: Total Cost of Ownership.

The air conditioning inside a BEV or HEV is very important, as this is the first impression for the user when driving an electric vehicle. Because of the very high energy efficiency of the electric drivetrain, not much energy remains for heating. On the other side, it would be even more efficient to heat or cool the battery system to an optimal temperature range. As for example presented by BMW in 2012 [3] different ways of active heating systems in the interior of electric cars have to be investigated.

Affordability is another very important aspect when discussing e-mobility today. One of the main things to be changed when discussing the practicability of e-mobility is the cost of the battery and the possible system reach [4]. The main reasons why electromobility has not yet found its way to the mass market are: system costs, modularity, and comfort. Therefore another focus in our research work is the battery and its power electronics.

2. Research on a system level

We at Fraunhofer IPA understand the importance of lowering system costs, developing modular systems for electric cars and also increasing the driving comfort when using electromobility. All these have to be considered from a systemic point of view. In the interdisciplinary research initiative "Fraunhofer Systemforschung Elektromobilität" (Fraunhofer System Research for Electromobility – FSEM I and II) our clear focus was on the manufacturing and infrastructure side. FSEM can be seen as one project divided in two parts. Part one was funded by the German federal government. Part II was the follow-up project where 16 institutes were pooling their expertise and were working in three clusters: "Drivetrain / Chassis", "Battery / Range Extender", and "Body / Infrastructure".

During the first project phase – FSEM I, from 2009 to 2011 – we developed among other projects a decentralized battery management system (DBMS). This battery management system enables a modular building block for electric and hybrid cars. Research efforts for the decentralized BMS were performed as part of a doctoral thesis [6] and led to the definition of the Intelligent Cell, described later in chapter 4.

During project part II we at Fraunhofer IPA focused more on the "comfort side" of e-mobility. We therefore developed a smart and fast surface heating system based on carbon nanotubes (CNT). The aim here was an easy to apply, comfortable, and fast heating system for passenger interiors.

3. Comfort and Electromobility

One of the most important factors why people use mobility systems like cars, trains or bicycles is comfort. In [5] it is discussed that the purchasing decision for or against electric vehicles is strongly dependent on the following factors: TOC, everyday suitability, reach, comfort and space. The first three properties are well addressed by a substantial amount of research projects all over the globe. However, comfort has never been discussed as one of the possible drivers for electromobility. During our research project we looked at possible technologies to bring more comfort into electric cars.

In FSEM II we developed a non-liquid heating system that can be integrated in an electric or hybrid vehicle. Fig. 1 shows the experimental electric heating system mounted in a test car. This system is manufactured using CNT technology. The system allows a very fast heating response time and is very easy to install. We used the CNTs as a paint coating applied on the inside of the car door.

Fig. 2 shows the heat measurement with sensors positioned at the door (T2 and T3) and at a reference position in the inside of the vehicle (T1 and T4). The x-axis represents the absolute time. The system was switched on at 18:30 on the day of testing, and subsequently switched off at 19:07. As shown in Fig. 2, the heating response of the system is very fast. Only 4 minutes after engaging the CNT heater (at 18:30) the full heating power can be measured at the surface of the door panel.

Fig. 1. CNT heating system integrated in a hybrid car.

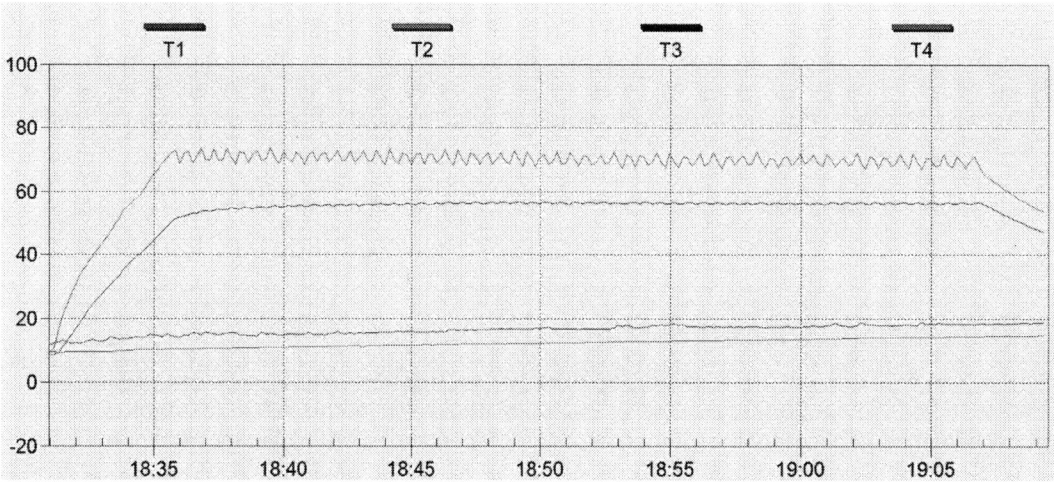

Fig. 2: Heat curve measured at different sensor positions.

4. Cost reduction and modularity

4.1. Developing the Intelligent Cell

The biggest part of the purchase price of electrified vehicles is represented by the traction battery and its subsequent lifetime and exchange costs. Also, currently available battery management systems require extended wiring efforts to communicate with every cell.

In an effort to improve both the TOC and range of these batteries, a concept of an Intelligent Cell as part of a modular structure down to a single energy cell has been developed at Fraunhofer IPA. It features a distributed BMS (DBMS) using powerline communication (PLC).

A prototypic setup has also been developed. It runs a decentralized communication and arbitration protocol over powerline, using the existing electrical wiring of the battery.

Each cell is fitted with "intelligence" consisting in a processing unit and circuit coupling components, allowing communicating freely on the PLC network using a proprietary protocol that enables data transfer without a master-slave connection. Temperature, voltage and current measurement are performed directly on the cell, directly connected to the microcontroller (intelligence).

Cell performance can thus be monitored using the additional intelligence. Parameters such as state of charge, health, or functioning (SoC, SoH, SoF) can be calculated, additional functioning history data can be saved or broadcasted, and a concept for an energy balancing system across the cells is also proposed. Safe functioning is granted by temperature monitoring and providing protection circuits, allowing to decouple defective cells from the battery itself. The Intelligent Cell prototype is shown in Fig. 3.

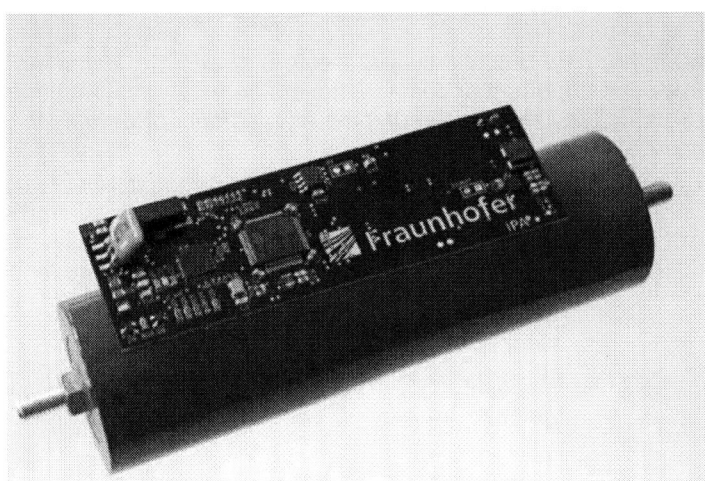

Fig. 3. Prototype of the Intelligent Cell. (image © Fraunhofer IPA)

Trials with the functional demonstrator have shown that battery cells can communicate autonomously and without a master unit on the powerline network. Laboratory trials and simulations of real-life driving patterns proved that the concept is functional, with a possible range improvement in the single-digit percentile area, and demonstrating that all cells of a traction battery can be monitored independently and – in case of low performance or damage – could be isolated from the battery pack [6].

Most electric vehicle manufacturers are achieving ranges between 120 and 400 km [8] and higher values are expected as improved cell designs appear in market products. The average distance covered in urban driving cycles is, on the other hand, between 40 and 60 km per day. This translates into a large share of daily commuters that would be pleased with a maximum range of 60 km for small electric cars, leading to a considerable price decrease for the traction battery of those vehicles. In this case the DBMS with Intelligent Cells could really offer a competitive advantage. Scalable, flexible and modular car batteries offer clear logistic, environmental and financial advantages for both the manufacturers and the customers, as shown in Fig. 4.

Currently, an obsolete traction battery is exchanged as a whole component after it ceases to meet certain functional parameters. This means that potentially good cells from the battery are also discarded at this point. Individual SoH monitoring can offer, for example, early warnings for defective or obsolete cells, allowing the DBMS to apply an alternative exploiting strategy, even removing defective cells from the array via designated one-way fusible circuits, and marking them for exchange at the next scheduled car inspection.

Fig. 4. Fixed sized batteries (left) are less practical than modular ones (right). These can be subject to targeted maintenance, optimized charging and are safer to use. (image © Fraunhofer IPA)

4.2. Improving the concept

Comfortable e-mobility also means that new technologies are safe and easy to implement. High-frequency (HF) compatibility is one important aspect that covers both of these attributes and is of high importance in the automotive field, considering the high amount of devices that communicate within the vehicle.

Further research now focuses on determining the HF behavior and compatibility of the developed PLC solution, resulting in a HF characterization of the PLC channel. A single automotive lithium-ion cell, and subsequently a serial connection of 10 cells were investigated using impedance measurements. Based on the obtained knowledge, a circuit based model of a battery cell has been developed, allowing the simulation of a traction battery. The model also takes the battery casing (or can), and the coupling values between the battery connectors into account.

Transfer channel analysis allowed for a behavior estimation of larger batteries. The simulation model was also used to determine different coupling methods to transmit and receive signals using PLC.

Additional conclusions were drawn regarding the dependence between the resonance frequency (f_R), cell separation and housing size, resulting in the graphic shown in Fig. 5. This is a useful representation when designing a new battery with known cells, and for choosing the appropriate separator medium in order to retain HF limitations.

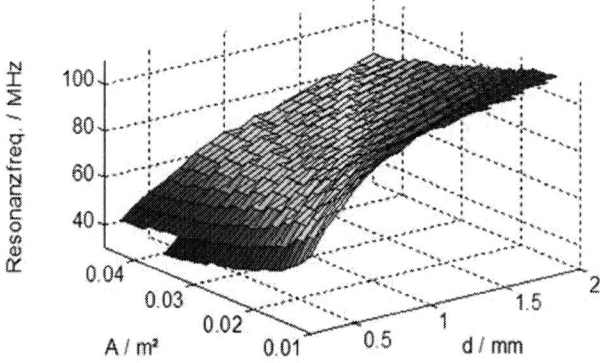

Fig. 5. f_R depending on housing size (A) and separation (d). (image © Fraunhofer IPA)

Further aspects of using PLC in a 48 V traction battery are currently being developed and improved within the EU-financed project 3Ccar that aims to gain complexity control of electrified cars by studying, improving, and integrating their components [7]. The BMS used for this project has a master-slave configuration, resulting in an alternative communication solution, but still based on the same principles. Fraunhofer IPA currently works on providing a project-specific PLC modem using an existing PSOC architecture for the master unit and the slaves (cells or cell arrays with integrated electronics). A different aspect will consist in adapting the communication channel simulation to the specification of the 3Ccar project and providing results that are relevant here.

4.3 Advantages and disadvantages

Taking it one step further, the Intelligent Cell aims to follow the success story of introducing the standard 18650 lithium-ion cell. This has led to rapid price drops of notebook accumulators. Prior to that, every manufacturer provided their own battery cells for their portable computers. By implementing a standard, the battery cells of the notebook battery became interchangeable. The market was stimulated, which finally helped make laptops affordable for everybody.

By proposing a standard intelligent energy cell with specific dimensions and functions, battery compatibility across all manufacturers of electrified vehicles could be established. Additionally, the quality of every standardized cell could be evaluated and would improve over time due to competition. Furthermore, newcomers could also enter the market by offering their product according to established standards.

Another advantage is the early identification of viable cells for their Second and Third Life. Battery cells for the automotive industry must meet higher standards than the ones for other applications such as stationary energy storage. Production batches with the highest quality results are therefore used for electric vehicles in their so called First Life. Still viable cells from discarded traction batteries are often reused in other less demanding scenarios. Based on its internal diagnosis, an Intelligent Cell could be singled out for future usage after it has been removed from an automobile. Besides the obvious environmental contribution, this also increases its residual value and thus effectively lowers purchasing costs for car owners.

The currently proposed discrete electronic circuit added to each battery cell is obviously unacceptable for any battery manufacturer standard. There is need to integrate this "cell intelligence" in a small and cheap microchip (ASIC or similar). Further investigations on HF interference and an increased PLC transfer rate are also required.

A further discussion point could also be about the granularity level of such "smart" batteries. Considerations must be made on whether it should be all the way down to a single cell, or rather if an intelligent cell group would be more feasible. Such cell modules equipped with a common intelligent circuit would possibly allow for even lower prices and assembly efforts.

5. Conclusion and Outlook

Looking at current developments in the car manufacturing industry, there seems to be an intermediate step on the way to full electromobility: the so called micro and mild hybrid cars and technologies. We at Fraunhofer IPA are working on new technologies and manufacturing ways for the electromobility of the near future and of the next generation of vehicles.

Comfort is and always will be one of the main drivers for customers to choose or, in the case their expectations are undercut, not to choose an e-mobility solution. Our surface heating system enables an easy to produce car interior heater that is easy to integrate on different kinds of surfaces. During research efforts, the main focus of our system was to provide a fast heating system for small travel distances. However one could imagine to also integrate this solution on larger areas in the car, for example on the interior roof or on the car floor and therefore to optimize the amount of radiated heating power.

We also presented a concept for a completely modular battery design. It offers new possibilities to customize the battery and match it to individual driving scenarios. A complete evaluation of the system in real driving scenarios must be performed. The additional circuitry raises the total price of a battery cell but we expect that a certain level of standardization will determine a price drop to compensate these extra costs. Thus the Intelligent Cell could become affordable compared to a conventional one.

As an outlook the combination of a modular battery design together with an easy to integrate surface heating system has to be taken into account for future automotive solutions. Such optimized battery systems could offer the possibility of interior heating within an optimal surrounding with optimal climate conditions.

6. References

[1] ZSW, Statista 2015: "Anzahl der zugelassenen Autos in Japan, China, Deutschland undden USA im Jahr 2014", http://de.statista.com/statistik/daten/studie/243993/umfrage/bestand-elektrofahrzeuge-nach-laendern/

[2] KBA, VDA, IHS, Statista 2015: „Anzahl der Neuzulassungen von Pkw in Deutschland von 1955 bis 2018 (in Millionen)", http://de.statista.com/statistik/daten/studie/74433/umfrage/neuzulassungen-von-pkw-in-deutschland/

[3] GoingElectric, ElektroAuto News, 2012: "BMW: Infrarot-Heizung und Wärmepumpe für mehr Effizienz im Elektroauto", http://www.goingelectric.de/2012/09/20/news/bmw-infrarot-heizung-und-waermepumpe-fuer-mehr-effizienz-im-elektroauto/, accessed on 1.03.2016.

[4] Fraunhofer IAO: „STRUKTURSTUDIE BWe MOBIL 2015 – Elektromobilität in Baden-Württemberg", Stuttgart (2015)

[5] Bozem, Karlheinz, et al. „Elektromobilität: Kundensicht, Strategien, Geschäftsmodelle: Ergebnisse der repräsentativen Marktstudie FUTURE MOBILITY", Wiesbaden (2013), Springer Vieweg.

[6] Bauernhansl, T.; Verl, A.; Westkämper, E.; (Hrsg.); Brix, J., „Entwicklung eines verteilten Energiemanagementsystems", Stuttgarter Beiträge zur Produktionsforschung, Band 46, ISBN: 978-3-8396-0909-5, Fraunhofer Verlag, 2015.

[7] 3Ccar homepage; http://3ccar.eu/; accessed on 10.10.2015, 15:25.

[8] http://www.plugincars.com/cars; accessed on 11.02.2016, 14:21.

Power Hardware-in-the-Loop Emulation of Permanent Magnet Synchronous Machines with Nonlinear Magnetics – Concept & Verification

Alexander Schmitt, Jan Richter, Michael Braun, Martin Doppelbauer
Institute of Electrical Engineering (ETI), Karlsruhe Institute of Technology (KIT)
Kaiserstraße 12, 76131 Karlsruhe, Germany, a.schmitt@kit.edu

Abstract

This paper presents a power hardware-in-the-loop emulation test bench (PHIL), based on a modular multiphase multilevel converter (MMPMC), to mimic arbitrary permanent magnet synchronous machines with nonlinear magnetics as they are used in automotive applications. Measurements in stationary operation as well as high dynamic torque steps are conducted at a real automotive machine and precisely reproduced at the PHIL system to demonstrate the excellent performance of the PHIL test bench. Moreover, the superiority of a PHIL test bench over conventional motor test benches is proven by the unproblematic emulation of a blocking rotor or a cracking shaft.

1. Introduction

In modern automotive drive inverter development the importance of simulation increases rapidly. In early stage, various simulation tools are used to simulate and validate the accurate function of inverters. Afterwards, real-time hardware-in-the-loop test benches are used to test the developed software in conjunction with the signal processing unit of the power converter [1, 2, 3]. Finally, the converter has to be connected to a motor test bench to test and improve the performance, reliability or the manufacturing of the device. Unfortunately, there are several drawbacks inherent to conventional motor test benches. The inverter can only be tested when the motor is already available, which is usually not the case. Moreover, exchanging motors is extensive and several test benches are needed to cope with different power demands. Thus, the space required for test beds can be large and additional costs and maintenance efforts are caused. Furthermore, conventional motor test benches are limited in their fault emulation capability. Faults like a crack of the shaft, a blocking rotor or winding short circuits are very difficult to test. Therefore, it is desirable to connect the converter to a power hardware-in-the-loop (PHIL) emulation test bench (see Figure 1) to evaluate its proper function in all possible operating conditions. Such a device can mimic any machine using parameters that can be easily calculated by measurement or finite element analysis. Changing the motor type or data set can be executed by software within seconds.

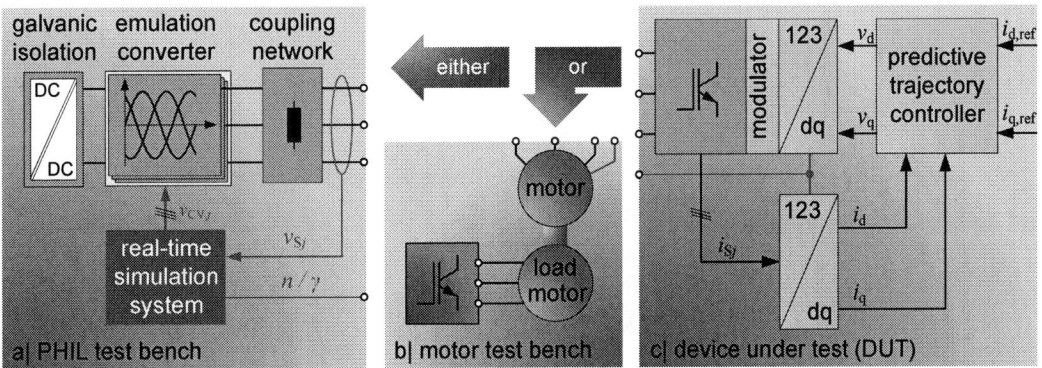

Figure 1: Either a motor test bench or a PHIL system can be used to test drive converters.

© VDE VERLAG GMBH · Berlin · Offenbach

PCIM Europe 2016, 10 – 12 May 2016, Nuremberg, Germany

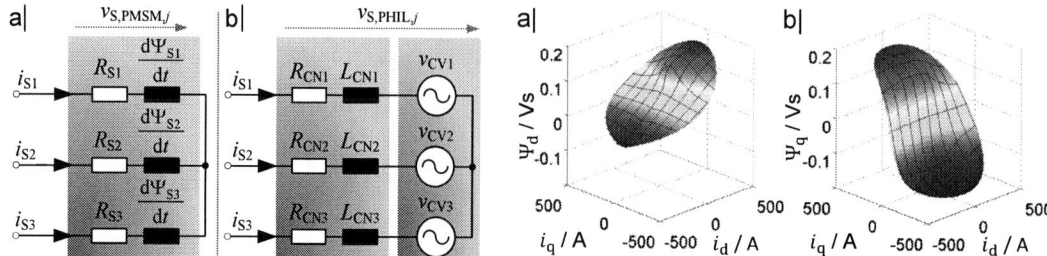

Figure 2: Equivalent circuit of a PMSM (a) and of the power hardware-in-the-Loop test bench (b).

Figure 3: Flux linkage $\Psi_d = f(i_d, i_q)$ (a) and flux linkage $\Psi_q = f(i_d, i_q)$ (b) of the PMSM [5].

This paper presents a PHIL test bench and its application to permanent magnet synchronous machines with nonlinear magnetics for automotive applications. Because the PHIL test bench was already introduced in [4], this paper focuses on the verification and accuracy of the PHIL system. Indeed, for comprehension issues the theory of the underlying machine model [5] is briefly described in section 2. Subsequently, Section 3 summaries the emulation concept and the used control scheme of the PHIL test bench [4]. Afterwards, the hardware setup for the verification is presented (Section 4) and the PHIL test bench is verified by means of stationary and dynamic measurements using a predictive trajectory dead-beat current controller (Section 5). Conclusions are stated in Section 6.

2. Theory

The proposed PHIL test bench consists of an emulation converter connected to the device under test (DUT) by means of an inductive coupling network (see Figure 1). The objective of a PHIL system is that the electrical behavior of the PHIL is identical to the real machine. Therefore, a machine model is required that describes the behavior of nonlinear permanent magnet synchronous machines considering the voltage drop at the coupling network. Figure 2 opposes the equivalent circuit of a PMSM and the equivalent circuit of the PHIL test bench. Nonlinear models of saturated, anisotropic permanent magnet synchronous machines are available and employed in this contribution [6]. The stator voltages v_{Sj}, with $j \in \{1,2,3\}$ as phase numbers, can be calculated by employing Ohm's law, Faraday's law of induction and Kirchhoff's laws to the machine coils. This leads to:

$$v_{S,\mathrm{PMSM},j} = R_{Sj} \cdot i_{Sj} + \frac{d\Psi_{Sj}}{dt} \tag{1}$$

Since the characteristics of the PHIL test bench should be identical to the real PMSM, the phase voltages of the PHIL system must be identical to the phase voltages of the real machine [5]. This can be obtained by subtracting the voltage drop of the coupling network from the derivatives of the stator flux linkages and leads to:

$$v_{S,\mathrm{PMSM},j} = v_{S,\mathrm{PHIL},j} = R_{CNj} \cdot i_{Sj} + L_{CNj} \cdot \frac{di_{Sj}}{dt} + \underbrace{\left(\frac{d\Psi_{Sj}}{dt} - L_{CNj} \cdot \frac{di_{Sj}}{dt} - (R_{CNj} - R_{Sj}) \cdot i_{Sj}\right)}_{v_{CV,j}} \tag{2}$$

Subsequent transformation to the rotor-fixed dq-reference frame yields:

$$v_d = R_{CN} \cdot i_d + L_{CN} \frac{di_d}{dt} - \omega \cdot L_{CN} \cdot i_q$$
$$+ \left(\frac{d\Psi_d}{dt} - \omega\Psi_q - L_{CN}\frac{di_d}{dt} + \omega \cdot L_{CN} \cdot i_q - (R_{CN} - R_S) \cdot i_d\right) \tag{3}$$

$$v_q = R_{CN} \cdot i_q + L_{CN} \frac{di_q}{dt} + \omega \cdot L_{CN} \cdot i_d$$
$$+ \left(\frac{d\Psi_q}{dt} + \omega\Psi_d - L_{CN}\frac{di_q}{dt} - \omega \cdot L_{CN} \cdot i_d - (R_{CN} - R_S) \cdot i_q\right) \tag{4}$$

© VDE VERLAG GMBH · Berlin · Offenbach

There, R_S denotes the stator resistance, ω the electric frequency, t the time and v_x, i_x and Ψ_x the voltages, currents and flux linkages in the direct and quadrature axes ($x \in \{d, q\}$). Furthermore, R_{CN} and L_{CN} are the resistance and the inductance of the coupling network. The required model parameters are the resistance R_S and the flux linkages Ψ_d and Ψ_q that depend nonlinearly on the currents i_d and i_q as defined by the function f and shown in Figure 3.

$$f \colon \mathbb{R}^2 \to \mathbb{R}^2, \ \left(i_d, i_q\right) \ \mapsto \left(\Psi_d, \Psi_q\right) \tag{5}$$

The model parameters can be obtained by finite-element method calculations [6] or by stationary measurements of the machine [7]. The machine model is used in a real-time simulator as illustrated in Figure 1 to calculate the counter voltages $v_{CV,j}$ of the MMPMC so that the PHIL behaves exactly like the real motor [4]. Therefore the output voltages v_x of the DUT inverter are measured. Using (3) and (4), the derivatives of the machine currents can be calculated to:

$$\frac{di_d}{dt} = \frac{v_d - R_S i_d + \frac{L_{dq}}{L_{qq}}(-v_q + R_S i_q + \omega\Psi_d) + \omega\Psi_q}{L_{dd} - \frac{L_{dq} \cdot L_{qd}}{L_{qq}}} \tag{6}$$

$$\frac{di_q}{dt} = \frac{v_q - R_S i_q + \frac{L_{qd}}{L_{dd}}(-v_d + R_S i_d - \omega\Psi_q) - \omega\Psi_d}{L_{qq} - \frac{L_{dq} \cdot L_{qd}}{L_{dd}}} \tag{7}$$

Therein L_{xy} are differential inductances and thus partial derivatives of the flux linkage function in the direct and quadrature direction ($x, y \in \{d, q\}$). The counter voltages of the MMPMC $v_{CV,d}$ and $v_{CV,q}$ can then be calculated by:

$$v_{CV,d} = \frac{di_d}{dt}(L_{dd} - L_{CN}) + \omega \cdot L_{CN} \cdot i_q + L_{dq} \cdot \frac{di_q}{dt} - \omega\Psi_q - (R_{CN} - R_S) \cdot i_d \tag{8}$$

$$v_{CV,q} = \frac{di_q}{dt}(L_{qq} - L_{CN}) - \omega \cdot L_{CN} \cdot i_d + L_{qd} \cdot \frac{di_d}{dt} + \omega\Psi_d - (R_{CN} - R_S) \cdot i_q \tag{9}$$

3. PHIL Concept

A precise emulation requires a complete identical behavior at the terminals of the PHIL test bench compared to the real machine. Therefore, the PHIL has to apply the counter voltage $v_{CV,j}$ at the coupling network very precise and with a minimal dead-time to ensure the correct current slopes $\frac{di_x}{dt}$ of arbitrary machines within the coupling inductance. For this reason, the basic challenge of PHIL emulation is the calculation and generation of the counter voltage. Modern FPGAs and A/D-converters allow the calculation of the machine model including the counter voltage with sample rates f_M of more than 1.5 MHz quasi continuous in real-time [5]. Indeed, the counter voltage generation is more challenging, especially for high power applications. The counter voltages are discontinuous functions since the current slopes $\frac{di_x}{dt}$ depend on the clocked output voltages v_x and are different in active as well as freewheeling states of the DUT [4]. Moreover, the phase inductance of the coupling network L_{CN} can not correspond to the differential inductances L_{xy} of the machine (8), (9) due to iron saturation or the magnetic anisotropy of the rotor. For this reason, modelling of the current slopes requires a converter topology that allows a high dynamic and very precise generation of the counter voltages $v_{CV,j}$. The modular multiphase multilevel converter (MMPMC) [8] and the associated modulation scheme [9] offers such a dynamic and precise voltage generation and is used in this PHIL test bench. The schematic diagram of the entire PHIL test bench is shown in Figure 4. A MMPMC with $n = 6$ branches per phase is used to generate a seven level output voltage waveform with a resulting PWM-frequency of $f_{PWM} = 120$ kHz [9]. The MMPMC has

PCIM Europe 2016, 10 – 12 May 2016, Nuremberg, Germany

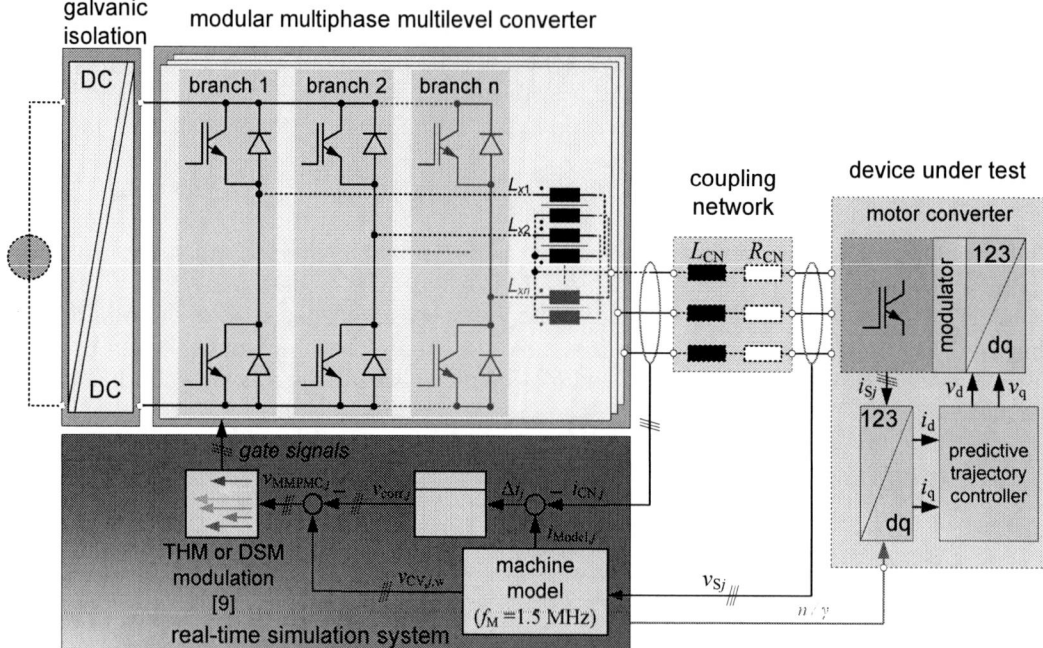

Figure 4: Detailed schematic diagram of the proposed PHIL test bench including the modular multiphase multilevel converter and the real-time simulation system as well as the coupling network and the device under test [4].

to be fed by a galvanic isolated DC-DC-converter because the real machine coils are also galvanically isolated. Furthermore, an external coupling network is necessary to connect the DUT inverter to the PHIL test bench since the MMPMC has the behavior of a voltage source. The real-time simulation system is based on an FPGA [5]. This FPGA contains the machine model, derived in Section 2, as well as the modulation of the MMPMC [9]. Since the counter voltage generation has unavoidable inaccuracies e.g. dead-times, forward voltages, zero current clamping etc. the real-time simulation system contains an additional P-controller. This P-controller is necessary to avoid a drift of the inner model currents $i_{\mathrm{Model},j}$ and the real currents $i_{\mathrm{CN},j}$. A difference between the model and the real currents would affect the calculation of the counter voltages and the inner torque. Hence, it would distort the behavior of the PHIL test bench compared to the real machine [4]. Indeed, a simple P-controller is sufficient and does not affect the stability of the current controller of the DUT. In addition, the real-time simulation system is able to emulate an incremental encoder as well as a resolver. Identically to real machines, this sensor signal is the only connection between PHIL and DUT besides the three power terminals (Figure 4) [4].

4. Experimental Setup

An interior permanent magnet synchronous machine for automotive traction applications of type *Brusa HSM1-6.1712-CO1* is used as a test motor (Figure 5 (b)). The machine has strongly nonlinear magnetics as can be seen in Figure 3, a maximum shaft power of 97 kW at a torque of 220 Nm and a rotor speed of 4200 min^{-1} (see. Table 1). Furthermore, Figure 5 (c) shows the DUT to control the test motor as well as the PHIL test bench. The motor converter is based on a *Semikron SkiiP (513GD122-3DUL)* six-pulse bridge and can be optionally connected to the motor (Figure 5 (b)) or the PHIL test bench (Figure 5 (a)). All measurements are carried out at a DC-link voltage of 300 V at the DUT and a DC-link voltage of 650 V at the MMPMC. Furthermore, a coupling inductance of $L_{\mathrm{CN}} = 1\,\mathrm{mH}$ is used inside

© VDE VERLAG GMBH · Berlin · Offenbach

PCIM Europe 2016, 10 – 12 May 2016, Nuremberg, Germany

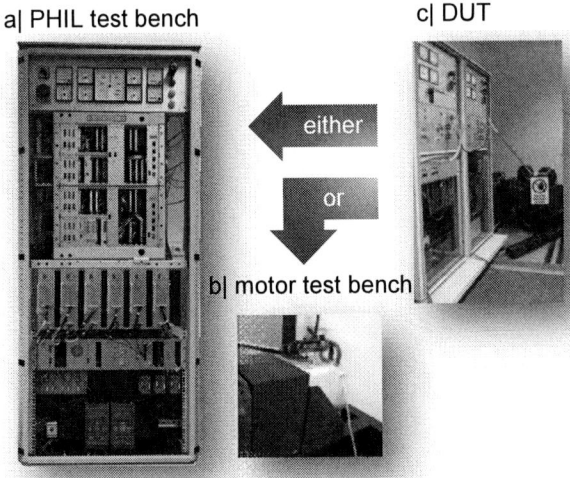

a| PHIL test bench   c| DUT

either

or

b| motor test bench

Figure 5: Test bench setup including the PHIL test bench (a), the real PMSM (b) and the DUT (c) [4].

Table 1

Machine Parameters	
Parameter	Value
line voltage nom.	212 V
current nom. /max.	169 A / 300 A
shaft power nom. / max.	57 kW / 97 kW
number of pole pairs	3
torque nom. / max.	130 Nm / 220 Nm
speed nom. / max.	4200 min-1 / 11000 min-1
$L_{dd}(0\,A, 0\,A)$ / $L_{dd}(-200\,A, 200\,A)$	410 µH / 204.5 µH
$L_{qq}(0\,A, 0\,A)$ / $L_{qq}(-200\,A, 200\,A)$	2.1 mH / 163.6 µH

the PHIL test bench. The test motor is set to a constant rotational speed of $n = 1000\,\mathrm{min}^{-1}$ by means of a speed controlled load motor which is connected to the PMSM. The rotational speed of the virtual machine was set to $n = 1000\,\mathrm{min}^{-1}$ by software. Furthermore, the current controller of the DUT is executed with a control and switching frequency of $f_c = 8\,\mathrm{kHz}$. The real-time simulation system calculates the machine model with a sampling rate of $f_M = 1.5\,\mathrm{MHz}$ which is why the counter voltages are first averaged over one modulation period T_{PWM}. Afterwards, the threshold modulation [9] generates the counter voltage with a PWM frequency of $f_{PWM} = 120\,\mathrm{kHz}$.

5. Results and Discussion

A predictive trajectory dead-beat controller as proposed in [7] is applied to control the motor converter. During the measurements, the integral component of the current controller was disabled. This ensures that dynamic as well as stationary variations between the PHIL test bench and the real motor are not compensated by the integral component. Thus, only the model accuracy of the PHIL test bench determines the accuracy of the current controller. Subsequently, the measurement results at the test motor and the PHIL test bench can be compared to precisely analyze the quality of the PHIL test bench.

5.1 Stationary Measurements

First, stationary measurements within the current plane were conducted at the PHIL test bench and at the test motor. Figure 6 shows the difference ε_x of the measured currents i_x.

$$\varepsilon_x = i_{x,\mathrm{PHIL}} - i_{x,\mathrm{PMSM}} \tag{10}$$

Note the current plane is currently limited to $|i_x| < 200\,A$ due to the hardware limits of the PHIL test bench. The differences ε_q in the q-axis are illustrated in Figure 6 (a) and the differences in the d-axis ε_d are depicted in Figure 6 (b). The measured points are marked by dots in the diagram and the remaining points are interpolated on the basis of the measured values. The plots depict the excellent stationary performance of the PHIL test bench since the differences between the measured currents are $\varepsilon_x < 2\,A$ in a wide operating range. Indeed, the difference between motor test bench and PHIL emulation system increases with increasing current caused by inaccuracies of the PHIL measurement and the counter voltage generation.

© VDE VERLAG GMBH · Berlin · Offenbach

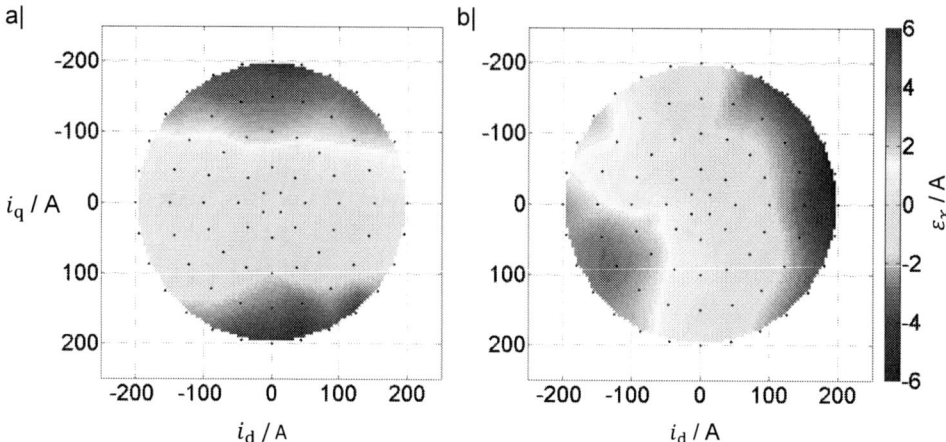

Figure 6: Difference between the q-currents (a) and the d-currents (b) at the PHIL test bench and the real motor during stationary operation.

5.2 Dynamic Measurements

The dynamic validation is conducted by a comparison of torque steps, shown in Figure 7. Thereby, a torque step from $M_i = 0\,\mathrm{Nm}$ to $M_i = 100\,\mathrm{Nm}$ is executed using three different control strategies at the DUT. The inductance of the d-axis is significant smaller than the inductance of the q-axis ($L_{\mathrm{qq,0A}} \approx 5 \cdot L_{\mathrm{dd,0A}} \approx 2\,\mathrm{mH}$) for this reason the d-axis offers significant higher current changes (see Table 1). This property can be used by high performance current controllers to optimize their strategies how a set value is reached. The left column of Figure 7 depicts the torque step for a direct current connection (DCC) of the set values (straight line), the middle column shows the short time to reference value strategy (STRV) which is the fastest way to reach the set value. Finally, the right column illustrates the fast torque response (FTR) trajectories. Therefore, the current jumps as fast as possible to the constant torque line to reach the requested torque and moves than along the constant torque line to the set value [7]. The torque steps are compared in the current plane (first row) as well as in their time response (second row). Finally the differences ε_x of the sampled values are calculated (third row). It can be seen that the PHIL precisely reproduces the dynamic q-current trend independent from the control strategy with variations of less than $\varepsilon_{\mathrm{q,max}} < 10\,\mathrm{A}$. In the d-axis the currents are at the beginning also very similar but differ dependent on the control strategy in maximum between $\varepsilon_{\mathrm{d,max,STRV}} \approx 10\,\mathrm{A}$ and $\varepsilon_{\mathrm{d,max,DCC}} \approx 25\,\mathrm{A}$. Indeed, these differences are not caused by modelling errors of the machine but by the limited output voltage of the PHIL test bench. A sufficient voltage reserve is essential since the machine inductance L_{dd} is significant smaller than the coupling inductance L_{CN}. Due to this, high counter voltages are necessary to generate the desired current slope $\frac{di_d}{dt}$ within the coupling network (see eq. (8)). Otherwise, the dynamic of the current $i_{\mathrm{d,PHIL}}$ is limited in case that the output voltage is limited. For this reason, the possible machine inductance, the precision of the counter voltage generation and the DC-link voltage of the emulation converter have to be considered for the design of the coupling inductance. However, if the PHIL system operates within its maximum voltage, it emulates the machine nearly perfect.

5.3 Fault Emulation

Finally, the PHIL test bench is used to emulate fault conditions which cannot be tested at a real motor test bench. Figure 8 (a) depicts the emulation of a blocking rotor. Therefore, the rotational speed is abruptly set from $n = 1000\,\mathrm{min}^{-1}$ to $n = 0\,\mathrm{min}^{-1}$ at $t = 0$. The set value for the load torque is held constant at $M_i = 75\,\mathrm{Nm}$. In contrast, Figure 8 (b) shows a step of

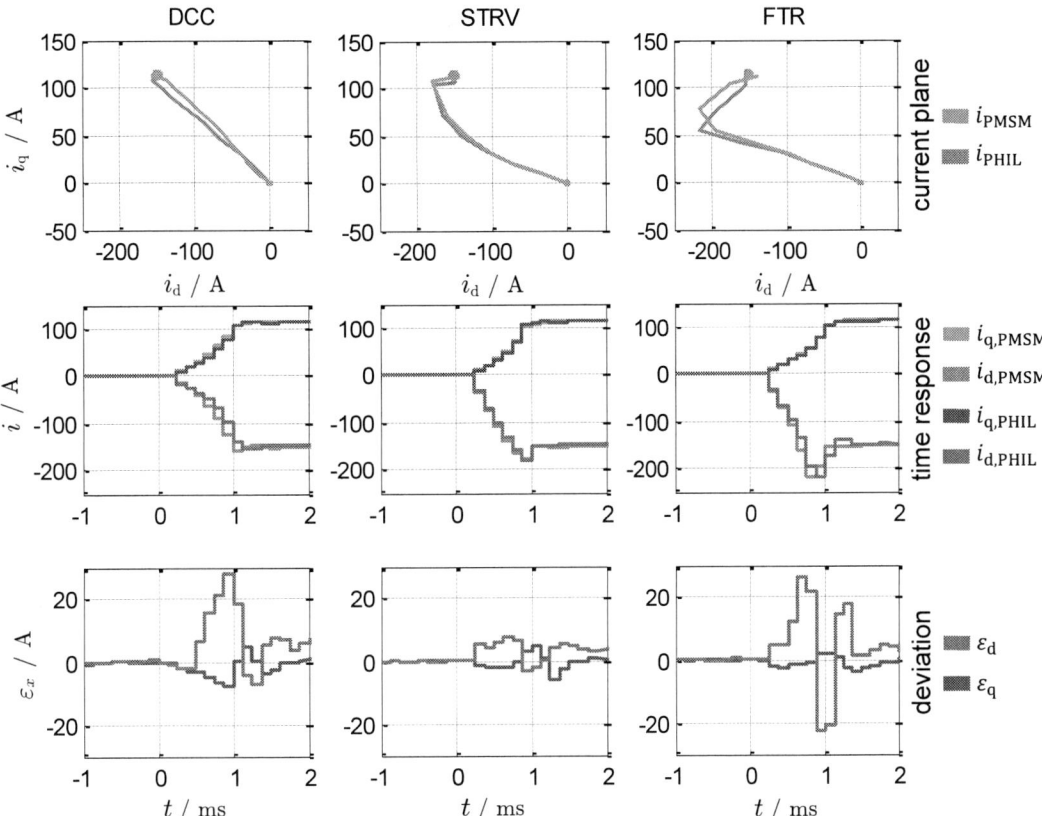

Figure 7: Torque step from $M_i = 0\,\text{Nm}$ to $M_i = 100\,\text{Nm}$ for three different control strategies: Direct Current Connection DCC (left column), Short Time to Reference Values STRV (middle column) and Fast Torque Response FTR (right column) [7].

the rotational speed from $n = 1000\,\text{min}^{-1}$ to $n = 2000\,\text{min}^{-1}$ at $t = 0$ whereas the load torque is $M_i = 25\,\text{Nm}$. Such a step can occur when the inertia torque suddenly decreases e.g. due to a cracking shaft. However, a blocking rotor or a change of the inertia torque cannot be tested with conventional motor test benches. Indeed, since the validity of the PHIL test bench is already proven, the PHIL test bench allows reliable tests of the DUT in operating points that are not possible on conventional motor test benches. This underlines the superiority of a PHIL test bench over conventional motor test benches in automotive drive development processes.

6. Conclusion

This paper has presented a power hardware-in-the-loop emulation test bench (PHIL), based on a modular multiphase multilevel converter (MMPMC), to mimic arbitrary permanent magnet synchronous machines with nonlinear magnetics. The underlying machine model as well as the PHIL concept using a seven level modular multiphase multilevel converter is introduced and verified. Therefore, the PHIL test bench is parametrized for an automotive PMSM and controlled by a DUT using a predictive trajectory dead-beat current controller. Measurements in stationary operation as well as high dynamic torque steps are conducted at a real automotive motor and precisely reproduced at the PHIL system and demonstrate the excellent performance of the PHIL test bench. Furthermore, the superiority of a PHIL test

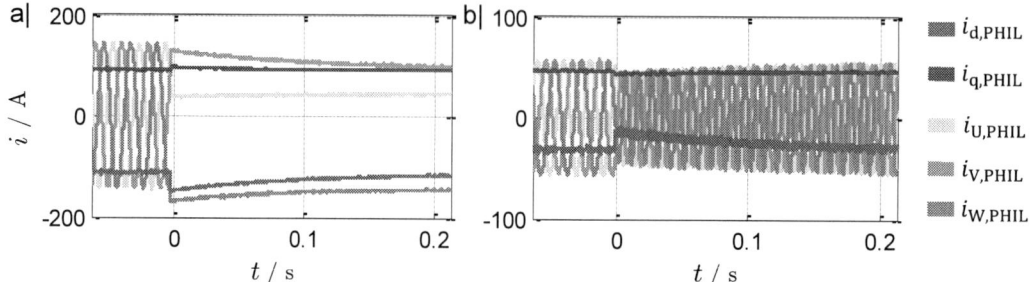

Figure 8: Emulation of a rotational speed step from 1000 min^{-1} to 0 min^{-1} (a) and from 1000 min^{-1} to 2000 min^{-1} (b)

bench over conventional motor test benches in automotive drive development processes is proven by the unproblematic emulation of a blocking rotor or a cracking shaft.

References

[1] C. Dufour, S. Cense, T. Yamada, R. Imamura, and J. Belanger, "Fpga permanent magnet synchronous motor floating-point models with variable-dq and spatial harmonic finite-element analysis solvers," in *15th International Power Electronics and Motion Control Conference (EPE/PEMC)*, 2012, pp. LS6b.2–1–LS6b.2–10.

[2] M. Matar and R. Iravani, "Massively parallel implementation of ac machine models for fpga-based real-time simulation of electromagnetic transients," *IEEE Transactions on Power Delivery*, vol. 26, no. 2, pp. 830–840, Apr. 2011.

[3] L. Herrera and J. Wang, "Fpga based detailed real-time simulation of power converters and electric machines for ev hil applications," in *IEEE Energy Conversion Congress and Exposition (ECCE)*, Sept. 2013, pp. 1759–1764.

[4] A. Schmitt, J. Richter, M. Gommeringer, T. Wersal, and M. Braun, "A novel 100 kw power hardware-in-the-loop emulation test bench for permanent magnet synchronous machines with nonlinear magnetics," in *8th IET International Conference on Power Electronics, Machines and Drives (PEMD 2016), Glasgow*, Apr. 2016, pp. 1–6.

[5] A. Schmitt, J. Richter, U. Jurkewitz, and M. Braun, "FPGA-based real-time simulation of nonlinear permanent magnet synchronous machines for power hardware-in-the-loop emulation systems," in *IECON 2014 - 40th Annual Conference of the IEEE Industrial Electronics Society*, Oct. 2014, pp. 3763–3769.

[6] X. Chen, J. Wang, B. Sen, P. Lazari, and T. Sun, "A high-fidelity and computationally efficient model for interior permanent-magnet machines considering the magnetic saturation, spatial harmonics, and iron loss effect," *IEEE Transactions on Industrial Electronics*, vol. 62, no. 7, pp. 4044–4055, Jul. 2015.

[7] J. Richter, P. Bauerle, T. Gemassmer, and M. Doppelbauer, "Transient trajectory control of permanent magnet synchronous machines with nonlinear magnetics," in *IEEE International Conference on Industrial Technology (ICIT)*, Mar. 2015, pp. 2345–2351.

[8] A. Schmitt, M. Gommeringer, J. Kolb, and M. Braun, "A high current, high frequency modular multiphase multilevel converter for power hardware-in-the-loop emulation," in *International Exhibition and Conference for Power Electronics, Intelligent Motion, Renewable Energy and Energy Management; Proceedings of PCIM Europe*, May 2014, pp. 1537–1544.

[9] A. Schmitt, M. Gommeringer, C. Rollbühler, P. Pomnitz, and M. Braun, "A novel modulation scheme for a modular multiphase multilevel converter in a power hardware-in-the-loop emulation system," in *IECON 2015 - 41st Annual Conference of the IEEE Industrial Electronics Society*, Nov. 2015, pp. 1276–1281.

Multimode Charging of Electric Vehicles

A combined concept with multiple use of components and strategies for decreasing power losses, weight and volume

Marco, Jung, Fraunhofer IWES, Germany, marco.jung@iwes.fraunhofer.de

Axel, Seibel, Fraunhofer IWES, Germany, axel.seibel@iwes.fraunhofer.de

Jonas, Steffen, Fraunhofer IWES, Germany, Jonas.steffen@iwes.fraunhofer.de

Georgios, Lempidis, Fraunhofer IWES, Germany, georgios.lempidis@iwes.fraunhofer.de

Jörg, Kirchhof, Fraunhofer IWES, Germany, joerg.kirchhof@iwes.fraunhofer.de

Roland, Gaber, Fraunhofer IWES, Germany, roland.gaber@iwes.fraunhofer.de

René, Marklein, Fraunhofer IWES, Germany, rene.marklein@iwes.fraunhofer.de

Abstract

In this paper, a complete multimode (wired and wireless) charging and discharging infrastructure and their implementation in an electric vehicle are presented. The overall system management and a control strategy for increasing the efficiency are discussed. In order to increase the grid connection of an electric vehicle, a highly flexible bidirectional battery charging concept has been developed which is capable of charging 1-phase grid wired, 3-phase grid wired as well as wireless with an inductive power transfer system. Ancillary services and self- consumption are possible due to the bidirectional power flow capabilities. The multiple use of components, for saving costs and overall installed size, requires a compromise in efficiency and dimensioning of power electronic components, since they must satisfy all requirements for a bidirectional 1-phase and 3-phase wired grid connection as well as a wireless grid connection of the battery. To reduce the negative aspects by reason of over dimensioning for different charging modes, a new operating control technique will be presented to find an acceptable solution. Measurements of the efficiency and electromagnetic compatibility document the performance and electromagnetic safety of the system. A proof of concept and the system integration in an electric vehicle is demonstrated.

1. Multifunctional bidirectional charging and discharging concept

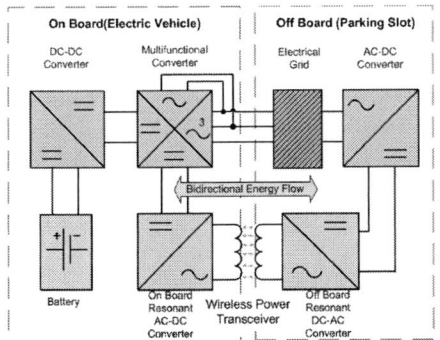

Fig. 1 Multifunctional bidirectional battery charging concept (left) [1, 2] and vehicle demonstrator with integrated multifunctional bidirectional battery charging system (right)

The block diagram of the presented multifunctional bidirectional charging concept for electric vehicles is depicted in Fig. 1 left. The concept allows the charging of the battery either wired with a 1-phase or 3-phase grid or/and wireless with an inductive power transfer system. The bidirectional concept provides a reactive and active power flow for grid ancillary services or self-consumption for households [1, 2].

In this concept, the primary side of the coil from the inductive power transfer system is integrated in the ground of the parking space and the secondary side is mounted underneath the car [2]. Additionally in Fig. 1, right the vehicle demonstrator is shown, where the components are integrated. In the following chapters the components will be described in detail.

2. Bidirectional wireless power transfer system

An inductive power transfer system for bidirectional charging can be described as an inductive coupled resonant coil system. In the vehicle the primary coil is mounted in the parking place and the secondary coil underneath the car.

As a result of the large distance between the primary and secondary coils, the low coupling requires a large amount of reactive power in establishing the magnetic field. The system efficiency can be increased by building a resonant circuit, including the leakage inductance Lr and a compensation capacitor Cr in series with the resonance frequency fr (see Fig. 2). Any misalignments can be partly balanced by adapting the switching frequency [3]. Greater misalignments as well as different primary and secondary coil geometries lead to a variable output voltage V_{OUT}.

Fig. 2 presents the power electronic topologies of a full (left) and half (right) bridge LLC resonant converter [4, 5], which can supply the resonant circuit. Using the mutual inductance Lm to discharge the parasitic capacitances during the dead time provides an optimum ZVS (Zero Voltage Switching) condition and thus significantly reduced switching losses.

If the full bridge configuration is selected, it is possible to operate the converter as a half bridge in partial load condition. For this the transistors T4 and T6 are switched on constantly and the transistors T3 and T5 are switched off constantly. This control strategy leads to an increase of the efficiency under partial load conditions, because of the decreased semiconductor and resonant circuit losses and is presented more detailed in [2].

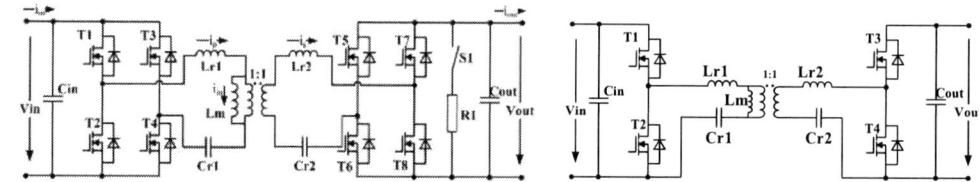

Fig. 2   Bidirectional full bridge (left) and half bridge (right) LLC resonant converter [2]

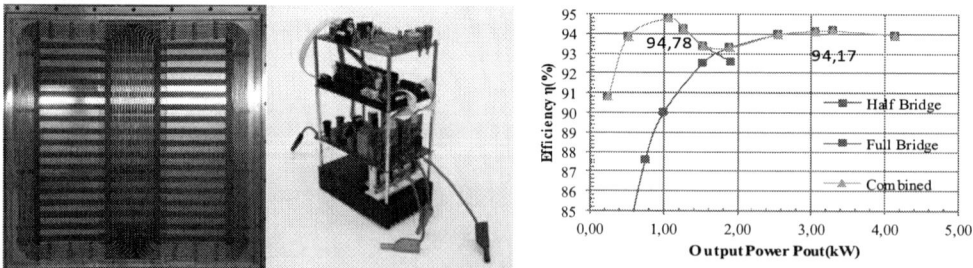

Fig. 3   Demonstrator of the LLC resonant converter (left: double D coil, middle: power electronics) and efficiency curves (right) with Vin = Vout = 400 V and air gap = 20 cm [2]

Fig. 3 depicts the prototype of the full bridge LLC resonant converter with a nominal output power of 3.3 kW and the double D coil system that were developed and additionally the efficiency curves of the system for half bridge operation, for full bridge operation and for combined half and full bridge operation. Detailed information about the inductive power transfer system is described in [2].

3. Multifunctional bidirectional charger

The power electronic topology of the multifunctional bidirectional battery charger is presented in Fig. 4. For 3-phase grid connection, all three half bridges and chokes are used. For 1-phase operation, only two of the three half bridges are connected with the grid as H4 bridge. The half bridges combined with the chokes enable the connection of a bidirectional inductive power transfer system with DC-output. In this operation mode, the power electronic components work as a bidirectional interleaved boost converter in order to achieve the claimed voltage (DC link voltage V_{ZW} or battery voltage V_{Batt}). Finally a DC/DC converter is connected in series and regulates the charging current and voltage. With such a flexible concept and thus with such a bidirectional charging device it is possible to use different battery systems with different battery voltages as well as different bidirectional inductive power transfer systems with different output voltages and system topologies [1, 6].

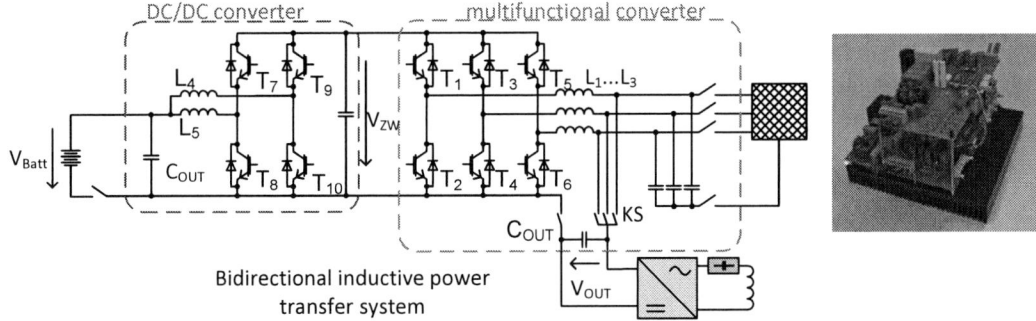

Fig. 4   Topology multifunctional bidirectional battery charger and realized demonstrator [1]

One big advantage of the converter concept is the multiple use of components such as cooling system, power semiconductors, connections, passive components, control and sensors. This allows reductions in cost, volume and weight of the overall system in comparison with other on board systems with the same flexibility of grid connection (wired 3-phase and 1-phase and wireless) as well as a bidirectional power flow. The result of the comparison is depicted in Fig. 5. More detailed information's are in [1, 6].

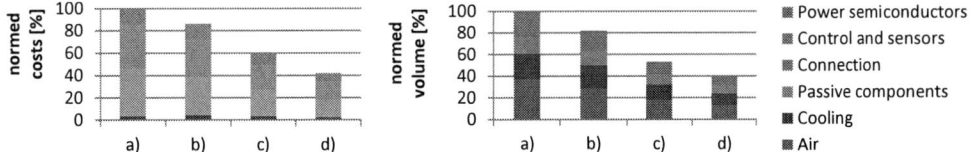

Fig. 5   Comparison between different on-board charging concepts and necessary bidirectional charger: a) 3 x 1-phase wired + DC/DC for wireless, b) 1-phase wired + 3-phase wired + DC/DC for wireless, c) combined 1- and 3-phase wired + DC/DC wireless, d) multifunctional [1, 6]

Typical charging devices work for the most part in partial load condition and therefore efficiency needs to be optimized in this operating range. A new optimization and control strategy for bidirectional multiphase buck converters (DC/DC converters) is presented in [7, 8] and is used for the proposed charging device. The benefit of this strategy is the increase in the overall system efficiency and the power density by adjusting switching frequency and

© VDE VERLAG GMBH · Berlin · Offenbach

phase adaption.

The multiple use of components requires a compromise in efficiency and dimensioning of power electronic components. Looking at the concept of the multifunctional converter, semiconductors with a blocking voltage of 1200 V must be used for the 3-phase grid connection (see Fig. 4). For the 1-phase and inductive grid connection a blocking voltage of 600 V is sufficient. In addition, the semiconductor and chokes can be designed for larger currents, which would lead to additional losses. Thus a compromise must be found to achieve a high level of efficiency despite the over dimensioning of the components.

To achieve interoperability, different coil geometries for the primary and secondary side of the inductive power transfer system must work together. This means that the voltage V_{OUT} (see Fig. 4) can be variable, because the voltage depends on the coil geometries, the input and output voltage of the resonant converter, the resonant frequency and the allowed frequency adaption, the misalignment of the coils etc. Additionally the battery voltage depends on the state of charge and leads to a wide voltage range. In order to handle these different voltage levels easily for a bidirectional power flow without a complex communication structure and control strategy, a bidirectional DC/DC converter is necessary, which can increase or decrease the voltage level in both power flow directions. Table 5.14 shows the required blocking voltage V_{Block}, inductor currents I_L and DC link voltages V_{ZW} based on the different charging mode as well as the selection for a 10 kW bidirectional battery charger.

Table 1 Requirements on the blocking voltage and current capability depending on the charging modus and the resulting impact on the demonstrator

	3-phase		1-phase		Inductive		Demonstrator	
P_{Grid}	10.8	kW	3.7	kW	3,7	kW	10.8	kW
V_{Block}	1200	V	600	V	600	V	1200	V
$I_{L.RMS}$	16	A	16	A	9	A	16	A
$\Delta I_{L.max}$	5.2	A	5.2	A	5.2	A	5.2	A
$V_{ZW.max}$	700	V	400	V	200 - 400	V	700	V

Equation (1) shows the switching losses and ensuing the dependency on switching frequency f_{sw} and voltages U_{DC} and $U_{DC.ref}$ for semiconductors:

$$P_{SW} = E_{SW}(I) \cdot \left[\frac{U_{DC}}{U_{DC.ref}} \right]^x \cdot f_{sw} \tag{1}$$

Equations (2) – (5) calculate the maximum inductor current ripple ΔI for the different charging modes. Especially for the 1-phase grid connection (H4 bride) two different modulation forms are possible. The bipolar modulation provides two voltage levels with the same frequency, in comparison with the switching frequency at the output. The unipolar modulation provides three voltage levels with the double frequency, in comparison with the switching frequency at the output [9]:

$$\Delta I_{L\_1-phase\_2-Level} = \frac{V_{ZW}}{4 \cdot L_{1,2} \cdot f_{sw}} \tag{2}$$

$$\Delta I_{L\_1-phase\_3-Level} = \frac{V_{ZW}}{16 \cdot L_{1,2} \cdot f_{sw}} \tag{3}$$

$$\Delta I_{L\_3-Phase} = \frac{V_{ZW}}{4 \cdot L_{1/2/3} \cdot f_{sw}} \tag{4}$$

$$\Delta I_{L\_DC/DC} = \frac{V_{ZW}}{4 \cdot L_{1,2} \cdot f_{sw}} \tag{5}$$

Considering Table 1 with the equations (1) –(4), it is possible to increase the efficiency for the 1-phase grid connection by decreasing the switching frequency and decreasing the DC link voltage depending on the modulation form.

For this demonstrator ($f_{sw\_3\text{-phase}}$ = 16 kHz), the switching frequency can be decreased to 9.2 kHz for the bipolar modulation and for the unipolar modulation to 2.2 kHz. The efficiency can be increased from 94.56 % to 96.3 % (bipolar) or 96.78 % (unipolar). The use of the unipolar modulation in combination with the non-galvanic isolated charger must be analysed system specifically, due to the parasitic capacitance to ground.

If the inductive power transfer system is switched on, a DC link voltage adaption (V_{ZW} = V_{BATT} or V_{ZW} = V_{OUT}) allows only one of both DC/DC converters to be activated. The other converter passively conducts the current thus drastically reducing the losses. Additionally, the switching frequency can be decreased because of the decreased DC link voltage and if the control strategy [7, 8] is used, the efficiency can be increased even more (consider Table 1 with the equations (1), (4) and(5)). The measured increasement is from 96.3 % up to 98.4 %.

Table 2 lists basic information concerning the components of the multifunctional bidirectional charger.

Table 2 Specifications of the multifunctional bidirectional battery charger

Inductance (L_1, L_2, L_3)	2.1 mH	Battery voltage (V_{Batt})	350 V
Inductance (L_4, L_5)	630 µH	Thermal resistance (R_{th})	0.66 K/W
IGBTs (T_1-T_{10})	IKW40N120H3	Ambient temperature (T_a)	30 °C
Output capacitance (C_{out})	45 µF, Film	Switching frequency (f_{sw})	16 kHz
Battery current (I_{Batt})	29 A	Input voltage (V_{ZW})	700 V

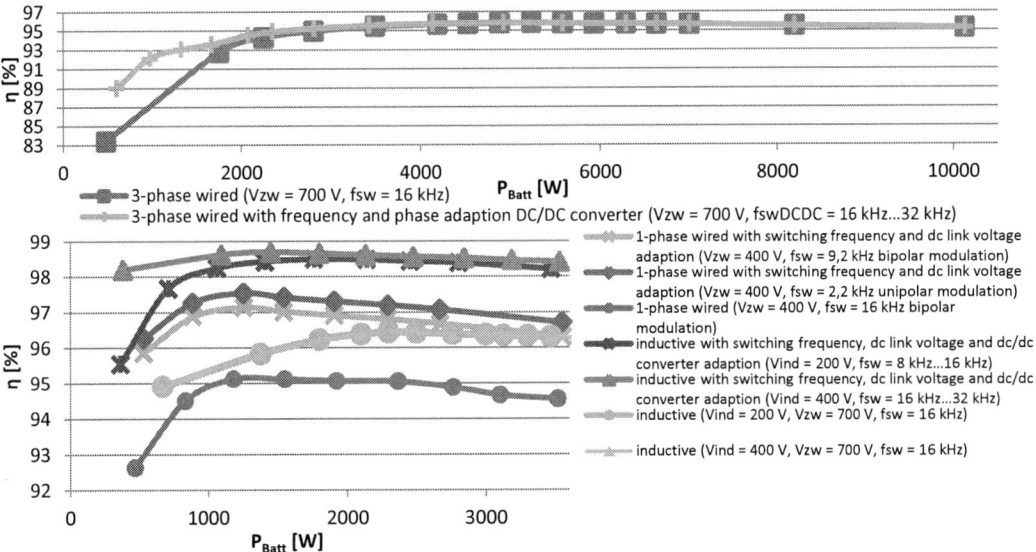

Fig. 6   Efficiency curves of the multifunctional bidirectional battery charger

Fig. 6 represents all efficiency curves of the multifunctional bidirectional battery charger. The highest efficiency level of around 96 % is achieved at partial load P_{Batt} ≈ 5 kW with the 3-phase operation mode. The switching frequency and phase adaption of the DC/DC converter (see [7, 8]) increases the partial load efficiency at 500 W about 3 percentage points and at 2.25 kW about 0.5 percentage points.

For 1-phase operation, the switching frequency can be decreased, because the DC link voltage is lower. Different modulation types for H4 bridges (bipolar and unipolar) were tested and as a result, the switching frequency with unipolar modulation can be decreased more in comparison to the bipolar modulation, because of the third output voltage level and the double output frequency and the fixed inductance value calculated for the 3-phase operation mode.

When using the inductive power transfer system operation mode, the converter achieves a high efficiency over the entire power range, because due to the DC link voltage adaption ($V_{ZW} = V_{Batt}$ or $V_{ZW} = V_{OUT}$) allows that only one of the two in series connected DC/DC converters must be activated.

4. Control strategy of the overall system vehicle demonstrator

All the described on-board components from the last chapters are included in a system demonstrator box depicted in Fig. 7. This system demonstrator box also contains a communication module for exchanging system data with an external charging station over wireless CAN. The external charging station includes a DC power supply, a LLC full bridge converter and the double D coil, which is integrated in the ground of the parking space. In this chapter, a practical control strategy for tracking the most applicable frequency for transferring energy is presented.

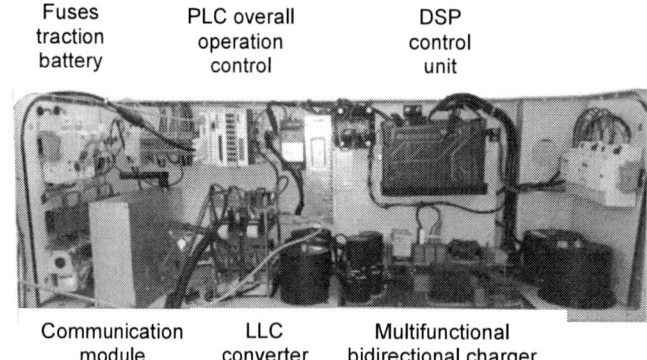

Fig. 7   System demonstrator box with integrated multifunctional bidirectional battery charging system

One set point for the LLC converter system (see Fig. 2) frequency can be a voltage transfer function unity gain of 1 [10]. That means the input voltage Vin is equal to Vout. Practical tests in the lab have shown that increasing the charging load of the traction battery results in a decrease of the output voltage Vout. So the transfer function is also related to the power (see Fig. 8). However, with such system behavior, continuous control of unity voltage gain depends on the actual load and can lead to dangerous states during system faults (significant power leap). To overcome this problem, the desired transfer voltage gain has to be tracked once unsing a defined gain. This transfer voltage gain should depend on the most commonly used charging power and can differ from 1.

The transfer function also depends on the position of the vehicle and the coil in relation to the charging station. In the lab, different characteristic lines were measured (see Fig. 8). By variing the pulse frequency and the power, the gain in the transfer function was measured. To ascertain the most practical transfer ratio, the position is also changed in each measurement:

1) x=0cm, y=0cm, z= 20cm
2) x=0cm, y=0cm, z=15cm
3) x=10cm, y=0cm, z=20cm
4) x=0cm, y=10cm, z=20cm

The optimal frequency here can be found by the most linear transfer function that is around 133 kHz - 135 kHz with y = 0 cm displacement (see Fig. 8). In this case and low power the transfer gain is between 1.03 - 0.98 and decreases by rising power up to 3 kW to 0.95 - 0.92. Using lower frequencies can lead to overvoltages [10].

The displacement in y direction has the strongest impact on the gain-power characteristic and here a transfer ratio of 1 is never achieved. So the exact positioning of the vehicle is imperative when using double D coil systems. Even a tracking algorithm must be limited in its frequency range.

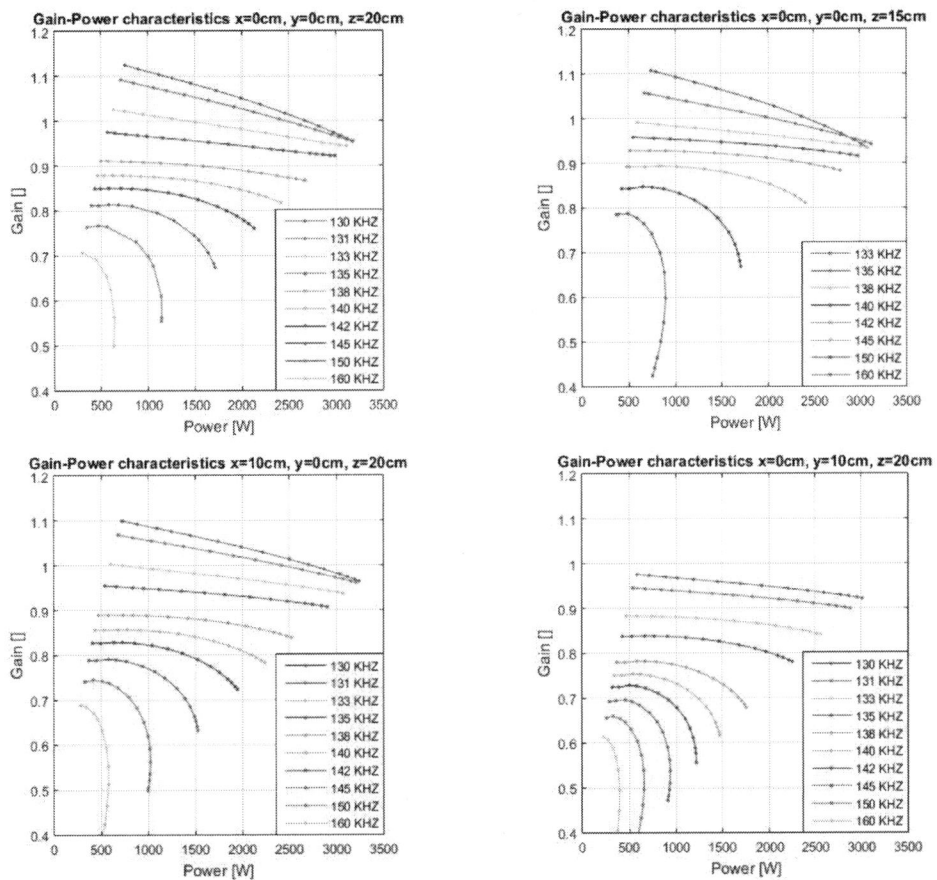

Fig. 8   Output Gain – Power characteristics of LLC converters with coil system at different positions (Input Voltage Vin= 400V) while charging the traction battery

To charge or discharge the battery, a defined sequence has to be started after positioning the electrical vehicle exactly (see Fig. 9).

From that position, a pragmatic approach is necessary to find the operation frequency of the LLC converter system. In order to track the wanted transfer gain (frequency), an additional minimum load is required at the beginning, because the multifunctional bidirectional battery charger is offline during this process and the capacitor Cout (see Fig. 2 left) must be continously discharged to obtain a stable voltage. This is realized by the discharging resistor R1 (normally used for protection purposes) of the LLC converter which can be connected parallel to Cout by closing the switch S1. The resistor R1 will be disconnected when the charging process of the traction battery starts.

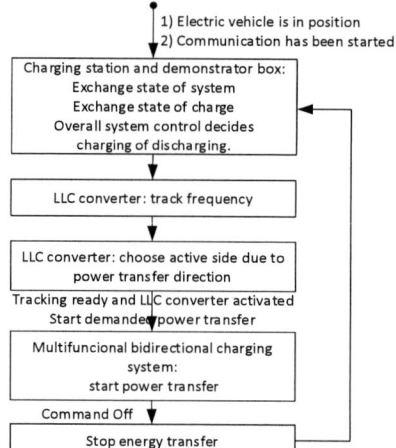

Fig. 9   Proposed charging/discharging progress of demonstrator system

The duration of the tracking process relates to the transfer speed of the communication module. A new frequency will be adjusted only if the external charging station on the parking space receives the voltage feedback from the vehicle. In this approach, the frequency tracker starts from a frequency much higher than the resonant frequency to avoid overcurrents at the beginning.

5. Automated system for electromagnetic compatibility (EMC) measurements

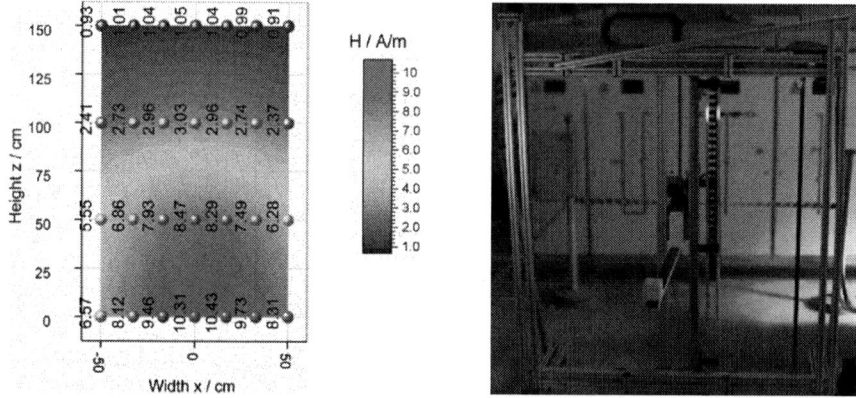

Fig. 10   Left: Magnetic field strength on a vertical plane 50 cm apart from the coil system. 28 individual points have been measured automatically. The contour plot is an interpolation between these points. Right: Positioning system with field probe mounted on a wooden bar.

At the Fraunhofer IWES a test setup according to IEC 61980 has been developed [11]. To measure the electromagnetic stray field, a field probe is used which can be moved in three orthogonal directions by a computer-controlled 3-D positioning system (see Fig. 10). With an additional system, it is possible to control mechanical offsets between both coils of the wireless power transfer (WPT) system to simulate the case where the car is not placed in the optimum position. An offset between the WPT coils reduces the coupling between them and leads to higher stray fields if the transmitted power is constant.

© VDE VERLAG GMBH · Berlin · Offenbach

All components are controlled by a LabView programme which runs on a standard PC

In contrast to broadband measurements, the Fraunhofer IWES system can deliver high resolution spectrum measurements of the magnetic field. This can help to investigate the source of distortions and losses. Additionally, tuned single frequency measurements with different resolution bandwidths are possible if time saving measurements with a large number of positions is necessary.

6. Conclusion

In this paper, a multifunctional bidirectional charging concept for electric vehicles enabling 1-phase wired, 3-phase wired as well as wireless charging of the vehicle battery has been presented. The multiple use of existing components in the vehicle allows a weight, volume and cost reduction compared to other charging concepts while still allowing the same flexibility. Furthermore, control strategies have been presented which increase the efficiency of the multifunctional bidirectional battery charger in different operation modes. Additionally a practical control strategy for tracking the most applicable frequency for the bidirectional inductive power transfer system and a system to perform automated electromagnetic compatibility (EMC) measurements are presented.

7. References

[1] M. Jung, H. Barth, and M. Braun, "Höher integrierter Stromrichter - Kombiniert kabelgebundenes und induktives Laden von Elektrofahrzeugen," in *Internationaler ETG-Kongress 2011: Umsetzungskonzepte nachhaltiger Energiesysteme - Erzeugung, Netze, Verbrauch ; Vorträge des Internationalen ETG-Kongresses vom 8. - 9. November 2011 in Würzburg ; Fachtagungen 1 bis 5*, Berlin [u.a.]: VDE-Verlag, 2011, pp. 638–643.

[2] G. Lempidis, Y. Zhang, M. Jung, R. Marklein, S. Sotiriou, and Y. Ma, "Wired and wireless charging of electric vehicles: A system approach," in *2014 4th International Electric Drives Production Conference (EDPC)*, pp. 1–7.

[3] *VDE-AR-E 2122-4-2 - Elektrische Ausrüstung von Elektro-Straßenfahrzeugen - Induktive Ladung von Elektrofahrzeugen - Teil 4-2: Niedriger Leistungsbereich*, 2011.

[4] J.-H. Jung, H.-S. Kim, J.-H. Kim, M.-H. Ryu, and J.-W. Baek, "High efficiency bidirectional LLC resonant converter for 380V DC power distribution system using digital control scheme," in *2012 IEEE Applied Power Electronics Conference and Exposition - APEC 2012*, pp. 532–538.

[5] D. Kürschner, *Methodischer Entwurf toleranzbehafteter induktiver Energieübertragungssysteme*. Aachen: Shaker, 2010.

[6] Marco Jung, Georgios Lempidis, René Marklein, Axel Seibel, and Jonas Steffen, "Kabelgebundenes und kabelloses bidirektionales Laden von Elektrofahrzeugen," in *Elektrik, Elektronik in Hybrid- und Elektrofahrzeugen*, Renningen: expert-Verl, 2015, pp. 459–481.

[7] M. Jung, G. Lempidis, D. Holsch, and J. Steffen, "Control and optimization strategies for interleaved dc-dc converters for EV battery charging applications," in *2015 IEEE Energy Conversion Congress and Exposition*, pp. 6022–6028.

[8] M. Jung, G. Lempidis, D. Holsch, and J. Steffen, "Optimization considerations for interleaved DC-DC converters for EV battery charging applications, in terms of partial load efficiency and power density," in *2015 17th European Conference on Power Electronics and Applications (EPE '15 ECCE Europe)*, pp. 1–9.

[9] F. Zare and A. Nami, "A New Random Current Control Technique for a Single-Phase Inverter with Bipolar and Unipolar Modulations," in *2007 Power Conversion Conference - Nagoya*, pp. 149–156.

[10] G.-C. Hsieh, C.-Y. Tsai, and S.-H. Hsieh, "Design Considerations for LLC Series-Resonant Converter in Two-Resonant Regions," in *2007 IEEE Power Electronics Specialists Conference*, pp. 731–736.

[11] J. Kirchhof, O. Strecker, R. Marklein, M. Jung, G. Lempidis, M. Wang und M. Z. Rahen, "Automatisierte Messung von räumlichen Streufeldern beim induktiven Laden," in *EMV 2016*

Design of a Contactless Energy Transfer System for an Electric Vehicle

Mike Böttigheimer, University of Stuttgart, Pfaffenwaldring 47, 70569 Stuttgart, Germany, boettigheimer@iew.uni-stuttgart.de

Nejila Parspour, University of Stuttgart, Pfaffenwaldring 47, 70569 Stuttgart, Germany, parspour@iew.uni-stuttgart.de

Marco Zimmer, University of Stuttgart, Pfaffenwaldring 47, 70569 Stuttgart, Germany, zimmer@iew.uni-stuttgart.de

Anna Lusiewicz, University of Stuttgart, Pfaffenwaldring 47, 70569 Stuttgart, Germany, lusiewicz@iew.uni-stuttgart.de

Abstract

This paper presents the design and dimensioning of a contactless energy transfer setup for an electric vehicle. The aim is to operate the setup in two different power levels with the same design for the secondary and primary side respectively. The resulting reactive power is compensated by capacitors in series on both sides. It is aimed to keep the battery voltage and the coupling factor variable in a defined range.

The adjustment of both power levels to the characteristic battery curve is realized by switching the power electronics on the secondary side between a half bridge and a full bridge rectifier. The system is dimensioned such that the transfer function has the same slope like the battery's characteristic curve. The greatest appearing coupling factor determines the inductivity on the secondary side, whereas the primary side's inductivity is fixed by the maximum voltage at the input inverter. Depending on the coil's current limitation and the capacitor's voltage limitation there are two alternatives for the primary side's coil size and the inverter topology.

1. Introduction

A contactless energy transfer in electric vehicles means transferring energy via coupled coils. Usually, the primary coil is fixed in the floor whereas the secondary coil is located in the car's underbody. In this paper a general dimensioning is presented for the contactless energy transfer setup. For the interoperable control it is necessary to consider the switching of the power electronics in the firmly embedded transfer setup. This aspect will be closer investigated and the limitations of the system setup will be presented.

2. Fundamentals of an Inductive Charging System

An inductive energy transfer setup can be split in two parts, the primary side and the secondary side. On each side there is an inductivity L, which is compensated for the resonant frequency with a capacitor C. The influence from the respectively other side is represented by the current I through the mutual inductance M. Fig. 1 shows an inductive energy transfer setup. The transfer function is presented in Fig. 2.

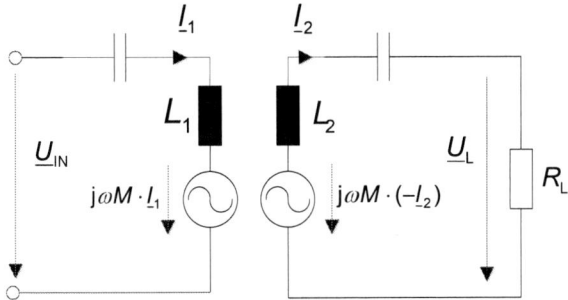

Fig. 1: An inductive energy transfer setup

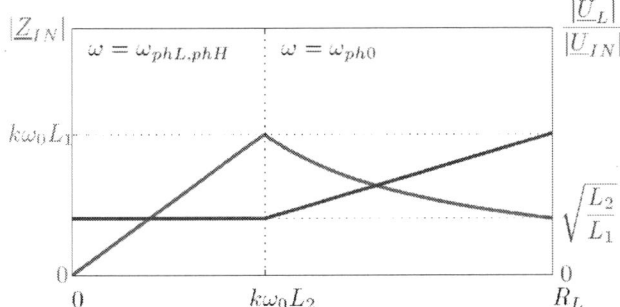

Fig. 2: Transfer function of an inductive energy transfer setup with series compensation on both sides [1]

Using Kirchhoff's laws, the transfer function on the right hand side of the discontinuity is shown in the following equation [1]:

$$\frac{U_L}{U_{IN}} = \frac{R_L}{\omega_{ph0} k \sqrt{L_1 L_2}}$$

(2.1)

The operation of a contactless energy transfer setup for electric vehicles requires two particularities: A power inverter with an alternating voltage in the range of a few kHz at the input and a rectifier (AC/DC) to obtain a direct voltage for charging the battery at the output (Fig. 3).

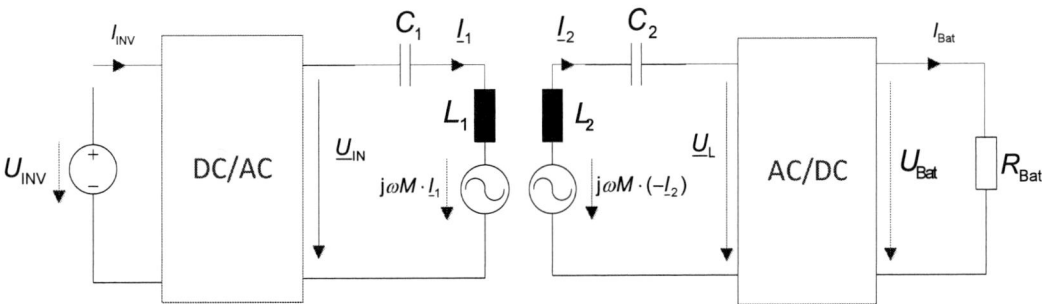

Fig. 3: The complete inductive charging setup

© VDE VERLAG GMBH · Berlin · Offenbach

The battery voltage and the battery current have to be transformed in order to obtain the correct calculations for the system left of the rectifier. With the transformed variables the calculation of the setup can be made. The following equations describe the transformation that contains a distinction between a full bridge rectifier (FB) and a half bridge rectifier (HB) [2]:

$$U_{Bat,FB} = \frac{\pi}{2\sqrt{2}} \cdot U_L \approx 1,11 \cdot U_L \tag{2.2}$$

$$U_{Bat,HB} = \frac{\pi}{\sqrt{2}} \cdot U_L \approx 2,22 \cdot U_L \tag{2.3}$$

$$I_{Bat,FB} = \frac{2\sqrt{2}}{\pi} \cdot I_L \approx 0,9 \cdot I_L \tag{2.4}$$

$$I_{Bat,HB} = \frac{\sqrt{2}}{\pi} \cdot I_L \approx 0,45 \cdot I_L \tag{2.5}$$

The transformed equivalent load resistance of the battery at the input of the rectifier can be calculated by using equations (2) - (5):

$$R_{L,FB} = \frac{8}{\pi} \cdot R_{Bat} \approx 0,81 \cdot R_{Bat} \tag{2.6}$$

$$R_{L,HB} = \frac{2}{\pi} \cdot R_{Bat} \approx 0,2 \cdot R_{Bat} \tag{2.7}$$

3. Boundary conditions, definitions and degrees of freedom

For the desired system the following presets are necessary:

Parameter	Symbol	Value
Battery voltage range	U_{Batt}	259 V..394 V
Resonant frequency	f_0	85 kHz
Compensation topology		1s2s
Coupling factor	k	0,2..0,3
Power level 1 for the battery	P_1	3 kW
Power level 2 for the battery	P_2	12 kW
Efficiency	η	> 0,9
Input voltage range of the inverter.	U_{INV}	0 V..650 V

Table 1: Design parameters

4. System Dimensioning and Operating Points

It is aimed to charge the battery with two different power levels. Either a full bridge rectifier or a half bridge rectifier can be used as a secondary side rectifier. If a half bridge rectifier is used instead of a full bridge rectifier, the equivalent load resistance is increased by the factor of 4, like shown in the equations (2.6) and (2.7). For dimensioning of the inductive transfer setup the two power levels P_1 and P_2 will be used. The dielectric strength of the compensation capacitors C_1, C_2 gives a restriction for the system's operation. Since the highest power occurring limits this influence, the complete dimensioning in the following equations is calculated with P_2. The calculation disregards ohmic losses.

The equivalent load resistance of the empty battery amounts to:

$$R_{\text{Bat,empty}} = \frac{U_{\text{Bat,empty}}^2}{P_{\text{Bat}}} \tag{4.1}$$

The equivalent load resistance of the full battery can be calculated as well:

$$R_{\text{Bat,full}} = \frac{U_{\text{Bat,full}}^2}{P_{\text{Bat}}} \tag{4.2}$$

Because of the rectifier at the output of the setup, the equivalent load resistance of the battery both for a full bridge and a half bridge rectifier calculates to:

$$R_{\text{L,empty}} = 0{,}9^2 \cdot R_{\text{Bat,FB,empty}} = 0{,}45^2 \cdot R_{\text{Bat,HB,empty}} \tag{4.3}$$

And for the full battery:

$$R_{\text{L,full}} = 0{,}9^2 \cdot R_{\text{Bat,FB,full}} = 0{,}45^2 \cdot R_{\text{Bat,HB,full}} \tag{4.4}$$

Parameter for the empty battery	Symbol	Value
Load resistance of the battery for $P_2 = 12\,\text{kW}$	$R_{\text{Bat2,empty}}$	5,6 Ω
Load resistance of the battery for $P_1 = 3\,\text{kW}$	$R_{\text{Bat1,empty}}$	16,9 Ω
Equivalent load resistance at the output of the transfer setup for $P_1 = 3\,\text{kW}$ and $P_2 = 12\,\text{kW}$	$R_{\text{L,empty}}$	4,5 Ω

Table 2: Calculated parameters for the empty battery

Parameter for the full battery	Symbol	Value
Load resistance of the battery for $P_2 = 12\,\text{kW}$	$R_{\text{Bat,full}}$	12,9 Ω
Load resistance of the battery for $P_1 = 3\,\text{kW}$	$R_{\text{Bat,full}}$	25,7 Ω
Equivalent load resistance at the output of the transfer setup for $P_1 = 3\,\text{kW}$ and $P_2 = 12\,\text{kW}$	$R_{\text{L,full}}$	10,5 Ω

Table 3: Calculated parameters for the full battery

At the characteristic ratio $r_c = 1$, the reactive power in the system is at its minimum [3]. With R_c as the characteristic load, the characteristic ratio is defined as:

$$r_c = \frac{k\omega_0 L_2}{R_L} = 1 \tag{4.5}$$

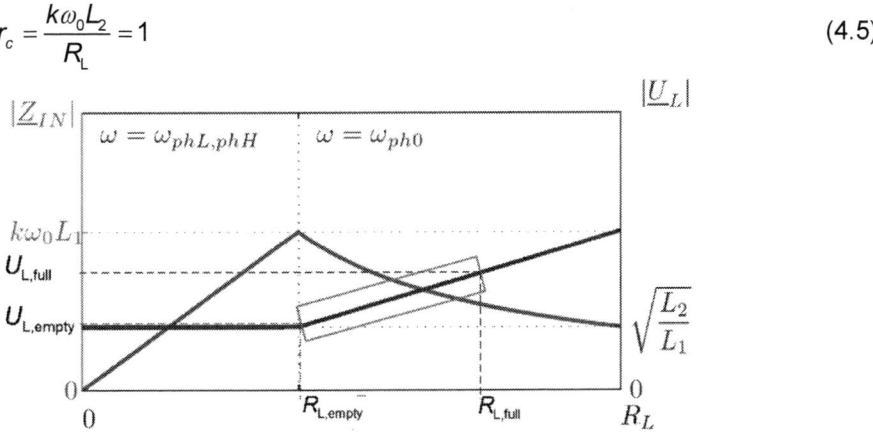

Fig. 4: Operating range of the inductive energy transfer setup (framed)

To properly use the increasing transfer function of a double side series compensated inductive transfer setup, the inductive transfer setup has to be adjusted to the increasing battery voltage. Ideally, the increasing line of the transfer function will be adjusted to the equivalent load resistance of the battery. This means for the dimensioning, that the following equation has to be valid:

$$R_c = R_{L,empty} \tag{4.6}$$

In Fig. 4 the operation range of the transfer setup is shown. Inserting the existing values into equation (4.5) and solving it, the inductivity of the secondary side is given to:

$$L_2 = \frac{R_{L,empty}}{k\omega_0} \tag{4.7}$$

The frequency is fixed for the dimensioning. During operation, the frequency can be changed in a little range up to $f_0 + 500\,\text{Hz}$ to adjust the switching behavior of the primary side inverter, but not to balance a change of the coupling factor. Changing the coupling factor by an adjustable frequency would require variable capacitors, which is not realistic regarding aspects of costs. The inductivity of the secondary side coil should be set to one fixed value. The coupling factor ranges between $k = 0,2..0,3$, which was determined by simulations of the two coils at the Institute of Electrical Energy Conversion [4]. To maintain the operating point above the characteristic load resistance R_C, the biggest possible coupling factor k_{max} of the system has to be considered for the dimensioning. The inductivity for the secondary coil is given to:

$$L_2 = \frac{R_C}{k_{max} \cdot \omega_0} \tag{4.8}$$

Parameter	Symbol	Value
Inductivity of the secondary side	L_2	28,1 µH
Dielectric Strength of the secondary side capacitor	$U_{C2,max}$	1092 V

Table 4: Parameters for the secondary side

The inverter on the primary side can also be chosen as either a full bridge inverter or a half bridge inverter. The maximum voltage at the input of the inverter is given by $U_{INV} = 650\,\text{V}$. In the following equations, the voltages at the output of the inverter are calculated for both full bridge and half bridge rectifiers:

$$U_{IN,FB} = 0,9 \cdot U_{INV} \tag{4.9}$$

$$U_{IN,HB} = 0,9 \cdot 0,5 \cdot U_{INV} \tag{4.10}$$

For each of the these two options one inductivity for the primary side can be determined:

$$L_{1,HB} = \frac{1}{L_2} \cdot \left(\frac{R_{L,empty} U_{IN,HB}}{\omega_0 k_{max} U_{L,empty}} \right)^2 \tag{4.11}$$

$$L_{1,FB} = \frac{1}{L_2} \cdot \left(\frac{R_{L,empty} U_{IN,FB}}{\omega_0 k_{max} U_{L,empty}} \right)^2 \tag{4.12}$$

The currents and voltages in the system have to be checked for both possible values. If one value leads to improper coil current or capacitor voltage, the dimensioning has to be adjusted or the other option has to be chosen. The input values to the inductive transfer setup for the lowest possible coupling factor of $k_{min} = 0,2$ are given to:

$$U_{\text{IN,HB,full}} = \sqrt{L_1 \cdot L_2} \cdot \frac{\omega_0 k_{\text{min}}}{R_{\text{L,full}}} \cdot U_{\text{L,full}} \tag{4.13}$$

$$I_{\text{IN,HB,full}} = \frac{P_{\text{IN}}}{U_{\text{IN}}} \tag{4.14}$$

$$U_{\text{IN,FB,full}} = \sqrt{L_1 \cdot L_2} \cdot \frac{\omega_0 k_{\text{min}}}{R_{\text{L,full}}} \cdot U_{\text{L,full}} \tag{4.15}$$

$$I_{\text{IN,FB,full}} = \frac{P_{\text{IN}}}{U_{\text{IN}}} \tag{4.16}$$

These values mark the worst case for the system's operation. With the aid of the currents from the equations (4.14) and (4.16), the highest voltages at the capacities can be calculated. The worst coupling factor for the operation is the lowest possible coupling factor $k_{\text{min}} = 0,2$:

$$U_{\text{C2,max}} = \sqrt{2} \cdot \frac{I_{\text{L,empty}}}{\omega \cdot C_2} \tag{4.17}$$

$$U_{\text{C1,max,HB}} = \sqrt{2} \cdot \frac{I_{\text{IN,HB,full}}}{\omega \cdot C_{\text{1,HB}}} \tag{4.18}$$

$$U_{\text{C1,max,FB}} = \sqrt{2} \cdot \frac{I_{\text{IN,FB,full}}}{\omega \cdot C_{\text{1,FB}}} \tag{4.19}$$

The maximum voltage $U_{\text{C2,max}}$ is independent of the primary side inductivity, if the same electrical power is transferred and ohmic losses are neglected. For the full bridge inverter and the half bridge inverter the voltage $U_{\text{C2,max}}$ reaches its maximum when the current I_{L} also reaches its maximum. This happens when the battery is empty. The voltage at the primary side capacitor for both operation points depends on the particular capacities $C_{\text{1,FB}}$ and $C_{\text{1,HB}}$ and on maximum input current into the inductive transfer setup. The maximum input current $I_{\text{IN,full}}$ appears with full battery.

The equations (4.17) and (4.21) show that at a half bridge inverter the current on the primary side is very high with $I_{\text{IN,HB,full}}$, whereas using a full bridge inverter the voltage at the primary side capacitor is very high with $U_{\text{C1,max,HB}}$. Both operation points are possible for the dimensioning, it depends on how the worst case values are going to be handled technically.

Parameter	Symbol	Value
Maximum input voltage of the transfer setup	$U_{\text{IN,HB}}$	292,5 V
Inductivity of the primary side	$L_{\text{1,HB}}$	44,2 µH
Input voltage on the transfer setup	$U_{\text{IN,HB,full}}$	127,7 V
Input current on the transfer setup	$I_{\text{IN,HB,full}}$	94,2 A
Dielectric strength of the primary side capacitor	$U_{\text{C1,max,HB}}$	3145 V

Table 5: Parameter for the half bridge inverter on the primary side

© VDE VERLAG GMBH · Berlin · Offenbach

Parameter	Symbol	Value
Maximum input voltage of the transfer setup	$U_{IN,FB}$	585 V
Inductivity of the primary side	$L_{1,FB}$	176,9 µH
Input voltage on the transfer setup	$U_{IN,FB,full}$	283,1 V
Input current on the transfer setup	$I_{IN,FB,full}$	47,1 A
Dielectric strength of the primary side capacitor	$U_{C1,max,FB}$	6291 V

Table 6: Parameter for the full bridge inverter on the primary side

5. Conclusion

In this paper the general design of a contactless energy transfer setup for electric vehicles is presented. The switching from half bridge to full bridge was achieved on the primary side and also on the secondary side. For the decision whether to choose a half bridge inverter or a full bridge inverter at the primary side it is important to know how to handle the voltages and the currents technically

A. References

[1] Zimmer, M.: "Blindleistungskompensation und Regelverfahren für kontaktlose Energie-übertragungsstrecken mit translatorischen Freiheitsgraden". Diploma Thesis, Institute of Electrical Energy Conversion, (2011)

[2] Abschlussbericht Bipol: "Berührungsloses, induktives und positionstolerantes Ladekonzept für elektrisch angetriebene Fahrzeuge." Verbundforschungsprojekt mit Förderung des Ministeriums für Finanzen und Wirtschaft Baden-Württemberg, (2011)

[3] Zimmer, M., Heinrich, J., Parspour, N.: "Design of a 3kW Primary Power Supply Unit for Inductive Charging Systems Optimized for the Compatibility to Receiving Units with 20kW Rated Power". ETEV (2014)

[4] Heinrich, J. and Parspour, N.: "Contribution to the Development of Positioning Tolerant Inductive Charging System". VDE Congress E-Mobility, Leipzig, (2010)

B. Acknowledgment

The results of the paper were developed in the project "B2LE – Betriebssicherheit von berührungsloser Ladeinfrastruktur in Elektrofahrzeugen", also called "CETeCAR", which is supported by the Vector Foundation in Stuttgart.

PCIM Europe 2016, 10 – 12 May 2016, Nuremberg, Germany

3300V HiPak2 modules with Enhanced Trench (TSPT+) IGBTs and Field Charge Extraction Diodes rated up to 1800A

Chiara Corvasce, Maxi Andenna, Sven Matthias, Liutauras Storasta, Arnost Kopta, Munaf Rahimo, Luca De-Michielis, Silvan Geissmann, Raffael Schnell
ABB Switzerland Ltd. Semiconductors, Fabrikstrasse 3, 5600 Lenzburg, Switzerland
e-mail: chiara.corvasce@ch.abb.com

Abstract

In this paper, we introduce the new generation 3300V HiPak2 IGBT module (130x190)mm employing the recently developed TSPT+ IGBT with Enhanced Trench MOS technology and Field Charge Extraction (FCE) diode. The new chip-set enables IGBT modules with improved electrical performance in terms of low losses, good controllability, high robustness and soft diode recovery. Due to the lower losses and the excellent SOA, the current rating of the 3300V HiPak2 module can be increased from 1500A for the current SPT+ generation to 1800A for the new TSPT+ version.

1. Introduction

Over the past two decades, the IGBT and antiparallel diode have experienced important performance breakthroughs due to improved device processes and design concepts. Nevertheless, further development work has shown the potential to achieve higher power densities, improved controllability and robustness. The trend for lowering the on-state losses of IGBTs has remained one of the main goals of this device concept development. After the introduction of the Field Stop (FS) / Soft-Punch-Through (SPT) buffer structures for providing thinner structures for low losses, the focus in recent years has shifted primarily towards further improvements on the MOS cell design with important steps made on both Planar and Trench cell concepts [1].

The fast diode also underwent similar developments to match the IGBT capabilities through new anode and buffer designs combined with optimized lifetime control for achieving low losses, soft recovery and high SOA capability.

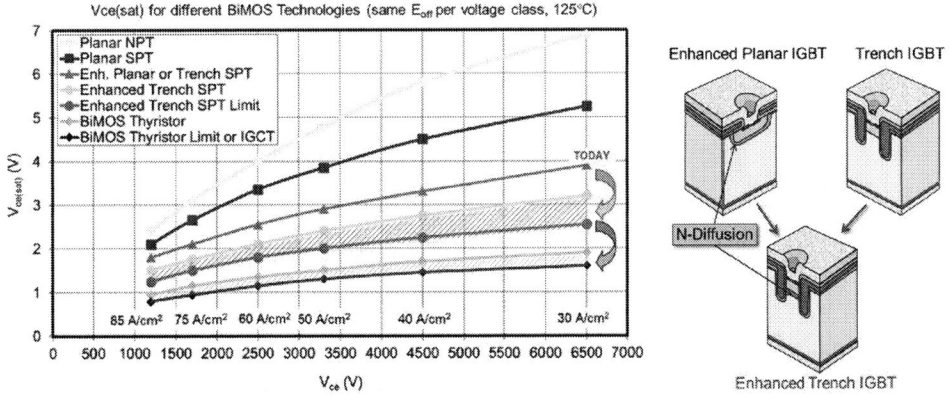

Fig. 1.   IGBT technologies and impact on on-state losses for different voltage ratings.

© VDE VERLAG GMBH · Berlin · Offenbach

Figure 1 shows an estimation of the on-state V_{ce_sat} loss reductions per voltage class with each generation of IGBT technology. For high voltage devices (>2kV), Enhanced Planar (EP) IGBT or Trench IGBT MOS cell concepts are employed on SPT structures demonstrating similar power ratings and loss performance. Nevertheless, for lower voltage devices (<2kV), the Enhanced Trench MOS cell employing an n-enhancement layer in the active trench region is already an established technology [2]. It is important to point out that trench based IGBTs, especially for higher voltages, exhibit an inherently higher effective gate input capacitance when compared to planar IGBTs which results in less controllability for optimum switching performance during IGBT turn-on [3][4]. The new TSPT+ presented in this paper is overcoming this negative aspect while offering a low loss performance and therefore provides an ideal solution for the next generation high voltage IGBTs [5].

For the anti-parallel diode, the losses and recovery softness remain the crucial performance targets in order to match the performance of the new TSPT+ IGBT. The FCE concept enables soft recovery under critical conditions in combination with low losses without compromising ruggedness [6].

2. New technologies and design elements

2.1. TSPT+ IGBT technology

In order to achieve the targeted enhanced carrier concentration near the trench emitter, aimed to reduced losses, the TSPT+ uses a striped architecture for the active Trench MOS Cell in combination with an n-enhancement layer encompassing the p-channel regions, as shown in Fig. 2.

A floating p+ well, deeper than the trench depth, is introduced in the region between active cell pairs and its capacitive coupling to the emitter potential node is optimized in order to achieve a low effective gate input capacitance compared to state-of-the-art trench IGBT designs while providing optimum reverse blocking capability. Moreover, the enhancement layer is used outside the trench MOS cell to separate the active trench gates from the floating deep p+ well. Such feature allows the reduction of the hole accumulation in the regions between active cells during turn-on switching, which plays a crucial role in defining the Miller capacitance and the consequent dI/dt.

The combination of the described enhanced trench cell architecture with the already well-optimized SPT buffer technology enables the TSPT+ to provide excellent losses performance, smooth switching behavior and high turn-off ruggedness.

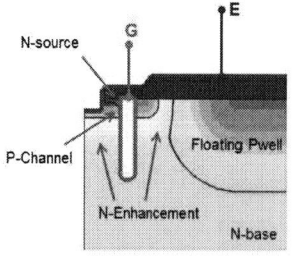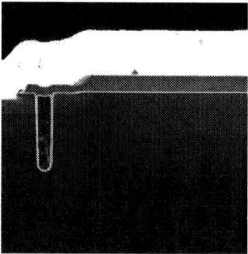

Fig. 2. Schematic and SEM cross section of half pitch TSPT+ IGBT cell

2.2. Field Charge Extraction Diode with Field Shielded Anode

Figure 3a shows the schematic cross-sections of the active area of the Field Shielded Anode (FSA) diode [7]. A high p+ doping forms the anode contact and a deep diffused p- buffer anode supports the electric field during blocking. An n-buffer layer is introduced at the cathode and the high n+ doping forms the contact. Deep levels generated by heavy ion-irradiation tailor the static and dynamic properties.

In contrast to the uniform backside contact of the FSA diode, the cathode of the FCE diode consists of a two-dimensional lattice of p+ islands embedded in the high n+ cathode doping (Fig. 3b and 3c). At the end of the recovery phase, when the electric field gets close to the cathode, the gain of the internal bipolar transistor increases and triggers an injection of holes from the p+ areas. This additional supply of holes enables a soft reverse recovery by preventing the stored charge to get too small to keep up a continuously decreasing reverse current as the electric field propagates towards the cathode, which would result in a current snap-off.

On the other hand, the forward voltage of the FCE diode depends on the buffer concentration, and on the area ratio of the p+ and n+ cathode regions: the larger the p+ area the more pronounced is the impact on the forward voltage V_F drop. By a thinner silicon and an optimum buffer design combined with a p+ area, which is kept less than 10% of the full diode area, the 3.3kV FCE diode is able to provide a forward voltage drop, which is only 50mV higher than the FSA diode platform for an improved softness behavior and with comparable blocking and SOA capability.

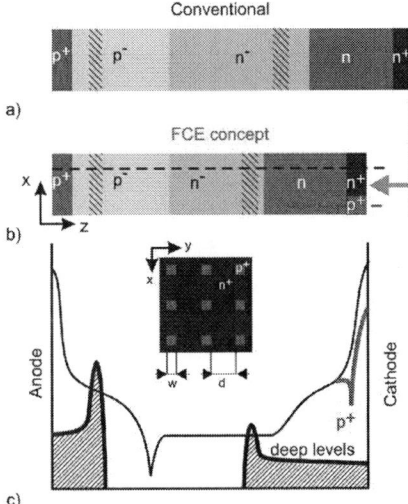

Fig. 3. Schematic drawing of the diode structure. a) Cross-section of the active area region of FSA diode b) Cross-section of the FSA with FCE concept using p+ areas at the cathode side. c) Schematic doping profiles with deep levels. In the inset: backside of the diode with the p+ islands in the n+ cathode contact. The width w and the distance d are critical design parameters for the performance.

3. Electrical performance of the 3.3kV HiPak2 module

The new 3.3kV HiPak2 module employing TSPT+ IGBT and FCE diode technologies is rated 1800A and specified for operating at a junction temperature of T_j=150°C. On module package side, the substrate layout is optimized to incorporate 20% more diode active area so that the diode performance can keep up with the increased current rating of the TSPT+ IGBT. This section illustrates the results of the static and dynamic measurements carried out to assess the module performance.

3.1.   TSPT+ IGBT characteristics and losses

In Fig. 4, the on-state curves of the 3.3kV TSPT+ IGBT are shown compared to the SPT+ characteristics. The typical on-state voltage drop at a nominal current of 1800A and T_j=150°C is 2.8 V. The TSPT+ IGBT shows a positive temperature coefficient for V_{ce_sat}, starting already at low currents, which enables a good current sharing capability between the individual chips in the module.

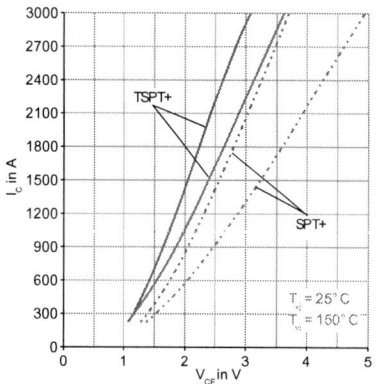

Fig. 4. HiPak2 Module conduction characteristics: 3.3kV TSPT+ IGBT compared to SPT+.

The turn-off waveforms shown in Fig. 5 are measured under nominal conditions, i.e. I_C=1800A and V_{CE}=1800V. In this test, the module is turned off using an external gate resistor $R_{g,off}$=1.5Ohm and a stray inductance L_S=100nH which results in a voltage rise with dV/dt=2230V/us. The optimized n-base region combined with the SPT buffer allows the collector current to decay smoothly, ensuring soft turn-off behavior without excessive voltage peaks or oscillations.

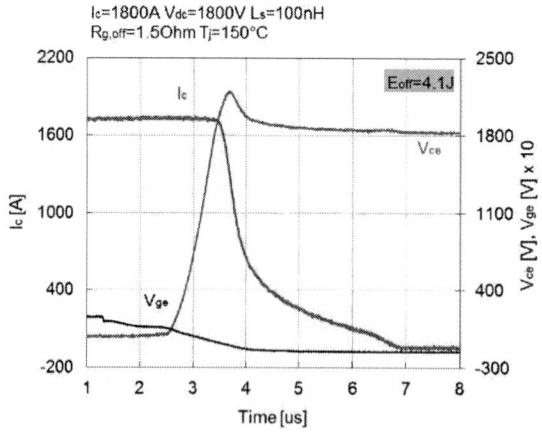

Fig. 5. 3.3kV TSPT+ IGBT turn-off under nominal conditions measured on module level.

The trade-off curves between the IGBT on-state voltage drop and the turn-off losses are presented in Fig.6 where the TSPT+ technology is compared to the SPT+ at the nominal current rate of the new module (1800A).The different sets of points on the technology curves correspond to IGBTs with different anode emitter efficiencies. The TSPT+ IGBT exhibits approximately 20% lower on-state voltage drop for the same turn-off losses as compared to the standard SPT+ technology.

In order to reduce the influence of the Miller capacitance on the turn-on speed, a gate emitter capacitor C_{ge}=330nF is used to characterized the turn-on switching. Since the external gate-emitter capacitance slows down the switching behavior of the IGBT, a gate resistors value $R_{g,on}$=1Ohm is used as an optimum driving condition.

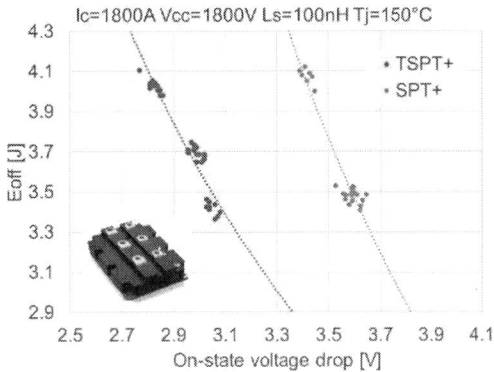

Fig. 6. 3.3kV IGBT technology curve: TSPT+ compared the SPT+ at I_c=1800A.

Figure 7 shows the turn-on waveforms under the mentioned gate driving conditions, which, in combination with the low FCE diode losses, brings the turn-on switching losses down to a typical value of 2.8J at T_j=150°C.

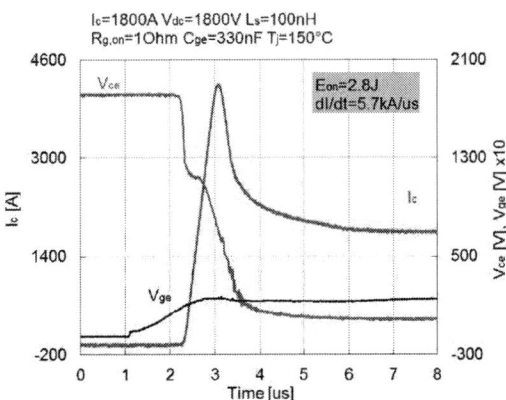

Fig. 7. 3.3kV TSPT+ IGBT turn-on under nominal conditions measured on module level.

© VDE VERLAG GMBH · Berlin · Offenbach

3.2. FCE diode characteristics and losses

In Fig. 8, the on-state characteristics of the 3.3kV FCE diode are shown. Due to the advanced plasma shaping utilizing double local lifetime control by irradiation, the diode has a positive temperature coefficient of V_F starting from I_F= 1200A. At rated current and T_j=150°C, the diode shows a typical on-state voltage drop of 2.3V.

The reverse recovery waveforms of the diode under nominal switching conditions is presented in Fig. 9. By carefully designing the cathode-sided lifetime irradiation peak, a short but still smoothly decaying current tail was achieved.

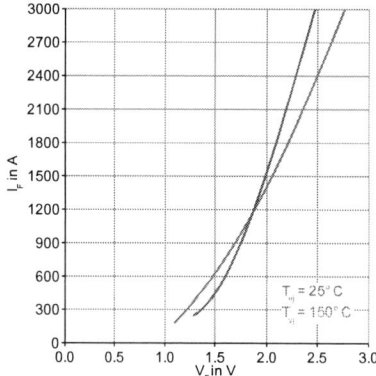

Fig. 8. Forward characteristics of the 3.3kV FCE diode (module level measurements).

Under nominal conditions and using $R_{g,on}$=1Ohm and C_{ge}=330nF, the diode recovery losses are 2.5J at T_j=150°C. Thanks to the high ruggedness and soft recovery behavior of the FCE technology, the diode can be switched with a high dI_F/dt of 5700A/us.

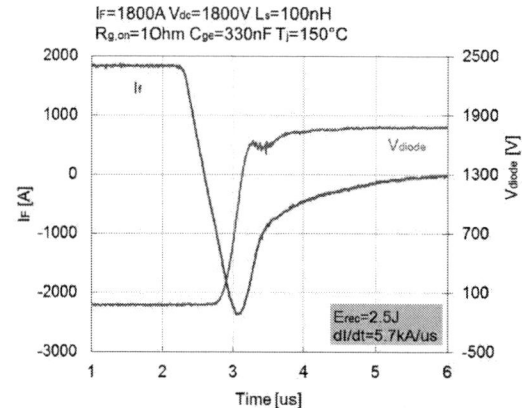

Fig. 9. 3.3kV FCE diode reverse recovery under nominal conditions measured on module level.

© VDE VERLAG GMBH · Berlin · Offenbach

3.3. IGBT turn-off and short circuit ruggedness

The cell architecture of the TSPT+ IGBT technology has been designed with respect to the cell cross section and in the direction perpendicular to the cell stripes in order to keep the high standard of turn-off ruggedness shown from the SPT+ generation. Figure 10a shows a turn-off waveform at module level, where a current of 10105A, which corresponds to more than five times the nominal current, is switched-off against a DC-link voltage of 2500V at a junction temperature of 150°C. The test is performed with an external gate resistance of 1.5Ohm and a stray inductance of 100nH without using any clamp or snubber.

The TSPT+ IGBTs are capable of sustaining a long period of strong dynamic avalanche during the turn-off transient showing an excellent SOA capability with a maximum turn-off peak power $P_{pRec}=26.8MW$.

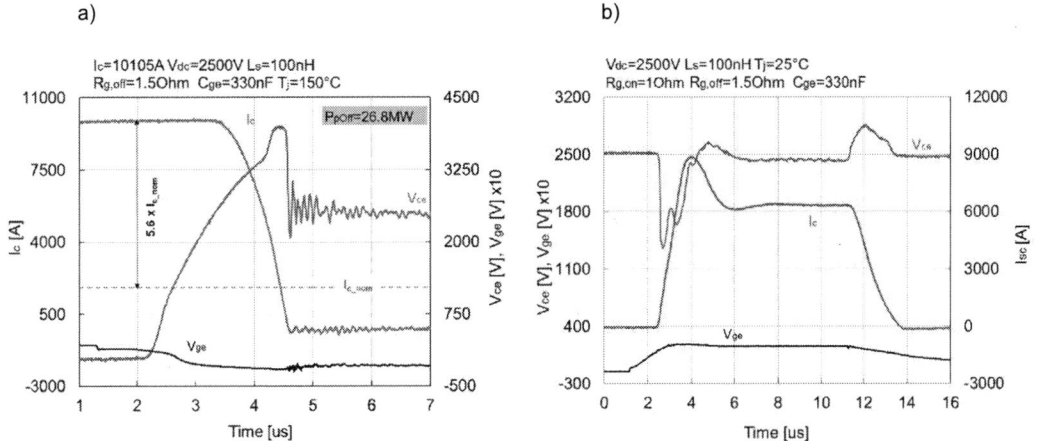

Fig. 10. 3.3kV TSPT+ IGBT module: a) turn-off under extreme SOA conditions b) short circuit under nominal SOA conditions.

The short circuit waveforms of the 3.3kV TSPT+ module measured at room temperature are shown in Fig. 10b. The SPT-buffer and the anode emitter efficiency of the IGBT have been optimized for the targeted short-circuit ruggedness to withstand a short circuit at $V_{GE}=15V$ for all DC-link voltages up to 2500V and junction temperatures up to $T_j=150°C$.

3.4. FCE diode softness and reverse recovery ruggedness

The diode recovery waveforms in Fig.11a show the comparison of the snap-off behavior at $T_j=150°C$ of the FSA and FCE diode when used as freewheeling diodes for the TSPT+ IGBT in the adverse conditions of very low current ($I_F=30A$) and very fast IGBT turn-on switching speed ($R_{g,on}=0.666Ohm$). A snap-off at the end of the tail phase is present for both diodes. However, the overvoltage for the FCE is 250V lower than for the FSA, which enables to operate it within the specification of the rated voltage. The beneficial effects of the p+ island hole injection is clear in the oscillations following the main peak which are immediately damped to the DC-link voltage in the case of the FCE.

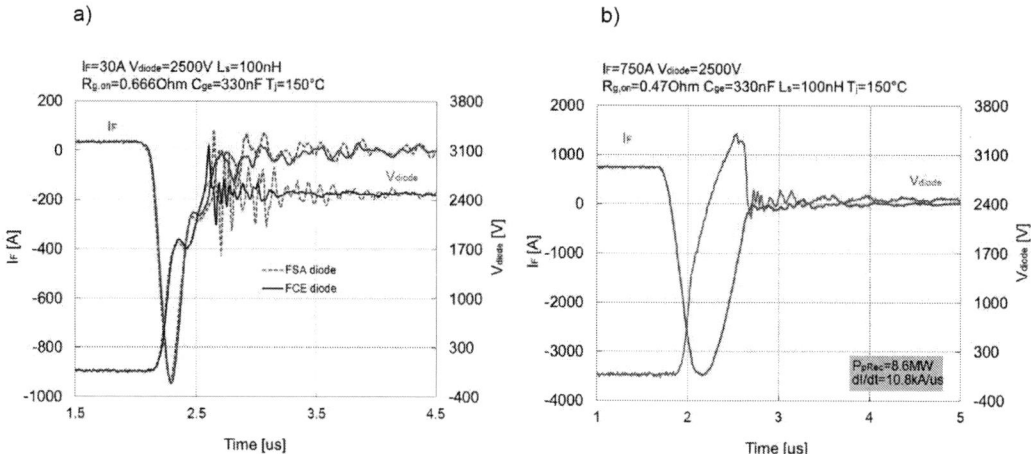

Fig. 11. 3.3kV HiPak2 1800A FCE diode: a) turn-off under snap-off conditions comparing FSA and FCE diode as auxiliary device for TSPT+ IGBT b) 3.3kV FCE diode turn-off under SOA conditions.

The diode reverse recovery SOA module test measured with a forward current of 750A and a DC-link voltage of 2500V is presented in Fig.11b. The diode is switched under extreme conditions at T_j=150°C using an external gate resistor of 0.47Ohm, which results in a switching speed of 10800A/us and a peak power of 8.6MW.

4. Conclusions

A new 3300V, 1800A HiPak2 IGBT module employing the TSPT+ IGBT with Enhanced Trench MOS technology and Field Charge Extraction FCE diode has been presented. Due to the advanced IGBT and diode device technologies, the module offers exceptionally low losses and high Safe Operating Area. Smooth switching characteristics and high dynamic ruggedness have also been demonstrated. The new module will provide high voltage system designers with enhanced current ratings and open possibilities for optimized designs. In the future, the new chip technologies will also be extended to higher voltage classes up to 6.5kV and employed in next generation modules, including the LinPak platform [8].

References

[1] Rahimo, M et al. "Novel Enhanced-Planar IGBT Technology rated up to 6.5kV for Lower Losses and Higher SOA capability" Proc. ISPSD'06, Italy.

[2] K. Nakamura et al, Proc. "Wide cell pitch 1200 V NPT CSTBTs with short circuit ruggedness" Proc. ISPSD'01, Japan.

[3] K. Nakamura et al, "Wide Cell Pitch LPT(II)-CSTBTTM(III) Technology Rating up to 6500V for Low Loss" Proc. ISPSD'10, Japan.

[4] Y. Toyota et al, "Novel 3.3kV Advanced Trench HiGT with Low Loss and Low dv/dt Noise" Proc. ISPSD'13, Japan.

[5] M. Andenna et al, "The Next Generation High Voltage IGBT Modules utilizing Enhanced-Trench ET-IGBTs and Field Charge Extraction FCE-Diodes" Proc. EPE`14, Finland.

[6] S. Matthias et al, "Field Shielded Anode concept enabling higher temperature operation of fast diodes" Proc. ISPSD'11, USA.

[7] S.Matthias et al, "Inherently Soft Free-Wheeling Diode for High Temperature Operation" Proc. ISPSD'13, Japan

[8] S.Hartmann et al. "The LinPak high power density design and its switching behavior at 1.7kV and 3.3kV" Proc. PCIM'16, Nürnberg.

PCIM Europe 2016, 10 – 12 May 2016, Nuremberg, Germany

Durable Design of the New HVIGBT Module

Nobuhiko Tanaka*, Shuichi Kitamura*, Kenji Ota*, Shinichi Iura*, Keiichi Nakamura*, Eugen Wiesner**, Eckhard Thal**

*Mitsubishi Electric Corporation, 1-1-1 Imajukuhigashi, Nishi-Ku, Fukuoka, Japan
**Mitsubishi Electric Europe B.V., Germany
E-mail: Tanaka.Nobuhiko@aj.MitsubishiElectric.co.jp

Abstract

Power semiconductors installed in high-reliable applications must maintain a certain quality in the field for periods up to thirty years. Using our proven high quality design approach, we developed a new High Voltage (HV) IGBT module (HVIGBT) called the X-Series. Based on realistic environmental conditions its robustness was verified to minimize the field failures. In this paper, we describe those design characteristics of the X-Series which are not generally indicated in the datasheets.

1. Introduction

Although power electronic applications such as railways and power distribution systems generally have a requirement to maintain the lowest possible failure rate, unexpected accidents sometimes do occur. Ideal power modules possess durability against unexpected conditions that are sometimes not defined in the specifications. However, since the possibility of predicting all unexpected application conditions is difficult, it is important to provide a suitable design margin for power modules without sacrificing performance. The following are our power module design approaches against the occurrence of unexpected accidents and tough environmental conditions.

2. Durable design concepts

2-1. Dynamic robustness against hard switching condition
Overload conditions

When using a power module, we have to consider operating environments such as circuit stray inductances, gate drive conditions, DC link voltage fluctuations, and ambient temperatures. The maximum turn-off switching current in a data sheet is commonly defined as twice the nominal current. The turn-off capability of an HVIGBT is mainly determined by the IGBT chip design. For example, by applying the partial P collector structure [1] [2], a high turn-off capability can be realized. Fig. 1 shows a cross-section schematic of the latest IGBT chip generation, including a partial P collector structure.

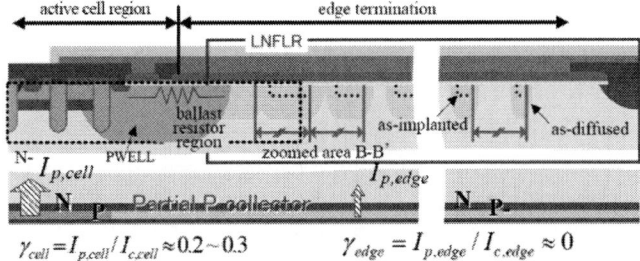

Fig. 1 Cross-section schematic of latest generation IGBT chip structure [1] [2].

© VDE VERLAG GMBH · Berlin · Offenbach

Short circuit conditions

Several types of short circuit modes exist, including during the IGBT's off-state (type 1), during its on-state (type 2), and during the diode's freewheeling mode (type 3) [3] [4]. There are principally two robust design approaches for high short circuit capability. One is the capability improvement of the IGBT chip, and the second is package design optimization to minimize the impact on the gate voltage by the magnetic flux due to the rise of the short circuit current. The short circuit capability of IGBT chips can be improved with respect to the latch-up immunity based on the design of the optimized MOS gate cell structure and the cell density (i.e., the design of a suitable saturation current). An electromagnetic field analysis was used to design an internal gate circuit to minimize the induced EMF that can be applied to the gate during a short circuit event [5].

Moreover, if a type 3 short circuit occurs, the diode might be destroyed by the reverse recovery phenomena. Thus, it is important that considering the type 3 short circuit, the module is designed with the corresponding diode ruggedness. By a employing the Relaxed Field of Cathode (RFC) diode [6] which possess a high-peak power ruggedness, the HVIGBT can obtain the capability to withstand the type 3 short circuit.

Driving limitation (di/dt during turn-on switching)

When a reverse recovery current flows through a conventional p-i-n diode, at high DC link voltage conditions, high surge voltages with diode ringing is generated by a snap-off phenomenon (Fig. 2). This can be suppressed by driving an IGBT with a low turn-on di/dt. Consequently, however this low di/dt increases the turn-on switching loss. Using an RFC diode, the snap-off phenomenon is suppressed even if the IGBT is driven at a high turn-on di/dt (without an increase in the turn on losses [6]). A cross-section of RFC diode chip structure is shown in Fig. 3.

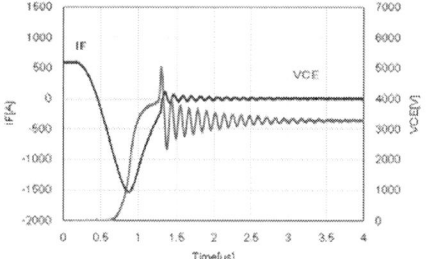
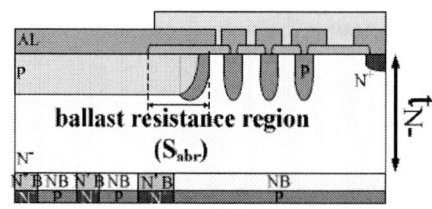

Fig. 2 (Left) Reverse recovery waveform example with high surge voltage by snap-off phenomenon.
Fig. 3 (Right) Cross-section schematic of RFC diode chip structure [6].

2-2. Thermal considerations and wider operation temperature ranges

Typically, the field failure rate of a semiconductor device increases with increase in operating temperature (based on the MIL standard (MIL-HDBK-217F), and the temperature factor of the field failure rate (π_T) at 150°C becomes 1.37 times higher than 125°C for a Si NPN transistor).

$$\pi_T = \exp\left\{-2114\left(\frac{1}{T_j + 273} - \frac{1}{298}\right)\right\} \qquad (1)$$

Table 1 Temperature factor of field failure rate

T_j	π_T	Ratio
125°C	5.94	1.00
150°C	8.14	1.37

Hence, understanding the limits and setting the design margins against high temperatures on semiconductor devices are important. The prevention of thermal runaway is a particularly crucial requirement under high temperature conditions for HVIGBT. In addition, we must pay sufficient attention to module materials and assembly processes that can withstand high temperature conditions.

Thermal runaway

Reducing the leakage current (I_{CES}) under high temperatures is an effective way for preventing thermal runaway events. The I_{CES} of a power semiconductor chip can be reduced by employing the following designing strategies:

1. The optimized N buffer layer profile (Fig. 4 Left)
2. Optimizing the lifetime control in the N⁻ drift region
3. Applying a suitable edge termination structure (Fig. 1) [7]
4. Strengthening the gettering wafer process to minimize the impurity-induced crystal defects in silicon (Fig. 4 Right).

By adopting the above technologies, the I_{CES} at 150°C is decreased to the same level as that of conventional module (considering 125°C).

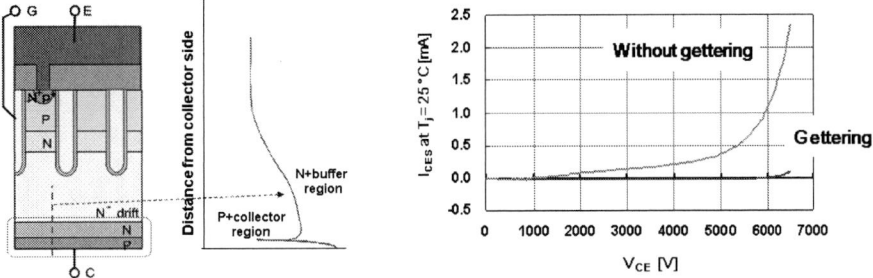

Fig. 4 (Left) Schematic of N buffer layer profile.
(Right) Example of I_{CES} with gettering process and I_{CES} without gettering process.

Insulation

One important characteristic of insulated power modules is the partial discharge characteristics for insulation stability. To maintain long-term, high insulation performance, a design approach is required that does not generate partial discharges after high temperature stresses. We adopted silicone gel, which has good insulation performance over the maximum operating temperature range and a suitable injection process was adopted to minimize the partial discharges due to air bubbles in the silicone gel.

2-3. Environments and reliabilities

For applications such as railways, controlling ambient temperature and humidity around power modules is difficult. Robust designs against high humidity during high voltage conditions are an effective measure to maintain the necessary power module quality.

Humidity and condensation

The capacity to endure conditions of high humidity can be improved by a chip edge termination design that is less likely to accumulate surface charges (Q_{ss}) on the guard ring (an occurrence when high voltages are being applied between the collector and the emitter). The amount of Q_{ss} on the guard ring mainly influences the chip's blocking voltage capability. By applying semi-insulating passivation film materials and processes that are less likely to accumulate Q_{ss} on the edge termination, we adopt suitable designs against I_{CES} increase while applying high voltage under tough environmental conditions (e.g., a dew condensation condition).

Long term DC stability

Long-term DC stability (LTDS), which is represented as cosmic ray ruggedness, is a significant aspect of an HVIGBT design. Our design ensures an HVIGBT's LTDS by Light Punch Through (LPT) structure on the collector side and I_{CES} reduction, as described previously [7].

3. Design verification results

We developed a new HVIGBT, called the X-Series, 6.5 kV, 1000A, with high robustness using durable design concepts described in the previous section. Next we describe our verification test results for durable designs.

3-1. Verification results of dynamic robustness

(1) Reverse Bias Safe Operating Area (RBSOA)

We verified the turn-off capability by gradually increasing the collector current (I_c) at a DC link voltage (V_{cc}) of 4500V and a maximum junction temperature (T_j) of 150°C for a half-bridge circuit with an inductive load. As shown in Fig. 5, we confirmed that the new HVIGBT can smoothly turn-off more than four times the nominal current (I_c=4000A) without being destroyed.

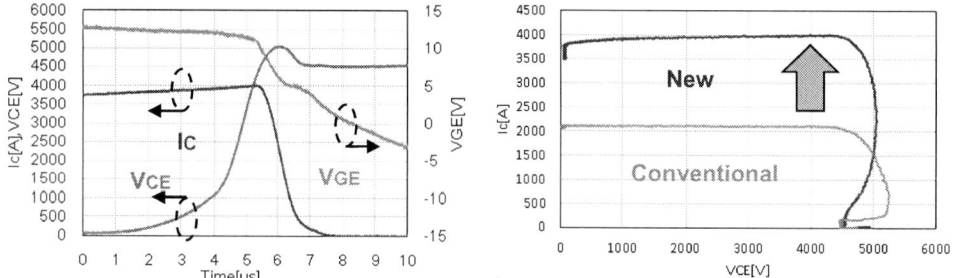

Fig. 5 (Left) Turn-off switching waveform at V_{cc}=4500V, I_c=4000A, T_j=150°C and L_s=150nH.
(Right) Comparison of I-V plots at identical V_{cc} and T_j conditions.

(2) Short circuit capability

The short circuit withstand capability is often indicated by the pulse width, i.e., the corresponding energy in the arm short circuit that is connected with low wiring inductance. However, various short circuit loops and modes actually exist. Below is shown an example of the verification result for short circuit modes during the IGBT on-state (type 2) and the freewheeling mode of an antiparallel diode (type 3). We carried out tests of short circuit types 2 and 3 by configuring the half-bridge circuit (Fig. 6) and the corresponding pulse sequences (Fig. 7). The switch (IGBT 3) was put on the short circuit loop. Because the DUT's short circuit current must not be limited by the saturation current of the IGBT 3, a segment of a module is used as the DUT (Fig. 6). The test conditions were V_{cc}=4200V and T_j=150°C. If no optimal short circuit capability is designed, the respective module might be destroyed after the short circuit current reaches its peak (Fig. 8). We confirmed that the new HVIGBT has sufficient robustness against short circuit types 2 and 3 (Fig. 9).

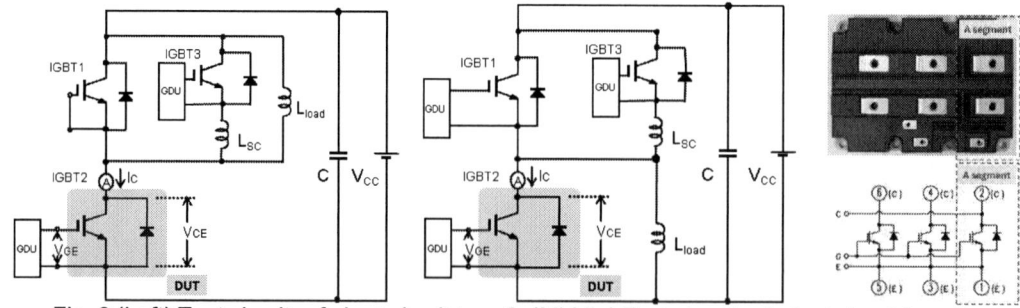

Fig. 6 (Left) Test circuits of short circuit type 2. (Middle) Test circuits of short circuit type 3.
(Right) DUT is a 1/3 module of a 6.5 kV HVIGBT (I_{nom}=333A).

PCIM Europe 2016, 10 – 12 May 2016, Nuremberg, Germany

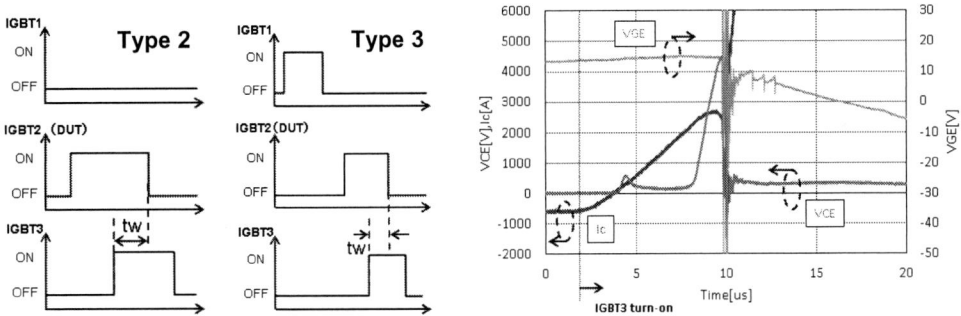

Fig. 7 (Left) Gate pulse sequences of short circuit types 2 and 3
Fig. 8 (Right) Example waveform at destruction by short circuit type 3

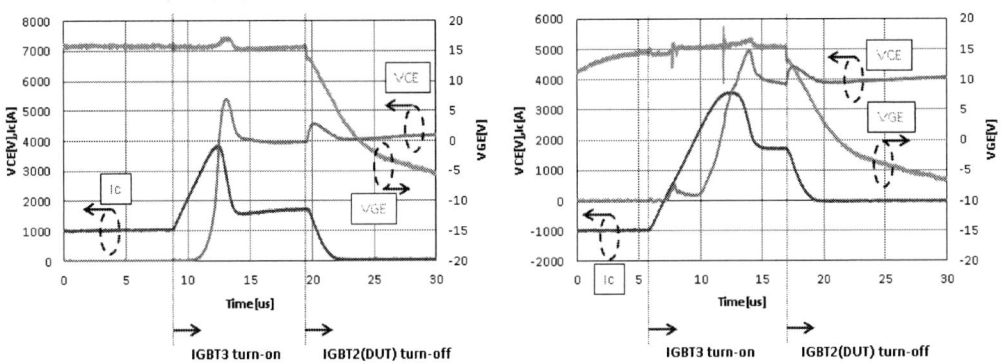

Fig. 9 (Left) A measured waveform of short circuit type 2. V_{cc}=4200V, V_{GE}=15V, T_j=150°C, tw=10us
and L_{SC}=4.2uH. Turn-on current before short circuit is 1000A.
(Right) Measured waveform at V_{cc}=4200V, V_{GE}=15V, T_j=150°C and tw=10us and L_{SC}=4.2uH.
Freewheeling current before short circuit is 1000A.

(3) Ruggedness of the diode

A significant characteristic for the robustness of a freewheeling diode is the peak power (P_{rr}) capability during a reverse recovery event. When a diode with insufficient P_{rr} capability is used at high voltages along with a high di/dt (diode) conditions, it might be destroyed. We confirmed that the new HVIGBT withstands harsh reverse recovery conditions with high P_{rr} =13.0MW by adopting the RFC diode. In addition, from Fig. 10, even if the new HVIGBT operates under a hard reverse recovery condition such as high temperature, high voltages and large currents, it is clear that the reverse recovery behavior is soft without the snap-off phenomenon.

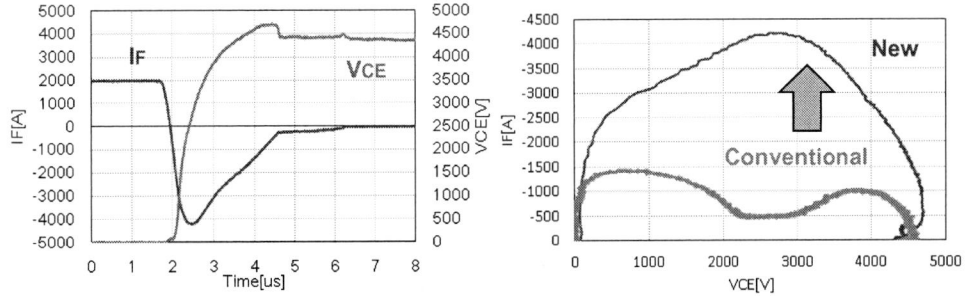

Fig. 10 (Left) Reverse recovery waveform at V_{cc}=4500V, I_F=2000A (2× I_{nom}), V_{GE}=15V, T_j=150°C and
L_s=150nH. (di/dt>5000A/us, P_{rr}=13.0MW)
(Right) Comparison of I-V plots at identical V_{cc}, I_F and T_j conditions.

© VDE VERLAG GMBH · Berlin · Offenbach

3-2. Verification results under tough environmental conditions

(1) High temperature

To verify the thermal runaway phenomena, we measured the I_{CES} of the new HVIGBT, which was mounted on a heat sink maintained at 150°C.Thereaftere, a DC voltage from 1 kV to 6.5 kV was applied (step-wise). The I_{CES} was less than 30mA at 150°C and the V_{CE}=6.5 kV (Fig. 11). Moreover, it was reduced to the same level as that of the I_{CES} of the conventional module at 125°C.

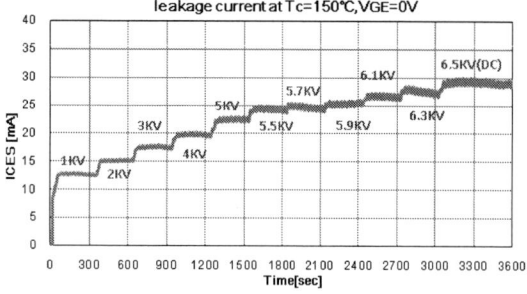

Fig.11 I_{CES} measurement result.

Table 2 I_{CES} specifications.

Product	I_{CES} at 125°C	I_{CES} at 150°C
New (X-Series)	5mA (TYP)	30mA (TYP)

The new HVIGBT's insulation performance was verified by evaluating the partial discharge characteristics after applying thermal stress. We measured the partial discharge behavior in the module by applying voltages up to 6.9 kV$_{rms}$ (Fig. 12) after thermal stress tests (Table 3). From the measurement results, since there is no difference in the partial discharge characteristics before and after the thermal stress, we confirmed that the insulation design of the new HVIGBT is robust enough.

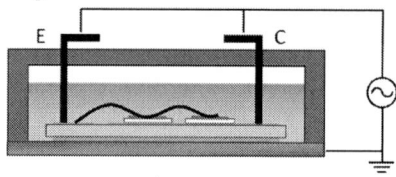

Fig.12 Illustration of the partial discharge test circuit.

Table 3 Partial discharge verification test results

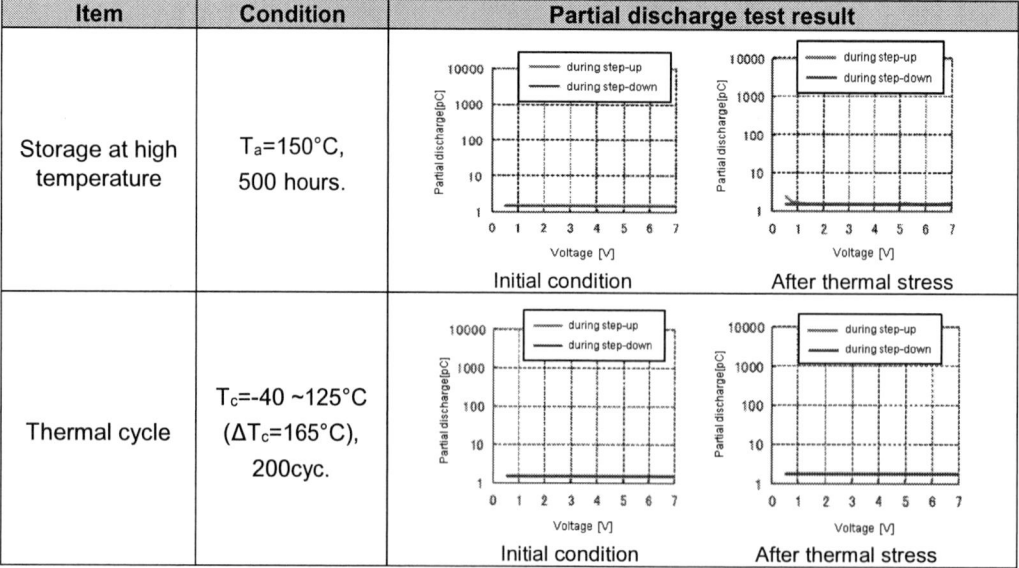

Item	Condition	Partial discharge test result
Storage at high temperature	T_a=150°C, 500 hours.	Initial condition / After thermal stress
Thermal cycle	T_c=-40 ~125°C (ΔT_c=165°C), 200cyc.	Initial condition / After thermal stress

(2) High humidity

The robustness of the power modules under high humidity conditions can be verified by the dew condensation test [9]. We measured the I_{CES} of the new HVIGBT during dew condensation at V_{CE}=5200V (DC), V_{GE}=0V, and T_a=25°C. The new HVIGBT has robustness against high humidity, because the I_{CES} characteristics do not change after repeating the dew condensation test five times (Fig.13).

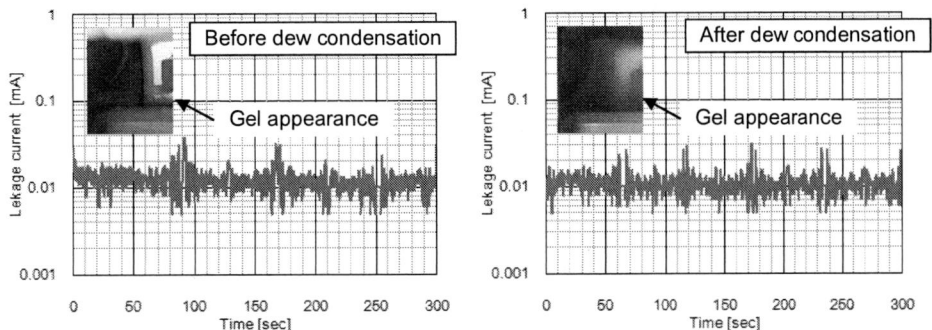

Fig. 13 Dew condensation test result of new 6.5 kV HVIGBT.

I_{CES} measurement conditions are V_{CE}=5200V (DC), V_{GE}=0V, and T_a=25°C.

(Left) I_{CES} at initial condition. (Right) I_{CES} after repeating dew condensation test five times.

4. Conclusion

This paper describes design approaches to maintain long-term stability. Power modules are key devices in power electronic equipments and must have robustness against disturbances without sacrificing performance. X-Series HVIGBT modules were developed based on high durability design concepts. The electrical robustness for overload and short circuit events was confirmed by evaluation. The module partial discharge requirement can be fulfilled after high temperatures (150°C) stresses. Finally the durability against tough environmental conditions like high humidity was also confirmed by testing.

5. References

[1] Z. Chen, et al., "LPT (II)-CSTBTTM (III) for High Voltage Application with Ultra Robust Turn-off Capability Utilizing Novel Edge Termination Design," Proc. ISPSD 2012, pp. 25, Belgium

[2] K. Hatori, et al., "The Next Generation 6.5 kV IGBT Module with High Robustness," Proc. PCIM 2014 Europe, pp. 28-33, Germany

[3] J. Luts, et al., "Short Circuit III in High Power IGBTs," Proc. EPE 2009, Barcelona

[4] S. Pierstorf, et al., "Different Short Circuit Types of IGBT Voltage Source Inverters," Proc. PCIM 2011 Europe, pp. 592-597, Germany

[5] K. Hatori, et al., "Wide temperature operation of high isolation HV-IGBT," Proc. PCIM 2010 Europe, pp. 470-475, Germany

[6] A. Nishii, K. Nakamura, F. Masuoka, and T. Terashima, "Relaxation of Current Filament due to RFC Technology and Ballast Resistor for Robust FWD Operation," Proc. ISPSD'11, pp. 112-115, San Diego, USA, 2011

[7] H. Uemura, et al., "Optimized Design against Cosmic Ray Failure for HVIGBT Modules," Proc. PCIM 2011 Europe, pp. 26-31, Germany

[8] Ze Chen, et al., "A Balanced High Voltage IGBT Design with Ultra Dynamic Ruggedness and Area-efficient Edge Termination," Proc. ISPSD 2013, pp. 37, Kanazawa, Japan

[9] N. Tanaka, et al., "Robust HVIGBT module design against high humidity," Proc. PCIM 2015 Europe, pp. 368-373, Germany

The 62Pak IGBT module range employing the 3rd Generation 1700V SPT^{++} chip set for 175°C operation

Sven Matthias, Chiara Corvasce, Athanasios Mesemanolis, Emilia Gustafsson, Charalampos Papadopoulos, Arnost Kopta, Silvan Geissmann, Martin Bayer, Raffael Schnell, Munaf Rahimo

ABB Switzerland Ltd. Semiconductors, Fabrikstrasse 3, 5600 Lenzburg, Switzerland

sven.matthias@ch.abb.com

Abstract

In this paper, we present a newly developed 62Pak with optimized IGBT and diode performance for high temperature operations. The latest generation of ABB's enhanced planar 1700V IGBT (SPT^{++}) has been improved by introducing a new termination concept and silicon design. Diode leakage current at 175°C has been significantly reduced by introducing the field shielded anode concept while still keeping the same electrical characteristics of the previous soft SPT^{+} diode platform. All mentioned features enable the development of a 1700V 62Pak module rated at 300A and operating at -40°C to 175°C junction temperature range with low losses and high Safe Operating Area (SOA).

1. Introduction

Throughout the past years the efforts in power semiconductor development were targeted to increase the power density for a given application. This performance target can be achieved by reduced losses, increased safe operating area and maximizing the allowable junction temperature during operation. A higher allowable junction temperature of the semiconductor is increasing the margin in the application when it comes to overload conditions. However, the design of power devices able to be operated at increasing temperatures bares certain challenges: Demonstration of the switching capability at the maximum specified junction temperature, proven thermal stability and high reliability at the full temperature range to fully utilize the potential. Here, we present the third generation of 1700V SPT^{++} IGBTs able to operate up to a maximum temperature of T_{vj}=175°C. The new IGBT design exploits the full potential of the optimized enhancement layer in combination with a novel termination technology and aggressive silicon design. Therefore offering outstanding performance for low to medium inductance applications as demanded in industrial or regenerative power applications. We used our 62Pak module platform to demonstrate the superior properties of our latest chip set (fig. 1).

	125°C	175°C	unit
I_{ces}			
I_{ces}	1.5	30	mA
V_{CEsat}	2.55	2.75	V
V_F	1.75	1.7	V
E_{on}	95	115	mJ
E_{off}	75	95	mJ
E_{rec}	75	110	mJ

300A version

Fig. 1. The 1.7kV 62Pak (standard footprint of 62mm x 106.4mm) module using the third generation SPT^{++} chipset. ABB SPT^{++} high temperature module (nominal conditions: T_{vj}=175°C, V_{DC}=900V, I_C=300A, R_{Gon} = 2.2 Ω, R_{Goff} = 2.2 Ω, V_{GE} = ±15 V, L_σ = 60 nH). Available in three current ratings of 300A, 200A and 150A.

2. Third generation chipset

The enhanced planar MOS cell concept has been successfully introduced for all voltage classes ranging from 1.2kV to 6.5kV during the last decade [1]. The major benefit of this SPT+ technology is the combination of minimized conduction losses, when compared to standard planar IGBTs, and excellent controllability of the enhanced planar design. This is achieved by the n-type enhancement layer surrounding the p-well in the IGBT MOS cell (fig. 2).

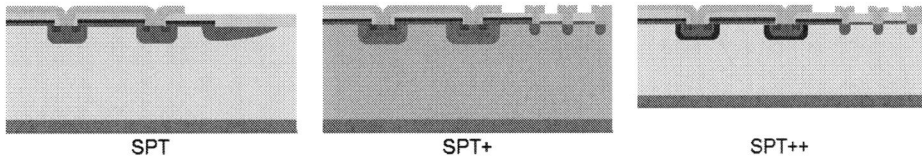

Fig. 2. Schematic drawing of the various planar IGBT generations.

The enhancement layer increases the carrier concentration at the cathode side of the IGBT and is lowering the on-state voltage drop without significantly increasing the turn-off losses. However, the inherent challenge is the reduction of the blocking capability of the device and its increasing sensitivity to dynamic avalanche failures resulting from the enhancement layer. Hence, the shape and the doping profile of this layer must be carefully optimized to maximize the benefit of the on-state losses and limit the negative impact on blocking margin and safe-operating area. This can be achieved by narrowing the doping profile and increasing the peak concentration as demonstrated in the latest SPT++ generation. As a consequence, the high electric field area is reduced with a minimum impact on the blocking and the plasma concentration is further enhanced. The beneficial effect of the blocking capability with the optimized enhancement layer enables a reduction of the n-base thickness by 10% compared to the previous generation. This results in minimized conduction and switching losses [2].

A reliable operation at high junction temperatures was enabled by the development of a new junction termination design based on the floating guard ring concept [3]. The termination consists of a number of diffused p-type rings contacted by metal plugs and interconnected by a semi-insulating layer. Such a termination design has been proven to be immune to inter-ring distance variations and interface states while offering very low leakage current levels. The achieved reduction of the leakage current by a factor of four compared to the previous generation allow us to expand the operating temperature to 175°C.

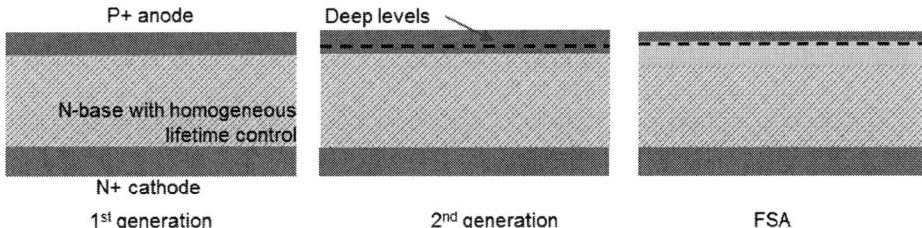

Fig. 3. Schematic drawing of the various diode generations.

The second generation diode uses locally incorporated deep levels by H+ irradiation for local lifetime control placed in the deep pn-junction and a homogeneous electron irradiation (fig. 3). The deep level allow us to tailor the plasma distribution and guarantee stable operation at T_{jmax}=150°C [4]. In this design, the electric field evolving during reverse blocking is penetrating into the zone of radiation defects already at very low reverse voltages. This generates a leakage current which does not allow a stable blocking operation of the chip at T_{jmax}=175°C. The newly developed Field Shielded Anode (FSA) design, previously applied for large area discrete diodes and 3.3kV fast recovery diodes [5], is characterized by a modified doping

© VDE VERLAG GMBH · Berlin · Offenbach

profile. The depth of the anode is maintained by introducing a deep profile having a reduced concentration and resembling a low p-doped buffer, preventing the electric field from reaching the zone of radiation defects during blocking. In addition, a shallow highly doped p-layer ensures good contact and good anode injection in the high-current regime to enable a good surge current capability. The radiation defects position and concentration are then tuned to tailor the plasma distribution to match the conduction and dynamic properties of the second generation diode. This FSA design has the inherent advantage of separating the radiation defects from the space charge region resulting in a significantly reduced high temperature leakage current. The leakage current distribution of the FSA diode chip when compared with the second generation diode and conventional platform is reduced by a factor of three (at same applied reverse voltage and temperature) and allows the FSA diode to be operated at $T_{jmax}=175°C$.

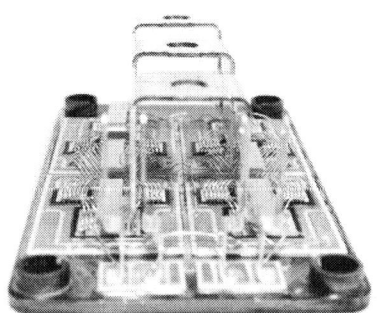

Fig. 4.   Picture of the 62Pak electrical layout.

3. Electrical Performance of the 1.7kV 62Pak module

Extensive measurements have been carried out to demonstrate the performance of the new 1700V chipset at the maximum junction temperature of 175°C. The device under test is a standard industrial 62Pak with a high and a low-side switch consisting of two parallel IGBT and four antiparallel diodes (fig.4). A stray inductance value of 60nH per module has been used. The results of this characterization are presented in this section.

3.1.   IGBT nominal switching characteristic

In Fig. 5 the switching waveforms of the new 1700V IGBT are shown as measured under nominal conditions i.e. at 300A and 900V at $T_{vj}=175°C$. In the turn-off test (Fig. 5a), the IGBT was switched off using an $R_{G,off}$ of 2.2Ω with a stray inductance of 60nH. The optimized n-base region combined with the SPT buffer allows the collector current to decay smoothly, ensuring a soft turn-off behavior without any disturbing voltage peaks or oscillations. The turn-off losses integrate to 95mJ at the maximum junction temperature. Figure 5b shows the turn-on waveforms under nominal conditions at $T_{vj}=175°C$. The low input capacitance of the planar SPT++ cell allows a fast drop of the IGBT voltage during the turn-on transient. This, combined with the low loss FSA diode brings the turn-on switching losses down to a typical value of 115mJ.

© VDE VERLAG GMBH · Berlin · Offenbach

PCIM Europe 2016, 10 – 12 May 2016, Nuremberg, Germany

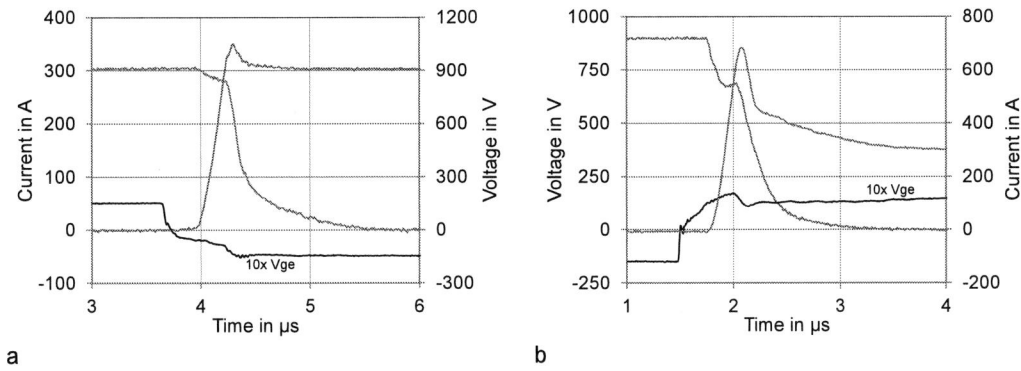

a   b

Fig. 5.   a) IGBT turn-off at nominal conditions: T_{vj}=175°C, V_{DC}=900V, I_C=300A, R_{Gon}=2.2Ω, R_{Goff}=2.2Ω, V_{GE}=±15V, L_σ=60nH. b) IGBT turn-on at nominal conditions.

3.2.   Diode nominal switching characteristic

The reverse recovery waveforms of the diode under nominal conditions at T_{vj}=175°C are shown in Fig. 6a. By carefully designing the anode profile and the local lifetime control peak, a short, but still smoothly decaying current tail was achieved resulting in total recovery losses of 110mJ. The reverse recovery softness test performed at 1/6th of the nominal rated current at a DC-link voltage of 900V and ambient temperature is shown in Fig. 6b. It confirms the soft recovery behavior showing only small oscillations and a peak overshoot voltage of 900V.

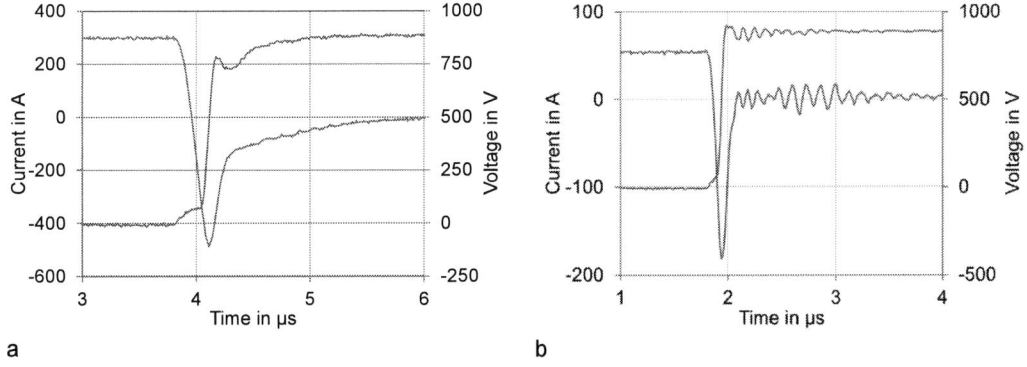

a   b

Fig. 6.   a) Diode turn-off at nominal conditions: T_{vj}=175°C, V_{DC}=900V, I_C=300A, R_{Gon}=2.2Ω, R_{Goff}=2.2Ω, V_{GE}=±15V, L_σ=60nH. b) Diode turn-off at low current 1/6 of the nominal current and T_{vj}=25°C.

3.3.   Maximum ratings

In order to evaluate the SOA performance of the third generation 1700V SPT^{++} IGBT technology, the chips have been subjected to a wide range of switching tests under extreme conditions in terms of current, voltage, temperature and stray inductance. Figure 7a shows the IGBT chip turn-off capability of two parallel chips in the 62Pak measured at room temperature and T_{vj}=175°C without an active clamp. The SOA capability of the IGBT chip at T_{vj}=175°C is proven from the waveforms shown in Fig. 7. The chip withstands a strong dynamic avalanche regime turning off a current exceeding the nominal value by a factor of five!

© VDE VERLAG GMBH · Berlin · Offenbach

PCIM Europe 2016, 10 – 12 May 2016, Nuremberg, Germany

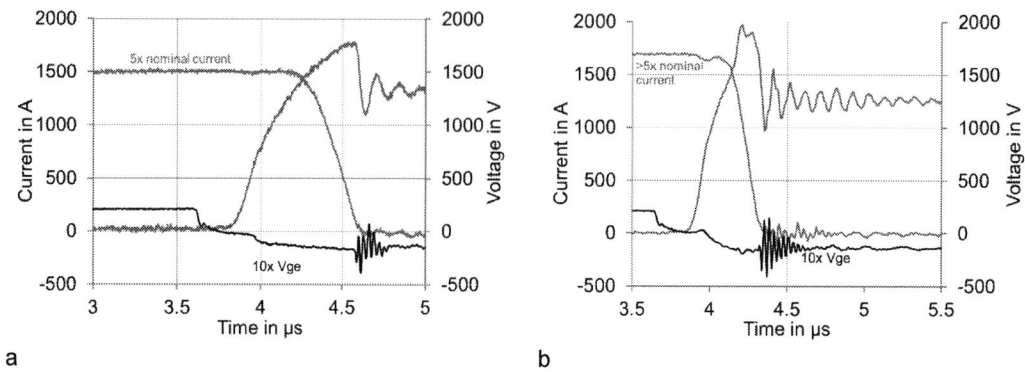

a   b

Fig. 7. a) IGBT turn-off at last pass conditions: T_{vj}=175°C, V_{DC}=1300V, I_C=1500A, R_{Gon}=2.2Ω, R_{Goff}=2.2Ω, V_{GE}=±20V, L_σ=60nH. b) IGBT turn-off at last pass conditions: T_{vj}=25°C, V_{DC}=1300V, I_C=1700A, R_{Gon}=2.2Ω, R_{Goff}=2.2Ω, V_{GE}=±20V, L_σ=60nH.

In the ambient temperature test shown in Fig. 7b, a current of 1700A, which is more than five times the nominal current, is turned off against a DC-link voltage of 1300V. A self-clamp overshoot voltage of 1970V can be observed during the turn-off period. The chip survives dynamic avalanche conditions and successfully withstands the switching self clamping mode (SSCM) of operation with a peak-power of more than 2300kW per module.

The reverse recovery safe operating area (SOA) of the new FSA technology was extensively investigated over the whole temperature range using a high DC-link voltage of V_{DC}=1300 V. The current was stepped up to 2 times the nominal value and after a successful pass, the reverse recovery di/dt was increased by lowering the gate-resistor value R_{Gon} until the diode failed. In Fig. 8 the last pass reverse recovery waveforms of the 1700V FSA diode can be seen. The diode manages to withstand a reverse recovery commutation speed of more than 3.4kA/µs. This high recovery robustness was achieved thanks to the optimization of the anode design in conjunction with the resistive extension of the active junction region as explained above.

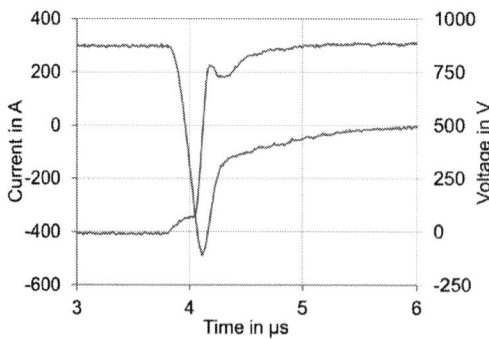

Fig. 8. Diode turn-off at last pass conditions: T_{vj}=175°C, V_{DC}=1300V, I_C=560A, R_{Gon}=0.47Ω, R_{Goff}=2.2Ω, V_{GE}=±15V, L_σ=60nH.

© VDE VERLAG GMBH · Berlin · Offenbach

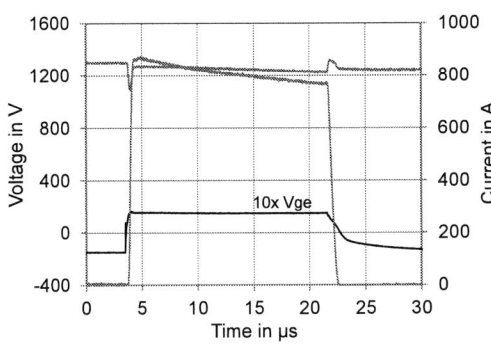 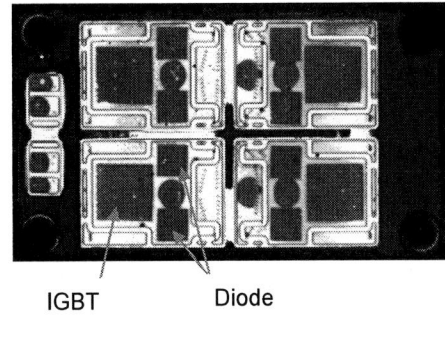

a   b

Fig. 9. a) Short-circuit last pass conditions: T_{vj}=175°C, V_{DC}=1300V, I_C=560A, R_{Gon}=2.2Ω, R_{Goff}=2.2Ω, V_{GE}=±15V, L_σ=60nH, pulse length t=18µs. b) Scanning acoustic microscopy image of the chip-to-substrate solder interface reflecting the very homogeneous soldering process.

The excellent short circuit capability of the new 1700V SPT++ IGBT is shown in Fig. 9a where the short circuit waveforms at T_{vj}=175°C and a DC-link voltage of 1300V can be seen. No thermal runaway after the test has been observed for pulse times up to 18µs with a short circuit current of more than 800A and a total dissipated energy of 18J. The SPT buffer and anode design employed in the SPT++ IGBT have been optimized in order to obtain a high short-circuit SOA capability, even at gate voltages exceeding the standard gate drive voltage of 15V over the whole junction temperature range from -40°C to 175°C. Figure 9b indicates the homogeneous soldering interface between the chips and the substrate resulting in an optimized heat transfer during operation.

4. Conclusion

In this paper we presented the latest generation of 1700V SPT++ IGBT and diode chip set targeting operation at a maximum junction temperature of 175°C in a standard industrial 62Pak package. Outstanding SOA turn-off and short circuit capability as well as considerable IGBT conduction losses reductions have been achieved whilst keeping the ultra low switching lossses by exploiting the full potential of the enhanced planar technology. A careful optimization of the IGBT termination design and process has enabled stable and reliable operation at T_{jmax}=175°C. By using the Field Shielded Anode concept, a soft and low losses recovery diode has been developed which can be operated at the same high temperature with a high recovery ruggedness.

References

[1] Rahimo M. et al., "SPT+, the Next Generation of Low-Loss HV-IGBTs", Proc. PCIM '05, Nürnberg, Germany 2005.

[2] Corvasce C. et al., "New 1700V SPT+ IGBT and Diode Chip Set with 175°C Operating Junction Temperature", Proc. EPE '11, Birmingham, UK 2011.

[3] Macary M.V. et al., "Comparison between biased and floating guard rings used as junction termination technique", Proc. ISPSD '92, Tokyo, Japan 1992.

[4] J. Lutz and U. Scheuermann, "Advantages of the new controlled axial lifetime diode," Proc. PCIM'94, pp. 163, 1994.

[5] Matthias et al., "Field Shielded Anode (FSA) Concept Enabling Higher Temperature Operation of Fast Recovery Diodes", Proc. ISPSD '11, San Diego, USA 2011.

PCIM Europe 2016, 10 – 12 May 2016, Nuremberg, Germany

Extended Power Rating of 1200V IGBT Module with 7G RC-IGBT Chip Technologies

M. Takahashi, D. Hofmann*, S. Yoshida, A. Tamenori, Y. Kobayashi, O. Ikawa

Fuji Electric Co. Ltd. / Japan, *Fuji Electric Europe GmbH / Germany

4-18-1 Tsukama, Matsumoto, Nagano, Japan 390-0821

takahashi-misaki@fujielectric.com

Abstract

Recently the main requirements for IGBT modules are increasing the power density and higher efficiency with sufficient reliability. To fulfill these requirements a new RC-IGBT (7G RC-IGBT) module based on Fuji Electric's latest 7th generation (7G) chip and package technology and RC-IGBT technology was developed. In this paper 7G RC-IGBT module was fabricated which achieved 40% smaller chip size compared to 6th generation (6G) IGBT+FWD with 12% smaller inverter loss. Therefore, the current rating of DualXT package was extended from 6G 600A to 1000A which reaches the region of PrimePACK2™. Additionally the DC lock and low frequency (Fo=1Hz) dTjc peak are 50% reduced compared to 6G module. In conclusion 7G RC-IGBT module enables further miniaturization and cost down and high reliability of power conversion systems.

Introduction

Currently energy savings are strongly desired to prevent global warming and ensure environmental sustainability. For these purposes low loss power semiconductor devices are widely used in power conversion systems. IGBT (Insulated-Gate Bipolar Transistor) modules are the most commonly used power semiconductors in various fields like industrial-, consumer-, automotive- and renewable energy applications. In most of the applications, customers are in need for more compact IGBT modules to reduce the cost and size of their product. The developed 7th generation (7G) chip and package technology, thereby achieved not only high compactness and high power density, but also high reliability and maximum operating chip junction temperature (Tjmax) of 175°C [1]. In addition the RC (Reverse Conducting) -IGBT using 6th generation (6G) chip technology has been developed and accomplished 21% smaller chip size compared to conventional 6G IGBT+FWD [2]. In this report we combined 7G chip and package technology with RC-IGBT technology to realize further increasing of power density with high reliability.

1. 7G RC-IGBT chip technology

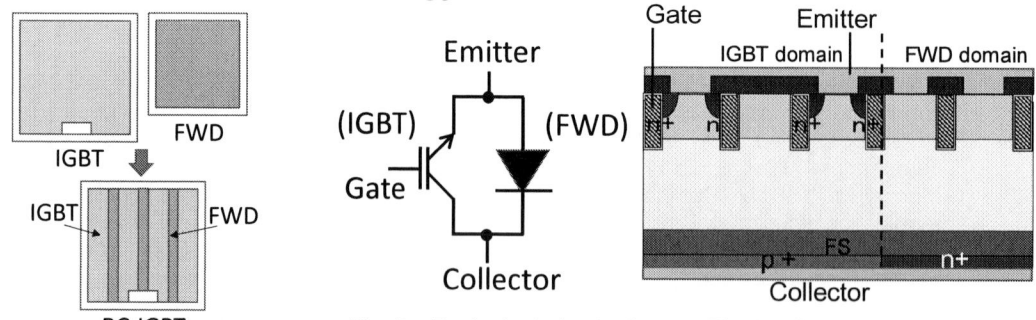

Fig. 1. Top view of 7G RC-IGBT

Fig. 2. Equivalent circuit of RC-IGBT

Fig. 3. Cross section of 7G RC-IGBT

Note: PrimePACK™ is registered trademark of Infineon Technologies AG, Germany.

© VDE VERLAG GMBH · Berlin · Offenbach

Recently many researchers reported about RC-IGBT [3]-[6] in the past, which is a type of IGBT having FWD function at the same time. Fig. 1 shows the top view of 7G RC-IGBT. 7G RC-IGBT has both IGBT and FWD domains which is arranged in a stripe pattern. Therefore it can operate as IGBT and FWD. The RC-IGBT can eliminate FWD chips, which leads to significant chip downsizing. The equivalent circuit of RC-IGBT is shown in fig. 2. RC-IGBT device is a parallel connection of IGBT and FWD. The cross section of 7G RC-IGBT is shown in fig. 3. The surface structure of 7G RC-IGBT is trench gate and has a Field-Stop (FS) layer on the back side similar to 6G IGBT and 7G IGBT. The 7G RC-IGBT applies the latest 7G IGBT chip technology which wafer thickness and surface structure are the same as 7G IGBT. 7G IGBT and 7G RC-IGBT applies thinner wafer compared to 6G IGBT, then achieves lower on-state voltage drop. Moreover 7G IGBT and 7G RC-IGBT has finer pitch trench gate than 6G IGBT which leads to increasing Injection Enhanced (IE) effects. Accordingly the trade-off relationship between conduction loss and turn-off loss was further improved. The fabrication process of 7G RC-IGBT is almost the same as 7G IGBT. Additional processes of the RC-IGBT production are a backside photo-etching ion-implantation process for the formation of backside P/N structure and a lifetime control process.

2. 7G RC-IGBT chip characteristics

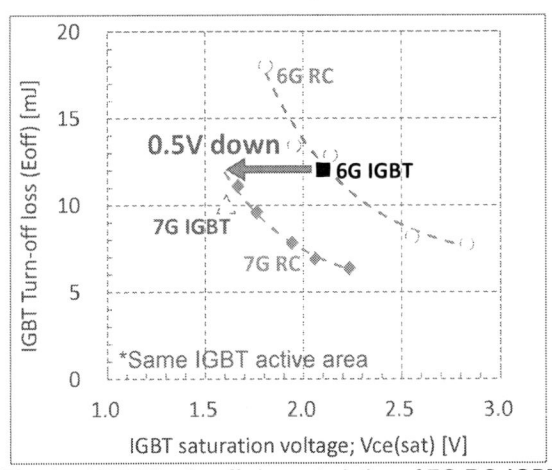

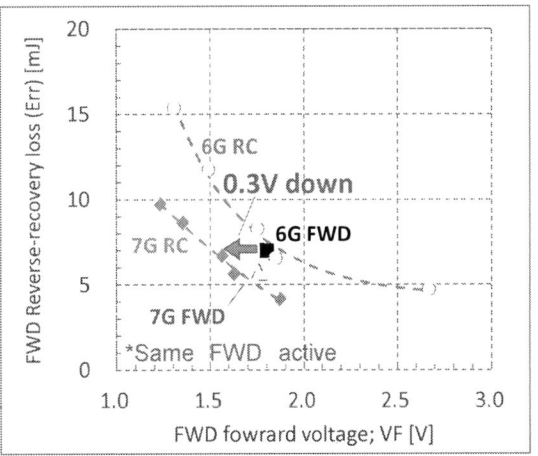

Fig. 4. IGBT Trade-off characteristics of 7G RC-IGBT
Measurement condition;

Fig. 5. FWD Trade-off characteristics of 7G RC-IGBT

Vce(sat); Ic=100A, Vge=15V   VF; If=100A, Vge=-15V(RC-IGBT)

Eoff, Err; Vcc=600V, Ic=100A, Vge=+15V/-15V, Max. reverse-recovery dV/dt=10kV/μsec

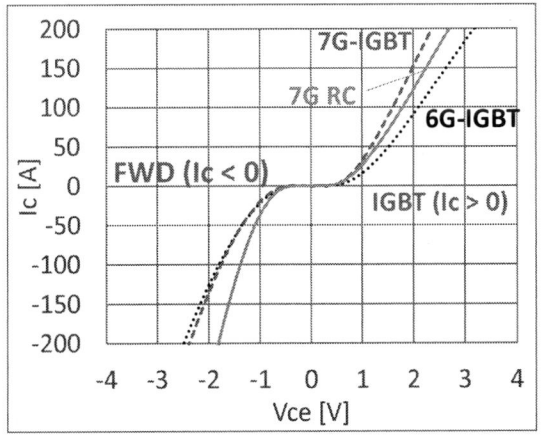

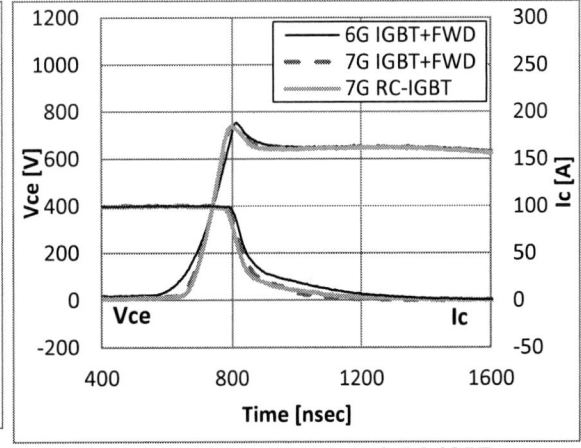

Fig. 6. Output characteristics of 7G RC-IGBT

Fig. 7. Turn-off waveform of 7G RC-IGBT

© VDE VERLAG GMBH · Berlin · Offenbach

PCIM Europe 2016, 10 – 12 May 2016, Nuremberg, Germany

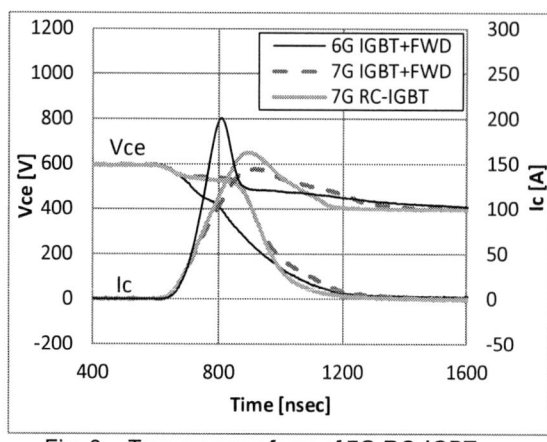

Fig. 8. Turn-on waveform of 7G RC-IGBT

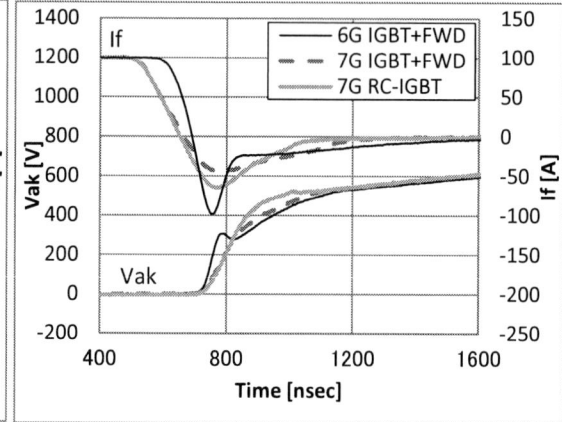

Fig. 9. Reverse-recovery waveform of 7G RC-IGBT

The IGBT loss trade-off characteristics of 6G IGBT, 7G IGBT and 7G RC with the same IGBT active area are shown in fig. 4. The 7G RC active area implies only IGBT domain active area. The FWD loss trade-off characteristics of 6G FWD, 7G FWD and 7G RC with the same FWD active are shown in fig. 5. The 7G RC active area contains only FWD domain active area. Each 7G RC results in fig. 4 and 5 are varied by lifetime control. The IGBT loss performance of 7G RC-IGBT are over 0.5V smaller compared to 6G IGBT. The FWD loss performance of 7G RC-IGBT are over 0.3V smaller compared to 6G FWD. The IGBT and FWD loss performance of 7G RC-IGBT are further improved compared to 6G IGBT+FWD. This great improvement was achieved by applying thinner wafer and finer pitch trench gate compared to 6G IGBT. The IGBT and FWD loss performance of 7G RC is equivalent to 7G IGBT+FWD. Generally RC-IGBT applies lifetime control therefore RC-IGBT loss performance is inferior to non-killer IGBT. However by optimizing the lifetime control 7G RC-IGBT accomplished equivalent loss performance to non-killer 7G IGBT.

The output characteristics of 7G RC-IGBT are shown in fig. 6. 7G RC-IGBT has reverse conducting function and no snapback behavior is observed. The turn-off, turn-on and reverse-recovery waveform of 7G RC-IGBT is shown in fig. 7, 8 and 9 respectively. The waveform of 7G RC is similar to 7G IGBT+FWD because 7G RC has the same IGBT structure as 7G IGBT. As shown in fig. 7, turn-off surge voltage of 7G RC is same as 6G and 7G IGBT. The tail current is much smaller than 6G IGBT hence turn-off loss is 23% smaller compared to 6G IGBT. As shown in fig. 8 and 9, 7G RC has very soft recovery behavior compared to 6G FWD and recovery current is smaller than 6G FWD. Therefore the reverse-recovery loss is reduced by 20%. Generally, it is known that thinner wafer and drift layer causes oscillation during turn-off and reduction of breakdown voltage. By an optimization of wafer resistivity and edge termination and FS layer, 7G RC-IGBT realizes the suppression of voltage oscillations and keeping its breakdown voltage.

3. 7G RC-IGBT module performance

Table 1. 1200V-100A module characteristics of 7G RC-IGBT

1200V-100A module comparison	6G IGBT+FWD IGBT/FWD (Total)	7G IGBT+FWD IGBT/FWD (Total)	7G RC-IGBT RC-IGBT
Chip size (a.u.)	0.62 / 0.38 (1.00)	0.48 / 0.31 (0.79)	0.61
Active area (a.u.)	0.64 / 0.36 (1.00)	0.48 / 0.33 (0.81)	0.65
Insulating substrate	Al2O3	Thin-AlN	Thin-AlN
Thermal resistance Rth(jc) [K/W]	0.29 / 0.44	0.22 / 0.31	0.16 / 0.16 (IGBT/FWD)
Max. operating Tj [°C]	150	175	175

© VDE VERLAG GMBH · Berlin · Offenbach

7G RC-IGBT shows excellent chip performance and it enables significant chip miniaturization. For example 1200V-100A rated module characteristics of 7G RC is demonstrated in table 1. Showing by fig. 10 calculation results, 7G RC-IGBT realizes 40% chip downsizing compared to 6G IGBT+FWD, and further 23% chip downsizing compared to 7G IGBT+FWD. 7G technology innovation includes not only chip innovation but also package innovation. Rth(jc) of 7G RC module is greatly reduced compared to 6G module by using a new thin and low thermal impedance AIN insulating substrate developed for 7G package technology. Moreover the IGBT Rth(jc) and FWD Rth(jc) of 7G RC module are 25% and 47% lower than 7G IGBT+FWD module. Nevertheless the total chip size of 7G RC is 23% smaller than 7G IGBT+FWD. As shown in fig. 1 IGBT and FWD domains of 7G RC is arranged in a stripe pattern and the distance of two adjacent regions is less than 0.6mm. Thus IGBT domain temperature and FWD domain temperature are equalized by thermal diffusion. In consequence generated loss is dissipated by whole chip area. 7G RC-IGBT chip size is 29% larger than 7G IGBT and 94% larger than 7G FWD, then Rth(jc) of 7G RC is much lower than 7G IGBT+FWD. Furthermore the operating chip junction temperature (Tj) of 7G RC module reaches 175°C by improvements of solder and silicone gel.

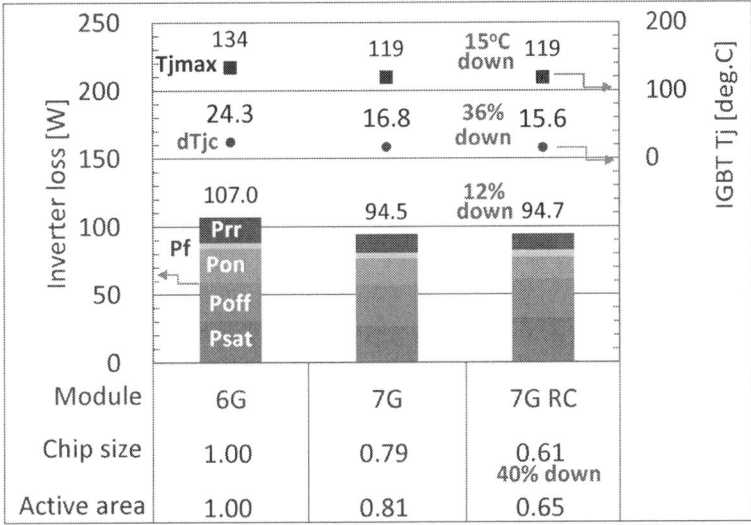

Fig. 10. Inverter loss and Tj of 7G RC-IGBT 100A module
Calculation condition;

Vcc=600V, Fo=50Hz, Fc=8kHz, Io=50Arms, cosφ = 0.9, modulation rate = 1.0

Tair=50°C, Rth(heatsink)=0.085K/W, Thermal paste; 50μm, 2W/m·K

Inverter loss and IGBT chip junction temperature (Tj) calculation results are shown in fig. 10. The inverter loss of 7G and 7G RC is also 12% reduced compared to 6G IGBT and the Tj of 7G and 7G RC is also 15°C lower than 6G IGBT. The performance of 7G and 7G RC module is greatly improved by 7G chip and package technology. At the same time the chip size of 7G RC-IGBT is 40% smaller than 6G IGBT+FWD and still 23% lower compared to 7G IGBT+FWD.

4. Extended power rating with 7G RC-IGBT

7G RC-IGBT achieves 40% chip downsizing compared to 6G IGBT+FWD. It enables significant expansion of module current rating. Table 2 shows the 1200V line-up of DualXT and PrimePACK2™ package. The line-up of DualXT was extended to 800A by 7G and moreover extended to 1000A by 7G RC-IGBT. It surpassed the lineup of PrimePACK2™. Table 3 shows the characteristics of 7G RC 1000A DualXT module. The current rating and current density of 7G RC DualXT is 67% higher compared to 6G DualXT at the same time Rth(jc) is 45% lower. In comparison with 6G PrimePACK2™ the current rating of 7G RC DualXT is 11% higher with 40% smaller footprint and 27% smaller Rth(jc). Therefore the current density of 7G RC DualXT is 83% higher than 6G PrimePACK2™. In conclusion 7G RC DualXT is able to cover the region of PrimePACK2™. The calculation results of maximum IGBT chip junction temperature (Tj) during continuous inverter drive are shown in fig. 11. 7G 800A DualXT can handle 37% larger output current compared to 6G 600A DualXT, on the top of that 7G RC 1000A DualXT can

PCIM Europe 2016, 10 – 12 May 2016, Nuremberg, Germany

handle 56% larger output current although in comparison with same cooling condition of heatsink.

Table 2. 1200V Line-up of DualXT and PrimePACK2™

		Rating Current [A]						
		225	300	450	600	800	900	1000
DualXT	6G							
	7G							★ 7G RC
PrimePACK2™	6G		7G IGBT+FWD					

Table 3. Characteristics of 7G RC 1000A DualXT module

Package	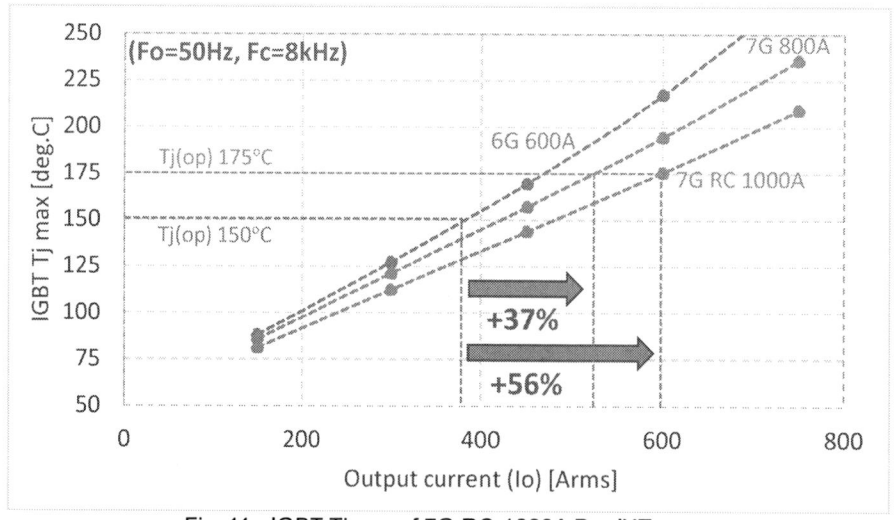 DualXT			PrimePACK™
Footprint [cm2]	93	93	93	153
Chip generation	6G	7G	7G RC	6G
Rating current [A]	600	800	1000	900
Current density [A/cm2]	6.45	8.60	10.75	5.88
Insulating substrate	SiN	Thin-AlN	Thin-AlN	Al2O3
Thermal resistance Rth(jc) [K/W]	0.04 / 0.06 (IGBT/FWD)	0.037 / 0.044 (IGBT/FWD)	0.022 / 0.022 (IGBT/FWD)	0.03 / 0.054 (IGBT/FWD)

Fig. 11. IGBT Tjmax of 7G RC 1000A DualXT

Calculation condition;

Vcc=600V, Fo=50Hz, Fc=8kHz, cosφ=0.9, modulation rate=1.0

Tair=50°C, Rth(heatsink)=0.015K/W, Thermal paste; 50μm, 2W/m·K

5. dTjc reduction of DC lock and low frequency drive (Fo=1Hz)

7G RC-IGBT achieves significant chip downsizing and enables to expand rating current and power density over 67% compared to 6G IGBT. In addition DC lock and low frequency AC drive are investigated which has large influence on a power cycling capability. In DC lock and low frequency drive larger Tj variations are applied to chips and a module compared to a normal continuous drive (Fo~50Hz) which leads to severe thermal stress for IGBT modules.

© VDE VERLAG GMBH · Berlin · Offenbach

PCIM Europe 2016, 10 – 12 May 2016, Nuremberg, Germany

Thereby power cycling capabilities of DC lock and low frequency drive mainly determined the reliability and lifetime of IGBT modules for NC servo application which DC lock and low frequency drives occur quite often.

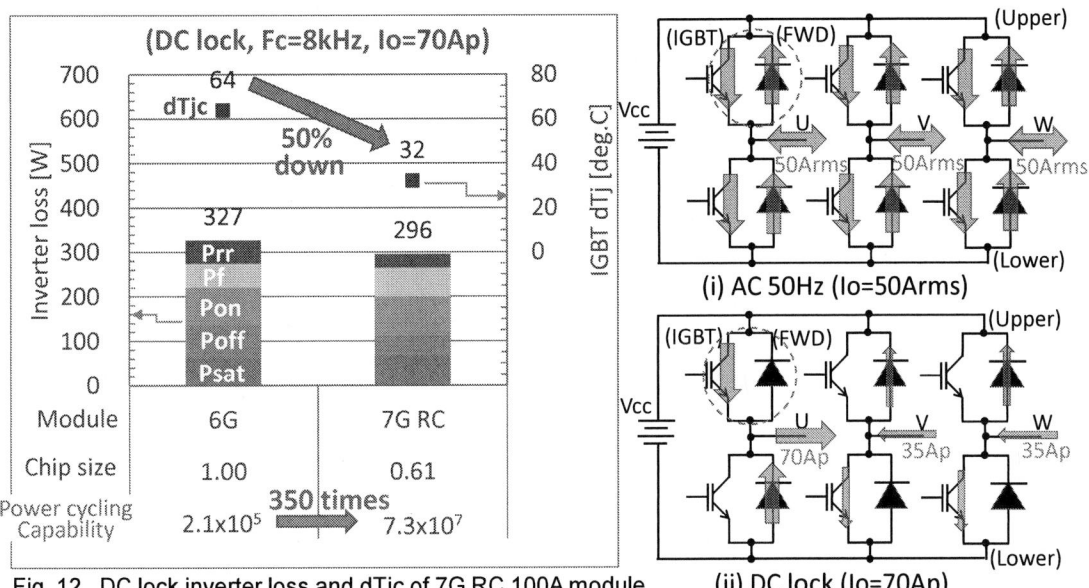

Fig. 12. DC lock inverter loss and dTjc of 7G RC 100A module
Calculation condition; Vcc=600V, Fc=8kHz, Io=70Ap, duty=0.5
Tair=50°C, Rth(heatsink)=0.085K/W
thermal paste; 50μm, 2W/m·K

(i) AC 50Hz (Io=50Arms)

(ii) DC lock (Io=70Ap)

Fig. 13. Inverter drive pattern of AC 50Hz and DC lock

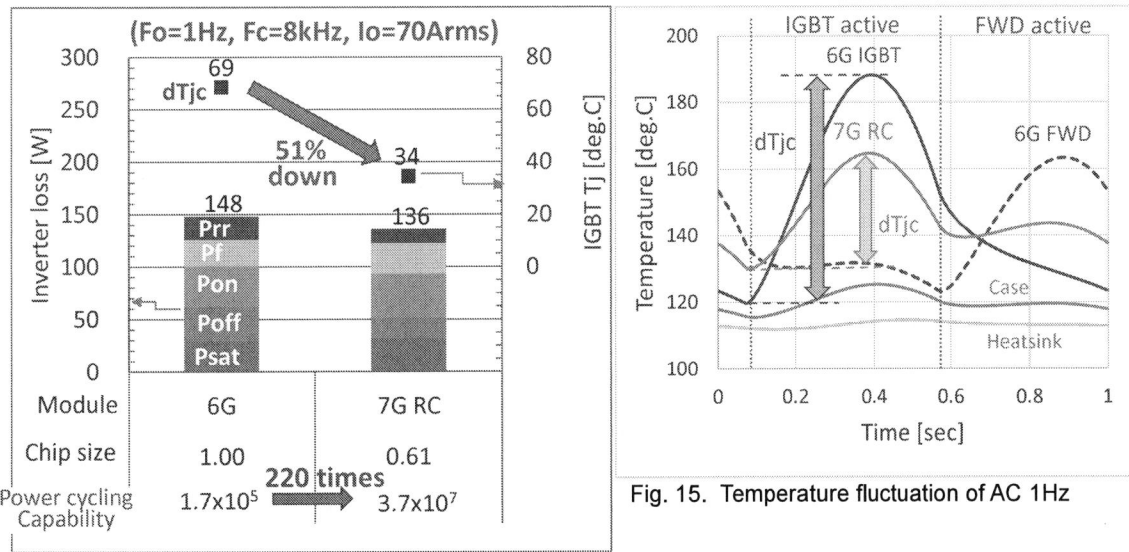

Fig. 14. AC 1Hz inverter loss and dTjc of 7G RC 100A module

Fig. 15. Temperature fluctuation of AC 1Hz

Calculation condition; Vcc=600V, Fo=1Hz, Fc=8kHz, Io=70Arms, cosϕ=0.9, modulation rate=0.02
Tair = 50°C, Rth(heatsink)=0.085K/W, thermal paste; 50μm, 2W/m·K

Fig. 12 shows the DC lock inverter loss and dTjc of 7G RC 100A module. In fig. 12, dTjc denotes Tj variation between before and after DC lock drive which determines a power cycling capability. As shown in fig. 12 the dTjc of 7G RC is 50% lower than 6G. In comparison with a power cycling capability, the lifetime of 7G RC is 350 times higher than 6G. The reliability of 7G RC-IGBT modules are drastically improved by 7G chip and package technology and RC-IGBT technology. Comparing fig. 10 and fig. 12, the AC 50Hz dTjc of 7G RC is 36% smaller than 6G, however DC lock dTjc of 7G RC is 50% smaller than 6G. The dTjc improvement of

© VDE VERLAG GMBH · Berlin · Offenbach

443

DC lock is higher compared to AC 50Hz. The inverter drive pattern of AC 50Hz drive and DC lock are shown in fig. 13. In AC 50Hz drive IGBT and FWD are active simultaneously in each arm, thus in case of RC-IGBT IGBT and FWD loss are applied to RC-IGBT because RC-IGBT includes FWD function. Therefore additional FWD loss increases the RC-IGBT dTjc of AC 50Hz drive. On the other hand in DC lock only IGBT is active in U-phase upper arm, hence only IGBT loss is applied to RC-IGBT. Consequently the dTjc of DC lock is further improved compared to AC 50Hz.

The inverter loss and dTjc of low frequency drive (AC 1Hz) are shown in fig. 14, dTjc denotes Tj difference between Tjmax and Tjmin among AC 1Hz drive. As shown in fig. 14 the dTjc of 7G RC is 51% lower than 6G. Fig. 15 shows the reason why dTjc of 7G RC is extremely small, which denotes the temperature fluctuation of AC 1Hz drive. The Tjmax of 7G RC is 23.6°C smaller compared to 6G IGBT because of lower loss and lower Rth(jc), and the Tjmin of 7G RC is 10.5°C higher than 6G since 7G RC temperature of FWD period is increased by FWD loss of RC-IGBT. In comparison with a power cycling capability the lifetime of 7G RC is 220 times higher than 6G. Similar to DC lock, in the case of AC 50Hz drive the reliability of 7G RC-IGBT modules are also drastically improved.

6. Conclusion

7G RC-IGBT module was fabricated by combination of 7G chip and package technology and RC-IGBT technology. 7G RC-IGBT module achieved 40% smaller chip size compared to 6G-IGBT+FWD module with 12% smaller inverter loss and 45% smaller IGBT Rth(jc). Therefore, the rated current of DualXT package was extended from 6G 600A to 1000A which reaches the region of PrimePACK2™. The power density was increased by 67%. Additionally the DC lock and low frequency (Fo=1Hz) dTjc of 7G RC module is 50% lower than 6G module, therefore the power cycling capability of 7G RC module is over 220 times higher compared to 6G module. Consequently 7G RC module is appropriate especially for NC servo application. In conclusion 7G RC module enables further miniaturization and cost down of power conversion systems.

7. Reference

[1] J. Kawabata et al. "The New High Power Density 7th Generation IGBT Module for Compact Power Conversion Systems", Proceeding of PCIM Europe 2015.

[2] K. Takahashi et al. "1200V Class Reverse Conducting IGBT Optimized for Hard Switching Inverter", in Proceeding of PCIM Europe 2014.

[3] H. Takahashi et al "1200V Reverse Conducting IGBT", pp. 133-136 Proceeding of ISPSD 2004.

[4] K. Satoh et al "A New 3A/600V Transfer Mold IPM with RC (Reverse Conducting)-IGBT", Proceeding of PCIM Europe 2006.

[5] H. Ruthing et al "600V Reverse Conducting (RC-) IGBT for Drives Application in Ultra-Thin Wafer Technology", pp. 89-92 Proceeding of ISPSD 2007.

[6] S. Voss "Anode Design Variation in 1200-V Trench Field-stop Reverse-conducting IGBTs", pp. 169-172 Proceeding of ISPSD 2008.

DC-DC Converter based on the Asymmetric Multistage Stacked Boost Architecture with Feed-Forward Control for Photovoltaic Plants

Georgios Mademlis, Aristotle University of Thessaloniki, GR-54124, Thessaloniki, Greece, mademlig@ece.auth.gr

Gina K. Steinke, École Polytechnique Fédérale de Lausanne, EPFL, CH-1015 Lausanne, Switzerland, gina.steinke@epfl.ch

Alfred Rufer, École Polytechnique Fédérale de Lausanne, EPFL, CH-1015 Lausanne, Switzerland, alfred.rufer@epfl.ch

Abstract

A new feed-forward control technique for a DC-DC step-up converter based on the asymmetric Multistage Stacked Boost Architecture (MSBA) is presented in this paper. The proposed closed loop control scheme can provide improved accuracy in the steady-state operation and high dynamic performance. The asymmetric MSBA converter is a single output high-voltage DC-DC step-up converter that comprises several in-series connected capacitors with active voltage balancing circuits. The converter can attain high voltage ratios and thus, it is a suitable solution for the high voltage transmission requirements of the large photovoltaic (PV) plants. Selective simulation and experimental results are presented in order to validate the effectiveness and the operational improvements of the proposed control system.

1. Introduction

In this paper, a problem of high concern for the efficient transfer of the produced electric energy in high power PV plants is addressed. Since PV plants of hundred MWs installed power cover a large area, the electric power generated by the solar panels should be collected over long distances. The transfer losses of the power system can be drastically reduced by applying medium voltage (MV) in the DC link between the PV installation and the point of common coupling to the grid, as it has initially been proposed in [1] and experimentally validated in [2]. The microgrid topology of such a PV plant is shown on Fig. 1. Output voltage of the each PV can be increased up to MV level by using an asymmetric MSBA converter, since it can provide very high voltage step-up ratios with high efficiency and low installation cost [1].

The asymmetric MSBA converter is a multilevel step-up DC-DC converter, where the voltage elevation is accomplished by using series connected stages. Each stage of the converter consists of one power switch, one diode, one capacitor and one inductor. These stages balance the voltage of the output capacitors and control the step-up ratio of the converter. The balancing in the voltages between the stages is very important and therefore the design of the control system for the power switches is crucial. An open loop control system has been examined, which was sensitive in oscillations, and a closed loop control that was based on PI controllers for controlling the voltage and current of each converter stage [1]. Although the performance of the MSBA converter with PI controllers can be considered as acceptable, investigations for further improvement have been conducted, particularly in the dynamic response and the stability of the system. These objectives can be accomplished by adding feed-forward of output currents to the current and voltage controllers of each converter stage [3]-[4]. The feasibility of this control technique has been examined on the symmetric MSBA converter [5] as well as on buck converters [6].

Aim of this paper is to present a feed-forward control scheme for an asymmetric MSBA converter with highly dynamic performance and improved accuracy in the steady-state operation. The developed converter comprises several in-series connected capacitors with active voltage balancing circuits. Since it can provide high voltage rations, it is suitable solution for the high voltage transmission requirements of the large photovoltaic (PV) plants.

The structure of this paper is as follows. The operation of an asymmetric MSBA converter with 4-stages is analyzed and the basic voltage and current equations are presented in Section 2. The suggested control technique is described and discussed in Section 3. Comprehensive collection of simulation and experimental results are presented in Sections 4 and 5, respectively, order to evaluate the feasibility and demonstrate the operational improvements of the proposed control technique.

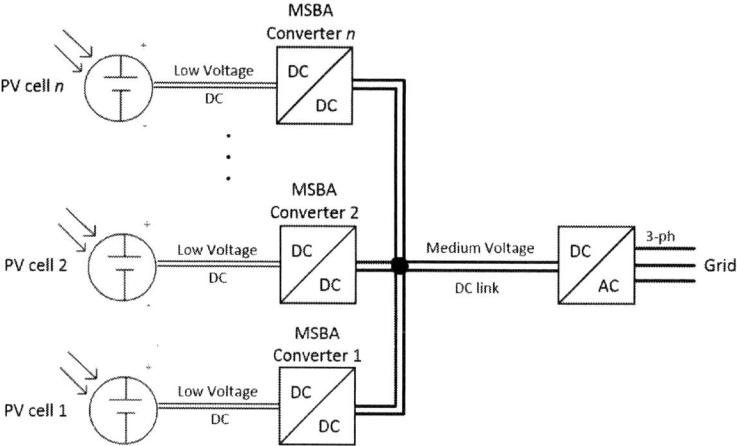

Fig.1. Schematic diagram of a PV plant with MSBA converters connected at the output of all the solar arrays and a central DC-AC inverter installed at the point of common coupling to the grid

2. The MSBA converter operation

The MSBA converter belongs to the multilevel non-isolated step-up DC-DC converters and its operation is based on the principle of balancing the voltage of series connected capacitors using a so-called *current diverter*. The current diverter is an active voltage balancing circuit and it has initially been proposed for the balanced charging of series connected batteries [7] and supercapacitors [8]. Figure 2(a) shows the elementary cell of an active voltage balancing circuit with unidirectional power flow in order to be suitable for a PV application.

The MSBA converters are divided into the *symmetric* and the *asymmetric* topologies. In the asymmetric topology the input boost-stage is connected to the lowest or the top level of the capacitor chain and the converter has one output at the high voltage side. This topology can attain a high voltage step-up ratio and therefore, it is suitable for increasing the output voltage of a PV cell. An asymmetric MSBA converter with 4 stages is analyzed in this paper and thus, a nominal voltage step-up ratio of 1:8 is attained [Fig. 2(b)].

In the 1st stage of the asymmetric MSBA converter, the relationship between the input and output voltage (U_{in} and U_{C_1}, respectively) is given by

$$\frac{U_{C_1}}{U_{in}} = \frac{1}{1-D_1}$$

(1)

where D_1 is the duty cycle of the power switch T_1. The relationship between the input voltage U_{C_2} and the output voltage U_{C_1} of the 2nd stage of the converter is described as follows

$$U_{C_2} = U_{C_1} \frac{D_2}{1 - D_2} \tag{2}$$

The voltage relationships of the 3rd and 4th stages are given by equations similar to (2). The average inductor current of the 2nd stage I_{L_2} can be calculated as follows

$$I_{L_2} = I_{L_1}\left(1 - D_1\right) + D_3 I_{L_3} \tag{3}$$

and the average inductor current of the 3rd stage I_{L_3} is

$$I_{L_3} = I_{L_2}\left(1 - D_2\right) + D_4 I_{L_4} \tag{4}$$

The average 4th stage inductor current (the last stage of the converter) is given by

$$I_{L_4} = \frac{I_{out}}{\left(1 - D_4\right)} \tag{5}$$

Finally, by using (1) and (2), and by considering that the duty cycles D of all the power switches are equal, the relationship between the input and output voltage of the whole 4-stage asymmetric MSBA converter is

$$\frac{U_{out}}{U_{in}} = \frac{1}{1 - D}\left[1 + \sum_{k=1}^{3}\left(\frac{D}{1 - D}\right)^k\right] \tag{6}$$

From (6) and with duty cycle $D = 0.5$, the voltage step-up ratio of the 4-stage converter is 1:8.

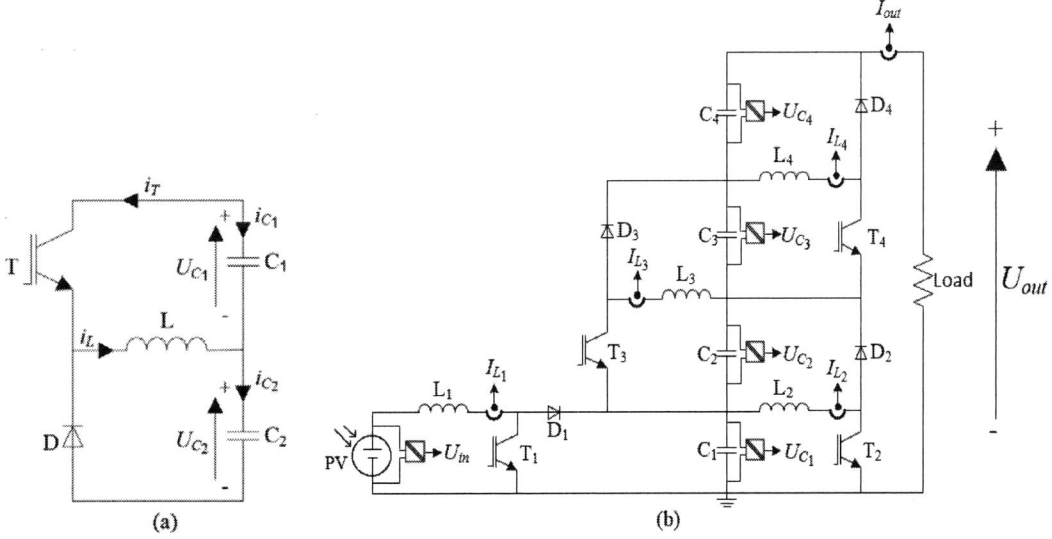

Fig.2. (a) Elementary cell of the current diverter with unidirectional power flow; (b) 4-stage asymmetric MSBA converter including the necessary voltage and current sensors for the implementation of the feed-forward control system

3. Control system

In a feed-forward based control system, the adjustment of the control variable is not based on the error between the reference and the measures value, but on the a priori knowledge of the performance from the mathematical model and measurements of the perturbations of the power system. The output currents of each stage of the MSBA converter are the perturbations that are taken into account in the feed-forward system. Thus, the feed-forward technique can improve the dynamic response of the control system, since it can respond very fast to any abrupt changes of the output currents, by changing the internal variables of the converter and also can successfully eliminate the steady state error.

The control system is closed loop and consists of cascaded voltage and current controllers for each stage of the converter. The voltage controller is the primary control loop and the current controller is the secondary control loop of the control scheme. The voltage controller determines the reference inductor current and the current controller provides the PWM pulses for the power switch. The feed-forward signal I_{FF} contributes to the elimination of the error between the inputs of the voltage controller, which is the reference capacitor voltage $U_C{}^*$ and the capacitor voltage to be controlled U_C. Since the static errors are eliminated by the feed-forward control, both controllers can be designed with simple proportional gains. The gains of the controllers for the 4-stage converter are reported in Table I.

The block diagram of the proposed control system for each converter stage is illustrated in Fig. 3, indicating the inputs to the voltage and the current controllers. Fig. 4 shows a more detailed view of the power circuit and the control system of the 4-stage asymmetric MSBA converter, including the feed-forward signals. The feed-forward signals for all the stages of the converter are calculated as follows

$$
\begin{aligned}
I_{FF_1} &= I_{L_2} - I_{L_3} D_3 \\
I_{FF_2} &= I_{L_3} - I_{L_4} D_4 \\
I_{FF_3} &= I_{L_4} \\
I_{FF_4} &= I_{out}
\end{aligned}
\tag{7}
$$

From the above analysis, it can be concluded that fast dynamic response can be achieved with the feed-forward control technique, since the voltage and current controllers of each stage are designed with simple proportional gains and, therefore, the delay that is caused by the integrator in a conventional system with PI controller has been considerably reduced. However, the control system is sensitive to the feed-forward signals and thus, the accurate measurements of the output current of each converter stage and all the capacitor voltages of the converter are of vital importance. Any inaccuracy in the measurements of the currents or of the voltages may result to a malfunction of the control in the whole system.

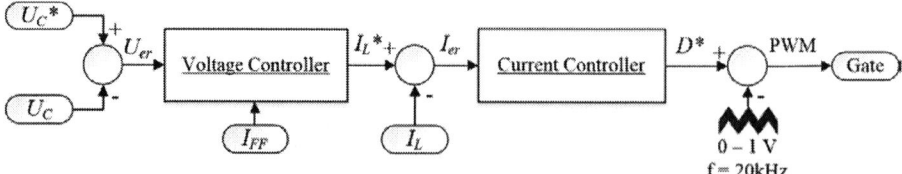

Fig.3. Block diagram of the control system in a stage of the asymmetric MSBA converter

PCIM Europe 2016, 10 – 12 May 2016, Nuremberg, Germany

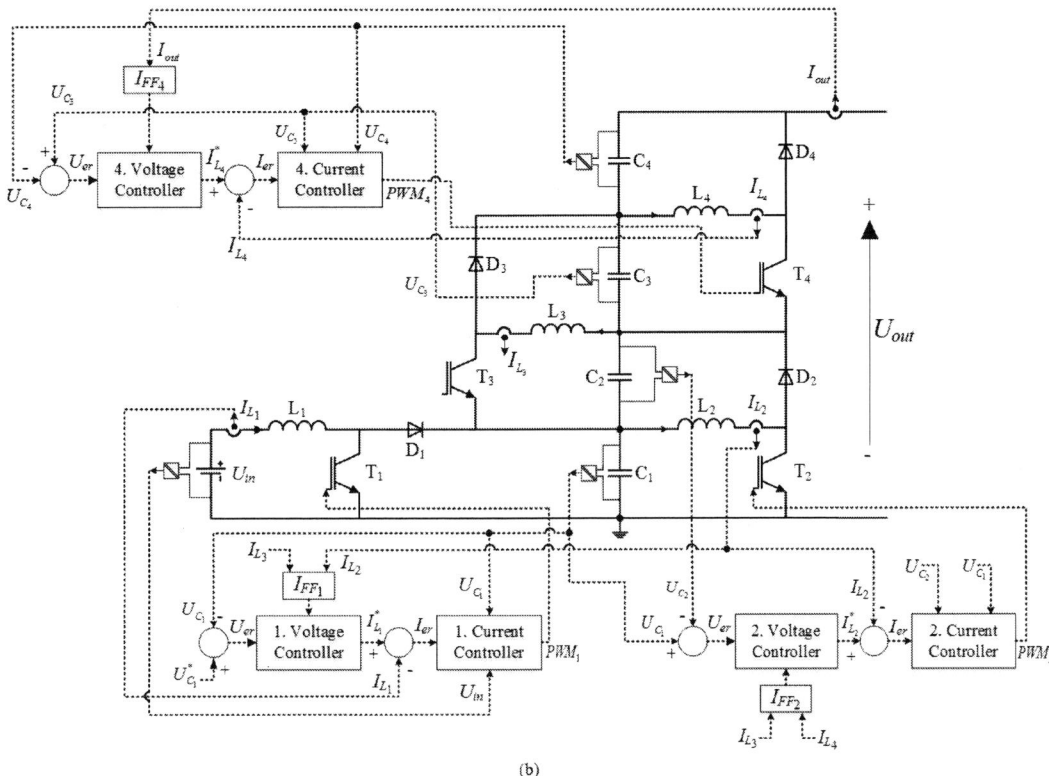

(b)

Fig.4. Power circuit and control diagram of the 4-stage asymmetric MSBA converter with feed-forward control

4. Simulation results

The Matlab/Simulink is used for the simulations of the steady state operation and the dynamic response of the asymmetric MSBA converter, in order to validate the effectiveness of the proposed control system. Table I shows all the parameters of the 4-stage converter that are used during the simulations and the experimental investigations. The gains of the voltage and current controllers are decided on the base of the magnitude optimum criterion [9], in order to accomplish the fastest dynamic response with an optimum balance between rise-time and overshoot.

The simulation results shown in Fig. 5 demonstrate the dynamic response of the developed MSBA converter. Specifically, Figs. 5(a) and (b) examine the response of the system in a step change of the reference capacitor voltage from 100 V to 160 V and the load change from 773 W to the nominal power of 1.17 kW, respectively. The input voltage in both cases is equal to the nominal value of 80 V. It can be seen that the inductor currents of all stages respond very fast and they reach steady-state without any oscillations. Also, the capacitor voltages of all stages get their new values with almost zero steady-state error. The fastest response corresponds to the 1st stage, while the slowest response corresponds to the 4th stage. This occurs, because the capacitors are connected in series and the power flows from the 1st to the 4th stage of the converter. Additionally, the reference capacitor voltage of each stage is the capacitor voltage of the preceding stage. From Fig. 5(b) it is also validated that the feed-forward based control system exhibits satisfactory rejection of the load disturbance by reaching equilibrium in less than 40 ms.

© VDE VERLAG GMBH · Berlin · Offenbach

PCIM Europe 2016, 10 – 12 May 2016, Nuremberg, Germany

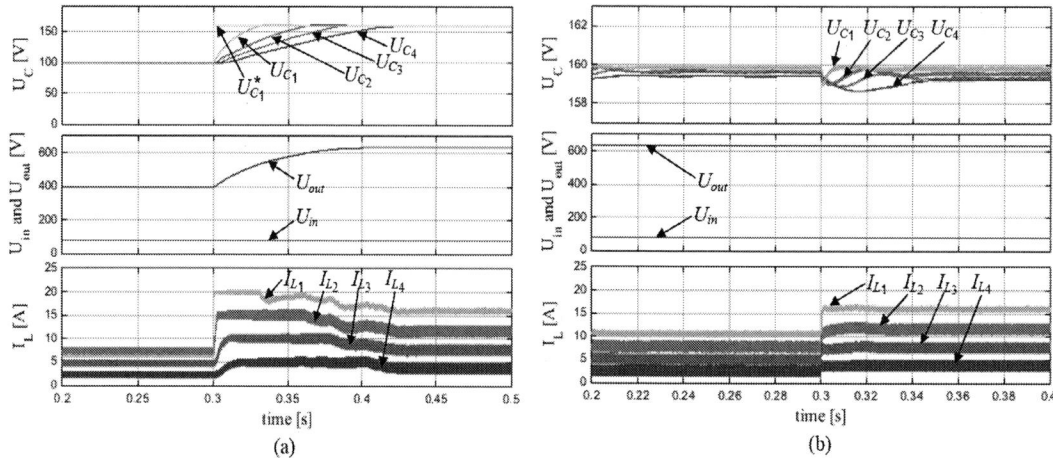

Fig.5. Simulation results with input voltage U_{in} = 80 V for a step change of: (a) the reference capacitor voltage from 100 V to 160 V and (b) the load from 0.77 kW to 1.17 kW

Table I: Parameters of the 4-stage Asymmetric MSBA Converter

Symbol	Quantity	Value
U_{in}	Nominal input voltage	80 V
U_{out}	Nominal output voltage	640 V
I_{out}	Nominal output current	1.83 A
U_{C1}	1st stage nominal capacitor voltage	160 V
$R\ (P_{load})$	Nominal load	350 Ω (1.17 kW)
$C_1 \div C_4$	Capacitances of stages 1 ÷ 4	1.5 mF
$L_1 \div L_4$	Inductances of stages 1 ÷ 4	1.7 mH
f_s	Switching frequency	20 kHz
G_{U_1}, G_{I_1}	1st stage voltage and current controller gains	1.1, 11
G_{U_2}, G_{I_2}	2nd stage voltage and current controller gains	0.35, 50
G_{U_3}, G_{I_3}	3rd stage voltage and current controller gains	0.25, 40
G_{U_4}, G_{I_4}	4th stage voltage and current controller gains	0.15, 30

5. Experimental results

An experimental prototype of the 4-stage asymmetric MSBA converter has been built in order to verify the theoretical considerations (Fig. 6). The parameters of the converter are the same with those used in the simulation analysis and have been reported in Table I. The components of the prototype converter are listed in Table II.

Similarly to the simulation analysis, the response of the system is examined at first with a step-change of the output reference voltage from 100 V to 160 V [Fig. 7(a)]. The converter operates with nominal input voltage and the load is an ohmic resistor of 350 Ω. As can be seen, the capacitor voltages of the 3rd and the 4th stage are balanced and the steady state error between the two of them is negligible before and after the change of the reference value. The dynamic response of the converter is also quite fast, since it takes 0.1s for the voltages to reach their new reference values.

© VDE VERLAG GMBH · Berlin · Offenbach

Figs. 7(b) and 7(c) illustrate the steady state performance of the converter with reference capacitor voltage 160V. As can be seen, the static errors of the feed-forward based voltage controllers are very small. Fig. 7(d) shows the experimental results for an abrupt change of the load from 530 Ω to the nominal value of 350 Ω. It can be seen that, the system responds satisfactorily to the load change and after a very short voltage drop of around 10 V, the capacitor voltages return balanced regardless of the new load conditions.

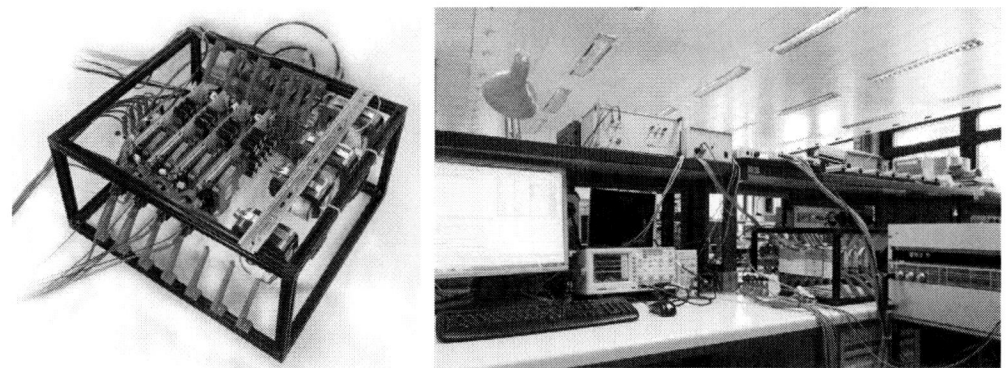

Fig.6. (a) Experimental prototype of the 4-stage asymmetric MSBA converter; (b) Real-time experimentation environment

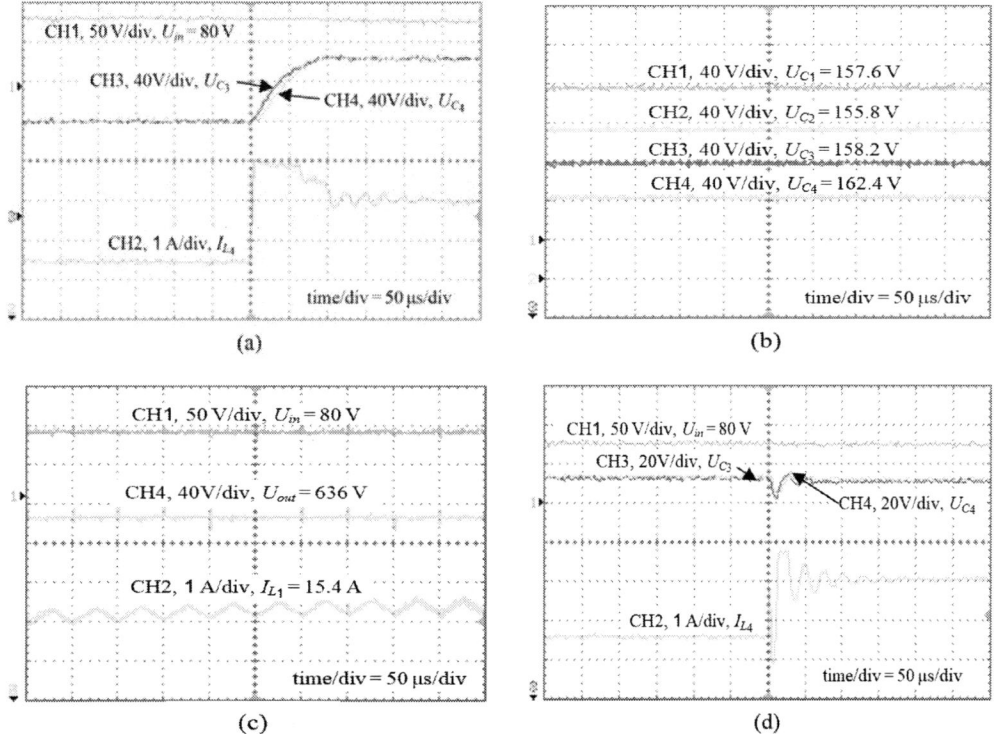

Fig.7. Experimental results of the asymmetric 4-stage MSBA converter: (a) the capacitor reference voltage step-changes from 100 V to 160 V, (b) and (c) steady state operation with capacitor reference voltage equal to 160 V and (d) abrupt change of the load from 530 Ω to 350 Ω

Table II: Component list of the experimental prototype

Name	Product Code	Quantity
IGBT	IXYS MMIX1X200N60B3H1	8
Output capacitor	Panasonic EEUEE2E151	45
Inductor core	Hitachi AMCC0160	4
PCB	Imperix PEB4250	4
Current sensor	LEM LAH 50-P	4
Current sensor	Imperix IX-LEM LAH 50-P	1
Voltage sensor	Imperix IX-ModuLink400V	5
Control platform	Imperix Boombox	1

6. Conclusions

The operation principle of a feed-forward based control system for an asymmetric MSBA DC-DC converter has been presented in this paper. The converter and the proposed control system can be used in high power PV plants in order to increase the output voltage of the PV cells up to the medium voltage level and thus, to reduce the transfer loss of the generated electric energy. The structure of the control scheme has been analyzed and the operation of the converter has been simulated on Matlab/Simulink. The proposed control system has also been tested on an experimental prototype. It has been validated from the simulation and experimental results that the proposed feed-forward based control technique provides fast dynamic response and eliminates the steady state error in the converter performance.

References

[1] A. Rufer, P. Barrade, and G. Steinke, "Voltage step-up converter based on Multistage Stacked Boost Architecture (MSBA)", in *Proc. Conf. IPEC-2014*, pp. 1081-1086.

[2] G. K. Steinke and A. Rufer, "Use of a DC-DC step up converter in photovoltaic plants for increased electrical energy production and better utilization of covered surface area", in *Proc. Conf. PCIM-2015*.

[3] R. Redl and N. O. Sokal, "Near-Optimum Dynamic Regulation of DC-DC Converters Using Feed-Forward of Output Current and Input Voltage with Current-Mode Control", *IEEE Trans. Power Electron.*, vol. PE-1, no. 3, pp. 181-192, July 1986.

[4] M. K. Kazimierczuk and A. Massarini, "Feedforward Control of DC–DC PWM Boost Converter", *IEEE Trans.* Circuits and Systems I: Fundamental Theory and Applications, vol. 44, no. 2, pp. 143-148, Feb. 1997.

[5] A. Rufer, "A Five-Level NPC Photovoltaic Inverter with Active Balanced Capacitive Voltage Divider", in *Proc. Conf. PCIM-2015*.

[6] A. Rufer and P. Barrade, "Non-Isolated DC-DC Converters for High Power Applications – Control of the Capacitive Voltage Divider", in *Proc. Conf. PCIM-2014*.

[7] N. H. Kutkut, "A modular non-dissipative current diverter for EV battery charge equalization", in *Proc. IEEE Conf. APEC '98*, 1998, vol. 2, pp. 686–690.

[8] P. Barrade, S. Pittet, and A. Rufer, "Energy storage system using a series connection of supercapacitors, with an active device for equalising the voltages", in *Proc. Conf. IPEC-2000*.

[9] H. Bühler, *Réglage de systèmes d'électronique de puissance - Volume 1: Théorie* (Control of power electronic systems – Textbook 1), Lausanne: Presses polytechniques et universitaires romandes, 1997, Chap. 3.

A Novel Submodule Concept for Modular Multilevel Converters

Benjamin Ruccius[1], Nicola Burani[1], Marek Galek[2], Dirk Malipaard[1]
[1] Fraunhofer IISB, Schottkystr. 10, 91058 Erlangen, Germany
[2] Siemens AG, Corporate Technology, Otto-Hahn-Ring 6, 81739 München, Germany

Abstract

Modular Multilevel Converter (MMC) topologies have been successfully employed for voltage source converters in high as well as in medium voltage applications. In the low voltage range, the success of this topology strongly depends on the capability of exploiting the best performing key components commercially available. A brief survey of the state of the art reveals that standard submodule concepts do not allow the employment of the best semiconductors and capacitors in a fully optimized design. In this paper, a novel submodule concept is presented which overcomes this constraint. The dimensioning of the capacitors and the possible modulation strategies are also discussed.

1. Introduction

Modular Multilevel Converter topologies (Fig. 1) have been already successfully used as self-commutated voltage source converters (VSC) in several HVDC as well as in medium voltage applications [1], [2]. A non-exhaustive list of advantages provided by an MMC solution includes:

- Low degree of total harmonic distortion (THD) avoiding filtering at the AC terminals.
- Use of industrially proven standard power modules.
- Reduction of conduction and switching losses through optimization of the semiconductor components, lower switching frequency and lower switching voltages.
- Modular construction of the system.
- Voltage scalability.
- Fault tolerance through redundant design.
- Flexible insulation design concept allowing, for example, low thermal resistance module construction [3].

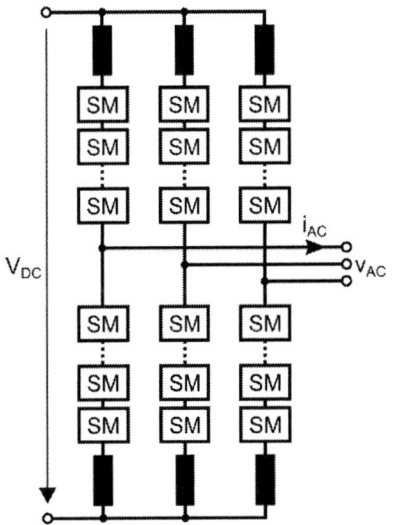

Fig. 1 - Structure MMC

Recently, increasing efforts have been made to prove that the advantages of a multilevel approach can be extended to applications in the low voltage range (1000 V_{rms}, below 1500 V_{dc}) [4]. It is nevertheless clear that key system parameters like efficiency, cost, volume and reliability, will be strongly influenced by the choice of the two main components of the converter: the semiconductors in the switching cells and the capacitors used for the energy storage.

Looking at commercially available semiconductors in the LV high power range, three possible solutions are easily identified:

- LV Silicon MOSFETs, from sub 20 V up to maximum 250-300 V breakdown voltage.
- Super-junction MOSFETs, from 500 V up to a maximum of 900 V breakdown voltage.
- Silicon IGBT, typically from 600 V upward (300 V parts are only sparely available).

To take advantage of the MMC converter topology, for example the low degree of total harmonic distortion (THD), a minimum of 4 switching levels in the output waveforms is commonly used. For low voltage applications, this requirement results in submodule capacitor voltages lower than 300V. Between the three semiconductor product categories identified above, the LV Si-MOSFETs are the most promising ones for this voltage range. They offer industrially proven chip designs, low conduction losses, and fast switching times for high system efficiencies. Moreover, external free-wheeling diodes and snubber circuits can be avoided by selecting components with an intrinsic diode optimized for hard-switching applications [5].

Considering the possible solutions for the energy storage there are:

- Electrolytic capacitors in a voltage range from sub 20 V up to 500-600 V.
- Polyethylene Terephthalate (PET) film capacitors, available in principle for voltages down to 200 V and up to the 650 V to 1 kV range.
- Polypropylene (PP) film capacitors, designed for conventional DC link applications with voltages typically not lower than 400 V.

Due to the relatively high ripple current specifications, the employment of electrolytic capacitors in industrial MMC front-end applications with extended lifetime requirements (typ. more than 100.000 h) at operating ambient temperatures up to 70°C, leads easily to a dimensioning of the DC-link which is mainly lifetime driven. To avoid over-dimensioning, film capacitors have to be preferred as they offer longer expected life-time and much higher ripple current handling capabilities. Table 1 shows a comparison of Polyethylene Terephthalate and Polypropylene as dielectric material according to their theoretically reachable maximum energy density at breakdown field strength E_{br} and at operative field strength E_{op} in metallized film capacitors [6].

Dielectric Material	Relative permittivity (ε_r)	Breakdown field strength E_{br} (V/µm)	Breakdown Energy density $W_{br} = \frac{1}{2}\varepsilon_r\varepsilon_0 E_{br}^2$ (J/cm^3)	Operative field strength E_{op} (V/µm) [7]	Operative Energy density $W_{op} = \frac{1}{2}\varepsilon_r\varepsilon_0 E_{op}^2$ (J/cm^3)
Polyethylene Terephthalate (PET)	3.2	570	1.3	80	0.1
Polypropylene (PP)	2.2	640	2	180	0.3

Table 1 - Energy densities for PET and PP dielectrics

According to the data presented here, polypropylene offers three times higher energy density than PET. Nevertheless, due to the higher price and the limited capabilities in the production of stable ultra-thin metallized PP films (2µm or lower), the operative energy densities obtainable for voltages below 400-450 V are practically reduced [8]. This makes PET a potentially competitive material for applications in the 200-250 V range and below. Other factors and parameters are important as well to derive the energy density practically obtainable in an optimized capacitor design.

Commercially available, compact, high-capacitive, and low-inductive DC-link solutions are typically offered for voltages not lower than 400 V and for capacities not higher than 10 mF. These solutions are typically based on polypropylene films, due to the higher temperature stability of the dielectric, its superior electric characteristics and in particular to the lower dielectric losses. Polyethylene Terephthalate is not commonly used for high-capacitive DC-link capacitors or bigger foil capacitor blocks in general. A custom solution for this application would be possible; nevertheless, due to the missing mass production of similar devices, a cost disadvantage is expected.

This brief survey of the commercially available state-of-the-art components shows, that traditional half or full bridge MMC submodule concepts are not suitable to match the overall best performing semiconductors and PP film capacitors in an optimized design (Fig. 2).

© VDE VERLAG GMBH · Berlin · Offenbach

A similar situation occurs in the high voltage range at 100-1050 kV DC-link voltages. In this range, a MMC solution allows the employment of fast switching industrial semiconductor power modules (IGBT), available up to 6.5 kV blocking voltage. For the energy storage, new dielectric materials like ultra-thin glass are opening perspectives in the design of capacitors operating at 10 kV and above offering very high energy density and high temperature capabilities [9], [10]. The matching of such devices with standard IGBT modules would not be possible by employing a traditional submodule concept.

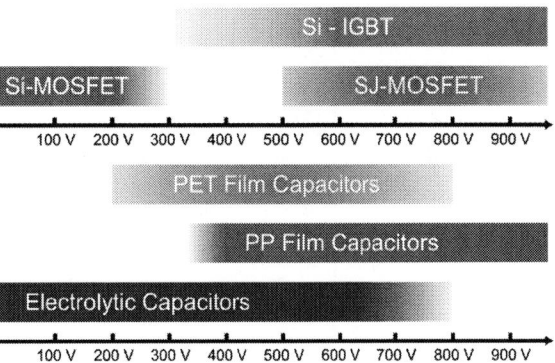

Fig. 2 - Voltage range of suitable semiconductors and capacitors

The work presented here is therefore focused on the development of a novel LV MMC submodule (SM) which allows the employment of best-in-market standard power semiconductors together with high performance capacitors in a fully optimized design.

2. Novel topology

The standard half-bridge MMC submodule is shown in Fig. 3. Fig. 4 shows a double cell submodule, which is commonly used as well [4]. The novel double cell submodule features a common storage capacitor and is shown in Fig. 5. By adding the voltage divider capacitors C_{Vd_I} and C_{Vd_II}, the voltage requirements regarding the semiconductors can be halved compared to the voltage of the common storage capacitor $C_{storage}$.

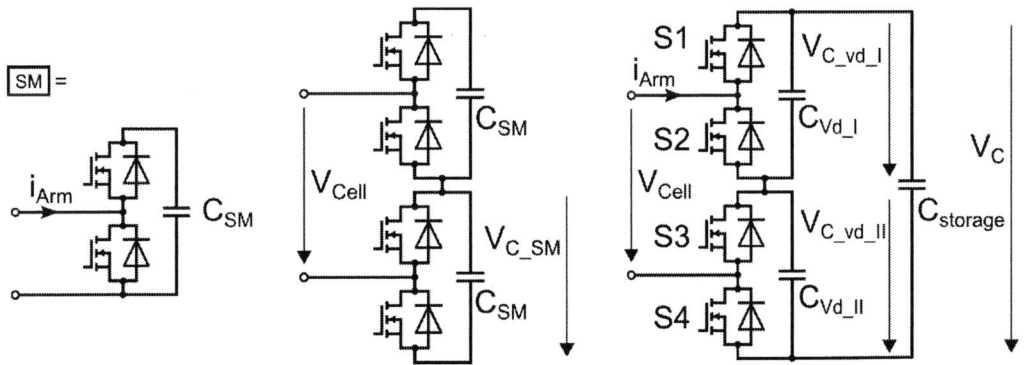

Fig. 3 – Half-bridge submodule

Fig. 4 – Double-cell-submodule with half-bridges

Fig. 5 – Novel submodule topology

2.1. Possible switching states

The possible cell output voltages V_{cell} of the newly proposed topology are the same as the ones of a double-submodule cell. There are two possible switching states 1a and 1b to set $V_{cell}=0.5 \cdot V_C$. The possible cell states and the corresponding current paths are summarized in Table 2. The current flow in state 1a, 1b and 2 is divided into two parallel connections, which are shown in the figures below the table.

© VDE VERLAG GMBH · Berlin · Offenbach

Voltage V_{cell}	0	$0.5 \cdot V_C$	$0.5 \cdot V_C$	V_C
No. of state	0	1a	1b	2
S1	Off	On	Off	On
S2	On	Off	On	Off
S3	On	On	Off	Off
S4	Off	Off	On	On
Current flow				

Table 2 - Possible switching states

2.2. Dimensioning of capacitors

In a Modular Multilevel Converter the sum of the submodules of one converter arm can be considered as a controlled voltage source. The capacitors C_{SM} serve as energy storage for this voltage source. They can supply energy, if the sign of i_{Arm} is negative and they can be recharged, if the sign of i_{Arm} is positive. The control of the MMC has to ensure, that the average energy in the capacitors remains constant over an output period $1/f_{sw}$. The minimum capacity of a submodule capacitor $C_{SM,min}$ is depending on the maximum energy lift per arm $\Delta W_{arm,max}$ during an output base period [11], [12], which depends on the output current i_{AC}, the DC-voltage V_{DC} the output power P_{AC} and the output base frequency f_{AC}.

$$\Delta W_{arm,max} = \frac{\hat{\imath}_{AC}}{2 \cdot \pi \cdot f_{AC}} \cdot \frac{V_{DC}}{2} \cdot \left(1 - \left(\frac{P_{AC}}{3 \cdot \hat{\imath}_{AC} \cdot \frac{V_{DC}}{2}}\right)^2\right)^{\frac{3}{2}} \tag{1}$$

The maximum available energy lift per arm $\Delta W_{arm,stored}$ is depending on the permissible maximum and minimum voltages of the capacitor of the submodule $V_{C\_SM,max}$ and $V_{C\_SM,min}$, as well as on the number of submodules per arm n_{SM} and the capacity of the submodule capacitors C_{SM}.

$$\Delta W_{arm,stored} = n_{SM} \cdot \frac{1}{2} \cdot C_{SM} \cdot \left(V_{C\_SM,max}{}^2 - V_{C\_SM,min}{}^2\right) \tag{2}$$

By equalizing (1) and (2), the minimum size of the submodules capacitor $C_{SM,min}$ can be calculated:

$$C_{SM,min} = \frac{2 \cdot \Delta W_{arm,max}}{n_{SM} \cdot \left(V_{C\_SM,max}{}^2 - V_{C\_SM,min}{}^2\right)} \tag{3}$$

The maximal voltage of the submodules capacitor $V_{C\_SM,max}$ has to be chosen considering the voltage rating of the capacitor and the semiconductors.

Due to the doubled capacitor voltage, the capacity of $C_{storage}$ can be halved, compared to C_{SM}, while still maintaining the same energy stored in the double cell submodule:

$$W_{double\_cell} = 2 \cdot \left(\frac{1}{2} \cdot C_{SM} \cdot V_{C\_SM}{}^2\right) = 2 \cdot \left(\frac{1}{2} \cdot C_{SM} \cdot \left(\frac{V_C}{2}\right)^2\right) = \frac{1}{4} \cdot C_{SM} \cdot V_C{}^2 \tag{4}$$

$$W_{new\_topology} = \frac{1}{2} \cdot C_{storage} \cdot V_C{}^2 \tag{5}$$

The capacitors C_{Vd_I} and C_{Vd_II} can be much smaller than the capacitors C_{SM} and $C_{storage}$, as they don´t have to store the energy, which has to be provided for the energy lift during the output base period [13].

The voltage over the voltage divider capacitors equals to half the voltage over the storage capacitor and an additional superimposed voltage ripple. The amplitude is determined by the arm current i_{Arm}, the switching frequency and the capacitor size. If states 1a and 1b are used by the modulation, the minimal capacity of the voltage divider is depending on the highest allowable voltage ripple over the voltage divider capacities ΔV_{max}, the maximum current through the cell I_{Arm_max} and the switching frequency f_{sw}. Assuming $C_{Vd} \ll C_{storage}$ and $C_{Vd\_I} \cong C_{Vd\_II}$:

$$C_{Vd,min} \approx \frac{0.5 \cdot I_{Arm,max} \cdot \Delta t_{state\_1}}{\Delta V_{max}} \leq \frac{0.5 \cdot I_{Arm,max} \cdot T_{sw}}{\Delta V_{max}} = \frac{0.5 \cdot I_{Arm,max}}{\Delta V_{max} \cdot f_{sw}} \tag{6}$$

with Δt_{state_1}: time applying state 1 in one switching period $T_{sw}=1/f_{sw}$.

While the size of $C_{storage}$ and the traditional submodules' capacitor C_{SM} depends inversely on the output base frequency (equations (3) and (1)), the size of the capacitors C_{Vd_I} and C_{Vd_II} depends inversely on the switching frequency.

2.3.   Modulation

A possible and widely applied modulation strategy for MMCs built of traditional half bridge submodules is shown in Fig. 6 [13]. The blue curve shows the target or desired arm voltage V_{target}, while the black curve shows the arm output voltage, which is the sum of the output voltages of the single cells. Within the control algorithm, a time period $T_A = n \cdot T_{PWM}$ (with typically n = 1) is defined, with T_{PWM} typically in the range of 0,05-1 ms. Within T_A, all the cells of an arm, except one, are continuously activated (Top semiconductor is switched on) or bypassed (Bot semiconductor is switched on). The one cell left is switched between active and bypass state following a high-frequency Pulse Width Modulated (PWM) signal with time period T_{PWM}. The cells which are activated, bypassed or driven by a PWM modulation in each period T_A are selected based on a sorting algorithm, ensuring that the capacitor voltages of each cell are equalized and balanced. If $i_{Arm} > 0$ A, cells with the lowest voltages within the same arm are activated to charge the capacitors. If $i_{Arm} < 0$ A, cells with the highest voltages are activated to discharge their capacitors. As T_A is normally selected to be much smaller than the period of the main output frequency, low levels of harmonic distortion are easily obtained.

With this type of modulation, the number of voltage levels that can be obtained at the AC output terminal of one phase is:

$$n_{Vac} = n_{SM} + 1 \tag{7}$$

with n_{SM}: number of submodules or cells per arm.

PCIM Europe 2016, 10 – 12 May 2016, Nuremberg, Germany

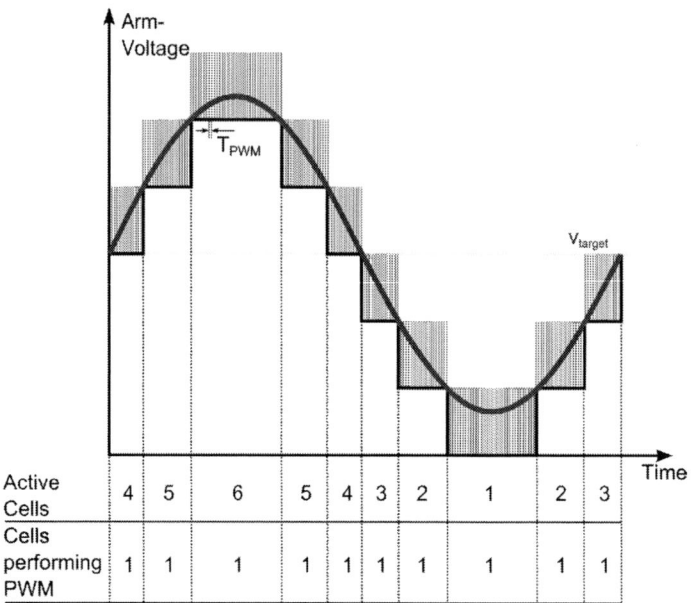

Fig. 6 - Modulation strategy for a MMC

For the new topology proposed, there are different types of modulation possible. The easiest possible modulation utilizes only two of the states from table 1: state 0 and state 2, where $V_{cell} = 0 \cdot V_C$ and $V_{cell} = V_C$ respectively. The modulation strategy is similar to the one previously presented for the MMC with traditional half bridge submodules, whereas the sorting algorithm is tuned to balance the voltage over the cells' storage capacitors. The number of voltage levels obtainable at the AC output terminal can be still calculated as previously presented (7). Nevertheless, as the target in the dimensioning of the novel submodule is an increase or doubling of the submodule voltage and therefore a reduction of the number of submodules per arm, the available output voltage levels per arm would be practically reduced.

With such a modulation strategy, the voltage divider capacitors could turn out to be smaller than calculated with equation (6), as they are only used as voltage dividers and won´t be used actively. Balancing of the voltage divider capacitors is in any case mandatory to avoid uneven voltage stresses due, for example, to components tolerances. This can be achieved passively with high value and low tolerance resistors or varistors connected in parallel to the capacitors. One drawback of such a passive balancing system is the occurrence of additional losses. Advantages of such a modulation strategy are its relatively low complexity and, when compared to the MMC with traditional half bridges submodule, a lower demand for measurement and communication between the cells and the central control; a reduced number of voltages has to be, for example, sampled and transmitted.

A modulation strategy for the proposed submodule concept, which utilizes all the possible cell output voltage levels, can be implemented as a structure with two stages. The desired output voltage for each cell is decided at a central level. At the local level, within each single submodule, switching between the states 1a and 1b is controlled to balance the voltage over the divider capacitors and, if required, the PWM-switching pattern is calculated. The central level of the modulation communicates with the local level and sets the desired output voltages ($V_{cell} = 0$ V, $0,5 \cdot V_C$ or V_C). PWM-switching output between these states can also be set as far as the duty factor is specified. If 0 V output voltage is the desired value, state 0 is applied and the cell is bypassed. If V_C output voltage is required, state 2 is applied for the whole time period. If $V_{cell} = 0,5 \cdot V_C$ is required, the cell-level modulation can switch between state 1a and 1b, depending on the polarity of i_{Arm} and the voltages $V_{C\_vd\_I}$ and $V_{C\_vd\_II}$. For example, if $V_{C\_vd\_I} > V_{C\_vd\_II}$ and $i_{Arm} > 0$ then state 1b will be used to charge C_{vd_II} and discharge C_{vd_I}. If a PWM output with a specified duty factor is demanded, the local modulator can calculate the PWM and drive the switches independently between state 0 and state 1a or 1b or between state 2 and state 1a or 1b.

© VDE VERLAG GMBH · Berlin · Offenbach

The advantage of this modulation strategy compared to the one previously presented is the preservation of all AC-voltage levels provided by a MMC built with traditional half bridge submodules. The number of output voltage levels per phase is in this case:

$$n_{Vac} = 2 \cdot n_{SM\_NT} + 1 \tag{8}$$

with n_{SM_NT}: number of submodules per arm with novel topology.

Another advantage is the reduction of the demand for communication between the submodules and the central control system compared to the MMC with standard half bridges, as only the voltage of the storage capacitor must be transmitted from the submodule to the central control unit. On the other hand, switching losses can be higher, as minimization of the capacity of C_{Vd} may lead to an increase of the switching frequency f_{sw} (see equation (6)).

A further modulation strategy without the previously presented division between a superimposed central and a local modulation is also possible. In this case, the central control would not only determine which cells should be active, bypassed, output half the storage capacitor voltage $V_{cell,desired} = 0{,}5 \cdot V_C$ or operate in PWM mode, but also specify which state should be used between state 1a and 1b for $V_{cell} = 0{,}5 \cdot V_C$ or in case of PWM mode. With this modulation, all voltage levels are maintained and local submodule intelligence can be reduced or avoided. A disadvantage, compared to the previously presented modulation, is an increased load of the communication system, as a higher number of signals have to be transmitted at higher data rate between the cells and the central control.

3. Summary and Outlook

This paper presents a novel submodule (SM) for Modular Multilevel Converters which contributes to the expansion of this topology into the low voltage (< 1 kV) but also high voltage (several 100 kV) range. The employment of best-in-market semiconductors and high-performance capacitors in a fully optimized design is made possible by increasing the energy-storage capacitor's voltage in each submodule and adding small capacitors acting as voltage dividers. Different possible modulation strategies for this novel topology together with the dimensioning of the capacitors are presented and compared to one of the modulation patterns widely applied to traditional MMC topologies.

As part of this research work, a prototype of the submodule is under construction to verify the concept proposed.

References

[1] Knaak, H.J. *Modular Multilevel Converters and HVDC/FACTS: a success story,* EPE 2011 Conference Birmingham, United Kingdom

[2] S. Rohner, S. Bernet, M. Hiller, R. Sommer *Modelling, Simulation and analysis of a Modular Multilevel Converter for medium voltage applications,* ICIT Conference 2010

[3] U. Waltrich, D. Malipaard, A. Schletz *Novel Design Concept for Modular Multilevel Converter Power Modules,* PCIM Europe 2014 Conference, Nuremberg, Germany.

[4] M. Galek *MOSFET-based Modular Multilevel Converter,* ECPE Workshop 2014 Advanced Multicell / Multilevel Power Converters Tolouse, France

[5] A. Huang *Hard Commutation of Power MOSFET OptiMOS*[TM] *FD 200/250 V,* Application Note AN 2014-03 Infineon Technologies Austria AG

[6] L. Qi, L. Petersson, T. Liu *Review of Recent Activities on Dielectric Films for Capacitor Applications,* ICEE Journal Vol. 4, No. 1 2014

[7] W. Grimm *Reliability of Film Capacitors,* ECPE Workshop: "Innovations in passive components for power electronics applications", Berlin 2014

[8] H. Vetter, *Miniaturized System Solutions for Innovative HEV Converters Based on PCC Technology,* APE 2006 Conference, Paris, France

[9] M. Letz et al. *Glass ceramics as dielectrics for high power capacitors,* IPMHVC, IEEE International 2014

[10] N. Tham, T. Erlbacher *Capacitors with High Energy Density,* ECPE Workshop Innovations in Passive Components for Power Electronics Applications 2014

[11] J. Kolb, F. Kammerer, M. Braun *Dimensioning and Design of a Modular Multilevel Converter for Drive Applications,* EPE-PEMC 2012 ECCE Europe, Novi Sad, Serbia

[12] A. Lesnicar *Neuartiger, Modularer Mehrpunktumrichter M2C für Netzkupplungsanwendungen,* Dissertation, Universität der Bundeswehr München, 2008

[13] J. Kolb *Optimale Betriebsführung des Modularen Multilevel-Umrichters als Antriebsumrichter für Drehstrommaschinen,* Dissertation, Karlsruhe 2014

PCIM Europe 2016, 10 – 12 May 2016, Nuremberg, Germany

Electro-Thermal Design of a Modular Multilevel Converter Prototype

Emilien Coulinge, Alexandre Christe, Drazen Dujic
Power Electronics Laboratory (PEL), École Polytechnique Fédérale de Lausanne (EPFL), Switzerland
emilien.coulinge@epfl.ch, alexandre.christe@epfl.ch, drazen.dujic@epfl.ch

Abstract

This paper describes the electro-thermal design of a medium voltage modular multilevel converter proto-type, from the submodule power semiconductor thermal requirements to the overall system level integration. The high number of semiconductors and capacitors involved are stressed differently depending on the actual operating conditions, as discussed in the paper. The presented design considers air cooling concept at the submodule level, with its enclosures used as air guide towards the chimney like structure. Forced air cooling is applied at the cabinet level. The numerical design is verified by means of 3-D finite element method simulations of different complexities.

1. Introduction

Ever since its introduction [1], the modular multilevel converter (MMC) and its derivatives gained in popularity both in academia and in industry, and is being considered for various applications. High voltage direct current (HVDC) bulk power transmission applications, were the first to commercially adopt this topology and successfully deploy in the field [2]. The benefits arise from the high modularity, scalability, reduced filtering effort compared to the competing solutions. At the medium voltage level, MMC is deployed for railway grid inter-ties and for STATCOMs [3], variable speed drives [4] and pumped hydro storage applications [5], albeit industrial wide acceptance, commercialization is still somewhat limited. Lower operating medium voltages require less submodules per arm, thus offsetting power quality benefits offered by MMC or requiring more care at the modulation level. Nevertheless, increased interest in MVDC power distribution networks requires new breed of conversion solutions for flexible high power DC-DC or DC-AC conversion, and MMC or its derivatives are viable solutions [6], [7].

Even though MMC is commercially available, it is far away from the mature stage and various technological improvements are still possible, from the submodule (SM) design, control, protection or overall system integration. While numerous MMC prototypes are reported, the majority of them are laboratory scale prototypes rated for low voltage and power, thus effectively removing problems related to the development of technologies that would normally have to be considered for medium voltage converters. Among others, these include: insulation coordination, auxiliary power supply, protection, communication, realistic switching frequencies, etc. Industrial data are often not easily accessible, but there are some notable exceptions, such as the work related to behaviour of MMC submodules during faults [8] or technology integration work reported in [9].

A part of the development of a medium voltage MMC [7] is presented in this paper, focusing on the electro-thermal design aspects of the basic SM, and its further integration into one MMC arm. The presentation is restricted to a single phase (two arms) arranged in one single cabinet, as this is found to be representative enough for a complete converter design. The power ratings of the MMC prototype are $0.5\,\text{MVA}$ with $10\,\text{kV}$ on the MVDC side and up to $6.6\,\text{kV}$ on the MVAC side. The design choice of $1200\,\text{V}$ IGBT voltage class has resulted in a need for 16 SMs per arm, or 32 per phase-leg organized inside the cabinet. Considering the rated system voltage, careful design approach is required when it comes to various electrical, thermal and mechanical considerations, to ensure safe operation. For simplicity, cost and reliability reasons, air cooling and air insulation (not discussed in the paper) are selected and design steps are described in the rest of the paper. This paper is organized as follows: Section 2 presents the submodule design, while Section 3 provides design data used for thermal design. FEM simulations for the SM and MMC cabinet are presented in Section 4, while conclusions are given in Section 5.

© VDE VERLAG GMBH · Berlin · Offenbach

461

PCIM Europe 2016, 10 – 12 May 2016, Nuremberg, Germany

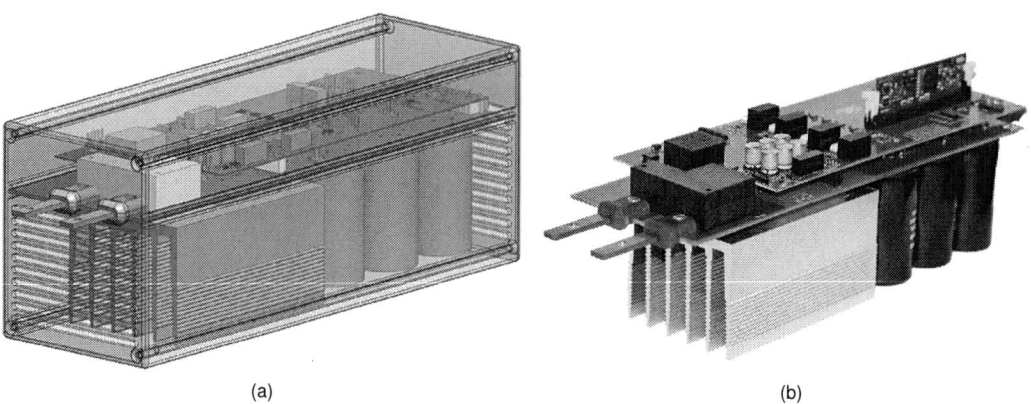

(a)   (b)

Fig. 1: MMC submodule: (a) 3D CAD rendering with its enclosure and (b) prototype.

2. Submodule Design

The SM and its metallic enclosure, normally at floating potential, are shown in Fig. 1a. Different PCBs are used for power and signal parts of the SM. The IGBT power module (*SEMITOP* from Semikron - *SK50GH12T4T*) and the bypass thyristor module are mounted on the common heatsink with electrical connections established on the power PCB, where the capacitor bank is connected as well (rated at $2.1\,\mathrm{mF}$, $650\,\mathrm{V}$). As the IGBT module is of full-bridge type, both unipolar or bipolar SM configurations are achievable by mechanical reconfiguration on the power PCB. Thus, current ratings are $50\,\mathrm{A}$ in full-bridge configuration and $100\,\mathrm{A}$ in half-bridge configuration. In addition to the fast SM thyristor protection, a slow permanent bypass (bi-stable relay) is installed as well. The upper PCB hosts the SM controller, gate circuits, measurement circuits, protection logic, inductive power transfer receiving coil and rectifier (not explicitly shown) and fibre optical communication interfaces. The actual SM prototype is shown in Fig. 1b.

Even though the fast bypass thyristors module is installed on the same heatsink with the IGBT module, there are no associated losses during normal converter operation of the MMC, and the thermal impact is negligible. The selection process for admissible temperature rise and fan airflow starts with the thermal resistance definition:

$$R_{th} \triangleq \frac{\Delta T}{\dot{Q}} \tag{1}$$

where ΔT [K] is the allowed temperature difference with respect to the ambient temperature and $\dot{Q}$ [W] the dissipated power. Based on this generic definition, the heatsink thermal resistance can be derived for the presented case. It is given in Eq. (2), where T_J refers to semiconductor junction temperature and T_H to the heatsink baseplate temperature. Considering *SEMITOP* modules, we are interested in modelling the equivalent thermal resistance from the junction to the heatsink.

$$R_{th_{J-H}} = \frac{T_J - T_H}{P_{tot}} \tag{2}$$

The maximum admissible junction temperature is defined by the manufacturer in the device data-sheet [10] to be $125\,°\mathrm{C}$. The best practices, safe operating areas, mechanical and thermal recommendations are collected in [11]. Keeping the junction temperature below this maximum value is critical and the heatsink selection is made accordingly. For the mechanical assembly, the IGBT module is directly mounted on the heatsink (separated by a thin layer of thermal grease). The maximum admissible temperature at the heatsink baseplate is computed in Eq. (3), where the thermal grease is taken into account in the equivalent thermal resistance $R_{th_{J-H}}$ as shown in Fig. 2.

$$\begin{aligned}T_{H_{\max}} = T_{J_{\max}} - \max(&P_{\mathrm{IGBT,up}} \times R_{th_{J-H_{\mathrm{IGBT}}}}, P_{\mathrm{IGBT,down}} \times R_{th_{J-H_{\mathrm{IGBT}}}},\\ &P_{\mathrm{Diode,up}} \times R_{th_{J-H_{\mathrm{Diode}}}}, P_{\mathrm{Diode,down}} \times R_{th_{J-H_{\mathrm{Diode}}}})\end{aligned} \tag{3}$$

© VDE VERLAG GMBH · Berlin · Offenbach

PCIM Europe 2016, 10 – 12 May 2016, Nuremberg, Germany

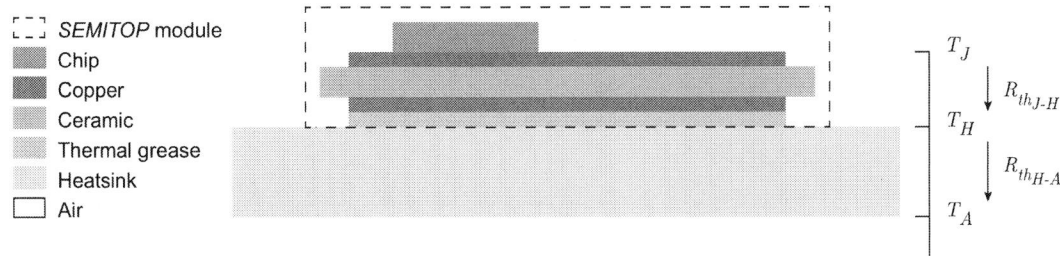

Fig. 2: *SEMITOP* module thermal model - baseplate free module.

The chip model represents either an IGBT or a diode inside the *SEMITOP* module. Subscripts used refer to the device and its position within the phase-leg. The module *SK50GH12T4T* features the following parameters: $R_{th_{J-H_{Diode}}} = 1.05\,\mathrm{K/W}$ and $R_{th_{J-H_{IGBT}}} = 0.65\,\mathrm{K/W}$. These values combined with the losses during the operation are used for the heatsink selection.

3. Electrical Losses

To evaluate the semiconductor and the capacitor losses of the SM, both DC and second harmonic circulating current injection operation of the MMC are considered [12]. These values are obtained from the closed-loop simulations with a $3\,\mathrm{kHz}$ apparent switching frequency (Figs. 3a and 3c), corresponding to a SM switching frequency of $f_{sw_{SM}} = 250\,\mathrm{Hz}$. The capacitor losses are evaluated as well and are presented in Figs. 3b and 3d, considering [13].

The worst-case losses (per component) used for the design are extracted and summarized in Table 1. It is noticeable that the global semiconductor losses are the highest for a power factor of 1, but the maximum is taken per device, regardless of the power factor or the circulating current.

Table 1: Maximum losses and temperature variation.

Device	$\max(P)$ [W]	Power factor [-]	Circulating current [-]	$R_{th_{J-H}}$ [K/W]	ΔT [K]
IGBT,up	13.10	−0.5	DC	0.65	8.52
IGBT,down	39.70	1	DC + 2nd harmonic	0.65	25.81
Diode,up	10.92	0	DC	1.05	11.47
Diode,down	35.13	−1	DC + 2nd harmonic	1.05	36.89
Capacitor	5.58	0	DC	-	-

According to Eq. (2) and the specifications of the module, Eq. (4) could be derived in order to get the maximum admissible heatsink temperature:

$$T_{H_{\max}} = 125\,^\circ\mathrm{C} - 36.89\,^\circ\mathrm{C} = 88.11\,^\circ\mathrm{C} \tag{4}$$

Considering a safety margin of $10\,\%$, the maximum limit is set to $80\,^\circ\mathrm{C}$. From there, the required thermal resistance to be matched by the heatsink is $0.7\,\mathrm{K/W}$. The final heatsink selection is made from standard extruded profile (SK 135 from Fischer-Elektronik) considering all the integration constraints of the boards and the enclosure.

Required SM capacitance, voltage ratings and current ripple constraints can not be reached using a single capacitor, and a capacitor bank is designed. It is realized with six $450\,\mathrm{V}$ capacitors, with three parallel branches of two series-connected capacitors. Balancing resistors are added to guarantee equal voltage sharing on all capacitors. Their maximum operating temperature is defined in [13] and its lifetime expectancy is subject to the ripple, frequency and temperature, as discussed in [14]. Numerically, we have derived a limit of $85\,^\circ\mathrm{C}$ for the capacitor inner core temperature, and this value is used as limit during the FEM simulations.

© VDE VERLAG GMBH · Berlin · Offenbach

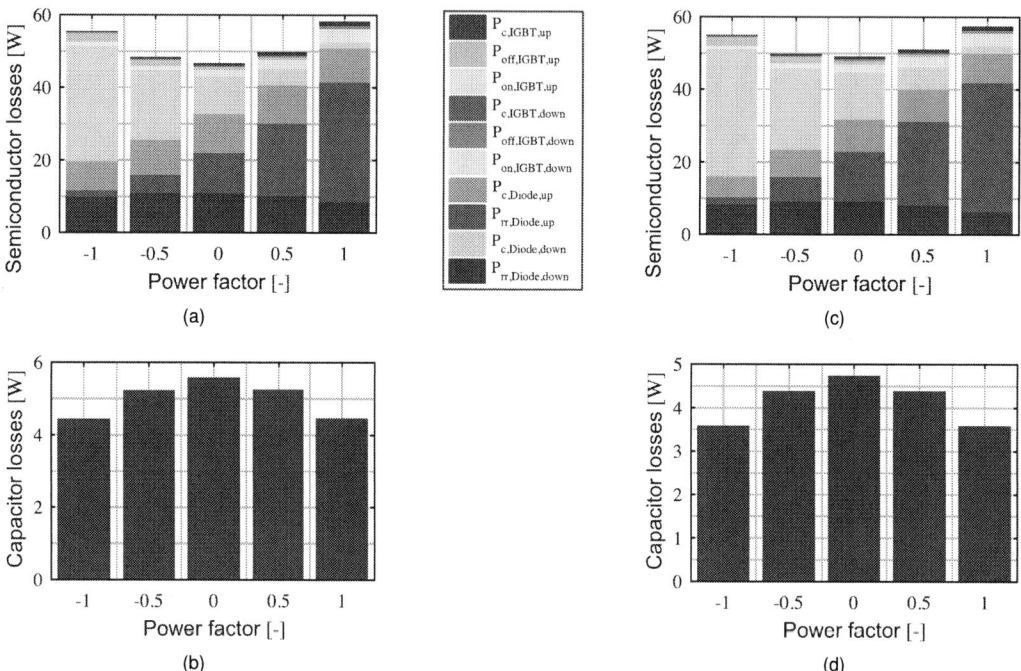

Fig. 3: Electrical losses (averaged): (a) semiconductor loss split for different power factors for the whole IGBT module with DC circulating current, (b) capacitor losses with DC circulating current, (c) semiconductor loss split for different power factors for the whole IGBT module with second harmonic circulating current and (d) capacitor losses with second harmonic circulating current.

4. FEM Thermal Simulations

The FEM simulations, carried with Ansys ICEpak, are used to verify the overall performance of the cooling system under maximum stress conditions. The maximum ambient air temperature is set as $40\,°C$, throughout the simulations.

4.1. Submodule FEM simulations

The role of the enclosure is to shape the electrical field, protect the SM and ensure a proper airflow path through the heatsink and around the capacitors. From a thermal point of view the influence of the enclosure is considered to be minimal, and the key elements to be considered are the heat sources (semiconductor base plate and capacitor windings), the heatsink and the opening grilles in the front and back panels of the enclosure. A simplified SM model considers only the lower part of the SM, where these relevant elements are located.

The IGBT baseplate is defined as a heat source with a power corresponding to the maximum semiconductor losses previously computed: $58\,W$ (Fig. 5a). The heatsink is modelled as an aluminium extrusion including a thermal interface facing the IGBT submodule baseplate, whose properties correspond to the ones of the thermal grease (Fig. 5a). Its length is defined by both mechanical constraints and thermal resistance. The length is maximized in order to reach the lowest achievable resistance with the selected profile. The electrolytic capacitors are modelled using the Parler model [15] by defining the windings and electrolyte, similar to the work presented in [16]. The windings are replaced by a hollow cylinder made of cylindrical orthotropic material featuring a high conductivity along z-axis ($100\,W/(m\,K)$) and a small one along the θ and r-axes ($0.21\,W/(m\,K)$), as suggested by Table II in [15]. Its dimensions are estimated as the supplier does not provide such informations. The windings are enclosed within the cylinder modelling the electrolyte. Its external radius corresponds to the capacitor diameter (Fig. 5b).

PCIM Europe 2016, 10 – 12 May 2016, Nuremberg, Germany

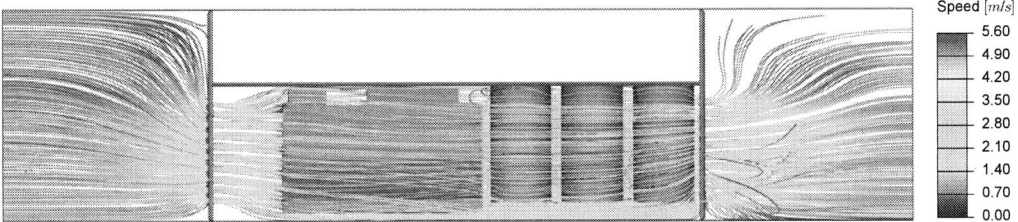

Fig. 4: Particle trace of the air speed across the SM.

The FEM offers the opportunity to determine the critical airflow that ensures an acceptable temperature across the SM. The airflow through the SM is depicted in Fig. 4, and defined at the outtake at $45\,\mathrm{m^3/h}$. Running the simulation for a given airflow at the outtake provides the temperature of the critical elements. Results are depicted in Fig. 5.

Based on this model, simulations are run to get the hottest spot of the heatsink and capacitors in order to obtain the optimal airflow across the SM: extracting the air too fast would lead to an oversized cabinet fan, extracting too slow would lead to hot components and their premature degradation. A volumetric outtake flow sweep is performed. The hottest point of both elements is outputted from the simulations and plotted in Fig. 6. This allows to determine the reference value for the airflow as $45\,\mathrm{m^3/h}$. One should note that the simulations are performed under the worst case conditions: losses corresponding the full power operation and maximum air temperature at the intake.

The volumetric resistance, also called loss coefficient, is computed through the FEM software during the simulations. The loss coefficient K is obtained as defined in Eq. (5).

$$K = \frac{P_{tot-in} - P_{tot-out}}{P_{dyn-out}} \tag{5}$$

This formula is implemented using the computation of the maximum pressure (P_{stat}) and the average speed (U_{avg}) over the SM air intake and outtake planes on the z axis, as detailed in Eq. (6), where P_{dyn} is expressed in Pascal and $\rho(@40\,^{\circ}\mathrm{C}) = 1.127\,\mathrm{kg/m^3}$.

$$P_{tot-in/out} = P_{stat-in/out} + P_{dyn-in/out}; \; P_{dyn} = 0.5 \times \rho \times U_{avg}^2 \tag{6}$$

Based on the same trials as illustrated in Fig. 6, the aforementioned values are computed and provided in Table 2. These results are used for the cabinet level simulations.

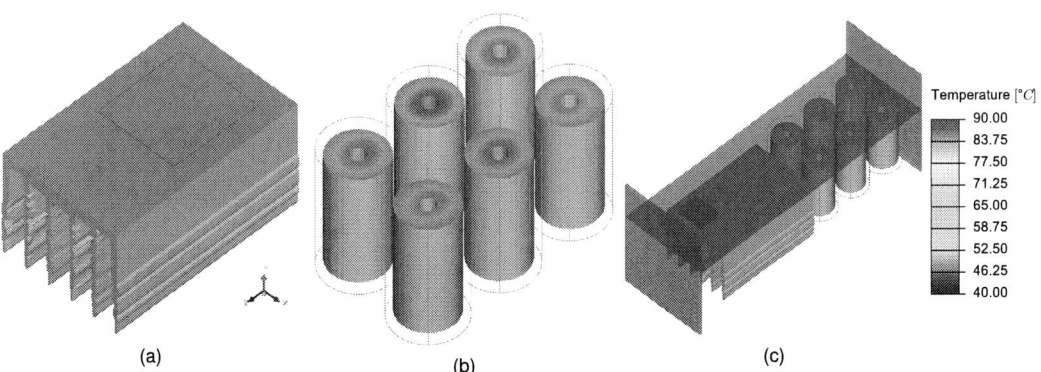

Fig. 5: SM level FEM results with $45\,\mathrm{m^3/h}$ airflow: (a) heatsink alone with semiconductor baseplate as heat source, (b) capacitors windings (solid) and electrolyte (frame) and (c) complete SM with the grille openings.

© VDE VERLAG GMBH · Berlin · Offenbach

465

PCIM Europe 2016, 10 – 12 May 2016, Nuremberg, Germany

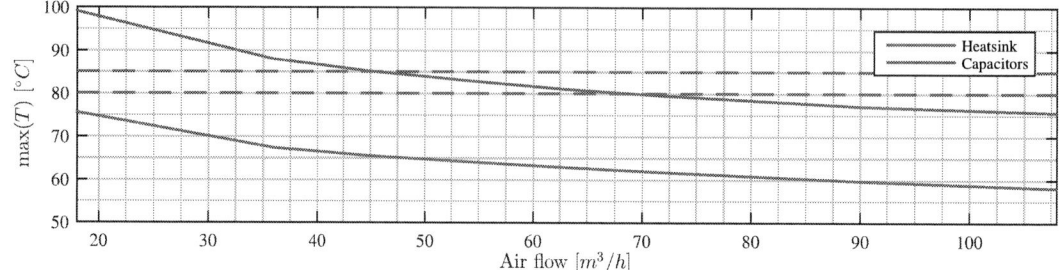

Fig. 6: Hottest spot for the heatsink and capacitors observed with FEM; the dashed lines represent the respective limits.

Table 2: FEM output for the various trials.

Q_{out} [m³/h]	$P_{stat-in}$ [Pa]	$P_{stat-out}$ [Pa]	U_{ave-in} [m/s]	$U_{ave-out}$ [m/s]	P_{dyn-in} [Pa]	$P_{dyn-out}$ [Pa]	K [-]
18	-1.748	-3.252	-0.1678	-0.1686	0.01587	0.01602	93.89
36	-7.09	-10.91	-0.3354	-0.337	0.0634	0.06401	59.73
45	-11.14	-15.73	-0.4188	-0.4201	0.09885	0.09944	46.16
54	-16.08	-22.37	-0.5032	-0.5036	0.1427	0.1429	43.98
63	-21.88	-30.13	-0.5859	-0.5871	0.1935	0.1942	42.46
72	-28.48	-38.79	-0.6701	-0.6701	0.253	0.253	40.74
90	-44.38	-60.71	-0.8373	-0.8382	0.395	0.3959	41.27
108	-63.78	-86.05	-1.004	-1.006	0.5683	0.5708	39.02

4.2. Cabinet FEM simulations

One MMC phase-leg consisting of 32 SMs is integrated in a standard 19" industrial cabinet. The mechanical arrangement is presented in Fig. 7a. Two cases are initially evaluated: (i) the complete front door as air intake (Fig. 7b) and (ii) a slot at the bottom of the cabinet as air intake (Fig. 7c). In both cases, the air is extracted out by one or more fans at a slot placed on the top of the cabinet. The results obtained in the second case (Fig. 7c) were producing unsatisfactory thermal performance, therefore the approach with the complete front door as air intake is selected and further evaluated.

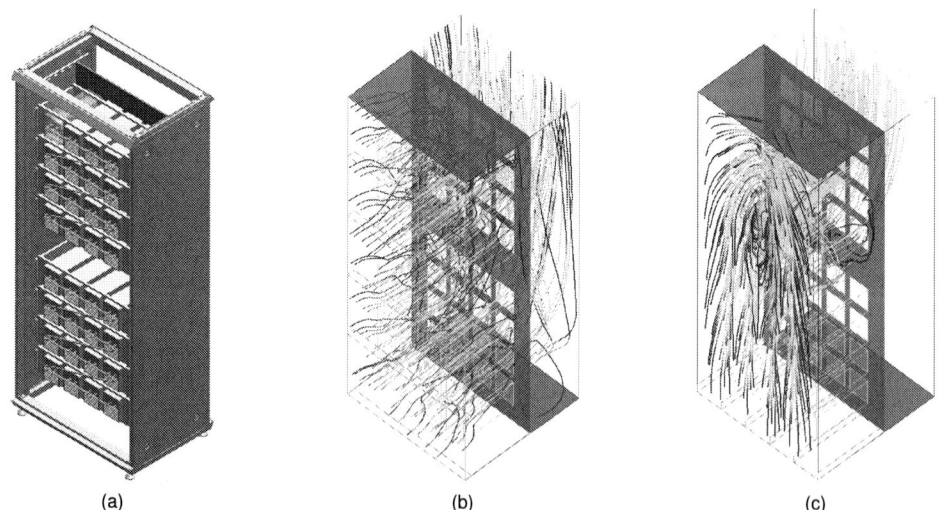

(a)   (b)   (c)

Fig. 7: Converter level FEM: (a) considered layout, (b) thermal simulation result for the case 1 and (c) thermal simulation result for the case 2.

© VDE VERLAG GMBH · Berlin · Offenbach

PCIM Europe 2016, 10 – 12 May 2016, Nuremberg, Germany

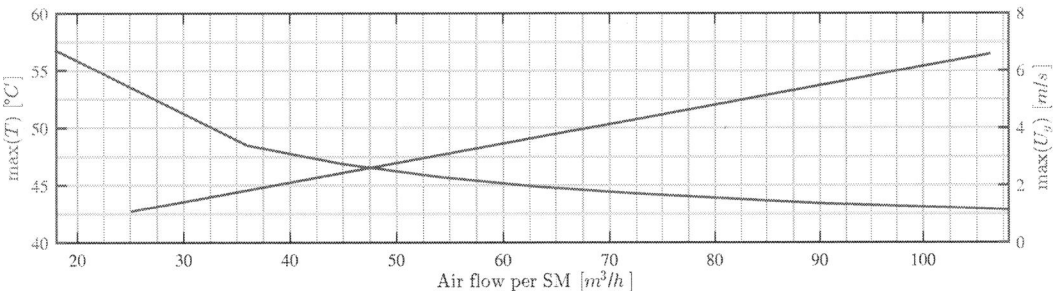

Fig. 8: Maximum temperature and airspeed inside the cabinet.

The SM level FEM simulations provided the required volumetric airflow at the SM intake. The complexity of the cabinet level FEM requires certain simplifications of the models for each of the 32 SMs: an equivalent macroscopic SM representation is necessary. This is achieved using a volumetric resistance, which corresponds to the equivalent volumetric resistance of the SM associated with an internal heat source equivalent to the overall losses inside the SM. The cabinet model is reduced to its walls, neglecting the structural parts.

To be consistent with the study at the SM level, the airflow across one SM is taken as a reference. By virtue of the mechanical arrangement, the total amount of air to be extracted from the cabinet is 32 times the volume of air that goes through one SM. The back sides of the SMs are all linked to a closed volume or "air chimney" which is located at the back of the cabinet. It serves to collect and guide the air from the SM outtakes to the cabinet roof. The fan, located at the top of cabinet, extract the air from this volume into the ambient. Trials are run based on the required volume flow through the cabinet. For each trial, the loss coefficient is modified according to the one indicated in Table 2. The minus sign refers to the normal orientation of the surfaces, with respect to the origin of the FEM coordinate system.

The results plotted in Fig. 8 show that the maximum airspeed is linear with the airflow, and as seen in Fig. 7b this speed is reached at the top of the air corridor. In practice this speed is limited by the capacity of the fan, placed on the cabinet roof (its exact specifications are not discussed in this paper). Nevertheless, the results provide requirements for the fan selection, responsible to maintain the temperature of all SM within specified limits.

5. Conclusions

The electro-thermal design of a phase-leg of a medium voltage MMC, composed of 32 SMs, has been presented. The high modularity of the converter topology allows for system partitioning and design optimization of different complexity. The losses during different operating conditions (influenced by the circulating current control strategy) can be easily calculated and used to size the SM accordingly. FEM simulations can aid, improve and speed-up the design process significantly, provided that the complexity of the models is properly adjusted with every simulation. Different mechanical arrangements can be easily evaluated and compared, before the actual prototyping, as shown from the cabinet level design.

Acknowledgement

This work is part of the Swiss Competence Centres for Energy Research (SCCER) initiative which is supported by the Swiss Commission for Technology and Innovation (CTI) with focus on Future Swiss Electrical Infrastructure (FURIES).

References

[1] R. Marquardt, A. Lesnicar, and J. Hildinger. Modulares Stromrichterkonzept für Netzkupplungsanwendungen bei hohen Spannungen. In *Proceedings of ETG Conference*, 2002.

© VDE VERLAG GMBH · Berlin · Offenbach

[2] H.-J. Knaak. Modular Multilevel Converters and HVDC/FACTS: a Success Story. In *Proceedings of the 2011-14th European Conference on Power Electronics and Applications (EPE 2011)*, August 2011, pages 1–6.

[3] H. Akagi, S. Inoue, and T. Yoshii. Control and Performance of a Transformerless Cascade PWM STATCOM With Star Configuration. *IEEE Transactions on Industry Applications*, 43(4):1041–1049, July 2007.

[4] M. Hiller, D. Krug, R. Sommer, and S. Rohner. A New Highly Modular Medium Voltage Converter Topology for Industrial Drive Applications. In *EPE 2009-13th European Conference on Power Electronics and Applications*, September 2009, pages 1–10.

[5] P.K. Steimer, O. Senturk, S. Aubert, and S. Linder. Converter-Fed Synchronous Machine for Pumped Hydro Storage Plants. In *2014 IEEE Energy Conversion Congress and Exposition (ECCE)*, September 2014, pages 4561–4567.

[6] S. Kenzelmann, A. Rufer, M. Vasiladiotis, D. Dujic, F. Canales, and Y.R. De Novaes. A Versatile DC-DC Converter for Energy Collection and Distribution Using the Modular Multilevel Converter. In *Proceedings of the 2011-14th European Conference on Power Electronics and Applications (EPE 2011)*, August 2011, pages 1–10.

[7] A. Christe and D. Dujic. On the integration of low frequency transformer into modular multilevel converter. In *2015 IEEE Energy Conversion Congress and Exposition (ECCE)*, 2015.

[8] U. Waltrich, D. Malipaard, and A. Schletz. Novel Design Concept for Modular Multilevel Converter Power Modules. In *Proceedings of PCIM Europe 2014; International Exhibition and Conference for Power Electronics, Intelligent Motion, Renewable Energy and Energy Management*, May 2014, pages 1–6.

[9] D. Cottet, F. Agostini, T. Gradinger, D. Velthuis R. Baumann, B. Wunsch, W. Gerig, A. Rüetschi, D. Dzung, H. Verfling, A. E. Vallestad, D. Orfanus, R. Indergaard, T. Wien, and W. van der Merwe. Integration Technologies for a Medium Voltage Modular Multi-Level Converter with Hot Swap Capability. In *2015 IEEE Energy Conversion Congress and Exposition (ECCE)*, September 2015.

[10] Semikron. *SK50GH12T4T Datasheet*. 27-04-2009 dil edition, 27-04-2009 dil edition, April 2009.

[11] M. Di Lella and R. Ramin. SEMITOP The Low and Medium Power Module for High Integrated Applications. Technical Information. Semikron, January 2008.

[12] M. Winkelnkemper, A. Korn, and P. Steimer. A Modular Direct Converter for Transformerless Rail Interties. In *2010 IEEE International Symposium on Industrial Electronics (ISIE)*, July 2010, pages 562–567.

[13] Exxelia Sic Safco. *SNAPSIC 4P Datasheet*.

[14] S.G. Parler. Deriving Life Multipliers for Electrolytic Capacitors. *IEEE Power Electronics Society Newsletter*, 16(1):11–12, February 2004.

[15] S.G. Parler. Thermal Modeling of Aluminum Electrolytic Capacitors. In *Industry Applications Conference, 1999. Thirty-Fourth IAS Annual Meeting. Conference Record of the 1999 IEEE*. Volume 4, 1999, 2418–2429 vol.4.

[16] T. Gradinger, F. Agostini, and D. Cottet. Two-phase cooling of hot-swappable modular converters. In *Proceedings of PCIM Europe 2014; International Exhibition and Conference for Power Electronics, Intelligent Motion, Renewable Energy and Energy Management*, May 2014, pages 1–8.

PCIM Europe 2016, 10 – 12 May 2016, Nuremberg, Germany

Non-isolated three-port DC/DC converter for ±380V DC microgrids

Yunchao Han, Fraunhofer IISB, Erlangen, Germany, yunchao.han@iisb.fraunhofer.de

Julian Kaiser, Fraunhofer IISB, Erlangen, Germany, julian.kaiser@iisb.fraunhofer.de

Leopold Ott, Fraunhofer IISB, Erlangen, Germany, leopold.ott@iisb.fraunhofer.de

Matthias Schulz, Fraunhofer IISB, Erlangen, Germany, matthias.schulz@iisb.fraunhofer.de

Fabian Fersterra, Fraunhofer IISB, Erlangen, Germany, fabian.fersterra@iisb.fraunhofer.de

Bernd Wunder, Fraunhofer IISB, Erlangen, Germany, bernd.wunder@iisb.fraunhofer.de

Martin, März, Fraunhofer IISB, Erlangen, Germany, martin.maerz@iisb.fraunhofer.de

The Power Point Presentation will be available after the conference.

Abstract

This paper describes a novel topology for a non-isolated three-port DC/DC-converter. Target application is the connection of various sources and loads, e.g. photovoltaic, batteries and fuel cells, to a three-wire, two-phase ±380 V_{DC} microgrid with center-tapped neutral.

The DC/DC-converter provides a bidirectional power transfer between the DC microgrid, energy sources or storages. The proposed topology can additionally perform power balancing between the positive and negative phase and thus omitting a second DC/DC-converter for balancing. The extended functionality is achieved by using only one coupled inductor.

A prototype DC/DC-converter with a nominal power of 6 kW is introduced which was realized to experimentally verify the proposed topology. Measurement results and performance of this prototype are presented.

1. Introduction

A local DC microgrid in a commercial or industrial facility offers several benefits. Those include higher efficiency and minimized cost for cables, as less electronic component count. Nominal operating voltage in commercial and industrial buildings is 380 V_{DC} [1]. A two-phase ±380 V_{DC} microgrid offers more power with less cable losses [2]. High power sources and loads can be connected between the +380 V_{DC} and -380 V_{DC} line wires to reduce the current and therefore the cable losses. Sources and loads with lower power can be connected between either the positive or negative phase and neutral.

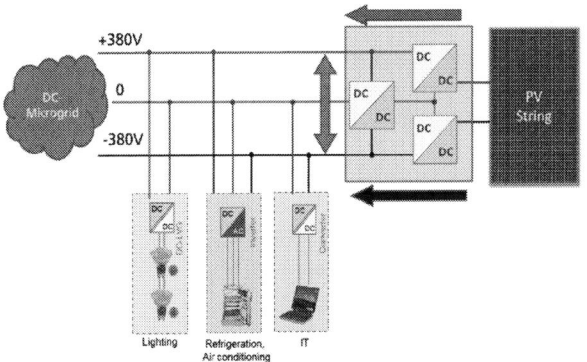

Fig. 1: Scheme of a ±380 V_{DC} microgrid with a DC/DC converter to balance the power between two phases.

Analog to a three-phase AC-grid, a power imbalance between the two DC phases must be compensated. In general, a DC/DC converter [3] is necessary to balance the power between the phases as shown in Fig 1. The proposed topology works as a bidirectional converter with integrated balancing capability. It is optimized for applications such as connecting a two-

© VDE VERLAG GMBH · Berlin · Offenbach

PCIM Europe 2016, 10 – 12 May 2016, Nuremberg, Germany

phase DC microgrid with different sources and loads like photovoltaic, batteries and fuel cells.

2. Circuit description

The proposed topology consists of two half-bridges with a coupled inductor as shown in Fig 2. During normal operation without balancing, it works like two independent buck- or boost converters, when switches T1, T2 respectively T3, T4 are switched at the same time. With a phase shift between T1 and T2 as well as T3 and T4, current flow builds up through the coupled inductor in order to transfer power between the two phases. It works like a bidirectional dual active half-bridge converter, which is presented in [4].

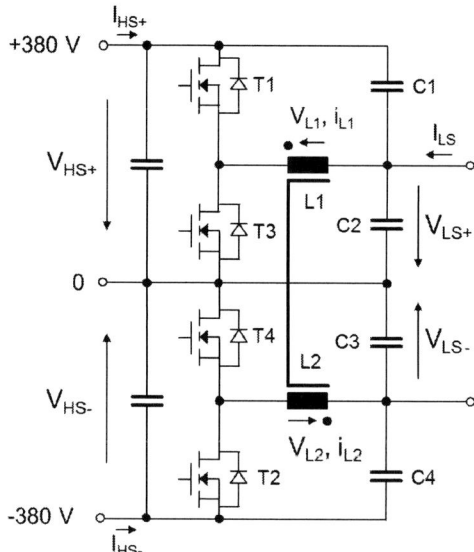

Fig. 2: The proposed topology with coupled inductor

The switches of both phases have the same duty cycle, which is controlled by a current controller as in a normal buck- or boost converter. To explain the principle of the topology, the following assumptions are made:

1) All the components (MOSFETs, inductors and capacitors) have ideal characteristics without losses.

2) The converter is in boost operation (I_{LS} positive). Input voltage at the low-voltage side is V_{LS} with $|V_{LS+}| = |V_{LS-}| = V_{LS}$. The voltage at the high-voltage side is $|V_{HS+}| = |V_{HS-}| = V_{HS}$. Under steady-state operation the switches T3 and T4 have an on time t_{on}. The switching period is T.

3) A short phase shift t_{delay} between T1/T3 and T2/T4 is applied to transmit power between the two phases.

4) Each of the capacitors C1, C2, C3 and C4 has the same capacitance value C, which is large enough to establish a constant voltage within a switching period.

5) The coupled inductor has a magnetizing inductance L_C and a leakage inductance L_s.

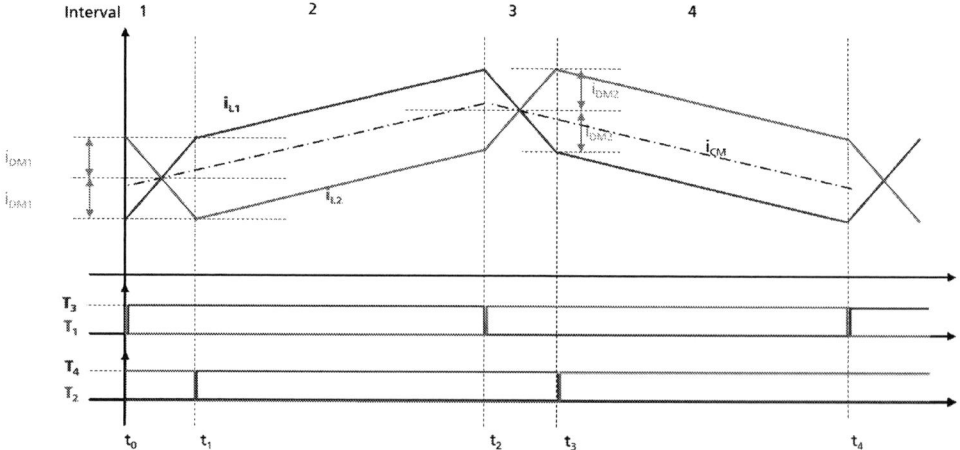

Fig. 3: The waveforms of two phases and gate signals of the switches.

With these assumptions, the waveforms of two phases are shown in Fig 3. Additionally the waveforms without phase shift are shown in the dashed line named i_{CM}.

A. Interval 1 ($t_0 < t < t_1$)

In the last interval of the previous cycle, both switches T1 and T2 were ON while T3 and T4 were OFF like in a normal boost converter. At the beginning of this interval, T1 is turned off but the current still flows through the body diode of T1. After a short dead time T3 is turned on and the current is commutated from T1 to T3. The duration of this interval is the phase shift $t_{delay} = t_1 - t_0$. In this interval a current difference between the two phases is established. It will be used in the next interval to transmit power between the two phases. The slopes of both currents are given as follows

$$L_c \cdot i'_{L1}(t) + L_s \cdot i'_{L1}(t) + L_c \cdot i'_{L2}(t) - V_{LS} = 0 \tag{1}$$

$$L_c \cdot i'_{L2}(t) + L_s \cdot i'_{L2}(t) + L_c \cdot i'_{L1}(t) + V_{HS} - V_{LS} = 0 \tag{2}$$

$$i'_{L1}(t) = \frac{L_c \cdot V_{HS} + L_s \cdot V_{LS}}{2 \cdot L_c \cdot L_s + L_s^2} \tag{3}$$

$$i'_{L2}(t) = -\frac{L_c \cdot V_{HS} + L_s \cdot V_{HS} - L_s \cdot V_{LS}}{2 \cdot L_c \cdot L_s + L_s^2} \tag{4}$$

The formula can be simplified with the assumption $L_c \gg L_s$

$$i'_{L1}(t) = -i'_{L2}(t) = \frac{V_{HS}}{2 \cdot L_s} \tag{5}$$

It is seen that the slopes are equal but the sign is opposite. The current difference magnitude is set by phase shift time t_{delay}.

B. Interval 2 ($t_1 < t < t_2$)

At the beginning of this interval, T2 is turned off and with a very short dead time T4 is turned on. The duration of interval 2 is $t_{on} - t_{delay}$. In this interval the current flow through the two inductors consists of two parts, one is magnetization current as in a normal boost converter, the other part is the current difference which was established in interval 1.

$$i_{L1}(t) = i_{CM}(t) + i_{DM1}(t) \tag{6}$$

$$i_{L2}(t) = i_{CM}(t) - i_{DM1}(t) \tag{7}$$

The current flow in this interval is shown in Fig 4. In the positive phase i_{CM} and i_{DM1} are in the same direction and in the negative phase i_{CM} and i_{DM1} are in different directions. The power transmission between the two phases is given by

$$P_{DM1} = \frac{1}{2} V_{HS} \cdot i_{DM1} \tag{8}$$

i_{DM1} is constant in this interval, because the voltages of both inductors are equal.

C. Interval 3 ($t_2 < t < t_3$)

At the beginning of this interval, T3 is turned off and after a very short dead-time T1 is turned on. The duration of this interval also is the phase shift t_{delay}. Similar to interval 1, the slopes of both currents in both inductors are given as follows

$$L_c \cdot i'_{L1}(t) + L_s \cdot i'_{L1}(t) + L_c \cdot i'_{L2}(t) - V_{LS} + V_{HS} = 0 \tag{9}$$

$$L_c \cdot i'_{L2}(t) + L_s \cdot i'_{L2}(t) + L_c \cdot i'_{L1}(t) - V_{LS} = 0 \tag{10}$$

$$i'_{L1}(t) = -\frac{L_c \cdot V_{HS} + L_s \cdot V_{HS} - L_s \cdot V_{LS}}{2 \cdot L_c \cdot L_s + L_s^2} \tag{11}$$

$$i'_{L2}(t) = \frac{L_c \cdot V_{HS} + L_s \cdot V_{LS}}{2 \cdot L_c \cdot L_s + L_s^2} \tag{12}$$

The formula can be simplified with the assumption $L_c \gg L_s$

$$i'_{L1}(t) = -i'_{L2}(t) = -\frac{V_{HS}}{2 \cdot L_s} \tag{13}$$

It is seen that the slopes are equal to the slopes in the interval 1 but again the sign is different.

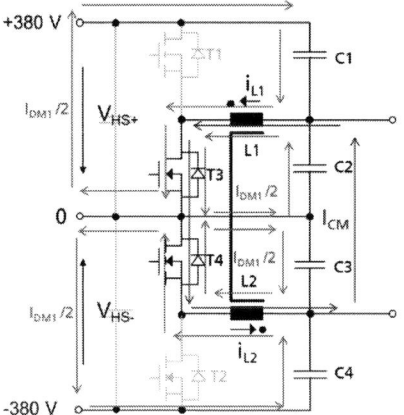

Fig. 4: Current-loop path in interval 2.

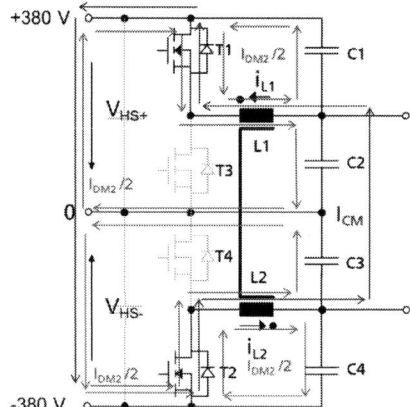

Fig. 5: Current-loop path in interval 4.

D. Interval 4 ($t_3 < t < t_4$)

At the beginning of this interval, T4 is turned off and with a very short dead time T2 is turned on. The duration of this interval is $T-t_{on}-t_{delay}$. In this interval the current flow of two inductors are analogous to the current flow in the interval 2.

$$i_{L1}(t) = i_{CM}(t) - i_{DM2}(t) \tag{14}$$

$$i_{L2}(t) = i_{CM}(t) + i_{DM2}(t) \tag{15}$$

The current flow in this interval is shown in Fig 5. In the positive phase i_{CM} and i_{DM} are in different direction and in the negative phase i_{CM} and i_{DM} are in same direction. The power transmission between the two phases is given by

$$P_{DM2} = \frac{1}{2}V_{HS} \cdot i_{DM2} \tag{16}$$

i_{DM2} is constant in this interval, because the voltages of both inductors are equal.

E. Summary

Current difference i_{DM1} and i_{DM2} were established in interval 1 and 3 to transfer power between the two phases. In interval 2, capacitors C_1 and C_3 are charged by i_{DM1} as well as C_2 and C_4 are discharged. In interval 4, capacitors C_2 and C_4 are charged by i_{DM2} as well as C_1 and C_3 are discharged. The DC current through the capacitor is zero, so that i_{DM2} is determined by

$$i_{DM2} = \frac{t_{on}-t_{delay}}{T-t_{on}-t_{delay}} \cdot i_{DM1} \tag{17}$$

With the assumption $t_{on} >> t_{delay}$ and $T >> t_{delay}$, the formula can be simplified

$$i_{DM2} = \frac{t_{on}}{T-t_{on}} \cdot i_{DM1} \tag{18}$$

i_{DM1} and i_{DM2} were established in interval 1 and 3 by following:

$$\frac{V_{HS}}{2 \cdot L_s} \cdot t_{delay} = i_{DM1} + i_{DM2} \tag{19}$$

And total power transmission between two phases is

$$P_{DM} = \frac{V_{HS}^2}{2 \cdot L_s} \cdot \frac{t_{delay} \cdot t_{on} \cdot (T-t_{on})}{T^2} \tag{20}$$

3. Design of the coupled inductor

The power transmission between the two phases is dependent on the leakage inductance L_S and phase shift t_{delay} and t_{on}. The on-time t_{on} is given by the high-side and low-side voltage and the phase shift must be much smaller than t_{on}. In order to maximize the power transmission the leakage inductance must be minimized. But, in reality the leakage inductance is also needed because of the PWM resolution. Normally, digital controllers are used to generate PWM signals with time resolution t_d down to 5 ns. The phase shift t_{delay} is discretized. The current slope in interval 1 and 3 was given in (5) (13). The current slope in t_d is expressed as follows:

$$\Delta i_{L1} = -\Delta i_{L2} = \frac{V_{HS}}{2 \cdot L_s} t_d \tag{21}$$

For example, a DC/DC converter with a high-side voltage of 380 V, a leakage inductance of 1 µH and a PWM resolution of 5 ns has a current slope of about 1 A. With a smaller leakage inductance, it is difficult to realize a current controller.

The magnetizing inductance is chosen as with a normal buck/boost converter. The leakage inductance is chosen with the compromise between the power transmission and PWM resolution. A magnetic saturation of the leakage inductance at peak current must be avoided.

A special structure was used to increase the leakage inductance and saturation current. An integrated leakage inductance for dual active bridge converter is presented in [5]. With the same principle the leakage inductance can be increased. Fig 6 (a) shows the structure of the coupled inductor. The winding is divided into two parts. 5 turns are wound around the middle leg and 2 turns around the right or left leg. Fig 6 (b) (c) shows a 2D simulation of the coupled inductor.

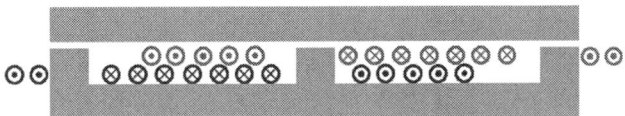

(a) The structure of the coupled inductor

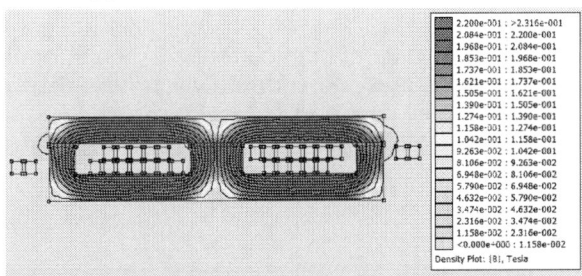

(b) Flux density with a symmetric current 15 A of two windings

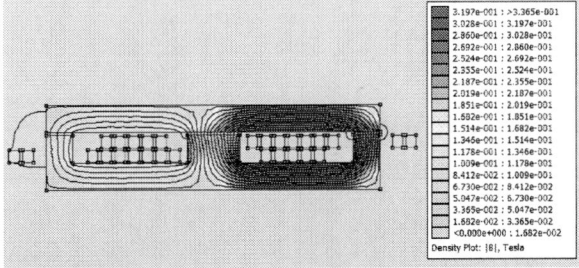

(c) Flux density with an asymmetric current in two windings (-5 A and 35 A)

Fig. 6: The structure of the coupled inductor and the simulation result.

4. Control strategy

The control strategy for the proposed topology is presented on the basis of a DC/DC converter for photovoltaic panels. The string of photovoltaic cells is connected to the low-side of the DC/DC converter. The high-side of the converter is connected to the $\pm 380\ V_{DC}$ microgrid. Therefore, the current of the PV-string must be controlled to realize maximum power point tracking. The power transmission between two phases is controlled through the difference of the voltage between the two phases. For example, if the voltage of the positive phase exceeds the voltage of the negative phase, more power will be transmitted to the -380 V phase.

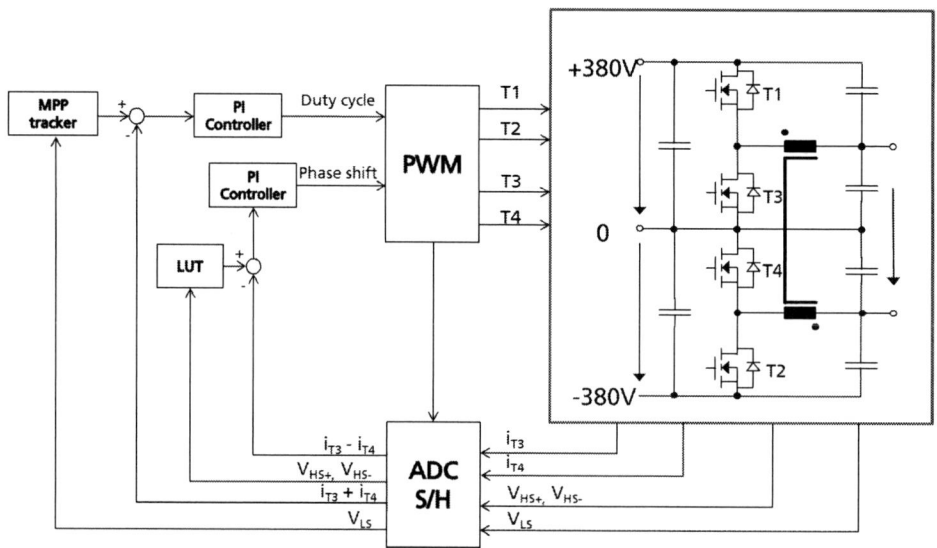

Fig. 8: The control scheme of the proposed DC/DC converter.

Fig. 8 shows the control scheme of the converter. Two current transformers are used to measure the current of the switches T3 and T4. The currents are sampled at the middle of the turn on time t_{on}. The average value of both currents is the current from low-side to high-side. The difference of both currents is the current, which flows between two phases. The turn on time t_{on} is controlled by a PI controller, which regulates low-side current. The phase shift t_{delay} is the manipulated variable of the other PI controller, which regulates the current difference between two phases.

5. Experimental results

A prototype DC/DC converter (Fig. 11) was built to verify the performance of the proposed topology. The specification of the prototype DC/DC converter is shown in Tab. 1.

Low-side voltage V_{LS}	100~300 V
High-side voltage V_{HS}	300~400 V
Switching frequency	200 kHz
PWM resolution	5 ns
Rated power (Low-side to High-side)	6 kW
Rated power (between phases)	2 kW
Magnetizing inductance	25 µH
Leakage inductance	3 µH
MOSFET	C3M0065090D

Tab. 1: The specification of the prototype DC/DC converter.

PCIM Europe 2016, 10 – 12 May 2016, Nuremberg, Germany

To emulate a two-phase DC microgrid, two electric loads were connected to the high-voltage side of the converter, which were set in constant voltage mode of 350 V_{DC}. The low-voltage side of the converter was connected to a DC power supply with a constant voltage of 350 V_{DC}. Fig. 9 shows the drain source voltage of T1 and T2, the drain source current through T3 and T4 and the output current of both phases. The output current and the drain source currents were almost identical without phase shift applied, as shown in Fig. 9 (a). When a phase shift was used, a current difference was built between T3 and T4 and the currents through both electric loads were different as shown in Fig. 9 (b, c). Then two efficiency measurement curves are shown in Fig. 10. The first efficiency curve was measured without phase balancing and the converter was working like a conventional boost converter. For the second measurement, the power transfer between low-voltage side and high-voltage side was constant at 2700 W. The balancing power between the two phases was varied from -1000 W to 1000 W. The efficiency of the converter is about 97.5% and it is almost independent from the transferred power, because the switching losses with a frequency of 200 kHz are dominating.

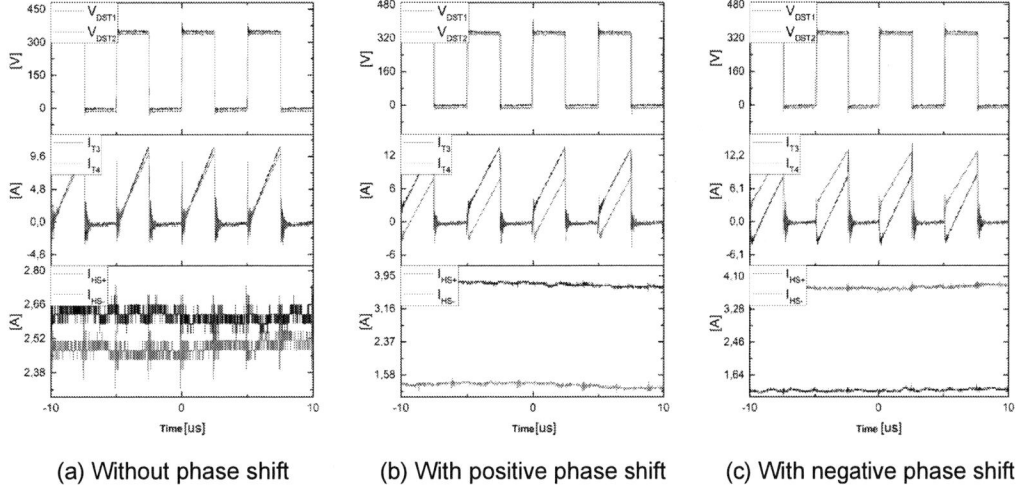

(a) Without phase shift   (b) With positive phase shift   (c) With negative phase shift

Fig. 9: The drain-source voltage of T1 and T2, the drain-source current through T3 and T4 and the output current of the both phases (V_{HS} = 350 V, V_{LS} = 250 V).

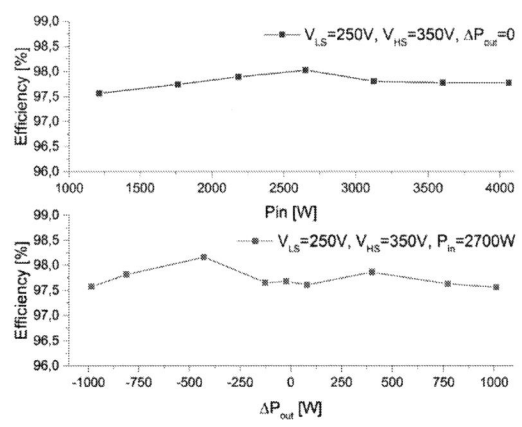

Fig. 10: Efficiency curves without phase balancing (above) and with phase balancing (below)

Fig. 11: The prototype DC/DC converter

© VDE VERLAG GMBH · Berlin · Offenbach

475

6. Conclusion

In this paper, a new topology for a DC/DC converter has been analyzed and a prototype converter was built. With a coupled inductor power can be transferred between the low-side and the high-side of the converter as well as between two phases of the high-side at the same time. The topology is optimized for a $\pm 380\,V_{DC}$ microgrid. The maximum power transmission between two phases is limited by the leakage inductance and the maximum duty cycle of the converter. A controller with very high PWM resolution is needed to regulate the current by a converter with a smaller leakage inductor. The ratio between high-side and low-side voltage also limits the power balancing capability. The converter has an optimized working point, at which low-side voltage is half the high-side voltage. For some application for photovoltaic cells the converter has always a nearly optimum working point, because the voltage in the maximum power point is always near to 60-70% of the open circuit voltage. The prototype converter has been built for PV application, in which the current flows from low-side to high-side. In principle the topology is bidirectional.

ACKNOWLEDGMENT

This contribution was supported by the Bavarian Ministry of Economic Affairs and Media, Energy and Technology as a part of the Bavarian project "Leistungszentrum Elektroniksysteme (LZE)".

REFERENCES

[1] Wunder, B.; Ott, L., Szpek, M.; Boeke, U.; Weiß, R., Energy Efficient DC-Grids for Commercial Buildings, 36th IEEE International Communications Energy Conference INTELEC 2014, Sep 28 - Oct 2 2014, Vancouver, Canada.

[2] Boeke, U.; Weiß, R.; Mauder, A.; Hamilton L.; Ott, L., Efficiency Advantages of ±380 V DC Grids in Comparison with 230 V/400 V AC Grids. DCC+G White Paper. [Online]. Available: http://dcgrid.tue.nl/files/2014-05-05_DCC+G-White_Paper_Efficiency_Advantages_of_DC_Power_Grids_v1-0.pdf (Accessed 07 March 2016).

[3] Zhang, X.; Gong, C., Dual-Buck Half-Bridge Voltage Balancer, IEEE Transactions on Industrial Electronics (Volume:60, Issue:8).

[4] Ngo, T.; Won, J.; Nam, K., A Single-Phase Bidirectional Dual Active Half-Bridge Converter, Applied Power Electronics Conference and Exposition (APEC), 2012 Twenty-Seventh Annual IEEE.

[5] Bernardo Cougo; Integration of leakage inductance in tape wound core transformers for dual active bridge converters, Integrated Power Electronics Systems (CIPS), 2012 7th International Conference.

Wide Voltage Input Range Insulated Current Fed Buck Flyback-Forward for HV/LV Power Conversion in Electric/Hybrid Vehicle

Reda CHELGHOUM, VALEO, Cergy Saint-Christophe, France, reda.chelghoum@valeo.com
Luis DE SOUSA, VALEO, Cergy Saint-Christophe, France
Larbi BENDANI, VALEO, Cergy Saint-Christophe, France
Daniel SADARNAC, GeePs | Group of electrical engineering-Paris, UMR CNRS 8507, CentraleSupélec, Univ.Paris-Sud, Sorbonne Universités, UPMC Univ Paris 06 3 & 11 rue Joliot-Curie, Plateau de Moulon 91192 Gif-sur-Yvette CEDEX, France

Abstract

This paper proposes a wide input voltage range 2kW isolated automotive DCDC converter. The converter is based on original Current Fed Buck Flyback-Forward topology. The presented topology combines optimization size, efficiency and low cost. These objectives are achieved thanks to a wide Zero Voltage Switching (ZVS) range for a wide input voltage range, naturally free output current ripple and reduced blocking voltage switches. We describe the topology characteristics, operating modes and the reducing of output EMC filter then we share some experimental results obtained on a 2kW 170V to 470V/14V prototype.

1. Introduction

DCDC in a hybrid (HEV) or electric (EV) vehicle replacing the alternator is a real technical challenge. Carmakers want to make it a smallest and cheapest as possible because of its fully transparent function to the user. Several automotive DCDC solutions were developed; they are often developed on the basis of conventional topologies (LLC, Phase-shit ...). These topologies show some limitations that prevent us from having a thermal autonomy of the DCDC and a high power density. The Phase-shift converter shows some limitations that have been developed in the literature such as the low-range ZVS operating mode, the generated losses by the freewheeling currents that are reflected in secondary and the overvoltage on the secondary rectification switches which increase the switch blocking voltage. Resonant LLC has the advantage of operating in ZVS and ZCS modes in its entire operating range [1] [2], but the output current is strongly rippled increasing the size of the output filter. In front of all these limitations an original topology based on a Current Fed Buck Flyback-Forward has been designed. Its main advantages are a ripple free DC output current and a wide range of ZVS operation for a wide range of input voltage (170V to 470V) and a lower blocking voltage on secondary switches. The combination of these benefits enables us to have a more compact DCDC with high efficiency. This paper presents the topology and its different operating modes then shows some results on a 2kW DCDC prototype HV/LV (170V to 470V /14V).

2. Current Fed Buck Flyback-Forward topology

Power stage of the proposed converter is illustrated in Figure 1. The first stage is a buck-boost converter that can buck or boost the capacitor C_{DC} voltage V_{DC} which can be expressed by: $V_{DC}=2.\alpha_{buck}.V_{in}$. The second stage is the Flyback-Forward or modified

Asymmetrical Half Bridge Converter (AHBC) [3] [4] operating at 50% duty cycle in open loop. The role of the Flyback-Forward is to operate like a "DC Transformer" converting a continuous DC Voltage into DC voltage with a fixed gain.

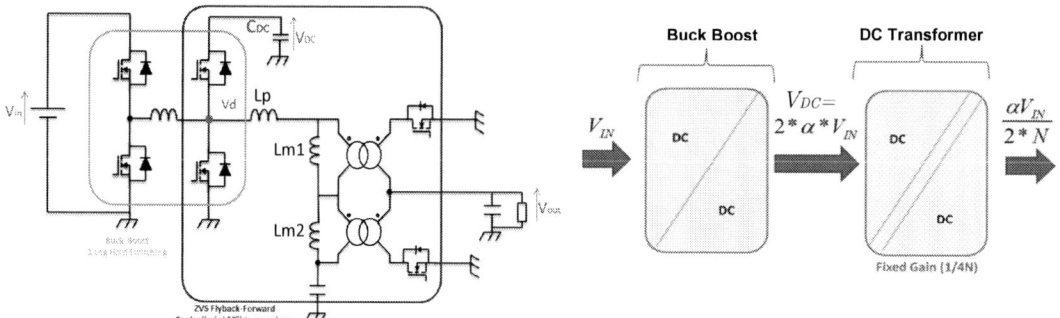

Figure 1: Configuration of proposed Current Fed Buck Flyback-Forward Converter

2.1. Converter operating modes

Operating modes of Flyback-Forward stage are shown in figures bellow:

t0-t1: L_{m1} is storing energy like a Flyback and L_{m2} is transferring energy to the secondary like a Forward (Figure 2.a).

t1-t2: C_L is discharged and C_H is charged by L_{m1}(Figure.2b) until $V_d=V_{DC}/2$ because at this moment the transformers are shorten (Figure 2.c). Then C_L is discharged by the leakage inductance L_p if the energy is enough (Figure 2.c). The voltage applied to the leakage inductance L_p makes current decrease and reverse its polarity. The capacitor must be discharged before the current inversion. The body diode of low side MOSFET is conducting when V_d try to go under zero volt (Figure 2.d).

t2- t3: L_{m1} is transferring energy to the secondary like a Forward and L_{m2} is storing energy like a Flyback (Figure 2.e)

Figure2.f shows the waveforms of primary current and magnetization inductors currents

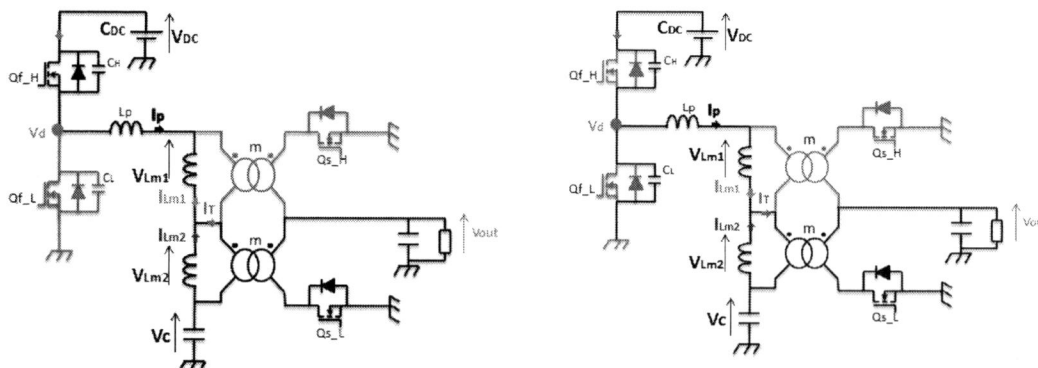

Figure 2.a: L_{m1} storing, L_{m2} transferring energy

Figure 2.b: ZVS transition

PCIM Europe 2016, 10 – 12 May 2016, Nuremberg, Germany

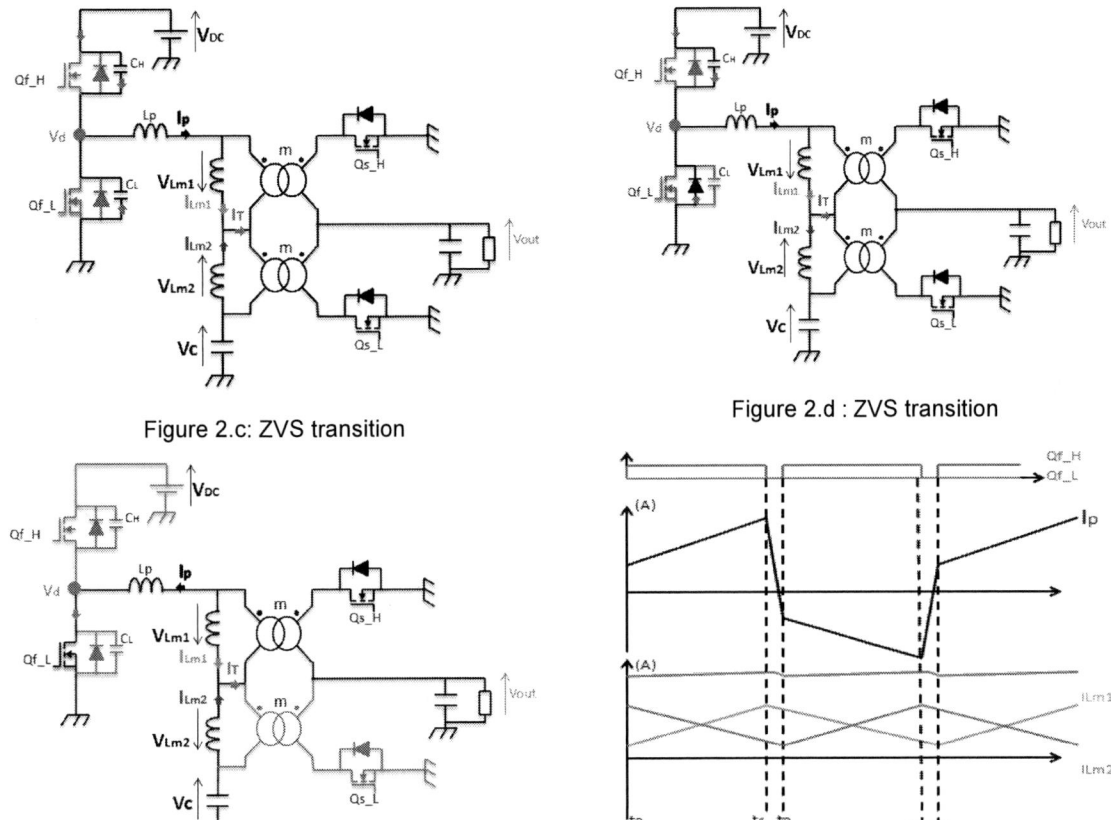

Figure 2.c: ZVS transition

Figure 2.d : ZVS transition

Figure 2.e:Lm1 transferring and Lm2 storing energy

Figure 2.f Primary currents waveforms

Figure 2: Operating Modes of Flyback-Forward stage and current waveforms

2.2. Zero ripple current

The Current transferred to secondary by transformers is expressed by $I_T=I_{Lm1}+I_{Lm2}$. This current is the result of magnetization current in one transformer and the demagnetization of the other transformer. Thanks to the identical design of transformers (ratio and magnetizing inductance) and 50% Duty cycle the currents I_{Lm1} and I_{Lm2} are opposed (Figure 2.f). Theoretically these two currents are perfectly interleaved which gives a continuous current; this current is reported to secondary by transformers and allow the minimization of the output filter. The dispersions in magnetizing inductance in the two transformers and stray inductance can cause an unbalance of transformers currents increasing slightly output current ripple. Figure 3 shows output current converter at 1.2kW and Vout=12v with 40µF ceramic filtering capacitors. These capacitors were mounted to compensate prospective imbalance in transformers and a slight ripple during dead times.

© VDE VERLAG GMBH · Berlin · Offenbach

PCIM Europe 2016, 10 – 12 May 2016, Nuremberg, Germany

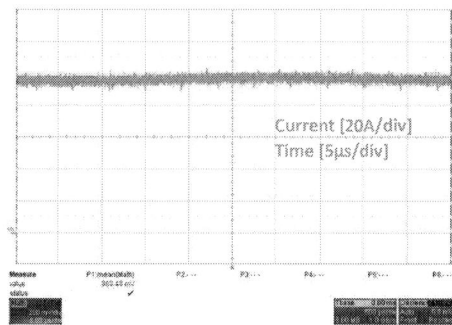

Figure3: Output converter current with 40µF ceramic filtering capacitors at 1.2kW/12V

2.3. Reduced blocking secondary switches

The Flyback Forward topology has the advantage of a low blocking voltage secondary rectifier switches. The constraints on secondary MOSFETs is reduced compared to some others topologies like a phase shift converter. In phase shift converter blocking voltage is depending on transformer ratio and input voltage $V_{breaking\_phaseshift} = V_{in\_max}*m$. In Flyback-Forward secondary switches breaking voltage is reduced from 150V to 60V, by the way R_{dson} is reduced from 7.2mOhms (IPB072N15N3) to only 1mOhms (IPB010N06N). Figure 4 shows blocking secondary rectifier MOSFETs.

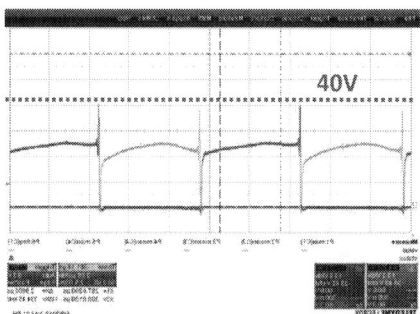

Figure 4: Secondary MOSFETs blocking voltage

2.4. Current Fed Flyback forward converter:

Output voltage of the Flyback forward is given by: $V_{out} = m*\alpha*(1-\alpha)*V_{DC}$ [3], with fixed duty cycle to 50% we obtain: $V_{out} = m*0.25*V_{DC}$. By adding an upstream current Fed or pre-regulator (Figure 5) it's possible to control V_{DC} Voltage. Then VDC will be expressed by $V_{DC} = 2*\alpha_{buck}*V_{in}$ and V_{out} by $V_{out} = 0.5*m*\alpha_{buck}*V_{in}$. This current Fed Flyback-Forward topology [7] can address à wide input voltage range like shown in figure 5.

© VDE VERLAG GMBH · Berlin · Offenbach

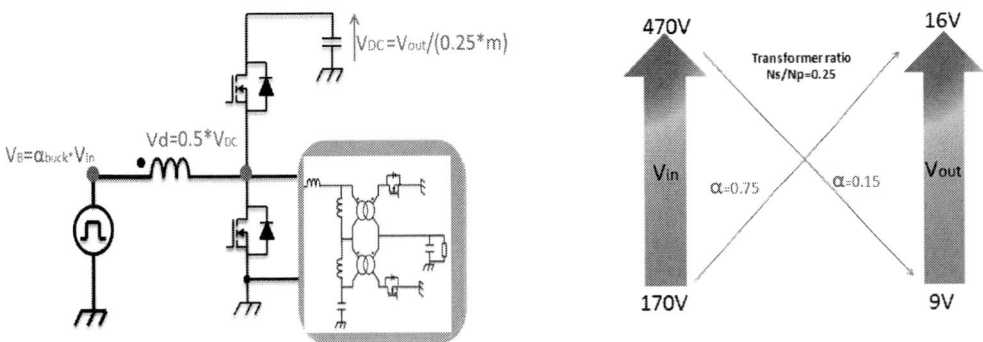

Figure 5: Wide input voltage range converter

3. Design considerations:

3.1. Primary MOSFETs choice

The choice of power components has been driven by our performance criteria and cost. In the pre-regulator leg we decided to use SiC devices (SCT2080KE, STPSC20H065CWY) in order to minimize switching losses and increase the frequency to 240 kHz. We use Superjonction Si MOSTETs (STW88N65M) in ZVS leg because of its low R_{dson} (25mOhms typical). By pressing at 70N and using 380µm gap pad we obtain $R_{th\ j\text{-}cooling}$=1,4°C/W.

3.2. Magnetic design

We choose planar technology for magnetics to simplify process assembly. Planar technology facilitates interleaving primary and secondary layers in the case of the transformer, reducing the alternatives losses in the windings [5] [6]. With our chosen interleaved windings we obtain inductor losses mapping shown in figure 6.

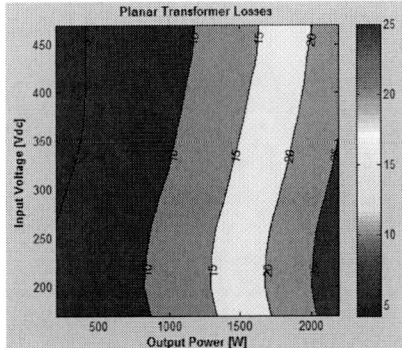

Figure 6: Transformer Losses

4. Experimental results

We realized a prototype of the topology to validate the concept and correlate with simulation results. Figure 7 shows some experimental measurements at 500W and Vin=200V on the prototype. Figure 8 shows a picture of the 2Kw realized Mockup which dimensions are 20cm*10cm*7cm (without input Filter). We also realized a thermal view of the prototype to identify hottest points; figure 9 shows a thermal capture at 1.5kW, The Flyback-Forward transformer is the most critical point in the design because the temperature reaches 98°C.

© VDE VERLAG GMBH · Berlin · Offenbach

PCIM Europe 2016, 10 – 12 May 2016, Nuremberg, Germany

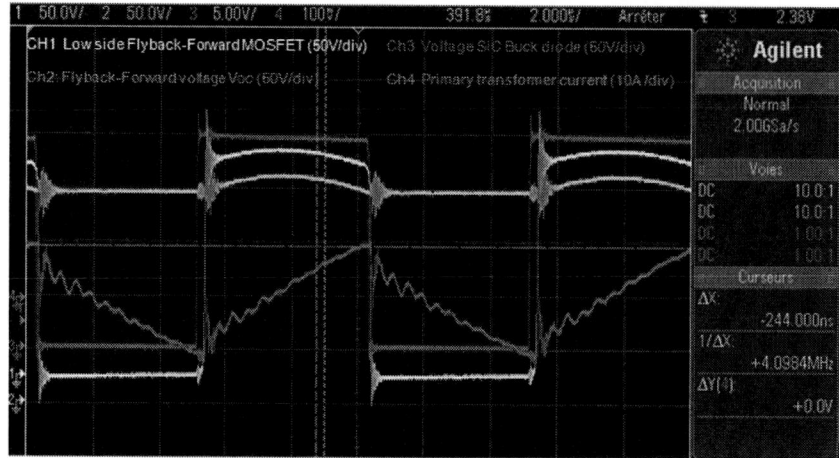

Figure 7: Prototype experimental measurements

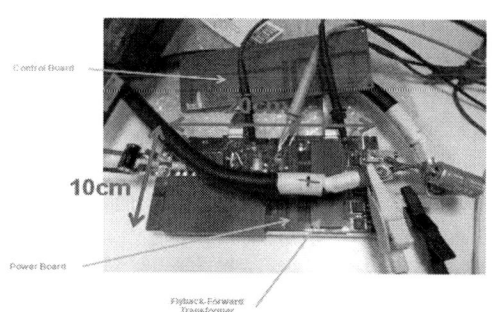

Figure 8: Picture of realized Converter

Figure 9: Thermal View at Vin 200V,1.5kW

An estimated Efficiency is given over the entire input voltage and power range to 2kw the converter is given in figure 10

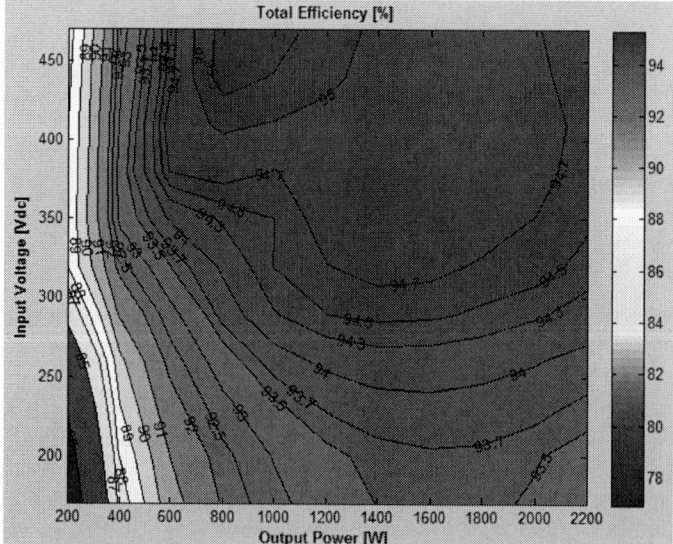

Figure 10: Estimated efficiency of the converter

© VDE VERLAG GMBH · Berlin · Offenbach

5. Conclusion

A Current Fed Buck Flyback-Forward topology was designed and validated. The ability to address a wide input voltage range with this topology has been demonstrated on the prototype. The zero output current ripple allows a drastically reduce the size of the output filter. The design of this topology is simple which reduces manufacturing costs. Some improvements like ZVS switching on Buck Leg could be realized in future designs in order to improve efficiency.

References

[1] G.Yang, P.Dubus, D.Sadarnac, "High efficiency parallel-parallel LLC resonant converter for HV/LV power conversion in electric/hybrid vehicles", PCIM Europe 2014, pp.1-8, May 2014

[2] G.Yang, P.Dubus, D.Sadarnac, "Design of a high efficiency, wide input range 500kHz 25W LLC resonant converter", Journal of Low Power Electronics, vol. 8, no.4, pp.1-8, August 2012

[3] S.Fraidlin, R.Miftakhutdinov, A.Nemchinov, V.Meleshin , "Modified asymmetrical ZVS half-bridge DC-DC converter", Applied Power Electronics Conference and Exposition, 1999.

[4] S.Korotkov, V.Meleshin, A.Nemchinov, S.Fraidlin, "Small-Signal Modeling of SoftSwitched Asymmetrical Half-Bridge DCDC Converter", APEC 1995, pp. 707-711.

[5] P.L.Dowell, "Effects of eddy currents in transformer windings", Proc Inst. Elect. Eng, vol.113, No.8, Aug,1966.

[6] Y,Ma "Detailed losses Analysis of High-Frequency Planar Power Transformer" Power Electronics and Drive Systems, 2007. PEDS '07. 7th International Conference on

[7] L.De Sousa, "Convertisseur DC/DC Isolé", Valeo patent, FR1459951

PCIM Europe 2016, 10 – 12 May 2016, Nuremberg, Germany

SiC JFET Cascode Enables Higher Voltage Operation in a Phase Shift Full Bridge DC-DC Converter

Jonathan Dodge, P.E., United Silicon Carbide, USA, jdodge@unitedsic.com

The presentation will be available after the conference.

1. Abstract

This paper presents a phase shift full bridge (PSFB) operating with 800 V input, 48 V output and utilizing SiC JFET cascode devices as the full bridge switches. The construction and characteristics of the cascode make it ideally suited for operation in PSFB converters. Greatly improved on-resistance and output capacitance per chip area, and low switching loss enable cost effective and efficient operation at higher operating voltages. Reliability is improved by eliminating the body diode failure mode in the PSFB. A gate drive compatible with conventional IGBTs and MOSFETs rejects noise and minimizes cost. Experimental results are included.

2. Introduction

SiC JFETs have very low $R_{DS(on)}$ per chip area, can support high blocking voltage, have low switching loss like a MOSFET, and short circuit and avalanche ruggedness. The major disadvantage is the depletion mode characteristic (normally-on characteristic with zero gate-source voltage) and the lack of an intrinsic anti-parallel diode. MOSFETs have low switching loss and good gate control. There is also an intrinsic anti-parallel diode, but in high voltage MOSFETs it has some undesirable characteristics including high recovery charge, snappy recovery if hard-commutated, and a possible latch-up failure mode in PSFB circuits [4, 5]. The main disadvantage of high voltage MOSFETs is a combination of high $R_{DS(on)}$ and output capacitance that makes it impractical to use them with ratings above about 900 V. IGBTs have reasonable cost at higher voltage ratings, but the turn-off switching loss severely limits the operating frequency, driving up system-level costs. What is needed is a device that combines the best features of a SiC JFET and a high voltage MOSFET, yet excludes the undesirable characteristics of each. This is precisely what the SiC JFET cascode provides, hereafter referred to simply as cascode.

3. Cascode Characteristics

From the perspective of the device terminals, the cascode is effectively an extremely high performance, high voltage MOSFET; but with an intrinsic diode that rivals the performance of a discrete SiC diode [1, 2]. A proprietary low voltage MOSFET design is the key to optimizing the performance of the cascode. The low blocking voltage of this MOSFET results in extremely low $R_{DS(on)}$, which is a fraction of the SiC JFET $R_{DS(on)}$. Being a silicon-based MOSFET, there is no problematic gate oxide and no threshold voltage drift with time and temperature. The reverse recovery charge is extremely low, has soft recovery, and changes very little with temperature. Leakage current is low, comparable to conventional MOSFETs or IGBTs. These characteristics make the cascode ideal to use in hard-switched topologies such as inverters and bridgeless totem-pole PFC circuits in addition to resonant and quasi-resonant converters such as the PSFB. Thus only one device type is needed to cover a full range of applications, and there is no need for special processing such as irradiation or the use of heavy metals [3].

Most low voltage MOSFETs have logic-level threshold for use in synchronous rectifiers. The

© VDE VERLAG GMBH · Berlin · Offenbach

cascode's MOSFET however has a typical threshold voltage of 4.5 V, making it immune to spurious turn-on in bridge circuits and eliminating the need for negative gate drive voltage. The cascode is fully on with a gate voltage of 10 V. The gate voltage range is ±20 V, and operates well with existing MOSFET or IGBT gate driver voltage ranges. The symmetric gate voltage range enables the use of gate drive transformers, which reduces the cost of a PSFB and other bridge circuits.

The switching slew rates of the cascode's MOSFET are controllable by adjusting the gate resistance, which in turn controls the JFET switching speed, aided by the resistive connection to the MOSFET source. Turn-off tends to be faster than turn-on, so using separate turn-on and turn-off gate resistors is the best practice.

The absence of a PN junction in the JFET conduction path and the design of the MOSFET result in very low capacitances. Of critical importance to the PSFB is the output capacitance C_{oss}. The product of $R_{DS(on)} \cdot C_{oss}$ is a good figure of merit for PSFB switch devices, and in this regard the cascode is in a class of its own compared to any other high voltage device type.

4. Intrinsic Anti-Parallel Diode Failure Mode Elimination

An intrinsic anti-parallel diode failure mode inherent in MOSFET-based PSFB circuits is due to the MOSFET drain-source voltage being insufficient to sweep out trapped minority carriers, especially during light-load operation [4, 5]. When the MOSFET subsequently turns off and drain-source voltage rises, current crowding of these minority carriers can cause the intrinsic NPN transistor to turn on (latch-up). According to [5], this can happen at high current as well. This failure mode is eliminated in the cascode for three reasons. First, the recovery charge of the cascode MOSFET is extremely low due to the low voltage rating of this MOSFET (less than 30 V) and a thin drift region. There are simply too few minority carriers present to cause current crowding. Second, there is very little resistance in the cascode MOSFET drift and body regions. The result is there is never enough voltage drop from charge-induced recovery current through the low resistance body region to turn on the parasitic NPN transistor. Finally, high voltage develops only across the SiC JFET, which has no minority carriers at all.

5. High Voltage PSFB Design

A demonstration PSFB DC-DC converter was designed with the following electrical requirements: 1500 W output power across an input voltage range of 700 to 820 V and an output voltage range of 40 to 60 V, maintaining resonance for zero-voltage turn-on (ZVS) over the full input and output voltages at an output power less than or equal to 750 W (half the rated power). These requirements are very ambitions, and as will be seen, one corner case must be given up. The design is based largely on a 400 V to 12 V PSFB reference design by Texas Instruments [6], but redesigned for higher voltage and higher power operation. New code for the TMS320F28027 microcontroller was written for voltage mode control (VMC), with the output voltage tracking at one-sixteenth the input voltage until the input voltage exceeds 700 V, providing a soft-start function.

Figure 1 shows a simplified schematic of the demonstration converter. The input connects to the H-bridge on the primary side of transformer T1, which provides galvanic isolation as well as voltage translation. The H-bridge switches Q1 through Q4 are SiC cascode devices from United Silicon Carbide, part number UJC1210K. These are rated at 1200 V, with a maximum $R_{DS(on)}$ of 0.1 Ω at 25 °C. Diodes D1 and D2 are clamp diodes, implemented with U2J1210T, a 1200 V, 10 A SiC diode from United Silicon Carbide. The output is a conventional center-tapped, single inductor, with synchronous rectifier (SR) MOSFETs attached to an RCD

© VDE VERLAG GMBH · Berlin · Offenbach

snubber. The SR MOSFETs are 200 V, 9.7 mΩ maximum $R_{DS(on)}$ from International Rectifier (Infineon), part number IRFP4668PbF.

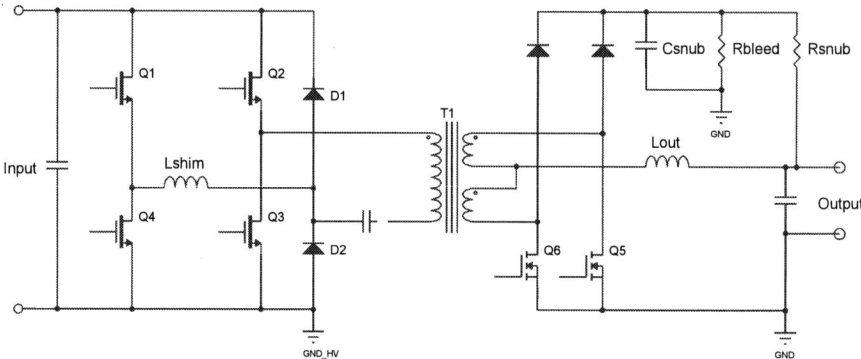

Figure 1 High voltage PSFB simplified schematic

5.1. Operation

Much has been written about the operation of the PSFB converter [4 - 7], yet a quick review will be helpful to understand differences relating to higher voltage operation. To begin, there are four operating states: power, freewheeling, active-to-passive (AtoP), and passive-to-active (PtoA). The power state is when diagonal switches in the H-bridge are on and power is transferred across the transformer to the secondary (output) side. The AtoP transitional state occurs after the power state when Q2 turns off while Q4 remains on, or when Q3 turns off while Q1 remains on. The Q2-Q3 leg is referred to as the AtoP leg because switching events in this leg relate to the AtoP transition. The freewheeling state is when Q1 and Q2, or Q3 and Q4 are simultaneously on, thus shorting the transformer primary, and no power is transferred to the output. The Q1-Q4 leg is the PtoA leg because switching events in this leg relate to the PtoA transition state following the freewheeling state.

Secondary-side duty cycle is defined as the time when diagonal switches in the H-bridge are on and power is transferred across the transformer to the output during the power state. Primary-side duty cycle is simply the time when diagonal pair switches are gated on regardless of whether power is transferred. This difference between primary and secondary-side duty cycle is due to the time required for current to change polarity through the transformer primary winding and is affected by the resonant inductance. Output power/voltage is controlled by the secondary-side duty cycle. Ignoring deadband time for the moment, each SR is turned off only when a corresponding diagonal in the H-bridge is on. During the freewheeling state both SR MOSFETs can be on.

5.2. Magnetics

The shim and output inductors and the transformer were custom designed by Triad Magnetics based on requirements derived as follows. The transformer turns ratio was set based on minimum and maximum input and output voltages respectively, and a maximum secondary-side duty cycle of 85 % (maximum primary-side duty cycle at 95 %, primary-secondary duty loss kept less than 10 %).

$$turns = \frac{V_{in\_min}}{V_{out\_max}} \cdot d_{sec} = \frac{700\,V}{60\,V} \cdot 0.85 = 9.92 \tag{1}$$

This must be checked against the voltage imposed across the SR MOSFETs, which is at its maximum when the output voltage is at its minimum value, and the input voltage is at its maximum.

$$d_{\text{sec\_min}} = \frac{V_{out\_min}}{V_{in\_max}} \cdot turns = \frac{40\,V}{820\,V} \cdot 9.92 = 0.48 \tag{2}$$

$$V_{SR\_max} = 2 \cdot V_{out\_min} \cdot \left[1 + \frac{1 - d_{\text{sec\_min}}}{d_{\text{sec\_min}}}\right] = 2 \cdot 40 \cdot \left[1 + \frac{1 - 0.48}{0.48}\right] = 165\,V \tag{3}$$

For a 200 V MOSFET this represents a 17 % voltage margin, which is quite narrow. Using a 250 V MOSFET of similar size and cost results in a 180 % increase in $R_{DS(on)}$, which would have a significant impact on efficiency. Increasing the turns ratio to 10.5 reduces the maximum SR voltage to 156 V worst case, corresponding to a still narrow but more acceptable 22 % voltage margin. It also increases the maximum duty cycle.

$$d_{\text{sec\_max}} = \frac{V_{out\_max}}{V_{in\_min}} \cdot turns = \frac{60\,V}{700\,V} \cdot 10.5 = 0.9 \tag{4}$$

If the primary-secondary duty cycle loss is 10 %, this would result in a worst-case primary-side duty cycle of 100 %, leaving no operating margin. It is possible to address this with the resonant inductance value by reducing the primary-secondary duty cycle loss, chosen to be 7 % with the primary-side maximum duty increased to 97 % to match the 10.5 turns ratio. Such duty cycle values are very optimistic, but it was decided that what is actually achievable would be determined by testing. The maximum resonant inductance is limited by this duty cycle loss.

$$L_{res\_max} = V_{in\_min} \cdot \frac{\Delta d_{max} \cdot \frac{T_{sw}}{2}}{2 \cdot \frac{1}{turns} \cdot I_{out\_min}} = 700\,V \cdot \frac{0.07 \cdot \frac{1}{2 \cdot 75\,kHz}}{2 \cdot \frac{1}{10.5} \cdot 25\,A} = 69\,\mu H \tag{5}$$

The minimum resonant inductance required is calculated by equating the energy stored in the resonant inductance and the H-bridge switch output and transformer capacitances. Only the PtoA transition is of concern here. There is always enough energy available from reflected output inductor current to achieve zero-voltage switching during the AtoP transition.

The output capacitances of the two switches in each leg of the H-bridge are effectively in parallel during switching transitions, so twice the energy-related output capacitance $C_{oss(er)}$ from the UJC1210K datasheet is added to the estimated transformer capacitance to calculate the resonant capacitive energy stored, which is C_{res}. The transformer capacitance is simply an estimate at design time.

$$C_{res} = 2 \cdot C_{oss(er)} + C_{transformer} = 2 \cdot 55\,pF + 30\,pF = 140\,pF \tag{6}$$

To calculate the stored inductive energy requires first estimating the magnitude of the primary-side current at the moment of switching. The minimum current available for the PtoA transition occurs at maximum output voltage, minimum full-load output current. This current is the peak output current reflected to the primary side, and neglecting losses is considered constant during the freewheeling state. Output inductor ripple current is 20 % peak-peak.

$$I_{PtoA\_min} = \frac{1}{turns} \cdot \left(I_{out\_min} + \frac{\Delta I_{Lout\_min}}{2}\right) = \frac{1}{10.5} \cdot \left(25\,A + \frac{0.2 \cdot 25\,A}{2}\right) = 2.62\,A \tag{7}$$

Now the minimum resonant inductance is calculated by equating capacitive and inductive energies.

$$\frac{1}{2} \cdot C_{res} \cdot V_{in\_max}^2 = \frac{1}{2} \cdot L_{res\_min} \cdot I_{PtoA\_min}^2 \; \therefore \; L_{res\_min} = \frac{C_{res} \cdot V_{in\_max}^2}{I_{PtoA\_min}^2} = \frac{140\,pF \cdot 820\,V^2}{2.62\,A^2} = 13.7\,\mu H$$

This is the minimum inductance required to maintain resonance under the worst-case full-load operating condition. Extending the resonant operating range to half-power results in a minimum resonant inductance of 55 µH.

Constructing the transformer to have such high leakage inductance would result in poor efficiency and regulation of the transformer. The transformer was designed to maximize efficiency, which tends to minimize its leakage inductance, so a shim inductor was used. Assuming the transformer has a leakage of about 5 µH (an educated guess at the time of the design) the maximum shim inductance can be 64 µH. The actual shim inductance value is 60 µH.

Note that the somewhat narrow range between the minimum and maximum resonant inductance values is due to both the high input voltage and the relatively wide input and output voltage ranges for which resonance is maintained down to half of rated power. The energy stored in the H-bridge switch output capacitance is proportional to the square of the input voltage, so doubling the input voltage from 400 V to 800 V quadruples the energy stored in the H-bridge switch output capacitances. More energy stored in output capacitances requires more resonant inductance, which is bounded by the loss of duty cycle and the switching frequency. Thus operating at high input voltage mandates a device with very low output capacitance in order to operate at a switching frequency where the size of the magnetics is cost effective. At the same time, $R_{DS(on)}$ must be low, otherwise efficiency would be degraded. The UJC1210K has an output capacitance of about 43 pF at V_{DS} = 800 V. Similar-rated SiC MOSFETs have roughly two times this much output capacitance, making it impossible meet the design specifications with them because the minimum and maximum required resonant inductance values would overlap. Therefore the SiC JFET cascode represents an enabling technology to operate the PSFB at high input voltage.

5.3. Gate Drive

The gate drive is transformer isolated, driven with a single 12 V power supply by UCC27324D dual gate driver ICs, as shown for one H-bridge leg in Figure 2. The SR MOSFETs are also driven by a UCC27324D gate driver supplied by 12 V, but without a transformer. The gate drive transformer provides substantial impedance by itself, so very low gate resistance values were used. In fact, after initial tests, the turn-on resistance at the cascode gate was reduced to zero, and the turn-off to only 3 Ω, which could probably be further reduced to zero. A 3 Ω resistor in series with the DC blocking capacitor is for damping and to limit power dissipation in the gate driver IC.

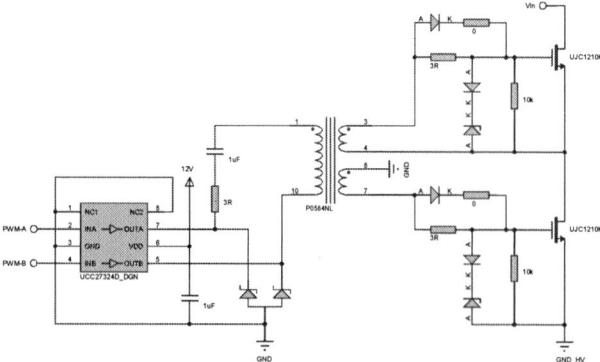

Figure 2 Gate drive for one leg of H-bridge

6. Experimental Results

Figure 3 (a) shows clean voltage waveforms across Q1 and the transformer primary winding, characteristic of a PSFB with its resonant operation. Figure 3 (b) is a photo of the PSFB demonstration board. The upper-right section of the board contains the microcontroller on a plug-in card and emulation circuitry. From left to right power flows to two film capacitors (only one is needed for the ripple current), the H-bridge with cascodes, the small shim inductor, down to the main transformer, through the SR MOSFETs, then up through the output inductor, and finally the output capacitors.

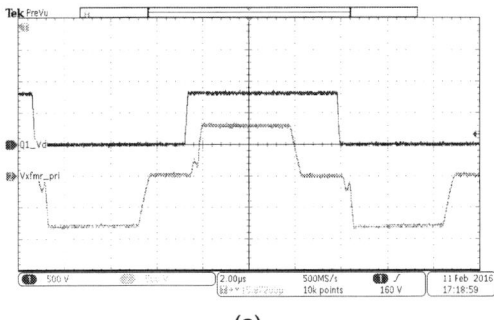

(a)   (b)

Figure 3 (a) Q1 drain-source and transformer primary voltages: 800 V in, 48 V out, 717 W out; (b) photo of the demonstration PSFB under test

Digital control facilitates adjusting the deadband times based on operating conditions. By experiment it was found that the PtoA deadband time could be left constant at 333 ns, regardless of the load current. This corresponds roughly to when the resonant voltage would peak. The AtoP deadband time however was dependent on the load current, and was adjusted as shown in Figure 4. A curve fit could be made and a simple lookup table could be used by the microcontroller to adjust the AtoP and SR deadband times. The SR deadband time was set equal to the AtoP deadband time plus a fixed offset of 350 ns.

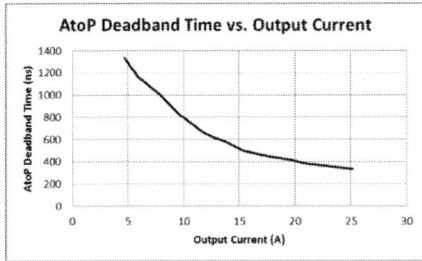

Figure 4 AtoP deadband time versus output current, 800 V in, 48 V out

Resonance was checked by looking for a "Miller plateau", which looks like a notch in the gate-source waveform of one of the PtoA cascodes, Q4. Presence of a Miller plateau corresponds to discharging of the reverse transfer capacitance through the gate terminal, meaning the voltage across the output capacitance is not completely discharged and therefore turn-on was not completely resonant. At output power between 714 W to 794 W the V_{GS} Miller plateau disappeared, so turn-on switching was fully resonant. Even though the load bank did not permit finer resolution at 60 V, it seems clear that the design goal of resonance at worst-case conditions of 750 W out at 820 V in, 60 V out was achieved. In fact the resonance limit was very close to calculated, meaning the combined transformer leakage and capacitance estimates were reasonable.

The PSFB was able to maintain 60 V out with 700 V in up to about 800 W output power. Above this power level the output voltage began to decrease slightly. This limitation is due to duty cycle loss and the increased voltage drop across circuit components with increasing output current. Increasing the primary-side duty (phase shift) above 96 % had no effect because there was no further duty cycle margin available. The design specifications are aggressive even for a 400 V PSFB, so it is not surprising that one of the corner requirements would be "clipped off". There are two alternatives. One would be to accept a lower maximum output power at 60 V out, 700 V in. In a battery charging application this would be completely reasonable since equalization is only done at low power. Alternatively, the shim inductance could be reduced to a value that maintains resonance under nominal input/output voltages instead of maximum. With 800 V in, 48 V out, resonance was maintained down to 511 W, or 34 % of the rated 1500 W output power, so there is room to decrease the shim inductance. This would increase the current slew rate during the PtoA transitions, reducing the duty cycle loss, and consequently increasing the maximum secondary-side duty cycle.

Efficiency calculations include all power loss, including magnetics and SR MOSFETS, and all gate drive and controller power consumption. Gate drive and controller power were measured as 0.88 W and 0.92 W respectively, with the PSFB operating at 800 V in. Note that controller power includes analog measurement, microcontroller, and serial communication over USB. Gate drive and controller power were added to the input power measurement. Four digital multimeters (Keysight 34465A) were used for input and output voltage and current measurements. Output current was sensed as voltage across a 100 mV / 100 A precision shunt from Deltec. The results for 700 V and 800 V, 48 V out in are shown in Figure 5.

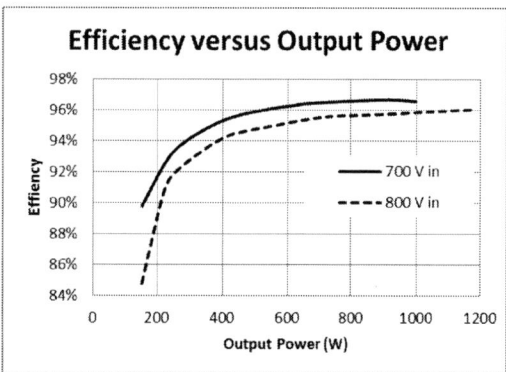

Figure 5 Efficiency with 48 V out, 75 kHz switching, calculation includes gate drive and control power

Noteworthy is the high efficiency achieved by this PSFB switching high voltages at 75 kHz. Peak efficiency with 700 V in is 96.7 %, and for 800 V in it is 96.0 % but still increasing with output power. (Testing at higher power has not yet been done due to a limitation of the high voltage power supply.) These efficiency results are what one would expect to achieve to date when switching 400 V with silicon-based switches. The 700 V efficiency curve is higher because switching loss dominates. At higher power, conduction loss increases, especially in the SR MOSFETs, which would cause the efficiency curves to slope down.

7.   Applications of High Voltage PSFB

The ability to operate at 800 V and high switching frequency with 1200 V rated full bridge switches opens new possibilities for lower cost and more efficient power conversion. In addition, the PSFB circuit can be operated in current-fed push-pull mode to transfer power in the reverse direction, creating opportunities in applications such as data centers, UPS, telecom, energy storage, and others [10]. One example of how to implement bidirectional power is by combining a PSFB/current-fed push-pull with a three-level active rectifier/inverter connected to 480 VAC, as shown in Figure 6. Considering a data center as an example, with such an arrangement a step-down transformer and boost PFC are eliminated. Power routed at higher voltage close to the end usage improves overall efficiency and reduces cable power loss and cost. There would be no reason to migrate to 380 VDC distribution with its bulky and expensive switchgear. Finally, operation in current-fed push-pull mode is very simple because the H-bridge cascodes can simply be left off, thanks to the excellent performance of the intrinsic anti-parallel diode.

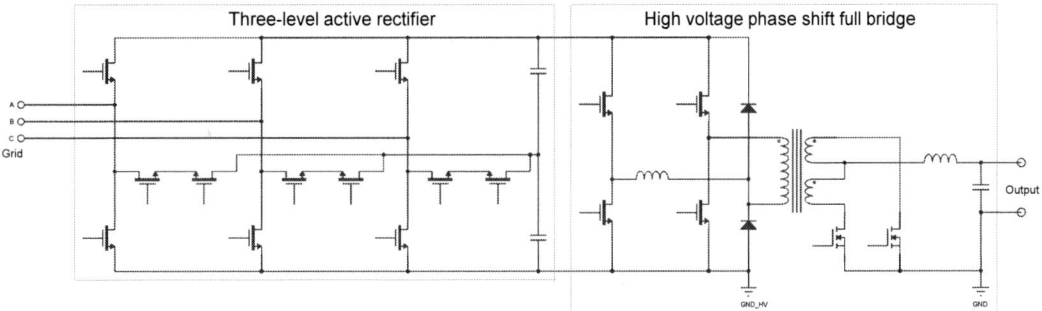

Figure 6 Three-level active rectification of 480 VAC supplying a high voltage PSFB

8.   Summary

With a low $R_{DS(on)} \cdot C_{oss}$ figure of merit combined with a high performance intrinsic diode, the SiC cascode is an enabling technology for the reliable and cost-effective implementation of an 800 V PSFB. Tests on a demonstration PSFB yielded 96 % peak efficiency at 800 V in, 48 V out, at 75 kHz switching frequency. Wide input/output voltage ranges of 700 V to 820 V in, 40 V to 60 V out were achieved with resonance down to half the rated 1500 W output power under all conditions. Operation at 60 V out with 700 V in was achieved at reduced output power. Many applications could benefit from this new high voltage capability.

9.   References

[1] A. Bhalla, X. Li, J. Bendel; "Switching Behavior of USCi's SiC Cascodes", Bodo's Power Systems, June 2015

[2] J. Bendel; "Cascode Configuration Eases Challenges of Applying SiC JFETs in Switching Inductive Loads", How2Power Today, August 2014

[3] J. Dodge; "Power MOSFET Tutorial", Microsemi application note APT-0403

[4] K. Dierberger, R. Redl, L. Saro; "High-Voltage MOSFET Behavior in Soft-Switching Converters: Analysis and Reliability Improvements", Intelec 1998

[5] Fiel, T. Wu; "MOSFET Failure Modes in the Zero-Voltage-Switched Full-Bridge Switching Mode Power Supply Applications", International Rectifier application report

[6] Texas Instruments, "Phase-Shifted Full Bridge DC/DC Power Converter Design Guide", TIDU248, May 2014

[7] Texas Instruments, "Phase-Shifted Full-Bridge, Zero-Voltage Transition Design Considerations", application report, September 1999

[8] D. Hamo; "A 50W, 500kHz, Full-Bridge, Phase-Shift, ZVS Isolated DC to DC Converter Using the HIP4081A", Intersil application note AN-9506, April 1995

[9] S. Abdel-Rahman; "Design of Phase Shifted Full-Bridge Converter with Current Doubler Rectifier", Infineon design note DN 2013-01, V1.0, January 2013

[10] Texas Instruments, "Bidirectional DC-DC Converter", TI Designs

Detailed Comparison of One Stage Topologies for LED Lighting Applications

A. Pawellek, T. Duerbaum
Friedrich-Alexander University Erlangen-Nürnberg
Chair of Electromagnetic Fields
Cauerstr. 7, Erlangen, Germany

Abstract

Light emitting diodes (LEDs) have great potential to replace existing light sources, like incandescent and fluorescent lamps due to their long life time and higher efficacy. Therefore, also the life time and efficiency of the LED-ballast are important in order to improve the overall system. In addition, size and costs are relevant factors in the highly competitive LED market. The paper compares three different realizations of a one stage topology for LED lighting applications. The highly efficient ballasts provide on the one hand a ripple-free LED current, while fulfilling the regulation for the line current harmonics on the other hand. The investigation and optimization is validated by practical setups.

Introduction

In general lighting applications, the LEDs are replacing more and more existing light sources due to their high efficacy and long life time. One the one hand, a constant and ripple free LED current is necessary for a high light quality. One the other hand, the input current of the ballast has to fulfill several standards, requiring a PFC functionality of the ballast in addition. The standard AC/DC input stage with a bridge rectifier and a huge electrolytic bus capacitor is cost effective but typically doesn't meet the PFC-requirements. This paper will show, how to choose the right value of the bus capacitor in order to fulfill the regulation for the line current harmonics according to the Class C of the DIN EN 61000-3-2. In addition, the proposed topology provides a constant, ripple-free LED current. With a buck-, a boost-converter and a linear current source, three different prototypes are optimized and compared to each other.

At the beginning of this paper, the special description of the input current waveform, given by the regulation, is introduced. The next paragraph deals with the design guidelines in order to fulfill the regulation. Based on the analysis, the prototypes of an LED lamp, replacing a 75 W incandescent lamp, will be designed and built for experimental verification.

Limits for harmonic current emissions

The regulation DIN EN 61000-3-2 must be applied for all electrical devices at the low voltage grid with an input current less or equal 16 A for each phase. The regulation sets limits for the harmonics in the input current for multiples of the mains frequency, which occur by a non-sinusoidal current consumption of the device. Lighting devices are categorized by the regulation in the Class C. The regulation gives a limit for each harmonic up to the 39th. But in the power range $P \leq 25$ W exists an alternative for the tabled limits. Instead of fulfilling the tabled limits, the regulation gives the possibility to fulfill the requirements of a described waveform:

- $I_3/I_1 \leq 86\,\%$ and $I_5/I_1 \leq 61\,\%$
- Current-threshold 5 % is reached before or at 60°
- Peak current is located before or at 65°
- Current doesn't drop down to the threshold 5 % before 90°

I_n is the rms-value of the current of the nth harmonic. The current-threshold 5 % refers to the maximum, absolute value in the measurement window. The reference for the phase angle is the zero crossing of the fundamental of the supply voltage at 0°. For illustration purposes, Fig. 1 shows a possible waveform with the relevant phase angles and current parameters.

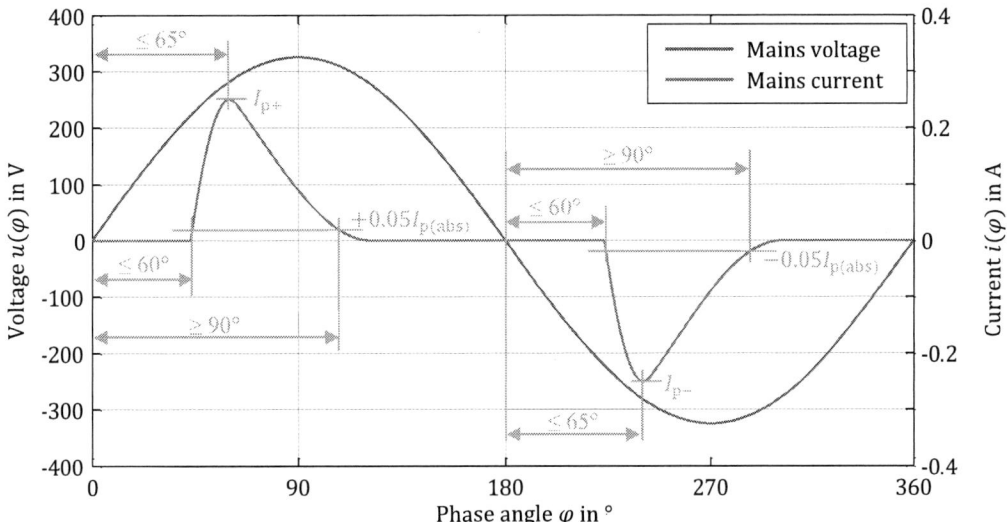

Fig. 1: Illustration of the phase angles and the current-parameters of the description of the waveform according to DIN EN 61000-3-2. The amplitudes are not relevant.

Structure in principle and operation mode

The typical input stage for an AC to DC conversion consists of a bridge rectifier and a bus capacitor. Since the bus voltage varies with the input voltage and exhibits a frequency component with twice the line frequency, a power stage next in line is necessary in order to supply the load with a constant voltage or constant current in case of an LED-ballast. The power stage can be either a switched DC/DC converter (Fig. 2) or a linear current regulator (Fig. 3).

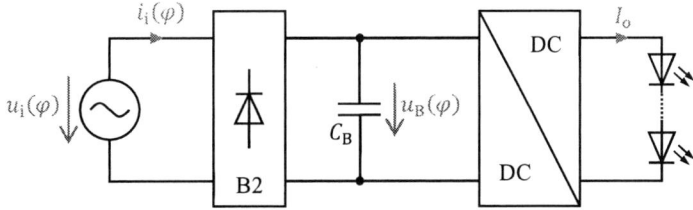

Fig. 2: Schematic of the complete system based on a switched DC/DC converter for supplying an LED-string with a constant current I_o from the mains.

The disadvantage of the rectification of the line voltage with B2-rectifier and bus capacitor is the high harmonic content of the input current. It can be shown, that it is not possible to fulfill the tabled limits of the DIN EN 61000-3-2 without further measures. One option to reduce the harmonic content and increasing the power factor is the use of passive filter components. The simplest realization is an inductance in the power line. However, for input powers in the interesting power range $P \leq 12.4\,W$ (on order to reach energy efficiency A+ for 940 lm), the needed inductor value for fulfilling the tabled limits is in the range of 500 mH and therefore excessive high.

PCIM Europe 2016, 10 – 12 May 2016, Nuremberg, Germany

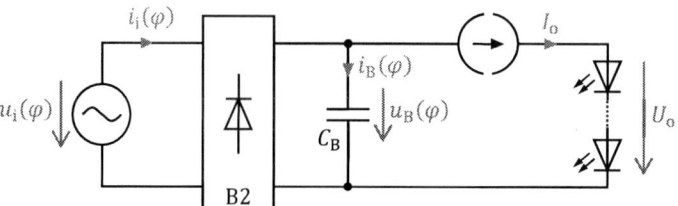

Fig. 3: Schematic of the complete system based on a linear current source for supplying an LED-string with a constant current I_o from the mains.

But with a proper design of the input stage, it is possible to fulfill the described input current waveform of the regulation in order to have a simple and cost effective solution.
The circuits of Fig. 2 and Fig. 3 are supplied by the sinusoidal input voltage

$$u_i(\varphi) = \hat{u}_i \sin(\varphi), \tag{1}$$

which represents the mains voltage. For the analysis of the topologies, the DC/DC converter with the LED-string in Fig. 2 can be modeled by a power sink, which draws a constant power P from the bus capacitor. In contrast, the current source in Fig. 3 draws a constant current I_o from the bus capacitor. The topology of Fig. 2 is analyzed in detail in a previous paper from the author ("Analysis and Design of a Cost Effective One Stage Topology for LED Lighting Application", *EPE'15 ECCE-Europe*, Sep. 2015), where the equations of the waveforms of bus voltage and input current are given. A corresponding analysis can also be done with the topology of Fig. 3. With the known waveforms, the influence of the power related bus capacitance C_B/P on the minimal bus voltage as well as the influence on the power factor can be calculated. Fig. 4 and Fig. 5 show the mentioned curves for the two schematics of Fig. 2 and Fig. 3. As expected, the minimal bus voltage decreases with smaller values of C_B/P respectively C_B/I_o. At the minimum value

$$\frac{C_B}{P} = 0.083 \, \frac{\mu F}{W} \tag{2}$$

in Fig. 4, it is not possible anymore to bridge the zero crossing of the nominal mains voltage. The has a maximum $\lambda = 0.594$ at $C_B/P = 0.164 \, \mu F/W$. It tends fast towards zero with smaller values of C_B/P and tends slowly towards zero for higher values of C_B/P. Fig. 5 shows a minimum value of

$$\frac{C_B}{I_o} = 9.8 \, \frac{\mu F}{A}. \tag{3}$$

Below this limit, the bus voltage always follows the rectified mains voltage and there is no bridging of the zero crossing. The power factor is monotonically decreasing. In addition to the traces, a green area is marked in both pictures. In this area, the corresponding waveforms of the input current fulfills the description of the regulation DIN EN 61000-3-2.
In order to fulfill the specification, the description of the waveform (DIN EN 61000-3-2) must be satisfied and the power factor has to be above 0.5 (European regulation 1194/2012) for the nominal input voltage. In addition, the bus voltage must not drop to zero at the lowest mains voltage (205 V) and mains frequency (47 Hz). Therefore, the span for the power related bus capacitance is given by

$$0.111 \frac{\mu F}{W} < \frac{C_B}{P} < 0.309 \frac{\mu F}{W}. \qquad\qquad 11.7 \frac{\mu F}{A} < \frac{C_B}{I_o} < 90.6 \frac{\mu F}{A}. \tag{4}$$

Therefore, there exists an upper limit for the value of the bus capacitor, in order to have a sufficient drop in the bus voltage.

© VDE VERLAG GMBH · Berlin · Offenbach

PCIM Europe 2016, 10 – 12 May 2016, Nuremberg, Germany

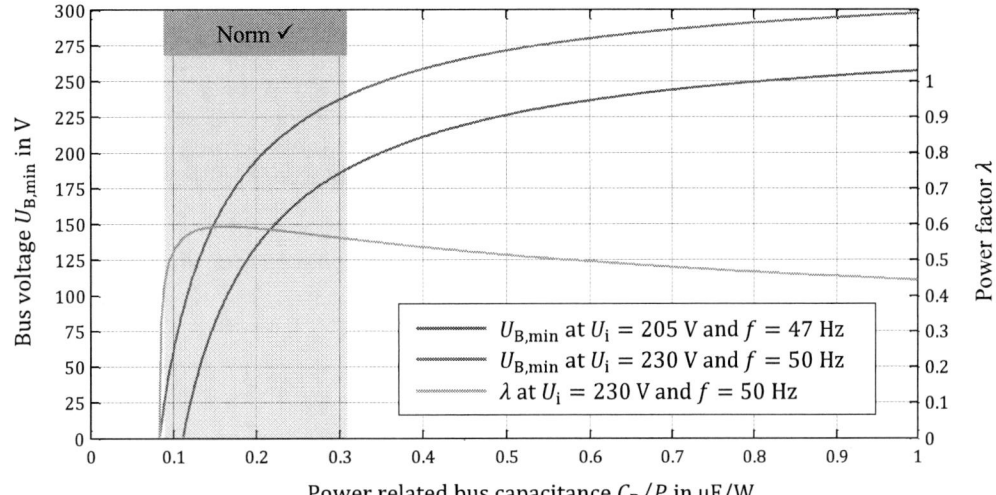

Fig. 4: Minimum voltage $U_{B,min}$ at the bus capacitor and power factor λ as a function of the power related bus capacitance C_B/P for the system of Fig. 2. The green marked area fulfills the description of the waveform according to DIN EN 61000-3-2.

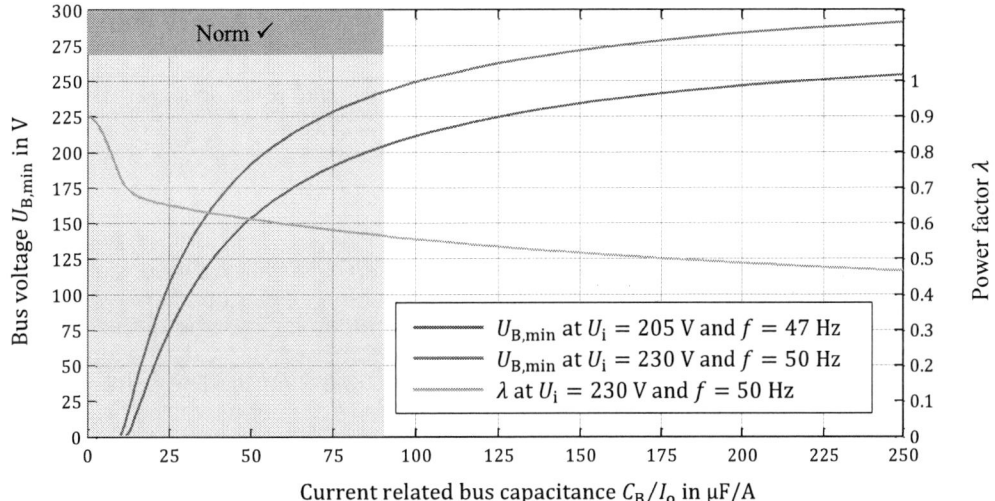

Fig. 5: Minimum voltage $U_{B,min}$ at the bus capacitor and power factor λ as a function of the current related bus capacitance C_B/P for the system of Fig. 3. The green marked area fulfills the description of the waveform according to DIN EN 61000-3-2.

In the considered power range from 9 W to 12 W, values for the bus capacitor around 2 µF are required. This value is low enough for using foil capacitors in the small housing of the LED-lamp. The volume of comparable electrolytic capacitors with regard to capacity and voltage rating is indeed smaller, however, foil capacitors have a higher lifetime. Since the lifetime of electrolytic capacitors often determines the lifetime of the whole LED-lamp, foil capacitors are preferable. Especially in this topology, an electrolytic capacitor would be highly stressed with the relatively high charging and discharging currents.

One degree of freedom for the systems of Fig. 2 and Fig. 3 is the selection of a suitable LED-string in order to define the output current and output voltage of the LED ballast. Fig. 6 shows possible LED-strings in blue dots by using LEDs from Cree. All of them provide a lu-

© VDE VERLAG GMBH · Berlin · Offenbach

minous flux of 940 lm. The product portfolio of other manufactures is similar and can also be used. The DC/DC-converter in Fig. 2 can be realized with different topologies. The most promising ones are the boost- and the buck-converter, since a safety isolation is not required. Each of these topologies and the linear regulator of Fig. 3 have a different area of output configuration, characterized by the output current (= LED-current) and output voltage. These areas are shown in Fig. 6. The areas are limited by a maximum or minimum voltage (orange traces) and a maximum input power (red traces). In order to reach an energy efficiency class of A+ or higher, the input power has to be lower than 12.4 W. With the realization of a buck-converter or a linear regulator, the output voltage has to be lower than the minimal bus voltage. In contrast, a boost-converter needs an output voltage, which is higher than the peak of the mains voltage.

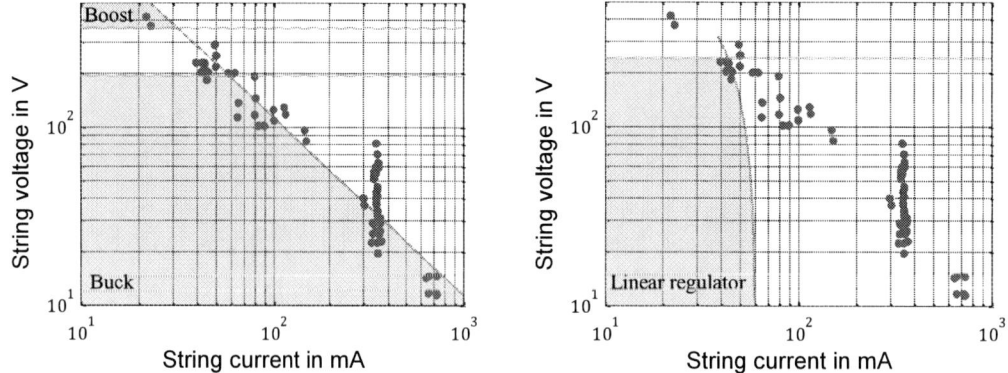

Fig. 6:   Possible area of output current and output voltage of the buck-converter and boost-converter (left) as well as for a linear regulator (right). Blue dots are possible LED-string configurations.

Specification

The LED-retrofit-lamp should replace a commercial 75 W incandescent lamp, having equivalent values in luminous flux (940 lm), color temperature (2700 K) with a proper color rendering index (>80) and form factor (A19). The specification of the values concerning the light is important for choosing the LEDs and will determine the output voltage and output current of the ballast. The LED-retrofit will be designed for the European mains, giving the input specification. The ballast has to fulfill the regulation DIN EN 61000-3-2 for the line current harmonics and the DIN EN 55015 for the RF interference voltage. An input power below 12.4 W is necessary in order to reach the desired energy efficiency class A+ (European regulation 874/2012)

Realization of a buck converter

The DC/DC converter in Fig. 2 is realized as a buck-converter. The schematic in principle is shown in Fig. 7. The MOSFET Q is put in the low side of the power path in order to have a simple drive of the gate. With this position of the MOSFET, it is reasonable to put the inductor L in the same path. Thus, the LED-string with its comparatively high parasitic capacitances towards earth is on a potential without high-frequency fluctuation. Therefore, additional common mode disturbances are avoided. The converter operates in BCM and controls the LED-current to a constant value. Compared to other topologies, this one has the advantage, that the output current is directly the mean value of the inductor current, respectively the half of the peak value of the inductor current. Therefore, an easy implementation of the control loop for the LED current is possible. In each high frequency cycle, a voltage signal proportional to the current through the MOSFET will be compared with a reference signal in order to determine the moment to switch off. No separate measurement of the output current or additional feedback structures are necessary.

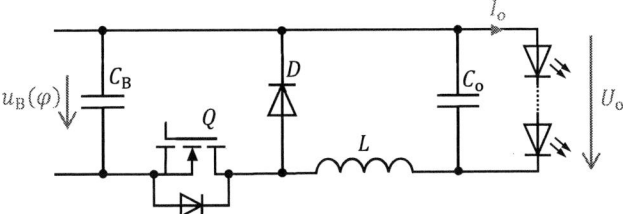

Fig. 7: Buck topology for the DC/DC-converter in Fig. 2 for supplying an LED-string.

For the optimization of the complete system, the luminous efficacy of the LED-string as well as the losses in the ballast must be considered. Fig. 8 shows the losses of the converter in the steady state as a function of the inductor value L. Three possible LED-strings with a high relative luminous efficacy and different string currents are chosen from Fig. 6. In addition to the three different output currents, the losses are calculated in each case for two different MOSFETs SPD01N60C3 and SPD02N60C3, with the last mentioned one having a smaller on-resistance, but higher capacitances. The traces for each case have a global and flat minimum of the losses, which is lower and located at higher inductor values for smaller output currents.

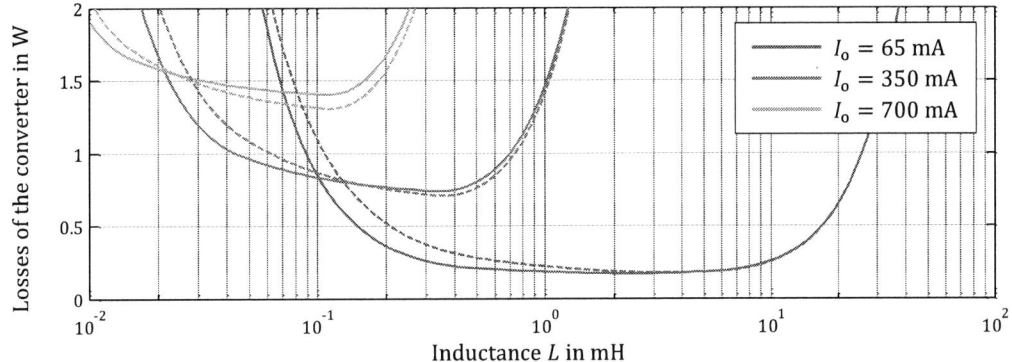

Fig. 8: Losses of the buck converter as a function of the inductance L for three different output currents I_o and in each case with two different MOSFETs (solid line: SPD01N60C3, dashed line: SPD02N60C3) at a power level $P = 9$ W

The hardware realization was done with an inductor value of 4.7 mH and a bus capacitance of 1.4 µF. Fig. 9 shows the top and the bottom layer of the PCB.

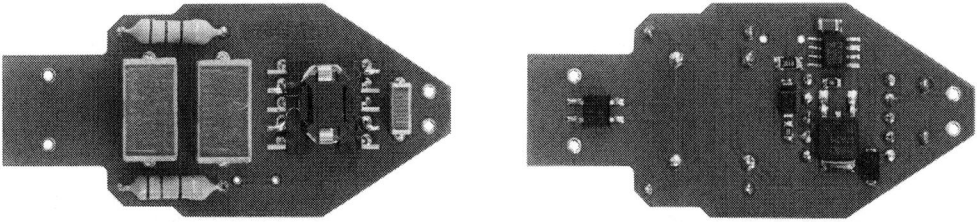

Fig. 9: Top (left) and bottom layer (right) of the realized LED-ballast with buck-converter.

The ballast draws an active power $P_i = 7.7$ W from the mains and delivers, with an output voltage of $U_o = 112$ V and an output current of 67 mA, the electrical power $P_o = 7.5$ W to the LED-string. Thus, the electrical efficiency of the ballast is 97 %. With this input power, the LED-lamp reaches the highest energy efficiency level A++. The output current is controlled to a constant value. The measured modulation in the output current

$$M = \frac{I_{o,\text{max}} - I_{o,\text{min}}}{I_{o,\text{max}} + I_{o,\text{min}}} = 0.4 \% \tag{5}$$

is very low, which yields a high light quality. A summary of more measurement results are given in Table 1.

Realization of a boost converter

In Fig. 10, the DC/DC-converter of Fig. 2 is realized as a boost-converter. With this topology, the bus voltage and the output voltage have the same ground potential. In contrast to the buck-converter, the information of the peak inductor current or switch current is not sufficient for controlling the LED-current to its nominal value. Therefore, a feedback control is necessary.

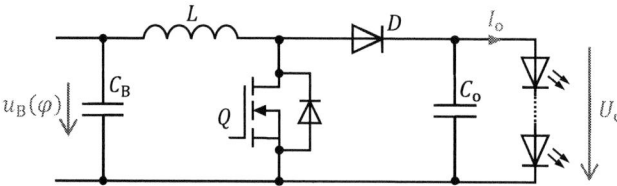

Fig. 10: Boost topology for the DC/DC-Converter in Fig. 2 for supplying an LED-string.

According to Fig. 2, only two LED-strings are in the possible area of the boost-converter, each with a string current around 22 mA. Fig. 11 shows a breakdown of the losses in the mains components as a function of the inductance value L, which determines mainly the switching frequency. At low inductor values, the switching losses are dominant, whereas the conducting losses in the inductor are dominant at higher values. A global minimum of the losses is around 8 mH.

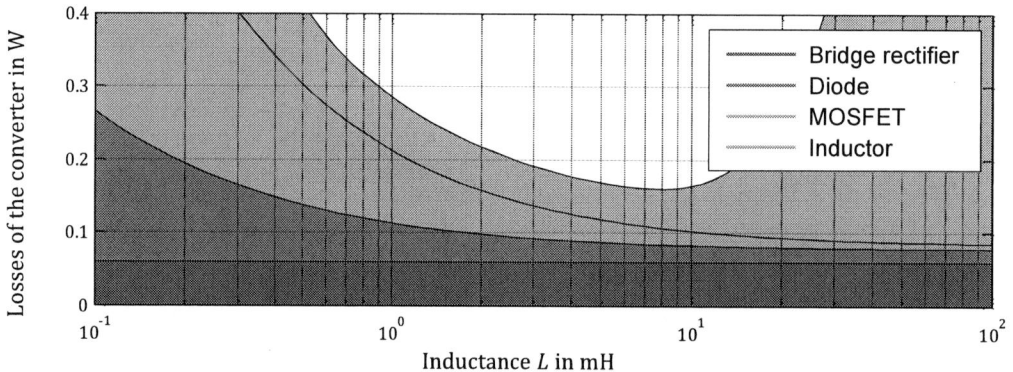

Fig. 11: Breakdown of the losses in the boost-converter of Fig. 10 for an LED-string with a current $I_o = 22$ mA and a power $P_o = 9$ W.

On the final hardware realization, the inductor value is with 7.7 mH near the optimal value of Fig. 11. The bus capacitance is with 1.4 µF the same as for the buck-converter. Pictures of the prototype are shown in Fig. 12. Comparing the bottom layer of the boost-prototype and the buck-prototype reveals the bigger component count of the boost-hardware, mainly due to the additional feedback structure. The LED-string needs an electrical power of $P_o = 9$ W for the specified luminous flux of 940 lm. At the mains side, the ballast draws an active power $P_i = 9.6$ W. Thus, the electrical efficiency of the ballast is 94 %. Controller and feedback structure have a relative high power consumption, which dominates the overall losses and therefore reducing the efficiency. More measurement results are listed in Table 1.

© VDE VERLAG GMBH · Berlin · Offenbach

PCIM Europe 2016, 10 – 12 May 2016, Nuremberg, Germany

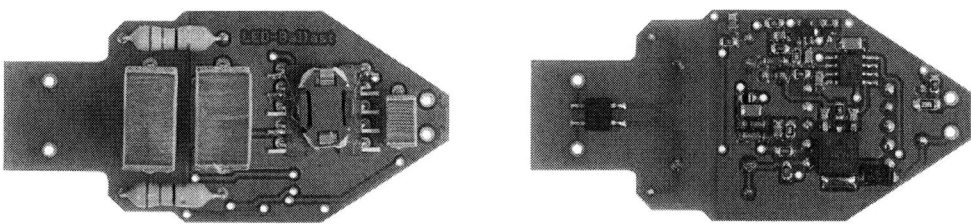

Fig. 12: Top (left) and bottom layer (right) of the realized LED-ballast with boost converter.

Realization of a linear current regulator

There are only a couple of LED-strings in the possible area of output current and output voltage in Fig. 6 for the design of the ballast with a current source. The current of the LED-strings are around 43 mA. Fig. 13 shows the schematic of the realization of the current source in Fig. 3. The constant output voltage of a linear regulator drops across a resistor, which results in a constant output current. Since the linear regulator can't handle high voltages at its input terminals, a depletion-mode power MOSFET is connect upstream. Nearly the complete voltage across the current source drops across the MOSFET. It should be mentioned, that driving a power MOSFET in the linear region can results in an internal thermal runaway. Therefore, the Safe Operating Area with the boundary to the thermal stability must be considered.

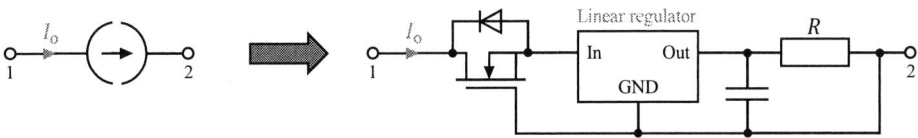

Fig. 13: Realization of the current source in Fig. 3 with a low-voltage linear regulator and a depletion-mode power MOSFET.

Fig. 14 shows the realized prototype. Only six components are necessary. The bus capacitor, a film capacitor of 3.3 µF, is on the top layer, the other components are in SMT on the bottom. A sufficient thermal management is essential, since the power MOSFET generate losses in the range around 3 W.

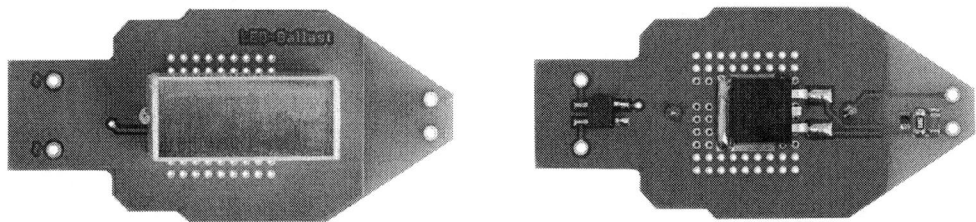

Fig. 14: Top (left) and bottom layer (right) of the realized LED-ballast with linear current source.

The ballast draws a power of 12.3 W and delivers 8.8 W to the LED-string. The difference of 3.5 W produces heat, mainly in the power MOSFET. Nevertheless, the ballast reaches the desired energy efficiency class A+. The electrical efficiency of the ballast is 72 %. The ballast controls the output current to a perfect constant value. No modulation can be measured. More measurement results are given in Table 1.

© VDE VERLAG GMBH · Berlin · Offenbach

499

Conclusion

The typical input stage, consisting of a bridge rectifier and a capacitor, is cost effective but doesn't fulfill the requirements of the regulation for the line current harmonics. Lighting devices with an input power below 25 W have the alternative option to fulfill the requirements of a described waveform. This paper presented, how to proper design the input stage in order to meet this requirements. Three prototypes with different topologies are realized and compared to each other.

Table 1: Summary of selected measurement results for the three realizations.

Characteristic value	Buck	Boost	Linear regulator
Input power	7.74 W	9.58 W	12.25 W
Output power	7.48 W	9.00 W	8.83 W
Efficiency	96.6 %	93.9 %	72.1 %
Power factor	0.57	0.58	0.58
String voltage	112.3 V	393.5 V	204.7 V
String current	66.5 mA	22.8 mA	43.15 mA
Modulation	0.4 %	2.5 %	0 %
Luminous efficacy	124 lm/W	97 lm/W	77 lm/W

Both ballasts with the switched topologies achieve efficiencies of more than 93 %, whereas as matter of principle the linear regulator reaches with 72 % a lower efficiency. The luminous efficacy of the overall system is mainly determined by the luminous efficacy of the LEDs. Thus, also the LED-lamp with the linear regulator as well as the LED-lamp with the boost-converter reaches the energy efficiency class A+. The LED-lamp with the buck-converter reaches even the currently highest class A++. Only six components are necessary for the linear ballast, but also the ballast based on the buck-converter has a low component count. This is especially with a simple implementation of the control loop, with available controllers on the market, possible. For the boost-converter, a part of the control loop must be done with discrete components, which results in a higher complexity and additional losses. This is mainly the reason for the lower efficiency. All of the three ballasts control the LED-current to a constant value, having only a negligible low modulation in the light output and therefore a high light quality. Fig. 15 shows one of the realized LED-retrofit-lamps with an opened housing.

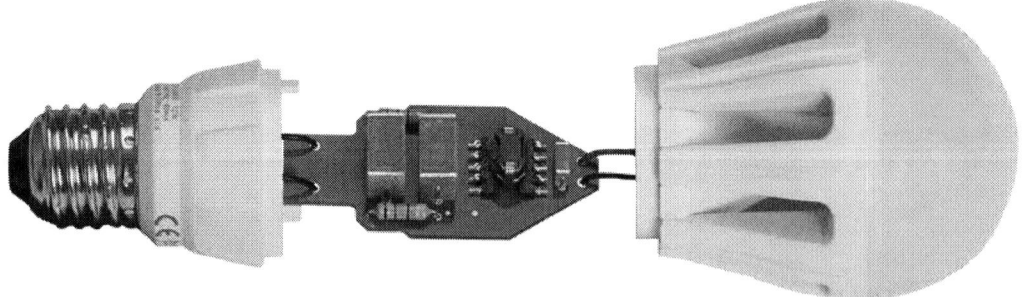

Fig. 15: Realized LED retrofit-lamp in fractional exploded view. The plastic housing and heat sink are from the commercial, available lamp "Parathom Pro Classic A 80" from Osram.

Low Cost High Density AC-DC Converter for LED Lighting Applications

Simon Nigsch, University of Applied Sciences NTB, Switzerland, simon.nigsch@ntb.ch
Janosch Marquart, University of Applied Sciences NTB, Switzerland, janosch.marquart@ntb.ch
Kurt Schenk, University of Applied Sciences NTB, Switzerland, kurt.schenk@ntb.ch

Abstract

Low cost, high power density designs of LED ballasts are not readily available on today's market. With every new generation the LED technology itself is becoming more efficient and the devices are becoming smaller. This leads to an increased demand for smaller lamps. Thus, a compact design of all components in a luminaire is becoming increasingly important. This is the main motivation to focus on miniaturizing the electronics, in order to facilitate small lighting fixtures. As this market is very competitive and thus price sensitive, the cost must be kept at a minimum. This paper presents an extensive topology evaluation and comparison followed by a size and cost optimized design of an AC-DC converter for LED lighting applications. The basis of this work is a comparison of more than twenty different topology combinations. On the set of the five most promising solutions more detailed analyses has been carried out. It turned out that for the given specifications a two-stage design achieves the best performance. It consists of a BCM (boundary conduction mode) Boost PFC as an AC-DC stage and an LLC converter providing galvanic isolation and output regulation. This solution allows high frequency operation, up to 1 MHz, which reduces the size while keeping the switching losses low, owing to zero voltage switching over the entire range of operation. On a prototype for an industrial application the analytical and simulation results were verified by measurements[1].

1. Introduction

Numerous single-stage and two-stage processing topologies have been presented in the past predicting high efficiency and low cost. A lot of industrial applications require more than that. Typically there is a list of specifications which have to be satisfied. Depending on these, different topologies can be used to meet the requirements best. The purpose of this paper is to find the optimal topology which fits the given specifications best. The method employed to compare the different solutions is by rating each topology with various weighted parameters which are defined later. Often, a reduced size of the converter is of great importance. The ever-present trend towards smaller size on power supplies requires an even higher efficiency to cope with the increased power density. Therefore, particularly the full load efficiency must be improved to achieve similar thermal performance with reduced size. The efficiency must be optimized over the entire input voltage range. For conventional LED converters, a two-stage approach is widely adopted to achieve a high efficiency design with universal input voltage, high power factor and a constant DC bus voltage. The DC-DC stage can be optimized for a wide output voltage range to cover different types of LED loads [1-3]. In contrast, single-stage solutions are mainly used for a cost efficient and small volume solutions. However, the disadvantage is to cope with wide input and output voltage ranges as well as a wide dimming range. In the next

---

[1]This project was generously supported and funded by Tridonic GmbH & Co KG

section a topology evaluation and comparison followed by a size and cost optimized design of an AC-DC converter for LED lighting applications is presented.

As a starting point for the investigation to determine the optimum topology, typical main specifications are given in Tab. 1.

Input Voltage:	90-265 V_{RMS}	Dimming Level:	1-100 %
Output Voltage:	30-60 V_{DC}	Power Density:	1 W/cm^3
Output Power:	35 W	Efficiency:	$\geq 90\%$

Tab. 1: Converter specifications

2. Single-Stage vs. Two-Stage Design

The motivation for a single-stage topology is driven by cost and volume. The circuit is required to allow tight regulation of the output current and to provide other control tasks (short-circuit protection, dimming, etc...) [4–7]. Moreover, its efficiency may be high as only one energy conversion stage is implemented. However, for the given specifications (wide input and output voltage range as well as the wide dimming range) the control of such a converter is a challenging task. In most single-stage solutions the necessary bulk capacitor is placed at the output. This implies that the energy is stored at a relatively low voltage and that the 100/120 Hz voltage ripple must be very small. In a two-stage approach, the bulk capacitor is decoupled from the output, which allows a much higher ripple at a higher DC-voltage. Both properties lead to a substantially reduced size and cost of the bulk capacitor. The second stage suppresses any low-frequency ripple present at its input. The output filter can be designed to filter only the switching ripple. It can be shown that one of the main advantages of a two-stage approach is a higher reliability and so the lifetime is considerably extended, whereas the main disadvantages are the higher number of components and its size. Often it is misunderstood that reducing the number of stages automatically achieves higher efficiency. Proper design and proper power processing achieve high efficiency. In general, low component stress can be translated into high efficiency, small physical size and low cost. In the low power range some of the alternative solutions may have an advantage in cost compared to the two-stage solution but some efficiency and control flexibility will be sacrificed. For the given specifications it turns out that a two stage approach is more favorable.

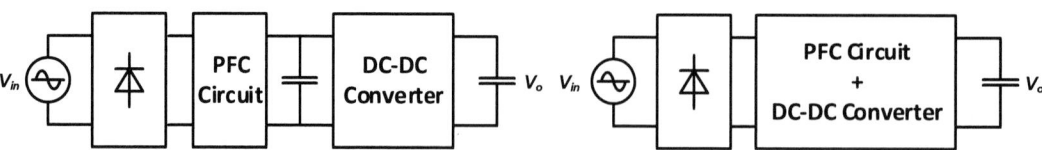

Fig. 1: Two-stage vs. single-stage topology.

3. Finding the Optimum Topology

To increase the power density a high switching frequency is indispensable. Typically, switching frequencies in the range of 100–200 kHz perform best in the terms of efficiency and size. To reach a power density in the range of 1 W/cm³ switching frequencies up to 1 MHz are required [8–10]. At this frequency the size of the magnetics and EMI-filter effort can be substantially reduced. To find the topology, which optimally fulfills the specifications, 23 different solutions have been compared with each other. In order to select five most promising versions for a more detailed investigation, 23 different weighted parameters have been defined which include economical as well as technical demands. To simplify the description of the method, these parameters were grouped to seven requirement groups as shown in Fig. 2(a). The breakdown and weighting of the parameters are shown in Fig. 2(b). Requirements like size, dimming and input/output voltage range are rated with higher importance than efficiency or cost. To preselect the best solution each parameter of all 23 different topologies were rated individually. The rating of each requirement was kept rather simple with three rating levels.

- **Good**   meets all requirements and is ideally suited for the considered purpose

- **Acceptable**   meets almost all requirements and is suitable for the considered purpose

- **Poor**   hardly meets the requirements and is not particularly suitable for use

| (a) Requirements List | (b) Weighting of the requirements |

Fig. 2: Requirements list and weighting.

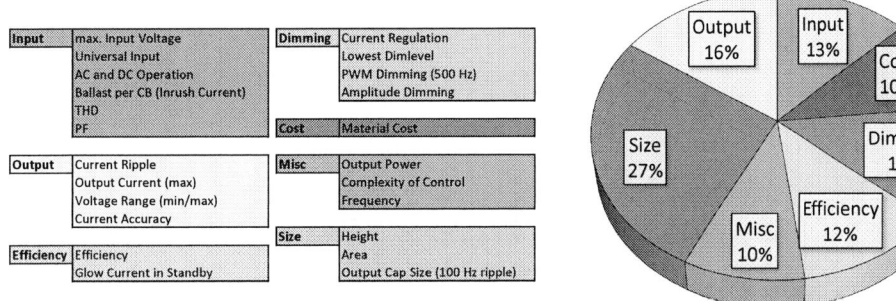

Topology Selection Ranking	Buck PFC LLC	BIFRED PFC	Totem Pole LLC	Double Clamp	Double Clamp Buck	S4CS	S4CS Buck	SS Bridgeless	SS Bridgeless Buck	Flyback	Fly-Forward	Cuk
Normalized Fulfillment	74%	49%	91%	53%	80%	29%	45%	60%	80%	36%	40%	30%
Rank Overall	5	10	2	8	3	23	12	6	4	18	14	22

	Sepic	Zeta	Boost PFC Flyback	Boost PFC Forward	Boost PFC LLC	TwoStage Buck Boost	Boost/Flyback	Boost/Three Level DCDC	Boost LLC	Buck/Flyback	Flyboost
Normalized Fulfillment	31%	31%	52%	57%	100%	37%	36%	35%	46%	37%	41%
Rank Overall	20	20	9	7	1	16	17	19	11	15	13

Fig. 3: Result of the topology comparison.

To get a preselection of the five best solutions, all ratings for each requirement with the weighted importance were summed up. This figure represents a scale of how well the topology is suitable for the given specifications. By adjusting the specifications or changing the weighting of the individual parameters, a different set of solution would be obtained, depending on the application. With this method a topology selection from a large pool of different topologies can easily be achieved. By doing this for the given specifications following result, shown in Fig. 3, is obtained. The green colored solutions were preselected for further investigations, given in Tab. 2.

Boost BCM PFC/LLC	Buck BCM PFC/LLC	Totem Pole/LLC
Double Clamp/Buck BCM	SS-Bridgeless/Buck BCM	

Tab. 2: Five selected solutions for a more detailed investigation [11–14].

For each of the five selected solutions a detailed analytical description has been carried out. This allows a proper converter and control design with an accurate loss model for all components facilitating an objective comparison. The result of the analytical calculation for all PFC-stages is shown in Fig. 4. As it can be seen, for every possible solution all currents and losses in each semiconductor as well as the losses in the magnetic components have been derived. The BCM Boost PFC and Totem Pole PFC achieve the highest efficiency since both converters can be operated with almost no switching losses. The same calculation was carried out with the DC-DC stage.

	Buck PFC			BCM Boost PFC			Totem Pole			Bridgeless PFC			Double Clamp PFC		
	Parameter	Current [A]	Losses [W]	Parameter	Current [A]	Losses [W]	Parameter	Current [A]	Losses [W]	Parameter	Current [A]	Losses [W]	Parameter	Current [A]	Losses [W]
Output Voltage [V]	90			400			400			400			400		
Output Power [W]	55			55			55			55			55		
Input Voltage [Vrms]	265			265			265			265			265		
MosFet															
Conduction Losses	Irms	0,4484	0,076	Irms	0,1085	0,004	Irms	0,4722	0,169	Irms	0,2001	0,030	Irms	0,8031	0,980
Switching Losses			1,379			0,576			0,244			8,261			0,425
Gate Drive Losses	Qg=39nC, Vcc=12V		0,123	Qg=39nC, Vcc=12V		0,167	Qg=39nC, Vcc=12V		0,135	Qg=39nC, Vcc=12V		0,187	Qg=39nC, Vcc=12V		0,749
Rectifier Diode															
Conduction Losses	Iavg	0,4263	0,469	Iavg	0,1376	0,151	Iavg	0,1375	0,002	Iavg	0,145	0,290	Iavg	0,1375	0,275
Bridge Diode	Vf=1V	0,1854	0,371	Vf=1V	0,187	0,374	Vf=1V	0	0,000	Vf=1V	0	0,000	Vf=1V	0	0,000
PFC Choke	L=78µH, ELP 14/3.5/5, R=100mR			L=360µH, ELP 14/3.5/5			L=150µH, ELP 14/3.5/5			L=503µH, ELP 14/3.5/5			Lm=30µH,Lr=2uH, ELP 14/3.5/5		
Inductance	L[uH]	78		L[uH]	360		L[uH]	150		L[uH]	903			85	
Winding	n	66		n	144		n	77			85				
Core Loss			0,800			0,372			0,436			0,386			0,643
Winding Loss		0,8491	0,721		0,2398	0,547		0,4722	0,607		0,2313	0,107		2,215	0,491
LI²		56,24			20,70			33,45			48,31			0,00	
Capacitor	dU=2%, Co=	2200		dU=2%, Co=	110		dU=2%, Co=	110		dU=2%, Co=	110		dU=2%, Co=	110	
THD	13,20%			0,60%			4,20%			1,50%			15,40%		
Total Losses			3,939			2,192			1,593			9,262			3,563
Efficiency			92,8%			96,0%			97,1%			83,2%			93,5%

Fig. 4: Analytical comparison of all PFC-Stages.

For the given application, cost and size are also of great importance. That means, each of these five solutions will also be rated by number, size, and cost of the individual circuit components, as well as the complexity of the control method. By doing this, the best combination of PFC and DC-DC stage can be found by minimizing size, cost, and complexity while maximizing the efficiency. Fig. 5 shows the final result of each solution itemized for each parameter group. As it can be seen, each solution has its advantage for particular requirements where a higher number represents a better solution. Using the weighted importance of each parameter, a Boost BCM PFC with a LLC converter as DC-DC stage with galvanic isolation, shown in Fig. 6, best fulfills the given requirements. The final ranking of this comparison is given in Fig. 5.

PCIM Europe 2016, 10 – 12 May 2016, Nuremberg, Germany

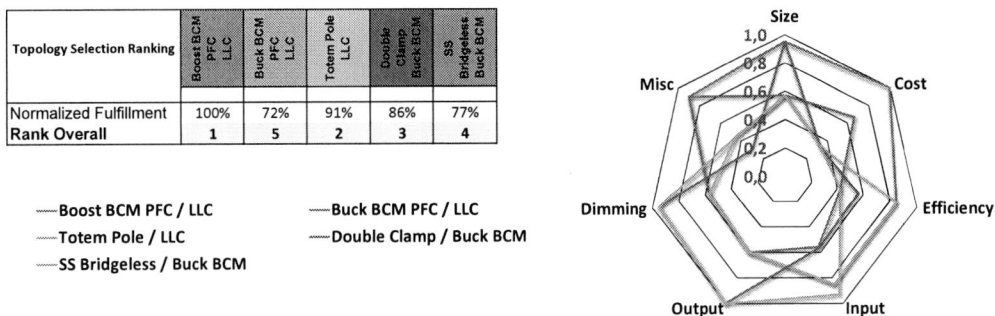

Topology Selection Ranking	Boost BCM PFC LLC	Buck BCM PFC LLC	Totem Pole LLC	Double Clamp Buck BCM	SS Bridgeless Buck BCM
Normalized Fulfillment	100%	72%	91%	86%	77%
Rank Overall	1	5	2	3	4

----Boost BCM PFC / LLC   ----Buck BCM PFC / LLC
----Totem Pole / LLC   ----Double Clamp / Buck BCM
----SS Bridgeless / Buck BCM

Fig. 5: Ranking of the five preselected solutions and its parameter comparison.

4. Design

The most widely used operation modes for the boost converter are continuous conduction mode (CCM) and boundary conduction mode (BCM). As the names indicate, the inductor current in CCM is continuous, while in BCM, the new switching period is initiated when the inductor current returns to zero, which is at the boundary between continuous conduction and discontinuous conduction operation. Even though the BCM operation leads to higher RMS currents in the inductor and switching devices, it provides better switching conditions for the MOSFET and the diode, as the reverse recovery is eliminated. The MOSFET is also turned on with zero current, which reduces the switching losses. A by-product of this operation mode is that the boost converter runs with variable switching frequency, which depends primarily on the selected output voltage, the instantaneous value of the input voltage, the boost inductor value, and the output power. The operating frequency changes as the input current follows the sinusoidal input voltage waveform. By limiting the switching frequency to a maximum of 700 kHz for all load situations the boost inductor is designed with L=280 µH. This inductor value is optimized for a high density PFC design, while keeping the losses of the PFC-stage low.

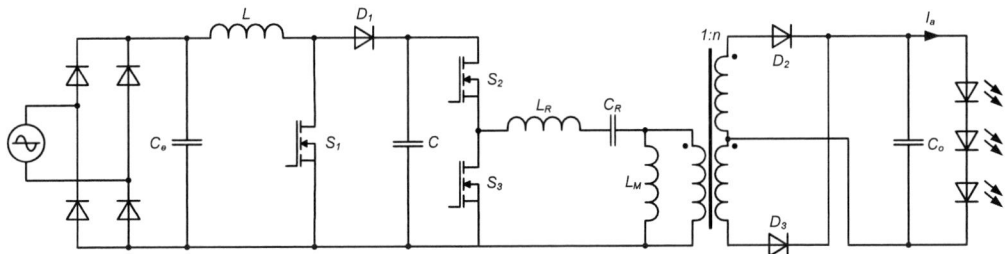

Fig. 6: Two-stage Boost BCM PFC with LCC converter.

An optimized design of the LLC converter is more challenging since a wide output voltage range and a dim-level from 1 % to 100 % is required. Considering the main goal of reducing the size of the whole LED converter, the optimization focuses on size and efficiency while maintaining ZVS for the whole area of operation including a dim-level down to 1 %. With a parameter variation algorithm, several hundred designs have been calculated. The design with the highest

© VDE VERLAG GMBH · Berlin · Offenbach

$L = 280\mu H / RM6 - Core$	$C = 10\mu F$	$S_1 : IPD65R1k4C6$
$L_R = 460\mu H$	$L_M = 450\mu H$	$C_R = 200pF$
$C_o = 10\mu F$	$n = 0.122$	$S_2, S_3 : IPD50R3K0CE$

Tab. 3: Converter component values

power density has been selected with the converter values given in Tab. 3. With these values, the maximum switching frequency reaches about 800 kHz and the lowest is about 450 kHz. This allows for a small magnetic design and optimized conduction losses due to the minimized magnetizing current, but still ensures proper ZVS for the whole area of operation.

With this design the two-stage LED converter reaches a calculated efficiency of about 91 %. The distribution of the losses is depicted in Fig. 7(a).

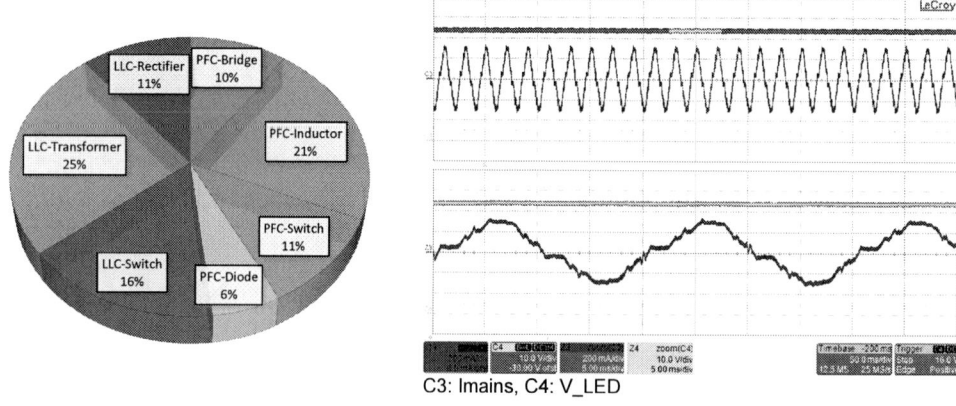

(a) Loss distribution of PFC and LLC

C3: Imains, C4: V_LED

(b) Input current measurement

Fig. 7: Analytical loss distribution and measured input current.

5. Measurements and Buildup

The designed two-stage circuit was built and packed in a housing with the outer dimensions of 32x86x21mm including an auxiliary power supply and dimming interface. The main specification is given in Tab. 1, whereas the prototype is shown in Fig. 9(a). With a thermal simulation software, the packaging and placement of the critical components have been simulated to get a homogeneous heat distribution, so that the housing of the fully potted converter stays below 60 °C. The thermal simulation and measurement results are shown in Fig. 8(a) and Fig. 8(b). The measured maximum temperature at full load reaches about 70 °C which is 10 °C more than the simulation predicted. Although, there is still some potential for improvement as the ferrite losses of the LLC transformer can be reduced by further optimizing the core structure.

The measured input current (blue) together with the output voltage (green) is shown in Fig. 7(b). As requested by the specifications and standards, the converter draws a sine-shaped input current. At full load, a THD and PF with values as low as 5 % and as high as 0.99, respectively,

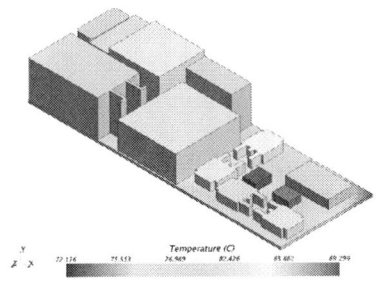
(a) Thermal simulation

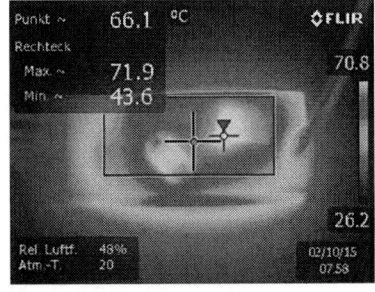
(b) Thermal measurement of the fully potted converter

Fig. 8: Thermal design: simulation and thermal measurement at full load.

are achieved. By utilizing a two-stage design, the 100/120 Hz ripple voltage can be filtered with the bulk capacitor and the following LLC stage supplies a constant current to the load. Thus, the measured output current ripple is below 10 % and with this easily meets the specification of $I_{pp} < 30$ %. The measured efficiency including auxiliary supply exceeds 90 %, as shown in Fig. 9(b), even with this low cost, high density design and at this high switching frequency. The measurement values are in good agreement with the calculated efficiency and loss-distribution presented in Section 4.

(a) Converter prototype

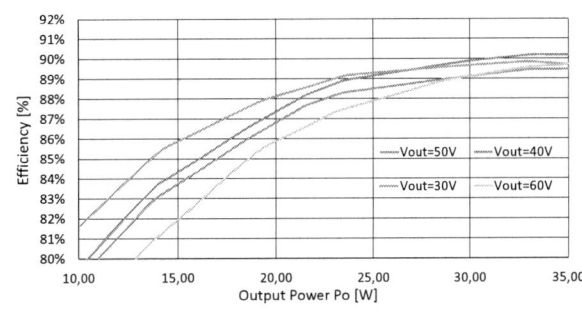
(b) Efficiency measurement

Fig. 9: Converter prototype and efficiency measurement.

6. Conclusion

In this work, a low cost high density LED driver has been designed and characterized through measurements on a hardware realization. By means of the presented topology selection process the best suitable topology solutions have been determined. This involves simulation tools combined with analytical description of the various circuits. A straightforward design-process delivers an optimized converter solution achieving a high power density while keeping the losses low. Including the auxiliary supply an efficiency of over 90 % was reached, even at this low power level of 35 W.

7. References

[1] H. van der Broeck, G. Sauerlander, and M. Wendt. Power driver topologies and control schemes for leds. In *Applied Power Electronics Conference, APEC 2007 - Twenty Second Annual IEEE*, pages 1319–1325, Feb 2007.

[2] A. Vazquez M. Arias and J. Sebastian. An overview of the ac-dc and dc-dc converters for led lighting applications. *Automatika - Journal for Control, Measurement, Electronics, Computing and Communications*, 53(2):156–172, May 2012.

[3] Seungbum Lim, D. M. Otten, and D. J. Perreault. New ac-dc power factor correction architecture suitable for high-frequency operation. *IEEE Transactions on Power Electronics*, 31(4):2937–2949, April 2016.

[4] C. Qiao and K. M. Smedley. A topology survey of single-stage power factor corrector with a boost type input-current-shaper. *IEEE Transactions on Power Electronics*, 16(3):360–368, May 2001.

[5] J. M. Alonso, J. Vina, D. G. Vaquero, G. Martinez, and R. Osorio. Analysis and design of the integrated double buck-boost converter as a high-power-factor driver for power-led lamps. *IEEE Transactions on Industrial Electronics*, 59(4):1689–1697, April 2012.

[6] D. Gacio, J. M. Alonso, A. J. Calleja, J. Garcia, and M. Rico-Secades. A universal-input single-stage high-power-factor power supply for hb-leds based on integrated buck-flyback converter. In *Applied Power Electronics Conference and Exposition, 2009. APEC 2009. Twenty-Fourth Annual IEEE*, pages 570–576, Feb 2009.

[7] Jae-Kuk Kim, Jae-Bum Lee, and Gun-Woo Moon. Isolated switch-mode current regulator with integrated two boost led drivers. *IEEE Transactions on Industrial Electronics*, 61(9):4649–4653, Sept 2014.

[8] C. Wang, M. Xu, B. Lu, and F. C. Lee. New architecture for mhz switching frequency pfc. In *Applied Power Electronics Conference, APEC 2007 - Twenty Second Annual IEEE*, pages 179–185, Feb 2007.

[9] M. P. Madsen, A. Knott, and M. A. E. Andersen. Very high frequency resonant dc/dc converters for led lighting. In *Applied Power Electronics Conference and Exposition (APEC), 2013 Twenty-Eighth Annual IEEE*, pages 835–839, March 2013.

[10] L. Corradini and G. Spiazzi. A high-frequency digitally controlled led driver for automotive applications with fast dimming capabilities. *IEEE Transactions on Power Electronics*, 29(12):6648–6659, Dec 2014.

[11] J. Zhang, J. Shao, P. Xu, F. C. Lee, and M. M. Jovanovic. Evaluation of input current in the critical mode boost pfc converter for distributed power systems. In *Applied Power Electronics Conference and Exposition, 2001. APEC 2001. Sixteenth Annual IEEE*, volume 1, pages 130–136 vol.1, 2001.

[12] B. Su, J. Zhang, and Z. Lu. Totem-pole boost bridgeless pfc rectifier with simple zero-current detection and full-range zvs operating at the boundary of dcm/ccm. *IEEE Transactions on Power Electronics*, 26(2):427–435, Feb 2011.

[13] S. Nigsch, S. Cuk, and K. Schenk. Analysis, modeling and design of a true bridgeless single stage pfc with galvanic isolation. In *Applied Power Electronics Conference and Exposition (APEC), 2015 IEEE*, pages 469–476, March 2015.

[14] M. Salato and P. Kowalyk. Double-clamp zvs converter interfaces high voltage traction batteries with 12v legacy system in hybrid and pure-electric vehicles. In *Vehicle Power and Propulsion Conference (VPPC), 2012 IEEE*, pages 667–670, Oct 2012.

PCIM Europe 2016, 10 – 12 May 2016, Nuremberg, Germany

Design Optimization for a High Power-Density, Wide Output, High Frequency LLC Resonant Converter for Lighting Applications

Janosch Marquart, University of Applied Sciences NTB, Switzerland, janosch.marquart@ntb.ch
Simon Nigsch, University of Applied Sciences NTB, Switzerland, simon.nigsch@ntb.ch
Kurt Schenk, University of Applied Sciences NTB, Switzerland, kurt.schenk@ntb.ch

Abstract

The ongoing development in the area of lighting applications demands increasing power density ($2\,W/cm^3$ as a goal) in power converters and a small and planar designs. Yet the functionality and the various specifications necessary to sell the products, such as wide output voltage range (30 V to 60 V in the presented example), have to be fulfilled. In this paper the optimization process for a fully functional LLC resonant converter for LED lighting applications is described. A very small, integrated magnetic with low parasitic capacitance between the winding enables high frequency operation and still provides zero voltage switching over the entire range of operation, including a dimming range from 1 % to 100 %. The presented design optimization introduces an optimal placement of the load independent point of operation and optimization parameters for an iterative design process. The design optimization is verified by simulation and measurements at a pre-series prototype[1].

1. Introduction

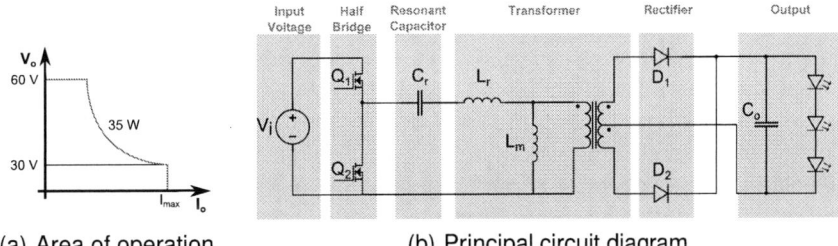

(a) Area of operation (b) Principal circuit diagram

Figure 1: LLC for lighting application

The LLC converter is a well-known topology frequently used in various applications ranging from front end DC-DC converters [1] over solar power conversion [2] and battery charger [3] to LED driver circuits [4]. For LED driver applications the half-bridge LLC converter is often used. Several resonant half-bridge topologies are known for almost 30 years [5]. LLC converters are best suited for fixed conversion ratios, high efficiency and small volumes. For the LED driver described in this paper, a wide output voltage range and a wide dim-level from 1 % to 100 % is required, demanding a more complex and challenging design process. Figure 1(a) shows the area of operation with a wide output voltage range from 30 V to 60 V.

Figure 1(b) shows the basic topology commonly used for LED lighting applications. A half-bridge drives the resonant tank consisting of a resonant capacitor C_r, a resonant inductor L_r and a parallel inductor L_m. A secondary side rectifier is driven by a center tapped transformer. For future lighting application the size (power density) and especially the height matters. The

---

[1]This project was generously supported and funded by Tridonic GmbH & Co KG

© VDE VERLAG GMBH · Berlin · Offenbach

findings of this paper are carried out for a 35 W converter[2]. The maximum height is specified as 21 mm, including casing, and the target power density for the LLC converter is 2 W/m³.

Section 2 covers the topic of the design optimization for the wide-output high-frequency LLC converter. Section 3 deals with the optimization process for the integrated magnetic transformer, focusing on size reduction and minimal parasitic effects using FEA simulations and mathematical verification. Measurement results for a pre-series prototype are presented in section 4.

2. Design Optimization for Wide Output Voltage Operation

2.1. State of the Art Design Process

The design process of an LLC converter is an iterative one ([6], [7] and [8]) and can mainly be divided in three parts. First designing the resonant circuit to cover the required input and output voltage ranges, then optimizing the magnetics considering parasitic effects, size and performance and finally optimizing the component selection.

The following analysis follows the well-known theory of the first harmonic approximation for circuit analysis.

The LLC design is governed by two resonant frequencies. The series resonance f_0 (or load independent point) and the load dependent peak resonance f_p. At the series resonance f_0 the series connection of L_r and C_r represent zero impedance. The output voltage is then directly determined by the input voltage and is therefore load independent. The magnetizing inductance L_m leads to a load dependent resonance f_{c0}, moving within the range of $f_p \leq f_{c0} \leq f_0$.

$$f_0 = \frac{1}{2\pi\sqrt{L_r C_r}} \qquad (1) \qquad f_p = \frac{1}{2\pi\sqrt{(L_r + L_m)C_r}} \qquad (2)$$

$f_{c0} = f_p$ at no load and moves towards f_0 as the load increases as shown in Figure 2(b), with equalty at load short circuit condition. Most of the time an LLC converter is designed to operate in the vicinity of f_0, limiting the load dependency to a minimum. For this reason a proper selection of the load independent point is a critical factor.

The output voltage of the converter can be regulated by M_g (normalized voltage gain function) through controlling f_n (normalized switching frequency defined as $f_n = f_{sw}/f_0$).

$$V_o = Mg(f_n, \lambda, Q_e)\frac{1}{n}\frac{V_{in}}{2} \qquad (3) \qquad M_g = \left| \frac{\lambda f_n^2}{[(\lambda + 1)f_n^2 - 1] + j[(f_n^2 - 1)f_n Q_e \lambda]} \right| \qquad (4)$$

The main design parameters are the switching frequency f_s, the transformer turns ratio n, the inductance ratio $\lambda = L_m/L_r$ and the quality factor of the series resonant circuit $Q_e = \sqrt{L_r/C_r}/R_e$. Generally, an LLC converter is designed based on the minimum and maximum voltage gain required for the application and then choosing a proper λ and Q_e. The peak gain is directly influenced by λ. From the definition of λ it is evident that a smaller λ means a smaller L_m and therefore higher peak gain (see Figure 2(a)) leading to a higher magnetizing current. Higher magnetizing current helps to ensure proper zero voltage switching (ZVS) at the cost of increased conduction losses. A smaller Q_e increases the peak gain while associated gain curves exhibit a larger frequency variation for a given gain range as shown in Figure 2(a) and 2(b).

---

[2]The goal is to design a fully functional LED driver, including dimming interface (e.g. DALI), power factor corrector (PFC) and auxiliary supply as a pre-series prototype. Input voltage range: 380 V to 420 V, output power 35 W and output voltage range 30 V to 60 V. This paper only covers the LLC converter used therefore.

PCIM Europe 2016, 10 – 12 May 2016, Nuremberg, Germany

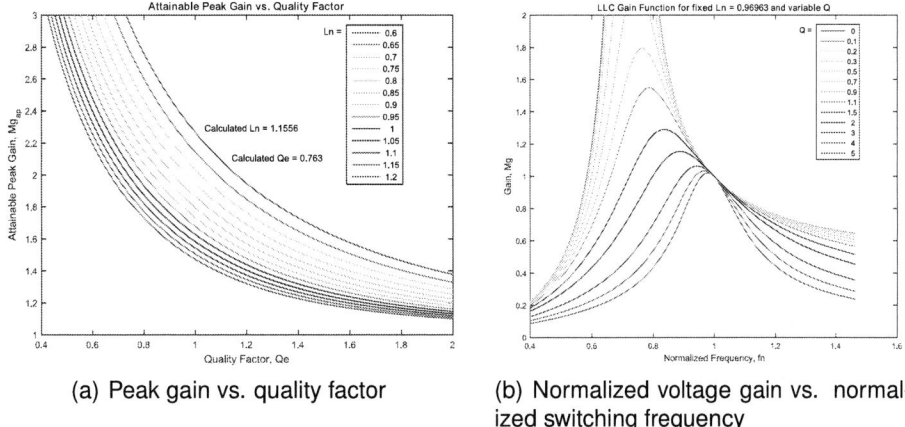

(a) Peak gain vs. quality factor

(b) Normalized voltage gain vs. normalized switching frequency

Figure 2: Peak gain and normalized gain curves for the given LLC converter design

The standard design process now iterates through changes of λ and Q_e and checking them against the required switching frequency range and gain.

2.2. Optimization

The placement of the load independent point $((f_n, M_g) = (1,1))$ within the range of operation[3] (see Figure 3) has a strong impact on how the losses are distributed between the primary and secondary side. The LLC converter should be operated in the vicinity of the load independent point as mentioned above. Often this point is placed at the highest frequency and therefore lowest output voltage [7] as the transfer function of the LLC converter behaves very differently below, at and above the series resonance[4]. This leads to a challenging task of implementing a fast and reliable controller. LED applications however do not generally demand a fast controller and thus, the selection of the load independent point is less restrictive.

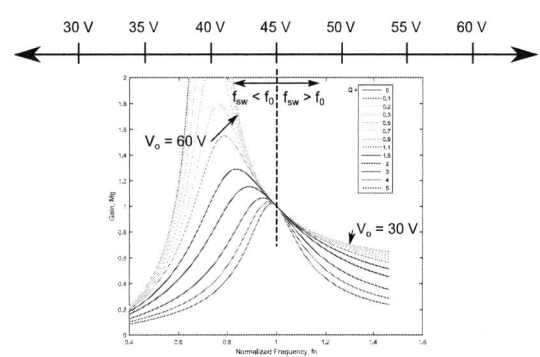

Figure 3: Load independent point of operation positioned at 45 V

The optimization for this design has been carried out using a parameter variation algorithm[5] that calculated several hundred designs. A validation check selects valid designs and the resulting set of designs is used for the optimization process. Considering the main goal of reducing the size of the whole converter, the optimization focuses on size and efficiency while maintaining ZVS for the entire area of operation including a dim-level down to 1 %. Two expressions can be used for proper design selection. The expression $L_m \cdot I_{m,rms}^2$ is proportional to the energy

---

[3]Area between lowest and highest output voltage.

[4]Operation above resonance frequency leads to lower crossover frequency with decreasing load, operation below leads to increasing crossover frequency with decreasing load.

[5]Different placements for load independent point at different output voltages; Various frequency ranges with a maximum frequency limit of 1 MHz.

© VDE VERLAG GMBH · Berlin · Offenbach

stored in the integrated magnetic and therefore proportional to the size of the transformer. Additionally, it can be shown that the product of $L_m \cdot f_{s,max}$ is related to the available energy in the magnetizing inductance for ZVS. The energy necessary for proper ZVS can be calculated with (5) where the worst case condition is $V_o = 30\,\text{V}$ and 1 % dim-level. With this information a proper design can be selected.

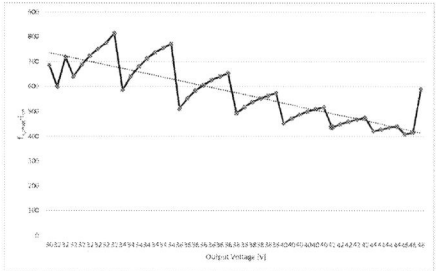

(a) Maximum switching frequency multiplied by magnetizing inductance

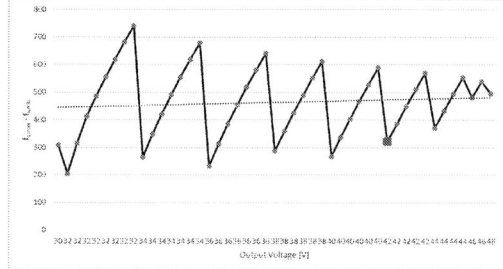

(b) Frequency variation for all valid designs

Figure 4: Optimal design selection.

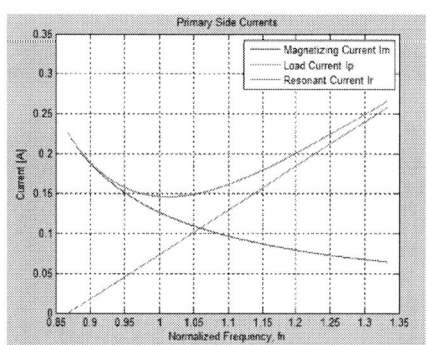

Figure 5: Resonant current (red), magnetizing current (blue) and load current (green) plotted over normalized switching frequency

The optimal design has been selected (red dots in Figure 4(a) and 4(b)) by minimizing the size of the integrated magnetic using the above mentioned expressions, while maintaining a small frequency band for the whole area of operation. Figure 4(b) shows the frequency variation ($f_{s,max} - f_{s,min}$) for all valid designs. Keeping this variation small results in reduced conduction losses for the primary side. The highest frequency determines the available energy for ZVS. Figure 5 shows that the magnetizing current dominates for lower frequencies leading to the necessity of keeping the frequency variation as small as possible. The maximum frequency now reaches about 800 kHz and the lowest frequency is about 450 kHz. This design enables a small magnetic device (high frequency operation and smallest possible frequency variation), and optimized conduction losses due to the smallest possible magnetizing current which still ensures proper ZVS over the whole area of operation.

3. Integrated Magnetics Design

3.1. Overview

A number of publications can be found on how to integrate magnetics for LLC converters, two examples are [9] and [10]. The main aspect is to integrate the separate series inductance L_r into the transformer and use the magnetizing inductance as the parallel inductor L_p. The main procedure, including a method to design the leakage inductance is presented in [11]. In this paper, the transformer is designed keeping in mind the main goal, namely the reduction of size, while still maintaining ZVS over the whole area of operation. As aforementioned, the most critical point of operation for ZVS is 1 % dim-level at $V_o = 30\,\text{V}$. At this point of operation the frequency reaches its highest value (up to 800 kHz in this design). High frequency operation has impacts on different levels, beginning at the device technology level, selecting magnetic materials and also the level of integration [12]. Under these considerations, a very low parasitic capacitance (from the integrated transformer) parallel to the output capacitor C_{oss} of the MOSFETs are necessary. According to [6] inequalities (5) and (6) must hold for proper ZVS. Increasing the parasitic capacitor leads to higher energy necessary for ZVS and therefore a higher dead-time for the switching transition to take place. Higher energy stored in the magnetizing inductor also means higher conduction losses due to higher currents.

$$\frac{1}{2}(L_m + L_r)I_{m,peak}^2 \geq \frac{1}{2}(2C_{oss} + C_{parasitic})V_{in}^2 \quad (5) \qquad t_{dead} \geq 16(C_{oss} + C_{parasitic})f_{sw}L_m \quad (6)$$

3.2. Integrated Magnetics Optimization

The output capcitance of the MOSFET is in the range of 10 pF per MOSFET. As aforementioned it is necessary to keep any capacitance in parallel to this output capacitance as small as possible. Many designs for integrated magnetics for LLC converters show an integration of the resonant inductance L_r on one leg and the transformer windings on the other leg. For this design the decision was made to separate the primary and secondary winding of the transformer to keep the parasitic capacitance in the range of a few pF and use the leakage inductance as the series resonant inductance L_r. An EI-shaped core (shown in Figure 6(a)) was used for a first design approach using a slightly altered E2010 core of the shelf (smaller center leg). The primary winding is wound on one of the outer legs, the secondary winding on the other one. This allows to individually set the magnetizing inductance (air gap in the outer leg) and the leakage inductance (air gap in the center leg) while minimizing the parasitic winding capacitance. Several transformer samples have been built for verification (not reported in this paper).

3.3. FEA Simulation and Verification

Different publications covering the topic of modeling and simulation of LLC converter and magnetics can be found [13] and [14]. For this design the magnetics have been simulated using modern FEA-simulation tools. The whole LLC converter has been simulated using PLECS[6]. The verification of the transformer model has been carried out using a reluctance based model simulation with PLECS.

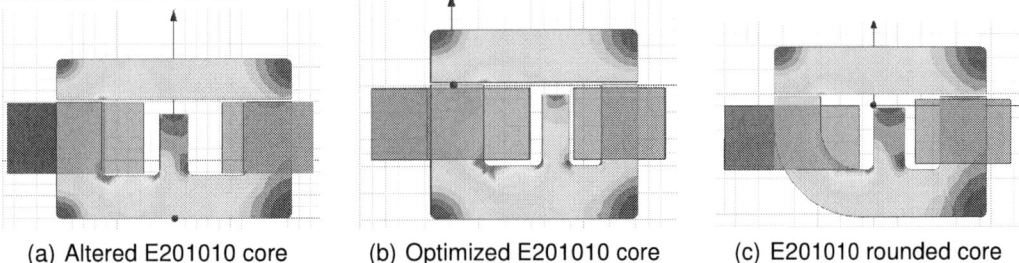

(a) Altered E201010 core   (b) Optimized E201010 core   (c) E201010 rounded core

Figure 6: FEA simulations for the magnetic design process

Figure 6 shows the magnetic-flux-density for three different core shapes. All three simulations use the same scale (blue: 10 mT, red: 100 mT). As shown in Figure 6(a) the core is not fully utilized as the magnetic flux density is higher on the left side (primary winding) and there is some space left for the winding. The optimized core shape (shown in Figure 6(b)) shows more equal magnetic flux density in both outer legs (primary and secondary winding). A third version was simulated (shown in Figure 6(c)) with an additional possibility for optimization. The rounded corner leads to a more homogeneous magnetic flux distribution in the leg[7].

---

[6]Piecewise Linear Electric Circuit Simulation tool from Plexim GmbH

[7]Not suitable for mass-production.

Transient FEA-simulations and measurements showed ferrite losses of 1 W. Figure 10(a) shows the different cores including the one used in the design (middle one without winding and most right one with bobbin and winding)[8]

3.4. Confirmation with Reluctance Model

FEA tools are not necessarily available at every work place. An alternative method to describe magnetic circuits is by means of a reluctance model. Each relevant magnetic path is represented by a reluctance. The reluctance $\Re$ corresponds to the ratio of the magneto motive force V_m to the magnetic flux ϕ in any given geometric structure. For cylindrical geometries, such as in ferrite cores with a defined length l_{Fe} and cross section A_{Fe} this can be expressed as: $\Re = l_{Fe}/\mu A_{Fe}$.

That means, each leg of a structure as shown in Figure 7(a) can be represented by a reluctance, where the effects of the air gaps in $\Re_1$ through $\Re_3$ are accounted for. The corresponding model is shown in Figure 7(b). The flux through $\Re_L$ accounts the leakage through the surrounding air.

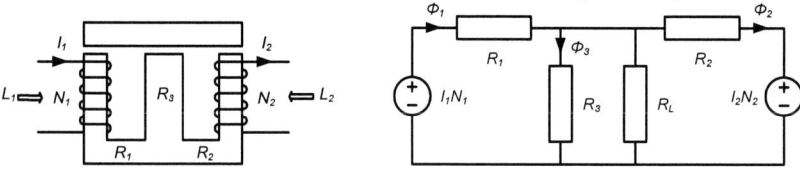

(a) Magnetic Structure   (b) Corresponding reluctance model

Figure 7: Confirmation with reluctance model

The values of the core reluctances could be estimated using the core geometry and air gap width. On an existing device it is easier to determine these values by inductance measurements on the windings, in this case with the following result[9]: $L_{1O} = 848\,\mu H$, $L_{1S} = 455\,\mu H$, $L_{2O} = 13.9\,\mu H$, $L_{2S} = 7.42\,\mu H$.

The subscript O stands for open and S for shorted opposite winding. The turns ratio $N_2 : N_1 = 9 : 70$. With these measurements the three independent reluctances $\Re_1$, $\Re_2$ and $\Re_3 || \Re_L$ can be determined. One measurement is redundant and can be used to increase accuracy. Here, without derivation, we have:

$$\Re_3 || \Re_L = N_1 N_2 \sqrt{\frac{1}{L_{2S}} \left(\frac{1}{L_{1S}} - \frac{1}{L_{1O}} \right)} \quad (7) \qquad \Re_3 || \Re_L = N_1 N_2 \sqrt{\frac{1}{L_{1S}} \left(\frac{1}{L_{2S}} - \frac{1}{L_{2O}} \right)} \quad (8)$$

Due to limited measurement accuracy (7) and (8) result in slightly different values. For further calculations the average value of the two is used, which is: $\Re_3 || \Re_L = 7392\,kH^{-1}$

The reluctance $\Re_1$ of the left leg and $\Re_2$ in the right leg, respectively:

$$\Re_1 = \frac{N_1^2}{L_{1S}} - \Re_3 || \Re_L = 3377\,kH^{-1} \quad (9) \qquad \Re_2 = \frac{N_2^2}{L_{2S}} - \Re_3 || \Re_L = 3524\,kH^{-1} \quad (10)$$

As the fluxes through the center leg and through the air are topologically in parallel, the values of $\Re_3$ and $\Re_L$ cannot be distinguished by measurements in the windings. However, it can be measured with a missing center leg on an otherwise identical core. A number of measurements

---

[8]LxWxH (including windings and bobbin): 22 mm x 20 mm x 13 mm

[9]Results measured at setup with the core shown in Figure 6(a)

have shown that on an E-type core the value of the leakage reluctance $\Re_L$ is relatively constant, irrespective of the size. Thus, here this typical value is used: $\Re_L = 12.5\,\mathrm{MH}^{-1}$

Either from above inductance measurements or from the reluctance model the components of the T-model of the transformer can be determined. Using the latter, the reluctances seen by the primary and secondary windings $\Re_P$ and $\Re_S$, as well as the respective k-coefficients are:

$$\Re_p = \Re_1 + \Re_2||\Re_3||\Re_L = 5764\,\mathrm{kH}^{-1} \quad (11) \qquad \Re_s = \Re_2 + \Re_1||\Re_3||\Re_L = 5842\,\mathrm{kH}^{-1} \quad (12)$$

$$k_p = \frac{\Re_3||\Re_L}{\Re_2 + \Re_3||\Re_L} = 0.677 \quad (13) \qquad k_s = \frac{\Re_3||\Re_L}{\Re_1 + \Re_3||\Re_L} = 0.686 \quad (14)$$

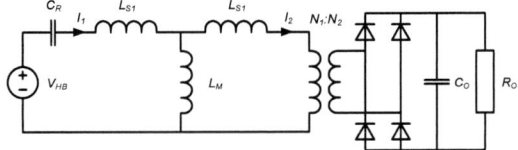

Figure 8: T-model of the LLC converter

From these values the components of the T-model (shown in Figure 8) can be calculated:

$$L_M = k_p \frac{N_1^2}{R_p} = 576\,\mu\mathrm{H} \tag{15}$$

$$L_{S1} = (1 - k_p)\frac{N_1^2}{R_p} = 275\,\mu\mathrm{H} \tag{16}$$

$$L_{S2} = (1 - k_s)\frac{N_1^2}{R_s} = 263\,\mu\mathrm{H} \tag{17}$$

The voltage source V_{HB} represents the square wave voltage generated by the half bridge. With this model the currents I_1 and I_2 can be determined by simulation for any point of operation. These values can be used directly in the reluctance model above (even concurrently in the same simulation) to determine the fluxes ϕ_1 through ϕ_3 in each respective leg. The flux density is then found by:

$$B_i = \frac{\phi_i}{A_{Fe,i}} \tag{18}$$

where the leg cross sections are: $A_{Fe,1} = A_{Fe,2} = 41.9\,\mathrm{mm}^2$ and $A_{Fe,3} = 26.5\,\mathrm{mm}^2$

For a given point of operation the following flux densities are achieved (with the transformer as specified above):

Bridge Voltage	$V_{HB} = \pm200\,\mathrm{V}$	Flux density in Leg 1	$B_1 = 59\,\mathrm{mT}$
Frequency	$f_s = 437\,\mathrm{kHz}$	Flux density in Leg 2	$B_2 = 27\,\mathrm{mT}$
Resonant capacitor	$C_r = 226\,\mathrm{pF}$	Flux density in Leg 3	$B_3 = 21\,\mathrm{mT}$
Output capacitor	$C_o = 100\,\mathrm{uF}$		
Load resistance	$R_o = 103\,\Omega$		

Table 1: Results form simulated reluctance model

These values correspond well to the values from the FEA simulation under the same conditions.

4. Pre-Series Protoype

Several prototypes have been build including fully functional pre-series prototypes featuring all necessary requirements [15], such as PFC and dimming interface. Figure 10(b) shows an example of a prototype for an input voltage range of 380 V to 420 V and the output specifications mentioned above. It is packaged into a case with a height of 21 mm and is thermally stable over the entire area of operation.

Figure 9 shows the efficiency of the LLC converter plotted for several output voltages. The LLC reaches almost 95 % efficiency. Although there is still some potential for improvement. As aforementioned the ferrite losses are too high. They can be reduced by increasing the cross-sectional-area of the core while reducing its volume with a thinner covering plate (I-shaped core).

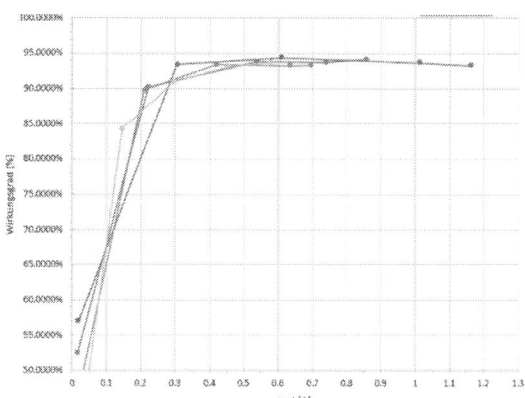

Figure 9: LLC efficiency for several different output voltages

It is a trade-off between the design optimization (mainly reducing conducted losses) mentioned in section 2 and the magnetic optimization (reducing size and ferrite losses).

(a) Different custom made magnetics

(b) Pre-series protoype of the whole LED driver

Figure 10: Protoype of the magnetics and the LED driver

5. Conclusion

The paper presents an optimized design for a fully functional pre-series LED driver protoype. Using two cost functions, an optimal design has been calculated with the focus on reducing the size and increasing the power density while still reaching an LLC efficiency of almost 95 % at 35 W. An optimized integrated magnetic design including reluctance model, FEA simulation and prototype is presented. As a result, a wide output voltage range (30 V to 60 V) fully dimmable (1 % to 100 %) high efficient and small (1.8 W/cm^3) LLC converter has been built and is presented in the paper.

© VDE VERLAG GMBH · Berlin · Offenbach

6. References

[1] Bo Yang, F. C. Lee, A. J. Zhang, and Guisong Huang. Llc resonant converter for front end dc/dc conversion. In *Applied Power Electronics Conference and Exposition, 2002. APEC 2002. Seventeenth Annual IEEE*, volume 2, pages 1108–1112 vol.2, 2002.

[2] Ching-Shan Leu, Pin-Yu Huang, and Wei-Kai Wang. Llc converter with taiwan tech center-tapped rectifier (llc-tct) for solar power conversion applications. In *Future Energy Electronics Conference (IFEEC), 2013 1st International*, pages 515–519, Nov 2013.

[3] G. Yang, P. Dubus, and D. Sadarnac. Double-phase high-efficiency, wide load range high-voltage/low-voltage llc dc/dc converter for electric/hybrid vehicles. *IEEE Transactions on Power Electronics*, 30(4):1876–1886, April 2015.

[4] H. Wu, S. Ji, F. C. Lee, and X. Wu. Multi-channel constant current (mc3) llc resonant led driver. In *Energy Conversion Congress and Exposition (ECCE), 2011 IEEE*, pages 2568–2575, Sept 2011.

[5] R. L. Steigerwald. A comparison of half-bridge resonant converter topologies. *IEEE Transactions on Power Electronics*, 3(2):174–182, Apr 1988.

[6] Hong Huang. Designing an llc resonant half-bridge power converter. *TI Power Supply Design Seminar*, 2010.

[7] Brent McDonald and Dave Freeman. Design and optimization of a high-performance llc converter. *TI Power Supply Design Seminar*, 2013.

[8] Bing Lu, Wenduo Liu, Yan Liang, F. C. Lee, and J. D. van Wyk. Optimal design methodology for llc resonant converter. In *Applied Power Electronics Conference and Exposition, 2006. APEC '06. Twenty-First Annual IEEE*, pages 6 pp.–, March 2006.

[9] H. Choi. Analysis and design of llc resonant converter with integrated transformer. In *Applied Power Electronics Conference, APEC 2007 - Twenty Second Annual IEEE*, pages 1630–1635, Feb 2007.

[10] J. Zhang, W. G. Hurley, W. H. Wolfle, and M. C. Duffy. Optimized design of llc resonant converters incorporating planar magnetics. In *Applied Power Electronics Conference and Exposition (APEC), 2013 Twenty-Eighth Annual IEEE*, pages 1683–1688, March 2013.

[11] G. Spiazzi and S. Buso. Effect of a split transformer leakage inductance in the llc converter with integrated magnetics. In *Power Electronics Conference (COBEP), 2013 Brazilian*, pages 135–140, Oct 2013.

[12] Q. Li, M. Lim, J. Sun, A. Ball, Y. Ying, F. C. Lee, and K. D. T. Ngo. Technology road map for high frequency integrated dc-dc converter. In *Applied Power Electronics Conference and Exposition (APEC), 2010 Twenty-Fifth Annual IEEE*, pages 533–539, Feb 2010.

[13] Seung-Hee Ryu and Byoung-Kuk Lee. Highly accurate analysis method for llc dc-dc converters with an improved transformer circuit model. *Electronics Letters*, 51(12):928–930, 2015.

[14] R. Yu, G. K. Y. Ho, B. M. H. Pong, B. W. K. Ling, and J. Lam. Computer-aided design and optimization of high-efficiency llc series resonant converter. *IEEE Transactions on Power Electronics*, 27(7):3243–3256, July 2012.

[15] G. Sauerlander, D. Hente, H. Radermacher, E. Waffenschmidt, and J. Jacobs. Driver electronics for leds. In *Industry Applications Conference, 2006. 41st IAS Annual Meeting. Conference Record of the 2006 IEEE*, volume 5, pages 2621–2626, Oct 2006.

PCIM Europe 2016, 10 – 12 May 2016, Nuremberg, Germany

Computationally efficient Anisotropy-Identification based on a Square-Shaped Injection Pattern

P. Landsmann, D. Paulus, S. Kühl and R. Kennel
Technische Universität München, Germany

Abstract

This paper proposes a discrete-time anisotropy-based sensorless control approach that compared to literature causes only a fraction of the computational load. While all known methods consist of two computation stages with vector transformations and/or matrix inversions, the proposed method identifies the anisotropy in a single stage with only 6 subtractions. The basis for this simple evaluation is a square-shaped injection pattern and a current control cycle of two PWM periods. The paper provides the mathematical derivation, a comparison to other methods and the experimental validation of the proposed approach.

1. Introduction

Today's highly efficient methods for driving electrical motors presume that the rotor position is known at all times. The position is normally measured during motor operation by means of a position sensor, mounted at the back end of the shaft. While allowing for efficient drive control, position sensors entail several drawbacks, such as increased system cost, reduced robustness, higher fault probability and increased construction space. These drawbacks explain the strong industrial demand to gain the position signal without using a position sensor. Methods realizing this are referred to as „sensorless control methods" and can be distinguished into two classes: 1. Fundamental model (FM) based methods evaluate the voltage induced by rotor movement and do, hence, fail at low and zero speed. 2. Anisotropy based methods evaluate the position dependence of the inductance from the current response to a high frequency voltage signal (so called "injection"), for which rotor speed is not necessary. Anisotropy based methods are still barely present in industry, which is inter alia explained by their higher complexity and computational effort compared to FM based methods. In order to meet the industrial requirements, Anisotropy based methods must be comprehensible for an engineer without extensive research and implementable on a micro controller with already high computational loading.

Anisotropy based methods can be further distinguished into *modulation based methods* [1] [2] (which entail several drawback and are, hence, not further discussed) and *discrete-time methods* [3] [4] [5] [6]. All discrete-time methods in literature are composed of two stages: In the first stage the inductive part of the current response is segregated by means of a few subtractions which only cause little computational effort. In the second stage the anisotropy of the inductance is resolved, which is realized differently in each method from literature: The most efficient methods [4] and [6] require in the best case inter alia 8 multiplications, the least efficient [3] and [5] additionally require up to 2 computationally expensive operations (divisions or angle calculation). Since these operations have to be executed thousands of times per second, anisotropy based methods for low speed sensorless control increase the computational load of a micro controller significantly.

This paper proposes a new method for anisotropy identification, referred to as Square-Injection, that presumes an injection pattern in which the injected voltage pulses are aligned with the axes of the stator fixed reference frame. Consequently, the evaluation equations can be simplified extensively, resulting in only a single computation stage wherein the anisotropy vector is calculated directly from the current response by a simple difference equation.

Firstly in Section 2 the anisotropy identification in literature is reviewed and analysed. Based on these insights the idea of Square-Injection is introduced and derived in Section 3 and experimentally validated in Section 4.

© VDE VERLAG GMBH · Berlin · Offenbach

518

2. Anisotropy Identification

Discrete-time methods for anisotropy identification rely on a discrete-time version of the machine voltage equation which is now given exemplarily for a permanent magnet (PM) synchronous machine

$$u_s^s = R_s i_s^s + \mathbf{L}_s^s \frac{\Delta i_s^s}{T_s} + \mathbf{J}\omega\psi_{pm}^s. \tag{1}$$

Herein u_s^s is the voltage (u) at the stator winding (subscript s) expressed in stator fixed reference frame (superscript s, i.e. a vector with two dimensions), R_s the resistance of the stator winding, i_s^s the stator current in stator frame and ψ_{pm}^s the PM-flux linkage in stator frame, and T_s is the sampling time. $\mathbf{L}_s^s(\theta)$ is the inductance of the stator winding in stator frame and because of its position dependent anisotropy it forms the source of information at low speed.

In order to identify the position dependence of the inductance, we analyse the current response to a high frequency voltage excitation (so called injection) by rewriting the voltage equation (1) for the current response

$$\Delta i_s^s = \mathbf{Y}_s^s \left(u_s^s - R_s i_s^s - \mathbf{J}\omega\psi_{pm}^s \right). \tag{2}$$

Instead of the inductance, there is now $\mathbf{Y}_s^s$, the product of its inverse times the sampling time T_s which is often referred to as "admittance". Like $\mathbf{L}_s^s$, $\mathbf{Y}_s^s$ is a 2x2 matrix with three variable entries (energy conservation demands equal mutual entries)

$$\mathbf{Y}_s^s(\theta) = \mathbf{L}_s^{s-1}(\theta)\, T_s = \begin{bmatrix} Y_{\alpha\alpha}(\theta) & Y_{\alpha\beta}(\theta) \\ Y_{\alpha\beta}(\theta) & Y_{\beta\beta}(\theta) \end{bmatrix} \tag{3}$$

This admittance $\mathbf{Y}_s^s$ shows a position dependence analogous to $\mathbf{L}_s^s$ and can therefore be employed for the estimation of the rotor position θ.

The magnetic anisotropy is comprised in the admittance matrix $\mathbf{Y}_s^s$ and can be segregated by separating its isotropic component $Y_\Sigma \mathbf{I}$

$$\mathbf{Y}_s^s = Y_\Sigma(\theta)\mathbf{I} + [\mathbf{y}_\Delta^s(\theta) \quad -\mathbf{J}\mathbf{y}_\Delta^s(\theta)] = \begin{bmatrix} Y_\Sigma + Y_{\Delta\alpha\alpha} & Y_{\Delta\alpha\beta} \\ Y_{\Delta\alpha\beta} & Y_\Sigma - Y_{\Delta\alpha\alpha} \end{bmatrix}. \tag{4}$$

$Y_\Sigma = \frac{T_s}{2}\left(\frac{1}{L_D} + \frac{1}{L_Q}\right)$ is the mean admittance and $\mathbf{y}_\Delta^s = [Y_{\Delta\alpha\alpha} \quad Y_{\Delta\alpha\beta}]^T$ is the anisotropy vector. In this representation, the three former variables $Y_{\alpha\alpha}$, $Y_{\alpha\beta}$ and $Y_{\beta\beta}$ are now transformed to Y_Σ, $Y_{\Delta\alpha\alpha}$ and $Y_{\Delta\alpha\beta}$, where for most machines the position dependence of Y_Σ is significantly weaker than that of $Y_{\Delta\alpha\alpha}$ and $Y_{\Delta\alpha\beta}$. This is why normally only the anisotropy is considered for position estimation.

The most simple position dependence (ideal, sinusoidal machine) is a circular rotation of the anisotropy vector with twice the electrical rotor angle

$$\mathbf{y}_\Delta^s = Y_\Delta \begin{bmatrix} \cos 2\theta \\ \sin 2\theta \end{bmatrix} \tag{5}$$

$$\mathbf{Y}_s^s = Y_\Sigma \begin{bmatrix} 1 & 0 \\ 0 & 1 \end{bmatrix} + Y_\Delta \begin{bmatrix} \cos 2\theta & \sin 2\theta \\ \sin 2\theta & -\cos 2\theta \end{bmatrix}, \tag{6}$$

where $Y_\Delta = \sqrt{Y_{\Delta\alpha\alpha}^2 + Y_{\Delta\alpha\beta}^2} = \frac{T_s}{2}\left(\frac{1}{L_D} - \frac{1}{L_Q}\right)$ is the anisotropy magnitude and the rotor angle can be obtained simply as half the angle of the anisotropy vector. For non-sinusoidal machines a more complex assignment of the rotor angle is necessary [7] [8].

However, in any case the goal of the anisotropy identification itself and, hence, the primary algorithmic challenge of anisotropy based methods in literature is to resolve the discrete-time voltage equation (2) such that the anisotropy vector $\mathbf{y}_\Delta^s$ is obtained. All methods in literature accomplish this by means of two computation stages:

2.1. Stage I: Segregating the Admittance

In order to analyse the admittance, all other terms of the voltage equation (2) must be eliminated. For this, methods from literature [3] [4] [5] [6] employ the assumption that these terms do only change to a negligible extent during the injection interval and can, hence, be eliminated by subtracting the current response of two consecutive intervals

$$\Delta^2 i_s^s = \Delta i_s^s[k] - \Delta i_s^s[k-1] \tag{7}$$

$$= Y_s^s u_s^s[k] - Y_s^s (R_s i_s^s + J\omega\psi_{pm}^s) - Y_s^s u_s^s[k-1] + Y_s^s (R_s i_s^s + J\omega\psi_{pm}^s) \tag{8}$$

$$= Y_s^s (u_s^s[k] - u_s^s[k-1]) = Y_s^s \Delta u_s^s. \tag{9}$$

As a result of this first stage we gain the direct relation between injection and injection response, solely described by the anisotropic admittance Y_s^s.

2.2. Stage II: Resolution of the Admittance

This relation described by the admittance Y_s^s is the basis for the second stage. According to (3) and (4), Y_s^s comprises three unknowns, which makes the two-dimensional vector equation (9) underdetermined. Hence, the admittance Y_s^s can only be resolved by employing at least two current responses. The particular solution of this problem is what distinguishes the methods of literature:

INFORM

In the so called „Indirect Flux detection by on-line Reactance Measurement" (INFORM) [9] [4] three voltage pulses aligned with the three phase directions form the voltage excitation Δu_s^s. In order to resolve the admittance, each current response is regarded in phase direction (i.e. injection direction) and separated into a parallel and an orthogonal component. The summation of all parallel component vectors in stator frame (and/or all orthogonal component vectors as well) yields the anisotropy vector $y_\Delta^s(\theta)$.

The computational effort of INFORM is generated mainly by the 8 to 24 multiplications required for the transformation of the current responses into voltage frame and back.

Thales' Theorem

Another discrete-time approach [5] [10] utilizes Thales's theorem to deduce the anisotropy angle $\theta_a = \arg y_a^s$ from two consecutive current responses in voltage in voltage frame. In contrast to INFORM, this approach allows to evaluate arbitrary voltage combinations, but does additionally require the voltage angle θ_u.

The computational effort of this approach is generated mainly by the transformation in an arbitrary voltage frame (variable multiplications and one division) and the additional angle calculation, which both belong to the most time consuming operations in micro controllers. Moreover, this approach only provides the anisotropy angle θ_a and is therefore not suited for certain decoupling methods [7] [8], since the anisotropy vector is not available as intermediary result.

Matrix Inversion

In [6] the underdeterminedness of (9) is solved by concatenating two consecutive voltages to matrix $U_s^s = [\Delta u_1^s \ \Delta u_2^s]$ and two current responses to matrix $I_s^s = [\Delta i_1^s \ \Delta i_2^s]$, leading to a matrix equation that is not underdetermined anymore

$$I_s^s = Y_s^s(\theta) \, U_s^s. \tag{10}$$

By using matrix inversion (10) is solved for Y_s^s

$$Y_s^s(\theta) = I_s^s \, U_s^{s-1} \tag{11}$$

the components of which reveal the anisotropy vector according to (4)

$$y_\Delta^s = \frac{1}{2} \begin{bmatrix} Y_{\alpha\alpha} - Y_{\beta\beta} \\ Y_{\alpha\beta} + Y_{\beta\alpha} \end{bmatrix}. \tag{12}$$

Regarding the computational effort, the inversion of a 2x2 matrix is just a signed swap of its components, finally scaled by its inverse determinant. In case of simple angle evaluation by $\arg \boldsymbol{y}_\Delta^s$ or in case of a fixed excitation voltage magnitude $|\Delta \boldsymbol{u}_s^s| = \text{const.}$ the final division is not necessary. In these cases the computational effort mainly consists of the 8 multiplications in (11). Hence, computation-wise Matrix Inversion is one of the most efficient methods.

Arbitrary Injection

The last discrete-time anisotropy identification method [3] is model-based, aiming to extract the rotor position from the current response to any voltage excitation (injection and/or current controller). The error $\boldsymbol{e}_{prd}^s$ between the current response of the machine and of a parameter-free isotropic model

$$\boldsymbol{e}_{prd}^s = \Delta^2 \boldsymbol{i}_s^s - Y_\Sigma \, \Delta \boldsymbol{u}_s^s = [\boldsymbol{y}_\Delta^s \quad -J\boldsymbol{y}_\Delta^s] \, \Delta \boldsymbol{u}_s^s. \tag{13}$$

is solved for the anisotropy vector

$$\boldsymbol{y}_\Delta^s = \begin{bmatrix} u_\alpha e_\alpha - u_\beta e_\beta \\ u_\alpha e_\beta + u_\beta e_\alpha \end{bmatrix} \frac{1}{u_\alpha^2 + u_\beta^2}. \tag{14}$$

The inverse voltage magnitude square in (14), that implies a time consuming division, is also required for the computation of Y_Σ. This mean admittance Y_Σ can be estimated in differntly expensive ways, which depend on the assumptions made regarding the injection pattern. In any case there must be a transformation into voltage frame and in the most expensive case a second division is required. Hence, computation-wise Arbitrary Injection belongs to the least efficient anisotropy based methods.

3. Square-Injection

According to the above state of the art, the computational effort of anisotropy based methods is caused mostly during the resolution of the admittance, since at this point voltage and current response are brought into relation. For this resolution at least a transformation into voltage frame or an inverse matrix multiplication is used.

The idea of Square Injection addresses exactly this point: The transformation into voltage frame is simplified if the underlying excitation voltages are aligned with the axes of the stator fixed reference frame. Then the transformation becomes a signed swap of the current response vector components.

But as will be shown in the following section, when strictly rewriting this simplification, the reduction in computational effort exceeds this expectation: finally the anisotropy is identified in a single computation stage directly from the current response using only 6 additions/subtractions. This is only a fraction of the computations required by state of the art methods and can, moreover, be implemented by a control engineer without having to understand anisotropy based sensorless control.

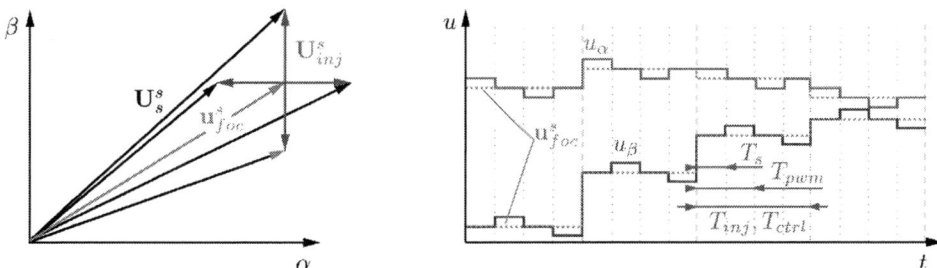

Fig. 1: Injection pattern in stator frame (left) and over time (right).

3.1. Derivation of the approach

Square-Injection presumes a voltage pattern as indicated in Fig. 1: the current controller is executed only once per injection interval T_{inj}, such that its output voltage is constant during this period. Upon this constant control voltage $\boldsymbol{u}_{foc}^s$ a square shaped injection pattern $\mathbf{U}_{inj}^s$ is superimposed

$$\mathbf{U}_{inj}^s = \begin{bmatrix} 1 & 0 & -1 & 0 \\ 0 & 1 & 0 & -1 \end{bmatrix} u_{inj} \tag{15}$$

$$\mathbf{U}_s^s = \mathbf{U}_{inj}^s + \boldsymbol{u}_{foc}^s \begin{bmatrix} 1 & 1 & 1 & 1 \end{bmatrix}, \tag{16}$$

where in style of (10) the column vectors of the voltage matrix represent voltage vectors of consecutive instants in time. Assuming the fundamental wave to be nearly constant during one injection interval (low speed), the excitation (16) results in the current response $\Delta\mathbf{I}_s^s$

$$\Delta\mathbf{I}_s^s = \mathbf{Y}_s^s \mathbf{U}_{inj}^s - \mathbf{Y}_s^s\left(\boldsymbol{u}_{foc}^s - R_s\,\boldsymbol{i}_s^s - \mathbf{J}\omega\boldsymbol{\psi}_{pm}^s\right)\begin{bmatrix} 1 & 1 & 1 & 1 \end{bmatrix} \tag{17}$$

$$= \Delta\mathbf{I}_u^s + \Delta\boldsymbol{i}_{FM}^s\begin{bmatrix} 1 & 1 & 1 & 1 \end{bmatrix} \tag{18}$$

$$= \begin{bmatrix} \Delta i_{\alpha 0} & \Delta i_{\alpha 1} & \Delta i_{\alpha 2} & \Delta i_{\alpha 3} \\ \Delta i_{\beta 0} & \Delta i_{\beta 1} & \Delta i_{\beta 2} & \Delta i_{\beta 3} \end{bmatrix}. \tag{19}$$

Since the Injection pattern is average-free, the right hand side term of (17), referred to as fundamental wave (FM), can be calculated by averaging $\Delta\mathbf{I}_s^s$

$$\Delta\boldsymbol{i}_m^s = \frac{1}{4}\Delta\mathbf{I}_s^s\begin{bmatrix} 1 & 1 & 1 & 1 \end{bmatrix}^T = \begin{bmatrix} \Delta i_{m\alpha} \\ \Delta i_{m\beta} \end{bmatrix} \approx \Delta\boldsymbol{i}_{FM}^s. \tag{20}$$

This fundamental wave is then subtracted from the total current response $\Delta\mathbf{I}_s^s$ in order to separate the response to the injection $\Delta\mathbf{I}_u^s$

$$\Delta\mathbf{I}_u^s = \Delta\mathbf{I}_s^s - \Delta\boldsymbol{i}_m^s\begin{bmatrix} 1 & 1 & 1 & 1 \end{bmatrix} \tag{21}$$

$$= \mathbf{Y}_s^s\mathbf{U}_{inj}^s \tag{22}$$

$$= \begin{bmatrix} \Delta i_{u\alpha 0} & \Delta i_{u\alpha 1} & \Delta i_{u\alpha 2} & \Delta i_{u\alpha 3} \\ \Delta i_{u\beta 0} & \Delta i_{u\beta 1} & \Delta i_{u\beta 2} & \Delta i_{u\beta 3} \end{bmatrix}. \tag{23}$$

This is the direct relation between voltage and current response: stage 1 is complete, the admittance segregated.

Stage 2 benefits from the fact that all injection voltages are aligned with the stator frame axes, as the current response in voltage frame $\Delta\mathbf{I}_u^u$ is then obtained by simply swapping the components of (23)

$$\Delta\mathbf{I}_u^u = \begin{bmatrix} +\Delta i_{u\alpha 0} & +\Delta i_{u\beta 1} & -\Delta i_{u\alpha 2} & -\Delta i_{u\beta 3} \\ +\Delta i_{u\beta 0} & -\Delta i_{u\alpha 1} & -\Delta i_{u\beta 2} & +\Delta i_{u\alpha 3} \end{bmatrix}. \tag{24}$$

By averaging (24) we obtain the mean injection response in voltage frame

$$\Delta\boldsymbol{i}_\Sigma^u = \frac{1}{4}\begin{bmatrix} \Delta i_{u\alpha 0} + \Delta i_{u\beta 1} - \Delta i_{u\alpha 2} - \Delta i_{u\beta 3} \\ \Delta i_{u\beta 0} - \Delta i_{u\alpha 1} - \Delta i_{u\beta 2} + \Delta i_{u\alpha 3} \end{bmatrix} = \begin{bmatrix} \Delta i_{\Sigma x} \\ \Delta i_{\Sigma y} \end{bmatrix}. \tag{25}$$

As indicated in Fig. 2, this mean injection response in voltage frame $\Delta\boldsymbol{i}_\Sigma^u$ represents the isotropic component of the current response, around which the injection response lies symmetrically distributed on the "anisotropy circle" known from literature.

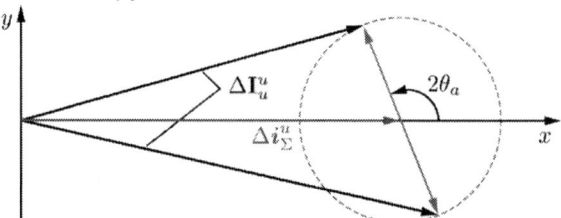

Fig. 2: Injection response in voltage frame.

Hence, a directed subtraction of the isotropic current response $\Delta\boldsymbol{i}_\Sigma^u$ from the stator frame injection response $\Delta\mathbf{I}_u^s$ will result in the anisotropic current response matrix $\Delta\mathbf{I}_\Delta^s$

$$\Delta \mathbf{I}_{\Delta}^{s} = \begin{bmatrix} \mathbf{y}_{\Delta}^{s}(\theta) & -\mathbf{J}\mathbf{y}_{\Delta}^{s}(\theta) \end{bmatrix} \mathbf{U}_{inj}^{s} \tag{26}$$

$$= \begin{bmatrix} \Delta i_{u\alpha 0} - \Delta i_{\Sigma x} & \Delta i_{u\alpha 1} + \Delta i_{\Sigma y} & \Delta i_{u\alpha 2} + \Delta i_{\Sigma x} & \Delta i_{u\alpha 3} - \Delta i_{\Sigma y} \\ \Delta i_{u\beta 0} - \Delta i_{\Sigma y} & \Delta i_{u\beta 1} - \Delta i_{\Sigma x} & \Delta i_{u\beta 2} + \Delta i_{\Sigma y} & \Delta i_{u\beta 3} + \Delta i_{\Sigma x} \end{bmatrix} \tag{27}$$

$$= \begin{bmatrix} +Y_{\Delta\alpha\alpha} & +Y_{\Delta\alpha\beta} & -Y_{\Delta\alpha\alpha} & -Y_{\Delta\alpha\beta} \\ +Y_{\Delta\alpha\beta} & -Y_{\Delta\alpha\alpha} & -Y_{\Delta\alpha\beta} & +Y_{\Delta\alpha\alpha} \end{bmatrix} u_{inj} \tag{28}$$

$$= \begin{bmatrix} \Delta i_{\alpha\alpha 0} & \Delta i_{\alpha\alpha 1} & \Delta i_{\alpha\alpha 2} & \Delta i_{\alpha\alpha 3} \\ \Delta i_{\alpha\beta 0} & \Delta i_{\alpha\beta 1} & \Delta i_{\alpha\beta 2} & \Delta i_{\alpha\beta 3} \end{bmatrix}. \tag{29}$$

From this matrix $\Delta \mathbf{I}_{\Delta}^{s}$ the targeted anisotropy vector is obtained with optimal SNR by adding/subtracting the current response of all matching entries of (28) respectively

$$4\, u_{inj}\, \mathbf{y}_{\Delta}^{s} = \begin{bmatrix} \Delta i_{\alpha\alpha 0} - \Delta i_{\alpha\beta 1} - \Delta i_{\alpha\alpha 2} + \Delta i_{\alpha\beta 3} \\ \Delta i_{\alpha\beta 0} + \Delta i_{\alpha\alpha 1} - \Delta i_{\alpha\beta 2} - \Delta i_{\alpha\alpha 3} \end{bmatrix} \tag{30}$$

In case of simple angle assignment by $\arg \mathbf{y}_{\Delta}^{s}$ the vector (30) does not need to be scaled. In case of a subsequent decoupling method [7] [8] the vector (30) may have to be scaled by the (possibly constant) factor $\frac{1}{4u_{inj}}$ in order to gain $\mathbf{y}_{\Delta}^{s}$ with correct magnitude.

3.2. Simplification

So far the anisotropy vector is obtained using additions, subtractions and multiplications with ¼ (fixed point bit shift). However, the computation can be simplified significantly when rewriting the terms $\Delta i_{\alpha xx}$ and Δi_{uxx} according to their definition.

Firstly, $\Delta i_{\alpha xx}$ in (30) is rewritten using its definition in (26)

$$4\, u_{inj}\, \mathbf{y}_{\Delta}^{s} = \begin{bmatrix} (\Delta i_{u\alpha 0} - \Delta i_{\Sigma x}) - (\Delta i_{u\beta 1} - \Delta i_{\Sigma x}) - (\Delta i_{u\alpha 2} + \Delta i_{\Sigma x}) + (\Delta i_{u\beta 3} + \Delta i_{\Sigma x}) \\ (\Delta i_{u\beta 0} - \Delta i_{\Sigma y}) + (\Delta i_{u\alpha 1} + \Delta i_{\Sigma y}) - (\Delta i_{u\beta 2} + \Delta i_{\Sigma y}) - (\Delta i_{u\alpha 3} - \Delta i_{\Sigma y}) \end{bmatrix} \tag{31}$$

$$= \begin{bmatrix} \Delta i_{u\alpha 0} - \Delta i_{u\beta 1} - \Delta i_{u\alpha 2} + \Delta i_{u\beta 3} \\ \Delta i_{u\beta 0} + \Delta i_{u\alpha 1} - \Delta i_{u\beta 2} - \Delta i_{u\alpha 3} \end{bmatrix} - \begin{bmatrix} \Delta i_{\Sigma x} - \Delta i_{\Sigma x} + \Delta i_{\Sigma x} - \Delta i_{\Sigma x} \\ \Delta i_{\Sigma y} - \Delta i_{\Sigma y} + \Delta i_{\Sigma y} - \Delta i_{\Sigma y} \end{bmatrix} \tag{32}$$

$$= \begin{bmatrix} \Delta i_{u\alpha 0} - \Delta i_{u\beta 1} - \Delta i_{u\alpha 2} + \Delta i_{u\beta 3} \\ \Delta i_{u\beta 0} + \Delta i_{u\alpha 1} - \Delta i_{u\beta 2} - \Delta i_{u\alpha 3} \end{bmatrix}. \tag{33}$$

Hence, the isotropic component in the injection response is eliminated implicitly, such that calculating the isotropic component $\Delta \boldsymbol{i}_{\Sigma}^{u}$ is not necessary and one can work directly with the (EMF free) injection response $\Delta \mathbf{I}_{u}^{s}$.

Secondly, the terms Δi_{uxx} in (33) are rewritten using their definition in (21)

$$4\, u_{inj}\, \mathbf{y}_{\Delta}^{s} = \begin{bmatrix} (\Delta i_{\alpha 0} - \Delta i_{m\alpha}) - (\Delta i_{\beta 1} - \Delta i_{m\beta}) - (\Delta i_{\alpha 2} - \Delta i_{m\alpha}) + (\Delta i_{\beta 3} - \Delta i_{m\beta}) \\ (\Delta i_{\beta 0} - \Delta i_{m\beta}) + (\Delta i_{\alpha 1} - \Delta i_{m\alpha}) - (\Delta i_{\beta 2} - \Delta i_{m\beta}) - (\Delta i_{\alpha 3} - \Delta i_{m\alpha}) \end{bmatrix} \tag{34}$$

$$= \begin{bmatrix} \Delta i_{\alpha 0} - \Delta i_{\beta 1} - \Delta i_{\alpha 2} + \Delta i_{\beta 3} \\ \Delta i_{\beta 0} + \Delta i_{\alpha 1} - \Delta i_{\beta 2} - \Delta i_{\alpha 3} \end{bmatrix} - \begin{bmatrix} \Delta i_{m\alpha} - \Delta i_{m\beta} - \Delta i_{m\alpha} + \Delta i_{m\beta} \\ \Delta i_{m\beta} + \Delta i_{m\alpha} - \Delta i_{m\beta} - \Delta i_{m\alpha} \end{bmatrix} \tag{35}$$

$$= \begin{bmatrix} \Delta i_{\alpha 0} - \Delta i_{\beta 1} - \Delta i_{\alpha 2} + \Delta i_{\beta 3} \\ \Delta i_{\beta 0} + \Delta i_{\alpha 1} - \Delta i_{\beta 2} - \Delta i_{\alpha 3} \end{bmatrix}. \tag{36}$$

Hence, using (36) with a square shaped injection pattern, the anisotropy vector $\mathbf{y}_{\Delta}^{s}$ (scaled by $4\, u_{inj}$) can be calculated directly from the current response – without eliminating the EMF and the isotropic component. I.e. in contrast to all methods in literature, with (36) the anisotropy is identified in a single computation stage. For simple anisotropy angle evaluation the overall necessary code is

$$\theta_a = \frac{1}{2} \operatorname{atan} \left(\frac{\Delta i_{\beta 0} + \Delta i_{\alpha 1} - \Delta i_{\beta 2} - \Delta i_{\alpha 3}}{\Delta i_{\alpha 0} - \Delta i_{\beta 1} - \Delta i_{\alpha 2} + \Delta i_{\beta 3}} \right). \tag{37}$$

4. Experimental Results

The Square-Injection method has been validated at a test bench with a DSP-based real time system (left), a 7.5kW inverter (middle) and a 1.6kW permanent magnet synchronous machine (right), indicated in Fig. 3.

 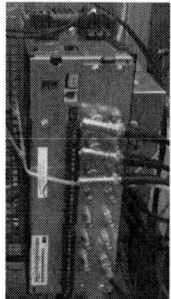 

Fig. 3: Test bench setup.

The switching frequency of the inverter is 8kHz, resulting in a maximum voltage actuation- and synchronous current measurement frequency of 16kHz. Hence, with Square-Injection the injection and current control frequency is 4kHz. The d- and q-axis inductances of the machine are $L_d = 8.3$mH and $L_q = 7.1$mH, resulting in an anisotropy ratio of only $\frac{L_q-L_d}{L_q+L_d} \approx 8\%$ - a relatively low value that makes the application of anisotropy based methods challenging.

The following experimental results have been carried out with an injection magnitude of 11V (at 560V DC-link voltage). Fig. 4 shows the measured electrical rotor angle (Ch1) and the α- and β-components of the (low pass filtered[1]) difference vector $4\,u_{inj}\,y_\Delta^s$ during closed-loop[2] low speed sensorless operation (Ch2 & Ch3).

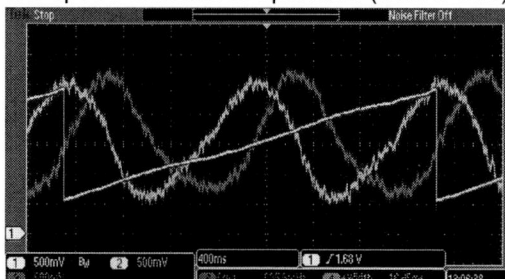 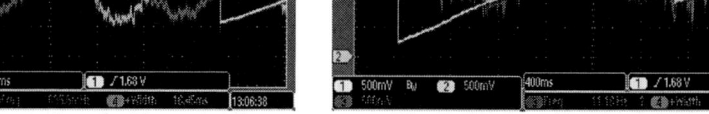

Fig. 4: Anisotropy vector y_Δ^s at low speed.   Fig. 5: Angle estimation at low speed.

For this relatively sinusoidal machine the difference vector can be evaluated simple by atan2(), the result of which is shown by the blue graph (Ch2) in Fig. 5. The purple graph (Ch3) shows the estimation error, i.e. the difference between the estimated and the measured angle with 5°/div (electrical). This error has a usual width of noise which could be reduced by increasing the injection magnitude (if necessary). Moreover the error contains a 2nd harmonic which is the result of a stator fixed anisotropy, and a slight 6th harmonic that can be found in most machines and is the result of geometrical effects (stator teeth, rotor pole shape, winding distribution etc.).

All in all the error shows properties that are usual for this type of machine and remains within a relatively narrow band.

---

[1] Due to the current measurement noise, all anisotropy based methods require some digital filtering for noise suppression. As the filtering does not depend on the method, it is has been excluded from the comparison of the computational effort. The noise shown in the figures results from a first order low pass filter with 125/s bandwidth.

[2] The estimated rotor angle has been used for the transformation of the field oriented current control and the rotor angle time derivative has been employed as a feedback signal for the speed controller.

5. Conclusion

„Square-Injection" has been introduced as a new method for the anisotropy identification of three phase alternating current machines. In contrast to all other methods in literature the anisotropy identification is realized in a single computation stage consisting of only 6 subtractions. Square-Injection does therefore only lead to a fraction of the computational effort of all other methods and can be implemented by a control engineer that is not familiar with the theory of anisotropy based sensorless control.

References

[1] P. L. Jansen und R. D. Lorenz, „Transducerless Position and Velocity Estimation in Induction and Salient AC Machines," *IEEE Trans. on Industrial Applications,* Bd. 31, pp. 240-247, 1995.

[2] M. J. Corley und R. D. Lorenz, „Rotor Position and Velocity Estimation for a Salient-Pole Permanent Magnet Synchronous Machine at Standstill and High Speeds," *IEEE Trans. on Industrial Applications,* Bd. 34, pp. 784-789, 1998.

[3] D. Paulus, P. Landsmann und R. Kennel, „Sensorless Field- oriented Control for Permanent Magnet Synchronous Machines with an Arbitrary Injection Scheme and Direct Angle Calculation," *IEEE Conf. SLED,* pp. 41-46, 2011.

[4] M. Schroedl, „Operation of the permanent magnet synchronous machine without a mechanical sensor," *Conf. Power Electronics and Variable-Speed Drives,* pp. 51-56, 1990.

[5] F. De Belie, T. Vyncke und J. Melkebeek, „Parameterless Rotor Position Estimation in a Direct-Torque Controlled Salient-Pole PMSM without Using Additional Test Signal," *IEEE Conf. ICEM,* pp. 1-6, 2010.

[6] S. Kim, Y.-C. Kwon, S.-K. Sul, J. Park und S.-M. Kim, „Position Sensorless Operation of IPMSM with Near PWM Switching Frequency Signal Injection," *IEEE Conf. ICPE - ECCE Asia,* pp. 1660-1665, 2011.

[7] R. Lorenz und M. Degner, „Using multiple saliencies for the estimation of flux, position, and velocity in ac machines," *IEEE Transactions on Industry Applications,* p. 1097–1104, 1998.

[8] D. Paulus, P. Landsmann, S. Kuehl und R. Kennel, „Arbitrary injection for permanent magnet synchronous machines with multiple saliencies," *IEEE Conf. ECCE,* pp. 511-517, 2013.

[9] M. Schroedl, „Verfahren und schaltungsanordnungen zur bestimmung maschinenbezogener elektromagnetischer und mechanischer zustandsgrössen an über umrichter gespeisten elektrodydynamischen drehfeldmaschinen". DE,EU,WO Patent DE59204692, 8 Apr 1992.

[10] J. M. Frederik De Belie, „Sensorless control of salient-pole machines". US,EP,WO Patent US20100327789, 13 02 2009.

[11] P. Landsmann, D. Paulus, P. Stolze und R. Kennel, „Saliency based encoderless predictive torque control without signal injection," *IEEE Conf. IPEC,* pp. 3029-3034, 2010.

[12] P. Landsmann, D. Paulus, A. Doetlinger und R. Kennel, „Silent Injection for Saliency based Sensorless Control by means of Current Oversampling," *IEEE Conf. ICIT,* pp. 398-403, 2013.

[13] P. Landsmann, D. Paulus, S. Kuehl und R. Kennel, „Dynamische geberlose Regelung im gesamten Drehzahlbereich für PM-Synchronmaschinen," *Conf. SPS Drives,* p. 229, 2013.

[14] P. Landsmann, D. Paulus und R. Kennel, „Online identification of load angle compensation for anisotropy based sensorless control," *IEEE Conf. SLED,* pp. 80-84, 2011.

Estimation of the Excitation Current and the Rotor Resistance of an Externally Excited Synchronous Machine with an Inductively Supplied Excitation Coil

Köhler, Stefan, TH Nürnberg GSO, Germany, stefan.koehler@th-nuernberg.de

Wagner, Bernhard, TH Nürnberg GSO, Germany, bernhard.wagner@th-nuernberg.de

Endres, Stefan, Fraunhofer IISB, Germany, stefan.endres@iisb.fraunhofer.de

The Power Point Presentation will be available after the conference.

Abstract

A nonlinear observer for an externally excited synchronous machine (EESM) with an inductively supplied excitation coil is presented, which estimates the excitation current and the rotor resistance. This observer is designed with the direct method of Lyapunov. Using the developed estimation algorithm a precision of over 96 % can be achieved.

1   Introduction

The progress of the energy turnaround impacts all areas of technology. Thereby, the automotive industry is particularly affected with regard to the topic electric mobility. The efficiency plays a major role for current and future technologies. Unlike the asynchronous machine and the permanently excited synchronous machine the externally excited synchronous machine (EESM) achieves a high to very high efficiency for the whole speed and torque range [1]. Nevertheless, it is still used rarely. One major disadvantage of the standard design is the mechanical slip ring, that supplies the rotor with excitation energy: mechanical wear can cause insulation problems and the motor design needs to support the replacement of the slip ring brushes. To overcome these disadvantages, an inductive energy transfer system (IETS) was designed

This publication presents a nonlinear observer for the rotor of the EESM as described in [2] with an IETS for the rotor excitation energy as presented in [3]. The investigations are done with a standard EESM with the following parameters:

- Max. excitation current: $I_{exc,max}$ = 16 A
- Rotor resistance range: R_{exc} = 7,5 Ω…11,5 Ω (over temperature)
- Max. excitation power: $P_{exc,max}$ = 2000 W
- Rotor inductance: L_{exc} = 0,3 H…0,8 H (under saturation)
- DC input voltage range: U_{in} = 250 V…400 V.

This estimation algorithm is important, because there is a huge effort to take measurements on the rotating rotor. The excitation current estimation is necessary for control purposes, whereas the estimation of the rotor resistance is needed to protect the EESM against overheating.

2   Inductive Energy Transfer System

2.1   Structure of the Inductive Energy Transfer System

Figure 1 shows the IETS consisting of a buck converter (BC) followed by a full bridge CLL resonant converter (FB CLL RC). Therefore, the IETS is a two stage variant, which operates as follows: The FB CLL RC operates on its upper resonance frequency (1:1 voltage transfer) for a minimization of the switching losses [4], whereas the BC adopts the voltage control.

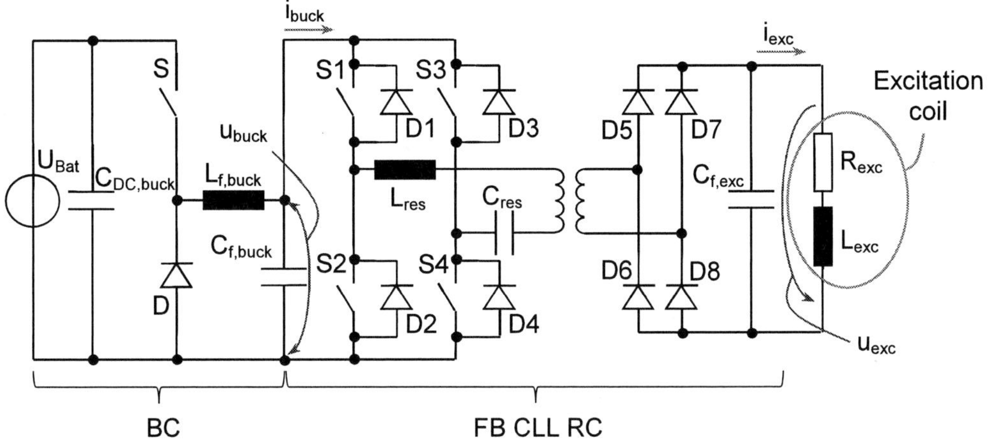

Figure 1: Equivalent circuit of the IETS consisting of a FB CLL RC with a BC as preliminary stage

2.2 Reason for the chosen Two Stage Variant

This two stage variant is chosen, because a FB CLL RC as single stage converter cannot cover the wide input and output voltage range (excitation voltage range) of 250 V...400 V as well as 7,5 V...120 V (R_{exc} = 7,5 Ω). The small negative slope of the voltage frequency response function of the FB CLL RC for high switching frequencies prevents the required reduction of the excitation voltage. Table I comprised diverse other concepts (without the chosen two stage variant) in conjunction with their exclusion criteria.

Table I: Other concepts for the realization of the IETS and their exclusion criteria

Concept	Exclusion Criteria
Phase Shift Converter	- Huge voltage overshoot on the secondary side - Snubber required → Additional components on the rotor (resistor and capacitor, preferred variant: only an additional capacitor is needed) → Efficiency decreases
Huge additional resonant inductance as described in [5]	- Significantly larger build space required
Operation concepts: CLL RC with phase shift operation, CLL RC with burst mode	- Compared to the two stage variant: Lower efficiency over the whole operational range (cf. Figure 2)

Figure 2 shows the comparison of the efficiency between the CLL RC with burst mode, the CLL RC with phase shift operation and the CLL RC with BC. As mentioned in Table I the CLL RC with BC achieves the best efficiency over the whole operational range. The maximum reached efficiency regarding to the preferred converter concept amounts 93 %, whereas the maximum efficiency of the other both concepts remains under 92 %.

PCIM Europe 2016, 10 – 12 May 2016, Nuremberg, Germany

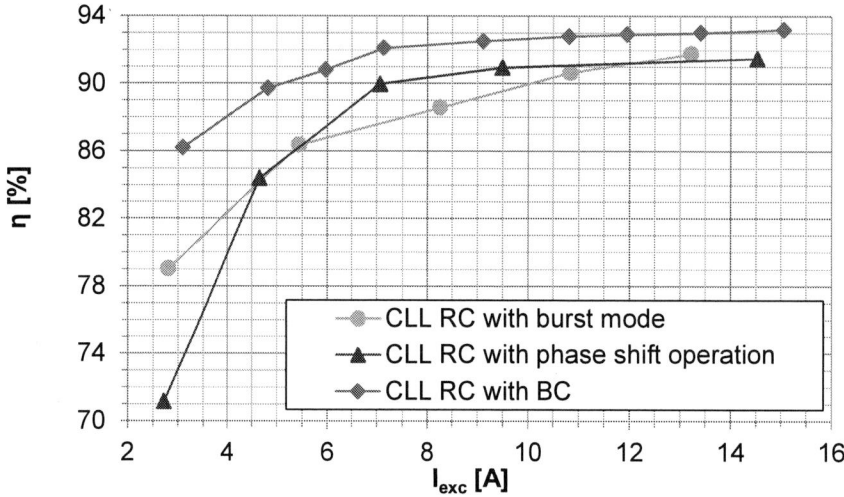

Figure 2: Comparison of the efficiency

2.3 Rotary Transformer

In Figure 3 a principle drawing, the manufactured ferrite cores (left side: secondary ferrite core, right side: primary ferrite core) and a FEM analysis of the rotary transformer is shown. The detailed description regarding the choice of the core geometry, the design and the measurement results are composed in [3].

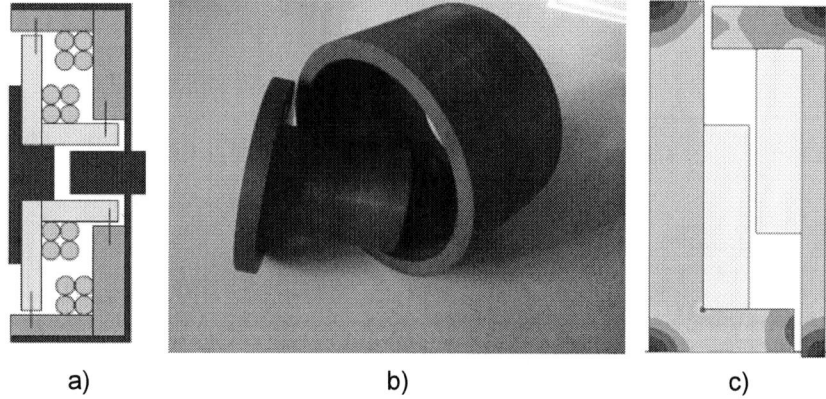

Figure 3: a) Principle drawing b) Manufactured ferrite cores c) FEM analysis of the rotary transformer

3 Lyapunov-based nonlinear observer

3.1 Structure of the Lyapunov-based nonlinear observer

The direct method of Lyapunov is used for the design of the nonlinear observer to ensure a stable estimation [6]. Figure 4 shows the simplified structure of the Lyapunov-based nonlinear observer. Its input signals are the output voltage u_{buck} and the output current i_{buck} of the BC. The output signals are the estimated excitation current $\hat{i}_{exc}$ and $\hat{R}_{exc}$.

© VDE VERLAG GMBH · Berlin · Offenbach

528

PCIM Europe 2016, 10 – 12 May 2016, Nuremberg, Germany

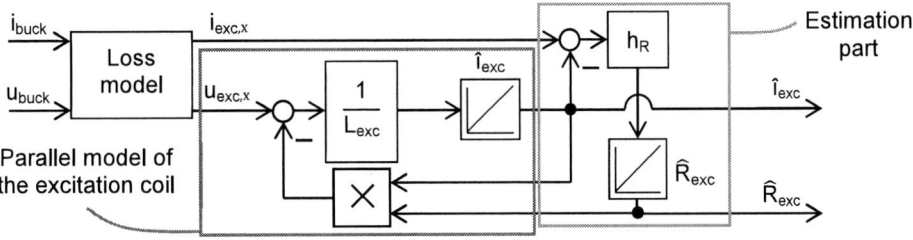

Figure 4: Simplified structure of the Lyapunov-based nonlinear observer

3.2 Design of the Nonlinear Observer with the Direct Method of Lyapunov

Following there is a detailed description for the design of a Lyapunov-based nonlinear observer for the estimation of the excitation current and the rotor resistance. The program flow chart, shown in Figure 5, Figure 6 and Figure 7, reflects a structured summary of this procedure.

The first step is to set up the state equations of the excitation coil (ohmic-inductive load). For the real and the estimated excitation current it follows

$$\frac{di_{exc}}{dt}=\frac{1}{L_{exc}(i_{exc})+\dfrac{dL_{exc}(i_{exc})}{di_{exc}}\cdot i_{exc}}\cdot(u_{exc}-R_{exc}\cdot i_{exc}) \quad (1)$$

and

$$\frac{d\hat{i}_{exc}}{dt}=\frac{1}{L_{exc}(\hat{i}_{exc})+\dfrac{dL_{exc}(\hat{i}_{exc})}{d\hat{i}_{exc}}\cdot\hat{i}_{exc}}\cdot(u_{exc}-\hat{R}_{exc}\cdot\hat{i}_{exc}). \quad (2)$$

It is assumed that the denominators of equation (1) and (2) are equal, because of the chosen control method. Here, the exact linearization and the trajectory planning are applied [7]. Therefore, the real excitation current and the estimated excitation current will follow the reference trajectory in a similar way. Consequently, the estimated excitation current is always near the real excitation current. To simplify the notation of the following equations the denominator of (1) and (2) is now written as f_L. The rotor resistance R_{exc} has to be defined as state variable, whereby its estimation gets possible. Since the rotor resistance changes very slowly, a simply integrator model for the real and the estimated rotor resistance is chosen. Thus, it follows

$$\frac{dR_{exc}}{dt}=0 \quad (3)$$

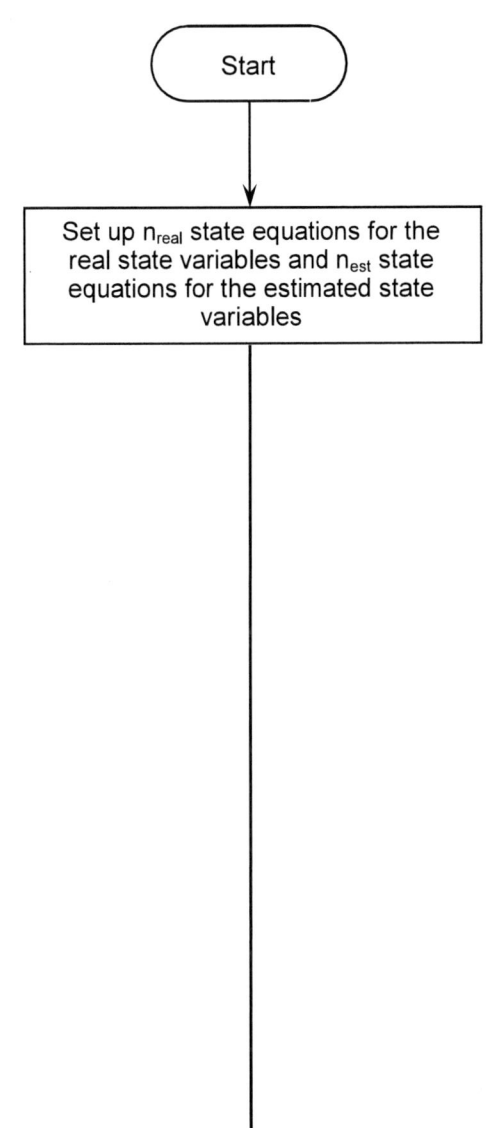

Figure 5: Program chart flow (Part 1)

© VDE VERLAG GMBH · Berlin · Offenbach

and

$$\frac{d\widehat{R}_{exc}}{dt}=h_R\cdot(i_{exc}-\hat{i}_{exc}).\qquad(4)$$

Here, the equilibrium point is reached, when the estimated values are equal to the real values. It is defined

$$e_i=i_{exc}-\hat{i}_{exc}\qquad(5)$$

and

$$e_R=R_{exc}-\widehat{R}_{exc}.\qquad(6)$$

The next step is the differentiation of equation (5) and (6) to insert equation (1) and (2) as well as equation (3) and (4), which result in

$$\frac{de_i}{dt}=-\frac{R_{exc}}{f_L}\cdot i_{exc}+\frac{\widehat{R}_{exc}}{f_L}\cdot\hat{i}_{exc}\qquad(7)$$

and

$$\frac{de_R}{dt}=-h_R\cdot e_i.\qquad(8)$$

Refer to the equations (5), (6), (7) und (8) an arbitrary function must be found, which fulfill the following stability conditions of Lyapunov according to [6]:

1. $V(e_i,e_R) > 0$ for $(e_i,e_R) \neq (0,0)$

   $\rightarrow$ positive definite

2. $V(0,0) = 0$

   $\rightarrow$ positive definite

3. $\dot{V}(e_i,e_R) < 0$

   $\rightarrow$ negative definite

For this investigation the function

$$V(e_i,e_R)=\frac{1}{2}\cdot a\cdot e_i^2+\frac{1}{2}\cdot b\cdot e_R^2\qquad(9)$$

is chosen. The lines of constant values of $V(e_i,e_R)$ are circles for $a = b$ and an ellipses for $a \neq b$. These two factors are almost freely selectable and serve for the tuning of the nonlinear observer. It must be ensured, that the stability conditions 1. and 2. are met.

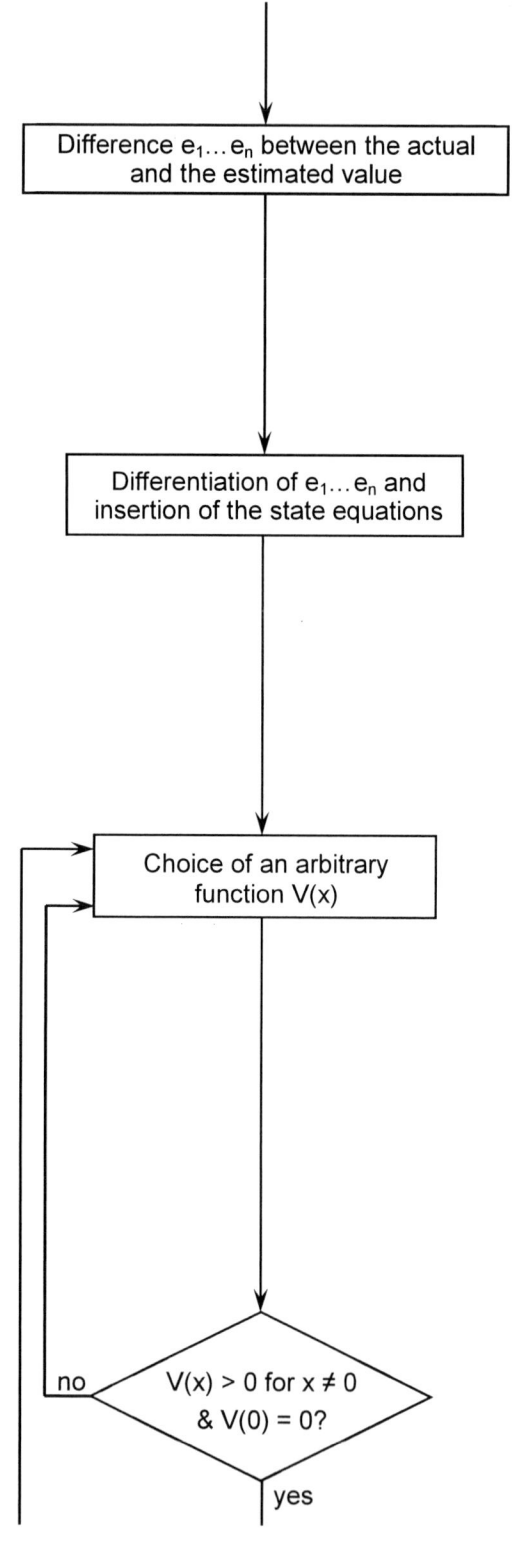

Figure 6: Program chart flow (Part 2)

Therefore, a and b have to be > 0. Stability condition 3. has to be checked with the aid of the differentiation of equation (9) and the insertion of equation (7) and (8). It follows

$$\dot{V}(e_i, e_R) = -\left(a \cdot \frac{i_{exc}}{f_L} + b \cdot h_R\right) e_i \cdot e_R - a \cdot e_i^2 \cdot \frac{\widehat{R}_{exc}}{f_L} \quad (10)$$

The correction factor h_R is chosen in such a way, that the first term of equation (10) is equal to 0. Thus, equation (10) is simplified to one term. This result in

$$h_R = -\frac{a}{b} \cdot \frac{i_{exc}}{f_L}. \quad (11)$$

Inserting equation (11) into equation (10) leads to

$$\dot{V}(e_i) = -a \cdot e_i^2 \cdot \frac{\widehat{R}_{exc}}{f_L} \quad (12)$$

Stability condition 3. is fulfilled for $\widehat{R}_{exc} > 0$ and a > 0. But it can be seen, that $\dot{V}(0) = 0$. Therefore, $\dot{V}$ is only negative semidefinite. This means, that the equilibrium point is stable, but not asymptotically stable according to Lyapunov. With the aid of LaSalle's invariance principle [8] the asymptotic stability of the equilibrium point is proved. This principle will not be discussed in detail here.

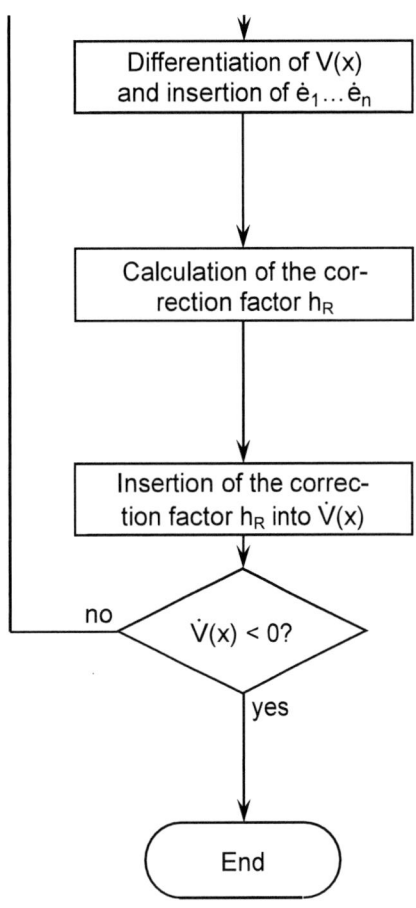

Figure 7: Program chart flow (Part 3)

4 Simulation Results

The integration of equation (11) into the observer for various initial values $\hat{i}_{exc}(0)$ and $\widehat{R}_{exc}(0)$ leads to the simulation results shown in Figure 8. It can be seen, that the estimated parameters (orange) converge to the real values (green) for all defined initial values with Lyapunov stability. The equilibrium point is asymptotically stable.

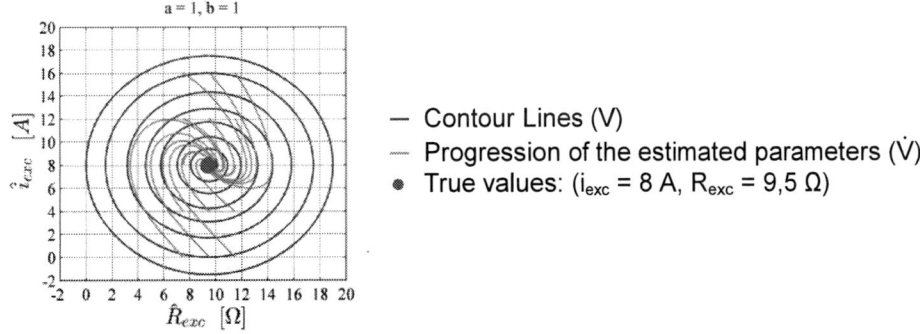

Figure 8: Contour lines and the progression of the estimated parameters to the equilibrium point

5 Measurement

5.1 Measurement Arrangement

Figure 9 shows the measurement arrangement for the tests of the nonlinear observer.

Figure 9: Measurement arrangement for the test of the nonlinear observer

In Figure 10 the power electronics of the IETS is shown.

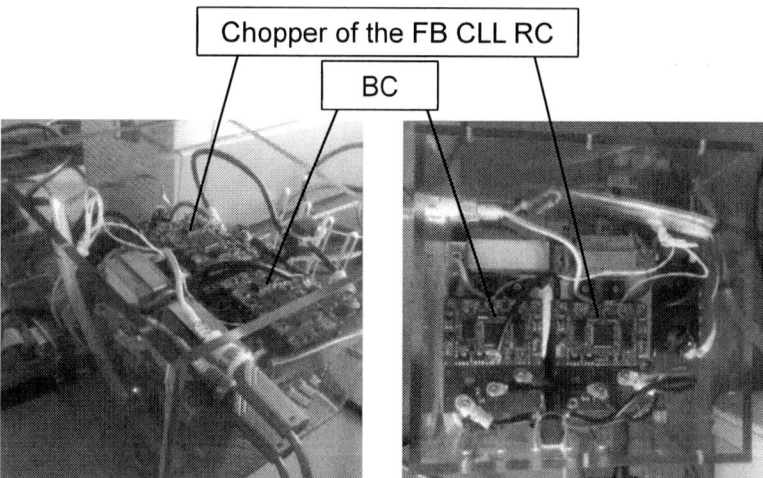

Figure 10: Power electronics of the inductive energy transfer system

5.2 Measurement Results

Figure 11 shows measurement results done with a dismounted rotor of a standard EESM. The relative deviations for $\hat{i}_{exc}$ and $\hat{R}_{exc}$ regarding the maximum excitation current of 16 A are 2,96 % and -3,75 %. Since these measurements were done with a dismounted rotor, the inductive value of that rotor is constant (no saturation).

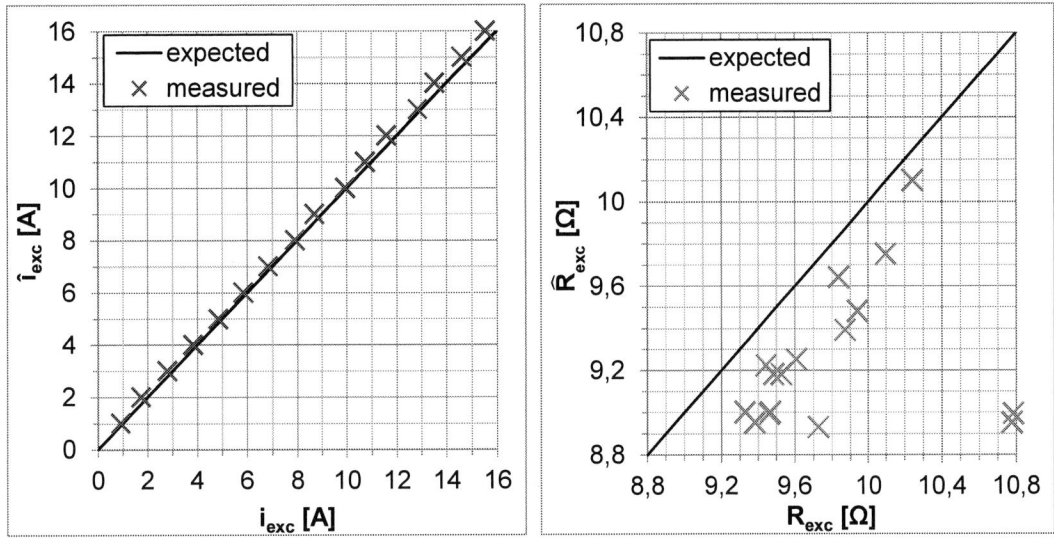

Figure 11: Test of the calculated nonlinear observer on the extended rotor of the EESM with IETS

6   Conclusion

A nonlinear observer is designed with the aid of the direct method of Lyapunov to estimate the excitation current and the rotor resistance of a rotor of an EESM with IETS. On the measurement arrangement the estimation algorithm achieves a precision of over 96 % regarding the maximum excitation current of 16 A (deviation: $\hat{\imath}_{exc} \rightarrow 2{,}96$ %, $\hat{R}_{exc} \rightarrow -3{,}75$ %).

References

[1] W. Hackmann, B. Wagner, R. Zwingel, I. Dziedzek, K. Welke, „Fremderregte Synchronmaschinen im Einsatz als Achshybridantriebe", International ETG Congress, Karlsruhe, Germany, pp. 55-64, Oct. 2007

[2] S. Köhler, „Erregerstrom- und Rotortemperaturschätzung einer durch einen DC-DC-Wandler gespeisten fremderregten Synchronmaschine", Master Thesis, TH Nürnberg GSO: Nuremberg, Nov. 2014

[3] A. Littau, B. Wagner, S. Köhler, A. Dietz, S. Weber, „Design of Inductive Power Transmission into the Rotor of an Externally Excited Synchronous Machine", CoFat, Munich, March 2014

[4] M.-K. Kazimierczuk, D. Czarkowski, „Resonant Power Converters", Hoboken: Wiley, pp. 448-457, 2011

[5] R. Beiranvand, M. R. Zolghadri, S. M. H. Alavi, „Using LLC Resonant Converter for Designing Wide-Range Voltage Source", IEEE Transactions on Industrial Electronics, Vol. 58, No. 5, pp. 1746-1756, May 2011

[6] A. N. Michel, L. Hou, D. Liu, „Stability of Dynamical Systems – Continuous, Discontinuous and Discrete Systems", Boston: Birkhäuser Boston, pp. 199-211, 2008

[7] A. Isidori, „Nonlinear Control Systems: An Introduction", Heidelberg: Springer-Verlag, pp. 178-253, 1985

[8] J. Adamy, "Nichtlineare Systeme und Regelungen", Heidelberg: Springer Vieweg, 2. Edition, pp. 101-104, 2014

High speed sensorless control of a synchronous motor with Kalman filter

Philipp Niedermayr, alpitronic, Italy, p.niedermayr@alpitronic.it
Luigi Alberti, Faculty of Science and Technology, University of Bolzano, Italy, luigi.alberti@unibz.it
Silverio Bolognani, Electric Drives Laboratory, University of Padova, Italy, silverio.bolognani@unipd.it
Reiner Abl, BMW Group, Germany, reiner.abl@bmw.de

Abstract

This paper presents the sensorless control of a permanent magnet synchronous motor operating at high speed (125000rpm). The motor drives a compressor for the air supply of a fuel cell cathode in a car. The control algorithm must fulfill both starting and dynamic performance, which are constrained by the particular application. For the sensorless control an algorithm based on the extended Kalman filter has been adopted. The system has been designed and implemented on two different prototypes.

1. Introduction

The interest in high-speed electrical drives has been increased in the last years in many applications: pumps, compressors, blowers, hybrid-electric vehicles and micro gas turbines [1, 2]. Some new technologies have been emerged and consolidated such as bearing-less operation. Among the advantages of high speed machines is the increase of the power per kilogram ratio of the machine. This is particularly well suited in mobile applications such as electrical vehicles where any saving in space and weight is a mandatory target in order to improve the system's consumption and emissions. Another advantage in other applications is the removal of an intermediate gearbox thanks to the adoption of a high speed motor: this yields an immediate reduction of the maintenance cost of the system and an improvement of its reliability. Nevertheless, operating the motor at high speed requires a particular design of the entire system. Traditional bearings are normally avoided for high speed machines and a bearing-less solution or air bearing configurations are adopted. The latter solution is particularly advantageous in high speed compressors, where the air bearing can be feed directly by the compressor outlet. Such solutions introduce additional requirements on the control system, e.g. when it comes to the design of the start-up procedure. Further, the mechanical constraints in the use of position resolvers for high speed permanent magnet synchronous motors (PMSM) call for efficient sensorless control algorithms. The electrical peculiarities of such high speed machines (such as low stator inductance and resistance) can cause a series of problems with the stability of the control, if the position and speed of the rotor is not known with high precision.
The Kalman filter is an efficient state observer also widely used in the sensorless control of PMSM since decades (see e.g.[3]). However, the Kalman filter still suffers from a number of problems if used for super high speed PMSM (for speeds > 50krpm), so that other sensorless control techniques were prefered so far (e.g. open loop constant V/f, I/f modes or hybrid modes [4], or different implementations of the back-EMF method [5, 6]). One main problem is the demanding calculation time of the Kalman filter, causing overruns at the high switching frequencies needed for the high speed applications. Another problem one has to deal with at high speed applications, are phenomena such as iron losses in the stator and eddy currents in

the rotor (which are not considered in the state equations normally used), which become more enhanced at high speeds.

This paper presents the control algorithm of the electrical drive for a high speed compressor (125000rpm) designed for an electrical vehicle. The performance of the drive is introduced starting from the constraints of the applications. Then the control scheme, based on a sensorless Extended Kalman filter (EKF) algorithm, is described and presented. The control has been implemented and tested in a real drive system with two different prototypes (which are called prototype 1 and prototype 2 in this paper). Experimental measurements are included in the paper.

2. Electrical drive requirements

Due to the particular application of the high speed air compressor for a vehicle's fuel cell supply, the performance in dynamics of the compressor is of high importance: it influences, in fact, the vehicle's performance directly, as the fuel cell can only deliver electrical energy at the rate the compressor delivers air to its cathode.

The start-up algorithm of the motor control has to account for the mechanical constraints of the compressor: the start-up from zero to idle speed (30000rpm) has to be completed in less than a second. An operation of the compressor below its idle speed for longer than 3s leads to irreversible hardware damage. The control algorithm is designed as speed control with a required control accuracy of 100rpm. A variable external load torque (depending on the operating point in the pressure-flow map of the compressor) represents a perturbation which has to be compensated efficiently. The DC link of the inverter is supplied by a battery and hence with a variable voltage. This voltage can vary in a range between 200V and 450V. The control algorithm has to compensate the variable DC voltage too. The table below gives a summarized overview of the requirements for the control algorithm.

start-up time 0krpm to 30krpm	< 1s
acceleration 25krpm to 110krpm	~ 1s
maximum speed	125krpm
control accuracy	± 100rpm
maximum overshoot for acceleration	1000rpm
DC link voltage	200V to 450V

Tab. 1: electrical drive requirements

3. The sensorless control algorithm

3.1. Modeling of the PMSM

The system equations of the non-salient ($L_d = L_q = L_s$) PMSM with the system matrix **A**, the input matrix **B**, the output matrix **C** (feed through matrix **D** $= 0$), the state vector **x**, the input vector **u** and the output vector **y** in the $\alpha - \beta -$ reference frame can be written as follows:

© VDE VERLAG GMBH · Berlin · Offenbach

$$\frac{\mathrm{d}}{\mathrm{d}t}\underbrace{\begin{pmatrix} i_\alpha \\ i_\beta \\ \omega \\ \theta \end{pmatrix}}_{\frac{\mathrm{d}}{\mathrm{d}t}\mathbf{x}} = \underbrace{\begin{pmatrix} -\frac{R_s}{L_s} & 0 & \frac{\lambda_{mg}p}{L_s}\sin\theta & 0 \\ 0 & -\frac{R_s}{L_s} & -\frac{\lambda_{mg}p}{L_s}\cos\theta & 0 \\ 0 & 0 & 0 & 0 \\ 0 & 0 & p & 0 \end{pmatrix}}_{\mathbf{A}} \cdot \underbrace{\begin{pmatrix} i_\alpha \\ i_\beta \\ \omega \\ \theta \end{pmatrix}}_{\mathbf{x}} + \underbrace{\begin{pmatrix} \frac{1}{L_s} & 0 \\ 0 & \frac{1}{L_s} \\ 0 & 0 \\ 0 & 0 \end{pmatrix}}_{\mathbf{B}} \cdot \underbrace{\begin{pmatrix} u_\alpha \\ u_\beta \end{pmatrix}}_{\mathbf{u}} \qquad (1)$$

with the stator phase inductance L_s, the stator phase resistance R_s, the number of pole pairs p and the stator permanent magnet flux linkage λ_{mg}. The state vectors entries correspond to the stator currents in the α - and β-axis, the rotor revolution speed ω and the rotor position θ. This model of the PMSM assumes an "infinite rotor inertia", i.e. the change of the rotor speed during one computation step is considered to be zero. This is manifested by zeros in the third row of the system matrix **A**. This approximation is proven to work fine for the particular application and has the advantage to speed up computations essentially by freeing the equations from all mechanical parameters (see e.g. [7, 8]). The output vector **y** can then be written as

$$\underbrace{\begin{pmatrix} i_\alpha \\ i_\beta \end{pmatrix}}_{\mathbf{y}} = \underbrace{\begin{pmatrix} 1 & 0 & 0 & 0 \\ 0 & 1 & 0 & 0 \end{pmatrix}}_{\mathbf{C}} \cdot \underbrace{\begin{pmatrix} i_\alpha \\ i_\beta \\ \omega \\ \theta \end{pmatrix}}_{\mathbf{x}} \qquad (2)$$

Hence, it is clear, that of the 4 dimensional state vector **x** we can only measure its first two entries, the phase currents (three phase currents in the a,b,c - reference frame reduce to two currents in the $\alpha - \beta -$ reference frame with the Clark-transformation).

3.2. The Kalman Filter

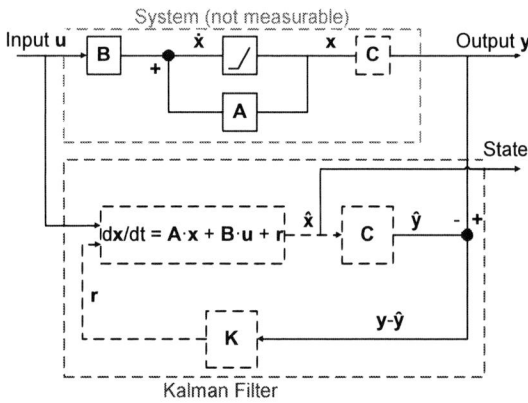

Fig. 1: System with Kalman Filter

The Kalman filter is a linear state observer that runs parallel to the actual system and provides an optimal estimation of the states of the system. Some of the states can be accessible by measurements (see Eq. (2)), but not all must be measurable. However, the Kalman filter provides an estimation of each state which is more accurate than the measurement alone. Figure 1 shows a system, with the input vector **u** and the state vector **x**, of which not all elements are measurable. Only the output states $\mathbf{y} = \mathbf{C} \cdot \mathbf{x}$ are measurable. The Kalman filter is running parallel to the system. It is provided with the input **u** of the system and estimates all the measurable and not measurable states $\hat{\mathbf{x}}$ of the system (in this paper we will denote estimated states of the Kalman filter with a hat).The Kalman filter compares its output $\hat{\mathbf{y}}$ to the output of the system **y**. As shown in section 3.1, the system equations of the PMSM are not linear (see system matrix **A** in Eq. (1)). An Extended Kalman filter (EKF) algorithm has to be implemented therefore. The EKF is a linearization of the system equations around the current estimate by means of a Taylor

PCIM Europe 2016, 10 – 12 May 2016, Nuremberg, Germany

expansion to the linear term. It has hence the same structure as the linear Kalman filter (see e.g. [9]). Starting from the system equations Eq. (1) and Eq. (2) we introduce the process noise σ (which represents inaccuracies in the model equations and disturbances in the system) and the observation noise ρ (which represents the noise present in the measurements) and rewrite the system equations as follows.

$$\frac{\mathrm{d}}{\mathrm{d}t}\mathbf{x}(t) = \mathbf{A} \cdot \mathbf{x}(t) + \mathbf{B} \cdot \mathbf{u}(t) + \sigma(t) \tag{3}$$

$$\mathbf{y}(t) = \mathbf{C} \cdot \mathbf{x}(t) + \rho(t) \tag{4}$$

Here $\sigma(t)$ and $\rho(t)$ represent zero mean Gaussian noises, which in general are time dependent. However their time dependency can be neglected and their covariance matrices $\mathbf{Q}$ and $\mathbf{R}$ can be written as time independent diagonal matrices (see for example [7]). A diagonal covariance matrix in this case means, that the variances of the single states are independent from each other. By writing the system Eq. (3) as difference equations (omitting the system noise σ, for which we will account in a separate predict step for the error covariance matrix $\hat{\mathbf{P}}_k$),

$$\frac{\hat{\mathbf{x}}_k - \hat{\mathbf{x}}_{k-1}}{T_s} = \mathbf{A} \cdot \hat{\mathbf{x}}_{k-1} + \mathbf{B} \cdot \mathbf{u}_k \tag{5}$$

one can derive the predict step of the Kalman filter, which calculates the predicted states $\hat{\mathbf{x}}_k^-$ out of the prior state $\hat{\mathbf{x}}_{k-1}$, the input vector $\mathbf{u}_k$, applied to the system during the period k, for which the state is predicted and with T_s the sampling time interval (**1** represents the unity matrix)[10]:

$$\hat{\mathbf{x}}_k^- = \underbrace{(\mathbf{1} + \mathbf{A} \cdot T_s) \cdot \hat{\mathbf{x}}_{k-1} + \mathbf{B} \cdot T_s \cdot \mathbf{u}_k}_{f(\hat{\mathbf{x}}_{k-1}, \mathbf{u}_k, 0)} \tag{6}$$

The predicted state estimation error covariance matrix of the state $\hat{\mathbf{x}}_k^-$ can be written as follows:

$$\hat{\mathbf{P}}_k^- = \mathbf{J}_k \cdot \hat{\mathbf{P}}_{k-1} \cdot \mathbf{J}_k^T + \mathbf{Q} \tag{7}$$

Here $\hat{\mathbf{P}}_{k-1}$ represents the state estimation error matrix of the previous step, and $\mathbf{J}_k$ is the Jacobian matrix of partial derivatives of $f(\hat{\mathbf{x}}_{k-1}, \mathbf{u}_k, 0)$ with respect to the state vector $\mathbf{x}$ [11]:

$$J_{k[i,j]} = \frac{\partial f_{[i]}}{\partial x_{[j]}}(\hat{\mathbf{x}}_{k-1}, \mathbf{u}_k, 0) = \begin{pmatrix} 1 - T_s\frac{R_s}{L_s} & 0 & T_s\frac{\lambda_{mg}p}{L_s}\sin\theta & T_s\frac{\lambda_{mg}p}{L_s}\omega\cos\theta \\ 0 & 1 - T_s\frac{R_s}{L_s} & -T_s\frac{\lambda_{mg}p}{L_s}\cos\theta & T_s\frac{\lambda_{mg}p}{L_s}\omega\sin\theta \\ 0 & 0 & 1 & 0 \\ 0 & 0 & T_sp & 1 \end{pmatrix} \tag{8}$$

Now the Kalman gain $\mathbf{K}$ can be calculated:

$$\mathbf{K}_k = \hat{\mathbf{P}}_k^- \mathbf{H}_k^T(\mathbf{H}_k\hat{\mathbf{P}}_k^-\mathbf{H}_k^T + \mathbf{R})^{-1} \tag{9}$$

Here $\mathbf{H}_k$ denotes the Jacobian matrix of partial derivatives of $h(\hat{\mathbf{x}}_k, 0)$ with respect to $\mathbf{x}$, where h represents the output function of Eq. (4):

© VDE VERLAG GMBH · Berlin · Offenbach

$$\mathbf{y}(t) = \mathbf{C} \cdot \mathbf{x}(t) + \rho(t) \quad \longrightarrow \quad \mathbf{y}_k = h(\hat{\mathbf{x}}_k, \rho) \quad \longrightarrow \quad H_{k,[i][j]} = \frac{\partial h_{[i]}}{\partial x_{[j]}}(\hat{\mathbf{x}}_k, 0) \quad \longrightarrow \quad \mathbf{H} = \mathbf{C} \quad (10)$$

The calculation of the Kalman gain contains a minimization of the square error of the states of the system (see [9]). The correct step can be calculated with the Kalman gain as follows:

$$\hat{\mathbf{x}}_k = \hat{\mathbf{x}}_k^- + \mathbf{K}_k(\mathbf{y}_k - \underbrace{\mathbf{H} \cdot \hat{\mathbf{x}}_k^-}_{\hat{\mathbf{y}}_k}) \tag{11}$$

$$\hat{\mathbf{P}}_k = (\mathbf{1} - \mathbf{K}_k \mathbf{H}_k)\hat{\mathbf{P}}_k^- \tag{12}$$

The Kalman gain can be seen as a sort of weighting factor: equation (11) reveals the correct step of the state vector $\hat{\mathbf{x}}$,where the measurable entires of the predicted system vector $\hat{\mathbf{x}}_k^-$ (see Eq. (6)) are compared to the measured states $\mathbf{y}_k$. The Kalman gain $\mathbf{K}$ is a measure of how certain the predicted states are in comparison to the measured states. In a one-dimensional system the Kalman gain would be a scalar between 0 and 1 (obviously updated at each calculation step). In extreme cases, a Kalman gain of 0 would mean that the information of the measurement is not considered at all and hence, the correct step would equal the predict step. Otherwise a Kalman gain of 1 would cancel the predict step out and take the only measurement states for the correct step.

3.3. Parametrization of the Kalman Filter

The difficult task in the implementation of a Kalman filter is the parametrization of the covariance matrices $\mathbf{R}$ and $\mathbf{Q}$, as well as the initial covariance matrix $\hat{\mathbf{P}}_0$ of the state vector. For the initial state vector $\hat{\mathbf{x}}_0$ the task is quite trivial, as one can be certain that the two currents $i_{\alpha,0}$ and $i_{\beta,0}$ are zero before the start of the PMSM, as well as the initial speed ω_0. However, as we do not know the initial rotor position θ_0 and every possible angle has the same probability, we keep things simple and choose $\theta_0 = 0$ as well. For the initial covariance matrix $\hat{\mathbf{P}}_0$ we choose a variance of 2 for the currents and a variance of 0.05 for the initial speed, as this is the state we are most certain in its initial value. Furhter $\hat{P}_{4,4}$ was set to 1. For the measurement noise a variance of 4 was adopted as this is a reasonable value for the current transducers used in the experiments.

For the parametrization of the covariance matrix representing the process noise, it was found that $Q_{1,1} = Q_{2,2} = 1$, $Q_{3,3} = 0.1$ and $Q_{4,4} = 0.00001$ [12] work very well for this application.

It is worth mentioning that we did the parametrization of the covariance matrices with a closed loop MATLAB/SIMULINK simulation with a very exact model of the PMSM and that here we also tried a method proposed by Bolognani et.al. in [13]. This method consists in a normalization of the motor parameters and the EKF algorithm. The choice of the covariance matrices is then straightforward, as the EKF equations are normalized and hence the very same covariance matrices can be used for different motor parameters. This method showed promising results in the simulation, however, its feasibility for this particular application has still to be proven in an experiment.

3.4. Open and closed loop control algorithm

Due to the particular mechanical restrictions at low speed of the PMSM (see Sec. 2) we decided not to start with a closed loop Kalman driven control algorithm (as the simulation showed, that the convergence of the Kalman Filter at low speeds is too slow or even fails). The adopted start-up algorithm is an open loop constant V/f mode as described e.g. in [4]. The Kalman filter is enabled at the start of the PMSM and compares its estimated position to the position given in the V/f mode. The switching to the closed loop Kalman filter algorithm is then made, if this difference between the Kalman estimated position and the V/f-mode given position is small enough and a certain speed is reached. Figure 2 shows the start up algorithm with the switch and a detailed scheme of the Kalman driven closed loop control algorithm.

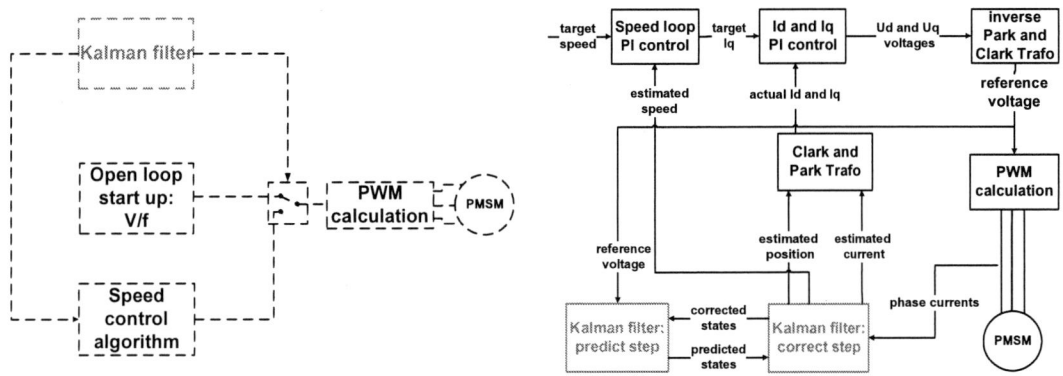

Fig. 2: Left: Schematic of the control algorithm with an open loop start-up V/f-control and a closed loop Kalman driven control algorithm. Right: Detailed schemtaic of the closed loop sensorless control algorithm.

4. Experimental setup and implementation

The experimental setup was realized in the high voltage laboratory of alpitronic. The DC power is supplied by a 1000V battery simulator, simulating the drive battery. The DC voltage was varied between 200V and 450V during the tests. The inverter used was developed by alpitronic. It is designed for a AC nominal current of 400A. The switching frequency was chosen to 20kHz, in order to have a minimum factor of 10 between the switching frequency and the fundamental electrical frequency at the maximum speed of the machine.

In order to reduce the risk of hardware damage at the high speed air compressor prototypes, a small 1.5kW commercial air blower was bought for the first tests of the control algorithm. Hence the control algorithm was designed such that it could be easily switched from one parameter set to another one. In a first step the consistency and functionality of the control algorithm was verified by means of a closed-loop SIMULINK model. The PMSM was simulated in a continuous model in the abc-reference frame: The model accounted for the IGBT on-resistance, the diode on-resistance and the DC-link voltage of the inverter, as well as the PWM centered current acquisition and the electrical parameters of the PMSM, such as the phase resistance and inductance and the emf-constant. The control algorithm was then coded in order to run on a dSpace PX10 System, which was connected to the gate driver circuit board of the inverter and the current measurement board. The parameters of the control algorithm were optimized in a series of experiments with the PX10 dSpace system. In the last step, the SIMULINK model with the optimized parameter set was coded again for the processor of a control circuit board developed and designed by alpitronic, which was integrated in the inverter.

5. Simulation and experimental results

Figure 3 shows the measured phase currents of the start-up (zero speed to idle speed) recorded with the prototype 1. The Figure reports also the result of the MATLAB/SIMULINK simulation of the phase currents for the same start-up. The consistency between the measured and simulated phase currents is very good. As reported in Sec. 3.4 for the start-up of the PMSM an open loop V/f control algorithm was used for the lower speed region prior to switching to the closed loop Kalman driven control algorithm. Both graphs show the phase currents during this open loop start up procedure from $t = 0$s to $t = 0.6$s. Due to the unknown initial rotor position, the first period length in the V/f mode is about 0.3s. The frequency then gets essentially increased (oscillations in the current amplitudes can be seen here in the measured as well as the simulated currents).The switching procedure to the closed loop algorithm occurs at $t = 0.6$s, which is manifested by a small decrease of the amplitudes of the phase currents for a few 10ms, as can be seen in both, the measured and the simulated currents.

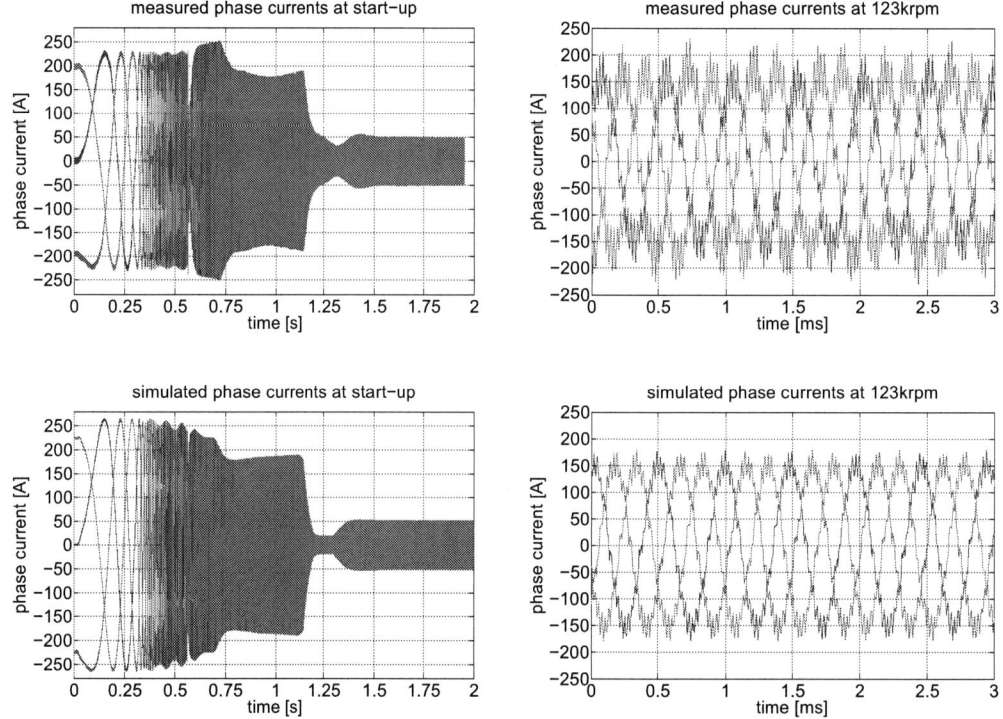

Fig. 3: Left: Comparison of measured and simulated phase currents during start-up procedure
Right: Comparison of measured and simulated phase currents ripple at 123krpm

Figure 3 also shows the phase currents at a speed of 123krpm measured and simulated with the prototype 2. Again, there is a good consistency between the measured and simulated currents, although the simulation underestimates the current ripple amplitude. This is mainly due to the fact, that in our model we did not include the effects of the eddy currents in the rotor, which can lower the inductance essentially.

The good reproducibility of the PMSM behavior in the simulation allows to evaluate certain aspects of the Kalman filter with the simulation rather than in an experiment. Also quantities which are not accessible during an experiment, such as the rotor position, can be accessed in the simulation and compared to the estimated ones. Figure 4 (left) reports the rotor position estimated by the Kalman filter compared to the real rotor position of the simulated PMSM. They are in good agreement.

Figure 4 (right) reports the accuracy of the speed (estimated by the Kalman filter) at the maximum revolution speed of 123krpm in an experiment performed with the prototype 2. The accuracy of the controlled speed lies within a band of ±100rpm, which is less than one part per thousand of the maximum speed. This clearly evidences the fact that the accuracy of the Kalman filter's estimated states is better than the accuracy of the single measurements (currents in this case).

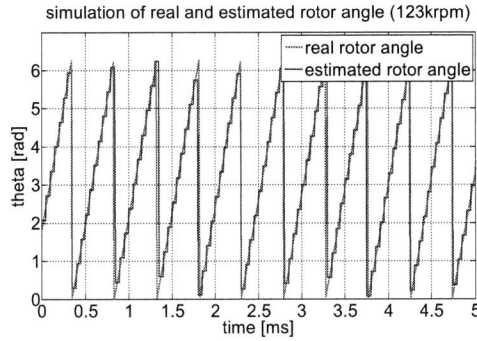 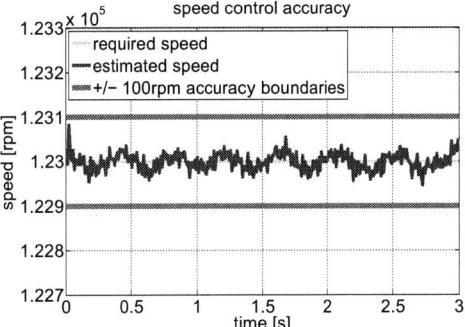

Fig. 4: Accuracy of the estimated position (left) and the estimated speed (right) of the Kalman filter. Please note: The graph showing the estimated speed as a function of the time was recorded in a real measurement with the prototype 2. The graph showing the estimated and the real rotor angle contains simulation and not measurement data, due to the fact, that there is no possibility to measure the real rotor position in an experiment on the test bench. However, as shown in Section 5, the simulation results are in good agreement with the measurement results. We therefore conclude that the comparison of the estimated and real rotor position in a closed loop SIMULINK simulation is a good measure for the accuracy of the Kalman position estimation.

6. Conclusion

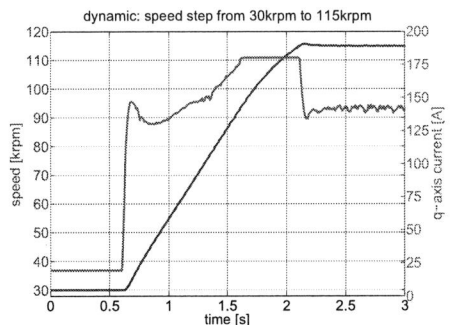

Fig. 5: acceleration from 30krpm to 115krpm

The adopted algorithm is well suited for this application, since it provides a very good speed estimation quality also in presence of measurement errors and noise, imprecise knowledge of the motor parameters and voltage non linearity due to the inverter. The implemented control algorithm has good dynamic performance, which is particularly important for its application.

Figure 5 shows the measurement of an acceleration from 30krpm to 115krpm performed with the prototype 2. The plot reports the speed and the corresponding i_q current as a function of the time. As the speed reference was handled by a rate limiter in the control algorithm to avoid instabilities, the i_q current does not reach its maximum value during the first part of the acceleration. Here it is possible to reach even better dynamic performance if the rate limiter is tuned

© VDE VERLAG GMBH · Berlin · Offenbach

further. During the experiments we did not observe any issues with the Kalman filter affecting the dynamic performance.

7. References

[1] D. Gerada, A. Mebarki, N.L. Brown, C. Gerada, A. Cavagnino, and A. Boglietti. High-Speed Electrical Machines: Technologies, Trends, and Developments. *Industrial Electronics, IEEE Transactions on*, 61(6):2946–2959, June 2014.

[2] Md Arifur Rahman, Akira Chiba, and Tadashi Fukao. Super high speed electrical machines-summary. In *Power Engineering Society General Meeting, 2004. IEEE*, pages 1272–1275. IEEE, 2004.

[3] A. Bado, S. Bolognani, and M. Zigliotto. Effective estimation of speed and rotor position of a PM synchronous motor drive by a Kalman filtering technique. In *Power Electronics Specialists Conference, 1992. PESC '92 Record., 23rd Annual IEEE*, pages 951–957 vol.2, June 1992.

[4] Longya Xu and Changjiang Wang. Implementation and experimental investigation of sensorless control schemes for PMSM in super-high variable speed operation. In *Industry Applications Conference, 1998. Thirty-Third IAS Annual Meeting. The 1998 IEEE*, volume 1, pages 483–489 vol.1, October 1998.

[5] Bon-Ho Bae, Seung-Ki Sul, Jeong-Hyeck Kwon, and Ji-Seob Byeon. Implementation of sensorless vector control for super-high-speed PMSM of turbo-compressor. *Industry Applications, IEEE Transactions on*, 39(3):811–818, May 2003.

[6] F. Genduso, R. Miceli, C. Rando, and G.R. Galluzzo. Back EMF Sensorless-Control Algorithm for High-Dynamic Performance PMSM. *Industrial Electronics, IEEE Transactions on*, 57(6):2092–2100, June 2010.

[7] S. Bolognani, R. Oboe, and M. Zigliotto. Sensorless full-digital PMSM drive with EKF estimation of speed and rotor position. *Industrial Electronics, IEEE Transactions on*, 46(1):184–191, February 1999.

[8] S. Bujacz, A. Cichowski, P. Szczepankowski, and J. Nieznanski. Sensorless control of high speed permanent-magnet synchronous motor. In *Electrical Machines, 2008. ICEM 2008. 18th International Conference on*, pages 1–5, September 2008.

[9] Peter S. Maybeck. *Stochastic models, estimation, and control*, volume 141 of *Mathematics in Science and Engineering*. 1979.

[10] Greg Welch and Gary Bishop. An Introduction to the Kalman Filter. 2006.

[11] A. Qiu, Bin Wu, and H. Kojori. Sensorless control of permanent magnet synchronous motor using extended Kalman filter. In *Electrical and Computer Engineering, 2004. Canadian Conference on*, volume 3, pages 1557–1562 Vol.3, May 2004.

[12] A. Cichowski, S. Bujacz, J. Nieznanski, and P. Szczepankowski. Sensorless startup of super high speed permanent magnet motor. In *Industrial Electronics (ISIE), 2010 IEEE International Symposium on*, pages 3101–3106, July 2010.

[13] S. Bolognani, L. Tubiana, and M. Zigliotto. Extended Kalman filter tuning in sensorless PMSM drives. *Industry Applications, IEEE Transactions on*, 39(6):1741–1747, November 2003.

Physical Modeling and High-Fidelity Simulation of the Transient Behavior of Multiply-Contacted Power Busbars

Vanessa Basler, Technical University of Munich, Germany, basler@tep.ei.tum.de
Wolfgang Hölzl, Technical University of Munich, Germany, hoelzlw@in.tum.de
Gerhard Wachutka, Technical University of Munich, Germany, wachutka@tep.ei.tum.de

Abstract

For optimizing the geometry and layout of power busbars it is indispensable to analyze in detail the transient electromagnetic field distribution inside and outside the busbars. The correct computation of these fields requires special boundary and interface conditions. In our work we developed a methodology to calculate the electromagnetic behavior even for complex structures with more than two contacts and with two or more electrically separated components of the busbar aggregate. A central concept is the generalized inductance matrix which can be extracted from the continuous-field model to optimize the shape of the busbars.

1. Introduction

The progressively faster switching transients and high power densities of semiconductor power modules and applications tighten the requirements on the commutation circuit. Among other, low-inductive passive components with high current capability such as DC busbars have to be employed. Furthermore, these components have to be optimized with respect to current crowding and local eddy currents. So far, a number of tools have been developed for the numerical analysis of field and current distributions in the frequency domain (see [1], [2] and [3] for example). In today's high power technology, we have to cope with fast pulse-shaped voltage and current waveforms encountered in power modules, which contain fast switching semiconductor devices connected by complex busbar structures. Therefore, a tool for the accurate numerical simulation of waveforms in the time domain is needed as well. The inductance extraction programs available from commercial distributors mostly rely on "Neumann's formula" for a static version of the inductance matrix (see e.g. [4]). However, using the concept of time-invariant self- and mutual inductance coefficients is not appropriate for modeling and optimizing the dynamic behavior of multiply-contacted busbars with complicated shape and composed of several electrically separated components. Instead, a generalized impedance operator has to be introduced, which is well-defined for transient waveforms and applicable to interconnects with more than two contact electrodes. For the proper description of transient current crowding, skin effect and displacement current, this generalized impedance matrix is extracted from the space and time dependence of the electro-magnetic field distribution in quasi-stationary approximation. To this end, it is necessary to model and to numerically simulate these physical effects in the time domain and not in the frequency domain.

2. Electrodynamic Simulation Model

For a detailed description of the current distribution and the electric and magnetic fields inside and outside the busbar structure, a transient approach must be pursued. To this end, a formulation based on the scalar and vector potentials φ and $\vec{A}$ (4-potential), is appropriate. The vector potential $\vec{A}$ generates the magnetic field as $\vec{B} = \nabla \times \vec{A}$ inside and outside the busbars, and the scalar potential φ together with $\vec{A}$ generates the electric field as $\vec{E} = -\nabla\varphi - \frac{\partial \vec{A}}{\partial t}$ in the interior of the busbars. For the situation and configuration considered, the quasi-stationary approximation is sufficient, in which the occurrence of electromagnetic waves is suppressed. Using Maxwell's equations the following governing equations can be deduced [5]:

$$\nabla \cdot (\varepsilon \nabla \varphi) + \frac{\partial}{\partial t}\left(\nabla \cdot (\varepsilon \vec{A})\right) = 0 \tag{1}$$

$$\nabla \times \frac{1}{\mu}\nabla \times \vec{A} + \sigma \frac{\partial \vec{A}}{\partial t} = -\sigma \nabla \varphi. \tag{2}$$

Here, ε denotes the permittivity, μ the permeability, and σ the conductivity of the respective material. These equations are valid inside the conducting material Ω_c. $\vec{A}$ and φ are not uniquely determined by equations (1) and (2) due to the fact that they can be altered by a gauge transformation [5]. In our approach, the Coulomb gauge is used to obtain a unique solution. If ε is constant inside the conducting domain, equation (1) becomes a Laplace equation for the scalar potential φ: $\Delta\varphi = 0$. The Finite Element Method implemented in our simulation tool uses Nédélec's edge elements as basis functions for the representation of $\vec{A}$. These basis functions have their degrees of freedom in the middle of the edges and satisfy the condition $\nabla \cdot \vec{A} = 0$ in the strong sense [6]. A complete solution requires the calculation of the magnetic field also outside the busbars in the non-conductive region Ω_n. For this purpose we use equation (1) with a formal conductivity σ_n set to zero, leading to the equation

$$\nabla \times \frac{1}{\mu}\nabla \times \vec{A} = 0. \tag{3}$$

This conforms with the approach reported in [7]. For a complete problem definition, boundary and interface conditions must be specified. These are:

$$\nabla\varphi \cdot \vec{n}_c = 0 \text{ on } \Gamma_{cn} \text{ and } \varphi(t) \text{ known at the contacts } C_j \text{ for all } j = 1, \ldots, N \tag{4}$$

$$\vec{A} \times \vec{n}_c = 0 \text{ on } \Gamma_n \tag{5}$$

$$\frac{1}{\mu}\nabla \times \vec{A} \times \vec{n}_c = \frac{1}{\mu}\nabla \times \vec{A} \times \vec{n}_n \text{ on } \Gamma_{nc} \tag{6}$$

$$\vec{A} \times \vec{n}_c = \vec{A} \times \vec{n}_n \text{ on } \Gamma_{nc} \tag{7}$$

The vector $\vec{n}_c$ denotes the outward unit normal vector along the boundary of domain Ω_c; evidently we have $\vec{n}_n = -\vec{n}_c$. Condition (4) implies that a voltage-controlled problem is considered. This means that the potential applied at the contacts is given and the potential-driven contribution to the current flow $\vec{j}_{pot} = \sigma \vec{E}_{pot} = -\sigma \nabla\varphi$ through the interface Γ_{cn} between the conducting domain Ω_c and the surrounding air is suppressed. This condition ensures the correct potential distribution and, hence, the correct electric field in the asymptotic stationary state attained for $t \to \infty$. The boundary condition (5) forces $\vec{A} \approx 0$ along Γ_n, the boundary of the domain Ω_n. The interface conditions (6) and (7) ensure continuity of the tangential component of $\vec{A}$ and $\vec{B}$ along the interface Γ_{cn}. Since current flow from the conducting material Ω_c into the non-conductive material Ω_n must be prevented, the interface condition $\vec{E} \cdot \vec{n}_c = 0$ on Γ_{cn} has to be posed. Together with the boundary condition $\nabla\varphi \cdot \vec{n}_c = 0$ along Γ_{cn} (cf. eq. (4)) we conclude that the additional condition

$$\frac{\partial \vec{A}}{\partial t} \cdot \vec{n}_c = 0$$

must hold along the interface Γ_{cn} as additional constraint. Having computed the field distributions as primary quantities, a generalized impedance matrix can be calculated in a post-processing step as described in [7], and then be used for the optimization of the shape of the busbar structure.

3. Results

An exemplary busbar structure made of copper was considered to demonstrate the capability and practicality of the implemented method. A voltage ramp up to $10\,\text{V}$ in $1\,\mu s$ as shown in Fig. 1(a) is applied. Fig. 2(a) shows the scalar potential inside the busbar, where the contacts are marked by labels with the values of the contact potentials applied ($\varphi = 0\text{V}$ and $\varphi = 1\text{V}$). The resulting electric field distribution at the end of the switching process is displayed in Fig. 3. Most of the electric field lines flow in the vicinity of the corners and edges of the conducting material. In the middle of the busbar, the potential-driven current is largely suppressed, and an induced current flowing in reverse direction prevails. Evidently, the skin effect dominates the current distribution inside the structure. In Fig. 2(b) the electric field of the stationary state attained after $1\,\text{ms}$ is depicted.

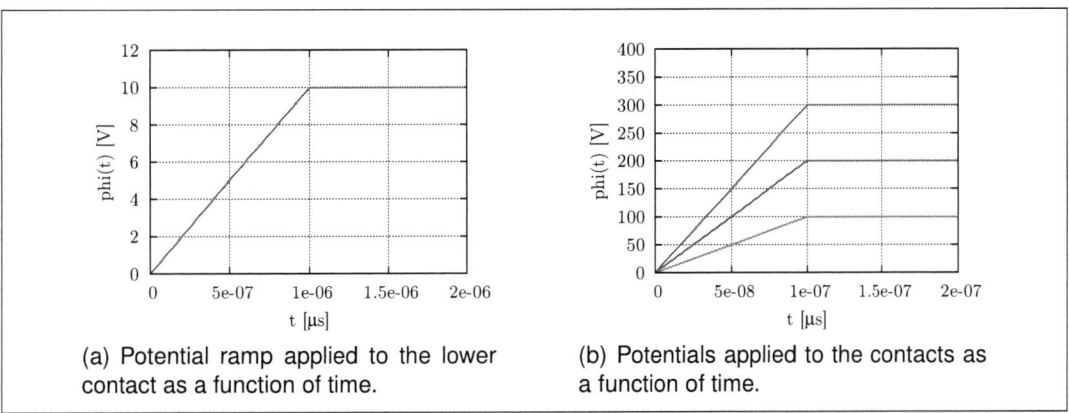

(a) Potential ramp applied to the lower contact as a function of time.

(b) Potentials applied to the contacts as a function of time.

Fig. 1: Applied potential over time.

The potential-driven current

$$i_{pot}(t) = -\int_C \sigma\,\nabla\varphi \cdot \mathrm{d}a$$

and the induced current

$$i_{rot}(t) = -\int_C \sigma\,\frac{\partial \vec{\mathbf{A}}}{\partial t} \cdot \mathrm{d}a,$$

where C denotes the contact with the positive potential, are displayed in Fig. 4(a) together with the total current

$$i(t) = i_{pot}(t) + i_{rot}(t).$$

The potential-driven current $i_{pot}(t)$ attains immediately to its final value, proportional to the waveform of the potential. The induced current, however, has opposite sign and rises as long as the potential at the contacts vary. When the switching process is finished and the potential attains its final value, the current density starts to redistribute and the induced current falls down until the stationary state is reached.

© VDE VERLAG GMBH · Berlin · Offenbach

PCIM Europe 2016, 10 – 12 May 2016, Nuremberg, Germany

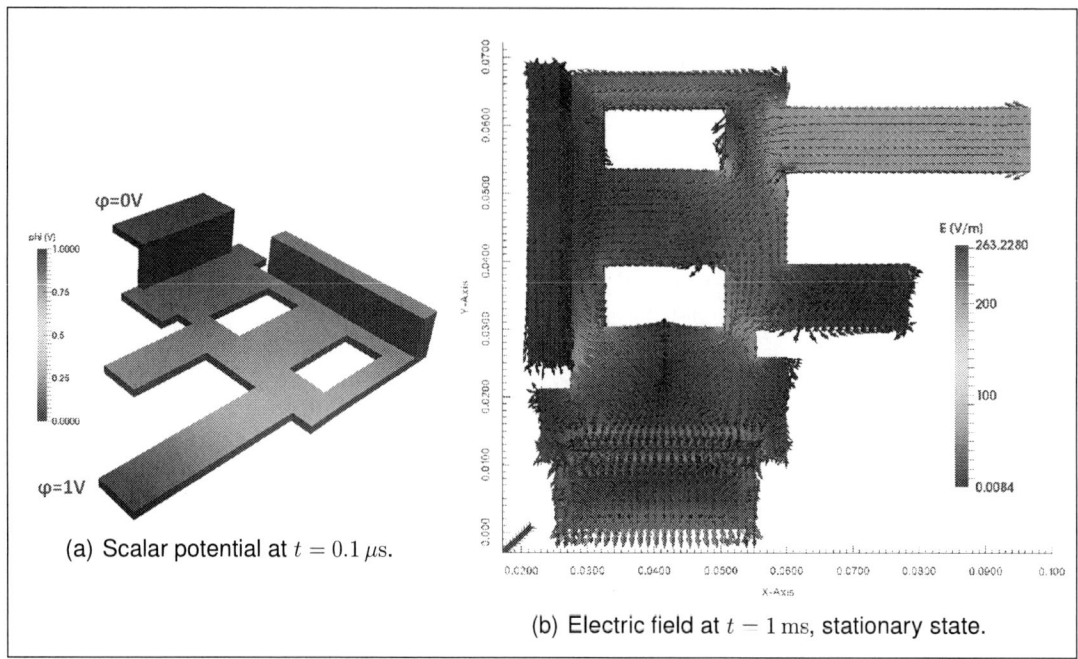

(a) Scalar potential at $t = 0.1\,\mu s$.

(b) Electric field at $t - 1\,ms$, stationary state.

Fig. 2: Examplary structure with 2 contacts.

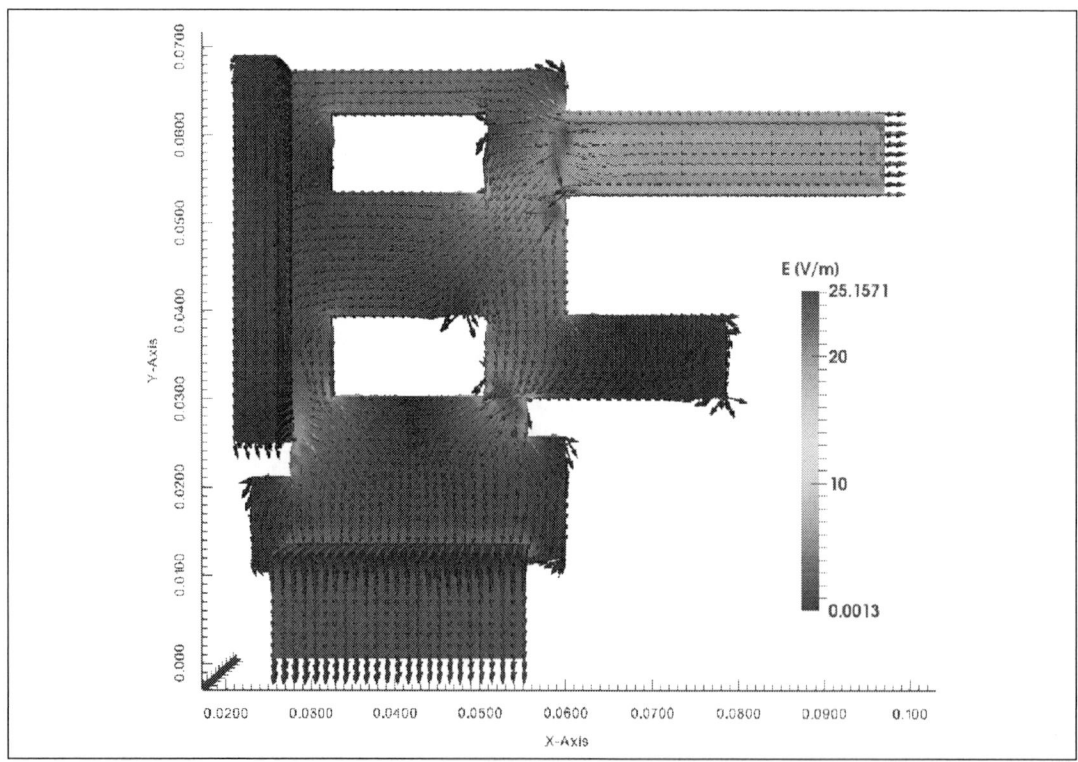

Fig. 3: Electric field in a structure with 2 contacts at $t = 0.1\,\mu s$, skin effect state.

Fig. 4(b) shows the apparent ohmic resistance of the busbar. In consequence of the skin effect

© VDE VERLAG GMBH · Berlin · Offenbach

PCIM Europe 2016, 10 – 12 May 2016, Nuremberg, Germany

the resistance is strongly enhanced at the beginning of the switching process. The transient inductance of the busbar structure is displayed in the same figure. At the beginning of the switching process the inductance is very small. With progressing time both the resistance and the inductance converge to their asymptotic stationary value. It is clearly visible that the time required for the busbar to reach its stationary state is much longer than the switching time of $10\,\mu s$.

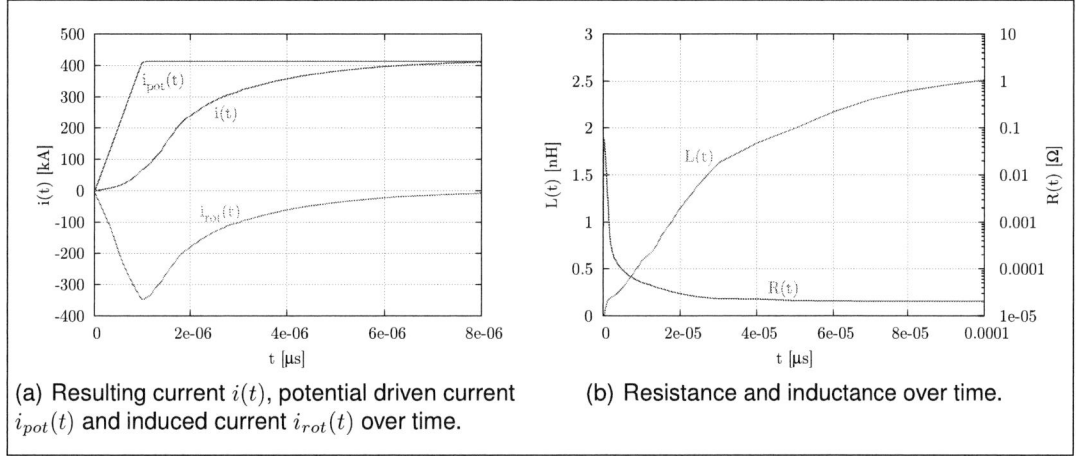

(a) Resulting current $i(t)$, potential driven current $i_{pot}(t)$ and induced current $i_{rot}(t)$ over time.

(b) Resistance and inductance over time.

Fig. 4: Transients during the switching process.

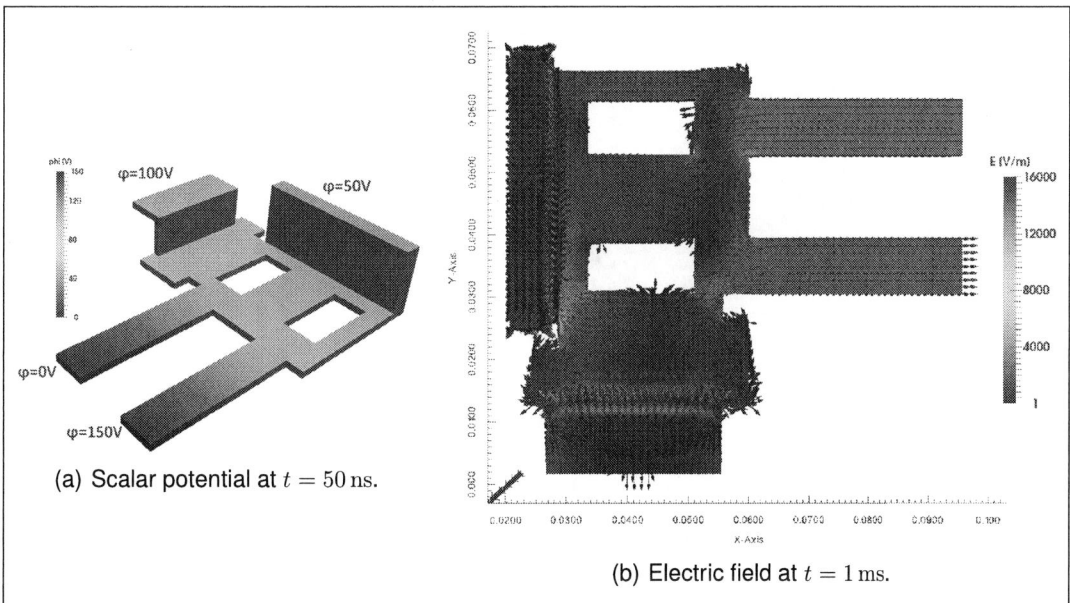

(a) Scalar potential at $t = 50\,ns$.

(b) Electric field at $t = 1\,ms$.

Fig. 5: Examplary structure with 4 contacts.

Using our method, structures with more than two contacts can be analyzed as well. An illustrative example is depicted in Figs. 5(a), 5(b) and 6. Fig. 5(a) shows the scalar potential φ in a similar busbar as in Figs. 2(a) to 3, but now four contacts are attached. In Fig. 1(b) the potentials applied to the four contacts as a function of time can be seen. Fig. 6 shows the current distribution inside the conducting busbar during the switching process. Also in this structure the

© VDE VERLAG GMBH · Berlin · Offenbach

547

skin effect is visible. In Fig. 5(b) the asymptotic stationary state is shown at a time long after the voltage reached its final value.

4. Conclusion

For the future optimization of power busbars it is decisive to gain a qualitative and quantitative insight into their transient behavior under realistic switching conditions. A complete simulation model using the $A - \varphi$-formulation has been presented. By using edge elements for the space discretisation, a fast and intelligible algorithm has been developed to solve the dynamic inductance problem. Our results show that the model is able to reproduce the involved physical effects and visualize the internal current density. The current distribution and the vector potential can be used to extract a generalised impedance matrix with a view to optimizing the busbar structure.

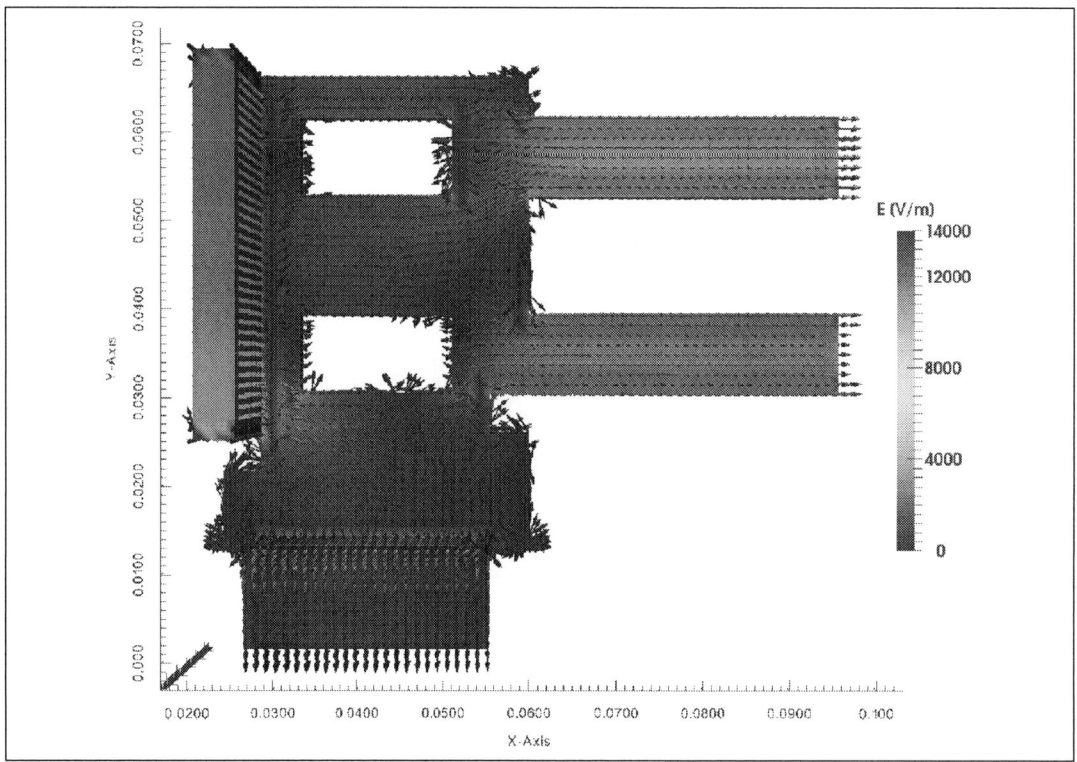

Fig. 6: Electric field in a busbar struture with 4 contacts at $t = 50\,\text{ns}$; skin effect state.

References

[1] Xin Hu. *Full-wave Analysis of Large Conductor Systems over Substrate*. PhD thesis, Massachusetts Institute of Technology, 2006.

[2] A.C. Cangellaris, J. Prince, and L.P. Vakanas. Frequency-dependent inductance and resistance calculation for three-dimensional structures in high-speed interconnect systems. *Components, Hybrids, and Manufacturing Technology, IEEE Transactions on*, pages 154–159, 1990.

[3] H.V. Jorks, E. Erion Gjonaj, and T. Weiland. Eddy current analysis of a pwm controlled induction machine. *COMPEL - The international journal for computation and mathematics in electrical and electronic engineering*, pages 1609–1619, 2013.

[4] M. Kamon, M.J. Tsuk, and J.K. White. Fasthenry: a multipole-accelerated 3-d inductance extraction program. *IEEE Transactions on Microwave Theory and Techniques*, pages 1750–1758, 1994.

[5] O. Biro, K. Preis, and K.R. Richter. On the use of the magnetic vector potential in the nodal and edge finite element analysis of 3d magnetostatic problems. *IEEE Transactions on Magnetics*, pages 651–654, 1989.

[6] J.C. Nedelec. Mixed finite element in 3d in h(div) and h(curl). *Lecture Notes in Mathematics, Springer Berlin Heidelberg*, pages 321–325, 1986.

[7] P. Böhm and G. Wachtuka. Transient electromagnetic behavior of multiply contacted interconnects. *2nd Int. Conf. on Modeling and Simulation of Microsystems, Semiconductors, Sensors and Actuators*, pages 301–304, 1999.

Small-Signal Output Impedance Modeling of Intersil's R4™ Technology

Yi Huang, Intersil Corporation, New Jersey, USA, Email: yhuang01@intersil.com
Chun Cheung, Intersil Corporation, New Jersey, USA, Email: ccheung@intersil.com

Abstract

In this paper, the small signal output impedance of a synchronous buck converter is modeled using Intersil's R4™ modulator. The describing function approach is employed in deriving the transfer function of the modulator. Good matching of the output impedance in s-domain is shown between the analytical model and the simulation results obtained from SIMPLIS. Design considerations and time-domain simulation results are also provided.

1. Introduction

The development of modern microprocessors has reinforced the need for fast transient response from voltage regulator modules (VRM). Over the past few years, significant research efforts have been devoted to developing novel control algorithms, such as ripple-based modulators [1], [2], that provide faster response than conventional voltage/current mode control. One of the most advanced ripple modulation technologies is the R4™ modulator invented by Intersil Corporation [3], which features linear control loop for optimal transient response, variable frequency and duty cycle control during load transient to achieve the fastest possible response, and inherent voltage feed-forward for wide range input [4], [5].

A simplified schematic of the R4™ modulator is shown in Fig. 1. To evaluate the fast transient response benefits of Intersil's R4™ technology, small-signal models are required, including the loop gain model and the output impedance model. The small-signal control-to-output and output impedance are very effective tools for analyzing the stability and load transient performance of the ripple-based controls. These tools have been thoroughly studied and examined in constant On-time V^2 control [6], [7], and constant frequency V^2 control [8].

Average modeling has not been effective for the dynamic characteristics of ripple regulators, so the describing function (DF) method [9] was employed to analyze this issue. The accuracy of DF method has been proven in peak-current-mode control [10], constant On-time current mode control [11], constant On-time V^2 control [6], [10], constant frequency V^2 control [8], and has been successfully used in obtaining the loop gain model of the R4™ modulator [12]. The purpose of this paper is to extend the loop gain model to calculate the small-signal output impedance of a buck converters using R4™ Technology, and to assist in the design of buck converters that provide optimal transient response.

The rest of the paper is organized as follows: in Section 2, the circuit configuration and the small-signal model of a simplified R4™ modulator are presented. In Section 3, the closed loop output impedance analysis and its verification are described. Design considerations and time domain verification are presented in Section 4, and conclusions are presented in Section 5.

© VDE VERLAG GMBH · Berlin · Offenbach

PCIM Europe 2016, 10 – 12 May 2016, Nuremberg, Germany

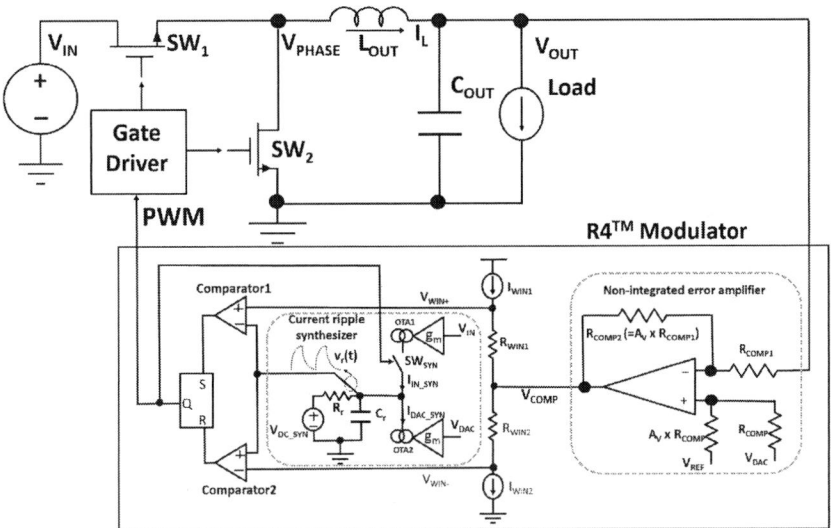

Fig. 1: A buck converter with Intersil's R4™ Technology.

2. Modeling of the R4™ Modulator

As shown in Fig. 1, a simplified R4™ modulator consists of two major blocks: the non-integrated error amplifier and the PWM generator based on a current ripple synthesizer [4], [5]. In the non-integrated error amplifier, the non-inverting input is generated from the reference voltage V_{DAC}, and the output voltage of the buck converter V_{OUT} is fed back into the inverting input node via a series resistor R_{COMP1}. The ratio of the resistor R_{COMP2} to the resistor R_{COMP1} equals a constant number A_V. Therefore, V_{COMP} contains the inverted amplified AC component as the AC signal of the V_{OUT}, and feeds into the PWM generator. In the synthetic current ripple generator, two identical currents I_{WIN1} and I_{WIN2} go through resistors R_{WIN1} and R_{WIN2} to form the hysteresis window, with its central level at V_{COMP}. The input voltage of the buck converter V_{IN} goes through a transconductance amplifier ($OTA1$) to generate a charge current I_{IN_SYN} to the capacitor C_r. A discharging current is formed via the V_{DAC} and $OTA2$. A DC voltage V_{DC_SYN} is connected with C_r via a series resistor R_r. V_{DC_SYN} sets the DC level of the generated synthetic current ripple signal $V_r(t)$, while R_r and C_r form its time constant. The synthetic current ripple signal $V_r(t)$ is then compared with the hysteresis window via two comparators, whose outputs set/reset the RS flip-flop to generate the PWM signal. The generated PWM signal also drives SW_{SYN} in the current ripple synthesizer to terminate the charging operation from I_{IN_SYN} at the end of the On time. The steady state waveform of this PWM generator is depicted in Fig. 2-a. To obtain the transfer function from control to duty cycle, the procedure of applying DF method in [12] is briefly repeated here: a pure DC voltage (V_{DC}, not shown in the figures) is applied at V_{COMP} node to keep the normal operation of the system. A sinusoidal signal perturbation is superposited, to have $V_{COMP}(t)$ expressed as:

$$V_{COMP}(t) = V_{DC} + \hat{r}\sin(2\pi f_m t) \tag{1}$$

where $\hat{r}$ is the magnitude of the perturbation signal. The perturbation frequency f_m and the switching frequency f_{SW} are commensurable ($Mf_m = Nf_{SW}$, where M and N are two positive integers) [9]. As shown in Fig. 2-b, the waveforms of the closed loop hysteresis window can be simplified to two constant upper and lower levels, with no perturbation, with the unchanged

© VDE VERLAG GMBH · Berlin · Offenbach

PCIM Europe 2016, 10 – 12 May 2016, Nuremberg, Germany

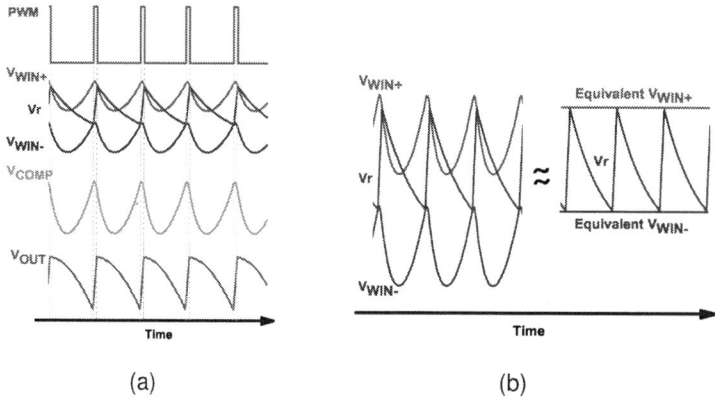

(a)   (b)

Fig. 2: (a) Steady state operation waveforms (b) Equivalent hysteresis window at no perturbation

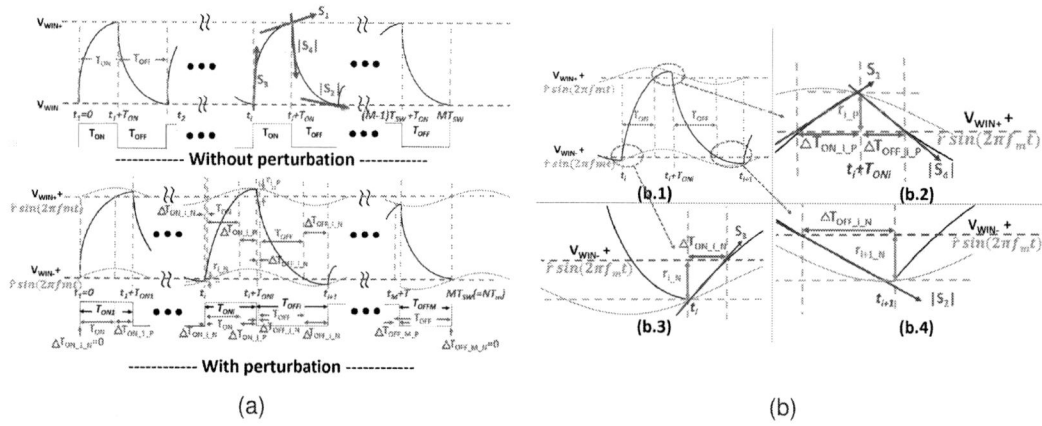

(a)   (b)

Fig. 3: (a) Perturbed hysteresis window and synthetic current ripple (b) T_{ON} and T_{OFF} modulation of the ith cycle (after [12])

synthetic current ripple signal $V_r(t)$. $V_r(t)$ can be described as:

$$
\begin{cases}
V_{DC\_SYN} + R_r g_m (V_{IN} - V_{DAC}) - \left(R_r g_m V_{IN} \left[1 - \dfrac{\left(1 - e^{\frac{-V_{DAC}}{V_{IN} R_r C_r f_{SW}}} \right) e^{\frac{-1}{R_r C_r f_{SW}}}}{e^{\frac{-V_{DAC}}{V_{IN} R_r C_r f_{SW}}} \left(1 - e^{\frac{-1}{R_r C_r f_{SW}}} \right)} \right] \right) e^{\frac{-t}{R_r C_r}} \\
\hspace{8cm} (iT_{SW} \le t \le iT_{SW} + T_{ON}) \\[4mm]
V_{DC\_SYN} - R_r g_m V_{DAC} + \left(R_r g_m V_{IN} \dfrac{1 - e^{\frac{-V_{DAC}}{V_{IN} R_r C_r f_{SW}}}}{e^{\frac{-V_{DAC}}{V_{IN} R_r C_r f_{SW}}} \left(1 - e^{\frac{-1}{R_r C_r f_{SW}}} \right)} \right) e^{\frac{-t}{R_r C_r}} \\
\hspace{8cm} (iT_{SW} + T_{ON} \le t \le (i+1)T_{SW})
\end{cases}
\tag{2}
$$

Based on (2), the absolute values of the slopes at the switching decision points of the PWM signal (S_3, S_1, S_4 and S_2 in Fig. 3-a) can be derived by differentiating (2) at the corresponding timing points. Once the expressions of the slopes are available, the perturbation signal is applied. In Fig. 3-a, the comparison between the steady state and the perturbed V_{WIN+} and

© VDE VERLAG GMBH · Berlin · Offenbach

PCIM Europe 2016, 10 – 12 May 2016, Nuremberg, Germany

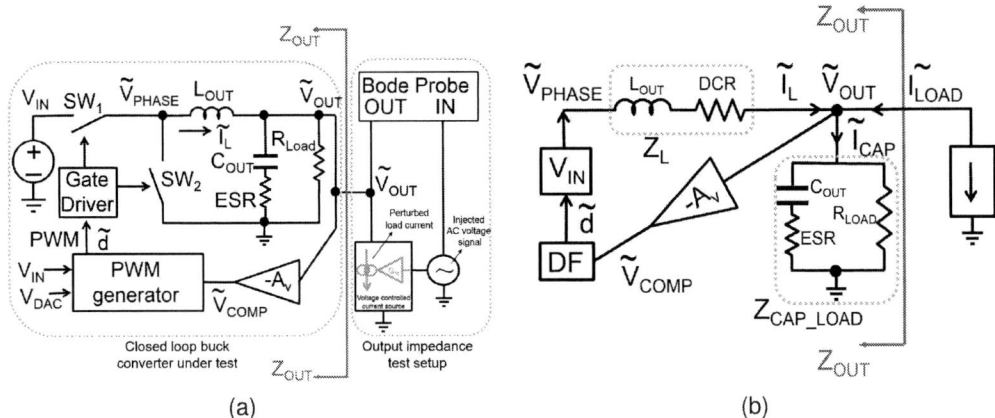

Fig. 4: (a) Output impedance test setup with closed loop buck converter (b) Equivalent block diagram for testing output impedance

V_{WIN-} are illustrated. In Fig. 3-b, geometry analysis is applied to the ith switching period, specifically at the regime around T_{ON} and T_{SW}. Following the derivation shown in [12], the detailed expressions can be described as:

$$\Delta T_{ON\_i} = -\frac{\hat{r}\sin(2\pi f_m t_i)}{S_3} + \frac{\hat{r}\sin(2\pi f_m(t_i + T_{ON\_i}))}{S_1}, \tag{3}$$

and

$$\Delta T_{OFF\_i} = \frac{\hat{r}\sin(2\pi f_m(t_i + T_{ON\_i}))}{S_4} - \frac{\hat{r}\sin(2\pi f_m t_{i+1})}{S_2}. \tag{4}$$

The duty cycle of this hysteresis mode control can be expressed as,

$$d(t)|_{0\leq t\leq MT_{SW}} = \sum_{i=1}^{M} \left[u(t - t_i) - u(t - t_i - T_{ON\_i}) \right], \tag{5}$$

where $u(t)$ is the step function. Thereafter, Fourier analysis is performed on (5), and the transfer function from V_{COMP} to PWM, $DF(s) = \frac{\hat{d}}{\hat{v}_{COMP}}$, is:

$$DF(s) = \frac{f_{SW}}{1 - e^{sT_{SW}}} \left(\frac{-1 + e^{sT_{OFF}}}{S_3} + \frac{-e^{sT_{SW}} + e^{sT_{OFF}}}{S_2} + \frac{-e^{sT_{SW}} + e^{sT_{ON}}}{S_1} + \frac{-1 + e^{sT_{ON}}}{S_4} \right) \tag{6}$$

3. Closed Loop Output Impedance Model and Its Verification

A closed-loop converter is shown in Fig. 4-a. To model its output impedance, an equivalent block diagram is shown in Fig. 4-b. The load current perturbation (shown in Fig. 4-b) at the output terminal of the buck converter ($\hat{i}_{LOAD}$) would cause the disturbance in the inductor current $\hat{i}_L$ and an additional AC current $\hat{i}_{CAP}$ from the parallel of the output capacitors and R_{LOAD}. Note that R_{LOAD} itself only represents the steady state DC load operation, and not interface with injected $\hat{i}_{LOAD}$. Due to the closed loop regulation, $\hat{i}_L$ is determined by the difference between two perturbed voltage sources $\hat{v}_{PHASE}$ and $\hat{v}_{OUT}$, while $\hat{i}_{CAP}$ is determined by $\hat{v}_{OUT}$ only. The frequency domain analysis can be conducted by applying KCL and KVL in Fig. 4-b:

$$i_L(s) = \frac{v_{PHASE}(s) - v_{OUT}(s)}{L \cdot s + DCR} = \frac{-A_V DF(s)V_{IN}v_{OUT}(s) - v_{OUT}(s)}{L \cdot s + DCR}, \tag{7}$$

© VDE VERLAG GMBH · Berlin · Offenbach

553

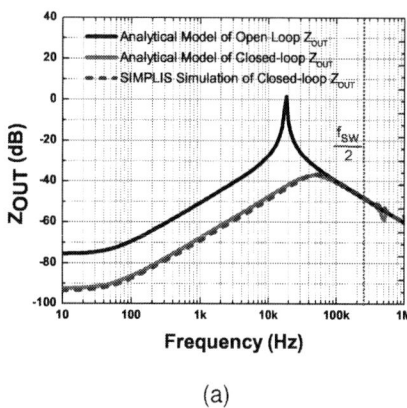 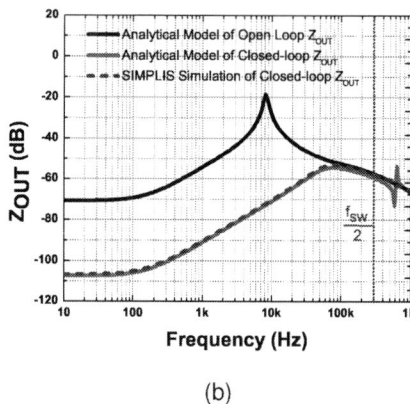

(a)   (b)

Fig. 5: Closed-loop output impedance model versus SIMPLIS simulation results: (a) case A (b) case B

$$i_{CAP}(s) = \frac{v_{OUT}(s)}{(\frac{1}{C_{OUT}\cdot s} + ESR)//R_{LOAD}}, \tag{8}$$

$$i_{LOAD}(s) = i_L(s) - i_{CAP}(s). \tag{9}$$

The small-signal output impedance can be obtained by solving (7) to (9), which yields

$$Z_{OUT}(s) = \frac{v_{OUT}(s)}{i_{LOAD}(s)} = \frac{1}{\frac{1}{(\frac{1}{C_{OUT}\cdot s}+ESR)//R_{LOAD}} + \frac{A_V DF(s)V_{IN}v_{OUT}(s)+1}{L\cdot s+DCR}}. \tag{10}$$

To verify the proposed model, the analytical model in (10) and the simulation results from SIMPLIS are compared. Two cases distinguished configurations are used, while the system requirements, the power stage parameters, and the controller are summarized in Table 1 and Table 2. The equivalent series resistances (ESR) of the $22\mu F$ ceramic capacitor and the 470 μF bulk capacitor are 4 $m\Omega$ and 6 $m\Omega$, respectively. For the sake of simplicity, the conducting resistances (R_{DS_ON}) of both switches, and the equivalent series inductance (ESL) of the output capacitors are ignored in the analysis.

	$V_{IN}(V)$	$V_{OUT}(V)$	$Load(A)$	$L_{OUT}(nH)$	$DCR(m\Omega)$	$C_{OUT}(\mu F)$
Case A	5	3.3	30	470	0.165	20×22
Case B	12	1	25	300	0.29	2×470+17×22

Tab. 1: System requirements and power stage configurations of two cases for model verification

	$f_{SW}(kHz)$	$R_r(k\Omega)$	A_V
Case A	500	800	7
Case B	600	200	14

Tab. 2: R4™ modulator configurations of two cases for model verification

The well matched plots in Fig. 5-a and Fig. 5-b validate the effectiveness of the model proposed in (10). By comparing the open loop output impedance, one can observe that in both cases, the buck converter's peaking at the resonance frequency of the output filter has been eliminated by the closed loop operation.

© VDE VERLAG GMBH · Berlin · Offenbach

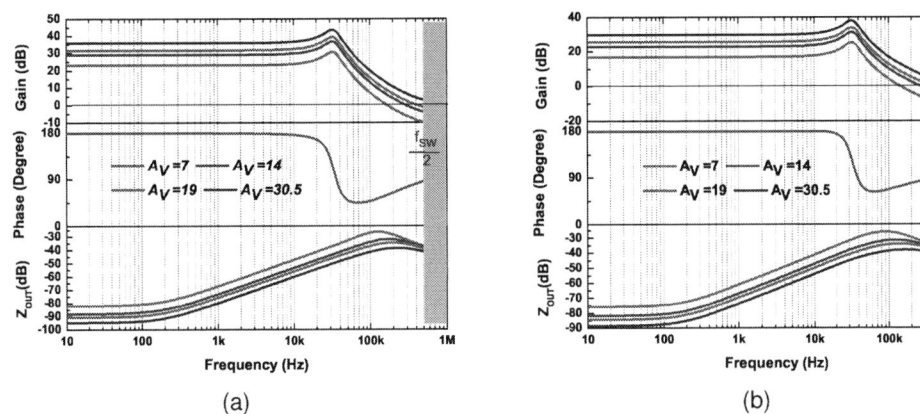

(a)   (b)

Fig. 6: Frequency domain analysis from the analytical models: (a) case 1 (b) case 2

4. Design Considerations and Time Domain Verification

In practical designs, most of the R4[TM] modulator parameters listed in Table 2 can be pro-grammed to optimize system transient response. In this section, design considerations of the two most important parameters, R_r and A_V , for optimizing the control loop are discussed, and the transient simulation waveforms are provided. In this example, the system f_{SW} is $1MHz$, and all the other parameters are summarized in Table 3.

System requirements	$V_{IN}(V)$	$V_{OUT}(V)$	$Load(A)$	
	12	5	10 (for frequency analysis)	
Power train	$L_{OUT}(nH)$	$DCR(m\Omega)$	$C_{OUT}(\mu F)$	$ESR(m\Omega)$
	1000	1.17	2×47	2.5/2

Tab. 3: Configurations of the case for time domain verification

In designing the control loop, users can start with a pre-defined R_r value since it determines the shapes of the synthetic current ripple signal $V_r(t)$ in Fig. 2. In the following analysis, two different R_r are defined as the initial conditions: $400k\Omega$ for case 1 and $800k\Omega$ for case 2. For each case, four different A_V are available for designing the loop: 7, 14, 19, and 30.5. The first step is the frequency analysis: by adapting the loop model proposed in [12], and the output impedance model (10), the bode plots and the output impedance of these two cases are plotted in Fig. 6. Because A_V is only used to tune the loop bandwidth, in each case, only one phase curve is shown for all subcases with different A_V. By probing the gain and phase plots in Fig. 6-a, one can observe that the increase of A_V from 7 to 19 results in a bandwidth boost within half of the switching frequency, without any zeros or poles showing up in the right half plane (RHP). Hence, it is reasonable to validate their stabilities, and move ahead to the output impedance plot. By analyzing the output impedance for these three subcases, the amplitude of the output impedance is attenuated as the increase of A_V, which indicates a better transient response in time domain. For the last subcase with A_V 30.5, according to the analytical model of the bode plot, the bandwidth is beyond half of the switching frequency, and the instability of the system is expected. At this point, the output impedance plot of this A_V is no longer valid. For Fig. 6-b, a similar analysis result can be obtained. Therefore, for these two cases, the transient response performance of the subcase with A_V=19 is expected to be the best, while the subcase with

PCIM Europe 2016, 10 – 12 May 2016, Nuremberg, Germany

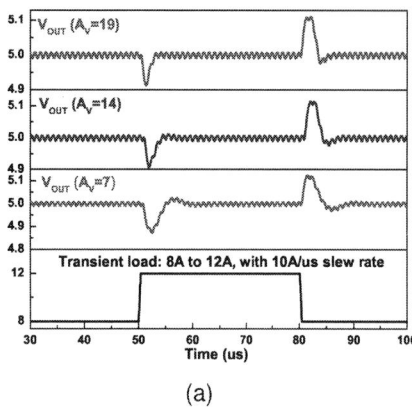
(a)

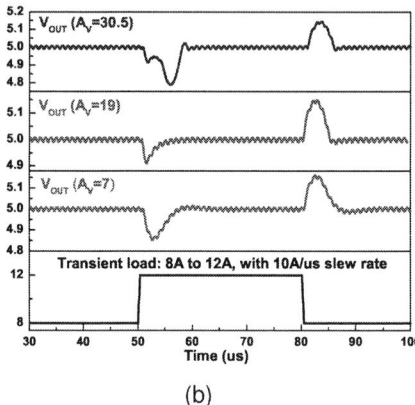
(b)

Fig. 7: Simulation results of transient tests: (a) Case 1 ($R_r = 400k\Omega$) (b) Case 2 ($R_r = 800k\Omega$)

A_V=30.5 is unfavorable.

With the frequency analysis concluded, the transient tests can be performed. Because the large-signal performance of the ripple-based regulator does not exactly follow the small-signal behavior, the phase margin shown in Fig. 6 is only useful as evidence of the stabilities. According to the analytical models, the small-signal behaviors of any DC load levels between 8A to 12A are very close to the 10A case shown in Fig. 6 and therefore the steady state 10A case can be used to partially predict transient performance when the 8A to 12A load step is applied. Therefore, the profile of the transient load is defined as an 8A to 12A load step, with $10A/\mu s$ slew rate.

The transient test simulated waveforms are plotted in Fig. 7-a and Fig. 7-b, both including three subcases with different A_V. In Fig. 7-a, V_{OUT} ringback is seen in both load insertion and release when $A_V = 7$ was chosen. The ringback at load insertion was alleviated at the subcase of $A_V = 14$, and successfully removed at the subcase of $A_V = 19$. Regarding the ringback at load release, the performance of the subcase of $A_V = 14$ and $A_V = 19$ is better than that of the subcase with $A_V = 7$. Increasing A_V also helps reduce the amplitude of the overshoot/undershoot. On the other hand, in Fig. 7-b, a similar dynamic performance improvement can be observed, by comparing the subcases with $A_V = 7$ and $A_V = 14$. For the subcase of $A_V = 14$, the V_{OUT} ringback at load insertion and load release is eliminated, revealing an optimally designed control loop. For the subcase with $A_V = 30.5$, the model prediction of a loop bandwidth higher than half of the switching frequency is still useful. Because in bench test operation, the theoretical bandwidth may be "bent" to be less than half of the switching frequency by the sidebands effect [7], therefore maintaining an acceptable steady state operation. However, the dynamic response in such cases could jeopardize the system's stability. Furthermore, high bandwidth can also cause severe PWM jittering, which degrades EMI performance during steady state operation.

In summary, to optimize the loop for the transient tests of a R4™ modulator, the user should start from the frequency analysis to locate the A_v levels within a reasonable range. The time domain tuning will also be necessary. The best dynamic performance can be achieved efficiently, considering the programmability of the control loop parameters. Compared with conventional voltage/current mode control schemes, the R4™ modulator provides a simple and elegant solution in designing the control loop, and does not require any effort in designing type II/III

© VDE VERLAG GMBH · Berlin · Offenbach

compensation networks. In addition, due to the low output impedance level at low frequency range (-90dB and -110dB in Fig. 5-a and Fig. 5-b), the DC regulation is tightly maintained, without introducing a zero at the origin in the loop, which would result in a more complicated design and a slower transient performance.

5. Conclusion

In this paper, the small-signal output impedance model of Intersil's R4™ technology was modeled using the describing function method. A simplified controller circuit configuration is introduced, with detailed modulation law and perturbed waveforms. This model is verified by comparing the SIMPLIS simulation results and Mathcad calculations, both in frequency and time domain. This model can be used to predict the buck converter's dynamic response to load transients in time domain.

Acknowledgement

The authors are grateful to Mr. Mark Alden of Intersil Corporation, Milpitas, CA, for his review and edit of the manuscript.

6. References

[1] Jian Sun. Characterization and performance comparison of ripple-based control for voltage regulator modules. *IEEE Trans. on Power Electronics*, 21(2):346–353, 2006.

[2] Kuang-Yao Cheng et al. Digital hybrid ripple-based constant on-time control for voltage regulator modules. *IEEE Trans. on Power Electronics*, 29(6):3132–3144, 2014.

[3] Rhys SA Philbrick et al. Switching regulator with balanced control configuration with filtering and referencing to eliminate compensation, April 10 2012. US Patent 8,154,268.

[4] ISL68200: Single-Phase R4™ Digital Hybrid PWM Controller with Integrated Driver, PMBus and PFM, Data Sheet, Intersil Corporation, 2016. Available: http//www.intersil.com.

[5] ISL68201: Single-Phase R4™ Digital Hybrid PWM Controller with PMBus/SMBus/I^2C and PFM, Data Sheet, Intersil Corporation, 2016. Available: http//www.intersil.com.

[6] Shuilin Tian et al. Small-signal analysis and optimal design of external ramp for constant On-Time V^2 control with multilayer ceramic caps. *IEEE Trans. on Power Electronics*, 29(8):4450–4460, 2014.

[7] Shuilin Tian et al. Unified equivalent circuit model and optimal design of V^2 controlled buck converters. *IEEE Trans. on Power Electronics*, 31(2):1734–1744, 2016.

[8] Shuilin Tian et al. Small-signal analysis and optimal design of constant frequency control. *IEEE Trans. on Power Electronics*, 30(3):1724–1733, 2015.

[9] Jian Li and Fred C Lee. Modeling of V^2 current-mode control. *IEEE Trans. on Circuits and Systems I*, 57(9):2552–2563, 2010.

[10] Jian Li and Fred C Lee. New modeling approach and equivalent circuit representation for current-mode control. *IEEE Trans. on Power Electronics*, 25(5):1218–1230, 2010.

[11] Shuilin Tian et al. Three-terminal switch model of constant On-time current mode with external ramp compensation. *IEEE Trans. on Power Electronics*, 2016.

[12] Yi Huang and Chun Cheung. Small signal modeling of the hysteretic modulator with a current ripple synthesizer. In *31th Annual IEEE Applied Power Electronics Conference and Exposition*, 2016.

An Approach of Reinforcement Learning Based Lighting Control for Demand Response

Xinxing Pan, Athlone Institute of Technology, Athlone, Ireland, xpan@research.ait.ie

Brian Lee, Athlone Institute of Technology, Athlone, Ireland, blee@ait.ie

Abstract

Lighting is a major contributor of building energy consumption. Lighting systems will thus be one of the important component systems of a smart grid for dynamic load management services such as demand response (DR). We consider the problem of autonomous control of multiple lighting systems in a building for providing DR Service, while keeping occupants' illuminance comfort. To achieve an online and adaptive control for lightings, we propose to use reinforcement learning (RL) rather than other intelligent control algorithms to learn the lighting system environments with consideration of both DR signals and users' illuminance requirements, for lighting control.

1. Introduction

Lighting energy consumption is one of the main electricity consumption contributor in commercial and residential buildings. Lighting systems are reported to consume 20-35% of the energy used in buildings, and 38% of the used electricity, more than any other end use [1]. For commercial buildings sector, lighting systems consume about 25% of the electrical energy used in US [2]. Lighting, therefore, has a major impact on electricity demand in commercial buildings. However, through recent publications the usage of lighting in buildings is not efficient enough. First, there are currently lots waste of lighting energy in buildings, which refers to occasions of lighting on with no occupancy at all. Second, over lighting exists in many occasions. Over-lighting or over-illumination is the presence of lighting intensity higher than that which is appropriate for a specific activity, which was commonly ignored, especially in office and retail environments [3]. Third, lighting harvest has not been used properly [4].

Accordingly, lighting systems have strong potential for dynamic load service such as Demand Response in building energy demand side management. With the advent of easily and accurately dimmable sources such as light emitting diodes (LEDs), lighting systems have become attractive as controllable loads to support demand response [5].

However, lighting DR is still slow on the uptake due to several reasons. First of all, lighting is much relevant to health, safety and productivity, which requires a careful plan for any changes on lighting visual comfort. Another obstacle is the limitation of current technologies, although the wireless network communication and dimming control technologies are already applicable, they are not cheap enough to take place of current lighting devices. Last but not the least, there's a lack of advanced design on demand response programs for lighting energy. In academic area, according to our study on up-to-date research, we found that there are many research work being done in general energy improvements and savings, or specifically on HVAC in buildings. Comparatively, very less research focus on demand response for lighting. Another issue is that traditional lighting controls depend much on manual work which makes lighting energy consumption involved with demand response control even more difficult. In order to solve the latter problem, intelligent controls for lighting have been drawn attentions nowadays. Both centralized control and distributed control integrated with intelligent algorithms such like fuzzy logic, genetic algorithm and artificial neural network are investigated [6]. However, these implementations are short at dealing with dynamic lighting environment like changing occupants' behavior, and furthermore, with adding demand response signals to the problem the lighting control become more complex.

For the reasons mentioned above, an online reinforcement learning (RL) is employed for the lighting control to implement demand response programs and occupants illuminance requirements. The solution comprises a Q-learning approach to learn a lighting control policy with considering both energy cost based on DR programs and occupant's preference on illuminance comfort. In summary the main contributions of this paper are

- A learning architecture which can support demand response programs, occupancy preferences and lighting controls, concurrently optimizing both user comfort and energy cost
- A novel state action space formalism for the environment of lighting controls, with daylighting, DR programs and user behaviors

2. Background Research

2.1 Lighting management system (LMS)

In-building lighting management system provides mainly interior illumination by using artificial lighting, aiming at a controllable lighting demand supply. Artificial lighting has three main influences on occupants in buildings, the visual impact, the health impact and the emotion impact [7]. Thus the daylighting and different work visual requirements have to be considered due to visual desire, the minimum illuminance level should be settled for consideration of health, and for the impact of occupants' emotion, the individual occupant's preference needs to be investigated.

2.1.1 Schedules

A schedule normally means a plan based on time frames. A basic control strategy for a LMS is to use lighting schedules to control the luminaires' status, on or off, based on the time. Proper schedules for lighting could help a LMS avoiding unnecessary usage of lighting and achieving an easy and safe lighting experience. For example, after working time all luminaires except emergency lightings are turned off automatically based on the schedule can avoid the waste of lighting energy due to that people might forget to turn the lights off. Another example is that turning on the lightings 10 minutes ahead of the working time would let people access their building easier and safer.

2.1.2 Zone function based standards

Not everywhere inside of the building demands the same illuminance level. For example, the illuminance in the corridors might not be needed as strong as in the meeting room. A specification on standard illuminance level in different type of rooms can be found in the European Committee for Standardization (CEN) [8].

2.1.3 Daylighting

The term of Daylighting or daylight harvesting, generally means the illumination of buildings by natural light. Daylighting has been recognized as an essential element in architecture for enhancing visual comfort, energy-efficiency and green building developments. Even a well daylighted building may have a high level of lighting energy use if the lighting controls are inappropriate. Hence, an appropriate lighting control linked with day-light can save electric lighting energy consumption and reduce peak electrical demands.

2.1.4 Occupancy and behavior

User occupancy and behavior can have a large impact on the energy consumption, including the lighting usage, of a building and hence on DR programs [9]. A number of field studies have been carried out over the years to investigate the relationship between occupancy and lighting. Mahdavi conducted a long term study – over the period of a year – of people's interactions with the buildings' environmental systems (lighting, shading, ventilation) in a number of office buildings in Austria [10]. He posits that, given sufficient observations, statistical methods can be used to generate individual occupancy patterns. Sadeghi [11] found that better user interfaces can encourage more use of daylighting and consequent energy savings. He also found a strong relationship between occupant perception of control and acceptability of a wider range of visual conditions.

2.1.5 Demand Response Programs

As mentioned in the introduction, there is great potential in lighting energy consumption for demand response programs to work. Accordingly, DR programs could be an important factor to affect the lighting use in buildings e.g. by elastic pricing. More details are discussed in the following section.

2.2 Lighting Energy Potential in Demand Response (DR) Programs

DR is a commercial program that requires communication and feedback between the electricity supply side and demand side in order to achieve a better grid balance. In simplest term, DR refers to the load shed during peak load time.

2.2.1 load based DR

DR is effected by either by shifting load or shedding load. While both approaches can be applied to thermal loads mechanisms such as HVAC or some forms of pluggable and production loads only the latter can be applied to lighting i.e. while lighting load can be reduced it cannot be postponed. Active lighting demand response involves acting upon the capability to specifically reduce load during peak demand periods or periods of high pricing (economic demand response) or upon utility request during times of grid stress (emergency demand response). This capability requires the ability to measure lighting load at any point in time, accept a utility signal to start, stop and measure a load shed event; and temporarily reduce lighting load by a significant amount while respecting occupant lighting needs. While it's possible to manually shut off non-critical loads, automating the process and incorporating dimming can be more reliable and less disruptive.

2.2.2 pricing based DR

In smart grid, price based programs are designed to lower consumption in high critical events or as a continuing incentive to reduce overall demand. Price signals are sent from the DR program servers to the LMS in the building, for example by the advanced meter. Roofegari proposed an algorithm for controlling consumption level of lighting loads in a smart home by real-time pricing signals [12]. His work depicted the relationship between the occupants' accepted illuminance level (from $E_{suggest}$ to E_{min}) and the electricity price σ_p, in one aspect. It is

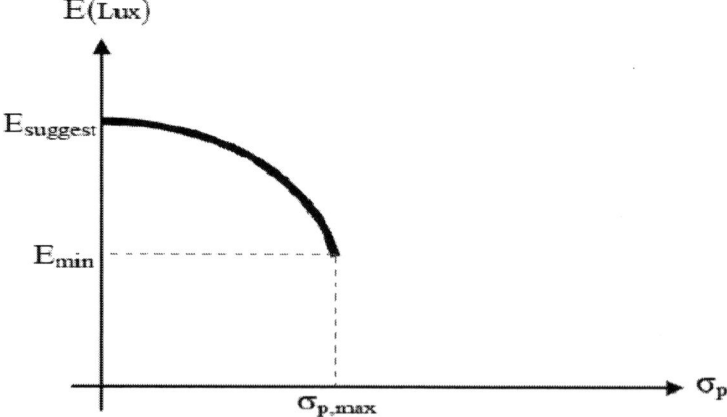

Figure 1: Illumination versus electric market price curve which schematically demonstrate the proposed controlling method [12]

presented in fig.1 that the deviation of illumination related to price described by a parabola curve with downward concavity, which makes dimmable lighting controls applicable in response to DR programs.

3. Reinforcement Learning Concept and Methods
3.1 Overview of RL

RL is a field of machine learning in which an agent learns to take actions in an environment in such a way that it maximizes its performance [13]. The performance is measured in terms of rewards that the agent gathers while acting in the world. The aim of a reinforcement learning algorithm is to find a policy that maps states of the world to actions that the agent should take when being in these states.

RL problems are often formulated as (Partially Observable) Markov Decision Processes ((PO) MDPs). Markov Decision Processes (MDPs) provide a mathematical framework for modelling decision-making under uncertainty. The MDP model assumes that the next state is solely determined by the current state (the Markov assumption). It also assumes that the state that the model is in is completely observable. This means that the current state has to be completely known at all times. A Markov Decision Process can be described as a tuple {S; A; T; R; Y} where:

- S is a finite set of world states.
- A is a finite set of available actions.
- T(s, a, s') is the state transition function, it denotes the probability of ending up in state s', given the current state s and action a.
- R(s, a) is the reward function. This function gives the expected reward gained by taking each action in each state.
- Y [0, 1] is the discount factor, weighing the reward function in such a way that rewards in the near future are of a greater influence than rewards later in time.

3.2 RL methods

There are three major categories of reinforcement learning methods, Dynamic Programming (DP) methods, Monte-Carlo (MC) methods and Temporal Difference (TD) methods. The DP is a method for solving a complex problem by breaking it down into a collection of simpler subproblems. The MC is a class of computational algorithms that relies on repeated random sampling to compute results. The TD learning methods combine the advantages of MC and DP methods by learning from interaction with the environment on a step by step basis, without requiring a model of the environment.

Q-learning algorithm, as one of TD methods, can be used to find an optimal action-selection policy for any given (finite) MDP. It works by learning an action-value function that ultimately gives the expected utility of taking a given action in a given state and following the optimal policy thereafter. One of the strengths of Q-learning is that it is able to compare the expected utility of the available actions without requiring a model of the environment. Additionally, Q-learning can handle problems with stochastic transitions and rewards, without requiring any adaptations.

3.3 RL applications in LMS

RL is much different from other machine learning approaches, for example the supervised learning which is based on leaning from examples provided by an external supervisor. RL is not dictated which actions to take, but need to interact with the environment and discover the actions which yield the most reward by trying them. This feature makes RL more and more popular as a solution to decision making problems with uncertainties in the electricity markets and consumer clients e.g. the dynamic pricing and consumers' preference.

It is a challenge of building DR for lighting energy consumption due to its non-shift-able load. There are still some approaches using RL for lighting controls. Tomoyuki [14] used the actor-critic algorithm to make the users to be able to set the brightness of the system through sensory operation, such as "much brighter" or "slightly darker". The RL agent is used to learn the users' sensory scale, such as "very" or "slightly" and finally would decrease the user burden by efficiently lighting controls. Bielskis [15] proposed a concept of using RL as one component in a multi-agent system for ambient lighting controls. The human comfort is expressed as an ambient lighting affect reward in the RL agent. Both of these approaches focus on lighting controlling without the consideration of DR features such as dynamic pricing and load shedding.

4. Lighting Control

This section discusses the specifics of applying each technology to the domain. We present a novel state action space formalism for Q-learning which enables it to effectively control lighting in an online manner. We formulate the problem as a Markov Decision Process and propose an optimization approach based on reinforcement learning (RL). An overview of our approach is described in fig. 5-2, where the basic concept of RL has applied in the lighting controls. The electricity price, daylighting and occupants' preference on the left form the environment the lighting controller should consider. On the right side, the set of actions comprises three options in this case. At each learning epoch or time step the agent, represented by the lighting controller in the picture, observes the state of the environment and chooses an action a, either to dim-up, dim-down or maintain the dimming level of the luminaires. After the action being executed, the reward and new states would update the knowledge the agent mastered. Eventually, the RL based lighting controller will be able to control the lighting automatically, with the consideration of both energy cost and occupants' comfort, which are illustrated in more details in the following sectors.

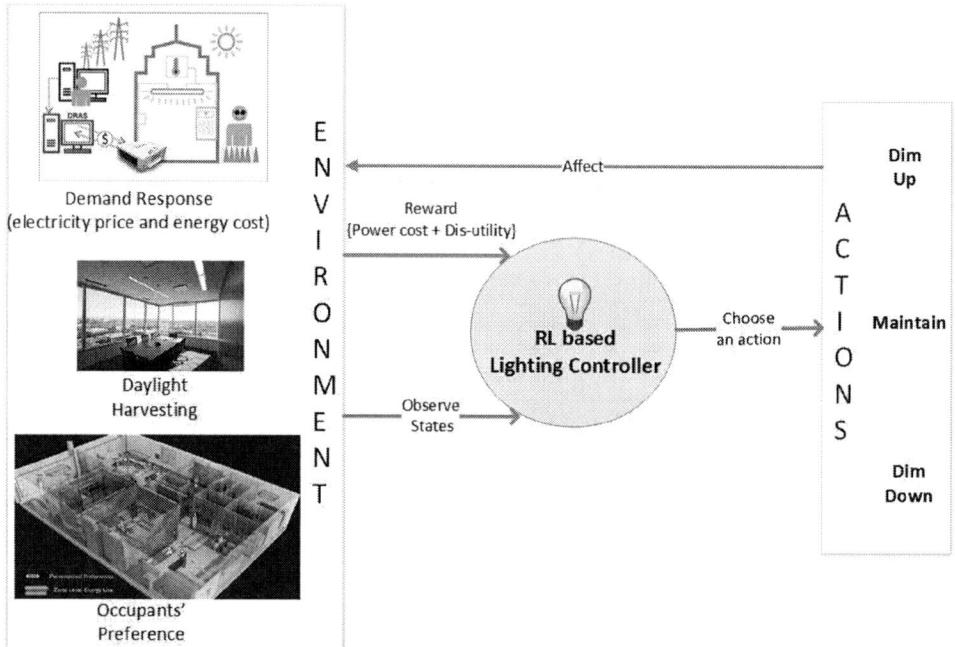

Figure 2: Structural schema of a RL model for lighting control

4.1 Electricity Pricing from DR Programs

To simplify our illustration on pricing based DR programs, we take real-time pricing as an example in our case study. Referring to [16], from customers' side, we consider that there is an advanced LMS which can receive real time variable pricing. Time is discrete t = 1, ..., and pricing signals are indicated by the pricing sequence p(t) $\in R_+$. In a LMS, the luminaires to be controlled have many dimmable levels, m = 1, ..., M, such as 0%, 37% or 100%, for example, take 1% as the interval, M will be 101. Each dimmable level requires a certain quantity γ_m of electrical energy to operate and complete its work. In some cases this quantity can be spread over several time slots to achieve, perhaps by controlling the luminaires with dim-up or dim-down over several times. The LMS controller selects a dimmable level of the luminaires and indicate the desire to run it by, for example, dimming up. The *controller reservation vector z(t)* $\in R_+^M$ captures this controlling action. If at time *t*, the controller selects the dimmable level *m*, then $z_m(t)$ = γ_m. However, this desire does not need to be achieved instantly as the controller would optimize its control strategies to minimize the energy cost but still satisfy the occupants' need, with consideration on pricing p(t) and other factors.

The sequence of controller reservation vectors, $z(t)$, captures the visible actions of the set of control strategies in a LMS. Frequently, there will be statistical relationships among controller's selections about what illuminance level to be achieved that can be exploited to reduce overall costs (*e.g.* using lower dimmable level when the price is high or daylighting is strong.) Thus, we model $z(t)$ as a Markov chain.

Similarly, the energy pricing sequence, $p(t)$, is assumed to be Markov, also with an unknown probabilistic structure. Note that the correlation between the processes $z(t)$ and $p(t)$ can be easily modeled by representing the overall system's state as the aggregate $(z(t), p(t))$. In this case, the algorithm will learn the statistics of the aggregate process.

4.2 Modelling User Preferences and Behaviours

User occupancy and behaviour, as one major influence factor to lighting usage, is hard to model or predict. However, the occupants' actions are the result of complex behaviour and are strongly dependent on the occupants' education, responsibility and other factors. That is possible to characterize occupant behaviour as a composition of observable states that represents actions, and hidden states that we cannot observe which represent the complex behaviours that have influenced these observed actions. Consequently, a hidden markov model (HMM) can be used to solve such a problem. One approach is to consider the whole building as a set of spaces and for each space the occupant occupies as a HMM [17]. This approach inspires us that we can model the occupants' behaviour as a HMM, where a RL can be applied without need to consider the hidden states as illustrated in [13].

Another issue needed to be clarified in this sector is the occupant's preference on illuminance, which can be formulised as utility functions. Utility functions are an abstraction used to numerically model that consumers prefer one product versus another or, as used here, that consumers have time preferences for things. Dainel chose to use a dis-utility function to represent the negativity of this utility function in solving users' preference in arranging devices usage plan [16]. Similarly, we use the dis-utility function $U_m(t) \in R_+$ to represent the dissatisfaction the occupants have on the illuminance. Here we make an assumption that, occupants in building offices would prefer to have the maximum comfort level of illuminance sooner rather than later, the longer the occupants stay in an un-satisfied illuminance level, the bigger value the dis-utility function would have. So that this preference can be expressed as a strictly concave utility function.

4.3 A Q-learning based Lighting Controller for Demand Response

The lighting controller employs Q-learning by framing the environment as a MDP. Based on the discussions we had in previous sectors, we must firstly define the set of states S and actions A.

States: let S be the States of the environment,

$$S = \{D_{out}, D_{in}, P_{rt}, L_{current}\}$$

- D_{out} : is the outside illuminance level (source: illuminance sensor, unit: lux)
- D_{in} : is the inside illuminance level (source: illuminance sensor, unit: lux)

- P_{rt} : is the real-time pricing based on DR programs (source: price signals from DR aggregator/agent, unit: $/kwh)
- $L_{current}$: is the current dimming level (source: lighting control panel, percent (%))

Actions: let A be the actions of the environment,

$$A = \{A_{up}(n), A_{down}(n), A_{keep}(n)\}$$

- $A_{up}(n)$: dim up the luminaires with n percent of the maximum lighting or n lux
- $A_{down}(n)$: dim down the luminaires with n percent of the maximum lighting or n lux
- $A_{maintain}$: maintain the dimming level of the luminaires

The idea is to try to keep the number of sates and actions low so that the problem remains within the bounds of tractability. However, even though there are only four state variables selected, each state variable can take on a wide range of values, or even actually infinite values in practice, which to create a relatively large state space. For instance the illuminance level of the daylighting is, or close to, infinite in values that can vary from 120,000 lux for direct sunlight at noon, to less than 5 lux for thick storm clouds with the sun at the horizon (even <1 lux for the most extreme case). In addition the price single based on the real-time pricing DR programs could vary from very low to pretty high value, according to the dynamic demand and supply in the smart grid market.

As lighting is related to human health, where the changing of illuminance level of lighting should be controlled within a certain frequency. Hence, lighting controller actions are executed at discrete time intervals known as epochs. At the end of each epoch the learning agent observes the current state of the environment and chooses whether or not to execute an automated lighting controlling action (dim-up, dim-down, maintain).

The transition probabilities T, is the likelihood of transitioning between states after executing particular actions, which is unknown, so that this problem cannot be solved using Dynamic Programming methods such as Policy or Value iteration. In addition we do not attempt to estimate T, instead we let Q-learning to observe the consequences of T and adjusts accordingly.

The reward function R, is defined as the combination of two types of costs: financial cost of consuming energy and the dis-utility of waiting for the luminaries to dim up to its maximum comfort level. The performance metric of the RL lighting controller based on DR programs is the average total discounted financial and dis-utility costs over an infinite horizon. The financial cost at time t is simply $\sum_{m=1}^{M} P_{rt}(t) L_{current,m}(t)$ and the dis-utility of delay at time t is $\sum_{m=1}^{M} U_m(D_{in},t)$. Because U_m is strictly convex, positive and increasing, every illuminance desire will eventually be satisfied. For a given state, Ω ($D_{out}(t)$, $D_{in}(t)$, $P_{rt}(t)$, $L_{current}(t)$), and energy control policy, $u(t)$, the cost incurred is the sum of the financial cost and dis-utility cost $R(\Omega(t), u(t)) = \sum_{m=1}^{M} P(t) L_{current,m}(t) + \lambda U_m(D_{in},t)$, where $\lambda \geq 0$ determines the trade-off between the financial cost and dis-utility cost and can be seen as the dollar cost of average delay. System performance is defined as the total discounted period costs over an infinite horizon $V_u(\Omega_o) = \lim_{T \to \infty} E\left[\sum_{t=o}^{T} \gamma^t R(\Omega(t), u(t))\right]$, where $\Omega 0$ is the initial state of the system and $0 \leq \gamma \leq 1$ is the discount factor. The discount factor reflects the fact that the consumer may give greater relevance to the immediate financial and dis-utility cost. The expectation operator is over *the time step*, given the system starts from Ω_0. The variable $V_u(\Omega_0)$ is the total discounted cost under the energy control policy u. Different policies can result in different total discounted costs. The objective is to find the best stationary energy control policy $u*$ that minimizes $V_u(\Omega_0)$. Policy u∗ is optimal if one can find a function $V(\Omega)$ such that $V(\Omega) = \min_u (R(\Omega, u) + E[V(\Omega_1)])$ where Ω_1 is the system state after using the energy control policy $u(\Omega)$. The function $V(\Omega)$ is called the value function and is the optimal cost when starting from state Ω. Unfortunately, Equation $V(\Omega)$ is very difficult to solve analytically [18]. Numerical methods can be used when the state space $|\Omega|$ is small and the transition probability distributions are known. However, here the Markov transition probabilities for states' transitioning and $P_n(t)$ are assumed unknown. Consequently, Q-learning is used in our approach. The relationship between the V value and Q-functions is that $V(\Omega) = \min_{u} Q(\Omega, u)$.

At state Ω, Q-learning samples the behaviour of the system in response to using control policy u and then updates an estimate of the Q-function $Q(\Omega, u)$.

Then Q-learning algorithm goes as follows:
1. set the gamma parameter, and environment rewards in matrix R.
2. Initialize matrix Q to zero.
3. for each episode:

    Select a random initial state.

    Do while the goal state (convergent) hasn't been reached.

- Select one among all possible actions for the current state.
- Using this possible action, consider going to the next state.
- Get maximum Q value for this next state based on all possible actions.
- Compute: $Q(s, a) = Q(s, a) + \alpha[r + \gamma \max Q(s', a') - Q(s, a)]$
- S $\leftarrow$ S', Set the next state as the current state.

    End Do

End For

5. Conclusions

In this paper, we proposed a novel approach, using RL agent as an automatic lighting controller, to implement DR programs for lighting management system. We also provided a fully understanding of influence factors which affect the lighting controls, discussed the lighting potential of implementing the DR programs, and illustrated the state action space in lighting control environment. In the future, the proposed state action space formalism could be extended further to give a greater observation over the environment. Hence, an approximation function might be in need for better presentation of large state action space. In addition, a field survey of occupants' illuminance preference will be conducted, to help a more practical design of trade-off between energy cost and the dis-utility.

Reference

[1] Behjat Hojjati and Steven H. Wade, "Commercial Buildings Energy Consumption and Intensity Trends", 2012
[2] Rubinstein, Francis, "Demand Responsive Lighting: A Scoping Study" LBNL, 2010
[3] M.D. Simpson, CIBSE, 1990
[4] Danny H.W.Li, "Study of daylight data and lighting energy savings for atrium corridorswith lighting dimming controls" Energy and Buildings, 2014
[5] Sri Andari husen, ""Lighting Systems Control for Demand Response", Innovative Smart Grid Technologies (ISGT), 2012
[6] Konstantinos Dalamagkidis, "Reinforcement Learning", book, 2008
[7] SCENIHR," Scientific Committee on Emerging and Newly Identified Health Risks: Health Effects of Artificial Light" , 2012
[8] EN 12464-1, standard
[9] Marcia Baptista, Anjie Fang, Helmut Prendinger, "Accurate Household Occupant Behavior Modeling Based on Data Mining Techniques", the Twenty-Eighth AAAI Conference on Artificial Intelligence, 2014
[10]Ardeshir Mahdavi and Claus Pröglhöf, "USER BEHAVIOR AND ENERGY PERFORMANCE IN BUILDINGS", (IEWT), 2009
[11] S.A. Sadeghi, P. Karava, "Stochastic Model Predictive Control of Mixed-mode Buildings Based on Probabilistic Interactions of Occupants with Window Blinds", International High Performance Buildings Conference, 7/15/2014
[12] R. Roofegari Nejad, "A New Method for Demand Response by real-time pricing signals for lighting loads", Power Engineering and Automation Conference (PEAM), 2012
[13] Richard S. Sutton and Andrew G. Barto, ""Reinforcement Learning: An Introduction", book, Second edition, 2015
[14] Tomoyuki Hiroyasu, Second World Congress on Nature and Biologically Inspired Computing, 2010
[15] A. A. Beilskis, E. Guseinoviene, "Ambient Lighting Controller Based on Reinforcement Learning Components of Multi-Agent", Electronics and Electrical Engineering, 2012
[16] Daniel O'Neill, Marco Levorato, "Residential Demand Response Using Reinforcement Learning", Smart Grid Communications (SmartGridComm), 2010
[17]JoãoVirote and Rui Neves-Silva, "Stochastic models for building energy prediction based on occupant behavior assessment", Energy and Buildings, 2012
[18] D. Bertsekas, "Dynamic Programming and Optimal Control". Massachusetts: Athena Scientific, 2005.

Cosmic Ray Failure Mechanism and Critical Factors for 3.3kV Hybrid SiC Modules

Tetsuya Nitta, Toshiba Corporation, Japan, tetsuya1.nitta@toshiba.co.jp
Yoko Sakiyama, Toshiba Corporation, Japan, yoko.sakiyama@toshiba.co.jp
Raita Kotani, Toshiba Corporation, Japan, raita.kotani@toshiba.co.jp
Tomoki Inoue, Toshiba Corporation, Japan, tomoki.inoue@toshiba.co.jp
Ryoichi Ohara, Toshiba Corporation, Japan, ryoichi.ohara@toshiba.co.jp
Kenya Sano, Toshiba Corporation, Japan, kenya.sano@toshiba.co.jp
Masakazu Yamaguchi, Toshiba Corporation, Japan, masakazu.yamaguchi@toshiba.co.jp
Georges Tchouangue, Toshiba Electronics Europe GmbH, Germany,
GTchouangue@tee.toshiba.de

Abstract

The cosmic ray failure mechanism for a 3.3kV IGBT has been investigated. It could be clarified that the dynamic avalanche current dominates the catastrophic self-heating, and the impact of the parasitic bipolar action is negligible for a 3.3kV IGBT unlike a lower voltage class IGBT. For practical use, the integral over the electric field with a specific threshold value was proposed as an index for the robustness for a High Voltage (HV) IGBT such as a 3.3kV class device. The cosmic ray robustness of a 3.3kV SiC Schottky Barrier Diode (SBD) was also investigated: it was found that the failure rate is almost the same as that of 3.3kV class Si devices.

1. Introduction

High Voltage (HV) IGBT modules have been widely used for industrial applications such as railway systems, HVDC (high-voltage direct current), renewable energy systems, etc. The improvement has been contributed to increase the energy efficiency of the systems. For the industrial applications, Long-Term DC voltage Stability (LTDS) of the IGBT is important from the point of view of maintenance costs. One influential factor for the LTDS is cosmic ray failure of an IGBT. On the ground level, it is well-known that neutrons originating from cosmic rays induce catastrophic failures called Single-Event Burnout (SEB). Therefore, considering robustness for neutrons in IGBT structure design is of high importance for LTDS [1].
The neutron induced SEB mechanism of an IGBT is generally considered as follows. The neutrons radiated into silicon generate heavy charged particles by nuclear reaction. The charged particles produce hole and electron pairs losing the energy along the trail. These hole and electron pairs make a high electric field in the reverse biased drift region, and induce dynamic avalanche. As an IGBT includes a parasitic NPN transistor and a PNP transistor (Fig.1), the impact ions produced by dynamic avalanche are amplified, which leads to catastrophic thermal runaway. The influence of the parasitic bipolar action was analyzed and reported in [2]. The paper reports that the SEB failure rate of a diode, which has no parasitic bipolar transistor, is lower than that of an IGBT, and a parasitic bipolar transistor has an important role on SEB for IGBTs. On the other hand, [3] reports that the robustness of IGBTs and diodes are almost identical.

We considered that the SEB failure mechanisms and the important design parameters of IGBTs are different for different IGBT voltage classes. This paper reports the cosmic ray failure mechanism and the critical design factors for a 3.3kV IGBT and diode by means of experimental measurement and simulation analysis. Moreover, in order to confirm the LTDS of a 3.3kV Si-IGBT/SiC-SBD (Schottky barrier diode) hybrid module, the SEB robustness of a 3.3kV SiC-SBD was measured and compared with Si-IGBT/Si PiN diode.

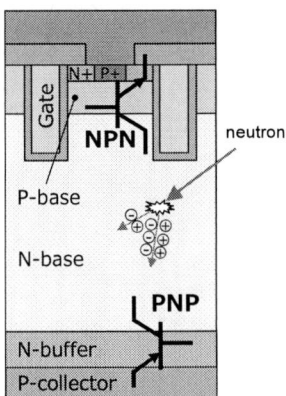

Fig.1. Schematic structure of IGBT and parasitic bipolar transistors.

2. Cosmic ray failure mechanism of HV-IGBT

2.1. Experimental Setting

We measured the neutron induced SEB failure rate by irradiating white neutron beams that were provided by the Research Center for Nuclear Physics (RCNP) at Osaka University in Japan [4]. The energy spectrum of the neutron beam is almost the same as the spectrum of natural neutron irradiation. The flux acceleration rate is about 1.5×10^8 times the natural neutron irradiation at ground level. The neutron beam was irradiated to reverse biased IGBTs or diodes at room temperature. The leakage currents of the devices were monitored during the neutron irradiation for 16 devices at once. The time to failure from the start of the irradiation was recorded individually for each device, and the SEB failure rate was calculated.

2.2. Experimental results

In order to investigate the influence of parasitic PNP action for 3.3kV IGBTs, the SEB failure rates were measured for different collector type IGBTs. The measured structures were standard IGBT, high N-buffer dose IGBT (low PNP gain), high p-collector dose IGBT (high PNP gain) and MOSFET (no PNP gain). The MOSFET type was fabricated by modifying the P-collector layer to an N-drain layer. Fig. 2 shows the measured failure rates as a function of biased collector voltage. The failure rates of the measured devices are almost identical, even for the MOSFET structure. From the result, it can be said that the PNP action hardly affects the SEB failure rate for 3.3kV IGBTs.

As an investigation of the influence of parasitic NPN, the failure rate of 3.3kV PiN diodes were measured and compared with that of the 3.3kV MOSFETs which were measured above. The comparison is shown in Fig.3. The failure rate of the diode is almost the same as that of the MOSFET that includes a parasitic NPN structure. It can be said the NPN action hardly affects the failure rate for 3.3kV class devices.

From the measurements, it is clear that the SEB failure rate of IGBTs, MOSFETs and diodes are identical, and neither PNP action nor NPN action affects SEB for 3.3kV IGBTs.

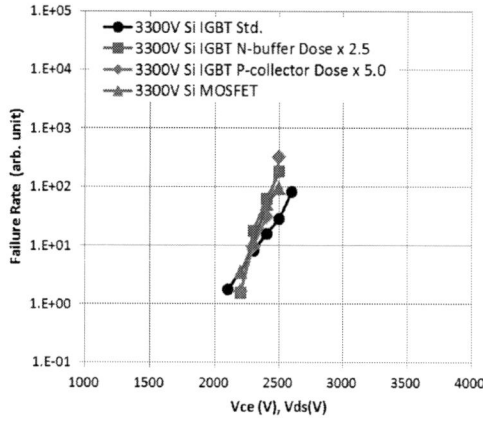

Fig.2. Measured failure rate of 3.3kV IGBTs for the parameters related to parasitic PNP gain.

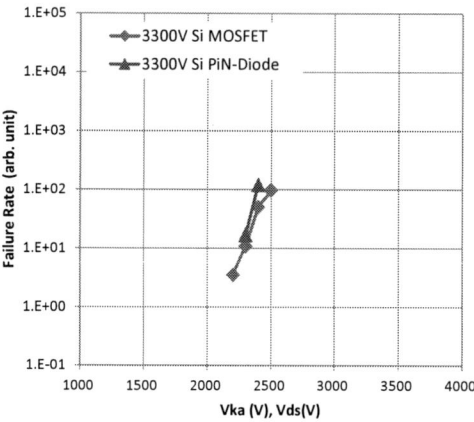

Fig.3. Measured failure rate of 3.3kV MOSFET and diode.

2.3. Simulation Method

We carried out device simulation to understand the experimental result that neither NPN nor PNP affects SEB failure rates for 3.3kV IGBTs. The simulation was performed with a Synopsys Sentaurus Device using the "heavy ion model" [5]. In the model, a heavy ion penetrates the device, and the energy loss creates a trail of electron/hole pairs. The SEB phenomena can be simulated by generating heavy ions into the reverse biased IGBT. We also took the self-heating effect into account in order to simulate device destruction due to thermal runaway. The simulation was performed with a simplified 2-dimensional IGBT structure with essential elements; N-emitter, P-base, N-base, N-buffer and P-collector. As the measurement was performed in off-state for all time, the gate structure was omitted from the simulation.

2.4. Simulation results

The simulation was performed for 1.2kV IGBTs and 3.3kV IGBTs in order to verify whether the SEB mechanism is the same or not depending on the voltage class. The difference in simulation settings is only the thickness and resistivity of the N-base layer. The response of collector current (Ice) and maximum junction temperature (Tmax) to a heavy ion penetration was simulated for various collector biases. The simulation results for the collector biases near a destructive condition for a 1.2kV IGBT and 3.3kV IGBT are shown in Fig.4. In Fig.4, the collector bias condition is indicated with a static peak electric field in the off-state.

For the 1.2kV IGBT with the condition that the electric field is less than 1.49×10^5 V/cm, the collector currents diminish in 60ns, and the peak of Tmax is not so high. On the other hand, for the condition that the electric field is 1.53×10^5 V/cm, the collector current runs away just after the current peak, and the Tmax rises to over 800K. In this case, the device may destruct (Fig.4(a)) [6].

For the 3.3kV IGBT, the collector peak currents are lower, but the collector currents continue longer than that of the 1.2kV IGBT. Although current runaway does not occur for all simulated conditions for the 3.3kV IGBT, peak Tmax exceeds 800K when the electric field is over 1.29×10^5 V/cm, accordingly the device may destruct in that case (Fig.4(b)).

Fig. 5 shows the simulated electric field distribution depending on times for a 1.2kV IGBT and 3.3kV IGBT for Case C in Fig.4(a) and Case C' in Fig.4(b) respectively. The heavy ion is set at the center of the N-drift layer for each (see the curve of 1.0×10^{-11} sec). Just after the heavy ion irradiation (~1×10^{-9} sec), the electric field at the emitter side increases, and causes impact ionization for both cases. For the 1.2kV IGBT, the electric field at the collector side intensely increases after parasitic NPN activation (7×10^{-9} sec~). The collector side high electric field causes reproduction of the base current of the NPN, and positive feedback of the self-heating. On the other hand, for a 3.3kV IGBT, self-heating occurs without an increase of the electric field at the collector side (2.5×10^{-8} sec~). Even for an 800K condition (at 1.07×10^{-7} sec), the electric field at the collector side is low.

From the results, we can conclude as follows: for the 1.2kV IGBT, the energy of the dynamic avalanche is small. The destruction is caused by the parasitic bipolar action at the current peak and following positive feedback of the collector side impact ionization. As for the 3.3kV IGBT, the avalanche current continues for a long period, which may due to the difference of N-base volume, and the destruction is induced by self-heating of the avalanche energy itself. For the 3.3kV IGBT, the impact of parasitic bipolar action for SEB should be small.

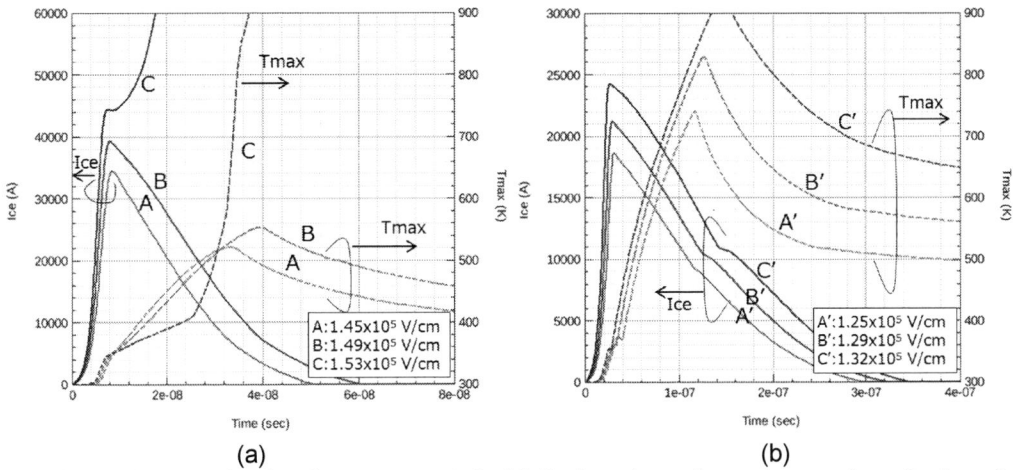

(a)   (b)

Fig.4. Simulation result of collector current (solid line) and maximum temperature (broken line) behaviour for heavy ion irradiation. (a) 1.2kV class IGBT. (b) 3.3kV class IGBT.

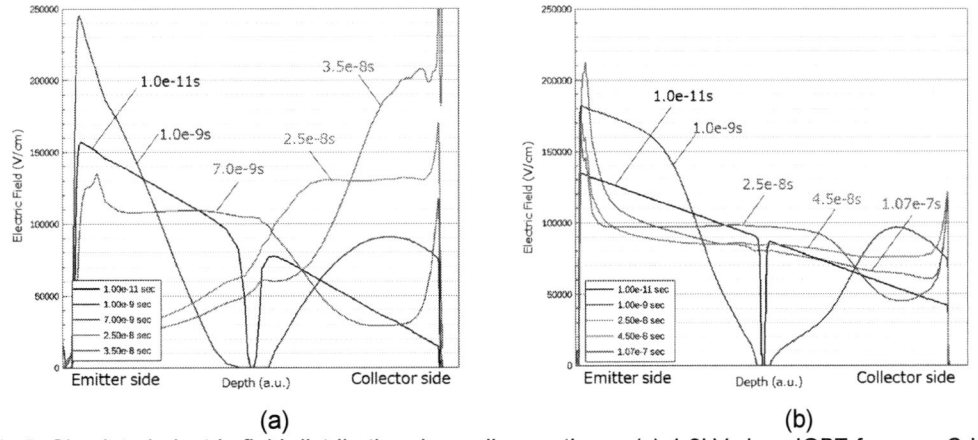

(a)   (b)

Fig.5. Simulated electric field distribution depending on times. (a) 1.2kV class IGBT for case C in Fig. 4(a). Tmax is about 800K at 3.5×10^{-8} sec. (b) 3.3kV class IGBT for case C' in Fig. 4(b). Tmax is about 800K at 1.07×10^{-7} sec.

3. Critical factor of cosmic ray failure for HV-IGBT

As the parasitic bipolar action does not affect the SEB failure rate for 3.3kV IGBTs, only the electric field should determine the SEB failure rate [7]. In order to evaluate the dependence between the electric field and the failure rate quantitatively, we measured the failure rates of 3.3kV diodes for various wafer thickness (Tsi) and resistivity (ρ). As the failure rate of a 3.3kV diode is proven to be identical to that of a 3.3kV IGBT, we can also apply this evaluation for a 3.3kV IGBT.

Fig.6 (a) shows the measured dependency of failure rates for wafer thickness and resistivity as a function of cathode bias (Vka). The failure rates are lower for thicker wafer and higher resistivity. In contrast, Fig.6 (b) shows the same data as a function of the maximum electric field in the N-base. The failure rates are lower for lower resistivity. We can see that the maximum electric field does not determine the failure rate. Fig. 7 shows the analytical electric field distribution at the same failure rate as indicated with broken line in Fig.6(a). The peak electric fields are clearly different for each. As an explanation to be the same failure rate, we propose an integral of the electric field with a specific threshold value (Eth), which can include the volume impact of the high electric region and neglect the contribution of low electric field regions. Fig. 8 shows the measured failure rates as the integral for Eth at 5.5×10^4 V/cm. We can see that integrals of the electric field are the same for each data point when setting the Eth to 5.5×10^4 V/cm. The failure rate can be quantitatively explained by the integral. Although the universality of Eth should be studied for more cases, we propose the integral as a practical and useful index for the robustness of SEB for high voltage IGBTs such as 3.3kV class.

From the analysis, we can also state an important point for high voltage IGBT design from the viewpoint of SEB robustness. Generally, the resistivity of the N-base is higher for higher voltage class IGBTs. Therefore, the high electric field volume is larger for higher voltage IGBTs. Accordingly, the SEB robustness is lower for higher voltage IGBTs if the design of the maximum electric field is same as for lower voltage IGBTs. As suppressing of parasitic bipolar gain has no impact for high voltage IGBT, it is very important to lower the electric field at the rated collector voltage to be less than the requirements from electrical SOA.

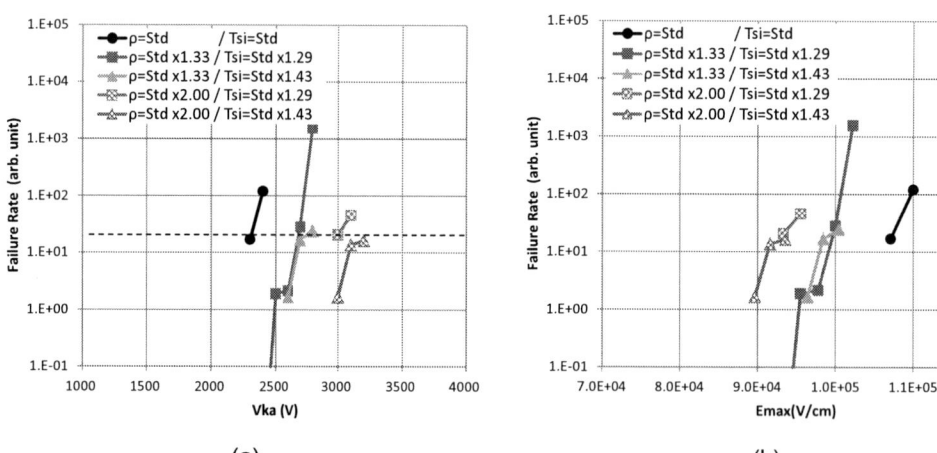

(a)   (b)

Fig.6. (a) Measured failure rate of 3.3kV diodes versus cathode voltage. (b) Measured failure rate of 3.3kV diodes versus maximum electric field.

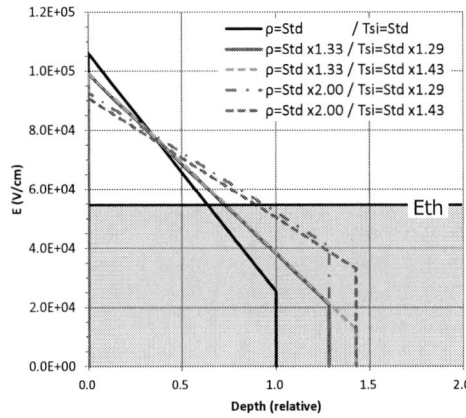

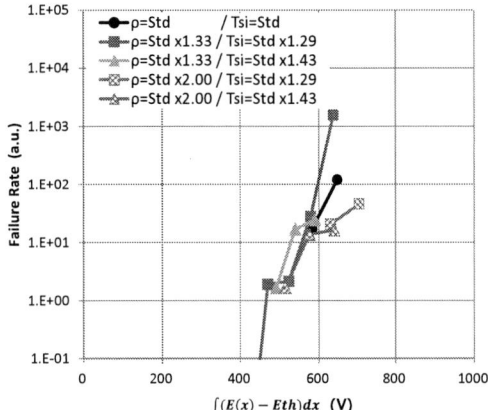

Fig.7. Analytical electric field of 3.3kV diode for the same failure rate condition (broken line in Fig.6 (a)).

Fig.8. Measured failure rate of 3.3kV diodes versus integral of electric field with specific threshold value. In this case, the threshold is set to 5.5×10^4 V/cm.

4. Cosmic ray robustness of 3.3kV Hybrid SiC Modules

The SEB failure rate of a 3.3kV SiC-SBD was also evaluated in order to confirm the LTDS of Si-IGBT/SiC-SBD hybrid modules. The 3.3kV hybrid module uses a 3.3kV SiC-SBD instead of the conventional Si PiN-Diode evaluated above. Fig.9 shows the 3.3kV /1500A hybrid module for railway systems. The SiC-SBD reduces the recovery loss to 3% compared with Si PiN-Diode, and improves turn-on loss of the IGBT. The AVAF (Adjustable Voltage Adjustable Frequency) inverter system introduced in the hybrid module contributes to improving inverter loss, efficiency of regenerative braking and motor loss of harmonic distortion.

The measured SEB failure rate of the SiC-SBD is shown in Fig. 10 and compared with that of the Si device shown in Fig.3. We can see that the failure rate of the SiC-SBD is almost at the same level as that of the Si-IGBT or Si-diode. Although further investigations of the failure mechanism of SiC-devices are needed [8], it can be said that the cosmic ray robustness of 3.3kV hybrid SiC modules is comparable to conventional Si modules.

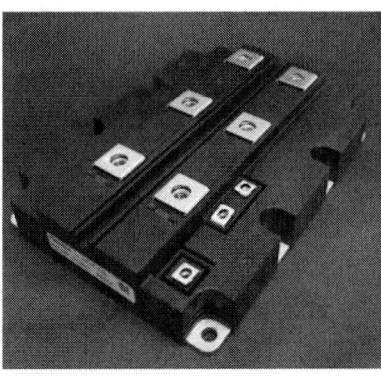

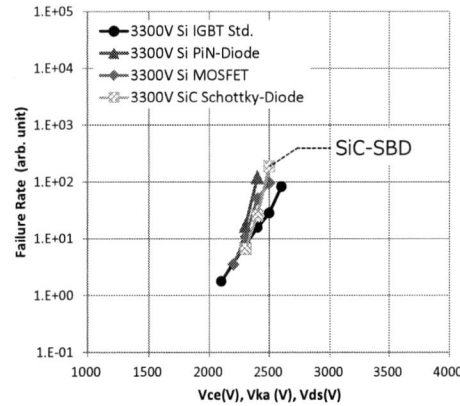

Fig.9. 3.3kV Hybrid SiC Module.
(1500A, 1in1 type, Si-IGBT + SiC-SBD)

Fig.10. Measured failure rate of 3.3kV SiC-SBD compared with 3.3kV Si-devices.

5. Conclusion

This work presented the cosmic ray failure mechanism and the critical factors that determine the failure rate of 3.3kV IGBTs. The measurements show that the failure rate of a 3.3kV IGBT is almost the same as that of a 3.3kV MOS-FET or 3.3kV diode, and the parasitic bipolar does not affect the failure rate. Simulation analysis reveals that the failure mode of the 1.2kV IGBT may be positive feedback of the parasitic bipolar, and that of the 3.3kV IGBTs may be self-heating of dynamic avalanche. The failure rate of high voltage IGBTs such as the 3.3kV is dominated by the electric field, and we propose an integral of the electric field with a specific threshold value as a practical and useful index for the robustness of SEB. The measurements also show that the failure rate of a 3.3kV SiC-SBD is almost the same as Si IGBTs and Si PiN diodes. The cosmic ray robustness of a 3.3kV Si-IGBT/SiC-SBD hybrid module is considered to be comparable to that of a conventional Si-IGBT/Si PiN diode module.

6. References

[1] H. Uemura, S. Iura, K. Nakamura, M. Kim, E. Stumpf, "Optimized Design against Cosmic Ray Failure for HVIGBT Modules", Proc. PCIM2011, pp.26-31, 2011.

[2] T. Shoji, S. Nishida, K. Hamada, "Triggering Mechanism for Neutron Induced Single-Event Burnout in Power Devices", Japanese Journal of Applied Physics 52 (2013) 04CP06, 2013.

[3] W. Kaindl, S. Soelkner, H.-J. Schulze, G. Wachutka, "Cosmic Radiation-Induced Failure Mechanism of High Voltage IGBT", Proc. ISPSD2005, pp.199-202, 2005.

[4] Y.Iwamoto, M. Fukuda, Y. Sakamoto, A. Tamii, K. Hatanaka, K. Takahisa, K. Nagayama, H. Asai, K. Sugimoto, I. Nashiyama, "Evaluation of the White Neutron Beam Spectrum for Single-Event Effects Testing at the RCNP Cyclotron Facility", Nuclear Technology, Vol.173, pp.210-217, 2011.

[5] S. Kato, E. Shimada, T. Yoshihira, A. Oyama, S. Ono, H. Ura, G. Ookura, W. Saito, Y. Kawaguchi, "Temperature Dependence of Single-Event Burnout for Super Junction MOSFET", Proc. ISPSD2015, pp.93-96, 2015.

[6] H. Hagino, J. Yamashita, A. Uenishi, H. Haruguchi, "An Experimental and Numerical Study on the Forward Biased SOA of IGBT's ", IEEE Transactions on Electron Devices, Vol. 43, pp. 490-500, 1996.

[7] F. Pfirsch, G. Soelkner, "Simulation of Cosmic Ray Failures Rates using Semiempirical Models", Proc. ISPSD2010, pp.125-128, 2010.

[8] J. Lutz, R. Baburske, "Some aspects on ruggedness of SiC power devices", Microelectronics Reliability 54 (2014), pp. 49-56, 2014.

Benefits of increased cosmic radiation robustness of SiC semiconductors in large power-converters

Christian Felgemacher, Samuel Vasconcelos Araújo, Christian Nöding, Peter Zacharias
Centre of Competence for Distributed Electric Power Technology (KDEE)
University of Kassel, Wilhelmshöher Allee 71, 34121 Kassel, Germany
c.felgemacher@uni-kassel.de

Abstract

Cosmic radiation induced single-event-burnout is a known failure mode for power semiconductors. To achieve high reliability the maximum voltage applied to power semiconductors in the application is commonly much less than their nominal voltage rating. Recent measurements of the voltage dependent failure rate of SiC MOSFETs have shown that SiC devices can be operated closer to their nominal voltage than comparable Si components. A case study to identify the failure rates expected in photovoltaic central inverters using different power semiconductors is conducted to investigate possible advantages arising from the increased robustness of SiC devices.

1. Introduction

Single-event-burnout (SEB) triggered by cosmic radiation was identified as a cause for failures of power semiconductor devices in the 1990s [1]. It is well known today that the failure rate of the devices depends strongly on the applied voltage stress. Below a certain threshold which usually lies at around 65% of the nominal voltage of a power semiconductor device no detectable amount of failures is seen. If this threshold voltage is exceeded the failure rate increases exponentially. The maximum voltage that is applied to the semiconductors in the application is therefore typically limited to 65 to 70% of the devices' rated voltage to ensure high reliability.

In the following, measurements of the failure rate due to SEB for a number of Si and SiC power semiconductor devices will be presented. After a brief qualitative discussion of the measurement results a case study using the example of two photovoltaic central inverters will be presented. The aim of this case study is to provide voltage stress profiles for the semiconductors in central inverters for 1,000V as well as 1,500V photovoltaic systems. Using the voltage profiles the importance of detailed knowledge of the voltage dependent failure rate of the semiconductors will be shown. Finally, the robustness data obtained for the SiC devices will be used to illustrate the advantage of SiC devices in terms of robustness against SEB in this application.

2. SEB failure rates of state of the art semiconductor devices

During a recent measurement campaign at the ANITA facility at the 'The Svedberg Laboratory' in Uppsala, Sweden the voltage dependent failure rate for a number of Si as well as SiC power semiconductor devices was measured. A picture of the measurement setup is shown in Fig. 1. Details of the measurement campaign including a detailed description of the experimental setup, the measurement methodology and all the results are presented in [2].

PCIM Europe 2016, 10 – 12 May 2016, Nuremberg, Germany

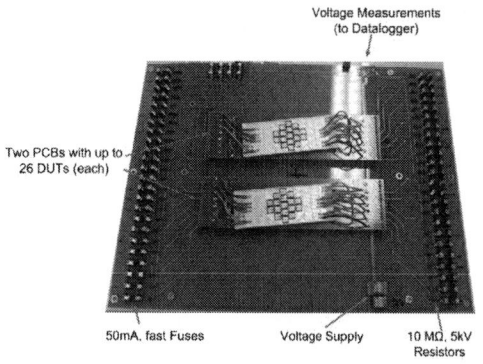

a) Printed-circuit-board used for SEB testing. Four of these were used in a measurement run each containing up to 52 DUTs on two smaller PCBs.

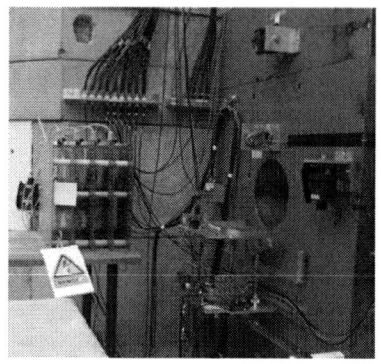

b) Measurement setup positioned in beam line. Up to 208 DUTs were tested at a time.

Fig. 1: Experimental setup that was used to test the SEB robustness of power semiconductors.

For the purpose of this investigation a selection of measurement results is provided in Fig. 2 and Fig. 3 alongside some data for state of the art Si diodes measured by Scheuermann and Schilling [3]. Using this data a comparison between inverter systems using different semiconductor devices will be conducted.

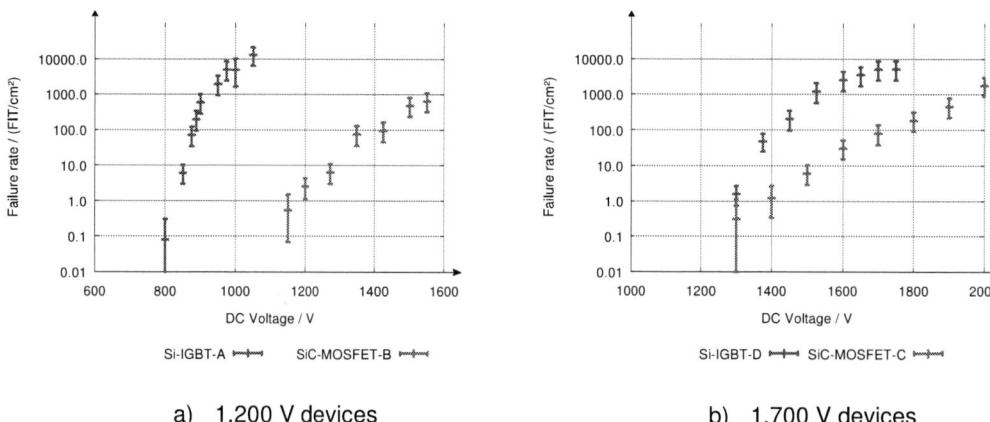

a)   1,200 V devices

b)   1,700 V devices

Fig. 2: Measured failure rates for 1,200 V / 1,700 V Si and SiC semiconductor switches.

The results in Fig. 2 show that the tested 1,200V SiC MOSFET is far more robust against cosmic radiation than the tested Si device. In fact, at the nominal voltage rating a failure rate of less than 10 FIT/cm² was measured. The tested 1,700V SiC MOSFET is also more robust than the tested Si IGBT. As can be seen a failure rate of less than 200 FIT/cm² was observed at the nominal voltage and the first failures were observed at a voltage of 1,300V, which was also the point at which the first failures were seen for the Si IGBT.

The results for the failure rate of the SiC diode (see Fig. 3a) show that the SiC diode is more robust than a state-of-the-art silicon diode (Si-Diode-C, data from [3]). In addition one can see that the Si CAL diodes (Si-Diode-A / Si-Diode-B, own measurements) are even more robust. The measured failure rates for the two Si CAL diodes are in agreement with the results presented by Scheuermann and Schilling [3], who also measured the failure rates of these devices.

© VDE VERLAG GMBH · Berlin · Offenbach

574

PCIM Europe 2016, 10 – 12 May 2016, Nuremberg, Germany

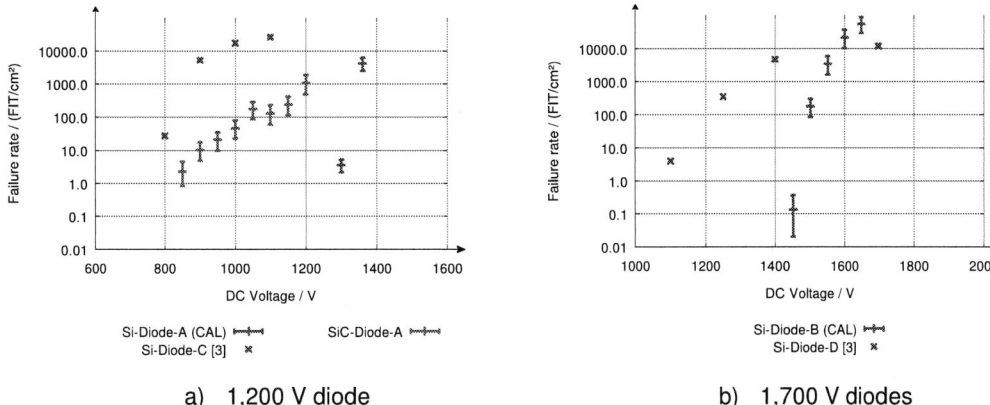

a) 1,200 V diode   b) 1,700 V diodes

Fig. 3: Measured failure rates for 1,200 V / 1,700 V diodes. Also shown are failure rates of state-of-the-art Si diodes published by Scheuermann and Schilling in [3].

3. Case-study: Photovoltaic central inverters in Sunbelt regions

To determine voltage profiles that can be used to estimate the failure rate due to single-event-burnout in the case of photovoltaic central inverters a system level simulation of two fictional photovoltaic systems was performed. For the simulation two exemplary photovoltaic central inverters using a standard 2-level bridge with specifications as illustrated in Table 1 were considered. Above the maximum MPP voltage the maximum output power is linearly reduced to 50% of the nominal power at the maximum DC voltage to provide sufficient headroom for switching transients and to avoid unnecessary thermal over-rating that would be required for operation at maximum input power at high DC voltages. In this investigation derating of the maximum inverter power output due to high ambient temperatures was not modeled, as such limits depend on the actual thermal design of an inverter system.

Table 1: Specifications for two fictional photovoltaic inverters.

	Option A (1,000 V system)	Option B (1,500 V system)
$P_{DC(max)}$	800 kW	800 kW
$V_{DC(min)}$	530 V	800 V
V_{MPP}	667 … 850 V	941 … 1,275 V
$V_{DC(max)}$	1,000 V	1,500 V
$I_{DC(max)}$	1,200 A	850 A
$V_{AC(nom)}$	360 V	540 V
$I_{AC(max)}$	1,280 A	850 A

The photovoltaic generator was modelled using the set of equations given by Alqahtani et.al. in [4]. This model is advantageous for this kind of investigation, as it provides an analytical description of a photovoltaic generator with N_{pp} parallel strings which each contain N_{ss} modules in series. This results in fast computation times even for long mission profiles in the range of an entire year at a time resolution in the range of seconds. The temperature of the photovoltaic modules was estimated using the following equation from [5].

$$T_{module} = T_{amb} + G \cdot \frac{NOCT - 20°C}{800(1 + \eta)}$$

T_{amb} is the ambient temperature, G is the solar irradiance on the module, NOCT is the normal operating cell temperature of the module and η the efficiency of the module at MPP.

The number of photovoltaic modules per string was selected to ensure that the open circuit voltage is not violated for the coldest expected temperature in a given location and the number of strings was computed to obtain a photovoltaic generator of the desired size. For this investigation it was assumed that the nominal power of the photovoltaic generator at STC (1000 W/m², 25°C) exceeds the power rating of the inverter by a factor of 1.5.

© VDE VERLAG GMBH · Berlin · Offenbach

575

For each time step in the simulation the maximum-power-point (MPP) of the inverter is computed and checked against the specifications of the inverter. If the MPP lies within the operational area (see Fig. 4) it is assumed that the inverter operates at the generator MPP. Otherwise, the DC voltage is increased to limit the power from the inverter until an acceptable operating point can be found. If no such point can be found (e.g. if the input voltage is too low) it is assumed that the inverter ceases to operate and that the open-circuit voltage of the photovoltaic generator is applied to the semiconductors.

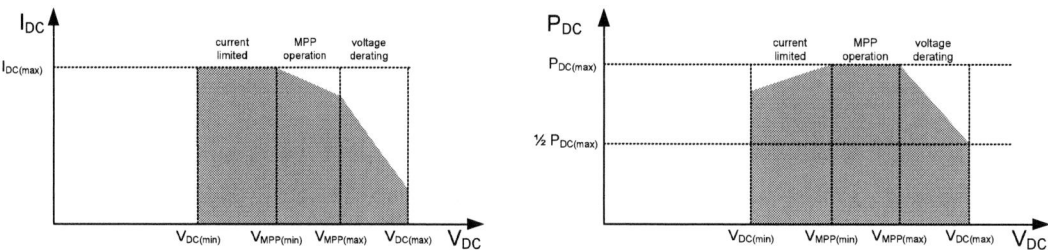

Fig. 4: Illustration of possible operating points of fictional photovoltaic inverters.

3.1. Voltage stress on semiconductors in 1,000 V / 1,500 V PV inverters

In Fig. 5 the simulated voltage stress on the semiconductors for four locations is shown. The inverter operates between 630V and 750V for most of the time; however, due to the high solar irradiance in locations in the Sunbelt (Bombay, Calama, Goldmud) the inverter also operates a certain amount of hours each year at up to 820V to limit the output power of the photovoltaic generator.

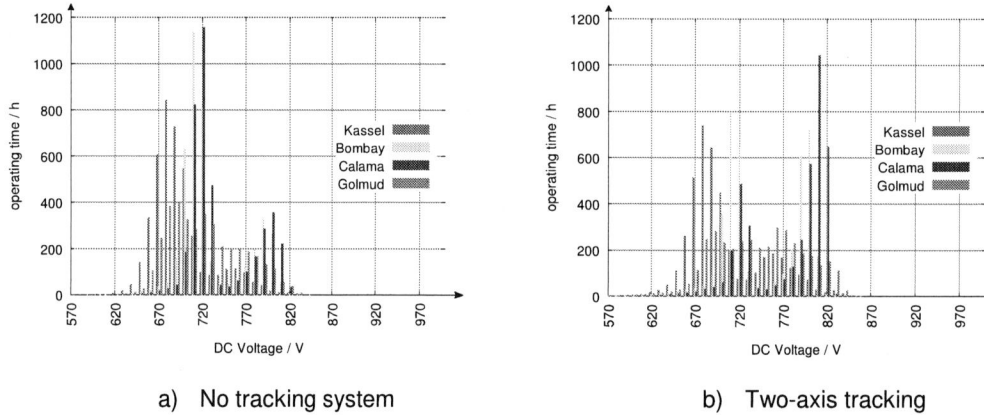

a) No tracking system   b) Two-axis tracking

Fig. 5: Distribution of operating hours across input voltage, sizing-ratio: 1.5, system voltage: 1,000 V

To illustrate the importance of considering all possible operating conditions that may impact the stress on the semiconductors the same system was simulated again under the assumption that the photovoltaic modules are placed on a two-axis tracking system. As this increases the irradiance on the photovoltaic modules and hence the available power from the photovoltaic generator the inverter has to limit the output power more often and therefore a higher voltage stress is applied to the semiconductors (see Fig. 5b).

Comparing the maximum voltages expected during normal operation of the inverter with the capabilities of the power semiconductors (see Fig. 2 / Fig. 3) one can see that very rarely voltages are reached at which failures are expected to occur. However, other operating conditions will be encountered in the field, such as the need to provide reactive power (depending on the sizing of the inverter this may result in reduced real power output) or requests by the distribution network operator to reduce power output (e.g. curtailment according to §14 EEG in Germany). The data in [6] shows that curtailment of renewable energy systems in Germany has increased to 1,581 GWh in 2014, which is a significant increase over previous years. For 2015 a further increase is expected according to [7].

© VDE VERLAG GMBH · Berlin · Offenbach

Output power curtailment has been evaluated by simulating the annual profiles for different levels of curtailment (reduction in maximum output power by a proportion of the system's rated power) and then averaging the occurring voltage stress profiles using the relative frequency of occurrence of each curtailment level in Table 2.

How much curtailment an actual photovoltaic installation is exposed to in the application cannot be predicted based on the available data, as every system can be affected to a different degree. Hence, the values in Table 2 have to be understood as exemplary values that were selected to highlight the effect a significant level of curtailment may have on photovoltaic inverters for the purpose of this case study. For actual sizing of a real converter these values need to be replaced by curtailment levels expected to occur in the application.

Table 2: Assumptions to estimate effect of curtailment.

Level of curtailment	Frequency of occurrence
0 % - (no restrictions)	19 h / 24 h
40 %	4 h / 24 h
70 %	30 min / 24 h
90 %	15 min / 24 h
100 % - (stop operating)	15 min / 24 h

The results in Fig. 6 show that the assumed curtailing results in significantly increased voltage stress. Without this effect the maximum voltage for Kassel was predicted to be approximately 820V whereas up to 940V are predicted if curtailment is considered.

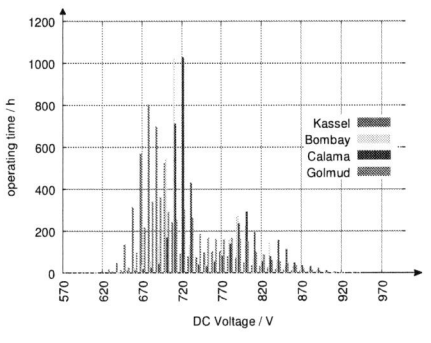
a) No tracking system

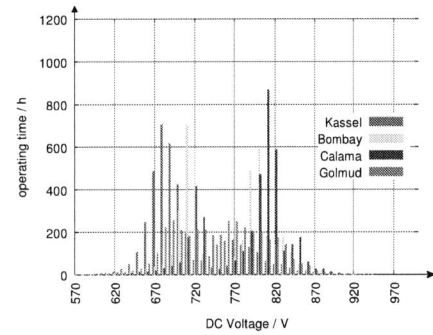
b) Two-axis tracking

Fig. 6: Distribution of operating hours across input voltage, sizing-ratio: 1.5, system voltage: 1,000 V, power curtailment as per Table 2.

The voltage stress for a 1,500V system is shown in Fig. 7 for the same level of power curtailment as for the 1,000V system. The maximum voltage in this case is around 1,430V.

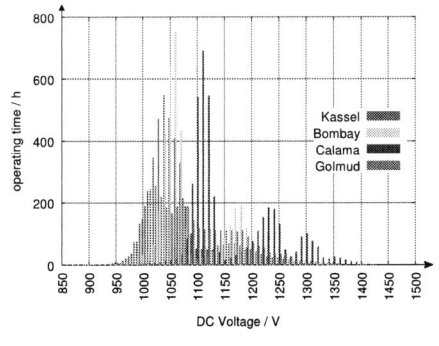
a) No-tracking system

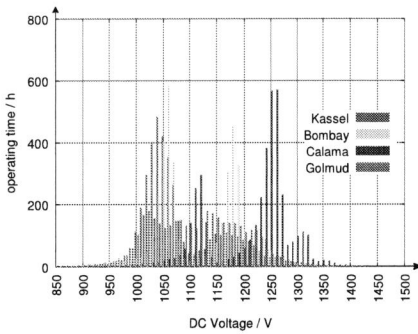
b) Two-axis tracking

Fig. 7: Distribution of operating hours across input voltage, sizing-ratio: 1.5, system voltage: 1,500 VOC, power curtailment as per Table 2.

3.2. Expected failure rate due to SEB for different semiconductors

In order to compare the failure rates for inverter systems that use different semiconductor devices it is necessary to relate the voltage stress expected to occur in the application (Fig. 6 / Fig. 7) to the voltage dependent failure rate for the selected semiconductor devices (Fig. 2 / Fig. 3). This gives the expected failure rate per unit area of the semiconductors. Additionally, the chip area needs to be estimated. For the purpose of this investigation the following chip area requirements were estimated by calculating the losses at all operating points using data for typical devices and choosing the minimum chip area which is required to maintain the chip temperatures at tolerable levels. The values shown here have to be understood as theoretical minimum values that may need to be exceeded in an actual design, e.g. due to non-ideal cooling or as a consequence of available module sizes.

Table 3: Assumptions made for absolute minimum chip area requirements under ideal conditions using data for typical devices (800 kW inverter / f_{sw} = 3 kHz)

	1,000V system (1,200V device)	1,500 V system (1,700V devices)
Si IGBT	83 cm²	82 cm²
Si diode	22 cm²	31 cm²
SiC MOSFET	28 cm²	30 cm²
SiC diode	12 cm²	12 cm²

In addition to the voltage dependent failure rates, the chip areas and the voltage stress profiles the local acceleration factors for the neutron flux at the four locations were calculated as per JESD89A [8] and used to calculate the failure rates in the following section.

3.2.1. System voltage: 1,000 V

In Table 4 the expected average failure rate for the eight voltage profiles shown in Fig. 6 is shown considering different combinations of power semiconductors. The failure rates computed using data for a 1,200V Si-IGBT (Si-IGBT-A) and a 1,200V Si-Diode (Si-Diode-C) show that problematic failure rates may occur if standard 1,200V devices are used in this application.

Table 4: SEB failure rate for 1,000 V inverter system with different semiconductor devices

voltage profile:	Kassel (no tracker)	Kassel (2-axis tracker)	Bombay (no tracker)	Bombay (2-axis tracker)	Calama (no tracker)	Calama (2-axis tracker)	Golmud (no tracker)	Golmud (2-axis tracker)
Si-IGBT-A + Si-Diode-C [3]								
Si-IGBT-A failure rate / (FIT/cm²)	0,84	0,84	2,87	2,61	5,45	4,67	3,12	2,89
Si-Diode-A failure rate / (FIT/cm²)	47,10	67,99	158,30	220,18	244,59	456,55	133,23	188,54
Total failure rate / FIT	618,94	876,24	983,00	1336,98	9873,80	17657,47	7830,16	10769,71
Failures in % / year	0,54%	0,76%	0,86%	1,16%	8,29%	14,33%	6,63%	9,00%
Si-IGBT-A +SiC-Diode-A								
Si-IGBT-A failure rate / (FIT/cm²)	0,84	0,84	2,87	2,61	5,45	4,67	3,12	2,89
SiC-Diode-A failure rate / (FIT/cm²)	0,04	0,05	0,12	0,13	0,23	0,23	0,11	0,12
Total failure rate / FIT	39,14	39,17	63,27	57,59	769,84	660,26	639,85	593,26
Failures in % / year	0,03%	0,03%	0,06%	0,05%	0,67%	0,58%	0,56%	0,52%
Si-IGBT-A +Si-Diode-A (CAL)								
Si-IGBT-A failure rate / (FIT/cm²)	0,84	0,84	2,87	2,61	5,45	4,67	3,12	2,89
Si-Diode-A failure rate / (FIT/cm²)	0,00	0,00	0,00	0,00	0,00	0,00	0,00	0,00
Total failure rate / FIT	38,88	38,86	62,89	57,19	765,11	655,50	636,47	589,60
Failures in % / year	0,03%	0,03%	0,06%	0,05%	0,67%	0,57%	0,56%	0,52%

To overcome this problem components of a higher voltage class (i.e. 1,700V IGBTs / diodes) are typically used, as these are readily available. With the 1,700V semiconductors measurable failure rates start to appear at approximately 1,300V for the IGBT (see Fig. 2) and at around 1,100V for the diode (see Fig. 3). These voltages are not reached in a 1,000V inverter system and hence the failure mode of single-event-burnout can be avoided by using devices of the higher voltage class.

A close look at the results in Table 4 shows that the dominant share of the failure rate if the selected 1,200V devices are used in a 1,000V system can be attributed to the diode. As pointed out by Scheuermann and Schilling [3] the robustness of the CAL diode is far superior to that of other state-of-the-art Si diodes. This is confirmed by the measurement results presented in Fig. 3. The measurement results for the SiC diode in Fig. 3 show that this device's robustness against SEB lies between that of the CAL diode and the state-of-the-art Si diode. The results in Table 4 show that by replacing the state of the art Si diode with a Si CAL diode or a SiC diode the failure rate is reduced significantly.

A further alternative would be the use of 1,200V SiC MOSFETs. These have shown no failures below 1,100V in the measurements (see Fig. 2) and would thus help to avoid failures due to single-event-burnout, if no extra freewheeling diode is used. However, this may need to be reevaluated if in the future the breakdown voltage of SiC MOSFETs, which is currently very high, gets reduced by the manufacturers and the robustness may change.

3.2.2. System voltage: 1,500 V

In the following table the expected average failure rate for the eight voltage profiles shown in Fig. 7 are computed using data for different combinations of power semiconductors. The results show that using standard 1,700V Si semiconductors in 1,500V photovoltaic inverters may be problematic. Unfortunately, very few semiconductors with an appropriately higher voltage rating (e.g. 2,500V) are available, which can lead to the use of multi-level topologies or a series connection of devices in this application.

The data in Table 5 further shows that replacing the standard 1,700V Si diode with a 1,700V CAL diode reduces the failure rate significantly. It would also be interesting to investigate the failure rates using 1,700V SiC diodes; however, no data was available for these devices.

Table 5: SEB failure rates for 1,500 V inverter systems with different semiconductor devices

voltage profile:	Kassel (no tracker)	Kassel (2-axis tracker)	Bombay (no tracker)	Bombay (2-axis tracker)	Calama (no tracker)	Calama (2-axis tracker)	Golmud (no tracker)	Golmud (2-axis tracker)
Si-IGBT-D + Si-Diode-D [3]								
Si-IGBT-D failure rate / (FIT/cm²)	0,10	0,11	0,06	0,06	1,08	1,02	0,08	0,07
1700V Si Diode failure rate / (FIT/cm²)	36,82	53,82	52,52	74,92	206,19	334,20	50,09	63,96
Total failure rate / FIT	574,97	838,62	816,58	1163,72	3240,36	5221,85	779,67	994,21
Failures in % / year	0,50%	0,73%	0,71%	1,01%	2,80%	4,47%	0,68%	0,87%
Si-IGBT-D + Si-Diode-B (CAL)								
Si-IGBT-D failure rate / (FIT/cm²)	0,10	0,11	0,06	0,06	1,08	1,02	0,08	0,07
Si-Diode-B (CAL) failure rate / (FIT/cm²)	0,00	0,00	0,00	0,00	0,00	0,00	0,00	0,00
Total failure rate / FIT	4,22	4,34	2,46	2,50	44,40	41,77	3,35	2,79
Failures in % / year	0,00%	0,00%	0,00%	0,00%	0,04%	0,04%	0,00%	0,00%

Finally, the high robustness of the 1,700V SiC MOSFET suggests that in the future it may be possible to build 1,500V systems using only 1,700V SiC MOSFETs, since for the tested 1,700V SiC MOSFET a failure rate of less than 10 FIT/cm² was measured at 1,500V (see Fig. 2). This device would therefore also allow failures due to SEB to be avoided. As stated before, this may need to be reevaluated if future generations of SiC MOSFETs are manufactured with lower breakdown voltages, as this may influence the robustness of the devices.

4. Conclusion

It was shown that in the application of photovoltaic central inverters the voltage stress on the semiconductor devices is significantly increased by curtailment of the inverter power output at the level that was assumed in this case study. As the results that consider the curtailment show, the use of standard 1,200V Si components can be problematic in 1,000V systems. For the same reason, the use of standard 1,700V components in 1,500V systems may be problematic.

The results in the case study have shown that with the assumptions that were made the failure rates are dominated by the state-of-the-art Si diodes which were considered. It was

© VDE VERLAG GMBH · Berlin · Offenbach

also shown, that replacing these diodes with more robust diodes, e.g. SiC diodes or Si CAL diodes, can reduce the expected failure rates.

Whether the use of the most robust 1,200V devices that are available may actually be viable in a 1,000V system (or 1,700V devices in a 1,500V system) depends on a multitude of factors (e.g. the robustness of the selected device, the location a system is operated in, the expected level of curtailment, tolerable failure rates, PV park operating strategies, etc.) and cannot be answered generally. For example park management strategies could implement curtailment by turning off individual inverters instead of reducing the output power, as suggested in [9].

The investigation has highlighted the need to consider all factors that influence the voltage stress in a given application to obtain stress profiles. Additionally, exact knowledge of the voltage dependent failure rates of the power semiconductor devices due to single-event-burnout is required.

Based on the measured failure rates the 1,200V SiC MOSFETs in a 1,000V inverter or 1,700V SiC MOSFETs in a 1,500V inverter would yield low failure rates, as the voltages at which the SEB failure rate starts to reach significant levels is very high. This makes the use of the tested SiC MOSFETs beneficial from the point of robustness against single-event-burnout. This needs to be reevaluated for future device generations if the breakdown voltage of SiC MOSFETs, which is currently very high, gets reduced by the manufacturers. A further benefit of SiC devices is that a smaller chip area can be used which also has a positive influence on the failure rate due to SEB. Finally, the use of SiC devices will likely allow the switching frequency to be increased significantly, which will reduce the size of passive components.

5. Acknowledgement

The work presented in this paper was undertaken as part of the research project "HHK", supported by the German Federal Ministry of Education and Research (BMBF). Project funding reference number: 16ES0096. Responsibility for the contents of this publication lies with the authors.

6. References

[1] H. Kabza, H.-J. Schulze, Y. Gerstenmaier, P. Voss, J. Schmid, F. Pfirsch and K. Platzöder: *Cosmic radiation as a cause for power device failure and possible countermeasures*, Proceedings of 6th ISPSD, pp. 9 – 12, 1994.

[2] C. Felgemacher, S. Araújo, P. Zacharias, K. Nesemann and A. Gruber: *Cosmic Radiation Ruggedness of Si and SiC Power Semiconductors*, Proceedings of 28th ISPSD, 2016. (accepted for publication)

[3] U. Scheuermann and U. Schilling: *Cosmic ray failures of power modules – the diode makes the difference*, PCIM Europe 2015, 19 – 21 May 2015.

[4] A. Alqahtani, M. Abuhamdeh, and Y. Alsmadi: *A simplified and comprehensive approach to characterize photovoltaic system performance*, 2012 IEEE Energytech, 2012.

[5] Marańda, W and Piotrowicz, M.: *Extraction of thermal model parameters for field-installed photovoltaic module*, 27th Intl. Conference on Microelectronics Proceedings (MIEL), pp. 153-156, 2010.

[6] Bundesnetzagentur, *EEG in Zahlen 2014* [Online]. Available: http://www.bundesnetzagentur.de/DE/Sachgebiete/ElektrizitaetundGas/Unternehmen_Ins titutionen/ErneuerbareEnergien/ZahlenDatenInformationen/zahlenunddaten-node.html

[7] Bundesnetzagentur / Bundeskartellamt, *Monitoringbericht 2015* [ONLINE]. Available: http://www.bundesnetzagentur.de/SharedDocs/Downloads/DE/Allgemeines/Bundesnetza gentur/Publikationen/Berichte/2015/Monitoringbericht_2015_BA.pdf?__blob=publicationFi le&v=3

[8] JEDEC Std. JESD89A, "Measurement and Reporting of Alpha Particle and Terrestrial Cosmic Ray-Induced Soft Errors in Semiconductor Devices", 2006.

[9] M. Morjaria and D. Anichkov.: *"Grid Friendly" Utlility-Scale PV Plants*, White Paper, First Solar, 2013. [Online]. Available: http://www.firstsolar.com/en/Technologies-and-Capabilities/Grid-Integration/Documents/Grid Integration White Paper.aspx?dl=1

© VDE VERLAG GMBH · Berlin · Offenbach

PCIM Europe 2016, 10 – 12 May 2016, Nuremberg, Germany

Passive IGBT turn-off during short-circuit type V

Jan Fuhrmann, University of Rostock, Germany, jan.fuhrmann@uni-rostock.de
Sebastian Klauke, Infineon Technologies AG, Germany, sebastian.klauke@infineon.com
Hans-Guenter Eckel, University of Rostock, Germany, hans-guenter.eckel@uni-rostock.de

Abstract

A new short-circuit type with a passive turn OFF of the IGBT is presented. It can occur in an active neutral point clamped (ANPC) 3L inverter, when the IGBT is turned OFF without taking blocking voltage. The plasma inside the IGBT is not removed until the short-circuit occurs. The di/dt of the plasma removal is given by the short-circuit inductance and results in a high dv/dt. An active-clamping interference in the gate-emitter voltage is impaired by the switched OFF gate. The influence of the DC link voltage, the current before the short-circuit and the delay between switching and short-circuit is analyzed with the help of measurements on a $3,3\,kV$-IGBT. The plasma removal can be explained with an equivalent capacitance. This effect is simulated and described with a causal chain and an equivalent circuit.

1. Introduction

Within an ANPC 3L inverter, especially the modified version in [1], a breakdown of a semiconductor can trigger a new short-circuit type (here called type V). In literature four different short-circuit types are described, depending on the ON-state of the IGBT or the diode [2, 3, 4, 5]. This paper describes a fifth short-circuit scenario, where the IGBT is turned OFF without taking blocking voltage. The plasma is not removed during turn OFF and remains in the IGBT until it is recombined. To take blocking voltage the plasma has to be removed. During the short-circuit type V a high di/dt and dv/dt occur with a limited active-clamping capability.

In Fig. 1 a schematic of an ANPC 3L phase is given. The red path in the left inlet shows a commutation path without a voltage source. During a commutation from the upper path (T12

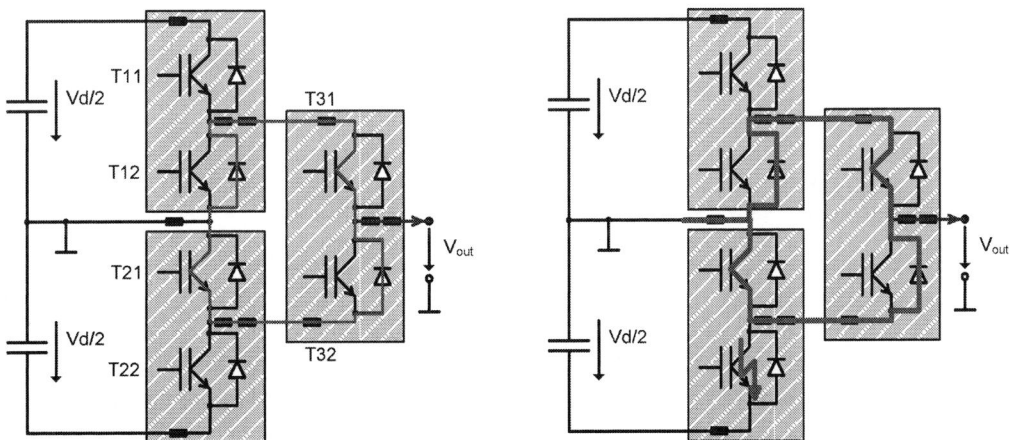

Fig. 1: Short-circuit type V after breakdown of T22 and zero voltage current commutation of T31

© VDE VERLAG GMBH · Berlin · Offenbach

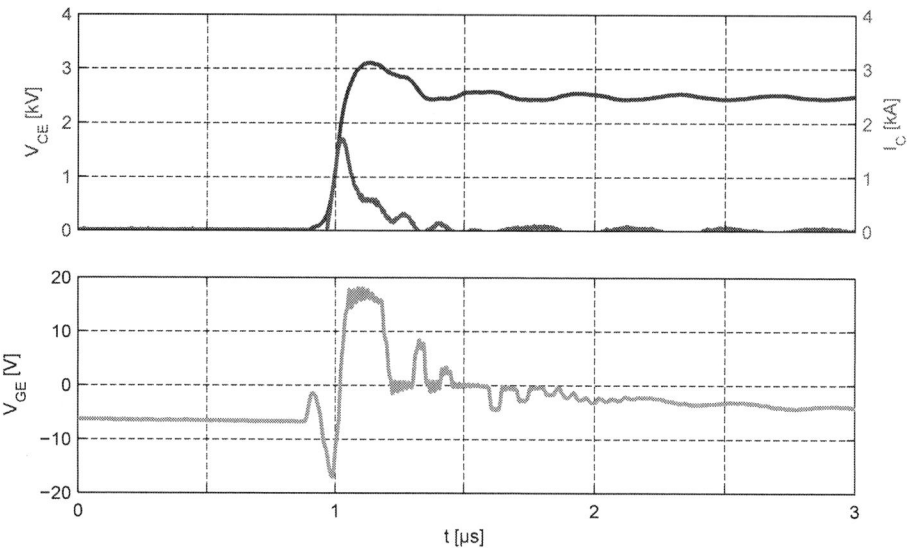

Fig. 2: Measurement: Short-circuit V with 3,3 kV-IGBT at V_D=2,5 kV, I_C=1,4 kA and t_{delay}=38 µs

and T31 switched ON) to the lower path (T21 and T32 switched ON) the IGBT T31 is switched OFF without taking blocking voltage. Only the parasitic inductances creating an overvoltage, which is zero after the switching. IGBT T31 is switched OFF with plasma remaining inside. A breakdown of T22, marked on the right hand side of Fig. 1, forces T31 to take blocking voltage, which is only possible, when the plasma is removed. It is like a zero voltage turn OFF of the IGBT [6] combined with hard switching techniques and not useful to combine.

2. Short-circuit type V

The described passive IGBT turn OFF during short-circuit is here defined as type V. The main condition for this short-circuit is a turned OFF IGBT, which has not taken blocking voltage, and a semiconductor breakdown, which forces the IGBT to take blocking voltage. A short-circuit type V is shown in Fig. 2, where a 3,3 kV-IGBT is switched OFF at V_D=2,5 kV and I_C=1,4 kA and 38 µs after turn-OFF the short-circuit occurs. The *di/dt* of the short-circuit current, in this case 24 kA/µs, is determined by the parasitic inductance and the DC link voltage. At the beginning of the short-circuit at 0,9 µs a forward-recovery of the IGBT cannot be observed. The turned-OFF IGBT can only carry the short-circuit current by removing the remaining plasma, which evoke a *dv/dt* of 28,5 kV/µs. When the collector-emitter voltage reaches the DC link voltage, the maximum short-circuit current is reached. The high collector-emitter voltage triggers the active-clamping which lifts the gate-emitter voltage and switches the IGBT ON. A current plateau at 1,1 µs and a current oscillation at the end of the short-circuit can be observed.

The gate-emitter voltage is at the beginning of the short-circuit at −7 V. The feedback of the Miller capacitance lifts the voltage to nearly 0 V until the current feedback reduces it to −15 V. During the *dv/dt* the gate channel is closed which has two big disadvantages. First, the active-clamping is inhibited. A triggered active-clamping needs more time and energy to lift the gate-emitter voltage over the threshold voltage to limit the *dv/dt*. Second, inside the space-charge region the electron current is missing. Normally the electrons decrease the electric field steepness which can avert an avalanche. Without the electrons an avalanche, which can destroy the

IGBT, is more probable. At the end of the short-circuit at $1,1\,\mu s$, the IGBT is switched ON and the collector-emitter voltage is limited until the short-circuit is switched OFF finally. Overall this short-circuit type V is very fast and lasts only several $100\,ns$.

3. Influence of different parameters on the short-circuit behavior

After showing the short-circuit type V itself, the influence of the DC link voltage, the load current and the delay between turn OFF and short-circuit shall be analyzed. For this, various measurements are made and presented. The first parameter is the DC link voltage, shown in the upper inlets of Fig. 3. If a typical DC link voltage is used, see Fig. 3(a), a maximum short-circuit current of $1,5\,kA$ is reached. The overvoltage is uncritical and the active-clamping time is rather low. If the DC link voltage is increased to $2,7\,kV$, displayed in Fig. 3(b), the overvoltage is critical and the active-clamping time is increased to $500\,ns$. The maximum short-circuit voltage reaches $2\,kA$. The influence of the active-clamping can be seen at the end of the short-circuit. The high negative di/dt is decreased by an additional electron current through the gate. The overvoltage was limited with a delay.

By changing the collector current before the short-circuit occurs, the short-circuit behavior is directly influenced. The turned OFF collector current influences the plasma inside the IGBT, which remains inside after turn OFF. Two measurements with varied collector current are displayed in center inlets in Fig. 3. The delay between turn OFF and short-circuit and the DC link voltage is constant. The increased collector current slightly decreases the dv/dt of the collector-emitter voltage. The maximum short-circuit current and the gate-emitter voltage is not influenced. The short-circuit duration is in both measurements with $500\,ns$ nearly constant.

Another parameter that influences the plasma inside the IGBT is the time between turn OFF and short-circuit. During the delay the plasma concentration reduces due to recombination. In the lower inlets in Fig. 3 this varied delay is displayed. In the lower left inlet the short-circuit occurs $38\,\mu s$ after the turn OFF. Most of the plasma is still inside the IGBT, which has to be removed before taking blocking voltage. The slope of the collector-emitter voltage is smaller compared to the lower right inlet with a delay of $158\,\mu s$. A longer delay decreases the maximum short-circuit current and the short-circuit duration. The dv/dt massively increases, while the maximum collector-emitter voltage is constant. The fast plasma removal reduces the feedback from the active-clamping. The gate is not turned ON again to decrease the collector-emitter slope. A snappy behavior with several oscillations can be seen.

The presented three parameters influence the short-circuit behavior of the IGBT. The collector current and the delay influence the plasma inside the IGBT, which influences the dv/dt. This behavior is explained in the following section with the help of an equivalent circuit of the IGBT.

4. Equivalent circuit for the short-circuit type V

The displayed behavior in the measurements can be explained with the help of the equivalent circuit displayed in Fig. 4(a). The equivalent circuit uses a voltage-controlled current source instead of the gate channel. The current through the collector consists of a hole and an electron current. During ON-state the electron current is carried by the voltage-controlled current source. The hole current, one fourth of the total current, is carried by the second current source. If the IGBT is turned OFF, the electron current is fed by the capacitance representing the plasma. This is the normal turn OFF which removes the plasma inside the IGBT.

PCIM Europe 2016, 10 – 12 May 2016, Nuremberg, Germany

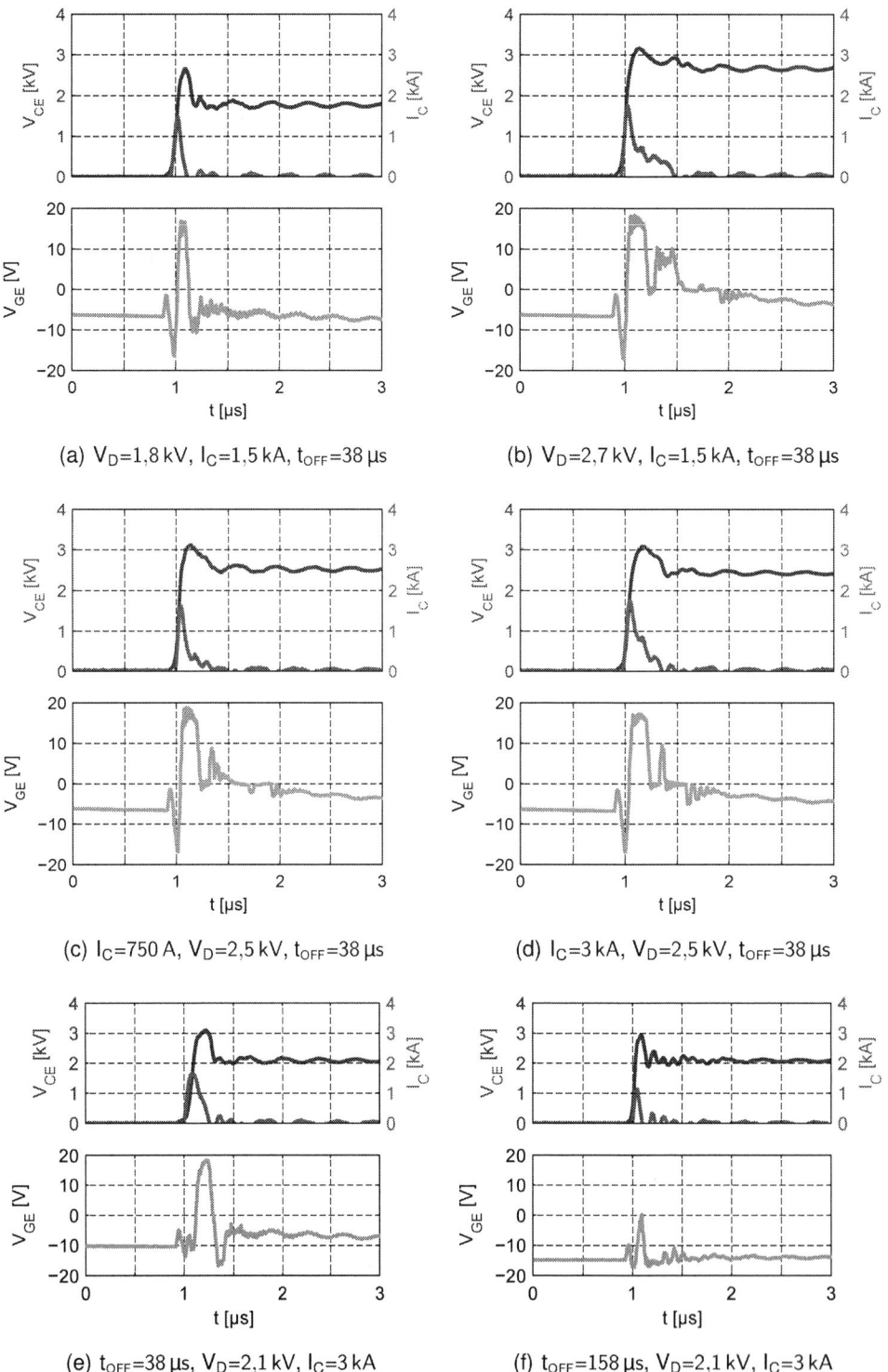

(a) V_D=1,8 kV, I_C=1,5 kA, t_{OFF}=38 µs

(b) V_D=2,7 kV, I_C=1,5 kA, t_{OFF}=38 µs

(c) I_C=750 A, V_D=2,5 kV, t_{OFF}=38 µs

(d) I_C=3 kA, V_D=2,5 kV, t_{OFF}=38 µs

(e) t_{OFF}=38 µs, V_D=2,1 kV, I_C=3 kA

(f) t_{OFF}=158 µs, V_D=2,1 kV, I_C=3 kA

Fig. 3: Measurement: Short-circuit V with 3,3 kV-IGBT, varied collector current, t_{delay} and V_D

© VDE VERLAG GMBH · Berlin · Offenbach

PCIM Europe 2016, 10 – 12 May 2016, Nuremberg, Germany

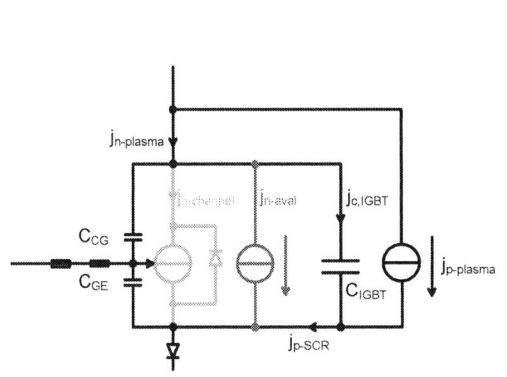

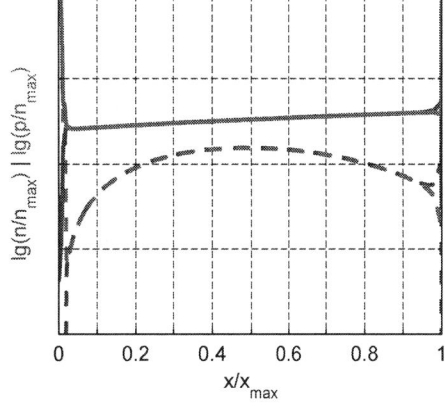

(a) Equivalent circuit of the IGBT during short-circuit type V with turned OFF gate, all capacitance are variable and non-linear

(b) Simulation of the plasma concentration during normal turn-OFF (solid lines) and before short-circuit type V, 100 ns after turn-OFF (dashed lines)

Fig. 4: Equivalent circuit and plasma concentration at the beginning of short-circuit type V

During short-circuit type V the IGBT is turned OFF without taking blocking voltage. The plasma remains inside the IGBT and has to be removed during short-circuit. The DC link voltage and the parasitic inductances determines a di/dt. This current is carried by the IGBT by removing its plasma. A dv/dt is the consequence. This behavior is described in the equivalent circuit by the plasma capacitance. A current through this capacitance result in a dv/dt. Compared with a normal turn OFF, two differences exist: Firstly, the gate is completely closed, no electron current through the gate occurs and the active-clamping is inhibited. An overvoltage protection is only with a delay possible. And secondly, the plasma distribution inside the IGBT is different. During a turn OFF the plasma concentration is increasing towards the collector, displayed in Fig. 4(b) with solid lines. If the IGBT is turned OFF without taking blocking voltage the diffusion of the charges changes the plasma concentration, which is displayed with dashed lines. This difference separates the short-circuit type V from a perfect intrinsic turn OFF.

5. Device simulation of short-circuit type V

A simulation of the short-circuit with the help of SYNOPSYS TAURUS MEDICI is presented in this section. With the help of the measurements and the equivalent circuit the dependence of the dv/dt and the plasma is shown. An increased collector current increases the plasma and decreases the dv/dt. In the simulation a good gate clamping without an active-clamping can be used without interference by other parameters.

If a short occurs, after the IGBT has not conducted any current, the dv/dt is very high and only a small displacement current occurs. A massive overvoltage occurs. With increasing current, the dv/dt reduces and the maximum short-circuit current increases. A massive avalanche occurs, which is different to the previous measurements.

The effect of the plasma, which shows capacitive behavior, can also be seen in the delay between short and turn OFF. A simulation is shown in Fig. 5(b). A short period leads to more plasma which has to be removed and resulting in a lower dv/dt. With increasing time more

© VDE VERLAG GMBH · Berlin · Offenbach

585

PCIM Europe 2016, 10 – 12 May 2016, Nuremberg, Germany

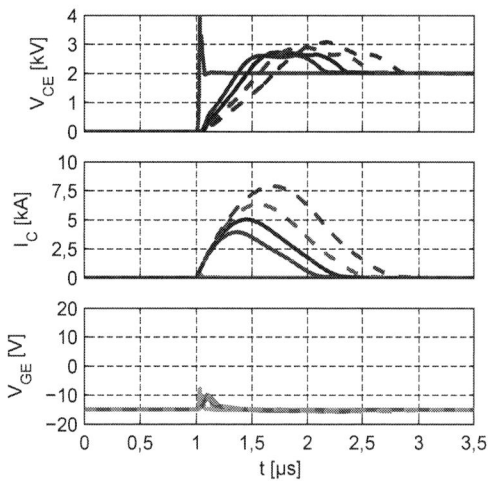

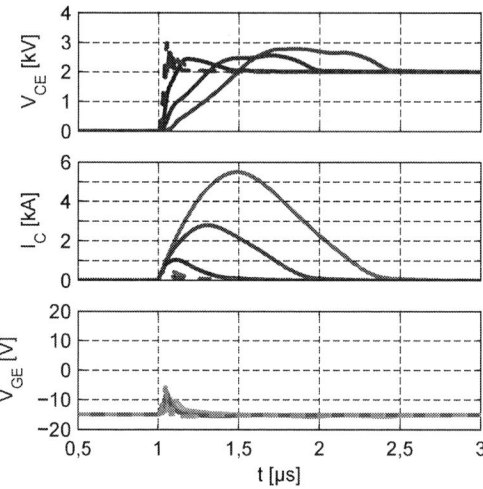

(a) Short-circuit type V simulation with varied current before turn OFF and short (0 A to 3 kA, increased current decreases *dv/dt*)

(b) Short-circuit type V simulation with varied delay between short and turn OFF (10 µs to 100 µs, increased time increases *dv/dt*)

Fig. 5: Short-circuit type V simulation with SYNOPSYS TAURUS MEDICI and a 3,3 kV-model

plasma has recombined and the equivalent capacitance reduces. The short-circuit current leads to a higher *dv/dt*. At 100 µs in the simulation nearly the complete plasma recombined and only a small amout of plasma is removed during short-circuit. The *dv/dt* is very high.

The simulation of the short-circuit type V shows the same dependence of the plasma capacitance like the previous measurements.

6. Causal chain during short-circuit type V

With the help of the equivalent circuit a causal chain, displayed in Fig. 6, for the short-circuit can be found. This causal chain explains with a closed feedback loop the progress of the short-circuit type V.

A short-circuit leads to a voltage drop over the parasitic inductance, which is at the beginning the complete DC link voltage. Later on it is reduced by the collector-emitter voltage of the IGBT. The integral of the resulting *di/dt* is the short-circuit current. With the perfectly turned OFF gate and no avalanche, this current is fed by the plasma. The current through the plasma, represented by a non-linear capacitance, leads to a *dv/dt*. Its integral gives the collector-emitter voltage. An increased collector-emitter voltage reduces the voltage drop over the parasitic inductance, which reduces the *di/dt* until the DC link voltage is reached and afterwards the *di/dt* is negative. This negative *di/dt* decreases the short-circuit current until it is turned OFF.

If an active-clamping turns ON the gate, an electron current through the voltage-controlled current source enables a parallel path to the plasma capacitance. The current through the capacitance is reduced, which reduces the *dv/dt*. Another parallel path can occur when an avalanche occur inside the IGBT. This avalanche creates additional charges which also reduces the capacitive current.

© VDE VERLAG GMBH · Berlin · Offenbach

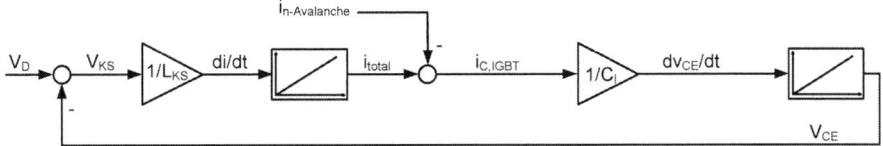

Fig. 6: Causal chain during the short-circuit type V, the capacitance is variable and non-linear

The plasma capacitance is non-linear and depending from the history of the device. A snappy behavior of the device, when the current abruptly stops, can hardly described. With this causal chain the complete short-circuit can be described, especially the *dv/dt* during taking blocking voltage.

7. Conclusion

In this paper a completely new short-circuit type is described. It can occur, when an IGBT is turned OFF without taking blocking voltage. After a semiconductor breakdown this IGBT has to take blocking voltage and remove the remaining plasma. A very high *dv/dt* is possible. This short-circuit is called type V and is completely different to a normal turn OFF. With the closed gate no active-clamping is possible and the missing electron current enhances the slope of the electric field. With the help of measurements, simulations and an equivalent circuit the influence of different parameters on the short-circuit behavior can be retrieved. The plasma, which depends on the previous current and the delay between short and turn OFF, determines the *dv/dt* during the short-circuit. A casual chain describes the complete short-circuit.

References

[1] J. Fuhrmann and H.-G. Eckel, "Advanced Active Neutral Point Three Level Inverter with Standard Half-Bridge Modules," in *PCIM Europe 2014. International Exhibition and Conference for Power Electronics, Intelligent Motion, Power Quality*, pp. 1311–1317, 2014.

[2] H.-G. Eckel and L. Sack, "Experimental investigation on the behaviour of IGBT at short-circuit during the on-state," in *20th Annual Conference of IEEE Industrial Electronics*, pp. 118–123, 1994.

[3] J. Lutz, R. Dobler, J. Mari, and M. Menzel, "Short circuit III in high power IGBTs," in *13th European Conference on Power Electronics and Applications*, pp. 1–8, 2009.

[4] J. Lutz and T. Basler, "Short-circuit ruggedness of high-voltage IGBTs," in *28th International Conference on Microelectronics*, pp. 243–250, 2012.

[5] S. Pierstorf and H.-G. Eckel, "Short-circuit behavior of diodes in voltage source inverters Short-circuit types Diode short circuits," in *PCIM Europe 2012. International Exhibition and Conference for Power Electronics, Intelligent Motion, Power Quality*, pp. 931–937, 2012.

[6] K. Chen and T. Stuart, "A study of IGBT turn-off behavior and switching losses for zero-voltage and zero-current switching," in *APEC '92 Seventh Annual Applied Power Electronics Conference and Exposition*, 1992.

© VDE VERLAG GMBH · Berlin · Offenbach

High-Current Power Cycling Test-Bench for Short Load Pulse Duration and First Results

Guang Zeng[1], Christian Herold[1], Menia Beier-Möbius[1], Sascha Kubera[2], Rodrigo Alvarez[2], Josef Lutz[1]

[1] Chemnitz University of Technology, Reichenhainer Str. 70, D-09126 Chemnitz, Germany

[2] Siemens AG, Günther-Scharowsky-Str. 2, D-91058 Erlangen, Germany

Abstract

A power cycling test-bench for short load pulse duration in the millisecond range has been developed. This new setup with a cross regulator is in addition capable to test IGBTs and diodes of different manufacturers with the same desired combination of load pulse duration t_{on} and junction temperature swing ΔT_j at the same time. Series of power cycling tests can be performed on this test-bench to build up new empirical lifetime models for IGBT modules. The focus lies on the impact of the short load pulse duration and small junction temperature swing on the power cycling capability.

1. Introduction

The power cycling test is a reliability test, which is used to characterize packaging design and lifetime expectation of power electronic devices. In [1], numerous power cycling tests were performed on various IGBT modules to investigate the impact of different parameters on power cycling capability, among which t_{on} and ΔT_j play significant roles. Special study for Al wire bonds was undertaken in [2] with minimum t_{on} = 70 ms, which confirmed the dependency of power cycling lifetime on t_{on} for module with Al wire bonds. On the contrary, other research groups at the same time showed no t_{on} dependencies of power cycling lifetime for wire bonds in their results [3], [4]. Furthermore, the majority of load pulse durations and junction temperature swings of the devices in many power electronic applications are below the test conditions from which existing empirical lifetime models were generated.

In [5], the temperature profile of an IGBT module in a hybrid car is analyzed with rain-flow method, whose junction temperature swing ΔT_j caused by short term power cycles are mainly below 40 K. Considering grid applications, for example converters for HVDC transmission or wind power application, ΔT_j of the semiconductor devices are also typically below 40 K [6], [7]. Power losses of the semiconductors, which are corresponding to their active temperature cycles, have two components, conduction losses and switching losses. In high power grid applications, the conduction losses dominate, which generate a grid frequency temperature swing in the devices with a duty cycle of 50%. For the mainly used grid frequency of 50 Hz, t_{on} of the dominant component in the spectrum analysis for junction temperature variation is 10 ms.

There is a lack of experimental results in the range of ΔT_j below 40 K and t_{on} of some ten-milliseconds, since the effort for a test is much higher. In addition, tests comparing power cycling capability of IGBT and diode in the same module are inadequate. In several applications, e.g. as active rectifier, the diode may limit the device lifetime. For a more precise lifetime prediction of the power semiconductors in aforementioned fields, further studies are necessary.

Due to the above mentioned reasons, a new high-current power cycling test-bench for short load pulse duration with the possibility to test different devices at the same time was developed.

PCIM Europe 2016, 10 – 12 May 2016, Nuremberg, Germany

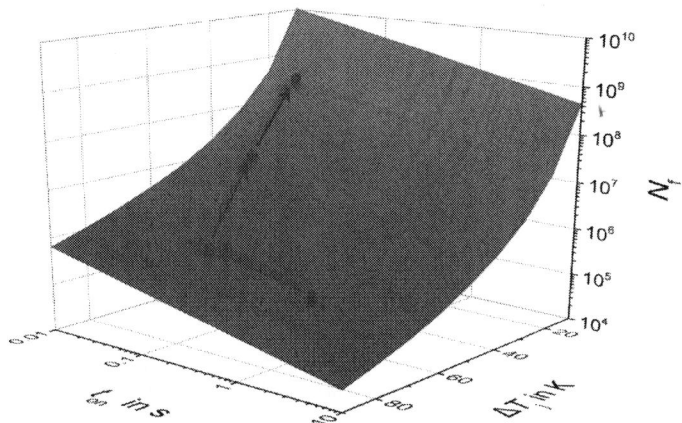

Fig. 1: Graphic demonstration of design of experiments (DOE)

Fig. 1 shows a possible test series to investigate the impact of small t_{on} and ΔT_j on power cycling lifetime. For a graphic demonstration of experiment design, reference lifetime cycles N_f in the figure are calculated by using the CIPS 08 model [1]. Test conditions for the first point are in the standard power cycling range with t_{on} = 1 s and ΔT_j = 70 K. The load current I_L could then be changed step by step to achieve a series of desired combinations of shorter t_{on} and smaller ΔT_j. In addition, the gate voltage V_{GE} of IGBTs can be adjusted separately for desired power loss.

2. Test-Bench Design and Setup

For testing 1200 A rated current IGBT modules, a 2600 A DC current source is chosen to create a relative large ΔT_j = 30 – 40 K within short t_{on} of some ten-milliseconds. A simplified circuit diagram of the test-bench is shown in Fig. 2.

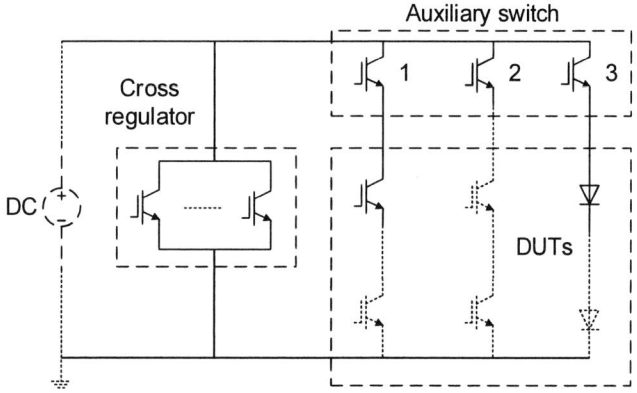

Fig. 2: Simplified electrical circuit diagram of the test setup

Devices under tests (DUTs) are kept in the forward conducting state during the test, while the auxiliary switches of power path 1, 2 and 3 are turned on and off alternately. Therefore the load current runs across the three power paths successively. The current is regulated by a cross regulator, which consists of four parallel connected IGBT modules. These IGBTs are working in linear mode to compensate the different operation points between the power paths.

© VDE VERLAG GMBH · Berlin · Offenbach

Because IGBTs and diodes from different manufacturers differ in their power losses and thermal behaviors at same load current and pulse duration, different load currents are needed for testing different devices at the same time. For example, if for a test the needed load currents of the three power paths are I_{L1} = 1800 A, I_{L2} = 1500 A and I_{L3} = 1700 A, then the desired current of the current source will be set to the maximum, i.e. 1800 A. During load pulse durations of power path 2, the unwanted difference of 300 A will be taken by the cross regulator. Similarly, the cross regulator works then with 100 A during load pulse durations of power path 3.

2400 A rated current IGBT modules FZ2400R17HE_B9 are used in the cross regulator for the consideration of overdesign. Due to the limit of cooling ability of one single heat sink, the maximum current per IGBT module is limited to 200 A, which lies far away from the nominal operation point of the used high power IGBT modules (see Fig. 3). In this working area of a multichip module, a very symmetric distribution of the conduction losses is normally quite unlikely. However the problem of inhomogeneous heating is expected to be uncritical due to the overdesign.

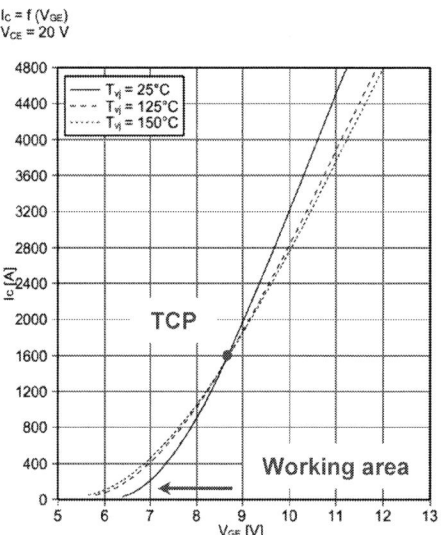

Fig. 3: Transfer characteristic of FZ2400R17HE_B9 used in cross regulator (Temperature compensation point and working area are marked) [8]

A PI control for the cross regulator is realized with a FPGA-based real-time system, which continuously calculates the difference between actual value and set point of load current in target power path and minimizes it accordingly by adjusting the gate voltage of the IGBTs in cross regulator. If the actual value of load current in the present conducting power path is higher than its set point, then the gate voltage of the IGBTs used in the cross regulator will be increased and thus more current will flow through cross regulator.

The difference between the actual value of current flowing out of the current source I_{CS} and the set point of current flowing through the power path I_{PP} will be divided by four (number of IGBT modules in cross regulator) and it will be taken as the set point of current I_{CR} for each single IGBT module used in the cross regulator. In this way, every IGBT module of the cross regulator will get a same set point of current, which leads to an identical current distribution of several modules in parallel operation. As it is shown in Fig. 4, the gate voltage V_{GE} of four IGBT modules is adjusted separately to achieve the desired set point of I_{CR}. Therewith an asymmetric operation of the modules is avoided, which could be caused by their positive temperature dependency of the collector current I_C in linear mode below temperature compensation point (TCP, see Fig. 3), if the modules are controlled with a same gate voltage V_{GE}.

PCIM Europe 2016, 10 – 12 May 2016, Nuremberg, Germany

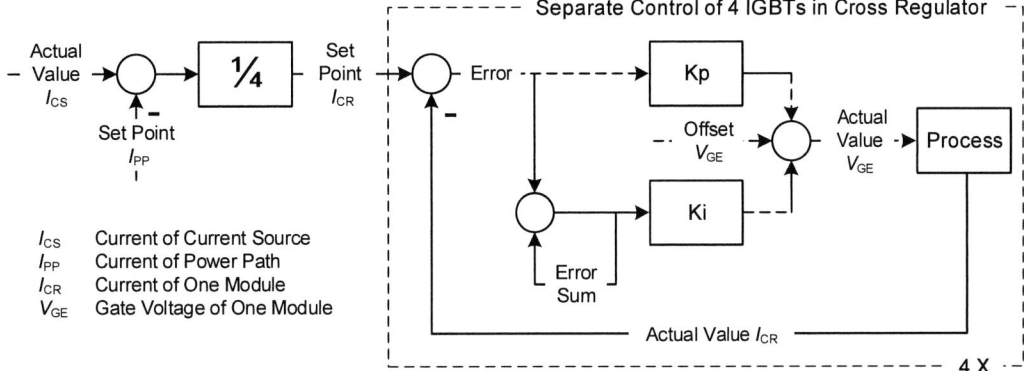

Fig. 4: Block diagram of the control loop for cross regulator

Between the conducting phases of two neighbor power paths, an overlap time of 100 μs is added to limit the inductive voltage peak, which is generated by the turn-off current slope di/dt at the parasitic inductance. Fig. 5 shows the current trends of two power paths and one module in the cross regulator around a switching event. Under this measurement condition, the load current through path 2 can approach ±5% of the set point of 1500 A within 6 ms, which is acceptable for tests with load duration t_{on} ≥ 40 ms. For tests with load duration of some millisecond, the current cannot be assumed to be constant during the whole load pulse.

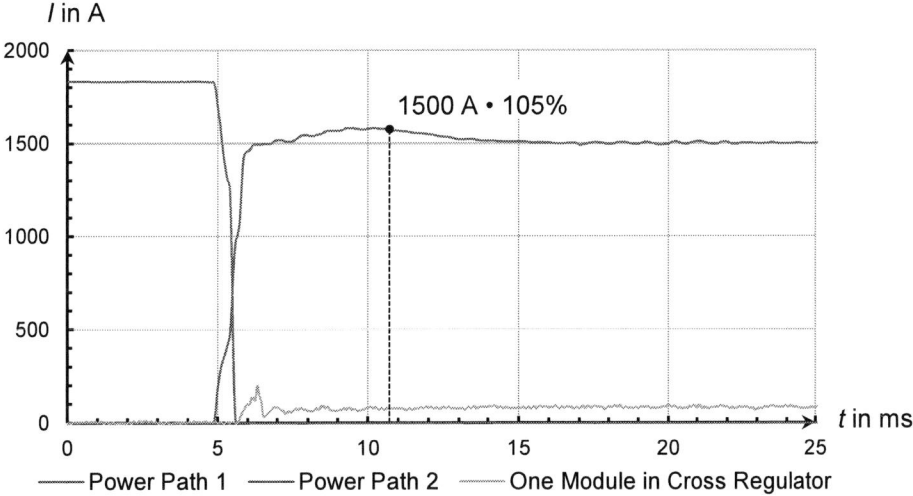

Fig. 5: Current developing of two power paths and one out of four module in the cross regulator around a switching event

The number of devices under test in each power path is variable. The maximum of devices in each power path is limited by the maximal output voltage of the current source. The sum of voltage drop of devices in each path should however be similar. The complete test-bench setup was designed in a way that the current source is exposed to constant current and its voltage jumps are limited when switching between power paths. Duty cycle of the test can be adjusted by varying the number of power paths. For the presented test setup with three power paths, the cooling phase is about twice as long as the heating phase.

The completed power cycling test-bench is shown in Fig. 6 with a cage with protection shields as test field, a 2600 A current source, a tempering device and the necessary supply,

© VDE VERLAG GMBH · Berlin · Offenbach

591

measurement and control systems. The cooling of the DUTs, auxiliary switches and cross regulator is realized with a water-operated tempering system. For each power path, a 500 mA current source is connected in parallel to the DUTs to determine the junction temperature by using the $V_{CE}(T)$-method [9]. The case temperature is measured by thermocouple in a small trench on the heat sink surface directly under the hottest chip.

(a)

(b)

Fig. 6: (a) Overview of the 2600 A short-pulse power cycling test-bench; (b) A more detailed picture of test field with protection shields open

3. First Power Cycling Test Results

A power cycling test was performed on 1200 A rated current IGBTs and diodes from two different manufacturers with load pulse duration t_{on} = 1 s, cooling phase duration t_{off} = 2 s, coolant inlet temperature T_{inlet} = 50°C and coolant flow rate per heat sink $\dot{V}$ = 6 - 7 l/min. The test conditions are shown in Table 1. Gate voltage V_{GE} of IGBTs from manufacturer B was reduced to 12.5 V to increase their power losses. Therewith, the difference of load current between three power paths was reduced. After the adjustment of test conditions in the start-up phase, the initial setting of test (t_{on}, t_{off}, I_L, V_{GE}, T_{inlet} and $\dot{V}$) was kept until the end of test.

Table 1: Test conditions of devices under first power cycling test

Manufacturer	Device	Label	ΔT_j in K	T_{jmin} in °C	I_L in A	V_{GE} in V
A	IGBT	A_T1	71.6	59.7	1554	15
		A_T2	70.8	60.0		
		A_T3	71.4	59.4		
		A_T4	72.8	59.6		
B	IGBT	B_T1	71.6	60.9	1806	12.5
		B_T2	70.9	60.5		
		B_T3	70.2	60.8		
		B_T4	71.0	60.2		
	Diode	B_D1	71.5	59.2	1782	-
		B_D2	69.8	59.0		
		B_D3	68.6	58.8		

The forward voltage V_{CE} (V_F) and thermal resistance (junction to case) R_{thjc} trends of IGBTs and diodes from two manufacturers during the first power cycling test are shown in the following Fig. 7, Fig. 9 and Fig. 10. The forward voltage V_{CE} (V_F) was measured short before turn-off in each cycle. To eliminate the influences of adjustment of test conditions in the start-up phase, the first 10% measurement data of V_{CE} (V_F) and R_{thjc} will not be used for the calculation of their initial reference values (100% values). The initial reference values of V_{CE} (V_F) and R_{thjc} are mean average of the corresponding measurement data from the 10 - 20% of the test procedure, when the test is considered to be already stable.

PCIM Europe 2016, 10 – 12 May 2016, Nuremberg, Germany

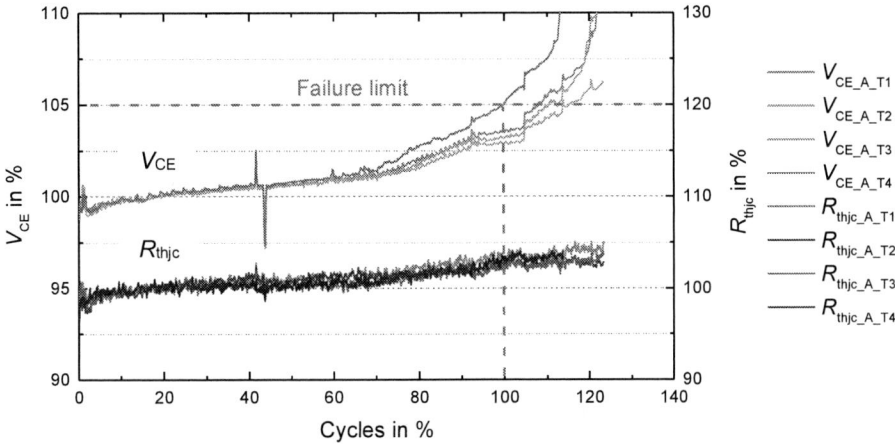

Fig. 7: V_{CE} and R_{thjc} trends of four IGBTs from manufacturer A during the test

Test cycles of all four IGBTs in Fig. 7 are normalized to the cycle of the first failure from manufacturer A (A_T4). The two peaks in the V_{CE} trends at around 40% of the test cycles were caused by operation errors, which were not related to the aging of the modules. The IGBTs in this group are defined to have reached their lifetime limit as an increase of forward voltage V_{CE} by 5%. The corresponding increases of thermal resistance R_{thjc} are less than 5%. At the end of the test, several wire bond lift-offs in IGBTs from manufacturer A are to be observed. Reconstruction of chip surface metallization after the power cycling test is also found under the microscope (see Fig. 8).

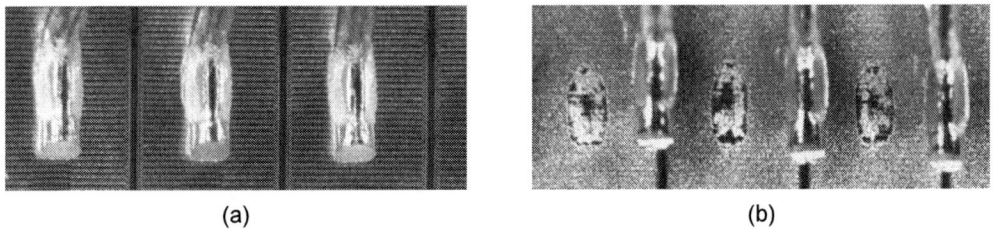

(a)   (b)

Fig. 8: Unstressed (a) and stressed IGBT chip with wire bond lift-offs (b) from manufacturer A

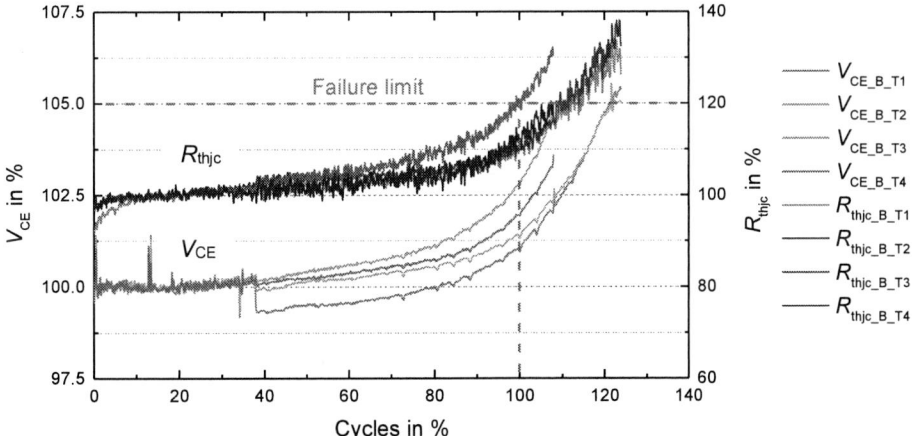

Fig. 9: V_{CE} and R_{thjc} trends of four IGBTs from manufacturer B during the test

© VDE VERLAG GMBH · Berlin · Offenbach

PCIM Europe 2016, 10 – 12 May 2016, Nuremberg, Germany

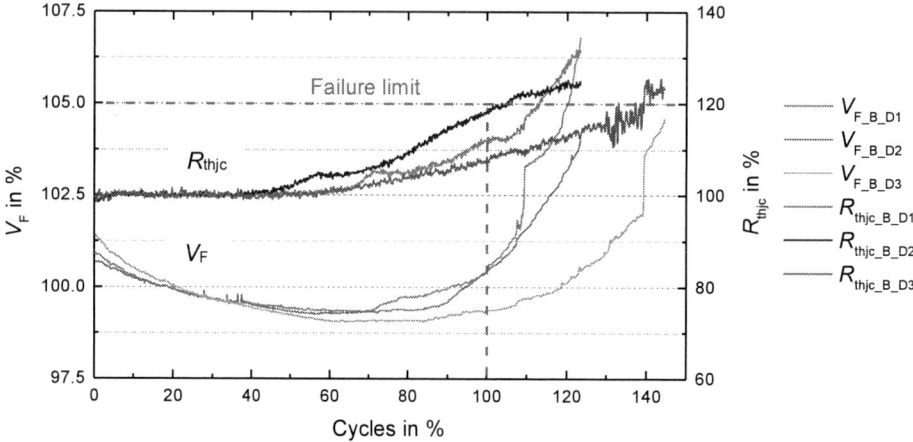

Fig. 10: V_F and R_{thjc} trends of three diodes from manufacturer B during the test

Test cycles of four IGBTs and three diodes in Fig. 9 and Fig. 10 are normalized to the cycle of the first failure from manufacturer B (B_T3). The small jumps in the V_{CE} trends of B_T1 and B_T2 at around 40% of the test cycles were caused by changing of current transducer. Despite the same set point of current, the offset difference between two current transducers led to the slight jumps in V_{CE} trends. The DUTs from manufacturer B (IGBT and diodes) are defined to have reached their lifetime limit as an increase of thermal resistance R_{thjc} by 20%, which is caused by the degradation of the chip solder. Discontinuous regions can be found at the chip solder edges between two stressed chips, where the stress of the chip solder is most severe during the test (see Fig. 11).

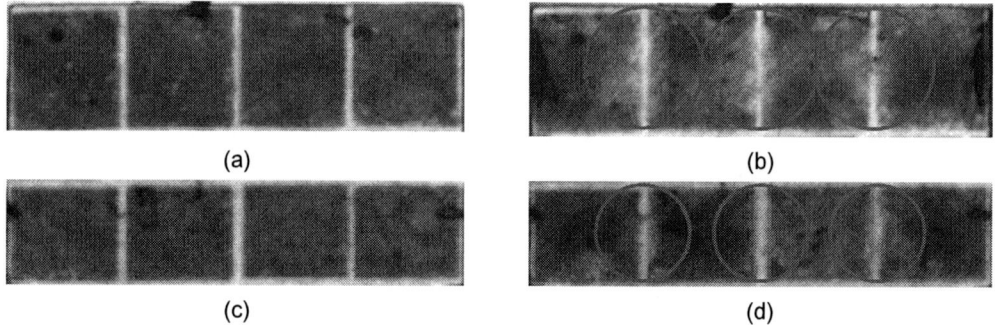

Fig. 11: Scanning acoustic microscopy of IGBT chip solder in B_T2 before (a) and after (b) the test, Scanning acoustic microscopy of diode chip solder in B_D3 before (c) and after (d) the test

At the end of the test, two clear jumps in the V_F trends of B_D1 and B_D3 are to be observed, which indicates wire bond lift-offs in the corresponding modules. However, the degradation of chip solder is the main and significant failure mechanism for manufacturer B in this test, which accelerated the aging of wire bonds. Compared to the IGBTs, the forward voltage of diodes first decreased slowly by about 2% due to the negative temperature coefficient of its forward voltage at this working point and then increased with dominating aging of wire bonds. Because of this effect, the aging of diode was compensated to some degree in comparison to IGBT. Even so, IGBT and diode from manufacturer B with same packaging technologies showed similar power cycling capability and failure mechanism.

© VDE VERLAG GMBH · Berlin · Offenbach

4. Summary

Devices from same manufacturer show similar lifetime and failure mechanism in the first power cycling test. All the devices have fulfilled the specification of power cycling capability from the manufacturers under the test conditions of the first test.

A cross regulator was developed to test different devices with the same desired combination of load pulse duration and junction temperature swing at the same time. The performance of the cross regulator regarding control speed and stability was verified by a demonstration power cycling test. Its suitability for the tests with extreme small t_{on} of some millisecond will be checked in further study.

For a large number of power electronic applications, short load pulse durations in millisecond range and small junction temperature swings below 40 K are of great interest. This paper presents a developed power cycling test-bench for this requirement. Power cycling tests based on the strategy as shown in Fig. 1 could be performed on this test-bench to build up new empirical lifetime models for high power IGBT and diodes, which will be actually and precisely studied for short load pulse durations.

References

[1] R. Bayerer, T. Herrmann, T. Licht, J. Lutz and M. Feller, "Model for Power Cycling lifetime of IGBT Modules - various factors influencing lifetime," in *Proc. CIPS 2008*, Nuremberg, 2008.

[2] U. Scheuermann and R. Schmidt, "Impact of load pulse duration on power cycling lifetime of Al wire bonds," *Microelectronics Reliability,* pp. 1687-1691, 2013.

[3] S. Hartmann and E. Özkol, "Bond wire life time model based on temperature dependent yield strength," in *Proc. PCIM 2012*, Nuremberg, 2012.

[4] Y. Wang, S. Jones, D. Chamund and G. Liu, "Lifetime Modelling of IGBT Modules Subjected to Power Cycling Tests," in *Proc. PCIM 2013*, Nuremberg, 2013.

[5] M. Denk and M. M. Bakran, "Comparison of counting algorithms and empiric lifetime models to analyze the load-profile of an IGBT power module in a hybrid car," in *International Electric Drives Production Conference*, Nuremberg, 2013.

[6] K. Ma, M. Liserre and F. Blaabjerg, "Lifetime Estimation for the Power Semiconductors Considering Mission Profiles in Wind Power Converter," in *IEEE Energy Conversion Congress and Exposition*, Denver, 2013.

[7] H. Liu, K. Ma, P. C. Loh and F. Blaabjerg, "Lifetime Estimation of MMC for Offshore Wind Power HVDC Application," *IEEE Journal of Emerging and Selected Topics in Power Electronics, Vol. PP, Issue 99,* p. 1, 07 Sept. 2015.

[8] Infineon, "Data sheet IGBT Modules FZ2400R17HE_B9," 2013.

[9] U. Scheuermann and R. Schmidt, "Investigations on the VCE(T)-Method to Determine the Junction Temperature by Using the Chip Itself as Sensor," in *Proc. PCIM 2009*, Nuremberg, 2009.

© VDE VERLAG GMBH · Berlin · Offenbach

Issues in Testing Advanced Power Semiconductor Devices

Gabor Farkas, Mentor Graphics MAD Division, Hungary, gabor_farkas@mentor.com
Zoltan Sarkany, BME Department of Electron Devices, Hungary, sarkany@eet.bme.hu
Marta Rencz,   BME Department of Electron Devices, Hungary, rencz@eet.bme.hu

The Power Point Presentation will be available after the conference.

Abstract

Thermal transient testing is a widely used tool in reliability testing of power semiconductors and structure integrity analysis of packages, cooling mounts etc.   The paper demonstrates novel concepts for powering and sensing in thermal tests when wide bandgap compound semiconductor components (GaN, SiC) replace silicon. We highlight problems like the non-linearity of devices in broad temperature ranges; and slow effects (surface charge, carrier absorption on traps) producing false transient signals. We propose methods to overcome these obstacles.

Keywords

Thermal transient testing, compound power semiconductor devices, Si, SiC, GaN, IGBT, HEMT

1. Introduction

Thermal transient testing has a special role among the techniques used for the thermal characterization of semiconductors. Using the inherent heat generated by the devices it can be used for reliability tests at high power [4] or also for non-destructive structural analysis [1]. A single device acting as heat source and sensor the same time can produce a one-dimensional map of the heat conducting path including the die, die attach, package, thermal interface layers, cooling mounts etc. In case of modules and subsystems multiple sources and sensors can give a more detailed picture of the structure measuring thermal transfer impedances between different points [1].

During the previous decades power electronics was mostly realized on silicon. It offered a very robust technology working fine even at 150 °C operating temperature. For this reason thermal testing standards (e.g. [5]) have taken for given such features of silicon like

- pn junctions available for powering and sensing
- mostly linear temperature dependence of parameters
- availability of normally-off devices for power electronics
- no slowly moving surface charges in MOS structures
- rather long lifetime of charge carriers in depleted regions

Material / Device	Si	SiC	GaN
BJT	St	E	X
MiSFET	St	Td	Td
JFET	St	E	Td
HEMT	X	X	Td
IGBT	St	Td	X
pn Diode	St	E	X
Schottky Diode	St	Td	E

Table 2 Thermal testing methods for devices realized in certain materials

Newly, compound semiconductor devices have been introduced into power applications where their excellent properties can justify their higher cost. They work well at extremes like at 77K in liquid nitrogen for minimum noise; or at high temperatures up to 350 °C. They show low channel resistance due the high carrier mobility and high blocking voltages thanks to their wide bandgap.

SiC having a wide and indirect IV–IV-bandgap has always been intended for power and high temperature devices.

GaN has a direct bandgap; minority carriers would recombine after an extremely short lifetime, they have no chance to reach the opposite bank (i.e. collector).   Accordingly, only unipolar transistors can be made of it. However, on an AlGaN/GaN interface a two-dimensional (2DEG) electron gas forms easily and offers low sheet resistances in a HEMT device.

In Table 2 we list the existing device types and also hint on the existing knowledge base available for their thermal testing. (St: standard testing method, E: the device exists, X: the device does not exist, Td: tests at Mentor Graphics laboratories done and published.)

2. Transient Thermal Testing

Thermal transient measurement is a widely accepted test method for checking device integrity of semiconductor components and systems. For any device category the main steps of transient testing are the following:

(1) **calibration**,

(2) **powering** the device,

(3) **switching** on/off and

(4) **recording** the temperature transient.

Steps (2), (3), (4) and subsequent evaluation of the results for distilling them into descriptive functions like *Zth* curves or *structure functions* are illustrated in Fig 1.

For all device categories the temperature measurement is based on the tempera-ture dependence of the voltage – current

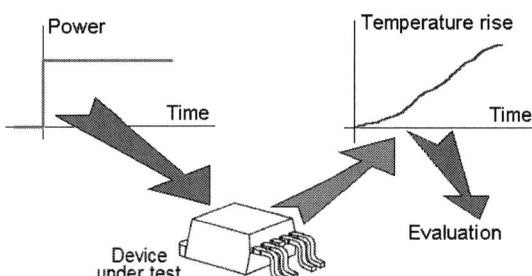

Fig 1 General scheme of the thermal transient testing

characteristics. One can fix either the voltage or the current and can use the change of the other quantity as temperature sensitive parameter (TSP).

Calibration and recording occur at a low power, which has to be maintained on the device for having a TSP at all. Standard testing methods for the devices in Table 2 typically use the same physical effect for powering and sensing/recording, e.g. the forward voltage of a particular junction at different current levels. In optimal case the powered state corresponds to the typical use of the device. The challenge in step (3) (switching) is just coping with the electric transients distorting the thermal signal of TSP.

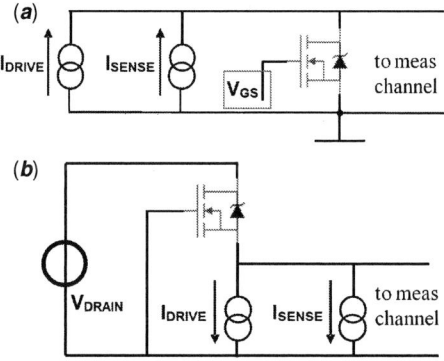

Fig 2: Insulated gate device powered and measured in "common gate" (**a**) and "common source" arrangement (**b**) . Control of the V_{GS} voltage defines the actual measurement mode

For the new device categories which show no monotonous temperature dependence of a certain parameter [8], or lack stability of it; a way around the problem can be using a parasitic effect as temperature sensor in the low powered state.

Some standard methods are elaborated for normally-off devices which dominate the power electronics, for normally-on devices at least a feasible modification of the method has to be introduced.

The standards define two basic ways of doing transient tests. Either after a prolonged heating period a *continuous cooling* is recorded capturing the change of the TSP. Alternatively, one can switch between a high powered and low powered state by applying a series of *power pulses*. Both methods have advantages and disadvantages; some hints will be given in following sections.

3. Testing of Three-Pin Devices

In case of three-pin devices which are the most characteristic in power electronics we have two usual ways for powering: common source/emitter (CS) or common gate/base (CG) arrangement (Fig 2). In both arrangements we can fix some currents, and measure a voltage, or vice versa. The measured quantity has to be a TSP showing a characteristic dependence on temperature.

The arrangements are often simplified for thermal testing. Connecting some pins we can reduce the arrangement to a two-pin measurement, resembling the measurement of diodes.

For MOSFET devices a frequent option is connecting the gate to the source in CS and applying a negative current. This makes the inherent bulk diode of the MOSFET powered. We will refer to this powering below as BD (bulk diode) powering.

An other common technique is connecting the gate to the drain in CS and applying a positive current (same as putting V_{DRAIN} to zero in CG). This defines an approximately quadratic curve in the MOSFET characteristic, often referred to as MOS diode (red curve in Fig 3, MD powering below).

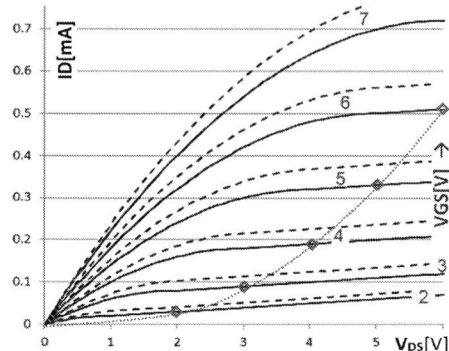

Fig 3: Characteristics of a normally-off MOSFET in common source arrangement, cold (solid), hot (dashed) state

Keeping V_{GS} at a proper fixed value makes the MOSFET behave as a positive or negative thermal coefficient resistor (R_{DSON}, RDS mode). An IGBT in the same arrangement exposes a pn junction of negative thermal coefficient (saturation mode, SM).

In all above powering modes the same structure can be used for powering and sensing. We can observe that a sudden power change can be realized either by switching from I_{DRIVE} to I_{SENSE} (current jump) or by switching V_{DRAIN} between two voltage levels (voltage jump). All these powering modes are discussed in detail in [5][7][11].

Although so far we treated cases where powering and sensing was established on the same structure, with proper programming of the sources also mixed modes can be realized, such as powering in MD or RDS mode, but sensing on the reverse bulk diode.

4. Reference Measurement on a Standard Device

Now we will present measurement techniques on a commercially available IXTH6N50D2 depletion (normally-on) power MOSFET. Details of this measurement are disclosed in [7].

As wide bandgap devices behave at room temperature resembling the operation of silicon at cryogenic temperatures, we carried out several very careful temperature-voltage calibrations of the device in a closed thermostat in a wide temperature range. The calibration of the forward voltage in BD mode was practically linear.

For RDS we shorted gate and source (V_{GS}=0 in Fig 2 is a plausible selection for normally-on device) and applied 0.5A measurement current.

We saw a clear proportional change between −70 °C and 150 °C (markers x). Moreover, selecting a subset of the measured points and extrapolating their "trendline" in an exponential way we got matching with the remaining points at low quadratic error. The voltage drop was proportional to the drain bias at all current levels and all temperatures, the device behaved as a true resistor in this current range.

We also submerged the sample (mounted on an appropriate heat sink) into liquid nitrogen for reaching low temperature (boiling point 77K, −196 °C).

We found that the channel resistance was much higher than at many points of higher temperature (♦ marker in the chart).

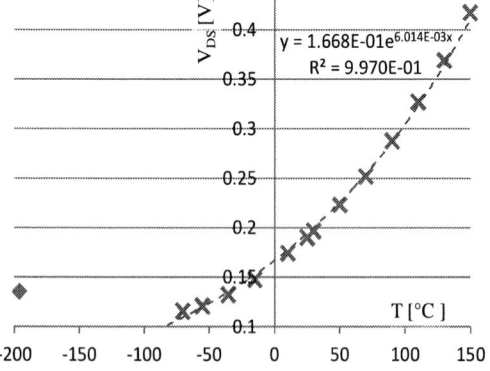

Fig 4: RDS calibration curve of the depletion device with exponential regression, at 0.5A measurement current

PCIM Europe 2016, 10 – 12 May 2016, Nuremberg, Germany

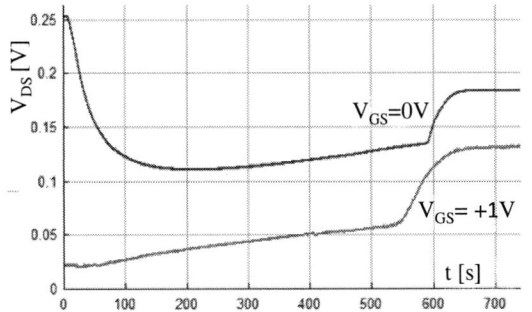

Fig 5: V_{DS} change in time, the device temperature grows from −196 °C to 25 °C. V_{DS} measured at I_D = 0.5A, V_{GS} = 0V and V_{GS} = +1V

As we had no instrument to fill the temperature gap between −196 °C and −70 °C, we measured the transient change of the drain voltage when we took out the mounted sample from the liquid nitrogen. This simple test ensured continuous and monotone growth of the device temperature.

Fig 5 demonstrates the "melting" of the channel resistance. The plot proves that the higher resistance in liquid nitrogen is not an artefact; R_{DSON} really grows towards low temperatures on the V_{GS}=0V plot.

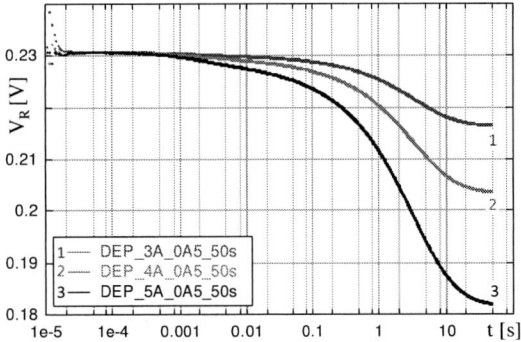

Fig 6: Recorded cooling transient of the device

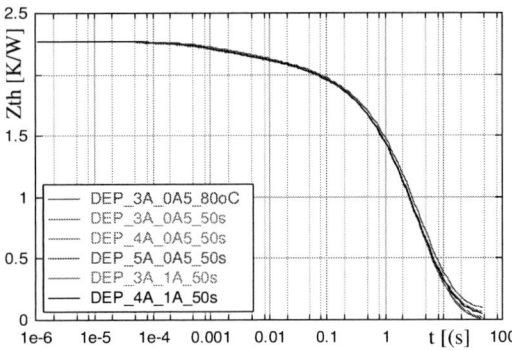

Fig 7: Zth curves, cooling normalized by power

After the calibration we recorded the voltage in a cooling test of the device. Applying heating current of 3A, 4A, 5A (curves 1,2,3) we saw the crisp transients of Fig 6.

Converting voltage to temperature upon the calibration curve of Fig 4 and normalizing it by the power measured on the device in its hot state we get the so called Zth curves. We see their perfect fit at all conditions after 80 µs (Fig 7: , RDS measurement, heating current 3A to 5A, measurement current 0.5A to 1A, base plate temperature 25 °C to 80 °C).   This means that we measure a true thermal effect, as expected at silicon.

5. Measurement of a power HEMT

As presented more in detail in [10], we tested a HEMT device (CLF1G0060) in all available modes.   We suggested in [9] a very stable measurement technique on the gate parasitic of Schottky gate HEMTs, but now for extending the methodology to insulated gate MISFETs, we restricted our choice to V_{GS} or R_{DSON} as TSP.

First we calibrated V_{GS} over a wide temperature range (−196 °C to 150 °C). Based on analytic equations presented in [8] we also calculated the Vth threshold voltage of the device.

Then we recorded the transient voltage signals at low sensor current in CG and RDS mode (1mA for CG, 20mA for RDS, Fig 9).

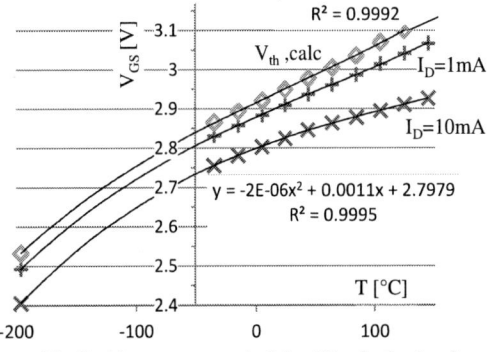

Fig 8: V_{GS} measured at I_D=10mA, I_D=1mA, V_{DS}=4V , −Vth calculated

The normalized transients were practically unusable because of strange changes lasting even in the second time range. The plausible explanation of the phenomenon is that while at standard MOS all physical changes *except* the heating of the die terminate in microseconds,

© VDE VERLAG GMBH · Berlin · Offenbach

PCIM Europe 2016, 10 – 12 May 2016, Nuremberg, Germany

the *carrier capture and emission on traps* in GaN are known to last really long.

The effects are described in the literature as surface or bulk trapping, forming a virtual gate, etc. The treatment of these is beyond the reach of this paper.

The reported time constants of these effects span from μs to minutes. We also experienced that after a large current or temperature shock the calibration curves were severely displaced and they relaxed in some minutes.

In order to overcome the problem we wrote a testing script using dozens of heating and sensing current and also ambient temperature values.

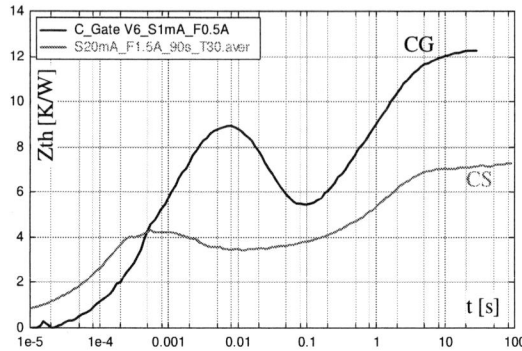

Fig 9: *Zth* curves of the HEMT,
in CG I_{sense}=1mA, ΔP=1W,
in CS I_{sense}=20mA, ΔP=1.8W

We found that selecting a high measurement current provides excellent results.
Fig 10 presents the *Zth* curves at 2A heating and growing sensor current (20 mA to 1A).

The key in the figure is of the format S<Isense>_F<Idrive>_<tmeas>_T<cold plate temp>.

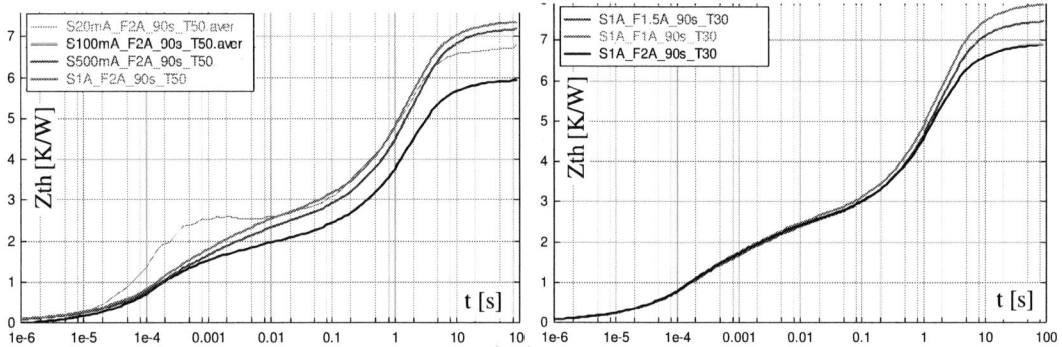

Fig 10: *Zth* curves at several I_{sense} values   Fig 11: *Zth* curves at I_{sense} =1A, several I_{drive} values

We see how the "bumps" belonging to time variant effects vanish. At higher I_{sense} we have a true thermal measurement (see Fig 11), the curves are proportional to heating, *Zth* fits at all I_{drive} and I_{sense} over an appropriate time range.

6. Measurement of a SiC module

In this experiment we measured the thermal characteristics of a commercially available Cree power module (CCS050M12CM2). Opening the package we found six separate SiC MOSFET transistors and six SiC Schottky diodes in it.   We decided to measure self-heating and transfer heating in the module.
First we measured the devices separately.

When measuring the diodes all Zth curves fit perfectly at different heating and sensor currents, as expected. The MOSFETs in MD mode showed good sensitivity but slow transients, this was acceptable in heat transfer measurements only.

RD mode was usable. The calibration curves are given in Fig 13, showing V_{DS} values at 7V gate-source voltage, at 0.5A, 1A, 2A sensor current. The voltages are here multiplied by 2 for I_{sense}=0.5A and by 0.5 for I_{sense}=2A, in such a way the chart corresponds more to the concept of R_{DSON}.

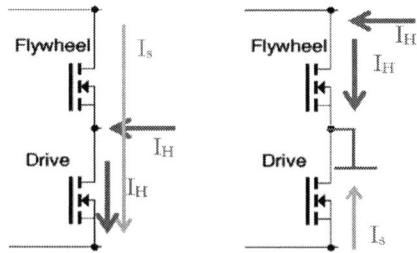

Fig 12: heating and sensing current directions for measuring self- and transfer heating in the module

© VDE VERLAG GMBH · Berlin · Offenbach

PCIM Europe 2016, 10 – 12 May 2016, Nuremberg, Germany

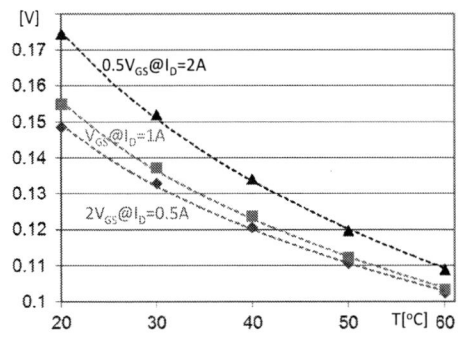

Fig 13 : V_{DS} at V_{GS}=7V, normalized on 1A

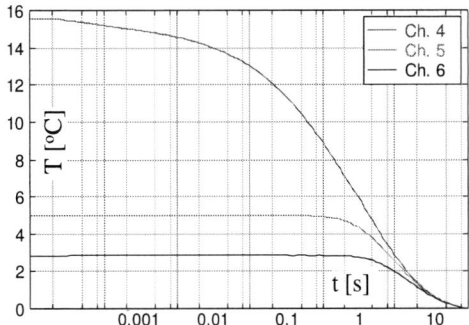

Fig 14: Cooling of the module, heat transfer effect

In Fig 14 we demonstrate the cooling of different devices in the module. The cooling was captured after one transistor at channel 4 of the tester was heated for 60 seconds.

7. Cascode HEMT measurement

In the next experiment we measured a device representing a real challenge, a cascode HEMT device (NTP8G202N). This popular device for high voltage power conversion contains the circuit encircled in Fig 15 and Fig 16 in a single package. The combination offers the normally-off operation and low charge storage of a low voltage MOSFET along with the high blocking voltage of the HEMT.

The thermal management of these packages is quite complex. As the channel resistance of the HEMT is typically an order of magnitude higher than that of similar size MOSFETs, the majority of the heat is dissipated on the HEMT channel in "on" state. Contrarily, with reversed current the bulk diode of the MOSFET chip will have the highest dissipation.

As we have no access to the internal point an iterative method was used for determining the individual electric characteristics and thermal resistances.

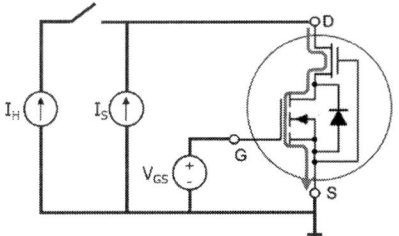

Fig 15: Measuring Rth on the channel resistances

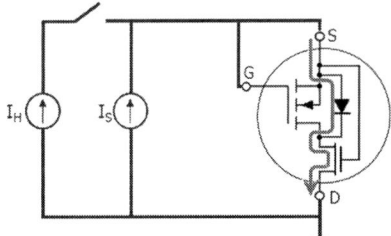

Fig 16 : Measuring Rth on the MOS bulk diode

First we applied different current levels from 1A to 4A (inducing 0.25W to 7.5W dissipation) on the arrangement in Fig 15. We did this in a pulsed operation, causing no significant heating of the channel. Using a regression scheme we measured 0.26 Ω channel resistance (data sheet value is 0.29 Ω).

Then we applied current levels from 1A to 4A on the reversed arrangement in Fig 16, again in pulsed operation. Subtracting the resistive portion from the measured characteristics we found the typical exponential diode characteristics, with a forward voltage of ~0.7V. The HEMT was biased by a positive V_{GS} voltage; which is now equal to the forward voltage of the diode, accordingly it was of a highly conductive channel.

In Fig 17 we see the transients measured (mainly) on the channel resistance of the HEMT at 1A sensing current and at 2A, 3A, 4A heating current, normalized as Zth. Fig 18 shows the structure functions condensed from the Zth curves (for more details on structure functions refer to [1][2][3]). In the spirit of the TDIM method defined by the JEDEC JESD51-14 standard [5] we repeated the measurement belonging to 2A heating with an additional TIM layer. We can read approximately 2.5 K/W junction to case thermal resistance for the HEMT.

© VDE VERLAG GMBH · Berlin · Offenbach

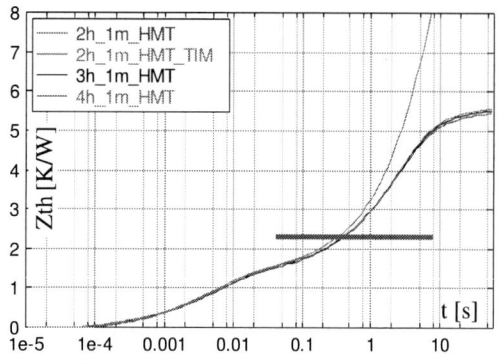

Fig 17 : *Zth* curves measured in the arrangement of Fig 15

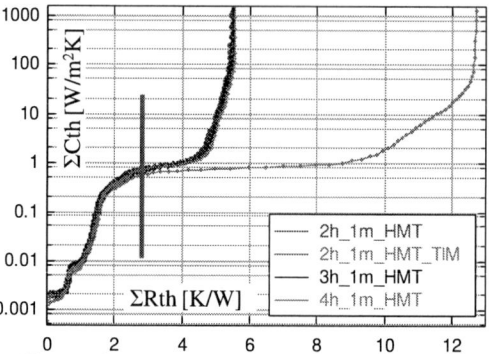

Fig 18: Structure functions measured in the arrangement of Fig 15

The *Zth* curves gained from similar transient measurements are shown in Fig 19. The accordion-like shrinking of the curves is a simple consequence of the definition of *Zth*, the actual temperature change on the diode as sensor is divided by the whole power measured on the device (diode plus HEMT channel). Accordingly, the distribution of the power between the two devices can be calculated from the factor of shrinking at different current level.

The MOSFET has some 4 K/W thermal resistance.

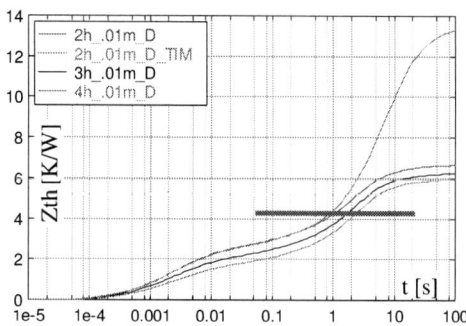

Fig 19 : *Zth* curves measured in the arrangement of Fig 16

8. Thermal transient testers

Present day thermal transient testers and their control and evaluation software are apt to test systems where the main heat source is a compound semiconductor. However, for RDS measurements a sensor current of several amperes and heating current of many amperes is needed.

It has to be noted that normally-on devices in CG need a tester which possesses 4-quadrant current sources: *Idrive* and *Isense* are negative but the voltage on them is above ground. We proposed a conversion from CG to CS in [6] using a simple external circuitry, it has to be extended to negative V_{GS} voltages for normally-on transistors.

Testers rather provide constant current than constant power as the latter needs regulation loops reacting faster than the early time constants of a system (e.g. die attach). This makes the use of pulsed ("dynamic") testing in RDS doubtful. Pulses of fixed current would result in quickly growing power level in subsequent pulses; and the power changes also during the pulse. Continuous methods (mistakenly called "static") cope with the problem by recording the power once, just before switching off and making no power pulses further on.

Reliability testers often apply repeated shorter cycles at high power and then carry out a structure analysis by transient testing at lower power [4]. The powering during reliability testing should resemble the actual operation of the device; otherwise the failure mechanism will differ from the expected. If this powering produces slow changes covering relevant time range needed for structure analysis, different sensing methods, also parasitic ones can be used for the latter.

9. Conclusions

Thermal testing of wide bandgap semiconductors can be done with available tester hardware/software but the measurement methodology has to be refined for relevant results.

The tests need stable operating points at high power and at low power; and a thermally sensitive device parameter (TSP) which can be calibrated.

Traditionally the TSP can be a forward voltage, threshold voltage or channel resistance.

Even if the calibration process proves that the TSP has a clear temperature dependence; slow effects influencing it (charge recombination, moving surface charge, absorption on traps) may cause false transient signals over a time range.

TSPs which were nearly linear in the typical operational temperature range of Si power devices are non-linear at wide bandgap. This has to be handled in the software tools.

The R_{DSON} channel resistance can show positive or negative temperature coefficient (PTC or NTC). In silicon we have experienced NTC only at cryogenic temperatures. In SiC we observed NTC at room temperature and above due to the wide bandgap.

Measurement on R_{DSON} makes the use of pulsed testing methods doubtful because of quadratic growth of the power level at fixed current. Continuous methods record the power once, just before switching off.

The physical effects for powering and sensing are not necessarily the same for reliability tests and structure analysis.

References

[1] V. Székely et al., "New approaches in the transient thermal measurements," Microelectronics J., vol. 31, pp. 727–733, 2000.

[2] M. Rencz, V. Székely, A. Poppe, G. Farkas, B. Courtois: "New methods and supporting tools for the thermal transient testing of packages."; Proceedings of the International Conference on Advances in Packaging (APACK'01). Singapore, 2001. pp. 407-411.

[3] O. Steffens, P. Szabo, M. Lenz, G. Farkas: "Thermal transient characterization methodology for single-chip and stacked structures", Proc. 21th SEMITHERM, San Jose,CA, pp. 313-321, 2005.

[4] Z. Sarkany, A. Vass-Varnai, M. Rencz: "Analysis of concurrent failure mechanisms in IGBT structures during active power cycling tests," Proc 16th EPTC, 3-5 Dec. 2014, pp.650-654, doi: 10.1109/EPTC.2014.7028349

[5] JEDEC Standard JESD51, "Methodology for the Thermal Measurement of Component Packages (Single Semiconductor Devices)". December 1995, www.jedec.org/sites/default/files/docs/Jesd51.pdf

[6] G. Farkas: "Thermal transient characterization of semiconductor devices with programmed powering", Proc. 21th SEMITHERM, San Jose,CA pp. 248-255, 2013.

[7] G. Farkas, T. Purak, G. Toth: "Thermal Transient Measurement of Insulated Gate Devices Using the Thermal Properties of the Channel Resistance and Parasitic Elements", Proc. 20th THERMINIC, London Sept. 24-26, 2014 doi:10.1109/THERMINIC.2014.6972517

[8] G. Farkas, G. Simon: "Thermal transient measurement of insulated gate devices using the thermal properties of the channel resistance and parasitic elements", Microelectronics Journal Vol. 46, Issue 12, Part A, Dec. 2015, pp 1185–1194, doi:10.1016/j.mejo.2015.06.027

[9] Z.Sarkany, G.Farkas: "Thermal transient characterization of pHEMT devices", Proc. 18th THERMINIC, 2012, pp. 225 – 228

[10] G. Farkas, G. Simon, Z. Sarkany: "Analysis of Advanced Materials Based on Measured Thermal Transients of Insulated Gate Devices in Broad Temperature Ranges", Proc. 21st THERMINIC, Paris Sept. 30-Oct.2, 2015

[11] G. Farkas, G. Simon: "Stability Criteria of the Thermal Characterization of Discrete Components", MIXDES Torun 2015, June 25-27, pp. 440-445

PCIM Europe 2016, 10 – 12 May 2016, Nuremberg, Germany

For information on the presentation

Requirements for Modern Lighting Systems and Expected Trends

please contact the speaker directly:

Mr. Diederik de Stoppelaar

LightingEurope
Boulevard Auguste Reyers 80
1030 Brussels
Belgium

Phone: +32 (0) 2 706 86 08
E-Mail: diederik.destoppelaar@lightingeurope.org

Special requirements on a SMPS for LED-lighting purposes by an example of an individual high-end LED-driver solution

Florian Müller, VS Lighting Solutions GmbH, Wasenstraße 25, 73660 Urbach, Germany, florian.mueller@vsu.vossloh-schwabe.com
Stefan Raithel, VS Lighting Solutions GmbH, Wasenstraße 25, 73660 Urbach, Germany, stefan.raithel@vsu.vossloh-schwabe.com

The Power Point Presentation will be available after the conference.

Abstract

Over the last years, the lighting market passed through substantial changes. When there was a variety of lighting technologies available on the market during the past years, suitable for different applications, there is now a clear trend to LED technology in nearly all areas of use. Meanwhile, there are 2nd-generation LED-drivers on the market which have improved concepts and improved performance values compared to the beginnings of the trend to LED, to fulfill the raised requirements stated by the luminaire manufacturers. This essay will explain the approach of Vossloh-Schwabe how to deal with these requirements and will name some features which will gain more and more importance in future applications.

1. Introduction

This paper will firstly give a short overview over common topologies used in LED-drivers for shoplight applications which generally have a high powerfactor, galvanic isolation and a constant current output.

Following this, an individual high-end Vossloh-Schwabe LED-driver solution will be presented. Naming all the challenges and requirements, which occur during the development of a state-of-the-art LED-driver, like realizing a high power factor, high energy density, reinforced isolation or accordance with several standards, would go beyond the scope of this abstract. So this concept is presented by means of three features which stand out between the competition and will become more and more relevant in future applications [1]. The first feature, which will be explained, is the constant current source which isn't realized with a buck converter, but with a linear concept. This concept shows some considerable advantages like excellent output current quality with lowest LF- and HF-ripple. Secondly, the regulation concept will be annotated to show the details of the entire LED-driver solution. Finally, the intelligent temperature protection is shown and it is explained, why such a concept is a remarkable advantage for the usage of such a driver in many applications. A diagram shows the different protection stages and thresholds.

© VDE VERLAG GMBH · Berlin · Offenbach

2. Topologies of a state-of-the-art LED-driver

There are several possible topologies or combinations of topologies to create a LED-driver concept which fulfill the requirements mentioned in chapter 1. Three functions have to be taken care of. The first function is the power factor correction to realize a high power factor and decrease the current harmonics values. These values have to stay below the limits given in EN 61000-3-2. The second function which is essential to reach SELV (safety extra low voltage) on the output is a reinforced galvanic isolation according to EN 61347-1/-2-13 and EN 60598. This isolation can be achieved by the usage of a galvanically isolated SMPS-topology (switched mode power supply topology). The third function is a constant current source on the output to drive the LEDs.

Figure 1 shows three examples, how these requirements can be fulfilled:

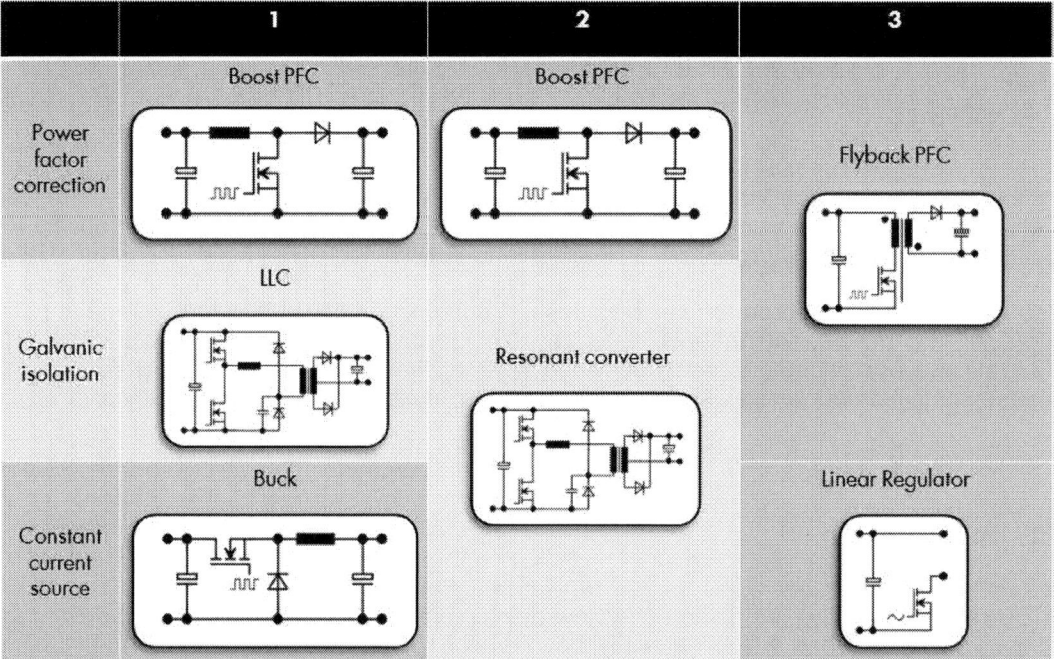

Fig. 1: Overview over topologies [2]

3. Vossloh-Schwabe LED-driver solution

The following chapters explain some of the features of the Vossloh-Schwabe LED-driver. A complete description of all functions and features of the driver would go beyond the scope of this abstract and so a summary was created of the functions which are special for lighting SMPS in contrast to SMPS for chargers or EDP systems (electronic data processing systems).

3.1. Topology

The topology of the described Vossloh-Schwabe LED-driver is topology 3, shown in figure 1. There are several advantages for this topology. Some of them are explained in the following chapters.

3.2. Constant current source

Due to the U-I-characteristic of a LED-module, it must be driven by a constant current source, as already mentioned in chapter 2. Vossloh-Schwabe chose a linear regulator as current source, since this solution benefits from a lot of features. The most significant advantage is that it is not a switched mode power supply topology which results in a very low LF- and HF-current ripple. This has a direct and positive effect on the health of people who are exposed to artificial light [3].

Another aspect of a linear regulator is its efficiency. Usually linear regulators are associated with a low efficiency, since in a standard application there is a considerably higher input voltage compared to the output voltage. But in contrast to a standard application, in the Vossloh-Schwabe driver, the 1st stage Flyback doesn't only guarantee a galvanic isolation and a correction of the power factor, but also a variable output voltage.

The secondary bus voltage can be controlled in small voltage steps to not only reduce the losses of this stage but also to realize a good impulse response. This way, efficiencies are reached which can be compared to efficiencies reached by a switched power supply, e.g. a buck-converter. But also a quick signal response is a crucial factor of the output stage, since load jumps from 0% to 100% at any output voltage without fading must be possible. Considering a quick signal response, also other parameters must be taken care of, like prevention of a voltage overshoot, since the requirements for SELV on the secondary side must not be violated and like prevention of oscillations which would result in visible flicker of the luminous flux.

In dimming mode, the linear regulator shows another advantage, since high PWM dimming frequencies are possible without further technical difficulties which are created, when using a SMPS as constant current source. This leads to improved perception and less interferences, when the LED-light is used e.g. in combination with cameras [4].

On top of that, there are several more advantages which are worth mentioning. One advantage is that a linear regulator topology possesses a much lower count of components which can be seen in table 1. This results in a smaller PCB.

Linear regulator topology	Buck Topology
Input capacitor	Input capacitor
Linear regulator	Diode
Output capacitor	Mosfet
	Inductor
	Control-IC
	Highside-driver
	Output capacitor

Tab. 1: Comparison – count of components

One more advantage is that by removing the switched constant current source, also a source of disturbances is removed which results in a much lower EMC noise level as can be seen in figure 2. Requirements according EN 55015 can be fulfilled with less effort compared to other topologies.

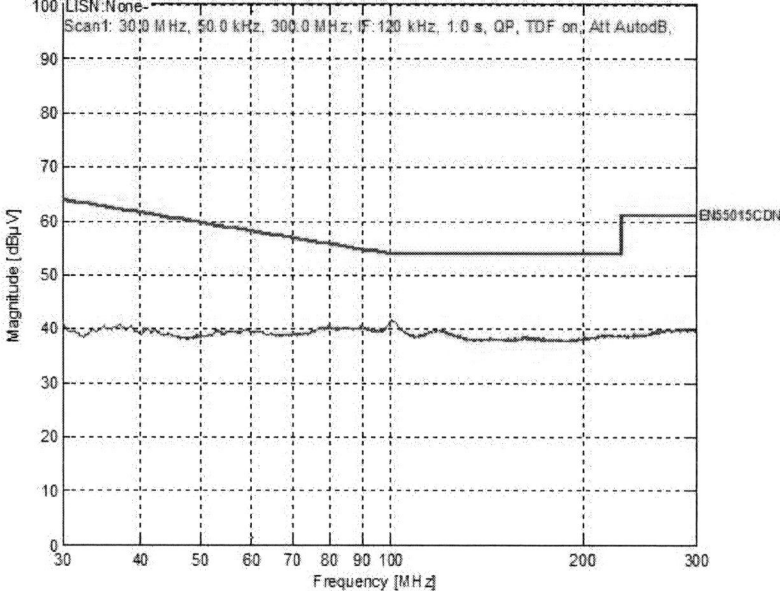

Fig. 2: EMC signature

Concluding, a linear regulator topology also makes it easy to adjust the output current in small steps to different levels. The only need is to influence the feedback voltage which changes the working point of the adjustable resistor.

3.3. Regulation concept

Figure 3 shows a simplified schematic of the regulation of the driver. It shows three regulation loops and the connection between each other.

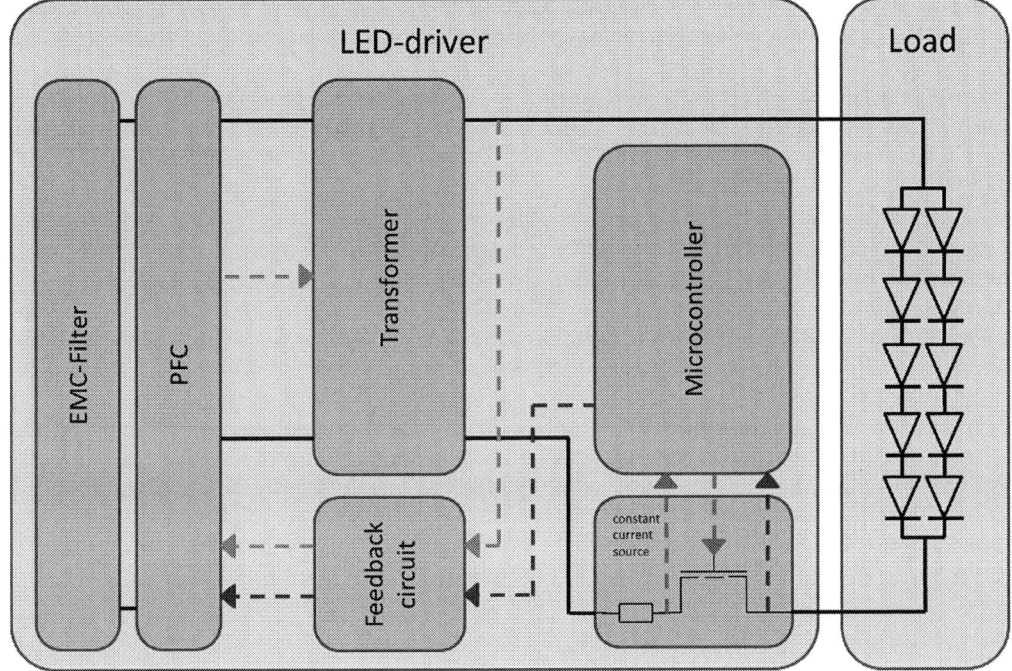

Fig. 3: Control loops

Regulation loop 1 (see red signals) is regulating the LED-current to the adjusted value. It compares the voltage at the shunt resistor with an internal reference voltage of the linear regulator. Additionally, the shunt voltage can be influenced by a voltage generated by the microcontroller. This way, the LED-current can be set to a value suitable for the connected LED-module. With this method, also multicurrent versions of the LED-driver can be realized which receive the desired output current level via a DALI (Digital Addressable Lighting Interface) command.

Regulation loop 2 (see blue signals) takes care of the voltage over the linear regulator, so that it can keep the set LED-current constant at any conditions. The controlled variable is the voltage at the linear regulator. The lower this voltage is, the higher is the efficiency of the constant current source. In normal operation this variable is set as low as possible to reduce losses. During power-up, this variable is set to a higher value to allow a turn-on-sequence from 0% to 100% without any drop of the output current which otherwise might occur due to the regulation latency of the PFC regulation loop. When the driver operated with a certain LED module for a defined time, the according bus voltage is stored. At every next start-up with the same load module, the bus voltage is only set up to exactly that value to reduce losses. The manipulated variable is the microcontroller-voltage setting the secondary bus voltage to the correct level via an adjustable Zener-Diode.

Regulation loop 3 (see yellow signals) is giving a feedback to the integrated circuit responsible for power factor correction and transforming energy to the secondary side. It compares the reference variable with the controlled variable to set up the correct operation point.

© VDE VERLAG GMBH · Berlin · Offenbach

The entire regulation procedure and schematic is subject to a patent to protect the approach of this concept for the future.

3.4. Advanced temperature protection

There are several concepts of temperature protections realized in a variety of LED-drivers in the market. The functionality reaches from simple turn-off-circuits through to more complex circuits which implement a power reduction.

In the LED-driver by Vossloh-Schwabe, one main focus was the safety of the driver which must not be impaired by excessive temperatures e.g. due to inadequate installation of the driver. The other main focus was a high convenience for the customer, whose main concern is a light output without interruption.

These targets have been reached by two separate thermistors which are placed close to two different hot spots within the housing. Two thermistors are necessary, since there are two independent overtemperature protections. The first step of the overtemperature protection is part of a software function which decreases the output current and therefore keeps the temperature at an adjusted maximum level. This way, it can be guaranteed that up to very high ambient temperatures, which are even out of specification, there will still be light from the according LED-module.

Since a software can be critical from safety point of view, another completely separated overtemperature protection is realized which is based solely on hardware and can turn off the Flyback stage at high temperatures that couldn't be reduced by decreasing the output current. With this additional circuit, it is realized that the safety of the driver will not be impaired at any ambient temperature which exceed the specified operating temperatures.

The following diagram (figure 4) shows the thresholds of the overtemperature protection.

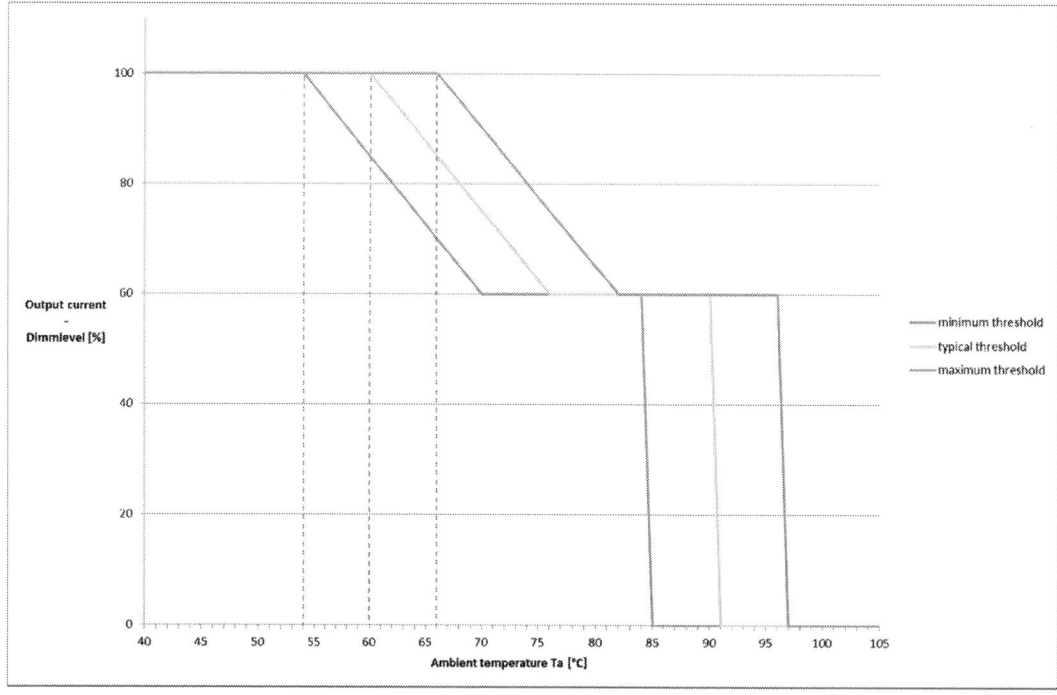

Fig. 4: Concept of overtemperature protection

4. Conclusion

Rather than proposing a 'best' solution for all applications, the target of this article is to show the advantages of the advanced high-end LED-driver concept which has some outstanding features compared to other driver concepts on the market and is suitable for many applications. The proposed concept has proven to be competitive on the market and will be continuously optimized and improved to also fit future demands by luminaire manufacturers.

5. Literature

[1] Magazine licht.wissen, edition 19, „Wirkung des Lichts auf den Menschen", Page 6-7

[2] Heinz Schmidt-Walter, „Übersicht: Schaltnetzteile", http://schmidt-walter-schaltnetzteile.de/snt/snt_deu/sntd_pdf.html, effective 14.10.15

[3] Brad Lehman and Arnold J. Wilkins, "Designing to Mitigate Effects of Flicker in LED Lighting: Reducing risks to health and safety", IEEE power electronica magazine, September 2014, pages 18-26

[4] Dipl.-Ing. Dmitrij Polin, technical university of Darmstadt, workshop for continued education „LED-Technik", „Zeitliche Wahrnehmungsaspekte"

PCIM Europe 2016, 10 – 12 May 2016, Nuremberg, Germany

The Seven Challenges of LED Lighting

Claudio Adragna, Francesco Ferrazza, Giovanni Gritti - STMicroelectronics s.r.l. - Italy

claudio.adragna@st.com, francesco.ferrazza@st.com, giovanni.gritti@st.com

Abstract

This paper focuses on some of the most important issues that Power Designers tackle when designing power conversion systems for LED lighting. Approaches and proposals to address them in terms of products, technology, control methodologies and power system architecture will be presented. Finally, looking ahead, an outline of how some of these challenges - and the relevant responses - are expected to evolve in the near future will be given.

1. Introduction

In the last decade, the development and diffusion of LED-based Solid-state Lighting (SSL) technology has gone through a tremendous boost, driven by the quest for solutions with higher energy efficiency, higher luminous efficacy and longer lifetime. This trend, supported by government policies and incentive campaigns worldwide, has resulted in the discontinuity in the lighting market brought by the transition from traditional light sources (incandescence, halogen, CFL, HID) to LED-based light sources.

Among the three major sectors of the lighting market (general lighting, backlighting and automotive), general lighting is a sector where a big growth is expected in the coming years: in 2020 more than 70% of general lighting solutions worldwide should be based on LED technology [1]. Perhaps, it is also the one posing the biggest technical challenges, if for no other reason, for its extreme diversification.

These challenges are located in different areas, including the LED technology itself (in short: the pursuit of higher lm/W, lm/cm^2 and lm/$) and the design of LED-based lamps and luminaires operated from the power line: thermal, optical, mechanical and power supply are the most important, with the cost challenge permeating all of them.

The focus of this paper is on the challenges of power supply. These are interrelated with the others, especially with mechanical and thermal. The link is so strong that the design of LED power supplies requires a system approach that accounts for and carefully balances the often contrasting requirements of electronics, mechanics and thermal management.

A list of seven (magic number!) challenges will be proposed, without claiming to be either exhaustive or of universal validity: in fact, each specific application has its own specific challenges. Also, every Power Designer could draft its own list based on her/his experience and find that not all the items in the proposed list are a real challenge to them, while other important items are missing. This is just meant to be a survey of the most frequent issues that, in authors' experience, are encountered when designing a power supply system for either an LED lamp, which we will refer to as an "embedded LED driver", or for a luminaire, which we will refer to as a "standalone LED driver".

A thorough analysis of these challenges and the possible ways to take them up is far beyond the scope of this paper: each of them is worth an essay. The approach that we will follow is to select a "case study" for each of them and describe a way to address it. The solution that will be proposed may concern products, technology, control methodologies or power system architectures, depending on the specific case. The reader will appreciate how some of these solutions address more than one challenge.

Finally, looking ahead, an outline of how some of these challenges - and the relevant responses - are expected to evolve in the near future will be provided.

© VDE VERLAG GMBH · Berlin · Offenbach

2. The challenges

2.1. Input current distortion (THD)

LED drivers, which must comply with class C harmonic emission limits of the IEC61000-3-2, are more and more often specified to meet market requirements that consider the Total Harmonic Distortion (THD) of the input current as well.

Standalone LED drivers are particularly challenging in this respect: in fact, these devices are often specified for a rated output current and for a range of output voltages (a 3:1 range is quite typical) to power different types/lengths of LED string. These devices must then comply with the IEC61000-3-2 and meet the THD targets even when operated at the specified minimum output voltage (e.g. at 1/3 of their rated maximum power).

Things get even tougher when dimming is in place, which further extends the power range under consideration for THD targets, and/or in drivers specified to operate over an extra-wide input voltage range that includes the 277 Vac power line as well (e.g. from 90 to 305 Vac).

The drivers where these issues are particularly challenging are those where the Power Factor Corrector (PFC) front-end uses a Quasi-resonant (QR) flyback converter (i.e. operated close to the boundary between continuous – CCM - and discontinuous conduction mode - DCM). This topology, often referred to as Hi-PF QR flyback converter, is extremely popular because it is very cost-effective and addressable with the same control ICs as boost topology.

However, Hi-PF QR flyback converters feature inherent distortion of the input current [2], unlike boost converters operated in the same way. Though this distortion is not an issue for the IEC61000-3-2, it makes some emerging THD targets (< 10% at full power) that are becoming a market requirement in some geographical areas [3]-[4] very difficult to achieve.

One solution to this issue is to switch from QR operation to fixed-frequency discontinuous mode (FF-DCM) operation [5]. However, there are some benefits in QR operation that are lost with FF-DCM operation: lower conducted EMI emissions, better current form factor and near zero-voltage switching, which result in higher conversion efficiency. This considering, an innovative control method able to combine the benefits of QR operation with the ability of FF-DCM operation to (ideally) get a sinusoidal input current has been proposed recently [6].

As shown in fig. 1 (left), this method is based on the insertion of a "current shaper" circuit in the control loop that counteracts the root cause of the distortion: the nonuniform chopping of the input current. With this method PFC stages based on QR flyback converters achieve the same THD levels as those based on boost converters, as shown by the green line in the plot of fig. 1 (right). Additionally, the THD is insensitive to frequency reduction methods aiming to maximize light load efficiency, thus the THD vs. efficiency trade-off at light load improves and converter's overall performance when dimming at low levels is enhanced.

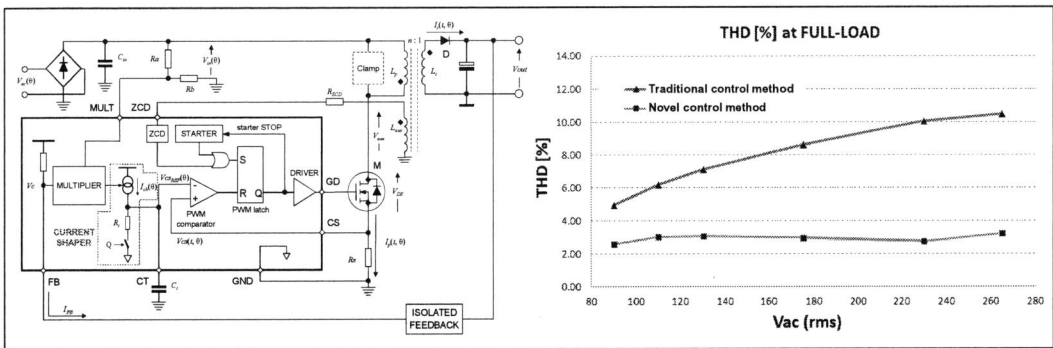

Fig. 1. New control of Hi-PF QR flyback [6]: principle schematic (left); THD performance and comparison with the traditional method (right).

2.2. Output current ripple

Any switching power converter for SSL operated from the power line has a ripple component in the output current they deliver to the LEDs. The ripple typically comprises a low-frequency component (at twice the ac frequency of the power line) and a high-frequency component at the switching frequency (typically above 30-40 kHz). The discussion will be focused on the low-frequency ripple. In fact, it may contribute to reducing the average LED current for a given peak value, to raising the operating temperature of the LEDs, thus shortening their lifetime [7], and generating light fluctuations (flicker and shimmer), which are annoying if perceptible and are reported to cause health problems even when imperceptible [8].

A two-stage conversion is the typical approach to address this challenge. Here two different ways to take it up will be considered. The first solution, depicted in fig. 2 along with a scope picture showing its performance, is a ripple canceling circuit. It works as a damper that absorbs the ac component of the output voltage and lets the LED string "see" the dc component $Vout$ - V_{LDO} only. There is a small penalty in efficiency depending on the V_{LDO}/$Vout$ ratio but if $Vout$ is large enough this loss may be acceptable. It is worth mentioning that this circuit acts as an inrush current limiter as well, thus allowing a hot insertion of the LED string.

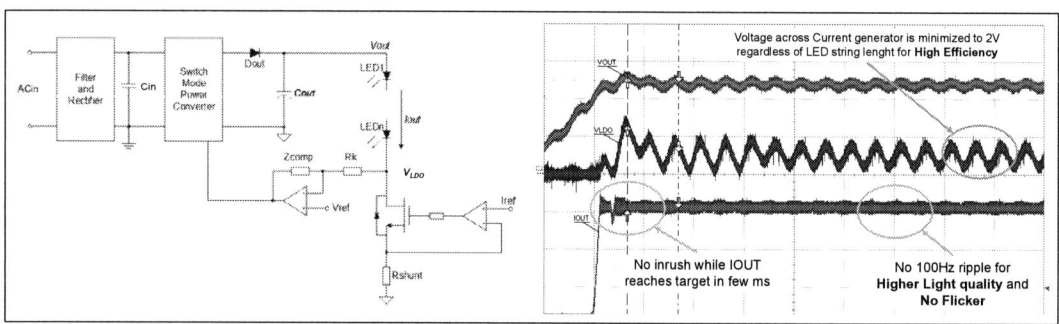

Fig. 2. Ripple canceling circuit: principle schematic (left); key waveforms (right)

The second solution makes use of a so-called S⁴ICS (single-stage, single-switch, input-current-shaping) topology. There are plenty of these topologies [9], but few of them are used in industry. One of the most interesting is the quadratic buck-boost (QBB) converter shown in fig. 3 (left), where two buck-boost stages work in a cascade sharing the only controlled switch. To authors' knowledge its application to LED lighting was first proposed in [10], although the topology had been previously treated as an isolated PFC stage in [11].

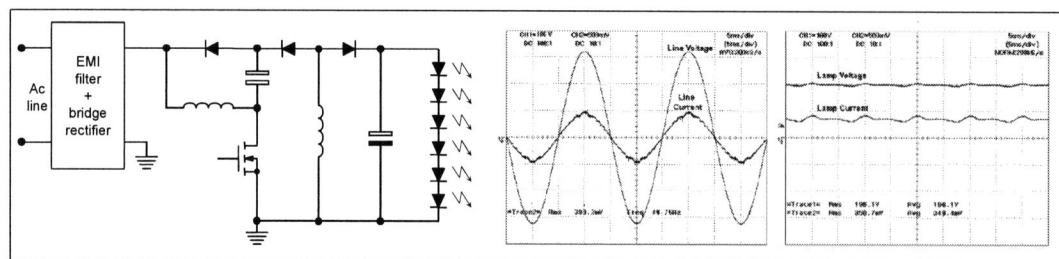

Fig. 3. QBB converter: principle schematic (left); typical key waveforms (right); source: [10]

The input stage may be operated either in FF-DCM or QR to achieve high power factor; the second stage may be operated either in DCM or CCM; the latter option permits a reduction of the output capacitor bank for a given high-frequency ripple through the LED string. This might give the opportunity to use only film capacitors and avoid electrolytic capacitors, which is beneficial for the rated life of the converter (see section 2.5).

The con of this approach is essentially in the size of the power switch: since this has to handle the current of two stages it will be considerably bigger if compared to that of a switch in a standard topology for the same power level. Therefore this approach might not be a viable solution for application rated more than ≈100 W with wide range input.

2.3. Dimming

Probably, the most challenging situation is encountered in embedded LED drivers specified to be compatible with triac-based phase-cut wall dimmers. These devices, thought to handle resistive loads such as incandescent and halogen lamps, when applied to LED lamps pose two big challenges. Firstly, it is difficult to operate them throughout the dimming range at power levels at least one order of magnitude lower than those they have been designed for, without flicker and/or emitting audible noise. Secondly, it is not easy to reproduce the characteristics of incandescent lamps in terms of light intensity at different dimmer settings, as desired by most manufacturers of dimmable LED lamps. This is made even more difficult by the huge number of different dimmer models, each with pretty unique characteristics.

A strategy that mitigates some of these issues is to turn the chopping style of dimming inherent in phase-cut dimmers into linear dimming. The control method [12], whose basic principle is illustrated in fig. 4 (left), is based on this strategy. It enables phase-cut dimming with a customizable current profiling, shown in fig. 4 (right), which eases the design of dimmable LED drivers compliant with industry standard dimming requirements.

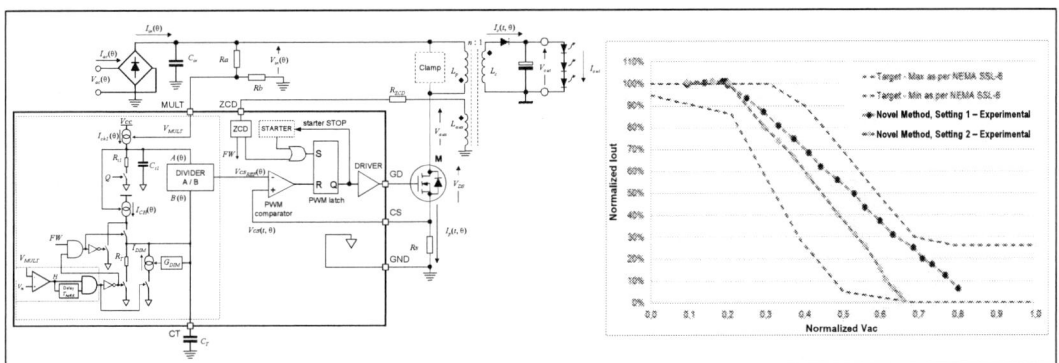

Fig. 4. Hi-PF QR flyback with PSR, low THD and optimized phase-cut dimming compatibility [12]: principle schematic (left); typical performance for two different current profile settings (right).

The method actually hits three targets simultaneously: besides providing phase-cut dimming capability, it is also able to accurately regulate the current delivered to the LED string using primary-sensing regulation (PSR), then with no optocoupler, and to draw a sinusoidal current from the ac line, for THD-sensitive applications like those mentioned in section 2.1.

It is worth reminding that, to make an LED driver actually compatible with a triac-based wall dimmer, additional circuit provisions to correctly operate the triac are needed [13].

2.4. Efficiency and Thermal

Efficiency in LED drivers is a matter of power consumption as well as heat generation: standalone drivers are generally boxed and sometimes sealed, then with scarce or no chance of convection cooling; embedded drivers receive the heat generated by the LEDs in addition to their self-heating. Perhaps the worst situation is found in downlights, where the driver receives heats from the LEDs by both conduction and convection. Heat is the number one enemy of power supply reliability (see next section) and is not friendly to LEDs either: a too high temperature shortens their useful life, reduces their lumen output and causes color shift.

The solution to these issues is a mixture of a careful electrical design that aims to maximize converter's efficiency and of thermal countermeasures able to take the heat out of the lamp as efficiently as possible within the mechanical and design constraints.

In this context we just want to highlight how in certain applications efficiency maximization can go through the selection of appropriate power supply architecture.

Offline power supplies of LED luminaires for street lighting applications (also for RGB or color tunable applications) normally includes three conversion stages: a PFC boost pre-regulator, a dc-dc converter (typically an LLC resonant half-bridge converter to achieve high efficiency and provide isolation from the power line) and a number of cascaded current regulators - linear or switching – driving each LED string separately and providing dimming capability.

With three stages, even if very efficient, the overall efficiency cannot be that high. For example, assuming three 96% efficient stages, the overall efficiency is still below 89%.

The system proposed in [14] and illustrated in fig. 5, enables an overall two-stage conversion from the ac line to the LED strings, PFC included. It is based on a soft-switched high-efficiency multiple current source able to drive multiple strings of LED in applications where electrical isolation is required. With this topology, LED current can be accurately controlled using PSR with excellent matching. Additionally, each LED string can be dimmed (either PWM or linear) independently from the others. In particular, PWM dimming can go down to very low duty cycles with excellent cross-regulation and dynamic response.

With this architecture an overall peak efficiency (line-to-LED) over 92% is feasible [15].

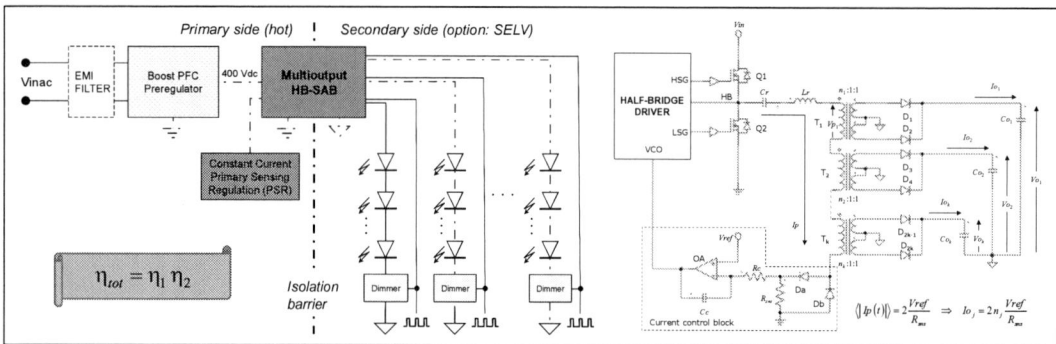

Fig. 5. Multistring LED driver [14]: overall architecture (left); multioutput soft-switched half-bridge converter with PSR and inherent current matching (right); dimming circuit is not shown

2.5. Reliability

The useful life of LEDs normally spans between 35,000 and 50,000 hours, and some manufacturers claim useful life L_{70} values (i.e. when the light output has declined to 70% of its initial lumens) greater than 100,000 hours. Their driver must be up to this level; otherwise the useful life of the lamp or luminaire will not be that of the LEDs (operated in the conditions allowed by the fixture) but the rated life of the driver, i.e. the time when the unit is expected to fail. The consequence is that the lamp is dead and one throws it away because the power supply failed, while the LEDs are still fully working.

The operating temperature is determinant for the rated life of the driver: this is why efficiency is important and in those situations where the heat generated by the losses of the converter is not the only one that keeps the operating temperature high, heat exchange becomes of paramount importance.

The weak link of the chain is typically represented by the aluminum electrolytic capacitors, which tend to dry up and lose their characteristics. Depending on where in the circuit the capacitor is used, the outcome of this wear may go from degrading driver's performance (which might still be acceptable if the lamp keeps on emitting light) to causing a catastrophic failure. The usage of special capacitors rated for high temperature operation (130 °C grade capacitors are now readily available) and where capacitors are placed on the PCB (away from heat generating parts and, when applicable, close to convection air flow) can be beneficial to increase their lifetime and that of the overall unit.

A good idea is not to use electrolytic capacitors at all or, at least, no electrolytic capacitors in those circuit locations where their reaching end-of-life causes a catastrophic failure.

The application proposed in [16], which is composed of a front-end boost PFC and an LLC resonant converter, is an example of a design with no electrolytic capacitors. In particular, the output capacitor of the PFC stage uses film capacitors. This requires a series of design trade-offs to properly handle the larger output ripple and prevent this low-frequency ripple to affect the output of the LLC stage. Component derating has been also carefully applied, decreasing the stress of the components as recommended by the MIL-HDBK-217D.

Another critical component that is at the origin of many field failures is the optocoupler used to close the feedback loop that regulates the output voltage (or current) in an isolated converter. In fact, this part is quite prone to fail as a consequence of lightning or may upset converter's operation as a result of large common mode transients. The usage of PSR, like in [12], [14], totally eliminates this possible mechanism of malfunction or failure.

2.6. Connectivity

Since the dawn of Smart Lighting, a growing number of luminaires have no longer been devices simply emitting light. With the aim of providing energy saving and a level of comfort and convenience, they are equipped with control systems and sensors and are able to communicate, so that they can be arranged in an intelligent networked system. Besides the more traditional wired control systems, such as 0-10V or DALI, the progress of wireless technologies has made systems such as Bluetooth or Zigbee quite popular in lighting as well.

All this does not strictly concern Power Designers directly but may have implications that cannot be neglected. As an example, the use of microcontrollers, needed to make a lighting device "smart" as above specified, has opened the door to digital control of power converters. For many Power Designers, that are essentially analog engineers, "going digital" has been a challenge in itself.

In this context an interesting solution is represented by the STLUX385A [17], a digital controller specific for lighting applications. This controller, based on an innovative peripheral called SMED (State Machine, Event Driven) [18], enables real-time signal processing even without a high-performance microcontroller or a DSP. It also embeds the hardware interface that provides full IEC 62386 slave DALI interface, thus significantly simplifying both the hardware and software implementation [15], [19].

2.7. Cost

Last but not least, all the previous challenges must be faced taking cost into account. Especially in some applications such as bulb replacement, cost is priority number one and sometimes even two or three. The fact is that in most cases those who specify an LED lamp or an LED-based luminaire do not care about the intricacies of the design, they care about the benefits that the product is expected to offer: good light quality, low consumption and long lifetime at the cheapest price. The initially mentioned system approach, which requires thorough understanding of the application and its pitfalls, will also help the designer to control the cost by not overdesigning the entire system. This, of course, applies to the driver as well.

It is worth noticing that cost is addressed by many of the solutions here proposed, [10], [12], [14], [15], [19], acting on both part count reduction and/or system level optimization.

3. Looking ahead

To figure out what power conversion requirements to anticipate in the next future, it may be useful to start from the driving forces that today it is possible to observe in the SSL sector: LED technology evolution, energy efficiency and energy saving, form factor and power density, regulations, total cost of ownership. Actually, the scenario is very complex and a few simplifications are needed to identify some possible directions.

The evolution of the LED technology, where the trend is to increase the luminous efficacy (lm/W) and decrease their cost, is definitely a major driving force. The good news for the electronics is that with more lm/W available the thermal issue is going to be mitigated. The bad news is that with more lm/$, the cost-down pressure on the electronics of the driver will increase. Lamps will be especially impacted: the BOM of luminaires includes other items, such as material costs and installation costs, where significant cost-down can be pursued too.

Today we are seeing a low-end lamp market, where cost is king and the electronics is expected to be simplified, e.g. with the ac-LED approach [20], or even disappear almost completely, and a high-end market where manufacturer are trying to put more value to this commodity product. Tunable lighting, both white point and color, and networked lighting – which paves the way for lighting to become part of the Internet of Things - is probably the mainstream both in luminaires and high-end lamps. Here the electronics is a key player but more as far as connectivity and controls are concerned. In this scenario, new challenges for power designers might derive from a broader diffusion of digital control of power converters, which calls for a cultural change of purely analog Power Designers, from new more severe form factors requiring higher power density or from more stringent standby requirements.

Standby requirements are related to another major driving force: regulations. It is reasonable to expect that new regulations will be issued in the coming years that will affect the way the LED drivers are designed. For example, the regulations concerning standby and off-mode losses might require more aggressive targets like those for the so called EPSs (External Power Supplies) [21], which especially standalone LED drivers are akin to. The solutions are already available: today it is not uncommon to find an EPS in the 40-50W range able to consume less than 30 mW from the power line when operated with no load.

In this context the "Zero Power" function [22] seems quite appealing, especially in luminaires. This function, along with the advanced power management features provided by modern microcontrollers, enables to build lighting apparatuses capable of being shut down with zero residual power consumption from the power line and turned on via any wired or wireless connectivity means. It is worth specifying that "zero power" actually means an input power level lower than 5 mW, which can be rounded down to zero as per IEC62301 clause 4.5.

4. Conclusion

SSL, as a disruptive technology, has created a series of challenges for the lighting industry. This paper has focused on those directed to Power Designers and has proposed a list of some of those most frequently found. It has been shown through examples that the weapons available are at different levels, from active and passive components to conversion topologies up to system architecture. The conclusion that can be drawn from this survey is that a system approach that considers and synthesizes the complexity of requirements and constraints is instrumental in carrying out a successful design.

How the challenges and the relevant responses will evolve in the near future appear to be mainly linked to the driving forces observable today. From the present trends, it seems that a mitigation of thermal issues (unless counterbalanced by more aggressive form factors), a broader usage of digital control of power converters as well as the introduction of advanced power management techniques to minimize standby consumption are to be expected.

© VDE VERLAG GMBH · Berlin · Offenbach

5. References

[1] Trevis Team "GE lighting sees brighter future with LED growth", http://www.forbes.com/ sites/greatspeculations/2013/06/11/ge-lighting-sees-brighter-future-with-led-growth/#659beae8 19ef

[2] C. Adragna, "Design equations of high-power-factor flyback converters based on the L6561", STMicroelectronics Application Note, AN1059, www.st.com.

[3] "India announces critical standards on LED lighting", http:// www.theclimategroup.org/what-we-do/news-and-blogs/india-announces -critical-standards-on-led-lighting/.

[4] "Lighting industry in India", www.globallightingassociation.org%2Fdocuments%2Fgla_ papers%2FELCOMA_presentation_to_GLA__April_24_2013.pdf.

[5] R. Erickson, M. Madigan, S. Singer "Design of a simple power-factor rectifier based on the flyback converter", Applied Power Electronics Conference and Exposition, 1990. APEC '90, Conference Proceedings 1990, pp. 792-801, March 1990.

[6] C. Adragna, G. Gritti "High-power-factor quasi-resonant flyback converters draw sinusoidal input current", Applied Power Electronics Conference and Exposition, 2015. APEC '15, Conference Proceedings 2015, pp. 498-505, March 2015

[7] Philips, "Understanding power LED lifetime analysis", Technology White Paper 2010

[8] A. Wilkins, J. Veitch, B. Lehman "LED Lighting Flicker and Potential Health Concerns: IEEE Standard PAR1789 Update", Energy Conversion Congress and Exposition , 2010. ECCE 2010 Conference Proceedings 2010, pp. 171-178, September 2010

[9] L. Huber, J. Zhang, M. M. Jovanović, F. C. Lee "Generalized topologies of single-stage input-current-shaping circuits", IEEE Transactions on Power Electronics, Vol. 16, No. 4, July 2001, pp. 508-513

[10] J. M. Alonso, et al. "Analysis and design of the quadratic buck-boost converter as a high-power-factor driver for power-LED lamps", IECON 2010 - 36th Annual Conference on IEEE Industrial Electronics Society, Conference Proceedings 2010, pp. 2541-2546, November 2010

[11] T. F Wu, Y. K. Chen, "Analysis and design of an isolated single-stage converter achieving power-factor correction and fast regulation", IEEE Transactions on Industrial Electronics, Vol. 46, No. 4, August 1999, pp. 759-767

[12] G. Gritti, C. Adragna "Primary-controlled constant current LED driver with extremely low THD and optimized phase-cut dimming compatibility", European Conference on Power Electronics and Applications, EPE ECCE 2015, Conference Proceedings 2015, Paper 0007, September 2015

[13] T. Stamm "120 Vac input-triac dimmable LED driver based on the L6562A", STMicroelectronics Application Note, AN2711, www.st.com.

[14] C. Adragna, E. Pastori "High-efficiency multiple LED-string driver for street Lighting and TV backlighting applications", AEIT Annual Conference - From Research to Industry: The Need for a More Effective Technology Transfer (AEIT), 2014, pp. 1-6, September 2014

[15] F. Ferrazza, A. D'Adda "100 W LED street lighting application using STLUX385A", STMicroelectronics Application Note, AN4461, www.st.com.

[16] C. Spini "48 V - 130 W high-efficiency converter with PFC for LED street lighting applications", STMicroelectronics Application Note, AN3106, www.st.com.

[17] "STLUX385A, Digital controller for lighting and power supply applications", datasheet, www.st.com

[18] A. Loidl, J. Hajek, G. Bosisio, I. S. Bellomo, "Innovative event driven state machine (SMED) peripheral for digital control of power conversion & lighting applications with 8 bit microcontrollers", PCIM Europe 2012 Proceedings, pp. 986 – 993, May 2012

[19] A. Loidl, F. Ferrazza, A. D'Adda "Digital Power Conversion: Easy, flexible and efficient way to decrease BOM complexity and increase feature set", PCIM Europe 2015 Proceedings, pp. 325 – 330, May 2015

[20] L. Peters, "AC-LED lighting products find niche, perhaps more (MAGAZINE)", LEDs Magazine, July/August 2012 issue

[21] European Commission – Renewable Energy Unit "Code of Conduct on Energy Efficiency of External Power Supplies" Ver. 5, October 2013, http://iet.jrc.ec.europa.eu/energyefficiency/ict-codes-conduct/efficiency-external-power-supplies

[22] "Viper0P, Zero Power offline high-voltage converter", datasheet, www.st.com

For information on the presentation

Highly Flexible Single Stage Flyback Quasi-Resonant Digital Controller for Advanced LED Applications

please contact the speaker directly:

Mr. Marc Fahlenkamp

Infineon Technologies AG
IFAG PMM SYS INO
Am Campeon 1-12
85579 Neubiberg

Phone: +49 89-234-25653
E-Mail: marc.fahlenkamp@infineon.com

Carrier-Based Modulation Technique to Reduce Low Frequency Ripple at the Partial Dc-Link Voltages of a Three-Level/-Phase/-Wire NPC Converter Applied to Future Dc Bipolar Active Distribution Networks

Joabel Moia[1] and Marcelo L. Heldwein[2]

[1] IFSC – Federal Institute of Santa Catarina, Eletronics Department, Florianópolis, SC – Brazil (e-mail: joabel.moia@ifsc.edu.br)
[2] UFSC – Federal University of Santa Catarina, Power Electronic Institute (INEP), P.O. box: 5119, 88040-970, Florianópolis, SC – Brazil (e-mail: heldwein@inep.ufsc.br)

Abstract

This work proposes a modulation technique for a three-phase/-wire/-level Neutral Point Clamped converter (NPC) aiming to reduce low frequency (LF) neutral point current components in future dc bipolar distribution network applications. The technique is based on the analysis of three-level converters in the *abc* reference frame, where the ac measured currents and duty-cycle signals are used to calculate the 0-axis duty-cycle, which ensures that the capacitor dc-link LF current components are eliminated, even when loads/microgeration units inject LF currents to the network central line. Simulation results validate the proposed technique.

1. Introduction

In recent years, many efforts have been made to improve the electric power system, due to increased consumption, need to ensure energy supply and search for better power quality indices [1-4]. In this way, smart grids and microgrids play a fundamental role. The dc microgrid configuration has some advantages compared to ac microgrids [5-8], especially the dc bipolar distribution network configuration [9-11]. Some works present comparisons that show the three-level/-phase/-wire Neutral-Point Converter (NPC-3L) PWM as a good solution to implement a dc bipolar active distribution system [12, 13].

Since its introduction [14], the three-phase three-level NPC converter (NPC-3L) has some significant advantages compared to conventional two-level converters, particularly for high power and medium-voltage applications. Nowadays, applying it in renewable energy and regenerative low voltage (LV) applications appears very promising [15, 16]. On the other hand, the balance of the central point voltage should be ensured by control and modulation strategy and it present a low frequency oscillation in the partial voltages (capacitors) for most modulation techniques industrially used [17].

A modulation technique is proposed here for the NPC-3L rectifier applied to dc bipolar active distribution networks in order to reduce the low frequency ripple at the partial dc-bus voltages of the system. This is achieved even when loads/microgeration units connected to the dc grid inject low frequency current components to the line center point of the network. The proposed technique is based on the analysis of three-level converters in the stationary *abc* reference frame, where the ac measured currents and duty-cycle signals are used to calculate the required 0-axis duty-cycle to zero the partial voltage ripple.

Due to the inherent limits of the proposed modulation, an additional circuit is also proposed to cancel the low frequency current at the center point of the NPC, when the modulation proposed here is not capable of doing so. This additional circuit comes into operation only when the proposed modulation cannot operate in the linear modulation range.

The main consequence of the application of the proposed compensation is that the rms current value is reduced in the dc bus capacitors, reducing capacitors losses and increasing life time. Moreover, high quality voltages will be achieved, which is an important subject in future dc distribution networks.

2. Partial Dc-Link Voltage Ripple Cancellation in the *abc* Reference Frame

Fig. 1 shows the system of a three-level/-phase/-wire bidirectional high power factor rectifier applied to dc bipolar microgrid with an additional circuit, which is a fourth NPC-3L leg. It is possible to connect loads and microgeneration units in the dc-side.

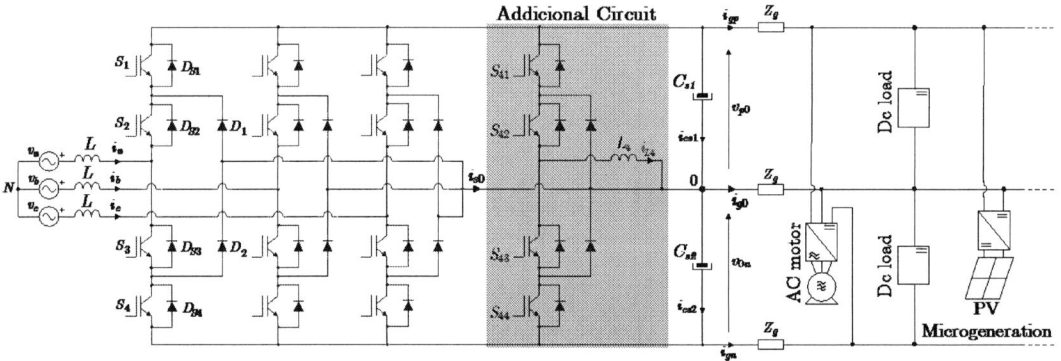

Fig. 1. Bidirectional three-phase/-wire/-level NPC converter with an additional circuit applied to a dc bipolar active network system.

The main causes of the low frequency oscillation on the partial bus capacitors of the NPC dc bipolar microgrid system is due to the three times the frequency of the modulation signals due to dc-link center point current (i_{s0}) from rectifier and due to low frequency center current (i_{g0}) inject by loads/microgeneration connected on dc side, specially the three-level/-phase inverter.

Based on PWM modulation, the duty-cycle applied to the NPC-PWM converter (Fig. 1) can be written according to equation (1)

$$d_x = d_{x0} + d_0$$
$$x = a,b,c \tag{1}$$

Equation (2) presents the neutral point current local average value $<i_{s0}>$ in a three-level/-phase/-wire NPC converter [18, 19]:

$$\langle i_{s0} \rangle = -\underbrace{\left| d_{a0} + d_0 \right| i_a}_{f_1} - \underbrace{\left| d_{b0} + d_0 \right| i_b}_{f_2} - \underbrace{\left| d_{c0} + d_0 \right| i_c}_{f_3} \tag{2}$$

which is a general expression defining the local average current value $<i_{s0}>$ in the *abc* reference frame. Functions f_1, f_2 and f_3 are used to define the 0-axis duty-cycle that cancels the low frequency voltage oscillation at the dc buses. The main objective here is to make the neutral point current local average value from the rectifier equal the low frequency center current from the loads, i.e:

$$\langle i_{s0} \rangle = \langle i_{g0} \rangle \tag{3}$$

By matching this condition the dc-link partial voltages will only present ripple generated by the switching frequency harmonics. Applying the definition of the absolute value in f_1, f_2 and f_3 in order to eliminate the module, the solution is found by solving (2) for d_0 with (3). This results in six solutions for the 0-axis duty-cycle d_0^{canc} that cancels the partial voltage low frequency oscillation on the system, i.e.

$$d_0^{canc} = \begin{cases} d_{0,1}^{canc} & se \ f_1 \geq 0, f_2 \geq 0, f_3 < 0 \\ d_{0,2}^{canc} & se \ f_1 \geq 0, f_2 < 0, f_3 \geq 0 \\ d_{0,3}^{canc} & se \ f_1 \geq 0, f_2 < 0, f_3 < 0 \\ d_{0,4}^{canc} & se \ f_1 < 0, f_2 \geq 0, f_3 \geq 0 \\ d_{0,5}^{canc} & se \ f_1 < 0, f_2 \geq 0, f_3 < 0 \\ d_{0,6}^{canc} & se \ f_1 < 0, f_2 < 0, f_3 \geq 0 \end{cases} \tag{4}$$

where,

$$d_{0,1}^{canc} = \frac{-i_a d_{a0} - i_b d_{b0} + i_c d_{c0} - \langle i_{g0} \rangle}{i_a + i_b - i_c} \quad d_{0,2}^{canc} = \frac{-i_a d_{a0} + i_b d_{b0} - i_c d_{c0} - \langle i_{g0} \rangle}{i_a - i_b + i_c} \quad d_{0,3}^{canc} = \frac{-i_a d_{a0} + i_b d_{b0} + i_c d_{c0} - \langle i_{g0} \rangle}{i_a - i_b - i_c}$$

$$d_{0,4}^{canc} = \frac{i_a d_{a0} - i_b d_{b0} - i_c d_{c0} + \langle i_{g0} \rangle}{-i_a + i_b + i_c} \quad d_{0,5}^{canc} = \frac{i_a d_{a0} - i_b d_{b0} + i_c d_{c0} + \langle i_{g0} \rangle}{-i_a + i_b - i_c} \quad d_{0,6}^{canc} = \frac{i_a d_{a0} + i_b d_{b0} - i_c d_{c0} + \langle i_{g0} \rangle}{-i_a - i_b + i_c} \tag{5}$$

The variables i_a, i_b, i_c and $<i_{g0}>$ are obtained from current measurements, while duty-cycles d_a, d_b and d_c are readily available at the output of the converter controllers. There is the need for

PCIM Europe 2016, 10 – 12 May 2016, Nuremberg, Germany

a sensor to measure the line grid central current (i_{g0}), and a low pass filter for only get the low frequency content of it might be necessary in case the loads are power converters without filters. Fig. 2 shows a control block diagram for implementing the proposed modulation. It is noticed that the dc-link low frequency ripple voltage cancelation operates in open loop, only calculating the 0-axis duty cycle according to equation (4). Thus, the modulation technique presented in this work is a carrier-based PWM scheme.

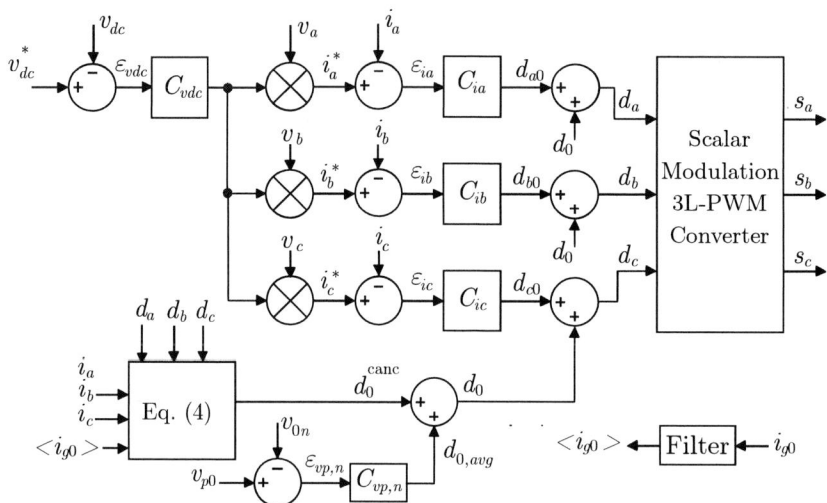

Fig. 2. Control block diagram implemented in the *abc* reference frame including the proposed modulation technique according to (4), ac currents control for unity power factor, dc buses partial and total voltage control.

The operation of the additional circuit is straightforward: it comes into operation when the proposed modulation is not able to cancel out the dc-link low frequency ripple voltage, due to its limitation, especially when there is a high amplitude low frequency current injected into the center point by the loads/microgeneration units. It also happens if the partial voltages control is not able to ensure its regulation, i.e., when there is a strong load unbalance on the dc-bus partial voltages. The additional fourth leg can provide such functions for both situations. In any other situation this circuit remains off in order to increase system efficiency. Reference [20] presents the operation and simulation results for the additional circuit in a dc bipolar microgrid application. It is possible to operate the fourth leg in two-level or three-level and provide fault-tolerance.

3. Simulation Results

The results presented in this section use a three-level converter employing a three-level/phase inverter load and two linear loads (pure resistive) connected to the two partial voltage dc-link in the dc bipolar active distribution network. Simulation parameters are:

- Switching rectifier/inverter frequency f_s^r = 20 kHz/ f_s^i = 20 kHz;

© VDE VERLAG GMBH · Berlin · Offenbach

PCIM Europe 2016, 10 – 12 May 2016, Nuremberg, Germany

- Mains/Inverter frequency f_g^r = 60 Hz/ f_g^i = 50 Hz;

- Rectifier ac rms line voltage $V_{ac,rms}$ = 380 V;

- Power of the three phase inverter load S_i = 11.04 kVA;

- Linear load power each partial dc-link bus P_l = 3 kW;

- Output total power P_o = 17.04 kW;

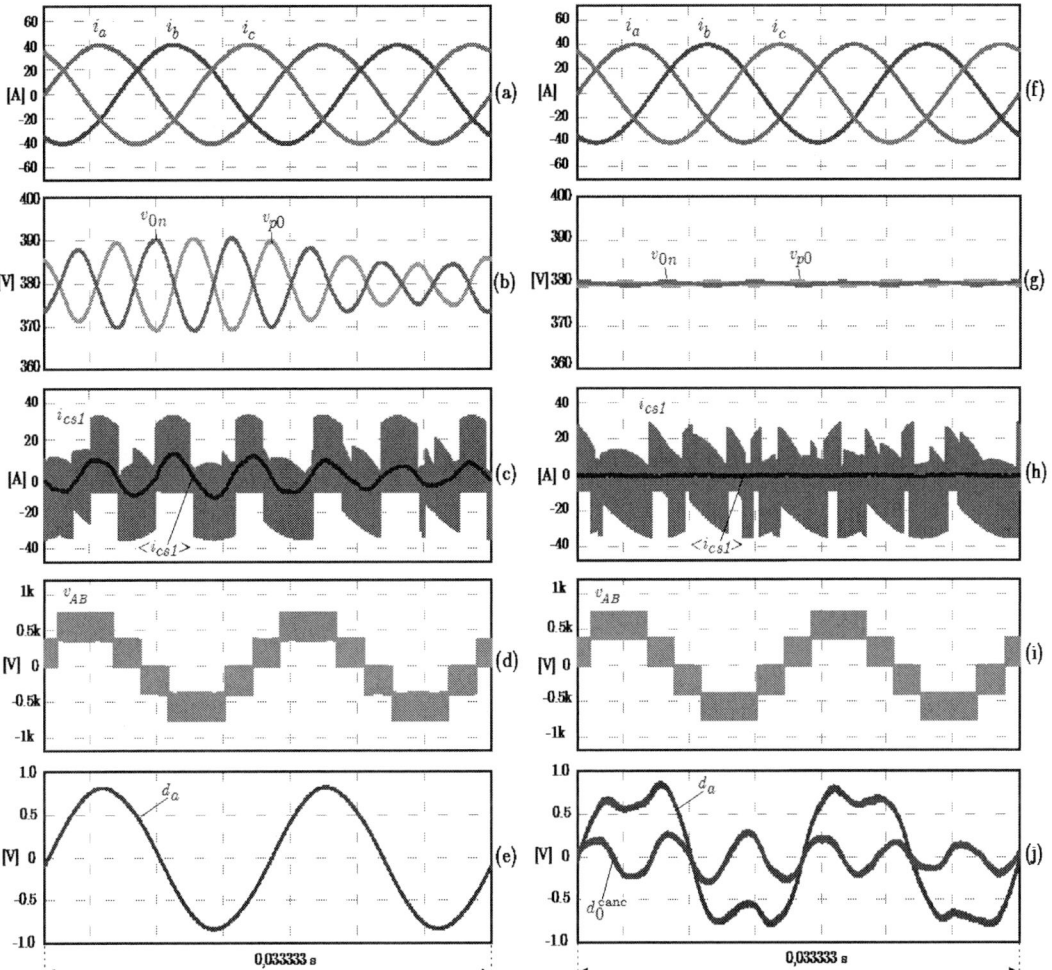

Fig. 3. Simulation results for balanced dc buses powers: (a),(b),(c),(d) and (e) show the simulation results for NPC rectifier with SPWM modulation without compensation. (f),(g),(h),(i) and (j) show the waveform for proposed modulation technique.

- Total dc-link voltage V_{dc} = 760 V;

- Rectifier power factor $PF^r \approx 1{,}0$;

- Dc-Link capacitors (each): C_s = 1000 µF.

The Fig. 3 leftmost column presents steady state simulation results for the NPC rectifier system with conventional SPWM modulation, while the rightmost column shows the same condition with the proposed modulation technique.

Fig. 3(c) and Fig. 3(h) show the instantaneous capacitor current i_{cs1} and its local average values $<i_{cs1}>$ for SPWM modulation and the proposed one, respectively. It can be seen that the proposed modulation is able to reduce the low frequency contents and the consequence is that no low frequency ripple appears at the partial dc-bus voltages and only the ripple at the switching frequency is observed (see Fig. 3(g)). The rectifier ac-side currents are clearly not affected by the proposed modulation, since the results for the proposed modulation is approximately the same as in the original SPWM.

Phase a modulation signals (d_a) are seen in Fig. 3(e) and Fig. 3(j) for SPWM modulation and the proposed one, respectively. Fig. 3(j) shows also the 0-axis duty cycle signal which is the result of the calculation of equation (4).

The dc-link capacitor current frequency spectra for the conventional SPWM modulation and for the proposed technique are presented in Fig. 4. It is noted that the low frequency harmonics contents with the proposed modulation is reduced to nearly zero. Nevertheless, increased harmonic current is present at the switching frequency. In typical electrolytic capacitors, the equivalent series resistance falls with frequency, the increased switching frequency current component generates lower losses and by consequence, there is an increase in the useful life of them.

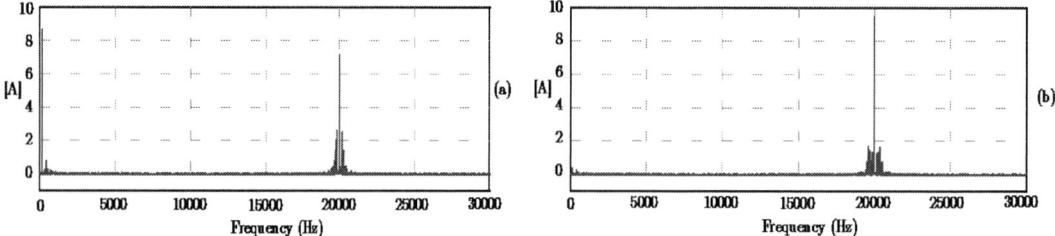

Fig. 4. Capacitor current i_{cs1} frequency spectra: (a) conventional SPWM modulation; (b) proposed modulation technique.

4. Conclusions

This work presented a novel modulation technique to reduce the low frequency oscillation of the bidirectional three-level/-phase/-wire NPC converter applied to future dc bipolar distribution networks. The analysis of the mathematical circuit model in the *abc* reference frame is used to implement the cancellation of the low frequency current contents of the line system center point even when loads/microgeration units inject low frequency current components to the center line of the microgrid

5. References

[1] J. V. Milanovic, J. Meyer, R. F. Ball, W. Howe, R. Preece, M. H. J. Bollen, S. Elphick, and N. Cukalevski. International industry practice on power-quality monitoring. *Power Delivery, IEEE Transactions on*, 29(2):934–941, April 2014.

[2] P.S. Wright, A. Bergman, A.-P. Elg, M. Flood, P. Clarkson, and K. Hertzberg. Onsite measurements for power-quality estimation at the sweden-poland hvdc link. *Power Delivery, IEEE Transactions on*, 29(1):472–479, Feb 2014.

[3] P. Bagheri and W. Xu. A technique to mitigate zero-sequence harmonics in power distribution systems. *Power Delivery, IEEE Transactions on*, 29(1):215–223, Feb 2014.

[4] C.-I. Chen and Y.-C. Chen. A neural-network-based data-driven nonlinear model on time- and frequency-domain voltage-current characterization for power-quality study. *Power Delivery, IEEE Transactions on*, 30(3):1577–1584, June 2015.

[5] Weixing Li, Xiaoming Mou, Yuebin Zhou, and Chris Marnay. On voltage standards for dc home microgrids energized by distributed sources. In *Power Electronics and Motion Control Conference (IPEMC), 2012 7th International*, volume 3, pages 2282–2286, June 2012.

[6] H. Kakigano, Y. Miura, T. Ise, and R. Uchida. Dc micro-grid for super high quality distribution; system configuration and control of distributed generations and energy storage devices;. In *Power Electronics Specialists Conference, 2006. PESC '06. 37th IEEE*, pages 1–7, June 2006.

[7] M.R. Starke, Fangxing Li, L.M. Tolbert, and B. Ozpineci. Ac vs. dc distribution: Maximum transfer capability. In *Power and Energy Society General Meeting - Conversion and Delivery of Electrical Energy in the 21st Century, 2008 IEEE*, pages 1–6, July 2008.

[8] D.J. Hammerstrom. Ac versus dc distribution systemsdid we get it right? In *Power Engineering Society General Meeting, 2007. IEEE*, pages 1–5, June 2007.

[9] H. Kakigano, Y. Miura, and T. Ise. Low-voltage bipolar-type dc microgrid for super high quality distribution. *Power Electronics, IEEETransactions on*, 25(12):3066–3075, Dec 2010.

[10] J. Rekola and H. Tuusa. Comparison of line and load converter topologies in a bipolar lvdc distribution. In *Power Electronics and Applications (EPE 2011), Proceedings of the 2011-14th European Conference on*, pages 1–10, Aug 2011.

[11] M. Brenna, E. Tironi, and Giovanni Ubezio. Proposal of a local dc distribution network with distributed energy resources. In *Harmonics and Quality of Power, 2004. 11th International Conference on*, pages 397–402, Sept 2004.

[12] J. Rekola, A. Virtanen, J. Jokipii, and H. Tuusa. Comparison of converter losses in an lvdc distribution. In *IECON 2012 - 38th Annual Conference on IEEE Industrial Electronics Society*, pages 1240–1245, Oct 2012.

[13] J. Moia, J. Lago, A.J. Perin, and M.L. Heldwein. Comparison of three-phase pwm rectifiers to interface ac grids and bipolar dc active distribution networks. In *Power Electronics for Distributed*

Generation Systems (PEDG), 2012 3rd IEEE International Symposium on, pages 221–228, June 2012.

[14] A. Nabae, I. Takahashi, and H. Akagi. A new neutral-point-clamped pwm inverter. *IEEE Transactions on Industry Applications*,IA-17(5):518 –523, sept. 1981.

[15] R. Teichmann and S. Bernet. A comparison of three-level converters versus two-level converters for low-voltage drives, traction, and utility applications. *IEEE Transactions on Industry Applications*, 41(3):855–865, 2005. 0093-9994.

[16] J. Rodriguez, S. Bernet, P.K. Steimer, and I.E. Lizama. A survey on neutral-point-clamped inverters. *IEEE Transactions on Industrial Electronics*, 57(7):2219 –2230, july 2010.

[17] N. Celanovic and D. Boroyevich. A comprehensive study of neutral-point voltage balancing problem in three-level neutral-point-clamped voltage source pwm inverters. *Power Electronics, IEEE Transactions on*, 15(2):242–249, Mar 2000.

[18] J. Moia, A.J. Perin, and M.L. Heldwein. Three-level/-phase pwm converters dc-link voltages ripple reduction technique in the alpha-beta reference frame. In *Applied Power Electronics Conference and Exposition (APEC), 2012 Twenty-Seventh Annual IEEE*, pages 740–747, Feb 2012.

[19] J. Zaragoza, J. Pou, S. Ceballos, E. Robles, P. Ibaez, and J.L. Villate. A comprehensive study of a hybrid modulation technique for the neutral-point-clamped converter. *Industrial Electronics, IEEE Transactions on*, 56(2):294–304, Feb 2009.

[20] J. Moia and M. L. Heldwein, "Three-level NPC-based bidirectional PWM converter operation for high availability/power quality bipolar DC distribution networks," *Power Electronics and Applications (EPE), 2013 15th European Conference on*, Lille, 2013, pp. 1-10.

PCIM Europe 2016, 10 – 12 May 2016, Nuremberg, Germany

FPGA Based Direct Model Predictive Power and Current Control of 3L NPC Active Front Ends

Zhenbin, Zhang, Technische Universität München, Germany, james.cheung@tum.de
Ralph, Kennel, Technische Universität München, Germany, ralph.kennel@tum.de

The Power Point Presentation will be available after the conference.

Abstract

Three-level neutral-point clamped (NPC) power converter is a viable candidate for high power grid-tied renewable energy generations. Direct model predictive control (DMPC) is a good alternative, in particular for multi-level converters. This work presents and experimentally evaluates two DMPC schemes for controlling a grid-tied Active-Front-End (AFE) with a fully FPGA based solution. The two methods are namely, *direct model predictive power control* (DMPPC) and *direct model predictive current control* (DMPCC). Both use a cost function based (nonlinear) control concept to regulate the control objectives and require no extra modulations. Their relationship has been analyzed. Both their steady state and transient phase performances are evaluated through experimental data, which confirm that, similar control performances are achieved by both, except that, slightly smaller current ripples are seen with DMPPC. The possible reasons are also given.

1. Introduction

Currently, 7.5 MW wind turbine systems are available in the market and numerous research activities aim at 10-12 MW level for offshore applications [1]. This increase in the power rating will require converter/inverter topologies with more than two voltage levels to meet grid codes and to guarantee a low total harmonic distortion [2]. In particular, the three-level neutral-point (diode) clamped (3L-NPC) converter seems promising. It allows for more than two voltage levels, but the required amount of components is drastically less than e.g. for five-level topologies [3, 4].

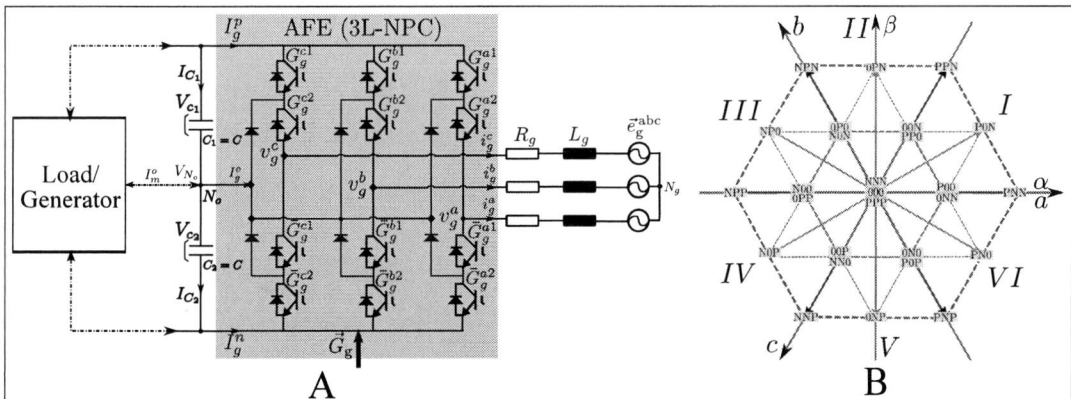

Figure 1: A: Simplified circuit of a 3L NPC Active Front End with choke ((R)L-filter) and B: its switching vectors in $\alpha\beta$ plane.

© VDE VERLAG GMBH · Berlin · Offenbach

Grid side converter control, i.e., the Active-Front-End (AFE) control, represents one of the key parts for a grid-tied renewable system (e.g., wind turbine system). A simplified grid-tied 3L-NPC AFE is shown in Fig. 1. It covers two of the three main control objectives for a PMSG wind turbine system [3], i.e., *i. grid side power or current control* and *ii. DC-link voltage control*.

Direct Model Predictive Control (DMPC), also named as Finite Control Set Model Predictive Control (FCS-DMPC) [5], takes the switching nature of a power converter into consideration and can easily include multiple nonlinear constraints into a customer designed cost function. It is conceptually straightforward and achieves good control dynamics. Therefore it has received increasingly much attention in the field of power electronics and drives since 1983 [6].

In this work, two DMPC schemes, namely, direct model predictive power control (DMPPC) and direct model predictive current control (DMPCC), dealing with a grid-tied 3L NPC converter AFE system, are presented and experimentally evaluated. Their control performances, both steady and transient phases, are compared with experimental data using a fully FPGA based solution. Moreover, their theoretical relationship is also derived.

2. Control algorithms

In this section the concept of direct model predictive control is briefly revisited and then the proposed DMPPC and DMPCC control schemes with neutral point voltage balancing control as the constraint target are presented.

2.1. Direct Model Predictive Control Concept

Classical DMPC schemes evaluate a given cost function J, which represents the control targets, for each switching state $\vec{u}$ of some admissible (finite) set $\mathcal{U}$ by using a prediction model. More specifically, for a grid-tied 3L NPC converter (See Fig.1-B), the switching vector $\vec{u}_{\mathrm{g}}$ is chosen from the set

$$\mathcal{U}_{27} := \{\mathrm{NNN}, \mathrm{NN0}, \ldots, 000, \ldots, \mathrm{PP0}, \mathrm{PPP}\} \tag{1}$$

of *27 admissible switching vectors*. The optimal gate signal vector $\vec{G}_g^* = \mathcal{G}(\vec{u}_{\mathrm{g}}^*)$ is found through minimizing the cost function, where $\mathcal{G}$ is a function maps the switching state to the gate signal. The application guidelines of the classical DMPC technique is given as pseudo code in **Algorithm I** [7].

Algorithm I: Direct Model Predictive Control for 3L-NPC power converter, with $\vec{u}_{\mathrm{g}} \in \mathcal{U}_{27}$

Step I: *State Prediction for all* $\vec{u}_{\mathrm{g}} \in \mathcal{U}_{27}$ with the measurements/feedback and the available system model.
Step II: *Cost enumeration and minimization.* The cost values with all the predicted states are calculated and the optimal switching vector is chosen as the one leading to smallest cost, i.e., $\vec{u}_g^* := \arg\min_{\vec{u}_{\mathrm{g}} \in \mathcal{U}_{27}} J(\vec{u}_{\mathrm{g}})$, where J is the costumer designed cost function representing the control targets (which usually contains both the *targeting control set* (e.g., current, torque and power tracking) and *the (nonlinear) constraint set* (e.g., switching frequency regulation, neutral point/common mode voltage reduction, etc.)).
Step III: Applying the optimal gate signal *for the whole control interval*: $\vec{G}_g^* = \mathcal{G}(\vec{u}_{\mathrm{g}}^*)$

Such technique requires the system discrete time models to "predict" the future system behaviors. The models of a 3L NPC AFE with (R)L filter will be presented in the next section.

PCIM Europe 2016, 10 – 12 May 2016, Nuremberg, Germany

2.2. Modeling of the three-level NPC AFE

The system discrete time models considering both the (i) current and (ii) power dynamics, applying the forward Euler method: $\frac{dx(t)}{dt} \approx \frac{x_{[k+1]} - x_{[k]}}{T_s}$ [7] and instantaneous power theory [8] can be described as follows [9]. Note that, all quantities $\vec{x}^{\alpha\beta}$ in the $\alpha\beta$ coordinate, and quantities $\vec{x}^{dq}$ in the dq coordinate can be derived by the corresponding quantities $\vec{x}^{abc}$ in the abc coordinate invoking (power invariable) Clarke- and Park-Transformation, as

$$
\vec{x}^{\alpha\beta}(k) = \underbrace{\sqrt{\tfrac{2}{3}} \begin{bmatrix} 1 & -\tfrac{1}{2} & -\tfrac{1}{2} \\ 0 & \tfrac{\sqrt{3}}{2} & -\tfrac{\sqrt{3}}{2} \end{bmatrix}}_{=: \mathbf{T}_c \ \text{(Clarke transformation)}} \vec{x}^{abc}(k), \ \vec{x}^{dq} = \underbrace{\begin{bmatrix} \cos(\phi_g) & \sin(\phi_g) \\ -\sin(\phi_g) & \cos(\phi_g) \end{bmatrix}}_{=: \mathbf{T}_P(\phi_g) \ \text{(Park transformation)}} \vec{x}^{\alpha\beta}. \tag{2}
$$

(i) Current discrete time models: A RL-filter connected to an ideal (balanced) grid in $\alpha\beta$ and dq are given by (see e.g. [7]):

$$
\vec{i}^{\alpha\beta}_{g[k+1]} = \left(1 - \tfrac{T_s R_g}{L_g}\right) \vec{i}^{\alpha\beta}_{g[k]} + \tfrac{T_s}{L_g} \left(\vec{v}^{\alpha\beta}_{g[k]} - \vec{e}^{\alpha\beta}_{g[k]} \right). \tag{3}
$$

$$
\vec{i}^{dq}_{g[k+1]} = \left(1 - \tfrac{T_s R_g}{L_g}\right) \vec{i}^{dq}_{g[k]} + \omega_g L_g T_s \begin{bmatrix} 0 & -1 \\ 1 & 0 \end{bmatrix} \vec{i}^{dq}_{g[k]} + \tfrac{T_s}{L_g} \left(\vec{v}^{dq}_{g[k]} - \vec{e}^{dq}_{g[k]} \right) \tag{4}
$$

where $R_g\,[\Omega]$ and $L_g\,[\text{Vs/A}]$ are the filter resistance and inductance, respectively, $\vec{z}^{xy}_g = (z^x_g, z^y_g)^\top [\text{A}]^2/[\text{V}]^2, (z \in \{i, v, e\}, x \in \{\alpha, d\}, y \in \{\beta, q\})$ can be the current, converter or source voltage vector in $\alpha\beta$ or dq frame.

(ii) Power discrete time models: According to the instantaneous power theory, *given the grid side voltage vector aligns with the d-axis using certain adequate phase-lock-loop (PLL)*, grid-side active and reactive power are given by (see [9, 10]):

$$
\left(P_{[k]}, \quad Q_{[k]} \right)^\top = \left(\Re\{\vec{i}^{\dagger}_{g[k]} \cdot \vec{e}_{g[k]}\}, \quad \Im\{\vec{i}^{\dagger}_{g[k]} \cdot \vec{e}_{g[k]}\} \right)^\top = \left(e^d_g \cdot i^d_g, \ e^d_g \cdot i^q_g \right)^\top \tag{5}
$$

where † represents conjugate of a vector. Defining the slope of a discrete signal $x_{[k]}$ at sampling point of k as $Sl_{x[k]} = \frac{x_{[k]} - x_{[k-1]}}{T_s}$, then the slopes of the active and reactive power for a given actuating converter voltage vector of $\vec{v}^i_{g[k]}$ (corresponding to switching state $\vec{u}^i_g$ at sampling point k are [10]

$$
Sl^{\vec{u}^i}_{P[k]} = \tfrac{1}{L_g} \left(||\vec{e}^{\alpha\beta}_{g[k]}||^2 - \Re\{\vec{e}^{\alpha\beta}_{g[k]} (\vec{v}^{\alpha\beta}_{g[k]})^\dagger\} - R_g P_{[k]} - \omega_g Q_{[k]} \right); \tag{6}
$$

$$
Sl^{\vec{u}^i}_{Q[k]} = \tfrac{1}{L_g} \left(\Im\{\vec{e}^{\alpha\beta}_{g[k]} (\vec{v}^{\alpha\beta}_{g[k]})^\dagger\} - R_g Q_{[k]} + \omega_g P_{[k]} \right). \tag{7}
$$

$\vec{Sl}^{\vec{u}^i}_{S[k]} = [Sl^{\vec{u}^i}_{P[k]}, Sl^{\vec{u}^i}_{Q[k]}]^\top$, is the apparent power slope vector corresponding to $\vec{v}^i_{g[k]}$ and switching state $\vec{u}^i$. Considering the active and reactive power evolving in a linear manner (i.e., its slope is a constant during a whole control interval), one can obtain the grid side active and reactive power at time instant $[k+1]$ as

$$
\vec{S}_{[k+1]} = \left(P_{[k+1]}, \quad Q_{[k+1]} \right)^\top = \left(P_{[k]}, \quad Q_{[k]} \right)^\top + \vec{Sl}^{\vec{u}^i}_{S[k]} \cdot T_s. \tag{8}
$$

The modeling of a 3L NPC power converter considering the converter phase voltage $\vec{v}^{abc}_g$, switching state $\vec{u}^{abc}_g$ and the neutral point voltage $V_{o[k+1]}$ has been described by Equations (4), (3) and (8)[1] in [11].

---

[1]Note that, the neutral point voltage prediction equation based on the current flow direction for this work shall be $V_{o[k+1]} = V_{o[k]} - \frac{T_s}{C}(\vec{i}^{abc}_{g[k]})^\top \cdot |\vec{u}^{abc}_{g[k]}|$, the converter phase voltage and switching state equations are the same as (4) and (3) in [11].

© VDE VERLAG GMBH · Berlin · Offenbach

PCIM Europe 2016, 10 – 12 May 2016, Nuremberg, Germany

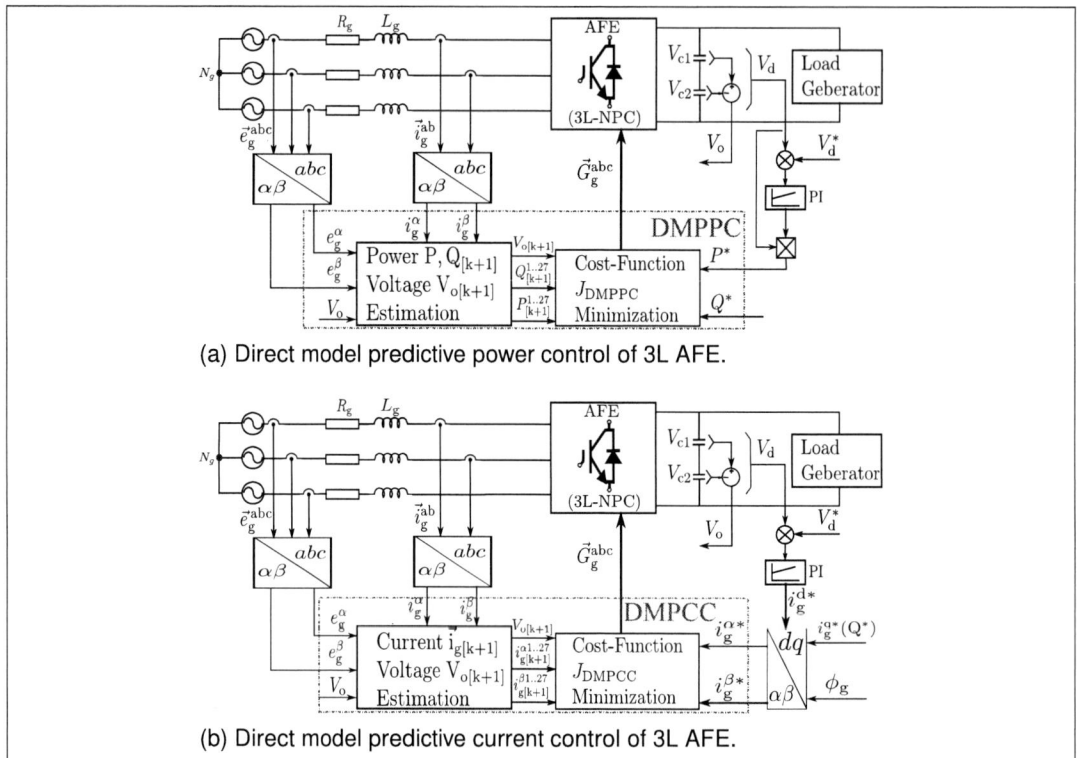

Figure 2: Structures of DMPPC and DMPCC for controlling a 3L AFE.

2.3. Direct Model Predictive Power Control

The control targets for a predictive power control method are active / reactive power control and DC-link voltage balancing. Therefore the cost function to fulfill these two control targets is defined as

$$J_{\text{DMPPC}}(\vec{u}_{\text{g}}) = \underbrace{\gamma_P \big(P^* - P_{[k+1]}(\vec{u}_{\text{g}})\big)^2 + \gamma_Q \big(Q^* - Q_{[k+1]}(\vec{u}_{\text{g}})\big)^2}_{=:J_{\text{PQ}}} + \underbrace{\gamma_{V_o}^{PPC} \big(V_o^* - V_{o[k+1]}(\vec{u}_{\text{g}})\big)^2}_{J_{V_o}} \quad (9)$$

with weighting factors $\gamma_P[\frac{1}{W}] = \gamma_Q[\frac{1}{\text{Var}}](= 1)$ and $\gamma_{V_o}^{PPC}[\frac{1}{V}](= 0.5)$. J_P and J_Q represent the targets for active and reactive power control, respectively. Whereas J_{V_o} shall assure voltage balancing. The predicted active and reactive power $P_{[k+1]}, Q_{[k+1]}$ and DC-link voltage difference $V_{o[k+1]}$ are obtained using the system models described in Sec. 2.2.

2.4. Direct Model Predictive Current Control

For direct model predictive current control of the grid-tied 3L NPC AFE, taking the current and voltage control as targets, the inner loop can be designed in the $\alpha\beta$ frame to eliminate the otherwise required synchronous frame transformations. Therefore, cost functions is defined as

$$J_{\text{DMPCC}}(\vec{u}_{\text{g}}) = \underbrace{\gamma_{i_{\text{g}}^{\alpha}} \big(i_{\text{g}}^{\alpha*} - i_{\text{g}[k+1]}^{\alpha}(\vec{u}_{\text{g}})\big)^2 + \gamma_{i_{\text{g}}^{\beta}} \big(i_{\text{g}}^{\beta*} - i_{\text{g}[k+1]}^{\beta}(\vec{u}_{\text{g}})\big)^2}_{=:J_{i_{\text{g}}^{\alpha\beta}}} + \underbrace{\gamma_{V_o}^{PCC} \big(V_o^* - V_{o[k+1]}(\vec{u}_{\text{g}})\big)^2}_{J_{V_o}}$$

$$(10)$$

© VDE VERLAG GMBH · Berlin · Offenbach

with weighting factors[2] $\gamma_{i_g^\alpha}[\frac{1}{A}] = \gamma_{i_g^\beta}[\frac{1}{A}](=1)$ and $\gamma_{V_o}^{PCC}[\frac{1}{V}](= \gamma_{V_o}^{PPC}/(e_g^d)^2)$. $J_{\vec{i}_g^{\alpha\beta}}, J_{V_o}$ are to control the current and DC-link voltage balancing, respectively. The predicted currents $i_{g[k+1]}^\alpha, i_{g[k+1]}^\beta$ and DC-link voltage difference $V_{o[k+1]}$ are obtained using the system models described in Sec. 2.2.

After minimizing the relevant cost of (9) and (10), the optimal switching sequence can be then obtained. The general overview of both schemes are given by Fig. 2. Note that, the outer DC-link voltage controller (for both a PI controller is used) is not the scope of this paper and is therefore not redundantly reported.

Parameter [unit]	Value
DC-Link cap. $C_1 = C_2 = C$ [F]	1000×10^{-6}
Sampling frequency $f_s = 1/T_s$ [kHz]	20
Grid-side phase voltages $\|\vec{e}_n\|$ [V]	250
Grid-side voltage frequency ω_n [rad/s]	100π
Grid-side resistance R_g [Ω]	1.56×10^{-3}
Grid-side inductance L_g [H]	16×10^{-3}
$\gamma_P = \gamma_Q = \gamma_{i_g^\alpha} = \gamma_{i_g^\beta}[1]$	1
$\gamma_{V_o}[1]$	0.5

Table 1: Implementation and system data.

3. Experimental results

To verify and compare their control performances, both schemes are implemented on a FPGA based real-time controller and tested at a self-constructed grid-tied regenerative three-level N-PC power converter system with a similar configuration[3] as shown in Fig. 1. Parameters are collected in Tab. 1. The relevant state predictions, cost function calculation/minimization procedures are divided into subroutines and optimized with the Single-Cycle-Timed-Loop method using NI-FPGA technique. The control interval is set as $T_s = 50e^{-6}$[s] for both for a fair comparison. The load side is connected to a controllable electrical drives, which can work at both generator and motor mode to generate and consume power to/from the AFE under test. Experimental results are given in Fig. 3 and Fig. 4.

Fig. 3 depicts the overall control performances for both methods. The testing scenarios are: a 80% of its rated torque is mounted to a generator, while its speed reference is changed (following a set sequence) which generates different active powers to the grid. For all the test, the reactive power/current references remain zero to achieve "unit power factor" control. As can be seen from Fig. 3, both methods achieve quite similar control dynamics, while, DMPPC shows slightly smaller ripples in the active and reactive power waveforms.

Fig. 4 shows the steady state performance comparison at a similar average switching frequency. As can be seen, the ripples of both the active and reactive powers using DMPPC are smaller than that of the DMPCC method. Reasons can be found from its less distorted current waveforms in comparison with the DMPCC method. Meanwhile, the synthesized command voltage

---

[2]The reason for $\gamma_{V_o}^{PCC}[\frac{1}{V}](= \gamma_{V_o}^{PPC}/(e_g^d)^2)$ is given by Sec. 4.1.

[3]Note that, for safety concerns, a variac is installed between the grid and the filter to lower down the line voltages, which is the different part from Fig. 1.

PCIM Europe 2016, 10 – 12 May 2016, Nuremberg, Germany

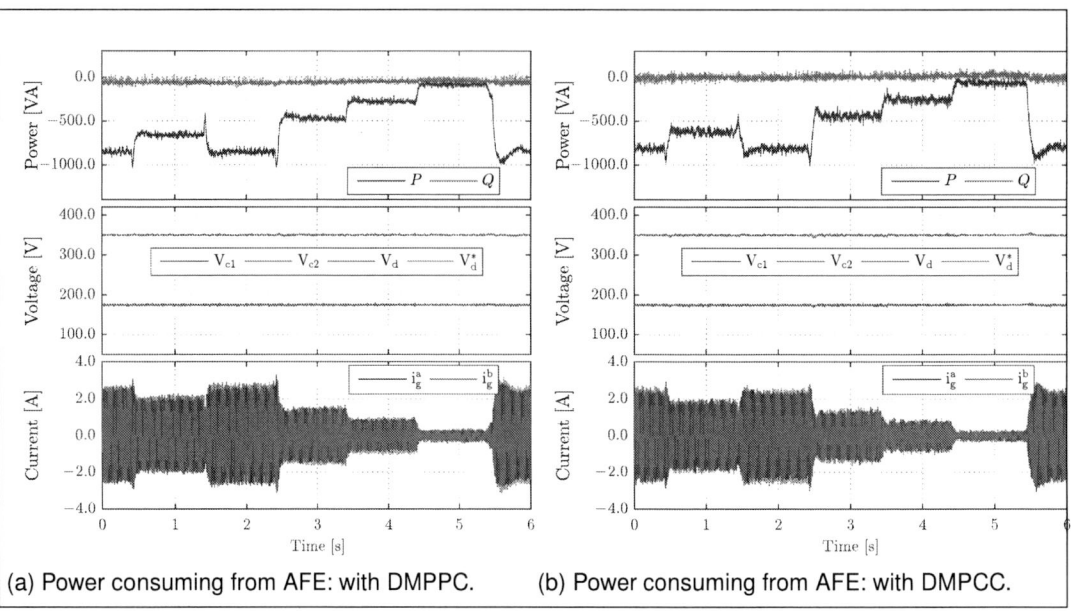

(a) Power consuming from AFE: with DMPPC.　　(b) Power consuming from AFE: with DMPCC.

Figure 3: Overall control performances comparison. For each sub-figure, from up to down are: active and reactive powers, DC-link and capacitor voltages, grid side currents in phase A and B.

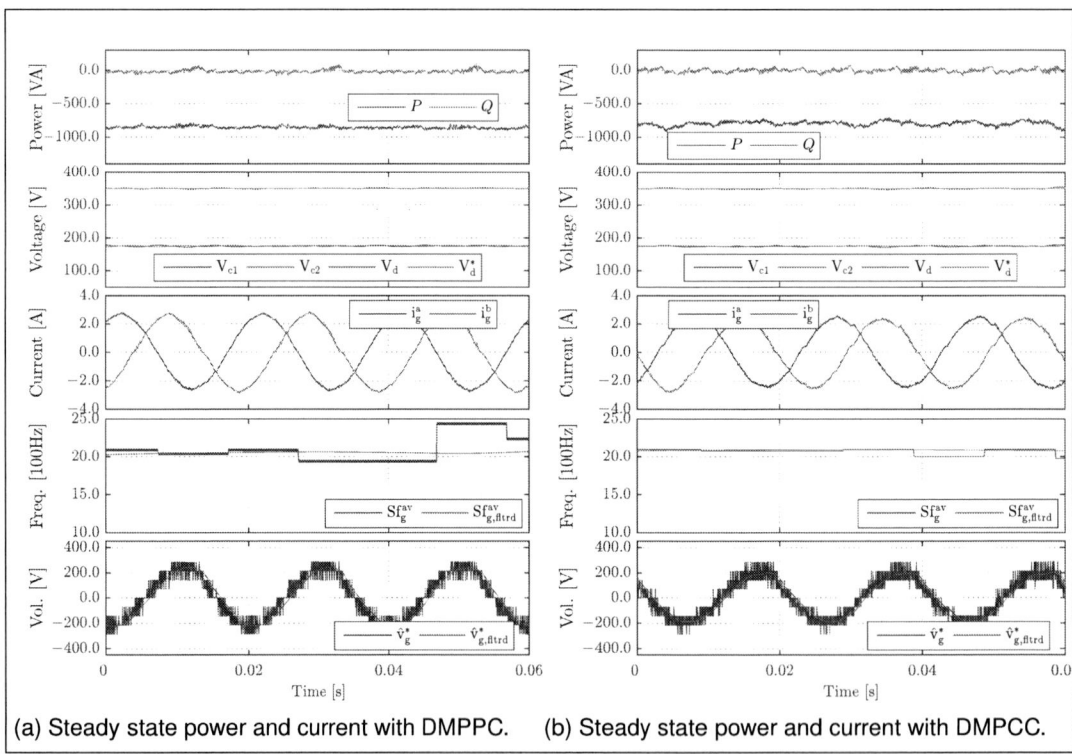

(a) Steady state power and current with DMPPC.　　(b) Steady state power and current with DMPCC.

Figure 4: Steady state control performance comparison. For each sub-figure, from up to down are: active and reactive powers, DC-link and capacitor voltages, grid side currents in phase A and B, switching (average) frequency Sf_g^{av} (and its filtered value $Sf_{g,fltrd}^{av}$, with a cut-off frequency of 5Hz), and the estimated command voltage $\hat{v}_g^*$ (and its filtered value $\hat{v}_{g,fltrd}^*$, with a cut-off frequency of 300Hz).

© VDE VERLAG GMBH · Berlin · Offenbach

(switching positions) of DMPPC method is also (slightly) smoother than the DMPCC method (see the last sub-figure in Fig. 4).

4. Discussion and summery

4.1. Discussion on the relationship between DMPPC and DMPCC

Based on the assumption that

A: the grid side voltage vector is aligned with the d-axis with an adequate PLL method,

B: within a sampling interval T_s, the grid side voltage vector does not change, i.e., $e^d_{g[k+1]} \approx e^d_{g[k]} = e^d_g$ and $\phi_{g[k+1]} \approx \phi_{g[k]}$,

invoking Eq. (2) and Eq. (5) yields

$$
\begin{aligned}
(P^* - P_{[k+1]})^2 &= (e^d_g)^2 \left(i^{d*}_g - i^d_{g[k+1]} \right)^2 = (e^d_g)^2 \left(\cos(\phi_{g[k]})(i^{\alpha*}_g - i^\alpha_{g[k+1]}) + \sin(\phi_{g[k]})(i^{\beta*}_g - i^\beta_{g[k+1]}) \right)^2 \\
&= (e^d_g)^2 \left(\cos^2(\phi_{g[k]})(i^{\alpha*}_g - i^\alpha_{g[k+1]})^2 + \sin^2(\phi_{g[k]})(i^{\beta*}_g - i^\beta_{g[k+1]})^2 \right) \\
&\qquad + 2(e^d_g)^2 \cdot \cos(\phi_{g[k]})(i^{\alpha*}_g - i^\alpha_{g[k+1]}) \cdot \sin(\phi_{g[k]})(i^{\beta*}_g - i^\beta_{g[k+1]}) \quad (11)
\end{aligned}
$$

$$
\begin{aligned}
(Q^* - Q_{[k+1]})^2 &= (e^d_g)^2 \left(i^{q*}_g - i^q_{g[k+1]} \right)^2 = (e^d_g)^2 \left(\sin(\phi_{g[k]}) \cdot (i^\alpha_{g[k+1]} - i^{\alpha*}_g) + \cos(\phi_{g[k]})(i^{\beta*}_g - i^\beta_{g[k+1]}) \right)^2 \\
&= (e^d_g)^2 \cdot \left(\cos^2(\phi_{g[k]})(i^{\alpha*}_g - i^\alpha_{g[k+1]})^2 + \sin^2(\phi_{g[k]})(i^{\beta*}_g - i^\beta_{g[k+1]})^2 \right) \\
&\qquad - 2(e^d_g)^2 \cdot \cos(\phi_{g[k]})(i^{\alpha*}_g - i^\alpha_{g[k+1]}) \cdot \sin(\phi_{g[k]})(i^{\beta*}_g - i^\beta_{g[k+1]}) \quad (12)
\end{aligned}
$$

Adding up Eq. (11) and (12) yields

$$
\underbrace{(P^* - P_{[k+1]})^2 + (Q^* - Q_{[k+1]})^2}_{J_{PQ}} = (e^d_g)^2 \cdot \underbrace{\left(i^{\alpha*}_g - i^\alpha_{g[k+1]}(\vec{u}_g) \right)^2 + \left(i^{\beta*}_g - i^\beta_{g[k+1]}(\vec{u}_g) \right)^2}_{=:J_{\vec{i}_g \alpha \beta}}. \quad (13)
$$

In steady state, e^d_g is a constant value; therefore, Eq. (13) will suggest that, DMPPC is equivalent to DMPCC for the *targeting set control*. Therefore, it is not difficult to understand, with a proper tuning of $\gamma^{PCC}_{V_o} = \gamma^{PPC}_{V_o}/(e^d_g)^2$, DMPPC and DMPCC shall achieve the same performances.

However, in practice, both assumptions (A) and (B) cannot (perfectly) hold for all time, because, assumption (A) might not hold for starting and transient phases or unbalanced grid situations due to the inevitable delay introduced by the PLL method, while (B) cannot hold when T_s is big. These might be the reasons why we see slightly different performances with DMPPC and DMPCC control in this work.

Apart from the slightly degraded performances in practice, DMPCC requires the voltage vector position to produce the reference signals. This process will rely on an outer loop PLL estimator. This might be taken as a shortcoming in comparison with DMPPC technique, although from the inner loop point of view, both methods can be realized in the $\alpha\beta$ frame.

© VDE VERLAG GMBH · Berlin · Offenbach

4.2. Summery

This paper has presented two FPGA based direct model predictive control schemes, dealing with a grid-tied three-level AFE control in renewable energy systems under balanced grid voltage situations. Both methods have been designed in the $\alpha\beta$ frames and require no extra modulation. From theoretical point of view, both can be equivalent under certain assumptions. However, experimental results confirm that, although both achieve quite similar control dynamics, DMPPC slightly outperforms DMPCC in terms of control variable ripples. Moreover, from realization point of view, DMPCC requires the grid voltage vector position information from its outer-loop (to generate the references), which might depend on the quality of the outer loop PLL estimator; while DMPPC can be purely realized without any PLL and the grid voltage vector position.

Acknowledgment: This work is supported by DFG founding (No.: KE817/32-1). Zhenbin Zhang is the corresponding author and would like to express his gratefulness to Mr. Marc Backmeyer from National Instruments and Dr. Martin Schulz from Infineon for their help during his test-bench construction.

References

[1] Frede Blaabjerg. Future on Power Electronics for Wind Turbine Systems. *IEEE Journal of Emerging and Selected Topics in Power Electronics*, 1(3):139–152, 2013.

[2] Frede Blaabjerg. Future on Power Electronics for Wind Turbine Systems. *IEEE Journal of Emerging and Selected Topics in Power Electronics*, 1(3):139–152, 2013.

[3] Zhenbin Zhang and Ralph Kennel. Direct Model Predictive Control of Three-Level NPC Back-to-Back Power Converter PMSG Wind Turbine Systems Under Unbalanced Grid. In *Predictive Control of Electrical Drives and Power Electronics (PRECEDE 2015), Valparaiso, Chile.*, 2015.

[4] M.P. Kazmierkowski, L.G. Franquelo, J. Rodriguez, M.A. Perez, and J.I. Leon. High-performance motor drives. *Industrial Electronics Magazine, IEEE*, 5(3):6–26, Sept 2011.

[5] S. Kouro, M. A. Perez, J. Rodriguez, A. M. Llor, and H. A. Young. Model predictive control: Mpc's role in the evolution of power electronics. *IEEE Industrial Electronics Magazine*, 9(4):8–21, Dec 2015.

[6] R Kennel and D Schöder. A predictive control strategy for converters. In *IFAC Control in Power Electronics and Electrical Drives*, pages 415–422, 1983.

[7] Zhenbin Zhang and Ralph Kennel. Fully fpga based direct model predictive power control for grid-tied afes with improved performance. In *Industrial Electronics Society, IECON 2015 - 41th Annual Conference of the IEEE*, Nov 2015.

[8] H Akagi, Yoshihira Kanazawa, and A Nabae. Instantaneous Reactive Power Compensators Comprising Switching Devices without Energy Storage Components. *IEEE Transactions on Industrial Applications*, IA-20(3):625–630, 1984.

[9] Zhenbin Zhang, He Xu, Ming Xue, Zhe Chen, Tongjing Sun, R. Kennel, and C.M. Hackl. Predictive control with novel virtual-flux estimation for back-to-back power converters. *Industrial Electronics, IEEE Transactions on*, 62(5):2823–2834, May 2015.

[10] Zhenbin Zhang, Fengxiang Wang, Tongjing Sun, J. Rodriguez, and R. Kennel. Fpga-based experimental investigation of a quasi-centralized model predictive control for back-to-back converters. *Power Electronics, IEEE Transactions on*, 31(1):662–674, Jan 2016.

[11] Zhenbin Zhang, Christoph Hackl, and Ralph Kennel. Fpga based direct model predictive current control of pmsm drives with 3l npc power converter. In *PCIM - 2016, Nurnburg*, May 2016.

PCIM Europe 2016, 10 – 12 May 2016, Nuremberg, Germany

Scalable insulated DC/DC converters for safe and efficient coupling of fuel cells, electrolyzers and DC grids

Bernd Seliger, Fraunhofer IISB, Erlangen, Germany, bernd.seliger@iisb.fraunhofer.de
Stefan Matlok, Fraunhofer IISB, Erlangen, Germany, stefan.matlok@iisb.fraunhofer.de
Stefan Zeltner, Fraunhofer IISB, Erlangen, Germany, stefan.zeltner@iisb.fraunhofer.de

Abstract

This paper describes scalable insulated DC/DC converters for safe and efficient coupling of fuel cells and electrolyzers with a 380 V single phase respectively +/-380 V split-phase DC grid. Energy storage systems in the MWh range, based on liquid organic hydrogen carriers (LOHC), can be realized this way. An 11 kW insulated fuel cell DC/DC as well as an 11 kW insulated electrolyzer converter, both based on novel universal building blocks, are presented together with efficiency measurements for different operating points. High efficiencies under partial load conditions are achieved by using a modular design approach and phase shedding. The modular power electronic design is described in detail.

1.   Introduction

The storage of electrical energy is the key element for a successful change in electricity generation [1]. A safe transition from a centralized electricity generation with only a few powerful plants to a decentralized system with a huge number of small renewable, volatile power sources is only possible by using energy storages. LOHC (Liquid Organic Hydrogen Carriers) storage systems are an auspicious approach whereat hydrogen is stored

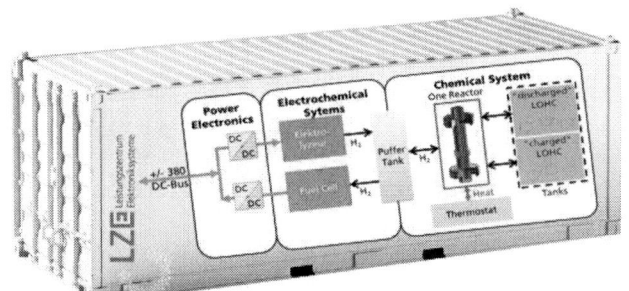

Fig. 1: Planned all-in-one LOHC 1 MWh storage system

in a liquid carbon carrier, as illustrated in Fig. 1. Due to a storage density of 1 MWh/m³ to 2 MWh/m³ it is possible to build up storages in the several MWh range [2]. For charging a LOHC system the electrical energy will be transformed to hydrogen by an electrolyzer. With the reversal process electrical energy can be generated. Discharging (dehydrogenation) the LOHC in a reactor produces hydrogen. That hydrogen can be used by fuel cells to generate electrical energy. Because the electrolyzer as well as the fuel cell needs DC voltages the LOHC storage system can be operated preferable within a DC microgrid. Therefore two DC/DC converters are required, one for the electrolyzer, and one for the fuel cell.

2.   Electrical System Requirements

Fig. 2 shows the entire electrical architecture which will be used for a prototype 1 MWh LOHC storage system. All necessary subsystems of the energy storage will be placed in a conventional 20 ft container. This means that a relative high power density for the power electronics is required, compared to typically stationary power equipment [3]. Another requirement is the galvanic isolation of the different grounded subsystems, which are operated in DC microgrids with different existing grounding concepts (IT grid, PE-grounded, weak

© VDE VERLAG GMBH · Berlin · Offenbach

grounded). Because DC microgrids are a future topic it should be possible to use a conventional bidirectional AC/DC topology to connect the storage system with the AC mains. Hence 11 kW subsystems are preferable and compatible with an alternative 3 phase 16 A (11 kW), 32 A (2x11 kW) or 64 A (4x 11 kW) AC grid connection.

3.   DC/DC system topology derived from electrical system requirements

To be prepared for high as well as for lower power applications it should be possible to connect the DC/DC converters to ± 380 V (760 V) split-phase DC grids as well as to 380 V_{DC} single-phase microgrids. This can be done easily if a series or parallel connection of the DC/DC high voltage sides is possible. A second reason for using a modular system approach is the possibility to implement phase shedding strategies to reach a high efficiency in partial load conditions. Finally, a third reason for a modular system approach is the limitation of high switch-off currents. Therefore both kind of DC/DC converters will be realized by 2.75 kW subsystems with a maximum current of 50 A at the low voltage side. For efficiency reasons a switching frequency of 47 kHz was chosen. A further important advantage of the shown modular approach is that also non symmetrical load conditions in a ± 380 V DC grid can be controlled.

The figures 3 and 4 show the modular approach at a glance. Each 11 kW unit can be

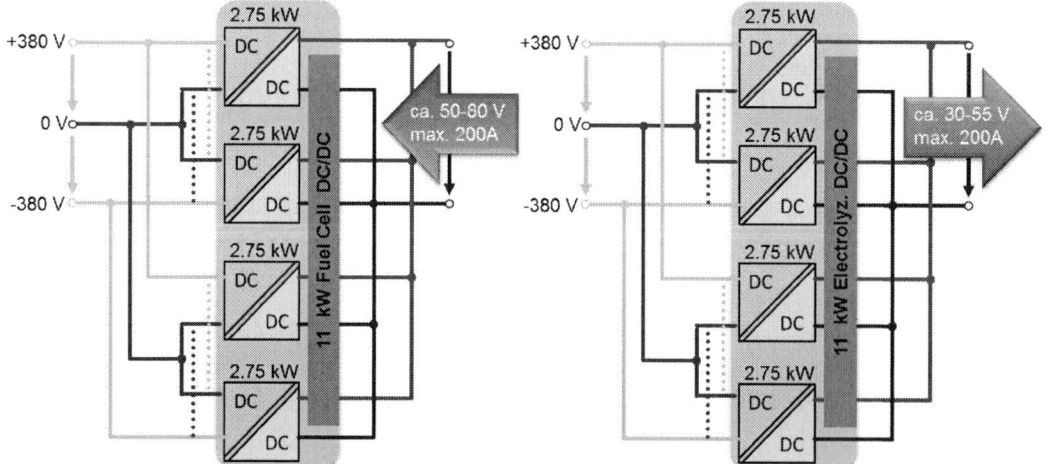

Fig. 2: System architecture for 600 A DC/DCs

operated by two 5.5 kW sub-units by using a HV-side series connection configuration. That means the 600 A system architecture shown in Fig. 2 can be operated by 6 sub-units via phase shedding to achieve a high efficiency in a wide load range (see 5.1).

Fig. 3: 11 kW / 200 A fuel cell DC/DC converter

Fig. 4: 11 kW / 200 A electrolyzer converter

© VDE VERLAG GMBH · Berlin · Offenbach

It has to be mentioned that in cases with higher fuel cell or electrolyzer voltages a series connection of two 2.75 kW sub-units on the LV-sides is possible. Due to the current controlability of the chosen full-bridge ZVS phase-shift topology down to 0 A no further changes are necessary.

4.   The two 50 A building blocks (2.75 kW)

A single stage two converter approach is chosen. Together with an optimization of both DC/DC converters (because input/output voltage ranges of electrolyzers and fuel cells are different in general), high efficiencies can be reached.

Fig. 5 shows the proposed insulated full bridge ZVS DC/DC converter topologies with phase shift control.

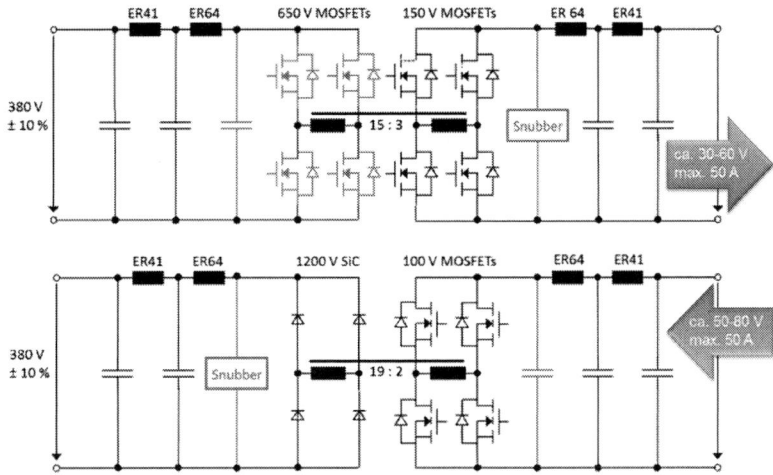

Fig. 5: Uniform circuit topology for electrolyzer and fuel cell DC/DC

It shows that only some minor changes are necessary to convert the electrolyzer topology to the fuel cell topology. Therefore similar submodules can be used for universal DC/DC building blocks (one for the electrolyzer and one for the fuel cell DC/DC converter).

Fig. 6 shows the universal 2.75 kW building block for the fuel cell DC/DC converter. The building block consists of different submodules like a full bridge MOSFET module, a rectifier module with 1200 V SiC diodes and a gate driver module. Additional snubber modules are necessary to avoid over voltages on the secondary side rectifiers for both topologies. All submodules have the same mechanical design what means they fit together in different combinations. E.g. the low voltage MOSFET module can be

Fig. 6: Universal building block with submodules

used as a synchronous rectifier module in the electrolyzer DC/DC as well as a full bridge module in the fuel cell converter.

© VDE VERLAG GMBH · Berlin · Offenbach

639

4.1. Transformer winding ratio design

The most important design parameter of the chosen isolated full bridge ZVS converter topology with phase shift control is the transformer winding ratio. It is given by the condition that the maximum output voltage $V_{out,max}$ must be reached with the minimum input voltage $V_{in,min}$ by choosing the right transformer ratio.

$$V_{out,max} = N_{eff} \cdot V_{in,min} \cdot \alpha \tag{1}$$

The additional factor α of about 0.9 is to keep margin because of parasitic effects and losses. Table 1 shows the results for the effective turns-ratio N_{eff} (equal N_2:N_1) for fuel cell as well for electrolyzer DC/DC.

Table 1:

DC/DC	$V_{in,range}$ (V)	$V_{in,min} \cdot \alpha$ (V)	$V_{out,range}$ (V)	$V_{out,max}$ (V)	N_{eff}
Fuel Cell	50 ... 80	45	380 ± 10 %	418	9.3
Electrolyzer	380 ± 10 %	308	30 ... 60	60	0.19

The real winding numbers N_1, N_2 are given by calculating the minimum primary side number of turns $N_{1,min}$ with:

$$N_{1,min} \geq \frac{V_{pri,max} \cdot T/2}{\Delta B_{max,pp} \cdot A_{min}} \tag{2}$$

$V_{pri,max}$ is the voltage amplitude on the primary side winding, T is the switching period, $\Delta B_{max,pp}$ is the maximum allowed peak-to-peak flux density and A_{min} is the minimum magnetic core area. With a chosen $\Delta B_{max,pp}$ = 0.6 Tesla (equal to ± 0.3 Tesla), a PM74 core (A_{min} = 630 mm²) and a switching frequency of f_s = 47 kHz, Table 2 shows the results for both converter types:

Table 2:

DC/DC	$V_{pri,max}$ = $V_{in,range,max}$ (V)	$T = 1/fs$ (us)	$\Delta B_{max,pp}$ (T)	A_{min} (mm²)	$N_{1,min}$
Fuel Cell	80	21.3	0.6	630	≈ 2
Electrolyzer	418	21.3	0.6	630	≈ 12

With the given effective turns-ratio N_{eff} from Table 1 and the minimum primary side windings from Table 2 the winding numbers N_1 (primary) and N_2 (secondary side) are given in Table 3:

Table 3:

DC/DC	$N_{eff} = N_2$:N_1	$N_{1,min}$	$N_{1,(chosen)}$	$N_{2,(chosen)}$	N_{real}
Fuell Cell	9.3	≈ 2	2	19	9.5
Electrolyzer	0.19	≈ 12	15	3	0.2

Table 3 shows that the chosen transformer winding ratio (2:19 for the electrolyzer DC/DC and 15:3 for the fuel cell DC/DC) is very close to the calculated values. Especially for the fuel cell DC/DC N_1 is close to $N_{1,min}$ what means, that the fuel cell DC/DC transformer will have more core losses under part load conditions.

PCIM Europe 2016, 10 – 12 May 2016, Nuremberg, Germany

4.2. Efficiency measurements

The results of the efficiency measurements for both 50 A (2.75 kW) DC/DC building blocks are shown in Fig. 7 (fuel cell DC/DC building block) and Fig. 8 (electrolyzer DC/DC building block) for an environment temperature of 25 °C and a cooling temperature of 65 °C.

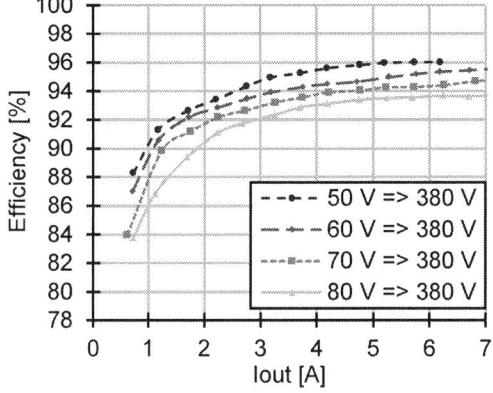

Fig. 7: Efficiency of fuel cell building block

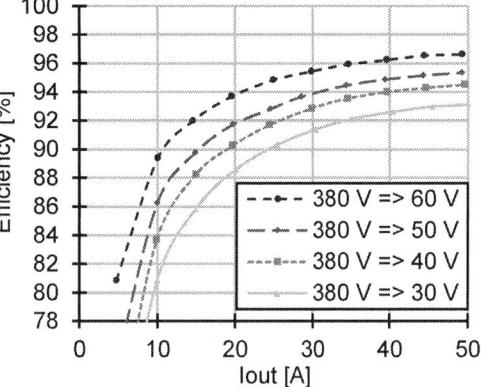

Fig. 8: Efficiency of electrolyzer building block

Fig. 7 shows that the fuel cell DC/DC building block reaches a maximum efficiency of 96 % at the minimum input voltage. Because the fuel cell voltage drops under full load (and with increasing life time) the fuel cell DC/DC was designed to reach its peak efficiency at maximum output power. Due to synchronous rectification the efficiency of the electrolyzer DC/DC is with 96.6 % a little bit higher, as shown Fig. 8. Also the electrolyzer DC/DC reaches the highest efficiency at the maximum output power.

4.3. Converter Control

As described in chapter 3 an important feature of the presented modular approach is the flexibility of possible series and parallel connections of submodules on the high-voltage side as well on the low-voltage side. Only a little additional effort (galvanic isolated voltage measurements and additional galvanic isolated auxiliary supplies) within the control architecture is necessary to provide this flexibility. Fig. 9 shows a basic block diagram to illustrate the necessary control blocks for a 2.75 kW building block.

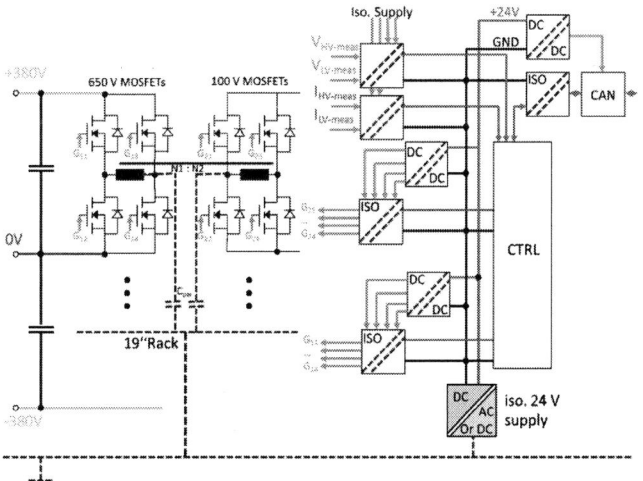

Fig. 9: Basic control architecture of a 50 A (2.75 kW) building block

© VDE VERLAG GMBH · Berlin · Offenbach

5. Scalable modular fuel cell and electrolyzer DC/DC converters

The insulation design allows stacking or parallelizing the converters on each side separately. This enables the phases and converters to be used in different applications, choosing voltages and currents. By using more converters in a rack, even the power can be adapted to a huge scale.

5.1. Phase shedding

To gain maximum efficiency, phase shedding can be used to reduce losses on a large power range down to low load conditions [4]. The efficiency shown in Figures 7 and 8, contain information of best phase shedding strategy for this converter type, not only depending on current and power, but although on the corresponding optimum voltage of a typical fuel cell.

$$\eta(x) = \frac{P_{out}(x)}{P_{out}(x) + P_{loss}(x)} = \frac{P_{out}(x)}{P_{out}(x) + (a_0 + a_1 \cdot x^1 + \cdots + a_N \cdot x^N)} \tag{3}$$

For analyzing the effects of phase shedding, the loss mechanisms are empirically fitted to analytical solvable equations. While efficiency curves need a high polynomic fitting and detailed data, calculations based on the loss curves are much more accurate, easier to handle and linked to physical effects. Based on the measurements of the fuel cell converter, the resulting losses for matching coefficients of quadratic fitting are shown in Fig. 10. As to see on this plot, the data matches the measurement quite precisely on 60 to 80 volt input curves. For 50 volt, there is too much deviation in the base data for a good fitting. So there is only a generic line shown, representing a rough estimation set of coefficients.

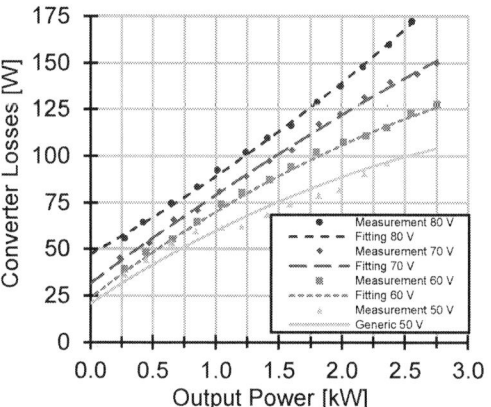

Fig. 10: Polynomial fitting of DC/DC losses

Based on the polynomial fitting, there can be done calculations on effects splitting the load on one or more phases. While the maximum power per phase is a hard limit, there is no restriction on symmetric or asymmetric phase usage for lower loads. The factor d in (4) defines the ratio of current on the first phase I_{PH1} to the sum of output current I_{OUT}.

$$d = \frac{I_{PH1}}{I_{OUT}} \tag{4}$$

The result of the overall loss reduction P_{LR} at a specific operations point of load (x) on different current distributions between the first and second phase is defined by:

$$P_{LR}(x) = a_0 - 2a_2 x^2 \cdot (d - d^2) \tag{5}$$

This shows the fact that activating a second phase on part load does only gain efficiency if the following conditions occur:

- The overall losses function does have a positive quadratic coefficient a_2
- The coefficient is significant enough to compensate initial losses of a_0 as shown in (5)

The best way to share the current is always at $d = 0.5$, caused by minimum of (5) by derivation of factor $(d-d^2)$.

© VDE VERLAG GMBH · Berlin · Offenbach

Following those calculations, the phase shedding strategy follows some simple rules for that specific converter based on its coefficients:

- Use as less phases as possible up to the maximum individual power
- On new phase activation, distribute load evenly between all active phases or converters

The resulting efficiency for 4 parallel phases and 3 converters at different voltages is shown in Fig. 11. The efficiency resulting in the final application is additionally linked to the typical voltage characteristics of the input power source. The result for a typical fuel cell voltage, at half aged lifetime condition showing nearly linear load current dependency, is shown in Fig 12.

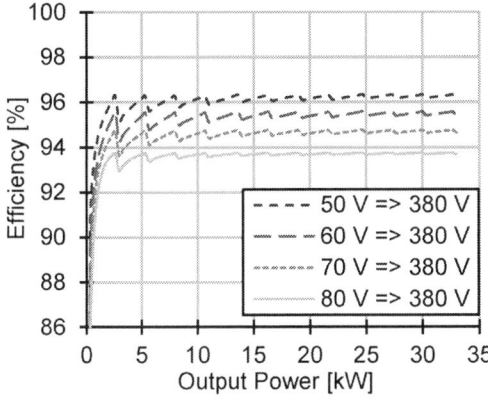

Figure 11: Efficiency with phase shedding

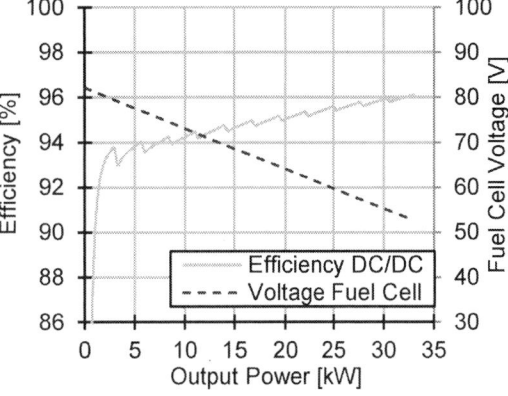

Figure 12: Final efficiency related to fuel cell

5.2. Design of the 600 A (3 x 11 kW) DC/DC converter systems

In Fig. 13 the CAD of the nearly identical 11 kW fuel cell DC/DC and 11 kW electrolyzer DC/DC is shown. Each is realized by four 2.75 kW building blocks (see Fig. 6) within a 19″ rack in two height units. The converters are water cooled with a maximum coolant temperature of 85 °C. All connectors are placed on the back side.

Measurements have shown that the temperature rise on the semiconductor packages are lower than 15 K at a cooling temperature of 65 °C.

Moreover all organic-based capacitors are placed beside the cooling plate what means lower temperature stress in environments with temperatures lower than the coolant temperature. Therefore a robust design was chosen to fulfill non-stop operation requirements.

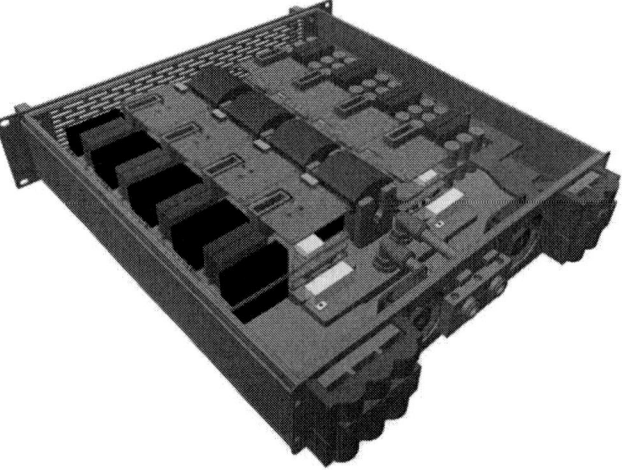

Fig. 13: CAD of the 11 kW DC/DC (fuel cell and electrolyzer)

ACKNOWLEDGMENT

This contribution was supported in part by the Bavarian Ministry of Economic Affairs and Media, Energy and Technology as a part of the Bavarian project "Leistungszentrum Elektroniksysteme (LZE)"

6.   Reference

[1]   IEA, „Technology Roadmap – Energy Storage", International Energy Agency, 2014

[2]   D. Teichmann, W. Arlt, P. Wasserscheid und R. Freymann, „A future energy supply based on Liquid Organic Hydrogen Carriers (LOHC)," Energy & Environmental Science, Bd. 4, Nr. 8, pp. 2767-2773, 2011

[3]   LZE, „Forschung in neuen Dimensionen", www.lze.bayern

[4]   B. Eckardt, A. Hofmann, S. Zeltner, M. Maerz, „Automotive Powertrain DC/DC Converter with 25kW/dm³ by using SiC Diodes" 4th International Conference of Integrated Power Systems (CIPS), 2006, pp.1-6.

SiC MW PV Inverter

Maja Harfman Todorovic, GE Research Center, USA, harfmanm@ge.com
Fabio Carastro, GE Research Center, Germany, Fabio.Carastro@ge.com
Tobias Schuetz, GE Research Center, Germany, schuetz@ge.com
Robert Roesner, GE Research Center, Germany, robert.roesner@ge.com
Ljubisa Stevanovic, GE Research Center, USA, stevanov@ge.com
Gary Mandrusiak, GE Research Center, USA, mandrusi@ge.com
Brian Rowden, GE Research Center, USA, rowden@ge.com
Fengfeng Tao, GE Research Center, USA, tao@ge.com
Philip Cioffi, GE Research Center, USA, cioffi@ge.com
Jeffrey Nasadoski, GE Research Center, USA, nasados@ge.com
Rajib Datta, GE Research Center, USA, Rajib.Datta@ge.com

Abstract

With utility-scale PV installations being built at an accelerated pace the need for highly efficient inverters is increasing. A critical step in enabling such solution is the introduction of SiC power devices which are now capable of handling megawatt-scale loads while operating at higher frequencies with significantly reduced losses. This results in a simple two-level air-cooled inverter design with improved system reliability in harsh desert conditions (typical to such large PV farm installations). In this paper, the design and demonstration of the world's most efficient megawatt-scale solar inverter built on GE's advanced SiC MOSFET technology is presented. Experimental results shown here were obtained from a fully operational, field deployed pilot megawatt-scale SiC inverter system.

1. Introduction

With the proliferation of utility-scale photovoltaic (PV) energy generation systems, deploying more efficient and robust inverters becomes increasingly important. Existing IGBT based solutions, while being above 98% efficient, suffer from stringent thermal management requirements, larger size, and insufficient reliability. The efficiency is particularly low during partial load conditions. These design challenges can be effectively addressed by the introduction of SiC power devices. The objective of the work presented in this paper is to look at the system design as a whole and optimize it in order to maximize the benefits of the SiC technology; and not to simply replace standard Si IGBTs with SiC switches. This approach allowed us to improve overall system performance, which resulted in higher power, higher efficiency, and smaller footprint.

Optimizing the system-level design is a multi-stage process that started with modifying the PV inverter topology from a typical three-level [1] to a simpler two-level design utilizing the higher voltage and higher switching frequency capability of SiC [2]. This reduced the count of components and simplified the control, which has positive implications on the higher long-term reliability of the inverter.

In addition to lower parts count, the fact that SiC can run at higher junction temperature and has lower losses compared to the Si IGBT [3], allowed us to replace liquid cooling with air cooling thereby simplifying the system and reducing costs.

As the final optimization step, use of the body diode of the SiC MOSFET in switching application, as discussed in our previous paper [4], showed similar room-temperature switching losses as compared to using antiparallel SiC Schottky diode. While a further optimization of this concept is needed due to the increase of MOSFET turn-on losses at elevated temperatures (originated from the upper body diode recovery), the use of the SiC MOSFET body diode provides significant cost and size benefits. Additionally, results of long-term switching converter testing suggest reliable SiC MOSFET body diode operation in hard switched applications.

Figure 1 and 2 shows the drain-source voltage versus diode or channel current of SiC MOSFET. Black curve on Figure 2 refers to the typical on-state characteristics in third quadrant of GE SiC MOSFET when body diode is conducting, and the blue curves correspond to the MOSFET channel current, as underlined in Figure 1. This device offers symmetric reverse conduction (RC) with positive gate bias in the third quadrant generally referred to as synchronous rectification. It is seen that when module operates in the third quadrant with gate voltage of 20V, the conduction loss is much lower than the body diode drop.

Since SiC-MOSFETs have reverse conduction capability [5], an inverter module can be constituted without freewheeling diodes (FWDs), resulting in a cost effective and power dense solution.

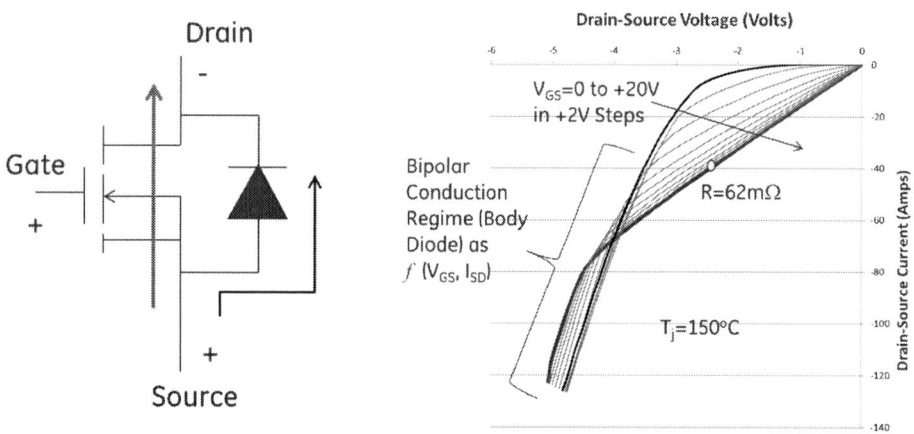

Figure 1: Diode and channel current

Figure 2: Typical on-state characteristics in third quadrant of GE SiC MOSFET chip

Another key SiC MOSFET characteristics is fast switching (which will be shown in the next section through the double pulse testing) thus enabling high switching frequency and reduced filter size and cost.

2. SiC multi-chip module

A half-bridge module is designed using GE's SiC MOSFET chips in an industry standard housing. In order to extract maximum switching performance, the module incorporates a low-inductance, laminated internal bus bar with patent-pending design. The DC terminal inductance is 4.5nH. An intelligent gate drive, optimized for fast switching, robust EMI/EMC performance and short-circuit protection, enables optimal switching performance of the module. Multiple modules can be switched synchronously with less than 1ns jitter, resulting in almost perfect dynamic sharing of currents and voltages. This enables easy scaling for higher power system applications.

a.   Main features
- Built on highly reliable GE SiC MOSFET devices
- Six die per switch
- Low Rds(ON) (6.7mΩ at 150°C) and low inductance (4.5nH) for ease of paralleling and fast switching
- Ultra-low switching losses up to 175°C
- Utilization of body diode during hard switching

b.   Benefits
- Very rugged performance
- Optimized internal design for low inductance and fast switching
- Excellent current sharing
- Easily integrated with liquid or air cooled systems
- Industry-standard footprint, flexible design

Double-pulse inductive switching tests at elevated temperature using the half bridge module are shown in Figure 3. The MOSFET's body diode is used here as the freewheeling diode. These characteristics depict fast switching with low losses (Eon=23.6mJ and Eoff=17.3mJ per module) with reasonable overshoots in voltage and current. Excellent static and dynamic current sharing with three modules in parallel is also observed, which indicates possibility of further scalability in current rating.

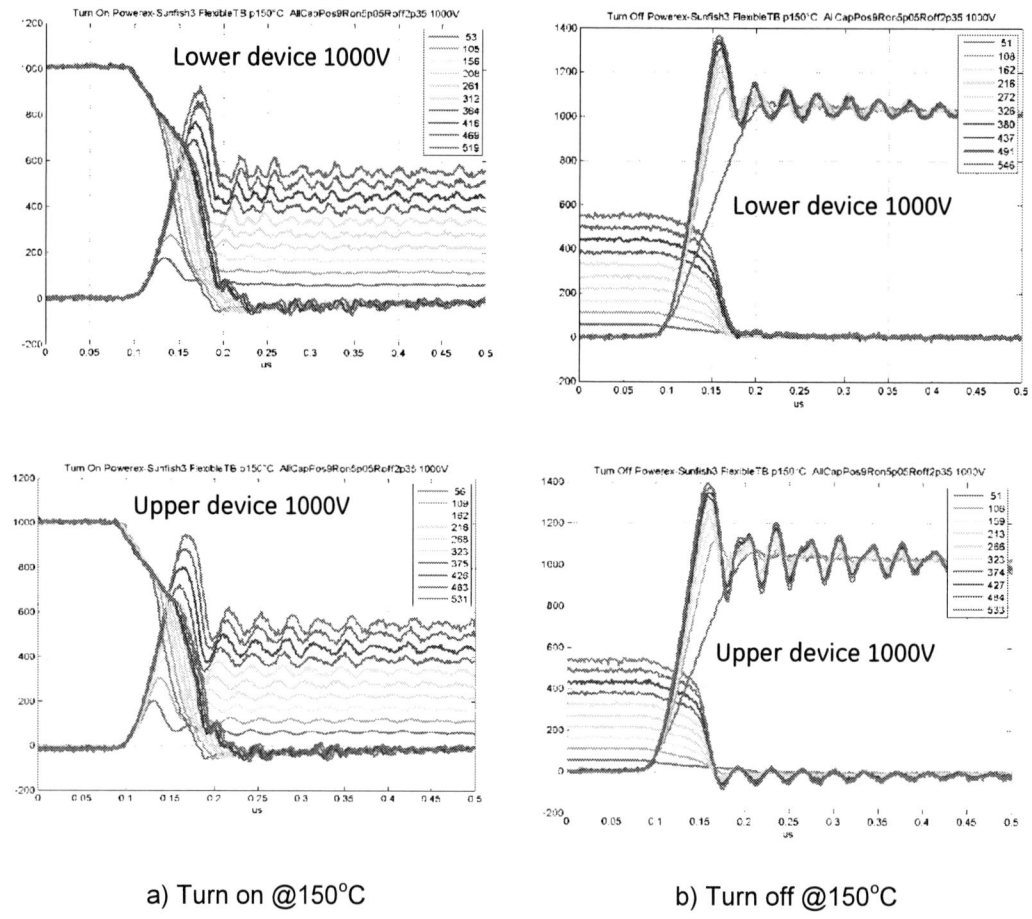

a) Turn on @150°C   b) Turn off @150°C

Figure 3: Double-pulse inductive switching of the 1.7kV, 500A SiC MOSFET module performed at 1000V, 150°C. The multi-plots show switching waveforms as a function of load current.

3. PWM operation in three-phase configuration

In order to demonstrate high switching frequency operation of SiC, three modules were configured to operate in parallel per phase. One inverter bridge comprising nine modules was installed in a forced air-cooled heat sink. Figure 4 shows the arrangement of the bridge.

© VDE VERLAG GMBH · Berlin · Offenbach

Figure 4: Three-phase solar converter layout

The testing was performed at room temperature in a typical pump-back configuration using a passive diode rectifier at the input and an AC source at the output terminals. At a current output of 600A rms, the device junction temperature was measured using a fiber optic sensor as shown in Figure 5. At this operating point a maximum temperature of only 120^0C was recorded, demonstrating a large margin to the maximum junction temperature rating of 175^0C. The measurement performed was also used to verify the thermal model.

Figure 5: Real time optical junction temperature

Besides the low switching loss, the low junction temperature observed was also achieved by excellent current sharing between the paralleled switches. The modules have been engineered for very low inductance and they exhibit outstanding current sharing, both static and dynamic, without any matching impedances. This is enabled by SiC MOSFET's positive temperature coefficient, symmetrical low inductance commutation loops comprising the modules and busbars, symmetrical impedances of individual gate layouts, matched propagation delay times for control and gate driver circuits.

Figure 6 shows the current sharing within two paralleled power modules.

© VDE VERLAG GMBH · Berlin · Offenbach

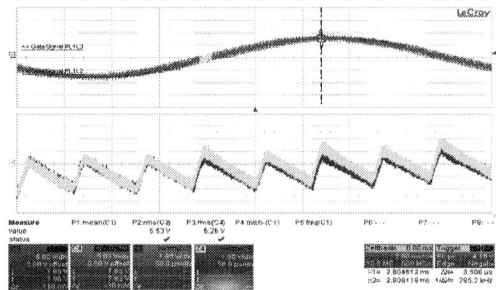

Figure 6: Current sharing

4. 1 MW Solar Converter Operation and Efficiency

In order to validate the performance in a 1MW solar inverter, two such bridges were integrated into a fully air cooled bridge assembly using a total of 18 modules. Figure 7 shows the whole cabinet assembly.

Figure 7: 1MW solar converter assembly

The next step was to measure total system efficiency which accounts for all electrical losses, including input and output filtering components, cooling fans and control power. The experimental results are highlighted in Figure 8, showing the system efficiency as a function of MPPT DC input voltage and output power. Efficiency curve for the lower voltage range exceeds the 99% point. The CEC efficiency results approach 99.0%. Such high efficiency dramatically simplifies the cooling system.

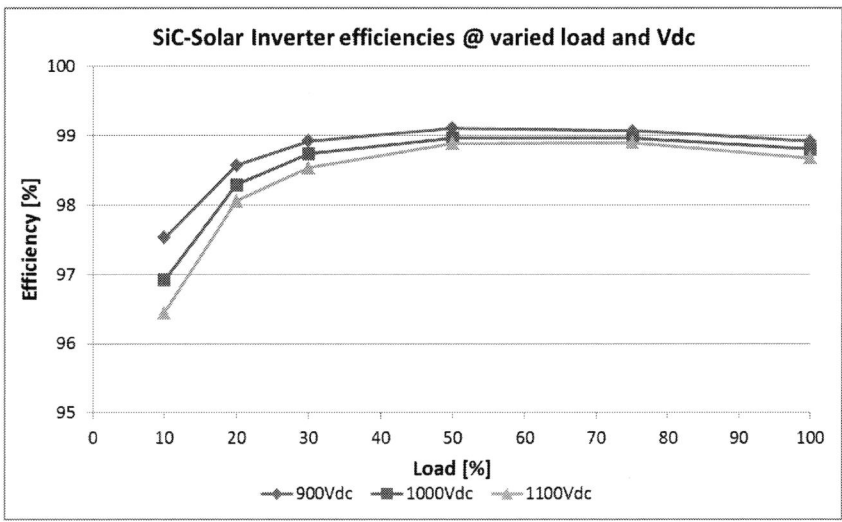

Figure 8: Efficiency test results

5. Future development

By utilizing the next generation of SiC modules (currently in production) the output power can be optimized to fit PV plant requirements. Figure 9 shows one of the SiC half-bridge modules with 6 chips per switch. As can been seen from a figure there is a room to place more die and increase the power ratting of the module resulting in smaller R_{dson} and smaller conduction loss of the converter in general. This will increase efficiency even further.

Figure 9: SiC half-bridge module

In summary, the 1MW solar inverter demonstrates the best-in-class efficiency using an air cooled design with simpler topology and controls. The filter size is reduced due to higher switching frequency, resulting in a smaller footprint. The overall system optimization also results in a lower cost solution.

6. References:

[1] J. Pou, R. Pindado, D. Boroyevich, P. Rodríguez, "Evaluation of the Low-Frequency Neutral-Point Voltage Oscillations in the Three-Level Inverter", IEEE Transactions on Industrial Electronics, Vol. 52, no. 6, Dec 2005.

[2] A. Bolotnikov, P. Losee, A. Permuy, G. Dunne, S. Kennerly, B. Rowden, J. Nasadoski, M. Harfman-Todorovic, R. Raju, F. Tao, P. Cioffi, F. J. Mueller, Lj. Stevanovic, "Overview of 1.2kV – 2.2kV SiC MOSFETs targeted for industrial power conversion applications", 30th Annual IEEE APEC 2015, pp. 2445 – 2452

[3] J. Glaser et al., "Direct Comparison of Silicon and Silicon Carbide Power Transistors in High-Frequency Hard-Switched Applications," 26th Annual IEEE APEC, March 2011, pp. 1049-1056.

[4] A. Bolotnikov, J. Glaser, J. Nasadoski, P. Losee, S. Klopman, A. Permuy, L. Stevanovic, "Utilization of SiC MOSFET body diode in hard switching applications," Materials Science Forum, Vols. 778-780, pp. 947-950, 2014.

[5] Zheng Chen, Yiying Yao, Wenli Zhang, D. Boroyevich, Khai Ngo, P. Mattavelli, and R. Burgos, "Development of a 1200V, 120 A SiC MOSFET module for high-temperature and high frequency applications," WiPDA, 2013, pp. 52–59.

PCIM Europe 2016, 10 – 12 May 2016, Nuremberg, Germany

A performance comparison of a 650 V Si IGBT and SiC MOSFET inverter under automotive conditions

Teresa Bertelshofer, Roman Horff, Andreas März, Mark-M. Bakran
University of Bayreuth, Department of Mechatronics, Center of Energy Technology
Universitätsstr. 30, 95447 Bayreuth, Germany
teresa.bertelshofer@uni-bayreuth.de

Abstract

This paper researches the performance benefits of replacing the Si IGBTs and Si PIN-diodes of a 650 V/ 100 kW 3-phase power module for automotive drives with SiC MOSFETs. For this purpose the maximum current density of the devices and their losses during a load cycle are evaluated. It will be shown, how much the chip area and the power losses, especially under partial load, can be reduced by substituting the Si power semiconductors with the SiC MOSFETs.

1. Introduction

There are already several papers which benchmark Si- and SiC-based power semiconductors, e.g. for 1.2 kV or 1.7 kV power modules or discrete devices [1-4]. With the recent commercial availability of 650 V SiC MOSFETs it becomes very interesting to establish their performance in comparison to a 650 V Si-inverter module [5] already used in electric and hybrid vehicles. The Si and SiC module should have the same target current rating, in this case 800 A. In the 1.2 kV voltage class there are several published works on the benefits of full-SiC modules with this current rating [6, 7].

The major benefit of SiC is its low switching losses. However for the main drive in an automotive inverter a relatively low switching frequency of f_{sw} = 10 kHz is applied. Therefore it will be shown, that the conduction losses can be significantly reduced as well. Especially in the partial-load operational area of an inverter it is expected, that the SiC MOSFETs grant a considerable saving in conduction losses.

Two different chip generations of 650 V SiC MOSFETs are used in this investigation, a planar SiC MOSFET [8] and a trench MOSFET [9]. These are compared to each other as well as to the Si IGBT.

2. Evaluation Basis

The most important step to fairly compare a Si-inverter module with a discrete SiC MOSFET is to convert the thermal and electrical characteristics of the discrete device to the chip area and package of a module. With the information of the active and total chip area of the discrete MOSFET and one switch of the inverter module (consisting of several paralleled IGBT and diode chips) the conduction and switching characteristics of the SiC MOSFET can be adjusted.

In case of the SiC MOSFET, all performance evaluations are conducted under the premise, that no additional free-wheeling SiC Schottky diode is used, only the body diode (in contrast to the Si IGBT, which inherently needs an external diode) [10]. To reduce losses during free-wheeling, synchronous rectification is used. With the applied switching frequency of f_{sw} = 10 kHz the necessary dead time to prevent a short circuit is small compared to the periodic time T_{sw}. Accordingly the high conduction losses occurring, when the current can only flow through the body diode, can be neglected.

For the IGBT and its diode the calculation of conduction and switching losses during SPWM (averaged over a sinusoidal current wave) in dependency on collector current I_C, DC voltage U_{DC} and junction temperature T_J is performed analog to [11].

© VDE VERLAG GMBH · Berlin · Offenbach

653

$$P_{\text{cond,IGBT}} = \left(\frac{1}{2\pi} + \frac{m \cdot \cos\varphi}{8}\right) \cdot U_{\text{CE0}}(T_{\text{J}}) \cdot \hat{I} + \left(\frac{1}{8} + \frac{m \cdot \cos\varphi}{3\pi}\right) \cdot r_{\text{CE}}(T_{\text{J}}) \cdot \hat{I_{\text{C}}}^2 \tag{1}$$

$$P_{\text{sw,IGBT}} = f_{\text{sw}} \cdot \left(E_{\text{on}}(T_{\text{J}}, I_{\text{C,eff}}, U_{\text{DC}}) + E_{\text{off}}(T_{\text{J}}, I_{\text{C,eff}}, U_{\text{DC}})\right) \cdot \frac{\sqrt{2}}{\pi} \tag{2}$$

$$P_{\text{cond,diode}} = \left(\frac{1}{2\pi} - \frac{m \cdot \cos\varphi}{8}\right) \cdot U_{\text{F0}}(T_{\text{J}}) \cdot \hat{I_F} + \left(\frac{1}{8} - \frac{m \cdot \cos\varphi}{3\pi}\right) \cdot r_{\text{F}}(T_{\text{J}}) \cdot \hat{I_F}^2 \tag{3}$$

$$P_{\text{sw,diode}} = f_{\text{sw}} \cdot E_{\text{rec}}(T_{\text{J}}, I_{\text{F,eff}}, U_{\text{DC}}) \cdot \frac{\sqrt{2}}{\pi} \tag{4}$$

For the IGBT this is a sufficiently accurate approximation, since its switching losses show a linear dependency on the collector current. However this is not the case for SiC MOSFETs. Fitting measurement data shows a quadratic dependency on the drain current as well as an offset caused by charging and discharging losses of the output capacity C_{oss}, which are higher than for Si IGBTs due to the thinner depletion layer. Therefore the calculation of the mean switching losses during a sinusoidal current wave $i(t)$ (with k samples) has to be adjusted by a current dependent correction factor cf. The turn-on losses include the recovery losses of the body diode, which have an almost negligible share in the overall losses, because the SiC body diode features a very small recovery charge. Furthermore, unlike the IGBT, the SiC MOSFET losses exhibit only a minor temperature dependency with a slightly negative temperature coefficient.

$$P_{\text{sw,MOS}} = \frac{1}{k}\sum_1^k E_{\text{sw}}(i(t_k)) \cdot f_{\text{sw}} = \left[\left(E_{\text{on+rec}}(I_{\text{D,eff}}, U_{\text{DC}}) + E_{\text{off}}(I_{\text{D,eff}}, U_{\text{DC}})\right) \cdot \frac{\sqrt{2}}{\pi} \cdot f_{\text{sw}}\right] \cdot cf(I_{\text{D,eff}}) \tag{5}$$

The conduction losses during forward and reverse conduction are:

$$P_{\text{cond,MOS}} = \frac{1}{4} \cdot r_{\text{DS,on}}(T_{\text{J}}) \cdot \hat{I_D}^2 \tag{6}$$

Since synchronous rectification is used the channel carries both the forward and the major share of the reverse current. Thus the conduction losses are not dependent on power factor $\cos\varphi$ and modulation index m. It must be kept in mind that the equations (1-6) are only valid, as long as the junction temperature T_{J} remains nearly constant during conducting the sinusoidal current (permissible with $f_{\text{el}} \geq 50$ Hz) and the dead time can be neglected.

The switching losses of the Si IGBT in the module and of the SiC MOSFETs are measurement results. The discrete SiC MOSFETs' losses were identified in an adapted test setup, which represents the situation in a power module with low stray inductance L_σ. This scaling of operational conditions lead to a substantial increase in switching losses compared to the datasheet values. However, the correct scaling and test setup parameters are dependent on the achievable current density.

3. Static performance analysis: maximum current density

As a first indication of the possible reduction of the chip area when using SiC MOSFETs, the maximum current density $J_{\text{eff,max}}$ for a given temperature of cooling fluid T_{cool}, maximum allowed junction temperature $T_{\text{J,max}}$ and specified switching frequency $f_{\text{sw}} = 10$ kHz is calculated.

As a first step the electrical characteristics of the MOSFET are adjusted to the same chip size as the IGBT in the module by arithmetically paralleling n discrete devices.

$$n = \frac{A_{\text{active; IGBT}}}{A_{\text{active; discrete SiC MOS}}} \tag{7}$$

This evidently reduces the on-state resistance compared to that of one discrete device.

$$P_{\text{cond,MOS}} = \frac{1}{4} \cdot \frac{r_{\text{DS,on,discrete}}}{n} \cdot \hat{I_D}^2 \tag{8}$$

Fig. 1 shows the temperature dependent output characteristics of a Si IGBT and both generations of SiC MOSFETs (all having the same active chip area). As a reference the values of the mean and maximum output current of a load profile, which will later be discussed in detail, are marked. The planar Gen2 SiC MOSFET shows a significantly higher temperature dependency than the Si IGBT, however it profits from its ohmic output characteristic. Even for high temperatures the SiC MOSFET provides a lower on-state voltage than the Si IGBT over a wide current range. The next generation of SiC MOSFETs (Gen3, which is not yet commercially available) alters the gate structure from planar to trench, which offers an even better output characteristic (reduction of R_{DSon} by 50%) and a low temperature dependency [9].

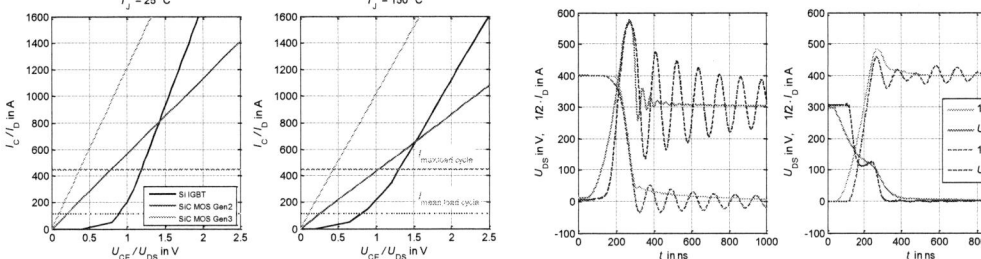

Fig. 1: Output characteristic of Si IGBT and SiC MOSFET of same chip size (two chip generations)

Fig. 2: Switching waveforms of Si IGBT and SiC MOSFET @ T_J = 25°C, U_{DC} = 300 V, I_C & I_D = 800 A

The influence of n on the switching losses is more complicated, since their dependence on the current is non-linear. For high module currents $I_{D,eff}$ the overall switching losses are reduced when using a higher number of paralleled devices, however for small currents they increase because of the constant offset k_3, representing the charging losses of C_{oss}.

$$P_{sw,MOS} = n \cdot E_{sw,discrete} \cdot \frac{\sqrt{2}}{\pi} \cdot f_{sw} \cdot cf = n \cdot \left[k_1 \cdot \left(\frac{I_{D,eff}}{n} \right)^2 + k_2 \cdot \left(\frac{I_{D,eff}}{n} \right) + k_3 \right] \cdot \frac{\sqrt{2}}{\pi} \cdot f_{sw} \cdot cf$$

$$P_{sw,MOS} = \frac{1}{n} \cdot k_1 \cdot I_{D,eff}^2 + k_2 \cdot I_{D,eff} + n \cdot k_3 \tag{9}$$

Additionally the switching losses are limited by the parasitics in the commutation loop and guidelines regarding the motor isolation. The datasheet values of the discrete devices inherently refer to single chip measurements with low di_D/dt and therefore low overvoltage peaks. This allows an aggressive gate circuit with zero external gate resistors (only the internal gate resistor limits the time constant of the gate drive loop).

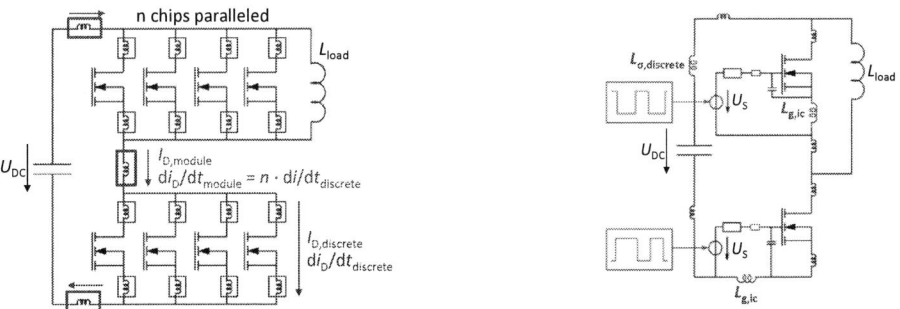

Fig. 3: Influence of parasitic inductances in a power module   Fig. 4: Adapted double pulse test

However, in a power module some parasitic inductances share the same commutation loop (see Fig. 3). These inductances induce overvoltage peaks according to:

$$\Delta U = n \cdot di_{D,discrete}/dt \cdot L_{\sigma,module} \tag{10}$$

To accurately measure the switching losses of one chip within a module, these overvoltage peaks must be emulated in a standard double pulse test. To that extent an external stray inductance is inserted in the commutation loop according to:

$$L_{\sigma,discrete} = L_{\sigma,module} \cdot \frac{I_{Dmax,module}}{I_{Dmax,discrete}} \triangleq n \cdot L_{\sigma,module} \tag{11}$$

© VDE VERLAG GMBH · Berlin · Offenbach

The gate resistors were chosen in a way, that at maximum current the overvoltage peak of the MOSFET at $T_J = 25°C$ and the overvoltage peak of the body diode at $T_J = 150°C$ do not exceed the maximum drain source voltage of $U_{BR,DSS} = 650$ V. To avoid the need to increase the gate resistors to a level, where the caused switching losses are too high, an inversely coupled gate loop is used (see Fig. 4). The gate inductance for the inverse coupling $L_{g,ic}$ decreases the switching speed the most, when the overvoltage could be critical, namely at high current gradients. At low current gradients the switching process is only slightly affected.

To model the thermal behaviour of the SiC MOSFETs in a module, the Foster network given in [5] is applied, since IGBT and MOSFET are scaled to the same chip size. This slightly penalises the SiC MOSFET, since in this calculation it is not included, that SiC has a higher thermal conductivity than Si.

In the following steady-state calculation of the maximum current density the thermal capacities can be omitted and only the overall thermal resistance from junction to cooling fluid is relevant. The maximum current density is defined as:

$$J_{max,eff}(f_{sw}) = \frac{I_{max,eff}(f_{sw})}{A_{chip}}; \quad P_{cond}(I_{max,eff}) + P_{sw}(I_{max,eff}) = \frac{T_{J,max} - T_{cool}}{R_{th}} \tag{12}$$

Since the MOSFET carries both forward and reverse current in its channel, whereas the IGBT carries only the forward current, the maximum current density of the IGBT is obtained by evaluating the maximum current and the total chip area of both IGBT and free-wheeling diode of the switch.

The operating point, where the current density is calculated, is derived from a load cycle (see Fig. 9): **$U_{DC} = 380$ V, $\cos\varphi = 0.9$, $m = 0.5$;**

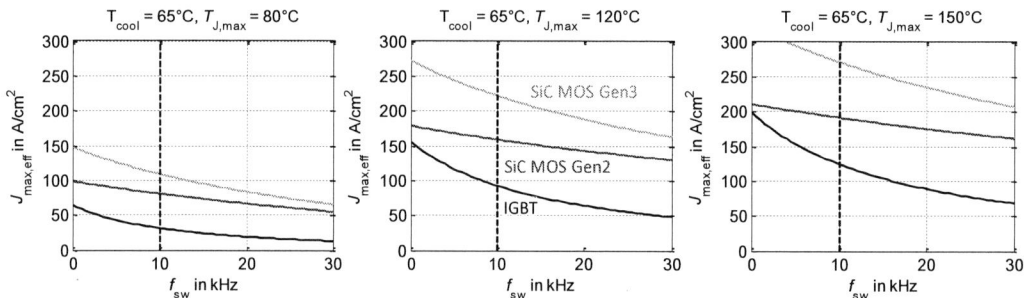

Fig. 5: Maximum current density $J_{max,eff}$ of SiC MOSFET (two chip generations) and Si IGBT for different maximum junction temperatures

Fig. 5 shows the maximum current density of Si IGBT and SiC MOSFETs for different maximum junction temperatures ($T_{J,max} = 80...150°C$). Knowing the ratio between $J_{max,rel}$ of Si IGBT and SiC MOSFET the necessary relative chip size of the MOSFETs to achieve the same current rating as the IGBT can be calculated in dependency of f_{sw} and $T_{J,max}$ (100 % ≙ $A_{ges,IGBT} + A_{ges,diode}$) (see Fig. 6).

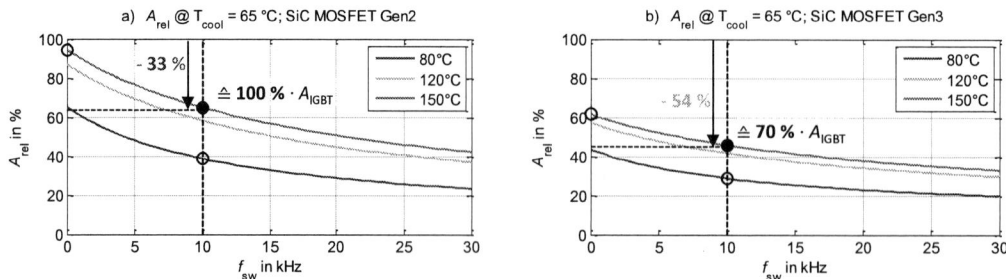

Fig. 6: Required relative chip area of SiC MOSFET compared to Si IGBT with free-wheeling diode

The aforementioned better performance of SiC during partial load is represented by a low junction temperature ($T_{J,max}$ = 80°C) in Fig. 5 and Fig. 6. At a switching frequency f_{sw} = 10 kHz the SiC MOSFETs exhibit a possible current density more than twice as high as the Si IGBT's. Even at high currents (full load) and without the advantage of smaller switching losses ($T_{J,max}$ = 150°C, f_{sw} = 0 kHz) the Gen2 MOSFET still features the same power density as the IGBT. Dimensioning the SiC chip area for full load ($T_{J,max}$ = 150°C, f_{sw} = 10 kHz), it equals the chip area of the Si IGBT without its diode (A_{rel} = 66% $\triangleq$ 100% · $A_{ges,IGBT}$ in Fig. 6). The Gen3 SiC MOSFETs permit an even higher reduction in chip area due to their low on-state resistance $R_{DS,on}$.

Dimensioning the MOSFET chip area in a way that both semiconductors can bear the same current as the IGBT at $T_{J,max}$ = 150°C results in a ratio of SiC MOSFET and Si IGBT losses at full load of 100% and 70% respectively for Gen2 and Gen3, since this is the necessary relative chip size found in Fig. 6. However in the following comparison not only the Si IGBT's losses are taken into account, but also the losses of its required free-wheeling diode. Additionally in this steady-state calculation it is included, that at low currents the junction temperature is set to T_{amb} = 25°C and at full load it rises to $T_{J,max}$ = 150°C.

Fig. 7a) shows the resulting ratio of losses of Si IGBT (with diode) and SiC MOSFET for the dimensioning according to Fig. 6. As a reference the mean and maximum currents $I_{eff,mean}$ and $I_{eff,max}$ in a load profile are marked. Again the better performance in partial load becomes obvious (P_{SiC}/P_{Si} @ $I_{mean,load\ cycle}$ ≈ 30%). Fig. 7b) analyses the distribution of switching and conduction losses: Whereas the switching and conduction losses are split approximately equally among the Si semiconductors ($P_{sw}/P_{sw+cond}$ ≈ 50-60%), the SiC MOSFETs are clearly dominated by conduction losses at full load, while the switching losses have the most influence in partial load. Reducing switching losses in order to improve the total performance becomes increasingly more important in newer chip generations, which feature lower on-state resistances $R_{DS,on}$.

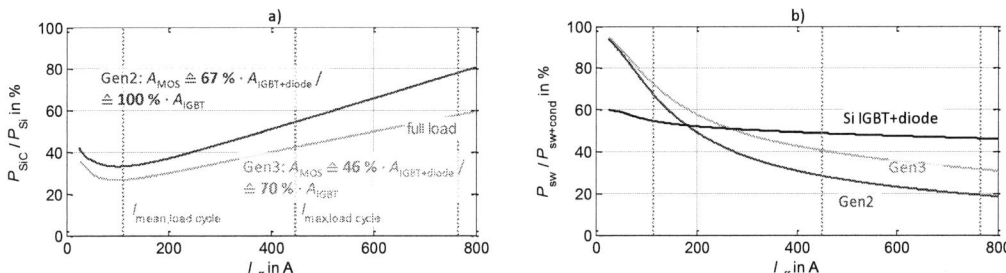

Fig. 7: a) Ratio of SiC MOSFET overall losses regarding Si overall losses (IGBT+diode); b) Ratio of switching losses regarding overall losses

Fig. 8a) shows the ratio of the conduction losses of the SiC MOSFET with regard to the conduction losses of IGBT and diode. Over a wide current range the SiC MOSFETs show lower conduction losses due to their ohmic output characteristic without the IGBTs knee-voltage of U_{CE0} = 0.7 V (see also Fig. 1). Fig. 8b) analyses the reduction in switching losses: Starting at $P_{SiC,sw}/P_{Si,sw}$ >> 100% for I_{eff} = 0 V (the SiC MOSFETs cause higher charging and discharging losses, because $C_{oss,SiC}$ > $C_{oss,Si}$) they quickly drop to 30...40%. This reduction is possible, because SiC MOSFETs are unipolar, show neither current nor voltage tail and their body diode has very little reverse recovery charge (see Fig. 2). A further reduction of E_{sw} by increasing di_D/dt and du_{DS}/dt is limited by the permissible voltage peak ΔU_{max} = $U_{BR,DSS}$ − U_{DC} and the maximum du/dt, if one wishes to meet the standard limit for motor isolation of about 10 kV/μs. Comparing SiC MOSFET Gen2 and Gen3 in Fig. 8b) one can clearly see the different influences of factors k_1, k_2 and k_3 (see (9)): Gen3 allows for a smaller chip size than Gen2, resulting in lower charging losses of C_{oss} (represented by k_3) and therefore smaller switching losses at low currents. At high currents however it is affected by the SiC MOSFETs' quadratic dependence of E_{sw} on I_D (represented by k_1): Each single chip has to bear more current which causes disproportionately higher switching losses.

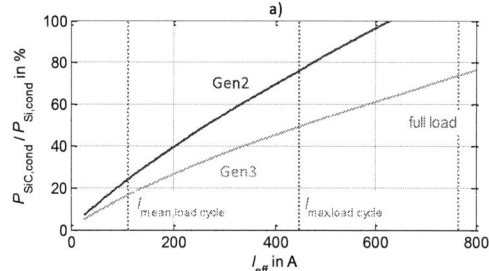 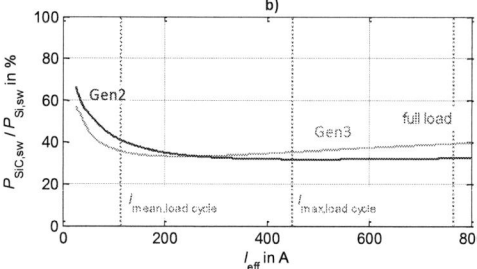

Fig. 8: a) Ratio of SiC MOSFET conduction losses regarding Si conduction losses (IGBT+diode); b) Ratio of SiC MOSFET switching losses regarding Si switching losses (IGBT+diode);

4. Performance during load cycle

Now the step from a steady-state performance to a dynamic loss and temperature analysis during a load cycle (Fig. 9) should be done. Important results in this analysis are the mean power loss, which indicates, how much cooling power must be installed, the maximum junction temperature, which specifies, whether the installed chip area is used to full capacity, and the resulting efficiency of the inverter.

The calculation of the power loss is still conducted with the aforementioned formulas in (1-9). Noteworthy is the simplification in this formulas, that the junction temperature remains approximately constant during conducting one sinusoidal current wave. This is valid only for electrical frequencies $f_{el} > 50$ Hz. However special operating points, where this assumption is not valid, e.g. during starting the vehicle, must be considered as well.

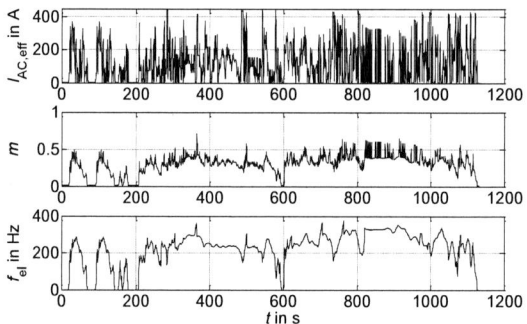

Fig. 9: Load profile of a hybrid vehicle

For this purpose the chip temperature is calculated with a Foster model with 5 R_{th}-C_{th}-elements. If the highest occurring junction temperature is lower than a given value for maximum T_J, e.g. 150°C, it is considered to minimise the semiconductor area. This has two effects: For one the thermal resistance R_{th} increases, reducing the possible heat flow and increasing the chip temperature. At the same time the thermal capacity C_{th} decreases, so the same heat flow into a layer causes a higher rise in temperature. Additionally a smaller chip area leads to a higher on-state resistance $R_{CE,on}$ or $R_{DS,on}$ causing higher conduction losses and therefore again a higher chip temperature.

Analog to Fig. 6 it is expected, that the chip area can be significantly reduced using SiC MOSFETs. However since the Si power module in these comparisons is also overdimensioned for this particular load profile, its chip size is optimised too and the necessary chip size of the SiC MOSFETs must be compared to this optimised value as well.

In Fig. 10 it is evaluated, how much chip area is necessary to drive the load cycle and how much reduction in mean power losses can still be achieved. The area optimisation of Si IGBT and its diode starts at 100% ($\triangleq A_{ges,IGBT} + A_{ges,diode}$), the optimisation of the SiC area starts at 67% and 46% for Gen2 and Gen3 respectively. Here the maximum occurring junction temperature during the load profile is about 100°C. Although the SiC MOSFETs' area is much smaller than the IGBTs and its diode, the mean power losses are reduced by 61 % (Gen2) or

69% (Gen3), since the load profile used in this comparison is dominated by partial load operating areas.

If a maximum junction temperature of 150°C is pursued the chip area of Si IGBT and diode can be reduced by 45%. Compared to this new chip area the SiC MOSFETs enable a further reduction by 27% and 47% for Gen2 and Gen3 respectively. These values deviate slightly from the prognosis of the static performance analysis, because $T_{J,max}$ occurs at a different operation point as was used in the calculation depicted in Fig. 5 to Fig. 8. The area reduction of Si IGBT and diode cause a relatively low rise in mean power loss, since the main part of the conduction losses are caused by the unchanged forward voltage drop U_{CE0} and the switching losses of the IGBT are approximately independent of the chip area. Compared to this the MOSFET losses are clearly influenced by the area reduction. But even then the mean power losses compared to Si IGBT and diode can still be reduced by 50% (Gen2) or 63% (Gen3).

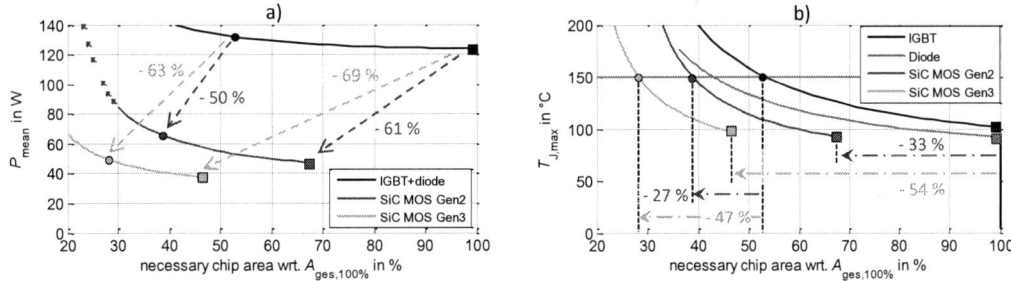

Fig. 10: Load cycle analysis: a) mean power losses and b) highest chip temperature in dependency of reduction of semiconductor area; 100 % $\triangleq A_{ges,IGBT}+A_{ges,diode}$; 66% (starting point MOSFET) $\triangleq A_{ges,IGBT}$;

As mentioned above the maximum junction temperature occurs at a different operation point than was used in the steady-state analysis, namely after a long current plateau caused by regenerative braking. Therefore the Si diodes temperature must be examined as well, but it can be found, that even at an operation point with $\cos\varphi < 0$ the diode is still colder than the IGBT (Fig. 10b). Thus the temperature of the IGBT is the limiting factor for scaling down the chip size. Even at this operation point with $\cos\varphi < 0$ the Si diode and the SiC MOSFETs' body diodes cause lower losses during free-wheeling, since the diodes' switching losses can be neglected compared to the IGBTs'/ MOSFETs' turn-on and turn-off losses. Since the SiC MOSFETs bear both forward and reverse current and dissipate lower losses than the Si IGBT the occurring temperature ripple of the chip is much smaller (Fig. 11). This could be a significant advantage regarding module lifetime [12].

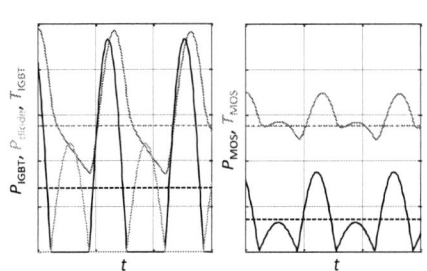

Fig. 11: Temperature ripple of Si IGBT and SiC MOSFET

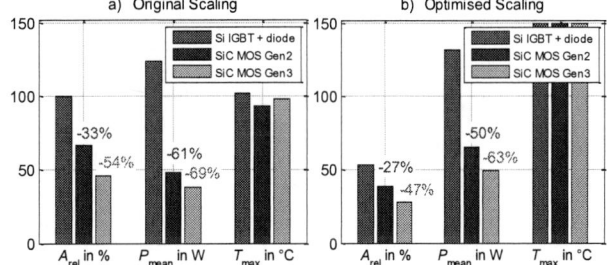

Fig. 12: Power loss and highest chip temperature during load cycle for Si IGBT (+diode) and SiC MOSFET (Gen2+Gen3)

The results of the investigation regarding loss and chip size reduction are summarised in Fig. 12 a and b. The resulting efficiency for the operation points in the load profile are shown in Fig. 13. Compared to the efficiency before optimising the chip area the Si-inverter's efficiency remains mostly the same (since P_{loss} increases only slightly as seen in Fig. 10b), and the SiC-inverters' efficiency only decreases slightly for high currents. Since the load profile is dominated by partial load operating points, using SiC MOSFETs increases the efficiency by ~3%.

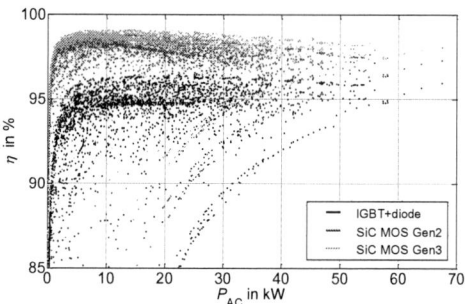

Fig. 13: Efficiency in the load profile after optimising the chip size

5. Conclusion

In this paper the advantages of replacing the Si IGBT and Si diode in a 650 V automotive inverter with SiC MOSFETs, that use their body diode for free-wheeling, were investigated, namely the possible reduction of semiconductor area and overall power losses. Different analyses to quantify the benefits were conducted: a steady-state calculation of the maximum power density and an investigation of efficiency, necessary chip area and occurring mean power losses during a load cycle. Additionally it was shown, how the performance of a SiC inverter can further be improved by using next generation SiC trench MOSFETs: Compared to the Si inverter the mean power losses can be reduced by 63% and the semiconductor area by 47%.

Acknowledgement

This project was supported by the ZF Friedrichshafen AG. Special thanks go to the department of Engineering Electric Mobility & Mechatronics, Auerbach i. d. Opf.

References

[1]   A. Maerz, R. Horff, M. Helsper and Mark-M. Bakran, "Requirements to change from IGBT to Full SiC modules in an on-board railway power supply ", EPE 2015.

[2]   S. Hazra, A. De, S. Bhattacharya, Lin Cheng, J. Palmour, M. Schupbach, B. Hull, and S. Allen, "High switching performance of 1.7kV, 50A SiC power MOSFET over Si IGBT for advanced power conversion applications", IPEC 2014.

[3]   J. Biela, M. Schweizer, S. Waffler and J. W. Kolar, "SiC versus Si-Evaluation of Potentials for Performance Improvement of Inverter and DC-DC Converter Systems by SiC Power Semiconductors", IEEE Transactions on Industrial Electronics, vol. 58, no.7, pp. 2872-2882, July 2011.

[4]   T. Zhao, J. Wang, A. Q. Huang and A. Agarwal, "Comparisons of SiC MOSFET and Si IGBT Based Motor Drive Systems", Industry Application Conference 2007.

[5]   Datasheet FS800R07A2E3_B31, www.infineon.com.

[6]   E. Wiesner, K. Masuda, and M. Joko, "New 1200V full SiC module with 800A rated current", EPE 2015.

[7]   R. Wood and T. Salem, "Evaluation of a 1200-V, 800-A All-SiC Dual Module", IEEE Transactions on Power Electronics, vol. 26, no. 9, pp.2504-2511, September 2011.

[8]   Datasheet SCT2120AF, www.rohm.com.

[9]   Press Release, "New Products Under Development 3rd Gen SiC MOSFET", 2014, www.rohm.com

[10] R. Horff , A. Maerz, M. Lechler and Mark-M. Bakran, "Optimised Switching of a SiC MOSFET in a VSI using the Body Diode and additional Schottky Barrier Diode ", EPE 2015.

[11] A. Wintrich, U. Nicolai and W. Tursky, "Applikationshandbuch Leistungshalbleiter", ISLE, 2010.

[12] R. Bayerer , T. Herrmann, T. Licht, J. Lutz and M. Feller, "Model for Power Cycling lifetime of IGBT Modules – various factors influencing lifetime ", CIPS 2008.

PCIM Europe 2016, 10 – 12 May 2016, Nuremberg, Germany

A generic topology for electrical energy storage systems

Christoph Marxgut, Helbling Technik, Germany, christoph.marxgut@helbling.de

Abstract

This paper presents an energy storage system which is aimed for energy recuperation of electrical drives. The topology is based on a combination of a multilevel converter (MLC) and a bidirectional boost converter (BBC). The MLC enables the application of low voltage energy storage components; thus super- or ultracapacitors with arbitrary voltage rating can be utilized. The BBC is applied to limit the MLC current when energy is stored in the cells and also allows to supply the output voltage almost independently on the state of charge of the ultracapacitors. Due to its generic structure, the topology can be adapted to different voltage, power, and energy levels. The paper highlights the benefits and potential drawbacks of the topology, details the design of the MLC, and presents the modulation for the MLC converter stage. The theoretical results are validated on a prototype employing 10 cells with 50F/5.4V each (i.e. 2 Wh in total). The system is designed to deliver a power of 500W for 5s to an 48V voltage bus. The system's efficiency for an entire store-release cycle is 80.1% at a volume of $1.05\,\mathrm{dm}^3$.

1.   Introduction

The increasing awareness and particularly the regulations regarding environmental issues have shifted the focus of almost any industry sector towards an increasingly green production and use of energy. This trend is based on the fact that energy costs, especially the oil prize, have never accounted for its entire life cycle, i.e. the processing and the recycling or disposal of energy have not holistically been included. CO2 taxes, energy standards, and efficiency regulations [1, 2] try to balance the energy prize which affect particularly industry sectors dealing with energy processing, e.g. power electronics. In that context, the trend of increasingly efficient power electronic converters has been apparent in academics and industry over the past decades and the achievable efficiency has been pushed above the 99% boundary [3]. Apart from high converter efficiencies, in modern drive systems, the utilization and recycling of kinetic energy is considered to boost the overall system efficiency. This idea has already been employed in electrical vehicles [4] and with increasingly stringent regulations, kinetic recycling systems could also find application in branches for which it has not been profitable so far, e.g. e-bike, e-scooter, cranes, or elevators.

[5] reports that for commercial cars fuel savings up to 17% are achievable with a 48V recuperation system consisting of a permanent magnet synchronous motor, an inverter, a Lithium-Ion battery, and a dc/dc converter which connects the system to the existing 12V bus. Although Li-Ion cells feature a high energy density (80-170Wh/kg) their power density is comparably low (0.8-2kW/kg) and hence they are not capable to store or deliver the high power peaks during acceleration and breaking of the motor [4]; mechanical breaks must still be employed in order to transform the residual kinetic energy into heat.

Ultracapacitors (UC) are prominent for a high power density (4-10kW/kg) and are available either as low voltage components (e.g. 2.7V, 3.3V) or as modules with voltage ratings up to 160V. Compared to battery cells, UCs do not suffer from limited charging cycles or flammability which identify the battery as a critical element in the life cycle and the safety of a drive system.

© VDE VERLAG GMBH · Berlin · Offenbach

PCIM Europe 2016, 10 – 12 May 2016, Nuremberg, Germany

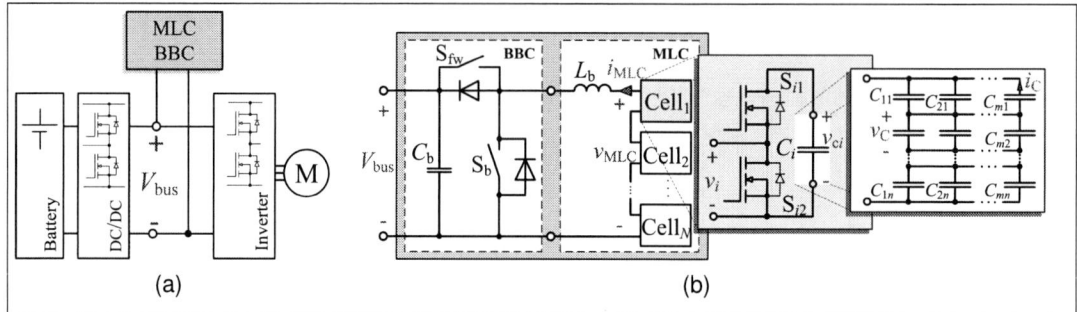

Fig. 1: (a) Overview of the drive system. (b) Schematic of the proposed energy storage system. m, n, and N are three parameters to scale the system regarding the voltage, power, and energy specifications.

However, since the energy density of UCs is rather low (2-30Wh/kg) a combination of Li-Ion batteries and UCs is a promising candidate for a recuperation system with both high power and high energy capability (cf. **Fig. 1(a)**).

[6] proposes a hybrid energy storage system (battery-UC combination) in which the battery is connected to a voltage bus employing a bidirectional dc/dc converter while the UC is directly clamped to the bus. This direct connection, however, limits the application of UCs to modules with the same voltage rating as the voltage bus. Furthermore, since the bus voltage varies in relatively small range around the nominal value (typ. $< \pm 20\%$) the UC is not fully utilized because the majority of the stored energy cannot be released. In order to have the flexibility of designing a customized UC-based energy storage system and to fully utilize the storage capability of the system, this paper presents a generic topology based on a multilevel converter (MLC) that allows to freely scale the system regarding stored energy, applied voltage, and required power rating without being dependent on the available UCs.

2. Topology

As already mentioned, energy recycling systems could be interesting for a variety of applications which might range from a few Wh to several hundred Wh. Since the range of available UCs is rather limited regarding voltage level or storage capacity and because UCs mainly determine size and costs of the system an MLC would be beneficial as it allows to design the system regardless of available storage devices and specified voltage, energy, or power requirements.

The MLC consists of a series connection of cells which are composed of a half-bridge and a capacitor C_i (cf. **Fig. 1(b)**) which can be a combination of several series and/or parallel connected low voltage UCs. Besides, the multicell arrangement enables the use of low voltage switches (S_{i1} and S_{i2}) which feature a low $R_{\mathrm{DS,on}}$ and thus low ON-state losses. Also, the reliability of the system can be improved by an inclusion of redundant cells.

A drawback of the MLC, however, is that the output voltage can only be controlled by enabling or disabling cells which results in an inherent tracking error of the output voltage, $v_{\mathrm{MLC}}(t)$. Furthermore, the MLC current $i_{\mathrm{MLC}}(t)$ cannot be limited and might damage the UCs. Both issues can be addressed by the employment of a bidirectional boost converter (BBC) which is connected between the bus and the MLC (see Fig. 1(b)).

Boost mode

The BBC operates as boost converter when energy needs to be delivered from the MLC to the bus (energy release mode). With it, V_{bus} can be tightly regulated regardless of the system's

© VDE VERLAG GMBH · Berlin · Offenbach

PCIM Europe 2016, 10 – 12 May 2016, Nuremberg, Germany

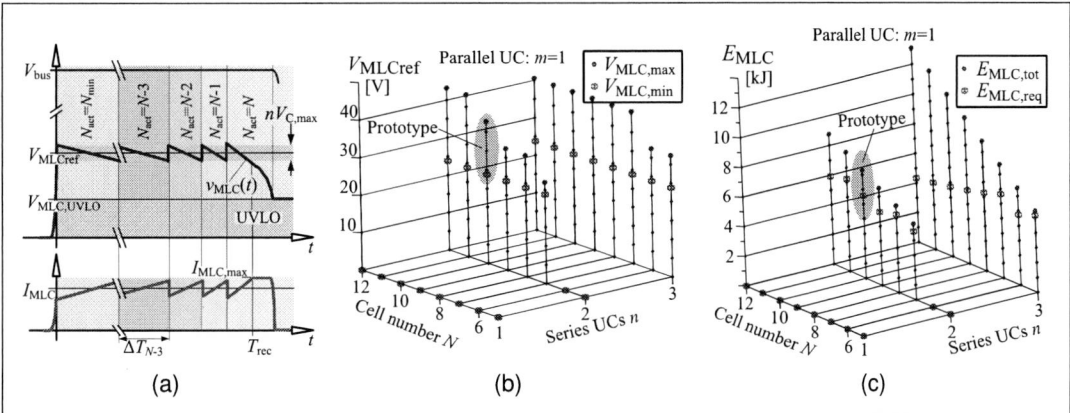

Fig. 2: (a) Management of the MLC output voltage $v_{\mathrm{MLC}}(t)$ when energy is fed back to the bus. (b) Calculation result of the allowed range for the MLC output voltage V_{MLCref}. (c) Minimal required storage capability $E_{\mathrm{MLC,req}}$ and totally stored energy $E_{\mathrm{MLC,tot}}$ as function of n and N ($m = 1$).

state of charge (SOC) and thus the majority of the energy in the cells can be utilized[1]. **Fig. 2(a)** illustrates the cell management when energy is released: $v_{\mathrm{MLC}}(t)$ tracks V_{MLCref} by activating additional cells, i.e. the number of active cells, N_{act}, increases. When all N cells are active $v_{\mathrm{MLC}}(t)$ cannot be kept constant and drops until the UVLO is reached to prevent the depletion of the cells[1].

Current limitation

When energy needs to be stored (energy storage mode), the BBC can be employed to limit the MLC input current $i_{\mathrm{MLC}}(t)$ and, hence, to avoid a damage of UCs or switches. In this mode, the BBC basically operates as current-controlled buck converter. Apart from that, the BBC also allows to disconnect the MLC from the bus in idle mode by turning-OFF S_{fw}.

Design considerations

The system shall be designed to provide the voltage bus with a defined maximum power P_{max} for a specified recuperation time T_{rec}. To meet the specifications, the MLC has several free parameters (cf. Fig. 1(b)):

- the number of cells, N,
- the number of series connected capacitors, n,
- the number of parallel connected UCs, m, and
- the applied UCs which define the maximum capacitor voltage $V_{\mathrm{C,max}}$, the maximum current per UC, $I_{\mathrm{C,max}}$, and the capacitance C.

In the MLC design, for each parameter set (m,n,N,UC) six equations need to be fulfilled and from the resulting solution space a solution can be chosen with respect to a certain goal (circuit complexity, cost, etc.). Also, a thorough system optimization regarding efficiency and/or power density which includes the BBC can be performed; this, however, is not in the scope of this paper.

Firstly, an ultracapacitor needs to be chosen yielding two obvious limitations which state that

---

[1] The cells are discharged until an undervoltage lockout level (UVLO) is reached in order to ensure that each cell can still supply its peripheral electronics (ADC, gate drive, communication).

© VDE VERLAG GMBH · Berlin · Offenbach

the maximum voltage and current of an UC must not be exceeded:

$$v_{\mathrm{C}}(t) \leq V_{\mathrm{C,max}}, \quad \text{and} \quad i_{\mathrm{C}}(t) \leq I_{\mathrm{C,max}}. \tag{1}$$

By consideration of Fig. 1(b), it is apparent that $v_{\mathrm{MLC}}(t)$ must be smaller than V_{bus} to take advantage of the BBC; thus, the reference voltage of $v_{\mathrm{MLC}}(t)$, V_{MLCref}, is limited to that voltage level

$$V_{\mathrm{MLCref}} < V_{\mathrm{bus}}. \tag{2}$$

Since the maximum power is determined by the maximum UC current $I_{\mathrm{C,max}}$, the number of parallel UCs, m, and a certain MLC voltage V_{MLCref}, a lower bound for V_{MLCref} can be defined with

$$P_{\mathrm{max}} \leq V_{\mathrm{MLCref}} \cdot m\, I_{\mathrm{C,max}} \quad \Longrightarrow \quad V_{\mathrm{MLCref}} \geq V_{\mathrm{MLC,min}} = \frac{P_{\mathrm{max}}}{m \cdot I_{\mathrm{C,max}}}. \tag{3}$$

Fig. 2(a) illustrates a simplified discharge profile (energy release mode) that assumes that all inactive cells are fully charged. Each time the number of active cells, N_{act}, is increased $v_{\mathrm{MLC}}(t)$ rises immediately with an offset of $n V_{\mathrm{C,max}}$ followed by the relatively slow discharge process. The discharge duration ΔT_i for a given N_{act} can easily be calculated with

$$C \cdot \frac{\mathrm{d}v(t)}{\mathrm{d}t} = i(t) \quad \Longrightarrow \quad \Delta T_i = C_{\mathrm{act}} \cdot \frac{n V_{\mathrm{C,max}}}{m \cdot I_{\mathrm{C}}} - C \cdot \frac{V_{\mathrm{C,max}}}{N_{\mathrm{act}} \cdot I_{\mathrm{C}}}. \tag{4}$$

In (4), C_{act} corresponds to the capacitance of the active MLC cells, i.e. $C_{\mathrm{act}} = m\,C/(n\,N_{\mathrm{act}})$ which causes ΔT_i to decrease with increasing N_{act}. Besides, $I_{\mathrm{C}} = I_{\mathrm{MLC}}/m$ is the discharge current per UC which is assumed to be constant for the sake of lucidity. In order to provide P_{max} during T_{rec} the discharge duration of all successively activated cells need to exceed T_{rec},

$$T_{\mathrm{rec}} \leq \sum_i \Delta T_i = C \cdot \frac{V_{\mathrm{C,max}}}{I_{\mathrm{C}}} \cdot \sum_{i=N_{\mathrm{min}}}^{N} \frac{1}{i}, \quad \text{with} \quad N_{\mathrm{min}} = \mathrm{ceil}\left(\frac{V_{\mathrm{MLCref}}}{n V_{\mathrm{C,max}}}\right). \tag{5}$$

Hereby, N_{min} is the minimal number of cells that need to be turned-ON to obtain a certain V_{MLCref}. Assuming a high V_{MLCref} then many cells need to be initially active whereas only few cells and thus discharge intervals are left for the summation in (5). Therefore, (5) yields another upper limit for V_{MLCref}.

The voltage ripple of $v_{\mathrm{MLC}}(t)$ (cf. Fig. 2(a)) also tightens the requirement for the minimal reference voltage (3)

$$V_{\mathrm{MLC,min}} \geq \frac{P_{\mathrm{max}}}{m \cdot I_{\mathrm{C,max}}} + \frac{n V_{\mathrm{C,max}}}{2}. \tag{6}$$

The final equation verifies that the system has enough energy storage capacity for the given parameter set. Thereby, the total storage capability $E_{\mathrm{MLC,tot}}$ must exceed the energy required to deliver P_{max} during T_{rec} plus the residual energy E_{res},

$$E_{\mathrm{MLC,req}} = \underbrace{\frac{1}{2}\frac{m\,C}{n\,N} \cdot \left(V_{\mathrm{MLCref}} - \frac{n V_{\mathrm{C,max}}}{2}\right)^2}_{E_{\mathrm{res}}} + \underbrace{P_{\mathrm{max}} \cdot T_{\mathrm{rec}}}_{E_{\mathrm{Pmax}}} \leq \underbrace{\frac{1}{2} n\,m\,N\,C \cdot V_{\mathrm{C,max}}^2}_{E_{\mathrm{MLC,tot}}}. \tag{7}$$

As can be concluded by inspection of (7), the requirement yields an additional upper bound for V_{MLCref} because for a given parameter set all variables in (7) but V_{MLCref} are constant. The derivation of E_{res} can easily be seen from Fig. 2(a): when T_{rec} has passed and all cells are active $v_{\mathrm{MLC}}(t)$ equals $V_{\mathrm{MLCref}} - n V_{\mathrm{C,max}}/2$ which defines E_{res}. This energy typically cannot be used

PCIM Europe 2016, 10 – 12 May 2016, Nuremberg, Germany

Parameter set for the analysis		System specifications	
Parallel UCs	$m = 1$	Parallel UCs	$m = 1$
Series UCs	$n = 1 \dots 3$	Series UCs	$n = 2$
Cell number	$N = 5 \dots 12$	Cell number	$N = 10$
Max. UC voltage	$V_{C,max} = 2.7V$	Max. energy	$E_{max} = 7.3kJ$
Max. UC current	$I_{C,max} = 21A$	Cell capacitance	$C_i = 50F/5.4V$
UC capacitance	$C = 100F$	BBC switching freq.	$f_{sw} = 100kHz$
Bus voltage	$V_{bus} = 48V$	Inductance	$L_b = 10\mu H$
Max. power	$P_{max} = 500W$	Min. MLC voltage	$V_{MLC,min} = 23.8V$
Recuperation time	$T_{rec} = 5s$	Max. MLC voltage	$V_{MLC,max} = 37.8V$

(a)   (b)

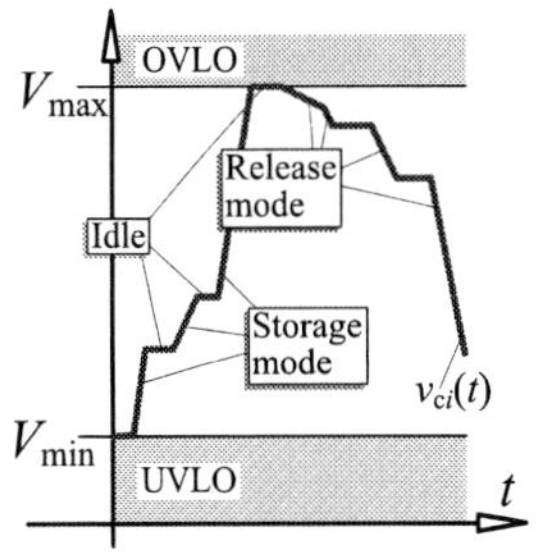

Fig. 3: (a) Parameter set used for the analysis and specifications of the realized prototype. (b) Operation modes of the cell voltage $v_{ci}(t)$.

to deliver P_{max} to the bus due to the current limit. However, it can be fed to the drive system with reduced power and thus almost[1] the entire energy stored in the MLC can be utilized.

The equations (1),(2), and (5)-(7) have been evaluated for the parameter set listed in **Fig. 3(a)** and the results, presented in **Fig. 2(b)** and **(c)**, show that for some parameters no solution can be found (indicated with dots at zero) as the equations are violated. Fig. 2(b) illustrates the range $[V_{MLC,min}, V_{MLC,max}]$ in which V_{MLCref} can freely be chosen while still meeting the requirements. A superimposed optimization could be performed to obtain a V_{MLCref} in the determined range which is optimal regarding some goal(s) (e.g. efficiency and/or power density). In Fig. 2(c), the total storage capability $E_{MLC,tot}$ and the required energy $E_{MLC,req}$ can be compared. Apparently, with an increasing number of UCs, the ratio between $E_{MLC,tot}$ and $E_{MLC,req}$ increases because more energy can be stored without being needed to meet the specifications. Nevertheless, for reliability reasons, it might be desired to include some degree of redundancy which, however, comes at the expense of an oversized system.

The BBC can be designed according to conventional boost converter models [7]. The small input voltage range of the BBC, $V_{MLCref} \pm nV_{C,max}/2$, allows for an optimal design and also simplifies the control compared to boost converters with a wide input voltage range.

3. Modulation and control

The modulation of the system can be divided into two parts: one is to control all cells and hence the MLC output voltage, $v_{MLC}(t)$, while the second part is the modulation of the BBC which is based on a conventional hard switching boost converter with constant switching frequency.

Fig. 3(b) illustrates the voltage operation range of a cell; when energy is stored from the bus (storage mode), the voltage of an active cell, $v_{ci}(t)$, is rising and vice versa if energy is needed at the bus (release mode). Each cell has an under- and an overvoltage lockout protection, UVLO and OVLO, respectively, which prevents the depletion or an overcharging of a cell. While the former case is less fatal the galvanically isolated cell could not be controlled anymore since it is supplied by the energy in the UCs.

For the modulation in storage mode, illustrated in **Fig. 4(a)** for $N = 5$ and $N_{act} = 3$, N_{act} active cells are turned-ON according to obtain a defined $V_{MLC,ref}$. The cell with maximum SOC is shorted while the cell with minimal SOC is always active to catch up regarding SOC. The other cells are cyclically turned-ON to achieve a balanced charging of all cells. The storage mode

© VDE VERLAG GMBH · Berlin · Offenbach

PCIM Europe 2016, 10 – 12 May 2016, Nuremberg, Germany

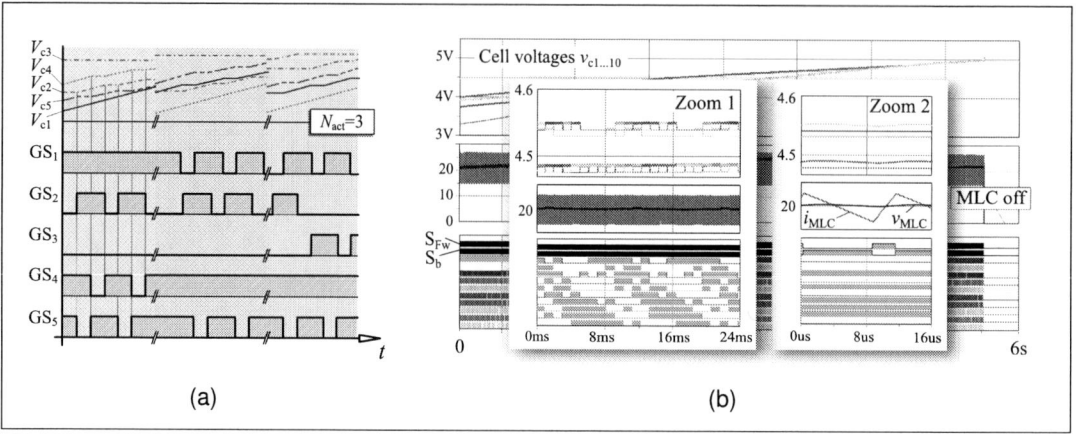

Fig. 4: (a) MLC cell management in energy storage mode. (b) Simulation of the system in storage mode. Zoom 1 shows the modulation of the cells and Zoom 2 illustrates the current control of the BBC.

ends either if all cells are fully charged or no more energy is being provided from the motor.

Fig.4(b) shows a simulation of the system in storage mode. As can be seen, due to the cyclical modulation scheme the cell voltages converge to each other and as soon as all cells are fully charged the modulation stops. Thereby, the switching frequency of the MLC can be low (e.g. 500Hz, cf. Zoom 1). The simulation further shows the limitation of $i_{MLC}(t)$ due to the BBC (cf. Zoom 2).

The modulation in energy release mode is identical except that the cell with the highest SOC is being active while the cell with lowest SOC is shorted.

4. System performance and loss analysis

The total system losses can be separated into BBC and MLC caused losses. The BBC losses for storage and release mode can be calculated using conventional loss models [7]. Due to the low switching frequency and the low ON-state resistance of the MLC switches ($<400\mu\Omega$, 5V/60A for PI5101 by Picor) the losses of the MLC are mainly determined by conduction losses caused by the ESR in UCs (ESR $< 6m\Omega$ for RSC2R7107SR by Ioxus).

Fig. 5(a) presents the balance of losses for a storage mode cycle of the system specified in Fig. 3(a). The BBC alone achieves an efficiency of over 95%; however, taking the capacitor losses into account, the efficiency drops down to 89.5%. Since the losses are almost identical for either mode the efficiency for an entire store-release cycle is $\eta = 80.1\%$ at an energy density of $1.93\,\mathrm{Wh/dm}^3$ and a power density of $\alpha = 0.48\,\mathrm{kW/dm}^3$.

A holistic optimization including different types of UCs with low ESR and/or an MLC arrangement which decreases I_{MLC} (e.g. increasing n, m, or N) could improve the efficiency. It comes, however, at the expense of an increased volume and a higher prize of the system:
Assuming $m = 2$ with all other parameters being kept constant, an efficiency for a store-release cycle of $\eta = 84.8\%$ could be achieved at an energy density of $2.34\,\mathrm{Wh/dm}^3$ and a decreased power density of $\alpha = 0.29\,\mathrm{kW/dm}^3$. The power density α corresponds to the specified 500W; the system, however, could actually provide more power which could be taken into consideration when several solutions are compared.

© VDE VERLAG GMBH · Berlin · Offenbach

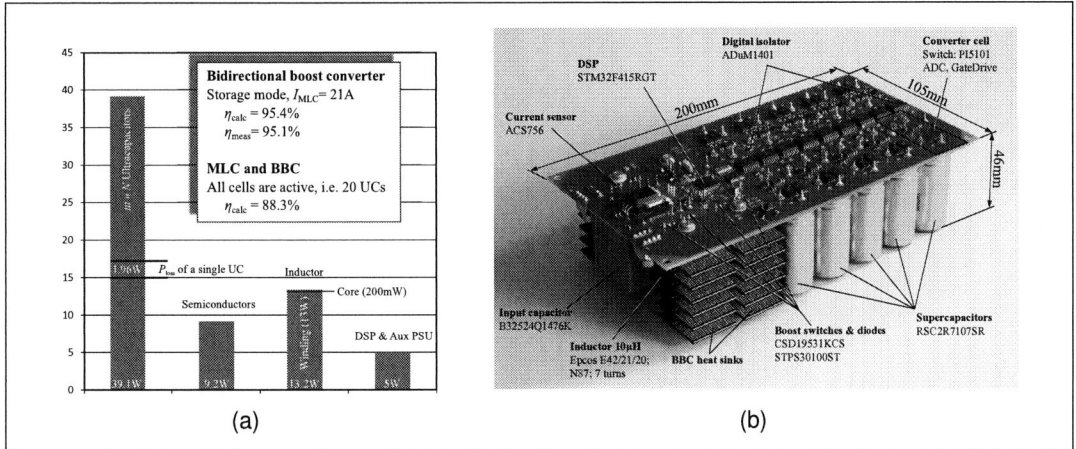

Fig. 5: (a) Balance of losses for one storage mode cycle. The total losses are 58.8W for a rated power of 500W. (b) Prototype of the proposed energy storage system.

5. Laboratory setup

In order to verify the analysis, a prototype for a small but highly dynamic drive system according to Fig. 3(a) is implemented and presented in **Fig. 5(b)**. Each cell is supplied by its UCs which avoids isolated auxiliary PSUs and only four isolated signal lines are required to monitor and control the cells (one gate signal and three SPI bus lines). Beside the measurement of each cell voltage, the bus voltage V_{bus}, the inductor current i_{MLC}, and the MLC voltage v_{MLC} are needed for the system control and to implement safety measures (e.g. overcurrent protection). An even voltage sharing between the two UCs is ensured with a high-impedance voltage divider.

Fig. 6 presents measurement results for both modes. Different views for release mode illustrate the discharge process (a), the proposed modulation scheme (b), as well as the current limitation (c) which confirm the simulations. Fig. 6(d) shows the measurement in storage mode with a current limit of 17.5A. When the cells are fully charged the MLC transitions into idle mode and is ready to release the energy when needed.

6. Conclusion

In pursuing the goal of a green use of electrical drives, recuperation systems are increasingly demanded for specific applications. Therefore, this paper presents the evaluation, analysis, and design of a scalable and flexible topology consisting of a multilevel converter and a bidirectional boost converter. In the system design, six equations determine the operation range in which the system meets the specifications for a given parameter set. A superimposed optimization could be performed on top to achieve a performance aimed for a certain goal. Also, optimized modulation schemes can be applied as the topology enables a high degree of freedom to individually modulate each cell and control the boost converter. The analytical results have been validated using measurements on a prototype for an 48V bus with an energy capacity of 2.04 Wh. An efficiency of 84.1% has been achieved for a store-release cycle at a volume of 1.05 dm^3. The topology could also feature a buck-boost converter instead of the BBC which yields even more flexibility concerning the bus voltage; this and other possibilities, however, are yet to be analyzed.

PCIM Europe 2016, 10 – 12 May 2016, Nuremberg, Germany

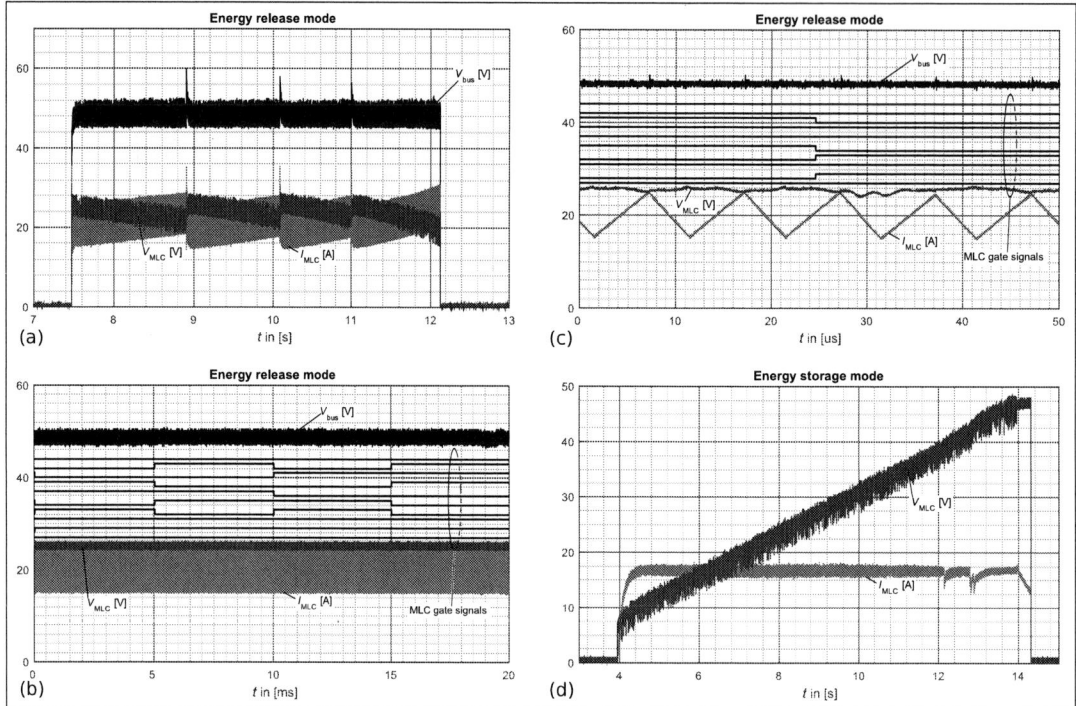

Fig. 6: (a)-(c) Measurement results for energy release mode at 500W. (d) Measurement result in energy storage mode with a constant current of $I_{\mathrm{MLC}} = 17.5$A.

7. References

[1] Energy Star©. Energy Star Program©; Requirements for uninterruptable power supplies. http://www.energystar.gov/, August 2012.

[2] International electrotechnical commission, IEC. IEC 60034, Global efficiency standards for AC motors, 2015.

[3] A. Stupar, T. Friedli, J. Miniböck, and J.W. Kolar. Towards a 99% Efficient Three-Phase Buck-Type PFC Rectifier for 400-V DC Distribution Systems. *IEEE Transactions on Power Electronics*, 27(4):1732–1744, April 2012.

[4] P. Spichartz, L. Bußmann, and C. Sourkounis. Comparison of Recuperation Strategies for Electric Vehicles Regarding Energy Efficiency. In *40th Annual Conference of the IEEE Industrial Electronics Society - IECON*, pages 2984–2990, October 2014.

[5] C.S. Kim, K. Park, H. Kim, G. Lee, K. Lee, H.J. Yang, H. Cho, M. Song, and Y. Son. 48V Power Assist Recuperation System (PARS) with a Permanent Magnet Motor, Inverter and DC-DC Converter. In *1st International Future Energy Electronics Conference (IFEEC)*, pages 137–142, November 2013.

[6] J. Cao and A. Emadi. A New Battery/UltraCapacitor Hybrid Energy Storage System for Electric, Hybrid, and Plug-In Hybrid Electric Vehicles. *IEEE Transactions on Power Electronics*, 27(1):122–132, January 2012.

[7] Robert Erickson and Dragan Maksimovic. *Fundamentals of Power Electronics*. Springer, 2nd edition, 2002.

© VDE VERLAG GMBH · Berlin · Offenbach

Pulse width- and frequency modulated DC/DC converter for hybrid- and electrical vehicles

Magnus Böh, Cologne University of Applied Sciences, Germany, magnus.boeh@th-koeln.de

Andreas Lohner, Cologne University of Applied Sciences, Germany, andreas.lohner@th-koeln.de

Christoph Engelhard, HELLA KGaA Hueck & Co., Germany, christoph.engelhard@hella.de

The Power Point Presentation will be available after the conference.

Abstract

An increasingly significant part of hybrid- and electric vehicles are DC/DC-converter which couples the traction battery with the DC-link of the inverter.

The chosen circuit is a multiple two-quadrant transducer topology. Each half bridge is phase shifted to the other half bridges. Due to this the separate phase currents are phase shifted, too. By the superposition of the individual currents, the total current has a less ripple. In opposite to a conventional pulse width modulation, a combination between frequency- and pulse width modulation is used. The frequency controls the current between the "continuous current mode" and the "discontinuous current mode". By operation in this work load, a zero current switching is realized, which decreases the "switch on" losses.

1. Context of the Converter

To preserve the individual traffic in future, it is necessary to make our society independently from fossil fuels like petrol, diesel or gas. An important intermediate step for the fossil fuel free traffic is the development of hybrid vehicles. Therefore, the Cologne University of applied Sciences started in 2012 a government founded project. This project, called DrEM-Hybrid, is done in cooperation with two medium sized companies, "Meta Motoren und Energie-Technik GmbH" from Herzogenrath, near to Aachen and "Centre for Concepts in Mechatronics" from Nuenen, near to Eindhoven, Netherlands.

The aim of the project was the design of a new powertrain. The advantage of this new powertrain is a "Double rotating Electrical Machine" – called DrEM. As the name implies, the machine has two rotors. The first rotor is connected to the internal combustion engine, which works as range extender, and the second is coupled to the front wheels via the differential gear. By suitable current feed of the electrical machine, a speed control of the range extender is possible. This offers to shift the range extender in his most efficient speed area. To decouple the combustion engines torque from which is needed for traction, two electric machines are installed at the rear axle. Through this both motors, the range extender works in its most efficient torque area.

For pure electrical drive, the torque shaft of the Range extender is clamped and the DrEM works as a conventional electrical machine. Further, the two rear axle machines increases the traction torque. A schema of the drive train is shown in figure 1.

PCIM Europe 2016, 10 – 12 May 2016, Nuremberg, Germany

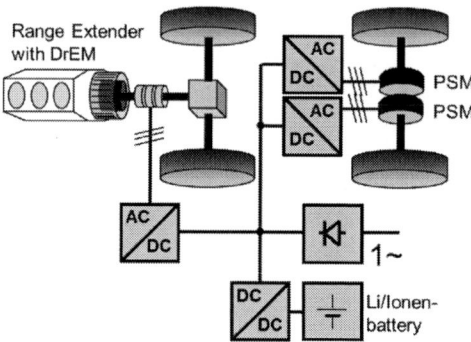

Fig. 1: Schema of the DrEM Hybrid [1]

To couple the low voltage traction battery to the high voltage DC-link, a bidirectional DC/DC converter is needed. The advantages by using DC/DC converter are less losses in the high voltage components (e.g. the traction inverter or the synchronous machines) by using a variable DC-link voltage, a traction battery with less number of cells and a simpler battery management system.

In this abstract the specialty of the frequency- and pulse width modulation for this application is described.

2. Requirements of the Converter

2.1. Requirements of the low voltage (battery) side

40 lithium polymer cells connected in series represent the traction battery. The used cells are SLPB100216216H from Kokam [2]. The major cell ratings are shown in Table 1.

Maximum continuous charge current	120 A
Maximum continuous discharge current	320 A
Maximum peak discharge current	480 A
Nominal cell voltage	3.7 V
Minimum cell voltage	2.7 V
Maximum cell voltage	4.2 V

Table 1: Ratings of the used cells

Considering table 1, a minimum, nominal, and maximum battery voltage of 108 V, 148 V and 168 V are given. Due to the series connection of the cells, the charge- and discharge currents of the battery are the same like the cell currents. The maximum charge and discharge power are shown in equals 1 to 3.

$$P_{charge} = I_{charge} \cdot U_{Batt\ nom} = 18\ kW \qquad \{1\}$$

$$P_{discharge} = I_{discharge} \cdot U_{Batt\ nom} = 47\ kW \qquad \{2\}$$

$$P_{discharge\ peak} = I_{discharge\ peak} \cdot U_{Batt\ nom} = 71\ kW \qquad \{3\}$$

© VDE VERLAG GMBH · Berlin · Offenbach

2.2. Requirements of the DC-link side

The requirements of the high voltage side are defined by the traction motors, the two rear axle machines and the DrEM. The calculation of the maximum DC-link voltage is shown with equals 4 to 9.

$$U_{DC-Link} = \sqrt{2} \cdot \sqrt{3} \cdot \sqrt{U_L{}^2 + U_0{}^2} \qquad \{4\}$$

$$U_L = L_{line} \cdot I_{nom} \cdot \omega_{electric} \qquad \{5\}$$

$$U_0 = U_0{}' \cdot \omega_{mechanic} \qquad \{6\}$$

$$\omega_{mechanic} = r_{differntial} \cdot \frac{v_{vehicle}}{3.6 \cdot \pi \cdot d_{wheel}} \qquad \{7\}$$

$$\omega_{electric} = n_{polepairs} \cdot \omega_{mechanic} \qquad \{8\}$$

Based on a maximum vehicle speed of $v_{vehicle}$ = 140 km/h, front wheel differential ratio of $r_{differential}$ = 3.05, tire diameter of d_{wheel} = 0.64 m, an motor inductance of L_{Line} = 730 µH and synchronous generated voltage of $U_0{}'$ = 840 mV/ω is a DC-link voltage calculated of

$$U_{DC-Link} = 635\,V \approx 650\,V. \qquad \{9\}$$

Based on the requirements of higher and lower voltage side, the DC/DC converter has a maximum power of 71 kW and a maximum DC-link voltage of 650 V. Both voltage areas, the battery and the DC-link are classified as high voltage areas. Therefore, a galvanic isolation isn't needed.

3. Circuit configuration

3.1. Basic topology

Due to the requirements in chapter 2, the decision fells on a two quadrant transducer for the circuit configuration. This topology allows a bidirectional energy flow between the battery and the DC-link and a variability in the DC-link voltage in combination with a simple structure. Figures 2 and 3 show the schematic picture of a two-quadrant transducer and its current profile in the boost mode.

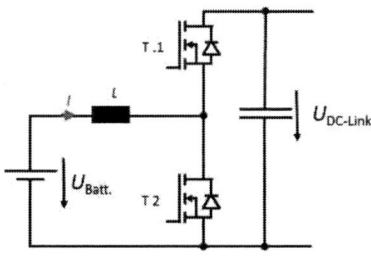

Fig. 2: Schematic of the two-quadrant transducer

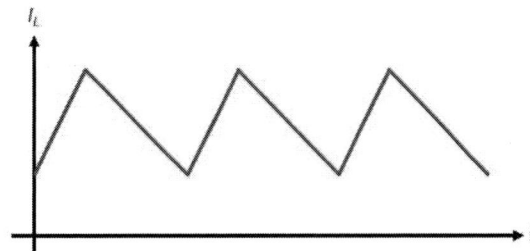

Fig. 3: current profile of the two-quadrant transducer

As seen in figure 3 the current has a high ripple. To reduce the current ripple by increasing the switching frequency is possible. The disadvantages of a higher switching frequency are the rising losses in the semiconductors and the inductance. Another method to reach a less current ripple is increasing the inductance. This is reflected in a higher volume, cost and weight. A further method to decrease the ripple is the usage of a multiphase DC/DC converter, which is described in chapter 3.2.

3.2. Multiphase topology

Due to the usage of further phases, a superposition is achieved by the individual phase currents (I_1 to I_3 in figure 4). With a phase shifted controlling of the half bridges current gradients of each half bridge cancels – in the ideal mode – to zero. The superposition, simulated for a converter with three phases, is shown in figure 5. In this ideal case is the DC-link voltage three times higher than the battery voltage. In this case, the magnetization is half as long as the demagnetization. The figure shows the converter in the boost mode.

Further is in figure 4 the schematic shown of a multiphase converter with three phases.

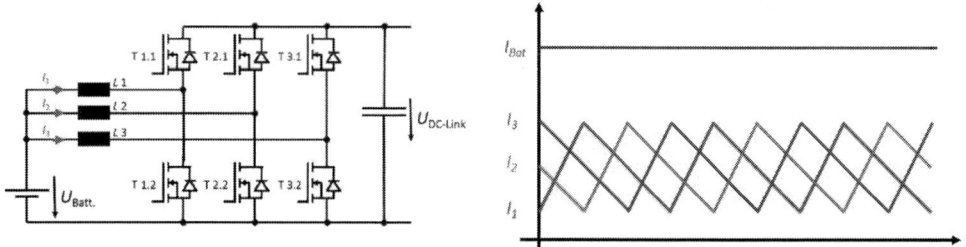

Fig. 4: Schematic of a multiphase converter   Fig. 5: Superposition of the individual currents

Based on a PQ 50/50 core the Matlab/Simulink simulation model shows, the ideal number of phases is 12. [3,4]. Figure 6 shows the maximum as a function of the phase numbers. In order not to saturate the core flux density is limited to 350 mT.

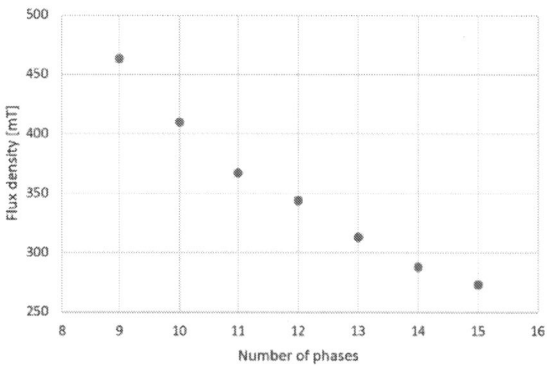

Fig. 6: Flux density vs. number of phases [4]

4. Control strategy

4.1. State of the technology

The most important control algorithm in power electronics are the PWM. By using a fix switching frequency, the pulse width is variable and set the current in an inner loop. An outer loop is for controlling a DC-Link voltage. In the most workloads the converter works in continuous current mode (CCM). For calculating the switching losses for a MOSFET the formulas 10 to 13 can be applied.

$$P_{switching\ losses} = P_{switch\ on} + P_{switch\ off} + P_{on} \qquad \{10\}$$

$$P_{switch\ on} = E_{on} \cdot f_{sw} \qquad \{11\}$$

$$P_{switch\ off} = E_{off} \cdot f_{sw} \qquad \{12\}$$

$$P_{on} = R_{DS\ on} \cdot I_C{}^2 \cdot a \qquad \{13\}$$

In equal 11 is shown, that the switching frequency is dependent from the "switch on energy" E_{on}. E_{on} increases with a rising collector current. By reducing the collector current to zero, the "switch on" losses also go nearly to zero. This switching behavior is known as "zero current switching" (zcs).

4.2. Frequency and pulse width modulation

A frequency modulation offers a "zero current switching" (zcs), which is described in chapter 4.1. Therefore, the pulse width of the generated switching signals is dependent of the ratio between battery- and DC-Link voltages. Further, the phase current is controlled by the switching frequency. For a higher current the switching frequency is decreasing. By a decreasing switching frequency the magnetization time is increasing. This leads to a higher battery current. To better explain and illustrate, in figure 7 and 8 are two different workloads shown.

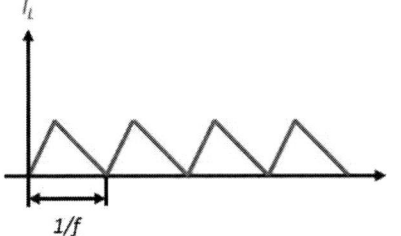 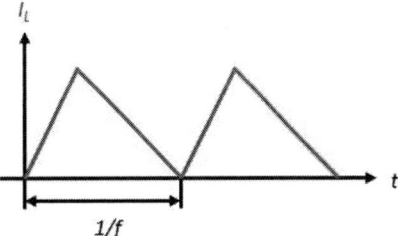

Figure 7: Lower phase current / higher switching frequency [5]

Figure 8: Higher phase current / lower switching frequency [5]

To range the switching frequency in the not hearable sphere, a minimum switching has a lower limit of 20 kHz. To realize bigger battery currents, the converter switch into a pure pulse width modulation without frequency modulation. In this area, the converter works in a continuous current mode (CCM).

A further switching frequency limit for less currents is at 100 kHz. Thus, the switching losses can be kept low, which would occur due to the high switching frequencies. In this area, the converter works in a discontinuous current mode (DCM). In figure 9 a family of characteristics for the switching frequency as a function of the battery current and the DC-Link voltage is shown. In the hatched areas the converter operates with a fix switching frequency in pulse

© VDE VERLAG GMBH · Berlin · Offenbach

width modulation mode. The color gradient goes from 20 kHz (blue) to 100 kHz (orange) with further frequency modulation.

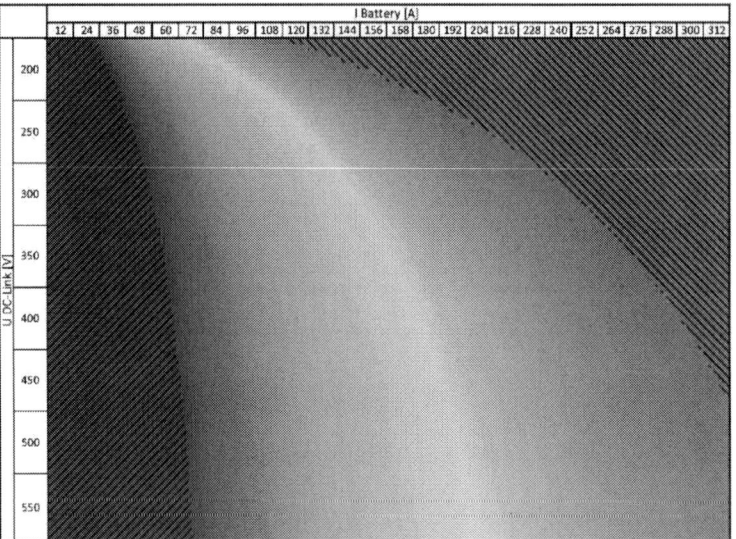

Fig. 9 Operating areas as a function of battery current and DC-Link voltage
(Left hatched: DCM; right hatched: CCM; other: frequency modulation) [5]

5. Constructed Hardware and Measurement Results

To verify the previous simulation, a test half bridge was built up to compare three different semiconductor types. Silicon (Si) IGBT with antiparallel Si diodes, Si IGBT in combination with silicon carbide (SiC) diodes and SiC MOSFET. In figure 11 is shown the simulated efficiency of the three different semiconductor combinations. Due to the highly efficiency of SiC MOSFET the decision was taken on this semiconductor type. In figure 10 is a picture of the first layout of a half bridge.

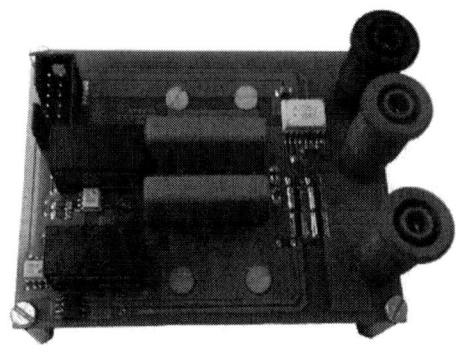

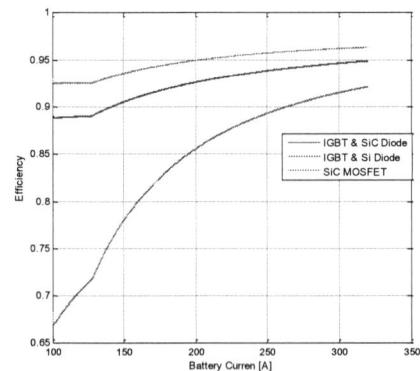

Figure 10 picture of the first prototype half bridge [5]

Figure 11 efficiency of Si IGBT, Si IGBT with antiparallel SiC and SiC MOSFET [6]

Based on this first prototype pcb, a compare between the efficiency of the pure pule width modulation and the pulse width combined with the frequency modulation is realized. The result of this measurement is given in figure 12.

© VDE VERLAG GMBH · Berlin · Offenbach

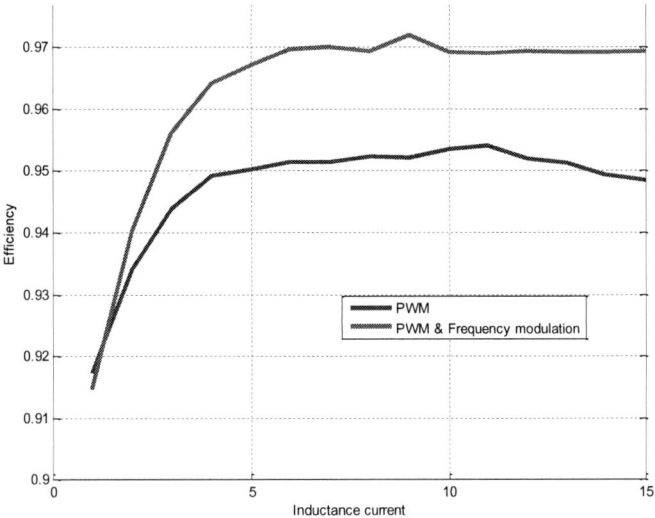

Fig. 12: Efficiency of PWM - and PWM & frequency modulation controller

In the second step, a pcb with a better footprint and included DC-Link capacitor was designed. For a better utilization of the DC-Link capacitor, on each pcb in one half bridge on the left, and one on the right side. This "double half bridge" is shown in figure 13. The overall CAD construction, which are designed with a forced air cooling, is displayed in figure 14. It includes the twelve inductances, six double half bridges, a further DC-Link capacitor designed in electrolytic technology and the controller unit in the middle of the converter. The case is for a better display not shown.

Figure 13 picture double half bridge pcb [5]   Figure 14 CAD rendering of the converter [5]

Abbreviations

P_{charge}	Charge power	[W]
I_{charge}	Charge current	[A]
$U_{Batt\ nom}$	Nominal battery voltage	[V]
$P_{discharge}$	Discharge Power	[W]
$I_{discharge}$	Discharge current	[A]
$I_{discharge\ peak}$	Discharge peak current	[A]
$U_{DC\text{-}Link}$	DC-Link voltage	[V]
U_L	Voltage drop across the DrEM inductance	[V]
L_{Line}	Line inductance of the DrEM	[H]
I_{nom}	Nominal Phase current of the DrEM	[A]
$\omega_{electric}$	Electric angular frequency	[rad]
U_0	Synchronous generated voltage	[V]
$U_0{}'$	Synchronous generated voltage per angular frequency	[V/rad]
$\omega_{mechanic}$	Mechanic angular frequency	[rad]
$v_{vehicle}$	Vehicle speed	[km/h]
$r_{differential}$	Differential gear ratio	[]
d_{wheel}	Wheel diameter	[m]
$n_{oolepairs}$	Polepais DrEM	[]
$P_{switching\ losses}$	Switching losses	[W]
$P_{switch\ on}$	Switch on losses	[W]
$P_{switch\ off}$	Switch off losses	[W]
E_{on}	Switch on energy	[J]
E_{off}	Switch off energy	[J]
f_{sw}	Switching frequency	[Hz]
P_{on}	On losses	[W]
$R_{DS\ on}$	Drain-source on resistance	[Ω]
I_C	Collector current	[A]
a	Duty cycle	[]

References

[1] SIGMUND, Daniel: *Matlab/Simulink-basierte Modellbildung und Simulation eines leistungsverzweigten Hybridantriebes für PKW*, Bachelor Thesis CUAS, Köln, 2012
[2] KOKAM: *KD07-RC23-01*, Datasheet, Gyeonggi-Do (Korea), 2011
[3] EPCOS: *PQ 50/50 Cores and coil former*, Datasheet, 2009
[4] BOEH, Magnus et al.: *Frequency and PWM controlled, not galvanic isolated multiphase DC/DC converter for hybrid- and electric vehicles*, EEVC, Bruessel, 2014
[5] ENGELHARD, Christoph: *Entwicklung eines FPGA-gesteuerten, mehrphasigen DC/DC-Wandlers für Elektro- und Hybrid PKW*, Master Thesis CUAS, Köln, 2015
[6] LOHNER, Andreas et al.: *Entwicklung eines kompakten, hocheffizienten, bidirektionalen multiphasigen DC/DC-Wandlers zur Anbindung einer Traktionsbatterie an einen Zwischenkreis mit integrierter Netzanbindung für Hybrid und Elektrofahrzeuge*, EEHE Conference, Bad Boll 2015

PCIM Europe 2016, 10 – 12 May 2016, Nuremberg, Germany

New High Power Density Modules for EV/HEV Applications

Seiichiro Inokuchi, Mitsubishi Electric Co., Japan, Inokuchi.Seiichiro@bk.MitsubishiElectric.co.jp

Shoji Saito, Mitsubishi Electric Co., Japan, Saito.Shoji@ak.MitsubishiElectric.co.jp

Arata Izuka, Mitsubishi Electric, Co., Japan, Iizuka.Arata@dr.MitsubishiElectric.co.jp

Yuki Hata, Mitsubishi Electric Co., Japan, Hata.Yuki@dx.MitsubishiElectric.co.jp

Shinji Hatae, Mitsubishi Electric Co., Japan, Hatae.Shinji@cw.MitsubishiElectric.co.jp

Khalid Hussein, Mitsubishi Electric Europe B.V., Germany, Hussein.Khalid@meg.mee.com

Abstract

This paper addresses a newly developed IGBT module dedicated for large Electric Vehicle (EV) and Hybrid Electric Vehicle (HEV) power-train inverter applications. The new IGBT module family adopts a 6-in-1 circuit configuration in one package, optimized internal wiring layout, Direct Lead Bond (DLB) structure and direct cooling structure with an integrated water-cooled Al fin. As a result, this new IGBT module family has simultaneously achieved high performance, low self-inductance, compact package and light weight. Compared to conventional products with similar power capabilities, the adoption of these innovative technologies has led to an improved thermal performance of 20%, and has reduced the module's self-inductance by 30%, the footprint by 50% and the weight by 70%.

1. High Power J1-Series

1.1. Introduction

The market for EV/HEV is growing by increasing global environment protection awareness. The power semiconductor module has become an important part to determine vehicle performance. Along with the increase of the market size, the system of the operation is also diversified to high power, and high power density and high capacity modules have become a requirement especially in recent years.

A new high-power-density modules family named "High Power J1-Series" has been developed in response to the automotive market essential requirements which are "high power", "high reliability", "compact size" and "high efficiency" [1, 2]. The "High Power J1-Series" has been implemented utilizing "6-in-1 circuit configuration", "Direct Lead Bond (DLB) structure", "Direct cooling structure" and "7th Generation CSTBT™ and Relaxed Field of Cathode (RFC)-diode" power chips [3]. The optimized combination of these technologies has brought successful improvement of the "High Power J1-Series" performance dedicated for EV/HEV applications. The High Power J1-Series power module's external appearance and circuit configuration are given in Fig.1 and Fig.2. Dimensions and corresponding current/voltage ratings are given in Table 1.

© VDE VERLAG GMBH · Berlin · Offenbach

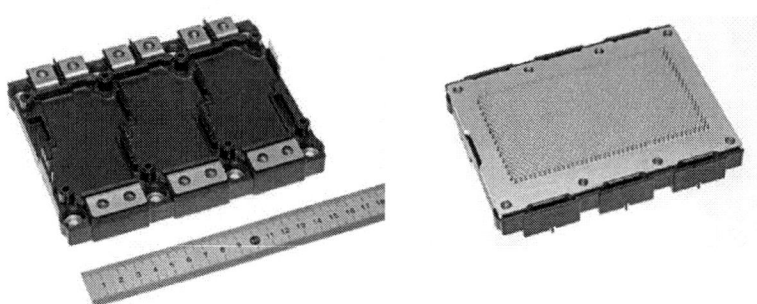

Fig. 1, High Power J1-Series Power Module (Left : front side, Right : Cooling fin side)

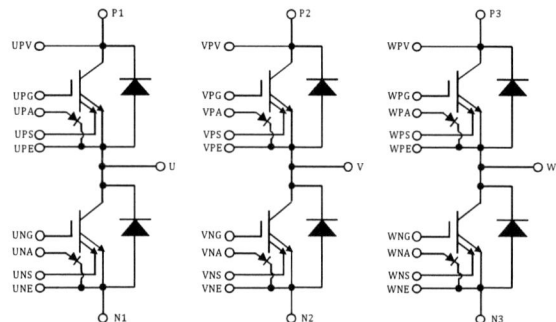

Fig. 2, High Power J1-Series block diagram

Table. 1. High Power J1-Series Line-up

Model	Ratings (Vces/Ic)	Package Size
CT1000CJ1B060	650V/1000A	163×124.5×33.6mm
CT600CJ1B120	1200V/600A	(including Control terminal and Pin-fin, 6-in-1)

1.2. PACKAGE TECHNOLOGY

Further advancing the capabilities of the previously reported J-Series and J1-Series [4, 5], the newest member of power modules with built-in Aluminum cooling-fin for EV/HEV applications is proposed in this paper as shown in Fig. 3. The new 6-in-1 module's package is characterized by several features including a highly reliable Direct Lead Bonding (DLB) structure, compact size, light weight, and high power handling capability. High power capacity modules (e.g. 1000A/650V or 600A/1200V) imply justified large size internal leads and power terminals hence increasing the module package size. Moreover, bigger packages have larger self-inductance than smaller ones which is quite critical problem for high power applications during high di/dt conditions. However, by adopting an optimized internal power leads and chips layout aiming at cancelling the magnetic flux between the PN terminals, the newly developed high-power J1-Series successfully achieved low overall self-inductance. Fig. 4 shows the inductance simulation results of the newly developed high-power J1-Series compared with conventional design.

© VDE VERLAG GMBH · Berlin · Offenbach

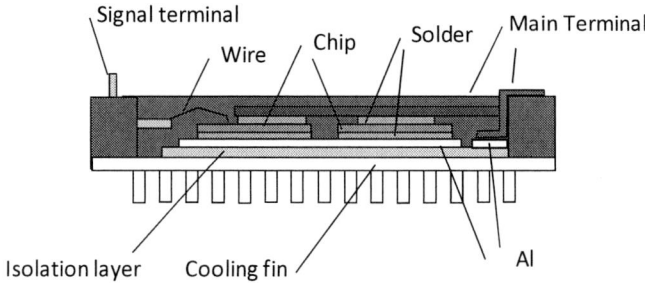

Fig. 3, Internal structure of High Power J1-Series

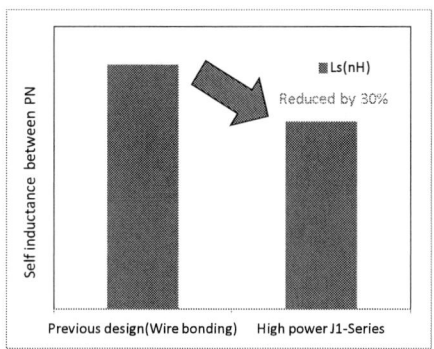

Fig. 4. Self inductance between P and N

In comparison with more conventionally packaged products (J-Series T-PM), the new power module family reduces the footprint more than 50% (Fig.5). The reduced size of the High Power J1-Series is the result of combining an optimized aluminum pin-fin structure with high efficiency 7th Generation CSTBT^{TM}/RFC-Diode chips. Despite the fact that Aluminum cooling-fins have lower thermal conductivity compared to copper cooling-fin structures, this selection provides several advantages to EV/HEV applications. Among these advantages the most prominent ones are corrosion resistance when Aluminum is exposed directly to coolants and light weight as much as 70% weight reduction when comparing 6-in-1 power module inverter solutions (Fig.6). Aluminum is not susceptible to galvanic corrosion like copper. If a copper fin is used it becomes necessary to apply thick nickel plating to prevent corrosion. In addition, the light weight of aluminum contributes to the reduction of electricity costs and fuel consumption in the EV/HEV.

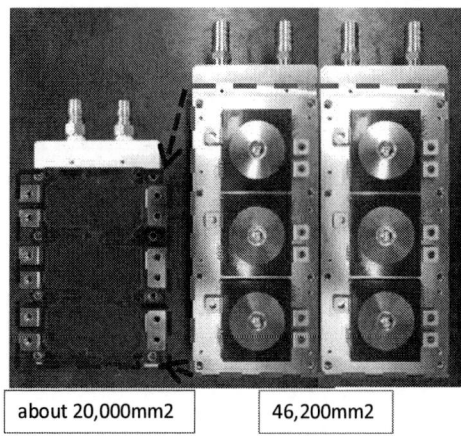

Fig. 5, High Power J1-Series footprint comparison with conventional T-PM solution.

© VDE VERLAG GMBH · Berlin · Offenbach

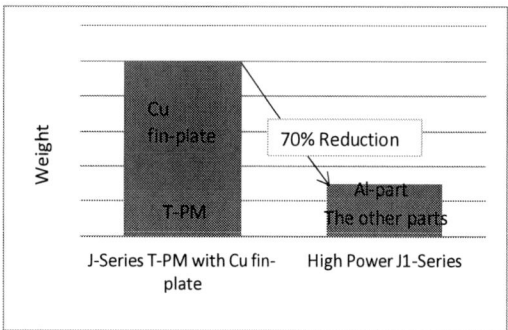

Fig. 6, High Power J1-Series weight comparison vs. conventional T-PM with Cu fin.

Additionally High Power J1-Series has eliminated two layers in the thermal path. One is the solder layer between the substrate and the baseplate, the other is the grease layer between the baseplate and water jacket. As a result a 20% thermal performance improvement is achieved when comparing 6-in-1 power module inverter solutions (Fig.7). At the same time, the reduction of layers contributes to improved thermal cycling capability.

These solutions compared in Fig.5, Fig.6 and Fig.7 are based on equivalent module current and voltage ratings for three-phase EV/HEV motor drives.

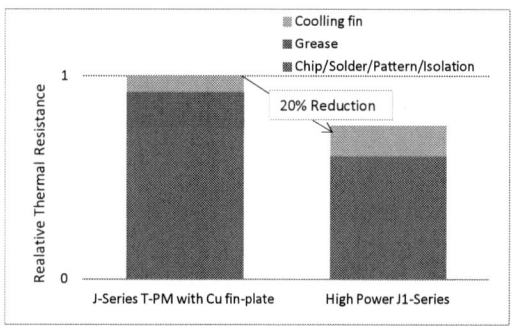

Fig. 7, High Power J1-Series thermal resistance comparison vs. conventional T-PM with Cu fin.

1.3. EXPERIMENTAL RESULTS WITH LATEST CHIP TECHNOLOGY

The High Power J1-Series power handling capability in conjunction with the performance of the thermal interface construction was experimentally verified under the following test conditions for the 650V/1000A module: Main battery voltage = 450V; PWM switching frequency (fc) = 5kHz, 10kHz; coolant temperature (Tw) = 65°C; coolant flow-rate = 10 l/min, IGBT_Rth(j-w)=max, IGBT_characteristics are representative example. Similarly, the test conditions for the 1200V/600A module: Main battery voltage = 600V; PWM switching frequency (fc) = 5kHz, 10kHz; coolant temperature (Tw) = 65°C; coolant flow-rate = 10 l/min, IGBT_Rth(j-w)=max, IGBT_characteristics are representative example.

Under these conditions the inverter output current of the 650V/1000A module can exceed 600Arms (corresponding to more than 120kW output power) at a maximum operation junction temperature less than 150°C. Moreover, the inverter output current of the 1200V/600A module can exceed 400Arms (corresponding to more than 120kW output power) at a maximum operation junction temperature less than 150°C. These results are shown in Fig.8 and Fig.9.

PCIM Europe 2016, 10 – 12 May 2016, Nuremberg, Germany

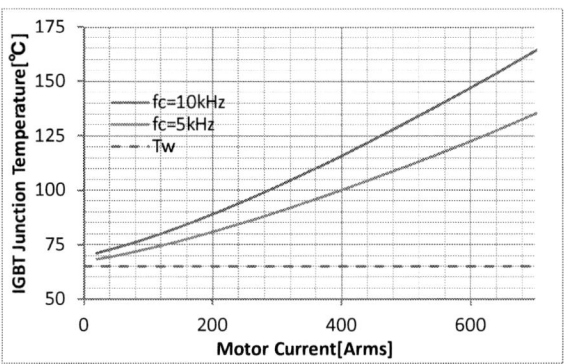

Fig.8.   Experimental performance of the High Power J1-Series (650V/1000A)

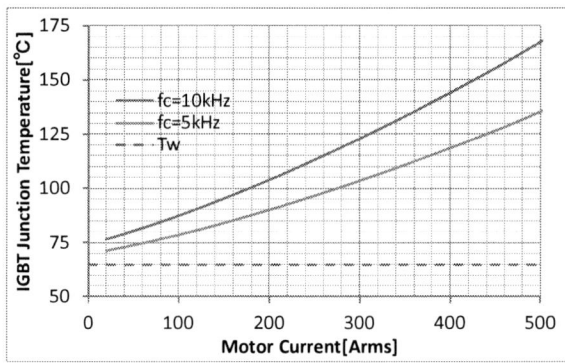

Fig.9, Experimental performance of the High Power J1-Series (1200V/600A)

This attractive and favorable result has been achieved by the utilization of the latest power chip technologies 7th Generation CSTBT^{TM} and RFC-diode. Advances in IGBT technology have always been driven by the continuing demand for higher power densities and higher efficiencies as reflected in the progress in IGBT generations with improved internal structures aiming at optimizing the well-known $V_{CE(sat)}$ vs Eoff tradeoff characteristic. By adding an extra layer of carriers within the IGBT structure, the CSTBT^{TM} concept achieves higher efficiency by reducing both the saturation voltage $V_{CE(sat)}$ and the switching loss Eoff. The 7th Generation IGBT further optimizes the CSTBT^{TM} $V_{CE(sat)}$ vs. Eoff trade-off characteristic as illustrated in Fig.10 which pictorially summarizes the continuous improvement in IGBT characteristics with advanced generations. Considering the very compact size of the J1-Series innovative package design (power module volume less than 0.68 liters), the very high power density level realized by the utilization of the 7th Generation IGBT is clearly evident.

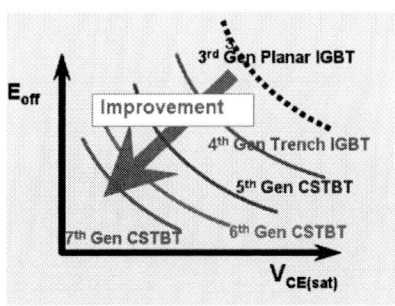

Fig.10, IGBT trade-off characteristics improvement

© VDE VERLAG GMBH · Berlin · Offenbach

2. Conclusion

New High Power Density Modules "High Power J1-Series" has been developed to meet the requirements of the evolving EV/HEV market. The High Power J1-Series achieves high performance, low self-inductance, compact size and light weight. These attractive features were made possible through the combination of optimized package structure technology and state-of-the-art chip technology (7th Generation CSTBT[TM] and RFC-diode). As conclusion, the High Power J1-Series can realize a wide range inverter operation and accommodate a variety of requests for EV and HEV applications.

3. References

[1] S. Inokuchi et al., "A new versatile high power Intelligent Power Module (IPM)", PCIM – ASIA 2015 pp.205-209

[2] K. Hussein, et al., "IPMs Solving Major Reliability Issues in Automotive Applications", IEEE-ISPSD 2004, Proceedings, pp. 89-92.

[3] S. Honda, et al., "Next generation 600V CSTBT[TM] with an advanced fine pattern and a thin wafer process technologies", ISPSD 2012, pp. 149-152.

[4] M. Ishihara et al., "New compact-package Power Modules for Electric and Hybrid Vehicles (J1-Series)", PCIM –Europe 2014, pp. 1093-1097

[5] K. Hussein, et al., "New compact, high performance 7th Generation IGBT module with direct liquid cooling for EV/HEV inverters", IEEE-APEC 2015, Proceedings, pp. 1343-1346.

PCIM Europe 2016, 10 – 12 May 2016, Nuremberg, Germany

Fault-Tolerant B6-B4 Inverter Reconfiguration with Fuses and Ideal Short-On Failure IGBT Modules

Michael Gleissner, Mark-M. Bakran, michael.gleissner@uni-bayreuth.de
University of Bayreuth, Department of Mechatronics, Centre of Energy Technology, Germany

Abstract

Fault-tolerance is a common measure to improve power electronics availability and to ensure functional safety. A six-switch three-phase voltage source inverter (B6) can be reconfigured to a four-switch three-phase voltage source inverter (B4) with DC link midpoint connection after semiconductor failure. This paper presents a new reconfiguration method without active switches for ideal short-on failure IGBT modules via diodes and fuses. The ideal short-on failure behaviour of modules with sandwich structure after current stress over a longer period is investigated and verified by cross-section analysis. Details on the inverter reconfiguration and post-fault operation strategy are explained in theory and with a prototype. The new fault-tolerant inverter is less complex than known equivalents.

1. Fault-tolerant inverters

Fault-tolerant systems enable an ongoing operation after failure. This is relevant for high availability and safety critical applications. Therefore, redundant structures on component or even system level are necessary. For inverters, two well-known strategies are an additional fourth leg or the reconfiguration to a B4 inverter with DC-link midpoint connection [1]. The fourth leg enables the same output power after failure, but requires more additional switches. The fourth leg can replace any of the three standard legs by firing the corresponding active reconfiguration switches (see Fig. 1a) or by passive reconfiguration switches combined with ideal short-on failure semiconductors (see Fig. 1b) [2]. Ideal short-on failure behaviour indicates that the switch is always turned-on after failure and can conduct nominal current without higher loss than in normal-on mode. The B4 reconfiguration reduces the post-fault output power, but needs less additional components [3–5]. Normally, the reconfiguration is achieved by turning on the corresponding bidirectional active reconfiguration switch (see Fig. 1c), e.g. a triac, depending on the failed phase, which has to be disconnected by fuses [6].

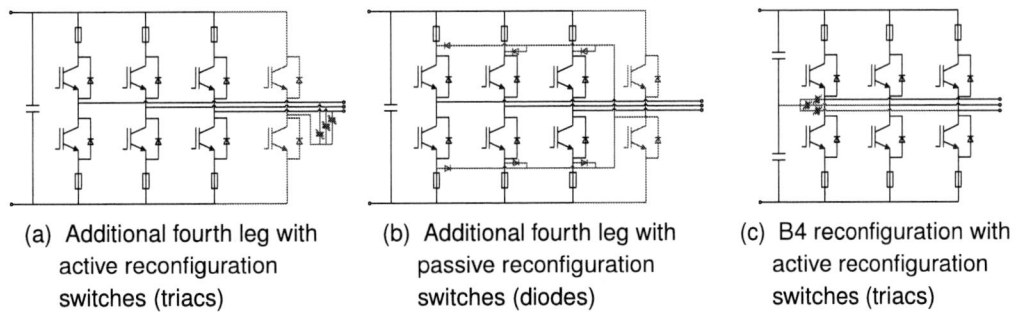

(a) Additional fourth leg with active reconfiguration switches (triacs)

(b) Additional fourth leg with passive reconfiguration switches (diodes)

(c) B4 reconfiguration with active reconfiguration switches (triacs)

Fig. 1: Known fault-tolerant inverters with fuses for isolation of faulty phase legs

© VDE VERLAG GMBH · Berlin · Offenbach

PCIM Europe 2016, 10 – 12 May 2016, Nuremberg, Germany

If the semiconductor switches fail always ideal short-on and are able to conduct nominal current for a sufficient time, passive reconfiguration switches like diodes are sufficient (see Fig. 2a). Thus, the reconfiguration is automatically performed by the diodes and simplified, because no active switches with drivers and control unit are necessary. A possible application for this new fault-tolerant inverter is electric steering or a limp home mode for vehicles with electrical drive-train. Details on the ideal short-on failure behaviour of modules with sandwich structure are provided in Section 2. The reconfiguration of a B6 to B4 inverter with these modules is described and verified with a prototype linked to an induction machine in Section 3.

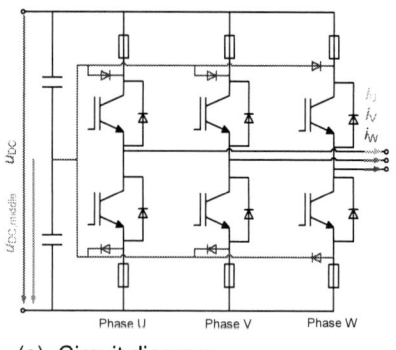

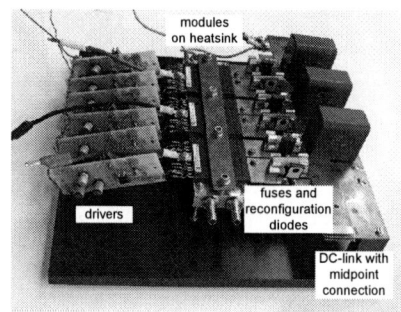

(a) Circuit diagram

(b) Prototype with ideal-short on failure modules

Fig. 2: Proposed new fault-tolerant B6 to B4 reconfiguration inverter with passive reconfiguration switches (diodes) for ideal short-on failure semiconductors, DC-link midpoint connection and fuses

2. Ideal short-on failure IGBT modules

2.1. Module design and failure characteristic

Modules with sandwich structure are applied to improve the electrical, thermal and reliability performance. The die is integrated between an upper and lower copper layer and there are no bond-wires in the collector-emitter path (see Fig. 3), which are a weak-point because of lift-off in case of material fatigue or a failure event. Several publications indicate that these sandwich structure modules from different manufactures offer an ideal short-on failure characteristic [2, 7, 8]. The analysis of a bond-wire free module from International Rectifier in [7] has shown, that melted material forms a low ohmic connection between the top and bottom copper structure. The same half-bridge module, including high- and low-side IGBTs with diodes each rated 680 V - 300 A, is investigated in this paper. The failure resistance is less compared to normal-on mode and the failed modules can conduct nominal current. Consequently, these ideal short-on failure modules can be applied in fault-tolerant converters with serial structures of semiconductors like in multilevel DC-DC converters or inverters. But they can also be implemented in 2-level structures to fire fuses and to enable a reconfiguration path in the failed always on mode.

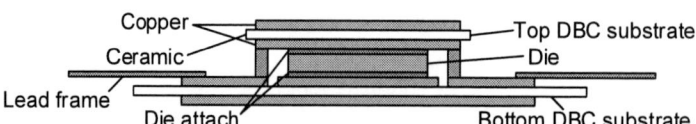

Fig. 3: Copper-die-copper sandwich structure of bond-wire free module

© VDE VERLAG GMBH · Berlin · Offenbach

2.2. Long-term behaviour of failed modules stressed with nominal current

The high-side and low-side switches in three modules have been destroyed by collector-emitter overvoltage. Directly after failure, the voltage-current output characteristic of all failed switches has been recorded in IGBT and diode direction at ambient temperature with a curve tracer.

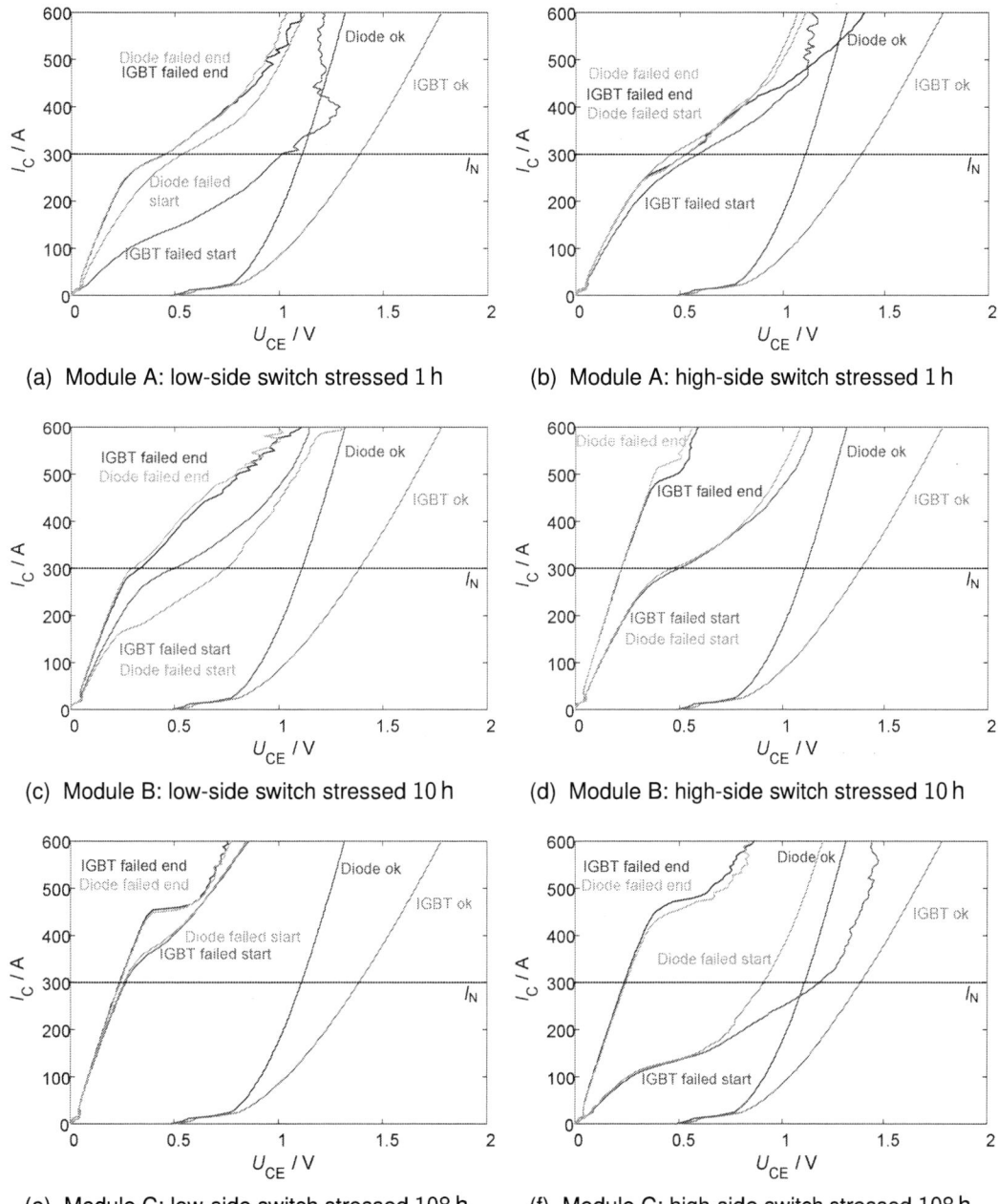

Fig. 4: Voltage-current output characteristic of three failed half-bridge modules in IGBT and diode direction at start and end of constant nominal current stress test with 300 A and varying time after collector-emitter overvoltage destruction compared to normal mode at ambient temperature

Afterwards, all switches have been stressed with DC nominal current of 300 A. Both switches in module A were stressed 1 h, in module B 10 h and in module C 108 h in order to see if the ideal short-on failure characteristic changes or even interrupts the current transfer depending on time. During the current stress test, the modules have been mounted on a heat-sink. The voltage-current output characteristic has been measured again at the end of the current stress test for all switches. The results are depicted in Fig. 4. Each figure also includes the voltage-current output characteristic of the IGBT and diode in normal not failed mode for comparison. The voltage-current output characteristic of all failed switches is always left-hand of the voltage-current output characteristic of a healthy IGBT. Thus, compared to normal continuous-on mode, the failed devices have a smaller resistance and less loss, which is the first precondition for usage in fault-tolerant converters. The failure resistance characteristic is not identical for all failed devices, but always consists of a linear ohmic region up to the nominal current of 300 A and increases above this value, but is still smaller compared to the OK characteristic. With higher current and local temperature the resistance behaviour changes from a linear ohmic characteristic to a drifting voltage zone, because high local temperatures can lead to change of the conduction channel. Moreover, the comparison of the measurements before and after the current stress test shows that the ideal short-on failure characteristic even improved for all switches by decreasing the resistance. The current stress test after switch failure indicates that the failed modules can conduct nominal current for a period of more than 100 h, which is the second requirement for the application in fault-tolerant converters. This time should be sufficient to achieve a safe state or to arrange a repair of the system.

2.3. Failure location analysis

After the current stress test, the modules have been examined by cross-section failure analysis. The failure location has been identified with X-ray. The position of the failure location varies from switch to switch. The cross-section view of the failure locations shows that solder, die and copper have melted and form a connection of the upper and lower copper layer and thus result in a short-circuit (see Fig. 5). There is no big difference between short- (1 h) and long-term (108 h) stressed switches obvious. The current density at the failure location can be

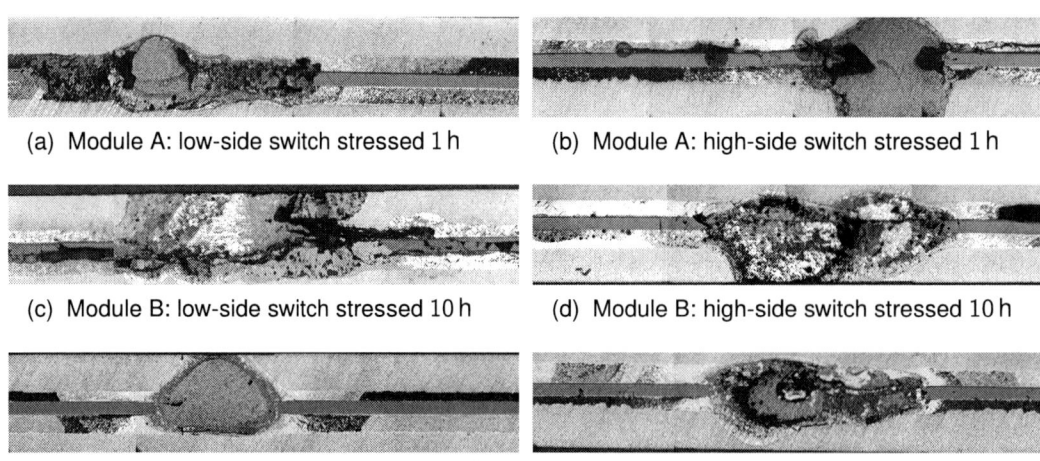

(a) Module A: low-side switch stressed 1 h   (b) Module A: high-side switch stressed 1 h

(c) Module B: low-side switch stressed 10 h   (d) Module B: high-side switch stressed 10 h

(e) Module C: low-side switch stressed 108 h   (f) Module C: high-side switch stressed 108 h

Fig. 5: Cross-section analysis of failure location centre of three half-bridge modules after collector-emitter overvoltage destruction and subsequent constant nominal current stress test with 300 A and varying time

very high and depending on current, temperature and previous stress result in material conversion. The material conversion is a good explanation for the decreased resistance at the end of the current stress test compared to the start (see Fig. 4). With increasing current and loss, the failure zone heats up and melts further material until the resistance decreases and a stable state is achieved. The material can hardly move away because of the sandwich copper plate construction. Thus, an open-failure state is unlikely.

Further analysis on the failure characteristic of modules with sandwich structure should focus on which failure reasons like e.g. overvoltage, surge current, avalanche robustness, mechanical stress or insufficient cooling consequence the ideal short-on behaviour. Especially, it is important to know which failure energy is required to achieve the ideal short-on failure state and which failure energy must not be exceeded to avoid severe destruction like an explosion of the module.

3. Passive B6 to B4 reconfiguration inverter

3.1. Prototype

A prototype with sandwich structure IGBT modules, reconfiguration diodes (1200 V, 60 A), fuses and DC-link midpoint connection is shown in Fig. 2b. The IGBT modules are rated 680 V, 300 A and mounted on a water cooled heat-sink. The test AC-load is a maximum 1 kW, $U_\Delta =230$ V induction machine with two pole pairs in delta connection coupled to a servo machine for adjusting the mechanical power without speed control. Thus, 10 A fuses are applied. The IGBT modules are clearly current oversized, but no other sandwich structures modules have been available. The DC-link is composed of six 100 µF film capacitors (each three parallel for upper and lower capacitor) resulting in 150 µF total capacitance. The control and Sinus PWM output is realized with a modular dSPACE DS1007, DS2004 and DS5101 system. The failure state is achieved by setting the control signal of one IGBT to always on. The DC-link voltage is 400 V.

3.2. Operation in B6 and B4 mode

Normally, the inverter semiconductors are rated to achieve nominal system power in normal mode. The six-switch inverter has six active switching vectors and two zero vectors whereas the four-switch inverter has only four active switching vectors, which have different lengths (see Fig. 6). The maximum output voltage in B4 mode is reduced by a factor of 2 compared to B6 mode, which limits the operation area after reconfiguration. Depending on the connected machine type and load characteristic, the total maximum output power is reduced. Moreover, the phase current flowing through the DC-link capacitors in B4 mode requires a large enough capacitance for a small voltage ripple at low output frequencies. An increasing capacitor voltage ripple consequences asymmetric phase currents and torque ripples. Alternatively, the modulation scheme can be adjusted with feedback from the capacitor voltage to achieve symmetric phase currents also with a small capacitance. However, the output voltage is further reduced by this adjusted modulation strategy [4, 5]. Fig. 7 illustrates the measured phase currents and DC-link voltages and in B6 and B4 mode without resp. with compensation of asymmetrical phase currents for varying loads.

But nevertheless, there is still an ongoing operation possible with this fault-tolerant converter compared to a standard B6 inverter, which has no fault-tolerance. This is beneficial for applications with high functional safety requirements, e.g. for safe-operational systems, where turning-off the power electronics consequences an unsafe state of the total system.

© VDE VERLAG GMBH · Berlin · Offenbach

PCIM Europe 2016, 10 – 12 May 2016, Nuremberg, Germany

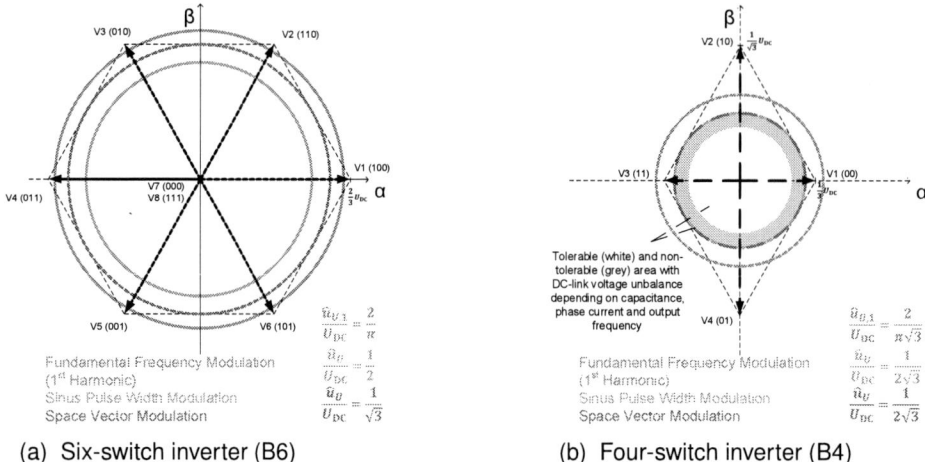

(a) Six-switch inverter (B6)

(b) Four-switch inverter (B4)

Fig. 6: Space vectors of B6 and B4 inverter and peak phase output voltages depending on modulation

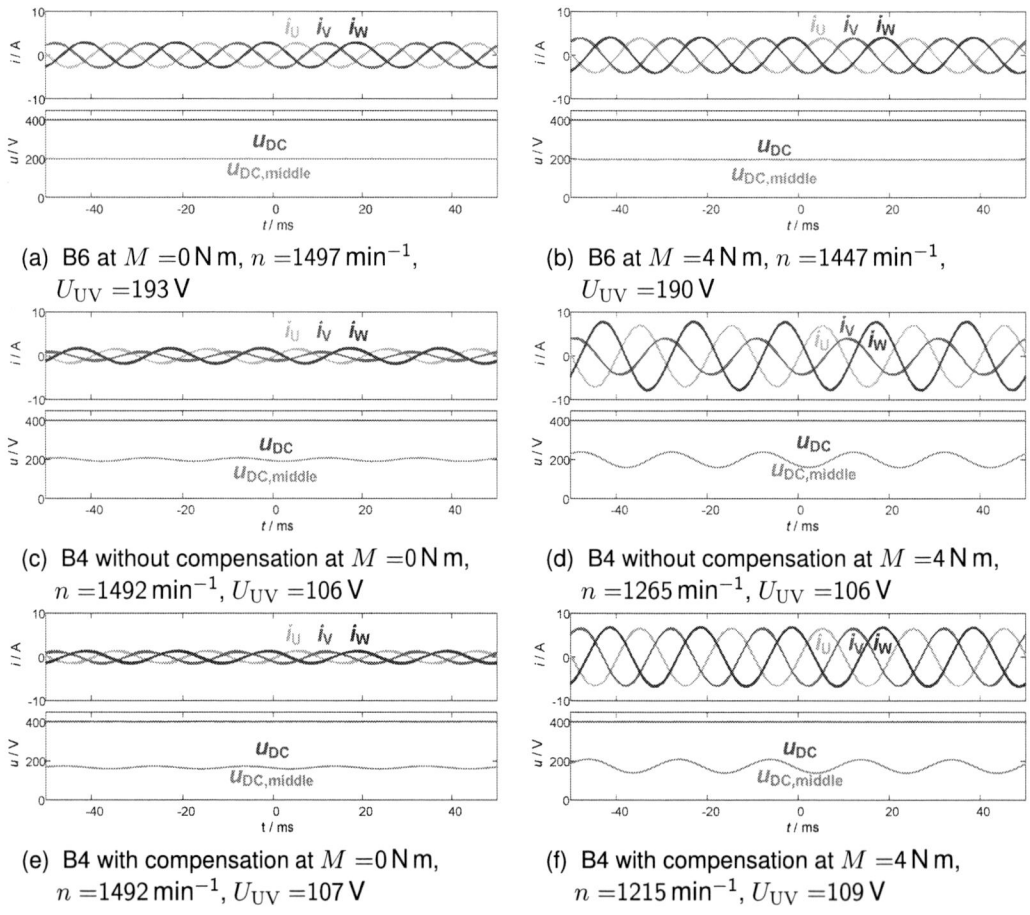

(a) B6 at $M = 0\,\mathrm{N\,m}$, $n = 1497\,\mathrm{min}^{-1}$, $U_{\mathrm{UV}} = 193\,\mathrm{V}$

(b) B6 at $M = 4\,\mathrm{N\,m}$, $n = 1447\,\mathrm{min}^{-1}$, $U_{\mathrm{UV}} = 190\,\mathrm{V}$

(c) B4 without compensation at $M = 0\,\mathrm{N\,m}$, $n = 1492\,\mathrm{min}^{-1}$, $U_{\mathrm{UV}} = 106\,\mathrm{V}$

(d) B4 without compensation at $M = 4\,\mathrm{N\,m}$, $n = 1265\,\mathrm{min}^{-1}$, $U_{\mathrm{UV}} = 106\,\mathrm{V}$

(e) B4 with compensation at $M = 0\,\mathrm{N\,m}$, $n = 1492\,\mathrm{min}^{-1}$, $U_{\mathrm{UV}} = 107\,\mathrm{V}$

(f) B4 with compensation at $M = 4\,\mathrm{N\,m}$, $n = 1215\,\mathrm{min}^{-1}$, $U_{\mathrm{UV}} = 109\,\mathrm{V}$

Fig. 7: Measured phase currents and DC-link voltages depending on operation mode and load torque of induction machine with output frequency $50\,\mathrm{Hz}$ and switching frequency $8\,\mathrm{kHz}$

© VDE VERLAG GMBH · Berlin · Offenbach

3.3. Reconfiguration process

After short-on failure of any switch and closing of its complementary switch, which is still working, the DC-link is short-circuited in the respective phase. The upper and lower fuses are blown and isolate the faulty phase, if the rated I^2t value is achieved by the DC-link energy. The new path of the phase current is via the two reconfiguration diodes, the short-on failed device and the complementary, always turned-on switch to the midpoint of the DC-link. The switch pattern of the four switches of the two healthy phases is adjusted to the B4 switch pattern. To identify the faulty phase, either feedback from the drivers or monitoring of the phase currents and comparison with expected values in normal mode can be employed [6]. Active reconfiguration switches like in Fig. 1c are superfluous, which reduces semiconductor cost and control effort.

3.4. Measurement results

Fig. 8 presents the measured reconfiguration process at constant 3 N m load torque of the induction machine after always-on failure of the low-side switch in phase W with a failure detection time of less than 7 ms and reconfiguration time of 150 ms until a constant speed is reached again. There is no speed control implemented in order to see the effect of the changed speed-torque characteristic at reduced output voltage. The failed phase is detected by comparing the three phase currents with expected values. The control pattern of the remaining two healthy phases is adjusted to B4 sinus PWM. The machine is working at part load when the failure occurs. Consequently, the mechanical output power is only reduced from 460 W to 420 W. A higher constant load torque can stop the machine, if the maximum torque of the induction machine at reduced voltage is exceeded. The reconfiguration depends on the load characteristic of the machine. The measured currents of upper and lower fuse as well as reconfiguration diodes and the respective I^2t values are shown in Fig. 9. There is an induced current flow through the reconfiguration diodes because of parasitic inductances and the high current slope during the DC-link capacitor short-circuit and fuse blowing. Thus, these diodes are stressed during reconfiguration and their rated I^2t value should not be exceeded. The diodes in the prototype are rated with $I^2t = 3.62\,\text{kA}^2\,\text{s}$ for $t = 10\,\text{ms}$. The induced I^2t is much less.

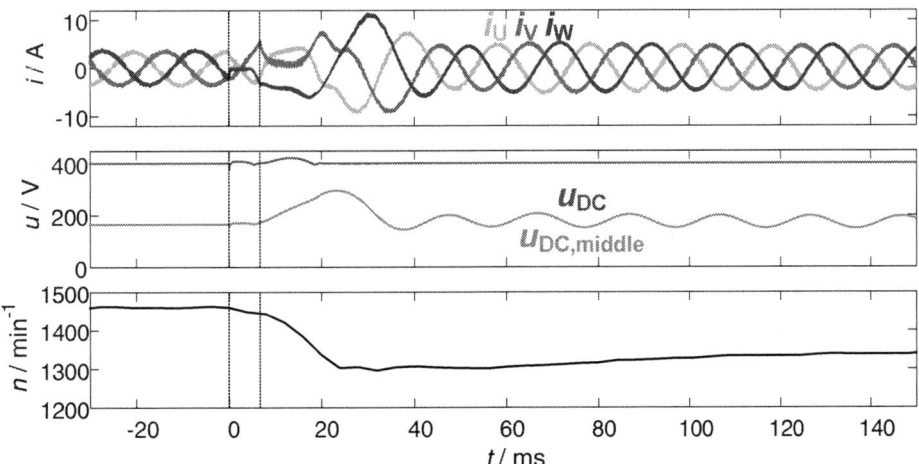

Fig. 8: Measured B6 to B4 reconfiguration with induction machine at 50 Hz output frequency and 3 N m load torque after always-on failure of low-side switch in phase W at $t = 0$. The fault is detected by comparing the phase currents in less than 7 ms.

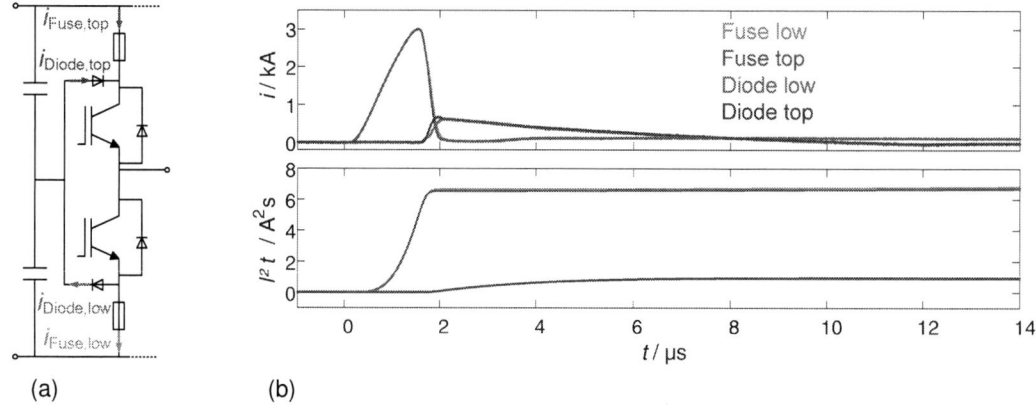

(a)   (b)

Fig. 9: Measured fuse and induced diode current after always-on failure of one switch and closing of the complementary switch (DC-link short-circuit). The fuses are blown after the rated I^2t value is achieved.

4. Conclusion

Adding six diodes and a DC-link midpoint connection in combination with ideal short-on failure switches is a simple and cost-effective method to integrate fault-tolerance in a standard B6 inverter. A new and less complex fault-tolerant B6 to B4 reconfiguration inverter is presented to enable a degraded, but continuous operation after switch failure. The characteristic of ideal short-on failure IGBT modules over a longer period with nominal current stress is analysed and offers a promising characteristic for fault-tolerant converters. Moreover, the reconfiguration strategy is presented and validated with a prototype. Design hints for the diode rating are given.

Acknowledgement

This work was supported by International Rectifier an Infineon Technologies Company by providing module samples and failure analysis and SIBA by supplying fuses.

References

[1] B. Welchko, T. Lipo, T. Jahns, and S. Schulz, "Fault Tolerant Three-Phase AC Motor Drive Topologies: A Comparison of Features, Cost, and Limitations," *IEEE Transactions on Power Electronics*, vol. 19, no. 4, pp. 1108–1116, 2004.

[2] W. Sanfins, F. Richardeau, D. Risaletto, G. Blondel, M. Chemin, and P. Baudesson, "Failure to short-circuit capability of emerging direct-lead-bonding power module. Comparison with standard interconnection. Application for dedicated fail-safe and fault-tolerant converters embedded in critical applications," in *Proceedings of the 17th European Conference on Power Electronics and Applications (EPE ECCE Europe)*, 2015, pp. 1–10.

[3] H. W. van der Broeck and J. D. van Wyk, "A Comparative Investigation of a Three-Phase Induction Machine Drive with a Component Minimized Voltage-Fed Inverter under Different Control Options," *IEEE Transactions on Industry Applications*, vol. 20, no. 2, pp. 309–320, 1984.

[4] R. Wang, J. Zhao, and Y. Liu, "A Comprehensive Investigation of Four-Switch Three-Phase Voltage Source Inverter Based on Double Fourier Integral Analysis," *IEEE Transactions on Power Electronics*, vol. 26, no. 10, pp. 2774–2787, 2011.

[5] M. B. d. Rossiter Correa, C. B. Jacobina, E. R. C. da Silva, and A. M. N. Lima, "A General PWM Strategy for Four-Switch Three-Phase Inverters," *IEEE Transactions on Power Electronics*, vol. 21, no. 6, pp. 1618–1627, 2006.

[6] J. O. Estima and A. J. M. Cardoso, "Fast fault detection, isolation and reconfiguration in fault-tolerant permanent magnet synchronous motor drives," in *Proceedings of the IEEE Energy Conversion Congress and Exposition (ECCE)*, 2012, pp. 3617–3624.

[7] M. Gleissner, M.-M. Bakran, and J. Marcinkowski, "First Fault-Resilient High-Power 5-Level Flying Capacitor DC-DC Converter with Ideal Short-On Failure IGBT Modules," in *Proceedings of the International Exhibition and Conference for Power Electronics, Intelligent Motion, Renewable Energy and Energy Management (PCIM Europe)*. VDE VERLAG, 2015, pp. 1338–1345.

[8] I. Yaqub, J. Li, and C. M. Johnson, "Dependence of overcurrent failure modes of IGBT modules on interconnect technologies," *Microelectronics Reliability*, vol. 55, no. 12, pp. 2596–2605, 2015.

Statistical Evaluation of Current Imbalance in Parallel Devices

Uwe Scheuermann, SEMIKRON Elektronik GmbH & Co. KG, Germany,
uwe.scheuermann@semikron.com

Abstract

The concept of paralleling plays a fundamental role in high power applications. A source of concern is the imbalance of current due to varying parameters of devices or components. As was already stated in [1], worst-case assumptions overestimate the consequences. However, the statistical method applied in this publication was incorrect. In this paper, the current and temperature imbalance of paralleled freewheeling diodes caused by a distribution in forward voltage drop will be analyzed on a correct statistical basis. The results confirm the general conclusions related to worst-case assumptions and give estimations for derating factors at defined statistical probabilities of occurrence.

1. Introduction

The concept of paralleling is inherent to power electronic applications; it is applied for parallel inverters in high power applications, parallel modules in high power inverters, parallel devices in high power modules and even for parallel operated cells in IGBTs or MOSFETs. However, the variation of parameters of components in parallel operation is a severe source of concern, since the resulting imbalance in the distribution of losses will compromise the system functionality and reliability.

A common practice to evaluate the impact of parameter variation is the worst-case assumption: components at the lower specification limit (LSL) are combined with components at the upper specification limit (USL) in the most unfavorable way and the imbalance in loss distribution and resulting component temperature is calculated.

The paper published in 2005 [1] investigated the problem of current imbalance resulting from the variation of forward voltage drop in parallel operated IGBT chips and diode chips. However, the statistical analysis of the problem was based on incorrect assumptions. This paper intends to give a correct statistical approach based on the same data base as published in 2005. A single parameter variation will be assumed: the variation of the on-state voltage. Since diodes in general exhibit a small positive or even negative temperature coefficient of the forward voltage drop dV_F/dT in the range of the nominal current – which will not mitigate imbalances as the typically pronounced positive temperature coefficient of IGBTs – this investigation is limited to diodes only.

In order to give a correct statistical analysis for the boundary conditions discussed in [1], all parameters remain unchanged and reflect the state of technology in 2005 – this should be kept in mind when transferring the results to up-to-date technologies.

The basis of this analysis is a statistical distribution of a month's production of Semikron CAL diodes SKCD61C120I. The data was collected in the production line end-test at I_F=50A and room temperature for 125,000 diodes produced in February 2004. The V_F values exhibit a normal distribution with the mean value V_{mean}=1816mV and a standard deviation σ=42.6mV as shown in Fig. 1. The first step is to approximate the on-state voltage drop of CAL-diodes as a function of current I_F and virtual junction temperature T_j. Additionally, a parameter s is introduced to identify the position of the device within the normal distribution. All dependencies are modelled by linear relationships as a first order approximation:

$$V_F(I_F, T_j, s) = V_{F0} + \beta_1 s + (\alpha + \beta_2 s)T_j + (r_F + \beta_3 s)I_F \qquad (1)$$

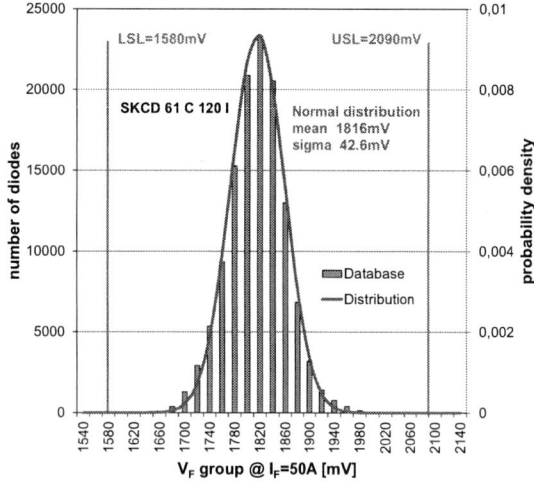

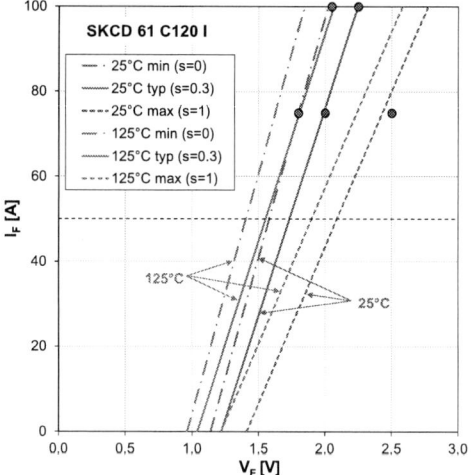

Fig. 1: Statistical distribution of on-state voltage at I_F=50A and T_j=25°C for 125,000 CAL diodes.

Fig. 2: On-state voltage of as a function of I_F, T_j and s according to eq. (1) together with data sheet values (full circles).

A set of coefficients was selected to match the data sheet values which results in a good correlation as shown in Fig. 2. It should be noted than eq. (1) contains no temperature dependence of the resistance parameter r_T. Furthermore, the scaling factor s equally affects V_{F0} and r_F which increases differences in on-state voltage with increasing currents.

To estimate the impact of variation of on-state voltage on the current imbalance, parallel operated power modules SKM100GB123 were considered. This module was selected, because it contains a single diode chip SKCD61C120I per switch which simplifies the statistical analysis. This selection also supports the assumption of thermally decoupled elements applied for the thermally and electrically consistent calculation of current distribution. Thus, an identical thermal resistance $R_{th(j-c)}$ and an identical constant case temperature T_c was assumed for all modules operated in parallel. For a group of n elements operated in parallel, the following boundary conditions must be fulfilled:

$$T_{j(i)} = T_c + R_{th(j-c)}V_{F(i)}I_{F(i)} \qquad i = 1 \dots n \qquad (2)$$

$$V_{F(1)} = V_{F(2)} = \cdots = V_{F(n)} \qquad (3)$$

$$\sum_{i=1}^{n} I_{F(i)} = n \cdot I_{F,nom} \qquad (4)$$

For compliance with the statistical distribution shown in Fig. 1, the nominal current was assumed as 50A. A case temperature T_c=85°C and a thermal resistance $R_{th,(j-c)}$=0.5K/W were chosen as constant values. These boundary conditions together with eq. (1) allow to calculate the current distribution for any given combination of elements in parallel operation.

2. Worst-Case Consideration

For the worst-case consideration, the most unfavorable combination of elements in a group of n elements within the specification limits is examined. The lower specification limit (LSL) and the upper specification limit (USL) for the discussed diode chips are also depicted in Fig. 1. The combination of elements with an on-state voltage given by LSL and USL will be analyzed.

For a single element (n=1) the worst-case is obtained for an element with an on-state voltage at the USL. For two elements (n=2) the worst-case is to combine one element with an on-state voltage at the LSL with another element with an on-state voltage at the USL. For this case, current is shifted from the diode at the USL to the diode at the LSL. This will increase the junction temperature of the diode at the LSL which further increases the current due to the negative temperature coefficient. As shown in Fig. 3, this will result for the diode at the LSL to a current $I_F{\sim}75A$ and a junction temperature $T_j{\sim}145°C$.

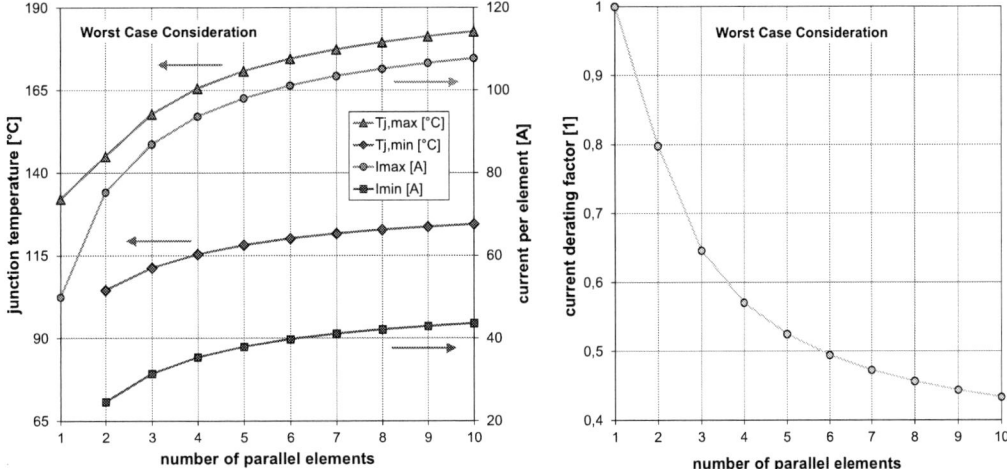

Fig. 3: Minimum and maximum values of T_j and I_F for worst-case combinations of n diodes (lines are only guide to the eye).

Fig. 4: Current derating factor to maintain the same maximum junction temperature as for a single element.

For n>2, diodes at the USL will be added to the group. This is the most unfavorable combination since more and more current is shifted form the devices with a high on-state voltage to the singe device with a low on-state voltage. Without derating, the current will rise to $I_F{\sim}108A$ and the temperature of the LSL diode will reach $T_j{\sim}183°C$ for n=10.

It is quite obvious that these worst-case consequences of parallel operation cannot be accepted for any application and a derating of the current is required to keep the maximum junction temperature inside specification limits ($T_j{\le}150°C$). In order to postulate a more general requirement which is less dependent on specific boundary conditions, the following definition was chosen for the current derating in this analysis: The maximum junction temperature in a group of parallel operated elements shall not be higher than for a single element operated at the same boundary conditions.

Thus, the maximum junction temperature was fixed at $T_j{\sim}132°C$ as calculated for a single diode with an on-state voltage at the USL and the boundary condition in eq. (4) was no longer applied. The ratio of the resulting total current to the ideal total current $n{\cdot}I_{F,nom}$ is defined as the derating factor as depicted in Fig. 4. The current must be derated to 80% for n=2 and to 43% for 10 parallel operated modules. This result makes parallel operation of power modules very unattractive.

However, the worst-case consideration does not take into account the probability of occurrence of the discussed combination of elements. Therefore, this probability is calculated for elements with $V_F{\le}LSL$ combined with elements with $V_F{\ge}USL$. This procedure assumes no selection process to eliminate devices out-of-specification. On the other hand, a good selection process will only reduce the probability by two orders of magnitude from the statistical point of view, which would even further reduce the probability of finding such elements.

The probability of selecting one diode with $V_F \leq LSL$ in the first draw and a second diode with $V_F \geq USL$ in a second draw can be calculated easily by multiplying the individual probabilities. However, selecting one diode with $V_F \geq USL$ in the first draw and a second diode with $V_F \leq LSL$ in a second draw would result in the same group after sorting the elements with respect to V_F. When specific combinations of elements in a group of samples are considered, the appropriate procedure is to sort the elements in each group and apply order statistics.

When groups of n elements $x_1,...,x_n$ are drawn from an entity described by the probability density function $f_x(x)$ and sorted to obtain $x_{[1]},...,x_{[n]}$ with $x_{[1]} \leq x_{[2]} \leq ... \leq x_{[n]}$ then the probability density function of any specific combination $(x_1,...,x_n)$ is defined by [2]:

$$f_{x_{[1]},...,x_{[n]}}(x_1, ..., x_n) = n! \; f_x(x_1) \cdot ... \cdot f_x(x_n) \quad where \quad x_1 \leq x_2 \leq \cdots \leq x_n \tag{5}$$

The factorial n! accounts for all permutations of elements that results in the same ordered sequence in order statistic. Applied to the discussed worst-case consideration of parallel operated diodes, this delivers:

$$F\big(x_{[1]} \leq LSL, x_{[2]} \geq USL, ..., x_{[n]} \geq USL\big)$$
$$= n! \cdot F\big(x_{[1]} \leq LSL\big) \cdot F\big(x_{[2]} \geq USL\big) \cdots F\big(x_{[n]} \geq USL\big) \tag{6}$$

The resulting probability calculated for the worst-case consideration is shown in Fig. 5. In the publication from 2005 [1], the probability for the worst-case consideration did not account for the permutations and is therefore not correct. However, the probability of occurrence of such a worst-case combination of elements is still decreasing rapidly with growing numbers of elements. Thus, the conclusions drawn are still valid, that combinations derived from worst-case consideration are very unlikely to occur in real life, especially for a larger number of elements.

It would be much more realistic to replace the worst-case consideration by a consideration of combinations of elements with a defined probability of occurrence. However, the statistically correct calculation is quite complex. In [1] it was assumed, that all (n-1) elements with the high V_F have the same value, but this assumption is not correct. For a correct calculation, all combinations of elements delivering the same maximum junction temperature have to be collected and then the limit of the junction temperature related to a defined probability of occurrence must be determined. This will involve calculating nested convolution integrals with a growing complexity for an increasing number of elements. Therefore, the decision was taken to investigate this problem by Monte-Carlo-Simulation.

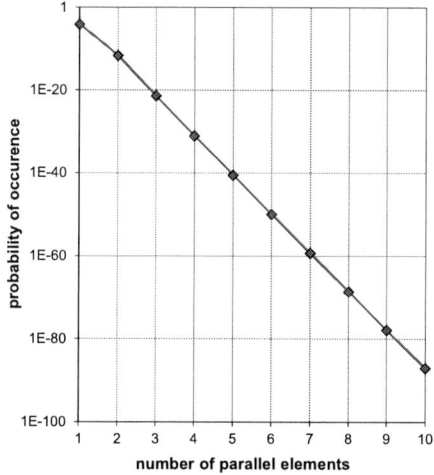

Fig. 5: Probability of occurrence for one diode with $V_F \leq LSL$ and (n-1) diodes with $V_F \geq USL$.

3. Monte-Carlo-Simulation

Microsoft Excel including Visual Basic for Applications (VBA) was chosen as platform for the Monte-Carlo-Simulation, since it is widely available and runs on any laptop computer. It was also used for the previously discussed calculations, which made use of the Solver tool to numerically solve the equations (1-4) by least square fit.

PCIM Europe 2016, 10 – 12 May 2016, Nuremberg, Germany

The goal is to create a large number of groups with n normally distributed elements each. Random values with a standard normal distribution can be obtained from uniform distributed random numbers in [0,1] by the Polar-method, which is a modified Box-Muller algorithm that uses polar coordinates instead of cartesian coordinates and thus avoids the evaluation of trigonometric functions [3].

The VBA function Rnd() delivers uniform distributed values in [0,1], but a test showed that the periodicity of this algorithm is only $2^{24} \sim 16.78$ million which is insufficient for the purpose of generating 1 million groups of up to 10 elements. Thus a Wichmann-Hill pseudo-random number generator [4] was implemented with an approximate period of $2^{43} \sim 8.8 \cdot 10^{12}$. The algorithm is seeded with the system timer value to generate independent sequences of uniform random numbers. Even though a periodicity of $>2^{60}$ is proposed [5] for high quality pseudo random number generators, the chosen algorithms should be adequate to give a good estimation of the current derating factors based on a statistical probability of occurrence.

To investigate the worst combination of elements with a defined probability of occurrence, a set of 1 million groups of standard normal distributed values was generated for each number of elements from n=2 to n=10. To check the quality of the values, the difference between the lowest value and the highest value in each set of n million values was determined which gives an estimation for a probability of 1ppm. Then the difference between the 10th highest value and the 10th lowest value was extracted which gives an estimation for a probability of 10ppm. Likewise, the differences for 100ppm and 1000ppm probability were identified. The result is shown in Fig. 6 along with the theoretical expectation for the associated number of elements. The visualization of the theoretical values as continuous lines only serves illustration purposes; the values are only valid for discrete number of elements.

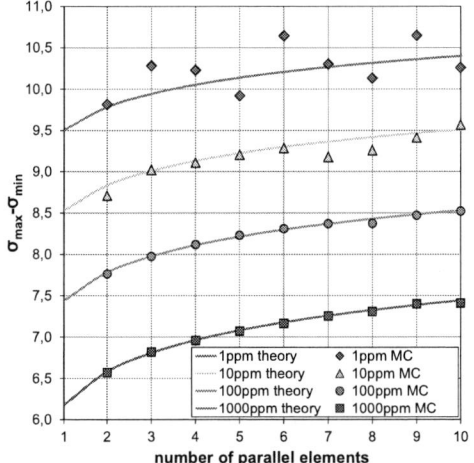

Fig. 6: Differences between highest and smallest standard normal values in 1 million groups of n elements.

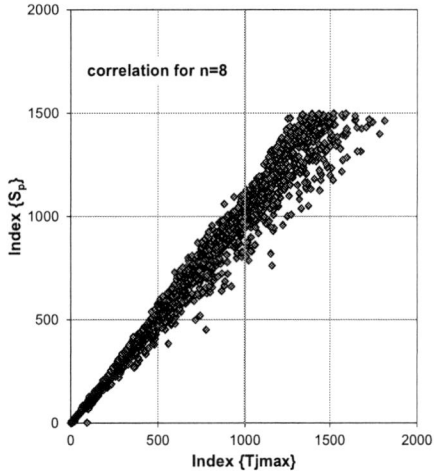

Fig. 7: Correlation between the indices of pre-sorting function S_p and maximum junction temperature $T_{j,max}$.

The result shows the expected behavior of the statistical samples. The differences for 1ppm probability exhibit a strong variation to both sides of the theoretical value because it is defined by single values. The variation decreases for 10ppm since it is determined by the 10 highest and 10 lowest values. It decreases further for 100ppm and correlates almost perfectly with the theoretical expectation for 1000ppm probability.

Now the maximum junction temperature must be determined. The values within each group were sorted in ascending order prior to further investigations. This allows to predict the elements which will exhibit the maximum junction temperature within each group, because

© VDE VERLAG GMBH · Berlin · Offenbach

the element with the smallest on-state voltage will always have the highest junction temperature in a group of parallel operated diodes. The creation of the set and the sorting within the groups were implemented in VBA and performed on a conventional laptop computer.

An algebraic solution for the equation system defined by eq. (1-4) can be found for n=2. This allowed to calculate the maximum junction temperature for each of the 1 million groups of two elements. After sorting the groups for descending maximum junction temperature, the highest value was selected for 1ppm probability, the 10th highest for 10ppm probability, and likewise for the probability of 100ppm and 1000ppm.

For n>2 the equation system must be solved numerically. This could not be done for 1 million groups in acceptable computation time, so a pre-sorting of groups is required. Based on the fact that the current imbalance will be more pronounced when the difference between the smallest element and the mean value of the rest of elements within a group is large, a first contributing factor can be identified. Secondly, the maximum temperature will increase if the mean value of the elements increases, which defines a second contributing factor. A parameter optimization was applied to define a pre-sorting value for the groups in the following form:

$$S_p\big(x_{[1]}, \ldots, x_{[n]}\big) = A\left(\frac{1}{(n-1)}\sum_{i=2}^{n} x_{[i]}\right) - x_{[1]} + B \quad with \quad A = 1.18\,, \; B = 0.36 \quad (7)$$

After the pre-sorting of groups for descending values of S_p the maximum temperature of the 2000 highest groups were calculated with the Excel Solver tool. Then these 2000 groups were sorted for maximum junction temperature. The correlation between the maximum junction temperature index and the pre-sorting function index for the case n=8 is displayed in Fig. 7 for index values of the pre-sorting function ≤1500. It validates that by numerically calculating 2000 pre-sorted groups, the 1000 groups with the highest maximum junction temperature will definitely be identified.

Now the maximum junction temperatures for different probability levels can be extracted with the result shown in Fig. 8. The fluctuation is high for 1ppm and decreases for increasing levels of probability. The maximum temperature for 10 parallel elements will not exceed T_j=147°C with a probability of 1ppm.

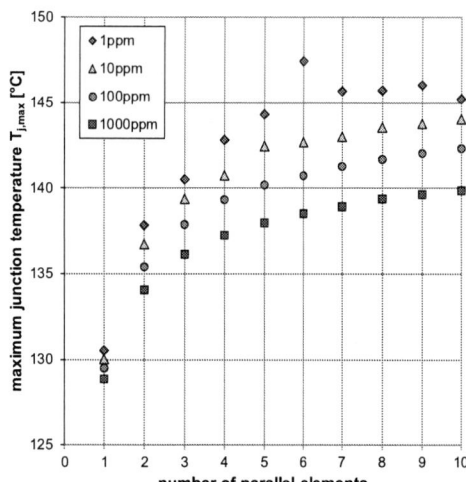

Fig. 8: Maximum junction temperature determined by Monto-Carlo simulation for defined probability levels.

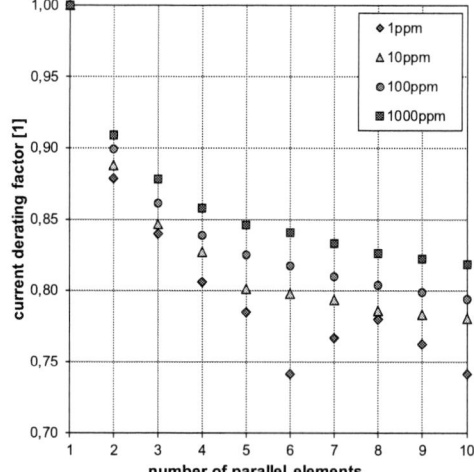

Fig. 9: Current derating factor to maintain the same maximum junction temperature as for a single element.

The current derating factor for defined statistical probability can be calculated based on the same definition as applied for the worst-case consideration. For each level of probability, the maximum junction temperature must be the same as for the case of a single element at the same probability level (shown in Fig. 8 for n=1). As expected, this derating factor is much lower than the values derived from the worst-case consideration (Fig. 4) The derating margin $1-I_{F,parallel}/(n \cdot I_{F,nom})$ is less than half the value predicted by worst-case consideration for n>2 even for a probability level of 1ppm.

The accuracy of the results from a single MC run is of course low and could be improved by multiple MC runs. However, since the results are only valid for the given statistical distribution and selected boundary conditions, this effort is not justified. Furthermore, the impact of asymmetric bus bar connections between modules – which is not considered here – can have a greater impact on the current distributions than parameter variations, as was shown in experimental investigations [6]. The results should merely be considered as an estimation for the impact of on-state voltage on current distribution.

4. Conclusion

The impact of parameter variations on the current imbalance in parallel operated diodes is investigated in this paper. Based on a distribution of the on-state voltage with LSL~-5.6σ and USL~6.4 σ in combination with realistic boundary conditions, the resulting maximum junction temperature is calculated with consistent thermal-electrical coupling for the worst-case consideration, as was already published in 2005 [1]. The probability of occurrence of worst-case combinations shows that this approach is not adequate.

Monte-Carlo-Simulation was applied to investigate the current imbalance with respect to defined levels of probability. In contrast to the publication from 2005 – which concluded that no current derating is required based on invalid statistical assumptions – current derating is necessary for parallel operation. The derating factors resulting from a statistical distribution of on-state voltage for the discussed example of parallel diodes can be estimated with Fig. 9 for different levels of probability.

The current rating for parallel operated diodes can be chosen much higher according to the statistical evaluation than predicted by the worst-case consideration. This will not prevent the occurrence of very unfavorable combinations of diodes, but it predicts that the number of such bad combinations will be acceptably small.

It should be noted that the computer resources to perform the Monte-Carlo-Simulation presented here were not available to the author 11 years ago.

5. References

[1] U.Scheuermann: Paralleling of Chips – From the Classical 'Worst Case' Consideration to a Statistical Approach, Proc. PCIM Europe 2005, 455-460.
[2] H.A.David: Order Statistics, Wiley, New York (1981).
[3] G.Marsaglia, T.A.Bray: A Convenient Method for Generating Normal Variables, Society for Industrial and Applied Mathematics (SIAM) Rev. 6(3) 1964, 260-264.
[4] B.A.Wichmann, I.D.Hill: Algorithm AS 183: An efficient and portable pseudo-random number generator, Applied Statistics 31(1982), 188-190.
[5] B.D.McCullough: Microsoft Excel's 'Not The Wichmann-Hill' random number generators, Computational Statistics and Data Analysis 52 (2008), 4587-4593.
[6] A.Wintrich, J.Nascimento, M.Leipenat: Influence of parameter distribution and mechanical construction on switching behaviour of parallel IGBT, Proc. PCIM Europe 2006, 511-516.

Batch purity in semiconductor power modules

Christian Aggen, Danfoss Silicon Power GmbH, Germany, christian.aggen@danfoss.com
Henning Ströbel-Maier, Danfoss Silicon Power GmbH, Germany, henning.stroebel-maier@danfoss.com
Matthias Mau, Danfoss Silicon Power GmbH, Germany, matthias.mau@danfoss.com
Jürgen Laue, Danfoss Silicon Power GmbH, Germany, jurgen.laue@danfoss.com
Marco Bäßler, Danfoss Silicon Power GmbH, Germany, marco.baessler@danfoss.com

Abstract

The influences of batch variations on temperature distribution of parallel connected IGBT chips are investigated. For this purpose the distribution of static and dynamic losses were determined by measurements and simulations to calculate the temperatures by using a Monte-Carlo simulation. In this paper the parameters V_{GEth} and V_{CE} of one chosen IGBT type were analyzed with the assumption it represents typical IGBT behavior. The results show the difference between a module with a single IGBT batch and multiple batches and the importance in comparison to package parasites.

1. Introduction

High power systems and devices often require connecting semiconductors in parallel to be able to switch higher currents than a single semiconductor device is capable of. Balancing of the current distribution and power dissipation between parallel connected semiconductor devices is crucial for maintaining equal temperature, achieving lifetime goals and avoiding overloads.

Variations in semiconductor characteristics of parallel connected semiconductor chips normally lead to unbalanced power distribution inside power modules. According to chip data sheets, chip parameters can vary between the given lower specification limit (LSL) and upper specification limit (USL). However, these limits are often theoretical and conservative. Experience shows that the measurable deviation of parameters within a single chip lot manufactured by reputable semiconductor manufacturers is considerably less than the specified limits indicate. That is why it is common use to restrict the power module production to single chip batches, it reduces the variation.

The difference between a small value span given by a single chip lot and the complete span between the specified limits by theoretical mixing different chip lots is evaluated by simulating temperature differences in a power module model with parallel connected IGBT chips. The resulting temperature difference between coolest and hottest chip per module is an important criteria to answer the question if single-batch production is a must to produce power modules.

2. Simulation

Evaluating the influence of chip lot variations would generally mean to regard all possible specific characteristics of the chip and crosscheck their impact to the module behavior. To reduce the complexity to a reasonable level it has been decided to define a few important parameters for simulations in this approach. Out of the many parameters which define the characteristics of an IGBT the dominant parameters for paralleling such devices are the gate-threshold-voltage V_{GEth} and the output characteristic or simply the V_{CE}, as also implied by other publications [1] [2]. While the V_{GEth} influences dynamic losses during switching the V_{CE} mostly determines static losses.

2.1. Monte-Carlo simulation – creation of random pairs of V_{CE} and V_{GEth}

The idea for comparing single-batch and mixed-batches characteristics by simulation is to generate pairs of V_{CE} and V_{GEth} values referring to their probability of occurrence within a single batch of chips on the one hand and within a normal distribution between lower and upper specification limits on the other hand. After the value pairs are generated they are put together in groups in the amount of parallel chips to be investigated for a chosen power module type which in turn are then used as input value for thermal simulations.

The information about the single lot can be gained from actual measurements while the specification limits can be taken from the corresponding chip datasheet, assuming a 6-sigma normal distribution between the limits. V_{GEth} and V_{CE} of several batches with thousands of chips each were analyzed to check if the assumption is suitable. In the left, Fig. 1 depicts the resulting distributions for these two parameters for a typical IGBT type (IGC136T170S8RH2) which show that the measured values lay reasonably well below the bell-curve with slight asymmetries and outliers. However, since the cumulated distributions are not too different from the Gaussians, the assumption is proved useful for this approach. The discrete value steps for analyzing the likelihood of appearance as well as for generating the value pairs afterwards were chosen to be not wider than 1% of the range between upper and lower specification limit, as depicted on the right of Fig. 1 for an exemplary single batch distribution.

2.2. Determining parameter influence with SPICE

Variation of IGBT model parameters

While differences of V_{CE} mostly have influence on the static losses by uneven current distribution between the chips, differences in V_{GEth} are more important during switching transitions and influence dynamic losses. Fig 2 shows a measured extreme example for this influence where one IGBT out of six in parallel had an about 1V higher V_{GEth} and switches off the load current I_C much earlier than the others. The offset was generated with a voltage source in series to the gate of this particular chip, subtracting 1 V from the driver voltage. Of course the switching losses of this IGBT are much lower than the others as depicted on the right of Fig. 2. The detailed effect on the switching losses was determined by simulating a double pulse scenario in a chopper circuit with 6 IGBT and diode models connected in parallel in a SPICE (Simulation Program with Integrated Circuit Emphasis) simulator, while electrical measurements with manipulated power modules only where used to check the simulations to certain extend.

The parameter V_{GEth} of one IGBT model at a time was tuned from lower specified limit to the upper while all other models were left the same. Observing the collector current and collector-emitter voltage of each IGBT then gives information about its static and dynamic losses depending on the chip parameters.

The simulation results were implemented in a thermal simulation model to convert the V_{GEth} and V_{CE} input values to power losses.

PCIM Europe 2016, 10 – 12 May 2016, Nuremberg, Germany

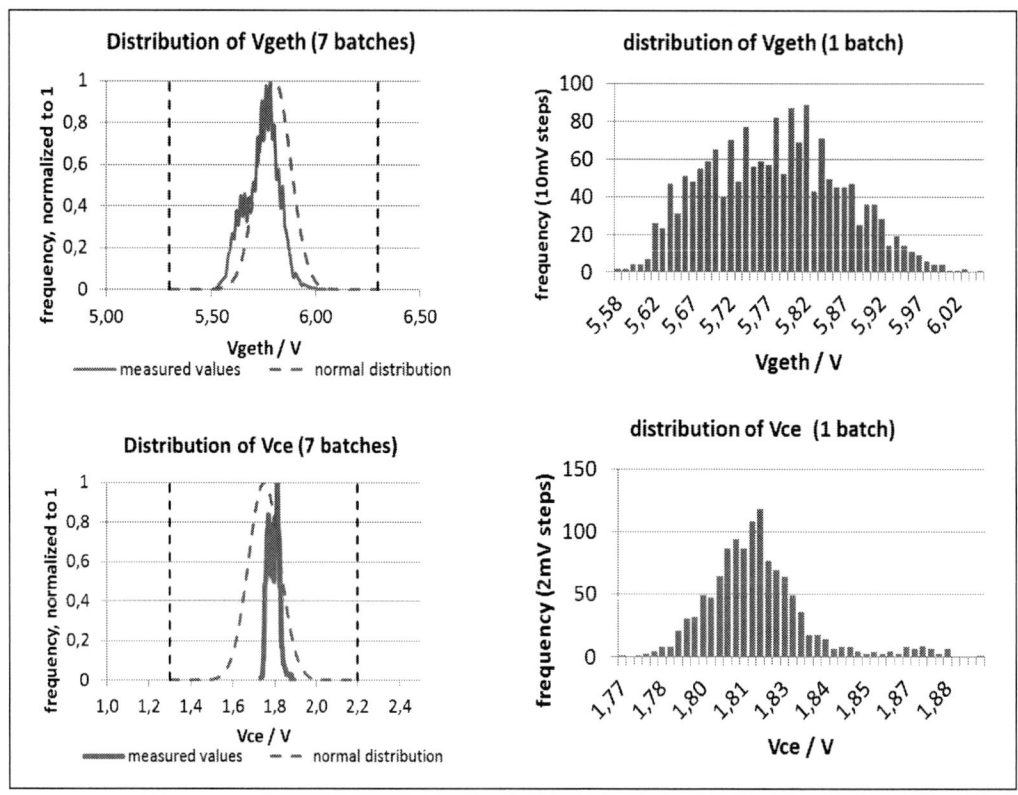

Fig. 1. Distribution of V_{GEth} and V_{CE} within 3 IGBT batches and normal 6 sigma distribution based on specification limits from chip supplier (left) and single batch (right).

Package parasites extraction with Q3D

Since losses and switching behavior of semiconductors in general are also affected by package parasites, a SPICE block containing a matrix of package resistances and inductances extracted from a 3D model of the power module of choice was added to the SPICE circuit (Fig 3). The extracted SPICE block represents a specific low-side chopper module of the P3 type with six chips per switch in parallel.

After implementation of the parasites block the SPICE simulation of the parameter influence was run again.

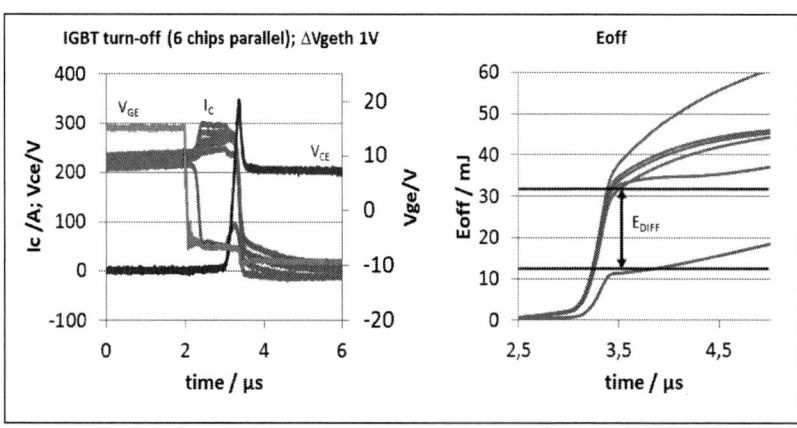

Fig. 2. Example illustration of a current distribution I_C during IGBT turn-off with different V_{GEth} and the effect on switching losses E_{DIFF}.

© VDE VERLAG GMBH · Berlin · Offenbach

700

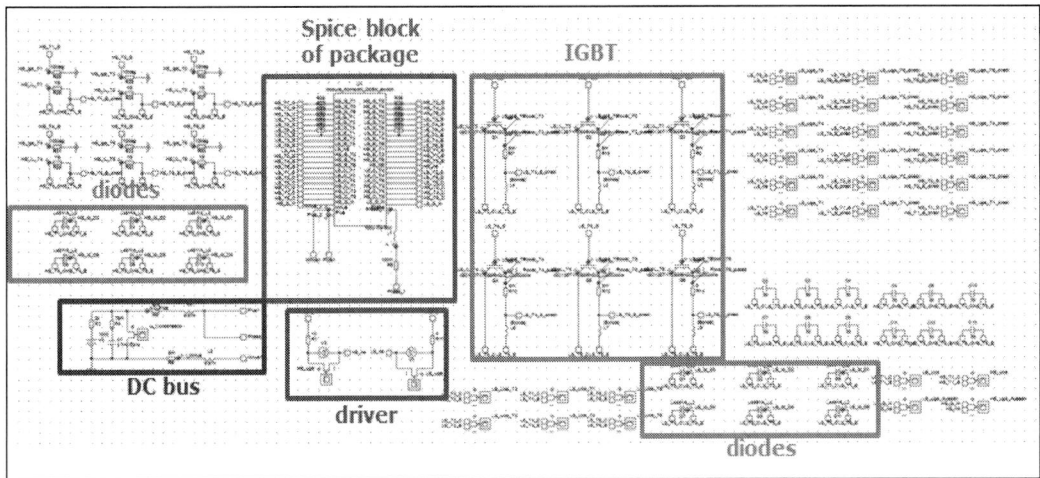

Fig. 3. SPICE model for switching simulations of a chopper module with package influence.

2.3. Thermal simulation

After having generated the database of V_{GEth} and V_{CE} values and their influence on thermal losses in a power module, the final thermal simulation was performed.

Overall 10.000 parameter sets were generated for a start to be fed into a thermal simulation tool. The procedure is depicted in Fig. 4. For specific operating conditions the simulation tool calculates the current, losses and temperature of each chip resembled by an equation. The conditions are overall load current, ambient temperature, thermal resistance for each chip and switching frequency. Since current capability, power losses and temperature are depending on each other the calculated values are fed back into another calculation loop. This way the tool iteratively solves the set of equations until quasi equilibrium between overall current is reached. Then the solver stops and outputs the maximum temperature difference ΔT_{max} between the hottest and coldest chip for this particular set of parameters.

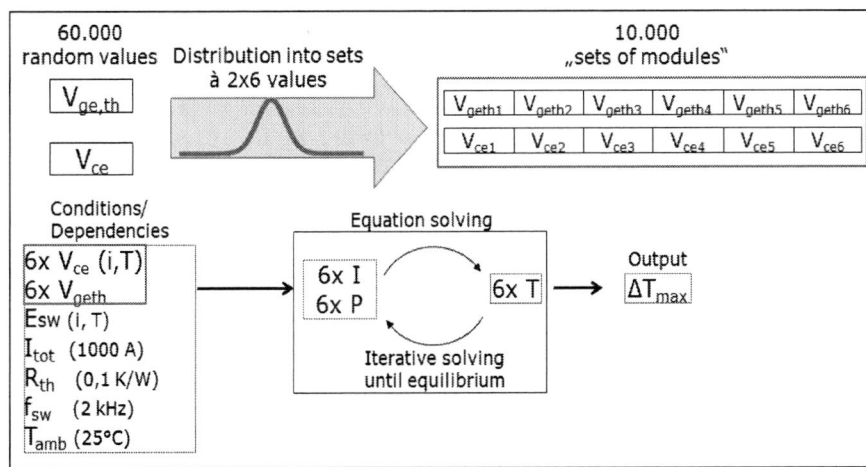

Fig. 4. Depiction of generating parameter sets via Monte-Carlo simulation (top) and principle of thermal simulation by iterative equation solving for a given operation condition (bottom)

2.4. Results

Parallel connected chips without package parasites

Fig. 5 shows the resulting frequency of occurrence of the maximum temperature difference between hottest and coldest chip for all simulated modules. As expected, the range of maximum temperature difference for the single batch is much narrower than for the normal distribution. Remarkable for the normal distribution results are the outliers which represent a high percentage of the total range. This huge difference to the single batch is a result of the much broader range of input values. For example, the range of possible values for the normal distributed V_{CE} is about 3 times wider as the typical distribution within a single batch as shown before. Also no extreme outliers were found in the measured values, hence the missing outliers in the related thermal simulation.

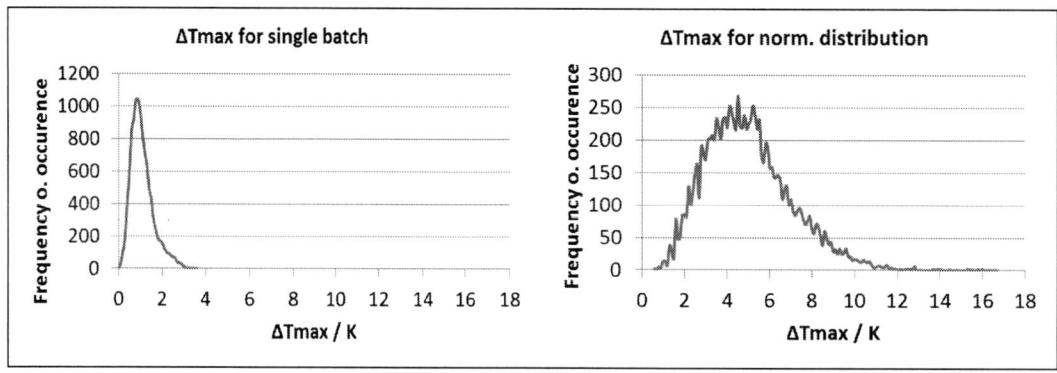

Fig. 5. Maximum temperature difference for one single chip batch (left) and for normal distributed parameters (right) for simulation without package parasites

Parallel connected chips with package parasites

The temperature distributions in Fig. 6 show most interesting results of the simulation now with package parasites but the same chip values as in the previous simulation. The range for mixed batches rises only 1 K in offset, but otherwise looks very similar to the results without package parasites. Whereas the range of the single batch rises about 2 K and gets wider to now reach up to about 8 K.

A difference to the simulation without package parasites was expected since most large power module packages have slightly asymmetric impedance current paths, leading to not completely even distributed currents and thus different chip temperatures. As these results show, the combination of mixing wide ranges of chip parameters and package parasites is not simply adding both effects. Obviously the package parasites can compensate a good part of a large distribution of chip parameters. It was also observed that some chips are more or less sensitive to parameter changes than others, depending on their position in the package. Certain positions predefined the range of switching speed the chip could cover by parameter variation. For example, a predefined fast and early switching chip was more sensitive, because it had to take some significant extra current of the neighboring recovering diode when switched even faster. This is a new aspect for future studies.

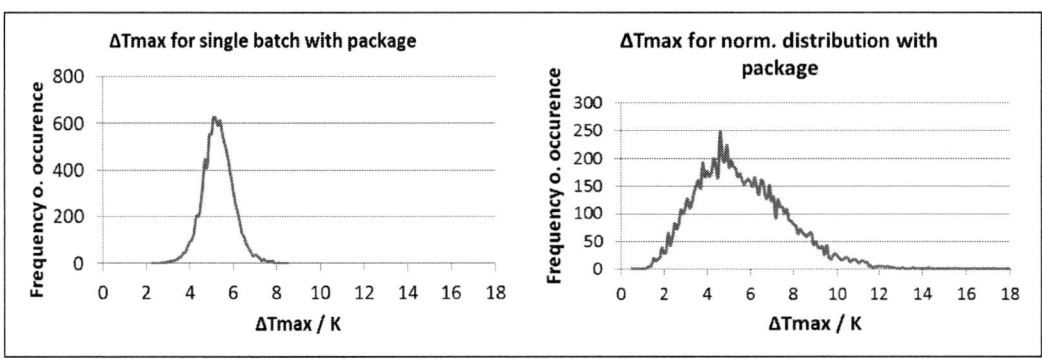

Fig. 6. Maximum temperature difference for one single chip batch (left) and for normal distributed parameters (right) with package parasites

3. Conclusion

The results show that using single IGBT batches lead to smaller temperature differences within a power module than using chips from several batches combined, provided the parameter distributions of multiple batches do not cover completely the same range.

Further, it can be recommended to narrow the parameter ranges beyond the typical spread of single batches to get even lower temperature differences, for instance by using classified chips.

It was also shown that the package of the chosen power module has quite a big influence on the temperature distribution. Therefore, it is reasonable to assume that depending on the choice of a package type the gain of the narrow parameter range is relativized to a certain degree.

4. References

[1] A. Wintrich; "Influence of parameter distribution and mechanical construction on switching behaviour of parallel IGBT"; PCIM Nuremberg; 2006

[2] U. Scheuermann; "Paralleling of Chips – From the Classical 'Worst Case' Consideration to a Statistical Approach"; PCIM Nuremberg; 2005

Novel Technique to Reduce Substrate Tilt & Improve Bondline Control between AlN Substrate and AlSiC Baseplate in IGBT Modules

J. Booth – Dynex Semiconductor Ltd - James_Booth@dynexsemi.com

K. Vijay – Indium Corporation - kvijay@indium.com

P. Mumby-Croft – Dynex Semiconductor Ltd - Paul_Mumby-Croft@dynexsemi.com

M. Packwood – Dynex Semiconductor Ltd - Matthew_Packwood@dynexsemi.com

K. Evans – Dynex Semiconductor Ltd - Kim_Evans@dynexsemi.com

A. Dai – Dynex Semiconductor Ltd - Andy_Dai@dynexsemi.com

Abstract

Bondline control, that is the use of spacers or standoffs applied to a power module's baseplate to achieve a homogenous solder layer, is a well understood technology employed by most power semiconductor manufacturers. It is understood that an inhomogeneous solder layer can lead to early device failure caused by cracking and delamination of the solder during thermal cycling, and that spacer technology such as stitched wirebonds for AlSiC baseplates or stamped 'bumps' on copper baseplates can inhibit this behaviour and increase joint lifetime.

This paper presents a novel alternative method of achieving bondline control on AlSiC baseplates whilst offering a drop in solution with no additional manufacturing steps or capital investment costs; a solder preform engineered with an embedded metal mesh across the area of the preform. Samples were made to evaluate the thermal fatigue resistance of this technology in comparison to both the traditional wirebond method and samples made without any bondline control. Active metal braze Cu-AlN-Cu substrates were soldered to 140x70mm AlSiC baseplates for these trials with each bondline variant. The samples were then temperature cycled at a ΔT of 200K, and analysed by scanning acoustic microscopy every 200 cycles. Cracking and solder layer delamination was observed at 600 cycles for the samples without bondline control and at 800 cycles for the samples with Al wirebonds; no signs of thermal fatigue was witnessed on the samples with the embedded metal mesh. The embedded metal mesh samples showed the least co-planarity deviation and superior reliability results to the Al wirebond method with the added advantage of no additional process steps.

1 Introduction

Large area solder joints in multi-chip power semiconductor packages experience fatigue caused by the periodic straining of the interconnection layers during thermal excursions as the device is operational. These stresses lead to delamination and cracks within the solder layer after many thermal cycles which increase the module to heat-sink thermal resistance and ultimately lead to early device failure [2].

Cracking and solder layer delamination occurs earlier in inhomogeneous solder joints due to stress concentration at thinner areas of the joint, Figure 1.0 shows how crack length within the solder joint increases greatly with solder layers thinner than 200µm. This figure illustrates how tilted samples where part of the joint is <200µm is more susceptible to cracking and delamination.

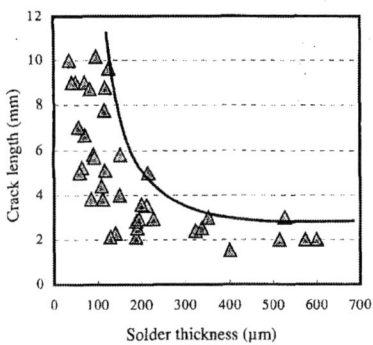

Figure 1.0: Correlation between solder joint thickness and induced crack length after thermal cycling [1]

The advent of spacer technology allows control of the solder joint thickness for a given solder volume by reducing substrate tilt to achieve a homogenous solder layer as Figure 1.1 demonstrates. This is most commonly done in power semiconductor modules by stitch bonding aluminium wire of a desired diameter to an AlSiC baseplate or for copper baseplate modules, copper 'bumps' can be stamped in the component (Figure 1.2).

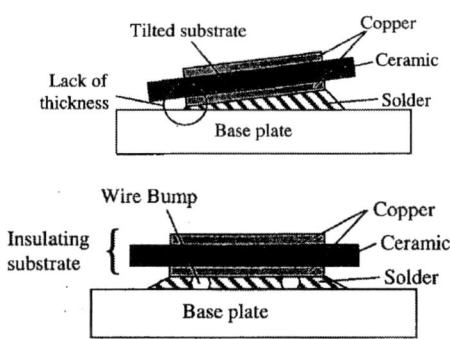

Figure 1.1: Substrate tilt example (top) and solution using wire bonds to achieve bondline uniformity (bottom) [1]

Figure 1.2: Traditional bondline control methods, aluminium wirebonds on AlSiC baseplate (left) & stamped 'bump' in copper baseplate (right)

The use of spacers in large area solder joints increases the joint lifetime by allowing for homogenous delamination, this occurs at a much slower rate than inhomogeneous delamination caused by substrate tilt [1-2]. This technology is well documented and employed today in power module assembly but this technique results in a high cost of ownership due to extra process steps and capital equipment costs.

© VDE VERLAG GMBH · Berlin · Offenbach

An alternative solution to achieving a homogenous solder layer is proposed by using a solder preform engineered with an embedded metal mesh. In the same way as the traditional wire-bond method, when the solder melts during reflow the metal mesh remains intact and serves to maintain a uniform bondline thickness. As well as the obvious reduction in manufacturing time by the advent of a drop in solution to achieving bondline homogeneity, the lifetime of this novel technique is evaluated against the traditional aluminium stitch bond method.

2 Sample Preparation

To evaluate the lifetime of the embedded metal mesh preform, sample modules were made and evaluated against the traditional wire-bond method for achieving a homogenous solder layer; these samples were also compared to reference modules with no bondline control; four of each variant was tested. The samples were then temperature cycled with a ΔT of 200K and cracking and delamination of the solder layer was monitored by scanning acoustic microscopy every 200 cycles.

Module assembly consisted of soldering ceramic AlN substrates (with Cu metalisation) to 140x70mm AlSiC baseplates using 200μm SnSb5 solder preforms. The samples with the embedded metal mesh consisted of a 200μm mesh with a 225μm net solder thickness. The samples with aluminium stitch bonds used 200μm diameter wire, the net height however is reduced to approximately 180μm due to the compression of the wire caused by the bond tool (Figure 2.0). These were compared to 200μm thick preforms with no bondline control.

200μm is the targeted bondline thickness as this offers the lowest thermal resistance without suffering from increased strain. Figure 2.1 illustrates how the normalised strain (non-dimensional strain normalised by the equivalent plastic strain) increases rapidly at thickness <200μm yet at thicknesses up to 600μm the normalised strain barely alters. This effect is also shown in Figure 1.0 as the increased strain on the solder joint promotes crack propagation [1].

Figure 2.0: Assembly images showing sample soldered with metal mesh preform (top) and sample with wirebond spacer (bottom)

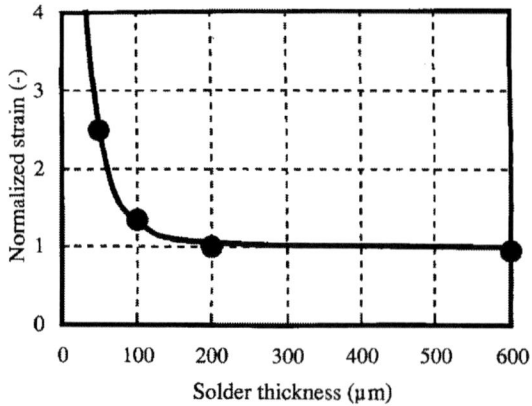

Figure 2.1: Correlation between solder thickness and the non-dimensional strain normalised by the equivalent plastic strain [1]

Following the assembly of the samples, one of each sample type underwent a laser surface profiling scan to determine the substrate tilt prior to thermal cycling tests. This was determined as the mean height variation across the top of the substrate at four points; the maximum deflection was also measured. The sample with the embedded metal mesh shows the least co-planarity deviation (smallest ΔZ) at 52.5µm and a maximum deflection of ~60µm (Figure 2.2), followed by the wire-bonded sample at 56.5µm with a maximum deflection at ~70µm (Figure 2.3) and finally the sample without bondline control at 67.5µm and a maximum deflection of ~90µm (Figure 2.4). It is noticed that on all samples the substrates are tilted inwards towards the baseplate, owing to the concave shape of the baseplate.

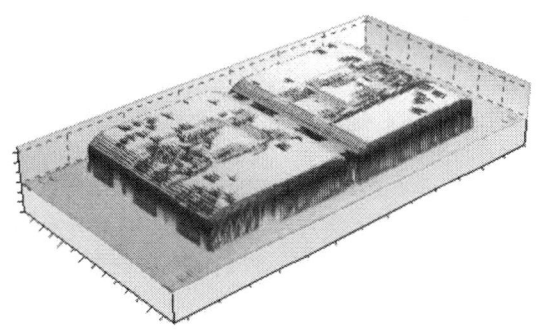

Figure 2.2: Sample with embedded metal mesh; Mean co-planarity deviation = 52.5µm, Max Deflection = 60µm

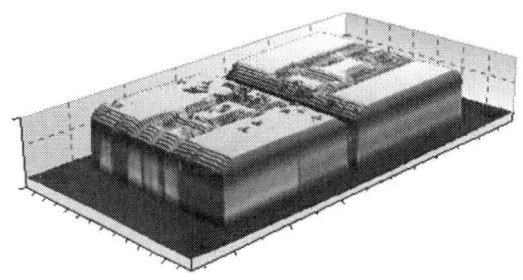

Figure 2.3: Sample with Aluminium Stitch bonds; Mean co-planarity deviation = 56.5µm, Max Deflection = 70µm

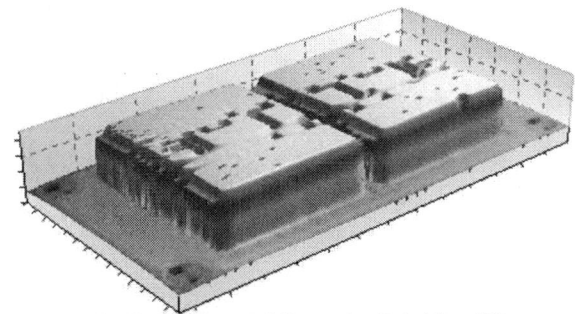

Figure 2.4: Sample with no bondline control; Mean co-planarity deviation = 67.5µm Max Deflection = 90µm

3 Thermal Cycling

Samples were thermal cycled using a Vötsch VT 7012 S3 chamber to chamber thermal cycler. The samples were cycled from -50°C to 150°C under the following conditions (figure 3.0).

t_{dwell} = 1 hour $\qquad$ $T_{s(max)}$ = 150°C

$t_{transition}$ = 30 seconds $\quad$ $T_{s(min)}$ = -50°C

ΔT = 200K

© VDE VERLAG GMBH · Berlin · Offenbach

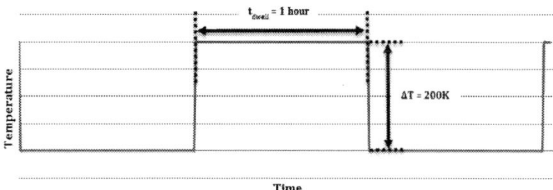

Figure 3.0: Representative temperature profile for thermal cycling test conditions

4 Results

Figure 4.0 below shows the initial SAM images at zero cycles of the baseplate/solder interface for all three techniques. Delamination of the solder layer is witnessed as bright reflections emanating from the edges of the solder layer. No delamination was witnessed at 200 and 400 cycles for all bondline variants however Figure 4.1 shows the metal mesh in the preform highlighted after 400 cycles. This is likely due to the expansion of the metal mesh within the solder; this expansion was not however noted to increase with further thermal cycling. Figure 4.2 (c) shows bright reflections appearing at the edges of the solder after 600 thermal cycles for the samples with no bondline control. The samples with the embedded metal mesh showed no signs of delamination or solder layer cracking.

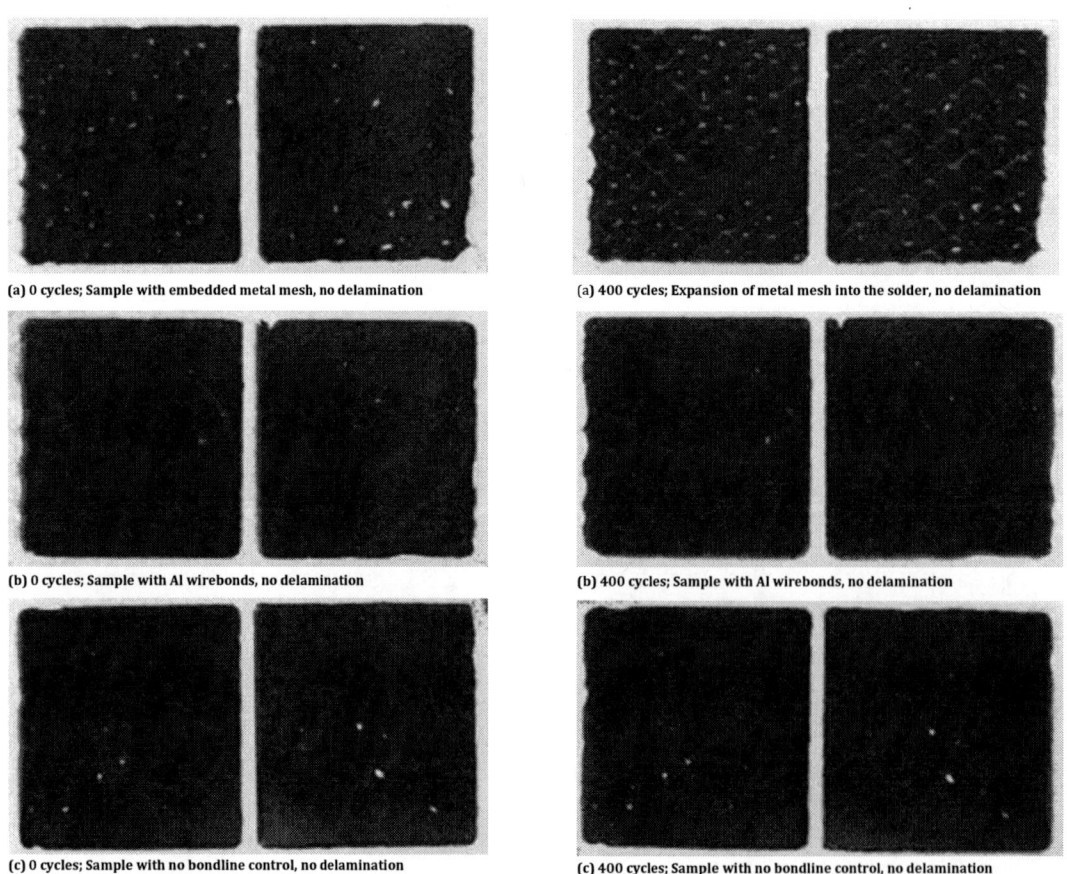

(a) 0 cycles; Sample with embedded metal mesh, no delamination

(a) 400 cycles; Expansion of metal mesh into the solder, no delamination

(b) 0 cycles; Sample with Al wirebonds, no delamination

(b) 400 cycles; Sample with Al wirebonds, no delamination

(c) 0 cycles; Sample with no bondline control, no delamination

(c) 400 cycles; Sample with no bondline control, no delamination

Figure 4.0: Initial SAM results at zero cycles

Figure 4.1: SAM images after 400 thermal cycles

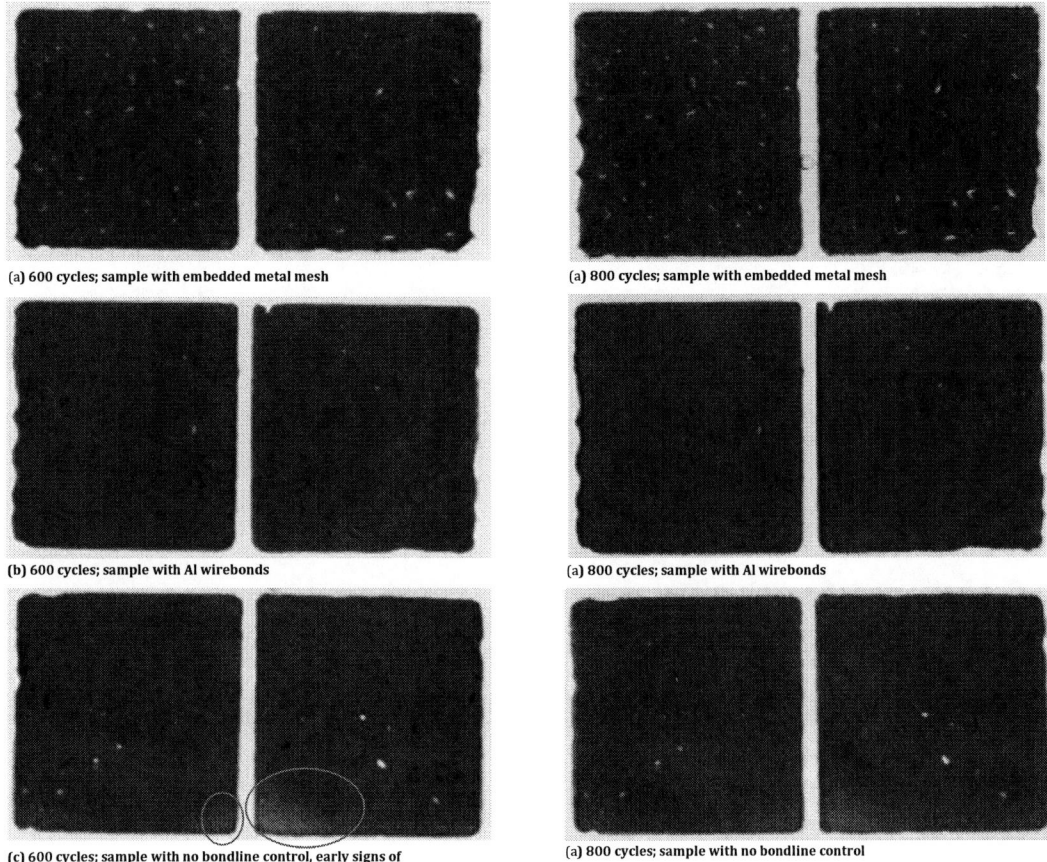

(a) 600 cycles; sample with embedded metal mesh

(a) 800 cycles; sample with embedded metal mesh

(b) 600 cycles; sample with Al wirebonds

(a) 800 cycles; sample with Al wirebonds

(c) 600 cycles; sample with no bondline control, early signs of inhomogeneous delamination occurring at corner of the substrate

(a) 800 cycles; sample with no bondline control

Figure 5.2: SAM results after 600 cycles

Figure 4.3: SAM results after 800 cycles

The sample shown in Figure 4.2 (c) and other samples made with no bondline control exhibited similar behaviour at the same number of thermal cycles. These defects however did not drastically increase at 800 cycles (Figure 4.3). Cracking was witnessed in the secondary SAM gate reflecting the substrate/solder layer by showed bright reflections indicating cracking. This was apparent with all samples tested without bondline control as shown in Figure 4.4. Figure 4.5 (a) shows signs of cracking with the samples made with Al wirebonds also, these appear at the same location where the solder layer is at its thinnest due to the concave nature of the baseplate. No signs of cracking or delamination after 800 cycles for samples with the embedded metal mesh is observed as shown in Figure 4.5 (b).

Figure 4.4: Secondary SAM gate highlighting cracks in the solder layer after 600 thermal cycles for samples made with no bondline control

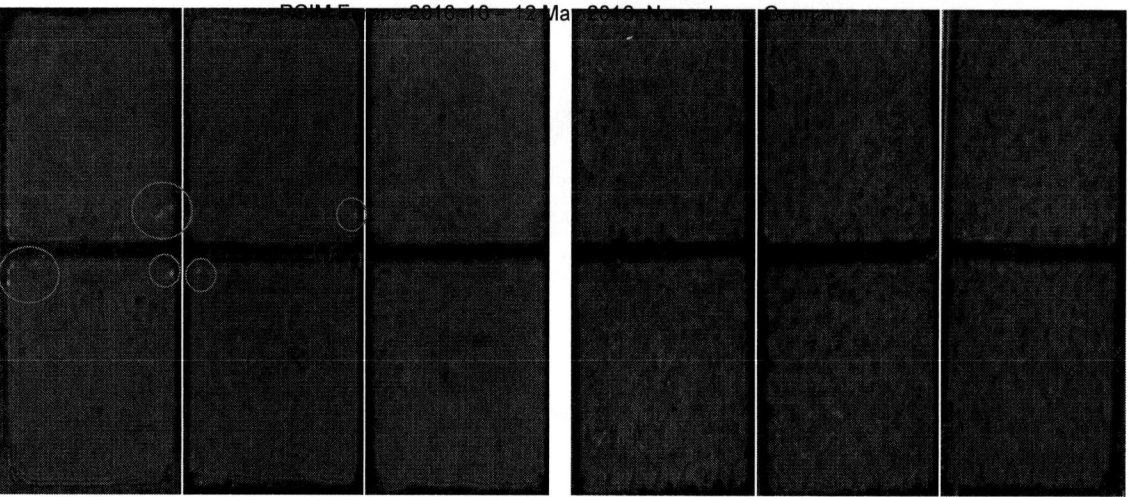

Figure 4.5 (a): 800 cycles; Secondary SAM gate highlighting cracks in the solder layer for samples with Al wirebonds

Figure 4.5 (b): 800 cycles; Secondary SAM gate showing no signs of cracking/delamination for the samples with the embedded metal mesh

5 Summary & Conclusion

A novel technique to prevent substrate tilt and maintain a homogenous 200μm solder layer using an embedded metal mesh was evaluated and compared to the traditional aluminium stitch bond technique for AlSiC baseplate modules. Samples were evaluated against aluminium stitch bonded modules and modules with no bondline control.

Following module assembly the top surface of the substrates underwent a surface profile scan to determine the substrate tilt. The mean co-planarity deviation revealed the sample with the embedded metal mesh had the lowest co-planarity deviation of 52.5μm and the sample with no bondline control had the highest deviation of 67.5μm. Additionally the maximum recorded values between two points over 40mm recorded the maximum tilt on the sample with no bondline control to be ~90μm whereas the values for the maximum deviation between two points for the other samples were ~60μm for the embedded metal mesh sample and ~70μm for the sample with aluminium wire-bonds.

Temperature cycling tests revealed solder layer delamination and cracking of the solder joint following the substrate tilt (as was measured before cycling). The samples with no bondline control showed cracking at 600 thermal cycles at the titled side with a difference of ~90μm; the lowest side of which emanated cracking and delamination. The samples with Al wirebonds showed signs of cracking at 800 thermal cycles with some but not all samples; again with cracks appearing at the thinner end of the solder joint, i.e. the centre of the baseplate. No cracks or solder delamination was seen for the samples with the embedded metal mesh.

Whilst the effect of bondline control has already been studied and shown to improve joint lifetime in power modules, this paper presents an alternative solution to the aluminium stitch bonding method. The metal mesh within the solder preform supressed solder fatigue up to

© VDE VERLAG GMBH · Berlin · Offenbach

800 cycles and shows a greater reliability to the traditional wire-bond method. The presence of the metal mesh became apparent after 400 cycles though this effect appeared to get no worse with an increasing number of cycles.

The embedded metal mesh preform offers a drop in and cost effective solution to achieving bondline homogeneity without additional process steps or costly capital investment for wire-bonding spacers (eliminates the need for dedicated wirebonders, maintenance, fixtures etc.). After 800 thermal cycles the embedded metal mesh preform showed no signs of cracking or solder layer delamination; in this test, this method shows improved reliability to the Al wirebond method.

Acknowledgments

The authors would like to thank the following for their contribution to this research: Dr Steve Jones.

References

[1] K. Hayashi & G. Izuta, "Improvement of Fatigue Life of Solder Joints by Thickness Control of Solder with Wire Bump Technique", ECTC 2002

[2] K. Guth & P. Mahnke "Improving the thermal reliability of large area solder joints in IGBT power modules", Integrated Power Systems (CIPS), 2006

[3] L. Mills & K. Vijay "InFORMS vs the Trimmed Wirebond Technique to Achieve Uniform Bondline Control Between Substrate and Baseplate", PCIM Europe 2015

PCIM Europe 2016, 10 – 12 May 2016, Nuremberg, Germany

Communicating Gate Driver for SiC MOSFET

Christophe Bouguet[1,2], christophe.bouguet@etu.univ-nantes.fr
Nicolas Ginot[2], IETR, France, nicolas.ginot@univ-nantes.fr
Christophe Batard[2], IETR, France, christophe.batard@univ-nantes.fr
[1] ECA GROUP, ZAC des hauts de Couëron, 24 rue Jan Palach, 44220 COUERON, France
[2] UBL Université - Université de Nantes, IETR, Rue Christian Pauc, BP 50609, 44306 Nantes cedex 3, France
This project is a joint effort between the IETR laboratory and ECA Group Company.

Abstract

Currently, the gate driver unit for power semiconductors provides the control of the components and ensures the safety of the switching cell. The integration of a communication function on a driver is described in this article. This new function sets up a communication channel between the primary side and the two control channels of the driver that are located on the secondary sides. The communication is done by using the CAN protocol. This paper introduces an innovative solution to provide a galvanic insulation to a CAN bus that fits the specific requirements of power electronics, especially a low parasitic capacity primary - secondary and the immunity to high dv/dt. A first prototype was designed and the experimental results are exposed.

Introduction

Silicon carbide (SiC) is a Wide Gap material which is nowadays used to produce power MOSFET [1] [2] [3]. This paper focuses on the transmission of monitoring information between the primary side and the secondary sides of a MOSFET SiC driver. In addition to gate control, the gate driver unit must properly supervise the voltage across the semiconductor when it is on-state. This monitoring enables to detect a quick and unexpected rise of the current in the semiconductor, especially in the case of a short circuit. In this case, the power semiconductor is properly switched-off and a binary error message (0 or 1) is sent towards the primary side. Other events, like low power supply voltage or an excessive temperature, can be at the origin of the error message. In this paper, we propose a communication solution based on CAN protocol that provides a bidirectional exchange of information between the primary and the secondary sides of the gate driver. If an error occurs, a quick basic binary message is sent to the primary side and a CAN frame giving details about the event is sent afterwards. This communication channel will eventually allow to monitor the evolution of parameters like temperatures (substrate, cold plate, passive components), power supplies voltage values or the DC bus voltage value. It will also be suitable to set dynamically some parameters of the driver such as the dead time value or the gate resistor.

The galvanic insulation of the communication channel is provided by a planar transformer for each channel. This technology was recently developed for gate drivers [4] [5], and « coreless » transformers too [6]. Using a common transformer for the communication CAN bus and for the

© VDE VERLAG GMBH · Berlin · Offenbach

transmission of the control orders was not considered for security reasons and also to get a fully dedicated media for the CAN bus. Our gate driver is then equipped with three planar transformers per channel, one for the power supply, a second one to transmit control orders and another one for the CAN bus. The scheme illustrated in Figure 1 shows the internal structure of the gate driver for SiC MOSFET.

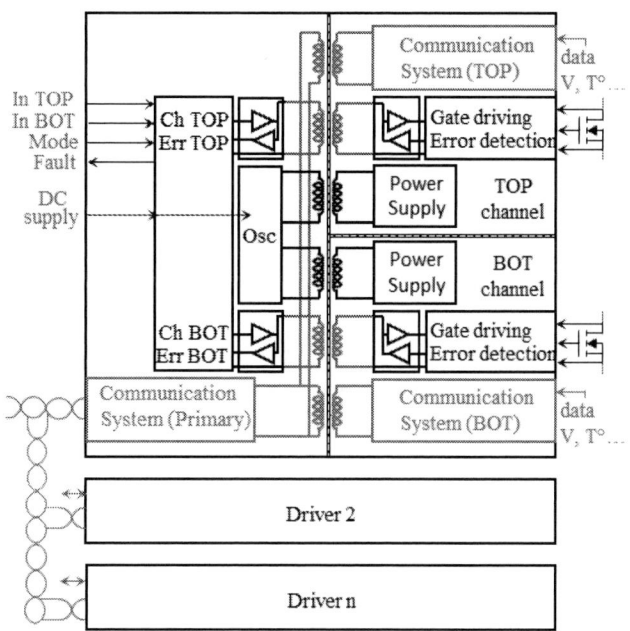

Figure 1 – Scheme including the communication system and several drivers

1 System description

The communication system shown in red in Figure 1 requires a galvanic insulation between the primary and the secondary sides. To provide this insulation, a planar transformer is used. The insulation is done thanks to the constitutive material of the PCB (FR4). The driver is designed for systems which require a 1250 V_{RMS} insulating voltage. The standard EN50178 then specifies that a double or reinforced insulation must be considered. Besides the functional aspect [7], the insulation also guarantees the users security.

The CAN protocol is widely widespread in the industry due to its robustness and its low cost. Moreover, its event triggered characteristic fully fits the need of the system. This protocol is based on dominant and recessive states of the communication channel. As a rule, the media is a twisted pair. Figure 2a shows an example of a conventional CAN network. The media is made of two conductors, CAN_L and CAN_H set to a potential equal to $V_{CC}/2$ when the network is in a recessive state. This twisted pair is ended with a 120 Ω resistor at both sides to restrict reflections at each end of the bus. When a node sets a dominant state on the media, the electrical potentials of CAN_L and CAN_H are modified so that a differential voltage (usually about 2 Volts) appears and is understood as a dominant state by all nodes of the network.

PCIM Europe 2016, 10 – 12 May 2016, Nuremberg, Germany

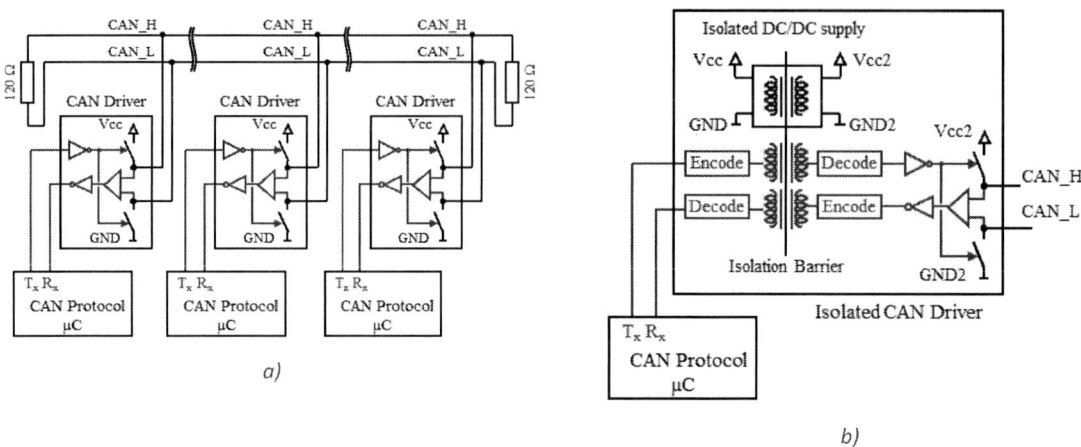

Figure 2 – Principle of CAN a) conventional CAN without insulation, b) CAN with insulation provided by transformers

As shown in Figure 1, the developed communication system permits keeping the advantages of CAN protocol with adding galvanic insulation. As a rule, each node is composed with a microcontroller and a transceiver. The microcontroller manages the protocol whereas the transceiver is an interface between the microcontroller and the media. A usual transceiver does not provide any galvanic insulation. The variety of electronic systems where such an insulation is needed have led manufacturers to design solutions based on capacitive or magnetic coupling transmissions. An example of magnetic coupling insulated CAN transceiver is shown in Figure 2b. The insulation barrier is established on the transmission and reception signals of the bus and not on the media itself. Moreover, the insulated CAN drivers are not designed for a galvanic insulation of several hundred volts. In addition, beyond the constraints inherent in the insulating voltage, some nodes of the communication network undergo repetitive variations of their reference potential. In the targeted applications, these cyclic variations are characterized by the voltage magnitude (some hundred volts) and by the dv/dt (10 kV/µs or more). The value of the capacity, between the primary and the secondary sides of a channel, has then a direct impact on the common mode currents value that flow through the galvanic insulation barrier. For example, a capacity of 5 pf with a dv/dt equal to 20 kV/µs lead a common mode current of 100 mA. To restrict the propagation of these currents, the parasitic capacity shall be the smallest. The low voltage solutions available on the market are based on two capacities or two transformers for each control channel, which increase the primary – secondary capacity value.

The proposed solution is shown in Figure 3. It requires only one transformer per control channel, which is located on the media itself instead of the R_X and T_X signals. The value of the parasitic capacity is thus reduced to the minimum. The principle is the following: a sinusoidal carrier is injected at the driver's primary side and is modulated according to the dominant or recessive state of the bus. A dominant state is conveyed by a reduced magnitude of the carrier whereas a recessive state does not have any effect on its magnitude. The carrier generator, the galvanic insulation barrier and the three VTI (Variable Termination Impedance) which allow each node to reduce the magnitude of the carrier are identified in Figure 3.

© VDE VERLAG GMBH · Berlin · Offenbach

PCIM Europe 2016, 10 – 12 May 2016, Nuremberg, Germany

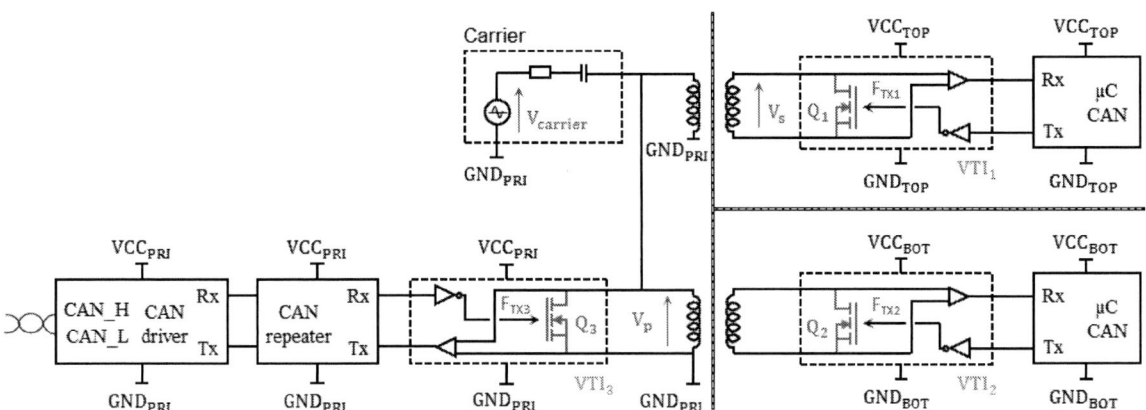

Figure 3 – Scheme of the insulated CAN solution with one transformer per channel

The choice of the carrier's frequency comes from a theoretical study which is not included in this paper. The communication channel modeling leads to the knowing of the recessive and dominant transfer functions for all possible arrangements of the VTI states. The magnitude of these transfer functions shown in Figure 4 concern two combinations of the states of the different VTI. The notation $F_{TXN} = 0$ means that VTI_N has no impact on the carrier magnitude and $F_{TXN} = 1$ means that VTI_N reduces the carrier magnitude. All possible values for the carrier frequency are located where the magnitude of the recessive state transfer function is higher than at the dominant state. By choosing a 5 MHz carrier, the magnitude difference between these two states is maximal.

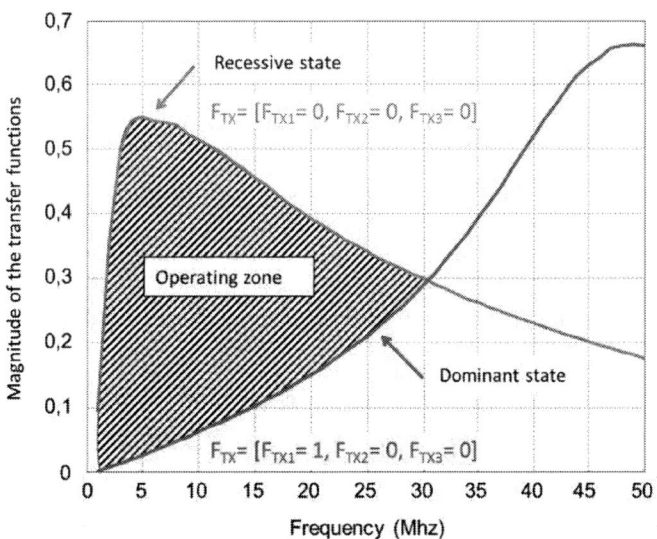

Figure 4 – Magnitude of the recessive and dominant transfer function

© VDE VERLAG GMBH · Berlin · Offenbach

2 Experimental tests of the communication function

These experimental tests are intended to verify that the communication between the primary side and the secondary TOP control channel of the driver is working as expected by monitoring R_X and T_X signals at both sides of the galvanic insulation barrier (Figure 3). At first, the communication is tested without any common mode voltage and so do not take into account the noisy environment. For it, the ground of the primary and TOP channel are set up at the same potential. BOT channel remains inactive on the bus by imposing a recessive state thanks to VTI_2 so that it does not lock the communication channel. To send a message, VTI3 controls the MOSFET Q_3 which is turned-on and turned-off. VTI_1 controls the MOSFET Q_1 and the MOSFET Q_2 of VTI_2 is permanently off. When Q_1 or Q_3 are turned-on, the carrier magnitude is reduced and the channel is set in a dominant state.

An example of data transmission using CAN protocol is described in Figure 5. The graph in Figure 5a focuses on just one frame whose data field is one-byte long. The blue curve corresponds to V_P and the red one to V_S as defined on figure 3. The green signal corresponds to the voltage at the R_X pin of the TOP channel microcontroller. It traces a frame which is being transmitted on the bus. The « Bit Time » is equal to 2.4µs, which corresponds to a gross bit rate of 417 kbits/s. Figure 5b focuses on the end of a data frame. In our system, the « ACK Slot » bit of the « ACK » field gives information about the emitting node. In fact, at this moment, the emitter lets the media idle by imposing a recessive state whereas at the same time, the receiver sets a dominant state on the bus only if the message transmission is a success. The blue curve allows to identify the emitting node thanks to the magnitude of V_P. Let's consider the beginning of the data frame. During a dominant bit slot, the carrier's magnitude is reduced by any VTI that turned-on its MOSFET. The V_P voltage magnitude (peak to peak) is about 0,4 Volts. During the « ACK Slot » bit, the V_P magnitude is twice smaller. This means that the VTI which is reducing the carrier magnitude is located nearer the carrier generator. As the latter is located at the primary side of the circuit, Figure 5b shows the end of a frame where TOP channel is emitting towards the primary side.

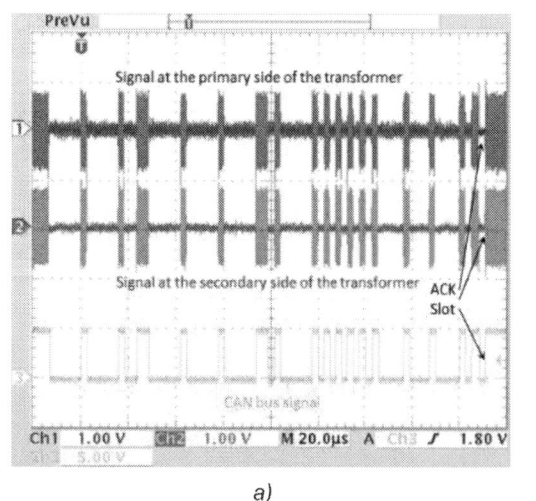 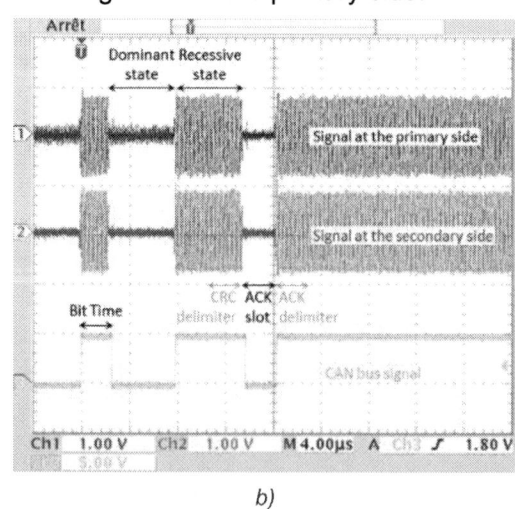

a)   b)

Figure 5 – Example of a data transmission based on CAN protocol with a galvanic insulation. a) A data frame and its "ACK slots" bits, b) End of a data frame emitted by TOP channel

3 Experimental trials with a common mode voltage

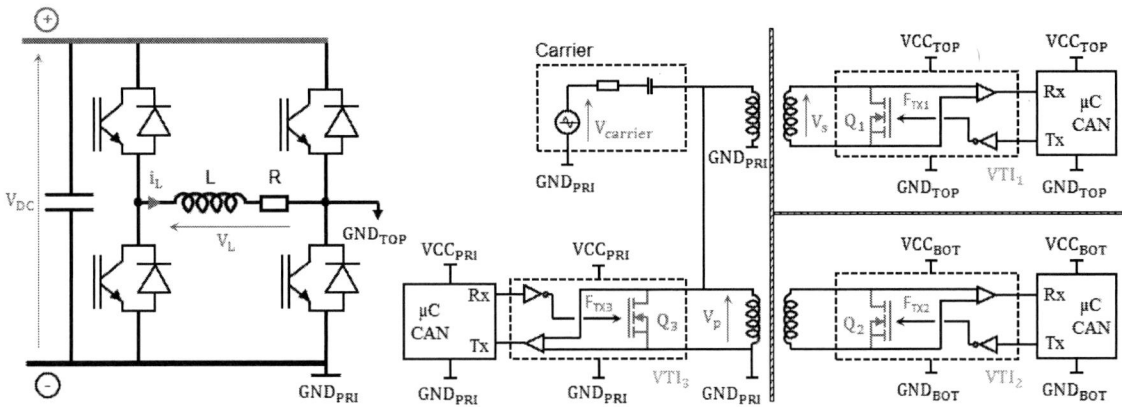

Figure 6 – Scheme corresponding to the trials with a common mode voltage

The experimental test with a common mode voltage between the primary and the TOP channel allows to generate disturbances that will be present in the final converter. The common mode voltage is obtained with a single phase inverter connected to an inductive load (L = 3 mH and R = 21 Ω) as specified in Figure 6. The electronic ground at the primary side of the communication board is linked to the negative polarity of the 500 Volts DC bus. The TOP channel power supply is done with a battery whose negative potential is connected to any leg of the single phase inverter.

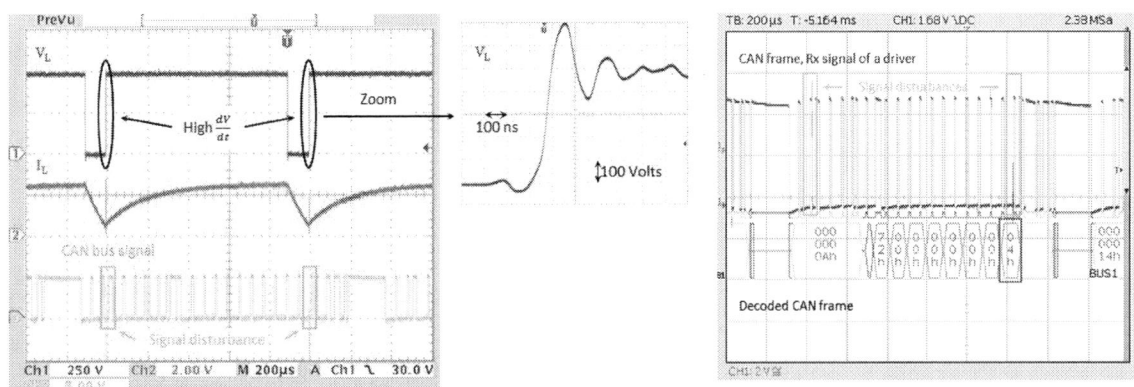

Figure 7 – Main traces

Figure 8 - Disturbance leading to a byte error value

Figure 7 illustrates a data transmission in a noisy environment. The green curve is an example of CAN frames going through the media of the communication system. We can notice the consequences of hard switching, painted in yellow. The signals distortion on the media can lead to an acquisition error value of some bits for the different nodes of the network. A problem occurs at a node when the amount of errors is too important. In this case, the receiving node is not able to recover the errors despite of the CAN protocol dedicated mechanism.

Figure 8 deals with the decoding of a data frame. The measurements were done with an oscilloscope (ROHDE & SCHWARZ HMO1002 Series) that is able to decode CAN frames circulating on the communication media. The error detected by the oscilloscope shows that the switching of power semiconductors perturbs the communication. This decoding device only observes the frames flowing through the communication bus and does not include any mechanism to recover from this kind of errors. However, such a mechanism is implemented on all the network nodes. It is clearly visible that the second switching produces an error in the CAN frame acquisition. In fact, the sampled value at the last byte of the data field is 0x04 whereas it is 0x00 actually. Even if a node sampled a wrong value for this byte, the original frame can be found thanks to the error recovery mechanism.

For a good functioning of the communication system, the number of hard switching during the transmission of a CAN frame must be limited. This involves that beyond a certain working frequency of the SiC MOSFET the functioning of the communication bus is no longer reliable. A data frame can be properly received and acknowledged by a receiving node if there is no more than five errors. An error is provoked by a dv/dt which is higher than 8.6 kV/µs. Furthermore, a triple sampling is possible to acquire the logic values of all bits forming a frame using CAN protocol.

Conclusion

A communication function added to a SiC MOSFET driver was presented. The communication system designed keeps the advantages of the CAN protocol while adding a galvanic insulation. The principle of the communication is the amplitude modulation of a 5 Mhz sinusoidal carrier according to the dominant or recessive state of the bus. A prototype was designed and the trial results concerning its immunity to high dv/dt were analyzed. Beyond the value of 8.6 kV/µs, the functioning of the communication system is no longer reliable.

References

[1] Joseph FABRE, Philippe LADOUX, Michel PITON, "Characterization and Implementation of Dual-SiC MOSFET Modules for Future Use in Traction Converters", IEEE transaction on Power Electronics, vol 30, pages 4079 to 4090, August 2015

[2] Paco Bogonez-Franco, Josep Balcells Sendra, "EMI comparison between Si and SiC technology in a boost converter", Electromagnetic Compatibility (EMC Europe), 2012 International Symposium on

[3] Ljubisa Stevanovic, Kevin Matocha, Zachary Stum, Peter Losee, Arun Gowda, John Glaser, Richard Beaupre, "Realizing the Full Potential of Silicon Carbide Power Devices", Control and Modeling for Power Electronics (COMPEL), 2010 IEEE 12th Workshop on

[4] Günter Schmitt, Wolf Kusserow and Ralph Kennel, "Power Supply for a IGBT-Driver with High Insulation Voltage based on a Printed Planar Transformers", Power Electronics and Motion Control Conference, 2008. EPE-PEMC 2008. 13th

[5] S. C. Tang, S. Y. Ron Hui, Henry Shu-Hung Chung, "A Low-Profile Power Converter Using Printed-Circuit Board (PCB) Power Transformer with Ferrite Polymer Composite", IEEE Transactions on Power Electronics, Vol 16, pages 493 to 498, 2001

[6] S. C. Tang, S. Y. (Ron) Hui, Henry Shu-Hung Chung, "Coreless Planar Printed-Circuit-Board (PCB) Transformers—A Fundamental Concept for Signal and Energy Transfer", IEEE Transactions on Power Electronics, Vol 15, pages 931 to 941, 2000

[7] Feng Luo, Jie Chen, Juexiao hen and Zechang Sun, "Controller Area Network Development for A Fuel Cell Vehicle", Vehicule Power and Propulsion Conference, 2009. VPPC '09. IEEE, 2009

New Ultra Fast Short Circuit Detection Method Without Using the Desaturation Process of the Power Semiconductor

Stefan Hain, Mark-M. Bakran
University of Bayreuth, Department of Mechatronics, Center of Energy Technology
Universitätsstraße 30, 95447, Bayreuth, Germany
www.mechatronik.uni-bayreuth.de, stefan.hain@uni-bayreuth.de

Abstract

This paper presents an ultra fast short circuit detection method for hard switching faults and fault under load short circuit conditions without using the desaturation process of the power semiconductor. The detection method is based on monitoring the simultaneous behaviour of the di/dt and the gate voltage of the power semiconductor, which is significantly different under short circuit conditions. Furthermore, a comparison with different state-of-the-art short circuit detection methods is shown which allows a classification of the presented method in terms of detection time and detection current. Therefore, it will be presented that the new di/dt-gate short circuit detection method is able to detect a fault behaviour of the IGBT close to earliest possible point in time a short circuit failure could be detected at all.

1. Introduction

The development of power semiconductor chips tends to reduce switching and conduction losses during operation which could induce other problems if the whole system is regarded. On the one hand, a reduced thickness of the power semiconductor chip will lead to lower conduction losses. On the other hand, the associated reduction of the thermal capacitance lowers the ruggedness against short circuit failures due to a lower withstand time in desaturation state. Using the desaturation process of a power semiconductor is a well-known and established method to detect a short circuit failure reliably. During desaturation condition, the device produces disproportionately high losses which could damage the device due to the necessary blanking time of the detection circuit (typical $5 - 10~\mu s$). Therefore, other short circuit detection methods were developed, which are able to detect a short circuit failure even before the IGBT reaches his desaturation current. The gate charge method e.g. is able to distinguish a normal turn-on process from a hard switching fault by measuring the flowed gate charge, but the detection of a high inductive hard switching fault or a fault under load also presupposes a desaturation process and the method is difficult to implement [1] [2]. Other short circuit detection methods which do not need the desaturation process for a detection have to measure the collector current of the device all the time. This can be done by integrating the induction voltage of the parasitic inductance between power and Kelvin emitter of the module or by the well-known and established current mirror [3] [4] [5]. Here, a short circuit detection signal (SCDS) is set, if the measured current exceeds a predetermined value. But using an integrated signal of a parasitic voltage drop can be very complex and costly. Furthermore, a current mirror needs a special chip design, which restrict the application to a selected field. This paper presents a new ultra fast short circuit detection method for hard switching faults and fault under load short circuit conditions by monitoring the simultaneous behaviour of the di/dt and the gate voltage of the power semiconductor, which is significantly different under short circuit conditions.

2. The di/dt-gate short circuit detection method

The new di/dt-gate short circuit detection method evaluates the appearance of two digital signals in order to distinguish between a normal turn-on process and a hard switching fault or between conducting mode and a fault under load. The first digital signal transmits information of the di/dt - behaviour of the dynamic current flow which can be measured with an inductive current sensor. If the di/dt value exceeds a predefined value, the output voltage of the inductive current sensor exceeds its reference value respectively and the digital signal is high. Therefore, every time the power semiconductor has a dynamic positive di/dt value, the event is marked with the corresponding digital signal. The second digital signal contains information if the actual gate voltage of the IGBT exceeds an other predefined reference voltage, which means high gate voltages are marked with the corresponding digital signal as well. Fig. 1 shows a turn-on process of an IGBT ($U_{ce,nom} = 1200$ V, $U_{ce} = 600$ V, $I_c = 240$ A, $L_\sigma = 12$ nH). Additional, the output voltage of an inductive current sensor can be seen. The area with a high di/dt value (output voltage of current sensor $u_{ind} > 0.3$ V) is highlighted in red, the area with a high gate voltage ($u_g > 12$ V) is highlighted in green. If the gate reference voltage is higher than the Miller plateau, both digital signals are separated in time. Fig. 2 presents the transients of both monitored signals under a low-inductive hard switching fault event. Since there is no Miller plateau present in this case, both values exceed their reference voltage simultaneously and the appearance of the two digital signals overlap in time, which is highlighted in blue. This coincidence can be used to distinguish between a normal turn-on process and a short circuit failure, which can be seen in fig. 3.

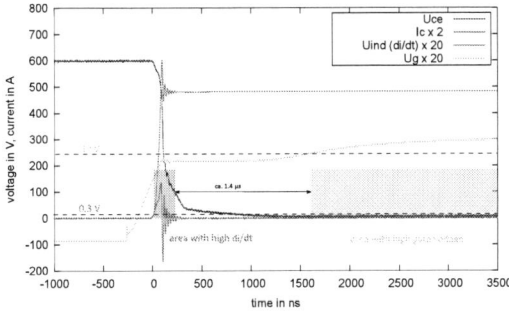

Fig. 1: Transients of the gate voltage, the collector-emitter voltage, the collector current and the di/dt-sensor during a turn-on process of an IGBT ($U_{ce,nom} = 1200$ V, $U_{ce} = 600$ V, $I_c = 240$ A, $L_\sigma = 12$ nH). The time intervals with a high di/dt value and a high gate voltage u_g are separated in time.

Fig. 2: Transients of the gate voltage, the collector-emitter voltage, the collector current and the di/dt-sensor during a low inductive hard switching fault of an IGBT ($U_{ce,nom} = 1200$ V, $U_{ce} = 600$ V, $L_\sigma = 12$ nH). The time intervals with a high di/dt value and a high gate voltage u_g overlap in time.

Here, the di/dt-sensor signal and the gate voltage are plotted against each other. By the precondition that both monitored signals exceed their reference values simultaneously, both reference voltages define a short circuit detection area (blue) in this di/dt - u_g phase diagram. During a normal turn-on process, transients of the di/dt-sensor signal and the gate voltage u_g move along the green trajectory and the two digital signals are never at logic high at the same time. If a low-inductive hard switching fault occurs, the transients move on the red trajectory and the short circuit detection area is entered. Therefore, the hard switching fault is detected even before the gate voltage could reach its maximum value. Fig. 4 shows the transient of a normal turn-off process and the transient of a fault under load event in the di/dt - u_g phase diagram. Since the turn-off process only generates negative di/dt values at high gate voltages, no short circuit detection signal is triggered. Even, strong oscillations after the turn-off process (periodic high di/dt values) can not trigger this short circuit sensor due to the low gate voltage. On the contrary, the transient of the fault under load event enters the short circuit detection area immediately with the occurance of the failure.

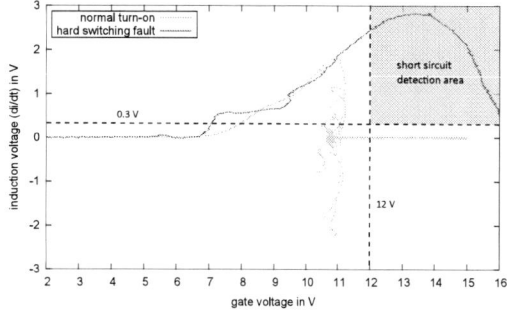 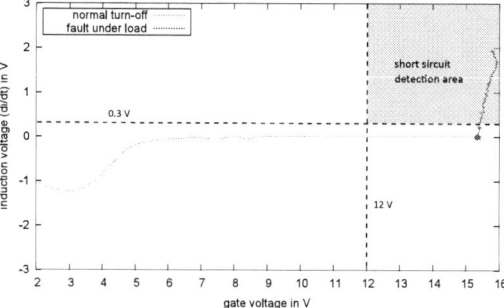

Fig. 3: di/dt-u_{ge} phase diagram with trajectories for a normal turn-on process and a hard switching fault. The short circuit detection area is defined by two reference voltages. If the gate voltage and the di/dt of the power semiconductor reaches the blue area, a short circuit failure will be detected.

Fig. 4: di/dt-u_{ge} phase diagram with trajectories for a normal turn-off process and a fault under load. The short circuit detection area is defined by two reference voltages. If the gate voltage and the di/dt of the power semiconductor reaches the blue area, a short circuit failure will be detected.

Monitoring the short circuit detection area can be realized by the circuit diagram shown in fig.5. As an inductive current sensor, the parasitic inductance between the Kelvin and the power emitter can be used. The digital signals can be generated by two comparators, which compares the di/dt and the gate signal with their corresponding reference values. The simultaneous appearance of both digital signals is observed by an AND logic gatter, which triggers a flip-flop circuit whose non inverting output signal is used as a short circuit detection signal (SCDS). In order to reduce the gate voltage immediately after a short circuit situation is detected, the SCDS is used to turn on an auxiliary MOSFET which is connected in parallel to the gate capacitance of the IGBT. Therefore, the auxiliary MOSFET turns off the IGBT and limits the maximum short circuit current although the signal voltage of the driver is still high.

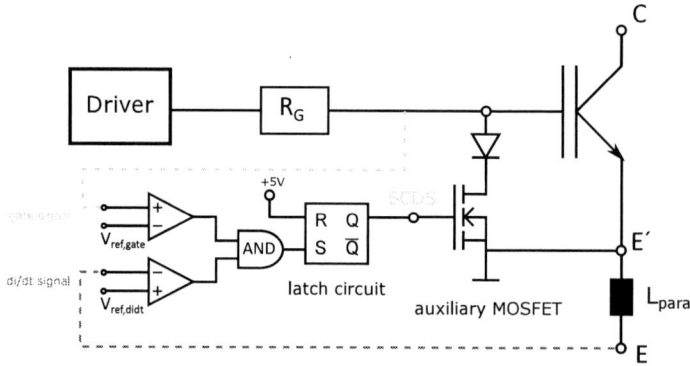

Fig. 5: Circuit diagram of the new short circuit detection method with the connection to the IGBT. As an inductive current sensor, the parasitic inductance between power and Kelvin emitter of the power semiconductor module is used. In order to react as fast as possible to short circuit failures, an auxiliary MOSFET is used to reduce the gate voltage immediately.

3. Experimental results

3.1. Adjusting the reference voltages

By defining the reference voltages of the di/dt and the gate comparator, the short circuit detection area can be adjusted to each application in order to guarantee no misdetection during normal operation. In the test setup, the reference voltage for the inductive voltage drop between Kelvin and power emitter is chosen to $v_{ref,di/dt} = 0.3V$ which is about ten times the induction voltage which appears due to the positive di/dt value during normal operation in conduction mode with a load inductance. The reference value for the gate comparator is chosen to $v_{ref,gate} = 12V$ which corresponds to a collector current of $I_{c,ref} \approx 550A$ according to the transconductance of the power semiconductor, which is about 2.3 times the nominal current of the module. Therefore, hard switching faults and fault under load situations can be detected up to a short circuit inductance of about one tenth of the normal load inductance. For even higher inductive short circuit events, slow but accurate current sensors, which were used to control the phase current, are able to detect these failures, which means there is no detection gap over the whole range of possible short circuit inductances.

3.2. Hard switching fault

Fig. 6 presents a normal turn-on process of an IGBT with the adapted di/dt-gate short circuit detection method. Additional, the appearance of both digital signals can be seen, which a have a solid time distance of about 750 ns. In comparison, fig. 7 shows the performance of the di/dt-gate method, detecting a low-inductive hard switching fault within about 50 ns after the hard switching fault can be distinguished from a normal turn-on process. The simultaneous appearance of both digital signals triggers a SCDS which immediately turns-on the auxiliary MOSFET. Due to the fact that the IGBT driver signal is still, the gate voltage drops to a constant plateau which is below the threshold voltage of the power semiconductor. After about 2.5 μs, the driver is turned-off by an external signal.

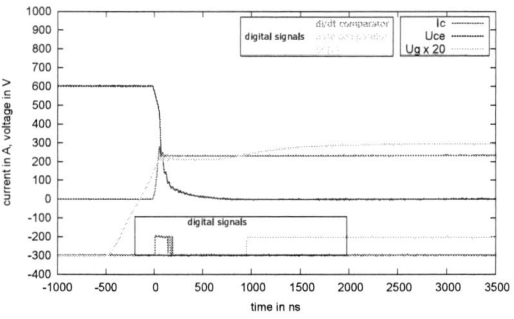

Fig. 6: Normal turn-on process of an IGBT ($U_{ce,nom} = 1200$ V, $U_{ce} = 600$ V, $I_c = 240$ A, $L_\sigma = 12$ nH). There is a solid time distance between the digital signals of the di/dt and gate comparator of about 750 ns if a gate reference voltage of 12 V is chosen.

Fig. 7: A low inductive hard switching fault of an IGBT ($U_{ce,nom} = 1200$ V, $U_{ce} = 600$ V, $L = 12$ nH) with the implemented new short circuit detection method. The auxiliary MOSFET is turned on by the SCDS which immediately leads to a decrease of the gate voltage.

In order to show more details of the performance of this detection method, fig. 8 presents a zoomed view of fig. 7. It can be seen that the low-inductive hard switching fault can not be

PCIM Europe 2016, 10 – 12 May 2016, Nuremberg, Germany

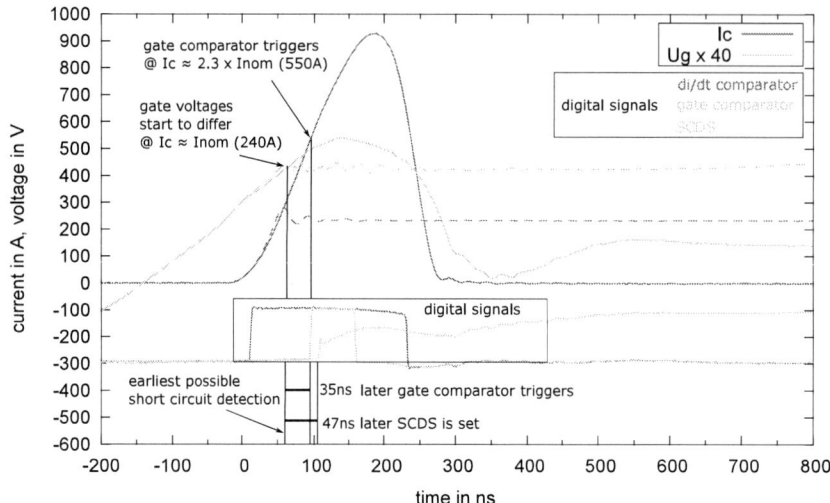

Fig. 8: Zoomed view of fig.7. The turn-on process and the hard switching fault can not be distinguished until the Miller phase appears during normal turn-on.

distinguished from a normal turn-on process until both gate voltages start to differ at a collector current of about $I_c = 240$ A because of the Miller effect. This is the earliest possible point in time a short circuit could be detected. About 35 ns later, the gate voltage exceeds its reference value of 12 V which corresponds to a current of about 550 A. This leads to a SCDS which appears after additional 12 ns due to the propagation delay time of the latch circuit. Therefore, after about 50 ns the short circuit waveforms started to differ from a normal turn-on process, the short circuit failure is detected by the di/dt-gate method.

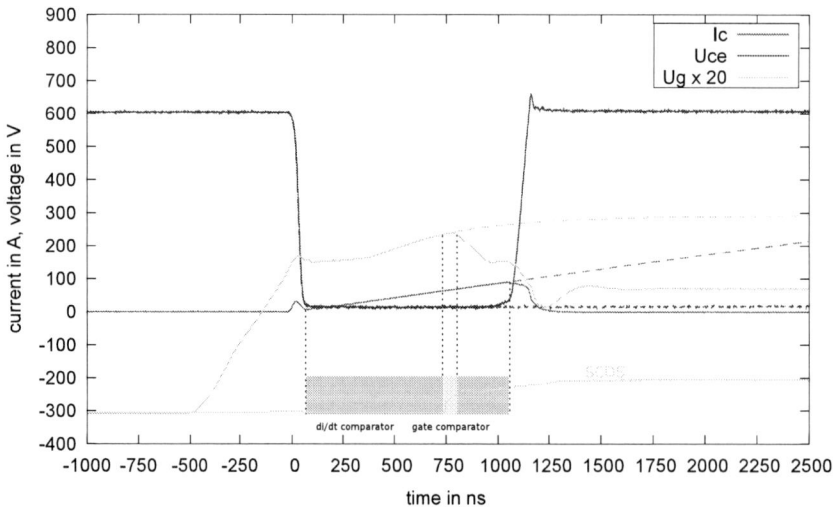

Fig. 9: A high-inductive hard switching fault of an IGBT ($U_{ce,nom} = 1200$ V, $U_{ce} = 600$ V, $L_{SC} = 3$ μH) with the implemented new short circuit detection method. The short circuit failure is detected at a collector current of about 160 A, which is even lower than the nominal current ($I_{nom} = 240$ A) of the device.

© VDE VERLAG GMBH · Berlin · Offenbach

If a high-inductive hard switching fault occurs, the di/dt is no longer controlled by the dynamic gate voltage but from the DC-link voltage and the remaining stray inductance in a short circuit situation. Fig. 9 shows the performance of the di/dt-gate method detecting a hard switching fault with a total stray inductance of about $L_{SC} = 3\ \mu\text{H}$. In this case, the fault can be detected at a collector current of about $160\ \text{A}$, which is even lower than the nominal current ($I_{nom} = 240\ \text{A}$) of the device.

3.3. Fault under load

Detecting a fault under load situation becomes very easy if the di/dt-gate detection method is used. After a normal turn-on process without any short circuit detection, the gate voltage remains at its high value and exceeds permanent its reference voltage in conduction mode. Therefore, every sufficiently high positive di/dt value is able to trigger the short circuit detection immediately, which only appears in a fault under load situation. Fig. 10 shows the performance of the di/dt-gate method, detecting a fault under load situation with a total stray inductance of about $L_{sc} = 1\ \mu\text{H}$. Here, the detection time is defined by the propagation delay time of the di/dt comparator ($\approx 4\ \text{ns}$) and the propagation delay time of the latch circuit ($\approx 12\ \text{ns}$). Therefore, only about $16\ \text{ns}$ after the failure occurred the SCDS is set in order to initiate a controlled turn-off process.

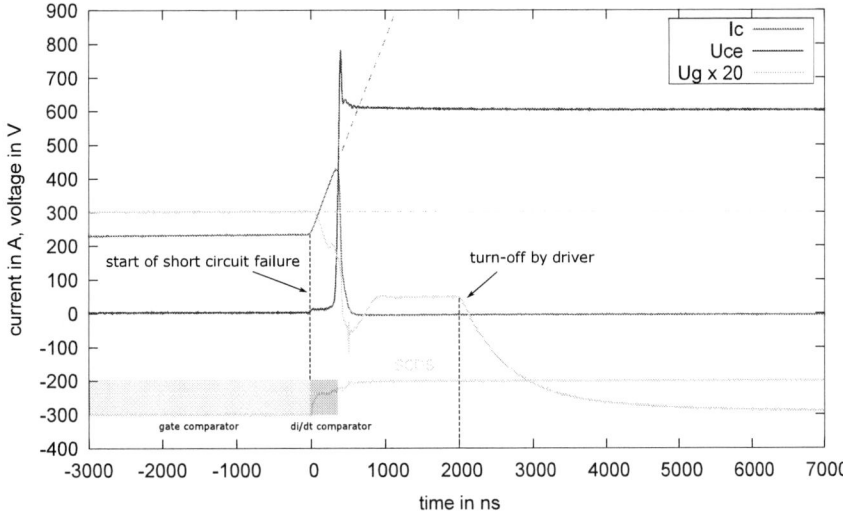

Fig. 10: Fault under load short circuit event for an IGBT ($U_{ce,nom} = 1200\ \text{V}, U_{ce} = 600\ \text{V},$ $I_c = 240\ \text{A}, L_{sc} = 1\ \mu\text{H}$). The short circuit can be detected about $16\ \text{ns}$ after the fault under load appears.

PCIM Europe 2016, 10 – 12 May 2016, Nuremberg, Germany

4. Comparison of the di/dt-gate method with different state-of-the-art detection methods

In the figures 11-13 a qualitative comparison of the di/dt-gate method with different state-of-the-art short circuit detection methods, such as $V_{ce,sat}$-method, gate-charge method and current mirror, is shown which allows a classification of the presented method in terms of detection time and detection current. The comparison is done by the transconductance characteristic of the power semiconductor which is suitable to show the detection performance of each method. In addition, the short circuit detection area of a current mirror is highlighted by the red area, which is defined by the reference current I_{sc}. The detection area of the di/dt-gate method is indicated by vertical red arrows, which means that trajectories which move in this marked area trigger the short circuit detection if the vertical component (positive di/dt- value) is too high. During a normal turn-on process, transients of the collector current and the gate voltage move along a trajectory which is similar to the green or blue curve, which represents turn-on processes with one and two times the nominal current. Since the red area is never reached (current mirror) and the trajectories move perpendicular to the red arrows (di/dt-gate method), no short circuit is detected.

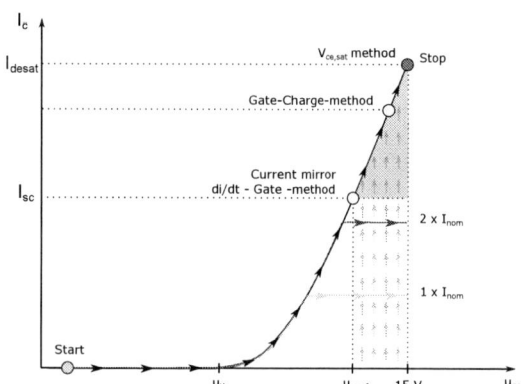

Fig. 11: Comparison of different state-of-the-art short circuit detection methods with the presented di/dt-gate method based on the transconductance of the IGBT for a low-inductive hard switching fault.

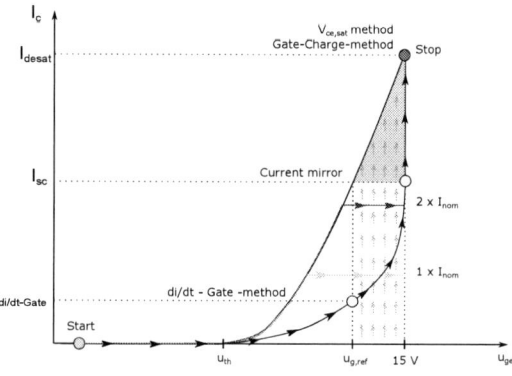

Fig. 12: Comparison of different state-of-the-art short circuit detection methods with the presented di/dt-gate method based on the transconductance of the IGBT for a high-inductive hard switching fault.

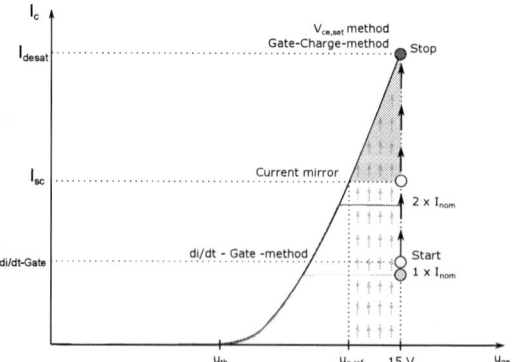

Fig. 13: Comparison of different state-of-the-art short circuit detection methods with the presented di/dt-gate method based on the transconductance of the IGBT for a fault under load.

© VDE VERLAG GMBH · Berlin · Offenbach

In figure 11 the performance in detecting a low-inductive hard switching fault of the different detection methods can be seen. The appearance of each SCDS signal is marked by a yellow or a red circle on the trajectory. Here, the current mirror and the di/dt-gate method have identical performances, if the reference voltage for the gate comparator corresponds to the reference current of the current mirror. Measuring the gate charge leads to a later detection, because more time and therefore higher gate voltages are needed in order to distinguish between a normal and a fault situation. Finally, the $V_{ce,sat}$-method presupposes the desaturation process for detection and therefore its SCDS signal is set always at the end of the fault trajectory. Figure 12 presents the different performances for a high-inductive hard switching fault. The di/dt-gate method is able to detect this failure even before the nominal current of the device is reached because the trajectory does not intersect the red arrows perpendicular, which was already discussed with figure 9. Since the current mirror detects a failure always at the same reference current, its SCDS signal does not change here. The SCDS signal of the gate-charge method is set at the end of the trajectory, because now a desaturation process is necessary to trigger this detection method, such as in the detection of a fault under load, which can be seen in figure 13. Due to the fact, that the trajectory starts with its movement within the red arrows, the SCDS signal is set immediately an therefore far ahead of the other detection methods.

5. Conclusion

This paper presents a new short circuit detection method which is based on monitoring the trajectory of the IGBT in a phase space which is defined by the di/dt value and the gate voltage. It can be shown that the simultaneous exceeding of both corresponding reference values is a solid indication for a fault situation, such as a hard switching fault or a fault under load. A comparison of this method with well known state-of-the-art detection methods illustrates that this method is able to detect a short circuit situation close to the earliest possible point in time, a failure could be detected at all. Furthermore, by evaluating the chronological order of the digital signals of the di/dt comparator and the gate comparator, a hard switching fault could be distinguished from a fault under load by using this method. This information could be used to initiate different turn-off strategies, which corresponds to the short circuit fault types respectively. Therefore, this method provides a fast, adaptable and easy to implement short circuit detection method for modern power semiconductors without the need of a desaturation process.

References

[1] Takeshi Horiguchi: *"A Short Circuit Protection Method based on a Gate Charge Characteristic",* IPEC, 2014.

[2] Byoung-Gun Park, Jun-Bae Lee, Dong-Seok Hyun: *"'A Novel Short-Circuit Detection Scheme Using Turn-On Switching Characteristics of IGBT'",* Industry Applications Annual Meeting, 2008.

[3] Ignacio Lizama, Rodrigo Alvarez, Steffen Bernet, Martin Wagner: *"A New Method for Fast Short Circuit Protection of IGBTs",* IECON, IEEE, 2014.

[4] Markus Oinonen, Matti Laitinen, Jorma Kyyr: *"Current measurement and short-circuit protection of an IGBT based on module parasitics",* EPE, IEEE, 2014.

[5] Frank Huang, Fred Flett: *"IGBT Fault Protection Based on di/dt Feedback Control",* IEEE PECS, 2007.

High power, high frequency gate driver for SiC–MOSFET modules

Gunter Königsmann, Gunter.Koenigsmann@semikron.com
Reinhard Herzer, Reinhard.Herzer@semikron.com
Sven Buetow, Sven.Buetow@semikron.com
Matthias Rossberg, Matthias.Rossberg@semikron.com
SEMIKRON Elektronik GmbH & Co. KG, Sigmundstraße 200, 90431 Nuremberg, Germany

Abstract

A high power, high frequency gate driver with high integration density using ASICs allowing to drive very low inductive 1200 V, 400 A SiC-MOSFET half bridge modules in both-side sinter technology (SKiN) is presented for the first time. Because of its low propagation delay, low dead times and very strong output stages of I_{peak} = +40/-78 A, the driver is excellently suited to switch the 400 A MOSFETs with a switching frequency up to 200 kHz with extremely low switching losses and low overvoltages.

1. Introduction

An increasing number of SiC switches (MOSFET, JFET) with voltage ratings between 600 V and 1700 V are hitting the market with decreasing $R_{DS(on)}$ and costs as well as sufficient reliability [1 - 4]. In order to make use of the outstanding performance of SiC devices it is extremely important to provide them with suitable application and system environments. This imposes the demand for extremely low inductivities and thermal resistances. Also high temperature operation (≥175°C) at increased reliability forces the need for improved packaging. On the driver side highly integrated gate drivers with driving and monitoring functions adapt for higher frequencies and higher operation temperatures are necessary as well as the possibility to use low inductive connections to the switches.

2. High Frequency Operation

The main benefit of using high switching frequencies for power inverters is that they allow the output-voltage and -current to fit more accurately to an ideal sinus form [see an example in Fig. 1: less ripple for 3-level, 8 kHz inverter (black curve) and especially for higher switching frequencies for 2-level, 50 kHz (grey curve) in comparison to 2-level, 8 kHz (red curve)]. The high ratio between switching frequency and output frequency and the reduced ripple current reduce filter size, cost and weight. Increasing the output frequency also allows to reduce the mechanical dimensions of actors like motors since their minimum inductivity can be lowered, as well. Both effects go hand in hand with an increasing market of applications which require high switching frequencies. Unfortunately switching losses increase linearly with the switching frequency which means using them can mean a significant drop of the achievable output current of the inverter. The solution that overcomes this problem is to drive a SiC-MOSFET in a way that only 20…25% of the switching losses of a Si-IGBT are produced. As seen in Fig. 2 the output current versus frequency is for the even 2.5 times smaller SiC-MOSFET die at frequencies > 6 kHz higher than for the Si-IGBT and already at 60 kHz the output current increases by a factor of roughly 3.

PCIM Europe 2016, 10 – 12 May 2016, Nuremberg, Germany

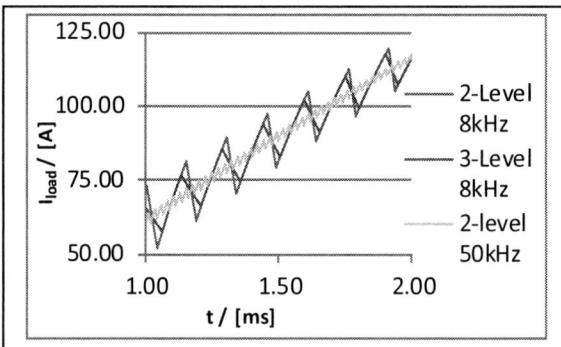

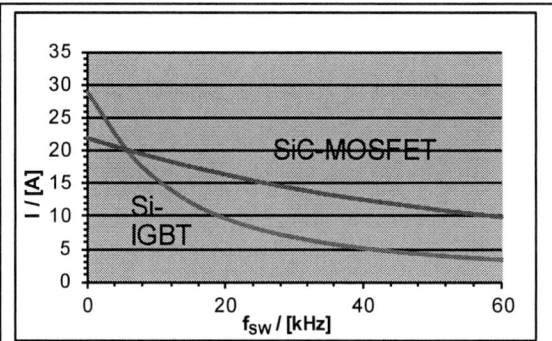

Fig. 1: Detail of simulated sinusoidal inverter output currents for 2-level inverter at 8 kHz switching frequency (red), 3-level inverter at 8kHz (black) and 2-level inverter at 50kHz (grey);

Fig. 2: Maximum output current over frequency for a 1200V, 25A IFX Si-IGBT4 (A_{Chip}= 25mm^2) and a 1200V, 20A SiC-MOSFET (A_{Chip}= 10mm^2; 80mΩ); air cooling[1], T_{jmax}=175°C, T_{am}=40°C

3. Low Inductive Module and System Design

Fast switching with extremely high dv/dt and di/dt has the potential of triggering many of the available oscillation modes. Furthermore a high di/dt will also increase the voltage overshoot (v = -L • di/dt) at the parasitic inductivities of the circuit. Therefore a low inductive module and system design is essential. Fig. 3 shows the newly developed half bridge module in which the Drain on the back side of the SiC-MOSFET chips is sintered to the DBC while their source and gate on the front side are sinter-contacted to a flexible circuit board (SKiN-technology [5]). This assembly technology allows dual-side cooling and leads to an outstanding performance regarding inductivity, reliability and thermal resistance. The optimized close commutation loops inside the module (see Fig. 4) guarantee a module inductivity lower than 1.5 nH per switch and the low inductive connection to a new developed DC link capacitor leads to a record-breaking overall inductivity of < 5 nH for the whole high power system.

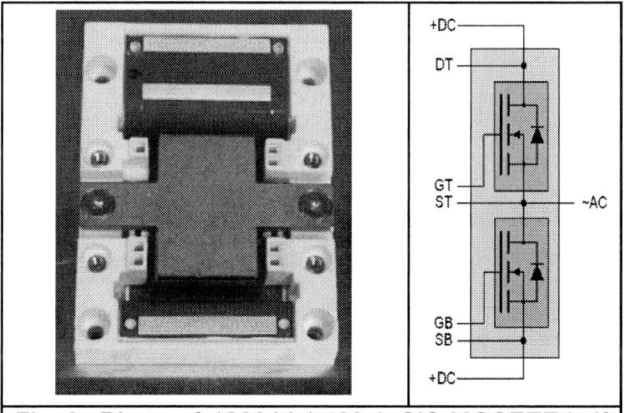

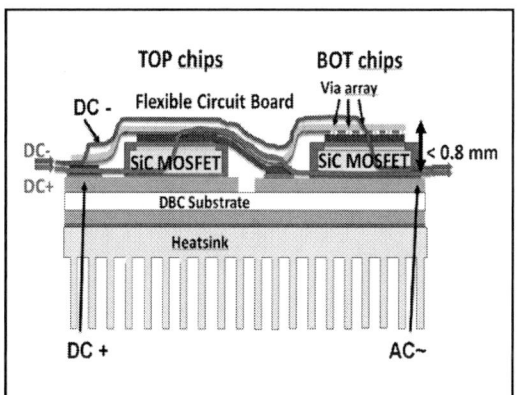

Fig. 3: Photo of 1200 V / 400 A SiC-MOSFET half bridge module assembled in both side sinter-technology (SKiN); one switch has 8 x 50 A MOSFET in parallel; module size is 69 x 92 mm

Fig. 4: Cross section of the module and simplified demonstration of the commutation loops of the very low inductive design

---

[1] SEMIKRON air cooler P14/120

© VDE VERLAG GMBH · Berlin · Offenbach

4. Gate Driver

The block circuit diagram of the gate driver with the implemented functions inside the newly developed driver ICs on the primary and secondary side is presented in Fig. 5 and the PCB of the driver is shown in Fig. 6. The gate driver for normally-off SiC-MOSFET [1; 2; 4] provides a nominal voltage of V_{GS}=20 V in the on-state which reduces the $R_{DS(on)}$. In the off-state negative voltages of V_{GS} = -3 V...-5 V are needed to prevent parasitic turn-on due to high dv/dt at the parasitic capacitance between gate and drain and in order to reduce the switching losses. The driver uses transformers for bidirectional signal transmission between primary and secondary side (Modem), can handle chips with breakdown voltages up to 1700 V and positive and negative offset voltages, respectively. A temperature and short circuit monitoring (using V_{DS}) of the MOSFET switch, as well as differential primary side inputs are used. A galvanic isolated power supply for the secondary sides and a monitoring of all operating voltages of the driver are implemented. The whole driver functionality is integrated in primary and secondary side Application Specific ICs (ASICs), for enhancing reliability and minimizing the area and therefore the EMC sensitivity of the whole system. As a result despite the powerful transformers, power supply and output stages only very few discrete components had to be placed on the PCB (see Fig. 6) whose back side could be completely reserved for shielding and cooling and whose interfaces have been optimized using electromagnetic field simulations.

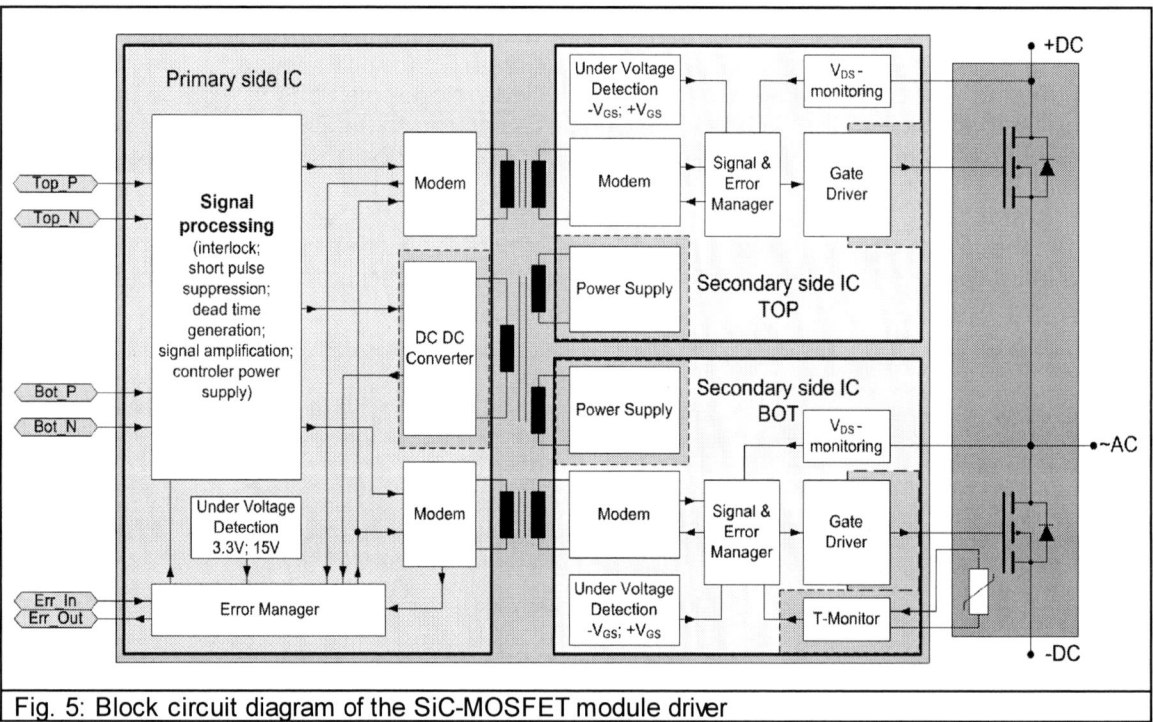

Fig. 5: Block circuit diagram of the SiC-MOSFET module driver

The enormous driver capability and speed with peak currents of +40 A / -78 A are needed to guarantee a high switching speed of the devices. The V_{DS}-monitoring based on SEMIKRON´s dynamic V_{DS}-reference is adapted to the SiC-MOSFET-switch and its high switching frequency and the resulting noise of the system.

Fig. 6: Front side of gate driver board with primary side IC, 2x secondary side ICs; secondary side power supply and gate driver output stages with V_{GS} = -5 V...+20 V, I_{Gpeak} = +40/-78 A

5. Electrical Behavior (Measurements)

Fig. 7 and Fig. 8 show the measured turn-on and -off behavior at nominal conditions (V_{DC} = 600 V and I_{DS} = 400 A). The switching losses (E_{SW} = E_{on} + E_{off} = 15.9 mJ + 1.8 mJ = 17.7 mJ) of the SiC-MOSFET switches are significantly reduced and are lower than 25 % of the losses of a conventional IGBT switch working under the same conditions. Thanks to the low inductivity of the commutation circuit the module can switch with very low gate resistances and so high dv/dt and di/dt are possible which causes the low losses of the MOSFET. Even at high di/dt there are very low over-voltages during turn-off (V_{DSmax} = 756 V). The turn-on losses are significantly higher (see Fig. 8), due to the Q_{rr} of the internal body diode of the MOSFET which is used as freewheeling diode (FWD). In combination with low module inductance, the fast ramp-up and high peak current provided by the driver the turn-on process can be made fast enough that the device is depleted of minority carriers first after the voltage commutation is complete. Therefore regardless of its relatively high speed a "snappy" behavior of the diode can be avoided [6] as can be seen in the smooth curve forms of V_{DS}, V_{GS} and I_{DS} during turn-on.

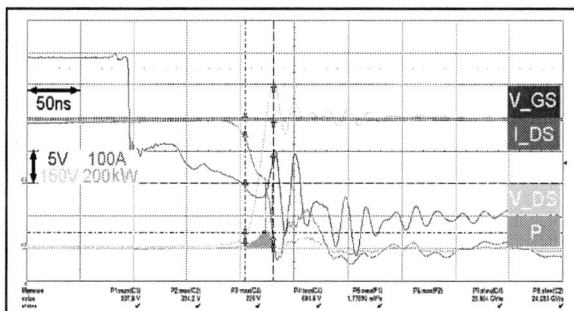

Fig. 7: Turn off behaviour of 1200 V SiC-MOSFET switch at V_{DC} = 600 V, I_{DS} = 400 A, R_G = 0.5 Ω, V_{GS} = +18 V...-5 V, dv/dt = 30 kV/µs, V_{DSmax} = 756 V, E_{off} = 1.8 mJ

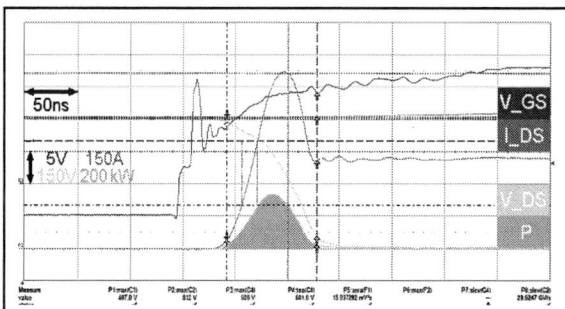

Fig. 8: Turn on behaviour of 1200 V SiC-MOSFET switch at V_{DC} = 600 V, I_{DS} = 400 A, R_G = 1 Ω, V_{GS} = -5V...+18 V, di/dt = 20 kA/µs, E_{on} = 15.9 mJ

Fig. 9 and 10 present the turn-off at maximum SOA condition (V_{DC} = 600 V, I_{DS} = 3 x I_{nom} and V_{DC} = 800 V, I_{DS} = 1.5 x I_{nom}). Because of the low inductive design of the module and the used setup switching speeds of over 40 kV/µs and 60 kA/µs are possible. Due to its optimized design the driver also works under such rough conditions. Even dv/dt of above 50kV/µs with its high magnetic field gradients which were reached above the traditional SOA conditions could be handled easily.

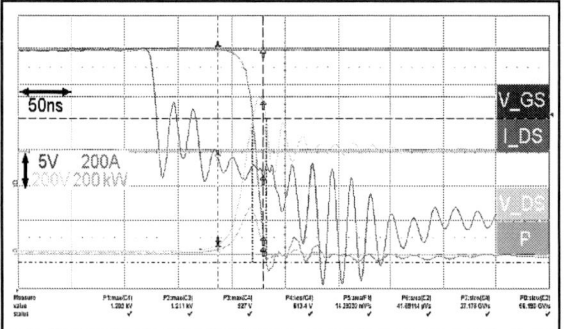

Fig. 9: Turn off under maximum conditions at V_{DC} = 600 V, I_{DS} = 1200 A, R_G = 1 Ω, V_{GS} = +18 V... -5 V, dv/dt = 37 kV/µs, di/dt = 66 kA/µs, V_{DSmax} = 927 V, E_{off} = 14.4mJ

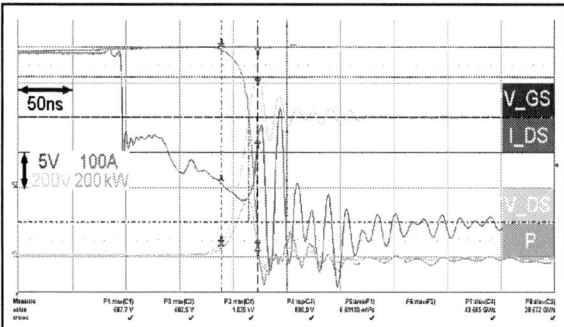

Fig. 10: Turn off under maximum conditions at V_{DC} = 800V, I_{DS} = 600 A, R_G = 0.5 Ω, V_{GS} = +18 V..-5 V, dv/dt = 43kV/µs, di/dt = 39 kA/µs, V_{DSmax} = 1035 V, E_{off} = 6.6mJ

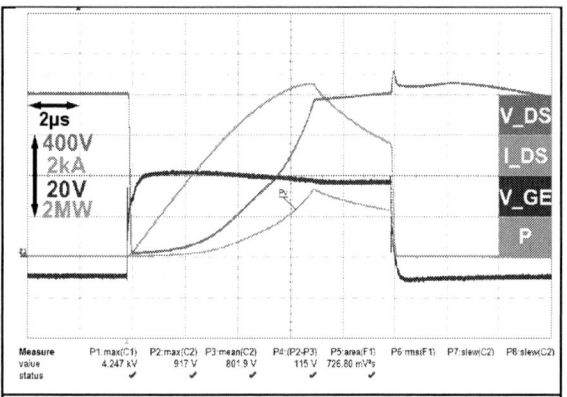

Fig. 11: Soft short circuit (L high) at V_{DClink} = 800V; T_A = 25°C; t_{sc} = 10µs and without V_{DS} detection, I_{DSmax}= 4247 A; V_{DSmax}= 917 V

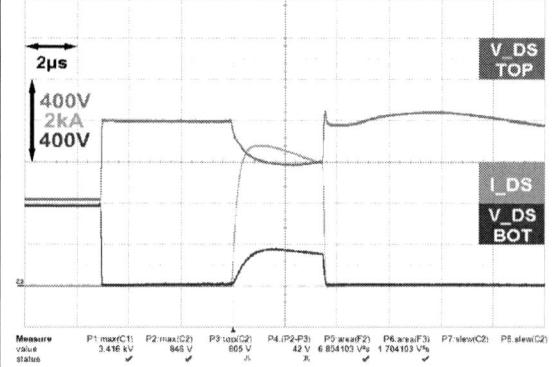

Fig. 12: Half bridge shot-through at V_{DClink} = 800V; 150°C with enabled V_{DS} detection (reaction within 4µs); I_{DSmax}= 3414A; V_{DSmax}= 846V

The SiC-MOSFETs were more and more optimized for low V_{DSsat}. Due to this optimization the MOSFETs and also the module have a very high short circuit current. Fig. 11 shows a soft short circuit at disabled V_{DS} detection. During the rising V_{DS} caused by the desaturation of the MOSFETs the gate voltage temporarily increases above 20V influenced by drain-gate capacitance and the I_{DSmax} increases to above 4200 A (525 A per chip) which is over 10 times the nominal current. Because of the high V_{DSsat} and high current of the SiC-MOSFETs during short circuit the device cannot withstand a hard short circuit or a short circuit at 150°C for 10 µs respectively. To prevent the thermal destruction of the module a V_{DS} detection is required which can handle this extreme rough conditions. Fig. 12 shows a half bridge shot-through which is turned off by the V_{DS} detection of the driver at about 4 µs. The I_{DSmax} is reduced to 3414 A (8.5× I_{nom}) and V_{DSmax} to 846 V.

Due to the use of the body diode of the MOSFET as a freewheeling diode also minor parameters of the gate driver like the dead time becomes more important for the total losses at high switching frequencies. At switching frequencies of 50 kHz and above the possible output current is limited to approximately the half of the nominal current determined by the cooling performance. After the dead time the off-switch is turned on and the voltage drop over the switch decreases from V_F of the body diode to V_{DSsat} of the MOSFET (see Fig. 13).

Fig. 14 shows this voltage drop during the commutation in a real dynamic measurement. In addition a variation of the dead time is shown in this picture. A reduction of the dead time from 800ns to 420ns reduces the static loses of the whole switch. The blue colored area is directly proportional to the reduction of the static losses which can be reached.

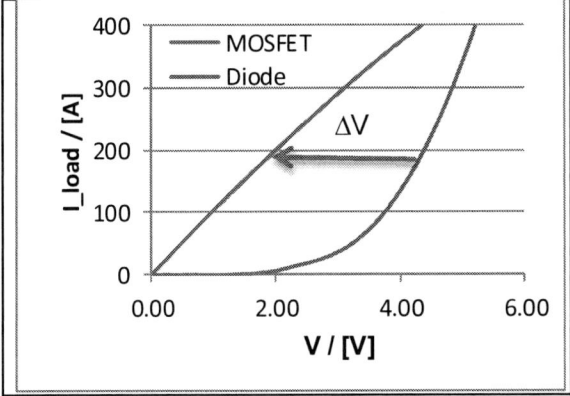

Fig. 13: Static measurement of V_{DSsat} of the MOSFET and V_F of the body diode

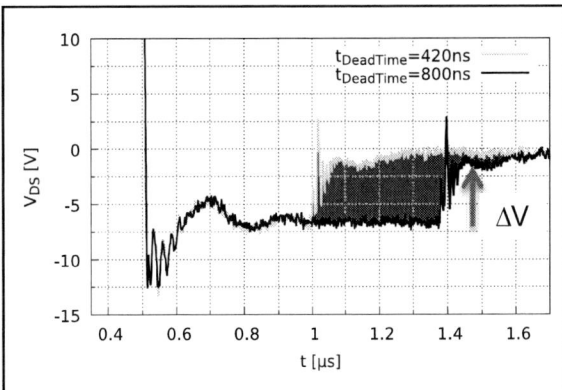

Fig. 14: Voltage drop at the switch during commutation depending on the dead time (V_{DClink} = 30 V; I_{load} = 83 A; T_J = 25°C)

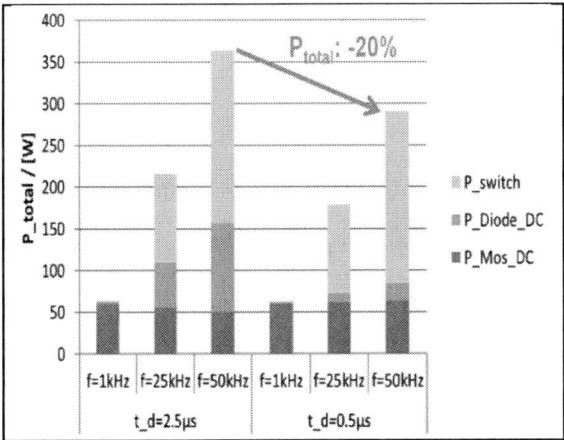

Fig. 15: Influence of driver controlled dead time between TOP and BOT switch on the different losses [static (DC) and dynamic] of MOSFET and internal FWD at 600V and 100 A_{rms}

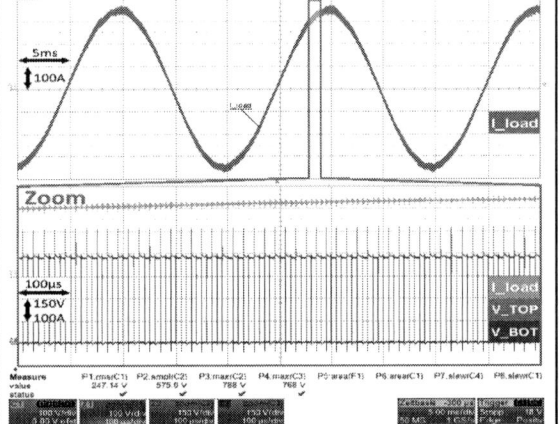

Fig. 16: Measurement of inverter output current (yellow) at high switching frequency (V_{DClink} = 600 V; f_{sw} = 50 kHz; I_{load} = 250 A_{rms} f_{out} = 50 Hz; $cos\varphi$ =1)

As consequence times where the MOSFETs aren't paralleled to the FWD increases the losses during each switching period. As shown in Fig 15 the total losses of the half bridge can be reduced at switching frequencies of 50kHz by 20% if the dead time decreases from 2.5 µs to 0.5 µs.). A further reduction of the dead time has to be handled carefully, because dead times that are too short will result in a cross current through the half bridge and subsequently increase the total losses.

Fig. 16 shows the behavior of two modules in a 4Q- single phase inverter test. Due to the high switching frequency of 50 kHz (limit of the controller of the test stand) the output current fits accurately to an ideal sinus. Fig. 17 shows the used test setup. The module is mounted on a water cooler and its connection to the DC link is realized with a four layer PCB. Each metallization has a thickness of 105µm to handle the high current. In addition due to the usage of a four layer design also a minimum spacing of 0.43 mm between +DC and -DC could reached to minimize the inductance of the setup. A further reduction of inductance is reached by mounting the DC-link on the PCB close to the module. The gate driver is placed on the top of the PCB direct above the module.

Test- PCB with DC-link

Gate driver

Adapter from SEMIKRON standard interface to gate driver

Module on water cooler

Fig. 17: Setup of 4Q- single- phase inverter test

6. Summary

A high power, high frequency gate driver with a high integration density by ASICs for very low inductive 1200 V, 400 A SiC-MOSFET half bridge modules in both side Sinter-technology (SKiN) is presented. The MOSFET module has an internal inductivity of < 1.5 nH, a very low size and only 25% of the thermal resistance of conventional modules. The module has a very close connection to the low inductive DC link and the gate driver which leads to a system inductivity of < 5 nH. Because of its low propagation delay, high robustness against di/dt and dv/dt, low dead times and very strong output stages of I_{peak} = +40/-78 A, the driver is able to switch the 400 A MOSFETs up to 200 kHz with low switching losses and a surprisingly low voltage overshoot.

References

[1] Data sheet S2307; Rohm 1200 V, 45 mΩ MOSFET

[2] Data sheet CPM2-1200-0025B; Cree 1200 V, 25 mΩ MOSFET

[3] Data sheet IJW120R100T1; IFX 1200 V, 80 mΩ SFET

[4] Q. Zhang, G. Wang, H. Doan et.al. "Latest Results on 1200 V 4H-SiC CIMOSFETs with $R_{sp,on}$ of 3.9 mΩcm² at 150°C";Proc.ISPSD 2015, pp. 89-92

[5] P. Beckedahl, S. Buetow, A. Maul, M. Roeblitz, M. Spang "400A, 1200V SiC power module with 1nH commutation inductance", CIPS 2016

[6] R. Bayerer "Power Circuits for Clean Switching and Low Losses"; ECPE Tutorial Feb 2016

Integrating a real-time T_{vj} calculation into an IPM

Stefan Schmies, Infineon Technologies AG, Germany, stefan.schmies@infineon.com
Dr. Peter Lahl, Infineon Technologies AG, Germany, peter.lahl@infineon.com
Wolfram Kruschel, Infineon Technologies AG, Germany, wolfram.kruschel@infineon.com
Matthias Lassmann, Infineon Technologies AG, Germany, matthias.lassmann@infineon.com

Abstract

In power semiconductors, the junction temperature is one of the main specification limits for the achievable output power. To maximize the power density, systems are designed corresponding to the maximum junction temperature allowed, taking overload conditions into account. Exceeding the specified maximum junction temperatures can damage the power semiconductor chips, reduce the lifetime or cause an immediate failure.

Usually only the DCB temperature in power modules and IPMs is directly measured by using an NTC. As a result, there either is a large uncertainty if the application relies on temperature monitoring based on the NTC temperature only, or the application needs to have detailed information on the thermal characteristics and the capability to perform a junction temperature calculation.

Looking at the integration trends, there is a large benefit of integrating a junction temperature determination into IPMs and hereby allowing the application to directly read-out the junction temperatures. Furthermore, the IPM is then able to monitor the junction temperatures and generate warning signals as well as shut-downs prior to exceeding specified limits.

Here, it is desirable to perform the junction temperature calculation at a kHz bandwidth to enable observing the specific limits in overload conditions like low-voltage ride-through in wind applications.

This paper presents an innovative approach for a precise real-time calculation of the junction temperatures of a half-bridge IPM based on a lateral thermal model and semiconductor loss data gained by applying a Design-of-Experiment approach.

Fig. 1. MIPAQ™ Pro half bridge IPM with implemented Tvj calculation [1]

1. IPM Data Acquisition

The input data for the calculation of diode and IGBT junction temperature (Tvj) is based on real-time measurements and fixed data gained from the IPM characterization.
DC-Link Voltage, AC output current, DCB temperature and PWM-pulse patters are measured in real time (see Fig.2). Diode and IGBT switching and static losses were characterized in a typical application setup.

This data is combined for the pulse by pulse power loss calculation for diode and IGBT.

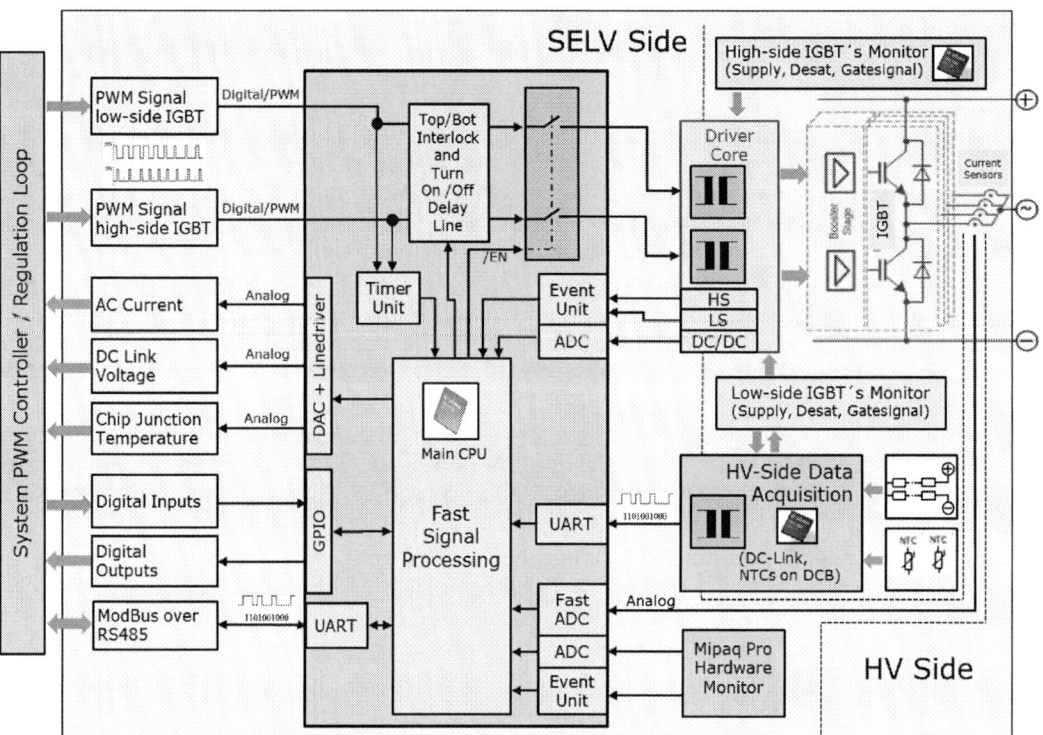

Fig. 2. Simplified MIPAQ™ Pro block diagram

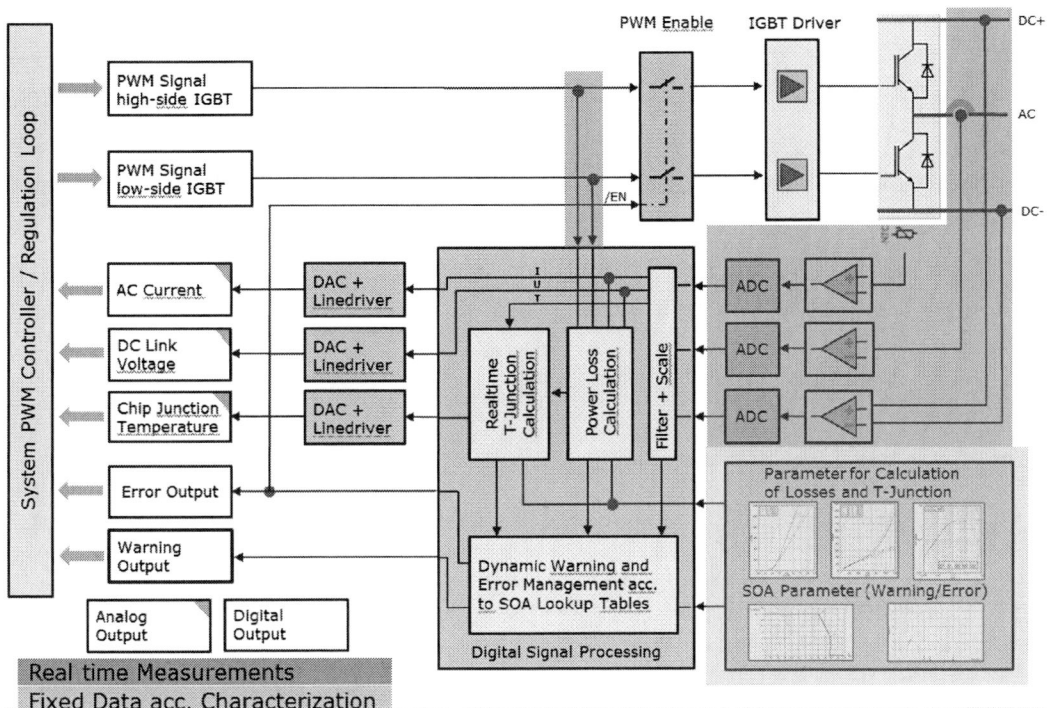

Fig. 3. Block Diagram of MIPAQ™ Pro data acquisition and processing

2.   Loss Calculation

The power losses in IGBT and diode are calculated for each switching operation and conduction phase. For this purpose, current-, voltage- and temperature dependencies of the dynamic and static losses have been determined using a Design-of-Experiments approach [2]. The acquired loss data is modelled by a cubic model including coupling terms. Using the determined cubic model, the losses can be calculated for specific operation conditions.

Calculating the losses at the maximum allowable junction temperature is sufficient for observing the specified limits as the temperature coefficients are positive. However, the cubic loss model can also be used for an accurate temperature calculation at lower temperatures depending on the microcontroller processing performance.

© VDE VERLAG GMBH · Berlin · Offenbach

PCIM Europe 2016, 10 – 12 May 2016, Nuremberg, Germany

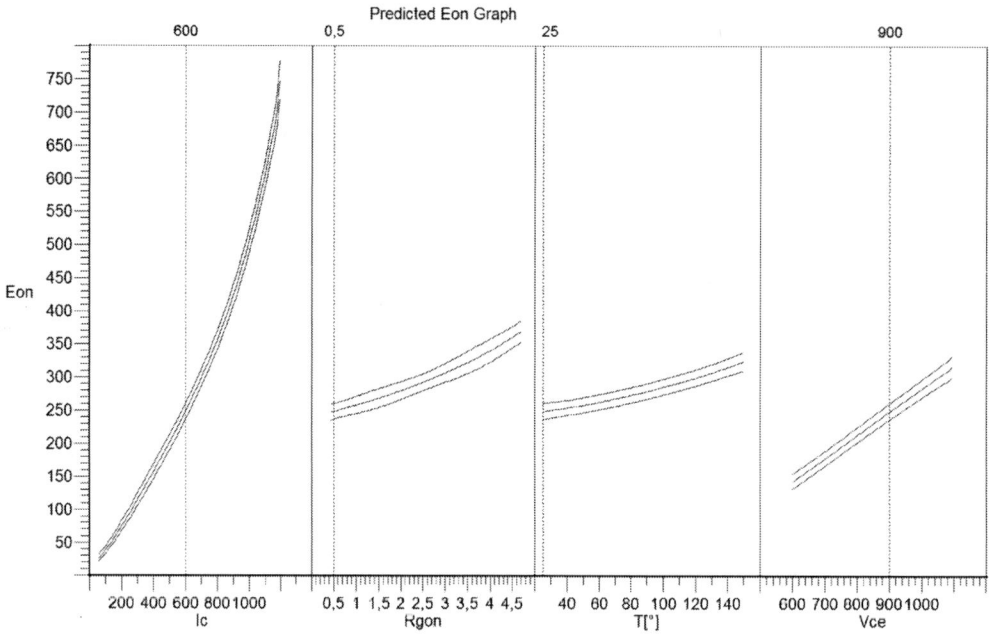

Fig. 3. DOE result created by "Cornerstone" software showing the dependencies for Eon

IGBT Eon	Cubic Model				

	Input				Output
	Rg	T	Ic	Vce	Eon
	[Ohm]	[°C]	[A]	[V]	[mJ]
	0,5	25	600	900	247,02

Min	0,45	25	60	600
Max	4,7	150	1200	1100
Midpoint	2,575	87,5	630	850
Term	-9,765E-01	-1,000E+00	-5,263E-02	2,000E-01

	Koeff.	Koeff * Term
Constant	1,790E+01	1,790E+01
T[°]	1,266E+00	-1,266E+00
Rg_on	2,574E+00	-2,514E+00
Vce	4,131E+00	8,263E-01
Ic	1,206E+01	-6,346E-01
T[°]^2	2,502E-01	2,502E-01
T[°] * Vce	4,579E-01	-9,159E-02
Ic * T[°]	2,398E+00	1,262E-01
Rg_on^2	4,487E-01	4,278E-01
Rg_on*Vce	5,475E-01	-1,069E-01
Ic * Rg_on	2,246E+00	1,155E-01
Vce^2	-3,381E-01	-1,352E-02
Ic * Vce	2,790E+00	-2,936E-02
Ic^2	2,559E+00	7,088E-03
Ic * Rg_on * T[°]	1,358E-01	-6,981E-03
Ic * T[°] * Vce	2,780E+00	2,926E-03
Ic^2 * T[°]	1,909E+00	-5,288E-03
Rg_off^3	-7,805E-01	7,267E-01
Ic * Rg_on * Vce	6,009E-01	6,176E-03
Ic^2 * Rg_on	4,893E-01	-1,324E-03
Ic * Vce^2	-3,107E-01	6,540E-04
Ic^2 * Vce	1,598E-01	8,856E-05
Ic^3	2,969E+00	-4,329E-04
	sum	1,572E+01
	sum ^ 2	2,470E+02

Fig. 4. Coefficients for Eon created by "Cornerstone" fed to an Excel sheet.

© VDE VERLAG GMBH · Berlin · Offenbach

To compute the IGBT and Diode switching losses, just like shown for Eon in Fig.4, a Microcontroller with Floating Point Unit is required. To handle the real-time measurements and to fulfil the demands on computing power a XMC4500 from Infineon was chosen. The Microcontroller has a 32Bit ARM Cortex M4 Core and is clocked at 120MHz. For the T_{vj} calculation at kHz bandwidth, the determined losses are summed up for each IGBT and diode per switch as a time series of discrete values.

3. Tvj Calculation

Thermal Model

The junction temperature calculation starting point is given by the NTC, which is situated on the DCB as can be seen in fig. 5a. Consequently, a lateral thermal model is used, which allows calculating the T_{vj} of IGBT and diode based on the measurement of the NTC temperature and the corresponding power losses.

To ensure a high accuracy, also the non-diagonal elements of the Z_{th} matrix, corresponding to the cross-coupling between IGBTs and diodes, are taken into account. These matrix elements are more important the larger the distance between the NTC and the chips is, compared to the inter-chip distance. Each element of the Z_{th} matrix as well as the power vector is time-dependent, representing the dynamic behavior.

Fig. 5b depicts the thermal model represented by an equivalent network. Fig. 6 gives the calculation formula using the complete Z_{th} matrix. The lateral nature of the model is contained in the matrix terms, which describe the temperature difference between NTC and IGBT or diode respectively as a result of the power loss vector. The temperatures are determined for IGBT and diode individually; the three indicated chips seen in fig 2a are combined within the thermal model.

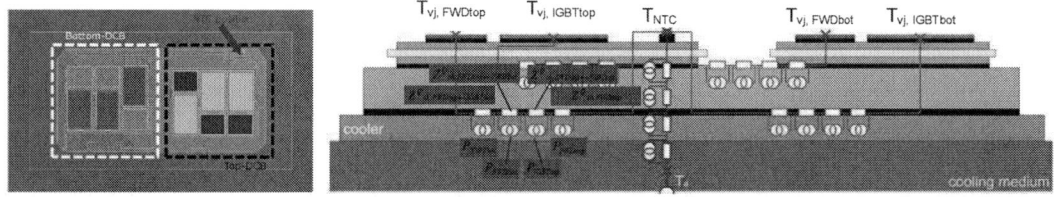

Fig. 5 (a) Top-view sketch of the power section with NTC position. (b) Lateral thermal model. The NTC temperature is measured and the junction temperature calculation for the diode of the top system is indicated.

$$
\begin{pmatrix} T_{IGBTbot} \\ T_{FWDbot} \\ T_{IGBTtop} \\ T_{FWDtop} \end{pmatrix} = \begin{pmatrix} Z^Q_{th,IGBTbot} & Z^Q_{th,IGBTbot\leftarrow FWDbot} & Z^Q_{th,IGBTbot\leftarrow IGBTtop} & Z^Q_{th,IGBTbot\leftarrow FWDtop} \\ Z^Q_{th,FWDbot\leftarrow IGBTbot} & Z^Q_{th,FWDbot} & Z^Q_{th,FWDbot\leftarrow IGBTtop} & Z^Q_{th,FWDbot\leftarrow FWDtop} \\ Z^Q_{th,IGBTtop\leftarrow IGBTbot} & Z^Q_{th,IGBTtop\leftarrow FWDbot} & Z^Q_{th,IGBTtop} & Z^Q_{th,IGBTtop\leftarrow FWDtop} \\ Z^Q_{th,FWDtop\leftarrow IGBTbot} & Z^Q_{th,FWDtop\leftarrow FWDbot} & Z^Q_{th,FWDtop\leftarrow IGBTtop} & Z^Q_{th,FWDtop} \end{pmatrix} \cdot \begin{pmatrix} P_{IGBTbot} \\ P_{FWDbot} \\ P_{IGBTtop} \\ P_{FWDtop} \end{pmatrix} + T_{NTC}
$$

$$
Z^Q_{th,Xy\leftarrow Ab} = \frac{T_{j,Xy} - T_{NTC}}{P_{Ab}}
$$

Fig. 6: T_{vj} calculation formula. Please note that the multiplication represents a convolution integral, as both Z_{th} and P are time-dependent. The Z_{th} matrix was determined and investigated using simulations and confirmed by measurements on the actual device.

Real-time Junction Temperature Calculation

For the junction temperature calculation, the time series of power loss vectors needs to be convolved with the Z_{th} matrix numerically.

This calculation was implemented on the integrated XMC4500 Microcontroller using a digital filter method. Here, the thermal impedances of each Z_{th} matrix element are approximated by one digital filter so that in total 16 filters need to be calculated. The calculation time must be significantly below 1ms for enabling the desired kHz bandwidth. The first implementation shows that the microcontroller is utilized less than 25% by calculating the Z_{th} matrix.

Detailed investigations regarding an efficient implementation of the real-time junction temperature calculation on the XMC4500 were carried out [3]. For these investigations, a MATLAB-based hardware-in-the-loop system was developed. Several filter types were evaluated with respect to calculation speed and accuracy. An IIR type filter was chosen for optimum performance and sufficient accuracy for the temperature calculation was demonstrated.

Application Impact

The application can affect the junction temperature calculation via the cooling conditions, namely the coolant thermal capacity and the flow rate. The coolant thermal capacity depends on the mixture of water with anti-freezing and corrosion protection additives.

The impact of these conditions on the T_{vj} calculation accuracy was investigated by a series of simulations. For this purpose, different temperature gradients representing application-relevant minimum and maximum cooling conditions for (a) low thermal capacity and low flow rate and (b) high thermal capacity and high flow rate were introduced into the heatsink from inlet to outlet. The lateral thermal models were determined for each case and compared. It was demonstrated that the application impact on the described T_{vj} calculation via the cooling conditions can be neglected.

Here, the lateral thermal model clearly has advantages compared to a vertical thermal model , because the temperature distribution in the heatsink down to the coolant is less relevant.

4. Summary and Conclusion

In this paper, the implementation of a junction temperature calculation using a XMC4500 microcontroller integrated into the high-power IPM MIPAQ™ Pro was demonstrated. The detailed loss determination and a lateral time-dependent thermal model are described.

The T_{vj} calculation is performed on the microcontroller using digital filters. Adequate calculation speed, accuracy and sufficient independence from the application's cooling conditions are demonstrated.

Implementing advanced functionalities into an IPM allows deep insight into the operating conditions like current, voltage and temperatures. Besides general electronics and DCB temperatures, the maximum power semiconductor's junction temperature is of profound interest, because it belongs to the main parameters limiting the output power.

Here, the implementation of a junction temperature calculation allows the IPM to monitor the system's operation and generate warning signals or shut down operation prior to violating specified limits. Furthermore, it enables a simple design-in into the application, especially when exploring the application limits during the design phase. Thus, the implementation of the junction temperature calculation into the IPM represents a consequent continuation of integrating advanced features into the IPM for the application's benefit.

5. References

[1] P. Lahl, S. Schmies, K. Schoo, M. Schulz: "Advanced Features in Sophisticated Inverter Design Supporting MW-Applications", PCIM Europe, 2015

[2] R. Zaazaa: "*Bestimmung und Modellierung der Schaltenergien eines High-Power-IPM nach dem Design of Experiments (DoE) Verfahren für den applikationsnahen Parameterraum*", Bachelor Thesis, Hochschule Düsseldorf, 2015

[3] P. Gräber: "*Entwurf, Ausarbeitung, Implementierung und Verifikation einer mikrocontrollerbasierten, echtzeitfähigen Sperrschichttemperaturberechnung für IGBT-Module*", Master Thesis, HTWK Leipzig, 2014

PCIM Europe 2016, 10 – 12 May 2016, Nuremberg, Germany

12 V lithium ion starter batteries

Schweiger, Hans-Georg, Technische Hochschule Ingolstadt, Germany,
hans-georg.schweiger@thi.de

Machuca-Garcia, Enrique, Technische Hochschule Ingolstadt, Germany,
enrique.machuca-garcia@thi.de

Löchel, Jonas, Technische Hochschule Ingolstadt, Germany, jonas@loecheledv.de

Abstract

The research focuses on the development of 12 V lithium-ion starter batteries for cold cranking. Electric Double Layer and lithium capacitors were connected in parallel to the lithium-ion battery to improve the power capability at low temperature. For controlling voltage, current and temperature of the cells, a Battery Management System was developed. Finally, lots of combinations were tested by applying cold cranking profiles at -28 °C. The goal was to find the best combinations that, fulfilling all the requirements, and saving weight, space and costs. This technology guarantees a longer life of the battery. The results show that the lithium capacitors failed at low temperature and only the Electrochemical Double Layer capacitors helped the lithium-ion cells to perform the cold cranking tests successfully. Lithium titanate cells proved good cold cranking performance, even without capacitors.

1. Introduction

Lead-acid batteries are the most extended technology used for the starting, lighting and ignition of the car, because they are cheaper than their competitors, but they have some disadvantages, like the low energy density, high weight, low cycle life and the emissions to the environment [1]. Lithium-ion are new batteries technology with high energy and power density, no heavy metals, long cycle life and light weight, resulting in a reduction of the battery size [2]. Electric double-layer capacitors (EDLC) are electrochemical capacitors that are able to store more energy than the conventional capacitors and to deliver more power than the lithium-ion batteries [3]. Finally, the lithium capacitors are an emerging technology that comprise the advantages of lithium-ion cells and EDLC. Compared with the lithium-ion cells, this hybrid system offers higher power density [4] and similar self-discharge. They also have higher cell voltage range and higher energy density if compared with EDLC [5].

Considering the advantages and disadvantages of the lithium-ion cells and the capacitors, the idea is to combine them in parallel, looking for the cheapest, lightest and smallest combination, getting an optimal starter battery [6]. Thanks to the capacitors, a better cold cranking power, a longer life time of the battery and a save of weight can be achieved. The lithium-ion cells will contribute with a very good energy density and, of course, with the capacity. Due to the 12 V nominal, there will be 4 Li-ion cells in series and 4 to 6 capacitors (depending if LiC or EDLC are employed) in parallel.

The main goals of the project are: 1) preselection of the cells by simulation; 2) simulate lots of combinations; 3) market research; 4) lithium-ion cells and capacitors selection; 5) electronics, hardware and software development to control the main battery parameters; 6) flexible prototype: build batteries connecting lithium-ion cells and capacitors in parallel, testing hundreds of combinations; 7) Select the best battery obtained in the lab tests.

2. Theoretical background

A 12 V starter battery combining in parallel lithium-ion cells and capacitors (EDLC and Li-capacitors) was developed for cold cranking tests at -28 °C. At low temperature, the lithium-ion cells have poor cold cranking capability because their internal impedance increases [7], and therefore, their power density decreases. As the proposed discharge profile demands a high current for several seconds, it was necessary to connect in parallel capacitors to increase

© VDE VERLAG GMBH · Berlin · Offenbach

the power capability of the battery [8]. Other benefits obtained from this combination are the weight saving (compared to the lead-acid batteries) and the increasing of the battery lifetime due to the higher cycle stability.

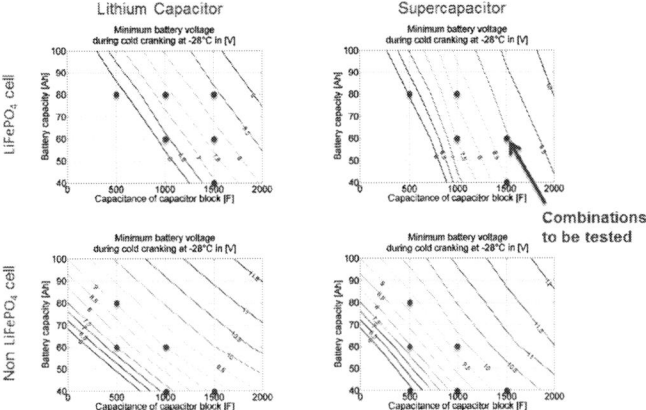

Figure 1 Simulation of all the possible combinations between the lithium-ion cells and capacitors. The colored lines represent different minimum voltages U_{MIN} of the battery and the red dots the combinations to be tested

To select the battery and capacitor cells simulations were carried out prior experiments. A battery model described in [9], [10] was used within the simulation software Matlab/Simulink R2013b (The Mathworks, Inc., USA). Parameters of the cells were derived from a market research and scaled to -28 °C [11]. The load profile shown in Figure 5 was applied different cell combinations and the minimum cell voltage was determined. In Figure 1 the results of these simulations are shown.

3. Experimental

Experimental setup

A homemade battery tester was used for the experiments. This tester is comprised of two GEN-60-85-3P400 power supplies (TDK-Lambda, Japan) and two electronic loads, EA-EL 9080-400 and EA-EL 9080-600 (Elektro-Automatik GmbH, Germany). This set-up provides a voltage range from 0-60 V, charging currents up to 170 A (limited to 10.2 W) [12] and discharging currents up to 1000 A (limited to 12 kW) [13]. A homemade autonomous safety switch-off box was implemented in the test bench, protecting battery from over voltage, deep discharge and over temperature. National Instrument LabView Version 2012 (Texas, USA) was used to automatize the test bench with a homemade virtual instrument. A VT 4021 temperature chamber (Vötsch Industrietechnik GmbH, Germany) was used to control the ambient temperature of the flexible prototype. And a VT 4011 temperature chamber from the same company used for the same purpose for the Demonstrator. Both chambers provide a stable temperature of -28 °C with an accuracy to ±0.5 K [14]. After assembly of the batteries, each single cell was charged with A NSP-2050 [15] power supplies (Manson Engineering Industrial Ltd., Hong Kong) to the same voltage, reducing the time need to balance the cells.

Flexible Prototype

The goal was to test different combinations of lithium-ion cells and capacitors forming a 12 V battery. The main focus of this prototype was the investigation of the best combination of battery and capacitor cells. So the main design goal was fast and convenient interchange of the cells modules instead of being a weight and space optimized battery. So, several lithium-ion cells chemistries (LiFePO$_4$, NMC, etc.) and shapes (18650, 26650, pouch, etc.) and different capacitors types (EDLC (cylindrical and prismatic) and lithium capacitor) were selected, and assembled in different combinations. Lithium-ion cells were arrange in 12 V / 20 Ah modules and four of these modules were build out of each cell type. So 20 Ah, 40 Ah,

© VDE VERLAG GMBH · Berlin · Offenbach

60 Ah and 80 Ah combinations were possible. Double layer and lithium capacitor cells were also arranged in 12 V and 0.5 kF modules. 3 of these modules were built for each cell type. So 0.5 kF, 1 kF and 1.5 kF combinations were possible.

By arranging the battery cell and capacitor modules, 102 combinations were possible. The assembled and tested combinations are shown in Table 1. According to the experience got from the first tests, some of the scheduled tests were skipped due to their chance of failure (if the biggest combination failed, smaller ones would fail too) or success (if the smallest combination passed, the bigger ones would pass too). In total, 47 tests were carried out.

Table 1 Selected lithium-ion cells [16 - 20], EDLC cells [21-22] and Li-capacitor cells [23] for the flexible prototype

Company	Model	Chemistry	Q [Ah]	C [kF]	R (AC) [mΩ]	Weight [g]	12 V 20 Ah / 0.5 kF module
A123	AMP20M1HD-A	LiFePO₄	20		1.0	496	4S1P
HETER	HTCF18650-1100-3.3	LiFePO₄	1.1		15	40.0	4S20P
LUMOS	HTCN26650-4.5Ah-3.6V	NMC	4.5		30	92.0	4S5P
LUMOS	HP-NP-3R2-200B	LiFePO₄	20		2.0	780	4S1P
KOKAM	SLPB 70205130P	Li-Po	12		1.5	354	4S2P
LS MTRON	LSUC 002R8P 3000F EA LR01	EDLC		3.0	0.25	650	6S1P
MAXWELL	BCAP3000	EDLC		3.0	0.29	510	6S1P
JSR MICRO	ULTIMO 2300F ULR	Li-Cap.		2.3	0.70	380	4S1P

For the mechanical design, despite fitting into the temperature chamber, there were no size restrictions. Heter (18650) and Lumos (26650) modules were assembled by using pure nickel strip (> 99.5%, 7x0.15 mm, 1.05 mm2, 9.35 g/m) welded on with resistance welding (ISQ20-6DC [24] weld controller with double pneumatic weld head, Miyachi Europe GmbH, Germany) the terminals and fixed with plastic holders. The modules of the screwed terminals cells (Lumos prismatic, LS Mtron, Maxwell and JSR Micro), copper bars were used for the connection. Pouch cells modules (A123 and Kokam) were assembled employing copper plates and plastic for pressing the cells terminals together [25]. The modules interconnection were carried out by using cables and cable shoes or metal bars.

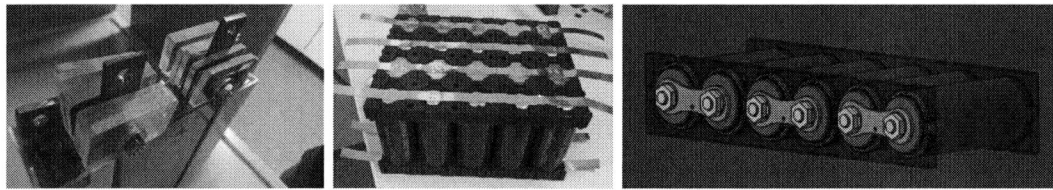

Figure 2 Different cell block assemblies, from left to right: Pouch cells assembly, cylindrical cells assembly, screwed terminal cells assembly.

Battery Management System (BMS)

To monitor the cells and capacitors (single cell voltages, currents distribution between the cell blocks, temperatures), as well as the system voltage and current of our batteries, a home build battery management system was used in both prototypes.

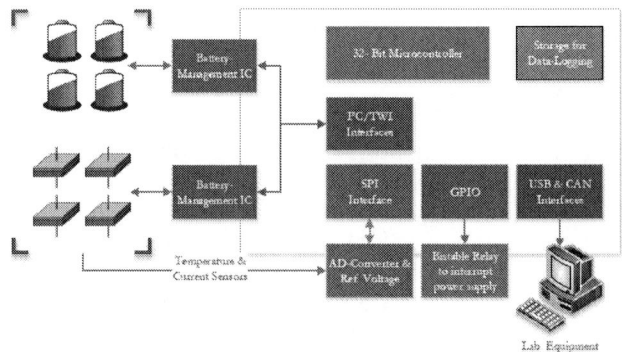

Figure 3 Architecture of the Battery Management System

© VDE VERLAG GMBH · Berlin · Offenbach

The BMS consists mainly of a microcontroller doing all kinds of calculations and data logging, two battery management ICs to monitor the cells and capacitors separately and an interface to obtain real-time data and access the data logged to a storage device. With the used battery management ICs the BMS is able to measure all cells and capacitors (with a voltage of up to 5 V) within 107 μs at an accuracy of up to 0.25 % [26].

Demonstrator

A more refined battery, the demonstrator was designed and built. This design had strong restrictions of size, weight, robustness and a cost-effective design, and it was oriented to the production. The eight best combinations of cells from the flexible prototype were preselected and ranked according to several features. The best rated batteries were selected for the Demonstrator. The evaluated features were volume, weight, assembly difficulty, U_{MIN} of the battery during the test, I_{MAX} of the lithium-ion cells block and compactness and power from the lithium-ion cells and capacitor block. Finally, a third battery only built with lithium-ion cells was selected as the best option from the market research carried out specifically for the demonstrator. In total, three batteries were designed, built and tested. The selected batteries for the Demonstrator were:

1) Kokam 60 Ah (4S5P) – Maxwell 1 kF (6S2P)

2) Kokam 60 Ah (4S5P) – LS Mtron 1 kF (6S2P)

3) Toshiba LTO 40 Ah (5S4P)

The main features, compared with the H9 [27] standard lead-acid battery are:

Table 2 Main parameters for the Demonstrator batteries. Comparison with a H9 standard lead-acid battery [27].

Feature	Battery 1	Battery 2	Battery 3	H9 lead-acid	Benefit
Capacity (Ah)	60	60	40	90	Less capacity
R_i AC (mΩ)	1.88	2.22	1.35	6 to 10	5 times smaller
R_i AC (mΩ) (-28°C)	2.1	1.66	2.32	9 to 12	5 times smaller
Dimension (mm)	465x295x195	460x285x195	394x175x190	394x175x190	Battery 3 same size
Weight (kg)	18	19.5	14	20 to 25	Always lighter

A CAD design of the three demonstrators are shown in Figure 4:

 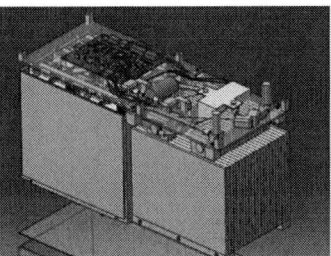 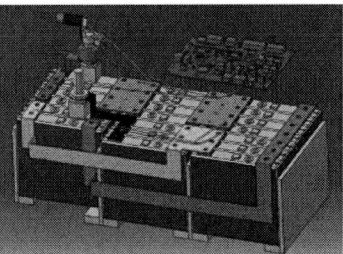

Figure 4 CAD drawings of the demonstrator. From left to right: 1) Kokam 60 Ah (4S5P) - Maxwell 1kF (6S2P); 2) Kokam 60 Ah (4S5P) - LS Mtron 1kF (6S2P); 3) Toshiba LTO 40 Ah (5S4P).

Tests of the Demonstrator were also performed by replacing the relay by a PSS-1 pyroswitch (Autoliv Inc., Sweden-USA) [28], a pyrotechnic safety switch that separates the electric connection irreversibly. This resulted in a smaller voltage drop between the main poles and the battery terminals.

Cold cranking tests

The profile used to test the battery was designed to simulate the following situation. The car is parked for 6 weeks and then the car is cold cranked at -28.0 °C. The procedure is shown in the left hand side of Figure 5. The battery was charged with 1 C with the constant current, constant voltage (CC/CV) method up to 15.5 V[1], the maximum voltage of the vehicles board net [29]. Due to quiescent currents 25 Ah [30] are removed from the battery during parking time. This was simulated by a 25 Ah 1 C CC discharge. While setting the charge of the battery was carried out at +25.0 °C, the battery was cooled down to -28.0 °C prior cold cranking. After a rest of 12 h at -28.0 °C to ensure thermal equilibration of the battery, the cold cranking profile, shown in the right hand side of Figure 5, was applied to the battery.

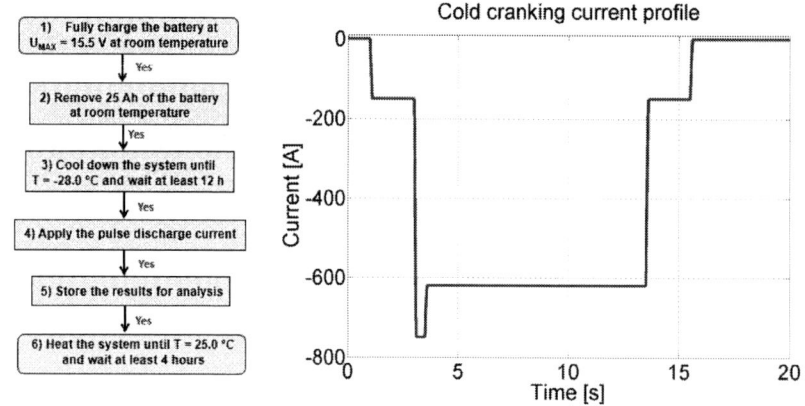

Figure 5 left hand side: Testing procedure of the demonstrators and the flexible prototypes
right hand side: Cold cranking profile

During all the tests the BMS was used to monitor and record each single cell voltage, battery voltage and current as well as the voltage and current of each single cell block. The flexible prototype, as well as the demonstrator we analyzed with the same profile.

A test was considered as successful if the minimum battery voltage during the cold cranking discharge was $U_{MIN} \geq 6.5$ V. If the battery voltage was below 6.5 V or the minimum voltage of a single cell was reached, the tests was stopped. After the test the battery was heated up to 25.0 °C. In case of the flexible prototype, the next combination of cells was assemble and the procedure was repeated. And in case of the demonstrator, the test was conducted with another model.

4. Results and discussion

Flexible Prototype

As shown in Figure 6, the most demanding part of the profile is the 620 A pulse applied during 10 s, especially because the battery is at -28 °C.

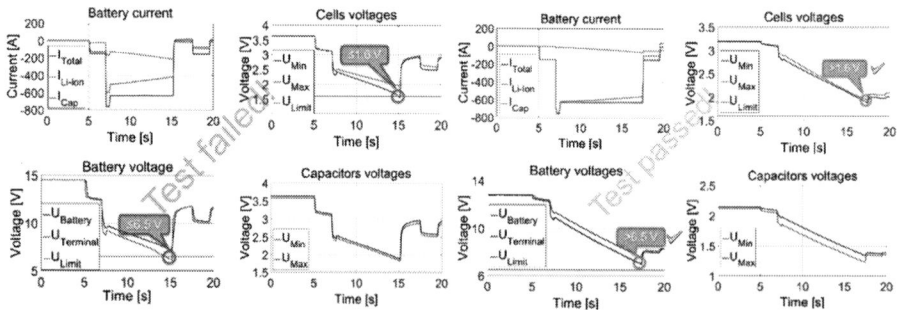

Figure 6 left side: a failed test with Kokam 84 Ah - JSR Micro 1.725 kF battery, where $U_{MIN} < 6.5$ V
right side: a succeeded test with A123 40 Ah - LS Mtron 1.5 kF battery, where $U_{MIN} > 6.5$ V

---

[1] 14 V in case of Toshiba battery. As the connection is 5S4P and $U_{MAX,cell}$ = 2.8 V, U_{MAX} = 2.8 V·5 = 14 V

A failed and a successful test cases including also U_{MIN} and U_{MAX} for the individual lithium-ion cells and capacitors are shown. After having performed all the scheduled tests, the most remarkable conclusions can be drawn. No combinations of lithium-ion cells and lithium capacitors worked. The lithium capacitors have a behavior between lithium-ion cells and EDLC, but at -28 °C the internal impedance (DC) of a single cell increases in a factor 7 in comparison with at 25 °C (from 0.7 mΩ [23] to 5 mΩ at -28 °C, obtained from the cold cranking tests). This value is much bigger than for the individual EDLC (factor 1.7, from 0.25 mΩ [21] to 0.41 mΩ at -28 °C approx.). This means that at these conditions, the lithium capacitors act similar to the lithium-ion cells, not delivering the requested power.

No combinations of lithium-ion cells and 0.5 kF EDLC modules worked. At -28 °C, taking a look at the battery current plots on Figure 6, an 80% approximately of the discharge current is absorbed by the EDLC. This means that the size of the capacitors system will be crucial for the demanded power. It can be concluded that at least 1 kF EDLC system was required to succeed, being hard to reach the standard H9 battery size.

Table 3 Tested combinations for the flexible prototype including U_{MIN} [V] / Weight [kg] / Volume [dm³]

		Combinations between Li-Ion cells and Capacitors								
		Capacitors								
		Maxwell 3000F EDLC			LS Mtron 3000 F EDLC			JSR Micro 2300 F LiCap		
	Modules	500 F	1000 F	1500 F	500 F	1000 F	1500 F	575 F	1150 F	1725 F
Heter 1.1 Ah 18650	40 Ah			8.9/15.8/11.4			7.2/18.1/11.1			
	60 Ah			7.2/18.8/12.8			7.4/21.5/12.5			
	80 Ah		<6.5/18.8/11.2	7.7/22/14.1	<6.5/16.7/8.2		7.2/24.5/13.9			<6.5/17.2/8.6
Lumos 4.5 Ah 26650	45 Ah		<6.5/9.8/7.2	7.6/12.9/10.1			7.9/15.4/9.8			
	68 Ah		<6.5/11.0/7.3	8.1/14.7/10.8						
	90 Ah		6.5/13.5/8.6	8.4/16.5/11.5	<6.5/13.3/8.1					
Kokam 12 Ah pouch	48 Ah		<6.5/11.4/5.5	7.9/14.8/12.4			8.1/17.4/12.1			
	60 Ah		6.9/13.2/10.4			6.7/14.9/10.2				
	84 Ah		7.1/16.1/12.3	8.8/19.1/15.2	<6.5/15.8/6.8	7.4/17.7/12.1	8.9/21.6/14.9			<6.5/18.3/6.9
A123 20 Ah pouch	40 Ah			5.8/13.2/11.1			7.0/15.7/10.8			
	60 Ah		<6.5/13.3/8.3	6.8/15.1/12.3	<6.5/13.8/9.2		7.2/21.6/13.1			
	80 Ah		<6.5/15.1/10.5	6.9/19.1/13.4	<6.5/14.3/7.5					
Lumos 20 Ah prismatic	40 Ah			7.1/15.4/12.2			7.2/18.9/11.9			<6.5/10.6/4.2
	60 Ah					<6.5/17.2/10.4				<6.5/13.8/3.7
	80 Ah	<6.5/18.7/9.4	<6.5/13.7/12.7	4.2/24.8/15.8		6.6/19.4/12.9		<6.5/18.5/8.1		

Test OK	Test failed	Test not scheduled
Skipped: expected to pass	Skipped: expected to fail	

All the test results are shown in Table 3. There, the U_{MIN} reached during the clod cranking discharge is included. Dark green means a succeeded test (above 6.5 V system voltage or above 1.6 V lithium-ion cell voltage) and dark red a failed one. The reasons why some scheduled tests were skipped (light green and red) are explained in chapter 3. Finally, light grey means not scheduled test. The behavior of Maxwell and LS Mtron EDLC was very similar as seen on the U_{MIN} values, having the last one a better performance (higher U_{MIN}). The small difference was due to the smaller value of the Ri (DC) of LS Mtron compared to Maxwell one (0.25 mΩ [21] vs 0.29 mΩ [22]).

Demonstrator

The results for the tests performed with the Demonstrator batteries including U_{MIN} and U_{MAX} for the individual lithium-ion cells and capacitors are shown in Figure 7 and Figure 8 shows all the individual lithium-ion cells voltages and also U_{MIN} and U_{MAX} values for the lithium-ion cells.

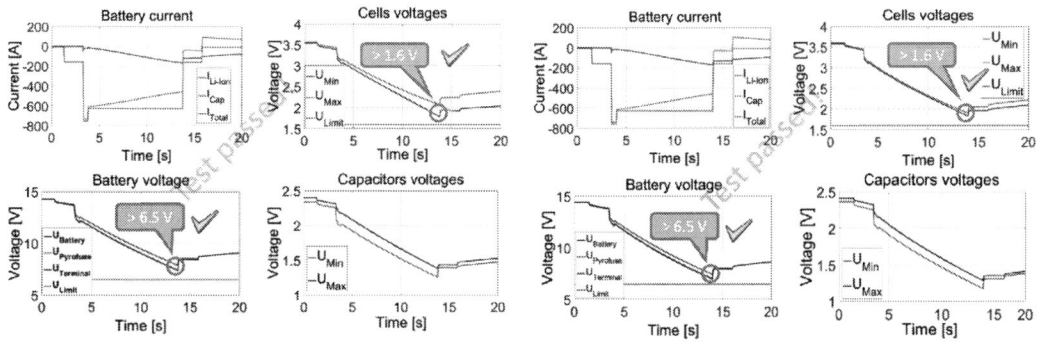

Figure 7 left side: Cold cranking test Battery 1 [Kokam 60 Ah - Maxwell 1 kF] with pyroswitch, U_{MIN} = 7.39 V right side: Cold cranking test Battery 2 [Kokam 60 Ah - LS Mtron 1 kF] with pyroswitch, U_{MIN} = 6.99 V

© VDE VERLAG GMBH · Berlin · Offenbach

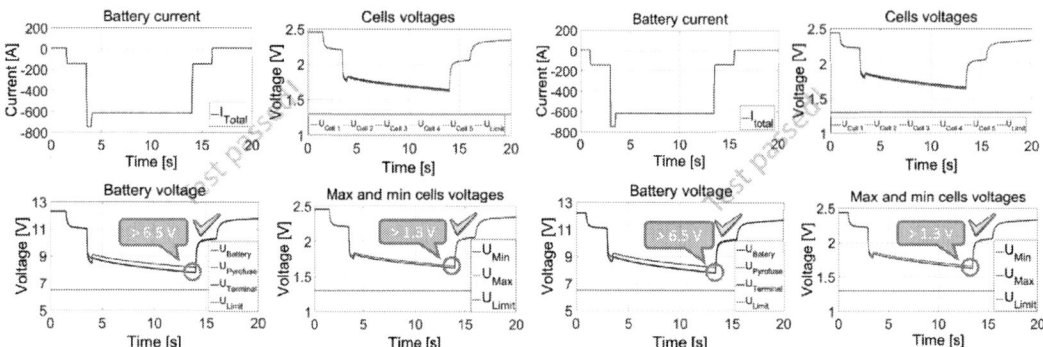

Figure 8 left side: Cold cranking test Battery 3 [Toshiba LTO 40 Ah] with pyroswitch; U_{MIN} = 7.74 V right side: Cold cranking test Battery 3 [Toshiba LTO 40 Ah] with pyroswitch after 5 minutes; U_{MIN} = 7.75 V

As expected, the batteries 1 and 2 succeeded the tests and can be reduced in terms of capacity [Ah] and/or capacitance [F] as U_{MIN} is near 1 V above the limit. The battery 3 passed also the test, even by applying a second discharge 5 minutes later. Thanks to the LTO cells, it's possible to avoid the use of capacitors, saving a lot of weight and space.

5. Conclusions

The lithium-ion cells contributed with the energy and the capacitors with the power. It was a challenge to select the number of each and therefore the battery size, so that the cold cranking tests could be performed successfully. The flexible prototype helped to discard the Li-capacitors. Despite their promising features at room temperature, they offer a bad cold cranking performance due to the high increase of the internal impedance at -28 °C.

Concerning the Demonstrator, the best option for building the 12 V lithium-ion starter battery are the LTO cells. No capacitors are required because the cells have a big power capability even at -28 °C. This allow to build a small, compact and light battery, improving the features of the lead-acid batteries even by being built only with lithium-ion cells.

The use of a pyroswitch decreases the voltage drop of the battery as well decreasing the internal impedance of the battery. As a consequence, a smaller battery could perform the test successfully. As the pyroswitch is 150 g lighter than the relay, also some weight was saved. More weight (255 g) can be also saved by using the pyroswitch to replace the relay and the fuse.

With the demanding discharge profile proposed, a working 12 V lithium-ion starter that could replace the standard lead-acid batteries was achieved.

6. Acknowledgements

The authors would like to thank to the Bundesministerium für Bildung und Forschung (BMBF) for funding and AUDI AG and EVA Fahrzeugtechnik GmbH for cofounding. Maxwell and Autoliv for providing free samples (DLC and pyro fuses), Miyachi Europe GmbH and VRI GmbH for assembling for us the cylindrical cells free of charge, and Marquardt GmbH for assembling the LTO cells for free. S. Barra (THI) for supporting us doing the tests, S. Vlasov for supporting BMS development and J. Winkler (Audi), D. Lamm (EVA) and O. Kanoun (TU Chemnitz) for discussion.

7. References

[1] D.A.R. Rand, J. Power Sources 64 (1997) 157–174

[2] M. Ceraolo, T. Huria, G. Pede, F. Vellucci, VPPC, 2011 IEEE, pp 1-6

[3] P. Sharma, T.S. Bhatti, Energ. Convers. Manage. 51 (2010) 2901–2912

[4] R. B. Sepe, A. Steyerl, S. P. Bastien, ECCE, 2011IEE, pp 1813-1818

[5] D. Porcarelli, D. Brunelli, L. Benini, INSS, 2012 9th Int. Conference on Networked Sensing, June 2012, pp 1-4

[6] H. Liu, Z. Wang, J. Cheng, D. Maly, Veh. Technol., Vol. 58, pp 1097-1105

[7] P. Suresh, A.K. Shukla, N. Munichandraiah, J. Appl. Electrochem., 32(3):267-273, 2002

[8] A. Burke, M. Miller, J. Power Sources 196 (2011) 514–522

[9] H. He, R. Xiong, J. Fan, Energies 2011, 4, 582-598

[10] Siqi Lin, Hengbing Zhao, Andrew Burke. Research Report – UCD-ITS-RR-12-12. Institute of Transportation Studies University of California-Davis

[11] Eduardo Cueva, Current State and the Future of the Energy Density and Specific Power of Lithium Ion Batteries (2014), Master thesis

[12] TDK-Lambda GEN-60-85-3P400 power supply datasheet, [Online]. Available: http://www.us.tdk-lambda.com/hp/pdfs/data%20sheets/93515001.pdf

[13] Elektro-Automatik GmbH EA-EL 9080-400 and EA-EL 9080-600 electronic loads datasheet, [Online]. Available: http://docs-europe.electrocomponents.com/webdocs/138c/0900766b8138cd88.pdf

[14] Vötsch Industrietechnik GmbH VT 4011 and VT 4021 datasheet, [Online]. Available: http://www.vt.com/sixcms/media.php/2335/VIT_Laboratory%20Temperature%20test%20chambers%5B1%5D.pdf

[15] Manson Engineering Industrial Ltd. NSP-2050 portable power supply, [Online]. Available: http://www.manson.com.hk/products/detail/31

[16] A123 systems, AMP20M1HD-A Nanophosphate® Lithium Ion Prismatic Pouch Cell, http://www.a123systems.com/prismatic-cell-amp20.htm

[17] Heter Electronics Group Co., Ltd. Technical Datasheet of Lithium Ion Cylindrical Cell, Model HTCF18650-1100-3.3

[18] Lumos Power & Electronics Co., Ltd. Technical Datasheet of Lithium Ion Cylindrical Cell, Model HTCN26650-4500mAh-3.6V.

[19] Lumos Power & Electronics Co., Ltd. Technical Data Sheet of Supercapacitor Li-ion Prismatic Cell, Model HP-NP-3R2-200B.

[20] Kokam Co., Ltd. Datasheet of Lithium Polymer Pouch Cell, Model SLPB 70205130P.

[21] LS MTRON Ltd., Datasheet of EDLC, Model LSUC 002R8P 3000F EA LR01.

[22] Maxwell Technologies, Inc., Technical Datasheet of EDLC, Model BCAP3000.

[23] JSR Micro, Inc. Technical Datasheet of Lithium Capacitor, Model ULTIMO 2300F ULR.

[24] Miyachi Europe GmbH, Technical datasheet of resistance welding machine, Model ISQ20-6DC http://www.amadamiyachieurope.com/cmdata/documents/TDS-AWS3-Servo-Motorised-RW-08-2015.pdf

[25] K. Afridi, S. Burhan, A. Escudero, T. Futshane, A. Gruber, A. Hackula, F. Ort, P. Ganti, R. Qayyum, A. Wild, 12 V lithium-ion battery (2013), IAE student project

[26] https://www.maximintegrated.com/en/products/power/battery-management/MAX11068.html

[27] DIN EN 50342-2 Lead-acid starter batteries - Part 2: Dimensions of batteries and dimensions and marking of terminals

[28] Autoliv, PSS-1 technical datasheet 5 (2015)

[29] Lastenheft zur 12V - Lithium- Starterbatterie Anforderungen, Spezifizierungen, Prüfungen V 1.0, Matthias Schneider (AUDI AG), Markus Mauerer (BMW Group), Thomas Binder-Leube (DAIMLER AG), Henning Kittel (Porsche AG), Sina Brunner (Porsche AG), Dr. Roland Kube (VW)

[30] Maxim Integrated, "Application note 3928: Automotive Linear Regulators Minimize Quiescent Current", 17th February 2015, available at https://www.maximintegrated.com/en/app-notes/index.mvp/id/3928

Electric Vehicles Batteries Modeling Analysis Based on a Multiple Layered Perceptron Identification Approach

Sender Rocha dos Santos, CPqD, Campinas, SP, Brazil, srocha@cpqd.com.br
Thais Tóssoli de Sousa, CPqD, Campinas, SP, Brazil, tsousa@cpqd.com.br
Alex Pereira França, CPqD, Campinas, SP, Brazil, afranca@cpqd.com.br

Abstract

A artificial neural network model of electric vehicles batteries to estimate the State of Charge (SoC) and State of Health (SoH) is presented. Two types of cells are investigated, lithium-ion and lead acid, confirming the versatility of the technique studied. Additionally, experimental case study results indicated outperforming, accuracy and fast convergence performance and robustness of the developed battery model free.

1. Introduction

Effective battery system for electric vehicles design is complicated by the fact that there is a wide range of battery chemistries to choose from Ragone Chart 2014 – each one with their own inherent characteristics, as example high density energy, high density power, and capacity. Modeling enables pack designers to select specific cells from specific manufacturers, and accurately and quickly simulate their behavior under specified conditions such as pack configuration, power delivery system and circuitry and load profile.

Sophisticated energy management strategies have been developed to provide better fuel economy in hybrid electric vehicles (HEVs) and high performance in pure electric vehicles (PEVs). The Battery Management System (BMS) is the key element to monitor and control the conditions and states of the traction battery pack, such as SoC and SoH. Therefore, the BMS is able to estimate the SoC, power and capacity fade and instantaneous available power, and it can adapt the cell characteristics over time as the cells in the battery pack age.

Conventionally, the estimation of parameters of battery is based on equivalent circuit models [1] that are mathematically represented by equation of space's state. In this type of model, the state of cell is function of SoC, of hysteresis and of constant's time of charge and discharge of battery.

Different from the aforementioned methods, the identification models represent the computational intelligent techniques for modeling of battery [2]. These classes of models ensure a model more global and with a computational cost more adequate. This paper presents an ANN for describe the dynamic of a lithium-ion battery and lead-acid battery (Fig. 1). The artificial neural network (ANN) is first trained offline to model the battery terminal voltages. Simulation results are validated against a great real database to demonstrate its validity of high accuracy.

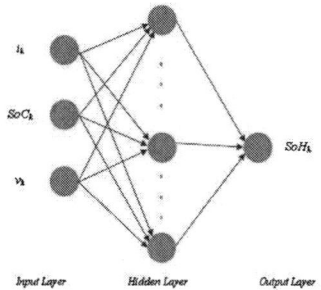

Fig. 1. Example of ANN model for battery SoH.

2. A dynamic system model for battery

To accurately estimate SoC and SoH, it is important to take into account the implicit nonlinear relationship between battery terminal voltage and input parameters, such as: SoC, current and capacity. However, precise modeling of this nonlinear relationship is challenging, due to complicated battery system dynamics. In this paper, an ANN is developed and employed to model this nonlinear relationship. The state-space model equations [3] can be written as:

$$x_k = F(x_{k-1}, u_{k-1}, \sigma_{k-1}) + w_{k-1}$$
$$y_k = G(x_k, u_k, \sigma_k) + \Omega_k$$

The equation above represents the state-space model of battery, where k is index of discrete time steps, x represents state of a dynamic system, y denotes outputs of system observations (or measurements), u is system input, σ represents system model parameter, w represents system process noises whereas Ω denotes measurement noise, and $F(\cdot)$ and $G(\cdot)$ represent system state and measurement functions, respectively.

The SoC can usually be calculated by the ampere hour counting technique. If an initial value of SoC is given, SoC at time can be computed by integrating current with respect to time. Using a rectangular approximation for integration and a small sampling period Δt, a discrete-time approximate recurrence may then be written as:

$$x_k = x_{k-1} + \frac{\eta \Delta t}{C_{k-1}} i_{k-1} + w_{k-1}$$

where x_k is the cell SoC, I is the instantaneous cell current (assumed positive for discharge, negative for charge) and C is the cell nominal capacity. Cell coulomb efficiency η is $\eta = 1$ for discharge and $\eta < 1$ for charge.

In the structure of the developed ANN, the network output is the battery terminal voltage, Voltage (k), and the input vector can be denoted as $v_{k_{Li}} = [soc_k, i_k, C_k]$.

In the developed approach, the designed ANN model can be updated at every time step, which indicates that, the values of neurons and the weights can vary with time. Thus, at a specific time step, the value for jth hidden node, h_k^j, can be calculated as:

$$h_k^j = \varphi\left(\sum_{i=1}^{I} v_k^i a_k^{ij} + a_k^{0j}\right)$$

where I is the index (i=1, 2, or 3) of the ith input node, j is the index (j = 1, 2, ..., J) of the node, a_k^{ij} is the weight connecting the ith input node and the jth hidden node at time step k, and $a_k^{\mathbf{0}j}$ is the bias of the jth hidden node at time step k. Thus, the value of the output node as z_k can be obtainded as

$$z_k = \varphi\left(\sum_{j=1}^{J} h_k^i b_k^j + b_k^{\mathbf{0}}\right)$$

where b_k^j is the weight connecting the jth hiddden node with the output node at time step k and $b_k^{\mathbf{0}}$ is the bias of output node at time step k. Finally, at time step k, the battery terminal voltage V_k as the output from the developed ANN can be analytically obtained based on previous equations:

$$V_k = \varphi\left(\varphi\left(\sum_{i=1}^{I} v_k^i a_k^{ij} + a_k^{\mathbf{0}j}\right)b_k^j + b_k^{\mathbf{0}}\right)$$

In the developed approach, first of all, the initial ANN model is trained by the collected experimental data. Then, the states of Li-ion batteries and the weights and biases of the ANN are all treated as the system states. Which the previous equations, the state transition and measurement equations for battery systems can then be obtained as

$$x_k = x_{k-1} + \frac{\eta\Delta t}{C_{k-1}}i_{k-1} + w_{k-1}$$

$$V_k = \varphi\left(\varphi\left(\sum_{i=1}^{I} v_k^i a_k^{ij} + a_k^{\mathbf{0}j}\right)b_k^j + b_k^{\mathbf{0}}\right) + \Omega_k$$

3. Experimental case study results

3.1. Validation of the ANN model for lithium-ion battery

In this subsection, the lithium-ion cell terminal voltage estimation using the developed ANN is validated by the experimental data provided by the Prognostics Center of Excellence (PCoE) at NASA Ames. Experimental data for a set of four batteries, numbered as 05, 06, 07 and 18, respectively, were available and employed in this study. There were three different operational profiles running alternately in this battery set, namely charge, discharge and impedance. In the charge cycle, the batteries were charged at a constant current of 1.5 A until the voltage reached 4.2 V, and then changed to a constant voltage until the charge current fell to 20 mA. In one regular discharge cycle, these four battery terminal voltage decreased to 2.7 V, 2.5V, 2.2 V, and 2.5 V for batteries 05, 06, 07 and 18, respectively. The experiments were forced to stop when the battery capacity degraded to 70% of the rated capacity. For the validation of the developed approach, the experimental data from the battery charging and discharging process are employed in this study.

In order to guarantee a valid generalization in different battery aging conditions, the experimental data set from battery 05, which consists of 20 discharge cycles and 20 charge cycles and 31,560 sample points, is employed for the training of the developed ANN. At each sample point, the instantaneous current, the terminal voltage, and the capacity have been measured, and the accumulated time at each sample point has also been recorded for calculating the SoC using the ampere-hour counting technique.

The three inputs of the developed ANN for lithium-ion battery model are the SoC, current, and capacity respectively. The number of hidden nodes is set to 100 in this study based on the experimental efficiency and accuracy. The output of the developed ANN is the battery terminal voltage estimation. The learning rule used in this study is the gradient descent algorithm, which is a first-order optimization algorithm being commonly used for ANN training. During the training process, the weights could be adjusted to the optimum values by running the gradient descent algorithm to reduce the training error. After training the developed ANN model using the data of battery 05, the other three batteries, 06, 07 and 18, are employed to validate the trained network model using experimental data from battery discharging process. Figure 2(a) shows training results of battery 05, while figure 2(b) shows the terminal voltage estimation performance using the trained ANN for battery 06. For the purpose of clear display, only the first 5500 samples have been plotted in each figure. Fig. 2, x-axis represents the time and y-axis denotes the terminal voltage, whereas the solid curve shows the value of the actual battery terminal voltage and the curve with circle signs shows the estimates provided by the trained ANN.

To quantify the accuracy of the trained ANN for the battery capacity estimation, the root mean squared (RMS) error and the mean absolute percentage error (MAPE) are used as error measures, defined respectively in the following equations as

$$RMS = \sqrt{\frac{1}{N}\sum_{i=1}^{N}[(A]_t - F_t)}$$

$$MAPE = \frac{1}{N}\sum_{i=1}^{N}\left|\frac{A_t - F_t}{A_t}\right| x100\%$$

where N is the total number of sample points, A_t is the true value of the battery capacity at time t, and F_t is the estimated battery capacity value obtained from the ANN at time t. The RMS and MAPE values for the battery terminal voltage estimation using the developed ANN for all the four batteries have been listed in Table 1.

Errors	Group			
	Battery 05	*Battery 06*	*Battery 07*	*Battery 18*
RMS	0.0404	0.1339	0.0621	0.0659
MAPE	0.43%	2.37%	0.66%	0.75%

Table. 1: Error of terminal voltage estimation by ANN.

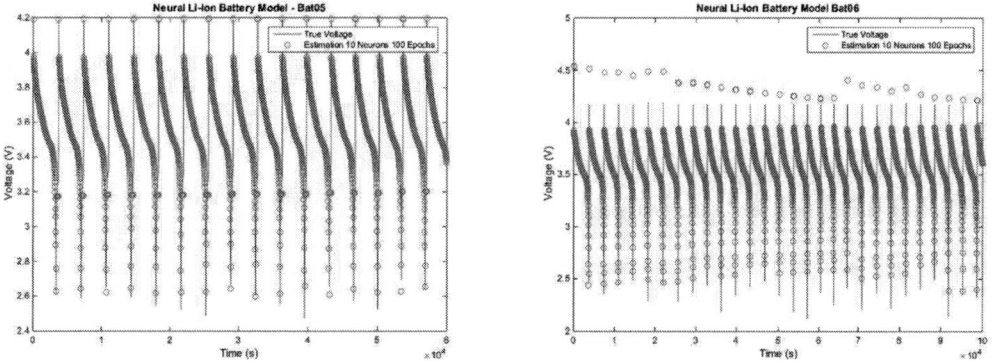

Fig. 2. Neural network estimation performance for lithium-ion battery (a) battery 05 (b) battery 06.

The developed ANN for lithium-ion battery model was compared with the commonly used empirical equivalent-circuit model based approach [4] (Fig. 3).

PCIM Europe 2016, 10 – 12 May 2016, Nuremberg, Germany

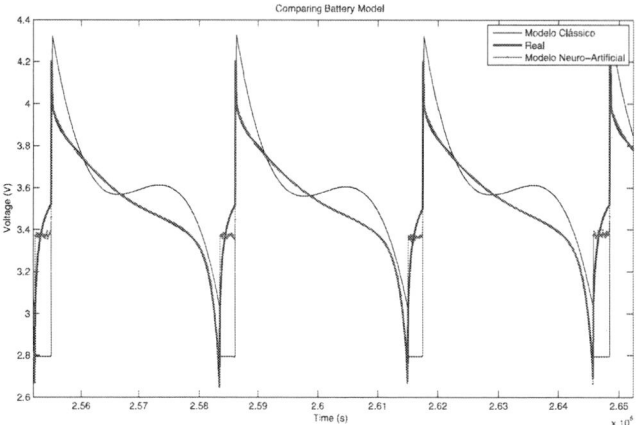

Fig. 3. Neuiral network model compared with empirical equivalent circuit model based approach

Errors	Classic Model	ANN Model
RMS	0.2394	0.0397
MAPE	4.75%	0.38%

Table. 2: Errors comparison between ANN model and classic model of battery.

3.2.   SoC and SoH estimation

This subsection presents the performance of SoC and SoH estimation using the developed approach in one discharge cycle. The estimation oft he SoC and SoH at each discharge cycle is carried out using the proposed model-free approach with the trained ANN as discussed previously above. Fig. 4 show the SoC estimation performance of the 10th discharge cycle for battery 06 to test the developed approach. The x-axis represents the voltage and the y-axis represents the SOC, whereas the solid curve shows the true SOC values calculated by the ampere-hour counting technique and the curves with circle marks show the estimated SOC by the developed model free approach.

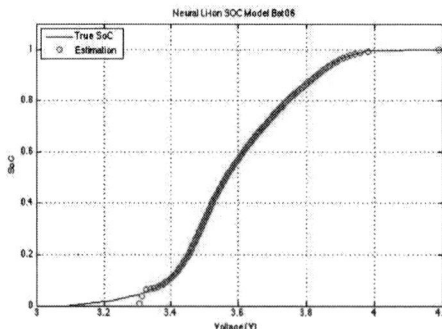

Fig. 4. SoC estimation for 10th discharge cycle of battery 06 (RMS: 0.0016)

Fig. 5 show the SoH estimation performance of the 10th discharge cycle for battery 05. The x-axis represents the time and the y-axis represents the SoH, whereas the solid curve shows the calculated SoH by capacity and the curves with circle marks show the estimated SoH by the developed model free approach.

© VDE VERLAG GMBH · Berlin · Offenbach

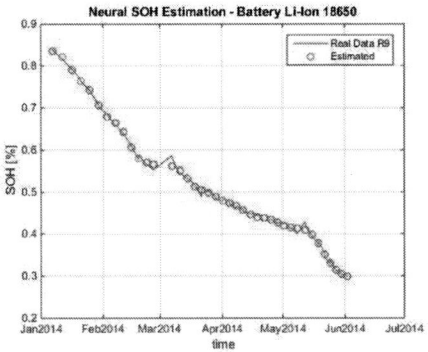

Fig. 5. Neural network model for SOH estimation. (a) RMS: 0.0066; MAPE: 0.94%.

3.3. Validation of the ANN model for lead acid battery

For validation the lead acid battery is used the experimental data based on the different cycle regime laboratory test of OPzS lead acid batteries. These data, provided by the certified laboratories CPqD, is used to validate the proposed approach for lead acid battery. Experimental data for four sets of three batteries (226 Ah, 2V), connected in series, were available and employed in this study. Each set of battery runs a different operational profile at 40ºC. The first group was charged during 1.5 hours with a current of 1.03 I_{10} and discharged during 1.5 hours with a current of I_{10} between 40% and 55% of SoC repeatedly. The second group was charged during 1.5 hours with a current of 1.03 I_{10} and discharged during 1.5 hours with a current of I_{10} between 55% and 70% of SoC repeatedly. The third group was charged during 1.5 hours with a current of 1.03 I_{10} and discharged during 1.5 hours with a current of I_{10} between 70% and 85% of SoC repeatedly. The fourth group was charged during 1.5 hours with a current of 1.03 I_{10} and discharged during 1.5 hours with a current of I_{10} between 85% and 100% of SoC repeatedly. The different cycle regime used in this study is shown in Table 4.

Group	Cycle Regime at 50ºC
1	Partial cycling between 40% and 55% of SoC
2	Partial cycling between 55% and 70% of SoC
3	Partial cycling between 70% and 85% of SoC
4	Partial cycling between 85% and 100% of SoC

Table. 4: Different cycle regime used in this study.

The developed ANN for lead acid battery used four inputs: current, i_k, state of charge, SoC_k, voltage, v_k, and battery SoH in the last time step, SoH_{k-1}, and one output, the battery SoH estimation, SoH_k. The number of hidden nodes is set to 25 and the number of epochs is set to 1000 in this study based on the experimental efficiency and accuracy. The learning rule used in this study is the gradient descent algorithm, which is a first-order optimization algorithm being commonly used for ANN training. During the training process, the weights could be adjusted to the optimum values by running the gradient descent algorithm to reduce the training error.
First, the lead acid ANN model was trained using the data from the group 01 of Table 4. After that, the other data from the groups 2-4 was used to quantitate the accuracy of the ANN developed for the battery SoH estimation.
The RMS and MAPE values for the battery SoH estimation using the developed ANN model for all the four lead acid groups have been list in Table 5.

Errors	Group			
	1	*2*	*3*	*4*
RMS	7.33e-7	6.97e-6	3.85e-6	1.04e-5
MAPE	6.03e-5%	5.23e-4%	2.76e-4%	6.28e-4%

Table. 5: Error of SoH estimation by lead acid ANN.

Fig. 6 shows the results of the ANN estimation performance for OPzS lead acid battery of group 1. As showed in Fig. 6, x-axis represents the time and y-axis denotes the battery SoH, whereas the solid curve shows the value of the actual battery SoH and the curve with circle signs shows the estimates provided by the trained ANN.

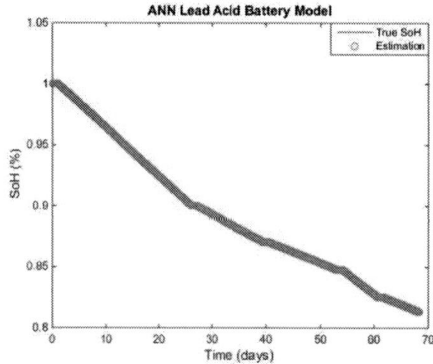

Fig. 6. ANN model for lead acid battery SoH of group 01.

4. Conclusion

A model-free battery for describes the dynamic of a lithium-ion and lead-acid battery is developed in this paper. The ANN here proposed shows superior learning capabilities, adaptation, contextual information, response to dynamics nonlinearities and failure tolerance. A big training set of real and reliable data are being concluded to enhance the performance of the ANN. Identification techniques are being analyzed for model validation.

5. Reference

[1] X. Hu et al.: A comparative study of equivalent circuit models for Li-ion batteries, J. Power Sources 198, 359–367, 2012.
[2] W. X. Shen et al.: Adaptive Neuro-Fuzzy Modeling of Battery Residual Capacity for Electric Vehicles, IEEE Transactions on Industrial Electronics, Vol. 49, No. 3, 2002.
[3] G. Bai et al.: A generic model-free approach for lithium-ion battery health management, Applied Energy, volume 135, Pages 247–260, 2014.
[4] B. Saha et al.: Comparison of prognostic algorithms for estimating remaining useful life of batteries, Transactions of the Institute of Measurement and Control, 2009.
[5] Hu, X. Li S. Peng. H. A comparative study of equivalent circuit models for Li-ion batteries. J. Power Sources 198, 359–367, 2012.
[6] Waagr, W. "Critical review of the methods for monitoring of lithium-ion batteries in electric and hybrid vehicles". Review Article Journal of Power Sources, Volume 258, Pages 321-339, 2014.

PCIM Europe 2016, 10 – 12 May 2016, Nuremberg, Germany

An Efficient Implementation of a Reconfigurable Battery Stack with Optimum Cell Usage

Martin Wattenberg,* and Martin Pfost**
*Reutlingen University, Germany, **University of Innsbruck, Austria
martin.wattenberg@reutlingen-university.de

Abstract

This paper presents an efficient implementation of a reconfigurable battery stack which allows full exploitation of the capacity of every single cell. Contrary to most other approaches, it is possible to electrically remove one or more cells from the battery stack. Therefore, the overall capacity of the system is not restricted by the weaker cells, and cells with very different states of health can be used, making the system very attractive for refurbished batteries.

For the required switches, low-voltage high-current MOSFETs are used. A demonstrator has been built with a total capacity of up to 3.5 kWh, a nominal voltage of 35 V, and currents up 200 A.

1. Introduction

To overcome the inherent fluctuations of renewable energy sources, energy storage systems, e.g. $LiFePo_4$ battery stacks, are proposed. Due to variations in battery cell quality and age (especially if refurbished batteries are used), the state of health of the cells in the battery stack will eventually drift apart. To compensate for these effects, battery monitors and balancers are commonly employed, see [1] for a comprehensive overview.

This paper presents an approach to increase battery stack performance by reconfiguring the connections between cells in the stack. Therefore the already fully charged or discharged batteries are bypassed during charging or discharging, respectively. The remaining batteries can be further utilized while disconnected cells are kept in a safe state. Systems with reconfigurable cell stacks with comparable functionality have been proposed by [2–4]. Contrary to them, this paper provides a more efficient implementation of the reconfigurable connections on the cell level using only two MOSFETs, one for including the cell in the stack and one for bypassing the cell.

Reconfigurable battery systems are introduced in Sec. 2 and our implementation in Sec. 3. The impact on battery stack performance due to increased resistance of the reconfigurable connections is investigated and an optimal cell usage scheme is described in Sec. 4. Experimental results of a prototype are presented in Sec. 5.

2. System Concept

To fully exploit the capacity of all cells in the stack, it is required that each cell be either included in the stack or bypassed, see Fig. 1. It should be noted that real switches have a certain resistance so it might be favourable to use several cells in series than just one, even though it might not be possible to fully exploit all cells. Such configuration aspects will be discussed in Sec. 4.

© VDE VERLAG GMBH · Berlin · Offenbach

PCIM Europe 2016, 10 – 12 May 2016, Nuremberg, Germany

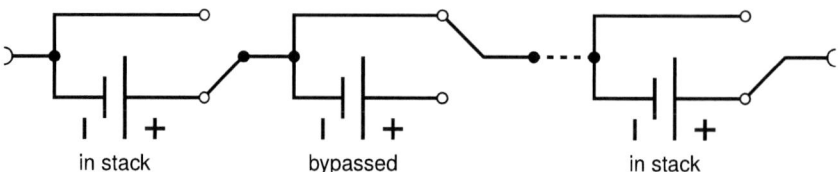

Fig. 1: Reconfigurable cell stack allowing full usage of all cell capacities.

To demonstrate the benefit of a reconfigurable cell stack for usable battery capacity, a Monte-Carlo simulation has been carried out. Fig. 2 shows that for a 12-cell battery stack, a 5.1% increase in usable capacity can be expected when up to six cells can be bypassed during discharge. Reconfigurable cell stacks also enhance the robustness of the system because bad cells can be bypassed, a failure of a single cell is not fatal to the entire system.

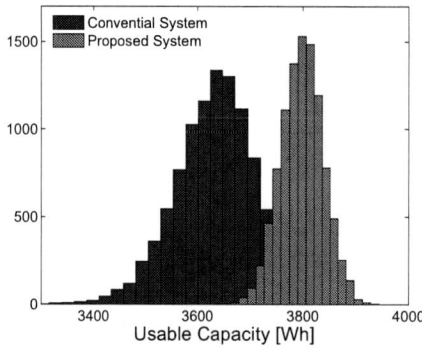

Fig. 2: Comparison of usable battery capacity of a fixed and a reconfigurable battery stack with 12 LiFePo$_4$ cells. Normally distributed batteries with a capacity of 100 Ah and $\sigma = 3.3$ Ah are assumed. The battery stack with fixed connections is considered depleted when one cell is completely discharged. The reconfigurable battery stack can be discharged as long as at least 6 out of 12 cells still contain energy.

3. Reconfigurable Battery Module

3.1. Switch Implementation

Fig. 3 (a) shows the simplified schematic of a reconfigurable battery module (RBM) and (b) its implementation. Every cell is connected to two MOSFETs which are switched complementary to each other. The orientation of the MOSFETs in Fig. 3 (b) allows bidirectional current flow through the cell so that it can be charged and discharged. During the (obligatory) deadtime, current can continue to flow through the body diode of either Q_1 during charge cycles, or Q_2 during discharge cycles.

The proposed system provides a very effective solution to achieve series or bypass connection using only two N-MOSFETs per RBM. Additional protection diodes as in [5] are not needed. An RBM designed for LiFePo$_4$ cells with 100 Ah and a peak current of 200 A is shown in Fig. 4.

© VDE VERLAG GMBH · Berlin · Offenbach

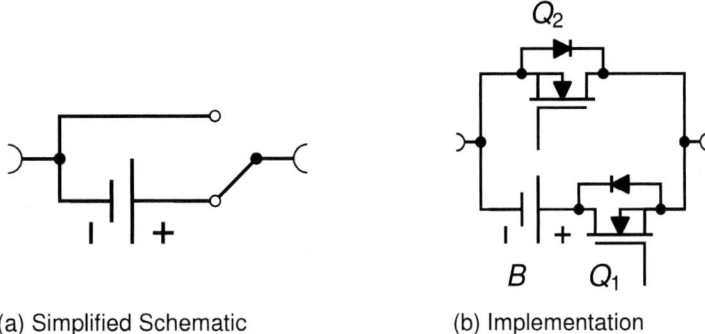

(a) Simplified Schematic   (b) Implementation

Fig. 3: Overview of a reconfigurable battery module consisting of two N-MOSFETs and one battery cell. Note that the diodes are the body diodes inherent to the MOSFETs.

Providing the supply voltage for the MOSFET gate drivers and for the cell monitoring circuitry is difficult due to the potential of an RBM with respect to ground depends on the state of charge of the battery stack and will change during operation. Previous solutions obtained the gate voltage V_{GS} either from the battery cell itself (which is especially problematic for discharged batteries), or used insulated DC/DC converters, which is an expensive solution. The presented approach takes advantage of the common source connection between two adjacent modules. Then, only one floating charge pump similar to [6] is sufficient per RBM.

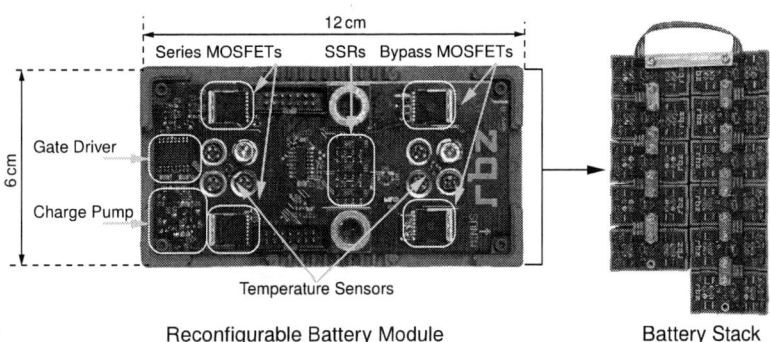

Reconfigurable Battery Module   Battery Stack

Fig. 4: Photography of the implemented RBM (left) and the entire battery stack with eleven cells (right).

3.2. Cell Monitoring

Keeping the battery cells within their safe operating area is crucial for the longevity of the cells, hence precise voltage measurements are required. A simple window comparator as presented in [7] does not provide enough information to predict the state of charge accurately (cf. [8]). In this work, a capacitor is utilized to sample the cell voltage. To select the cell to be measured, optically insulated solid state relays (SSR) are used. By this approach, level shifting is straightforward, and the same high precision ADC can be used for all cells. By a careful selection of the sample and hold capacitor and the ADC, an accuracy of ± 1 mV is achieved.

The approach for level-shifting is also well suited to sample the battery temperature. Two temperature sensors are used, one for each cell terminal. With this, the sensor that is further apart from the currently turned-on MOSFET can be used, thus reducing the influence of MOSFET power dissipation on the cell temperature measurement.

4. Optimum System Configuration

The proposed system is aimed at batteries with large variation in capacity, as refurbished batteries do. While a preselection could be feasible, it has to be taken into account that the cells age differently over time. Hence, it is not possible to predict cell capacities over the entire operating time of the battery stack. This implies that each cell should have its own RBM for full flexibility. Each RBM however, introduces an additional series resistance to the stack. This effectively reduces the available capacity, and it could be advisable to group cells nonetheless. For this reason, it was proposed in [5] to use several cells per RBM.

Clearly, the best choice of how many cells should be used in each RBM depends on the standard deviation σ of the cell capacity. Fig. 5 explores this with a Monte-Carlo simulation. The investigated battery stack is composed of 12 cells. The number of cells was assumed to be the same for all RBMs and ranges from 1 to 12. Note that 12 cells per RBM corresponds to the complete stack, so an RBM is not required in this case.

Fig. 5 (a) shows the achievable capacity for different numbers of cells per RBM and for different variation in cell capacity. Here, the losses over the RBM resistances were considered for a discharge current of 25 A. For σ = 5 Ah, the presented system provides an increase in usable capacity by over 200 Wh, which is 7%.

Fig. 5 (b) shows cell numbers per RBM for different discharge currents. The simulation assumes σ = 2 Ah. It can be clearly seen that while one RBM per cell provides the highest flexibility and makes optimal use of each cell, the ohmic losses (especially at high currents) affect the usable capacity in a significant way.

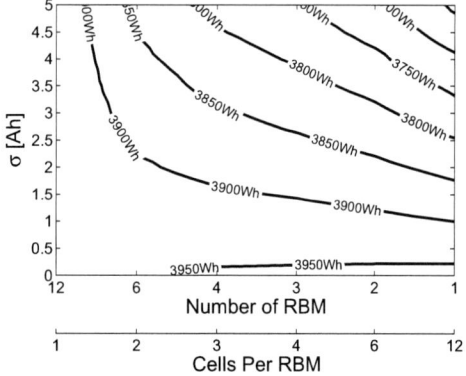

(a) Increase of usable capacity depending on battery cells per module and std. deviation σ.

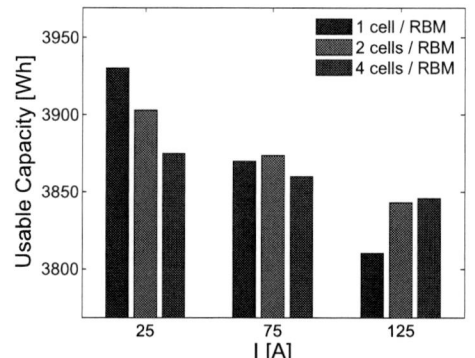

(b) Usable capacity of battery stack as a function of the current.

Fig. 5: Monte-Carlo simulation results for a system with 12 cells, showing the impact of cell capacity standard deviation and current on usable capacity.

5. Experimental Results

To demonstrate the performance of the proposed reconfigurable battery stack, a prototype has been built using eleven 100 Ah batteries. Because they were new, the standard deviation σ was only 4.5 Ah. Nevertheless, an increase in usable capacity of 11% was observed with 6 out of

12 cells being bypassed. Using all cells up to their maximal capacity (i.e. bypassing up to 10 cells), a further increase of 0.5% was achieved. Moreover, it is expected that with increasing cycle count the cells will age differently, and σ will increase over time.

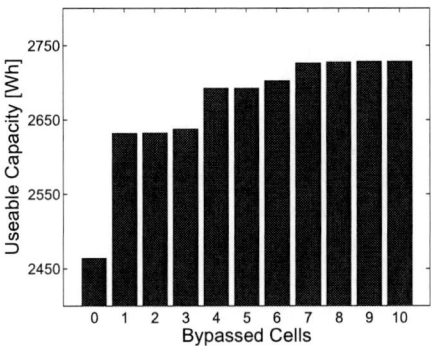

Fig. 6: Usable capacity depending on the number of cells which can be bypassed. A total of eleven batteries were discharged with 10 A constant current.

The resistance of the PCB itself amounts to 0.3 mΩ for the bypass connection and 0.7 mΩ for the series connection. A significant reduction of the PCB resistance can be expected with thicker copper layers.

Because cooling relies solely on passive heat removal, the thermal capabilities of the system have been investigated. For this, a charge current of 50 A (twice the nominal current) was used. Fig. 7(a) shows the temperature profile of the PCB after 30 min of operation. It can be seen that only a minimum amount of heat is generated.

To evaluate the impact of the MOSFET resistance, a (dis-)charge current of 25 A was chosen. The resulting voltage drop across the transistors Q_1 and Q_2, see Fig. 3 (b), can be seen in Tab.1. Even though the series MOSFET reduces the output voltage of the cell, the effect is minimal in practice. Tab. 1 shows that the voltage drop is less than 0.2%, even for the worst case scenario a of low cell voltage like $V_{Bat} = 3$ V. Compared to the energy efficiency of LiFePo$_4$ batteries of typically 93% to 97% [9], the impact of MOSFET voltage drop is small. It should

	Charge	Discharge
Bypass	$V_{DS(Q2)} = 4.5\,mV$	$V_{DS(Q2)} = -4.5\,mV$
Series	$V_{DS(Q1)} = -4.6\,mV$	$V_{DS(Q1)} = 4.6\,mV$

Tab. 1: Voltage drop across Q_1 and Q_2 for different cell connections. The (dis-)charge current was set to 25 A.

be noted that during battery charging, Q_1 is operated in third quadrant operation mode, i.e. $I_{DS} < 0$ A and $V_{DS} < 0$ V. The same applies for Q_2 during discharge. Despite the asymmetrical design of source and drain in power MOSFETs, the measurements confirm that third quadrant operation is not a problem. $V_{DS(Q1)}$ is well below the forward voltage of the intrinsic body diode, therefore it can be concluded that the total current is carried by the channel. Note that this operating condition can also be found elsewhere i.e. in synchronous buck-boost converters.

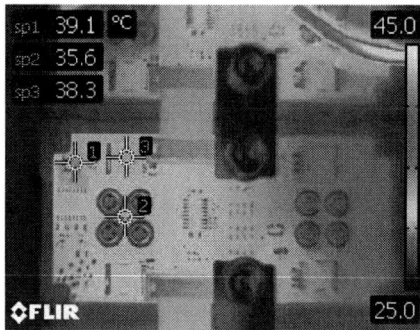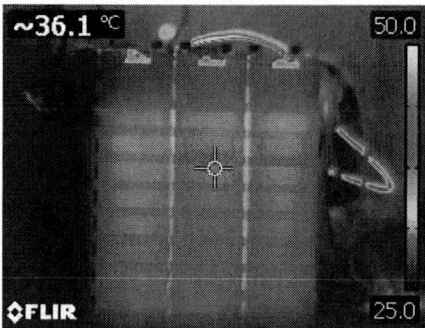

(a) Thermal top view of a RBM during a charge cycle with 50 A

(b) Thermal side view of three batteries during a charge cycle with 50 A

Fig. 7: Investigation of the thermal performance of the RBM mounted on top of a cell. Some heat is generated inside the MOSFET as well as in the PCB traces. The connection terminals reflect the room temperature due to their low emissivity. The image was taken 30 min after charging started. No forced cooling was used.

6. Conclusion

An efficient implementation of a reconfigurable battery stack has been proposed, achieving a significant increase of the usable capacity of the battery stack. Using a simple MOSFET-based switch, each cell can be electronically removed, i.e. bypassed from the battery stack. This allows to allowing to fully exploit the capacity of each cell, even if they differ significantly. Measurements with 100 Ah batteries and low-voltage high-current MOSFETs have shown negligible impact on the overall battery efficiency. The authors are confident that the additional cost of the proposed system is well compensated by the benefits of larger usable capacity and higher robustness to cell failures, but most notably by the possibility to use lower-grade or even refurbished cells.

7. References

[1] J. Cao, N. Schofield, and A. Emadi. Battery balancing methods: A comprehensive review. In *Vehicle Power and Propulsion Conference*, pages 1–6, September 2008.

[2] Taesic Kim, Wei Qiao, and Liyan Qu. Series-connected self-reconfigurable multicell battery. In *Applied Power Electronics Conference and Exposition*, pages 1382–1387, March 2011.

[3] A. Manenti, A. Abba, A. Merati, S. M. Savaresi, and A. Geraci. A new BMS architecture based on cell redundancy. *IEEE Transactions on Industrial Electronics*, vol. 58(9):4314–4322, September 2011.

[4] Yuang-Shung Lee and Ming-Wang Cheng. Intelligent control battery equalization for series connected lithium-ion battery strings. *IEEE Transactions on Industrial Electronics*, vol. 52(5):1297–1307, October 2005.

[5] F. Baronti, G. Fantechi, R. Roncella, and R. Saletti. Design of a module switch for battery pack reconfiguration in high-power applications. In *IEEE International Symposium on Industrial Electronics*, pages 1330–1335, May 2012.

[6] Shihong Park and T. M. Jahns. A self-boost charge pump topology for a gate drive high-side power supply. In *Applied Power Electronics Conference and Exposition*, pages 126–131, February 2003.

[7] Ui Seong Kim, Chee Burm Shin, and Chi-Su Kim. Effect of electrode configuration on the thermal behavior of a lithium-polymer battery. *Journal of Power Sources*, vol. 180(2):909–916, June 2008.

[8] F. Codeca, S. M. Savaresi, and G. Rizzoni. On battery state of charge estimation: A new mixed algorithm. In *IEEE International Conference on Control Applications*, pages 102–107, September 2008.

[9] Pei Zhang, Changqing Du, Fuwu Yan, and Jianqiang Kang. Influence of practical complications on energy efficiency of the vehicle's lithium-ion batteries. In *International Conference on Electric Information and Control Engineering*, pages 2278–2281, April 2011.

PCIM Europe 2016, 10 – 12 May 2016, Nuremberg, Germany

List of Authors of PCIM Europe 2016

Author	Institution	Pages
A		
Abbatelli, Luigi	STMicroelectronics, I	1536
Abdel-Rahman, Sam	Infineon Technologies Americas, USA	1615
Abe, Yasushi	Fuji Electric, J	188
Abl, Reiner	BMW, D	534
Abouchi, Nacer	Institut des Nanotechnologies de Lyon, F	1707
Abronzini, Umberto	University of Cassino and South Lazio, I	379 1217
Adachi, Shinichiro	Fuji Electric, J	1956
Adelmund, Melanie	University of Bremen, D	1232
Adler, Johannes	University of Bremen, D	1324
Adragna, Claudio	STMicroelectronics, I	612
Adrien, Mercier	ENS Cachan – SATIE, F	1647
Aggen, Christian	Danfoss Silicon Power, D	698
Agrawal, Binod	CREE India Private Limited, IN	2114
Ahmad, Jawad	Politecnico di Torino, I	2058
Aizpuru, Iosu	Mondragon University, ES	763 2074
Alberti, Luigi	University of Padova, I	534
Albertin, Thomas	Atherm, F	1132
Ali, Esmail A.	University of Science and Technology Yemen, YE	1046
Allen, Scott T.	Wolfspeed, USA	34
Almai, Farshid	University of Paderborn, D	232
Altstadt, Jochen	Kortec, D	1979
Alvarez, Rodrigo	Siemens, D	588
Alves-Rodrigues, Luis Gabriel	Commissariat à l'énergie atomique et aux énergies alternatives, F	128
Ambacher, Oliver	Fraunhofer-Institute IAF, D	319
Ammann, Ulrich	Esslingen University of Applied Sciences, D	780
An, Bao Ngoc	Karlsruhe Institute of Technology, D	1979
Andenna, Maxi	ABB Switzerland, CH	417
Antivachis, Michael	ETH Zürich, CH	1622
Appel, Tobias	STS, D	1434
Aragues-Penalba, Monica	CITCEA-UPC, ES	2211

PCIM Europe 2016, 10 – 12 May 2016, Nuremberg, Germany

Author	Institution	Pages
Aranzabal, Itxaso	University of the Basque Country, ES	1970
Araujo Vasconcelos, Samuel	University of Kassel, D	573
Arnold, Martin	ABB Switzerland, CH	917
Arpino, Matilde	University of Cassino and South Lazio, I	379 1217
Asada, Shinsuke	Mitsubishi Electric Corporation, J	326
Asfaux, Pascal	Airbus Operation, F	1814
Ashley, Grant	Vxl Power, GB	1455
Askari, Vahid	Teesside University, GB	1439 1685
Attaianese, Ciro	University of Cassino and South Lazio, I	379 1217
Augoustidis, Alex	GrafTech International, USA	1126
Austermann, Johann	University of Applied Sciences Ostwesfalen-Lippe, D	1639
Avenas, Yvan	G2ELAB, F	1494
Azcondo, Francisco	University of Cantabria, ES	1194 1631
Azuma, Katsunori	Hitachi Power Semiconductor, J	348
B		
Baghadadi, Mehdi	University of Cambridge, GB	1845
Baghaie Yazdi, Mehrdad	Fairchild Semiconductor, D	813 2231
Baier, Thomas	Friedrich-Alexander-University Erlangen, D	264
Bakran, Mark-M.	University of Bayreuth, D	272 355 653 683 720 1055 1400 1503 2181
Balke, Christian	Maccon, D	1353
Banerjee, Sujit	Monolith Semiconductor, USA	984
Barater, Davide	Università degli Studi di Parma, I	1073 1584
Barrios, Ernesto L.	Public University of Navarre, ES	1209
Barruel, Franck	Commissariat à l'énergie atomique et aux énergies alternatives, F	128
Barwig, Markus	Friedrich-Alexander-University of Erlangen, D	1693 1701
Baschnagel, Andreas	ABB Switzerland, CH	945

PCIM Europe 2016, 10 – 12 May 2016, Nuremberg, Germany

Author	Institution	Pages
Basler, Thomas	Infineon Technologies, D	180
Basler, Vanessa	Technical University of Munich, D	543
Bäßler, Marco	Danfoss Silicon Power, D	698
Batard, Christophe	University of Nantes, F	712
Bauer, Pavol	Delft University of Technology, NL	1201 1383
Baumann, Michael	SUMIDA Components & Modules, D	1728 1764
Bauwens, Filip	On Semiconductor, BE	1479
Bayer, Martin	ABB Switzerland, CH	432
Bazzano, Gaetano	STMicroelectronics, I	804
Beaucarne, Guy	Dow Corning Europe, B	992
Beaurenaut, Laurent	Infineon Technologies, D	1377 2027
Becker, Karl-Friedrich	Fraunhofer Institute IZM, D	1040
Beichler, Johannes	Vacuumschmelze, D	1471
Beier-Möbius, Menia	Chemnitz University of Technology, D	172 588
Bellingen, Jörg	Technical University of Vienna, AT	2219
Bellini, Marco	ABB Switzerland, CH	911
Bendani, Larbi	Valeo, F	477
Benkendorff, Berthold	Christian-Albrechts-University, D	1942
Berger, Hubert	FH Joanneum, AT	296 2151
Bergogne, Dominique	CEA Leti, F	312
Beringer, Sebastian	Leibniz University Hannover, D	107
Bernardinis, Gabriele	Analog Devices, USA	195
Bernd, Martin	Karlsruhe Institute of Technology, D	1979
Bertelshofer, Teresa	University of Bayreuth, D	272 653 1503
Beuermann, Max	Siemens, D	1888
Beyer, Harald	ABB Switzerland, CH	945
Bhalla, Anup	United Silicon Carbide, USA	1511 1549
Bier, Anthony	Commissariat à l'énergie atomique et aux énergies alternatives, F	128
Billet, Cyrille	Calyos, BE	1139
Bingham, Chris	University of Lincoln, GB	1769
Birkel, Andre	University of Bayreuth, D	2181

PCIM Europe 2016, 10 – 12 May 2016, Nuremberg, Germany

Author	Institution	Pages
Blaabjerg, Frede	Aalborg University, DK	831 1837
Blank, Mathias	Technical University of Vienna, AT	1339
Blank, Thomas	Karlsruhe Institute of Technology, D	1487 1979
Blum, Manuel	Siemens, D	1912 1919
Böcker, Joachim	University of Paderborn, D	232 780 1662 1992
Bödeker, Christian	University of Bremen, D	1232
Böh, Magnus	Technical University of Cologne, D	669
Bolognani, Silverio	University of Padova, I	534
Bolte, Sven	University of Paderborn, D	232 1992
Booth, James	Dynex Semiconductor, GB	704
Borcherding, Holger	University of Applied Sciences Ostwesfalen-Lippe, D	1639
Borecki, Jacek	University of Bremen, D	248
Borghetti, Giovanni	Nidec ASI, I	2175
Boroyevich, Dushan	CPES/Virginia Tech, USA	1286
Bortis, Dominik	ETH Zürich, CH	1622
Bosch, Swen	Heinrich Steinhart, HTW Aalen, D	2129
Böttigheimer, Mike	University of Stuttgart, D	410
Bouguet, Christophe	University of Nantes, F	712
Bourdon, Jeremy	Airbus Operation, F	1814
Boureghda, Monnir	Alpha Assembly Solutions, USA	1027
Bourner, David	Vicor Corporation, USA	1253 1822
Braband, Matthias	ISW – University of Stuttgart, D	256
Bräckle, Dennis	Karlsruhe Institute of Technology, D	788
Bramanpalli, Ranjith	Würth Elektronik eiSos, D	99 1757
Branas, Christian	University of Cantabria, ES	1631
Braun, Michael	Karlsruhe Institute of Technology, D	393 788
Brinkfeldt, Klas	Swerea IVF AB, SE	1063
Brix, Jonathan	Fraunhofer Institute IPA, D	386
Brubaker, Michael	SBE, USA	1391
Budaker, Bernhard	Fraunhofer Institute IPA, D	386

PCIM Europe 2016, 10 – 12 May 2016, Nuremberg, Germany

Author	Institution	Pages
Buetow, Sven	Semikron Elektronik, D	728
Burani, Nicola	Fraunhofer Institute IISB, D	453
Burger, Michael	SEW Eurodrive, D	144
Burgos, Rolando	CPES/Virginia Tech, USA	1286
Burns, Chris	AB Mikroelektronik, AT	969
Buschhorn, Stefan	Infineon Technologies, D	850
Buschkühle, Marc	Infineon Technologies, D	1800
Buttay, Cyril	INSA de Lyon, F	1792
C		
Cabezuelo, David	University of the Basque Country, ES	1970
Camus, Stephane	Aperam Alloys Amilly, F	84
Canales, Jose Mari	Mondragon University, ES	763 2074
Cao, Zhiyu	AEG Power Solutions, D	1662 1677
Carastro, Fabio	GE Research Center, D	645
Carcouet, Sébastien	University Grenoble Alpes CEA LITEN, F	2098
Carvalho, Adriano	Nomad Tech, PT	953
Casady, Jeffrey	Wolfspeed, USA	34
Catalisano, Giuseppe	STMicroelectronics, I	1536
Catellani, Stéphane	Commissariat à l'énergie atomique et aux énergies alternatives, F	128
Cavallaro, Daniela	STMicroelectronics, I	804
Cellier, Remy	Institut des Nanotechnologies de Lyon, F	1707
Cerezo, Jorge	Infineon Technologies Americas, USA	819
Chamaret, André-Philippe	SNCF, F	1103
Chandra Mouli, Gautham Ram	Delft University of Technology, NL	1383
Chatroux, Daniel	University Grenoble Alpes CEA LITEN, F	2098
Chaudhuri, Toufann	ABB Sécheron, CH	1103
Chelghoum, Reda	Valeo, F	477
Chen, Jingxuan	University of Toronto, CA	895
Chen, Zhiyang	ON Semiconductor, USA	1777
Cherix, Nicolas	EPFL – Ecole Polytechnique Fédérale de Lausanne, CH	1949
Cheung, Chun	Intersil Corporation, USA	550
Chinthavali, Madhu S.	Oak Ridge National Laboratory, USA	1391
Choi, Hanna	KCC Corporation, ROK	961
Choo, Byoungho	Infineon Technologies Power Semitech, ROK	863

PCIM Europe 2016, 10 – 12 May 2016, Nuremberg, Germany

Author	Institution	Pages
Choudhary, Vijay	Texas Instruments, USA	2019
Christe, Alexandre	EPFL – Ecole Polytechnique Fédérale de Lausanne, CH	461
Chun, Ho Tae	Hyundai Motors Company, ROK	1262
Chung, Daewoong	Infineon Technologies Power Semitech, ROK	858 863
Ciocia, Alessandro	Politecnico di Torino, I	2058
Cioffi, Philip	GE Research Center, USA	645 1853
Comola, Marco	STMicroelectronics, I	804
Concari, Carlo	Università degli Studi di Parma, I	1584
Conrad, Detlef	Synopsys, D	813
Consentino, Giuseppe	STMicroelectronics, I	804
Corvasce, Chiara	ABB Switzerland, CH	417 432
Costa, Francois	ENS Cachan – SATIE, F	288 1494
Coulinge, Emilien	EPFL – Ecole Polytechnique Fédérale de Lausanne, CH	461
Cousineau, Marc	University of Toulouse, F	371
Crisafulli, Vittorio	ON Semiconductor, D	1178 1865
Curea, Octavian	ESTIA, F	2066
D		
Dadanema, Gnimdu	ENS Cachan – SATIE, F	1494
Dai, Jian	GE Research Center, USA	1853
Dai, Xiaoping	Dynex Semiconductor, GB	704 845 938
Dalessandro, Luca	Schaffner, CH	148
Daly, Michael	Analog Devices, USA	195
Datta, Rajib	GE Research Center, USA	645 1853
Dauchy, Julien	University Grenoble Alpes CEA LITEN, F	2098
Davidson, Colin	Alstom Grid, GB	2197
Davidson, Jonathan	University of Sheffield, GB	1095
de Alegría, Iñigo Martinez	University of the Basque Country, ES	1785 1970
De Bernardinis, Alexandre	French Institute of Science and Technology for Transport, Development and Networks, F	2159
De Michielis, Luca	ABB Switzerland, CH	417

PCIM Europe 2016, 10 – 12 May 2016, Nuremberg, Germany

Author	Institution	Pages
De Monchy, Michiel	Alpha, RUS	957
de Rooij, Michael	Efficient Power Conversion Corporation, USA	304
De Sousa, Luis	Valeo, F	477
de Stoppelaar, Diederik	LightingEurope, BE	604
Deboy, Gerald	Infineon Technologies, AT	1622
Declercq, Frederick	On Semiconductor, BE	1479
Dede, Ercan	Toyota Research Institute North America, USA	72
Delsuc, Vincent	Dow Corning, BE	1017
Demattio, Horst	Karlsruhe Institute of Technology, D	1487 1979
Denk, Fabian	Karlsruhe Institute of Technology, D	1721
Denk, Marco	University of Bayreuth, D	1055
Deviny, Ian	Dynex Semconductor, GB	845
Di Leo, Paolo	Politecnico di Torino, I	2058
Di Monaco, Mauro	University of Cassino and South Lazio, I	379 1217
Diaz Reigosa, Paula	Aalborg University, DK	831
Dieckerhoff, Sibylle	Technical University of Berlin, D	1186
Dietrich, Lothar	Fraunhofer-Institute IZM, D	1155
Dincan, Catalin Gabriel	Aalborg University, DK	2050
Dinis, Corina Maria	Politehnica University Timisoara, RO	2106
Dinulovic, Dragan	Würth Elektronik eiSos, D	107
Djallel, Kerdoun	GLEC Constanine 1, DZ	1785
Dobusch, Julian	Friedrich-Alexander-University of Erlangen, D	1857
Dodge, Jonathan	United Silicon Carbide, USA	484 1549
Domb, Moshe	Infineon, USA	1555
Domes, Daniel	Infineon Technologies, D	53
Doppelbauer, Martin	Karlsruhe Institute of Technology, D	393
Dragan, Mihai	Fraunhofer Institute IPA, D	386
Draghici, Mihai	Infineon Technologies, AT	180
Draugedalen, Eirik	Valeo, NO	113
Dudin, Andrey	Technical University of Ilmenau, D	1896
Dujic, Drazen	EPFL – Ecole Polytechnique Fédérale de Lausanne, CH	461
Dupont, Vincent	Calyos, BE	1139
Dürbaum, Thomas	Friedrich-Alexander-University of Erlangen, D	492 1693 1701 1857

PCIM Europe 2016, 10 – 12 May 2016, Nuremberg, Germany

Author	Institution	Pages
Durham, Jeffrey	Alpha Assembly Solutions, USA	1027
Dworakowski, Piotr	SuperGrid Institute, F	1592
E		
Eberlin, Michael	Fraunhofer Institute ISE, D	1733
Ebli, Michael	Universtiy of Reutlingen, D	210
Eckel, Hans-Günter	University of Rostock, D	581 924 1888
Edelmoser, Karl	Technical University of Vienna, AT	1741 2137
Efika, Ikenna	Alstom Grid, GB	2197
Ehrhardt, Christian	Fraunhofer-Institute IZM, D	1155
Ellinger, Thomas	Technical University of Ilmenau, D	1896
Enami, Hiroji	Dow Corning Toray, J	1017
Endres, Stefan	Fraunhofer Institute IISB, D	526
Engel, Günter	CeraCap, AT	2151
Engelhard, Christoph	HELLA KGaA Hueck, D	669
Epp, Nikolai	AEG Power Solutions, D	1677
Ertl, Hans	Technical University of Vienna, AT	1339 1669 1846
Evans, Kim	Dynex Semiconductor, GB	704
F		
Fahlbusch, Sebastian	Helmut Schmidt University Hamburg, D	1528
Fahlenkamp, Marc	Infineon Technologies, D	620
Fahnert, Holger	AEG Power Solutions, D	1662
Fairweather, Andrew	Vxl Power, GB	1455 2090
Farkas, Gabor	Mentor Graphics,HU	596
Fazio, Gianluca	On Semiconductor Italy, I	1178
Felgemacher, Christian	University of Kassel, D	573
Fernandez, Manuel	Universty of Oviedo, ES	1479
Fernández-Serantes, Luis Alfonso	University of Applied Sciences Joanneum, AT	296
Ferrazza, Francesco	STMicroelectronics, I	612
Fersterra, Fabian	Fraunhofer Institute IISB, D	469
Filho, Faete	Eaton Corporation, USA	1518

PCIM Europe 2016, 10 – 12 May 2016, Nuremberg, Germany

Author	Institution	Pages
Fink, Karsten	Power Integrations, D	48
Fischer, Aaron	Technical University of Ilmenau, D	1896
Fischer, Fabian	ABB Switzerland, CH	945
Fischer, Hermann	Fairchild Semiconductor, D	813
Fortuna, Stefania	STMicroelectronics, I	804
Foster, Martin	University of Sheffield, GB	1095 1455 1769 2090
Franceschini, Giovanni	Università degli Studi di Parma, I	1073
Franco, Augusto	Nomad Tech, PT	953
Frangieh, Tony	GE Research Center, USA	1853
Frank, Wolfgang	Infineon Technologies, D	1615
Franke, Toke	Danfoss Silicon Power, D	1942
Frankeser, Sophia	Technical University of Chemnitz, D	1063
Freitas, Nuno	Nomad Tech, PT	953
Frick, Florian	ISW – University of Stuttgart, D	256
Fröhleke, Norbert	University of Paderborn, D	232 1662 1992
Fuchs, Friedrich W.	Christian-Albrechts-University, D	1942
Fuhrmann, Jan	University of Rostock, D	581 1888
Fujimoto, Mitsunao	ALPS Green Devices, J	91
Funk, Dustin	Technical University of Ilmenau, D	1749
G		
Gaber, Roland	Fraunhofer Institute IWES, D	401
Gafford, James	Mississippi State University, USA	1294
Galai Dol, Lilia	Efficacity, F	2159
Galea, Michael	University of Nottingham, GB	1073
Galek, Marek	Siemens, D	453 1912 1919
Gambino, Giusy	STMicroelectronics, I	1987
Gant, Levi	The University of Alabama, USA	984
Garnier, Laurent	University Grenoble Alpes CEA LITEN, F	2098
Gautier, Cyrille	ENS Cachan – SATIE, F	288
Geissmann, Silvan	ABB Switzerland, CH	417 432

PCIM Europe 2016, 10 – 12 May 2016, Nuremberg, Germany

Author	Institution	Pages
Gekeler, Manfred W.	HTWG Konstanz University of Applied Sciences, D	136
Gerada, Chris	University of Nottingham, GB	1073
Gerlach, Rolf	Infineon Technologies, D	180
Ghossein, Layal	SuperGrid Institute, F	1592
Gierschner, Sidney	University of Rostock, D	1888
Ginot, Nicolas	University of Nantes, F	712
Gleißner, Michael	University of Bayreuth, D	683
Gleitner, Volker	Electronicon Kondensatoren, D	1408
Glose, Daniel	Technical University of Munich, D	1270
Glück, Tobias	Technical University of Vienna, AT	1339
Godbold, Gerald W.	Hyperion Technology, USA	1294
Goikoetxea, Ander	Mondragon University, ES	763 2074
Gomez Suarez, Carlos	Aalborg University, DK	831
Gomez, German	On Semiconductor, BE	1479
Gony, Bashar	Aperam Alloys Amilly, F	84
Gonzalez, Diego	University of Oviedo, ES	1479
González, Manuela	University of Oviedo, ES	1631
Götz, Stefan	Duke University, USA	1904
Grady, Matt	United Silicon Carbide, USA	1549
Grasel, Bernhard	Dewesoft, AT	2204
Greca, Gustavo	Alpha Assembly Solutions, USA	1027
Green, James	University of Sheffield, GB	1095
Grégoire, Luc-André	University of Toulouse, F	371
Grider, Dave	Wolfspeed, USA	34
Grigans, Linards	Riga Technical University, LV	1880
Grimaud, Louis	Safran Group, F	1286
Gritti, Giovanni	STMicroelectronics, I	612
Grohmann, Rolf	HTWK Leipzig, D	2145
Guerre, Vincent	Local Energy Alternative & Fair, F	2066
Guillaume, Herault	ENS Cachan – SATIE, F	1647
Gumaan, Mohammed	Mansoura University, AE	1046
Gustafsson, Emilia	ABB Switzerland, CH	432
H		
Haaf, Peter	Fairchild Semiconductor, D	2231
Hacala, Amélie	ESTIA, F	2066
Hackl, Christoph	Technical University Munich, D	1926

PCIM Europe 2016, 10 – 12 May 2016, Nuremberg, Germany

Author	Institution	Pages
Hähre, Karsten	Karlsruhe Institute of Technology, D	1721
Hain, Stefan	University of Bayreuth, D	720
Hammes, David	University of Rostock, D	1888
Han, SeungHyun	Hyundai Motors Company, ROK	1262
Han, Yunchao	Fraunhofer Institute IISB, D	469
Harel, Jean Claude	Renesas Electronics America, USA	1027
Harfman Todorovic, Maja	GE Research Center, USA	645 1853
Harper, Jonathan	ON Semiconductor, D	877
Hartmann, Michael	Schneider Electric Power Drives, AT	1669
Hartmann, Samuel	ABB Switzerland, CH	945
Hase, Nobuhiro	Rohm Semiconductor, USA	42
Hatae, Shinji	Mitsubishi Electric Corporation, J	677
Haug, Martin	Würth Elektronik eiSos, D	107
Hausberger, Thomas	Technical University of Vienna, AT	1339
Hayakawa, Seiichi	Hitachi Power Semiconductor, J	348
He, Hongtao	Mitsubishi Electric & Electronics, CN	871
He, Weikun	Mentor Graphics, UK	1027
Heer, Daniel	Infineon Technologies, D	53 1800
Heering, Wolfgang	Karlsruhe Institute of Technology, D	1721
Heinemann, Lothar	AEG Power Solutions, D	1677
Heinzel, Thomas	Fuji Electric Europe, D	824 1956
Heldwein, Marcelo	Federal Institute of Santa Catarina – IFSC, BR	621
Helling, Florian	Universität der Bundeswehr Munich, D	1904
Helsper, Martin	Siemens, D	272
Heredero-Peris, Daniel	Universitat Politécnica de Catalunya, ES	2035
Hernandez Gutiérrez, Diego	CH	1178
Hernes, Magnar	SINTEF Energy Research, NO	1408
Herold, Christian	Technical University of Chemnitz, D	588
Herzer, Reinhard	Semikron Elektronik, D	728
Heseding, Johannes	Leibniz University Hannover, D	1421
Hewitt, David	University of Sheffield, GB	1095
Hill, Julia	Aperam Alloys Amilly, F	84
Hiller, Marc	Karlsruhe Institute of Technology, D	788
Himmelstoss, Felix	Technikum Vienna, AT	1741 2137

PCIM Europe 2016, 10 – 12 May 2016, Nuremberg, Germany

Author	Institution	Pages
Hinken, Reiner	Danfoss Silicon Power, D	1934
Hirao, Akira	Fuji Electric, J	1001
Hirayama, Tomohisa	Hitachi Power Semiconductor, J	348
Hochberg, Martin	Karlsruhe Institute of Technology, D	1305 1578
Hoffmann, Klaus F.	Helmut Schmidt University Hamburg, D	202 1300 1528
Hofmann, Daniel	Fuji Electric Europe, D	438
Hofmann, Norbert	University of Applied Sciences Nordwestschweiz, CH	917
Hofmann, Viktor	University of Bayreuth, D	355
Höltgen, Markus	Technical University of Cologne, D	796
Hölzl, Wolfgang	Technical University of Munich, D	543
Homann, Michael	Technical University of Braunschweig, D	216
Honsberg, Marco	Mitsubishi Electric Europe, D	889
Horff, Roman	University of Bayreuth, D	272 653 1503
Hori, Motohiro	Fuji Electric, J	188
Horiuchi, Keisuke	Hitachi, J	1110
Hosking, Terry	SBE, USA	1391
Hossein Khani, Milad Mohammad	Fraunhofer Institute ISE, D	1733
Houzouji, Hiroshi	Hitachi, J	78
Hruska, Miroslav	Skoda Electric, CZ	1808
Hu, Bo	Mitsubishi Electric & Electronic, CN	932
Huang, Jianwei	Dynex Semconductor, GB	845
Huang, Yi	Intersil Corporation, USA	550
Hudoffsky, Boris	PMK Mess- und Kommunikationstechnik, D	1463
Hull, Brett	Wolfspeed, USA	34
Hüning, Felix	University of Applied Sciences Aachen, D	1963
Hunziker, Christoph	University of Applied Sciences Northwestern Switzerland, CH	2167
Hussein, Khalid	Mitsubishi Electric Europe, D	677
I		
Iagar, Angela	Politehnica University Timisoara, RO	2106
Iannuzzo, Francesco	Aalborg University, DK	831
Ichikawa, Hiroaki	Fuji Electric, J	824

PCIM Europe 2016, 10 – 12 May 2016, Nuremberg, Germany

Author	Institution	Pages
Ikawa, Osamu	Fuji Electric, J	438 824
Ikeda, Osamu	Hitachi, J	78
Ikeda, Yoshinari	Fuji Electric, J	188
Immovilli, Fabio	Raw Power, I	1073
Ino, Kazuhide	Rohm, J	42
Inokuchi, Seiichiro	Mitsubishi Electric Corporation, J	677
Inoue, Daisuke	Fuji Electric, J	1956
Inoue, Tomoki	Toshiba, J	566
Iraola, Unai	Mondragon University, ES	763 2074
Irifune, Hiroyuki	Kaga Toshiba Electronics Corporation, J	839
Ishibashi, Hidetoshi	Mitsubishi Electric Corporation, J	342
Ito, Yoichi	Sanken Electric, J	2083
Izuka, Arata	Mitsubishi Electric Corporation, J	677
J		
Jacob, Mathew	Texas Instruments, USA	2019
Jagau, Martin	Technologienetzwerk Allgäu, D	1171
Jang, Hyosang	Infineon Technologies Power Semitech, ROK	863
Jang, JiWoong	Hyundai Motors, ROK	1600
Jang, KiYoung	Hyundai Motors, ROK	1600
Jara, Martin	West Bohemian University, CZ	1808
Jaschke, Reinhard	Helmut Schmidt University Hamburg, D	1300
Jeong, JeeHye	Hyundai Motors Company, ROK	1262
Jeong, KangHo	Hyundai Motors, ROK	1600
Jerinic, Vladan	Danfoss Silicon Power, D	1934
Jeyaprakash, Arun	Technical University of Munich, D	1904
Jiu, Jinting	Osaka University, J	1021
Jones, Steve	Dynex Semiconductor, GB	938
Jonke, Peter	AIT-Austrian Institute of Technology, AT	1846
Joo, JeongHong	Hyundai Motors Company, ROK	1262
Jorge-Ques, Fernando	Universitat Politécnica de Catalunya, ES	2035
Joshi, Shailesh	Toyota Research Institute North America, USA	72
Josso, Stieven	Henkel Electronics, BE	1035
Joubert, Charles	Ampere Laboratory, F	1286 1792
Jun, ChangHan	Hyundai Motors Company, ROK	1262

PCIM Europe 2016, 10 – 12 May 2016, Nuremberg, Germany

Author	Institution	Pages
Jun, Kisoo	KCC Corporation, ROK	961
Jung, Jaehoon	KCC Corporation, ROK	961
Jung, JinHwan	Hyundai Motors Company, ROK	1262 1600
Jung, Marco	Fraunhofer Institute IWES, D	401
Jungwirth, Herbert	SUMIDA Components & Modules, D	1764
K		
Kähr, Christian	University of Applied Sciences Nordwestschweiz, CH	917
Kaiser, Julian	Fraunhofer Institute IISB, D	469
Kaji, Yusuke	Mitsubishi Electric Corporation, J	326
Kakiki, Hideaki	Fuji Electric, J	188
Kako, Naotsugu	Toshiba Corporation Semiconductor, J	839
Kamal, Mustafa	Mansoura University, AE	1046
Kaminski, Nando	University of Bremen, D	1232 1324
Kammerer, Felix	Karlsruhe Institute of Technology, D	788
Kampen, Dennis	BLOCK Transformatoren-Elektronik, D	144
Kanai, Naoyuki	Fuji Electric, J	188
Kaneko, Satoshi	Fuji Electric, J	188
Kapaun, Florian	Universität der Bundeswehr Munich, D	1570
Kapels, Holger	Fraunhofer Institute ISIT, D	202
Kasper, Matthias	ETH Zürich, CH	1622
Kato, Koji	Sanken Electric, J	2083
Kaulfersch, Eberhard	Berliner Nanotest & Design, D	1063
Kawamoto, Noriaki	Rohm, J	42
Kawase, Daisuke	Hitachi Power Semiconductor, J	348 1110
Kawashima, Tetsuya	Fuji Electric, J	895
Ke, Maolong	Dynex Semconductor, GB	845
Kennel, Ralph	Technical University of Munich, D	120 518 629 1248 1270 1926
Keuck, Lukas	University of Paderborn, D	232
Keyse, Richard	Dynex Semiconductor, GB	903
Khani, Toktam	Technologienetzwerk Allgäu, D	2122
Khaselev, Oscar	Alpha Assembly Solutions, USA	1027

PCIM Europe 2016, 10 – 12 May 2016, Nuremberg, Germany

Author	Institution	Pages
Khatri, Danish	Infineon Technologies North America, USA	882
Killeen, Peter	Wolfspeed, USA	34
Kim, Hyunwoo	KCC Corporation, ROK	961
Kim, Taehyun	Infineon Technologies Power Semitech, ROK	858
Kimura, Takashi	Hitachi Automotive Systems, J	331
Kimura, Yoshitaka	Mitsubishi Electric Corporation, J	342
Kinzer, Dan	Navitas Semiconductor, USA	31
Kirchhof, Jörg	Fraunhofer Institute IWES, D	401
Kitamura, Shuichi	Mitsubishi Electric Corporation, J	425
Kizu, Naoyuki	Rohm, J	42
Kjaer, Philip C.	Aalborg University, DK	2050
Klarenbach, Christoph	Beckhoff Automation, D	796
Klauke, Sebastian	Infineon, D	581
Klein, Axel	Technical University of Braunschweig, D	216
Kling, Rainer	Karlsruhe Institute of Technology, D	1721
Klötzer, Sebastian	Helmut Schmidt University Hamburg, D	1528
Kobayashi, Hideto	Fuji Electric, J	1956
Kobayashi, Yasuyuki	Fuji Electric, J	438
Koch, Nelson	EPFL – Ecole Polytechnique Fédérale de Lausanne, CH	1949
Koga, Shunsuke	Osaka University, J	1021
Köhler, Stefan	Technical University Nuremberg Georg Simon Ohm, D	526
Köhnlechner, Benjamin	Technical University of Ilmenau, D	1147
Koike, Yoshihiko	Hitachi, J	78
Koini, Markus	Epcos, D	164
Kolar, Johann Walter	ETH Zürich Power Electronic Systems Laboratory, CH	32 1622
Kolb, Johannes	Schaeffler Technologies, D	1979
Kondo, Satoshi	Mitsubishi Electric Corporation, J	326
Königsmann, Gunter	Semikron Elektronik, D	728
Konishi, Yuichiro	Hitachi, J	1110
Konishide, Masaoki	Yasuhiko Kohno, Hitachi, J	280
Konno, Akitoyo	Hitachi, J	78
Kopta, Arnost	ABB Switzerland, CH	417 432
Körner, Julian	Karlsruhe Institute of Technology, D	1721
Kotani, Raita	Toshiba, J	566
Kouge, Takuma	Fuji Electric, J	1956
Krecek, Tomas	ON semiconductor, CZ	1089

PCIM Europe 2016, 10 – 12 May 2016, Nuremberg, Germany

Author	Institution	Pages
Krenn, Markus	FH Joanneum, AT	2151
Krishna Vytla, Rajeev	Infineon Technologies North America, USA	882
Kroics, Kaspars	Riga Technical University, LV	1880
Kropp, Jürgen	University of Bayreuth, D	1400
Kruschel, Wolfram	Infineon, D	735
Kubera, Sascha	Siemens, D	588
Kübrich, Daniel	Friedrich-Alexander-University of Erlangen, D	1693
Kugi, Andreas	Technical University of Vienna, AT	1339
Kühl, Sascha	Technical University of Munich, D	518
Kusano, Dai	Japan Fine Ceramics, J	64
Kwiecien, Marcin	Magneto, PL	1428
L		
Labrousse, Denis	ENS Cachan – SATIE, F	1647
Lacombe, Bertrand	Safran Group, F	1792
Ladoux, Philippe	University of Toulouse, F	371 2189
Laeuffer, Jacques	Dtalents, F	1541
Lagier, Thomas	SuperGrid Institute, F	2189
Lahl, Peter	Infineon, D	735
Lamo, Paula	University of Cantabria, ES	1194
Lampke, Thomas	Technical University of Chemnitz, D	976
Landsmann, Peter	Technical University of Munich, D	518
Lang, Klaus-Dieter	Fraunhofer Institute IZM, D Technical University of Berlin, D	1040 1155
Langhals, David	Technical University of Cologne, D	796
Larousse, Sebastien	Institut des Nanotechnologies de Lyon, F	1707
Larrañaga, Jon Andreu	University of the Basque Country, ES	1785
Larson, Kent	Dow Corning Corporation, USA	992
Lassmann, Matthias	Infineon, D	735
Laud, Satyavrat	Renesas Electronics America, USA	1027
Laue, Jürgen	Danfoss Silicon Power, D	698
Lautner, Jennifer	Friedrich-Alexander-University of Erlangen, D	1278
Leach, John	Castle, GB	1769
Leary, Alex	Carnegie Mellon University, USA	1518
Lechler, Armin	ISW – University of Stuttgart, D	256
Lee, Brian	AIT, IE	558
Lee, JaeWon	Hyundai Motors Company, ROK	1262

PCIM Europe 2016, 10 – 12 May 2016, Nuremberg, Germany

Author	Institution	Pages
Lee, JeongYun	Hyundai Motors Company, ROK	1262
Lee, Junbae	Infineon Technologies Power Semitech, ROK	858 863
Lee, KiJong	Hyundai Motors, ROK	1600
Lee, Minsub	Infineon Technologies Power Semitech, ROK	858 863
Lefebvre, Stéphane	ENS Cachan – SATIE, F	1647
LeHenaff, Francois	Alpha Assembly Solutions, D	1027
Leibfried, Thomas	Karlsruhe Institute of Technology, D	363
Lemke, Michael	AEG Power Solutions, D	1677
Lemmon, Andrew	The University of Alabama, USA	984
Lempidis, Georgios	Fraunhofer Institute IWES, D	401
Lenz, Kevin	Danfoss Silicon Power, D	1934
Leon-Ramirez, Christian Alejandro	University Simon Bolivar, VE	2211
Leszczynski, Jacek	AGH University of Science and Technology, PL	1428
Lexow, Daniel	University of Rostock, D	924
Leyrer, Benjamin	Karlsruhe Institute of Technology, D	1979
Li, Daohui	Dynex Semiconductor, GB	938
Li, Xueqing	United Silicon Carbide, USA	1511
Lifton, Anna	Alpha Assembly Solutions, USA	957 1027
Lindemann, Andreas	Otto-von-Guericke-University, D	1081 1829
Lindseth, Roar	Valeo, NO	113
List, Hans	FH Joanneum, AT	2151
Liu, Cheng	Chenyang Technologies, D	1248
Liu, Guoyou	Dynex Semconductor, GB	845
Liu, Hui	Valeo, NO	113
Liu, Ji-Gou	Chenyang Technologies, D	1248
Liu, Tao	Lenze SE, D	2007
Liu, Wenduo	Infineon Technologies Americas, USA	1655
Liu, Xiaoshan	ENS Cachan – SATIE, F	288
Lledó-Ponsati, Tomàs	TeknoCEA, ES	2035
Löchel, Jonas	Technical University of Ingolstadt, D	742
Loges, Werner	Vacuumschmelze, D	1471
Loh, Poh Chiang	Aalborg University, DK	1837
Lohner, Andreas	Technical University of Cologne, D	669

PCIM Europe 2016, 10 – 12 May 2016, Nuremberg, Germany

Author	Institution	Pages
Longo, Giuseppe	STMicroelectronics, I	804 1987
López, Felipe	University of Cantabria, ES	1194
López, Jesús	Public University of Navarre, ES	2043
Luo, Haihui	Dynex Semconductor, GB	845
Lura, Shinichi	Mitsubishi Electric Corporation, J	425
Lusiewicz, Anna	University of Stuttgart, D	410
Lutz, Josef	Chemnitz University of Technology, D	172 588
M		
Ma, Xiankui	Mitsubishi Electric & Electronics, CN	871
Machuca, Enrique	Technical University of Ingolstadt, D	742
Mackay, Laurens	Delft University of Technology, NL	1201
Mademlis, Georgios	Aristotle University of Thessaloniki, GR	445
Madiwale, Subodh	Analog Devices, USA	195
Madzharov, Nikolay	Technical University of Gabrovo, BG	1999
Makoschitz, Markus	Technical University of Vienna, AT	1669
Malipaard, Dirk	Fraunhofer Institute IISB, D	453
Mandrusiak, Gary	GE Research Center, USA	645 1853
Mangal, Navneet	CREE India Private Limited, IN	2114
Manier, Charles-Alix	Fraunhofer-Institute IZM, D	1155
Mansour, Madaci	University of the Basque Country, ES	1785
Mantzanas, Panagiotis	Friedrich-Alexander-University of Erlangen, D	1693
Marklein, René	Fraunhofer Institute IWES, D	401
Marlino, Laura D.	Oak Ridge National Laboratory, USA	1391
Marquardt, Rainer	Universität der Bundeswehr Munich, D	1570
Marquart, Janosch	University of Applied Sciences NTB Buchs, CH	501 509
Marroyo, Luis	Public University of Navarre, ES	1209
Martin, Christian	Ampere Laboratory, F	1286 1792
Martin, Jérémy	Commissariat à l'énergie atomique et aux énergies alternatives, F	128
Marxgut, Christoph	Helbling Technik, D	661
März, Andreas	University of Bayreuth, D	272 653 1503

PCIM Europe 2016, 10 – 12 May 2016, Nuremberg, Germany

Author	Institution	Pages
März, Martin	Fraunhofer Institute IISB, D	469 1315
Masana, Francesc	Barcelona Semiconductors, ES	1009
Masayoshi, Nakazawa	Fuji Electric, J	188
Matallana, Asier	University of the Basque Country, ES	1970
Mathieu, Olivier	Rogers Germany, D	1027
Matlok, Stefan	Fraunhofer Institute IISB, D	637
Matocha, Kevin	Monolith Semiconductor, USA	984
Matsushita, Akira	Hitachi Automotive Systems, J	331
Matthias, Sven	ABB Switzerland, CH	417 432
Mau, Matthias	Danfoss Silicon Power, D	698
Mauder, Anton	Infineon Technologies, D	1324
Maurer, Wilhelm	Infineon Technologies, D	1155
Mazzola, Michael	Mississippi State University, USA	1294
McHenry, Michael E.	Carnegie Mellon University, USA	1518
McPherson, Brice	Wolfspeed, USA	34
Meany, Tom	Analog Devices ERDC, IE	1361
Meisser, Michael	Karlsruhe Institute of Technology, D	1487 1979
Meradji, Moudrik	Harbin Institute of Technology, CN	2014
Merlo, Christophe	ESTIA, F	2066
Merten, Jens	National Solar Energy Institute, F	33
Mertens, Axel	Leibniz University Hannover, D	1421
Mesbahi, Tedjani	Ecole Centrale de Lille, F	2014
Mesemanolis, Athanasios	ABB Switzerland, CH	432
Middelstaedt, Lars	Otto-von-Guericke-University, D	1829
Mii, Kenji	Toshiba Corporation Semiconductor, J	839
Mikulla, Michael	Fraunhofer-Institute IAF, D	319
Millington, Alan	Dynex Semiconductor, GB	903
Misra, Sanjay	Henkel, USA	60
Mitic, Gerhard	Siemens, D	2226
Mitova, Radoslava	Technical University of Berlin, D	1155
Miura, Mineo	Rohm, J	42
Miyahara, Satoshi	Mitsubishi Electric, D	996
Miyata, Hiroshi	Fuji Electric, J	1956
Miyazaki, Takaaki	Hitachi, J	78
Mizushima, Takao	ALPS Green Devices, J	91

PCIM Europe 2016, 10 – 12 May 2016, Nuremberg, Germany

Author	Institution	Pages
Mochizuki, Eiji	Fuji Electric, J	188 1001
Moia, Joabel	Federal Institute of Santa Catarina – IFSC, BR	621
Momose, Fumihiko	Fuji Electric, J	1001
Montesinos-Miracle, Daniel	Universitat Politécnica de Catalunya, ES	2035
Morel, Florent	Ampère, F	1592
Morel, Hervé	Ampère, F	1592
Mori, Mutsuhiro	Hitachi, J	78 1110
Morita, Toshiaki	Hitachi, J	78
Motto, Eric R.	Powerex, USA	889
Müller, Christian	Infineon Technologies, D	850
Müller, Florian	Vossloh-Schwabe Lighting Solutions, D	605
Müller, Georg	Karlsruhe Institute of Technology, D	1305 1578
Mumby-Croft, Paul	Dynex Semiconductor, GB	704
Münster, Patrick	University of Rostock, D	924
Müter, Ulf	Helmut Schmidt University Hamburg, D	1528
N		
Nabhani, Farhad	Teesside University, GB	1439 1685
Naeberle, Norbert	Schaffner, CH	148
Nagahara, Teruaki	Mitsubischi Electric Corporation, J	889
Nagao, Shijo	Osaka University, J	1021
Nagashima, Kazuhito	Hitachi Power Semiconductor, J	348
Nagaune, Fumio	Fuji Electric, J	1956
Naitoh, Yutaka	ALPS Green Devices, J	91
Nakagawa, Ryosuke	Mitsubishi Electric Corporation, Japan	1311
Nakamura, Keiichi	Mitsubishi Electric Corporation, J	425
Nakamura, Masato	Hitachi, J	78
Nakanishi, Masaharu	Rohm Semiconductor, D	42
Nakano, Hiroshi	Hitachi, J	78
Nakatsu, Kinya	Hitachi, J	331
Nasadoski, Jeffrey	GE Research Center, USA	645
Nashiki, Masato	Toshiba Corporation Semiconductor, J	839
Nate, Satoru	Rohm, J	42
Neumaier, Klaus	Fairchild Semiconductor, D	2231
Ng, Chiu	Infineon Technologies Americas, USA	819

PCIM Europe 2016, 10 – 12 May 2016, Nuremberg, Germany

Author	Institution	Pages
Ng, Wai Tung	University of Toronto, CA	895
Nicolle, Thomas	Calyos, BE	1139
Niedermayr, Philipp	Alpitronic, I	534
Nigsch, Simon	University of Applied Sciences NTB Buchs, CH	501 509
Nishimura, Yoshitaka	Fuji Electric, J	1001
Nishio, Haruhiko	Fuji Electric, J	895
Nishiura, Akira	Fuji Electric, J	1956
Nitta, Tetsuya	Toshiba, J	566
Nöding, Christian	University of Kassel, D	573
Nogawa, Hiroyuki	Fuji Electric, J	1001
Nojima, Geraldo	Eaton Corporation, USA	1518
Norambuena, Margarita	Technical University of Berlin, D	1186
O		
O'Sullivan, Dara	Analog Devices, IE	1369
Oeder, Christian	Friedrich-Alexander-University of Erlangen, D	1701 1857
Ohara, Ryoichi	Toshiba, J	566
Ohodnicki, Paul	DOE-National Energy Technology Laboratory, USA	1518
Ohta, Hiroshi	Kaga Toshiba Electronics Corporation, J	839
Okinori, Chihiro	IWATSU Test Instruments, J	1463
Olejniczak, Kraig	Wolfspeed, USA	34
Oñederra, Oier	University of the Basque Country, ES	1970
Onno Krah, Jens	Technical University of Cologne, D	796
Onozawa, Yuichi	Fuji Electric, J	824
Oppermann, Hermann	Fraunhofer-Institute IZM, D	1155
Orlik, Bernd	University of Bremen, D	248
Ota, Kenji	Mitsubishi Electric Corporation, J	425
Otsubo, Yoshitaka	Mitsubishi Electric Corporation, J	342 996
Ott, Leopold	Fraunhofer Institute IISB, D	469
Otto, Alexander	Fraunhofer-Institute ENAS, D	1063
Owzareck, Michael	BLOCK Transformatoren-Elektronik, D	1447
P		
Pacas, Mario	University of Siegen, D	224 240 1873

PCIM Europe 2016, 10 – 12 May 2016, Nuremberg, Germany

Author	Institution	Pages
Packwood, Matthew	Dynex Semiconductor, GB	704 938
Pai, Ajay Poonjal	Infineon Technologies, D	1315
Pala, Vipindas	Wolfspeed, USA	34
Palmer, Patrick	University of Cambridge, GB	1845
Palmour, John W.	Wolfspeed, USA	34
Pan, Xinxing	AIT, IE	558
Papadopoulos, Charalampos	ABB Switzerland, CH	432
Parry, John	Mentor Graphics, UK	1027
Parspour, Nejila	University of Stuttgart, D	410 1447
Passmore, Brandon	Wolfspeed, USA	34
Patt, Michael	Technologienetzwerk Allgäu, D	1171 2122
Paulus, Dirk	Technical University of Munich, D	518
Paulwitz, Christian	Epcos, D	156
Pawellek, Alexander	Friedrich-Alexander-University Erlangen, D	492
Pellicone, Devin	Advanced Cooling Technologies, USA	1118
Peng, Mingkai	Shenzhen Zeasset Electronic Technology, CN	1416
Penzel, Michael	Technical University of Chemnitz, D	976
Pereira França, Alex	CPqD, BR	750
Perez, Angel Luis	University of the Basque Country, ES	1785
Perrin, Remi	INSA de Lyon, F	1286 1792
Persson, Eric	Infineon Technologies Americas, USA	1655
Peters, Dethard	Infineon Technologies, D	53
Petkov, Valeri	Technical University of Gabrovo, BG	1999
Petzoldt, Jürgen	Technical University of Ilmenau, D	1896 2145
Pezet, Francois	Nidec ASI, I	2175
Pfost, Martin	University of Innsbruck, AT	210 757
Phung, Van Trang	University of Siegen, D	240
Piepenbreier, Bernhard	Friedrich-Alexander-University of Erlangen, D	264 1278
Pierfederici, Serge	University de Lorraine, F	772
Pietkiewicz, Andrzej	Schaffner, CH	148
Pietrini, Giorgio	Università degli Studi di Parma, I	1073 1584

PCIM Europe 2016, 10 – 12 May 2016, Nuremberg, Germany

Author	Institution	Pages
Pigazo, Alberto	University of Cantabria, ES	1194
Pignataro, Gaetano	STMicroelectronics, I	804
Plumpton, Ashley	Dynex Semiconductor, GB	903
Pluta, Wojciech	Czestochowa University of Technology, PL	1428
Popa, Gabriel Nicolae	Politehnica University Timisoara, RO	2106
Postiglione, Gianluca	Nidec ASI, I	2175
Pottier, Frederic	Aperam Alloys Amilly, F	84
Proulx, Joe	Mentor Graphics, UK	1027
Puff, Markus	Epcos, AT	164
Q		
Qi, Fang	Dynex Semiconductor, GB	938
Qian, Peng	Technical University of Munich, D	1270
Qiu, Mianwei	Shenzhen Zeasset Electronic Technology, CN	1416
Quay, Rüdiger	Fraunhofer-Institute IAF, D	319
Quentin, Nicolas	SAGEM, F	1286 1792
R		
Rädel, Uwe	Technical University of Ilmenau, D	2145
Rahimo, Munaf	ABB Switzerland, CH	417 432 917
Raithel, Stefan	Vossloh-Schwabe Lighting Solutions, D	605
Ramirez Figueroa, Fernando David	University of Siegen, D	224
Ramirez-Elizondo, Laura	Delft University of Technology, NL	1201
Raso, Antonio	Nidec ASI, I	2175
Rathbone, Kevin	Robotae, GB	1845
Razik, Hubert	University Claude Bernard Lyon, F Laboratoire Ampere, F	312 1707
Reddig, Manfred	University of Applied Sciences Augsburg, D	120
Reger, Martin	Rogers Germany, D	1027
Rehm, Markus	IBR Ingenieurbüro Rehm, D	1332
Reimann, Tobias	Technical University of Ilmenau, D	1147 1749
Reiner, Richard	Fraunhofer-Institute IAF, D	319
Reinhold, Andreas	HTWK Leipzig, D	2145
Reiter, Tomas	Infinoen Technologies, D	1315 1391

PCIM Europe 2016, 10 – 12 May 2016, Nuremberg, Germany

Author	Institution	Pages
Remaci, Ahmed	ESTIA, F	2066
Rencz, Marta	BME, HU	596
Revol, Bertrand	ENS Cachan – SATIE, F	288
Richmond, Jim	Wolfspeed, USA	34
Richter, Dennis	Otto-von-Guericke-University, D	1829
Richter, Jan	Karlsruhe Institute of Technology, D	393
Rocha dos Santos, Sender	CPqD, BR	750
Rodriguez, Jose	University Andres Bello, CL	1186
Roesner, Robert	GE Research Center, D	645
Roig, Jaume	On Semiconductor, BE	1479
Rossberg, Matthias	Semikron Elektronik, D	728
Rossmanith, Hans	Friedrich-Alexander-University of Erlangen, D	1434
Rout, Colin	Dynex Semiconductor, GB	903
Rowden, Brian	GE Research Center, USA	645 1853
Ruccius, Benjamin	Fraunhofer Institute IISB, D	453
Rufer, Alfred	EPFL – Ecole Polytechnique Fédérale de Lausanne, CH	445 1713 1949
Rupp, Roland	Infineon Technologies, D	180
Ryu, Sei-Hyung	Wolfspeed, USA	34
Rzepka, Sven	Fraunhofer-Institute ENAS, D	1063
S		
Sack, Martin	Karlsruhe Institute of Technology, D	1305 1578
Sadarnac, Daniel	CentraleSupelec, F	477
Sahli, Nizar	Helmut Schmidt University Hamburg, D	1528
Saito, Katsuaki	Hitachi Europe, GB	348 1110
Saito, Ryuichi	Hitachi Automotive Systems, J	331
Saito, Shoji	Mitsubishi Electric Corporation, J	677
Saitou, Takashi	Fuji Electric, J	1001
Sakai, Shinji	Mitsubishi Electric Corporation, J	336
Sakiyama, Yoko	Toshiba, J	566
Sakurai, Naoki	Yasuhiko Kohno, Hitachi, J	280
Salerno, Paul	Alpha Assembly Solutions, USA	1027
Sanchis, Pablo	Public University of Navarre, ES	1209
Sander, Rene	Karlsruhe Institute of Technology, D	363

PCIM Europe 2016, 10 – 12 May 2016, Nuremberg, Germany

Author	Institution	Pages
Sano, Kenya	Toshiba, J	566
Sarkany, Zoltan	Mentor Graphics, UK	1027
Sárkány, Zoltán	Budapest University of Technology and Economics, HU	596 1155
Sasaki, Masahiro	Fuji Electric, J	895
Sawada, Mutsumi	Fuji Electric, J	824
Schanen, Jean-Luc	G2ELAB, F	1494
Schäning, Björn	Helmut Schmidt University Hamburg, D	1528
Scharwitz, Christian	Vacuumschmelze, D	1471
Schefler, Stefan	Epcos, D	164
Schenk, Kurt	University of Applied Sciences NTB Buchs, CH	501 509
Scherbaum, Markus	University of Applied Sciences Augsburg, D	120
Scheuermann, Uwe	Semikron Elektronik, D	691
Schiele, Jürgen	AEG Power Solutions, D	1662
Schijffelen, Jos	Power Research Electronics, NL	1383
Schilling, Marco	Technical University of Ilmenau, D	1147
Schlenk, Manfred	Infineon Technologies, D	120
Schliewe, Jörn	Epcos, D	156
Schmeller, Markus	SUMIDA Components & Modules, D	1728 1764
Schmidhuber, Michael	SUMIDA Components & Modules, D	1728 1764
Schmies, Stefan	Infineon, D	735
Schmitt, Alexander	Karlsruhe Institute of Technology, D	393
Schmitt-Landsiedel, Doris	University Rosenheim, D	1562
Schnarrenberger, Mathias	Karlsruhe Institute of Technology, D	788
Schnell, Raffael	ABB Switzerland, CH	417 432 945
Schreitmüller, Stefan	HTWG Konstanz University of Applied Sciences, D	136
Schubert, Andreas	Technical University of Chemnitz, D	976
Schuetz, Tobias	GE Research Center, D	645
Schulte-Overbeck, Christian	AEG Power Solutions, D	1677
Schulz, Matthias	Fraunhofer Institute IISB, D	469
Schulz, Nicola	University of Applied Sciences Northwestern Switzerland, CH	2167
Schumacher, Walter	Technical University of Braunschweig, D	216
Schupbach, Marcelo	Cree, USA	2114

PCIM Europe 2016, 10 – 12 May 2016, Nuremberg, Germany

Author	Institution	Pages
Schwalbe, Ulf	ISLE Steuerungstechnik und Leistungselektronik, D	1147 1749
Schweiger, Hans-Georg	Technical University of Ingolstadt, D	742
Schwenk, Holger	Vacuumschmelze, D	1471
Scibilia, Roberto	Texas Instruments, D	1224
Scrimizzi, Filippo	STMicroelectronics, I	804 1987
Seibel, Axel	Fraunhofer Institute IWES, D	401
Seldrum, Thomas	Dow Corning, BE	1017
Seleme Jr., Seleme Isaac	Federal University of Minas Gerais – UFMG, BR	371
Seliger, Bernd	Fraunhofer Institute IISB, D	637
Seliger, Norbert	University Rosenheim, D	1562
Sequeira, Luis	Nomad Tech, PT	953
Shalaby, Rizk	Mansoura University, AE	1046
Shang, Ming	Mitsubishi Electric & Electronics, CN	871
She, Xu	GE Research Center, USA	1853
Sheehan, Cathal	Bourns Electronics, IE	1224
Shekhar, Aditya	Delft University of Technology, NL	1201
Shelton, Ed	Silicon Contact, GB	1845
Shin, SangChul	Hyundai Motors, ROK	1600
Shorten, Andrew	University of Toronto, CA	895
Shousha, Mahmoud	Würth Elektronik eiSos, D	107
Sieweke, Nico	Beckhoff Automation, D	796
Simco, David	Wolfspeed, USA	34
Singer, Arthur	Universität der Bundeswehr Munich, D	1904
Sinn, Peter	Robert Bosch, D	1377
Sirmelis, Ugis	Riga Technical University, LV	1880
Slatter, Rolf	Sensitec, D	1240
Slawinski, Maximilian	Infineon Technologies, D	1800
Soinski, Marian	Magneto, PL	1428
Sokolovs, Alvis	Riga Technical University, LV	1880
Solanki, Jitendra	University of Paderborn, D	1662
Soldati, Alessandro	Università degli Studi di Parma, I	1073 1584
Song, Gaosheng	Mitsubishi Electric & Electronic, CN	871 932
Sorensen, Jens	Analog Devices, USA	1369
Sorrentino, Elmer	University Simon Bolivar, VE	2211

PCIM Europe 2016, 10 – 12 May 2016, Nuremberg, Germany

Author	Institution	Pages
Sorsdahl, Torbjorn	Valeo, NO	113
Spence, Michael	Dynex Semiconductor, GB	903
Spertino, Filippo	Politecnico di Torino, I	2058
Spro, Ole Christian	SINTEF Energy Research, NO	1408
Starks, Ann	ON Semiconductor, USA	1777
Steffen, Jonas	Fraunhofer Institute IWES, D	401
Stegmeier, Stefan	Siemens, D	2226
Steinbring, Manuel	University of Siegen, D	1873
Steinke, Gina	EPFL – Ecole Polytechnique Fédérale de Lausanne, CH	445 1713
Stevanovic, Ljubisa	GE Research Center, USA	645 1853
Stiasny, Thomas	ABB Switzerland, CH	917
Stiegler, Karlheinz	ABB Switzerland, CH	911
Stöckl, Johannes	AIT-Austrian Institute of Technology, AT	1846
Stocksreiter, Wolfgang	FH Joanneum, AT	296 2151
Stone, David	University of Sheffield, GB	1095 1769 2090
Storasta, Liutauras	ABB Switzerland, CH	417
Strauss, Bastian	Otto-von-Guericke-University, D	1081
Streb, Fabian	Infineon Technologies, D	976
Ströbel-Maier, Henning	Danfoss Silicon Power, D	698
Strzalkowski, Bernhard	Analog Devices, D	195
Stubenrauch, Franz	University Rosenheim, D	1562
Stuckmann, Christoph	Maccon, D	1345
Stuckmann, Tim	University of Applied Sciences Ostwesfalen-Lippe, D	1639
Subramanian, Prasanth	GrafTech International, USA	1126
Suganuma, Katsuaki	Osaka University, J	1021
Sumper, Andreas	CITCEA-UPC, ES	2211
Sun, Brian	Infineon Technologies North America, USA	882
Sun, Hui	Chenyang Technologies, D	1248
Suriyah-Jaya, Michael	Karlsruhe Institute of Technology, D	363
Surma, Alexey	Proton-Electrotex JSC, RUS	957
Swieboda, Cezary	Magneto, PL	1428
Szczesny, Paul	GE Research Center, USA	1853

PCIM Europe 2016, 10 – 12 May 2016, Nuremberg, Germany

Author	Institution	Pages
T		
Tadikonda, Ramakrishna	Infineon Technologies Americas, USA	819
Takahashi, Misaki	Fuji Electric, J	438
Takahashi, Takuya	Mitsubishi Electric Corporation, J	342
Takahashi, Yoshikazu	Fuji Electric, J	1001
Takamiya, Yoshikazu	Fuji Electric, J	1956
Takorabet, Noureddine	University de Lorraine, F	772
Tamai, Yuuta	Fuji Electric, J	1001
Tamenori, Akira	Fuji Electric, J	438
Tan, Xiaoya	Schaffner, CH	148
Tanabe, Gen	Japan Fine Ceramics, J	64
Tanaka, Nobuhiko	Mitsubishi Electric Corporation, J	425
Tanioka, Toshikazu	Mitsubishi Electric Corporation, J	336
Tao, Fengfeng	GE Research Center, USA	645
Tchouangue, Georges	Toshiba Electronics Europe, D	566 839
Teixeira Pinto, Rodrigo	CITCEA-UPC, ES	2211
Thal, Eckhard	Mitsubishi Electric Europe, D	48 425
Thesseling, Matthias	Lenze SE, D	2007
Thomas, Tina	Technical University of Berlin, D	1040
Titushkin, Dmitry	Proton-Electrotex JSC, RUS	957
Tiwari, Amit	Newcastle University, GB	2197
Tokuyama, Takeshi	Hitachi Automotive Systems, J	331
Tomasso, Giuseppe	University of Cassino and South Lazio, I	379 1217
Tóssoli de Sousa, Thais	CPqD, BR	750
Trainer, David	Alstom Grid, GB	2197
Treier, Christian	ABB Switzerland, CH	945
Trintis, Ionut	Aalborg University, DK	831
Trüller, Jürgen	HF Instruments, D	1463
Tsang, Chi Wa	University of Lincoln, GB	1769
U		
Uchida, Yoshiyuki	Japan Fine Ceramics, J	64
Ura, Hideyuki	Toshiba Corporation Semiconductor, J	839
Urtasun, Andoni	Public University of Navarre, ES	1209

PCIM Europe 2016, 10 – 12 May 2016, Nuremberg, Germany

Author	Institution	Pages
V		
Van Brunt, Edward	Wolfspeed, USA	34
Vanlathem, Eric	Dow Corning Europe, BE	992 1017
Velasco, David	Public University of Navarre, ES	2043
Vemulapati, Umamaheswara Reddy	ABB Switzerland, CH	917
Verl, Alexander	ISW – University of Stuttgart, D	256
Vershinin, Konstantin	Alstom Grid, GB	2197
Victory, James	Fairchild Semiconductor, D	813
Viera, Juan C.	University of Oviedo, ES	1631
Vijay, Karthik	Indium Corporation, GB	704
Villbusch, Tim	Infineon Technologies, D	1800
Vobecky, Jan	ABB Switzerland, CH	911 917
Vogelsberger, Markus	Bombardier Transportation Austria, AT	1339 2219
Voigt, Gunter	HTWG Konstanz University of Applied Sciences, D	136
Volay, Philippe	Centralp, F	1707
Volke, Andreas	Power Integrations, D	48
Vollaire, Christian	Laboratoire Ampere, F	1494
von Essen, Tobias	Berliner Nanotest & Design, D	2226
Vu, Trong Tue	Eisergy, IE	1163
W		
Wachutka, Gerhard	Technical University of Munich, D	543
Wagner, Bernhard	Technical University Nuremberg Georg Simon Ohm, D	526
Wallscheid, Oliver	University of Paderborn, D	780
Waltereit, Patrick	Fraunhofer-Institute IAF, D	319
Wanderoild, Yohan	CEA Leti, F	312
Wang, Chi-Ming	Toyota Motor Engineering & Manufacturing, USA	1608
Wang, Gang-Yao	Wolfspeed, USA	34
Wang, Gaolin	Harbin Institute of Technology, CN	2014
Wang, Xiongfei	Aalborg University, DK	1837
Wang, Yangang	Dynex Semiconductor, GB	938
Wang, Yazhe	Mitsubishi Electric Corporation, J	336
Warnakulasuriya, Kapila	Carroll & Meynell Transformers, GB	1685 1439

PCIM Europe 2016, 10 – 12 May 2016, Nuremberg, Germany

Author	Institution	Pages
Watabe, Kiyoto	Mitsubishi Electric Corporation, J	336
Wattenberg, Martin	Universtiy of Reutlingen, D	210 757
Weber, Marc	Karlsruhe Institute of Technology, D	1979
Weber, Stefan	Epcos, D	156 164
Wegelin, Viktor	Karlsruhe Institute of Technology, D	1979
Weis, Gerald	FH Joanneum, AT	296 2151
Weiss, Beatrix	Fraunhofer-Institute IAF, D	319
Weiß, Helmut	Technical University of Leoben, AT	2204
Wendt, Hans-Joachim	Lenze Drives, D	2007
Wendt, Michael	Infineon Technologies, D	1615
Werner, Quentin	Daimler, D	772
Wespel, Matthias	Fraunhofer-Institute IAF, D	319
Weyant, Jens	Advanced Cooling Technologies, USA	1118
Weyh, Thomas	Universität der Bundeswehr Munich, D	1904
Wiedemann, Simon	Maccon, D	1353
Wiesner, Eugen	Mitsubishi Electric Europe, D	48 425
Wintrich, Arendt	Semikron, D	1829
Wohlstreicher, Manfred	SUMIDA Components & Modules, D	1728
Wolbank, Thomas	Technical University of Vienna, AT	2219
Wood, John	Silicon Contact, GB	1845
Wright, Nick	Newcastle University, GB	2197
Wu, Xiaomin	Fairchild Semiconductor, D	2231
Wunder, Bernd	Fraunhofer Institute IISB, D	469
Wunderle, Bernhard	Technical University of Chemnitz, D	1155
Würfel, Alexander	University of Bremen, D	1324
Wurz, Marc C.	Leibniz University Hannover, D	107
Wüthrich, Martin	Schaffner, CH	148
X		
Xu, Dianguo	Harbin Institute of Technology, CN	2014
Xu, Yanan	Chenyang Technologies, D	1248
Y		
Yamada, Junji	Mitsubishi Electric Corporation, J	996
Yamaguchi, Masakazu	Toshiba, J	566

PCIM Europe 2016, 10 – 12 May 2016, Nuremberg, Germany

Author	Institution	Pages
Yamashita, Hiroaki	Toshiba Corporation Semiconductor, J	839
Yang, Gang	Valeo, F	113
Yao, Wenli	Northwestern Polytechnical University, CN	1837
Yasuda, Yuusuke	Hitachi, J	78
Yin, Hang	Technical University of Berlin, D	1186
Yoo, Inpil	Infineon Technologies, D	2027
Yoshida, Hiroshi	Mitsubishi Electric Corporation, J	326 342
Yoshida, Souichi	Fuji Electric, J	438 1956
Yoshiwatari, Shinichi	Fuji Electric, J	824
Young, George	Eisergy, IE	1163
Yu, Kezhuang	Shenzhen Zeasset Electronic Technology, CN	1416
Yu, Zhe	Fraunhofer Institute ISIT, D	202
Yuki, Hata	Mitsubishi Electric Corporation, J	677
Z		
Zacharias, Peter	University of Kassel, D	573
Zeidler, Henning	Technical University of Chemnitz, D	976
Zeltner, Stefan	Fraunhofer Institute IISB, D	637
Zeman, Miro	Delft University of Technology, NL	1383
Zeng, Guang	Technical University of Chemnitz, D	588
Zhang, Hao	Osaka University, J	1021
Zhang, Shirley	United Silicon Carbide, USA	1511
Zhang, Xiaobin	Northwestern Polytechnical University, CN	1837
Zhang, Yuancheng	Mitsubishi Electric & Electronics, CN	871
Zhang, Yuanzhe	Efficient Power Conversion Corporation, USA	304
Zhang, Zhenbin	Technical University of Munich, D	629 1926
Zhou, Haihua	Infineon Technologies Americas, USA	1655
Zhou, Lu	Dow Corning, CN	992
Zhou, Wei	Dynex Semiconductor, GB	938
Zhu, Ke	United Silicon Carbide, USA	1549
Zimmer, Marco	University of Stuttgart, D	410
Zippelius, Bernd	Infineon Technologies, D	180
Zöller, Clemens	Technical University of Vienna, AT	1339 2219
Zschieschang, Olaf	Fairchild Semiconductor, D	1063

Mesago PCIM GmbH
Rotebuehlstrasse 83-85
70178 Stuttgart Germany

ISBN 978-1-5108-2530-7

International Exhibition & Conference for Power Electronics, Intelligent Motion, Renewable Energy and Energy Management (PCIM Europe 2016)

Nuremberg, Germany
10 - 12 May 2016

Volume 2 of 3

International Exhibition & Conference for Power Electronics, Intelligent Motion, Renewable Energy and Energy Management (PCIM Europe 2016)

Nuremberg, Germany
10 - 12 May 2016

Volume 2 of 3

ISBN: 978-1-5108-2530-7

Printed from e-media with permission by:

Curran Associates, Inc.
57 Morehouse Lane
Red Hook, NY 12571

Some format issues inherent in the e-media version may also appear in this print version.

Copyright© (2016) by Mesago PCIM GmbH
All rights reserved.

Printed by Curran Associates, Inc. (2016)

For permission requests, please contact Mesago PCIM GmbH
at the address below.

Mesago PCIM GmbH
Rotebuehlstrasse 83-85
70178 Stuttgart Germany

Phone: 49 711 619 460
Fax: 49 711 619 4690

info@mesago.com

Additional copies of this publication are available from:

Curran Associates, Inc.
57 Morehouse Lane
Red Hook, NY 12571 USA
Phone: 845-758-0400
Fax: 845-758-2633
Email: curran@proceedings.com
Web: www.proceedings.com

PCIM Europe 2016, 10 – 12 May 2016, Nuremberg, Germany

Table of Content PCIM Europe 2016

Keynotes

Keynote: Welcome to the Post-Silicon World: Wide Band Gap Powers Ahead.................................... 31
Dan Kinzer, Navitas Semiconductor, USA

Keynote: Smart Transformers – Concepts-Challenges-Applications.............................. 32
Johann Walter Kolar, ETH Zürich Power Electronic Systems Laboratory, CH

Keynote: Trends of Solar Systems and their Integration in Electricity Networks................................. 33
Jens Merten, National Solar Energy Institute, F

SiC Devices

Ultra-low (1.25 mΩ) On-Resistance 900V SiC 62 mm Half-Bridge Power Modules Using New 10 mΩ SiC MOSFETs.. 34
Jeffrey Casady, Vipindas Pala, Edward Van Brunt, Brett Hull, Sei-Hyung Ryu, Gang-Yao Wang, Jim Richmond, Scott T. Allen, Dave Grider, John W. Palmour, Peter Killeen, Brice McPherson, Kraig Olejniczak, Brandon Passmore, David Simco, Wolfspeed, USA

Evolution of SiC Products for Industrial Application 42
Naoyuki Kizu, Satoru Nate, Mineo Miura, Noriaki Kawamoto, Kazuhide Ino, Rohm, J; Masaharu Nakanishi, Rohm Semiconductor, D; Nobuhiro Hase, Rohm Semiconductor, USA

Advanced Protection for Large Current Full SiC-Modules.. 48
Eugen Wiesner, Eckhard Thal, Mitsubishi Electric Europe, D; Andreas Volke, Karsten Fink, Power Integrations, D

Switching Performance of a 1200 V SiC-Trench-MOSFET in a Low-Power Module.............................. 53
Daniel Heer, Daniel Domes, Dethard Peters, Infineon Technologies, D

Module Materials

Beyond Thermal Grease, Enhancing Thermal Performance and Reliability 60
Sanjay Misra, Henkel, USA

High Thermal Conductivity Silicon Nitride substrate for Power Semiconductor Applications 64
Dai Kusano, Gen Tanabe, Yoshiyuki Uchida, Japan Fine Ceramics, J

Thermal Management of Future WBG Devices using Two-Phase Cooling 72
Shailesh Joshi, Ercan Dede, Toyota Research Institute North America, USA

Highly Reliable and Lead-Free High Power IGBT Modules Using Novel Copper Sintering Die Attachment 78
Akitoyo Konno, Takaaki Miyazaki,Yuusuke Yasuda, Osamu Ikeda, Hiroshi Nakano, Toshiaki Morita, Hiroshi Houzouji, Mutsuhiro Mori, Masato Nakamura, Yoshihiko Koike, Hitachi, J

PCIM Europe 2016, 10 – 12 May 2016, Nuremberg, Germany

Magnetics & Inductors

A New Generation of Nanocrystalline Magnetic Cores with Very Low Magnetic Losses 84
Bashar Gony, Stephane Camus, Julia Hill, Frederic Pottier, Aperam Alloys Amilly, F

The Fe-based Glassy Alloy Powder Core Inductor for the Boost Converter by GaN HEMT and SiC SBD 1 MHz Operation 91
Mitsunao Fujimoto, Yutaka Naitoh, Takao Mizushima, ALPS Green Devices, J

Accurate Calculation of AC Losses of Inductors in Power Electronic Applications 99
Ranjith Bramanpalli, Würth Elektronik Eisos, D

Thin-Film Based Microtransformer Suitable for High Switching Frequency Power Applications 107
Dragan Dinulovic, Mahmoud Shousha, Martin Haug, Würth Elektronik eiSos, D; Sebastian Beringer, Marc C. Wurz, Leibniz University Hannover, D

DC/AC and AC/DC Converters

Design of High Efficiency High Power Density 10.5 kw Three Phase On-Board-Charger for Electric/Hybrid Vehicles 113
Gang Yang, Valeo, F; Eirik Draugedalen, Torbjorn Sorsdahl, Hui Liu, Roar Lindseth, Valeo Powertrain Energy Conversion, NO

A Bridgeless, Quasi-Resonant ZVS-Switching, Buck-Boost Power Factor Correction Stage (PFC) 120
Markus Scherbaum, Manfred Reddig, University of Applied Sciences Augsburg, D; Ralph Kennel, Technical University of Munich, D; Manfred Schlenk, Infineon Technologies, D

A high efficiency 5.3 kW Current Source Inverter (CSI) Prototype using 1.2 kV Silicon Carbide (SiC) Bi-Directional Voltage Switches in hard Switching Mode 128
Jérémy Martin, Anthony Bier, Stéphane Catellani, Luis Gabriel Alves-Rodrigues, Franck Barruel, Commissariat à l'énergie atomique et aux énergies alternatives, F

Comparison of the EMC and Efficiency Characteristics of Hard and Soft Switching Three-Level Inverters 136
Manfred W. Gekeler, Stefan Schreitmüller, Gunter Voigt, HTWG Konstanz University of Applied Sciences, D

Special Session "Passive Components"

Drive System Loss Reduction by Allpole Sine Filters 144
Dennis Kampen, BLOCK Transformatoren-Elektronik, D; Michael Burger, SEW Eurodrive, D

Harmonic Filtering in Variable Speed Drives 148
Luca Dalessandro, Xiaoya Tan, Andrzej Pietkiewicz, Martin Wüthrich, Norbert Naeberle, Schaffner, CH

A just Comparison of Ferrite and Nanocristalline Common Mode Chokes 156
Jörn Schliewe, Christian Paulwitz, Stefan Weber, Epcos, D

Modelling an Anti-Ferroelectric Ceramic Capacitor for Time- and Frequency-Domain Simulations of Power Systems 164
Stefan Schefler, Markus Koini, Stefan Weber, Epcos, D; Markus Puff, Epcos, AT

PCIM Europe 2016, 10 – 12 May 2016, Nuremberg, Germany

SiC Reliability

Breakdown of Gate Oxide of 1.2 kV SiC-MOSFETs Under High Temperature and High Gate Voltage 172
Menia Beier-Möbius, Josef Lutz, Chemnitz University of Technology, D

Avalanche Robustness of SiC MPS Diodes 180
Thomas Basler, Roland Rupp, Rolf Gerlach, Bernd Zippelius, Infineon Technologies, D; Mihai Draghici, Infineon Technologies, AT

Compact, Low Loss and High Reliable 3.3 kV Hybrid Module 188
Satoshi Kaneko, Naoyuki Kanai, Motohiro Hori, Nakazawa Masayoshi, Hideaki Kakiki, Yasushi Abe, Yoshinari Ikeda, Eiji Mochizuki, Fuji Electric, J

DC/DC Converters I

Isolated Synchronous Forward Controller with Integrated Feedback Loop and Adjustable Dead Time for High Efficiency DC/C Converter 195
Bernhard Strzalkowski, Analog Devices, D; Subodh Madiwale, Gabriele Bernardinis, Michael Daly, Analog Devices, USA

Extreme High Efficiency Non-Inverting Buck-Boost Converter for Energy Storage Systems 202
Zhe Yu, Holger Kapels, Fraunhofer Institute ISIT, D; Klaus F. Hoffmann, Helmut Schmidt University, D

A High-Efficiency Bidirectional GaN-HEMT DC/DC Converter 210
Michael Ebli, Martin Wattenberg, Universtiy of Reutlingen, D; Martin Pfost, University of Innsbruck, AT

Control Converters

Direct Delta Sigma Signal Processing for Control of Power Electronics 216
Michael Homann, Axel Klein, Walter Schumacher, Technical University of Braunschweig, D

Finite Control Set Model Based Predictive Control of a PMSM with Variable Switching Frequency and Torque Ripple Optimization 224
Fernando David Ramirez Figueroa, Mario Pacas, University of Siegen, D

Frequency- and Mode-Adaptive Control of DC-DC Converter for Efficency Improvement 232
Lukas Keuck, Farshid Almai, Sven Bolte, Norbert Fröhleke, Joachim Böcker, University of Paderborn, D

Control Techniques in Intelligent Motion Systems I

Load Torque Estimation in Repetitive Mechanical Systems by Using Fourier Interpolation 240
Van Trang Phung, Mario Pacas, University of Siegen, D

Fast Current Waveform Calculation Algorithm for a Six Phase Switched Reluctance Machine 248
Jacek Borecki, Bernd Orlik, University of Bremen, D

Process Requirements-Based Adaptive PWM for Improved Efficiency of Machine Tool Feed-Drives 256
Matthias Braband, Florian Frick, Armin Lechler, Alexander Verl, ISW – University of Stuttgart, D

PCIM Europe 2016, 10 – 12 May 2016, Nuremberg, Germany

DC/AC Converters

Comparison of Bidirectional T-Source Inverter and Quasi-Z-Source Inverter for Extra Low Voltage Application 264
Thomas Baier, Bernhard Piepenbreier, Friedrich-Alexander-University Erlangen, D

Benchmarking of SiC JFET and SiC MOSFET Modules for the Application in Medium Power Traction Converters 272
Andreas März, Roman Horff, Teresa Bertelshofer, Mark-M. Bakran, University of Bayreuth, D;
Martin Helsper, Siemens, D

New Bus-bar Topology to Suppress the Current Imbalance of Parallel-connected IGBT Modules for High Power Railway 280
Naoki Sakurai, Masaoki Konishide, Yasuhiko Kohno, Hitachi, J

GaN Converters

EMI Investigation in a GaN HEMT Power Module 288
Xiaoshan Liu, Francois Costa, Bertrand Revol, Cyrille Gautier, ENS Cachan- SATIE, F

Ultra-High Frequent Switching with GaN- HEMTs using the Coss-Capacitances as non-dissipative Snubbers 296
Hubert Berger, Luis Alfonso Fernández-Serantes, Wolfgang Stocksreiter, Gerald Weis, University of Applied Sciences Joanneum, AT

eGaN® FET based 6.78 MHz Differential-Mode ZVS Class D AirFuel™ Class 4 Wireless Power Amplifier 304
Michael de Rooij, Yuanzhe Zhang, Efficient Power Conversion Corporation, USA

High Frequency, High Temperature designed DC/DC Coreless Converter for GaN Gate Drivers 312
Yohan Wanderoild, Dominique Bergogne, CEA Leti, F; Hubert Razik, University Claude Bernard Lyon, F

Monolithic GaN-on-Si Half-Bridge Circuit with Integrated Freewheeling Diodes 319
Richard Reiner, Patrick Waltereit, Beatrix Weiss, Matthias Wespel, Michael Mikulla, Rüdiger Quay, Oliver Ambacher, Fraunhofer-Institute IAF, D

Module Design

Resin Encapsulation Combined with Insulated Metal Baseplate for Improving Power Module Reliability 326
Shinsuke Asada, Satoshi Kondo, Yusuke Kaji, Hiroshi Yoshida, Mitsubishi Electric Corporation, J

An Experimental Study on the Thermal Performance of Double-Side Direct-Cooling Power Module Structure 331
Akira Matsushita, Ryuichi Saito, Takeshi Tokuyama, Takashi Kimura, Hitachi Automotive Systems, J;
Kinya Nakatsu, Hitachi, J

New Transfer Mold DIPIPM™ Utilizing Silicon Carbide (SiC) MOSFET 336
Yazhe Wang, Kiyoto Watabe, Shinji Sakai, Toshikazu Tanioka, Mitsubishi Electric Corporation, J

A 1700 V-IGBT module and IPM with new insulated metal baseplate (IMB) featuring enhanced Isolation Properties and Thermal Conductivity 342
Takuya Takahashi, Yoshitaka Kimura, Hiroshi Yoshida, Hidetoshi Ishibashi, Yoshitaka Otsubo, Mitsubishi Electric Corporation, J

PCIM Europe 2016, 10 – 12 May 2016, Nuremberg, Germany

NHPD² (Next High Power Density Dual) with Next Generation Chip Suitable for Low Internal Inductance Package 348
Daisuke Kawase, Kazuhito Nagashima, Tomohisa Hirayama, Katsunori Azuma, Seiichi Hayakawa, Hitachi Power Semiconductor, J; Katsuaki Saito, Hitachi Europe, GB

Power Electronics in Transmission Systems in Smart Grids

An Optimized Hybrid-MMC for HVDC 355
Viktor Hofmann, Mark-M. Bakran, University of Bayreuth, D

Selective HVDC Transmission Line Breaking for Bus Bar Applications under Reduced Expenses 363
Rene Sander, Michael Suriyah-Jaya, Thomas Leibfried, Karlsruhe Institute of Technology, D

Decentralized Controller for Modular Multilevel Converter 371
Seleme Isaac Seleme Jr., Federal University of Minas Gerais – UFMG, BR; Luc-André Grégoire, Marc Cousineau, Philippe Ladoux, University of Toulouse, F

Power Converters for PV Systems with Energy Storage: Optimal Power Flow Control for EV's Charging Infrastructures 379
Mauro Di Monaco, Umberto Abronzini, Ciro Attaianese, Matilde Arpino, Giuseppe Tomasso, University of Cassino and South Lazio, I

Special Session "E-Mobility"

Modular and Comfortable Electromobility 386
Mihai Dragan, Bernhard Budaker, Jonathan Brix, Fraunhofer Institute IPA, D

Power Hardware-in-the-Loop Emulation of Permanent Magnet Synchronous Machines with Nonlinear Magnetics – Concept & Verification 393
Alexander Schmitt, Jan Richter, Michael Braun, Martin Doppelbauer, Karlsruhe Institute of Technology, D

Multimode Charging of Electric Vehicles – A combined Concept with Multiple Use of Components and Strategies for Decreasing Power Losses, Weight and Volume 401
Marco Jung, René Marklein, Georgios Lempidis, Jonas Steffen, Axel Seibel, Jörg Kirchhof, Roland Gaber, Fraunhofer Institute IWES, D

Design of Contactless Energy Transfer System for an Electric Vehicle 410
Mike Böttigheimer, Nejila Parspour, Marco Zimmer, Anna Lusiewicz, University of Stuttgart, D

High Power Semiconductor

3300 V HiPak2 modules with Enhanced Trench (TSPT+) IGBTs and Field Charge Extraction Diodes rated up to 1800 A 417
Chiara Corvasce, Maxi Andenna, Liutauras Storasta, Sven Matthias, Arnost Kopta, Munaf Rahimo, Luca De Michielis, Silvan Geissmann, Raffael Schnell, ABB Switzerland, CH

Durable Design of the New HVIGBT Module 425
Nobuhiko Tanaka, Kenji Ota, Shuichi Kitamura, Shinichi Lura, Keiichi Nakamura, Mitsubishi Electric Corporation, J; Eugen Wiesner, Eckhard Thal, Mitsubishi Electric Europe, D

PCIM Europe 2016, 10 – 12 May 2016, Nuremberg, Germany

The 62Pak IGBT Module Range Employing the 3rd Generation 1700 V SPT++ Chip Set for 175 °C Operation... 432
Sven Matthias, Chiara Corvasce, Athanasios Mesemanolis, Emilia Gustafsson, Charalampos Papadopoulos, Arnost Kopta, Silvan Geissmann, Martin Bayer, Raffael Schnell, Munaf Rahimo, ABB Switzerland, CH

Extended Power Rating of 1200 V IGBT Module with 7G-RC-IGBT Chip Technologies 438
Misaki Takahashi, Souichi Yoshida, Akira Tamenori, Yasuyuki Kobayashi, Osamu Ikawa, Fuji Electric, J; Daniel Hofmann, Fuji Electric Europe, D

Multi Level Converters

DC-DC Converter based on the Asymmetric Multistage Stacked Boost Architecture with Feed-Forward Control for Photovoltaic Plants 445
Georgios Mademlis, Aristotle University of Thessaloniki, GR; Gina Steinke, Alfred Rufer, EPFL – Ecole polytechnique fédérale de Lausanne, CH

A Novel Submodule Concept for Modular Multilevel Converters 453
Benjamin Ruccius, Nicola Burani, Dirk Malipaard, Fraunhofer Institute IISB, D; Marek Galek, Siemens, D

Electro-Thermal Design of a Modular Multilevel Converter Prototype 461
Emilien Coulinge, Alexandre Christe, Drazen Dujic, EPFL – Ecole Polytechnique Fédérale de Lausanne, CH

DC/DC Converters II

Non-isolated Three-Port DC/DC-Converter for ±380VDC Microgrids 469
Yunchao Han, Julian Kaiser, Leopold Ott, Matthias Schulz, Fabian Fersterra, Bernd Wunder, Martin März, Fraunhofer Institute IISB, D

Wide Voltage Input Range Insulated Current Fed Buck Flyback-Forward for HV/LV Power Conversion in Electric/Hybrid Vehicle 477
Reda Chelghoum, Luis De Sousa, Larbi Bendani, Valeo, F; Daniel Sadarnac, CentraleSupelec, F

SiC JFET Cascode Enables Higher Voltage Operation in a Phase Shift Full Bridge DC-DC Converter 484
Jonathan Dodge, United Silicon Carbide, USA

Lamp Ballasts Lighting Systems

Detailed Comparison of One Stage Topologies for LED Lighting Applications 492
Alexander Pawellek, Thomas Dürbaum, Friedrich-Alexander-University Erlangen, D

Low Cost High Density AC-DC Converter for LED Lighting Applications 501
Simon Nigsch, Janosch Marquardt, Kurt Schenk, University of Applied Sciences NTB Buchs, CH

Design Optimization for a High Power-Density, Wide Output, High Frequency LLC Resonant Converter for Lighting Applications 509
Janosch Marquart, Simon Nigsch, Kurt Schenk, University of Applied Sciences NTB Buchs, CH

Sensorless Motor Control

Computationally efficient Anisotropy-Identification based on a Square-Shaped Injection Pattern 518
Peter Landsmann, Dirk Paulus, Sascha Kühl, Ralph Kennel, Technical University of Munich, D

PCIM Europe 2016, 10 – 12 May 2016, Nuremberg, Germany

Estimation of the Excitation Current and the Rotor Resistance of an Externally Excited Synchronous Machine with an Inductively Supplied Excitation Coil 526
Stefan Köhler, Bernhard Wagner, Technical University Nuremberg Georg Simon Ohm, D;
Stefan Endres, Fraunhofer Institute IISB, D

High Speed Sensorless Control of a Synchronous Motor with Kalman Filter 534
Philipp Niedermayr, Alpitronic, I; Silverio Bolognani, Luigi Alberti, University of Padova, I;
Reiner Abl, BMW, D

Software Tools and Applications

Physical Modeling and High-Fidelity Simulation of the Transient Behavior of Multiply-Contacted Power Busbars 543
Vanessa Basler, Wolfgang Hölzl, Gerhard Wachutka, Technical University of Munich, D

Small-signal Output Impedance Modeling of Intersil's R4TMTechnology 550
Yi Huang, Chun Cheung, Intersil Corporation, USA

An Approach of Reinforcement Learning Based Lighting Control for Demand Response 558
Xinxing Pan, Brian Lee, AIT, IE

Cosmic Ray & Ruggedness

Cosmic Ray Failure Mechanism and Critical Factors for 3.3 kV Hybrid SiC Modules 566
Tetsuya Nitta, Yoko Sakiyama, Raita Kotani, Tomoki Inoue, Ryoichi Ohara, Kenya Sano,
Masakazu Yamaguchi, Toshiba, J; Georges Tchouangue, Toshiba Electronics Europe, D

Benefits of Increased Cosmic Radiation Robustness of SiC Semiconductors in large Power-Converters 573
Christian Felgemacher, Samuel Araujo Vasconcelos, Christian Nöding, Peter Zacharias,
University of Kassel, D

Passive IGBT Turn-off during Short-circuit Type V 581
Jan Fuhrmann, Hans-Günter Eckel, University of Rostock, D; Sebastian Klauke, Infineon, D

High-Current Power Cycling Test-Bench for Short Load Pulse Duration and First Results 588
Guang Zeng, Christian Herold, Menia Beier-Möbius, Josef Lutz, Technical University of Chemnitz, D;
Sascha Kubera, Rodrigo Alvarez, Siemens, D

Issues in Testing Advanced Power Semiconductor Devices 596
Gabor Farkas, Mentor Graphics, HU; Zoltan Sarkany, Marta Rencz, BME, HU

Special Session "Smart Lighting"

Smart Lighting – Requirements for Modern Lighting Systems and Expected Trends 604
Diederik de Stoppelaar, LightingEurope, BE

Special requirements on a SMPS for LED-Lighting Purposes by an Example of an Individual High-End LED-Driver Solution 605
Florian Müller, Stefan Raithel, Vossloh-Schwabe Lighting Solutions, D

The seven challanges of LED lighting 612
Claudio Adragna, Francesco Ferrazza, Giovanni Gritti, STMicroelectronics, I

PCIM Europe 2016, 10 – 12 May 2016, Nuremberg, Germany

Highly Flexible Single Stage Flyback Quasi-Resonant Digital Controller for Advanced LED Applications 620
Marc Fahlenkamp, Infineon Technologies, D

New and Renewable Energy Systems

Carrier-Based Modulation Technique to Reduce Low Frequency Ripple at the Partial Dc-Link Voltages of a Three-Level/-Phase/-Wire NPC Converter Applied to Future Dc Bipolar Active Distribution Networks 621
Joabel Moia, Marcelo Heldwein, Federal Institute of Santa Catarina – IFSC, BR

FPGA Based Direct Model Predictive Power and Current Control of 3L NPC Active Front Ends 629
Zhenbin Zhang, Ralph Kennel, Technical University of Munich, D

Scalable Insulated DC/DC Converters for Safe and Efficient Coupling of Fuel Cells, Electrolyzers and DC Grids 637
Bernd Seliger, Stefan Matlok, Stefan Zeltner, Fraunhofer Institute IISB, D

SiC MW PV Inverter 645
Maja Harfman Todorovic, Ljubisa Stevanovic, Gary Mandrusiak, Brian Rowden, Fengfeng Tao, Philip Cioffi, Jeffrey Nasadoski, Rajib Datta, GE Research Center, USA; Fabio Carastro, Tobias Schuetz, Robert Roesner, GE Research Center, D

Power Electronics in Automotive

A Performance Comparison of a 650 V Si IGBT and SiC MOSFET Inverter under Automotive Conditions 653
Teresa Bertelshofer, Roman Horff, Andreas März, Mark-M. Bakran, University of Bayreuth, D

A Generic Topology for Electrical Energy Storage Systems 661
Christoph Marxgut, Helbling Technik, Dli

Pulse Width- and Frequency Modulated DC/DC Converter for Hybrid- and Electrical Vehicles 669
Magnus Böh, Andreas Lohner, Technical University of Cologne, D; Christoph Engelhard, HELLA KGaA Hueck

New High Power Density Modules for EV/HEV Applications 677
Seiichiro Inokuchi, Shoji Saito, Arata Izuka, Hata Yuki, Shinji Hatae, Mitsubishi Electric Corporation, J; Khalid Hussein, Mitsubishi Electric Europe, D

Module Technology

Fault-Tolerant B6-B4 Inverter Reconfiguration with Fuses and Ideal Short-On Failure IGBT Modules 683
Michael Gleißner, Mark-M. Bakran, University of Bayreuth, D

Statistical Evaluation of Current Imbalance in Parallel Devices 691
Uwe Scheuermann, Semikron Elektronik, D

Batch Purity in Semiconductor Power Modules 698
Christian Aggen, Henning Ströbel-Maier, Matthias Mau, Jürgen Laue, Marco Bäßler, Danfoss Silicon Power, D

PCIM Europe 2016, 10 – 12 May 2016, Nuremberg, Germany

Novel Technique to Reduce Substrate Tilt & Improve Bondline Control between AlN Substrate & AlSiC Baseplate in IGBT Modules 704
James Booth, Paul Mumby-Croft, Matthew Packwood, Kim Evans, Andy Dai, Dynex Semiconductor, GB; Karthik Vijay, Indium Corporation, GB

Drive Strategies in Power Converters

Communicating Gate Driver for SiC MOSFET 712
Christophe Bouguet, Nicolas Ginot, Christophe Batard, University of Nantes, F

New Ultra Fast Short Circuit Detection Method Without Using the Desaturation Process of the Power Semiconductor 720
Stefan Hain, Mark-M. Bakran, University of Bayreuth, D

High Power, High Frequency Gate Driver for SiC-MOSFET Modules 728
Gunter Königsmann, Reinhard Herzer, Sven Buetow, Matthias Rossberg, Semikron Elektronik, D

Integrating a real-time Tvj calculation into an IPM 735
Stefan Schmies, Peter Lahl, Wolfram Kruschel, Matthias Lassmann, Infineon, D

Energy Storage

12 V Lithium Ion Starter Batteries 742
Hans-Georg Schweiger, Enrique Machuca, Jonas Löchel, Technical University of Ingolstadt, D

Electric Vehicles Batteries Modeling Analysis Based on a Multiple Layered Perceptron Identification Approach 750
Sender Rocha dos Santos, Thais Tóssoli de Sousa, Alex Pereira França, CPqD, BR

An Efficient Implementation of a Reconfigurable Battery Stack with Optimum Cell Usage 757
Martin Wattenberg, Reutlingen University, D; Martin Pfost, University of Innsbruck, AT

Comparative Study and Evaluation of Passive Balancing Against Single Switch Active Balancing Systems for Energy Storage Systems 763
Iosu Aizpuru, Unai Iraola, Jose Mari Canales, Ander Goikoetxea, Mondragon University, ES

Control Techniques in Intelligent Motion Systems II

Voltage Levels Comparison and System Optimization for Electric Drives in Hybrid Vehicles 772
Quentin Werner, Daimler, D; Serge Pierfederici, Noureddine Takorabet, University de Lorraine, F

Real-Time Capable Model Predictive Control of Permanent Magnet Synchronous Motors Using Particle Swarm Optimisation 780
Oliver Wallscheid, Joachim Böcker, University of Paderborn, D; Ulrich Ammann, Esslingen University of Applied Sciences, D

A Modular Multilevel Matrix Converter for High Speed Drive Applications 788
Dennis Bräckle, Felix Kammerer, Mathias Schnarrenberger, Marc Hiller, Michael Braun, Karlsruhe Institute of Technology, D

Smart Supercapacitor based DC-link Extension for Drives offers UPS Capability and acts as an Energy Efficient Line Regeneration Replacement 796
Jens Onno Krah, Markus Höltgen, David Langhals, Technical University of Cologne, D; Nico Sieweke, Christoph Klarenbach, Beckhoff Automation, D

PCIM Europe 2016, 10 – 12 May 2016, Nuremberg, Germany

MOSFET, IGBTs, Freewheeling Diodes

New LV Wide SOA Power MOSFET Technology for Linear Mode Operation 804
Filippo Scrimizzi, Gaetano Bazzano, Daniela Cavallaro, Marco Comola, Giuseppe Consentino,
Stefania Fortuna, Giuseppe Longo, Gaetano Pignataro, STMicroelectronics, I

Field Stop Trench IGBT Process Parameter Calibration for Advanced Predictive Prototyping 813
Mehrdad Baghaie Yazdi, Hermann Fischer, James Victory, Fairchild Semiconductor, D;
Detlef Conrad, Synopsys, D

Best-in-class 1200 V IGBT for High Frequency Applications 819
Ramakrishna Tadikonda, Jorge Cerezo, Chiu Ng, Infineon Technologies Americas, USA

**Extra Electro-Thermal Performance of 1700V IGBT with the Latest 7th Generation Chipset/
Package Technologies** 824
Thomas Heinzel, Fuji Electric Europe, D; Mutsumi Sawada, Shinichi Yoshiwatari, Hiroaki Ichikawa,
Yuichi Onozawa, Osamu Ikawa, Fuji Electric, J

**Parameter Extraction for PSpice Models by Means of an Automated Optimization Tool –
An IGBT model Study Case** 831
Carlos Gomez Suarez, Francesco Iannuzzo, Paula Diaz Reigosa, Ionut Trintis, Frede Blaabjerg,
Aalborg University, DK

**800 V Super Junction MOSFET (HV-DTMOS IV) with Better Trade-Off Between Switching Loss
and dVDS/dt** 839
Hiroyuki Irifune, Hiroshi Ohta, Kaga Toshiba Electronics Corporation, J; Hiroaki Yamashita, Hideyuki Ura,
Kenji Mii, Masato Nashiki, Naotsugu Kako,Toshiba Corporation Semiconductor, J; Georges Tchouangue,
Toshiba Electronics Europe, D

Highly Robust 1700 V Diodes Fabricated on 8"Line Using Optimized Proton Implanted Buffer 845
Maolong Ke, Haihui Luo, Ian Deviny, Xiaoping Dai, Jianwei Huang, Guoyou Liu, Dynex Semconductor, GB

Loss and Softness Optimized IGBT-Diode System for Fast-Switching Applications 850
Christian Müller, Stefan Buschhorn, Infineon Technologies, D

Intelligent Power Modules

Protection Features of Intelligent Power Module against Transient State 858
Taehyun Kim, Minsub Lee, Junbae Lee, Daewoong Chung, Infineon Technologies Power Semitech, ROK

**New High Level Integrated Intelligent Power Module with Three Phase Inverter and Power Factor
Correction Topologies Optimized for Home Appliance** 863
Hyosang Jang, Byoungho Choo, Junbae Lee, Minsub Lee, Daewoong Chung, Infineon Technologies
Power Semitech, ROK

New DIPIPM+TM Series Module with All-in-one Integrated 871
Yuancheng Zhang, Xiankui Ma, Hongtao He, Gaosheng Song, Ming Shang, Mitsubishi Electric &
Electronics, CN

Improvement of System Level Power Density of 15 A / 600 V Intelligent Power Modules 877
Jonathan Harper, ON Semiconductor, D

Optimization of FREDFET-based µIPMTM for very Low Power Motor Drive Applications 882
Rajeev Krishna Vytla, Danish Khatri, Brian Sun, Infineon Technologies North America, USA

A novel Transfer Molding Intelligent Converter Inverter Brake IGBT Module (DIPIPM+) with Integrated Level Shifting Control ICs 889
Marco Honsberg, Mitsubishi Electric Europe, D; Teruaki Nagahara, Mitsubischi Electric Corporation, J; Eric R. Motto, Powerex, USA

An Automatic IGBT Collector Current Sensing Technique via the Gate Node 895
Jingxuan Chen, Andrew Shorten, Wai Tung Ng, University of Toronto, CA; Masahiro Sasaki, Tetsuya Kawashima, Haruhiko Nishio, Fuji Electric, J

High Voltage Devices

Design and Characterisation of Optimised Protective Thyristors for VSC Systems 903
Michael Spence, Ashley Plumpton, Colin Rout, Alan Millington, Richard Keyse, Dynex Semiconductor, GB

Cathode Emitter vs. Carrier Lifetime Engineering of Thyristors for Industrial Applications 911
Jan Vobecky, Marco Bellini, Karlheinz Stiegler, ABB Switzerland, CH

Experimental Results of a Large Area (91 mm) 4.5 kV "Bi-Mode Gate Commutated Thyristor" (BGCT) 917
Thomas Stiasny, Umamaheswara Reddy Vemulapati, Martin Arnold, Munaf Rahimo, Jan Vobecky, ABB Switzerland, CH; Christian Kähr, Norbert Hofmann, University of Applied Sciences Nordwestschweiz, CH

 **Effect of Self Turn-On during Turn-On of HV-IGBTs** 924
Patrick Münster, Daniel Lexow, Hans-Günter Eckel, University of Rostock, D

An Innovative 6500 V HVIGBT with High Robustness 932
Bo Hu, Gaosheng Song, Mitsubishi Electric & Electronic, CN

New High Power 3.3 kV / 1500 A IGBT Module Packaging 938
Daohui Li, Wei Zhou, Fang Qi, Matthew Packwood, Yangang Wang, Steve Jones, Xiaoping Dai, Dynex Semiconductor, GB

The LinPak High Power Density Design and its Switching Behaviour at 1.7 kV and 3.3 kV 945
Samuel Hartmann, Fabian Fischer, Andreas Baschnagel, Harald Beyer, Raffael Schnell, Christian Treier, ABB Switzerland, CH

Power Converter GTO to IGBT Upgrade – a New Life for Traction Converters 953
Luis Sequeira, Augusto Franco, Adriano Carvalho, Nuno Freitas, Nomad Tech, PT

Packaging Technologies and Materials

Aspects of Reliability Improvement for Large Area Power Semiconductor Devices through Sintering 957
Dmitry Titushkin, Alexey Surma Proton-Electrotex JSC, RUS; Michiel De Monchy, Anna Lifton, Alpha, RUS

Analysis of Interface Structure and Composition of Cu/Al_2O_3 for the High Stability of DBC (Direct Bonded Copper) 961
Hyunwoo Kim, Jaehoon Jung, Hanna Choi, Kisoo Jun, KCC Corporation, ROK

Power Stack – Advantages and Reliability of an Aluminum Based Stacked Power Module 969
Chris Burns, AB Mikroelektronik, AT

Evaluation of Metal-Matrix composites Baseplates with anisotropic thermal Conductivity Inserts 976
Fabian Streb, Infineon Technologies, D; Henning Zeidler, Michael Penzel, Andreas Schubert, Thomas Lampke, Technical University of Chemnitz, D

PCIM Europe 2016, 10 – 12 May 2016, Nuremberg, Germany

Analysis of Packaging Impedance on Performance of SiC MOSFETs 984
Andrew Lemmon, Levi Gant, The University of Alabama, USA; Sujit Banerjee, Kevin Matocha, Monolith
Semiconductor, USA

Low-Stress Silicone Encapsulant for Reliable Power Conversion Devices 992
Guy Beaucarne, Eric Vanlathem, Dow Corning Europe, B; Lu Zhou, Dow Corning, CN; Kent Larson,
Dow Corning Corporation, USA

Pumping out Failure Free Package Structure 996
Junji Yamada, Yoshitaka Otsubo, Mitsubishi Electric Corporation, J; Satoshi Miyahara, Mitsubishi
Electric, D

High power IGBT Module with New AlN Insulated Substrate 1001
Hiroyuki Nogawa, Akira Hirao, Yoshitaka Nishimura, Takashi Saitou, Yuuta Tamai, Fumihiko Momose,
Eiji Mochizuki, Yoshikazu Takahashi, Fuji Electric, J

Nanosilver Paste for Low Pressure Die Attach: A Turn Key Process 1009
Francesc Masana, Barcelona Semiconductors, ES

New Silicone Gel Enabling High Temperature Stability for next Generation of Power Modules 1017
Thomas Seldrum, Eric Vanlathem, Vincent Delsuc, Dow Corning, BE; Hiroji Enami, Dow Corning Toray, J

**A New Ag Paste Composed by Nano and Micro-Ag Particles prepared Simultaneously and
Application as Die-attachment Materials** 1021
Katsuaki Suganuma, Jinting Jiu, Hao Zhang, Shunsuke Koga, Shijo Nagao, Osaka University, J

Packaging and Reliability

Reliability of Double Side Silver Sintered Devices with various Substrate Metallization 1027
Francois LeHenaff, Alpha Assembly Solutions, D; Gustavo Greca, Paul Salerno, Oscar Khaselev,
Monnir Boureghda, Jeffrey Durham, Anna Lifton, Alpha Assembly Solutions, USA; Olivier Mathieu,
Martin Reger, Rogers Germany, D; Zoltan Sarkany, Weikun He, Joe Proulx, John Parry, Mentor Graphics,
UK; Jean Claude Harel, Satyavrat Laud, Renesas Electronics America, USA

New Interconnect Materials: For Future High Reliable Power Module Assembly 1035
Stieven Josso, Henkel Electronics, BE

**Encapsulation of Smart Power Electronic Devices – Thermal Degradation and Dielectric
Behavior** 1040
Tina Thomas, Technical University of Berlin, D; Karl-Friedrich Becker, Klaus-Dieter Lang, Fraunhofer
Institute IZM, D

**Improvement of the Mechanical Properties of Sn-Ag-Sb Lead-Free Solders: Effects of Sb
Addition and Rapidly Solidified** 1046
Mohammed Gumaan, Rizk Shalaby, Mustafa Kamal, Mansoura University, AE; Esmail A. Ali, University
of Science and Technology Yemen, YE

Health-Monitoring of IGBT Power Modules using repetitive Half-sinusoidal Power Losses 1055
Marco Denk, Mark-M. Bakran, University of Bayreuth, D

Reliability Investigation on SiC BJT Power Modules 1063
Alexander Otto, Sven Rzepka, Fraunhofer-Institute ENAS, D; Eberhard Kaulfersch, Berliner Nanotest &
Design, D; Sophia Frankeser, Technical University of Chemnitz, D; Klas Brinkfeldt, Swerea IVF AB, SE;
Olaf Zschieschang, Fairchild Semiconductor, D

PCIM Europe 2016, 10 – 12 May 2016, Nuremberg, Germany

Investigation of the Influence of Ageing Processes on Thermal Characteristics of an IGBT Power Module by Means of Transient Thermal Analysis 1072
Tobias von Essen, Berliner Nanotest & Design, D; Stefan Stegmeier, Gerhard Mitic, Siemens, D

Test Setup for Multistress Characterization of Insulation Degradation Mechanisms in Electric Drives 1073
Davide Barater, Alessandro Soldati, Giorgio Pietrini, Giovanni Franceschini, Università degli Studi di Parma, I; Chris Gerada, Michael Galea, University of Nottingham, GB; Fabio Immovilli, Raw Power, I

Integration of a Measurement Circuit to determine Junction Temperatures of IGBTs in a Three-phase Converter 1081
Bastian Strauss, Andreas Lindemann, Otto-von-Guericke-University, D

A Recuperation Topology for Power Device Testing 1089
Tomas Krecek, ON semiconductor, CZ

Electrolytic Capacitor Age Estimation using PRBS-Based Techniques 1095
David Hewitt, James Green, Jonathan Davidson, Martin Foster, David Stone, University of Sheffield, GB

On-line Monitoring for Diagnosis on Traction Transformer for Rolling Stock Application 1103
André-Philippe Chamaret, SNCF, F; Toufann Chaudhuri, ABB Sécheron, CH

Cooling

Direct-Water-Cooled Next High Power Density Dual (nHPD2) Considering Inverter Layout 1110
Keisuke Horiuchi, Yuichiro Konishi, Mutsuhiro Mori, Daisuke Kawase, Hitachi, J; Katsuaki Saito, Hitachi Europe, GB

Heat Pipes used as Heat Flux Transformers and for Remote Heat Rejection 1118
Devin Pellicone, Jens Weyant, Advanced Cooling Technologies, USA

New Class of Graphite TIMs provide Performance and Reliability 1126
Prasanth Subramanian, Alex Augoustidis, GrafTech International, USA

Heat Pipe System Development for Railway Application working with speed Motion Convection 1132
Thomas Albertin, Atherm, F

High Performances Passive Two-Phase Loops for Power Electronics Cooling 1139
Vincent Dupont, Cyrille Billet, Thomas Nicolle, Calyos, BE

Thermal Modelling and Management for increasing the Power Density in High Current Power Electronic Systems 1147
Marco Schilling, Benjamin Köhnlechner, Ulf Schwalbe, Tobias Reimann, Technical University of Ilmenau, D

Packaging and Characterization of Silicon SiC-based Power Inverter Module with Double Sided Cooling 1155
Charles-Alix Manier, Hermann Oppermann, Lothar Dietrich, Christian Ehrhardt, Fraunhofer-Institute IZM, D; Zoltán Sárkány, Budapest University of Technology and Economics, HU; Bernhard Wunderle, Technical University of Chemnitz, D; Wilhelm Maurer, Infineon Technologies, D; Radoslava Mitova, Klaus-Dieter Lang, Technical University of Berlin, D

PCIM Europe 2016, 10 – 12 May 2016, Nuremberg, Germany

Sensors, Control and Protection

Digital Adaptive Control Approach to Dynamic Response Improvement for Compact PFC Rectifiers 1163
Trong Tue Vu, George Young, Eisergy, IE

Nonlinear Output Characteristic of DAB Converter caused by ZVS Transition 1171
Martin Jagau, Michael Patt, Technologienetzwerk Allgäu, D

Efficiency Maximization for Half-Bridge LC Converter through Automatic Dead Time Tuning 1178
Vittorio Crisafulli, ON Semiconductor, D; Gianluca Fazio, On Semiconductor Italy, I;
Diego Hernandez Gutiérrez, CH

Improved Finite Control Set Model Predictive Control with Fixed Switching Frequency for Three Phase NPC Converter 1186
Margarita Norambuena, Hang Yin, Sibylle Dieckerhoff, Technical University of Berlin, D;
Jose Rodriguez, University Andres Bello, CL

Current Sensorless Totem-pole Bridgeless Power Factor Corrector 1194
Felipe López, Francisco Azcondo, Paula Lamo, Alberto Pigazo, University of Cantabria, ES

State Space Model for n-Parallel Connected DC-DC Converters with Predictive Current Control Strategy 1201
Aditya Shekhar, Pavol Bauer, Laurens Mackay, Laura Ramirez-Elizondo, Delft University of Technology, NL

Parameter-Independent Battery Voltage Control Based on Virtual Capacitor Emulation 1209
Andoni Urtasun, Ernesto L. Barrios, Pablo Sanchis, Luis Marroyo, Public University of Navarre, ES

FPGA Digital Control for VSI Nonlinearity Effect Compensation 1217
Mauro Di Monaco, Umberto Abronzini, Ciro Attaianese, Matilde Arpino, Giuseppe Tomasso, University of Cassino and South Lazio, I

Offline Non Isolated Converter Protection 1224
Cathal Sheehan, Bourns Electronics, IE; Roberto Scibilia, Texas Instruments, D

Optimisation of Shunt Resistors for Fast Transients 1232
Melanie Adelmund, Christian Bödeker, Nando Kaminski, University of Bremen, D

High Bandwidth Current Sensors as an Enabler for Advanced Control Techniques 1240
Rolf Slatter, Sensitec, D

Rotational Speed Measurement Based on Avago ADNS-9800 Laser Mouse Sensor 1248
Cheng Liu, Yanan Xu, Ji-Gou Liu, Hui Sun, Chenyang Technologies, D; Ralph Kennel, Technical University of Munich, D

Low EMI high efficiency converters

Practical EMI Control in a Power Component Design Space 1253
David Bourner, Vicor Corporation, USA

Converter Switching Noise Reduction for Enhancing EMC Performance in HEV and EV 1262
Ho Tae Chun, SeungHyun Han, ChangHan Jun, JeongYun Lee, JaeWon Lee, JeeHye Jeong,
JeongHong Joo, JinHwan Jung, Hyundai Motors Company, ROK

Efficiency and Vibration Observations of a Symmetrical Six-Phase Drive applying Interleaved Space Vector Modulation 1270
Daniel Glose, Peng Qian, Ralph Kennel, Technical University of Munich, D

PCIM Europe 2016, 10 – 12 May 2016, Nuremberg, Germany

High Efficiency Three-Phase-Inverter with 650 V GaN HEMTs 1278
Jennifer Lautner, Bernhard Piepenbreier, Friedrich-Alexander-University of Erlangen, D

A Large Input Voltage Range 1 MHz Full Converter with 95 % Peak Efficiency for Aircraft Applications 1286
Nicolas Quentin, SAGEM, F; Remi Perrin, INSA de Lyon, F; Christian Martin,
Charles Joubert, Ampere Laboratory, F; Louis Grimaud, Safran Group, F;
Rolando Burgos, Dushan Boroyevich, CPES/Virginia Tech, USA

Integrating Depletion-Mode SiC VJFETs into Production Motor Drives 1294
Michael Mazzola, James Gafford, Mississippi State University, USA; Gerald W. Godbold, Hyperion Technology, USA

Higher Light Efficacy in LED-Lamps by Lower LED-Current 1300
Reinhard Jaschke, Klaus F. Hoffmann, Helmut Schmidt University Hamburg, D

Synchronized Switching and Active Clamping of IGBT Switches in a Simple Marx Generator 1305
Martin Sack, Martin Hochberg, Georg Müller, Karlsruhe Institute of Technology, D

High Efficient and Lightweight Auxiliary Power supply with new SiC Power Device 1311
Ryosuke Nakagawa, Mitsubishi Electric Corporation, Japan

A New Behavioral Model for Accurate Loss Calculations in Power Semiconductors 1315
Ajay Poonjal Pai, Tomas Reiter, Infineon Technologies, D; Martin März, Fraunhofer Institute IISB, D

High Speed Electronic Over Current Breaker for DC-Grids without Additional Sensing 1324
Alexander Würfel, Johannes Adler, Nando Kaminski, University of Bremen, D;
Anton Mauder, Infineon Technologies, D

Wireless Power Transmission with High Efficiency for Extensive Applications 1332
Markus Rehm, IBR Ingenieurbüro Rehm, D

Motors and Motor Drives

Prevention of Traction Drives Stator Insulation Faults Based on Overvoltage Reduction Utilizing Active Edge Shaping 1339
Clemens Zöller, Thomas Hausberger, Mathias Blank, Tobias Glück, Hans Ertl, Andreas Kugi,
Technical University Vienna, AT; Markus Vogelsberger, Bombardier Transportation Austria, AT

Noise & Vibration Levels of modern Electric Motors 1345
Christoph Stuckmann, Maccon, D

Development Platform and Techniques for the Rapid Implementation of High Performance Drives 1353
Christian Balke, Simon Wiedemann, Maccon, D

Functional Safety for Integrated Circuits used in Variable Speed Drives 1361
Tom Meany, Analog Devices ERDC, IE

A Sytem Approach To Understanding The Impact of Non-ideal Effects In A Motor Drive Current Loop 1369
Jens Sorensen, Analog Devices, USA; Dara O'Sullivan, Analog Devices, IE

Gate Driver as Part of the Inverter Safety Concept: Optimizing Inverter's Design 1377
Laurent Beaurenaut, Infineon Technologies, D; Peter Sinn, Robert Bosch, D

PCIM Europe 2016, 10 – 12 May 2016, Nuremberg, Germany

Passive Components

Estimation of Ripple and Inductance Roll off when using Powdered Iron Core Inductors 1383
Gautham Ram Chandra Mouli, Pavol Bauer, Miro Zeman, Delft University of Technology, NL;
Jos Schijffelen, Power Research Electronics, NL

Optimized DC Link for Next Generation Power Modules 1391
Michael Brubaker, Terry Hosking, SBE, USA; Tomas Reiter, Infinoen Technologies, D;
Laura D. Marlino, Madhu S. Chinthavali, Oak Ridge National Laboratory, USA

**In-Circuit-Characterization of Ceramic Capacitor with Anti-Ferroelectric Material for Voltage
Source Inverters** 1400
Jürgen Kropp, Mark-M. Bakran, University of Bayreuth, D

Operability of Metallized Polypropylene Capacitors under High Pressure 1408
Magnar Hernes, Ole Christian Spro, SINTEF Energy Research, NO; Volker Gleitner, Electronicon
Kondensatoren, D

Application of High-Voltage 750 V Aluminum Electrolytic Capacitor in Inverter 1416
Kezhuang Yu, Mingkai Peng, Mianwei Qiu, Shenzhen Zeasset Electronic Technology, CN

Analytic Loss Calculation for E-Core Inductors including the End Windings 1421
Johannes Heseding, Axel Mertens, Leibniz University Hannover, D

The Applicability of Nanocrystalline Stacked and Block Cores for Power Electronics 1428
Cezary Swieboda, Marian Soinski, Marcin Kwiecien, Magneto, PL; Wojciech Pluta, Czestochowa
University of Technology, PL; Jacek Leszczynski, AGH University of Science and Technology, PL

The Benefit of Formed or Compacted Litz-Wire Coils 1434
Tobias Appel, STS, D; Hans Rossmanith, Friedrich-Alexander-University of Erlangen, D

**Development of a 100 kW, 20 kHz Nanocrystalline Core Transformer for DC / DC Converter
Applications** 1439
Kapila Warnakulasuriya, Carroll & Meynell Transformers, GB; Farhad Nabhani, Vahid Askari,
Teesside University, GB

Simulation of a 3-Phase Common- and Differential Mode Inductor on a Four-Limb Core 1447
Michael Owzareck, BLOCK Transformatoren-Elektronik, D; Nejila Parspour, University of Stuttgart, D

Design Procedure for Pot-Core Integrated Magnetic Component 1455
Martin Foster, University of Sheffield, GB; Andrew Fairweather, Grant Ashley, VxI Power, GB

Investigation of Core Losses under Different Conditions Applying the Cross Power Method 1463
Boris Hudoffsky, PMK Mess- und Kommunikationstechnik, D; Chihiro Okinori, IWATSU Test Instruments, J;
Jürgen Trüller, HF Instruments, D

A Finite Element Simulation of Nanocrystalline Tape Wound Cores 1471
Christian Scharwitz, Holger Schwenk, Johannes Beichler, Werner Loges, Vacuumschmelze, D

SiC and GaN

**An Insightful Evaluation of a 650 V High-Voltage GaN Technology in Cascode and Stand-Alone
Transistors** 1479
Jaume Roig, German Gomez, Frederick Declercq, Filip Bauwens, On Semiconductor, BE;
Manuel Fernandez, Diego Gonzalez, University of Oviedo, ES

PCIM Europe 2016, 10 – 12 May 2016, Nuremberg, Germany

Static Characterization of Discrete State-of-the-Art SiC Power Transistors 1487
Michael Meisser, Horst Demattio, Thomas Blank, Karlsruhe Institute of Technology, D

Analytical Losses Model for SiC Semiconductors dedicated to Optimization Operations 1494
Gnimdu Dadanema, Francois Costa, ENS Cachan - SATIE, F; Jean-Luc Schanen, Yvan Avenas,
G2ELAB, F; Christian Vollaire, Laboratoire Ampere,F

**Current Measurement and Gate-Resistance Mismatch in Paralleled Phases of High Power SiC
MOSFET Modules** 1503
Roman Horff, Teresa Bertelshofer, Andreas März, Mark-M. Bakran, University of Bayreuth, D

Gate Drive Strategies of SiC Cascodes 1511
Anup Bhalla, Xueqing Li, Shirley Zhang, United Silicon Carbide, USA

State-of-the-art of HF Soft Magnetics and HV/UHV Silicon Carbide Semiconductors 1518
Geraldo Nojima, Faete Filho, Eaton Corporation, USA; Paul Ohodnicki, DOE-National Energy
Technology Laboratory, USA; Alex Leary, Michael E. McHenry, Carnegie Mellon University, USA

**Comparison of Unipolar Silicon Carbide Power Transistors Used in High Switching Frequency
Inverter Topologies** 1528
Sebastian Fahlbusch, Nizar Sahli, Sebastian Klötzer, Ulf Müter, Björn Schäning, Klaus F. Hoffmann,
Helmut Schmidt University Hamburg, D

ST SiC MOSFETs in 1 MHZ DC-DC Converter 1536
Luigi Abbatelli, Giuseppe Catalisano, STMicroelectronics, I

Towards a One Nano-Henry Power Module for SiC and GaN 1541
Jacques Laeuffer, Dtalents, F

Scalable SiC Cascode Power Blocks 1549
Jonathan Dodge, Matt Grady, Ke Zhu, Anup Bhalla, United Silicon Carbide, USA

**High Power Density, High Efficiency 380v to 52v LLC Converter Utiliziing Emode GaN
Switches** 1555
Moshe Domb, Infineon, USA

Gate Drive Units

**A Low Impedance Drive Circuit to Suppress the Spurious Turn On in High Speed Wide
Band-Gap Semiconductor Halfbridges** 1562
Franz Stubenrauch, Norbert Seliger, Doris Schmitt-Landsiedel, University Rosenheim, D

Isolated Gate Driver for High Current/ High Speed FET-Converters 1570
Florian Kapaun, Rainer Marquardt, Universität der Bundeswehr Munich, D

Simple Gate-boosting Circuit for Reduced Switching Losses in Single IGBT Devices 1578
Martin Hochberg, Martin Sack, Georg Müller, Karlsruhe Institute of Technology, D

**Stability and Performance Analysis of a Voltage Controlled Resistor Circuit for Wide Band-gap
Device Gate Drivers** 1584
Alessandro Soldati, Giorgio Pietrini, Davide Barater, Carlo Concari, Università degli Studi di Parma, I

State of the Art of Gate-Drive Power Supplies for Medium and High Voltage Applications 1592
Layal Ghossein, Piotr Dworakowski, SuperGrid Institute, F; Hervé Morel, Florent Morel, Ampère, F

PCIM Europe 2016, 10 – 12 May 2016, Nuremberg, Germany

The Optimized Gate Driver Design Techniques for IGBT Properties and Downsizing in Eco-Friendly Vehicle 1600
KangHo Jeong, SangChul Shin, KiYoung Jang, JinHwan Jung, KiJong Lee, JiWoong Jang, Hyundai Motors, ROK

A Revisit to Resonant Gate Driver and a New Driver to Improve EMI vs. loss Tradeoff for SiC MOSFET 1608
Chi-Ming Wang, Toyota Motor Engineering & Manufacturing, USA

Application and Design Considerations of CoolMOS™ CFD2 and EiceDRIVER™ IC in Motor Drive Application 1615
Wolfgang Frank, Michael Wendt, Infineon Technologies, D; Sam Abdel-Rahman, Infineon Technologies Americas, USA

AC-DC Converters and Power Supplies

4D-Interleaving of Isolated ISOP Multi-Cell Converter Systems for Single Phase AC/DC Conversion 1622
Matthias Kasper, Michael Antivachis, Dominik Bortis, Johann Walter Kolar, ETH Zürich, CH; Gerald Deboy, Infineon Technologies, AT

Battery Charger Based on a Triple-LCp Resonant Converter 1631
Christian Branas, Francisco Azcondo, University of Cantabria, ES; Juan C. Viera, Manuela González, University of Oviedo, ES

High Efficient Flyback Converter with SiC-MOSFET 1639
Johann Austermann, Tim Stuckmann, Holger Borcherding, University of Applied Sciences Ostwesfalen-Lippe, D

PCB Integration of a Magnetic Component dedicated to a Power Factor Corrector Converter 1647
Herault Guillaume, Mercier Adrien, Stéphane Lefebvre, Denis Labrousse, ENS Cachan-SATIE, F

Evaluation of TCM and CrCM modulation for Totem Pole PFC 1655
Haihua Zhou, Wenduo Liu, Eric Persson, Infineon Technologies Americas, USA

System Concept and Model-Based Optimization of High-Current Variable-Voltage Chopper-Rectifiers 1662
Zhiyu Cao, Holger Fahnert, Jürgen Schiele, AEG Power Solutions, D; Jitendra Solanki, Norbert Fröhleke, Joachim Böcker, University of Paderborn, D

Evaluation of a Unidirectional Three-Phase Rectifier Based on the Third Harmonic Injection Concept in Comparison to a VIENNA Rectifier 1669
Markus Makoschitz, Hans Ertl, Technical University of Vienna, AT; Michael Hartmann, Schneider Electric Power Drives, AT

SiC Improves Switching Losses, Power Density and Volume in UPS 1677
Nikolai Epp, Christian Schulte-Overbeck, Zhiyu Cao, Michael Lemke, Lothar Heinemann, AEG Power Solutions, D

Optimization of 12 Pulse and 18 Pulse Rectifier Systems by the Selection of Optimum Parameters for Magnetics 1685
Kapila Warnakulasuriya, Carroll & Meynell Transformers, GB; Farhad Nabhani, Vahid Askari, Teesside University, GB

PCIM Europe 2016, 10 – 12 May 2016, Nuremberg, Germany

DC-DC Converters I

Analysis of the Flyback Converter Utilizing a Transformer with Stepped Air-Gap 1693
Panagiotis Mantzanas, Daniel Kübrich, Markus Barwig, Thomas Dürbaum, Friedrich-Alexander-
University of Erlangen, D

Novel Method for the Estimation of Switching Losses in Resonant Converters 1701
Christian Oeder, Markus Barwig, Thomas Dürbaum, Friedrich-Alexander-University of Erlangen, D

Active Dead-Time Optimization for wide Range Flyback Active-Clamp Converter 1707
Sebastien Larousse, Nacer Abouchi, Remy Cellier, Institut des nanotechnologies de Lyon, F;
Hubert Razik, Laboratoire Ampere, F; Philippe Volay, Centralp, F

**Energetic Macroscopic Representation (EMR) and Control Scheme for the Asymmetric 4-Stage
MSBA** 1713
Gina Steinke, Alfred Rufer, EPFL - Ecole Polytechnique Fédérale Dde Lausanne, CH

**Adjustable 20 kW Full-SiC Electronic Load with Energy Recovery for Medium-frequency
Inverter** 1721
Fabian Denk, Karsten Haehre, Julian Koerner, Rainer Kling, Wolfgang Heering, Karlsruhe Institute
of Technology, D

A New High Frequency Transformer for UPS 1728
Michael Schmidhuber, Manfred Wohlstreicher, Michael Baumann, Markus Schmeller, SUMIDA
Components & Modules, D

DC/DC-Converter for Modular Coupling of 48 V Battery Packs to a High Voltage DC Bus 1733
Michael Eberlin, Milad Mohammad Hossein Khani, Fraunhofer Institute ISE, D

High Dynamic Current Source for LED Light and Data Transmission Applications 1741
Karl Edelmoser, Technical University of Vienna, AT; Felix Himmelstoss, Technikum Vienna, AT

**Design Methods for LLC Converter considering Buck and Boost Mode with Limited Frequency
Range for Wide Input Voltage Range** 1749
Dustin Funk, Tobias Reimann, Technical University of Ilmenau; Ulf Schwalbe, ISLE Steuerungstechnik
und Leistungselektronik, D

The Behavior of Electro-Magnetic Radiation of Storage Inductor in DC-DC Converters 1757
Ranjith Bramanpalli, Würth Elektronik Eisos, D

DC-DC Converters II

A New High Frequency Ferrite Material for Gan Applications 1764
Herbert Jungwirth, Michael Schmidhuber, Michael Baumann, Markus Schmeller, SUMIDA Components
& Modules, D

Multi-Stage LLC Resonant Converters designed for Wide Output Voltage Ranges 1769
Chi Wa Tsang, Chris Bingham, University of Lincoln, GB; Martin Foster, Dave Stone, University of
Sheffield, GB; John Leach, Castle, GB

Application Advantages and Disadvantages of Modern Fast Switching MOSFETs in VRM 1777
Zhiyang Chen, Ann Starks, ON Semiconductor, USA

**Medium to Low Voltage DC/DC Resonant Converter with SiC SCRs and Nanocrystalyne
Magnetic Core Transformer** 1785
Iñigo Martinez de Alegria, Angel Luis Perez, Madaci Mansour, Jon Andreu Larrañaga, University of
the Basque Country, ES; Kerdoun Djallel, GLEC Constantine 1, DZ

PCIM Europe 2016, 10 – 12 May 2016, Nuremberg, Germany

GaN Active-Clamp Flyback Converter with Resonant Operation Over a Wide Input Voltage Range 1792
Nicolas Quentin, SAGEM, F; Remi Perrin, Cyril Buttay, INSA de Lyon, F; Christian Martin, Charles Joubert, Ampere Laboratory, F; Bertrand Lacombe, Safran Group, F

Demonstration of superior SiC MOSFET Module performance within a Buck-Boost Conversion System 1800
Maximilian Slawinski, Tim Villbusch, Daniel Heer, Marc Buschkühle, Infineon Technologies, D

High Efficiency and High Power Density Boost / Buck Converter with SiC JFET Modules for Advanced Auxiliary Power Supply in Trolleybuses 1808
Miroslav Hruska, Skoda Electric, CZ; Martin Jara, West Bohemian University, CZ

Development of a 12 kW isolated and bidirectional DC-DC Converter dedicated to the More Electrical Aircraft: The Buck Boost Converter Unit (BBCU) 1814
Pascal Asfaux, Jeremy Bourdon, Airbus Operation, F

Reverse Mode Application of Sine Amplitude Converters 1822
David Bourner, Vicor Corporation, USA

DC-AC Converters

Influence of the Configuration of the Load Cable on Switching Characteristics of IGBTs 1829
Lars Middelstaedt, Dennis Richter, Andreas Lindemann, Otto-von-Guericke-University, D; Arendt Wintrich, Semikron, D

Improved Power Decoupling Scheme for Single-Phase Grid-Connected Differential Inverter with Realistic Mismatch in Storage Capacitances 1837
Wenli Yao, Xiaobin Zhang, Northwestern Polytechnical University, CN; Xiongfei Wang, Poh Chiang Loh, Frede Blaabjerg, Aalborg University, DK

Technical Approach: Interleaved, Folding, Interpolating Dual-Path Adiabatic Autotransformer Based Power Converter 1845
John Wood, Ed Shelton, Silicon Contact, GB; Kevin Rathbone, Robotae, GB; Mehdi Baghadadi, Patrick Palmer, University of Cambridge, GB

Design and Performance Evaluation of a Three Phase AC Power Source with Virtual Impedance for Validation of Grid Connected Components 1846
Peter Jonke, Johannes Stöckl, Hans Ertl, AIT-Austrian Institute of Technology, AT

Design and Testing of a Modular SiC based Power Block 1853
Maja Harfman Todorovic, Rajib Datta, Ljubisa Stevanovic, Xu She, Philip Cioffi, Gary Mandrusiak, Brian Rowden, Paul Szczesny, Jian Dai, Tony Frangieh, GE Research Center, USA

A Novel Method to simulate the Control-to-output Transfer Function of Resonant Converters 1857
Julian Dobusch, Christian Oeder, Thomas Dürbaum, Friedrich-Alexander-University of Erlangen, D

A Study of the Thermal and Parasitic Optimization of a Large Current Density Highly Parallelized Three-Phase Reference Board for Motor Drive Applications 1864
Mehrdad Baghaie Yazdi, Xiaomin Wu, Peter Haaf, Klaus Neumaier, Fairchild Semiconductor, D

PCIM Europe 2016, 10 – 12 May 2016, Nuremberg, Germany

AC-AC and Multilevel Converters

Trends in Residential and Industrial Induction Cooking: Topologies and Power Devices for High Efficiency 1865
Vittorio Crisafulli, ON Semiconductor, D

Direct Power Control for a Grid Connection of a Three Phase Z-Source Inverter 1873
Manuel Steinbring, Mario Pacas, University of Siegen, D

Interleaved Series Input Parallel Output forward Converter with Simplified Voltage Balancing Control 1880
Kaspars Kroics, Alvis Sokolovs, Linards Grigans, Ugis Sirmelis, Riga Technical University, LV

Fault-Tolerant Behaviour of the Three Level Advanced-Active-Neutral-Point-Clamped Converter ... 1888
Sidney Gierschner, David Hammes, Jan Fuhrmann, Hans-Günter Eckel, University of Rostock, D;
Max Beuermann, Siemens, D

Cell Voltage Balancing Controller for the Modular Multilevel Converter Arm using Symmetrical Transformation 1896
Andrey Dudin, Aaron Fischer, Thomas Ellinger, Jürgen Petzold, Technical University of Ilmenau, D

Isolated low-power multi-output DC-DC Converters with Heterogeneous Loads for an Efficient Supply of Modular Power Electronics Systems 1904
Arthur Singer, Thomas Weyh, Florian Helling, Universität der Bundeswehr Munich, D; Arun Jeyaprakash,
Technical University of Munich, D; Stefan Götz, Duke University, USA

A wire based communication interface for Medium and High-Voltage Converters 1912
Marek Galek, Manuel Blum, Siemens, D

An Auxillary Power Supply with integrated Communication Capability for Medium and Highvoltage Applications 1919
Manuel Blum, Marek Galek, Siemens, D

FPGA Based Direct Model Predictive Current Control of PMSM Drives with 3L-NPC Power Converter 1926
Zhenbin Zhang, Christoph Hackl, Ralph Kennel, Technical University Munich, D

IGBT Power Module in Three-Level Neutral Point Clamped Type 2 (NPC2, T-NPC, Mixed Voltage) Topology in Short Circuit Modes 1934
Vladan Jerinic, Kevin Lenz, Reiner Hinken, Danfoss Silicon Power, D

Efficiency Verification Power Circulation Method of a High Power Low Voltage NPC Converter for Wind Turbines 1942
Berthold Benkendorff, Friedrich W. Fuchs, Christian-Albrechts-University, D; Toke Franke, Danfoss
Silicon Power, D

Control of the Actively Balanced Capacitive Voltage Divider for a Five-Level NPC Inverter-Estimation of the Intermediary Levels Currents 1949
Alfred Rufer, Nelson Koch, Nicolas Cherix, EPFL – Ecole Polytechnique Fédérale de Lausanne, CH

Automotive Applications

Automotive Power Module Technologies for High Speed Switching 1956
Shinichiro Adachi, Takuma Kouge, Souichi Yoshida, Hiroshi Miyata, Daisuke Inoue, Yoshikazu Takamiya,
Hideto Kobayashi, Akira Nishiura, Fumio Nagaune, Fuji Electric, J; Thomas Heinzel, Fuji Electric Europe, D

PCIM Europe 2016, 10 – 12 May 2016, Nuremberg, Germany

Power Semiconductors for the Automotive 48 V Board Net 1963
Felix Hüning, University of Applied Sciences Aachen, D

Status and Advances in Electric Vehicle's Power Modules Packaging Technologies 1970
Itxaso Aranzabal, Asier Matallana, Oier Oñederra, Iñigo Martinez de Alegría, David Cabezuelo,
University of the Basque Country, ES

A Highly Integrated Full SiC Six Pack Power Module for Automotive Applications 1979
Bao Ngoc An, Viktor Wegelin, Martin Bernd, Benjamin Leyrer, Michael Meisser, Horst Demattio,
Thomas Blank, Marc Weber, Karlsruhe Institute of Technology, D; Johannes Kolb, Schaeffler Technologies, D;
Jochen Altstadt, Kortec, D

**Automotive-grade P-channel Power MOSFETs for Static, Dynamic and Repetitive Reverse
Polarity Protection** 1987
Filippo Scrimizzi, Giuseppe Longo, Giusy Gambino, STMicroelectronics, I

Isolated On-Board DC-DC Converter for Power Distribution Systems in Electric Vehicles 1992
Sven Bolte, Joachim Böcker, Norbert Fröhleke, University of Paderborn, D

Innovative Solution of Static and Dynamic Contactless Charging Station for Electrical Vehicles 1999
Nikolay Madzharov, Valeri Petkov, Technical University of Gabrovo, BG

Combining an External Rotor Motor with Vernier Concept for Drives in Intralogistics 2007
Matthias Thesseling, Tao Liu, Lenze SE, D; Hans-Joachim Wendt, Lenze Drives, D

**Dynamic Modeling and Optimal control for Series-parallel Drivetrain based on Lithium-ion
battery** 2014
Tedjani Mesbahi, Ecole Centrale de Lille, F; Moudrik Meradji, Gaolin Wang, Dianguo Xu, Harbin
Institute of Technology, CN

**Smart Diode and 4-Switch Buck-Boost Provide Ultra High Efficiency, Compact Solution for
12-V Automotive Battery Rail** 2019
Vijay Choudhary, Mathew Jacob, Texas Instruments, USA

On-Chip Temperature Measurement: A new Approach for Optimizing Automotive Inverters 2027
Laurent Beaurenaut, Inpil Yoo, Infineon Technologies, D

Renewable Energy Systems

Resonant load Emulator for Distributed Energy Resources to test Anti-islanding Algorithms 2035
Daniel Heredero-Peris, Fernando Jorge-Ques, Daniel Montesinos-Miracle, Universitat Politécnica de
Catalunya, ES; Tomàs Lledó-Ponsati, TeknoCEA, ES

Low Voltage Ride Through (LVRT) Capability of an Enhanced DFIG System 2043
David Velasco, Jesús López, Public University of Navarre, ES

Control and Modulation for Loss Minimization for Dc/Dc Converter in Wind Farm 2050
Catalin Gabriel Dincan, Philip C. Kjaer, Aalborg University, DK

**A Variable Step Size Perturb and Observe Method Based MPPT for Partially Shaded
Photovoltaic Arrays** 2058
Jawad Ahmad, Filippo Spertino, Paolo Di Leo, Alessandro Ciocia, Politecnico di Torino, I

**Renewable Electricity Conversion and Storage: Focus on Power to Gas process, EMR
Modelling and Simulation** 2066
Ahmed Remaci, Octavian Curea, Christophe Merlo, Amélie Hacala, ESTIA, F; Vincent Guerre, Local
Energy Alternative & Fair, F

PCIM Europe 2016, 10 – 12 May 2016, Nuremberg, Germany

Balancing Current and Efficiency Modelling of Single Switch Active Balancing Systems for Energy Storage Systems 2074
Iosu Aizpuru, Unai Iraola, Jose Mari Canales, Ander Goikoetxea, Mondragon University, ES

A Control Strategy for Multiple Energy Storage Devices for Power Leveling of Renewable Energy Systems 2083
Koji Kato, Yoichi Ito, Sanken Electric, J

Examining Contrasting Excitation Modes within Battery Characterisation using Maximum Length Sequences 2090
Andrew Fairweather, VxI Power, GB; David Stone, Martin Foster, University of Sheffield, GB

Comparison Between Standard and Innovative Solutions to exchange Energy between High Energy Storage Systems 2098
Laurent Garnier, Daniel Chatroux, Sébastien Carcouet, Julien Dauchy, University Grenoble Alpes CEA LITEN, F

EMI, Harmonics, Filters

Simulation and Experimental Analysis of Non-Linear Loads from Residential and Educational Buildings 2106
Gabriel Nicolae Popa, Angela Iagar, Corina Maria Dinis, Politehnica University Timisoara, RO

A Digital Predictive Constant Frequency Controller for High Frequency 3-Phase Silicon Carbide PFC Rectifier 2114
Marcelo Schupbach, Cree, USA; Binod Agrawal, Navneet Mangal, CREE India Private Limited, IN

DC-link Harmonic Content in Double Two-Level Inverter for Permanent Magnet Synchronous Motor Drive Systems – Comparison and Analysis 2122
Toktam Khani, Michael Patt, Technologienetzwerk Allgäu, D

Hybrid Filter With an Optimized Switching Method of the Compensation Capacitors and Predictive Active Filter Control 2129
Swen Bosch, Heinrich Steinhart, HTW Aalen, D

Active Mains Filters with Combined Feed-Forward and Feed-Back Control 2137
Felix Himmelstoss, Technikum Vienna, AT; Karl Edelmoser, Technical University of Vienna, AT

Influence of the Zero Sequence Voltage on the Design of a Series Active Filter 2145
Andreas Reinhold, Rolf Grohmann, HTWK Leipzig, D; Uwe Rädel, Jürgen Petzoldt, Technical University of Ilmenau, D

Electromagnetic Emissions in High Density and Fast GaN Switched Half Bridges with Resonance Filter Structures 2151
Wolfgang Stocksreiter, Hans List, Hubert Berger, Gerald Weis, Markus Krenn, FH Joanneum, AT; Günter Engel, CeraCap, AT

Energy Transmission and Grid

AC or DC Grid for Railway Stations? 2159
Lilia Galai Dol, Efficacity, F; Alexandre De Bernardinis, French Institute of Science and Technology for Transport, Development and Networks, F

Solid-State Transformer Modeling for Analyzing its Application in Distribution Grids 2167
Christoph Hunziker, Nicola Schulz, University of Applied Sciences Northwestern Switzerland, CH

PCIM Europe 2016, 10 – 12 May 2016, Nuremberg, Germany

Hybrid Reactive Power Compensation System for Grid Code Compliance in Renewable Energy Power Plants 2175
Gianluca Postiglione, Antonio Raso, Giovanni Borghetti, Francois Pezet, Nidec ASI, I

A new shunt connected HVDC Tap Based on a Highly Efficient Resonant Cascade Converter 2181
Andre Birkel, Mark-M. Bakran, University of Bayreuth, D

Analysis of Voltage and Current Unbalance in a Multi-Converter Topology for a DC-Based Offshore Wind Farm 2189
Thomas Lagier, SuperGrid Institute, F; Philippe Ladoux, University of Toulouse, F

Experimental Demonstration of a Solid-StateDamping Resistor for HVDC Applications 2197
Konstantin Vershinin, Ikenna Efika, David Trainer, Colin Davidson, Alstom Grid, GB; Nick Wright, Amit Tiwari, Newcastle University, GB

High Precision Loss Measurement at HVDC Converter 2204
Helmut Weiß, Technical University of Leoben, AT; Bernhard Grasel, Dewesoft, AT

A Fast Methodology for Solving Power Flows in Hybrid AC/DC Networks: The European North Sea Supergrid Case Study 2211
Rodrigo Teixeira Pinto, Monica Aragues-Penalba, Andreas Sumper, CITCEA-UPC, ES; Christian Alejandro Leon-Ramirez, Elmer Sorrentino, University Simon Bolivar, VE

Panel Discussion "The smart future of power electronics"

Using Smart Converter to obtain Traction-Machine Insulation Health State Information 2219
Markus Vogelsberger, Bombardier Transportation Austria, AT; Clemens Zöller, Jörg Bellingen, Thomas Wolbank, Technical University of Vienna, AT

Manuscripts which were handed in late

Investigation of the Influence of Ageing Processes on Thermal Characteristics of an IGBT Power Module by Means of Transient Thermal Analysis 2226
Tobias von Essen, Berliner Nanotest & Design, D; Stefan Stegmeier, Gerhard Mitic, Siemens, D

A Study of the Thermal and Parasitic Optimization of a Large Current Density Highly Parallelized Three-Phase Reference Board for Motor Drive Applications 2231
Mehrdad Baghaie Yazdi, Xiaomin Wu, Peter Haaf, Klaus Neumaier, Fairchild Semiconductor, D

Comparative Study and Evaluation of Passive Balancing Against Single Switch Active Balancing Systems for Energy Storage Systems

Iosu, Aizpuru, Mondragon University, Loramendi 4 20500 Arrasate-Mondragón, Spain, iaizpuru@mondragon.edu
Unai, Iraola, Mondragon University, Loramendi 4 20500 Arrasate-Mondragón, Spain, uiraola@mondragon.edu
Jose Mari, Canales, Mondragon University, Loramendi 4 20500 Arrasate-Mondragón, Spain, jmcanales@mondragon.edu
Ander, Goikoetxea, Mondragon University, Loramendi 4 20500 Arrasate-Mondragón, Spain, agoikoetxeaa@mondragon.edu
The Power Point Presentation will be available after the conference.

Abstract

Series connection of energy storage cells implies the need of a BMS and a balancing system to control and improve the performance of the battery pack. Nowadays passive balancing is the most used balancing system in industrial applications, basically due to its simplicity and low price. Active balancing systems are mostly reserved to research articles and experimental prototypes. During this research article, single switch active balancing systems will be presented as a real option of passive balancing substitution. For that purpose during the article most important characteristics of balancing systems will be presented regarding to the impact on the final battery performance, behavior and price. After detecting most important comparison characteristics a single switch active balancing systems will be compared with a passive balancing system prototypes under different working situations.

1. Introduction

Energy storage applications are high demand and popular applications specially in portable technologies. One single cell is used in mobile phone applications and low number of series connected cells (3-4 cells) in laptops and other small portable devices.

High number of cells connected in series/parallel configuration are necessary for renewable energy and electro mobility applications. Energy storage systems permit to increase the impact and penetration of renewable energy in the electric grid [1]–[6] and are the key factor for future success of the electric vehicles [7]–[9].

High power applications require series connection of the cells to obtain high voltage working voltages reducing the power losses due to joule effect losses. Series connection of cells decreases the total energy of the battery pack and reduces the performance of the system [10], [11]. The performance reduction is due to little manufacturing and environment differences that induce a mismatching between single cells characteristics as SOC, capacity and internal resistance differences [12]–[15].

In order to improve the available battery pack energy and performance, a balancing system is connected to the Battery Pack to reduce the differences and mismatch effect of series connected cells [16].

Energy storage balancing systems are divided in passive balancing systems and active balancing systems [17]–[19]. The main difference between both topologies is that passive balancing systems balance the series connected cells burning the extra energy of the most charged cells and the active balancing systems redistribute the energy of the strong cells to the weak cells. Passive balancing systems are widely used in industry applications due to simplicity, reliability and low cost characteristics.

Single switch active balancing systems present good characteristics as low complexity, low component number and the ability to balance the series connected cells without in open loop control mode [20], [21].

The main goal of this article is to compare passive and single switch active balancing systems behavior under different conditions. 3 main comparison topics are evaluated:

- *Energy considerations:* Charging energy W_C, discharging energy W_D, efficiency η and standby energy consumption W_S.

- *Temperature behavior:* Temperature gradient ΔT between series connected cells, maximum temperature of cells T_{Max} and temperature and losses generated in the balancing system T_{BS}.

- *Cost:* The total cost of the balancing system is evaluated C_ϵ.

Even though passive balancing systems are the most used and popular balancing systems in industrial and commercial applications, this paper will demonstrate the performance improvement of single switch active balancing systems, and will give the industry an interesting point of view of the benefits of active balancing vs. passive balancing systems, not only in terms of behavior, but also in terms of simplicity and cost.

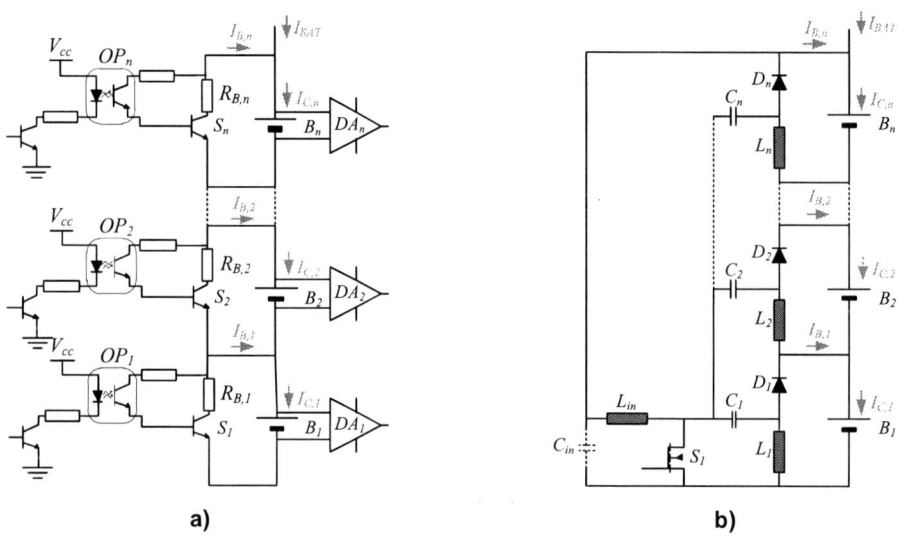

a)   **b)**

Fig. 1. Energy storage balancing systems under comparison. a) Passive balancing system. b) Sepic based single switch active balancing system.

TABLE I

ADVANTAGES AND DISADVANTAGES OF PASSIVE BALANCING SYSTEMS AND SEPIC BASED SINGLE SWITCH ACTIVE BALANCING SYSTEMS FROM LITERATURE KNOWLEDGE

	Passive balancing system	Sepic single switch active balancing system
Advantages	• Simple. • Cost.	• Simple. • Efficient. • Open loop balancing.
Disadvantages	• Energy wasting. • Useless in discharge.	• Only voltage balancing.

TABLE II

WEAK CELL $I_{B,W}$ BALANCING CURRENT AND STRONG CELL $I_{B,S}$ BALANCING CURRENT OF BALANCING SYSTEMS UNDER STUDY. 4S1P BATTERY PACK N=4 ONE WEAK CELL $V_{B,W}$=2V AND 3 STRONG CELLS $V_{B,S}$=3.65 V

	PB	AB
$I_{B,W}$ [mA]	0	365
$I_{B,S}$ [mA]	-304	-67

The paper starts with the main impact characteristics related to balancing systems under a theoretical point of view. Section 3 presents the experimental results of the balancing systems, and the comparison of their behavior related to the main impact characteristics. Last section presents the most important conclusions about the behavior and characteristics of the balancing systems.

2. Impact characteristics of balancing systems

Balancing systems are connected to energy storage systems with the main target of improving battery pack characteristics. The effect of the balancing system can be presented in

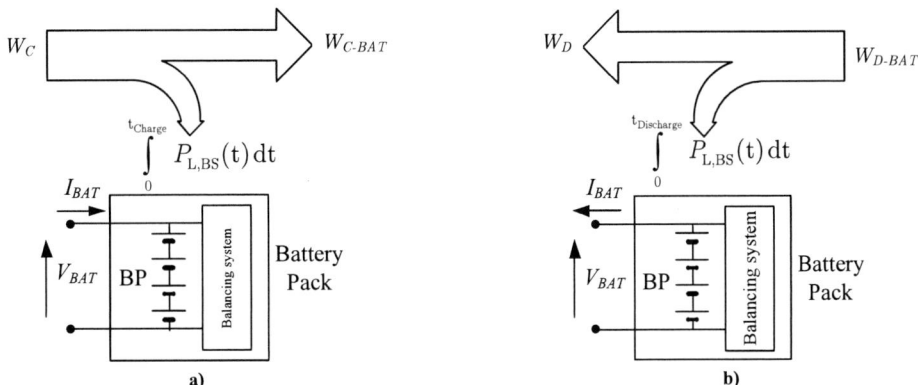

Fig. 2. Arrow diagram for the energetic evaluation of the battery pack cycling performance. $P_{L,BS}$ always considered positive losses a) Charge process (I_{BAT} positive. W_C and $W_{C\text{-}BAT}$ positive). b) Discharge process (I_{BAT} negative. W_D - $W_{D\text{-}BAT}$ negative)

improvements of power and energy specifications, energy efficiency of the system and extension of the cycle life.

2.1. Energy considerations

The energy evaluation of the battery pack is usually evaluated under two different criteria, considering cycle life and calendar life issues.

Cycle life issues will be related to the energy inserted to the battery pack during a charging process W_C (W_C considered positive), the energy obtained during discharge W_D (W_D considered negative) and the efficiency of the battery pack respect to the energy inserted and the energy discharged η.

Charging energy W_C is the total energy inserted from the point of view of a battery charger. The total energy inserted to the battery pack $W_{C\text{-}BAT}$ is measured by the integration of the voltage $V_{B,n}$ and the current $I_{C,n}$ of each cell. It is also the total energy measured by the charger W_C subtracting the energy lost by the power losses of the balancing system $P_{L,BS}$.

$$W_C = \int_0^{t_{charge}} V_{BAT}(t)I_{BAT}(t)dt \qquad W_{C-BAT} = \int_0^{t_{charge}} \sum_1^n V_{B,n}(t)I_{C,n}(t)\,dt = W_C - \int_0^{t_{charge}} P_{L,BS}(t)dt \qquad (1)$$

During the discharge process, active balancing redistributes energy to the weak cell improving the total discharge energy W_D by increasing the weak cell current $I_{C,w}$. As during charge, the battery discharged energy $W_{D\text{-}BAT}$ is reduced due to power losses of the balancing system $P_{L,BS}$. The term reduced means that it is more negative

$$W_D = \int_0^{t_{discharge}} V_{BAT}(t)I_{BAT}(t)dt \qquad W_{D-BAT} = \int_0^{t_{discharge}} \sum_1^n V_{B,n}(t)I_{C,n}(t)\,dt = W_D - \int_0^{t_{discharge}} P_{L,BS}(t)dt \qquad (2)$$

The efficiency η of the battery pack is the ratio between the energy inserted and the energy obtained from a battery pack.

$$\eta = \left| \frac{W_D}{W_C} \right| \cdot 100 \qquad (3)$$

This ratio defines the performance of the battery pack and the behavior due to the balancing system.

Calendar life behavior of the battery pack is represented by the aging process due to battery pack time degradation. For this issue the self-discharge current of the battery pack will be measured due to stand by energy discharge W_{SB}. The effect of the balancing system during the stand-by process should be analyzed to estimate their influence in the calendar life process.

2.2. Temperature behaviour

The balancing system influences in the temperature behaviour of the battery pack, in parameters as the maximum temperature T_{Max} and the deviation in temperature ΔT between different cells of the battery pack [22]. The power losses of the balancing system also increase the temperature of the balancing system board T_{BS} that could generate a hot spot for nearby positions cells.

T_{Max} and ΔT of the single cells composing a battery pack are directly dependent on the power losses of each single cell. The power losses of each cell are proportional to the internal resistance of the cell R_{in} and increase quadratically due to cell current $I_{C,n}$. The internal resistance could also be approached to the difference between the cell voltage V_B and the open circuit voltage V_{OC}.

$$P_{L,n} = R_{in}I_{C,n}{}^2 \approx \left|V_{B,n} - V_{OC}\right| I_{C,n} \tag{4}$$

Balancing systems equalize the voltage level of each cell, so they improve the temperature gradient between cells ΔT respect to no balancing systems, assuming equal aging of cells.

Maximum temperature T_{Max} is improved in series connected cells thanks to active balancing systems during discharge processes. Passive balancing systems do not deliver any balancing current to the weak cell, however active balancing systems insert current to the weak cell decreasing the total current that flows through the cell during discharge.

$$P_{L,W} = R_{in}I_{C,W}{}^2 = R_{in}\left(I_{BAT} + I_{B,W}\right)^2 \quad AB \quad P_{L,W} \downarrow\downarrow$$
$$P_{L,W} = R_{in}I_{C,W}{}^2 = R_{in}I_{BAT}{}^2 \qquad PB \quad P_{L,W} \approx cte \tag{5}$$

During charge both passive and active balancing systems decrease the maximum temperature of the cell due to decrease in current of the strong cell $I_{C,S}$.

$$P_{L,S} = R_{in}I_{C,s}{}^2 = R_{in}\left(I_{BAT} + I_{B,S}\right)^2 \quad PB, AB \quad P_{L,S} \downarrow\downarrow \tag{6}$$

High temperatures in the balancing system T_{BS} could induce a mismatch in the temperature of different cells of a battery pack due to heat concentration next to the balancing system. Passive balancing systems burn all the energy in the balancing resistor, while active balancing systems only generate heat due to power losses in the balancing converter. Passive balancing systems can generate a hot spot in the battery pack.

2.3. Cost

Balancing system cost is one of the most important parameters, if not the most important one, why passive balancing is the principal balancing architecture in industrial applications.

Active balancing systems usually are complex architectures requiring several active switches. Each active switch requires a high frequency isolated driver increasing the cost, and decreasing the reliability of the whole system.

Passive balancing systems typical architecture presented in Fig. 1 a) require an optocoupler OP_n to isolate the system and a power switch S_N in conjunction with a balancing resistor $R_{B,n}$ to discharge the excess energy of the cells. The voltage measurement of each cell is made via differential amplifiers DA_n for voltage balancing control, even if specific ICs for BMS operation have decreased the complexity and the price of passive balancing systems.

Single switch balancing topologies, as the Sepic architecture presented in Fig. 1 b) only need one single switch to balance one series string of cells. Voltage measurement is not necessary as the balancing process is made naturally and the switching of the active switch is made in open loop, decreasing the complexity of the driver and the control system.

The specifications of single switch open loop systems could be a key point as the main architecture to change the industry position from passive configurations to active balancing systems.

3. Experimental results and comparison

The evaluation and comparison of passive and active balancing systems is made under 3 different environments. The comparative evaluation is made for a 4S1P battery pack with the PB and AB systems presented in section II. The comparative results are presented in Fig. 4.

- *High power fresh cells (HPF)*: a 4S1P battery pack of 6.5 Ah LiFePO$_4$ fresh cells has been connected to prove the balancing systems behavior under a newly assembled battery pack.
- *High power aged cells (HPA)*: An 80% *SOH* 6.5Ah nominal capacity LiFePO$_4$ cell

TABLE III
TOTAL CHARGE, DISCHARGE AND EFFICIENCY BEHAVIOR OF 3 DIFFERENT 4S1P BATTERY PACKS UNDER CCCV CHARGE AND CC DISCHARGE PROCESSES.

	HPF		HPA		LPF	
	W_C	$W_{C\text{-}BAT}$	W_C	$W_{C\text{-}BAT}$	W_C	$W_{C\text{-}BAT}$
PB	97.89	97.21	70.68	68.76	13.79	13.64
AB	96.96	96.9	66.96	66.64	13.60	13.53
	W_D	$W_{D\text{-}BAT}$	W_D	$W_{D\text{-}BAT}$	W_D	$W_{D\text{-}BAT}$
PB	-92.28	-92.61	-63.25	-65.22	-12.59	-12.7
AB	-92.48	-92.5	-63.95	-64.02	-12.66	-12.69
	η		η		η	
PB	94.28		89.49		91.4	
AB	95.39		95.51		93.19	

has been connected with 3 fresh cells making a 4S1P battery pack with one cell in an advanced aging stage.

- *Low power fresh cells (LPF)*: A 4S1P battery pack of 1,1 Ah low capacity LiFePO4 cells has been implemented to view the effect in lower capacity cells, where the balancing current $I_{B,n}$ is closer to the nominal capacity of the cell.

3.1. Energy considerations

In order to evaluate energy characteristics related to cycle life and cycling behavior due to balancing systems a charge discharge cycle has been designed.

- *Charge cycle*: A 1C Constant Current CC until $V_{B,n}=3.65$ Constant Voltage CV until $I_{BAT}=C/20$ charge cycle has been defined to evaluate charging energy W_C.

- *Discharge cycle*: A 1C CC discharge cycle until $V_{B,n}=2V$ is designed to evaluate the total discharge energy W_D.

Both cycles are repeated 10 times to evaluate the balancing system performance during repetitive cycling. The balancing systems are controlled by voltage difference. The threshold voltage for balancing switching on is set to 10 mV difference between two cells. The 4S1P battery packs are cycled inside a temperature chamber with a constant 25°C ambient temperature.

During the charge process passive balancing consumes more energy from the charger to charge the battery pack. Even that energy excess consumption, the battery pack is more charged with *PB* than with the *AB*. However the charging efficiency is much lower in the *PB* system than in the *AB* due to higher power losses of the *PB*. The power losses in the *PB* system are easily calculated knowing the balancing resistor $R_{B,n}$, the cell voltage $V_{B,n}$ and the balancing time when *PB* is connected. The *AB* power losses are measured experimentally as 0.195 W for an unbalance situation of 2 V for the weak cell $V_{B,W}$ and 3.65 V for the strong cells (presented in Fig. 3). When AB is active constant 0.195 W power losses are considered, even the real losses will be smaller. However, even if the power losses are overestimated the *AB* power losses are much lower than the *PB* system power losses.

Discharge process is greatly improved by the *AB*. The passive balancing only burns energy in order to decrease the voltage difference between cells reducing the total energy of the battery pack. The *AB* redistributes the energy to the lowest voltage cell, increasing the discharge time and the total discharge energy. The power losses during discharge are higher in *PB* than in *AB*.

TABLE IV
TOTAL DISCHARGED ENERGY W_D AND ENERGY WASTED DURING STANDBY OPERATION W_{SB} FOR A 4S1P LiFePO4 BATTERY PACK WITHOUT BALANCING SYSTEM, WITH *PB* SYSTEM AND WITH *AB* SYSTEM.

	W_D	W_{SB}
No balancing	-91.33 (100%)	0 (0%)
PB	-91.13 (99,78%)	0,2 (0,22%)
AB	-91.24 (99,9%)	0,09 (0,1%)

The cycling results presented in Table III, conclude that -92.48 Wh are discharged W_D with the *PB* system, 0,2 Wh less than with the *AB* system although 97.89 Wh are charged, 0,93 Wh more than with the *AB* for *HPF* cells.

The results with *HPA* cells are even better for the *AB* system. W_C=70.68 Wh are charged with the PB system, 3.72 Wh more than with the *AB* system. However the *AB* system discharges W_D=-63.95 Wh 0.6 Wh more than the *PB* system.

The results for the *LPF* cells are also superior for *AB* system compared with *PB* system. The charged energy W_C is decreased 0.19 Wh for the *AB* system, and the discharged energy increased W_D is increased 0.07 Wh in *AB* system.

The *AB* system charges less energy in the battery pack and discharges more energy being superior than the *PB* system.

Due to the excessive energy wasted, during charge and discharge process in *PB*, the battery pack efficiency is greatly improved with the *AB*. +1,11% with *HPF* cells, 6,02% for *HPA* cells and 1,79% for *LPF* cells. The *AB* system is much superior than the *PB* system in efficiency requirements. For continuous charge discharge applications as electro mobility, the efficiency requirement is primordial.

For standby energy consumption W_{SD}, 4S1P battery pack has been stored fully charged at 25 °C ambient temperature, during the standby period *PB* and *AB* balancing systems have been connected to evaluate the leakage current of the balancing systems. The balancing systems have been connected during 3 weeks consecutive periods. After the storage time a full discharge of the battery pack is evaluated, with 1C constant current, to compare the energy decrease with a non-balancing system battery pack.

The results presented in Table IV conclude that the balancing system do not contribute in an accelerated self-discharge process of the battery pack. The PB system increases the self-discharge only 0.22% and the AB system 0.1% respect to a 4S1P battery pack without balancing system.

3.2. Temperature behaviour

To compare the temperature improvement and behavior of the 4S1P battery packs due to balancing systems, 2 K type thermocouples have been connected to each cell to evaluate the maximum temperature T_{max} in the surface and the temperature distribution ΔT between the 4S1P cells.

In order to avoid temperature influence between nearby positioned cells, the cells are distanced 2 cm between them. Distancing the series connected cells isolates the impact of each cell from the heat generation of adjacent cells, taking only into account the performance of the balancing system for the temperature behavior.

The maximum temperature T_{max} and the maximum temperature gradient ΔT, take place at the end of the discharge process. The temperature behavior has been evaluated during the cycles presented for the energy considerations during the previous subchapter. The temperature values are the mean value of the 10 consecutive discharge cycles. The mean value permits to filter temperature measurement errors and dispersion.

Table V presents the temperature behavior results for 10 consecutive charge discharge cycles. For *HPF* cells *PB* reaches lower maximum temperature than *AB*. This result conflicts with equation (5). However the temperature increase in *AB* could be generated due to a deeper SOC reached (Lower V_{OC}), and higher R_{in} reached. The temperature dispersion between cells is reduced from 0.4 °C on the *PB* to 0.3 °C on the *AB*.

The results for the *HPA* configuration present that the PB reaches higher temperature than AB, if an aged cell is presented in the battery module. The temperature dispersion increases more than 1.5 °C from the *HPF* case, so it is also concluded that low SOH dispersion cells

TABLE V

TEMPERATURE BEHAVIOR DURING END OF DISCHARGE OF 3 DIFFERENT 4S1P BATTERY PACKS UNDER CCCV CHARGE AND CC DISCHARGE PROCESSES.

	HPF		HPA		LPF	
	T_{max}	ΔT	T_{max}	ΔT	T_{max}	ΔT
PB	28.61	0.4	29.59	2.07	27,86	0,59
AB	29.23	0.3	29.53	1.95	27,64	0,54

TABLE VI
COST DISTRIBUTION OF BALANCING SYSTEMS.

	PB			AB	
	Part ref.	€/pcs		Part ref.	€/pcs
OP_n	TLP523	0.71	D_n	SL13-E3/61T	0.26
$R_{B,n}$	ER7412RJT	0.28	L_n	74459168 Würth	2.18
S_n	BD437	0.233	C_n	C1206C106K3PA CTU	0.19 5
DA_n	INA 148	3.49	L_{in}^{*}	74459247 Würth	2.02
			S_1^{*}	IRF8721PBF	0.6
4S1P	18.852 €		4S1P	13.16 €	

PB		AB	
$P_{L,BS}$	T_{BS}	$P_{L,BS}$	T_{BS}
3.5 W	119.3 °	0.195 W	38.2 °

Fig. 3. Thermography camera photo for temperature measurement of the *PB* and *AB* under maximum unbalancing situation. $P_{L,BS}$ and T_{BS} are measured under unbalancing conditions..

Price of components for minimum order of 10 pieces in www.farnell.com
* Only one element per battery pack

have lower temperature dispersions. The temperature dispersion ΔT is 0.12ºC lower for *AB* than the 2.07 ºC dispersion presented for the *PB* system configuration.

For the *LPF* battery module configuration the *AB* presents better characteristics regarding T_{Max} and ΔT compared to the *PB* system. The balancing current rating of the *AB* system is closer to the nominal capacity of the *LPF* configuration than for the *HPF* configuration, presenting better results than the *PB* system for fresh cells. 0.22 ºC less respect to T_{Max} and 0.05 ºC less temperature dispersion ΔT.

The power losses $P_{L,BS}$ of the balancing system could generate a temperature increase near the battery pack decreasing the life span of the battery by accelerating aging mechanisms. In order to evaluate the temperature generated in the balancing system T_{BS} a test bench with 4 power supplies connected in series has been designed. The power supplies are bidirectional with power sinking capability. 3 power supplies are set to 3.65 V and the weak cell is simulated to 2V to evaluate the extreme unbalancing situation.

The power losses of the balancing circuits are experimentally measured and the temperature hot spots in the balancing system are measured by a thermography camera Fig. 3.

The *AB* system has a hot spot of 38.2 ºC, with total power losses of 0.195 W. The hot spot is presented in the mosfet driver and in the diode which inserts energy in the simulated 2 V weak cell. The *PB* system dissipates 3.5 W in 3 balancing resistors, generating 3 dangerous hot spots of 119.3 ºC. The excess temperature in the *PB* could increase the temperature of the battery pack significantly, or even could make a temperature unbalance between cells of a battery pack. The losses are independent of the battery pack cell, only depend on the battery pack cell voltage.

3.3. Cost

Cost issue is the biggest constraint, why *PB* is the most used balancing system and why is widely used in industry applications. However, single switch balancing systems are good candidates to deal with cost issues. A voltage measurement system is not required. Single switch systems operate in open loop, reducing the complexity of the control system.

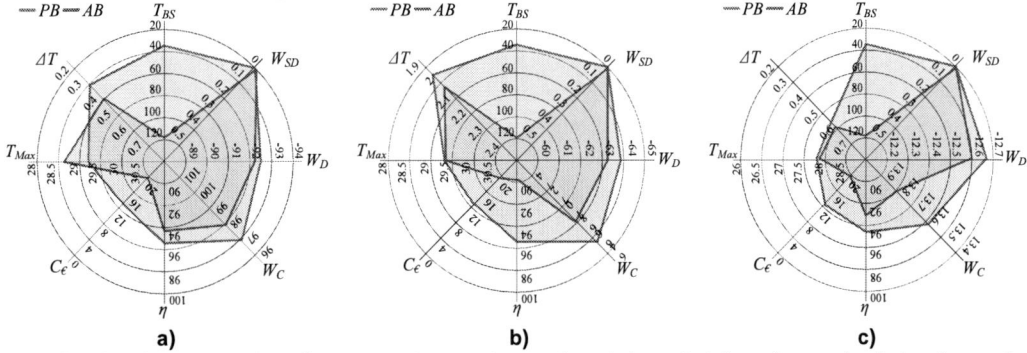

Fig. 4. Spyder chart comparison for energy, temperature and cost characteristics of a passive balancing system and an active single switch Sepic balancing system. a) High Power Fresh cells *HPF* b) High Power Aged cells *HPA* c) Low power fresh cells *LPF*.

To evaluate the cost of the *PB* and *AB* systems, the cost of each cell balancing unit will be evaluated, taking into account the price per each element.

- *PB*: Optocoupler OP_n, Balancing resistor $R_{B,n}$, power switch S_n and differential amplifier DA_n per each cell.

- *AB*: Diode D_n, inductor L_n and capacitor C_n per each cell, and one input inductor L_{in} and one power switch S_1 per battery pack.

The results presented in Table VI present that the *AB* is cheaper 13.16 € than the *PB*, mainly due to the high price of differential amplifiers DA_n. *AB* open loop control strategy avoids the use of a measurement system for each cell voltage, however due to the natural unsafe behavior of Li-ion cells voltage monitoring is indispensable. If a specific Battery Management System IC is used for *PB* systems the *PB* cost could be reduced.

4. Conclusions

A comparative study has been presented where a single switch Multi stacked Sepic active balancing system has been compared with a passive balancing system. Single switch active balancing systems are a good candidate to substitute passive balancing due to low complexity and open loop control in DCM.

The main conclusion claims that active single switch active balancing are good candidates to replace passive balancing systems in industrial applications.

The passive balancing system and the active balancing system are compared in 3 different scenarios with high power fresh cells, high power aged cells and low power fresh cell in a 4S1P configuration. The 3 different configurations are compared for energetic considerations temperature behavior and cost issues.

Regarding to energetic results, active balancing is much superior to the passive balancing system. Inserts less energy from the battery charger (reducing the electric bill), and increases the discharged energy (increasing battery energetic availability). The efficiency of the battery pack is greatly increased. 1% for high power fresh cells, 6 % for high power aged cells and nearly 2 % for low power fresh cells. Active balancing and passive balancing systems connection to a battery pack do not affect in self-discharge increase.

Temperature behavior results present a better behavior of the passive balancing system only regarding maximum temperature in high power fresh cells. Maximum temperature in aged and low power cells is reduced in active balancing system respect to passive balancing system. The temperature dispersion between cells is always lower in the active balancing system than in the tested passive balancing system. The balancing system temperature is dramatically reduced from 119.3 °C in the passive balancing system to 38.2 °C in the active balancing system, reducing the possibility of generating hot spots inside the battery pack.

The cost issue presents a lower cost for the active balancing system 13.16 € than for the passive balancing system 18.852 €. This is because single switch balancing systems can balance the cells without the need of a voltage measurement system. However the natural unsafety behavior of Li-ion cells forces to use a voltage monitoring system. Passive balancing system price could also be reduced by the use of commercial ICs for battery management systems.

References

[1]   G. Xu, L. Xu, D. J. Morrow, and D. Chen, "Coordinated DC Voltage Control of Wind Turbine With Embedded Energy Storage System," *Energy Conversion, IEEE Transactions on*, vol. 27, no. 4. pp. 1036–1045, 2012.

[2]   C. Abbey, K. Strunz, and G. Joos, "A Knowledge-Based Approach for Control of Two-Level Energy Storage for Wind Energy Systems," *Energy Conversion, IEEE Transactions on*, vol. 24, no. 2. pp. 539–547, 2009.

[3]   F. Giraud and Z. M. Salameh, "Steady-state performance of a grid-connected rooftop hybrid

wind-photovoltaic power system with battery storage," *Energy Conversion, IEEE Transactions on*, vol. 16, no. 1. pp. 1–7, 2001.

[4]   T.-Y. Lee and N. Chen, "The effect of pumped storage and battery energy storage systems on hydrothermal generation coordination," *Energy Conversion, IEEE Transactions on*, vol. 7, no. 4. pp. 631–637, 1992.

[5]   S. Schoenung and C. Burns, "Utility energy storage applications studies," *Energy Conversion, IEEE Transactions on*, vol. 11, no. 3. pp. 658–665, 1996.

[6]   D. Wu, F. Tang, T. Dragicevic, J. C. Vasquez, and J. M. Guerrero, "Autonomous Active Power Control for Islanded AC Microgrids With Photovoltaic Generation and Energy Storage System," *Energy Conversion, IEEE Transactions on*, vol. 29, no. 4. pp. 882–892, 2014.

[7]   W. A. Lynch and Z. M. Salameh, "Realistic electric vehicle battery evaluation," *Energy Conversion, IEEE Transactions on*, vol. 12, no. 4. pp. 407–412, 1997.

[8]   K. Thirugnanam, J. T. P. Ezhil Reena, M. Singh, and P. Kumar, "Mathematical Modeling of Li-Ion Battery Using Genetic Algorithm Approach for V2G Applications," *Energy Conversion, IEEE Transactions on*, vol. 29, no. 2. pp. 332–343, 2014.

[9]   C. Zhou, K. Qian, M. Allan, and W. Zhou, "Modeling of the Cost of EV Battery Wear Due to V2G Application in Power Systems," *Energy Conversion, IEEE Transactions on*, vol. 26, no. 4. pp. 1041–1050, 2011.

[10]   Y. Barsukov, "Battery cell balancing: what to balance and how," *Texas Instruments*, 2005.

[11]   C. Martinez, "Cell Balancing Maximizes The Capacity Of Multi-Cell Li-Ion Battery Packs," *Intersil. Inc.*

[12]   J. R. Belt, C. D. Ho, T. J. Miller, M. A. Habib, and T. Q. Duong, "The effect of temperature on capacity and power in cycled lithium ion batteries," *J. Power Sources*, vol. 142, no. 1–2, pp. 354–360, Mar. 2005.

[13]   M. Uno and K. Tanaka, "Influence of High-Frequency Charge-Discharge Cycling Induced by Cell Voltage Equalizers on the Life Performance of Lithium-Ion Cells," *Vehicular Technology, IEEE Transactions on*, vol. 60, no. 4. pp. 1505–1515, 2011.

[14]   S. Santhanagopalan and R. E. White, "Quantifying Cell-to-Cell Variations in Lithium Ion Batteries," *Int. J. Electrochem.*, vol. 2012, pp. 1–10, 2012.

[15]   I. Aizpuru, U. Iraola, J. M. Canales, E. Unamuno, and I. Gil, "Battery pack tests to detect unbalancing effects in series connected Li-ion cells," *Clean Electrical Power (ICCEP), 2011 International Conference on*. 2013.

[16]   M. Einhorn, W. Roessler, and J. Fleig, "Improved Performance of Serially Connected Li-Ion Batteries With Active Cell Balancing in Electric Vehicles," *Veh. Technol. IEEE Trans.*, vol. 60, no. 6, pp. 2448–2457, 2011.

[17]   S. Moore, "A review of cell equalization methods for lithium ion and lithium polymer battery systems," 2001.

[18]   B. Lindemark, "Individual cell voltage equalizers (ICE) for reliable battery performance," *Telecommunications Energy Conference, 1991. INTELEC '91., 13th International*. pp. 196–201, 1991.

[19]   W. C. Lee, D. Drury, and P. Mellor, "Comparison of passive cell balancing and active cell balancing for automotive batteries," *Vehicle Power and Propulsion Conference (VPPC), 2011 IEEE*. pp. 1–7, 2011.

[20]   M. Uno and K. Tanaka, "Single-Switch Cell Voltage Equalizer Using Multistacked Buck-Boost Converters Operating in Discontinuous Conduction Mode for Series-Connected Energy Storage Cells," *Vehicular Technology, IEEE Transactions on*, vol. 60, no. 8. pp. 3635–3645, 2011.

[21]   M. Uno and K. Tanaka, "Single-Switch Multi-Output Charger Using Voltage Multiplier for Series-Connected Lithium-Ion Battery/Supercapacitor Equalization," *Industrial Electronics, IEEE Transactions on*, vol. PP, no. 99. p. 1, 2012.

[22]   U. Iraola, I. Aizpuru, L. Gorrotxategi, J. M. C. Segade, A. E. Larrazabal, and I. Gil, "Influence of Voltage Balancing on the Temperature Distribution of a Li-Ion Battery Module," *Energy Conversion, IEEE Transactions on*, vol. PP, no. 99. pp. 1–8, 2014.

PCIM Europe 2016, 10 – 12 May 2016, Nuremberg, Germany

Voltage levels comparison and system optimization for electric drives in hybrid vehicles

Quentin Werner, Daimler AG / Univ. Lorraine, France, quentin.werner@daimler.com
Serge Pierfederici, Univ. Lorraine, France, serge.pierfederici@univ-lorraine.fr
Noureddine Takorabet, Univ. Lorraine, France, noureddine.takorabet@univ-lorraine.fr

Abstract

The main focus of this paper is on the global system optimization of hybrid electric vehicle drivetrain and the voltage level of the hybrid system. Utilizing the previous research on voltage levels [1], this paper shows the different methods to globally and simultaneously evaluate the integration and efficiency of the electric machine and inverter. A comparison and evaluation of typical automotive voltage levels is completed and potential for further research is suggested.

1. Introduction

1.1. Motivation

The hybrid vehicle market has quickly grown and extensive research on power levels and topologies have been done as in [2] and [11]. Besides vehicle optimization, deep analyses have been individually done for electric machines, power electronics and battery optimization as in [3]. However, an important question remains: what is the optimal voltage level for hybrid electric vehicles? Despite contributing interesting results about the voltage dependency of the efficiency, the previous research in [1] only superficially investigate the integration issues and the interactions with the electric machine characteristics.

1.2. Boundary conditions and automotive industry requirements

All the investigations in this paper are done for the following drivetrain topology: a combustion engine on the front axle and an electric machine on the rear axle.

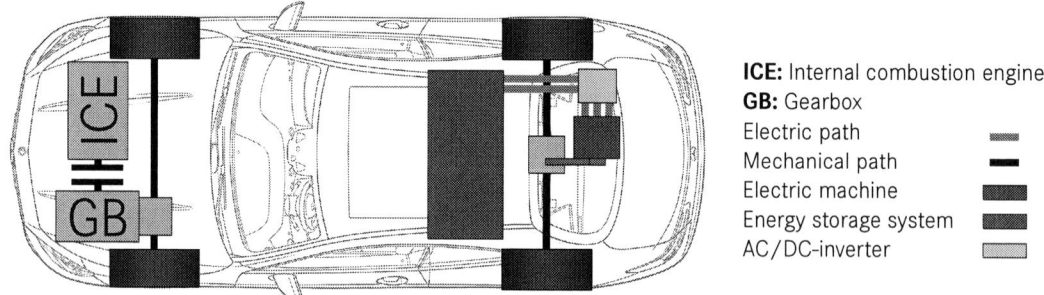

ICE: Internal combustion engine
GB: Gearbox
Electric path
Mechanical path
Electric machine
Energy storage system
AC/DC-inverter

Fig. 1. Hybrid drivetrain investigated

The investigation in this paper consists in a voltage levels analysis and is performed in two independent parts: First for a constant apparent power (systems A and B) and secondly with

© VDE VERLAG GMBH · Berlin · Offenbach

a constant mechanical power (Systems A and D or systems B and C) as shown in Table 1. The focus is set on the electric drive components: electric machine and the AC/DC-inverter. The aim is to optimize the global system by analyzing the different voltage levels. No investigation of the electric machine design is performed and the different results are deduced from a single, fixed induction machine design, whose characteristics were previously calculated with FEM-software. Thanks to the chosen architecture, the adaptation to mechanical requirements is done through variations of the gear ratio value.

System	A	B	C	D
Apparent power [kVA]	111	111	185	80
Mechanical power [kW]	80	100	100	80
AC current [A$_{rms}$]	300	150	500	105
DC voltage [V]	350	700	350	700

Table 1: Systems under investigation

The following criterions are evaluated: efficiency, dimensions and weight. Contrary to previous investigations in [1], the voltage levels are limited to two values: 350V and 700V, which correspond to the nominal voltage for systems using current 600V- and 1200V automotive certified semiconductor technology. These two voltage levels were chosen because they represent available components under automotive safety norms as shown in [4] and [8] and thus depict the potential of commercially available technology.

2. Adapted approach and modelling for system comparison

The analysis of global systems requires specific approach and adapted modelling method in order to produce consistent results. The aim is to find a compromise between the following requirements: simulation effort (complexity and time), parameter range and accuracy. In this part, the resulting approaches for the estimation of both weight and volume as well as efficiency evaluation is discussed.

2.1. Components volume and weight evaluation

When considering the integration aspects, the influence on the global electric drive volume needs to be evaluated. Since the electric machine design is fixed, only the influence of voltage and current of the AC/DC-inverter is studied in this paper. The aim is to enhance the works previously done in [1] by proposing an approach that can evaluate the whole component volume and weight. To achieve this goal, the development of an adapted method is required. For each component, weight and dimensions are estimated based on both electric requirements and safety norms. In this paper, the method is described in detail for the DC link capacitor and the power module (as examples) before detailing the approach for the estimation of the entire system (AC/DC-inverter).

DC link capacitor volume and weight estimation

An important parameter for the design of the DC link capacitor is the stress current. In [7], the authors developed an analytic method which enables calculating this stress current (I_{Crms}). It can be estimated based on the inverter output current (I_{Nrms}), the modulation index (M) and the power factor ($cos\varphi$) of the system as depicted by the following formula:

$$I_{Crms} = I_{Nrms}\sqrt{\left[2M\left\{\frac{\sqrt{3}}{4\pi} + cos^2\varphi\left(\frac{\sqrt{3}}{\pi} - \frac{9}{16}\right)\right\}\right]}$$

According to [7], the results can simplified based on the electric machine technology and can be approximated in the case of an asynchronous motor by the following equation:

$$I_{Crms} \approx \frac{1}{2} I_{Nrms}$$

The influence of the voltage is expressed in the calculation of the capacitance as described by the following formula, where the capacitance (C) is a function of the stress current, the switching frequency ($f_{switching}$) and the ripple voltage (ΔU):

$$C = \frac{I_{Crms}}{2\,\pi\,f_{switching}\,\Delta U}$$

In [8], the ripple voltage (peak-to-peak) is limited to 8V for 350V-systems and thus it is supposed to be less or equal to 2% of the maximum system voltage in this work. The requirements are then derived for the 700V-systems which results in a value of 16V. The voltage level influences the capacitor design when estimating the required capacitance. Based on the analysis in [9], a film capacitor technology is chosen and the C4AE series from the firm KEMET is taken as example for the DC-link capacitor estimation [10]. As it can be seen on the specification sheet, the capacitors are designed for different values of rated DC-voltage. The density of the capacitor is inversely proportional to voltage and thus higher voltage level will directly impact the DC-link capacitor volume and weight. Based on the previous formulas and capacitor specifications, the complex relations between current, voltage, the capacitor volume and weight can be calculated. The results for the capacitor volume are represented in the following diagram.

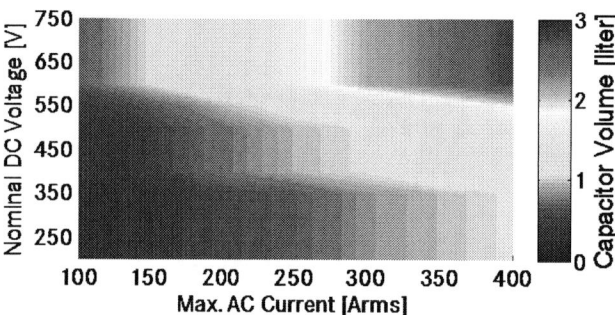

Fig. 2. Capacitor volume investigation for different voltage and current levels

The discontinuities in the capacitor volume are due to the capacitor specifications; the suppliers do not develop components for a specific voltage value but for pre-defined ranges. Each discontinuity represents a change in the capacitor rated voltage and thus a decrease of its capacitance per volume. This analytic approach, despite only representing an approximation, enables providing initial trends to analyze the voltage and current level influences during the system design process. Similar methods can be used on other passive elements but not for the power modules which impose considering additional safety requirements.

Power module choice under different voltage level requirements

The determination of adapted power module for a hybrid vehicle application depends on electrical and safety requirements. The safety requirements as explained in [4] impose defining voltage ranges in order to achieve the required reliability as depicted in Fig. 3. The application of these requirements on the IGBT-modules results in the use of 600V IGBTs for a 350V nominal voltage system and 1200V IGBTs for a 700V nominal voltage system.

© VDE VERLAG GMBH · Berlin · Offenbach

PCIM Europe 2016, 10 – 12 May 2016, Nuremberg, Germany

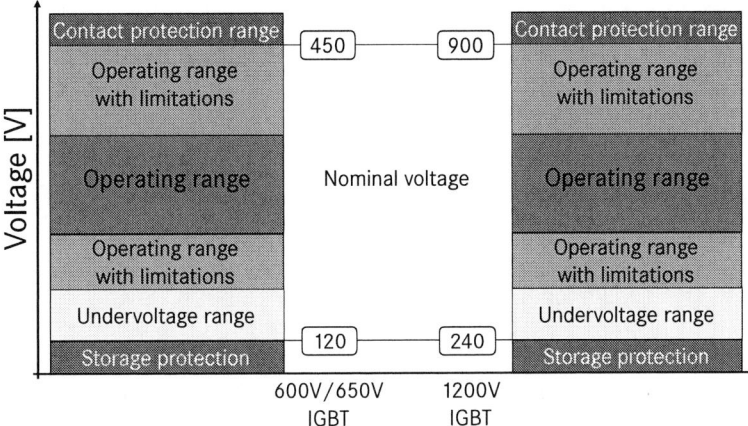

Fig. 3. Voltage range for 600/650V IGBT and 1200V IGBT adapted from [4] and [8]

AC/DC-inverter volume and weight estimation

Based on the two previous methods and on present inverter design and construction, an algorithm is created to estimate the entire inverter volume. The principle of this algorithm is as follows: using three architectures, the different components are spatially organized in a geometrical frame in order to estimate their position relative to the others. These positions are calculated using safety requirements concerning the creepage and clearance distances which are dependent on the voltage level. Then the busbar dimensions are calculated based on the current. Finally, weight is calculated based on materials density or the specified weight for the passive components, the connectors and the control boards.

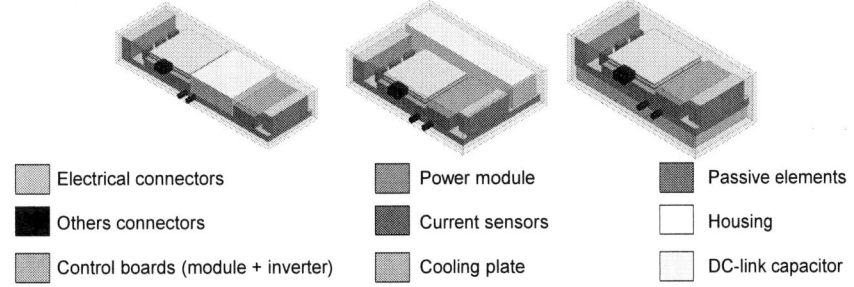

▢ Electrical connectors	▢ Power module	▢ Passive elements
■ Others connectors	▢ Current sensors	▢ Housing
▢ Control boards (module + inverter)	▢ Cooling plate	▢ DC-link capacitor

Fig. 4. Architectures for the AC/DC-inverter volume and weight estimation

The method was validated based on current benchmark automotive components and the validation is presented for two components:

		Length [mm]	Width [mm]	Height [mm]	Volume [Liter]	Weight [kg]
Automotive component 1	**Reference**	250	165	155	6.4	8.1
	Simulation	236	158	150	5.6	8.2
Automotive component 2	**Reference**	240	155	135	5.0	7.5
	Simulation	236	150	126	4.5	7.6

Table 2: Validation of the AC/DC-inverter dimensions and weight estimation

© VDE VERLAG GMBH · Berlin · Offenbach

The simulation results show a light deviation from the reference but the deviation remains under 10%. Using this method, the voltage and the influence on the whole component design can be evaluated. Doing this requires some approximations of the component's sub-parts. This first approximation is however sufficient to address initial system integration plausibility in the early development phase.

2.2. Efficiency evaluation

The second part of the voltage and current investigation consists in the evaluation of the system efficiency. The evaluation is based on the New European Driving Cycle and its mean value on the cycle is considered as the criteria. The electric machine characteristics are given and thanks to a well-defined interface, the interactions between the machine and its associated inverter can be accurately modelled. The losses are estimated using the approach defined in [5] by the following formulas:

$$\boldsymbol{P_{CT}} = U_{C0} \cdot I_0 \cdot \left(\frac{1}{2\pi} + \frac{m_i \cdot \cos\varphi}{8}\right) + R_C \cdot I_0^2 \cdot \left(\frac{1}{8} + \frac{m_i \cdot \cos\varphi}{3\pi}\right)$$

$$\boldsymbol{P_{CD}} = U_{D0} \cdot I_0 \cdot \left(\frac{1}{2\pi} - \frac{m_i \cdot \cos\varphi}{8}\right) + R_D \cdot I_0^2 \cdot \left(\frac{1}{8} - \frac{m_i \cdot \cos\varphi}{3\pi}\right)$$

$$\boldsymbol{P_{SW}} = \frac{1}{\pi} \cdot \frac{U_{DC}}{U_{nom}} \cdot \frac{I_{AC}}{I_{nom}} \cdot \left(E_{on} + E_{off} + E_{rec}\right) \cdot f_{sw}$$

$$with: I_0 = I_{ACrms} \cdot \sqrt{2}$$

P_{CT} and P_{CD} : Conduction losses
P_{SW}: Switching losses
U_{C0}, R_C: Transistor parameters
U_{D0}, R_D: Diode parameters
m_i: Motor displacement factor
$\cos\phi$: Power factor
U_{DC}: Input voltage
I_{AC} and I_{ACrms}: Output current and its rms value
E_{on}, E_{off} E_{rec}: Switching energy for the value Unom, Inom

Considering the requirements for the power modules, two of them are considered in this work: One for each voltage level. The parameters for the losses estimation are extracted from the FS400R07A1E3_H5 and FF600R12ME4_B11 IGBT-modules from INFINEON [6]. Semiconductors losses are enhanced with the addition of the bus-bar losses which are evaluated as copper losses (for both AC and DC bus-bar):

$$\boldsymbol{P_{bus-bar}} = R_{bus-bar} \cdot I_{bus-bar}^2$$

From with the total losses can be deduced:

$$\boldsymbol{P_{Total}} = P_{CT} + P_{CD} + P_{SW} + P_{bus-bar}$$

On the following diagram, the inverter efficiencies for the different system are shown:

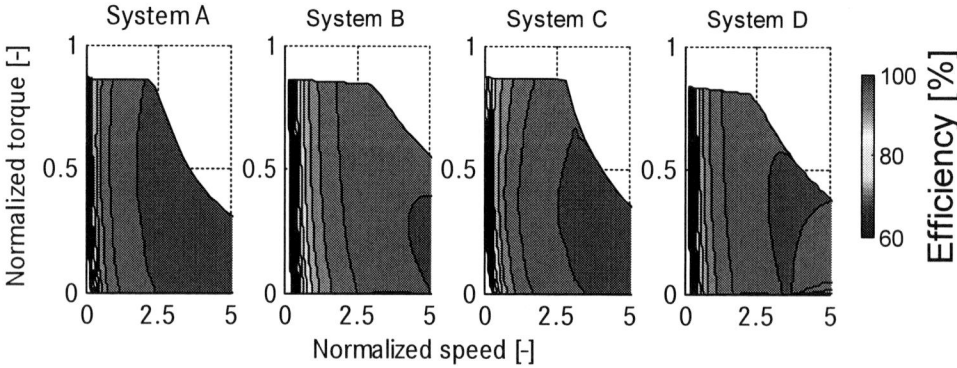

Fig. 5.   Efficiency AC/DC-inverter for two different IGBT-modules (600/650V and 1200V)

© VDE VERLAG GMBH · Berlin · Offenbach

As shown on Fig. 5, the efficiency shows globally a similar behavior over the speed range and the modelling was previously validated based on test-bench measurements. The validation was done for different voltage level for the 600/650V module and then applied in a same manner to estimate the losses of the 1200V module. The results of the validation are presented on the following diagram where the modelling results are shown with and without the bus-bar losses.

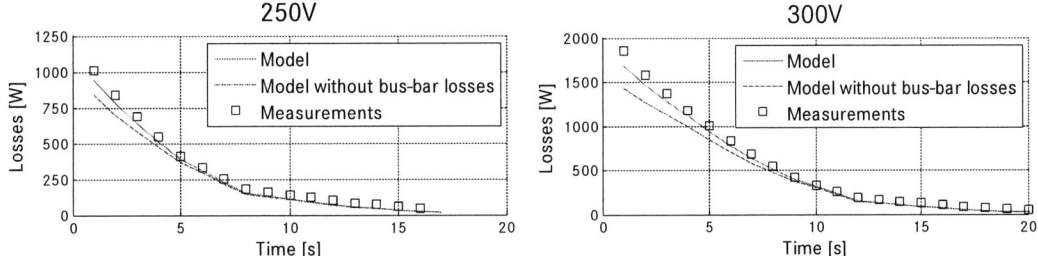

Fig. 6. Validation of the modelling with measurements for two different voltage levels

The importance of the bus-bar losses plays a significant role especially in the high stress areas. This enhancement of the losses plays a non-negligible role for the evaluation of the global system because the modelling can evaluate the component behavior as well as its interactions and its connection with the rest of the system.

3. Results of the voltage level investigation

By choosing and developing methods which shows a good compromise between accuracy and parameter range, an investigation of the global system was possible. The voltage and current influence on the whole electric drive can be evaluated and the results are shown on the following figure and table.

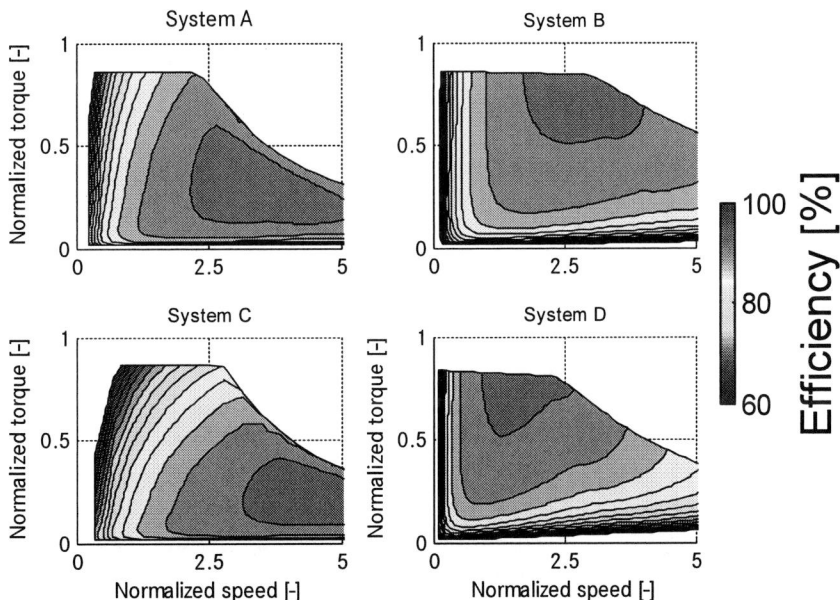

Fig. 7: System efficiency for the solutions under consideration

System	A	B	C	D
AC current [A$_{rms}$]	300	150	500	105
DC voltage [V]	350	700	350	700
NEDC mean efficiency [%]	88.3	75,8	88.1	71.6
Inverter volume [dm³]	5.9	7.0	6.2	6.7
Inverter weight [kg]	9.5	10	9.9	9.8
Power density [kW/dm³]	13.6	14.3	16.1	12.0

Table 3: Integration and efficiency evaluation of the electric drives

Under the boundary conditions defined for this paper, the 700V-systems show globally a decrease of the efficiency for a comparable integration. When comparing the systems with the same apparent power, the system B reaches higher machine utilization and a better power density as the system A. However for a hybrid electrical system in automotive industry, parameters such as the volume, the weight and the efficiency are far more important than the power density. Reaching a higher power density is always an improvement but can be insignificant if only 6.5dm³ is available. When comparing the systems with the same mechanical power (A and D or B and C), the higher machine utilization potential of the 700V-systems was no longer the case and the 350V-systems show higher potential.

4. Conclusion and overview

This paper enhances the previous investigations on the voltage level by introducing methods and modelling to evaluate both integration and efficiency characteristics of electric drives. As presented in this article, higher voltages demonstrate potential for electric machine utilization; however the examples in this paper do not improve the integration or the efficiency of the electric drive. The higher machine utilization of the 700V-systems needs to be further investigated and is limited in this paper Due to the use of a singular, fixed electric machine design and the research focus on the evaluation of integration and efficiency.

This work presents the method to globally and simultaneously evaluate the integration and efficiency of the electric machine and inverter. By doing the suitable approximations, the voltage level can be investigated and first indication can be done. Despite the advantages shown by the 350V-systems (lower volume and higher efficiency), no final recommendations can be made concerning the voltage level. A more detailed investigation of the machine parameter could lead to efficiency improvements and thereby increase the appeal of 700V-systems.

5. References

[1] U. Schwalbe and M. Schilling,
"Topology comparison and system optimization for a modular 25 kW motor–inverter drive train system", in PCIM Europe, Nuremberg (Germany), 2015.

[2] L. Fang, S. Qin, G. Xu, T. Li and K. Zhu;
"Simultaneous Optimization for Hybrid Electric Vehicle Parameters Based on Multi-Objective Genetic Algorithms", in Energies 2011, 4 (Journal), ISSN 1996-1073

[3] U. Vollmer;
"Entwurf, Auslegung und Realisierung eines verlustoptimierten elektrische Antriebs für Hybridfahrzeuge", Phd. thesis, 29.06.2012

[4] R. Foley, R. Nagappala, G. Ressler, P. Andres and B. Maretel,
"Application of Insulation Standards to High Voltage Automotive Applications", in SAE International, 04/08/2013 (date of publication)

[5] J. W. Kolar, H. Ertl and F. C. Zach,
"Influence of the Modulation Method on the Conduction and Switching Losses of A PWM Converter System", in IEEE transaction on industry applications, Vol. 27, No. 6, 1991.

[6] Infineon IGBT module specification sheets:
http://www.infineon.com/cms/de/product/power/igbt/automotive-igbts/automotive-igbt-module/channel.html?channel=db3a30432ba3fa6f012be33e87b75915

[7] J.W. Kolar and S. D. Round,
"Analytical calculation of the RMS current stress on the DC-link capacitor of voltage PWM converter systems", in IEE Proc.-Electr. Power Appl., Vol. 153, No. 4, July 2006

[8] VDA LV123 norm;
Electrical characteristics and electrical safety of high voltage components in road vehicles;

[9] M. März, A- Schletz, B. Eckardt, S. Egelkraut and H. Rauh;
"Power Electronics System Integration for Electric and Hybrid Vehicles", in Integrated Power Electronics Systems (CIPS), 2010 6th International Conference on, Fraunhofer Inst. of Integrated Syst. & Device Technol., Erlangen, Germany,

[10] KEMET capacitors specification sheets:
http://www.kemet.com/Lists/ProductCatalog/Attachments/366/KEM_F3046_C4AE_RADIAL.pdf

[11] W. Gao and C. Mi;
"Hybrid vehicle design using global optimisation algorithm", Int. J. Electric and Hybrid Vehicles, Vol. 1, No. 1, 2007

Real-Time Capable Model Predictive Control of Permanent Magnet Synchronous Motors Using Particle Swarm Optimisation

*Oliver Wallscheid, **Ulrich Ammann, *Joachim Böcker
*Paderborn University, Power Electronics and Electrical Drives, D-33095 Paderborn, Germany
**Esslingen University of Applied Sciences, D-73037 Göppingen, Germany
wallscheid@lea.upb.de, ulrich.ammann@hs-esslingen.de, boecker@lea.upb.de

Abstract

Permanent magnet synchronous motors (PMSM) are widely used in many industrial applications due to their attractive power and torque densities. To fully exploit their torque dynamics, non-linear model predictive control (MPC) approaches have been investigated in the last years. Besides the requirement of an accurate motor model solving a suitable cost function under real-time conditions is a challenging task, since highly-utilised PMSM exhibit a strongly non-linear behaviour due to (cross-)saturation effects. In this contribution a particle swarm optimisation (PSO) based MPC using multiple sub-swarms which are implemented on a multi-core processor platform is investigated. Thanks to parallel program execution, a suitable turnaround time ($<100\ \mu$s) is achieved. In contrast to conventional MPC approaches the proposed PSO-based technique is not limited to convex cost functions or linear plant models and therefore extends the MPC application scope. Measurement results based on a $60\ \mathrm{kW}$ prototype motor fed by a 2-level-IGBT inverter demonstrate the general proof of concept. In comparison to a state-of-the-art linear feedback controller (PI-type) a slightly improved control performance is observed.

1. Introduction and State-of-the-Art

Model predictive control (MPC) has attracted considerable attention during recent years. A huge number of publications have been issued, addressing different modulation methods, a variety of applications, implementation issues, computation time reduction, integration of non-linear boundary conditions, cost function design and other aspects. An overview of published work related to MPC has recently been given in [1]. In its basic form, MPC is a modulator-less, discrete-time control method, based on a strictly sequential algorithm:

- Based on actual measurement values, the behaviour of the system is predicted using system linear equations or even non-linear models. This prediction is repeated for every possible switching state within the power electronics. The prediction horizon can extend from only two (including one prediction to compensate for calculation delay and one prediction for the future) up to many steps ahead.

- The different predictions are compared with the aid of a cost or quality function, where different control targets can be combined into one single cost function with weighting factors $w_1,...,w_n$, e.g.

$$J = w_1 f_1(x_1) + ... + w_n f_n(x_n) \tag{1}$$

In most cases, the primary control target is the current error ($x_1 = i^* - i$) and $f_2,...,f_3$ are arbitrary functions.

- The switching state or switching state sequence which leads to the smallest value of the cost function is selected for the following sampling interval.

The freedom that lies within the design of the cost function offers a variety of different control targets to be integrated. In the majority of the published work, the absolute value of the control error (formulated as the Euclidean norm in the cost function) is the main optimisation goal (e.g. [2]). However, other mathematical formulations of the cost function can be found as well. In [3], a method to evaluate different cost functions with respect to stability and weighting factor design is presented. Even non-linear boundary conditions can be integrated into the cost function, e.g. voltage limitation conditions in power inverters, as shown in [4]. Although there are still some stability issues to be clarified, it can be stated that quadratic cost functions result in stable control loops for most topologies as long as the weighting factors are tuned properly.

Due to the discrete-time optimisation nature of the original MPC algorithm, the switching frequency of the converter is variable. This behaviour can be undesirable, e.g. with respect to the thermal layout of the power stage or EMC filter issues. As a solution to this problem, MPC can be combined with linear modulators, such as the well-known PWM approach [5].

Computational effort for the prediction is another major concern in MPC. Different strategies for calculation load reduction (e.g. sphere decoding [6], branch and bound [7] and others) have been investigated. This paper addresses several of the above mentioned drawbacks of MPC by combining a modulator-based scheme, a parallelised implementation on a multi-core controller, a non-linear plant model and non-linear boundary conditions. To solve such control problems under real-time conditions a particle swarm optimisation (PSO) based approach is investigated. As a prove of concept the proposed method is experimentally evaluated regarding the current control of a permanent magnet synchronous motor (PMSM). In contrast to other PSO-based MPC approaches, this paper is not limited to simulative studies (e.g. [8][9]) or large time constant applications (e.g. [10][11]).

2. Control Structure and Implementation

The proposed MPC approach is designed in a field-oriented control (FOC) environment as shown in Fig. 1. In comparison to classical, linear FOC methods (PI-type, [12]) only the current controller is realised in a model predictive way to demonstrate the real-time feasibility of the proposed PSO implementation on a multi-core processor platform. However, the MPC can be extended to include the operating point selection or the inverter modulation scheme by adapting the plant model and the cost function, as well. The operation point selection in Fig. 1 is realised as a look-up table (LUT) based maximum efficiency strategy without any additional time delay [13]. The inverter model is required to compensate inverter non-linearities, i.e. deviations between the desired and actual output voltage due to non-ideal switching characteristics and resistive voltage drops [14]. In addition, a precise flux observer compensates other plant parameter deviations (e.g. temperature-related permanent magnet flux linkage) to improve the prediction accuracy [15]. These two measures are particularly important because otherwise steady-state errors can occur due to deviations between the predicted state variables and the measured ones. The steady-state disturbance rejection is an interesting drawback of the MPC approach in comparison with standard feedback controllers (e.g. PI-type) which is not often addressed in literature. However, the mentioned elements of the entire control structure will not

PCIM Europe 2016, 10 – 12 May 2016, Nuremberg, Germany

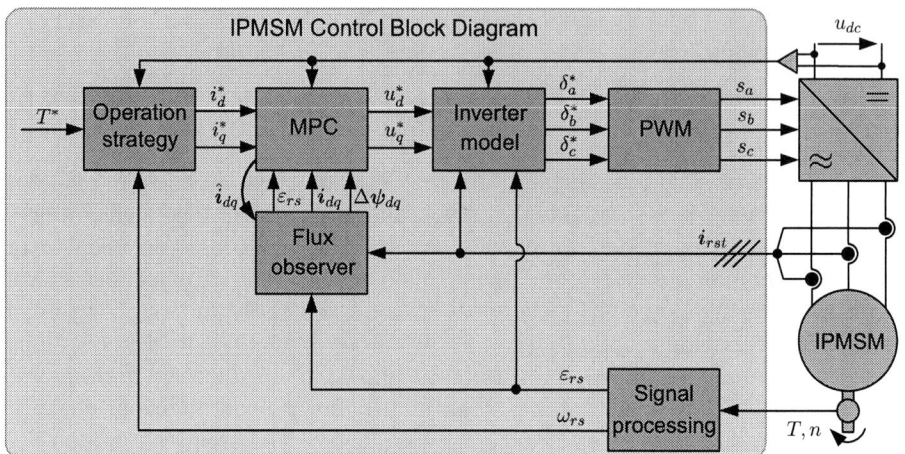

Figure 1: Proposed model predictive control approach embedded in a field-oriented setting

be covered within this contribution and therefore interested readers are referred to the above cited literature.

The first-order IPMSM discrete-time model in field-oriented dq-coordinates is given by [16]:

$$\frac{1}{T_s}(\boldsymbol{\psi}_{dq}[k+1] - \boldsymbol{\psi}_{dq}[k]) =$$

$$\underbrace{\boldsymbol{Q}(-\Delta\varepsilon_{rs}[k])\boldsymbol{u}_{dq}[k]}_{\text{terminal voltage}} - \underbrace{\boldsymbol{Q}(-\Delta\varepsilon_{rs}[k])R_s\boldsymbol{i}_{dq}[k]}_{\text{ohmic voltage drop}} - \underbrace{\frac{1}{T_s}(\boldsymbol{I} - \boldsymbol{Q}(-\Delta\varepsilon_{rs}[k]))\boldsymbol{\psi}_{dq}[k]}_{\text{induced voltage}} \quad (2)$$

Here, bold symbols denote matrix quantities. $\boldsymbol{\psi}_{dq}$ is the flux linkage, $\boldsymbol{Q}$ is the rotation matrix, $\boldsymbol{I}$ is the identity matrix, ε_{rs} is the electrical rotor angle, T_s is the sampling time, $\boldsymbol{u}_{dq}$ is the stator voltage, R_s is the ohmic winding resistance and $\boldsymbol{i}_{dq}$ is the stator current, respectively. To take (cross-)saturation effects into account the relationship between flux and current is given by the non-linear function $\boldsymbol{\psi}_{dq} = \boldsymbol{f}(\boldsymbol{i}_{dq})$, which can be identified experimentally or by finite elements analysis [17]. For implementation purpose look-up tables or polynomial approximation can be used, depending on the computational and memory-related constraints of a certain hardware platform. Rotor angle dependent flux harmonics, due to slotting or winding scheme influences, are neglected for the sake of model simplicity and to reduce the computational burden. For this case study and according the the overall control concept in Fig. 1, the MPC cost function is chosen as

$$J = \sum_{j=2}^{N_p} w_1 \left\| \boldsymbol{i}_{dq}^*[k+j] - \widehat{\boldsymbol{i}}_{dq}[k+j] \right\|_2 + w_2 \Theta(\left\| \boldsymbol{u}_{dq}^*[k+j-1] \right\|_2 - \frac{u_{dc}}{\sqrt{3}}) \quad (3)$$

with w_1 and w_2 as weighting factors, $\|\cdot\|_2$ is the Euclidean norm and $\Theta(\cdot)$ is the Heaviside function to penalise voltage demands outside the classical PWM 2-level-inverter limits. The prediction horizon is chosen to $N_p = 2$ and the control horizon to $N_u = 1$. Here, it should be noted that $N_p = 2$ is the minimum feasible prediction horizon due to the time delay between $\boldsymbol{u}_{dq}^*[k]$ and $\boldsymbol{i}_{dq}^*[k+1]$ (regular-sampling principle). The cost function (3) reduces the least squares current error and the resulting optimisation task is multidimensional as well as non-linear due to the influence of (cross-)saturation effects.

© VDE VERLAG GMBH · Berlin · Offenbach

782

To solve (3) a PSO scheme embedded in a regular-sampled FOC framework was designed and is shown in Fig. 2. Regular-sampling is chosen to ensure a fixed switching frequency and to blank out the high-frequency current ripple regarding the measured current signal. The proposed PSO-based MPC utilises modern processor designs by parallelising the particle swarm iterations on multiple cores. One master core is required to process the auxiliary control parts and to initialise the PSO parameters on every slave core. Every slave core is independent from the others to reduce communicational load and therefore the best optimisation result is selected at the end of each sampling period.

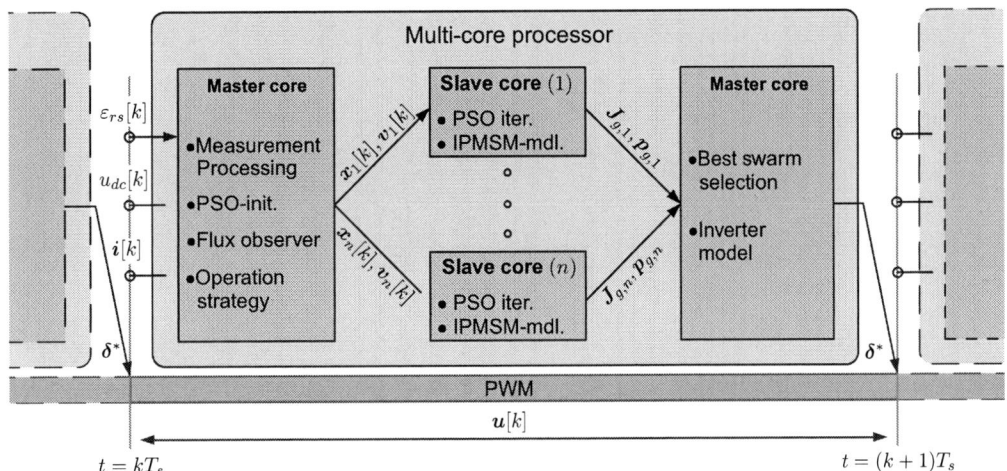

Figure 2: Simplified sequence diagram of PSO-based MPC approach on multi-core platform

PSO is a stochastic global optimisation tool inspired by the social behaviour of animals, such as bird or fish flocking. PSO exploits a population of potential solutions to probe the search space and relies on the exchange of information between individuals of the population. The basic entities of the PSO are called particles, which are connected among each other in one or several swarms. Each particle consists of

$$\boldsymbol{x}_i = [x_{i0}, ..., x_{id}], \quad \boldsymbol{v}_i = [v_{i0}, ..., v_{id}], \quad \boldsymbol{p}_i = [p_{i0}, ..., p_{id}], \quad J_{i,opt} = J(\boldsymbol{p}_i) \tag{4}$$

where x is the current particle position inside the search space, v is the particle velocity, $\boldsymbol{p}_i$ is the particle's historic best position till the current iteration and $J_{i,opt}$ is the respective cost function value. For this application the particle position corresponds to the reference voltage and the velocity is equal to a voltage difference inside the voltage plain:

$$\boldsymbol{x}_i \overset{!}{=} \boldsymbol{u}_{dq}^*[k+1], \quad \boldsymbol{v}_i \overset{!}{=} \Delta \boldsymbol{u}_{dq}^*[k+1] \tag{5}$$

At the beginning of each sampling period the particles are initialised on the master core according to Fig. 3. For this contributation a uniform distribution according to polar coordinates inside the dq-plane were chosen – a Cartesian grid or even a random distribution are possible, too. For simplification reasons particles were only placed inside the inner voltage plain circle of a 2-level inverter. For this example 37 particles are placed with respect to polar coordinates and additionally the global best particle position of the last iteration $p_g[k]$ is evaluated, as well. Among this configuration the best particle position $\tilde{p}_g[k]$ after the master core evaluation is used to initialise the sub-swarms on the slave cores. Each slave core is configured individually with 6 new particles randomly positioned in the reduced search room and v_i is initialised individually

PCIM Europe 2016, 10 – 12 May 2016, Nuremberg, Germany

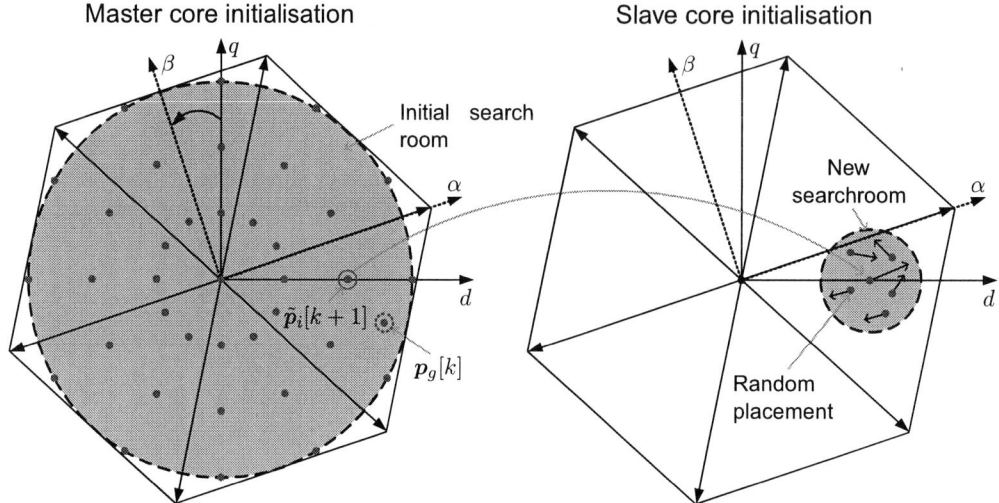

Figure 3: Concept of master core initialisation to reduce the PSO searchroom of every slave core

on each sub-swarm. After that the master core is send to idle mode and the following PSO algorithm is executed on each slave core independently in an iterative fashion:

$$\boldsymbol{v}_i[k] = \boldsymbol{v}_i[k-1] + \varphi_1(\boldsymbol{p}_i - \boldsymbol{x}_i[k-1]) + \varphi_2(\boldsymbol{p}_g - \boldsymbol{x}_i[k-1])$$

$$\boldsymbol{x}_i[k] = \alpha\boldsymbol{x}_i[k-1] + \boldsymbol{v}_i[k]$$

$$\boldsymbol{p}_i = \boldsymbol{x}_i[k], \qquad\qquad \text{if} \quad J(\boldsymbol{x}_i[k]) < J_{i,opt}$$

$$\boldsymbol{p}_g = \boldsymbol{p}_i, \qquad\qquad \text{if} \quad J(\boldsymbol{p}_i) < J_g \qquad\qquad (6)$$

$$\varphi_1 = c_1 \cdot r(0,1), \quad \varphi_2 = c_2 \cdot r(0,1) \qquad \text{(learn law)}$$

$$c_1 + c_2 \approx 4 \quad \text{(learn rates)}, \qquad \alpha < 1 \quad \text{(inertia weight)}$$

$$r(a,b) = \text{uniformly distributed random number within } [a,b]$$

Here, J_g is the global best cost value of each swarm. The tunable optimisation parameters are the learning rates c_1 and c_2, which emphasise the movement focus to the individual best $\boldsymbol{p}_i$ or global best position $\boldsymbol{p}_g$, and the inertia weight factor α which adjusts the transition between exploration of the search room and local convergence. The implementation (6) describes a global neighbourhood so that each particle can inform all other particles inside the swarm regarding $\boldsymbol{p}_g$. However, other neighbourhood topologies (e.g. ring or star) can be used instead, but due to the limited particle number per slave core the global neighbourhood offers best convergence characteristics. Hard constraints regarding particle's position and velocity can be added to (6), but for this example the voltage and current constraints are implicitly modelled in (3). The algorithm (6) is iterated on each slave core until a certain computation time has expired. Then, all best particle positions of each slave core are transferred to the master core and among these the one with the minimum cost value is send to the PWM register. More PSO basics or information concerning alternative PSO implementations can be found in [18].

3. Experimental Implementation and Results

The proposed PSO-based MPC was implemented on a laboratory test bench equipped with a dSPACE DS1006 (2.8 GHz AMD Opteron) quad-core processor board. Hence, $n = 3$ sub-

© VDE VERLAG GMBH · Berlin · Offenbach

swarms on the slave cores were generated each consisting of 6 particles. To realise a constant switching time $T_s = 100~\mu$s of the entire control sequence, 7 PSO iterations on each slave core could be evaluated. Thus, a trade-off between the particle and the iteration number has to be chosen. In general, a large amount of particles reduces the risk to converge into sub-optimal local minima whereas many iterations improve the convergence characteristics around the global optimum. All calculations are based on double floating-point accuracy and thus the computational burden could be reduced when using fixed-point variables. However, the iteration amount for this implementation example led to feasible convergence results and therefore a fixed-point transformation was not necessary. The test motor is a highly-utilised 60 kW interior PMSM prototype designed for traction application and a standard 3-phase 2-level inverter equipped with SKiiP 1242GB120-4DW Semikron half bridges were used. Also, a load motor ensured constant speed operation of the test device. The PSO learning rates were chosen $c_1 = 1.5$ and $c_2 = 2.5$, the inertia weight factor was $\alpha = 0.5$.

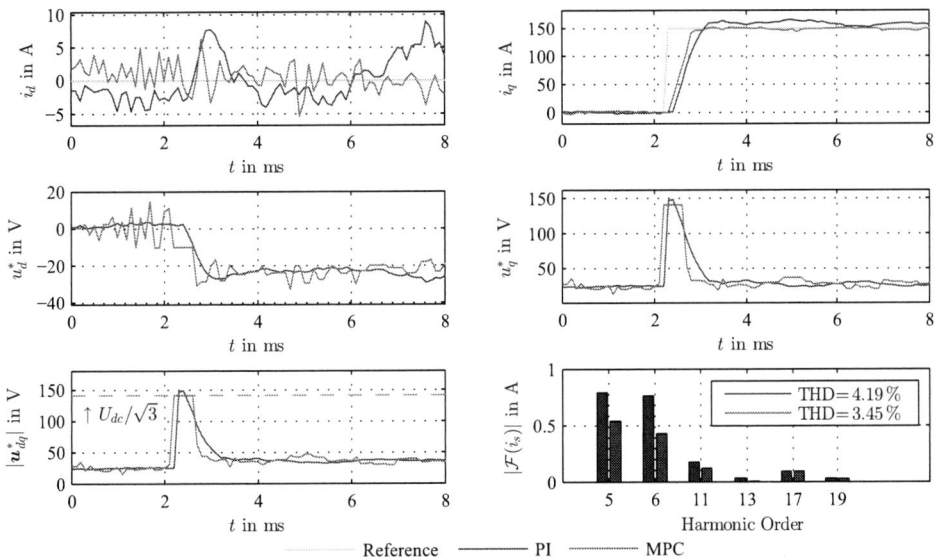

Figure 4: Measurement results: step response from $i^*_{dq} = \{0~\mathrm{A}, 0~\mathrm{A}\}$ to $i^*_{dq} = \{0~\mathrm{A}, 150~\mathrm{A}\}$ at constant speed $n = 500~\mathrm{min}^{-1}$ and current harmonics in steady state

In Fig. 4 the PSO-based MPC is validated and a first comparison to a time-discrete PI controller, designed according to [12] for highest dynamics, is shown. The PI-FOC design from [12] is fully parameter-adaptive and equipped with a sophisticated anti-reset windup measure to guarantee a fair comparison. In Fig. 4 it can be seen that the MPC performance is slightly better, i.e. faster current dynamics without overshoot and reduced current harmonics could be achieved. In addition, the currents are decoupled almost perfectly in the presence of significant saturation and the voltage constraint is strictly adhered.

In Fig. 5 another measurement result is depicted. Due to the increased motor speed the voltage constraint is more important and the resulting current step responses are slower. Consequently, the PI-FOC massively violates the voltage constraint and therefore a sub-optimal voltage choice results in an inferior control performance compared with the proposed method. Especially, the clear overshoot of the PI-FOC could be avoided and the steady-state current harmonics are better suppressed by the PSO-based MPC, as well.

© VDE VERLAG GMBH · Berlin · Offenbach

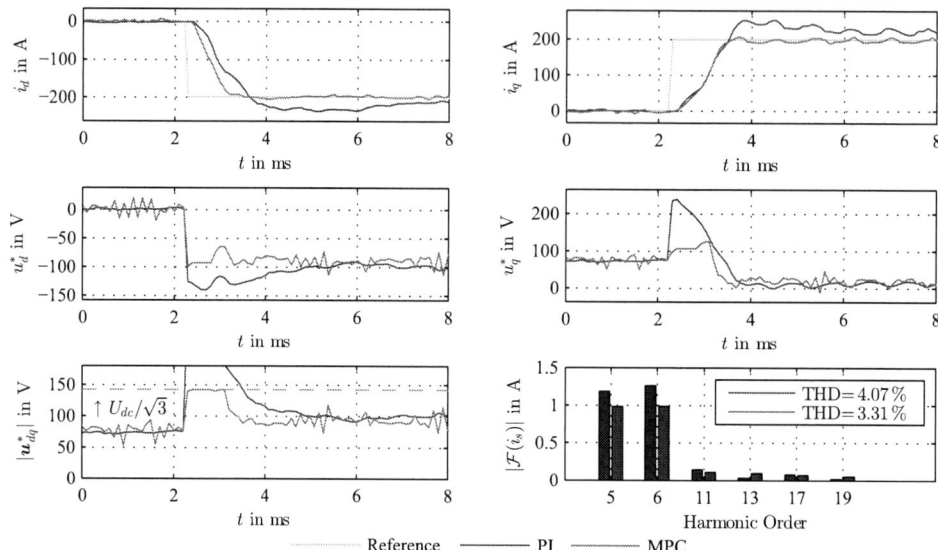

Figure 5: Measurement results: step response from $i_{dq}^* = \{0\,\text{A}, 0\,\text{A}\}$ to $i_{dq}^* = \{-200\,\text{A}, 200\,\text{A}\}$ at constant speed $n = 1500\,\text{min}^{-1}$ and current harmonics in steady state

4. Conclusion

The proposed PSO-based MPC concept exhibits two main key features:

- It utilises modern multi-core processor platforms since the control algorithm can be easily parallelised as well as scaled in terms of the number of available cores

- Due to the meta-heuristic PSO approach multidimensional and non-linear optimisation task are addressed appropriately reducing the risk to converge into sub-optimal local minima compared with classical gradient-based optimisation

The implementation and measurement results proved the real-time capability of the concept even for the relatively small sampling time $T_s = 100\ \mu\text{s}$. However, it is obvious that PSO-based MPC requires significantly more computational resources compared with standard linear feedback controllers or even gradient-based MPC approaches. Therefore, the proposed concept is considered as a long-term perspective regarding non-linear control techniques in the confidence that the computational power of state-of-the-art industrial processor platforms still increases in the next years and decades.

Regarding the demonstrated working example for the current control loop of a PMSM drive a slightly improved control performance compared with a state-of-the-art PI-FOC approach could be realised, i.e. the control dynamics as well as the harmonic suppression were increased. In future works other parts of the motor control, e.g. an efficiency-optimal operation point selection or the switching signal modulation task for the feeding inverter, can be realised within PSO-based MPC. In addition, the proposed concept is likely to be adaptable for many other power electronics or drive related control problems.

5. References

[1] S. Kouro, M.A. Perez, J. Rodriguez, A.M. Llor, and H.A. Young. Model Predictive Control - MPC's Role in the Evolution of Power Electronics. *IEEE Industrial Electronics Magazine*, vol. 9, no. 4:pp. 8–21, 2015.

[2] J. Rodriguez, J. Pontt, C. A. Silva, P. Correa, P. Lezana, P. Cortes, and U. Ammann. Predictive Current Control of a Voltage Source Inverter. *IEEE Transactions on Industrial Electronics*, vol. 54, iss. 1:pp. 495–503, 2007.

[3] U. Ammann. Using Assignment Figures to Evaluate Cost Functions in Predictive Inverter Control. In *IEEE International Symposium on Predictive Control of Electrical Drives and Power Electronics (PRECEDE)*, 2015.

[4] T. Geyer, R.P. Aguilera, and D.E. Quevedo. On the Stability and Robustness of Model Predictive Direct Current Control. In *IEEE International Conference on Industrial Technology (ICIT)*, 2013.

[5] S. Mariéthoz and M. Morar. Expllicit Model-Predictive Control of a PWM Inverter With an LCL Filter. *IEEE Transactions on Industrial Electronics*, vol. 56, no. 2:pp. 389–399, 2009.

[6] T. Geyer and D.E. Quevedo. Multistep Direct Predictive Control for Power Converters – Part 1: Algorithm. In *IEEE Energy Conversion Congress and Exposition (ECCE)*, 2013.

[7] A. Ayad, P. Karamanakos, and R. Kennel. Direct Model Predictive Current Control of Quasi Z-Source Inverters. In *IEEE International Symposium on Predictive Control of Electrical Drives and Power Electronics (PRECEDE)*, 2015.

[8] J. Thomas. Particle Swarm Optimization Based Model Predictive Control for Constrained Nonlinear Systems. In *International Conference on Informatics in Control, Automation and Robotics (ICINCO)*, 2014.

[9] F. Xu, H. Chen, X. Gong, and Q. Mei. Fast Nonlinear Model Predictive Control on FPGA Using Particle Swarm Optimization. *IEEE Transactions on Industrial Electronics*, vol. 63, no. 1:pp. 310–321, 2016.

[10] P. Arpaia, S. Manfredi, F. Donnarumma, and C. Manna. Model Predictive Control Strategy Based on Differential Discrete Particle Swarm Optimization. In *IEEE Workshop on Environmental Energy and Structural Monitoring Systems (EESMS)*, 2010.

[11] K. Van Heerden, Y. Fujimoto, and A. Kawamura. A Combination of Particle Swarm Optimization and Model Predictive Control on Graphics Hardware for real-Time Trajectory Planning of the Under-Actuated Nonlinear Acrobot. In *IEEE International Workshop on Advanced Motion Control (AMC)*, 2014.

[12] W. Peters and J. Böcker. Discrete-Time Design of Adaptive Current Controller for Interior Permanent Magnet Synchronous Motors (IPMSM) with High Magnetic Saturation. In *Annual Conference of the IEEE Industrial Electronics Society (IECON)*, 2013.

[13] W. Peters, O. Wallscheid, and J. Böcker. Optimum Efficiency Control of Interior Permanent Magnet Synchronous Motors in Drive Trains of Electric and Hybrid Vehicles. In *European Conference on Power Electronics and Applications (EPE)*, 2015.

[14] N. Seilmeier, C. Wolz, and B. Piepenbreier. Modelling and Model Based Compensation of Non-Ideal Characteristics of Two-Level Voltage Source Inverters for Drive Control Application. In *Electric Drives Production Conference (EDPC)*, 2011.

[15] A. Specht, O. Wallscheid, and J. Böcker. Determination of Rotor Temperature for an Interior Permanent Magnet Synchronous Machine Using a Precise Flux Observer. In *International Power Electronics Conference (IPEC)*, 2014.

[16] Andreas Specht, Sina Ober-Blöbaum, Oliver Wallscheid, Christoph Romaus, and Joachim Böcker. Discrete-Time Model of an IPMSM Based on Variational Integrators. In *IEEE International Electric Machines and Drives Conference*, 2013.

[17] M. Meyer and J. Böcker. Optimum Control for Interior Permanent Magnet Synchronous Motors (IPMSM) in Constant Torque and Flux Weakening Range. In *International Power Electronics and Motion Control Conference (EPE-PEMC)*, 2006.

[18] M. Clerc. *Particle Swarm Optimization*. ISTE, 2006.

© VDE VERLAG GMBH · Berlin · Offenbach

PCIM Europe 2016, 10 – 12 May 2016, Nuremberg, Germany

A Modular Multilevel Matrix Converter for High Speed Drive Applications

Dennis Bräckle, Felix Kammerer, Mathias Schnarrenberger, Marc Hiller, Michael Braun
Elektrotechnisches Institut (ETI) - Power Electronic Systems
Karlsruhe Institute of Technology (KIT), Kaiserstraße 12, 76131 Karlsruhe, Germany
Fon: +49 721 608 42922, E-Mail: dennis.braeckle@kit.edu

Abstract

The Modular Multilevel Matrix Converter (M3C) performs a direct three-phase AC to AC power conversion and is highly suitable for medium voltage high power drive applications. One area of application are high speed drives such as compressors. However, additional balancing power components which reduce the output power capability of the M3C when input and output frequencies are similar occur. This paper analytically examines the operation behavior and power capability in these operation points in order to assess whether the M3C can generate these additional components without oversizing the converter's components. Subsequently, the theoretical evaluation is verified by a laboratory scaled prototype with a rated power of 15 kW.

1. Fundamentals of the M3C topology

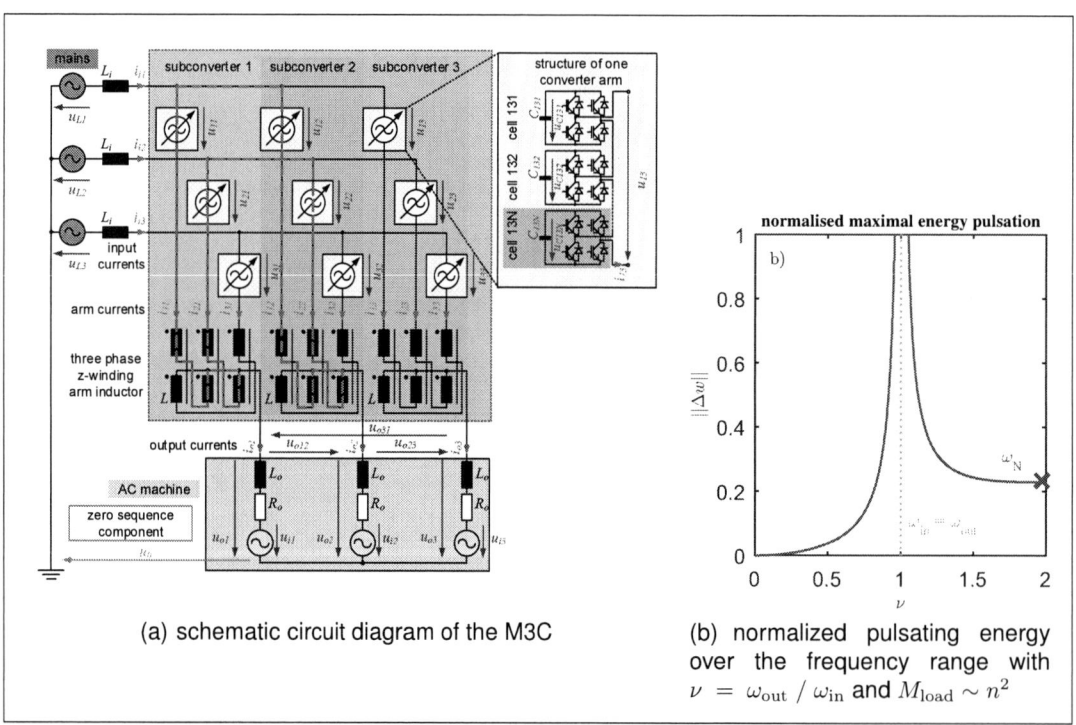

(a) schematic circuit diagram of the M3C

(b) normalized pulsating energy over the frequency range with $\nu = \omega_{\text{out}} / \omega_{\text{in}}$ and $M_{\text{load}} \sim n^2$

Fig. 1

Figure 1 a) shows the schematic structure of the overall system with the AC-grid, the Modular Multilevel Matrix Converter (M3C) and an AC machine. The M3C is composed of 3 subcon-

© VDE VERLAG GMBH · Berlin · Offenbach

verters. Each subconverter connects the three input phases to one of the output phases, using converter arms and arm inductances L. The converters consist of N series connected cells in H-bridge configuration. Each cell has a capacitor C_{xyz} (x: input phase, y: output phase, z: cell number) and is able to generate a voltage of $-\sum_{z=1}^{N} u_{Cxyz} \leq u_{xy} \leq \sum_{z=1}^{N} u_{Cxyz}$.

The M3C has been proven suitable for low frequency high torque drive applications [1] and for high speed drive applications [2]. Figure 1 (b) shows the normalized total energy pulsations $\|\Delta w\|$ over the positive frequency range when driving a machine with $\omega_N = 2\,\omega_i$ and a quadratic load characteristic. The converter's components are designed for the nominal operation point of the driven machine. The maximum energy pulsation at $\omega_o \approx \omega_i$ is a criterion for dimensioning the cell capacitors. However, when $\omega_o \approx \omega_i$, additional low frequent power components occur. They have to be compensated in order to reduce the total Δw. Using the control strategies presented in [3] and [4], additional balancing currents and a zero sequence voltage are generated to ensure stable operation at $\omega_o \approx \omega_i$. The maximum current in each arm is limited by semiconductors and designed for the nominal operation point of the machine. In order to determine the available performance at $\omega_o \approx \omega_i$, the maximum balancing power and current limits are calculated and verified with a laboratory scaled prototype with a rated power of $P = 15\text{kW}$ and $N = 5$ cells per converter arm.

2. Fundamentals of the M3C control strategy

The M3C is a highly complex topology with several degrees of freedom that need to be considered adequately when implementing a control scheme. First, the 9 arm currents need to be controlled in order to drive the load machine and exchange power with the feeding grid. There are 8 degrees of freedom and the constraint $\sum i_{xy} = 0$. Secondly, the converter's energy must be distributed equally among all capacitors. The control scheme presented in [3] separates the modulation of each arm from the converter control. The equivalent circuit diagram is shown in Figure 1(a). With this analysis, the important power components are identified and examined when driving a high speed machine.

2.1. Arm converter power control

In [3] a new control scheme based on a transformed arm power analysis is presented. The measured currents and capacitor voltages can be transformed so that a decoupled control of the input side, the output side and two inner components is possible. The power for each arm can be calculated with the arm voltage and arm current $p_{xy} = u_{xy} \cdot i_{xy}$. For converter arm 11 the resulting power component is

$$p_{11} = u_{11} \cdot i_{11} \approx (u_{i,\alpha} - u_{o,\alpha} - u_0) \cdot \left(\frac{i_{i,\alpha}}{3} + \frac{i_{o,\alpha}}{3} + i_{d1,\alpha} + i_{d2,\alpha}\right) \tag{1}$$

The resulting arm power consists of the α-components of the input voltage u_i, the output voltage u_o and the zero sequence voltage u_0 multiplied by the sum of the α-components of the input current $\frac{i_i}{3}$, the output current $\frac{i_o}{3}$ and the internal currents i_{d1} and i_{d2}. The calculated arm powers are also transformed with respect to the rules in [3]. Due to the different frequencies occurring in the power components, some must be considered separately during operation with similar input and output frequencies (see section 2.2).

2.2. Operational management when feeding a high frequency drive

The M3C preforms a direct 3AC-3AC energy conversion. It is suitable for driving machines whose nominal frequency is larger than the input frequency [2]. When running up the machine, the point of similar input and output frequency has to be overcome. At that point the frequency of inner reactive power components become low frequent and cause a large energy fluctuation in the cell capacitors. These components have to be compensated in order to assure stable operation of the M3C. According to the transformation in [3] and assuming sinusoidal functions with $\gamma = \omega t$ for all current and voltage components the resulting inner power components $p_{d1,\alpha\beta}$ and $p_{d2,\alpha\beta}$ are

$$p_{d1\alpha} = \frac{1}{2}\hat{U}_i\hat{I}_{d2}\cos(\gamma_i + \gamma_{d2}) - \frac{1}{2}\hat{U}_o\hat{I}_{d2}\cos(\gamma_o + \gamma_{d2}) + \frac{1}{2}\hat{U}_0\hat{I}_{d1}\cos(\gamma_0 \pm \gamma_{d1})$$
$$+ \frac{1}{6}\hat{U}_i\hat{I}_o\cos(-\gamma_i + \gamma_o + \varphi_o) - \frac{1}{6}\hat{U}_o\hat{I}_i\cos(\gamma_o - \gamma_i - \varphi_i) \tag{2}$$

$$p_{d1\beta} = \frac{1}{2}\hat{U}_i\hat{I}_{d2}\sin(\gamma_i + \gamma_{d2}) - \frac{1}{2}\hat{U}_o\hat{I}_{d2}\sin(\gamma_o + \gamma_{d2}) + \frac{1}{2}\hat{U}_0\hat{I}_{d1}\sin(\gamma_0 \pm \gamma_{d1})$$
$$+ \frac{1}{6}\hat{U}_i\hat{I}_o\sin(-\gamma_i + \gamma_o + \varphi_o) - \frac{1}{6}\hat{U}_o\hat{I}_i\sin(\gamma_o - \gamma_i - \varphi_i) \tag{3}$$

$$p_{d2\alpha} = \frac{1}{2}\hat{U}_i\hat{I}_{d1}\cos(-\gamma_i + \gamma_{d1}) - \frac{1}{2}\hat{U}_o\hat{I}_{d1}\cos(\gamma_o + \gamma_{d1}) + \frac{1}{2}\hat{U}_0\hat{I}_{d2}\cos(\gamma_0 \pm \gamma_{d2})$$
$$+ \frac{1}{6}\hat{U}_i\hat{I}_o\cos(\gamma_i + \gamma_o + \varphi_o) - \frac{1}{6}\hat{U}_o\hat{I}_i\cos(\gamma_o + \gamma_i - \varphi_i) \tag{4}$$

$$p_{d2\beta} = \frac{1}{2}\hat{U}_i\hat{I}_{d1}\sin(-\gamma_i + \gamma_{d1}) - \frac{1}{2}\hat{U}_o\hat{I}_{d1}\sin(\gamma_o + \gamma_{d1}) + \frac{1}{2}\hat{U}_0\hat{I}_{d2}\sin(\gamma_0 \pm \gamma_{d2})$$
$$+ \frac{1}{6}\hat{U}_i\hat{I}_o\sin(\gamma_i + \gamma_o + \varphi_o) - \frac{1}{6}\hat{U}_o\hat{I}_i\sin(\gamma_o + \gamma_i - \varphi_i) \tag{5}$$

The second line in each of the equations (2)-(5) is defined by the feeding grid and load machine. The power components depend on the input voltage $\hat{U}_i$, current $\hat{I}_i$ and phase shift φ_i as well as the output voltage $\hat{U}_o$, current $\hat{I}_o$ and phase shift φ_o. These components become low frequent and cause energy fluctuations during operation with $\omega_i \approx \omega_o$. Therefore, they have to be compensated. Otherwise stable operation of the M3C can not be guaranteed. The first line of the equations can be controlled separately. Using the internal currents $\hat{I}_{d1}$ and $\hat{I}_{d2}$ and the zero sequence voltage $\hat{U}_0$, these components compensate the low frequent reactive power(see section 3.2) and assure stable operation of the M3C.

3. Operation points with similar input and output frequencies

3.1. Compensation of power components

As discussed in 2.2, low frequent reactive power components occur during operation with similar input and output frequencies. An analytical discussion for compensating these components is presented. The following discussion is only focusing on inner power component 1 $p_{d1,\alpha\beta}$ (eq. (2) and (3)) with corresponding results to component 2 $p_{d2,\alpha\beta}$. The occurring low frequent components are

$$p_{\mathrm{d1,lf},\alpha} = \frac{1}{6}\hat{U}_{\mathrm{i}}\hat{I}_{\mathrm{o}}\cos(-\gamma_{\mathrm{i}}+\gamma_{\mathrm{o}}+\varphi_{\mathrm{o}}) - \frac{1}{6}\hat{U}_{\mathrm{o}}\hat{I}_{\mathrm{i}}\cos(\gamma_{\mathrm{o}}-\gamma_{\mathrm{i}}-\varphi_{\mathrm{i}})$$

$$p_{\mathrm{d1,lf},\beta} = \frac{1}{6}\hat{U}_{\mathrm{i}}\hat{I}_{\mathrm{o}}\sin(-\gamma_{\mathrm{i}}+\gamma_{\mathrm{o}}+\varphi_{\mathrm{o}}) - \frac{1}{6}\hat{U}_{\mathrm{o}}\hat{I}_{\mathrm{i}}\sin(\gamma_{\mathrm{o}}-\gamma_{\mathrm{i}}-\varphi_{\mathrm{i}})$$

$$\hat{p}_{\mathrm{d1,lf}} = \sqrt{p_{\mathrm{d1,lf},\alpha}^2 + p_{\mathrm{d1,lf},\beta}^2}$$

$$= \frac{1}{6}\cdot\sqrt{\left(\hat{U}_{\mathrm{i}}\hat{I}_{\mathrm{o}}\right)^2 + \left(\hat{U}_{\mathrm{o}}\hat{I}_{\mathrm{i}}\right)^2 - 2\hat{U}_{\mathrm{i}}\hat{I}_{\mathrm{o}}\hat{U}_{\mathrm{o}}\hat{I}_{\mathrm{i}}\cdot\cos(\varphi_{\mathrm{i}}+\varphi_{\mathrm{o}})} \tag{6}$$

Several assumptions are made:

- losses are neglected
- the ratio between input voltage $\hat{U}_{\mathrm{i}}$ and output voltage $\hat{U}_{\mathrm{o}}$ is k_{u}
- the ratio beween the output power S_{o} and the nominal output power $S_{\mathrm{o,N}} = \frac{1}{6}\hat{U}_{\mathrm{o,N}}\hat{I}_{\mathrm{o,N}}$ is s
- the worst-case-scenario of the maixmal balancing power is considered

With these assumptions the input current can be written as

$$P_{\mathrm{in}} = P_{\mathrm{out}}$$

$$\frac{\sqrt{3}}{2}\hat{U}_{\mathrm{i}}\hat{I}_{\mathrm{i}}\cos(\varphi_{\mathrm{i}}) = \frac{\sqrt{3}}{2}\hat{U}_{\mathrm{o}}\hat{I}_{\mathrm{o}}\cos(\varphi_{\mathrm{o}})$$

$$\hat{I}_{\mathrm{i}} = \hat{I}_{\mathrm{o}}k_{\mathrm{u}}\frac{\cos(\varphi_{\mathrm{o}})}{\cos(\varphi_{\mathrm{i}})} \tag{7}$$

Combining equation (6) and (7), the resulting, normalized power component p_{bal} needing to be compensated is

$$p_{\mathrm{bal}} = \frac{\hat{p}_{\mathrm{d1,lf}}}{S_{\mathrm{o,N}}} = \frac{\frac{1}{6}\left(\hat{U}_{\mathrm{o}}\hat{I}_{\mathrm{o}}\right)}{\frac{1}{6}\left(\hat{U}_{\mathrm{o,N}}\hat{I}_{\mathrm{o,N}}\right)}\cdot\sqrt{\frac{1}{k_u^2} + \frac{k_u^2\cos^2(\varphi_{\mathrm{o}})}{\cos^2(\varphi_{\mathrm{i}})} - 2\frac{\cos(\varphi_{\mathrm{o}})}{\cos(\varphi_{\mathrm{i}})}\cos(\varphi_{\mathrm{o}}+\varphi_{\mathrm{i}})}$$

$$p_{\mathrm{bal}} = \frac{s}{\cos(\varphi_{\mathrm{i}})}\cdot\sqrt{\frac{\cos^2(\varphi_{\mathrm{i}})}{k_u^2} + k_u^2\cos^2(\varphi_{\mathrm{o}}) - 2\cos(\varphi_{\mathrm{i}})\cos(\varphi_{\mathrm{o}})\cos(\varphi_{\mathrm{i}}+\varphi_{\mathrm{o}})} \tag{8}$$

Equation (8) shows the power component that has to be compensated in order to overcome the point of similar input and output frequency to accelerate a load machine to nominal frequencies beyond the input frequency. According to equation (2) to (5) the superposition of the low frequent components results in a sinusoidal reactive power component with an amplitude p_{bal} and a frequency ω_{bal}, which is not considered because of the worst-case consideration. p_{bal} provides information whether the pulsating power components can be compensated without oversizing the converter. In addition, the generation of the balancing power must be considered with respect to the current and voltage capability of the converter (see section 3.2).

3.2. Generation of the balancing power components

[3] and [5, 4] show control techniques that allow the compensation of the balancing power p_{bal} from equation (8). In [3] the balancing power is generated with an balancing current $i_{\mathrm{bal},[3]}$ and a zero sequence voltage u_0. The zero sequence voltage is a degree of freedom when the neutral

points of the input and output side are not connected and a common mode voltage is allowed on both sides. [4] presents a method where the power is generated with an internal current $i_{\mathrm{bal},[4]}$ and the output voltage u_{o}. The arm currents of the M3C are composed of $i_{\mathrm{arm}} = \frac{i_{\mathrm{i}}}{3} + \frac{i_{\mathrm{o}}}{3} + i_{\mathrm{bal}}$. The semiconductors of the converter arms are designed for nominal operation of the machine. Neglecting losses, the maximum arm current is $i_{\mathrm{arm,max}} = \frac{i_{\mathrm{i,N}}}{3} + \frac{i_{\mathrm{o,N}}}{3}$. Normalised on the nominal output power, the maximum amplitude of the arm current is given by $|\hat{I}_{\mathrm{arm,max}}| = 0.67$. With both proposed controlling methods, the balancing power from equation (8) can be generated. Depending on the operational conditions, either method [3] or [4] may be used. The resulting rated balancing current amplitudes are

$$i_{\mathrm{bal},[3]} = \sqrt{\frac{1}{36} \cdot \frac{S_{\mathrm{o}}^2 k_{\mathrm{u}} \cos^2(\varphi_{\mathrm{o}})}{\cos^2(\varphi_{\mathrm{i}}) u_0^2} + \frac{1}{36} \cdot \frac{S_{\mathrm{o}}^2}{k_{\mathrm{u}}^2 u_0^2} - \frac{1}{18} \cdot \frac{S_{\mathrm{o}}^2 \cos(\varphi_{\mathrm{o}}) \cos(\varphi_{\mathrm{o}} - \varphi_{\mathrm{i}})}{\cos(\varphi_{\mathrm{i}}) u_0^2}} \tag{9}$$

$$i_{\mathrm{bal},[4]} = \frac{1}{3} \cdot \sqrt{\frac{i_{\mathrm{o}}^2 k_{\mathrm{u}}^2 \cos^2(\varphi_{\mathrm{o}})}{\cos^2(\varphi_{\mathrm{i}})} + i_{\mathrm{o}}^2} \tag{10}$$

The balancing current is an additional current that reduces the output current capability when driving a load with similar input and output frequency. A converter, designed for nominal operation of a high speed drive, is able to accelerate this machine without oversizing the semiconductors if the maximum arm current is not exceeded during operation with similar input and output frequencies.

4. Experimental setup and final results

Fig. 2: low voltage laboratory prototype of the M3C with $P_{\mathrm{N}} = 15\mathrm{kW}$, signal processing unit and load machine/resistive load

Figure 2 (a) shows the laboratory setup for measurement and verification of the theoretical discussion. The low voltage prototype has a rated power of $P_{\mathrm{N}} = 15\mathrm{kW}$. It consists of $N = 9 \cdot 5 = 45$ cells with a local microcontroller (dsPIC from Microchip) in each cell. They are connected via optical fiber to a FPGA (Altera) as central modulation unit. The control scheme presented in [3, 6, 7] is programmed on a TMS320C6748 digital signal processor unit from Texas Instruments. The period of the control algorithm is $T_{\mathrm{ctrl}} = T_{\mathrm{PWM}} = \frac{1}{f_{\mathrm{PWM}}} = 125\mu\mathrm{s}$. The prototype can feed an AC-machine (depicted in figure 2) or a resistive load. The prototype

© VDE VERLAG GMBH · Berlin · Offenbach

operates over the whole frequency range[3] and is used to verify the results presented in this paper.

4.1. Experimental results of the balancing power analysis

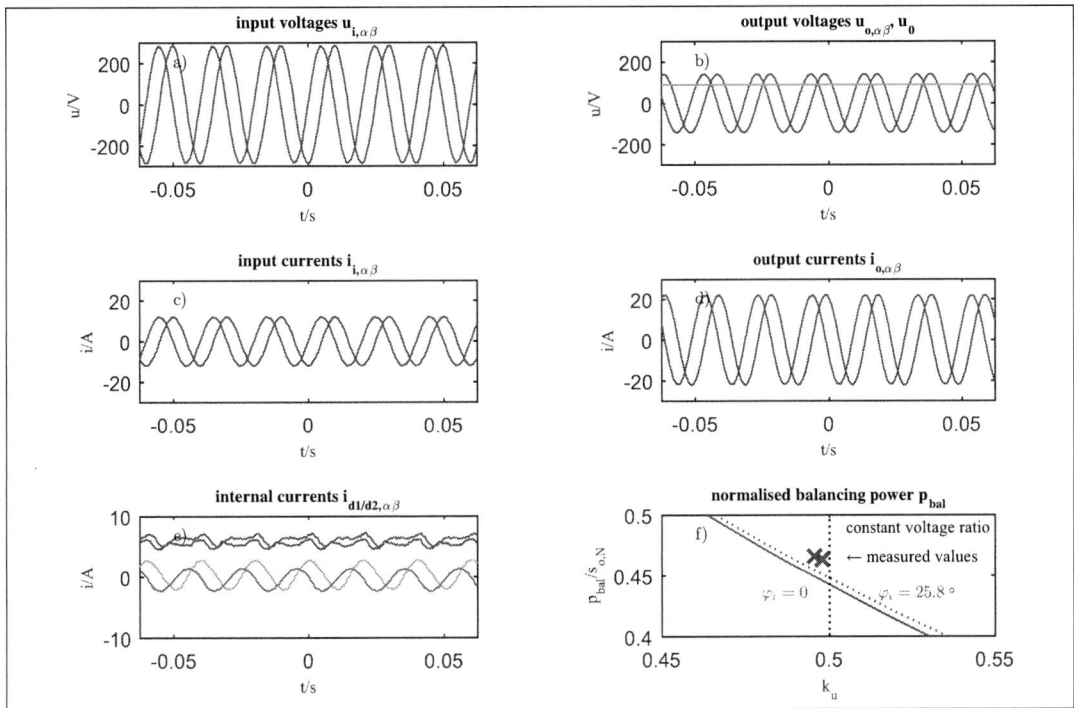

Fig. 3: a)-e) measurements of the laboratory prototype with a passive load at $\omega_o = 50Hz$, $s \approx 0.29$, $\varphi_i = 0$ and $k_u \approx 0.5$, short time averaged values at $T_A = \frac{1}{f_{PWM}} = \frac{1}{8kHz}$, f) compares the calculated and actual balancing power components at different $\varphi_i = 0$

Figure 3 shows experimental results from a laboratory prototype of the M3C [1] to verify the results from equation (8). The phase shift on the output side φ_o is defined by the load. The phase shift on the input side φ_i is a degree of freedom. In order to generate the balancing power, additional voltages and internal currents are necessary. The sum of the normalised output power s_o and balancing power p_{bal} must not exceed 1 to avoid an oversizing of the converter components. Figure 3 a)-d) depict the measured input and output values of the M3C laboratory prototype at $\omega_o = 50$Hz and a passive load. The output phase shift is $\varphi_o = 0$. The output power is $P_o = 0.29 \cdot P_{o,N} = 4.35kW$. Figure 3 e) shows the internal balancing currents. The additional balancing power is generated with the DC zero sequence voltage u_0 and the internal balancing currents. Figure 3 f) compares the theoretical balancing power p_{bal} in equation (8) and the measured power. Additionally, the effect of different phase shifts φ_i on the input side are examined.

The M3C is a symmetric system. Reactive power, either on the input or the output side, results in an additional balancing power when $\omega_i \approx \omega_o$. The discrepancy between the theoretical and measured balancing power in Figure 3 f) is only approx. 4% and results from the assumption of ideal conditions, e.g. neglecting losses. The results therefore verify the analytical discussion of the balancing power.

© VDE VERLAG GMBH · Berlin · Offenbach

4.2. Maximum output current capability with additional balancing current

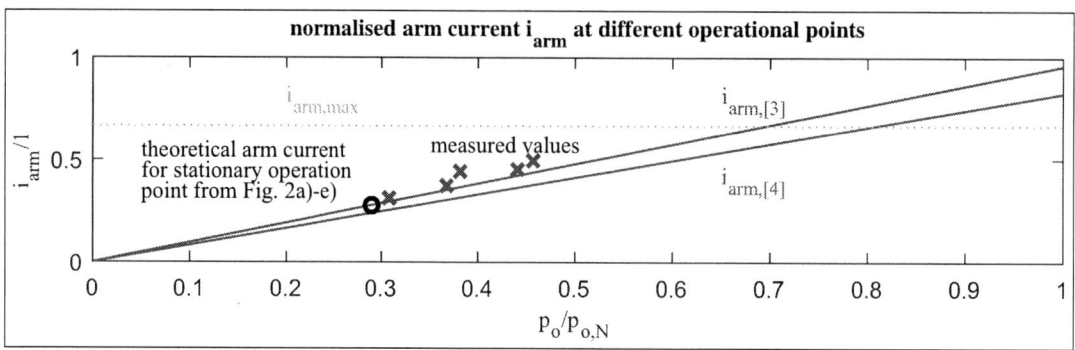

Fig. 4: normalised arm currents ($i_{\text{arm}} = \frac{i_i}{3} + \frac{i_o}{3} + i_{\text{bal}}$) according to [3] and [4] at constant output voltage ratio $k_u \approx 0.5$ and different output powers.

Figure 4 depicts the normalized arm currents $i_{\text{arm}} = \frac{i_i}{3} + \frac{i_o}{3} + i_{\text{bal}}$. The output voltage ratio is constant $k_u \approx 0.5$. Assuming a quadratic load characteristic, the maximum arm current $|\hat{I}_{\text{arm,max}}|$ (dashed yellow line) is not reached. Therefore the remaining arm current capability can be used to generate additional output current when operating at similar input and output frequencies. Transferred to a drive application, the additional current can generate an acceleration torque e.g. for accelerating the AC machine to its nominal speed without oversizing the semiconductors. Measurements with different output currents verify that additional output power can be provided.

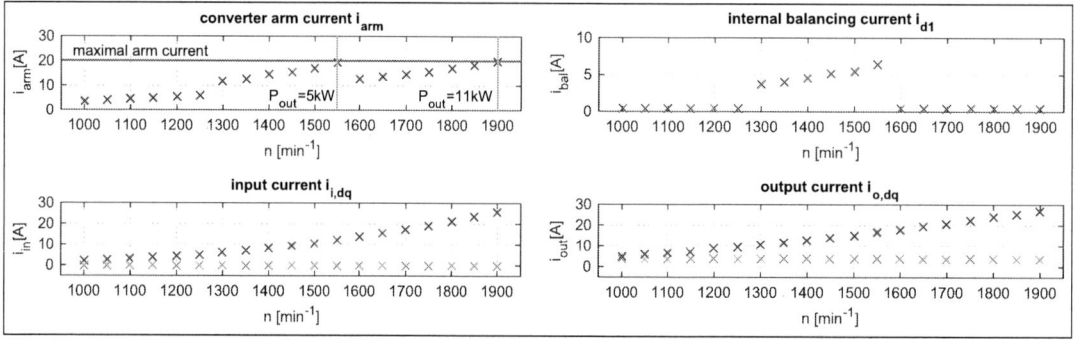

Fig. 5: short time averaged measurements with the prototype when driving a machine and a quadratic load profile. In addition the balancing, input and output currents

In Figure 5, an acceleration of an induction machine from $n_0 = 1000\text{min}^{-1}$ to $n_1 = 1900\text{min}^{-1}$ is shown. The load is a converter-fed DC machine with a quadratic load profile. During acceleration, the point of similar input and output frequency must be overcome. Therefore, additional internal currents are produced to compensate the power components from equation (8). The maximum arm current is never exceeded. It reaches its maximum value ($i_{\text{arm,max}} = 20\text{A}$) where the output frequency is similar to the input frequency and where $n_{\text{IM}} = n_1$. The machine can be accelerated without oversizing the converters semiconductors.

5. Conclusion

This paper presents an analytical solution for the performance of the Modular Multilevel Matrix Converter (M3C) in operation points for similar input and output frequencies. The necessary balancing power and the effect on the limit of the arm currents are calculated. A laboratory scaled prototype with a rated power of 15 kW demonstrates the outstanding performance of the M3C. With this results, the M3C is suitable for high speed drive applications where $\omega_N = 2\omega_i$ and a quadratic load characteristic. The current limits allow additional output currents to accelerate a load machine to its nominal speed. Subsequently, the cell capacitors and semiconductors of the M3C do not have to be oversized in order to accelerate a machine with a high nominal frequency e.g. for compressor applications.

6. Acknowledgment

The authors would like to thank the DFG (German Research Foundation) which finances this research project under grant BR 1780/8-2.

7. References

[1] F. Kammerer, D. Braeckle, M. Gommeringer, M. Schnarrenberger, and M. Braun. Operating performance of the modular multilevel matrix converter in drive applications. In *PCIM Europe 2015; International Exhibition and Conference for Power Electronics, Intelligent Motion, Renewable Energy and Energy Management; Proceedings of*, pages 1–8, 2015.

[2] K. Ilves, L. Bessegato, and S. Norrga. Comparison of cascaded multilevel converter topologies for ac/ac conversion. In *Power Electronics Conference (IPEC-Hiroshima 2014 - ECCE-ASIA), 2014 International*, pages 1087–1094, 2014.

[3] F. Kammerer, M. Gommeringer, J. Kolb, and M. Braun. Energy balancing of the modular multilevel matrix converter based on a new transformed arm power analysis. In *Power Electronics and Applications (EPE'14-ECCE Europe), 2014 16th European Conference on*, pages 1–10, 2014.

[4] W. Kawamura, M. Hagiwara, and H. Akagi. A broad range of frequency control for the modular multilevel cascade converter based on triple-star bridge-cells (mmcc-tsbc). In *Energy Conversion Congress and Exposition (ECCE), 2013 IEEE*, pages 4014–4021, 2013.

[5] W. Kawamura, M. Hagiwara, and H. Akagi. Control and experiment of a 380-v, 15-kw motor drive using modular multilevel cascade converter based on triple-star bridge cells (mmcc-tsbc). In *Power Electronics Conference (IPEC-Hiroshima 2014 - ECCE-ASIA), 2014 International*, pages 3742–3749, May 2014.

[6] F. Kammerer, J. Kolb, and M. Braun. Fully decoupled current control and energy balancing of the modular multilevel matrix converter. In *Power Electronics and Motion Control Conference (EPE/PEMC), 2012 15th International*, pages LS2a.3–1–LS2a.3–8, Sept 2012.

[7] F. Kammerer, J. Kolb, and M. Braun. A novel cascaded vector control scheme for the modular multilevel matrix converter. In *IECON 2011 - 37th Annual Conference on IEEE Industrial Electronics Society*, pages 1097–1102, Nov 2011.

PCIM Europe 2016, 10 – 12 May 2016, Nuremberg, Germany

Smart Supercapacitor based DC-link Extension for Drives offers UPS Capability and acts as an Energy Efficient Line Regeneration Replacement

Jens Onno Krah[1], Nico Sieweke[2], Christoph Klarenbach[2], Markus Höltgen[1], David Langhals[1]

[1] TH Köln, Campus Deutz, Betzdorfer Straße 2, 50679 Köln, Germany
[2] Beckhoff Automation, Hülshorstweg 20, 33415 Verl, Germany

Email: Jens_Onno.Krah@TH-Koeln.de

Abstract

Today most machines in industrial automation are build without UPS functionality. In case of a mains failure, a DC-link extension for drives can be used to maintain the voltage of the DC-link at a specified level to bypass the failure without disturbance and/or to bring the machine to a controlled stop.

Supercapacitors provide peak power and back-up power. A smart supercapacitor (SC) based DC-link extension can offer uninterruptable power supply (UPS) capability and acts also as an energy efficient line regeneration replacement. Peak power requirements for acceleration and deceleration can be provided by the SC based energy storage system. The size of the AC power supply can be reduced to lower power ratings.

1. Introduction

A notebook with rechargeable battery offers inherent UPS functionality. In case of an unexpected mains failure the battery keeps the PC operating for several hours. Some industrial PCs (IPC) are utilizing SC technology to operate the IPC for several seconds. The capacity is high enough to bridge a short power failure. In case of longer mains failures the UPS functionality allows to save persistent data on a solid-state-drive (SSD).

Low power servo systems – single-digit kilowatt range – used in robotic and handling applications are typically based on a common DC-link with a reasonable electrolytic capacity C. Mains failures up to 20 ms can be bridged by this technology. The storable recuperation energy E_{rec} depends on the rectified voltage v_{rect}, the maximal allowed DC-link voltage v_{max} and the DC-link capacity C.

$$E_{rec} = \tfrac{1}{2} C \left(v_{max}^2 - v_{rect}^2\right) \qquad (1)$$

In larger drives film capacitors are common due to their higher power density and their optional higher voltage ratings compared to electrolytic capacities. Disadvantageous is the even lower energy density.

The dynamic energy storage DES from Michael Koch can be operated in conjunction with most frequency converters and servo drives operating with a maximum DC-link voltage of 850 V. The unit can store up to 0.5 Wh by using an electrolytic capacity and a DC-DC-converter. It is expandable with additional capacity modules [1].

A machining center typically needs about 3 kW power in average and up to 50 kW peak power are used to accelerate a high speed spindle. The kinetic energy of a fast rotating machine tool spindle is typically about 20 Wh. Without energy storage the necessary power supply rating corresponds to the required peak power.

© VDE VERLAG GMBH · Berlin · Offenbach

Using an active front end for line regeneration is not always economical:

- An active front end inverter requires more space and increases the initial cost compared to a simple diode based rectifier.
- The rectifying efficiency drops from about 99% to 98% or less due to semiconductor switching losses and losses of the current smoothing inductors.

Therefore most servo systems typically do not use line regeneration. A brake chopper limits the DC-link voltage by switching the recuperated energy to a resistor where it is converted to heat.

Machining centers can benefit from Uninterruptable Power Supplies (UPS):

- An UPS can prevent expensive work-pieces from getting destroyed by a power outage especially in case of long processing times.
- Continuously running diesel generators are available at reasonable cost, but they are not designed for fast load changes or power regeneration. High dynamic load changes can result in overvoltages, which can be critical for sensible devices.

Machine internal energy storage offers several advantages:

- UPS functionality; short term bridge power for orderly shut down
- Power buffering for large momentary power surges allows lower power supply ratings
- Higher efficiency due to no or less braking resistor heat

2. Selection of a suitable supercapacitor

The maximum voltage of a single SC-cell is limited by design to around 2.7 V. In order to achieve higher voltages quantities of cells are connected in series. The capacity of an individual SC-cell is typically between 1000 and 3000 F. Several suppliers are offering preassembled SC-modules with integrated voltage balancing. A SC-module with a rated voltage of 81 V and a capacity of 40 F for example is built with thirty 1200 F, 2.7 V cells. Usually SC-modules also offer monitoring signals for over-voltage and/or over-temperature for protection [2].

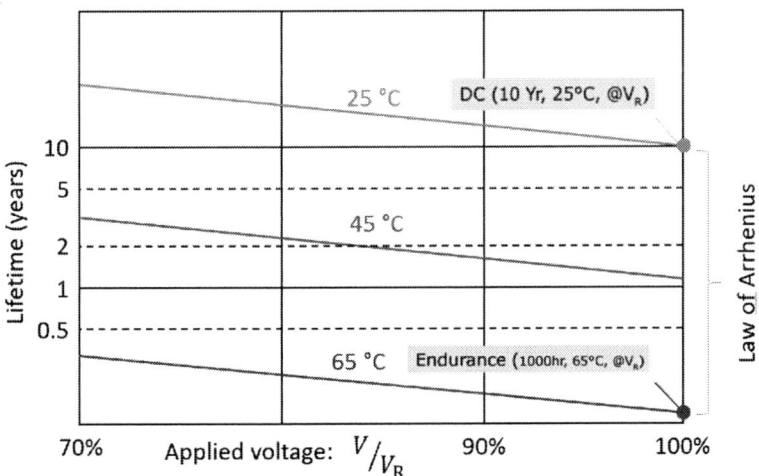

Fig. 1: Lifetime of a SC is basically a function of operating voltage and temperature according to the low of Arrhenius [5].

Unlike batteries, supercapacitors do not have a hard end of life criteria. Supercapacitors degradation is apparent by a gradual loss of capacitance C and a gradual increase in equivalent serial resistance *ESR*. End of life is when the capacitance and resistance is out of the application range and can differ depending on the application [3, 4]. Therefore life prediction is easily possible by condition monitoring of C and R_{ESR}.

Most SC are specified with an operating temperature range of -40°C ... 65°C. The specified lifetime depends strongly on temperature and voltage, fig 1. A rule of thumb – related to the law of Arrhenius – states that a reduction in temperature of 10 K, doubles the lifetime. In addition a voltage decrease of 0.1 ... 0.2 V at voltages beyond 2 V also doubles the lifetime [5].

The maximal environmental temperature depends on the internal losses P_{loss} and the thermal resistance R_{th}. Self-heating is produced almost exclusively by current related losses:

$$P_{loss} = I_{rms}{}^2 \cdot ESR \tag{2}$$

$$\Delta T = \overline{P_{loss}} \cdot R_{th} \qquad \text{(thermal time constant is here not counted in)} \tag{3}$$

Where *ESR* (R_i) is the equivalent series resistance of the SC-module. To ensure a reasonable lifetime it is recommended to operate significantly below the maximal specified temperature and also below the rated voltage. For example an environment temperature of maximal 30°C is required to keep the SC internal temperature below 40°C in conjunction with 10°C increase due to the rms. current.

Key parameters for sizing the SC are the rms. current in conjunction with the environment temperature and of course the amount of energy to store.

3. Operating the supercapacitor

There are a number of reasons why a DC-DC-converter should be used to utilize a SC as efficient DC-link extension, fig. 2.

- The SC current is in all operation modes under control
- Allowing a notable SC voltage span makes a significant part of the total storable energy available
- In lower power systems it is more efficient to use SC-modules with rated voltages much lower than the inverter DC-link voltage
- Pre-charging and shutting down is without difficulties and can be used for condition monitoring

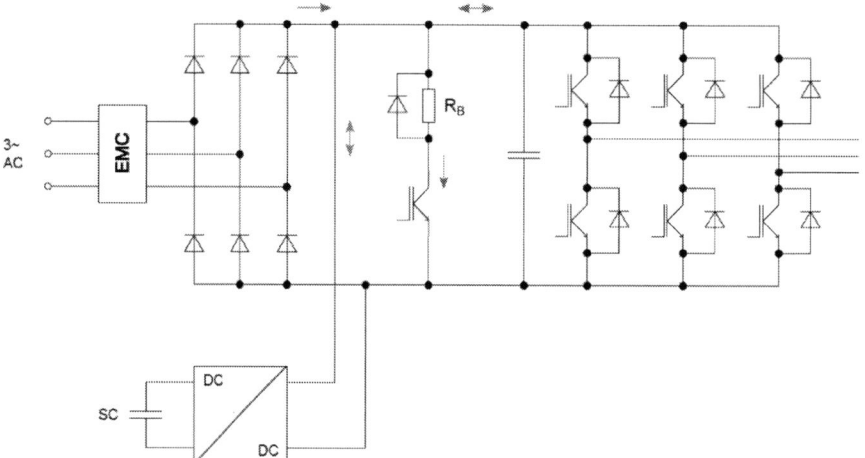

Fig. 2: The supercapacitor (SC) is physically connected via a DC-DC-converter to the DC-link of a drive

4. Considering the operational conditions

UPS applications are requiring much less power – peak and rms. – than the SC maximum peak power. Therefore in UPS applications SC internal heating is not an issue because of the thermal mass in combination with the rare occurrence of power faults.

Energy storage use-cases are different. As an example a machine tool spindle has to be cyclically accelerated and decelerated. Typically this is performed with a constant acceleration rate α. This leads to a triangular power profile.

$$P = t_{motor} \cdot \omega = \alpha \cdot J \cdot \omega \tag{4}$$

Peak power is usually defined by max torque t_{max}, max speed ω_{max} and efficiency η.

$$P_{Peak,acc} \approx t_{max} \cdot \omega_{max} \cdot {}^1/_\eta \tag{5}$$

$$P_{Peak,dec} \approx t_{max} \cdot \omega_{max} \cdot \eta$$

Assuming a nearly constant SC voltage the current profile is also triangular. The crest factor C of a triangle wave is:

$$C_{Triangle} = \frac{|I_{peak}|}{I_{rms.}} = \sqrt{3} \tag{6}$$

In case of constant power acceleration/deceleration the current is also nearly constant which leads to a crest factor of $C = 1$.

Also the pulse (t_1) / period (T_P) relation of the application has to be considered:

$$C_{pulse-period} = \sqrt{\frac{T_P}{t_1}} \tag{7}$$

In an example the peak power is 30 kW and the SC voltage is 300 V. The efficiency is neglected. This leads to a peak SC current of

$$I_{peak} = \pm \frac{30\ kW}{300V} = \pm 100\ A$$

During acceleration or deceleration the rms. current is

$$I_{rms,acc} = \frac{|I_{peak}|}{C_{Triangle}} = \frac{100A}{\sqrt{3}} \approx 57.7\ A$$

Considering a pulse-period relation of $t_1 = 1$ s and $T_P = 10$ s the SC rms. current is

$$I_{rms} = \frac{I_{rms,acc}}{C_{pulse-period}} = \frac{57.7\ A}{\sqrt{\frac{T_P}{t_1}}} = \frac{57.7\ A}{\sqrt{10}} = 18.3\ A$$

With knowledge of the SC-module ESR, the internal power loss (averaged) can be estimated:

$$P_{loss} = I_{rms}{}^2 \cdot ESR = (18.3\ A)^2 \cdot 12m\Omega \approx 4\ W \tag{8}$$

By multiplying with the thermal resistance R_{th} the SC internal temperature increase can be projected:

$$\Delta T = P_{loss} \cdot R_{th} \approx 4W \cdot 0.5\frac{K}{W} = 2K \tag{9}$$

Even at the end of the SC lifetime – typically characterized by a doubled ESR – the temperature rise is here only 4 K which keeps the SC temperature low, as required.

5. Sizing of inverter and choke

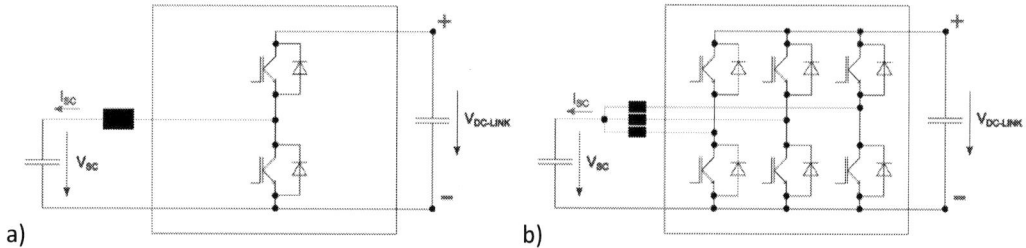

Fig. 3: Realization of the DC-DC-Converter   a) with one choke and one inverter leg
b) with 3 chokes and 3 inverter legs and 120° phase shifted PWM

Fig. 3a shows the schematic of the SC-based DC-link extension using a two-quadrant DC/DC converter. This configuration is for example used in [1]. Storing energy in the SC is utilizing the step-down mode; feeding back energy into the DC-link is using the step-up mode. Due to cost, size and losses of the choke a high current ripple is considerable. By dividing the two-quadrant chopper into three individual inverter lags – each with its current smoothing choke – with phase shifted PWM the current ripple can be reduced significantly to keep the SC rms. current low, fig. 3b. Fig. 4 illustrates the current waveforms.

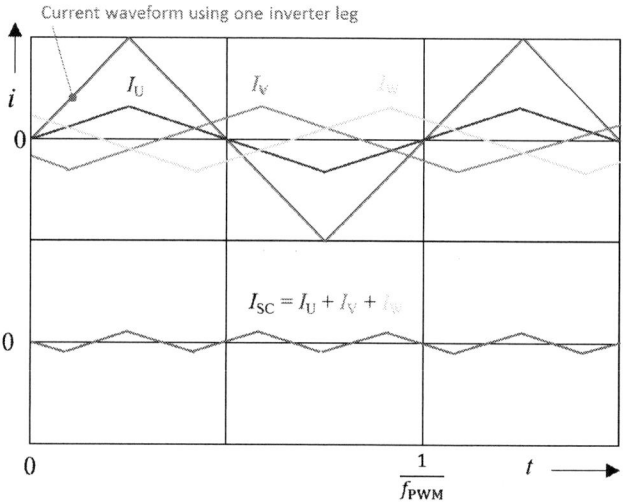

Fig. 4: Waveform of the SC current. By utilizing 3 inverter legs with 120° phase shifted PWM instead of only one the SC current ripple is reduced by a factor of nine.

6. Control strategy

Using a chopper with a ballast resistor to limit the DC-link voltage the regeneration energy will be converted into heat after the voltage is increased above the specified level. This level is usually configured to approximately 850 V to protect the power stage from damage. It is advantageous to operate several inverters by one common DC-link voltage. The DC-link current distribution – charge and discharge – is related to the individual capacity C_k in relation to the total capacity C_Σ:

$$\frac{I_k}{I_\Sigma} = \frac{C_k}{C_\Sigma} \tag{10}$$

A patent for power distribution between a plurality of parallel-connected ballast circuits is presented in [6].

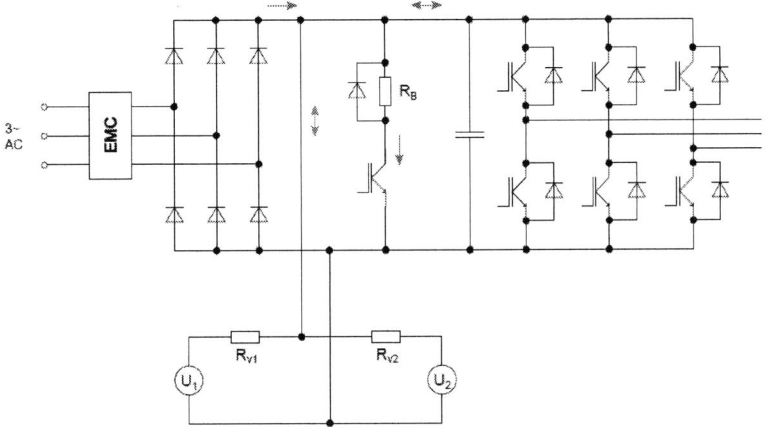

Fig. 5: By using more than one voltage source with similar rated voltages in parallel the current distribution is defined by the shown (internal) resistances of the voltage sources.

Fig. 5 shows two voltage sources with similar voltages v_1, v_2 connected in parallel with the DC-link. The resulting internal resistance R_Σ is just the parallel connection of the individual resistors R_i.

$$R_\Sigma = \left(\sum_i \frac{1}{R_i} \right)^{-1} \tag{12}$$

The current distribution is related to the individual internal resistances:

$$\frac{I_k}{I_\Sigma} = \frac{R_\Sigma}{R_k} \tag{13}$$

A method for power distribution between a plurality of parallel-connected active front end inverters is presented in [7]. Here a virtual resistor takes care of the current distribution. Due to this control strategy the active front end converter can also benefit from the DC-link capacity for energy storage due to the (within limits) inconstant DC-link voltage.

The control of the SC based DC-link extension uses also this virtual resistor technology for current distribution. The resulting current from the DC-link i_{ES} can be easily calculated:

$$i_{ES} = \frac{v_0 - v_{dc}}{R_v} \tag{14}$$

If v_0 is configured to 600 V, R_v is configured to 1 Ω, and the DC-link voltage is actually 590 V the current command will be $i_{ES} = +10$ A.

The calculated current loop command value is limited according to the SC voltage and also to system restrictions. At high SC voltage charging is prohibited, at low SC voltage further discharging is prevented.

This control method has several advantages:

© VDE VERLAG GMBH · Berlin · Offenbach

- The current command is inherently generated with a slew rate limitation due to the DC-link capacity which limits the DC-link voltage slew rate.
- Two or more energy storage systems can be operated independent of each other in parallel with in-built current distribution. The systems are always stable and they need no parameter adaption.
- The energy storage capability of the connected DC-link capacity can be used due to the limited DC-link floating capability.

7. Experimental results

Experimental results are created with a 25 kW / 100 Wh supercapacitor based DC-link extension in conjunction with a high speed spindle during periodically acceleration and deceleration.

For energy storage Ultra-Capacity modules form Ioxus are used:

$$C = 40 \text{ F}, \ U_{max} = 81 \text{ V}, \text{ESR} = 10.5 \text{ m}\Omega \qquad [2]$$

Resulting capacity of four serial connected modules to build the energy storage system:

$$C = 10 \text{ F}, \ U_{max} = 324 \text{ V}, \text{ESR} = 42 \text{ m}\Omega$$

Operational energy storage capacity:

$$E_{usable} = \frac{1}{2}C\left(U_{max}^{2} - U_{min}^{2}\right) = \frac{1}{2} \, 10\text{F} \, (324\text{V}^2 - 162\text{V}^2) = 393\,660 \text{ Ws} \approx 100 \text{ Wh}$$

A standard AC-servo drive inverter from Beckhoff [8] is used as 3 phase buck / boost converter. Only the control firmware is modified and the max. phase current was configured to 30 A. This results in 90 A energy storage peak current. The IGBT switching frequency is configured to 16 kHz. Three 120° phase shifted carriers are used. One toroid inductor per phase with $L = 0.5$ mH is chosen to realize the buck/boost converter with reasonable current ripple.

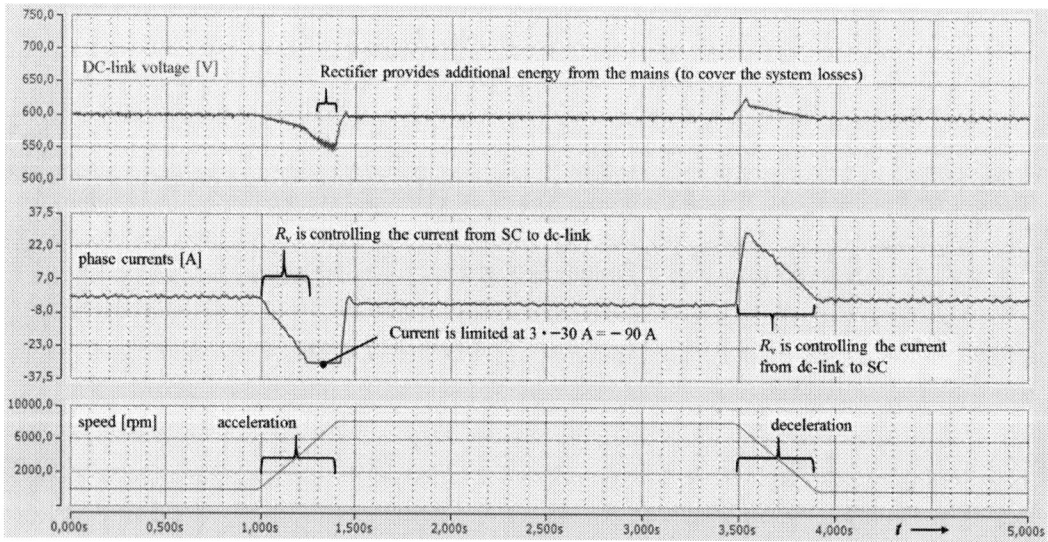

Fig. 6: High speed spindle cycle – TwinCAT scope plot
Top: DC-link voltage [Volt]   Middle: phase currents [Ampere] (the 3 phases are similar)
Bottom: spindle speed [rpm]:   spindle acceleration starts at $t = 1$ s,   dec. starts at $t = 3.5$ s

Figure 6 shows the operation of the energy storage system in conjunction with a machining spindle. At t = 1.0 s the spindle starts to accelerate to 9000 rpm. At t = 3.5 s the spindle deceleration starts. The three inductors current are also plotted. Due to scaling and high performance control the phase current waveforms are appearing in the figure almost as a single plot.

During time periods without acceleration and deceleration the DC-link voltage is stabilized to 600 V. In case of UPS operation it is important that not too much energy is requested. Usually is not necessary to accelerate the spindle in case of a mains failure.

For the period of spindle acceleration the voltage is ramping down which correlates – due to the virtual resistor – with an increasing current from the SC storage to the DC-link. At t = 1.3 s the rectifier starts to provide additional current/energy from the mains to the DC-link which covers the system losses.

During spindle deceleration the DC-link voltage rises which correlates – due to the virtual resistor – with an increasing current from the DC-link to the SC storage.

8. Conclusion

The contribution shows that using virtual resistors to control the energy flow between DC-link and supercapacitor energy storage results in a smart control strategy. The energy storage system needs no tuning and is always stable – even if more than one system is operated at a common DC-link circuit.

9. References

[1] European Patent Office EP 2 372 892 B1; "Vorrichtung und Verfahren zur Zwischenspeicherung elektrischer Bremsenergie eines an einem Wechselrichter betriebenen Motors".

[2] Datasheet iMOD X-SERIES, www.ioxus.com/english/products/modules/

[3] H. Gualous, H. Louahlia, R. Gallay: "Supercapacitor Characterization and Thermal Modelling With Reversible and Irreversible Heat Effect", IEEE Transactions on Power Electronics Volume 26, 2011, p. 3402 – 3409.

[4] G. Alcicek, H. Gualous, P. Venet, R. Gallay, A. Miraoui: "Experimental study of temperature effect on ultracapacitor ageing", European Conference on Power Electronics and Applications, 2007.

[5] P. Kreczanik, P. Venet, A. Hijazi, G. Clerc: „Study of Supercapacitor Aging and Lifetime Estimation According to Voltage, Temperature, and RMS Current" IEEE Transactions on Industrial Electronics, 2014, pp: 4895 – 4902.

[6] Patent: DE 37 35 621: "Method for power distribution between a plurality of parallel-connected servo drive amplifier ballast circuits, and a ballast circuit operating in accordance with this method".

[7] Patent: EP2820752 "Semi-active front end converter with vector control of reactive power"

[8] AX5140 Servo drive www.beckhoff.com

PCIM Europe 2016, 10 – 12 May 2016, Nuremberg, Germany

New LV Wide SOA Power MOSFET technology for Linear Mode operation

Filippo Scrimizzi, Gaetano Bazzano, Daniela Cavallaro, Marco Comola, Giuseppe Consentino, Stefania Fortuna, Giuseppe Longo, Gaetano Pignataro,

STMicroelectronics, Stradale Primosole 50, Catania, Italy,

filippo.scrimizzi@st.com, gaetano.bazzano@st.com, daniela.cavallaro@st.com, marco.comola@st.com, giuseppe.consentino@st.com, stefania.fortuna@st.com, giuseppe-mos.longo@st.com, gaetano.pignataro@st.com.

Abstract

Many applications, especially in the Telecom (hot-swap + PoE) and Automotive field, require Power MOSFET robustness in linear mode. Modern devices have been designed for ever-decreasing on-resistance and very high current capability at the expense of reduced FBSOA (Forward-biased Safe Operating Area).

Linear Mode operation highlights a different scenario than a switched-mode application where the design is usually driven by dynamic and static power losses. If the device is operated in Linear Mode, the power dissipation is quite high because it works with high voltage drop and high current that could result in rapid increase of the junction temperature that may lead the Power MOSFET to thermal runaway or in other words to an unstable condition that occurs when the junction temperature increases without control until device failure occurs.

The objective of the following paper is to evaluate the key requirements of a Power MOSFET working in a hot-swap application and to compare the latest advanced LV technology, not suitable for it, with the new tailored technology designed by ST to satisfy the harsh requirement of the Linear Mode Operation.

© VDE VERLAG GMBH · Berlin · Offenbach

Introduction

High-availability systems, such as servers, network switches, redundant-array-of-independent-disk (RAID) storage, and other forms of communications infrastructure, need to be designed for near-zero downtime throughout their useful life. If a component of such a system fails or needs updating, it must be replaced without interrupting the rest of the system so it means that the board or module will have to be removed and its replacement plugged in while the system remains up and running. This process is known as hot swapping

When a line card is plugged into a live backplane, the card's discharged power supply filter capacitors present a low impedance and demand a large, sudden "inrush" current. This sudden high load can cause the backplane's power supply to collapse.

Hot-swap controllers, provide inrush current limiting when the card is first inserted, and short-circuit protection while the card is in operation.

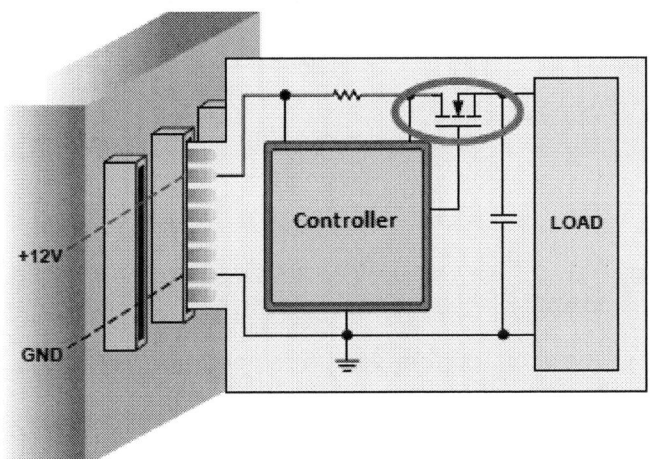

Figure 1: Hot-swap controller

A hot-swap controller limits the inrush current by slowly increasing the on-state resistance of an N-channel MOSFET. When the board is first plugged in, the controller slowly enhances the MOSFET, allowing the voltage at the MOSFET's drain to rise from zero volts (or fall from zero volts for PC boards powered by a negative supply).

Following figure shows the typical SOA curve included in most power MOSFET datasheets. The constant power curves, shown to the right of constant current line within the SOA boundary, are extracted from the thermal data with the assumption that the junction temperature is essentially uniform across the power MOSFET die. The dissipated power does not cause a catastrophic failure to the device, but brings its junction temperature up to the maximum guaranteed temperature when the applied power pulse is evenly distributed on the die surface.

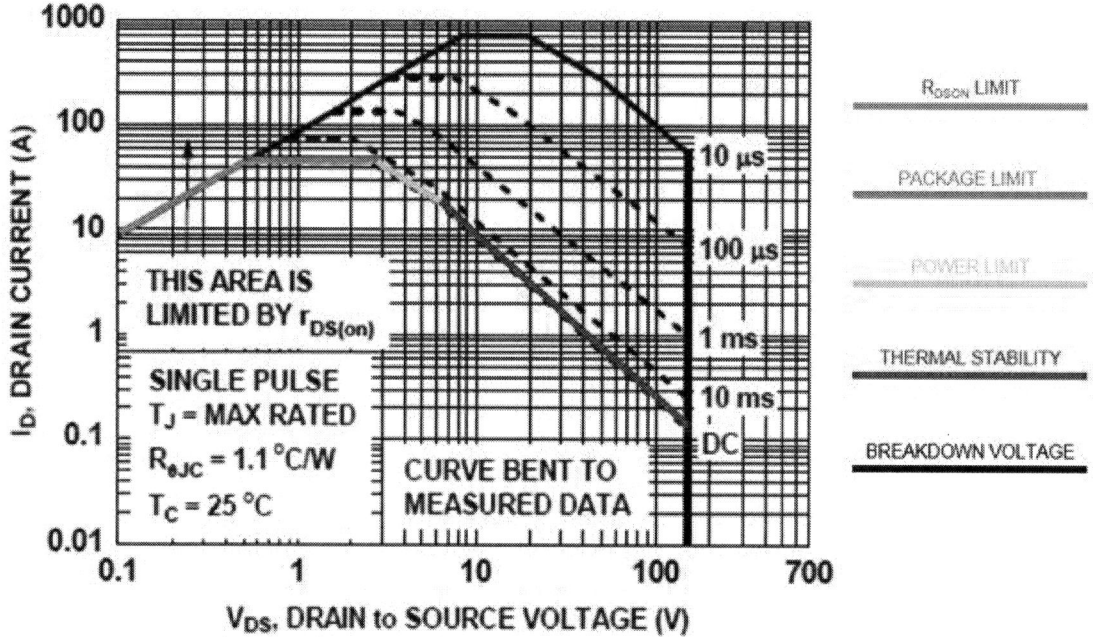

Figure 2: SOA limits

The parameter that measures such thermal instability is the thermal coefficient (TC). The thermal coefficient is defined as the derivative from the drain current versus temperature $\left(\frac{\partial I_D}{\partial T}\right)$ and is generally normalized by the active area or device perimeter in order to compare the performances of different devices or structures.

When TC is zero, or negative, by increasing the temperature, the drain current decreases thus, the device works in thermal stability conditions.

Whenever the TC is positive, the device may fail. That depends on the capability of the whole die thermal system to catch away the heat per unit area and time developed by the electrical power pulse. If the heat produced by unit time can be totally extracted from the device, then the power MOSFET works in safety conditions.

The above reported conditions are set by the external application conditions that together with the phisical behaviour linked to the thermal stability of the mosfet elementary cells, define the thermal stability of the device working in linear mode.

The heat developed inside the junction is due to the electrical power dissipation in linear mode:

$$P_D = V_{DS} \cdot I_D = \frac{\Delta T}{R_{thj-a}}$$

When the temperature rises the drain current changes depending on the thermal coefficient; the device can manage power till reaching of the failure point

$$V_{DS} \cdot \frac{\partial I_D}{\partial T} < \frac{\partial \Delta T}{\partial T} \cdot \frac{1}{R_{thj-a}}$$

Sometimes even if the structural features of the device don't match this requirement, it doesn't fail because the R_{thj-a} is enough low to dissipate the generated heat due to the applied power, the failure happen when

$$\left| \frac{\partial I_D}{\partial T} \right| = \frac{1}{V_{DS}} \cdot \frac{1}{R_{thj-a}}$$

Here below are reported the transcharacteristics at different temperatures with the crossing point at 0 TEMPCO and the thermal coefficient of the MOSFET. The lower is the 0 TEMPCO in term of Vgs and Id, the better is the MOSFET in self-limiting the current during the linear mode operation.

The crossing among the limits conditions defined by the application (Vds voltage drop and thermal budget/Rth) and the thermal coefficient define the thermal instability current window. Therefore at a fixed Vds and Rth the MOSFET cannot withstand for long time any current inside this forbidden window.

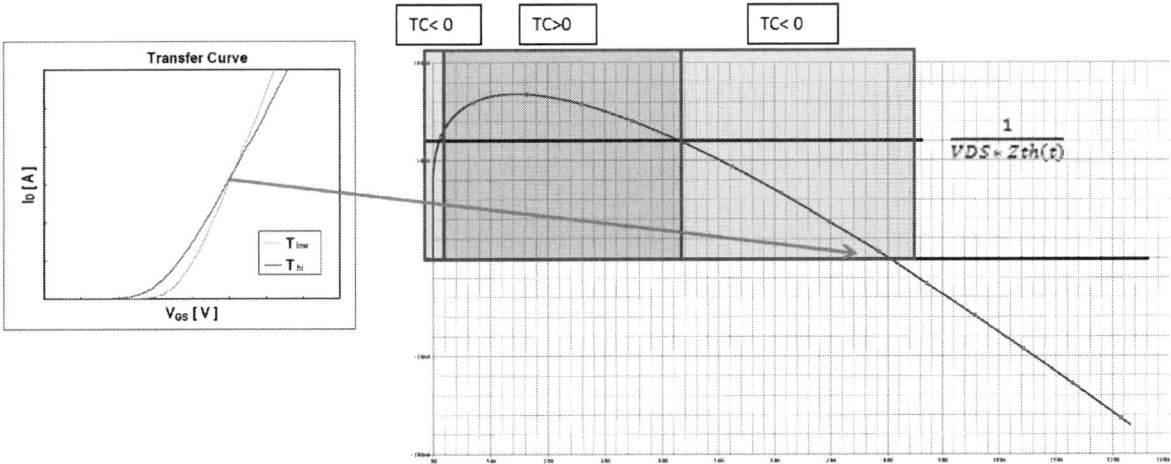

Figure 3: Thermal coefficient and Thermal instability current window

$$\frac{\partial P_{generated}}{\partial T} > \frac{\partial P_{dissipated}}{\partial T} \quad \rightarrow \quad V_{DS} \cdot \frac{\partial I_{DS}}{\partial T} > \frac{1}{Z_{thJC}\,(t_{pulse})}$$

So practically:

$$Instability = \frac{1}{V_{DS} \cdot Z_{thJC}\,(t)} < \frac{\Delta I_{drain}}{\Delta T}$$

To avoid this unstable condition, ST has introduced a new technology able to meet a wider FBSOA capability in conjunction with an extremely low on-state resistance (trade-off between Linear Zone operation and switching operation efficiency).

Here below are reported the comparison in terms of thermal coefficient, at high Vds and Tc, among a standard low Rds(on) trench technology and its equivalent die-size wide SOA one.

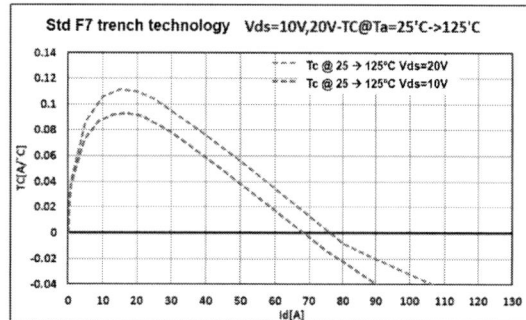 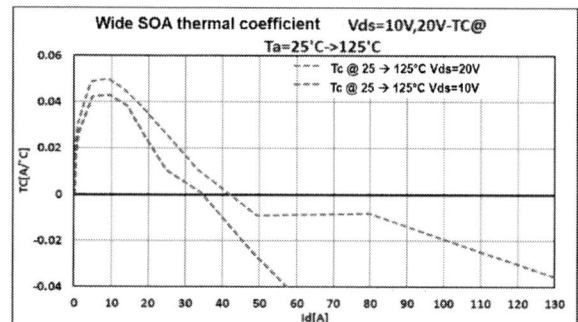

Figure 4: TC vs Id in STD F7 trench technology and Wide SOA

Comparing those two curves is evident the maximum value of the two parabolic curves and the current value starting from which the thermal coefficient becomes negative (thermal stability region for the MOSFET).

The wide SOA technology has lower maximum thermal coefficient and narrower positive coefficient area (TC<0 when Ids > 43A at Vds=20V).

The slope of the Trans-characteristic shows a controlled current gain. The lower is the slope the better is the linear mode capability.

Basing on a detailed characterization at high temperature we run a simulation trial to define the thermal stability behavior of our wide SOA technology and the best competition.

Thanks to a dedicated silicon structure elementary cells and package features, the current trend vs the working time in linear mode at high Vds shows an almost flat current curves.

© VDE VERLAG GMBH · Berlin · Offenbach

This remarks the capability of the device to manage that working conditions (see below reported simulation curves).

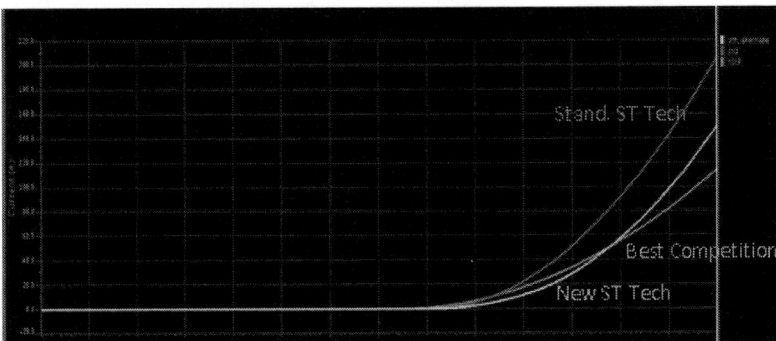

Figure 5: Trans-characteristics

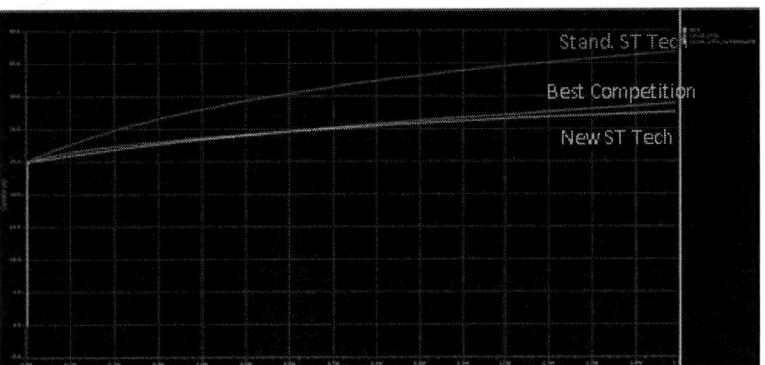

Figure 6: Id vs Time

The benchmark

To see the performances of our devices in a Hot Swap application we used the below simple board schematic:

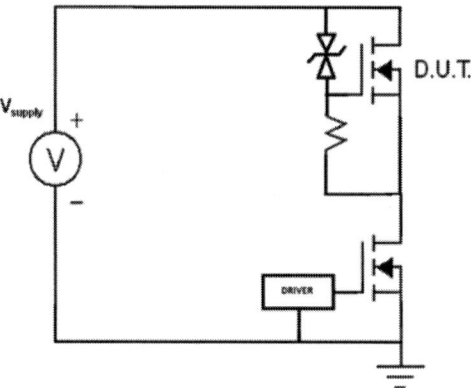

Figure 7: Board schematic

Once the driver transistor is on, the pre-charged capacitance connected to the power supply will discharge through the DUT. The external gate-drain zener diodes will clamp the Vds voltage at the selected value and the gate-source resistor will provide the relevant Vgs to flow the fixed Ids current.

A STD trench technology device fixing the board at Vds=30V, Ids=25A (typical working condition in TELECOM application) fails at around 800µs. Here below is reported the relevant scope waveform:

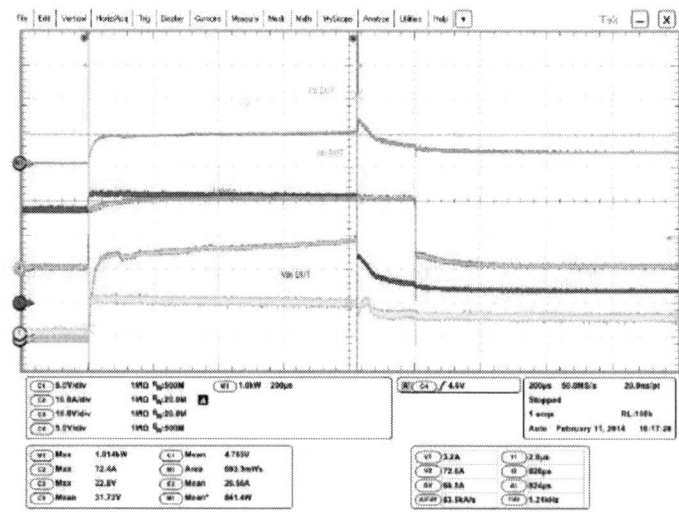

Figure 8: 25A/30V test results using STD Trench techonology

PCIM Europe 2016, 10 – 12 May 2016, Nuremberg, Germany

Applying the same testing condition to a new WIDE SOA technology device, we are able to survive at least 2 ms that represents the minimum time gap required from the main CTMs.

Wide SOA device can withstand even Vds=40V, Ids=25A, T_{pulse}=1ms in linear mode condition.

Here below are reported the relevant scope waveforms:

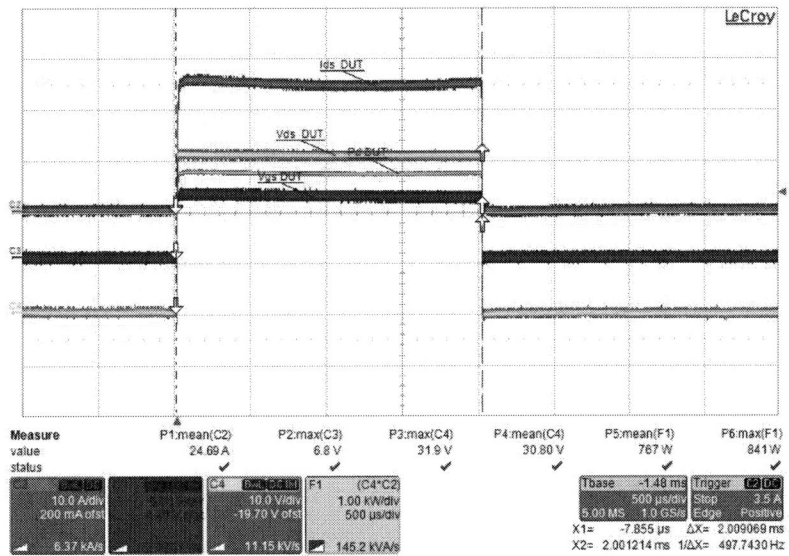

Figure 9: 25A/30V test results for Tp=2ms using Wide SOA Trench technology

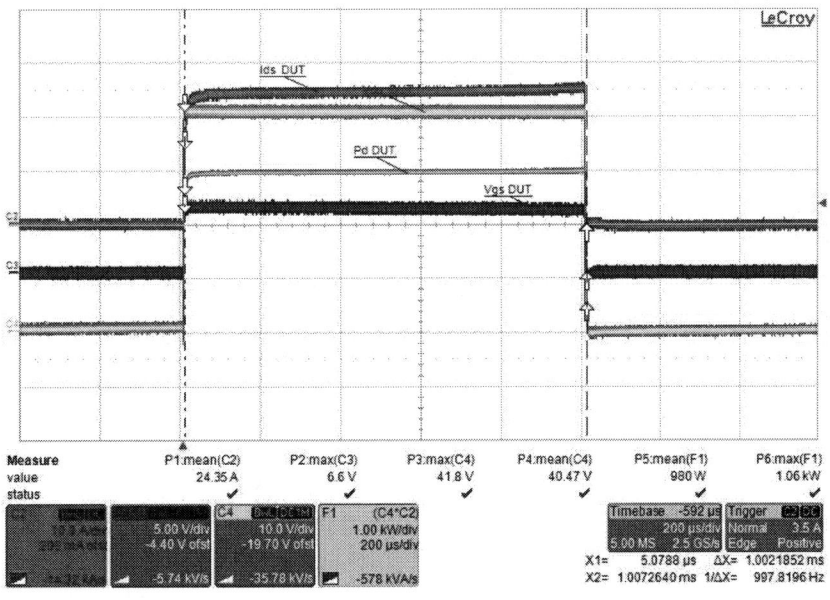

Figure 10: 25A/40V test results for Tp=1ms using Wide SOA Trench technology

© VDE VERLAG GMBH · Berlin · Offenbach

Conclusions

The new wide SOA technology ST got for 100V MOSFETs shows good results both in terms of thermal coefficient and current gain control providing the thermal stability in the high voltage and current SOA area becoming less sensitive to the FBSOA drift (Spirito effect).

The already available samples are now able to withstand STD CTM's hot swap condition.

The process and design features of the new «WIDE SOA» technology allow to achieve a good linear mode ruggedness;

References

[1]   A. Consoli, F. Gennaro, A. Testa, G. Consentino, F. Frisina, R. Letor, A. Magri' "Thermal Instability of Low Voltage Power MOSFET's", IEEE Transaction on Power Electronics, Vol. 15, NO. 3 May 2000

[2]   G. Consentino, G. Bazzano "Safe Operating Limits in Linear Mode for the latest generation of Low Voltage Power MOSFETs: a mathematical model and experimental results" PCIM Europe 2005, Nuremberg, Germany

Field Stop Trench IGBT Process Parameter Calibration for Advanced Predictive Prototyping

Mehrdad, Baghaie Yazdi, Fairchild Semiconductor, Germany,
mehrdad.baghaie@fairchildsemi.com
Hermann, Fischer, Fairchild Semiconductor, Germany, hermann.fischer@fairchildsemi.com
James, Victory, Fairchild Semiconductor, Germany, james.victory@fairchildsemi.com
Detlef, Conrad, Synopsys, Germany, detlef.conrad@synopsys.com

Abstract

This paper advances the state of the art through a first time implementation of a multi-stage optimizer procedure within the Synopsis Sentaurus TCAD environment to achieve precise capacitance calibration between a real and a simulated Field Stop Trench IGBT. The methodical approach is detailed, starting from a raw, non calibrated process to the novel optimizer results. The simulations are compared to measurements of the real device. In addition, an analytic model that governs the theory and aids the modeling of capacitances is derived.

1. Introduction

Prototyping and development of power semiconductor technologies such as Insulated Gate Bipolar Transistors (IGBT) or power MOSFETs can prove to be time consuming and costly without robust simulation methodologies. TCAD tools present a well known mechanism to qualitatively design new technologies. However in order to be able to use TCAD as a quantitative prototyping tool, one must assure that the base technology that is to be changed is modeled to the best precision possible. This requires a so called fully calibrated Process deck, which then enables accurate predictive simulation. In order to truly consider a process deck as calibrated, not only static, but also dynamic and AC characteristics of the resulting device need to match between simulation and real device performances. Often the more complicated AC characteristics in particular the capacitances are only marginally calibrated due to the difficulty of achieving a good physical match to real data. However, device capacitance and its derivatives contain a wealth of device physical information that can be exploited to achieve physically based TCAD calibration. In this paper we present an innovative method of using both analytical and software automated capacitance optimization to achieve the best possible match between TCAD simulation and real device measurement. The methods can then be further linked to physical SPICE models for optimized application specific device development [1].

For this paper the Fairchild Semiconductor Field Stop 3 (FS3) Trench IGBT with the designation FGH40T65SPD was used. This is a third generation trench IGBT with a current rating of 40 A and a break down voltage of 650 V.

2. Procedure and Results

In the following we will show, how starting from the initial capacitance simulations obtained by implementing the process flow of the FGH40T65SPD in a TCAD deck, one can finish with a highly matched result between simulation and measurement. First the initial simulations are shown, followed by the derivation of an analytic equation in order to fine tune process parameters. Using the results obtained from the analytic equation the Sentaurus Optimizer is used to finalize the process parameter adjustment and finding the best match to measurements.

2.1. Initial Simulation Results based on Process Run sheet

The initial point for generating the FS3-IGBT model in TCAD was to implement the run sheet of the device, as is typically done in setting up a TCAD deck. After a complete calibration according to the static characteristics, the initial Capacitance simulation vs. real data (data is acquired using the methods presented in [2]) is as seen in figure 1.

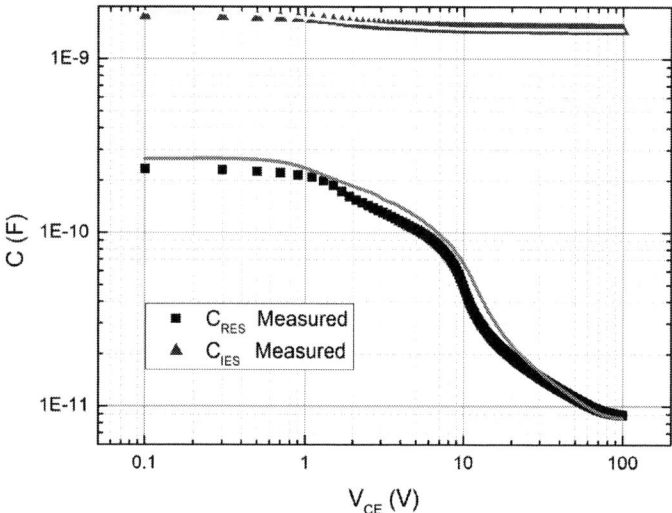

Figure 1 Simulated (solid lines) vs. measured (symbols) capacitance of IGBT. Simulations are performed with TCAD after implementing process run sheet and static calibration of process deck.

2.2. Manual Optimization based on Physical Equations

In order to improve matching between measured and TCAD simulated device, a better understanding of the influence of the doping profile on the various capacitances needs to be established. For an IGBT the three capacitances of interest are C_{RES}, C_{OESS} and C_{IES}, their definition with respect to the device contacts, gate, collector and emitter are shown in figure 2. The close relation of capacitances and physical properties of the devices is shown in figure 2 (right), it becomes clear that the epi, the JFET, the oxide thickness and the field stop layer all convolute to form the capacitance values which dictate the devices switching behavior.

A simple starting point, which we will not discuss in depth, is to check the calibration of C_{GG} (the total gate capacitance). Comparing measured and simulated total gate capacitance is a direct means to verify the calibration of the oxide thickness and channel doping parameter.

Under the assumption that the inversion layer forming during the sweeping of C_{GG} is the plate of the parallel-plate capacitor and the gate oxide the dielectric, the thickness of the oxide can be easily calculated with $C = \frac{\varepsilon_0 * \varepsilon_r * A}{A * t_{gox}}$.

Looking at the relation between the IGBT capacitances, shown in figure 2, it becomes evident that the most important to have well adjusted is C_{RES}. The transfer capacitance (C_{RES}) is only dependant on the gate-collector capacitance, and all both C_{IES} and C_{OES} have C_{RES} as a component. Furthermore the miller-plateau, seen in gate-charge measurements and switching tests, is a result of the gate-collector capacitance. A good understanding of this parameter is critical for gate driver design and for any optimization of switching characteristics.

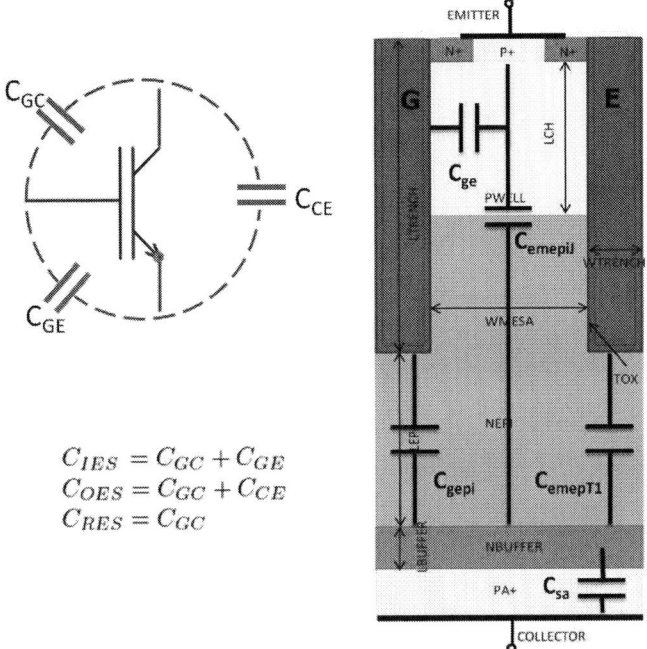

$$C_{IES} = C_{GC} + C_{GE}$$
$$C_{OES} = C_{GC} + C_{CE}$$
$$C_{RES} = C_{GC}$$

Figure 2 Left, simple schematic of IGBT capacitances and their relationships. Right side Field Stop IGBT cell with gate-emitter Trench and the correspondence of structural regions to capacitances within a cell.

In order to calibrate the transfer capacitance first knowledge of its relation to the doping profile of the IGBT needs to be established. Inspecting figure 2, it can be seen that C_{RES} can be considered as the series connection of the oxide, the JFET and the epi capacitors. Since the oxide capacitance C_{ox} has already been adjusted with C_{GG}, we are left with the junction and the epi capacitance. These two parameters are adjusted by determining the appropriate doping profiles within the TCAD deck. Under the assumption of basic physics resulting from the Maxwell equations (eq. 1) and the correlation between charge and capacitance (eq. 2) it can be concluded that depending on the location of the electric field, we are probing a different doping region of the IGBT.

$$\vec{\nabla} \bullet \vec{E}(\vec{r}) = \frac{\rho(\vec{r})}{\varepsilon} \tag{1}$$

$$Q = C * U \tag{2}$$

Under the assumption of the field dependence and the derived relationship of C_{RES} and the physical structure of the IGBT, two voltage ranges are chosen for calibrating the required doping parameters: the low voltage regime which is ranging from 0.01-10 V, and corresponds to the JFET doping and 13 – 70 V, which is correlated to the epi doping.

Following the first physical assumption, a second one is made, in order to derive an analytic equation that helps finding a calibration coefficient between measured device and TCAD generated deck derived from the process flow.

Starting with eq. 3 for the specific capacitance of the semiconductor depletion region, and substituting the expression for the depletion width (eq. 4), one can derive an equation that correlates doping, applied voltage and the first derivative of the measured capacitance.

$$C_{dpl} = \frac{\varepsilon_0 * \varepsilon_{si}}{W dpl} \tag{3}$$

$$C_{dpl} = \frac{\varepsilon_0 * \varepsilon_{si}}{\sqrt{\frac{2 * \varepsilon_0 \varepsilon_{si} * V_{CE}}{q * N_D}}} \tag{4}$$

$$\frac{dC}{dV} = -0.5 * \varepsilon_0 * \varepsilon_{si} * \sqrt{\frac{q * N_D}{2 * \varepsilon_0 \varepsilon_{si}}} * \frac{1}{(-V_{CE})^{3/2}} \tag{5}$$

From equation 5 one can derive a doping ratio between the measured and simulated capacitance that can be used to calibrate the TCAD deck. This ratio is given by $= \sqrt{\frac{m_s}{m_m}}$, where m_s and m_m are the simulated and measured slopes respectively of the capacitance. Applying a numeric iteration process for adjusting the slopes[1] of the simulated and measured transfer capacitance, the results shown in figure 3 are achieved.

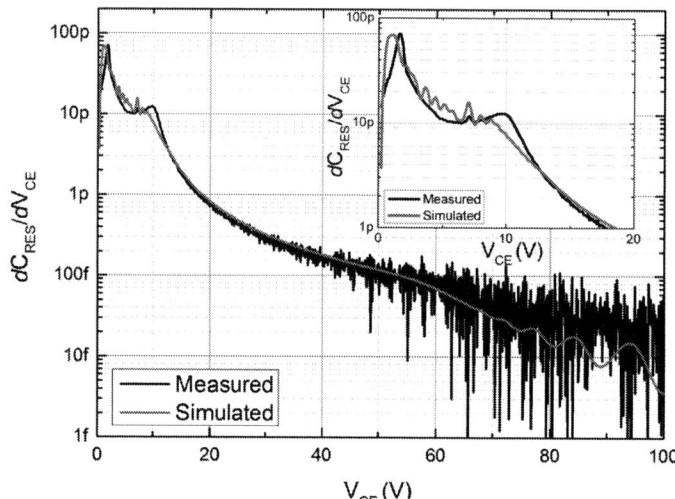

Figure 3 First derivative of measured and simulated transfer capacitance used to analytically correct TCAD deck.

---

[1] Since the series capacitor C_{janode} is always forward biased and considered constant, it has minimal influence on the derivative studies. A fully analytical solution is ongoing.

The resulting improvement in C_{RES} and C_{IES} are shown in figure 4. These encouraging results are a validation of the approach and physical assumptions made in the derivation of the structure to capacitance dependencies. This knowledge can be transferred in a next step to make use of the build in optimizer in Synopsis Sentaurus to further improve the results.

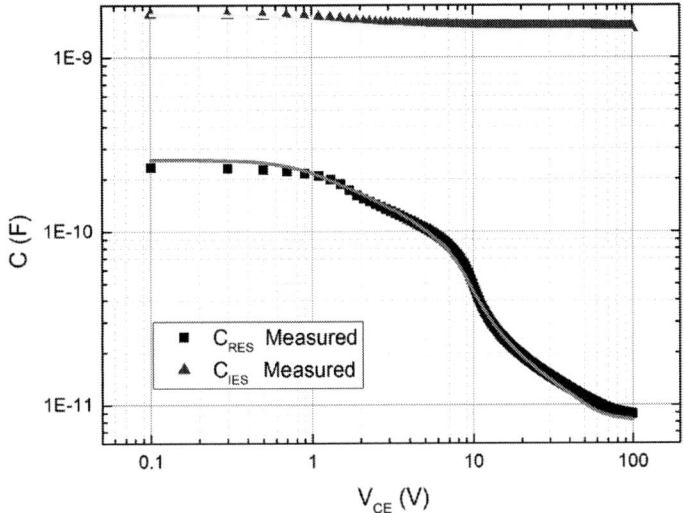

Figure 4 Simulated (solid lines) vs. measured capacitance of IGBT using analytic function for doping profile.

2.3. Using the Optimizer to automatically calibrate the Deck

In the final step we present the optimizer which is used to modify the appropriate process parameters that have been identified in the previous section. We have shown that the JFET and epi doping are the crucial process parameters that require optimization for a perfect fit of the C_{RES}. Within the TCAD deck, the actual process parameter is not the doping itself, but rather the dose and implant energy that define the final doping profile. Using a multi-point two variable Parameter-Response optimization within the Sentaurus environment is used to achieve the best possible result. This complex optimization is a unique method to use the TCAD simulator for calibrating process parameters in respect to expected values from measurement.

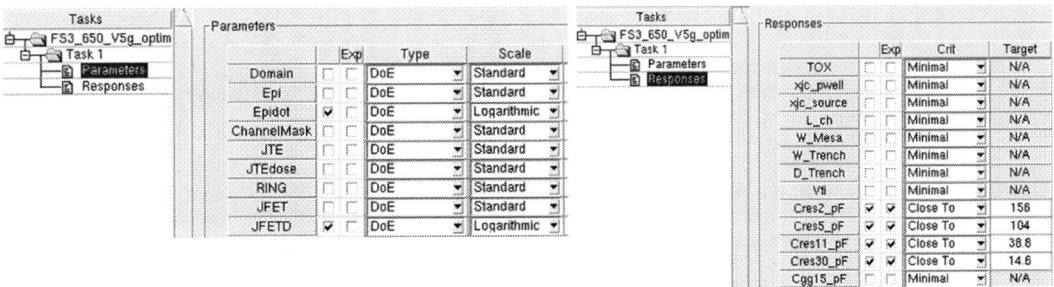

Figure 5 Synopsis Sentaurus optimizer GUI, showing the Parameter-Response Matrix

In figure 5 we show the Parameter-Response setting GUI, where the *JFETdose* and the *EPIdoping* are defined as process parameters influencing the response, and measured C_{RES}

values at four different voltages, namely (2, 5, 11 and 30 V), as response. The number of optimization points can be increased for even higher accuracy, however given the complexity of the simulation, calculation time is heavily impacted. Making the proper choice of the voltage points, namely within the physically crucial range (defined in the previous section), results in sufficient resolution to obtain highly accurate matches. The optimizer then, given our starting points for dose and energy derived from the analytic methods, iterates the chosen process variables until a match with a maximum of 10 % error is achieved. The final capacitance curves can be seen in figure 6.

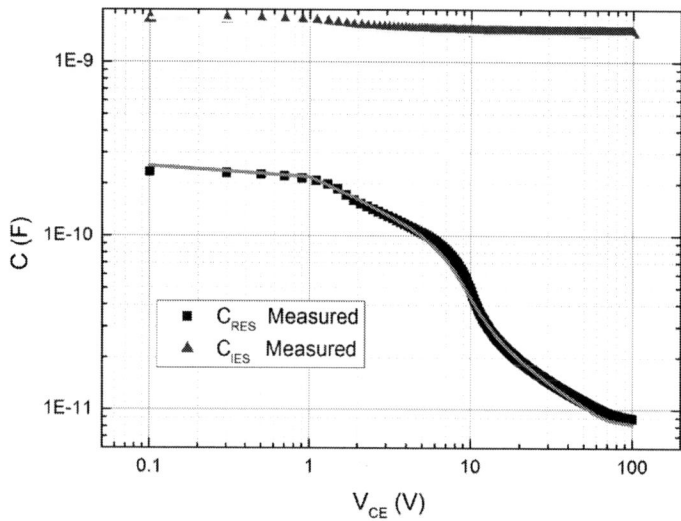

Figure 6 Final results of measurement vs. TCAD simulation using the optimizer to fine tune

3. Conclusion

In conclusion, we have shown a unique approach in calibrating a TCAD device deck for a field stop trench IGBT to an unpreceded accuracy using a combination of analytic derivation of process parameter dependence to capacitance and multi-variable optimization within the Synopsis Sentaurus TCAD environment. This method, overcomes the difficulty of properly calibrating doping profiles which are a challenging to measure and monitor in production.

References

[1] J. Victory, D. Son, T. Neyer, K.Lee, E. Zhou, J. Wang, and M. Baghaie Yazdi "A Physically Based Scalable SPICE Model for High-Voltage Super-Junction MOSFETs," *PCIM Europe 2014; International Exhibition and Conference for Power Electronics, Intelligent Motion, Renewable Energy and Energy Management; Proceedings of*, Nuremberg, Germany, 2014, pp. 1-8.

[2] M. Baghaie Yazdi, J. Victory and T. Neyer, "On-Wafer high voltage discrete device capacitance characterization for parameter extraction of physically scalable electro-thermal SPICE models," *PCIM Europe 2015; International Exhibition and Conference for Power Electronics, Intelligent Motion, Renewable Energy and Energy Management; Proceedings of*, Nuremberg, Germany, 2015, pp. 1-6.

PCIM Europe 2016, 10 – 12 May 2016, Nuremberg, Germany

Best-in-class 1200V IGBT for High Frequency Applications

Ramakrishna Tadikonda, Infineon Technologies Americas Corp, El Segundo, USA,
Ramakrishna.tadikonda@infineon.com

Jorge Cerezo, Infineon Technologies Americas Corp, El Segundo, USA,
Jorge.cerezo@infineon.com

Chiu Ng, Infineon Technologies Americas Corp, El Segundo, USA, chiu.ng@infineon.com

Abstract

Modern Power Converters demand increase in efficiency and power density while keeping the costs low by increasing the switching frequency of Power devices. A major challenge to increase switching frequency of Insulated Gate Bipolar Transistor (IGBT) is due to large presence of minority carriers during turn-off event. A 1200V ultra-fast Trench IGBT based on Field Stop technology has been developed and optimized exclusively for high switching frequency applications in the range of 30kHz to 100kHz. This paper describes the feature of this ultra-fast switching IGBT in a critical comparison with equivalent products available in the market today.

1. Introduction

The trend in Power conversion is constantly treading towards higher switching frequency to improve the overall system performance, cost and size. IGBTs are being widely used power device in various power conversion systems from medium to high power range [1]. However, it has been a challenge for IGBT to evolve and achieve faster switching speeds due to the presence of minority carriers which leads to a long tail current during the turn-off event [2]. This tail current increases the turn-off losses and enforces an increase in the dead time between the conduction of two devices in a converter circuit. Thus the IGBT has to be optimized for extremely low turn-off losses for its applicability at higher frequencies. If achieved, such an IGBT can be very attractive in applications such as Welding, Battery chargers, UPS, PFC etc. Since switching losses dominate over conduction losses at higher frequencies, a new ultra-fast 1200V IGBT was developed by applying new innovative technologies to reduce turn-off switching losses significantly while achieving better conduction-turn-off switching losses (V_{CEON}-E_{OFF}) tradeoff. This new IGBT enables to operate at much higher frequencies than previous generations of IGBT.

2. New Ultra-fast IGBT device structure and Fabrication

1200V high speed Trench IGBTs [3] are fabricated on thin wafer using Field-Stop technology as shown in Fig.1. The new IGBT adapted trench-gate structure and N-enhancement layer that was similar to IR's previous generation IGBT [3 & 5]. The presence of N-enhancement layer enhances the conductivity modulation and helps to reduce on-state losses. The N-base layer is thinned further to reduce V_{CEON} and E_{OFF} further and show reduction with thickness by ~15mV/µm and

© VDE VERLAG GMBH · Berlin · Offenbach

~10µJ/µm respectively. N-base and N-buffer layers are optimized to secure requisite breakdown voltage capability.

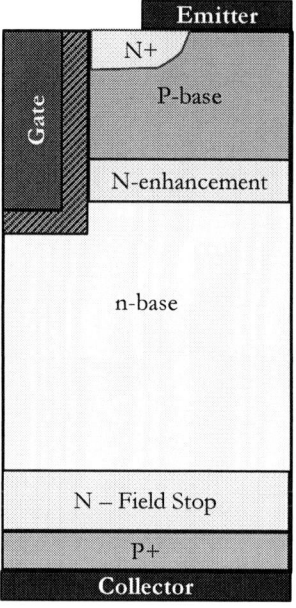

Fig.1 Ultra-Fast Field-Stop IGBT structure

Switching losses in IGBTs result from the slow dispersal of holes in the drift region after the Gate-Emitter voltage falls below threshold voltage to turn the device off. Either the holes recombine or a voltage gradient sweeps them out. Until this process completes, the IGBT exhibits a tail current which slows the switching speed and increases the switching losses. The use of thin wafers allows a lightly doped collector which reduces stored charge during turn-off, which means shorter tail current. These IGBTs are not processed with lifetime killing techniques, such as electron irradiation or metal doping, since they limit the operating temperature of the device.

This new device has more than 50% lower E_{OFF} in comparison to the previous generation devices [4-5]. A summary of the key device characteristics is shown in Table.1.

$T_J = 150°C$	V_{CEON} (V)	E_{OFF} (mJ)
IRGP20B120U	3.7	1.05
IRG7PH35U	2.3	1.2
New Fast IGBT	2.8	0.59

Table.1 20A, 1200V IGBT characteristic comparison at 150°C

3. Performance comparison in Buck Converter Topology

All device performance models have been obtained using a typical double pulse switching test setup (as shown in Fig.2) and a Semiconductor Automated Test Equipment (ATE). Then, with these models, the total system power losses have been estimated [6] using a typical Buck converter topology, focusing mainly on the Turn-off switching losses performance. As it is well known, for hard switching applications, the total power losses incurred in each power transistor consists of conduction and switching losses, i.e. turn-on and turn-off losses. Since turn-on losses are strongly dependent on reverse recovery characteristics of the diode, this comparison has been disregarded here.

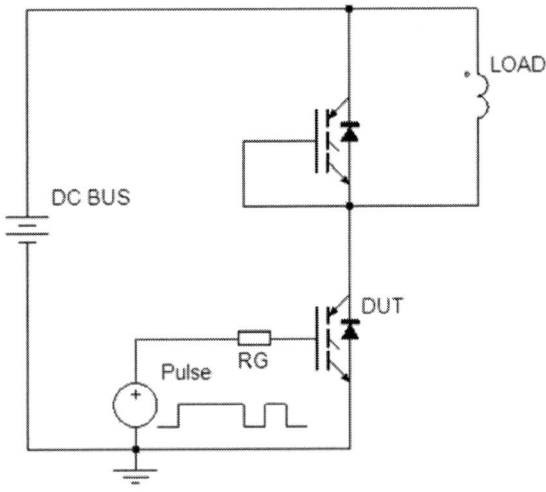

Fig.2 Typical Double pulse Switching Test Setup

The turn-off switching waveforms of new ultra-fast IGBT and competitors are shown in Fig.3 and can be observed from the switching waveforms that the new IGBT is much faster than the competitors. The device optimization has led to significantly shorter tail without causing high over-voltage and oscillations. The system power losses are simulated using the conduction (PSW Cond) and turn-off switching losses (PSW OFF) information of all the devices. Fig.4 shows that new IGBT is 34% lower system losses in comparison to its competitors at 30 kHz. Competitor B couldn't be tested at 9A @ 30kHz, hence not displayed in Fig.4(b). In addition, the performance comparison in terms of maximum current capability versus switching frequency, as shown in Fig.5, the new ultra-fast IGBT is performing much better than its competitors above 30 kHz.

© VDE VERLAG GMBH · Berlin · Offenbach

PCIM Europe 2016, 10 – 12 May 2016, Nuremberg, Germany

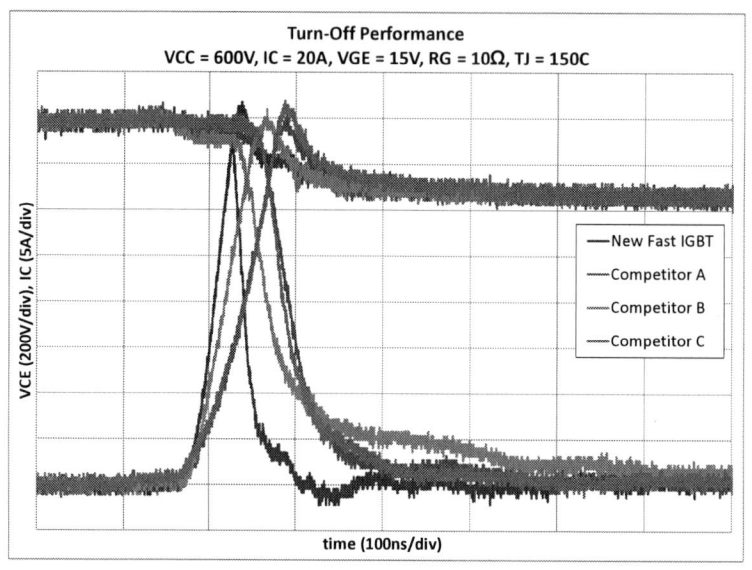

Fig.3 Comparison of Turn-off switching waveforms with competitors

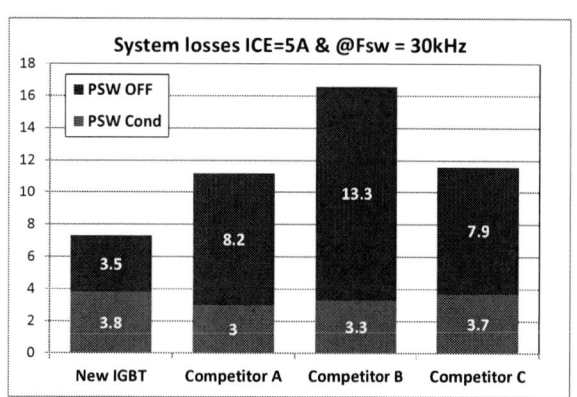
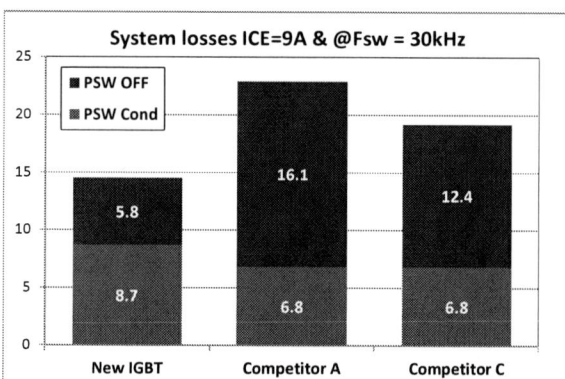

Fig.4 System Power losses at 30kHz (a) at 5A and (b) at 9A currents.

© VDE VERLAG GMBH · Berlin · Offenbach

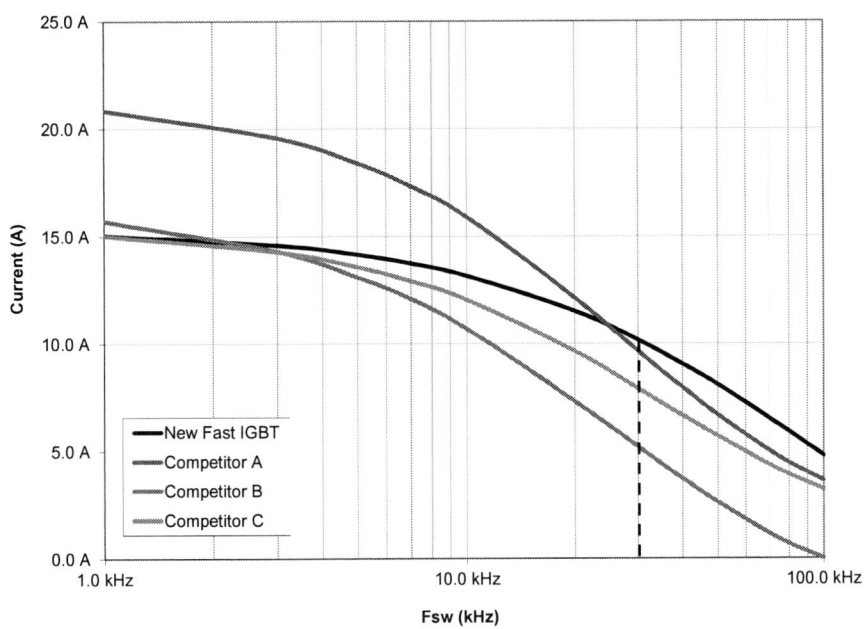

Fig.5 Maximum current capability versus Switching Frequency

4. Conclusion

A new ultra-fast 1200V Trench IGBT optimized for ≥30kHz frequency applications has been presented. The new IGBT is optimized with low V_{CEON} and much lower switching losses than earlier generation IGBTs. The new IGBT provides better E_{OFF}-V_{CEON} tradeoff, lower system power losses and higher current capability at >30kHz frequencies in comparison to its competitors.

5. References

[1] http://www.irf.com/technical-info/whitepaper/choosewisely.pdf

[2] B. J. Baliga; "Fundamentals of Power Semiconductor Devices" Springer, 2008th Edition

[3] Chiu Ng & Yuan-Heng Chao; "Variable Threshold Trench IGBT with Offset Emitter Contacts",

US 8,067,797 B2, 2008

[4] http://www.irf.com/product-info/datasheets/data/irgp20b120u-e.pdf

[5] http://www.irf.com/product-info/datasheets/data/irg7ph35upbf.pdf

[6] http://www.irf.com/product-info/igbt/tool.pdf

© VDE VERLAG GMBH · Berlin · Offenbach

Extra Electro-Thermal Performance of 1700V IGBT with the latest 7th generation chipset/package technologies

Mutsumi Sawada, Shinichi Yoshiwatari, Hiroaki Ichikawa, Yuichi Onozawa, Osamu Ikawa,
Fuji Electric Co., Ltd., Japan,
Thomas Heinzel, Fuji Electric Europe GmbH, Germany,
thomas.heinzel@fujielectric-europe.com

Abstract

This paper describes 1700V IGBT module new product family with extended power range up to 1700V-1800A in PrimePACK[TM]3. The 7th generation IGBT technologies were presented in the last PCIM2015 for 1200V rated devices [1]. The 1700V IGBT/FWD chipset has also been developed with similar technological background and design concept.

In addition, an extra thermal performance of lower thermal impedance was achieved with successful implementation of novel enhanced strength AlN isolation substrate. The integration of new chip and package technologies made it possible to achieve significant increase of watts density of about 30% ($11.2kW/cm^2$ (6G) -> $14.4kW/cm^2$ (7G))

1. Introduction

Insulated-Gate Bipolar Transistors (IGBTs) are widely used power semiconductors in various fields. Especially 1700V IGBT modules have been utilized mainly for large industrial drives, renewable energy applications (solar and wind), and rail transportation traction systems. In those markets, IGBT modules are required to reduce the switching energy loss as well as lower on-state voltage drop. In addition, one of the recent trends is to upgrade the operation temperature (maximum junction temperature) up to 175 °C.

The 7G IGBT/Free Wheeling Diode (FWD) technologies which were presented in the last PCIM2015 for 1200V rated devices developed in order to improve the trade-off relationship between the on-state voltage drop and the turn-off energy. The 1700V IGBT/FWD chipset has also been developed with similar technological background and design concept.

In addition to the improvement of chipset, the packaging technologies of the 7G improved thermal cycling capability with new alumina insulating substrate of higher thermal conductivity.

The integration of new chip and package technologies made it possible to achieve significant increase of watts density of about 30%.

2. The 7th generation device technologies for 1700V

The 7G-IGBTs have achieved remarkable reduction in switching energy loss as well as lower on-state voltage drop. As a result, high conversion efficiency is expected even with an extra die shrink in IGBTs and FWDs.

In addition, the 7G chipset and package integration were done with the concept of upgrading the continuous operation temperature of 175°C, the critical device capabilities, such as Short Circuit Safe Operation Area (SCSOA), Reverse Bias Safe Operation Area (RBSOA) for IGBTs, reverse recovery ruggedness for FWD and so on, are carefully considered so that the devices have consistent withstand capabilities when they are compared to the conventional devices of Tj-op (max) 150°C.

PrimePACK[TM] is registered trademark of Infineon Technologies AG, Germany

© VDE VERLAG GMBH · Berlin · Offenbach

2.1. IGBT chip technologies

Fig. 1 shows the device cross section of the 7G-IGBT with comparison to that of the 6G [2] [3]. The trench gate Field-Stop (FS) IGBTs are well known as the state-of-the-art IGBT which have good trade-off relationship between the on-state voltage drop and the turn-off dissipation. The drift layer thickness of the 7G-IGBT was reduced due to optimization of the surface and backside structure compared to the 6G. This thinner drift layer and the fine pitch of the surface design achieved the extra improvement in overall performance of IGBTs.

By the optimization of the FS layer the 7G-IGBT realized the soft switching behavior and enough high breakdown voltage even with thinner bulk thickness. The leakage current of the 7G-IGBT was also reduced for more than 37% in comparison to that of the 6G-IGBT at Tj=175 °C, VCE=1700V.

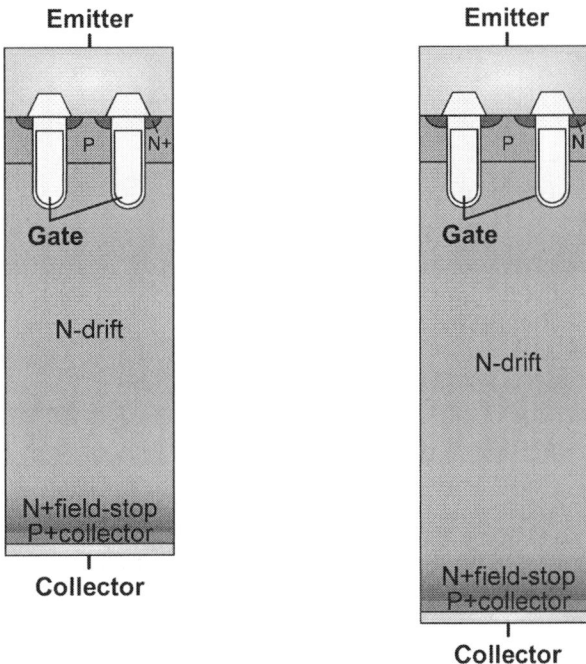

(a)7th generation IGBT   (b) 6th generation IGBT

Fig. 1. Comparison of the device cross-section

The output characteristics of the 7G 1700V 150A rated IGBTs are shown in Fig. 2. The on-state voltage drop of the 7G-IGBT is about 1.8V and 1.9V at 150°C and 175°C, respectively, which are about 0.4V lower Von when it compared to those of the 6G at the same collector current. The thinner drift layer and the optimized surface structure contributed to achieve much higher performance.

The trade-off relationship between the on-state voltage drop and the turn-off losses is shown in Fig. 3. The measurement was done with the inductive load switching at Vcc=900V and the load current was 150A. At the same turn off loss of 40mJ/A, the on-state voltage drop of the 7G-IGBT is 0.4V lower than that of the 6G-IGBT. On the other hand, when they are compared at the same on-state voltage drop of at 2.5V, the 7G-IGBT will have as much as 30% lower turn-off energy than that of the 6G-IGBT.

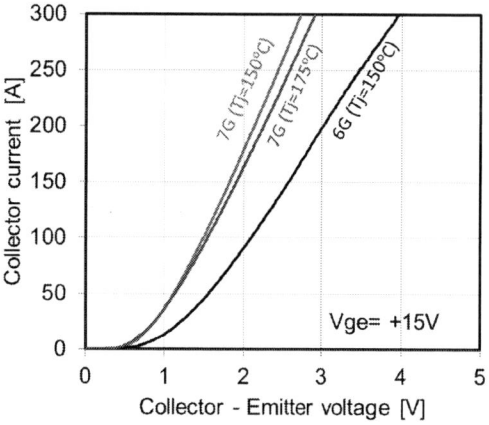
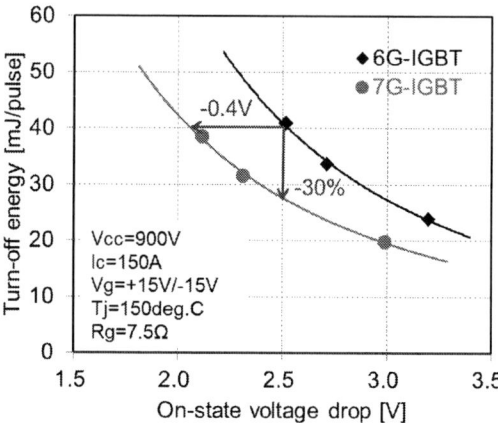

Fig. 2 Output I-V characteristics of
1700V150A IGBTs

Fig. 3 Eoff-Von trade-off relationship of
1700V150A IGBTs

2.2. FWD chip technologies

The performance improvement in the 7G-1700V chipset is not just in IGBTs, but also in FWDs. The forward voltage drop of the 7G-FWD was reduced by the thinning of the drift layer. The trade-off relationship between reverse recovery loss (Err) and forward on voltage of the 7-FWD is shown in Fig. 4. Comparing to the 6G-FWD, the forward voltage was reduced by 0.3V at the same reverse recovery loss (Err) at Tj=175 °C.

The reverse recovery switching waveforms are shown in Fig. 5. The 7G-FWD has soft recovery behavior, which is similar to the 6G-FWD.

In general, special considerations are necessary in FWD design when trying the thin drift layer. In case the depletion layer punchthrough happens during the reverse recovery process [4], the FWD easily gets snappy behavior, which leads the serious ringing and/or high surge voltage issues. Because the backside structure of the 7G-FWD was well-optimized, the depletion punchthough effect can be pushed out from the safe operating area of the IGBT module. As a result, the 7G 1700V FWDs were successfully developed with soft and fast recovery as well as much lower forward voltage drop.

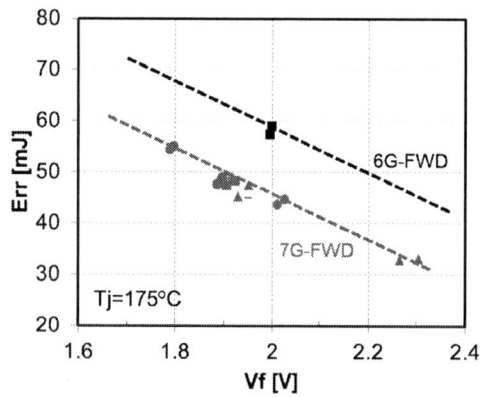
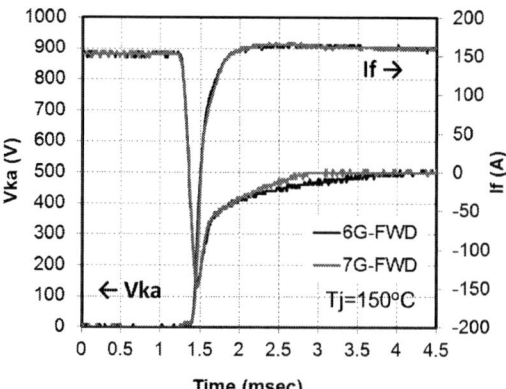

Fig. 4 Err-Vf trade-off relationship of
1700V 150A FWD

Fig. 5 Reverse Recovery characteristic of
1700V 150A FWD

3. The 7th generation package technologies

The trend of higher power density challenges are not just small/medium power converters, but also large power applications, in which the 1700V IGBTs are installed, have also strong requirement of higher output power and/or reduction in size and volume of the power conversion systems. The new generation IGBTs of smaller footprint size or extended power rating in the same footprint size are expected to higher output power, reduce number of parallel of the IGBTs, smaller thermal management systems and so on.

In order to upgrade the IGBT module power ratings, the thermal management is the most important to compensate power density increase and the thermal network. The smaller die size, the higher thermal impedance if the dies are packed into the same power module structure. The new package must have extra thermal performance which is able to compensate thermal impedance degradation (Rthj-c) to prevent the junction temperature increase.

In addition, another trend of 175°C continuous operation temperature upgrading eventually requires the extra improvement in the packaging integration, especially at material interfaces, such as bond wired, solders, insulating ceramic, in order to have even longer reliability and/or product lifetime even with 25°C Tjmax increase from the previous generation of Tjop(max)=150°C.

3.1.   New high heat dissipation AlN insulated substrate

Basic idea of heat compensation is: for example, 30% die shrink needs 30% thermal impedance improvement when the target junction temperature stays the same value. Because the die shrink were done with 7G-IGBTs and FWDs, the thermal performance of the DCB ceramic has to be improved since it has the highest contribution in the total thermal performance of the module.

The practically available ceramic are: Aluminum oxide (Al_2O_3), Aluminum Nitride (AlN) and Silicon Nitride (Si_3N_4). The Al_2O_3 DCBs are the most popular in various power modules, however, it is not thermally high performance since the heat conductivity is lower than other materials. On the other hand, AlN has been known as high thermal conductivity material. The mechanical weak feature of the conventional AlN needs thicker ceramic to have affordable temperature cycling (heat cycling, heat shock) capability. Alternatively, Silicon Nitride DCBs are installed in some high performance power module since the Si_3N_4 DCB has better balance in the thermal performance and mechanical features.

Recently, the new AlN ceramic which have outstanding improvement in the mechanical strength were developed. The new innovation made it possible to install thin-AlN into IGBT module [5]. The mechanically strong feature of the new-AlN has overcome the temperature cycling issues. The high reliability as well as significant improvement in the thermal impedance has been achieved simultaneously. The junction to case thermal impedance Zth (j-c) with new AlN and the conventional Al_2O_3 substrate are shown in Fig. 6. Comparing to the Al_2O_3 substrate, the thermal impedance of new AlN substrate was reduced as much as 45%. By applying the new AlN substrate in the 7G-IGBT modules the temperature rise due to the shrinking of the chip size was successfully solved.

PCIM Europe 2016, 10 – 12 May 2016, Nuremberg, Germany

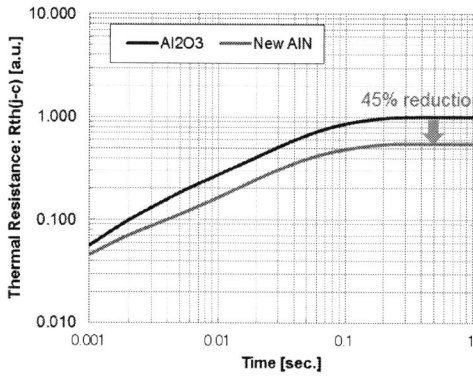

Substrate	New AlN	Conventional AlN	Al₂O₃
Thermal Conductivity [W/m·K]	170	170	20
Thickness [a.u.]	1.2	2.0	1.0
Bending Strength [MPa]	500	450	500
Thermal Cycling Capability	Very good	good	Very good

Fig. 6 Comparison of the thermal resistance between junction and case at the same die size (left). Comparison of principal characteristics between new AlN and conventional substrates (right).

3.2. Improvement of delta-Tj power cycling capability

Fig. 7 shows the estimated thermal cycling capabilities of the 6G and the 7G-IGBT for two different temperature conditions. The curves show so-called the delta-Tj power cycling of two different types of the IGBT generation of the 6G and the 7G. The curves show different maximum junction temperature conditions of at 150°C and at 175°C.

At the same delta-Tj of 50deg, the estimated thermal cycling capabilities are:

1.2E6 cycles 7G@Tjmax=150°C   6.0E5 cycles 7G@Tjmax=175°C
5.5E5 cycles 6G@Tjmax=150°C   3.0E5 cycles 6G@Tjmax=175°C

This means two different ways of upgrading will be possible. One is the reliability upgrade. When the Tjmax stays the same temperature, for example 175°C, 2times enhancement of 3.0E5 to 6.0E5 cycles can be expected by replacing the IGBT module from 6G to 7G.

Another way of upgrading is to have more power output with new generation. When the estimated cycles stay the same number, for example, 3E5cycles, about 8deg of delta-Tj, 58deg, will be possible with the 7G.

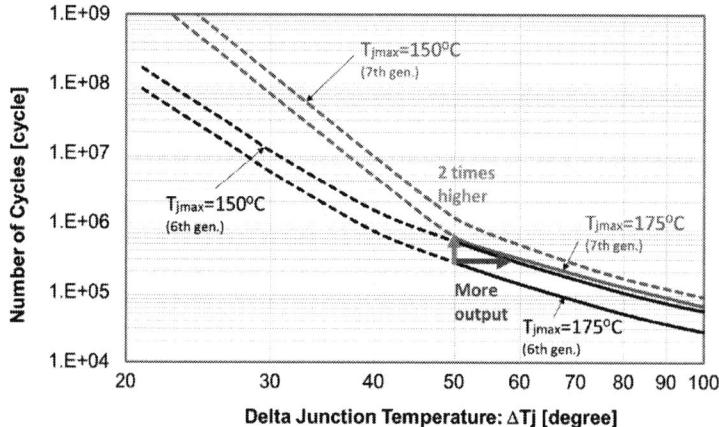

Fig. 7 Thermal cycling(delta-Tj) curves of the 7G-IGBT at 150°C and 175°C and those of 6G

© VDE VERLAG GMBH · Berlin · Offenbach

4. The 7th generation IGBT module

Above performance improvement of the chipset as well as novel integration of the extra-thermal impact of new-AIN isolation substrate, it was successfully achieved to pack 1700V-1800A rating into PP3 foot print size. This means the 7G will have about 30% more power density when compared to 6G-PP. Fig. 8 shows the estimated total power dissipation and the maximum junction temperature calculation in typical AC690V power converters. The calculation was done with 1700V-300A module of output current of 150Arms.

At the carrier frequency of 4kHz, the estimated total watt of the 7G is about 460W which is about 9% lower than that of 6G. The maximum junction temperature swing estimation also indicates that 7G will have 29deg compared to 38deg. of 6G. These results indicate that the 7G-IGBT provides more flexibility in power converter design options of 1) lower price centric deign of less thermal management efforts, 2) higher power challenges, 3) Longer lifetime expectation by less temperature swing, and so on.

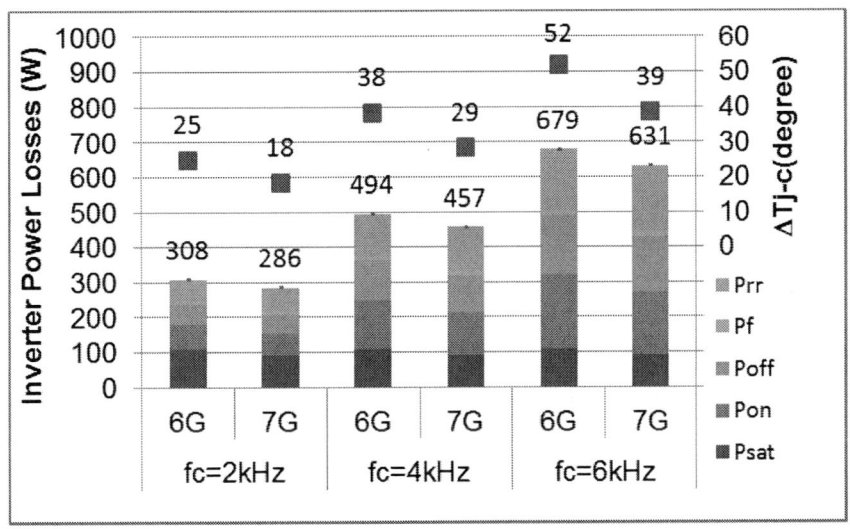

Fig. 8 Loss and ΔTj-c of 1700V300A modules

Fig. 9 shows the product family plan of the newly developed 1700V rated X-series IGBT (the 7th generation). The bars with "Current" mean the current ratings which are already available in the previous generation of 6G-family. The bars with "Enhanced" are exclusive to the 7th generation X-series IGBT family, which are now possible to extend the power range with new integration of the chip and package technologies.

The 34mm, 62mm modules will have new rating of 150A, 400A, respectively. 1700V-600A DualXT and 600A EP+ are also new ratings. The grate upgrading is also planned in PP2 and PP3. PP2 will have 1700V/1200A. The new rating of 1700V-1800A in PP3 will have the highest power density of 14.4kW/cm^2, which is about 30% higher than 11.2kW/cm^2 of the 6G.

© VDE VERLAG GMBH · Berlin · Offenbach

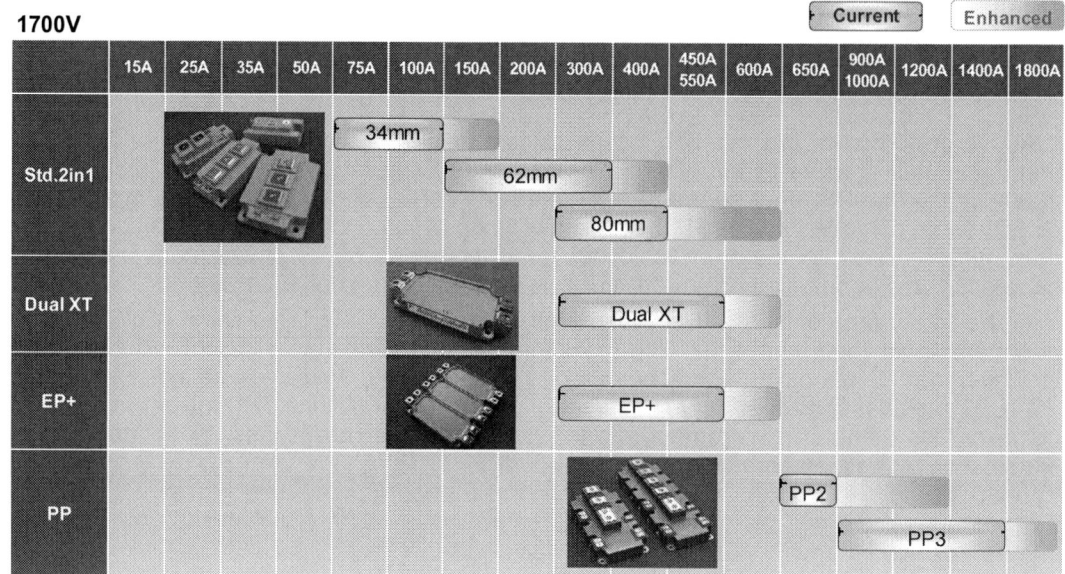

Fig. 9 The 7th generation 1700V IGBT product family plan.

Note : EconoPACK^{TM}+ is registered trademark of Infineon Technology AG , Germany.

Note: Prime PACK^{TM} is registered trademark of Infineon Technologies AG, Germany.

5. Conclusions

X-series, the 7th generation 1700V IGBT new family, have been introduced in this paper. The optimized new trench-field stop IGBT and ultra-soft recovery FWD chipset promises great improvement in switching performance as well as reduction in the on-state voltage drop. These results indicate that the additional die shrinking is possible compared to the previous generation of V-series IGBT.

New thin-AIN substrate, which has thinner but mechanical stronger feature compared to the conventional AIN material, also provides much lower thermal impedance while it keeps or even better thermal cycling capability.

The integration of the chip and package new technologies, the X-series 1700V IGBT family will be soon available. Each package will have extended power rating compared to V-series, such as 1700V-1800A new rating in PP3 or 1700V-650A in DualXT.

References

[1] T.Heinzel et al. "The New High Power Density 7th Generation IGBT Module for Compact Power Conversion Systems", PCIM Europe 2015

[2] M.Otsuki et.al., "The 6th generation 1200V advanced Trench FS-IGBT chip technologies achieving low noise and improved performance", Proceeding of PCIM Europe 2007.

[3] Y. Kobayashi et al. "The New IGBT-PIM with the 6th generation V-IGBT chip technology", Proceeding of PCIM Europe 2007.

[4] Y. Onozawa et al. "Development of the 1200V FZ-Diode with Soft Recovery Characteristics by the New Local Lifetime Control Technique", pp.80-83 Proceeding of ISPSD 2008.

[5] F. Momose et al. "The New High Power Density Package Technology for the 7th Generation IGBT Module", PCIM Europe 2015

Parameter Extraction for PSpice Models by means of an Automated Optimization Tool – An IGBT model Study Case

Carlos Gomez Suarez, Paula Diaz Reigosa, Francesco Iannuzzo, Ionut Trintis, and Frede Blaabjerg, Department of Energy Technology, Aalborg University, Pontoppidanstræde 101, 9220 Aalborg, Denmark. carlosgs91@gmail.com, pdr@et.aau.dk, fia@et.aau.dk, itr@et.aau.dk, fbl@et.aau.dk

Abstract

An original tool for parameter extraction of PSpice models has been released, enabling a simple parameter identification. A physics-based IGBT model is used to demonstrate that the optimization tool is capable of generating a set of parameters which predicts the steady-state and switching behavior of two IGBT modules rated at 1.7 kV / 1 kA and 1.7 kV / 1.4kA.

1. Introduction

Reliability of power semiconductor devices has become an increasingly important factor in power electronics systems [1]. For the reliability design and analysis of power semiconductors, physics-based models for circuit simulators are a must to better understand the complex failure mechanisms as well as the prediction of DC and switching characteristics. However, their use has been accompanied by a common challenge, i.e. the model parameter identification.

Most of the prior-art works focus on parameter extraction techniques based on the datasheet information and/or switching characteristics [2-4], which are suitable for initial parameter extraction. A more efficient approach is to apply an optimization algorithm capable of modelling the device performance by a set of parameters which accurately characterizes the device, such as presented in [5-7]. However, these optimization tools are kept as proprietary, making the parameter identification a tedious task.

To overcome the problem this paper proposes a freely available optimization-based parameter extraction methodology to model the device performance. The automated tool has been implemented through a user-friendly Graphical User Interface (GUI). Such interface, named Model Based Parameter Identifier (MBPI), has been applied to predict the behavior of two IGBT modules rated at 1.7 kV / 1 kA and 1.7 kV / 1.4 kA.

2. Operational principle

Fig. 1 illustrates the basic operational principle of the tool. A set of experiments is performed in a device of which parameters p need to be extracted. The same experiments are replicated under simulations with different models. Both experiment and simulations are subjected to the same inputs (i.e., voltages) and testing conditions (i.e., temperature, load) and the outputs (i.e., currents) are compared. Then MBPI will modify iteratively the set of parameters p used in the simulations to minimize the difference between both until the stopping criteria are met.

An initial guess of the parameters, p_0, is needed. It can be taken from some rough estimations from datasheets or approximations. Then those estimated parameters can be refined through optimization techniques. The proposed tool, MBPI, uses an optimization algorithm to refine the parameters based in the comparison between experimental waveforms denoted as y_e in Fig. 1 with the simulation waveforms, y_s, from PSpice.

PCIM Europe 2016, 10 – 12 May 2016, Nuremberg, Germany

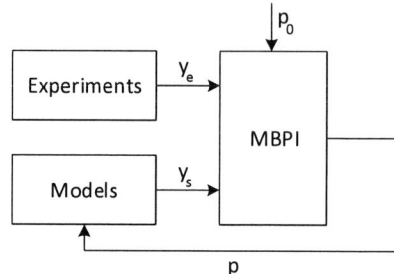

Fig. 1. Basic operational principle of MBPI, where p_0 represent the initial set of parameters, p the set of parameters in each iteration and y_e and y_s the experimental and simulation data respectively

3. Configuration file creation

MBPI works by using configuration files that contain all the data required to perform the optimizations between experiments and simulations. This file is created using a separate tool named Data Filler. The required data is presented below and the GUI from the Data Filler depicted in Fig. 2.

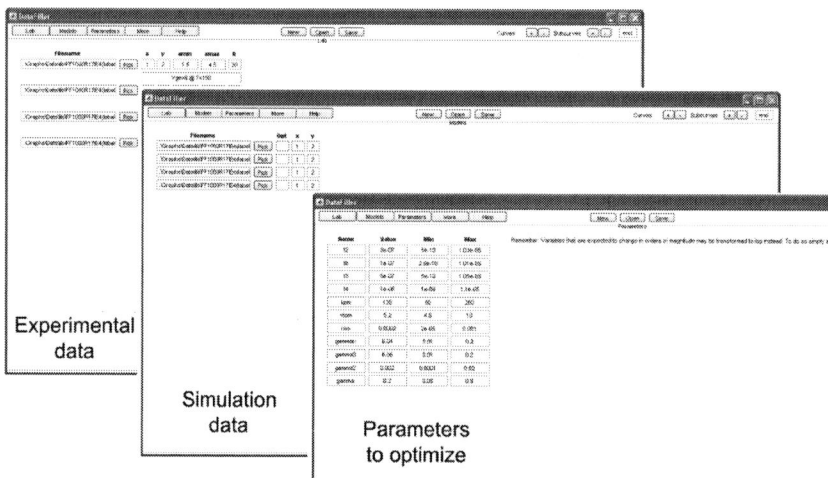

Fig. 2. Configuration file creation with Data Filler

- Experimental curves: The data recorded from experiments is saved in space-separated txt files where each variable is stored in a new column. The different text files are then selected from Data Filler and the x and y variables used to compare with the simulations are selected.
- Simulations: Each simulation file from PSpice is selected as well as the x and y variables to use in the comparison against the experiments. Those simulation files must be modified to contain a set of "variable parameters" that are then changed within MBPI. This is performed by rewriting any parameter as *varParam(#number)*.
- Parameters: The parameters to estimate are added with their respective names and initial values.
- Constraints and optimization method: Linear constraints can be added for the parameters and the optimization method can be selected.

© VDE VERLAG GMBH · Berlin · Offenbach

832

PCIM Europe 2016, 10 – 12 May 2016, Nuremberg, Germany

Once all the data is fed the configuration file can be exported and loaded into MBPI where the only action required after is to click on the RUN button. The configuration file can be modified later on from Data Filler if any change needs to be performed.

4. Parameter Optimization Procedure

The proposed tool MBPI has been used for extracting the parameters of a physics-based IGBT model implemented in PSpice [8]. MBPI includes four major steps as shown in Fig. 3.

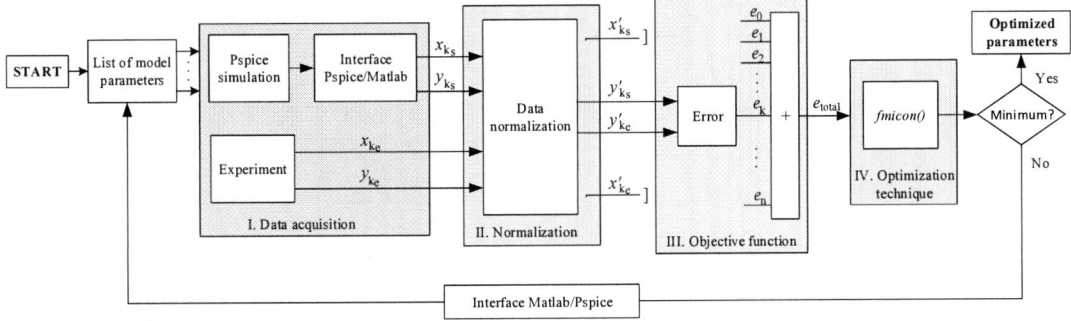

Fig. 3. Flowchart of MPBI

A) Data evaluation (step I)

The configuration file created is read by MBPI and the experimental waveforms are acquired (i.e., in the IGBT case static and dynamic). Simulation data is obtained in each iteration with the new parameters to use. The connection between PSpice models and MBPI is performed trough an interface that can be extended to support other simulation software.

B) Data normalization (step II)

In the configuration file both the experimental and simulation data is attached and they are compared in pairs (the first experimental curve with the first simulation, the second with the second and so on). However as the sampling of the experimental data and the simulation files is usually different the registered data points need to be normalized so they can be compared one by one. This step is represented in Fig. 4.

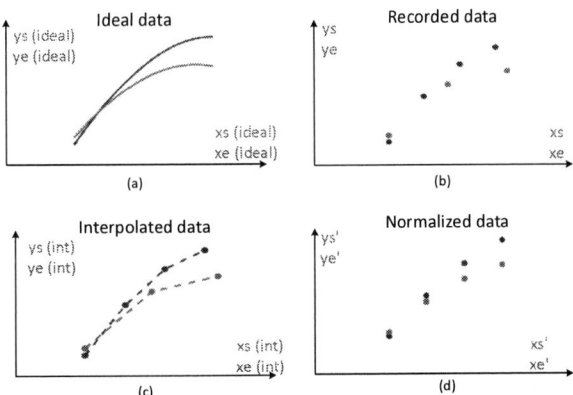

Fig. 4. Normalization procedure of recorded data. The ideal continuous data is shown in (a). An example of how the actual recorded data from simulations (red) and experiments (blue) looks is shown in (b). In (c) a linear fit is made with the recorded points. In (d) a new set of points is generated based in the previous interpolation so experimental and simulation points can be easily compared as pairs.

© VDE VERLAG GMBH · Berlin · Offenbach

833

In Fig. 4a the ideal data is shown (in continuous time). However the data recorded in experiments and the one from simulations is sampled as shown in Fig. 4b. The samplings are usually different so they can be normalized to perform comparison between each curve-pair to create an error function to optimize. The proposed method is to create an interpolated curve as shown in Fig. 4c with a linear regression and use it to create a new set of points at the same instants as illustrated in Fig. 4d.

C) Objective function (step III)

Each pair of experiment-simulation curves will generate an error. This error can be calculated as the mean absolute error between experiment and simulation, scaled by the mean of the experiment values as given in (1). N_k is the number of points, y'_{k_e} and y'_{k_s} are the experiment and simulation y-axis values (i.e., voltage and current) after normalization.

$$e(k) = \frac{1}{N_k} \sum_{j=1}^{j=N_k} \left| \frac{\overrightarrow{y'_{k_e}}(j) - \overrightarrow{y'_{k_s}}(j)}{mean(\overrightarrow{y'_{k_e}})} \right| \tag{1}$$

The total error is calculated by adding all the individual errors scaled by the number of experiment-simulation curves, N_c, as written in (2) for the IGBT case. The individual errors corresponding to the IGBT static and dynamic experiment-simulation curves are denoted by e_{static} and $e_{dynamic}$.

$$e_{total} = \frac{1}{N_c} \sum_{k=1}^{k=N_c} e(k) = \frac{1}{N_c} \left(\sum e_{static} + \sum e_{dynamic} \right) \tag{2}$$

D) Optimization technique (step IV)

The optimization procedure consists of minimizing the objective function previously defined in (2) by fulfilling the parameter boundaries and linear constraints. Based on an initial parameter estimation [5], a local gradient-based method, such as the *fmincon()* from MATLAB™ can be applied to refine the initial parameters. This method starts with the initial list of parameters defined by the user. A gradient of the error is estimated based on small perturbations in each parameter and is used to generate a new set of parameters, whose boundary values can be set by the user. The algorithm is repeated until the stopping criteria is met.

When a gradient method is used typically the algorithm will stop at the closest local minima of the error function and thus the solution is dependent of the initial parameters set. The tool also allows the use of other optimization methods designed to find the global minimum such as *globalsearch()* or *ga()* more suitable if the trust in the initial parameters is low. However the computation time required for those methods to converge is highly increased when compared with *fmincon()*.

Lastly, MBPI can also be used as an interface between experiments and simulation software to create the objective function and let the user implement the optimization strategy of choice which may be more suitable for the particular problem.

5. Tool description

The user may start using Data Filler described in Section 3 to generate the configuration file. Once it is loaded into MBPI the optimization can begin. The program is separated in different tabs as shown in Fig. 5. In the first tab a text-based version of the configuration file pops up so the user can check that the correct file is loaded. The run button can then be clicked to begin the optimization. In the second tab the evolution of the estimated parameters and error function over iterations is shown. This can be used to see the evolution of the optimization and if the user is satisfied can be stopped at any moment. In the third tab the user has access to the experiment-simulation comparison of the curves. Finally when an optimization is completed the results can be exported into an HTML file where all the information presented in the tabs is saved. Moreover a copy of the simulation files with the estimated parameters is also given.

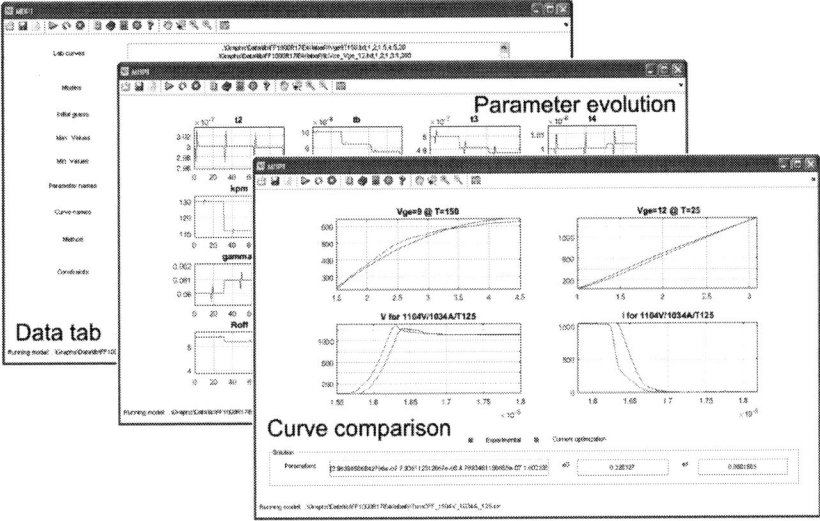

Fig. 5. MBPI tabs during an optimization execution

6. Validation

A model for high-voltage IGBTs based on the lumped-charge approach has been used. This model has been previously presented by Iannuzzo et. al in [8], and therefore, the equations for defining the physics-based IGBT model are not further explained. A commercially available IGBT module rated at 1.7 kV / 1 kA [9] and another one at 1.7 kV / 1.4 kA [10] have been selected to validate that the optimization tool is capable of obtaining the list of parameters which best predict the real behavior of the device.

6.1 IGBT model and experiments

MBPI requires the acquisition of experimental waveforms to compare with the simulations for the parameter optimization. For the IGBT model case study, two types of experimental measurements are done in order to characterize the behavior of the system and are depicted in Fig. 6.

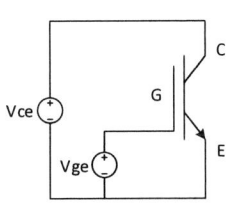   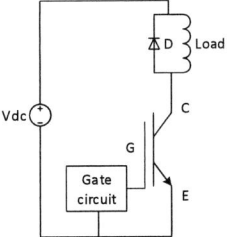

Fig. 6a. Static test schematics   Fig. 6b. Dynamic test schematics

In Fig. 6a the static test is shown. For different gate voltages, V_{GE}, the potential V_{CE} is swept and the steady-state value of the current I_C is measured. The test can be performed at different temperatures and characterizes the steady-state behavior of the device. The dynamic response can be obtained under the test depicted in Fig. 6b. The load current can be controlled with the inductor and once the desired value is met the DUT can be turned off and on. The evolution of I_C and V_{CE} are acquired as a function of time. The models in PSpice used in MBPI

© VDE VERLAG GMBH · Berlin · Offenbach

replicate the same experiments. The same circuits with the same inputs (voltages) are applied and the time-domain simulations are synchronized (the firing time is set the same in experiment and simulation). Then the different curves between experiment and models can be compared in pairs. Those are summarized in Table 1.

Test	States	x-axis	y-axis
Static	V_{GE} and T	V_{CE}	I_C
Dynamic	I_{LOAD}, T and turn on/off waveform and	t	V_{CE}, I_C

Table 1. Summary of tests performed

6.2 Optimized parameters results

In order to validate the capability of MBPI, the IGBT model with the initial set of parameters (i.e., non-optimized) and the final set of parameters (i.e., optimized) have been compared with the steady-state and dynamic measurement waveforms of two modules rated at 1.7 kV / 1 kA and 1.7 kV / 1.4 kA. Fig. 7 and Fig. 9 show the static and dynamic responses of both modules before and after optimization. The conduction and turn-off losses are depicted in Fig. 8 and Fig. 10.

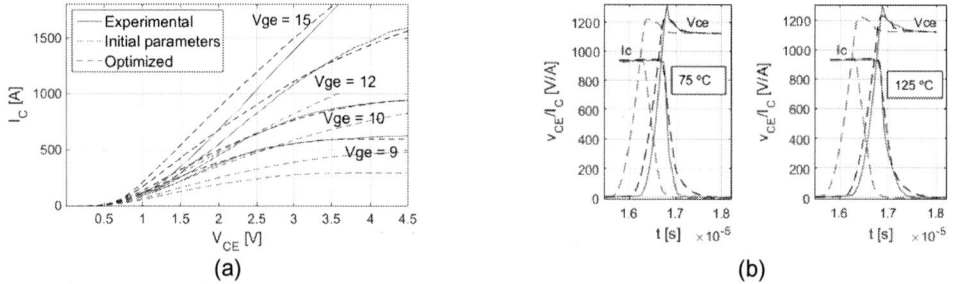

Fig. 7. IGBT model of 1.7 kV / 1 kA IGBT module: (a) I-V static output characteristics at T = 150°C, and (b) turn-off switching characteristics at V_{CE} = 1125 V, T = 75°C and V_{CE} = 1115 V, T = 125°C

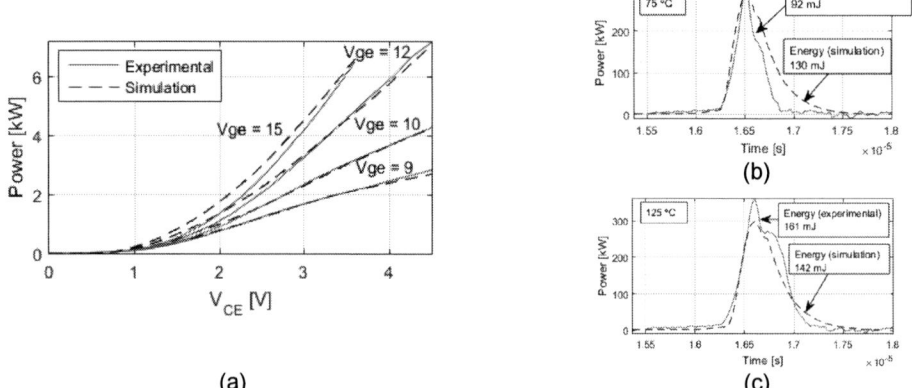

Fig. 8. IGBT model of 1.7 kV / 1 kA IGBT module: (a) I-V conduction losses at T = 150°C, (b) turn-off switching losses at V_{CE} = 1125 V, T = 75°C and (c) turn-off switching losses at V_{CE} = 1115 V, T = 125°C

© VDE VERLAG GMBH · Berlin · Offenbach

PCIM Europe 2016, 10 – 12 May 2016, Nuremberg, Germany

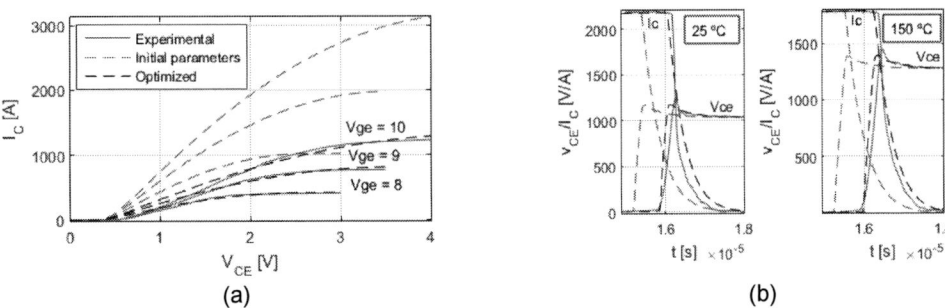

(a)   (b)

Fig. 9. IGBT model of 1.7 kV / 1.4 kA IGBT module: (a) I-V static output characteristics at T = 150°C, and (b) turn-off switching characteristics at V_{CE} = 1034 V, T = 25°C and V_{CE} = 1279 V, T = 150°C

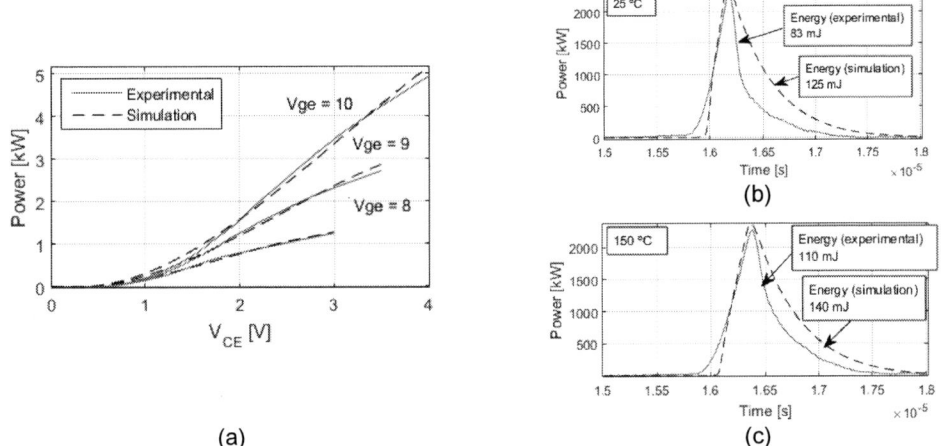

(a)   (c)

Fig. 10. IGBT model of 1.7 kV / 1.4 kA IGBT module: (a) I-V conduction losses at T = 150°C, (b) turn-off switching losses at V_{CE} = 1034 V, T = 25°C and (c) for V_{CE} = 1279 V, T = 150°C.

7. Conclusion

An original automated tool for parameter optimization of PSpice models is presented in this paper. This tool can be applicable for different models and two examples are included to validate that MBPI can give a set of parameters that can be implemented in the IGBT model from [8]. The tool is released for free through the author's website (www.igbtmodel.org). Due to its open-source nature it provides the basis for future research platforms where interfaces to other simulation software can also be implemented.

8. Literature

[1] H. Wang, M. Liserre, and F. Blaabjerg, "Toward reliable power electronics: Challenges, design tools, and opportunities," IEEE Industrial Electronics Magazine, vol. 7, no. 2, pp. 17–26, June 2013.

[2] R. Withanage, N. Shammas, S. Tennakoorr, C. Oates, and W. Crookes, "Parameter extraction for the Hefner IGBT model," in Proc. Of the 41st Internationa lUniversities Power Engineering Conference (UPEC), vol. 2, Sept 2006, pp. 613–617.

© VDE VERLAG GMBH · Berlin · Offenbach

[3] G. Sfakianakis, M. Nawaz, and F. Chimento, "A temperature dependent simple spice based modeling platform for power IGBT modules," in Proc. Of the 2014 IEEE Energy Conversion Congress and Exposition (ECCE), Sept 2014, pp. 2873–2879.

[4] M. Miyake, D. Navarro, U. Feldmann, H. Mattausch, T. Kojima, T. Ogawa, and T. Ueta, "HiSIM-IGBT: A compact Si-IGBT model for power electronic circuit design," IEEE Transactions on Electron Devices, vol. 60, no. 2, pp. 571–579, Feb 2013.

[5] A. Bryant, X. Kang, E. Santi, P. , and J. Hudgins, "Two-step parameter extraction procedure with formal optimization for physics-based circuit simulator IGBT and p-i-n diode models," IEEE Transactions on Power Electronics, vol. 21, no. 2, pp. 295–309, March 2006.

[6] M. Saadeh, H. Mantooth, J. Balda, E. Santi, J. Hudgins, S.-H. Ryu, and A. Agarwal, "A unified silicon/silicon carbide IGBT model," in Proc. of the 2012 Twenty-Seventh Annual IEEE Applied Power Electronics Conference and Exposition (APEC), Feb 2012, pp. 1728–1733.

[7] D. Cavaiuolo, M. Riccio, G. De Falco, G. Romano, A. Irace, G. Breglio, D. Dapra, L. Merlin, R. Carta, C. Sanfilippo, and F. Crudelini, "An effective parameters calibration technique for Pspice IGBT models application," in Proc. Of the 2014 International Symposium on Power Electronics, Electrical Drives, Automation and Motion (SPEEDAM), June 2014, pp. 133–138.

[8] F. Iannuzzo and G. Busatto, "Physical CAD model for high-voltage IGBTs based on lumped-charge approach," IEEE Transactions on Power Electronics, vol. 19, no. 4, pp. 885–893, July 2004.

[9] Infineon, "FF1000R17IE4," technical information datasheet, Rev 3.2.

[10] Infineon, "FF1400R17IP4," technical information datasheet, Rev 2.4.

800V Super Junction MOSFET (HV-DTMOS IV) with better trade-off between switching loss and dV_{DS}/dt

Hiroyuki, Irifune, *1, Japan, hiroyuki.irifune@toshiba.co.jp
Hiroshi, Ohta, *1, Japan, hiroshi3.ohta@toshiba.co.jp
Hiroaki, Yamashita, *2, Japan, hiroaki7.yamashita@toshiba.co.jp
Hideyuki, Ura, *2, Japan, hideyuki.ura@toshiba.co.jp
Kenji, Mii, *2, Japan, kenji.mii@toshiba.co.jp
Masato, Nashiki, *2, Japan, masato.nashiki@toshiba.co.jp
Naotsugu, Kako, *2, Japan, naotsugu.kako@toshiba.co.jp
Georges, Tchouangue, *3, Germany, GTchouangue@tee.toshiba.de

*1 Kaga Toshiba Electronics Corporation
*2 Toshiba Corporation Semiconductor & Storage Products Company,
*3 Toshiba Electronics Europe GmbH

Abstract

A new generation 800V-class MOSFET, HV-DTMOS IV, is developed using super junction (SJ) technology. The proposed device shows high speed and low loss switching characteristics. By optimizing the drift region and MOS structure, the specific on-state resistance $R_{DS(on)}*A$ and the switching F.O.M. $R_{DS(on)}*Q_G$ could be reduced by 27% and 69% respectively compared to the competitor. The trade-off between switching loss and the drain-source voltage change rate (dV_{DS}/dt) could also show better results compared to a competitor's SJ-MOSFET.

1. Introduction

Power MOSFETs are used as the main switching devices for power supply systems. Superior characteristics such as low on-resistance and efficient switching are always demanded for higher efficiency and reduction of system volume. At high voltage ratings, SJ-MOSFETs realize ultra-low specific $R_{DS(on)}*A$ below the Si-limit. With this technology, we can reduce their package size and parasitic capacitances. Adopting these devices means that power systems could be made much smaller and cooler through the use of smaller packages and faster switching.

Simple flyback topology is used to reduce costs and size of power systems in designs such as the power conversion circuits of LED lighting and AC/DC adapters. In these applications, high voltage class MOSFETs are required to accommodate voltage spikes during switching transients. SJ-MOSFETs are suitable for these applications because of their good trade-off between breakdown voltage and $R_{DS(on)}*A$. At the same time, electromagnetic interference (EMI) noise performance of MOSFETs is important in power supply system design, because the EMI noise generated by MOSFET switching has a significant affect on the application's noise performance. In switching operations, switching noise depends on dV_{DS}/dt during the switching period. This is because high dV_{DS}/dt induces voltage overshoot and generates EMI noise due to the existence of parasitic components in the power circuit. Designers can select the value of external gate resistance (R_G) and supress EMI noise by reducing dV_{DS}/dt. However, as R_G is set to higher values, the switching loss of the MOSFET increases. Therefore, a trade-off between switching loss and dV_{DS}/dt exists.

This paper reports the following two advantages of our newly developed SJ-MOSFET with higher voltage rating. By adopting SJ technology, ultra-low $R_{DS(on)}*A$ at 800V-class is realized. The trade-off between switching loss and dV_{DS}/dt is improved by optimizing internal gate resistance (r_G).

2. Structure and process

We optimized our manufacturing process to extend our SJ-MOSFET line-up. Fig. 1 illustrates three general MOSFET cell design structures. Fig.1 a) shows a conventional process with uniform n-type epitaxial drift region. Fig. 1 b) and c) show the SJ structure utilizing a charge compensating principle by forming p- and n-type columns in the drift region. There are two main fabrication processes; multi-epitaxial process (Fig.1 b) and deep-trench filling process (Fig.1 c). The multi-epitaxial process consists of multiple ion-implantations and n-type epitaxial growth. To improve the trade-off between breakdown voltage and $R_{DS(on)}*A$, the number of process steps must be increased to form p- and n-type columns with high aspect ratio [1]. On the other hand, the deep-trench filling process requires only one deep trench etching and p-type filling epitaxial growth. Thereby, this process enables the reduction of process steps compared to the multi-epitaxial process [2].

In order to extend our line-up to higher voltage ratings, in this work, we optimized the deep-trench filling process by controlling doping concentration and thickness of the drift region.

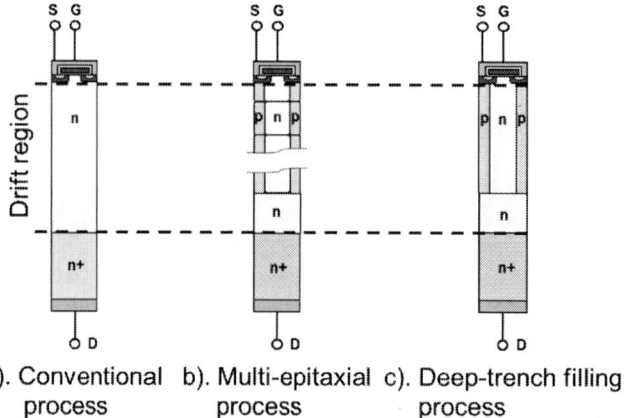

a). Conventional   b). Multi-epitaxial c). Deep-trench filling
    process   process   process

Fig. 1.   Cross sectional view of each structure: a). Conventional process, b). SJ with multi-epitaxial process and c). SJ with deep-trench filling process. The deep-trench filling process is optimized for HV-DTMOS IV without increasing the number of process steps.

3. Device characteristics

With our optimized deep-trench process, we realized an 800V rating SJ-MOSFET with superior characteristics. Fig. 2 compares the $R_{DS(on)}*A$ of the conventional process, HV-DTMOS IV and an SJ-MOSFET from a competitor. HV-DTMOS IV shows a 79% and 27% reduction of $R_{DS(on)}*A$ against the conventional process and the competitor respectively. This is due to the reduction of the resistance of the drift region by optimizing the deep-trench filling process. In addition, HV-DTMOS IV shows good switching figure of merit. By optimizing the MOS structure, $R_{DS(on)}*Q_G$ and $R_{DS(on)}*Q_{GD}$ could be reduced by 69% and 77% respectively compared to the competitor (Fig. 3). The reduction of $R_{DS(on)}*A$ also contributes to the reduction of the energy stored in output capacitance during switching (E_{oss}). Due to the smaller chip area realized by the lower $R_{DS(on)}*A$, E_{oss} is reduced by 20% compared to the competitor (Fig. 4). From these results, HV-DTMOS IV shows not only low $R_{DS(on)}*A$ but also excellent high-speed switching performance.

We are planning to launch an 800V class line-up of HV-DTMOS IV series as shown in Fig. 5. The lower $R_{DS(on)}*A$ supports ultra-low $R_{DS(on)}$ with small packages (e.g. TO-252 or TO-251) which cannot be realized using a conventional process.

PCIM Europe 2016, 10 – 12 May 2016, Nuremberg, Germany

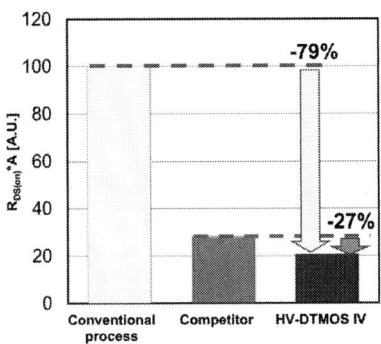

Fig. 2. $R_{DS(on)} \cdot A$ for a conventional process, competitor and HV-DTMOS IV. HV-DTMOS IV shows a dramatic switching reduction of $R_{DS(on)} \cdot A$ compared to both the conventional and competitor.

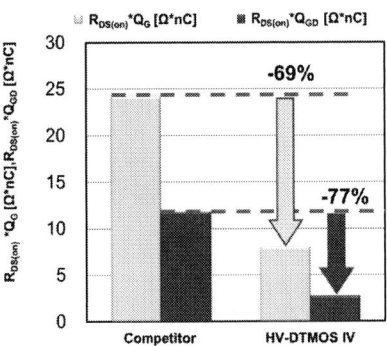

Fig. 3. $R_{DS(on)}{}^*Q_G$ and $R_{DS(on)}{}^*Q_{GD}$ of HV-DTMOS IV and competitor. These low F.O.M. realizes low loss.

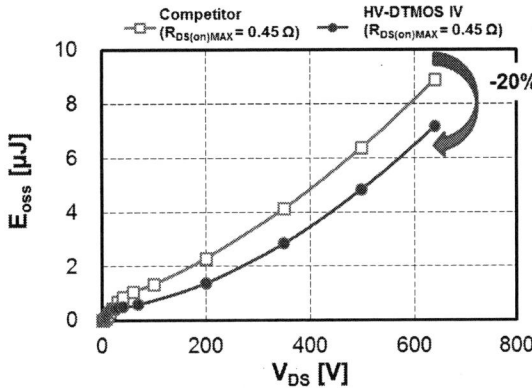

Fig. 4. E_{oss} dependence on drain-source voltage in each device. HV-DTMOS IV shows 20% lower E_{oss} at V_{DS}=640V compared to the competitor.

$R_{DS(on)}$ [Ω] MAX	ID [A] MAX	Q_G [nC] Typ	TO-252 DPAK	TO-251 IPAK	TO-220	TO-220 Full PAK
4.9	3.0	12.0			Conventional product areas	
(2.25)	(2.5)	(4.3)	Under Development	Under Development		
1.05	5.5	12.0	Sample available	Sample available		
(0.95)	(6.5)	(13.0)			Under Development	Under Development
0.55	9.5	19.0			Sample available	Sample available
0.45	11.5	23.0			Mass production	Mass production
0.29	17.0	32.0			Sample available	Mass production

*(): Planning

Fig. 5. Product line-up of HV-DTMOS IV with 800V-rating.

© VDE VERLAG GMBH · Berlin · Offenbach

4. Switching characteristics

In order to improve the trade-off between switching loss and dV_{DS}/dt, we optimized the MOS structure. dV_{DS}/dt can be modified by changing the characteristics of parasitic capacitances and the gate resistance. During the change of V_{DS}, two capacitances - gate-drain capacitance (C_{GD}) and drain-source capacitance (C_{DS}) - affect the switching behavior. The C_{GD} charging process depends on gate resistance. Therefore, we can control the charging speed of C_{GD}.

On the other hand, the charging process of C_{DS} only depends on dV_{DS}/dt. However, if the process of the charging of C_{DS} dominates the switching behavior, the switching speed cannot be controlled by the gate resistance [3].

In the development of HV-DTMOS IV, we focused on the internal gate resistance (r_G). By optimizing r_G, switching controllability is improved. In this section, we focus on the turn-off switching behavior. The switching characteristic with an inductive load was evaluated with the circuit shown in Fig. 6. Fig. 7 shows the turn-off waveforms with the external gate resistance (R_G) of 10 Ω. With each device, the gate-source voltage waveform oscillates as a result of the high dV_{DS}/dt and parasitic components in the circuit. This noisy behavior is undesirable because large V_{GS} oscillation has the potential to cause unintended turn-on and the destruction of the MOSFET. Fig. 7b) and c) show turn-off waveforms of HV-DTMOS IV with different r_G. The peak-to-peak value of gate-source voltage ($V_{GSp.p}$) is reduced to 10 V with optimized r_G. This is due to the reduction of the switching speed that is achieved through r_G optimization.

Fig.8 shows the relationship between turn-off switching loss (E_{off}) and dV_{DS}/dt in each device. By optimizing r_G, the HV-DTMOS IV technology shows smaller dV_{DS}/dt at the same R_G without a large increase of E_{off}. With the same dV_{DS}/dt, HV-DTMOS IV exhibits a lower E_{off} than the competitor, which means a better trade-off between switching loss and dV_{DS}/dt. This is due to the smaller E_{oss} realized by shrinking the chip size.

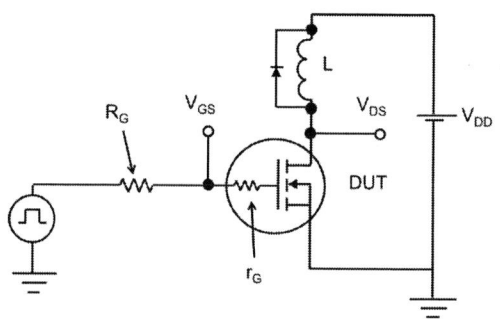

Fig. 6. Inductive load switching evaluation circuit.
(L = 1 mH, V_{DD} = 640 V)

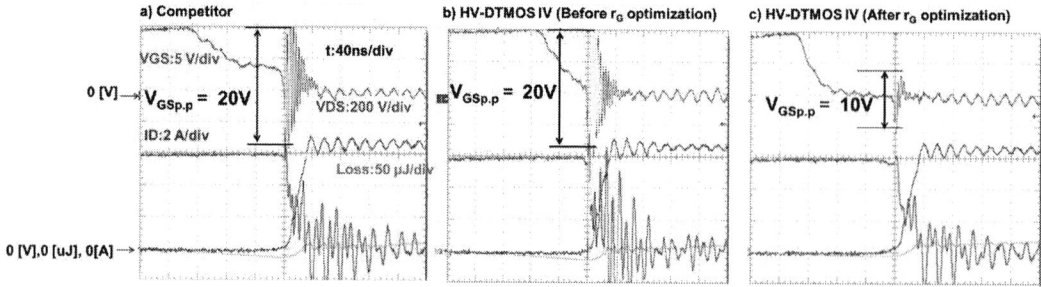

Fig. 7. Turn off switching waveform with an external gate resistance R_G of 10 Ω.

© VDE VERLAG GMBH · Berlin · Offenbach

PCIM Europe 2016, 10 – 12 May 2016, Nuremberg, Germany

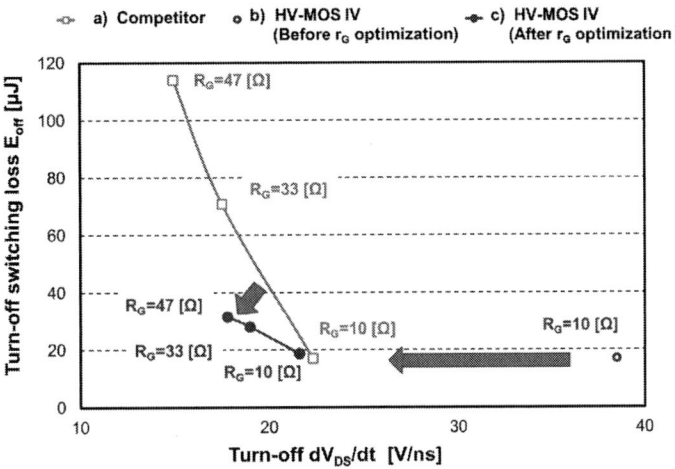

Fig. 8. Trade-off characteristic between turn-off switching loss E_{off} and dV_{DS}/dt.
HV-DTMOS IV (after r_G optimization) shows a better trade-off than the competitor.

5. Application evaluation

In order to evaluate the performance of the technology in an actual application, we tested our device in a power supply circuit with an 85W rating. A schematic of the circuit is shown in Fig. 9. Fig. 10 shows the component of the loss generated by the MOSFET when using an off-side external gate resistance ($R_{G\ (off)}$) of 4.7 Ω. The turn-off switching loss is larger than the turn-on switching loss and the conduction loss in this configuration. To compare the switching characteristics between HV-DTMOS IV and the competitor, we analyzed the loss dependence on $R_{G(off)}$ (Fig. 11). The total and turn-off loss of the competitor's MOSFET becomes larger as $R_{G(off)}$ increases. HV-DTMOS IV, however, shows a negligible dependence on $R_{G(off)}$. The better trade-off between switching loss and dV_{DS}/dt with HV-DTMOS IV discussed in the previous section contributes to improved efficiency in actual usage (Fig. 12).

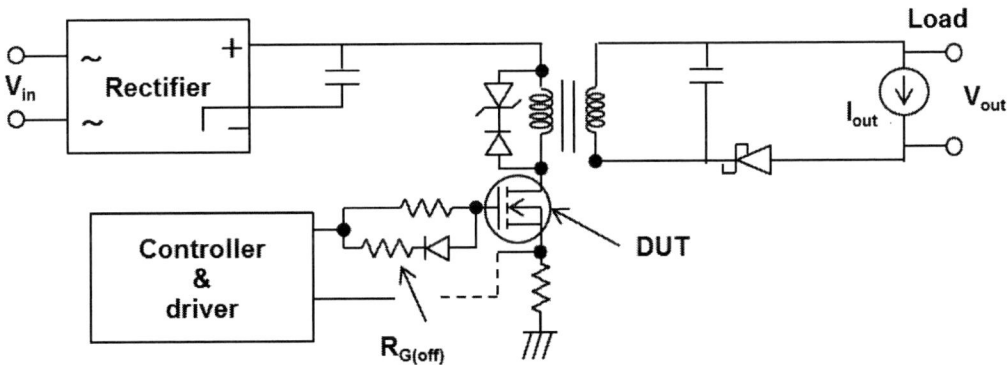

Fig. 9. Evaluation circuit of 85W class power supply.
Condition: V_{in} = 100 V_{AC}, V_{out} = 20 V, I_{out} = 4.25 A, Switching frequency f = 50 kHz, Ta = 298 K

© VDE VERLAG GMBH · Berlin · Offenbach

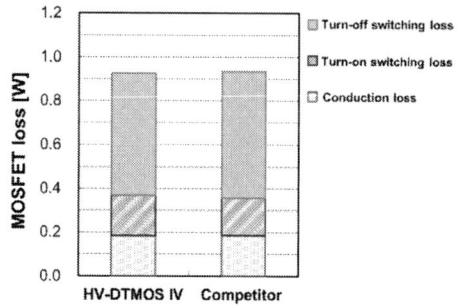

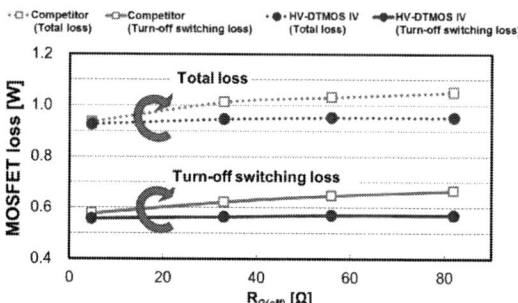

Fig. 10. The component of the loss with an $R_{G (off)}$ = 4.7 Ω. Turn-off loss is larger than other, smaller components in this application.

Fig. 11. The MOSFET loss dependence of $R_{G (off)}$. HV-DTMOS IV shows better loss with each $R_{G (off)}$.

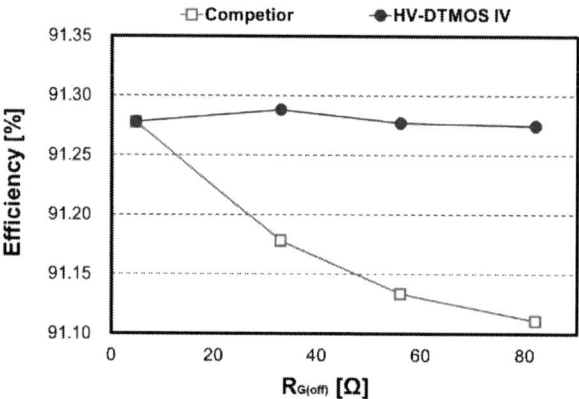

Fig. 12. Efficiency dependence on $R_{G(off)}$.
HV-DTMOS IV shows better efficiency compared to the competitor.

6. Conclusion

Utilizing our deep-trench filling technology allows a higher voltage SJ-MOSFET with low on-resistance and switching loss to be realized. Better trade-off between switching loss and dV_{DS}/dt could be also achieved by optimizing internal gate resistance.

[1] Wataru Saito et al, "**A 15.5mQcm2-680V Superjunction MOSFET reduced On-Resistance by Lateral Pitch Narrowing**", ISPSD 2006
[2] Syotaro Ono et al, "**DTMOS-IV : RDS(ON) innovation by deep-trench filling SJ technology**", PCIM Europe 2012
[3] Syotaro Ono et al, "**Improvement of gate controllability for new generation Superjunction MO SFETs : DTMOS-III series**", PCIM Europe 2011

PCIM Europe 2016, 10 – 12 May 2016, Nuremberg, Germany

Highly robust 1700V diodes fabricated on 8" line using optimized proton implanted buffer

Maolong Ke, Haihui Luo, Ian Deviny, Xiaoping Dai, Jianwei Huang, Guoyou Liu
Power Semiconductor R&D Centre, CSR Times Electric Co.,Ltd,
Doddington Road, Lincoln, United Kingdom. Tel. +44-1522502865, FAX: +44-1522502747
Emails: Maolong_Ke@dynexsemi.com, liugy@csrzic.com

Abstract

Migrating high power device production from 6" to 8" wafers can significantly reduce unit cost. However, for 1700V freewheeling diodes, it is not practical to continue using the diffused buffer on 8" wafer production line as it has been the norm for 6" wafer production line. This is because much thicker starting wafers are typically used in 8" production line for frontside processing due to the requirement of mechanic stability. Therefore, very large thermal budget would be required to produce a very deep buffer for 8" wafer, leading to significant wafer bow, breakage and cost. Therefore an alternative method of producing sufficiently strong buffer has to be found.

1. Proton implantation and activation

Standard production for high power diodes starts from backside buffer formation. This is because high temperature and longtime drive are required to form a deep and soft buffer for smooth switching operation. If the frontside of the diodes were processed first, then the driving or activation temperature would have to be limited to 500^0C, well below the temperature required for typical buffer diffusion.

Proton implantation and subsequent thermal activation have been reported to be capable of producing a n-type doping profile, and more importantly, the activation temperature was found to be typically below 500^0C [1-2], making it potentially a very suitable tool for buffer formation after the frontside processing. Unfortunately, the carrier concentration can be achieved from each proton implantation is quite limited because of the difficulty in donor formation from protons [1] and, to our best knowledge, it has only been used in commercial production for up to 1200V rated diodes so far [3]. Many obstacles need to be overcome in order to for it to be used in higher voltage rated diodes. Firstly, proton to donor conversion rate and the peak carrier concentration need to be increased. Secondly, the dynamics of the activation and donor formation process need to be better understood. Thirdly, the behaviour and impact of deep level defects created by the proton implantations need to be studied. Finally, hydrogen passivation to phosphorous at the back surface needs to be minimized.

In order for proton implantations to be used in 1700V rated high powder diodes, detailed study was carried out for properties of proton implantations and activation. The study included different thermal annealing, spread resistance profiling (SRP), secondary ion mass spectroscopy (SIMS), deep level transient spectroscopy (DLTS).

Figure 1 displays typical SRP profiles obtained at various annealing temperatures. The curve for "no anneal" displayed much lower carrier concentration at the left hand side than the background doping at the right hand side. This is probably mainly because the left hand side was damaged by the proton implantation and the carrier mobility is much lower than the right hand side, which manifests itself as much lower carrier concentration in SRP measurements. At the low temperature anneal (350^0C or lower), a pronounced peak at the end of the range appeared quickly, but at the left hand side of the peak the carrier concentration was still

© VDE VERLAG GMBH · Berlin · Offenbach

lower than the background doping at the right hand side, suggesting crystal damages wasn't fully healed. Further increase in annealing temperature (to 400^{0}C or 450^{0}C) has gradually removed the damages and increased the carrier concentration at the left side. The SRP profile suggested the implantation damages were fully healed at around 450^{0}C anneal as the concentration profile at the left hand side was smooth and no sudden dip ("concentration valley") again. Also, further increase in annealing temperature to 490^{0}C did not increase the carrier concentration. In fact, the overall integrated carrier concentration peaked at the annealing temperature of around 400^{0}C and slowly decreases as the annealing temperature increases further. Another interesting observation was the existence of a "concentration valley" at the left side when the annealing temperature was below 420^{0}C (not shown here). The valley decreased in depth and shifted left towards the back surface as the annealing temperature increased from room temperature upwards. The formation of the concentration valley under the 420^{0}C anneal may have suggested the activation/healing happened quicker at the both sides of the valley. But the shift of the valley towards the leftside with the annealing temperature indicates the activation/healing processing is strongly linked to propagation of something. It is either the propagation of defects/vacancies towards the back surface or protons and proton complexes from the peak regions towards the back surface. We believe it is likely to be a combination of both. As the vacancies/defects propagate away, the region's mobility recovers. At the same time, protons will propagate backwards as well to form donor complexes and lift the carrier concentration along the way even further. The peak concentration at the end of the implantation range was found to decrease with annealing temperature, and at the same time the peak position shifted leftwards as well. All that seem to confirm our above hypothesis. On the other hand, the proton propagation towards back surface has one serious drawbacks. As the back surface is heavily phosphorous doped for n$^+$ contact purpose for high power diodes, the hydrogen ions (protons) can form a complex bond with phosphorous ions and passivate the n$^+$ contact layer. We have seen an increase in Vf (or the contact resistance) just by increasing the proton annealing temperature, confirming the back contact layer was passivated by increased hydrogen propagation towards the back surface. Also, Secondary Ion Mass Spectroscopy (SIMS) measurements (not shown here) have shown strong presence of hydrogen at the back surface.

In order to understand the activation and healing of proton implantation further, we have performed deep level transient spectroscopic (DLTS) measurements to investigate the deep level defects behaviour. Similar to the results obtained by Leveque [4-7] and others, we have also seen 5 deep level defect signals related to implanted protons in DLTS spectroscopy. The origin of the peaks has been well identified for some. For example, the vacancy-oxygen (VO) complex at 0.17eV, the double vacancies defects at 0.23 and VOH at 0.32 are all well identified. But the defect origins for both 0.42eV and 0.45eV positions are still to be confirmed, although they were believed to be vacancy and hydrogen (H) related complexes [5-6]. We found that the concentration of those deep level defects were very high (~10^{13}cm^{-3}) at 350^{0}C temperature anneal, especially those at the deeper end of the spectrum (>0.3eV), but decreased markedly as the annealing temperature increased. Since these deeper level defects are all believed to be H-related, hence the sharp drop in their signals further suggests that the hydrogen propagation and formation of hydrogen-related donor complexes are the important mechanisms in the activation and healing process. As the annealing temperature increased to 490^{0}C, the concentration of deep level defects reduced to around 10^{12}cm^{-3}, and some signals, 0.45eV for example, disappeared completely. In an ideal world, we would like to anneal at close 500^{0}C to get rid of the deep level defects. But, as discussed early, too higher temperature anneal can lead to both a drop in the buffer concentration and an increase in Vf due to the H-passivation of phosphorous contact layer at the back surface. Hence a careful compromise is needed in the proton annealing/activation temperature.

In fact, as similar deep level defects will be introduced anyway at a later stage of the diode processing for lifetime control purpose, we can argue that there is no need to completely get rid of them now and re-introduce them later. Moderate amount of these defects left at this stage of processing should not present any problem for the diode performance. More detailed SIMS measurements to be carried out shortly in order to study the hydrogen propagation throughout.

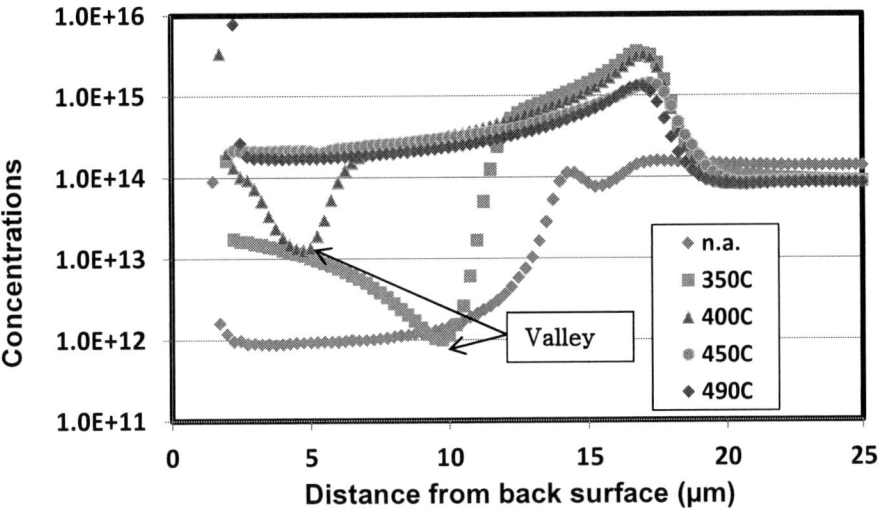

Fig. 1 Measured carrier concentration against annealing temperature for proton implanted buffer.

2. Diode fabrication

High power diodes operating at 1700V were fabricated through front-side processing, grinding, backside n-type buffer formation through multiple proton implantations and careful thermal annealing as discussed above, back metal contact and sintering. The diode wafers are then undergone frontside helium implant for local lifetime killing to control carrier injection, followed by electron irradiation to reduce the carrier lifetime across the n⁻ region. The optimization of the lifetime control processes is critical for device performance. Once the wafer fabrication is completed, the wafers are probe-tested and then diced into individual devices to build into substrates along with suitable IGBTs for further tests. Finally F-outline modules were built for comprehensive tests and qualifications.

Figure 2 shows two representative F- modules assembled for the tests, one was built with our 1700V diodes produced on 8" production line used the above mentioned methods of buffer formation through multiple proton implantations and activation, the other with diodes from a leading supplier, which were fabricated on a 6" wafer production line with diffused buffer, to our best knowledge. The diodes' voltage and current rating were exactly the same. The IGBT dies used in both types of modules were exactly the same as well for comparison.

Fig. 2 F-module outlines used for extensive testing and benchmarking purpose

3. Diode testing

Firstly, static tests were carried out for the two types of modules and Vf was found to be very similar and both are within the range of 1.8 ~1.85V at 1600A at room temperature. Then extensive dynamic tests were carried out. Figure 3 shows the comparison of switching waveforms between the two types of modules. As expected, both modules passed typical standard operating conditions at 900V, 1600A and 900V, 2400A (3a) without oscillation, but the modules with our diodes inside had significantly lower Irr than the competitor's. The Qrr and Erec were quite close between the two types of modules because our diodes had a longer current tail, which on the one hand indicates the smoothness of recovery from our diodes, but on the other, increases both Qrr and Erec from our diodes to the same level as our competitor's.

The Fig. 3b shows the switching performance at 1000V and 2400A. The recovery for our diodes was still smooth as the voltage increased, but that for our competitor's oscillated seriously. Further increase in line voltage to 1100V destroyed the modules with competitor's diodes inside while ours still smoothly recovered (not shown here). This clearly demonstrated the robustness in switching performance of our diodes against our competitors. Figure 4 displays the room temperature low current (15% nominal) operation. It is well known that room temperature and low current operation is a very harsh testing condition for diodes, and it can easily lead to oscillation and failure of diodes. It is another important test for diode's robustness. As seen from Fig. 4, the module with our diodes inside recovered smoothly again while the modules with competitor's diodes inside oscillated strongly. Further increase in line voltage to 1000V with the current remaining low destroyed the modules with competitor's diodes again while ours survived.

The technology trade-off between Vf and Erec are shown in Fig.5 and they are at very similar position for both type of modules, which means the total losses will be very similar for both on state (Vf) and switching (Erec) operations. As our diodes are made on 8" wafer production line against 6" from competitors, there will be cost advantages. So our diodes are potentially both lower cost and more robust than competitor's.

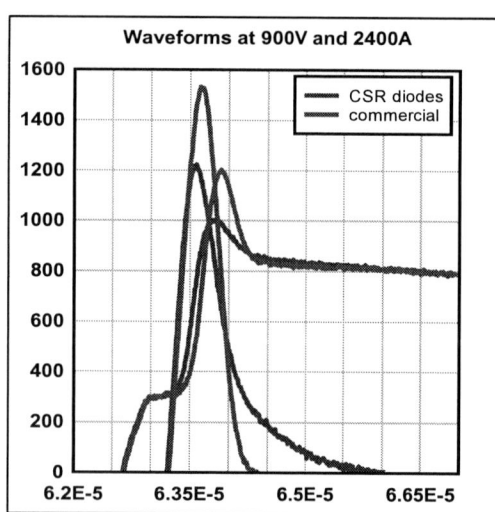 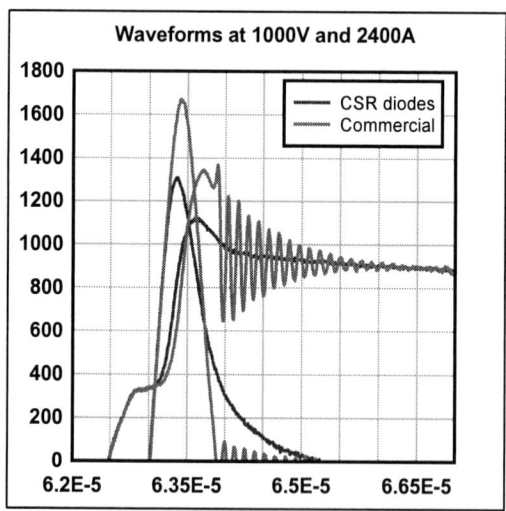

Fig. 3 Switching waveform comparison between modules made from our newly fabricated diodes and those from a leading commercial suppliers. 3a Nominal switching condition of 900V 2400A, and 3b 1000V and 2400A condition.

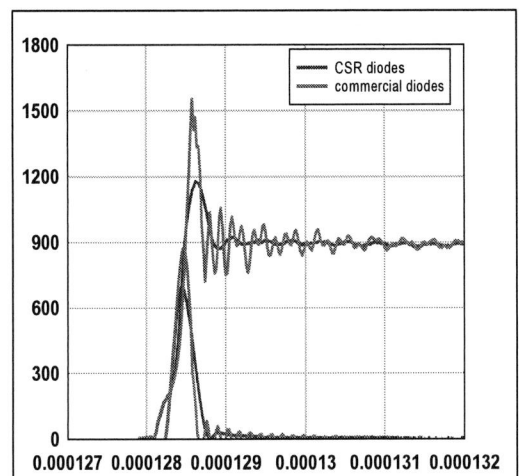

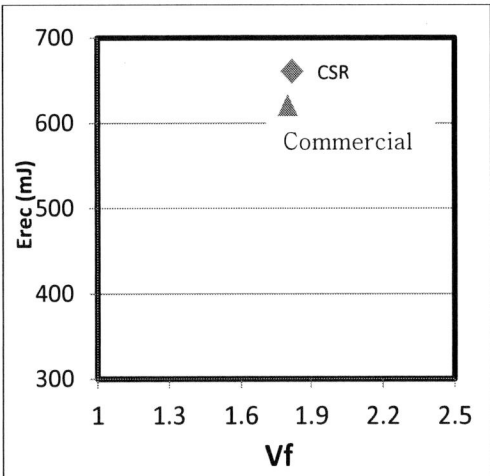

Fig. 4 Switching waveforms at low current (~15% nominal) and room temperature operation. Our diodes recovered smoothly while the leading commercial diodes oscillated seriously

Fig. 5 Technology trade off points between two types of diodes.

4. Conclusions

Extensive study of proton implantation and annealing behaviour have been carried out with a view of producing highly robust 1700V how power diodes on 8" wafer production line. The study led us to find an optimal condition for proton activation and defects healing, which subsequently enabled us to produce highly robust 1700V rated diodes with potentially much lower cost. Benchmarking tests were carried out between our newly fabricated diodes with similarly rated diodes from a leading commercial supplier. The results of the benchmarking tests confirmed that our new diodes are more robust and potentially lower cost.

We greatly acknowledge the help and assistance from Simon Leonard and Professor Bruce Hamilton at Manchester University for performing the DLTS measurements and subsequent discussions. We are also grateful to the support received from colleagues both at Dynex Semiconductor Ltd at Lincoln and CSR Times Electric Co. at Zhuzhou.

References:

[1] Klug J. N., Lutz J., Meijer J. B., "n-type doping of silicon by proton Implantation" Proceedings of the 14th European conference on Power Electronics and Applications (EPE2011), p.1-7 (2011)

[2] Ohmura, Y.; Zohta, Y.; Kanazawa, M. (1972): Electrical properties of n-type Si layers doped with proton bombardment induced shallow donors, Solid State Communications, vol. 11/1 ,pp. 263–266.

[3] Mauder A., Schulze H. J., Hille F., Schulze H., Pfaffenlehner M., Schaffer C., Niedernostheide F. J., "Semiconductor Device and Fabrication Method Suitable Therefor" US Patent No: US 7514750 B2 (2009)

[4] Leveque P., Pellegrino P., Hallen A., Svensson B. G., Privitera V., Nucl. Instrum. Methods B 174, 297-303 (2001)

[5] Leveque P., Hallen A., Svensson B. G., Wong-Leung J., Jagadish C., Privitera V., Eur. Phys. J. AP 23, 5–9 (2003)

[6] Watkins G. D., Corbett J. W., Phys. Rev. 121, 1001 (1961)

[7] Svensson B. G., Hallen A., Sundqvist B. U. R., Mater. Sci. Eng. B 4, 285 (1989)

Loss and softness optimized IGBT-diode system for fast-switching applications

Christian R. Müller and Stefan Buschhorn, Infineon Technologies AG, Max-Planck-Str. 5, 59581 Warstein, Germany

Abstract

The interaction between IGBT and diode is analyzed with focus on fast-switching operation. By addressing intrinsic design parameters, the softness of both devices is varied independently and a broad range of design combinations is investigated. The usage of a statistical approach allows identifying the mutual interdependency in-between the design combinations comprehensively. Loss and softness optimized IGBT-diode combinations are presented and their impact on the application is discussed.

1.   Fast switching in applications and power modules

Within the last decade, high-speed devices emerged as one of the major topics in power electronics due to the enormous benefit arising from either higher switching frequencies or reduced losses [1-4]. Based on this approach, it is essential to provide an optimized environment for operation, i.e., low stray inductance, symmetric designs, and good device control [5-11]. In addition to this, the performance of power semiconductors can be optimized for a given system if the identified and known constraints are taken into account [12]. It is essential to consider not only the switch itself but also the diode in the free-wheeling path due to their mutual interdependency during the switching event [13].

Focusing on this interdependency in-between switch and diode, additional degrees of freedom have to be considered in the development of power modules as well as for the application itself. These degrees of freedom correlate with the device design, i.e., the device physics and technology-inherent trade-off relations. In contrast to operating conditions, which can be addressed in the application itself and, hence, offer the option to be dynamically adjusted during operation, the optimization degree of the IGBT-diode system is fixed after its definition and implementation. Therefore, a deep understanding of the interplay between diode and switch under operating conditions is essential. For clarity: On module level, the design of both switch and diode can be altered, e.g., by addressing the softness of the devices. This leads to a variety of power-module design scenarios. In the system, typical parameters like stray inductance and gate resistor are defined in order to set switching slopes and limit the resulting peak voltages for a given power module. Finally, i.e. in the respective application, the target operating conditions like switching frequency and output power are set with respect to the environmental restrictions like, e.g., cooling. The performance of both system and power module is based on this interplay.

Ultimately, the performance of the power module is determined by power losses and switching slopes. The latter ones are typically characterized in terms of dv/dt and di/dt at the respective devices. In addition to this, peak voltages during the IGBT turn-off and diode recovery can occur and are directly related to the addressable di/dt values. To overcome these challenges, the usage of low-inductively and symmetrically designed setups is recommended. For given designs, system parameters like the maximum operating voltage can be lowered in order to avoid violations of the safe-operating area as a countermeasure. Finally, it is about the trade-off between fast switching and low switching losses on the one hand, and maximum values of dv/dt and peak voltages, on the other hand. These

PCIM Europe 2016, 10 – 12 May 2016, Nuremberg, Germany

parameters are typically related to the device softness and also go hand in hand with oscillations and steep slopes.

2. Interplay between diode design and switching performance

Due to the fact that the switching parameters of both, i.e. switch and diode, influence the switching characteristics significantly, the analysis of the interplay in-between diode design and switching performance is based on using similar switching characteristics for each investigated diode design. Therefore, 650-V TRENCHSTOP™ 5 IGBTs with nominal current of 50 A were implemented in a half-bridge topology with a single IGBT per system. The investigated diode designs were operated in the free-wheeling path, i.e. added as anti-parallel diode to the high-side and low-side IGBTs. The target operation mode was diode recovery.

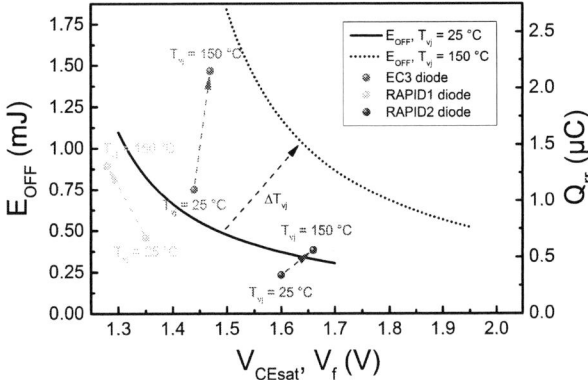

Fig. 1. Trade-off curves E_{OFF} versus V_{CEsat} and Q_{rr} versus V_f of the 650-V TRENCHSTOP™ 5 IGBT and the investigated diode types, respectively.

Here, three different diode designs were used: an emitter-controlled diode (EC3), a RAPID1 diode, and a RAPID2 diode. All diodes provide a maximum blocking voltage of 650 V and a nominal current of 30 A. Fig. 1 shows the trade-off diagram for diodes and IGBT. With focus on the diode, the reverse-recovery charge (Q_{rr}) of the diode designs varies significantly. A low Q_{rr} is typically related to larger dv/dt, a more snappy switching behavior, and low recovery losses (E_{rec}). The forward voltage (V_f) directly corresponds to the static losses of the diode in conducting mode. Based on these two characteristic parameters, i.e. Q_{rr} and V_f, the diode design can be distinguished as follows at room temperature ($T_{vj} = 25\ °C$): The EC3 diode provides the largest Q_{rr} and moderate V_f which indicates an ideal operation at lower switching frequencies. Compared to the EC3 diode, the RAPID1 diodes offers a 100 mV smaller V_f and a Q_{rr} which is only about of 60 %. This allows diode operation at higher switching frequencies as well as lower static losses. The RAPID2 diode shows by far the lowest Q_{rr} of all three diode designs, but, in turn, the largest V_f. Therefore, the RAPID2 diode can be operated at highest switching frequencies. The implemented IGBT provides a collector-emitter saturation voltage (V_{CEsat}) of 1.65 V at $T_{vj} = 25\ °C$. This indicates the ability for operating at highest switching frequencies as well.

For the IGBT, a clear interdependency between the turn-off energy (E_{OFF}) and V_{CEsat} exists [12]. For clarity, a decreasing V_{CEsat} results in softer switching and, in turn, leads to less overvoltage (V_{CEmax}) and oscillations during turn-off.

The measurements were performed in a setup with a stray inductance of 35 nH. For operating the IGBT, a gate-emitter voltage $V_{GE} = \pm15\ V$ and a gate resistor of $R_G = 20\ \Omega$ was used. Fig. 2 displays characteristic switching curves of all diode designs for a DC-link voltage $V_{DC} = 400\ V$ and at $T_{vj} = 25\ °C$.

© VDE VERLAG GMBH · Berlin · Offenbach

PCIM Europe 2016, 10 – 12 May 2016, Nuremberg, Germany

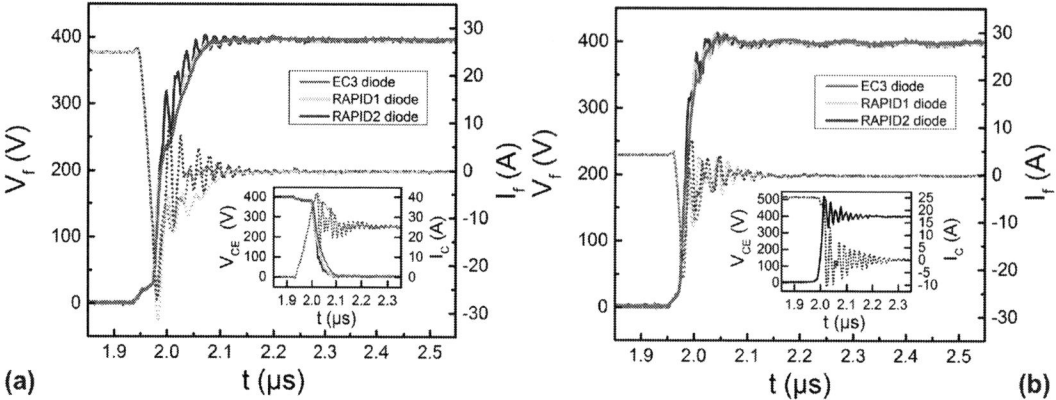

Fig. 2.   Switching characteristics of the different diode designs for I_f = 25 A (a) and I_f = 4 A (b). Solid lines correspond to V_f (and V_{CE}) and dotted lines to I_f (and I_C). Insets: IGBT turn-on (a) for the diode designs at I_C = 25 A. IGBT turn-off (b) for I_C = 25 A.

In Fig. 2b, the diode recovery is shown at a small current (I_f = 4 A). Here, oscillations on both V_f and I_f occur for all diode designs. Again, the oscillations are more pronounced for RAPID1 diode and the use of a RAPID2 diode is leading to a higher dv/dt. The insets of Fig. 2a and Fig. 2b visualize the turn-on and turn-off characteristics of the IGBT for a collector current I_C = 25 A. As expected, the diode design interacts with the turn-on characteristics of the IGBT and the same findings as above described are observed. For the turn-off characteristics, no interactions in-between IGBT and diode design occur. Therefore, I_C and V_{CE} are displayed only in combination with the RAPID2 diode. The observed oscillations are attributed to the switching characteristics of the IGBT.

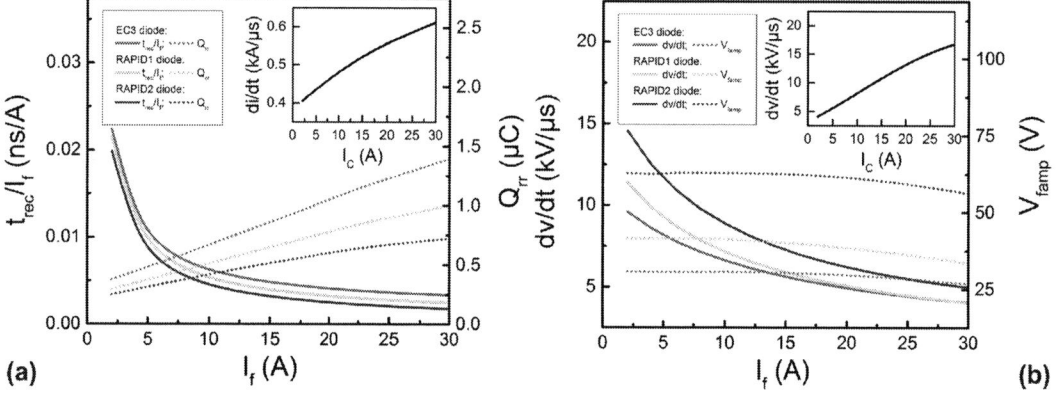

Fig. 3.   Characteristic switching parameters of the different diode designs during the diode recovery. (a) t_{rec}/I_f and Q_{rr} of the different diode designs versus I_f. Inset: di/dt of the IGBT versus I_C during turn-on. (b) dv/dt and V_{fmax} versus I_f. Inset: dv/dt of the IGBT versus I_C during turn-off.

The characteristic switching parameters are depicted in Fig. 3. The left-hand side shows Q_{rr} and t_{rec}/I_f versus I_f for the three diode designs. The recovery time (t_{rec}) corresponds to the average time which is required for clearing out the diode and is proportional to Q_{rr}/I_{rmax}. For a given stray inductance and switching speed, the usage of t_{rec} allows comparing the different designs with respect to their snappiness. Lower t_{rec} corresponds to a faster clearing out and, hence, leads to a higher likeliness of oscillations. Independent of the diode design, Q_{rr} increases for larger forward currents due to the fact that the devices provide a higher carrier

© VDE VERLAG GMBH · Berlin · Offenbach

density. By investigating t_{rec}/I_f, the interplay between clear-out rate and carrier density is taken into account. The RAPID2 diode provides both, low t_{rec}/I_f and small Q_{rr}, which emphasizes the ability of high-frequency operation as well as a snappy switching behavior. For the RAPID1 diode, larger values for t_{rec}/I_f and Q_{rr} are observed compared to the RAPID2 diode, whereas the EC3 diode provides the largest values. Shown as inset of Fig. 3a, the di/dt during the diode recovery illustrates that all diode designs were operated under equal switching conditions.

Fig. 3b displays the dv/dt and V_{famp} versus I_f for the different diode designs. V_{famp} corresponds to the initial amplitude of the snap-off oscillations which is superimposed on V_f. Independent of the diode design, larger I_f leads to a reduction of dv/dt which is attributed to the slower propagation of the device's intrinsic electric field with an increasing amount of stored charges. It has to be pointed out that all diode designs achieve dv/dt values up to 10 kV/µs for very small I_f. With focus on dv/dt, the EC3 diode provides the smallest values of all investigated diode designs. In comparison, the RAPID2 diode provides the highest values of dv/dt and also V_{famp}. Whereas for the RAPID1 diode significant values of V_{famp} are observed, almost no oscillation amplitude is found for the EC3 diode. For the RAPID2 diode, dv/dt values up to 15 kV/µs are reached and minimum values are still above 5 kV/µs, which is a critical value in distinct applications.

The dv/dt during the turn off of the IGBT is plotted in the inset of Fig. 3b. For the IGBT, also high values of dv/dt in the same range as for the diodes are reached. At low currents, the highest dv/dt originates from the switching behavior of the diodes and not of the IGBT. Hence, in low-current applications, the dv/dt of the diode design will be the limiting factor instead of the IGBT softness.

RAPID1 diode and RAPID2 diode provide the smallest Q_{rr} and t_{rec}/I_f. With respect to the trade-off curve, at least the RAPID1 diode offers an V_f that is comparable to the EC3 diode. For fast-switching applications, RAPID1 diode and RAPID2 diode should be taken into further consideration whereas the EC3 diode is not optimized for this purpose. Therefore, the upcoming analysis will focus only on RAPID1 diode and RAPID2 diode.

3. Design criteria for optimized IGBT-diode systems

The behavior of an IGBT-diode system is mainly determined by the design criteria of each single element, i.e. IGBT and diode. In addition, the interplay needs to be considered sufficiently. For a proper understanding of this interplay, the influence of external and internal factors is considered.

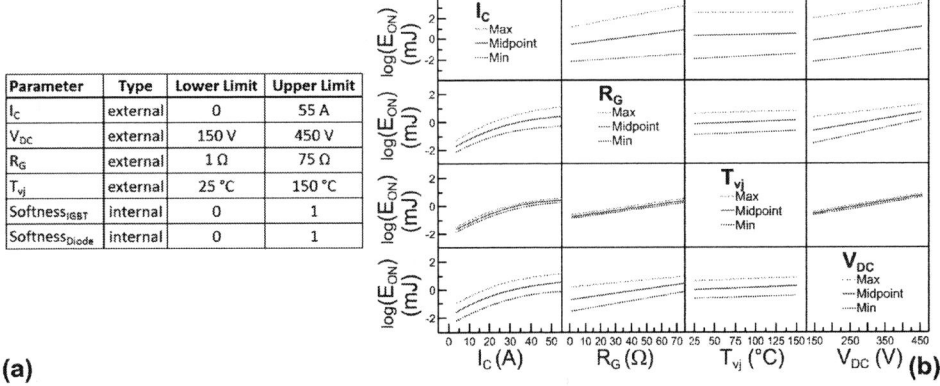

Parameter	Type	Lower Limit	Upper Limit
I_C	external	0	55 A
V_{DC}	external	150 V	450 V
R_G	external	1 Ω	75 Ω
T_{vj}	external	25 °C	150 °C
Softness$_{IGBT}$	internal	0	1
Softness$_{Diode}$	internal	0	1

(a)

(b)

Fig. 4. (a) Addressed parameters within the performed DoE approach. For external parameters, a broad range is accessible within the parameter space, whereas internal parameters are limited by design. (b) The interaction plot visualizes the main effects and interactions for all external parameters.

PCIM Europe 2016, 10 – 12 May 2016, Nuremberg, Germany

Fig. 4a illustrates the identified factors and their ranges which were taken into account for the experiment. The external factors are the collector current and forward current, respectively, the DC-link voltage, the virtual-junction temperature (T_{vj}), and the gate resistor. All these factors can be controlled and modified in the experimental setup as well as in the application easily. The internal factors are the softness of IGBT and diode, which are defined by the device design.

For planning the experiments, a statistical approach (DoE) was used to analyze the system behavior. Unless otherwise specified, all shown results are derived under reference conditions: V_{DC} = 400 V, I_C = 25 A, $1/R_G$ = 1 1/Ω, and T_{vj} = 25 °C. In addition, the same setup and gate driver as mentioned above was used. For the analysis, the switching speed is related to $1/R_G$. Here, low values of $1/R_G$ characterize a slower switching whereas an increasing $1/R_G$ is associated with a higher switching speed.

First, the interaction of the external factors is investigated. Fig. 4b shows the interaction plots of the turn-on losses (E_{ON}) for all external factors. Based on these interaction plots, main effects and interaction effects are easily distinguished. It can be clearly seen that there is no significant interaction between T_{vj} and other external factors. Also only a very weak interaction is observed in-between I_C and V_{DC}. In general, no effects were identified which provide a reversed interaction in the considered range. In addition, all external factors lead to a monotonically increasing E_{ON}. Consequently, no reduction of E_{ON} can be realized by interaction of the external factors. The same observations and conclusions are valid for E_{OFF} as well as for E_{rec} (interaction plots not shown here).

As internal factors, $Softness_{IGBT}$ and $Softness_{Diode}$ were identified and are represented in arbitrary units (a.u.). The $Softness_{IGBT}$ was defined as the normalized ratio of E_{OFF} and V_{CEmax}. With respect to Fig. 1, $Softness_{IGBT}$ covers the range of the shown trade-off curve. For $Softness_{Diode}$, the designs of RAPID1 diode and RAPID2 diode correspond to the maximum and minimum value, respectively.

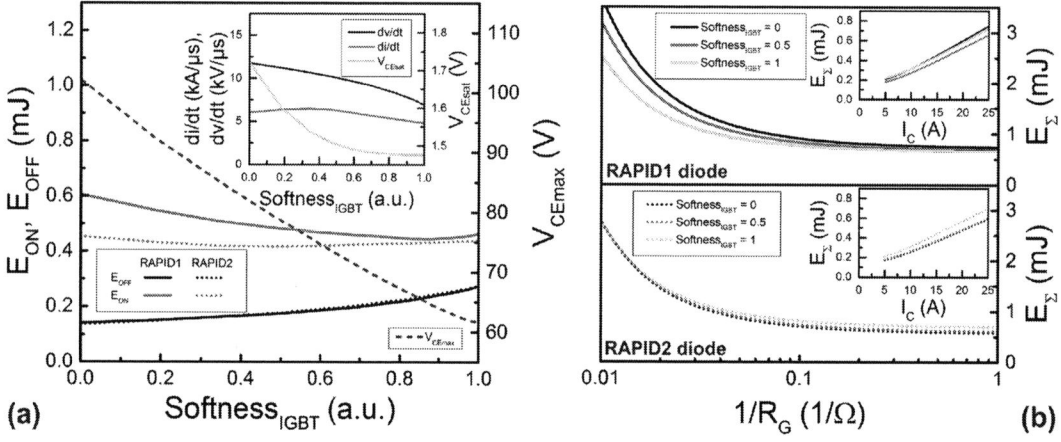

Fig. 5.   (a) E_{OFF}, E_{ON}, and V_{CEmax} versus $Softness_{IGBT}$. Inset: di/dt during turn-on, dv/dt during turn-off, and V_{CEsat} versus $Softness_{IGBT}$. (b) For RAPID1 diode (upper part) and RAPID2 diode (lower part), E_Σ versus $1/R_G$ and $Softness_{IGBT}$. The insets show E_Σ versus I_C and $Softness_{IGBT}$, respectively.

Fig. 5a displays E_{ON}, E_{OFF}, and V_{CEmax} in dependency on the IGBT softness. With increasing $Softness_{IGBT}$ and independent of the used diode design, E_{OFF} monotonically increases and doubles its value. In addition, higher IGBT softness reduces V_{CEmax} to about 60 % of its initial value. The inset of Fig. 5a shows dv/dt during IGBT turn-off and V_{CEsat} versus the IGBT softness. Also here, monotonically decreasing dv/dt and V_{CEsat} are found with increasing IGBT softness. In contrast to this, larger $Softness_{IGBT}$ does not consequently lead to a lowering of E_{ON}. If a RAPID1 diode is used in the system, E_{ON} decreases with increasing

© VDE VERLAG GMBH · Berlin · Offenbach

IGBT softness but slightly rises again for $Softness_{IGBT} \geq 0.8$. For the RAPID2 diode, a minimum value of E_{ON} is reached at $Softness_{IGBT} = 0.5$. The temporary decrease of E_{ON} with increasing IGBT softness correlates to a higher di/dt during the IGBT turn-on (inset of Fig. 5a). For $Softness_{IGBT} \geq 0.5$, the di/dt reduces again. Thus, any further decrease of E_{ON} does not originate from the IGBT softness itself but is related to the interplay of diode and IGBT.

The sum of losses ($E_\Sigma = E_{ON} + E_{OFF}$) is depicted in Fig. 5b. With increasing switching speed, E_Σ lowers and reaches its minimum value at maximum switching speed. Larger I_C leads to higher E_Σ values (see insets). These observations are independent of the used diode design. In contrast to this, a higher IGBT softness leads to a reduction of E_Σ only for the RAPID1 diode. For the RAPID2 diode, the lowest E_Σ is provided for $Softness_{IGBT} = 0$. These findings are directly related to the stronger impact of E_{ON} on E_Σ for using the RAPID1 diode compared to the RAPID2 diode. Therefore, a higher IGBT softness leads to lower E_{ON} which almost compensates the increase of E_{OFF} in case of the RAPID1 diode.

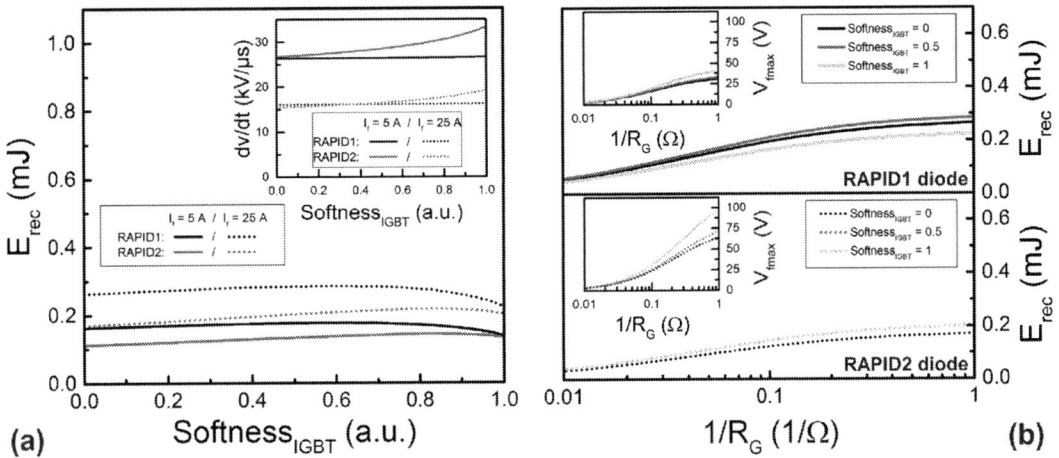

Fig. 6. (a) E_{rec} versus $Softness_{IGBT}$ and I_f. Inset: dv/dt during diode recovery versus $Softness_{IGBT}$ and I_f. (b) For RAPID1 diode (upper part) and RAPID2 diode (lower part), E_{rec} versus $1/R_G$ and $Softness_{IGBT}$. The insets show V_{fmax} versus $1/R_G$ and $Softness_{IGBT}$.

In Fig. 6a, E_{rec} is shown versus the IGBT softness for the two diode designs. As expected, the RAPID2 diode provides smaller recovery losses than the RAPID1 diode. With increasing $Softness_{IGBT}$, E_{rec} grows slightly but shows a reduction for large values of $Softness_{IGBT}$. This behavior of E_{rec} is in good agreement with the observed di/dt-$Softness_{IGBT}$ dependency, as higher di/dt values lead to larger E_{rec}. At this point the mutual interdependency in-between IGBT and diode comes to the fore. Whereas the RAPID1 diode is limiting the switching speed of the IGBT-diode system, the RAPID2 diode offers more potential for fast switching, and, in this case, the IGBT is the limiting factor (inset of Fig. 6a). With increasing IGBT softness, the IGBT's ability to drive higher dv/dt during turn-on is enhanced (not shown). In turn, if the dv/dt is not limited by the diode design, a higher possible dv/dt leads to higher recovery losses for a given di/dt. Due to this effect, the recovery losses will increase although the di/dt is decreasing.

Fig. 6b displays the recovery losses in dependency of the switching speed. One can see clearly that a higher switching speed results in larger recovery losses. With respect to the di/dt-$Softness_{IGBT}$ dependency (inset of Fig. 5a), the RAPID1 diode provides the highest E_{rec} for medium IGBT softness. For the RAPID2 diode, E_{rec} raises with the IGBT softness in general. It has to be pointed out that the recovery losses of the RAPID2 diode are significantly smaller compared to the RAPID1 diode independent of the IGBT softness. On the one hand, this observation supports the expected improved switching performance of the RAPID2 diode in terms of losses. On the other hand, the insets of Fig. 6b emphasize the upcoming consequences in the system. Under equal switching conditions, the RAPID2 diode provides more than twice as large overvoltage (V_{fmax}) compared to the RAPID1 diode. This additionally points out that an overvoltage reduction at the IGBT due to an increased IGBT softness does not consequently provide a reduction of the system's overvoltage. A balancing between losses and overvoltage is done by using either RAPID1 diode or RAPID2 diode.

For a further analysis of the IGBT-diode interplay, Fig. 7 shows the temperature increase of IGBT and diode (ΔT_{vjIGBT} and $\Delta T_{vjDiode}$) at a fixed output current (I_{rms}) based on a boost-converter setup. For the calculation, a heat-sink temperature (T_{HS}) of 90 °C and a thermal resistance in-between junction and heat sink of 1.9 K/W for the IGBT and 3 K/W for the diode is assumed. With respect to the specification, T_{vj} must not exceed 150 °C.

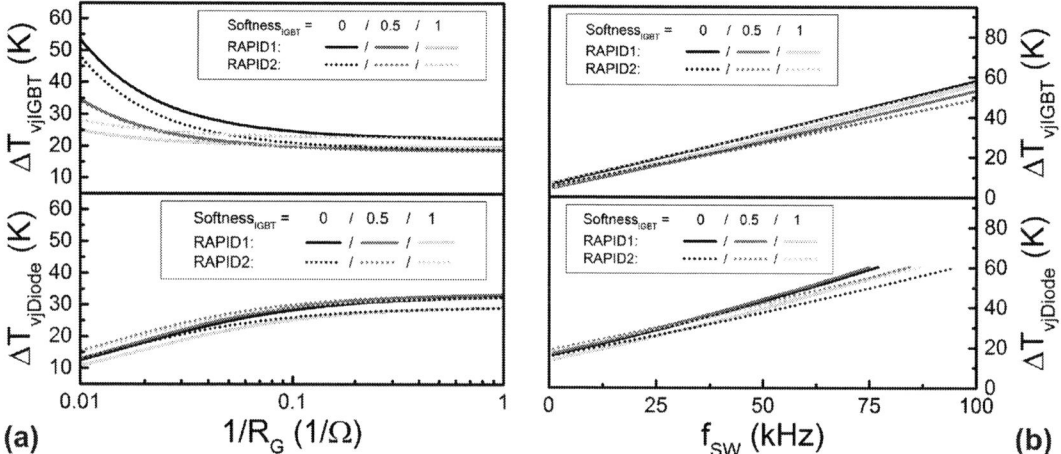

Fig. 7.   (a) ΔT_{vjIGBT} and $\Delta T_{vjDiode}$ versus $1/R_G$ for $I_{rms} = 5$ A, $f_{SW} = 32$ kHz, and $T_{HS} = 90$ °C. (b) ΔT_{vjIGBT} and $\Delta T_{vjDiode}$ versus f_{SW} for $I_{rms} = 5$ A, $1/R_G = 1$ 1/Ω, and $T_{HS} = 90$ °C.

In Fig. 7a, ΔT_{vjIGBT} and $\Delta T_{vjDiode}$ are shown in dependency on the switching speed at a fixed switching frequency $f_{SW} = 32$ kHz. At this switching frequency, neither the IGBT nor the diode is limiting the system. Anyhow, it becomes obvious that the IGBT reaches its maximum temperature if operated at very low switching speed. Fig. 7b visualizes the temperature increase in dependency of the switching frequency. The IGBT can be operated up to 100 kHz without exceeding its maximum temperature. In contrast to this, both diode designs are limiting the system operation. For the investigated operating conditions, the RAPID1 diode is limited to a maximum switching frequency of about 80 kHz ($Softness_{IGBT} = 1$), whereas the RAPID2 diode can be operated up to 95 kHz ($Softness_{IGBT} = 0$). For switching frequencies up to 32 kHz, both diodes can be operated although every diode design reaches its optimum in combination with a different IGBT softness. For higher switching frequencies, a RAPID2 diode together with a low-softness IGBT provides the best results. In addition, an increase of the output current will lead to lower achievable maximum switching frequencies and, in turn, the benefits of the RAPID2 diode are pointed out at high f_{SW}.

4.   Conclusion

The article guides the reader through the variety of interactions and supports a proper choice of IGBT-diode device combinations. The interplay between three different 650-V diode designs, i.e. one emitter-controlled diode (EC3) and two RAPID-diode types (RAPID1 and RAPID2), and an IGBT (650-V TRENCHSTOP™ 5 IGBT) was analyzed. Although the IGBT is able to operate at high switching speed, it is found that high switching frequencies can be reached only in the combination with RAPID diodes. The high Q_{rr} of the EC3 diode limits the operating frequency significantly. Therefore, the EC3 diode is out of scope for fast-switching applications.

Focusing on the RAPID diodes, the performance of IGBT and diode is characterized for differently switching IGBT designs. The RAPID2 diode is designed for faster-switching applications and, consequently, provides lower recovery losses compared to the RAPID1 diode. In turn, the same conclusion stands for the IGBT and the lowest overall losses are provided in combination with the RAPID2 diode. However, this conclusion is only valid if the devices can be operated at highest switching speed. A reduced switching speed due to, e.g., limitations emerging from stray-inductance induced overvoltages, might lead to a different conclusion for a given and not optimized setup.

By taking the system performance into account and by varying the intrinsic IGBT design, it is shown that an increasing IGBT softness clearly interacts with the diode performance and can strongly affect the system performance. Whereas softer IGBT designs provide a benefit for the operation of a RAPID1 diode and lead to reduced overall losses, the best performance of the RAPID2 diode is only reached in combination with the lowest IGBT softness. Also here, the best overall losses are provided only in combination of a fast-switching IGBT and the RAPID2 diode. In turn, it is pointed out that this combination requires low-inductive and symmetric designs, due to the inherent more snappy behavior of IGBT and diode.

5.　References

[1]　A. Huang, L. Cheng, J. W. Palmour, and C. Scozzie, "Ultra high voltage SiC power devices and its impact on future power delivery system", PCIM Europe, Nuremberg, Germany, 2014.

[2]　D. Ueda, T. Fukuda, S. Nagai, H. Sakai, N. Otsuka, T. Morita, N. Negoro, T. Ueda, and T. Tanaka, "Present and future of GaN power devices", CIPS, Nuremberg, Germany, 2014.

[3]　G. Miller, "New semiconductor technologies challenge package and system setups", CIPS, Nuremberg, Germany, 2010.

[4]　J. Popovic, J. A. Ferreira, J. D. van Wyk, and F. Pansier, "System integration of GaN converters – paradigm shift", CIPS, Nuremberg, Germany, 2014.

[5]　R. Bayerer and D. Domes, "Power circuits design for clean switching", CIPS, Nuremberg, Germany, 2010.

[6]　R. Bayerer, "Parasitic inductance - a problem in power electronics", ISiCPEAW, Stockholm, Sweden, 2013.

[7]　R. Bayerer and D. Domes, "Parasitic inductance in gate driver circuits", PCIM Europe, Nuremberg, Germany, 2012.

[8]　P. Beckedahl, M. Spang, and O. Tamm, "Breakthrough into the third dimension – Sintered multi layer flex for ultra low inductance power modules", CIPS, Nuremberg, Germany, 2014.

[9]　E. Hoene, A. Ostmann, and C. Marczok, "Packaging very fast switching semiconductors", CIPS, Nuremberg, Germany, 2014.

[10]　C. R. Müller and R. Bayerer, „Low-inductive inverter concept by 200 A / 1200 V half bridge in an EasyPACK 2B – following strip-line design", CIPS, Nuremberg, Germany, 2014.

[11]　C. R. Müller and S. Buschhorn, „Impact of module parasitics on the performance of fast-switching devices", PCIM Europe, Nuremberg, Germany, 2014.

[12]　C. R. Müller and S. Buschhorn, "Power-module optimizations for fast switching – a comprehensive study", PCIM Europe, Nuremberg, Germany, 2015.

[13]　S. Buschhorn and C. R. Müller, "Performance of power semiconductor devices and the impact on system level", PCIM Europe, Nuremberg, Germany, 2015.

© VDE VERLAG GMBH · Berlin · Offenbach

PCIM Europe 2016, 10 – 12 May 2016, Nuremberg, Germany

Protection Features of Intelligent Power Module against Transient State

Taehyun Kim, Infineon Technologies Power Semitech, Korea, Taehyun.Kim@Infineon.com
Minsub Lee, Infineon Technologies Power Semitech, Korea, Minsub.Lee@Infineon.com
Junbae Lee, Infineon Technologies Power Semitech, Korea, Junbae.Lee@Infineon.com
Daewoong Chung, Infineon Technologies Power Semitech, Korea,
Daewoong.Chung@Infineon.com

Abstract

For motor drive applications, various protection functions are necessary to enable safe operation of all motor drive systems under the conditions of the short circuit, over current, and over temperature. This paper presents protective methods under the specific transient states such as abnormal operation of micro controller, disconnection of 15V power supply and short circuit between adjacent pins of Intelligent Power Module (IPM). They effectively contribute to the enhancement of system reliability in low power motor drives.
In this paper, the specific transient states are described and remedial methods are provided with actual experimental results.

1. Introduction

Energy-saving is being very important in the world these days. Therefore, inverter technology with IPM is being increasingly accepted and used by a wide range of users such as refrigerators, washing machines and air conditioners [1]. In general, an IPM provides general protection functions under the conditions of short circuit, over current and over temperature [2]. The protection of IPM against failures caused by short circuit transients and over temperature is of concern in many applications and especially in the low power motor drives, where devices may be subject to several fault types [3].
In this paper, various transient states in the low power motor drive applications are introduced. Input signals of one phase high side and another phase low side in inverter topology have kept full turn-on state by abnormal operation of micro controller. For inverter operation, 15V power supply is suddenly disconnected by weak control board condition. And there can be short circuit condition between adjacent pins of IPM. These transient states make a severe damage of IPM in case there are not proper protective methods. This paper introduces detailed transient states and describes optimal protective methods with actual experimental results.

2. Protective Features

2.1 Abnormal operation of micro controller

It is possible to have an unexpected and incorrect operation of the micro controller which transfers input signal to the IPM. Figure 1 shows input signals of W phase high side and U phase low side have kept full turn-on state by abnormal operation of micro controller. This situation makes repetitive over current and fault output signal of IPM. When it is continue for a long time, temperature of both IGBTs are extremely increased by over current, and then,

© VDE VERLAG GMBH · Berlin · Offenbach

PCIM Europe 2016, 10 – 12 May 2016, Nuremberg, Germany

one or both IGBTs in IPM are damaged by extreme high temperature. The case temperature of IPM is increased up to 142°C as displayed in Figure 2.

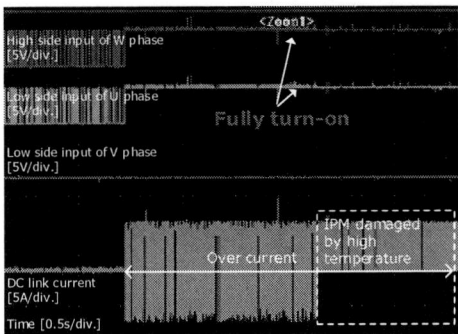

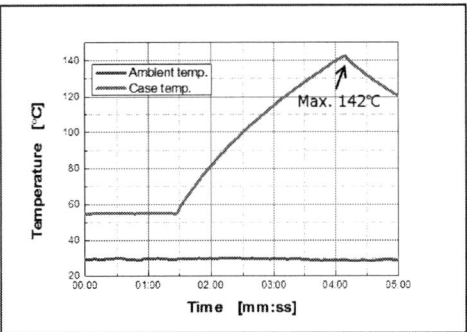

Fig. 1. Waveform of micro controller malfunction

Fig. 2. IPM case temperature by over current

An optimized protection function of an IPM at overcurrent is the sleep function. As shown in Figure 3, new edge input signal is mandatory to activate gate drives after fault duration time.

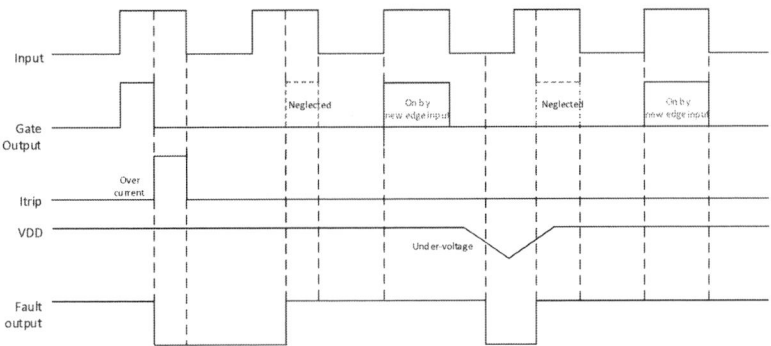

Fig. 3. Sleep function timing diagram

When over current is happened as shown in Figure 1, fault output is activated to shutdown all gate drives. After the reset of fault output, full turn-on input signal is neglected by the sleep function. Therefore, repetitive over current is blocked by the sleep function as shown in Figure 5.

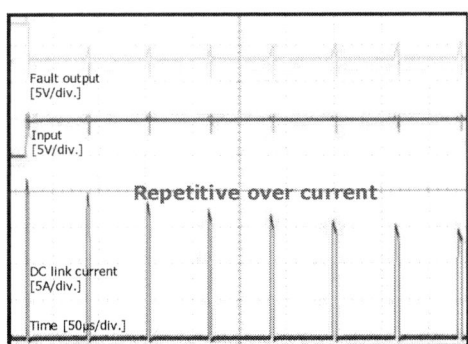

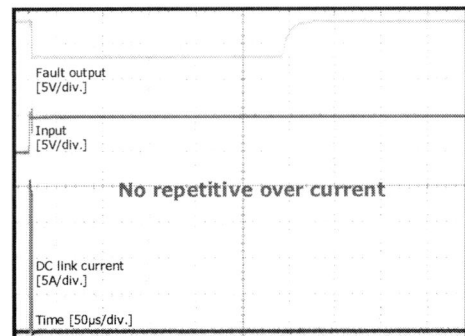

Fig. 4. Waveform without sleep function

Fig. 5. Waveform with sleep function

© VDE VERLAG GMBH · Berlin · Offenbach

2.2 Disconnection of 15V power supply

The integrated Driver IC in an IPM offers Under-Voltage Lock Out (UVLO) function to prevent high conduction loss of the IGBT by low supply voltage. There are two kinds of UVLO for high side (V_{BSUV}) and low side (V_{DDUV}). When the V_{DDUV} is detected, driver IC is stopped with fault output. However, when the V_{BSUV} is detected, fault output is not generated. Normally, driver IC has similar UVLO trigger level for both V_{BSUV} and V_{DDUV}.

High side supply voltage is slightly lower than low side caused by the forward voltage (V_F) of the bootstrap diode. One high side IGBT can be turned off earlier by V_{BSUV} without any fault output even under the operation of other 5 IGBTs in inverter topology.

Figure 6 shows an unbalance of operation of IPM under disconnection of 15V power supply. In this case, IPM operates only with charged VDD capacitor. It is gradually discharged until under-voltage level, and V_{BSUV} is triggered earlier than V_{DDUV} due to V_F of bootstrap diode. Then, IPM is operating under unbalance state due to suddenly turning off of one high side IGBT by V_{BSUV}. It happens that the loss of specific low side IGBT is increased, and eventually, IPM is damaged by this unbalanced operation.

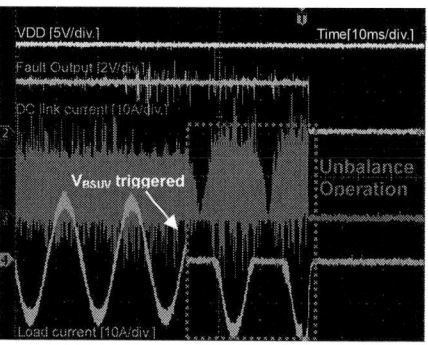

Fig. 6. Unbalanced operation ($V_{DDUV} = V_{BSUV}$)
Condition: VDC=300V, VDD=15V, Ipeak=15A, Rshunt=20mΩ, Tair=25°C, Output freq.=60Hz, Power factor = 0.99, Switching freq.=5kHz

It is important that each of V_{BSUV} and V_{DDUV} has different value for preventing unbalanced operation. When V_{DDUV} has higher trigger level than V_{BSUV}, V_{DDUV} will be detected earlier than V_{BSUV}. It is strongly required to stop driver IC operation. In Figure 7, driver IC is stopped by V_{DDUV} triggered, and fault output is activated. Also, load current has stable sine wave. That means IPM is safely turned off without any damages under normal operating. As a result, higher trigger value of V_{DDUV} than V_{BSUV} is effective solution preventing unbalance operation.

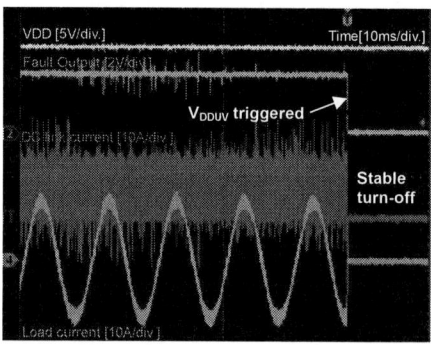

Fig. 7. Normal operation and stable turn-off ($V_{DDUV} > V_{BSUV}$)
Condition: VDC=300V, VDD=15V, Ipeak=15A, Rshunt=20mΩ, Tair=25°C, Output freq.=60Hz, Power factor = 0.99, Switching freq.=5kHz

2.3 Short circuit between adjacent pins of IPM

Short circuit test between adjacent pins during system operation is one of the product liability tests in home appliances. Specially, short circuit case between Fault Output (VFO) and VDD pins of IPM has possibility to cause malfunction of entire system. Normally, VDD (15V) is used for IPM, and 5V bias is used for VFO and micro controller with pull up resistor.

In Figure 8, VFO is connected to 5V through pull-up resistor, and VDD is connected to 15V directly. When VFO and VDD pins are shorted for the operation of inverter, 5V bias is connected to 15V through pull-up resistor. Then, VFO is rising to 15V as shown in Figure 9.

IPM itself doesn't take a problem for internal clamp diode between VFO and VDD. It shows maximum specification range to VDD+0.5V. However, 15V can make over-stress and damage to micro controller. It is normally designed to use 5V or 3.3V bias. Therefore, there can be abnormal operation of micro controller in Figure 9, and IPM can be damaged by it as well.

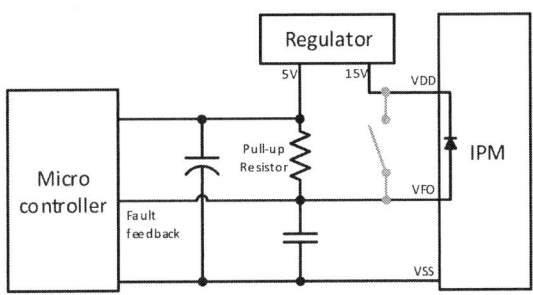

Fig. 8. Diagram of short circuit test

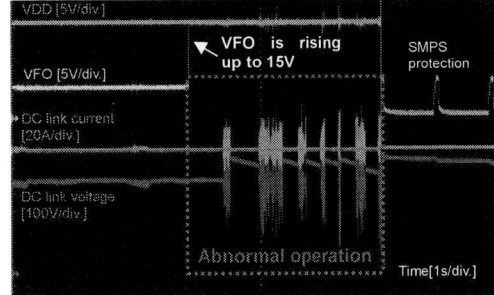

Fig. 9. Short circuit test waveform

In this short circuit test, zener diode in Figure 10 is effectively used for protecting micro controller and IPM. When zener diode is connected with VFO and VSS, it prevents VFO from 15V rising by zener effect as shown in Figure 11. Therefore, VFO and micro controller maintain stable voltage without abnormal operation, and also, 5V and 15V bias are fluctuated by SMPS protection in Figure 11.

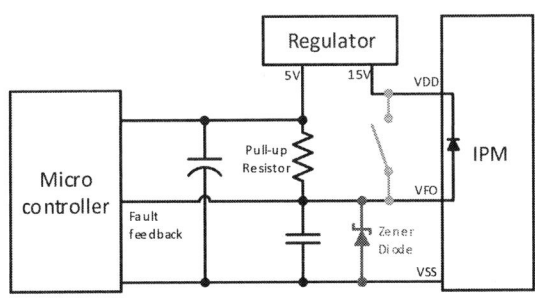

Fig. 10. Zener diode for short circuit test

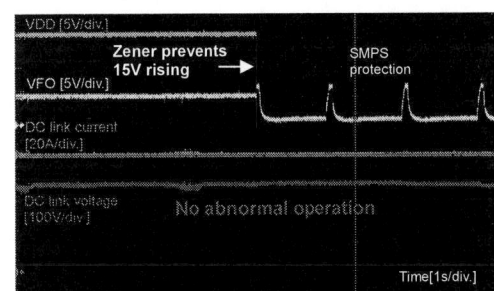

Fig. 11. Waveform of short circuit test with zener diode

It is required that zener voltage should be lower than V_{DDUV} and higher than 5V. Then, 15V is decreased below V_{DDUV} by zener diode, and IPM is not operated by under voltage function of driver IC without any damage.

© VDE VERLAG GMBH · Berlin · Offenbach

3. Conclusion

There are various protection functions in low power motor drives under the specific transient conditions. They are very important to guarantee safety of low power motor drive systems. In this paper, three kinds of transient states are introduced and protection functions for stable system operation are described. Most of protection features in this paper are constructed for total system, which can be effectively used to enhance system reliability. Activities for the improvement of entire system reliability will be kept continuously.

4. Reference

[1] J.Song. J.Lee, D.Chung, B.Suh, Wolfgang.F: A New Intelligent Power Module with Reverse Conducting IGBTs for up to 2.5kW Motor Drives, IPEC2010, pp.156 – 158.

[2] EiceDRIVER™, 6ED family, Application note, Infineon Technologies AG

[3] J.Lee, H.Hyun: Gate Voltage Pattern Analyzer for Short-Circuit Protection in IGBT Inverters," IEEE PESC, 2007.

New High Level Integrated Intelligent Power Module with Three Phase Inverter and Power Factor Correction Topologies Optimized for Home Appliances

Hyosang Jang, Infineon Technologies Power Semitech, Korea, Hyosang.Jang@infineon.com
Byoungho Choo, Infineon Technologies Power Semitech, Korea,
Byoungho.Choo@infineon.com
Junbae Lee, Infineon Technologies Power Semitech, Korea, Junbae.Lee@infineon.com
Minsub Lee, Infineon Technologies Power Semitech, Korea, Minsub.Lee@infineon.com
Daewoong Chung, Infineon Technologies Power Semitech, Korea,
Daewoong.Chung@infineon.com

Abstract

This paper introduces a new high level integrated IPM (Intelligent Power Module) with a three phase inverter and a power factor correction (PFC) topology optimized for home appliance applications.

A three phase Inverter and a single boost PFC stage are integrated in a single miniaturized DIL (Dual-In-Line) transfer molded type package. The inverter stage is built with six 600V rated TRENCHSTOP™ IGBTs, six Emitter Controlled Diodes and a SOI (Silicon On Insulator) gate driver. The PFC stage consists of a 650V rated TRENCHSTOP™ IGBT and a 650V rated Rapid Switching Emitter Controlled Diode which has fast and soft switching characteristics. 650V of voltage rating provides higher reliability and ruggedness for a PFC topology [1]. DCB (Direct Copper Bond) substrate is adopted for good thermal performance considering small package size.

This paper provides an overall description of the new IPM as well as performance, power ratings and characteristics of semiconductors built in the module. Using this new IPM, reduced PCB and heatsink size and simplified assembly process can also be achieved.

1. Introduction

In these days the importance of energy saving with regards to environmental issues has been grown bigger and bigger. Inverters are increasingly accepted and used for home appliance applications such as room air conditioners in order to improve efficiency, reliability and controllability of the system.

PFC (Power Factor Correction) function is also becoming more important in low power major home appliance applications in order to avoid negative effect to the power grid and many home appliance companies are trying to improve power factor (PF) of their system to meet the tightened regulations [2].

In case of room air conditioner, one of the widely used structures is the discrete solution as PFC stage which consists of a discrete IGBT and a discrete boost diode together with an IPM solution as an inverter stage for compressor driving. A single phase bridge rectifier is also used to make DC power source from AC power grid. In this case, all of the power semiconductor components (i.e. a bridge rectifier, a discrete IGBT, a discrete boost diode, and an IPM) are mounted on a heatsink for their heat dissipation. Integrating discrete power semiconductors and drivers into one package allows system designers to reduce the time and effort spent on development. It enables to reduce the PCB and heatsink size and as well as simplified the assembly process. Fig. 1 shows the space and assembly comparison between the discrete PFC solution with inverter IPM and the new IPM solution. The number of devices which are mounted on a heatsink is reduced so the heatsink size and the space which is occupied with these devices can be reduced and the assembly process can also be simplified.

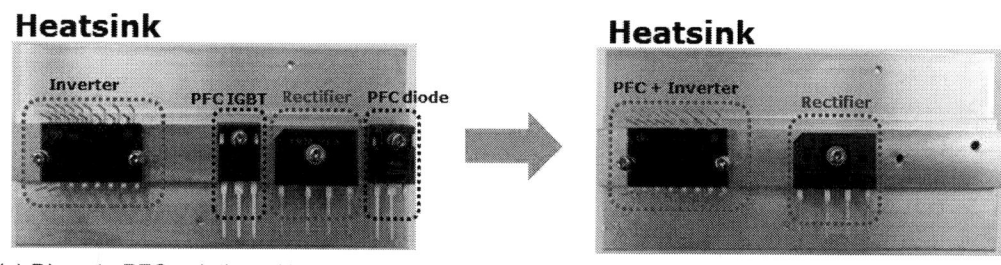

(a) Discrete PFC solution with inverter IPM   (b) New IPM solution

Fig. 1. Space and assembly comparison of discrete PFC solution and new IPM solution.

To meet strong demand for small size, higher power density and simple assembly process, Infineon Technologies has developed a new family of highly integrated intelligent power modules that contain almost all of the semiconductor components required for electronically controlled variable-speed electric motor drives.

This new IPM incorporates a three-phase inverter stage and PFC stage in one package with DCB (Direct Copper Bond) technique which is for reducing thermal resistance causes energy loss increase by heat. It contributes to increase better thermal performance and dissipation capability.

This paper describes the feature of internal components as well as the package structure and thermal performance.

2. External view and circuit configuration

Package outlines of new high level IPM are shown in Fig. 2. The new IPM build in a compact size of Infineon Technologies CIPOS^{TM} (Control Integrated POwer System) MINI package of 21mm X 36mm. Due to its compactness, it can radiate heat effectively by adopting DCB and integrate nearly all of semiconductors into one package.

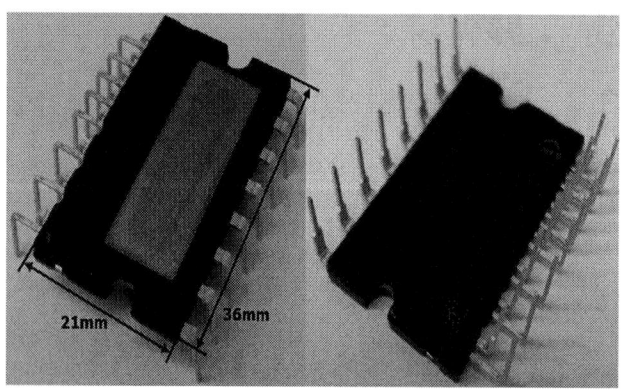 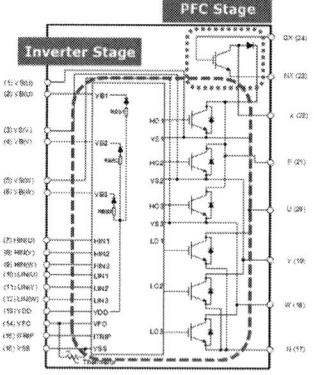

Fig. 2. External view and internal circuit of the new IPM.

The internal circuit of a new IPM is composed of six 600V rated TRENCHSTOP™ IGBTs and six Emitter Controlled Diodes in a three phase inverter structure together with one SOI gate drive IC which provides integrated bootstrap circuit and thermistor for temperature monitoring. The PFC stage consists of a 650V rated TRENCHSTOP™ IGBT and a Rapid Switching Emitter Controlled Diode which has fast and soft switching characteristics (Fig. 2).

3. PFC Stage

3.1 650V rated TRENCHSTOP™ IGBT

The trench-field-stop technology is the most common concept for modern IGBTs with blocking voltages in the range of 600 V to 1200 V. Within this technology, the device performance is mainly controlled by design parameters like cell geometry, chip thickness, and doping profile. These parameters determine the carrier distribution in the device and, hence, influence the static and dynamic characteristics of the IGBT significantly [3].

Infineon Technologies has developed two kinds of products according to their PFC IGBT characteristics. High Speed 3 (HS3) for 20kHz switching frequency is included in IFCM15S60GD and TRENCHSTOP™ 5 (TS5) for 40kHz switching frequency is included in IFCM15P60GD.

3.1.1   High Speed 3 IGBT

The performance of an IGBT is determined by a trade-off curve which describes the correlation between the turn-off energy (E_{off}) and the collector-emitter saturation voltage (V_{CEsat}). Therefore, the performance of an IGBT can be either optimized for high-frequency applications which need devices with low dynamic losses, or for low-frequency applications which benefit from low static losses.

The High Speed 3 IGBT is optimized for high-frequency applications compared with IGBT 3 due to its high collector-emitter saturation voltage V_{CEsat} = 1.85V and low turn-off energy E_{off} = 0.6mJ. The IGBT3 provides a more than a factor of 2 larger turn-off energy E_{off} = 1.4mJ, whereas the V_{CEsat} is reduced to 1.45V [3].

Fig. 3 shows I-V curve and turn-off waveform of High speed 3 IGBT characteristics which is built in IFCM15S60GD as a PFC IGBT. The V_{CEsat} is about 1.81V at nominal current (30A) and the turn-off energy loss is about 400uJ because of its small tail current during turn-off. So the energy loss can be reduced to be optimized for 20kHz switching operating.

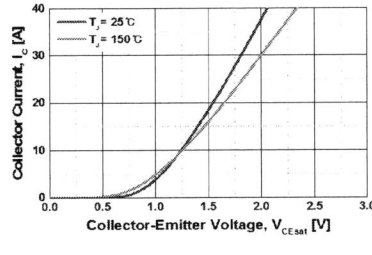

(a) I-V curve

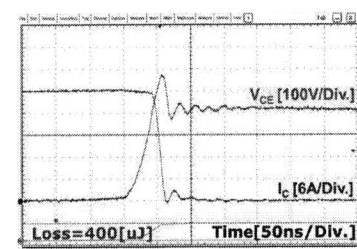

(b) Turn-off Waveform

Fig. 3. I-V curve and turn-off waveform of High Speed 3 IGBT

Conditions: DC-link voltage V_{DC} = 400V, Bias supply voltage V_{DD} = 15V, Collector current I_C = 30A, Gate driver IC = IRS44273 and Gate resistor Rg=5.1Ω.

3.1.2 TRENCHSTOP™ 5

The TRENCHSTOP™ 5 device is a forceful advancement of Infineon's field stop technology to gain lowest switch off losses due to drift zone reduction to < 50μm. This results in a considerable reduction of charge carriers during device-on operation which need to be removed from the device at switch off. Furthermore, the device is characterized by a significantly increased channel width compared to Infineon's TRENCHSTOP™ series in order to reduce simultaneously V_{CEsat} and switch off losses [4].

Fig. 4 shows I-V curve and turn-off waveform of TRENCHSTOP™ 5 IGBT characteristics built in IFCM15P60GD as a PFC IGBT. The V_{CEsat} is about 1.75V at nominal current (30A) and the turn-off energy loss is about 176uJ because of its MOSFET-like switching performance. So the energy loss can be dramatically reduced to be optimized for 40kHz of switching operation.

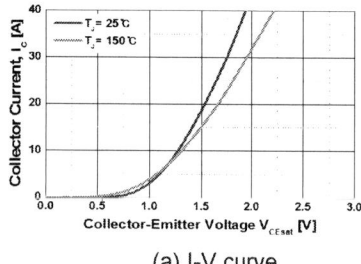

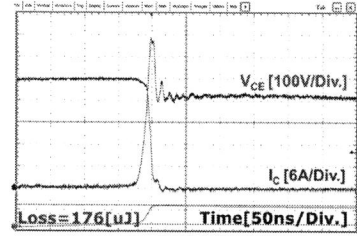

(a) I-V curve

(b) Turn-off Waveform

Fig. 4. I-V curve and turn-off waveform of TRENCHSTOP™ 5 IGBT

Conditions: DC-link voltage V_{DC} = 400V, Bias supply voltage V_{DD} = 15V, Collector current I_C = 30A, Gate driver IC = IRS44273 and Gate resistor Rg=5.1Ω.

3.2 Rapid Emitter Controlled Diode

The Rapid Emitter Controlled Diode of Infineon is optimized to operate with TRENCHSTOP™ IGBT as a boost diode in PFC topology when the switching frequency is less than 40kHz because conduction losses dominate switching losses at lower switching frequency operation. Rapid diode advancement in thin wafer technology helps to maintain a stable V_F over temperature. Fig. 5 is the I-V characteristics of the Rapid diode. It shows that V_F of Rapid diode is very low and less dependent on junction temperature.

The Rapid diode combines low V_F for lower conduction losses and low Irr to reduce E_{on} of the TRENCHSTOP™ IGBT. Increased efficiency, with the additional benefit of having a 650V breakthrough voltage can be achieved [5].

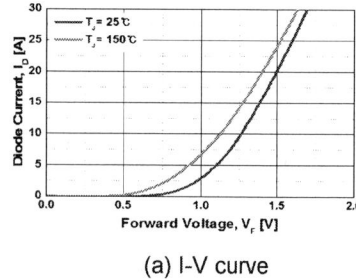

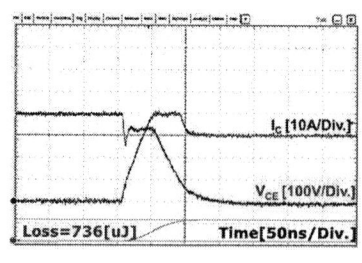

(a) I-V curve

(b) Turn-off Waveform

Fig. 5. I-V curve and reverse recovery characteristics of Rapid Emitter Controlled Diode

Conditions: DC-link voltage V_{DC} = 400V, Bias supply voltage V_{DD} = 15V, Collector current I_C = 10A, Gate driver IC = IRS44273 and Gate resistor Rg=5.1Ω.

© VDE VERLAG GMBH · Berlin · Offenbach

4. Inverter stage

4.1 Inverter power devices

The internal circuit of inverter devices is composed of six 600V rated TRENCHSTOP™ technology IGBTs and anti-parallel diode for free-wheeling in a three phase inverter structure. The assembled IGBT-chips in the inverter part are employed Infineon´s TRENCHSTOP™ technology. This is improved by implementing an additional layer between the substrate region and the collector layer. This is called field-stop layer and has also an n-doping. The combination of both the field-stop and the trench gate technology lead to highly improved conduction and switching characteristics. Especially, the TRENCHSTOP™ IGBTs and anti-parallel diodes are designed for fast switching without excessive ringing and low overall losses [6].

Fig. 6 shows I-V curve and turn-off waveform of TRENCHSTOP™ IGBT characteristics built in the new IPM for the inverter stage. The switching performance is smooth and soft even though the VCEsat is low.

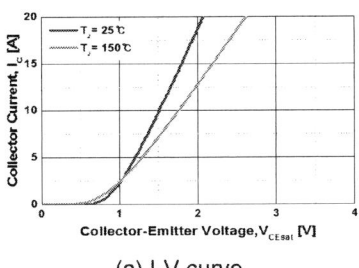 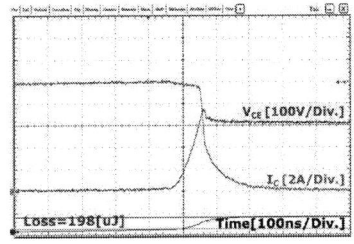

(a) I-V curve

(b) Turn-off Waveform

Fig. 6. I-V curve and turn-off waveform of TRENCHSTOP™ IGBT

Conditions: DC-link voltage V_{DC} = 300V, Bias supply voltage V_{DD} = 15V, Collector current I_C = 10A

4.2 SOI Gate driver IC

A silicon-on-insulator (SOI) gate drive IC depicted in Fig.7 is built in the CIPOS™ modules to get better integration, reliability and performance. SOI is an advanced technique for MOS/CMOS fabrications. The silicon is separated by a buried silicon oxide layer to one layer on the top and the other on the bottom. The buried silicon oxide provides an insulation barrier between the active layer and silicon substrate and hence reduces the parasitic capacitance tremendously. Moreover, this insulation barrier disables leakage or latch-up currents between adjacent devices. This prevents the latch-up effect even in case of high dv/dt switching and surge under elevated temperature and hence provides improved robustness. No latch-up will happen when surge voltage is applied between VDD (control supply) and VSS. Besides the thin-film SOI technology provides additional benefits like low power consumption and higher immunity to radioactive radiation or cosmic rays. This 6 channel single SOI gate drive IC has a build in inter-lock function for shoot through protection, and shut down all outputs when a fault situation such as over current and under voltage. Additionally, it provides an integrated bootstrap circuit. And it is possible to observe temperature by thermistor on the internal PCB. SOI gate drive IC is designed with fully dedicated and minimum necessary functions. It enables optimized IGBT switching and provides inter-lock, under voltage lock out (UVLO) of drive supply and over circuit protection functions [7].

© VDE VERLAG GMBH · Berlin · Offenbach

PCIM Europe 2016, 10 – 12 May 2016, Nuremberg, Germany

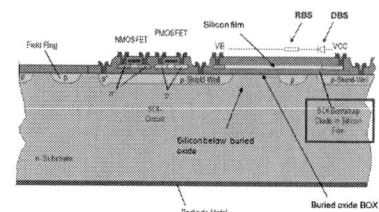

Fig. 7. Cross section of HV SOI technology.

5. Package

Fig. 8 shows that the package structure is composed of PCB part for driver devices and DCB part for power devices in DIL (Dual-In-Line) transfer molded package internally. The compact intelligent power module DCB type IPM is applied Al wire bonding technology for mechanical connection method. The compact intelligent power module applies a thermal substrate of DCB (Direct Copper Bond) for high thermal performance and Al wire bonding method for electrical and mechanical connection. The thermal substrate, DCB is composed of a ceramic plate with high thermal conductivity and copper plates. Power devices are attached on the one side of DCB and the other side is exposed on the surface of the package. This technique contributes to increase better thermal performance to improve thermal dissipation capability, and achieve an excellent solution for up to 2.0kW motor drives [8].

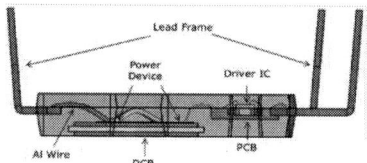

Fig. 8. Vertical structure of the new IPM

6. Thermal performance

Fig. 9 and Fig. 10 are the test circuit and measured waveforms which show the test system's operating status for evaluating thermal performance of the new IPM of an input power of 2kW. The input current is controlled as sinusoidal waveforms under continuous current mode. The input power factor is higher than 0.99 and THD is below 10% at 40kHz as PFC switching frequency. The similar performance is achieved at the switching frequency of 20kHz. Infineon Technologies has developed two kinds of products according to their PFC IGBT characteristics. High Speed 3 (HS3) for 20kHz switching frequency is included in IFCM15S60GD and TRENCHSTOP™ 5 (TS5) for 40kHz switching frequency is included in IFCM15P60GD.

© VDE VERLAG GMBH · Berlin · Offenbach

PCIM Europe 2016, 10 – 12 May 2016, Nuremberg, Germany

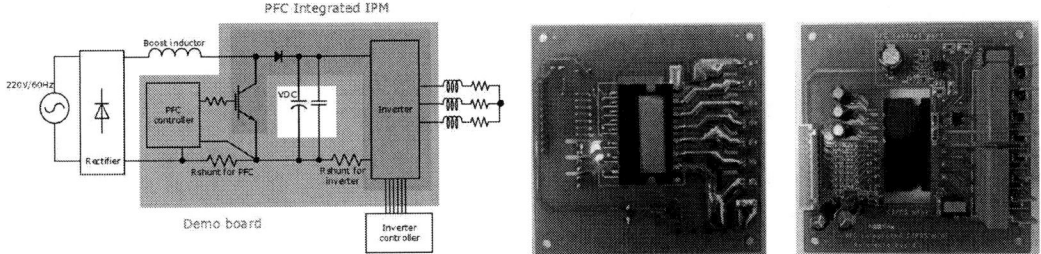

Fig. 9. Test circuit and evaluation board for thermal performance evaluation

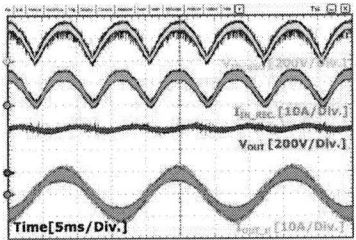

(a) IFCM15S60GA(@20kHz)

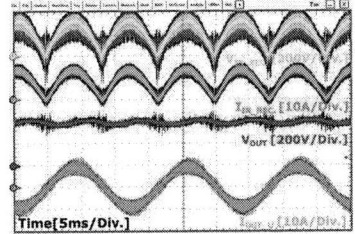

(b) IFCM15P60GA(@40kHz)

Fig. 10. Operating waveforms of test system for thermal performance evaluation

Conditions: PFC controller = ICE2PCS05G, Input source V_{IN} = 220V/60Hz, DC-link voltage V_{DC} = 400V, Switching frequency of inverter = 5kHz, SVPWM, RL load (R = 13.75Ω, L = 2.96mH), MI = 0.69, power factor of load = 0.99, Bias supply voltage V_{DD} = 15V, Ta = 25°C

DUT	Fsw of PFC	V_{IN}	I_{IN}	P_{IN}	P.F.	THD[%]
IFCM15S60GD	20kHz	220V/ 60Hz	9.06 Arms	1.99 kW	0.995	9.78
IFCM15P60GD	40kHz	220V/ 60Hz	9.04 Arms	1.99 kW	0.997	8.11

Table 1. Input power and power factor of the new IPMs

To confirm the thermal performance, case temperatures in Fig. 11 are measured. The points in Fig. 11 represent the hottest points in a module. Tc_inverter is the low side U phase IGBT temperature in inverter and Tc_PFC is the PFC IGBT temperature.

At 20kHz PFC switching frequency, Tc_PFC is about 67.5°C as the highest point and it is higher than Tc-inverter about 2.6°C with 2kW handling. In case of 40kHz, the temperatures are 71.6°C and 68.0°C. These results show the good performance of this new IPM.

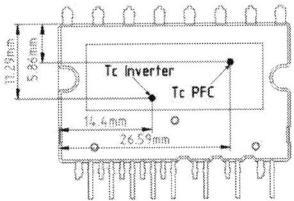

Fig. 11. Temperature measurement point

DUT	Fsw of PFC	Tc_Inverter[°C]	Tc_PFC[°C]	Ta[°C]
IFCM15S60GD	20kHz	64.9	67.5	26.8
IFCM15P60GD	40kHz	68.0	71.6	27.2

Table 2. Case temperatures of the new IPMs

© VDE VERLAG GMBH · Berlin · Offenbach

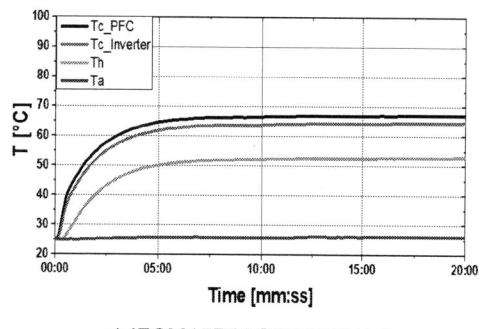
a) IFCM15P60GD(@20kHz)

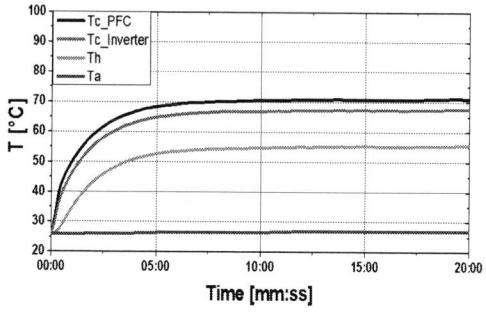
(b) IFCM15P60GD(@40kHz)

Fig. 12. Thermal performance of new integrated IPMs

Conditions: PFC controller = ICE2PCS05G, Input power PIN = 2kW, Input source V_{IN} = 220V/60Hz, DC-link voltage V_{DC} = 400V, Switching frequency of inverter = 5kHz, SVPWM, RL load (R = 13.75Ω, L = 2.96mH), MI = 0.69, power factor of load = 0.99, Bias supply voltage V_{DD} = 15V, Gate resistor Rg = 5.1Ω, Ta = 25°C

7. Conclusion

New Intelligent Power Modules are the optimized solution with inverter and PFC topologies for variable speed motor drive such as room air conditioner. The thermal performance up to 2kW is evaluated with 20kHz and 40kHz of PFC switching frequency. They offer high efficiency enabling the development of increasingly energy efficient appliances while they provide small size to miniaturize a system. Infineon Technologies owns all necessary technologies and is committed to support its customers to realize compact and more efficient solutions. Strong demands to reduce the system size, total cost and time to market will be satisfied with this new high level integrated Intelligent Power Module.

References

[1] S Shim, B Choo, J Lee, D Chung: "A New High Efficient 2-Phase and 3-Phase Interleaved Power Factor Correction Boost Converter typed Intelligent Power Module with high switching capability for low power home appliances", ICPE 2015 Asia, Seoul, Korea.

[2] IEC 61000-3-2:2014: Limits for harmonic current emissions

[3] Christian R. Müller: "Using 650-V High Speed 3 IGBTs in Power Modules for Solar Inverter Performance Improvement", PCIM Europe 2013, Nuremberg, Germany.

[4] Thomas Kimmer, Erich Griebl: "TRENCHSTOP[TM] 5: A new application specific IGBT series", PCIM Europe 2012, Nuremberg, Germany.

[5] "650V Rapid Diode for Industrial Applications", Application note, Infineon Technologies AG V2.0, Jul. 2014

[6] Wolfgang Frank: "TrenchStop-IGBT-Next Generation IGBT for Motor Drive Application", Application Note, Infineon Technologies AG V1.0, Oct. 2004

[7] R. Keggenhoff, Z.Liang, A. Arens, P. Kanschat, R. Rudolf: "Novel SOI Driver for Low Power Drive Applications", Power Systems Design Europe Nov.2005

[8] S Park, J Cho, H Kwon, J Lee, D Chung: "A Compact Intelligent Power Module with High Thermal Performance for up to 4kW Power Motor Drives", PCIM Europe 2014, Nuremberg, Germany.

New DIPIPM+^{TM} Series Module with All-in-one integrated

Yuancheng Zhang, Xiankui Ma, Hongtao He, Gaosheng Song

Mitsubishi Electric & Electronic (Shanghai) Co., Ltd, China

Ming Shang

Mitsubishi Electric GEM Power Device (Hefei) company Limited

Email: ZhangYC@mesh.china.meap.com

Abstract

This paper presents a new series of DIPIPM+^{TM} power module with all-in-one internal circuit and compact size for industrial applications. To achieve lower power losses, the 7th generation IGBT chip is employed into the DIPIPM+. With highly integrated and compact size, it can be used easier in the industrial field, such as inverter, servo, motor driver, etc.

Synopsis

1. Introducing

DIPIPM product family, for inverter application, has been more and more popular in the industrial market. The characteristics of high integrated, high power density, compact size, easy assembly are advantages which be preferred by engineers.

Now, combined with rectifier, inverter unit and brake unit have been integrated into a single package, DIPIPM+ makes it much easier to utilize then shortening the period from R&D to mass production.

And, with the newest 7th generation IGBT technology has been employed, DIPIPM+ has lower power losses.

2. Description of the DIPIPM+^{TM}

2.1 Chip Technology

DIPIPM+ has employed the 7th generation IGBT chip with CSTBT^{TM} technology. The new chip uses the new process to achieve power losses reducing in which conducting and switching simultaneously; thinner wafer for lower $V_{CE\,(sat)}$, and fined cell pith for lower E_{off}.

To compare with 6th generation IGBT chip, the characteristics of $V_{CE\,(sat)}$ and E_{off} has reduced for about 7% (Figure 1) and 25% (Figure 2) respectively. Therefore, high current density can be realized in a compact package.

PCIM Europe 2016, 10 – 12 May 2016, Nuremberg, Germany

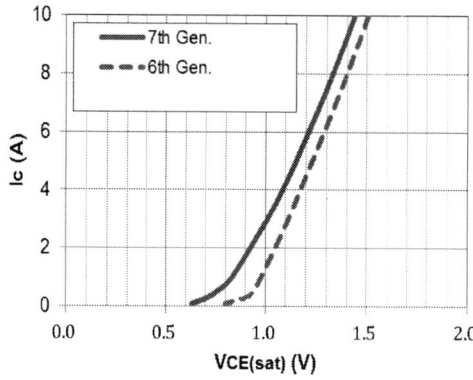

Figure 1 $V_{CE\,(sat)}$ comparison – 6th vs. 7th

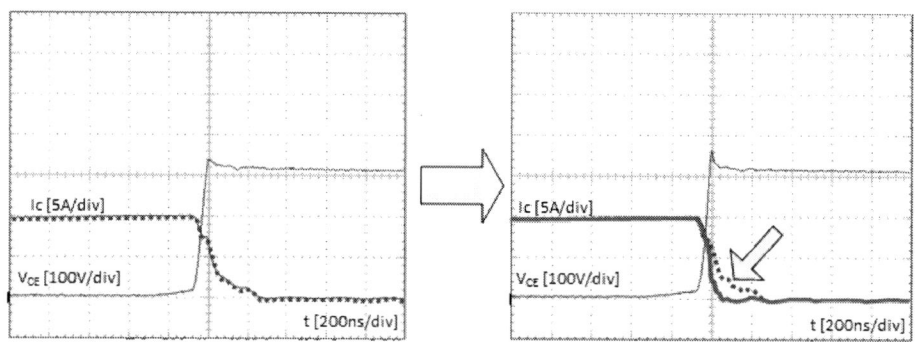

Figure 2 IGBT chip turn-off loss comparison - 6th vs. 7th

2.2 High integration

To designers, ones prefer that, the power devices which for structuring a three-phase inverter are integrated into a single package (Fig.3). By high integration, the works for circuit design and module assembly become simple and easy. Then, CIB (Converter-Inverter-Brake) and PIM module came into being as market demand. Further, adds on with gate drive and protection circuit, DIPIPM+ is a convenient choice for designers.

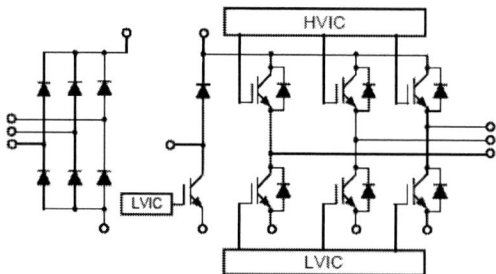

Figure 3 Illustration for internal connection of DIPIPM+

© VDE VERLAG GMBH · Berlin · Offenbach

PCIM Europe 2016, 10 – 12 May 2016, Nuremberg, Germany

2.3 Optimized control part

2.3.1 Block diagram of Inverter unit

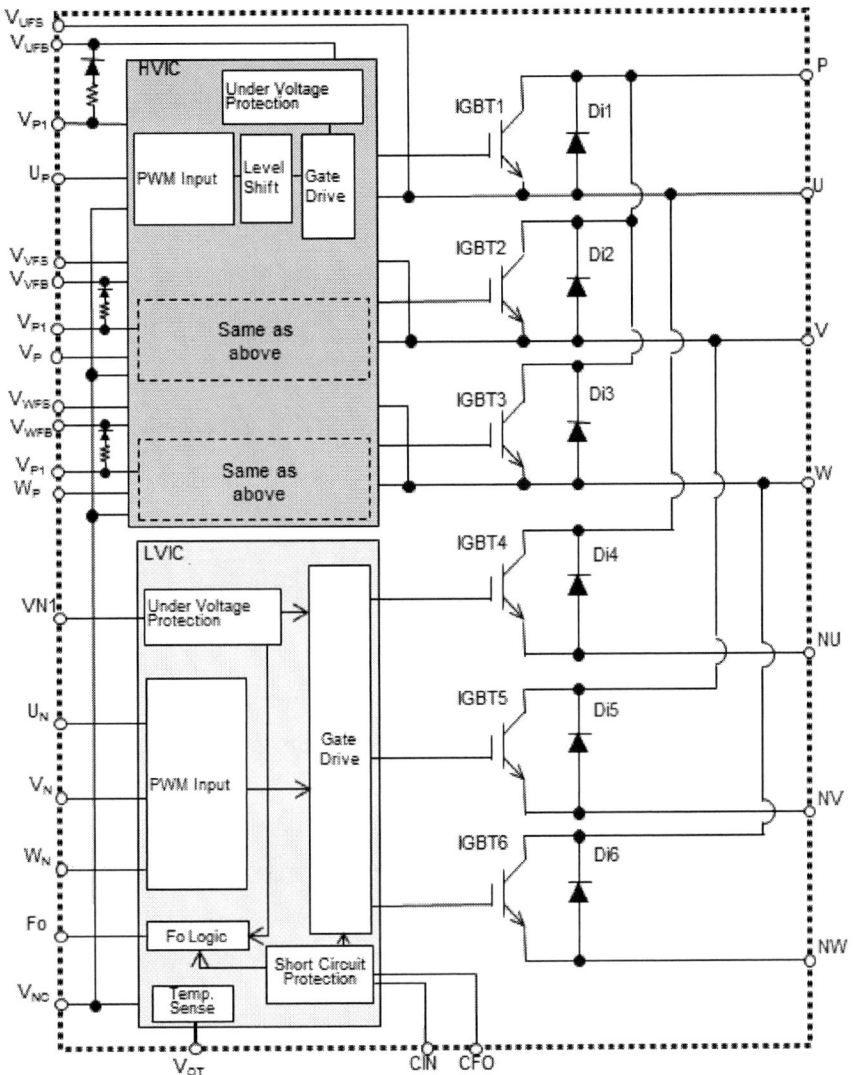

Figure 4 Block Diagram of Inverter unit

According to the block diagram, control part includes HVIC and LVIC in which to adapt up-arm and low-arm respectively. Refer to the function block; PWM signal transmitting circuit, under voltage protection, short circuit protection and fault signal output circuit are integrated. And more, analog temperature voltage output (V_{OT}) function is integrated so that the temperature of module can be monitored in real time.

2.3.2 Internal bootstrap diode and current limiting resistor

In purpose that it can be used with single +15V power for control part of DIPIPM+, HVIC is

© VDE VERLAG GMBH · Berlin · Offenbach

873

employed. Corresponding that, the bootstrap circuit is needed for supporting to drive up-arm. Benefited by utilizing of new processes, power consumption of HVIC is lower so that the bootstrap diode and the current limiting resistor can be integrated (Figure 4).

2.3.3 Temperature-analog output (V_{OT}) Function

Based on the Statistic Analysis of failure data, OT is a main cause of power chip destruction; thus, it should be avoided when the module is operating under abnormal conditions. Therefore, the temperature of power chip or case of module should be sensed.

In practical, the NTC resistor is installed into the power module as the sensor to monitor the related temperature in purpose. DIPIPM+ has the similar design; the thermal sensor is embedded on the LVIC to monitor the case temperature (Figure 5).

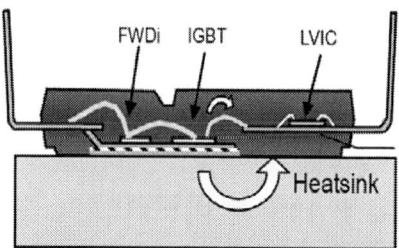

Figure 5 Illustration for the theory of thermal detecting

However, the thermal sensors of which be used in the power modules are the component with non-linear characteristics versus temperature. It may cost some resource of software on logarithmic computing or table look-up so as to get the corresponding value when it would be used. To simplify this procedure, the LVIC of DIPIPM+ has built in a compensation circuit for translating the temperature on LVIC to voltage which with linear characteristic (Figure 6). The voltage can be utilized by MCU easily with analog-digital conversion program. Meanwhile, programmer can adjust the control algorithm according to the real time temperature.

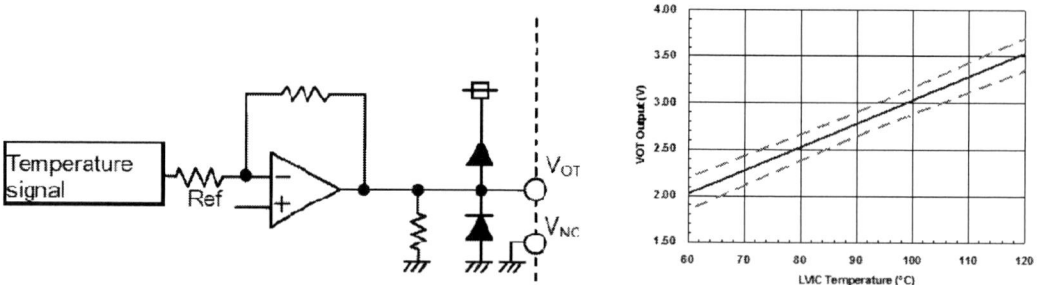

Figure 6 Internal circuit for V_{OT} Function & output voltage characteristics

3. Power losses simulation

To describe the performance of DIPIPM+, the power losses simulation has been done; and the comparison between DIPIPM+ and large type DIPIPM has been done also (Figure 7).

Figure 7 Power Losses comparison – Large DIPIPM vs. DIPIPM+

According to the figure 7 shown, DIPIPM+ has about 10% reduction of power losses than the large DIPIPM.

4. PCB Layout example

Another advantage for DIPIPM+ is the reasonable pin assignment. It is easy to divide the pins into power pin portion and signal pin portion (Figure 8).

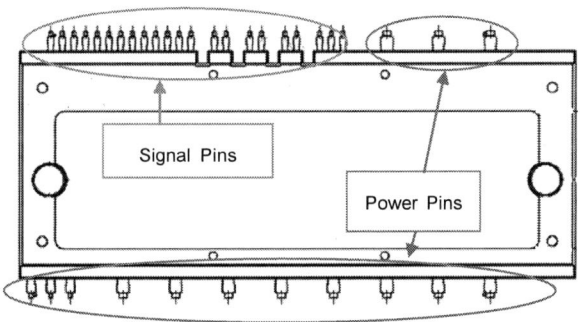

Figure 8 Pin assignment

So, it seems to design the PCB layout easier (Figure 9).

It's easy to find that, the power terminals such as input terminal, output terminal and brake resistor terminal are all set onto the same side without cross wiring. And, low count of peripheral component makes the design simple, more, PCB size is saved.

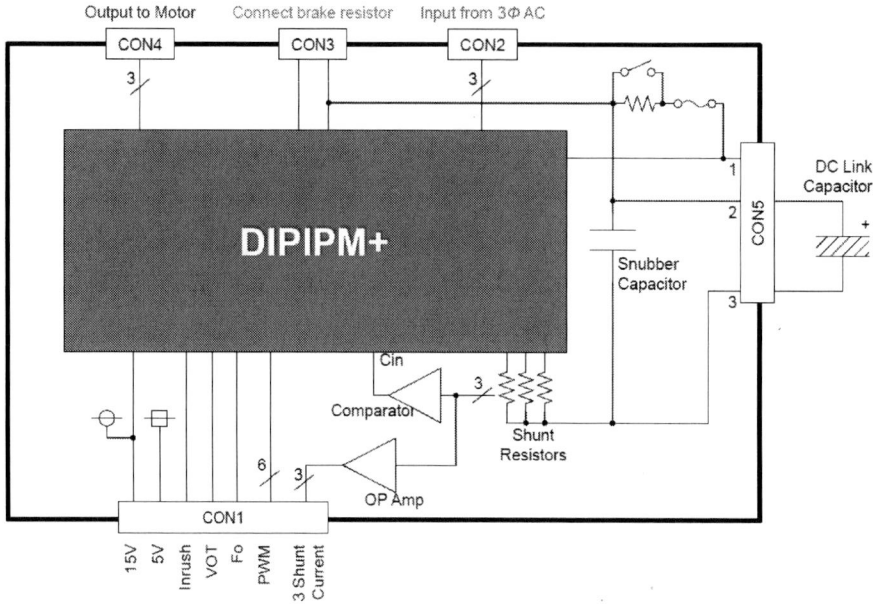

Figure 9 PCB layout example

5. Conclusion

The new DIPIPM+ series module shows more advantages; high integration, lower power losses and lower count of peripheral component make the application simple and compact. However, high integration and power density mean the heating is concentrated, so, it is needed to be careful on cooling system design. And, absence of short circuit protection in HVIC is a disadvantage which may cause failure by large current which in case of the load would be shorted to N point. Furthermore, it's an unknown risk for the situation in which fault info of HVIC cannot be passed out.

References

[1]. Eric R. Motto, "Standard packages and features for DIPIPM[TM] in small motor driver applications", ECCE 2013.

[2]. L. Xiaoguang, M. Shiramizu, A, Yamamoto, C. Tadokoro, T, Nakano "BSD embedded new series super mini DIPIPM[TM] Ver.4" 2010 PCIM China.

Improvement of system level power density of 15A/600V intelligent power modules

Jonathan Harper, ON Semiconductor GmbH, Zamdorferstrasse 100, 81677 Munich, Germany, jonathan.harper@onsemi.com

Abstract

A novel approach to implementing a 15A/600V intelligent power module using a new type of DBC substrate is presented. The module is 40% smaller than other intelligent power modules having a similar current and voltage specification which use IMS, lead frames or other combinations of substrates in their construction. Despite the reduction in size, the thermal resistance of the package is 5% lower. Careful optimization of the IGBT and diode reduces the overall power losses by 5%.

1. Introduction

1.1. Energy Saving Initiatives driving Module Use

The European Union ErP programme is forcing the adoption of energy-saving techniques in industrial and consumer end markets. Replacing a universal motor with a permanent magnet synchronous (PMSM) motor or a brushless DC (BLDC) motor saves energy but requires a more complex drive stage using an inverter. Inverters generate more heat and are significantly larger than the simple triac drive needed for a universal motor. A transfer-molded module is generally used for the inverter power stage as this has higher reliability, lower electromagnetic interference (EMI) and a smaller footprint than a discrete solution. As more motors in a given system transition from universal to PMSM or BLDC, the number of inverters on a given board increase, driving the requirement for modules which are smaller and easier to cool.

1.2. Module Components

The components of the inverter power stage consist of drivers, thermistors, IGBT transistors and rectifier diodes. There are a number of ways of constructing such modules. One method is to mount the drivers, thermistor, IGBT's and rectifier diodes onto an insulated metal (IMS) substrate [1]. Another option is to mount the power devices onto one lead frame and the drivers onto a second lead frame [2]. A third option is to combine use of DBC for the power components and use lead frames for the control components [3].

In the presented module (Compact Intelligent Power Module) all components including the drivers are mounted onto a single substrate. This technique has been used on IMS substrates for over a decade. With suitable manufacturing process techniques, this can be applied to a modified direct bonded copper (DBC) substrate, combining the integration density of IMS with the excellent thermal performance of DBC substrates. The resulting module has a smaller size and a better thermal performance than modules using lead frame or IMS-based solutions.

© VDE VERLAG GMBH · Berlin · Offenbach

PCIM Europe 2016, 10 – 12 May 2016, Nuremberg, Germany

2. Compact Intelligent Power Module

2.1. Package

Figure 1 shows the compact intelligent power module with rated current specifications up to 15A and voltage rating of 600V. The peak current rating goes up to 30A. The DBC substrate is visible on the top of the package and provides an excellent cooling path for the integrated IGBTs and rectifiers. The package can be mounted directly onto the case of the motor which it is driving. Alternatively, a heat sink can be mounted on top of the package. The dimensions of the molded part of the package are 29.6mm by 18.2mm with a height of 4.3mm.

Fig. 1.   15A/600V Compact Power Integrated Module Package

2.2. Construction

Figure 2 shows a cross-section of the module. The module is a transfer molded module using a DBC substrate. The DBC substrate consists of an insulating ceramic with an excellent thermal conductivity with a copper layer on each side. The external copper plate is covered with a finishing layer. The internal copper layer is used to make a printed circuit layout. The components are soldered onto the DBC. Interconnect bond wires are used to connect between the dies and the printed circuit layout, or between different dies. Thicker wires and printed circuit patterns are used for the power traces. Using techniques developed for IMS designs, and applying them in the right way to modules using DBC technology, allows the use of thinner wires and fine circuit patterns for low current traces than is possible with other approaches. This leads to a smaller overall layout, resulting in a high power density.

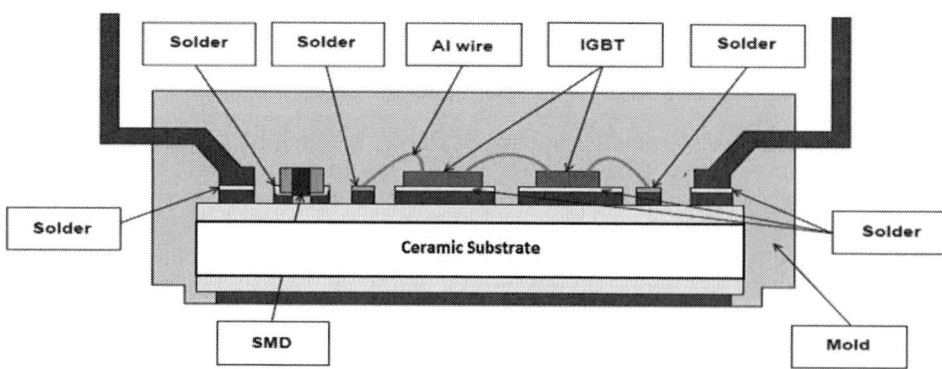

Fig. 2.   15A/600V Compact Power Integrated Module Package

© VDE VERLAG GMBH · Berlin · Offenbach

2.3. Integration

Figure 3 shows the components integrated into the module. The compact intelligent power module contains six IGBTs, six rectifiers, a thermistor and a three channel half-bridge driver with protection functions for cross-conduction, undervoltage lockout on both high and low sides, overvoltage protection and over-current protection. The high density routing approach discussed in the previous section allows for the integration of an inverter driver stage as shown for current levels of up to 15A.

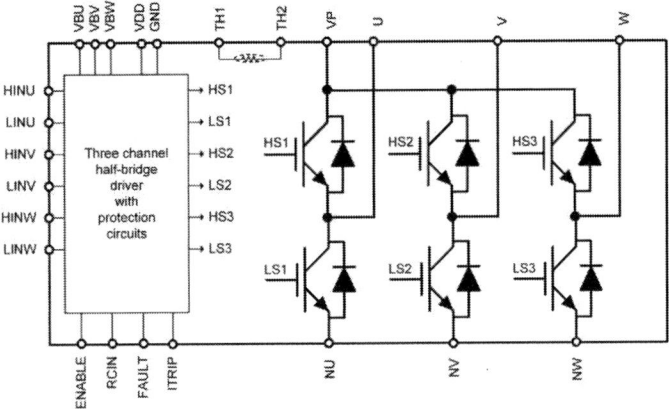

Fig. 3.   Components integrated into 15A/600V Compact Power Integrated Module

3. Performance Characteristics

3.1.   Size and Thermal Resistance

Figure 4 shows the size of the module compared to other DIP modules available on the market. The compact IPM module has a thermal resistance of 2.6K/W and a size of 450mm2. This is 40% smaller than the other DIP modules. Despite the much smaller package size, the thermal resistance is 5% lower due to the use of DBC substrate.

Assuming that the power losses in the modules under comparison were identical, for example by using the same drivers and power components, the temperature difference between the junction and ambient would be 5% lower.

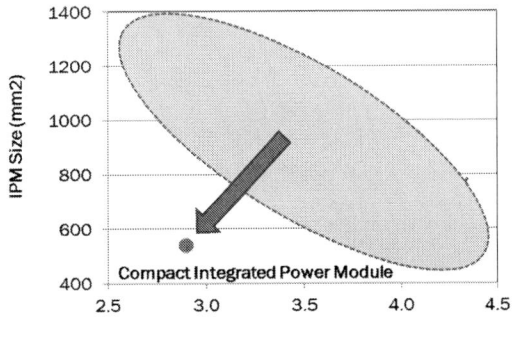

Fig. 4.   Comparison of size and thermal resistance of compact intelligent power module

3.2. Power Losses

Figure 5 shows the losses in the module as a function of current measured at both 25 and 100 degrees Centigrade. The blue and black lines refer to the power losses of standard DIP modules. The losses of the compact intelligent power module are around 5%-10% lower than that of the standard DIP modules. This is however related to the improved performance of the IGBTs, the optimization of the IGBT and rectifier tradeoffs and is not related to the packaging. The module shown here has a rated current of 10A and a peak current of 20A.

Condition : Vcc=300V, f=15kHz, PF=80%, Duty cycle=96%

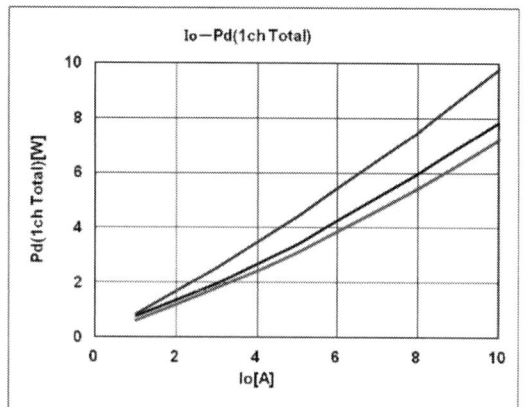 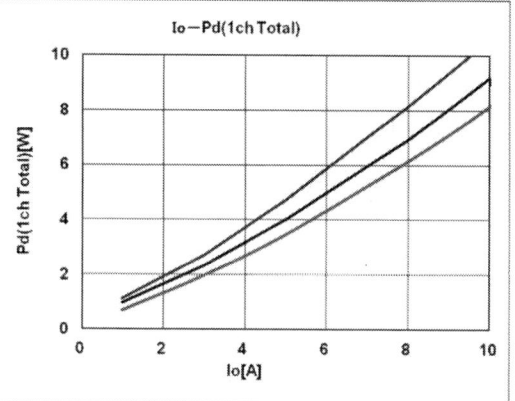

Fig. 5. Loss comparison of compact intelligent power module with standard IPM

3.3. EMI

Reduced power losses can often be achieved by faster switching, which in turn worsens conducted and radiated EMI. Additionally fast voltage transitions with high dv/dt applied to the motor windings can generate large displacement currents in the motor bearings and windings. These currents affect long term motor reliability.

Figure 6 shows the switching transitions of the 10A compact intelligent power module from the previous section. During turn on, the ringing caused by the diode reverse recovery damps quickly. The di/dt is approximately 125A/us. The dv/dt is approximately 5V/ns. During turn off, the dv/dt is approximately 8V/us. System level EMI tests of different inverters using the compact intelligent power module have shown that the EMI emissions generated by the module are within the limits.

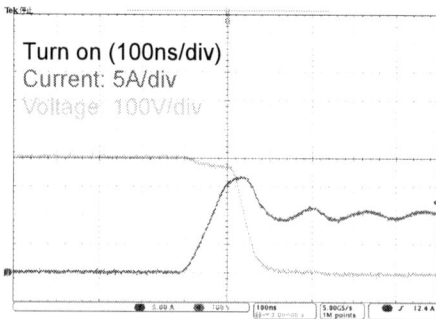 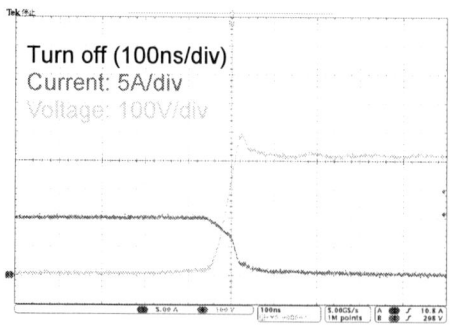

Fig. 6. Turn-on and turn-off transitions for 10A compact intelligent power module at 100°C

© VDE VERLAG GMBH · Berlin · Offenbach

4. Conclusion

The benefits of lower package size, lower thermal resistance and lower power losses extend beyond improved system efficiency. Lower power losses and lower thermal resistance result in lower junction temperature at full power, and therefore lower change in junction temperature when load cycling between low power and full power conditions. Long term module reliability is improved by lower changes in junction temperature during power cycling and improved by lower average junction temperature. The compact intelligent power module approach presented thus results in improved power cycling reliability.

In summary, using the compact intelligent power module in an inverter results in a more compact, more efficient and more reliable solution.

5. References

[1] T. S. Kwon, J. H. Song, J. B. Lee, S. H. Paek, S. I. Yong, "A new smart power module for low power motor drives", ICPE, p. 695-699, 2007.

[2] Y. C. Son, K.Y. Jang, B. S. Suh, "Integrated MOSFET Inverter Module for Low-Power Drive System", IEEE Transactions on Industry Applications (Volume: 44, Issue: 3), pp 878 – 886, 2008

[3] T. S. Kwon, S. H. Hong, J. H. Baek, S. I. Yong, "Development of New 1200V SPM smart power module for Industrial Motor Drive Applications", PCIM, p1-6, 2014.

PCIM Europe 2016, 10 – 12 May 2016, Nuremberg, Germany

Optimization of FREDFET-based µIPM™ for Very Low Power Motor Drive Applications

Rajeev Krishna, Vytla, Infineon Technologies AM, USA, Rajeev-Krishna.Vytla@infineon.com
Danish, Khatri, Infineon Technologies AM, USA, Danish.Khatri@infineon.com
Brian, Sun, Infineon Technologies AM, USA, Brian.Sun@infineon.com

Abstract

This work presents a new Infineon µIPM™ module with optimized 500V FREDFET for very low power (<200W) motor drive applications operating in a PWM frequency range of 15 – 20 kHz. The improvement in the new module performance is due to the improvement of the body diode characteristics using optimized electron irradiation dose. The results showed that the new improved module helped in improving the thermal performance of the system by up to five degrees compared to the reference.

1. Introduction

µIPM™ is a family of Infineon low power Intelligent Power Modules (IPM) [1] designed to efficiently power small motors rated up to 200W for air conditioning and pump applications. In a majority of these applications, the motors spend a vast majority of their operating life running under light load conditions at a fraction of the maximum load they were designed for. For overall efficiency, it is thus important to optimize the performance of the IPMs under light load conditions. Compared to IGBTs, FREDFETs offer greater light load efficiency advantages and more economical packaging solutions since they do not require any freewheeling antiparallel diodes. In the 15 - 20kHz frequency range, turn on switching losses dominate over the conduction losses.

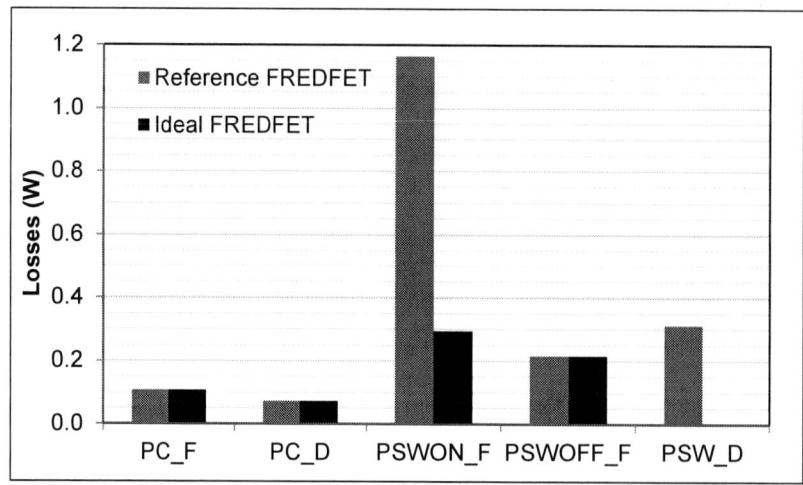

Fig. 1. Simulated loss components of an inverter operating at V_{BUS} = 320 V, I_{motor} = 200mArms, f_{sw} = 20kHz under space vector modulation for a reference FREDFET and an ideal FREDFET (zero recovery body diode). P_{C_F} = FET conduction loss, P_{C_D} = Body diode conduction loss, P_{SWON_F} = FET turn-on loss, P_{SWOFF_F} = FET turn-off loss, P_{SW_D} = Body diode recovery loss

© VDE VERLAG GMBH · Berlin · Offenbach

Figure 1 shows the simulated loss components of an inverter operating at bus voltage V_{BUS} = 320 V, motor phase current I_{motor} = 200mArms, and PWM frequency f_{sw} = 20kHz under continuous symmetrical space vector modulation for a reference FREDFET and an ideal FREDFET with zero recovery body diode. Under these conditions, reverse recovery losses of the reference FREDFETs makeup 63% of the total inverter losses. Therefore, it is very important to reduce the body diode recovery losses. This work is aimed at further optimizing the FREDFET body diode to reduce the recovery-related inverter losses.

2. FREDFET Optimization

2.1. Body diode optimization

It is important to note that in order to minimize the switching losses to a level comparable to a typical freewheeling diode normally used with IGBTs, the FET body diode needs to be designed and processed to reduce the switching recovery time and recovery charge [2]. Faster reverse recovery characteristics of the body diode can be achieved by simply using room temperature high energy electron irradiation followed by a proper thermal annealing process. Electron irradiation creates lattice damage, and thus introduces recombination energy levels within the bandgap to help reduce the carrier lifetime during reverse recovery. The result is greatly reduced reverse recovery charge (Q_{rr}), reverse recovery time (T_{rr}), and peak reverse recovery current (I_{rr}). Fig. 2. shows the reduction of Q_{rr} and T_{rr} as electron dose increases. It is found that there is an optimum dose beyond which there is no further improvement in body diode performance [3]. For the selected optimum dose, other DC parameters (R_{DSon}, V_{th}, V_f, I_{DSS}) are not affected significantly and are in the acceptable range. For example, R_{DSon} increase from higher irradiation is only increased by 5% compared to the reference.

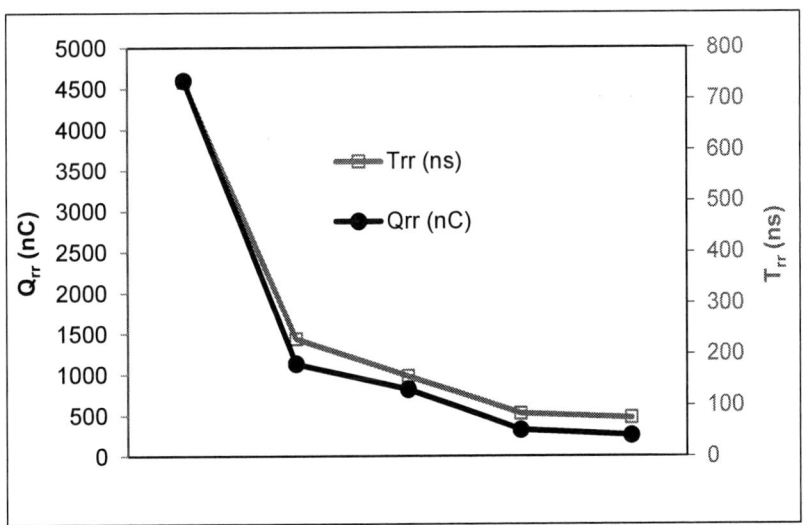

Fig. 2. Variation of Q_{rr} and T_{rr} with variation of Electron irradiated dose

PCIM Europe 2016, 10 – 12 May 2016, Nuremberg, Germany

3. Electrical Results

3.1. Discrete package test

Devices with nominal (reference FREDFET) and optimized (new FREDFET) electron irradiation doses were produced and assembled into TO-220 packages. Body diode performance was evaluated for the various doses using a double pulse test setup, shown in Figure 3.

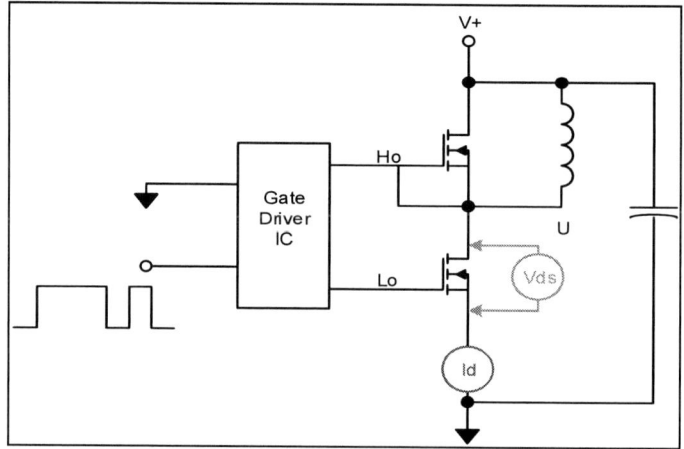

Fig. 3. Double pulse test setup

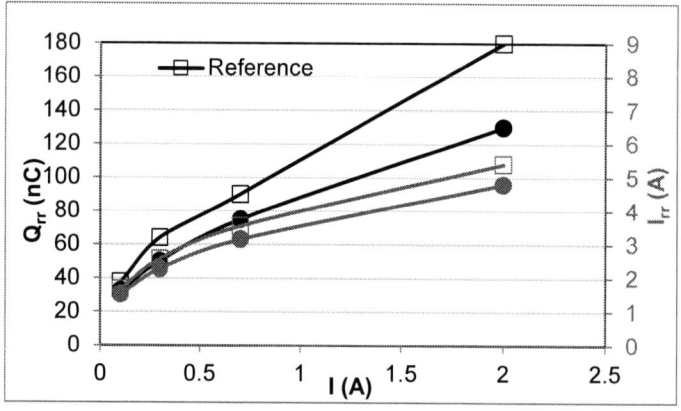

Fig. 4. Measured Q_{rr} and I_{rr} for reference and new FREDFET vs current (V_{BUS}=320V, T_J=25C, L=7mH)

Body diode parameters including T_{rr}, I_{rr} and Q_{rr} were measured. All the measurements were performed at V_{BUS} = 320V, T_J = 25°C and 125°C, L = 7mH. Figure 4 shows Q_{rr} & I_{rr} of the new and reference FREDFETs when switched with various forward test current up to 2A at

© VDE VERLAG GMBH · Berlin · Offenbach

V_{BUS}= 320V, T_j= 25°C, L= 7mH. It can be noted that the Q_{rr} and I_{rr} is proportional to the test current and increases due to increase in minority carrier charge with increase in current.

There is an improvement in the body diode performance of the new FREDFET in terms of T_{rr}, I_{rr}, and Q_{rr} compared to the reference. The improvement in Q_{rr} is more evident as the current level goes up.

A motor test was then performed to determine the effect of FREDFET body diode optimization at a system level. Figure 5 shows the EMI spectrum of the system with the new FREDFET vs the reference one under the following conditions: I_{motor} = 200mArms, f_{sw} = 20kHz, V_{BUS} = 320V, continuous symmetrical space vector modulation. There was no significant difference in conducted emissions compared to the reference.

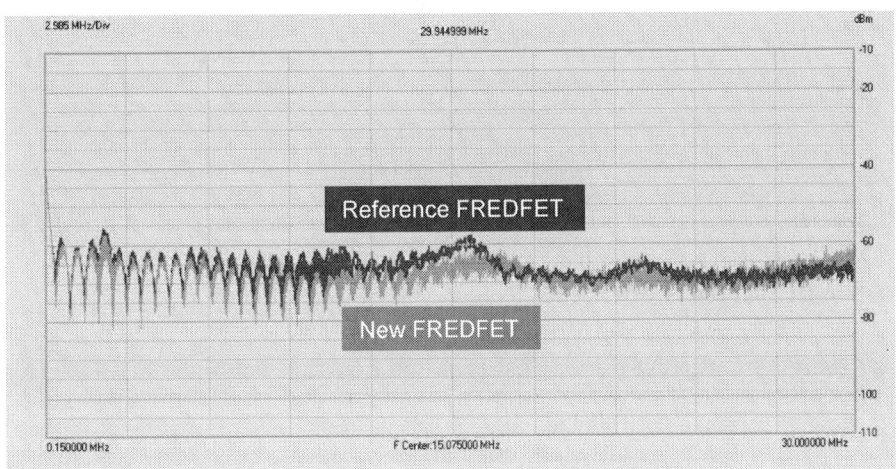

Fig. 5. Measured EMI for a system with new and reference FREDFETs (I_{motor} = 200mArms, f_{sw} = 20kHz, V_{BUS} = 320V)

3.2. Module test

Modules with assembled new and reference FREDFETs were tested using the double pulse test circuit shown in Figure 3 (L = 40mH, V_{BUS} = 320V). Figure 6 (left) shows the variation of FREDFET turn-on (E_{ON}) and recovery (E_{REC}) energy under various test currents. There is a reduction in E_{ON} and E_{REC} of about 17% and 23% respectively for the module with the new FREDFET compared to the reference at 300mA. Figure 6 (right) shows E_{ON} and E_{REC} at 150°C.

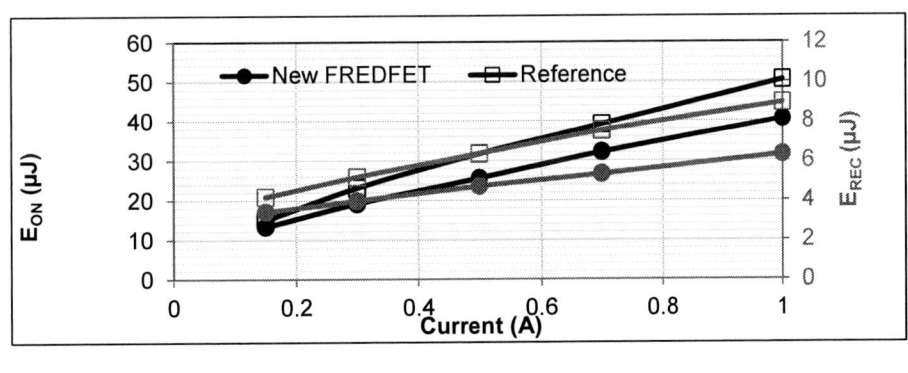

(a)

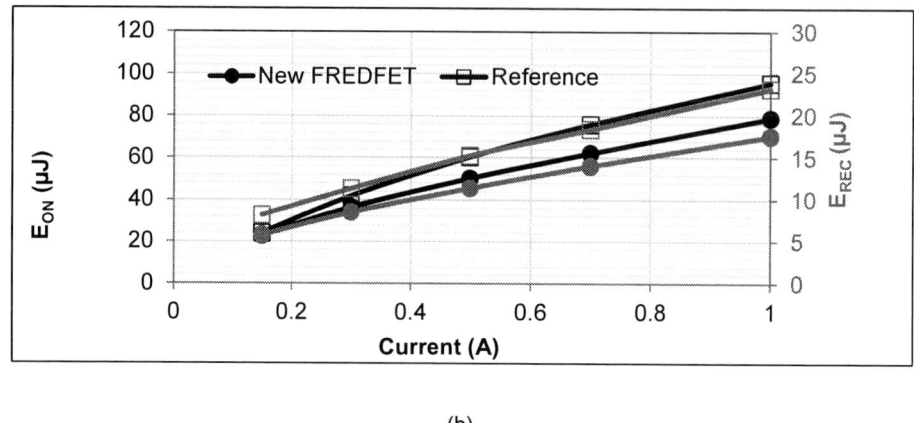

(b)

Fig. 6. Measured E_{ON} and E_{REC} at 0.15 – 1.0 A at 25°C (a) and 150°C (b).

4. Motor test performance

Motor tests were performed with modules using the new and reference FREDFETs and the thermal performance was compared. Figure 7 shows the test board used, a complete reference board capable of running any 3-phase motor via sensorless space vector modulation. In order to have consistent comparison, modules built with reference and new FREDFETs were tested under identical conditions (V_{BUS} = 320 V, f_{sw} = 20 kHz, I_{motor} = 200mArms and 300mArms, continuous symmetrical space vector modulation). The modules' case temperature was measured using an infrared camera with the hottest point on the top surface of the module recorded. Figure 8 shows thermal images of the modules running at identical conditions. It can be observed that the module with the new FREDFET runs 1.7°C cooler than the reference at 300mArms and 5°C cooler at 400mArms compared to the reference (Table I).

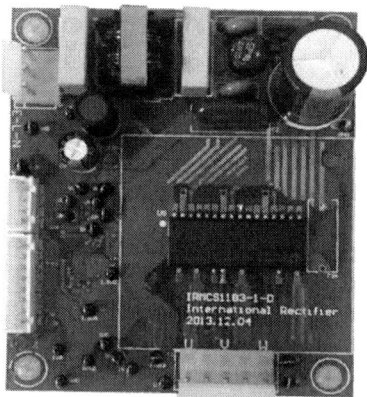

Fig. 7. Test PCB board with µIPM™

300mArms, Reference **300mArms, New FREDFET**

400mArms, Reference **400mArms, New FREDFET**

 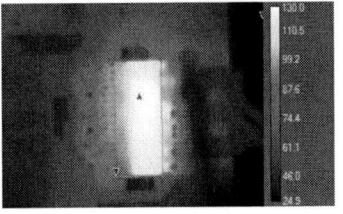

Fig. 8. Thermal images for reference (left) and new (right) FREDFET based modules at 300mArms (top), and 400mArms (bottom)

Part	Max Temp (°C)	I_{motor} (mArms)
Reference	97.4	300
New Module	95.7	
Reference	124	400
New Module	119	

Table I: Measured maximum case temperatures at 300mArms and 400mArms current for modules with reference and new FREDFETs.

5. Conclusions

Reverse recovery performance of the new FREDFET body diode is improved using optimized electron irradiation dose without significantly affecting the other DC parameters. EMI performance of the new and reference devices were compared with no significant difference in conducted emissions. An improvement in E_{ON} of about 17% and about 23% improvement in E_{REC} for the new FREDFET compared to the reference at 300mArms test current. Thermal performance of the modules during motor test showed that the module with

© VDE VERLAG GMBH · Berlin · Offenbach

the new FREDFET runs 1.7 degrees cooler than the reference at 300mArms current and 5 degrees cooler at 400mArms current compared to the reference.

References

1. http://www.irf.com/product-info/datasheets/data/irsm505-065.pdf
2. H. Yilmaz, K. Owyang, P. O. Shafer, and C. Borman, "Optimization of power MOSFET body diode for speed and ruggedness," IEEE Trans. Ind. Appl., vol. 26, no. 4, pp. 793–797, Jul. 1990.
3. Brian Sun, IRF Internal communication, International Rectifier.
4. Polenov, Dieter, et al. "The influence of turn-off dead time on the reverse-recovery behavior of synchronous rectifiers in automotive DC/DC-converters. *Power Electronics and Applications, 2009. EPE'09. 13th European Conference on.* IEEE, 2009.

PCIM Europe 2016, 10 – 12 May 2016, Nuremberg, Germany

A novel Transfer Molding Intelligent Converter Inverter Brake IGBT module (DIPIPM+) with integrated level shifting control ICs

Marco Honsberg, Mitsubishi Electric Europe B.V., Mitsubishi Electric Platz 1, 40882 Ratingen, Germany, marco.honsberg@meg.mee.com
Teruaki Nagahara, Mitsubishi Electric Power Device Works, 1-1-1 Imajukuhigashi, Nishi-Ku, Fukuoka 819-0192 Japan
Eric R. Motto, POWEREX Inc. 173 Pavilion Lane, Youngwood, PA 15697 USA, emotto@pwrx.com

Abstract

This paper introduces a new family of a compact intelligent CIB module and the specific design features of the used IGBT, Diodes and the High Voltage Integrated Circuit (HVIC). The module is manufactured in Transfer Mold technology which provides a good matching of mechanical expansion coefficients especially between the Copper lead frame and the mold material. The optimization of internal bond wiring provided better compactness of the module now being only slightly bigger than the standard DIPIPM™ and including a 3-phase input bridge rectifier and brake chopper topology. Test results of the switching performance and the related impact on the EMI are shown in a comparative analysis of EMI behavior of a previous generation DIPIPM versus the new DIPIPM+™. As a test platform for this new intelligent power module with converter and brake the Evaluation Board for the new DIPIPM+™ module is introduced in this paper.

1. Introduction

The Transfer Mold package of the DIPIPM+™ comprises the 3 main power electronic functional blocks to build compact stand-alone inverters supplied by the 3-phase 400V/480V AC line. The inverter, converter and brake parts are integrated along with low-voltage (LV) and high voltage (HV) control ICs and their corresponding bootstrap circuit to drive and to protect the IGBTs. Hence, the package contains six IGBT and Di's (Free Wheeling Diodes) assigned to the inverter output and furthermore a B6 diode rectifier bridge circuit and for most current ratings optionally a brake chopper circuit to dissipate the energy generated into the DC-link bus by motors in regenerative "braking" operation. The supply of the floating High side portion of the employed HVICs is realized by embedded 1200V Bootstrap Diode (BSD) chips that permit using only a simple single 15V power supply for the entire power stage of inverter system.

Fig.1 shows the photo of the new DIPIPM+™ and Fig.2 the internal block diagram.

© VDE VERLAG GMBH · Berlin · Offenbach

2. Internal construction and functional blocks

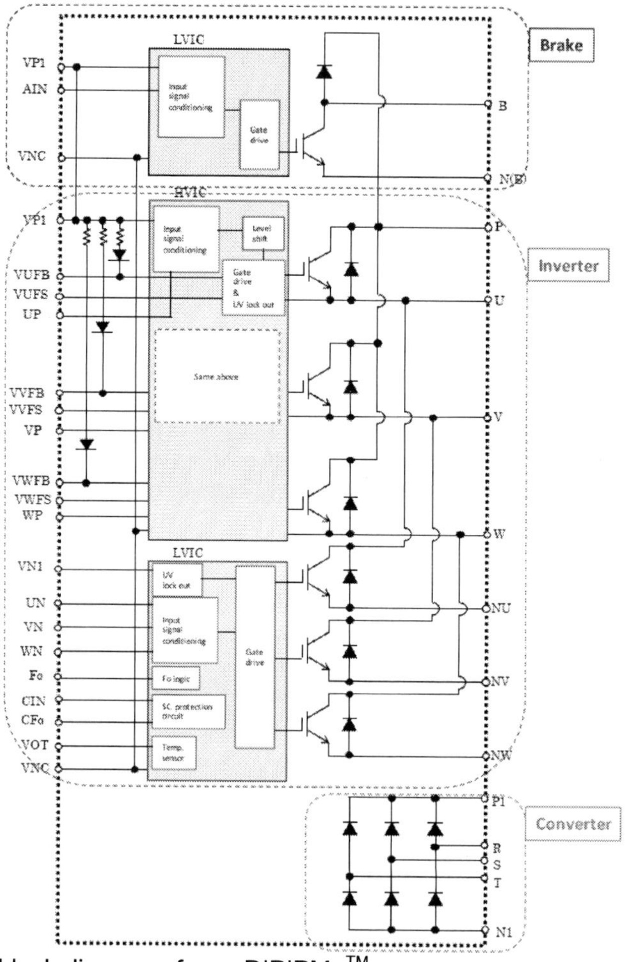

Fig.2 Internal block diagram of new DIPIPM+^{TM}

The driving HVIC and LVIC (Low Voltage IC) are placed as bare chips on the lead frame and besides the fundamental driver function these control ICs employ multiple protection functions like under voltage (UV) protection, short circuit (SC) protection and furthermore they provide a precise linear analog output signal (VOT) corresponding to the case temperature of the device. In contrast to simple NTC solutions this reference diode based output voltage VOT is much more precise in higher temperature range. Indeed the circuit's accuracy is optimized for a case temperature of approximately 85℃. Besides the over temperature protection the essentially important short circuit protection has been implemented into the DIPIPM+^{TM} referring to the voltage drop across externally placed shunt resistors at the open Emitter structure. The internal short circuit protection function will protect the IGBT chips against destruction at a short circuit condition and safely shut-down before the critical short circuit energy is reached under the precondition that the shunt resistor in conjunction with the internal reference voltage determined desaturation current has been set to 1,7 times the rated current of the IGBT and the externally set short circuit delay time is selected as less than 2µsec. For this purpose a built-in comparator with a very tight tolerance reference voltage of 0,48V±0,025V is sensing the CIN terminal and will turn-off the IGBTs as previously described when the threshold level is exceeded at this terminal.

3. Compactness improvement

In order to increase the compactness of the module the internal construction of the module package has been further developed and new approaches to connect the control ICs to the IGBTs have been developed. In the conventional DIPIPM™, the ICs and IGBTs are indirectly connected, e.g. through an additional island as part of the lead frame. The space consuming island facilitated the transition from gold wire bonding utilized between the chip and the island and on the other side the connection from the same island to the power chip by an Aluminum bond wire. Figure 3a and 3b illustrate the elimination of the space consuming intermediary island:

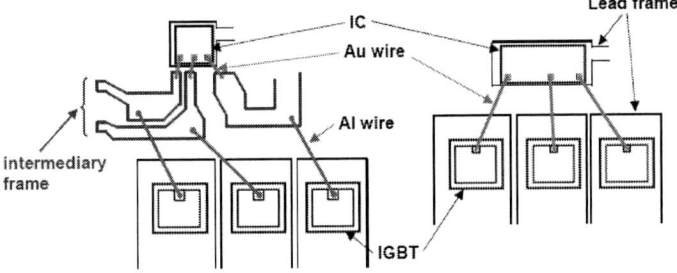

Fig.3a and 3b Wire bonding with and without intermediary island realized by the new lead frame shape

The advantages of the developed direct-wire-bonding technology are apparently contributing to the newly reached compactness, e.g. a module size reduction of about 35% comparing with today's state-of-the-art 1200V DIPIPM™s and simplification of the bonding process by utilization of only a gold wire bond has become possible.

PCB layout optimized pin terminal assignment

The described internal construction and especially the technological capability to manufacture a transfer mold device with long Aluminum and gold wire bonds provides a pin out of the DIPIPM+™ module that eases the routing of the traces of a Printed Circuit Board (PCB). Especially the implemented BSD and the clear separation of the control part from the power and high voltage related part of the module allows an optimized and compact design of the entire inverter. Fig. 4 indicates the pin assignment of the new DIPIPM+™:

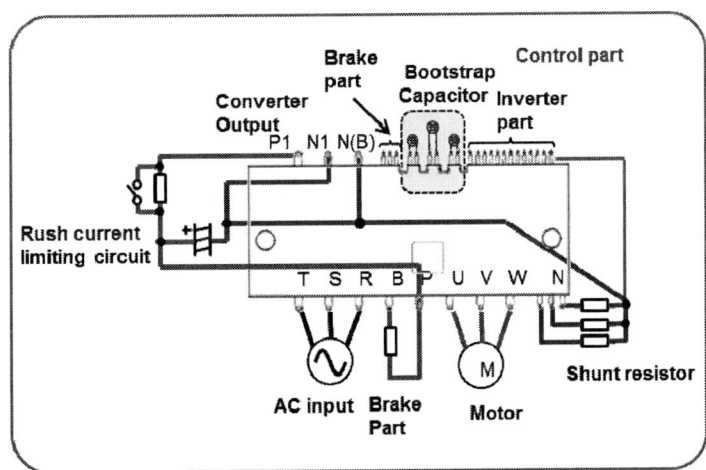

The well-organized pin out of the DIPIPM+^{TM} permits designing low inductive connection of the power stage to the DC-link even by double layer PCBs and, furthermore, the Dual – Inline structure itself provides a control and power signal separation, e.g. the control side terminals on one side and the high voltage terminals on the other side.

4. Switching loss performance and 7th generation chipset

In an Intelligent Power Module (IPM) the switching performance versus static loss tradeoff of the performance is fixed by internal settings of the built-in driver stage of the IGBTs. Thus, it is the manufacturer's choice to provide an optimized setup which is matching with most commonly used application requirements and to set a suitable operation point in the trade-off between switching loss corresponding to switching speed and the impact on EMI behavior.

A simple reduction of IGBT turn-off loss by increasing the switching speed would result in a disadvantageous EMI behavior and a too slow turn-on behavior would compromise the total loss too much in certain applications. In fact the driver's inherent settings need optimization for the desired chip behavior of both the IGBT and the Di's and the utilized new 7th generation CSTBT^{TM} chipset features an improved controllability of switching speed di/dt and dV/dt and loss performance whereas the new diodes are tuned for fast but soft recovery behavior.

The new chips of Di and IGBT acting together in the DIPIPM+^{TM} reveal an unprecedented performance and provide further improvement of the generated loss versus EMI performance trade-off compared with previous generations of chipsets. The switching operation at rated current of a DIPIPM+ with employed state-of-the-art 7th generation CSTBT^{TM} chips and new Diode structure is shown in fig.5 and fig.6 respectively.

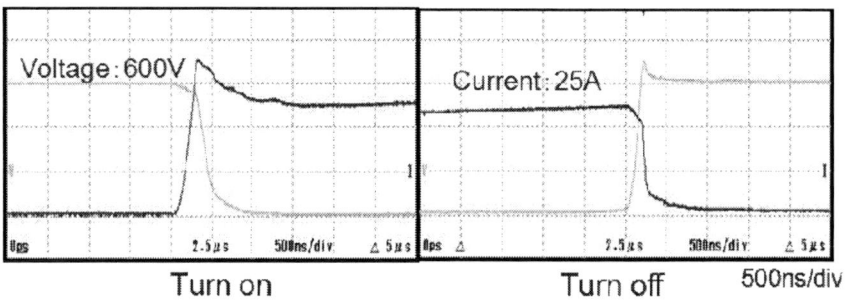

Fig.5: Turn on and turn-off behavior of the new DIPIPM+^{TM} (Tj=125℃, 25A device)

The figures of voltage in green colour and the current in blue colour reveal that there aren't any oscillations during the indicated switching operation and furthermore that the tail current is smoothly declining but short in its duration. As a consequence the implied switching energy in the tail current duration becomes comparatively small and that in turn reduces the total dynamic loss of the new DIPIPM+^{TM}. Apparently the switching behavior and its contained frequency spectrum is linked to the generation of EMI noise and since fig. 5 indicates a widely smooth waveform an advantageous EMI behavior of the DIPIPM+^{TM} can be expected.

5. EMI impact of the new chipset setup

This switching behavior as mentioned before shall correspond to a rather low EMI noise generation. A comparative analysis has been carried out and under similar conditions the frequency spectrum from 30MHz to 100MHz has been recorded for the DIPIPM^{TM} version 4 and the new DIPM+^{TM}. The results are indicated in fig. 6 and they reveal a superior EMI behavior of the DIPIPM+^{TM}:

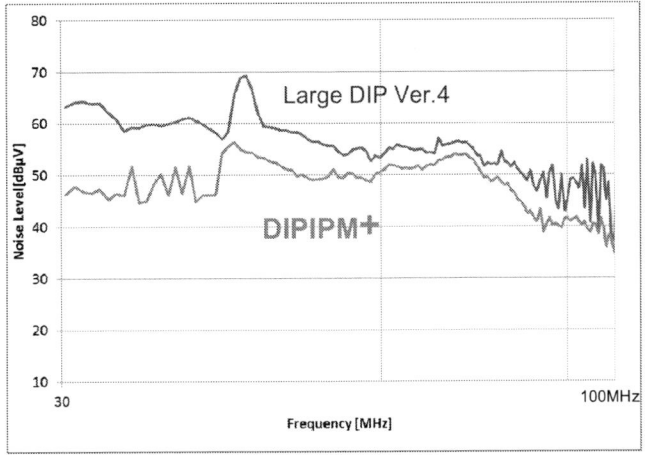

Fig. 6: Radiated noise emission comparing DIPIPM+^{TM} with a previous chip generation (DIPIPM 1200V version 4)

6. Evaluation platform

A platform to commission the DIPIPM+^{TM} has been developed providing the required peripheral components to basically operate this new IPM and to understand the simplicity to facilitate a power stage of an inverter based on the DIPIPM+^{TM}. The developed board provides numerous test points to acquire the signals and a non-isolated interface for micro controller signals referring to the negative voltage rail of the DC-link. The evaluation board provides a snubber capacitor, a power supply and even an inrush current limiter that can be bypassed by a relay circuit. All power connections can be done solder-less and the control signals are applied through an industrial standard JST XH connector with 16 pins. Fig. 7 shows a photo of the evaluation platform.

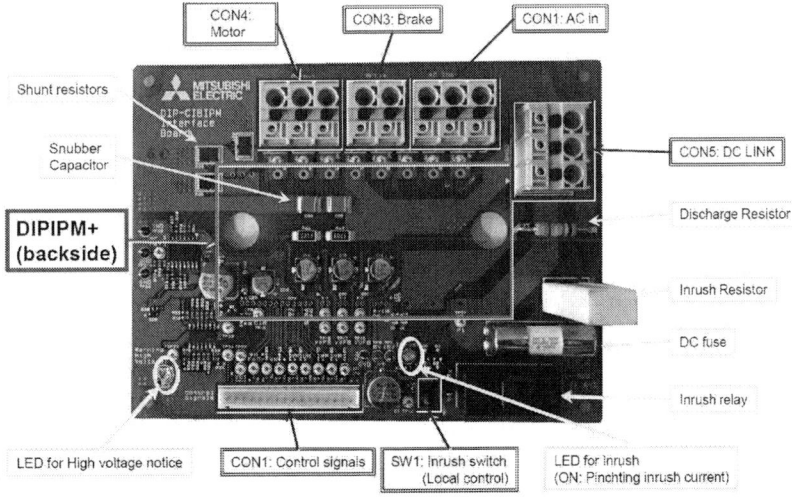

Fig. 7: DIPIPM+^{TM} evaluation board – position of the DIPIPM+^{TM} indicated with red colour.

The DIPIPM+[TM] is mounted from bottom side on the Printed Circuit Board which provides a degree of freedom to experiment with different heatsink structures since it is the only component mounted from the bottom side of the PCB. The comprehensive signal description on the silk screen of the PCB allows an easy identification of the most important signals as referred to in the schematic of the corresponding evaluation board's documentation. Hence, a verification of the control signals, the bootstrapping operation as well as the switching performance can be facilitated by the evaluation board for this new DIPIPM+[TM].

7. Conclusion

Based on Transfer Mold technology a new Converter Inverter Brake (CIB) module has been developed employing level shifting HVICs including BSDs and LVICs with complete protection function for SC, OT and UV. Furthermore the new DIPIPM+[TM] is an easy-to-use device by its high integration and compact design and the pin terminal assignment that simplifies the PCB routing. The internal chip setup provides an optimized trade-off between loss performance and the generated EMI that will contribute to the overall cost of an inverter by optimizing the EMI/EMC filtering effort as well as the heatsink requirement.

8. References

[1] S. Shirakawa, T. Iwagami, H.Kawafuji, M. Seo, K. Satoh "A New Version Transfer Mold-Type IPMs with Compact Package" 2005 PCIM China.

[2] M. Honsberg et al.: "A Novel Family of 1200V Transfer Mold Converter - Inverter - Brake (CIB) Modules Driven by a New 1200V High Voltage Integrated Circuit (1200V HVIC)", Proc. of PCIM 2005, pp. 461-468

[3] Toru Iwagami, Katsumi Satoh, Kou Shomei, Hisashi Kawafuji, Shinya Shirakawa, Tomofumi Tanaka " A Development of 30A/600V Super mini DIPIPMTM " 2006 ipemc,

[4] T.Takahashi, Y.Yoshiura " The 6th-Generation IGBT & Thin Wafer Diode for New Power Modules" 2011 Mitsubishi Electric ADVANCE June 2011

[5] K.Yamaguchi, Shinya Nakagawa, Tomofumi Tanaka, Yazhe Wang: "A New Version Transfer Mold-Type DIPIPMTMs with built-in Converter and Brake function", Proceedings PCIM China 2015

An Automatic IGBT Collector Current Sensing Technique via the Gate Node

J.X. Chen, A. Shorten, M. Sasaki[+], T. Kawashima[+], H. Nishio[+], W.T. Ng

The Edward S. Rogers Sr. Department of Electrical and Computer Engineering, University of Toronto, Toronto, Ontario, Canada, e-mail: chenjx@vrg.utoronto.ca

[+]Power Semiconductor Development Division, Fuji Electric Co., Ltd. Matsumoto, Nagano, Japan

Abstract

In this paper, a novel IGBT collector current sensing technique via the gate current and its physical implementation are presented. The proposed technique utilizes the Miller plateau phenomenon observed in power MOSFETs and IGBTs. A Miller plateau sensing circuit has been designed such that I_G can be measured in the plateau region during the turn on and turn of phase of the IGBT. The value of the I_G is digitized using an ADC and a distinct relationship with I_C can be observed for the IGBT. The proposed current sensing method only monitors I_G from the low voltage side of the gate drive circuit and allows a cycle-by-cycle measurement of I_C with high accuracy. The maximum deviation is ±0.8A within the range of 0~25A for turn-on and 0~30A for turn-off. This technique is also suitable for integration into a gate driver IC.

1. Introduction

In modern high voltage and high current power systems, current measurement for switching devices is critical for output regulation and for safety purposes. Conventional IGBT collector current sensing methods usually utilize external series current sensing resistor; which is typically placed between the emitter of the IGBT and ground [1]. The considerable power loss across this resistor reduces the system efficiency. In addition, any voltage drop produced across the sensing resistor during the on state of the switch reduces the actual gate-to-emitter voltage of the IGBT and increases the on-state voltage drop of the IGBT [2].

Another common collector current sensing method uses the embedded sense structure inside the IGBT. The sensed current, I_S can be deduced through an external sense resistor. There is a current sensing ratio between the sense current I_S and the actual collector current I_C. Due to the unique characteristics of the IGBT, there are three options for the embedded current sensor: 'active current sense', 'MOS current sense' and 'bipolar current sense'. The linearity of these sensors has been studied in [3, 4]. However in practical applications, it is very difficult to achieve a constant current sensing ratio across all the entire operating condition of the IGBT. In addition, the sense resistor must be sufficiently small to keep the potential at the sense electrode to be less than 50mV. As a result, the sensing circuit has poor noise immunity [2, 5, 6]. Another IGBT current sensing technique recently proposed by International Rectifier is based on the measurement of $R_{DS(on)}$ or V_{CE} of the IGBT [7]. This approach involves a HVFET which connects directly to the switching node for V_{CE} sensing; therefore, a discrete component is necessary.

All the above mentioned current monitoring methods involve extra discrete components which typically reduce overall system efficiency. Furthermore, the measurement data must cross an electrical isolation boundary between the gate and the collector terminals as the IGBTs are usually operating in a high voltage environment. This problem could be eliminated by using the proposed technique where the collector current can be deduced by measuring I_G from the low voltage gate drive circuit [8].

2. Proposed Technique

It is well known that for IGBTs, the V_{GE} or I_G exhibits a Miller plateau due to the charging (discharging) of C_{GD} during turn-on (turn-off), as shown in Fig. 1 (a) and Fig. 1 (b) [8-10]. The gate voltage at which the V_{GE} plateau occurs always corresponds to the I_C waveform [10]. For a fixed R_G, a distinct relationship between I_G and I_C during turn-on and turn-off can be observed [8], as shown in Fig. 2(a) and Fig. 2(b). The Miller plateau values for I_G during turn-on and turn-off were extracted and plotted against I_C for difference V_{DD} voltages on the load side, from $25V$ to $250V$. It can be seen that the measurement results remain consistent regardless of different V_{DD} levels. The plateau values of I_G was extracted by averaging all the values during the plateau interval, and the worst case maximum discrepancy between the actual value and the average value is 1.5% for turn-on plateaus and 1.8% for turn-off plateaus.

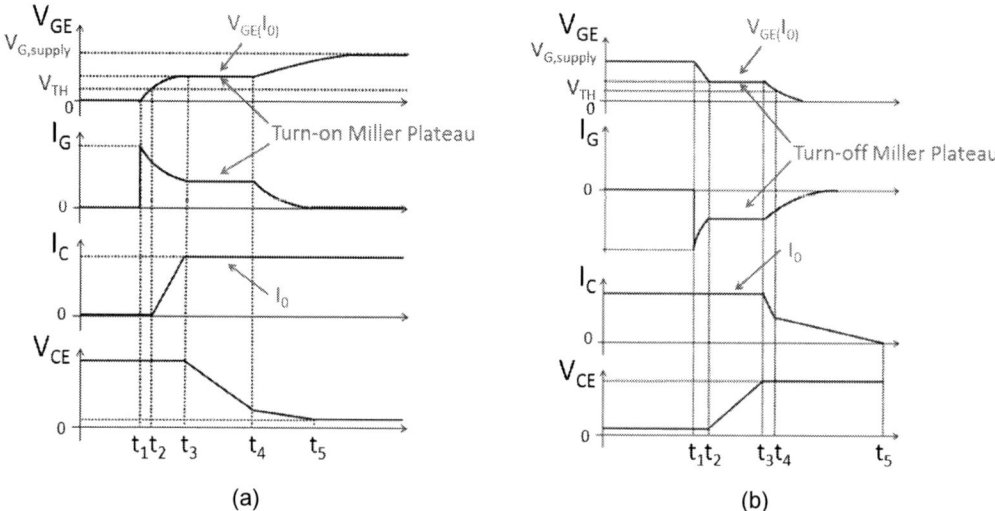

Figure 1: Typical switching waveforms for an IGBT at (a) turn-on and (b) turn-off.

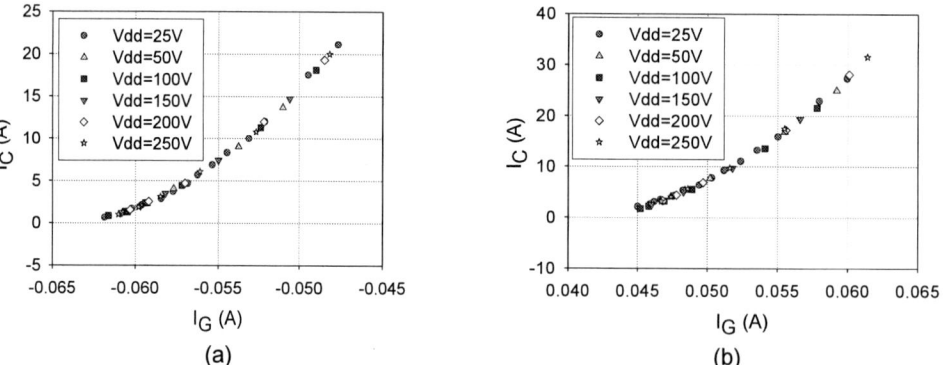

Figure 2: (a) Measured I_C vs. I_G (turn-on plateau values) [1]. (b) Measured I_C vs. I_G (turn-off plateau values) [1].

The proposed technique is an extension of the previously reported work [8] with an automated measurement current sensing circuit using discrete components. The block diagram of the circuit is as shown in Fig. 3. The proposed circuit senses the onset of the

Miller plateau region and utilizes a differential integrator to integrate the voltage across the gate resistance in the plateau region. Based on the integrator parameter R and C, the integrator output can be obtained as

$$V_{out\_diff} = \frac{V_{in\_diff}}{RC}t = \frac{I_G \times R_G}{RC}t \tag{1}$$

The differential output of the integrator is proportional to the gate current for a fixed time integration. V_{out_diff} is digitized at the end of the integration time to produce a digital representation of the collector current.

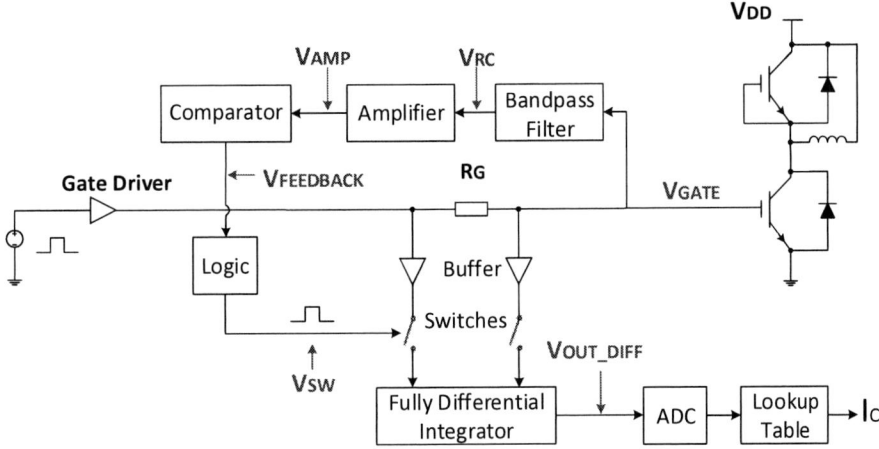

Figure 3: Block diagram of the proposed current sensing circuit.

The reason for using an integrator instead of sensing the gate voltage directly is due to that the plateau region is usually very short and noisy. The integration provides an averaging function that strongly suppresses noise. Another reason is the fact that the change in gate current is relatively small when compared to the changes in the collector current (as it can be seen from Fig. 2(a) and Fig. 2(b)). Therefore, the integrator can expand the change in gate current to provide higher sensitivity and higher accuracy.

3. Experimental Results

3.1 Test Bench Setup

The experimental test setup to verify the current sensing circuit is as shown in Fig. 4(a). Three levels of customised PCBs are cascaded onto the base level IGBT power module. The power module only includes three phase IGBTs and freewheeling diodes. The first level PCB only contains connection headers which allows for easy addition and removal of the control and current sensing circuits. Various terminals for all the IGBTs are accessible via the first level PCB. The second level PCB allows access to a Field Programmable Gate Array (FPGA) board that plugs directly onto the board. It includes two segmented gate drivers constructed using discrete components. It also provides access to important nodes via the third level PCB that contains all the analog components. The third level board is the actual I_C current sensing system, which is connected to the second level PCB via headers to provide access to the FPGA digital control and gate signal generation circuits.

A double gate pulse technique is used to first set the collector current of the IGBT. The IGBT turn on and off transients are then measured as illustrated in Fig. 4(b). Measurements are performed at the rising and falling edge of the second gate pulse. The width of the first gate pulse is adjusted in order to set the level for I_C. The voltage at the switching node, V_{SW} is the pulse used to trigger the fully differential integrator. It is basically a representation of the turn-off plateau of the IGBT. The ADC samples the output of the integrator at around the end of the Miller plateau. The output of the ADC can be read directly through the SPI interface of the oscilloscope.

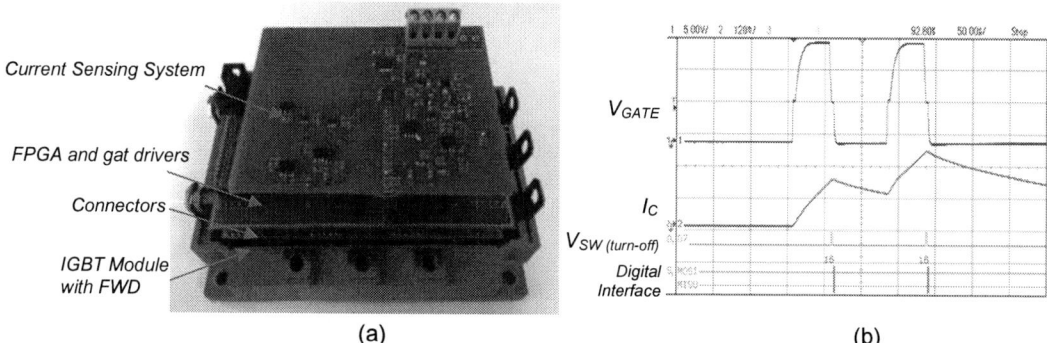

(a)   (b)

Figure 4: (a) Experimental testbed for the current sensing circuit. (b) Typical waveforms for the proposed testing methodology.

3.2 Turn-on and Turn-off Plateau Sensing

The Miller plateau sensing block is designed based on a passive band pass filter. Fig. 5(a) illustrates the schematic for the turn-on Miller plateau sensing block. A high pass filter is used to sense the gate node of the IGBT, where the time derivative of the plateau region equals zero. This is followed by a low pass filter which filters out the high frequency noise. The resulting V_{RC} is then amplified and compared with a constant reference voltage. The digitized $V_{FEEDBACK}$ is sent to a FPGA for digital processing. The FPGA uses a state machine to generate V_{SW} which corresponds to the plateau regions. To sense the turn-off plateau, a similar circuit is used except that V_{RC} needs to be inverted, as shown in Fig. 5(b).

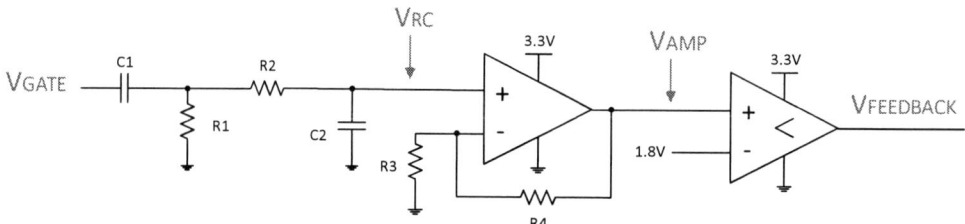

Figure 5(a): Schematic for the turn-on Miller plateau sensing circuit.

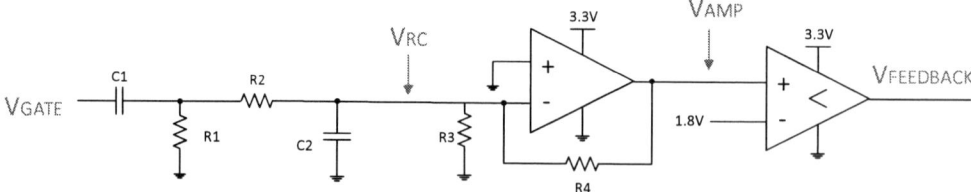

Figure 5(b): Schematic for the turn-off Miller plateau sensing circuit.

The V_{SW} signal is used to trigger the differential integrator. This signal corresponds to either the turn-on or turn-off Miller plateau of the IGBT. Fig. 6 shows the typical waveforms for the turn-on and turn-off Miller plateau sensing.

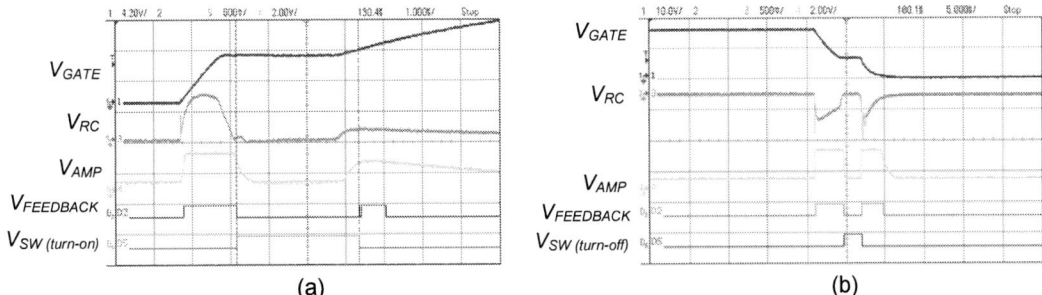

Figure 6: (a) Sample waveforms for sensing the turn-on Miller plateau. (b) Sample waveforms for sensing the turn-off Miller plateau.

3.3 Integrator Design and ADC Interface

The fully differential integrator is used to integrate the voltage across the gate resistor R_G when the switching node voltage V_{SW} is turned on during the IGBT turn-on and turn-off plateau period. The schematic for the integrator is illustrated in Fig. 6(a). The common mode voltage needs to be carefully selected to accommodate both the common mode input range of the integrator Op Amp as well as the ADC. The integrator R and C values are determined based on the integration time t in order to obtain reasonable resolution from the output voltage. C should be no less than 200pF to avoid excess charge injection from the reset switches. The trigger signal and output waveforms for the fully differential integrators during turn-off are shown in Fig. 6(b).

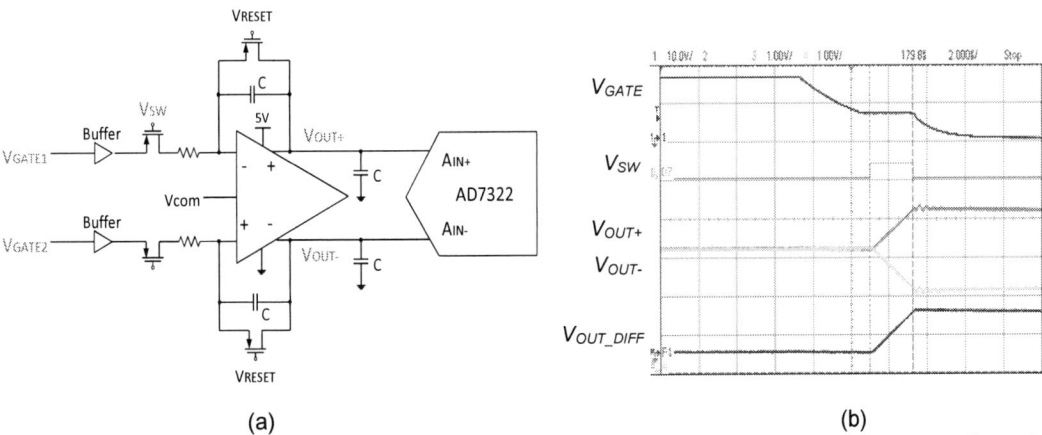

Figure 6: (a) Schematic for the fully differential integrator. (b) Output waveforms of the differential integrator during turn-off.

The ADC is configured to accept fully differential inputs in the range of ±10V for turn-off and ±5V for turn-on. This is based on the fact that the effective plateau length for turn-on is shorter than turn-off for the same R_G value. Therefore, the ±10V input range setting could not provide enough resolution for the turn-on plateau sensing. The ADC samples V_{OUT+} and V_{OUT-} signals when the chip selection signal V_{CS} goes low and needs 16 clock cycles to complete the conversion, as shown in Fig. 7. Since V_{CS} is set to be enabled at the first rising edge of a clock cycle after the falling edge of $V_{SW\ (turn-off)}$, therefore, the real integration time is not necessarily the exact duration of the $V_{SW\ (turn-off)}$ signal; this may cause variation of the integration time t. Fig .7.illustrates all the interfacing signals of the fully differential integrators with the ADC for turn-off.

© VDE VERLAG GMBH · Berlin · Offenbach

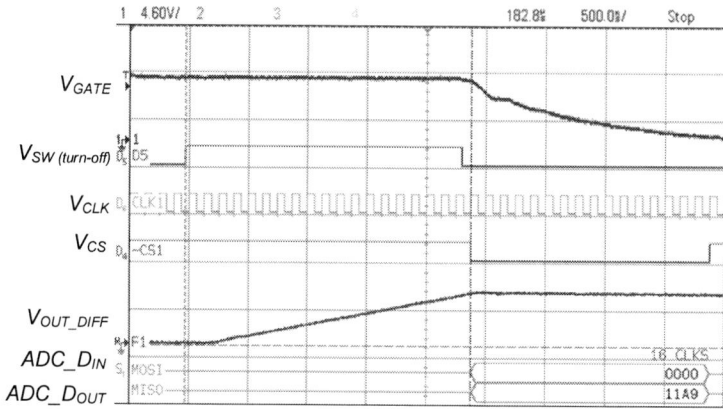

Figure 7: Switching waveforms of V_{GATE}, $V_{SW\,(turn\text{-}off)}$, V_{OUT+}, V_{OUT-} and V_{OUT_DIFF} during turn-off.

3.4 Fixed Time Integration

In order to eliminate the error source mentioned in the previous section, a fixed time integration method was proposed and tested. Based on extensive testing results, it is found that this fixed time integration could achieve much higher accuracy. A new signal $V_{SW\text{-}CLK}$ is generated from V_{SW} through a chain of D Flip-Flops as shown in Fig. 8(a). This fixed time integration method ensures that a multiple cycle of the master clock (8 MHz) are used. The integration time is determined based on the graph in Fig. 8(b); where I_G is plotted as a function of time for different I_C levels and with V_{CE}. In practical applications, V_{CE} is normally fixed at very high voltages. When V_{CE} is higher, the plateau length will be slightly longer [8]. In this case, the plateau length varies from 2.7 to 2.75 µs when I_C changes from 2 to 30 A with V_{CE} at 20 V. The variation is less than 2%. Therefore, the fixed time integration is practical and the time should be chosen to be as long as possible but shorter than the minimum plateau length. In this case, the integration time is chosen to be 1.6 µs for turn-on and 2.4 µs for turn-off.

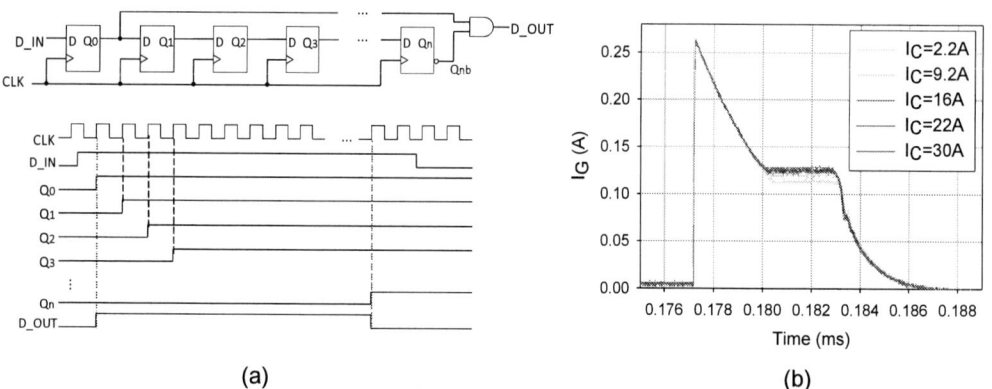

(a)   (b)

Figure 8: (a) D Flip-Flop chain for fixed time integration (b) I_G as a function of time at different I_C levels for turn-off when V_{CE} = 20 V.

Fig. 9 (a) and (b) illustrate the switching waveforms for the fixed time integration for turn-on and turn-off. The gate signal at turn-on is purposely zoomed in to show the initial ringing at the plateau region which is due to the Electromagnetic Interference of the switching. Therefore, the starting point of the $V_{SW\text{-}clk\,(turn\text{-}on)}$ signal is purposely delayed for several clock cycles.

© VDE VERLAG GMBH · Berlin · Offenbach

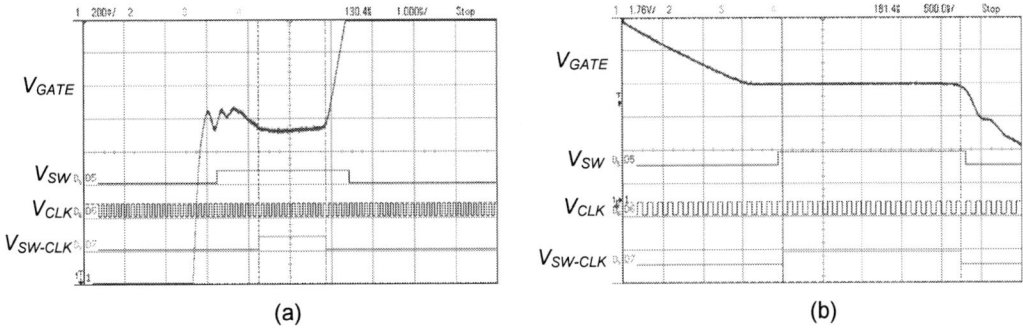

(a)   (b)

Figure 9: (a) Typical waveforms illustrating the fixed time integration for turn-on and (b) turn-off.

The extracted I_C versus ADC digital output for both turn-on and turn-off are as shown in Fig. 10(a) and Fig. 10(b), matching the previously obtained I_C vs. I_G relationship very well. Regression analysis is performed on the actual I_C and the ADC outputs in order to evaluate the accuracy of the technique. The maximum deviation is less than 0.8 A for both turn-on and turn-off over the entire testing range, as shown in Fig. 11. Therefore, it is demonstrated that the IGBT collector current can be measured digitally from the low voltage gate drive circuit.

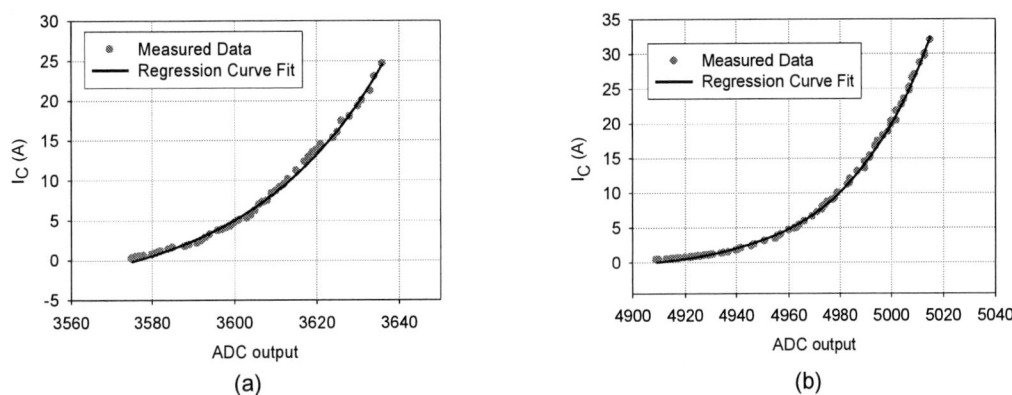

(a)   (b)

Figure 10: (a) I_C vs. digital ADC output for turn-on. (b) I_C vs. digital ADC output for turn-off.

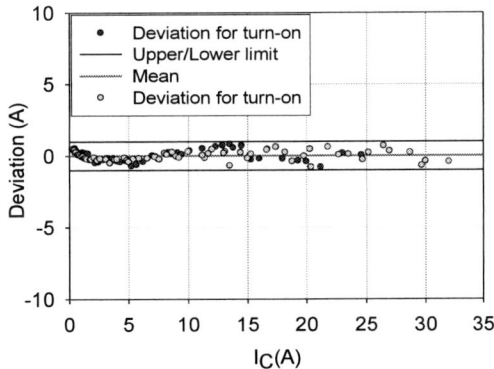

Figure 11: Deviation of the measured I_C to the regression fitted curve vs. I_C.

4. Conclusions

In this paper, a unique method is introduced to measure the IGBT collector current (I_C) indirectly from the gate node. The proposed technique leverages the Miller effect to allow the monitoring points to be situated entirely on the low voltage gate drive circuit. As this measurement can only occur at turn-on and turn-off, the proposed technique allows for a cycle-by-cycle measurement of I_C. An automatic gate current sensing system constructed with discrete components was designed and tested, the IGBT collector current could be known through a lookup table during each turn-on and turn-off cycle with good accuracy. Future work will be carried out to integrate the current sensing system to the gate driver IC. Also, it is well known that the transfer characteristics of the IGBT will change with the temperature. Therefore, additional circuitry will be designed to compensate for the temperature effect.

ACKNOWLEDGEMENTS

The author would like to thank Fuji Electric and NSERC for their generous support.

References

[1]   D. Tam, "ICs Protect IGBTs and Sense Currents in Motor Drives ", Power Electronics Technology, April 2005.

[2]   B. J. Baliga, *The IGBT Device*, 1st ed., 2015. ISBN: 978-1-4557-3143-5.

[3]   T. P. Chow, D. N. Pattanayak, E. J. Wildi, J. M. Pimbley, B. J. Baliga, and M. S. Adler, "Design of current sensors in IGBTs," *Electron Devices, IEEE Transactions on*, vol. 39, no. 11, pp. 2673, 1992.

[4]   T. P. Chow, Z. Shen, D. N. Pattanayak, E. J. Wildi, M. S. Adler, and B. J. Baliga, "Modelling and analysis of current sensors for N-channel, vertical IGBTs." pp. 253-256. Dec 1992.

[5]   H. Yamagiwa, T. Saji, S. Kaneko, S. Takahashi, T. Uno, and K. Sawada, "Sense Device Structure in Hybrid IGBT with Constant Current Sense Ratio for Entire Collector Current Range." pp. 217-220. May 2008.

[6]   "IGBT Sense Emitter Current Measurement Using the Keysight B1505A", 2014. URL: http://literature.cdn.keysight.com/litweb/pdf/5991-4543EN.pdf

[7]   T. Ribarich, "Enhancing switching-mode power performance with RDS(on) current sensing" Power System Design: Empowering Global Innovation, 2014.
URL: http://www.irf.com/pressroom/articles/1099psd1404.pdf

[8]   A. S. J.X. Chen, W.T. Ng, "IGBT Collector Current Sensing Using Gate Current," ISPS 2014. Prague, 27-29 August 2014.

[9]   R. Chokhawala, J. Catt, and B. Pelly, "Gate drive considerations for IGBT modules." pp. 1186-1195 vol.1. 06 August 2002

[10] J. Karlsson, "The concept of IGBT modeling and the evaluation of the PSPICE IGBT model," Master Thesis, Conducted at ALSTOM Power, Växjö, 2002.

Design and Characterisation of Optimised Protective Thyristors for VSC Systems

Spence M., Plumpton A., Rout C., Millington A., Keyse R.

Bipolar R&D Group, Research and Development Centre, Dynex Semiconductor Ltd, Doddington Rd, Lincoln, United Kingdom LN6 3LF, email: michael_spence@dynexsemi.com

Abstract

Voltage Source Converter (VSC) technology provides a number of advantages over traditional HVDC including self commutation, small footprint and black start capability, and is becoming increasingly popular in applications such as offshore wind [1,2]. One major drawback of VSC systems is their vulnerability to DC faults, which cause a short circuit current to pass through the IGBT diode path. The magnitude of the fault current can be many times the rated current, can last for a number of cycles and can destroy the diode before an AC circuit breaker is able to clear the fault. A common solution is to include a thyristor in parallel with the IGBT diode to divert the majority of the fault current from the diode in such an event. The correct design of such a device can increase fault current handling capability of the VSC and also enable a more compact and cost effective sub-module. This paper discusses the design methodology adopted to produce an optimised protective thyristor and presents a method for characterising and comparing differing thyristor designs for VSC protective applications.

1. Introduction

In each sub-module the IGBT diode is protected by a bypass thyristor (SW1) and mechanical circuit breaker (SW2), both in parallel with the diode (figure 1). During normal operation the thyristor will be switched off and, in the forward direction, is only required to block a small voltage. During DC faults the diode will be exposed to a large fault current typically lasting up to hundreds of milliseconds. Once this fault is detected and the voltage across the thyristor is sufficient, it will be triggered, diverting a large proportion of the current until the mechanical breaker is tripped.

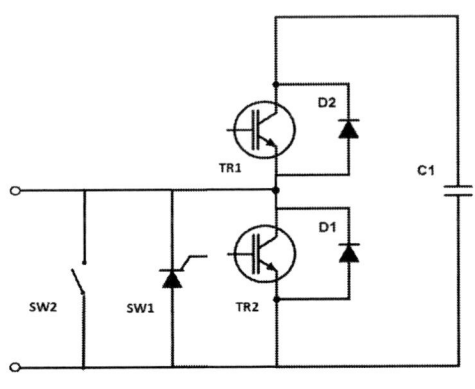

Fig1. Topology of VSC half bridge sub-cell

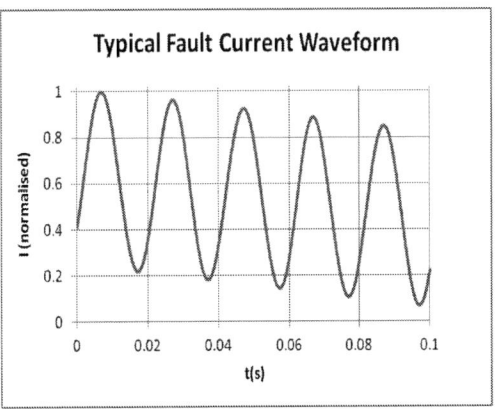

Fig2. Typical short-circuit fault current 50Hz

2. Design

2.1 Requirements

There are a number of requirements on the performance of protection thyristors in VSC applications:

- Survivability – A primary requirement on the thyristor is to survive the fault current which it diverts. The device must also be able to block the forward and reverse voltage waveforms.
- Pickup voltage – This describes the anode-cathode voltage at which the thyristor is able to latch when supplied with a given gate signal. This voltage is generated by the on-state voltage of the diode as the fault current rises. A reduced pick-up voltage would allow the device to be fired earlier.
- Plasma Spreading – Another important factor is how long the full area of the thyristor takes to turn on. Along with the pickup voltage this will help determine the maximum current experienced by the diode and the total energy (I^2t) received in the first half-sine of the fault.
- On-state voltage – The Vt of the fully turned on thyristor will affect the total I^2t received by the IGBT diode.
- Reliability – Cosmic ray induced failures may become problematic due to the duty cycle, which means that such devices often must be overrated in voltage.

2.2 Survivability

The application requires a device which will survive the specific type of fault current. This can either be determined by reproducing the current directly or the response can be simulated using single shot surge values. Survivability is primarily dependent on the surge capability which is a factor of the current rating and the thermal properties of the device and packaging. During normal operation, the thyristor is also subjected to a fast rate of voltage rise, produced by the dynamic V_F of the IGBT diode. The device must be able to withstand a much higher dV/dt than is commonly quoted in datasheets and therefore careful emitter design is required.

2.3 On-State Voltage

On-state voltage can most easily be reduced by increasing device diameter or decreasing silicon thickness. The former is not desirable, as compactness is a prime consideration in VSC converters; however contact area can be increased without increasing the external footprint of the package. This can be achieved by a double positive bevel which gives a very good utilisation of the active area, giving a larger active area for the same diameter [3]. Thickness reductions can be achieved by a number of methods. The bevel can be designed so that the device is not limited by edge breakdown allowing optimisation of the start silicon. As the only forward voltage seen by the device is from the dynamic V_F of the diode, the forward junction of the thyristor is only required to support a small proportion of the rated voltage of the reverse. As a consequence of this the forward blocking capability of the device can be de-rated and modifications can be made to the diffusion profile allowing a reduction in silicon thickness

Where the bypass thyristor diverts sufficient current, these improvements in on-state voltage can mean that a smaller SCR wafer can be used. This translates to a smaller device encapsulation, allowing for a more compact sub-module and potentially reducing total build costs.

2.4 Emitter/Gate Design and Current Spreading

Improved current spreading in the bypass thyristor will allow the device to turn on more quickly. In other words, it will divert a greater amount of the fault current from the IGBT diode at the beginning of the fault. In addition careful emitter design will allow high di/dt withstand capability.

Good current spreading can be realised by selecting an appropriate gate geometry and suitable emitter shorting [1]. Simulation shows that modification of the shorting density close to the gate region can strongly affect the spreading of the electron-hole plasma. Figure 3 shows the improvement in uniformity of plasma spreading achieved by moving the shorting spots away from the gate arms.

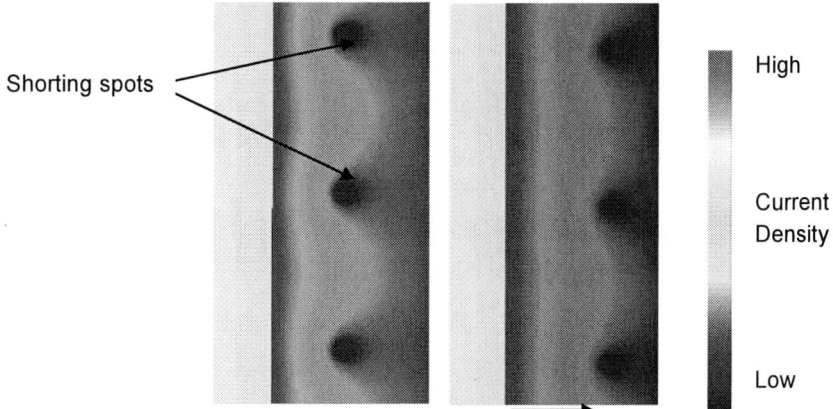

Fig 3. Simulation comparing shorting layouts

As described earlier, the rate of change of voltage can be quite fast; therefore when designing the shorting, improvements in current spreading must be balanced with deterioration in dV/dt withstand capability.

2.5 Reliability

In general, the wear on the thyristor is much lower than in many applications such as rectification as the thyristor is only required to conduct only a few times per year to clear faults. One potential issue for this application is the voltage waveform experienced by the thyristor during normal operation of the circuit. Thyristor devices are normally designed for blocking sinusoidal voltage waveforms, whereas in VSC thyristors are blocking square waves. This can make the devices susceptible to cosmic ray induced failures [4]. The FIT rating is strongly dependent on blocking voltage and n-base resistivity [5]. For obvious reasons, reducing the voltage is not desirable; however the resistivity of the device can be modified. Increasing resistivity sufficiently can provide dramatic improvements to the cosmic failure rate, making it negligible when compared to other failure modes. It is necessary, particularly in non-punch through structures, to ensure that increases in the n-base resistivity do not degrade the reverse blocking too much.

3. Device performance

3.1 Device Characteristics

Three device types were included. The first generation device was a standard SCR optimised for on-state voltage, with the voltage rating selected to ensure a lower cosmic ray failure rate than the IGBT diode. The second generation devices had reduced on-state voltage and further improved cosmic ray robustness. The third generation development exhibits the best improved cosmic ray robustness and an even further reduced on-state voltage. In addition, two different emitter patterns were compared on the second and third generation devices, both designed to turn-on at a low anode-cathode voltage. The first of these emitter masks (Em1) featured an improved shorting pattern and a greater turn-on length compared to the second (Em2).

The figures below show the device characteristics of first, second and third generation bypass thyristors.

Device Type	3300V companion		4500V companion	
	Vdrm (V)	Vrrm (V)	Vdrm (V)	Vrrm (V)
Generation 1	4200V	4200V	5200V	5200V
Generation 2	>1000V	3300V	>1500V	4500V
Generation 3	>1000V	3300V	>1500V	4500V

Fig 4. Forward and reverse voltage ratings for each generation of bypass thyristor

Device Type	Typical Pick Up Voltage	
	3300V companion	4500V companion
Generation 1	<3V	<3V
Generation 2	<2V	<2V
Generation 3	<2V	<2V

Fig 5.Typical anode-cathode voltage (pick-up voltage) at which the thyristor can be fired. Example pulse 3A, 100µs

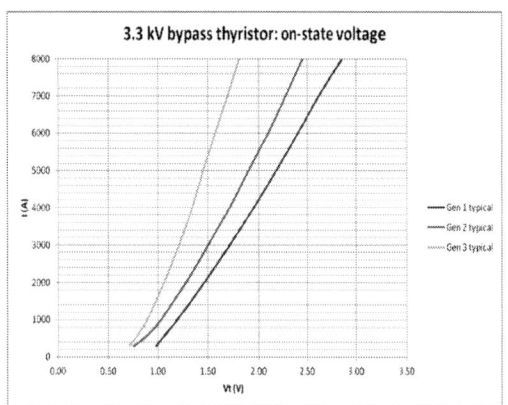 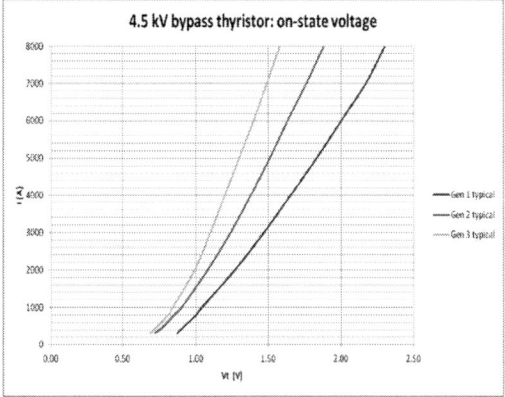

Fig 6. Comparison of on-state voltage for each generation of bypass thyristor

3.2 Comparison of Bypass Thyristor Sharing

A direct measurement approach is used to examine the fault current through the IGBT diode and thyristor whilst in the application configuration. Direct measurement provides a method of assessing the total impact of different design factors on the effectiveness of the protective thyristor at diverting the current.

Samples of the three different generations of devices described in section 3.0 were included in the test, with different emitter masks (Em1 and Em2) compared on the latter. In order to explore the above aspects and compare different designs, it is not necessary to use multiple cycles of fault current. To incorporate all the aspects of turn-on, spreading and the response of a fully turned on thyristor to an increasing current, the system only needs to be exposed to the first half sine of the current.

3.3 Sharing Test Circuit

A simplified version of a VSC sub cell was recreated in the laboratory (figure 7). A major consideration of this circuit was to minimise the total impedance. In commercial applications, designers are very careful to minimise the impedance of the system to improve the efficiency of the converter. While these measurements are comparative and the circuit generalised, the aim is to represent a real VSC system as closely as possible. Additionally, without careful consideration of circuit layout, the impedance of the circuit becomes the dominant factor in current sharing.

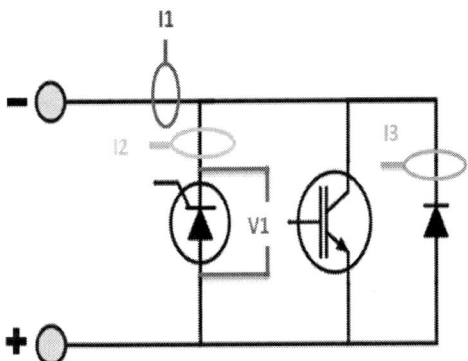

Figure 7. Test circuit

The IGBT modules were controlled at a temperature of 125°C to reflect typical operating conditions. Current was measured using three current transducers along the IGBT diode path (I3), the bypass thyristor path (I2) and across the transformer terminal (I1) to monitor total input current. Voltage was also measured across the thyristor pole pieces (V1). The circuit described in figure 7 was then connected to a transformer and capacitor bank capable of producing the required half sine current pulse of 50Hz, >15kA.

PCIM Europe 2016, 10 – 12 May 2016, Nuremberg, Germany

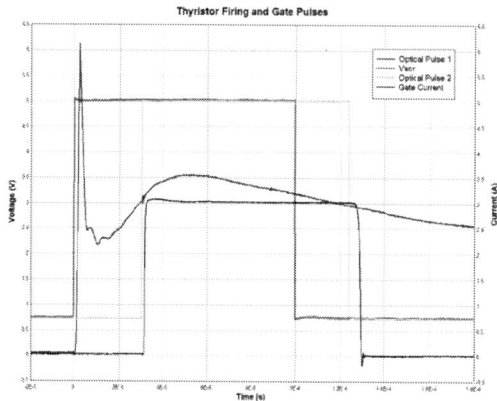

Fig 8. Bypass thyristor firing and anode-cathode voltage

Control of the circuit was achieved by a pair of linked optical pulses (figure 8). An initial pulse discharged the capacitor bank and the following pulse then sent a 3A, 100µs gate signal to fire the thyristor switch. The time delay of these pulses could be selected to fire the thyristor at the chosen anode-cathode voltage, ensuring that the device is triggered above its pick-up voltage. It is also worth noting that the on-state voltage of the diode rises quickly, after the initial peak. Thus, the ability to reliably fire the thyristor at an anode-cathode (pick-up) voltage of 2V rather than 3V would only allow the device to be fired a few tens of microseconds earlier, yielding a minute difference in the total current received by the IGBT diodes.

3.4 Sharing Test Parameters

The test was carried out using devices from each of the three generations as bypass switches for the IGBT diode. The 3.3kV companion devices were tested with a 3300V, 1500A Dynex IGBT module and the 4.5kV devices with a 4500V, 1200A single switch Dynex IGBT module, both in a single switch configuration. A comparison was also carried out on two different emitter variants to explore the effect of plasma spreading.

3.5 Current Sharing Results

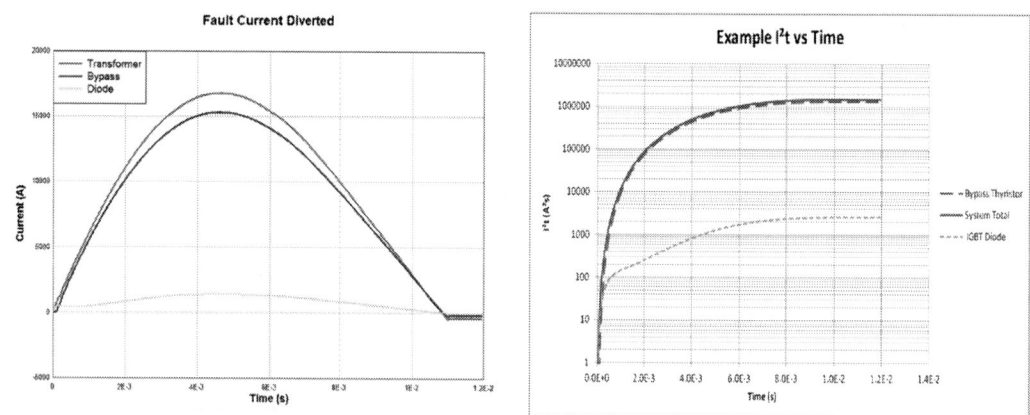

Fig 9. I²t experienced by diode compared to that seen by the thyristor

© VDE VERLAG GMBH · Berlin · Offenbach

Figure 9 compares the I²t experienced by the thyristor an IGBT diode through the half sine wave. Figure 10 shows the ratio of fault current diverted by an example bypass thyristor. As we can see the thyristor diverts a considerable proportion of the current.

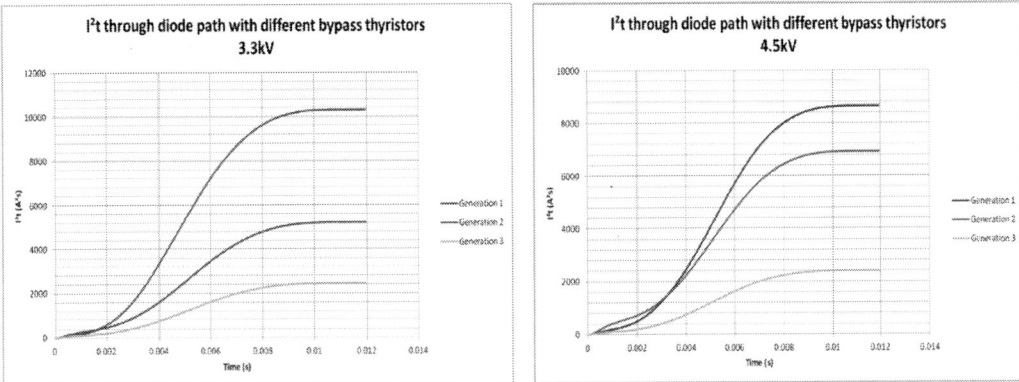

Fig 10. Reduction in total current experienced by the IGBT diode with improvements in bypass thyristors

Figure 10 shows the average I²t through the IGBT diode using a generation 1, generation 2 and generation 3 bypass thyristors. As we can see, there is a substantial reduction with each generation. Due to the long fault duration, the most dominant factor in current diversion is the on-state voltage of the thyristor. With the second generation devices, the I²t experienced by the diode is reduced by between 25-50%. For the third generation devices, where the on-state voltage is lowest, the voltage reduction is 70-75%. In a commercial system, where bus bars and clamps are specifically designed to reduce impedance; improvements in performance from the newer generations of devices are likely to be amplified.

3.6 Thyristor Turn-on

The emitter design can have an effect on plasma spreading in the thyristor, as described earlier. An improved emitter design (Em2), exhibits a more rapid diversion of current from the IGBT diode when compared to a more simplistic design (Em1) as gate arm and emitter shorting design produce improved current spreading.

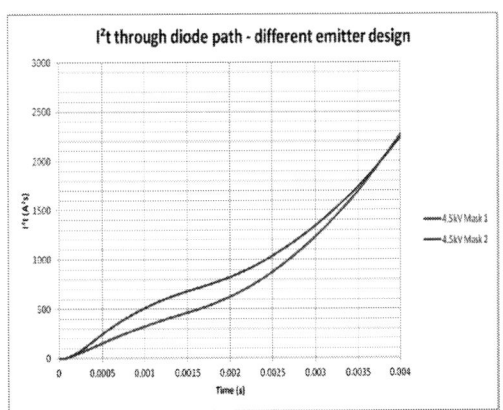

Fig 11. Comparison of plasma spreading with different emitter designs.

Figure 11 shows the difference in spreading for an example device type with the two emitter designs. When considering the total I²t through the diode, this is secondary to changes in on-state voltage of the bypass thyristor, which have a much more pronounced effect.

4. Conclusion

The requirements placed on a bypass thyristor in a VSC system are numerous and often atypical of those involved in the design of a phase control thyristor. Cosmic ray robustness must be considered, as the thyristor is required to block a square wave. The device must be able to withstand large voltage transients, necessitating careful emitter design. The on-state voltage of the device must also be kept to a minimum to ensure effective fault current diversion from the IGBT diode.

Devices with an improved on-state voltage are able to divert a greater amount of a given fault current from the IGBT diode. This can present advantages to the system designer for fault currents of higher magnitudes and for higher voltage systems, where the thyristors must be up-rated in voltage. In existing systems, where an off the shelf rectifier device will divert sufficient current, reductions in on-state voltage, achieved by factors such as increased contact area for a given device size and optimised n-base width allow reduction in package size aiding the system designer in creating a compact converter.

5. References

1) Lin W., Jovcic D. *LCL and L-VSC Converters with DC Fault Current-Limiting Property and Minimal Power Losses*. IEEE Transactions on Power Delivery, vol. 29, no. 5, pp. 2359-2368, (Oct 2014)
2) Bucher MK, Franck CM. *Contribution of Fault Current Sources in Multi-Terminal HVDC Cable Networks*. IEEE Transactions on Power Delivery, vol. 28, no. 3, (Jul 2013)
3) Ghandi, SK. *Semiconductor Power Devices*. John Wiley & Sons, Inc., (1977).
4) Zeller HR. *Cosmic ray induced failures in high power semiconductor devices*. Solid-State Electronics, vol. 38, no.12 p2041–2046, (1995)
5) Bauer FD. *Calculation of cosmic ray limited maximum DC blocking voltages of high voltage silicon PIN diodes*. Solid-State Electronics vol. 52, no. 8, p1052-1057, (2008)

Cathode Emitter versus Carrier Lifetime Engineering of Thyristors for Industrial Applications

J. Vobecký, ABB Switzerland Ltd, Semiconductors, jan.vobecky@ch.abb.com

M. Bellini, ABB Corporate Research Center, Switzerland, marco.bellini@ch.abb.com

K. Stiegler, ABB Switzerland Ltd, Semiconductors, karlheinz.stiegler@ch.abb.com

Abstract

We experimentally demonstrate that the traditional way of reducing the recovery charge Q_{rr} by electron irradiation can be replaced by appropriate density of cathode shorts of an appropriate size. Functionality of this approach is demonstrated for the thyristors of 1.8 and 2.8 kV voltage classes, for which an improvement of technology curve $Q_{rr} - V_T$ was achieved. In addition, the higher density of cathode shorts improves the dV/dt capability and circuit commutated recovery time t_q. The *Irradiation Free Concept* provides thyristors with improved ratings, while it eliminates the fear of annealing the radiation defects from electron irradiation under overload conditions, when thyristors are repeatedly exposed to ON-state currents close to the nominal value of surge ON-state current I_{TSM}.

1. Introduction

A Phase Control Thyristor (PCT) is the device of choice for the applications, where the highest current handling capability per unit area (low ON-state voltage drop), easiness of control and low cost are required, while the missing turn-off capability can be substituted by circuit commutation. This is the reason, why PCTs continuously maintain a significant market share, though they belong to the oldest semiconductor device concepts [1].

Electron irradiation is the state-of-the-art technique for setting the position at the technology curve $Q_{rr} - V_T$ of PCTs [2]. It is practical in volume production, because we can not only set the required reverse recovery charge Q_{rr} after the device is completely processed including metallization, but also because we can repeat the irradiation to further reduce the Q_{rr}, if the required specification was not reached. In this paper, we go beyond these undoubted advantages by designing the PCT, which has a proper Q_{rr} value already after leaving the production line, i.e. without the electron irradiation. The demonstrated technique is suitable for the industrial applications, where PCTs are not operating in parallel and precise tuning of Q_{rr} into a very narrow band is not necessary [4].

2. Thyristor Design

Cathode emitter shorting (shorts) is the design technique widely accepted to reduce the current gain of the internal NPN transistor at low currents under forward blocking [1, 2, 3]. This allows us to achieve an equally high blocking voltage under forward bias (break-over voltage) as under the reverse bias, where the multiplication of leakage current is significantly lower due to the lower current gain of the PNP transistor and no amplification from the NPN one. Analogically, the increased density of shorts reduces the amplification of the displacement current $C_{SCR}.dV/dt$ generated during a steep rise of anode voltage dV/dt hereby increasing the so-called dV/dt capability. C_{SCR} is the capacity of space charge region of the thyristor, which typically amounts to units or tens of nanofarads at low voltage depending on device area.

The above mentioned higher density of cathode shorts lowers the active cathode area used for carrier injection, which can increase the ON-state voltage drop V_T. This trade-off is therefore considered during design optimization. The relevant parameters are short diameter, short separation, relative shorted area (given by the number of shorts), and the shape of shorting array, which is the triangular one in this work. At glance, the doping profiles of the shorts, P-base and cathode are also important items in the design.

The existing level of surface patterning in the silicon technology for bipolar power devices allows us to push the minimal short diameter below the size, which would be considered impractical in the past. Also the cleanliness of starting material and diffusion processes have improved so, that the required levels of leakage current (blocking voltage) can be achieved for any short design and without the electron irradiation at low voltage devices presented below. All these aspects represent a good reason for reviewing the state-of-the-art thyristor design.

Fig.2.1a) shows simulated distribution of the ON-state current density for small and big cathode shorts while the total cathode area consumed by the shorts is equal. The device simulation using Sentaurus Device from Synopsys was used. Fig.2.1b), which is cut out of the Fig.2.1a), indicates that bigger shorts affect the ON-state current density deeper in the bulk and that the smaller shorts result in more homogeneous current distribution. This feature provides a degree of freedom for further optimization of device performance.

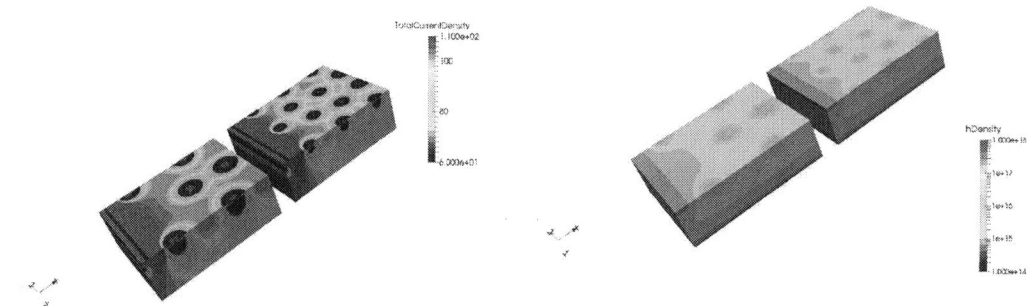

Fig.2.1a): ON-state distribution of current density of *Classical* (left) and *New* (right) devices with equal cathode area covered by shorts (3D device simulation).

Fig.2.1b): ON-state distribution of current density of devices from Fig.2.1a) with top layer 50 μm below the cathode surface.

The device simulation further confirms, that by increasing the number of shorts we can create the technology curve $Q_{rr} - V_T$ likewise we normally do by increasing the dose of electron irradiation (see Fig.2.2). The simulation, which is based on drift-diffusion approximation of electron-hole transport, includes standard settings for silicon with Shockley-Read-Hall model of generation-recombination, and models of Auger recombination, impact ionization and bandgap narrowing.

We show experimentally below that, in addition to improved technology curve, we can improve the dV/dt capability and circuit commutated recovery time t_q while keeping the rest of parameters unchanged. The smaller short diameter used in this concept also facilitates its placement around the regions with complex shape like the amplifying gate structures, which allows us to keep the gate triggering likewise the dV/dt and di/dt capability at required level.

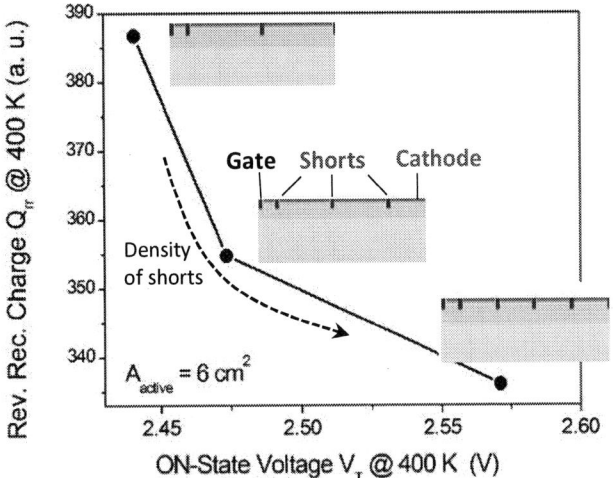

Fig.2.2: Simulated trade-off curve Q_{rr}-V_T by changing the cathode area coverage by shorts of similarly small size as in Fig.2.1b). V_T is simulated at I_T = 4 kA, T = 400 K. Q_{rr} is simulated at I_T = 2 kA, T = 400 K, with snubber parameters R_s = 20 Ω, C_s = 5 µF.

3. Experimental Results and Discussion

Figs.3.1 and 3.2 compare the measured trade-off curves of *Classical* and *New* PCTs with active area of about 6 cm^2 for 1.8 kV and 2.8 kV voltage classes. The relative shorted area in % is shown as a parameter. Figs.3.2 and 3.4 show that the similar curves for larger devices (three times in this case). The advantage of having a smaller short diameter in the *New* device is that it improves the cathode utilization for current conduction in agreement with the Fig.2.1 and gives us a lower V_T for the same shorted area. This feature allows us to place more shorts to reduce Q_{rr} while keeping the V_T equally low.

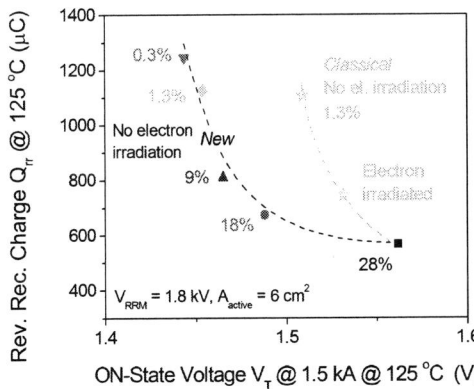

Fig.3.1: Trade-off curve Q_{rr} – V_T of *Classical* and *New* 1.8 kV PCTs. Active area ≈ 6 cm^2.

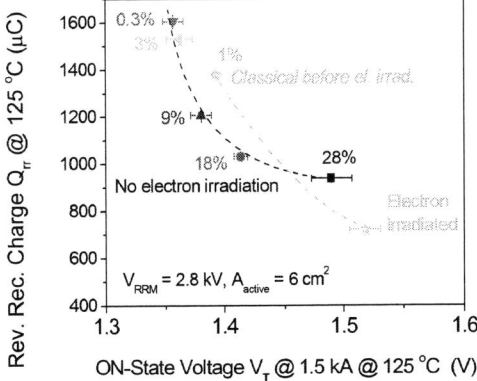

Fig.3.2: Trade-off curve Q_{rr} – V_T of *Classical* and *New* 2.8 kV PCTs. Active area ≈ 6 cm^2.

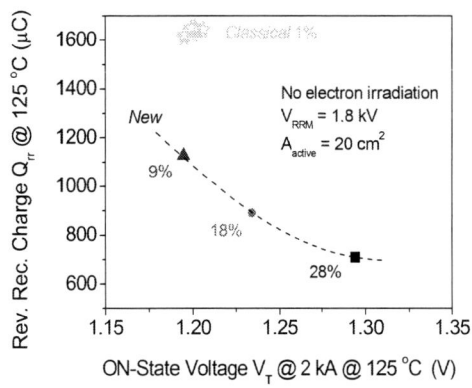 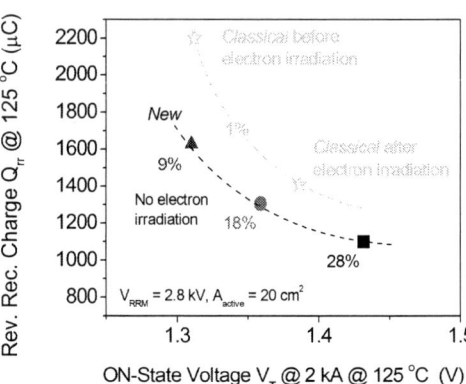

Fig.3.3: Trade-off curve $Q_{rr} - V_T$ of *Classical* and *New* 1.8 kV PCTs. Active area ≈ 20 cm².

Fig.3.4: Trade-off curve $Q_{rr} - V_T$ of *Classical* and *New* 2.8 kV PCTs. Active area ≈ 20 cm².

The possibility to replace the electron irradiation by the densely placed shorts is obvious from the figures above. On one hand, it is clear that the *Irradiation Free Concept* brings a better technology curve. On the other hand, it is clear that the beneficial effect vanishes at certain percentage of shorted area at which the technology curve of the *Irradiation Free concept* crosses that of the *Classical* one achieved by electron irradiation. This limit depends on voltage class (wafer thickness) and device specification for Q_{rr} and V_T. However, for the two voltage classes and device sizes the mentioned crossing does not appear before the devices run out from their typical $Q_{rr} - V_T$ specifications for industrial PCT.

At glance, the *Irradiation Free Concept* (cathode short engineering) is not functional at high-voltage PCTs processed at thicker wafers. This is illustrated for the 8.5 kV PCT in Fig.3.5, which illustrates the limited control of Q_{rr} by changing the short density between 0.3 and 12 % in comparison with the electron irradiation using the short density of about 1 %. The Q_{rr} values achieved by increasing the shorted area are not much different from the Q_{rr} values of unirradiated ones with low shorted area.

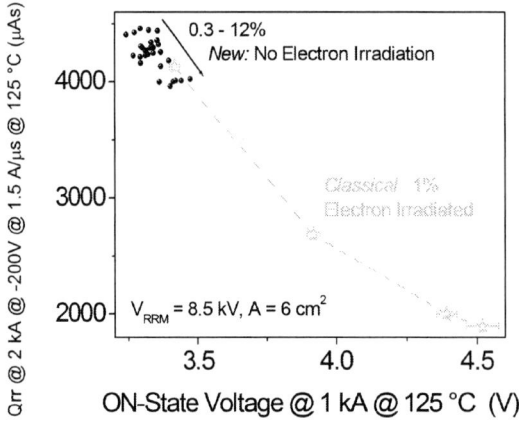

Fig.3.5: Trade-off curve $Q_{rr} - V_T$ of *Classical* and *New* 8.5 kV PCTs. Active area ≈ 6 cm².

Fig.3.6 and 3.7 illustrate how can the dV/dt capability benefit from the New concept. The graphs take into account the PCTs with both single gate and amplifying gate structures and they utilize an equal cathode area. The test was performed at harder conditions in comparison with the ones specified at typical data sheets to obtain a better sorting of device capability. This consists in increasing the maximal applied forward voltage V_D from the standard specification test level at 2/3 of the V_{DRM} value to the nominal value of V_{DRM} = 1.8 and 2.8 kV.

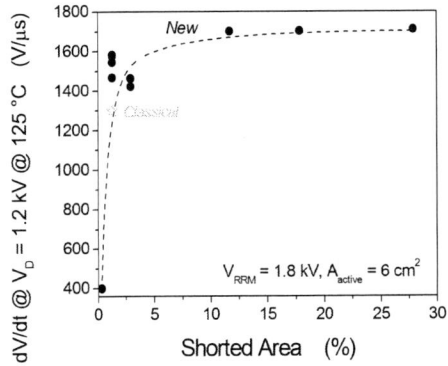 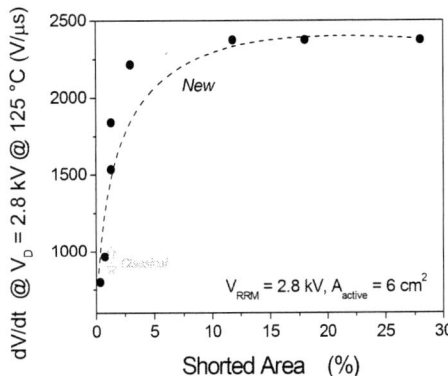

Fig.3.6: dV/dt capability of *Classical* and *New* 1.8 kV PCTs. Active area ≈ 6 cm^2.

Fig.3.7: dV/dt capability of *Classical* and *New* 2.8kV PCTs. Active area ≈ 6 cm^2.

The increased short density increases the dV/dt capability likewise does the electron irradiation. The measurements show that the dV/dt capability saturates at about the relative shorted area of ≈ 10 %. This value already lowers the reverse recovery charge Q_{rr} to a typical datasheet value obtained by electron irradiation. Moreover, this shorting density brings the benefit of lower V_T compared to the electron irradiation.

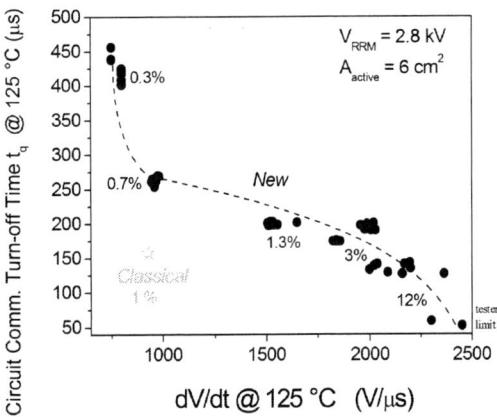

Fig.3.8: dV/dt capability vs. circuit commutated recovery t_q of *Classical* and *New* 2.8 kV PCTs.

Fig.3.8 brings the circuit commutated recovery time t_q into play. Note that the t_q for the *Classical* and *New* devices in Fig.3.8 do not appear at the same curve, because they have different gate structures in this example. For the density of shorts equal or higher than ≈10 %, we hit the limit of our tester for both the dV/dt and t_q. For the density of shorts below

1 %, the dV/dt capability and circuit commutated recovery time t_q get outside a typical datasheet specification. On the other hand, the leakage current under both forward and reverse biases remains equally low regardless of short density. This indicates the unrivalled importance of cathode shorts for the dynamic performance of contemporary PCTs.

4. Conclusions

The increasing shorted area of the *Irradiation Free Concept* has a similar effect as the increasing dose of electron irradiation of the *Irradiation Concept*. The cathode shorts of state-of-the-art PCTs are indispensable for the dynamic performance (circuit commutation time t_q and dV/dt), while they also improve the technology curve. It is proved for phase controlled thyristors of 1.8 and 2.8 kV voltage classes with sizes between 5 and 20 cm^2.

In addition to the reduced cost of processing, it allows one to reduce the safety margin on surge current I_{TSM} rating, because there is no fear of annealing the radiation defects from electron irradiation during repetitive overload conditions at the ON-state.

5. Reference

[1] N. Holonyak, The Silicon p-n-p-n Switch and Controlled Rectifier (Thyristor), IEEE Transactions on Power Electronics, Vol.16, 2001, pp. 8 – 16.

[2] B. J. Baliga, E. Sun, Lifetime Control in Power Rectifiers and Thyristors using Gold, Platinum and Electron Irradiation, International Electron Device Meeting Technical Digest, 1976, pp. 495 -498.

[3] P. D. Taylor, Thyristor Design and Realization, John Wiley & Sons Ltd., Chichester, 1987

[4] ABB Technology AG, Zurich, Switzerland, Rückwärtssperrender Thyristor ohne strahlungsinduzierte Effekte, Utility model No. DE 20 2015 101 476 U1 2015.05.21

Experimental results of a Large Area (91mm) 4.5kV "Bi-mode Gate Commutated Thyristor" (BGCT)

Thomas Stiasny[+], Martin Arnold[+], Umamaheswara Reddy Vemulapati[++], Munaf Rahimo[+], Jan Vobecky[+], Christian Kähr[*], Norbert Hoffmann[*]

[+] ABB Switzerland Ltd., Semiconductors, Lenzburg, Switzerland

[++] ABB Switzerland Ltd., Corporate Research, Baden-Dättwil, Switzerland

[*] Fachhochschule Nordwestschweiz, Windisch, Switzerland

thomas.stiasny@ch.abb.com

Abstract

In this paper, we present the experimental results of the 91mm, 4.5kV Bi-mode Gate Commutated Thyristor (BGCT). The BGCT is a new type of Reverse Conducting Integrated Gate Commutated Thyristor (RC-IGCT) [1]. In this work, we have also compared the results of the BGCT with that of the 91mm, 4.5kV conventional RC-IGCT both electrically (i.e. on-state and turn-off characteristics) and thermally (i.e. temperature distribution during the conduction of the device). The experimental results show that the BGCT has a better technology curve, improved safe operation area, soft reverse recovery behavior and lower leakage current compared to conventional RC-IGCT. The thermal simulation results performed with Abaqus show that BGCT has a better thermal distribution i.e. more uniform temperature distribution throughout the wafer in both GCT- and diode-modes of operation.

1. Introduction

The IGCT has been established as the device of choice for high power applications such as medium voltage drives, pumped hydro, STATCOMs, railway interties and power quality applications [2-3]. The main features of the IGCT are:

- It conducts like a Thyristor in the on-state resulting in low conduction losses.
- Turns-off like an Insulated Gate Bipolar Transistor (IGBT) in open base pnp transistor mode (i.e. hard switching turn-off capability due to the integration of the low inductive gate unit).
- Good device scalability for increased current rating [4-5] and increased blocking voltage rating [6].
- Its hermetic press pack design for good reliability in the field with respect to the power-semiconductor device protection and load cycling capability.

Today, IGCTs have been optimized for Voltage Source Inverter (VSI), Current Source Inverter (CSI) and event switching (solid state circuit breaker) applications and are available as Asymmetric, Symmetric (Reverse Blocking), and Reverse Conducting devices. The paper [7] reports the major developments of the IGCT technology, such as high power technology (to increase safe operation area), high temperature operation (140°C), and lower losses towards 1V on-state.

For VSI topologies, the Asymmetric IGCT has the highest power level for a given wafer size while the RC-IGCT provides compactness by integrating a diode on the same GCT wafer [8]. However, in the state-of-the-art RC-IGCTs, the GCT and diode are integrated into a single wafer but they are fully separated from each other as shown in Fig. 1a. Consequently, in the RC-IGCT, the utilization of the silicon area is limited in the GCT region when operating in GCT-mode and in the diode region when operating in the diode-mode. The BGCT, on the other hand, features an interdigitated integration of diode- and GCT-areas as shown in Fig. 1b to Fig. 1d. The interdigitated integration of the GCT- and diode- in the BGCT offers the following advantages over conventional RC-IGCT:

- An improved diode as well as GCT area.
- Better thermal distribution.
- Soft turn-off/reverse recovery behavior
- Lower leakage current.

The BGCT concept has been demonstrated first experimentally with 38mm, 4.5kV devices [9]. The experimental results show that the BGCT has a better technology curve in GCT-mode and almost matches the technology curve of the RC-IGCT in diode-mode while maintaining other advantages such as soft reverse recovery behavior and significantly lower leakage current (about 4 times lower leakage current at 2.8kV, 125°C) due to the distributed anode shorts at the backside as shown in Fig. 1d.

In this paper, we demonstrate the BGCT concept with a large area i.e. 91mm, 4.5kV BGCT for high power electronics applications (the current handling capability increases with enlarged device area).

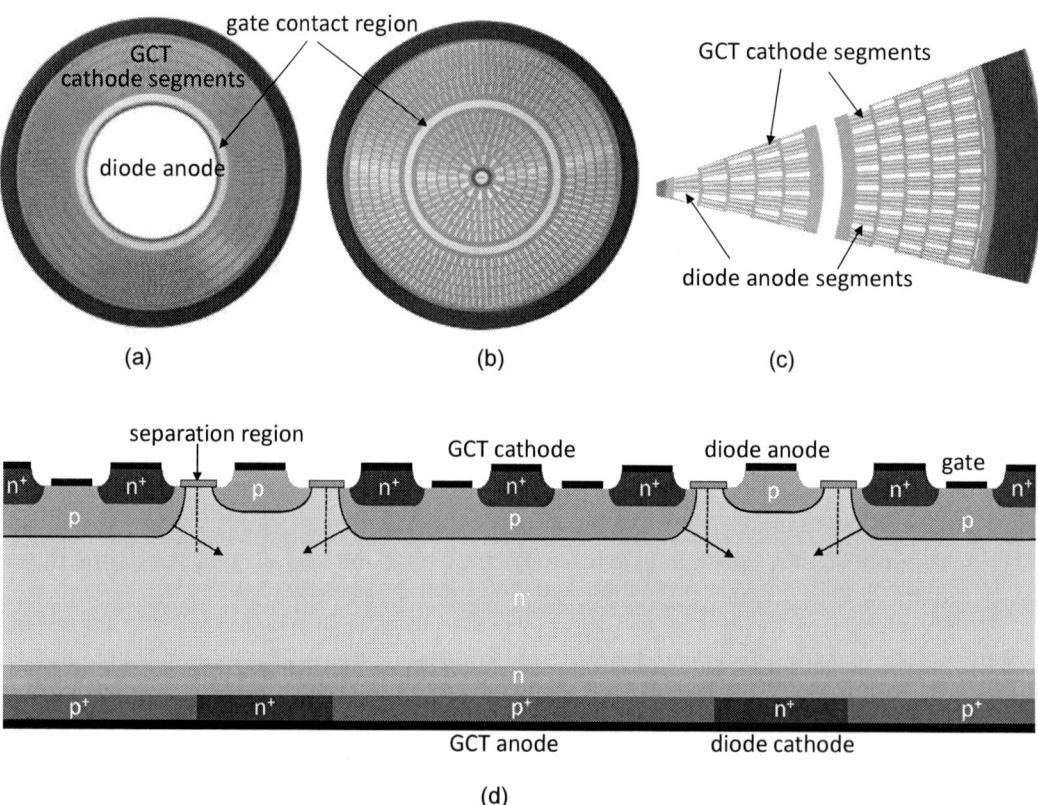

Fig. 1. (a) top-view of conventional 91mm, 4.5kV RC-IGCT, (b) top-view of 91mm, 4.5kV BGCT, (c) zoomed part of the BGCT; the wide white rectangular/triangular regions are the diode anode segments and the ones between the diode anode segments are the GCT cathode segments, (d) schematic cross section of a BGCT with shallow diode anode.

2. Experimental Results in GCT-mode

2.1. On-state Characteristics

The on-state characteristics of a 91mm, 4.5kV BGCT at 25°C & 125°C after carrier lifetime engineering are as shown in Fig. 2. For the carrier lifetime engineering, the combination of both homogenous lifetime control and local lifetime control techniques is used to improve the technology curves of the BGCT both in GCT- and diode-modes. It is worth mentioning that, the local lifetime control is applied only at the backside of the device (diode cathode side).

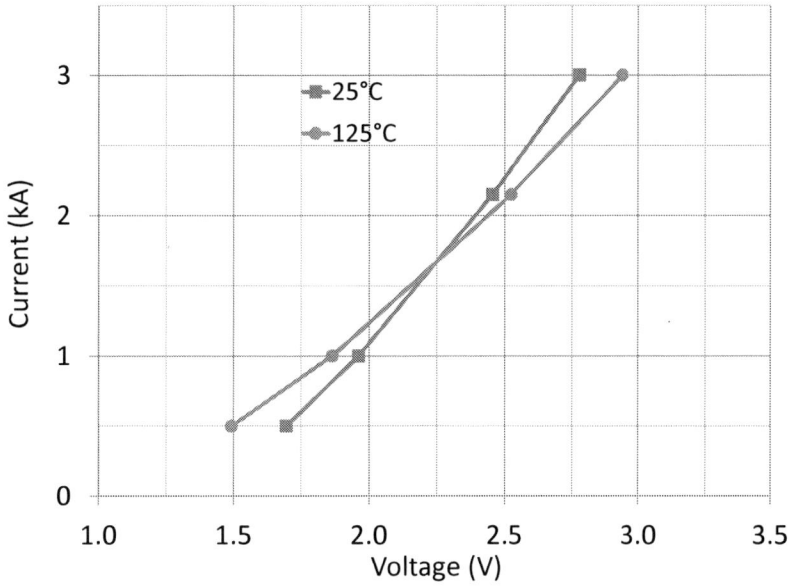

Fig. 2. The measured on-state characteristics of a 91mm, 4.5kV BGCT in GCT-mode at 25°C & 125°C.

2.2. Turn-off Characteristics

The switching circuit shown in Fig. 3 is used to test the turn-off behavior of the BGCT and RC-IGCT. We have used the same gate control unit for both cases. The comparison of the turn-off waveforms between BGCT and RC-IGCT is shown in Fig. 4. It can be seen from Fig. 4 that the BGCT shows soft turn-off behavior (around 4µs) compared to RC-IGCT.

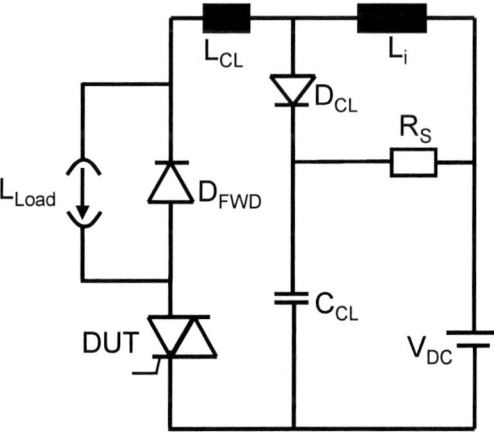

Fig. 3. The switching circuit used to test the turn-off behavior of the BGCT and RC-IGCT.

© VDE VERLAG GMBH · Berlin · Offenbach

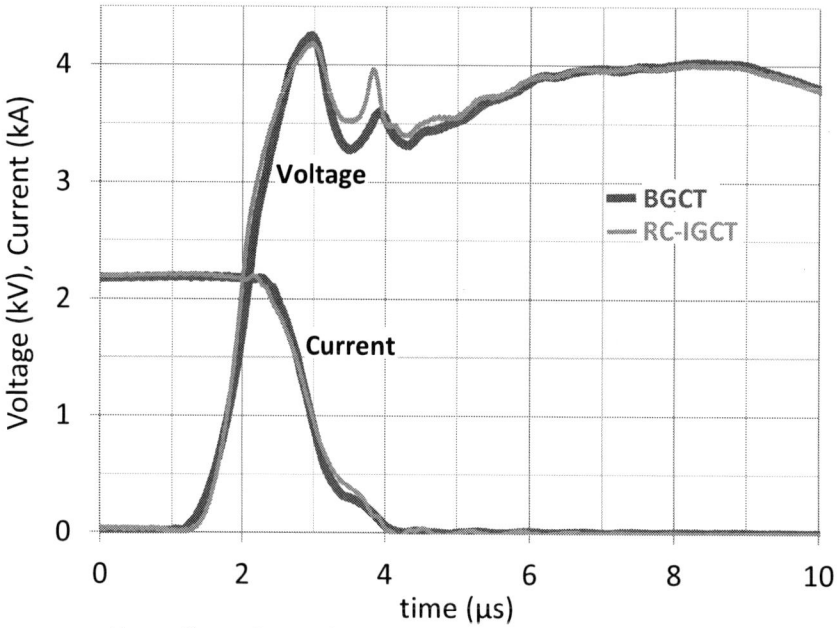

Fig. 4. The measured turn-off waveforms of a 91mm, 4.5kV BGCT and 91mm, 4.5kV RC-IGCT at 2.2kA, 2.8kV and 115°C.

2.3. Maximum Turn-off Current Capability

The 91mm, 4.5kV BGCT offers significantly higher safe operation area (SOA) i.e. higher maximum controllable turn-off current capability compared to the conventional RC-IGCT at a given dc-link voltage. Fig. 5 illustrates that the BGCT is able to successfully turn-off the currents as high as 4.4kA at 2.8V and 115°C, which is more than 1.3 times higher than the maximum turn-off current capability of the RC-IGCT.

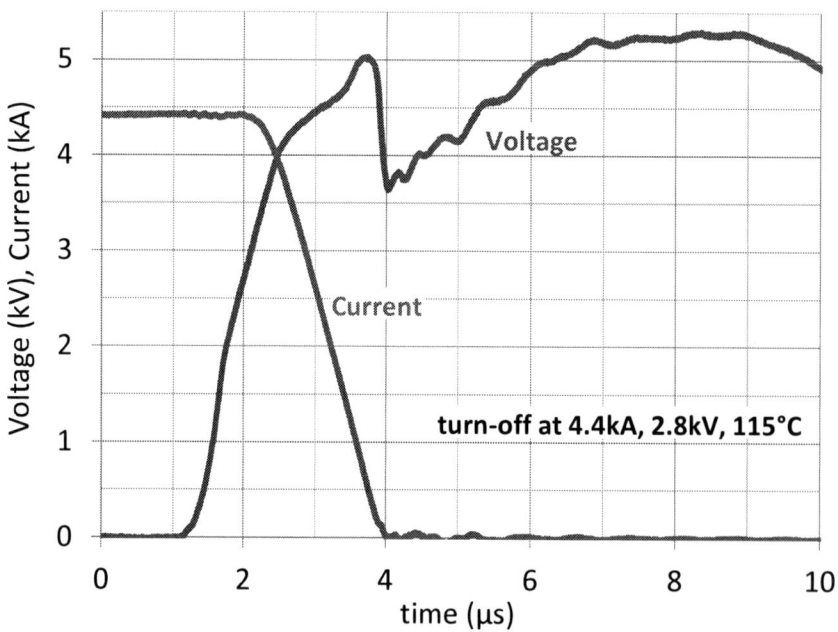

Fig. 5. The measured SOA turn-off waveforms of a 91mm, 4.5kV BGCT. The measurement results showing successful turn-off current capability up to 4.4kA at 2.8kV and 115°C, which is 1.3 times higher than the RC-IGCT.

© VDE VERLAG GMBH · Berlin · Offenbach

3. Experimental Results in Diode-mode

3.1. On-state Characteristics

The on-state characteristics of a 91mm, 4.5kV BGCT at 25°C & 125°C after carrier lifetime engineering are as shown in Fig. 6. As explained before, for the carrier lifetime engineering, the combination of both homogenous lifetime control and local lifetime control techniques is used in BGCT. In the RC-IGCT case, local lifetime control is used on both sides of the device only in the diode part (i.e. anode and cathode sides of the diode) along with the homogeneous lifetime control. The main reason to use local lifetime control at the anode side of the diode in conventional RC-IGCT is to reduce the reverse recovery peak current, I_{RM} and hence limit the high power dissipation ($I_{RM} * V_{DC}$) and thereby protect the diodes from high power failures [10].

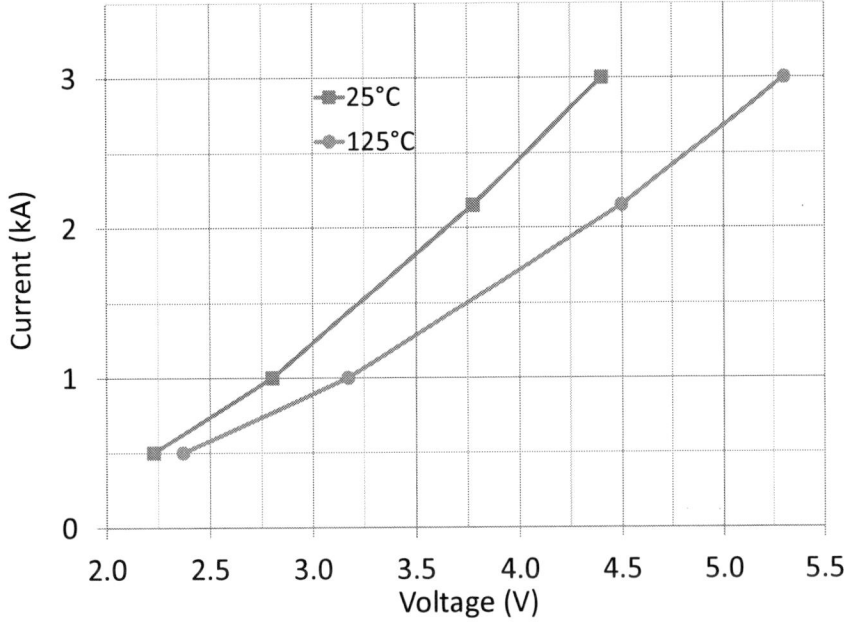

Fig. 6. The measured on-state characteristics of a 91mm, 4.5kV BGCT in diode-mode at 25°C & 125°C.

3.2. Reverse Recovery Characteristics

In this section, the reverse recovery characteristics of a 91mm, 4.5 kV BGCT are compared with that of the conventional RC-IGCT in diode-mode. It can be seen from Fig. 7 that, the BGCT shows a softer reverse recovery behavior compared to RC-IGCT. The main reason is the distributed p+- and n+-regions on the backside i.e. the diode cathode side (see Fig. 1d) as explained in [9]. It is worth noting that the BGCT in diode-mode is tested only up to dc-link voltages of 2kV to avoid destruction of the device from high power failure as no local lifetime is used at the front side of the device (i.e. diode anode side).

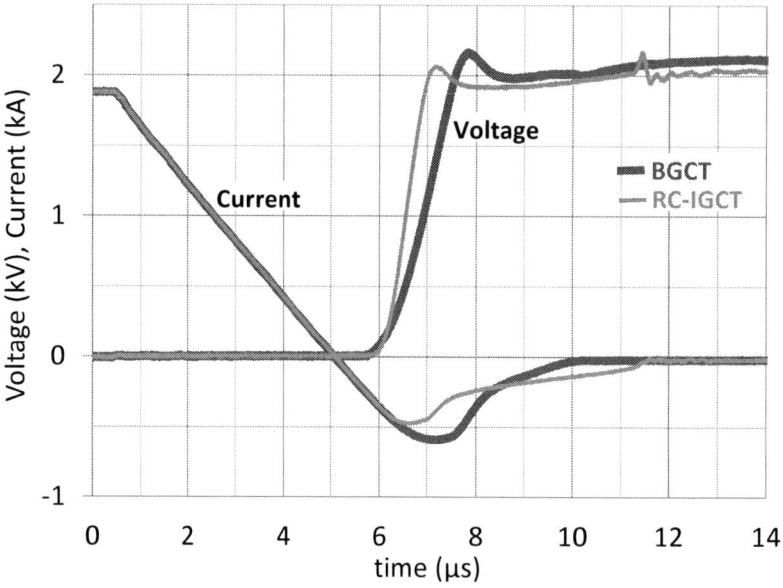

Fig. 7. The measured reverse recovery characteristics of a 91mm, 4.5kV BGCT and 91mm, 4.5kV RC-IGCT at 2kA, 1.9kV and 115°C.

4. Technology Trade-Off of BGCT and RC-IGCT

4.1. GCT-mode

Like in 38mm, 4.5kV BGCT [9], the 91mm, 4.5kV BGCT has a better technology curve (turn-off losses vs. on-state voltage drop) than that of the RC-IGCT in GCT-mode. For the same turn-off losses, the on-state voltage drop of the BGCT is more than 100mV (5%) lower compared to that of the RC-IGCT. This is likely due to more available area in BGCT compared to that of the GCT in RC-IGCT.

4.2. Diode-mode

The 91mm, 4.5kV BGCT has a better technology curve (reverse recovery losses vs. on-state voltage drop) than that of the RC-IGCT also in diode-mode. Like in 38mm, 4.5kV BGCT [9], the technology curve of the 91mm, 4.5kV BGCT in diode-mode improves with increasing diode anode efficiency. For the same reverse recovery losses, the on-state voltage drop of the BGCT is more than 200mV (6%) lower compared to that of the RC-IGCT.

5. Thermal behavior of BGCT and RC-IGCT

Fully coupled (i.e. electrical, thermal, mechanical) simulations have been performed with Abaqus to analyze the thermal behavior of the BGCT and RC-IGCT in both GCT- and diode-modes of operation. As expected, the simulation results show that BGCT has a better thermal distribution (i.e. more uniform temperature distribution throughout the wafer) due to the interdigitated integration of GCT- and diode-areas in the BGCT (utilization of whole silicon volume in both GCT- and diode-modes of operation), compared to that of the RC-IGCT as shown in Fig. 8. The better thermal distribution leads to both lower losses and enhanced reliability.

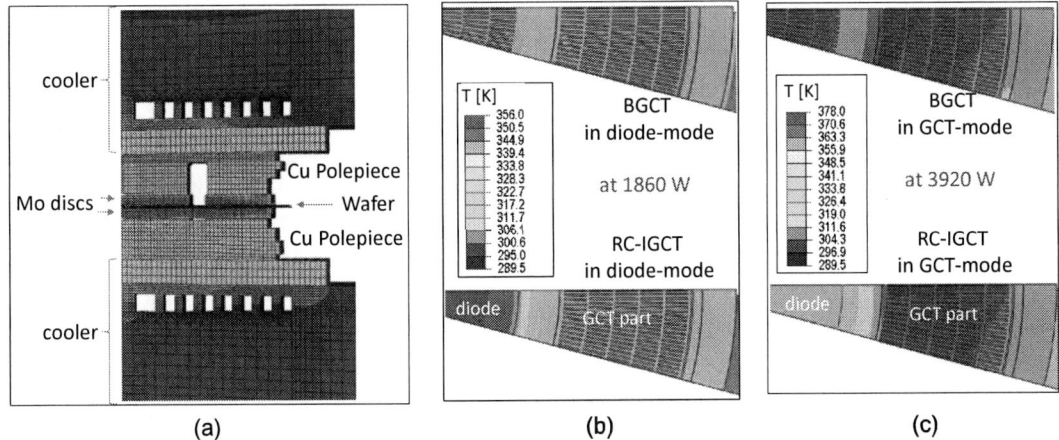

(a)   (b)   (c)

Fig. 8. Thermal simulations of BGCT and RC-IGCT (a) simulation setup (b) temperature distribution in diode-mode (c) temperature distribution in GCT-mode. The temperature is more uniformly distributed throughout the wafer in BGCT and as a result, the maximum junction temperature is lower in BGCT.

6. Conclusions

We have presented the experimental results of a 91mm, 4.5kV BGCT. The BGCT is a new type of reverse conducting IGCT with improved performance (i.e. better technology curve, soft reverse recovery/turn-off behavior and lower leakage current) than that of the conventional RC-IGCT. Furthermore, the experimental results show that the 91mm, 4.5kV BGCT has significantly higher turn-off current capability (i.e. >1.3 times) compared to RC-IGCT. The thermal simulation results show that the BGCT has a better thermal distribution compared to RC-IGCT due to utilization of whole silicon volume in both GCT- and diode-modes of operation.

References

1. Vemulapati, U., Bellini, M., Arnold, M., Rahimo, M. and Stiasny, T.: The concept of Bi-mode Gate Commutated Thyristor-a new type of reverse conducting IGCT, Proc. ISPSD, Bruges, 2012, pp. 29-32

2. Gruening, H., Oedegard, B., Weber, A., Carroll, E., Eicher, S.: High power hard-driven GTO module for 4.5 kV/3kA snubberless operation, Proc. PCIM Europe, Nürnberg, Germany, 1996, pp.169-183

3. Klaka, S., Frecker, M. and Grüning, H.: The Integrated Gate Commutated Thyristor: A new high efficiency, high power switch for series or snubberless operation, Proc. PCIM Europe, Nürnberg, 1997

4. Yamaguchi, Y. et al.: A 6 kV/5 kA reverse conducting GCT, IEEE IAS, Chicago 2001

5. Wikström, T. et al.: The 150 mm RC-IGCT: a device for the highest power requirements, Proc. ISPSD, Waikoloa, HI, 2014, pp. 91-94

6. Nistor, I., Wikström, T., Scheinert, M., Rahimo, M. and Klaka, S.: 10kV HPT IGCT rated at 3200A, a new milestone in high power semiconductors, Proc. PCIM Europe, Stuttgart, Germany, 2010, pp. 467-471

7. Vemulapati, U., M., Rahimo, M., Arnold, Wikström, T., Vobecky, J., Backlund, B., Stiasny, T.: Recent Advancements in IGCT Technologies for High Power Electronics Applications, EPE'15 ECCE Europe, Geneva, 2015

8. Linder, S., Klaka, S., Frecker, M. and Zeller, H.: A new range of reverse conducting gate-commutated thyristors for high voltage, medium power applications, Proc. EPE, Trondheim, 1997

9. Vemulapati, U. et al.: An experimental demonstration of a 4.5 kV "Bi-mode Gate Commutated Thyristor" (BGCT), Proc. ISPSD, Hong Kong, 2015, pp. 109-112

10. Linder. S.: Power Semiconductors, EPFL Press, 1st edn. 2006.

© VDE VERLAG GMBH · Berlin · Offenbach

Effect of *Self Turn*-ON during turn-ON of HV-IGBTs

Patrick Münster, University of Rostock, Germany, patrick.muenster@uni-rostock.de
Daniel Lexow, University of Rostock, Germany, daniel.lexow@uni-rostock.de
Hans-Günter Eckel, University of Rostock, Germany, hans-guenter.eckel@uni-rostock.de

Abstract

During the turn-ON of a high-voltage IGBT a *Self Turn*-ON can be observed, which is analysed in this paper. This positive feedback, which has not been observed in HV-IGBTs before, is caused by a *Self-Charging Displacement Current*. This effect influences the turn-ON process and can lead to oscillations of the gate-emitter voltage. The results are based on high-voltage measurements on single-chip modules and semiconductor simulations.

1. Introduction

For the development of high performance IGBT drivers, it is essential to understand IGBT switching behaviour. Especially if there are any feedbacks through the IGBT itself, which can be further amplified through an insufficient driver design. Such effects, like the *parasitic turn*-ON *due to stray inductances* [1] or *Self Turn*-OFF [2] have been described before. The consideration about these effects becomes more important for the control of paralleled IGBTs to avoid current imbalance [3, 6]. Concerning the turn-ON the *Self-Charging Displacement Current* [4, 5] has been introduced for IGBTs with low blocking voltage, but not for HV-IGBTs.

2. Self Turn-ON during turn-ON

In this paper the *Self-Charging Displacement Current* is investigated regarding positive feedbacks on the IGBT turn-ON switching behaviour. Therefore the IGBT is turned ON with a charge pulse. During the switching transient, the IGBT gate is high-resistant disconnected from the gate-driver unit.

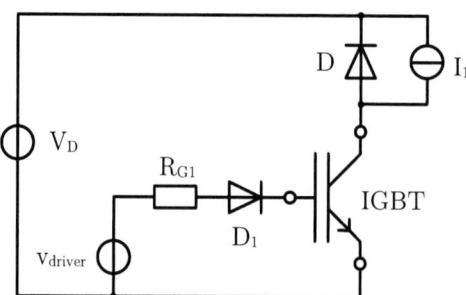

Fig. 1: Self Turn-ON simulation circuit

PCIM Europe 2016, 10 – 12 May 2016, Nuremberg, Germany

2.1. Simulation results

Simulations are carried out with SYNOPSYS TAURUS MEDICI. The circuit is displayed in Fig. 1. V_D represents the DC link voltage and I_1 the load current, which is paralleled to a free-wheeling diode.

The gate-drive unit consists of a diode D_1, a gate resistor R_{G1} and a voltage source, which applies v_{driver}. Regarding the physical simulation a 6.5 kV/600 A planar IGBT model and a 6.5 kV/750 A trench IGBT model are used. For both models different gate resistors are chosen to achieve almost the same diode stress at normal turn-ON. R_{G1} is set to 10 Ω for the planar respectively 2.5 Ω for the trench model. To eliminate other effects on the switching behaviour and to evaluate the positive feedbacks through the IGBT, no parasitic inductances in the commutation path are used. For an easy comparison of both IGBT simulations, their collector-emitter voltages v_{ce} are normalised to V_D, their collector currents I_{c1} to their nominal current I_{nom} and their gate-emitter v_{ge} as well as their driver voltage v_{driver} to $v_{driver,on}$.

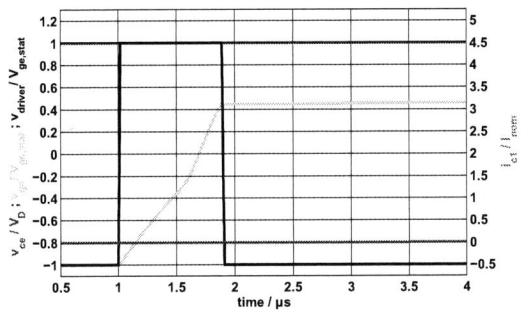

(a) Charge time = 890 ns; v_{ge} below the level to trigger *Self Turn*-ON

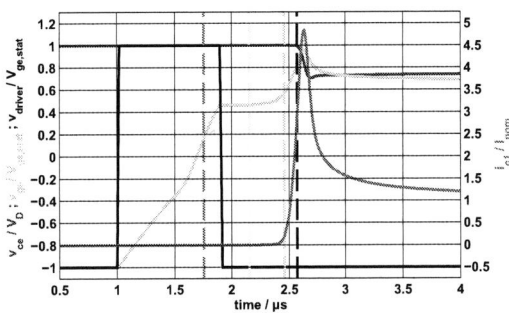

(b) Charge time = 900 ns; slightly increased v_{ge} triggers *Self Turn*-ON; v_{ge} increases even with decoupled gate-drive unit

Fig. 2: Simulation: Turn-ON of a planar field-stop IGBT with decoupled gate-drive unit (R_{G1} = 10 Ω) at V_D=3.6 kV and I_1=600 A

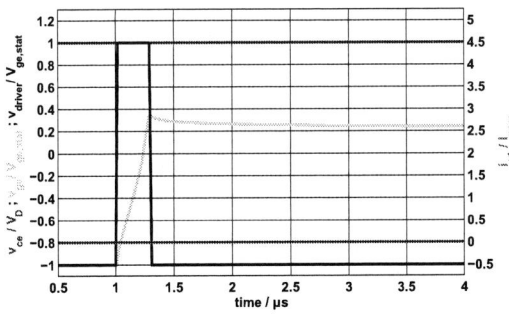

(a) Charge time = 290 ns; v_{ge} does not exceed threshold voltage; thus *Self Turn*-ON is not triggered

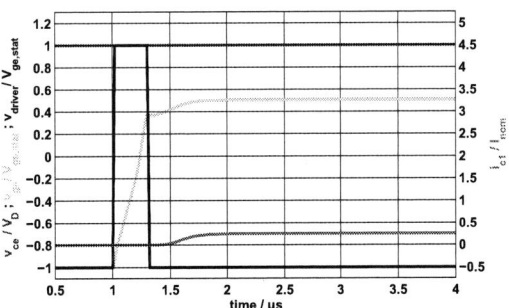

(b) Charge time = 300 ns; v_{ge} exceeds threshold voltage and triggers *Self Turn*-ON; again v_{ge} increases with decoupled gate-drive unit

Fig. 3: Simulation: Turn-ON of a trench field-stop IGBT with decoupled gate-drive unit (R_{G1} = 10 Ω) at V_D=3.6 kV and I_1=750 A

© VDE VERLAG GMBH · Berlin · Offenbach

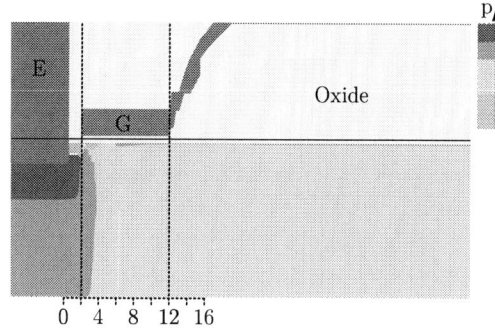
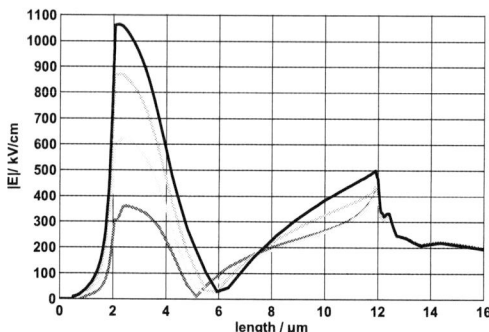

(a) Structure of the planar 6.5 kV field-stop IGBT gate including black cut line with illustrated hole concentration at the moment of the dashed cyan line in Fig. 2(b)

(b) Magnitude of the electric field along the cut line marked in 4(a) referring the instants of time illustrated in Fig. 2(b)

Fig. 4: Results of the physical simulation of the planar field-stop IGBT at V_D=3.6 kV and I_1=600 A with decoupled gate-drive unit in Fig. 2(b)

During the simulation v_{driver} (black) is switched from -15 V to 15 V ($v_{driver,on}$) at t=1 μs and v_{ge} (green) increases, as shown in Fig. 2(a), 2(b), 3(a) and 3(b). v_{driver} is switched back to -15 V after 890 ns displayed in Fig. 2(a) and after 290 ns displayed in Fig. 3(a). Hence, v_{ge} remains at the previous voltage level, because of the blocking diode D_1. In both cases the IGBTs are still switched OFF.

In Fig. 2(b) the applied voltage is switched back to -15 V after 900 ns. As a result v_{ge} exceeds the threshold voltage, the n-channel is built and electrons flow from the emitter to the backside p^+ collector. There holes are injected and accumulated below the gate. In Fig. 4(a) the area of the hole accumulation (below the gate) is orange highlighted and leads to an increasing potential in this region [4, 5]. After the gate is high-resistant disconnected, the n-channel is still open, electrons flow from the emitter to the collector and holes accumulate below the gate oxide. The potential in this region is further raised. As a consequence the magnitude of the electric field, which is shown in Fig. 4(b) along the cut line in the gate-oxide (Fig. 4(a)), increases in the region of the hole accumulation between 8 μm and 12 μm. The instants of time of the cuts in Fig. 4(b) are pursuant marked in Fig. 2(b) with the coloured dashed lines. Thus, a displacement current through the gate-oxide is caused, which charges the gate and ends in the *Self Turn*-ON process. The increasing v_{ge} leads to a decreasing resistance of the n-channel and i_{c1} rises. After i_{c1} exceeds the load current, the reverse-recovery charge of the diode D gets removed and the diode starts to take over the DC link voltage. Now the displacement current, caused by the hole carriers, gets superposed through a negative displacement current caused by negative dv_{ce}/dt at the Miller capacitance. Thus, v_{ge} decreases. Finally the IGBT carries the load current desaturated.

Compared with Fig. 3(a) the trench IGBT is switched OFF after 300 ns, illustrated in Fig. 3(b). Again v_{ge} rises as a result of the effects and processes mentioned above. As a consequence i_{c1} increases also, but compared with the planar IGBT, the intensity of the *Self Turn*-ON is reduced. Obviously the gate structure of the trench IGBT attenuates the influence of the *Self Turn*-ON on the gate-emitter voltage.

2.2. Experimental results

If an usual 6,5kV-IGBT module is used to validate the simulation results, the effect of the *Self Turn*-ON is superposed by other effects. Compared to the simulation the influence of the commutation inductance cannot be neglected. During the switching transient the di_{c1}/dt leads to a voltage drop over the parasitic inductance, which reduces the collector-emitter voltage of the IGBT. This reduced collector-emitter voltage causes a discharge of the Miller capacitance, which superposes the mechanisms of the *Self Turn*-ON. The effective stray inductance was reduced of approximately 24 times by the use of a corresponding single-chip IGBT module, which leads to a less inhibited *Self-Charging Displacement Current*. A double-pulse test bench with a load inductance of 10 mH was used to create turn-ON transients with a nominal current of I_{nom}=31.25 A. The gate circuit consists of a gate driver with three output states. Hence, the gate can get charged with a scaled turn-ON resistor $R_{G,on}$=19.5 Ω up to 15 V. The driver has two turn-OFF states. The first switches the IGBT OFF with a scaled turn-OFF resistor $R_{G,off}$=119 Ω and the second consists of a resistor $R_{G,disc}$=100 kΩ, which is used to disconnect the gate high-resistant.

Fig. 5(a) and 5(b) illustrate the measuring results. At t=0.75 μs the gate gets charged through the applied voltage v_{driver} (black) and v_{ge} (green) starts to rise. After 400 ns the gate is disconnected and finally turned OFF after further 5 μs. The dashed line in Fig. 5(a) marks the zero crossing of the gate current (cyan), which was measured by a Pearson transducer. According to Fig. 5(b) the gate is disconnected after 500 ns. Under these conditions v_{ge} exceeds the threshold voltage and the IGBT starts to conduct. As well as in Fig. 5(b), the first dashed line marks the zero crossing of i_g. Although the gate current ends, v_{ge} rises. This is affected by *Self-Charging Displacement Current* and demonstrates the *Self Turn*-ON mentioned above. In comparison to the simulation, the commutation inductance leads to a slight Miller feedback through the drop of v_{ce} (blue), which inhibits the *Self-Charging Displacement Current*. The second dashed line marks the point, when i_{c1} exceeds the load current. Now the charge of the diode D should be removed to take over the DC link voltage, but hence to the lower v_{ge} the IGBT remains desaturated and takes the DC link voltage.

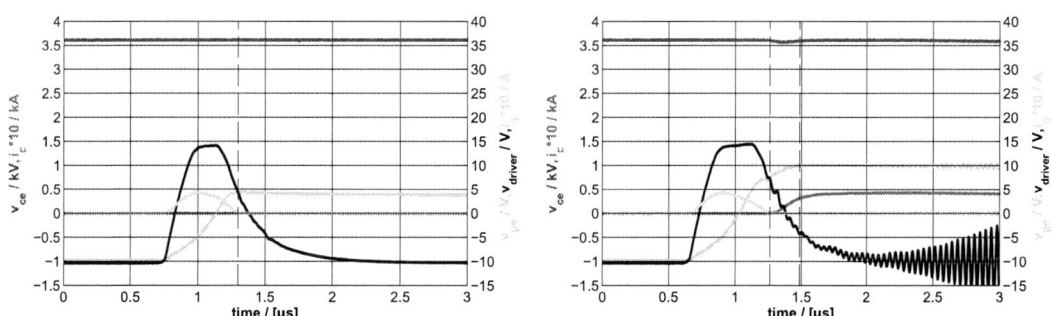

(a) Charge time = 400 ns; compared to simulation Fig. 2(a) and Fig. 3(a); *Self Turn*-ON is not triggered

(b) Charge time = 500 ns; v_{ge} exceeds the threshold voltage; *Self Turn*-ON is triggered

Fig. 5: Measurement: Turn-ON trench field-stop IGBT with high-resistant disconnected gate ($R_{G,on}$ = 19.5 Ω, $R_{G,off}$ = 119 Ω, $R_{G,disc}$ = 100 kΩ) at V_D=3.6 kV and I_1=31.25 A

3. Self Turn-ON during low-inductive short circuit type 1

Due to the current dependency of the *Self-Charging Displacement Current* [4, 5], the short circuit behaviour of the *Self Turn*-ON is investigated. Compared to section 2, the IGBT is turned ON with a charge pulse, but turned OFF with a low-ohmic resistor during the switching transient to avoid further failure mechanisms.

3.1. Simulation results

Again simulations are made with SYNOPSYS TAURUS MEDICI and for the physical simulation the 6.5 kV/750 A trench IGBT model is used. Fig. 6 illustrates the circuit to simulate a low-inductive short circuit type 1. Compared to Fig. 1 the use of V_D, $v_{driver,1}$, $R_{G1} = 2.5\,\Omega$ and D_1 remains. Additionally $v_{driver,2}$, R_{G2} and D_2 are used and R_{G2} is set to $22.5\,\Omega$, which value fits to an usual application. The stray inductance $L_\sigma = 100\,nH$ is added and corresponds to the stray inductance of the measurement mentioned in the following section. For a simplified comparison of the simulations v_{driver} (black), v_{ge} (green), i_{c1} (red) and v_{ce} (blue) are normalised as mentioned in the previous section. Concerning to the simulation, displayed in Fig. 7, the rising edge of v_{driver} marks the time, when $v_{driver,1}$ and $v_{driver,2}$ are switched ON. Due to the diodes D_1 and D_2, only R_{G1} is effective and turns ON the IGBT.

A typical low-inductive short circuit at usual operational conditions ($V_D = 3.6\,kV$) is shown on Fig. 7(a). $v_{driver,1}$ is switched from $-20\,V$ to $20\,V$ at $t=1\,\mu s$ and v_{ge} starts to rise. After v_{ge} exceeds the threshold voltage, i_{c1} increases and a voltage drop at the stray inductance L_σ occurs. The negative dv_{ce}/dt leads to a Miller feedback, which reduces the dv_{ge}/dt. When the IGBT desaturates, v_{ce} starts to increase again and the positive dv_{ce}/dt charges the Miller capacitance, so that v_{ge} increases higher than the applied driver voltage. Summarising, no effect of the *Self-Charging Displacement Current* can be observed.

As mentioned earlier, the Miller feedback during turn-ON attenuate the mechanisms, which lead to the *Self Turn*-ON. Hence, a simulation with reduced DC link voltage of 0.5 kV is displayed on Fig. 7(b). Compared with the simulation of a typical short circuit type 1, the switching conditions of $v_{driver,1}$ are equal, but now $v_{driver,1}$ is switched OFF after $t=1.65\,\mu s$. During the turn-ON pulse v_{ge} increases as a reason of a charge current limited by R_{G1}. After the turn-OFF, the gate gets low-ohmically discharged. Additionally the dv_{ce}/dt is negative, which should cause a discharge of the gate through the Miller capacitance. But again, the *Self Turn*-ON appears and v_{ge} increases. Thus, i_{c1} increases also until the IGBT starts to desaturate and the influence of the *Self Turn*-ON decreases, so that the IGBT gets switched OFF.

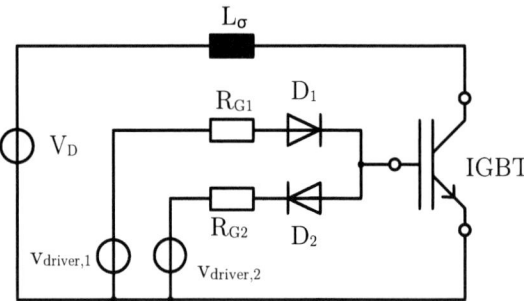

Fig. 6: Short circuit type 1 simulation circuit

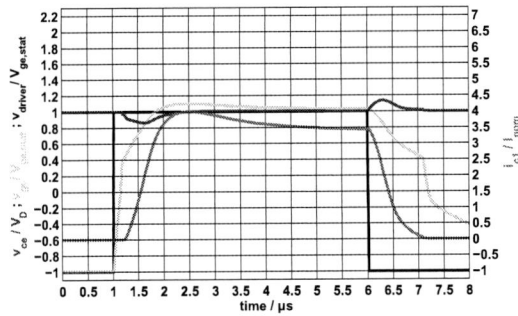 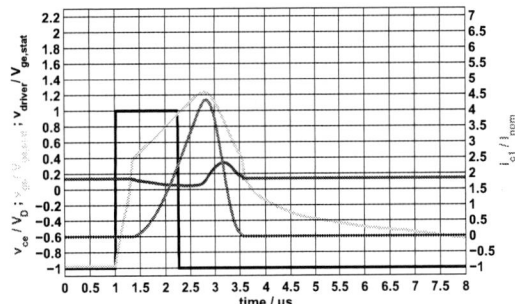

(a) $V_D = 3.6\,kV$, $t_{on} = 5\,\mu s$, typical low-inductive short circuit type 1

(b) $V_D = 0.5\,kV$, $t_{on} = 1.65\,\mu s$; even with a turned OFF gate-drive unit, the gate-emitter voltage increases for further $0.5\,\mu s$ until the *Self Turn-*ON is overcome and the short circuit is turned OFF

Fig. 7: Simulation: Short circuit type 1 trench field-stop IGBT with $R_{G1,on} = 2.5\,\Omega$, $R_{G2,off} = 22.5\,\Omega$ and $L_\sigma = 100\,nH$

Further simulations are made with $V_D = 3.6\,kV$, which fit the usual operational conditions of the IGBT. But as a result of the higher V_D, the dv_{ge}/dt is higher. A higher voltage drop at the stray inductance is caused by the higher di_{c1}/dt and results in an amplified Miller feedback. The effects, which lead to the *Self Turn-*ON, get superposed by the Miller feedback, therefore no *Self Turn-*ON can be observed, even with a turn-ON pulse width lower than $1.65\,\mu s$ compared with Fig. 7(b).

3.2. Experimental results

For the investigation of the *Self Turn-*ON during a low-inductive short circuit type 1, a trench field-stop IGBT ($6.5\,kV/750\,A$) has been used. But compared with section 2.2. a full IGBT module is utilised instead of a single chip IGBT module to meet usual operational conditions. The test circuit consists of a double-pulse test bench. But the upper module is shorted at its bus bar connections. Hence, a test bench with a short circuit type 1 is built, which stray inductance is approximately $100\,nH$. A common IGBT driver with a $R_{G1,on} = 2.5\,\Omega$ at $v_{driver,on} = 15\,V$ and $R_{G2,off} = 22.5\,\Omega$ at $v_{driver,off} = -15\,V$ is used. Again for a simplified comparison with the simulation mentioned above all values of Fig. 8(a) and 8(b) are normalised equally.

Compared with the simulation in Fig. 7(a), the measured low-inductive short circuit type 1, shown in Fig. 8(a), results in an obviously higher collector-current peak, but a smoother dv_{ge}/dt. The smoother dv_{ge}/dt occurs because of parasitic inductances in the gate path, like the gate inductance or the chip-auxillary-emitter inductance. Both were not considered in the simulation. Additionally, at $t = 1.2\,\mu s$ there are some oscillations on v_{ge} detectable, which seem to occur because of the interaction between the positive feedback through the *Self-Charging Displacement Current* and the negative feedback through the Miller capacitance related to the negative dv_{ce}/dt. Similar to the simulation the IGBT turns OFF after $t = 5.5\,\mu s$.

One result of the simulation is that the *Self Turn-*ON is more intense at low DC link voltages. Hence, measurements are done with $V_D = 0.5\,kV$ to investigate the *Self Turn-*ON with reduced Miller feedback. Fig. 8(b) illustrates the results and can be compared with Fig. 7(b). In comparison v_{driver} (black) is switched to $+15V$ at $t = 0.5\,\mu s$. As a result v_{ge} start to rise until v_{driver} is switched OFF $2\,\mu s$ later. Through the turn-OFF v_{ge} decreases.

© VDE VERLAG GMBH · Berlin · Offenbach

PCIM Europe 2016, 10 – 12 May 2016, Nuremberg, Germany

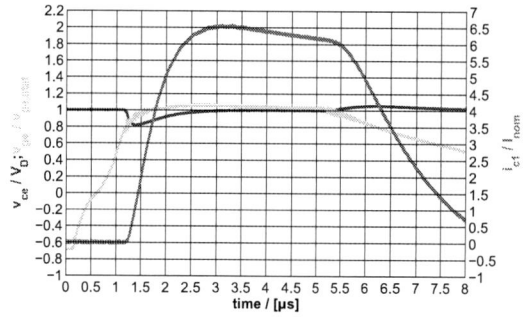

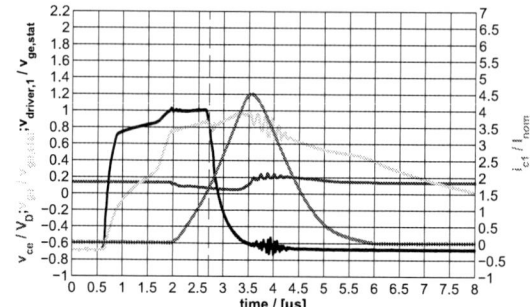

(a) $V_D = 3.6\,kV$, $t_{ON} = 5\,\mu s$, typical low-inductive short circuit type 1

(b) $V_D = 0.5\,kV$, $t_{ON} = 2\,\mu s$, again v_{ge} increases even with turned OFF gate-drive unit and the *Self Turn-ON* occurs until the IGBT desaturates and the short circuit is turned OFF

Fig. 8: Measurement: Short circuit type 1 trench field-stop IGBT with $R_{G1,on} = 2.5\,\Omega$, $R_{G2,off} = 22.5\,\Omega$ and $L_\sigma = 100\,nH$

The reason is the measuring point of v_{ge}. The voltage probe is directly connected to the contacts outside of the module. But inside the module is an additional internal gate resistor. During turn-ON the gate current results in a voltage drop at this resistor, which is also measured. When the gate-drive unit is turned OFF, the gate current and consequently the voltage drop over the internal resistor changes its directions.

On Fig. 8(b) the dashed line marks the point, when v_{ge} and v_{driver} are equal. Although v_{driver} is less than v_{ge} and the IGBT should turn OFF, v_{ge} increases, because of the mechanisms of the *Self Turn-ON*. Similar to the section 3.1. the IGBT turns OFF after it desaturates as a reason of the weaken *Self Turn-ON* mechanisms.

4. Conclusion

In this paper the *Self Turn-ON* is analysed and described for high-voltage IGBTs. With the help of semiconductor simulations and measurements on high-voltage IGBTs the effect of the *Self Turn-ON* can be found.

The *Self Turn-ON* is caused by a *Self-Charging Displacement Current*, where an accumulation of holes under the gate oxid occurs. This *Self Turn-ON* can overcome the negative turn-OFF voltage of the gate drive unit, resulting in a delayed IGBT turn-OFF.

This effect is dependent on the load current and the gate structure of the IGBT. A planar gate structure seems to be more susceptible to the *Self Turn-ON*. The Miller feedback caused by a voltage drop over the parasitic inductance at turn-ON can compensate this effect.

Further investigations, especially with the help of more measurements, are needed to describe the presented effect more in detail and to understand the emerged v_{ge} oscillations. An additional problem could be the parallel connection of IGBTs. A *Self Turn-ON* would lead to an asymmetric stress of the devices and a possible breakdown. Moreover an analyse in other voltage classes can improve the knowledge of the *Self Turn-ON* effect.

© VDE VERLAG GMBH · Berlin · Offenbach

Acknowledge

This work is supported by the Federal Ministry for Economic Affairs and Energy on the basis of a decision by the German Bundestag.

References

[1] P. Baginski. Driving IGBTs with unipolar gate voltage. 49(0):1–8, 2006.

[2] J. Böhmer, J. Schumann, and H.-G. Eckel. Negative differential miller capacitance during switching transients of IGBTs. *Proceedings of the 2011 14th European Conference on Power Electronics and Applications*, pages 1–9, 2011.

[3] J. Böhmer, J. Schumann, K. Fleisch, and H.-G. Eckel. Current mismatch during switching due to the self-turn-off effect in paralleled IGBT. *Power Electronics and Applications (EPE), 2013 15th European Conference on*, 2013.

[4] H. Feng, W. Yang, Y. Onozawa, T. Yoshimura, A. Tamenori, and J. K. O. Sin. Transient Turn-ON Characteristics of the Fin p-Body IGBT. *IEEE Transactions on Electron Devices*, 62(8):2555–2561, 2015.

[5] I. Omura, W. Fichtner, and H. Ohashi. Oscillation Effects in IGBTs Related to Negative Capacitance Phenomena. *IEEE Transactions on Electron Devices*, 46(1):237–244, 1999.

[6] U. Schlapbach. Dynamic Paralleling Problems in IGBT Module Construction and Application. *CIPS 2010*, 2010.

An innovative 6500V HVIGBT with high robustness

Bo Hu, Gaosheng Song

Mitsubishi Electric & Electronics (Shanghai) Co., Ltd, China

E-mail: HuBo@mesh.china.meap.com

Abstract

The paper introduces a new type 1000A/6500V HVIGBT (High Voltage Insulated Gate Bipolar Transistor). The new generation Mitsubishi Electric HVIGBT modules, X-series HVIGBT, use the latest IGBT chip and diode chip technology. As a result, it can significantly reduce the conduction saturation voltage, the switching losses and the thermal resistance. In this paper, the electrical characteristics are presented. Due to its excellent performance, it is very suitable for the traction application, and the detailed applications in 4QC (4-Quadrant Converter) and PWMI (Pulse Width Modulation Inverter) are talked to express the advantages.

1. Introduction

As the development of power converters, it puts forward higher requirements of the system reliability, the efficiency and the volume. To meet the requirements, as a key component in power converters, the IGBT modules should have the high robustness, low power loss and high current density. Similarly, low losses, high robustness and high reliability are just the main three concepts in Mitsubishi Electric HVIGBT design. Mitsubishi Electric has developed the new generation 1000A/6500V HVIGBT: CM1000HG-130XA.

The package adopts the standard package, and the current rating is boosted from 750A to 1000A, so the system output current capability is improved correspondingly. The IGBT chip adopts the third generation CSTBT^{TM} (Carrier Stored Trench Gate Bipolar Transistor) technology, while the diode chip adopts RFC (Relaxed Field of Cathode) technology, so it can realize a lower saturation voltage and lower switching power losses. The paper pays more attention to traction application and makes some comparisons with existing HVIGBT. The comparison results show that CM1000HG-130XA is a very good switching device for high power traction converters.

2. Features of CM1000HG-130XA

2.1 A brief descriptions

Simple descriptions of CM1000HG-130XA are as below:
- V_{CES} (Collector-emitter voltage) =6500V;
- I_C (Collector current) =1000A;
- Maximum operation junction temperature T_{jop}=150℃;

PCIM Europe 2016, 10 – 12 May 2016, Nuremberg, Germany

- High isolation voltage: V_{iso}=10.2kV;

Fig.1. Outline of CM1000HG-130XA

We can see the package as shown in Fig.1: CM1000HG-130XA also adopts the 140mm*190mm standard package to maintain the mechanical compatibility with the existing H-series and R-series HVIGBT.

2.2 The IGBT chip technology

X-series HVIGBT adopts the third generation CSTBT technology. We can see its cross-section in Fig.2. Compared to conventional planer and trench structure, an additional carrier enhanced N layer is injected. The carrier stored layer restrains the holes flow from N⁻ substrate layer to P⁺ layer. As a result, the carrier concentration is increased, and then the conduction impedance is reduced, finally lower $V_{CE(sat)}$ under on-state is realized.

The $V_{CE(sat)}$ comparison with R-series HVIGBT was shown in Fig.3, we can see X-series HVIGBT has a notable decrease. In other words, the conduction power loss of X-series HVIGBT will be lower than R-series HVIGBT under the same working conditions. For low switching frequency applications, such as traction, HVDC, SVG and so on, the conduction power loss takes a relatively high proportion of the total power loss, so low $V_{CE(sat)}$ is very worthwhile.

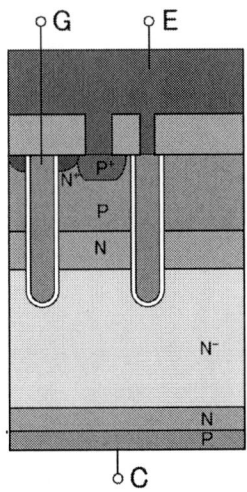

Fig.2. Cross-section of IGBT chip

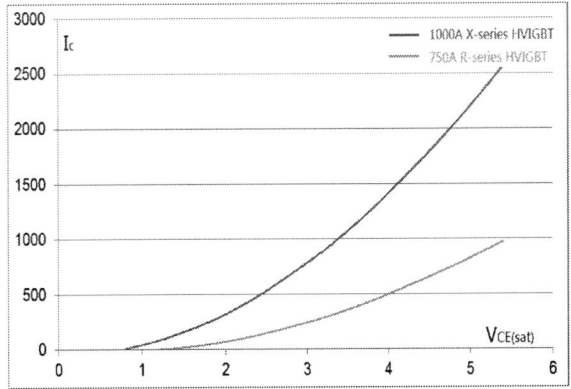

Fig.3. $V_{CE(sat)}$ comparison with R-series HVIGBT

© VDE VERLAG GMBH · Berlin · Offenbach

For FWD (Free Wheeling Diode) chip, by adopting the RFC technology, the forward on voltage V_{EC} is also decreased. Fig.4 is the cross-section of FWD chip, and we can see Partial P is injected on the cathode backside. The benefits are that the diode has a wide SOA tolerance and high snap-off tolerance. Sometimes, when an IGBT was turn on very rapidly, another diode in the same bridge will have an extremely high dv/dt in reverse recovery procedure, in such conditions, commonly referred to as "snappy" recovery, can cause very high transient voltage of V_{CE}, which may damage HVIGBT modules. X-series HVIGBT use the latest technology and can virtually eliminate this issue. Under such conditions: T_j=150℃, V_{CC}=3600V, I_E=2000A, L_s=150nH, we can see the reverse recovery waveforms of freewheeling diode in Fig.5. The diode performs well and the reverse recovery curve is also very smooth and no snappy recovery.

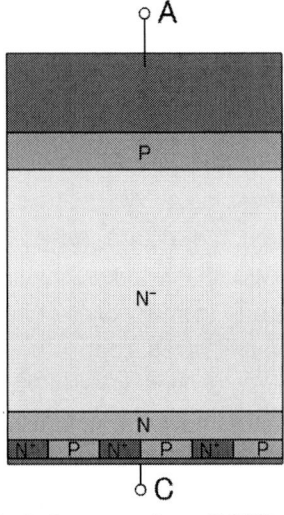

Fig.4. Cross-section of IGBT chip

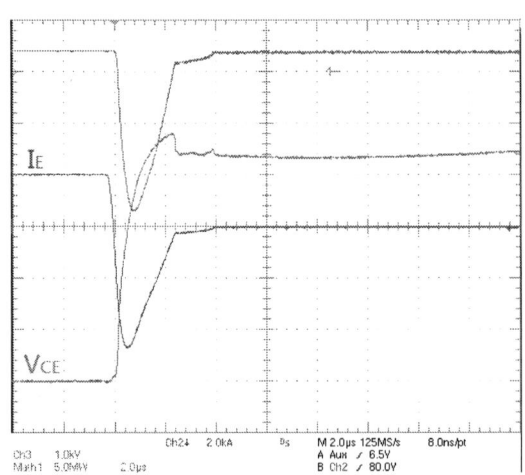

Fig.5. Reverse recovery waveform

X series HVIGBT adopts new edge termination structure which uses the Linearly Narrowed Field Limiting Ring (LNFLR) technology. As showed in Fig.6, the active area is increased about 28% by shrinking edge termination area. As a result, the junction to case thermal resistance decreased. Table.1 shows the difference, in which, $R_{th(j-c)Q}$ is the IGBT thermal resistance between chip and case, and $R_{th(j-c)D}$ is the diode thermal resistance between chip and case.

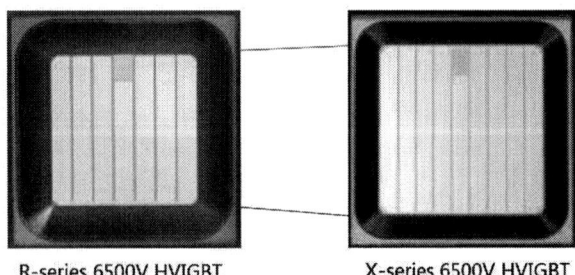

R-series 6500V HVIGBT   X-series 6500V HVIGBT

	$R_{th(j-c)Q}$	$R_{th(j-c)D}$
R-series	12.0K/kW	22.0K/kW
X-series	11.0K/kW	17.0K/kW

Fig.6. Chip comparison   Table.1 Thermal resistance comparison

All the above latest chip technologies assure CM1000HG-130XA is the first 6500V HVIGBT module in the world, which guarantees 150℃ junction operation temperature.

3. CM1000HG-130XA application in traction converters

Compared to other transportation ways, railway transportation has many advantages, such as fast speed, large capacity, low energy consumption, low pollution, good safety, etc. With the development of the economy, railway transportation system will play a more and more role in future.

3.1.   Traction converter principle

The traction converter is a main part of the traction system, and it controls the train speed by the adjustment of output voltage and frequency. As a switching device, HVIGBT is a key component in the converters. Fig.7 is a common topology of high speed railway traction converter. It is mainly composed of 4QC, brake circuit and PMW inverter.

To illustrate the advantages of CM1000HG-130XA in traction converters, the following example is cited here. The topology is as Fig.7, the rated output power is 1100kW and 2 pieces of 550kW motor are driven, and the operation parameters are shown as the table 2.

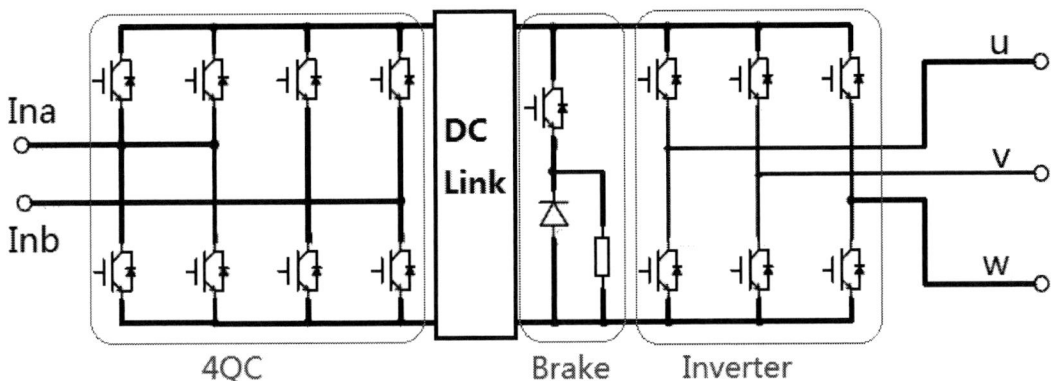

Fig.7. An example of traction converter

Table.2 Operation parameters

4QC		PWM Inverter	
Input voltage	2100V AC	Input voltage	3600V DC
Output voltage	3600V DC	Output voltage	0-2000V AC
Input current	600A	Output current	350A
Switching frequency	450Hz	Switching frequency	450Hz
HVIGBT module	CM600HG-130H	HVIGBT module	CM750HG-130R

3.2.   Discussion on 4QC

By using the latest chip technology in CM1000HG-130XA, low losses, low thermal resistance and high operation junction temperature are achieved. In the above 4QC, 2 pieces of CM600HG-130H in parallel were used. As we all know, there is static and dynamic current imbalance among the paralleled IGBTs. To restrain the current imbalance, we need to pay more attention to the circuit design and IGBT selection, so it will greatly increase the difficulty of system design. To simplify the design and increase the system reliability, let's discuss if we can use 1 piece of CM1000HG-130XA to substitute 2 pieces of CM600HG-130H in parallel. Based on the conditions in table.2, the power loss and thermal rise simulation are taken. Besides, the maximum junction temperature $T_j=125℃$, power factor PF=0.85, and we ignore the current imbalance and set the heat sink temperature $T_s=80℃$(The measuring point is just under the chip). Here, $\Delta T_{(j-c)Q}$ is the thermal rise between IGBT junction to case, $\Delta T_{(j-c)D}$ is the thermal rise between FWD junction to case, T_{jQ-ave} is the average IGBT operation junction temperature, T_{jD-ave} is the average FWD operation junction temperature.

Table.3 Comparison of power loss and thermal rise

Converter capacity	HVIGBT module	Power loss	T_{jQ-ave}	Margin	T_{jD-ave}	Margin
1200kW	CM600HG-130H	1674W×2 =3348W	105.6℃	19.4℃	102.4℃	22.6℃
1200kW	CM1000HG-130XA	2902W	118.4℃	31.6℃	115℃	35℃

We can see from table.3: the total power loss decreased about 15%. Although the operation junction temperature of CM1000HG-130XA is higher than CM600HG-130H, however, CM1000HG-130XA has a higher maximum junction operation temperature(150℃), and there is still a big margin.

Through the above simulation, we can get the result that 1 piece of CM1000HG-130XA can substitute 2 pieces of CM600HG-130H in parallel under certain conditions. By using single CM1000HG-130XA, the current imbalance issue can be avoided. Meanwhile, another advantage is that the IGBT quantity is reduced, so the heat sink volume and the converter volume can be reduced correspondingly.

Of course, the above simulation was done under the specific conditions. If we want to use 1 piece of CM1000HG-130XA to substitute 2 pieces of CM600HG-130H, we need to simulate the actual working conditions and evaluate it comprehensively.

3.3. Increase converter capacity

According to the distribution of the traction power, EMU (Electric Multiple Units) can be divided into power-centralized EMU and power- distributed EMU. Because power-centralized EMU has the advantages of low noise and simply maintenance, it receives increasing concern in some countries. Meanwhile, power-centralized EMU raised requirements to high voltage and large current power device. Compared to H-series and R-series HVIGBT, the rated current of CM1000HG-130XA is increased, so output power of the traction converters can also be increased by using CM1000HG-130XA.

Let's take an example to explain it. As mentioned above, the inverter output power is 1100kW by using CM750HG-130R, if we use CM1000HG-130XA, can the inverter be operated at 1600kW or higher power?

Similarly, the power loss and thermal rise simulation were taken. The working conditions had been listed in table.2 and the output current is about 500A when output power is 1600kW. The simulation result can be seen as table.4.

Table.4 Power loss and thermal rise simulation

Inverter capacity	HVIGBT module	Power loss	T_{jQ-ave}	Margin	T_{jD-ave}	Margin
1600kW	CM750HG-130R	2582W	120.6℃	4.4℃	106.2℃	18.8℃
1600kW	CM1000HG-130XA	2569W	116.9℃	33.1℃	105.8℃	44.2℃

From the table, we can see there is also a big margin with the maximum junction temperature. Compared to CM750HG-130R, CM1000HG-130XA can boost the converter capacity obviously.

4. Conclusion

The paper introduce an innovative 6500V HVIGBT CM1000HG-130XA, the features and the chip design technology are briefly explained. Because of its good characteristics, it's very suitable for mega power converters, especially for traction application. By applying CM1000HG-130XA, below benefits can be reached: (1)Compared to conventional 4QC topology with 600A/6500kV HVIGBT, no paralleling can be achieved with new 1000A/6500V X-series HVIGBT; (2)The total converter capacity can be boosted to 1.6MW or higher under normal operation conditions.

5. References

[1] M. Rahimo, A. Kopta, and S. Linder, "Novel Enhanced-Planar IGBT Technology Rated up to 6.5kV for Lower Losses and Higher SOA Capability," Proc. ISPSD '06, pp. 33-36, Naples, Italy, 2006

[2] A. Nishii, K. Nakamura, F. Masuoka, and T. Terashima, "Relaxation of Current Filament due to RFC Technology and Ballast Resistor for Robust FWD Operation," Proc. ISPSD' 11, pp. 112-115, San Diego,USA, 2011

[3] K. Nakamura, K. Sadamatsu, D. Oya, H. Shigeoka, and K. Hatade, "Wide Cell Pitch LPT(II)-CSTBTTM(III) Technology Rating up to 6500V for Low Loss," Proc. ISPSD ' 10, pp. 387-390, Hiroshima, Japan,2010

[4] Mitsubishi Electric, 'General Considerations for IGBT and Intelligent Power Modules', Sep.1998.

[5] Mitubishi Electric, "CM1000HG-130XA specification", http://www.mitsubishielectric.com/semiconductors/products/powermod/hvigbtmod/index.html.

New High Power 3.3kV/1500A IGBT Module Packaging

Daohui Li, Wei Zhou, Fang Qi, Matthew Packwood, Yangang Wang, Steve Jones, Xiaoping Dai

Power Semiconductor R&D Center, Dynex Semiconductor Ltd, Lincoln LN6 3LF, United Kingdom

Email: daohui.li@dynexsemi.com

Abstract

High power IGBT modules have been widely utilised in modern power electronics applications, i.e. traction, renewable energy, electrical vehicle and other industry application areas. A 3.3kV/1500A IGBT module with 140mmX190mm footprint is one of the key standard module packages that have been widely used. In this paper, a new internal design and assembly process have been applied during the development of the high power IGBT module package. A newly designed 3D busbar-substrate assembly has >40% lower inductance than the traditional 2D structural busbars, and ohmic losses from both metal busbars and metal layer of substrates are much lower for the new design. The pre-bent 3D busbars and ultrasonic welding (USW) processes provide better reliability than the usual busbars based on soldering process. During the development of the module package, different 3D finite element method (FEM) simulation packages have been utilised to verify the component design and optimise the overall design with respect to electromagnetic, electrical, thermal/mechanical aspects。

I. Introduction

With the fast development in power electronics applications in traction, renewable energy, electric vehicle and related industry fields, and the demands for more and more switching power, users require higher power density, higher voltage and current ratings, higher switching frequency, high operation/storage temperature with relative low cost. All these requirements are driving forces pushing power module package suppliers to utilise new packaging materials, new assembly processes, novel designs together with corresponding voltage/current rating chips to provide the industry with increased power, improved performance, enhanced reliability standard modules [1-3].

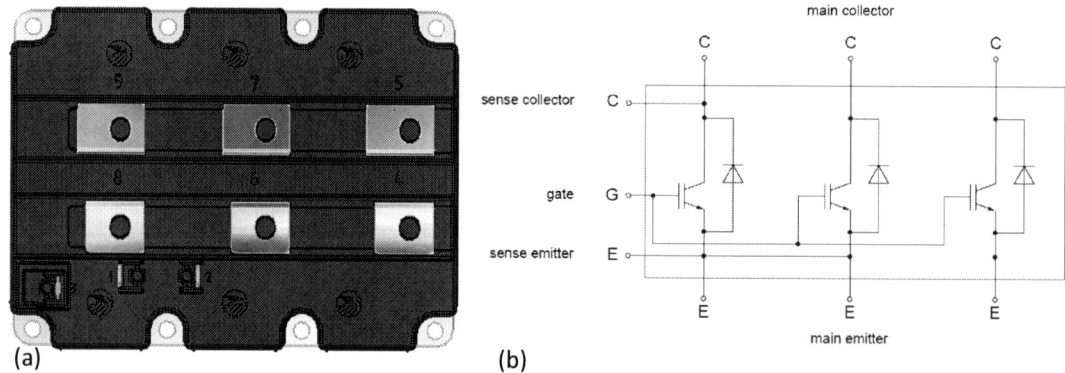

Fig. 1 Single switch IGBT module with 140X190 footprint: (a) Outlook of IGBT module, (b) Single switch schematic circuit diagram.

The standard 140X190 dimension single-switch IGBT module with schematic circuit, as shown in Fig. 1(b), is one of several key standard modules used in high power applications for different voltage levels, i.e. 1.7kV, 3.3kV, and 4.5kV.

II. Design Simulation and New Assembly Processes

A cross-section of a module can be seen in Fig. 2, which includes chips, chip solder layer, substrate with AIN ceramic, substrate solder layer, and AlSiC baseplate. The different types of material have different electromagnetic, thermal, mechanical properties. The parts design, material selections and assembly processes are key challenges for module packaging design. 3D FEM computational simulation technologies have been used in the design and optimization processes. Based on ANSYS EM simulation, some comparison between the old design and the new structure are shown in Table 1 [3].

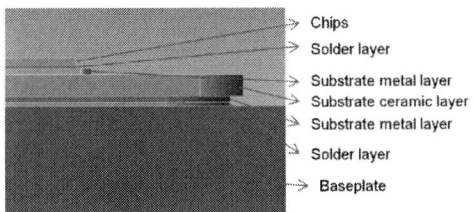

3.3kV/1500A Module		*Older version*	*New version*
Stray Inductance *(nH)*		22.7	15.6
Busbar Loss *(W)*	Emitter	18.4	8.7
	Collector	18.7	12.5
Substrate Loss *(W)*	Left	17.2	15.5
	Right	16.6	15.9
Busbar Highest Temperature *(°C)*	Emitter	112.3	93.0
	Collector	112.7	93.5

Fig. 2 Layer structure of module. Table. 1 Comparison on busbar-substrate assembly.

Fig. 3 steady-state thermal/mechanical FEM simulation. (a) Layer temperature distribution, and (b) thermal deformation.

Steady state thermal-mechnical simulation has been carried out to show the juntion temperature of the multi-layer structure and their corresponding mechanical deformation in Fig. 3 (a) and Fig. 3(b), respectively [3].

Some new assembly technologies compared to traditional 140X190 modules have been used to increase module reliablities, such as UltraSonic Welding (USW) of 3D prebent busbars instead of soldering, formic acid fluxless soldering process to reduce traditional soldering post process with flux residue which is rather difficult to be cleaned, stand-off spacer has been used to provide even soldering layer between substrates and baseplate to improve reliablity of solder as shown in Fig. 4(a), plastic frame mounting by using glue together with screw from backside of baseplate as shown in Fig. 4(b).

© VDE VERLAG GMBH · Berlin · Offenbach

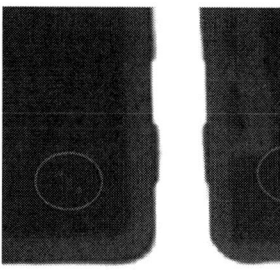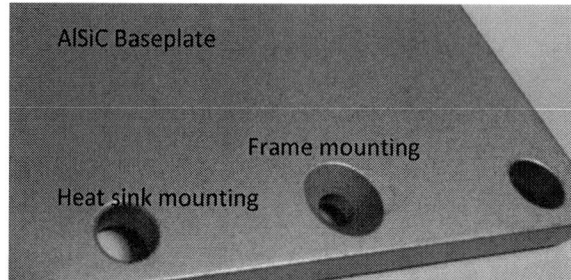

Fig. 4 (a) X-ray image of stand-off bonding wires, (b) Mouting hole from backside of baseplate.

III. Inductance Measurement, Static/Dynamic Tests

A. Inductance measurement:
New type of 3D prebent busbars have been designed and tested in the development process of the IGBT module packaging. The inductance measurement by using a HP4192A LF impedance analyzer, has been carried out. One set of busbar-substrate assemblies has ~17nH inductance as shown in Fig. 5(a)-(c): (a) is the USW bonded busbar feet, (b) is the inductance measurement probe, and (c) is the inductance measurement display.

For one single switch module, there are three set of busbar-substrate assemblies in parallel, that the module inductance can be calculated as 17nH/3=5.7nH. The simulation results and measurement results have shown less than 8% difference, that give us confidence in future low inductance power module design [2-4].

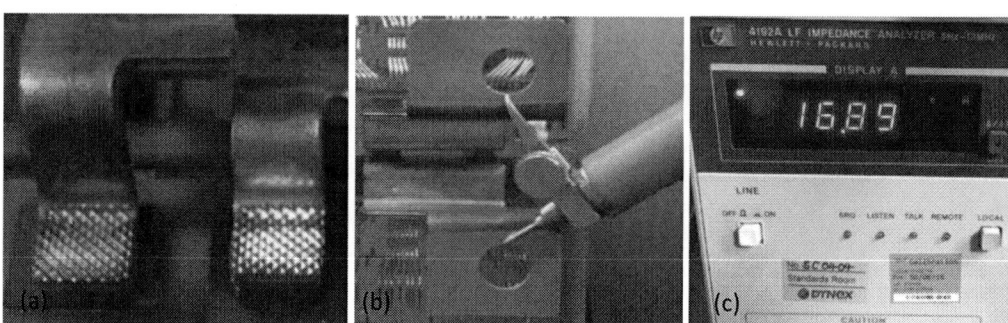

Fig. 5 Measurement of USW welded prebent busbars by using Impedance Analyzer.

B. Static test:
The static electrical test at substrate level has shown the Vce voltage difference between substrate level and IGBT at chips level is about ~0.2Volts, which agrees well with the EM simulation results as shown in Fig. 6(a) and (b). The Vce difference from current conduction voltage drop via topside copper layer of the substrate is about 0.15Volts and that from the parallel Al wires is about 0.07Volts.

Three types of 140X190 modules with same voltage/current ratings 3.3kV/1500A have been compared via static test. The redesigned new module is so-called E2 module, and the other two modules are called Dynex-E (E) and CSR-E (CSR) modules, respectively [5].

High temperature static test at module level has been carried out to show the conduction power loss difference for the three types of moduels. The comparison for collector-emitter voltage $V_{ce(sat)}$ vs. collector current I_c of three modules can be seen in Fig. 7(a). One can see that $V_{ce(sat)}$ of E2 module has shown as the smallest one for the same Ic value amongst all the three modules. The comparison for foward voltage V_f vs. forward current I_f of three modules can be seen in Fig. 7(b). One can see that the V_f of E2 module is the smallest at same I_f value of all the three moduels. Those two static test comparisons can lead to smallest conduction power loss for E2 module in the actual application compared to the other two E-modules from Dynex and CSR. Thus E2-module could reduce conduction power and thermal dissipation.

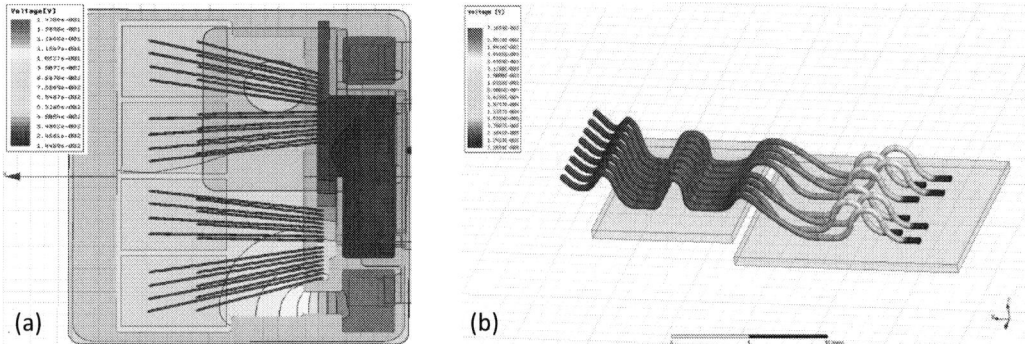

Fig. 6 Conduction DC voltage (a) substrate copper layer, (b) Al wire conducting

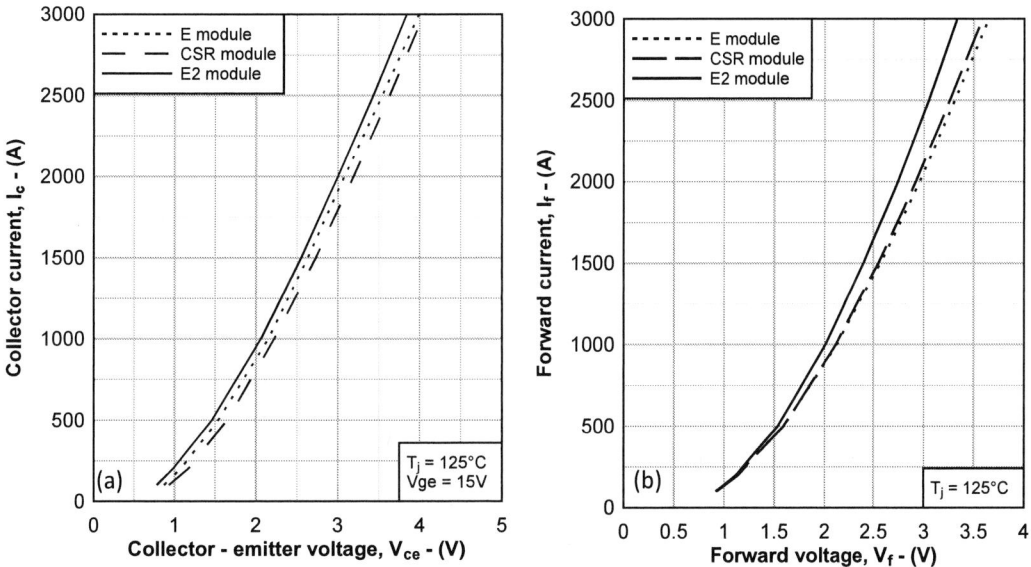

Fig. 7 Static performance of Dynex-E, Dynex-E2, CSR-E modules at 125°C. (a) Vce(sat) vs. Ic, and (b) V_f vs. I_f

During the development stage of E2 module, temperature plastic frames by using selective laser sintering (SLS) technique were used. The frame material can not afford too high temperature up to 150^{0}C, that high temerpature tests at module level were carried out at 125^{0}C only. The 150^{0}C test for module level will be done when proper plastic frames received.

C. Dynamic test :

Some dynamic tests have been carried out at both substrate level and module level. The E2 substrate level dynamic tests have been carried out at room temperature and high temperature 150^0C, module level double pulse test at 125^0C due to SLS frame used at developing stage. Some of the dynamic test results and test comparisons between three types of modules have been introduced in this session [5].

(i). Substrate level dynamic test:

The chopping test, short-circuit test and double pulse test have been carried out with nominal voltage and current at substrate level, respectively. Due to paper length limit, only high temerature at 150^0C of the substrate dynamical test results can be seen from Fig. 8(a)-(d).

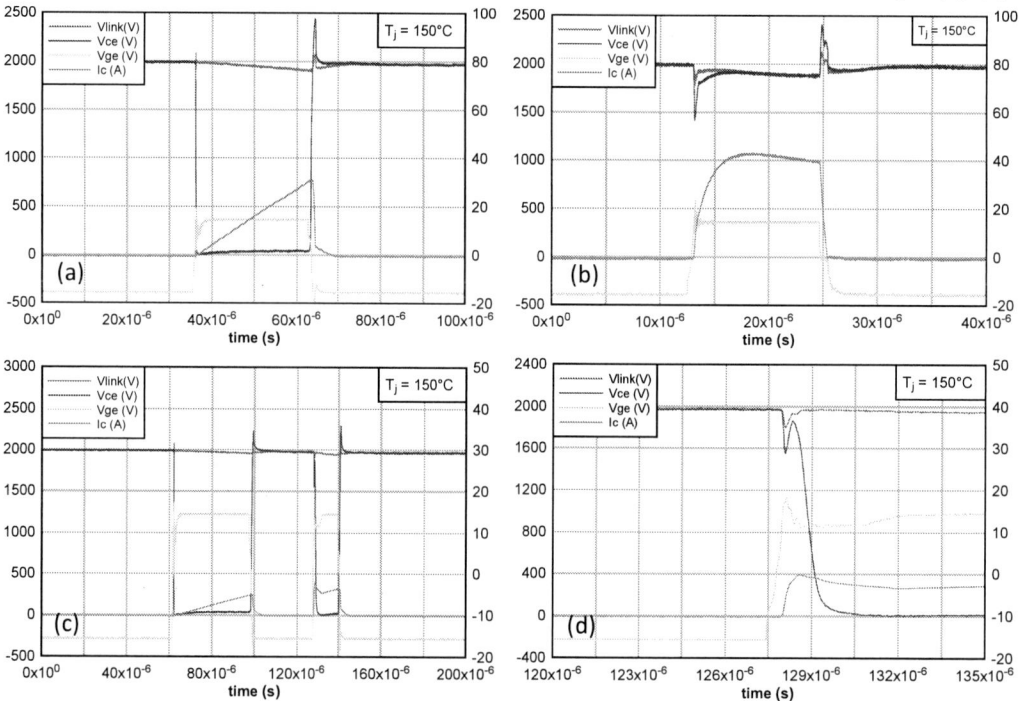

Fig. 8 E2-substrate dynamical test at 150^0C: (a) single pulse test, (b) short-circuit test (T_{on}=12us), (c) double pulse test, and (d) the second turn-on transient

For substrate level test, some test conditions are under high temperature 150^0C together with 2kV line voltage, 6Ω for R_{on} and R_{off}. In Fig. 8(a), the collector current increases upto 750A, which is three times of nominal current rating for substrate, which is 250A; short-circuit test result is shown in Fig. 8(b) with ~1000A maximum current, which is four times of nominal current rating for substrate; typical double pulse substrate level test can be seen in Fig. 8(c) and the second turn-on transient details can be seen in Fig. 8(d). The low temperature and high temperature dynamic test have shown that the E2-substrate can fit the typcial performance requirements.

(ii). Module level dynamic test comparison:

Different types of module level tests have been carried out in the developing processes of E2-module. Due to paper length limitation, only the dynamic test comparisons between three types of IGBT modules, i.e. E2, Dynex-E, and CSR-E module, will be mainly introduced here. The double pulse dynamical module test conditions are set as 1.8kV line voltage and 1500A current rating at high temperature of 125^0C. The gate resistor has been set as the same for all the three modules. Module level double pulse test results of three modules have been shown in Fig. 9(a) with their gate drive, collector-emitter voltage V_{ce}, collector current I_c. Via the comparison, some of E2 module's design advantage can be seen after checking details of the comparisons.

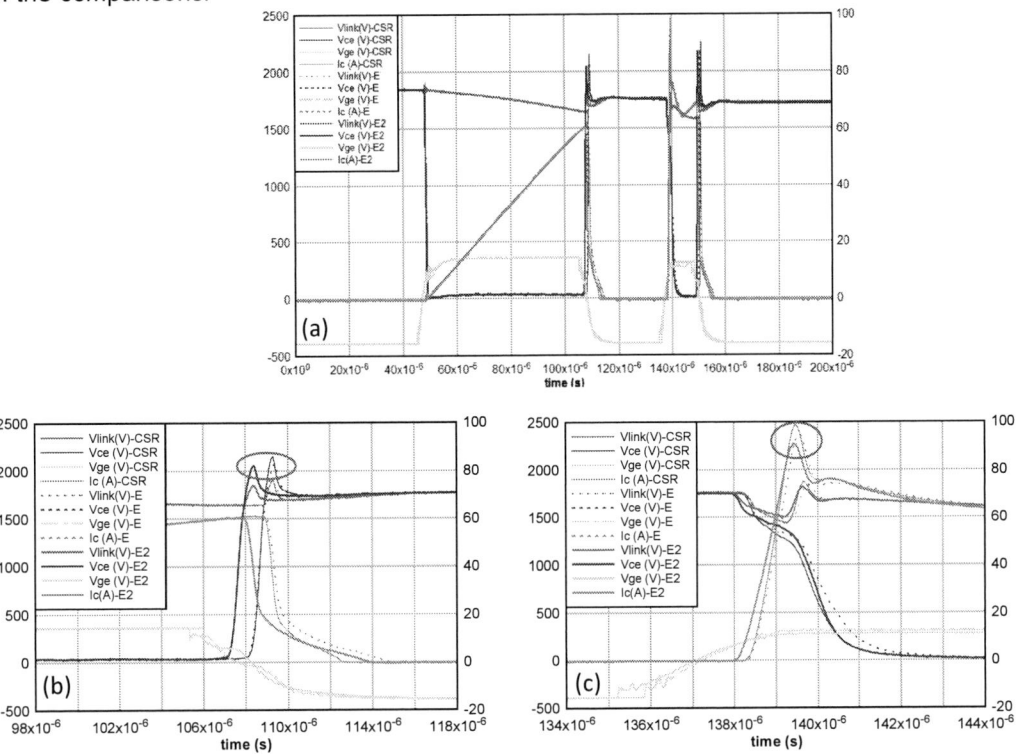

Fig. 9 Double pulse test comparison for three types of modules: (a) double pulse curves, (b) the first turn-off transient, and (c) the second turn-on transient

For double pulse test, two stages are very important to reflect the performance of IGBT module, i.e. the first turn-off transient and the second turn-on transient processes. At the first turn-off transient stage, gate drive signal reduces from +15V to -15V, IGBT is switched off with Ic current change from high current rating down to zero with tail current, Vce changes from $V_{ce(sat)}$ to high voltage. There is so-called overshooting voltage happened during the transient due to module parasitic inductance (L) and current change slope (di/dt). One can see from Fig. 9(b) from the highlighted circle that the overshooting voltage of E2 modules was smaller than both Dynex-E and CSR modules with same gate resistor. This can verify that the internal stray inductance of E2 IGBT module is lower than both Dynex-E and CSR module, which is one of key advantages of E2 module structure based on 3D prebent busbar. Consequently, E2 module has wide safe operating area (SOA) which can use smaller turn-off gate resistor $R_{g(off)}$ to reduce Eoff in actual application.

The di/dt and reverse recovery current Irr of freewheeling diode (FRD) affected by E2-module has shown smaller values than those of both Dynx-E and CSR modules as highlighted in Fig. 9(c), which indicates more details for the second turn-on transient in time domain. Via the comparison, E2-module can use smaller turn-on gate resistor $R_{g(on)}$ to expand reverse bias safe operating area (RBSOA) of FRD meanwhile to reduce Eon.

The power loss comparison of E, E2 and CSR modules under the same test conditions is shown in Fig.10, the E_{on} of E2 module is between the E and CSR modules and E_{off} of E2 module is similar as E modules which was bigger than CSR modules, E_{rec} of three modules were almost the same.

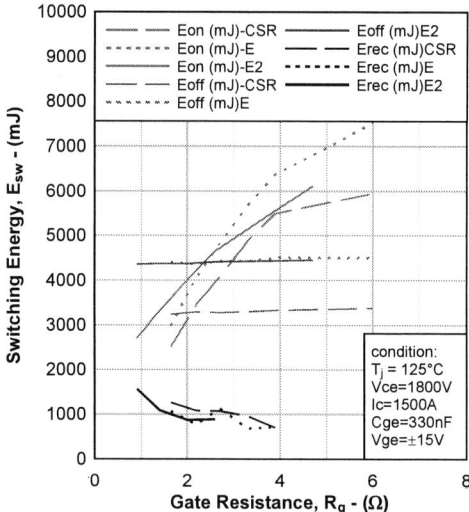

Fig. 10 Power loss comparison of E, E2 and CSR modules

According to the dynamic test results, we can say that E2 module has low inductance, low total power loss and wide SOA so that it has enough margin to tolerate the strict working conditions. Then, whole efficiency of power electronics system by using E2-module can be increased.

IV. Summary

A new low inductance single switch 3.3kV/1500A IGBT power module has been designed, tested and verified with new assembly material and processes to enhance the reliability. The good agreement between simulation and measurement on inductance values gives us confidence for future low inductance module designs.

The static and dynamic experimental tests at substrate level and module level for different temperatures have been carried out. Via the tests, E2 module package can provide low inductance, low total power loss and wide safe operating area. More static and dynamic tests are ongoing together with optimisation of the design, update will be reported during the conference.

Reference:

[1] A. Volke, M. Hornkamp, "IGBT Modules-Technologies, Driver and Applications", published by Infineon Technologies AG, 2nd Edition, 2012.

[2] Z. Tang, "High Power Electronics Design", ANSYS presentation, 3rd June, 2014.

[3] D. Li, M. Packwood, F. Qi, Y. Wu, Y. Wang, S. Jones, X. Dai, G. Liu, "3D Multiphysics Modelling of High Voltage IGBT Module Packaging", ICEPT 2015, Changsha, China, Aug, 2015.

[4] H. Li, Z. Beczkowski, S. Munk-Nielsen, R. Maheshwari, "Circuit mismatch and current coupling effect influence on paralleling SiC MOSFETs in multichip power modules", PCIM Europe 2015, 19-21 May 2015, Nuremberg, Germany.

[5] W. Zhou, D. Li, Dynex R&D Center internal test report, Nov. 2015.

The LinPak high power density design and its switching behaviour at 1.7 kV and 3.3 kV

Samuel Hartmann, Fabian Fischer, Andreas Baschnagel, Harald Beyer, Raffael Schnell, Christian Treier
ABB Switzerland Ltd., Semiconductors, Fabrikstrasse 3, 5600 Lenzburg, Switzerland
e-mail: samuel.hartmann@ch.abb.com

Abstract

The LinPak IGBT power module is designed to meet the requirements of demanding high power applications. Robust switching behaviour is shown experimentally on a set-up with only 20 nH of commutation loop inductance. The resulting overvoltage at IGBT turn-off stays far below the module's voltage rating, which makes the LinPak suited for fast switching chip sets with low losses. Several design variants are discussed considering possible current rating and module reliability. Further a method is proposed for estimating the chips' temperature from the reading of the LinPak's integrated NTC temperature sensor.

1   Introduction

Today's high power IGBT modules with the typical foot print of 140 mm x 190 mm in single switch configuration reach its limit. New fast switching chip sets like the 1700V SPT[++] IGBT [3], that allow switching with low losses require low commutation loop inductance to keep the voltage overshoot small. Additionally the available power terminal connection area limits the continuous rms current possible to draw from these modules.

The new LinPak power module is developed as a new open standard to overcome these limits. The half-bridge module is designed for easy paralleling as shown on the right of Figure 1. Different converter output power can be achieved simply by using the appropriate number of modules. With its low module stray inductance of 10 nH and the possibility to implement a low inductive bus bar equipped with DC link capacitors, the voltage overshoot during IGBT turn-off can be reduced typically by a factor of 5. The amount of terminal connection area per amp of current rating is doubled, which avoids the need of bus bar cooling. The combination of the well established AlSiC base plate and AlN substrates with a reliable particle free ultrasonic terminal welding process makes the LinPak ideally suited for applications with demanding power cycling requirements like traction and CAV (commercial, construction and agricultural vehicles).

A new feature to standard high power IGBT modules is the built-in NTC temperature sensor. It can be used to detect failures in the cooling system or to derive the chips' temperature using the mathematical model provided in this paper.

Figure 1: The LinPak IGBT power module and its low inductive bus bar connection.

2 Module design

For a pioneering power module design, many aspects need to be considered: high current rating, high reliability, good manufacturability, low cost, low thermal resistance, low inductance, stable electro-magnetic behaviour and many more. The optimum arrangement of the power terminals for easy paralleling and low inductance bus bar design is already given [1] [2]. The LinPak power module shown on the left in Figure 1 implements this layout. For realization of the internal power circuit layout, three concepts have been considered for the LinPak. The three layouts are shown in Figure 2. For each layout only two of four AlN substrates used within the module are shown. The further two substrates are identical and in parallel to the shown two substrates.

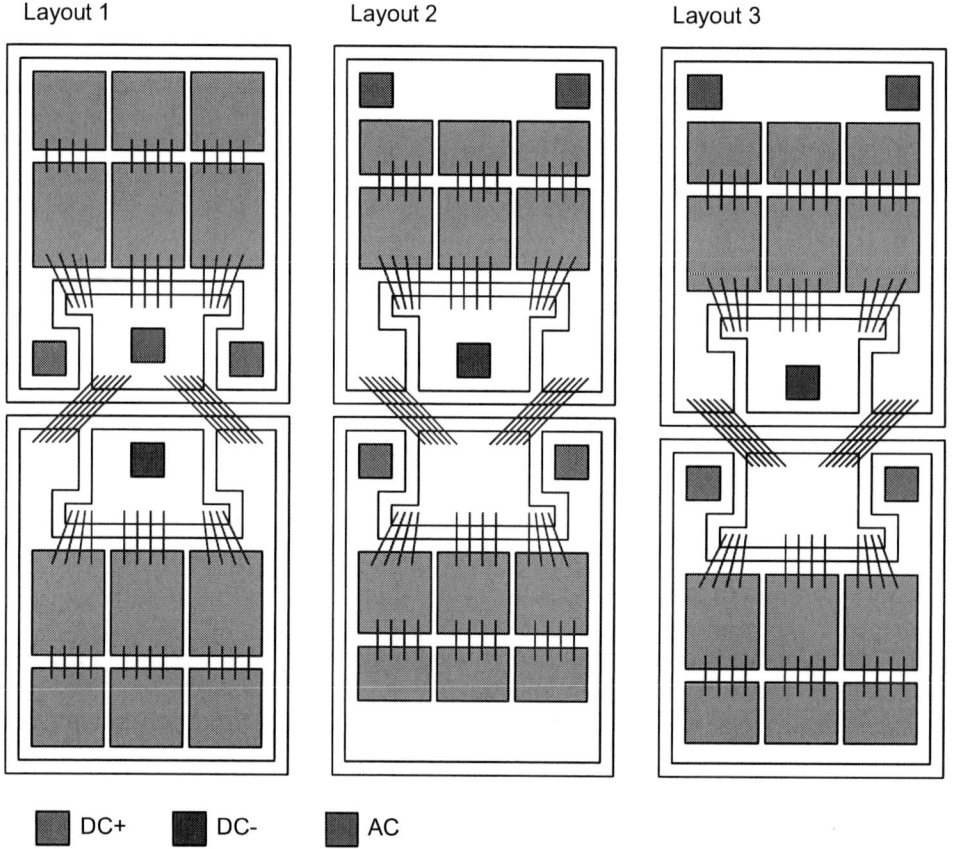

Figure 2: Three design concepts for a low inductive half-bridge module. Only two of four substrates are shown per variant.

The first layout is used in the LinPak module. The high side and low side switches are realized with identical substrates. All terminal plates are connected to the substrates in the middle of the module. For the second and third layout, the AC terminal plates are connected on the AC side of the module. This allows a shorter AC path with lower resistance, but additional area is needed for the AC terminal connection. Also the second layout is realized with identical substrates, therefore the area for the AC connection is lost two times. The third layout is realized using two different substrates for the high and low side switches, and the AC terminal connection area is needed only once.

The difference in AC terminal connection area needed, strongly influences the area remaining for the power semiconductors and consequently the achievable current rating. A comparison of the silicon area available is shown Figure 3. In comparison with Layout 1, Layout 2 and 3 take up only 77% and 89% silicon area. With Layout 1, the LinPak achieves a current rating of 1000 A based on SPT[++] IGBT technology [3], future modules using enhanced trench IGBT technology have the potential to reach a rating of up to 1300 A [6].

The silicon area also influences the thermal resistance. Using finite element analysis the three layouts are compared to each other (Figure 3). Layout 1 with 27.2 K/kW has a considerably lower thermal resistance junction to case than Layout 2 with 33.6 K/kW.

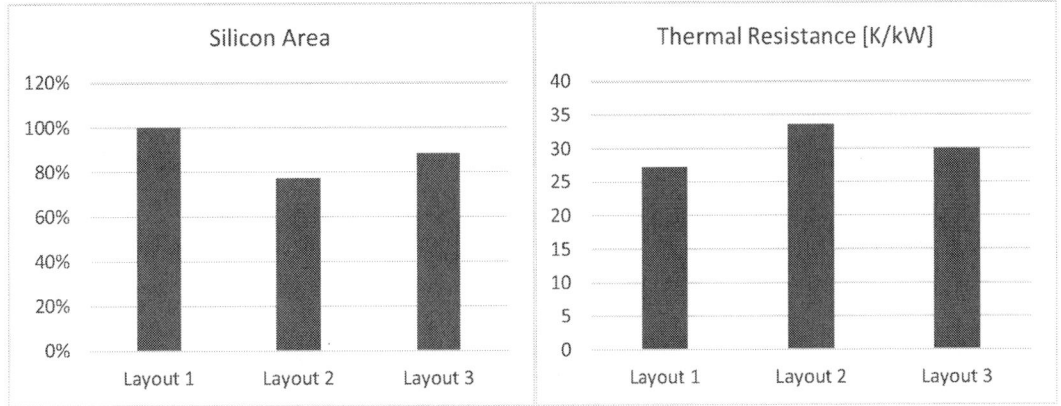

Figure 3: Comparison of the three layouts: silicon area (left) and thermal resistance for IGBT, junction to case (right).

A further advantage of layout 1 is seen in a reliable ultra sonic welding process for connecting the power terminals to the substrates [4]. Ultrasonic welding is known as a reliable connection when it comes to thermal cycling load, but the technology also poses two challenges: the generation of particles during the welding process and the possibility of crack formation in the ceramic insulation. The particle challenge is addressed by having all welding connections located at some distance to the semiconductor chips, which are very sensitive to particle contamination. In addition particle removal is implemented in the welding process. The low level of particles in the LinPak would not have been achievable with layout 2 or 3, where the AC terminal feet are located very close the semiconductor chips. The insulation integrity challenge is mastered by having the welding pad on the substrate large enough which is reasonably possible only with layout 1. The distance between the welded terminal foot and the copper edge of the welding pad is found to be a critical design parameter. Together with carefully chosen process parameters, cracks in the aluminum nitride ceramic are successfully prevented in the LinPak module.

3 Measured switching behaviour

Prototype modules of the LinPak are built and tested. The switching waveforms for the 1.7 kV LinPak at nominal conditions are shown in Figure 4. With the low stray inductance of 20 nH achieved in the commutation loop, the overvoltage at IGBT turn-off remains far below the maximum device voltage rating. The high side switch (continuous line) and the low side switch (dashed line) show almost identical switching behaviour.

The safe operating area (SOA) curves are shown in Figure 5. The IGBT can turn off safely a current of 2000 A when the DC link voltage is 1300 V whilst the overvoltage still staying safely below 1700 V. Both switches can be safely turned off after 10 μs in a phase short circuit (SCSOA).

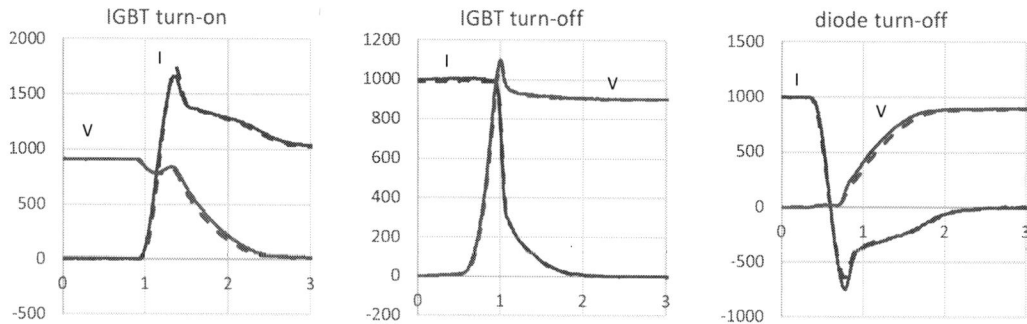

Figure 4: Measured waveforms of the 1.7 kV LinPak at nominal conditions.
I_c = 1000 A, V_{cc} = 900 V, T_j = 125°C, L_σ = 20 nH, R_{Gon} = 1.2 Ω, R_{Goff} = 2.2 Ω.
Continuous line is high side switch and dashed line is low side switch. Units are V, A and μs.

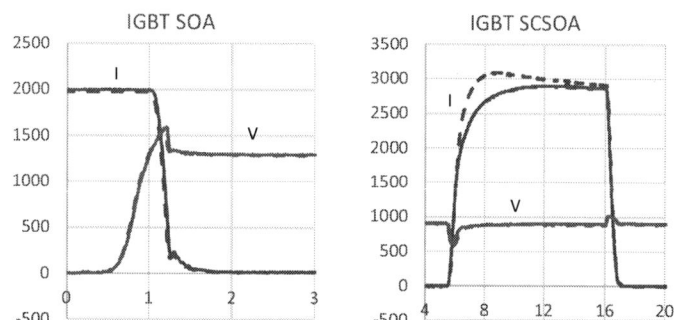

Figure 5: Measured SOA curves of the 1.7 kV LinPak at 125°C.
IGBT SOA with I_c = 2000 A, V_{cc} = 1300 V, L_σ = 20 nH, R_{Goff} = 2.2 Ω.
IGBT SCSOA with V_{cc} = 900 V, L_σ = 20 nH, R_{Goff} = 12 Ω.
Continuous line is high side switch, dashed line is low side switch. Units are V, A and μs.

The switching behaviour of the 3.3 kV LinPak is shown in Figure 6 at nominal conditions and in Figure 7 under SOA conditions. The measurements are done on a production equipment. Therefore the commutation loop inductance was rather high at 80 nH. In a low inductive application set-up, the same low commutation loop inductance will be achieved as with the 1.7 kV LinPak. Also the 3.3 kV LinPak shows smooth switching behaviour, almost identical switching speed of high side and low side and robust turn-off capability under SOA conditions.

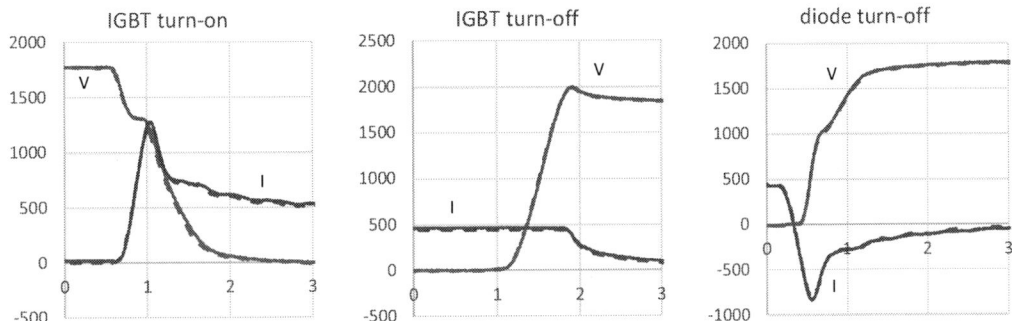

Figure 6: Measured waveforms of the 3.3 kV LinPak at nominal conditions.
I_c = 450 A, V_{cc} = 1800 V, T_j = 125°C, L_σ = 80 nH, R_{Gon} = R_{Goff} = 3.3 Ω.
Continuous line is high side switch and dashed line is low side switch. Units are V, A and µs.

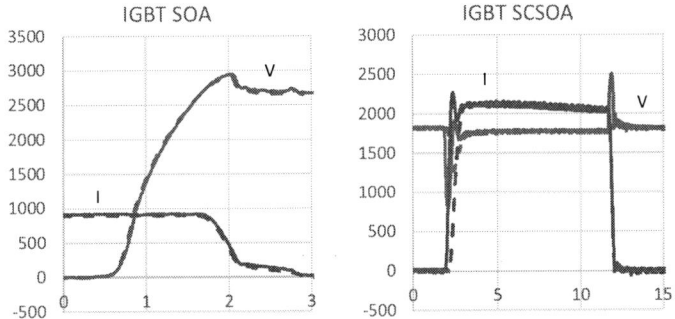

Figure 7: Measured SOA curves of the 3.3 kV LinPak at 125°C.
IGBT SOA with I_c = 900 A, V_{cc} = 2600 V, L_σ = 80 nH, R_{Goff} = 3.3 Ω.
IGBT SCSOA with V_{cc} = 1800 V, L_σ = 80 nH, R_{Goff} = 10 Ω.
Continuous line is high side switch, dashed line is low side switch. Units are V, A and µs.

4 Integrated temperature sensor

4.1 Different type of sensors

A temperature sensor integrated into the LinPak power module allows monitoring the temperature of the power module during operation. The feature is mainly used to detect failures of the cooling system. Typically, NTC resistors (negative temperature coefficient) are used as sensor within power modules. Often a round MELF type NTC resistor is used which is soldered on both ends to two separated copper pads. While the MELF type features a hermetical glass encapsulation, which makes it reliable in corrosive environment, the soldering process with solder preforms, as typically used for high voltage modules, is difficult. Furthermore, it will not be possible to use advanced die bonding techniques such as silver sintering or diffusion soldering. In addition, the high voltage insulation might suffer from bridging two conductor pads with the resistor. Partial discharges could occur below the resistor. Therefore, an alternative chip type resistor was verified for use within the LinPak module. The chip type resistor can be attached using the same technique as for attaching the IGBTs. The top side of the resistor chip is contacted by thick aluminum wire bonding.

Figure 8: NTC resistors tested, chip type (left) and MELF type (right).

To verify the reliability of the NTC resistors, the following tests are done: high temperature storage 168 h @ 175°C, low temperature storage 168 h @ -48°C, 100 temperature shocks from -50°C to 150°C, HAST (highly accelerated stress test) 168 h @ 125°C, 85% humidity. While the chip type NTC passed all tests, the MELF type NTC failed the shock cycling test. Cracks in the solder connections and considerable drift in resistance values were found. However since the shock cycling was done without the protective silicone gel, the test might be too harsh.

4.2 Model to derive the chips' temperature

The integrated NTC temperature sensor of the LinPak module can easily be evaluated during operation. The sensor is spaced apart from the IGBT chips and even further apart from the diode chips. The sensor therefore is not measuring the chips' temperature directly, but from the NTC temperature signal T_N the chips' temperature can be calculated. In steady state, this can be done using the following equations:

IGBT Temperature: $\quad T_{j,I} = T_N + P_I R_{th,I} + P_D R_{th,D \to I}$

diode temperature: $\quad T_{j,D} = T_N + P_I R_{th,I \to D} + P_D R_{th,D}$

Where the values of P_I and P_D are the power dissipation of the IGBTs and diodes per switch. The R_{th} values used in the equation are all with reference to NTC temperature. The definitions are the following:

IGBT heats IGBT: $R_{th,I} = \frac{T_{j,I} - T_N}{P_I}$ $\qquad$ IGBT heats diode: $R_{th,I \to D} = \frac{T_{j,D} - T_N}{P_I}$

diode heats diode: $R_{th,D} = \frac{T_{j,D} - T_N}{P_D}$ $\qquad$ diode heats IGBT: $R_{th,D \to I} = \frac{T_{j,I} - T_N}{P_D}$

The above equations are valid if the power is constant for some seconds. After this time the temperature difference between the chips and the NTC saturates. However, if the power is changing fast and the impact on the chips' temperature is needed to be known, the above equations need to be adjusted for the influence of time. This is done using thermal impedance (Z_{th}) curves instead of thermal resistances. The equations for getting the temperatures will turn into convolution integrals:

$$T_{j,I}(t) = T_N(t) + \int_{\hat{t}=0}^{t} \dot{P}_I(\hat{t})Z_{th,I}(t-\hat{t})d\hat{t} + \int_{\hat{t}=0}^{t} \dot{P}_D(\hat{t})Z_{th,D\to I}(t-\hat{t})d\hat{t}$$

$$T_{j,D}(t) = T_N(t) + \int_{\hat{t}=0}^{t} \dot{P}_I(\hat{t})Z_{th,I\to D}(t-\hat{t})d\hat{t} + \int_{\hat{t}=0}^{t} \dot{P}_D(\hat{t})Z_{th,D}(t-\hat{t})d\hat{t}$$

The Z_{th} curves for the above equations are obtained from finite element simulation. The curves are shown in Figure 9. For simplify the calculation, the curves can be fitted using a Foster network. For each curve, this is done using the equation below. The values from the fitting are given in Table 1.

$$Z_{th}(t) = \sum_{i=1}^{n} R_{th,i}\left(1 - e^{-\frac{t}{\tau_i}}\right)$$

	IGBT	diode	I → D	D → I
$R_{th,1}$	4.99 K/kW	12.9 K/kW	43.37 K/kW	-7.63 K/kW
$R_{th,2}$	10.58 K/kW	26.0 K/kW	-40.02 K/kW	29.0 K/kW
$R_{th,3}$	29.47 K/kW	48.9 K/kW		
$R_{th,tot}$	45.1 K/kW	87.8 K/kW	3.34 K/kW	21.4 K/kW
τ_1	2.15 ms	4.86 ms	1.97 s	0.513 s
τ_2	44.7 ms	94.2 ms	2.52 s	1.55 s
τ_3	0.597 s	1.38 s		

Table 1: Foster model parameters for the 1.7 kV LinPak NTC temperature sensor.

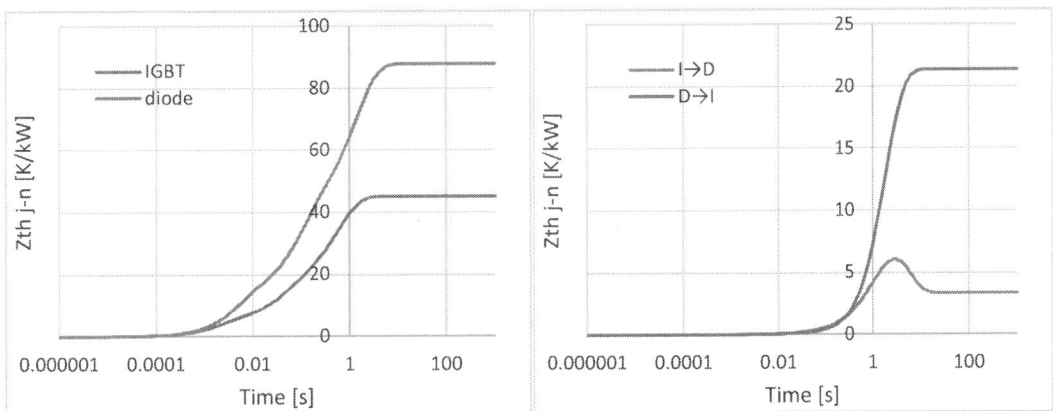

Figure 9: Zth curves for IGBT and diode junction temperatures with respect to NTC temperature. Self-heating of IGBT and diode on the left and on the right the curves for IGBT heats diode (I→D) and diode heats IGBT (D→I).

The accuracy of the described model has been estimated considering the following error sources: inaccuracy of the simulation assumptions (e.g. thermal conductivities), power terminal as additional heat source, NTC inaccuracy, changes in the cooling path. The overall accuracy for the chip temperature estimation obtained is 9.6°C for the IGBT and 20.4°C for the diode. The considerably high error for the diode temperature is due to its larger distance to the temperature sensor. Also for the IGBT temperature, the error is quite large. However, for some applications the accuracy might be sufficient.

A major contributor to the inaccuracy are within the simulation assumptions. These can be avoided by measuring the NTC response curves instead of using simulated curves. The IGBT chips or the diode chips of LinPak module can be heated up to steady-state using DC current. After turning off the current source, the cooling curve of the diodes and the IGBTs can be measured using the $V_{CE}(T)$ or $V_F(T)$ method respectively [5]. At the same time the NTC can be read. Finally the needed curves are obtained for the temperature model. When using measured NTC temperature response curves, the error of the model is expected to come down to 6.1°C for the IGBT and 12.6°C for the diode. This is already a quite reasonable accuracy for many applications. For instance, the model can be used to estimate the remaining power cycling lifetime of the modules used in a traction application in the field.

5 Conclusions and Outlook

Measurements on the 1.7 kV and 3.3 kV LinPak have shown stable switching behaviour under nominal and safe operating area conditions. The robust switching together with the exceptionally low stray inductance and the possibility of easy paralleling make the module ideally suited to become the standard in demanding high power applications. To meet the high reliability requirements of these applications, the module design has been done carefully following safe design rules and introducing a reliable particle free ultra sonic welding process.

An integrated NTC temperature sensor allows detecting cooling failures in the application and estimating the chips' temperature by a method proposed in this paper. An experimental verification of the temperature estimation will be done later this year.

The LinPak IGBT power module is currently in qualification. Market introduction for the 1.7 kV, 1000 A LinPak will be end of 2016/beginning 2017. The 3.3 kV, 450 A LinPak will be released soon after.

References

[1] Raffael Schnell et al.: *LinPak, a new low inductive phase-leg IGBT module with easy paralleling for high power density converter designs.* PCIM Europe 2015, Nuremberg

[2] Georg Borghoff: *Implementation of low inductive strip line concept for symmetric switching in a new high power module.* PCIM Europe 2013, Nuremberg

[3] Chiara Corvasce et al.: *New 1700V SPT⁺ IGBT and Diode Chip Set with 175°C Operating Junction Temperature.* EPE 2011, Birmingham, UK, Aug. 2011

[4] Roman Tschirbs et al.: *Ultrasonic metal welding as contact technology for state-of-the-art power modules.* PCIM Europe 2008, Nuremberg

[5] Uwe Scheuermann et al.: *Investigations on the $V_{CE}(T)$-Method to Determine the Junction Temperature by Using the Chip Itself as Sensor.* PCIM Europe 2009, Nuremberg

[6] Chiara Corvasce et al.: *3300V HiPak modules with Enhanced Trench (TSPT⁺) IGBTs and Field Charge Extraction Diodes rated up to 1800A.* PCIM Europe 2016, Nuremberg

Power Converter GTO to IGBT upgrade – a new life for traction converters

Luis Sequeira, Nomad Tech, Lisbon, Portugal, luis.sequeira@nomadtech.pt

Adriano Carvalho, FEUP, Porto, Portugal, asc@fe.up.pt

Augusto Franco, Nomad Tech, Porto, Portugal, augusto.franco@nomadtech.pt

Nuno Freitas, Nomad Tech, Porto, Portugal, nuno.freitas@nomadtech.pt

Abstract

The power traction converter with thyristor GTO (Gate Turn Off) technology was, in the last 30 years, the most widely used in railways traction applications. Currently, owners and operators find themselves in great difficulties due to the fact of the technological performance of GTO technology compared with new ones.

This GTO to IGBT(Insulated Gate Bipolar Transistor) upgrade/modernization project was launched as a pioneer solution in this type of application, resulting in huge benefits for the owners and operators of the rolling stock, such as, but not limited to: increase in the lifetime of the rolling stock and consequently reduced life-cycle cost; higher efficiency either operational or energy-wise; embedded remote condition monitoring, with no need for major and intrusive changes on the vehicle or the need for special approvals or re-homologation. The modernization installation process can be performed on an existing immobilization short period of the rolling-stock.

Introduction

The main factors that led to the implementation of this project, referred as Lusogate, were the increase of traction malfunction situations related to the GTO technology and the rise of the costs associated with maintaining this technology. Owners, operators and maintainers of such systems are struggling to deal with the technological obsolescence of the GTO technology, that has induced operational and maintenance problems, and restrained them to a very limited supply chain - sometimes they have even to stop valid working units as spare units are difficult to manage in real-time conditions.

The research, study and laboratory development phase started in 2010, and consisted on the design of new power modules for the traction converters based on IGBT technology and compatible drive system for existing railways traction control, incorporating new features for remote condition monitoring and online supervision. Figure 1 illustrates the triggering of the power module in transition from off-state to on-state.

Clearly and grounding the project approach, the technical and operational IGBT technology causes an immediate increase of the rolling stock lifetime as well as introduced reliability and availability increases and improvements.

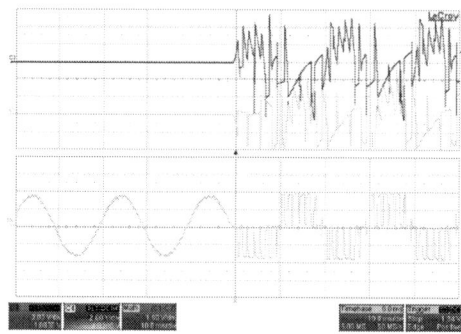

Figure 1 - Illustration of the first triggering for the first traction power test. Collector current top and bottom chopper voltage.

Main steps of the GTO to IGBT upgrade project – chronological summary of the project

<u>Study and analysis of the problem</u> – the reduction of the traction converter reliability, train disruptions in commercial service and the negative impact on the life-cycle cost were the main triggers for the project's kick-off. The study was initially conducted for the characterization of the traction converter operational running conditions in real commercial service, for the identification of the best technical solutions and for analyzing the performance of the new devices when compared to previous ones.

<u>Specification of the physical limits of intervention</u> – for the success of the project it was necessary to study and understand the existing solution (implemented with GTO technology) on traction converters. The limits of intervention were based on the power equipment safe operation area and keeping transparent the train traction control for the modernized traction converter control electronical interface.

<u>Lab essays in Power Test Bench</u> – this phase shows to be a necessary step, as it allows at establishing conditions similar to real operational conditions. The power bench is composed of two phase modules and an external circuit (snubber for GTOs), allowing to essay 400A switching current in a 600V DC bus.

Figure 2 shows a picture of the power test bench.

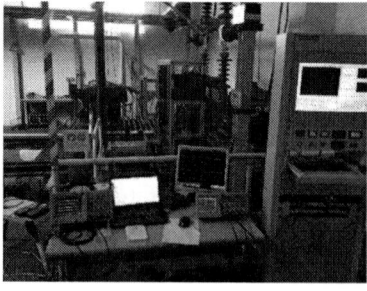

Figure 2 – Power test bench.

Prototyping, testing and proof of concept – to validate the solution was chosen as appropriate to replace two phase modules with GTO by IGBT, in the chopper converter in the traction converter. The results of the static/dynamic tests confirmed the perfect compatibility of the Lusogate system with the GTO system. Figure 3 illustrates a power module in essay, an upper view of power module arrangement and the monitoring and control desk in an essay condition.

Figure 3 – Final assembly stages.

Final assembly – After the proof of concept some functional improvements were carried out (mechanical and electrical), then they were built the final 10 IGBT phase modules as well as installed in the train. Figure 4 illustrates the final assembling in the train.

Figure 4 – Final validation stages.

Final validation - Testing and EMC approval - the first commercial unit has been performing regular commercial services as from March 2015 in Portugal. This unit has been operating an average of 20 hours per day in a train line with the hardest service either in driving profile or load profile.

Real operational results have shown:

- Good performance in terms of harmonic voltages and currents pattern;
- Improved behavior when facing power transients, namely due to inherent automatic handling by the IGBT driver of very short high power currents, what avoids frequent actuations of the DC Bus protection breakers and leads to almost inexistent "stop in line" situations.
- Complete technical compatibility with GTO based solution;

© VDE VERLAG GMBH · Berlin · Offenbach

- Appropriate cooling system perfect response (previously based on Freon circulation, now on forced air)
- Finally, in terms of energy efficiency, savings are currently monitored at 12%, by comparison between one EMU half equipped with GTO and the other half with IGBTs.

Conclusions

The paper summarizes the framing and motivation to analyze a movement from GTO technology to IGBT technology for traction power modules, showing the project steps developed by the authors.

The present technological status of IGBT technology as well as its technical advantages when they improve reliability and availability of the power converters operating according to rolling stock lifetime, with consequent reduction of maintenance costs clearly becomes the analyzed upgrade an issue.

The operational results carried out with a IGBT prototype running one year in real commercial service what allowed a precise performance comparison allow at concluding that the upgrade must be implemented always the GTO technology fails demanding for maintenance actuation. Besides the authors' conclusion it should be added that the upgrade costs are similar to maintenance costs.

References

1. Sequeira, Luis. "Estudo das Funções de Auto-protecção e Diagnóstico de Estado para Módulos de Disparo de IGBT num Ondulador para Tracção Ferroviária", Dissertação de Mestrado de Engenharia Electrotécnica do Instituto Superior de Engenharia de Lisboa 2013.
2. Motto Eric R, Yamamoto M., "New High Power Semiconductors: High Voltage IGBTs and GCTs", Proc. PCIM98 Power Electron. Conf., 1998
3. Volke Andreas, Hornkamp Michael, "IGBT Modules Technologies, Driver and Application", Munich, Infineon Technologies, 2012.

Aspects of reliability improvement for large area power semiconductor devices through sintering

Dmitry Titushkin, «Proton-Electrotex»,JSC, d.titushkin@proton-electrotex.com,
Alexey Surma, «Proton-Electrotex»,JSC, a.surma@proton-electrotex.com,
Michiel de Monchy, «Alpha Alent», MdeMonch@alent.com
Anna Lifton, «Alpha an Alent plc Company», alifton@alent.com

Abstract

Present article covers aspects of reliability improvement for power semiconductor devices of 80+ mm chip diameter by low temperature of silicon die- molybdenum disc junction with silver nanoparticles. The article studies the connection between the porosity of the joint and the cycling endurance in experimental samples and shows the area of optimal pressure and temperature for the junction procedure.

Introduction

Sintering technology is widely used in power semiconductor electronics [1,4]. One of the perspective areas of sintering application is joining semiconductor crystals and molybdenum discs of high and ultra-high power thyristors and diodes with 80-150 mm silicon die.

Experience with use of sintering in manufacturing of single-crystal power thyristors and diodes proves the advantages of this technology, including thermal cycling [3, 4], thermal resistance decrease [2, 3, 4] and surge current value increase [1]. Another important advantage of sintering is decreased area spread of emitter layer injection coefficient [6] as there is no dissolving of silicon die surface layers. Transition to silver nanoparticle materials is a common tendency nowadays, as they provide pressure and temperature lowering at sintering thanks to higher free energy. In turn, high free energy may result into inner mechanic damage during sintering. This damage may be reduced by increasing the porosity of the joint, which, however, causes a number of problems, discussed further in this work. That is why it is important to find balance in manufacturing of semiconductor devices with 80mm+ silicon die, in order to secure right values of all characteristics mentioned above.

Sintering Mechanisms

Sintering technology is based on diffusion welding principle and plastic deformation of silver particles as free energy reserve on their surface is the driving force of sintering process. Basically sintering is described by Mackenzie Shuttleworth model [5]:

$$\frac{d\rho}{dt} = \frac{3\gamma}{2r_0\eta(T)}(1 - \rho_0),$$

Where γ - surface energy;
r_0 – initial radius of the spherical pores;
ρ_0 – initial porosity;
$\eta(T)$ – temperature dependent shear viscosity.

As the silver particle size decreases, its free energy value goes up.

Table. 1. Dependence free energy

Silver Particle Size	2.0 µm	100 nm	30 nm
Driving Force	2.0 MPa	40 MPa	143 MPa

Free energy increase allows temperature lowering at sintering which in its turn positively affects residual deformation values and mechanical distortions in silicon die-molly disc structure.

However, due to free energy excess, baking speed can be very high leading to internal mechanical damages and cracking of the seam. This can be avoided by raising the junction porosity as it improves its elasticity (Young's modulus decreases), but with porosity increase thermal conductance of the seam degrades and can negatively affect device's thermal resistance.

Also, there is a negative effect caused by the decrease of Young's modulus, which is specific only for thyristors and diodes with large area die mounted in press-pack capsules. During thermal cycling of such devices, tangential mechanical force is transmitted by friction from upper and lower copper base to the semiconductor element joined with the molybdenum disc. The molybdenum disc serves precisely to compensate for the impact.

However, the decrease of Young's modulus in the junction causes the molybdenum temperature compensator to stop functioning and the silicon wafer to break.

For semiconductor devices with 80+ mm chip it is important to provide the joint porosity value, which eliminates internal stress during sintering but excludes critical decrease of thermal resistance and mechanical robustness of the joint.

Experimental samples

To get the dependence of thermal cycle endurance on the joint porosity and determine the optimal area for sintering temperature and pressure, we produced experimental samples of TFI393-2500-28 thyristors with 100 mm semiconductor element. These samples were aimed at repetitive peak reverse voltage of 2800 V and average current of 2500 A. For sinter material we used silver film Argomax 8010 by Alpha an Alent plc Company. The samples were produced in the following sintering regimes: temperature range from 195 to 235 °C and pressure range from 5 to 20 MPa.

Possibilities of improving the reliability of power bipolar devices

Figure 1 shows the dependence of porosity on low-temperature joint regimes that we have determinded.

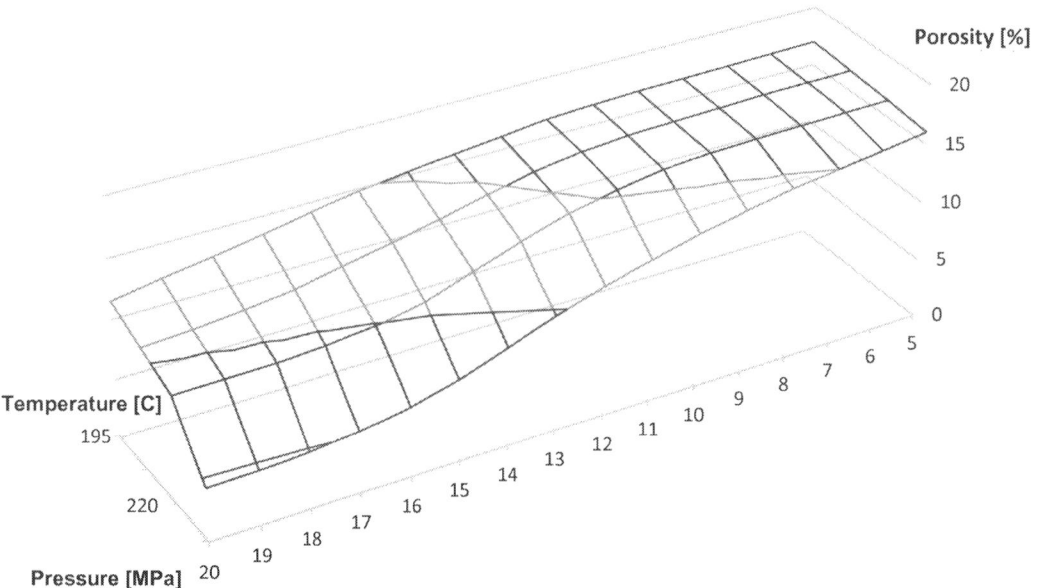

Figure 1. Dependence of silver joint porosity on sintering process parameters

PCIM Europe 2016, 10 – 12 May 2016, Nuremberg, Germany

Experimental samples were tested for thermal cycling endurance in the following regime: temperature range from 25 to 150 °C (125 °C), 100 cycles. Test results are below (Figure 2):

- Thermal cycling endurance of the samples produced at temperature no less than 220 °C did not exceed 10-15 cycles, i.e. the porosity of the resulting joint was clearly insufficient;
- Thermal cycling endurance of the samples produced at 220-235 220-235 °C and 10MPa pressure also did not exceed 10-15 cycles;
- Sintering pressure rise over 12 MPa causes drastic increase of thermal cycling endurance
- Samples produced in 20MPa/235°C showed no disruption or decrease in characteristics after the cycling;

The results we received are well combined with the joint porosity changes in different sintering regimes (Figure 3).The figure shows that the acceptable level of cycling endurance is possible if the porosity of the joint does not exceed 7%.

The results allow us to forecast the pressure and temperature area where sintering of the used nano material gives the silicon-molybdenum junction for elements of thyristors and diodes with large area crystal (Figure 4).

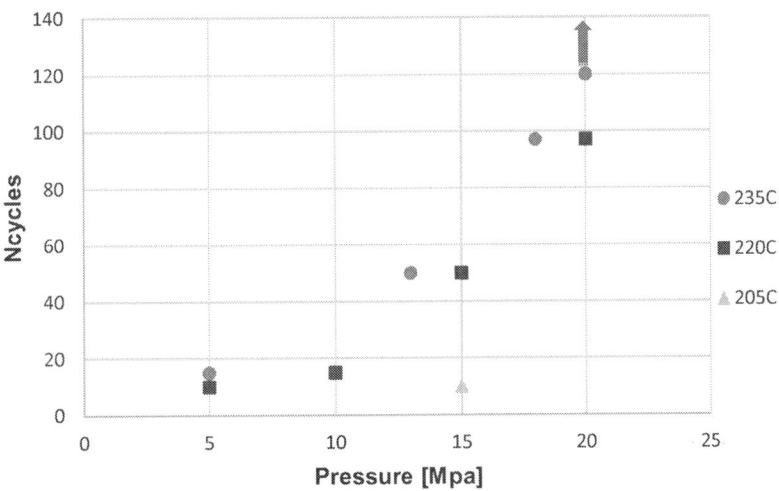

Figure 2. Thermal cycling endurance in experimental samples obtained in different sintering regimes

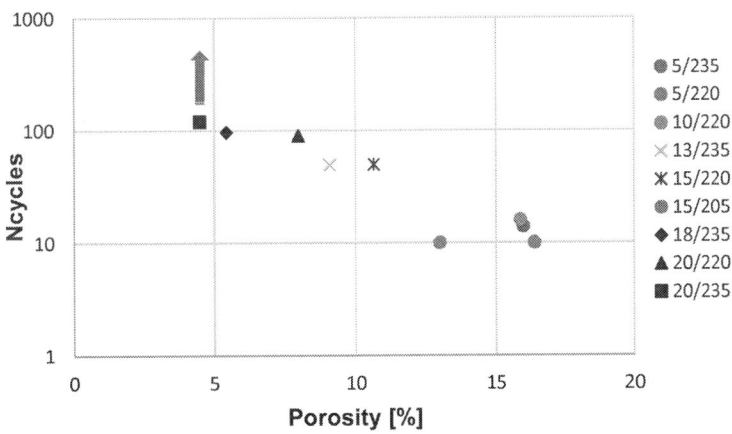

Pic. 3. Dependence of thermal cycling endurance on joint porosity.

© VDE VERLAG GMBH · Berlin · Offenbach

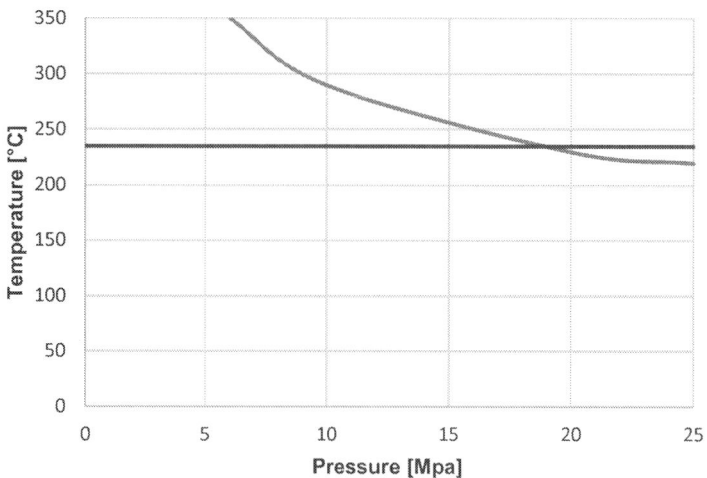

Figure 4. Area thermal cycling endurance (above the boundary line)

Conclusion

The article studies the influence of silver nanoparticle sintering regime on the porosity and thermal cycling endurance of the large area silicon die/molybdenum disc joint. It also shows the connection between the thermal cycling endurance of the experimental samples and the porosity of the joint. In addition, the article explains, that in order to provide thermal cycling endurance of press-pack capsule thyristors and diodes with large area silicon crystal, the porosity must not exceed 5%. We have determined the temperature and pressure rate for nano film sintering which enables achieving high cycling endurance of the joint in the devices mentioned above.

References

1. H. Schwarzbauer. Novel Large Area Joining Technique for Improved Power Device Performance. - IEEE Transactions on Industrial Applications, 27 (1), 1991, p. 93- 95.
2. A. Chernikov, A. Stavtsev, A. Surma. Features of wafer - Mo joining by sintering of silver paste for large area silicon devices. – Proc. EPE'2013, Lille, 2013
3. U. Scheuermann, P.Wiedl. Low Temperature Joining Technology-A High Reliability Alternative to Solder Contacts. - Workshop on Metal Ceramic Composites for Functional Application, Vienna, 1997, p 181-192.
4. C. Göbl, P. Beckedahl, H. Braml. Low temperature sinter technology Die attachment for automotive power electronic applications. – Proc. Automotive Power Electronics, Paris ,2006.
5. M.O.Prado, C. Fredericci, E.D. Zanotto. Glass sintering with concurrent crystallisation. Part 2.Nonisothermal sintering of jagged polydispersed particles. – Physics and Chemistry of Glasses, 43 (5), 2002.
6. Dmitriy Titushkin, Alexey Suma. «New ways to produce fast power thyristors» - Bodo's Power Systems 08, 2015, p. 28- 29.

© VDE VERLAG GMBH · Berlin · Offenbach

PCIM Europe 2016, 10 – 12 May 2016, Nuremberg, Germany

Analysis of interface structure and composition of Cu/Al₂O₃ for the high stability of DBC (direct bonded copper)

Hyunwoo Kim[1], Jaehoon Jung[1], Hanna Choi[1], Kisoo Jun[2]

[1]KCC Corporation, 17-3, Mabungro240Beon-gil, Giheung-gu, Yongin-si, Gyeonggi-do, Korea
[2]KCC Corporation, 19, Wanjusandan 4-ro, Bongdong-eup, Wanju-gun, Jeollabuk-do, Korea

Topic Number: 1.5; Topic Name: Power Semiconductors; Power Modules and Power Hybrids

Abstract

Copper and aluminum oxide have been bonded using two procedures; solid state bonding and liquid state bonding by the eutectic method. As a consequence of these treatments, the $CuAlO_2$ and $CuAl_2O_4$ phases are formed at the interface. The phases transformations of delafossite $CuAlO_2$ and spinel $CuAl_2O_4$ structures shows a particular orientation with respect to oxygen content during the joining process leading to improve the mechanical properties. In this study, eutectically bonded copper-aluminum oxide joints are subjected to oxygen potential. The interface phases are of important characteristics for manufacturing DBC substrates. Especially, the bonding strength is closely related to the different phases of eutectic conditions. One of these conditions, we used a copper powder in order to control the concentration of oxygen. The amount of copper powder to reduce the oxygen concentration increased, the bonding strength after firing increased due to the increase of $CuAlO_2$ as well as $CuAl_2O_4$. In this paper, we identify the phase characteristics using electron backscatter diffraction (EBSD) for phase created by varying the components of the interface.

Introduction

When copper and Al_2O_3 are bonded, CuO layers or components are formed by supplied oxygen. This oxygen is provided by several paths such as low oxygen atmosphere in the furnace, pre-oxidation of copper foil, and copper oxide powder. The joint is formed through the synthesis of copper oxides with Al_2O_3 in the presence of the molten Cu-Cu_2O eutectic which well wets the ceramic substrate. It is known that the oxygen promotes wetting of liquid copper to Al_2O_3 substrate and is necessary to achieve good spreading of the liquid along the bond interface. Oxygen-rich liquid copper is known to form a $CuAlO_2$ layer contacting with Al_2O_3. However, it is not certain if the formation of the complex oxide is indispensable to make strong bonding strength or is simply a result of good wetting. The synthesis of $CuAlO_2$ and $CuAl_2O_4$ at the interface is promoted by the oxygen released during the reduction of copper oxide. By this CuO layer or components, $CuAlO_2$ and $CuAl_2O_4$ layers are formed in the bonding interface of copper/Al_2O_3. The reaction for forming $CuAlO_2$ and $CuAl_2O_4$ layers are simply described by following reactions [1, 14].

$$2\,CuO \rightarrow Cu_2O + \tfrac{1}{2}\,O_2 \qquad (1)$$

$$Cu_2O \rightarrow 2Cu + \tfrac{1}{2}O_2 \qquad (2)$$

$$Cu_2O + Al_2O_3 \rightarrow 2\,CuAlO_2 \qquad (3)$$

$$Cu_2O + Al_2O_3 + 2Cu + 2O_2 \rightarrow 2CuAl_2O_4 \quad (4)$$

Namely, for bonding between copper and Al_2O_3 by forming eutectic liquid, the formation of $CuAlO_2$ and $CuAl_2O_4$ layers must be formed. For the formation of $CuAlO_2$ layer, the reduction

© VDE VERLAG GMBH · Berlin · Offenbach

to Cu_2O from CuO is necessary. The phase change to Cu_2O from CuO happens at about 600~900℃ because Cu_2O is better stable phase than CuO. The formation of $CuAlO_2$ from Cu_2O and Al_2O_3 at about 1075°C is the reaction for better stable phase. These reactions are necessary for bonding process between copper and Al_2O_3.

As following reaction (5), we can notice that the phase changed to Cu_2O from CuO is noticeable by the reaction between CuO and copper directly.

$$2\,CuO + 2\,Cu \rightarrow 2\,Cu_2O \qquad (5)$$

For this direct reaction between CuO and copper, interface phases which can promote the bonding strength by the increase of $CuAlO_2$ without adjusting other eutectic conditions such as pressure, temperature, reductive atmosphere. However, at higher oxygen concentration the spinel $CuAl_2O_4$ may be formed according to the results of the equilibrium experiments carried by Jacob and Alcock[9]. The aim of the present study was to examine how the phase transformations that occur at the interface between copper and Al_2O_3 during joining process in the presence of oxygen and mixed copper and CuO powder affect the mechanical strength of the copper alumina joints.

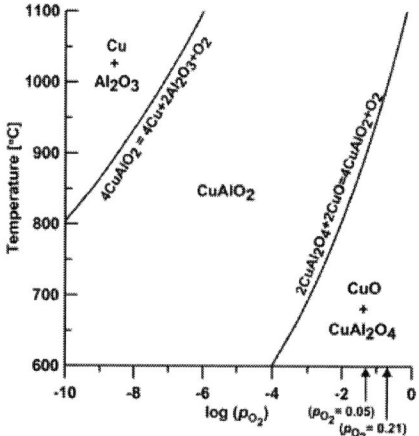

Fig.1: Equilibrium phase relationships in the ternary Cu-Al-O system in a 1:1 molar mixture of CuO and Al_2O_3 at 1 atm total pressure based on the Gibbs energy data reported by Jacob an Alcock [9]

Method
Experiments

In order to examine the bonding characteristics of copper/Al_2O_3 depending on the interface phases various compositions of copper and CuO powders were prepared. Copper and cupric oxide(CuO) are used in the commercial powders with a purity of 99.9 wt% and 98.5 wt% and average particle size of 10 μm and 20 μm, respectively. The procedure consisted of the spray of mixed powder (slightly covered with constant amount of power) on the Al_2O_3 substrate followed by the annealing at elevated temperature in the tube type furnace. The pressure in the furnace was sustained about 2~3 torr and the inflow of oxygen from outside were blocked by using argon gas. The argon gas inflow was kept at 3L/min and the temperature inside of the furnace was kept at 1075°C for 6 minutes. The size of Al_2O_3 substrates was 40 mm x 40 mm and the size of copper sheets was 38 mm x 38 mm. Thermal oxidation of copper sheet is usually performed in a temperature range of 200 to 300 ℃ in the DBC process. We employed pre-oxidized copper sheet without changing its property that may be optimized in the DBC process. It is reported that the grain size of oxidized copper surface is attributable to lower yield strength of the copper metallization.[10] The mixture of copper and CuO powder balanced were sprayed evenly on the Al_2O_3 substrate and then cover the copper foil etched by hydrofluoric acid. To ensure the uniform distribution of the mixed powders, the thickness of the oxide layer after firing is restricted to 2~3 µm. Samples intended for examining the interface phases and joint strength were prepared by depositing paste, made of Cu and CuO mixing compositions according to following table.

Tab.1: Composition of the mixtures used in the experiments (wt%)

Compositions no.	Cu	CuO
1	-	100
2	10	90
3	20	20
4	50	50
5	100	-

In order to analyze the interface phase and compositions, electron backscatter diffraction (EBSD)[15] followed by ion milling on the target surface and focused ion beam(FIB) with energy dispersive x-ray spectroscopy(EDAX) are used.

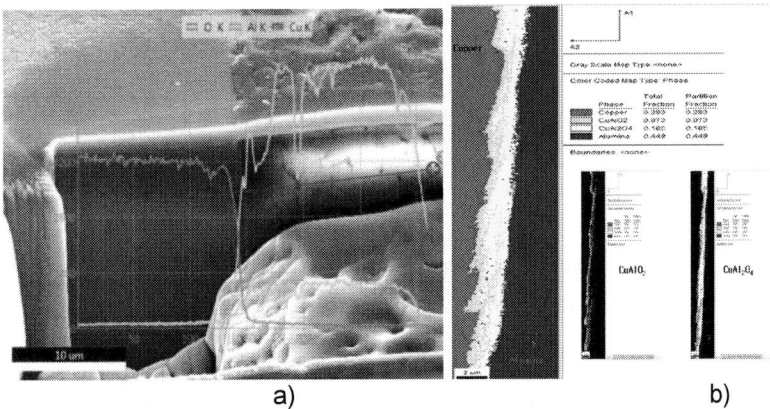

a)   b)

Fig.2. a) Cross section of Copper and Al_2O_3 and the element (Al, O, Cu) profile of the DBC bonding, b) $CuAlO_2$ and $CuAl_2O_4$ distribution at the interface of Copper and Al_2O_3, and the dominant phase showing yellow area at the interface is the spinel structure of $CuAl_2O_4$.

Structural properties

Amsterdam density functional (ADF)[12] is used to predict and understand the structural properties of interlayer materials at 0 K. Generalized gradient approximation (GGA) functional was used which is parameterized by Perdew, Bruke, and Ernzerhof(PBE)[13]. All calculations were preformed spin-polarized with DZ, triple-zeta quality basis set. The Brillouin zone was sampled by a 1x1x1 mesh of k-points for the calculation of density of states and band structure of $CuAlO_2$ and $CuAl_2O_4$.

The $CuAlO_2$ as well as $CuAl_2O_4$ are formed at the interface between copper and Al_2O_3 as shown in Fig.2. The spinel structure of $CuAl_2O_4$ is a little dominant phase, but the delafossite structure of $CuAlO_2$ which is in the vicinity of copper face also increased when the copper content increased. This means that the Cu atom from Cu powder and CuO powder was considered to belong to the spinel $CuAl_2O_4$ preferably.

Mechanical properties

The specimens were subjected to a 90° peel test after being positioned in a testing jig and the test was conducted at a pull rate of 0.1 mm/min. Peel strength was defined as the average load per unit width of copper expect for the initial transient. The mechanical strength of the ceramic/copper joints strongly depends on the phase change that occurs in the interface layers of alumina and copper during annealing with mixed copper and copper oxide, the bonding of alumina substrates by copper with these two compositions. The measured strength values are very varied, especially in the joints bonded directly through the ratio of $CuAlO_2$ and $CuAl_2O_4$. The copper's addition to copper oxide causes a substantial increase of the joints strength and repeatability of the results. The higher bonding strength was obtained in the joints bonded directly through the metallic copper addition composed of copper powder with cupric oxide (CuO).

Results

Metal and ceramics bonding is difficult because of the difference in their electrical, thermal, mechanical, and chemical characteristics. Table2 illustrated physical properties for the copper and Al_2O_3 system which is representative in the direct bonded copper (DBC) process.

Brief description of the bulk structure and physical properties of copper and copper derivative compounds enable us to understand the interfacial properties which may cause phase transformation. Bonds performed under oxygen pressure atmosphere lead to the growth of interfacial oxides as a result of a diffusion of copper metal in the Al_2O_3 by the addition reaction leading to form spinel $CuAl_2O_4$ as well as delafossite $CuAlO_2$ reacted from the Cu_2O and Al_2O_3. Table summarized the crystallographic parameter of interfacial oxide phases.

Tab. 2: Physical properties of copper and Al_2O_3 [3]

	Tm(℃)	Expansion coeff.(C^{-1})	Young's modulus (GPa)	Diffusion Coeff. (cm^2/s)	Thermal cond. (Wm/s)	Surface energy (mJ/m2)
Cu	1,083	17×10^{-6}	125	2×10^{-9} (at 1000℃)	153	1,200
Al_2O_3	2,037	6.5×10^{-6}	340	Al: 10^{-19} O: 10^{-23} (at 1000℃)	1.4	Mono:748 Poly:1,560

Tab. 3: The crystallographic parameters of interfacial phases between Copper and Al_2O_3

	Cu_2O	$CuAlO_2$	$CuAl_2O_4$	Al_2O_3
Structure	FCC,BCC	Delafossite R-3m	Spinel	Corundum
Group	Cubic	Rhombohedral	Cubic	Trigonal
Lattice parameters (theoretical, Å)	a=8.069	a=2.8604 c=16.953	a=8.08	a=4.76 c=12.99
Volume of unit cell(theoretical, $Å^3$)	77.85	119.76	131.88	84.93

The crystalline structure of Rhombohedral delafossite $CuAlO_2$ was found that the lattice parameters of $CuAlO_2$ are observed to be a=2.802 Å, c=16.704 Å (experiment, a=2.858 Å, c=16.958 Å [4]) and the internal parameter is u=0.1097. The Cu cations are linearly coordinated by oxygen and the Cu_2O dumbbells are separated by a layer of edge sharing AlO_6 octahedra. It is also known that copper planes which has a 2D-structure as shown in Fig.3 a) show the conduction characteristic.[5] The structure of $CuAl_2O_4$ has a closed-packed face-centered-cubic structure and in its unit cell Cu ions are tetrahedrally coordinated and Al ions occupy octahedral sites as illustrated in Fig.3 b). The lattice parameters of $CuAl_2O_4$ are observed to be 8.08 Å (exp. 8.0778 Å [6]) after geometry optimization.

© VDE VERLAG GMBH · Berlin · Offenbach

PCIM Europe 2016, 10 – 12 May 2016, Nuremberg, Germany

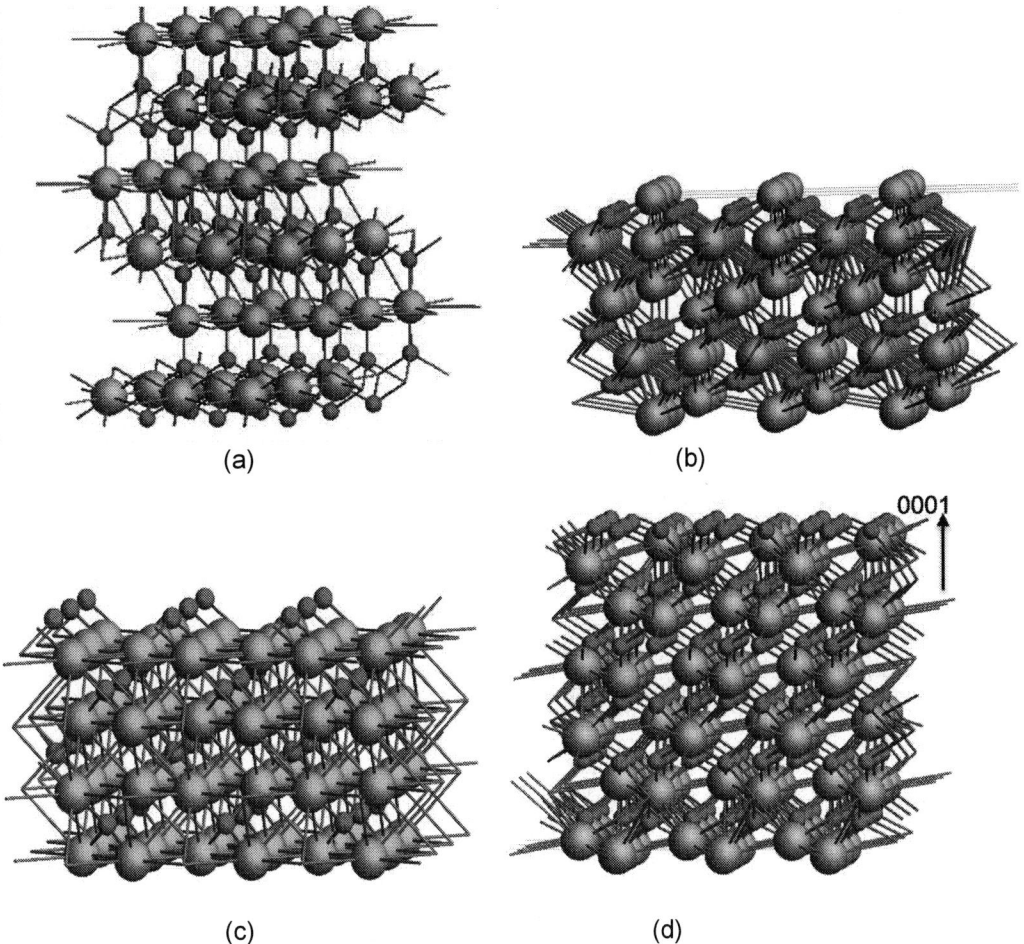

(a)   (b)

(c)   (d)

Fig. 3: Bulk Structures of a) delafossite $CuAlO_2$ b) spinel $CuAl_2O_4$, c) Cu_2O, and surface structure of d) Corundum $Al_2O_3(0001)$, yellow, gray, and red balls represent copper, aluminum, and oxygen, respectively.

The spinel $CuAl_2O_4$ is observable at the Al_2O_3 surface by the EBSD analysis, which is depended on the crystallographic orientation of the surface of alumina, especially if the system was not in thermodynamic equilibrium. At the transitional range where $CuAl_2O_4$ and $CuAlO_2$ coexist, copper prefers to form $CuAl_2O_4$ by reacting with the (0001) plane of alumina, and $CuAlO_2$ by reacting with the (1120) plane [7]. There can be two Al^{3+} ions for every three O^{2-} ions to maintain electrical neutrality. Thus, the cations occupy only two thirds of the octahedral sites of the basic array, which make different types of aluminum cations layer. The Oxygen-terminated surface is more preferable to be adsorbed by the metal atom than the Al-terminated surface because the surface polarity.[8]

Interfacial phase of Cu_2O has the simple cubic structure where the oxygen atoms occupy tetrahedral interstitial positions relative to the copper sublattice and the copper is linearly coordinated by two neighboring oxygens. The Cu_2O is observable in the region of copper sheet. Thermodynamically, the Cu_2O may be in contact with $CuAl_2O_4$ layer because these two phase have the same symmetry point group and the lattice parameter of the $CuAl_2O_4$ phase is almost twice that of the Cu_2O (a = 8.08 Å and 4.25 Å, respectively). However, Cu_2O grain is found in the Cu sheet probably due to the non-equilibrated state of the process in our test.

© VDE VERLAG GMBH · Berlin · Offenbach

PCIM Europe 2016, 10 – 12 May 2016, Nuremberg, Germany

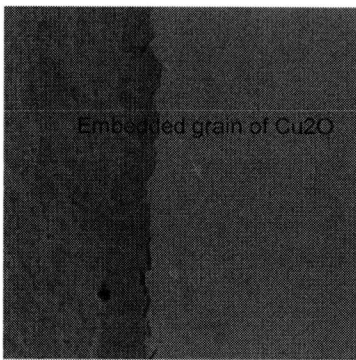

Fig. 4: The cross sectional view of alumina and copper and Cu2O grain embedded in the copper region.

Cross-sectional microstructure and the EDS analysis show that Al, Cu, O concentration profiling and the roughness of layers at the interface are obvious and there are two adjacent layers of $CuAlO_2$ and $CuAl_2O_4$ as shown in Fig.2. The spinel structure of $CuAl_2O_4$ is preferably formed at the mixed copper and CuO with 1~2 times greater than the forming of the delafossite structure of $CuAlO_2$. EBSD measurement indicated that the thickness of $CuAlO_2$ layer increased as copper powder increased as shown in Fig.6. Similarly, the relative area percent are calculated based on the specific cross-sectional area as shown in Table4. The rest part at the specific area includes pure copper and Al_2O_3 regions. Irrespective of whether only copper powder added or cupric oxide added, $CuAlO_2$ and $CuAl_2O_4$ phases are obtained. In the presence of only copper corresponding to the lowest concentration of oxygen, the spinel $CuAl_2O_4$ as well as delafossite $CuAlO_2$ is obtained favorably.

Tab. 4: Results of the EBSD phase identification and bonding strength for the DBC.

Type of powder (wt%)			EBSD		Bonding Strength
			Relative values (wt% on specific cross-sectional area)		Averaged (N/mm)
No.	copper powder	cupric oxide (CuO) powder	$CuAlO_2$	$CuAl_2O_4$	
0	-	-	5.40	4.80	10.63
1	0	100	3.30	3.20	9.36
2	10	90	4.90	7.40	NA
3	20	80	5.70	12.40	10.95
4	50	50	5.20	4.20	NA
5	100	0	7.30	18.50	12.58

Fig. 6 shows the oxygen effect on the $CuAlO_2$ and $CuAl_2O_4$ area which increase as the copper content increase, since the oxygen is deficient in reaction into the $CuAlO_2$ from the CuO and $CuAl_2O_4$ from the equation (4) with the aid of copper powder. At the highest Cu content, the highest $CuAlO_2$ and $CuAl_2O_4$ phases can be obtained as shown in Fig.5. This implies that the free oxygen may be surplus probably due to the existence of oxygen in the atmosphere, copper power, pre-oxidized copper sheet, and Al_2O_3. In addition, as the amount of copper increase, the $CuAlO_2$ increases according to the reaction (3).

© VDE VERLAG GMBH · Berlin · Offenbach

PCIM Europe 2016, 10 – 12 May 2016, Nuremberg, Germany

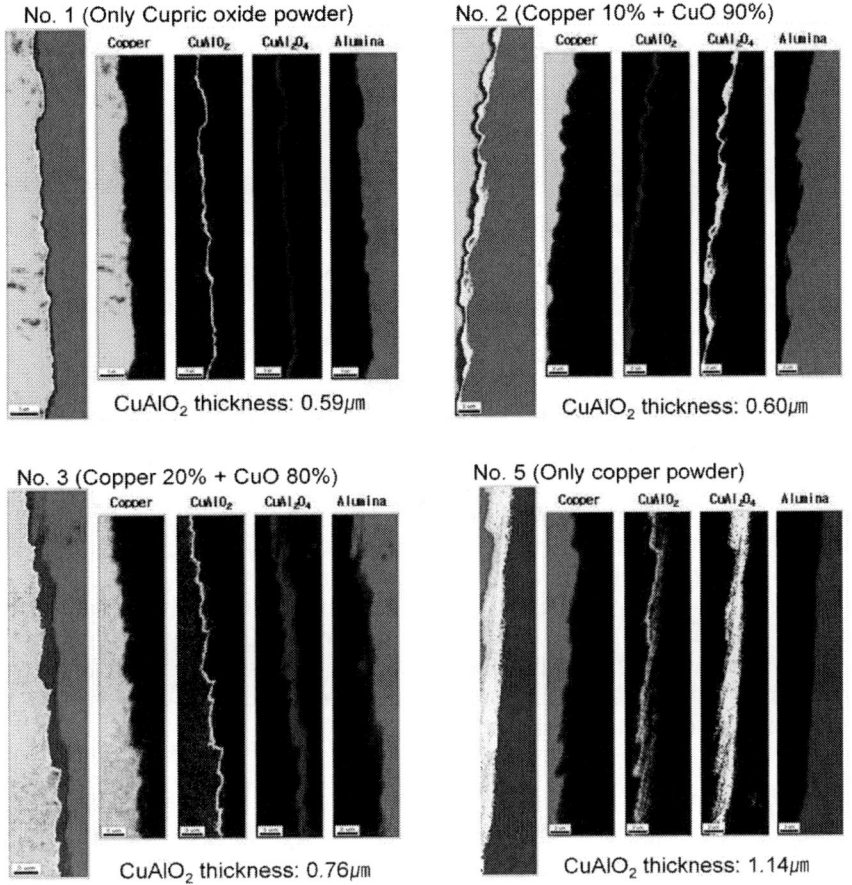

Fig.5: Image map showing the two reaction phases of $CuAlO_2$ and $CuAl_2O_4$ layers between copper and alumina.

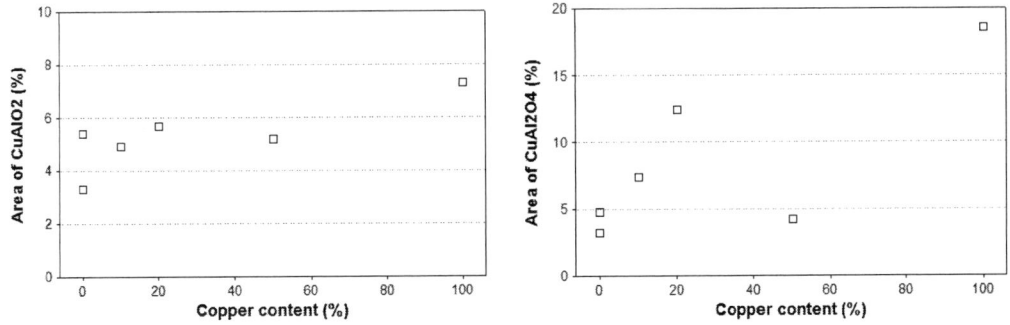

Fig. 6: The area percentage of (a) $CuAlO_2$ and (b) $CuAl_2O_4$ depending on the copper content in the interface of copper and alumina.

The results of the bonding strength influenced by the copper content were shown in Fig. 7. In the absence of copper powder, the bonding strength shows the lowest 9.4 N/mm, which is corresponded to the lowest amount, 3.2 wt% of $CuAlO_2$ relative area in the interface. As the copper concentration increases, bonding strength reach to 12.6 N/mm. This result is due to

© VDE VERLAG GMBH · Berlin · Offenbach

the increase of the $CuAlO_2$ as well as $CuAl_2O_4$. We found that the bonding strength can be accomplished in the case of the highest copper powder added and the smallest amount of oxygen in the interface.

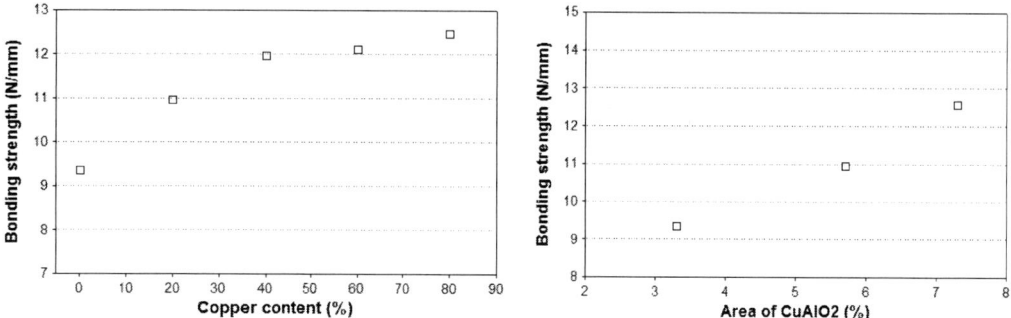

Fig. 7: Bonding strength depending on the copper content in the interface of copper and alumina.

Summary

This paper presents the copper and copper oxide interface phases for the bonding of copper sheet and Al_2O_3 substrate. As the copper content increases, the $CuAlO_2$ phase increase and resultantly the bonding strength increase. It is noticed that $CuAlO_2$ phase at the interface have a critical role of bonding property between copper and Al_2O_3. In addition, the formation of reactive oxide layer which can enhance the bonding property and conformity without adjusting the eutectic conditions such as temperature and reduction atmosphere can be achievable by replacing CuO with Cu powder.

Reference

[1] Kim, S.T. et al., Interfacial reaction product and its effect on the strength of copper to alumina eutectic bonding, *J. Materials Science*, 27, 2061-2066, 1992.

[2] J. Hong et al., Direct Bonding of Cu/AlN using $Cu-Cu_2O$ Eutectic Liquid, *J. Kor. Powd. Met. Inst.*, 20, 2, 2013.

[3] Interfaces in New Materials, P. Gange, B. Delmon, edited, Springer Science & Business Media, 171, 2012.

[4] Nie, S.-H. Wei, and S. B. Zhang, *Phys. Rev. Lett.* 88, 066405, 2002.

[5] D.B. Rogers et al., Chemistry of noble metal oxides. III. Electrical transport properties and crystal chemistry of ABO2 compounds with the delafossite structure. *Inorganic Chemistry* 10(4), pp. 723-727,1971.

[6] http://www.fiz-karlsruhe.de/icsd.html

[7] David W. Susnitzky, C. Barry Carter, *J. Mater. Res.*, 6,1991,1958.

[8] R. Cavallotti et al. Role of surface hydrolxy groups on zinc adsorption characteristics on a-Al_2O_3(0001) surfaces: First-principles study, *J.Phys.C*, 118, 13578-13589, 2014.

[9] Jacob, K. T.; Alcock, C. B. Thermodynamics of CuAlO2 and CuAl2O4 and phase quilibria in the system $Cu_2O-CuO-Al_2O_3$, *J. Am.Ceram. Soc.*, 58 (5−6), 192−195, 1975.

[10] A. Roth et al, Improved thermal cycling reliability of direct copper bonded substrates by manipulating metallization properties, *PCIM* 2013.

[11] Kirfel, A. et al., Accurate structure analysis with synchrotron radiation. The electron density in Al2O3 and Cu2O. *Acta Crystallogr.,Sect. A,* 46 271-284, 1990.

[12] http://www.scm.com

[13] Perdew, J.P., et al., Generalized gradient approximation made simple," *Phys. Rev. Lett.,* 77(18), 3865- 3869, 1996.

[14] M. Diemer, et al., *Amer. Ceram. Soc.*, 82, 10, 2825-32, 1999.

[15] Michael, J.R., in *Electron Backscatter Diffraction in Materials Science*, 75, Kluwer Academic/Plenum Publishers, New York, 2000.

Power Stack – Advantages and Reliability of an Aluminum Based Stacked Power Module

Robert Christopher, Burns, AB Mikroelektronik GmbH, Austria, c.burns@ab-mikro.at

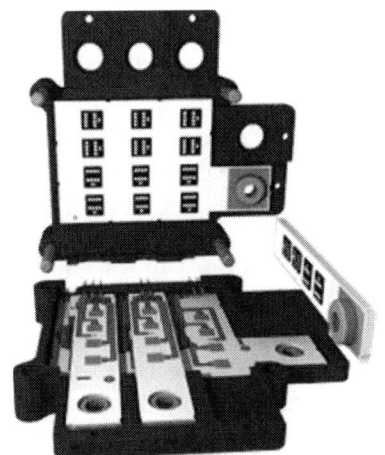

Abstract

Electromobility cannot be stopped. As battery technology improves year after year more and more electric cars will take to the streets. At the same time, as energy sources such as batteries or fuel cells improve, power electronics also will need to improve to most efficiently convert energy from its source to its output, example given to the wheels, board electronics, heating and cooling instruments etc. Because the use of electronics in an automobile is also increasing with every new model, inverters and converters will need to handle higher currents, higher voltages, higher self-induced temperatures and at the same time must be smaller and lighter weight to improve performance and increase the drivers overall driving experience and satisfaction. AB Mikroelektronik part of the TT Electronics plc, has developed a new power module technology, Power Stack, that is easily customizable and is aimed at solving the challenges that the electromobility future poses. The Power Stack is a solid solution for automotive applications due to the many thermal, electrical and footprint advantages as well as maintaining the performance during standard automotive environmental testing such as thermal cycling, humidity, and power cycling described in the LV 324.

Introduction

The Power Stack is a vertically stacked module that allows for cooling on both the top and bottom side which effectively cuts the thermal resistance in half and doubles the power density. This module is built using low cost materials, possible with various aluminum alloys and thick film dielectrics and conductors that can be selectively printed where needed. There are no extra etching processes that create dangerous waste and added processing time. Screen printing is used for electronic patterning, much like a t-shirt factory, which allows for quick and easy design changes which will and do occur when designing a new module for a new application. By customizing a module, the exact requirements and demands can be met which help to optimize the overall performance and design needed for an application. The Power Stack 48V module has been realized for a full B6 bridge for inverter applications; however, single switches or single phase applications can easily be adapted or scaled down. A brief explanation of the advantages as well as the design of experiments for determining the best bill of materials to withstand the extreme automotive testing posed by the LV 324 reliability standard will be presented.

© VDE VERLAG GMBH · Berlin · Offenbach

Advantages of the Power Stack

The Power Stack has many advantages over standard power modules such as having a smaller footprint, being light weight, low resistance and inductance, while at the same time having better thermal resistance and impedance characteristics. Because the Power Stack concept is so different from standard modules it is sometimes hard to compare the two technologies. However, compared to a standard power module with chip and wire technology on DCB substrates mounted on Cu Baseplate, with the same power class, the power stack was able to reduce the size by 60% and the weight by more than half of a kilogram or from 700g to 40g. The layout from the Power Stack has also been optimized to reduce electrical and thermal properties. By using a parallel switching configuration, the R_{dson} and the R_{th} are drastically reduced to 1.1 mohm and 0.04 K/W respectively. This correlates to less temperature created with in each Mosfet, but also evenly spreads the heat over the entire Al substrate giving the module a lower and more homogeneous temperature.

The Power Stack also excels in dynamic performance. Due to the laws of self and mutual inductance, a stacked module will also have much lower inductance which directly translates into a reduction of the voltage overshoot during switching. The Power Stack that has been tested has an inductance of 6nH. This delivers a much more responsive power module during operation. The thermal characteristics of the Power Stack are also much better when compared to a standard power module. Firstly, the Power Stack has heat conduction on both the top and bottom of the transistor effectively doubling the contact area and allowing heat to be transferred in both directions. By using large Aluminum bus bars as the current carrier, the benefits of high thermal capacitance are present due to the fact that the chips are directly attached to the busbars eliminating the need for large and heavy Cu baseplate or heat spreaders. Through thermal transient testing, it is seen that at 20kHz current pulses, the chip temperature doesn't rise. As current pulses increase to 3 minutes, the absolute Rth is finally reached. This allows standard applications to be downsized saving cost and space, but also allows higher peak currents when longer overdrive times are necessary.

Reliability Testing LV 324

The Power Stack is a solid solution for automotive applications due to the performance during standard reliability testing demanded by the LV 324. To start out, different die attach materials, underfill materials, and layouts were used to find the best bill of materials for the power module. The first and hardest tests, thermal shock cycling and humidity testing, were used to "weed" out the material systems that cannot perform and to find the best material set that not only performs well again the LV 324, but also aim to optimize price. All modules were initially characterized to quantify the modules in the initial or nominal state. The current vs. voltage curves were taken from 0 to 470A, using a 50µs pulse to prevent heat development in the module and changing the measured Rdson at room temperature.

PCIM Europe 2016, 10 – 12 May 2016, Nuremberg, Germany

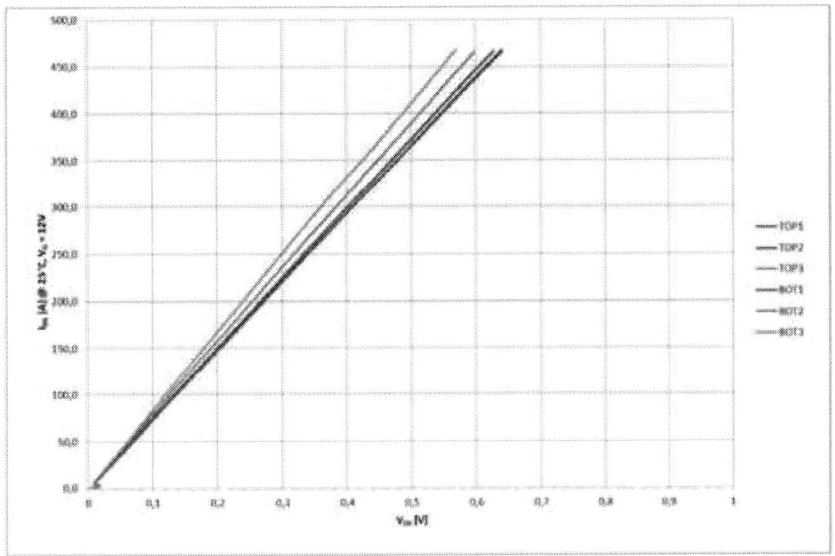

Figure 1 current vs. voltage plot from the power stack in initial condition at 25°C.

Each switch has a slightly different layout and therefore there are slightly different resistances in the measurement pfad. Using Ohm's law, the Rdson from all modules tested has a maximum of 1.4 mohm for the high side switches.

Important to monitor is also the breakdown voltage. Because the chips are sandwiched in between two large bus bars, the stresses present could damage the chips during reliability testing. The reverse bias curves are shown below.

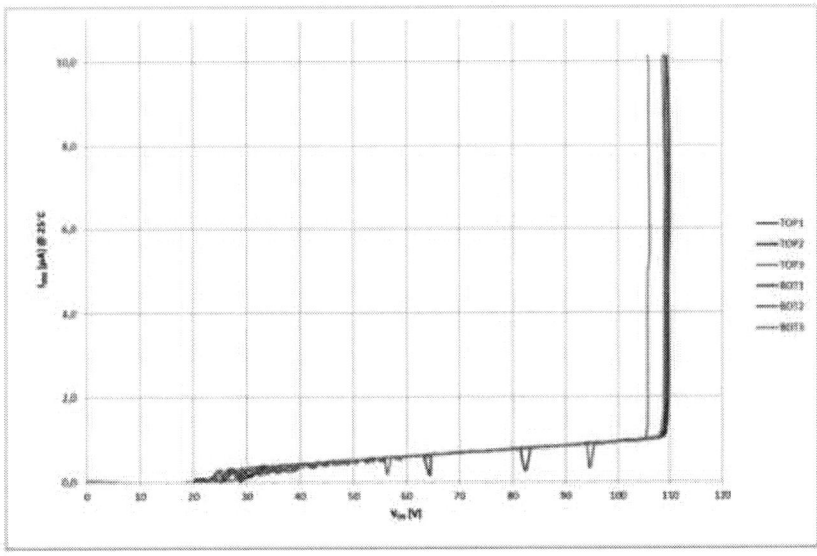

Figure 2 shows the reverse bias breakdown voltage. At 105-110V the current breaks through.

© VDE VERLAG GMBH · Berlin · Offenbach

The reverse bias breakdown voltage is important because it will give added information on how the chip ages under thermal shocks as well as humidity storage. To round out the initial testing the inverse diode charachteristic curves were also measured. This is helps to calculate the losses that occur during the "free-wheeling" state or when current flows back through the Mosfet when it is switched off, but the induced current from the motor needs to be sent back to the battery.

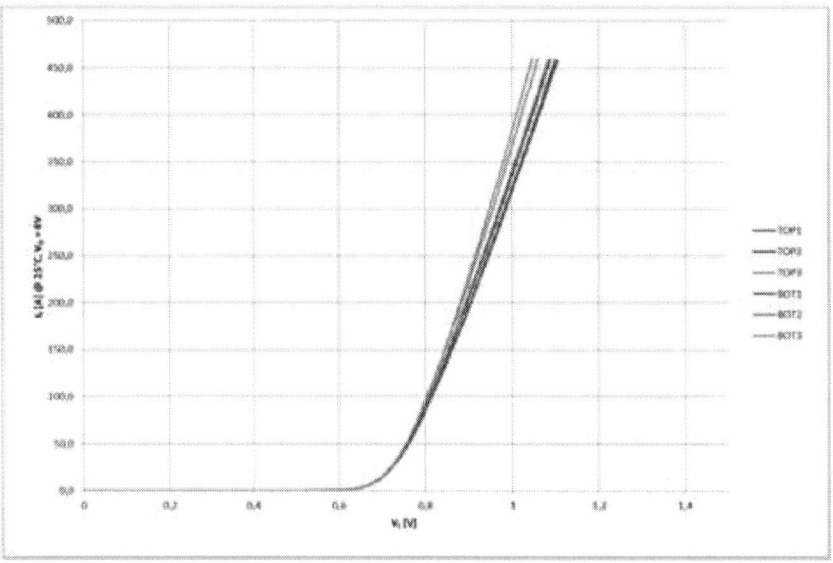

Figure 3 shows the charachteristics of the inverse body diode within the Mosfets.

Once the modules were characterisized, the modules were subjected to thermal cycling and humidity testing to determine the best bill of materials. For the thermal cycling test, each module was tested as in the initial state after 250 thermal cycles upto 1000 thermal cycles which is taken directly from the LV 324 standard. Ultrasonic pictures were also taken because it is a non-desctructive test that gives a good indication of the aging process of the module. The best and worst cases are presented below in figures 4 and 5.

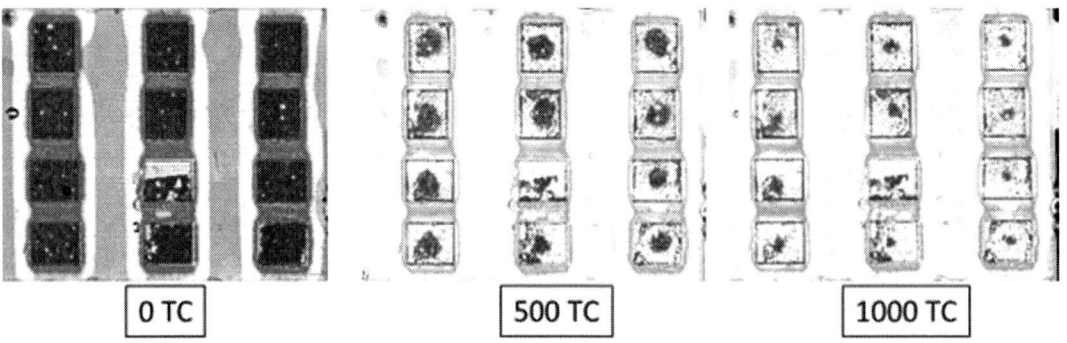

Figure 4 shows the powerstack "worst case" without any optimization in die attach, underfill or mechanical modifications. After 500 thermal cycles these modules have failed.

PCIM Europe 2016, 10 – 12 May 2016, Nuremberg, Germany

Figure 5 shows the best case with the best bill of materials that also passes the LV 324 requirements.

The 4 Mosfets aligned in columns are for each of the 3 individual phases. The dark areas demonstrate the intial die attach to bus bar connection is completely intact. In figure 4, the solder has already deteriorated after 500 thermal cycles indicated in the white areas. When air from a crack, void, or delamination in solder joints is present, the acoustic signal will be dampened and the signal turns that part of the image white In figure 5, the best case shows initially a couple of voids, however the voids do not increase or delamination doesn't increase even after 1000 thermal cycles. As seen, by optimizing the die attach, the underfill material, and the 3D thick film layout tremendous improvement can be accomplished.

The entire test matrix was designed in order to take advantage of the Anova and Regression analysis. After the thermal cycling tests were finished, all modules were tested again indicating which bill materials is acceptable for a series production. A typical Rdson value is presented in figure 6.

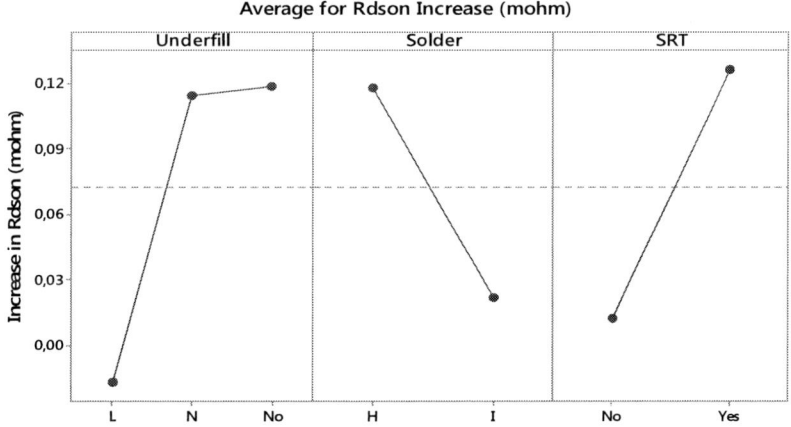

Figure 6 presents the main effects diagram for obtaining the best material combination to pass the LV 324 requirements.

© VDE VERLAG GMBH · Berlin · Offenbach

The modules were tested initially and after testing and then compared to the pass/fail criteria. After being subjected to 1000 thermal cycles in a temperature range between -40°C and 125°C, the modules were tested for source-drain resistance and breakdown voltage. Initially the module had an average Rdson between 0 and 400 Ampere of 1.1mohm and a breakdown voltage of 100V. The pass-fail criteria according to the LV 324, is set at a 5% increase in resistance after thermal cycling, which is an absolute value of 55µohm. All modules with the best solder and encapsulant remained below the required pass-fail level.

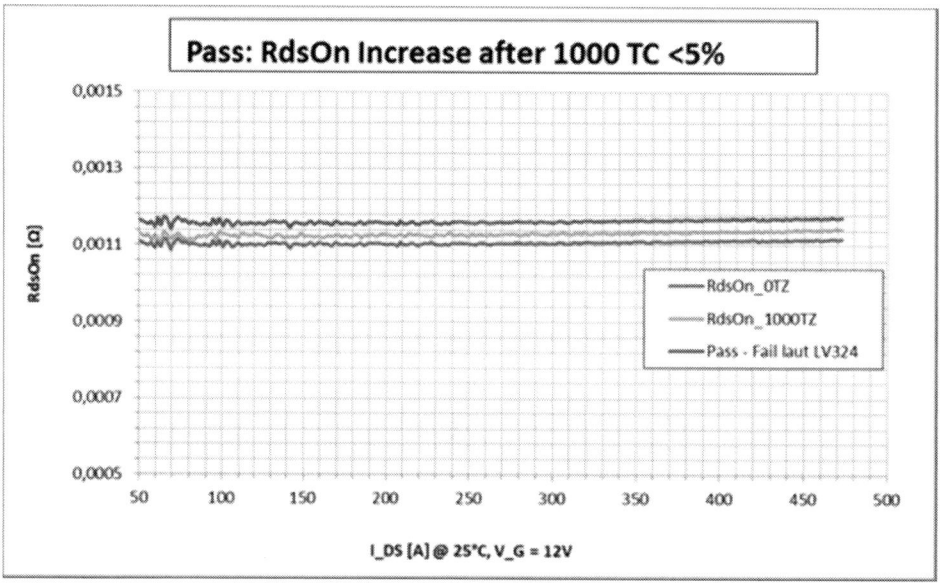

Figure 7 shows the initial R_{dson}, R_{dson} after 1000TC, and the limit for failure. Upto 470A, the R_{dson} remains stabile.

The break down voltage did pose some failures or had a reduction of voltage. This has been found to be due to flux residues trapped inside the Power Stack before the encapsulation process. This issue has been addressed and fixed. Humidity testing showed no problems with an increase in R_{dson} or with break down voltage. All modules remained above the required 100V niveau.

Single switches were subjected to the power cycling test, and after 100,000 power cycles the test was completed with all modules operating as in initial condition. The next step is to finish the power cycling test required by the LV 324, tested till end of life at 2 different temperatures. What has been learned from the initial power cycling tests is that in order to achieve the required temperature delta of, example given, 80°C at the chip junction to baseplate, the power stack requires 30% more current when compared to a standard power module of the same class. A new power cycling standard should be created for double sided cooling modules because there is big difference between switching with 300A or with 400A. With work done at a sister company Semelab Ltd. within the TT electronics plc, it was shown that even with the power cycling current being raised by 30% the stacked module still outlasts a standard module more than five times.

Conclusion

There are many advantages for a double sided or stacked power module that will enable higher current densities and longer lifetimes that high power electromobility applications require. Due to better performance electrically and thermally, while at the same time reducing the overall package size, the Power Stack when compared to standard modules allows larger current densities as well as a reduction in footprint. Not only is the Power Stack architecture better for performance but it has proven to be robust against the harsh testing required by the automotive industry and LV 324 Standard.

References

1. A.B. Mikroelektronik GmbH. 2012. "Verfahren zur Herstellung eines metallisierten substrats" Patent Application A 527/2012. 04 May 2012.
2. Chris Burns, PCIM Europe 2014; International Exhibition and Conference for Power Electronics, Intelligent Motion, Renewable Energy and Energy Management, VDE, May 20-22 2014 pp. 1-8.
3. Mario Kubik, B.S. (2013). *Aufbau- und Zuverlässigkeitsuntersuchung für Leistungsumrichter in Vertical-Stacked Technologie.* Bachelor. Thesis. Fachhochschule Technikum Vienna: Austria.

Evaluation of metal-matrix composites baseplates with anisotropic thermal conductivity inserts

Fabian Streb[1,2], Henning Zeidler[3], Michael Penzel[3], Andreas Schubert[3], Thomas Lampke[4]
[1] Corresponding author, PhD student at Institute of Materials Science and Engineering, Professorship of Materials and Surface Engineering, Technische Universität Chemnitz, Germany
[2] Infineon Technologies AG, Germany
[3] Professorship Micromanufacturing Technology, Technische Universität Chemnitz, Germany
[4] Chair of Materials and Surface Engineering, Technische Universität Chemnitz, Germany

Abstract

In order to fulfill the increasing demands on thermal management of high performance power modules we investigated the usability of heat spreading via materials with anisotropic thermal conductivity λ. As a vivid example we took a look at macroscopic anisotropic inserts integrated into metal baseplates. We used annealed pyrolytic graphite (APG) as insert material with $\lambda = 1600$ Wm^{-1}K^{-1} in two and $\lambda = 10$ Wm^{-1}K^{-1} in one spatial direction. First we evaluated the influence of geometry and orientation of those inserts on the heat dissipation of typical power modules and compared them to isotropic baseplates using steady-state thermal FEM simulations. Common heat spreading was shown to be detrimental and we suggest the use of increased thermal conductivity in the vertical instead of the horizontal direction. Next we investigated the long term stability of cast aluminum-APG composite baseplates via thermal cycling and transient plane source as well as scanning acoustic microscope measurements. The interfaces between matrix and inserts show deterioration over time while the effective lateral thermal conductivity remains constant.

1 Steady-state FEM simulations

A common method to improve heat dissipation in electronic devices is the usage of anisotropic materials. These materials, like annealed pyrolytic graphite (APG), have a very high thermal conductivity λ of 1600 Wm^{-1}K^{-1} in two dimensions at the cost of reduction in the remaining dimension down to 10 Wm^{-1}K^{-1} [1,2]. Consequently, it is of outmost importance to optimize the geometry, position and orientation of those inserts. The usage of such anisotropic materials with the two directions of high thermal conductivity in the horizontal direction (perpendicular to the direction from heat source to heat sink) is quite common and results in heat 'spreading'. This allows more material to take part in the heat dissipation and hence lowers the chip temperatures. In this work we apply steady-state FEM simulations to evaluate the influence of geometry and orientation of macroscopic APG inserts on the heat dissipation of typical power modules and compare them to isotropic baseplates.

1.1 Simulation model

We created a FEM steady-state thermal simulation of a power module (Ansys Workbench 15.0) and evaluated the chip temperatures and the thermal resistance of the module. The references model uses an isotropic copper baseplate ($\lambda = 385$ Wm^{-1}K^{-1}), which we compare to a module using a baseplate with an APG plate-like insert. Fig. 1 (b) shows the design of the power module. It is based on a real-life power module yet simplified to minimize the calculation times. The mounting holes are omitted and an artificial 'substrate' layer is created, representing the thermal properties and geometry of the direct copper bonded substrate (DCB) as well as the chip and the system solder within one body. These simplifications allow to shift the focus of the mesh towards the baseplate and the insert. Fig. 1 (a) shows an APG plate-like insert within the baseplate and the mesh. The dimensions of the insert in X&Y direction, its position along the Z direction and the orientation of the two directions with high thermal conductivity are varied.

Within the X&Y direction the insert is always centered and is completely enclosed by the baseplate matrix (at least 10 µm). Overall the model consists of 6 switches, 2 IGBTs and 2 FWDs with 220 W per switch. Despite the simplifications the resulting chip temperatures are qualitatively comparable to state-of-the-art power modules. As a heat source we apply an internal heat generation of 73.905 Wmm^{-3} on all IGBTs. As a heat sink the baseplate is mounted onto an aluminum cooler (not shown in Fig. 1) with a convection surface on the bottom side, a heat transfer coefficient of 4000 Wm^{-2} and an ambient temperature of 40 °C. For the interface between matrix and APG inserts as well as any other contact areas we assume perfect thermal contact conductance (TCC). The resulting chip temperatures are proportional to the chip lifetime. Fig. 1 (c) and (d) show typical simulation results. The thermal resistance R_{th} (lower implies better heat dissipation) is calculated from the temperature difference between the chip temperature and the temperature on the bottom side of the baseplate perpendicular below the respective chip center. The reference module has a maximum chip temperature of 124.7 °C, minimum chip temperature of 110.4 °C and an averaged thermal resistance over all IGBTs of $<R_{th}>$ = 0.1182 K/W.

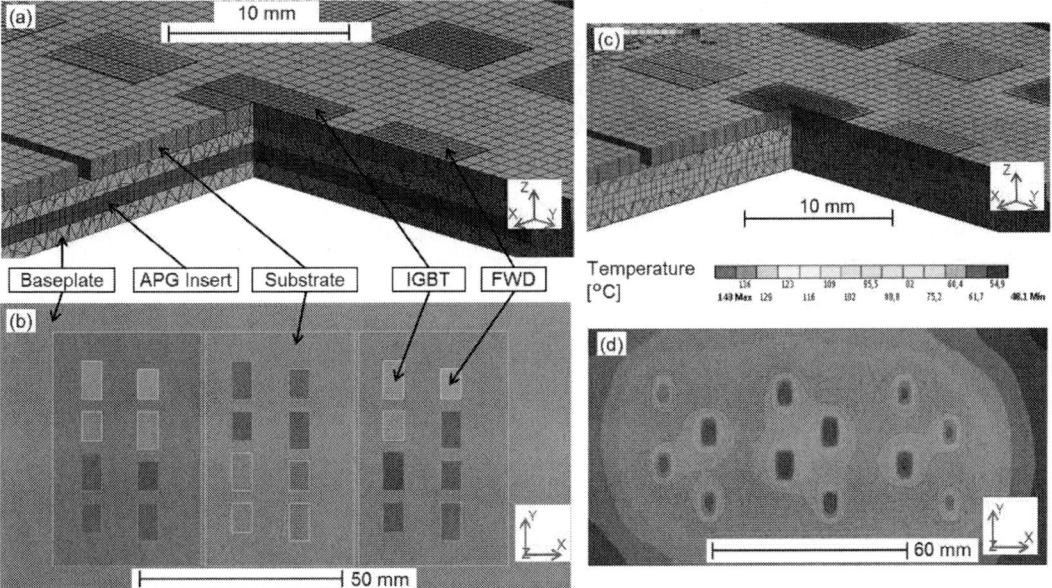

Fig. 1.: (a) Magnified cross-section of the simulation model including mesh, (b) top view of the whole simulated power module, (c) & (d) show typical simulation results in cross-section and top view respectively.

1.2 Heat spreading: Results & Discussion

First we investigate the usability of the insert with high thermal conductivity in the horizontal (X&Y) direction. We vary the following parameters of the insert: The length in X direction from 60 to 130 mm, the length in Y direction from 30 to 60 mm, the insert height from 1 to 3 mm and its position in vertical direction within the metal matrix (either centered in Z direction, bottom or top side of the baseplate). Fig. 2 shows the maximum chip temperature and $<R_{th}>$ depending on the volume. The steps within one data set origin from different shapes (ratio between length and width) of the inserts. Fig. 2 (a) shows the results for a 1 mm thick APG insert with different positions in Z direction and (b) for a centered insert with different heights. Overall the insert is impairing the heat dissipation. The higher the volume of the insert is the lower is the heat dissipation. Despite the improved heat spreading in horizontal direction the reduction of the thermal conductivity in Z direction (from heat source to sink) reduces the performance of the composite baseplate. For the same volume and shape the position of the insert on the bottom side yields the best results. If heat spreading would be beneficial one would expect the exact opposite. The closer to the heat source the heat spreading occurs the more it would enhance the heat dissipation. This confirms that the increased conductivity in horizontal direction does not remedy the decrease in vertical direction.

© VDE VERLAG GMBH · Berlin · Offenbach

For the same volume and position but different shapes the chip temperatures and the average thermal resistance contradict each other, see Tab. 1: The $<R_{th}>$ and the minimum chip temperature suggests that with increased thickness and reduced area the heat dissipations improves, while the maximum chip temperature suggests the exact opposite. While the minimum chip temperature origins from chips at the edge of the baseplate, the maximum chip temperature origins from chips in the center of the baseplate. Since the insert impairs the heat dissipation and a thicker insert at the same volume has a smaller area, the heat of chips at the edge of the baseplate is dissipated without disturbance. This reveals that despite the common usage of the $<R_{th}>$ to evaluate isotropic materials, using it to evaluate anisotropic materials can lead to wrong conclusions. Instead of looking at the average value the maximum and minimum value of both chip temperature and R_{th} should be used.

Tab. 1: Selected examples of chip temperatures and average thermal resistance for inserts positioned centered in relation to the Z direction with different insert dimensions. High conductivity in X&Y direction, lowest values are marked.

Insert dimensions			$<R_{th}>$	Max. chip temp.	Min. chip temp
X [mm]	Y [mm]	Z [mm]	[K/W]	[°C]	[°C]
120	50	1	0.245	**142.4**	125.6
100	30	2	0.216	149.4	114.9
60	50	2	**0.205**	152.7	**111.9**

Overall the APG insert in horizontal orientation does not improve the heat dissipation. Instead of spreading the heat to the side of the baseplate, heat congestion at the center of the baseplate occurs. With APG insert the thermal fields of the centered chips overlap stronger than with a plain isotropic baseplate and the chip temperatures are increased. The lifetime of all chips is decreased. This could be prevented by increasing the total area of the baseplate, allowing the insert to spread the heat to more material but also increasing the cost of the whole module. Next we take a look at the usage of the APG insert with one direction of high thermal conductivity in the Z direction, from heat source to heat sink.

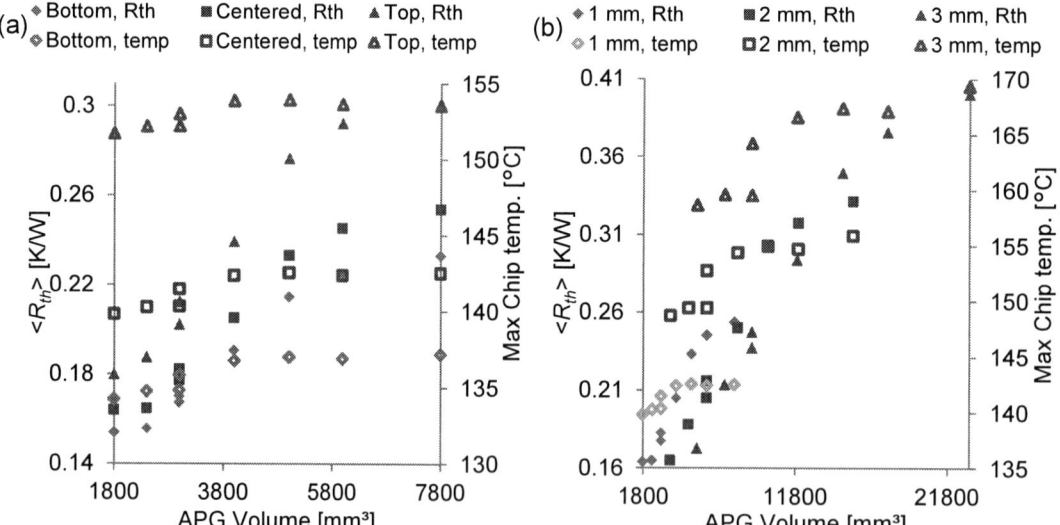

Fig. 2: Average thermal resistance and maximum chip temperature depending on the overall APG volume using an insert with X&Y direction and (a) using a 1 mm thick APG insert for different insert positions, (b) using an APG insert centered within the baseplate for different insert thicknesses. Without APG the max. chip temperature is 124.7 °C and the $<R_{th}>$ = 0.1182 K/W.

1.3 Vertical heat dissipation: Results & Discussion

We repeated the simulations from 1.2 with a different orientation of the APG insert; the high thermal conductivity is now in Y&Z direction. Fig. 3 shows the results. For X&Z orientation the results are nearly identical. $<R_{th}>$ as well as maximum chip temperature are reduced the bigger the insert volume is. The APG significantly improves the heat dissipation. For the given geometries the chip temperature can be decreased by more than 8 °C (≈7 %), nearly doubling the chip lifetime.

Tab. 2 shows similar inconsistencies between $<R_{th}>$, maximum and minimum chip temperatures for a constant insert volume. All three propose a different geometry for the best heat dissipation. More detailed transient simulations are necessary to come to a conclusive result. However the differences between geometries of the same insert volume are small. Overall this shows the importance of the heat dissipation in vertical direction over heat spreading, for the given ratio between heat source area and heat sink area.

In order to investigate the influence of the thermal contact conductivity (TCC) on the heat dissipation we perform a simulation with a 1 mm thick insert, centered in relation to the Z axis within the baseplate and an area of 120x50 mm². Tab. 3 shows maximum chip temperature and $<R_{th}>$ depending on the TCC between insert and matrix. For a TCC of 400 kWm^{-2}K^{-1} and better the maximum chip temperature is smaller than using a plain copper baseplate. We expect that for metal matrix with a lower thermal conductivity than copper, like Al or AlSiC, the necessary TCC for improvement over a plain metal baseplate to be smaller, depending on the difference between matrix and insert λ.

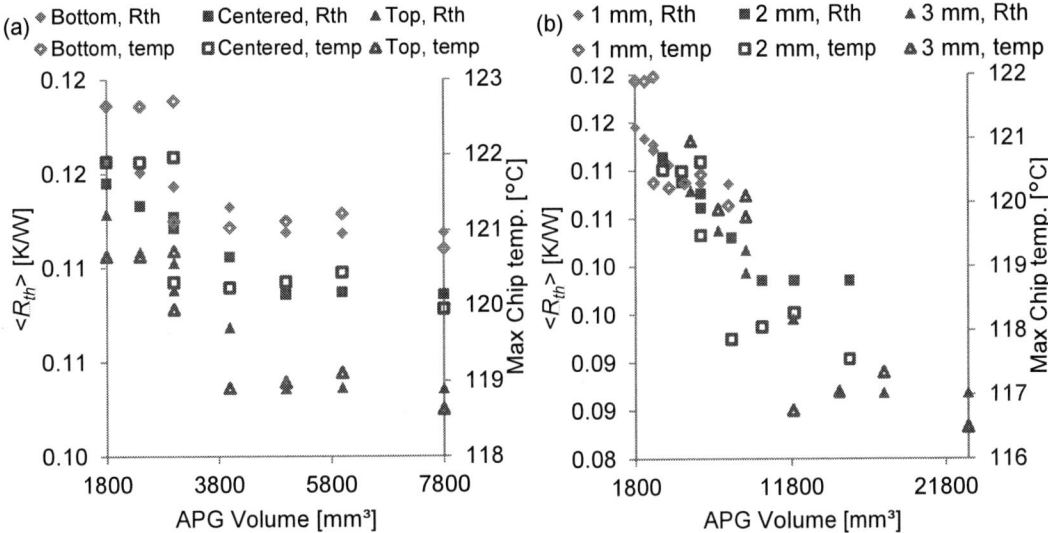

Fig. 3: Average thermal resistance and maximum chip temperature depending on the overall APG volume using an insert with Y&Z direction and (a) using a 1 mm thick APG insert for different insert positions, (b) using an APG insert centered within the baseplate for different insert thicknesses.

Tab. 2: Selected examples of maximum chip temperature and average thermal resistance for inserts positioned centered in relation to the Z direction with different insert dimensions. High conductivity in Y&Z direction, lowest values are marked.

Insert dimensions			$<R_{th}>$	Max. chip temp.	Min. chip temp
X [mm]	Y [mm]	Z [mm]	[K/W]	[°C]	[°C]
120	50	1	0.109	120.4	**108.8**
100	30	2	**0.106**	120.6	110.0
60	50	2	0.108	**119.5**	109.9

Tab. 3: Influence of the thermal contact conductance between matrix and insert on chip temperature and thermal resistance.

TCC [kWm^{-2}K^{-1}]	Max. chip temp [°C]	$<R_{th}>$ [K/W]
'infinite'	120.4	0.1087
800	122.0	0.1187
400	123.5	0.1277
200	126.0	0.1430
100	130.2	0.1670
10	170.4	0.3519

1.4 Summary

With steady-state FEM simulations we could show that macroscopic inserts with anisotropic thermal conductivity integrated into metal baseplates can be used to improve the heat dissipation of high performance power modules. It is necessary to adjust the insert geometry and orientation to achieve an improvement compared to plain metal baseplates. If the insert is optimized the chip temperature can be reduced by more than 8 °C, nearly doubling the chip lifetime. Without optimization of the insert orientation heat spreading is shown to be detrimental for the given example because of the increased heat congestions at the inner IGBTs. Improvement compared to plain metal baseplates is only possible for inserts with high thermal conductivity in vertical direction. Modern power modules and their baseplate are shaped in such a way that only a very small amount of material doesn't take part in the heat dissipation. Thus, an increased heat spreading in horizontal direction only leads to an increased overlap and therefore to a heat accumulation. The conventional method to evaluate heat dissipation via an average thermal resistance value fails whenever anisotropic materials are involved. It is necessary to take a look at individual chip temperatures to fully comprehend the influence of the insert. The thermal contact conductance plays a major role in the performance of baseplate. If the TCC between insert and matrix is below 400 kWm^{-2}K^{-1} the insert is not able to absorb heat and acts as a void instead.

2 Long term stability

Another point of interest concerning the usability of APG as inserts in metal baseplates is the brittle behavior of graphite. During the assembly process of power modules, the different coefficients of thermal expansion (CTE) between insert, baseplate, chip and DCB substrate lead to internal mechanical strains. To compensate for the difference in CTE the baseplate needs to be bent before the assembly process. Furthermore, during usage of the power module the baseplate is deformed periodically because of the pulsed load. It is unclear if the brittle APG insert within a metal baseplate can withstand these deformations. Tab. 4 shows selected CTEs, 'in-plane' refers to the directions of high thermal conductivity. A European funded project 'THERMACO' is presently ongoing in order to make use of the anisotropic thermal conductivity of graphite despite its brittle behavior. As proposed above the THERMACO project [3,4] uses thinner graphite stripes with high thermal conductivity in the vertical direction in contrast to the state-of-the-art heat spreading. The strain on the inserts is reduced using insert geometries which neither cover the whole length nor width of the baseplate. We investigate the long term stability of these cast aluminum metal matrix baseplates via thermal cycling and transient plane source as well as scanning acoustic microscope measurements.

Tab. 4: CTE of selected materials.

Material	CTE [10^{-6} K^{-1}]	References
Silicon	2.6	[5]
Copper	17	[6]
Aluminum	23	[6]
APG		
in-plane	-1	[2]
Thru-plane	26	

2.1 Measurement setup

Two baseplates per type have been measured. Fig. 4 shows the three different baseplate types we investigated, plain Al baseplates and two composite baseplates: one with discontinuous APG inserts ('short' APG, center) and one with continuous APG inserts ('long' APG, right side). On the backside of the composite baseplates (the side the DCB would be soldered onto) the inserts are not visible, they are covered by several hundred μm of aluminum. Using a scanning acoustic microscope (Sonoscan S-CAM D-9000) the interface between insert and the aluminum matrix in Z direction have been investigated.

The effective lateral thermal conductivity λ of the baseplates have been investigated using a transient plane source technique (Hot Disk TPS 3500 by Hot Disk AB, Sweden). For details of the thermal conductivity measurements please see [7–9]. The sensor acts as a heat source and as a temperature sensor at the same time. The thermal conductivity is calculated from the temperature-time graph. The λ values cannot be compared quantitatively to each other, since the volume fraction of APG and Al responsible for the heat conduction differ from sensor position to sensor position and from baseplate type to baseplate type. Yet they allow a qualitative investigation of the resistance of the matrix-insert interface to thermal degradation and of the influence of sensor position and baseplate type on the heat dissipation. The measurements are performed before thermal cycling and after 1, 10, 110, 510 and 1010 thermal cycles. Thermal cycling is performed according to the JEDEC JESD22-A104D standard (Vötsch VT 7012 S2), 10-12 minutes per chamber at -55 °C to +150 °C with 2-3 min ramp time.

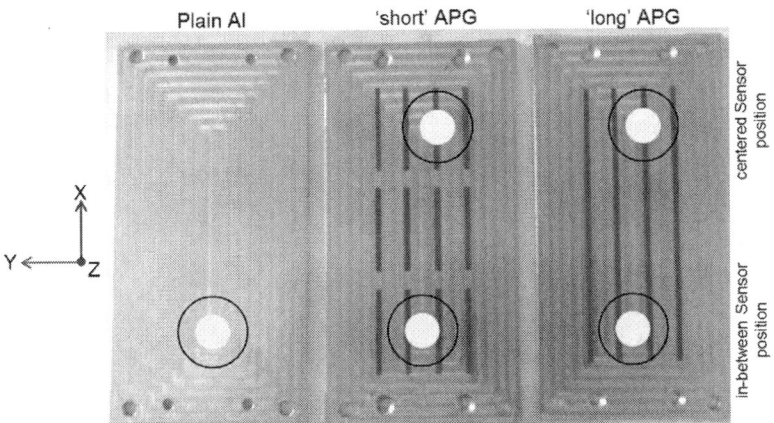

Fig. 4: Image of the three different baseplate geometries. Yellow dots illustrate the position and size of the used sensor. The circle illustrate how far the energy would dissipate into the baseplate if the material is plain, isotropic aluminum for the used measurement parameters.

2.2 Results and Discussion

The sensor position plays a vital role for the thermal conductivity measurements. Fig. 5 shows the effective lateral thermal conductivity depending on the number of thermal cycles for different baseplate geometries and sensor positions. If the sensor is positioned directly on the APG insert ('centered' position) λ is up to 32.3 % higher compared to the plain Al baseplate for the 'long' inserts and 34.3 % for the 'short' inserts respectively (after 510 cycles). The plain Al baseplates outperform the composite baseplates for all other sensor positions. For the centered sensor position the difference between the 'short' and the 'long' geometry is negligible (<10 %). The 'short' insert type outperforms the 'long' type by up to 17 % for the 'between' position and by up to 49.6 % for measurements on the Al backside. The difference between the 'short' and 'long' type can be explained with the thermal contact conductance (TCC) between inserts and matrix. If the TCC is very low no heat is exchanged between insert and matrix. If the heat is introduced mainly into the baseplate matrix and not into the insert (which is the case for the 'between' and Al backside position) the inserts act as voids and the heat dissipation is hampered. This shows that the TCC is a major influence for the overall heat dissipation performance of a composite baseplate. After a drop in λ between 0 and 10 thermal cycles it stabilizes between 10 and 1010 cycles for all evaluated baseplate types. The plain Al baseplate shows a change in λ after 1010 cycles of -3 %. Depending on sensor position the 'short' type shows a change in λ of -5.5 % down to -7.1 % and the 'long' type of +1.6 % down to -1.7 %. The plain Al baseplate has no internal interfaces which could be damaged during the thermal cycling test. The temperatures used are very low and hence prevent any changes in the Al microstructure. Therefore, for the plain Al baseplate the changes in λ can only be explained with measurement noise. Compared to this measurement noise the 'long' type shows no and the 'short' type only very limited degradation with increasing number of thermal cycles.

The SAM images demonstrate a similar trend as the thermal conductivity measurements. Fig. 6 shows parts of the 'long' baseplate type before and after 1, 10 and 1010 thermal cycle(s). The interface between matrix and insert gets sharper with increasing number of thermal cycles. The interface deteriorates slightly. This very small trend is similar for all investigated baseplate types and in agreement with the HotDisk measurements.

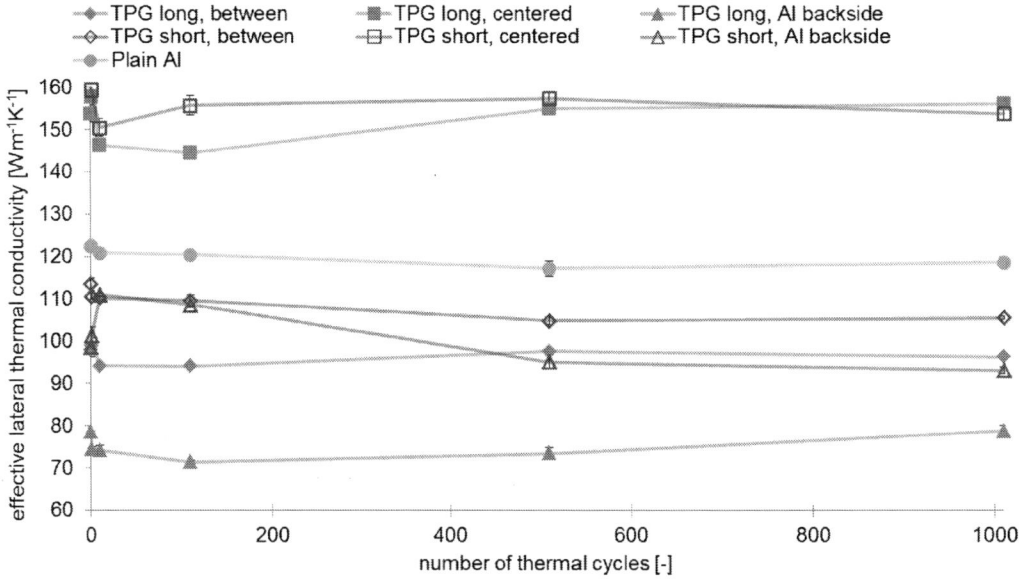

Fig. 5: Effective lateral thermal conductivity depending on the number of thermal cycles for different baseplate geometries and sensor positions. (Tested cycles: 0, 1, 10, 110, 510, 1010)

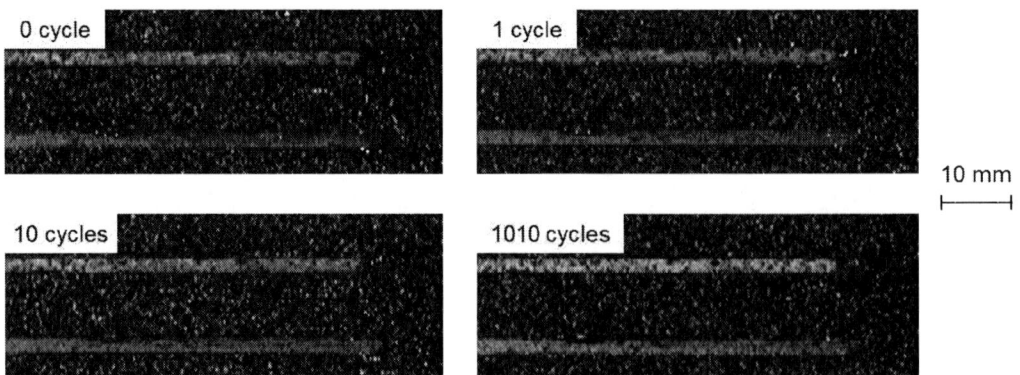

Fig. 6: Typical SAM results before and after 1, 10 and 1010 thermal cycle(s). Image shows 2 insert halves of the 'long' baseplate type. From 0 to 1010 thermal cycles the interfaces become sharper and less diffuse.

2.3 Summary

Using transient plane source measurements, scanning acoustic microscopy and thermal cycling experiments we could show that the interface between Al matrix and APG inserts degraded only slightly over time. The degradation is so small that 1010 thermal cycles (corresponding to 505 h of thermal cycling) have no measurable effect on the lateral thermal conductivity of the composite baseplates. We could also demonstrate that the measurement position plays a major role in the heat dissipation properties of the baseplate. Only if the sensor (acting as heat source) is positioned centered on the inserts an improvement compared to the plain metal baseplate could be accomplished. For all other positions the thermal contact conductance between insert and matrix is too low to exchange heat between insert and matrix. Without further optimization of the interface and without correct placement of the heat source the inserts act as voids and have a detrimental influence on the heat dissipation.

Acknowledgments

This work is supported by the ongoing project THERMACO (Smart Thermal conductive Al MMCs by casting), which receives funding from the European Union's Seventh Framework Programme for research, technological development and demonstration under grant agreement no. 608978. For details please see www.thermaco.eu.

References

[1] R.J. Lemak, Y. Nogami, R.J. Moskaitis, K. Chikuba, High Performance Pyrolytic Graphite Composite Heat Spreaders, (2011).

[2] High-Thermal Conductivity Graphite and Composites - Momentive Performance Materials, http://www.momentive.com/Categories/Ceramics/High-Thermal-Conductivity-Graphite-and-Composites.aspx (accessed February 12, 2016).

[3] H. Zeidler, A. Schubert, S. Flemmig, M. Penzel, S. Essel, Graphene Aluminium Matrix Nano-Composites for Heat Transfer Applications, in: Proc. 4th Int. Conf. Nanomanufacturing NanoMan2014, Germany, 2014.

[4] H. Zeidler, A. Schubert, S. Flemmig, M. Penzel, Simulation of the anisotropic thermal conductivity generated by graphene integration in aluminium matrices, in: Proc. 14th Euspen Int. Conf., Dubrovnik, 2014.

[5] H. Watanabe, N. Yamada, M. Okaji, Linear thermal expansion coefficient of silicon from 293 to 1000 K, Int. J. Thermophys. 25 (2004) 221–236.

[6] D.D.L. Chung, Materials for thermal conduction, Appl. Therm. Eng. 21 (2001) 1593–1605.

[7] M. Gustavsson, H. Wang, R.M. Trejo, E. Lara-Curzio, R.B. Dinwiddie, S.E. Gustafsson, On the Use of the Transient Hot-Strip Method for Measuring the Thermal Conductivity of High-Conducting Thin Bars, Int. J. Thermophys. 27 (2006) 1816–1825. doi:10.1007/s10765-006-0072-z.

[8] Y. Ma, J.S. Gustavsson, Å. Haglund, M. Gustavsson, S.E. Gustafsson, Pulse transient hot strip technique adapted for slab sample geometry to study anisotropic thermal transport properties of μm-thin crystalline films, Rev. Sci. Instrum. 85 (2014) 044903. doi:10.1063/1.4871589.

[9] F. Streb, G. Ruhl, A. Schubert, H. Zeidler, M. Penzel, S. Flemmig, I. Todaro, R. Squatritio, T. Lampke, Simulations and measurements of annealed pyrolytic graphite–metal composite baseplates, in: 18. Werkstofftechnisches Kolloquium Chemnitz, 2016.

Analysis of Packaging Impedance on Performance of SiC MOSFETs

Andrew Lemmon[1], Sujit Banerjee[2], Kevin Matocha[2], Levi Gant[1]

[1]The University of Alabama, 245 7TH AVE, Tuscaloosa, AL 35487, USA

[2]Monolith Semiconductor Inc., 408 Fannin Ave, Round Rock, TX 78664, USA

Abstract

This paper presents an analysis of the impact of packaging impedances for high-performance SiC MOSFET's, building on previous packaging analysis performed by the author [1] and others (e.g., [2]-[3]). Devices from the same fabrication lot were assembled into two standard discrete package types (TO-247 and TO-263), and subjected to a range of frequency-domain and time-domain analysis procedures. The objective of this work is to provide an understanding of the performance limitations imposed by standard package types, and to evaluate whether the TO-263 package provides a measurable advantage over the larger TO-247 package in the context of high-edge-rate-capable SiC MOSFET's. Frequency-domain analysis was utilized to extract parasitic impedance estimates for each of these two package types; and the TO-263 package is shown to have 32% lower inductance in the drain-source path compared to the TO-247 package. In addition, empirical switching experiments were carried out with identical conditions for devices in these two packages, and a comparison is presented on the basis of switching dynamics and switching energy calculated from time-domain waveforms. Empirical results reported here indicate that the total per-cycle switching losses are reduced by 29% in the TO-263 package as compared to the TO-247 package at a load current of 20 A, all other conditions being equal. This difference is found to be attributable to a small reduction in the common source inductance of the TO-263 package compared to the TO-247 package.

1. SiC MOSFET Design and Process

1200V, 80 mΩ SiC MOSFETs manufactured with advanced processing techniques in a state-of-the-art 150mm silicon CMOS fab were used in this work. The devices are designed with minimal gate resistance (typical 2.5 Ω) by optimizing the internal gate bus design and fabrication process. The objective of this design optimization is to enable rapid charge and discharge of the input capacitance in order to maximize the signal edge rates achievable in hard-switched converter design. The device design also ensures robustness to high dV/dt in excess of 100V/ns allowing circuit designers to take full advantage of inherent capability of SiC MOSFETs. Further details on the device performance and reliability of these MOSFETs are provided in [4]. In this work, devices from the same wafer lot were assembled in different packages to minimize any process dependence on the switching characteristics.

2. Frequency-Domain Analysis

The first step in the analysis procedure was to perform a detailed package impedance analysis using a 120 MHz precision impedance analyzer and a set of custom-developed adapter/interface PCB's. These PCB's provide a means for accurately measuring circuit components of arbitrary geometry such as the components of the clamped-inductive-load (CIL) test stand used in the time-domain portion of this analysis. The frequency-domain test setup used in this work is shown in Fig. 1, along with a sample result for measurement of the drain-source loop inductance of the TO-247-packaged SiC MOSFET device. The impedance data captured by this instrument was imported into MATLAB and matched to a series LC network frequency response, which agrees with the empirical data shown in Fig. 1. The lumped inductance of the TO-247 package drain-source terminal pair was determined to be 13.6 nH; while this value was 9.2 nH for the TO-263 package (32% lower).

PCIM Europe 2016, 10 – 12 May 2016, Nuremberg, Germany

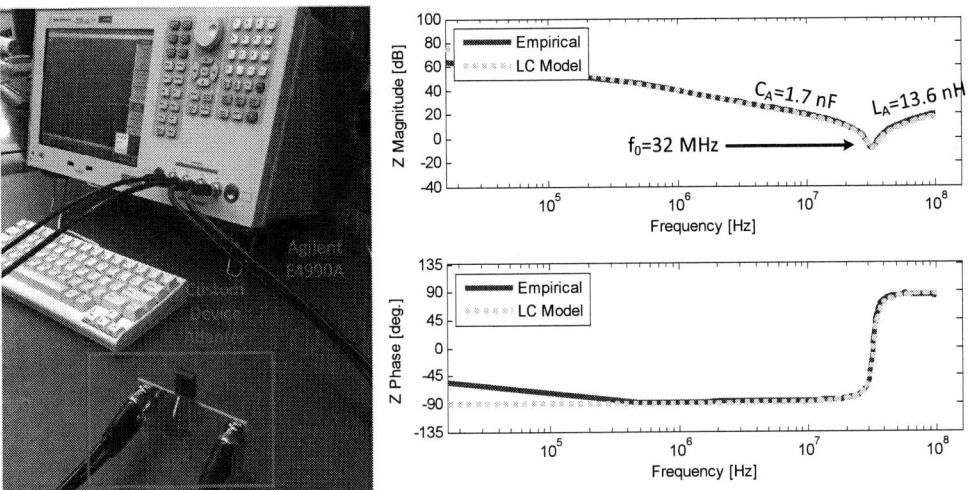

Fig. 1: Frequency-domain packaging analysis setup and results. Left subplot: impedance analyzer with custom interface adapter used to characterize TO-247 lead inductance; Right subplot: results of TO-247 packaging impedance analysis, demonstrating series LC-resonance at 32 MHz.

3. Time-Domain Analysis

3.1. Test Apparatus and Empirical Procedures

The second step in the analysis procedure was to perform a set of pulsed switching experiments using a clamped-inductive-load (CIL) test fixture. The purpose of these experiments is to enable a quantitative analysis of the transient behavior variations caused by subtle differences in the parasitic impedances of the two packaging options considered. A schematic diagram of the circuit used for this analysis is shown in Fig. 2. It should be noted that parasitic elements appear in this circuit in red. The semiconductor packaging inductances are represented in this figure as per-terminal lumped equivalent elements: drain inductance L_D, source inductance L_S, and gate inductance L_G. In addition, the stray inductance of the DC bus is identified separately as L_{BUS}. Although the influence of the bus inductance is convolved with that of the packaging inductances, it must be identified separately in order to enable the parametric analysis which follows. In addition, the circuit quantities which will be compared in the following analysis are shown in this diagram in blue.

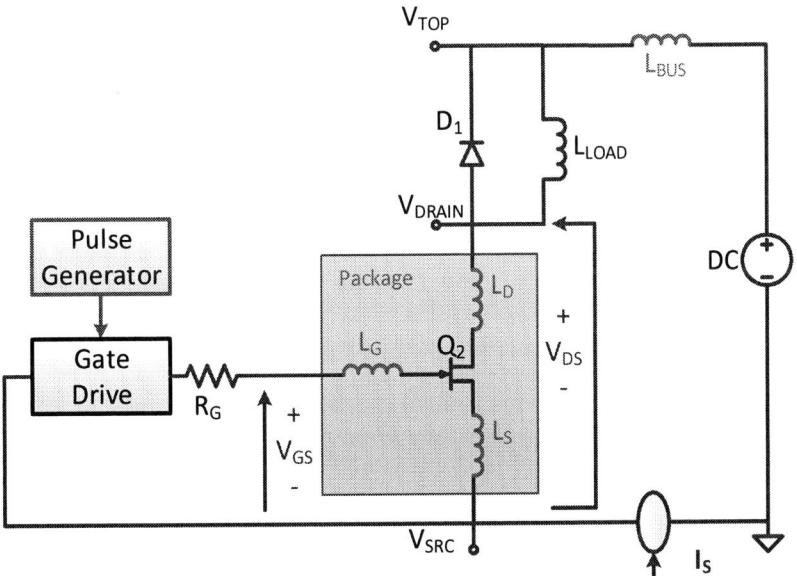

Fig. 2: Schematic diagram of CIL test circuit used for transient analysis, showing packaging parasitics

© VDE VERLAG GMBH · Berlin · Offenbach

An annotated picture of the physical realization of the CIL test circuit is presented in Fig. 3. The details of this test stand design have been previously described in [5]. However, there are a few updates to the test stand which were made during the course of the evaluation presented here. Notably, a new custom gate-drive circuit capable of 14 A peak drive current was designed and built to maximize the switching performance achievable with the SiC MOSFET's used in this work. This gate-drive circuit incorporates galvanic isolation for both the power and signaling input paths, even though it is used in the low-side configuration for the current analysis. Ensuring sufficient common-mode transient immunity and minimizing the parasitic capacitance across the isolation boundaries were among the primary considerations during the design of this circuit. The importance of these design principles for high-slew-rate gate-drive circuits is described in [6].

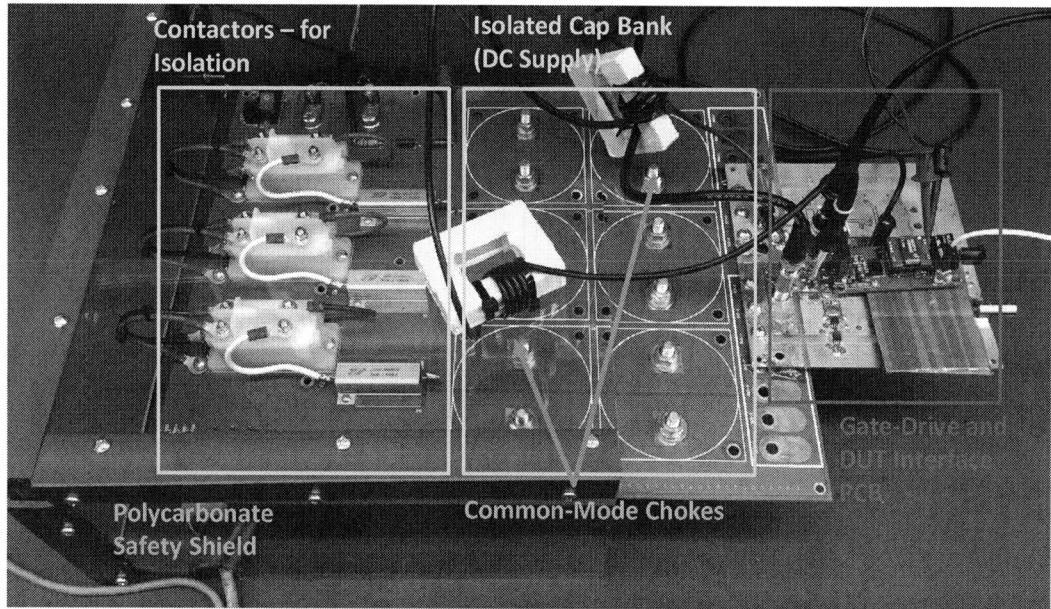

Fig. 3: Annotated picture of CIL test stand used for transient analysis

To evaluate the impact of the different values of parasitic inductance of the TO-247 and TO-263 packages, a set of identical experiments was carried out using these two package types, with all other factors kept constant. For all experimental results described here, the custom gate-drive circuit was fitted with a 2Ω gate resistor. The load inductor used in this work was a helical-wound inductor bank with a value of 370 µH and a current handling capability of more than 80 A (without saturating). The bus voltage used for all experiments was 600 V. The arbitrary waveform generator used to produce the gate signal pulses was programmed to deliver a range of load currents from 5.0 A to 20.0 A.

3.2. Metrology Considerations

One significant difference in the configuration of the test stand used in this work compared to the previous description in [5] relates to the metrology used. In the previous work, a 200 MHz Pearson current transducer [7] was used to make measurements of the device drain current. However, it was discovered that although this instrument has sufficient Bandwidth and rise-time ratings to capture the essential characteristics of the drain current signal, it is very difficult to remove the time misalignment between the current and voltage sensors when using this device. As a result, a 2.0 GHz coaxial current viewing resistor from R&M research [8] was used in this work. A fast rise-time, high-voltage, passive, ground-referenced voltage probe was used to monitor the drain-source voltage of the SiC MOSFET. Prior to taking any measurements, the de-skew technique described in [2] was used to ensure proper alignment of the current and voltage sensors. A synopsis of the instrumentation used in this work is presented in Table 1.

Table 1: Instrumentation used in empirical analysis procedures

Instrument Type	Instrument Model	Measured Quantity	Band-width	Rise-Time
Oscilloscope	Tektronix MD04054B-3	-	500 MHz	N/A
High-Voltage Probe	Tektronix TPP0850	V_{DS}	800 MHz	700 ps
Low-Voltage Probe	Tektronix TPP0500	V_{GS}	500 MHz	?
Current Probe	T&M Research SSDN-10	I_S	2.0 GHz	180 ps
Current Transducer	*Pearson Electronics 2877*	*Not Used*	*200 MHz*	*2 ns*

3.3. Transient Results Example

An overlay plot of the time-domain waveforms for the identical TO-247 and TO-263 pulsed switching experiments at a load current of 20 A is shown in Fig. 4. From this comparison, it can be seen that while the switching dynamics of both devices are similar, the TO-263 package switches more quickly at turn-on. At turn-off, when a low value gate resistor is utilized, the switching dynamics are usually governed by the external circuit rather than by the device turn-off speed as described in [9]. Under the parameters used for the current work, this condition is observed to hold, which explains why the two devices behave very similarly at turn-off. However, at turn-on, the drain-source voltage of the TO-263 device falls about 5-8 ns before it does for the identical part in a TO-247 package. In addition, the "shelf" which is present in the drain-source voltage waveform occurs at approximately 200 V for the TO-263 package, whereas it is centered at approximately 400 V for the TO-247 package. This "shelf" results from the L*di/dt drop which appears across L_{BUS} during the time that the device current is rapidly changing from zero up to the maximum peak value. The difference in the shelf voltage in these two cases results from the difference in the current slew rate observed in the two examples rather than from the difference in the packaging inductance. Although there is voltage drop across both the bus inductance (L_{BUS}) and the parasitic packaging inductance ($L_D + L_S$) during this interval, only the influence of L_{BUS} is observable in determining the value of the "shelf" voltage. This is because the drain-source voltage shown here was measured on the outside of the package; from this point the voltage drop across the SiC MOSFET (V_{DS}) is summed with the voltage drop across the parasitic packaging inductance ($L_D + L_S$).

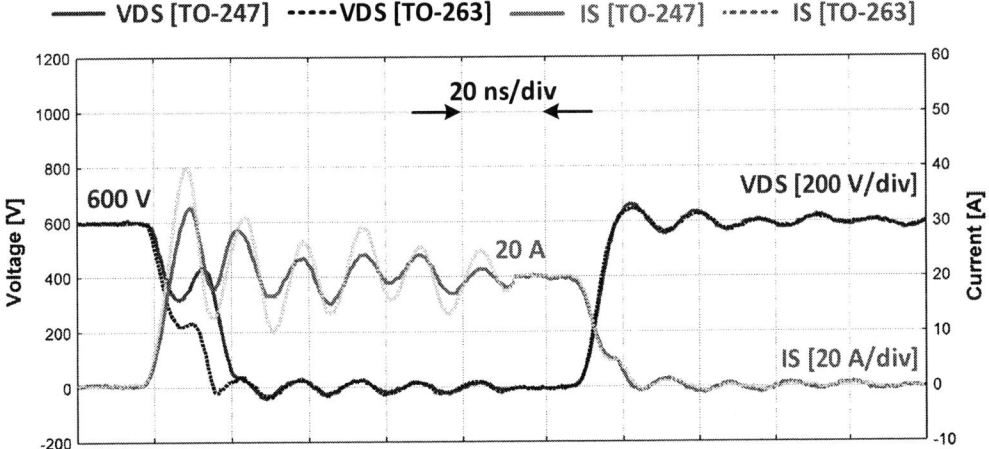

Fig. 4: Comparison of TO-247 and TO-263 packaged SiC MOSFET's switching at 600V and 20A

3.4. Simulation-Based Packaging Impact Analysis

In order to better understand the underlying reasons for the difference in the transient behavior illustrated in the previous section, a detailed simulation study was conducted. This simulation study was based on the use of a SPICE model for a commercially-available SiC MOSFET with similar characteristics to the one used in this study. Since the goal of this study was to understand the origin of the significant difference in the empirical behavior observed for the TO-247 and TO-263 package types, it was not necessary to precisely match the simulation output to the trajectory of the empirical waveforms. However, very good agreement with the empirical waveforms shown in the previous section was nevertheless achieved. By performing a parametric sweep of the three parasitic packaging inductances shown in Fig. 2, it was possible to identify the source inductance L_S as the parameter of sensitivity which produced the transient behavior difference described previously. Since neither the TO-247 nor the TO-263 has a gate-return (or "Kelvin source") terminal, the source inductance L_S also serves as the "common" source inductance (CSI), which is present both in the gate-loop and the power loop. The influence of CSI has been extensively studied in the literature (e.g. [10]-[11]). CSI is known to substantially influence the speed of switching transitions due to the negative-feedback effect it invokes by introducing a voltage drop in the gate-loop which militates against the action of the gate-drive circuit. In this case, it is recognized that even a very small change in the CSI value can produce a substantial change in the resulting transient behavior of SiC MOSFET's.

For the comparison of the TO-263 and TO-247 packages, this paper previously identified that the TO-263 package has 32% lower drain-source parasitic inductance than the TO-247. However, it should be noted that this quantity represents ($L_D + L_S$), and it was not possible to distinguish which portion of this value is attributable to the individual terminals within the package. Therefore, the simulation exercise described here was utilized to determine the difference in L_S which could be expected to produce the difference in observed transient behavior. Accordingly, Fig. 5 presents a simulation result similar to the measured behavior of the TO-263 package. This example incorporates 0.7 nH of source inductance, with the remainder of the 9.2 nH measured for ($L_D + L_S$) appearing in series with L_{BUS}. This configuration produces a current overshoot of 42 A, and a V_{DS} voltage shelf of approximately 200 V at turn-on, both of which are in good agreement with the empirical results. On the other hand, Fig. 6 presents a simulation result similar to the measured behavior of the TO-247 package. This example incorporates 2.0 nH of source inductance, with the remainder of the 13.6 nH measured for ($L_D + L_S$) appearing in series with L_{BUS}. This configuration produces a current overshoot of 32 A, and a V_{DS} voltage shelf centered at 400 V at turn-on, which is also in good agreement with the empirical results.

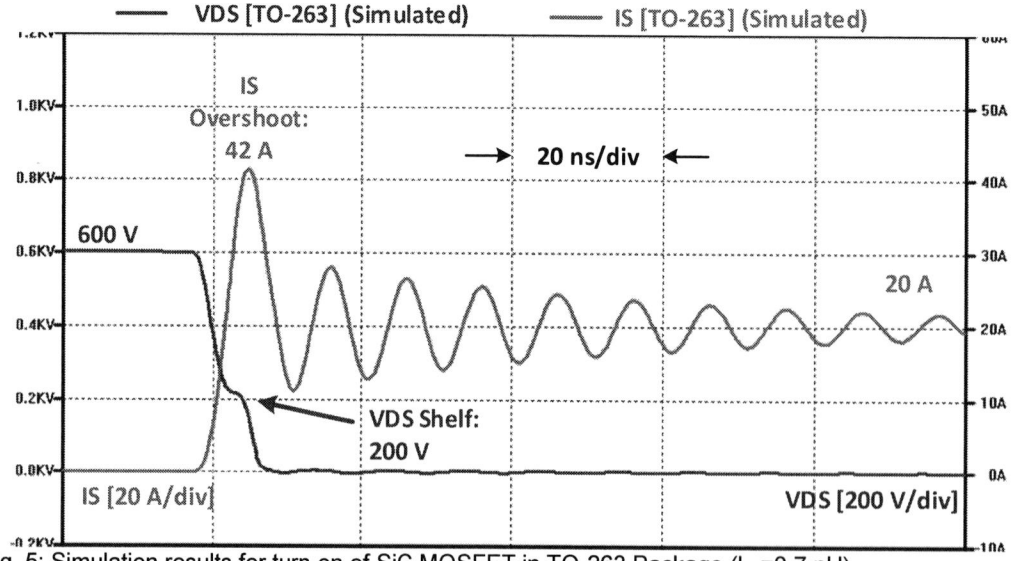

Fig. 5: Simulation results for turn-on of SiC MOSFET in TO-263 Package (L_S=0.7 nH)

PCIM Europe 2016, 10 – 12 May 2016, Nuremberg, Germany

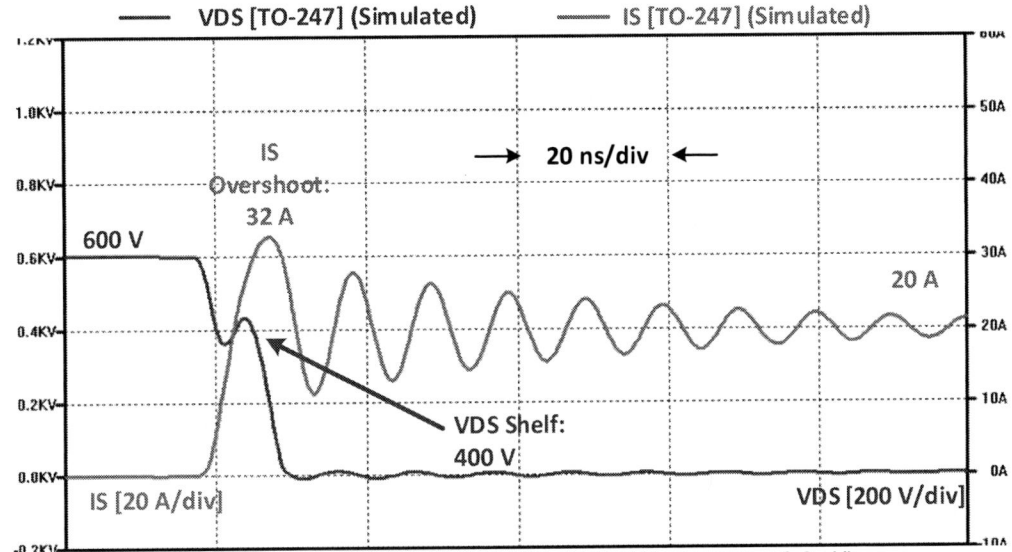

Fig. 6: Simulation results for turn-on of SiC MOSFET in TO-247 Package (L_S=2.0 nH)

3.5. Transient Results Summary

While the qualitative difference in transient behavior observed for the TO-263 and TO-247 package types may be more obvious than expected, the results themselves do not suggest a clear reason to prefer one package over the other. As a matter of fact, the TO-247 results demonstrate improved transient dynamics in terms of reduced overshoot and increased damping, when compared to the TO-263 results. However, some further quantifiable metric is needed to make a quantified assessment of the practical implications of this difference in qualitative behavior. For this reason, the results of the previously-described pulsed switching experiments were utilized to quantify the switching losses associated with the behavior of this SiC MOSFET in these two packages across a range of load currents from 5.0 to 20.0 A. The results of this analysis are presented in Fig. 7. From this figure, several observations can be made. First, the switching loss of the two package types at turn-off is virtually indistinguishable, as might be expected from the close transient agreement demonstrated in the previous section. Turn-off loss for both packages is also independent of the load current.

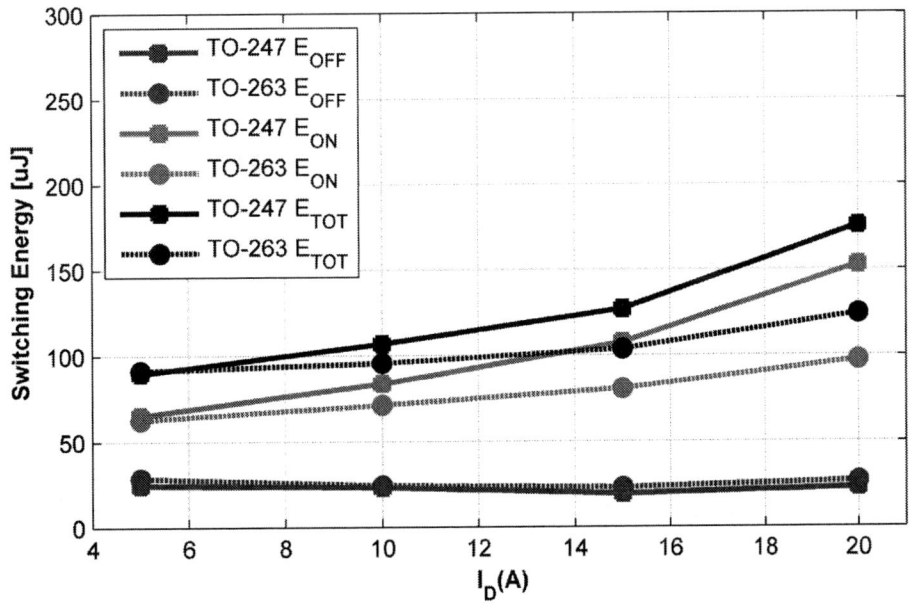

Fig. 7: Summary of switching energy results obtained for the two package types with R_G=2.0 Ω

© VDE VERLAG GMBH · Berlin · Offenbach

Second, the turn-on loss of the two package types is similar at light load, but an increasing difference is apparent as the load current is increased, in favor of the lower-inductance TO-263 package. This difference reaches a maximum of 55 uJ at a load current of 20 A, resulting in a reduction in total switching losses of 29% for this operating point. It is also suspected that operating at higher load currents would yield increasing disparity in the two package types. Third, the total switching losses for the SiC MOSFET in both package types are significantly lower than would be expected for alternative 1200V devices such as Silicon IGBT's.

4. Conclusion

This paper has presented an empirically-based evaluation of identical, high-performance SiC MOSFET's in two standard discrete packages: TO-263 and TO-247. As part of this work, a frequency-domain analysis was performed to quantify the differences in parasitic inductance of the two packages types. Subsequently, a set of pulsed switching characterization experiments was carried out to determine the implications of this difference in packaging parasitics. The results of this analysis indicate that there is a substantial difference in the behavior of devices in these two package types at turn-on, which results from a modest difference in the common source inductance of the two packages. The TO-263 package is characterized by slightly lower common source inductance than the TO-247 package, which results in a 29% reduction in total per-cycle switching losses at a load current of 20 A. Based on this analysis, it is projected that additional packaging improvements aimed at reducing the common source inductance would be expected to further improve the switching capabilities of high-performance SiC MOSFET's.

5. References

[1]   A. Lemmon, M. Mazzola, J. Gafford, and C. Parker, "Gate-Drive Considerations for Silicon Carbide FET-Based Half Bridge Circuits," In *Proc. Power Conversion and Intelligent Motion Europe,* 2013, pp. 311-318.

[2]   G. Laimer and J. Kolar, "Accurate measurement of the switching losses of ultra high switching speed CoolMOS power transistor/SiC diode combination employed in unity power factor," in *Proc. Power Conversion Intelligent Motion Europe, 2002.*

[3]   Josifovic and J. A. Ferreira, "Improving SiC JFET Switching Behavior Under Influence of Circuit Parasitics," *IEEE Transactions on Power Electronics*, vol. 27, no. 8, pp. 3843–3854, 2012.

[4]   K. Matocha, S. Banerjee and K. Chatty, "Advanced SiC Power MOSFETs Manufactured on 150mm SiC wafers", in *Proc. International Conference on Silicon Carbide and Related Materials (ISCRM)*, Sicily, Italy, 2015, in press.

[5]   A. Lemmon, R. Graves, and J. Gafford, "Evaluation of 1.2 kV, 100A SiC Modules for High-Frequency, High-Temperature Applications," in *Proc. IEEE Applied Power Electronics Conference and Exposition (APEC)*, 2015, pp. 789-793.

[6]   X. Zhang, N. Haryani, Z. Shen, R. Burgos, and D. Boroyevich, "Ultra-Low Inductance Phase Leg Design for GaN-Based Three-Phase Motor Drive Systems," in *Proc. Workshop on Wide Bandgap Power Devices and Applications*, 2015.

[7]   "Pearson Current Monitor Model 2877," Pearson Electronics, Inc., Palo Alto, CA., Available: http://www.pearsonelectronics.com/pdf/2877.pdf

[8]   "SSDN Series - 2 Watt Unit," T&M Research Products, Inc., Albuquerque, NM, Available:
http://www.tandmresearch.com/uploads/images/Products/2W_SSDN_DRAW.PDF

[9]   J. Hancock, F. Stueckler, and E. Vecino, "CoolMOS C7: Mastering the Art of Quickness," Infineon, Inc., Villach, Austria, 2013.

[10]   B. Yang and J. Zhang, "Effect and Utilization of Common Source Inductance in Synchronous Rectification," in *Proc. Applied Power Electronics Conference (APEC)*, 2005, pp. 1407–1411.

[11] Y. Xiao, H. Shah, T. P. Chow, and R. J. Gutmann, "Analytical Modeling and Experimental Evaluation of Interconnect Parasitic Inductance on MOSFET Switching Characteristics," in *Proc. Applied Power Electronics Conference (APEC)*, 2004, pp. 516–521.

Low-stress silicone encapsulant for reliable power conversion devices

Guy Beaucarne, Dow Corning Europe S.A., Rue Jules Bordet, Parc Industriel Zone C, B-7180 Seneffe, Belgium, guy.beaucarne@dowcorning.com

Eric Vanlathem, Dow Corning Europe S.A., Rue Jules Bordet, Parc Industriel Zone C, B-7180 Seneffe, Belgium, eric.vanlathem@dowcorning.com

Lu Zou, Dow Corning(China) Holding Co.Ltd, Zhangjiang Hi-Tech Park, Pudong District, Shanghai 201203, China, lu.zou@dowcorning.com

Kent Larson, Dow Corning Corporation, 2200 W Salzburg Rd, Midland, MI, United States of America, k.larson@dowcorning.com

Abstract

The emergence of distributed energy production and vehicle electrification has created the need for power conversion devices that remain highly reliable even when exposed to outdoor climatic conditions for decades. We describe a new encapsulant that contributes to reaching this kind of reliability. It is a tacky and soft silicone material that applies minimal stress on electronic components during thermal cycling and shows thermal conductivity to evacuate heat, while enabling fast assembly thanks to low viscosity and high cure speed. Material properties are described and compared to those of conventional power device encapsulants. Device testing is also reported, including stress measurement in encapsulated devices.

1. Introduction

Power conversion and conditioning devices such as inverters and DC-DC converters play a crucial role in the current energy transition. Many of those devices are exposed to widely varying and sometimes extreme weather conditions, for instance micro-inverters and power optimizers at the back of photovoltaic modules [1,2], or DC to DC converters and traction inverter modules in electric vehicles [3]. Demands on reliability are however extremely stringent for those devices. Their lifetimes need to match the lifetime of the systems they are part of, while acceptable failure rates are very low. At the same time, the manufacturers of those devices need to meet high productivity and low capex requirements for their products to be cost-effective. A key component in those devices is the circuit protection material.

2. Material requirements

The primary function of this material is to prevent exposure of the electronics to liquid water in order to avoid corrosion. In some electronic devices a thin conformal coating is sufficient to provide adequate protection. For power devices meant to operate outdoors or in other severe conditions there are additional requirements which lead to a different material choice. First the constant outdoor exposure present a non-negligible risk of water standing in the device enclosure at some point during its lifetime. In such event, a coating does not give appropriate protection. Conformal coating therefore can only be used in combination with a completely watertight enclosure to prevent liquid water from entering. Designers will therefore often choose potting of the circuit (complete immersion in the encapsulant) so that contact with liquid water is excluded and water tightness requirements are relaxed. Another factor in the choice of material is thermal management. Power conversion devices generate heat during operation, which results in a temperature increase that may come on top of a high ambient temperature. To limit this temperature increase it is necessary to enable heat dissipation by providing a media that can conduct heat. For this reason as well potting is favored, preferably with a high thermal conductivity encapsulant material. Moreover, as the devices are exposed to large temperature variations but consist of different materials with dissimilar coefficients of thermal expansions, it is important to limit temperature cycling-induced stress by choosing the right encapsulant material. It has been recognized in the past that stiff materials should be avoided in power device packages to allow thermal deformation and avoid stress build-up [4]. Finally, a low viscosity is preferred to enable quick filling of the enclosures, enabling high productivity, and avoid trapping air or creating voids, which are detrimental to reliability.

Dow Corning recently launched an encapsulant specifically designed for this type of applications called *Dow Corning*® EE-3200 Silicone Encapsulant.

3. Material properties

Table 1 shows material properties of EE-3200 and other encapsulant materials considered for potting of power conversion devices

	Viscosity at 10 s^{-1} shear rate (Pa.s)	Thermal conductivity (W/mK)	Linear CTE (10^{-6} /°C)	Hardness	Young's modulus (MPa)
Established silicone encapsulant	2.1	0.48	275	40 type A	5.2
Typical Polyurethane encapsulant	1.2	0.58	142	60 type D	55
EE-3200	1.7	0.5	360	20 type 00 43 type 000	0.09

Viscosity and thermal conductivity of EE-3200 is comparable to the established silicone encapsulant and the typical polyurethane encapsulant. The major difference however is the much lower hardness and Young's modulus. The thermal expansion coefficient is higher than for polyurethane encapsulant and that will tend to increase thermal stress. However this effect is expected to be overcompensated by far by the much lower modulus, resulting in much lower stress on electronic components.

Another material property that was studied is water/moisture absorption. The encapsulants listed Table 1 were immersed in water, and the weight change was studied in function of immersion time. It was observed that polyurethane encapsulant had a high percentage of water adsorption whereas silicone encapsulants showed only very low absorption. Interestingly, water absorption in the polyurethane encapsulant strongly increased with higher immersion temperature (0.3 % weight change for 1000 h at 25 °C, 1.1 % at 50 °C and 6 % at 85 °C), whereas it remained well below 0.1 % for EE-3200 at all temperatures. The water absorption content did reduce to zero when encapsulants were vacuum backed dried at 85 °C, for both all encapsulants tested. This indicates that it is a reversible process, and that water is physically absorbed and not chemically bound. Nevertheless, a high level of water absorption next to and around electronic components in some operating conditions is a hazard for circuits. It should be mentioned that tests involving short time exposure at room temperature may not reveal large differences between polyurethanes and EE-3200. This is because polyurethanes take a longer time to get saturated than silicones. However a relevant test has to run long enough to observe saturation, to take into account some extreme conditions that may occur over the lifetime of the device.

4. Device testing

The stress caused by the encapsulant during temperature changes was tested experimentally with the apparatus depicted in Figure 1. When temperature is increased, the expanding encapsulant will be constrained by the walls of the cylindrical container resulting in a pressure increase. This pressure increase per degree of temperature increase is measured by a pressure transducer. Figure 2 compares measurement results for EE-3200 and a leading polyurethane material. As anticipated based on material properties, thermal stress is much lower for EE-3200. The soft material will therefore result in much lower forces on the electronics in thermal cycling, which is beneficial for device reliability.

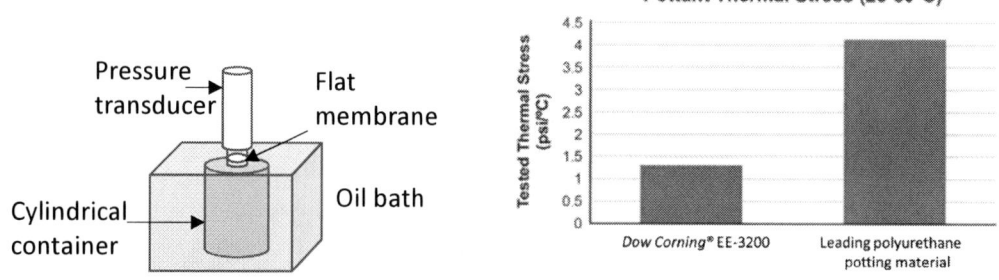

Fig. 1.: Apparatus to measure thermal stress   Fig. 2 : thermal stress measurements

The impact on device reliability was tested in a combined accelerated aging program. It consists of several rounds for subsequent humidity freeze, thermal cycling and damp heat tests. The devices that survive one round undergo a second, identical round of tests. In Figure 3, we show the results for three devices of the same type, but with three different circuit protection materials. The first one is just a conformal coating solution (no potting), the second one is the PU encapsulant mentioned earlier, and the last one is EE-3200. As can be seen in the graph, the device with only circuit coating fails during the first round. The device with PU encapsulant survives the first round, but fails during the initial humidity freeze cycles of the 2nd round. Only the device with EE-3200 encapsulant survives two full testing rounds. This results in accelerated aging will translate in lower rate of failure in the field.

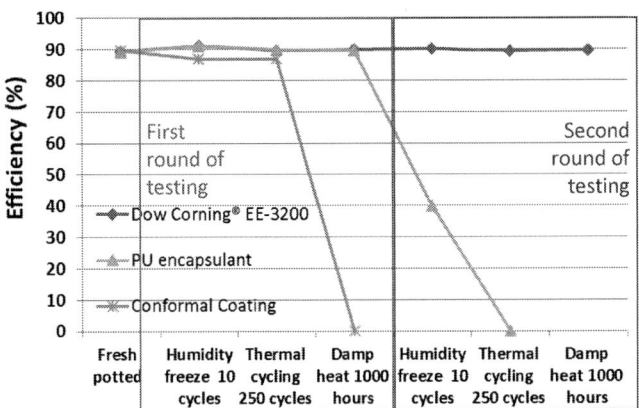

Fig. 3 : Combined accelerated aging tests on power conversion devices potted with different encapsulants

5. Conclusions

The new silicone encapsulant EE-3200 for power conversion devices is a very soft material, which leads to low stress on electronic components, but features appropriate thermal conductivity and a low viscosity in the uncured state. The low-stress aspect of this material is very important for devices that are exposed to outdoor conditions or generally experience a high temperature cycling stress, as lower stress leads to slower failure and improves reliability.

6. References

[1] Hadeed Ahmed Sher, Khaled E. Addoweesh, Micro-inverters — Promising solutions in solar photovoltaics, Energy for Sustainable Development 16 (2012) 389-400

[2] Perry Tsao, Sameh Sarhan, Ismail Jorio, Distributed max power point tracking for photovoltaic arrays, Proceedings 34th IEEE Photovoltaic Specialists Conference (2009) 2293-2298

[3] Young-Joo Lee, Alireza Khaligh, Ali Emadi, Advanced Integrated Bidirectional AC/DC and DC/DC Converter for Plug-In Hybrid Electric Vehicles, IEEE Transactions on Vehicular Technology 58 (2009) 3970 - 3980

[4] H. de Lambilly, H.O. Keser, Failure analysis of power modules: a look at the packaging and reliability of large IGBT's, IEEE Trans. Components Hybrids Manuf. Technol. 16 (4) (1993) 412–417.

Pumping out failure free package structure

Junji YAMADA, Yoshitaka OTSUBO

Mitsubishi Electric Power Device Works, 1-1-1 Imajukuhigashi, Nishi-Ku, Fukuoka 819-0192 JAPAN

Satoshi MIYAHARA, Mitsubishi Electric Europe B.V., Mitsubishi-Electric-Platz 1, 40882 Ratingen

Abstract

This paper presents an advanced technology and its benefit for new power module structures which enable preventing pumping out failures of the thermal interface material between base plate of the module and the cooling fin. This benefit is realized by developing a packaging technology with a new structure which has less warpage of the base plate with change in temperature.

1. Introduction

Nowadays, all kinds of electric power converter systems are required to be compact in size and yet deliver higher output power. Therefore, power modules as a key part of power converters are also required to have smaller package size along with higher output power. Power density of power modules have been improving accordingly, and it makes the heat dissipation design attain greater significance. For better thermal dissipation, one crucial factor is the reduction of the thermal resistance between the base plate and cooling fin. In general, thermal interface materials are applied between base plate and cooling fin to get an enhanced thermal contact, and its parameters (thickness, performance, material property, etc.) have an impact on heat dissipation capability and power module reliability. On the other hand, it is well known that the shape of the base plate of the power module is deformed by temperature changes generated by power loss in IGBTs and Diodes. This small but repetitive deformation of the base plate pushes out the thermal interface material, and is referred-to as the "pumping out failure". For assuring a long-term stability in the thermal interface material, it is required to develop an advanced package structure which is capable of preventing the warpage of the base plate under continuous temperature cycles.

2. Mechanism of "Pumping out failure"

The conventional structure of the power module is shown in Figure 1 and their components along with their coefficients of thermal expansion (CTE) are described in Table 1. In this structure, DBC, solder and copper base plate, which are consist of module base, have different CTEs. When the case temperature changes with heat generated by the IGBTs / FWDs operation, each component expands and contracts to different extents due to the different CTEs. Finally, among the various layers, the differential strain results in differential deformation, such as typical bimetal structures. This phenomenon is the cause of the base plate warpage.

Figure 2 shows the image of this phenomenon. The repetitive temperature change in a power module creates repetitive warpage of the base plate as shown in Figure 2 (a) and (b). This warpage pushes out the thermal interface material. This pumping out failure results in an ineffective thermal contact between the base plate and heatsink. This causes the degradation of Rth(c-s) and heat dissipation capability of the power module. In the worst case, it will lead to the thermal destruction of the power module when the junction temperature exceeds the absolute maximum rating.

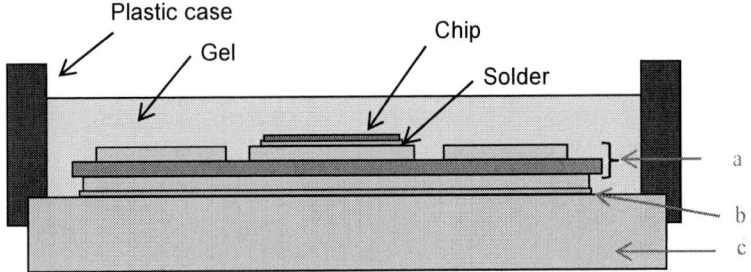

Figure 1. Conventional structure of power module

Symbol	a	b	c
Material	DBC	solder	copper
CTE(10^{-6}/K)	6.9	26	16.7

Table 1. CTE of component materials in conventional structure

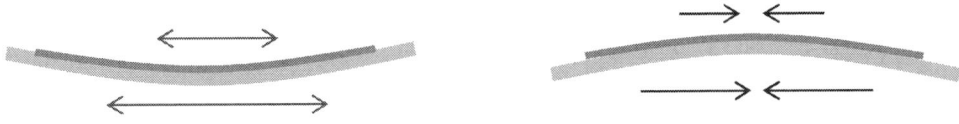

a) Plus bend (convex) by high temperature   b) Minus bend (concave) by low temperature

Figure 2. Image of base plate bending phenomenon

3. Measures against pumping out phenomenon

The new package structure can dramatically reduce the differences of strain between component materials of the module by matching their CTEs. The internal structure of the package and their CTEs are described in Figure 3 and Table 2 respectively. The new package structure employs solid resin as a potting material instead of the gel which is used in the conventional structure. Therefore, not only base materials but also the potting resin has been developed so that each component has a CTE compatible to the component adjoining it.

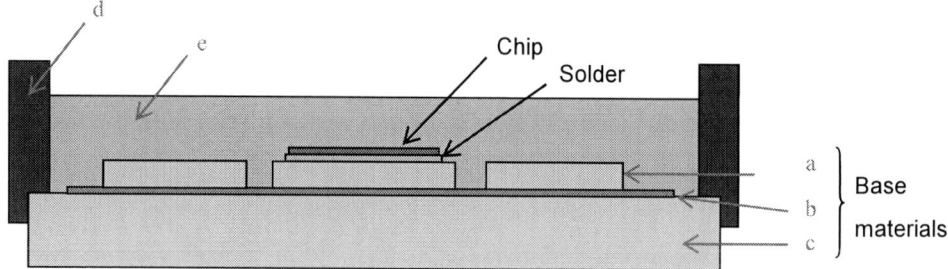

Figure 3. New structure of power module

Symbol	a	b	c	d	e
Material	Copper	Insulator	Copper	Plastic	Resin
CTE (10^{-6}/K)	16.7	16.7	16.7	16.5	16.0

Table 2. CTEs of component materials in new structure

This well-balanced structure in terms of CTEs creates much less deformation of the base plate caused by temperature change. Figure 4 and 5 shows the result of the stress analysis (finite element method) for the base plate components (conventional and new structure respectively), when the ambient temperature (Ta) changes from 25°C to -40°C. The amount of displacements shown in Figure 4 and 5 are five times the actual value in order to make the differences more visible.

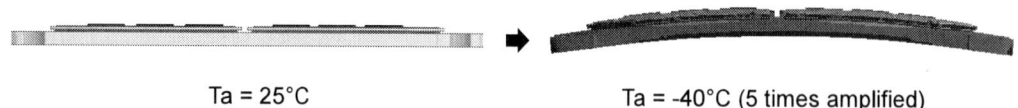

Ta = 25°C   Ta = -40°C (5 times amplified)

Figure 4. Simulation result for base plate deformation in conventional structure at Ta=25°C and -40°C

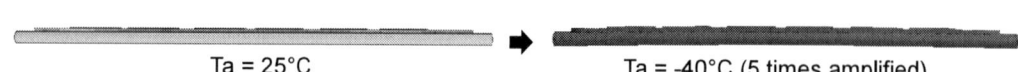

Ta = 25°C   Ta = -40°C (5 times amplified)

Figure 5. Simulation result for base plate deformation in new structure at Ta=25°C and -40°C

The simulation results depict that the new structure with the resin material improves the impact of temperature change on base plate deformation (providing less chances of warpage).

4. Measurements with completed prototypes

The temperature dependency of base plate warpage in the new structure is indicated in Figure 6 and Table 3. The vertical displacement of the measurement point shown in Figure 6 was measured at three different ambient temperatures controlled by an environmental chamber. The reference length is 91mm in longitudinal direction. Table 3 shows the measurement results of the average of nine completed prototypes.

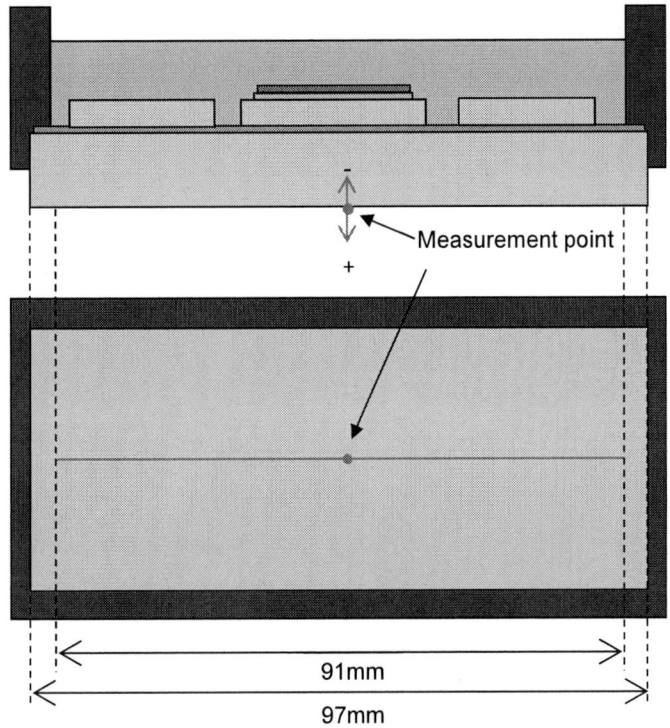

Figure 6. Measurement point of vertical displacement

Temperature (°C)	25	100	125
Displacement (µm)	+49.6	+39.6	+36.2

Table 3. Measurement result of vertical displacement

The gap of measured displacement in temperature change from 25°C to 125°C results in merely 13.4µm, which means there is no significant warpage of the base plate. This brings reduced pumping out failures to our newly developed modules. As an actual evaluation results with typical thermal interface material, no pumping out was found after 300 cycle heat cycle test (-40°C ~ +125 °C).

5. Conclusion

It is confirmed that the new power module structure, which has a matched coefficient of thermal expansion between adjacent layers, is able to control the warpage of base plate during temperature cycling and thereby provide a substantial reduction of the possibility of a pumping out failure event.

PCIM Europe 2016, 10 – 12 May 2016, Nuremberg, Germany

High Power IGBT Module with New AlN Substrate

H. Nogawa, A. Hirao, Y. Nishimura, Y. Tamai, F. Momose, T. Saito, E. Mochizuki,
Y. Takahashi

Fuji Electric Co., Ltd.
4-18-1, Tsukama, Matsumoto, Japan,
nogawa-hiroyuki@fujielectric.com

Abstract

This paper presents the packaging technologies for high power insulated gate bipolar transistor (IGBT) module which applied new thin aluminum nitride (AlN) insulated substrates with high heat dissipation and reliability to achieve higher power density. To apply new thin AlN insulated substrates to the high power IGBT module, we developed thin AlN ceramic substrate with high strength for higher reliability in thermal cycle test. In addition, applying the terminal with high current capability and reliability is necessary for high power IGBT module. For the purpose of that, the ultrasonic copper welding technology for AlN insulated substrates are developed. These technologies lead to increase the output power density compared to the conventional packaging technologies of high power modules.

1. Introduction

Recently, the market of inverter systems in a variety of area, such as general inverter, factory automation, renewable energy, automobile and electric train are growing with increasing of worldwide requirement to suppress global warming and energy consumption. Especially, the usage of renewable energy is expanding for the purpose of carbon dioxide emission reduction and environmental conservation. In the inverter systems for renewable energy and electric train area, high power IGBT modules are mainly applied. In these areas, the inverter systems are required to have higher power efficiency and reliability, and the size of inverter systems tend to be decreasing in contrast to improving their power density. To meet this trend, high power IGBT modules are required to achieve higher current density and reliability. For the packaging technology of power modules, improving the heat dissipation of the package is one of the solutions for increasing current density of IGBT modules.

Application of ceramics insulated substrates with high thermal conductivity is effective for improving heat dissipation of IGBT modules. Table 1 shows the material properties of a variety of ceramics insulated substrates. Ceramics insulated substrates are selected depending on the application of the inverter systems. For instance, Si_3N_4 insulated substrates have been applied to IGBT modules for automotive due to their high strength as shown in table 1. This is because IGBT modules for automotive mainly apply direct liquid cooing structure with aluminum heatsinks and large coefficient of thermal expansion (CTE) deference between aluminum fin and ceramics of insulated substrate cause large stress[1][2]. On the other hand, the most of power modules in industry area have air cooling structure with a copper base plate and it leads to the smaller CTE mismatching compared to the automotive ones. Therefore, Al_2O_3 insulated substrates are generally used for the power modules in industry area because of their low price. However, the thermal conductivity of Al_2O_3 insulated substrates is lower than those of AlN and Si_3N_4. In order to achieve higher current density, AlN substrates are applied for IGBT module in industrial area which need higher current density because the thermal conductivity of AlN is higher than that of Al_2O_3 and Si_3N_4. However, AlN insulated substrates exhibit low ceramic strength and need to be thicker to prevent ceramic layer breaking when IGBT module is screwed to heatsink.

© VDE VERLAG GMBH · Berlin · Offenbach

Conventional thick AlN insulated substrates lead to weakness in temperature cycle test because thicker ceramics substrates generate higher stress due to their high stiffness.

Table 1. Material properties of the variety of ceramics insulated substrates

	Thermal conductivity (W/mK)	Strength(MPa)	CTE (x 10^{-6}/K)
Al_2O_3	20	500	7.0
Si_3N_4	80	700	3.4
AlN	170	450	4.6

High power modules apply Cu terminals for a part of wiring inside module instead of wires in contrast to small and middle power modules that the all of their wiring are composed of Aluminum wires. These days, some high power IGBT modules apply ultrasonic welded Cu terminal instead of soldering. Because the welded Cu interfaces exhibits longer lifetime than solder layer of soldered Cu terminals. However, ultrasonic welding technique is likely to break ceramics by ultrasonic vibration. Therefore ceramics with high strength, and optimization of Cu pattern designs and bonding condition are required for ultrasonic welding technique. Especially, ultrasonic Cu welding to AlN insulated substrates has not been reported because AlN insulated substrates exhibit low strength.

As described above, applying AlN ceramics insulated substrate and ultrasonic welded Cu terminal is effective for increasing current density of high power IGBT modules. However, it has never been realized because of less reliability of solder under conventional thick AlN insulated substrates and breaking AlN by ultrasonic welding. This paper presents the new thin AlN insulated substrates with high strength that can prevent decreasing the reliability of power modules in temperature cycle tests, and the technology for ultrasonic Cu welding to AlN insulated substrates.

2. New thin AlN substrate

Applying conventional thick AlN insulated substrates lead to decrease the reliability of the solder layer under the insulated substrate in thermal cycle test compared to Al_2O_3 insulated substrates because the CTE mismatching between AlN and Cu is more than those between Al_2O_3 and Cu. One of the solutions to increase the reliability of solder layer under the insulated substrate is reducing the stiffness property of insulated substrate in order to decrease the stress that generate in solder layer. For the smaller stiffness, AlN insulated substrates with thin ceramics layer were applied to IGBT modules. The problems of conventional thick AlN insulated substrate when ceramic layer become thinner are as follows: (a) Decrease isolation capacity of IGBT module, (b) Ceramics breaking in power module mounting process on heatsink.

The new thin AlN insulated substrates were developed by using following three technologies, (1) Strengthening ceramic with the optimization of the sintering condition, (2) Improvement of copper pattern design to reduce the stress, (3) Optimization of design to maintain the insulating capacity. Table 2 shows the comparison of characteristics new AlN, conventional AlN and Al_2O_3. As shown in table 2, the strength of new AlN is raised to 500 MPa from 450 MPa of that of conventional AlN. Consequently, the thickness of AlN insulated substrates can be reduced by strengthening AlN and optimizing the Cu pattern design to decrease the stress.

Fig.1 shows the measurement results of the crack lengths in the solder under a variety of insulated substrates in temperature cycling test. For these measurements, same solder and base materials are used for each test samples. The condition of thermal cycle test (T/C test) was from -40 °C to 150 °C and crack lengths are measured by using ultrasonic microscopy at different time points in temperature cycle test. The solder crack length under the new thin AlN insulated substrates about 40 % smaller than those under conventional thick AlN insulated substrates. From these results, it is clarified that reducing shear stress of a solder layer under an AlN insulated substrate cause to suppress the propagation of solder crack.

To investigate the insulating capacity of IGBT module with new thin AlN insulated substrates, the dielectric strength test was performed. The test voltage was 4.0 kV and it was loaded for 1 minute. The number of test samples was five for each ceramics. Table 3 shows the dielectric strength test results of IGBT modules with new thin AlN or conventional Al_2O_3 insulated substrates before and after thermal cycle test. The condition of thermal cycle test was from -40 °C to 150 °C. As a result, none of IGBT module with new thin AlN insulated substrates is broken down before and after 300 cycles in thermal cycle test.

Fig.2 shows the thermal impedance of IGBT module with new thin AlN or Al_2O_3 insulated substrates between junction and case [3]. As shown in Fig.2, the thermal resistance of IGBT module with new thin AlN insulated substrates reduces by 45% compared to those with Al_2O_3 insulated substrates. Therefore, it is assumed that new thin AlN insulated substrates make it possible to increase the current density of high power modules.

Table 2. Material properties of the variety of ceramics insulated substrates.

	Thermal conductivity (W/mK)	Strength(MPa)	CTE (x 10⁻⁶/K)	Thickness (a.u.)
Al_2O_3	20	500	7.0	1.0
Conventional thick AlN	170	450	4.6	2.0
New thin AlN	170	500	4.6	1.2

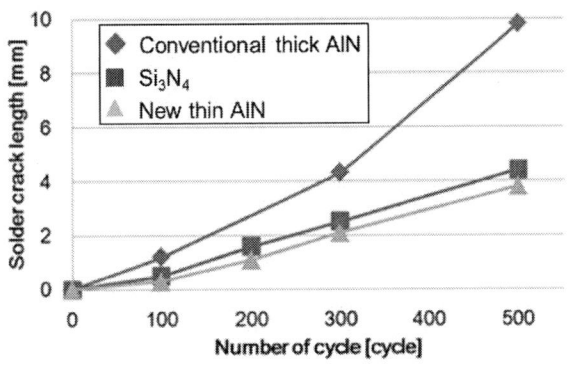

Fig. 1. Crack lengths in T/C test

Table 3. The results of the dielectric strength test: 4000V 1min.
The number of test samples was five for each ceramics.

	Initial	T/C 300 cyc.
Al_2O_3	5/5 pass	5/5 pass
New thin AlN	5/5 pass	5/5 pass

© VDE VERLAG GMBH · Berlin · Offenbach

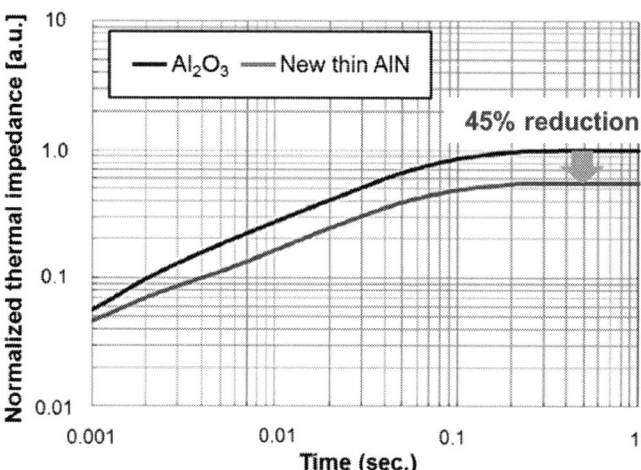

Fig. 2. Comparison of the thermal impedances [3]

3. Ultrasonic welding technology for AlN substrates

Table.4 shows the examples of the structure of small and middle power modules, and high power module. In high power modules, Cu terminals are applied as the interconnect structure to connect the output terminals in order to achieve high current density. These days, some high power IGBT modules apply ultrasonic welded Cu terminal instead of soldered copper terminal because of high reliability and high current capability. Fig.3 shows the result of the finite element method (FEM) simulation of the thermal stresses at an ultrasonic welded Cu interface and a soldered Cu terminal. As shown in fig.3, in power modules with soldered Cu terminal, high stress generate in the solder layer because of CTE mismatching between Cu and ceramics and that cause solder cracks and reducing the reliability of the solder layer under Cu terminal. On the other hand, less stress generate in Cu interface between Cu terminal and plate compared to soldered Cu terminal because the Cu terminal and Cu plate made from same Cu materials.

In ultrasonic welding technique for Cu terminal, surface Cu oxide layers on a Cu terminal and a Cu plate on ceramics insulated substrate are physically removed by ultrasonic vibration. After that, both of Cu surfaces of the terminal and Cu plate are bonded by solid phase diffusion. Therefore, ultrasonic welding can be performed at room temperature. High power modules have large Cu base plate and their camber size by soldering become larger than small IGBT modules. In other words, soldering Cu terminals cause large camber of high power IGBT modules by heating in soldering and make it difficult to assemble itself.Therefore, ultrasonic welding have advantage in assembly of high power modules compared to soldering.

To achieve high current density and high reliability, the Cu terminal ultrasonic welding technique for new thin AlN insulated substrates is developed. Cu materials exhibit a high hardness and the recrystallization temperature because of their high melting point, and solid phase diffusion of Cu need high energy for ultrasonic welding. Moreover, the shear stress in a horizontal direction in ultrasonic welding concentrate on the edge of the Cu pattern on ceramics insulated substrate. Therefore it is assumed that the position of the ultrasonic welded Cu terminal and optimum design of Cu pattern on insulated substrates strongly influence breaking ceramics [4]. Cu terminal ultrasonic welding for the new thin AlN substrates was realized by following technique: (A) high strength AlN ceramic, (B) optimization of copper pattern design, (C) location of terminals to reduce the damage and the optimum process condition of ultrasonic welding.

Table 4. The IGBT module structure small power module vs high power module

	Small Power Module		High Power Module	
Interconnect Technology	Al Wire	Cu Wire	Cu Terminal	Cu Terminal
Bonding	Ultrasonic Bonding	Ultrasonic Bonding	Solder	Ultrasonic Welding
Cross Section				
Appearance				
Current	1.00	1.67	3.54	3.54
Reliability (T/C)	> 2000 Cy.	> 2000 Cy.	> 300 Cy.	> 2000 Cy.

(a)   (b)

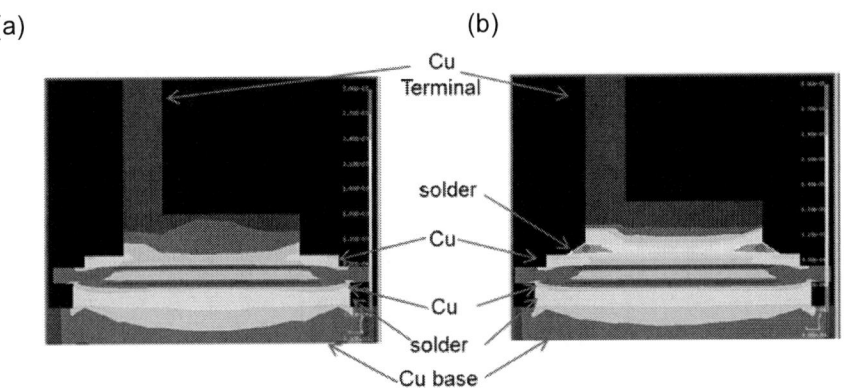

Fig. 3. Results of simulation of the thermal stresses
(a) Ultrasonic welded Cu terminal (b) Soldered Cu terminal

Fig.4 shows the appearance cross-sectional pictures with SEM (Scanning Electron Microscopy) of ultrasonic welded Cu terminal on new thin AlN insulated substrate bonded under the optimum welding condition and design for new thin AlN. As shown in fig.4(b), non-bonded area is little observed in the interface between Cu terminal and a Cu plate on the AlN insulated substrate. Cu welded layer with smaller Cu grain size than those of Cu terminal and Cu plate is observed as shown in fig.4(c). The small Cu grains in the welded interface indicate that both metallic surface of Cu terminal and plate are atomically closed each other, and the Cu atoms near the surfaces form new small Cu grains. Furthermore, it is assumed that the heat dissipation in ultrasonic welding proceed promptly due to high thermal conductivity of Cu and therefore the temperature of the welded interface in ultrasonic welding do not exceed crystallization temperature of Cu. Therefore, the size of Cu grain is assumed to be remained small during ultrasonic welding process. Thus, it is expected that the strength of welded interface are stronger than parent Cu materials because the grain size of welded Cu interface is fine.

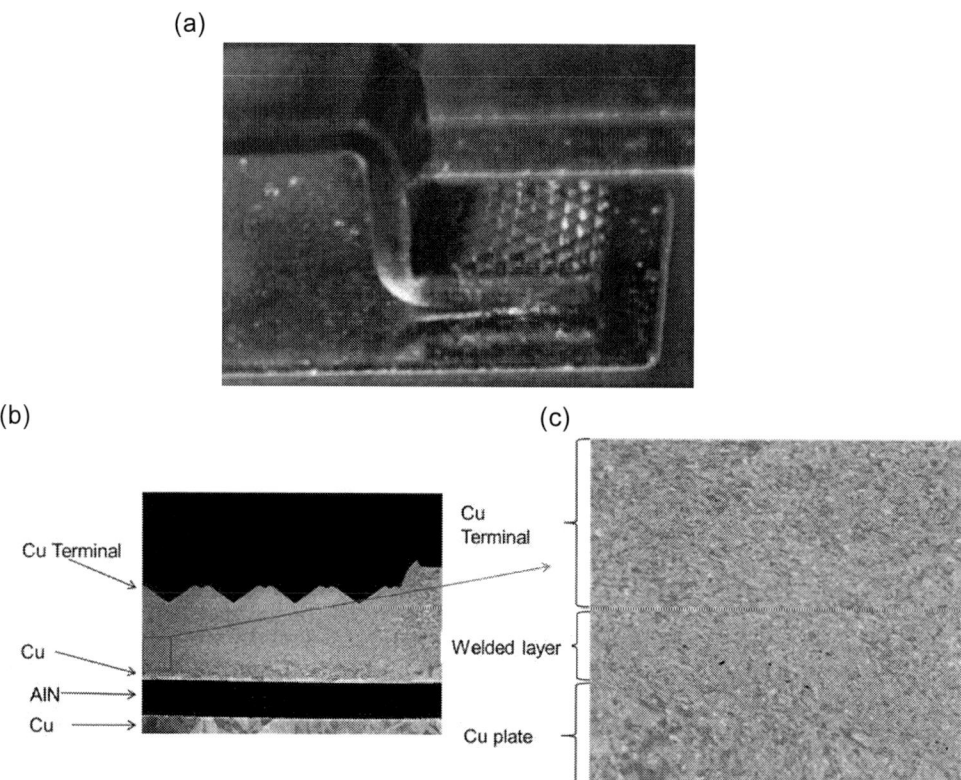

Fig. 4. Appearance and Cross sectional pictures of ultrasonic welded Cu terminal

(a)Appearance (b)Overall view (c)Extended view

To investigate the reliability of the ultrasonic welded Cu interface under the optimum welding condition, a thermal cycle test was performed. Fig.5 shows the tensile strength test results of ultrasonic welded Cu terminals on new thin AlN or conventional Al_2O_3 insulated substrates before and after thermal cycle tests. The condition of thermal cycle test was from -40 °C to 150 °C. The red line in fig.5 indicates the peeling strength of the interface between Cu and ceramics with same width as Cu terminal. In the case that tensile strength of ultrasonic welded Cu terminal is greater than the peeling strength of the interface between Cu and ceramics, the welded Cu terminal is not peeled before breaking the interface between Cu and ceramics which has the weakest tensile strength among Cu, solder, ceramics, and interface layers between Cu and ceramics. As shown in fig. 5, although the initial tensile strength of Cu terminal welded on new thin AlN insulated substrates is 14% lower than that on conventional Al_2O_3 insulated substrates, the tensile strength of Cu terminal that ultrasonically welded on new thin AlN insulated substrates greatly exceed the peeling strength between Cu and ceramics. The reason for the differences of the initial tensile strengths between ultrasonic welded Cu terminals of AlN and Al_2O_3 insulated substrates are assumed to be that welding conditions and Cu plate material of each ceramics substrates are different. In addition, the tensile strength of Cu terminal on new thin AlN insulated substrates almost remain unchanged after 300 cycles in thermal cycle test as well as that on Al_2O_3 insulated substrates. These results indicate that ultrasonic welded Cu terminal on AlN exhibit adequate strength against thermal cycle. Fig.6 shows the cross-sectional picture of ultrasonic welded Cu terminal before and after the thermal cycling test. After 300 cycles, no breaking of AlN and flaking of Cu terminal are observed. From these results, we concluded ultrasonic welded Cu terminal welded on new thin AlN under the optimum condition exhibits adequate reliability in thermal cycle test.

© VDE VERLAG GMBH · Berlin · Offenbach

PCIM Europe 2016, 10 – 12 May 2016, Nuremberg, Germany

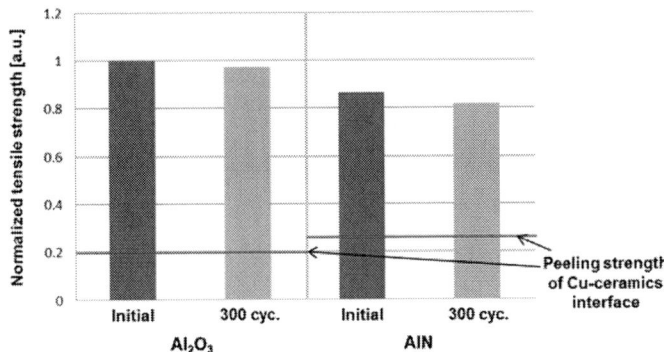

Fig. 5. Tensile strength test of ultrasonic welded Cu terminal in T/C test

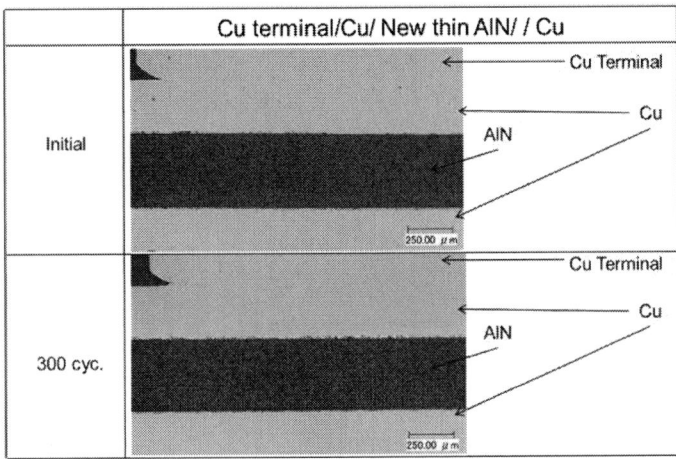

Fig. 6. Cross-sectional view of ultrasonic welded Cu terminal in T/C test

4. Conclusion

In order to apply AlN insulated substrates to high power IGBT module, it is necessary to prevent the crack propagation in solder under the insulated substrate and increasing the lifetime of terminal joint. As a solution of these problems, we developed following technologies:

(1) Development of new thin AlN substrate with high strength: it leads to improve the lifetime of the solder under insulated substrate by 65% compared to conventional thick AlN substrates.

(2) Optimized ultrasonic welding process for new thin AlN insulated substrates.

Using these technologies enable to increase the output current density of high power IGBT modules compared to those of conventional ones.

5. References

[1] A. Morozumi, H. Hokazono, Y. Nishimura, Y. Ikeda, Y. Nabetani and Y. Takahashi, Direct Liquid Cooling Module with High Reliability Solder Joining Technology for Automotive Applications, Proceedings of the 25th ISPSD & ICs, Kanazawa, May 26-30, 2013.

© VDE VERLAG GMBH · Berlin · Offenbach

[2] H. Gohara, Y. Nishimura, A. Morozumi, P. Dietrich, E. Mochizuki and Y. Takahashi, "Next-gen IGBT module structure for hybrid vehicle with high cooling performance and high temperature operation", pp. 1-8, Proceedings of PCIM Europe 2014, Nuremberg.

[3] T. Heinzel, J. Kawabata, Y. Kusunoki, Y. Onozawa, Y. Nishimura, Y. Kobayashi, O. Ikawa, "The New High Power Density 7th Generation IGBT Module for Compact Power Conversion systems", pp.359-367, Proceeding of PCIM Europe 2015.

[4] K. Kido, et al "Development of Ultrasonic Welding for IGBT Module Structure (Report 1)", pp.335-338 Proceeding of Mate 2010.

Nanosilver Paste for Low Pressure Die Attach: A Turn Key Process

Francesc Masana, Barcelona Semiconductors S.L.U., Sta Agnes 18, Pol. Ind. Sud, 08440 Cardedeu, Spain. e-mail: fmasana@bcnsem.com

Abstract: In this work, we present a new solder paste for die attach based on a bimodal nanosilver particle blend, with good mechanical, electrical and thermal characteristics. Also, in order to simplify processing while achieving a good repeatability and improved reliability, a sintering furnace has been developed providing the required control of sintering atmosphere, temperature, pressure and time in a programmed sequence with just the push of a button.

The sintering proceeds under a vacuum of ≈ 100HPa in order to help removal of organics from underneath the die by lowering its boiling point and collapsing the air or solvent bubbles that may form during evaporation while still leaving some oxygen for organics burnout. Also, a given amount of mechanical pressure, in the order of 1-5MPa, is applied onto the die surface through a compliant pad, by means of a pneumatic cylinder with controlled air pressure. This helps in two ways: it provides an intimate contact between silver particles and also with die and substrate surfaces, and helps collapse large voids eventually present in the paste that may impair sintering.

In the course of this work, we undertake parameter optimization in order to set up a die attach process for power semiconductor devices that is well suited to industrial applications. We also present some power cycling results that demonstrates the ability of both the paste and the process as good candidates for high temperature, high reliability die attach.

1. Introduction

The profusion of wide band gap semiconductor devices with their ability to perform at high temperatures is pushing packaging materials to meet the challenge. Also, silicon devices improved performance at high temperatures demands for tougher assembly materials. Besides of this, stricter reliability constraints also put increasing requirements on solder materials.

A critical point for high temperature operation is die attach. New high temperature materials have jumped on stage in the last few years that can be grouped mainly in two categories: the so called Low Temperature Transient Liquid Phase (LTTLP) bonding and the Sintered Silver die attach. As couldn't be otherwise, both of them have advantages and drawbacks:

LTTLP bonding rely on the transformation of a eutectic alloy of low melting point into non-eutectic intermetallics with a higher melting point [1-3]. It involves both an added material, usually a preform, and the chip and substrate metallization. This fact can limit the choice of suitable metals that may be compatible with common die and substrate metallization. Moreover, intermetallics use to be brittle materials and this can also impair reliability. On the other side, the fact that solder melts during process helps wetting and consequent adhesion to die and substrate. Homologous temperatures of 0.6 at 150ºC operation can be achieved, compared with 0.9 for conventional die attach materials.

Sintered Silver joints rely on the ability of nanosized particles to sinter into a solid at low temperatures, due to the high internal energy of the nanoparticles because of its small curvature radius [4-7]. The material usually comes as a paste containing silver nanoparticles and solvents. Before sintering can start solvents have to evaporate and/or burn, which means an oxidizing ambient. Pressure is generally required in order to obtain a good joint [8]. Adhesion to die and substrate is achieved by interdiffusion, making interface contact critical to adhesion. Ag sintering is directly compatible with the most common die backside metallization, namely Ag and Au, but is not so well suited to the Ni finish of some lead-frames. Homologous temperature is 0.35 at 150ºC because die attach material is pure Ag. This makes it the best candidate for high temperature, high reliability die attach.

© VDE VERLAG GMBH · Berlin · Offenbach

Accordingly, sintered silver die attach has received a good deal of attention in the recent years, not only for its potential as high temperature material but also for its improved reliability in front of its soft solder and LTTLP counterparts. However, the scarcity of commercial equipment availability and the relatively high cost of nanosilver solder pastes, along with a certain reluctance of semiconductor manufacturers to a not yet well established technology cast some shadows to its otherwise successful progression.

In the following paragraphs we will focus on such limitations. On the one hand, we will present a bimodal nanosilver paste that helps to reduce overall cost by using larger particle size while still maintaining or even improving material properties, and on the other we introduce a piece of equipment that enormously simplifies processing by including all that is needed into a single enclosure. On top of that, a series of experiments will demonstrate the good performance of the couple and will also establish the optimum process conditions.

2. Nanosilver Paste

The nanosilver paste is formulated by dispersing two different size nanosilver particles in a high temperature solvent. Particle size distribution is 40-90nm and 100-500nm and particle morphology is crystalline for both. They are mixed in a 50-50 proportion to a total solids content of ≈ 85wt%. Figure 1 shows the aspect of the paste. Its texture is smooth and suitable for both dispensing and stencil printing.

Figure 1: Aspect of nanosilver paste

The bimodal structure increases ductility of the final solid layer without much impairing strength [8,9]. Ductility is a very convenient property for die attach that significantly improves power cycling capability of the joint. Moreover, crystalline morphology along with particles of many different sizes increases packing density and intergrain contact that facilitates sintering at low temperatures [10].

The resulting solder paste can be dispensed or stencil printed down to 25μm thickness. Thin printed layers are convenient for ultrathin (100μm or less) die to avoid paste overflow over die edges when pressure is applied, but it may require a low roughness substrate to achieve a reasonable bondline thickness.

The relatively large grain size allows to substantially lowering the cost of the paste, which is still one of the main obstacles to spread its use. Finally, shelf life also compares favourably to other similar products. If stored in a fridge at 5°C, its shelf life is well over one year.

3. Description of the Process and Process Equipment

Although the paste is suitable to bond die on Ag and Au coated substrates, the process and its related equipment is designed to get good results on standard DCB Cu substrates, the main reasons being cost and compatibility with well established processes.

Process begins by stencil printing the paste on the DCB substrate to the required thickness (Fig.2-1). For die of 200μm or more in thickness, 50μm stencil is considered optimum. For very thin die, 25μm stencil may be better, although 50μm can also be used if stencil aperture is

© VDE VERLAG GMBH · Berlin · Offenbach

made slightly smaller than die size. For small die (less than 2mm in side) dispensing may also work well.

Because no previous drying is necessary, die can be placed on the paste print by means of standard vacuum pick and place equipment (Fig.2-2). Slight pressure is applied on die until paste begins to show at die edges.

Substrates with die are then placed on the loading tray (Fig.2-3) of sintering equipment and the compliant pad plate is gently placed on top of substrates through alignment pins (Fig.2-4). Tray with pad is then placed on the door alignment pins and against the edge of heating plate (Fig.2-5) and door is pushed in to close the sintering chamber.

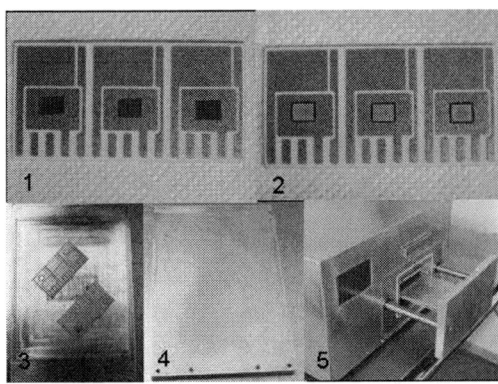

Figure 2: Process steps

From that point on, the process is automatically sequenced by the sintering equipment. Once the required sintering program is selected, the "RUN" key is pressed and door is pushed against the chamber edge until vacuum holds it in place. Sintering sequence is as follows:

- ✓ Chamber is first evacuated to a vacuum level of ≈ 150HPa
- ✓ Once this vacuum level is reached, cylinder is lowered to apply the set force to the pressing pad
- ✓ Heaters are switched on and pump continues to a vacuum level of ≈ 100HPa
- ✓ Temperature rises to the set point and stays there for the set process time
- ✓ Heaters are switched off and temperature begins to drop
- ✓ Once temperature is low enough, say around 100ºC, vacuum is switched off and chamber is flushed to atmosphere
- ✓ Chamber can then be opened and tray with substrates removed

In the above sequence, some of the process steps deserve further attention to fully understand its usefulness.

3.1. Sintering Atmosphere

Two apparently contradictory situations have to be resolved, i.e., copper oxidation and organics burn-out. Although a certain oxygen partial pressure is needed for the second, this pressure can be low enough to keep the first under control.

In our case, considering oxygen partial pressure as 21% of that of air, its value is approximately 21HPa. This value is lower than the optimum proposed in [11] but this can be due to the different organics used in paste formulation. Moreover, this lower value helps to keep copper oxidation to a minimum.

Apart from the above, vacuum also helps to remove organics from underneath the die by lowering its boiling point and collapsing the air or solvent bubbles that may form during evaporation. This, along with applied mechanical pressure increases compactness of solder layer and, as such, improves sintering.

3.2. Sintering Pressure

The effect of pressure in sintering is well known [8,12] to the point that, despite of some attempts to leave it out, results hardly reach acceptable values except for small die sizes [12, 13].

In a sintering process under ideal conditions, the ratio of final to initial volume of pores is constant [14]. This means that if we start with a low density (high volume of pores) compound, the final density will be lower than it would be if we had started with a higher density (lower volume of pores) compound, for the same sintering conditions. Then, reducing the initial volume of pores is of paramount importance to get a high density layer after sintering. Moreover, large pores are difficult to close, because the principal mechanism operating in them is surface diffusion that may reconstruct its internal surface but will not decrease its size [15].

Pressure is the best way to achieve this [15]. Uniaxial unrestricted pressure is applied normal to die (and solder layer) surface. Because paste can overflow past die edges, a shear force is in fact applied to the solder layer and the effect of this shear force is higher because the silver grains inside the paste are still "lubricated" by the solvent and they are able to flow with less force. This tends to close large pores, improve the inter-particle contact and enhance adhesion with die and substrate surfaces, and it can be achieved with lower pressure values, that greatly simplify the construction of sintering equipment as well as its operation.

Summarizing, sintering under (moderate) pressure and in a vacuum (oxygen partial pressure $\approx$ 20Hpa) atmosphere are the main features of the proposed sintering process, while temperature and time will be adjusted to optimize final results.

4. Experimental Procedure

In the following experiments we will use a silicon die of 5.7 x 4.5mm (25.65mm^2) and 240μm thick. Die backside metallization is standard Ti-Ni-Ag. The substrate is a standard DCB substrate with bare copper 300μm thick.

Paste is applied to the substrate through a 50μm Kapton$^\circledR$ stencil. Then, die is positioned on top of paste imprint with a pick and place equipment and slightly pressed until paste shows at die edges. No previous drying step is used.

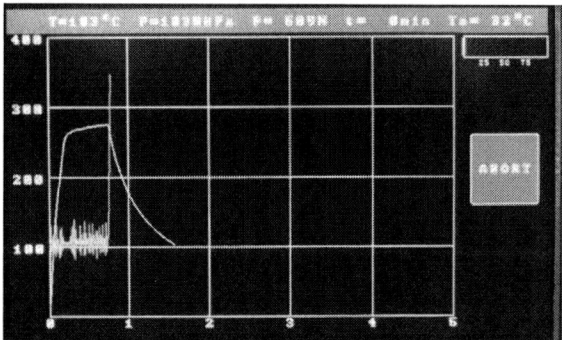

Figure 3: Typical Temperature profile (yellow) for a 250ºC, 45' process.

Die shear strength criteria follows the MIL-STD 883E. However, the interpretation of the standard for large die is somewhat ambiguous. Although it can be interpreted as if the maximum applied force for any die larger than 80x80mil ($\approx$2x2mm) should be 5Kg ($\approx$50N), this would lead to impractically low shear strength for large die. Instead, we interpret it as if the slope of the plot given should be constant. This slope is ≈12MPa for the x2 curve so we set the threshold to this value, irrespective of the failure mode, i.e., the amount of silicon remaining after shear. This yields a threshold force of ≈300N for the test die used.

In order to determine the optimum process parameter values, a set of experiments is designed where process temperature and time vary and the rest remain fixed. At least 3 die per temperature-time point are measured. Temperatures tested are: 200, 225, 250 and 275ºC and times 20, 30, 45, 60 and 120 minutes. Not all the possible combinations are tested. Low temperatures are mostly combined with long times and vice-versa. Applied force is set to ≈400N and vacuum level to ≈100HPa. Test times include the ramp up so short process times become more affected.

Figure 3 shows a typical temperature profile. As can be seen, ramp up time is 10 to 15 min. Green line is chamber vacuum level in HPa.

The test results are shown in Figure 4 both in tabular and graphic form.

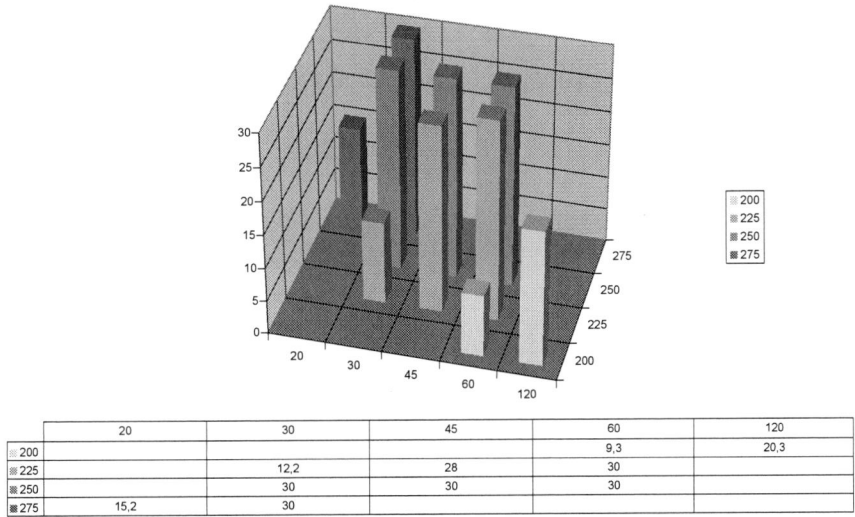

	20	30	45	60	120
200				9,3	20,3
225		12,2	28	30	
250		30	30	30	
275	15,2	30			

Figure 4: Temperature-Time test results

The experiment shows that a temperature of 225-250ºC and a process time of 45-30 min. yield the best results, although if a lower temperature is needed for, i.e., substrate material compatibility, 200ºC along with a longer process time of 90-120 min. can still be used with acceptable results. In all cases, shear strengths above 20MPa can be obtained, which are considered adequate for good die attach performance.

5. Power cycling

Besides the shear strength of die attach, reliability of the joint is the final target. Power cycling test is the best and most used way to ascertain this.

Power cycling test has been performed on an IGBT die 6x6mm sintered to a DCB substrate with 300µm bare copper. The sintering process used is: T=250ºC, t=45min., force F=400N and chamber pressure p=100HPa.

For the test, the IGBT has its gate shorted to its collector and a current of 11A is passed through the device, giving a power dissipation of ≈ 65W. The current is on for 1s and off for 5s in a 6s cycle that produces a temperature excursion of $\Delta T \approx 86\text{-}115ºC$ at each cycle.

The value of the maximum and minimum temperature as well as the current through the IGBT and its collector-emitter drop are measured and stored on a cycle basis. Thermal impedance to check the state of die attach is measured approximately after every 32Kcycles.

Thermal impedance vs. time measurement can give valuable information about die attach, as long as it is interpreted correctly. In fact, there is a direct correspondence between time and length [16] through the thermal diffusivity of the involved materials so, if we take the value of thermal resistance for a given elapsed time since the start of the measurement, and this time corresponds to a diffusion length a bit in excess of die thickness, then the value of thermal impedance will be the total thermal resistance between junction and DBC substrate and as such it will offer reliable information about the integrity of die attach. For the die used in the test ($\approx 350\mu$m thick) the time chosen is 2ms.

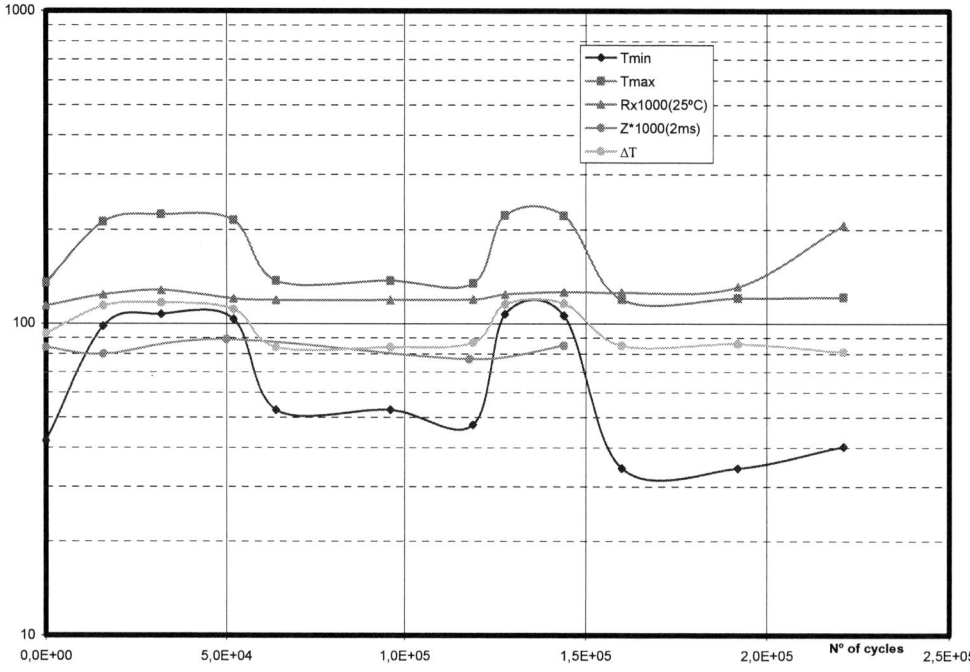

Figure 5: Power cycling results

Figure 5 summarizes the power cycling results. There are three curves related to temperature, i.e., T_{max}, T_{min} and ΔT. We can see that temperature excursion goes between 85 and 115ºC and the maximum temperature goes as high as 215ºC during the course of the test.

The other two curves are the slope of the IGBT output (CE) voltage drop (x1000 to fit the scale) at 25ºC, an indication of the (among other effects) state of collector electrical contact to the substrate plus the state of wire bond contact to emitter pad, and the thermal impedance at 2ms (also x1000 to fit the scale) that is a clear indication of the state of die attach.

The test finally failed after 221.000 cycles due to wire bond lift. At about 200.000 cycles it can be already seen a steady increase in output voltage drop till the final failure that interrupted the measurements.

The last measurement for thermal impedance was after 144.000 cycles. An excess confidence made us to postpone the measurement and failure came before we did it again. However, after wire bond failure we ran a shear strength test on the die, that gave a value over 30MPa, putting it crystal clear that die attach was still in very good condition.

6. Conclusions

Along this work we have introduced a new solder paste for die attach based on a bimodal nanosilver particle blend, with good mechanical, electrical and thermal characteristics. Also, a sintering furnace providing the required control of sintering atmosphere, temperature, pressure and time in a programmed sequence has been presented.

The sintering process has two significant characteristics: it runs under vacuum of ≈ 100HPa and a given amount of mechanical pressure in the order of 1-5MPa is applied onto the die surface. Both of these features are implemented to easy the evaporation of solvents and to reduce initial paste porosity in order to improve the mechanical, electrical and thermal properties of the final sintered layer.

The optimal process parameters are established in a range of 225-250ºC and 30-45 min. with a vacuum level of 100HPa and an applied force of 400N for the die tested.

The sintering equipment is developed to simplify the processing and can be operated with a single button, with full programmability of process parameters that makes it very convenient for production use.

Finally, both shear strength and power cycling results clearly demonstrates that the presented paste and process equipment are good candidates for research and production in the area of die attach for power devices.

7. References

[1] C.C. Lee, C.Y. Wang and G.S. Matijasevic. "A New Bonding Technology using Gold and Tin Multilayer Composite Structures". IEEE Trans. On Comp. Hybrids and Manuf. Tech, Vol 14, Nº 2, pp 407-412, June 1991

[2] J.W. Roman. "An Investigation on Low Temperature Transient Liquid Phase Bonding of Silver, Gold and Copper". Thesis for the Master of Science in Materials Engineering. MIT, 1991

[3] S.W. Yoon, K. Shiozaki, S. Yasuda and M.D. Glover. "Highly reliable nickel-tin transient liquid phase bonding technology for high temperature operational power electronics in electrified vehicles". 27th Annual IEEE Applied Power Electronics Conference and Exposition, Orlando FL, pp 478-482, 2012

[4] Z.Z. Zhang and Guo-Quan Lu. "Pressure-Assisted Low-Temperature Sintering of Silver Paste as an Alternative Die-Attach Solution to Solder Reflow". IEEE Trans. On Electronics Packaging Manuf. Vol 25, Nº 4, pp 279-283, Oct. 2002

[5] J.G. Bai, Z.Z. Zhang, J.N. Calata and Guo-Quan Lu. "Characterization of Low-Temperature Sintered Nanoscale Silver Paste for Attaching Semiconductor Devices". Conference on High Density Microsystem Design and Packaging and Component Failure Analysis, pp 1-5, June 2005

[6] Z.Z. Zhang. "Semiconductor Device Attachment by Silver Paste Sintering: Theory, Approaches and Applications". VDM Verlag Dr. Müller Publisher. ISBN: 978-3-639-15979-0, 2009.

[7] T.G. Lei, J.N. Calata, Guo-Quan Lu, X. Chen and S. Luo. "Low Temperature Sintering of Nanoscale Silver Paste for Attaching Large-Area (>100mm^2) Chips". IEEE Trans. On Components and Packaging Tech. Vol 33, Nº 1, pp 98-104, March 2010.

[8] K.S. Siow. "Mechanical properties of nano-silver joints as die attach materials". Journal of Alloys and Compounds, 514 (2012) pp 6-19.

[9] I.A. Ovid'ko and A.G. Sheinerman. "Plastic Deformation and Fracture Processes in Metallic and Ceramic Nanomaterials with Bimodal Structures". Rv.Adv.Mater.Sci. 16 (2007) pp 1-9.

[10] C. Fruh, M.Gunther, M. Rittner, A. Fix and M. Nowottnick. "Characterisation of silver particles used for the Low Temperature Joining Technology". Electronic System-Integration Technology Conference (2010) pp 1-5.

[11] H. Zheng, L. Xu, J.N. Calata, K. Ngo and Guo-Quan Lu."Effect Of Oxygen Partial Pressure On Sintering Nanoscale Silver Die-attachment On Copper Substrate". CPES Conference 2011, Blacksburg, VA (April 3-5, 2011) pp D4.2.

[12] K.S. Siow. "Are Sintered Silver Joints Ready for Use as Interconnect Material in Microelectronic Packaging?". Journal of Electronic Materials, Vol. 43, Nº 4, 2014.

[13] S.T. Chua, K.S. Siow and A. Jalar. "Effect of Sintering Atmosphere on the Shear Properties of Pressureless Sintered Silver Joint". IEEE 36th International Electronic Manufacturing Technology Conference (IEMT) 2014, pp 1-6.
[14] V.A. Ivensen. "Densification of Metal Powders During Sintering". Plenum Publishing Co. N.Y. 1973. ISBN : 0-306-10881-X
[15] C.C. Koch. "Nanostructured Materials: Processing, Properties and Applications". Noyes Pub. N.Y. ISBN: 0-8155-1451-4
[16] F. Masana. "Extraction of Structural Information from Thermal Impedance Measurements in Time Domain". 2011 MIXDES Conference, 2011

New silicone gel enabling high temperature stability for next generation of power modules

Thomas, Seldrum, Dow Corning Europe S.A., Belgium, thomas.seldrum@dowcorning.com
Eric, Vanlathem, Dow Corning Europe S.A., Belgium
Vincent, Delsuc, Dow Corning Europe S.A., Belgium
Hiroji, Enami, Dow Corning Toray Co. Ltd., Japan

The Power Point Presentation will be available after the conference.

Abstract

The power electronics industry is continuously seeking for higher operating temperature of power devices and modules, in order to increase the current density and decrease the cooling requirements in power systems. This is driven by the market trends, together with the emergence of new technologies allowing higher temperature of operation (such as silicon carbide and gallium nitride-based devices). The higher temperature of devices implies that the packaging materials (housing and encapsulant materials) also need to follow the same trend. As a leader in the silicone industry, Dow Corning developed encapsulant solutions for power modules to meet the industry requirements: soft and though silicone gels have been formulated in order to sustain a continuous temperature of 200°C. The integrity of the gel in terms of hardness, dielectric strength and adhesion to commonly used substrates is presented in this paper.

1. Introduction

Silicone gels are already used in the power electronics industry for decades. The main purpose of these materials is to protect the power chips from the environment (moisture in particular) and to provide electrical insulation for high voltage operation. Majority of today's power modules are specified for junction operating temperatures up to 150°C and slowly rising on the market are power modules rated at 175°C. Increasing the junction temperature allows a higher power density, allowing size and weight reduction. This increase in junction temperature is also driven by the spread of new devices based on SiC and GaN technologies which can intrinsically tolerate a higher operating temperature compared to the Si counterparts.

In order to pass the reliability requirements at higher temperatures, accelerated tests are performed by power module manufacturers in order to simulate the 20 years or longer lifetime of power modules. These tests consist in storing the modules at higher temperature than operating temperature (typically 25°C higher according to UL1557) and measuring the mechanical and electrical performances after the required hours of aging. The higher the aging temperature the shorter the time of aging can be.

Dow Corning is a major player in the silicone gel industry since decades. We developed a broad range of silicone-based solutions covering an operating temperature from -40°C to +175°C. In order to meet the industry requirements of higher temperature of operation, it has been required to develop silicone gels with resistance up to +200°C to simulate the aging conditions of power modules to be operated at +175°C. Moreover, low temperature resistance for particular areas in the world is required and chemistry suitable for operations down to -60°C was also investigated. As a result, Dow Corning is currently developing a new generation of silicone gel for operating temperatures between -60°C and +200°C in continuous mode.

Two different polydimethyl-based formulations have been developed in order to meet the temperature requirements of next generation power modules; these include a tough gel version and a soft gel version. These materials have a low viscosity (<700cP) in order to ensure proper filling of cavities within the power module architecture and high dielectric strength (>15kV/mm) to ensure required dielectric insulation properties.

2. Development of High Temperature Resistant Silicone Gels

2.1. Oxidative Degradation of Regular Silicone Gels

The temperature has a strong impact on the silicone gel properties in terms of hardness and integrity. Thermally driven oxidative degradation mechanisms occur when a polydimethylsiloxane silicone gel is exposed to elevated temperature for an extended period of time. This is demonstrated in Figure 1 where we can observe that silicone gels rated for operating temperature of 150°C (Sylgard 527) and 175°C (EG-3896) are hardening over time when exposed to a temperature of 200°C. The consequence is the cracking of the silicone material, starting with the creation of a hard crust on the surface directly exposed to the oxygen. This cracking occurs after 800 hours for the Sylgard 527 and after 1750 hours with the EG-3896.

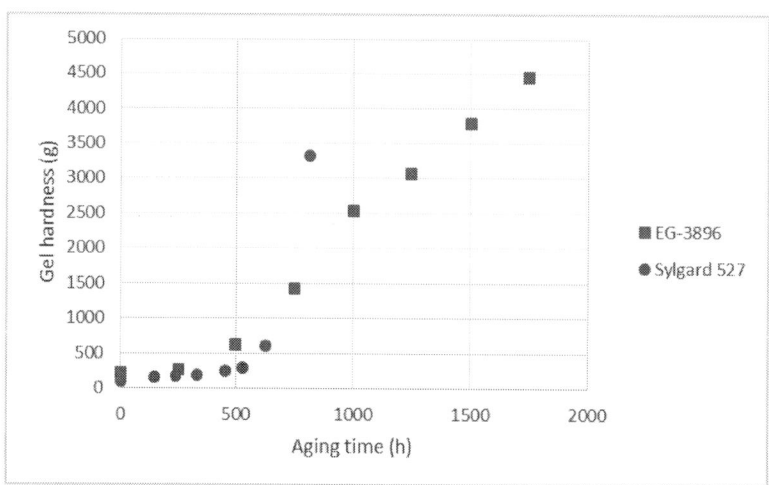

Fig. 1. Hardness of silicone gels Sylgard 527 (circles) and EG-3896 (squares) after aging in air at 200°C

The effect of oxygen during the ageing at elevated temperature is well-understood and acts on four different pillars:

- Increase of the cross-linking density (increased hardness) combined with desorption of CO_2 and water in already cross-linked material;

- Reaction of initially un-reacted Si-H groups with vinyl groups in the silicone matrix (increased hardness is observed);

- Even though silicone is an inorganic material, it contains a set of carbon containing units such as CH3 (methyl), CH=CH2 (vinyl) and CH2-CH2 (post-reacted vinyl), that are the most sensitive to oxidative degradation;

- Catalysis of cyclics by reaction with oxygen.

2.2. High Temperature Resistant Silicone Gels

The oxidation process cannot be avoided in air environment due to the high permeability of silicones to gases. However, it is possible to modify the formulation using additives in order to delay this degradation process, creating a silicone gel suitable for operation at higher temperatures.

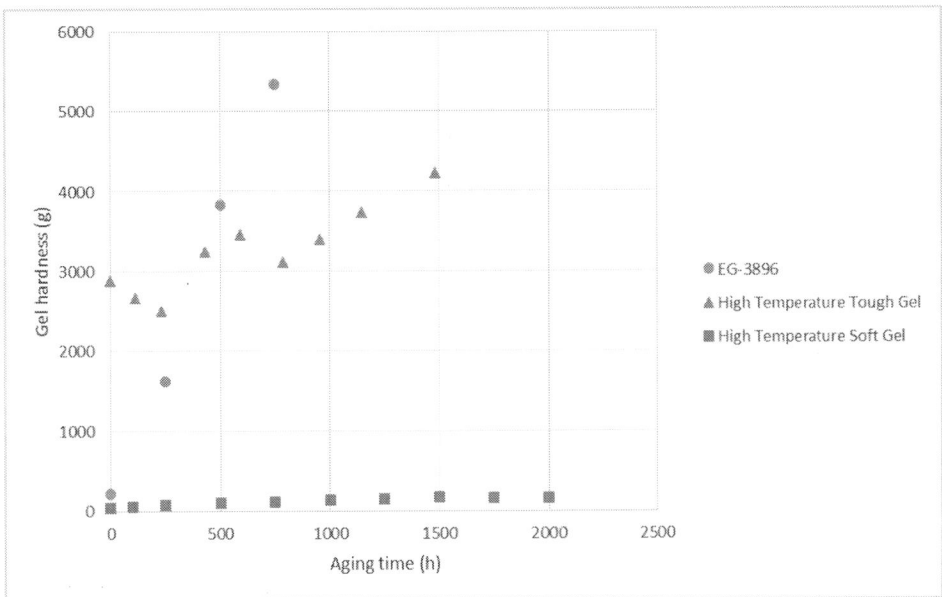

Fig. 2. Hardness of silicone gels EG-3896 (circles) and new formulations (soft gel with triangles and tough gel with squares) after aging in air at 225°C.

Dow Corning developed a high temperature resistant silicone gel suitable for continuous operation at 200°C. In order to meet the customer requirements in terms of power module lifespan, accelerated aging at 225°C has been used for the qualification of the products. Two new formulations were developed with different initial hardness to address the market needs. As depicted in Figure 2, both options could survive more than 2000 hours of aging at 225°C in air without cracking (squares and triangles for the soft and hard versions respectively). An increase in the hardness is still observed (driven by oxidation degradation mechanism), but the speed of degradation was considerably reduced when compared to the EG-3896 (circles).

The dielectric strength stability of the gel was also investigated to confirm its stability after the aging conditions at 200°C and 225°C and despite the hardening observed over time the dielectric strength of the soft and hard new formulations could be maintained (Figure 3). For this experiment, the dielectric breakdown was measured on a 2mm sheet of gel immerged in dielectric mineral oil. The test was done using a HIPotronics 775-5D149-5-B.

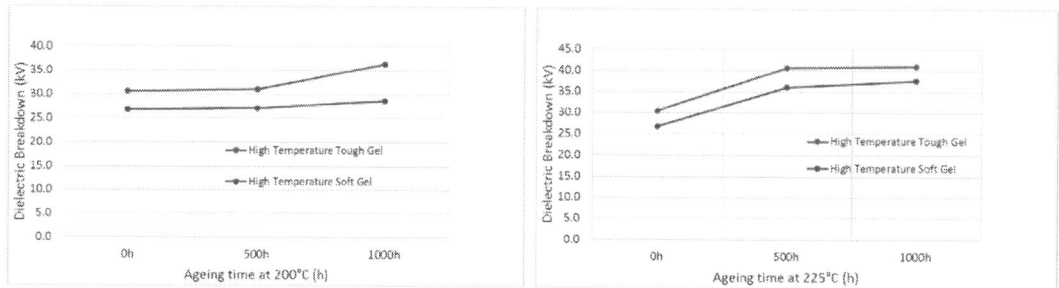

Fig. 3. Dielectric breakdown using 2mm sheets of Tough and Soft Gels under 200°C (left) and 225°C (right) ageing conditions.

We also investigated the adhesion and absence of delamination of the gel on different substrates commonly used in the power electronics industry. We could demonstrate that the adhesion on the substrate survived the aging condition at 225°C for more than 2,000 hours on gold, copper, aluminum, ceramic and silver.

The stability of the new formulations up to 200°C in continuous mode can also be used as a reliability enhancer for applications operating at a temperature below 200°C. This new family of silicone products can henceforth be extended to the entire range of power modules for reliability benefits.

3. Conclusion

Dow Corning developed a silicone gel suitable for operation at high temperature in order to meet the market trends. The silicone gel retains its integrity in terms of flexibility, dielectric strength and adhesion to substrates after several thousands of hours of aging up to 225°C. This new generation of silicone gels meets all requirements to allow usage of SiC and GaN devices at higher temperature and to increase the reliability of existing power modules rated at lower temperatures.

© VDE VERLAG GMBH · Berlin · Offenbach

A new Ag paste composed of nano- and micro-Ag particles prepared simultaneously and its application as die-attachment materials

Katsuaki Suganuma*, Hao Zhang, Shunsuke Koga, Jinting Jiu*, Shijo Nagao
Institute of Scientific and Industrial Research, Osaka University, Ibaraki, Osaka 567-0047, Japan. suganuma@sanken.osaka-u.ac.jp; jiu@eco.sanken.osaka-u.ac.jp;

Abstract

Although nano-Ag was developed to attach a die to substrates, and high performance in joint strength and electronic conductivity were realized, the complicated synthesis process and low yield of nano-Ag, and high joint pressure and temperature largely limited the application of nano-Ag in the industry. Here, a simple two-step polyol process was used to simultaneously synthesize nano and micro-Ag particles in large-scale and high-yield. The size and size distribution of Ag particles depended on the reaction parameters. These Ag particles were used to make hybrid Ag pastes to joint Cu sheets. Shear strengths over 40 and 52 MPa were achieved by using an Ag paste composed of nano-Ag particles and micro-Ag particles under a small sintering pressure of 0.4 MPa at 200 and 250 °C, respectively. These joints realized high thermal stability with shear strength over 30 MPa even aging at 250 °C for 500 h. These results are superior to those obtained with only nano-Ag pastes and suggest that the Ag paste composed of both nano-Ag and micro-Ag with suitable ratio and size distribution will be an excellent die-attachment material for high power semiconductor device.

1. Introduction

Wide-bandgap (WBG) high power semiconductor devices based on silicon carbide (SiC) and gallium nitride (GaN) have attracted much attention since the first commercialization of silicon carbide (SiC) Schottky diodes in 2001 [1], because these devices have lower power losses, higher conversed efficiency, higher thermal conductivity, much higher operation temperature, higher current density, and higher blocking voltages than normal silicon-based power devices [2-4]. Recently, several SiC-based power devices, such as JFETs (Junction Field Effect Transistor), MOSFETs (metal-oxide-semiconductor field-effect transistor), and BJTs (bipolar junction transistor), are beginning to enter the power electronics marketplace, which are expected that the global market for these high-power semiconductor devices will be over $3 billion in 2020. However, the WBG device market is still in an early stage. There are still many issues to be resolved, such as how to reduce the price of SiC/GaN dies which is still far higher than its Si counterparts, how to design the gate drive circuit by considering the unique operating characteristics of SiC/GaN in MOSFETs, how to join the SiC/GaN dies to the chip-carrier or leadframe under a suitable jointing temperature and pressure ranges. Among these issues, reliable joint and packaging technologies must be developed firstly to meet the demands of high temperature operating environments, which will further fulfill the potential of WBG devices. For example, the maximum junction temperature of SiC is 600 °C which is far higher than the maximum allowable junction temperature of standard silicon technology (150°C). Hence, the SiC device can be operated well even over 400°C, which is very good to be applied in high-end devices, such as server and telecom power supplies, solar power conversion, hybrid vehicles, and space and military monitoring systems. Die-attachment materials for joint, which physically connect the SiC dies to the rest of the system and also provide electrical and thermal pathways for semiconductor devices, also must be able to operate at such high temperatures.

Traditional die-attachment materials including Sn-based solders and Pb-free solders are ideal joint materials in silicon technologies because of their desirable joining process and temperatures. However, the low melt point of about 210-250 ºC for these solders is unsuitable for high power semiconductor devices [5-6]. Thus, looking for low-temperature bonding and high-temperature service die-attached material was an urgent issue. Some alternative die-attach materials that can withstand high temperatures have been reported, which included an off-eutectic gold-based alloy, silver-indium-based, nano-Ag, micro-Ag and hybrid silver die-attach materials [7-9]. Among these, hybrid Ag paste is a promising die-attach material because of its high melting temperature, intrinsically excellent electrical and thermal conductivity, dense filling property and long-term reliability [10-11].

Normally, hybrid Ag paste included nano-Ag and macro-Ag particles, which were bought from different companies and simply mixed with suitable solvent. The surface state of Ag particles and organics coating-agent in these Ag particles kept unclear and stochastic, which have proved to largely affect the sintering behavior of these Ag pastes and further determine the performance of die-bonding. In the present work, therefore, a new hybrid Ag paste composed of nano-Ag and micro-Ag were prepared by polyol synthesis with controlling organics amount and suitable size distribution, and used as die-attached materials to joint copper sheets to elucidate the sintering mechanism.

2. Experimental

Hybrid Ag pastes were prepared with a simple polyol process. Firstly, Ag seeds were synthesized by mixing poly(vinylpyrrolidone) (PVP, MW = 360000), ethylene glycol (EG) and silver nitrate ($AgNO_3$) and the mixture was reacted at 150°C for 1 h. After that the fresh Ag seeds were washed and mixed with EG and $AgNO_3$ again and reacted at 130 and 200°C to obtain the last Ag particles, which were washed with lots of ethanol to remove organics as much as possible, and then mixed with suitable EG to make hybrid Ag pastes with the concentration of Ag particles over 90 wt%. The morphology and size of the Ag particles were measured using a JEOL JSM-6700F electron microscope. The thermal degradation of the Ag particles was measured using a NETZSCH 2000SE/H/24/1 thermogravimetric (TGA) analyzer (NETZSCH, Selb, Germany) under a pure N_2 flow rate of 60 mL·min^{-1} and at a heating rate of 10°C·min^{-1}. To evaluate the electrical conductivity of the Ag paste, pastes were printed on glass substrates and annealed at 150, 180, and 200°C for 30 min to make Ag tracks. The Ag tracks resistivity was measured using a four-probe method with a surface resistivity meter (Loresta GP T610, Mitsubishi Chemical Analytech Co. Ltd.). The bonding strength was measured by a die shear test (DAGE, XD-7500).

3. Results and discussion

Chemical methods are simple and easy processes for the synthesis of metal particles with controlling size and size distribution by adjusting the reaction parameters, such as temperature, time, molar ratio of reactants and additives and so on. Polyol synthesis is a kind of chemical methods for the preparation of metal nanoparticles since the first development by Fievet et al. in 1989 to fabricate fine and monodisperse submicrometer-size metal nanopartilces [12-13]. The morphology, size and size distribution of metal particles in the polyol process can be easily tailored by adjusting the inject speed of reactants, using various growth agents and capping agents, seeds and so on [14-15]. In order to minimize the size and narrow the size distribution of the nanoparticles, normally, large amounts of organic dispersant was added to ensure monodisperse and prevent agglomeration. Here, in order to balance and optimize the size of the Ag particles for die attachment purposes, a two-step method was used to tailor the diameter distribution of the Ag particles and cut down dispersant as much as possible. Fig. 1a shows the SEM images of the Ag seeds formed in the first step. The Ag seeds were irregular and had a broad size distribution from below 100 nm to several hundred nanometer when $AgNO_3$ was continuously added into the reaction solution during preparation process. These Ag ions were reduced into Ag nucleus and further grew into Ag particles. New Ag nucleus was formed or deposited on the formed Ag particles when new Ag source was added, which led to the broad size distribution seen in Fig.1a. However, these Ag particles did not grow indefinitely even with continuous supply of Ag source due to the existence of PVP capping agent , which always played a role to prevent those Ag particles from aggregation. In order to form Ag particles with large size in the polyol process, these Ag particles were completely washed with ethanol to remove capping agent. After that, these particles was used as the seeds and nucleus in the second step to grow the Ag particles. Fig.1b and 1c shows the SEM images of Ag particles prepared at 130 and 200°C, respectively. It is clear that these small Ag seeds have grown into large submicrometer particles with different size distributions depending on the reaction temperatures. At 130°C, some micro-Ag mixed with nano-sized ones are always seen. These micro-Ag particles are always surrounded by small ones, which was expected to make excellent packed structure in die-attach bonding. At 200°C, these seeds also developed into large Ag particles which trended to have a same diameter comparing to those Ag particles formed at low temperature. Some small Ag particles were also included (Fig. 1c). The above two Ag particles formed at different temperatures were washed with ethanol for three times and dry at 100 °C for one week to determine the content of organics. When the Ag particles were heated from room temperature to 300 °C, the weight loss was about 0.03 and 0.08 wt% in the two Ag particles prepared at 130 and 200 °C, respectively. It indicated that the organics in the Ag particles are very few comparing to those nanoparticles with thick organic shell which prevent Ag nanoparticle from self-cohesion and agglomeration during preparation and storage [16].

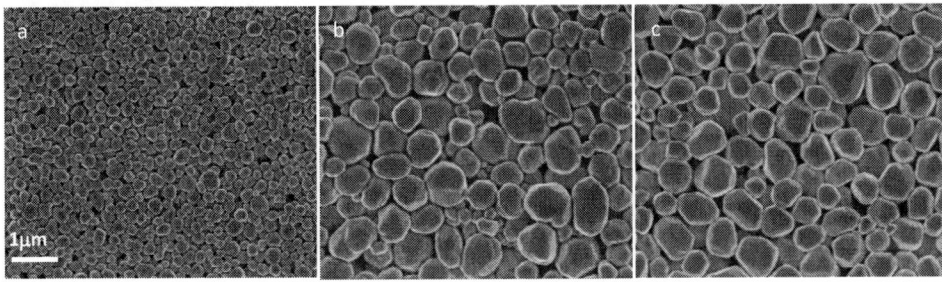

Fig. 1. The SEM images of Ag seeds (a) and Ag particles prepared at 130 (b) and 200 (c) °C.

As the die-attach materials, high electronic conductivity is an important indicator. Fig. 2a shows the electrical resistivity evolution of Ag tracks with different sintering temperatures. The two Ag pastes were screen printed on slide glasses with a width of 3 mm, length of 76 mm, and thickness of 200µm. Sintering was carried out in air at temperatures from 150 °C to 250 °C for 30 min. The resistivity was measured by four point probe method for five samples per dataset. The resistivity decreased with the increase in sintering temperature. At 150 °C, the resistivity of the two pastes were about 27 and 33 µΩ·cm, and then drastically decreased to 5 and 5.2 µΩ·cm at 200 °C, which is lower than the value of Ag nanoparticle paste [17-20]. The reasons can be attributed to the following: (1) the excellent packaging structure formed by Ag particles with different size in the two pastes, which made a dense Ag track by filling those voids among large particles with small particles, leading to low resistivity (Fig 2b and 2c show the surface of Ag tracks annealed at 200 °C for 30 min. It is clear that these Ag particles have been superbly arranged according to the size to make a dense film); (2) the trace amount of organics in the pastes due to the big size comparing to those nanoparticles pastes with thick protecting organic layer to avoid aggregation; (3) the multiplying effect of these Ag particles with various diameters in which small particles allowing for easy sintering might enhance the diffusion

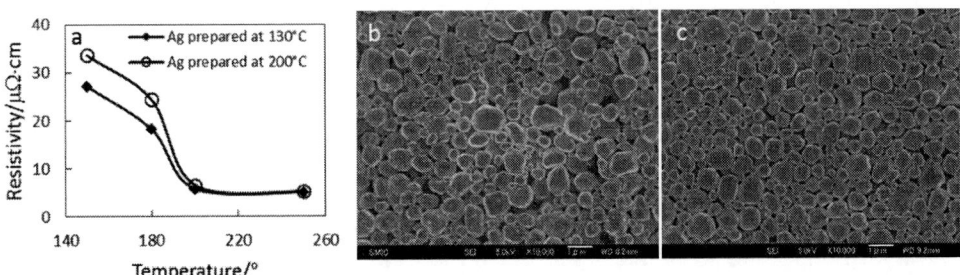

Fig. 2. The resistivity evolution of Ag tracks (a) and the surface SEM images of Ag tracks sintered at 200°C for 30 min for Ag pastes prepared at 130 (b) and 200 (c) °C.

growth between Ag particles because grain boundary occurs readily between small and large particles [21]. Moreover, with low-temperature sintering, the resistivity of Ag tracks from Ag paste prepared at 130 °C was always lower than that prepared at 200 °C, which might contribute to the matching of Ag particles in diameter. High-temperature sintering over 200 °C gave same resistivity. These results implied that the tailoring of Ag size and size distribution might be beneficial for the low-temperature sintering.

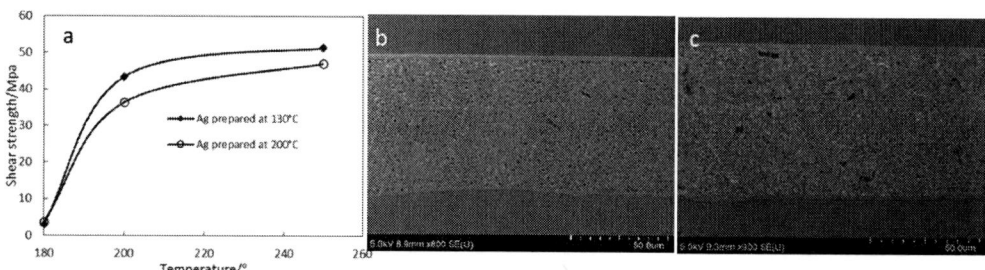

Fig. 3. The shear strength of Ag pastes (a) and the cross-section SEM images of Ag joints sintered at 200°C for Ag pastes prepared at 130 (b) and 200 (c) °C.

In most die-bonding processes of Ag nanoparticles pastes, applied pressures of about 1-10 MPa are always needed to improve the bonding performance, and to increase the bonding strength of joints. For example, Ag nanoparticles with the diameter of about 40 nm achieved joint strength of 25 and 40 MPa at 300 °C under 1 and 5 MPa bonding pressure, respectively [16]. The strength was decreased substantially to only 10 Mpa which is far below the requirements of industrial applications, when the sintering temperature was reduced to 200 °C without any bonding pressure [22]. To improve the bonding performance, hybrid Ag pastes including micro-Ag sheets and submicron Ag particles have been developed to achieve high performance joints [23-25]. Here, considering the excellent packaging structure of these Ag pastes seen in Ag tracks, the Ag pastes were used to join two Ag-plated copper sheets at 200 and 250 °C for 1 h under a low bonding pressure of 0.4 MPa. In order to remove the EG solvent including in the two pastes, the temperature was firstly increased to 85°C and kept at 85°C for 10 min to relax most of the low-molecular-weight organic compounds, and then increased to 185°C for 10 min to remove EG, finally, went to 200°C for 1 h to complete the sintering of the Ag pastes. Thereafter, the joining strength was measured after the temperature was cooled naturally to room temperature. Fig. 3 shows the shear strength of joints fabricated with the two Ag pastes. The strength was generally increased with sintering temperatures. Although the resistivity of Ag tracks formed at 180 °C gave a high electrical conductivity comparing to Ag nanoparticles, the shear strength was not achieved at this low temperature. Interestingly, the average shear strength of five samples was drastically increased to 43.5 and 36.3 MPa at sintering temperature of 200 °C, and then went up to 51 and 47 MPa when the sintering temperature was increased to 250°C. The strength is equal or even higher than that realized by only Ag nanoparticles pastes [16-20], which suggests that large Ag particles with suitable size can realize high shear strength for joints. Fig. 3b and 3c show the cross-section SEM images of the two joints sintered at 200 °C. Ag pastes evolved into a porous Ag structure and was tightly connected with the Ag-plate Cu substrates for realization of high strength. Comparing to the porous structure formed with Ag paste prepared at 200 °C, another Ag paste prepared at 130 °C seems to give a more uniform porous structure with few voids (Fig. 3b). Some big voids seen in figure 3c might be related to the size distribution of Ag particles. In the former paste, Ag particles had a similar and large diameter including a fraction of small particles, but the latter included much smaller particles seen in Fig. 1. When Ag particles were sintered and grown by surface and grain-boundary diffusion, some voids between big particles are always formed due to natural stacking. When sufficient small particles were arranged around these big particles, it could fill these voids to make a uniform and dense structure seen in Fig. 3b. Moreover, the grain size formed in the Ag paste with similar diameter Ag particles seen in Fig. 3c might imply that the growth by grain-boundary diffusion was rapid in these Ag particles, however, it is still unclear. The slightly low strength was most likely to be related to

© VDE VERLAG GMBH · Berlin · Offenbach

those big and irregular voids seen in Fig. 3c. These results furtherly suggest that the design and tailor of Ag particles diameter are crucial in the die-attached materials to meet the low-temperature bonding and high-temperature application. The joints prepared at 250 °C were used to confirm the stability of high-temperature storage. After stored at 250°C for 500h, the jointing shear strength of Ag paste prepared at 130°C slightly decreased from 51 to 43 MPa. However, the shear strength of Ag paste prepared at 200°C was increased from 48 to 65 MPa after aging for 500h. Whether the increase in strength is related to grain-boundary diffusion growth mentioned above is still unclear. However, these results indicated that joints exhibited higher reliability in high temperature storage test and implied that the size and size distribution of Ag particles were important for the die-attached materials.

4. Conclusions

Hybrid Ag pastes composed by nano-Ag and micro-Ag were synthesized by a simple polyol process in large-scale and high yield. The size and size distribution of Ag pastes were tailored by modifying the reaction temperature. These Ag pastes were used to joint Ag-plate Cu sheets at a low sintering temperature of 200 and 250°C under a very small pressure of 0.4 MPa. A shear strength of over 40 MPa were achieved, and the thermal cycle test indicated excellent long term reliability. These results are superior to those obtained with nanoparticle pastes and open a new way for the design of Ag particles to realize low-temperature bonding and high-temperature application in high-power devices.

Acknowledgement

The work is partly supported by the Japan Society for the Promotion of Science Grant-in-Aid for Scientific Research (Grant No. 24226017). The authors would like to thank Yasuha Izumi, Tsukasa Takahashi for their help and support in experimental work.

References

1. R. Rupp, C. Miesner, I. Zverev, In: Proceedings of PCIM 2001 Power Electronics Conference, 210-220 (2001).
2. K. Shenai, M. Dudley, R.F. Davis, ECS J. Solid State Sci. Tech. 2, N3055-N3063 (2013).
3. J. Millan, In: Proceedings of Semiconductor Conference (CAS), 2012 International, 1, 57-66 (2012).
4. J. Li, C. M. Johnson, C. Buttay, W. Sabbah, and S. Azzopardi, J. Mater. Pro. Tech. 215, 299–308 (2015).
5. R. W. Johnson, J. L. Evans, P. Jacobsen, J. R. R. Thompson, and M. Christopher, IEEE Trans. Electron. Packag. Manufact. 27, 164-176 (2004).
6. P. Godignon, X. Jorda, M. Vellvehi, X. Perpina, V. Banu, D. Lopez, J. Barbero, P. Brosselard, and S. Massetti, IEEE Trans. Ind. Electron. 58, 2582-2590 (2011).
7. M.J. Palmer, R.W. Johnson, In: Proceedings of the international high temperature electronics conference. 118-24 (2006)
8. F. Le Henaff, S. Azzopardi, J. Y. Deletage, E. Woirgard, S. Bontemps, and J. Joguet, Microelectron. Reliab. 52, 2321-2325 (2012).
9. S. Wang, M. Li, M. Ji, C. Wang, Scr. Mater. 69, 789–792 (2013).
10. K. Suganuma, S. Sakamoto, N. Kagami, D. Wakuda, K. S. Kim, and M. Nogi, Microelectron. Reliab. 52, 375-380 (2012).
11. H. Zhang, S. Nagao, K. Suganuma, J. Electron. Mater. 44, 3896-3903 (2015).
12. F. Fievet, J.P. Lagier, M. Figlarz, Mater. Res. Bull. 14, 29 (1989)
13. C. Ducamp-sanguesa, R. Herrera-urbina, M. Figlarz, J. Solid State Chem. 100, 272(1992).
14. J. Jiu, K. Murai, D. Kim,K. Kim, K. Suganuma K, Mater. Chem. Phys. 114: 333-338 (2009).
15. S.E.krabalak, L. Su, X. Li, Y. Xia, Nat. Protoc. 2, 2182-2190 (2007)
16. E Ide, S. Angata, A. Hirose, K. Kobayashi, Acta. Mater. 53, 2385-2393 (2005)
17. C. Huang, M.F. Becker, J.W. Keto, and D. Kovara, J. Appl. Phys. 102, 054308 (2007).
18. T. Wang, X. Chen, G.O. Lu, G.Y. Lei, J. Electron Mater. 36, 1333–1340 (2007).
19. M. Jakubowsk, M. Jarosz, K. Kiełbasinski, A. Mozniak, Microelectron. Reliab. 51, 1235-1240 (2010)
20. K. Moon, H. Dong, R. Maric, S. Pothukuchi, A. Hunt, Y. Li, C.P. Wang, J. Electron. Mater. 34, 168-

175 (2005)

21. L. Ding, R. L. Davidchack, J. A. Pan, Comput. Mater. Sci. 45, 247−256 (2009).

22. J.Yan, G. Zou, A. Wu, J. Ren, J. Yan, A. Hu, Y. Zhou, Scr. Mater. 66, 582−585 (2012)

23. A. Oestreicher, T. Ro¨hricha, J. Wildenb, M. Lerchc, A. Jakobd, and H. Langd, Appl. Surf. Sci. 265, 239-244 (2013)

24. M.Kuramoto, S. Ogawa,M. Niwa,K.S.Kim, K. Suganuma, IEEE Trans. Compon. Packag. Manuf. Technol. 33, 801-808 (2010)

25. K. Suganuma, S. Sakamoto, N. Kagami, D. Wakuda, K. Kim, M, Nogi, Microelecton. Relia. 52, 375-380 (2012).

Reliability of Double Side Silver Sintered Devices with various Substrate Metallization

Francois Le Henaff[1], Gustavo Greca[2], Paul Salerno[2], Olivier Mathieu[3], Martin Reger[3], Oscar Khaselev[2], Monnir Boureghda[2], Jeffrey Durham[2], Anna Lifton[2], Jean Claude Harel[4], Satyavrat Laud[4], Weikun He[5], Zoltan Sarkany[5], Joe Proulx[5], John Parry[5]

[1] Alpha Assembly Solutions, Elisabeth-Selbert-Strasse 4, 40764 Langenfeld, Germany
[2] Alpha Assembly Solutions, 109 Corporate Boulevard South Plainfield NJ 0890, USA
[3] Rogers Germany GmbH, Am Stadtwald 2, 92676 Eschenbach, Germany
[4] Renesas Electronics America (REA), 2801 Scott Blvd, Santa Clara, CA 95050 USA
[5] Mentor Graphics, 81 Bridge Road, East Molesey, KT8 9HH UK

Abstract

Silver sintering technology is one of the most promising high performance lead-free die-attach technologies. The work presented focuses on the thermo-mechanical and electrical performance evaluation of double side sintered modules. Die and clip attachments are processed with pressure assisted Argomax® silver sintering technology developed by Alpha Assembly Solutions. Power cycling tests following automotive requirements (200A, 5s ON and 10s OFF, until failure) and thermal cycling tests (liquid to liquid, -55°C/+165°C, 3 minutes dwell time for 1000 cycles) were applied to the devices. Thermal impedance measurements along with optical observations (metallographic and CSAM analysis) showed a significant increase of the 650V/200A IGBT device reliability when compared with best in class solder technology.

1. Introduction

The continual evolution of electric and hybrid automotive vehicles have increased the requirement for high power component reliability in smaller packages. Lead-free technologies are widely investigated in academic research [2, 3 and 4] and private consortium; such as the DA5 composed by Infineon, Bosch, NXP, ST Microelectronics and Freescale, as thermal, electrical and mechanical efficiency of standard solder based interconnection is approaching its limit [1] Several technologies have emerged over the past decade to achieve high temperature power modules (175°C and beyond junction temperatures) with high reliability. Some examples of such technology include gold base, high cost solders such as AuGe and AuSn, self-gripping assemblies using Van der Waals forces as the gecko, electrically conductive adhesives (ECAs), transient liquid phase bonding, SnSb alloys, high lead solder, SAC systems and silver sintering.

The commercialization of new technology has proven to be a complex process. Silver sintering has attracted a lot of interest over the last decade and has shown the ability to fulfill the electronics device reliability requirements. Argomax® silver sintering paste and film technologies are developed by Alpha Assembly Solutions and allow a fast (< 2 min), low-pressure (5-10 MPa) process for a wide range of applications including power modules, power discretes, thyristors, high power LEDs, and power RF devices.

Additionally, wire-bondless technology allows further miniaturization and significant reduction in parasitic inductance related losses. This design also improves the thermal management of power modules as heat dissipation is possible from both sides of the die [5, 6]. An internal simulation study has shown IGBT junction temperature decreases approximately 33% when using silver sintering as die-attach material in place of conventional solders (Fig. 1). This assumes nominal bond lines of 20 µm for sintering, 150 µm for soldering, and same die configurations and power applied (2kW). Specific process conditions were investigated to confirm the results obtained during simulation.

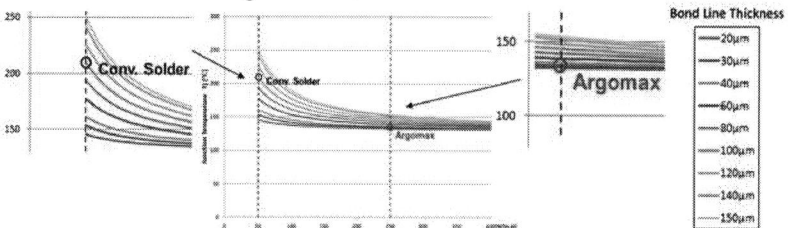

Figure 1: Junction Temperature in °C (Y Axis) for a 2kW power applied to 90um Si chip, using Argomax® sinter material and solder at different bond line thicknesses (BLT). Thermal conductivity in [W/(m.K)] (X Axis)

In this paper, we propose to evaluate the impact of double side sintering (die and clip attach) on the thermo-mechanical performances of 650V/200A IGBTs on Si_3N_4 AMB substrates with three different metallization: raw copper, spot silver and full electroless silver, when exposed to thermal shock and power cycling. Thermal impedance measurements throughout the power and thermal cycling tests were used to characterize the thermomechanical behavior of the sintered modules. Metallographic and CSAM (Scanning acoustic microscopy) analysis were used to correlate those failures to a defect inside the assembly.

2. Experimental approach

A 650V/200A 8.80 x 8.80 x 0.090 mm IGBT from Renesas Electronics America (REA) was assembled on a Si_3N_4 AMB substrate from Rogers (Curamik® Performance) with 3 types of metallization: raw copper, spot silver and full electroless silver (Fig. 2). Argomax® 8020 film technology was used for the die-attach process on electroless silver and spot silver finish substrates; whereas, Argomax® 8050 film technology that is specifically developed for copper finishes was used for attachment on the raw copper substrates. Copper clips were attached to the die and substrates using Argomax® film technology as well. This design replaced the wire bonding as the top connection to create the electrical interface for the power cycling and thermal impedance measurement using the Mentor Graphics MicReD 1500A tester. A four steps process was used to manufacture seven sintered modules for each substrate metallization. Three were used for power cycling tests, three more for thermal shock tests and the remaining one for baseline metallographic analysis. One of the modules used in power and thermal cycling tests were analyzed optically for any failure occurring in the assembly. CSAM analysis was completed on all sintered modules to detect delamination throughout the tests.

Assembly process:
1. Die placement step (Fig. 2)
   i. Laminate the dies with the Die Transfer Film (DTF) process using die-bonder equipment capable to apply heat and pressure via the bond-head.
   ii. Place the die on the bonding position at room temperature
   iii. Transfer to the sintering press
2. Die sintering step: Sintering of the assembly with the following parameters
   - 10 MPa, 2 minutes at 250°C
3. Laminate the clip using Argomax® 8020 film
4. Sinter the clip onto the assembly (die + substrate) (Fig. 3)

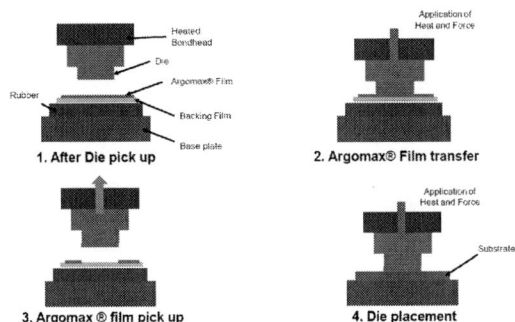

Figure 2: Die Transfer Film (DTF) process and die placement using a Die Bonder equipment

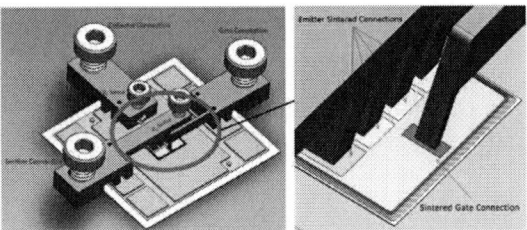

Figure 3: Test vehicle design for all configurations, using Si$_3$N$_4$ AMB substrate and a clip to provide power to the IGBT and substrate. Detail shows the full sintered top connection. Clip design courtesy of Renesas Electronics America (REA)

The MicReD Industrial 1500A Power Tester from Mentor Graphics measures the variation in V$_{CE}$ (voltage between collector and emitter of the IGBT) during each relaxation period (cooling period) and calculates the variation in the thermal resistance to help indicate if the die- or clip-attach interfaces of the assembly are failing. The devices were continually subjected to 200A for 5s and then to a 10s relaxation period. The double side sintered modules were placed on the cooling sink with a thermal grease (G751 from ShinEtsu Group) filled with silver particles to improve the thermal path. A torque wrench and consistent application of thermal grease was used to ensure the same pressure was placed on all modules to the cooler sink. The cooling liquid was kept at 30°C at all-time throughout the tests. When carrying 200A, the max junction temperature was 130°C with little variation in function of the finish observed. Junction temperature reached 125.5°C for the raw copper substrates. 3 soldered pieces with wire bonding used as electrical connection for top face were also cycled under same conditions to create a baseline for comparison.

The power cycling tests were conducted until the sintered modules failed. Thermal impedance was conducted throughout the experiment as thermo-mechanical failures in power modules are typically due to cracks between attachment interfaces. Coefficient of Thermal Expansion (CTE) mismatches between the different layers of a power module contributes to the expansion and shrinkage phenomena created during thermal variations (table 1); especially between the substrates materials and the dies (Silicon) as well as the die-attach layer. Thermal dissipation performance in a power modules becomes compromised when the ratio of cracking becomes too high in the module layers. Therefore, thermal impedance measurements, taken throughout power cycling, are used to find failure in power modules. The measurements were conducted continuously throughout the power cycling tests with the same equipment. An increase of 15% of the thermal resistance or max delta T°C or on the V$_{ce}$ (voltage between collector and emitter) of the module was defined as the failure criteria. Optical analysis using Scanning Acoustic Microscopy (CSAM) and metallographic analysis were conducted after

© VDE VERLAG GMBH · Berlin · Offenbach

initial processing the parts (time t=0) and at the end of the power cycling tests to correlate any failure to the sintered modules.

Table 1: CTE mismatches between power module materials

Material	Si_3N_4	Copper	Silicon	Sintered silver
CTE ($10^{-6}.°C$)	3.3	16.9	2.8	19.5

The sintered modules were submitted to -55/+165°C (liquid-to-liquid, three minutes dwell time) thermal shock testing for 1000 cycles. Profiles were continually adjusted to ensure temperature measured on the boards is in line with test specifications. Thermal impedance was measured prior to the test and at the end of 1000 cycles. If the thermal impedance presented an increase of 15% the modules were considered to have failed. CSAM and metallographic analysis were used to identify delamination and crack that lead to failure in the sintered module assembly.

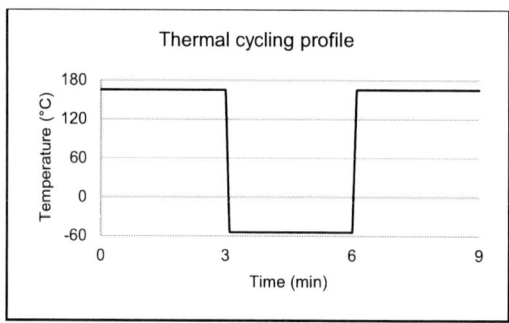

Figure 4: Thermal cycle profile - Liquid-to-liquid, -55°C/+165°C and 3 minutes dwell time

3. Tests results and analysis

3.1. Thermal shock test analysis

Nine sintered modules were submitted to liquid-to-liquid thermal cycling (three for each substrate metallization). The tests were conducted for 1000 cycles between -55°C and +165°C with three minutes dwell time (Fig. 4). Liquid-to-liquid equipment was used to ensure fast transition between the extreme temperatures. The thermal impedance variations for all tested sintered modules were below the failure criteria: 15% increase from the initial measurements (Table 2). This illustrates the encouraging properties of sintered modules compare to soldered modules. Indeed, previous internal tests have shown that similar soldered modules with Innolot (*SnAgCu+Sb+Bi alloy*) 150 µm preform cannot withstand more than 200 thermal cycles without a drastic increase of the thermal impedance. The CSAM analysis confirmed the behavior of the sintered modules compared to the soldered modules. No delamination was observed on the sintered modules compare to those observed on the soldered modules (Fig. 5 and 6).

Metallographic analysis was conducted on two sintered modules for each substrate metallization: one at t = 0 cycle and one at t = 1000 cycles. The cross-sections were taken on one edge and in the middle of each module. The sintered silver layer exhibits an average density of 90% for all three substrate metallization types. The sintered modules present well-formed bonds between the die and sintered silver layer as well as between the substrates and sintered silver layer (Fig. 7). As depicted in the microscope pictures below (Fig. 8), no defect was found on all of the sintered modules. The absence of defect was expected due to the intermediate test results for thermal shock. The sintered layers and interfaces with the die and substrates did not exhibit any structural modifications throughout thermal cycling test.

Table 2: Average thermal impedance variation throughout thermal cycling for the three types of substrates metallization - Three modules were tested for each substrates configuration

Module type	Thermal Impedance (°C/W) @ t = 0 cycle	Thermal Impedance (°C/W) @ t = 1000 cycles	Variation (%)
Ag AMB	0.350	0.355	+1.40%
Spot Ag AMB	0.338	0.343	+1.46%
Bare Cu AMB	0.326	0.322	-1.22%

Module type	Sintered modules @ t = 0 cycle	Sintered modules @ t = 1000 cycles
Ag AMB	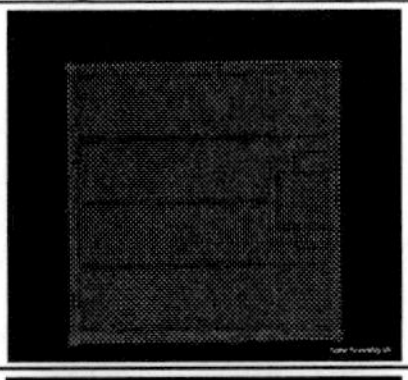	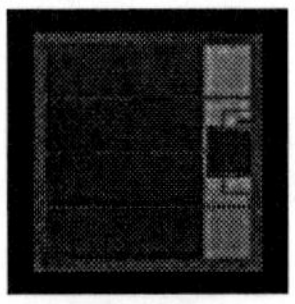
Cu AMB	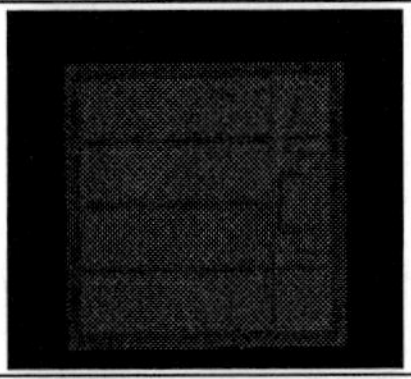	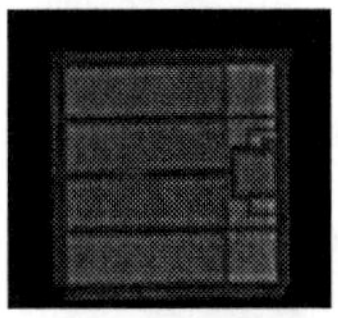
Spot Ag AMB	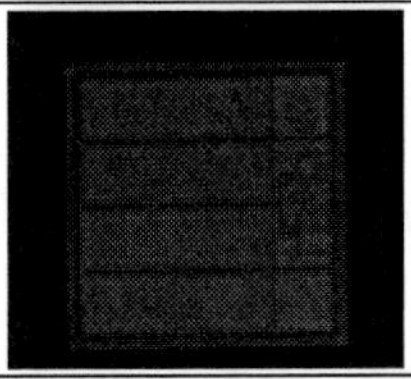	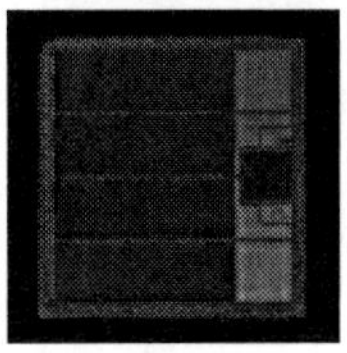

Figure 5: CSAM pictures for sintered modules @ t = 0 and t = 1000 thermal cycles (-55/165°C)

| 0 cycles | 200 cycles | 400 cycles |

Figure 6: Delamination on the soldered modules (Innolot (SnAgCu+Sb+Bi alloy) soldered interface thickness 150um, die size 8.80 x 8.80 x 0.090 mm) observed during the CSAM analysis after 200 and 400 thermal shocks. At 200 cycles, the thermal impedance already reached the increasing of 15% failure criteria, at 400 cycles the assembly is severe delaminated. Internal study.

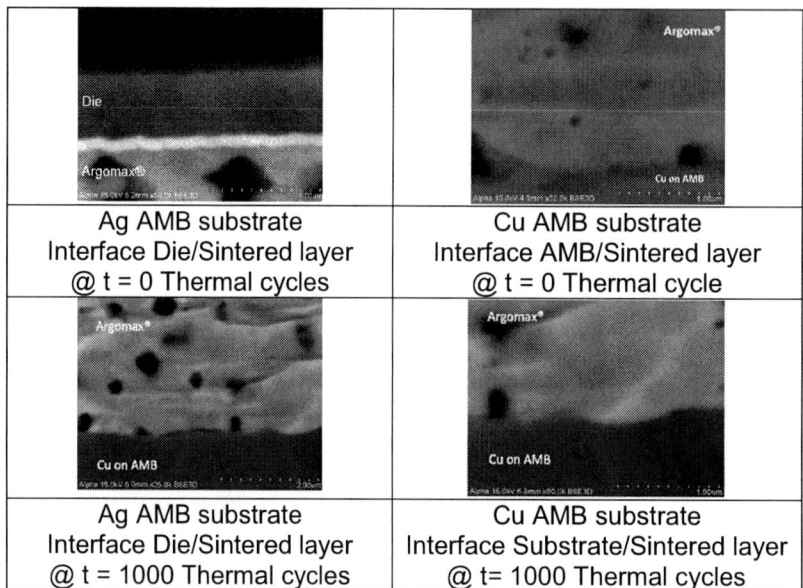

Figure 7: SEM analysis of the interfaces Die/Silver layer and Silver layer/AMB @ t = 0 and t = 1000 thermal cycles

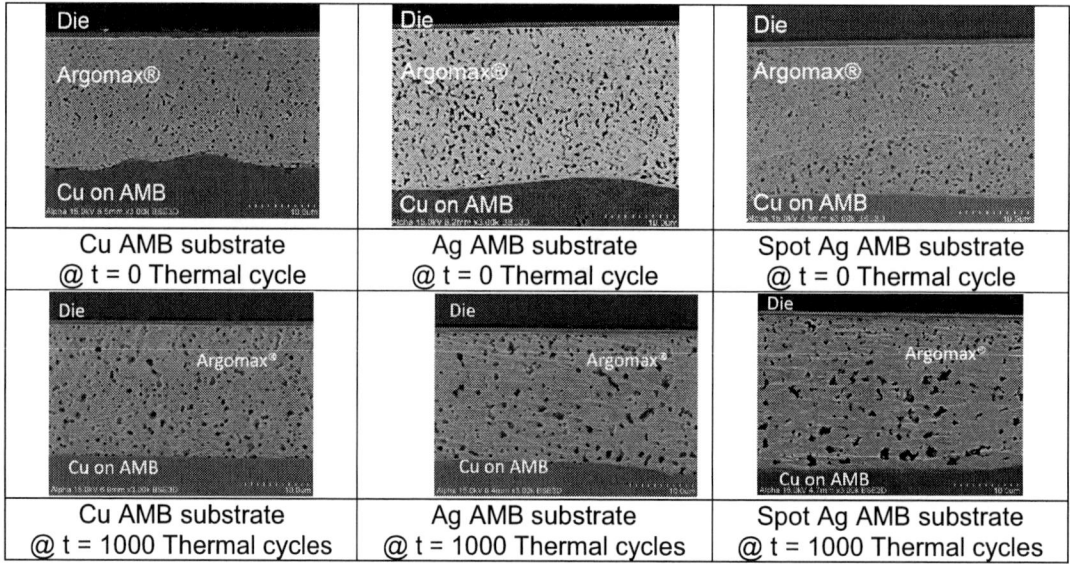

Figure 8: SEM analysis of the sintered modules @ t = 0 and t = 1000 thermal cycles

3.2. Power cycling test analysis

Nine sintered modules, three for each substrate metallization, were subjected to power cycling using the MicReD Industrial 1500A Power Tester from Mentor Graphics. It was agreed that the test were to end when the failure criteria of 15% increase of the thermal impedance, max delta T°C or V_{ce} was reached for each module. After 65k cycles (200A, 5s ON and 10s OFF, Delta T= 100°C), none of the nine sintered modules presented an increase of 15% of the thermal impedance (Table 3). Nevertheless, the power cycling tests were stopped to proceed with the

CSAM and metallographic analysis. Analysis of the aged sintered modules did not present any defect (delamination or cracks) inside the assemblies (Fig. 9, 10 and 11). The low thermal impedance of the silver layer creates a better thermal path inside the assembly and reduces the overall thermomechanical stress. The improved thermal performance of the sintered module permits a reduction of the maximum junction temperature to 130°C subjected to 200A current. For solder, the junction temperature reaches 172.6°C due to the high thermal impedance of the solder layer (SnAgCu). Soldered power modules failed at 45.6K cycles due to cracking on the die attach that caused the wire bond to lift off.

Table 3: Average thermal impedance variation throughout power cycling for the three types of substrates metallization - Three modules were tested for each substrates metallization

Module type	Thermal Impedance (°C/W) @ t = 0 cycle	Thermal Impedance (°C/W) @ t = 65000 cycles	Variation (%)
Ag AMB	0.345	0.320	- 7.25
Spot Ag AMB	0.338	0.330	- 2.37
Bare Cu AMB	0.326	0.295	- 9.51

Module type	Sintered modules @ t = 0 cycle	Sintered modules @ t = 65000 cycles
Ag AMB		
Cu AMB		
Spot Ag AMB		

Figure 9: CSAM pictures @ t = 0 and 65000 power cycles for sintered modules, clips were detached from the top for analysis.

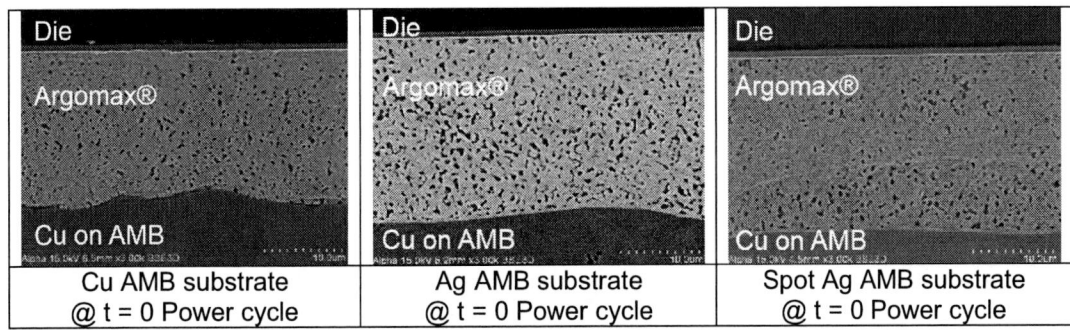

Cu AMB substrate @ t = 0 Power cycle	Ag AMB substrate @ t = 0 Power cycle	Spot Ag AMB substrate @ t = 0 Power cycle

Figure 10: SEM analysis of the sintered modules @ t = 0 Power Cycle

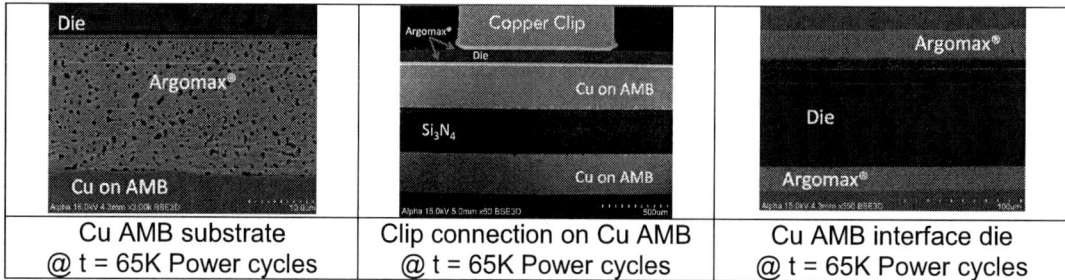

| Cu AMB substrate @ t = 65K Power cycles | Clip connection on Cu AMB @ t = 65K Power cycles | Cu AMB interface die @ t = 65K Power cycles |

Figure 11: SEM analysis of the Cu finish sintered modules @ t = 65000 Power cycles

4. Summary

The findings on double sided sintered power module for automotive applications in presented in this work. Silver sintering was chosen because of its promising properties for die and clip attach applications for power electronic modules and other devices. Silver sintering has demonstrated superior thermo-mechanical performance compared to conventional solder. The double side sintered module increased the performance of the assembly due to the high thermal conductivity of the silver layer and subsequent heat path (top and bottom side of the die) created to extract the heat from the semiconductor. The sintered modules did not exhibit any defect after thermal cycling (1000 cycles, liquid to liquid, 3 minutes dwell and -55°C/+165°C) and power cycling (65000 cycles, 200A, 5s ON and 10s OFF). The CSAM and metallographic analysis confirmed the thermal impedance measurements done throughout the cycling tests. The failure criteria was not met for all tested sintered modules; whereas, soldered samples failed at 45k cycles. The work presented confirmed the significant improvement that sintered silver technology offers for the thermomechanical performance of power module. The lower thermal impedance and junction temperature of the sintered assembly when compared to that of solder offers the possibility to operate semiconductors at a higher current level without reaching the 200°C Silicon barrier where its performance will be reduced.

This work is part of a consortium between Alpha (electronic interconnection materials supplier), Rogers Germany GmbH (former Curamik - electronics' substrate manufacturer), Renesas Electronic America (semiconductor manufacturer) and Mentor Graphics (electronic design automation company).
The authors would like to thank Dr. Aoki from Keio University, Akihiro Mochizuki, Yoshio Murakami and Goro Yoshinari from Alpha Japan for their valuable help on the Finite Element Modelisation work.

5. References

[1] U. Scheuermann et al. "The road to the next generation power module: a 100% solder free design", CIPS'2008.
[2] F. Le Henaff et al. "A preliminary study on thermal and mechanical performances of sintered nano-scale silver die-attach technology depending on the substrate metallization", ESREF'2012
[3] Ikeda, Y. et al. "Investigation on Wire-bondless Power Module Structure with High-Density Packaging and High Reliability, International Symposium on Power Semiconductor Devices and IC's." 2011, Proceeding CD, p.272-275.
[4] Horio, M. et al. "New Power Module Structure with Low Thermal Resistance and High Reliability for SiC Devices." PCIM Europe. 2011, Proceeding CD, p.229- 234.

PCIM Europe 2016, 10 – 12 May 2016, Nuremberg, Germany

New interconnect materials:
For future high reliable power module assembly

Stieven Josso, Henkel Electronic materials N.V., Belgium, Stieven.josso@henkel.com

Abstract

The trend in power electronics to move to higher operating temperatures and higher current densities has an effect on all components used inside the power module. By using different semiconductor IGBT and wire bond materials higher operating temperatures or enhanced power cycle life time can become reality. Also enhanced thermal interface materials are developed to guarantee a good thermal management during higher reliability requirements. For the chip interconnect material, standard SAC solder can be replaced by other Pb-free solder solutions using either AuSn, SnSb, Bi or Zn alloyed solders. Also Ag sinter materials -either with or without organic resin still present after sintering- or TLPS (transient liquid phase sintering) materials can be seen as possible alternatives for performing the interconnection. In this paper different interconnection materials are presented as a high reliable solution for IGBT die interconnection to cope with higher temperature requirements. Main focus will be on new developments showing Ag sintering during the die interconnection process. High dense materials with higher thermal conductivities -which either require pressure or no pressure during the sinter process- are presented.

1. Performance of Interconnection materials

1.1. Interconnection properties

In the picture below different interconnection possibilities are mapped based on thermal performance (thermal conductivity) in function of their homologous (possible working-) temperature. These are bulk properties of the materials but will in reality not only depend on the material used but, certainly with sintering materials, also on the assembly process.

The selection of which of the interconnection materials to use will depend on the substrate, substrate finish, the die, its BSM (backside metallization), die size and thickness, wire bond selection…

Further requirements like thermal cycling, active power cycling, temperature and adhesion requirements will also determine what material to use but, especially with sintering materials, also the process that needs to be used with these materials.

© VDE VERLAG GMBH · Berlin · Offenbach

PCIM Europe 2016, 10 – 12 May 2016, Nuremberg, Germany

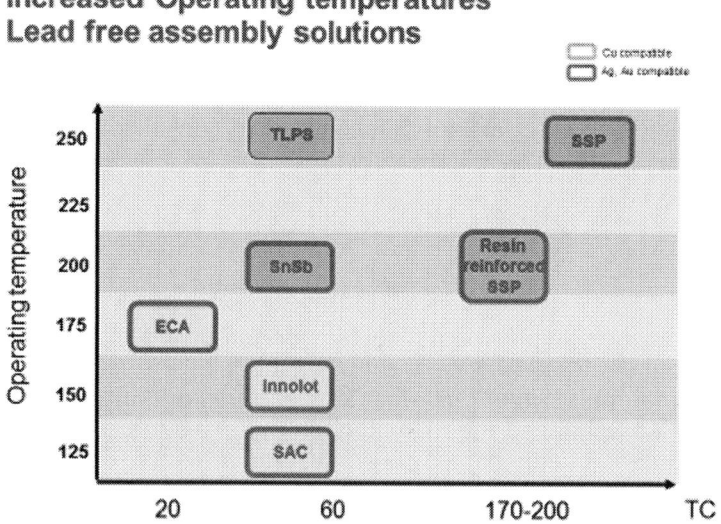

Fig. 1. Operating temperatures relation to thermal conductivity

1.2 Interconnection solutions

We will map some of the possible lead free interconnection solutions for high power applications and their properties.

1.2.1 Die attach paste – ECA

High Conductivity Die Attach Adhesive are designed for wide variety of Medium Power and High Reliability applications up to 8x8mm die size.

These are highly (up to 85%) Ag stacked and are based on low stress hybrid resins.

Thermal conductivity up to 19 W/mK bulk conductivity and 0.000019 ohm-cm volume resistivity can be reached.

These are temperature stable up to 175°C.

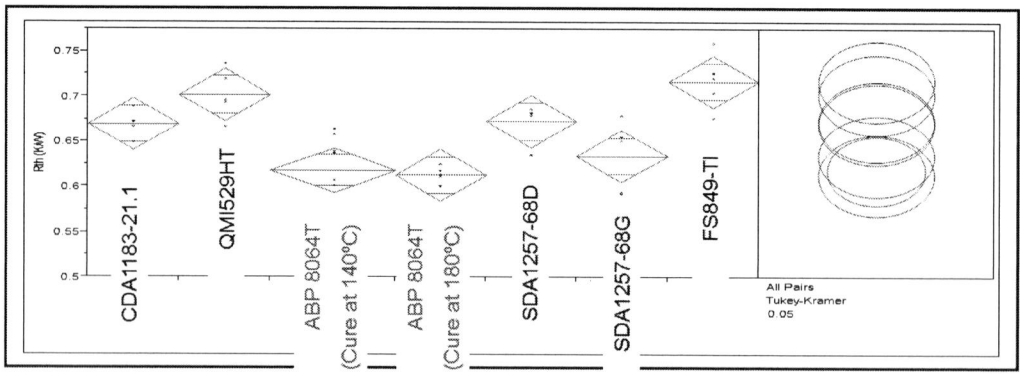

Fig. 2. Thermal resistance values of resin based die attach materials

© VDE VERLAG GMBH · Berlin · Offenbach

1036

1.2.2 Solder Materials

Different lead free solder alloys are available with operating temperatures up to 175°C and thermal conductivity in the 50-60 W/mK range. These can be used as pastes or preforms.

Solder fatigue will be one of the challenges to overcome for high power applications given the critical homologous temperature of solder in these applications.

1.2.3 Ag Sinter materials

Ag Sinter properties

The benefit of sintering materials compared to other solutions is shown in the table below.

Thanks to its high melting temperature, possible operating temperatures are a lot higher and will remain below the critical homologous temperature so operating temperatures up to 300°C are no problem. Besides this, the high electrical and thermal conductivity give a clear benefit over traditional interconnection materials.

		Ag Pure Silver	Ag Sinter Layer	SnAg Solder Layer	Factor
Liquidous	°C	961	961	221	4
Electric Conductivity	MS/m	68	41	7,8	5
Thermal Conductivity	W/mK	429	250	70	4
Density	Gr/cm	10,5	8,5	8,4	1
CTE		19,3	19	28	1
Tensile Strength	Mpa	139	55	30	2

Table 1.   Properties of an Ag sinter layer.

Thermal conductivity

Based and depending on process conditions (time-pressure-temperature), thermal conductivity of 100-200 W/mK can be reached.

New development materials already show over 200W/mK thermal conductivity when applied pressureless and even up to 270 W/mK when applied with pressure thanks to the dense silver structure.

Material Name	SSP2020	ISS20696-30A	ISS20696-75C
Recommend substrate	Ag plated		
Dry condition(Hot plate)	90°C30min		90°C60min
Sintering condition	300°C 5min 10MPa		
Viscosity (mPa · s)	19,000	10,000	70,000
Thixotropic index	3.7	4.2	4.5
Specific gravity (g/cm³)	5.4	6.6	7.40
Thermal conductivity (Laser flash, W/mK)	200	270	270
Memo	Commercial product	Printing type	Dispense type

Table 2.   New ag sinter development materials.

There is a clear relationship between thermal performance and density of the interconnection layer. This density can be altered/achieved by changing process parameters or using different interconnection materials with the same process.

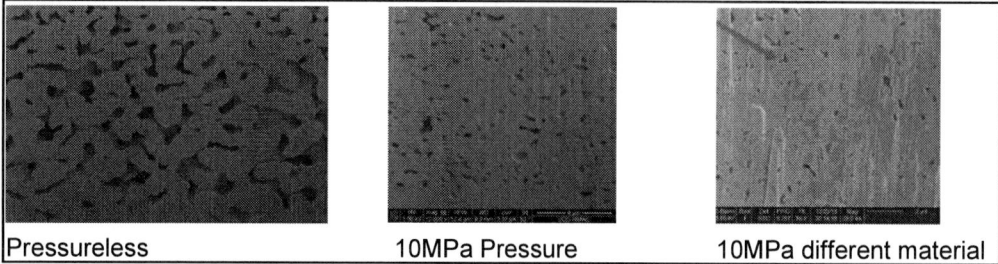

| Pressureless | 10MPa Pressure | 10MPa different material |

Fig. 3.   Cross sections of sinter layers showing different sinter densities.

Adhesion

An easy to check property is adhesion.

Adhesion tests like peel, die shear, bend performance are executed.

Bend performance becomes more and more a standard in the power electronics market when it comes to adhesion testing. This test is representative when it comes to warpage that can occur in Power modules in operation.

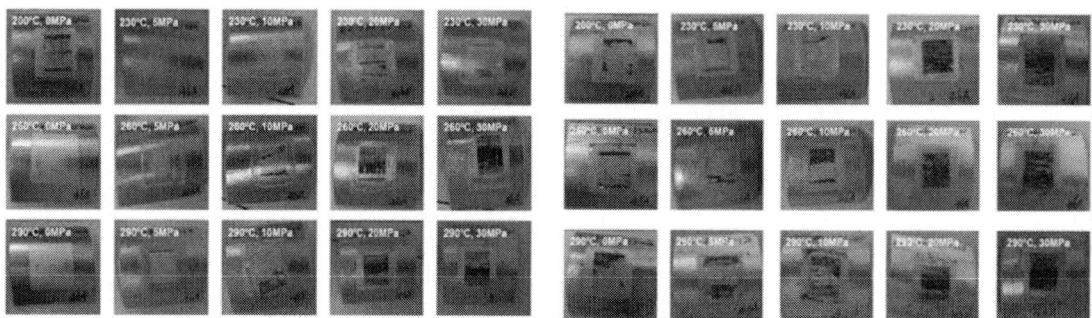

Fig. 4.   Mandrell bend test results.

It is clear that based on different assembly conditions (time – temperature – pressure) adhesion strengths can and will vary as well.

It is not clear whether improved adhesion passed a certain point will also give improved performance in thermal cycling and or active power cycling.

The selection of substrates, substrate finish, die, die backside metallization, wire bonding etc. will play a role, not only on adhesion but on the overall performance and on the quality of the assembly over time.

1.2.4 Sintering film

For small to medium die sizes, polymer assisted sinter layer (fig 2) in film format is available in both laminate and pre-cut.
Thermal and electrical performance is in the range of Pb solder.
Temperature stability up to 175°C has been proven and it is passing MSL1 and automotive tests like:

- PCT (96h @ 121C),
- HTS (1000h @ 175C)
- TCT 1000 cycles (-65C to 150C)

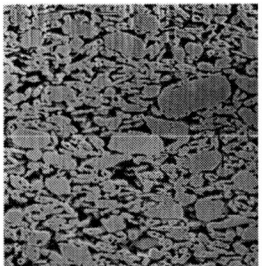

Fig 5 – sintering film cross section

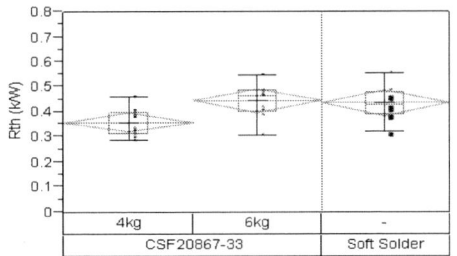

Fig 6 – thermal resistance of sintering film

Summary

There is a range of products available that need to be selected based on requirements in working temperature, thermal performance and adhesion, and based on the other assembly materials used.
Selection of material and process can be made based on the different resulting properties of these materials.

Encapsulation of Smart Power Electronic Devices – Thermal Degradation and Dielectric Behavior

Tina Thomas*, S. Gineiger, K.-F. Becker, L. Georgi*, K.-D. Lang*

*Technical University Berlin, Microperipheric Center
Fraunhofer Institute for Reliability and Microintegration [FhG IZM]
Gustav-Meyer-Allee 25, 13355 Berlin, Germany
phone: +49-30/464 03 625; fax.: +49-30/464 03 254
e-mail: tina.thomas@izm.fraunhofer.de

Abstract

This paper gives an overview about current processes and materials used for encapsulation of power electronics modules. The implications of the material classes on process and reliability will be discussed and special focus is put on thermal ageing of the encapsulants and the effect on dielectric strength and thus on functional reliability.

Latest results of thermo-oxidation and thermo-degradation of encapsulants will be presented, showing a minor influence of oxygen compared to effects of just thermal degradation. Besides the influence of ageing on thermo-mechanical behavior of mold compounds, dielectric strength is also an essential material property and this paper will discuss also the high voltage stability of epoxy material, depending on test conditions.

1. Introduction

Today power electronics packaging is one of the key enablers for a wide variety of applications ranging from automotive & e-mobility via industrial electronics drive control units to solar energy converters. In the past packaging for these applications was typically done by die attach on a ceramic substrate, wire bonding and subsequent silicone gel potting, yielding bulky packages with a heat spreader on bottom and a thermoplastic housing with connectors on top. Today, with the emergence of compact smart power modules, i.e. modules integrating both, powerICs and the necessary driver plus passives, the classic molding process is also a way to protect power modules from the environment and provide the necessary electrical insulation. For this "classic" transfer molding is used as well as compression molding – typically used for FO WLP [1], but also applicable to voluminous packages. Latest trend for power electronics is the embedding of power ICs into printed circuit boards, providing maximum miniaturization, shortest wiring length and low inductance [2]. In this case, encapsulation is provided by the substrate material itself – including the necessary dielectric performance over a wide temperature range.

Within the last years a constant increase of application temperatures for electronic packaging is observable, this is true for sensor applications, for power electronics packages and also for the Smart Power Modules, as described above. Drivers for high temperature [HT] use are on the one hand the introduction of GaN/SiC power electronic devices. These need packaged modules to operate at temperatures higher than 200 °C, at least for short time, due to the possibility to reach Tj > 250 °C using the wide band gap semiconductors [3,4]. On the other hand electronic devices will be implemented considerably closer to or even inside external heat sources as engines, gear boxes etc. with ambient temperatures ≥ 175 °C.

To select suitable materials for the encapsulation of smart power modules the determination of both thermal stability and dielectric strength is of great importance. When using standardized test methods to determine these values often a large amount of material is needed – often far beyond the economical focus of the project. So a test program was set up to evaluate the thermal/thermo-oxidative ageing of such compounds and to evaluate test methods for dielectric strength determination, that allows to gain insight in material performance with only small amounts of encapsulant.

In summary this paper discusses encapsulation issues for the manufacturing of compact power modules with a strong focus on dielectric material ageing performance at elevated temperatures and will give advice on the reliability potential of encapsulated power modules.

2. Thermal Degradation of Epoxy Mold Compounds

Molded bars have been manufactured and examined regarding Charpy impact resistance, which is a measure of material toughness. The two used Epoxy Mold Compounds (EMCs) were commercially available materials suggested for high temperature use with T_g of 130 °C and 235 °C, respectively. Samples have been divided into three groups: initial state (blue), thermal storage in air (red), and thermal storage in nitrogen (green), respectively. Each group comprised 10 EMC bars. Storage condition was 500 h at 200 °C.

Cross sections of both EMCs reveal a similar optical appearance; therefor just cross sections of one EMC are depicted: Figure 1, which shows a cross section of an initial state EMC-bar, and Figure 2, which shows an EMC-bar stored in nitrogen, resemble each other. Optical appearance was not changed by thermal storage. This was confirmed by SEM images as well.

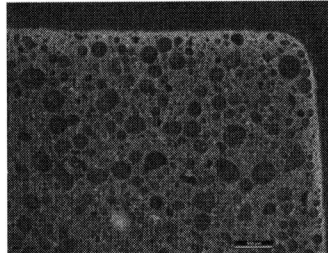 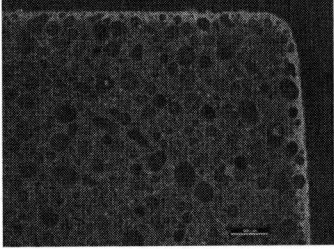 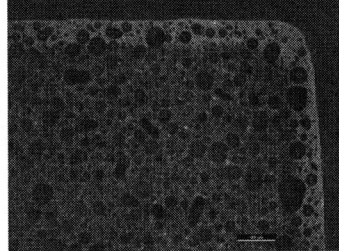

Figure 1: cross section of molded material in initial state

Figure 2: cross section of molded material after 500 hours of storage at 200 °C in N₂

Figure 3: cross section of molded material after 500 hours of storage at 200 °C in air

Only Figure 3, showing a cross section of an EMC bar, which was stored in air, indicates thermo-oxidation of EMC in the oxygen exposed area. An about 100 µm thick degraded layer is visible, which covers in approx. 7% of overall cross section area. Inner material presents itself as initial material – at least in optical appearance.

The Impact-Test is a standard method of determining the impact resistance of materials. The Charpy-impact resistance a_{cU} [J/m²] is the impact energy E_c absorbed during a break referring to the sample cross section h x b. This test determines brittleness of material. Test procedure is described in standard EN-ISO 179-1. The samples used for this test were of rectangular geometry (V = 80 x 10 x 4 mm³) and without v-groove. Analyzing impact resistance values of both EMCs reveals that not only the visible thermo-oxidized layer is responsible for change of impact resistance.

Also the thermally aged samples – done under N₂-atmosphere – experience a value drop of about 13 %, meaning the material gets more brittle. That embrittlement is not caused by thermo-oxidation, as O_2 was not involved. However, the value drop of thermo-oxidized samples is more pronounced, about 50 % of initial value. The thin degraded layer, visible in Figure 3 cannot be the only reason for embrittlement, its volume share is too low and furthermore we found thermal degradation without attendance of O_2 as well.

© VDE VERLAG GMBH · Berlin · Offenbach

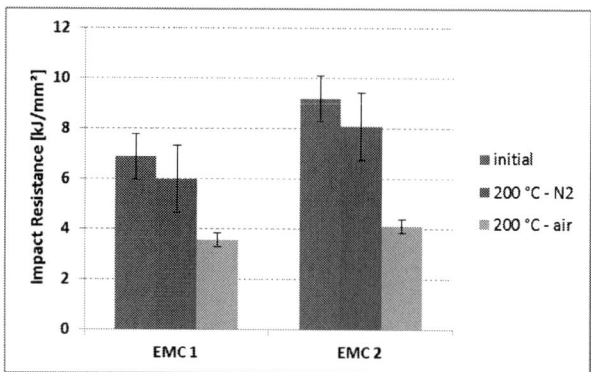

Figure 4: impact resistance of 2 EMCs in initial state and after thermal storage (500 h at 200 °C) in N_2 and air, respectively.

3. Dielectric Strength of Mold Compounds

As shown in the preceding chapter - high temperature storage does change mechanical properties of encapsulation material; but for use as an encapsulation material not only mechanical behavior, but also dielectric performance is of particular interest. Amongst others the test of dielectric strength is a common test procedure for assessment of dielectric performance. If dielectric breakdown occurs, usually the insulating material is destroyed along the discharge channel, which is an undesired scenario during application.

Applied standard for the test is DIN EN 60243-1/2 [5, 6] (comparable with ASTM D149). Although that standard test is quite useful for comparison of different materials, due to the specific geometry and manufacturing of test samples the results are usually not transferable to real application[7]. Measurement of dielectric strength in dependence of various material parameters like filling degree, test temperature and others is already described in literature[8,9,10]. But not only the insulating material itself – different test conditions give different results as well. Usually the insulating samples are clamped between electrodes, the more complex test setup with enclosed electrodes was not found very often. As influence of these two different test setups is described, but not quantified in literature, it will be examined more detailed in this study.

3.1 Test setup

Measurement is usually conducted on thin material samples as depicted in Figure 5.

Equipment used is a high voltage electrical breakdown tester with heating option up to 250 °C. A maximum of 50 kV can be applied; measurement can be done with DC and with AC up to 50 Hz. As material for test sample manufacturing, an unfilled epoxy was selected. The clear material allows detecting voids, which form weak points under electrical field and enable electric breakdown, quite easily. Samples for clamping between electrodes are poured 200 mm epoxy wafers. For enclosing the electrodes a dedicated casting mold was manufactured and epoxy was cured in direct contact to the test electrodes. As dielectric breakthrough shall not be provoked by local weak points as voiding, entrapped air bubbles or even cracks; sample manufacturing needs to be done carefully. Voidfree enclosure of electrodes was achieved by temperature assisted mold filling and cure under vacuum.

While literature describes influence of test temperature, ageing time and share of filler particles, here intrinsic test parameters will be examined. These parameters are connection of sample to electrode (clamped sample or enclosed electrodes), kind of current (AC or DC), and sample thickness. Combination of these three parameters results in eight test groups. Each test group with clamped insulation sample consists of ten test samples. Each test groups of enclosed electrodes consist of three test samples.

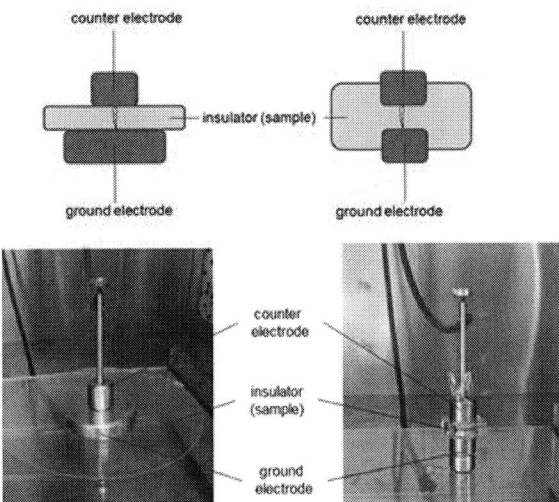

Figure 5: high voltage test setup – schematic and photograph; left: insulating sample (unfilled epoxy resin, wafer Ø 200 mm) clamped between electrodes; right: insulating sample (unfilled epoxy resin) encloses electrodes, which are covered with PI tape additionally

Sample thickness is 300 µm (the thinnest producible wafers) and 500 µm (thickest sample, where breakdowns are expected), respectively. AC frequency is 50 Hz; measurements are conducted at room temperature.

3.2 Test Results and Discussion

First measurements with clamped epoxy wafers did not result in any dielectric breakthrough through the material. A high speed camera rather observed flash overs as shown in Figure 6**Fehler! Verweisquelle konnte nicht gefunden werden.** on the left side. To detect real dielectric breakthrough these flash overs needed to be avoided. That was achieved by a silicone-made cover of the counter electrode – as depicted in Figure 6 (right). However, partial discharges along the epoxy surfaces were observed in measurements with clamped samples.

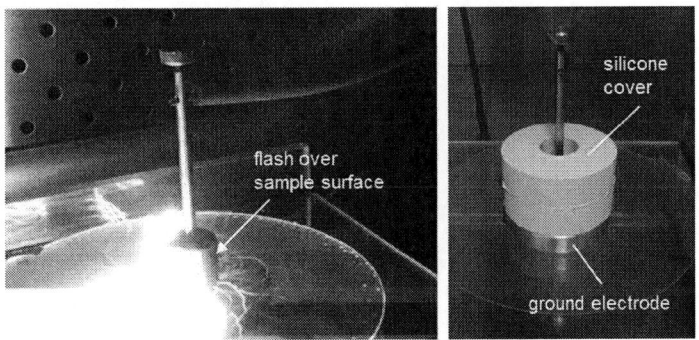

Figure 6: left - flash over sample surface - caused by ionized air; right – silicone covers counter electrode and prevents flash arcs over sample surface

Dielectric strength of epoxy wafers of 300 µm and 500 µm thickness in dependence of AC and DC is shown in Figure 7. As expected, dielectric breakthrough occurs at lower voltages for AC. Dielectric strength of clamped epoxy wafers in DC is about 80 % higher than in AC.

Breakdown of isolation in an AC field is caused by fixed charge carriers, which are shifted or realigned. In contrast breakdown in a DC field is caused by flexible charge carriers. The experiments with clamped insulating samples could quantify the difference of dielectric strength under AC versus DC.

In contrast, with enclosed electrodes, the significant differences of dielectric strength in AC and DC field cannot be observed. Just depending on electrode distances, dielectric strength was found to be ≥100 kV/mm for 500 µm and ≥150 kV/mm for 300 µm. For 300 µm electrode distance, sometimes U_{max} of 50 kV was not high enough to provoke a breakthrough. These values are significantly higher than for clamped samples: dielectric strength of theses insulating samples – carried out as electrode enclosure - is more than twice the value for samples, which are just clamped between electrodes in an AC field.

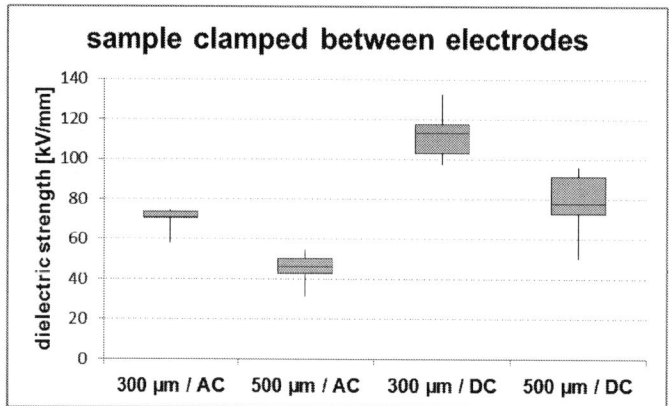

Figure 7: dielectric strength of four test groups: test setup with clamped 300 µm and 500 µm samples; test current AC and DC – significant higher dielectric strength for thinner samples and DC condition

While partial discharges on the surface were observed via high-speed camera with clamped insulating samples, no partial discharges of any kind were found for enclosed electrodes. These partial discharges –as shown in Figure 8 weaken the insulator regarding breakdown strength and the absence is assumed to be the reason for significantly higher dielectric strength.

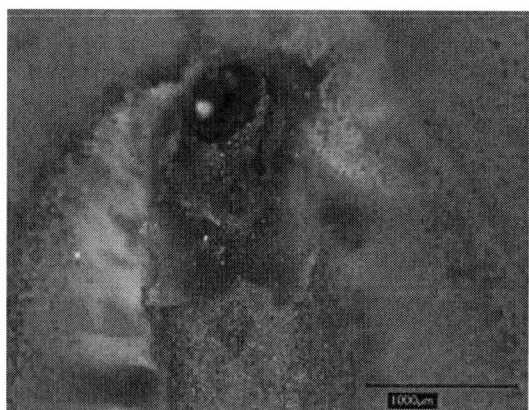

Figure 8: photograph of a dielectric breakthrough with surficial partial discharges

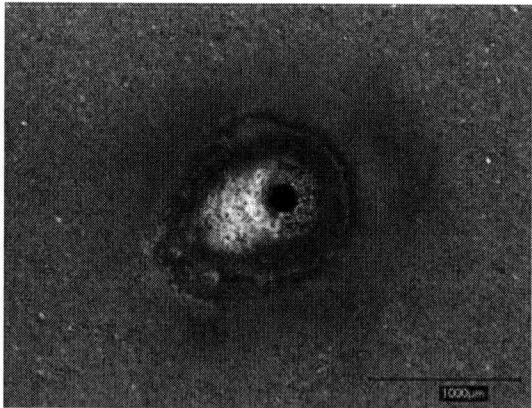

Figure 9: photograph of a dielectric breakthrough without surficial discharge

4. Summary and Outlook

Within this split study, the thermal aging behavior of epoxy mold compounds in air and in N_2 was examined. It was shown that even the absence of oxygen during thermal storage increases brittleness of material.

Second part showed significant influence of test setup measuring dielectric strength. Gaining insight in dielectric material performance with only small amounts of encapsulant was possible and even revealed strong dependence of test setup. Depending on connection of insulating sample and electrodes, dielectric strength can be twice as much for the same insulating material. Future work will include gathering more data of surficial partial discharges, and the classification of those. One target is to correlate these partial discharges with corresponding breakthrough.

5. References

[1] T. Braun et al.: Large Area Mold Embedding Technology with PCB Based Redistribution, Proc. EMTC 2012

[2] L. Böttcher et al.: Next Generation High Power Electronic Modules Based on Embedded Power Semiconductors, Proc. IMAPS 2014

[3] G. Feix et al.: Embedded Very Fast Switching Module for SiC Power MOSFETs, Proc. PCIM 2015

[4] A. Ostmann et al.: Embedded Power Electronics for Automotive Applications, Proc. Impact 2012

[5] DIN EN 60243-1: Electric strength of insulating materials - Test methods - Part 1: Tests at power frequencies (IEC 60243-1:2013); German version EN 60243-1:2013

[6] DIN EN 60243-2: Electric strength of insulating materials - Test methods - Part 2: Additional requirements for tests using direct voltage (IEC 60243-2:2013); German version EN 60243-2:2014

[7] H. Schmiedel; Handbuch der Kunststoffprüfung; Carl Hanser Verlag München Wien, 1992

[8] M.-H. Kim, I.-H. Son; D.-J. Lee, K.-H. Kim, C.-O. Park, J.-H. Kim; DC Breakdown Properties of Interpenetrating Polymer Networks based on Epoxy/MA/PU, IEEE Proceedings of the 5th International Conference on Properties and Applications of Dielectric Materials, Seoul Korea; 1997

[9] R. Kotte, E. Gockenbach, H. Borsi; Influence of the Filler on the Breakdown and Partial Discharge Behavior of Heat-resistant Cast Resins, IEEE Conference Record of the 2000 IEEE International Symposium on Electrical Insulation; USA; 2000

[10] T. Andritsch, R. Kochetov, Y. T. Gebrekiros, P. H. F. Morshuis, J. J. Smit; Short term DC Breakdown Strength in Epoxy based BN Nano-and Microcomposites; IEEE International Conference on Solid Dielectrics; Germany 2010

Improvement of the Mechanical Properties of Sn-Ag-Sb Lead-Free Solders: Effects of Sb addition and rapidly solidified

Mohammed, S. Gumaan[1,2], Esmail A. Ali[2], Rizk M.Shalaby[1] and Mustafa Kamal[1]

[1]Metal Physics Lab., Physics Department, Faculty of Science, Mansoura University, Mansoura, Egypt.

[2]Basic Science Department, Faculty of Engineering, University of Science and Technology, Yemen.

m.gumaan1@gmail.com, doctorrizk2@yahoo.co.uk and kamal42200274@yahoo.com.

Corresponding author: m.gumaan1@gmail.com

Abstract

The melt-spun processes of Sn-3.5 wt.%Ag and Sn-3.5 wt.%-Sb_x (x = 0.1, 0.3, 0.5, 0.7 and 1.0 wt.%) where analyzed using an x-ray diffractometer, a differential scanning calorimetry (DSC), and scanning electron microscopy (SEM). The results revealed that super saturated solid solution and intermetallic compounds (IMCs) were produced during melt-spun processes. The results revealed that there are more changes in thermal properties of these alloys due to structural reasons which are producing from micro-structural changes after Ag_3Sn and AgSb Intermetallic compounds formation. The mechanical properties of the ternary Sn-Ag-Sb alloys have been improved after small alloying additions of Sb dramatically and rapidly solidified due to microstructure refining. The solder alloys have been withstand creep deformation as show in their mechanical properties and creep results where the values of stress exponent and activation energy of Sn_{96}-$Ag_{3.5}$-$Sb_{0.5}$ were 7.87 and 66.51 KJ/mol respectively which indicate to dislocation climb controlled by dislocation pipe diffusion. It is known that even$\geq$ 0.1wt% level additions of Sb have significant effects on the microstructure of Sn-Ag solder alloys. Sb suppresses the growth of β-Sn dendrites in favour of eutectic formation.

1. Introduction

Researchers already turned from the producing Sn-Pb solder alloys towards Pb-free solder alloys as a result of the Pb health problems. The development of Pb-free solders became one of the most important research areas of the electronic industry, Pb-free solders should be controlled by many factors such as acceptable cost, adequate span life and performance as a solder paste, good fatigue resistance, low melting temperature and no poorer than those of Sn-Pb solder, so tin is far utilize as one of the primary constituent in Pb-free solder alternatives; recent research has been mainly focused on eutectic Sn-$Ag_{3.5}$ alloy due to its good mechanical properties and creep resistance, they got rid of the Pb health problems but some other problems such as high melting temperature, high cost and low interfacial bonding reliability have arisen. Some transition metals were added to these solders for improving the microstructural features of the Sn-$Ag_{3.5}$ solders such as Cu, Co and Fe [1, 2]. Currently, SnAgCu (SAC) solders are most widely used for Pb-free applications. In more papers published the formation of Ag_3Sn Intermetallic compound (IMC) enhances the creep resistance. Ag_3Sn intermetallic particles play two different roles. They may strengthen the alloy matrix and prevent the formation of large dislocation pile-ups at grain boundaries according to [3]. Adam J. Boesenberg et.al reported that the Al addition (>0.20Al) to the (SAC3595) solders had good thermal stability and led to spontaneous of Cu_3Sn appear with Ag_3Sn hiding [4]. Other articles observe that the formation of Ag_3Sn IMC in β-Sn matrix, cause to increase the creep rate, thus decreasing the creep resistance due to micro cracks initiated and then failure process. Although, the higher number of particles in a given matrix, the more matrix/intermetallics interfaces it contains, leading to a higher possibility of microcrack nucleation that can speed up the failure process. Adversely effecting of more amount of Ag_3Sn can occur on the plastic-deformation and restricts the formation of β-Sn phase [5]. Some alloying elements such as (Cu, Zn and Ni) have been added to the system to form ternary solders [6–8]. Rare researches using Sb as alloying element and used normal techniques, solders usually contain a maximum 0.5% Sb to eliminate the allotropic transformation. Hwa-Teng Lee et al [9]. reported that the In addition to SAS (Sn-Ag-Sb) alloys decreases the melting point and increased the pasty range temperature due to Ag_3Sn transformation to $Ag_2(In, Sn)$. Good creep resistance and mechanical strength of eutectic Sn-Ag alloys occur when it's doped with Zn, Cu, or Sb due to refined microstructure. Enhancement of the creep resistance was occurred by Sb addition to Sn-$Ag_{3.5}$ alloy, thus ternary alloys exhibited creep resistances higher than the binary Sn-$Ag_{3.5}$ alloy [10]. Hwa-Teng Lee et al [11] reported that the fatigue life of the as-soldered joint increased with the amount of Sb additions due to less plastic strain the hardened solders produced. Binary phase diagram for Sn-$Ag_{3.5}$ eutectic system indicate that there is little solid solubility of either minority constituent in the β-Sn phase near room temperature, thus microstructural coarsening is appear, since uniform phase distribution and small effective grain size forming have been attaining by quenching [12]. So this paper as a part of my Ph.D thesis contribute to solve the problem of creep deformation through refining Ag_3Sn IMC formed by Sb small additions to Sn-$Ag_{3.5}$ alloy using rapidly solidified (melt-spinning technique) reaching to superior mechanical properties, and present new information concerned the structural stability, thermal behavior, mechanical properties and creep resistance obtained.

2. Experimental methods:

2.1. Sample preparation:

Six Pb-free alloys of compositions $Sn_{96.5} - Ag_{3.5} - Sb_x$ (x = wt.% (0, 0.1, 0.3, 0.5, 0.7 and 1) were prepared from pure Sn, Ag and Sb (purity $> 99.99\%$). These alloys were putted in a porcelain crucible and melted by an electric furnace at 450 °C. After 25 min from heating the alloys become in a molten state then they are minutely agitated to increase the homogenization and again put in the furnace for 20 min. the molten alloys are shooting on the rotating copper wheel which has a linear speed of 31.4 m/s of the melt-spinning technique. The resulting alloys have long ribbons form of about 93 μm in thickness and 4 mm width.

2.2. Sample characterization:

A variety of technique was used to characterize the crystallographic, and transformation features of the melt-spun ribbons Pb-free alloys including x-ray diffraction. The x-ray diffraction study was carried out using CuKα radiation at room temperature. The microstructure analysis was carried out on a scanning electron microscope (SEM) of type (JEOL JSM-6510LV, Japan) operate at 30 kV with high resolution 3 nm. The melting temperature of these alloys was determined by differential scanning calorimetry, with a heating rate 10 k/min [13]. The mechanical properties were examined in air atmosphere with a dynamic resonance method [14–16]. The produced samples were tested in a Vickers microhardness tester, where a diamond pyramid indenter with square base is used and the Vickers hardness number is given by using $H_v = 0.185F/d^2$ where F is the applied load in gram force (gf) and d, is the average diagonal length in mm. Micro-creep measurements as described elsewhere [17] were also carried out using a Vickers hardness tester using the different loads 10, 25 and 50 (gf) for dwell time up to 99 s.

© VDE VERLAG GMBH · Berlin · Offenbach

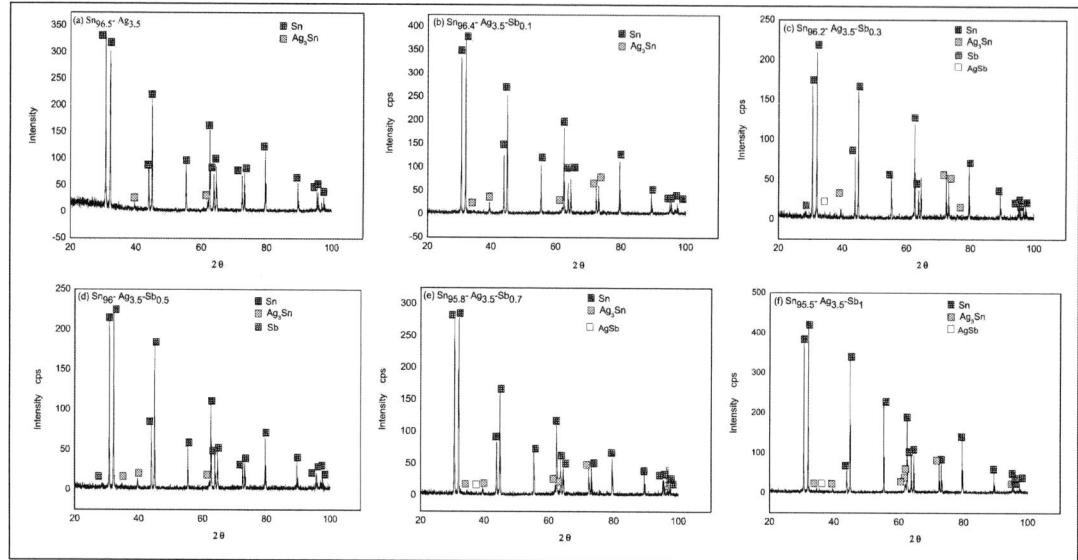

Fig. 1: The XRD patterns of as quenched melt-spun alloys

3. Result and discussion

3.1. Structural analysis:

In a detailed experimental study for structural properties where the rapid quenching of metallic alloys from melt was first carried out by Pol Duwez et al [18]. They found that the rapid quenching extends the solid solubility limits and produce non-equilibrium phase or amorphous alloys [19]. The patterns of X-ray diffraction obtained from the melt-spun ribbon for alloys are shown in fig.1. For $Sn_{96.5} - Ag_{3.5}$ is shown in Fig.1a contains pure β-Sn phase and only two peaks for Ag_3Sn (IMC) phase embedded in the Sn matrix. For $Sn_{96.4} - Ag_{3.5} - Sb_{0.1}$ alloy Fig.1b the number of peaks and intensity due to Ag_3Sn increases. According to the Hume-Rothery Rule, Sn atoms in the Ag_3Sn compounds can be easily replaced by Sb atoms to the phase transformation mechanism induced within the $Sn - Ag - Sb_x$ system via increasing Sb addition can be summarized as: $Ag_3Sn \longrightarrow AgSb$ AgSb IMC phase indicated by a single line appeared at 2θ= 37.54°, 2θ= 37.59° and 2θ= 37.62° for $Sn_{96.2} - Ag_{3.5} - Sb_{0.3}$ Fig.1c, $Sn_{95.8} - Ag_{3.5} - Sb_{0.7}$ Fig.1e and $Sn_{95.5} - Ag_{3.5} - Sb_1$ Fig.1f respectively, in addition to the presence of the above phases. It should be noted disappearance of the AgSb IMC phase in $Sn_{96} - Ag_{3.5} - Sb_{0.5}$ Fig. 1d. Pure rhombohedra Sb phase only exist at 0.3 wt.% and 0.5 wt.% additions. It's found that generally the Sb addition enhancement the Ag_3Sn phase. No ternary phases are formed in the system nor SnSb phase formed. The details of the XRD analysis are shown in Table 1. Estimating of the particle size was implemented using Scherrer equation t = (0.9λ/ B cosθ_B) , Where: B is the broadening of diffraction line measured at half its maximum intensity (radians), θ_B is the diffraction angle, t is the diameter of crystal particle and λ is the wavelength of x-ray. There are clearly decrease in the particle size of β-Sn as Sb increasing whereas fluctuation in values of Ag_3Sn IMC phase which is generally bigger than that of all other phases. The particle size of AgSb IMC phase is between 26.64nm-93.24nm, finally the lowest size was for Sb phase at 0.3 wt.% 13.01nm. Dominant phase is the phase of base metal. Refining of the particle size implemented by small amount additions of Sb where Sb suppresses the growth of coarse β-Sn dendrites in favour of eutectic formation. The variation of axial ratio c/a with different compositions is shown in table 1. Maximum value of axial ratio

Solder	Phases	Particle size(nm)	a($\overset{\circ}{A}$)	c($\overset{\circ}{A}$)	(c/a) of β-Sn	V($\overset{\circ}{A}$)^3	N	$\varepsilon \times 10^{-4}$
$Sn_{96.5} - Ag_{3.5}$	β-Sn	71.89	5.830	3.182	0.545	108.17	3.3	7.56
	Ag_3Sn	98.41						4.09
$Sn_{96.4} - Ag_{3.5} - Sb_{0.1}$	β-Sn	70.42	5.833	3.178	0.544	108.16	2.4	6.56
	Ag_3Sn	82.38						5.08
$Sn_{96.2} - Ag_{3.5} - Sb_{0.3}$	β-Sn	72.36	5.836	3.178	0.544	108.24	2.02	6.23
	Ag_3Sn	96.78						4.16
	Sb	13.01						
	$AgSb$	26.64						
$Sn_{96} - Ag_{3.5} - Sb_{0.5}$	β-Sn	63.02	5.825	3.182	0.546	108.01	2.3	6.92
	Ag_3Sn	96.44						4.41
	Sb	91.04						
$Sn_{95.8} - Ag_{3.5} - Sb_{0.7}$	β-Sn	67.33	5.832	3.185	0.546	108.01	2.4	7.50
	Ag_3Sn	87.73						5.21
	$AgSb$	93.24						
$Sn_{95.5} - Ag_{3.5} - Sb_1$	β-Sn	66.74	5.826	3.184	0.546	108.08	2.4	6.45
	Ag_3Sn	51.55						4.61
	$AgSb$	30.46						

Tab. 1: The details of the XRD analysis

c/a = 0.5463 at 0.5 wt% of Sb, due to expanding c-axis and contracting a-axis, this meaning decreasing in cell volume to become 108.014$\overset{\circ}{A}$. From XRD analysis it's found the relatively refining of Ag_3Sn as alloying element (Sb) increased. decreasing the number of atoms per cell volume which is due to existing point defects whereas maximum lattice distortion (ε) was at no Sb addition for Sn-matrix [20, 21].

3.2. Stabilized and refined microstructure:

Generally Pb-free solder surfaces tend to be dull, matte, rough, and grainy, contrary to the typical shiny appearance of a Sn-Pb solder.The properties of the Sn-based solder matrix will be similar to the physical properties of tin, because the tin has relatively low solubility for alloying elements, the mechanical properties of the solders alloys are greatly affected by the microstructural changes [22]. The Sb element is preferred as alloying element in $Sn_{96.5} - Ag_{3.5}$ due to solid solution hardening of Sb in the Sn matrix. The microstructres of $Sn - Ag_{3.5} - Sb_x$ ($x =$ 0, 0.1, 0.3, 0.5, 0.7, and 1 wt%) are shows in fig.2. For $Sn_{96.5} - Ag_{3.5}$, fig.2a The dark contrast dendritic globules are the β-Sn phase and the lighter contrast interdendritic regions contain the eutectic dispersion of Ag_3Sn precipitates within a β-Sn matrix. The large grains microstructure coarsening with two-phase eutectic colony solidification structure occurred. Cracks propagated at the solder/IMC interface and through the solder. By eliminating the large Ag_3Sn and uniform dispersion of Ag_3Sn precipitates, the solidification microstructure of $Sn_{96.5} - Ag_{3.5}$ alloy can be improved. With small amount of Sb addition, a more refined and uniform microstructure of $Sn_{96.4} - Ag_{3.5} - Sb_{0.1}$ solder alloy has been obtained Fig.2b. Fig.2c, illustrated that there is a competition in the Sn-Ag-Sb system to form Ag_3Sn vs AgSb or Sb and finest morphology has been occurred. At 0.5 wt% Sb fig.2d. the ductility in this ternary alloy is increased compared to that in the $Sn_{96.5} - Ag_{3.5}$ binary alloy, thus increasing in creep resistance and activation energy. Increased ductility should result from a smaller effective grain size [12]. Approximately the microstructure and the formation of precipitate-free β-Sn dendritic globules has been completely suppressed. At 0.7wt% Sb fig.2e there are increasing in β-Sn denetrites and a void has been

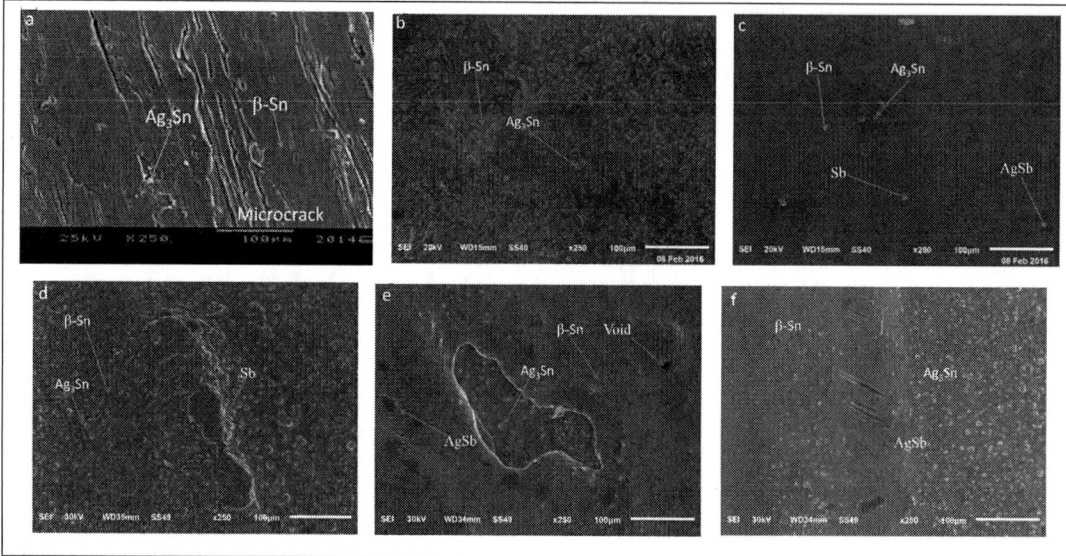

Fig. 2: Micrographs of SEM for solder alloys: (a) $Sn_{96.5} - Ag_{3.5}$ (b) $Sn_{96.4} - Ag_{3.5} - Sb_{0.1}$ (c) $Sn_{96.2} - Ag_{3.5} - Sb_{0.3}$ (d) $Sn_{96} - Ag_{3.5} - Sb_{0.5}$ (e) $Sn_{95.8} - Ag_{3.5} - Sb_{0.7}$ (f) $Sn_{95.5} - Ag_{3.5} - Sb_1$

occurred. At 1wt% Sb fig.2f the cracks initiated. The microstructural changes that occur as a result of 0.1wt% Sb addition have a clear effect on the mechanical properties, as shown in the comparative creep curves in Fig.5. Approximately with Sb addition alloys have over 30 % more stress exponent and over 50 % in activation energy which due to microstructural modifications and the microstructure then become to resemble.

3.3. Melting and thermal behavior:

The most challenge in Pb-free solders manufacturing is the high temperature, which it makes them difficult to be reflowed with devices on board and fabrication into existing physical forms of solder, i.e. wire, preforms, ribbon, spheres, powder, paste etc. Sb additions don't reduce the melt temperature of Sn-based solders; they go to enhance some properties. The optimal Sb content depends upon the particular properties of interest, to obtain the compromise of properties maximum limit of Sb addition is 1 wt.% which is provides a higher strength and fatigue life compared to eutectic Sn-Ag without a detectable increase in melt temperature. The DSC curves obtained from the six alloys during heating with heating rate 10 K/min are shown in Fig.3. The melting point (T_m), solidus temperature (T_s), Liquids temperature (T_l), enthalpy of fusion (ΔH_f), pasty range (ΔT), heat capacity (c_p) and thermal energy for melting (q) of these alloys were calculated in Table 2. There is only one endothermic peak for all solder alloys except at the first addition (0.1 wt.%) has been separated to two endothermic peaks due to β-Sn and Ag_3Sn phases. ΔT is highly sensitive to the antimony content. Normally at eutectic composition ΔT is zero, but in our case it is not equal to zero, the reasons may be due to the effects of rapidly solidification from melt. The solidus temperature (T_s) and liquids temperature (T_l) behavior shows in fig.4 where the drop in (T_l) to minimum value at 0.1 wt.% Sb and then slightly decrease to 1 wt.% Sb while the maximum value for (T_s) at the same point which will release low pasty range value for this system as show in table 2. Other solder alloys > 0.1 wt.% showing an increase up to 14.86 °C. The different properties for solder alloy in this range temperature have been occurred which clarified it easy to form in different shapes. From table 2 it's clear that the (T_m) and ΔH_f values are sensitive to Sb atoms concentration, the maximum heat flow at 0.5 wt.% Sb which due to maximum enthalpy 66.52 (J/g). It's indicated that c_p is important factor for the solder material as intrinsic property. The lowest c_p was for an eutectic $Sn_{96.5} - Ag_{3.5}$ alloy, suddenly c_p up to maximum value at 0.1 wt.% Sb which may be due to increasing Ag_3Sn content. After that continuous slightly decrease until 1 wt.% Sb as show in table 2. The low c_p means low energy needed for temperature changing hence lower cooling rate [23, 24]. By using this equation (q = m c_p Δt) for the same mass (1.75 mg) of all solder alloys it's obvious that maximum energy needed for melting at 0.1 wt.% Sb and minimum energy needed at eutectic alloy as a function for Ag_3Sn content and their contribution (phase intensity) as show in XRD results.

Solder	T_s (°C)	T_l (°C)	T_m (°C)	ΔH_f (J/g)	ΔT (°C)	c_p (J/g.°C)	q (J) for (1.75 mg)
$Sn_{96.5} - Ag_{3.5}$	212.5	237.5	217.4	1.391	25	0.0556	0.018
$Sn_{96.4} - Ag_{3.5} - Sb_{0.1}$	215.18	219.27	216.96	31.58	4.09	7.721	2.526
$Sn_{96.2} - Ag_{3.5} - Sb_{0.3}$	213.63	226.21	220.83	66.33	12.58	5.272	1.760
$Sn_{96} - Ag_{3.5} - Sb_{0.5}$	213.12	227.98	222.22	66.52	14.86	4.476	1.505
$Sn_{95.8} - Ag_{3.5} - Sb_{0.7}$	213.21	226.96	221.32	65.8	13.75	4.785	1.602
$Sn_{95.5} - Ag_{3.5} - Sb_1$	213.4	226.83	222.38	58.78	13.43	4.376	1.473

Tab. 2: Thermal analysis

3.4. Enhancement in mechanical properties:

Young's modulus and microhardness:

The microstructure is strongly influence the mechanical properties of a Pb-free solder, which is controlled by its cooling rate and alloying elements. The effective grain sizes and uniform dispersion of fine precipitates benefit the mechanical properties of solders and acquisition the ability to withstand the applied stresses. Mechanical observations reveal that Young's modulus and hardness depend on concentration, distribution and size of Ag_3Sn formed. Clear increasing in Young's modulus and hardness values due to the addition Sb as shown in Table 3; however the creep resistance was increased as shown in Fig. 5.

Stress exponent and activation energy of ordering:

Improvements in creep properties in the Pb-free alloys are desired to avoid rupture. The physical metallurgy of the solders is an important facet of mechanical deformation. Stress exponent strongly affected by the microstructural changes. Microstructural features such as grain boundaries and dislocation structures determine the strain response of the material to the applied stress. The creep tests were performed at room temperature. Deformation behavior of solder alloys was studied by micro-indentation creep experiments. In both alloys, micro-indentation creep rate was found to increase as a power-law function of indentation load. Fig.6 shows the plots of the quantity $\ln\Delta c_p T^2$ against the $(1/T)\times 10^{-3}$ give a straight line for each alloy. The slope of each straight line gives the activation energy (Q) of ordering for each alloy [25]. The exponents of the power-law were above 7 for lower creep rates. From this equation [26]:

$$n = \left[\frac{\partial \ln \dot{d}}{\partial \ln H_v} \right]_d \tag{1}$$

where H_v is the Vickers hardness number, d is the indentation diagonal length, and $\dot{d}$ is the rate of variation in indentation diagonal length, if $\dot{d}$ is plotted against H_v on double logarithmic scale, a straight line would be obtained whose slope gives the stress exponent as show in fig.7. The observation that both creep resistance and activation energy increased with more uniform distribution and finer microstructure in SEM micrographs which is significant, due to high activation energy required to break the covalent bonds of IMCs in this solder alloy as shown in (energy needed for melting) work done for melting values in DSC data in table 2. Results indicate that at both stress levels, the activation energy increases with Sb addition. There is a strong indication that the improvement of creep behavior of Pb-free solder with Sb addition is due to an increase in activation energy of ordering. Finally the creep behavior of the $Sn_{96.5} - Ag_{3.5}$ solder alloy has been improved due to the fine dispersive particles and the active properties of Antimony. It was found that the lowest creep for $Sn_{96} - Ag_{3.5} - Sb_{0.5}$ and $Sn_{95.8} - Ag_{3.5} - Sb_{0.7}$. The stress exponent, activation energy and strain rate sensitivity values which are scheduled in table 4 indicates that dislocation climb controlled by dislocation pipe diffusion with stress exponent of ≈ 7.8. Such change in creep behavior could be reflected by the microstructural changes observed in Fig.2, and XRD analysis Fig.1. However, the enhancement of creep resistance with increasing loading force is in good agreement with the result reported for Pb-free solders after Sb additions [8].

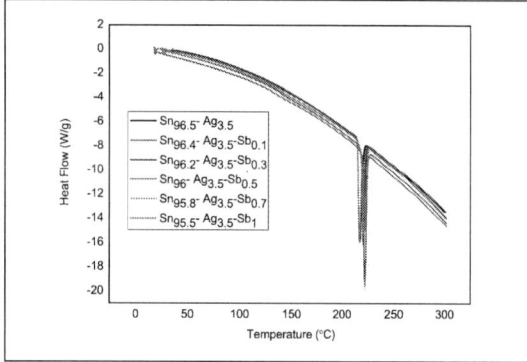

Fig. 3: Differential Scanning Caliometry (DSC) melting curves for solder alloys

Fig. 4: Variation of liquidus temperature T_l and solidus temperature T_s with Sb addition

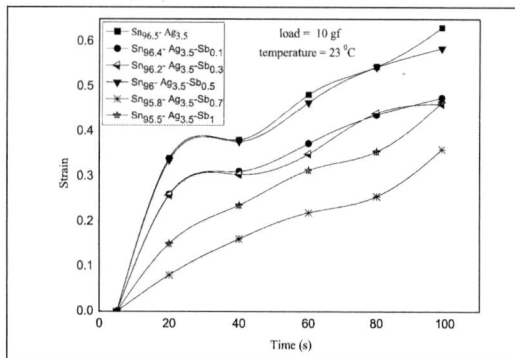

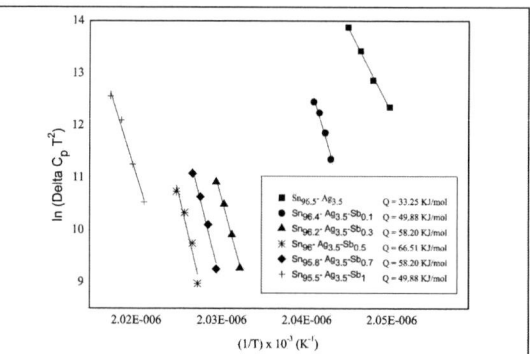

Fig. 5: The creep behavior of as-quenched solder alloys

Fig. 6: Relation between $\ln\Delta c_p T^2$ values and $(1/T)\times10^{-3}$

Comparison of the indentation creep behaviors indicated that both Pb-free solder alloys have high creep resistance at room temperature, and the $Sn_{96} - Ag_{3.5} - Sb_{0.5}$ has the highest creep resistance among the sex solder alloys. Results from as-quenched samples showed that producing tended to reduce the indentation creep resistance of the alloy by promoting micro-crack spreading in the eutectic microstructure. The stress exponent n measurement was implemented at three different loads 10, 25, and 50 gf, the average is presented herein. As a compromise for thermal, mechanical, electrical and creep properties of the produced solder

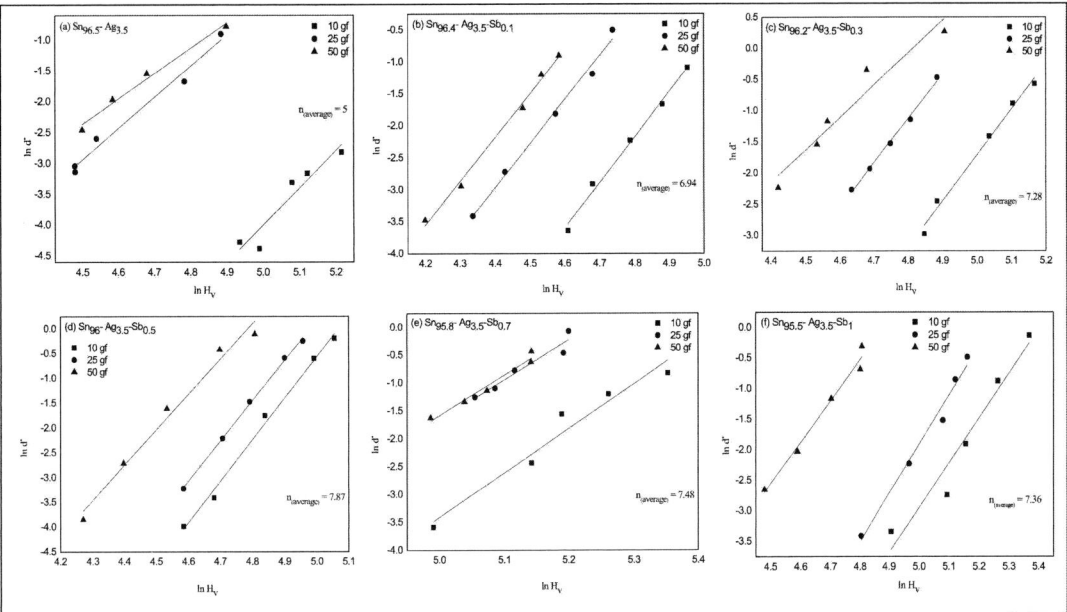

Fig. 7: $\ln-\ln$ plot of the rate of diagonal variation against the Vickers hardness numbers for solder alloys

Solder	E (Gpa)	H_v (Mpa)
$Sn_{96.5}-Ag_{3.5}$	212.5	237.5
$Sn_{96.4}-Ag_{3.5}-Sb_{0.1}$	215.18	219.27
$Sn_{96.2}-Ag_{3.5}-Sb_{0.3}$	213.63	226.21
$Sn_{96}-Ag_{3.5}-Sb_{0.5}$	213.12	227.98
$Sn_{95.8}-Ag_{3.5}-Sb_{0.7}$	213.21	226.96
$Sn_{95.5}-Ag_{3.5}-Sb_1$	213.4	226.83

Tab. 3: Young's modulus and Vickers micro-hardness

Solder	n	Q (KJ/mol)	m
$Sn_{96.5}-Ag_{3.5}$	5	33.25	0.166
$Sn_{96.4}-Ag_{3.5}-Sb_{0.1}$	6.94	49.88	0.141
$Sn_{96.2}-Ag_{3.5}-Sb_{0.3}$	7.28	58.20	0.136
$Sn_{96}-Ag_{3.5}-Sb_{0.5}$	7.87	66.51	0.119
$Sn_{95.8}-Ag_{3.5}-Sb_{0.7}$	7.48	58.20	0.131
$Sn_{95.5}-Ag_{3.5}-Sb_1$	7.36	49.88	0.131

Tab. 4: stress exponent (n), activation energy (Q) and strain rate (m)

alloys, the $Sn_{96}-Ag_{3.5}-Sb_{0.5}$ is a suitable alloy.

4. Conclusions:

Enhancement, refine, and more uniform distribution for Ag_3Sn were observed after Sb addition and rapidly solidified. Sb addition does not reduce the melt temperature of Sn-based solders, but enhance the mechanical properties. The values of stress exponent and activation energy of eutectic Sn-Ag after Sb addition were 7.87 and 66.51 KJ/mol respectively which indicates to dislocation climb controlled by dislocation pipe diffusion. Hardness and Young's modulus of solder depend on the size, distribution and the concentration of the Ag_3Sn. As a compromise for suitable soldering temperature and time, lower energy needed for soldering process and more resistance deformations of the produced solder alloys, it's concluded that the $Sn_{96}-Ag_{3.5}-Sb_{0.5}$ is a suitable solder alloy for interconnects in electronic packages.

References

[1] B. A. Cook, I. E. Anderson, J. L. Harringa, and R. L. Terpstra, "Effect of Heat Treatment on the Electrical Resistivity of Near- Eutectic Sn-Ag-Cu Pb-Free Solder Alloys," vol. 31, no.

11, 2002.

[2] M. Kamal and E. Gouda, "Effect of zinc additions on structure and properties of Sn-Ag eutectic lead-free solder alloy," J. Mater. Sci. Mater. Electron., vol. 19, pp. 81-84, 2008.

[3] C. M. L. Wu. Æ. Y. W. Wong, "Rare-earth additions to lead-free electronic solders," Lead-Free Electron. Solder., vol. A Special , pp. 77-91, 2006.

[4] A. J. Boesenberg, I. E. Anderson, and J. L. Harringa, "Development of Sn-Ag-Cu-X Solders for Electronic Assembly by Micro-Alloying with Al," vol. 41, no. 7, pp. 1868-1881, 2012.

[5] R. Kumar, "Effect of Ag on Sn-Cu Lead Free Solders Effect of Ag on Sn-Cu Lead Free Solders," no. 212, 2014.

[6] A. Gyenes, E. Nagy, P. Lanszki, and Z. Gácsi, "Investigation of Ni-Microalloyed Sn-0.5Cu Lead-Free Solders," Adv. Mater. Res., vol. 1120-1121, pp. 466-472, 2015.

[7] J. P. Jing, F. Gao, J. Johnson, F. Z. Liang, R. L. Williams, and J. M. Qu, "Brittle Versus Ductile Failure of a Lead-Free Single Solder Joint Specimen Under Intermediate Strain Rate," Ieee Trans. Components Packag. Manuf. Technol., vol. 1, no. 9, pp. 1456-1464, 2011.

[8] M. Pourmajidian, R. Mahmudi, A. R. Geranmayeh, S. Hashemizadeh, and S. Gorgannejad, "Effect of Zn and Sb Additions on the Impression Creep Behavior of Lead-Free Sn-3.5Ag Solder Alloy," J. Electron. Mater., 2015.

[9] H. Lee, F. Lee, T. Hong, and H. Chen, "En '," pp. 191-194, 2008.

[10] R. Mahmudi, M. Pourmajidian, A. R. Geranmayeh, S. Gorgannejad, and S. Hashemizadeh, "Materials Science and Engineering A Indentation creep of lead-free Sn-3.5Ag solder alloy: Effects of cooling rate and Zn / Sb addition," Mater. Sci. Eng. A, vol. 565, pp. 236-242, 2013.

[11] H. Lee, H. Lin, C. Lee, and P. Chen, "Reliability of Sn - Ag - Sb lead-free solder joints," vol. 407, pp. 36-44, 2005.

[12] M. Mccormack and S. Jin, "Improved mechanical properties in new, Pb-free solder alloys," J. Electron. Mater., vol. 23, no. 8, pp. 715-720, 1994.

[13] Y. A. GELLER and A. G. RAKHSHTADT, "Science of Materials," Sci. Mater, vol. 138, pp. 138-141, 1977.

[14] J.M. Ide, "Rev. Sci," Instrum, vol. 6, no. 296, 1935.

[15] S. Sppinert, and W.E. Teffit, "ASTM Proc," vol. 61, no. 1221, 1961.

[16] N. S. E. Schreiber, O.L. Anderson, "Elastic Constants and Their Measurements," McGraw Hill, New York, p. 82, 1973.

[17] T. El-Ashram and R. M. Shalaby, "Effect of rapid solidification and small additions of Zn and Bi on the structure and properties of Sn-Cu eutectic alloy," J. Electron. Mater., vol. 34, no. 2, pp. 212-215, 2005.

[18] P. Duwez, R. H. Willens, and W. Klement, "Metastable electron compound in Ag-Ge alloys," J. Appl. Phys., vol. 31, no. 6, p. 2006, 1960.

[19] M. Kamal, A. El-Bediwi, and M. Jomaan, "Rapid Quenching of Liquid Lead Base Alloys for High Performance Storage Battery Applications," 2012.

[20] B. D. Cullity, Elements of X-ray Diffraction. USA: Wesely, 1959.

[21] G. K. WILLIAMSONt and W. H. HALLt, "X-ray line broadening from filed aluminium and wolfram," Acta Metall., vol. 1, no. 1, pp. 22-31, Jan. 1953.

[22] R. M. Shalaby, "Indium, chromium and nickel-modified eutectic Sn-0.7 wt% Cu lead-free solder rapidly solidified from molten state," J. Mater. Sci. Mater. Electron., vol. 26, no. 9, pp. 6625-6632, 2015.

[23] Y. K. Wu, K. L. Lin, and B. Salam, "Specific Heat Capacities of Sn-Zn-Based Solders and Sn-Ag-Cu Solders Measured Using Differential Scanning Calorimetry," vol. 38, no. 2, 2009.

[24] J. N. Harris, MECHANICAL WORKING OF METALS, First edit., vol. 36. NEW YORK: MATERIALS SCIENCE AND TECHNOLOGY, 1983.

[25] R. M. Shalaby, "Correlation between thermal diffusivity and activation energy of ordering of lead free solder alloys $Sn_{65-X}Ag_{25}Sb_{10}Cu_X$ rapidly solidified from molten state," vol. 6, pp. 187-191, 2005.

[26] R. Roumina, B. Raeisinia, and R. Mahmudi, "Room temperature indentation creep of cast Pb-Sb alloys," Scr. Mater., vol. 51, no. 6, pp. 497-502, 2004.

PCIM Europe 2016, 10 – 12 May 2016, Nuremberg, Germany

Health-Monitoring of IGBT power modules using repetitive half-sinusoidal power losses

Marco Denk, Mark-M. Bakran

University of Bayreuth, Department of Mechatronics, Center of Energy Technology
Universitätsstraße 30, 95447 Bayreuth, Germany
marco.denk@uni-bayreuth.de

Abstract

In view of ever-increasing power densities and harsh operating conditions manufacturers of power electronic systems are demanding concepts to monitor the ageing process of IGBT power-modules in the field. This paper presents a diagnostic concept to identify the thermal resistance from chip to substrate and to monitor its increase due to the gradual delamination of the IGBT chip-solder. For this purpose the IGBTs are operated with repetitive half-sinusoidal power losses and a modified gate driver measures the amplitude ΔT_J and the minimum temperature $T_{J,Min}$ of the resulting junction temperature cycles. The focus of this paper is on the implementation of the diagnostic concept within a voltage source inverter of a hybrid transmission. It describes an inverter control strategy to realize the periodic heating current in a regular operating point and without generating any drive torque. Moreover, the additional increase of the temperature cycle amplitude ΔT_J due to the increased power losses at higher minimum cycle temperatures $T_{J,Min}$ is investigated. The diagnostic concept is tested with artificially aged power modules and enables a very robust and selective identification of a delaminated chip-solder and a contaminated cooling system with little implementation effort.

1. Introduction

The lifetime calculation of an IGBT power module in the hybrid transmission given in figure 1 suffers particularly from the inaccurate knowledge of the inverter mission profile. To verify the theoretical profiles with data from the field the recorder algorithm presented in [1] identifies the temperature cycles of the semiconductors during the regular inverter operation. In order to gain additional experience about the ageing process of the power module an in-situ diagnostic function ought to identify and monitor the thermal resistance $R_{th,1}$ of an IGBT semiconductor and its increase due to the gradual delamination of the IGBT chip-solder.

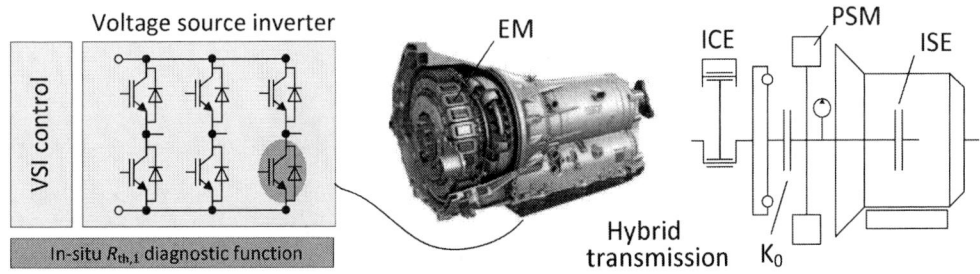

Fig. 1 The health-state of an IGBT power module that is integrated into the voltage source inverter of a hybrid transmission should be identified and monitored with small additional effort.

© VDE VERLAG GMBH · Berlin · Offenbach

The main challenge is to identify this rather small $R_{th,1}$-increase with little additional effort and without the need of highly sensitive sensors and laboratory equipment. However, the state of the art methods presented in figure 2 always measure the cooling curve of the semiconductor in the short-time domain and calculate the $R_{th,1}$-resistance by means of a thermal Cauer network [2, 3, 4]. Due to the small thermal capacity of the semiconductor $C_{th,1}$ these measuring methods are inevitably linked to fast sensors. Moreover, the heating of the IGBTs with a step- or impulse shaped power-profile requires additional current sources, so today there's no suitable solution to identify the thermal $R_{th,1}$-resistance of a power module within a voltage source inverter. In [5] a new diagnostic concept was presented that uses the characteristic thermal properties of the power module structure and identifies an increasing $R_{th,1}$ during the periodic heating of the IGBTs with repetitive, half-sinusoidal power losses and the measurement of the temperature cycle amplitude ΔT_J and the minimum cycle temperature $T_{J,Min}$.

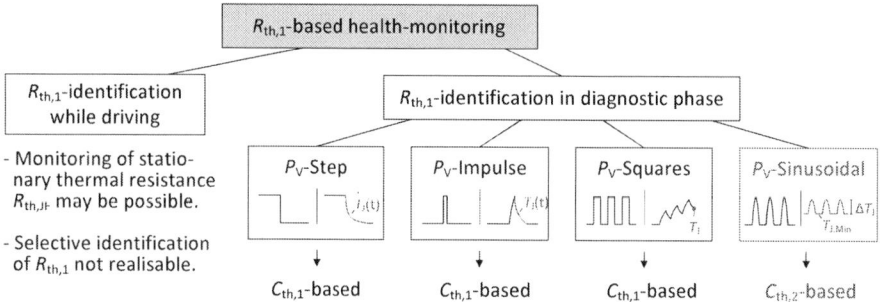

Fig. 2 Structuring of known methods to identify an increase of the thermal resistance $R_{th,1}$.

2. Implementation of the diagnostic concept

Prior to the development of an inverter control strategy the impact of the temperature dependent power losses on the temperature cycle parameters is examined. The aim is to establish whether the control of a constant phase current I_{AC} is sufficient to adjust the half-sinusoidal power losses with an approximately constant power losses amplitude ΔP_V.

2.1 Realization of half-sinusoidal power losses

The identification of a degraded chip-solder or cooling system by means of the temperature cycle parameters ΔT_J and $T_{J,Min}$ requires the periodic heating of the semiconductors with constant power cycles. However, the switching and conducting losses of an IGBT increase with temperature so that a higher minimum cycle temperature results in higher power losses and an additional increase of the cycle amplitude ΔT_J. To investigate this added increase ΔT_{ad} the

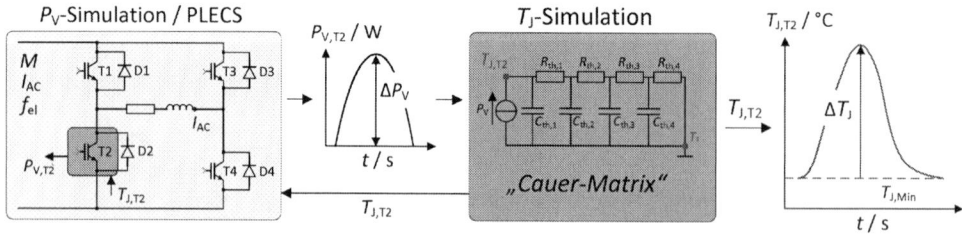

Fig. 3 Thermal simulation of the temperature dependent power losses $P_{V,T2}$ of the IGBT under test T2 and it's junction temperature $T_{J,T2}$ during inverter operation with a low-frequency phase current I_{AC}.

© VDE VERLAG GMBH · Berlin · Offenbach

thermal model in figure 3 was used. It consists of the IGBT T2 that is conducting the half-sinusoidal phase current I_{AC} with $\hat{\imath}_{AC}$ = 200 A and f_{el} = 1 Hz. The power losses $P_{J,T2}$ were applied to a thermal Cauer model that calculates the IGBT junction temperature $T_{J,T2}$. In order to consider temperature dependent power losses the junction temperature is returned after each calculation step. Figure 4 shows the simulated temperature cycles of an IGBT in the initial state and in case of an increased thermal resistance $R_{th,1}$ or $R_{th,5}$ by 20% of the stationary thermal resistance of the power module. As it was outlined in [5] the resulting increase of the temperature cycle parameters ΔT_J and $T_{J,Min}$ by 28,7% and 6,6% enables the identification of the Cauer element, which thermal resistance has increased. If both resistances $R_{th,1}$ and $R_{th,5}$ have increased, the cycle parameters ΔT_J and $T_{J,Min}$ were affected by 29,8% and 6,9%. This 0,1% larger increase of ΔT_J is called additional increase ΔT_{ad} and initiated by the higher minimum cycle temperature $T_{J,Min}$ and the equivalent higher power losses $P_{V,T2}$. Compared to the increase that is caused by the chip-solder delamination (28,7%) this additional increase is negligible and the periodic heating of the IGBTs with approximately con-stant power losses can be realized with a constant amplitude of the phase current I_{AC}.

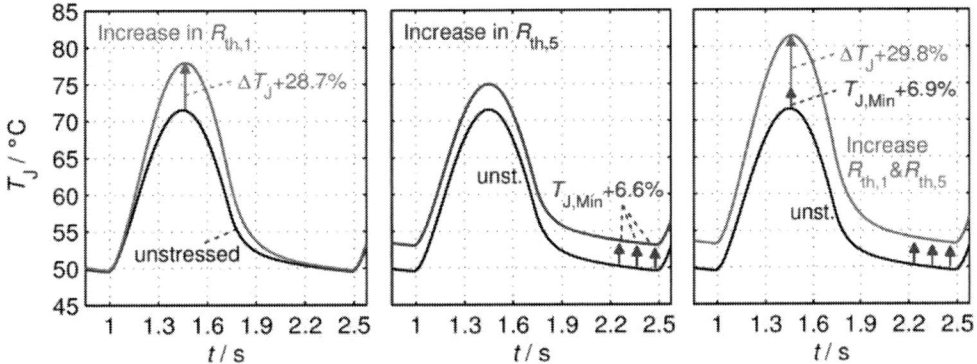

Fig. 4 Simulated temperature cycles in thermal equilibrium of an unstressed IGBT power module and in case of an increased thermal resistance $R_{th,1}$ or/and $R_{th,5}$.

2.2 Control strategy of the voltage source inverter

The special feature of the following inverter control strategy is the realization of the half-sinusoidal power losses in a regular inverter operating point and without generating any drive torque. For this purpose the IGBT under test T2 is controlled by the regular PWM signal of the inverter control while the opposite IGBTs T3 and T5 are permanently switched on.

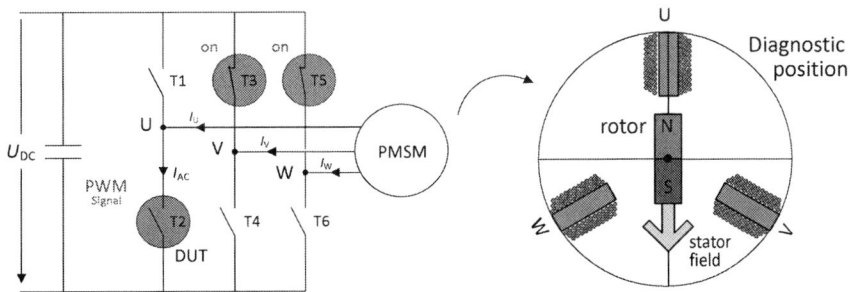

Fig. 5 Control strategy of the voltage source inverter to heat-up the IGBT under test T2 with repetitive half-sinusoidal power losses and without producing any drive torque.

© VDE VERLAG GMBH · Berlin · Offenbach

In the diagrammed diagnostic position the resulting phase currents I_U, I_V and I_W lead to a half-sinusoidal increase of the stator field, but not to a rotating field and any movement of the rotor (see figure 6). Prior to the periodic heating of T2 the rotor of the electric machine has to be moved into the diagnostic position. For this purpose the separating clutch K_0 and the integrated starting element (ISE) were released so that the rotor is able to rotate (see figure 1). Now the ramped increase of the phase current I_{AC} causes a slow rotation of the machine until the rotor reaches the diagnostic position. Figure 6 shows the simulated phase currents of the electric machine during the positioning of the rotor and the subsequent heating of the IGBT T2 with the half-sinusoidal diagnostic current $I_{AC} = I_U$. In this view the inverter control strategy uses the mechanical conditions of the hybrid transmission to exclude any drive torque on its output shaft. Consequently, the periodic heating of the IGBTs with half-sinusoidal power losses can be realized very easy and doesn't require any additional heating current sources.

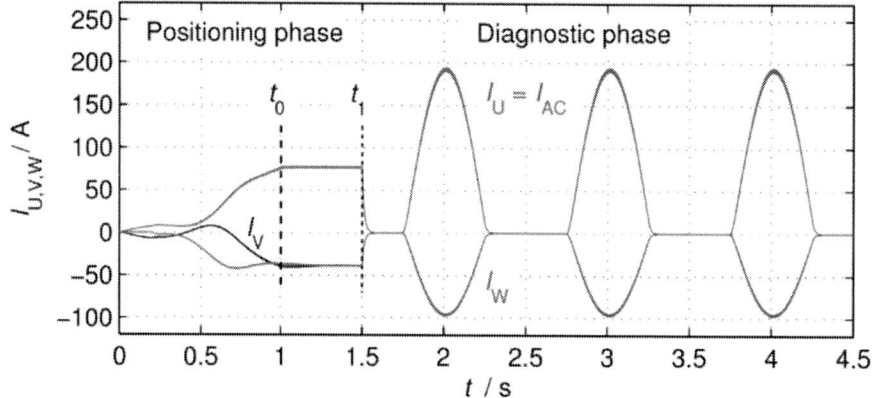

Fig. 6 Simulated phase currents of the synchronous machine while the positioning of the rotor and the periodic heating of the IGBT T2 due to the half-sinusoidal phase current I_{AC}.

3. Experimental verification

To verify the diagnostic concept within a real voltage source inverter the control strategy is implemented into a hardware-in-the-loop test setup [6] and tested with artificially aged power modules. Once again a particular focus is on the additional increase of the cycle amplitude ΔT_J due to the higher power dissipation in case of an increased minimum cycle temperature $T_{J,Min}$. In order to improve the identification accuracy, an automatic diagnostic function that determines the temperature cycle parameters ΔT_J and $T_{J,Min}$ of several half-cycles and a correction function to eliminate the additional increase were presented.

3.1 HiL setup with real-time T_J-measurement

As shown in figure 7, the HiL test setup consists of two half-bridge power modules PM1 and PM2 that imitate the loading of one inverter phase with a virtual electric machine. During the inverter operation the T_J-IGBT-Driver measures the junction temperature of the IGBT T2 and sends it to the VSI control. Additionally the phase current I_{AC} and the temperature of the coolant T_F is measured with the available sensors of the inverter. All IGBTs were controlled by a dSpace® system and a MATLAB/Simulink® model that consists of the inverter control strategy and an automatic identification function to determine the temperature cycle parameters ΔT_J and $T_{J,Min}$. Once the diagnostic function has started the inverter control increases the amplitude of the half-sinusoidal phase current by switching on T3 and the pulse-width modulated control of T2 (see section 2.2). The electric frequency of the phase current is set to the optimum excitation frequency of $f_{el} = 1$ Hz (see [5]).

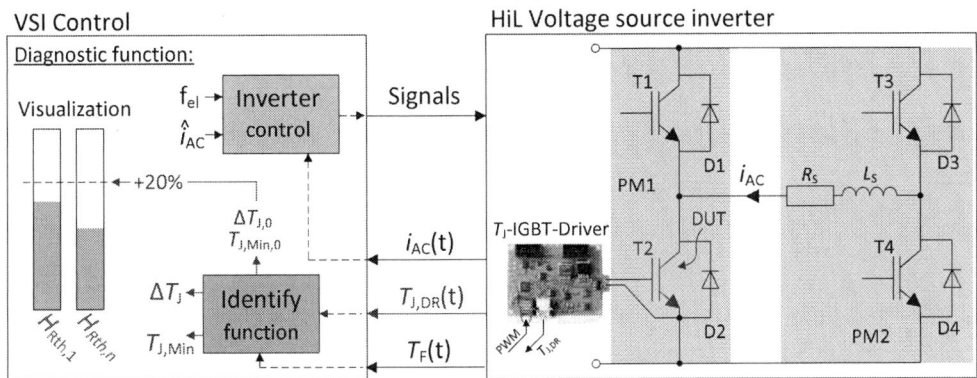

Fig. 7 Laboratory HiL test setup of a voltage source inverter with integrated real-time junction temperature measurement ("T_J-IGBT-Driver", [6,7]) to investigate the diagnostic concept.

Figure 8 shows the resulting junction temperature of T2 that is measured with the T_J-IGBT-Driver presented in [6, 7]. The identification function of the VSI control waits until the temperature reaches a thermally stable state. Afterwards it measures and averages the temperature cycle parameters ΔT_J and $T_{J,Min}$ over 30 temperature cycles. To further improve the measuring accuracy the identification function is repeated three times. In conjunction with a digital infinite impulse filter this averaging enables the identification of the temperature cycle amplitude with a random measurement error of $\sigma_{\Delta TJ} \leq 0,2$ K ($\leq 1\%$). The minimum cycle temperature can be measured with an accuracy of ±1K. To estimate and visualize the health-state of the power module $H_{Rth,1}$ and the cooling system $H_{Rth,n}$ the measured cycle parameters were set in relation to the reference values $\Delta T_{J,0}$ and $T_{J,Min,0}$ that were determined in the delivery state. Since an increase of the thermal resistance $R_{th,1}$ causes an approximately linear increase of the temperature cycle amplitude ΔT_J the health-state of the power module can be calculated from $H_{Rth,1} = (\Delta T_{J,0}-\Delta T_J)/\Delta T_{J,0} \cdot 100\%$.

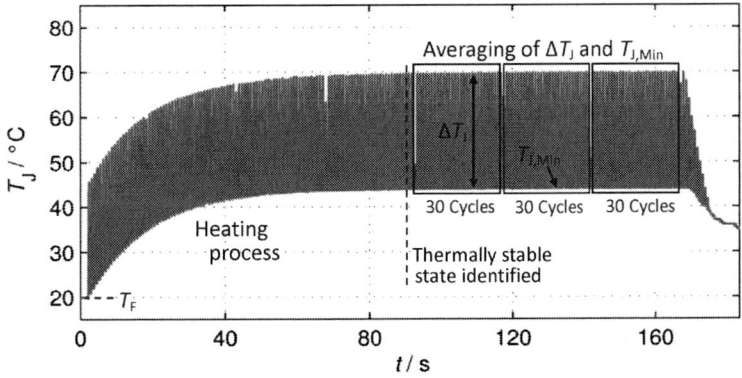

Fig. 8 Measured junction temperature cycles during the inverter operation with a phase current of 200 A and a frequency of 1 Hz. For T_J-measurement a modified IGBT gate driver was used.

© VDE VERLAG GMBH · Berlin · Offenbach

3.2 $R_{\text{th},1}$-identification within a voltage source inverter

Figure 9 shows a section of three temperature cycles that were measured within the middle averaging window of the identification function (see figure 8). The raw data points $T_{\text{J,TR}}$ were sent by the T_{J}-IGBT-Driver [6] with a sample rate of 2,5 ms and a temperature resolution of $A_{\text{TJ}} = 1$ K. The smoothed temperature curves $T_{\text{J,F}}$ were calculated with the online infinite impulse response filter. It can be seen that the measured temperature cycles are very similar so that the sensor properties of the modified gate driver are sufficient to satisfy the requirements of the diagnostic concept. In case of a delaminated chip-solder an increase of the amplitude ΔT_{J} by 3,8 K (15%) can be observed while $T_{\text{J,Min}}$ remains constant.

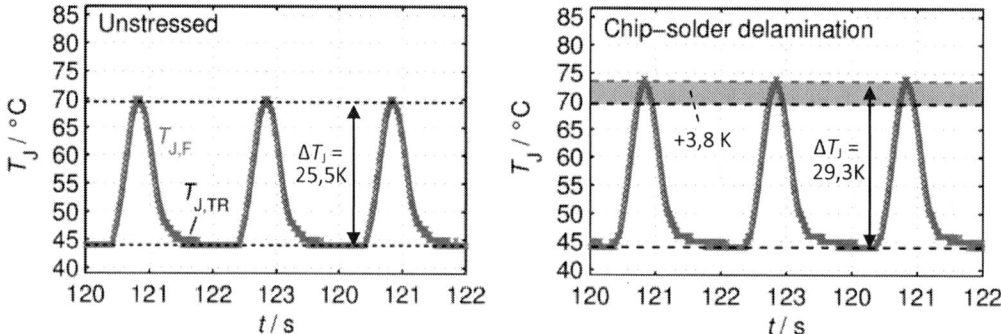

Fig. 9 Measured junction temperature cycles in thermal equilibrium of an unstressed power module (left) and an artificially aged power module with delaminated chip-solder (right).

In contrast, the contamination of the cooling system in figure 10 results in an increase of the minimum cycle temperature $T_{\text{J,Min}}$ by 4,2 K (14%) and an approximately unaffected temperature cycle amplitude of $\Delta T_{\text{J}} = 25,6$ K. The additional increase of the amplitude by 0,1K (0,4%) due to the higher minimum cycle temperature and the temperature dependent power losses in T2 is negligible and disappears in the random measurement error.

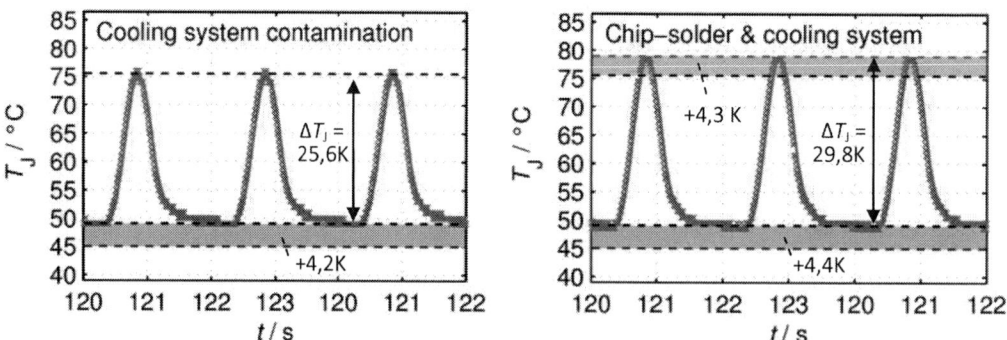

Fig. 10 Measured junction temperature cycles of an unstressed power module in case of a halved flow-rate of the coolant (left) and an additional delamination of the chip-solder (right).

If chip-solder and cooling system have deteriorated an increase of both cycle parameters ΔT_{J} and $T_{\text{J,Min}}$ can be observed. Compared to figure 9 the additional increase of the cycle amplitude is $\Delta T_{\text{ad}} = 29,8$ K $- 29,3$ K $= 0,5$ K (+2%) so the identification of a delaminated chip-solder (+3,8 K; 15%) remains possible. However, in practice the minimum cycle temperature $T_{\text{J,Min}}$ may vary in a wider temperature range so the additional increase of the amplitude has to be investigated in more detail. For this purpose table 1 shows the measured minimum cycle temperatures $T_{J,Min}$ and the cycle amplitudes ΔT_J at a gradual reduction of the coolant flow-rate dV/dt. Since the measurements were made with a new power module a delamination of the chip-solder can be excluded and the increase of the temperature cycle amplitude ΔT_J can be solely attributed to the increased power losses due to the increased minimum cycle temperature $T_{J,Min}$. It can be seen that in a temperature range from 40°C to 80°C the additional increase of ΔT_J is still below 2,5% so that the periodic heating of the IGBTs with a constant phase current is sufficient to identify a critical delamination of the IGBT chip-solder (+15%).

Flow-rate dV/dt / lmin^{-1}	Minimum $T_{J,Min}$ / °C	Cycle amplitude ΔT_J / K	Additional increase $\Delta T_{ad} = \Delta T_J - \Delta T_{J,8l/min}$
8,0	43,4	25,50 $= \Delta T_{J,8l/min}$	---
6,2	46,0	25,54	+0,04 (+0,2%)
4,1	49,6	25,60	+0,10 (+0,4%)
2,0	56,0	25,69	+0,19 (+0,7%)
1,0	69,4	25,95	+0,45 (+1,7%)
0,2	81,2	26,14	+0,64 (+2,5%)

Tab. 1 Additional increase ΔT_{ad} of the temperature cycle amplitude ΔT_J due to an increase of the temperature dependent power losses at higher minimum cycle temperatures $T_{J,Min}$.

In order to monitor the delamination process of the chip-solder in the field the accuracy of the diagnostic concept can be improved by correcting the impact of the temperature dependent power dissipation. As a basis for this, figure 11 shows the additional increase ΔT_{ad} of the temperature cycle amplitude as a function of the minimum cycle temperature $T_{J,Min}$. In the relevant temperature range from 40°C to 80°C the additional increase ΔT_{ad} can be fitted with a regression line. The linear relationship can be explained by the approximately linear increase of the power losses at higher junction temperatures. To eliminate the $T_{J,Min}$-dependency of the measured temperature cycle amplitude ΔT_J the diagnostic function calculates the corrected temperature cycle amplitude $\Delta T_{J,CS} = \Delta T_J - \Delta T_{ad}$ using the correction function $\Delta T_{ad} = f(T_{J,Min})$. The resulting amplitude $\Delta T_{J,CS}$ gives accurate information about the health-state of the IGBT chip-solder, independently from the minimum cycle temperature $T_{J,Min}$ and the additional increase of ΔT_J due to temperature dependent power losses.

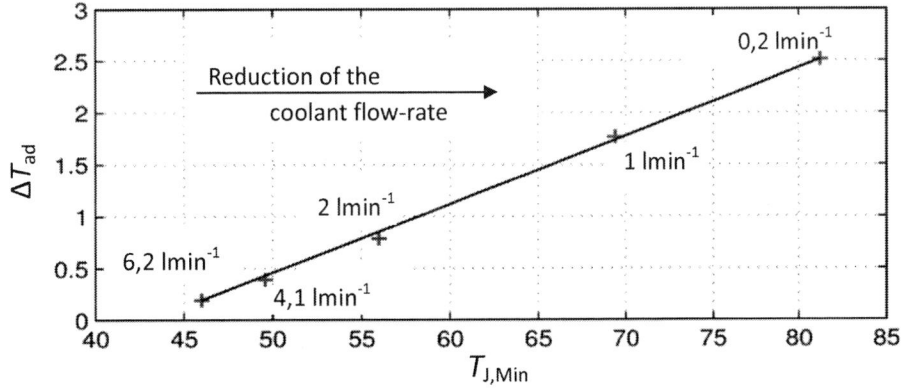

Fig. 11 Additional increase ΔT_{ad} of the temperature cycle amplitude due to increased power losses at higher minimum cycle temperatures $T_{J,Min}$. Values related to $\Delta T_{J,8l/min}$ at 8 lmin^{-1}.

© VDE VERLAG GMBH · Berlin · Offenbach

4. Conclusions

This paper presented a working diagnostic concept to monitor the delamination of the IGBT chip-solder and the deterioration of the cooling system of an IGBT power module that is integrated within a real voltage source inverter. In the diagnostic concept the IGBTs were heated with repetitive half-sinusoidal power losses and a modified gate driver measures the amplitude ΔT_J and the minimum temperature $T_{J,Min}$ of the resulting junction temperature cycles. It is shown that the implementation of the diagnostic function within a hybrid transmission only requires the installation of the modified gate driver and the operation of the voltage source inverter with a low-frequency phase current. Experimental investigations with artificially aged power modules revealed that the additional increase of the temperature cycle amplitude ΔT_{ad} due to the increased power dissipation at higher minimum cycle temperatures $T_{J,Min}$ is still below 2,5% so the identification of a critical delamination of the chip-solder (+15%) remains possible. For this reason, the periodic heating of the IGBTs with a constant phase current amplitude is sufficient to realize approximately constant power losses. In order to accurately monitor the chip-solder delamination of a power module in the field the accuracy of the diagnostic concept could be improved by a linear correction function that eliminates the impact of the temperature dependent power dissipation. In this view, the diagnostic concept enables an accurate and very robust identification and monitoring of a delaminated chip-solder and a contaminated cooling system. Moreover, the concept doesn't require any precise measurement equipment and additional current sources, so it holds great potential to be implemented within real voltage source inverters.

Acknowledgement

This project was supported by the ZF Friedrichshafen AG. Special thanks go to the department of Engineering Electric Mobility & Mechatronics, Auerbach i. d. Opf.

References

[1] Denk, M; Mark-M. Bakran: Efficient online-algorithm for the temperature cycle recording of an IGBT power module in a hybrid car during inverter operation, CIPS 2014

[2] Hensler, A.; Herold, C.; Lutz, J.; Thoben, M.: Thermal Impedance Monitoring during Power Cycling Tests, PCIM 2011, pp. 241-246

[3] Hiller, S.; Beier-Möbius, M.; Frankeser, S.; Lutz, J.: Using the Zth(t) - power pulse measurement to detect a degradation in the module structure,   PCIM 2015

[4] Tian, B.; Qiao, W.; Wang, Z.; Gachovska, T.; Hudgins, J.L.: Monitoring IGBT's Health Condition via Junction Temperature Variation, APEC 2014

[5] Denk, M; Mark-M. Bakran: Case Sensitive Condition Monitoring of an IGBT Inverter in a Hybrid Car, CIPS 2016

[6] Denk, M.; Bakran, M-M.: Junction Temperature Measurement during Inverter Operation using a TJ-IGBT-Driver, PCIM 2015, pp. 818-825

[7] Denk, M.; Bakran, M-M.: An IGBT Driver Concept with Integrated Real-Time Junction Temperature Measurement, PCIM 2014, pp. 214-221

PCIM Europe 2016, 10 – 12 May 2016, Nuremberg, Germany

Reliability Investigation on SiC BJT Power Module

Alexander Otto, Fraunhofer ENAS, Germany, alexander.otto@enas.fraunhofer.de
Eberhard Kaulfersch, Berliner Nanotest und Design GmbH, Germany, eberhard.kaulfersch@nanotest.org
Sophia Frankeser, Technische Universität Chemnitz, Germany, sophia.frankeser@etit.tu-chemnitz.de
Klas Brinkfeldt, Swerea IVF AB, Sweden, klas.brinkfeldt@swerea.se
Olaf Zschieschang, Fairchild Semiconductor GmbH, Germany, olaf.zschieschang@fairchildsemi.com
Sven Rzepka, Fraunhofer ENAS, Germany, sven.rzepka@enas.fraunhofer.de

Abstract

In this paper reliability investigation results for a power module fully based on silicon carbide (SiC) devices are presented. The module comprises four SiC bipolar junction transistors (BJT) and four SiC diodes in half-bridge configuration and is part of a newly developed 3-phase inverter for construction vehicles as well as for passenger car applications. The reliability investigations include electro-thermal and thermo-mechanical finite element simulations as well as power cycling tests with subsequent failure analyses. Furthermore, a double-sided cooling approach for the SiC BJT power module will be described and its thermal performance compared to the single-sided cooling version.

1. Introduction

In the frame of the European project COSIVU a novel system architecture for electric drive-train systems has been developed, consisting of a modular, compact and smart inverter including control and sensing electronics directly attached to the e-motor and gearbox [1]-[3]. The developed inverter comprises three equal inverter building blocks (IBB), a power supply, a central control unit and an inverter housing including the liquid cooling system. Each IBB in turn consists of a cooling plate, DC-link capacitors, current sensor, base drivers and three paralleled half-bridge SiC power modules (Fig. 1). Furthermore, health monitoring of the e-motor by structure-borne sound analysis [4] as well as of the power modules by means of thermal impedance monitoring [5] is implemented in order to improve the system's reliability and uptime, as highly required for the targeted heavy commercial vehicle applications.

SiC BJT power module   Inverter building block (IBB)   Inverter system

Fig. 1. SiC power module as part of the newly developed inverter for electrically powered heavy commercial vehicles and passenger cars

The developed inverter system has successfully run on a Volvo CE test bench in order to prove its functionality and ability to run in the targeted commercial vehicle application. A power loss reduction of up to 50% could be demonstrated compared to conventional solutions. Furthermore, the system has also been successfully implemented and tested on a converted passenger car (VW Sharan) by project partner Elaphe Propulsion Technologies [3].

© VDE VERLAG GMBH · Berlin · Offenbach

Reliability investigations have been performed in parallel to the design and realization phase in order to assure a high maturity of the developed system and thus allow a faster commercialization after the course of the project. Focus has been set in particular on the mechanical robustness of the inverter system [6] and on the novel SiC power module. The performed work for the latter case will be presented in the following sections.

2. SiC BJT power module

The investigated power module (Fig. 2) is based on the automotive qualified module package APM19 and is designed and manufactured by Fairchild Semiconductor. It contains four integrated SiC bipolar junction transistor devices (FSICBH017A120) from Fairchild and four anti-parallel SiC Schottky diodes (CPW2-1200S050) from Wolfspeed (former Cree) connected in half-bridge configuration. They are specified for 1200 V and 50 A (BJT) / 54 A (diode) and paralleled in order to drive motor currents up to 300 A at inverter system level.

 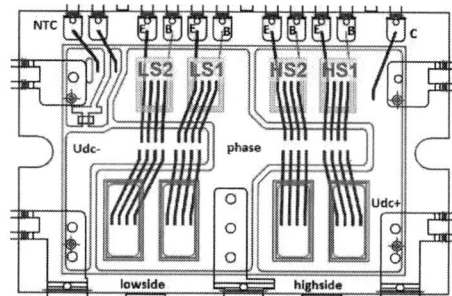

Fig. 2. Photographic image of SiC bipolar junction transistor (left) and layout of power module with BJT device numeration

The module construction, having an overall dimension of 44 mm by 29 mm by 5 mm, is based on a direct bonded copper (DBC) substrate with aluminium nitride (AlN) used as the ceramic isolator due to its superior thermal conductivity. As die attach, a lead-free SAC solder (SAC305) has been used. Electrical connection of the chip topside pads is realized with aluminium wire-bonds (300 µm for emitter, 150 µm for base). For the encapsulation an epoxy moulding compound (EMC) was applied. It has openings for threaded fasteners to assure an appropriate clamping force for an optimal heat transfer to the cooler. Furthermore, a thermistor for temperature indication is integrated into the module.

3. Finite element analysis

Electro-thermal and thermo-mechanical simulations have been performed to investigate, i.e. to locate and monitor, mechanical stress concentrations as well as accumulating plastic and creep strains (serving as failure criterion for low cycle solder fatigue) induced by manufacturing processes and during operation (power cycling) due to internal and external thermal loads. By this, the corresponding physics behind the relevant failure mechanism can be replicated, analysed and compared to experimental power cycling test results in order to assess the reliability and to derive design optimization guidelines.

The global geometry model of the module as well as the local wire-bond model (with cohesive zone approach) is shown in Fig. 3. The geometry data have been calibrated with real module dimensions by employing (3D) microscopic and computed tomography analyses. Prior to modelling, all relevant materials were analysed regarding temperature and process dependencies in order to derive proper material data. Furthermore, warpage measurements for different temperatures have been performed on the module to calibrate and validate the simulation model.

© VDE VERLAG GMBH · Berlin · Offenbach

PCIM Europe 2016, 10 – 12 May 2016, Nuremberg, Germany

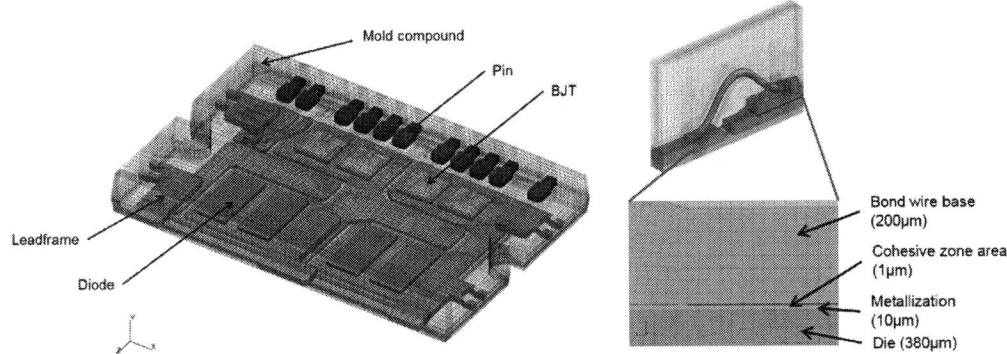

Fig. 3. Geometry model of power module including wire-bond model with cohesive zone approach

As a first step, 3D thermal simulations on the power module were conducted and calibrated with thermal measurements (see also Fig. 7) in order to assess the temperature evolution in the module during power cycling. This is not only important for gathering knowledge about the maximum and minimum temperature of the junction, but also to assess the total time and spatial dependence of the solder joint temperature $T_{solder}(t)$. The frame conditions have been oriented on the power cycling tests performed in parallel (section 4), i.e. with $t_{on} = t_{off} = 10$ s, $T_{heatsink} = 40°C$ and powering one high-side and one low-side BJT (see also Fig. 2, right).

In a second step thermo-mechanical process simulation was performed with individual materials added analogous to the real technological process to take into account intrinsic stresses induced by thermal mismatch at various steps of production, complemented by simulating a temperature profile reflecting the heating and cooling process during power cycling. In this procedure the stress-strain response was calculated for the simulated temperature profiles according to the experimental cycling conditions. In order to take into account creep and plastic time independent deformations of the SAC solder joint material, an in-house developed constitutive description was implemented in the calculations.

Primarily from the high CTE mismatch between DBC substrate and SiC dies significant strains and stresses evolve in the die attach. Accumulating die attach creep will lead to solder fatigue and therefore increased thermal resistance to the substrate. Fig. 4 is showing the incremental increase in creep deformation in the solder die attach of the SiC BJTs after one cycle with ton/toff of 10 s each and cooling down to cooling bench temperature (40°C). There is roughly 1.6‰ maximum rise in creep per cycle at the powered devices in test.

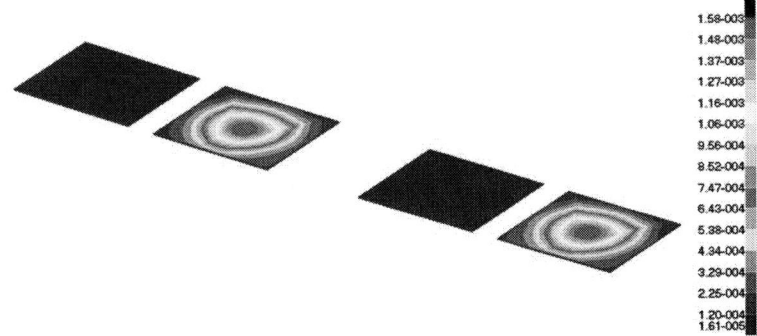

Fig. 4. Equivalent creep strain increment of die attach (solder) accumulated during one power cycle (ton/toff = 10 s)

The wire-bond based interconnection to the SiC devices is expected to be the main failure cause in power cycling due to the high stiffness of the semiconductor material. Sub-modelling has been employed to investigate stressing of the bond under thermo-mechanical loading during power cycling testing. Due to on-off cycles of 10 s, high cycle fatigue effects of the wire could be neglected. The simulation results of the plastic strain increment in Fig. 5 indicate that cracks are most prone to propagate either into the transition area between wire-bond material

© VDE VERLAG GMBH · Berlin · Offenbach

and chip metallization or in the heel region of the wire-bond. Thereby, a maximum increase of creep per cycle of approx. 0.55‰ was simulated. Although this is a lower value compared to the die attach (solder) value, based on this simulation results it is expected that the wire-bonds will fail earlier than die attach due to its much smaller footprint.

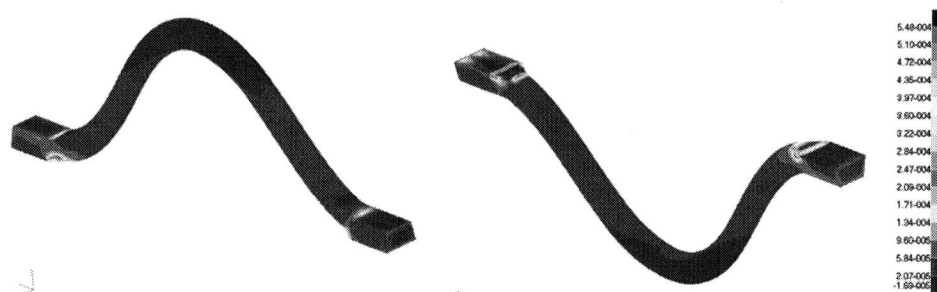

Fig. 5. Equivalent plastic strain increment accumulated in the bond wire during one power cycle with ton/toff of 10 s

Further simulation results, e.g. for the die attach as wells as for passive thermal cycles, are described in [7],[8] and [9] (latter one includes also information on followed lifetime assessment methodology and on performed warpage measurements).

4. Power cycling tests

4.1. Test setup

For performing power cycling tests the available test bench has been adapted to the specific needs of the SiC BJT power modules with the possibility to test single-sided as well as double-sided cooled power modules (Fig. 6, left) [9].

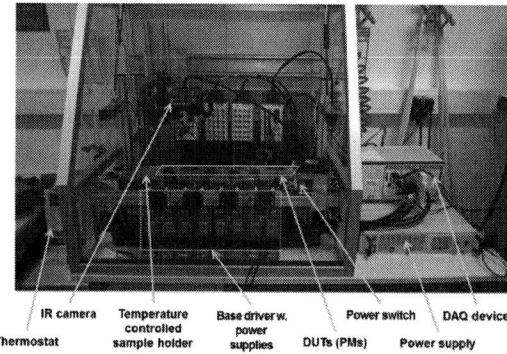

Test parameter	Value
Load current I_L	44 A (1st run)
Timing t_{on} / t_{off}	10 s (each)
Heatsink temp. $T_{heatsink}$	40°C (const.)
Min. junction temp. $T_{j,min}$	50...55°C
Max. junction temp. $T_{j,max}$	120...150°C
Junction temp. swing ΔT_j	71...91 K
Test strategy	Constant test parameters
Failure criterion	+5% U_{ce} +20% $T_{j,max}$

Fig. 6. Power cycling test bench (left) and applied test conditions (right)

In a first test run four modules, each involving one high-side and one low-side BJT (e.g. HS1 and LS1, see also Fig. 2, right), have been tested in series connection by applying a DC current load profile. In case one of the two BJTs of a certain module has failed, the whole module was temporary by-passed and the test continued on the remaining modules. After all modules were tested that way, the test on the remaining (and not failed) BJTs was continued by adapting the current level (from 44 A to approx. 50 A) in order to reach approximately the previous junction temperature profiles. This was necessary since a high mutual thermal influence of the different BJTs could be observed due to the small intermediate distances and the high thermal conductivity of the AlN substrate. For example, when powering HS1 and LS1 BJT together, a maximum junction temperature of approx. 142°C was measured for each BJT for a load current

© VDE VERLAG GMBH · Berlin · Offenbach

of 44A. In contrast, powering the HS1 BJT alone with the same load current resulted in a reduced maximum junction temperature of 108°C only. A third and fourth test run was performed further on to test the remaining BJTs (e.g. HS2 and LS2). This way, it was possible to test all individual BJTs of a power module. However, analysing the test results needs to be done carefully by considering the increased current levels for the second and fourth test runs and by taking possible pre-aging, due to the forced passive thermal cycling of the respective two remaining and by-passed BJTs during the first two test runs, into account. This is especially true for the in-between placed HS2 BJT when first testing HS1 and LS1 BJT, where a maximum (passively induced) junction temperature of e.g. 110°C for the HS2 BJT was measured (with HS1 ≈ LS1 ≈ 142°C).

Virtual junction temperature determination, being the most critical value for lifetime estimation, was done by means of TSEP (Temperature Sensitive Electrical Parameter) method [10] using the off-state V_{be} voltage [11]. For this reason, the voltage drop at the base-emitter junction, induced by a small sense current of 5 mA, was started to be measured approx. 100 µs after switching off the load current. To do so, first the base-current, provided by dedicated and within the COSIVU project developed base driver circuitry (1,5 A, 3V) [12],[13], was switched off followed by switching over an external IGBT power module with a time delay of 100 µs. For extrapolating the virtual junction temperature to t = 0 s the square-root-t method was used [14]. Verification of the calculated temperature values was done with IR camera on prepared power modules, where the die areas were opened with acid and blackened in order to get a homogenous emissivity.

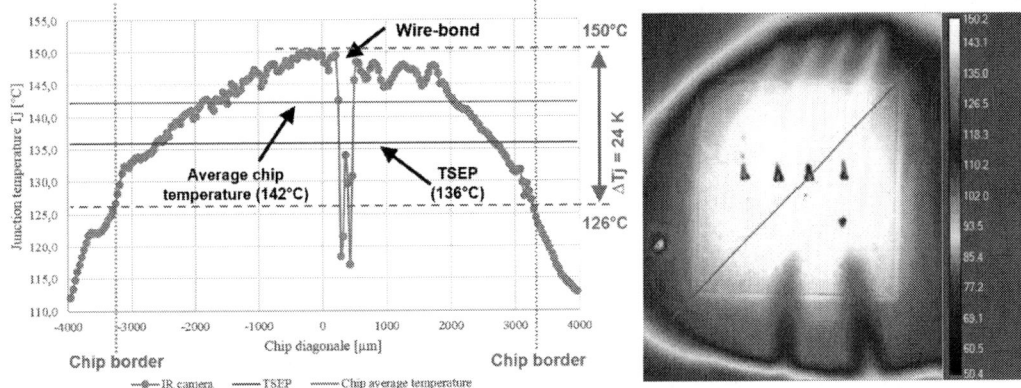

Fig. 7. Temperature profile over chip surface during power cycling at $T_{j,max}$ (269 W/cm²)

Fig. 7 shows the IR image of an HS1 BJT device during power cycling at $T_{j,max}$ for a power density of 269 W/cm² together with the corresponding temperature profile and the measured virtual junction temperature T_{vj} based on the TSEP method. For thermal characterisation as well as for testing, a thermal foil (R_{th} = 0,57 K/W) was used as thermal interface material in order to achieve the targeted maximum junction temperature of 150°C within the rated current range. The temperature difference between the center and the corner of the chip along the chip diagonal (6,27 mm) was in this case measured to be 24K. According to literature the TSEP based virtual temperature value is expected to be in the region of the average chip temperature value [15],[16], which was also roughly confirmed here.

4.2. Test results

First test results are available for 12 BJT devices in four power modules. Furthermore, in the case of two more modules electrical failures were observed after only 6235 and 8761 cycles, respectively, and the modules were removed from the test bench for further failure root cause analyses.

For all cases step increases of the collector-emitter voltage U_{ce} were observed indicating wire-bond failures as the dominating failure mechanism, whereas no significant increases of thermal resistances R_{th} could be detected, which is in line with the simulation findings. An exemplary

© VDE VERLAG GMBH · Berlin · Offenbach

$T_{j,max}$ and U_{ce} evaluation plot over the testing duration is shown in Fig. 9, left. In this particular case the test was continued until three out of four wire-bonds had failed and thus a temperature of almost 190°C was reached due to increased power loss at the remaining wire-bond.

In Fig. 9 (right) a microscopic cross section with a crack at the interface between chip metallization and wire-bond is shown. From the cross section it cannot be concluded, whether this crack has fully developed through the interface region and thus caused a disruption, since a lift-off might effectively be hindered by the mould compound.

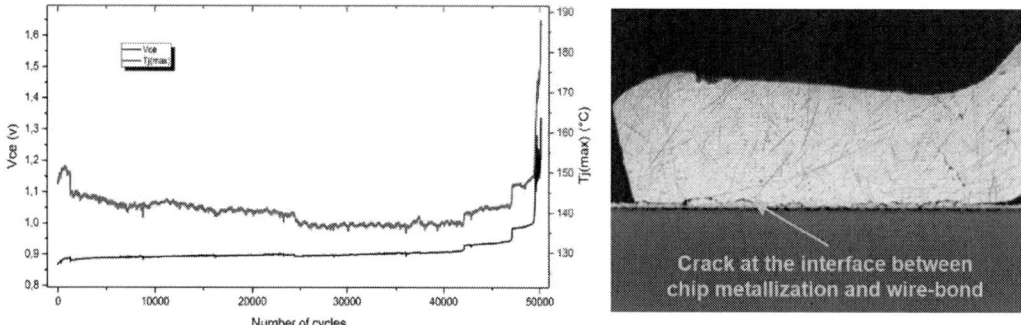

Fig. 9. Example of $T_{j,max}$ and V_{ce} evolution during power cycling (left) and microscopic cross section of failed wire-bond due to crack at interface region between chip metallization and wire-bond (right)

A further cross section image of the particular BJT tested to a temperature of 190°C is shown in Fig. 10, where changes and cracks in mould compound indicate increased temperatures. However, further analyses are necessary to obtain more information on the wire-bond failure modes. First 3D computed tomography analyses also showed small cavities in the solder material below the wire-bond interface locations, which are not assumed to originate from the soldering process, but probably from increased temperatures of cracked, but not completely disrupted wire-bonds.

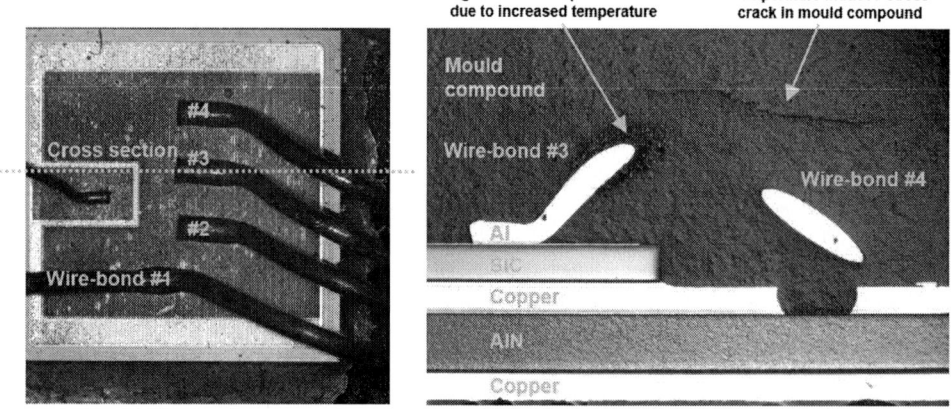

Fig. 10. Cracking and material changes of mould compound due to increased junction temperature (>200°C)

All test results are summarized and plotted in Fig. 11 (right). End-of-life was defined by an increase of V_{ce} by 5%, which usually was the case after the first wire-bond failure. Due to the limited number of data points it was not yet possible to derive differences between HS1 / LS1 and pre-aged (due to passive cycles) HS2 / LS2 BJTs. Furthermore, the effect of increased current level for the second and fourth test run was assumed to be not that critical with respect to the statistical analysis, since in all cases the modules had failed within a short time period after current level adjustment. Based on these first lifetime data, a very preliminary lifetime curve based on the LESIT model [17] was estimated (Fig. 11, left). However, further tests, in particular for different temperature swings to achieve further supporting points, are indispensable in order to derive a proper lifetime curve and to compare them to respective Si-based modules.

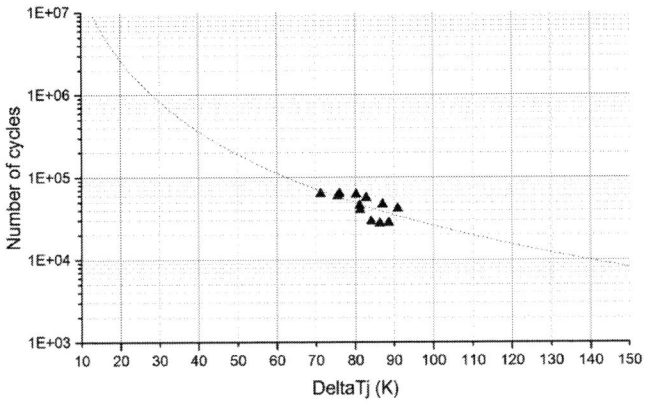

DUT	$T_{j,max}$ [°C]	ΔT_j [K]	EOL [cycles]
Module 1, HS1	137	84	29240
Module 1, HS2	138	87	47130
Module 1, LS1	134	81	45600
Module 1, LS2	141	91	42140
Module 2, HS1	126	76	63600
Module 2, HS2	139	89	28180
Module 2, LS1	119	71	63490
Module 2, LS2	126	80	62470
Module 3, HS1	139	86	27580
Module 3, LS1	134	81	40270
Module 4, HS1	133	83	57000
Module 4, LS1	126	76	59900

Fig. 11. Power cycling test results and preliminary, roughly estimated lifetime curve fitting based on LESIT model

5. Thermal benchmarking of double-sided cooled power module

A further subject of investigation was the thermal characterization of a novel double-sided cooled power module, as shown and compared to the single-sided cooled power module (the same one being described and investigated in the previous sections) in Fig. 12 [7],[8]. Both versions are based on exactly the same SiC components and half-bridge circuitry.

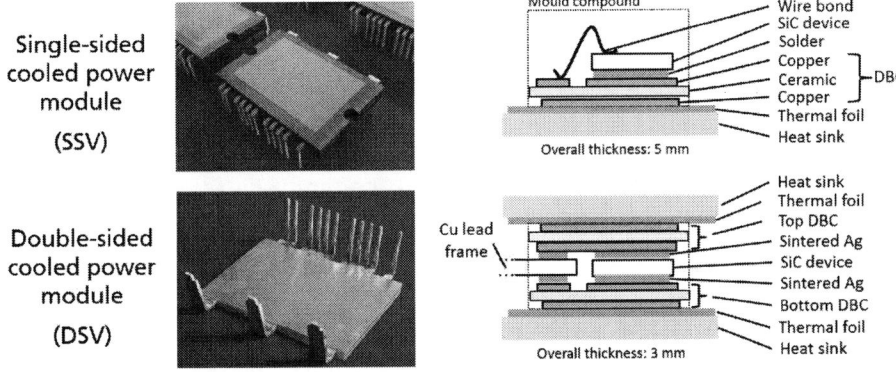

Fig. 12.　　　Image and cross section of the novel double-sided cooled SiC BJT power module in comparison with the single-sided version with both versions featuring the same SiC devices

For the measurements the power cycling test bench was equipped with an additional top cooling body (Fig. 13, right). The temperature of the cooling medium during the measurements was set to 25°C. Load currents of 15 A, 30 A, 40 A, and 50 A were applied for both versions. A maximum current value of 51 A was used for the single-sided cooled version (SSV), while the double-sided cooled version (DSV) was also measured at 60 A, 65 A 67 A and 67.5 A (until a maximum junction temperature of approx. 180°C was reached). The same timing (t_{on} = t_{off} = 10 s) as for the power cycling tests has been used. For each measurement level approximately 15 cycles were performed and the average values of the final cycle were used in order to ensure that a steady state status was reached. The measurement results in terms of maximum junction temperature, measured with the TSEP method, versus load current as well as versus

© VDE VERLAG GMBH · Berlin · Offenbach

power loss are shown in Fig. 13. They show a clearly increased thermal performance for the DSV with respect to the SSV: The maximum junction temperature was reduced by 18% – 55% for a respective load current range 15 A – 50 A. In other words, the maximum current and power loss capacity can with the double-sided cooling approach be increased by approx. 20% and 75%, respectively. Further details on the thermal characterisation results as well as on corresponding thermal simulations can be found in [18]. Further reliability investigations including power cycling tests are planned in order to investigate the robustness of the new DSV modules.

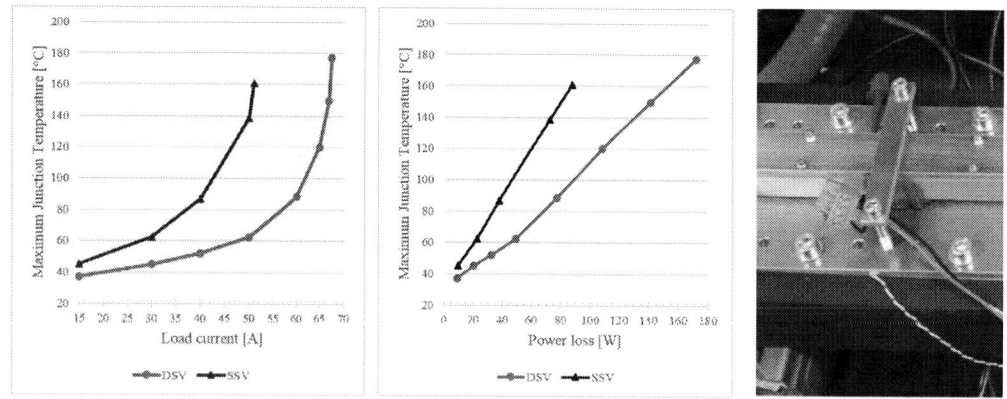

Fig. 13.   Thermal measurement results of the SiC-based power modules plotted against load current and power loss (left) and image of double-sided cooling test setup (right)

6. Conclusion

SiC based power modules are able to considerably enhance the energy efficiency of inverter systems due to their improved electrical and thermal properties in comparison to silicon based solutions.

The paper describes comprehensive reliability investigations on a SiC BJT power module, which constitutes an essential part of a newly developed 3-phase inverter for electric vehicles. FE analyses have been performed to assess the thermo-electro-mechanical stresses and strains induced during manufacturing and service operation in order to predetermine the main reliability failure mode. In parallel, power cycling tests of the new SiC BT power modules have been performed. The test results confirm the simulation findings that end-of-life is mainly determined by wire-bond failures.

Furthermore, a comparison between the investigated single-sided cooled power module and a new double-sided cooled version with the same SiC BJT and diode devices showed the ability to lower the total thermal resistance by nearly 50% and by this to decrease the maximum junction temperature from 139°C to 62°C (-45%) at a rated current of 50 A. This will not only allow to increase the power density, but potentially allow for increased lifetime.

Acknowledgement

The Authors would like to acknowledge the European Commission for supporting these activities within the project COSIVU under grant agreement number 313980.

References

[1]   Rzepka, S., Otto, A., COSIVU – Compact, Smart and Reliable Drive Unit for Fully Electric Vehicles, Micromaterials and Nanomaterials, issue 15, 116-121, 2013.

[2] Gustafsson, T., Nord, S., Andersson, D., Brinkfeldt, K., Hilpert, F., COSIVU - Compact, Smart and Reliable Drive Unit for Commercial Electric Vehicles, Proceedings of AMAA 2014, Advanced Microsystems for Automotive Applications, (Lecture Notes in Mobility, Springer 2014), 191-200.

[3] Andersson, R., D., et al., "COSIVU – Compact, Smart and Reliable Drive Unit for Fully Electric Vehicles", Proceedings of SMTA Pan Pacific Microelectronics Symposium, Big Island, HI, USA, 25-28 January 2016.

[4] Kern, J., Thun, C., Herman, J., "Development of a Solid-Borne Sound Sensor to Detect Bearing Faults Based on a MEMS Sensor and a PVDF Foil Sensor", in Advanced Microsystems for Automotive Applications 2014: Smart Systems for Safe, Clean and Automated Vehicles, Fischer-Wolfarth, J. and Meyer, G., Eds., Springer International Publishing, 2014, pp. 201-211.

[5] Frankeser, S., Hiller, S., Wachsmuth, G., Lutz, J., "Using the on-state-Vbe,sat-voltage for temperature estimation of SiC-BJTs during normal operation", PCIM Europe 2015, 19-21 May 2015, Nuremberg, Germany, p. 132-139. - Berlin Offenbach: VDE VERLAG GMBH, 2015.

[6] Otto, A., Kaulfersch, E., Gadhiya, G., Hilpert, F., Brabandt, I., Rzepka, S., "Reliability Assessment of a Smart and Compact Inverter Developed for Electrically", SSI 2016, 9-10 March 2016, Munich, Germany.

[7] Brinkfeldt, K., Edwards, M., Ottosson, J., Neumaier, K., Zschieschang, O., Otto, A., Kaulfersch, E., Andersson, D., "Thermo-mechanical simulations of SiC power modules with single and double sided cooling". 16th International Conference on Thermal, Mechanical and Multi-Physics Simulation and Experiments in Microelectronics and Microsystems, EuroSimE 2015.

[8] Brinkfeldt, K., Ottosson, J., Neumaier, K., Zschieschang, O., Kaulfersch, E., Edwards, M., Otto, A., Andersson, D., "Design and Fabrication of a SiC-Based Power Module with Double-Sided Cooling for Automotive Applications", in Advanced Microsystems for Automotive Applications 2015: Smart Systems for Green and Automated Driving, Schulze, T., Müller, B. and Meyer, G., Eds. Springer International Publishing, 2014, pp. 157-171.

[9] Otto, A., Kaulfersch, E., Brinkfeldt, K., Neumaier, K., Zschieschang, O., Andersson, D., Rzepka, S., "Reliability of New SiC BJT Power Modules for Fully Electric Vehicles", in Advanced Microsystems for Automotive Applications 2014: Smart Systems for Safe, Clean and Automated Vehicles, Fischer-Wolfarth, J. and Meyer, G., Eds., Springer International Publishing, 2014, pp. 235-244.

[10] Avenas, Y., Dupont, L., Khatir, Z., "TSEP paper Temperature Measurement of Power Semiconductor Devices by Thermo-Sensitive Electrical Parameters—A Review", IEEE Transactions on Power Electronics, vol. 27, no. 6, pp. 3081–3092, Jun. 2012.

[11] Frankeser, S., Herold, C., Franke, J., Lutz, J., On-line temperature measurement of SiC-BJTs using Vbe thermal sensitive electrical parameters, ISPS '14, 27 - 29 August 2014, Czech Technical University in Prague, Czech Republic, p.140-145.

[12] Frankeser, S., Muhsen, H., Lutz, J., "Comparison of drivers for SiC-BJTs, Si-IGBTs and SiC-MOSFETs", PCIM Europe 2015, 19 - 21 May 2015, Nuremberg, Germany, page 1096 - 1104. - Berlin - Offenbach: VDE VERLAG GMBH, 2015.

[13] Frankeser, S., Hiller, S., Lutz, J., Domes, K., "Proportional Driver for SiC BJT's in electric vehicle inverter application", PCIM 2014, S. 82 - 89. - Berlin Offenbach : VDE Verlag GMBH, 2014.

[14] Herold, C., Beier, M., Lutz, J., Hensler, A., "Improving the Accuracy of Junction Temperature Measurement with the Square-Root-t Method", Thermal Investigations of ICs and Systems (THERMINIC), 19th International Workshop on, 25-27 September 2013, Berlin, Germany 92-94.

[15] Goehre, J.-M., " Entwicklung und Implementierung einer verbesserten Lastwechseltestmethode zur experimentellen Bestimmung der Zuverlässigkeit von Dickdrahtbonds in Leistungsmodulen", Dissertation, Berlin, 2013, p. 57.

[16] Scheuermann, U., Schmidt, R., „Investigation on the VCE(T)-Method to Determine the Junction Temperature by Using the Chip itself as Sensor," International Exhibition and Conference on Power Electronics and Intelligent Motion (PCIM), 2009

[17] Held, M. ; Jacob, P. ; Nicoletti, G. ; Scacco, P. ; Poech, M.-H.: Fast power cycling test of IGBT modules in traction application.In: Power Electronics and Drive Systems Bd. 1, 1997.

[18] Brinkfeldt, K., Ottosson, J., Otto, A., Mann, A., Zschieschang, O., Frankeser, S., Andersson, D., "Thermal Simulations and Experimental Verification of Power Modules Designed for Double Sided Cooling", 66th The Electronic Components and Technology Conference (ECTC), Las Vegas, NV, 31 May - 03 June, 2016

© VDE VERLAG GMBH · Berlin · Offenbach

PCIM Europe 2016, 10 – 12 May 2016, Nuremberg, Germany

For information on the presentation

Investigation of the Influence of Ageing Processes on Thermal Characteristics of an IGBT Power Module by Means of Transient Thermal Analysis

Mr. Tobias von Essen

Berliner Nanotest & Design GmbH
Thermal Characterization
Volmerstr. 9B
12489 Berlin
Germany

Phone: +49 30 6392 3615
E-Mail: vonessen@nanotest.eu

Please note that the following manuscript
has been handed in late.

→ please go to the following page 2226

© VDE VERLAG GMBH · Berlin · Offenbach

Test Setup for Multistress Characterization of Insulation Degradation Mechanisms in Electric Drives

Davide Barater, University of Parma, Parma, Italy, davide.barater@unipr.it

Alessandro Soldati, University of Parma, Parma, Italy, alessandro.soldati@studenti.unipr.it

Giorgio Pietrini, University of Parma, Parma, Italy, giorgio.pietrini@studenti.unipr.it

Giovanni Franceschini, University of Parma, Parma, Italy, Giovanni.franceschini@unipr.it

Fabio Immovilli, Raw Power s.r.l., Reggio Emilia, Italy, fabio.immovilli@rawpowergroup.it

Michael Galea, University of Nottingham, Nottingham, UK, michael.galea@nottingham.ac.uk

Chris Gerada, University of Nottingham, Nottingham, UK, chris.gerada@nottingham.ac.uk

Abstract

In this paper a special test setup able to assess lifespan models of electrical motor insulation system under various stress conditions is proposed. The test bed will allow characterizing insulation degradation mechanisms under variable ambient and power supply parameters, from simple models, such as twisted pairs, up to coil form and complete machine operated at rated load.

1. Introduction

Studies have revealed that approximately 30% of the failures in electromechanical actuators are related to electrical faults [1]. Since the majority of electrical fault happens in stator or rotor windings insulation, the design of the insulation system of the windings and their structure play a major role on preventing possible failure mechanisms.

The detection of a change in the insulation state provides the possibility to track insulation deterioration and, in field applications, intervene with strategic maintenance to prevent an abrupt breakdown.

So far offline insulation monitoring methods, together with standardized aging and accelerated lifetime tests, have been used to assess the impact of a specific stress factor on the insulation system of electric motor [2][3]. Tests are usually performed on a formette or a portion of coils considering single aging mechanisms at once.

In this paper it is described an under development experimental setup, that will allow to experimentally investigate the effects of four major sources of stress in winding insulation (thermal, ambient, electrical and mechanical), both as single effects as well as combined.

Air pressure variations, which are normally neglected in everyday motor applications, are considered for realizing a reliable test setup even for characterization of electric drive used in harsh environment such as, aircraft applications or drives for high altitude mining [4].

The ultimate goal is to identify new standards for electric drive used in harsh environment, resulting from more reliable procedures that allow accurate estimation of a system lifetime starting from the design stage.

© VDE VERLAG GMBH · Berlin · Offenbach

2. Architecture of the Proposed Test Setup

The test setup will be able to reproduce an elevated number of environmental and operational conditions at which electric drives and motors, used in harsh environment, are subject. The system will also permit to stimulate the electric motors under test with different or combined environmental and electric stress factors, acquire data from the motor/drive during the test procedures, and support the experimental validation and fine tuning of life-time models predicting fault mechanisms in electric motors.

Apart from validating the models, the setup will provide the facility to test different degradation mechanisms at the same time, with the main advantage being the ability to evaluate the cross-couplings between the various mechanisms.

Moreover, the test setup will allow the execution of very specific accelerated life-time tests, tailored on the selected mission profile.

A functional scheme of the proposed test bed is reported in Figure 1 along with the installation footprint in Figure 2. The test setup comprises four main components:

- a Thermal vacuum chamber, housing the Motor Under Test (MUT), to apply environmental stress factors (temperature and pressure);

- a reconfigurable test bench that can be embedded in the test chamber;

- a custom inverter based on SiC devices capable to realize PWM commutation with different dv/dt ratio, thus applying different levels of electrical stress to the MUT.

- a data acquisition and control system to monitor an acquire relevant operating parameters

The test bench frame (7) will host the motor under test (2), the dynamometer (4), the torque meter and the encoder (3). In order to facilitate the set up procedures for different types of motors under test, the access flange of the thermal chamber (8) will be integrated into the structure of the test bench. Rails will guide the insertion\extraction of the motor under test into\from the thermal chamber (1). In this way, it will be possible to cabling and setting up the motor under test outside the thermal chamber, allowing for easy handling by the technicians, favoring the safety and speeding up the entire process.

The custom test inverter (5), and the dynamometer's drive (6) are housed in the cabinets, whereas the data control and acquisition unit (11) will collect all the data from the transducers and permit the control of the thermal chamber, the test and the brake inverters, allowing an easy configuration of the system from a unique control station. The component specifications for the test bed are resumed in Table I.

The entire system will be configurable from an unique control station, to which all data, collected during the experiments, will be addressed and made available for the qualified users.

The users will have the possibility to set-up the test procedure by means of a graphical user interface program (GUI) running on a dedicated desktop PC. The GUI will give access to 3 different subsystem blocks, each one governing a specific component of the test-setup (Figure 3).

PCIM Europe 2016, 10 – 12 May 2016, Nuremberg, Germany

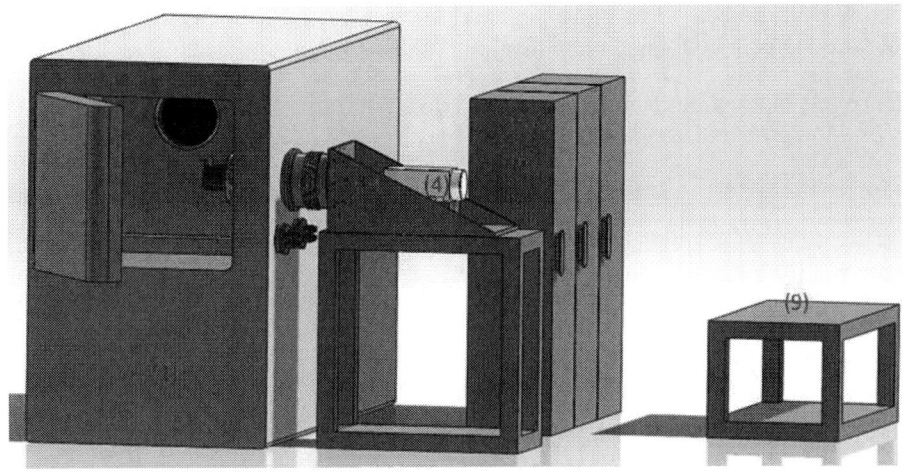

Figure 1 Deployed test setup rendering (MUT inserted).

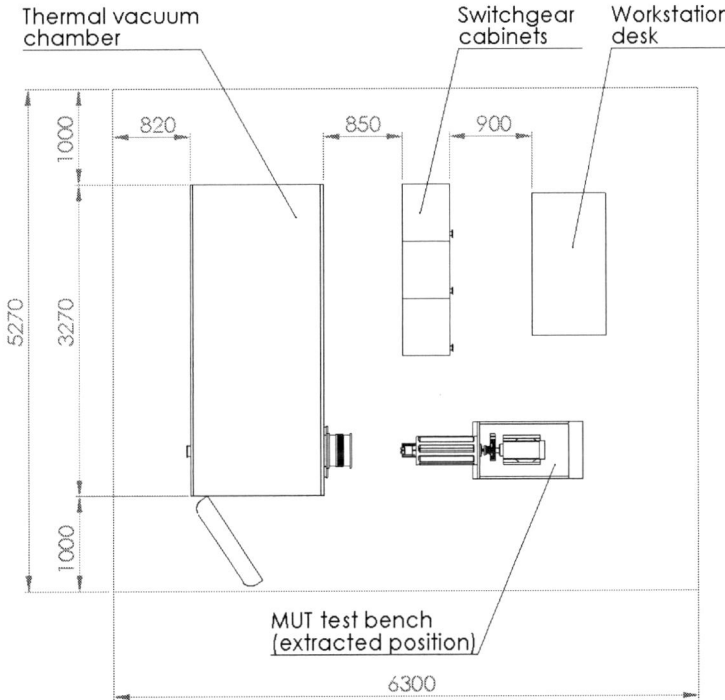

Figure 2 Installation footprint proposal of the complete test setup.

In particular, one subsystem block will be dedicated to the motor control, and will permit to control and acquire information from both the brake inverter and the MUT inverter. The communication between the host PC and the brake/MUT drive will be based on the standard protocol Modbus RTU used in industrial applications. Another subsystem block will be dedicated to the control of the thermal vacuum chamber. It will give the possibility to set the test profile of temperature and pressure and to constant monitoring the quantities of interest.

© VDE VERLAG GMBH · Berlin · Offenbach

The third block is dedicated to the data acquisition system for collecting and organizing all the electromechanical quantities monitored during the test procedure.

For each of this block a detailed description is given in the following, describing the technical aspects and the practical implementation.

Table I Test Setup Components Summary

No.	Name	Specifications
1	Vacuum thermal chamber	– Temperature: -40 +180 °C – Pressure: 1000 mBar down to 30 mBar (from sea level to 50000 ft altitude) – Internal useful dimensions 1000x1000x1000mm – Power handling capability (thermal): 5 kW (in the range -40 / 85°C and with P > 300 mBar) – Maximum leak rate @ 30 mBar: 0,5 kg/h – Remote interface for control of operating parameters and logging of environmental data
2	Motor/generator under test	– Nominal power: 40 kW @ 10 kRPM – Constant power operation up to 20 kRPM – Fixture: frontal (drive-end) flange – Maximum dimensions: 280mm OD X 300mm length – Shaft coupling: male involute splined shaft, max diameter 30mm
3	Torque meter	– High speed contactless torque meter, flanged construction (e.g. Kistler Ki Torq System type 4550A series) – Nominal torque range ±100Nm – Max torque: ±200 Nm (breaking torque > ±400Nm) – Accuracy class 0,05 (0.05% of the nominal full scale reading)
4	Test bench brake/dynamometer	– Commercial high speed synchronous spindle motor – Specifications to match the motor under test – Liquid cooled
5	Custom test inverter	– SiC MOSFET based (up to 100 kHz PWM) – 40 kW @ 10 krpm – Nominal voltage 550 V DC, current 58A – Air cooled – Hardware safety interlocks – Modbus 485 interface for remote control of operating parameters – Common DC Bus with the Brake inverter to allow regenerative operation
6	Brake inverter + Auxiliary MUT inverter	– Two commercial air cooled units (Emerson/Control Techniques) rated 270 A / 132 kW – Hardware safety interlocks – Modbus 485 interface for remote control of operating parameters – Common DC Bus with the Test inverter to allow regenerative operation
7	Test bench Frame	– Comprises fixture for the torque meter, test & brake motors – Self-contained cantilevered construction to avoid transmission of force/torque to the thermal vacuum chamber – Flange seals to mate the access port on Thermal vacuum chamber – Mounted on rails for insertion in the test chamber
8	Test Bench bulkhead	– Comprises mating flange for the thermal vacuum chamber MUT access port – Comprises an high speed face seal for the passage of the motor shaft
9	Feedthrough	– MUT Power Cables

		– MUT Encoder / Resolver cables – Thermocouples – Accelerometers
10	Closed circuit chiller	– Cooling of the ambient heat exchanger of the Thermal vacuum chamber and brake/dynamometer (part of installation requirements)
11	Data acquisition and control	– Control of the Vacuum thermal chamber parameters – Control of the test inverter – Control of the brake inverter – Data acquisition from the transducers (electrical, thermal and mechanical)

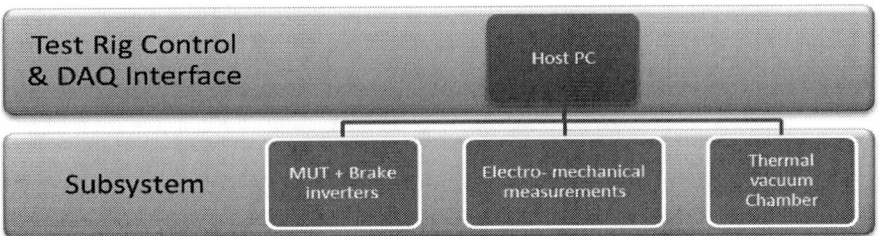

Figure 3 Test setup control and DAQ summary

3. Thermal vacuum Chamber and test bench coupling

The test bench structure is completely self-contained: it features a cantilevered construction, to support the MUT and allow it's placement inside the thermal vacuum chamber without placing stresses on the chamber frame/walls (see). Access to the test volume is given by an ISO-K flange DN320 that can accommodate MUT with body diameter up to 250 mm. The flange and mating metal bellows ensure air tightness. The test bench bulkhead incorporate a high speed shaft seal, that provide an airtight passage for the shaft from the test volume inside the thermal vacuum chamber and the dynamometer section housed outside at room pressure & temperature. Another two ISO-K flanges DN100 and two DN160 provide feedthrough for the shielded power cable to the MUT, together with various feedback (encoder/resolver) and transducer couplings (thermocouple, accelerometers) by means of MIL spec connectors (Figure 4).

4. Custom MUT Converter

The custom inverter, driving the motor under test MUT, is based on SiC power devices and capable to apply PWM commutation of different dv/dt ratio. The adoption of Silicon-Carbide power devices will give the possibility to reach both high frequency of the PWM carrier and fast rise and fall time of the inverter output voltage.

In this way it will be possible to vary the electrical stress intensity that the MUT has to withstand, and performing tests considering different levels of electrical stimulation.

CAS120M12BM2 by Cree was chosen as power devices, as it fits the requirements for the MUT inverter reported in table I.

© VDE VERLAG GMBH · Berlin · Offenbach

Figure 4 Detail of the test bench and chamber interface.

The device's datasheet reports a turn-on time ton=72ns and turn-off time toff=92ns, with DC voltage VDD = 600V, gate-source voltage VGS = -5/+20V, drain current ID = 120 A, and an additional external gate resistor RG(ext) = 2.5 Ω.

Obviously commutation time depends on the value of the external gate resistor adopted. Considering the dependency of delay turn-on time td(on), delay turn-off time td(off), rise time tr and fall time tf in function of RG(ext) value as reported in the datasheet, the total turn-on can be calculated as ton = td(on)+tr, where toff= td(off)+tf.

Therefore, according to the datasheet and considering three different gate resistor values of 6.66, 10, 20 Ω, and a DC voltage of 600 V the possible pre-set dv/dt ratio that the inverter can provide, considering the commutation time as the maximum between ton and toff, are reported in the table below.

Table II Summary of pre-settable dv/dt values

RG(ext)	td(on)	td(off)	tr	tf	ton	toff	dv/dt
6.66 Ω	50 ns	100 ns	50 ns	37 ns	100 ns	150 ns	4 kV/µs
10 Ω	75 ns	163 ns	63 ns	50 ns	138 ns	213 ns	2.8 kV/µs
20 Ω	130 ns	230 ns	90 ns	77 ns	220 ns	307 ns	2 kV/µs

5. Data Acquisition System

The data acquisition and control system will be tailored to the relevant operating parameters to be monitored and acquired:

- High Speed Acquisition (> 2 MS/s) for capturing induced voltage transients (dV/dt) and induced (di/dt)

- Medium speed acquisition e.g. for vibrations (up to 50 kS/s)

- Low speed acquisition: thermal transients (1-10 S/s)

© VDE VERLAG GMBH · Berlin · Offenbach

The data acquisition and control system will be interfaced with the load drive to enable testing to be carried out in a controlled and programmable manner with relevant test data acquired, stored and presented. This control system will also interface with both the test chamber temperature system and the vacuum arrangement for control of test temperature and simulated altitude level. The test system will also incorporate a range of monitoring and safely systems to ensure the security of personnel and equipment during setup and operation.

The control interface will be developed using NI Lab VIEW, as it is a commonly recognized programming environment for acquiring and manipulating engineering and scientific technical data. This will allow for a flexible, custom user interface to display live data, manage operator inputs and data log operations. As reported in Figure 5, the interface purpose is to:

- Control of the test inverter set points (MODBUS RTU 485)

- Control of the brake inverter set points (MODBUS RTU 485)

- Manage and display data acquisition from the transducers/instruments (electrical, thermal and mechanical)

- Manage and display data acquisition of thermal vacuum chamber environmental parameters

- Schedule and manage datalog operation for the acquired variables

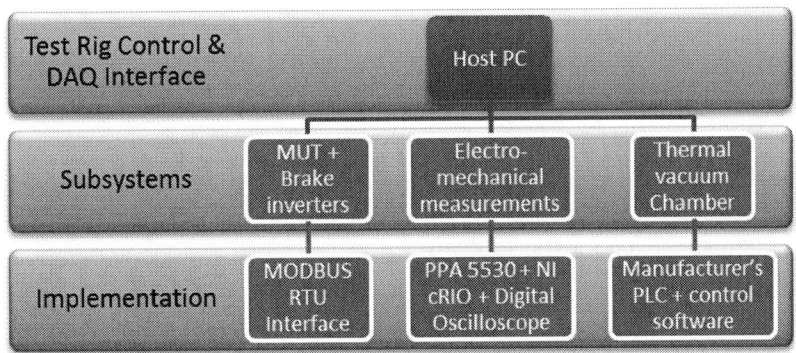

Figure 5 GUI specifications.

In any case, the user can select the quantities to be monitored and logged from a configuration panel. More in detail, datalogging can be set to three different modes of operation:

- Manual start/stop of acquisition, or single shot with a presettable acquisition buffer

- Automatic- cyclic recording, with a presettable repetition interval and data acquisition interval

- Event triggered with a presettable pre-trigger: (trigger sources examples: overcurrent, vibration level, over temperature)

The following table summarizes the data acquisition requirements for the test setup.

Table III Summary of DAQ requirements

Measured Quantity	Transducer description	Target sample Frequency	Channel No.
Temperature	K type thermocouple	1 S/s	8-16
Vibration	ICP Piezoelectric accelerometers	50 kS/s	4
Torque	Torque meter output ±10V	50 kS/s	1
Speed	Torque meter integrated encoder	50 kS/s	1
MUT Currents*	Shunt / Hall effect	2 MS/s	3
MUT Voltages*	Insulated differential probe	2 MS/s	3
DC bus Current*	Shunt / Hall effect	2 MS/s	1
DC bus Voltage*	Insulated differential probe	2 MS/s	1

* Galvanic insulation between channels required

6. Conclusion

This paper describes an experimental setup for investigating the fault mechanisms of the winding insulation system in electric motors. The proposed setup will be able to apply different stress factors even at the same time, enabling to assess the cross correlations between them. The ultimate goal is to identify new standards for electric drive used in harsh environment, and allow accurate estimation of a system lifetime starting from the design stage.

7. Acknowledgements

This work is carried out under the funding of the EU Clean Sky Joint Undertaking, in the framework of the Clean Sky call SP1-JTI-CS-2013-03 (topic: SGO-02-088) with the proposal no. 641496: ALEA - Accelerated Life tests for Electrical drives in Aircraft.

References

[1] Paoletti, G.; Golubev, A., "Partial discharge theory and applications to electrical systems," Pulp and Paper, 1999. Industry Technical Conference Record of 1999 Annual, vol., no., pp.124,138, 21-25 June 1999.

[2] Greg C. Stone, Ian Culbert, Edward A. Boulter, Hussein Dhirani, "Electrical Insulation for Rotating Machines: Design, Evaluation, Aging, Testing, and Repair," 672 pages, August 2014, Wiley-IEEE Press, ISBN: 978-1-118-05706-3.

[3] IEC Standard 60610 - Principal Aspects of Functional Evaluation of Electrical Insulation Systems: Aging Mechanisms and Diagnostic Procedures.

[4] Pontt, J.; Rodriguez, J.; Rebolledo, J.; Martin, L.S.; Cid, E.; Figueroa, G., "High-power LCI grinding mill drive under faulty conditions," in Industry Applications Conference, 2005. Fourtieth IAS Annual Meeting. Conference Record of the 2005 , vol.1, no., pp.670-673 Vol. 1, 2-6 Oct. 2005.

PCIM Europe 2016, 10 – 12 May 2016, Nuremberg, Germany

Integration of a measurement circuit to determine junction temperatures of IGBTs in a three-phase converter

Bastian, Strauß, Otto-von-Guericke-University Magdeburg, GER, bastian.strauss@ovgu.de
Andreas, Lindemann, Otto-von-Guericke-University Magdeburg, GER, andreas.lindemann@ovgu.de

Abstract

In order to realize an overtemperature protection for converter modules, an online determination of the junction temperatures is advisable. By comparing the measured junction temperatures with model based reference values, a condition monitoring of the converter module can be realized. By evaluating the long-term behaviour of the junction temperatures, deteriorated thermal interfaces or wearout can be detected. However, since the junction temperature can not be measured directly in power electronic standard applications, suitable measurement methods have to be defined. In this paper, the system integration of such a measuring concept into a three-phase converter will be described.

1. Introduction

Power electronic modules such as traction converters are often exposed to strong thermal stresses under real operating conditions. With increasing dynamic load these stresses become stronger, especially due to high temporal temperature gradient. They lead to variable inter-laminar thermomechanical stresses in the module structure. In long-term behaviour, ageing is the result. By this, an increase of the thermal resistance $R_{th,jh}$ between chip and heatsink is caused. The deteriorated thermal interfaces lead to a reduced heat dissipation. Thus the junction temperature increases, [1], [2].

Besides simulative and computational approaches [3], [4], the case-temperature measurement is used for monitoring the thermal behaviour in standard applications, [5]. In order to determine the junction temperatures within an converter module, various indirect measurement methods are recommended. Among other temperature sensitive electrical parameters (TSEP), previous studies have already shown that the threshold voltage V_{th} can be used as a suitable parameter to determine the junction temperature [6], [7]. Hereinafter, for threshold voltage and junction temperature the additional notation "virtual" is used. As mentioned in [9], this is due to the different behaviour of these physical magnitudes between their real and theoretical characteristics. Within this work, the system integration of a defined measurement concept to determine the junction temperatures of IGBTs in a converter module was realized. The measuring circuits are used as supplements to the drivers. By measuring the virtual threshold voltage and referring to pre-recorded calibration curves, the corresponding virtual junction temperatures ϑ_{vj} of each power semiconductor switch will be determined. This paper illustrates the integration of the measuring circuit into a converter to determine the virtual junction temperatures of its semiconductor switches indirectly. Furthermore, the central data acquisition in real online operation of the converter will be discussed in detail. Finally, a principal exemplary time profile of the virtual junction temperature — determined by using the virtual threshold voltage of one IGBT — is shown.

© VDE VERLAG GMBH · Berlin · Offenbach

PCIM Europe 2016, 10 – 12 May 2016, Nuremberg, Germany

2. Measurement concept and calibration of a V_{vth}-measurement system for determining the virtual junction temperature of an IGBT

2.1. Measurement concept

While the first results of a feasibility study [7] confirmed the suitability of the V_{vth}-measurement concept for detecting the virtual junction temperature, optimizations have been realized with regard to the duration of the measuring trigger pulse V_{trig} and the time accuracy. In figure 1(a) the principle measurement concept is shown. Figure 1(b) illustrates the behaviour of the respective measurement and logical signals. As described in [7], the time mask defines the time range

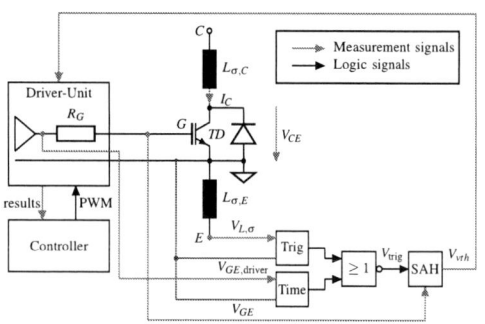

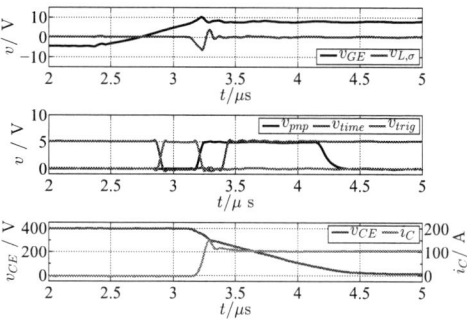

(a) Measurement concept for determining the virtual junction temperature, based on [8]

(b) Switching behaviour of an IGBT considering the defined variables of figure 1(a)

Fig. 1: Measurement concept for electro thermal deterining the virtual junction temperature and switching behaviour of an IGBT

during turn-on of an IGBT when the measurement should plausibly happen. The time mask begins if the gate-emitter voltage V_{GE} exceeds a minimum voltage level of above 2 V. At this point of time, also the measuring trigger pulse V_{trig} begins. By the output of a PNP-trigger circuit the ending of the measuring trigger pulse will be defined. This trigger circuit detects the beginning of turn-on by evaluating the parasitic voltage $V_{L,\sigma}$ between main emitter E and auxiliary emitter E' of one semiconductor power switch. If the measuring trigger pulse is HIGH, the IGBT-side gate-emitter voltage will be sampled by a sample-and-hold circuit (SAH), whereby the output voltage of the SAH follows the IGBT-side gate-emitter voltage. During the LOW state, the last sampled value is kept constant. The analog output voltage of the V_{vth}-measurement circuit is available for further signal processing as well as a centralized evaluation by the higher-level control unit.

2.2. Calibration of a V_{vth}-measurement system for determining ϑ_{vj} of an IGBT

In order to determine the corresponding virtual junction temperatures, during the initial system setup a calibration of the $V_{vth}(\vartheta_{vj})$-measurement has to be performed for each semiconductor switch. To evaluate ϑ_{vj} as accurately as possible, a reference temperature characteristic – illustrated in figure 2(a) – has been determined. For this, the internal NTC sensors of the converter module were used. In normal operation, these sensors are implemented to monitor the module temperatures. The calibration of the internal NTC sensors was performed using a climatic chamber. This ensures a uniform temperature distribution within the investigated

© VDE VERLAG GMBH · Berlin · Offenbach

PCIM Europe 2016, 10 – 12 May 2016, Nuremberg, Germany

module. In thermal steady state the measurement was performed using steps of $5\,\mathrm{K}$ over a temperature range of $25\,°\mathrm{C} < \vartheta_{vj} < 140\,°\mathrm{C}$.

For calibrating the temperature characteristic of $V_{vth}(\vartheta_{vj})$ a double pulse test setup was used. The boundary conditions of the power circuit were defined by a DC-link voltage of $V_Z = 400\,\mathrm{V}$ and a collector current of $I_{C,max} = 130\,\mathrm{A}$. In figure 1(b) an exemplary switching behaviour is documented. In temperature steps of $5\,\mathrm{K}$ the individual double pulse measurements were also performed in thermal steady state. So, the reference temperature determination using the NTC sensors corresponds to a measurement of the virtual junction temperature. The temperature curve $V_{vth}(\vartheta_{vj})$ of the investigated IGBTs – shown in figure 2(b) – was performed using a calibration system. During each temperature step 50 individual measurements of $V_{vth}(\vartheta_{vj})$ were determined. By averaging these data the temperature characteristic, shown in figure 2(b) was generated. The coefficients of a cubic approximation according to the measured temperature curve of V_{vth} are summarized in table 1. Thus, the approximate temperature characteristic of V_{vth} can be expressed as follows:

$$V_{vth}(\vartheta_{vj}) = A_3 \cdot \vartheta_{vj}^3 + A_2 \cdot \vartheta_{vj}^2 + A_1 \cdot \vartheta_{vj} + A_0 \qquad (1)$$

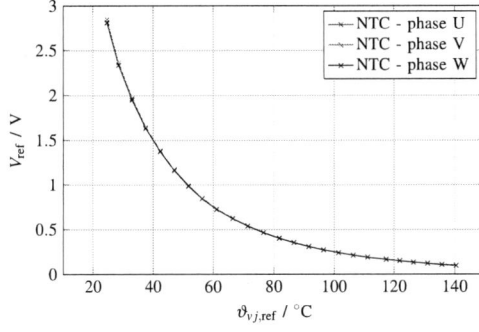

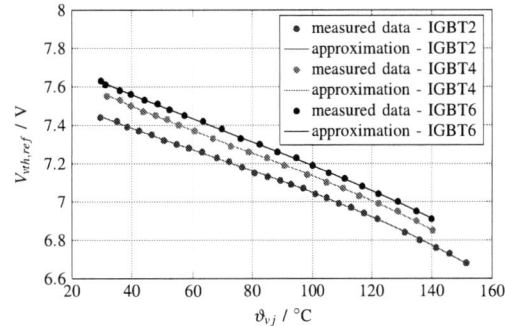

(a) Calibration curves for determining the reference virtual junction temperatures using NTC-sensors loaded with $0.5\,\mathrm{mA}$

(b) Measured temperature characteristic $V_{vth}(\vartheta_{vj})$ of three exemplary IGBTs

Fig. 2: Calibration curves of the intrinsic NTC sensors of the converter module and the measured virtual threshold voltage characteristics of three exemplary IGBTs

Tab. 1: Coefficients of the polynomial approximation according to the calibration measurement of $V_{vth}(\vartheta_{vj})$

Semiconductor	$A_3 \left[\frac{V}{K^3}\right]$	$A_2 \left[\frac{V}{K^2}\right]$	$A_1 \left[\frac{V}{K}\right]$	$A_0 \,[V]$
IGBT2	$-1.67 \cdot 10^{-7}$	$3.41 \cdot 10^{-5}$	$-7.78 \cdot 10^{-3}$	7.651
IGBT4	$-2.19 \cdot 10^{-7}$	$5.14 \cdot 10^{-5}$	$-9.87 \cdot 10^{-3}$	7.826
IGBT6	$-1.23 \cdot 10^{-7}$	$2.70 \cdot 10^{-5}$	$-8.01 \cdot 10^{-3}$	7.843

The temperature characteristics of V_{vth}, shown in figure 2(b) and the corresponding coefficients of the approximations in table 1 illustrate that each semiconductor chip has its individual temperature characteristic of the virtual threshold voltage. To achieve high accuracy it is thus advantageous to calibrate each semiconductor separately. As already described in [9] and [10], the temperature characteristics of the IGBT virtual threshold voltages are nearly linear. So a linear

© VDE VERLAG GMBH · Berlin · Offenbach

or cubic interpolation of the characteristic is sufficient. For a three-phase two-level converter, these six individual temperature characteristics can be stored in the higher-level controller to monitor the virtual junction temperatures in online mode operation. The following section illustrates the integration of six measurement circuits into a three-phase two-level converter. For this purpose the potential differences between the transistors need to be considered; the proposed bus architecture can advantageously deal with those.

3. Integration of the measurement concept into a three-phase two-level converter application

For a centralized data acquisition a suitable bus architecture is used. However, during pulsed mode operation of the converter, conducted as well as radiated disturbances will be generated on the bus lines. To avoid misinterpretation of information signals a Controller Area Network (CAN-bus) topology is used for data transmission. Here, the voltage difference between CAN-High and CAN-Low line is interpreted by each receiver. By evaluating of voltage differences, common mode interferences are directly eliminated. In figure 3 the voltage signals of CAN-Low and CAN-High line are shown in comparison to the evaluated logic signal. Depending on the

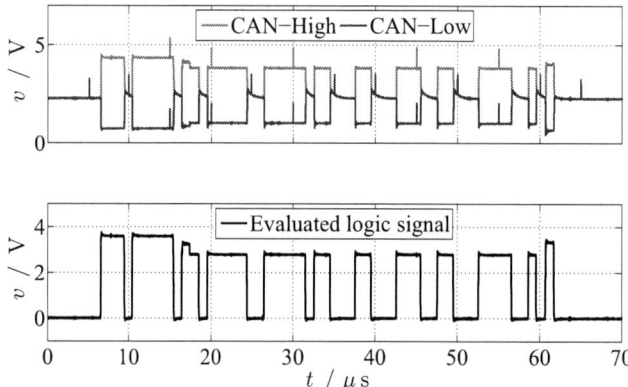

Fig. 3: Time behaviour of one CAN data frame in comparison to the evaluated logic signal

CAN-bus system-impedance, the voltage signals of CAN-High and CAN-Low show a delayed settling with first order. Furthermore, to demonstrate the active suppression of interferences by differential voltage analysis, the voltage signals of CAN-High and CAN-Low line were applied with a 200 kHz common mode interference signal. By differential voltage analysis, coupled interferences will be eliminated. Influences stemming from the impedance of the CAN-bus system, will also be repressed from the information signal. Thus, the pure information signal consists of clearly distinguishable voltage levels. So the interference-suppressed information signals can be evaluated by the DSP without additional misinterpretations of the CAN-protocol.

Because of the interference resistance of this bus topology, in the automotive industry the CAN-bus has become well established in recent years. Furthermore, the use of high-speed CAN-bus (Baudrate: 1 MBit/sec) allows a nearly real time control of various components within the automobile system. It is appropriate to embed the online monitoring of the virtual junction temperatures for a drive converter into the central CAN bus based control when implemented in an automobile.

For embedding the control unit as well as the V_{vth}-measurement systems into a CAN-bus, the respective CAN-interfaces must be electrically isolated from the associated power semiconductor switches. In figure 4 the integration of one V_{vth}-measurement circuit into a CAN-bus node is documented. The analog output magnitude of the V_{vth}-measurement circuit is delivered directly to the CAN controller. Using a 12-Bit analog digital converter (ADC), the sampled voltage signal will digitized. The on board CAN-controller implements the digitized measurement value into a standard CAN-protocol. The electrical isolation between one V_{vth}-measurement system and the higher level CAN-bus system is realized by using a CAN transceiver. Furthermore, this CAN transceiver allows a level adjustment of the signal voltages at the BUS level of CAN-High and CAN-Low line. The entirety of measurement board, CAN controller and CAN transceiver defines a CAN node of the CAN-bus system.

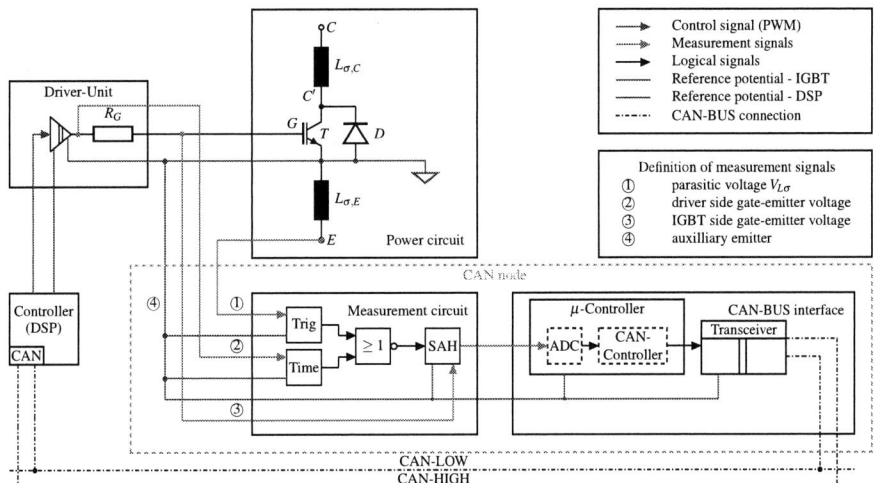

Fig. 4: Integration of the V_{vth}-measurement circuit into a CAN-node for connecting the measurement system to a CAN-bus architecture

For powering the respective CAN nodes the associated IGBT drivers are used. Here, the reference potential is characterized by the corresponding auxiliary emitter. Thus, for each semiconductor switch the same configuration of driver, measurement circuit an CAN-bus interface can be used. On the secondary side of the CAN transceiver, the reference potential of the CAN-bus – here the digital ground potential – is provided by the higher level control unit. The potential relationships are correspondingly presented in figure 4.

Based on the defined signals and potential relationships, visible in figure 4, the integration of each individual V_{vth}-measurement system into a three-phase converter can be carried out as shown in figure 5. At both ends of the CAN-bus system a termination using terminating resistances of $R_{term} = 120\,\Omega$ is necessary. During one measurement interval the central acquisition of each determined virtual threshold voltage will be realized by the higher level controller in interrupt mode.

In figure 6 the CAN-bus protocol for transmitting of six individual digitized measurement values is shown. The transmission will be initiated by the DSP using a request frame. This is followed by the serial responses of the individual V_{vth}-measurement systems in the form of data frames. Each data frame is led by an identifier which identifies individual objects of the CAN-bus system.

For serial transmission of all six measured values using a high-speed CAN bus topology with a baudrate of $1\,\text{MBit/sec}$, a total time of $500\,\mu s$ is required. In consideration of the computational

© VDE VERLAG GMBH · Berlin · Offenbach

PCIM Europe 2016, 10 – 12 May 2016, Nuremberg, Germany

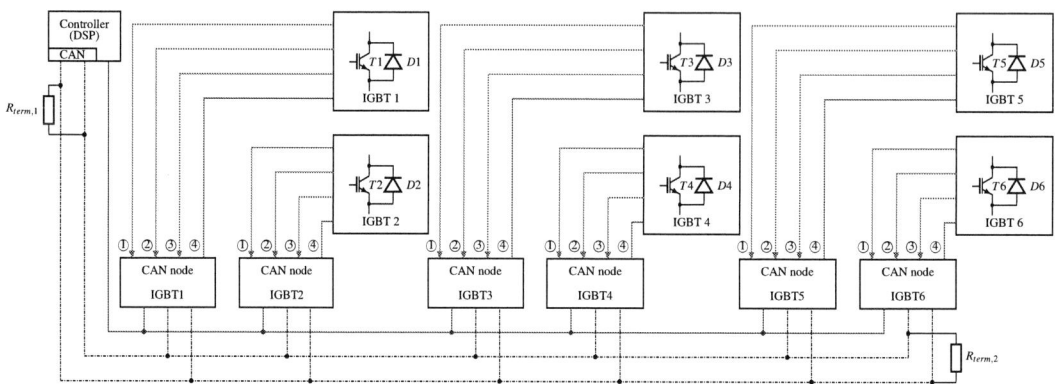

Fig. 5: Overall structure of a three-phase two-level converter system with integrated measurement of the six individual virtual junction temperatures in online mode; measurement signal definition (① to ④) for each individual IGBT based on figure 4

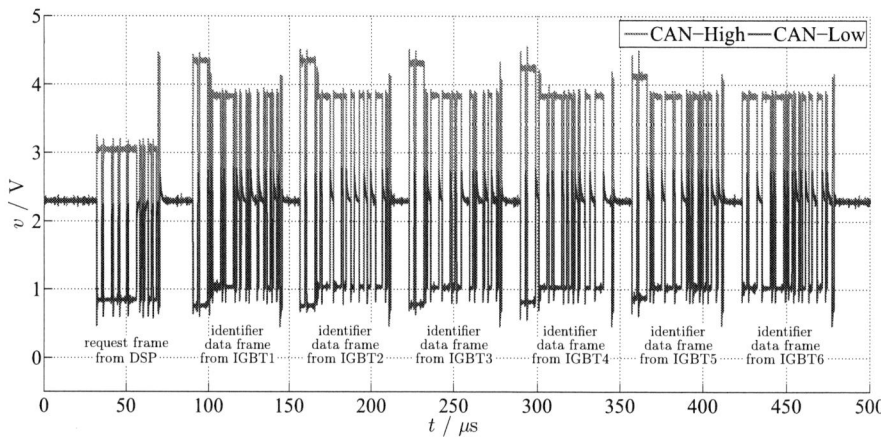

Fig. 6: CAN-bus protocol for transmitting of six determined and digitized measurement values from the individual IGBTs to the higher level controller (DSP)

load according to the higher level controller, an updating of the individual IGBT virtual junction temperatures of every $1.8\,\mathrm{ms}$ is sufficient. Furthermore, the thermal time constant of a standard power electronic module is in the range of a second, so a higher temporal resolution of the temperature measurement is not necessary for most purposes, see [9]. The digitization and transmission of the sampled analog output voltages must not generate any additional systematic errors in the assessment of measured data. These systematic errors are mainly influenced by the manufacturing tolerances of the used electronic components. In order to minimize influences of the signal electronics, a correction could be implemented in the CAN controller of each measurement system. Figure 7(a) illustrates the transfer characteristic of data transmissions between the measuring systems and the higher level controller. Apart from the slope C_1 all other polynomial coefficients have a negligible value, see table 2. Thus, between each analog output voltage of the measurement circuits and the DSP a transfer characteristic of exactly 1 can be assumed. Further results with respect to the accuracy of the realized measurement concept can be seen in [9]. In figure 7(b) the measured time dependent behaviour of the virtual junction temperature is documented exemplarily for IGBT4 in converter mode operation. It should be noted that IGBT4 will heat up during the negative half-wave of the load current

© VDE VERLAG GMBH · Berlin · Offenbach

Tab. 2: Coefficients of the polynomial approximation according to the transfer characteristics between analog output voltages of the V_{vth}-measurement circuits an the digitized values evaluated by the DSP

IGBT	$C_3 \left[\frac{V}{V^3}\right]$	$C_2 \left[\frac{V}{V^2}\right]$	$C_1 \left[\frac{V}{V}\right]$	$C_0 [V]$
1	$-1.8 \cdot 10^{-17}$	$2.3 \cdot 10^{-16}$	1.0	$1.7 \cdot 10^{-16}$
2	$2.2 \cdot 10^{-18}$	$9.4 \cdot 10^{-17}$	1.0	$1.1 \cdot 10^{-15}$
3	$-9.8 \cdot 10^{-18}$	$1.4 \cdot 10^{-16}$	1.0	$6.6 \cdot 10^{-16}$
4	$-1.7 \cdot 10^{-17}$	$3.5 \cdot 10^{-16}$	1.0	$2.8 \cdot 10^{-15}$
5	$-3.3 \cdot 10^{-17}$	$3.5 \cdot 10^{-16}$	1.0	$5.9 \cdot 10^{-16}$
6	$4.2 \cdot 10^{-18}$	$4.7 \cdot 10^{-17}$	1.0	$1.7 \cdot 10^{-15}$

period. While in the positive half-wave of the load current IGBT3 becomes warmer, IGBT4 will cool down.

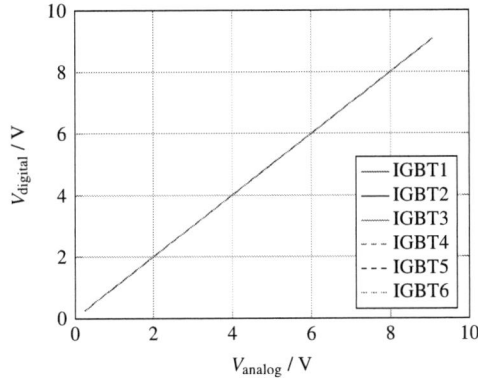

(a) Transfer characteristic between the analog output of the measurement circuits an the digitized values evaluated by the DSP

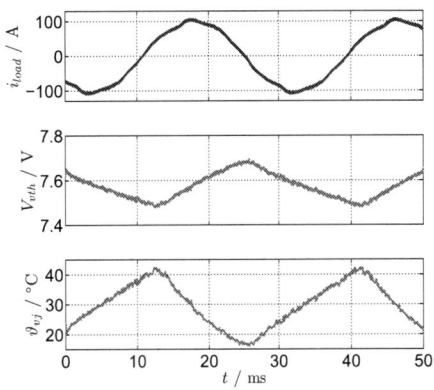

(b) Load current of phase V and time behaviour of ϑ_{vj} in converter mode operation exemplarily shown for IGBT4

Fig. 7: Transfer characteristic between measurement circuit and DSP as well as time behaviour of the virtual junction temperature in converter mode operation

4. Conclusion

The measurement concept to determine the virtual junction temperature ϑ_{vj} by using the virtual threshold voltage V_{vth} of semiconductor devices — already described in [7] — has been successfully implemented in a three-phase two-level converter of medium power class. Using an appropriate calibration procedure, the temporal behaviour of virtual junction temperatures for IGBTs within an exemplary converter module has been determined in online mode operation. Thus, by monitoring the virtual junction temperatures, a timely shut down of the converter is possible if defined temperature limits would be reached. Furthermore, it is possible to investigate the influence of ageing mechanisms onto the long term behaviour of virtual junction temperatures. So it is possible to develop an approach for life time prediction based on the real virtual junction temperature behaviour.

© VDE VERLAG GMBH · Berlin · Offenbach

5. References

[1] J. Lutz; H. Schlangenotto; U. Scheuermann; R. D. Doncker. *Semiconductor Power Devices; Physics, Characteristics, Reliability.* Springer Verlag, 2011.

[2] A. Middendorf. *Lebensdauerprognostik unter Berücksichtigung realer Belastungen am Beispiel von Bondverbindungen bei thermomechanischen Wechselbeanspruchungen.* PhD thesis, TU Berlin, 2009.

[3] P.M. Igic; P.A. Mawby; M.S. Towers. *Physics-based dynamic electro-thermal models of power bipolar devices (PiN diode and IGBT).* Proceeding of the 13th International Symposium on Power Semiconductor Devices & ICs. IPSD '01 (IEEE Cat. No.01CH37216), (3):381-384, 2001.

[4] C. Yun; P. Malberti; M. Ciappa; W. Fichtner. *Thermal component model for electrothermal analysis of IGBT module systems.* IEEE Transaction on Advanced Packaging, 24(3):401-406, 2001.

[5] B. Ji; V. Pickert; B. Zahawi. *In-situ Bond Wire and Solder Layer Health Monitoring Circuit for IGBT Power Modules.* Integrated Power Electronics Systems (CIPS), 2012 7th Int. Converence on, 9:1-6, 2012.

[6] J. A. Butrón Ccoa; B. Strauß; Dr. G. Mitic; A. Lindemann. *Investigation of Temperature Sensitive Electrical Parameters for Power Semiconductors (IGBT) in Real-time Applications.* PCIM 2014 Europe, Nuremberg, 2014.

[7] B. Strauß; A. Lindemann. *Indirect measurement of junction temperature for condition monitoring of power semiconductor devices during operation.* PCIM 2015 Europe, Nuremberg, 2015.

[8] I. Bahun; N. Čobanov; Ž. Jakopović. *Real-Time Measurement of IGBTs Operating Temperature.* AUTOMATIKA 52(2011) 4, 295305, 2011.

[9] B. Strauß; A. Lindemann. *Measuring the junction temperature of an IGBT using its threshold voltage as a temperature sensitive electrical parameter (TSEP).* International Multi-Conference on Systems, Signals and Devices 2016, Leipzig, 2016.

[10] B. Strauß; A. Lindemann. *Measurement of the junction temperature during operation of a drive converter.* 13. Braunschweiger Symposium Hybrid- und Elektrofahrzeuge, Braunschweig, 2016.

A Recuperation Topology for Power Device Testing

Tomas Krecek, ON semiconductor, 1.maje 2594 (C12 Building), 756 61, Roznov pod Radhostem, Czech Republic, Europe, tomas.krecek@onsemi.com

Abstract

The aim of this paper is to present versatile and modular power electronics architecture which provides a tool to implement and evaluate any power conversion topology, semiconductor devices or Power Integrated Modules (PIMs). The key goal is to design a dedicated bench test setup that emulates the real conditions such as current and voltage levels, $cos\,(\varphi)$, switching frequency etc. For this purpose, vector control technique has been chosen and developed. The adopted platform includes recuperation of energy so no extra expensive equipment such as power supply or power loads are needed. In addition, the implemented system avoids problems with huge heat dissipation. At the same time the system makes it possible to measure precise relative power loss.

Introduction

Power conversion in low and middle power ranges is getting much more demanding in terms of efficiency requirements. This is mainly due to the increase of energy demand and need for power density. This is especially true in applications where renewable sources, such as wind and solar, have taken a primary role as new energy suppliers. Same efficiency requirements impact other power conversion categories such as Uninterruptible Power Supply (UPS), which needs to improve the utilization of backup battery bank. The market also demands integration of the power converter into the overall system. This leads to enhanced efforts in designing new power topologies or utilization of existing but complex structures such as multilevel topologies, precisely optimized active and/or passive components or accurate customer designs optimization. In that process, it is very convenient to expose all parts of design to real conditions as in final application.

Proposed recuperation topology

In the *Fig. 1* we can see proposed topology which is using all fundamental converters. One operates always as DC to AC converter and second one as AC to DC stage. In *Fig. 1* we see T-type and Half-bridge topologies for instance but it is fully arbitrary what type of converter is used as depicted on the *Fig. 1*. The converters have common DC link and common output power filter. The idea is that DC to AC converter is supplying the AC to DC converter through the output power filter. In that process, the DC link works as an energy recuperation medium. The output and input DC link current must be the same. DC link voltage is naturally regulated and given by external power supply. This external power supply can be form by low-power variac for ultra-low cost version or by external high voltage low power DC source. The variac or external power source covers total power loss generated by active and passive components of the proposed system. This power loss can be used for relative power loss evaluation of the power devices, topologies or/and passive components such as inductors and capacitors. Relative expression for given device under test giving us information about total power loss generated by the device itself plus its effect on power loss in auxiliary passive and/or active components.

Three phase rectifier is used as AC-DC converter in the case usage variac or for DC link voltage separation (protection) for commercial HVDC source. The DC link capacitor divider is used for accommodation three-level converters so voltage balancing strategy must be implemented.

© VDE VERLAG GMBH · Berlin · Offenbach

PCIM Europe 2016, 10 – 12 May 2016, Nuremberg, Germany

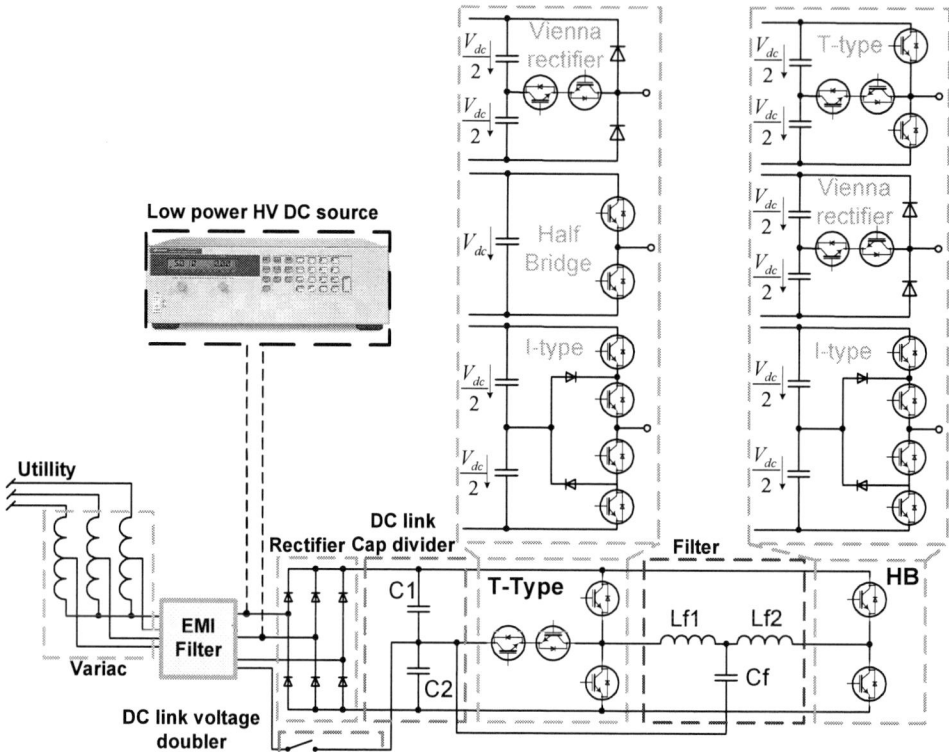

Fig.1 The proposed recuperation topology

The proposed topology utilizes two converters where the first one operates as inverter and the second one as rectifier. In that case $\varphi_1 = 0°$ and $\varphi_2 = 180°$ where φ_1 is phase displacement between V_{Cf} (voltage across C_f) and I_{Lf1} (L_{f1} current-carrying) and φ_2 is phase displacement between V_{Cf} and I_{Lf2} (L_{f2} current-carrying). The phase difference $180°$ between first and second converter must be hold accurately to ensure an equilibrium state.

The power filter is composed by two identical inductors and one capacitor. If one converter operates as inverter the closer inductor and capacitor works as filter and remaining inductor as boost inductor for second converter. In real application there is extra power resistance in series with capacitor works as dumping element (not depicted).

Total power loss dissipating as heat from passive and active components is covered by external source which can be DC or AC (this is the case in *Fig. 1*). This effect is utilized as another opportunity to get precise power loss measurement, which is very difficult for high efficiency topologies and advanced power device technologies. This technique brings to us very precise relative measurement.

Control strategy

In terms of control algorithm, this is most important point in the recuperation system. For semiconductor power devices evaluations it is critical to expose all designing components to all possible voltage/current levels and phase displacements. This is the reason why oriented vector control chosen. The *Voltage Oriented Control* (VOC) and *Virtual Flux Oriented Control* (VFOC) are close to Field Oriented Control for electric machines. The method is based on the transformation between stationary coordinate $\alpha\beta$ and synchronous rotating coordinates *dq*. *Fig. 2* shows simplify control structure based on *Virtual Flux Oriented Control* [1].

© VDE VERLAG GMBH · Berlin · Offenbach

PCIM Europe 2016, 10 – 12 May 2016, Nuremberg, Germany

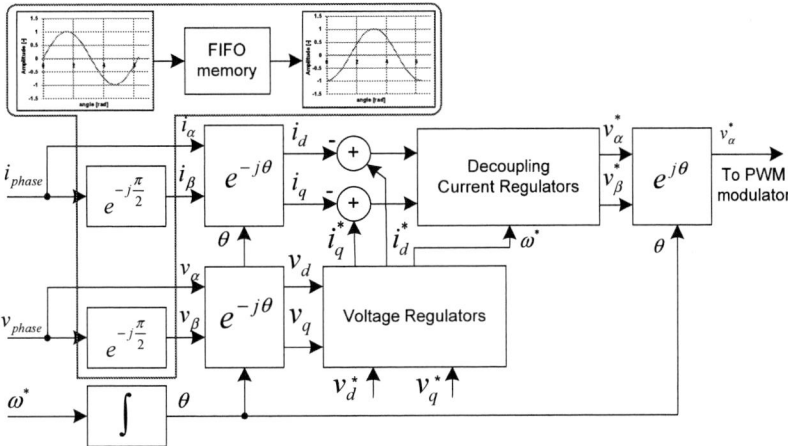

Fig. 2. The Virtual Flux Oriented Control as control strategy of the proposed topology. The structure is implemented for both converters

This strategy guarantees: fast transient response and high static performance via internal current control loop, phase displacement control and reactive or active power control. In a single-phase system, the use of a rotating frame is not possible unless a virtual system is coupled to the real frame in order to emulate a two-axis environment. There are several techniques to time shift the time dependent signal while FIFO memory features were used in this paper. Consequently, the performance depends on the quality of the current control loop. We can find several strategies that can be applied for current control. A widely used scheme for high performance current control is the DQ synchronous controller, where the regulated currents are DC quantities. This eliminates steady-state errors precisely.

Basically for successful implementation is need to have information about currents and voltages I_{Lf1} (Iphase in *Fig. 2*), I_{Lf1} (I_{phase} in *Fig. 2*), V_{DC} (or V_{C1}), V_{C2} and V_{Cf} (*Vphase* in *Fig. 2*).

The proposed structure contains transformation of the measured voltage and currents to $\alpha\beta$ which is easily implemented by FIFO memory. Further, the transformation to synchronous coordinates follows where angle of rotation is taken from required output line frequency ω^*. Due to the stabilization of converters cooperation is recommended to maintain one converter in voltage mode (for that the proposed structure is valid) and second converter in current mode (proposed structure without voltage regulators). This strategy gives us naturally constant power operation.

Because for $\alpha\beta$ transformation is used simple time shift by the help of FIFO memory feature there is no information about i_β for the first $\pi/2$ interval. This situation makes the system unstable during start-up operation. The unstable behavior can be easily resolved by slow down current regulators which means choose lower current regulator's bandwidth for example. Very often the design of PI regulators is not for specific bandwidth however IMC technique [2] implements that successfully. In term of bandwidth of current regulators there is a good practice to choose same for both implemented converters.

Another aspect which is not incorporated to the structure in *Fig. 2* is voltage capacitor balancing. It is clear that balancing of the particular voltages across C_1 and C_2 must be same for smooth converter operation. This problem is solved by adding to the actual current value is an offset signal proportional to the capacitor's voltage difference [3]. For this approach one extra voltage sensor needed but there is one cost-effective way uses V_{Cf}. The simple strategy exploits mean value output voltage (V_{Cf}) computation. Because the output line period is known exactly the computation is very simple and accurate. Result mean voltage is then used by adding to the actual current value as offset signal proportional to the capacitor's voltage difference or the signal can affect PWM modulation signal directly.

© VDE VERLAG GMBH · Berlin · Offenbach

PCIM Europe 2016, 10 – 12 May 2016, Nuremberg, Germany

Experimental results

The proposed power topology and control strategy were verified. In *Fig. 3* shows steady state condition of most important output/input quantities such as output and input currents and voltages for respective converters. In our case there are waveforms for half - bridge and T-type topologies. Because one of the converters operates as rectifier ($\varphi_2 = 180°$) and second one as inverter ($\varphi_2 = 0°$) the current should by exactly opposite but *Fig.3* shows inverted current of the rectifier for better comparison.

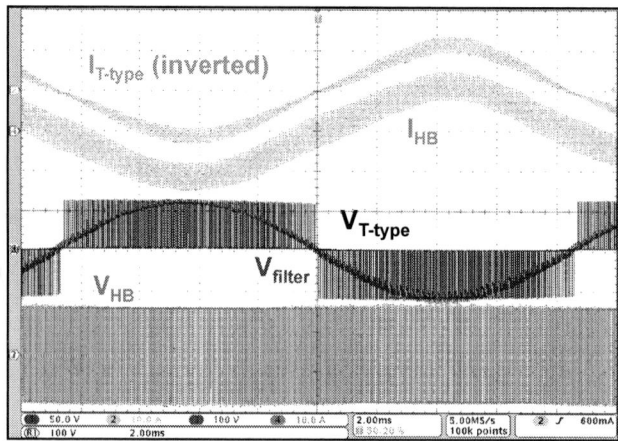

Fig.3. Output currents and voltages for both tested topologies

Fig. 3 shows immediately, online and at the same time that standard half bridge produces only two levels of output voltage which means: (a) high *dv/dt* stresses passive and active components, (b) high *dv/dt* produces high switching loss, (c) high *dv/dt* makes gate drive more difficult, (d) voltage pattern produces higher ripple current and high *dv/dt* produces higher EMI in comparison with three-level T-type topology.

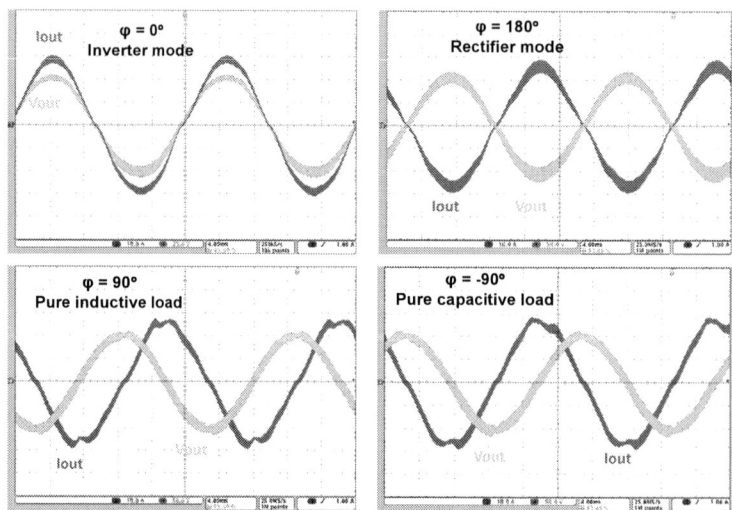

Fig.4. The proposed concept makes possible to control any phase displacement angle, voltage and/or current

© VDE VERLAG GMBH · Berlin · Offenbach

PCIM Europe 2016, 10 – 12 May 2016, Nuremberg, Germany

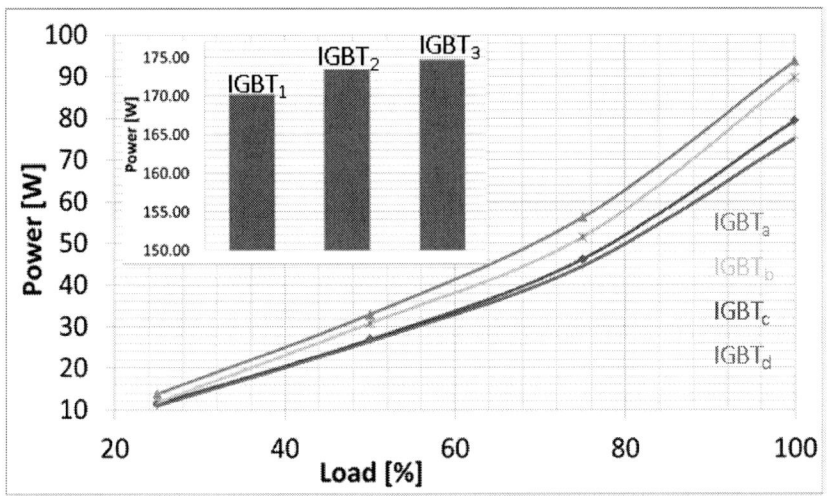

Fig.5. Power loss covered by external source represents total power loss inside the system

Another experimental verification of the proposed system can be seen in the *Fig. 4*. The waveforms show output voltage (measured on the capacitor C_f) and current (measured in series with inductance L_{f1}). Scope snapshots number 1 and 2 show inverter and rectifier mode respectively for T-type converter. As can be seen for case 1 the power is positive and for case 2 negative. The equivalent load has a purely resistive character so the reactive power is zero. Scope snapshots number 3 and 4 show pure inductive and capacitive load where the active power is zero and converter is loaded by reactive power only. The pictures show us four totally different extreme operation modes which mean that the all power devices or parts of a developing topology can be tested in four distinct conditions in that case.

Fig.6. Practical realization of the proposed system

© VDE VERLAG GMBH · Berlin · Offenbach

As mentioned above an external power source covers total power loss generated inside the recuperation system. This power loss can be used for relative power loss evaluation of the power devices, topologies or/and passive components such as inductors and capacitors. The common results from evaluation new IGBT technologies are shown in *Fig. 5*. The figure contains power loss for different levels of load and one chart for maximum power level. Because of the external power source is covering only power loss and in addition the input voltage/current are DC the power measurement is very accurate in relative point of view.

The practical realization of the proposed system is shown in *Fig. 6*. The depicted system is ready to test PIMs or discrete IGBTs in half-bridge (discrete) and T-type power module. The depicted system is developed as modular system to easily accommodate various customer requirements.

Conclusion

The paper presents a system level bench topology for evaluation of any power scheme (T-type, Vienna I-type or half-bridge for instance) or any semiconductor technology such as IGBTs, MOSFETs or PIMs. Proposed strategy makes possible to operate with any displacement angle ($cos(\varphi)$), with very wide range of switching frequencies and independent voltage and current control which means constant current power (reactive or/and active) regulation which is important for power components evaluation. Besides that, it brings us very cost effective and precise relative power loss measurement. To get these advantages, it is important use the most modern control methods such us Virtual Flux Oriented Control with time shifting methods in the case single phase system. The developed algorithm operates in closed loop for precise evaluation, tests, measurements, condition setting and handling. By applying minor rule of thumb to the control algorithm, very high power evaluations can be handled at low operations and equipment costs. This paper demonstrates these benefits through a lot of experimental results and analysis.

References

[1] Remus Teodorescu, Marco Liserre, Pedro Rodríguez. January 2011. *Grid Converters for Photovoltaic and Wind Power Systems*. 1st edition. John Wiley & Sons.

[2] P Brandstetter, T Krecek, *Speed and current control of permanent magnet synchronous motor drive using IMC controllers*, Advances in Electrical and Computer Engineering Volume 12, Number 4, 2012

[3] Rashid, H. M.. *Power Electronics Handbook*. Academic Press, 2001. ISBN 0125816502.

Electrolytic Capacitor Age Estimation Using PRBS-based Techniques

David A. Hewitt*, James E. Green, Jonathan N. Davidson, Martin P. Foster, David A. Stone

Department of Electronic and Electrical Engineering, University of Sheffield, Mappin Street,

Sheffield, S1 3JD, United Kingdom

*Corresponding author. Email: David.Hewitt@sheffield.ac.uk

Abstract

Electrolytic capacitors form a major part of most power electronic converters. System failures which can be directly attributed to electrolytic capacitors account for a particularly large proportion. In normal service, rather than failing instantly, capacitors tend to deteriorate through reduced capacitance and increased equivalent series resistance. Early detection of these changes permits mitigating action to be taken prior to failure. In this paper, a method of characterising these parameters using a pseudorandom binary sequence (PRBS) system identification approach is proposed as part of the development of a prognostic system for passive components in power electronic systems.

1. Ageing and failure of capacitors

Capacitors are an important component in most power converters and drives and therefore their reliability has a considerable influence on the overall system reliability. The failure of electrolytic capacitors accounts for a greater proportion of total failures than any other component in a power converter [1]. An accurate and easy-to-implement method of determining the state of health of a capacitor is of value as it allows the dynamic loading on a failing capacitor to be reduced, extending its remaining lifetime. Alternatively, preventative maintenance can be performed prior to failure, for example, through replacing aged capacitors. The system we propose requires a method of characterising the real-time capacitance and equivalent series resistance (ESR) of a capacitor, allowing its state of health to be determined. In this paper, we propose a pseudorandom binary sequence (PRBS) based system which can be used to identify the critical ageing parameters of an electrolytic capacitor which could be used to estimate their state of health.

An electrolytic capacitor has two failure states: open- or short-circuit. Before reaching one of these states, its performance will deteriorate. This deterioration is manifested as a reduction in capacitance and an increase in ESR. The mechanisms which drive both of these changes are the loss of electrolyte through evaporation [2] and chemical changes within the electrolyte [3]. A standard approach is to assume that that a capacitor should be considered 'failed' for practical purposes if the capacitance has fallen by 15 - 20 % from its value when new, or when its ESR has increased by 200 – 300% [3].

To develop an understanding of the ageing process and the trends involved, sample capacitors were placed into an environment chamber and aged at 115 °C. During this time the ESR and capacitance were calculated from an impedance measurement at 1 kHz. The results of this work can be observed in Fig. 1 where the measured data has been extrapolated to determine the likely progression of the capacitor parameters. The results are normalised to the values measured when new and the failure criteria for the capacitance and ESR value are shown.

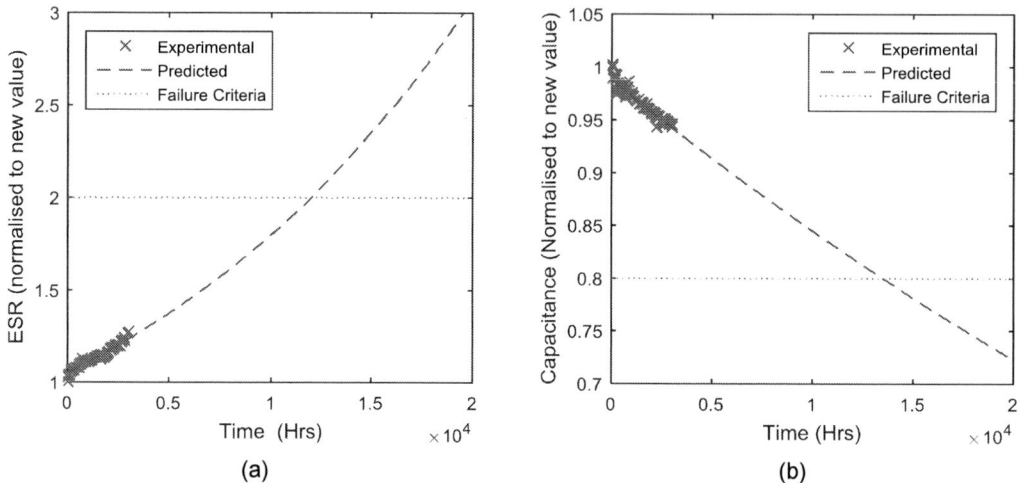

(a)   (b)

Fig. 1 – Experimental ageing values for capacitor measured at 115 °C and 1 kHz. (a) ESR values; (b) capacitance values

The rate of ageing for a capacitor is proportional to the operating temperature [2] and the heat generation within an electrolytic capacitor is related to the power losses within it. The power loss can be calculated using (1) [4], where P is the power dissipated within the capacitor; I_R is the RMS ripple current and ESR is the equivalent series resistance of the capacitor at the ripple frequency.

$$P = I_R{}^2 ESR \qquad (1)$$

Consequently, as the capacitor ages the power losses within it increase due to the increase in ESR. This leads to an increased operating temperature within the capacitor, further expediting the ageing. A useful prognostic system would detect the rise in ESR and therefore reduce the magnitude of the ripple current counteracting the increase in P by reducing I_R.

The ESR and capacitance are temperature-dependent. The temperature dependency of these values, measured at 1 kHz, can be seen in Fig. 2. It is important that any system designed to predict the age of a capacitor possesses some method of accounting for the temperature dependency. From Fig. 2(a) the optimum ESR is approximately 80 °C.

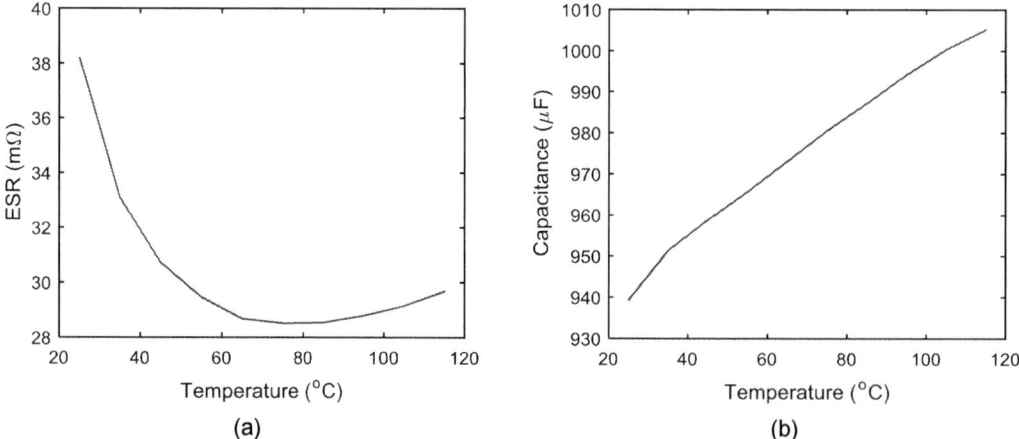

(a)   (b)

Fig. 2 – Temperature dependency of capacitor properties (measured at 1 kHz). (a) ESR with respect to temperature; (b) capacitance with respect to temperature

2. Pseudorandom binary sequences (PRBS)

2.1. PRBS-based parameter estimation

PRBS signals can be used to excite a system with band-limited white noise. By measuring the system response, its impedance spectrum can be calculated. Similar approaches have been used for a variety of different applications including the thermal characterisation of systems [5]; parameter estimation in electrical generators [6] and the modelling of batteries [7]. Using this technique it is possible to determine the impedance of a system over a pre-selected range of frequencies using a single test sequence.

An advantage of using PRBS techniques is that they are amenable to 'on-line' measurements, aiding the implementation of this prognostics into a functional drive system.

2.2. PRBS generation

A maximum length sequence PRBS can be generated using linear feedback shift registers. The sequence length (N) and clock rate (f_c) of a PRBS sequence are determined by the desired bandwidth and frequency limits of the measurements. A longer sequence length will increase the usable bandwidth of the signal, but will also take an increased length of time to complete. The clock rate of the system determines the bandwidth covered. By way of example, a 4-bit PRBS generator is illustrated in Fig. 3(a), with its corresponding PRBS signal (generated with a 1 Hz clock frequency) shown in Fig. 3(b). It is possible to determine the bandwidth of a PRBS system by using (2) and (3) [5]; where n is the bit length of the shift register and BW is the bandwidth (with f_1 being the minimum frequency and f_2 being the maximum). In the example shown here $n = 4$, therefore $N = 15$ and the bandwidth of the system is 0.368 Hz ($f_1 = 0.067$ Hz, $f_2 = 0.435$ Hz).

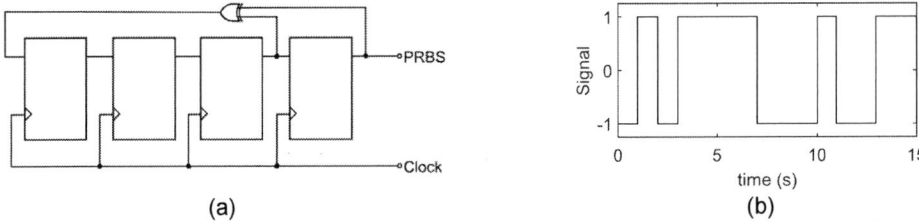

| (a) | (b) |

Fig. 3 – (a) Example 4-bit PRBS generator constructed using shift registers; (b) Example 4 bit PRBS sequence

$$N = 2^n - 1 \qquad (2)$$

$$BW = f_2 - f_1, \qquad f_1 = \frac{f_c}{N}, \qquad f_2 = \frac{f_c}{2.3} \qquad (3)$$

2.3. Use of PRBS to detect capacitor deterioration

To obtain the impedance spectrum of a capacitor, a bipolar current waveform in the form of a PRBS will be applied and the resulting voltage waveform on the capacitor terminals measured. The impedance spectrum is generated by taking the quotient of the Fourier transforms of the voltage and current. The whole process is summarised in Fig. 4

PCIM Europe 2016, 10 – 12 May 2016, Nuremberg, Germany

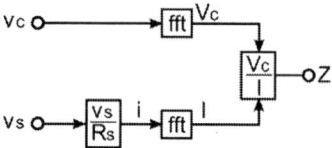

Fig. 4 – Schematic of data processing steps to obtain component impedance from PRBS data

A Simulink model of the PRBS capacitor tester using the configuration shown in Fig. 5 was used to validate the measurements obtained using PRBS. Here the voltage across the capacitor is denoted as v_c and the current is calculated from the voltage across a current sense resistor R_s.

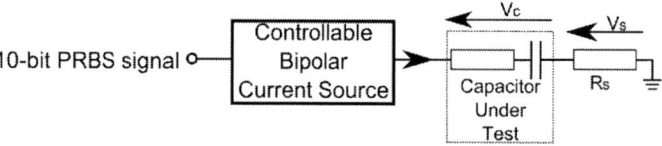

Fig. 5 - Schematic of test procedure

Simulation results are shown in Fig. 6 and were obtained from the Simulink model which employed a 10-bit PRBS sequence and a clock frequency of 100 kHz. For the purpose of comparison, theoretical impedances, obtained using (4), are also included in the figure. The results obtained by using PRBS are in close agreement with the theoretical values. Here the values selected for the 'New' capacitor are those measured for the capacitor which will be used later in the paper to test the practical system and the 'aged' values are obtained by multiplying the 'New' capacitance value by 0.8 and the ESR by 2, mimicking the failure criteria discussed previously. This is done to illustrate the fact that the changes in these parameters during ageing are measurable using this technique.

$$Z = \sqrt{\left(\frac{1}{2\pi f C}\right)^2 + R^2}$$

(4)

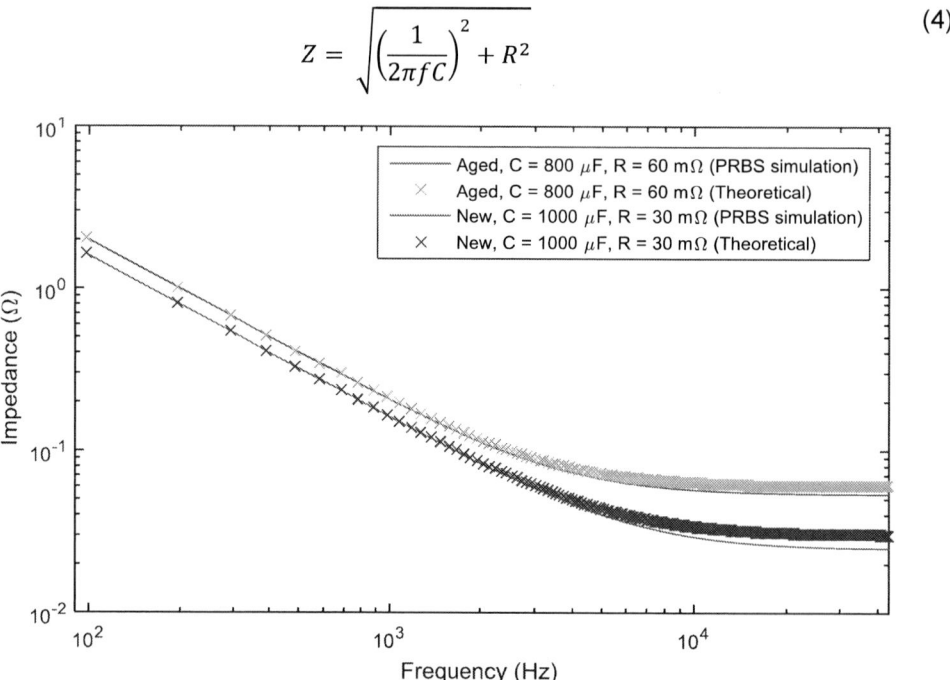

Fig. 6 - Simulated frequency spectrum for capacitor excited by 10-bit PRBS signal for aged and new capacitors compared to theoretical impedances

© VDE VERLAG GMBH · Berlin · Offenbach

Clearly the new and aged capacitors have a significant difference in impedance and so the impedance could be used to identify the failure of a capacitor. The complex impedance is obtained using this technique and therefore it is possible extract the resistive and reactive values and determine the ESR and capacitance values. This this technique will be demonstrated later in this paper, when the experimental results are considered.

3. Practical implementation of system

In the development of a practical system in order for a PRBS system to be effective there are three factors which must be considered. These are:

- the PRBS clock frequency;
- the PRBS bit length;
- the magnitude of the driving current.

These parameters are interconnected and depend on the frequency range of interest. As f_2 is determined solely by the clock frequency, the clock frequency should be selected so as to satisfy the upper bandwidth requirements of the system. Once this upper frequency has been set, the number of bits within the PRBS sequence can be selected to determine lower bandwidth frequency of the system.

Excitation current level within a practical system is determined by a trade-off between two parameters:

- The maximum voltage the current source can provide
- The minimum voltage the analogue-to-digital converter (ADC) can resolve.

It should also be noted that the capacitor voltage will vary proportionally to the magnitude of the current used, meaning that the signal to noise ratio of the measurements will be smaller for a smaller excitation current.

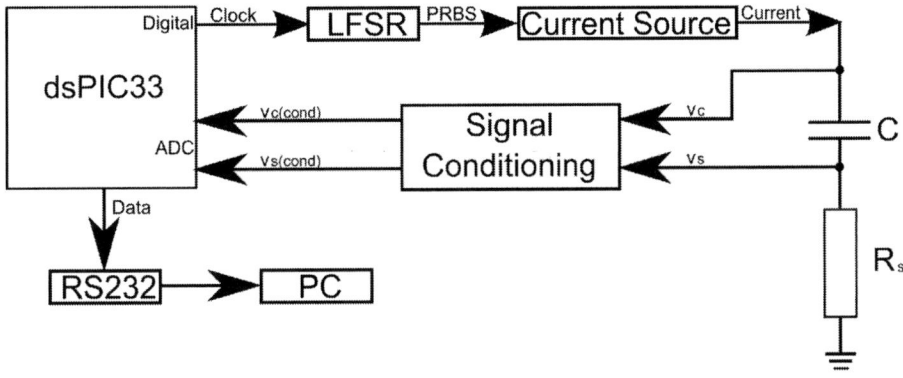

Fig. 7 – Block diagram of system implementation

A block diagram of the PRBS capacitor characterisation system is illustrated in Fig. 7. This system is based around a dsPIC33 microcontroller. This chip incorporates 4 sample-and-hold buffers and a 10-bit ADC. The availability of multiple sample-and-hold buffers is particularly useful for this application as it allows the two voltage signals to be sampled simultaneously thereby removing any ambiguities associate with time delays and phasing of signals. A 10-bit PRBS is generated using a discrete linear-feedback shift register (LFSR) driven by a clock generated by the dsPIC (f_c = 20 kHz). The current source provides a PRBS signal amplitude of +/- 1A. The dsPIC ADC is designed to operate with inputs within its power supply rails (0 - 3.3 V). The current source is operated from a +/- 10 V supply however. To ensure that that the ADC is not damaged by excessive voltage a signal conditioning circuit scales the voltage

© VDE VERLAG GMBH · Berlin · Offenbach

appearing across the device under test, in response to the PRBS current signal, to 0 – 3.3 V. The ADC data is sent via RS232 to a PC where the data is processed to obtain the capacitor impedance.

4. Experimental results

To evaluate the performance of the PRBS system described previously a 1000 μF capacitor manufactured by *Forever* was selected as a test device. The capacitance of the test capacitor was measured using an LCR meter at 1 kHz. C = 1062 μF and ESR = 40 mΩ. The impedance plot obtained from PRBS testing this capacitor is shown in Fig. 8. The impedance shows a good correlation to the simulation results. Fig. 9 shows plots of the ESR and capacitance obtained by considering the real and imaginary components of the impedance using (5) and (6).

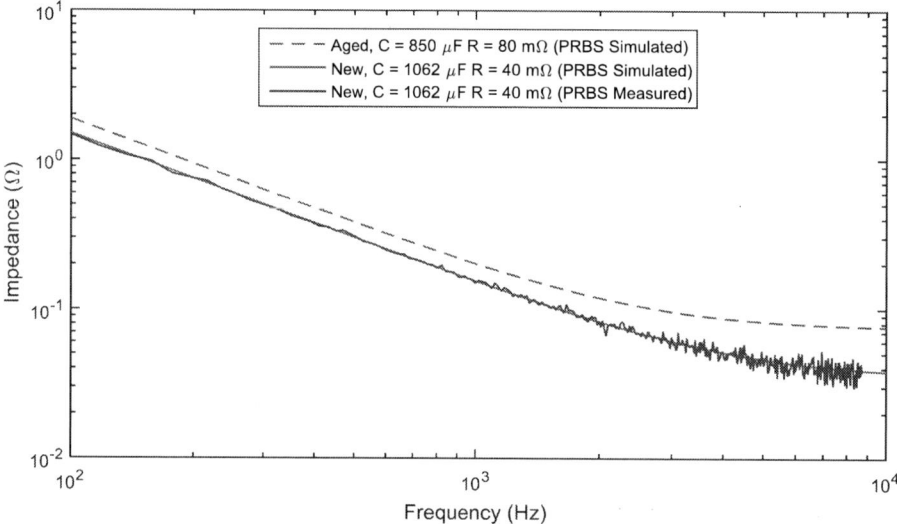

Fig. 8 – Experimentally obtained impedance of a 1062 μF capacitor with an ESR of 40 mΩ compared to new simulated results and aged simulated results measured at 25 °C

$$ESR = \Re(Z) \tag{5}$$

$$C = -\frac{1}{2\pi f\,\Im(Z)} \tag{6}$$

The PRBS results show reasonable agreement with the 'New' capacitor simulation results with the following exceptions. Firstly, the experimentally obtained ESR values shown here are higher than the predicted values for the capacitor at frequencies below 500 Hz. In simulation the ESR is assumed constant with no frequency dependence; this is not the case, especially at lower frequencies. Secondly, the capacitance data is noisy at frequencies above 2 kHz, meaning that distinguishing between a new and aged capacitor is difficult. The increase in noise is due to reactance becoming smaller as frequency increases, resulting in a falling signal to noise ratio with increasing frequency. Considering both the ESR and capacitance together, the most suitable frequency at which both parameters can be determined is approximately 1 kHz. If the system were to be used to characterise capacitors of different values the optimum frequency will vary. This highlights the importance of tuning the PRBS system parameters to ensure that it is possible to suitably measure the capacitor under test over its entire ageing range.

© VDE VERLAG GMBH · Berlin · Offenbach

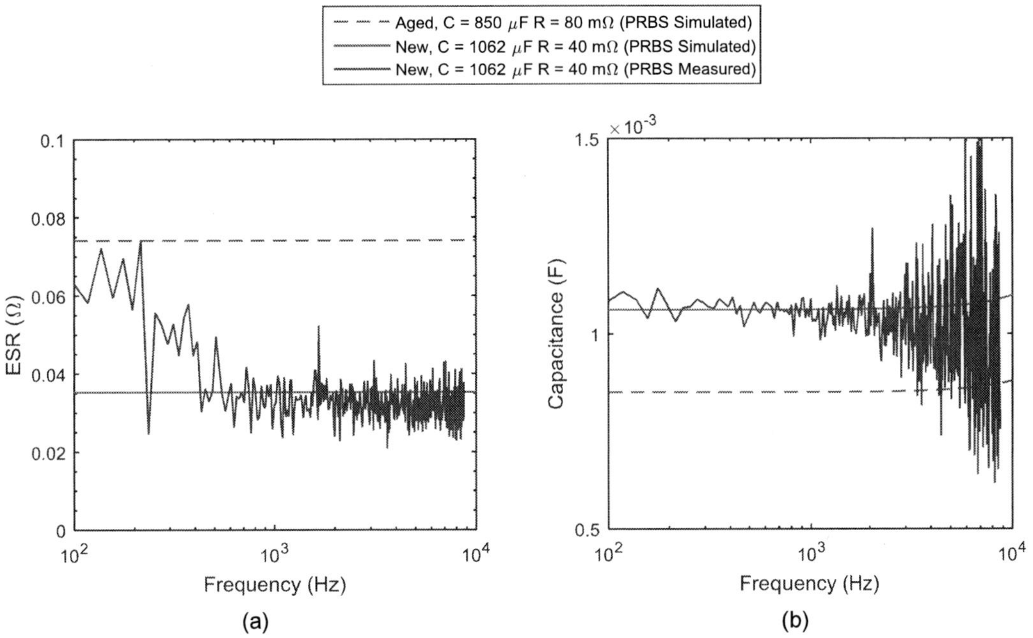

Fig. 9 – ESR and capacitance values extracted from impedance results. (a) ESR against frequency; (b) Capacitance against frequency

5. Conclusions

This paper proposed a PRBS-based method to measure the equivalent series resistance and bulk capacitance of electrolytic capacitors for state-of-health prediction. The PRBS technique generates impedance data over a range of frequencies from which the ESR and bulk capacitance are generated. The technique has been demonstrated in simulation and for practical devices, with both exhibiting good correlation with direct measurement of the parameters.

The influence of capacitor temperature on the ESR and capacitance was also highlighted, stressing the importance of any system which determines capacitor age to account for measurement temperature. By comparing up-to-date ESR and bulk capacitance at a given temperature to nominal values, it is possible to determine the age of the capacitor. Therefore its remaining useful life can be estimated.

References

[1] Department of Defence, Reliability prediction of electronic equipment, Military Handbook (MIL-HDBK-217F), Washington DC: US Department of Defence, 1995.

[2] K. Harada, A. Katsuki and M. Fujiwara, "Use of ESR for deterioration diagnosis of electrolytic capacitor," *Power Electronics, IEEE Transactions on,* vol. 8, no. 4, pp. 355-361, 1993.

[3] Emerson Network Power, "Capacitors age and capacitors have an end of life," Emerson Network Power, Columbus, 2008.

[4] Nichicon Corporation, "Technical Notes on Aluminum Electrolytic Capacitors /Cat. 8101E," Nichicon Corporation.

[5] J. Davidson, D. Stone, M. Foster and D. Gladwin, "Improved Bandwidth and Noise Resilience in Thermal Impedance Spectroscopy by Mixing PRBS Signals," *Power*

Electronics, IEEE Transactions on, vol. 29, no. 9, pp. 4817 - 4828, 2014.

[6] A. Saavedra-Montes and J. Ramirez-Scarpetta, "Identification of excitation systems with the generator online," *Electric Power Systems Research,* vol. 87, pp. 1 - 9, 2012.

[7] A. Fairweather, M. Foster and D. Stone, "Modelling of VRLA batteries over operational," *Journal of Power Sources,* vol. 207, pp. 56-59, 2012.

On-line monitoring for diagnosis on traction transformer for rolling stock application

André-Philippe CHAMARET[1], Toufann CHAUDHURI[2],
[1] SNCF Centre d'Ingénierie du Matériel, 4 allée des gémeaux, Le Mans, France
[2] ABB SECHERON, 4-6 rue des Sablières, 1217, Genève, Switzerland

Abstract

An on-board traction transformer with online monitoring systems for predictive and diagnosis maintenance is presented. Today, no tele diagnosis functions are connected with this component on rolling stock. Maintenance operations are made according to the suppliers' advice and return of experience. In power transformer area, the monitoring on the transformer and its accessories is essential, due to criticality for the exploitation in case of major failures. That's why we studied the possibility and the interest to monitor traction transformer for online diagnosis, to simplify the maintenance and to improve the availability of the transformer and also on the traction drive.

1. "French Intelligent Traction Transformer" Project description

1.1. Why work on IOT for Traction Transformer?

The on-board traction transformer is one of the main components of the traction drive for a rolling stock. Its function is to convert energy from the pantograph to the power electronics converter for traction and auxiliaries.

Its cost is very important, especially due to raw materials for the production (copper, tank, magnetic core, dielectric oil, etc.). So, the risks of failure need to be controlled to minimize the maintenance costs for the railway operator. For example, an on-board traction transformer cost on an EMU for regional trains or mass transit is close to 200k€ and can be multiplied by two for a TGV high speed train. Today, the fleet of traction transformers in SNCF's rolling stocks is close to 4500 units.

That's why the interest to use the digitalization on traction transformers to optimize the costs and the availability of the rolling stock is quite interesting for railway operators. In 2015, SNCF launched the first studies for the R&D project "FITT" for "French Intelligent Traction Transformer"

First, to improve the maintenance of traction transformers, it's important to begin by a list of t he traction transformers maintenance operations to be realized during a full life on rolling sto cks.
Traction transformer maintenance is divided according different levels:
- Light maintenance : realized in daily maintenance workshop
    o Silica gel substitution for the air dryer
    o Visual checks (oil leakages, radiator clogging, etc.)
    o Dielectric oil measurement (voltage breakdown, acidity, moisture)
    o Pressure relief valve control
    o Oil circulation detector control
    o Temperature sensors control (PT100, thermometer)
- Medium maintenance : realized in daily maintenance workshop

- o DGA (Dissolved Gas Analysis)
- o Auxiliary motors measurement (vibrations levels)
- Heavy maintenance : realized in specialized workshop for midlife control or after major failures
  - o Gaskets substitution
  - o Dielectric oil replacement
  - o Auxiliary motors replacement (oil pump and fans)
  - o Visual inspection of the active part and cleaning
  - o Routine test

Then, one of the major parts in this project was to analyze the root cause of the failures for a traction transformer and to determine the criticality.

Function	Failure type	Method control	Risks
Tank	Cracks	Visual check	Oil leakage
Air dryer	Humidity => silica saturation	Visual check	Moisture, dielectric rigidity breakdown
HV Bushing	Cracks	Visual check	explosion of the bushing
Dielectric oil	Humidity, oil overheated, production of gas	PT100, dielectric rigidity tester, oil sample send to external laboratoty	Insulation failure, cooling performance reduction, burning
Electrical connections	Looseness	Visual check	overvoltage, burning
Radiator	Pollution, cracks	Visual check	Oil leakage, low performance
Valves	Wrong position of the valve	Visual check	No oil circulation
Oil pump	Motor failure, bearing	vibrations test + test bench	No oil circulation
Fans	Motor failure, bearing	vibrations test + test bench	No cooling
Oil level	Oil level under the limit	Visual check	Major failure of the transformer (dielectric insulation)
Safety relief valve	No opening of the contact	Visual check	Major failure of the transformer without protection (overpressure inside the transformer)

Fig 1: Table of potential failures for on-board traction transformer

According to this list and the return of experience on the field, thanks to maintenance supervision, we decided to focus on:
- Dielectric oil analysis
- Auxiliary motors
- Oil leakages

1.2. State of the art for online monitoring on transformer

In fact, a lot of information about monitoring and diagnosis have been brought from the power and distribution transformer area. The experience in this domain is today very consistent with a high number of potential solutions available on the market (DGA system with one gas or more, bushing monitoring, electrical parameters, etc.). The analysis of the different components and a comparison list of advantages / inconvenients have been established.

It's easy to define electrical default as one of the major failures. This type of failure can cause important troubles on the transformer (thermal rise, mechanical movement, etc.). The dielectric oil inside the transformer plays the role of insulation and cooling for the active part. As dielectric oil is an important parameter to control, we decided to install a Dissolved Gas Analyzer system. The dielectric oil can be compared to the blood in a human body. The interest to analyze the chemical aspects of the oil, , is to follow the evolution of the different type of failures. Each type of failure will produce a common gas (H_2) and more specific gases (CO_2, C_2H_4, etc.).

CO_2: carbon dioxide → overheated solid insulation

H_2: hydrogen → general failures

C_2H_2: acetylene → arcing

C_2H_4: ethylene → overheated oil

Etc.

We can also monitor the humidity inside the oil. The interest of this control is to have a "picture" of the dielectric rigidity. The H_2O level (in ppm) inside the oil will have an important impact on the dielectric rigidity, especially for mineral and silicone oils. For ester oil, the relation is more complex, due to the specificity of the ester oil to accept a large proportion of water without any hazardous phenomena. Water appears also after non maintenance on the transformer, as silica gel replacement or leaks. The water has an important impact on the ageing of the solid insulation (H_2O stored inside paper insulation) and creates moisture. [1]

The monitoring of the dielectric breakdown strength of the oil is more complex. The conventional method uses 2 electrodes of bronze or steel, distant of 2,5mm. A voltage is applied and grows up until the apparition of an electrical discharge. This operation is realized 6 times to obtain an average value of dielectric breakdown. The problem is, this test will be destructive for the oil sample, and so, it's impossible to use this method for online motoring on a transformer. Whereas, another method developed by WEIDMANN, can give an online measurement of the time before voltage breakdown. This information can offer an interesting approach to determine the gap until maintenance operations without risks of dielectric insulation failures of the oil.

The other parameter to control is the auxiliary motor of the oil pump or of the fans.

The objective is to determine the performance of the auxiliary motor, to evaluate the risks of failures. A default on the auxiliary motor for an oil pump or on fans can be catastrophic:

- For oil pump: the cooling of the transformer is off, so a rise of the temperature inside the transformer will reach the thermal limits and protect the transformer (no authorization for the traction drive).
- For fans: the cooling of the transformer decreases and the thermal performances are reduced.

To control the oil pump, we could monitor vibrations and/or current on the auxiliary motor directly connected to the auxiliary motor.

Thanks to this system, we can monitor:

- Operating hours
- Energy consumption
- Loading
- Vibrations levels
- Etc.

The vibrations level analyses give a frequency spectrum of the motor. With these analyses, we can observe on different ranges if a trouble happens on the motors (bearing, rotor, stator, etc.).

The example bellow shown:

- C1: Cage frequency
- F0: Rotation frequency
- E1: External Bearing Ring frequency
- B1: Ball-bearing frequency
- I1: Internal Bearing Ring frequency

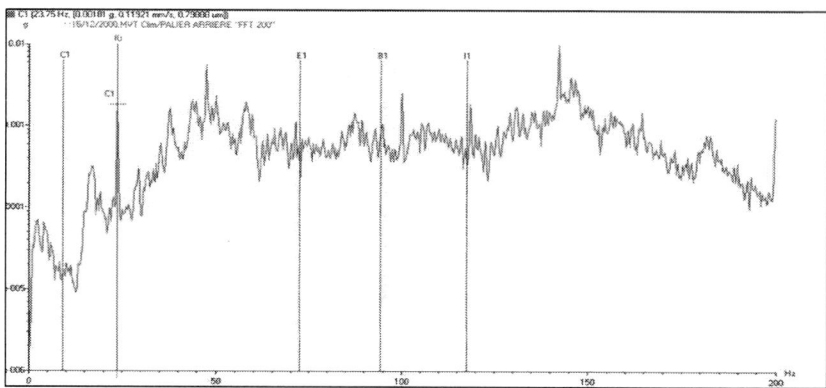

Fig 2: Spectrum of the different frequency of the bearings in an auxiliary motor

This sensor is especially interesting because you can obtain these data from the auxiliary motors, without loss of time due to problems of accessibility you could have when placing sensor into the motors.

Moreover, even if this study concerns auxiliary motors of the traction transformer, the results could be used for others auxiliary motors on trains (air compressor, doors, air cooling system, etc.). On recent EMU rolling stocks, we can find more than 200 auxiliary motors per train set.

With these two solutions, we will analyze and predict the main scope of potential failures on a traction transformer. The goal is to reduce the cost of maintenance around 50% and to avoid 75% of the major failures, as expected for power transformer sector. [3]

2. Tests and case of study with "TGV 603"

2.1. Selection of the sensors and integration on train

To realize a first proof of concept, we decided to install a dissolved gas analyzer system on the traction transformer, in the power car of a TGV high speed train. The tests are performed on a TGV with on-board Internet equipment, to connect the different system together.

For this prototype, we decided to test 3 systems:
- Mono gas H2 and moisture system in the oil (DGA system)
- Auxiliary motor analyzer system
- Oil leakage cable detection

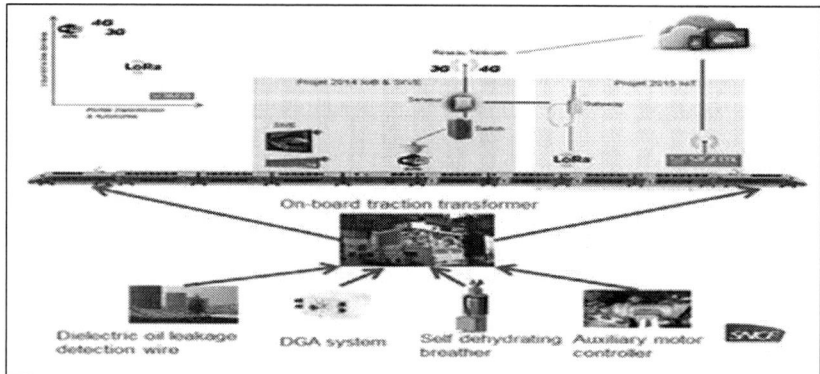

Fig 3:   Scheme of integration of the sensors in the train

The DGA system is installed on the traction transformer and connected to the tank by a valve. The DGA system is linked to a communication box with SIM card and a 3G connection. Thanks to this connection, the different data will be collected and transferred to an on-ground maintenance computer.

We selected a mono gas DGA system for our application because, as explained in the previous part, the hydrogen is the only gas to be monitored for covering all the scope of failure types. The other gases offer possibilities to have a better view on the different failure types, but, the monitoring system is far more complex (sensors, mass and volume, etc.) and expensive by the same time. So, the hydrogen monitoring for on-board traction transformer application represents the best compromise for this prototype.

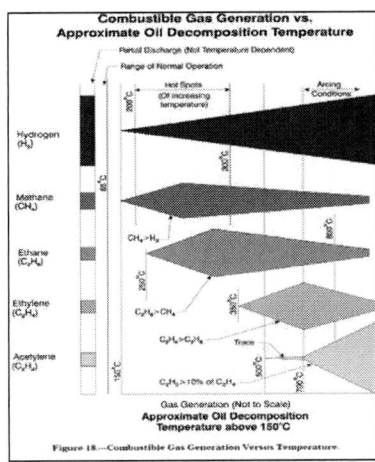

Fig 4: graph of combustible gas generation vs approximate oil decomposition temperature

For the auxiliary motors monitoring, we decided to test a prototype of monitoring system dedicated to low voltage motors. This system, mounted on the auxiliary motor transmits the data collected from the motor to a smartphone by Bluetooth communication. With a dedicated application, we can observe the evolution of the measurements and check if limits we can program have been exceeded.

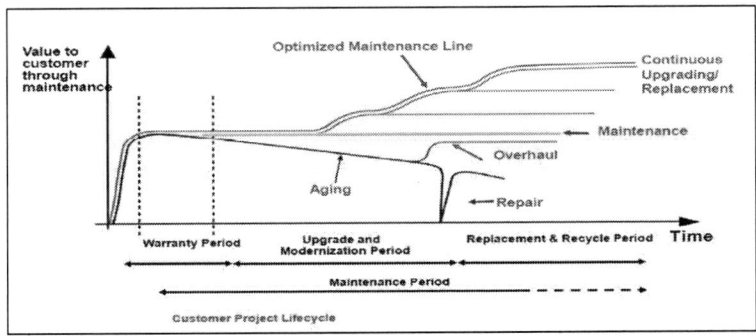

Fig 5: Life Cost Cycle of an electrical motor according to different scenarii

The auxiliary motor monitoring system can monitor different parameters according to values registered in the system. The parameters are:
-   Current
-   Vibrations

- Time
- Power
- Temperature

For the oil leakage detection, one of the major failures during the life of an on-board traction transformer, we decided to focus on the oil leak position and alert. The traction transformer, especially for high speed train TGV, is installed in the power car. Its volume is important and protections are mounted on the transformer for the security of maintenance operators circulating in the power car to repair technical components. So, some parts of the transformer are not very accessible, and the visual check to control the potentials leakages on the transformer can take a lot of time.

Moreover, today, the oil level of a traction transformer is only controlled by the visual operation of a maintenance operator. This operation is realized regularly, but unfortunately, it's too late and the oil level in the transformer is probably critical.

To monitor this failure type, we selected a flexible cable, able to detect an oil leakage of the transformer. This cable, installed on the different critical parts of the transformer (welding area, valves, bushing, etc.), can be linked with other cables. This structure offers the possibility to define different "zones" on the transformer (Radiator 1, Cover, Valve 2, etc.).

When an oil leakage appears, the oil drops on the detection cable and will modify the impedance of the cable. This modification of the impedance will create an alarm in the supervision system of the cables linked to it. The alarm information is then sent by MODBUS to the on-ground maintenance computer, which gives the vision and the localization of the alert.

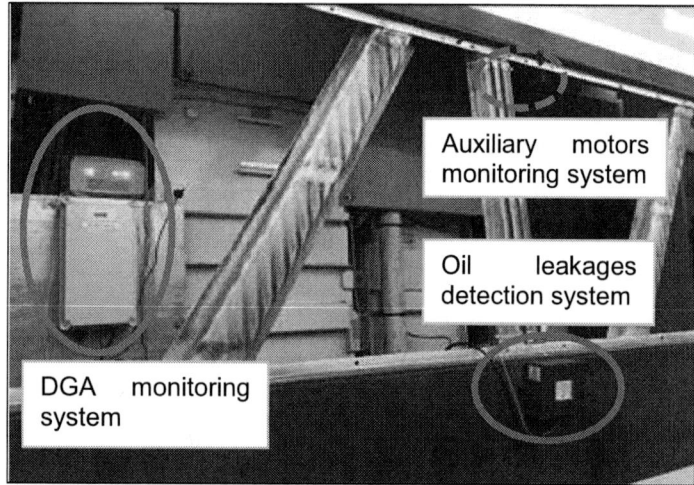

Fig 6: Integration of the different systems on the traction transformer in the power car of TGV 603

2.2. Tests and Results on TGV 603

To realize a first test of integration of the systems on a traction transformer on train, a high speed train, TGV 603, was modified with an internet box. Thanks to this box, the IOT can be connected and data are sent to the on-ground maintenance computers.

The big difficulties was first to install the systems on the traction transformer. For the DGA monitoring system, we had to mount the sensor close to the transformer to connect it with the tank. The system is the same as used for power transformer application, so, with no constraints of volume and mass for the installation.

Moreover, an additional box with 3G modem was necessary to share the data with the application of the manufacturer on the laptop. The computer analyses the data to have a live record of the hydrogen and moisture inside the transformer. Thanks to limits defined according our return of experience on this transformer series, we can supervise the evolution and predict potential failures.

For the auxiliary motor monitoring system, the installation was easier, due to the space available around the motor to install the sensor. The system was powered by a battery and used a Bluetooth connection to send the data to a smartphone.

Finally, for the oil leakage cable detector, a central system was installed on the transformer to obtain the link with the different cables. Four cables had been installed on the transformer radiator 1, bottom valve 1, welding of the cover of the main tank and bottom pipe 1.

A beacon was necessary too, to transfer the data in MODBUS to the on-ground maintenance homepage on computer.

Fig. 7: On-ground maintenance computers and smartphone

3. Conclusion

To conclude this first part of R&D project "FITT", a first step of validation of the different monitoring systems on the traction transformer has been realized. The results are interesting to predict the risks of failures. This new method of maintenance supervision will offer more capabilities to have a greater reliability and more trains available for the exploitation.

The second step of this project will be to perform running tests on the tracks and see if all the conditions are conform to our requests. This step will be also the possibility to design new products with the specifics requests of on-board rolling stock applications (vibrations, fire & smoke, volume and mass, costs). These new prototypes are probably the future of the traction transformer, but can also fit on existing rolling stocks, if the possibilities of integration and costs savings are clever.

Direct-water-cooled next High Power Density Dual (nHPD^2) considering inverter layout

Keisuke Horiuchi [1], Yuichiro Konishi [1], Mutsuhiro Mori [1], Daisuke Kawase [2], Katsuaki Saito [3]

[1] Hitachi, Ltd. Research & Development Group, Japan, keisuke.horiuchi.oa@hitachi.com
[2] Hitachi Power Semiconductor Device, Ltd., Japan
[3] Hitachi Europe Ltd. Power Device Division, United Kingdom

Abstract

Three types of cooling methods were compared for Hitachi's latest low inductive housing 2in1 module called "next High Power Density Dual (nHPD^2)" considering inverter layout. The first is an air-cooled type, the second is a conventional water-cooled type (called 'indirect-water-cooled' type), and the third is a water-cooled type without using thermal grease (called 'direct-water-cooled' type). The trade-off curve between thermal resistance and pressure drop was smallest when the direct-water-cooled nHPD^2 was used. The thermal resistance of direct-water-cooled type was half which resulted in electric current output doubling comparing to indirect-water-cooled type.

1. Introduction

The potential advantages of Hitachi's latest housing power module called next Power High Power Density Dual (nHPD^2) (designed to cover from 1.2 to 6.5 kV) with low stray inductance have been known for years [1-3]. This technological trend occurred because reducing stray inductance results in minimising the induced surge voltage and oscillation during high-speed switching of next-generation semiconductor devices. On the other hand, many researchers have reported that the direct-water-cooled power module enables both the module size and thermal resistance to be reduced since it eliminates the need for thermal grease [4-6]. However, few studies have compared the direct-water-cooled, indirect-water-cooled, and air-cooled power modules from the viewpoint of inverter layout. In this paper, we studied the parallel use of the direct-water-cooled nHPD^2 considering inverter layout and clarified its benefits.

2. Direct-water-cooled module for high voltage application

2.1. Reduction in both thermal resistance and pressure drop

Figure 1 shows the outline of the direct-water-cooled type. This module has an integrated base formed the pinfin heatsink. Note that the footprint size of this newer power module is compatible with already released standard indirect-water-cooled power modules. The open windows on top of the housing are manufactured by casting along with water channel, and the windows are covered with the flange space on the backside of the module base, as shown in Fig.1. The pinfin configuration is optimised to provide relatively small pressure drop and thermal resistance. An O-ring is used to seal the flange space of the pinfin heatsink. The housing has the same footprint size for the two other modules used for this particular study. To the best of our knowledge, the bolt pitch of [direct-water cooled power modules with nHPD^2 is the longest among other direct-water cooled power modules, so there should be no water leakage after careful sealant design.

PCIM Europe 2016, 10 – 12 May 2016, Nuremberg, Germany

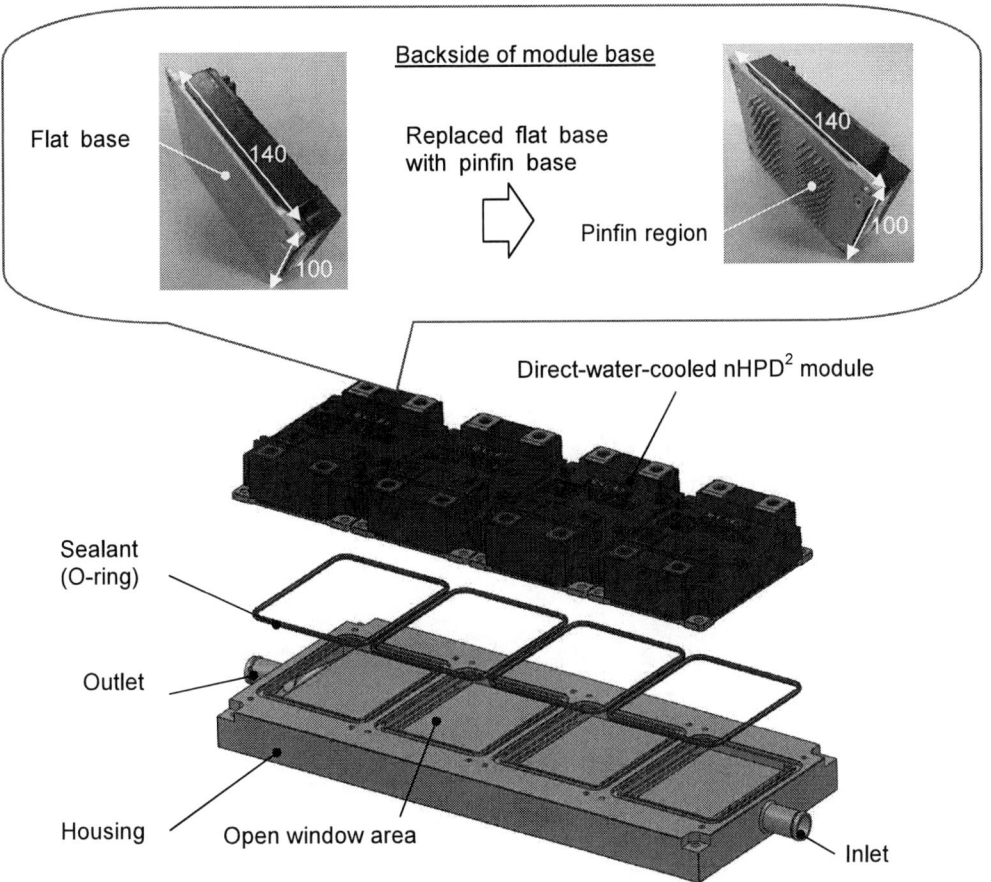

Fig. 1   Exploded view of direct-water-cooled nHPD^2 power module

2.2.   Delivering both high sealing reliability and high power density

Figure 2 represents Hitachi's previous direct-water-cooled power module called ECOBLOC for 1.7-kV application [5]. It sustains high water flow pressure of about 500 kPa because there are enough bolts on the base plate. On the other hand, it is challenging to increase the power density by increasing the size of dies since several bolts obstruct the die mounting area, especially at the centre of the module.

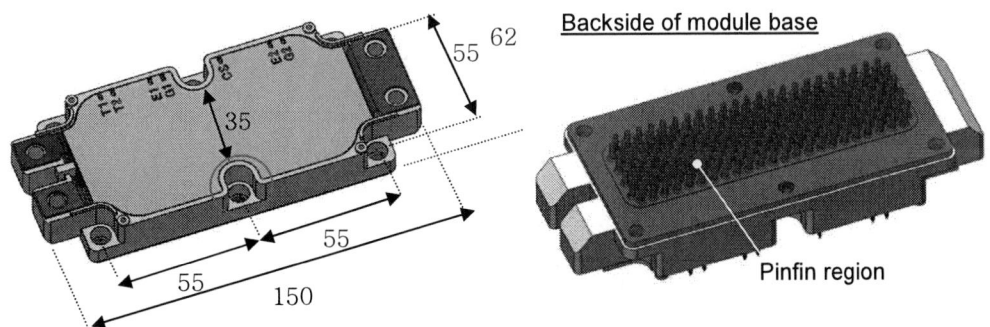

Fig. 2 Hitachi's previous direct-water-cooled power module called ECOBLOC for 1.7-kV application [5]

© VDE VERLAG GMBH · Berlin · Offenbach

We solved the above problem of delivering both high sealing reliability and high power density by optimizing the O-ring size and location as follows. Figure 3 is a schematic of the important parameters considering sealing design. We first need to determine the smallest compression ratio (denoted as "E") of the O-ring, which is typically 8% for flat-face-type seal [7]. The compression allowance can be calculated as

$$\sigma = E \times d - (H + \alpha),\qquad(1)$$

where d is the diameter of the O-ring, H is the depth of the O-ring groove, and α is machining tolerance. In practice, we need to take the initial warpage (denoted as "β") into consideration, so the allowable displacement, γ, is

$$\gamma = \sigma - \beta = E \times d - (H + \alpha) - \beta.\qquad(2)$$

We calculated displacement under various water-pressure conditions. For instance, if the required water pressure is 600 kPa, then the target allowable displacement should be larger than 0.3 mm. Table 1 lists γ corresponding to d. In Table 1, the flange width (a) and distance values from the die edges (b) are also listed. Figure 4 indicates the relationship between the ratio of thermal resistance increase (Rjw/Rjw_b = 7) and distance b, where "j" and "w" stands for junction and water, respectively. We can see that a preferable design area exists within $b > -2.0$ mm. From the above analysis, we achieved a sealing area as well as die mounting area at the same time to satisfy both high sealing reliability and high power density.

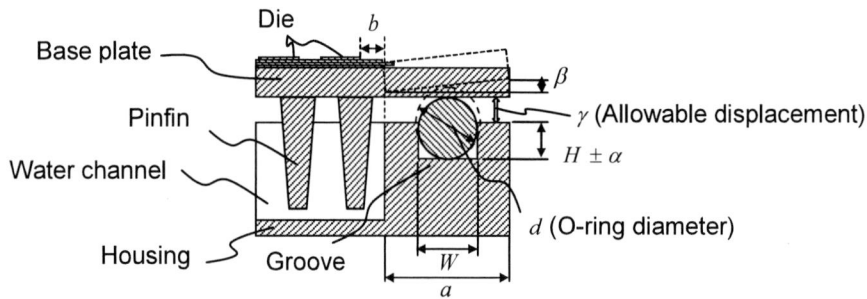

Fig. 3   Schematic of important parameters considering sealing design

Table 1 List of allowable displacements (γ) and distances from die edges (b) corresponding to several O-ring diameters (d).

O-ring diameter	Groove width	Groove depth	Machining tolerance	Initial warpage	Allowable displacement	Flange width	Distance from die edge
d	W	H	α	β	γ	a	b
[mm]	[mm]	[mm]	[mm]	[mm]	[mm]	[mm]	[mm]
1.9	2.5	1.4	0.1	0.1	0.14	6.5	6.7
2.4	3.2	1.8	0.1	0.1	0.20	7.2	0.2
3.5	4.7	2.7	0.1	0.1	0.32	8.7	-0.5
5.7	7.5	4.6	0.1	0.1	0.44	11.5	-2.0

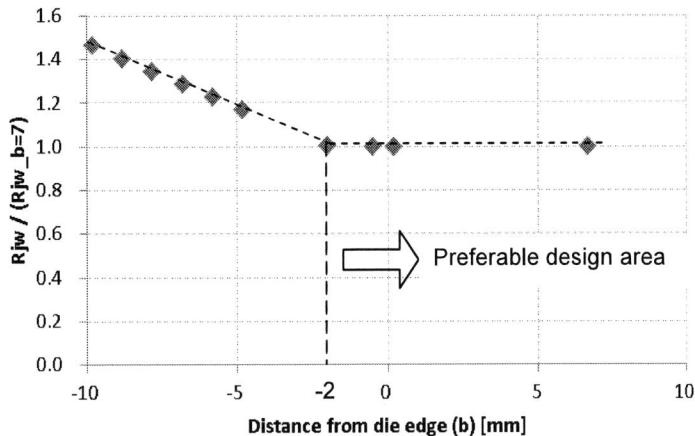

Fig. 4 Ratio of thermal resistance increase corresponding to distance from die edge (b)

3. Consideration of inverter layout

One of the advantages of nHPD^2 is its flexible scalability. Depending on the total electric current output required, the number of parallel modules can be varied properly. Table 2 lists four scenarios (possible module arrangements considering inverter electric circuit and coolant flow direction). For our study, we placed four modules in parallel and shared the same AC busbar (denoted as "U, V, and W" in the table). Note that the adjoining modules of the same AC phase current were paralleled in the direction of the module short side. By doing this, the gate driver substrate can be mounted on the four modules, and the main circuit inductance can be minimised by sharing the PN busbar of all modules over a short distance. Table 3 defines the symbols corresponding to several cooling methods and the inverter layout, results of which are shown in Fig. 6. Indirect-water-cooled type signifies one that has a conventional aluminium water housing that consists of fine pitch plate fins and grease (Shinetsu, G747) with a thermal conductivity of 0.9 W/(mK) .

We first calculated the thermal resistance of the insulated gate bipolar transistor (IGBT) and pressure drop for each scenario in Table 2. For the water-cooled type, the total volume flow rate of 40 L/min and the coolant condition was ethylene glycol solution (50%) at an inlet liquid temperature of 65 degrees C. With regard to the air-cooled type, the total air flow rate and inlet air temperature were assumed to be 4800 L/min and 40 degrees C. Note that the total flow rate was divided by the number of flow channels. For example, each module was cooled at a flow rate of 10(=40/4) L/min for Scenario #3 since there were four flow channels.

To experimentally verify cooling performance (thermal resistance and pressure drop), we fabricated a single module mounted on a small water housing, which is defined by the rectangular region in Fig. 5(a). Both indirect-water-cooled and direct-water-cooled nHPD^2 modules were evaluated using the setup shown in Figs. 5(b) and (c). The experimental results in Fig. 5, drawn from the calculation results for all scenarios in Table 2, are depicted in Fig .6. According to this graph, the direct-water-cooled module provided smaller thermal resistance and pressure drop than the other modules in all scenarios. The experimental results also suggested slightly higher pressure drop and smaller thermal resistance compared to the calculation results, but it is still predictable within an error of 20%.

Finally, we analysed both the IGBT thermal resistance shown in Fig.7 and diode thermal resistance. The interaction between IGBT and diode and coolant temperature increase at the outlet was also taken into account. Figure 7 compares the root mean square (RMS) phase current (I_{ce}) ratios under the same inverter operating conditions. The results suggest that the direct-water-cooled module produced twice as much current output as the conventional indirect-water-cooled module which enabled reduction in number of paralleled modules.

© VDE VERLAG GMBH · Berlin · Offenbach

Table 2 Possible inverter layout scenarios using nHPD^2 and definition of symbols used in Fig. 6

Scenario #1	Scenario #2	Scenario #3	Scenario #4
One serial water channel (40 L/min)	12 series water channels (3.33 L/min per channel)	4 series water channels (10 L/min per channel)	3 series water channels (13.33 L/min per channel)

Table 3 Symbols corresponding to cooling methods and the inverter layout used in Fig. 6

Cooling methods	Symbols used in Fig.6			
	Scenario #1	Scenario #2	Scenario #3	Scenario #4
Direct-water-cooled type	■	●	▲	◆
Indirect-water-cooled type	□	○	△	◇
Air-cooled type	▥	◍	◬	◈

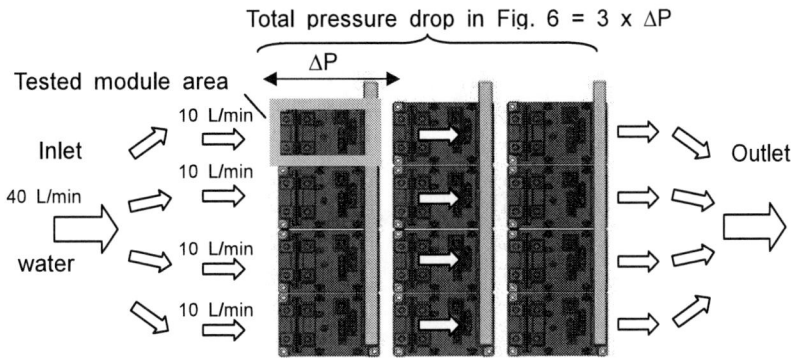

(a) Tested module area in Scenario #3

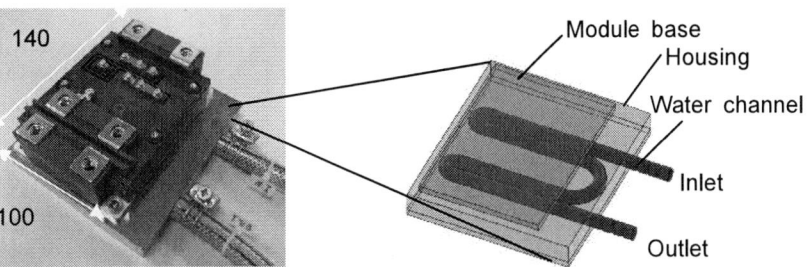

(b) Experimental setup for evaluating indirect-water-cooled nHPD^2 module

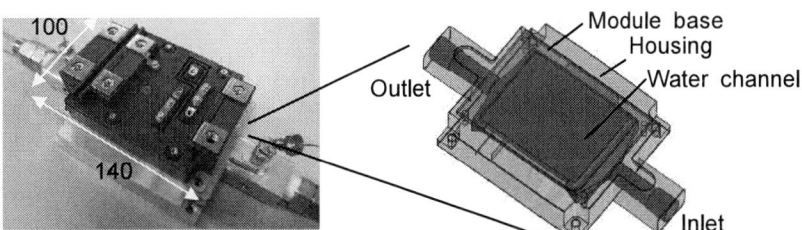

(c) Experimental setup for evaluating direct-water-cooled nHPD^2 module

Fig. 5 Experimental apparatus to validate cooling performance

PCIM Europe 2016, 10 – 12 May 2016, Nuremberg, Germany

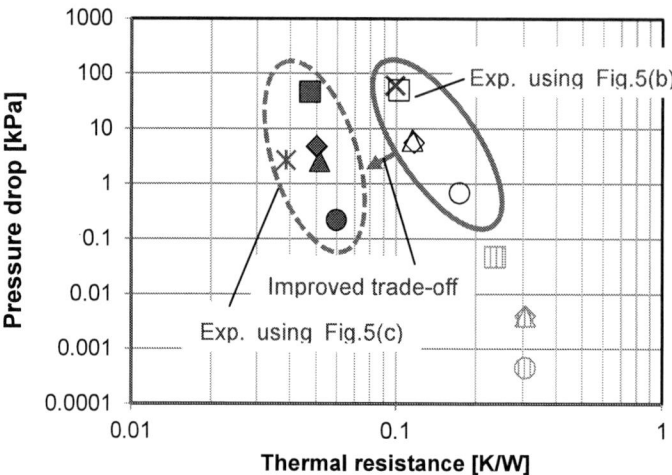

Fig. 6   Thermal resistance and pressure drop for each cooling method in Table 2 and Table 3

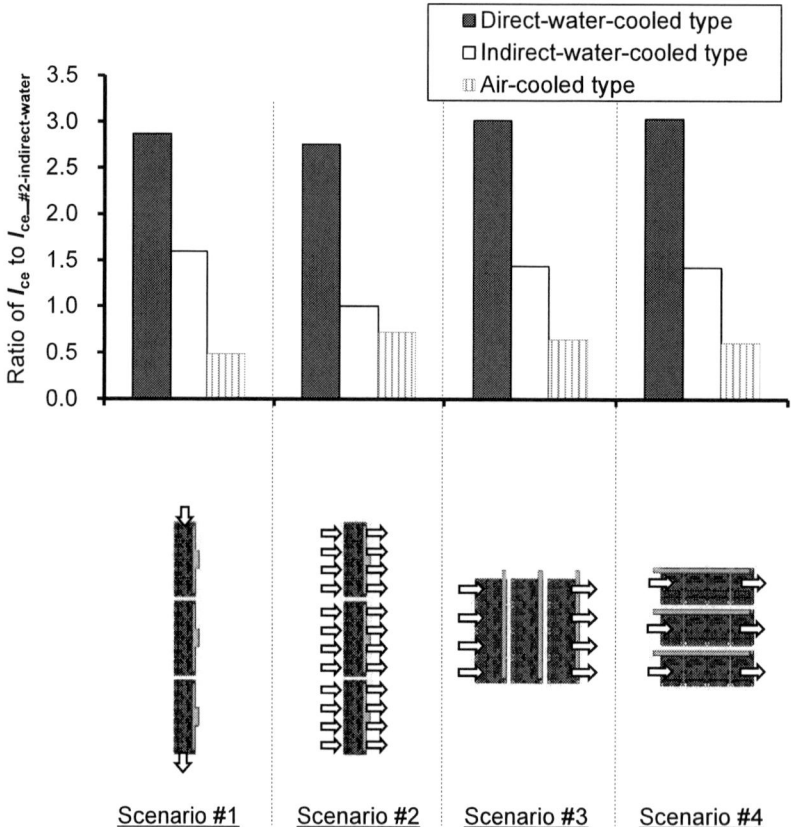

Fig. 7   RMS phase current ratio under conditions of junction temperature = 150 degrees C,
DC Voltage = 1500 V, fundamental frequency = 50 Hz, PWM switching frequency = 575 Hz,
modulation = 0.75, and power factor = 0.75 for each scenario

© VDE VERLAG GMBH · Berlin · Offenbach

1116

4. Conclusions

We developed a direct-water-cooled nHPD^2 module, which provides smaller thermal res istance and pressure drop than other cooling modules in all scenarios. The experiment al results suggested slightly higher pressure drop and smaller thermal resistance comp aring to the calculation results, but they are still predictable within an error of 20%. W e analysed both IGBT thermal resistance and diode thermal resistance. The interaction between IGBT and diode and coolant temperature increase at the outlet was also tak en into account and the RMS phase current ratios under the same inverter operating c onditions were compared. The results suggested that the direct-water-cooled type produ ced twice as much current output as the conventional indirect-water-cooled type in all module scenarios. Therefore, both the number of paralleled modules and the footprint size of the resulting inverter can be reduced in case of direct-water-cooed type.

5. References

[1] D. Kawase et al., High voltage Module with low internal Inductance for next Chip Genera-tion – next High Power Density Dual (nHPD^2), PCIM Europe 2015

[2] R. Schnell et al., LinPak, a new low inductive Phase-Leg IGBT Module with easy parallel-ing for high Power density Converter Designs, PCIM Europe 2015

[3] G. Borghoff, Implementation of low inductive strip line concept for symmetric switching in a new high power module, PCIM Europe 2013

[4] T. Kurosu et al., Packaging technologies of direct-cooled power module, IPEC 2010

[5] K. Sasaki et al., Small Size, Low Thermal Resistance and High Reliability Packaging Technologies of IGBT Module for Wind Power Applications, PCIM Europe 2010

[6] K. Horiuchi et al., Advanced Direct-water-cool Power Module having Pinfin Heatsink with Low Pressure Drop and High Heat Transfer, ISPSD 2013

[7] NOK Corporation, O-ring brochure, http://kythuataba.vn/wp-content/uploads/2015/12/O-ring-NOK.pdf

Heat Pipes used as Heat Flux Transformers and for Remote Heat Rejection

Devin Pellicone, Jens Weyant

Advanced Cooling Technologies, Inc, Lancaster, PA USA, info@1-ACT.com

Abstract

Heat pipes are commonly used as a tool to improve thermal performance when conduction through solid metal (aluminum and/or copper) alone is unable transfer heat within the temperature drop budget. Heat pipes can be embedded into aluminum spreaders to significantly improve thermal performance by accepting the high heat flux generated by the electronics and spreading it to a level that can be managed by the ultimate heat sink. Heat pipes can also be used to transfer heat to remote fin stacks when geometry and environmental constraints do not allow for local heat rejection. These applications of heat pipes are becoming more valuable as electronics become more powerful and compact packaging becomes a primary design goal.

1. Background

High performance electronics are increasing the amount of waste heat generated and the trend toward smaller packages amplifies the need for thermal solutions capable of rejecting high heat fluxes. Reliable operation and optimum performance depend on a thermal management solution capable of meeting system demands. Typical heat spreaders are made from aluminum or copper since they are low cost, easily machined and have high thermal conductivities of 180 W/m-K and 380 W/m-K, respectively. As heat flux requirements increase, conduction gradients become the largest contributor in the thermal resistance network. Heat pipes can be used as a cost effective enhancement to traditional spreaders without adding significant weight or volume.

1.1 Heat Pipes

Heat pipes transport heat by two phase flow of a working fluid [1,2]. Shown in Fig. 1, a heat pipe is a vacuum tight device consisting of a working fluid and a wick structure. The heat input vaporizes the liquid working fluid inside of the wick in the evaporator section. The vapor, carrying the latent heat of vaporization, flows towards the cooler condenser section. In the condenser, the vapor condenses and gives up its latent heat. The condensed liquid returns to the evaporator through the wick structure by capillary action. The vapor space of the heat pipe is at saturated conditions, so the temperature difference within the vapor space is driven be the pressure difference between the evaporator and condenser ends of the heat pipe. This means that the end-to-end heat pipe temperature difference is driven largely by the conduction losses through the pipe wall which is typically on the order of a few degrees Celsius [3,4].

Copper/water heat pipes are standard for electronics cooling. Water and copper are known to be compatible for long term operation. In the temperature range of typical electronics cooling environments (25°C to 125°C), water has the best combination of physical properties (surface tension, latent heat, viscosity, etc.) for heat pipe performance. Copper also has the highest thermal conductivity of any engineering metal, making it ideal for heat transfer applications. Copper's flexibility makes it ideal for conforming to different desirable geometries, allowing them to be bent (at a bend radius as tight as 3x the pipe O.D.) and flattened (up to 2/3 of the pipe O.D.) to conform to flat input surfaces or avoid structures.

Since heat pipes are essentially isothermal along their length, their effective thermal conductivity can range from 10,000 to 200,000 W/m-K [5].

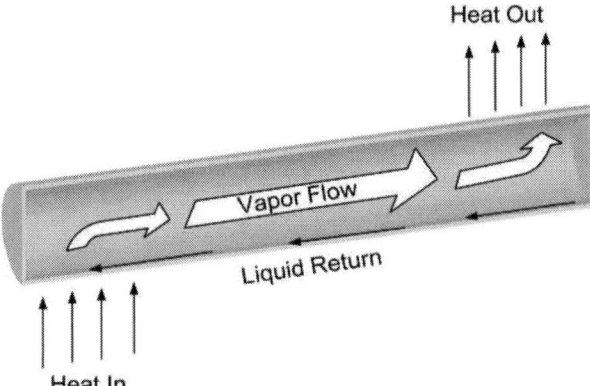

Fig. 1. The operation of a heat pipe is illustrated. Heat applied to one end of the heat pipe evaporates liquid off a wick. The vapor carrying its heat of vaporization moves toward the colder end of the heat pipe where it condenses. The wick returns fluid to the evaporator.

1.2 Heat Pipe Application Guidelines

A properly designed heat pipe will exhibit the following properties:

- Adverse gravity operation
- Freeze/thaw tolerance
- Shock/vibration tolerance

Operation in an adverse gravity orientation refers to any configuration in which the heat input area of the heat pipe is above the heat rejection area with respect to gravity. In this orientation the liquid return must overcome gravitational forces proportional to the vertical height difference between the evaporator and condenser sections. This is achieved by utilizing capillary action through a porous wick structure.

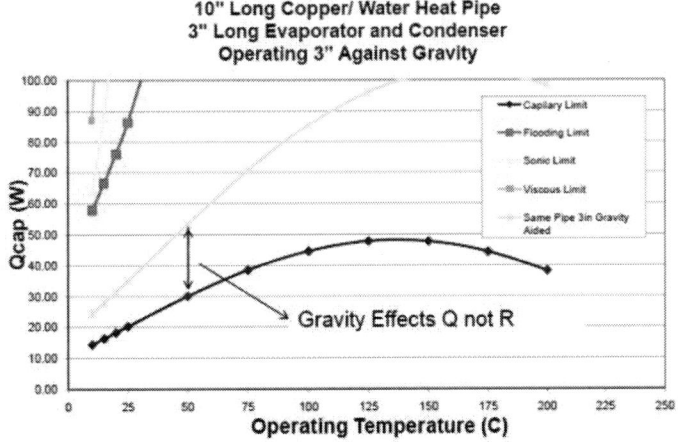

Fig. 2. Performance curves for a nominal heat pipe operating horizontal and against gravity show a difference in the maximum power handling capability. However, the thermal resistance of the heat pipe is unchanged and the expected DT across the pipe will be identical.

When the wick structure is designed properly, the heat pipe will continue to operate at the same thermal resistance (R) regardless of the orientation. However, it should be noted that adverse gravity orientations do affect the maximum amount of heat (Q) that the heat pipe can carry, as seen in Fig. 2. This simply means that when designing a heat pipe for a particular application the maximum power and worst case orientation must be considered. If the pipe is designed properly to perform at the worst case power and orientation it is virtually guaranteed to perform equally or better at all other conditions.

A common misconception about heat pipes that utilize water as the working fluid is that they are limited to applications in which the environment never drops below freezing (i.e. >0°C). The concern in this situation is that the water inside of the heat pipe will expand during freezing and result in bulging or complete failure of the heat pipe envelope. This is not actually the case when a heat pipe is properly manufactured. The working fluid inside of a heat pipe should fully saturate the wick structure without making a puddle of excess fluid. With the fluid completely contained within the wick, it is not able to bridge the gap across the inside diameter of the heat pipe. This allows multiple freeze thaw cycles to occur without any signs of deformation.

The authors routinely subject heat pipes to thermal cycling and typical freeze thaw tests are conducted from temperatures ranging from -20 to +20°C and -40 to +80°C. A typical cycle for such tests is shown in Fig. 3 which illustrates a thermal ramp rate of about 120°C per hour. The authors have tested heat pipes up to 1,200 cycles, but 50-300 cycles are a more standard practice. Heat pipes may be thermally cycled prior to installation into assemblies or the entire heat pipe assemblies can be thermally cycled as units to assure system level performance.

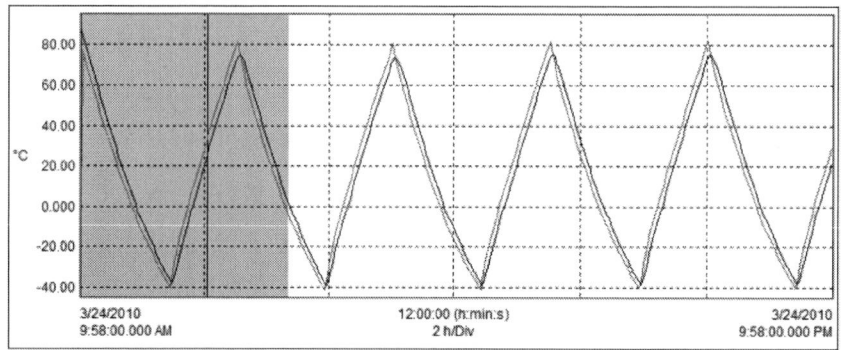

Fig. 3. Typical heat pipe freeze thaw testing criteria showing numerous cycles between 80°C and -40°C at a ramp rate of around 120°C per hour.

Another common concern regarding the use of heat pipes in some applications is in regards to the shock and vibration tolerance of the devices. It is often thought that the passive nature of heat pipes (i.e. there are no moving parts) makes them susceptible to failure under random shock or vibrational loading. In fact, heat pipes exhibit little or no change in performance when exposed to shock loads up to 9,000 lbf and 4,500 lbf of sustained vibration loads. The authors have tested numerous types of heat pipes in shock and vibration environments similar to those presented in Fig. 4 and have witnessed no degradation in thermal performance.

PCIM Europe 2016, 10 – 12 May 2016, Nuremberg, Germany

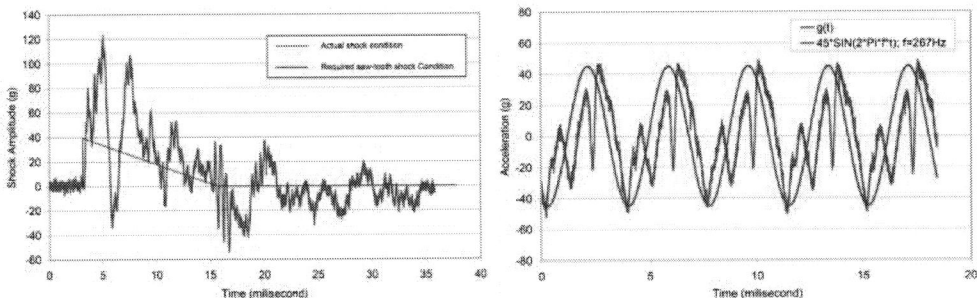

Fig. 4. Typical heat pipe testing curves for Shock Amplitude vs. Time and Acceleration vs. Time.

The reason for the sustained performance under these harsh conditions can be attributed to the wick structure. The working fluid inside of the heat pipe is protected from the external perturbations by the capillary pressure of the porous material. The surface tension forces are strong enough that the shock and vibration loads cannot dislodge the working fluid from the pores which is known as depriming the wick. It is worth noting that if an acceleration vector is sustained in a direction that opposes the flow of the working fluid back to the evaporator section the performance of the heat pipe could be hindered. This is very similar to the effect that gravity has on the operation of a heat pipe and the device would have to be designed accordingly to withstand this operating condition.

2. Remote Heat Rejection

In applications where local heat rejection is limited, heat pipes can be used to transfer heat to a location where there is sufficient volume. A common example is a luminaire design where the ceiling or wall fixtures are based on a pre-existing design using non-LED technologies. These designs commonly have both restricted space for heat dissipation through conduction and limited air flow to remove heat via convection. In cases where there is space to remotely dissipate the heat, heat pipes can be used to transport the heat from the device to a remote heat sink. Fig. 5 shows a photograph and an infrared (IR) image of a heat pipe transporting heat to a remote sink, clearly demonstrating the transport of heat isothermally from the heat source to the heat sink and the even distribution of heat to the heat sink.

Fig. 5. IR image and photograph of remote cooling with a heat pipe embedded radial heat sink. The temperature distribution clearly demonstrates that the heat pipe can transport heat almost isothermally, and then deliver it uniformly to the heat sink.

© VDE VERLAG GMBH · Berlin · Offenbach

Heat pipes can also be bent and manipulated to clearance multiple obstructions in a given assembly as shown in the two examples in Fig. 6. In this case the form factor that must be packaged in between the heat source and heat sink is simply the outer diameter of the heat pipe. This enables more compact packaging options and minimizes the number of components that may need to be displaced when implementing a thermal solution. The ultimate heat sink does not necessarily need to be an air-cooled heat sink. As shown in the figure on the right in Fig. 6, the ultimate heat rejection medium could be a central cold rail that could be cooled by a circulated liquid.

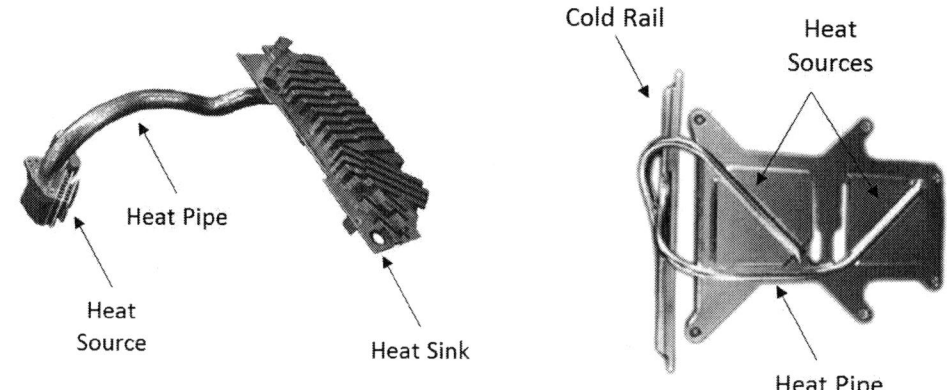

Fig. 6. (left) A heat pipe and heat sink assembly with multiple bends used to remove the waste heat from a thermoelectric device. (right) Heat pipes embedded in aluminum spreader plates and connected to a cold rail for heat rejection.

3. Heat Flux Transformers

3.1 HiK™ Plates

Heat pipes can be embedded into conventional spreaders to take advantage of their superior heat transport capabilities and increase the overall effective thermal conductivity (k). When heat pipes are embedded into aluminum, they are known as HiK™ plates. Heat pipes are typically embedded into aluminum instead of copper for several reasons. First, since the heat pipes are the primary means of heat transfer the copper provides marginal thermal improvement at a significant weight penalty. Secondly, aluminum has superior mechanical properties when compared to copper. These are maintained after heat pipes are installed. Finally, aluminum is a low-cost material that is easily machined and extruded.

Fig. 7 shows a design challenge where electronics mounted to a conventional aluminum heat spreader were overheating. Here large thermal gradients in the heat spreader results in a maximum base plate temperature of 91°C, which is above the 80°C max case temperature levied by many electronics. By embedding heat pipes within the plate the thermal gradient of the spreader was reduced to 70°C. The high heat flux components located in the upper corners of the plate are already located close to the edge of the spreader where heat is rejected. Using heat pipes to spread heat along the edge transformed the heat flux to a manageable level and minimized the thermal bottleneck at the clamping interface. The hot spot located at the bottom center is a result of high conduction gradients. The high effective thermal conductivity of the heat pipes reduced this gradient to maintain a safe base plate temperature. With the design complete the heat pipe solution was fabricated and is shown in the image to the right of Fig. 7.

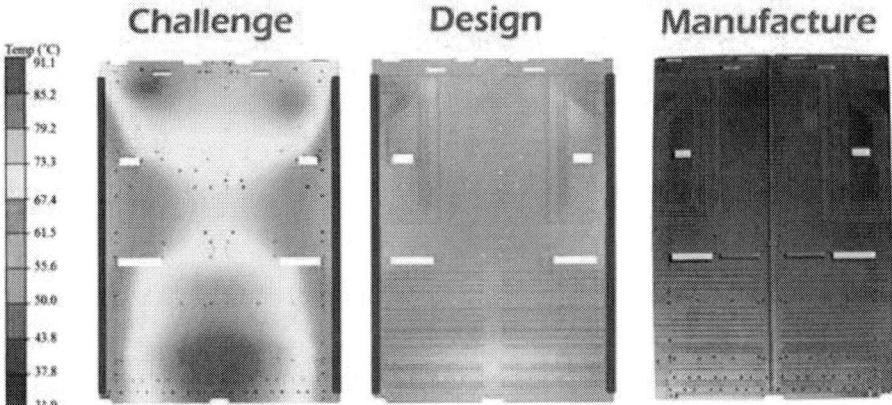

Fig. 7. The thermal model on the left shows the design "Challenge", a heat spreader made of aluminum. The thermal model in the center shows the heat pipe enhanced "Design". The image on the right shows the "Manufactured" heat spreader.

3.2 Vapor Chambers

Vapor chambers are effectively flat heat pipes which operate under the same principles by evaporating and condensing a working fluid inside of a vacuum tight enclosure. The main difference between a heat pipe and vapor chamber is the geometry. Heat pipes are typically cylindrical devices while vapor chambers are flat with parallel evaporator and condenser sections. The benefit to this configuration is that very effective heat spreading can be achieved. As shown in Fig. 8, the vapor spreads to the entire inner volume and condenses over a much larger, cooler surface of the vapor chamber. The condensed liquid is transported back to the heat input area in the wick structure lining the vapor chamber inner wall. In some cases, vapor chambers are referred to as a "heat flux transformers" because of this ability to convert higher heat fluxes into lower heat fluxes. Depending on the design of the wick structure, heat fluxes up to 1000 W/cm^2 can be achieved.

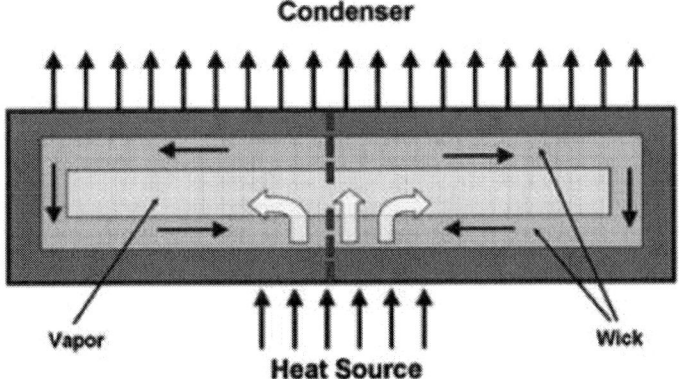

Fig. 8. A vapor chamber is essentially a flat heat pipe where the heat input area is substantially smaller than the heat rejection area. Image courtesy of [6].

The authors have fabricated advanced vapor chambers that are manufactured with low coefficient of thermal expansion (CTE) materials that enable direct solder attachment and high heat flux capability. The advanced wick structure is shown in Fig. 9, which allows for efficient liquid return to the evaporator section while maintaining low thermal resistance. Most vapor chambers also contain an internal support structure to provide rigidity and allow for more robustness (shown in Fig. 9). Typical vapor chamber thicknesses are on the order of 3-5mm which provides highly efficient and compact heat spreading for a wide variety of applications.

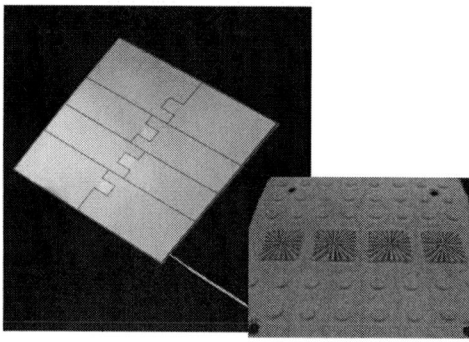 

Fig. 9. (left) CTE-matched vapor chamber with advanced wick structure capable of handling 1000 W/cm^2. (right) Internal support structure to provide structural integrity to the otherwise hollow vapor space.

4. Summary

Heat pipes transfer heat very efficiently and can be used to enhance the performance of conventional heat spreaders without compromising mechanical properties or significantly adding weight. Heat pipes are capable of accepting high heat fluxes and spreading them to a manageable level. Their high effective thermal conductivity allows designers to incorporate heat sinks of sufficient volume remotely when other geometry constraints don't allow for local heat rejection. Heat pipes are freeze/thaw and shock and vibration tolerant and can operate in any orientation when designed appropriately.

5. References

[1] Faghri A., "Heat Pipe Science and Technology", Taylor & Francis, Washington, DC. 1994

[2] Dunn P.D., Reay D.A., "Heat Pipes", Pergamon, Tarrytown, NY. 1994

[3] Collier J.G. Thome J.R., "Convective Boiling and Condensation", Oxford University Press, NY. 1994

[4] Incropera F.P., Dewitt D.P., "Fundamentals of Heat and Mass Transfer", John Wiley & Sons, NY. 2002

[5] Eastman Y.G.; "The Heat Pipe" Scientific American, Vol 218, No 5, pp 38-46, 1968

[6] http://www.qats.com/cms/2010/12/10/vapor-chambers-and-their-use-in-thermal-management-part-2-of-2/

This page intentionally left blank.

New Class of Graphite TIMs provide Performance and Reliability

Prashanth Subramanian, GrafTech International, USA, Prashanth.Subramanian@Graftech.com

Alex Augoustidis, GrafTech International, USA, Alex.Augoustidis@Graftech.com

The Power Point Presentation will be available after the conference.

Abstract

Thermal interface materials (TIMs) have one sole purpose, to displace the air between a power module and the baseplate with a higher thermally conductive material. An ideal thermal interface will minimize thermal impedance between the device and heat sink and without significant degradation in performance for the life of the inverter. Some of the key challenges include reliability and repeatability over the life and between modules while ensuring no significant additions to the processing steps. eGRAF® HITHERM™ HT-C3200 Compressible graphite (HT-C3200) addresses both the initial performance and maintains the same levels over the life of the modules while significantly simplifying the application process.

1. Thermal performance

1.1. Device temperature:

The temperature of the device ($T_{Resistor}$ aka junction temperature) during the test is regarded as the key criteria to determine performance of the TIM solution. Thermal impedance is defined as the opposition to the flow of heat within an assembly. Fig 1. Shows the performance of HT-C3200 in comparison with dry joint (No TIM), eGRAF® Hi-Therm HT-1210, "Competitive Graphite" is a commercially available 100μ thickness graphite solution and "Competitive Grease" is a commercially available and popular silicone based solution and is widely used in the power electronics industry. The HT-C3200 shows comparable performance to grease while clearly outperforming both the HT-1210 and the dry joint. The Temperature delta across the TIM shows the effectiveness of the heat transfer between the case plate and the cold plate and in turn the performance of the TIM. Fig 2. , shows the performance of the thermal distribution across the assembly using HT-C3200. Temperatures were measured at four locations as described below in Fig 3.

PCIM Europe 2016, 10 – 12 May 2016, Nuremberg, Germany

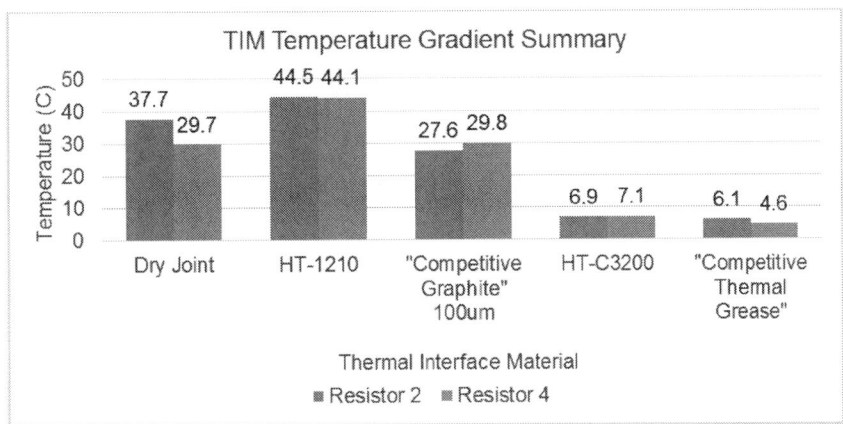

Fig. 1. TIM Temperature gradient across TIM

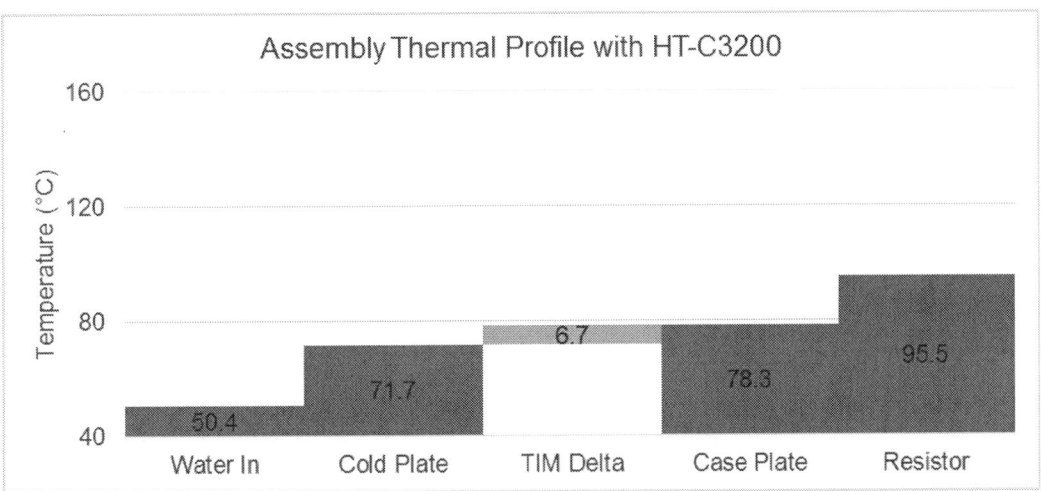

Fig. 2. Temperature gradient across assembly

1.2. Assembly

The application assembly defined in this case are the Resistor (device), TIM 1, the case plate, TIM 2 and the cold plate. The interfacing surfaces of the TIM in the study are the case plate and the cold plate ("contacting surfaces"). A separate assembly following the ASTM D5470 standard was used to measure material property data and is published in the TDS and will not be discussed in this paper. The thermal difference between the Device (Junction temperature) and the cold plate is seen as a representation of the total thermal impedance in the assembly.

© VDE VERLAG GMBH · Berlin · Offenbach

PCIM Europe 2016, 10 – 12 May 2016, Nuremberg, Germany

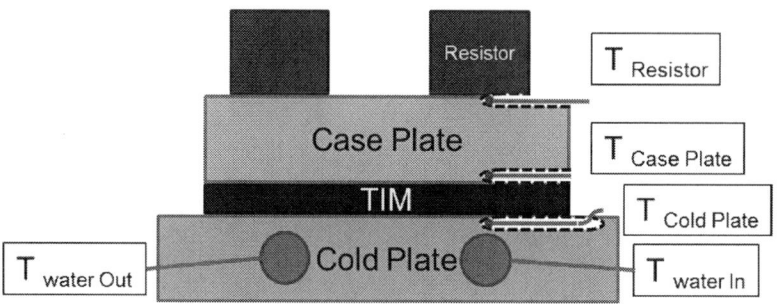

Fig. 3. Temperature measurement locations in the assembly

The final set up consists of four resistors connected to form the final assembly as shown in Fig 4. This mimics a 6 IGBT device module (Infineon FF450R12ME4) [1] on its thermal power and enables tighter control over heat dissipated at or above expected module temperatures. This also enables adding thermocouples to accurately measure temperatures.

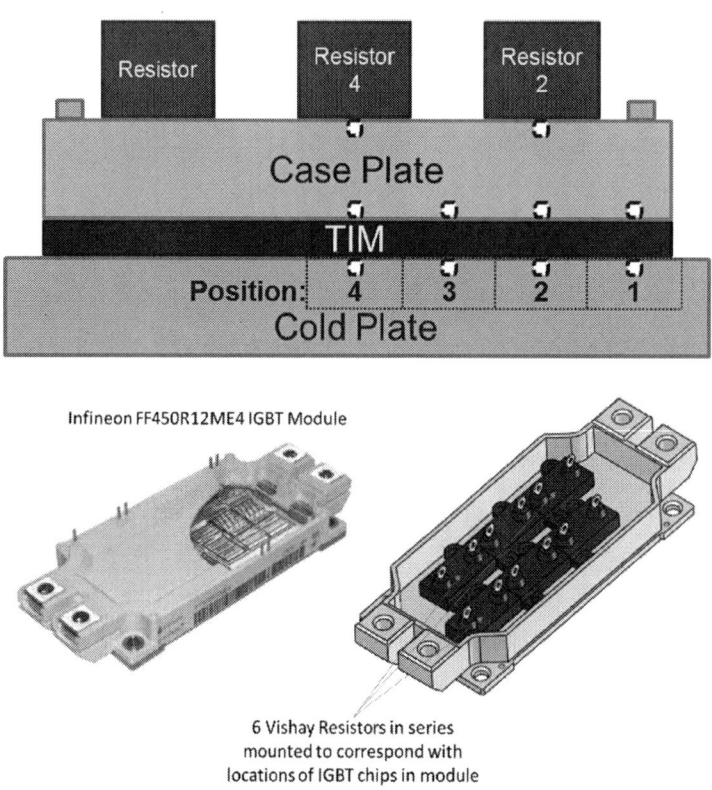

Fig. 4. Final assembly setup

© VDE VERLAG GMBH · Berlin · Offenbach

1.3. Compressibility

One of the key advantages of using a grease like solution over a non-compressible foil like solution is the increased effective wetting aided by the ability of the material to flow and adjust to the flatness variations of the metal surfaces. Most rigid foil based solutions (including some graphite and metals) typically have less than 10% compressibility and followability limiting their effectiveness in contacting both mating surfaces, hence allowing for air gaps that act as insulators. For the foil based solutions to work more effectively, they must be able to effectively fill the variations in the bond lines between the mating surfaces. The HT-C3200 is an engineered graphite foil that can be compressed to about 70% of its initial thickness under pressure. This compressibility helps mimic the followability of the grease like substance while maintaining its ability to not pump-out under pressure. This purely graphite based solution does not pump out or dry out during thermal cycling or while applying pressure. This significantly improves the life of the material both during storage and during operation. Fig. 5., shows the thickness of the material under compression. The difference in pressures between the edges and the center of the cold plate and case plate creates a bond line variation that can be between 50 and 100μ based on the flatness of the metal surfaces, the torque applied and the thickness of the metal plates. The thickness was measured using a vacuum controlled thickness gauge.

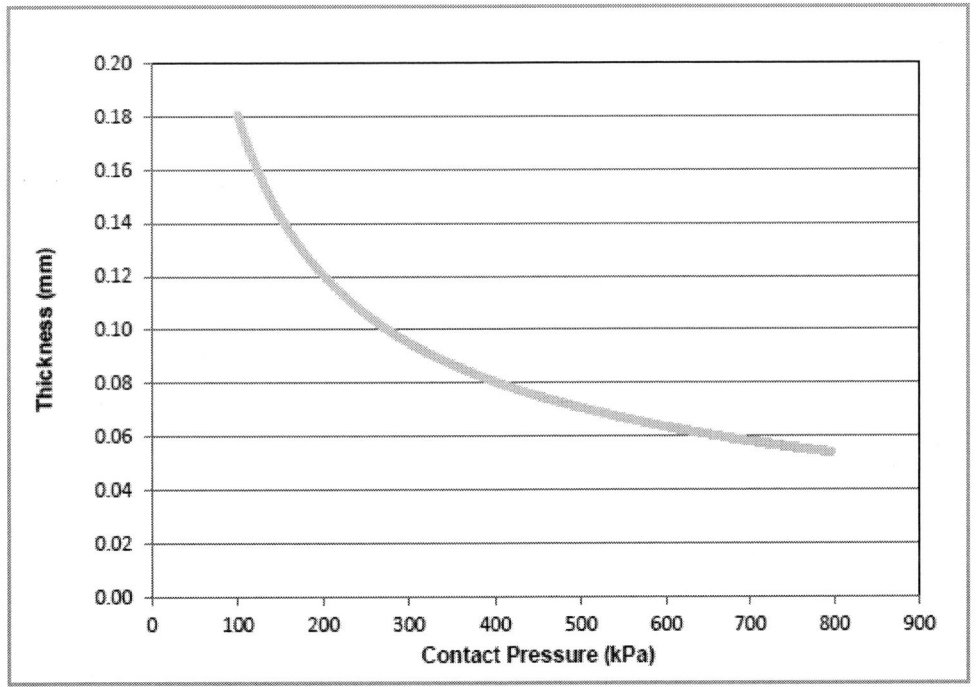

Fig. 5. Thickness vs. Pressure HT-C3200

(http://www.graftech.com/wp-content/uploads/2015/03/TDS319-HITHERM-HT-C3200.pdf)

1.4. Operating temperature

Most grease or Phase Change Material based TIMs have effective operating temperatures <= 150C, the higher temperatures cause the material to dry out and lose its performance significantly. With the gradual but definite proliferation of Wide Band Gap material based devices, the devices can operate comfortably in the 180C to 220C range, rending the current grease like solution inadequate. The HT-C3200 can operate in temperatures up to 400C with no noticeable difference in performance. This graphite solution does not have the challenges like grease to pump out or dry out either during assembly or during operation.

2. Test setup

2.1. Test assembly

This test assembly consists of four parts (Fig 6.)

a. Water Chiller: A ThermoCube 400L Liquid Cooled, Recirculating Chiller from Solid State Cooling Systems [2] was used to maintain water temperature at 50 C.
b. Assembly under Test: The assembly consists of the set up described in Fig 3. , thorough cleaning of the cold and case plates with alcohol, DI water and let dry before changing the TIMs for testing
c. DAQ: An Agilent 34972A LXI Data Acquisition / Data Logger Switch Unit [3] was used to record temperatures from the various thermocouples
d. Power Supply: An Agilent 6655A DC Power Supply 500W [4] (Max 4 A) was used to provide power to the assembly

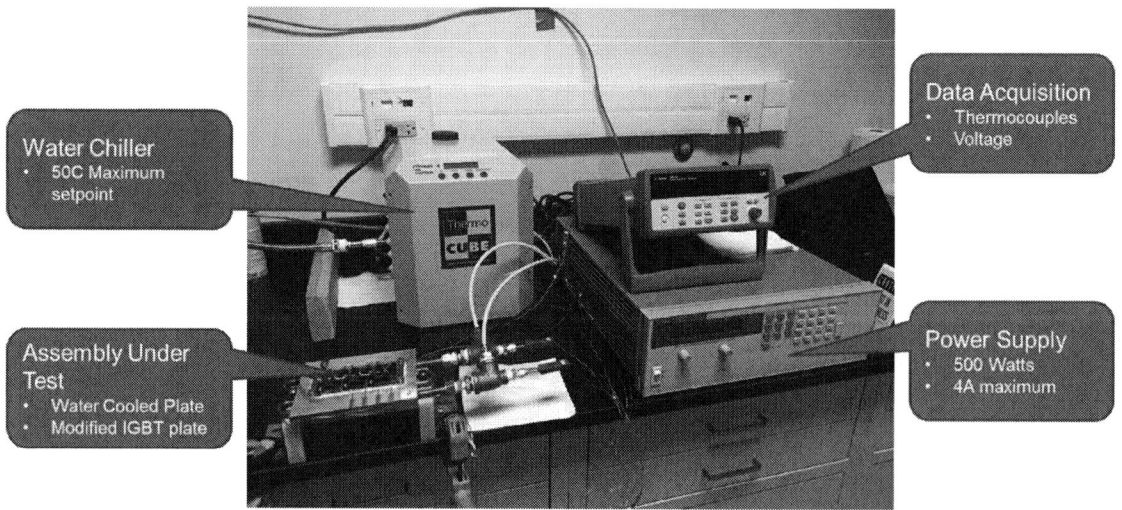

Fig. 6. Test setup

© VDE VERLAG GMBH · Berlin · Offenbach

The test assembly was used to provide 480W of constant power and the chiller temperature was maintained at 50C. The testing sequence was randomized and the tests repeated multiple times under the various TIM conditions. The data reported here is an accurate representation of the collected data. This assembly enables repeatable test conditions and enables tighter control of the power supplied.

Conclusion

The HT-C3200 is a highly engineered compressible graphite based TIM that provides both the surface wettability like grease while not having the same challenges with pump out and dry out that plague the long term performance of these paste like materials including Phase Change Materials. The HT-C3200 also does not require any dispensing equipment and is not messy in its application unlike the paste like materials, enabling clean and easy assembly and maintenance options. The performance of the HT-C3200 is comparable to the popular grease based solutions unlike the non-compressible foil based solutions including both graphite and metal based solutions.

Acknowledgement

We would like to thank Bret Trimmer, Gerald F Hoffert, Greg Kramer, Jon Taylor, Marty Smalc for their help in designing and execution of the various tests. We also would like to thank the team at GrafTech AET and ITC for their invaluable contribution and support.

Key:

[1]: http://www.infineon.com/dgdl/Infineon-FF450R12ME4-DS-v03_01-en_de.pdf?fileId=db3a30431a5c32f2011a769c4a506c4b

[2] : (http://sscooling.com/products/thermocube/thermocube-400l-600l/item/thermocube-liquid-cooled-400l-to-600l-liquid-recirculating-chiller-for-lab-laser-semiconductor)

[3]: (http://www.keysight.com/en/pd-1756491-pn-34972A/lxi-data-acquisition-data-logger-switch-unit?cc=US&lc=eng)

[4] : (http://www.keysight.com/en/pd-839376-pn-6655A/500-watt-system-power-supply-120v-4a?cc=US&lc=eng)

PCIM Europe 2016, 10 – 12 May 2016, Nuremberg, Germany

Heat Pipe system development for Railway application working with speed motion convection

Thomas ALBERTIN, Atherm, 1 rue Charles Morel 38420 Domène, France,
t.albertin@atherm.com

Abstract:

Atherm has been developing Heat Pipes systems for railway applications since 1988 and recently developed a "cold wall" heat pipe system in collaboration with a final customer producing locomotives. This "cold wall" is a speed motion cooling system based on a Heat Pipes assembly for railway application, aiming at extracting more than 6kW from assembled IGBTs and able to work from -40°C, both in forced and natural convection, with the same performances.

1. Atherm presentation and context:

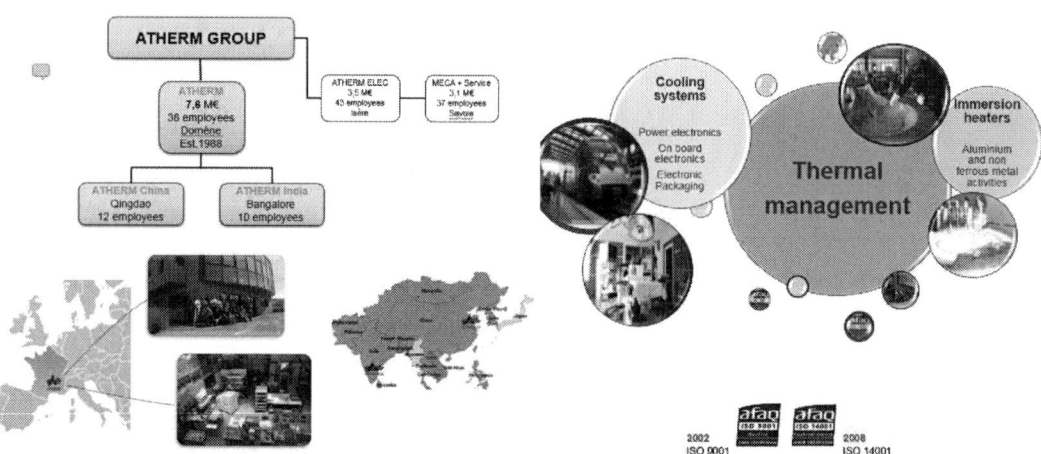

Fig. 1: Atherm presentation

Atherm is the central entity of a holding company named "Atherm group". The Company was founded in 1988. Its turnover in 2014 was around 7.6M€. Approximately 60% of our the over is realized on export and most of our products are manufactured in our French facility and sold directly all around the world. Atherm has 2 subsidiaries abroad, one in China and the other in India. Those facilities are only dedicated to supply local markets.

Generally speaking, Atherm is involved in the field of thermal management. 80% of the turn over is made by serial production of cooling systems: for Railway applications: with clients like Alstom and Bombardier, for Medical applications: with clients like Trixell or General Electrics, for Embedded electronics and electronic packaging in the field of aeronautics: with customers like Thales, Zodiac or Airbus. The 20 other percent of our turnover are realized by the manufacture of immersion heaters dedicated to the nonferrous metal industry. Atherm is also ISO 9001 and 14001 certified

© VDE VERLAG GMBH · Berlin · Offenbach

PCIM Europe 2016, 10 – 12 May 2016, Nuremberg, Germany

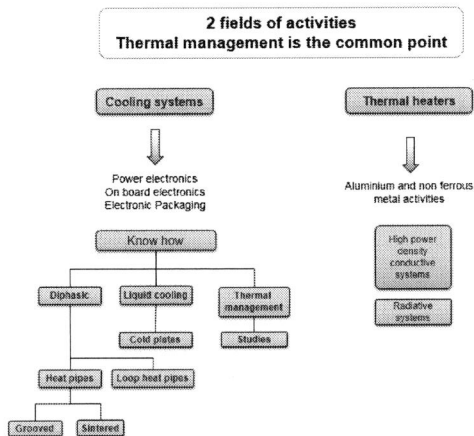

Fig. 2: range of products

In terms of cooling technologies, Atherm is able to work on liquid cooling systems (such as cold plates) as well as diphasic systems. Diphasic systems include grooved heat pipes, sintered heat pipes and loop heat pipes. All those systems and technologies are manufactured in our French facility. Atherm tries to codesign the solutions with its clients. It means that we try to be involved at the beginning of projects, when customers are developing their specifications. Then, we work on all the value chain from design and simulation up to the serial production, passing via prototypes, testing and qualification of pre-series. All that job is mainly done in France. Our design office gathers a great part of our competences. It is our clients preferred entry point for new developments. R&D represents 6.5% of our turnover. 4 permanent persons are able to work on the development of our products. We have numerical and analytical calculation possibilities. Our designers are competent in the field of: mechanical design, thermal design and thermal management, thermal calculation. They also commonly work on prototypes manufacturing and testing. One of the design office jobs also consists in developing test benches and tools for our serial production workshop. We like to say that one of our strength is the possibility to extend the theory to the real life.

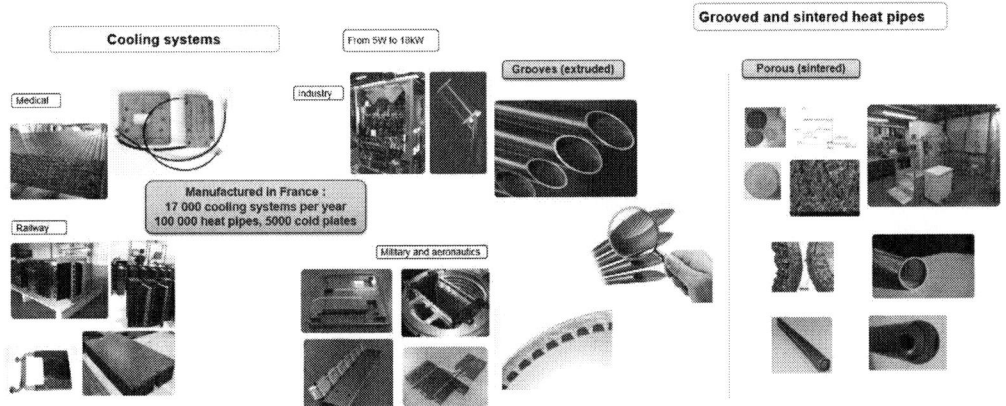

Fig. 3: heat pipes

© VDE VERLAG GMBH · Berlin · Offenbach

Grooved heat pipes are produced from extruded commercial copper tubes and sintered ones are realized with our know-how in our vacuum furnace.

2. development demarche, 2 practical examples:

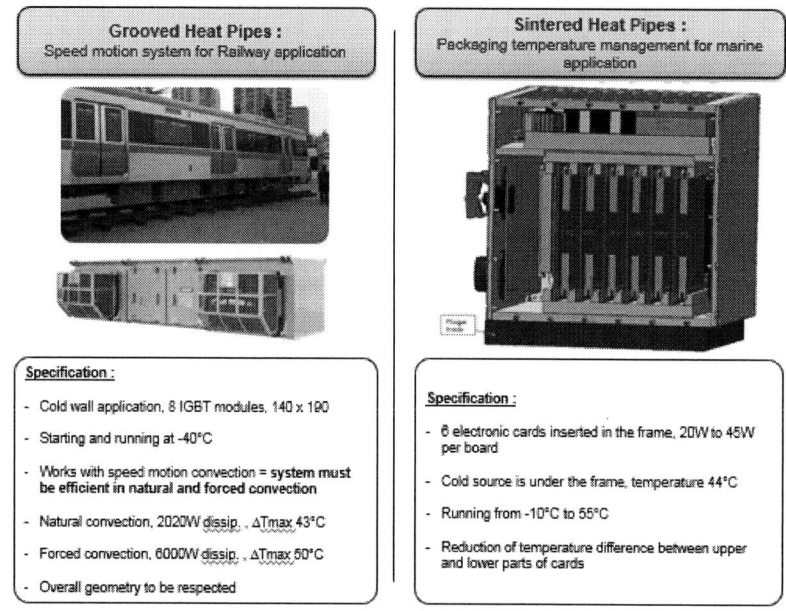

Fig. 4: characteristics of developed systems

The first development concerns a grooved heat pipe system for railway, the second one is a sintered heat pipes system for marine application.

The cooling system for an IGBT application has to dissipate up to six kilowatts and has to work either in natural convection and forced convection. It is a huge system weighting approximately 40 kilograms.

The naval system uses heat pipes to homogenize temperature on electronic boards which are vertically racked between 2 frames.

PCIM Europe 2016, 10 – 12 May 2016, Nuremberg, Germany

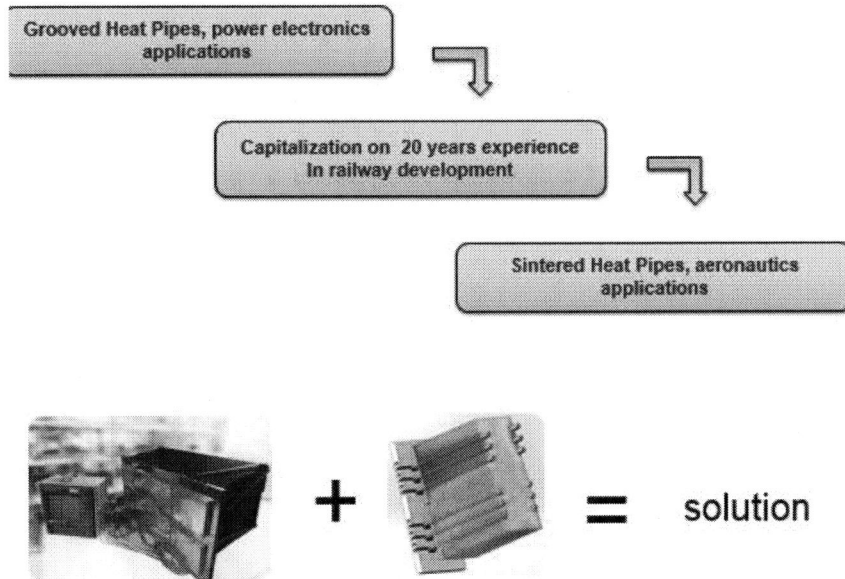

Fig. 5: demarche

For a long time, Atherm has been working in the field of railway applications, developing heat pipes systems for power electronic components such as IGBT.

In this field, we generally associate theoretical works with practical works, I mean prototypes and calculation. It really helps us to reach the solution.

Thus, we have simply decided to extend this methodology to aeronautics developments.

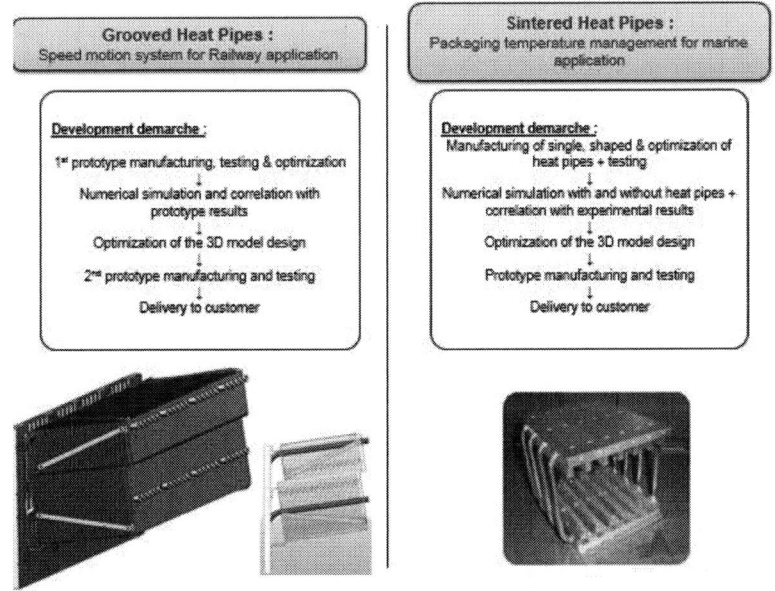

Fig. 6: demarche

© VDE VERLAG GMBH · Berlin · Offenbach

1135

When we start a new development for a client, we generally associate theoretical and practical works in a step by step demarche, mixing 3D models and prototypes.

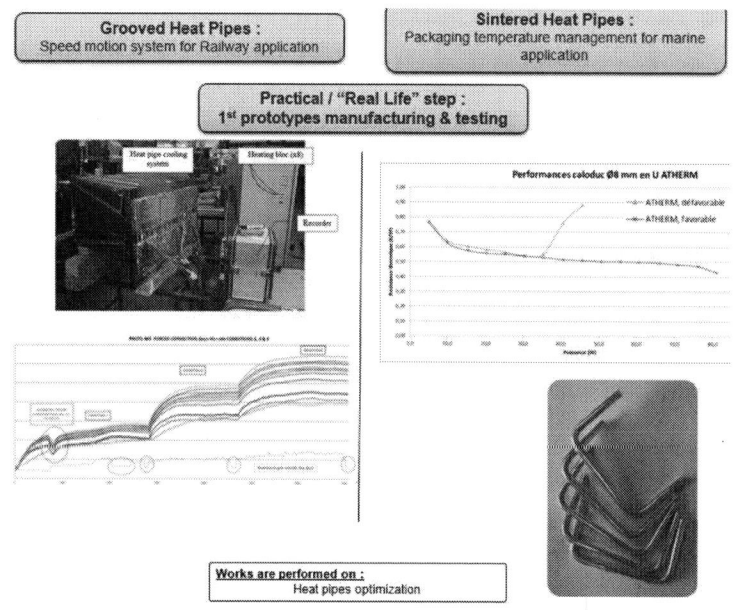

Fig. 7: results

First step consists in first tests. We can observe some typical testing installation and typical results on the fig. 7. We often estimate thermal resistance vs power. First step starts by a first prototype or mock up in order to make a first approach of the system in real life situation. Results of tests are gathered and recorded.

Those preliminary tests helps us to have materials to feed our models with results.

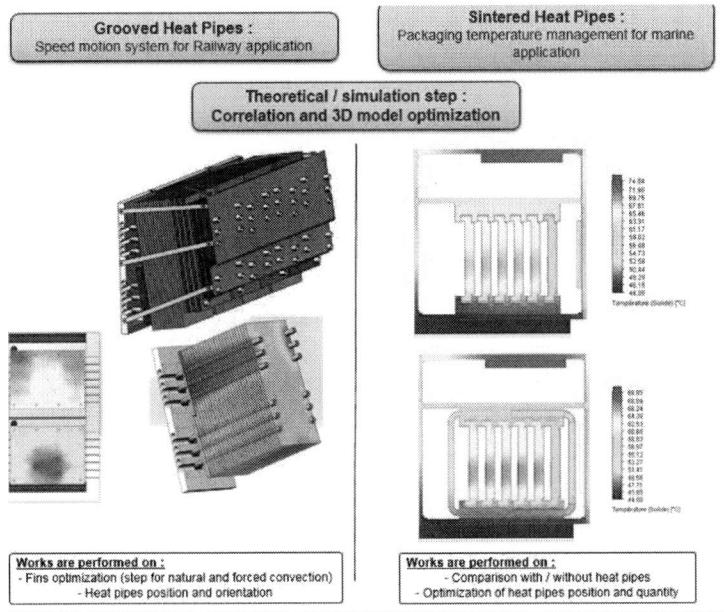

Fig. 8: results

Then a Numerical campaign can begin and before all, the model is correlated with real life results. Optimization can be driven on the numerical support until a solution is found. Concerning heat pipes themselves, we work on macro simulation, working with equivalent conductivities. Efforts are generally focused on the orientation of heat pipes, optimization of their location and quantity and optimization of the exchange surfaces.

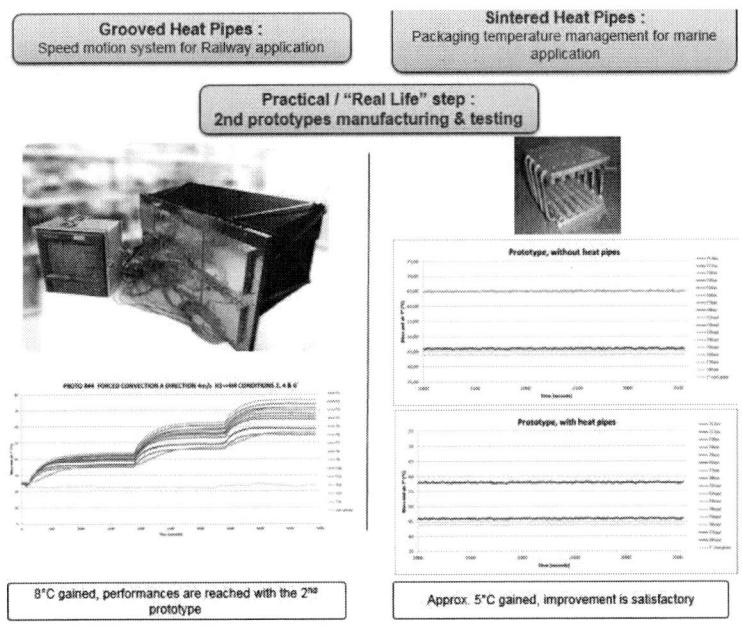

Fig. 9: results

After the numerical step, we generally manufacture a second set of prototype and test it before delivery. We can sometimes consider a second loop of simulation if performances are not reached yet.

3. Conclusion:

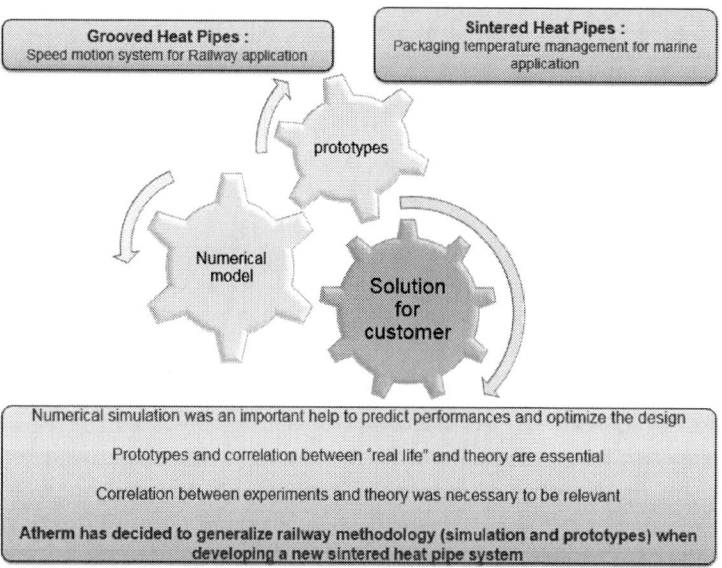

Fig. 10: demarche generalization

Atherm has decided to generalize the railway methodology mixing prototypes and simulation when developing new products. We estimate this helps us to reduce delivery times and development prices. We also think that using preliminary results coming from testing to feed the simulation helps to keep a strong link between theory and real life, and it is essential for a supplier of physical products.

High Performances Passive Two-Phase Loops for Power Electronics Cooling

DUPONT, Vincent, CALYOS S.A., Belgium, vincent.dupont@calyos-tm.com.
BILLET, Cyrille, CALYOS S.A., Belgium, cyrille.billet@calyos-tm.com.
NICOLLE, Thomas, CALYOS S.A., Belgium, thomas.nicolle@calyos-tm.com.

Abstract

Capillary Loops (CPL, LHP and Thermosiphon Loop) can be used for passive cooling of high power electronics. The present paper describes the recent advances in the field of high heat flux evaporator technology (up to 100 W/cm²) applied to IGBT modules cooling that dissipate several kilowatts Several application cases permit to demonstrate the benefits of such a passive system on existing and competing cooling systems in terms of junction temperature and life time.

1. Power Module Heat Dissipation

1.1. Thermal Resistance Chain

A power electronic module is basically an assembly of chips (Silicon, Silicon Carbide, etc.) for example IGBT and diodes, on several layers of materials with at the end a base plate clamped to the cooling device. Figure 1 shows an example of such a stacking. In a bipolar junction transistor (BJT) the collector-emitter voltage Vce, allows to turn-on and turn-off the base current in an efficient way. Thus, the thermal power losses can be divided in two parts [1]:

$$Q_{\mathrm{mod}\,ule} = Q_{conduction\ losses} + Q_{switching\ losses} \tag{1}$$

where, the conductive losses are due to joule effect inside the BJT (directly linked to the number of parallel switches available) and the switching losses during the rising and falling time of the collector current (directly linked to the switching frequency i.e. on the efficiency of the control architecture).

The thermal power sharing between the two kinds of losses depends on the application of the power module. For a motor drive, the thermal power depends on the load profile type (torque versus shaft speed) and the operation duty cycle. For example, for a metro motor drive case, the full thermal power is rejected in during less than 10 seconds (acceleration phase) with respect to a complete cycle duration of 68 seconds i.e. the time between two stations [2]. Thus, these losses are time dependent (and can also be slightly temperature dependent if the thermal amplitude is large enough). During continuous or fast transient behavior the heat is spread from the chip to the cooling system, through the multiples layers in a complex way, where thermal inertia, contact resistance between layers and bulk thermal resistance of the materials play a role. Different models are used to capture this transient evolution. The Foster-model is the easier to implement from the datasheet [3] but can be transformed in the Cauer-model more physically correct [1]. These models allows both to calculate the response time of the module. But Cauer-model allows to plot the thermal capacitance as a function of the thermal resistance in order to detect the location of a defect among the multiple resistances and to better understood the failures modes of the complete system i.e. the module and its cooling system [4].

Generally, the datasheet is the only source of information for the end-user and it gives the global thermal resistance between the junction of the chip (Rth IGBT and Rth diode) and the thermal capacitance of the whole module (Cth). Power module manufacturer might exchange drawings and informations with major end-users in order to perform complete 3D Finite Element Modeling (FEM). The thermal and electrical designs are strongly coupled in order fulfill two major targets. First, to maintain the virtual junction temperature T_j of each chip below the value specified in the datasheet. The definition of the junction temperature is susceptible to change from a manufacturer to another (average or maximum on the silicon) and engineers might take margins on this value based on the return of experience of the application. Improvement of chips technology lead to an increase of Tj typically from 125°C to 175°C [5]. Second, to achieve the lifetime of the power module specified by the end user of the equipment. In this case the thermal swing i.e. the amplitude of the temperature evolution of the module base plate under each chip is critical. Here, the failure mode and the failure cycle are the major law to know. For standard automotive power module, these failure laws are well known with for example the CIPS 2008 model [1], for large IGBT module, for motor drives, is more difficult to obtain these informations and in-house characterization are often necessary.

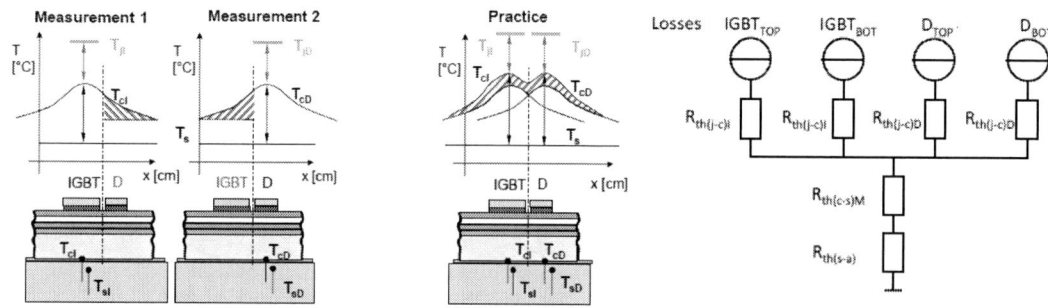

Fig. 1. Crosstalk and thermal resistance chain between the layers inside the power module [6].

The global system resistance given in Figure 1 is valid for a heatsink (air cooling) packaging. The cooling system of the present paper is based on the fluid evaporation and condensation inside a loop. Thus, the resistance chain is completed:

$$Rth_{global} = \frac{T_j - T_{cold\ source}}{Q_{source}} = Rth_{IGBT\ or\ diode} + Rth_{TiM} + Rth_{evaporator} + Rth_{condenser} + Rth_{heat\ exchanger} \qquad (2)$$

where, TiM is Thermal Interface Material (thermal grease or other) and T cold source is the inlet temperature of the final cooling fluid i.e. air or water-glycol for example.

Thermal engineer using cooling fluid systems prefers to take into account the local performances of the heat transfer thanks to the heat transfer coefficient (HTC), in W/m²K. The relation between the HTC at cooling surface level and Rth is given by:

$$Rth_{evaporator} = \frac{T_{evap} - T_{sat}}{Q_{source}} = \frac{1}{HTC_{evaporator}\ S_{source}} \quad i.e. \quad HTC_{evaporator} = \frac{q_{I/F\ vap}}{T_{evap} - T_{sat}} \qquad (3)$$

In a first approach the contact surface of the module base plate can be used to determine S_{source}. But in fact thermal analysis shows that the spreading resistance inside the module is not sufficient to suppress the hot spots [2] and a local value of the heat flux q I/F vap must be considered. T_{sat} is the saturation temperature i.e. the temperature of the fluid.

1.2. Thermal Interface Material (TiM) issues

The global trend in power electronics is the densification. For example, the power density reaches 160 kW/dm³ for a 60kW full SiC drive inverter for a high speed motor drive application [5]. This goal leads to a decrease of the thicknesses of the layers, the number of layers, an increase of the conductivity of the material, etc. Among these improvements the TiM issue is critical [7]. Figure 2 describes the HTC of the cooling alone as a function of the equivalent HTC of the cooling system including the TiM. A poor TiM performance can completely flatten the better cooling system. This well know phenomenon is due to the serial thermal resistance connection: the bottle neck is, often, the TiM. Of course the complete removing of the material, by integrating the cooling exchanger inside the IGBT module baseplate itself is an efficient way to improve the overall thermal performances (TiM "perfect" line of the figure). This can be achieved with water cooling cold plate [5] but also with passive two-phase evaporators [8].

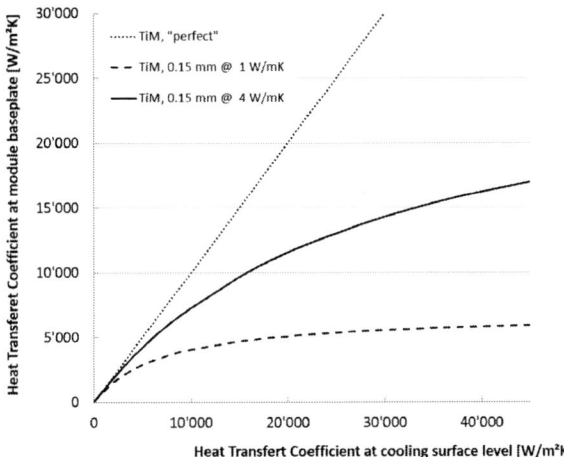

Fig. 2. Influence of the TiM thermal performances on the cooling system performances.

1.3. Heat flux at module saddle issues

The power sharing between diodes and IGBT chips has a direct influence on the heat flux entering inside the cooling system [6]. Figure 1 gives an example of the influence of two neighbor's chips. This crosstalk issue can be studied with FEM simulations [3] thanks to the knowledge of the internal architecture of the module. Figure 3 (left part), shows FEM simulations of different high voltage IGBT modules, where the maximum heat flux is plotted as a function of the average heat flux, both at module base plate level (TiM contact). The sharing between diode power and IGBT power, the module footprint, cooling HTC are different for each point. As the rule of thumb, there is a factor 2 between the average and the maximum local heat flux at base plate. Thus, TiM and cooling performances must be chose with respect to this maximum local value.

TiM performances impact the thermal gradient between the cold plate and the virtual junction temperature but, and, on a less obvious way, the heat flux at module saddle level. In order to illustrates this phenomenon, several FEM simulations have been performed with a standard high voltage IGBT module with Q_{IGBT} = 1.3 kW and Q_{diode} = 0.4 kW and a HTC (including the cooling system and the TiM) ranging from 2'500 to 25'000 W/m²K. Figure 3 (right) shows that for the 3 combinations of chip powers there is an increase of the maximum heat flux of 23%. It cannot be neglected in the thermal design. For this case, at 25'000 W/m²K, the crosstalk i.e. the activation of the diodes induces an increases of 10% of the local maximum heat flux under the IGBT chip. As described in § 2.1, the thermal losses are strongly linked to the complete control and duty cycle. Unfortunately, often during the thermal acceptance tests of the cooling system, this complete architecture is not available and the performances are evaluated in a resistive way (IGBTs alone and diodes alone) without the combined charge.

© VDE VERLAG GMBH · Berlin · Offenbach

FEM calculations are time consuming and the structure of the power module itself cannot be communicated by the end-user. Thus, CALYOS has validated a method based on the development of Fourier series expansions of N discrete sources located over 2 virtual layers proposed in [9]. The thickness and the thermal conductivity of the 2 layers are identified thanks to at least four FEM calculations. Figure 3 (right) shows the good agreement between FEM and semi-analytical approaches. Thanks to this 2-layers model, it is possible to simulate the crosstalk phenomenon and the iterative coupling between the evaporator performances and the local heat flux and, finally, the lifetime of the power module.

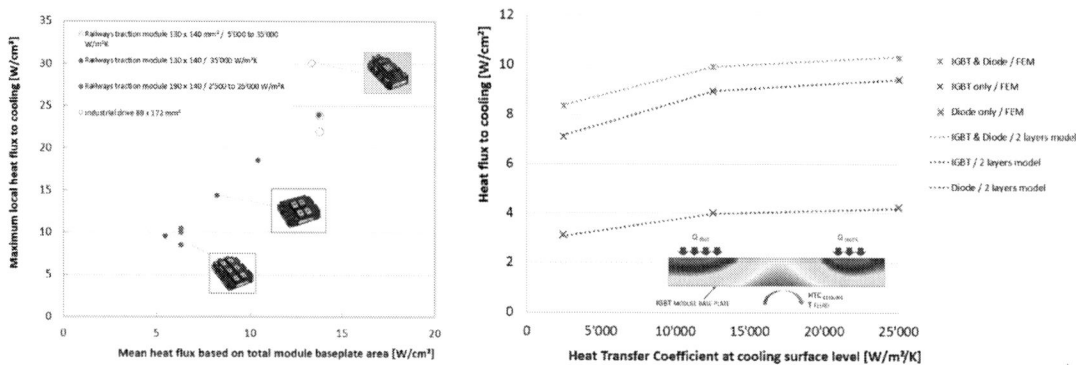

Fig. 3. FEM simulations of the IGBT module with different heat load between IGBT and diodes.

2. Passive Two Phase Cooling systems

2.1. Capillary Loops working principle

Capillary Pumped Loop (CPL) and Loop Heat Pipe (LHP) are heat transfer systems perfectly tight and filled with a working fluid [11]. The fluid is chosen thanks to the temperature operating range among ammonia (space applications), water (in door applications), methanol, ethanol, R245fa, etc. As described in Figure 4 (left), capillary loops are composed of one, or several, evaporator(s) in direct contact with the power module base plate, a condenser, in contact with the cold source and a reservoir to manage fluid thermal expansion. The cold source can be a heat sink, a finned tubes air exchanger, a structure with a large thermal inertia, etc. The evaporator contains an inner capillary structure made of a metallic foam (nickel, copper, titanium, etc.), the porous wick, with pores diameter ranging from 2 to 10 microns. The fluid vaporization at the surface of the wick creates a capillary pressure head that makes the fluid circulates inside the loop without any mechanical additional pump nor external energy. Transport capability and thermal resistance are much better than simple heat pipes thanks to the capillary structure located only at the evaporator wall level (cf. Figure 4, right). The main benefits of capillary loops are summarized in the following sections.

High heat flux values at power module saddle level.
As described in Figure 4 (right), the porous wick separates efficiently the vapor flow from the liquid flow. Thus, the Critical Heat Flux (CHF) i.e. the heat corresponding to the wall dry-out, and the power module destruction, is larger than a classical liquid cooling system and also classical heat pipes (HP) for example embedded inside heat sinks. Large heat flux value are a critical issue for an air heat sink cooling due to conduction gradient inside the saddle. Even with embedded HP, the relative location of the hot spots with respect to the HP evaporator tube is a concern. Moreover, the CHF of an heat pipe decreases drastically at low temperature [10]. In the case of liquid cooling, the CHF requires subcooling liquid and/or large flow in order to avoid boiling incipience inside the cold plate. The HTC has the same value at any place of the active surface of the evaporator thank to the nearly constant saturation temperature inside the system. This performance uniformity prevents derating and "thermal runaway". Heat Pipes and liquid cooling cold plates need custom designs to adapt the active surface to the hot spots.

© VDE VERLAG GMBH · Berlin · Offenbach

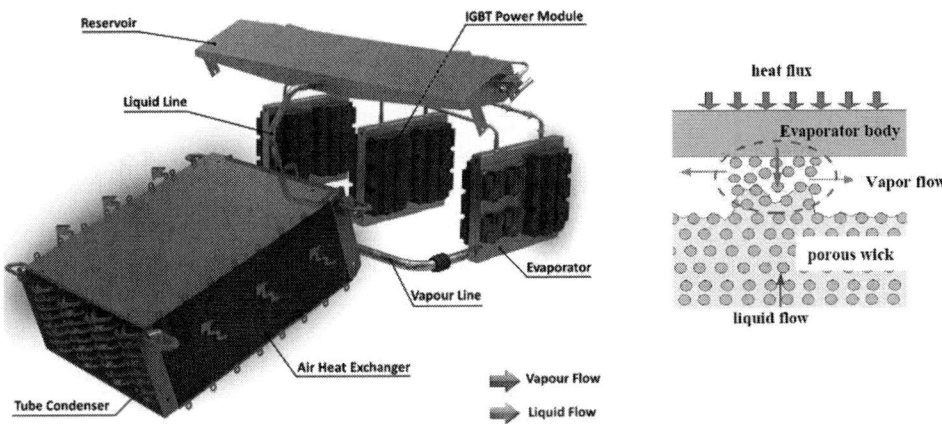

Fig. 4. Schematic view of a Capillary Pumped Loop (left) and the vaporization interface inside the capillary evaporator porous wick (right).

The following table summarize the heat flux value achieved during characterization tests.

Capillary Loop type	Working fluid – porous wick	Heat source footprint [cm²]	Maxi. Heat flux [W/cm²]	Ref.	Comments
CPL	Methanol Nickel	15.2 (special tool)	100	[12]	Demonstrator (TRL4)
CPL	Methanol Nickel	19.0 (special tool)	88	[13]	Product (TRL-8) with startup at -50°C
LHP	R245fa Nickel	4.6 (GPU)	73	[13]	Datacenter product (TRL-8)
CPL	Water Copper	4.0 (ceramic)	111	[8]	Development (TRL-3)
Loop Thermosiphon	Methanol Nickel	57.0 (special tool)	26	[14]	Natural convection heat sink (TRL-4)

Figures 5 shows the evaporator wall HTC as a function of the contact heat flux for a product evaporator [13]. These performances are far above the standard heat pipe systems and comparable to the better mechanically pumped liquid cooling systems. This graph illustates how the power electronic interact with the vaporization via the heat flux level as discussed in §1.3.

Transport capabilities against acceleration of gravity

CPL systems, as presented in Figure 4, has the ability to be equipped with a large number of evaporators in parallel way (typically 4). Thus, the transferred power can reach large values, up to 27 kW [12].

Thanks to the pressure head developed by the capillary wick, the heat exchanger can be placed on the side - or below - the evaporators and is able to operate with inclination and acceleration of a vehicle [2] and [10]. For example, typically with methanol, the wick is able to pump 1.5 m against gravity, half of this budget is used to move the fluid inside the loop. This capability allows to use smaller height than heat pipes thermosiphon systems.

© VDE VERLAG GMBH · Berlin · Offenbach

The final exchanger (air) can be located several meter away from the power electronics. In the limits of the pumping capability, this distance has no impact on the saturation condition inside the loop but increases the size of the reservoir (to stock the liquid located in the vapor line before startup).

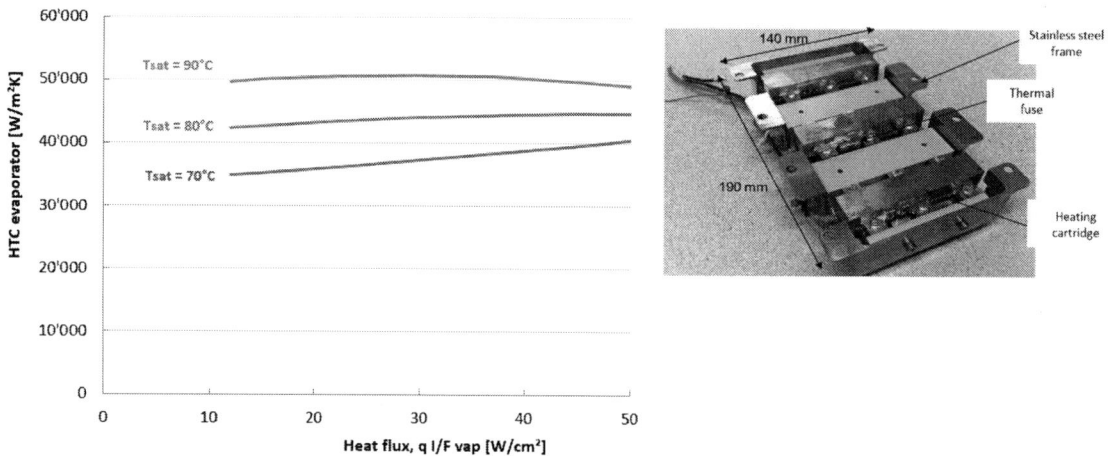

Fig. 5. HTC of the capillary evaporator as a function of the heat flux. The custom heating tool used to characterize high heat flux behavior (57 cm²) is presented on the right [13].

No losses, no vaporization

Contrary to the mechanically liquid cooling system i.e. equipped with a mechanical pump (or the some air heat sink) the heat transfer of the evaporator naturally stops when the thermal losses tend to zero. In other word, the thermal swinging is less pronounced with phase change than with constant cooling (designed for the maximum heat load in the hot case). This unique behavior, directly improves the lifetime of the power module.

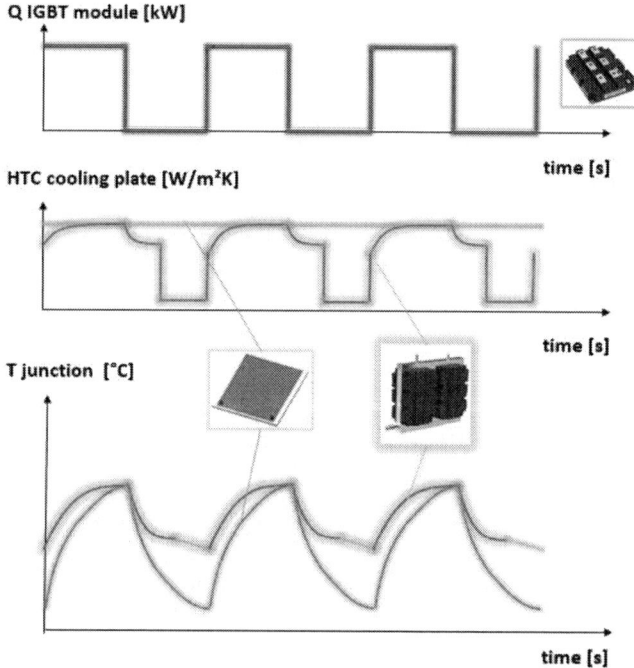

Fig. 6. Thermal swinging effect. The capillary loop (yellow line) exhibits a lower thermal amplitude than the active Liquid Cooling System (blue line) thanks to the extinction of the vaporization inside the evaporator (the wall decrease below the saturation temperature).

© VDE VERLAG GMBH · Berlin · Offenbach

3. Application examples

3.1. Railways Inverter for a Metro application

The first example is the world's first use of a CPL inboard a train. It is an inverter located under the floor of a tire mounted RATP vehicle operating on the line n°1 in Paris that has been successfully cooled during commercial service [2]. The inverter controls two 260 kW motors. The complete railways qualification of the technology has been performed including mechanical and environmental test. Maximum thermal resistance was lower than 5 K/kW at 1.3 kW at IGBT module level (130x140 mm²) and a transport capability of 7.7 KW have been reached in steady state, in natural and forced convection mode (to simulate the train movement). The maximum power tested during the metropolitan cycle was 10.4 kW. For the presented application, the heat fluxes were limited to 20W/cm². Operating temperature i.e. saturation condition saturation inside the CPL have been tested from 59°C to 96°C with methanol as a working fluid. Air temperature effect has been studied between –50°C to +70°C inside a climatic chamber.

Fig. 7. Location of the capillary loop below the roof of the train (left) and the dedicated air exchanger able to operate by natural cooling i.e. air speed and natural convection in station (right) [2].

3.2. Capillary Thermosiphon for cabinet cooling

This system is an innovative coupling between a CPL and a loop thermosiphon in order to suppress the system reservoir and allow to place the heat load anywhere on the complete height of the thermosiphon without significate change in the overall performances.

The demonstrator, described in Fig. 8, is based on a single capillary evaporator that use methanol as a working fluid with a nickel wick (cf. §3.1.). The thermosyphon loop rejects the heat to the air by natural convection thanks to fifteen enhanced copper tubes embedded inside an aluminum heat sink (0.55 x 0.26 x 1.20 m³). The thermal inertia can be used to smooth power cycle overload. The total system power ranges was up to 3 kW (heat flux ranging from 5 to 26 W/cm²) and a saturation temperature between 45°C to 99°C. A new evaporator configuration with a single active face will be available in 2016 with a nickel or a copper wick. This new configuration will decrease the cost of the system when double side active evaporator is not necessary.

4. Conclusions

Capillary Loops are efficient systems able to transfer up to 27 kW on several meters. There are fully passive, maintenance free, energy free like a standard heat pipe but capable to pump the liquid against gravity (or other accelerations) like a mechanically pumped liquid cooling system. Well adapted to power electronics cooling with high heat flux capability, up to 100 W/cm², it has been successfully qualified and operated on a power inverter inboard a train on the metro line n°1 in Paris.

Thanks to the extinction of vaporization below the saturation temperature and a better performance at hot spot level, the thermal swinging is less pronounced for a capillary evaporator than for a constant cooling system. This unique behavior, directly improves the lifetime of the power module. The evaporator technology can also be tailored to any power electronic footprint or directly integrated into the power module base plate.

© VDE VERLAG GMBH · Berlin · Offenbach

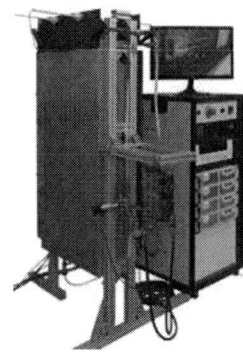 

Fig. 8. Capillary Thermosiphon cooled by natural convection with a close view of the heating tool used to mimic high voltage IGBT module [14].

5. References

[1] Lutz J., (2014), Reliability of IGBT Power Modules, PCIM Workshop 10, PCIM Europe, Nuremberg, Germany, 20-22 May.

[2] Dupont V., Van Oost S., Barremaecker L, S. Nicolau (2013), Railways qualification tests of Capillary Pumped Loop on a train, Proc. of the 17th IHPC, Kanpur, India, pp. no printed version.

[3] Hunger T. and Schilling O. (2008), Numerical investigation on thermal crosstalk of silicon dies in high voltage IGBT modules, Proceedings PCIM Europe, Nuremberg, Germany 27–29.

[4] Von Essen T. (2015), Transient thermal measurement and analysis techniques of high power modules for determination of heat transfer coefficient at different ambient conditions, 10th European Advanced Technology Workshop on Micropackaging and Thermal management, February 4 & 5, La Rochelle, France, pp. no printed version.

[5] Eckardt B. (2015) Cool Systems with SiC and GaN, ECPE Workshop, Advances in Thermal Materials and Systems for Electronics, Nuremberg, December 8 & 9.

[6] Wintrich A. (2014) Thermal resistance of IGBT Modules - specification and modelling, Application Note Semikron AN-1404.

[7] Saums D. L. (2015) Thermal Interface Material, Purpose, Classification, Applications, ECPE Workshop, Advances in Thermal Materials and Systems for Electronics, Nuremberg, Dec.8 & 9.

[8] Dupont V. (2015) High Performance Passive Two-Phase Cooling systems, ECPE Workshop, Advances in Thermal Materials and Systems for Electronics, Nuremberg, December 8 & 9.

[9] Muzychka Y.S. (2006), Influence Coefficient Method for Calculating Discrete Heat Source Temperature on Finite Convectively Cooled Substrates, IEEE Transactions on components and packaging technologies, Vol. 29, No. 3.

[10] Dupont V., J.C. Legros Van Oost S., Barremaecker L., (2013), Experimental investigations of a CPL pressurized with NCG inside a centrifuge up to 10g, Proc. of the 17th IHPC, Kanpur, INDIA, pp. no printed version.

[11] Maidanik Y.F. (1999), State-of-Art of CPL and LHP Technology, Proc. of the 11th IHPC, Tokyo, Japan, 1999, pp. 19-30.

[12] Dupont V., Van Oost S., Barremaecker L., S. Nicolau (2010), Experimental investigations on a methanol capillary evaporator equipped with four flat evaporators, Proc. of the 15th IHPC, Clemson, pp. no printed version.

[13] Dupont V. (2015), Thermal Performances of a fully passive capillary pumped loop with heat flux up to 90 W/cm² at evaporator, 10th European Advanced Technology Workshop on Micropackaging and Thermal management, February 4 & 5, 2015, La Rochelle, France, pp. no printed version.

[14] Dupont V., Nicolle T., Billet C. (2016), Capillary thermosyphon loop for high power electronic cooling by natural convection, Proc. of the 18th IHPC, Jenju, South Korea, to be published.

© VDE VERLAG GMBH · Berlin · Offenbach

PCIM Europe 2016, 10 – 12 May 2016, Nuremberg, Germany

Thermal modelling and management for increasing the power density in high current power electronic systems

Marco Schilling, TU Ilmenau, Germany, marco.schilling@tu-ilmenau.de
Benjamin Köhnlechner, TU Ilmenau, benjamin.koehnlechner@tu-ilmenau.de
Ulf Schwalbe, ISLE Steuerungstechnik und Leistungselektronik GmbH, u.schwalbe@isle-ilmenau.de
Tobias Reimann, TU Ilmenau, tobias.reimann@tu-ilmenau.de

Abstract

A good thermal management is the key for a reliable and compact design of high current integrated power electronics. Different thermal models are required to compute the heating of the system at the nominal operating point or for a complete load cycle. This paper presents the development of thermal models for a high current MOSFET with a finite - element - method (FEM) tool, the model simplification and verification as well as the development of design guidelines for high current power electronics. With the developed thermal models it is possible to design and optimize the thermal management.

1. Introduction

Today's power electronic systems becomes more and more complex. For design engineers it is necessary to have appropriate system models to solve various design problems prior to the development of the first reference demonstrator. A major problem is the system design with a controlled heat flux. This controlled heat flux or also called thermal management gains more and more importance caused by the predefined system geometry, the predefined load cycle or the optimization of the power density.

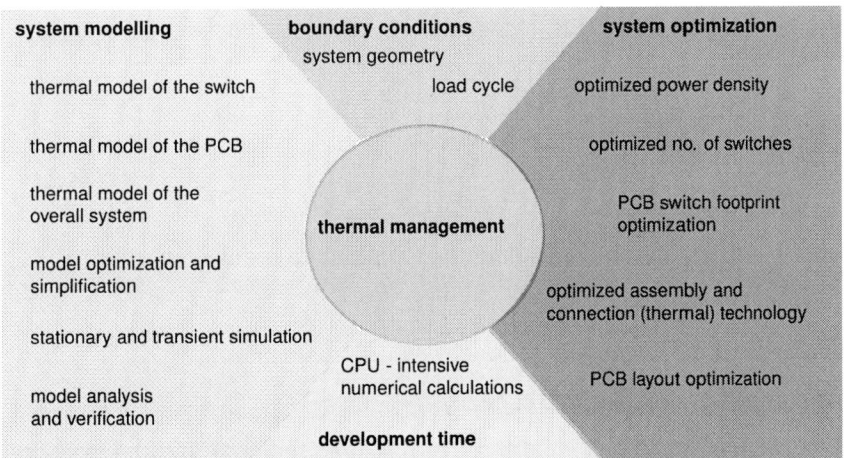

Fig. 1: Key issues in thermal design of power electronic systems

The key issues of thermal management are shown in Fig. 1 and divided in the groups: boundary conditions, system modeling, system optimization and development time. The focus of this

© VDE VERLAG GMBH · Berlin · Offenbach

(a) MOSFET mounted on PCB (b) inner view of the MOSFET

Fig. 2: Different views of the investigated MOSFET with the TO-Leadless Package [2]

paper is on the system modeling, model analysis and verification. Based on simulation and measurement results some printed circuit board (PCB) layout design rules will be proposed. The following investigation is for the low voltage, high current MOSFET "IPT015N10N5" (Infineon Technologies) with an improved low resistance package (TO - Leadless) and small package dimensions. Typical parameters of the MOSFET are shown in Tab. 1. The MOSFET mounted on a PCB is shown in Fig. 2(a) and the view of an open TO-Leadless Package with 5 bond-wires is shown in Fig. 2(b).

Parameter	V_{DS-max}	R_{DS-on} @ 25°C	L_σ **	R_{th-JC}	Chip Size	Footprint
Value	100V	1.5 mΩ	3 nH	0.4 K/W	(6.7 x 4.5) mm^2	(12 x 15) mm^2

Tab. 1: MOSFET IPT015N10N5 parameters [1], [2]; **datasheet parameter of the MOSFET package

2. Basics of thermal design

2.1. Modeling methods

The behavior of temperature fields is described by the partial differential equation (PDE) 1 for potential fields.

$$\lambda \cdot \Delta \vartheta = -p_v + c_p \frac{\partial \vartheta}{\partial t}, \vec{q} = -\lambda \nabla (\vartheta) \tag{1}$$

The used symbols are:

- ϑ - temperature (potential)

- $\vec{q}$ - heat flux density: is the result of the temperature gradient of ϑ

- p_v - internal heat source

- λ - material property: thermal conductivity

- c_p - material property: thermal capacity

With the boundary condition $p_v = 0$ (no internal heat sources) and the investigation of the steady state temperature field ($\frac{\partial \vartheta}{\partial t} = 0$) the PDE can be simplified to equation 2. For the heat flow through a homogeneous material (constant material properties) with the boundary conditions $\vartheta(x_E) = \vartheta_{ref}$ and $q|_{x_A} = \lambda \frac{\partial \vartheta}{\partial x} = \frac{P_v}{A}$ the solution of the PDE (equation 2) yield to the linear equation 3 (solution in 1-dimensional case). The equation 3 is the basis for the definition

© VDE VERLAG GMBH · Berlin · Offenbach

of the thermal resistance according to formula 4.

$$\lambda \cdot \Delta\vartheta = 0, \vec{q} = -\lambda\nabla(\vartheta)$$ (2)

$$\vartheta(x) = \frac{P_v}{\lambda A} \cdot x + \vartheta_{ref}$$ (3)

$$R_{th} = \frac{\vartheta(x_A) - \vartheta(x_E)}{P_v} = \frac{x_A - x_E}{\lambda A} = \frac{l}{\lambda A}$$ (4)

For simple geometries or for the series connection of different homogeneous materials with the same effective heat transfer area, a equivalent circuit can be determined (see Fig. 3).

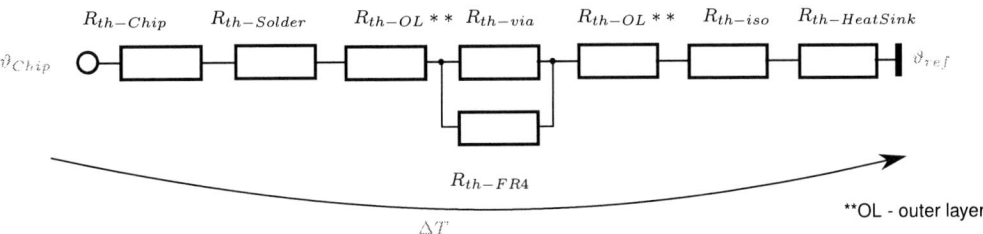

Fig. 3: Thermal equivalent circuit for the heat transfer between the chip and the heat sink in accordance with Fig. 6

For complex geometries like a power electronic PCB, the analytical approach only delivers approximate solutions. For more accurate solutions a finite element method (FEM) simulation is necessary. The basic idea of FEM is to devide the complex 3D geometry in a grid with discrete nodes. The result of the discretization process of the PDE is the state-space equation (SSE) 5. In this SSE the matrix C is the capacity-matrix, K is the heat conduction matrix, B is the input matrix, c^T is the output matrix and Q is the heat flux matrix. The temperature profile will be interpolated between the discrete calculated temperature nodes with a simple function approach (e.g. power function).

$$C\dot{\vartheta}(t) + K\vartheta(t) = Q = Bp(t) \quad y = c^T\vartheta(t)$$ (5)

For further investigation, e.g. a coupled electrical / thermal simulation with a load cycle, a model simplification is necessary. One approach is to simplify the SSE (from the FEM tool) with the arnoldi process into a SSE with reduced order. This procedure is described in paper [3]. A further method is the model parameter optimization based on the simulated heating process (transient thermal simulation). The model is for example the thermal impedance (see equation 6) which is described in [4, page 302 ff]. The model order will be determined for a adequate accuracy. The parameters can be determined by the leas mean square error method. In the following investigation this model simplification method will be preferred.

$$Z_{th}(t) = \sum_{\nu=1}^{n} R_{th\nu} \cdot \left(1 - e^{\frac{-t}{\tau_{th\nu}}}\right)$$ (6)

2.2. Design process for the thermal design of a high current system

The whole design process for the thermal management is summed up in the flow chart Fig. 4. The key steps are step one, three and four. The results of these steps will be shown in the following chapters for a PCB with a single high current MOSFET. In the first step the thermal design will be determined by "hand calculations" of the equivalent circuit in accordance with Fig. 3. In the third step the PCB - layout will be optimized by the using of the FEM tool "ANSYS - Workbench". The optimization parameter is the number of thermal vias. In the fourth step the model simplification will be done.

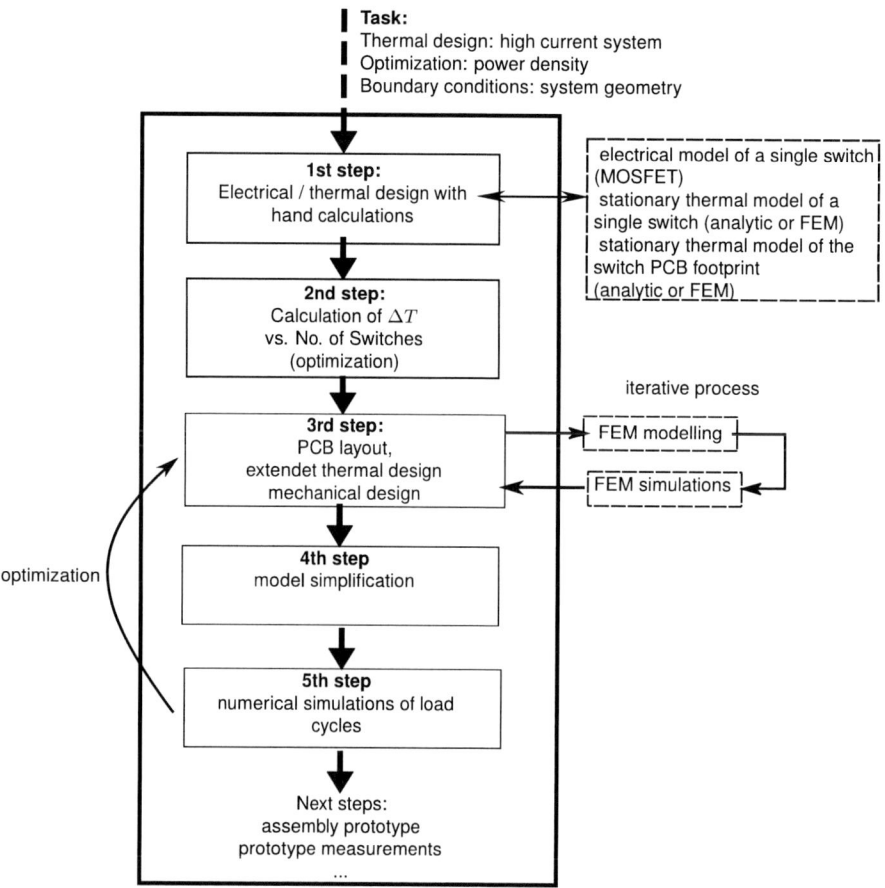

Fig. 4: Flow chart: thermal design for a high current system

3. Steady state thermal modeling of a single MOSFET soldered on a PCB

3.1. Model properties

Parameter	power loss	@ heat conduction paste	nodes	elements	mesh
Value	30W	Bergquist 3500S35 (3.6W/mK)	269606	128366	element size = 0.2 mm

Tab. 2: FEM model properties and measurement setup

The parameters of the system simulation and measurement arrangement are shown in Tab. 2. The appropriate footprints for 42 thermal vias (left) and 97 thermal vias (right) depicts Fig. 5. The cross section of the investigated system with the layer setup - 4 copper layers a $70\mu m$ thickness and 0.875 mm PCB thickness - is presented in Fig. 6. The constant reference temperature is at the bottom side of the isolation.

Fig. 5: Footprints for the measurement and simulation arrangement (left - 42 vias; right - 97 vias)

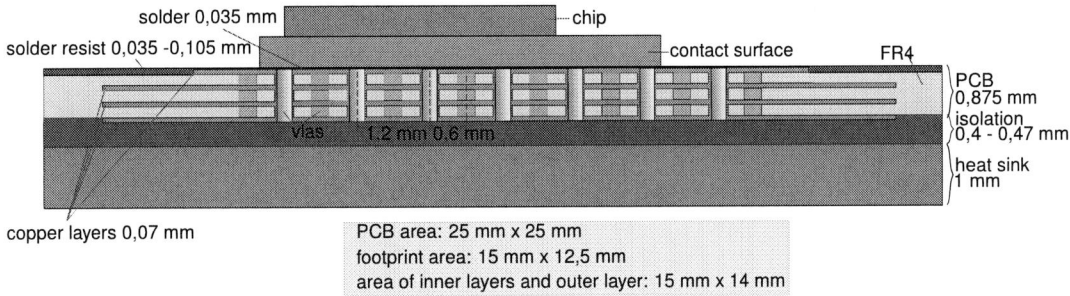

Fig. 6: PCB model cross section and model properties

3.2. Simulation and measurement results

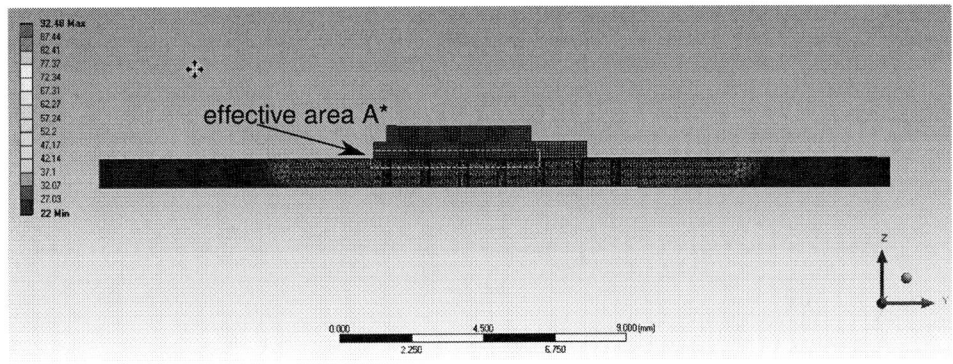

Fig. 7: PCB model cross section and steady state thermal simulation result for 97 vias

The simulation results for the system with 97 thermal vias including the preprocessed meshgrid is presented in Fig. 7. It can be seen that the overtemperature is 70.5 K and the thermal resistance results in 2.35 K/W. Additionally it can be seen that the MOSFET contact area is not

PCIM Europe 2016, 10 – 12 May 2016, Nuremberg, Germany

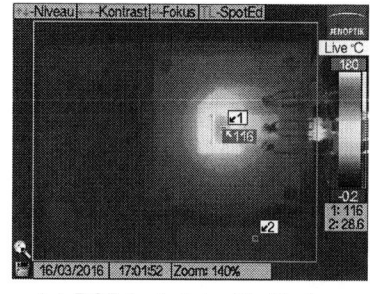
(a) PCB footprint with 42 vias

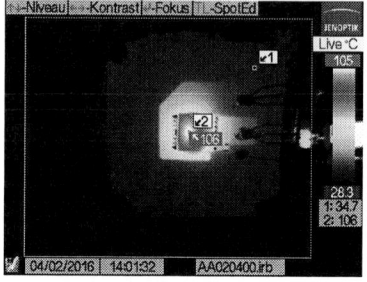
(b) PCB footprint with 97 vias

Fig. 8: Measurement results: single MOSFET with 30W power loss mounted on a PCB
yellow (2 in a), 1 in b) - heat sink temperature, red- maximum temperature $\approx$ chip temperature

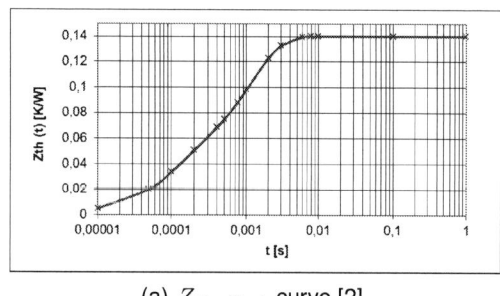
(a) $Z_{th-Pack}$ curve [2]

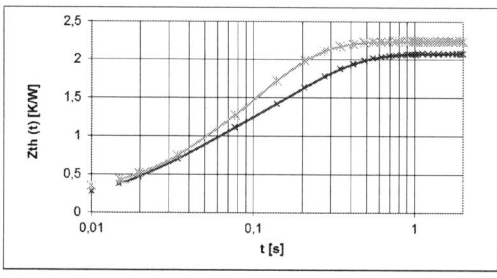
(b) simulated (blue) and analytic calculated (orange) $Z_{th-Pck+PCB+Gap}$ curves

Fig. 9: Z_{th} curve of the MOSFET Package (left) and MOSFET + PCB + Gapfiller (right)

completely effective for heat transfer. The results of the thermal simulations are usable for the calculation of an effective heat transfer area $A*$. In this way the combination of the approximate thermal resistance calculation in accordance with Fig. 3 and the results for the effective area A* should be helpfully for the investigation of systems with different properties. This point will be worked out more precisely in chapter 5.

The measuring arrangement was built up with an air cooled heat sink with a very low thermal resistance - so the assumption of a constant reference temperature at the bottom side of the isolation layer is nearly is given. For slight heating of the PCB contacts, the MOSFET operates in the linear active mode (not in the on- state). In the thermography pictures of the measurement setup (see Fig. 8(a) and 8(b)) the thermal resistances results in $R_{th}(42\text{vias}) = 2.91K/W$ and $R_{th}(97\text{vias}) = 2.38K/W$. All in all the simulated results are consistent with the measurement results. The simulation results and the measured thermal resistances are plotted versus the number of vias in Fig. 11(a).

4. Transient thermal modeling of a single MOSFET soldered on a PCB

The measured $Z_{th-Pack}(t)$ function for the TO-Leadless package (see paper [2]) is plotted in Fig. 9(a). This function is the base for the analytical calculation of $Z_{th-Pack+PCB+Gap}(t)$. $Z_{th-PCB}(t)$ and $Z_{th-Gap}(t)$ are calculated with a first order approach ($\nu = 1$, see equation 6). The parameters are determined by the system geometry according to equations 4, 7, 8. A*

© VDE VERLAG GMBH · Berlin · Offenbach

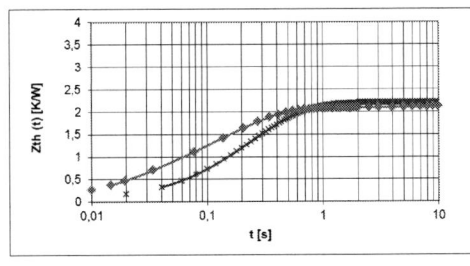
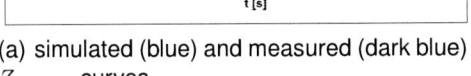

(a) simulated (blue) and measured (dark blue) Z_{th-OS} curves

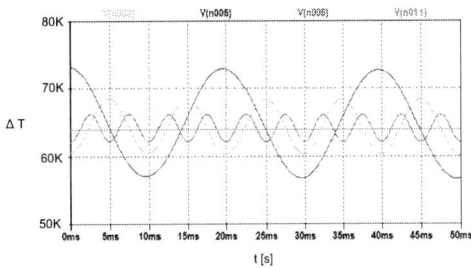

(b) temperature simulation of a 3-phase inverter @ $I_{ph-peak} = 130A$, blue - 50Hz, green - 100Hz, red - 200Hz, cyan - steady state

Fig. 10: measurement results of the Z_{th} measurement and simulation results of a 3-phase inverter

is the effective area for the heat transfer and is a function of different system properties (see equation 9). Because of this fact, the effective area was determined by the analysis of the simulated temperature field in the first step.

$$R_{th-PCB} = f(\text{no of vias, via-geometry}, A*), \quad R_{th-Gap} = f(\lambda_{Gap}, A*, h_{Gap}) \tag{7}$$

$$C_{th-PCB} = \rho_{Cu} \cdot A * \cdot h_{PCB} \cdot c_{th-cu}, \quad C_{th-Gap} = \rho_{Gap} \cdot A * \cdot h_{Gap} \cdot c_{th-Gap} \tag{8}$$

$$A = f(\text{no of vias, via-geometry, layer geometry, gapfiller, ...}) \tag{9}$$

In Fig. 9(b) the simulated and calculated $Z_{th-Pack+PCB+Gap}$ lines for the overall system are compared. It can be seen that the approximately approach delivers usable solutions if the effective heat transfer area A* is known. The measurement and simulation results of the overall system Z_{th-OS} can be seen in Fig. 10(a). The thermal impedance was determined by the temperature log at the cooling down process of the overall system arrangement. The usual printed system heat up curve is created with the estimated model parameters.

The measurement and simulation results are consistent - the FEM model calculates usable solutions for further investigations such as the calculation of the temperature oscillation during a load cycle of a 3-phase inverter.
The simulation results for a drive train inverter ($f_{nom} = 50Hz$) with a peak phase current of 130 A , different output frequencies and a single MOSFET IPT015N10N5 per switch can be seen in Fig. 10(b). The maximum overtemperature for a low inverter output frequency of 50Hz is about 10K higher than the steady state value. This point has to be considered in the thermal design of a power electronics system.

5. Development of design guidelines

The simulations show that an equivalent circuit calculation $Z_{th-OS} = Z_{th-Pack} + Z_{th-PCB} + Z_{th-Gap}$ (see also equation 7) yields usable results if the thermal effective area A* is known. In Fig. 11(a) the simulated R_{th} versus the number of thermal vias is plotted. For a high number of vias, the effective R_{th} of the system is limited. This effect is caused by the limited heat spreading. The idea is to calculate the ratio of the ideally calculated $R_{th-OS-ref}$ (the whole area is effective for heat transfer) and the simulated $R_{th-OS-sim}$ to yield the ratio of the effective area A* and the total area A_{ref}. This ratio is plotted in Fig. 11(b) versus the number of vias.

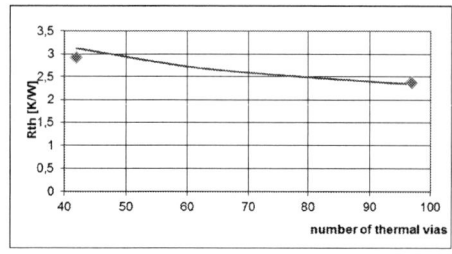 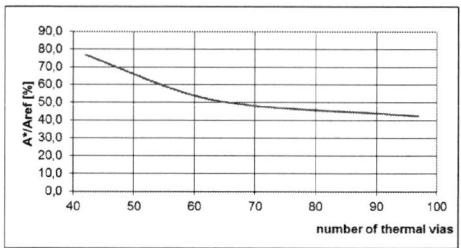

(a) R_{th} vs. number of thermal vias, blue line - simulation, red points - measurement

(b) effective area A* vs. number of thermal vias

Fig. 11: R_{th} and effective area A* vs. number of thermal vias

---

For the system design the following recommendations are derived:
- use a number of vias between 60 and 90.
- calculate the ideal heat transfer area A_{ref}
- a thumb value for the effective area A* can be taken from Fig. 11(b)
- calculate the thermal impedance with $Z_{th-OS} = Z_{th-Pack} + Z_{th-PCB} + Z_{th-Gap}$
- use the equations 4, 7, 8 for parameter calculation

So far the results are only validated for the investigated TO-Leadless package. Further package investigations and the development of a power electronics thermal design software tool will be performed in future

6. Conclusion

The thermal behavior of a single MOSFET with a TO-Leadless package, mounted on a PCB, was analyzed in this paper. Steady state and transient thermal simulations were performed and verified with measurements and analytical calculations. It was shown that the simulations and the analytical calculations yield usable results. The key issue in the analytical way is to determine the effective thermal area. The results can be used for the design of a high current power electronic system. Against this backdrop, some easy to use design guidelines were developed for the investigated high current package. In further investigation other packages and system configurations will be analyzed. Finally, the goal is the development of a power electronics thermal design software tool.

7. References

[1] Infineon. *Datasheet IPT015N10N5*. Munich, 2015.

[2] Ralf Walter, Ralf Siemieniec, and Marion Hoja. *An Improved and Low-Resistive Package for High-Current Mosfet*. Power Electronics and Applications (EPE'15 ECCE-Europe), Geneva, 2015.

[3] Uwe Franke, Thomas Ellinger, and Jürgen Petzold. *Thermal modelling of power electronic systems using model order reduction of large scale nite elemente models*. Power Electronics and Intelligent Motion (PCIM) - Europe, Nuremberg, 2008.

[4] Arendt Wintrich, Ulrich Nicolai, Werner Tursky, and Tobias Reimann. *Application Manual Power Semiconductors*. ISLE Verlag, Ilmenau, 2nd edition edition, 2015.

PCIM Europe 2016, 10 – 12 May 2016, Nuremberg, Germany

Packaging and Characterization of Silicon and SiC-based Power Inverter Module with Double Sided Cooling

Charles-Alix Manier[1], Hermann Oppermann[1], Lothar Dietrich[1], Christian Ehrhardt[1], Zoltán Sárkány[2], Marta Rencz[2], Bernhard Wunderle[3], Wilhelm Maurer[4], Radoslava Mitova[5], Klaus-Dieter Lang[6]

[1] Fraunhofer IZM, Berlin, Germany
[2] Budapest University of Technology and Economics, Budapest, Hungary
[3] Chemnitz University of Technology, Chemnitz, Germany
[4] Infineon Technologies AG, Munich, Germany
[5] Schneider Electric, Grenoble, France
[6] TU Berlin, Berlin, Germany

Abstract

Thermal management and especially cooling attracts more and more attention during phases of conception and fabrication of power modules to increase their field operation time and/or power density. Indeed thermal power losses impact the reliability of power systems, which are in a same time converging towards always higher power density and performances. Since standard solder die attach and wire bonding are reaching their limits, a novel concept of assembly has been realized with double sided cooling (DSC), as fully integrated part of the power module. Half-bridge modules either based on IGBTs and silicon diodes or based on SiC- devices have been assembled by combining Transient Liquid Phase Bonding (TLPB) and Transient Liquid Phase Soldering (TLPS). SiC and silicon wafers were prepared by deposition of a high density area array of Cu posts. In order to switch 1200V / 25A copper posts were electroplated 80 μm in height with a thin tin cap for flip chip bonding by TLPB. To attach a second DCB substrate on the backside, a TLPS paste was used. All interconnects of the modules are formed by intermetallic bonds for the target purpose of increased capability in heat dissipation, improved reliability and higher operation temperatures. Results on fabrication, thermal characterization and active power cycling (ΔTj = 125 K) are here also reported.

1. Introduction and general motivations for double sided cooling

Emerging applications, growing market segments and upcoming breakthrough of wide-band gap semiconductor technologies like e.g. SiC or GaN [1]-[3] for higher cost and global energetic efficiencies (i.e. from device fabrication to field application), are inevitably leading to higher requirements in terms of current densities, thermal management and packaging technologies. In order to address increasing power density of bare semiconductors and simultaneously higher module compactness, novel concepts, including new materials and joining technologies must be developed, tested and implemented [4] for extracting the highest potential of WBG semiconductors.

Silver sintering is one promising candidate for the next generation of die attach materials, with major technical advantages combining higher microstructural stability and higher thermal conductivity than free lead solder materials. In this regard, the thermal resistance of the die interconnects can be lowered and reliability increased. Nevertheless increasing the die attach robustness does not solve the global problematic but rather shift it further to a weaker interface. In the case of a traditional topology with wire bonds, it rather tends to intensify the reliability issues towards the wire-bond die interconnects, hereafter emphasizing the "Achilles heel" of power modules. Copper or bi-metallic wire-bonds with larger diameter for handling

© VDE VERLAG GMBH · Berlin · Offenbach

higher current loads, wire redundancy or even large ribbons are considerable remedial measures but traditional die metallization must be first adapted for example by wafer backend processing. To exploit the joint synergy between thermal management and packaging reliability, double sided cooling gains in attraction, since an additional heat flow path is provided to the top side and wire bonds eliminated with related EMC and reliability concerns.

2. Innovative Packaging

2.1. Module presentation

In the present approach, all the electrical interconnects rely on the formation of intermetallics phases, using an alternation of Transient Liquid Phase Bonding (TLPB) and Transient Liquid Phase Soldering (TLPS). These processes are based on low temperature metal fusion; the liquid phase wets the surfaces to be bonded but solidifies at the constant soldering temperature by transformation of the liquid alloy into solid intermetallic phases. In this way, high-temperature interconnections are achieved, the solder material cannot remelt when operation temperature exceeds the soldering temperature. Using this approach, assembly by stacking is feasible and high reliability is targeted [5]-[7] also under power cycling.

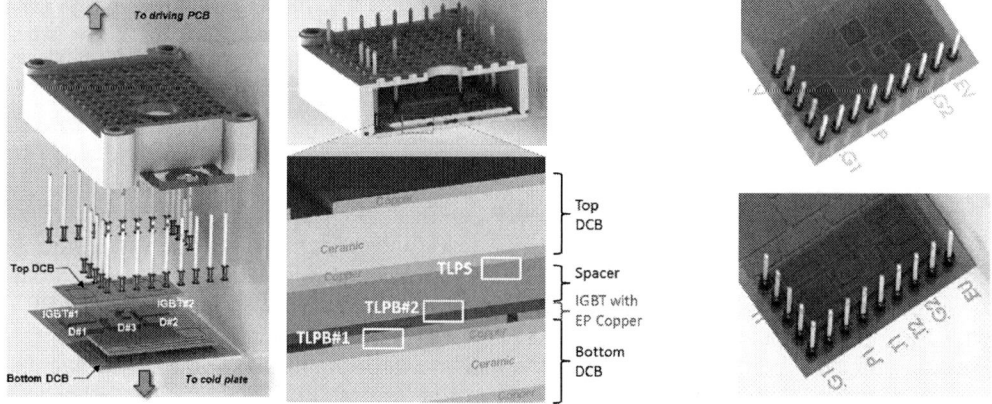

Fig. 1. Overview of the packaging concept for Double Sided Cooling (DSC) with Silicon half bridge inverter module (bottom right) and SiC-JFET based module (top right)

For housing and feeding the contacts out of the package, a standard casing based on the EasyPIM-1B from Infineon is used. With a housing bottom surface of ca. 38x32 mm², the DSC module is very compact. As presented in Figure 1 it can be screwed to a cooler or cold plate beneath and the pins soldered to a board placed above. The module and casing can be further tuned to add a single massive Thermo-Electric Cooler (TEC) or multiple small TECs on the top DCB and a thermal buffer above the TEC to cope with harsh cyclic power overload transients [8]. The assembled half-bridge module requires only half of an EasyPIM-1B housing. Two variants with the same bottom and top DCB designs were conceived. In one case, the module comprises two IGBTs and three Si-diodes, one dedicated to thermal sensing (Fig. 1, bottom right). In the other case, three SiC JFET were used instead of the Si devices, one being used for balance in assembly (Fig. 1, top right). In both cases conductive joiners are also required to close the electric circuit, i.e. to bring back the current from the top to the bottom DCB substrate, placed in the middle of the modules.

2.2 IGBT, Diode and SiC JFET wafer level preparation

Prior to assembly, the devices need to be adapted in respect of the assembly concept. The thin power devices (150 mm IGBT and Si-Diode wafers and 100 mm SiC-JFET wafers, voltage class of 1200V/35A, provided by Infineon) necessitate back-end processing. The

110 µm thin active device wafers are first mounted on temporary carrier wafer to enable a safe handling during processing. The front side aluminum finish metallization of the semiconductor devices is modified by addition of an 80 µm thick structured copper plating and a thin tin layer on top. The wafers are first sputtered with a seed layer consisting of a diffusion barrier and a Cu galvanic starting layer. Then a thick photoresist is lithographically structured, and the wafers are electroplated using specially tuned electrolytes to redistribute and pattern the contacts with a low stress Cu layer and fined grain Sn depots (fig. 2).

Since the plating speed is normally limited by the maximum copper concentration itself also being limited by the copper salt solubility in electrolytic solutions, a high pressure electroplating cell with a very strong bath agitation was used. Different designs were considered for the frontside contacts, i.e. anode of the diodes and for the IGBT and JFET the Gate and the Emitter/Source. They were either fully plated or structured in form of copper pillars (Fig. 2, right). During electrodeposition, current densities ranging from 3-4 up to 10 Adm were stepwise applied leading to a process time of ca. 80 min.

After plating, the deposited copper is smoothed by mechanical planarization. The height uniformity is here levelled over the entire wafer. The resulting plating uniformity is very accurate (+/- 1 µm). Not only all the contacts formed on the component are made planar with all the same height, but also all the components present on the wafer present a similar total thickness. After thin-film processing, the wafers are de-mounted from their carrier and diced.

Fig.°2. Back-End wafer processing with thick copper plating technology at Fraunhofer IZM [9]

The resulting devices present in the end a highly thermo-electrical conductive and solderable frontside interconnect. Ampacity (especially for the emitter) is also increased, compared to solder or wire bond technologies. As shown, a high aspect ratio structured copper metallization is formed, with a smooth thin tin surface, making the components highly suitable for assembly by TLPB and for dissipating heat from their frontside during operation.

2.3 Assembly technologies applied

The assembly starts with the bonding of the post-processed thin power devices. They are first placed per flip-chip on the bottom Aluminium Nitride (AlN) direct copper bonded (DCB) substrate having a smooth copper finish. Thick Joiners (~500µm) are also assembled during this step. A reflow is then performed with the help of a dedicated jig.

The second main step is the assembly of standoffs/posts over each of the devices and covering the device backside. The posts are first cut from plates of precise thickness with a tin finish These parts are then assembled on the backside metallization of all the respective active devices by transient liquid phase bonding for formation of an intermetallic interconnect [10]. This is followed by die underfilling, increasing the mechanical robustness and the electrical isolation between gate and emitter. The posts fulfil three distinct functions. First, they enlarge the gap between top and bottom DCBs for improved electrical isolation. Their thickness must be complementary to the assembled 500 µm joiners, for latter capping with the top DCB. Secondly, the posts contact electrically the backside of the vertical components with the top DCB. The third function is the dissipation of the thermal losses towards the top DCB. So they must present good electrical and thermal properties.

PCIM Europe 2016, 10 – 12 May 2016, Nuremberg, Germany

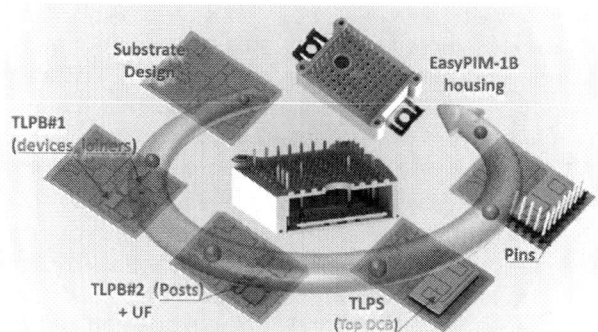

Fig.°2. Simplified illustration of assembly

After that the top DCB is assembled over the module. The module is designed to deal with a top DCB size as reduced as possible to limit warpage. However, the different elements cannot inevitably match perfectly the same total thickness over the entire module. Therefore capping of the top DCB can only be performed with an interconnecting technology tolerating the differences in thickness. The top DCB substrate is here assembled using a process based on a stencil printable paste [11]-[12] suitable for Transient Liquid Phase Soldering, "capping" and contacting together all the elements of the module.

The combination of these different assembly technologies present the advantage of employing a rather low temperature assembly sequence with in the end a potential operation of the module for high temperatures.

2.4 Resulting Package

The resulting package is presented in Figure 3. The bottom DCB and the top DCB are recognizable as well as the pins bonded on the bottom DCB. X-Ray imaging performed from the top of the module is also depicted in Figure 3. The different elements composing the half-bridge module can be recognized. Due to the stack and the multiple copper layers, X-Ray absorption is high and only the TLPS can be distinguished. The TLPS paste tends to spread slightly out of the bond areas, furthers optimizations on the bonding process might be here required. An optical side view shows also the underfill at the edges of the devices.

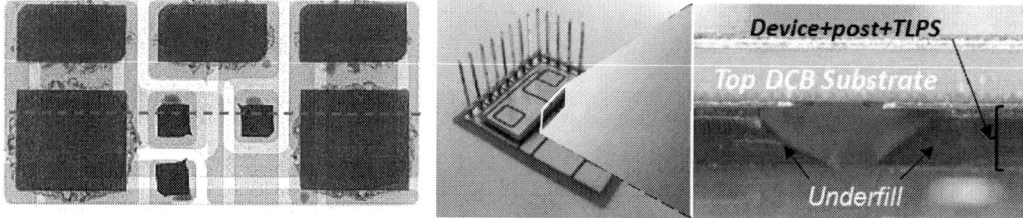

Fig.°3. Different views of the half bridge package

Figure 4 presents a metallographic cross-section across the full module done along the red dot line of Figure 3. The complete cross section is shown in the middle and the diverse interconnects above (TLPS) and below (TLPB) as well as the structure of the stack are more precisely shown.

On the bottom side the TLPB interconnect joining the Cu plated devices with the DCB substrates is exclusively formed of Cu_3Sn (η) and Cu_6Sn_5 (ϵ) intermetallics phases. Since both mating surfaces have been smoothed, the resulting bond interface is very thin and correspondingly its thermal resistance. In combination with the high thermal (300 – 400 W/mK) and electrical conductivity of the plated copper, the thermal losses generated by the components can be dissipated from the device frontside into the bottom substrate. The 80 μm thick copper plated on the frontside allows also rising the distance between the device and the bottom substrate, inhibiting any risk of flashover between both. The electrical

© VDE VERLAG GMBH · Berlin · Offenbach

isolation is also reinforced by the presence of the underfill. The TLPB bond between the device backside and the posts is also represented. Here the bondline is thicker, due to the waviness of the post surface. As a result, the bondline cannot be entirely transformed and some tin-rich rests in form of "Pockets" or thin interlayer in the middle of the bond can be found,the waviness allowing only locally the fine mating of both joined parts. The TLPS bond is constituted of Cu particles in a CuSn intermetallic matrix.

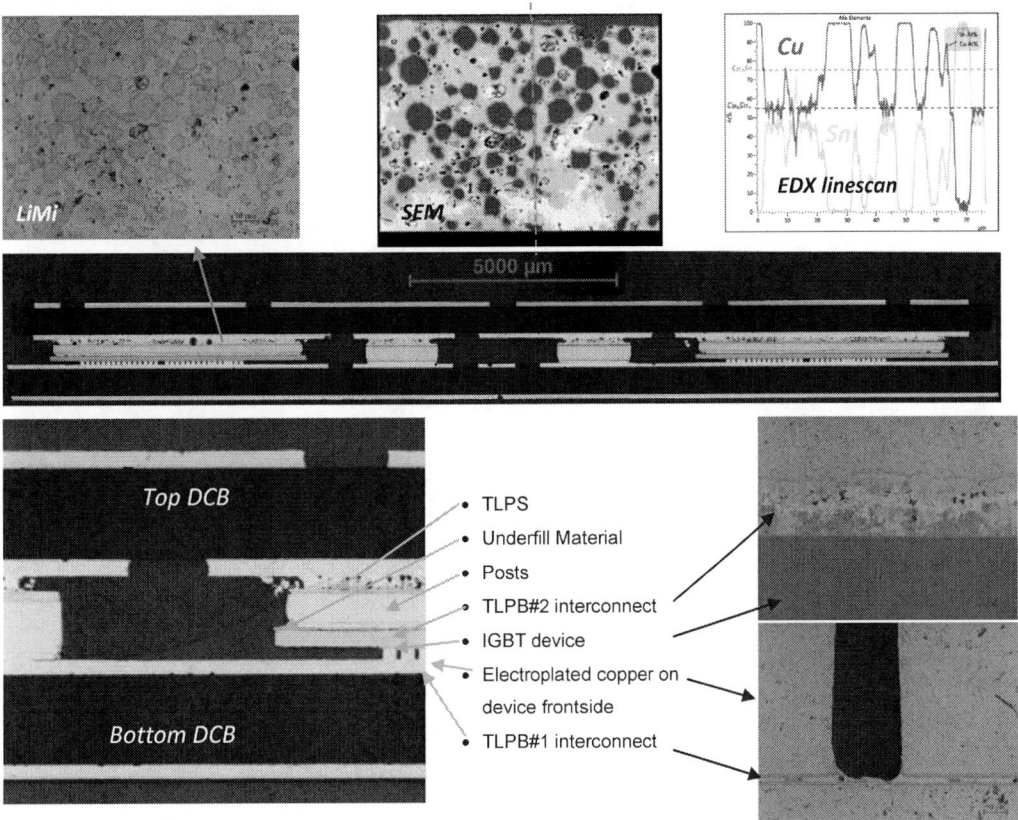

Fig. 4. Cross-section of a half-bridge module after assembly of the DSC structure

3. Electrical and thermal Characterization

3.1   DC characterization of DSC power modules and first switching results on SiC-Module

Static electrical characterization was carried out to verify the electrical functionality of the double side cooled power modules. The classical output characteristics of diode, IGBT and SiC JFET are plotted and proves the devices good performances after packaging (Fig.5). The electrical isolation of the IGBT based power modules was also measured and compared to the original EasyPIM module (DS_FP25R12W2T4). The measurement was carried out by applying high voltage between IGBT drain and source and by measuring the leakage current. The behavior of the double side module differs from the original, with a current leakage 5x lower for the DSC module, probably due do the use of isolating underfiller around the power devices. Both show a breakdown voltage nearing 1400V.

The SiC JFET double side cooled power modules were tested with DC/AC half-bridge power converter developed for this purpose. In figure 6 is shown the schematic principle of the converters. The JFET were controlled with the dedicated DS_FP25R12W2T4 gate driver. Figure 6 shows also the switching waveforms measured across one of the JFET, the gate and control signal across the complementary JFET and the load current. Due to limitation of the power supply the JFET were tested up to 400V and 1.5A input current. The oscillations of I_{Load} are caused by the low current used for this functionality test.

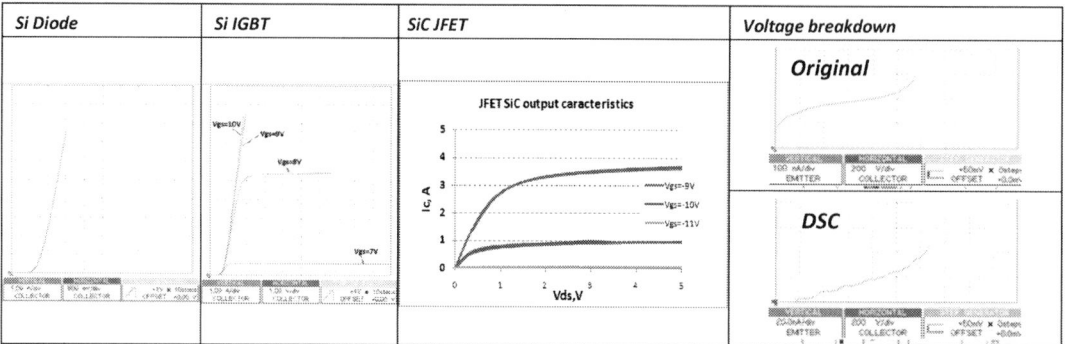

Fig. 5: output characteristics of diode I(V) and IGBT Ids(Vds) and SiC JFET Ids(Vds) and Voltage withstand of original EasyPIM module (top left) and the DSC IGBT based module (bottom)

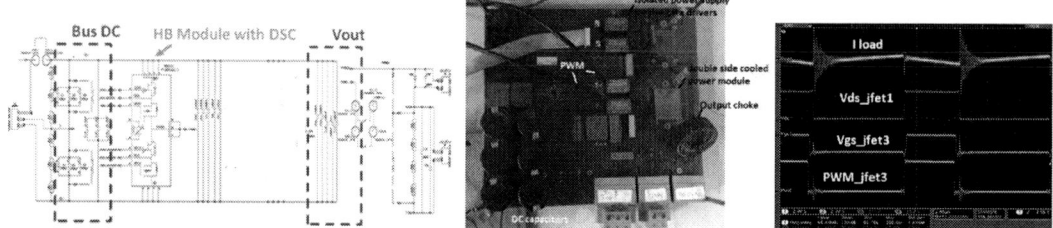

Fig. 6. Schematic of half-bridge inverter (left),and inverter PCB board (center) and switching of DSC packaged SiC-JFET (right)

3.2 Description of thermal transient characterization and results

A stepwise power excitation is applied to the semiconductor device and the time function of the junction temperature change is recorded. After proper evaluation of the thermal transient curve not only the junction to ambient thermal resistance can be calculated but the thermal resistances corresponding to the different material layers can also be identified [12]. For the identification and measurement of the thermal resistance of the package in both directions, i.e. towards the bottom (R_{thJC}, Junction-to-case) and top of the module where TECs can be mounted (R_{thJTEC}, junction-to-TEC) the Transient Dual Interface Method [13] was used, in this special case with 2x2 different boundary conditions. For R_{thJC}, the transient curve was captured with the module pressed to a cold plate with and without thermal interface material applied inbetween, while the top side was thermally insulated from the environment. Until the heat propagates inside the package the two transient curves are in perfect fit. The thermal resistance can be read at the separation point of the two curves where the effect of the different boundary conditions changes the shape of the transient responses. In a similar manner, R_{thJTEC} was measured by pressing the top surface of the modules to a cooling surface but with a copper spacer inbetween due to the pin length. The transient measurement was done with and without thermal paste applied between the cooling surface and the top surface of the module (the bottom surface being thermally insulated) and the R_{thJTEC} thermal resistance was read at the separation point of the latter two transient curves.

All chips of three Si half bridge modules and JFET transistors of two SiC half bridge modules were first characterized. The silicon based modules provided relatively low and consistent results with about 0.3 K/W thermal resistance for the transistors and 0.4 K/W resistance for the diodes both towards the top (R_{thJTEC}) and the bottom (R_{thJC}) side with low average deviation. The individual transistors of the silicon carbide modules provided different results.

© VDE VERLAG GMBH · Berlin · Offenbach

T2 transistor of both modules had relatively low thermal resistance in both directions, but the high thermal resistance of the *HB#81 T1* transistor indicates needs of assembly improvements. The *HB#82 T1* transistor was not measurable (no contact).

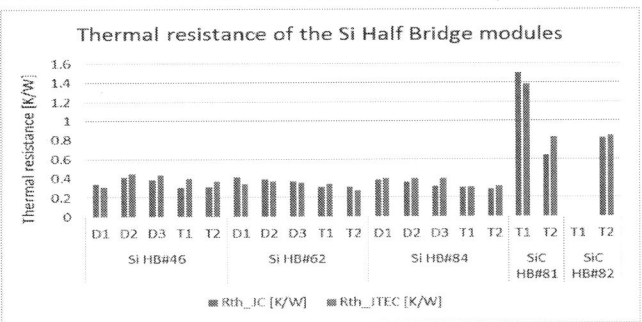

Fig. 9. Results of thermal resistance measurements of Si and SiC half bridge modules

4. Power Cycling, methodology and results

Active power cycling tests were carried out on two IGBT chips in two modules with 125 K temperature swing (ΔTj). The device is here also heated up by powering, and switched off to capture the cooling transient, thus periodically. The cooling transient of the DUT and corresponding structure function monitor the degradation of the package structure. During power cycling tests, the bottom side of the devices was fixed to a temperature stabilized cold plate, without cooling connected to the top side. To achieve large power dissipation with limited heating current, gates and collectors of the IGBTs were connected together. A constant current powering was applied to achieve 125 °C temperature change during a 30 second heating period and in case of change of the thermal resistance of the structure adjusted to keep the temperature change close to the target value. The cooling time was also set by 30 sec. A thin thermal insulator layer must be here put between the device cooling surface and the cold plate to increase the overall thermal resistance of the structure for achieving the desired high temperature swing. The heating power was ca. 90 W (about 10A). The cooling transient response was captured after every 1000 cycles. The IGBTs were selected so to avoid cross heating between adjacent transistors.

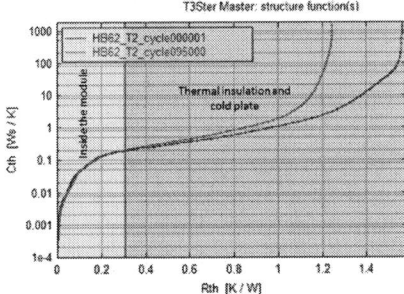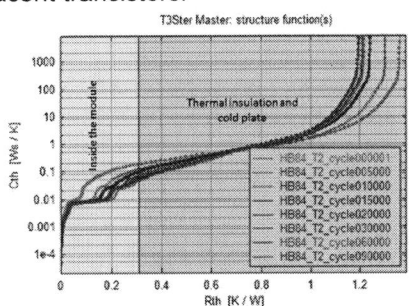

Fig. 10. Cumulative structure functions of device HB#62 T2 at the initial state and after 95000 power cycles (left) and of device HB#84 T2 after different number of power cycles (right)

More than 95000 power cycles have elapsed so far and both devices are still functional. Although none of the transistors failed up to now, the two devices show significantly different behavior. The junction to bottom thermal resistance of HB#62 T2 was initially measured to be 0.3K/W. No change is visible in the inner structure of the module; the structure functions (Fig. 10) captured at the beginning of the power cycling and after 95000 cycles both fit perfectly in this region (green background). Change was only experienced in the cooling environment outside of the module; the thermal resistance of the insulation layer put between the module and the cold plate decreased during the test significantly. This change of the overall thermal resistance was compensated during power cycling by increasing the heating current. However there is a significant degradation visible at the die attach level in case of HB#84 T2 transistor. There has been a gradual degradation. It starts already at the very beginning of cycling, but surprisingly this has slowed down as the cycles elapsed and stabilized after 60000 cycles, indicating a stop of the degradation. It has to be noted that already at the initial state the die attach layer had much higher resistance than at the other measured sample. Further analysis will be performed to understand the root cause of the structural evolution.

5. Conclusion

A compact packaging integrating double sided cooling for power modules was assembled with materials already in use in manufacturing of traditional power module packages. A new wafer level process including thin wafer handling was carried out for the adaptation of power devices for flip-chip assembly. The double sided cooled structure is completely built using emerging interconnect processes based on the transformation of the bond into intermetallic phases, making it suitable for high temperature applications. First results on half Bridge modules were demonstrated regarding functionality and thermal characterisations. Promising power cycling results are also reported with some evolution and stabilization of the internal package structure, notably without failure of the devices, testifying a first complete proof of concept. The origin of the structural evolution is probably related to the process developed for assembly and will be further investigated. The assembly process will be further extended towards a compact three phase Full Bridge configuration integrating in a similar manner double sided cooling.

6. Acknowledgements

The authors gratefully acknowledge their funding by the EU FP7 Project "Smartpower" *Smart Integration of GaN & SiC high power electronics for industrial and RF applications*, Grant #288801.

7. References

[1] J. Millán and P. Godignon, "Wide Band Gap power semiconductor devices," Electron Devices (CDE), 2013 Spanish Conference on, Valladolid, 2013, pp. 293-296.

[2] U.S. Department of Energy, Quadrennial Technology Review 2015, Chapter 6: "Innovating Clean Energy Technologies in Advanced Manufacturing", http://energy.gov

[3] R. Khazaka, L. Mendizabal, D. Henry and R. Hanna, "Survey of High-Temperature Reliability of Power Electronics Packaging Components," in IEEE Transactions on Power Electronics, vol. 30, no. 5, pp. 2456-2464, May 2015.

[4] P. Dietrich, "Joining and package technology for 175 °C Tj increasing reliability in automotive applications", Microelectronics Reliability, Volume 54, Issues 9–10, September–October 2014, Pages 1901-1905.

[5] S. A. Moeini, H. Greve and F. P. McCluskey, "Reliability and failure analysis of Cu-Sn transient liquid phase sintered (TLPS) joints under power cycling loads," WiPDA, 2015 IEEE 3rd Workshop, Blacksburg, VA, 2015, pp. 383-389.

[6] C. Ehrhardt, M. Hutter, C. Weber, K.-D. Lang, "Active power cycling results using copper tin TLPB joints as new die-attach technology", proceedings of PCIM 2015, Nuremberg, Germany, p. 1268-1275 .

[7] H. Greve, S. A. Moeini and F. P. McCluskey, "Reliability of paste based transient liquid phase sintered interconnects," Electronic Components and Technology Conference (ECTC), 2014 IEEE 64th, Orlando, FL, 2014, pp. 1314-1320.

[8] B. Wunderle, M. Springborn, D. May, R. Mrossko, M.A. Ras, C.-A. Manier, H. Oppermann, R. Mitova, "Phase change based thermal buffering of transient loads for power converter," THERMINIC 2014, vol., no., pp.1-7, 24-26 Sept. 2014.

[9] Annual report 2013Fraunhofer IZM, p. 38, http://www.izm.fraunhofer.de/content/dam/ izm/en/documents/Publikationen/Jahresberichte/AR_2013_14/AR_2013_14_EN.pdf

[10] T.-Y. Tsai, Y.-J. Chang, K.-N. Chen, "Quality and reliability investigation of Ni/Sn transient liquid phase bonding technology," Physical and Failure Analysis of Integrated Circuits (IPFA), 2015 IEEE 22nd International Symposium on the, Hsinchu, 2015, pp. 492-495.

[11] C. Ehrhardt, M. Hutter, H. Oppermann and K. D. Lang, "A lead free joining technology for high temperature interconnects using Transient Liquid Phase Soldering (TLPS)," ECTC, 2014 IEEE 64th, Orlando, FL, pp. 1321-1327.

[12] V. Szekely, S. Torok, E. Nikodemusz, F. Farkas, M. Rencz, "Measurement and Evaluation of Thermal Transients", In: Proceedings of the 18th IEEE IMTC 2001.Budapest, Hungary, pp. 210-215.

[13] JESD51-14, "Transient Dual Interface Test Method for the Measurement of the Thermal Resistance Junction-to-Case of Semiconductor Devices with Heat Flow Through a Single Path"

Digital adaptive control approach to dynamic response improvement for compact PFC rectifiers

Trong Tue Vu, Icergi Ltd., Dublin, Ireland, ttrongvu@icergi.com
George Young, Icergi Ltd., Dublin, Ireland, georgeyoung@icergi.com

Abstract

A simple digital controller capable of fast output voltage regulation of power factor correction (PFC) rectifiers with minimal output capacitance is presented in this paper. The method is based on the use of two controllers for the outer voltage loop: an inter-line-cycle controller with slow dynamic responses takes care of steady state operation and small deviations from the set point while a fast intra-line-cycle corrector observes any abrupt change in the load current, and accordingly force the power command level to avoid significant overshoot or undershoot at the output voltage. Unlike existing studies, both controllers are derived based on the idea of balancing the input and output power rather than minimizing output voltage errors, and as a consequence allow faster responses and globally stable operation. The practicality of the proposed controller is shown on a 200W four-level bridgeless PFC prototype with a switching frequency of 143kHz. Experimental results confirm that performance is significantly improved as compared to conventional controllers.

1. Introduction

PFC is a near universal requirement for AC-DC power converters with input power above 75W in the case of general power applications, and with a lower power threshold in the case of lighting applications. A typical control approach for a PFC stage involves a current loop being used to program the input current and a voltage loop being used to regulate the output voltage. The communication between two loops is via a control variable, called emulated resistance R_e, which defines the power transfer profile of the PFC stage.

There is considerable attraction in being able to reduce the size of the electrolytic capacitor, which can account for 20% plus of the volume utilization in many power designs. A practical limitation on usage of smaller values of electrolytic capacitance may come to control challenges. In particular, a fast voltage loop can result in gross distortion of the current waveform due to R_e varying widely across the line cycle while a slow voltage loop is unable to maintain transient output voltages within desired bands. A number of approaches has been proposed to bypass the compromise associated with conventional control approaches for PFC stages including output ripple elimination through discrete-time sampling [1] and comb filters [2], dead-zone approaches [3], digital control [4, 5], nonlinear control [6], and dual-voltage-loop controllers [7]. However, many of them [2, 3, 6] are too complex to be implemented in a low-cost microcontroller while the others [5, 7, 1] are not fast enough to deal with reduced capacitance applications in which any severe step load can cause dramatic overshoot or undershoot on the bus voltage,in turn, shortening lifespan of electronics components or causing device failures. Therefore, this paper proposes simple yet effective multiple voltage loop control based on a power-balance principle allowing fast dynamic responses with globally stable operation.

2. Proposed digital control scheme

Given the challenges of lifting the bandwidth of PFC stages without significant effect on current waveforms, this paper proposes an adaptive control structure as illustrated in Fig. 1. Fundamentally, the converter is still controlled by two loops: an inner current loop that uses a digital predictive approach to input current programming similar to that presented in [4], and an outer loop that regulates the output voltage through emulated resistance adjustment. However, the outer loop is now advanced by unique combination of two controllers, including a slow inter-cycle voltage compensator and a fast intra-cycle voltage corrector, operating in a manner that the emulated resistance R_e is periodically updated every half line cycle by the slow controller, and is only corrected by the fast controller whenever major disturbances in the operating conditions are detected within half line cycles.

In order to avoid bumpy transfers on control actions, the output of the intra-cycle controller, R_{ecor}, is used as a correction term penalising under/over voltage transients rather than completely replace the signal output by the inter-cycle one. This arrangement requires memoryless behaviours from the fast controller and consistency in control objectives for both compensators, which can be simply achieved by equating the averaged input and output power of the PFC such that the averaged capacitor voltage can be brought back to a desired level within a half/quarter line cycle. Information required for power conditioning includes only the bulk capacitor voltage measured at zero-crossing points, crests, and troughs of the input voltage waveform, which should be available if digital implementation of the current loop is of interest. Therefore, the

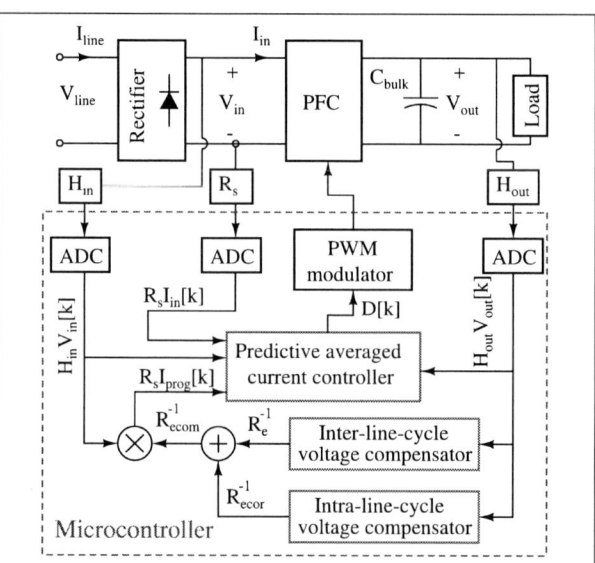

Fig. 1: Proposed digital control structure for PFC rectifiers.

proposed control scheme can be fully implemented in any microcontroller facilitating an Analog-to-Digital Converter (ADCs) with at least three channels for input voltage, output voltage, and inductor current measurements, and timer chains for PWM modulation. Designing compensators for the current loop and voltage loop is discussed in more detail in Sections 3 and 4.

3. Predictive current loop controller

The purpose of this loop is to keep the input current I_{in} proportional to the input voltage V_{in}, where the gain is controlled by the voltage loop through the emulated resistance R_e. Therefore, if the power stage is chosen, just for demonstration in this paper, to be a four-level bridgeless totem pole boost converter as outlined in Fig. 2, the control goal is to program the averaged inductor current I_{Lavg} to follow the reference governed by

$$I_{ref} = \frac{V_{in}}{R_{ecom}} \tag{1}$$

Generally, the four-level boost rectifier operates in a similar manner as conventional ones except that a multiple-phase modulation scheme is deployed to allow significant reduction in voltage stresses on the switching devices and inductor current ripples, see [8] for more details. Hence, the well-known state-space averaging technique as comprehensively presented in [9] can be applied to calculate the evolution of the averaged inductor current over time

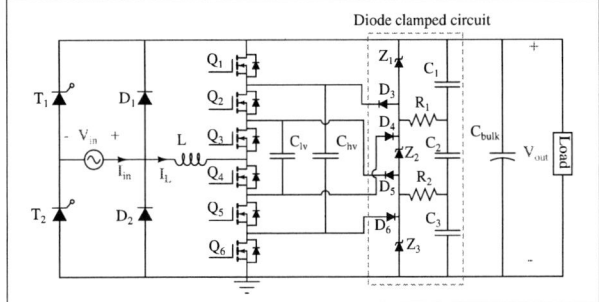

Fig. 2: Four-level bridgeless totem pole boost rectifier used as a prototype for validation of the proposed controller

$$L\frac{dI_{Lavg}}{dt} = -\left(1 - D_{bot}\right)V_{out} + V_{in}, \tag{2}$$

where D_{bot} is the duty ratio of the three bottom MOSFETs Q_4, Q_5 and Q_6. The averaged current I_{Lavg} is assumed to be periodically sampled every switching cycle, so the relation between samples in two consecutive cycles can be obtained by discretising (2) to

$$I_{Lavg}[n] - I_{Lavg}[n-1] = \frac{T_{pwm}}{L}\left(-\left(1 - D_{bot}[n-1]\right)V_{out} + V_{in}\right), \tag{3}$$

$$I_{Lavg}[n+1] - I_{Lavg}[n] = \frac{T_{pwm}}{L}\left(-\left(1 - D_{bot}[n]\right)V_{out} + V_{in}\right). \tag{4}$$

From the predictive control perspective [4], it is desired to set the duty ratio $D_{bot}[n]$ for the switching cycle n such that the averaged inductor current will equal the reference level at $t = (n+1)T_{pwm}$, i.e. $I_{Lavg}[n+1] = I_{ref}$. Substituting the control target into (4) and adding the corresponding sides of (3) and (4) gives

$$I_{ref} - I_{Lavg}[n-1] = \frac{T_{pwm}}{L}\left(-\left(2 - D_{bot}[n] - D_{bot}[n-1]\right)V_{out} + 2V_{in}\right). \tag{5}$$

If the duty ratio is assumed to change gently between two successive cycles, the sum of $D_{bot}[n]$ and $D_{bot}[n-1]$ can be approximated by $D_{bot}[n] + D_{bot}[n-1] \approx 2D_{bot}[n]$. Substituting the result equation into (5) and solving for $D_{bot}[n]$ gives

$$D_{bot}[n] = 1 - \frac{V_{in}}{V_{out}} + \frac{L}{2V_{out}T_{pwm}}\left(I_{ref} - I_{Lavg}[n-1]\right). \tag{6}$$

Equation (6) should be theoretically sufficient for inductor current programming, however, due to the lack of an integrator, there exists a disparity between I_{ref} and I_{Lavg}, causing distortions around the tail parts of the current waveform. To overcome such an issue, an integral term is introduced to (6), which results in

$$D_{bot}[n] = 1 - \frac{V_{in}}{V_{out}} + \frac{L}{2V_{out}T_{pwm}}\left(I_{ref} - I_{Lavg}[n-1] + k_I\sum_{n=1}^{\infty}I_{error}[n-1]\right), \tag{7}$$

where $I_{error}[n-1] = I_{ref} - I_{Lavg}[n-1]$ denotes the deviation from the current set-point. The integral gain k_I is practically set to be in the order of 0.04.

PCIM Europe 2016, 10 – 12 May 2016, Nuremberg, Germany

4. Voltage loop controller synthesizer

4.1. Inter-line-cycle control

It is assumed that the current loop is ideal, and power losses and energy storage in magnetic devices are negligible, suggesting that the instantaneous power appearing at the output of the PFC stage can be well approximated by

$$P_{out}(t) \approx P_{in}(t) = V_{in}(t)I_{in}(t) = \frac{V_m^2}{2R_e}\left(1 - cos\left(2\pi f_{line}t\right)\right), \tag{8}$$

where V_m denotes the amplitude of the line input voltage while f_{line} is the line frequency.

The PFC typically feeds one or several downstream stage driving loads with slow variations in voltage and current. Hence, the PFC load can be modelled by a constant power sink which should equals $\frac{V_m^2}{2R_e}$ at steady state. Therefore, the variable term of (8) will be absorbed by the bulk capacitor, which results in double-line-frequency ripples at the output voltage, as illustrated in Fig. 2.

The output voltage ripples are not desirable in the context of power factor correction as feeding them back will pollute the power demand controlled by R_e, and as a consequence distorts the inductor current. Therefore, processing the feedback voltage before using it for control purposes is generally required. For digital implementation, the ripples can be effectively removed by sampling the output voltage at the time instants where $P_{in}(t)$ is either zero or maximum. Such a sampling scheme will be used in this paper to obtain feedback signals for both the inter-line-cycle and intra-line-cycle controllers. Particularly, the output voltage is captured every quarter line cycle with sampling points highlighted by red dots as illustrated in Fig. 3. For inter-cycle control, only information at zero-crossing points, e.g. t_{n-2}, t_n , t_{n+2}, etc..., is used while all voltage samples are made use by the intra-cycle controller.

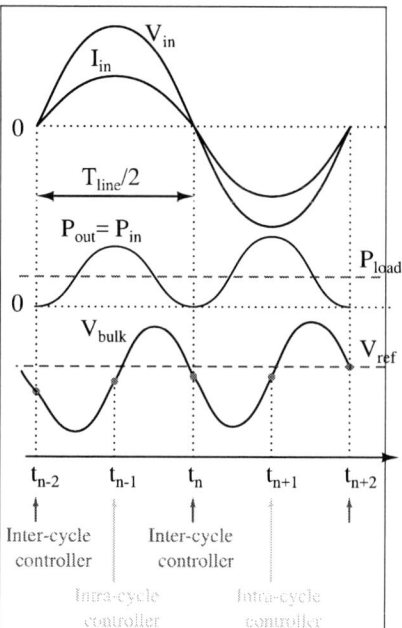

Fig. 3: Operational waveforms of a PFC stage.

The averaged output voltage is assumed to deviate from the set-point V_{ref} at time $t = t_n$, so it is desired to command a value of R_e such that the output voltage gets back to the reference level at $t = t_{n+2}$. This control objective can be achieved if the power balance condition during the period $[t_n, t_{n+2}]$ as described by (9) is satisfied

$$P_{in\_avg}^{[t_n,t_{n+2}]} = P_{load}^{[t_n,t_{n+2}]} + P_{adj}^{[t_n,t_{n+2}]}, \tag{9}$$

where $P_{in\_avg}^{[t_n,t_{n+2}]}$ and $P_{load}^{[t_n,t_{n+2}]}$ denotes the averaged power entering the PFC stage and power sunk by the load during $[t_n, t_{n+2}]$, respectively, while $P_{adj}^{[t_n,t_{n+2}]}$ is the extra power required to charge the bulk capacitor voltage from $V_{out}[t_n]$ to V_{ref}, which can be simply expressed by

$$P_{adj}^{[t_n,t_{n+2}]} = \frac{C_{bulk}}{T_{line}}\left(V_{ref}^2 - V_{out}^2[t_n]\right). \tag{10}$$

© VDE VERLAG GMBH · Berlin · Offenbach

The load power is typically not available at the time of calculation; therefore, an estimation must be carried out by some means. One interesting solution is to rely on the past averaged input power and capacitor voltage as given by (11)

$$P_{load}^{[t_n, t_{n+2}]} = P_{in\_avg}^{[t_{n-2}, t_n]} + \frac{C_{bulk}}{T_{line}} \left(V_{out}^2[t_{n-2}] - V_{out}^2[t_n] \right).$$ (11)

Sine R_e is only adjusted at zero-crossing points and kept fixed for the whole half line cycle, the averaged input power during the intervals $[t_{n-2}, t_n]$ and $[t_n, t_{n+2}]$ can be approximated by

$$P_{in\_avg}^{[t_{n-2}, t_n]} = \frac{V_m^2}{2R_e[t_{n-2}]}, \quad P_{in\_avg}^{[t_n, t_{n+2}]} = \frac{V_m^2}{2R_e[t_n]}.$$ (12)

Substituting (12) and (11) into (9) and solving for $R_e[t_n]^{-1}$ gives

$$R_e[t_n]^{-1} = R_e[t_{n-2}]^{-1} + \frac{2C_{bulk}}{T_{line}V_m^2} \left(V_{ref}^2 + V_{out}^2[t_{n-2}] - 2V_{out}^2[t_n] \right).$$ (13)

In theory, the proposed control law as described by (13) should be able to bring the PFC output to the set-point within a half line cycle if a step load is committed right after zero crossing points. However, in practice, performance is affected by non-deterministic load transition, component tolerances, and measurement errors, which pushes the transient responses up to two or three half-line cycles, which still a lot faster than conventional controllers [6]. Other advantages of (13) are simplicity benefiting digital implementation and inherent integral actions allowing zero steady state error.

As compared to the linearised algorithm developed by [5], (13) is an exact non-linear solution to the control problem which does not rely on any small signal approximation during the development. Therefore, the proposed controller is, in theory, globally stable. On top of that, the averaged capacitor voltage instead of the peak value is controlled, and so any concerns regarding variations in mean V_{out} as noted in [5] should disappear here.

4.2. Intra-line-cycle control

The intra-cycle controller can be derived in a similar manner as the inter-cycle controller; however, the updating interval now reduces to a quarter cycle, which boosts the closed-loop bandwidth of at least twice as compared to that of (13). Given the desire to force the averaged output voltage to reach the reference level within a quarter line cycle, one can find the expression governing the emulated resistance as

$$R_e[t_{n-1}]^{-1} = R_e[t_{n-2}]^{-1} + \frac{4C_{bulk}}{T_{line}V_m^2} \left(V_{ref}^2 + V_{out}^2[t_{n-2}] - 2V_{out}^2[t_{n-1}] \right).$$ (14)

By comparing (13) and (14) one may argue that there is no need to complicate the design with dual controllers for the voltage loop as the intra-line-cycle controller should inherent all features of the inter-line-cycle one but possess a lot fast dynamic responses. Unfortunately, experimenting with the intra-cycle controller only shows that control performance is quite sensitive to switching noises, and inductor current distortion happens occasionally at certain operating conditions. For this reason, (14) is adapted to operate in combination with the inter-cycle controller. Particularly, the second term on the right hand side of (14), named ΔR_e^{-1}, is employed to detect any significant sag in the averaged capacitor voltage and correspondingly correct the emulated resistance. In order to avoid racing conditions, the threshold for ΔR_e^{-1} is set to a high value and two controllers should be activated at different operating points as reflected in Fig. 3.

5. Results

A 200W prototype of a four level boost rectifier is built according to the topology suggested in Fig.2 to verify the performance of the proposed controller. The rectifier is designed with the following specifications: universal line input voltage = 85 - 265V$_{rms}$, output voltage V_{out} = 175 - 420V, maximal flying capacitor voltage ripples ΔV = 10V, and switching frequency f_{pwm} = 150kHz, which suggests the component values as follow: the PFC choke L = 165μH, output capacitor C = 68μF, flying capacitors C$_{lv}$ = C$_{hv}$ = 400nF, diode clamped circuitry C$_1$=C$_2$=C$_3$ = 100nF, R$_1$ = R$_2$ = 50Ω, and MOSFETs BSZ22DN20NS3G for main switches Q1, ..., Q6. A comprehensive discussion on the component selection can be found in [8]. The proposed controller is fully implemented in digital environment using low-cost, fixed-point, 48MHz ARM Cortex M0 STM32F051 microcontrollers.

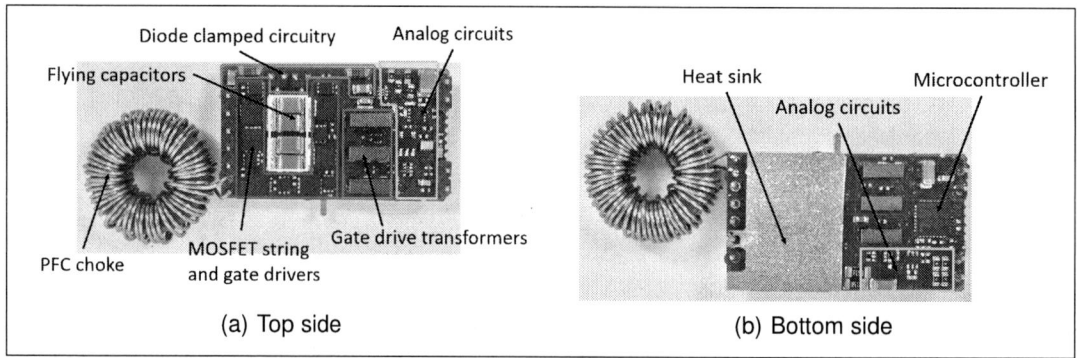

(a) Top side   (b) Bottom side

Fig. 4: Hardware prototype of a 200W bridgeless four-level PFC rectifier without bulk capacitors, thyristors, and EMI filters. The dimension of the board excluding the PFC choke is about 20mm x 35mm.

Figure 4 gives a snapshot of the hardware which has been used for final testing. Most components are amounted on the top side and the PCB board except the microcontroller and a part of the analogue circuitry, which allows some room for the mechanical part. Thanks to the multi-level approach, the whole design can be miniaturised into a small area of just 20mm x 35mm without penalising converter efficiency and control performance.

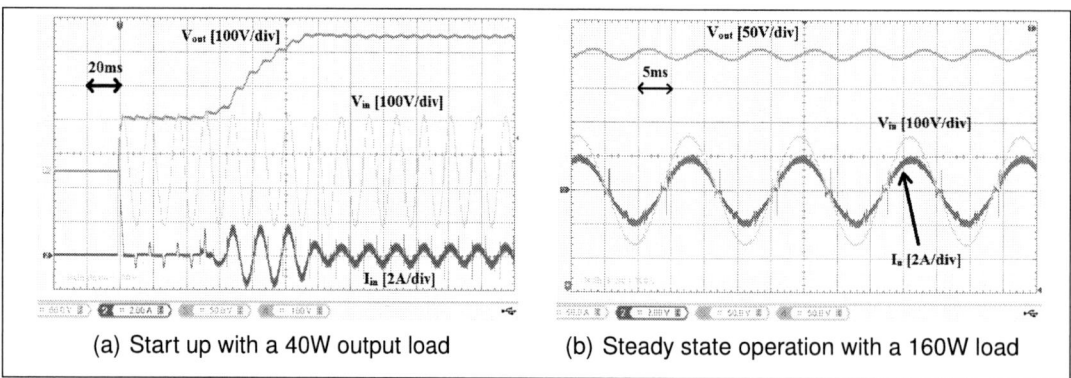

(a) Start up with a 40W output load   (b) Steady state operation with a 160W load

Fig. 5: Operating waveforms during startup and steady state conditions at low line V_{in} = 115V$_{rms}$: CH2 - Input current I_{in}, CH4 - Output voltage V_{out}, Math channel - Input voltage V_{in}

Figure 5 shows the startup and steady state operation of the four-level boost rectifier. For startup current management, the controller operates with limited demand power - in other words, putting constraints on the emulated resistance. One can realize from Fig.5(a) that the startup happens with no overshoot and the current is well controlled.

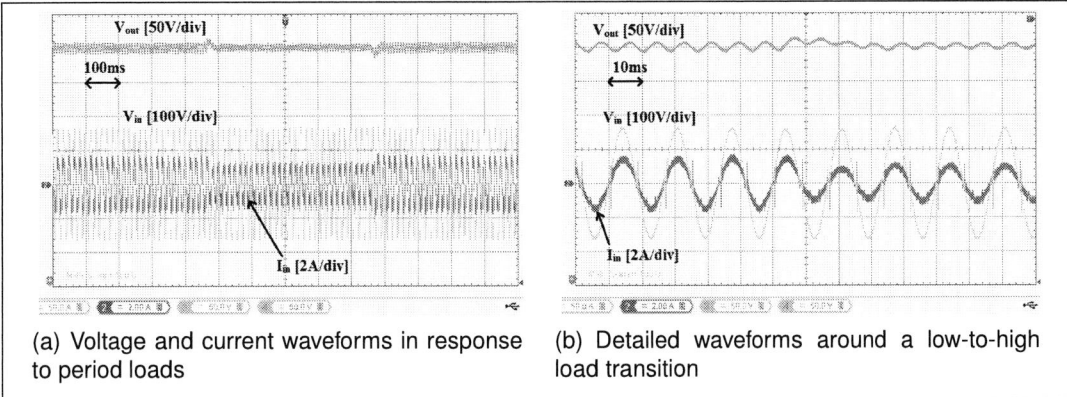

(a) Voltage and current waveforms in response to period loads

(b) Detailed waveforms around a low-to-high load transition

Fig. 6: Dynamic responses of the proposed controller under 80W-120W periodic load testing conditions: CH2 - Input current I_{in}, CH4 - Output voltage V_{out}, Math channel - Input voltage V_{in}

Figure 6 illustrates the line input current and output voltage of the PFC stage in response to a periodic load varying between 80W (40% nominal load) and 120W (60% nominal load) every 500ms. Examining the detailed transient waveforms in Fig.6(b) confirm that the controller manage to get the right value for R_e well within two line cycles with no significant overshoot. Due to the benign transient conditions, the inter-cycle controller can sort everything out with itself and no need to trigger the intra-cycle compensation.

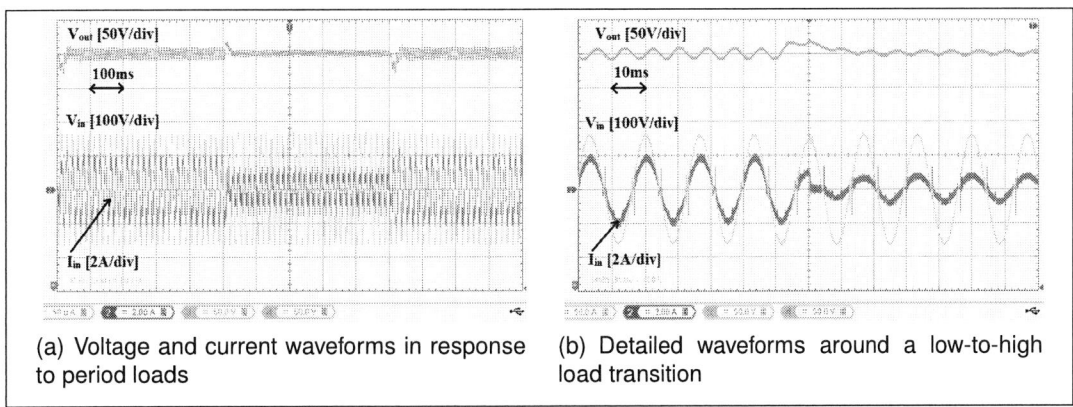

(a) Voltage and current waveforms in response to period loads

(b) Detailed waveforms around a low-to-high load transition

Fig. 7: Dynamic responses of the proposed controller under severe transient loads stepping between 60W and 160W: CH2 - Input current I_{in}, CH4 - Output voltage V_{out}, Math channel - Input voltage V_{in}

With a more severe loading condition varying between 60W and 160W as reflected in Fig. 7, the intra-cycle controller now gets kicked in at the peak input voltage to suppress the power transfer to the PFC output as detailed in Fig.7(b), which prevents any further overshoot and allow shorter settling time. Once again, a two-line-cycle achievement is still maintained. One should note that we use only $68\mu F$ capacitance at the output which is only a fourth of a conventional design [10], so the performance is expected to be a lot better with conventional applications.

© VDE VERLAG GMBH · Berlin · Offenbach

6. Conclusions

A simple yet effective digital control technique based on the principle of balancing the input and output power of the PFC stage is proposed to improve dynamics responses of low-harmonic rectifiers, which benefits high efficiency and high density power supply designs in which minimizing the cost and the physical volume of the products is the main driving factor. The proposed solution allows at least 50% reduction in the output capacitance while ensuring the output voltage well within allowable dynamic ranges even in the worst case of 80% to 20% step loads. The implementation is relatively simple with low hardware demand. While the power-balance-based concept is only demonstrated through the study of a four-level bridgeless rectifier, one can easily confirm the validity of such a principle for other PFC topologies.

7. References

[1] S. Wall and R. Jackson. Fast controller design for single-phase power-factor correction systems. *IEEE Trans. Industrial Electron.*, 44(5):654–660, Oct. 1997.

[2] A. Prodic, Jingquan Chen, D. Maksimovic, and R. W. Erickson. Self-tuning digitally controlled low-harmonic rectifier having fast dynamic response. *IEEE Trans. Power Electron.*, 18(1):420–428, Jan. 2003.

[3] A. Prodic, D. Maksimovic, and R. W. Erickson. Dead-zone digital controllers for improved dynamic response of low harmonic rectifiers. *IEEE Trans. Power Electron.*, 21(1):173–181, Jan. 2006.

[4] S. Bibian and H. Jin. Digital control with improved performance for boost power factor correction circuits. In *16th Annu. IEEE Applied Power Electron. Conf. and Expo.*, volume 1, pages 137–143, 2001.

[5] A. D. Castro, P. Zumel, O. Garcia, T. Riesgo, and J. Uceda. Concurrent and simple digital controller of an AC/DC converter with power factor correction based on an FPGA. *IEEE Trans. Power Electron.*, 18(1):334–343, Jan 2003.

[6] P. Das, M. Pahlevaninezhad, J. Drobnik, G. Moschopoulos, and P. K. Jain. A nonlinear controller based on a discrete energy function for an AC/DC boost PFC converter. *IEEE Trans. Power Electron.*, 28(12):5458–5476, Dec. 2013.

[7] M. Rathi, N. Bhiwapurkar, and N. Mohan. Dual voltage controller based power factor correction circuit for faster dynamics and zero steady state error. In *29th Annu. IEEE Industrial Electron. Society Conf.*, volume 1, pages 238–242, Nov. 2003.

[8] T. T. Vu and G. Young. Implementation of multilevel bridgeless PFC rectifiers for mid-power single pphase applications. To be appeared at IEEE Applied Power Electron. Conf. and Expo. (APEC), Mar. 2016.

[9] R. W. Erickson and D. Maksimovic. *Fundamental of Power Electronics*. Kluwer Academic Publishers, 2nd edition, 2004.

[10] M. Wierich. Design review: Power stage design for a 200w off-line power supply. https://www.fairchildsemi.com/technical-articles/, July 2015.

Nonlinear Output Characteristic of DAB Converter caused by ZVS Transition

M.Sc. Martin Jagau, Technologie Netzwerk Allgäu, Germany, martin.jagau@tn-allgaeu.de

Prof. Dr. Michael Patt, Technologie Netzwerk Allgäu, Germany, michael.patt@tn-allgaeu.de

Abstract

In order to incorporate renewable energies modern power converters with high efficiency and power density are essential. Resonant power converter topologies are implemented due to the reduction of switching losses, yielding higher switching frequencies which ultimately reduce the size of the magnetic components. Resonant topologies are known for several decades and main goal remains to optimize them in terms of efficiency and dynamics. It is essential to implement dynamic and robust control schemes over a wide operating range. In order to guarantee robust control the transfer function of the plant has to be known. In this paper the plant of a Dual Active Bridge Converter (DAB) was measured using a Bode 100 from Omicron LAB. A nonlinear phenomenon in the output characteristic can be observed, when a transition from soft switching (ZVS) to hard switching occurs. In this region, the gain of the transfer function of the plant drops significantly, which leads to a high drop in bandwidth and thereby decreasing the dynamics of the converter. Furthermore a technique is described by which the nonlinear region can be cancelled.

1. Dual Active Bridge Converter

A DAB-Converter [1] shown in Fig. 1 is used when a bidirectional and galvanic isolated power flow is needed. Further advantages are the low amount of passive components, the resonant switching behavior and the fact, that the parasitic effects can be used [2] [3] [4]. To implement stable control schemes with high bandwidth the exact transfer function of the plant has to be known. Power converters often have a wide input voltage range. To guarantee stability over the whole operating range, the transfer function of the plant has to be known at all the operating points. When the transfer function of the plant is known, high bandwidth compensators can be designed via pole and zero placements.

The DAB converter consists of two active H-Bridges interfaced by a transformer and a coupling inductor. The power transfer can be controlled by a phase shift between the two H-Bridges. This allows a constant switching frequency and rectangular bridge voltage waveforms (Fig. 3). The lossless model of a DAB converter is shown in Fig. 2, where the active bridges are replaced by square wave sources interfaced by the coupling inductor.

PCIM Europe 2016, 10 – 12 May 2016, Nuremberg, Germany

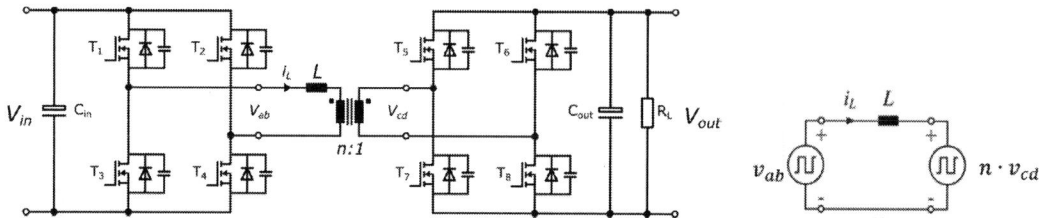

Fig. 1. Dual Active Bridge (DAB) Converter   **Fig. 2** Lossless Model

The output power when all losses are neglected is derived as [1]

$$P = \frac{n \cdot V_{in} \cdot V_{out} \cdot \varphi \cdot (\pi - |\varphi|)}{2 \cdot \pi^2 \cdot f_S \cdot L} \tag{1}$$

where V_{in} is the input Voltage, V_{out} is the output voltage, n is the transformer turns-ratio, f_S is the switching frequency, L is the coupling inductor and φ is the phase-shift angle in radians.

1.1. Soft Switching

The highest efficiency of a DAB Converter can be achieved if both active H-Bridges incorporate soft switching. Fig. 3 shows the ideal transformer Voltages V_{ab} and V_{cd} and the inductor current i_L. In the ideal model no dead time is present and therefore an instantaneous current commutation is assumed. The boundaries which enclose the resonant switching region can be defined by the following constraints:

$$i_L(0) \leq 0 \tag{2}$$

$$i_L(T_\varphi) \geq 0 \tag{3}$$

Defined by constraints (2) and (3) three different switching states are possible.

Tab. 1 Possible switching states of active bridges

	Primary Bridge	Secondary Bridge	Condition
Mode 1	ZVS	HS	$i_L(0) \leq 0 \cup i_L(T_\varphi) \leq 0$
Mode 2	ZVS	ZVS	$i_L(0) \leq 0 \cup i_L(T_\varphi) \geq 0$
Mode 3	HS	ZVS	$i_L(0) \geq 0 \cup i_L(T_\varphi) \geq 0$

Violating constraints (2) or (3) will lead to hard switching of the primary or the secondary bridge respectively.

Fig. 4 shows the output power as a function of the phase shift φ with the input voltage as a parameter. Furthermore the boundaries for the primary and the secondary bridge enclosing the resonant switching region are shown. It can be seen, that for $V_{in} = n \cdot V_{out}$ the resonant switching region is valid over the whole operation area.

It is evident, that a transition from hard switching to soft switching of the secondary bridge and a transition from soft switching to hard switching of the primary bridge occurs when the input voltage is varied. Especially at low output power these transitions occur, when a wide

© VDE VERLAG GMBH · Berlin · Offenbach

input voltage range is demanded, as the resonant switching region narrows with output power decreasing.

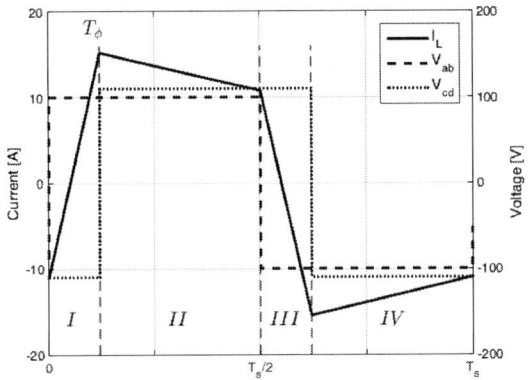

Fig. 3. Transformer Voltages and Inductor Current with $V_{in} = 100V$, $V_{out} = 100V$, $P = 1000W$ and $n = 1$

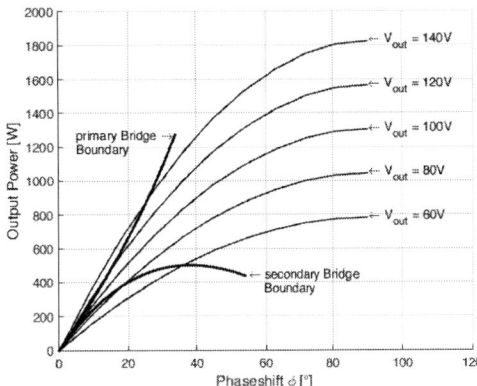

Fig. 4 Output Power versus φ and ZVS boundary with $V_{in} = 100V$ and $n = 1$

2. Output Characteristic of DAB Converter

Transforming formula (1) yields the ideal output voltage of a DAB converter

$$V_2 = \frac{n \cdot V_1 \cdot \varphi \cdot (\pi - |\varphi|)}{2 \cdot \pi^2 \cdot f_S \cdot L} \cdot R_L \tag{4}$$

with R_L being the output load resistance. The ideal output voltage in relation to the control parameter φ is plotted in Fig. 5 with different load resistances. However, when the real output voltage is measured, a flattening in the output characteristic can be observed. Fig. 6 illustrates the measured output voltage versus φ for a load resistance of 16Ω and different dead times. Fig. 7 shows the measured output voltages with a load resistance of 32Ω. Fig. 8 illustrates, that at high output power only one flattening of the output voltage is present (16Ω). With a decrease in output power, additional changes in the output characteristic occur (32Ω and 48Ω).

At low output power if $n \cdot V_{out} < V_{in}$ constraint (3) (secondary bridge boundary in Fig. 4) is violated, meaning the primary bridge is soft switched and the secondary bridge is hard switched (Mode 1 in Tab. 1). An increase in output voltage increases the output power and dependent on the output resistance at one point constraint (3) will start to be fulfilled. The starting point of region one is shown in Fig. 8. At this point a transition of the secondary bridge from hard switching to soft switching occurs. During this transition an increase of phase shift leads to no increase in output voltage and the output characteristic is flattened for the first time. If the transition is complete both bridges incorporate soft switching (Mode 2 in Tab. 1).

If the input voltage is increased further constraint (2) is no longer fulfilled and a transition from soft switching to hard switching of the primary bridge occurs. The starting point of region

© VDE VERLAG GMBH · Berlin · Offenbach

two can be seen in Fig. 8. This transition leads to another flattening of the output characteristic and after completion the DAB converter operates in Mode 3 (Tab. 1).

If the output resistance is low enough, the converter will not violate constraint (2) and the converter will not enter operating mode 3. In this case, only one flattening of the output characteristic takes place, which can be seen in Fig. 6.

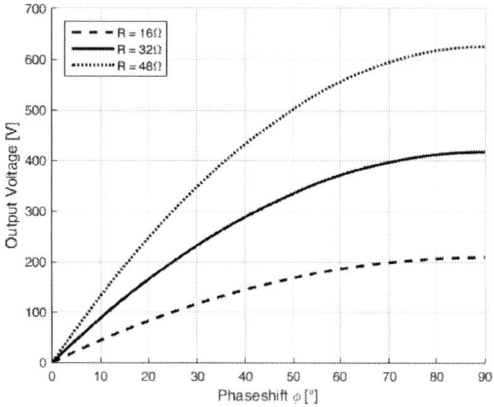

Fig. 5. Output Voltage versus φ with $V_{in} = 100V$, and $n = 1$

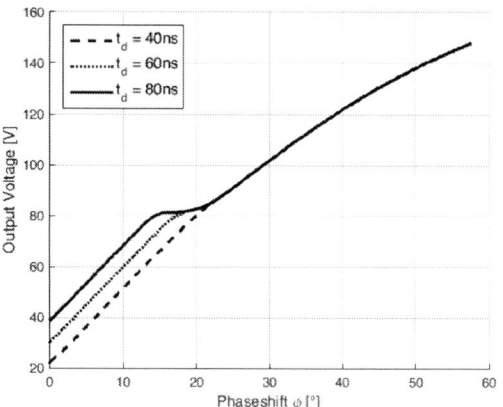

Fig. 6 Measured Output Voltage versus φ with $R = 16\Omega$, $V_{in} = 100V$ and $n = 1$

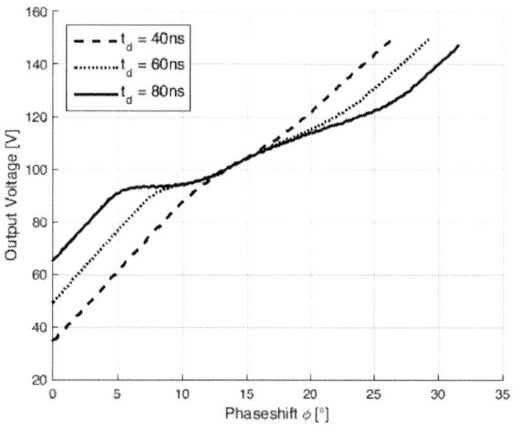

Fig. 7. Measured Output Voltage versus φ with $R = 32\Omega$, $V_{in} = 100V$ and $n = 1$

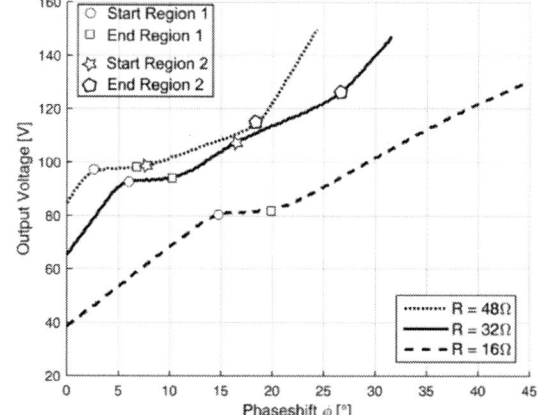

Fig. 8 Measured Output Voltage versus φ with $t_d = 80ns$, $V_{in} = 100V$ and $n = 1$

2.1. Effect of Dead Time on transfer function

The nonlinear regions are caused by the transition from soft switching to hard switching. It is visible, that the dead time has an effect on the output characteristic and that with an increase of dead time the length of the flattening increases. Furthermore it is apparent, that the length of the first flattening is not dependent on the output power and stays constant with rising output power (Fig. 8). At low output power, a second change in the output characteristic occurs and again the length of this region is dependent on the dead time, which is seen in Fig. 7.

A closer look at the nonlinear region in Fig. 8 shows, that an increase in phase shift leads to no increase in output voltage. In other words the gain of the plant is significantly reduced (Fig. 9 (b)) and the compensator is nearly ineffective in this region. Due to the wide input voltage range, the converter will have an operating region, at which this phenomenon occurs and dynamic and stable control cannot be certain.

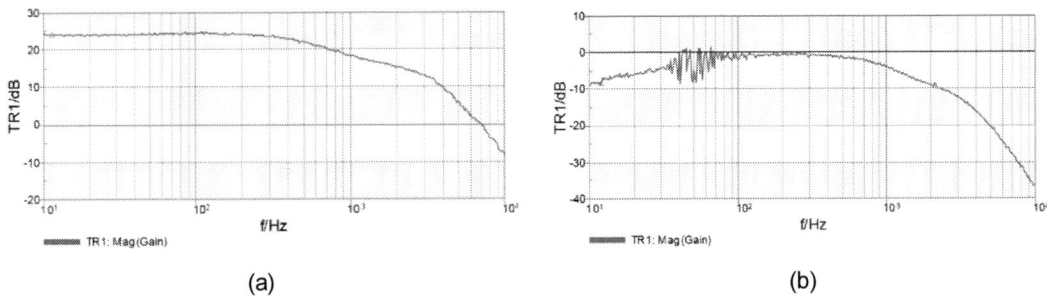

(a)   (b)

Fig. 9. Measured gain of DAB converter, (a) linear region, (b) flattened region

2.2. Eliminating flattened regions of output characteristic

In order to guarantee control with high bandwidth in the nonlinear region, the gain of the plant has to be increased in the relevant regions. Two possible transition regions occur which have to be eliminated, in order to omit a significant drop in gain of the DAB converter. Increasing the phase shift from zero the first flattening occurs, when $i_L(T_\varphi)$ rises above zero and the secondary bridge enters soft switching. The starting point of the region (Fig. 8) is constant at fixed load and is independent of the input voltage. The starting point is defined by $i_L(T_\varphi) = 0$ and the length of the region is proportional to the dead time.

Increasing the phase shift further a second transition occurs, when $i_L(0)$ falls below zero and the primary bridge loses soft switching. The start point of the second flattened region (Fig. 8) is defined by $i_L(0) = 0$ and the slope of this region is inversely proportional to the output resistance. Again the length of the second nonlinear region is a function of dead time.

Depending on the above mentioned characteristics the implemented nonlinear cancellation algorithm is based on:

- Calculating the starting points of flattened region one and two by constraints (2) and (3)
- Cancelling the first flattened region by adding the length of the region to the control parameter φ.
- Increasing the slope of the second flattened region by a factor based on the load's resistance.

The result of the nonlinearity cancelation algorithm is shown in Fig. 10 and it can be seen, that no more flattening of the output characteristic is present and no more drop in the gain of

© VDE VERLAG GMBH · Berlin · Offenbach

the transfer function of the plant occurs. It is now possible to design stable compensators with high bandwidth and performance over the whole operation range is guaranteed.

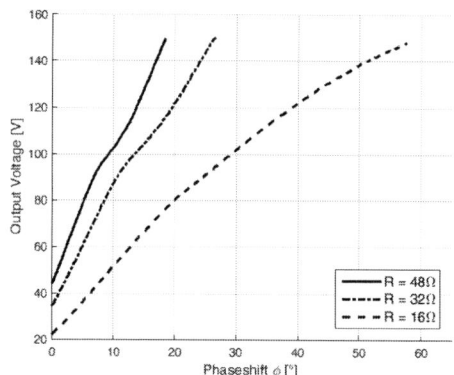

Fig. 10. Measured Output Voltage versus φ with $t_d = 60ns$, $V_{in} = 100V$ and $n = 1$

Conclusion

The effect of dead time on the output characteristic and thereby on the transfer function of a DAB converter is shown. It is shown, that the nonlinear regions occur by a transition of soft switching to hard switching or vice versa of the active bridges. It is important to tune the dead time to the resonant circuit during commutation, in order to keep the effect of dead time a small as possible. Furthermore a nonlinear cancelation algorithm was implemented and it is shown, that a cancellation of the nonlinear region can be achieved. By increasing the gain in the flattened regions the flattened output characteristic can be eliminated and the gain of the plant does not change significantly over the whole operating range. This makes it possible to design stable and dynamic compensators by proper poles and zero placement over a wide input voltage operating range.

References

[1] D. D. M. K. R.W.A.A. De Doncker, "A Three-phase Soft-Switched High-Power-Density DC-DC Converter for High-Power Applications," *IEEE Tran. on Industry Applications,* vol. 27, no. 1, pp. 63-73, Jan/Feb 1991.

[2] J. W. K. F. Krismer, "Accurate small-signal model for an automotive bidirectional Dual Active Bridge converter," *Proc. IEEE 11th Control Model. Power Electron. Workshop,* pp. 17-20, Aug 2008.

[3] J. K. F. Krismer, "Efficiency-optimized high current dual active," *IEEE Trans. Ind. Electron,* 2011.

[4] R. D. D. J. Walter, "High-power galvanically isolated dc/dc converter topology for future automobiles," *2003 IEEE 34th Annual Power Electronics Specialist Conference,* vol. 1, pp. 27-37, June 2003.

[5] "Phase-Shifted Full-Bridge, Zero-Voltage Transition Design Considerations," 2011.

[6] R. W. G. D. M. D. E. D. B. M. H. Kheraluwala, "Performance characterization of a high-power dual active bridge dc-to-dc converter," 1992.

Efficiency Maximization for Half-Bridge LC Converter through Automatic Dead Time Tuning

Vittorio, Crisafulli, ON Semiconductor Germany GmbH, Munich, Germany,
vittorio.crisafulli@onsemi.com
Diego Hernandez Gutierrez, Switzerland, diegongh9@gmail.com
Gianluca, Fazio, ON Semiconductor S.R.L, Milan, Italy, gianluca.fazio@onsemi.com

Abstract

Induction heating (IH) is becoming the preferred technology for domestic cooktop market. This kind of systems offer a higher efficiency than the standard solutions: electric and gas stove. IH operating condition can vary grandly making it difficult for the power converter to maintain soft switching operating mode. The required dead time for operating zero voltage switching (ZVS) condition, changes too, due to the different load conditions. In this paper a new method for automatic dead-time compensation is presented. Using this new method, soft switching is achieved for every load conditions even under worst case. This allows the converter to always operate in ZVS mode. The main benefits of it are improved effective SOA and efficiency. The method does not require any additional external components since current sensing functions are already implemented for power management purposes. Experimental measurements performed on a commercial Induction cooktop, with ON Semiconductor IGBTs, demonstrate the effectiveness of this method and its robustness with regard to component variations.

1. Introduction

In the early 80's the Induction Heating technology was introduced in the domestic market, ever since becoming the most appealing technology in food-prep applications. According to the U.S. Department of Energy the efficiency of energy transfer in these systems is about 90%, compared to 71% for a smooth-top non-induction electrical unit, providing an approximate 20% saving in energy for the same amount of heat transfer.

Recently Induction heating systems for cooking applications have become increasingly popular owing to their merits of very high thermal conversion efficiency, rapid heating, local spot heating, direct heating, high power density, high reliability, low running cost and not-acoustic noise.

Basically, an Induction cooker transfers electrical energy by induction from a coil of wire into a pot made of a material which must have high magnetic permeability. The heat generated is analogous to the unwanted heat dissipated in an electric transformer; most of the heat is due to the eddy currents generated in the pot bottom layer.

In literature several topologies have been presented, but the most used are those containing resonant tanks [1]. [2] [3] shows block diagram of an inverter for Induction Heating, which is composed by a filter block, resonant converter and the coils.

The main requirements for IH converter are as follows:
- High frequency switching
- Power factor close to unity
- Wide load range

PCIM Europe 2016, 10 – 12 May 2016, Nuremberg, Germany

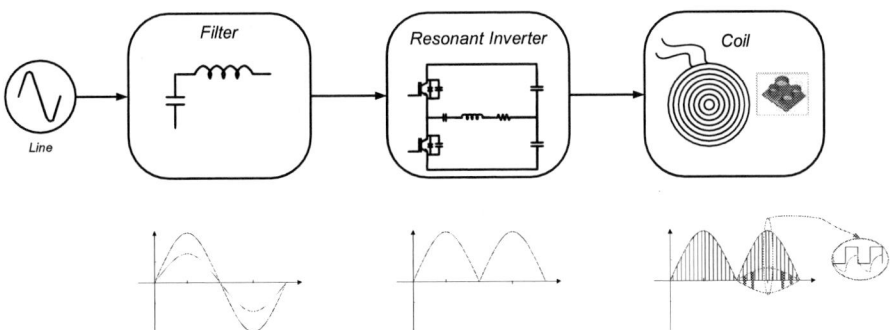

Figure 1: Common block structure of Inverter for Induction Heating

The most common output power control for induction heating applications is based on a variable frequency scheme Figure 1. This is a basic method that is applied against the variation of load or line frequency. The major disadvantage of this method is the large frequency variation required for output power control over a wide range.

The main advantage of resonant converters is that they can work in a very high switching frequency range with very low power losses. Several control techniques, like zero current switching (ZCS) or zero voltage switching (ZVS), can be used to reduce power loss in resonant converters.

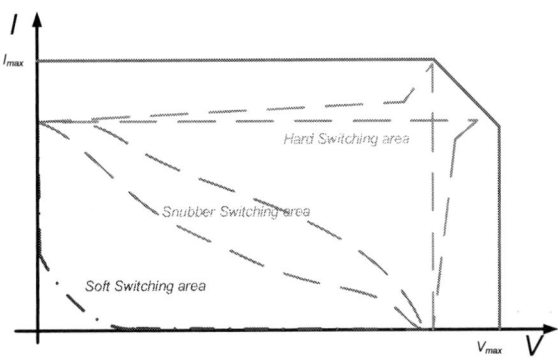

Figure 2: Switching Area

Figure 2 shows switching area for hard-switched condition, snubber assisted commutation and soft switching. During the hard switching turn-on and turn-off, the power device has to withstand high voltage and current simultaneously, resulting in high switching losses and stress. Dissipative passive snubbers are usually added to the power circuits so that the dv/dt and di/dt of the power devices can be reduced. However, the switching loss is proportional to the switching frequency, thus limiting the maximum switching frequency of the power converters.

The most popular topology for IH is the Half-Bridge (HB) series-resonant converter, as shown in Figure 4. In this topology an important parameter is the so-called dead time, which is the time between the two PWM signals that the power transistors; this time interval is necessary to avoid power transistor cross-conduction.

In order to simplify the design and control most of the converters have a fixed dead-time regardless of operating conditions, assuming that this constant value will always, guarantee ZVS/ZCS mode. However, a fixed value does not always match the ZVS/ZCS condition, due to the fact that the load can assume a very wide range of value (depend on the pot bottom layer EM characteristics) [9], [10] and [11].

© VDE VERLAG GMBH · Berlin · Offenbach

PCIM Europe 2016, 10 – 12 May 2016, Nuremberg, Germany

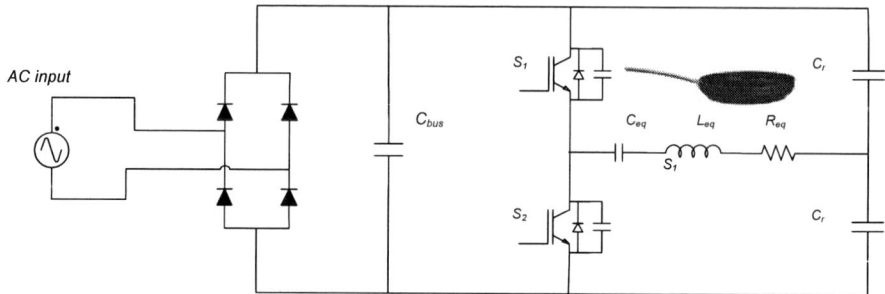

Figure 3: Half-Bridge series-resonant topology used in standard IH cooktop.

This paper presents a new method for dead time automatic compensation in HB topology. The proposed method can enable the converter to achieve soft switching operation over a wide output power regulation range with a high efficiency. The proposed method can be used in single phase home-cooking applications with a high efficiency and high reliability at each power levels.

2.   LOSS OF SOFT SWITCHING CONDITION

As explained above, avoiding hard switching condition is very important, especially in those applications where reliability is a must. The main causes for losing soft switching conditions can be either requesting very different power levels with two or more pot or working in continuous mode at lower power levels.
There are two different types of hard switching conditions that can be observed in standard power converters. They are shown in Figure 5.

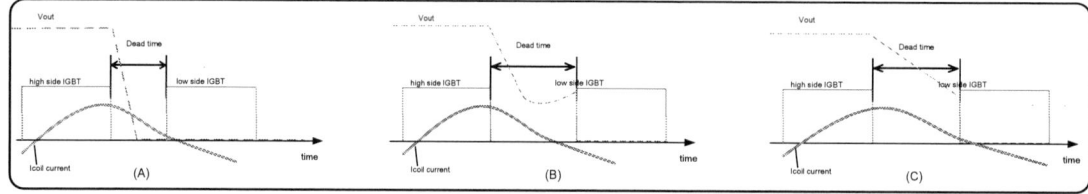

Figure 4: Waveforms of different hard switching conditions in resonant HB.

There are two types of hard switching differentiated on the basis of the operating frequency:
*   Operating frequency close to the resonant frequency
*   Operating frequency far from to the resonant frequency
The former condition occurs when the IH inverter operates close to the resonant frequency and the output current during the switch off of either high side IGBT or low side IGBT is too low and becomes zero before being able to fully charge/discharge both snubber capacitors, as shown in Fig. 5 B. The combination of output current and output voltage (indicated as Vout in the Figure 5) prevents the freewheeling diode from turning on, and thus prevents ZVS at the next turn-on IGBT. The output voltage reaches a minimum value coinciding with the coil current zero crossing.
At this point, the coil current changes direction and starts charging the snubber capacitors in reverse mode, meanwhile the low side IGBT is not turned on and the freewheeling diode cannot operate. The hard switching occurs since the IGBTs are essentially short-circuiting the still charged snubber capacitors.
The latter condition occurs when the IH inverter operates far from the resonant frequency where and the load coil-pot assembly is highly inductive, as shown in Fig. 5 C. In this case the output current is too low and its phase is almost equal to 90° with respect to the first

© VDE VERLAG GMBH · Berlin · Offenbach

harmonic. The zero crossing of the coil current occurs after the end of the dead time and the snubber capacitors cannot be fully charged/discharged.

It is important to note that these cases are more frequent whenever the converter control makes use of asymmetric duty cycle PWM drive signals.

3. PROPOSED METHOD FOR DEAD-TIME AUTO-TUNING

Standard PWM switching frequency for HB topology, used in IH cooktop, is in the range of 20 KHz-50 KHz with a fixed dead-time between PWM signals is set for to avoid power transistors cross-conduction Figure 6 and Figure 7.

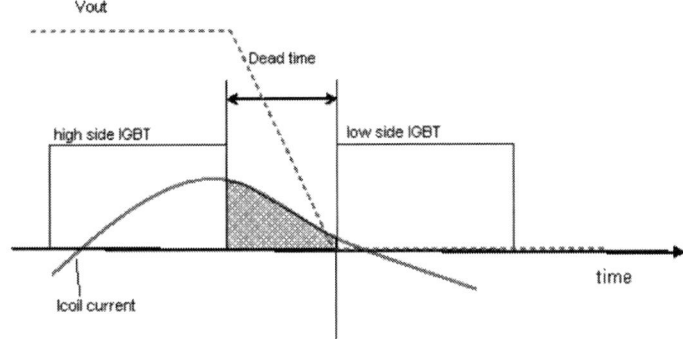

Figure 5: Soft switching conditions in resonant HB-1

For different reasons, the desired soft-switching ZVS operation mode cannot be achieved on frequencies that are either close to the LC tank resonance frequency (at maximum output power) or very far from the resonance frequency (where output power is minimum). A fixed dead time will be too long or too short, respectively, in order to ensure the required soft-switching operating mode. This will reduce the maximum delivered power as well as hasten the power transistors wear out.

The proposed method aims at guaranteeing soft switching operations by dead-time auto-tuning, thus maximizing the power converter efficiency. Fig. 8 shows the flow chart of the proposed method, according to the zero-crossing of the coil current and the effective calculated discharge of snubber capacitor during the dead time, the algorithm changes the dead time value (increasing/decreasing it) within min/max dead time values (which are fixed in the design stage) before the next power IGBT turn-on.

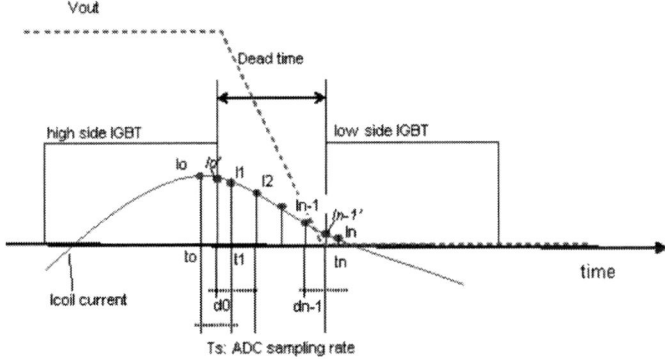

Figure 6: Soft-switching conditions in resonant HB-2.

3.1. Hard switching detection mechanism

© VDE VERLAG GMBH · Berlin · Offenbach

For simplicity we proceed to describe the detection mechanism of the hard switching conditions on those operating mode when the converter is working with an operating frequency close to resonance of the LC resonant tank. The same mechanism can be applied for detecting the hard switching conditions when working with frequency far from resonance.

The detection of the loss of soft switching is achieved by comparing the charge stored in the snubber capacitors with the effective charge being supplied by the coil current during the dead-time interval. As a result, if the first charge value exceeds the second one then it can be said that the IGBTs have lost soft switching conditions.

The charge stored in the snubber capacitors can be calculated with following equation:

$$Q_{snub} = 2C_{snub}Vout(turnoff)$$

Equation. 1

where:

C_{snub} is the snubber capacitor value (defined in the design specification)

V_{out} (during turn off) is the voltage at the middle point of the Half–bridge at the IGBT turn off. The max value of the V_{out} coincides with the peak of the Mains voltage after rectification.

The charge being actually supplied by the coil current that charges and discharges the snubber capacitors can be calculated assuming that the coil current varying linearly during the dead time interval (as dead time interval is small compared with the overall IGBT drive period).

The area below the current during the dead time interval (colored in pink in Figure 6) represents the effective charge/discharge due to coil current flow and its value can be calculated using the equation:

$$Q_{curr} = \frac{1}{2}I_{off} \cdot t_{snub}$$

Equation. 2

where:

t_{snub} is the actual interval where snubber capacitors are being charged or discharged.

I_{off} is the calculated averaged coil current value during the dead time interval

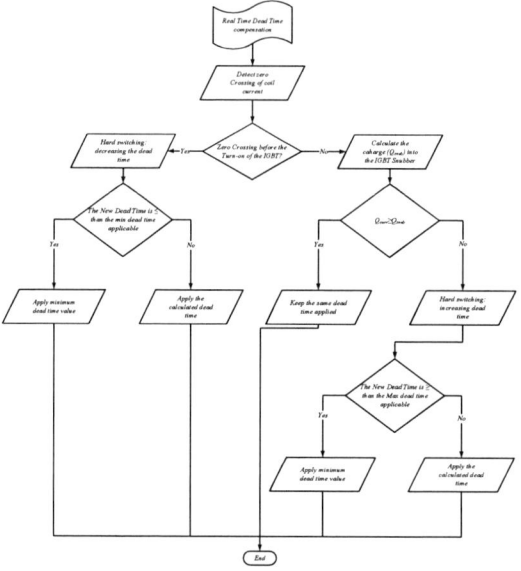

Figure 7 Flow Chart of the proposed method

In order to determine Q_{curr} it is necessary to first calculating I_{off} and t_{snub}. In the standard power converter used for IH cooktop the coil current is sampled synchronously with the IGBT drive signals (the sampling rate and resolution of the ADC output signals are determined by the technical specification of the microcontroller mounted onto the power converter board and have an important impact on the overall precision of the hard switching conditions).

As the ADC sampling of the coil current is synchronous with IGBT drive signals it is possible to identify those samples that coincides with the dead time interval, as shown in the Figure 7. The I_{off} value is calculated with a trapezoidal weighed integral as shown below:

$$I_{off} = \frac{d_0 * I_0' + \sum_{J=1}^{N-2} I_J * ts_j + d_{n-1} * I_{n-1}'}{d_0 + \sum_{i=1}^{n-2} ts_i + d_{n-1}}$$

Equation. 3

where

$$I_j = \frac{I_J + I_{J+1}}{2}$$

Equation. 4

I_1 and I_{n-1} : coil current samples that falls in the dead time interval.

I_0' , I_{n-1}' : coil current at high side IGBT turn off and low side IGBT turn on, respectively, that have been approximated with linear interpolation.

$t_1.. t_{n-1}$: ADC sample period

d_0, d_{n-1}: partial sample period of first and last sample within the dead time interval.

t_{snub} is determined as follows:

$$t_{snub} = t_{deadtime} - t_{delayIGBT}$$

Equation. 5

where:
- $t_{dead\ time}$: correspond to the actual dead time applied by the converter control.
- t_{delay} IGBT: correspond to the delay for the actual HW IGBT turn-off.

Once both charge values are calculated the ZVS soft switching condition is guaranteed if the following condition is true:

$$Q_{curr} \geq Q_{snub}$$

Equation. 6

3.2.   Dead Time Tuning method

In the last decades, many techniques have been proposed in literature [16-17]. The proposed method does not require any extra hardware.

First the algorithm waits until a coil current zero crossing is detected. After that, instant of zero crossing and PWM signal rising edge and compared.

a) If the zero crossing happens before the turn on IGBT, the system is in hard-switching. The algorithm calculates a new dead time, the new value is compared with the minimum value applicable for the system. If the new dead time is larger than the minimum value it is kept, otherwise the minimum value is used.

b)   If the zero crossing happens after the turn on of the IGBT, the algorithm compares the charge into the snubber capacitor Q_{snub} with the integral of the coil current during the dead time Q_{curr}. If $Q_{curr} \geq Q_{snub}$, the algorithm keeps the same dead time. Otherwise a new dead time is calculated by increasing the old value, the new value is compared with the maximum value applicable for the system. If the new dead time is smaller than the maximum value it is kept, otherwise the maximum value is used.

4.   Experimental results

A half-bridge series resonant inverter, with new Field Stop II ON Semiconductor IGBTs, has been used to test the proposed control scheme. Experimental tests were carried out on an IH commercial platform operating at variable switching frequency, in the range of 20 KHz – 50 KHz.

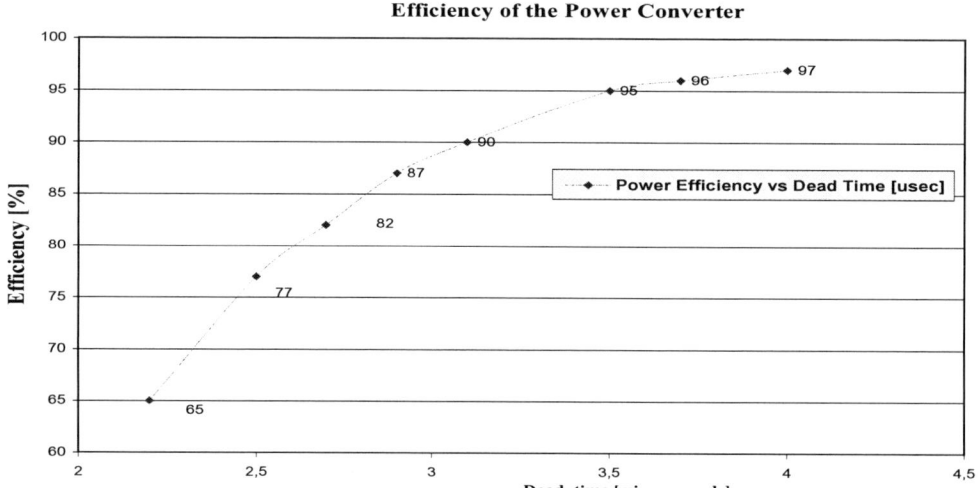

Figure 8 Power efficiency versus dead-time

A standard Microcontroller has been used to implement the digital control of the induction heating cooktop and the real time optimized automatic dead time tuning.

This new method has been tested under different conditions. Figure 9 shows the power efficiency versus dead-time from experimental data when the IH converter operates far from the resonance frequency. It is quite evident that the efficiency is hardly depending on the dead-time values. The figure shows the evolution of the new method, starting from a dead time of 2,2 μs as it reiterate the calculation until the optimal value of 4μs. In this region the dead-time calculated by using the new method results in max efficiency of 97%.

As we expected, our system automatically tunes the programmed delays in order to compensate the variations and it restores the maximum efficiency operation, as shown in Figure 10.

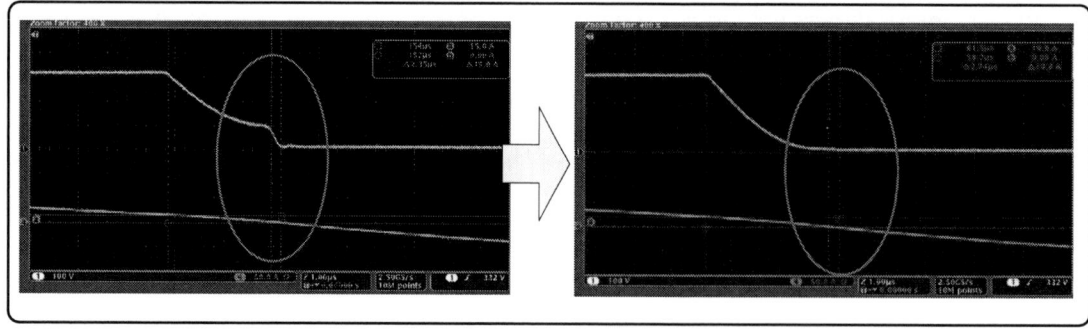

Figure 9: algorithm detect the hard switching and the period after come back in soft switching.

5. Conclusion

In this paper a new method has been presented for auto-tuning the dead-time of the PWM signal that drives the power-transistor in a Half Bridge series resonant topology.

This method does not require any extra hardware components and it can be implemented directly into the control of the IH power converter. This cost-free solution can be easily implemented inside a routine, based on iterative steps and which doesn't need fast execution time, and it will find straightforward the optimum dead-time at any operating point. It also should have a limited impact on the current architecture of low-level closed-loop control algorithm of the today IH cooktop.

Experimental results carried out in Half Bridge converter of IH domestic cook-top demonstrate the soft-switching ZVS mode can be achieved. A comparison of power efficiency between a standard control with fixed dead-time and a control with the new method shows the advantage of the new method.

References

[1] N. Mohan, T. Undeland, W.P. Robbins, "Power Electronics: Converters, Application, and design 3rd", Wiley; 3rd edition, October 10, 2002

[2] Henry W. Koertzen, Jacobus D. van Wyk and Jan A. Ferreira. "Design of the Half-Bridge Series Resonant Converter for Induction Cooking", in Proceedings of the1995 Power Electronics Specialist conference, 95CH35818, pp.729-735.

[3] M. K. Kazimierczuk, T. Nandakumar and Shang Wang,. "Analysis of series-parallel resonant converter", IEEE Transaction of AES, Vol. 29, No1, Jan 1992, pp. 88-97.

[4] J. M. Burdio, F. Monterde, J. R. Garcia, L. A. Barragan, and A. Martinez, "A two-output series-resonant inverter for inductionheating cooking appliances", IEEE Transactions on Power Electronics, vol. 20, no. 4, pp. 815-822, July 2005.

[5] H. W. Koertzen, J. D. v. Wyk, and J. A. Ferreira, "Design of the Half-bridge Series Resonant Converters for Induction Cooking", in IEEE Power Electronics Specialist Conference Records, 1995, pp. 729-735.

[6] S. Wang, K. Izaki, I. Hirota, H. Yamashita, H. Omori, and M. Nakaoka, "Induction-heated cooking appliance using new quasiresonant ZVS-PWM inverter with power factor correction", Industry Applications, IEEE Transactions on, vol. 34, no. 4, pp. 705-712, July/August 1998.

[7] J. M. Leisten and L. Hobson, "A parallel resonant power supply for induction cooking using a GTO", in Power Electronics and Variable-Speed Drives Conference, 1990, pp. 224-230.

[8] O. Lucía, I. Millán, J. M. Burdio, S. Llorente, and D. Puyal, "Control algorithm of half-bridge series resonant inverter with different loads for domestic induction heating", in International Symposium on Heating by Electromagnetic Sources, 2007, pp. 107-114.

[9] Millán, I.; Burdío, J.M.; Acero, J.; Lucía, O.; Palacios, D.;.: "Resonant inverter topologies for three concentric planar windings applied to domestic induction heating", IEEE Electronics Letters, Vol. 46 no 17, August 2010, pp. 1225- 1226.

[10] A.K. Paul.: "Comparative study of functional integrity on major topologies for induction heating equipment" 2010 Joint International Conference on Power Electronics, Drives and Energy Systems (PEDES) & 2010 Power India 6 pp 1-6.

[11] H. Ogiwara, M. Itoi andM. Nakaoka, 'PWM-controlled soft-switching SEPP high-frequency inverter for induction-heating applications' IEE Proc.-Electr. Power Appl., Vol. 151, No. 4, July 2004

[12] A. Pizzutelli, A. Carrera, M. Ghioni, S. Saggini, "Digital Dead Time Auto-Tuning for maximum Efficiency Operation of Isolated DC-DC Converters" Power Electronics Specialists Conference, 2007. PESC 2007 June 2007 Orlando FL pp 839 – 845

[13] Beiranvand, R.; Rashidian, B.; Zolghadri, M.R.; Alavi, S.M.H. "Optimizing the Normalized Dead-Time and Maximum Switching Frequency of a Wide-Adjustable-Range LLC Resonant Converter Example of fast switching component", IEEE Transaction on Power Electronics Vol. 26 no 2, February 2011, pp. 462- 47.

PCIM Europe 2016, 10 – 12 May 2016, Nuremberg, Germany

Improved Finite Control Set Model Predictive Control with Fixed Switching Frequency for Three Phase NPC Converter

Margarita, Norambuena, Technische Universitaet Berlin, Germany, margarita.norambuena@gmail.com
Hang, Yin, Technische Universitaet Berlin, Germany, hang.yin@win.tu-berlin.de
Sibylle, Dieckerhoff, Technische Universitaet Berlin, Germany, diecker@win.tu-berlin.de
Jose, Rodriguez, Universidad Andres Bello, Chile, jose.rodriguez@unab.cl

Abstract

This paper presents a new Finite Control Set Model Predictive Control (FCS-MPC) scheme with fixed switching frequency that is implemented in a Neutral Pointed Clamping (NPC) converter. The proposed control algorithm consists in setting a fixed switching frequency and dividing it in smaller evaluation steps. The proposed method fixes the switching frequency using commutation limitations in the structure of the FCS-MPC without weighting factors in the cost function. Experimental results show that the proposed method is practicable and has comparable performance both in transient and steady state with respect to the standard solution. It should be highlighted that this new method has a fixed frequency and a better loss distribution than the standard FCS-MPC for the NPC converter.

1. Introduction

In recent years, with the development of renewable energy integration, voltage sources based on power converters are gaining increasing importance. Furthermore, Voltage Source Converters (VSC) are commonly adopted in electric drives, power storage, power transmission and distribution, UPS and traction applications [1]. With regard to wind energy, the power of the individual wind turbine is increasing especially in offshore applications. Therefore, medium voltage technologies are becoming of interest. In this case, multilevel topologies are often replacing the standard 2-level VSC. The Neutral Point Clamped Converter (NPC) is a commonly adopted solution, that is also well known from medium voltage drives.

Model Predictive Control (MPC) is designed to track the error in the reference variables through the minimization of a cost function taking explicitly into account the state of the power switches [2, 3]. Some challenges with MPC are: variable switching frequency, high calculation effort, the weighting factor design, steady state errors and the quality of disturbance rejection [4]. Variable switching frequency makes the filter design difficult. Especially for high power applications, low and fixed switching frequency are critical issues. A variable switching frequency is a menace for the stability of the system.

In recent years, MPC has been improved to avoid variable switching frequency. Reference [5] presents a dual MPC with linear controller and Pulse Width Modulation (PWM), where the linear controller is activated during steady state and MPC is used in transient periods to achieve the new references value. Reference [6] proposes a MPC strategy with a long prediction horizon selecting the output that minimized the commutations over time. This is a good idea but has a high computational cost, and it is necessary to have the prediction of the references and the whole system for several prediction horizons. Reference [7] proposes a MPC with constant

© VDE VERLAG GMBH · Berlin · Offenbach

switching frequency and low sampling frequency applying a Discrete Space Vector Modulation technique. The drawback of this method is that it needs an external modulator to generate the firing pulses, and for low switching frequency the computational effort to achieve high performance is increased. Reference [8] presents a novel MPC technique with a fixed switching frequency by selecting a fixed switching period and dividing it into smaller evaluation steps. The switching behavior is obtained by using the analogy of PWM with triangular carriers to define regions during which only specific switching transitions are allowed. Reference [9] presents a new MPC scheme that consists in three steps: 1) Standard MPC to define the state in the next sampling instant; 2) The resulting switching function is filtered to eliminate high frequency components; 3) The resulting filtered signal is modulated to generate the converter switching signals. This algorithm needs a modulation stage and the design of a low pass filter. It has a trade-off between the cutoff frequency of the filter and the dynamic response of the system. Finally, reference [10] proposes a modulated MPC. The advantages of the standard MPC are maintained, but it is necessary to have a modulation stage, and the final switching frequency is higher than the average switching frequency in the standard MPC.

In this paper, a new algorithm is proposed to implement the standard MPC without additional modulation stages or linear controller, to achieve a fixed switching frequency. The proposed control will be implemented and validated in a Neutral Point Clamped converter.

2. Converter Model

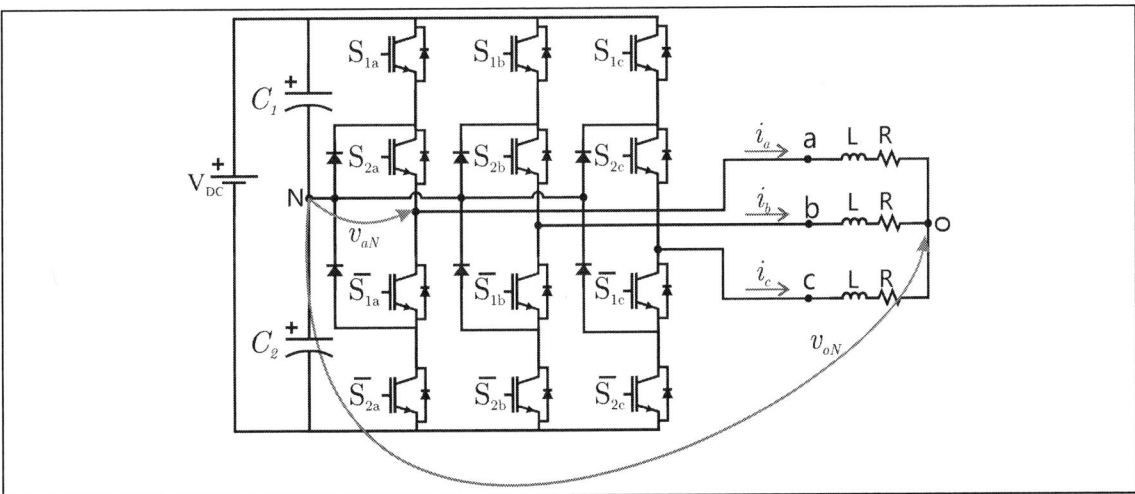

Fig. 1: Topology of the 3-phase NPC converter.

Fig. 1 shows the power circuit of the three-phase NPC inverter with RL-load. Each power switch can take only two possible states: $S_{ix} = \{0, 1\}$, where $i \,\epsilon\, \{1, 2\}$ and $x \,\epsilon\, \{a, b, c\}$, and the operation of switches S_{ix} and $\bar{S}_{ix}$ must be complementary. $3^3 = 27$ different switching states are possible. The output voltage v_{xN} delivered by the NPC is given by eq. (1).

$$v_{xN} = (S_{1x} + S_{2x} - 1) \frac{V_{DC}}{2} \tag{1}$$

In addition, the equations for the load side are:

PCIM Europe 2016, 10 – 12 May 2016, Nuremberg, Germany

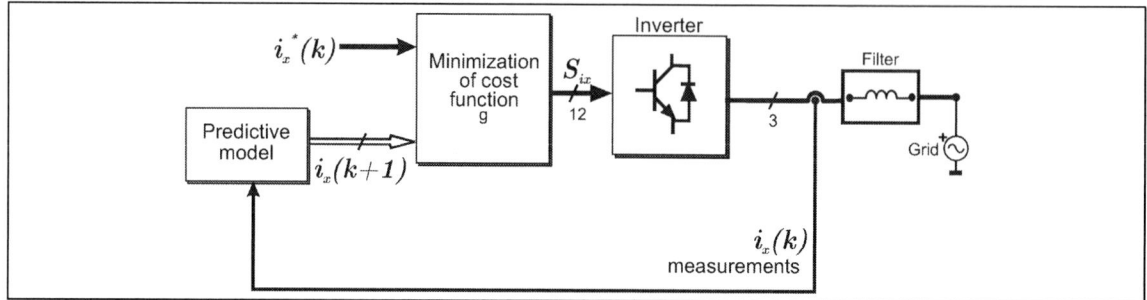

Fig. 2: Control scheme for MPC.

$$v_{xN} = Ri_x + L\frac{di_x}{dt} + v_{oN} \tag{2}$$

$$v_{oN} = \frac{1}{3}(v_{aN} + v_{bN} + v_{cN}) \tag{3}$$

3. Finite Control Set Model Predictive Control

Finite Control Set Model Predictive Control (FCS-MPC) explicitly takes the power switches in the optimal problem into account and thus the modulation stages are not necessary.

To implement FCS-MPC it is necessary to obtain a discrete time model of the converter and the load, either active or passive. The state variables in this case are i_a, i_b and i_c. The switches states S_{ix} are considered as control inputs. Since each switch can take only two values, i.e., $S_{ix} = \{0, 1\}$, there is a finite number of switch combinations that can be triggered. FCS-MPC chooses the switching states to be implemented at instant $k + 1$, S_{ix}^{k+1}, through minimization of a cost function.

$$g^k = \left(i_a^* - i_a^{k+2}\right)^2 + \left(i_b^* - i_b^{k+2}\right)^2 + \left(i_c^* - i_c^{k+2}\right)^2 \tag{4}$$

Eq. (4) shows the cost function used in this work, where variables with superscript $*$ are reference values corresponding to sampling time $k + 2$. The cost function is evaluated for all 27 possible switching states of the 3-phase NPC inverter. Fig. 2 shows the block diagram of FCS-MPC strategy.

3.1. Mathematical Model

The discrete-time model of the system for sampling instant $k + 1$ can be easily derived from eq. (1)-(3), being:

$$v_{xN}^k = \left(S_{1x}^k + S_{2x}^k - 1\right)\frac{V_{DC}}{2} \tag{5}$$

$$i_x^{k+1} = i_x^k e^{-T\frac{R}{L}} + \left(v_{xN}^k - v_{oN}^k\right)\frac{\left(1 - e^{-T\frac{R}{L}}\right)}{R} \tag{6}$$

$$v_{oN}^k = \frac{1}{3}(v_{aN}^k + v_{bN}^k + v_{cN}^k) \tag{7}$$

© VDE VERLAG GMBH · Berlin · Offenbach

1188

Where i_x^k denote the measured values of the output currents at instant k, and T is the sampling period. Notice from the model that the NPC converter is non-linear, since the system state is multiplied with the control input.

4. Proposed Algorithm

In power electronic converters with high and medium power ratings, losses due to switching often become the limiting factor. The control method that is presented in this paper tries to improve this disadvantage, which is even more apparent in MPC, by fixing and evenly distributing the switching actions.

4.1. Algorithm to fix the switching frequency

The switching frequency (F_{sw}) is defined as the number of cycles ($S_{ix} = \{0, 1, 0\}$ or $S_{ix} = \{1, 0, 1\}$, see fig. 3-(a)) that the switch carries out in a definite period. This means that during a switching period (T_{sw}), the switches change their states twice and thus, during half switching period, the switches change one time their states. The aim of the proposed control scheme is to guarantee one commutation during the defined half switching period.

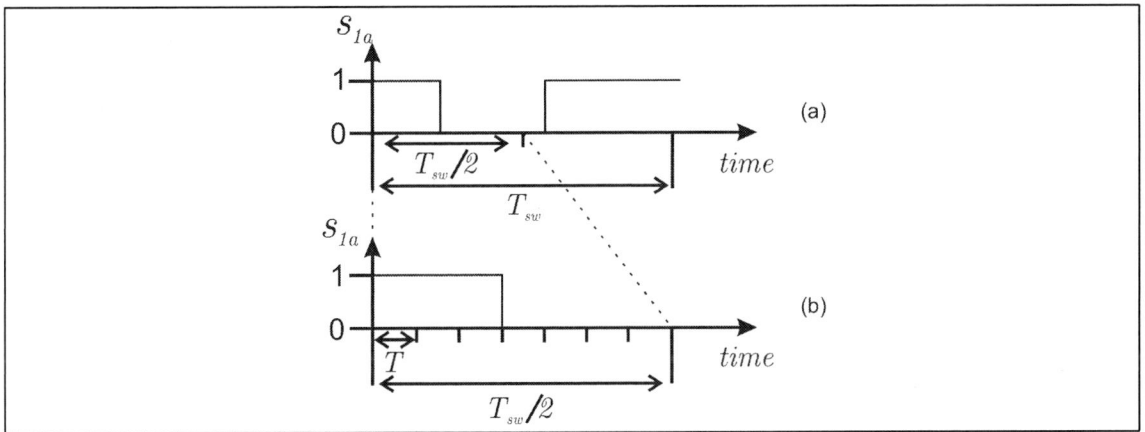

Fig. 3: Representation of switching pulse: (a) Cycle $S_{1a} = \{1, 0, 1\}$; (b) Sampling with period T faster than switching frequency to evaluate the FCS-MPC.

The proposed FCS-MPC algorithm is derived by evaluating the controller at a faster rate than the desired switching frequency and using the standard FCS-MPC with a strong constraint in the possible switching states to define the output of the control. Fig. 3 shows the representation of one cycle of switch S_{1a} and the associated sampling period T.

The switch state must change once during the half switching period (see Fig. 3-(b)), therefore, is necessary to determine the sampling frequency as N times of the double switching frequency ($2F_{sw}$), with N an integer number. To get satisfactory results, N must be the highest possible number. However, this number is limited due to the computing power of the microprocessor.

Once N is defined, the sampling frequency of the control is fixed. For the realization of the control algorithm, it is necessary to define a variable which saves the number of commutations of the different switches during the half of the switching period. $Comm_{1x}$ and $Comm_{2x}$ are the variables to monitor the states of the switches in phase x.

© VDE VERLAG GMBH · Berlin · Offenbach

PCIM Europe 2016, 10 – 12 May 2016, Nuremberg, Germany

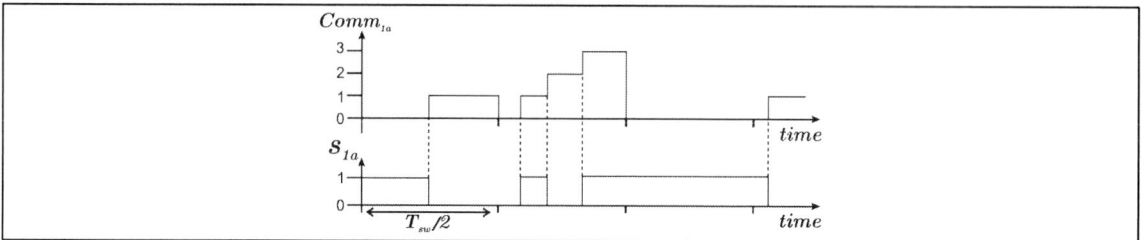

Fig. 4: Possible values of the $Comm_{1a}$ variable.

Fig. 4 exemplary shows states of the variable $Comm_{1a}$ when switch S_{1a} takes different values during several half switching periods. If the switches change their states many times during the half switching period, the switching frequency is higher than the proposed value and definitely this condition cannot be allowed. Therefore, the proposed control algorithm fixes the number of commutations to one for every switch in the inverter, this means $Comm_{1x} = 1$ and $Comm_{2x} = 1$ at the end of the half switching period.

4.2. Consideration with NPC inverter

The NPC converter has an unsymmetrical switching performance and this means that during a half period of the fundamental output variable, i.e. the load voltage, a pair of switches does not change their states. For a positive output voltage, switch $S_{2x} = 1$, and for negative output voltage, switch $S_{1x} = 0$.

Consequently, during the corresponding half period, the restriction on $Comm_{1x} = 0$ in case of negative half cycle, and the restriction on $Comm_{2x} = 0$ in case of positive half cycle is effective.

5. Experimental Results

This section shows the experimental results for the proposed control applied to a 3 phase NPC converter supplying an RL load. The *dc-link* is set to 200 V, and the load is $L = 8mH$ and $R = 10\Omega$. The switching frequency is set at 2500Hz and the average is 1250Hz, the sampling period is $T = 50\mu s$. For comparison purposes, is implemented the standard FCS-MPC according to chapter 3 with a sampling period $T = 200\mu s$ to have a similar switching frequency as the proposed control.

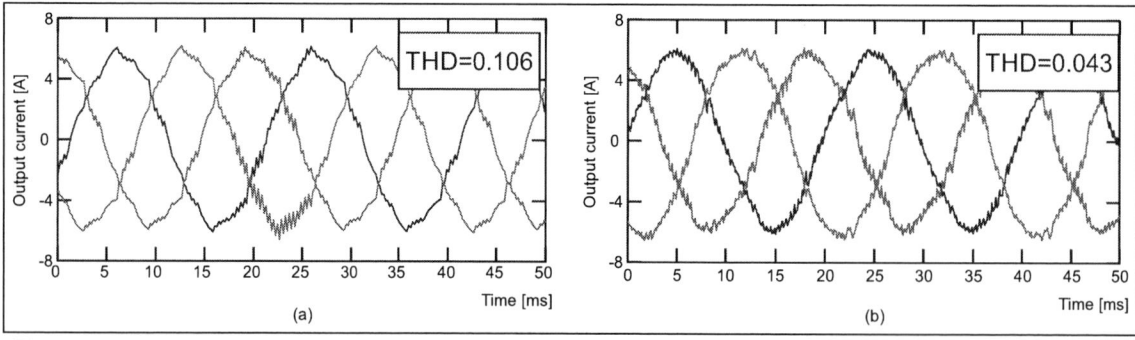

Fig. 5: Output current waveform in steady-state: (a) Standard FCS-MPC; (b) Proposed FCS-MPC.

© VDE VERLAG GMBH · Berlin · Offenbach

PCIM Europe 2016, 10 – 12 May 2016, Nuremberg, Germany

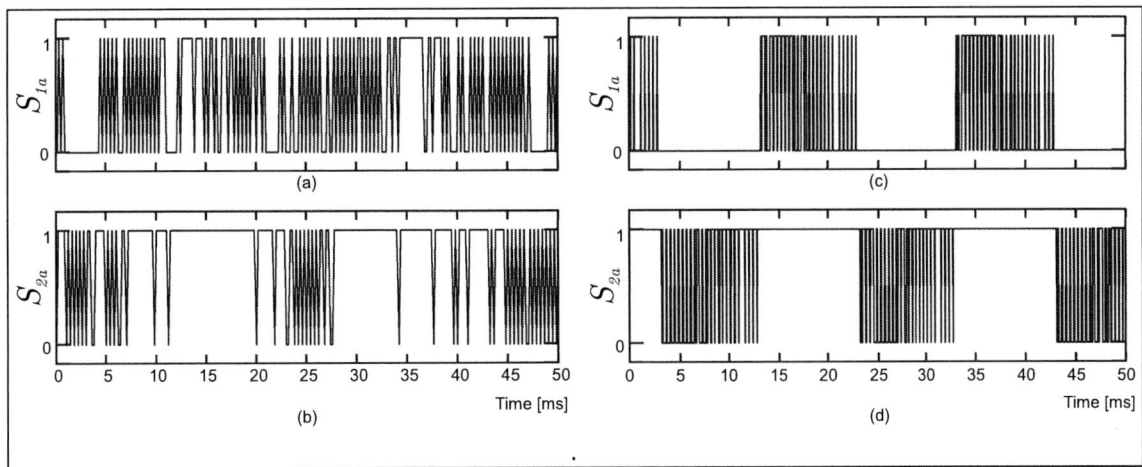

Fig. 6: Switching states with standard FCS-MPC: (a) S_{1a}; (b) S_{2a}. Switching states with proposed FCS-MPC: (c) S_{1a}; (d) S_{2a}

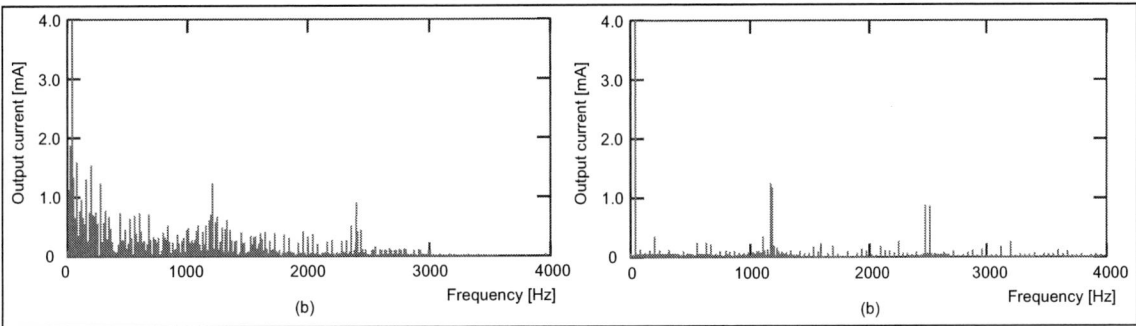

Fig. 7: Spectrum in the output current: (a) Standard FCS-MPC; (b) Proposed FCS-MPC.

Fig. 5 shows the output current waveform for the standard FCS-MPC and the proposed FCS-MPC in steady-state. It can be seen that the two control strategies achieve a three-phase balanced output current. Nevertheless, as the switching frequency average is set to 2500Hz, the standard FCS-MPC has a worse performance according to the numbers and magnitude of harmonics. The proposed FCS-MPC control has lower Total Harmonic Distortion (THD) than the standard control, 4.3% versus 10.06%. This is a consequence of the fast sampling time, which is N times faster than the switching period.

Fig. 6-(a) and fig. 6-(b) show the switching states of phase a for the standard FCS-MPC and the proposed control, respectively. It can be seen that the standard FCS-MPC generates a variable switching frequency, and the stress -in addition to the unsymmetrical loss distribution in the NPC- is not equal for all switches. The proposed FCS-MPC presents a fixed switching frequency and an even distribution of switching actions in all power switches with a good performance in steady-state. It has a fixed switching frequency at 2500Hz in a half period of the current, and 1250Hz in the whole period, while the standard FCS-MPC has an average switching frequency of 1570Hz in one period of the output current.

Fig. 7 shows the spectrum of the output current of the standard and the proposed FCS-MPC. It can be seen that the standard FCS-MPC presents a wide spread spectrum, while the proposed FCS-MPC produces a concentrated spectrum around the switching frequency and its multiples.Fig. 8 shows the output current waveform for the standard and the proposed FCS-MPC when a step change form -6A to 4A is applied in the reference currents. It can be seen

© VDE VERLAG GMBH · Berlin · Offenbach

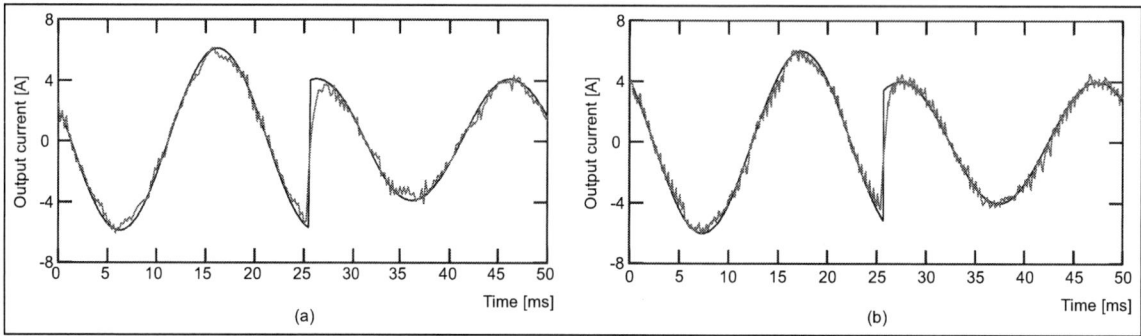

Fig. 8: Dynamic response of the control: (a) Standard FCS-MPC; (b) Proposed FCS-MPC.

that both FCS-MPC schemes have a fast dynamic response, in this case 2ms to achieve the new reference value.

6. Conclusions

In this work, a Model Predictive Control scheme with a fixed switching frequency is applied to a 3-level NPC inverter. The resulting strategy is easy to understand and to implement and does not need the use of linear controllers with PWM or an additional term in the cost function. Furthermore, the cost function does not need weighting factors to obtain a fixed switching frequency. The proposed strategy is simple but powerful and produces an effective control of the load currents and a fast dynamic response.

Acknowledgement

The authors acknowledge the support of the Chilean Research Council (CONICYT) under grant "Doctorado Nacional 2014" (21140574) and FONDECYT under grant Advanced Center for Electrical and Electronic Engineering (AC3E) and FONDECYT under grant 1150829. The authors acknowledge the support of German Academic Exchange Service (DAAD) under grant "Bi-nationally Supervised Doctoral Degrees" (57129430).

7. References

[1] J.M. Carrasco, L.G. Franquelo, J.T. Bialasiewicz, E. Galvan, R.C.P. Guisado, Ma.A.M. Prats, J.I. Leon, and N. Moreno-Alfonso. Power-Electronic Systems for the Grid Integration of Renewable Energy Sources: A Survey. *IEEE Transactions on Industrial Electronics*, 53(4):1002–1016, 2006.

[2] J. Rodriguez, J. Pontt, C. A. Silva, P. Correa, P. Lezana, P. Cortes, and U. Ammann. Predictive current control of a voltage source inverter. 54(1):495–503, February 2007.

[3] J. Rodriguez, M.P. Kazmierkowski, J.R. Espinoza, P. Zanchetta, H. Abu-Rub, H.A. Young, and C.A. Rojas. State of the Art of Finite Control Set Model Predictive Control in Power Electronics. *IEEE Transactions on Industrial Informatics*, 9(2):1003–1016, 2013.

PCIM Europe 2016, 10 – 12 May 2016, Nuremberg, Germany

[4] R.P. Aguilera, P. Lezana, and D.E. Quevedo. Finite-control-set model predictive control with improved steady-state performance. *Industrial Informatics, IEEE Transactions on*, 9(2):658–667, May 2013.

[5] P. Lezana, M. Norambuena, R.P. Aguilera, and D.E. Quevedo. Dual-stage model predictive control for Flying Capacitor Converters. In *Industrial Electronics Society, IECON 2013 - 39th Annual Conference of the IEEE*, pages 5794–5799, 2013.

[6] J. Scoltock, T. Geyer, and U.K. Madawala. Model Predictive Direct Power Control for Grid-Connected NPC Converters. *IEEE Transactions on Industrial Electronics*, 62(9):5319–5328, 2015.

[7] S. Vazquez, J. I. Leon, L. G. Franquelo, J. M. Carrasco, O. Martinez, J. Rodriguez, P. Cortes, and S. Kouro. Model predictive control with constant switching frequency using a discrete space vector modulation with virtual state vectors. In *Proc. IEEE International Conference on Industrial Technology ICIT 2009*, pages 1–6, 10–13 Feb. 2009.

[8] M. Tomlinson, T. Mouton, R. Kennel, and P. Stolze. Model Predictive Control with a Fixed Switching Frequency for a 5-level Flying Capacitor Converter. In *ECCE Asia Downunder (ECCE Asia), 2013 IEEE*, pages 1208–1214, 2013.

[9] R.O. Ramirez, J.R. Espinoza, F. Villarroel, E. Maurelia, and M.E. Reyes. A Novel Hybrid Finite Control Set Model Predictive Control Scheme With Reduced Switching. *IEEE Transactions on Industrial Electronics*, 61(11):5912–5920, 2014.

[10] M. Rivera, F. Morales, C. Baier, J. Munoz, L. Tarisciotti, P. Zanchetta, and P. Wheeler. A Modulated Model Predictive Control Scheme for a two-level Voltage Source Inverter. In *Industrial Technology (ICIT), 2015 IEEE International Conference on*, pages 2224–2229, 2015.

© VDE VERLAG GMBH · Berlin · Offenbach

PCIM Europe 2016, 10 – 12 May 2016, Nuremberg, Germany

Current Sensorless Totem-pole Bridgeless Power Factor Corrector

Felipe Lopez-Vidal, University of Cantabria, Spain, lopezvfe@unican.es
Paula Lamo, University of Cantabria, Spain, paula.lamo@unican.es
Alberto Pigazo, University of Cantabria, Spain, pigazoa@unican.es
Francisco J. Azcondo, University of Cantabria, Spain, azcondof@unican.es

Abstract

Compared with traditional PFC converters, Bridgeless PFC reduces the conduction losses since the power conversion is performed by only one stage. However, the input current measurement is floating, increasing the circuit complexity and cost. This paper proposes a solution that avoids the use of the sensor of the input current. Instead, the current is rebuilt inside a digital device (FPGA) by integrating the voltage across the input inductor over time. Transistors switching time unbalance are assessed to match the estimated and actual current.

1. Introduction

DC-loads are daily increasing in the domestic environment (laptops, TV, etc...) and, as a consequence, so does the number of AC/DC converters. Years ago, AC/DC conversion was based only on a bridge diode stage. Nowadays, the most commonly adopted solution is a high-frequency DC/DC boost converter in cascade with the bridge diode to reduce the pollution due to harmonic injection. Several advantages of that topology are known [1]. However, it is a two-stage converter, so the efficiency is lower compared to other PFC circuits [2], [3]. Therefore, a solution based on only one stage is preferred to increase the efficiency. Based on the number of the devices used and the EMC issue [4], the totem-pole bridgeless rectifier is one of the most promising topologies.

One of the drawbacks of the totem-pole bridgeless rectifier is that the current measurement is floating and, even knowing that several methods are available [5], all of them present at least one of the following drawbacks: the operation depends on the power rate, signal conditioning and isolating circuit, together with a filter to improve the noise immunity, are required, and power associated with the current sensor is dissipated. The paper proposes a current sensorless solution to shape the line current as required for PFC applications.

2. System Configuration

A current control is applied to the totem-pole bridgeless rectifier working as PFC, shown in Fig. 1, following [6], [7]. However, in this case, the grid current (i_g) is not being measured but estimated based on the line-voltage, the DC-link voltage and the switches states (S1 and S2).

© VDE VERLAG GMBH · Berlin · Offenbach

PCIM Europe 2016, 10 – 12 May 2016, Nuremberg, Germany

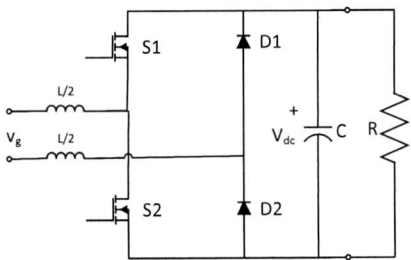

Fig. 1. Single phase totem-pole bridgeless topology

The totem-pole bridgeless rectifier can be split into two circuits. During the positive half-line cycle, two states are possible:

- Switch S2 is ON (Fig. 2a)
- Switch S2 is OFF (Fig. 2a)

When S2 is ON, the equations that represent this state are:

$$v_L = v_g - i_L(r_L + 2R_{DS}) \tag{1}$$
$$i_C = -i \tag{2}$$

In the same way, when S2 is OFF, the equations that represent this state are:

$$v_L = v_g - i_L r_L - V_F - V_{DC} - i_L R_{DS} \tag{3}$$
$$i_C = i_L - i \tag{4}$$

As it can be seen in the above equations, there are terms that represent the main parasitic elements in the converter. To better model the behavior of the bridgeless converter, those terms must be taken into account:
- r_L represents the parasitic resistance of the input inductor
- R_{DS} represents the parasitic resistance of the MOSFET
- V_F represents the voltage drop across the upper diode

Meanwhile, during the negative half-line cycle, two states are possible as well:

- Switch S1 ON (Fig. 2b)
- Switch S1 OFF (Fig. 2b)

When S1 is ON, the equations that represent this state are:

$$v_L = v_g - i_L(r_L + 2R_{DS}) \tag{5}$$
$$i_C = -i \tag{6}$$

In the same way, when S2 is OFF, the equations that represent this state are:

$$v_L = v_g - i_L r_L - V_F + V_{DC} - i_L R_{DS} \tag{7}$$
$$i_C = i_L - i \tag{8}$$

To sum up, the equations that models the behavior of the converter are shown below. Eq. (9) models the converter during the ON-state whereas (10) models the converter during the OFF-

© VDE VERLAG GMBH · Berlin · Offenbach

state:

$$v_L = v_g - i_L(r_L + 2R_{DS}) \tag{9}$$
$$v_L = v_g - sign(v_g)V_{DC} - i_L(r_L + R_{DS}) - V_F \tag{10}$$

Where:

$$sign(v_g) = \begin{cases} 1, v_g > 0 \\ -1, v_g > 0 \end{cases} \tag{11}$$

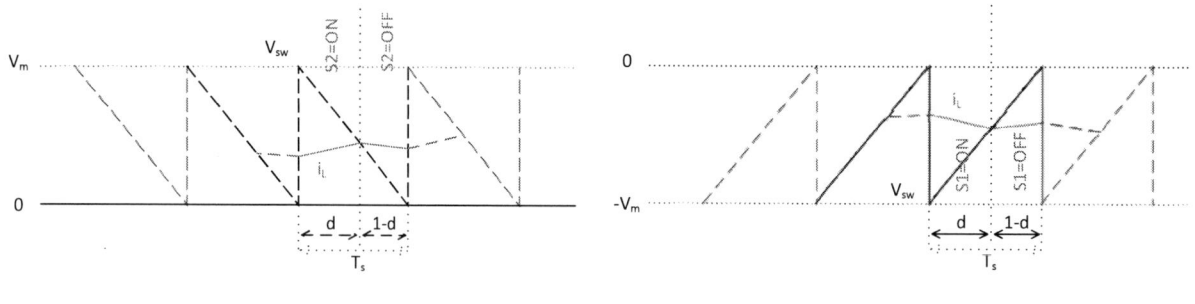

a) Positive half-line cycle   b) Negative half-line cycle

Fig. 2. Switching pattern followed

The duty cycle, d, is defined in the positive line semi period as the time in which transistor S2 is in the ON-state over the switching period. Similarly, d, is defined in the negative line semi period as the time in which transistor S1 is in the ON-state over the switching period.

Averaging through the whole duty cycle, with d obtained by comparing the estimated current with a carrier signal, using the Non-Linear Carrier Control (NLC) technique [8], as shown in Fig. 2, the following equation is obtained, where $<v_L>_{Ts}$ is the averaged voltage across the inductor over the switching period.

$$\langle v_L \rangle_{T_S} = \langle v_g \rangle_{T_S} - sign(v_g)V_{DC}(1-d) - i_L(r_L + R_{DS}) - i_L R_{DS}d - V_F(1-d)) \tag{12}$$

Based on this, it can be observed that to determine which active device (S1 or S2) is switching at high frequency, the phase of the input voltage must be known. A PLL is used to determine the sign of the input voltage. To do so several choices are available [9]–[11]. Due to their accuracy against low frequency harmonics and high switching noise, the PLL (Phase-Locked Loop) has been selected. The version implemented was the simplest one, called TD-PLL (Transport Delay-PLL), based on a FIFO memory [12]. The input voltage is measured through a sigma-delta analog-to-digital converter (ΣΔ ADC) and rebuilt by means of a Cascaded Integrator–Comb (CIC) filter, from where an α component is obtained directly and a virtual β component is generated through a T/4 delay buffer (FIFO memory), as shown in Fig. 3.

PCIM Europe 2016, 10 – 12 May 2016, Nuremberg, Germany

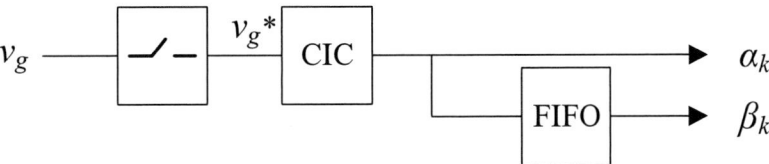

Fig. 3. Transport-Delay Phase-Locked Loop (TD-PLL)

The last step is the current rebuilt computation. To do so, the equations (9) and (10) are implemented inside the FPGA to obtain the voltage across the inductor (v_L). The current through the inductor is calculated with a digital version of the integral operator.

$$i_L(t) = \frac{1}{L} \int_0^t v_L(\tau) d\tau \tag{13}$$

Finally, the circuit block diagram that implements the current rebuilt and the control of the aforementioned converter working as PFC is shown in Fig. 4 and consists of:

- grid voltage phase detection based on a PLL algorithm (TD-PLL)
- current rebuilder algorithm following equations (1), (2) and (4)
- outer voltage control loop to keep constant the DC-link voltage (PI)
- inner current control based on non-linear carrier control (NLC)

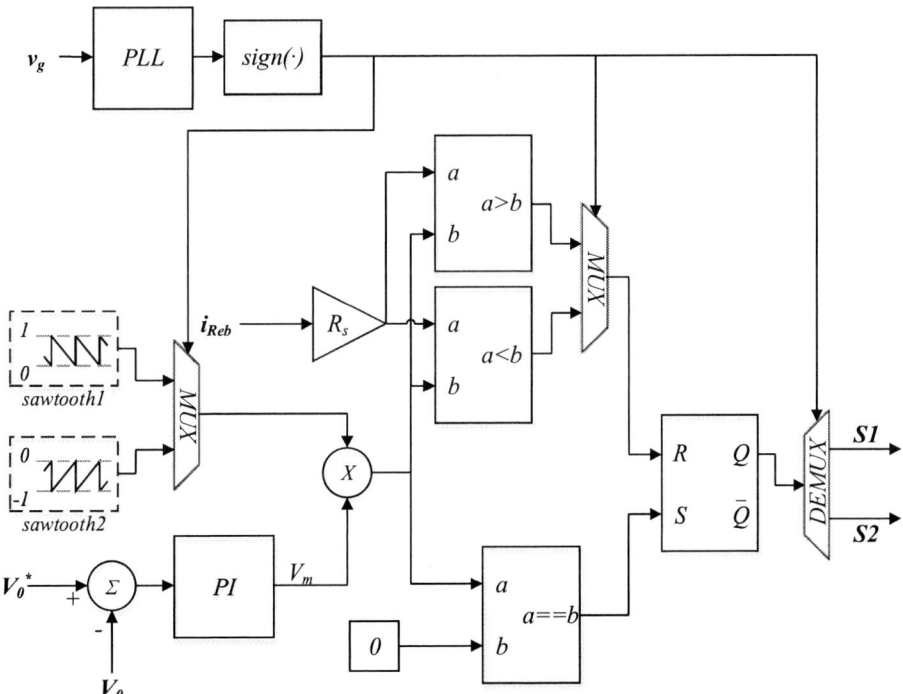

Fig. 4. Control scheme block diagram

© VDE VERLAG GMBH · Berlin · Offenbach

1197

3. Results

To validate the proposed control, simulations and experimental results have been carried out under the conditions shown in Table 1.

Input Voltage	Output Voltage	Load (R)	Grid Frequency	Switching Freq.
50 Vrms	150 V	500 Ω	50 Hz	48 kHz
V_F	R_{DS}	r_L	L	DC-Link
1.5 V	400 mΩ	1 Ω	1 mH	940 µF

Table 1. Parameters used throughout the simulations and experimental results

As explained in the previous section and expressed in (12), still two variables need to be measured: v_g and v_0. To do so, a simple resistor divider and an optically isolated ΣΔ ADC is used for each one. In the same way, to drive the MOSFETs from the FPGA, and optically isolated driver is used. The scheme described above is shown in Fig. 5.

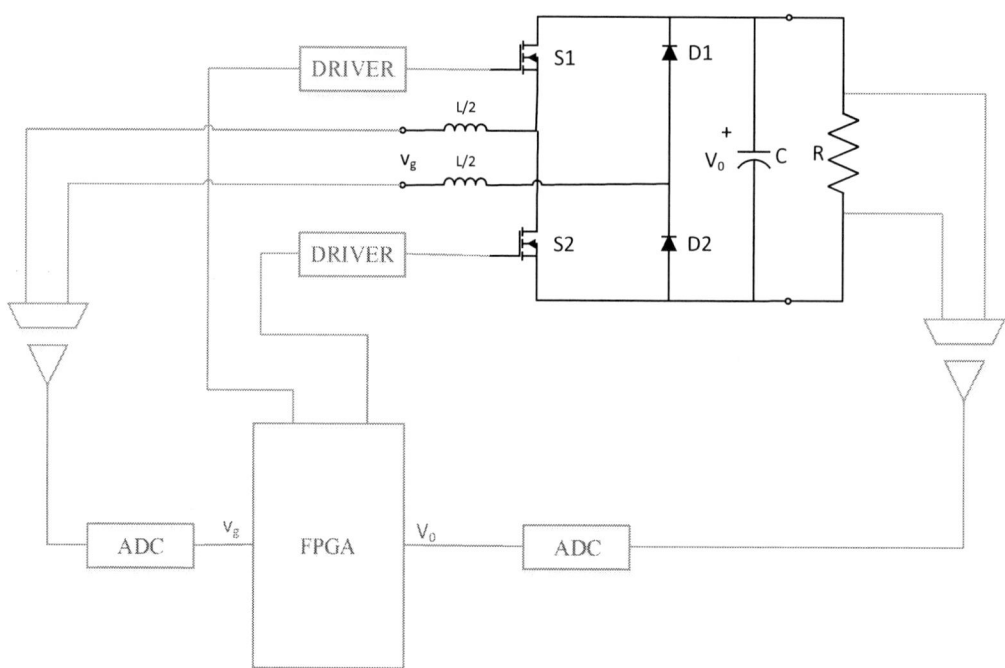

Fig. 5. Scheme used in the laboratory setup

One of the effects not included in the model up to this point is the delay in between the FPGA turn-on and turn-off events of the FPGA signals and the actual turn-on and off of the active switching devices (Fig. 6). The fact that turn-on and turn-off delays are different ($\Delta t = 1\mu s$), make the actual ON time in each switching cycle higher than the calculated by the FPGA algorithm. This effect is also considered as a parasitic element. In the case of the Totem-pole topology, the delays are different for both active switches (S1 and S2). As a consequence, a negative offset appears in the grid current, as it can be seen in Fig. 7 and Fig. 8. As a solution, the voltage v_L across the input inductor obtained during the negative half-line cycle was scaled down by 0.6 until the DC-offset disappears, as seen in Fig. 8b.

© VDE VERLAG GMBH · Berlin · Offenbach

PCIM Europe 2016, 10 – 12 May 2016, Nuremberg, Germany

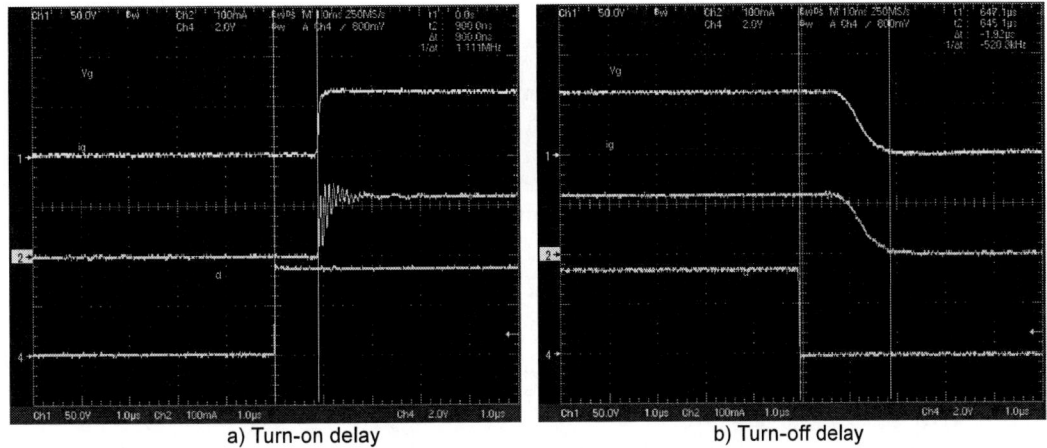

a) Turn-on delay   b) Turn-off delay

Fig. 6. Turn-on and turn-off delays of the active switching devices.

asdfas

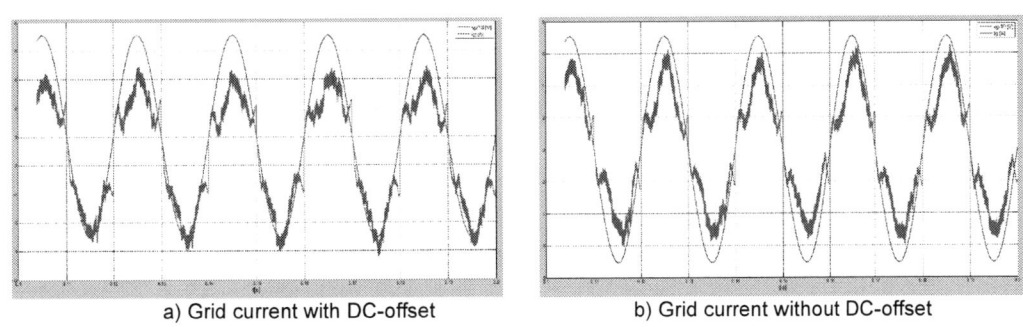

a) Grid current with DC-offset   b) Grid current without DC-offset

Fig. 7. Simulations results obtained using MATLAB/Simulink and PLECS

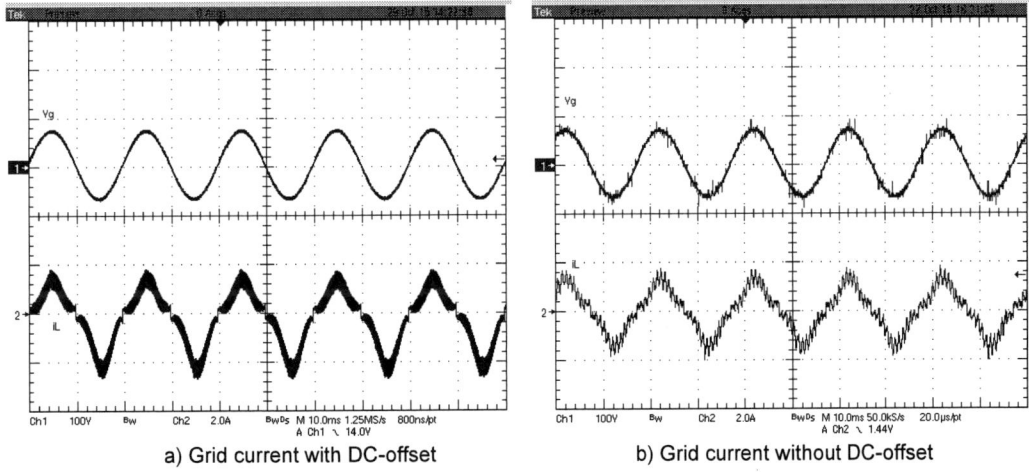

a) Grid current with DC-offset   b) Grid current without DC-offset

Fig. 8. Experimental results obtained using the laboratory setup

© VDE VERLAG GMBH · Berlin · Offenbach

4. Conclusions

A contribution to the current sensorless control for bridgeless totempole PFC converter has been presented. Key elements to make this technique feasible and robust are: accurate input and output voltage acquisition, modeling of the parasitic resistance, noise immunity line phase detection and active correction of the signal propagation and switches delays. The proposed technique has been verified are proved by simulation and experimental results.

5. Bibliography

[1]   A. De Bastiani Lange, T. B. Soeiro, M. Silveira Ortmann, and M. Lobo Heldwein, "Three-Level Single-Phase Bridgeless PFC Rectifiers," *Power Electron. IEEE Trans.*, vol. 30, no. 6, pp. 2935–2949, 2015.

[2]   K. Shi, M. Shoyama, and S. Tomioka, "A study of common mode noise current of bridgeless PFC circuit considering voltage change in Y-capacitors," in *Electromagnetic Compatibility, Tokyo (EMC'14/Tokyo), 2014 International Symposium on*, 2014, pp. 73–76.

[3]   L. Huber and M. M. Jovanovic, "Performance Evaluation of Bridgeless PFC Boost Rectifiers," *IEEE Trans. Power Electron.*, vol. 23, no. 3, pp. 1381–1390, May 2008.

[4]   Q. Li, M. A. E. Andersen, and O. C. Thomsen, "Conduction losses and common mode EMI analysis on bridgeless power factor correction," in *Power Electronics and Drive Systems, 2009. PEDS 2009. International Conference on*, 2009, pp. 1255–1260.

[5]   Y.-W. Cho, J.-M. Kwon, and B.-H. Kwon, "Single Power-Conversion AC--DC Converter With High Power Factor and High Efficiency," *IEEE Trans. Power Electron.*, vol. 29, no. 9, pp. 4797–4806, Sep. 2014.

[6]   F. Javier Azcondo, A. de Castro, V. M. Lopez, and O. Garcia, "Power Factor Correction Without Current Sensor Based on Digital Current Rebuilding," *IEEE Trans. Power Electron.*, vol. 25, no. 6, pp. 1527–1536, Jun. 2010.

[7]   F. Lopez, A. Pigazo, and F. J. Azcondo, "Bidirectional Current-sensorless High Power Factor Corrector," in *Power Conversion and Intelligent Motion (PCIM) Europe*, 2015.

[8]   D. Maksimovic and R. W. Erickson, "Nonlinear-carrier control for high-power-factor boost rectifiers," *IEEE Trans. Power Electron.*, vol. 11, no. 4, pp. 578–584, Jul. 1996.

[9]   O. Vainio and S. J. Ovaska, "Adaptive lowpass filters for zero-crossing detectors," in *IECON 02 [Industrial Electronics Society, IEEE 2002 28th Annual Conference of the]*, 2002, vol. 2, pp. 1483–1486.

[10]   D. Yazdani, M. Pahlevaninezhad, and A. Bakhshai, "Single-phase grid-synchronization algorithms for converter interfaced distributed generation systems," in *Electrical and Computer Engineering, 2009. CCECE '09. Canadian Conference on*, 2009, pp. 127–131.

[11]   F. Blaabjerg, R. Teodorescu, M. Liserre, and A. V. Timbus, "Overview of Control and Grid Synchronization for Distributed Power Generation Systems," *IEEE Trans. Ind. Electron.*, vol. 53, no. 5, pp. 1398–1409, Oct. 2006.

[12]   S. Golestan, J. M. Guerrero, A. Vidal, A. G. Yepes, J. Doval-Gandoy, and F. D. Freijedo, "Small-Signal Modeling, Stability Analysis and Design Optimization of Single-Phase Delay-Based PLLs," *Power Electron. IEEE Trans.*, vol. 31, no. 5, pp. 3517–3527, May 2016.

PCIM Europe 2016, 10 – 12 May 2016, Nuremberg, Germany

State Space Model for n-Parallel Connected DC-DC Converters with Predictive Current Control Strategy

Aditya Shekhar, Delft University of Technology, The Netherlands, a.shekhar@tudelft.nl
Laurens Mackay, Delft University of Technology, The Netherlands, l.j.mackay@tudelft.nl
Laura Ramírez-Elizondo, Delft University of Technology, L.M.RamirezElizondo@tudelft.nl
Pavol Bauer, Delft University of Technology, The Netherlands, P.Bauer@tudelft.nl

Abstract

In this paper, a state space model for n-parallel connected dc-dc buck converters is developed, and a predictive current control strategy is derived. A better dynamic response by eliminating the transient overshoot in the output current is incorporated in the prediction algorithm. Finally, a corrective term for output converter resistance is introduced for reduction in reference tracking error.

1. Introduction

With the emerging notion that opting for a dc distribution system can be beneficial, integration of dc-dc converters will gain more momentum in controlling the optimal required voltage level (1) and dc ready devices (2). Assuming that a dc distribution grid maintains a rigid, more or less stable voltage, such dc-dc converters will have voltage source behaviour at input side, thereby, a current controlled behaviour at the output side (3).

Several control strategies with PI controller based pulsed width modulation methods, peak current control and hysteresis control exist in the literature (4)-(6). With peak current control, the buck dc-dc converters switched off whenever the output current exceeds the set reference current value (5; 7). This strategy is simple and no system parametric values such as output inductance and resistance must be known for control. However, peak control is a responsive strategy and dependent on the measurement device error and delay. Furthermore, the transient overshoot when a step change in reference current value is commanded cannot always be eliminated, as the controller responds in subsequent switching cycle.

Experience with PI controller shows that it is difficult to tune the constants, particularly if the system configuration such as the output inductance can be changed through tapping. It is difficult also to respond adequately to the non-linearities of the system.

With information on the system physics and parameters, a predictive current control strategy can be developed. For control of two and three level voltage source inverters, this strategy is explored in (8)-(10). These look into the performance of the predictive current control against the conventional control strategy and highlight its robustness by including the non linearities of the system, thereby modelling and influencing the system variables in subsequent switching instances in order to obtain desired functionalities.

Section 2 describes the equivalent circuit and the state space model of the system for which the control strategy is described. Section 3 presents the algorithm developed for predictive control

© VDE VERLAG GMBH · Berlin · Offenbach

of the output current of the dc dc converter based on the derived system equations and the simulation results are depicted for specific system parameters. Section 4 refines the algorithm to eliminate the overshoot observed in the output current. Section 5 incorporates a corrective term for converter output resistance and therefore depicts the reduction in reference tracking error. Section 6 highlights the limitations of the presented control strategy. Finally, the results and conclusions are presented in Section 7.

2. System Description

The equivalent circuit of n buck dc-dc converters connected in parallel and supplying power to an inductive load is shown in Fig. 1.

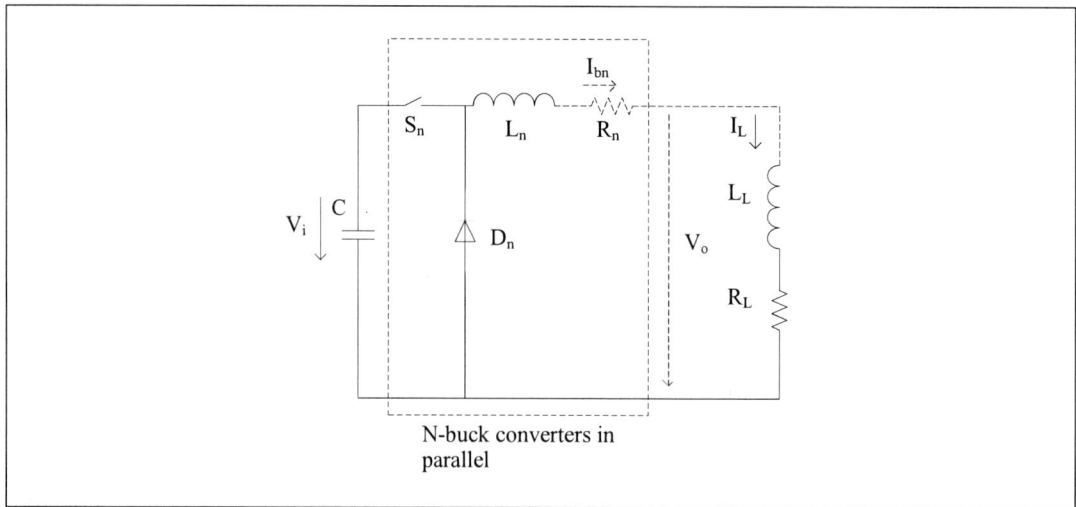

Fig. 1: Equivalent circuit of predictive current controlled buck converter.

Parameter values corresponding to the variables depicted in the diagram, for which the simulation results in this paper are presented are given in Tab. 1. The input voltage V_i can be a rigid

Tab. 1: Parameter Values for the Equivalent Circuit

Parameter	Value
Switching Frequency ($1/T_s$)	10.0 kHz
Buck Input Capacitance (C_{nn})	5.0 mF
Buck Output Inductance (L_n)	1.5 mH
Buck Output Resistance (R_n)	0.1 Ω
Load Inductance (L_L)	1 μH
Load Resistance (R_L)	1 Ω
Capacitor Voltage ($V_i(0)$)	1 kV
Initial Currents ($I_{bn}(0)$)	0.0 A

dc microgrid voltage. Herein, it is considered as an initial voltage to which the input capacitor is charged to, reducing with supplied energy to the load, in order to simulate the dynamic variation in input voltage.

The switching pulse parameter P_n for nth buck converter is defined such that,

$$S_n = 1 \quad \text{if switch is closed} \tag{1}$$

$$= 0 \quad \text{if switch is open} \tag{2}$$

The state space equations corresponding to the space variables V_i, I_{bn} and I_L associated with the energy storage elements of the circuit, derived to develop the simulation model are presented in (3)-(5).

$$\frac{dV_i}{dt} = \frac{-1}{C} \left(\sum_{n=1}^{n} S_n I_{bn} \right) \tag{3}$$

$$\frac{dI_{bn}}{dt} = \begin{cases} 0, & \text{if } I_{bn} = 0 \ \& \ S_n = 0 \\ \frac{S_n V_i - V_0 - R_n I_{bn}}{L_n}, & \text{otherwise} \end{cases} \tag{4}$$

$$\frac{dI_L}{dt} = \frac{V_0}{L_L} - \frac{R_L I_L}{L_L} \tag{5}$$

The output converter voltage V_0 must be represented in terms of the space variables. The circuit time constants are defined as $\tau_n = \frac{L_n}{R_n}$ and $\tau_1 = \frac{L_1}{R_1}$. In order to represent the derived expressions concisely, we define $\frac{1}{L_{in}} = \sum_{n=1}^{n} \frac{1}{L_n}$ (corresponding change in L_{in} based on the zero current flow through component inductances is incorporated in the model). From Kirchoff's law, it follows that,

$$\frac{dI_L}{dt} = \sum_{n=1}^{n} \frac{dI_{bn}}{dt} \tag{6}$$

Substituting (4) in (6) for $I_{bn} > 0$,

$$\frac{dI_L}{dt} = V_i \sum_{n=1}^{n} \left(\frac{S_n}{L_n} \right) - \frac{V_0}{L_{in}} - \sum_{n=1}^{n} \left(\frac{I_{bn}}{\tau_n} \right) \tag{7}$$

Substituting (5) in (7) and rearranging,

$$V_0 = \left(\frac{V_i \sum_{n=1}^{n} \left(\frac{S_n}{L_n} \right) - \sum_{n=1}^{n} \left(\frac{I_{bn}}{\tau_n} \right) + \frac{I_L}{\tau_L}}{\frac{1}{L_{in}} + \frac{1}{L_L}} \right) \tag{8}$$

Using the described state space model, n-parallel connected buck converter system can be simulated.

3. Predictive Current Control Algorithm

In this section, the expression for the duty cycle D_{k+1} in the $(k+1)^{th}$ switching instant for a constant switching time period (T_s) is derived, based on the predictive current control strategy. This strategy uses the measurements of the output inductor current $(I_{out,k})$ and the input voltage $(V_{in,k})$ at the end of kth switching instant. The rate change of the output inductor current that is required to reach the reference current magnitude (I_{ref}) in the $(k+1)^{th}$ switching instant is given by (9).

$$\left(\frac{dI_{bn}}{dt} \right)_{req} = \frac{I_{ref} - I_{bn,k}}{D_{k+1} T_s} \tag{9}$$

© VDE VERLAG GMBH · Berlin · Offenbach

Since the buck converter is a voltage source converter supported by a input capacitance (3), the variation in input voltage over consecutive switching instances can be assumed to be constant ($V_{i,k} \approx V_{i,k+1}$). Therefore, the duty cycle D_{k+1} is given by (10),

$$D_{k+1} = \frac{V_{o,k+1}}{V_{i,k}} \tag{10}$$

The actual rate change of the output inductor current in the $(k+1)^{th}$ switching instant described by the physics of the system is given by (11),

$$\left(\frac{dI_{bn}}{dt}\right)_{k+1} = \frac{V_{i,k} - V_{o,k+1}}{L_n} \tag{11}$$

In order to predict the duty cycle from the variables measured in the $(k)^{th}$ step, it is necessary to eliminate $V_{o,k+1}$ from the expression. This is because the output voltage can vary, particularly if any of the n switches change their switching state. Thereby, an approximation $V_{o,k} \approx V_{o,k+1}$ may not be accurate. Moreover, we reduce one measured variable in the control strategy by eliminating $V_{o,k+1}$.

From (9)-(11), the quadratic equation for D_{k+1} is given by (12),

$$D_{k+1}^2 - D_{k+1} + \frac{L_n(I_{ref} - I_{bn,k})}{V_{i,k}T_s} \tag{12}$$

The roots of the equation are given by (13),

$$D_{k+1} = \frac{1 \pm \sqrt{1 - \underbrace{\frac{4L_n(I_{ref} - I_{bn,k})}{V_{i,k}T_s}}_{x}}}{2} \tag{13}$$

In order to build an algorithm to predict the duty cycle from (13), we must take into account the following:

- With increasing value of error signal $I_{ref} - I_{bn,k}$, indicated by an increase in x, the predicted value of duty cycle D_{k+1} must increase.

- When the error signal is considerably large, indicated by x>1, the term under the square root can become negative, giving a complex value for duty cycle, which must be avoided.

- When the error signal is negative (x<0), the duty cycle should be zero.

Hence, the predicted duty cycle constrained by $0 \leq D_{(k+1)} \leq 1$ is given by (14),

$$D_{k+1} = \begin{cases} \frac{1 - \sqrt{1-x}}{2}, & \text{if } 0 \leq x \leq 1 \\ \frac{1 + \sqrt{x-1}}{2} & x > 1 \\ 0 & x < 0 \end{cases} \tag{14}$$

Based on the derived algorithm for the predictive current control strategy, the simulation results for load current, input voltage and duty cycle for a single buck converter supplying an inductive load is shown in Fig. 2. The reference current is first set to 100 A and then increased to 150 A.

The deviation from the reference current value depends on the switching frequency and the output inductance of the buck converter. It can be observed that when the reference value is

PCIM Europe 2016, 10 – 12 May 2016, Nuremberg, Germany

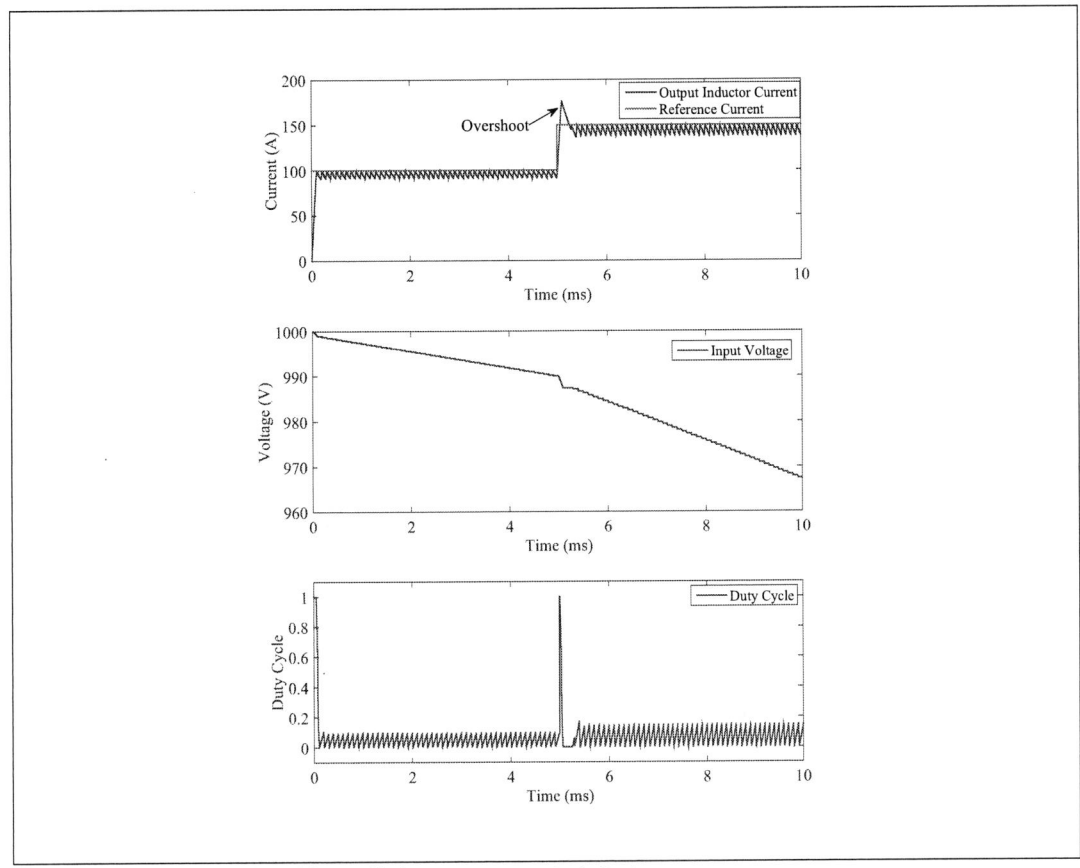

Fig. 2: Simulation results with predictive current control strategy.

changed to 150 A in step at 5 ms, the actual output current undergoes an overshoot to 175 A as the system can respond only in subsequent switching instant.

It is here that the developed algorithm by understanding the systems physics can aid in controlling the transient behaviour in a predictive way.

4. Transient Overshoot Elimination

The predictive correction in the $(k+1)^{th}$ duty cycle computation to eliminate the transient overshoot in the controlled output dc-dc converter current is given by (15).

$$\text{if, } x > 1 \text{ and } D_{k+1} > \frac{x}{4}, \text{ then,}$$
$$D_{k+1} = \frac{x}{4} \tag{15}$$

The mathematical expression describes that if the reference error $I_{ref} - I_{bn,k}$ that is reflected in variable x is higher than 1, which is true during a transient situation, the predicted duty cycle will be limited to a quarter of x. Fig. 3 depicts the simulation result for the same system with better dynamic response with the corrective algorithm for transient overshoot incorporated in the predictive current control strategy.

© VDE VERLAG GMBH · Berlin · Offenbach

PCIM Europe 2016, 10 – 12 May 2016, Nuremberg, Germany

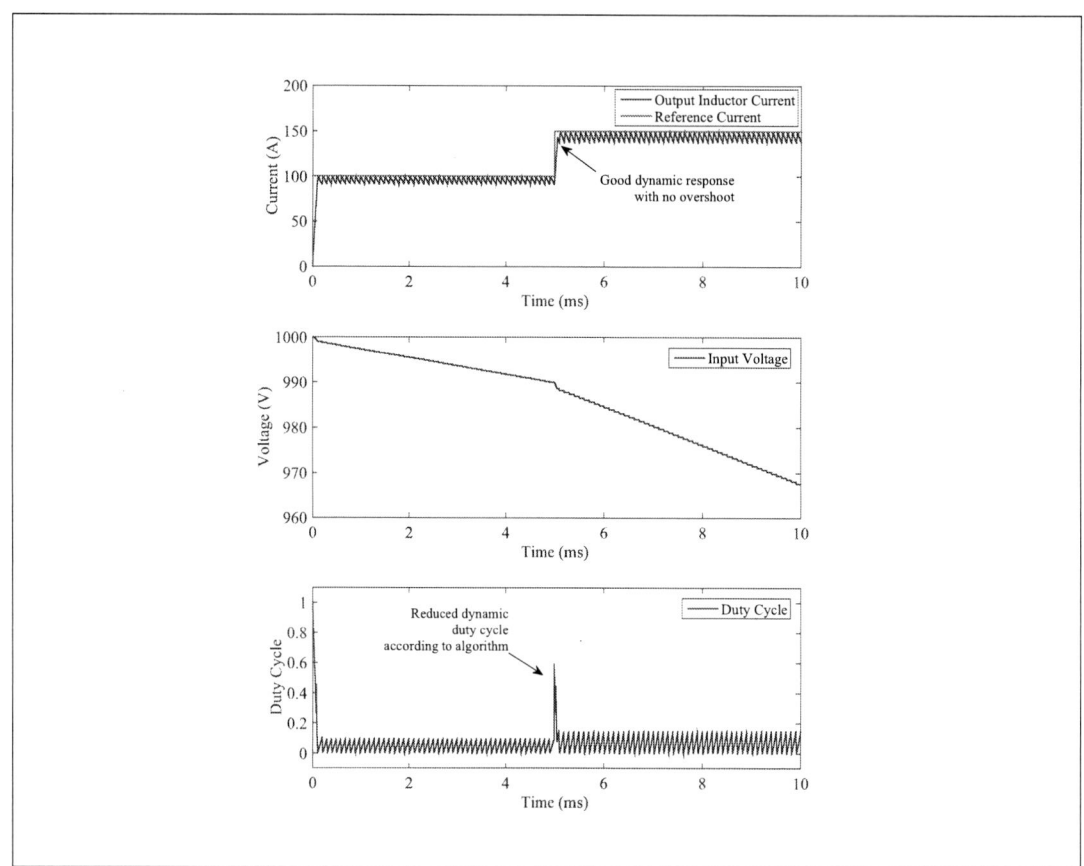

Fig. 3: Simulation results with elimination of transient overshoot.

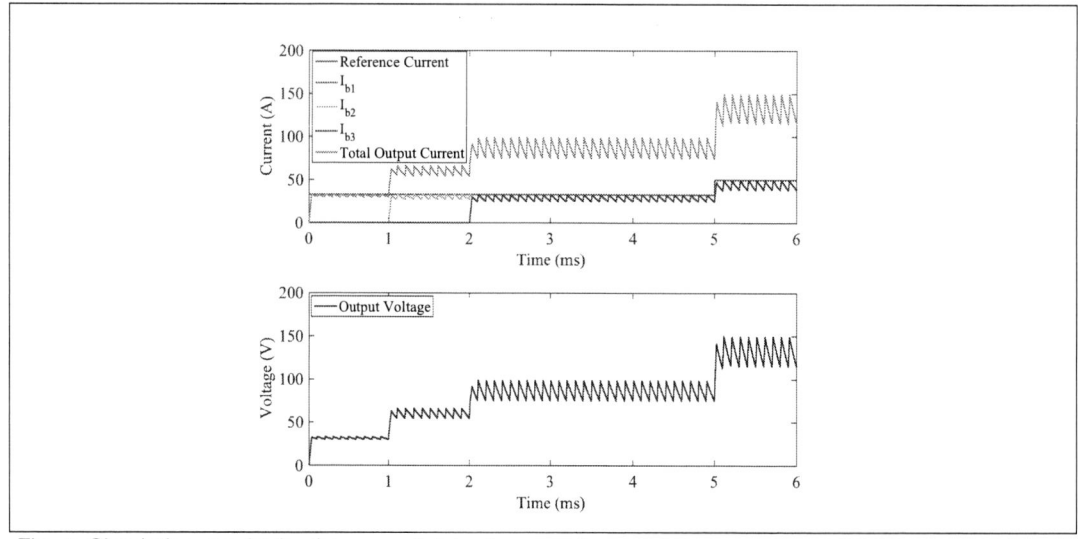

Fig. 4: Simulation results for three parallel dc-dc converters.

Fig. 4. shows the simulation results for output currents of three parallel dc-dc buck converters switched on with specified delay and reference current. The total load current I_L and the output voltage V_o at the point of common coupling is also depicted.

It can be observed that the ripple in the controlled current also gradually increases with the total output current because the output voltage is increasing. This can be minimized by asynchronously operating the three buck converters such that peak of each converter current occurs at different times.

5. Error Reduction with Output Resistance

The reference tracking error is due to the ripple component in the converter output current dependent on the inductance and switching frequency. With increasing switching frequency and inductance, the ripple reduces but the converter efficiency drops.

In the equations presented before, the output resistance is neglected in duty cycle computation. A minute, but finite reduction in error can be achieve by incorporating a corrective term corresponding to the converter resistance. The term 'x' in the predictive current control algorithm described in the preceding term along with this corrective term is given by (16).

$$x = \frac{4L_n(I_{ref} - I_{bn,k})}{T_s(V_{i,k} - I_{bn,k}R_n)} \tag{16}$$

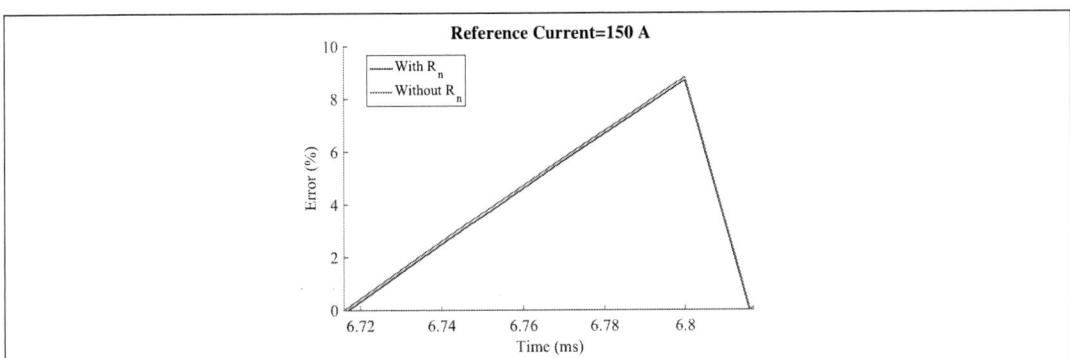

Fig. 5: Error reduction with output resistance correction.

Fig. 5 depicts the reference tracking error in percentage with and without the converter resistance corrective term at steady state for a reference current of 150 A.

The reduction in error is greater with increasing resistance and converter output current.

6. Limitations

The limitations of the control strategy include the following:

- Unlike peak current control, the algorithm is complex and input voltage measurement is also required apart from current measurement.

- The system parameters, particularly the output converter inductance must be known.

- The algorithm cannot effectively control the current as the output converter voltage becomes close to input voltage.

- With increasing load inductance, error in reference current tracking increases.

7. Conclusion

Therefore, a state space model of n-parallel connected buck dc-dc converters supplying power to an inductive load was developed. The output current was controlled by computing the duty cycle in for the next switching cycle in a predictive way. Modification in the control algorithm to eliminate the transient overshoot was incorporated. It was shown through simulations that the output current is able to track the set reference current with good dynamic response. Finally, a corrective term for converter output resistance is introduced and the reduction in reference current tracking error is depicted.

8. References

[1] S. Anand and B. G. Fernandes, "Optimal voltage level for DC microgrids," " *IECON 2010 - 36th Annual Conference on IEEE Industrial Electronics Society,* Glendale, AZ, 2010, pp. 3034-3039.

[2] L. Mackay, L. Ramirez-Elizondo and P. Bauer, "DC ready devices - Is redimensioning of the rectification components necessary?," *Mechatronics - Mechatronika (ME), 2014 16th International Conference on,* Brno, 2014, pp. 1-5.

[3] F. Z. Peng, "Revisit power conversion circuit topologies-recent advances and applications," *Power Electronics and Motion Control Conference, 2009. IPEMC '09. IEEE 6th International,* Wuhan, 2009, pp. 188-192.

[4] N. Mohan, T. M. Undeland, and W. P. Robbins, Power Electronics, 2nd ed. New York: Wiley, 1995.

[5] Shekhar, Aditya. "Detection, characterization and extinction of electric arcs in DC Systems "Diss. TU Delft, Delft University of Technology, 2015.

[6] L. K. Wong and T. K. Man, "Steady state analysis of hysteretic control buck converters," *Power Electronics and Motion Control Conference, 2008. EPE-PEMC 2008.* 13th, Poznan, 2008, pp. 400-404.

[7] M. M. Walter, C. M. Franck "Flexible Pulsed DC-Source for Investigations of HVDC Circuit Breaker Arc Resistance," XVIII International Conference on Gas Discharges and Their Applications, Greifswald, Germany, 2010.

[8] Wu, R.; Dewan, S.B.; Slemon, G.R., "Analysis of a PWM AC to DC voltage source converter under the predicted current control with a fixed switching frequency," in *Industry Applications, IEEE Transactions on* , vol.27, no.4, pp.756-764, Jul/Aug 1991.

[9] J. Rodriguez, J. Pontt, C. Silva, P. Cortes, U. Amman and S. Rees, "Predictive current control of a voltage source inverter," *Power Electronics Specialists Conference, 2004. PESC 04. 2004 IEEE 35th Annual,* 2004, pp. 2192-2196 Vol.3.

[10] R. Vargas, P. Cortes, U. Ammann, J. Rodriguez and J. Pontt, "Predictive Control of a Three-Phase Neutral-Point-Clamped Inverter," in *IEEE Transactions on Industrial Electronics,* vol. 54, no. 5, pp. 2697-2705, Oct. 2007.

Parameter-Independent Battery Voltage Control Based on Virtual Capacitor Emulation

Andoni Urtasun, Ernesto L. Barrios, Pablo Sanchis, and Luis Marroyo
Department of Electrical and Electronic Engineering, Institute of Smart Cities
Public University of Navarre (UPNa), Pamplona, Spain

Abstract

An adequate regulation of the battery voltage is fundamental to extend the battery lifetime and prevent it from overvoltage. However, due to the battery impedance variability, the voltage response becomes dependent on the operating point and battery characteristics. This paper proposes to emulate a large capacitor in parallel with the battery, making it possible to achieve a battery voltage control which is almost independent of the parameters. The proposed method is verified by simulation, showing how the voltage response is similar for batteries with completely different impedances.

1. Introduction

Due to the ongoing expansion of electrical transportation and renewable-based systems, batteries are currently in the spotlight. In order to reduce the cost of energy, main concerns are related to the battery lifetime, which can be extended thanks to the interfacing converter provided that a fine battery voltage control is performed. However, the battery voltage regulation involves many complications as a result of the battery impedance variability, whose value depends on the battery State-Of-Charge (SOC), state-of-health, temperature, battery technology, and the number of cells connected in series and parallel.

In order to make the control robust against parameter variations, some authors have proposed an adaptive voltage control [1], [2]. However, this method is complex because a complicated algorithm is required in order to estimate the battery internal resistance, R_{bat}, and then the controller parameters need to be modified.

In this paper, a simple and robust control is proposed for the battery voltage regulation. A large virtual capacitor in parallel with the battery is emulated, in such a way that the control becomes almost independent of the battery impedance variability. As a result, the converter becomes very versatile because a fine voltage regulation is carried out with no need of any information from the battery except for the voltage and current limits.

2. Battery Voltage Control

2.1. Description of the Problem

The analyzed system is shown in Fig. 1, consisting of a battery connected to a boost converter. The output can vary depending on the application, it could be a voltage-fed inverter or an electric vehicle DC bus, for example. A small capacitor is often included at the input to reduce the high-frequency current provided by the battery. In any case, this capacitor is very small and does not affect the voltage regulation, making it possible to be neglected.

© VDE VERLAG GMBH · Berlin · Offenbach

PCIM Europe 2016, 10 – 12 May 2016, Nuremberg, Germany

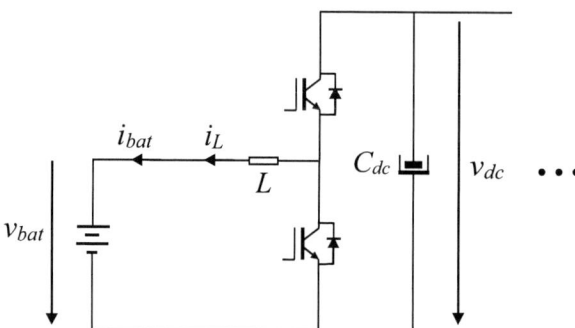

Fig. 1. Battery connected to a boost converter.

When the battery is fully charged, its voltage must be controlled to the limit value. Thanks to its advantages for this application [1], a cascaded control is adopted here, where the inner inductor current loop can be easily made independent of the input impedance by using the measured variables v_{bat} and v_{dc} as feedforward compensations [3]

In contrast, it is obvious that the battery impedance does have an influence on the voltage regulation. The small-signal battery model usually considers an internal resistance in series with one or more RC branches [4]. However, for the low frequencies around the cutoff frequency of the voltage regulation, the model can be considered as a pure resistance [1], R_{bat}. As a result, the small-signal plant to be controlled is:

$$G_{bat} = \frac{\hat{v}_{bat}}{\hat{i}_L} = \frac{\hat{v}_{bat}}{\hat{i}_{bat}} = R_{bat} \qquad (1)$$

The R_{bat} value depends on the battery technology, operating point, aging, and series-parallel cells connection. In order to evaluate the performance of the battery voltage control, it is important to obtain the resistance variation range. For this purpose, it will be considered that the converter can interface different battery technologies (such as lead-acid, lithium-ion and nickel-cadmium), the nominal voltage must be within 60 V and 240 V, and the maximum input current is 50 A.

From the various possibilities, the minimum battery resistance, $R_{bat,min}$, appears for a new lithium-ion battery, with low nominal voltage and high maximum current (60 V, 50 A), operating at high temperature (50°C) [5]–[7]. Taking into account commercial lithium-ion batteries and the mentioned conditions, one obtains $R_{bat,min}$ = 5 mΩ.

On the other hand, the maximum battery resistance, $R_{bat,max}$, can be found for an overused lead-acid battery, with high nominal voltage and low capacity (240 V, 75 Ah), operating at low temperature (-10°C) [8], [9]. In this situation, the maximum resistance for a commercial lead-acid battery becomes $R_{bat,max}$ = 1 Ω. As a result, it can be observed that the battery resistance can change from 5 mΩ to 1 Ω, i.e. within a rate of 200, making it difficult to achieve a parameter-independent voltage regulation.

The simplest method to control the battery voltage is by directly using an integral or proportional-integral (PI) controller [4], as shown in Fig. 2, where v^*_{bat} is the reference battery voltage, $v_{bat,m}$ the measured battery voltage, i^*_L the reference inductor current, C_v represents the voltage controller, G_{cl} the inductor current closed-loop and H_v the voltage filter.

© VDE VERLAG GMBH · Berlin · Offenbach

PCIM Europe 2016, 10 – 12 May 2016, Nuremberg, Germany

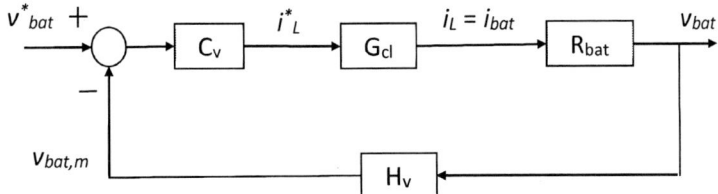

Fig. 2. Battery voltage control loop for the conventional control.

In this scheme, the parameter of the integral controller is calculated by assuming that the system plant is (1) for a certain battery resistance value. In this case, this resistance is chosen as the geometric mean between $R_{bat,min}$ = 5 mΩ and $R_{bat,max}$ = 1 Ω, that is $R_{bat,medium}$ = 70 mΩ. According to the battery features, a low cutoff frequency is required for the voltage regulation, and it is set to f_c = 0.5 Hz. As a result, the compensated open-loop for three different battery resistances is shown in Fig. 3, where the inner current loop and the voltage filter are modeled as first order transfer functions with time constants equal to 400 µs and 80 µs, respectively. As it can be observed in the figure, the voltage response is as designed, with f_c = 0.5 Hz, only for $R_{bat,medium}$ = 70 mΩ, while it reaches 7 Hz for $R_{bat,max}$ = 1 Ω and 36 mHz for $R_{bat,min}$ = 5 mΩ.

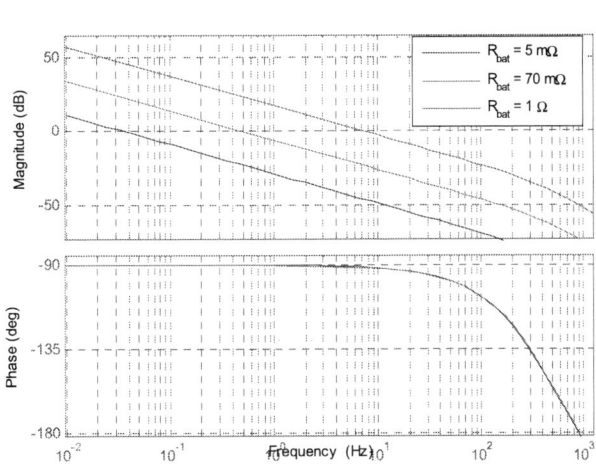

Fig. 3. Bode plot of the compensated open-loop for three different battery resistances, using the conventional control.

The conventional control is tested by using the simulation software PSIM. The model includes the battery, a boost converter, and a single-phase inverter connected to the grid. The simulation results for different battery resistances are shown in Fig. 4. It can be observed that the response is very variable depending on R_{bat}, specifically the rise time (from 10% to 90%) is $t_{r,5m\Omega}$ = 10.5 s, $t_{r,70m\Omega}$ = 750 ms, $t_{r,1\Omega}$ = 50 ms. In effect, the control is too distant from the desired control performance and is not suitable for this application.

In order to compensate the resistance variation, some authors have proposed an adaptive control [1], [2]. Although this control succeeds in improving the robustness against parameter variations, it is complex since the resistance needs to be continuously estimated, and then the control parameters need to be modified.

© VDE VERLAG GMBH · Berlin · Offenbach

1211

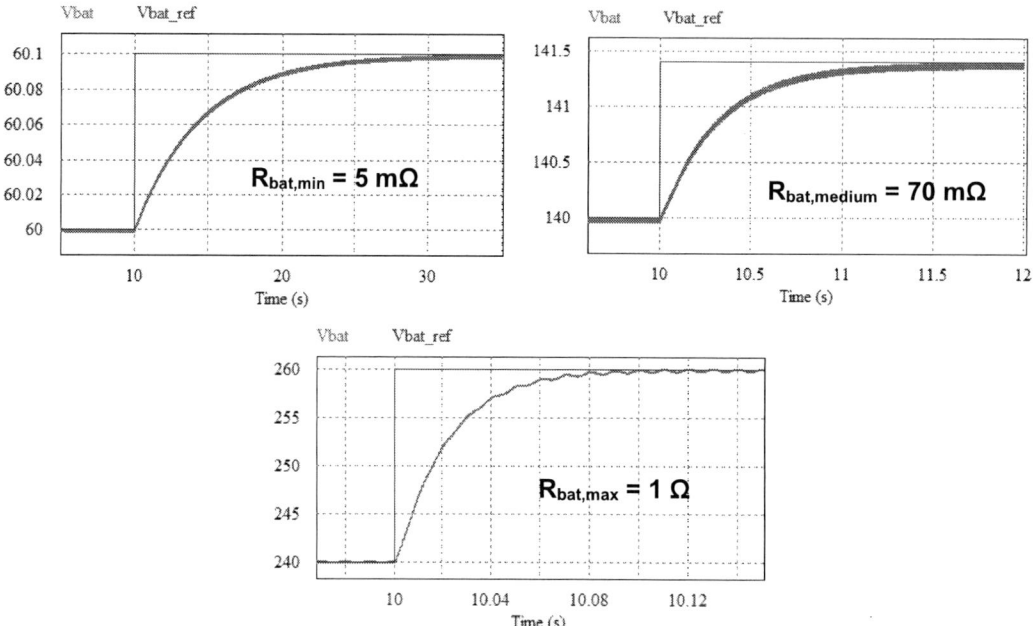

Fig. 4. Simulation results for the conventional battery voltage control for $R_{bat,min} = 5$ mΩ, $R_{bat,medium} = 70$ mΩ, $R_{bat,max} = 1$ Ω.

2.2. Proposed Method

A straightforward means to reduce the influence of the battery resistance variation would be to add a large capacitor, C_{real}, in parallel with the battery. If the capacitor impedance is small enough around the frequencies of concern, the system would behave as that known capacitor, and thus the plant variability would be completely removed. In doing so, the plant becomes:

$$G_{bat} = \frac{\hat{v}_{bat}}{\hat{i}_L} = \frac{R_{bat}}{C_{real} \cdot R_{bat} \cdot s + 1} \tag{2}$$

$$G_{bat} = \frac{\hat{v}_{bat}}{\hat{i}_L} \approx \frac{1}{C_{real} \cdot s}, \quad if \ C_{real} \cdot R_{bat} \cdot \omega \gg 1 \tag{3}$$

By using this capacitor, the battery voltage control loop is represented in Fig. 5(a), where i_C is the capacitor current. According to the battery features, a low cutoff frequency for the voltage regulation is required, and it is set to 0.5 Hz. The capacitor is chosen equal to 20 F, as a tradeoff between the capacitor value and the voltage response variability.

Although this capacitor is obviously too large for a practical application, ideally it could be emulated by calculating the capacitor current as $C_{virtual} \cdot dv_{bat}/dt$, where $C_{virtual}$ is the virtual capacitor. By using this virtual capacitor, the voltage control is represented in Fig. 5(b), where i_{Cv} is the virtual capacitor current, and $i_v = i^*_L + i_{Cv}$ is the virtual current. By comparing the figures, it can be observed that, in reality, the capacitor emulation is not exact because the measured voltage is used instead of the real voltage, and the reference current is modified instead of the real battery current. However, if one assumes that the current closed-loop and the voltage measurement are fast, both loops are equivalent. As a result, the system plant also becomes (2) when emulating a virtual capacitor, with the advantage that a real capacitor is not required.

© VDE VERLAG GMBH · Berlin · Offenbach

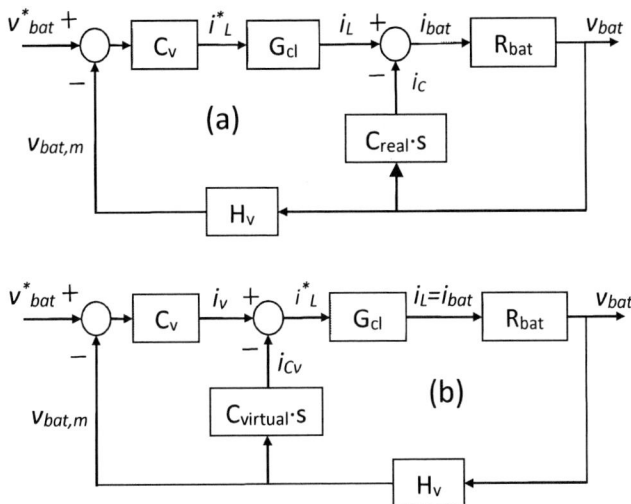

Fig. 5. Battery voltage control loops: (a) With a real capacitor, (b) With a virtual capacitor.

In order to safely implement the digital derivate, in this case H_v is a second-order digital filter with f_n=7 Hz (ω_n=2π·7 Hz) and ζ=0.7, which removes the high frequency noise as well as the ripple at twice the grid frequency present in single-phase systems. It is also worth noting that, in order to avoid a resonance for high R_{bat} values, the current reference $\overset{*}{i}_L$ should be filtered, which can be naturally obtained by implementing a slower inner loop or using the voltage loop sampling, or alternatively by implementing a digital first-order filter. In any case, in this work that transfer function will be included within G_{cl} and a time constant τ=80 ms is used.

As a result, in the scheme of Fig. 5(a), due to the filter H_v and the inner current loop, represented by G_{cl}, the system plant is in reality different from (2), and the non-compensated open-loop can be obtained as:

$$ G_{bat} \cdot H_v = \frac{\hat{v}_{bat}}{\hat{i}_v} \cdot H_v = \frac{G_{cl} \cdot H_v \cdot R_{bat}}{1 + G_{cl} \cdot H_v \cdot C_{virtual} \cdot R_{bat} \cdot s} \tag{6} $$

The bode plots of the non-compensated open-loop, represented by (6), are shown in Fig. 6 for three different battery resistances, $R_{bat,min}$ = 5 mΩ, $R_{bat,medium}$ = 70 mΩ, $R_{bat,max}$ = 1 Ω, and for the searched plant involving the pure virtual capacitor, $1/C_{virtual}$·s. As it can be observed, around 0.5 Hz, the plants with $R_{bat,medium}$ = 70 mΩ and $R_{bat,max}$ = 1 Ω behave as the virtual capacitor, while the plant with $R_{bat,min}$ = 5 mΩ behaves as the virtual capacitor in parallel with the battery resistance. Thus, thanks to the proposed method, the impedance variation at the frequencies of concern has been greatly reduced, particularly from a ratio of 200 to 3.3.

© VDE VERLAG GMBH · Berlin · Offenbach

PCIM Europe 2016, 10 – 12 May 2016, Nuremberg, Germany

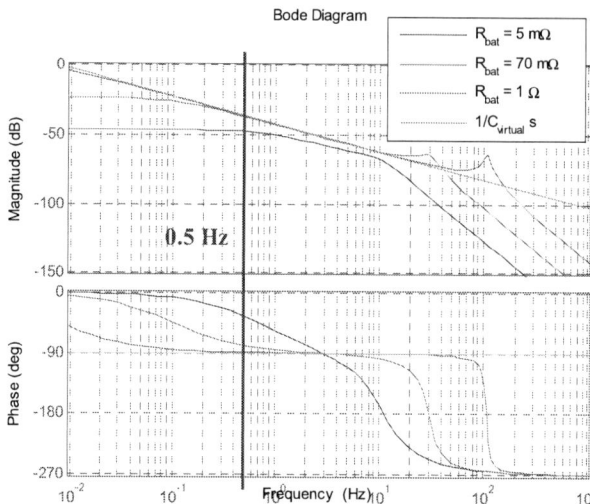

Fig. 6. Bode plot of the non-compensated open-loop for three different battery resistances, using the proposed method with a virtual capacitor.

Once the impedance variation has been drastically decreased, a simple PI controller can be used. Its parameters are calculated by assuming that the system plant is $1/C_{virtual} \cdot s$ for $C_{virtual}$ = 20 F, and for a cutoff frequency f_c equal to 0.5 Hz and a phase margin equal to 55°. Considering this controller, the bode plots of the compensated open-loop are represented in Fig. 7 for different battery resistances, $R_{bat,min}$ = 5 mΩ, $R_{bat,medium}$ = 70 mΩ, $R_{bat,max}$ = 1 Ω. It can be observed that the voltage response is as designed for high R_{bat} values and becomes slower for very low R_{bat} values. In particular, the voltage response slows down from 0.5 Hz for $R_{bat,max}$ = 1 Ω up to 0.1 Hz for $R_{bat,min}$ = 5 mΩ, greatly reducing the plant variability effect. It is worth noting that, if required, the voltage response variability can be further reduced by increasing the virtual capacitor, at the cost of deteriorating the noise immunity. Concerning the phase margin, it increases as R_{bat} decreases, making the response more damped.

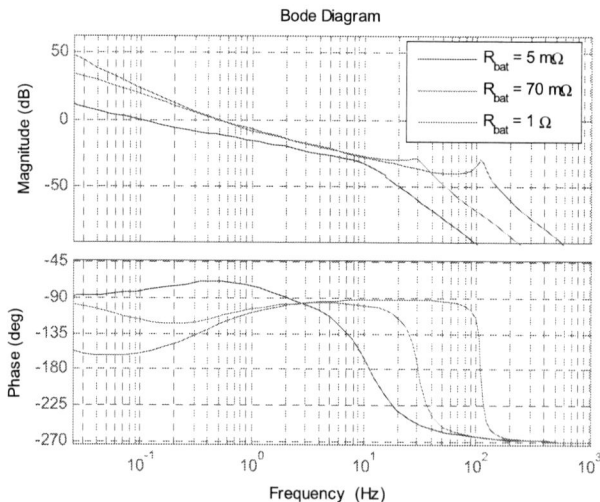

Fig. 7. Bode plot of the compensated open-loop for three different battery resistances, using the proposed method with a virtual capacitor.

© VDE VERLAG GMBH · Berlin · Offenbach

3. Simulation Results

The proposed control is tested by using the simulation software PSIM. The model includes the battery, a boost converter, and a single-phase inverter connected to the grid. The simulation results for different battery resistances are shown in Fig. 8, where the battery voltage, its reference and the battery voltage filtered by the second-order filter H_v are plotted. It can be observed that the voltage response is as designed for $R_{bat,medium} = 70$ mΩ and $R_{bat,max} = 1$ Ω, and it becomes slower and more damped for $R_{bat,min} = 5$ mΩ, as predicted by the previous analysis. More precisely, the rise time (from 10% to 90% for overdamped systems and from 0 to 100% for underdamped systems [10]) is $t_{r,5m\Omega} = 4.1$ s, $t_{r,70m\Omega} = 610$ ms, $t_{r,1\Omega} = 860$ ms, greatly improving the results of the conventional control. All in all, the proposed control is stable and fast enough for every situation, including the scenario with $R_{bat} = 5$ mΩ, where the operating conditions, battery technology and voltage and capacity levels represent the worst case.

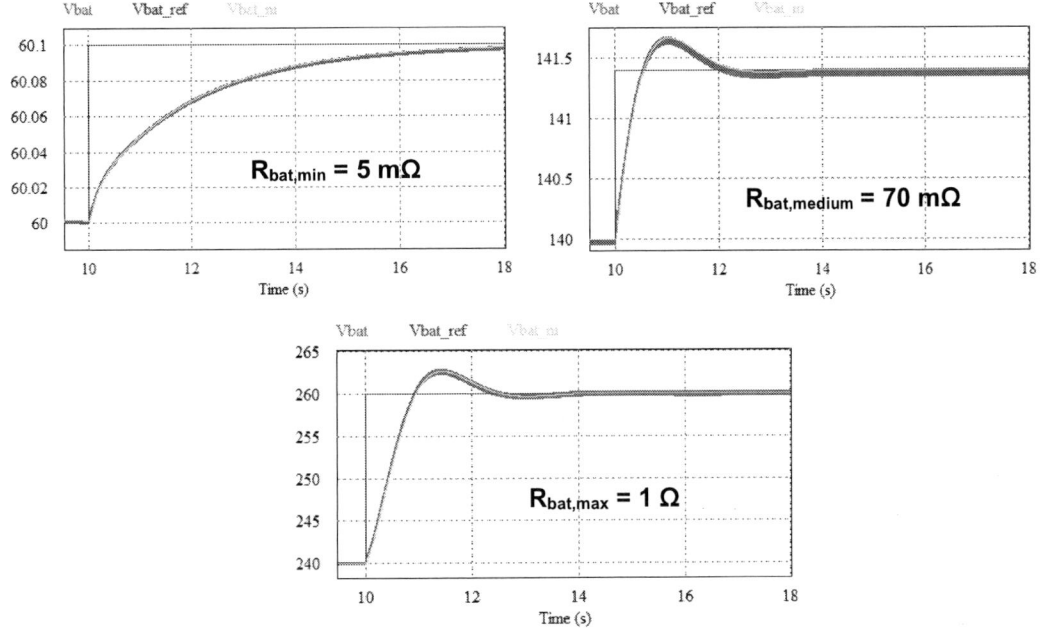

Fig. 8. Simulation results for the proposed battery voltage control
for $R_{bat,min} = 5$ mΩ, $R_{bat,medium} = 70$ mΩ, $R_{bat,max} = 1$ Ω

4. Conclusion

An adequate battery voltage regulation is a fundamental requirement to extend the battery lifetime. However, it is difficult to obtain a parameter-independent voltage control since the battery impedance can drastically vary depending on the battery technology, operating point, aging, and series-parallel cells connection.

This paper proposes a battery voltage regulation which emulates a large capacitor in parallel with the battery, making it possible to obtain a similar voltage response for completely different battery systems and conditions. In particular, the results show how an adequate voltage control is achieved for batteries with impedances between 5 mΩ and 1 Ω.

5. Acknowledgment

This work was supported in part by the Spanish Ministry of Economy and Competitiveness under Grant DPI2013-42853-R.

The authors gratefully acknowledge INGETEAM POWER TECHNOLOGY for their financial and permanent support.

References

[1] D. Pavkovic, M. Lobrovic, M. Hrgetic, A. Komljenovic, and V. Smetko, "Battery Current and Voltage Control System Design with Charging Application," in *2014 IEEE Conference on Control Applications (CCA)*, pp. 1133 – 1138, 2014.

[2] H. R. Eichi and M. Chow, "Adaptive Parameter Identification and State-of-Charge Estimation of Lithium-Ion Batteries," in *38th Annual Conference on IEEE Industrial Electronics Society (IECON)*, pp. 4012 – 4017, 2012.

[3] A. Urtasun, P. Sanchis, and L. Marroyo, "Adaptive Voltage Control of the DC/DC Boost Stage in PV Converters With Small Input Capacitor," *IEEE Transactions on Power Electronics*, vol. 28, no. 11, pp. 5038 – 5048, 2013.

[4] S. G. Tesfahunegn, P. J. S. Vie, O. Ulleberg, and T. M. Undeland, "A simplified Battery Charge Controller for Safety and Increased Utilization in Standalone PV Applications," in *2011 37th IEEE Photovoltaic Specialists Conference (PVSC)*, pp. 2441 – 2447, 2011.

[5] N. Somakettarin, and T. Funaki, "Parameter Extraction and Characteristics Study for Manganese-Type Lithium-Ion Battery," *International Journal of Renewable Energy Research*, vol. 5, no. 2, pp. 464 – 475, 2015.

[6] Y. Jing, W. Xuezhe, D. Haifeng, Z. Jiangong, and X. Xudong, "Lithium-ion Battery Internal Resistance Model Based on the Porous Electrode Theory," in *2014 IEEE Vehicle Power and Propulsion Conference (VPPC)*, pp. 1 – 6, 2014.

[7] J. D. Dogger, B. Roossien, and F. D. J. Nieuwenhout, "Characterization of Li-Ion Batteries for Intelligent Management of Distributed Grid-Connected Storage," *IEEE Transactions on Energy Conversion*, vol. 26, no. 1, pp. 256 – 263, 2011.

[8] S. Schaeck, A. O. Stoermer, F. Kaiser, L. Koehler, J. Albers, and H. Kabza, "Lead-Acid Batteries in Micro-Hybrid Applications. Part I. Selected Parameters," *Journal of Power Sources*, vol. 196, no. 3, pp. 1541 – 1554, 2011.

[9] T. Dragicevic, J. M. Guerrero, J. C. Vasquez, and D. Skrlec, "Supervisory Control of an Adaptive-Droop Regulated DC Microgrid With Battery Management Capability," *IEEE Transactions on Power Electronics*, vol. 29, no. 2, pp. 695 – 706, 2014.

[10] W. S. Levine, The Control Handbook, CRC Press, ISBN 0-8493-8570-9, 1996.

PCIM Europe 2016, 10 – 12 May 2016, Nuremberg, Germany

FPGA Digital Control for VSI Nonlinearity Effect Compensation

Umberto Abronzini, University of Cassino and Southern Lazio, Italy, u.abronzini@unicas.it
Ciro Attaianese, University of Cassino and Southern Lazio, Italy, attaianese@unicas.it
Matilde D'Arpino, University of Cassino and Southern Lazio, Italy, m.darpino@unicas.it
Mauro Di Monaco, University of Cassino and Southern Lazio, Italy, m.dimonaco@unicas.it
Giuseppe Tomasso, University of Cassino and Southern Lazio, Italy, tomasso@unicas.it

Abstract

This paper deals with the implementation on FPGA of a new compensation method of the non-linearities for grid-connected Voltage Source Inverters (VSIs). In steady state conditions, the compensation technique evaluates the distortion voltage error on the basis of the error between reference and actual output currents, within a fundamental period. This voltage error is, hence, used to correct the reference voltage space vector at the following fundamental periods.

1. Introduction

Voltage Source Inverter is widely used in several application fields, where high performances can be achieved by means of proper modulation techniques. However, the output voltage distortion due to non-ideal behaviour of VSI can considerably decrease the overall performances in the power conversion process in terms of harmonic distortion and loss in the fundamental [1]. Non-linearities are essentially due to turn-on/off delays, switches and diodes saturation voltages and the adoption of dead time delay. Methods proposed to compensate this undesirable behavior of VSI [2–6] can be divided into two main categories: averaging methods and pulse-based methods. In the former, the voltage error is averaged over an entire cycle and added to the reference voltage according to the direction of the load current. In the latter, the voltage error is evaluated for each PWM pattern modulation and compensated in the next PWM period. Averaging methods are usually characterized by slow compensation action, whereas the pulse-based ones are faster but they require a high performance control units. In [6] authors presented a recursive non-linearities compensation method for grid-connected current-controlled Photovoltaic (PV) power converters. The method is based on a step-by-step evaluation of the voltage distortion as a function of the error between reference and measured currents. This algorithm works in steady state conditions and it allows compensating non-linearities inside one single sampling interval during each period of the fundamental waveform; hence, several consecutive evolution periods of the current waveform are needed to achieve a full compensation.
In this paper an improvement of this technique and its FPGA implementation are proposed. In particular, the time interval needed to perform the full compensation is considerably reduced to the period of the fundamental voltage by the adoption of a suitable mathematical model of the power conversion system. Once the steady state condition is reached within a fundamental period, the proposed method calculates step by step the voltage error due to non-linearities on the basis of the error between reference and sampled output currents. The voltage compensation is kept until the working conditions change. Furthermore, thanks to the flexibility and fast computation time of FPGA, it is possible to implement on a single-chip both the modulation algorithm and the compensation method, achieving high speed and performance control

© VDE VERLAG GMBH · Berlin · Offenbach

PCIM Europe 2016, 10 – 12 May 2016, Nuremberg, Germany

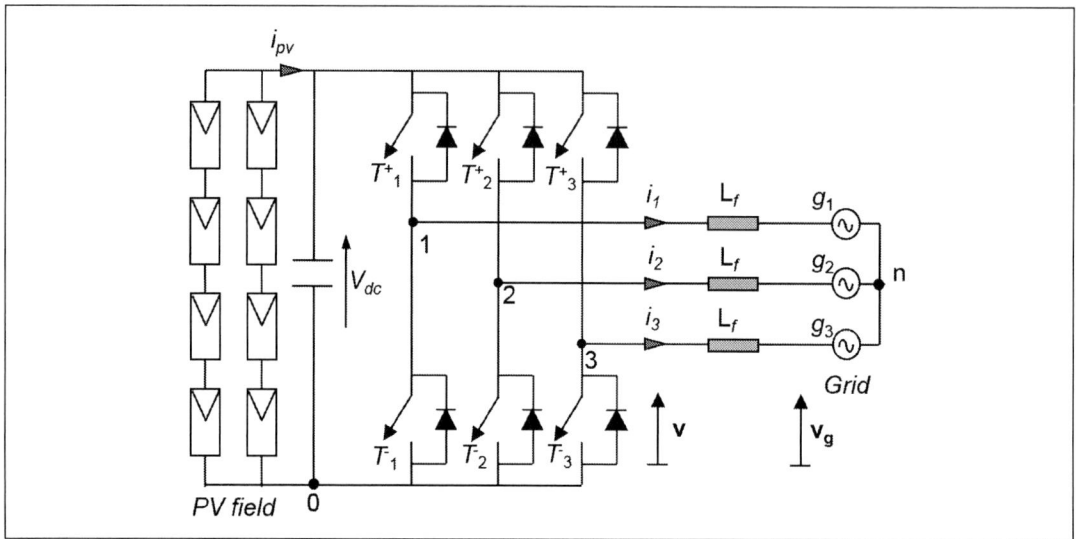

Fig. 1: Grid-connected PV power conversion system.

platform. In order to integrate modulation and compensation, a FPGA based architecture has been used for experimental implementation.

2. Compensation algorithm

With reference to fig. 1, the mathematical model of the considered grid-connected VSI for PV application is represented by the following equation:

$$\mathbf{v} = \mathsf{L}_f \frac{d}{dt} \mathbf{i} - \mathbf{v}_g \tag{1}$$

where:
- $\mathbf{v}$ voltage space vector of VSI;
- $\mathbf{v}_g$ grid voltage space vector;
- $\mathbf{i}$ grid current space vector;
- L_f inverter output inductance.

The reference voltage space vector to be applied at the n-th control interval (T_s) to achieve the $(n+1)$-th reference current $(\mathbf{i}^{*(n+1)T_s})$ can be carried out with one-step-ahead predictive control as a function of the actual current space vector $(\mathbf{i}^{nT_s})$ and the average value over a T_s of the grid voltage space vector $(\hat{\mathbf{v}}_g^{nT_s})$. In particular, it yields:

$$\mathbf{v}^{*nT_s} = \frac{\mathsf{L}_f}{T_s} \left(\mathbf{i}^{*(n+1)T_s} - \mathbf{i}^{nT_s} \right) + \hat{\mathbf{v}}_g^{nT_s} \tag{2}$$

Superscripts n and $(n+1)$ the quantities evaluated at the n-th and $(n+1)$-th sampling times respectively indicate . The magnitude of the reference current space vector $\mathbf{i}^*$ is usually provided by means of a Maximum Power Point Tracking (MPPT) algorithm in PV application, while its phase is function of the requested power factor [7].
The application of $\mathbf{v}^{*nT_s}$ allows achieving the value of the desired current $(\mathbf{i}^{*(n+1)T_s})$ at the end

© VDE VERLAG GMBH · Berlin · Offenbach

PCIM Europe 2016, 10 – 12 May 2016, Nuremberg, Germany

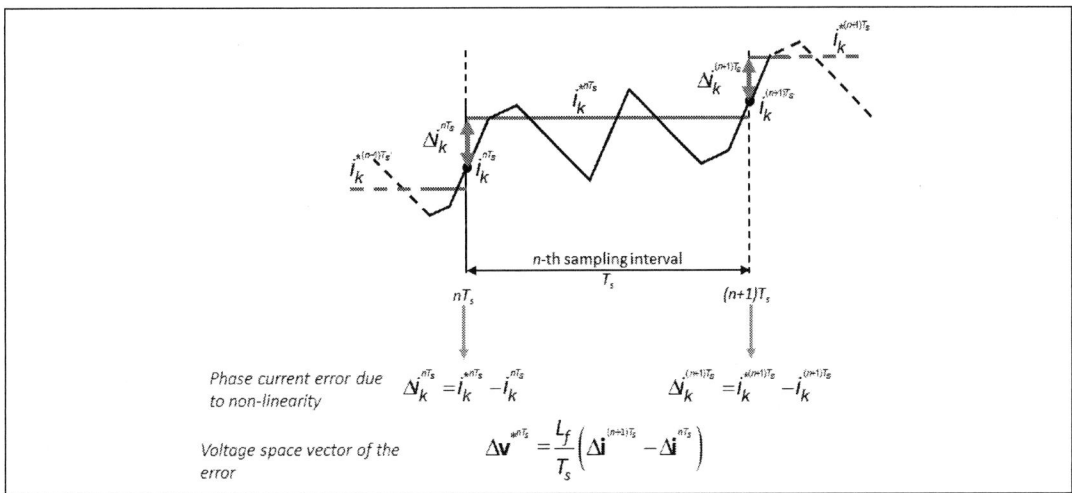

Fig. 2: Working principle of the non-linearities compensation algorithm.

of the n-th control interval only in the case of ideal behaviour of the power conversion system. In real operating conditions, the quality of the current decreases due to the voltage distortions. Considering a non-ideal semiconductor switches [6], the pole voltages of a VSI are characterized by an error with respect to the ideal ones, which is function of the sign of the phase currents. As shown in [6], the voltage distortion for the generic k-th phase is mainly due to the dead time that is a delay added on the switch-on command to avoid the short circuit of the dc link during the switches commutation over a leg. However, the turn-on/off delays, saturation voltage of active switches and diodes also play an important role on the non-linear behaviour of the output voltage. With reference to the n-th control interval, the generic k-th phase current does not reach the desired value ($\mathbf{i}^{*(n+1)T_s}$) due to the voltage distortion, as shown in fig. 2. The current error ($\Delta\mathbf{i}^{(n+1)T_s}$) gained at the end of n-th control interval is not only due to voltage error of the considered control period, but also due to the error voltage of the previous ones. In fact, the current value at the beginning of the analysed control period is different from the ideal one equal ($\mathbf{i}^{*nT_s}$) as well.

In the proposed compensation technique, the error between reference and measured currents is evaluated at the beginning and at the end of each control period T_s and it is used to carry out the real voltage compensation to be applied at the same T_s of next period of the fundamental waveform, as shown in fig. 3. In particular, by indicating with $\mathbf{v}^{*nT_s}$ and $\mathbf{v}^{nT_s}$ the ideal and real space vectors of the inverter output voltage at the n-th sampling time, it yields:

$$\begin{cases} \mathbf{v}^{*nT_s} = \frac{L_f}{T_s}\left(\mathbf{i}^{*(n+1)T_s} - \mathbf{i}^{*nT_s}\right) + \hat{\mathbf{v}}_g^{nT_s} \\ \mathbf{v}^{nT_s} = \frac{L_f}{T_s}\left(\mathbf{i}^{(n+1)T_s} - \mathbf{i}^{nT_s}\right) + \hat{\mathbf{v}}_g^{nT_s} \end{cases} \tag{3}$$

Thus, the voltage space vector $\Delta\mathbf{v}^{*n}$ related to the error due to the non-linear behaviour of the VSI during the only n-th interval can be expressed by the following equation:

$$\Delta\mathbf{v}^{*n} = \mathbf{v}^{*nT_s} - \mathbf{v}^{nT_s} = \frac{L_f}{T_s}\left[\left(\mathbf{i}^{*(n+1)T_s} - \mathbf{i}^{(n+1)T_s}\right) - \left(\mathbf{i}^{*nT_s} - \mathbf{i}^{nT_s}\right)\right] \tag{4}$$

In steady state condition and with reference to fig. 3, the real voltage error is evaluated during the 1-st fundamental period for each T_s by (4) and added to voltage reference, worked out by

© VDE VERLAG GMBH · Berlin · Offenbach

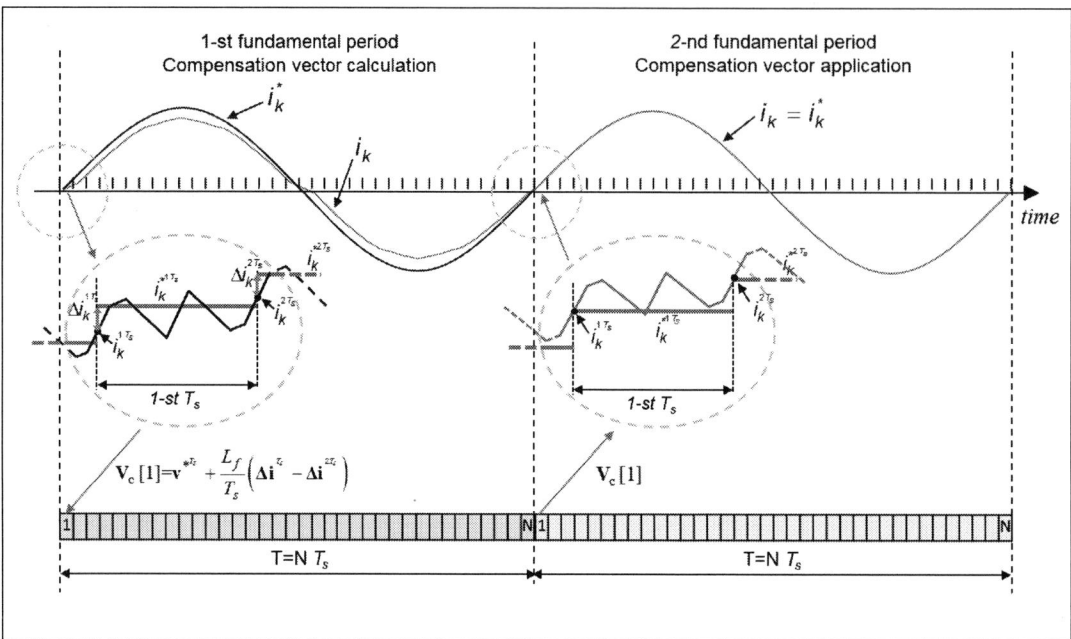

Fig. 3: Time evolution of the non-linearities compensation algorithm.

predictive control, in order to built up a new reference voltage array. Each element of this array is adopted as reference voltage space vector to be applied at the same control intervals within the 2-nd fundamental period. The operating principle is represented for the first control interval in fig. 3. This action is performed for all N control intervals and, hence, a full non-linearity compensation is achieved after one fundamental period. The voltage compensation is kept in the following fundamental periods until the working conditions change.

3. FPGA based implementation of the compensation method

A control unit based on low-cost Cyclone III FPGA has been designed and implemented to perform the predictive current control with the proposed compensation technique. Fig. 4 shows the architecture of the whole control system. FPGA chip has been used to perform an embedded microprocessor for the control algorithms implementation, to control the 6-channel acquisition board and to implement the Space Vector Modulation (SVM) with related dead time generator. The predictive control and the proposed compensation technique have been implemented by means a interrupt routine inside the embedded microprocessor, configured by means of Altera® Qsys tool. A soft-core processor Nios II/f with an external DDR memory has been adopted. This memory is used as cache to store with high-speed the reference voltage array that is built up during the compensation procedure of the non-linearities (fig. 3).

The FPGA chip has been interfaced with a ADC board for the acquisition of the control variables, such as phase currents and voltages and panel current and voltage. The number of ADC channel has been minimized considering that for a three-phase symmetrical power system, only two currents and voltages are linearly independent. The third current and voltage have been expressed as a linear combination of the other two into the control algorithm. Six SAR ADC converters TLV 1572 have been used, which are interfaced with FPGA board by 6 wire SPI bus.

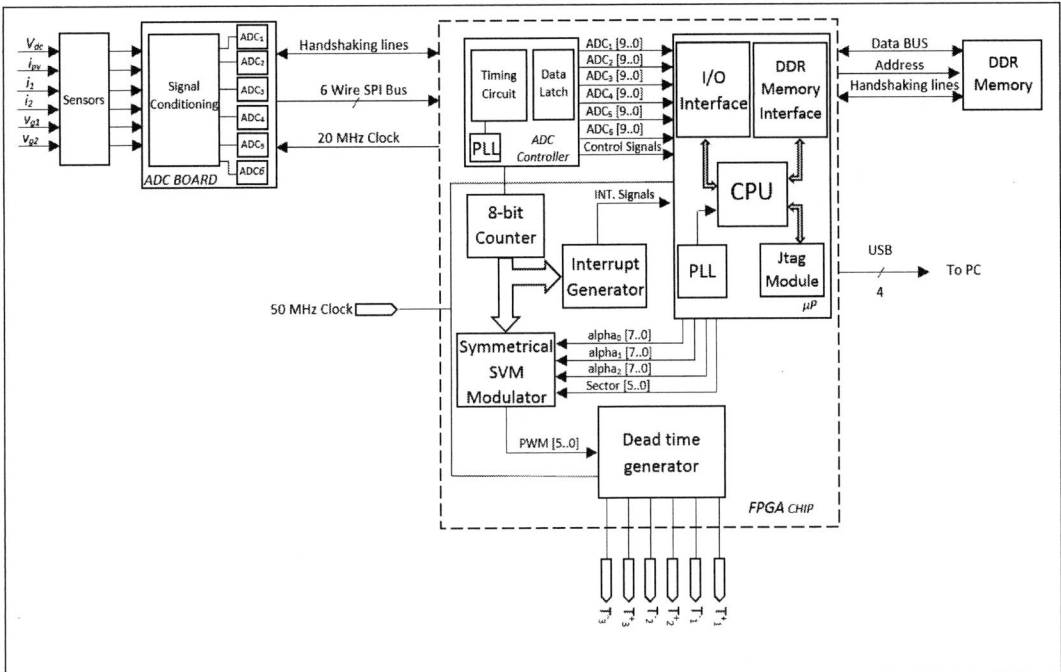

Fig. 4: Architecture of the FPGA based control unit.

To improve the performance of the control board, the ADC controller and the digital symmetrical SVM modulator with its dead time generator have been implemented in VHDL (VHSIC Hardware Description Language). ADC controller manages the handshake with ADC converters and stores the results of the conversion process.

The main advantage of the proposed architecture is represented by the possibility to develop the embedded processor and the other components of the control unit on a single chip. Furthermore, thanks to adopted architecture a perfect timing synchronization of all cyclic processes, from the fastest in microsecond cycle to the slower ones, has been performed by means of a suitable generation of 5 kHz interrupt signal. In particular, at the beginning of the each control interval, the microprocessor sends to ADC controller a start signal and waits until the results of conversion are available. Then, it runs the predictive current control described in the previous paragraph. Moreover, starting from the reference voltage space vector the microprocessor works out the application times ($alpha_i$ with $i = 0, 1, 2$) of the voltage vectors to be applied during the modulation period [5]. The symmetrical SVM modulator generates the PWM signals on the basis of these information and the dead time generator block adds the delay on the switch-on commands of power switches. Moreover, the current error space vector is calculated at the start of each sampling interval and stored to calculate the voltage compensation space vector of the proposed compensation method (eq. 4). The PWM signals are successively converted into optical signals, and transmitted over a fibre optic to the power converter drivers. Fig.s 5 (a) and (b) the pictures of the ADC board and of the whole control unit of the power converter respectively show .

PCIM Europe 2016, 10 – 12 May 2016, Nuremberg, Germany

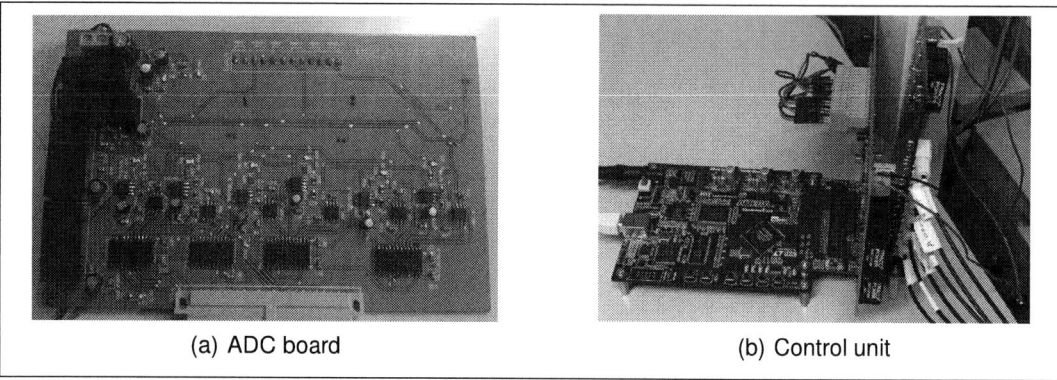

| (a) ADC board | (b) Control unit |

Fig. 5: Pictures of the FPGA based control unit.

4. Implementation results

To verify the performance of the FPGA based compensation system, both hardware and software, a three-phase VSI has been implemented by means of Mitsubishi® PM100DSA120 intelligent power modules and a 10.8 mH three-phase inductance filter has been used to connect it to the utility grid. For this system the control interval and the dead time have been set to 200 μs and 4 μs respectively.

First, a complete simulation of the power conversion system has been designed to test the proposed control algorithm and FPGA implementation. In particular, an accurate numerical model including PV field (V_{MPP}=800 V, i_{MPP}=3.4 A), real VSI, three phase voltage source (230 V, 50 Hz) and control section has been developed in Matlab-Simulink®. The non-linear model of the VSI has been performed, by taking into account current-dependent switches and diode voltage drops and on-off delay times of Mitsubishi PM100DSA120 power modules. The timing of the simulation has been designed on the basis of FPGA control unit. Fig. 6(b) shows the phase currents with and without the application of the compensation technique when a reference current of 5 A (peak) is adopted. In particular, the compensation algorithm starts at t=0.010s and within a fundamental period, it builds and stores the compensated reference voltage space vector array. The appropriate application of the elements of the array in the following fundamental period, which starts at t=0.030s, allows achieving a full-compensation of the currents. In detail, current THD decreases from 11.56% to 8.80% and RMS increases from 2.75 A to 3.53 A. Thanks to proposed compensation technique, a considerable improvement of the system performance has been achieved just after one fundamental period. In fact, the values of current THD and RMS are very close to the ones of the equivalent ideal power converter, which are respectively equal to 8.60% and 3.54 A.

Fig. 6 shows the experimental result achieved considering the same working condition of the numerical analysis: the proposed solution allows a strong reduction of the non-linearities effects on the current waveform. In particular, current THD decreases from 12.83% to 9.74% and RMS increases from 2.75 A to 3.56 A, confirming the validity of the proposed compensation method and architecture.

5. Conclusion

In this paper, a new steady-state recursive algorithm for the compensation of the non-linearity of the VSI and its implementation on a FPGA chip is presented in this paper. On the basis

© VDE VERLAG GMBH · Berlin · Offenbach

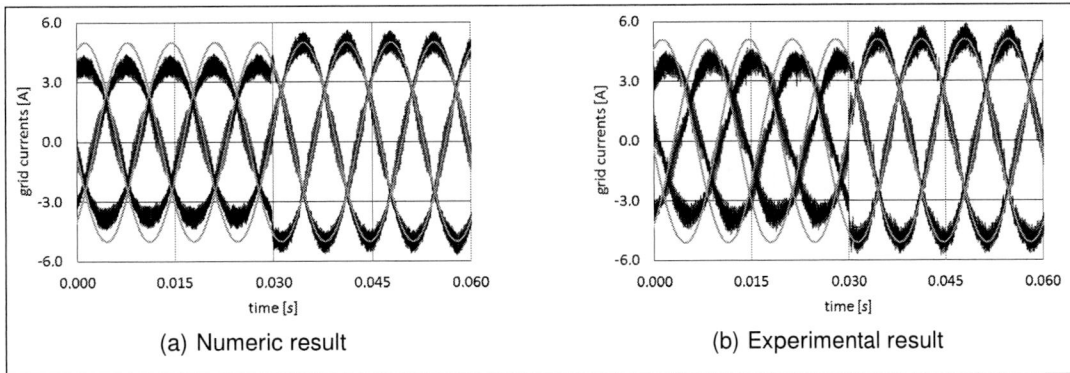

(a) Numeric result   (b) Experimental result

Fig. 6: Comparison between the numeric and experimental results for the phase currents without (for time $<0.030\ s$) and with (for time $>0.030\ s$) compensation action when a reference current of 5 A (peak) is considered.

of predictive control, the proposed system allows reducing the voltage drop and the current distortion due to the non-linear behaviour of VSI. Simulation and experimental results show a valuable improvement on the ac current waveforms, fully confirming the validity of the proposed control structure.

6. References

[1] Hengbing Zhao, Q.M.J. Wu, and A. Kawamura. An accurate approach of nonlinearity compensation for vsi inverter output voltage. *Power Electronics, IEEE Transactions on*, 19(4):1029–1035, July 2004.

[2] Seon-Hwan Hwang and Jang-Mok Kim. Dead time compensation method for voltage-fed pwm inverter. *Energy Conversion, IEEE Transactions on*, 25(1):1–10, 2010.

[3] Mario A Herrán, Jonatan R Fischer, Sergio Alejandro González, Marcos G Judewicz, and Daniel O Carrica. Adaptive dead-time compensation for grid-connected pwm inverters of single-stage pv systems. *Power Electronics, IEEE Transactions on*, 28(6):2816–2825, 2013.

[4] L. Idkhajine, E. Monmasson, and A. Maalouf. Ac drive system on chip controller with nonlinearity errors compensation. In *Industrial Electronics, 2008. IECON 2008. 34th Annual Conference of IEEE*, pages 2381–2386, Nov 2008.

[5] C. Attaianese, V. Nardi, and G. Tomasso. A novel svm strategy for vsi dead-time-effect reduction. *Industry Applications, IEEE Transactions on*, 41(6):1667–1674, Nov 2005.

[6] Ciro Attaianese, Matilde D'Arpino, Mauro Di Monaco, and Giuseppe Tomasso. Recursive dead time compensation techniques for pv system power converters. In *PCIM Europe 2014; International Exhibition and Conference for Power Electronics, Intelligent Motion, Renewable Energy and Energy Management; Proceedings of*, pages 1–8. VDE, 2014.

[7] C. Attaianese, M. Di Monaco, V. Nardi, and G. Tomasso. Dual inverter for high efficiency pv systems. In *Electric Machines and Drives Conference, 2009. IEMDC '09. IEEE International*, pages 818–825, May 2009.

© VDE VERLAG GMBH · Berlin · Offenbach

PCIM Europe 2016, 10 – 12 May 2016, Nuremberg, Germany

Offline Non Isolated Converter Protection

Cathal Sheehan, Bourns Electronics, Ireland Cathal.Sheehan@bourns.com

Roberto Scibilia, Texas Instruments GmbH, Germany R-Scibilia@TI.com

Abstract

This paper examines the use of resettable polymer fuses for protecting offline Flyback converters. It uses a thermal model of the resettable fuse, surrounding solder pads and copper to optimize the trip time so that the converter is protected during overloads. There are two potential positions for the PTC in the circuit which are considered. One position is directly on the winding and the other position is beyond the control loop. Results are taken from the converter and compared with a simulation.

1. Introduction

Polymer PTC resettable fuses (PTCs) are used for protecting circuits from overloads albeit with the following drawbacks:

A) Difference between rated hold current and trip current. Typically the trip current is twice the hold current with trip times of greater than ten seconds. This paper shows how this trip time can be reduced significantly.

B) Poor resistance stability over temperature leading to significant de-rating. The resistance of a resettable fuse at 100°C can be 220% of its nominal value at 25°C. High temperature PTCs however are now available and exhibit an increase of 150% of their resistance which is comparable to high power MOSFETs at 100°C as shown in Figure 1.

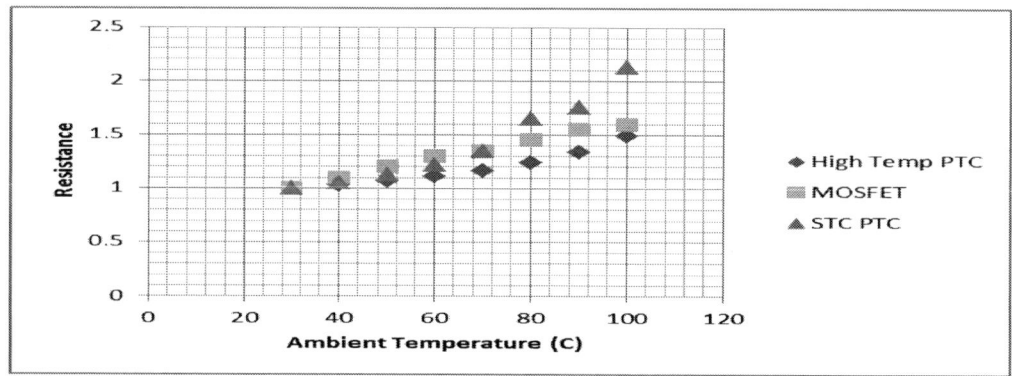

Figure 1: Normalized Resistance of a High Temperature PTC and a MOSFET

2. Description of PTC Model

The behaviour of a polymer PTC can be modelled using the laws of thermal dynamics. Polymer PTCs like fuses react to temperature and will change from low to high impedance at a certain trip temperature. The time to trip depends on the power generated in the component which increases the rate of change of temperature as well as the surroundings which can dampen the rate of change. We can define a PTC as a thermal 3 body model consisting of a power source which generates heat in PTC chip which in turn dissipates through packaging and surrounding solder pads and copper tracks.

© VDE VERLAG GMBH · Berlin · Offenbach

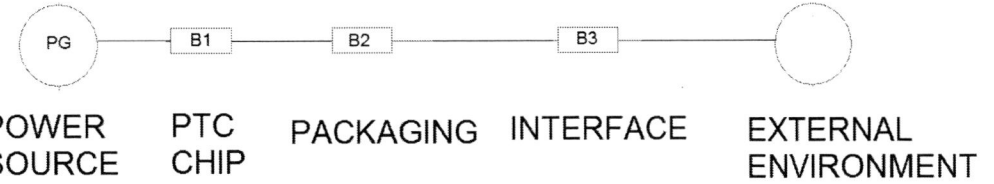

POWER SOURCE   PTC CHIP   PACKAGING   INTERFACE   EXTERNAL ENVIRONMENT

Figure 2: Three Body Thermal Model of a PTC

The equations for all 3 bodies B1, B2 and B3 are as follows:

$$\frac{dt_{B1}}{dt} = \theta_{12} k_1 p_G - k_1 (t_{B1} - t_{B2}) \tag{1}$$

$$\frac{dt_{B2}}{dt} = \frac{\theta_{23}}{\theta_{12}} k_2 (t_{B1} - t_{B2}) - k_2 (t_{B2} - t_{B3}) \tag{2}$$

$$\frac{dt_{B3}}{dt} = \frac{\theta_{3A}}{\theta_{23}} k_3 (t_{B2} - t_{B3}) - k_3 (t_{B3} - t_A) \tag{3}$$

Where:

- θ_{12} is the thermal resistance between bodies B1 and B2.
- θ_{23} is the thermal resistance between bodies B2 and B3
- θ_{3A} is the thermal resistance between body B3 and the environment
- p_G is the input power
- t_a, t_{B1} t_{B2}, t_{B3} is the temperature of the various bodies.
- k_1, k_2, k_3 are constants of proportionality

We turn to Spice to solve these differential equations. A RC network as shown in Figure 3 with a current source IS has the same differential equations. IS would represent the power generated in the circuit. V_{cth1} represents the temperature on the chip while V_{cth2} represents the temperature on the packaging and V_{cth3} is the temperature on the solder interface. The corresponding differential equations are now:

$$\frac{dV_{cth1}}{dt} = \frac{I_s}{C_{th1}} - \frac{(V_{cth1} - V_{cth2})}{R_{th1} C_{th1}} \tag{4}$$

$$\frac{dV_{cth2}}{dt} = \frac{(V_{cth1} - V_{cth2})}{R_{th1} C_{th2}} - \frac{(V_{cth2} - V_{cth3})}{R_{th2} C_{th2}} \tag{5}$$

$$\frac{dV_{cth3}}{dt} = \frac{(V_{cth2} - V_{cth3})}{R_{th2} C_{th3}} - \frac{(V_{cth3} - V_{ta})}{R_{th3} C_{th3}} \tag{6}$$

© VDE VERLAG GMBH · Berlin · Offenbach

PCIM Europe 2016, 10 – 12 May 2016, Nuremberg, Germany

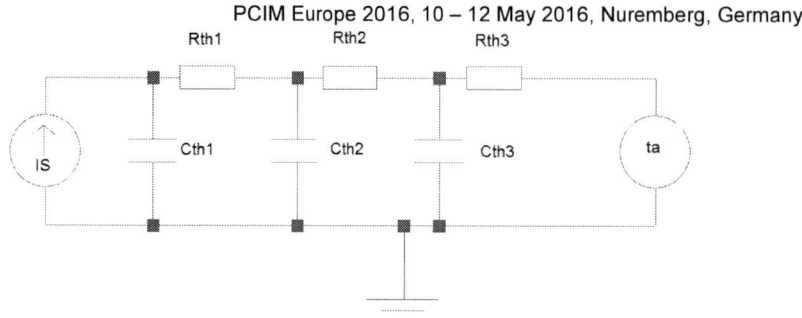

Figure 3 RC Network Equivalent of 3 Body Model

We can now use curve fitting to determine the correct vales of R_{th1}, R_{th2}, R_{th3}, C_{th1}, C_{th2}, C_{th3}. The thermal resistance of the system is calculated using the power dissipated by the component as well as the ambient temperature and the temperature that the component trips at. The thermal resistance is divided between R_{th1}, R_{th2}, R_{th3}.

The model can be used for predicting times to trip and for evaluating the effect of thermal resistance on trip times. Figure 4 shows modelled times for a 0.75A rated PTC superimposed on measured times taken from the actual datasheet.

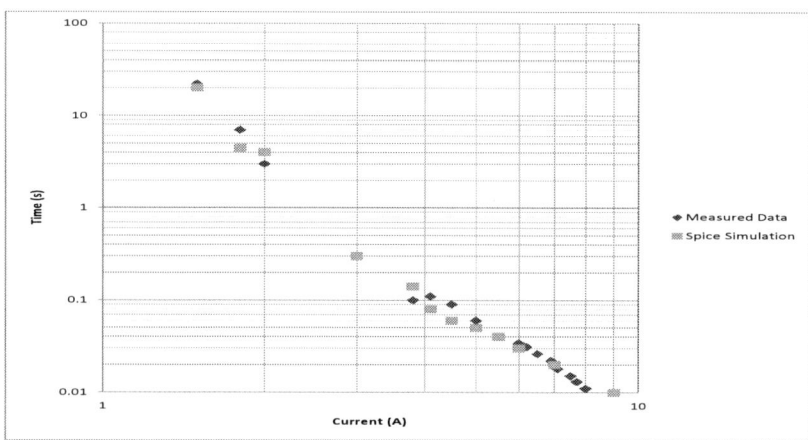

Figure 4: Modelled Times to Trip Compared with Data Sheet

If the PTC is mounted on a circuit board, then the 3rd body (B3) would be the output solder pad and connecting track drawn as shown in Figure 5 where W1, L1, W2, L2, W3, L3 represent 3 separate thermal resistances which form R_{th3}.

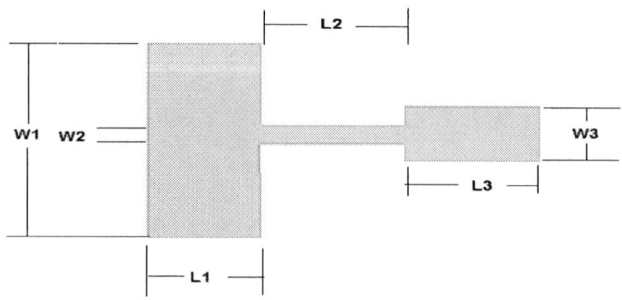

Figure 5 Representation of R_{th3} as a copper plane

© VDE VERLAG GMBH · Berlin · Offenbach

1226

The thermal resistance θ_{cu} of a copper plane can be expressed as:

$$\theta_{cu} = \frac{L}{W*t*\beta} \tag{7}$$

Where β is the thermal conductivity of copper (4W/(cm.°C)).

The thermal resistance of a plane as shown in Figure 5 of thickness t consisting of a pad plus copper trace can be represented by the following equation:

$$\theta_{plane} = (\frac{L1}{W1} + \frac{L2}{W2} + \frac{L3}{W3})(\frac{1}{\beta t}) \tag{8}$$

Let θ_{plane} be R_{th3}, as this represents the third body in the thermal model. Assuming we adjust L2 and W2 and assuming $\frac{L1}{W1}$ and $\frac{L3}{W3}$ are much smaller than $\frac{L2}{W2}$, we can express R_{th3} as:

$$R_{th3} = \frac{L2}{W2*\beta*t} . \tag{9}$$

Hence R_{th3} can be increased by adjusting W2 downwards.

3. Design Considerations for the Flyback Converter with PTC

An offline Flyback converter using a UCC28880 monolithic controller with was designed to operate in Continuous Conduction Mode with 5V +/-5% output and a maximum load of 0.5A from an input voltage range of (90-275Vac). The UCC28880 uses a high voltage MOSFET of 700V. It switches at 62kHz and has a typical peak current limit of 0.21A. The primary inductance was selected based on the fact that the controller has a maximum current at worst case -40°C of 0.3A . The minimum inductance required to keep the power supply in CCM mode is as follows:

$$L_p = \frac{VDCmin*Dmax}{I_{peak}*Fsw} \tag{10}$$

Lp was selected as 5mH based on a minimum input of 90V and a switching frequency of 62kHz as well as a worst case peak current of 0.3A.
The turns ratio N is calculated as:

$$N = \frac{Dmax*VDCmin}{Vout*(1-Dmax)} \tag{11}$$

Being able to operate the controller at the maximum duty cycle of at least 45% requires therefore a higher turns ratio but this also increases the stress on the output diode. Secondary detection of the current allows for automatic adjustment of the primary current limit. The controller protects itself from short circuit currents or overloads by entering a "run-away" protection mode whereby the switching frequency is reduced allowing the secondary side more time to discharge. Under worst case conditions the current limit could be 0.3A. The rms current in the secondary is given by the following equation:

$$I_{rmsout} = I * \sqrt{(1-D)} * \sqrt{1 + \frac{1}{3}(\frac{\Delta I}{I})^2} \tag{12}$$

© VDE VERLAG GMBH · Berlin · Offenbach

Where I represents the dc value of the current.

The duty cycle during overload will be very low so (12) can be simplified to:

$$I_{rmsout}=I^*\sqrt{1+\frac{1}{3}(\frac{\Delta I}{I})^2} \tag{13}$$

ΔI is calculated as 0.2A on the secondary side. This gives I_{rmsout}= 3.25A. If we ignore the ripple we can use the following formula:

$$I_{out}=I_{limit}^*N^*(1-D) \tag{14}$$

Using (14), I_{out} is 3.4A.

It is an overkill using a diode rated to withstand this short circuit current when the circuit is designed for 0.5Amperes. The secondary winding would also have to be chosen so it would not overheat during such a short circuit.

3.1 Location of PTC

A resettable fuse can be placed in two locations as shown in Figure 6. In position A the PTC could be directly assembled inside the winding. The voltage across the PTC in this position will be at least $\frac{VDCmin}{N}$ during the on time and Vout*N during the Flyback time where N is the number of turns and VDCmin is the minimum DC input voltage. The PTC therefore must be rated to this voltage. During an overload the PTC will reduce the feedback voltage to zero which in effect which creates a potential open loop leaving the output capacitor unprotected.

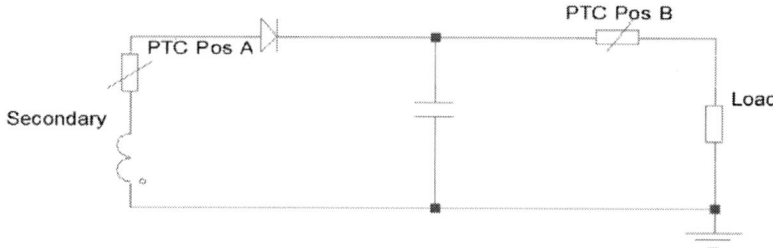

Figure 6: Illustration of 2 Locations of PTC

An alternative location B is located after the control loop and the output capacitor. In an overload situation the controller will regulate as before with the PTC acting as a high ohmic load. Furthermore the PTC can have a voltage rating equal to the output voltage which in our case was 5V instead of 32V if connected in position A. Location B therefore is judged as being the best location for the PTC.

4. Results and Findings

The PTC in position B was a surface Mount device with a resistance of 0.2 ohms on a board with 70µm of copper. We can use Spice to resolve the correct values for R_{th3} and C_{th3} in order to reduce the time to trip of the device when it is conducting 1.5A. Using a track of length 5mm, width 2.0mm and of normal thickness for power boards (70µm) we obtain a value for Rth_3 of 71.4 °C/W. This closely approximates our curve fitting of 69°C/W for Rth_3. The time to trip at 1.5A closely matches the simulation. The overload test was repeated with a track width of 1mm and the time reduced significantly to 3.5 seconds (Fig 8). The thermal resistance was recalculated to have increased to 178.5°C/W. By stepping the thermal resistance in increments of 60°C/W in Spice we were able to confirm the same measurements as shown in Figure 7.

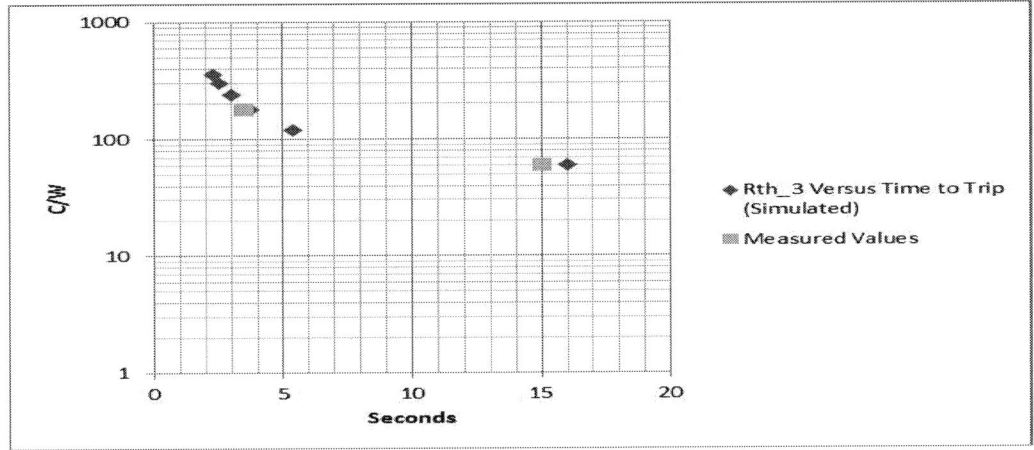

Figure 7: Comparison of Simulated Time to Trip Compared with Actual Values

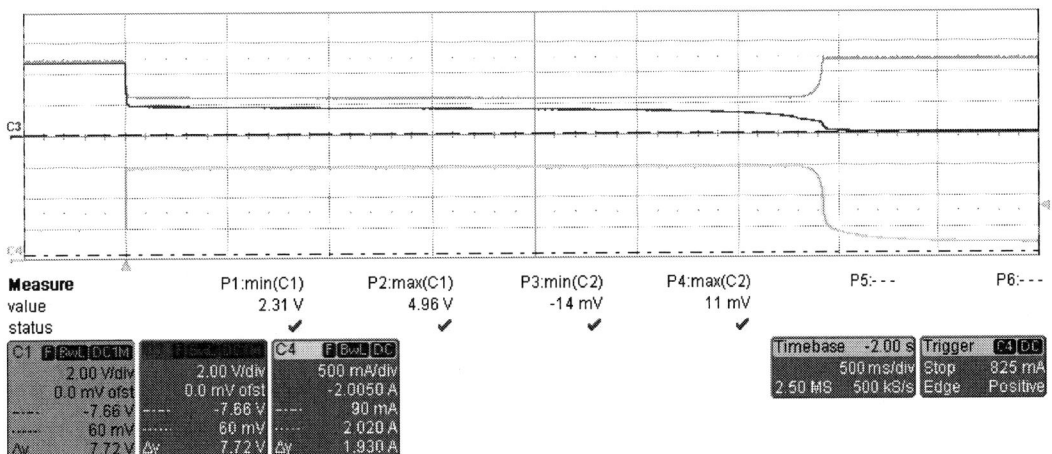

Figure 8: Time to Trip of PTC at 1.5Amps of current (starting from continuous 0.5A)

PCIM Europe 2016, 10 – 12 May 2016, Nuremberg, Germany

Figure 9: Thermal Image of Board during Short Circuit Test

The board with the PTC was connected to a 33mF capacitor. The capacitor charged in 0.3 seconds with a load of 0.5Amps (Fig 10). A power supply of similar output voltage and current (5V at 0.5A) without a PTC for protection but with integrated secondary overcurrent protection was also connected to the capacitor. The secondary current limit was set to 0.65A. Figure 10 also shows that the protection circuit remains tripped in this condition due to the very high initial charging currents. These currents are not long enough in duration to trouble the PTC. This short experiment illustrates one benefit of the PTC for circuits charging super capacitors.

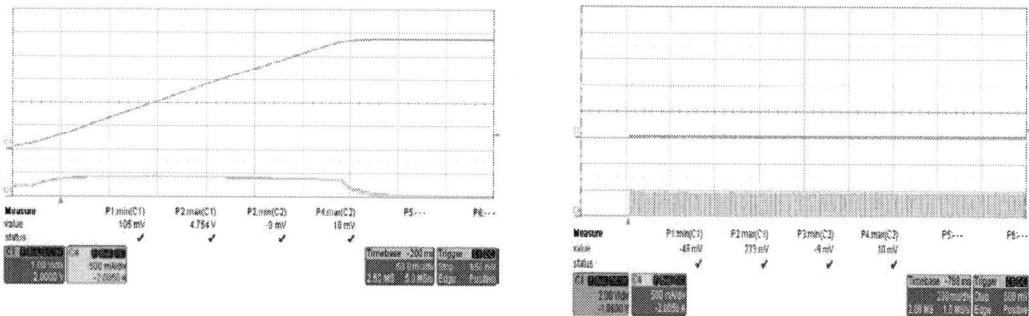

Figure 10 : Charging Voltage of 33mF with PTC and with Secondary Overcurrent Protection

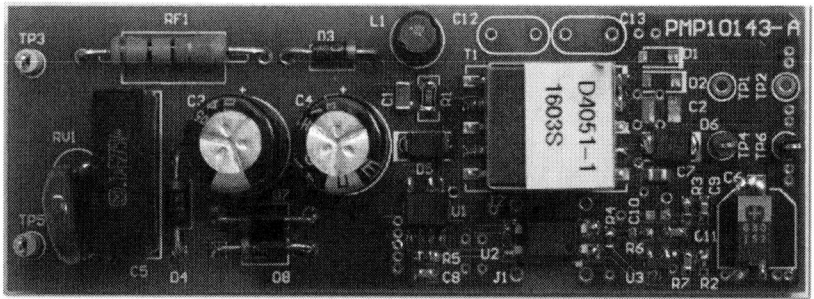

Figure 11: Photograph of Converter Used for this Paper

5. Conclusion

A PTC resettable fuse can be used to provide short circuit protection to a Flyback converter. It is possible using thermal dynamics to model the PTC and the environment and to calculate

© VDE VERLAG GMBH · Berlin · Offenbach

the required external copper traces to obtain the necessary trip time. High Temperature PTCs demonstrate a comparable resistance drift over temperature to MOSFETs and could be considered for circuits where there is not a secondary overcurrent protection mechanism or where the initial inrush is too much for the in built short circuit protection circuit. The best location for the PTC is on the output after the control loop. Putting the PTC on the secondary leaves the circuit vulnerable to open circuit conditions.

6. References

[1] Gupta, Weng Use thermal analysis to predict and IC's transient behaviour and avoid overheating *Maxim Engineering Journal Volume 68 pages 9-15*

[2] Erikson, Maksimović: Fundamentals of Power Electronics *Springer Science and Business Media LLC 2001*

[3] Texas Instruments AN-2020 Thermal Design By Insight, Not Hindsight, *SNVA419C–April 2010–Revised April 2013*

[4] Advanced Power Technology Power Mosfet Tutorial, *Application Note APT-0403 Rev B March 2, 2006*

[5]: Texas Instruments UCCC28880 High Voltage Switcher for Non Isolated AC/DC Conversion, *SLUSC05A –July 2014–Revised October 2014*

[6]: Bourns Electronics MF-RHT Series PTC Resettable Fuses *Datasheet Rev. H, 11/14*

© VDE VERLAG GMBH · Berlin · Offenbach

PCIM Europe 2016, 10 – 12 May 2016, Nuremberg, Germany

Optimisation of Shunt Resistors for Fast Transients

Melanie Adelmund, University of Bremen, Germany, melanie.adelmund@uni-bremen.de
Christian Bödeker, University of Bremen, Germany, christian.boedeker@uni-bremen.de
Nando Kaminski, University of Bremen, Germany, nando.kaminski@uni-bremen.de

Abstract

The development of fast power semiconductors increases the requirements for current sensors. Due to the higher switching frequencies and the higher gradients of the current and voltage transients, the capacitive and inductive coupling can affect the measurement signal considerably. The requirements increase especially in the characterisation process, but also for mass applications of fast semiconductors. In addition, some current sensor types introduce an inductance into the load circuit, which affects the switching process and, thus, has to be avoided. Therefore, the aim is to build a shunt with a minimum inductance and a high signal fidelity simultaneously.

1. Introduction

Precise current measurements are required in most areas of power electronics. There is a variety of sensor principles: Closed- and open-loop hall-effect sensors, Pearson current transformers, Rogowski coils, and shunts. Due to the inherent isolation and the still adequate bandwidth the Rogowski coil is widely used in power electronics. However, the maximum di/dt is limited by the integrator circuit and the measurement signal is affected by capacitive coupling into the windings. Recent publications show good results with a differential Rogowski coil, in which the capacitive component of the signal can be eliminated [1]. The other standard measurement devices are shunts, especially coaxial shunts, due to their very high achievable bandwidth. However, also this device has its drawbacks: There is no galvanic isolation between the shunt and the measurement instrument. Furthermore, the shunt has to be inserted into the circuit to be measured. Thus, the inductance of the shunt adds to the circuit and can affect the switching transient under investigation. In this work the focus is on optimising the inductances of shunts to get a minimum influence on the switching process and to measure without inductive coupling effects at the same time.

2. Shunt Structures

2.1. Coaxial Shunt

The coaxial shunt (Fig. 1a and Fig. 2a) offers the advantage to place the measurement taps in the field free region of the inner conductor [2] [3]. Due to the cylindrically symmetrical shape the space in the inner conductor is not affected by magnetic fields caused by the current through inner and outer conductor. The only influence on the measurement signal is caused by the effective inductance of this circuit, i.e. the inner inductance of the inner conductor, which is in the low picohenry range. Therefore, also high di/dt values will not affect the measurement significantly. However, the construction of the shunt is demanding and the cooling of the actual shunt material, i.e. the inner conductor is not easy to handle. To improve the heat dissipation from the inner conductor, the space between inner and outer conductor needs to be filled with

© VDE VERLAG GMBH · Berlin · Offenbach

PCIM Europe 2016, 10 – 12 May 2016, Nuremberg, Germany

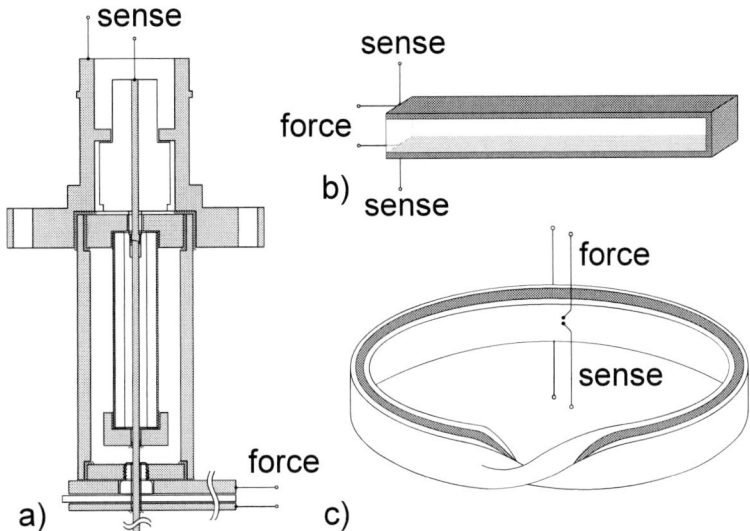

Fig. 1: Shunt structures: a) coaxial shunt b) hair-pin shunt c) Möbius shunt

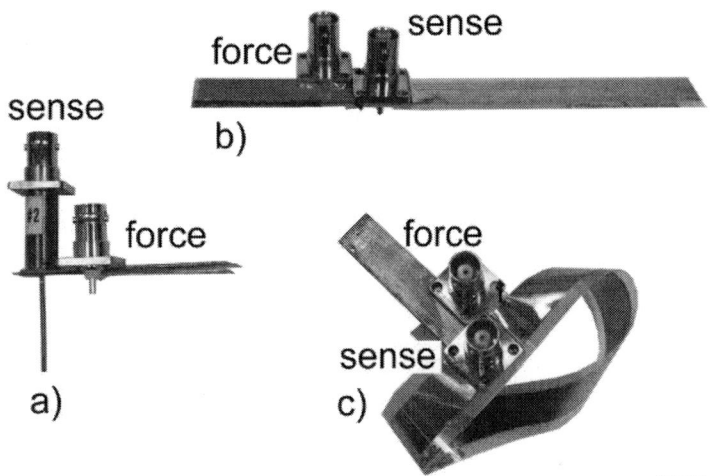

Fig. 2: Self-made shunts: a) coaxial shunt b) hair-pin shunt c) Möbius shunt. The additional BNC-connector is for the evaluation of the DC-resistance

a compound material. Resistance alloys like Manganin® offer the advantage of low temperature coefficients and, therefore, low temperature drifts. However, also these alloyed conductors can be damaged due to overheating. Thus, coaxial shunts can only be used with low continuous power or only in pulse operation. Furthermore, the inductance introduced to the load circuit by commercial coaxial shunts is at least in the range of 3 - 5 nH. Pipe or squirrel cage shunts are "a kind of" coaxial shunt with a better continuous power rating but without the advantage of an absolutely field free measurement in the inner conductor due to the openings in the walls to avoid eddy currents. Also the inductance brought to the load circuit will be much higher due to the bigger shape.

© VDE VERLAG GMBH · Berlin · Offenbach

2.2. Hair-pin Shunt

With a planar shunt structure called hair-pin shunt [4] (Fig. 1b and Fig. 2b) very low inductances can be achieved. Compared to the coaxial shunt the cooling is much easier due to the accessibility of the resistance material. Test structures with the resistance part made of Manganin® (Fig. 3) show inductances in the range of 1 nH or even lower in the load circuit. Thus, the inductances of the constructed hair-pin shunts are anyway lower than those of commercial coaxial shunts. However, the inductance also affects the measurement signal and, thus, the measured voltage V_{shunt} contains a large inductive component V_L (Fig. 5, blue line). The inductive component can be compensated by an RC-network across the shunt (Fig. 4) and its parasitic inductance [2] [5]. The resulting voltage V_C and, therefore, the current signal is measured across the capacitor C_C (Fig. 5, red line). It is easy to see that the signal is well compensated and shows similar curves like the best available reference, i.e. the coaxial shunt by T&M Research Products, Inc.

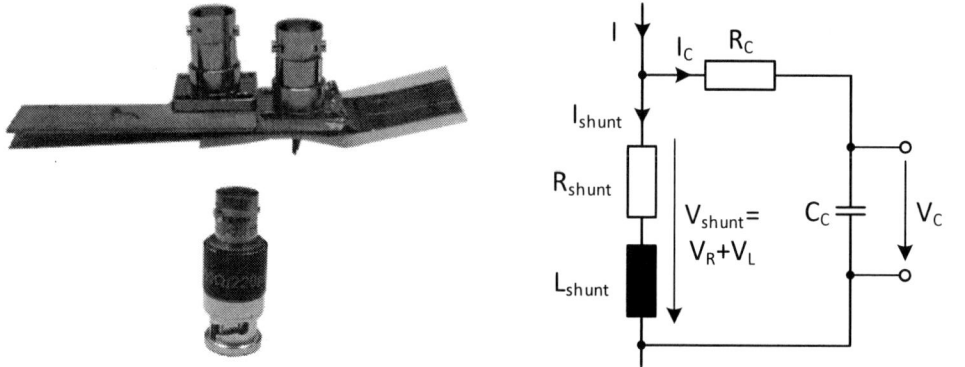

Fig. 3: Hair-pin shunt and compensation element

Fig. 4: RC network for signal compensation [2] [5]

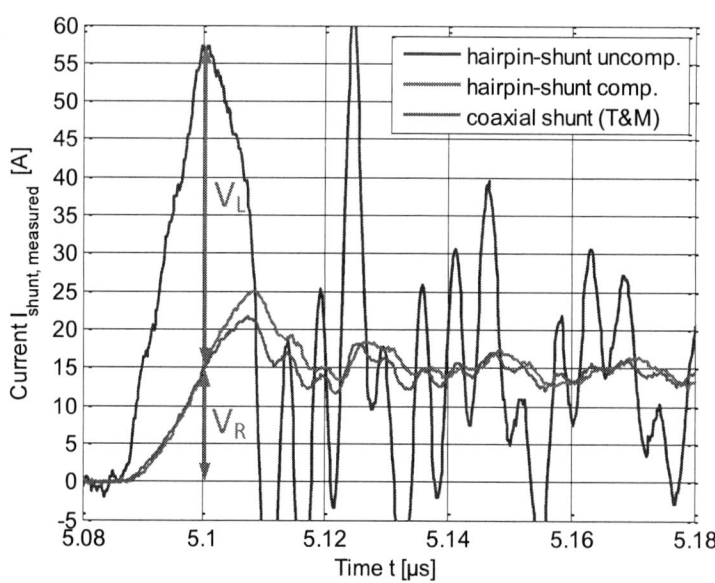

Fig. 5: Different shunt signals: reference signal of a coaxial shunt (green), uncompensated signal of a hair-pin shunt (blue) and compensated signal of the hair-pin shunt (red)

To determine the values for the compensating RC-network [5] [6] the voltage drop V_{shunt} across the uncompensated shunt can be expressed in the Laplace domain:

$$V_{shunt}(s) = R_{shunt} \cdot I_{shunt}(s) + L_{shunt} \cdot s \cdot I_{shunt}(s) = (R_{shunt} + s \cdot L_{shunt}) \cdot I_{shunt}(s) \tag{1}$$

The current I_C into the compensation branch consisting of the resistor R_C and the capacitor C_C can be calculated by:

$$I_C(s) = \frac{V_{shunt}(s)}{R_C + 1/s \cdot C_C} = \frac{V_{shunt}(s) \cdot s \cdot C_C}{s \cdot R_C \cdot C_C + 1} \tag{2}$$

Thus, the compensated measurement signal V_C across the capacitor C_C is given by:

$$V_C(s) = I_C(s) \cdot \frac{1}{s \cdot C_C} = R_{shunt} \cdot I_{shunt}(s) \cdot \frac{s \cdot L_{shunt} / R_{shunt} + 1}{s \cdot R_C \cdot C_C + 1} \tag{3}$$

With the matching condition of $L_{shunt} / R_{shunt} = R_C \cdot C_C$ the voltage V_C equals the voltage V_R across the resistor R_{shunt}.

$$V_C(s) = R_{shunt} \cdot I_{shunt}(s) \tag{4}$$

Fig. 5 shows the compensated signal V_C (red line), which is similar to the signal of the commercial coaxial shunt (green line).

For the compensation network, the boundary condition $R_C >> |R_{shunt} + s \cdot L_{shunt}|$ [5] has to be fulfilled to avoid a significant current, which would affect the measurements. Therefore, the resistance R_C is chosen to be about 1000 times higher than the shunt resistance R_{shunt}. Recommended values for the resistor R_C in the compensation network are in the range 30 Ω ... 1000 Ω and for the capacitor C_C within the range 47 pF ... 1 nF [2]. Thus, the compensation network (100 Ω / 220 pF) is in the recommended ranges. However, it has to be taken into account, that elements of the compensation network (Fig. 4) also have additional parasitic elements and, hence, are frequency depended. Also the resistance and capacitance of the connected probe have to be considered (here: 10 MΩ / 8 pF).

2.3. Möbius Shunt

Another type of shunt is the Möbius shunt (Fig. 1c and Fig. 2c), which was patented in 1966 as a "non-inductive electrical resistor" [7]. Investigations showed that this shunt does not offer a lower inductance or capacitance than planar shunt structures. The term "non-inductive" seems to be related to bifilar wired shunts and not to planar constructions. In addition, the measurement of the resulting voltage cannot be performed in a field free space like in the coaxial shunt and inductive coupling will affect the measurement. Also the Möbius shunt needs to be four times longer to get the same resistance like a hair-pin shunt due its parallel kind of structure. Therefore, the inductance in the load circuit is for sure larger than that of a hair-pin shunt of the same resistance. Thus, it does not offer any advantage over the hair-pin shunt.

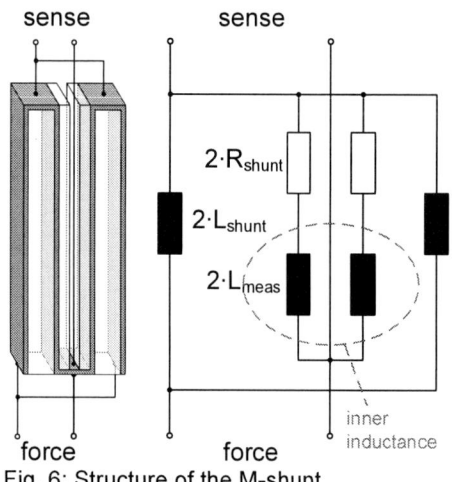

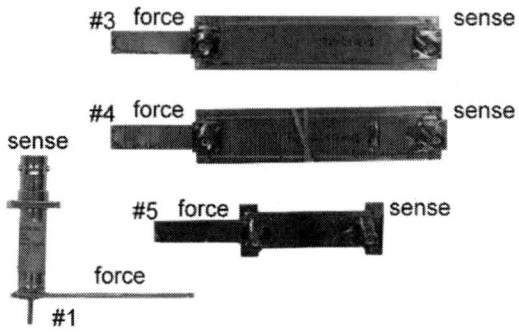

Fig. 6: Structure of the M-shunt

Fig. 7: Prototypes of the M-shunt and the coaxial shunt by T&M Research Products, Inc. used for comparison. The additional BNC-connectors were used for determination of the DC-resistances and the inductances the shunts add to the load circuit.

2.4 M-Shunt

A novel shunt structure, which should combine the advantages of coaxial and hair-pin shunts is a doubled hair-pin shunt. This planar M-shaped structure corresponds to the cross section (Fig. 6) of the coaxial shunt. The taps for the measurement signal are in a field free space, like in the case of the coaxial shunt, if both halves of the M-shaped structure are identical and the connection to the measuring taps are between them. In this case, only the inner inductances of the inner conductors will affect the measured signal. Furthermore, like the hair-pin shunt this kind of shunt introduces only a low inductance between its force connectors and, therefore to the load circuit.

3. Measurements and Results

All transient measurements were performed on a double pulse test bench, which is configured as a buck converter. A printed circuit board (PCB) was designed specifically for the tests (Fig. 8). As switching device, a SiC-MOSFET (Cree C2M0160120D) was used. The load inductance was a handmade spider web coil (L_L = 500 µH, C_{par} = 4.1 pF), which freewheels over a SiC-Schottky-Diode (Cree C4D05120E). An oscilloscope was used to capture the measurement data (Tektronix MSO4104). The test current was I_{test} = 14.5 A and the voltage of the DC-link was V_{test} = 150 V.

The investigated shunts are the coaxial shunt by T&M, a hair-pin shunt, and three different M-shunts (all Fig. 7 except the hair-pin shunt). The signal of the coaxial shunt was used as reference current i_{ref} for the calculation of the inductances of each shunt.

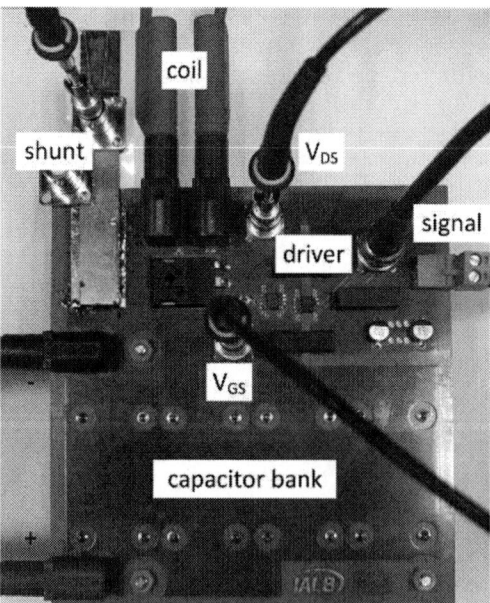

Fig. 8: PCB used during the double pulse tests for the investigation the parasitic inductances of the shunts (L_{PCB} = 31.8 nH)

© VDE VERLAG GMBH · Berlin · Offenbach

All shunts were measured with the PCB (L_{PCB} = 31.8 nH) shown in Fig. 8. The inductance, which each shunt adds to the load circuit was determined by eq. 5. An assumption for the calculation of the inductances from the shunt voltages V_{shunt} is that the inductances of the shunts are all similar and quite low compared to the stray inductance of the PCB ($L_{shunt} \ll L_{PCB}$) and, therefore, will not affect the switching transients by themselves. This method was used for the shunts #1 - #3. For even better precision, the coaxial shunt #1 and the M-shunts #4 and #5 were connected in series to guarantee an identical di/dt during the measurements. The investigated values are shown in Tab. 1.

$$L_{shunt} = \frac{V_{shunt} - R_{shunt} \cdot i_{ref}}{di_{ref}/dt} \qquad (5)$$

The hair-pin shunt (shunt #2) adds only one third of the inductance of the coaxial shunt to the load circuit. For the M-shunt, the evolution can be pursued by the results of the shunts #3 - #5. The resistance value of shunt #3 is 30.7 mΩ and the stray inductance effective in the load circuit is about 5.2 nH. The part of the inductance affecting the measurement (inner inductance of the inner conductor) is about 177 pH higher than that of the coaxial shunt, which is the reference. From these initial results further improvements have been derived: By using a resistance alloy instead of brass and optimising the shunt structure the inductance in the load circuit has decreased due to the reduced length of the resistance material and less space between the layers of the shunt, respectively. Also the influence on the measurement signal has decreased due to these improvements. By using Manganin© as resistance alloy (shunt #4) the inductance, which adds to the load circuit is only 2.5 nH. Also the additional inductance affecting the measurement signal is much lower with only 29 pH, whereby the measurement signal is obviously improved. Shunt #3 and #4 came in the same housing, which was designed for brass as resistance material. A further M-shunt (shunt #5) was again redesigned and, thus, has much smaller structures specifically for Manganin©. Additionally, a thinner isolation of only 75 µm Mylar© was used. Therefore, this shunt only adds an inductance of 0.4 nH to the load circuit. Even if the resistance would be four times higher and, thus, the shunt four times longer, this shunt will have a lower inductance than the 3.3 nH of the coaxial shunt (shunt #1). Also the inductance affecting the measurement signal is only 39 pH higher than that of the reference.

Fig. 9 shows the actual measurements of the M-shunts during turn-off of the transistor. The coaxial shunt is measured for comparison. All shunts were connected in series. Thus the di/dt

Shunt	Type	Resistance alloy	Isolation material	Resistance R_{shunt}	Inductance L_{shunt}	Additional inductance L_{meas} related to reference
#1	Coaxial-shunt	-	-	102.4 mΩ	3.3 nH	Reference
#2	Hair-pin shunt	Manganin© (17.5 µm)	Mylar© (125µm)	83.4 mΩ	1 nH	---[1]
#3	M-shunt	Brass (10 µm)	Mylar© (125µm)	30.7 mΩ	5.2 nH	177 pH
#4	M-shunt	Manganin© (17.5 µm)	Mylar© (125µm)	21.4 mΩ	2.5 nH	29 pH
#5	M-shunt	Manganin© (17.5 µm)	Mylar© (75µm)	22.4 mΩ	0.4 nH	39 pH

Tab. 1: Results of the measured shunts

[1] same inductance as L_{shunt} in the uncompensated case

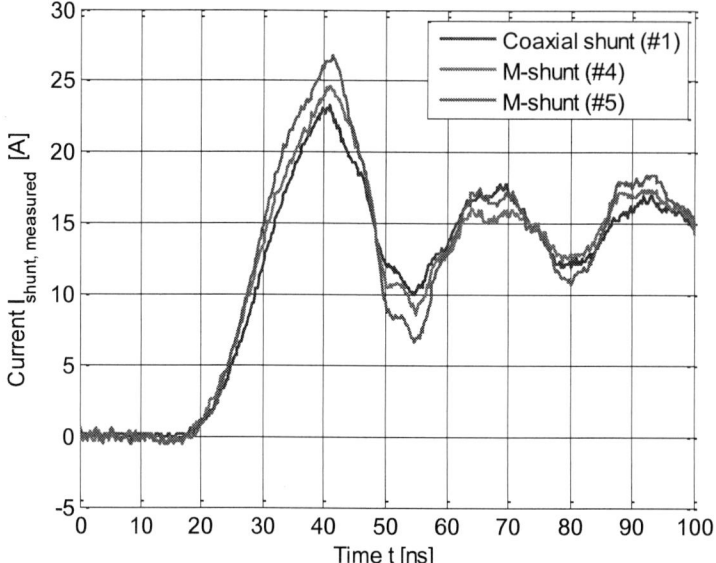

Fig. 9: Measurements of the M-shunts compared to the coaxial shunt

at all shunts is identical and the only influence on the measurement signal can appear due to the inner inductance of the inner conductors, which affects the measurement signal. The higher inductance L_{meas} of the M-shunt #5 compared to the coaxial shunt and M-shunt #4 may be due to deviations from the intended arrangement of the resistance alloy in the inner structure of the shunt. Further improvement is on its way.

4. Conclusion

The coaxial shunt is a high bandwidth shunt with measurement taps in the field free region of the inner conductor but commercially available types add an inductance of 3 - 5 nH to the load circuit. A hair-pin shunt offers a lower inductance compared to the coaxial shunt and is easy to cool. However, the measurement taps are not placed in the field free space introducing a large inductive component into the measured voltage. The proposed M-shunt combines the advantages of the coaxial shunt and the planar hair-pin shunt. The inner inductance of the inner conductors, which affects the measurement signal is quite low but not yet lower than that of the coaxial shunt. All signals measured with Manganin© M-shunts differ only slightly from the signal of the coaxial shunt and, therefore, are already sufficient for many applications. Probably, the inductance, which affects the measurement signal can be further minimised by a more precise arrangement of the inner structure. The inductance brought into the load circuit by the best M-shunt is much lower than the inductance a commercial coaxial shunt adds to the circuit. Another benefit of the M-shaped structure is the potentially better transfer of heat from the resistive alloy to the ambient compared to the considered coaxial shunts.

A further advantage of the M-shunt is the simple construction with only plane parallel surfaces. Therefore, the shunt is much easier and also cheaper to assemble than a comparable coaxial shunt.

5. Acknowledgement

The authors would like to thank Isabellenhütte Heusler GmbH & Co. KG for providing samples of their Manganin® foil.

6. References

[1] S. Hain and M.-M. Bakran, "New Rogowski coil design with a high DV/DT immunity and high bandwidth," presented at the 15th European Conference on Power Electronics and Applications (EPE), 2013.

[2] S. A. Dyer, *Wiley Survey of Instrumentation and Measurement*. John Wiley & Sons, 2004.

[3] A. J. Schwab, "Low-Resistance Shunts for Impulse Currents," *IEEE Trans. Power Appar. Syst.*, vol. PAS-90, no. 5, pp. 2251–2257, Sep. 1971.

[4] R. Davis, "Design Formulas for Nonreactive High-Voltage Pulse Resistors," *IEEE Trans. Parts Mater. Packag.*, vol. 1, no. 2, pp. 3–23, Sep. 1965.

[5] D. Schröder, *Leistungselektronische Bauelemente*, 2nd ed. Springer Berlin Heidelberg New York: Springer-Verlag, 2006.

[6] B. Hudoffsky, "Berührungslose Messung schnell veränderlicher Ströme," Dissertation, University of Stuttgart, Stuttgart, 2014.

[7] R. L. Davis, "Non-Inductive Electrical Resistor," Patent, US 3,267,405, Aug. 1966.

PCIM Europe 2016, 10 – 12 May 2016, Nuremberg, Germany

High bandwidth current sensors as an enabler for advanced control techniques

Dr. Rolf Slatter, Sensitec GmbH, Germany, rolf.slatter@sensitec.com

Abstract

New control techniques for electric drives pose new problems for the manufacturers of current sensors. In particular the sensorless control of permanent magnet synchronous machines places new demands on the current sensors used to measure the motor phase currents. The latest control techniques demand sensors with an extremely high bandwidth. Oversampling of the motor current allows a low noise calculation of the rate of change of current (di/dt), which improves the measurement resolution and allows smooth operation down to zero speed without negative side-effects, such as acoustic noise. The magnetoresistive (MR) effect offers a unique combination of high bandwidth, high resolution, miniaturization and robustness, and is particularly well-suited for the development of compact, fast and accurate current sensors. The power losses are significantly lower than for shunt resistors and the response time is almost an order of magnitude faster than for hall-effect based current sensors. Furthermore, MR-based current sensors have an extremely high bandwidth, in the range of several MHz, which allows the full exploitation of new power electronic technologies, such as Silicon Carbide switches and diodes.

1. Trends in Power Electronics

The precise, dynamic and low-loss measurement of electrical current is a basic, but decisive function in numerous power electronics devices for electromobility applications. The requirements in this field are different and partially more complex than for power electronics applications in the industrial field. The market-pull exerted by the trend to electromobility is reinforced by requirements from the emerging field of power electronics for "smart" applications, such as the smart grid or smart home. Technology-push in the form of wide bandgap semiconductors, digital control and advanced packaging techniques is helping to provide the basis for power electronics solutions with higher power density, higher efficiency and improved price/performance ratio as required by these emerging markets [1].

The fast growing number of applications for driving electric motors and for battery management for electromobility, specifically, is leading to a demand for sensors for measuring currents in the range of several hundreds of amps, with high bandwidth, low hysteresis, compact dimensions and simultaneously high isolation strength [2], [3], [4]. This complex set of requirements, combined with strong price pressure, is a big challenge for the manufacturers of current sensors. It is leading to the development of completely new products, deliberately designed for this application area, based on the latest R&D results.

2. New Requirements for Current Sensors

There is a wide variety of different applications in the field of power electronics where electrical currents need to be measured. However, the requirements for these measurement devices are becoming steadily more demanding regarding accuracy, size and especially bandwidth.

In order to increase the power density of power electronics, as particularly important for electromobility, there is a clear causal chain. Soft switching leads to higher efficiency and higher frequencies, which enable a smaller size for a given power output. Higher switching frequencies allow the size of magnetic and inductive components to be reduced significantly, resulting in

© VDE VERLAG GMBH · Berlin · Offenbach

more compact and lighter designs. This trend is now being reinforced by use of new semiconductor materials like silicon carbide (SiC) and gallium nitride (GaN), as their low on-resistances and low parasitic capacitances reduce switching losses.

One application field for high-speed current sensing arises for DC/DC converters [4]. The requirements for the bandwidth is rising into the MHz range in order to increase the power density. But also in safety critical applications, in which overcurrent situations need to be detected in the ns-range, extremely fast current sensors are necessary. To fulfill all these requirements, compact current sensors are necessary that detect currents highly dynamically, accurately and that are cost-effective.

Conventional current sensor solutions, e.g. hall- or shunt based sensors exhibit a limited bandwidth, typically less than 250 kHz. Other current sensors, like those based on the Rogowski-Coil, are capable of highly dynamic current measurement, but are significantly more expensive, larger, and hence not suitable for large series applications. Furthermore, Rogowski-Coils are only capable of measuring alternating currents (AC), which prevents their use in applications where DC currents must also be measured.

In order to meet the above mentioned requirements, magnetoresistive (MR) current sensors are ideally suited due to the fact that the bandwidth of the magnetoresistive effect extends up into the GHz-range.

Fig. 1. High bandwidth current sensors (Source: Sensitec GmbH)

3. High-bandwidth MR Current Sensors

The magnetoresistive effect is best known from the read heads of computer hard discs or from magnetic memory (MRAM) applications, but it is also well suited to uses in sensor technology. It has a long history, the anisotropic magnetoresistive (AMR) effect being first discovered in 1857 by Lord Kelvin. The AMR effect occurs in ferromagnetic materials, such as nickel-iron layers structured as strip elements, whose specific impedance changes with the direction of an applied magnetic field. Due to a special structure of the strips the resistance change is proportional to the applied magnetic field over a wide range. This means that by adept design of the sensor structure very small magnetic fields can be detected with very high accuracy. However, the MR-effect did not experience widespread use until the early 1980s, when the first MR-based read heads were implemented in hard disc drives. The first industrial applications for MR-based sensors followed at the beginning of the 1990s, since when the number of applications has increased dramatically. The applications are not only limited to terrestrial use – MR sensors are used to control the electric drives used on "Curiosity", the Planetary Rover that landed successfully on Mars in August 2012. MR sensors are also used extensively in safety-critical automotive applications, for example in wheel speed sensors for the ABS-system or in steering angle sensors for the ESC-system.

The magnetoresistive effect is particularly attractive in the field of electrical current measurement. The very high sensitivity means that there is no need to use an iron core to concentrate the magnetic field generated by the conductor carrying the current. This means that MR-based current sensors do not suffer from hysteresis and that they have a significantly higher bandwidth, enabling current sensors with bandwidths in the MHz area.

© VDE VERLAG GMBH · Berlin · Offenbach

Compared to shunt resistors MR-based sensors have the benefit of galvanic isolation and dramatically lower power losses. This is particularly important in high voltage applications and where overall power efficiency is a major design driver.

Sensitec has a long experience of developing MR-based current sensors for industrial applications. The increased demand for very dynamic current sensors generated by the recent trend to electromobility was the driver for the development of a highly integrated current sensor (CMS3000) comprising an AMR sensor chip, a signal conditioning circuit and two biasing permanent magnets (Fig. 1). The latter are necessary for maintaining the initial magnetization direction of the AMR structures in the case of overcurrent situations.

The quantity to be measured is a differential magnetic field, also referred to as field gradient that is generated by two currents with opposed current flow directions. The primary current conductor is typically U-shaped, with its straight parallel parts positioned underneath the sensor. For current measurement four AMR "resistors" are connected to form a Wheatstone bridge. The resistors on the silicon chip are placed so that they constitute a differential field sensor. This is necessary because interference fields can be eliminated this way. Combined with a signal conditioning circuit the chip is assembled on a ceramic substrate, incorporating a hybrid circuit (Fig. 2).

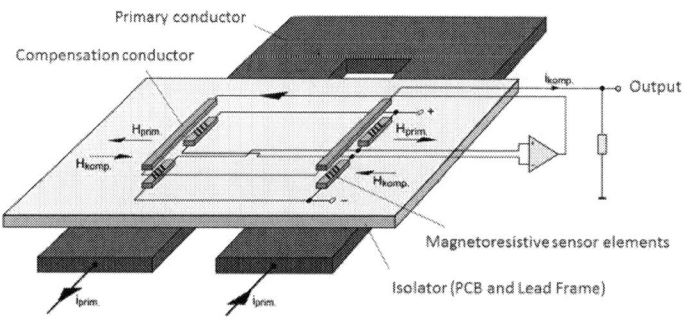

Fig. 2. Principle of operation (Source: Sensitec GmbH)

Furthermore, a compensation conductor is integrated on the chip with which a magnetic field can be generated close to the resistors. On the opposite side of the substrate the primary current conductor is attached below the MR chip in a U-shape. The geometry of the primary conductor defines the measurement range of the current sensor. Based on the output signal from the MR chip, the signal conditioning circuit (shown in Fig. 2 schematically as an operational amplifier) generates a current i_{comp} in the compensation conductor, which compensates the magnetic field generated by the primary conductor in the plane of the AMR resistors. With this method the signal achieves a high linearity (0.1 %) and is largely independent of temperature. This compensation current is directly proportional to the primary current to be measured and is used to generate the output signal from the current sensor.

This "closed-loop" principle results in an extremely compact sensor that is largely insensitive to homogeneous interference fields and temperature changes, with a low power consumption and very high efficiency. The AMR-based current sensor exhibits no hysteresis as observed in iron core based Hall-sensor solutions and no remaining magnetic offset after overcurrent events. Due to the high sensitivity of the AMR sensor chip, a flux concentrator is not necessary. The sensor is designed for high accuracy and very fast electronic measurement from DC up to 2 MHz AC. Fig. 3 shows a typical step response of a CMS3000 compared to a 200 kHz current sensor, demonstrating a response time of just 40 ns.

PCIM Europe 2016, 10 – 12 May 2016, Nuremberg, Germany

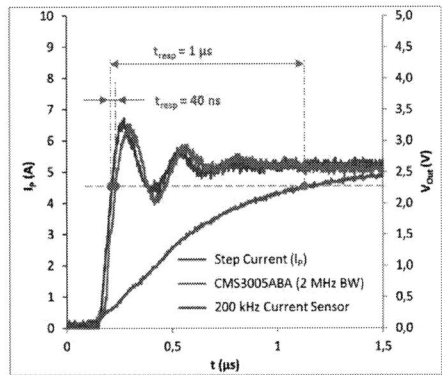

Fig. 3. Step response of CMS3000 current sensor (Source: Sensitec GmbH)

Contrary to Hall-effect based sensors, the described system enables differential magnetic field measurement by means of the advanced geometry of the magnetoresistive elements described above [5]. Due to this construction the sensor is immune to homogeneous interference fields and hence needs no magnetic shielding as required for surface mounted Hall-effect current sensors with external current bar. Fig. 4 summarizes the differences between MR-based current sensors and previous technologies.

Technology	Size	Bandwidth	Accuracy	Power losses	Cost	Ease of design-in
Hall-effect, Fluxgate (closed loop)	- -	-	+	-	- -	+
Hall-effect (open loop)	+ +	-	- -	+	+ +	-
Shunt	+	-	+ +	- -	-	-
Magnetoresistive	+	+ +	+	+	+	-

Fig. 4. Comparison of current sensor technologies (Source: Sensitec GmbH)

4. Application Examples

4.1. Sensorless control of permanent magnet motors

To control synchronous permanent magnet motors many important control parameters are referenced to a co-ordinate system that rotates with the rotor. This relies on knowledge of the absolute rotational position of the rotor at all times with high accuracy. Typically motor feedback systems in the form of resolvers or optical encoders are used to directly measure the rotor position. These motor feedback systems necessitate an additional cable between the motor and the frequency converter as well as the application of additional hardware to ensure the interference-free transmission and subsequent processing of the position signals.

There is an increasing number of applications where so-called sensorless (that is rotor position sensor-less) drives are desirable:

- In applications where the requirements regarding dynamic performance or positioning accuracy are not particularly high. Here the price of the complete drive system is increasingly under pressure and the removal of the motor feedback system means that the costs of the additional cabling and control hardware described above can be saved.

- The increasing use of high-pole-number direct drives, in which the motor is connected directly to the output machine element rather than via a gearbox, requires new methods for determining the rotor position to avoid the often complex integration of a motor feedback system

© VDE VERLAG GMBH · Berlin · Offenbach

1243

- Particularly in electromobility applications the compactness of the drive is particularly important. Removing the motor feedback system reduces the volume of the drive, so enabling more compact configurations.

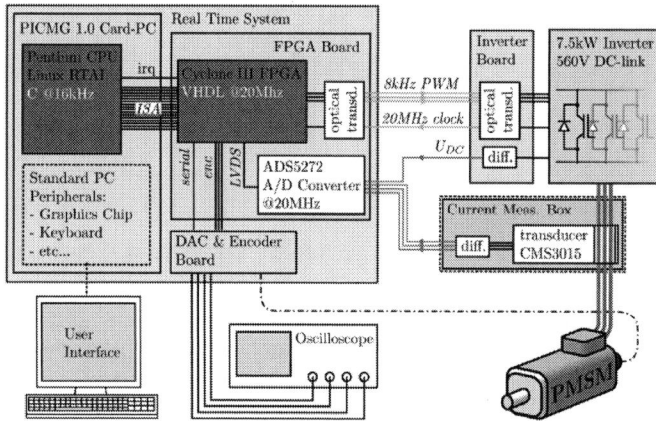

Fig. 5. System schematic (Source: TU Munich)

Due to these potential benefits there has been a lot of research into methods of indirectly determining the rotational position and speed of the rotor via electrical parameters. These methods are typically referred to as "sensorless" control techniques, although the term is somewhat misleading, because current sensors are still required to utilize these techniques effectively.

Fig. 6. Application of CMS3000 current sensors (Source: Peter Landsmann)

Despite the considerable research efforts these methods have not yet achieved a true industrial breakthrough. The main reason for this is that the physical effects used to determine the rotor position are highly dependent on the actual operating point of the motor. The determination of rotor position is particularly difficult at low speeds or at stand-still, because the back-EMF is low or zero respectively. Under these operating conditions other methods have to be used that allow the rotor position to be determined independent of rotor speed [6].

For speeds over ca. 10 – 20 % of rated speed the back-EMF is used to determine the rotor position passively. At low speeds or stand-still so-called injection techniques are used to evaluate inductive changes, generated by voltage pulses applied to the motor, that are dependent on rotor position [7]. Both approaches have been extensively studied, but neither has gained wide acceptance, either because they require more expensive control hardware, which negates the savings generated by removing the motor feedback system, or because the technique require more commissioning effort. Last, but not least, the injection technique is associated with disturbing acoustic noise.

© VDE VERLAG GMBH · Berlin · Offenbach

At the Institute for Electrical Drives and Power Electronics of the Technical University Munich a new technique has been proven that avoids the problem of noise and also much improves the accuracy of the estimated rotor position [8]. High bandwidth magnetoresistive current sensors allow an "oversampling" of the current signal with a frequency of 2 MHz (Fig. 5, Fig. 6). This oversampling is used for the low-noise (electrical) calculation of the rate of change of current (di/dt). This value can be used to estimate the inductance which, in turn, can be used to determine the rotor position. The low electrical noise content of the di/dt signal allows the injection voltage to be much reduced. This leads to a significant reduction in the acoustic noise generated by the injected voltage. The new technique not only allows operation down to stand-still, but also provides much better estimates for the rotor speed value, thus enabling a much improved dynamic performance (Fig. 7)

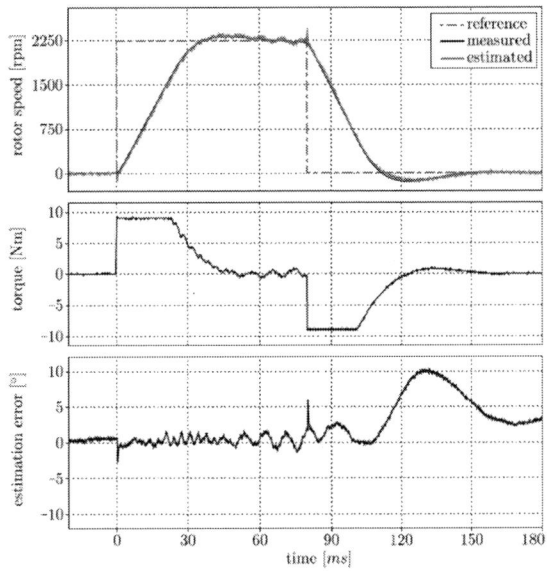

Fig. 7. Speed step-response (Source: TU Munich)

4.2 Integrated power electronics for hybrid commercial vehicle

Current sensors based on the MR-effect also played an important role in the EU-funded project COSIVU, where a number of suppliers and universities have worked together with the vehicle manufacturer Volvo to develop extremely compact power electronics and motors for a hybrid truck (Fig. 8).

The main objectives and highlights of the COSIVU project, which ran from 2012 to 2015, can be summarized as follows:

- Decentralized drive-train system, managed by a central vehicle computer, for reduced weight and cooling complexity and thus improved energy efficiency
- One compact system package due to increased level of mechatronic integration for the electric motor
- Development of next generation of highly integrated inverter modules based on novel SiC technology (1200V, 500A) with innovative cooling concepts (e.g. double sided cooling), being capable of reducing energy losses by 50% and more
- Benchmarking of wireless communication between drive units and central computer vs. wired based solutions with respect to EMC issues, costs and durability
- Implementation of innovative functional and health monitoring (SoF/SoH) features, like thermal impedance spectroscopy for SoF/SoH determination of the inverter module or structure-borne sound analysis for SoF/SoH determination of the motor/gear module

© VDE VERLAG GMBH · Berlin · Offenbach

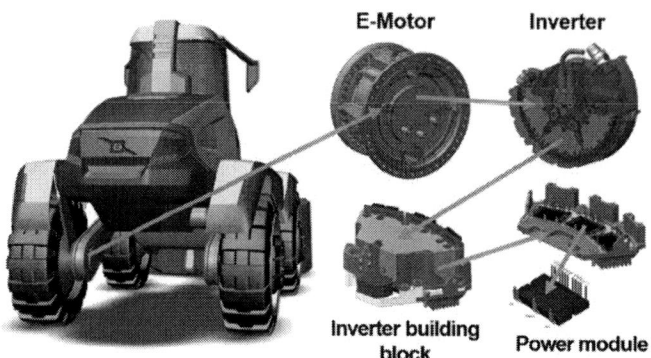

Fig. 8. COSIVU Drive system (Source: Swerea IVF)

Fig. 8 shows the basic concept for the decentralized drive-train, where the inverter for each actuator is mounted directly within the motor housing. This concept, called "site-of-action integration" greatly reduces the complexity and weight of the high voltage wiring harness. It also reduces the effort necessary to cool the various power modules and furthermore reduces the effort for EMI filtering. Fast, accurate current sensors are an important enabling technology for the high power density inverters essential for the implementation of this concept [9], [10].

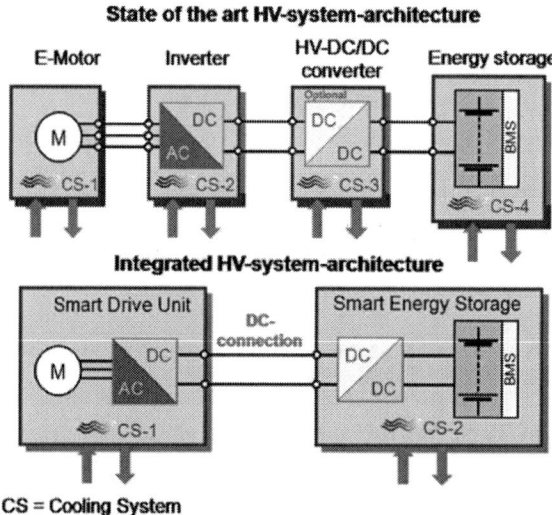

Fig. 9. Topology of the drive unit (Source: Fraunhofer IISB)

A particularly interesting result of the COSIVU project was the development of "building blocks" for the modular inverter. This low-impedance commutating cell provides the required robustness, EMC-compatibility and high efficiency through low switching losses (Figure 10). The inverter building block (IBB) incorporates the DC-link capacitor, base driver, current sensor and direct cooled baseplate.

This concept greatly simplifies the power, signal and thermal interfaces for the power electronics. The current sensor is used to measure and monitor the phase currents of the electric motor. To withstand the extremely harsh EMC environment in this application Sensitec developed a shielded current sensor with a specially shaped bus bar to allow easy integration within the IBB.

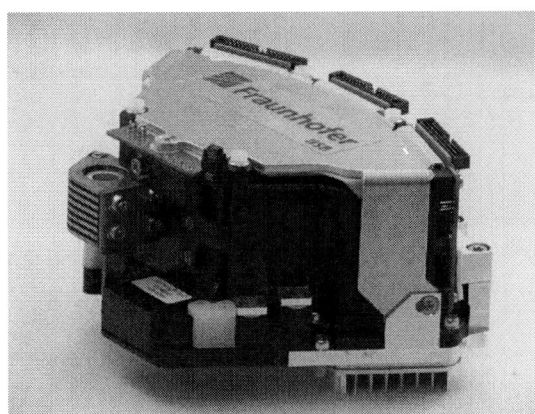

Figure 10: Inverter Building Block (IBB) for Hybrid Commercial Vehicle (Source: Fraunhofer IISB)

Summary and Outlook

The actual and potential electromobility applications for high bandwidth magnetoresistive current sensors range from DC/DC converters to rectifiers, inverters, gate drivers, switched power supplies and power electronics for inductive charging devices for electric vehicles. The high bandwidth not only enables higher switching frequencies and provides improved short-circuit protection, but also opens opportunities for the improved condition monitoring of power electronics as well as for so-called sensorless control [11],[12].

References

[1] J.W. Kolar et al, "What are the big challenges in Power Electronics?", 8th International Conference on Integrated Power Electronics Systems (CIPS), Nuremberg, 25.-27.2.2014

[2] J. Bockstette et al, "Bidirectional current controller for combination of different energy systems in HEV / EV", Proc. of Power Electronics, Machines and Drives Conf., Bristol, 2012

[3] G. Galzin et al, "Electrical Environmental Control System", Proc. of More Electric Aircraft Forum, 2009

[4] A. Kaiser et al. "Design of a lightweight DC/DC converter providing fault tolerance by series connection of low voltage sources", Proc. of 1st Aerospace Sensors Conference, Frankfurt / Main, 2012

[5] S. Scherner & R. Slatter, „New applications in power electronics for highly integrated high-speed magnetoresistive current sensors", 15th European Conference on Power Electronics and Applications, Lille, Frankreich, 3.- 5.9.2013

[6] J.C. Gamazo-Real, E. Vázquez-Sánchez & J. Gómez-Gil, „Position and Speed Control of Brushless DC Motors Using Sensorless Techniques and Application Trends", Sensors 2010, 10, 6901-6947

[7] M. Schrödl, "Dynamik und Überlastfähigkeit von sensorlosen Antrieben mit PM Synchronmaschinen einschließlich Stillstand und tiefen Drehzahlen", VDE/VDE Tagung Antriebssysteme 2013, Nürtingen, Deutschland, 17.- 18.9.2013

[8] P. Landsmann, D. Paulus, A. Dötlinger & R. Kennel, "Silent Injection for Saliency based Sensorless Control by means of Current Oversampling", Proc. of ICIT 2103 – IEEE International Conference on Industrial Technology, Cape Town, South Africa, 25.- 27.2.2013

[9] A. Otto et al, "Reliability for new SiC BJT Power Modules for Fully Electric Vehicles", 18th International Forum on Advanced Microsystems for Automotive Applications (AMAA), Berlin, 23.-14.6.2014

[10] A. Andersson, "COSIVU – Compact, Smart and Reliable Drive Unit for Commercial Electrical Vehicles", 18th International Forum on Advanced Microsystems for Automotive Applications (AMAA), Berlin, 23.-14.6.2014

[11] R. Slatter, "Hochdynamische, hochintegrierte Stromsensoren für die Elektromobilität", ETG-Fachtagung Forschung und Entwicklung für die Elektromobilität, Berlin, 5.-6.11.2013

[12] R. Slatter, R. Buss, „The Role of Magnetic Sensors in the Hybridization and Electrification of Commercial Vehicles", Proc. of 3rd International Commercial Vehicle Technology Symposium, Kaiserslautern, 11.-13.3.2014

© VDE VERLAG GMBH · Berlin · Offenbach

Rotational Speed Measurement Based on Avago ADNS-9800 Laser Mouse Sensor

Cheng Liu[1], Yanan Xu[1/2], Ji-Gou Liu[1], Hui Sun[1], Ralph Kennel[2]

[1] ChenYang Technologies GmbH & Co. KG, Markt Schwabener Straße 8, 85464 Finsing, Germany, Email: cheng.liu@chenyang-ism.com, john.liu@cy-sensors.com

[2] Institute for Electrical Drive Systems and Power Electronics, Technische Universität München, Arcisstraße 21, 80333 Munich, Germany, Email: ralph.kennel@tum.de

Abstract

In this paper, a rotational speed measuring system is proposed on the basis of a laser mouse sensor. The rotational speed of a motor can be determined by measuring the air gap change of a target disc mounted on the motor shaft with the laser mouse sensor ADNS-9800 of Avago / PixArt. By applying a Fast Fourier Transform algorithm, the fundamental frequency of the sampled data from the sensor can be extracted, which is directly proportional to the rotational speed. With this method, a measuring accuracy of 0.5% in speed range from 50 RPM to 6000 RPM can be achieved.

1. Introduction

Rotational speed measurement is an important operation for applications in industry and automation. Especially in electrical drive systems, rotational speed sensors are essential for a proper functionality of the whole machine [1].
In the past, analogue precision tachometers or tachogenerators did a quite good job for drive systems [4]. Nowadays the commonly used rotational speed sensors in electrical drive systems are mainly optical and magnetic encoders. For some high precision applications, encoders with high resolution and accuracy are needed. However, these encoders are very expensive.
A better solution is to develop a cheaper rotational speed sensor with comparable accuracy like high resolution encoders. Laser mouse sensors are widely used in high performance mice for computer gamers. They are cheap in production and can reach high resolution up to 8200 CPI (Counts Per Inch), which can also be expressed as DPI (Dots Per Inch).
So it is motivated to develop a rotational speed measuring system based on the laser mouse sensor like ADNS-9800 of Avago / PixArt.

2. Working Principle of ADNS-9800

The laser mouse sensor ADNS-9800 offers a resolution up to 8200 CPI, a moving speed up to 150 IPS (Inches Per Second), a processing speed of 12000 FPS (Frames Per Second) and a maximum acceleration up to 30 G [2].
The sensor is based on the Laser-Speckle interferometry and CMOS imaging technology. The sensor package contains a vertical-cavity surface-emitting laser (VCSEL), the sensor based on CMOS or CCD technology, an imaging lens and a collimating lens (see Fig. 1).
By optically acquiring sequential surface images (known as frames) via an Image Acquisition System (IAS), a Digital Signal Processor (DSP) can calculate the displacement by comparing the contrast of two images in a row (see Fig. 2). So the corresponding displacement in X and Y directions of the surface can be determined mathematically by the DSP [3]. By using Serial Peripheral Interface (SPI), the DSP can establish a communication interface with an external microcontroller for sensor data output.

© VDE VERLAG GMBH · Berlin · Offenbach

PCIM Europe 2016, 10 – 12 May 2016, Nuremberg, Germany

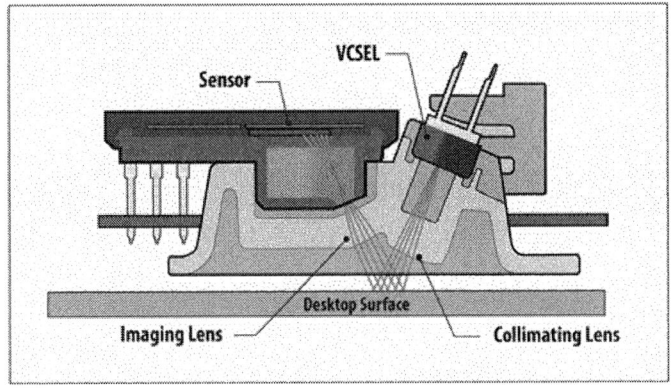

Fig. 1. Internal Structure of an ADNS-9800 Laser Mouse Sensor [5].

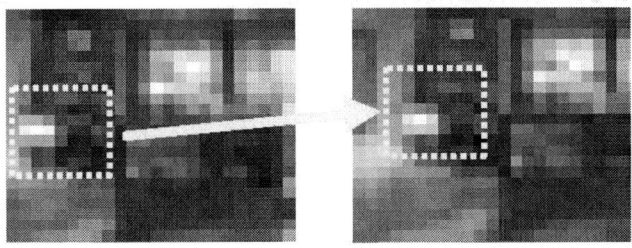

Fig. 2. Working Principle of ADNS-9800 Laser Mouse Sensor [5].

3. Test System

Our rotational speed measuring system consists of an ADNS-9800 laser mouse sensor, a microcontroller unit (MCU) and a Host-PC. Fig. 3 shows the functional block diagram of this system.

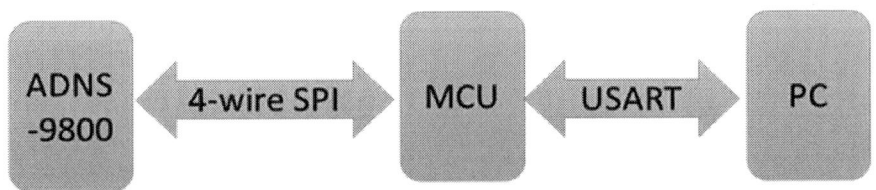

Fig. 3. Functional Block Diagram of the Rotational Speed Measuring System
with ADNS-9800 [5].

The communication between ADNS-9800 is realized with a four-wire SPI connection, whereas the MCU communicates with the Host-PC through a serial port connection with USART and RS-232.

The ADNS-9800 is mounted on a U-shaped plate above the target disc which is driven by a servo motor. Therefore, the laser mouse sensor can measure the displacement in the tangential direction of the target disc. This structure is shown in Fig. 4.

The rotational speed of the servo motor can be set in the Host-PC by software and a servo controller. For a rotational speed reference, the data from the built-in encoder is used.

© VDE VERLAG GMBH · Berlin · Offenbach

Fig. 4. Rotational speed of a servo motor is measured with a Laser Mouse Sensor [5].

4. Test Results

Due to an eccentricity of the target disc mounted on the motor shaft, the light signal amplitude changes periodically, as result of the periodic air gap change between sensor and target disc (see Fig. 5). Therefore, the laser mouse sensor gives out a sinusoidal output signal, whose fundamental frequency can be determined by a Fast Fourier Transform (FFT). Fig. 6 shows the amplitude spectrum of the measured sensor signal at a rotational speed of 100 RPM. The highest peak in amplitude spectrum marks the fundamental frequency of the periodic signal.

The rotational speed w can be easily calculated from this fundamental frequency f according to the following formula:

$$w = 60f \qquad \text{in RPM} \qquad (1)$$

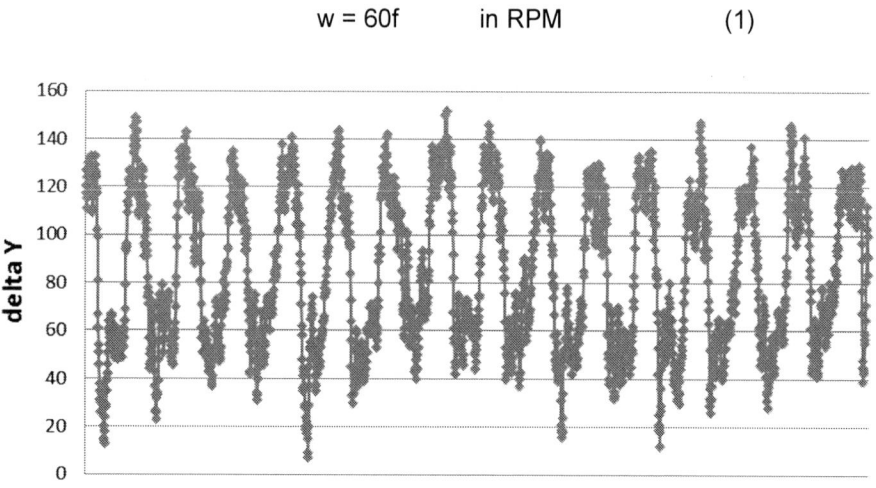

Fig. 5. Measured Sensor Signal at Rotational Speed of 100 RPM [5].

Normally, the ADNS-9800 only supports speed up to 150 IPS, which means 3.81 m/s and 1814 RPM with the target disc radius of 2 cm. But by measuring the air gap change, rotational speed up to 6000 RPM can be detected. The relative errors from the measurements are shown in Fig. 7. The accuracy can be controlled within 0.5% in the measuring range from 50 RPM to 6000 RPM.

© VDE VERLAG GMBH · Berlin · Offenbach

PCIM Europe 2016, 10 – 12 May 2016, Nuremberg, Germany

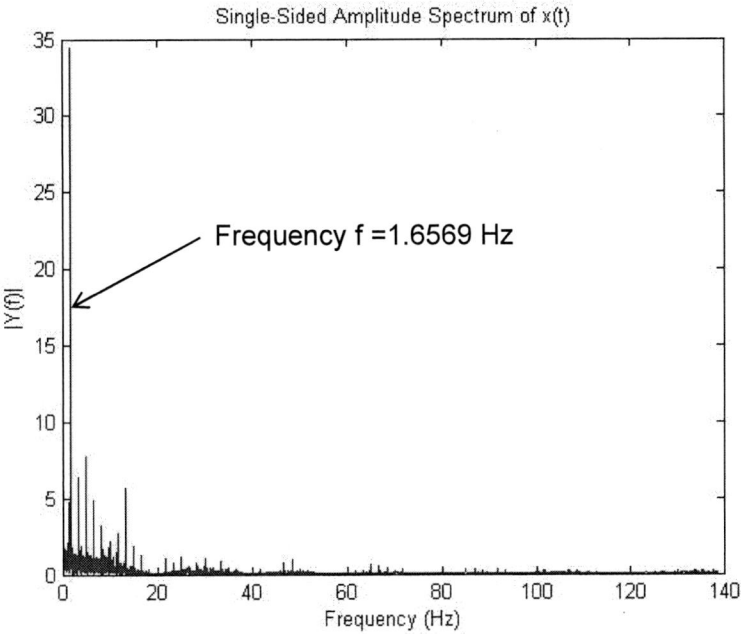

Fig. 6. Amplitude Spectrum of Measured Sensor Signal at Rotational Speed of 100 RPM [5].

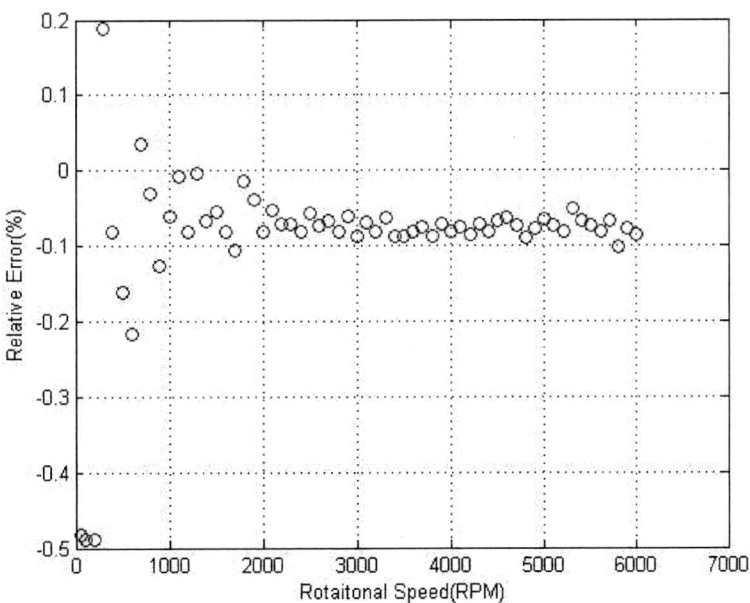

Fig. 7. Relative Errors of the Speed Measurement from 50 RPM to 6000 RPM [5].

5. Conclusions

This paper presents a rotational speed measuring system which is based on laser mouse sensor ADNS-9800.
From the results one can draw the following conclusions:

- The laser mouse sensor ADNS-9800 works with the Laser Speckle and Image Acquisition Principle. Two images are taken in a very short time and compared towards each other in order to determine the displacement.
- The laser mouse sensor provides a resolution of 8200 CPI, a moving speed of 150 IPS, and a maximum acceleration up to 30G. So it can be used for rotational speed measurement.
- By measuring the air gap change of the rotational target disc, the laser mouse sensor ADNS-9800 gives out a periodic output signal, which comprises the fundamental frequency, and therefore the rotational speed can be determined. A measuring accuracy of 0.5% is realizable for the range from 50 RPM to 6000 RPM.

Further research works should be done in order to improve the measuring accuracy of the rotational speed sensors by optimization of the sensor parameters.

References

[1]   D. Schröder, *Elektrische Antriebe - Regelung von Antriebssystemen*, 2nd Edition, Springer-Verlag, Berlin, Heidelberg, New York, 2001.

[2]   PixArt Imaging Inc.: *Datasheet ADNS-9800*.

[3]   B. A. Wandell, A. El Gamal, B. Girod, *Common Principles of Image Acquisition Systems and Biological Vision*, Proceedings of the IEEE, Vol. 90, No. 1, January 2002.

[4]   R. Kennel, *"Why Do Incremental Encoders Do a Reasonably Good Job in Electrical Drives with Digital Control?"*, 41th IEEE IAS 2006 Annual Meeting, Tampa, Florida, 2006.

[5]   Y. Xu, *Study of Rotational Speed Measurement Based on Laser Mouse Sensor with Applications to Electric Driving Systems*, Master Thesis, Technische Universität München, ChenYang Technologies GmbH & Co. KG, August 27, 2015.

[6]   P. Drabarek, R. Kennel, *"Are Interferometric Encoders a Reasonable Alternative in Servo Drive Applications?"*, PEMD 2008, IET 4th International Conference on Power Electronics, Machines & Drives, York, United Kingdom, Apr. 2008.

[7]   E. Schrüfer, *Elektrische Messtechnik – Messung elektrischer und nichtelektrischer Größen*, 6th Edition, Carl Hanser Verlag, Munich, Germany, 1995.

[8]   E. Schrüfer, *Signalverarbeitung: Numerische Verarbeitung digitaler Signale, 1st Edition*, Carl Hanser Verlag, Munich, Germany, 1990.

PCIM Europe 2016, 10 – 12 May 2016, Nuremberg, Germany

Practical EMI Control in a Power Component Design Space

David Bourner: Vicor Corp, 25 Frontage Road Andover MA 01844 USA   dbourner@vicr.com

Abstract

The control of electromagnetic interference (EMI) within switched mode power systems is a perennial topic. This article attempts to address the notion of control of conducted emissions in the context of applying Vicor Power Components in a customer application. Vicor has developed quasi-resonant topologies that mitigate noise to a great extent by design. Although there are significant noise reductions to be had using resonant topologies in SMPS, no converter is ever noise-free. Applying power modules in a way to assure compliance with CE engineering standards can often prevent unforeseen, costly delays in bringing products to market.

1. Managing Conducted Emissions from Concept through Implementation

1.1 Introduction

Conducted emissions control must be a consideration at the outset of a power system design, made well before final integration with other parts of a complete application. The noise mitigation schemes used must be developed alongside the power and signal processing pathways. The implement of the system should be routinely subject to a series of pre-qualification tests, which although subjective, quickly bear results which indicate whether or not the final product will be adequately arranged for successful qualification outcomes. We look for minimal emissions that are below those levels that represent adequate performance.

1.2 Model of an Embedded Power Component

The block diagram shown in figure 1 is the author's attempt to show all the features of a power system that address aspects that go beyond a simple DC input/output power specification. Identifying power support functions in this way allows us to be able to look at tradeoffs and interactions between these various functions. Knowledge of interactions between required functions allows the designer to adopt a design sequence that effectively addresses all aspects of performance of the finalized design.

© VDE VERLAG GMBH · Berlin · Offenbach

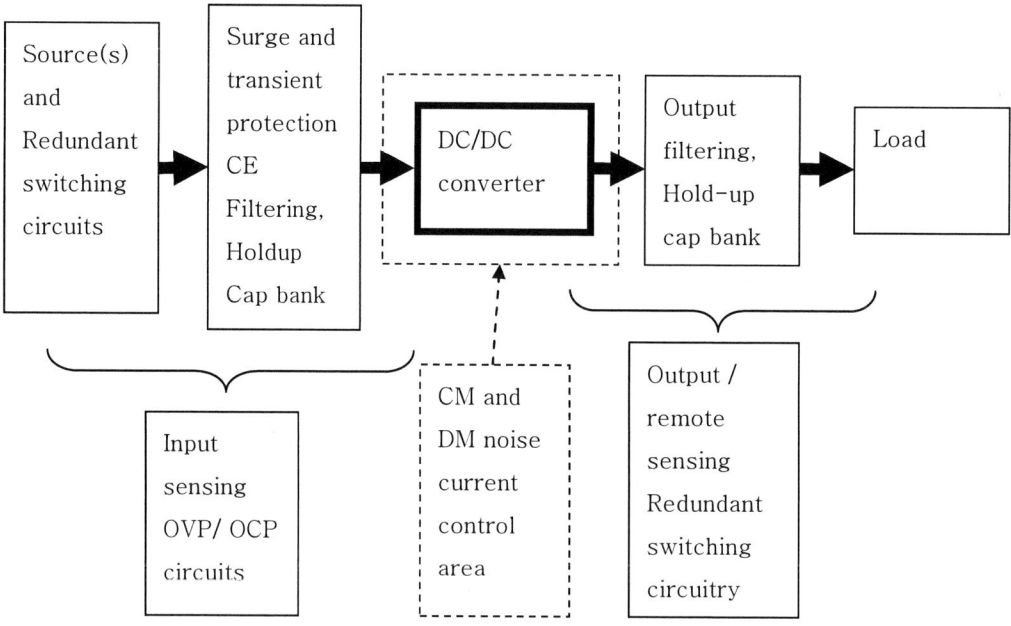

Figure 1 - Generic System Block Diagram for a Switching Power System. DC power flow on the busses are represented with bold arrows. The DC/DC converter is generally arranged with galvanically isolated input and output power ports.

1.3 Comments on the Model

In the context of noise, analog and digital systems are thought of respectively as receivers and transmitters of electrical noise. Switched-mode power converters present a *mixed mode* environment. Controllers, be they digital or analog in nature, are set alongside switched power devices. Power Components are modular: To a great extent the OEM (original engineering manufacturer) will have taken great pains to mitigate the effect of noise sources internal to the module, sometimes with mixed results. If not countered, self-noise of a power module is often sufficient to affect signaling affecting control of power within the Component.

Implementing a system with such products involves introducing measures that the OEM should provide in applications literature. Module self-noise can be reduced with the introduction of suitable external components that are selected and arranged in a board layout designed to introduce minimal parasitics. The components will interact with the printed circuit layout, which means that noise performance is not something that can be automatically guaranteed. In almost every case that such provisions are offered, there will be reference made to engineering standards for CE (conducted emissions) such as CISPR22 or EN61000-4-6. These two standards contain clear definitions and detailed test arrangements that can be used to authenticate a power system's emission and susceptibility characteristics.

CE standards focus on conducted electrical noise that propagates to the input power source. Noise appearing at the output side of the system can also be attenuated, with either CM (common mode) or DM (differential mode) filtering or a combination of the two, just as for input source terminated CE. It can be argued that using Power Components facilitates noise control design in the integrated application. Although it is generally not possible to integrate all the suppression elements in a power component, the manufacturer will recommend the minimum effective external measures required in an optimal board layout. Given the nature of the varied types of applications being worked on by customers, applications engineers at the OEM can often be called upon to determine adequacy of current noise control strategies and suggest ways to suppress noise. This paper illustrates such an example.

1.4 Applying the Model

To illustrate an important interaction example using figure 1, the inductance of the input power bus and the capacitors situated at the input port of the DC-DC converter (be they part of the holdup or the filtering function) interact with the dynamic (negative) input impedance the converter presents. This introduces the prospect of input power bus instability [2]. Clearly, the choice of the holdup capacitors with their attendant ESR is critical. These components, along with necessary surge and transient protection elements, impact the differential filtering of conducted noise travelling from the DC-DC converter's input port to the source.

Besides DM noise, there are CM noise currents emanating from each of the converter's input and output ports. Controlling such noise mandates the use of highly localized common-mode filtering components. These components will offer HF currents very small loops which are formed going from the power terminal, back through the coupling capacitance, afforded by a specially devised shield plane, back into the converter. Figure 2 shows the equivalent circuit of the DC-DC converter expressed in noise terms, as input-referenced noise current sources with X and Y capacitors set in place.

2. Noise Control Development Strategy

A full qualification test is expensive and time-consuming. This process should only be carried out once, at the pre-production stage of development, before freezing the design for production.

In the meantime, many simpler tests can be performed using equipment that is accessible to most designers and technicians working on hardware in a laboratory setting. It will be possible to set up and tailor noise control networks so as to minimize noise. This is often an iterative process, based on engineering intuition and a detailed understanding of parasitics in

PCIM Europe 2016, 10 – 12 May 2016, Nuremberg, Germany

the layout, possibly exploiting them to aid in noise control [8].

Test references should be established at the outset. It is important to run some "zero-signal" tests to establish the instrumentation's noise floor and to establish whether there are other noise sources that unexpectedly contribute energy to the system. These sources, if present need to be noted; If it is possible to eliminate them with modest measures, then a small investment in shielding and grounding of the lab bench may be in order.

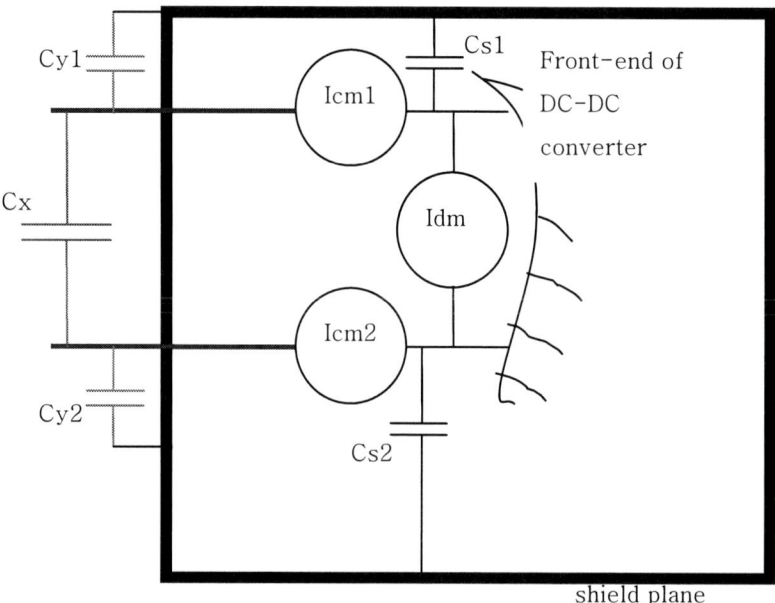

Figure 2 – input power port noise equivalent circuit of a DC-DC converter expressed with differential and CM noise current sources, along with noise suppression X and Y capacitors, installed and connected to an underlying shield plane which offers HF capacitive coupling (Cs1, Cs2 shown as lumped equivalents) back to the source at low impedance.

Once a *zero energy input* profile has been secured and collected, data should be developed for a signal with known characteristics. This test data set will be used as a background template which can be used to assess noise sources associated with the application's CE noise profile. With the veracity of the oscilloscope instrument and its coaxial probe established, it is now possible to exploit featured FFT processing running on the wideband oscilloscope as a rapid indicator of the conducted noise signal spectrum. It is easy to quickly establish read-back of the effectiveness of differing arrangements of CM and DM filtering and to iterate on topologies and discrete component selections for CE control.

© VDE VERLAG GMBH · Berlin · Offenbach

2.1 Example of Noise Control Assessment

The block diagram in figure 3 shows a system that was set up and augmented using networks of CE control elements. Some experimentation on component technology options and optimal placement of component networks (such as for the setups shown in figures 4 and 5) carried out to maximize noise control. At the completion of the iterative phase of establishing and testing noise control arrangements highlighted with figures 6 through 9 inclusive, the customer had a full qualification test done on their previously non-compliant target system, modified in accordance with the measures outlined. It was established that the CE noise profile of the system was robustly attained with a good margin to spare [8].

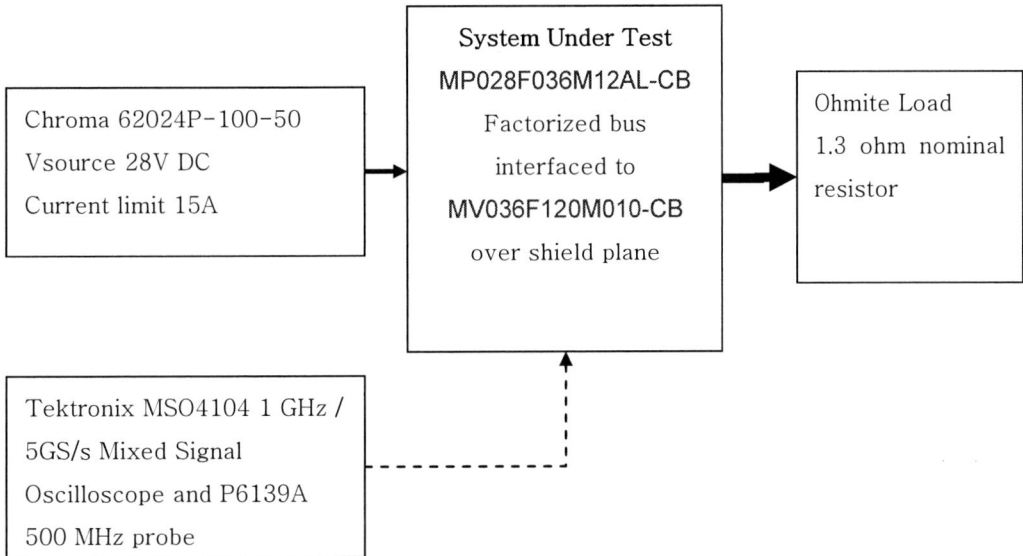

Figure 3 - Block diagram showing basic source / power converter and load arrangement for the unsuppressed system.

Figure 4 – Intermediate bench setup shows coaxially grounded scope probe, inboard shield ground contacts made at measurement stations out of adapted pieces of shielding cages of Johnson jacks. Shield ground plane is a solid copper layer. Y-cap provisional placement was changed to gain better oscilloscope acquired noise spectra.

Figure 5 – View on the VTM evaluation test fixture board's edge. Shows the noise reference voltage point for the system which is located at VTM's {–OUT} terminal.

PCIM Europe 2016, 10 – 12 May 2016, Nuremberg, Germany

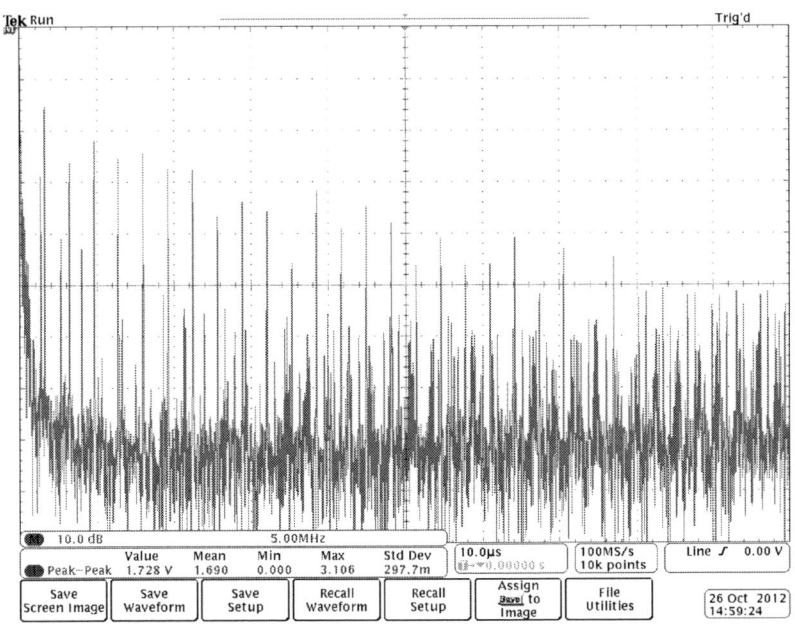

Figure 6 – Unsuppressed CE noise characteristic for the PRM {+IN} power terminal

MP028F036M12AL / MV036F120T100 TEST FIXTURE CE CONTROL ARRANGEMENT

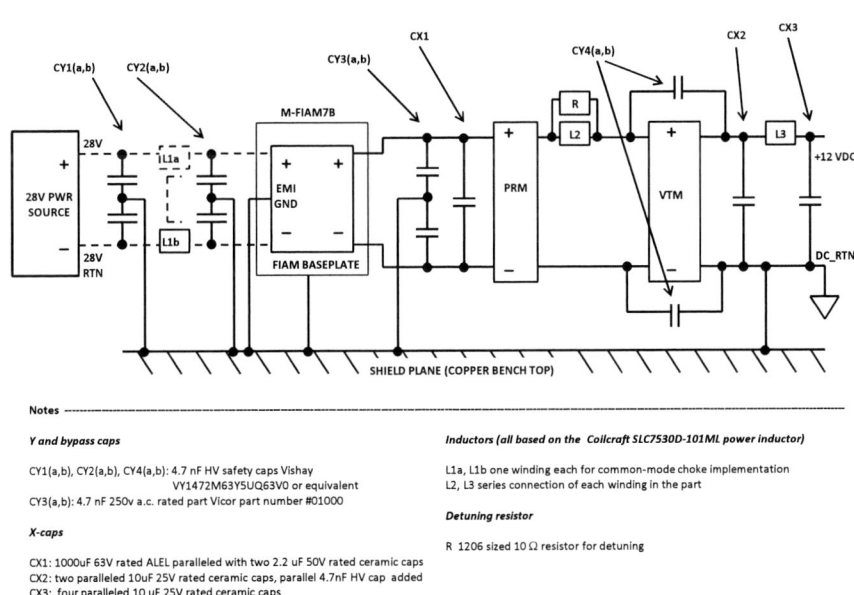

Notes ---

Y and bypass caps

CY1(a,b), CY2(a,b), CY4(a,b): 4.7 nF HV safety caps Vishay
VY1472M63Y5UQ63V0 or equivalent
CY3(a,b): 4.7 nF 250v a.c. rated part Vicor part number #01000

X-caps

CX1: 1000uF 63V rated ALEL paralleled with two 2.2 uF 50V rated ceramic caps
CX2: two paralleled 10uF 25V rated ceramic caps, parallel 4.7nF HV cap added
CX3: four paralleled 10 uF 25V rated ceramic caps

Inductors (all based on the Coilcraft SLC7530D-101ML power inductor)

L1a, L1b one winding each for common-mode choke implementation
L2, L3 series connection of each winding in the part

Detuning resistor

R 1206 sized 10 Ω resistor for detuning

Figure 7 – Outline of CE noise suppression arrangement. The M-FIAM is a filter and input attenuator module offering CE compliance in conformance with MIL-STD-461E.

© VDE VERLAG GMBH · Berlin · Offenbach

PCIM Europe 2016, 10 – 12 May 2016, Nuremberg, Germany

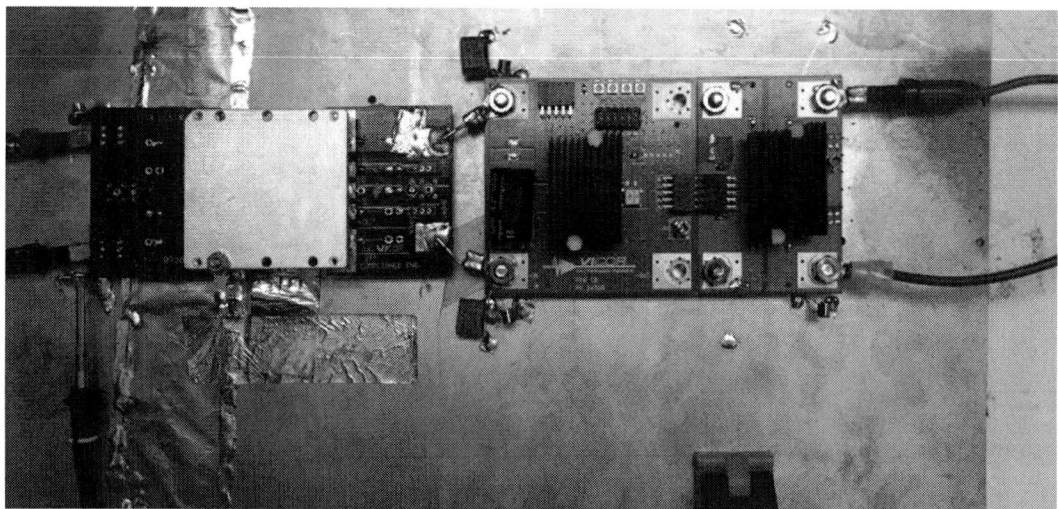

Figure 8 – Partial view of adapted noise control prototype: See the probe attached for the measurement of CE at the M-FIAM 's {-IN} terminal. Note that the "bypass style" Y caps {C4a, C4b} across the top and bottom of the VTM isolation barrier are mounted underneath the VTM evaluation board.

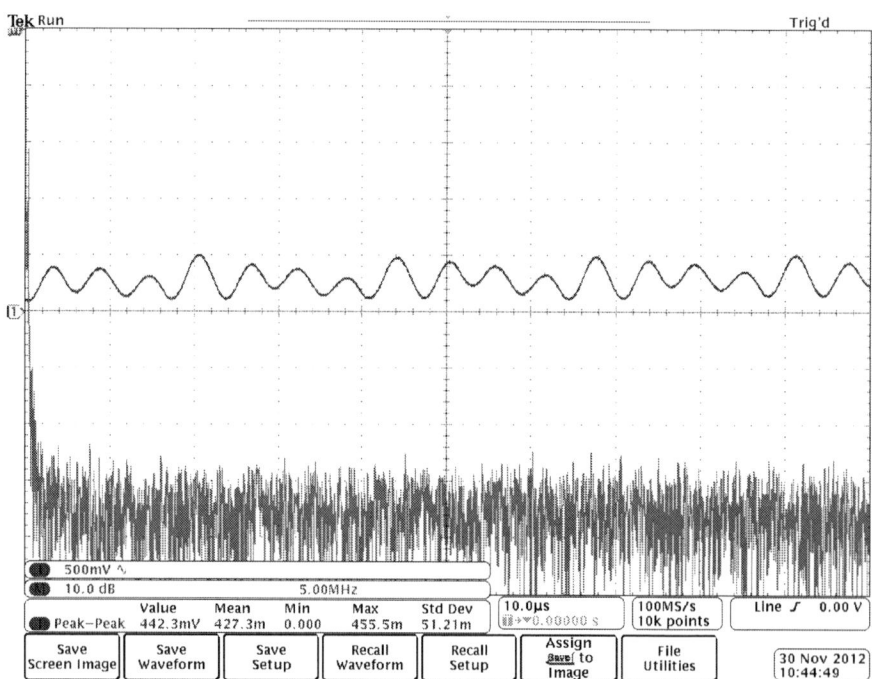

Figure 9 - Outcome of the CE noise suppression arrangements measured at the 28V DC source {+} terminal, location of which is shown in figure 8.

DC Voltages, Currents and Output Power Bus Inductance for Conducted Emissions for PRM, VTM power train based Test Rig		
V_SOURCE	27.993 V	Voltage measured at Chroma's output power terminals
V_IN_PRM	27.093 V	Voltage measured across the PRM's input power terminals {+IN, -IN}
R_LOAD	1.37 Ω	Load set across VTM's output power bus terminals {+OUT, -OUT}
I_SOURCE	4.507 A	Current delivered by the Chroma supply
I_LOAD	9.72A	Current in the load
L_OUTPUT	24 uH	Inductance presented by Ohmite load (1 kHz test frequency)

Table 1 - Summary of DC power figures for System Under Test

3. Summary

A suggested method for resolving a functional design whilst testing and providing for control of CE noise early on in the design and development of a power system has been outlined. This forms a paradigm that is easily followed. The noise reductions arising from exercising this approach have been successfully demonstrated, using equipment commonly available in most electronic development lab settings.

4. References

[1] Vicor Application Note AN: 005 "FPA Printed Cicuit Board Layout Guidelines" P Yeaman Rev 1.2 Nov 2013

[2] Vicor Application Note AN:023 "Filter Network Design for VI Chip DC to DC Converter Modules" Xiaoyan Yu Vicor Applications Engineering Aug 2012

[3] "Power System – CE EMI Topology Review". Vicor Internal document

[4] "Introduction to Electromagnetic Compatibility" P Clayton, Wiley Interscience ISBN-13 978-0471755005

[5] "The Circuit Designer's Companion" T Williams 2nd Edition EDN Series for Design Engineers Newnes ISBN-13: 978-0750663700

[6] "Back to Basics -- What are Y-Capacitors?" Vicor PowerBlog, June 5, 2013

[7] "Capacitor Characteristics Impact Power Supply Decoupling" D Bourner PCIM May 2001

[8] "CE EMI Topology Review" D Bourner, Vicor Internal Report

Converter Switching Noise Reduction
for Enhancing EMC Performance in HEV and EV.

HoTae, Chun, Hyundai Motor Company, Korea Rep, 6005588@hyundai.com
SeungHyun, Han, Hyundai Motor Company, Korea Rep, brian.han@hyundai.com
ChangHan, Jun, Hyundai Motor Company, Korea Rep, ch.jun@hyundai.com
JeongYun, Lee, Hyundai Motor Company, Korea Rep, jeongyooni@hyundai.com
JaeWon, Lee, Hyundai Motor Company, Korea Rep, leej1@hyundai.com
JeeHye, Jeong, Hyundai Motor Company, Korea Rep, jeehye88@hyundai.com
JeongHong, Joo, Hyundai Motor Company, Korea Rep, juinkong@hyundai.com
JinHwan, Jung, Hyundai Motor Company, Korea Rep jhjung10@hyundai.com

Abstract

In Eco-Friendly Vehicle like Hybrid Electric Vehicle (HEV) and Electric Vehicle (EV), the electric power conversion system performs the electric motor operating and 12V power supply, using the high voltage battery. The LDC, Low-Voltage DC-DC Converter, is the DC-DC Converter converting high voltage (usually 200~400V) to low voltage (usually 12~15V). In order to maximize the efficiency, the SMPS method is used for LDC. The SMPS includes the PWM switching which generates the severe EMC noise in the form of switching noise. This is because DC-DC Converter in HEV and EV is operated using the High-Voltage and High-Frequency. Since the power semiconductor device like FET has the nonlinear property, this intensive noise is expanded into harmonics of switching frequency. This switching noise harmonics contains the huge energy in particular frequency so it may create serious EMC problem in HEV and EV.

The RFI, radio frequency interference, is the test method by which we can measure the switching noise in the vehicle level. The RFI test measures the amount of EMC noise interference from the electric device to the radio antenna. Since the general switching frequency of LDC is about 100kHz, the AM (510kHz~1.8MHz) is the closest band to the switching noise harmonics. As a result, the switching noise from LDC appears strongly in AM RFI test. Therefore, reducing the switching noise is important feature for the robust design in terms of EMC performance. In this paper, several solutions of switching noise reduction of LDC output is suggested. The solutions include the noise path improvements and noise source reduction. In addition to that, the simulation and experiment are performed for each solution as well. Finally, the best effective way of switching noise reduction is proposed through the comparison analysis.

1. The Structure of LDC

The structure of general LDC is shown below (see Fig. 1). The transformer is located in the center of the LDC. With transformer as the center, the primary side is high voltage operating region and the secondary side is low voltage operating region respectively. The resistive divider may be used to step down the voltage. In order to maximize the efficiency, however, SMPS with power semiconductor device should be used for LDC. The full-bridge of power semiconductor device (FET) alternates the current direction of the transformer repeatedly so that the transformer can transfer the energy from primary to secondary, stepping down the voltage.

PCIM Europe 2016, 10 – 12 May 2016, Nuremberg, Germany

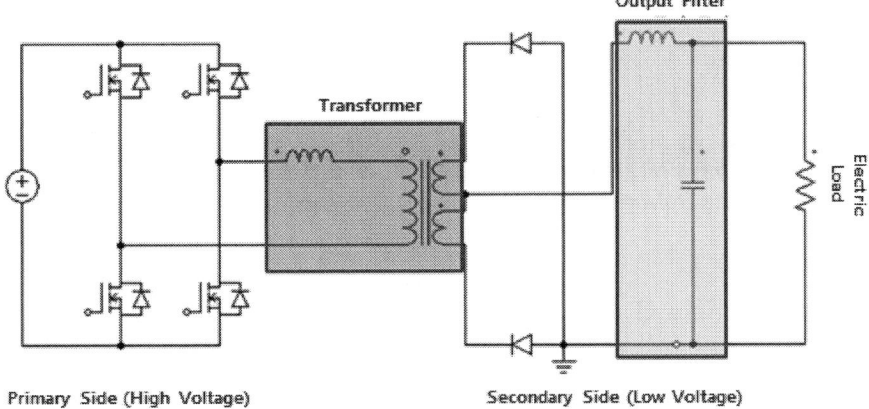

Fig. 1. The structure of LDC

FET in LDC operates at high speed in general. The LDC using in this paper has 100kHz switching frequency of FET. The reason of having such a high speed switching frequency is that passive elements like capacitor and inductor can be reduced in size with high switching frequency. However, unintended switching noise which includes the intensive EMC energy in particular frequency is generated because of 100kHz switching. Since power semiconductor like FET has the nonlinear property, this switching noise is expanded into harmonics of switching frequency. These switching noise harmonics are transferred from primary to secondary by noise coupling mechanism although the transformer is electrically isolated between primary and secondary side.

2. The measurement of switching noise

The switching noise of the converter like LDC can be measured distinctly by conducted emission measurement referring to CISPR 25 which is the international EMC standard. The measurement result is shown at Fig. 2. The measurement has been conducted at LDC output which is the secondary (low-voltage operation) side. From the measurement result, three important facts can be derived.

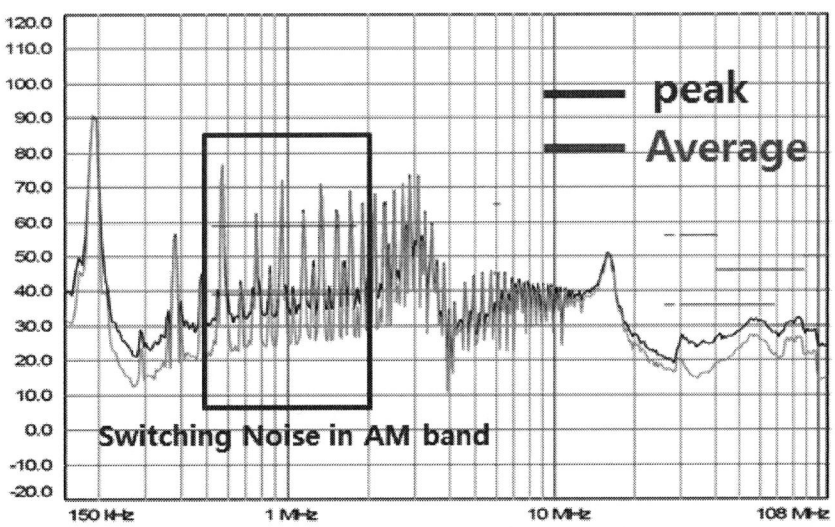

Fig. 2. LDC output CE(Conducted Emision)

© VDE VERLAG GMBH · Berlin · Offenbach

2.1. The Intensive energy in switching noise

In Fig. 2, the average and peak noise level of switching noise is almost equal to each other. The average noise actually represents the energy in resolution bandwidth. Thus this indicates that huge energy is concentrated in that particular frequency. If we take a closer look at Fig. 2, it is shown that the noise level of switching noise is decreasing gradually as the frequency is increasing. Since the AM (520~1.8MHz) is low frequency band closed to the harmonics of switching frequency, the switching noise level in AM is relatively high compared to the other frequency.

2.2. The mechanism of noise transferring from the primary to secondary

In fig. 2, switching noise harmonics have 100kHz interval correspond to switching frequency. However, If we take a closer look at fig 2, we can find out that the relatively high level noise elements appear in every other switching harmonics. To find out the reason for this, it is necessary to analyze "Phase-Shift Full-Bridge ZVS" which is the topology of the LDC.

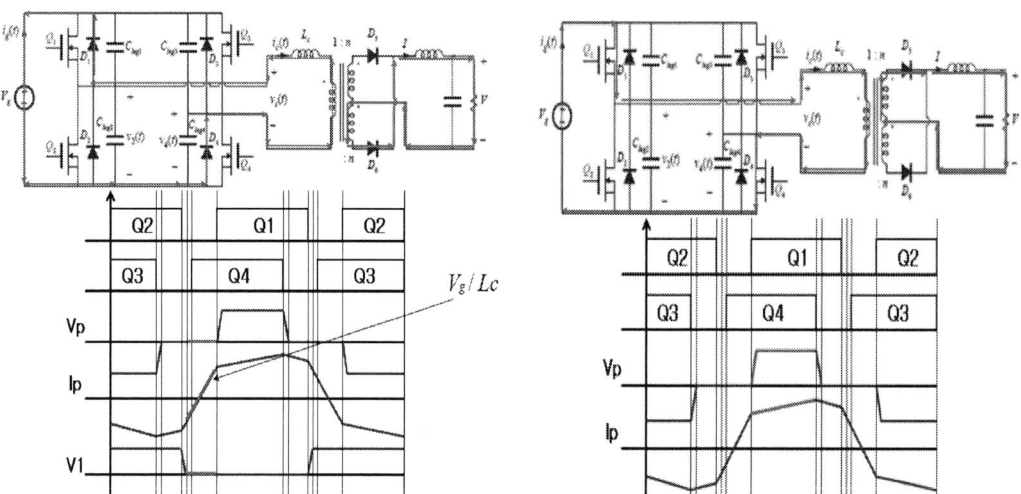

Fig. 3. Phase-Shift Full Bridge ZVS [Non-Powering Mode (Left) / Powering Mode (Right)]

In "Phase-Shift Full-Bridge ZVS LDC" (See Fig. 3), there are two legs that each one has one pair of FETs. Thus, LDC has four switches consist of Q1, Q2, Q3 and Q4. As illustrated in Fig. 3, Q1 (top) and Q2 (bottom) makes one pair and the rest of Q3 (top) and Q4 (bottom) makes the other. Although it has several operation modes in detail, the operation can be largely divided into two states, which are the powering mode and non-powering mode. The operation state that Q1 and Q4 are turned on at the same time is "powering mode" where the energy transferring from the primary to secondary actually takes place. Whereas the state that Q1 and Q4 are not turned on at the same time is "non-powering mode" where no energy is transferred from the primary to secondary at all. Adding more detailed explanation, the "powering mode has current of "Lm" which is the large inductance in transformer so that energy in the primary is big enough to be delivered into the secondary. Otherwise, the "non-powering mode has current of leakage inductance which is the pretty small value in transformer so that the energy in the primary is not big enough to be transferred into the secondary. Since the leakage inductance is really small value, current in the primary has the sharp gradient leading to quick discharge of parasitic capacitor between drain and source of FET. As a result, the voltage between drain and source goes down to zero quickly in "non-powering mode" and now the FET is ready for the "zero voltage switching (ZVS)". The ZVS is important feature for not only efficiency enhancement but also EMC noise reduction anyway.

However, the "powering mode" takes place twice in one period since the LDC has four switches and each pair (Q1 and Q4 / Q2 and Q3) performs the energy transferring. When the energy transfer occurs in LDC, the switching noise is also transferred to LDC output unintentionally. Finally, it is possible that relatively strong noise of 200kHz harmonics appears at LDC output by "Conducted Emission" measurement.

2.3. Switching Noise: The potential EMC risk to the xEV vehicle

As described above, the high level of switching noise harmonics can be transferred to the LDC output. However, LDC output is connected to the 12V battery and its electrical devices in xEV vehicle. Since 12V system in vehicle uses the chassis ground, the switching noise at LDC output may lead to the serious EMC problem in xEV vehicle.

The influence to the vehicle level can be measured by RFI (Radio Frequency Interference). The RFI test result for Electrical vehicle is shown at Fig. 4.

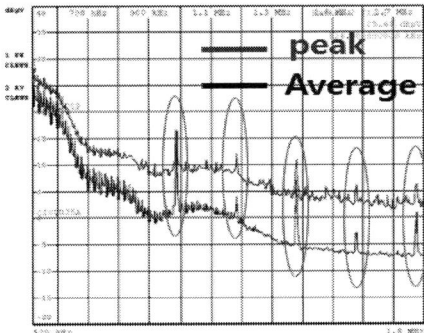

Fig. 4. RFI Test Result in AM Band

As described above, the switching noise of 200kHz harmonics which is twice the switching frequency appears in RFI AM band test result. Definitely, the AM band radio reception problem shall take place at switching frequency harmonics. Furthermore, this may cause other potential EMC risk as well. Thus, reducing the switching noise in xEV vehicle is really important feature to make the robust EMC design.

3. The switching noise reduction in xEV vehicle

To reduce the switching noise, this paper developed the noise path improvement and noise source reduction method.

3.1. The Noise path improvement method

The idea of the noise path improvement is shown in Fig. 5.

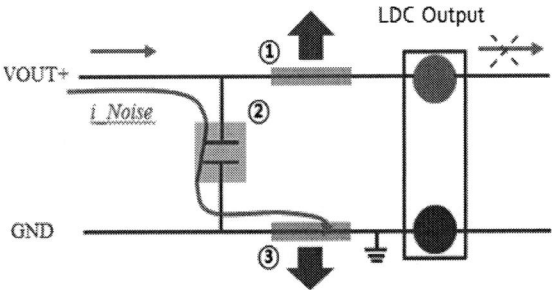

Fig. 5. The idea of the noise path improvement

© VDE VERLAG GMBH · Berlin · Offenbach

Since the noise should takes lower impedance path, pursuing higher differential impedance between output terminal and chassis ground is important. There are three ways to make differential impedance increase.

Adding inductance to LDC output terminal (+)

In order to increase the line inductance of LDC output (+), adding inductor is the most general method. The impedance of inductor increases with the frequency. Thus, the absolute value of S-parameter (S21) of the inductor also inceases with frequency so that the inductor can block the noise. Since noise should take the lower impedance path, switching noise doesn't get out of LDC output (+) and make a detour to the chassis ground instead. In this paper, the inductor of 100uH has been added and its result is shown at Fig. 6.

Adding capacitance to LDC output terminal between (+) and (-)

In order to perform the noise filtering, adding capacitor is the most general method as well. The impedance of capacitor decreases with the frequency. Thus, the absolute value of S-parameter (S21) of the capacitor also decreases with frequency so that the noise can pass through the capacitor. Since noise is filtered by capacitance, switching noise doesn't get out of LDC output (+) and make a detour to the chassis ground instead. In this paper, the capacitor of 1500uF has been added and its result is shown at Fig. 6 as well.

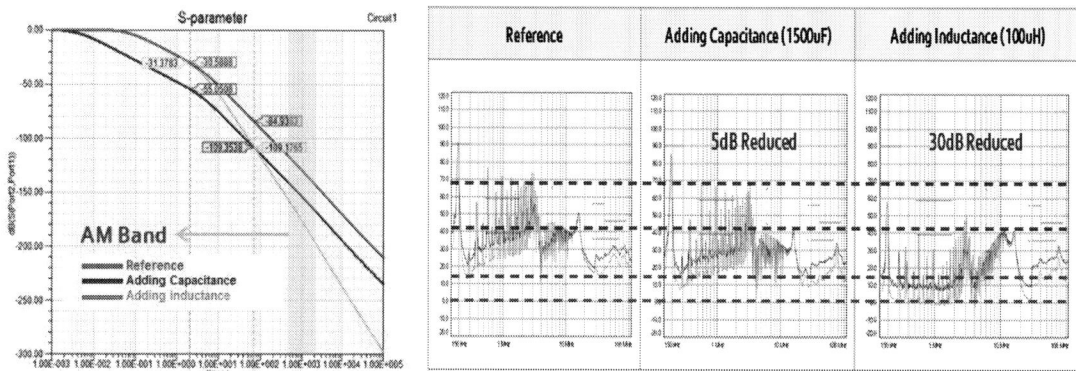

Fig. 6. Simulation (Left) and Test Results(Right) Rerarding Capacitance and Inductance

The simulation and test results (LDC output CE) of comparison between inductor and capacitor are shown at Fig. 6. The test results are consistent with simulation that the switching noise level is actually reduced by adding capacitor and inductor. If we take a closer look at Fig. 6, adding inductor has larger noise reduction than adding capacitor in both simulation and CE test.

However, the switching noise has intensive energy in particular frequency. Therefore, pretty big inductor or capacitor is needed to manage the switching noise reduction. In terms of cost and size, these are not the effective solution for reducing the switching noise definitely.

Reducing the plane-impedance of output filter board (PCB)

Adding inductance or capacitance is pretty general method to reduce the EMC noise. However, these traditional methods are not the best way to reduce the switching noise since the cost rise and size increase are too much.

To find out better solution for the noise path improvement, this paper implemented the plane-impedance reduction on the output filter board. For example, the lay-out of LDC is like below (See Fig. 7). In order to maximize the filter performance, the output filter board should be installed at end location of LDC.

PCIM Europe 2016, 10 – 12 May 2016, Nuremberg, Germany

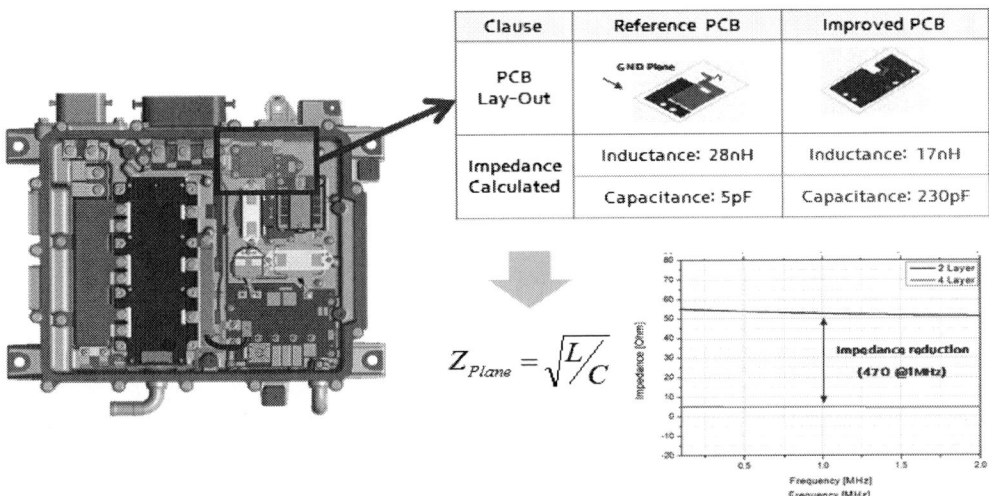

$$Z_{Plane} = \sqrt{L/C}$$

Fig. 7. Simulation Rerarding Plane Impedance

Generally, the PCB of the output filter board has multi-layers and trace pattern on it. The plane impedance can be calculated using inductance and capacitance which are extracted parameters from the PCB. It is possible to minimize the plane impedance by maximizing the pattern overlap and the number of PCB layers. The simulation result is shown at Fig. 7 as well. From the result, it is expected that the switching noise in vehicle may decrease. The switching noise in RFI test is shown at Fig. 8. The switching noise in RFI is reduced as the plane impedance of filter board is minimized. Since there is no additional element like capacitor or inductor for reducing the plane impedance, cost rise and size increase should be negligible. Thus, it should be more effective method compared to before.

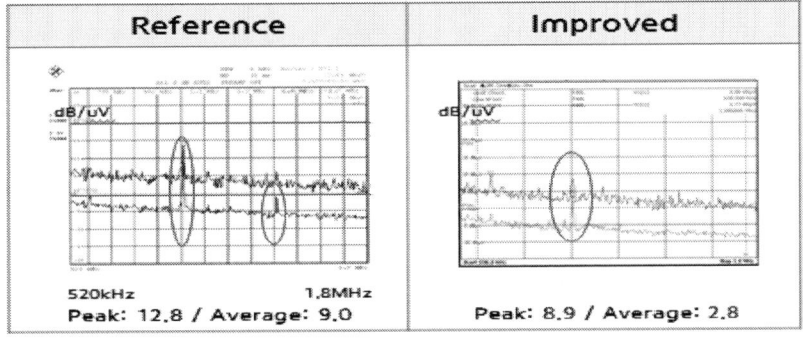

Fig. 8. Test Results (RFI of AM Band) Rerarding Plane Impedance

3.2. The noise source reduction method

SSFM (Spread Spectrum Frequency Modulation)

However, the switching noise still exists keeping its identical shape in narrow bandwidth even though the plane impedance reduction is implemented. Therefore, this paper implemented the noise source reduction method called SSFM. The SSFM, Spread Spectrum Frequency Modulation, is the noise source reduction method since it actually lowers the energy of the switching noise. The concept and simulation of SSFM are shown at Fig. 9.

© VDE VERLAG GMBH · Berlin · Offenbach

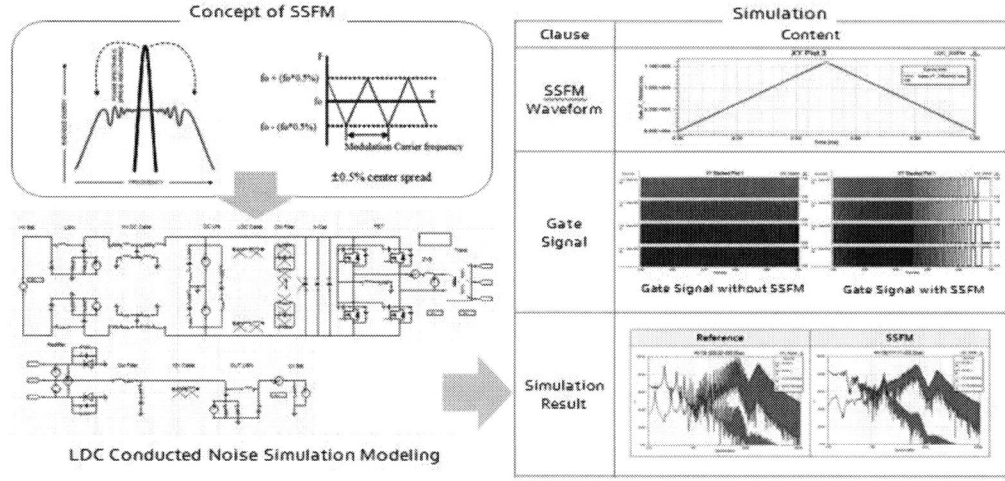

Fig. 9. SSFM Simulation Modeling

In SSFM, the switching frequency varies so that the intensive energy in narrow bandwidth should be spread over a large bandwidth. Therefore, the noise energy dispersion takes place leading to the switching noise source reduction. The test result is shown at Fig. 10. The switching noise level in RFI is reduced by 10dB. If take a closer look, we can find out that the only average level goes down whereas the peak level is unchanged. It makes sense since the average level means the calculated energy in bandwidth and SSFM performs the energy dispersion over a larger bandwidth respectively. The peak level is the measurement result in bandwidth so that the peak value should remain unchanged.

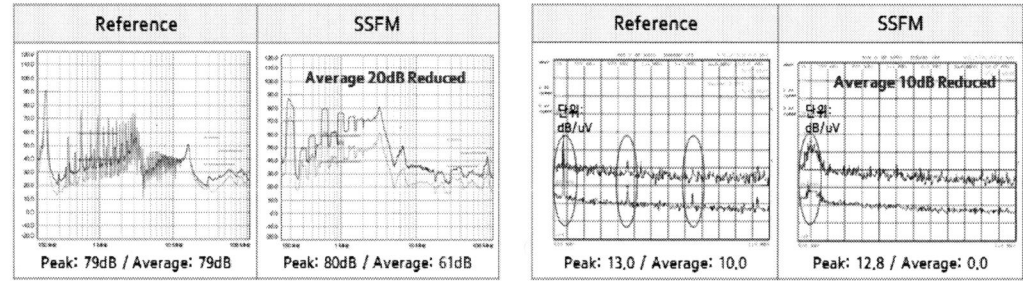

Fig. 10. LDC output CE (Left) and RFI (Right) reragding SSFM

Implementation method of SSFM (Spread Spectrum Frequency Modulation)

The implementation method of SSFM can be divided into random modulation and linear modulation largely.

Modulation	Random Modulation	Linear Modulation
Component	LTC690X	Modulation Circuit
Modulation Period	Fixed	Variable
Ripple	High	Low
NVH	High	Low
Modulation Signal		

© VDE VERLAG GMBH · Berlin · Offenbach

In this paper, "Linear Modulation" method is used since it has advantage of low ripple and low NVH. The PWM IC, UCC2895, has the function of adjusting switching frequency. Thus, it is possible to implement SSFM by adding linear modulation circuit which has only few elements like resistor, capacitor and comparator (See Fig. 11).

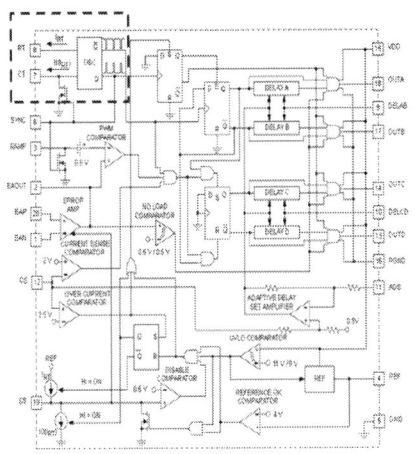 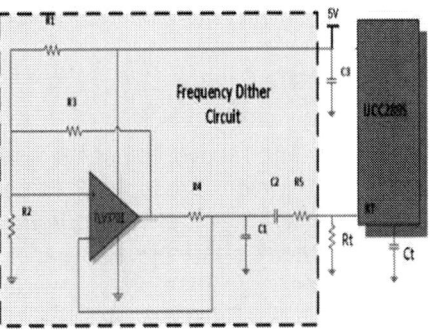

Fig. 11. PWM IC "UCC2895" (Left) and Linear Modulation Circuit (Right)

Since the circuit that added is really small, the cost and size increase is negligible.

4. Conclusion

For the robust EMC design, this paper introduces the switching noise reduction methods including noise path improvement and noise source reduction. The huge EMC filter is needed to reduce the switching noise since switching noise exhibits large amount of spectral energy at the switching frequency harmonics. In order to make the best effective solution for the switching noise reduction, the plane impedance reduction and SSFM method are proposed. Throughout the proposed methods, the RFI noise in xEV vehicle is reduced by more than 10dB with low cost and no size-up.

5. Reference

[1] X. Fang, S. Li, D. Jiandong, "Prediction Model of Conducted Common-mode EMI in PWM Motor Drive System"
[2] Y. Lee, A. Nasiri, "Conductive CM and DM noise analysis of power electronic converters in electric and hybrid electric vehicles
[3] T. Van. Doren, Grounding and Shielding of Electromanetic System, UMR.1999.

© VDE VERLAG GMBH · Berlin · Offenbach

PCIM Europe 2016, 10 – 12 May 2016, Nuremberg, Germany

Efficiency and Vibration Observations of a Symmetrical Six-Phase Drive applying Interleaved Space Vector Modulation

Daniel, Glose, TUM, Germany, daniel.glose@tum.de
Peng, Qian, TUM, Germany, peng.qian@tum.de
Ralph, Kennel, TUM, Germany, ralph.kennel@tum.de

Abstract

In this paper, a novel interleaved space vector modulation (SVM) scheme for a symmetrical six-phase drive is presented. It's capable of reducing current distortions and machine vibrations caused by the switching behavior of inverters. The six-phase machine is composed of two symmetrical three-phase winding sets, spatially shifted by 60 electrical degrees and fed by two independent three-phase inverters with a common DC-link.
To validate the proposed modulation technique, a series of experiments are carried out. According to the results, the efficiency improvement is significant compared to the traditional SVM and there is a notable reduction in vibration.

1. Introduction

The multi-phase and in particular the six-phase drives are a subject of increasing interest in the last few years and have found wide applications, especially in transport. They offer several advantages over the conventional three-phase drives, such as reduced torque pulsation, lower harmonic currents and improved reliability on the system level.

 Six-phase machines can be divided into two types according to the phase shift angle of the two three-phase sets: the symmetrical and the asymmetrical machines (see Fig. 1). With an angle between the two sets of $\gamma = j \cdot \frac{\pi}{3}$ ($j \in 1, 3, 5, ...$) the symmetrical machine allows the number of pole pairs to be changed during operation [1]. The asymmetrical type has an arbitrary displacement between the two subsets (30 electrical degrees are very common), which enables the control of the odd spatial harmonics [2].

The drive-setup considered in this paper is composed of a symmetrical six-phase machine with a phase shifting of 60 degrees and two common voltage source inverters (VSI) with two levels,

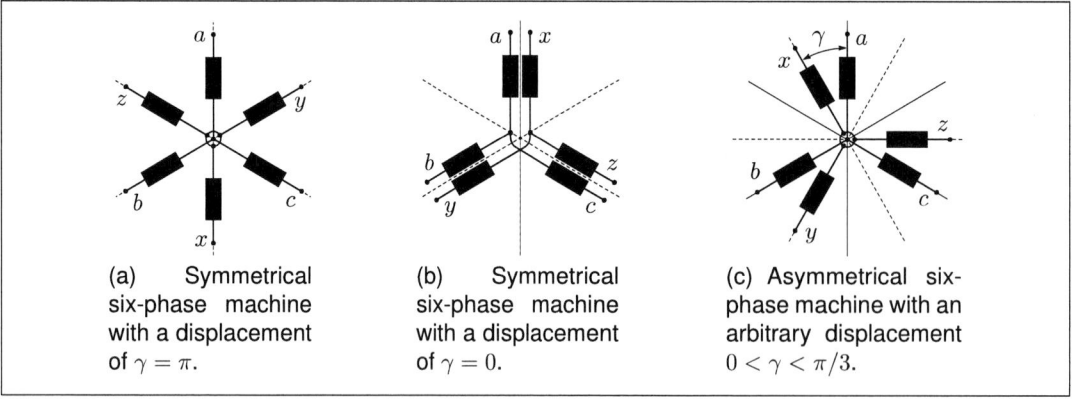

(a) Symmetrical six-phase machine with a displacement of $\gamma = \pi$.

(b) Symmetrical six-phase machine with a displacement of $\gamma = 0$.

(c) Asymmetrical six-phase machine with an arbitrary displacement $0 < \gamma < \pi/3$.

Fig. 1: Typical phase arrangements of the machine.

© VDE VERLAG GMBH · Berlin · Offenbach

PCIM Europe 2016, 10 – 12 May 2016, Nuremberg, Germany

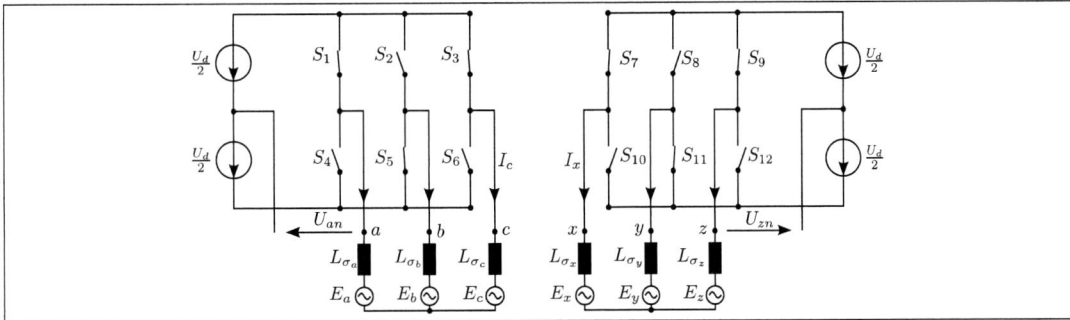

Fig. 2: Six-phase drive setup, consisting of two three-phase, two-level VSIs and a high frequency model of a six-phase machine with two neutrals

as illustrated in Fig. 2.

Due to the inverter-switching behaviors, current distortions occur, incurring extra losses in the machine. There exist a variety of articles proposing methods to suppress these harmonics. For the conventional SVM scheme an optimal switching strategy leads to minimal harmonics as derived mathematically in [3]. For the six-phase drive, however, the most attentions are given to the asymmetrical setup and corresponding optimal SVM schemes are described in [2, 4, 5]. For the symmetrical counterpart, in [6] interleaved carrier-based modulation techniques are given. A comprehensive analysis for an interleaved SVM technique is still missing.

In the present paper a novel interleaved SVM technique is derived and its performance regarding efficiency and machine vibration is evaluated. In section 2, a model of the six-phase machine is presented, which is the basis of further analysis. The novel control scheme is described in section 3 and the formulas of the optimal time shifting between the two three-phase SVM is derived in detail in section 4. In order to validate the interleaved SVM, a series of experiments have been carried out and the results are shown in section 5. Finally in section 6, conclusions and perspectives are given.

2. Modeling and Decomposition of a Symmetrical Six-Phase Drive

According to [6], the high-frequency model of a six-phase machine can be greatly reduced to a combination of a series-connected stray inductance and a voltage source (see Fig. 2).

The well known "Clark-Transformation" with the matrix $\mathbf{T_{C3}}$ [7] can be extended for the six-phase machine and the system is decomposed into two isolated star-connected three-phase circuits. The novel transformation matrix is given as follows:

$$\begin{bmatrix} X_{\alpha_I} & X_{\beta_I} & X_{0_I} & X_{\alpha_{II}} & X_{\beta_{II}} & X_{0_{II}} \end{bmatrix}^T = \begin{bmatrix} \mathbf{T_{C3}} & \mathbf{0} \\ \mathbf{0} & \mathbf{T_{C3}} \end{bmatrix} \cdot \begin{bmatrix} X_a & X_b & X_c & X_x & X_y & X_z \end{bmatrix}^T \quad (1)$$

Another decomposition technique, called "Generalized Two-Phase Real Component Transformation" [8] is introduced here, which offers a great insight for further analysis. The transformation of arbitrary phase states is defined as:

$$\begin{bmatrix} X_{\alpha_1} \\ X_{\beta_1} \\ X_{\alpha_2} \\ X_{\beta_2} \\ X_{z_1} \\ X_{z_2} \end{bmatrix} = \mathbf{T_{pp}} \cdot \begin{bmatrix} X_a \\ X_b \\ X_c \\ X_x \\ X_y \\ X_z \end{bmatrix} = \frac{1}{3} \cdot \begin{bmatrix} 1 & -\frac{1}{2} & -\frac{1}{2} & -1 & \frac{1}{2} & \frac{1}{2} \\ 0 & \frac{\sqrt{3}}{2} & -\frac{\sqrt{3}}{2} & 0 & -\frac{\sqrt{3}}{2} & \frac{\sqrt{3}}{2} \\ 1 & -\frac{1}{2} & -\frac{1}{2} & 1 & -\frac{1}{2} & -\frac{1}{2} \\ 0 & -\frac{\sqrt{3}}{2} & \frac{\sqrt{3}}{2} & 0 & -\frac{\sqrt{3}}{2} & \frac{\sqrt{3}}{2} \\ \frac{1}{2} & \frac{1}{2} & \frac{1}{2} & -\frac{1}{2} & -\frac{1}{2} & -\frac{1}{2} \\ \frac{1}{2} & \frac{1}{2} & \frac{1}{2} & \frac{1}{2} & \frac{1}{2} & \frac{1}{2} \end{bmatrix} \cdot \begin{bmatrix} X_a \\ X_b \\ X_c \\ X_x \\ X_y \\ X_z \end{bmatrix} \quad (2)$$

© VDE VERLAG GMBH · Berlin · Offenbach

The first two rows of $\mathbf{T_{pp}}$ map the machine states onto the first subspace (α_1, β_1), which corresponds to the harmonic content of the first order. In the second subspace (α_2, β_2), the states describe the harmonic content of the second order. The fifth and sixth rows correspond to the zero-sequence components, which are not under consideration here, since the two neutral points are isolated.

Combining equation (2) with (1), a relation between the transformed three-phase sets and the two subspaces is achieved:

$$
\begin{bmatrix} X_{\alpha_1} \\ X_{\beta_1} \\ X_{\alpha_2} \\ X_{\beta_2} \end{bmatrix} = \frac{1}{2} \begin{bmatrix} 1 & 0 & 0 & -1 & 0 & 0 \\ 0 & 1 & 0 & 0 & -1 & 0 \\ 1 & 0 & 0 & 1 & 0 & 0 \\ 0 & -1 & 0 & 0 & -1 & 0 \end{bmatrix} \cdot \begin{bmatrix} X_{\alpha_I} \\ X_{\beta_I} \\ X_{0_I} \\ X_{\alpha_{II}} \\ X_{\beta_{II}} \\ X_{0_{II}} \end{bmatrix}
\tag{3}
$$

The machine is usually operated in either one of the two subspaces, i.e. the design intended (α_1, β_1)- or (α_2, β_2)-subspace. If the fundamental currents in phase a and x are equal, i.e. $I_a = I_x$, the machine is operated in the second subspace only (parallel mode). However, for an operation in the first subspace, it is $I_a = -I_x$ (anti-parallel mode). The overall harmonics are calculated using the square sum of the harmonic currents (RMS-values) of both subspaces.

3. Interleaved SVM

For the generation of the control signals, the synchronized and continuous SVM is used for each three-phase set [9]. According to the space-vector concept, the on-time of the switches is computed from a reference voltage vector. For symmetry reasons, this paper only focuses on the first sector. In order to simplify the calculations, the on-time t_R of a certain voltage vector V_R is expressed by the duty cycle d_R, i.e. $d_R = t_R/T_S$ for $R \in \{0, 1, 2, 7\}$, where T_S is the switching period. The duty cycles of two active space vectors can be computed with the modulation index M and the space vector angle Θ [10]:

$$
\begin{bmatrix} d_1 \\ d_2 \end{bmatrix} = \frac{\sqrt{3}M}{\pi} \begin{bmatrix} -1 & \sqrt{3} \\ 2 & 0 \end{bmatrix} \begin{bmatrix} \sin(\Theta) \\ \cos(\Theta) \end{bmatrix},
\tag{4}
$$

where M is defined as the ratio of the output voltage U^* to the maximal fundamental magnitude $U_{SS} = 2U_{dc}/\pi$, i.e. $M = U^*/U_{SS}$. Thus, the two duty cycles of the zero vectors can be expressed by d_1 and d_2:

$$
d_0 = d_7 = (1 - d_1 - d_2)/2
\tag{5}
$$

Since the machine is fed by two common three-leg VSIs, an extra degree of freedom appears for the modulation compared to the traditional three-leg case, namely the time shifting Δd between the two pulse patterns. As illustrated in Fig. 3(a), Δd leads to an asymmetrical switching behavior of the two sets and results in an interleaved harmonic flux characteristic of both three-phase sets (see Fig. 3).

4. Optimal Time Shifting

According to [11], applying an interleaved SVM to the six-phase drive leads to a decreased harmonic current in one subspace (e.g. α_2, β_2) but to an increased value in the other one (e.g. α_1, β_1). In order to find the optimal Δd, an expression of the overall distortion must be derived.

PCIM Europe 2016, 10 – 12 May 2016, Nuremberg, Germany

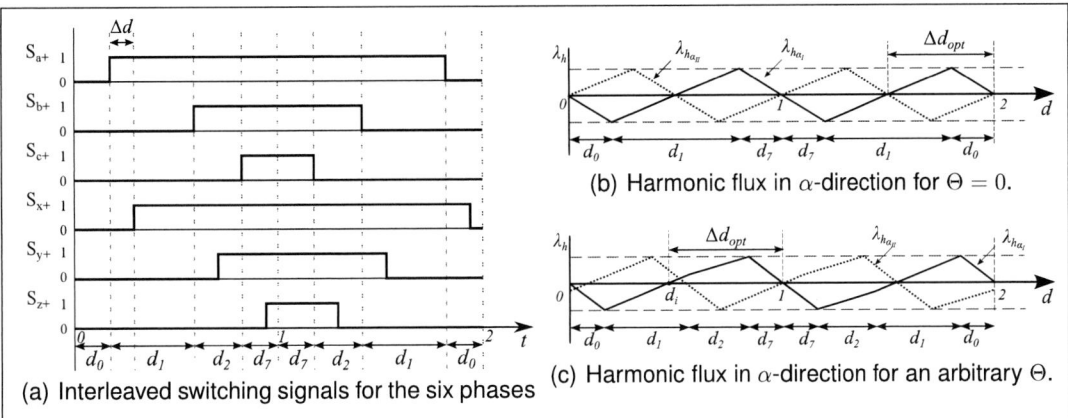

(b) Harmonic flux in α-direction for $\Theta = 0$.

(c) Harmonic flux in α-direction for an arbitrary Θ.

(a) Interleaved switching signals for the six phases

Fig. 3: Interleaved SVM and waveforms of the optimal time-shifted harmonic flux.

4.1. Overall Distortion

With the assumption of high switching frequency, the conceptual harmonic flux λ_h can be used to estimate the current distortion [9]. The harmonic flux over a half switching period of an arbitrary space vector is calculated as follows:

$$\lambda_h(M, \Theta) = \int_0^1 (v_k - v^*)\, dd \tag{6}$$

where the reference voltage vector v^* can be assumed to be constant during a switching cycle and v_k is the inverter output-voltage vector of the k-th state.

Employing equation (6), the RMS-value of the harmonic flux can be calculated as:

$$\lambda_{rms}^2(M, \Theta) = \int_0^1 \lambda_h^2\, dd \tag{7}$$

In order to estimate the overall losses, different leakage inductances have to be considered for both subspaces. Therefore, the inductance ratio κ is introduced:

$$\kappa = L_{\sigma_1}/L_{\sigma_2} \tag{8}$$

An expression of the overall distortion over a half switching period can then be derived according to equations (6), (7) and (8):

$$\lambda_{rms}^2 = \int_0^1 \left[\left(\lambda_{h_{\alpha_1}}^2 + \lambda_{h_{\beta_1}}^2 \right) \frac{1}{\kappa^2} + \left(\lambda_{h_{\alpha_2}}^2 + \lambda_{h_{\beta_2}}^2 \right) \right] dd \tag{9}$$

Combining equation (3) with (9), the overall distortion can be rearranged as a sum of a Δd-dependent and an independent term:

$$\lambda_{rms}^2 = \frac{\kappa^2 + 1}{4\kappa^2} \underbrace{\int_0^1 \left(\lambda_{h_{\alpha_I}}^2 + \lambda_{h_{\beta_I}}^2 + \lambda_{h_{\alpha_{II}}}^2 + \lambda_{h_{\beta_{II}}}^2 \right) dd}_{\neq f(\Delta d)} + \frac{\kappa^2 - 1}{2\kappa^2} \underbrace{\int_0^1 \left(\lambda_{h_{\alpha_I}} \lambda_{h_{\alpha_{II}}} + \lambda_{h_{\beta_I}} \lambda_{h_{\beta_{II}}} \right) dd}_{= f(\Delta d)} \tag{10}$$

The Δd-independent term of equation (10) is constant for a certain space vector and inductance ratio. Therefore, the objective function of the optimization is:

$$\text{Min.} \frac{\kappa^2 - 1}{2\kappa^2} \int_0^1 \left(\lambda_{h_{\alpha_I}} \lambda_{h_{\alpha_{II}}} + \lambda_{h_{\beta_I}} \lambda_{h_{\beta_{II}}} \right) dd \to \Delta d_{opt} \tag{11}$$

© VDE VERLAG GMBH · Berlin · Offenbach

1273

4.2. Operation Strategies

According to equation (11), improved operational strategies should depend on Δd as well as the inductance ratio κ. For symmetry reasons, it is only necessary to estimate the optimal Δd_{opt} in the first half sector, i.e. $\Theta \in [0, \pi/6]$.

Optimal Δd for $\kappa = 1$

Setting $L_{\sigma_1} = L_{\sigma_2}$ results in $\kappa = 1$, which sets the Δd-dependent term to zero. With the control of Δd, the overall distortion of the six-phase drive cannot be influenced and there is no difference between the interleaved and the common SVM.

Optimal Δd for $\kappa < 1$

Under this condition, $\left(\kappa^2 - 1\right)/2\kappa^2 < 0$, hence equation (11) can be rewritten as:

$$\int_0^1 \left(\lambda_{h_{\alpha_I}}\lambda_{h_{\alpha_{II}}} + \lambda_{h_{\beta_I}}\lambda_{h_{\beta_{II}}}\right) dd \to \text{Max.} \tag{12}$$

Equation (12) can be maximized if the harmonic components in the α- and β-axis of both equivalent circuits are equal, i.e. $\lambda_{h_{\alpha_I}} = \lambda_{h_{\alpha_{II}}}$ and $\lambda_{h_{\beta_I}} = \lambda_{h_{\beta_{II}}}$. This is achieved with $\Delta d = 0$, which is again equal to the common SVM.

Optimal Δd for $\kappa > 1$

For $L_{\sigma_1} > L_{\sigma_2}$, the optimal solution based on equation (10) is very complex to calculate and not applicable in real-time. In order to reduce the complexity, the following approximation is made to achieve an analytical solution anyway: Since $\Theta \in [0, \pi/6]$, the harmonic flux distortion in the β-direction can be neglected compared to the one in α-direction. With this assumption, the square sum of the harmonic flux according equation (10) can be simplified:

$$\lambda_h^2 = \frac{1}{4}\int_0^1 \left(\lambda_{h_{\alpha_I}} + \lambda_{h_{\alpha_{II}}}\right)^2 dd = \frac{1}{4}\underbrace{\int_0^1 \left(\lambda_{h_{\alpha_I}}^2 + \lambda_{h_{\alpha_{II}}}^2\right) dd}_{\neq f(\Delta d)} + \frac{1}{2}\underbrace{\int_0^1 \lambda_{h_{\alpha_I}}\lambda_{h_{\alpha_{II}}} dd}_{=f(\Delta d)} \tag{13}$$

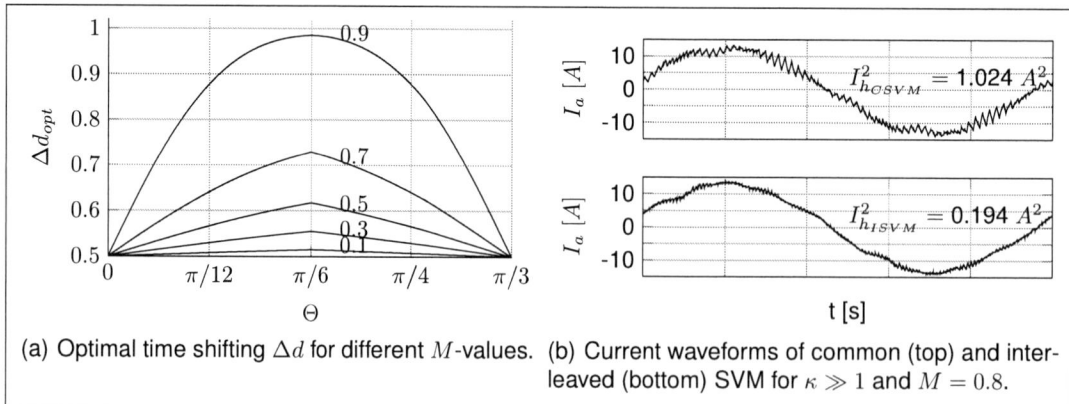

(a) Optimal time shifting Δd for different M-values. (b) Current waveforms of common (top) and interleaved (bottom) SVM for $\kappa \gg 1$ and $M = 0.8$.

Fig. 4: Distribution of Δd and current waveforms over one fundamental cycle

The left integral of equation (13) is determined by the system parameters and cannot be changed during operation. Only the multiplication, within the right integral of equation (13), can be minimized.

As an example, for $\Theta = 0$, the optimal time-shifted $\lambda_{h_{\alpha_I}}$ and $\lambda_{h_{\alpha_{II}}}$ trajectories are given in Fig. 3(b). It can be seen that the harmonic flux $\lambda_{h_{\alpha_I}}$ or $\lambda_{h_{\alpha_{II}}}$ is shifted with $\Delta d = 0.5$, its product $\lambda_{h_{\alpha_I}} \lambda_{h_{\alpha_{II}}}$ is negative for any d and the minimal value of the overall distortion is achieved.

In general, the harmonic flux should be shifted in such a way, that the negative quarter wave of $\lambda_{h_{\alpha_I}}$ for $d > 1$ is covered by the positive one of $\lambda_{h_{\alpha_{II}}}$. This can be achieved by moving the first zero-point d_i of $\lambda_{h_{\alpha_I}}$ to the next one, as shown in Fig. 3(c).

According to equation (6), d_i can be estimated with:

$$\int_0^{d_i} (v_k - v^*) \, dd = 0 \ (d_0 < d_i < d_0 + d_1) \tag{14}$$

which yields an explicit solution $d_i = 2d_0 / (4d_0 + d_2)$.

Employing equations (4) and (5) the optimal time shift Δd_{opt} can be expressed as:

$$\Delta d_{opt}(M, \Theta) = 1 - d_i = 1 - \frac{2d_0}{4d_0 + d_2} = \frac{1}{2} + \frac{\sqrt{3}M \sin(\Theta)}{2(\pi - 3M \cos(\Theta))} \tag{15}$$

The approximated Δd_{opt} depends on Θ, as illustrated in Fig. 4 for several values of M. Note that Δd_{opt} is mirrored at $\Theta = \pi/6$.

5. Experimental Results

To evaluate the validity of the theoretical considerations discussed previously, a variety of experiments was carried out on a $15kW$ test bench. As shown in Fig. 5, the rig consists of a diode-rectifier, two three-phase inverters, an optional transformer and a six-phase machine. The latter is mechanically connected to a load.

Two different cases were examined: $\kappa \approx 1$ (transformer disconnected) and $\kappa \gg 1$ (transformer connected). The transformer features three coils on the primary side and three on the secondary. The inverter legs a, b and c are connected to the first terminal of each primary coil, the second terminal being linked with the machine phases a', b' and c' respectively. The same is done with the legs x, y and z and the phases x', y' and z' using the coils on the secondary side. This leads to a strong coupling between phases a' and x', b' and y' as well as c' and z'. Under anti-parallel operation (i.e. $i_a = -i_x$), a high mutual flux is established in the transformer resulting in a high L_{σ_1}-value, whereas under parallel operation (i.e. $i_a = i_x$) only very low mutual flux occurs, resulting in a nearly unmodified L_{σ_2}. Hence, using the transformer leads to $\kappa \gg 1$.

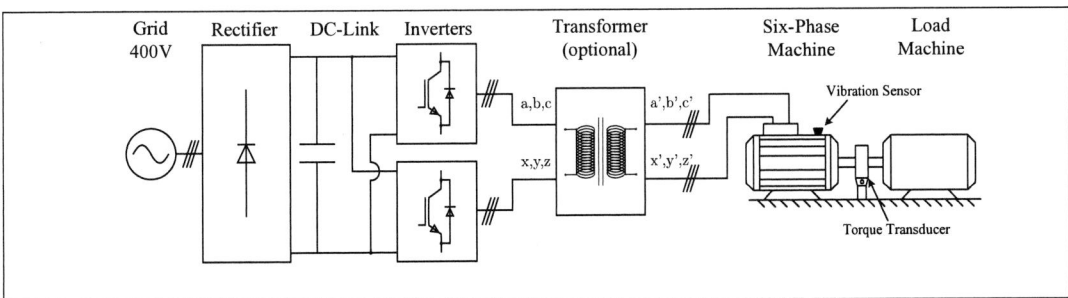

Fig. 5: Experimental setup.

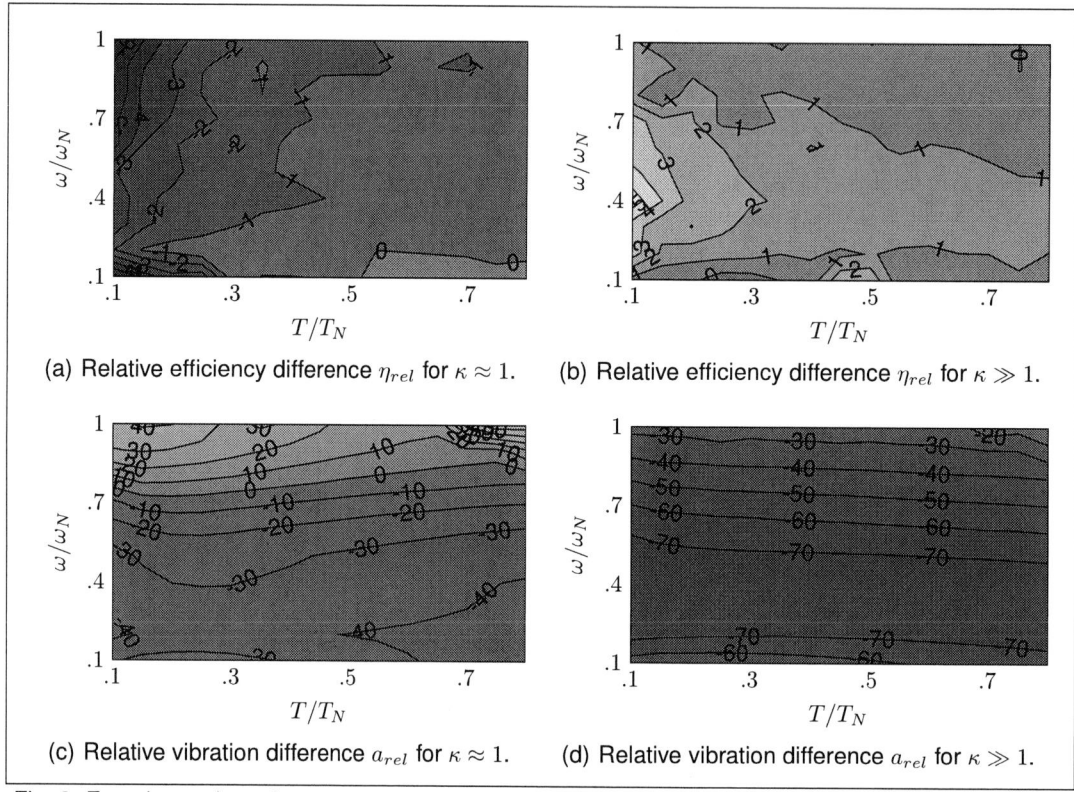

(a) Relative efficiency difference η_{rel} for $\kappa \approx 1$. (b) Relative efficiency difference η_{rel} for $\kappa \gg 1$.

(c) Relative vibration difference a_{rel} for $\kappa \approx 1$. (d) Relative vibration difference a_{rel} for $\kappa \gg 1$.

Fig. 6: Experimental results.

The efficiency of the six-phase drive η can be defined as the ratio of the mechanical power P_m and the electrical power P_e. To estimate the input power P_e, the DC-link voltage U_{dc} and current I_{dc} were measured. For the calculation of the output power P_m, the torque T_m and the angular speed ω_m were also recorded.

Beside the efficiency, the vibration of the machine, another criterion for the modulation quality was measured with an acceleration sensor mounted on the machine casing, as shown in Fig. 5. The RMS-value of acceleration is used as a figure of merit here. For a better comparison, the relative efficiency and vibration gain between the common SVM and the interleaved technique (CSVM and ISVM) are estimated:

$$\eta_{rel} = \frac{\eta_{ISVM} - \eta_{CSVM}}{\eta_{CSVM}} \times 100\% \qquad (16)$$

$$a_{rel} = \frac{a_{ISVM} - a_{CSVM}}{a_{CSVM}} \times 100\% \qquad (17)$$

For $\kappa \approx 1$, the η_{rel} values are depicted in Fig. 6(a). It can been seen that the efficiency decreases slightly compared with the common SVM in the results, especially at the low-torque range. Actually without the transformer, the inductance ratio of this test machine is a little bit smaller than 1 and applying the interleaved SVM leads to a decrease in efficiency. However if the transformer is connected, i.e. $\kappa \gg 1$, the current waveforms for CSVM and ISVM and the squares of the harmonic currents $I^2_{h_{ISVM}}$ and $I^2_{h_{CSVM}}$ are given in Fig. 4(b). The efficiency is improved significantly, specifically at the mid-speed range, as shown in Fig. 6(b). Although connecting the transformer increases the resistance slightly, the overall losses are decreased due to lower current distortions. Efficiency improvements of up to 5% at mid speeds and low torques are achieved. Although, at low and high speed, the vibration is increased for $\kappa \approx 1$, a decrease in vibration is observed in the mid-speed range, as shown in Fig. 6(c). For $\kappa \gg 1$, the decline in vibration is tremendous throughout the whole operating range and reaches values of up to -70%. It can be stated that the acoustic/hearable noise is also significantly reduced.

6. Summary and Outlook

An interleaved SVM strategy for a symmetrical six-phase drive is presented in this paper and a mathematical way of calculating the optimal time shifting Δd_{opt} for minimizing the overall losses of the system is given in detail. Significant efficiency improvements are observed if $\kappa \gg 1$. The experiments proved the validity of this theory, especially at high torques and low angular speeds. In addition, the vibration of the machine is significantly suppressed, which also proves the performance of the modulation strategy. There is still no explanation for the reduced vibrations, which may be the focus of further works.

7. References

[1] K. Mizuno, T. Tsuboi. Basic principle and maximum torque characteristics of a six-phase pole change induction motor for electric vehicles. *Electrical Engineering*, 118:78–91, 1997.

[2] Y. Zhao and T.A. Lipo. Space vector pwm control of dual three-phase induction machine using vector space decomposition. *Industry Applications, IEEE Transactions on*, 31(5):1100–1109, Sep 1995.

[3] G. Narayanan, D. Zhao, H. K. Krishnamurthy, R. Ayyanar, and V. T. Ranganathan. Space vector based hybrid pwm techniques for reduced current ripple. *IEEE Transactions on Industrial Electronics*, 55(4):1614–1627, April 2008.

[4] R. Bojoi, A. Tenconi, F. Profumo, G. Griva, and D. Martinello. Complete analysis and comparative study of digital modulation techniques for dual three-phase ac motor drives. In *Power Electronics Specialists Conference*, volume 2, pages 851–857 vol.2, 2002.

[5] Lin Chen and Lijun Hou. A novel space vector pwm control for dual three-phase induction machine. In *Power Electronics and Motion Control Conference, 2004. IPEMC 2004. The 4th International*, volume 2, pages 724–729 Vol.2, Aug 2004.

[6] D. Glose and R. Kennel. Carrier-based pulse width modulation for symmetrical six-phase drives. *Power Electronics, IEEE Transactions on*, PP(99):1–1, 2015.

[7] E. Clarke. Circuit analysis of a-c power systems, ser. general electric series. wiley. 1950.

[8] D. White and H. Woodson. Electromechanical energy conversion. Wiley, 1950.

[9] A.M. Hava, R.J. Kerkman, and T.A. Lipo. Simple analytical and graphical methods for carrier-based pwm-vsi drives. *Power Electronics, IEEE Transactions on*, 14:49–61, 1999.

[10] H.W. van der Broeck, H.-C. Skudelny, and G.V. Stanke. Analysis and realization of a pulsewidth modulator based on voltage space vectors. *Industry Applications, IEEE Transactions on*, 24(1):142–150, Jan 1988.

[11] D. Glose and R. Kennel. Continuous space vector modulation for symmetrical six-phase drives. *IEEE Transactions on Power Electronics*, 31(5):3837–3848, May 2016.

© VDE VERLAG GMBH · Berlin · Offenbach

High Efficiency Three-Phase-Inverter with 650 V GaN HEMTs

Jennifer Lautner and Bernhard Piepenbreier
Friedrich-Alexander-University of Erlangen-Nuremberg, Germany, Jennifer.Lautner@fau.de

Abstract

GaN power devices offer great performance improvements compared to Si transistors. The lower conduction and switching losses of GaN transistors enable to build high efficiency power converters with high switching frequencies. An increase of the PWM frequency has many advantages in motor drive applications like reduced motor current ripple, lower motor losses and reduced filter size and cost. This paper presents the evaluation of a normally-off 650 V, 30 A GaN transistor in a motor drive application. The switching performance of the GaN HEMT is investigated by using a double pulse tester. Furthermore, design and experimental results of a high efficiency 1.5 kW three-phase-inverter with sine-output filter and a PWM frequency of 100 kHz are shown. The GaN inverter with sine-output filter achieves 97 % efficiency.

1. Introduction

The trend of modern power converters is towards higher efficiency, higher power density and lower size and cost. To meet these different requirements better power devices are needed. Wide bandgap power devices such as gallium nitride (GaN) transistors are promising candidates. GaN has a higher bandgap, breakdown field and electron mobility compared to silicon (Si) which offers great performance improvements [1, 2]. The lower switching and conduction losses of GaN transistors enable to build highly efficient power converters with high switching frequencies. Using GaN transistors in a motor drive application has many advantages [3]. Due to the low switching losses the pulse width modulation (PWM) switching frequency of the inverter can be about 10 times higher. A higher PWM frequency allows to build a sine-output filter with a higher corner frequency which leads to smaller values of the filter components. Therefore, the size and cost of the filter can be reduced. Furthermore, the current and voltage output waveforms of a three-phase-inverter with sine-output filter are close to pure sinusoidal which reduces the motor losses and increases the efficiency of the total system [4].

This paper presents the evaluation and application of a commercially available 650 V, 30 A enhancement mode GaN high electron mobility transistor (HEMT) of *GaN Systems*. In Section 2 the switching performance of the GaN HEMT in a phase-leg configuration under influence of different parameters is evaluated. In Section 3 the printed circuit board (PCB) and thermal design of a 1.5 kW GaN three-phase-inverter with sine-output filter in a motor drive application is discussed. Experimental results including waveforms and efficiency of the GaN inverter are shown.

2. Evaluation of 650 V GaN HEMTs Switching Behavior

When using a new power device it is important to evaluate the switching behavior of the transistor in order to optimize the driving parameters like gate resistance and dead time and to estimate the switching losses and possible frequency for a future power converter (e.g. for thermal dimensioning).

© VDE VERLAG GMBH · Berlin · Offenbach

2.1. Measurement test setup

The switching behavior of the 650 V, 30 A GaN HEMT of *GaN Systems* (Type GS66508P) is investigated by using a double pulse tester with clamped inductive load as shown in Fig. 1(a). A typical phase-leg configuration is used and both transistors are switched complementary. The lower transistor is the active device and the upper transistor is used in reverse mode as freewheeling path without additional diode. Since in a phase-leg configuration the parasitic elements have a significant impact on both devices [5], the switching transients of the upper and lower transistor are evaluated. The gate resistances R_{on} and R_{off} determine the turn-on and turn-off switching speed, respectively, and can be adjusted separately by using the gate driver LM5114. Due to the application of the GaN phase-leg in a three-phase-inverter the gate resistance values have to be the same for lower and upper transistor, whereby different values for R_{on} and R_{off} are recommended for GaN devices.

The experimental test setup of the double pulse tester is shown in Fig. 1(b). The parasitic inductances of the gate and power loop are minimized by proper PCB layout using small ceramic dc link capacitors placed directly below the GaN phase-leg. In order to evaluate the switching behavior of the GaN HEMT the drain current is measured with a 100 mΩ coaxial shunt and the drain-source and gate-source voltages are measured with high bandwidth passive probes. The measurements were performed with a constant dc link voltage of 300 V and different load current levels. The same circuit board is used to measure the switching transients of the upper device but the inductive load is clamped over the lower transistor instead of the upper transistor and the gate signals are reversed.

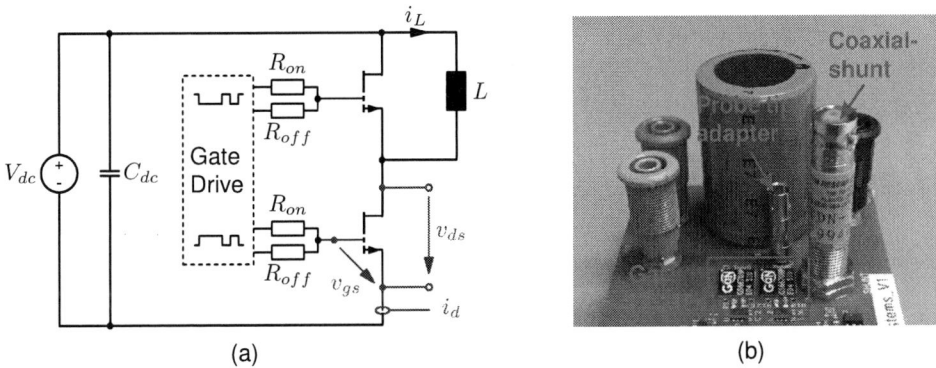

Fig. 1: (a) Schematics and (b) experimental test setup of the double pulse tester with GaN HEMTs.

2.2. Switching behavior of 650 V GaN HEMT

Fig. 2 shows the measured hard switching waveforms of the upper and lower GaN HEMT at 300 V and 10 A and with different gate resistance values. As described in [5], the switching transients of both devices in a phase-leg configuration are mainly determined by the gate resistance of the active (lower) transistor. However, picking a high turn-off gate resistance for the upper transistor causes a visible peak in the gate source voltage (see Fig. 2(c)) which may result in an unintentional turn-on of the upper device and should be avoided. Nevertheless, Fig. 2 shows clean switching waveforms with less ringing at all gate resistance values. The maximum turn-off voltage (see Fig. 2(a)) is about 360 V. A very small turn-off gate resistance value leads to slight ringing. Therefore, R_{off} is chosen to be 4.7 Ω. The turn-on current peak (see Fig. 2(b)) is not a reverse recovery phenomenon (because GaN HEMTs have no recovery

© VDE VERLAG GMBH · Berlin · Offenbach

PCIM Europe 2016, 10 – 12 May 2016, Nuremberg, Germany

charge) but the consequence of charging and discharging the parasitic capacitances. This also causes a turn-off overvoltage at the upper transistor (see Fig. 2(c)) which can be reduced by using a higher turn-on gate resistance value of the lower transistor. Therefore, a resistance value of $18\,\Omega$ is chosen for R_{on}.

Fig. 3 shows the switching transients at 300 V with $R_{off} = 4.7\,\Omega$ and $R_{on} = 18\,\Omega$ and for load current levels between 5 A and 15 A. Fig. 3(a) and Fig. 3(d) depict that the turn-off switching speed of the lower device and also the turn-on switching speed of the upper device depends on the load current level. This speed determines the dead time between turn-off of the active (lower) transistor and turn-on of the upper transistor. In this case, the dead time is about 20 ns.

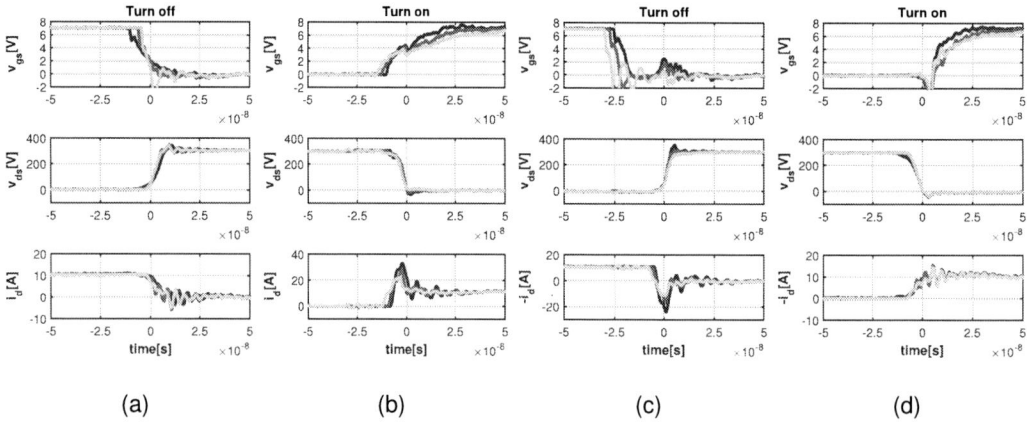

Fig. 2: (a)-(b) Measured switching transients of the lower transistor and (c)-(d) measured switching transients of the upper transistor at 300 V, 10 A with $R_{off} = [10\,\Omega, 4.7\,\Omega, 1\,\Omega]$ and $R_{on} = [10\,\Omega, 18\,\Omega, 27\,\Omega]$.

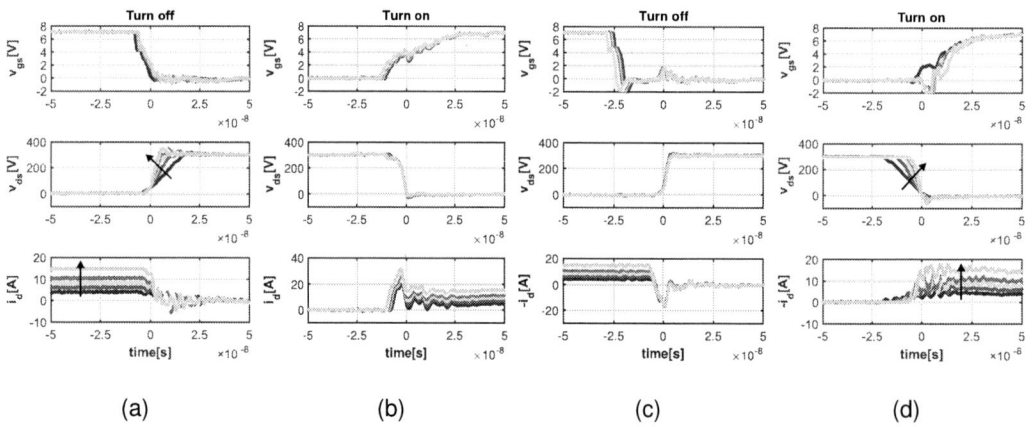

Fig. 3: (a)-(b) Measured switching transients of the lower transistor and (c)-(d) measured switching transients of the upper transistor at 300 V with $R_{off} = 4.7\,\Omega$ and $R_{on} = 18\,\Omega$ and for load current levels between 5 A and 15 A.

© VDE VERLAG GMBH · Berlin · Offenbach

2.3. Switching loss analysis

Based on the measured turn-on and turn-off switching transients of Sec. 2.2 the switching energies as functions of the drain current and the drain-source voltage can be calculated by

$$E_{on/off} = \int_{t_1}^{t_2} v_{ds}(t) \cdot i_d(t) \; dt. \tag{1}$$

The integration limits are chosen as depicted in Fig. 4(a). Fig. 4(b) shows the calculated switching energies of the GaN HEMT phase-leg based on the measured waveforms of the upper and lower device. The turn-on process of the lower transistor causes the dominant switching energy $E_{on,l}$ which has values between 20 µJ and 60 µJ depending on the load current. The other calculated switching energies are about 5 µJ for all load current levels. The latest datasheet of the used GaN HEMT also provides values of the switching energies [6] which is in well accordance with the presented results in this paper.

Based on the calculated energies the switching losses of a GaN three-phase-inverter can be estimated by

$$P_S = \frac{6}{\pi} \cdot f_s \cdot E_{tot}(v_{ds}, i_d), \tag{2}$$

where f_s is the PWM switching frequency and E_{tot} is the sum of all switching energies of the phase-leg [7]. The conduction losses of the forward operating GaN HEMTs in a three-phase-inverter can be calculated by

$$P_C = 6 \cdot R_{DS,on} \cdot \hat{i}_d \cdot \left(\frac{1}{8} + \frac{M \cos(\phi)}{3\pi} \right), \tag{3}$$

where $R_{DS,on}$ is the on-resistance of the HEMT and M is the modulation factor. The conduction losses of the GaN HEMTs, which operate in reverse mode, can be estimated in the same way. The losses of the "'body diode'", which occurs in the dead time between turn-off and turn-on, are neglected. Fig. 4(c) shows the calculated losses of a three-phase-inverter using GaN HEMTs with a PWM frequency of 100 kHz. The conduction losses are considerably higher as the switching losses at higher current levels. Furthermore, the on-resistance $R_{ds,on}$ increases with the junction temperature which is not considered in Eq. 3 and Fig. 4(c). Therefore, it is expected that the conduction losses are underestimated and become even more dominant in the practical test setup. This has to be taken into account for thermal dimensioning.

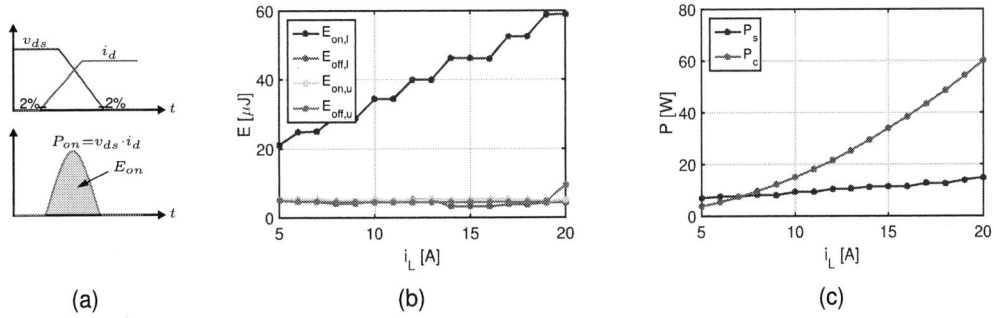

(a) (b) (c)

Fig. 4: (a) Switching loss calculation, (b) calculated switching energies of the phase-leg with $R_{off} = 4.7\,\Omega$ and $R_{on} = 18\,\Omega$ for different load current levels and (c) calculated switching and conduction losses of the GaN three-phase inverter with $f_s = 100\,\text{kHz}$ and $R_{ds,on} = 50\,\text{m}\Omega$.

3. Application of 650 V GaN HEMT in a Three-Phase-Inverter

A three-phase voltage source inverter for a motor drive application using GaN HEMTs was developed. The schematic of the test setup is shown in Fig. 5. The inverter drives a three-phase 2.2 kW induction motor which is loaded by a 4.2 kW DC motor. The GaN three-phase-inverter was operated in V/f control with a PWM switching frequency of 100 kHz. Thus, the GaN inverter can be tested in different operating points. Since the high dv/dt of the GaN transistors can cause motor insulation breakdown issues at the AC motor a sine-output filter should be used. The inverter is fed by an external DC voltage source with 300 V. The efficiency calculations were carried out by a power analyzer LMG500 which measures the electrical input power and the AC output power, as shown in Fig. 5.

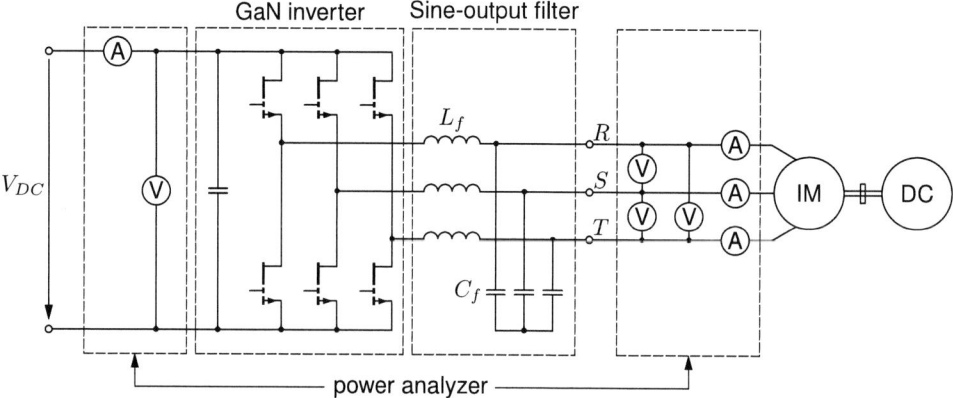

Fig. 5: Schematic test setup of three-phase-inverter with GaN HEMTs and sine-output filter for a motor drive application.

3.1. Experimental test setup

Three-phase-inverter with GaN HEMTs

The three-phase-inverter consists of six 650 V, 30 A enhancement mode GaN HEMTs of GaN Systems (Type GS66508T) and is diode-free because the HEMTs can conduct in both directions without reverse recovery [6]. Fig. 6(a) shows the experimental test setup of the GaN inverter. The GaN transistors, small ceramic DC-link capacitors and the isolated gate drivers are mounted on the upper circuit board. This board can easily be replaced in order to evaluate different types of GaN transistors. The parasitic inductances of the gate loop and the power loop are minimized by proper PCB layout. The power loop PCB design of one GaN HEMT phase-leg is schematically shown in Fig. 6(b). The return path of the power loop is directly under the the top layer on the first inner layer. This provides a small physical loop size and a good field self cancellation. Every GaN half bridge has the same pcb design, as shown in Fig. 7(a). The six GaN HEMTs are forced air cooled by one small heat sink. The used GaN transistors have a thermal pad on the top-side which is attached to the heat sink. Due to the internal connection between source and the thermal pad of the GaN HEMT a thermal interface material has to be used. The gate resistance values are 4.7 Ω and 18 Ω as discussed in Sec. 2.2. The second circuit board contains the logic circuit, DC-link capacitors, monitoring circuits and connectors. The GaN HEMTs can be operated in hard switching mode up to frequencies of 100 kHz using a Xilinx FPGA. The measured line voltages referred to the negative rail are shown in Fig. 7(b). The measurements were performed with a 400 MHz differential probe of PMK and a 1 GHz Oscilloscope of LeCroy (Type 610Zi).

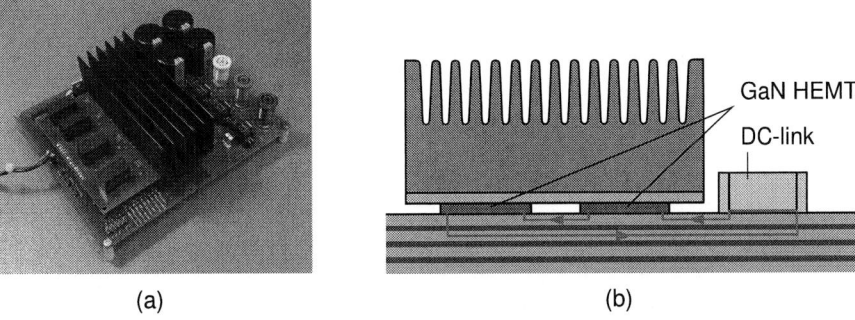

(a)   (b)

Fig. 6: (a) Experimental test setup of the three-phase-inverter with 650 V GaN HEMTs and (b) schematic of the PCB design for one GaN phase-leg whereby the power loop is highlighted in red color.

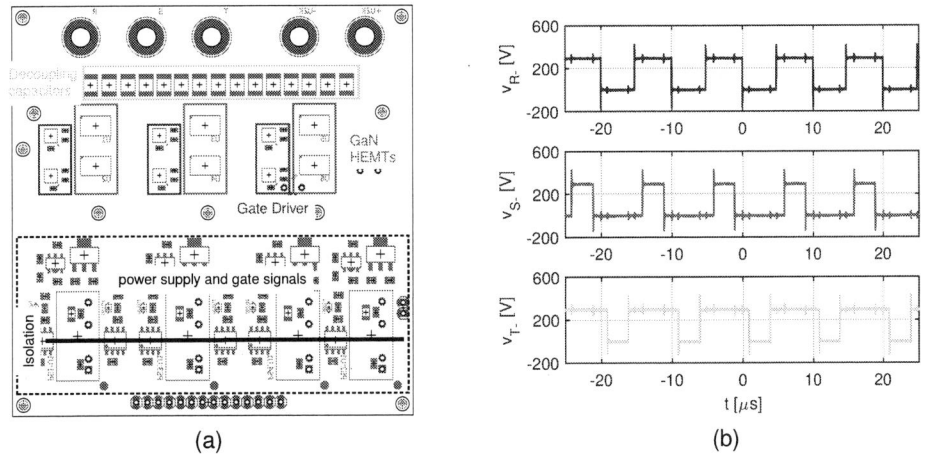

(a)   (b)

Fig. 7: (a) PCB design of the three-phase-inverter with 650 V GaN HEMTs and (b) measured line voltages referred to negative rail at 100 kHz switching frequency.

Sine-output filter

The sine-output filter is designed as shown in the schematic in Fig. 5. Due to the high PWM frequency of the GaN inverter the output filter can have a high corner frequency f_f which reduces the size and cost of passive filter components. The corner frequency of the filter can be calculated by

$$f_f = \frac{1}{2\pi\sqrt{C_f L_f}}. \tag{4}$$

The filter inductances are chosen with $L_f = 100\,\mu H$ and the filter capacitances are chosen to be $C_f = 0.68\,\mu F$. This leads to a filter corner frequency of 19.3 kHz which is five times smaller than the PWM switching frequency of 100 kHz. Additional damping resistors are not required because of $f_s >> f_f$ and the core losses of the filter inductor provides some damping. Fig. 8

© VDE VERLAG GMBH · Berlin · Offenbach

shows the measured inverter output waveforms of the line currents and the line-to-line voltages with and without sine-output filter at 1 kW output power. It can be seen that both the line current and the line-to-line voltage are pure sinusoidal with sine-output filter which reduces the motor losses and increases the total system efficiency.

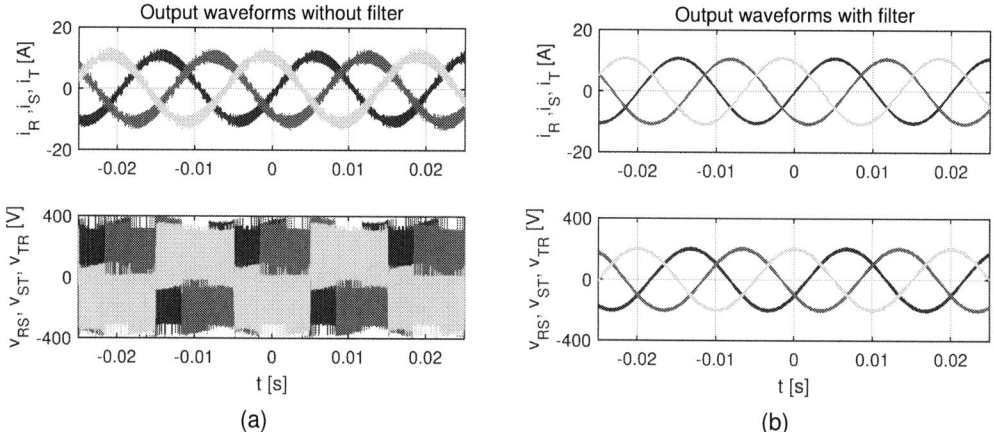

Fig. 8: (a) Measured output line currents i_R, i_S, i_T and line-to-line voltages v_{RS}, v_{ST}, v_{TR} without sine-output filter and (b) with sine-output filter at 1 kW output power.

3.2. Experimental results

The GaN inverter was tested in different operating points by changing the motor frequency respectively the line to line output voltages and by changing the load in steps between no load and full load. Due to the DC-link voltage of 300 V the maximum line to line output voltage of the GaN inverter is 212 V, which is half of the rated motor voltage. The efficiency of the GaN inverter with sine-output filter at a motor frequency of 25 Hz and 20 Hz respectively a line to line output voltage of 200 V and 160 V is shown in Fig. 9(a). The GaN inverter including output filter achieves a maximum efficiency of 97 % at 1.5 kW output power. The maximum power losses of the GaN inverter including the sine-output filter are 52 W. Fig. 9(b) shows the heatsink of the GaN inverter captured with a thermal image camera at 1.5 kW. The heatsink temperature is about 40 °C. As expected, the efficiency is lower by operating at a smaller motor voltage but under the same load conditions. On the other side, by using a higher DC-link voltage a higher inverter output voltage could be generated which may further improves the efficiency. However, due to the higher DC-link the switching losses also increases in this case.

4. Conclusion

In this paper the evaluation and application of a new 650 V, 30 A enhancement mode GaN HEMT was discussed. The switching behavior of both GaN transistors in a phase-leg configuration was evaluated. With proper PCB layout and suitable gate resistance values clean switching waveforms with less overshoots can be achieved. Furthermore, the advantages of using GaN HEMTs in motor drive applications was pointed out. The low switching losses of the GaN HEMTs enable a PWM switching frequency of 100 kHz which allows to design a small sine-output filter with a high corner frequency. A highly efficient 1.5 kW GaN three-phase-inverter

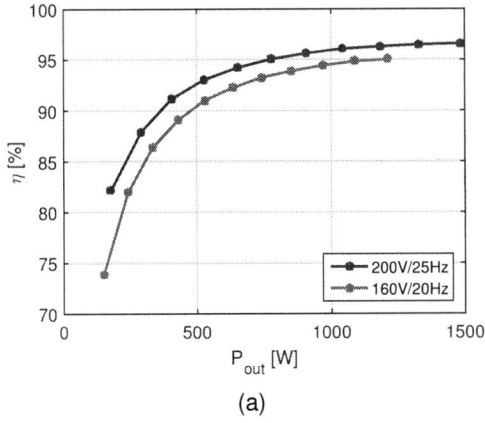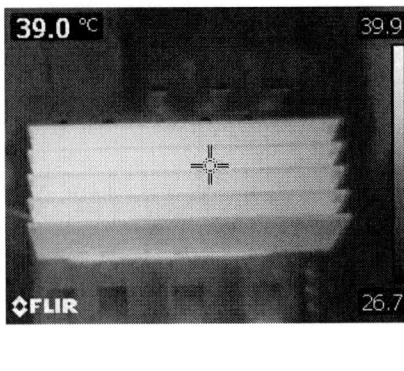

(a)   (b)

Fig. 9: (a) Measured GaN inverter efficiency including losses of sine-output filter at different output power and (b) thermal image of inverter heatsink at 1.5 kW output power

with pure sinusoidal output waveforms was successfully developed. The functionality of the GaN inverter at different operation points was shown in experimental results. The maximum efficiency of the system, consisting of GaN inverter and sine-output filter, is 97 %.

References

[1] N. Kaminski. State of the art and the future of wide band-gap devices. In *Power Electronics and Applications, 2009. EPE '09. 13th European Conference on,*, pages 1–9, 2009.

[2] J. Millan, P. Godignon, X. Perpina, A. Perez-Tomas, and J. Rebollo. A Survey of Wide Bandgap Power Semiconductor Devices. *Power Electronics, IEEE Transactions on*, 29(5):2155–2163, 2014.

[3] K. Shirabe, M. Swamy, Jun-Koo Kang, M. Hisatsune, Yifeng Wu, D. Kebort, and J. Honea. Advantages of high frequency PWM in AC motor drive applications. In *Energy Conversion Congress and Exposition (ECCE), 2012 IEEE*, pages 2977–2984, 2012.

[4] K. Shirabe, M. Swamy, Jun-Koo Kang, M. Hisatsune, Yifeng Wu, D. Kebort, and J. Honea. Efficiency Comparison Between Si-IGBT-Based Drive and GaN-Based Drive. *Industry Applications, IEEE Transactions on*, 50(1):566–572, 2014.

[5] J. Lautner and B. Piepenbreier. Analysis of GaN HEMT switching behavior. In *Power Electronics and ECCE Asia (ICPE-ECCE Asia), 2015 9th International Conference on*, pages 567–574, 2015.

[6] GaN Systems. GS66508P: 650V E-mode GaN transistor; Preliminary Datasheet. 2016.

[7] A. Wintrich, U. Nicolai, W. Tursky, and T. Reimann. *Application manual power semiconductors*. ISLE-Verlag, 2nd edition, 2015.

© VDE VERLAG GMBH · Berlin · Offenbach

PCIM Europe 2016, 10 – 12 May 2016, Nuremberg, Germany

A Large Input Voltage Range 1 MHz Full Converter with 95% Peak Efficiency for Aircraft Applications

Nicolas Quentin, University of Lyon/Ampere/Safran Group, France, nicolas.quentin@sagem.com
Remi Perrin, INSA Lyon/Ampere, France, remi.perrin@insa-lyon.fr
Christian Martin, University of Lyon/Ampere, France, christian.martin@univ-lyon1.fr
Charles Joubert, University of Lyon/Ampere, France, charles.joubert@univ-lyon1.fr
Louis Grimaud, Safran Group, France, louis.grimaud@sagem.com
Rolando Burgos, CPES, USA, rolando@vt.edu
Dushan Boroyevich, CPES, USA, dushan@vt.edu

Abstract

This paper presents a design methodology of a $50\ W$ isolated DC/DC converter serving as a power supply for aircraft equipment like a FADEC (Full Authority Digital Electronics Control). The particularity of this work is the design of a full converter regarding two antagonistic requirements which are a wide input voltage with an high efficiency. To fulfill those requirements an optimization regarding three different levels is done: 1- A suitable topology for wide input voltage range, 2- Proper technologies for high frequency operation and 3- A converter architecture to minimize the surface. A $50\ W$, $18 - 80\ V$ input voltage, $1\ MHz$ switching frequency converter is built using GaN transistors and a planar transformer. The proposed converter operates under optimized soft-switching conditions in order to minimize switching losses and reached a peak efficiency of $95\ \%$. This converter is compared with 2 other converters: an hard-switching $400\ kHz$ Flyback topology which is the industrial standard equipment and an $1\ MHz$ soft-switching topology without the architecture optimization.

1. Introduction: Industrial problem in an aircraft application

The power supply system in an onboard aeronautical application represents an important contribution to the recurring cost, global efficiency and the volume of the system. A conventional design approach [1], [2] demonstrates the significant impact of the variation of the input voltage on the design. In avionics applications, the regulator has to face large input voltage variations by a factor of six to eight, whereas this variation is more as a factor of two to three for other industrial applications.

The variation of the power supply input voltage is a major constraint because as a first approach, the size of the converter reflects

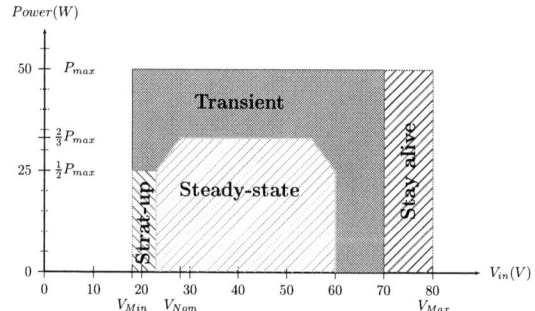

Fig. 1: Operating area of the wide input voltage converter

the envelope of the maximum current and maximum voltage it has to deal with. As shown in Fig. 1, for safety reason most of the converters have an operating range oversized to withstand

© VDE VERLAG GMBH · Berlin · Offenbach

the different operating modes: start-up, steady-state, transients, stay alive (extreme transients). The regulation purpose makes it very difficult to improve the steady-state efficiency. This fluctuation creates a technical obstacle that manufacturers usually circumvent by using an association of converters in order to limit electrical and feedback control constraints on each power stage of the system. This combination, if done adequately, allows to limit power losses over the full operating range but increases the size and control complexity of the overall system.

All the equipments located inside an aircraft network have to respect several requirements imposed by international standards such as the DO160 or manufacturer customs like the derating/stress on a component. One of the most constraining standard is within the Electromagnetic Interference (EMI) where the converter does not disturb or is disturbed by other equipments on the electrical network.

To reduce the converter size, the general trend is the switching frequency increase but this implies losses increase. Besides, power supplies are installed in an confined environment. They cannot be cooled down by forced convection in the case of the most extreme applications (operating temperature from $-55°C$ to $+110°C$). In this condition, power supply losses impact the converter volume and weight to prevent it from overheating. Therefore, a significant gain in efficiency is the main purposes, since it results in a reduction of weight and volume of the equipment.

2. Overall Description: A full converter

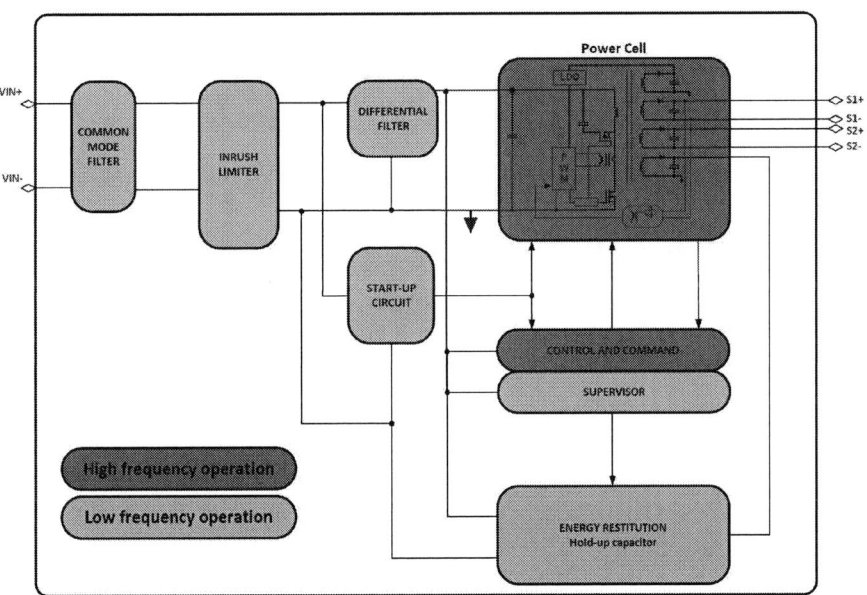

Fig. 2: Synoptic of the overall converter including autonomous and safety functions

When fed by a DC bus in the 18 to $80\ V$ input voltage range, the power cell provides two $15\ V$ isolated outputs to power all standard electronic equipments. As presented in Fig. 2, the work is not limited to the power cell and filters which represent only 44.5% of the components size area. This paper deals with the design and implementation of the full converter. It includes all auxiliary functions required for the safety and autonomous operation. The table below summarizes the main functions.

Tab. 1: Functions description and size

Functions	Description	Components number / area ratio
Power cell	Provide two isolated voltages from primary power bus to user.	67 / 27.1%
Start-Up Circuit	Provide energy to the PWM controller in order to start the DC-DC converter.	17 / 5 %
Filters: CM, DM and output	Limit voltage ripple on the power bus.	47 / 17.4 %
Inrush limiter	Limit inrush current during the start-up.	15 / 6.7%
Energy Restitution	Withstand input power interruption with an hold-up capacitor.	55 / 20 %
Supervisor	Protect against output overload and short-circuit and input under voltage.	57 / 10.6 %
Control and command	Control the duty cycle of the switches in order to control the outputs voltages. A current mode PWM and a driver stage compose this function.	79 / 13.2%

3. Design consideration

As mentioned previously, in order to provide a compact and efficient converter, 3 lines of work have been investigated.

3.1. The soft-switching topology suitable for wide input voltage

The goal is to use a topology which creates the capacity to increase the switching frequency in order to reduce the passive elements. Soft-switching is a reasonable technique to increase the switching frequency and limit the power losses at the same time. Fig. 3 shows the simplified schematics (power cell only) of the Flyback Active-Clamp circuit [3]. This topology is suitable for wide input voltage applications with a step-down and step-up transfer function depending only on the duty cycle α:

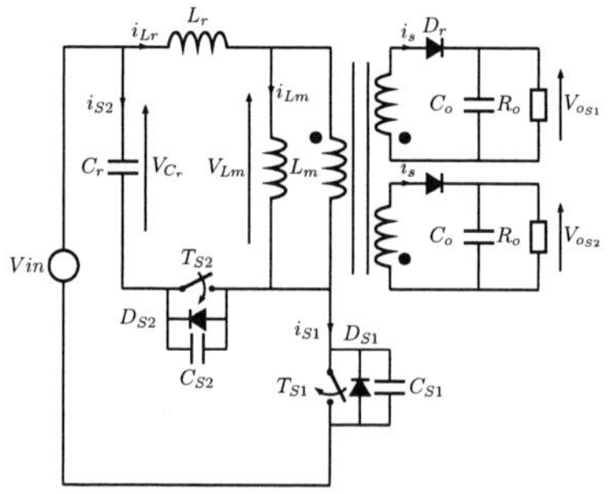

Fig. 3: Double Outputs Flyback Active-Clamp topol[o]gly

$$\frac{V_o}{V_{in}} = \frac{n\alpha}{1-\alpha} \qquad (1)$$

when α is low $\frac{n\alpha}{1-\alpha} \approx n\alpha$, when α is high $\frac{n\alpha}{1-\alpha} \approx \frac{n}{1-\alpha}$. This topology has two additional benefits for a low power application: a low number of components and a simple control (duty cycle at fixed frequency).

Regarding the efficiency, as shown in Fig. 4 this topology can achieve Zero Voltage Switching (ZVS) at the primary side allowing the converter to increase the switching frequency compared to a conventional hard-switching converter such as the classical Flyback converter.

ZVS condition consists in discharging the parasitics capacitances of both $S1$ and $S2$ switches to allow the body diode to naturally be forward biased and then conducts the power current. In the Flyback Active-Clamp topology, the energy necessary to provide ZVS on $S1$ and $S2$ come from 2 different sources. For $S1$, ZVS occurs by using the leakage inductance energy, $E_{L_r} = \frac{1}{2}L_r(\Delta i_{Lr})^2$, to discharge C_{S1} and charge C_{S2} during the dead time $[t_3\text{-}t_4]$. ZVS condition for $S2$ is satisfied when the magnetizing current, $i_{Lm}(t_1)$, is large

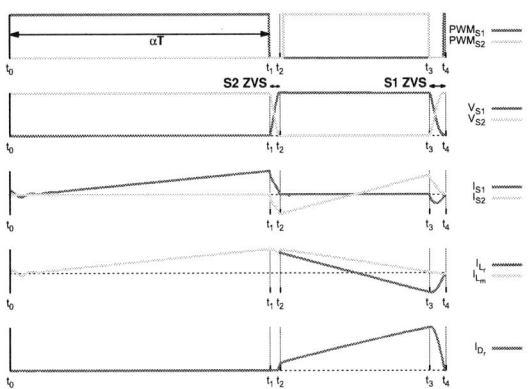

Fig. 4: *Main waveforms in steady-state operation*

enough to charge and discharge the drain-source parasitics capacitances during the deadtime $[t_1 - t_2]$. Considering that the power flow is positive, the current $i_{Lm}(t_1)$ is always able to discharge C_{S2}. Therefore the switching of $S2$ is always soft.

3.2. The technologies dedicated to an high frequency operation

This converter includes advanced technologies dedicated to high frequency and high efficiency operation with the use of GaN transistors from EPC (EPC2010C; $200\ V$ - $22\ A$) and planar transformer.

In high frequency applications, it seems clear that using GaN transistor creates a real improvement of the efficiency [4, 5] due to its low output charge, Q_{oss}. In a soft-switching topology the device output charge has an important impact on the energy required to achieve ZVS condition providing a larger power storage and transfer period means naturally a higher efficiency.

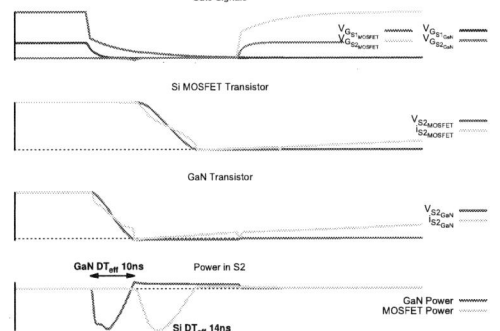

Fig. 5: GaN transistor and Si MOSFET comparison in the Flyback Active-Clamp on S2 commutation

The other GaN transistor benefits are its low total gate charge, Q_g, and its low drain to source on-state resistance, $R_{DS_{On}}$, reducing the drive power (losses) and conduction losses respectively. All these features are implemented in a smaller packaging with less parasitics inductances compared to a Si MOSFET with similar power characteristics.

To illustrate that Fig. 5 shows the current, voltage and the power trajectory of the auxiliary switch ($S2$) in the Flyback Active-Clamp topology for a traditional Si MOSFET (IPD320N20N3 From Infineon) and the selected GaN transistor. Fig. 5 highlights that for the same operating point, due to a lower drain to source capacitance, gate charge and voltage. GaN transistor reduces the effective dead-time DT_{eff} and thus it increases of 2% (1% for each commutation) the effective duty cycle allowing an higher amount of power transfered to the secondary. Nevertheless, this figure also shows one of the GaN transistor drawback in a soft-switching topology which is a higher dead time reverse voltage with a drop of about 2 volts due to the reverse conduction. Where the conduction losses during the dead time are more important than for a Si MOSFET.

In an isolated converter, the transformer is the key component which has to be designed properly because of its important impact on the efficiency and on the global volume of the power cell. For a soft-switching converter, the planar transformer allows to increase the efficiency and the reliability [6] of the converter due to its very low profile, excellent thermal management and the good reproducibility. This last criterion is very important since both magnetizing (L_m) and leakage (L_r) inductances are used to achieve the power transfer and the ZVS conditions.

In a full converter, the transformer has several outputs: two $15 \ V$ - $25 \ W$ power outputs, one $15 \ V$ - $0.75 \ W$ to supply the control and command part and one $45 \ V$ - $3 \ W$ to charge the hold-up capacitor. Fig. 6(a) shows the transformer architecture and Fig. 6(b) the main PCB characteristics.

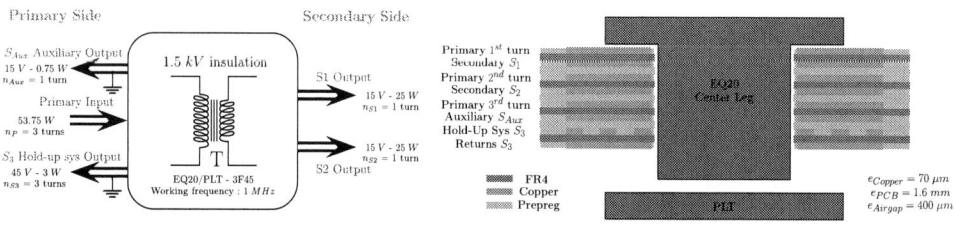

(a) Windings distribution   (b) PCB characteristics

Fig. 6: Planar transformer specifications

8 layers $70 \ \mu m$ PCB have been used to design windings and specific care has been taken on DC resistance in order to limit self-heating. To reduce the leakage inductance, the primary turns are interleaved with the 2 secondaries. Note that the interleaving windings (primary and secondaries) requires to increase the dielectric layers (so the PCB thickness) to still insure the $1.5 \ kV$ insulation requirement. To keep as low as possible parasitic capacitance outside transformer (due to layout considerations), the primary and auxiliary outputs (referred from the primary ground, S_3 and $_{Saux}$) are on one side of the PCB and the 2 power outputs on the other. Regarding the magnetic part, a 3F45 ferrite EQ20/PLT-core from Ferroxcube has been optimized for the application. Fig. 7 presents the shape of the transformer including winding layout used in finite element simulation.

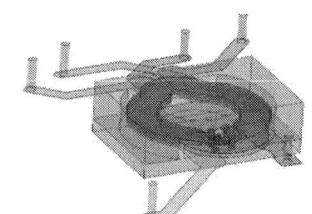

Fig. 7: Planar transformer layout design

3.3. The architecture using 3D assembly strategy

Tab. 1 shows that there are a lot of components in an aircraft full converter where for safety and environmental/thermal requirements all the functions have to be implemented with analog components. This important number of components increases the PCB area and layout complexity.

However, PCB size is not directly dependent on the number of components. For example, the supervisor has a lot of small components whereas the filter has less components but larger like the common and differential mode chokes. However in general, the height of the converter is imposed by the highest component (most of the time the hold-up capacitor or the common mode choke). In order to decrease the surface of the converter, the idea is to realize a 3D module on 2 stages (2 PCB). The goal is to use the free space around the highest component to implement a part of the converter on a second PCB above the first one. As shown in Fig. 8, the separation of the functions on the 2 PCB is done regarding the operating frequency; the low frequency functions (power interruption management system, supervisor, current limiter and soft-start) will be on the mother board whereas the upper stage is dedicated to the high frequency functions (power cell, driver, control/command) with custom PCB.

This modular strategy allows to adapt the PCB requirements for each part with less connection issues. For instance, the power stage needs thick copper layers ($35 - 70~\mu m$) in order to decrease parasitics resistances. Whereas the load (digital components) needs high density implementation that imply very thin track (about 5 mils). PCB choice is a trade-off between layer thickness and track width. This approach helps the designer to

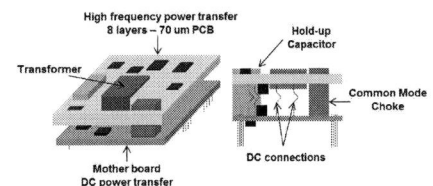

Fig. 8: Schematic representation of the modular PCB Strategy

create a suitable PCB for each function (especially for the planar transformer). Another benefit is the layout simplification of the full converter. To summarize, the goal of this modular architecture is to increase the power density. For a $50~W$ DC/DC converter the frequency limit is imposed by the magnetic components and $1~MHz$ seems to be the upper value for ferrite material (at $50~W$). Therefore, the easiest way to continue to increase the power density is a better use of the space, especially the height available.

4. Experimental results

The $1~MHz$ - $50~W$ prototype of the proposed full converter is shown in Fig. 9(a). The Fig. 9(b) depicts also the operating waveforms with the primary voltages and currents at $28~V$ - $30~W$.

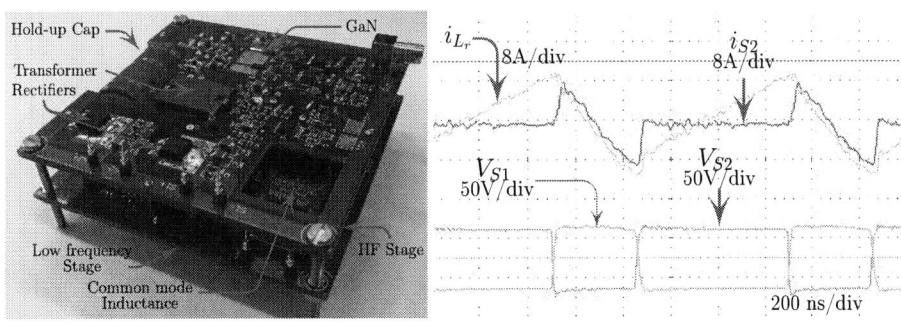

(a) Prototype with GaN: EPC2010C, Transformer: EQ20/PLT 3F45

(b) Experimental waveforms, V_{S1} and V_{S2} (20 V/div and 200 ns/div)

Fig. 9: Prototype of the proposed full converter

According to Fig. 10(a) which represents the efficiency regarding the load variation for several

PCIM Europe 2016, 10 – 12 May 2016, Nuremberg, Germany

input voltages, the efficiency approaches 92% at the nominal input voltage ($28\ V$) and the peak efficiency is 95.5% for an input voltage of $18\ V$.

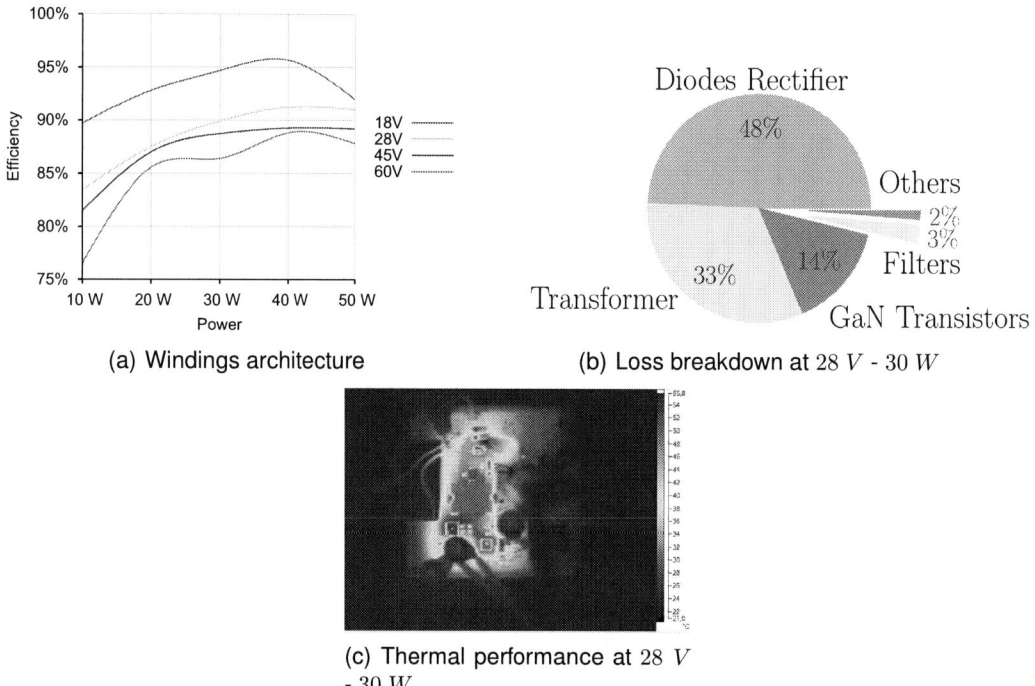

(a) Windings architecture

(b) Loss breakdown at $28\ V$ - $30\ W$

(c) Thermal performance at $28\ V$ - $30\ W$

Fig. 10: Experimental results of the $1\ MHz$ - $50\ W$ Flyback Active-Clamp converter

The thermal analysis has been done at the nominal operating point $28\ V$ - $30\ W$ power after half an hour of operation. The thermal behavior shown in Fig. 10(c) is good with a small temperature increase of about $30\ °C$. The hottest component is the main GaN transistor, $S1$. The temperature increase is due to an important thermal resistance.

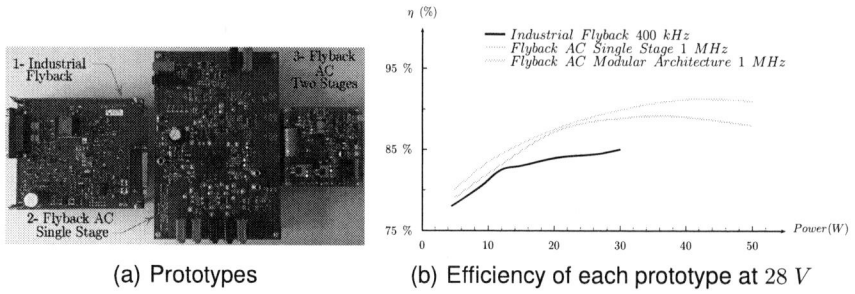

(a) Prototypes

(b) Efficiency of each prototype at $28\ V$

Fig. 11: Prototypes comparison

5. Aircraft full converters comparison

Tab. 2 and Fig. 11 present a comparison between the proposed converter (board # 3) with a conventional industrial Flyback converter (board # 1) and the same converter without the

© VDE VERLAG GMBH · Berlin · Offenbach

modular architecture (board # 2). Regarding this comparison, Fig. 11(b) highlights that the efficiency is improved with a soft-switching topology and the surface is reduced with the modular architecture.

Tab. 2: Prototypes comparison

	Flyback 400kHz-30W Board #1	Flyback AC 1MHz-50W Board #2	Flyback AC 1MHz-50W Board #3
Surface (without connector)	$168\ cm^2$	$184\ cm^2$	$72,25\ cm^2$
Cost	+	++	+++
efficiency between $\frac{P_{Max}}{3}$-P_{Max}	82-84 %	87-90 %	88-95 %

6. Conclusion

This paper presents a $50\ W$ full converter operating at $1\ MHz$ switching frequency for a wide input voltage range in aircraft applications. The use of the soft-switching topology coupled with suitable technologies allows the converter to increase the efficiency and frequency at the same time. At the end, the converter is efficient and compact due to the components and modular architecture. This solution creates a reduction of $60\ \%$ of the surface with a better use of the space and an easier layout compared to a single stage converter with the same schematic.

7. References

[1] Carsten Bruce. Smps topology selection and circuit design tricks. In *PCIM Europe 2002 Seminar notes, Seminar 5*, 2002.

[2] Carsten Bruce. Converter component load factors: performance limitation of various topologies. In *PCIM Europe 88, Munich*, 1988.

[3] R. Watson, F. C. Lee, and G.-C. Hua. Utilization of an active-clamp circuit to achieve soft switching in flyback converters. In *Power Electronics Specialists Conference, PESC '94 Record., 25th Annual IEEE*, pages 909–916 vol.2, Jun 1994.

[4] A. Lidow D. Reusch, J. Strydom. Improving system performance with egan fets in dc-dc applications. In *International Microelectronics Assembly and Packaging Society (IMAPS)*, 2013.

[5] A. Lidow D. Reusch, J. Strydom. A new family of gan transistors for highly efficient high frequency dc-dc converters. In *Applied Power Electronics Conference and Exposition (APEC), 2015 IEEE , vol., no., pp.1979,1985*, 2015.

[6] M.A.E. Ziwei Ouyang; Thomsen, O.C.; Andersen. Optimal design and tradeoffs analysis for planar transformer in high power dc-dc converters. In *Power Electronics Conference (IPEC), 2010 International , vol., no., pp.3166,3173*, 2010.

PCIM Europe 2016, 10 – 12 May 2016, Nuremberg, Germany

Integrating Depletion-Mode SiC VJFETs into Production Motor Drives

Michael S. Mazzola, Dept. of Electrical and Computer Engineering, Mississippi State University, USA, mazzola@ece.msstate.edu

James R. Gafford, Center for Advanced Vehicular Systems, Mississippi State University, USA, gafford@cavs.msstate.edu

Gerald W. Godbold, Hyperion Technology Group, USA, ggodbold@hyperiontg.com

Abstract

A 75-kW three-phase IGBT based motor drive in current production is analyzed and found to have native reliability features similar to that needed to accept normally on SiC JFETs. With modest changes to the OEM drive, vertical-junction SiC JFETs from United Silicon Carbide are successfully integrated into the drive and experimental results show significant improvement in efficiency.

1. Introduction

Much has been written concerning the challenges of integrating normally on (i.e., depletion-mode) SiC junction field effect transistors into half-bridge power electronic converters, such as motor drives. Many efforts focus on making the normally on switch functionally equivalent to the normally off switch by using the cascode technique [1]. Many other methods accept purely normally on SiC JFETs but pair them with self-powered gate drives [2]. And perhaps the most elaborate is a combination of the two called "direct-drive" gate-drive technology from Infineon [3] for use with the lateral-junction normally on SiC JFET.

What is largely overlooked in this literature is that a motor drive designed for normally off devices, principally the silicon IGBT, must have commercial reliability features that prevent closure of a DC contactor during start up if the internal DC bus is shorted. This is because normally off devices can and often do fail closed. Conventional design identifies this single-point of failure as a reliability concern requiring a fail-safe response. Analysis of a 75-kW Yaskawa A1000 commercial variable frequency motor drive in current production reveals an internal DC bus contactor that will not close unless the DC bus capacitors charge. This prevents a shorted inverter from being connected to the main power bus. It can be easily adapted to make a similar motor drive using normally on SiC JFETs have similar reliability and cost as the IGBT drive. This can unlock rapid adoption of SiC JFETs at near cost parity with IGBTs that remains out of reach for SiC MOSFETs.

This paper reports recent work to demonstrate this novel approach including experimental results with a Yaskawa motor drive before and after drop-in replacement of the original equipment manufacturer (OEM) silicon IGBTs with normally on SiC vertical junction JFETs from United Silicon Carbide, Inc. Significant improvement in efficiency was achieved despite the overall high efficiency of the original IGBT motor drive.

2. Description of Drive Modifications

SiC JFET drop-in replacement of IGBTs in an industrial motor drive is accomplished through multiple tasks. These include baseline evaluation of the targeted motor drive platform,

© VDE VERLAG GMBH · Berlin · Offenbach

development of a discrete component based module, a half-bridge module gate drive system, integration of SiC JFET modules into the motor drive, and testing that observed an expected increase in efficiency due to the lower conduction losses resulting from the use of UJN1205K SiC JFETs from United Silicon Carbide. In the course of this research, it was recognized that the existing hardware built into the OEM drive could be exploited to make the motor drive with the drop-in normally on silicon carbide JFET power modules have equivalent reliability as the motor drive with the original normally off silicon IGBT power modules. To see how this is so, the following two sections compare the unmodified drive functional architecture with the same drive modified to accept the SiC JFETs.

2.1. The unmodified drive

The OEM motor drive in its original functional architecture is shown in Fig. 1. Note the mechanical contractor after the three-phase rectifier which acts as a bus short-circuit

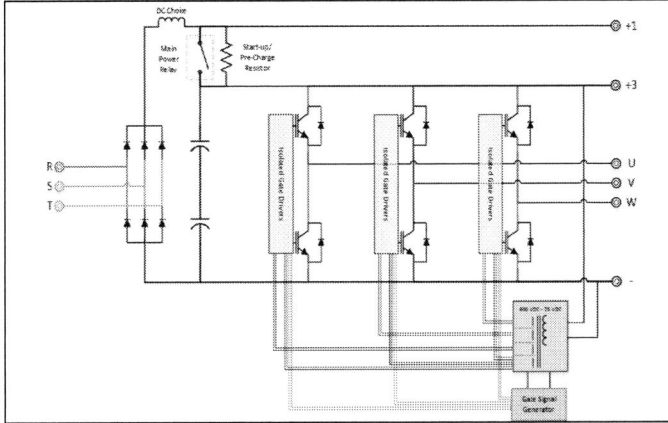

Fig. 1. Unmodified motor drive functional architecture.

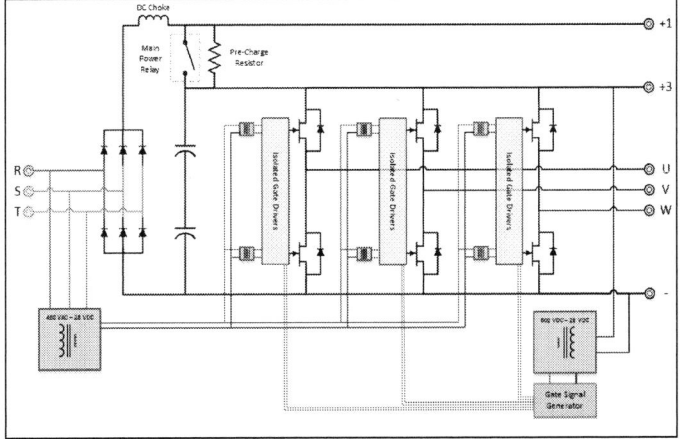

Fig. 2. Modified motor drive functional architecture.

protection device. The power supply provisioning the supervisory controller shown in Fig. 1 is supplied by the 600-V DC bus and does not power up until the bus capacitors charge above a minimum voltage level. This occurs due to a bypass resistor shown across the contactor in Fig. 1. Only after the bus capacitors charge to above the under-voltage lock-out level will the power supply start and close the contactor. At this time a low-impedance path is available to the bus capacitors from the 480-V AC line to supply power to the motor through the three-phase IGBT bridge. If an IGBT should fail, the contactor will open when the bus voltage again falls below the under-voltage lock-out level, thus providing a fail-safe response to a single unit failure of the IGBT bridge. This feature was exploited in the motor drive modification described in the next section.

2.2. The modified drive

Moving the main power supply connection to a second low-current and low-cost three-phase bridge rectifier allows the controller to power up nearly instantly while the bus capacitors are charging. Figure 2 shows this change. When the 480-VAC mains are closed, the JFET gate drivers power up while the main DC contactor remains open and the main bus capacitors charge through the pre-charge resistor. The normally on SiC JFETs quickly assume the correct blocking state before they see a bus voltage. Only when the bus voltage rises due to this blocking action will the drive controller close the main DC contactor, which is the normal function of this controller. In this way, an inherent self-protection feature against a shorted IGBT switch is adopted to allow startup of the drive with normally on SiC JFETs, while retaining the same inherent self-protection of the original drive in case of a failed power transistor.

© VDE VERLAG GMBH · Berlin · Offenbach

3. Experiment

If the motor drive were to be built according to the functional architecture shown in Fig. 2 by the OEM, full integration of normal systems specific to the SiC JFETs would be used. In this work, however, only a one-off modified drive was needed to test the feasibility of modification and to demonstrate the improved efficiency expected from the SiC JFETs. Therefore, a hybrid drive control architecture to accommodate drop-in replacement of the SiC JFETs for the convenience of this laboratory demonstration was employed as described next.

3.1. Laboratory modifications

In the OEM system the gate drive board, in addition to conditioning gate control signals to properly drive the gates of IGBT modules, is responsible for generating dead-time control and feedback of input and output potentials. If the gate drive board is removed from the gate control signal chain the possibility of commanding both devices in a half-bridge module into the on-state, resulting in catastrophic failure due to shoot-through, is likely. The OEM gate drive is also bonded to Kelvin connections to measure DC bus input potential and output potentials. Rather than adding to the complexity and risk of interpreting the distribution level control signals and incorporating dead-time control into the JFET module gate drive, the OEM gate drive was incorporated into the drive signal chain. However, this decision requires that the OEM gate drive outputs be conditioned for the JFET module gate drive input.

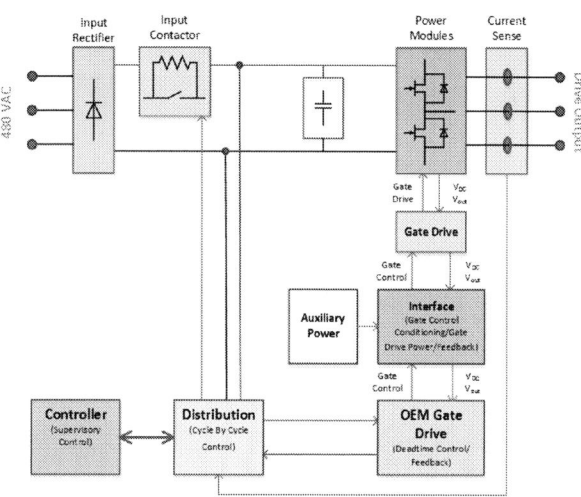

Fig. 3. Modified motor drive with hybrid functional architecture. The auxiliary power subsystem is powered by the 480-VAC bus through a parallel connection as shown in Fig. 2.

As shown in Fig. 3, the laboratory modified A1000 drive includes all OEM subsystems while adding three additional subsystems. A simple signal interface subsystem was developed to integrate the OEM gate drive with the JFET gate drive. This subsystem is powered by an auxiliary power supply. This system ensures that each of the three JFET gate drives provides adequate reverse bias to all modules when the motor drive is in a quiescent state. Since critical system control and protection features are not managed by a single subsystem but are distributed in the original system this integration of SiC JFET modules was successful due to maintaining the native signal chain to a high degree. The approach highlights the requirements imposed by actual products such as a production industrial motor drive.

Figure 4 shows drop-in replacement modules being installed into the laboratory modified A1000 motor drive. The new power modules are based on two normally on UJN1205K SiC JFETs plus one UJ2D1230K Schottky diode per switch position, both supplied by United Silicon Carbide, Inc. In the left image, two modules are shown installed over a thermally conducting but electrically isolating silicon pad which is not needed in the OEM configuration because the IGBT modules are internally electrically isolated from the heat sink, whereas the

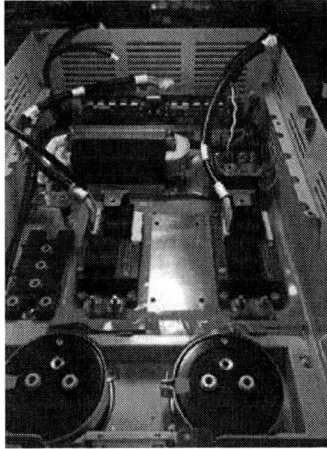  

Fig. 4. Installation of the SiC discrete modules with JFET gate drives into the A1000 Motor Drive

cases of the TO247 JFET and SBD are not. (Phase II of this project will use custom modules supplied by United Silicon Carbide that were designed to replace the IGBT modules without need of external electrical insulation.) The center image shows all three modules installed on the A1000 heat sink. The right image shows the A1000 OEM control cards installed in their normal location over the power semiconductor modules. As described earlier, the OEM gate drive boards are still used in this laboratory modification.

3.2. Experimental Setup

The SiC-switched A1000 motor drive was reinstalled into the motoring dynamometer test stand shown in figure 5. The A1000 powers a 50-HP induction motor that drives a brushless

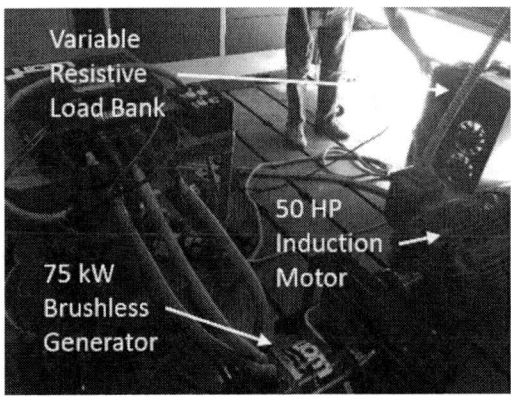

Fig. 5. Motoring dynamometer test stand.

DC machine operated as a generator through a 1.8:1 gear box. An adjustable resistive load bank dissipates the electrical power produced by the induction motor with the DC generator, which permits fixed, repeatable control of the electrical load seen by the A1000 at the output terminals. A Yokogawa WT1600 power analyzer utilizing the two-watt meter method measures both input and output power to the A1000 so that overall efficiency of just the A1000 can be accurately measured to within the metrological precision of the power analyzer and its associated transducers. The estimated resolution is limited to efficiencies up to about 99%. A Tektronix TDS5054B digital oscilloscope monitors the output voltage and current waveforms to confirm proper operation of the motor drive.

3.3. Experimental Results

Although the capacity of the discrete modules shown in figure 4 is up to four 50-mΩ SiC JFETs in parallel (12.5 mΩ equivalent switches at room temperature), the results given in this summary are for the case where two SiC 50-mΩ JFETs in parallel per switch position were installed in the A1000 motor drive. Thus, the efficiency results reported here represent the use case in which a low-cost option is obtained by reducing the number of SiC switches. As will be seen, the 25-mΩ equivalent switches at room temperature outperformed the more heavily rated IGBT modules at up to about half rated power of the motor drive. (In contrast, the Phase II SiC JFET integrated power modules exceed the rating of the original IGBT modules and are expected to significantly improve efficiency at all power levels.)

Figure 6 shows the efficiency curves for the SiC-switched A1000 motor drive measured at constant speed for three different induction motor synchronous speeds. The control of the A1000 implements a V/f type relation where the higher the speed, the higher the RMS output line-to-line voltage driving the stator of the induction machine. So, for the same power the line currents are lower at higher shaft speed and the efficiency of the A1000 is expected to be correspondingly higher. This is born out in both the SiC-switched and the IGBT-switched versions of the A1000. This observation reinforces an original assumption when planning this project, namely, that the conduction losses of the power semiconductor switches would dominate the semiconductor-related loss budget due to the low switching frequency of the motor drive. As a result, it was expected that at light load the SiC switched A1000 would likely show a higher efficiency and this is clearly the case with the experimental results. In figure 6, the scatter of the data points is more than with the baseline IGBT results probably due to the fact that the computed efficiency typically exceeds 98.5%, which is approaching the resolution of the power analyzer. In contrast, the best curves with the unmodified motor drive switched with silicon IGBTs

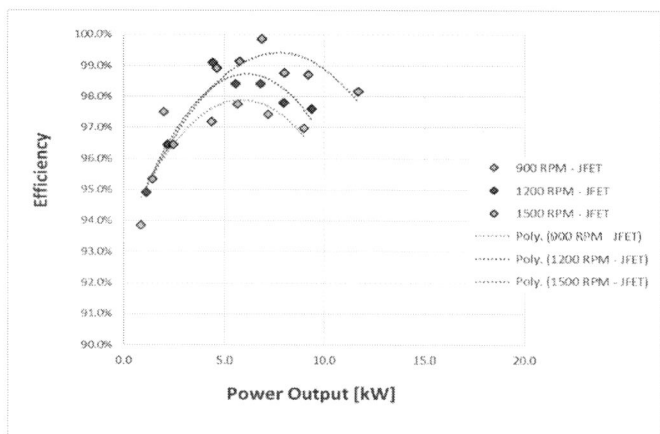

Fig. 6. Experimental results with the SiC-switched A1000 motor drive in terms of steady state average power efficiency measured from the three-phase input terminals to the three-phase output terminals of the A1000 drive at constant synchronous motoring speed.

(1500 and 1800 RPM) peak at about 98%, and at the lower power levels fall to significantly less than 98%. At 900 RPM, the peak efficiency of the motor drive modified with SiC JFETs is nearly 98%, as compared to about 96.5% in the unmodified drive under similar conditions.

A second feature of both the SiC and the IGBT switched curves is a classic single-peaked efficiency profile. The peaks of the curves in figure 6 are not as easily identified as they are for the IGBT curves due to the scatter. But an estimate was made using the quadratic polynomial trend lines shown in Fig. 6. The location of the peak as a function of output power for each of the three speeds is increasing with speed, as expected from the V/f control imposed by the motor drive. (It is also expected that with the Phase II modules, these peaks will be pushed out to higher power giving the Phase II SiC switched drive the likelihood of higher efficiency over all power ranges.)

4. Conclusions

Inherent to the design of Si IGBT based commercial motor drives are features that protect against the possibility of applying input power into a short circuit. These features can be adopted to safely integrate normally on SiC JFET power modules with little change to system architecture and while maintaining the same commercial standards of reliability as the other options (such as SiC MOSFETs). As a result, performance enhancements can be realized by adopting SiC power JFETs without negatively impacting unit cost by significantly redesigning the motor drive architecture.

5. References

[1] John Bendel, "*Cascode Configuration Eases Challenges of Applying SiC JFETs in Switching Inductive Loads*," How2Power Today, available on-line in issue August 2014.

[2] Michael S. Mazzola and Robin L. Kelley, "*Half-Bridge Circuits Employing Normally ON Switches and Methods of Preventing Unintended Current Flow Therein*," U.S. Patent 7,602,228 issued 13 Oct. 2009, and related divisional patents 7,907,001 and 8,456,218.

[3] Karl Norling, Christian Lindholm, and Dieter Draxelmayr, "*1st Commercial SiC JFET Driver for DirectDrive JFET Topology*," Proc. PCIM Europe, paper 57, pp. 452-457, 8-10 May 2012.

6. Acknowledgement

This work was sponsored by the Air Force Research Laboratory under contract FA8650-13-C-2349 (prime contractor: United Silicon Carbide, Inc.). The author's gratefully acknowledge the provision of silicon carbide devices by United Silicon Carbide, Inc.

Higher Light Efficacy in LED-Lamps by lower LED-Current

R. Jaschke; K.F. Hoffmann
Helmut Schmidt University / University of the Federal Armed Forces Hamburg
Faculty of Electrical Engineering, Power Electronics
Holstenhofweg 85, 22043 Hamburg, Germany
Email: reinhard.jaschke@hsu-hh.de

Abstract

About 20% of worldwide electric energy is consumed by lighting systems. LED-lamps in future have the biggest part and so it is important that they work with highest efficacy [3]. With LED the energy wasting of traditional lighting technologies is obsolete. In 2015 in the households most LED-lamps less 25 watt at AC line voltage produce only 70 Lumen per Watt. Some small power filament LED lamps in bulb shape until 6 Watt give 120 Lumen per Watt. If we reduce the LED-current on the halve value we get a bigger efficacy of 150 Lumen per Watt. In the laboratory we get measurements of luminous intensity in an Ulbricht sphere in a temperature chamber from -20°C to 60°C. The presented measurement results verify this theory.

1. Introduction

In the 2014 news we can read "Cree First Break 300 Lumens-Per-Watt Barrier". In figure 1 we show the Cree leading innovations from 2006 with ε_{2006} = 130 lm/W until 2014 with ε_{2014} = 303 lm/W at a color temperature of 5150 K and current 350 mA by standard room temperature in research laboratory. In 2010 CREE announces that one LED produced 208 lm with ε_{2010} = 208 lm/W at a blue color temperature 4579 K and current 350 mA at room temperature. The low LED-voltage is calculated by U_{LED} =1W/0.35A =2.857 V. The luminous-current factor is κ_{2014} = Φ_{LED} / I_{LED} = 303lm/350mA = 0.86 lm/mA. For a monochromatic light of 555 nm (green) it gives the maximum luminous efficacy ε_{max} =683 lm/W.

The reality market efficacy values in 2006 with 50 lm/W and 2015 with 120 lm/W at color temperature 2700K include the losses of power electronics, reflectors and encapsulation.
In [4] gives a single-die warm white LED: 200lm, 0.7A, 2.9V with a luminous-current factor κ_{Cree}=Φ_{LED} / I_{LED} = 200lm/700mA=0.29 lm/mA and a luminous efficacy ε_{Cree}=Φ_{LED} /($I_{LED}U_{LED}$) = κ_{Cree} /U_{LED} = 290 lm/(2,9V*A) = 100 lm/W and specified tolerance 7%.

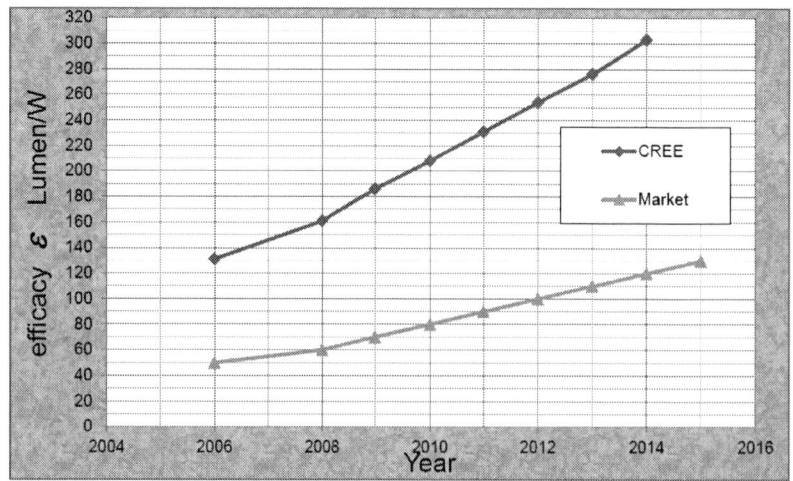

Fig. 1: LED-light innovation in laboratory and market in 10 year range

2. Measurements with filament LED

In figure 2 we see a filament LED lamp in historic light bulb optic with 108 filament LEDs in four strings. The driver circuit can be a linear regulator transistor [1] in figure 3 to produce a constant LED current. The electronic circuit can also be a flyback converter in figure 4. The environmental chamber in figure 5 can artificially replicate the conditions of temperatures, relative humidity or cyclic variations.

In figure 6 we measured only the light emitting diode characteristic $I_{LED} = f(U_{LED}, \vartheta_{LED})$ of one filament LED string of the 0,8W series with 27 LEDs on COB. We see the typical NTC-characteristic. In figure 7 we measured only the electrical LED power $P_{LED}=U_{LED}*I_{LED}$ from one filament LED string of the 0,8W series. Simultaneous measured is the luminous flux Φ_{LED} in an Ulbricht sphere [2] with the maximal efficacy $\varepsilon_{LED0.4W} = \Phi_{LED}/P_{LED} = 130$lm/W. For a LED-Power $P_{LED} = 0.8$W we have an efficacy $\varepsilon_{LED0.8W}=120$lm/W at 35°C. In figure 8 we show the luminous flux Φ_{LED} over the electrical LED current I_{LED}. There is an interesting current factor κ_{LEDn} for one filament LED $\kappa_{LEDn}=\Phi_{LED}* U_{LED} /(n\ P_{LED}) = \varepsilon_{LED}\ U_{LED} /n = 130$lm/W*71V/27) = 0.34 lm/mA. This current factor is nearly the same in CREE first milestone LED dies $\kappa_{Cree2010} = 0.29$ lm/mA in the introduction. In the year 2014 it is three times higher $\kappa_{Cree2014}= 0.86$ lm/mA, but we don´t know the lifetime and junction temperature. The important rule more efficacy with less LED power is shown in figure 9 and figure 10. With half LED power we get 8.3% more light energy. The decrease temperature with less LED power in figure 10 is not the mean reason for this efficacy droop, but quantum physics have an Auger recombination theory.

Fig. 2: Historic light bulb optic with 108 filament LEDs in four strings at 230 V AC

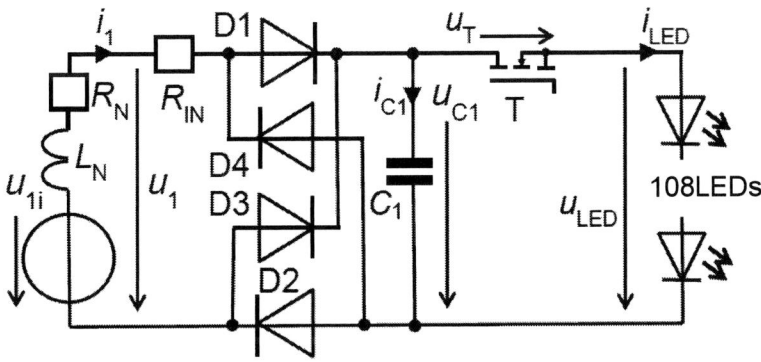

Fig. 3: Circuit of a 4 Watt filament LED lamp with linear regulator transistor at 230V AC

PCIM Europe 2016, 10 – 12 May 2016, Nuremberg, Germany

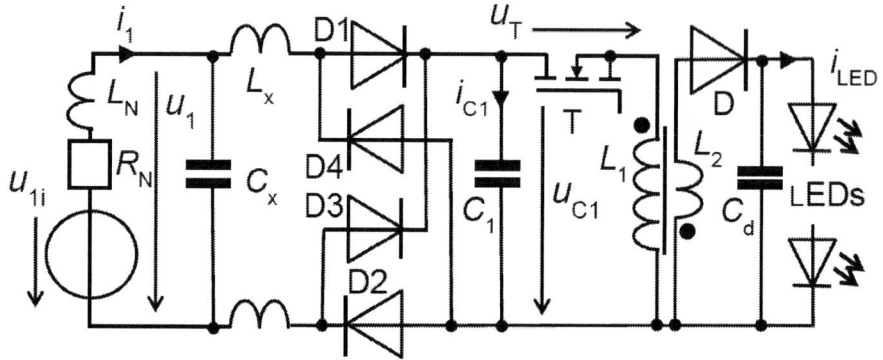

Fig. 4: Circuit of a 5 Watt filament LED lamp with flyback converter

Fig. 5: Chamber for measurement the luminous efficacy depending on temperature

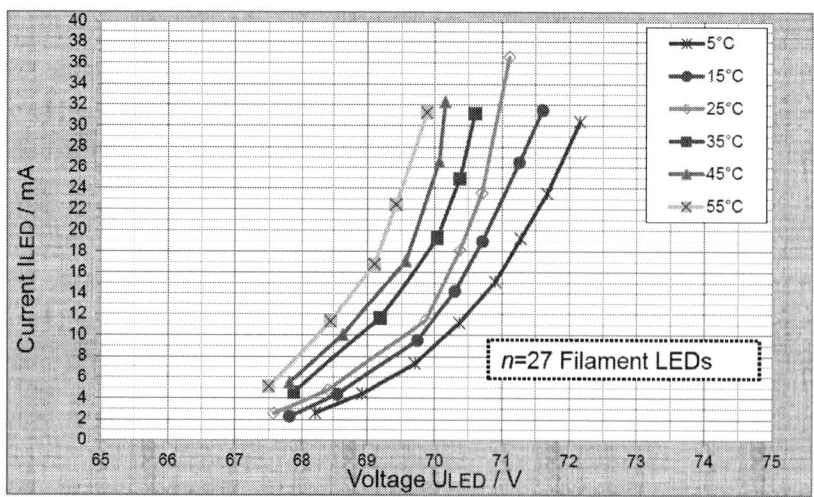

Fig. 6: LED-characteristic of a filament string with 27 LEDs and different temperatures

© VDE VERLAG GMBH · Berlin · Offenbach

PCIM Europe 2016, 10 – 12 May 2016, Nuremberg, Germany

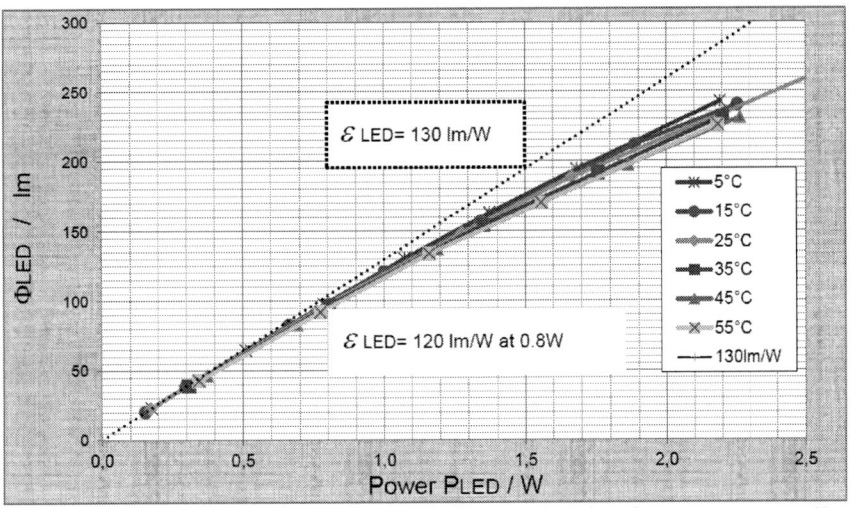

Fig. 7: Light measurement of a LED filament string lumen over watt

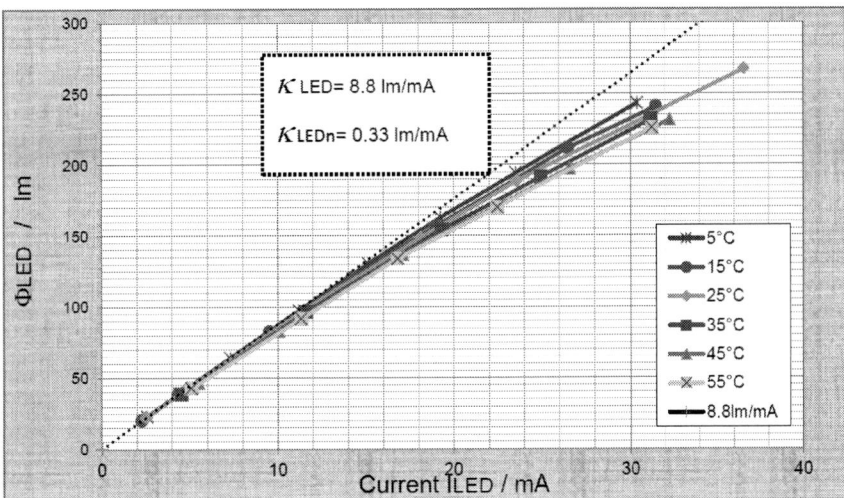

Fig. 8: Light measurement of a LED filament string lumen over current in mA

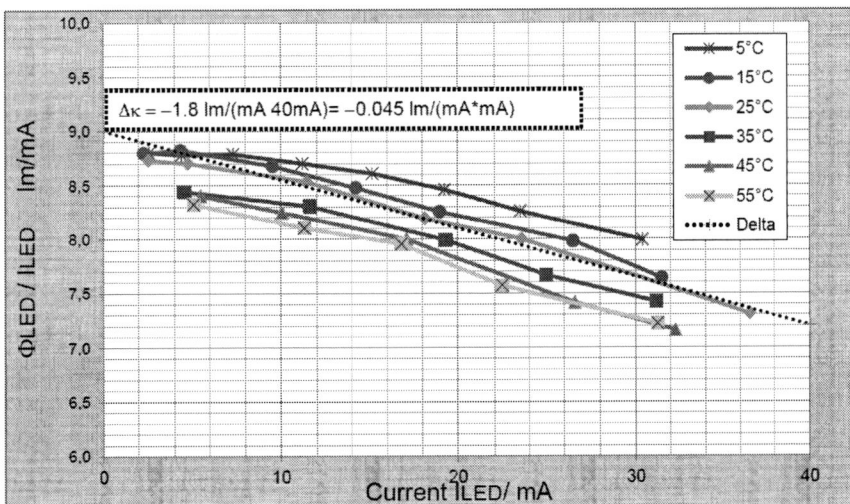

Fig. 9: Luminous efficacy lm/W of a LED filament string lumen over current in mA

© VDE VERLAG GMBH · Berlin · Offenbach

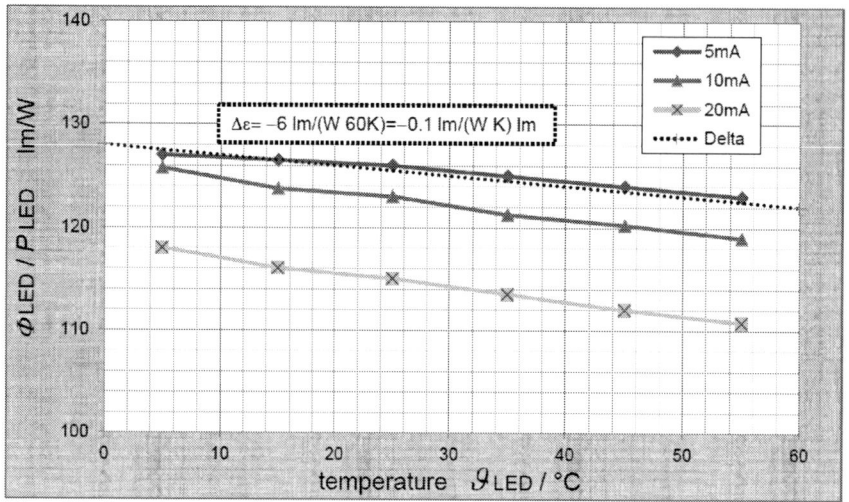

Fig. 10: Luminous efficacy lm/W of a LED filament string over LED-temperature

3. Conclusion

We modified filament LED lamps with 4 strings. Powered it by half LED-current and got nearly a luminous efficacy from 120 lm/W to 130 lm/W. Also we can work with parallel LED-strings and more LED in one string for a better adaption of the linear regulator to 230V AC. The measurements show, that lamps from different distributors with different binning specifications have efficacies up to 150 lm/W.

The aim to save energy with higher light efficacy we have reached with more LED filaments. Also an important contribution is an intelligent light management [5] with motion detectors. An automatic shut-down of unused lamps can save up to 70% energy.

4. References

[1] Jaschke, R.: Better Efficiency with Linear-Regulator compared to Flyback-Converters in LED Lamps at AC Grids. PCIM Asia 2015, 24-26 June 2015, Shanghai, China

[2] Jaschke, R.: Wirkungsgrad und Energieeffizienz-Kennzeichnung für LED-Lampen. NEIS 2015, 10.-11. September 2015, Hamburg, Germany

[3] Jaschke, R.: Höhere Energie-Effizienz bei LED-Licht-Anwendungen durch Kompensation der negativen LED-Temperatur-Koeffizienten. NEIS 2013, 12.-13. September 2013, Hamburg, Germany

[4] Cree XLamp XM-L LEDs, Datasheet, 2010-2015 Cree, Durham, USA

[5] Commission Delegated Regulation (EU) No 874/2012 of 12 July 2012 supplementing Directive 2010/30/EU of the European Parliament and of the Council with regard to energy labelling of electrical lamps and luminaires.

Synchronized Switching and Active Clamping of IGBT Switches in a Simple Marx Generator

Martin Sack, Martin Hochberg, Georg Mueller
Karlsruhe Institute of Technology, Institute for Pulsed Power and Microwave Technology,
Hermann-von-Helmholtz-Platz 1, 76344 Eggenstein-Leopoldshafen, martin.sack@kit.edu

Abstract

For an experimental setup a simple Marx-type pulse generator for pulses in the microsecond range equipped with IGBT switches and a charging path with current-compensated inductors has been designed. To protect the switches against over-voltage active clamping based on combined capacitive and avalanche-diode coupling between collector and gate has been employed. Power supply for the gate driver circuitry and transmission of the trigger signal are inductively coupled to the stages. A current source based on inductors boosts the rise of the trigger signal and, hence, fosters synchronized switching of the stages. So far, a stack of three stages has been assembled and successfully tested.

1. Introduction

For experiments on the application of a pulsed electric field to biological tissue a modular pulse generator has been designed and is currently set up. The application of pulses of sufficiently high electric field strength, pulse length, and energy causes an irreversible permeabilization of the cell membranes [1]. Subsequently, substances can be extracted from the biological tissue more easily. To cause permeabilization, charging the cell membranes to a voltage in the order of 0.5...1 V is required. For pulse application, the material is immersed into a conductive liquid, for example water or juice, to establish the electric contact to an electrode system inside a treatment chamber. Due to the combined series and parallel connection of many cells in a tissue and surrounding liquid, the application of high-voltage is required invoking also a high current flow. Hence, pulses are applied, which are just long enough for membrane charging and subsequent permeabilization.

Pulses can be generated by means of a capacitor discharge. The pulse circuit can be described as a RLC resonant circuit in series connection. For the experiments an aperiodically damped or strongly damped oscillating voltage shape across the load is required [2]. It is achieved by appropriate selection of the circuit elements. Although in most cases a complete discharge of the pulse capacitor is desired, in the case of a short circuit due to a flash-over inside the treatment chamber immediate opening of the pulse switch is required to switch off the current fast. The inductive component of the pulse circuit on one hand reduces the current rise giving the control circuitry and the switch more time for detection of a short-circuit condition and breaking the circuit. However, on the other hand when opening the switches the inductive energy needs to be removed without causing excessive over-voltage across the switches.

2. Design of the Pulse Generator

For the pulse generator a Marx-type topology has been selected. It allows for an easy voltage multiplication at a comparably low inner impedance of the generator. Fig. 1 shows a simplified schematic of one stage of the pulse generator and fig. 2 a photo of one stage module. The charging path comprises current-compensated chokes (L_C) enabling a parallel configuration of the capacitors with comparably low inductance for fast charging. Compared to a charging circuit set up with diodes and semiconductor switches, a charging circuit with passive elements does not require any additional gate-drive circuitry for switches in the charging path. Moreover, it can be designed to have a low DC resistance for low losses during charging. A stage voltage of 1 kV has been selected. The pulse capacitor C_{Stage} has been set up as parallel connection of six capacitors of 1 µF each. The pulse switch T_3 consists of a parallel connection of six IGBTs. Each IGBT is driven by an inexpensive integrated gate driver Dr capable of delivering a driving voltage of either 0 V or 15 V. The gate-drive circuitry of each stage is powered inductively via a transformer consisting of one turn of high-voltage cable as primary. For the photo the high-voltage cable has been removed. When neglecting the transformers' main inductance, such a configuration results in a series connection of the stages' power supplies across the transformers' stray inductance. Hence, the primary winding is powered by a current source.

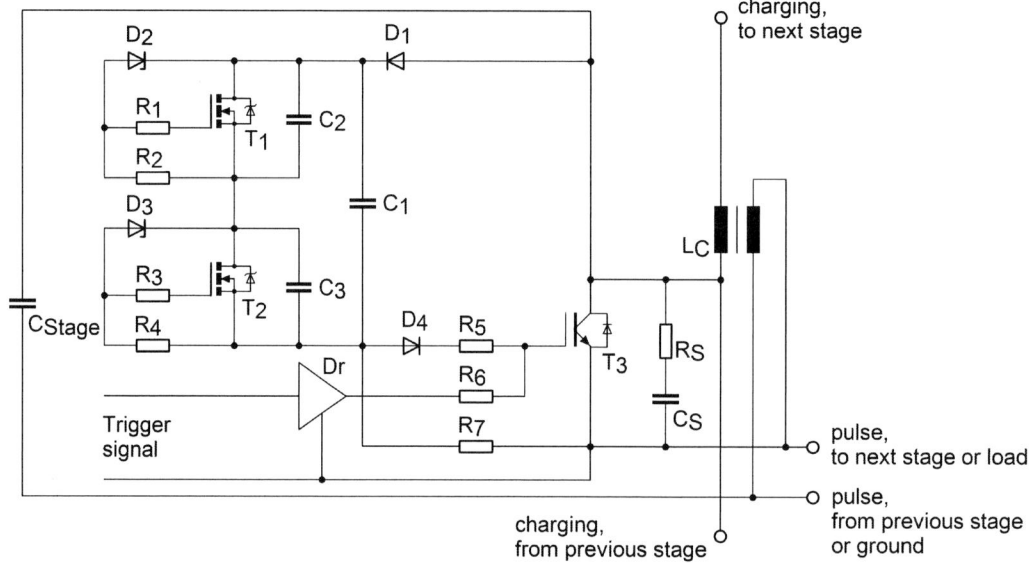

Fig. 1. Simplified schematic of one stage of the Marx generator.

Fig. 2. Photo of one stage module of the Marx generator.

© VDE VERLAG GMBH · Berlin · Offenbach

For pulse generation all stages need to switch synchronized to each other at the same time. The trigger signal is transferred via the same transformer used for supplying the gate drive circuitry. For triggering a single pulse of approximately twice the amplitude of the supply pulses is used. To foster synchronized switching, a steep rise of the trigger signal is of advantage. It reduces the time to reach the threshold level for detecting the trigger signal and, therefore, also the uncertainty for threshold detection caused by parameter variations between the stages and low-level interference. Hence, the current source has been equipped with inductances to compensate for the voltage drop across the stray inductance of the transformers and the loop inductance of the trigger circuit.

Fig. 3 shows the current through the transformers' primary windings I_{Supply} during supply and the application of a 50 µs trigger pulse. Additionally, the voltage at the output of the current source V_{Supply} and the trigger signal V_{Signal} at the stage is shown. For supplying the stages a 50 kHz rectangular current of 1 A peak value is used. For switching a unipolar current of 3 A is fed to the cable. For the duration of this trigger signal the IGBTs are switched on. An inductive voltage drop across the loop of up to 280 V for the trigger pulse has been measured, which is beyond the boundaries of the diagram in fig. 3. As a consequence, especially the trigger signal exhibits a steep rise and decline.

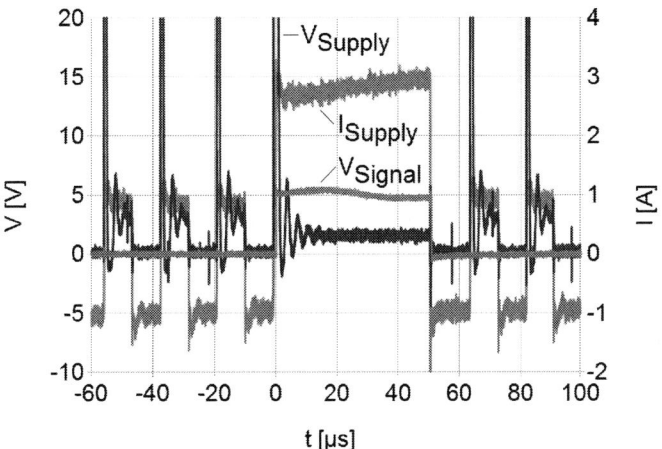

Fig. 3. Supply current and trigger signal.

The lack of free-wheeling diodes across the stages involves the risk of an over-voltage across the switch either, if one switch closes delayed with respect to the other switches or in the case of an inductively driven over-voltage when opening the switch. Therefore, active clamping has been used. The clamping circuit comprises a voltage source in series with a diode between collector and gate of the IGBT switch. It is shown in Fig. 1. The voltage source is set up by the capacitor C_1 in parallel to a stack of avalanche diodes (D_2, D_3) boosted by MOSFETs (T_1, T_2). If the gate driver Dr is off, the gate is connected to emitter potential via the damping resistor R_6. This connection provides fast discharge of the gate, also during voltage clamping. However, the voltage source for clamping needs to deliver a continuous discharge current in addition to the gate drive current. The voltage source might be set up using a capacitor only [3]. However, the capacitor needs to be such large that its voltage does not rise significantly during clamping. Using a stack of avalanche diodes only might be not fast enough [4]. Moreover, the clamping voltage of avalanche diodes exhibits some variation with the current. The use of MOSFETs for boosting the avalanche diode current lowers the current dependency of the clamping voltage significantly. For a parallel connection of six IGBTs, each with a damping resistor of 10 Ω, a gate voltage of approximately 8 V results in a current of 4 A. In the parallel connection of a stack of boosted avalanche diodes with a capacitor the capacitor can have much lower capacitance than without parallel path, as only the higher frequency components pass through it.

PCIM Europe 2016, 10 – 12 May 2016, Nuremberg, Germany

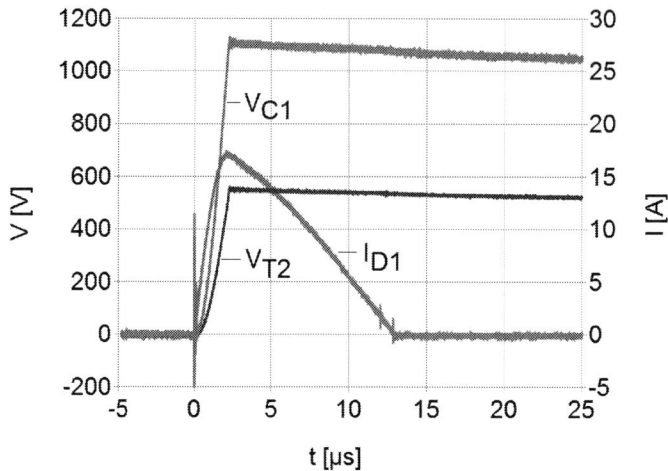

Fig. 4. Voltage and current measurement at the voltage source for voltage clamping.

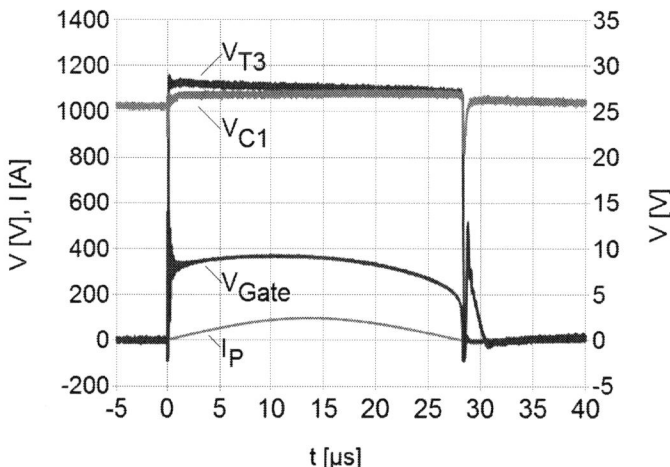

Fig. 5. IGBT connected to a clamping circuit during voltage clamping.

To validate voltage stabilization, resistor R_7 has been shorted and a transient current has been fed through D_1 into the voltage clamp. Fig. 4 shows the voltage across C_1 and T_2 together with the current through D_1. After charging the capacitor C_1 the voltage across it remains nearly constant at 1.1 kV. The voltage is distributed equally between T_1 and T_2. Fig. 5 shows voltage clamping of a single IGBT connected to the clamping circuit. Its gate voltage rises to approximately 9 V for a pulse current I_P of approximately 100 A. The voltage across the IGBT differs from the voltage across C_1 by the sum of gate-voltage and voltage across D_1. For the experiment C_1 has been charged continuously up to the clamping voltage via an external resistor from an external voltage source. For operation in a Marx circuit no additional voltage source is required. C_1 is charged in parallel to the stage capacitor C_{Stage} via D_1 and R_7 in parallel to D_4, R_5 and R_6. The charging current is such low, that the voltage drop across R_6 is far below the IGBT's threshold voltage.

3. Testing the Pulse Generator

A stack of three modules has been tested in a pulse circuit. Fig. 6 shows the voltage at each stage with respect to ground (V_1, V_2, V_3) and the pulse current for the stack connected to a wirewound 5 Ω resistor and operated at a charging voltage of 1 kV per stage when switching on. A sufficiently well synchronization of the switching moment of the stages has been

© VDE VERLAG GMBH · Berlin · Offenbach

achieved. The output voltage of the third stage with respect to ground V_3 rises within 100 ns to 2.5 kV. In comparison to the total charging voltage of the Marx circuit of 3 kV it is reduced by the inductive voltage drop across the inner inductance of the generator, mainly the inductance of its leads. The inductance of the leads of the whole circuit and additionally of the wirewound resistor causes a slower rise of the current. The current exhibits a rise time of approximately 900 ns and reaches its crest value of 540 A after 1.6 µs. This delay is a desired effect. Although as already mentioned the inductance of the pulse circuit is primarily designed for pulse shaping, it allows also for an energy-efficient soft switching of the semiconductors near zero current. Moreover, it allows for a reduced size of the IGBTs' heat sink. This is especially important for repetitive operation of the pulse generator.

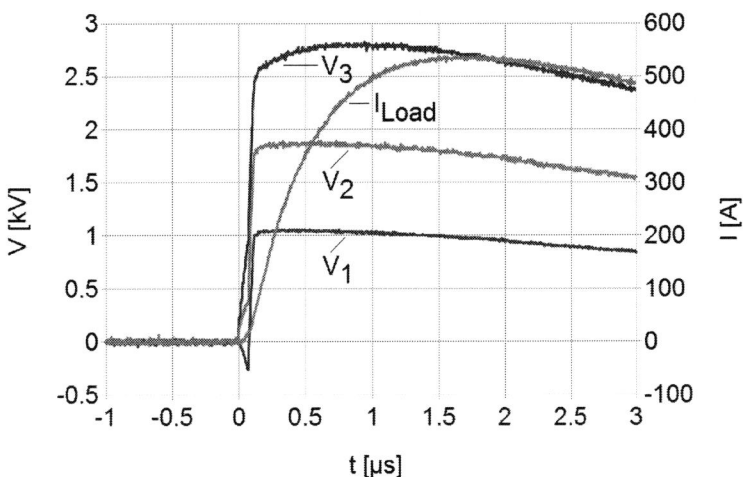

Fig. 6. Stage voltages and load current when switching on.

The first stage exhibits a slightly delayed switching. The reason for this delay is a slightly modified driver circuit. However, delayed switching may also occur due to parameter variations. In series connection of the stages, a RC circuit (R_S, C_S according to fig. 1) in parallel to each IGBT switch helps for transient voltage balancing during switching. Without the RC circuits oscillations have been observed.

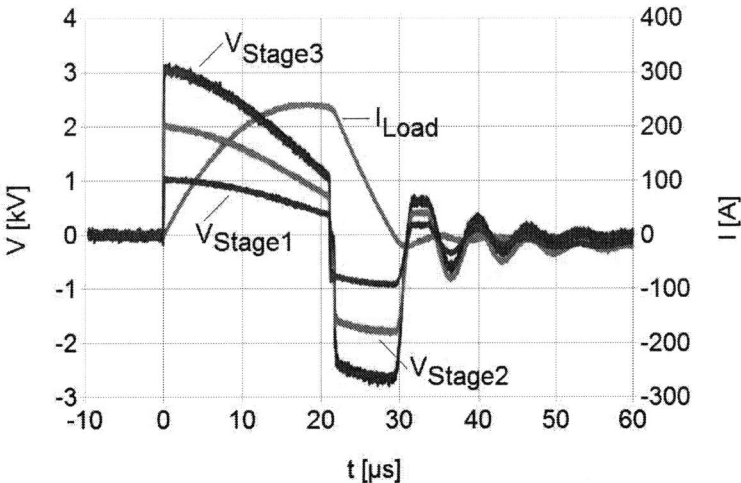

Fig. 7. Stage voltages and load current during discharge and in clamping operation.

Fig. 7 shows active voltage clamping when switching off the load current through a combined resistive and inductive load 20 µs after beginning of the pulse just after the current has passed its maximum. When switching on the IGBTs the current rises sinusoidally and the output voltage of each stage decays accordingly. After opening the semiconductor switches the inductance of the circuit still drives the current. Voltage clamping across the IGBT switches occurs. The voltage across each IGBT is clamped to approximately 1.1 kV. Due to the series connection of stage capacitor and switch, the clamping voltage of opposite polarity with respect to the voltage across the stage capacitor is superimposed to the capacitor voltage. As the stage capacitors are already considerably discharged and are still continued to be discharged by the load current, the output voltage of each stage becomes negative. However, an equal voltage distribution between the stages for both discharging and clamping has been observed. The inductive energy stored inside the inductance of the circuit is transferred to the IGBTs. The load current exhibits a nearly linear decay governed by the essentially constant clamping voltage. A thin aluminum plate serves a transient heat sink during this discharge process. Moreover, it couples the IGBTs thermally.

4. Conclusion

A three-stage Marx generator with IGBT switches and current compensated charging coils has been set up and tested successfully. Synchronized switching of stages and voltage clamping across each stage of the stack has been demonstrated. When softly switching on the stack near zero current a voltage rise within 100 ns to 2.5 kV has been measured. Voltage clamping at a load current of 240 A has been demonstrated. In both cases equal voltage distribution among the stages has been observed. Future work will be devoted to a setup of a larger stack.

References

[1] Weaver J C, Chizmadzhev, Yu A: Theory of electroporation: A review, Bioelectrochemistry and Bioenergetics Vol. 41, 135-160, 1996

[2] Sack M, Keipert S, Herzog D et al.: Design of a PEF Treatment Device for Experiments on Food Preparation, Proc. PPC 2015, May 31-June 4, 2015, Austin, Tx, USA.

[3] Lefranc P, Bergogne D, Planson D et al.: Active clamping of IGBT: capacitor replaces TRANSIL diodes, Proc. EPE 03, 2-4 Sep. 2003, Toulouse.

[4] Hong T, Pfirsch F, Thoben M, Bayerer R: Robustness improvement of high-voltage IGBT by gate control, Proc. PCIM Europe 2008, 27 – 29 May 2008, Nuremberg.

High efficient and lightweight auxiliary power supply with new SiC power device

R.Nakagawa, Y.Fukuda, H.Takabayashi, T.Kobayashi, T.Tanaka
Mitsubishi Electric Corp., Itami Works, 8-1-1, Tsukaguchi-Honmachi, Amagasaki City, Hyogo, Japan
Nakagawa.Ryosuke@cb.MitsubishiElectric.co.jp

Abstract

Efficiency is one of the most important factors for the power supply system. In addition, size and weight are important factors for the onboard railway applications. On the other hand, silicon carbide (SiC), which is characterized by its "low-loss" and "high temperature operation" features, has recently started being used in large power electronics equipment. The latest version of high power Full SiC modules consists of SiC MOSFET and SiC Schottky Barrier Diode (SBD). Authors developed compact auxiliary power supply (APS) for the railway vehicle that has more than 97% efficiency using new 1.2kV SiC power device that includes two features mentioned above.

1. Design concept of auxiliary power supply system

Fig.1 shows the schematic diagram of APS system. APS consists of MOSFET-rectifier, inverter and several filter circuits. It is important that both ACL reactor and ACC capacitor supply precise sine wave voltage to the output of APS. The LC filter passes 50 Hz or 60Hz voltage wave and cuts switching frequency. Power loss in power devices is around 50% and in ACL reactor is also around 30% of total loss of APS.

By replacing conventional type of power device to the latest SiC type, high efficiency is obtained only in MOSFET-rectifier and inverter itself. Furthermore, cut-off frequency of the LC filter can be increased by higher switching frequency with lower loss performance of SiC device. This means capacity of the LC filter can be reduced. Smaller values of ACL and ACC can result in loss reduction and weight reduction.

Table 1 shows basic specification of APS applying new design concept with the latest SiC power module. Fig.2 shows appearance of the developed APS.

2. Concept of APS with 1.2kV Full SiC

With higher switching frequency based on above APS design concept, the ratio of dead time in PWM switching pulse becomes high and the availability of voltage becomes lower. At least 700VDC is necessary for DC link voltage to output 3 phase 415Vrms.

On the other hand, surge voltage is generated between terminals of power module caused by parasitic inductance of power circuit during power module switching. Therefore, the 1.7kV hybrid SiC module consisting of IGBT and Schottky Barrier Diode has been used to decrease loss of power modules in conventional APS system. However, the lower voltage rating device such as 1.2kV has lower voltage drop for same current value.

The new 1.2kV Full SiC power module which can be applied in large power APS has recently become available. Authors developed to use this new 1.2kV Full SiC module to obtain much lower loss by improvement of lower parasitic inductance of power circuit and decreasing surge voltage. Furthermore, higher switching frequency of Full SiC power module results in

© VDE VERLAG GMBH · Berlin · Offenbach

better stability of DC link voltage of the system. It becomes easier to obtain lower DC link voltage against the output voltage without output transformer.

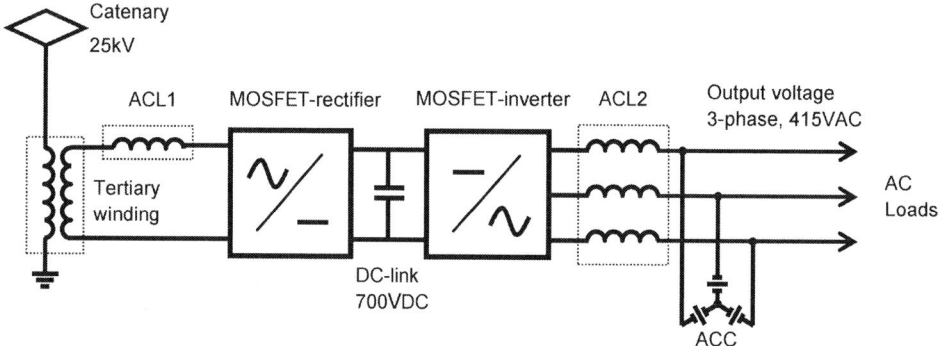

Fig.1 Schematic diagram of auxiliary power supply system

Table 1 Specification of auxiliary power supply

Main circuit	2-level voltage source PWM MOSFET-rectifier + inverter
Output capacity	136kVA
Input voltage	380Vrms
DC link voltage	700VDC
Output voltage	415Vrms
Output frequency	50Hz
Cooling system	Natural air cooling

Fig.2 Appearance of the developed APS

Fig.3 shows the appearance of SiC power module applying to the developed APS. This module has 1.2kV rated voltage and two arms that consist of SiC MOSFET and SiC Schottky Barrier Diode (SBD). By applying 7 modules, that mean 4 modules in the single phase MOSFET-rectifier and 3 modules in the three phase inverter, the developed APS has 136kVA output capacity.

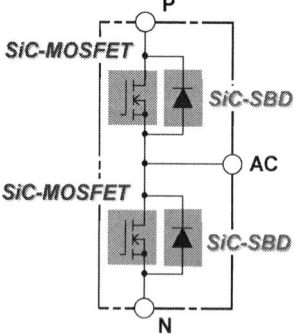

Fig.3 1.2kV Full SiC power module consisting of SiC-MOSFET and SiC-SBD

3. Reduction of volume and mass of filter circuit

Table 2 shows comparison between 1.7kV Hybrid SiC system and 1.2kV Full SiC system regarding of carrier frequency, volume and mass of filter reactor and total harmonic distortion (THD) of MOSFET-rectifier and inverter.

Authors apply the carrier frequency for MOSFET-rectifier and inverter as approximately 5kHz. Volume and mass of the cooling unit for MOSFET-rectifier and inverter increase when carrier frequency for them is set higher. However, considering a trade-off of power loss and the filter circuit, authors optimized the most suitable carrier frequency.

With higher carrier frequency, filter reactor inductance of MOSFET-rectifier, filter reactor inductance and capacity of AC filter capacitor of inverter can be decreased. Furthermore, THD of MOSFET-rectifier input current can be reduced to the approximately 1/3 of THD of the conventional APS.

On the other hand, THD of 3-phase 415VAC output voltage of inverter was designed to be of a low value. It is due to requirement from auxiliary load of a train. THD of the developed APS, with less inductance and less capacity of filter circuit, remains on the same low level compared with conventional APS.

Fig.4 shows the comparison of waveform of MOSFET-rectifier input current between conventional APS and developed APS. The waveform of the developed APS has less ripple than the waveform of the conventional APS.

Table 2 Volume and mass of filter circuit

Power module of Auxiliary power supply			Applying 1.7kV Hybrid SiC	Applying 1.2kV Full SiC
Converter	Carrier frequency		1.05 kHz	5.3 kHz
	Filter reactor	Volume	1	0.31
		Mass	1	0.29
Inverter	Carrier frequency		1.95 kHz	5.5 kHz
	Filter reactor	Volume	1	0.34
		Mass	1	0.17
	Filter capacitor	Volume	1	0.33
		Mass	1	0.39

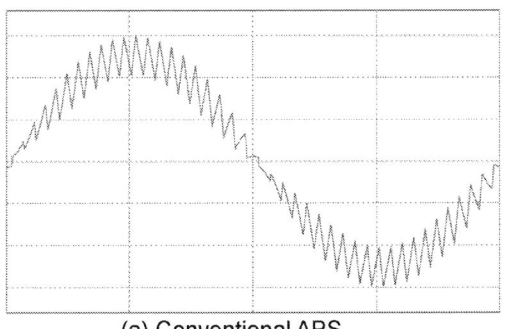 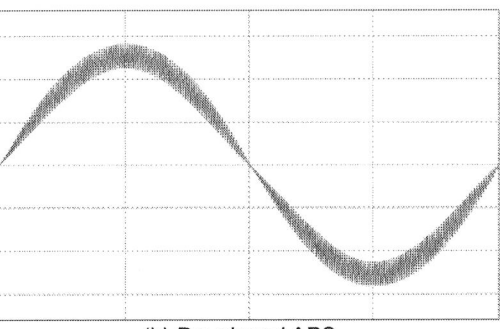

(a) Conventional APS   (b) Developed APS

Fig.4 Waveform of MOSFET-rectifier input current

4. Improvement of power consumption and total efficiency

Fig.5 shows loss comparison between the conventional and the developed system. Though the switching frequency becomes higher from approximately 1kHz to 5kHz, total loss of APS is reduced by 40% and total efficiency is improved from 95.6% to 97.3% by applying 1.2kV Full SiC power devices. Reduction of power loss of reactor is an effect of the decreased ripple current that is achieved by high frequency switching of Full SiC module.

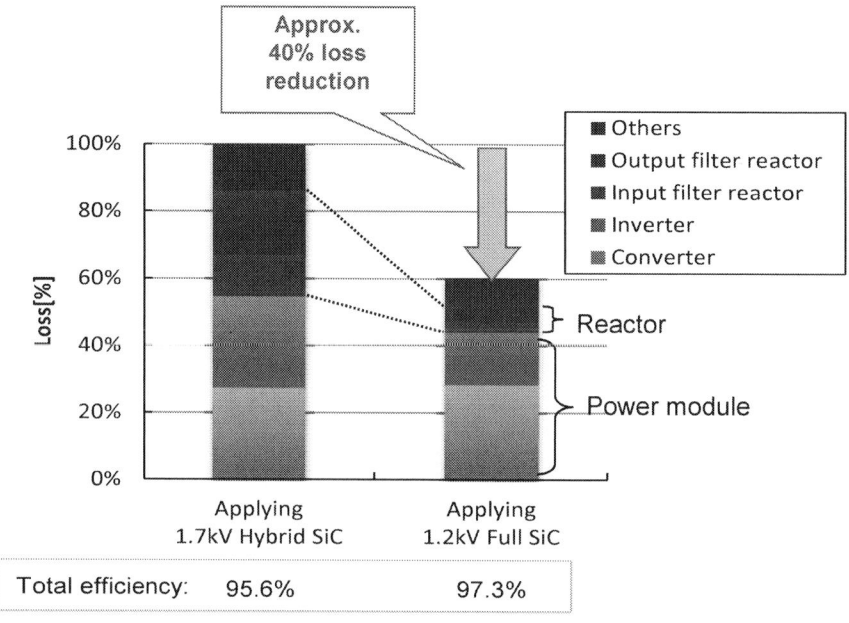

Fig.5 Comparison of power consumption

5. Conclusion

By using Full SiC module, which has low loss and high temperature operation characteristic, higher carrier frequency can be applied to railway applications. Authors designed new auxiliary power supply system (APS) considering the trade-off of the increase of the loss of the switching device, and volume and mass reduction of the filter circuit. In newly developed APS, power loss is reduced by 40%, and volume and mass of filter circuit is 70% lower compared with conventional APS.

6. Literature

[1] T. Kobayashi, Y. Nakashima, K. Kaneko, Y. Yamashita, A. Murahashi: Energy Saving by Railway Inverter System with SiC Power Module, PCIM Europe (2012) p.547 - 553

[2] A. Furukawa, S. Kinouchi, H. Nakatake, Y. Ebiike: Low on-resistance 1.2 kV 4H-SiC MOSFETs integrated with current sensor, ISPSD (2011) p.288 - 291

PCIM Europe 2016, 10 – 12 May 2016, Nuremberg, Germany

A New Behavioral Model for Accurate Loss Calculations in Power Semiconductors

Ajay Poonjal Pai, Infineon Technologies AG, Germany, AjayPoonjal.Pai@Infineon.com
Tomas Reiter, Infineon Technologies AG, Germany, Tomas.Reiter@Infineon.com
Martin Maerz, Fraunhofer IISB, Germany, Martin.Maerz@IISB.Fraunhofer.de

Abstract

In this paper, a new behavioural model is proposed for calculating power losses in power semiconductor switches. In contrast to models existing in literature which mostly model losses only in terms of $I_{c/f}$, V_{dc} and T_j, this paper also takes into account R_g and V_{ge} which heavily affect the losses but are generally neglected. Moreover, the model also calculates losses as a function of the chip area per switch, which makes this model ideal for calculating the optimum chip area for a given application. The accuracy of this model is experimentally demonstrated on a HybridPACK Drive FS820R08A6P2 power module from Infineon, and the model is found to offer significantly better accuracy compared to the existing models.

1. Introduction

As power semiconductors continue to evolve, and with more players entering the market, the current density of modern devices is increasing dramatically and the devices are becoming more efficient. As a consequence, the estimation of losses in semiconductors with higher accuracy is becoming extremely important in order to benchmark the devices. Loss calculations serve three major purposes. The first one is for dimensioning the power semiconductors and for facilitating the most efficient choice of the semiconductors. Second, for estimating the reliability and lifetime. The third purpose is for carrying out the design of the cooling system, to ensure that the junction temperatures do not exceed the maximum values specified by the device manufacturers.

Often, semiconductor companies are required to estimate the losses of their devices at the system level. A typical example is that of an inverter for Hybrid Electric Vehicle (HEV) application, which has to be thoroughly simulated for several thousand hours of its lifetime, as per various mission profiles or drive cycles. Running detailed physics-based simulations in such cases would be impractical as the simulation run time would be too long. Therefore, most semiconductor companies resort to simple behavioural models where they characterize the devices under different operating conditions and use these characterizations to develop simple equations to model the losses. Depending on the desired accuracy and the permissible levels of complexity, most models estimate losses as functions of one or more of the following parameters: device currents $I_{c/f}$, dc-link voltages V_{dc} and junction temperatures T_j. However, in practice, losses also depend on the gate-resistance R_g and the gate-voltage V_{ge}. Moreover, these models are parametrized for a certain device, and need to be re-parametrized for a device of a different chip area, even if it is of the same family. A single model that takes into account all the above dependencies is missing in literature. In this paper, a new model for losses in power semiconductors is proposed. It takes into account $I_{c/f}$, V_{dc}, T_j, R_g, V_{ge} and also the IGBT chip area (A_I) and diode chip area (A_D). It offers a significantly higher accuracy compared to the state-of-the-art models. This model is simple to implement and is in the form of closed analytic equations.

© VDE VERLAG GMBH · Berlin · Offenbach

Therefore, it can be used to derive simple analytic expressions for losses in standard topologies such as the Two-Level Voltage Source Converter (2L-VSC), which is the most commonly used topology in automotive inverters. Another important feature of this model is that it is backward compatible to the models commonly used in industry. Since the insulated-gate bipolar transistor (IGBT) is the most widely used power transistor in the medium- and high-voltage range [1], this paper focuses only on the IGBT. However, with a little modification, the model can be made suitable to other power devices as well, and will be demonstrated on a SiC MOSFET in a future paper.

The organisation of this paper is as follows. In the first part, the various models available in literature and used by major semiconductor companies are introduced, and their pros and cons are studied. Then, the proposed model is discussed in sufficient detail, and its application on a HybridPACK Drive Module [2] from Infineon is explained. Finally, the model is validated with respect to simulation data as well as with experimental data.

2. Review of Literature

Model	$V_{ce/f}$	E
1989 [3]		-
1991 [4]	$V_{ce/f} = V_{ce0/f0} + R_{c/d} \cdot I_{c/d}$	$E = E_{nom} \frac{I_c}{I_{c,nom}} \frac{V_{dc}}{V_{dc,nom}}$
1994 [5]		Calculated by integrating the product of approximations of current and voltage transients curves
1995 [6]	$V_{ce/f} = V_{ce0/f0} + R_{c/d} \cdot I_{c/d}^K$	$E = E_{nom} (\frac{I_c}{I_{c,nom}})^A$
1996 [7]	$V_{ce/f} = (V_{ce0/f0} + A \cdot T_j) + (R_{c/d} + B \cdot T_j) \cdot I_{c/d}^K$	$E = E_{nom} (\frac{I_c}{I_{c,nom}})^A (\frac{V_{dc}}{V_{dc,nom}})^B (\frac{T_j}{T_{j,nom}})^C$
2004 [8]	$V_{ce/f} = V_{ce0/f0} + R_{c/d} \cdot I_{c/d}$	$E = E_{nom} \frac{I_c}{I_{c,nom}} \frac{V_{dc}}{V_{dc,nom}}$
2005 [9]		$E = E_{nom}(A \cdot I_c^2 + B \cdot I_c + C) \frac{V_{dc}}{V_{dc,nom}}$
2013 [10]		$E = E_{nom} \frac{I_c}{I_{c,nom}} (\frac{V_{dc}}{V_{dc,nom}})^K$
Infineon	$V_{ce/f} = (V_{ce0/f0} + A \cdot T_j + B) + (R_{c/d} + C \cdot T_j + D) \cdot I_{c/d}$	$E = E_{nom} \frac{I_c}{I_{c,nom}} \frac{V_{dc}}{V_{dc,nom}} \cdot (E \cdot T_j + F)^G$
ABB [11]	$V_{ce/f} = V_{ce0/f0} + R_{c/d} \cdot I_{c/d}$	$E = E_{nom}(A \cdot I_c^2 + B \cdot I_c + C) \frac{V_{dc}}{V_{dc,nom}} \cdot (E \cdot T_j + F)^G$
Semikron [12]	$V_{ce/f} = (V_{ce0/f0} + A \cdot T_j + B) + (R_{c/d} + C \cdot T_j + D) \cdot I_{c/d}$	$E = E_{nom}(\frac{I_c}{I_{c,nom}})^A (\frac{V_{dc}}{V_{dc,nom}})^B (C \cdot T_j + D)^K$

Tab. 1: Comparison of different loss-calculation models

Tab. 1 provides a summary of the several behavioral models existing in literature and industry, and it is clear that all the models consider the variation of only $I_{c/f}$, V_{dc} and T_j, but not R_g and V_{ge}. Also, most models assume a linear relationship between the losses and the different parameters, which is not an accurate assumption. Furthermore, the coefficients of the model would have to be re-determined for a chip of different area, even if it belonged to the same family/technology as the reference chip. This would be a serious shortcoming if these models have to be integrated into algorithms that estimate the optimum chip area for a given application. To overcome all these shortcomings, a new model is proposed in this paper.

3. The Proposed Model

The following goals are set for the proposed model:

1. The model equations have a closed form and can be integrated for sinusoidal values of the load current, in order to derive simple expressions for common applications such as the 2L-VSC.

2. The model offers significantly higher accuracy than the existing models.

3. The model takes into account $I_{c/f}$, V_{dc}, T_j, R_g, V_{ge}, A_I and A_D to determine the losses.

4. The model does not need too much characterization data.

© VDE VERLAG GMBH · Berlin · Offenbach

PCIM Europe 2016, 10 – 12 May 2016, Nuremberg, Germany

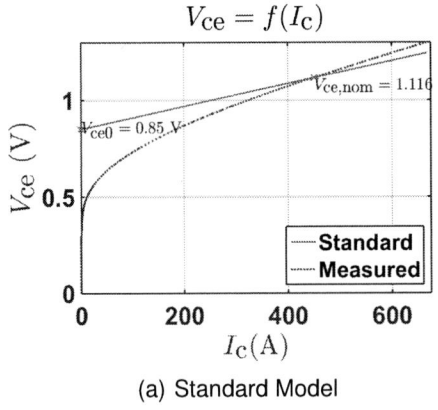

(a) Standard Model

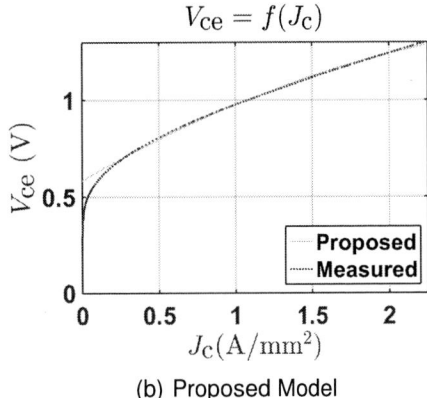

(b) Proposed Model

Fig. 1: V_{ce} v/s I_c/J_c- Comparison of the models

For assessing the performance of the proposed model, it will be compared against the Infineon Model mentioned in tab. 1 which is referred to as the 'standard model' in this publication. While comparing the accuracy of the proposed model against the standard model, a look-up table based model would be taken as the benchmark and is thus called the 'benchmark model'. For all measurement purposes, a HybridPACK Drive module FS820R08A6P2* from Infineon will be used. It is necessary to define base values for all the parameters in this model, hereafter named as 'nominal values'. The model has its highest accuracy at these values, and therefore, it is recommended to choose the most commonly occurring values in typical applications as the nominal values. In this paper, the nominal values chosen are given in tab. 2.

$I_{c/f}$ (A)	T_j (°C)	V_{dc} (V)	$R_{g,on}$ (Ω)	$R_{g,off}$ (Ω)	$t_{M,on}$[†] (µs)	$t_{M,off}$[†] (µs)	V_{ge} (V)	A_I (mm²)	A_D (mm²)	J_c[†] ($\frac{A}{mm^2}$)	J_f[†] ($\frac{A}{mm^2}$)
450	25	400	0.5	2.8	0.803	0.2672	15	300	150	1.5	3

Tab. 2: Nominal values for all parameters used in the model

3.1. Conduction Losses

Almost all models, including the standard model, decompose V_{ce} linearly into an offset component V_{ce0} and a drop across an incremental resistance R_c as given below:

$$V_{ce} = V_{ce0} + R_c \cdot I_c \qquad \text{[Standard Model]} \qquad (1)$$

The characterization process for determining the coefficients of the model is demonstrated for HybridPACK Drive as follows. V_{ce} values are measured at different values of I_c, keeping all the other parameters equal to their nominal values mentioned in tab. 2. The measured values are shown by the dotted blue curve in Fig. 1(a). For fitting the linear curve, the offset V_{ce0} is typically taken as 0.85V for an IGBT at room temperature. The voltage drop across R_c is represented by drawing a line connecting $V_{ce,nom}$ with V_{ce0}. R_c is equal to the slope of this line. The resulting curve is shown in solid red. It can be seen that the fitted curve matches the

---

*FS820R08A6P2 is a B6-bridge IGBT module with an implemented current rating of 820 A and blocking voltage of 750V. It has three IGBTs of 100 mm² each in parallel per switch (in total 300 mm² per switch) and three anti-parallel diodes of 50 mm² each in parallel per diode (in total 150 mm² per switch).

[†]$J_{c/f}$ and $t_{M,on/off}$ will be explained in equations 2 and 8 respectively.

measurement data exactly at the nominal point. However, as we calculate at points besides the nominal point, there is an increasing error. At light-load conditions, the error is significantly high. It has to be noted that automotive inverters operate in light-load conditions most of the time, and not in nominal conditions. Thus, using the standard model for estimating losses in automotive inverters leads to significant inaccuracy. Therefore, this paper proposes to use a better function for fitting the curve. An exponential function would have been a more intuitive and, beyond doubt, a better fit. But, while deriving an equation for the average conduction losses in an inverter application, it is necessary to carry out integration of the product of $I_c(t)$, which is a sinusoid, and $V_{ce}(t)$. If V_{ce} is also modelled as an exponential function, integrating a product of a sinusoid and an exponent would result in a final expression containing both the sinusoidal and the exponential terms making it complicated, thereby defeating the very purpose of having a behavioural model. Therefore, this paper avoids exponential functions and chooses to model in terms of quadratic functions instead. Moreover, choosing quadratic functions makes this model backward compatible with the standard models which make use of linear functions. Since we shall be dealing with different chip areas in the proposed model, it makes more sense to talk in terms of the current per chip area rather than the absolute current. Therefore, the proposed model will use the current densities J_c and J_f instead of I_c and I_f respectively. The nominal values for J_c and J_f would be 1.5 A/mm^2 ($= \frac{450A}{300 \text{ mm}^2}$) and 3 A/mm^2 ($= \frac{450A}{150 \text{ mm}^2}$) respectively, as given in tab. 2. The equation for V_{ce} now becomes:

$$V_{ce} = (A_{11} \cdot J_c^2 + A_{12} \cdot J_c + A_{13}) \tag{2}$$

where the coefficients A_{11}-A_{13} are obtained by fitting a quadratic curve to the measured values of V_{ce} v/s J_c. This is depicted in fig 1(b), where it can be seen that the quadratic curve (green) of the proposed model fits better than the standard model discussed previously. Weighting equation 2 with a quadratic function of T_j extends the model to T_j as follows:

$$V_{ce} = (A_{11} \cdot J_c^2 + A_{12} \cdot J_c + A_{13}) \cdot (A_{21} \cdot T_j^2 + A_{22} \cdot T_j + A_{23}) \tag{3}$$

To obtain the coefficients A_{21}-A_{23}, characterization measurements of V_{ce} are carried out as previously, but this time varying T_j and keeping all the other parameters at their nominal values mentioned in tab. 2. Additionally, as the multiplied term is only intended to be a weighting function, the V_{ce} values must be normalised to its value at $T_{j,nom}$

Although datasheets generally provide curves of V_{ce} v/s V_{ge}, none of the investigated models take into account the dependency of conduction losses on V_{ge}. In the proposed model, the dependency of V_{ce} on V_{ge} is taken into account by further weighting equation 3 with a quadratic function of V_{ge}:

$$V_{ce} = (A_{11} \cdot J_c^2 + A_{12} \cdot J_c + A_{13}) \cdot (A_{21} \cdot T_j^2 + A_{22} \cdot T_j + A_{23}) \cdot (A_{31} \cdot V_{ge}^2 + A_{32} \cdot V_{ge} + A_{33}) \tag{4}$$

A_{31}-A_{33} can be obtained as discussed previously, by fitting a quadratic curve to the measurement data of the normalised V_{ce} v/s V_{ge}.

For a diode, the above discussion applies too, except for the fact that, naturally, there is no dependency on V_{ge}. Thus, the following equations can be written:

$$V_f = (B_{11} \cdot J_f^2 + B_{12} \cdot J_f + B_{13}) \cdot (B_{21} \cdot T_j^2 + B_{22} \cdot T_j + B_{23}) \tag{5}$$

PCIM Europe 2016, 10 – 12 May 2016, Nuremberg, Germany

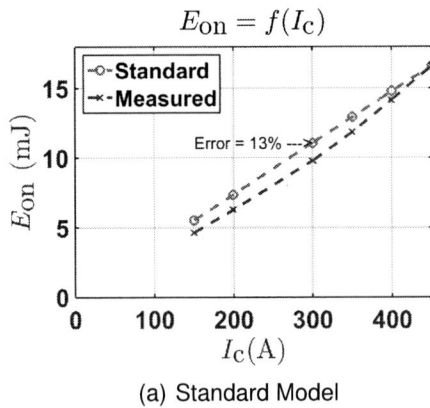
(a) Standard Model

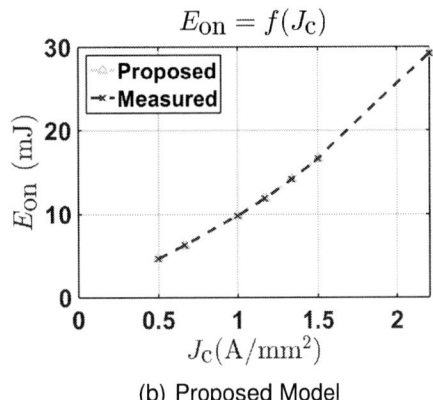
(b) Proposed Model

Fig. 2: E_{on} v/s $I_{\mathrm{c}}/J_{\mathrm{c}}$- Comparison of the models

3.2. Switching Losses

Most semiconductor companies normally specify the turn-on energy E_{on} and the turn-off energy E_{off} for IGBTs as a function of I_{c}, V_{dc} and T_{j}. For brevity, we choose to explain the model only for E_{on} in this paper, and the same discussion applies also to E_{off}. For diodes, the turn-off energy due to reverse recovery E_{rec} is generally specified. As per most models, including the standard model, the turn-on energy E_{on} at any operating point I_{c}, V_{dc}, T_{j} is modelled as:

$$E_{\mathrm{on}} = E_{\mathrm{on,nom}} \cdot \frac{I_{\mathrm{c}}}{I_{\mathrm{c,nom}}} \cdot \frac{V_{\mathrm{dc}}}{V_{\mathrm{dc,nom}}} \cdot (A \cdot T_{\mathrm{j}} + B) \qquad \text{[Standard Model]} \qquad (6)$$

Figure 2(a) shows the measured values of E_{on} v/s I_{c} for HybridPACK Drive. It can be seen that the linear curve matches the measurement only at the nominal point I_{c}=450 A. At all other points, there is a significant error, with the maximum error being 13%. Moreover, as E_{on} is scaled proportionally with I_{c}, the standard model calculates $E_{\mathrm{on}} \approx 0$ at $I_{\mathrm{c}} \approx 0A$. But in reality, there are losses due to capacitive effects even at zero currents, as long as $V_{\mathrm{dc}} \neq 0$. This is a drawback of the standard model, particularly for automotive inverter applications which operate at low currents most of the time. As discussed in the case of conduction losses, we replace I_{c} by J_{c}, and propose to model E_{on} as a product of quadratic functions of the parameters. The model equation taking into account the commonly used parameters I_{c}, T_{j} and V_{dc} is given below:

$$E_{\mathrm{on}} = (C_{11} \cdot J_{\mathrm{c}}^2 + C_{12} \cdot J_{\mathrm{c}} + C_{13}) \cdot (C_{21} \cdot T_{\mathrm{j}}^2 + C_{22} \cdot T_{\mathrm{j}} + C_{23}) \cdot (C_{31} \cdot V_{\mathrm{dc}}^2 + C_{32} \cdot V_{\mathrm{dc}} + C_{33}) \quad (7)$$

To determine the coefficients C_{11}-C_{13}, characterization measurements are carried out for E_{on} at different values of I_{c}, with the other parameters being maintained at their nominal values. Figure 2(b) shows that the quadratic curve is a better fit to the measurement data. In contrast to the standard model, the proposed model clearly offers much higher accuracy. Similarly, C_{21}-C_{23} and C_{31}-C_{33} are obtained by varying T_{j} and V_{dc} respectively. As mentioned previously, all coefficients other than C_{11}-C_{13} are to be determined at normalised values of E_{on}

In addition to the parameters described above, the switching losses also depend on the switching speed $\frac{di_c}{dt}$ of the IGBT which is controlled by means of the external gate resistor R_{g}. However, $\frac{di_c}{dt}$ measurements are quite complicated to obtain as they require probes with a very high bandwidth, which can also affect the measurement set-up. It can be experimentally observed that for the same R_{g}, two identical devices would not switch at the same speed. This is because the characteristics of the gate driver circuit, the stray inductances of the gate- and the

© VDE VERLAG GMBH · Berlin · Offenbach

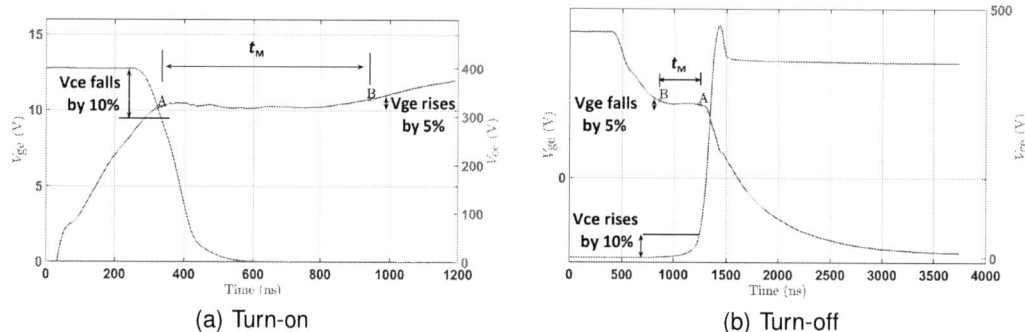

(a) Turn-on   (b) Turn-off

Fig. 3: Definition of t_{M} for indication of IGBT's switching speed

power-circuit, the internal gate resistances are not identical, which results in a different $\frac{di_c}{dt}$ for the different set-ups. Therefore, taking into account the $\frac{di_c}{dt}$ by means of R_{g} would lead to heavy dependencies on the measurement set-up and result in significant inaccuracies. Instead, this paper chooses the duration of the miller plateau of V_{ge} which does not need a high bandwidth current probe, to represent the switching speed. As shown in figures 3(a) and 3(b), the duration of the miller plateau t_{M} is defined as the time (in µs) between the point 'A' at which V_{ce} changes by 10 % of its off-state value to the point 'B' at which V_{ge} is 5% higher than its value at A.

Measurements are carried out at different values of R_{g} and the corresponding readings of t_{M} and E_{on} are noted. Eqn. 7 can now be updated to take into account t_{M}:

$$E_{\mathrm{on}} = (C_{11} \cdot J_{\mathrm{c}}^2 + C_{12} \cdot J_{\mathrm{c}} + C_{13}) \cdot (C_{21} \cdot T_{\mathrm{j}}^2 + C_{22} \cdot T_{\mathrm{j}} + C_{23}) \tag{8}$$
$$\cdot (C_{31} \cdot V_{\mathrm{dc}}^2 + C_{32} \cdot V_{\mathrm{dc}} + C_{33}) \cdot (C_{41} \cdot t_{\mathrm{M}}^2 + C_{42} \cdot t_{\mathrm{M}} + C_{43})$$

C_{41}-C_{43} are determined by fitting a quadratic curve to the function E_{on} v/s t_{M}.

It is a well known fact that E_{on} depends on the IGBT chip area A_{I} as well as the diode chip area A_{D}[‡]. Most models in literature consider E_{on} to scale linearly with A_{I}. In practice, however, this is not fully true. This is because E_{on} is highly dependent on the stray inductance which depends on the wire bonds, layout etc., and does not scale linearly with the chip area. To take this into account, the model can be updated to:

$$E_{\mathrm{on}} = (C_{11} \cdot J_{\mathrm{c}}^2 + C_{12} \cdot J_{\mathrm{c}} + C_{13}) \cdot (C_{21} \cdot T_{\mathrm{j}}^2 + C_{22} \cdot T_{\mathrm{j}} + C_{23}) \cdot (C_{31} \cdot V_{\mathrm{dc}}^2 + C_{32} \cdot V_{\mathrm{dc}} + C_{33}) \tag{9}$$
$$\cdot (C_{41} \cdot t_{\mathrm{M}}^2 + C_{42} \cdot t_{\mathrm{M}} + C_{43}) \cdot (C_{51} \cdot A_{\mathrm{I}}^2 + C_{52} \cdot A_{\mathrm{I}} + C_{53}) \cdot (C_{61} \cdot A_{\mathrm{D}}^2 + C_{62} \cdot A_{\mathrm{D}} + C_{63})$$

It was quite straightforward to obtain measurements for E_{on} at different I_{c}, V_{dc}, T_{j} and R_{g}. But obtaining readings at different values of A_{I} and A_{D} requires that we are able to change the dies of the modules, which is not practical. But there is a workaround. The HybridPACK Drive has three IGBT dies per switch each of 100 mm², and three diode dies of 50 mm². A die can be disconnected from the module by cutting off the bond wires of that die. In this way, by cutting off one die at a time, we can obtain measurements for E_{on} at three different chips areas, viz., 300, 200, 100 mm² for IGBT and 150, 100, 50 mm² for the diode. One must keep in mind that

---

[‡]It is here that the model for E_{off} would slightly differ from that of E_{on}. As E_{off} does not depend on A_{D}, E_{off} would not be a function of A_{D}.

PCIM Europe 2016, 10 – 12 May 2016, Nuremberg, Germany

the gate resistance has to be compensated to maintain the same reference switching speed in the device, modelled in terms of t_{M} in this paper. This is because when one of the dies is cut off, for the same R_{g}, the gate current per die would now be higher. Thus, the IGBTs would switch faster, which would lead to discrepancies. Finally, C_{51}-C_{53} are determined from the measurements of E_{on} v/s A_{I} and C_{61}-C_{63} are determined from E_{on} v/s A_{D}.

For a diode, we can similarly write:

$$E_{\mathrm{rec}} = (E_{11} \cdot J_{\mathrm{f}}^2 + E_{12} \cdot J_{\mathrm{f}} + E_{13}) \cdot (E_{21} \cdot T_{\mathrm{j}}^2 + E_{22} \cdot T_{\mathrm{j}} + E_{23}) \qquad (10)$$
$$\cdot (E_{31} \cdot V_{\mathrm{dc}}^2 + E_{32} \cdot V_{\mathrm{dc}} + E_{33}) \cdot (E_{41} \cdot A_{\mathrm{D}}^2 + E_{42} \cdot A_{\mathrm{D}} + E_{43})$$

The coefficients A_{11}-E_{43} obtained from the characterization process described so far are summarised in tab. 3, along with the number of points at which measurements have been obtained. It is to be noted that these 60 coefficients completely describe the conduction and switching losses in the device, for any values of $I_{\mathrm{c/f}}$, T_{j}, V_{ge}, V_{dc}, R_{g} and $A_{\mathrm{I/D}}$. Using this model, it is possible to calculate the losses at any combination of values of the parameters. As a minimum of three points are required to fit a quadratic curve, it is worth mentioning that the minimum no. of measurements for this model to work is three per parameter. Naturally, more the number of measurements, higher will be the accuracy. Moreover, in case of previous designs where it may be necessary to run the standard model, the proposed model can be still used by setting the quadratic parameters as zero, and refitting the curves. This makes the model backward compatible, which is an attractive feature.

Measured Parameter	Varied Parameter	No. of points	Coefficients Determined		
V_{ce}	J_{c}	213	$A_{11} = -6.32 \cdot 10^{-2}$	$A_{12} = 4.55 \cdot 10^{-1}$	$A_{13} = 5.84 \cdot 10^{-1}$
	T_{j}	9	$A_{21} = -1.25 \cdot 10^{-9}$	$A_{22} = 1.15 \cdot 10^{-4}$	$A_{23} = 9.83 \cdot 10^{-1}$
	V_{ge}	6	$A_{31} = 1 \cdot 10^{-2}$	$A_{32} = -3.35 \cdot 10^{-1}$	$A_{33} = 3.76$
V_{f}	J_{d}	16	$B_{11} = -2.39 \cdot 10^{-2}$	$B_{12} = 2.95 \cdot 10^{-1}$	$B_{13} = 6.57 \cdot 10^{-1}$
	T_{j}	4	$B_{21} = -4.98 \cdot 10^{-6}$	$B_{22} = 1.39 \cdot 10^{-4}$	$B_{23} = 1.09$
E_{on}	J_{c}	7	$C_{11} = 3.5322$	$C_{12} = 4.869$	$C_{13} = 1.386$
	T_{j}	12	$C_{21} = -1.264 \cdot 10^{-6}$	$C_{22} = 2.745 \cdot 10^{-3}$	$C_{23} = 9.3217 \cdot 10^{-1}$
	V_{dc}	8	$C_{31} = 2.273 \cdot 10^{-6}$	$C_{32} = 1.58 \cdot 10^{-3}$	$C_{33} = 3.975 \cdot 10^{-3}$
	$t_{\mathrm{M,on}}$ $(R_{\mathrm{g,on}})$	5	$C_{41} = 0.427$	$C_{42} = 0.673$	$C_{43} = 0.185$
	A_{I}	3	$C_{51} = -1.0842 \cdot 10^{-5}$	$C_{52} = 0.0066$	$C_{53} = -0.0014$
	A_{D}	3	$C_{61} = -1.13 \cdot 10^{-6}$	$C_{62} = 0.0016$	$C_{63} = 0.7877$
E_{off}	J_{c}	5	$D_{11} = 3.6537$	$D_{12} = 7.3424$	$D_{13} = 2.7006$
	T_{j}	12	$D_{21} = 1.555 \cdot 10^{-5}$	$D_{22} = 3.2552 \cdot 10^{-4}$	$D_{23} = 0.9821$
	V_{dc}	9	$D_{31} = -3.386 \cdot 10^{-6}$	$D_{32} = 0.00375$	$D_{33} = 0.0421$
	$t_{\mathrm{M,off}}$ $(R_{\mathrm{g,off}})$	4	$D_{41} = 0.947$	$D_{42} = 0.387$	$D_{43} = 0.829$
	A_{I}	3	$D_{51} = 3.3068 \cdot 10^{-6}$	$D_{52} = 0.0022$	$D_{53} = 0.0342$
E_{rec}	J_{f}	14	$E_{11} = -0.08983$	$E_{12} = 1.919$	$E_{13} = 0.755$
	T_{j}	12	$E_{21} = 4.93 \cdot 10^{-5}$	$E_{22} = 0.00305$	$E_{23} = 0.893$
	V_{dc}	7	$E_{31} = 1.0097 \cdot 10^{-6}$	$E_{32} = 0.0021$	$E_{33} = -0.0066$
	A_{D}	3	$E_{41} = -7.146 \cdot 10^{-6}$	$E_{42} = 0.0077$	$E_{43} = 0.0039$

Tab. 3: Coefficients of the proposed model for HybridPACK Drive

4. Validation of the Model

The developed model was used to evaluate $V_{\mathrm{ce/f}}$ and $E_{\mathrm{on/off/rec}}$ at various operating points at which no measurements were taken during the characterization. For instance, as per tab. 3, the characterization process included only one measurement of E_{on} for A_{I}=200 mm^2 i.e, with

© VDE VERLAG GMBH · Berlin · Offenbach

PCIM Europe 2016, 10 – 12 May 2016, Nuremberg, Germany

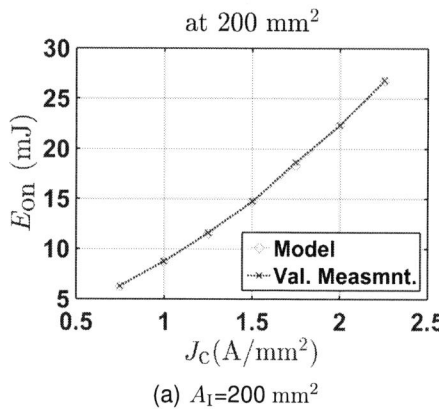

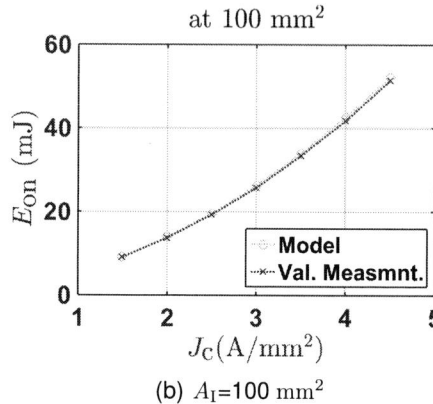

(a) A_I=200 mm^2 (b) A_I=100 mm^2

Fig. 4: Comparison of E_on values calculated by the proposed model against the validation measurements.

one IGBT die cut off from the module. Using the model, E_on was calculated as a function of J_c for A_I=200 mm^2. The calculated values are indicated by the green solid curve in fig. 4(a). Now, in order to check the accuracy of these values, measurements of $E_\mathrm{\ddot{o}n}$ were carried out as a function of J_c at A_I=200 mm^2, as can be seen from the blue dotted curve in fig. 4(a). These 'validation measurements' are not to be confused with the 'characterization measurements' described previously. The validation measurements were taken only for validating the model and are not necessary for the characterization process. It can be seen that there is a good overlap between the estimated values and the measured values, thereby confirming the validity of the developed model. The same conclusion can be arrived at from fig. 4(b) which depicts the case for A_I=100 mm^2.

5. Application Example

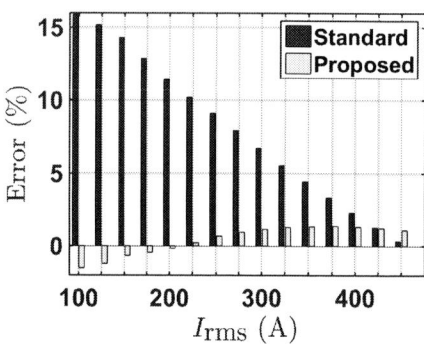

Fig. 5: Comparison of error in the standard and proposed models

For demonstrating the model at the system-level, an automotive inverter (V_dc= 400 V, $I_\mathrm{rms,nom}$= 450 A) with the above module is considered. Loss calculations are performed with the standard and the proposed models at several I_rms points, and the errors are compared with reference to the benchmark model as shown in Fig. 5. As the standard model is designed to be accurate at the nominal point, it offers close to zero error at full-load conditions ($I_\mathrm{rms,nom}$= 450 A). At all other points, it has a higher error, with the worst case error being 16% at I_rms= 100 A. It can be concluded that the standard model is suitable only for applications operating mostly at full-load

© VDE VERLAG GMBH · Berlin · Offenbach

conditions. However, for applications such as automotive inverters which operate mostly at light-load conditions, the standard model results in more than 15% error. This is not acceptable for efficiency calculations, especially when comparing across device technologies, e.g., Si vs SiC, or across chip generations, where the difference in the compared technologies itself may be in the range of 10-15%. In contrast, the proposed model consistently offers less than 2% error for the entire range of I_{rms}.

6. Conclusions

A new behavioural model was proposed for calculating power losses in power semiconductors. This model takes into account not only $I_{\mathrm{c/f}}$, V_{dc} and T_{j}, like most other models in literature, but also R_{g}, V_{ge}, A_{I} and A_{D}. The step-by-step procedure to experimentally determine the model coefficients was demonstrated on HybridPACK Drive module. The developed model was used to evaluate the losses for the module at several operating points, with good accuracy. The model was also used to simulate the losses for an automotive inverter application. The standard model was found to have around 15% error over a wide operating range, making it unsuitable for efficiency calculations. In contrast, the proposed model had less than 2 % error over the entire range. This makes it ideal for efficiency calculations, especially when the device technologies considered differ by less than 15% in terms of losses.

7. References

[1] B Jayant Baliga. Analytical modeling of igbts: challenges and solutions. *Electron Devices, IEEE Transactions on*, 60(2):535–543, 2013.

[2] Product brief hybridpack drive. *Infineon AG*, 2014.

[3] LK Mestha and PD Evans. Analysis of on-state losses in pwm inverters. In *IEE Proceedings B (Electric Power Applications)*, volume 136, pages 189–195. IET, 1989.

[4] Johann W Kolar and Hans Ertl. Influence of the modulation method on the conduction and switching losses of a pwm converter system. *Industry Applications, IEEE Transactions on*.

[5] F Casanellas. Losses in pwm inverters using igbts. *IEE Proceedings-Electric Power Applications*, 141(5):235–239, 1994.

[6] Frede Blaabjerg and Stig Munk-Nielsen. Power losses in pwm-vsi inverter using npt or pt igbt devices. *Power Electronics, IEEE Transactions*, 10(3):358–367, 1995.

[7] Frede Blaabjerg and John K Pedersen. An extended model of power losses in hardswitched igbt-inverters. In *IEEE Industry Applications Conference, 1996*.

[8] Michael H Bierhoff. Semiconductor losses in voltage source and current source igbt converters based on analytical derivation. In *IEEE Power Electronics Conference, 2004*.

[9] Sibylle Dieckerhoff. Power loss-oriented evaluation of high voltage igbts and multilevel converters in transformerless traction applications. *Power Electronics, IEEE Transactions*.

[10] Volodymyr Ivakhno. Estimation of semiconductor switching losses under hard switching using matlab/simulink subsystem. *Electrical, Control and Communication Engineering*.

[11] Björn Backlund, Raffael Schnell, Ulrich Schlapbach, Roland Fischer, and Evgeny Tsyplakov. Applying igbts. *ABB, Lenzburg, Switzerland, Appl. Note 5SYA2053-01*, 2007.

[12] Arendt Wintrich, Ulrich Nicolai, Werner Tursky, and Tobias Reimann. Application manual power semiconductors. ISLE, 2011.

© VDE VERLAG GMBH · Berlin · Offenbach

PCIM Europe 2016, 10 – 12 May 2016, Nuremberg, Germany

High Speed Electronic Over Current Breaker for DC-Grids without Additional Sensing

Alexander Würfel, University of Bremen, Germany, wuerfel@ialb.uni-bremen.de
Johannes Adler, University of Bremen, Germany, adler@ialb.uni-bremen.de
Anton Mauder, Infineon Technologies AG, Germany, anton.mauder@infineon.com
Nando Kaminski, University of Bremen, Germany, nando.kaminski@uni-bremen.de

Abstract

An electronic over current breaker (OCB) for DC-grids without the need of additional sensing circuitry is presented. The OCB consists of standard Si-MOSFETs and is cascoded with SiC-JFETs for increased blocking capability. The over current detection threshold can be adjusted in a wide range. Due to the lack of additional sensing circuitry, the detection and clearing of a fault current is extremely fast and fail-safe compared to conventional OCBs at least with respect to the control.

1. Motivation

The use of direct current offers several advantages especially for the use in battery powered systems. A crucial factor is the ability to switch off the DC current in case of a failure or short circuit. Especially in onboard applications the cable lengths between battery and OCB are very short, resulting in very small parasitic wire inductances. Assuming a wire inductance in the fault current path of 10 µH in a 400 V DC grid leads to a di/dt of 40 A/µs. Common electro-mechanical OCBs need approximately 5 to 10 ms to switch off a fault current. With the given di/dt of 40 A/µs, the current would theoretically reach values of up to 400,000 A until the fault is cleared. Such current values are far above the maximum current capability of any battery system and the fault current would be limited by the maximum short circuit current of the battery in that case. Each very high over current stresses the battery and will reduce the overall lifetime significantly or can even lead to unexpected burning of the battery. The only remedy is detecting the over current and switching it off within several microseconds.

Switching speeds of several microseconds can only be achieved by solid state circuit breakers. Even when using SiC devices with additional detection circuits, the resulting switching is not very fast ([1], [2]). A very fast OCB based on SiC devices switching off in a few microseconds, is shown in [3]. However this OCB uses a normally-on JFET devices and is, therefore, not fail-safe with respect to the gate control.

2. Functional principle of the proposed electronic OCB

The proposed OCB is based on the so-called dual thyristor principle ([4], [5]) in which the sources of a p- and a n-channel semiconductor are linked together and the gates are connected to the corresponding drains. The simplest way to realise this principle is a circuit out of two JFETs (Fig. 1a). When current flows through the p-JFET, the voltage between S and Dp ($V_{S,Dp}$) rises. Because of $V_{S,DP} = -V_{Gn,S}$, the n-JFET starts to turn off and vice versa with current through the n-JFET. If the current exceeds a certain value, the JFETs enter a positive feedback loop. The gate voltages rise, resulting in an increased resistance, which increases the gate voltage again until the JFETs pinch-off and turn off the current completely. The OCB will not conduct again until the voltages $V_{Gn,S}$ and $V_{S,Gp}$ fall below the pinch-off voltages.

Unfortunately, this version can only be realised for very small currents and voltages due to the lack of commercially available JFETs (especially p-type) for high power. Instead, MOS-

© VDE VERLAG GMBH · Berlin · Offenbach

PCIM Europe 2016, 10 – 12 May 2016, Nuremberg, Germany

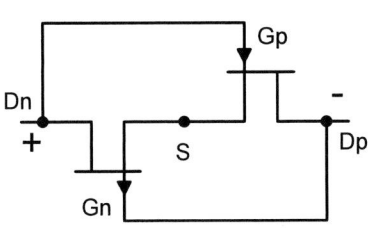

(a) The easiest way is the use of two normally-on devices like JFETs.

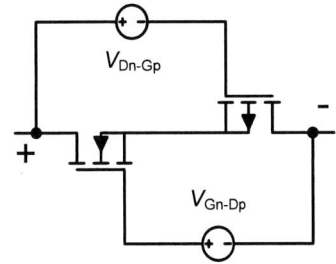

(b) Two normally-off MOSFETs need additional voltages.

Fig. 1: Simple realisations of the dual thyristor based OCB each with one p- and n-type device.

FET devices can be used, of which normally-off devices are the most common and advanced ones. When using normally-off devices, additional gate voltages need to be supplied as shown in Fig. 1b, in order to switch on the OCB. In case of a failure in the gate supply voltage, the OCB will switch off and is therefore fail-safe. The applied gate voltages can be used to adjust the trip value of the OCB and define the maximum current.

The result of a spice simulation of an OCB with MOSFETs is shown in Fig. 2. Here the models of the Infineon MOSFETs used for the experiment were used and the off-set voltages were varied. The voltage was ramped without a load inductance and any over voltage protection. It can be seen, that the current rises with the on-state voltage until the trip value is reached. Then the positive feedback begins and the OCB stops conducting with higher voltage.

When the OCB is switched off, the whole DC supply voltage is applied across the OCB. Assuming two devices with identical impedances half of the DC supply voltage is applied to each gate. With a maximum gate voltage of 20 V, the maximum overall voltage must not exceed 40 V. In order to switch higher voltages, the OCB can be cascoded using a device with a higher blocking capability. As the OCB is already fail-safe, a normally-on device can be used for the cascode switch requiring no additional offset voltage source.

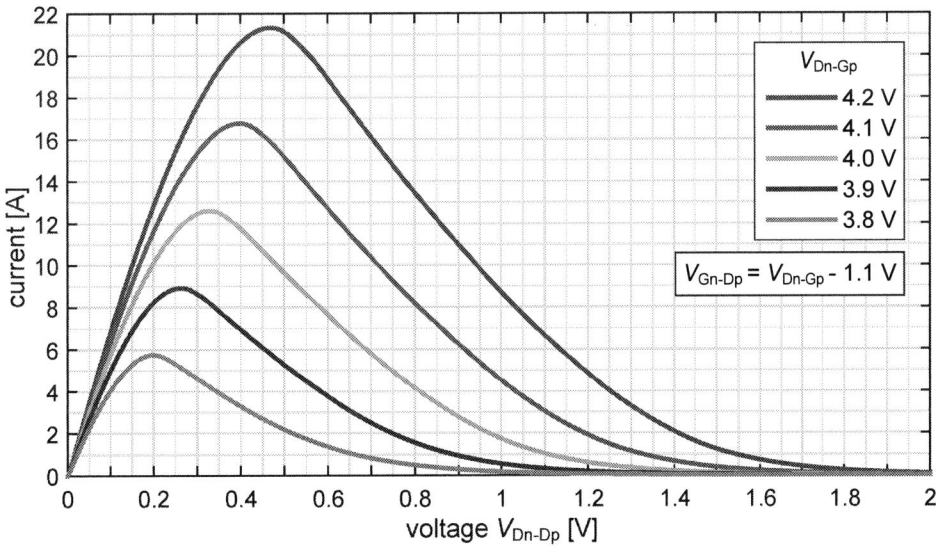

Fig. 2: Current versus voltage curves of a spice simulation of the circuit of Fig. 1b with models of the MOSFETs in Tab. 1. The trip value is defined by the offset voltages. A higher voltage leads to a higher current. The offset voltage for the n-MOSFET is set 1.1 V lower than for the p-type to get a symmetric turn-off and voltage distribution.

© VDE VERLAG GMBH · Berlin · Offenbach

3. Experimental setup

3.1. Cascoded OCB

A printed circuit board (PCB) test module was constructed (Fig. 3b). For the OCB part low voltage power MOSFETs of Infineon's OptiMOS 3 series are used, which both have a steep transfer characteristic. The cascode is build with silicon carbide (SiC) 1200 V JFETs having a gate threshold voltage of about -14 V also from Infineon. Detailed information about the devices can be found in Tab. 1.

Fig. 3a shows the circuit diagram of the cascoded OCB. Each of the MOSFETs has a pull-up or pull-down resistor respectively at the gate to preserve defined $V_{GS} = 0$ V and, therefore, ensure a turn-off of the OCB when the external voltage sources are disconnected. With respect to this the circuit is fail-save.

To reduce the overall on-state resistance we use three JFETs parallel for the over current switching tests, each with its own gate resistor. The JFETs have a positive temperature coefficient, such that the on-resistance rises with higher temperature, so that the device is very suitable for parallelisation.

The required offset voltages for the MOSFETs are generated externally and are supplied to the PCB (Fig. 3b) over pin headers with buffer capacitors close to the gates. By adjusting these voltages a control of the trip value of the over current is possible.

Because of the inductance in the load the switch need an over voltage protection. This is also realised with add-on boards, so one can easily change between different snubbers and varistors.

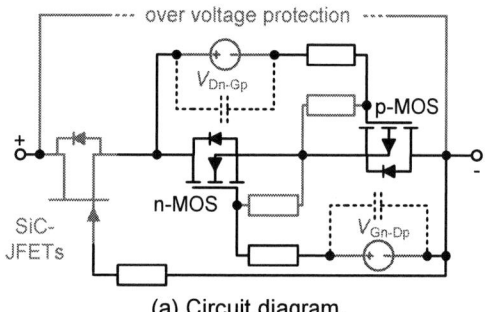

(a) Circuit diagram.

(b) Photo of the PCB realization with 3 JFETs.

Fig. 3: The cascoded OCB. The two voltage sources and the over voltage protection are connected to the pin headers.

		n-MOSFET BSC030N03MS G	p-MOSFET BSC030P03NS3 G	SiC JFET IJW120R070T1
Drain-source breakdown voltage	V_{DS}	30 V	-30 V	1200 V
Continuous drain current	I_D	100 A	-100 A	35 A
Drain-source on-state resistance	$R_{DS(on)}$	3.8 mΩ @ $V_{GS} = 4.5$ V	4.6 mΩ @ $V_{GS} = 6$ V	70 mΩ @ $V_{GS} = 0$ V
Gate-source voltage	V_{GS}	-20 V – 20 V	-25 V – 25 V	-19.5 V – 2 V

Tab. 1: Datasheet maximum values at $T_C = 25$ °C. All devices are from Infineon.

© VDE VERLAG GMBH · Berlin · Offenbach

3.2. Over current switching circuit

For testing the main function of the OCB an over current has to be applied to the DUT. Fig. 4 shows the schematic of the used test setup.

The source can provide up to 600 V and is buffered with capacitors for the high current peak. For over voltage protection RCD-snubbers with changeable capacitors or different varistors are used.

For the test, the offset voltages are provided by laboratory DC power supplies. The voltages are set to $V_{Gn\text{-}Dp} \approx 3$ V and $V_{Dn\text{-}Gp} \approx 4$ V. In order to make the measurements comparable these voltages were kept almost constant.

First the load draws a base current between 3 A and 7 A defined by the base load wire wound resistor R_{L1} with L_{L1}. This current is lower than the tripping current of the OCB. Optional a 100 µH air coil L_{Lo} was used as load. To emulate an over current failure a second low inductive and low resistive load (R_{L2}, L_{L2}) is switched in parallel, using a fast IGBT switch (S_1). The resistance and inductivity are reduced to $R_L = R_{L1} \| R_{L2}$ and $L_L = L_{L1} \| L_{L2} + L_{Lo}$, so the current rises and trips the OCB. When the voltage over the MOSFETs reaches the threshold voltage of the cascoded JFETs, the JFETs pinch-off and take the rest of the voltage.

To reactivate the tripped OCB the voltage has to be removed. This can be done by opening the switch S_2 and then discharge the OCB over a resistor with the push-button S_3.

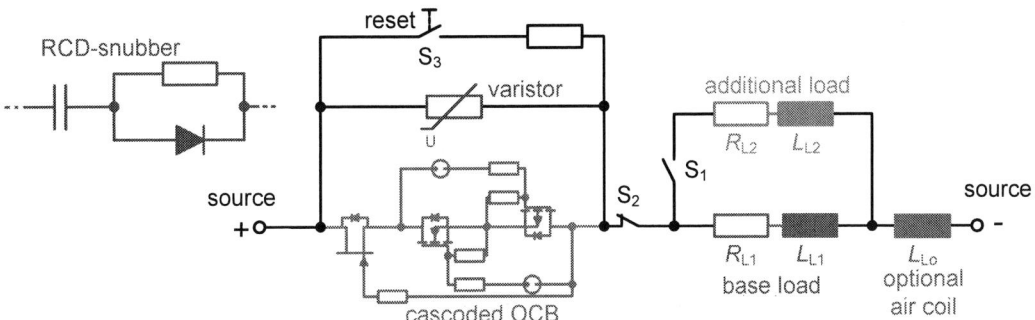

Fig. 4: Schematic of the over current switching circuit. To the base load an additional load can be switched in parallel with S_1 to simulate the over current failure. S_3 is to reset the OCB. Instead of the varistor one can also use a RCD-snubber for over voltage protection.

4. Experimental results

4.1. On-state resistance

The on-state resistance of the whole cascoded OCB was measured at a base load of 4 A with one to three parallel JFETs. The offset voltages were set to a tripping current of about 12 A. The results are listed in Tab. 2. With the assumption, that all JFETs have the same on-state resistance at all three measurements, for one JFET the on-state resistance is calculated to 48 mΩ and the two MOSFETs together have 17 mΩ in this setup.

The MOSFETs here have higher resistance compared to the datasheet but this is because the tripping current is only 8 A above the base current and the gate voltage according to this is lower than in the datasheet, so that the MOSFETs have lower channel conductance. The spice simulation for the inner OCB part (Fig. 2) also shows the increasing on-state resistance but slightly less than in the experiment.

© VDE VERLAG GMBH · Berlin · Offenbach

Number of JFETs	1	2	3
Forward voltage	260 mV	165 mV	133 mV
On-state resistance	65 mΩ	41 mΩ	33 mΩ

Tab. 2: On-state resistance of the whole cascoded OCB at 4 A forward current as a function of the number of parallel used JFETs.

For higher tripping currents the on-state forward voltage at low currents will be less than the values here.

4.2. Over current turn-off

The over current turn-off performance was investigated with RCD-snubbers and varistors. Due to the limited cooling of the MOSFETs in this test bench it is not possible to determine the exact trip value like in the spice simulations in Fig. 2. To emulate different grids, the load inductance was varied. In measurements with $L_L \geq 100$ µH the air coil L_{Lo} was used.

RCD-snubber

The turn-off behaviour was studied with different capacitances in the RCD-snubber at a source voltage of 600 V. The optimum value for this experimental setup is 0.33 µF. A lower capacitance results in too high over voltage and a higher capacitance reduces the voltage slope and so increases the over current.

Fig. 5 shows the voltage over the complete cascode with three JFETs and the current during the over current turn-off. At ① the switch S1 is closed and the current begins to rise. At ② the MOSFETs have already switched off due to the forward voltage feedback and the cascode begins to take significant voltage. This is the tripping point, which is also marked by the orange lines to indicate the tripping current level. Due to the capacitance of the snubber the switching is decelerated and the current can rise until the maximum ③ where the time axis is set to zero. This is because the current rises as long as the capacitor voltage (integral of the load current over time divided by the capacitance) has not reached the supply voltage.

The load inductance is the main influence for the current slope. This can be seen by comparing the two graphs in Fig. 5, on the left an inductance of 140 µH and right 40 µH. The tripping

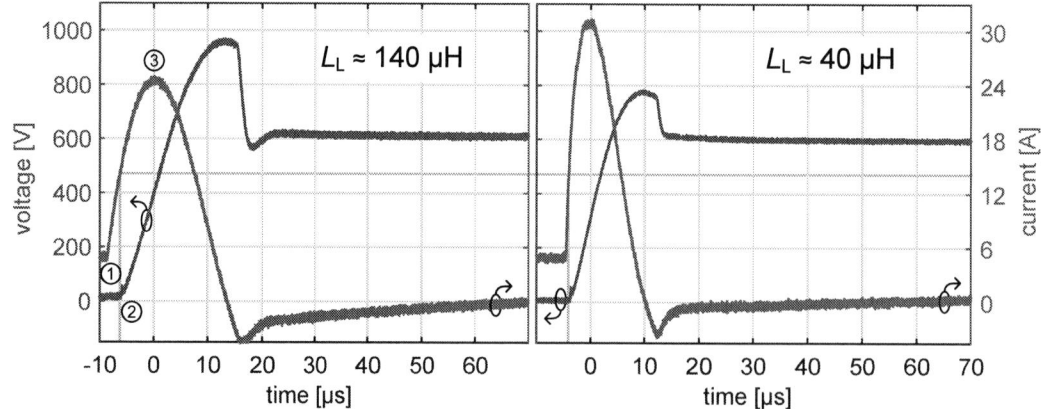

Fig. 5: Turn-off measurement with RCD-snubber with two different load inductances. The y-axis ticks are the same for both graphs. Lower inductance leads to less over voltage but more over current. The orange lines indicate the tripping current and time.

current is the same in both cases. With lower inductance the switching becomes faster and the over voltage is lower, but the over current gets higher. So the RCD-snubber is not suited for fast switching and low inductances but the turn-off is very smooth. The snubber circuit has the advantage to be more reliable than varistors presented below, which may show aging effects with each clamping event.

Varistor

Alternative to the RCD-snubber different metal-oxide varistors (MOV) were used to dissipate the energy of the inductivity. Fig. 6 shows turn-off curves for two different varistors with two different loads (100 µH and 8 µH with the additional load only a cable) at a source voltage of 400 V. The varistors both are from EPCOS. Varistor A is a S20K300 with a varistor voltage of about 510 V and Varistor B a S20K420 with a varistor voltage of about 680 V.

Compared with the RCD-snubber the voltage rises much faster and the current slope is only limited by the load inductance. Due to the very fast switching the over current is much lower. This can be seen especially at the 100 µH graphs where the maximum current reached is negligibly higher than the tripping current. For the very low inductance of only 8 µH, the current is so steep, that there is an overshoot. However, it does not reach critical values.

Varistor B allows higher voltages so that the energy is dissipated very fast and the time elapsed from tripping of the OCB to the first zero crossing of the current is less than 1 µs when switching 8 µH (bottom right in Fig. 6). A zoom into this curve is shown in Fig. 7. Here the time scale is in nanoseconds. The tripping current is about 13 A so that the "reaction time" from tripping to the time when the current starts to decrease is under 200 ns.

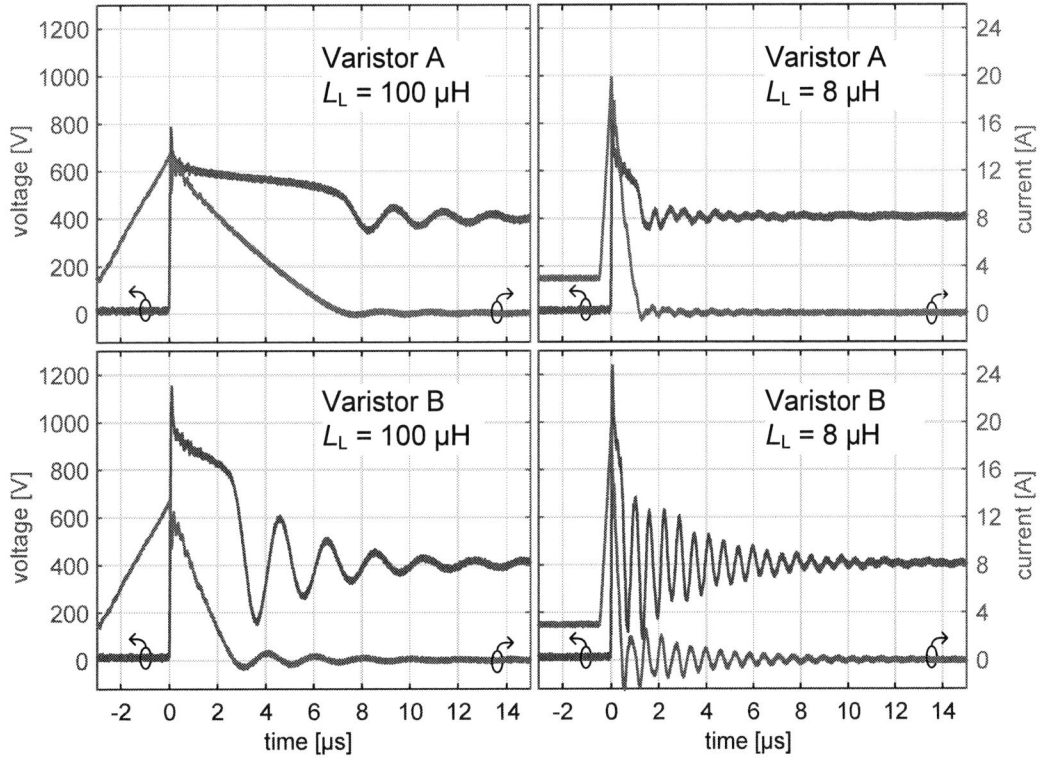

Fig. 6: Turn-off measurement with two different varistors and two different load inductances. All graphs have the same axis ticks. The varistor B with the higher clamping voltage results in faster switching with higher over voltage. Low inductance leads to a very fast current transient.

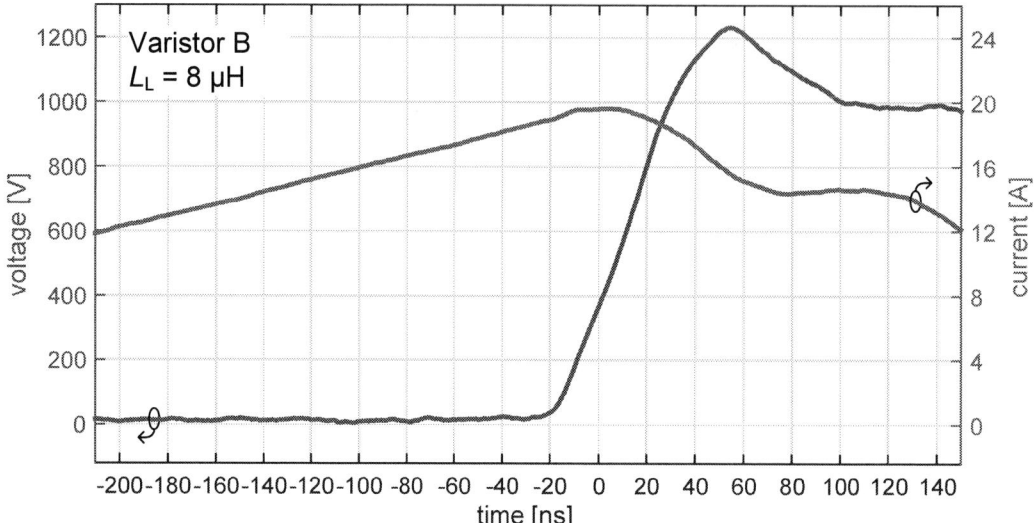

Fig. 7: Zoom in the turn-off curve with a load of 8 µH and Varistor B. After taking voltage it lasts only about 20 ns to the point where the maximum current is reached.

The very fast switching leads to oscillations but especially for the current the amplitude is very low and not critical.

Over all, the investigated cascoded OCB switches off over-currents safely and reproducibly and is fail-safe against the loss of the necessary offset voltages.

5. Conclusions

A novel high speed over current breaker based on the dual thyristor principle has been proposed. The realisation is based on advanced Si low voltage MOSFETs, which are cascoded with SiC-JFETs or alternatively with super junction MOSFETs for higher blocking capability. With the required gate offset voltage sources the threshold current can be adjusted. By using a varistor, the OCB turns off within microseconds and, thus, can cope with the variety of stray inductance down to a nearly short circuit in DC circuits. This way, the OCB is a promising candidate for a general purpose DC breaker.

6. Acknowledgements

This work is funded by the Federal Ministry of Education and Research of the Federal Republic of Germany under grant 16ES0116.

7. References

[1] D. P. Urciuoli, V. Veliadis, H. C. Ha, and V. Lubomirsky, "Demonstration of a 600-V, 60-A, bidirectional silicon carbide solid-state circuit breaker," in *Applied Power Electronics Conference and Exposition (APEC), 2011 Twenty-Sixth Annual IEEE*, 2011, pp. 354–358.
[2] Y. Sato, Y. Tanaka, A. Fukui, M. Yamasaki, and H. Ohashi, "SiC-SIT Circuit Breakers With Controllable Interruption Voltage for 400-V DC Distribution Systems," *Power Electron. IEEE Trans. On*, vol. 29, no. 5, pp. 2597–2605, May 2014.
[3] Z. J. Shen, G. Sabui, Z. Miao, and Z. Shuai, "Wide-Bandgap Solid-State Circuit Breakers for DC Power Systems: Device and Circuit Considerations," *Electron Devices IEEE Trans. On*, vol. 62, no. 2, pp. 294–300, Feb. 2015.

© VDE VERLAG GMBH · Berlin · Offenbach

[4] J. L. Sanchez, M. Breil, P. Austin, J.-P. Laur, J. Jalade, B. Rousset, and H. Foch, "A new high-voltage integrated switch: the «thyristor dual» function," in *Power Semiconductor Devices and ICs, 1999. ISPSD '99. Proceedings., The 11th International Symposium on*, 1999, pp. 157–160.

[5] B. Rosensaft, U. R. Vemulapati, and D. Silber, "Circuit Breaker and Safe Controlled Power Switch," in *Power Semiconductor Devices and IC's, 2007. ISPSD '07. 19th International Symposium on*, 2007, pp. 169–172.

Wireless Power Transmission with High Efficiency for Extensive Applications

Rehm, Markus, IBR Ingenieurbuero Rehm, Germany, rehm@ib-rehm.de

Abstract

Wireless power transmission has been known for many years, with inductive near field proximity coupling being the most commonly used technology. Among the publications, there are the typical technical papers describing the physical behavior of the system but there is also a definite trend attempting to impose specific features on standards (WPC [1], A4WP [2] and PMA [3]).

After an enthusiastic start the industry seems to have realized that this is a difficult technology grow and as a result the development of wireless power is hampered. The demand for more power, scalable systems and federal emission constraints are the main barriers to development.

The „universal Wireless Power" („uniWP") presented here represents a new solution to overcome the existing technology barriers.

1. Known arts and their drawbacks

Traditionally, wireless power transmission uses coupled resonance circuits, where the power transmission reaches its maximum when the resonance frequency of the two circuits are identical and the transmitter operates at this resonance frequency.

Unfortunately the resonance frequency of this coupled arrangement changes due to variations or drifts in the components (tolerances, aging and temperature) and in the coupling (mis-positioning- or geometric changes between transmitter and receiver). Lastly, even load changes on the receiver side cause the resonance frequency to change as well.

- The solution adopted by WPC´s QI [1] and others senses the resulting resonance frequency when a receiver becomes coupled with a transmitter. The operation frequency is then set to the identified resonance frequency under this static coupling condition. This system is limited in its flexibility e.g. the system is sensitive to a change of the number of coupled receivers while operating and dynamic coupling conditions (variable geometry due to vibrations or shaking etc.) are not considered at all. Additionally, the tight standardization down to circuitry- and component level severely limits adaptability to future evolutions.

- Other solutions adjust a generator frequency that drives the resonant circuit in a closed loop manner to the actual resonance frequency and makes the system insensitive to dynamic coupling conditions. Unfortunately, the loop stability is largely dependent on the resonant circuit quality and there is no possibility to determine the resonance frequency actively. In fact, the generator is the slave of the various parameters of the resonant circuit.

- The solution according to [4] can control the resonance frequency of the resonator actively but linear operation is not possible because the resonance frequency is not only dependent on a control input but also on the current/voltage amplitude in the resonator. This requires a costly loop which is sensitive to the quality of the resonance network (load dependent loop behavior) and therefore slows down the loop response dramatically.

Last but not least, all known approaches to wireless power transmissions limit their operation to a more or less predetermined coupling factor. In fact, generally a system is designed to operate in either a loose- or tight coupling condition. Further, operation near the physical limit representing an approximated wired connection between transmitter and receiver was avoided due to instability problems (risk of bifurcation i.e. over coupling).

© VDE VERLAG GMBH · Berlin · Offenbach

2. Coupled resonant circuits equivalent to a wired connection?

Figure 1a depicts an inductively coupled wireless power transmission link using leakage compensation networks on its primary and secondary side.

Figures 1b and 1c depict simplified equivalent circuits using different resonant circuit topologies for the leakage compensation.

Depending on the chosen resonant circuit topology, different behaviors result on the quality factors (Qs) in the secondary circuit. Actually there are four possible variants; the ones not shown behave similarly [5].

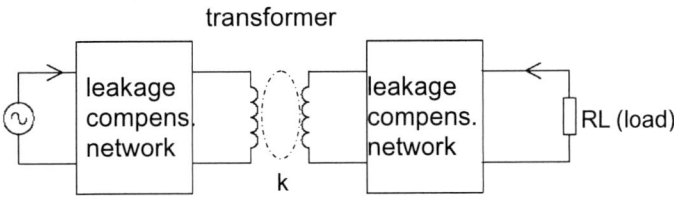

Fig. 1a: Inductively coupled wireless power transmission link using leakage compensation networks

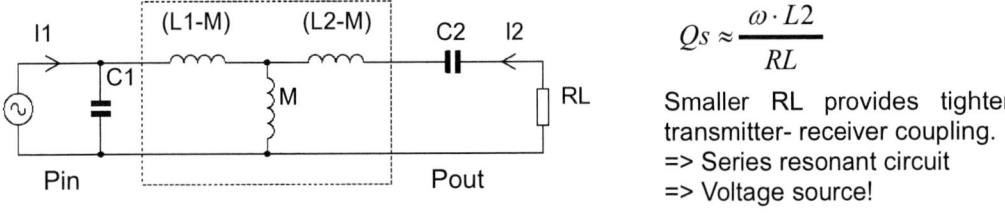

$$Qs \approx \frac{\omega \cdot L2}{RL}$$

Smaller RL provides tighter transmitter- receiver coupling.
=> Series resonant circuit
=> Voltage source!

Fig. 1b: Transformer equivalent circuit with parallel- series resonant circuits

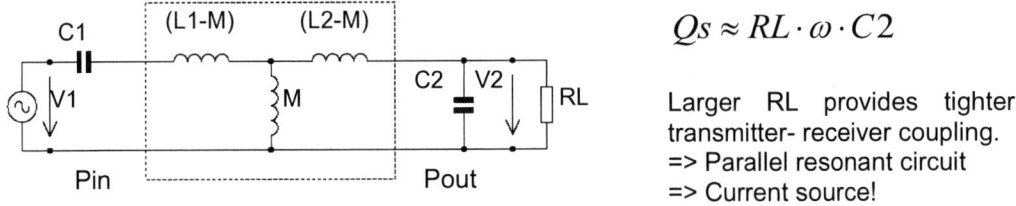

$$Qs \approx RL \cdot \omega \cdot C2$$

Larger RL provides tighter transmitter- receiver coupling.
=> Parallel resonant circuit
=> Current source!

Fig. 1c: Transformer equivalent circuit with series- parallel resonant circuits

Generally, the higher Qs is, the more coupling factor (k) can be compensated (see Figure 2). One defines: Undercritical coupling, critical coupling and overcritical coupling.
The critical coupling (green curve) is the equivalent of a wired connection. A further increase in Qs moves the system into an over coupled condition state (bifurcation), wherein there is no stable resonance frequency any longer (red). Operation in this state should be avoided in wireless power transmission because it increases losses and allows the frequency spectrum to become uncontrollable. To avoid overcritical operation common systems work in undercritical coupling (yellow) which results in very low efficiency.

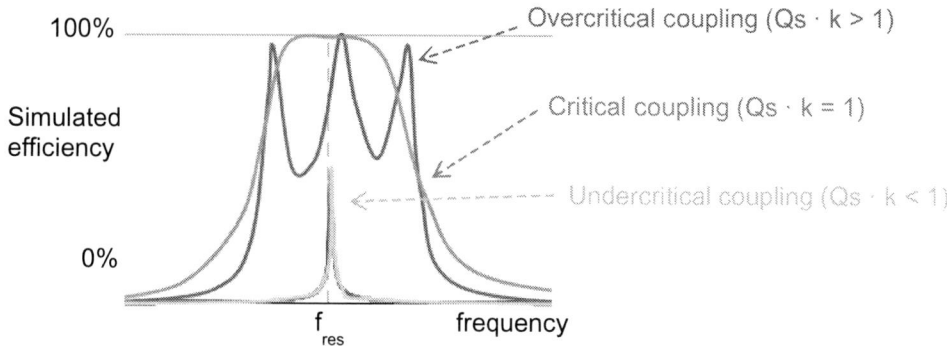

Fig. 2: Theoretical efficiency in different coupling conditions

3. uniWP offers a wide dynamic range for wide range applications

The new technology „uniWP" is a layered implementation whose power transmission layer comprises several essential blocks, see figure 3.

One is a large signal VCO, that is itself part of a PLL. In uniWP the transmitter operates in a frequency- resp. phase control loop that is closed over the resonant circuit. Consequently the resonance frequency is tightly bound to a reference frequency. This results in an output spectrum characteristic whose fundamental frequency has the accuracy of the reference frequency. This accuracy can be chosen to be arbitrarily high. Any detuning effects in the primary- and / or secondary resonant circuits, or changes in the coupling- and / or load conditions immediately trigger a retuning action in the resonant circuit until the resonance frequency is returned to the original value, and any phase difference is eliminated.

uniWP comprises over coupling detectors (det) that are implemented in the transmitter and in the receiver for redundancy.

Another essential block (VCC Cont) controls the transmitted power and also the load (RL) coupling in the receiver. When an over coupling condition is detected by "det", the transmitter interrupts power transmission. After a pause it retransmits power until "det" detects the next over coupling condition. The response time of "det" in the transmitter determines the minimal duration of a transmission session. Similarly, the load will become immediately decoupled in the receiver as soon as "det" signals an over coupling condition. Consequently, according to the explanation of fig. 1b, Qs drops and the over coupling condition is inhibited. After a pause the receiver couples the load again until det signals another over coupling condition.

Essentially, the system defines a longer response time for det in the transmitter than for det in the receiver. In this manner it is guaranteed that the load always becomes decoupled before the transmitter interrupts power transmission. This strategy allows uniWP to become coupling factor- and Qs tolerant. The maximum power transmission is limited to the critical coupling condition.

© VDE VERLAG GMBH · Berlin · Offenbach

Another essential block comprises a load filter, whose main task is energy storage and decoupling the resonant circuits from rapid load changes. This prevents spectral disturbances.

Another essential block comprises the status communication means. These are basically similar to those known in the art.

The high level of flexibility in the system is apparent and since there are only a few predefined constraints there is a great freedom for implementation:
A small coupling factor in the receiver can be compensated with higher Qs.
The load control in the receiver can be part of a post output regulator (see Figure 3). This provides superior load transient behavior.

It is also worth mentioning that all four variants of coupled resonant circuits are possible.

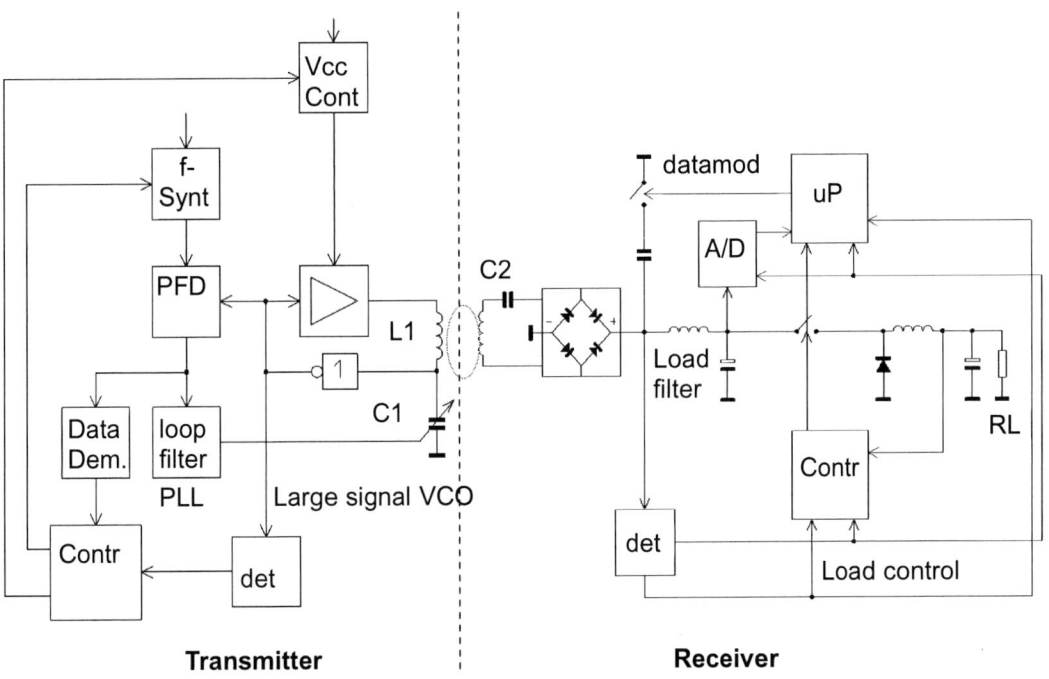

Figure 3: Block diagram of a wireless „uniWP"- transmission link example implemented as a voltage source (transmitter according to figure 1c, receiver according figure 1b)

4. New solution „uniWP" with high efficiency

If the resonant circuit is driven precisely at its resonance frequency, the wireless transmission link behaves like a real transformer. This automatically results in the optimum matched condition since the real load is directly transformed to the transmitter side. This is exactly what occurs in „uniWP" technology.

The maximum efficiency is achieved under the critical coupling condition as explained above. Additionally, the efficiency is also dependent on ohmic losses. The transmitter efficiency and the harmonics behavior are determined by the total quality (Qtot) and the cross current in M. If Qtot is high, there is a high level of kinetic energy circulating in the resonant circuit and this results in lower efficiency. But the harmonics suppression is higher due to the higher filter selectivity. On the other hand, if Qtot is low, there is a low level of kinetic energy circulating in the resonant circuit and this results in higher efficiency. But the harmonics suppression is lower due to the lower filter selectivity.

In this manner „uniWP" offers great flexibility to designers who can prioritize the benefits. The „uniWP" concept was verified with a regulated output voltage on the secondary side and a parallel push-pull large signal VCO in the transmitter using an arbitrary loop antenna (Figure 4). A common PLD circuit provides almost all non power functions in Figure 3.

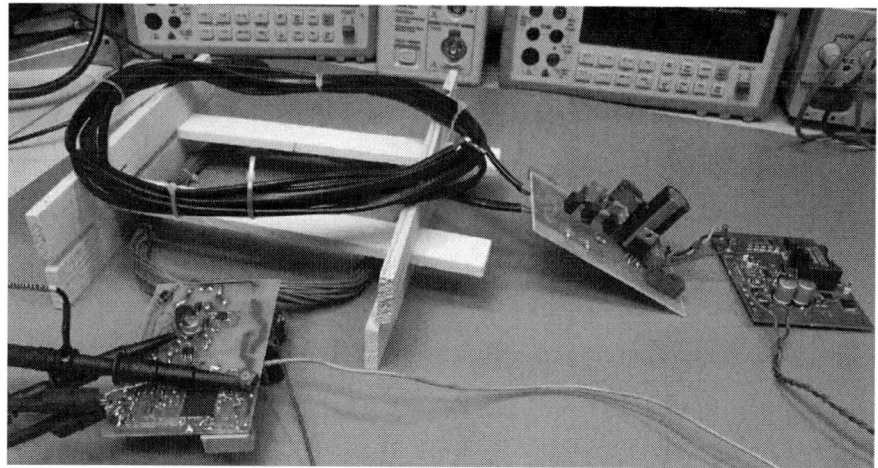

Figure 4: One example of „uniWP": Transmitter with blue loop antenna. Receiver with black loop antenna, synchronous rectification und buck-boost-converter for a controlled output, here 24V.

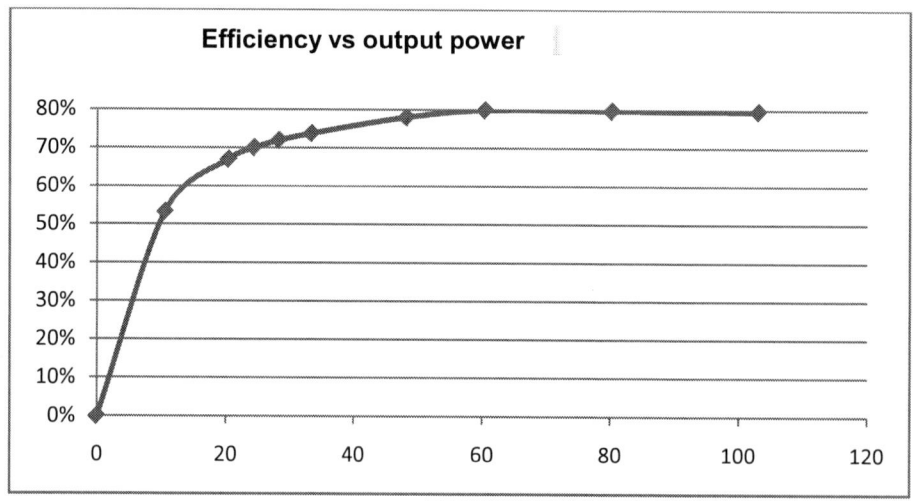

Figur 5: Efficiency of example as shown in figur 4. There is no auxilliary supply! The distance between transmitter and receiver is approx. 4cm. The output voltage is a controlled 24V.

© VDE VERLAG GMBH · Berlin · Offenbach

5. New solution „uniWP" improves EMI

Thanks to the high dynamic in the new „uniWP" large signal resonance frequency control loop, arbitrary frequency spectrums can be generated through software (frequency synthesizer).
Figure 6a depicts a discrete output transmission frequency peak (134kHz) and Figure 6b shows sweep operation (120 … 134kHz).
The „uniWP" frequency spreading feature allows a reduction in the spectral density of 10dB for the same transmitted power!

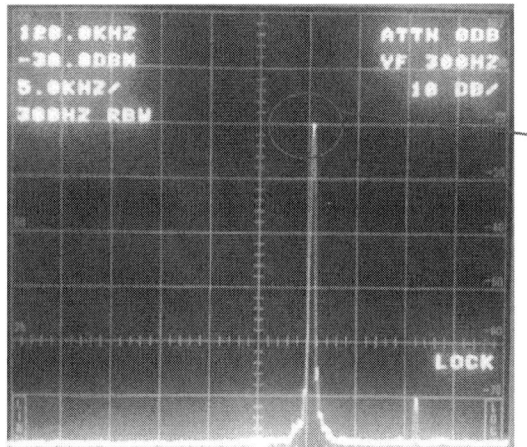

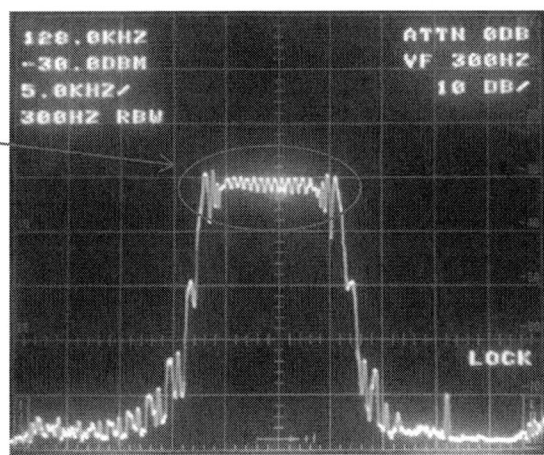

Figure 6a): Transmitter output spectrum in discrete frequency operation

Figure 6b): Transmitter output spectrum in sweep frequency operation

In this manner federal frequency standard EN300330 (Figure 7) can easily be achieved by software without any hardware change (synthesizer frequency data). E.g. frequency notches are generated at excluded frequencies in the swept frequency hopping mechanism. EN300330 and others specify the maximum levels in the specific frequency range.

Consequently, thanks to the frequency spreading feature, „uniWP" can transmit higher power levels than all other wireless transmission solutions.

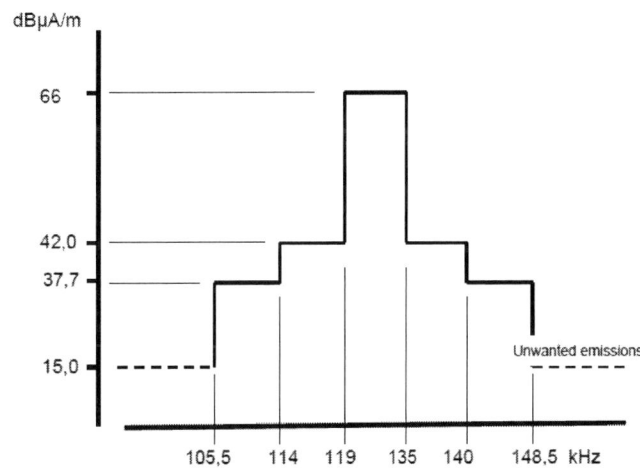

NOTE: The limit at 129,1 kHz ± 500 Hz is maximum 42 dBµA/m at 10 m.

Figure G.1: Boundary for LF RFID systems

Figure 7:   Excerpt from EN300330

© VDE VERLAG GMBH · Berlin · Offenbach

6. Summary

The concept to transmit power over a wireless link according to „uniWP" allows multiple new applications and features. The advantageously insensitive operation even in harsh and dynamic coupling environments allows the system to be operated at the maximum physical limit. The simple power scalability far outperforms existing solutions. The research activities driven by cost, reliability and practical constraints demonstrate uniWP is a good alternative for future product implementations and is well suited for future evolution.

	Known arts	uniWP
Resonant frequency is not inside the allowed band	no power transfer	maximum power transfer, because of large signal VCO
Dynamic load and coupling condition	no power transfer	maximum power transfer, due to fast resonance tracking
Efficiency	low	High, due to critical coupling
Power	a few watts	up to 200W (1kW planned)
Area of application	different for each region	worldwide (frequency management sw.)
Range	small range	long range due to high quality factor
Number of receiver	one	several
Coupling	tight	tight or loose
EMI performance	low	high
Flexibility	no (standard)	high
Data transfer included	yes	yes
Cost	high, small tolerance components	low, standard components

References

[1] WPC: http://www.wirelesspowerconsortium.com/
[2] A4WP: http://www.a4wp.org/index.html
[3] PMA: http://www.powermatters.org/
[4] Si, Hu, etc: A Frequency Control Method for Regulating Wireless Power to Implantable Devices, IEEE 1.3.2008
[5] S. Chopra, P. Bauer: Analysis and Design Considerations for a Contactless Power Transfer System, IEEE 2011

Prevention of Traction Drives Stator Insulation Faults Based on Overvoltage Reduction Utilizing Active Edge Shaping

C. Zoeller[1], M.A. Vogelsberger[3], T. Hausberger[2], M. Blank[2], T. Glueck[2], H. Ertl[1], A. Kugi[2]

TU Wien, [1] Institute of Energy Systems and Electrical Drives / [2] Automation

and Control Institute, Gusshausstraße 25-29, 1040 Vienna, Austria

[3] Bombardier Transportation PPC-Drives Development Center 1, Hermann Gebauer Straße 5

1220 Vienna, Austria; E-Mail: markus.vogelsberger@rail.bombardier.com

Abstract

Modern inverter-fed traction drives frequently operate near and even above their rated values leading to high strains on the machine and especially on its winding insulation system. With new emerging semiconductor technologies higher inverter switching frequencies will be possible and high inverter dv/dt rates appear, resulting in transient overvoltages at the machine stressing the insulation system. A strategy to avoid/minimize such overvoltages is proposed by active edge shaping of the inverter. The semiconductor switches (IGBTs) are driven by a feedforward gate current profile which is adaptively modified cycle-by-cycle such that finally a Gaussian-shaped reference switching profile of the inverter's output voltage is achieved. First, experimental results of this concept tested at a 1.4MW induction machine for railway application are presented. Additionally, a brief overview of the characteristics and parameters influencing the overvoltage oscillation at the stator winding of inverter-fed machines will be given.

1. Introduction

The fast switching transients of modern voltage source inverters used in today's traction drive applications lead to transient overvoltages of the machine's insulation system. Additionally, with new emerging wide-bandgap semiconductor technologies, the increased switching speed of the inverter (high dv/dt rates) causes additional stress on the motor winding and may lead to a deterioration of the insulation strength and finally to a fault. The insulation breakdown usually is a slowly proceeding process, starting with the deterioration of the insulation and then leading to severe turn-to-turn, phase-to-phase and finally phase-to-ground short circuit. To increase the reliability different strategies like fault tolerant design, condition monitoring as well as protection systems (e.g., electrical filters) can be implemented. A survey of additional stress factors, e.g., thermal, mechanical etc. is given in Fig.1. These main causes responsible for a machine breakdown have been analyzed in [1]-[6].

In this paper, an enhanced method for preventing motor overvoltages in drive systems based on the usage of a smart inverter power switch and active control of the switching transients is presented. The desired switching transition control is based on adaptive gate current profiles implemented with an Iterative Learning Control (ILC) strategy. Instead of simply controlling the maximum slew rate of the inverter output voltage, which will still excite voltage ringing in the winding system, a limitation also of the higher-order derivatives of the switching transient is introduced. In accordance with [7], the best approach to accomplish these requirements is a Gaussian switching profile. The performance of the control strategy is demonstrated by a series of measurements conducted on a 1.4MW induction machine for railway application which are presented in a later section.

2. Drive System Components and Transient Overvoltage

A traction drive system can be described in principle by its three main components, the inverter, the cabling and the machine. These components define a complex impedance system that can be represented by the equivalent electrical components depicted in Fig.2. For the inverter (yellow box) the ground impedance, in particular its capacitive coupling $C_{Inv\text{-}Gnd}$ to ground is considered. The cabling system (green box) is modeled by per unit length inductance L_{Cable} and resistance R_{Cable}. Furthermore, phase-to-phase and phase-to-ground parasitic capacitances ($C_{Ph\text{-}Ph}$, $C_{Ph\text{-}Gnd}$) are considered, which substantially influence the behavior of the system at higher frequencies. Finally, for the motor (blue box) the basic parameters are its stator resistance R_S and the stator inductance L_S. In addition parasitic winding capacitances, i.e., $C_{Ph\text{-}Gnd}$, $C_{Ph\text{-}Ph}$ and $C_{Turn\text{-}Turn}$ are included which largely influence the high frequency behavior and consequently the transient overvoltages at the machine.

Due to the mismatch of the cable and machine impedance and high dv/dt-rates of modern semiconductor switching devices, transient overvoltages at the machine terminals in a traction drive system appear. According to [8] and [9], this phenomenon can be described by traveling wave theory. The machine's impedance is by far higher than the cable's characteristic impedance. Hence, the reflection coefficient is close to +1. A voltage step generated by the inverter and showing a rise time being small in comparison to the cable's propagation delay therefore is theoretically doubled at the machine terminals [10]. For practical traction systems, these idealized relations are valid merely approximately. Nevertheless, overvoltage oscillations, decaying in amplitude, will result at and even within the motor showing typical frequency components in the range of tens kHz to tens MHz [11]. For increased cable length, the oscillation frequency is decreased [9]-[10] in contrast to the magnitude of the overvoltage, which is increased [11]. The increased resistance of the conductors (skin effect) improves the damping of the ringing voltages at higher frequencies [9], [11]. On the contrary, if the resonance frequency of the cable and the machine are close, additional overvoltages may occur.

3. Test Bench and Control Strategy to Prevent Overvoltages

The experimental analyses are carried out on a 1.4MW (4-pole) squirrel-cage induction machine. To verify the effect of the applied edge-shaping control strategy on the voltage transient and voltage peaks, respectively, the machine is equipped with specific fiber-insulation wires with winding taps accessible at the terminal connection block. Figure 3 depicts the test bench and shows the used wiring of the machine (blue shaded area) and the voltage measurement positions taken from the 1st, 2nd, 6th and 12th coil of phase L1 (accessible through the taps) with respect to the terminal connection point of phase L3. The length of the machine cabling is 10m. Both ends of phase L2 and L3 are connected to the positive DC-link potential. Phase L1 is connected to the collector of the power-semiconductor switch T1. The overvoltage at different machine winding positions is recorded and analyzed in several

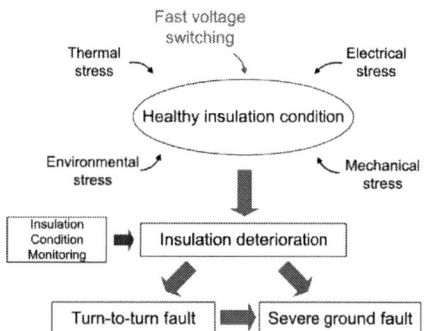

Fig.1 Root causes of insulation deterioration and resulting failure.

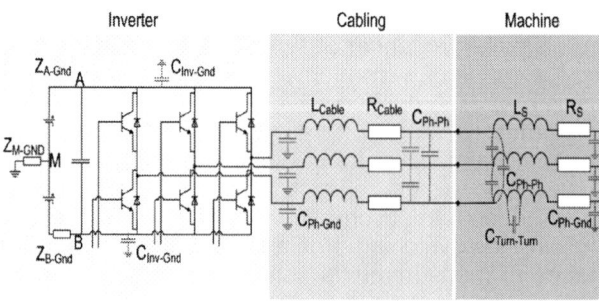

Fig.2 Schematic view of the complex system inverter, cabling and machine with parasitic capacitances.

© VDE VERLAG GMBH · Berlin · Offenbach

PCIM Europe 2016, 10 – 12 May 2016, Nuremberg, Germany

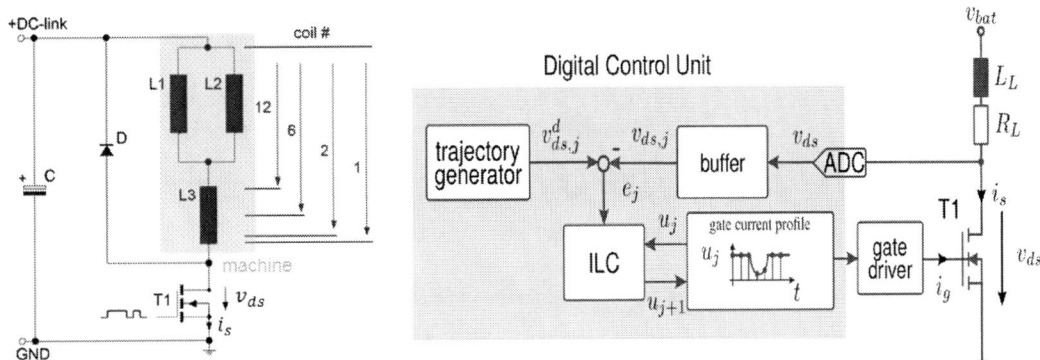

Fig.3 Test bench arrangement and
voltage measurement positions.

Fig.4 Block diagram of the digital control strategy.

experiments with changed test setups, e.g., different gate resistors for the gate driver to analyze the effect of dv/dt-rate and different control strategies for the iterative switching control.

Figure 4 depicts the block diagram of the used digital control. The digital control strategy tracks a desired output voltage $v_{ds}^d(t)$ using an adaptive feedforward gate current profile. The gate driver is built as a digitally controllable current source which provides a gate current profile for turn-on and turn-off of T1. The depicted elements L_L and R_L represent the load (machine of Fig. 3). The resulting switch current $i_s(t)$ and the drain-source voltage $v_{ds}(t)$ are sampled and discretized by an ADC unit (implemented on an FPGA extension board) at high sampling rates T_s. Based on the set of N samples which are taken for each switching instant of T1 and which subsequently are compared to a predefined desired $v_{ds}^d(t)$ or $i_s^d(t)$ reference switching trajectory, an adaptive controller (iterative learning controller (ILC)) iteratively adapts the gate current profile until a minimum tracking error between the desired and the actual shape of $v_{ds}(t)$ is achieved. For the sake of brevity, further implementation details and descriptions of the iterative learning control strategy are omitted here and given in [12], [13].

4. Experimental Results

In Fig. 5, the results of the measured line-to-line voltage at the 1st coil (first subfigure), 2nd coil (second subfigure), 6th coil (third subfigure) and 12th coil (fourth subfigure) of phase L1 with respect to phase L3 for a gate resistor value of R_G=44Ω (blue traces) compared to active edge shaping control (Gaussian-shape switching, "Gauß"; green traces) are depicted.

A voltage overshoot at the beginning of the switching transition is clearly visible with the highest amount visible at the 1st coil of phase L1 (first subfigure). The line-to-line voltage shows about 20% overshoot (referred to as $U_{DC\text{-}Link}$) for R_G=44Ω at a dv/dt-rate of 1kV/µs. The waveforms with the active control strategy show a very smooth and well damped behavior and no overvoltage occurs, which in particular also holds true for the inner winding positions. This achievement will clearly reduce the additional stress level for the motor winding insulation (at converter supply) and will therefore strengthen the lifetime of the traction motor.

Measurement results with the boundary condition of the same switching losses for both switching configurations (standard switching operation and active Gaussian-shaped switching) are presented in Fig. 6. The figure depicts the voltage traces for the 1st coil (first subfigure), 2nd coil (second subfigure) and 6th coil (third subfigure) for standard switching operation and gate resistance configuration of R_G=100Ω (red traces) compared to the traces of the active Gaussian-shaped switching trajectory (gate resistance is still R_G=44Ω). The switching losses are determined by evaluating $v_{ds}(t)$ and $i_s(t)$ to 49mWs. As can be seen, the overvoltage is significantly reduced in all waveforms with the active control strategy.

© VDE VERLAG GMBH · Berlin · Offenbach

PCIM Europe 2016, 10 – 12 May 2016, Nuremberg, Germany

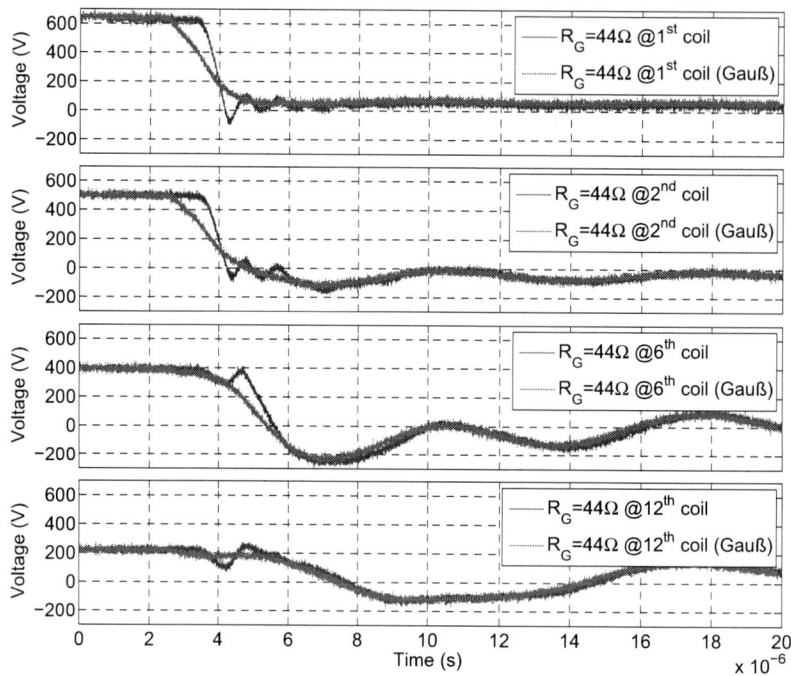

Fig.5 Line-to-line voltage traces at standard switching (blue traces; $R_G=44\Omega$) and active Gaussian edge shaping (green traces; $R_G=44\Omega$).

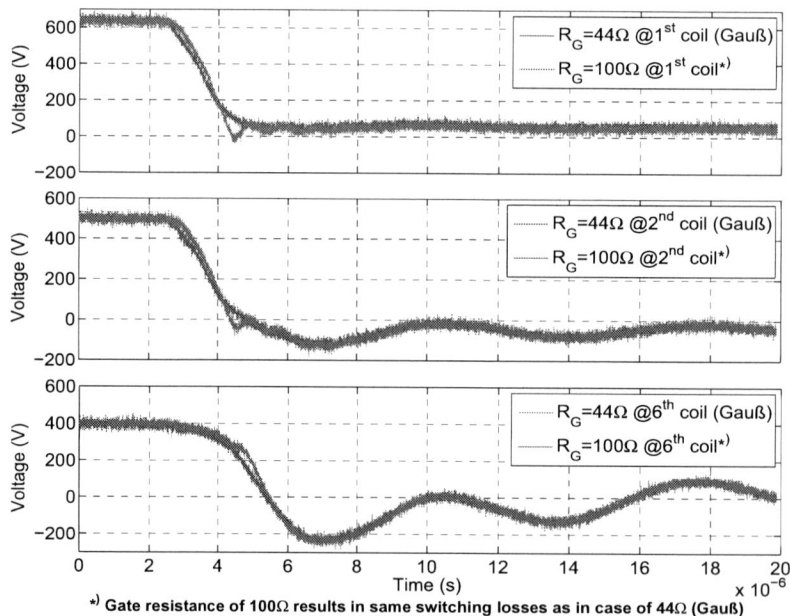

*) Gate resistance of 100Ω results in same switching losses as in case of 44Ω (Gauß)

Fig.6 Line-to-line voltage traces at standard switching (red traces; $R_G=100\Omega$) and active Gaussian edge shaping (green traces; $R_G=44\Omega$).

© VDE VERLAG GMBH · Berlin · Offenbach

The proposed method shows an approach to reduce the overvoltage and the high frequency resonances occurring at the switching transition of inverter-fed drives to inhibit the insulation degradation process. Although the investigations are conducted at low dv/dt-rate ($\leq$ 1kV/µs), which is due to the low sampling rate of the ADC in the digital control unit, a significant reduction of the switching transients is achieved. This effect is expected to become even more significant at higher dv/dt-rates. Furthermore, the application of the suggested active control strategy enables a reduction of the overvoltages at the machine compared to the standard switching operation with increased gate resistance at approximately the same switching losses.

5. Conclusion

It is shown, based on an initial experimental setup, that the machine-terminal switching overvoltages and the high frequency resonances of inverter-fed traction drives can be effectively reduced by active control of the switching transition. A Gaussian-shape switching transition seems to be a good approach to achieve this goal. Hence, the transient overvoltage stressing of the traction machine stator winding insulation system will be clearly reduced. A certain trade-off between switching losses and overvoltage reduction has to be accepted leading most likely to a substantial improvement of the reliability of the insulation system of modern traction drives, which, however, has to be verified by additional material-oriented research running in parallel at present.

6. References

[1] IEEE Committee Report; "Report of large motor reliability survey of industrial and commercial installation," *IEEE Transaction on Industry Applications*, vol.21, pp.853–864, 1985.

[2] Farahani, M.; Gockenbach, E.; Borsi, H.; Schaefer, K.; Kaufhold, M.; "Behavior of machine insulation systems subjected to accelerated thermal aging test," *IEEE Transactions on Dielectrics and Electrical Insulation*, vol.17, no.5, pp.1364-1372, 2010.

[3] Nussbaumer, P.; Wolbank, T.M.; Vogelsberger, M.A.; "Separation of disturbing influences on induction machine's high-frequency behavior to ensure accurate insulation condition monitoring," *28th Annual IEEE Applied Power Electronics Conference and Exposition (APEC)*, pp.1158-1163, 2013.

[4] Nussbaumer, P.; Vogelsberger, M.A.; Wolbank, T.M.; "Induction Machine Insulation Health State Monitoring Based on Online Switching Transient Exploitation," *IEEE Transactions on Industrial Electronics,* vol.62, no.3, pp.1835-1845, 2015.

[5] Yang, D.J.; Cho, J.; Lee, S.B.; Yoo, J.Y.; Kim, H.D.; "An Advanced Stator Winding Insulation Quality Assessment Technique for Inverter-Fed Machines," *IEEE Transaction on Industrial Applications*, vol.44, no.2, pp.555-564, 2008.

[6] Kaufhold, M.; Aninger, H.; Berth, M.; Speck, J.; Eberhardt, M.; "Electrical stress and failure mechanism of the winding insulation in PWM-inverter-fed low-voltage induction motors," *IEEE Transactions on Industrial Electronics*, vol.47, no.2, pp.396-402, 2000.

[7] Patin, N.; Vinals, M.L.; "Toward an optimal Heisenberg's closed-loop gate drive for Power MOSFETs," *IECON 2012 - 38th Annual Conference on IEEE Industrial Electronics Society*, pp.828-833, 2012.

[8] Nussbaumer, P.; Zoeller, C.; Wolbank, T.M.; Vogelsberger, M.A.; "Transient distribution of voltages in induction machine stator windings resulting from switching of power electronics," *IECON 2013 - 39th Annual Conference of the IEEE Industrial Electronics Society,* pp.3189-3194, 2013.

[9] Kerkman, R.J.; Leggate, D.; Skibinski, G.L.; "Interaction of drive modulation and cable parameters on AC motor transients," *IEEE Transactions on Industry Applications*, vol. 33, pp. 722-731, (1997).

[10] Persson, E.; "Transient effects in application of PWM inverters to induction motors," *IEEE Transactions on Industry Applications*, vol. 28, pp. 1095-1101, (1992).

[11] Peroutka, Z. and Kus, V.; "Adverse effects in voltage source inverter-fed drive systems," 17th *Annual IEEE Applied Power Electronics Conference and Exposition (APEC)*, pp. 557-563, 2002.

[12] Blank, M.; Glueck, T.; Kugi, A.; Kreuter, H.-P.; "EMI Reduction for Smart Power Switches by Iterative Tracking of a Gaussian-shape Switching Transition," *International Conference and Exhibition on Power Electronics, Intelligent Motion, Renewable Energy and Energy Management (PCIM Europe)*; pp.1361–1368, 2015.

[13] Blank, M.; Glueck, T.; Kugi, A.; Kreuter, H.-P.; "Digital Slew Rate and S-Shape Control for Smart Power Switches to Reduce EMI Generation," *IEEE Transactions on Power Electronics*, vol.30, no.9, pp.5170-5180, 2015.

Noise & Vibration Levels of modern Electric Motors

Christoph, Stuckmann, MACCON GmbH, Germany, c.stuckmann@maccon.de

ABSTRACT

The performance demands on modern electric motors are steadily increasing. In addition to high expectations regarding efficiency, power-density and torque ripple, the user expects a low noise and vibration level. Safety considerations are also important when selecting the best motor technology for an application.

In this paper the absolute and relative performance characteristics of six different types of electric motors are comprehensively analyzed:
- AC-induction motors
- PM-excited, brushless synchronous motors
- PM-excited, brushless synchronous motors with embedded magnets
- Switched reluctance motors
- Synchronous reluctance motors
- Wound-field synchronous motors

This comparison considers noise and vibration in particular. In addition to a theoretical consideration using analytical methods also simulation techniques are presented. Finally recommendations have been developed, which can help design optimization with respect to noise and vibration, once the choice of motor technology has been finalized.

1. Electric motors for E-Mobility

The electric motor is increasingly used as a vehicle drive, both for passenger cars and commercial vehicles with short and medium range. Ignoring the issue of primary energy storage, such motors and inverters are much smaller and lighter than internal combustion engine. They are also much more flexible to control; not to be forgotten is the efficiency at full power well above 90%.

1.1 Summary of different e-motor Technologies

There are various electric motor technologies that are available for this type of application:
- AC induction motors (AC)
- PM-excited, brushless synchronous motors (PSM)
- PM-excited, brushless synchronous motors with embedded magnets (IPM)
- Switched reluctance motors (SR)
- synchronous reluctance motors (SYR)
- Wound field excited synchronous motors (WSy)

An initial evaluation of these technologies is summarized in Table 1. The particular advantages for IPM, WSy and SyR technologies for e-Mobilty can immediately be recognized. Therefore these three technologies will be focused on in this paper.

PCIM Europe 2016, 10 – 12 May 2016, Nuremberg, Germany

Techno-logy	Maturity / costs	Efficiency / power factor	field weakening capability	power density	safety on short circuit	Control complexity	Sensitivity noise excitation
AC	1	3	2	3	1	3	4
PSM	2	1	5	1	5	2	1
IPM	2	2	2	1	4	3	2
SR	3	3	2	2	1	4	5
SyR	3	3	2	2	1	4	2
WSy	3	1	1	2	2	2	1-2

Table 1. Comparison of different motor types

1.2 Detailed description of the IPM, WSy and SyR- motor technologies

1.2.1 IPM motors

A major problem with the use of magnets for generating a magnetic field is the fact that this field is permanently present. This has the disadvantage in inducing high voltages at high speed, which are reflected to the battery. To avoid that the battery is subjected to reverse current, it is possible to impose a short circuit at the motor terminals. This short circuit at the motor terminals has as a strong braking effect, which may itself be a safety issue. Also the electrical braking energy has to be dissipated; one possibility hare is to use large regeneration resistors. Regardless of which mechanism is used to solve this problem, safety issues have to be addressed.

Permanent magnet machines have the advantages that they are compact and efficient. The IPM- version with embedded magnets in the rotor has the advantages that it can be field-weakened very efficiently.

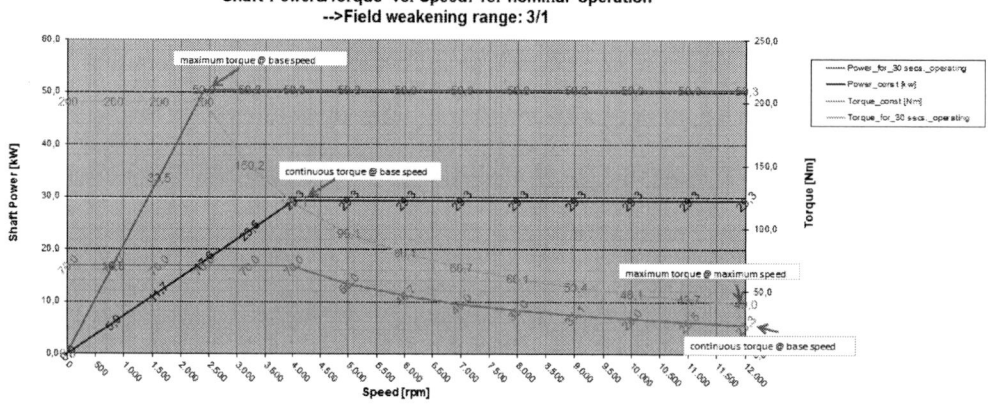

Figure 1: Example for Typical shaft power / shaft torque diagram characteristic of a IPM / SrY
motor with field weakening range of 3 to 1 (Source: MACCON)

This is prerequisite for an enhanced speed characteristic (see Fig. 1.).

Field weakening is achieved by phase advancing current w.r.t. the back-emf voltage. In practice, the armature current is split into two components (d/q). As shown in figure 2 the d-axis is the axis of symmetry which is in the center in the permanent magnet axis. The q-axis is placed in the inter-polar gap and is therefore electrically orthogonal, oriented at 90° to the permanent magnet axis. Both axes are mathematical symmetrically. In this particular example a salient-pole machine with different inductive properties in the d- and q-axes is shown. Salient-pole machines are ideally suited for field weakening operation.

© VDE VERLAG GMBH · Berlin · Offenbach

PCIM Europe 2016, 10 – 12 May 2016, Nuremberg, Germany

For the synchronous speed and steady state operation a phasor diagram can be generated, which is described in the d- and q-axis formulation as shown in Figure 3. In equation (1), the motor torque is described as a function of the rotor excitation current and the d-and q-axis current of the stator.

$$Ti(Ie, Id, Iq) = 1.5 * p * (LM * Ie * Iq + (Ld - Lq) * Id * Iq) \quad (1)$$

The inductances itself are depending on the respective current as shown in equation (2).

$$Ld, Lq, LM = f(Id, Iq, If) \quad (2)$$

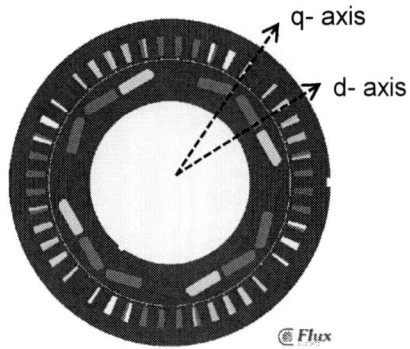

Fig. 2: dq- axis definition for the stator and rotor of an 8-pole IPM machine with 48 slots (Source: MACCON)

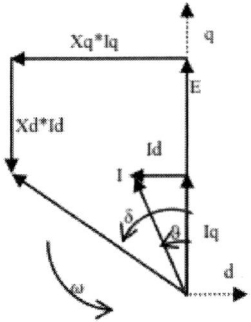

Fig. 3: the stator and rotor of an 8-pole IPM machine (Source: MACCON)

To allow the rotor field to be influenced the motor current has to be controlled by splitting it into a q-axis current and a d-axis component; the control algorithm is usually designated as current vector control. The q-current component in this operation mode is responsible for generation of motor torque; the d-current component for the field weakening.

The design of an IPM motor can be seen in Fig. 4. The stator is conventional and similar to AC and PSM motors. The rotor magnets are embedded in the laminated rotor.

In Table 1 it was shown that the IPM design has more dominant noise than PSM. This is generally due to saturation effects which cause harmonics induced in the motor voltage and consequently in the current. This effect can be avoid by a optimizing the shape for the pole design. For high speed applications special pole shaping may not be needed; such pole configurations have a negative impact on rotor robustness.

© VDE VERLAG GMBH · Berlin · Offenbach

Therefore the induced harmonics in current and voltage in IPM motor types may have the effect of a greater the noise excitation.

Fig. 4: Stator und Rotor of an 8-pole IPM- motor (Source: MACCON)

1.2.2 WSy engines

The externally excited synchronous machine has the big advantage that the magnetic field can be influenced over a wide range by varying the current in a rotor winding. Although some energy is indeed needed for the field generation, the efficiency can even better than permanent magnet machines under boundary conditions (for example, for low power at high speeds).

In order to ensure the maximum efficiency of a WSy motor under all operating conditions, a complex control algorithm is required: Depending on the specific operating point, there is a corresponding optimum ratio between the armature and field power. The geometry of the rotor poles can be optimized in order to achieve a favorable relation between the inductance in the d and q-axis. The position d and q- axis can be seen in Fig. 4.

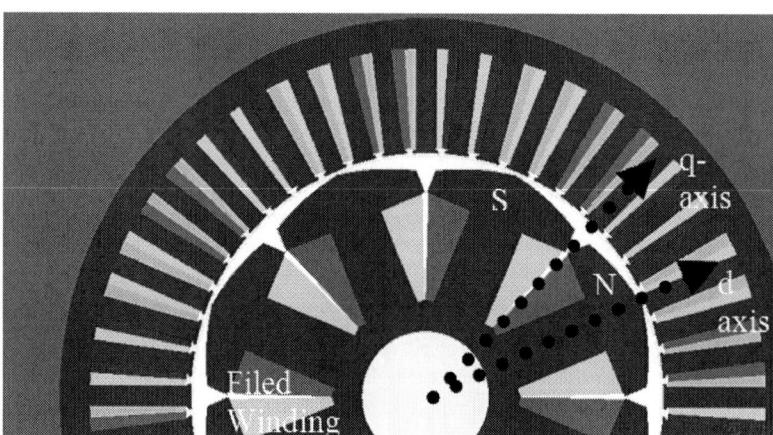

Fig. 4: Stator und Rotor of an 8-pole wound field synchronous motor
(Source: MACCON)

The production of WSy-motor stators is similar to the permanent magnet excited machines. In both cases, distributed windings are in use with a relatively high number of stator slots as compared to the rotor poles.

However, the use of separately excited synchronous motors causes additional electronics and cabling requirements for providing the driving power to the rotor winding. This can be done by means of slip rings (2 channels) or inductively via a rotary transformer. Nevertheless the FSy engine thanks to its ultra-flexible controllability is a very interesting candidate for e-mobility applications.

As listed in table 1 the sensitivity of the noise excitation compared to PSM variant is as well as for the IPM generally more critical. The reason of this can generally be found in the pronounced salient pole compared.

1.2.3   SyR- motors

The synchronous reluctance motor type (SyR) is characterized in Table 1. The SyR uses no magnets; the electromagnetic motor torque is produced by the reluctance principle. The rotor is mirror-symmetric w.r.t. the q-axis. Non-magnetic rotor slits with a banana shape are employed; a gap is directly implemented in the q-axis. The legs of the slots are oriented parallel to the d- axis. The physical operating principle of the synchronous reluctance motor is similar to that of the IPM- motor except that no permanent magnet flux is needed. Only the reluctance torque component is responsible for inducing the electromagnetic torque.

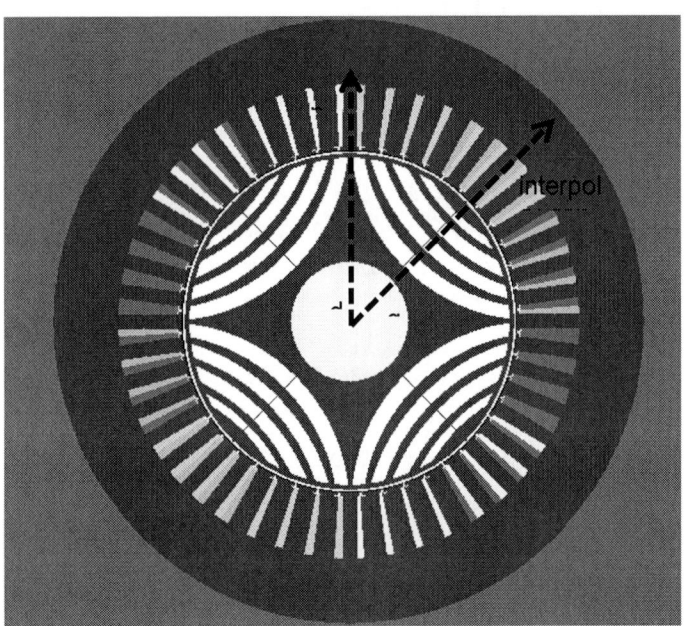

Fig. 5: Synchronous reluctance / motor cross section (48 slots and 4 rotor poles; distributed winding scheme)
(Source: MACCON)

The stator of the Synchronous reluctance motor can be driven with a classical 3-phase power stage. Consequently a similar controller solution may be used as for IPM motor control.

However the control of a SyR- motor demands exact knowledge of the rotor position as the rotor field has to be precisely controlled w.r.t. to the magnetic axis. Although the synchronous reluctance motor tends to higher torque ripple as the reluctance torque varies depending on the position. To minimize these effects special designs are needed. The power density of this variant is similar to the WSy-variant.

It should be pointed out that this variant is a real alternative solution to the IPM machine. It does not depend on rare earth magnets and is not subject to the safety issues described above. Furthermore this motor type has the potential of a lower level of noise excitation.

© VDE VERLAG GMBH · Berlin · Offenbach

2. Noise Generation Phenomena

2.1 Basic mechanisms of noise excitation

Noise excitation occurs if electromagnetic forces act at the stator teeth, causing the stator yoke to deform. This deformation is transmitted to the motor housing and surrounding structure and noise is induced. Magnetic conductivity deviations occur in the magnetic circuit due to changes in the flux path through the stator teeth compared to their path through the stator slots; thus winding harmonics and current ripple are induced in the motor current.

2.2 Mathematical description for the working mechanism of noise excitation

To evaluate the forces, which are induced by the magnetic fields in the air gap and interacting with the stator tooth, the equations for the Maxwell stress tensor are needed. If the axial components of the magnet field at the tooth tips are neglected, The radial stress component can be deduced:

$$F_{rad} = \iint_A \left(\mu_0 H_{rad}^2 - \frac{\mu_0}{2} \vec{H}^2 \right) \cdot dA \qquad (3)$$

Further the tangential component can be deduced, if the axial components of the magnet field at the tooth tips are neglected:

$$F_{tan} = \iint_A \left(\mu_0 H_{tan} H_{rad} \right) \cdot dA \qquad (4)$$

Following the computation method for the electromagnetic forces of the individual stator teeth as implemented in FLUX® is rapidly shown.

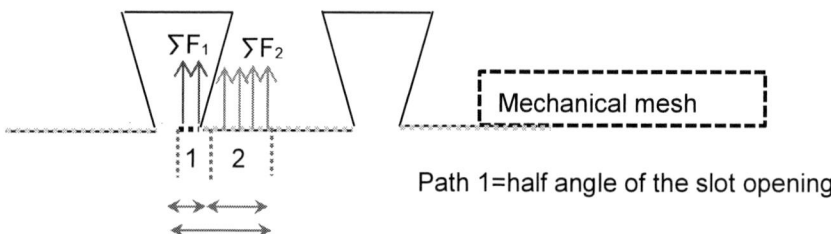

F1+F2= summation of the electromagnetic forces

The electromagnetic forces on the evaluation areas on the section A and B are computed by the magnetic pressure again derived from the Maxwell stress tensor but now described by differential equations.

Magnetic pressure in the normal direction:

$$\left(\frac{F}{A_{1+2}} \right)_{normal} = \frac{B_r{}^2 - B_t{}^2}{2\mu_0} \qquad (5)$$

Magnetic pressure in the tangential direction:

$$\left(\frac{F}{A_{1+2}} \right)_{tangential} = \frac{B_r \times B_t}{\mu_0} \qquad (6)$$

The corresponding forces are deduced by multiplying the pressure by the area of the mechanical mesh element. For each node of the mechanical mesh the FFT is done in the time domain; the amplitude and phase of each harmonic can then be evaluated. The calculation is done over one mechanical period. Therefore the fundamental frequency corresponds to the mechanical period.

3. Comparison of noise excitation of IPM and SyR- motors

As described in section 1.2.3 the SyR motor type has a smaller potential for noise excitation than the IPM motor type; in this motor the embedded magnets lead to higher noise excitation, as described in chapter 1.2.1.

The main reason for the lower noise excitation of the SyR motor type is due to its physical operation principle; the rotor field has no intrinsic source. The induction of the rotor field is based only on the reluctance principle. No direct interaction between a rotor pole and the stator tooth occurs. The saturation effects are smaller; therefore lower harmonics are induced. This results in a lower sensitivity to noise excitation for the SyR motor type.

4. Comparative analyzing of the Noise level of IPM vs. SyR-Motors

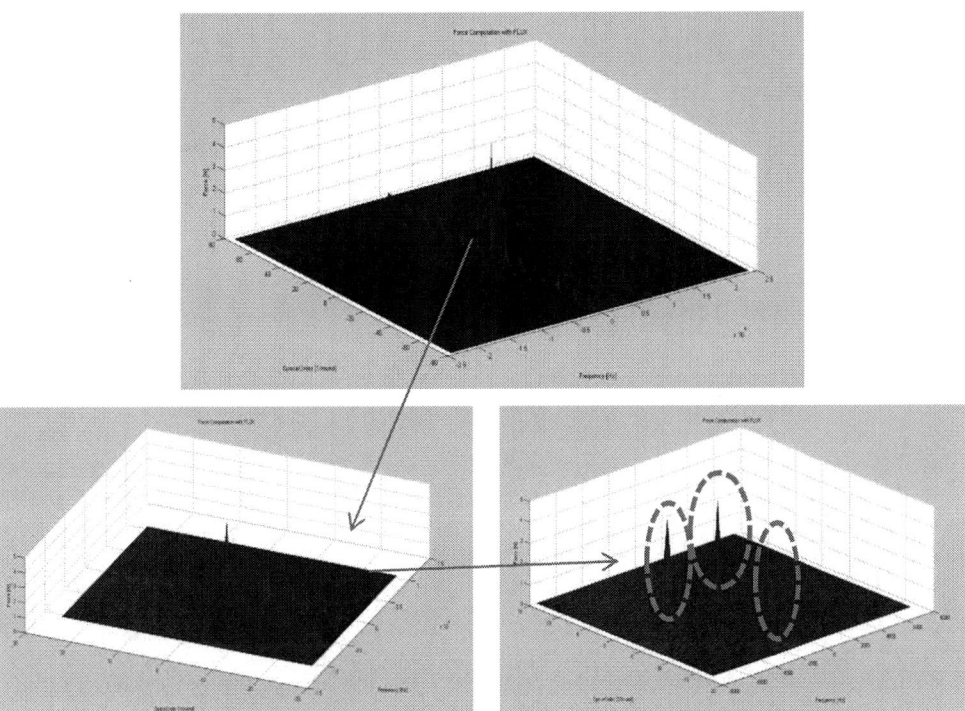

Fig. 6: IPM motor / space harmonics and frequency distribution of the tooth forces
→ successively zoomed in the relevant parts / (each force component has to be multiplied with the number of FEM- nodes on the stator tooth which is 30 in this case) (Source: MACCON)

In the figure 6 the space harmonics and frequency distribution of an IPM motor are shown for a particular design example. As can be seen there is one main component at zero frequency with a space harmonic of zero order. This DC- component is caused by the main force vector acting on the tooth tips as static tension. This component is not relevant for the analysis of the sensitivity of force components with the risk of noise excitation. The more interested harmonics for analysis of noise excitation are the both shown in Figure 6 in red. They are induced by the 8 poles and have the frequencies of -533Hz and +533Hz (double the fundamental frequency as the considered operating point is as shown in Figure 1: 4000rpm for the continuous operating torque). These components can cause noise excitation if the surrounding mechanics have Eigen-frequencies in this range. The harmonics with of 16th order are not relevant as these frequencies are too to excite the surrounding mechanics.

© VDE VERLAG GMBH · Berlin · Offenbach

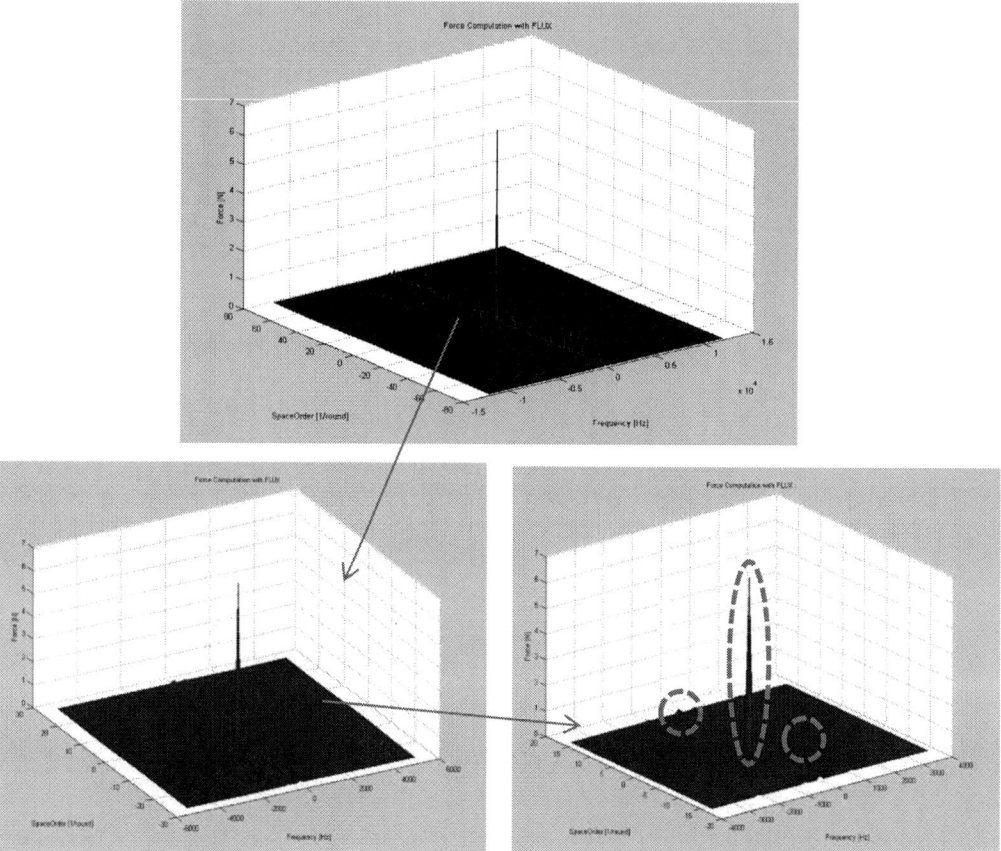

Fig. 7: Synchronous reluctance / space harmonics and frequency distribution of the tooth forces
→successively zoomed in the relevant parts / (each force component has to be multiplied with the number of
nodes on the stator tooth which is 30 in this case) (Source: MACCON)

Also for the SrY the more interested harmonics for analysis of noise excitation are the both shown in Figure7 is red. They are induced by the 4 poles and have the frequency of -266Hz and +266Hz (double the fundamental frequency as the considered operating point is as shown in Figure 1: 4000rpm for the continuous operating torque).

These components increase the risk of noise excitation if the surrounding mechanics have Eigen frequencies in this range.

What can be seen in Figure 7 is that the amplitude for the frequency of -266Hz and +266Hz is much lower. Therefore the level of noise excitation is less for this motor type.

5. Summary

In the paper different kinds of motor designs have been reviewed as options for use as traction motors. The IPM, WSy and SrY seem to be the most promising solutions.

The IPM version has issues related to safety - in the case of id-failure at high speeds and thus field weakening failure. The WsY is better in this respect as field excitation can be influenced by the field current. However in order to optimize its operation the WsY motor also needs id-current control. There still a safety issue but this is reduced.

The SrY seems to be the most promising solution. Furthermore it was shown that the SrY has the advantage that noise excitation is lower.

The efficiency and torque density of the SrY lies between between that of IPM and WSy.

© VDE VERLAG GMBH · Berlin · Offenbach

PCIM Europe 2016, 10 – 12 May 2016, Nuremberg, Germany

Development Platform and Techniques for the Rapid Implementation of High Performance Drives

Author: Dipl. Ing. Christian Balke, MACCON, Deutschland, c.balke@maccon.de
Co-author: M.Sc. Simon Wiedemann, MACCON, Deutschland, s.wiedemann@maccon.de

Abstract

For most drive applications, a wide range of standard converters is available on the market. In some special cases, a custom drive solution is required. Sometimes, for a customized solution, a standard converter can be modified, that a special version of the standard product can be offered.

But sometimes, a complete custom design of the converter is required. This is time consuming, expensive and needs a lot of expertise in this area. A complete custom development is required for topologies different from three phase half-bridge power stages, for complex mechanical geometries or for special regulation or control requirements. These specialized converters can use different pulse control techniques for different working points of the motor, which can be important for high performance drive solutions.

For a custom converter design the project requirements are:
- Short Development Times
- Less Expertise required for successful Design
- Minimum Redesign Cycles
- Cost Efficiency

Solution for these requirements can be a modular, scalable platform, because nearly all converters use the same functional basic hardware- and software blocks, independent from the physical seize or the required power. The development of these blocks takes a lot of effort and expertise and some redesign cycles for a successful product. Once successfully qualified, the blocks can be easily plugged together to a very specialized custom design.

Critical reusable hardware blocks are:
- Control-Box for Implementation of Customized Software and Algorithms
- Isolated DC-Link Voltage Sensor
- Isolated Half-Bridge Gate-Driver capable of driving large IGBTs at high PWM Frequency
- Complete Half-Bridge Power-Block with IGBT, Driver, Current- and Temperature Sensor, Cooling and Snubber Capacitor
- Three-Phase B6 Rectifier-Block with Cooling, compatible to the IGBT-Block

On the software and controller side, proven reusable blocks are:
- Regulating Blocks for Current and Speed
- Multiphase PWM-Block
- Control Block for a Brake-Chopper Switch
- Software Drivers for different Motor Feedback Systems, Communication Protocols and Auxiliary I/Os

Based on these theoretical findings, a modular platform has been developed and will be introduced block by block on the following pages.

© VDE VERLAG GMBH · Berlin · Offenbach

1. Control-Box

1.1. Description

The control-box is a highly flexible controller-platform for drive-applications in a tiny 180mm x 60mm x 245mm housing. Combined with other platform standard modules for converters, it offers a plug and play hardware concept for unusual topologies or difficult mounting space requirements. This concept allows designing customized converters with minimum engineering-effort in short delivery times.

Application software can be easily generated with the HDL-coder from a proven Matlab/Simulink simulation without deeper knowledge of internal hardware details or VHDL programming skills.

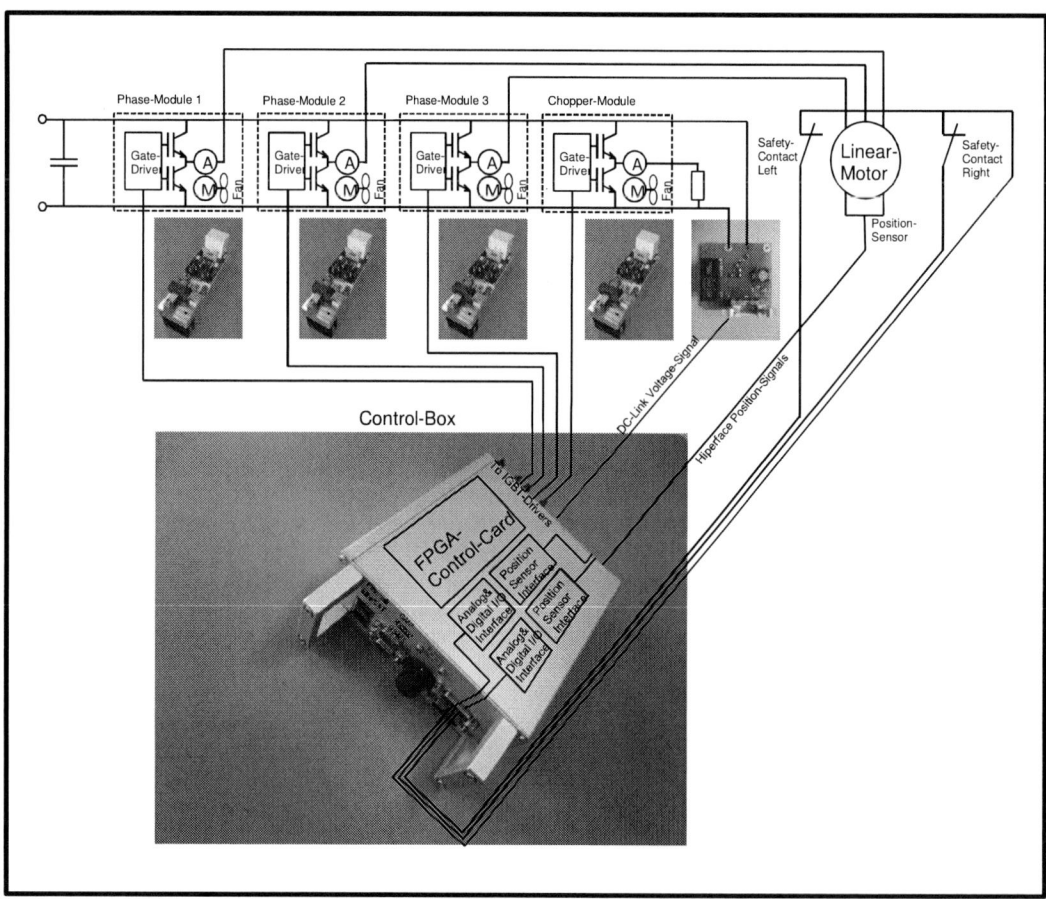

Fig. 1. Application Example: Control-Box with three Motor Phase Half-Bridge Power Blocks, one Chopper Half-Bridge Power Block and an Isolated Voltage Sensor in a Linear Motor Application.

1.2. Control-Box Features

- **Large FPGA core** combined with 64MByte parallel flash and 32MByte SDRAM offers a powerful and flexible solution, sufficient for independent dual-motor-drive applications.

- **Multiple communication interfaces** for Ethernet/EtherCAT, CAN, RS232, RS422/RS485 cover most requirements for industrial and automotive applications.

- **Eight half-bridge connectors** can be used for complex single-motor applications or for medium complex dual motor applications. These connectors are compatible with the gate driver for EconoDUAL IGBT-modules and the air cooled half-bridge power block. Each connector supports signals for driver supply, half-bridge gate drive, IGBT failure, temperature signal, phase current signal and fan supply on a small 10-pole ribbon-cable connector. The fan supply for all eight connectors comes from the same source, which can be regulated depending on the IGBT-module with the actual highest baseplate temperature. This feature reduces the noise and the current consumption of the fans to a minimum.

- **Two position sensor interface slots** are designed as independent plug-in modules. For a dual motor application, different position sensors can be internally factory-mounted, if required. The first available universal sensor interface card can read resolvers, encoders and Hiperface sensors. Also two PT100 and one PTC thermal-switch can be connected. All sensor signals need to be double or reinforced isolated from high voltage parts inside or around the motor!

- **Two universal I/O-interface slots:** Available are plug-in cards with 2 digital inputs, 2 digital outputs, 2 differential analog inputs, 1 analog +/-10V output and 2 PT100 inputs combined on one card. Maximum of two cards can be factory-configured.

- **Signal connector for DC-Link voltage and current** is compatible with the isolated voltage sensor module to measure the DC-link voltage and offers also the possibility to connect an external 5V current sensor to measure the DC-link current.

- **Strong isolated 24V supply** can prevent ground loops or significant voltage drop on signal ground when the negative supply input gets a separate cable to the source. The large input voltage range fulfills 60Vpk railway requirements and is reverse polarity protected. The box provides sufficient power to drive eight half-bridges with 4W/Gate drivers plus the fans from eight half-bridge blocks. For air-cooled systems, there is nearly no limitation from the supply or gate driver side to reach 10kHz or higher PWM frequency, because the switching losses inside the IGBTs for these frequencies drive the air-cooling system into the limit. The controller has enough power to drive a liquid cooled power stage at higher PWM frequencies.

- **Speed-potentiometer and LED-display** for initial bring-up or manual debug. In these phases, manual control and one-digit error code display can be very helpful.

2. Rapid Prototyping Software Platform

For the software concept, the platform requirements for rapid prototyping designs are also valid. Hardware and software need to merge to a uniform platform. The software development can also be done without detailed basic knowledge of all hardware details. Software development can be done in three steps:

- Simulation in Matlab/Simulink: Like in hardware, also for the simulation model proven blocks can be configured like type of motor, power-stage, position feedback-sensor, PWM-model and other. This ensures easy development of the control-model combined with the high confidence of simulated functionality in an early stage of the project.

- Transformation of the proven simulation model into VHDL-code with an automatic tool called HDL-Coder ensures efficient code transformation with minimum effort and a minimum failure rate. Based on the architecture of the proven simulation model, executable control software is generated.

- Tests on a real hardware system to ensure reliable functionality under all circumstances. This task takes a lot of effort to deliver a high quality system and must not be underestimated. Combined with the preparation of the two previous development steps, the total software development effort is very low.

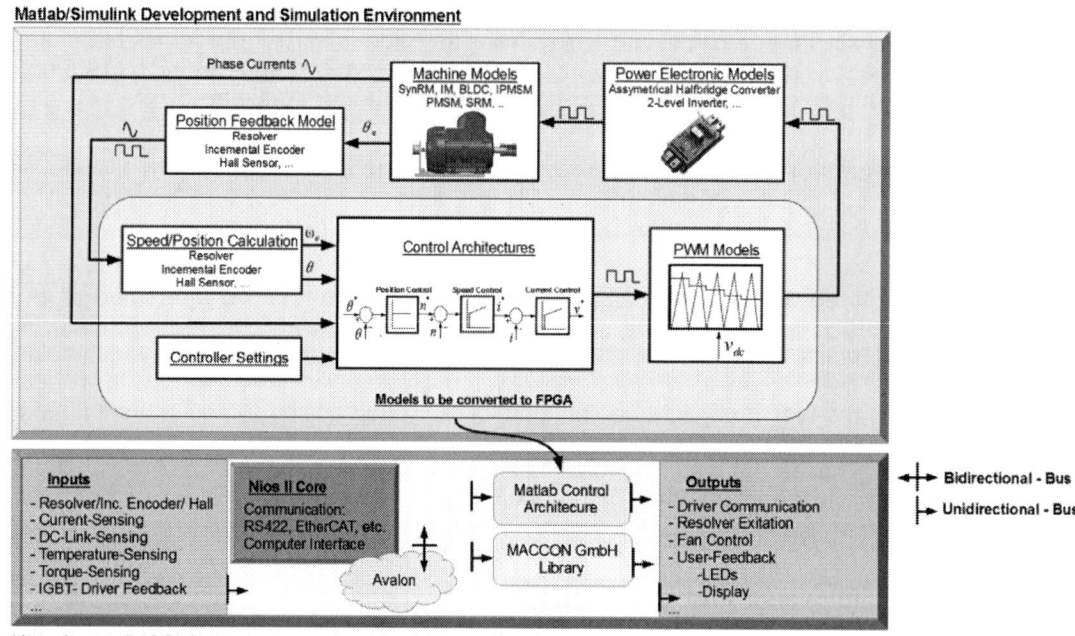

Fig. 2. Rapid Prototyping Architecture of Control-Algorithm using Matlab/Simulink

The grey area shows the whole Matlab/Simulink simulation area without any real hardware. The embedded green area will be transformed into VHDL-code and programmed into the FPGA on the real hardware platform. The NIOS II Core is an embedded standard microcontroller core, available for FPGAs.

3. Isolated Voltage Sensor

- Up to 1200V DC Primary Voltage

- Galvanic Isolation up to 1200VDC

- High 1.2 MOhm Input Resistance

- Wide Supply Range 3V...16VDC

- Integrated Y-capacitor can be enabled

- -40°C to 85°C at Free Convection

Fig. 3. Isolated Voltage Sensor

The isolated voltage sensor module has been developed to measure isolated or non-isolated DC-link voltages. It offers interesting features to build compact and efficient motor-drives.

- **High impedance voltage sensing** with integrated primary side sense resistors allows voltage measurements at all high-voltage DC-links up to 1200V in a compact 59mm x 68mm package.

- **Wide auxilary supply voltage range** from 3V to 16V allows connection of the supply to a logic supply, to an analog supply or directly to the gate-driver supply. A 3.3V supply is only recommended in a low-noise environment.

- **Galvanic isolation** up to 1200V operating voltage and 8mm creepage distance offer safe voltage-sensing in high voltage circuits, even in harsh environments.

- **Integrated Y-Capacitor** can be enabled by using metal spacers to mount the module on the chassis. The capacitor can be disabled by mounting the module (PE terminal) with isolated spacers with a minimum length of 8mm. The Isolated Voltage Sensor needs DC-link terminals and controller-GND on the same module, so using the Y-Capacitor inside the module is the most space-efficient way for the motor drive design.

- **Status LEDs for fast debug:** The green LED "PWR" indicates supply voltage on the high-voltage side. The red LED "HV" indicates measured voltages > 55VDC.

4. Isolated IGBT Half-Bridge Driver for EconoDUAL Packages

- Dual 4W/15A Drivers

- DESAT Protection

- Isolated NTC Temperature Sensing

- Up to 1200V Peak DC-Link Voltage

- Temperature Range -40°C to 100°C, (Full Load up to 50°C)

- Reserved Space for Snubber Capacitor

Fig. 4. Driver-PCB, IGBT-Module, Snubber Capacitor

The gate driver module is a highly integrated, powerful gate driver for standard EconoDUAL IGBT-modules. It offers an interesting combination of many features to build high-performance motor drives.

- **Powerful ready-to-use half-bridge gate-driver** (2x4W DC/DC converter combined with 2x15A driver and integrated gate resistors) ensures a compact, high-performance solution for large gates up to 100°C ambient temperature at free air convection, full load up to 50°C (see derating curve).

- **DESAT protection circuitry** can protect modules which are specified to withstand short circuit times larger than 6µs at the specified DC-link voltage. The DESAT protection circuit produces insignificant leakage on the DC-link compared to the IGBT-leakage current.

- **Galvanically isolated temperature sensing circuit** to safely measure the module's NTC temperature. The continuous baseplate temperature sensing offers the possibility to build converters working close to the limits to reach high power-densities at pulsed loads.

- **Special driver output stage minimizes gate-ringing** during the Miller-Plateau Phase.

- **Efficient 10-pole ribbon cable interface** offers a high density signal interface in a standard format. All signals like 15V driver supply, driver control and error signals, temperature signal, 12V supply for optional external fan and current signal from optional external current sensor are combined in this interface. The driver offers a 5V supply for an optional external current sensor, which is produced from the 15V single driver supply.

Based on this highly flexible and powerful gate-driver block, a whole modular IGBT half-bridge power-block has been developed.

© VDE VERLAG GMBH · Berlin · Offenbach

5. Modular IGBT Half-Bridge Power-Stage

- Plug and Play IGBT Half-Bridge Solution up to 1200V

- 10-Pole Standard Ribbon Cable Connector for all Signals

- Strong Gate-Driver (2x4W/15A)

- Current Sensor

- Isolated Module NTC Temperature Sensing

Fig. 5. Modular IGBT Half-Bridge Power-Stage

The modular power-stage is a highly integrated, ready to use IGBT half-bridge solution. It integrates all components like IGBT-module, gate-driver, current sensor, temperature sensor, snubber capacitor and heatsink with fan. There is also a compatible B6 rectifier module on the same type of heatsink available.

The IGBT half-bridge can be delivered with different types of IGBT-modules from different manufacturers. The NTC-sensing circuit is compatible for Infineon and Fuji modules.

The key component of the power-stage is the gate driver module. In this application, the driver module uses all of its features:

- Powerful Ready-To-Use Half-Bridge Gate-Driver 2x4W/15A
- DESAT Protection Circuitry
- Galvanically Isolated Temperature Sensing Circuit
- Interface and Supply for External Current Sensor
- Supply Connection for External Fan
- All Signals in a tiny 10-Pole Standard Ribbon Cable Interface

- **Snubber capacitor:** The snubber capacitor is a critical component in converter power-stages, because it has to take high current pulses. It is strongly recommended to connect a snubber capacitor close to the DC-Link terminals of the IGBT-module. A matching high performance snubber capacitor, which fits to the IGBT terminals, is part of the power-stage package to simplify the choice of this critical part.

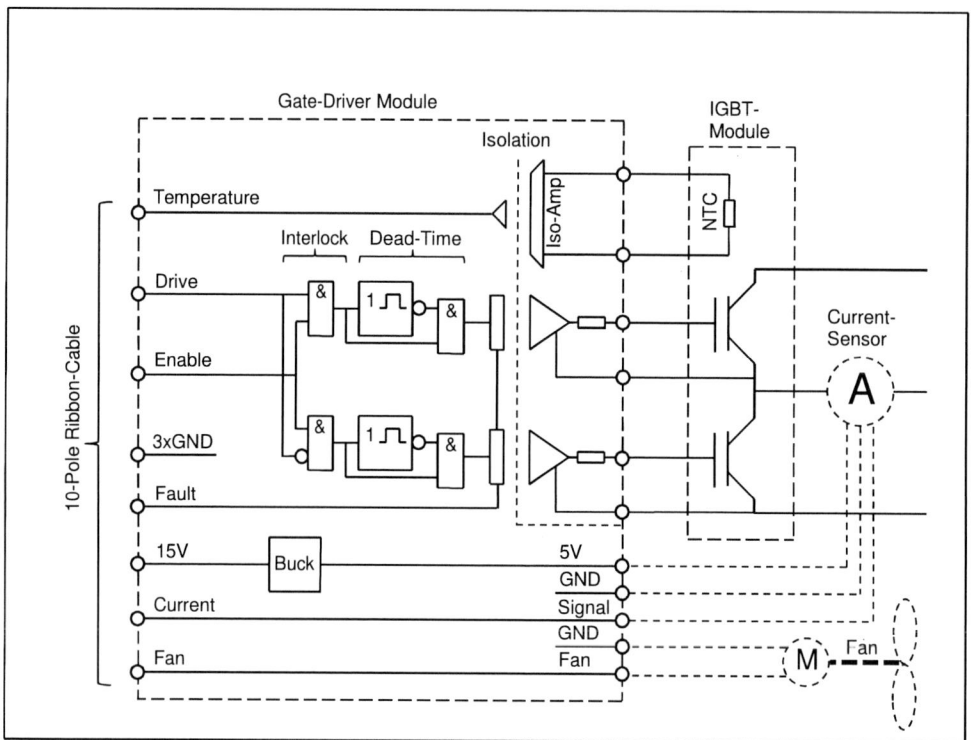

Fig. 6. Block Diagram of the Modular Half-Bridge Power-Module

6. Summary

The combination of powerful and flexible controller hardware, proven hardware blocks for efficient power-stage design and easy software implementation of simulated regulation algorithms is forming a solid rapid prototyping platform concept.

The requirements for rapid prototyping projects of

- Short Development Times
- Less Expertise required for successful Design
- Minimum Redesign Cycles
- Cost Efficiency

can be fulfilled. Designers can use only single blocks of this platform to cover partial problems of their own design, they can use the whole platform to build their own converter or they can order a tested and verified custom converter according to their specifications based on this platform concept.

Nearly no system designer for complex projects like a custom converter has the expertise for all details in analog and digital hardware, power electronics, software, DIN/ISO-standards for isolation, distances, materials... If even possible, combining of proven blocks or subsystems is the most efficient way to develop a complex system.

Engineering is fun. Inventing the wheel twice is no fun.

Functional safety for integrated circuits used in variable speed drives

Tom Meany
Analog Devices
Lurraga road
Raheen Business Park
Limerick
V94 RT99
Ireland
tom.meany@analog.com

Topic Number 9.4 Functional Safety
Preferred Presentation Form: Oral

Abstract

Functional safety is the branch of safety related to correct functioning of electrical and electronic systems. Variable speed drives now play an important part in implementing functional safety. Previously functional safety for motor control applications was realized using safety relays and contactors external to the drive. But with safety integrated in the drive safety functions such as STO and SLS can be implemented within the drive offering productivity enhancements on the factory floor. Integrated safety requires the use of integrated circuits but interpreting the functional safety requirements for integrated circuits used in variable speed drives is challenging. Ideally all such ICs would be assessed to IEC 61508 but this would be expensive and is not demanded by the standards. This paper will attempt to summarize what guidance is available for integrated circuits being used in the design of variable speed drives. A goal of this paper is to provide an overview of the topics without the use of jargon.

Synopsis

At the machinery level ISO 13849 and IEC 62061 set out the functional safety requirements for control systems in machinery. At the variable speed drives level the standard IEC 61800-5-2 sets out the functional safety requirements to design a drive. However when it comes to integrated circuits only the generic standard IEC 61508 is available for guidance.

None of the system level standards state that only certified components must be used. Instead it is the responsibility of the system or module designer to satisfy themselves that the components they choose are suitable for use in their system. Having a fully certified component would be the easiest way for the designer to be satisfied but components developed to IEC 61508 are largely confined to dual core microcontrollers.

The key 3 requirements of Functional Safety

Functional safety has three key requirements.

Requirement 1 – To use reliable components. This means ICs with a sufficiently low FIT rate. FIT rates are often calculated according to standards such as IEC 62380 or SN 29500 which base their results on the "average" failure rate seen in the field for various types of components. Alternatively data can be based on accelerated life testing such as that found at analog.com/ReliabilityData. One important consideration is that the PFH (probability of failure per hour dangerous) figures given in IEC 61508 and similar standards are for an entire safety function and not just for a single IC. Therefore the PFH figure of 1e-7 h^{-1} for a SIL 3 safety function (100 FIT) might give an error budget of only 1 FIT for a given IC. It is also worth noting that the term PFH actually means the probability of dangerous failures per hour. It can be argued that at least 50% of failures are safe and that the reliability limit for the IC can be doubled.

Requirement 2 – Implement a set of measures that have been shown in the past to design products with high safety. This is referred to it the standards as systematic integrity. Rather than random hardware failures systematics failures are built into a system and only a design change can eliminate them. Software bugs are an example of systematic failures. So also are EMC failures.

Requirement 3 – Be fault tolerant in that you accept that faults due to random hardware failures or systematic faults will occur no matter how reliable the components or how good the development process followed. Two ways to then cope with the faults are through diagnostics and redundancy. Diagnostics detect the faults and take the system to a safe state. For motor control the safe state is generally to bring the motor to a stop with a safety sub function such as STO from IEC 61800-5-2. The other alternative is to implement redundancy so that there are two or more items either one of which can detect an unsafe state and bring the system to a safe state when required to do so. Standards generally allow a tradeoff between diagnostics and redundancy. Measures of the effectiveness include SFF from IEC 61508, DC from ISO 13849 and the single point fault metric from ISO 26262.

IEC 61800-5-2

IEC 61800-5-2 is a C-type standard. This means that this standard sets out the requirements for a particular machine category in this case a variable speed drive. Having a type C standard is very valuable because it interprets the generic standard IEC 61508 for that equipment type and only keeps what is relevant for that machine. A generic standard by its nature has to cope with many different types of equipment and situations which means that it contains lots of information and requirements which are not relevant to a specific design. IEC 61800-5-2 boasts that "By applying the requirements from this part of the IEC 61800 series, the corresponding requirements of IEC 61508 that are necessary for a *PDS (SR)* are fulfilled." However in the event of there being topics on which the C-type standards such as IEC 61800-5-2 gives no guidance then IEC 61508 is the fallback.

Within IEC 61800-5-2 safety sub functions such as STO (safe Torque off) and SLS (safely limited speed) are defined and a functional safety life cycle outlined.

PCIM Europe 2016, 10 – 12 May 2016, Nuremberg, Germany

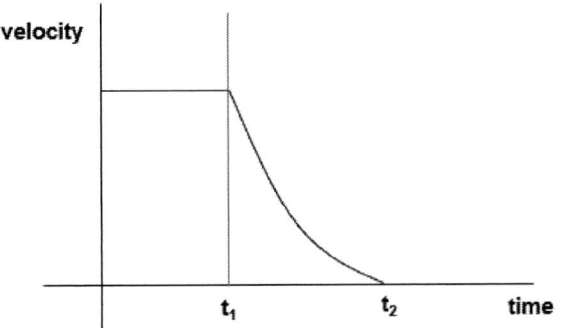

Figure 1 - STO safety function

With the STO safety sub function a safe state can be achieved by preventing "force-producing power from being provided to the motor". Typically this will be done using pulse blocking or power removal at the gate driver when a guard is open. Since total power to the drive is not removed a quick restart can be facilitated once the guard is closed.

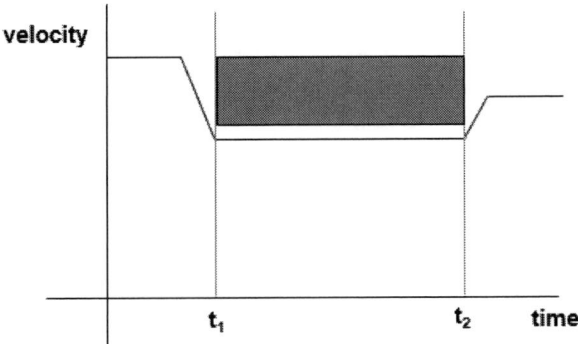

Figure 2 - Safely limited speed

With the SLS safety sub function the speed of the motor is monitored and if a set level is exceeded the drive takes the motor to a safe state, most often STO. A typical use of this safety sub function might be during the cleaning of a roller in conjunction with a three position grip switch. The above diagram shows SLS engaging at t_1 and disengaging at t_2. The red block indicates a speed region which if entered will cause the drive to go to a safe state.

While IEC 61800-5-2 does not enforce a requirement for two channel safety most drive manufacturers will also want to claim a performance level according to ISO 13849 and therefore two channels are common.

ISO 13849

ISO 13849 is the machinery standard based on the now redundant EN954 standard. In contrast to IEC 61800-5-2, IEC 61508 and IEC 62061 it uses performance levels (PL) instead of SIL levels. The levels are PLa through PLe. ISO 13849 also has a clear preference for two channel systems in that for the higher performance levels a category 3 or 4 system must be used. ISO 13849 uses DC (diagnostic coverage) as a metric for diagnostic effectiveness as opposed to the SFF from other standards. On the assumption that faults are 50% safe/50% dangerous SFF and DC are related using the equation below.

© VDE VERLAG GMBH · Berlin · Offenbach

$$SFF = 0.5 + 0.5 * DC$$

IEC 62061

IEC 62061 is the machinery interpretation of IEC 61508. It is in effect a parallel standard to ISO 13849 and in fact there is an effort to combine both machinery standards in ISO/IEC 17305.

In the scope of IEC 62061 in states in note 2 that "*In this standard, it is presumed that the design of complex programmable electronic subsystems or subsystem elements conforms to the relevant requirements of IEC 61508. This standard provides a methodology for the use, rather than development, of such subsystems and subsystem elements as part of a SRECS*".

IEC 61508

IEC 61508-2:2010 contains significant IC requirements but they can be easy to miss on a casual or incomplete reading of the standard. The requirements include an ASIC development V model, see IEC 61508-2:2010 figure 3. The V model is targeted at digital ASICs in that it references "synthesis placement and routing" along with "final coding" but the V model can be interpreted for analog or mixed signal ASICs using a little imagination.

The preference for digital ASICs continues into Annex F which is entitled "*Techniques and measures for ASICs – avoidance of systematic failures*" and in note 1 states "*The following techniques and measures are related to digital ASICs and user programmable ICs only. For mixed-mode and analogue ASICs no general techniques and measures can be given at the moment*". Despite the limitations however it is quite feasible to complete the checklist for the digital portion of a mixed signal ASIC and with some use of "not applicable" to even a purely analog IC.

Annex E is entitled "*Special architecture requirements for integrated circuits (ICs) with on-chip redundancy*". Once again a digital limitation is put on the annex when it states in E.1 that "*The following requirements are related to digital ICs only. For mixed-mode and analogue ICs no general requirements can be given at the moment*". Another restriction on Annex E which seems to be widely ignored when the annex is quoted in other standards is that "*on-chip redundancy as used in this standard means a duplication (or triplication etc.) of functional units to establish a hardware fault tolerance greater than zero*". The word duplication means identical redundancy and the author of this paper believes the target was dual core micros possibly using the lockstep technique. While most of the techniques are good they may be excessive when applied to separation between diverse redundant blocks or between a block and another on-chip block used as a diagnostic on the first block. Duplicated blocks are subject to common cause failures such as temperature, ESD, power supply failures and others which are less likely to affect diverse blocks in the same way at the same time. An example of how annex E is quoted is in section D.2.4 of ISO 13849-2:2012 where it states "*Consequently, it is highly unlikely that the multi-channel functionality necessary for the fault tolerance and/or detection requirements of category 2, 3 or 4 can be achieved using a single integrated circuit, unless it satisfies the special architecture requirements of IEC 61508-2:2010, Annex E*". IEC 61800-5-2 FDIS(Autumn 2015) allows a possible exclusion for on-chip short circuits based on the requirements of Annex E of IEC 61508-2:2010 but on examining Annex E you find that only items f) and g) refer to directly to on-chip shorts. Item f) requires spacing between separate blocks of at least 10x the minimum design rule for the process and item g) talks only about "adjacent lines of separate physical blocks".

© VDE VERLAG GMBH · Berlin · Offenbach

Tables A.1 of IEC 61508-2:2010 gives faults or failures to be assumed when calculating the SFF. Tables A.2 to A.14 give examples of typical DC which can be claimed for typical diagnostics but the tables can sometimes need interpretation for integrated circuits. Annex H of IEC 62380 and the related Appendix A of UL 1998 are more detailed especially for digital microcontrollers and similar.

In terms of calculating FIT rates for integrated circuits IEC 62380 and SN29500 are both referenced along with other sources.

The requirement to consider soft errors was added in the 2010 revision of the standard and has implications for the addition of ECC and parity to volatile memories (e.g. RAM) in order to detect and control soft errors which affect RAMs in particular.

ISO26262 requirements

ISO 26262 is the automotive interpretation of IEC 61508. In was developed in parallel to revision 2 of IEC 61508 and contains some requirements related to integrated circuits not found in IEC 61508, some clarification of items in IEC 61508 but omits other requirements. For instance ISO 26262-10:2012 contains an automotive version of IEC 61508-2:2010 Annex F and table D.1 of ISO 26262-5:2011 clarifies the position for automotive on how to consider on chip shorts with "It *is not intended here to require an exhaustive analysis, for example to require the exhaustive analysis of bridging faults that can affect any theoretical combination of any signal inside a microcontroller or in a complex PCB. The analysis focuses on main signals or on very highly coupled interconnections identified with a layout level analysis*".

Part 10 in particular contains nuggets such as "if a CPU area occupies 3% of the whole microcontroller die area, then its failure rate could be assumed to be equal to 3% of the total microcontroller failure rate". While such a process is part of the "custom and practice" of IEC 61508 it is good to see it written down.

An integrated circuit interpretation of ISO 26262 is being worked on as ISO/AWI PAS 19451-1 under ISO/TC 22/SC32.

Assistance in designing in integrated circuits

Having reviewed the standards the author has a number of recommendations on how IC manufacturers can assist drive manufacturers in designing in integrated circuits into their drives.

Firstly a safety manual for integrated circuits should be of benefit to drive designers. This can be produced even if the ASIC or device was not developed to IEC 61508.

Items which could be available in the safety manual could include
- The development process and lifecycle model used
- A completed Annex F checklist from IEC 61508-2:2010
- The assumed mission profile
- FIT rate predictions according to IEC 62380 and SN29500 at a reasonable average operating temperature, for instance 55'c with thermal cycling of 10'C over a 24 hour period
- Die size, number of die, number of RAM cells and transistor counts to allow drive designers to calculate their own FIT rates using SN29500 and IEC 62380 (better still if the calculations are pre-done and details of the calculations given)

- Evidence to support claims of on-chip separation
- Evidence to support the claiming of any relevant fault exclusions
- Details of the on-chip diagnostics
- Details of assumed system level diagnostics
- Results of a pin FMEA giving λ_{DU}, λ_{DD}, λ_S and calculated SFF and DC for an assumed set of diagnostics looking at expected package failure modes
- Results of an FME(D)A giving λ_{DU}, λ_{DD}, λ_S and calculated SFF and DC for an assumed set of diagnostics looking at expected die failure modes
- A FIT rate for the various blocks shown on the datasheet to allow the drive manufacturer to redo the FME(D)A

Given the nature of the data, safety manuals may only be available under NDA (non-disclosure agreement).

Parts relevant to motor control safety for which safety manuals are currently being developed at Analog Devices include the AD7403 isolated ADC and the ADuM4135 isolated gate driver.

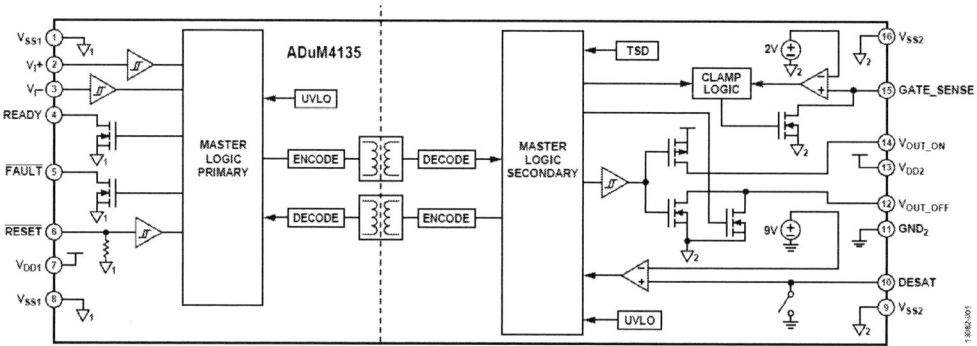

Figure 3 - ADuM4135 isolated gate driver

Secondly, the IC manufacturer, having an understanding of the system level design can assist in designing it the required features for functional safety. For instance
- Knowing that only a fraction of the PFH, perhaps just 1% is available to the IC
- Knowing that while in general for functional safety simpler is better but that transistors on chip are extremely reliable and if increasing the number of transistor on-chip by a factor of 10 leads to less components on a PCB the overall PFH will come down.
- Knowing that on-chip diagnostics can react much faster than system level diagnostics and can help in preventing an accumulation of errors.
- Knowing that the typical lifetime of a drive is 20 years and data should be available to prove the IC can match this lifetime under a given mission profile.

- Knowing that the addition of hardware accelerators such as CRC engines to reduce the software burden

Thirdly a set of recommended architectures showing how ICs could be combined to implement the safety functions from IEC 61800-5-2. This could involve
- Recommendations on system level diagnostics
- Recommendations on suitable components
- Recommendations on meeting the independence requirements between different channels
- Recommendations on software independence between safety and non-safety software which could reduce the number of required processors from three to two if

control and safety can be combined in at least one of the processors. If sufficient independence cannot be shown then everything must be treated as safety related.

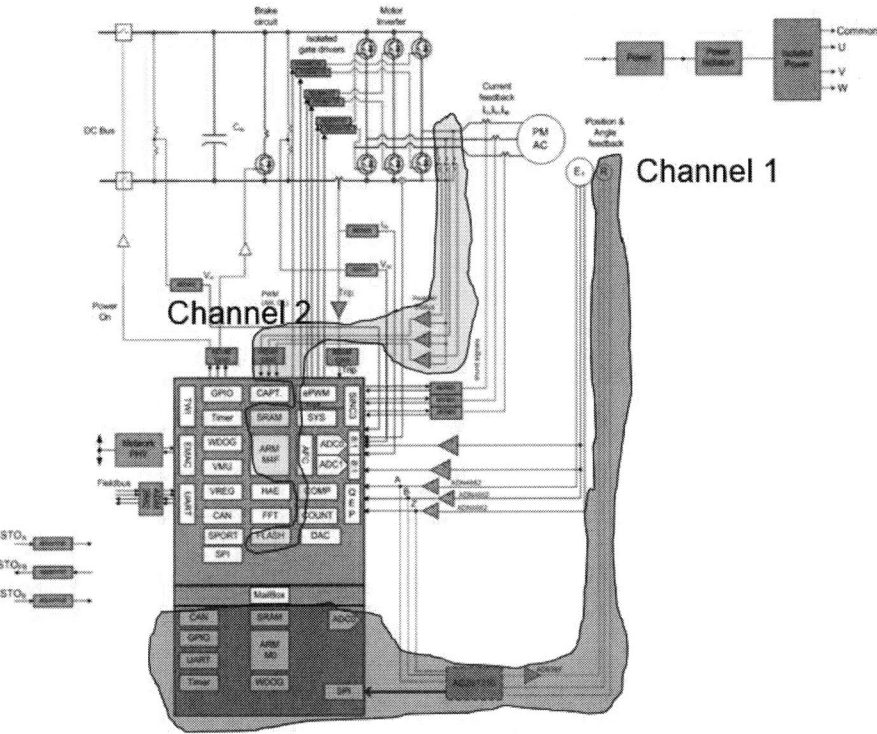

Figure 4 – a concept 2 channel architecture for implementation of SLS safety sub function from IEC 61800-5-2 using the ADSP-CM419(/8/7/6) DSP core

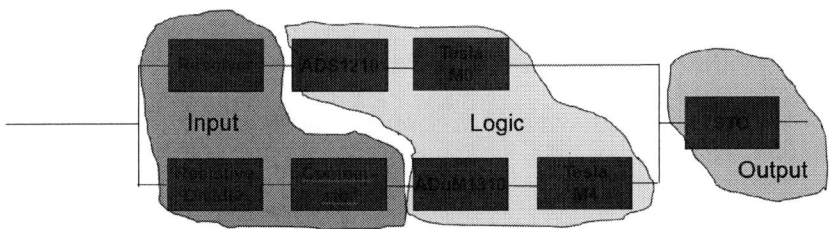

Figure 5 - Reliability block diagram for the SLS interpretation

Fourthly – the standards should be influenced to clarify requirements. For instance

- What protection against data corruption should be placed on an SPI interface connecting an ADC to a microcontroller or DSP on the same PCB? Standards such as IEC 61800-5-2:2006 refer the reader back to IEC 61508 which in turn refers to rail standards. The next version of IEC 61800-5-2 has added text to clarify that the requirements of IEC 61784-3 do not apply to such interfaces but when the author read his own words in the new standard the clarification is not as clear as he had hoped. A better clarification is contained in the new draft standard of EN 50402 where it distinguishes between "*Signal-transmission between spatially separated modules*" and "*Signal-transmission between modules not spatially separated*".

- Clarification on the on-chip separation requirements for ICs implementing diverse redundancy

- Clarification on the on-chip separation requirements for analog and mixed-signal ICS

Fifthly - Removing references particular solutions from the standards which leads some readers to believe these are the only solutions to a problem. For instance opto-couplers are an old and well known means to achieve signal isolation but have a number of disadvantages in terms of reliability, power and speed compared to newer digital isolators. Editing standards such as ISO 13849 and IEC 61800-5-2 and replacing references to opto-couplers with more generic terms such as "galvanic isolators" will also help the adoption of newer more reliable digital isolators. This has been done in the latest FDIS (final draft) of IEC 61800-5-2 in 2015.

Conclusions

This paper presents a review of the main functional safety standards related to machines and variable speed drives in particular. From this review conclusions on the requirements related to integrated circuits have been drawn. One conclusion is that to aid in meeting the requirements of functional safety IC manufacturers can supply additional information and features. This paper lists some of the most important of that information. A second conclusion is that semiconductor manufacturers need to know more about the system level requirements and Analog Devices have embarked on an analysis of their own, non-functional safety, motor control demo system design. The goal is to uncover how that architecture can be modified to meet the requirements of functional safety and to discover what information is missing to allow our customers to design in our products into a drive with functional safety requirements.

References
[1] IEC 61800-5-2:2007: Adjustable speed electric power drive systems, Safety Requirements, Functional Safety
[2] IEC 61508-2:2010, Functional safety of electrical/electronic/programmable electronic safety-related systems – Part 2: Requirements for electrical/electronic/programmable electronic safety related systems
[3] ISO 13849-1:2006, Safety of machinery – Safety-related parts of control systems – Part 1: General principles for design
[4] ISO 13849-2:2012, Safety of machinery – Safety-related parts of control systems – Part 2: Validation
[5] IEC 62061:2005 – Safety of machinery – Functional safety of safety-related electrical, electronic and programmable electronic control system
[6] ISO 26262 -
[7] BGIA Report 2/2008e – Functional safety of machine controls – Application of EN ISO 13849
[8] ADuM4135 datasheet on http://www.analog.com/media/en/technical-documentation/data-sheets/ADuM4135.pdf
[9] Analog Devices motor control web page at http://www.analog.com/motorcontrol
[10] ADSP-CM408F datasheet at http://www.analog.com/en/products/processors-dsp/cm4xx-mixed-signal-control-processors/adsp-cm408f.html#product-overview
[11] AD7403 datasheet at http://www.analog.com/en/products/interface-isolation/isolation/isolated-ad-converters/ad7403.html

© VDE VERLAG GMBH · Berlin · Offenbach

PCIM Europe 2016, 10 – 12 May 2016, Nuremberg, Germany

A System Approach to Understanding the Impact of Non-ideal Effects In A Motor Drive Current Loop

Jens Sorensen; Analog Devices; Wilmington, MA, USA; Jens.Sorensen@analog.com

Dara O'Sullivan; Analog Devices; Cork, Ireland; Dara.Osullivan@analog.com

Abstract

An essential part of any digitally controlled motor drive is phase current feedback. The quality of the measurement directly relates to system parameters such as torque ripple and torque settling time. While there is strong correlation between system performance and phase current measurement, it is difficult to translate into hard requirements to the feedback system. From a system point of view, this paper discusses how to design a feedback system optimized for motor control. Error sources are identified and mitigation effects are discussed.

1. Introduction

The performance of the current loop in a motor drive or servo, see Figure 1, directly impacts the torque output from the motor, which is critical for smooth response as well as accurate positioning and speed profiles. A key measure of smooth torque output is torque ripple. This is particularly important for profiling and cutting applications where torque ripple directly translates into achievable end application precision. Parameters related to the current loop dynamics, such as response time and settling time, are very important for automation applications where production efficiency is directly impacted by the available control bandwidth. Apart from the motor design itself, multiple factors within the drive directly impact these performance parameters.

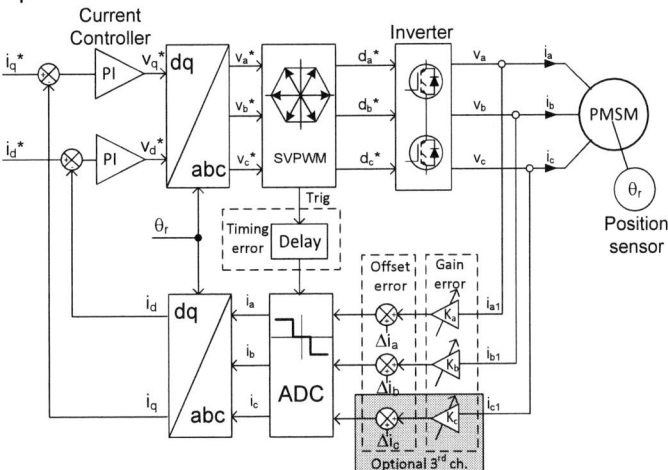

Figure 1: Current loop in a motor drive with non-ideal elements in the feedback path.

There are several sources of torque ripple within a motor drive. Some are from the motor itself, such as cogging torque due to stator winding and slot arrangement and rotor EMF harmonics [1]. Other sources of torque ripple are related to offset and gain error in the phase current feedback system [2], see Figure 1.

Inverter dead time also directly impacts torque ripple in that it adds low frequency (predominantly 5th and 7th[3]) harmonic components of the stator electrical frequency to the PWM output voltage. The impact on the current loop in this case is related to the disturbance rejection of the current loop at the harmonic frequencies.

This paper will focus on torque ripple due to phase current measurement. Each of the errors are analyzed and approaches to minimizing the effect of measurement error are discussed.

2. Torque Ripple Due To Current Measurement Error

The electro-magnetic torque equation of a 3 phase permanent magnet motor is:

© VDE VERLAG GMBH · Berlin · Offenbach

$$T_e = \frac{3}{2}PP\big(\lambda_{PM} \cdot i_q + (L_d - L_q) \cdot i_d \cdot i_q\big) \tag{1}$$

T_e is the electromagnetic torque, PP is the number of pole pairs, λ_{PM} is the permanent magnet flux, L_d and L_q are the stator inductances in the synchronous rotating reference frame, i_d and i_q are the stator currents in the synchronous rotating reference frame. In steady state and under ideal conditions, i_d and i_q are DC quantities and consequently, the produced torque will also be a DC quantity. A torque ripple will be present when there is an AC component in i_d or i_q. Because of the direct relationship between i_{dq} and the produced torque the approach used in this paper is to analyze how various measurements errors affect i_d and i_q. As a basis for this analysis consider current feedback for a 3-phase motor:

$$i_a = i_{a1} + i_{ae}$$
$$i_b = i_{b1} + i_{be} \tag{2}$$
$$i_c = i_{c1} + i_{ce}$$

Where i_x is the measured phase current (x=a,b,c), i_{x1} is the actual phase current and i_{xe} is the measurement error. No assumptions are made about the nature of the error; it can be an offset, a gain error, or an AC component. Using the Clarke transform the currents are projected on to stationary 2-phase quantities i_α and i_β:

$$i_\alpha = \frac{2}{3}\left(i_a - \frac{i_b + i_c}{2}\right)$$
$$i_\beta = \frac{2}{3}\left(\frac{\sqrt{3}}{2}i_b - \frac{\sqrt{3}}{2}i_c\right) \tag{3}$$

Using the Park transform the currents are projected on to rotating 2-phase quantities i_d and i_q:

$$i_d = i_\alpha \cdot \cos\theta + i_\beta \cdot \sin\theta$$
$$i_q = i_\beta \cdot \cos\theta - i_\alpha \cdot \sin\theta \tag{4}$$

where θ is the angle of the rotor. For field oriented control of a 3-phase motor it is necessary to know all 3 phase currents. One common approach is to measure all 3 currents which naturally requires 3 sensors and 3 feedback channels. Another common approach is to only measure 2 channels and calculate the 3rd current. For cost and complexity reasons it is desirable to have fewer sensors and measurement channels but as will be explained, measuring all 3 currents makes the system much more robust towards measurement errors.

2.1. Two Phase Measurement

First consider a 3-phase drive that has 2 phase current measurements. The 3rd phase current is calculated assuming the currents sum to 0. If i_a and i_b are measured, i_c is calculated as:

$$i_c = -i_a - i_b \tag{5}$$

Using equation (2) and (5):

$$i_c = -i_{a1} - i_{b1} - i_{ae} + i_{be} \tag{6}$$

In the stationary reference frame the currents are:

$$i_\alpha = \frac{2}{3}\left(i_a - \frac{i_b - i_a - i_b}{2}\right) = i_a = i_{a1} + i_{ae}$$
$$i_\beta = \frac{2}{3}\left(\frac{\sqrt{3}}{2}i_b + \frac{\sqrt{3}}{2} \cdot (i_a + i_b)\right) = \frac{\sqrt{3}}{3}i_a + \frac{2\sqrt{3}}{3}i_b = \frac{\sqrt{3}}{3}(i_{a1} + i_{ae}) + \frac{2\sqrt{3}}{3}(i_{b1} + i_{be}) \tag{7}$$

In the rotating reference frame the currents are:

$$i_d = (i_{a1}+i_{ae}) \cdot \cos\theta + \frac{\sqrt{3}}{3}(i_{a1}+i_{ae}) \cdot \sin\theta + \frac{2\sqrt{3}}{3}(i_{b1}+i_{be}) \cdot \sin\theta$$
$$= i_{a1}\left(\cos\theta + \frac{\sqrt{3}}{3}\sin\theta\right) + \frac{2\sqrt{3}}{3}i_{b1} \cdot \sin\theta + i_{ae}\left(\cos\theta + \frac{\sqrt{3}}{3}\sin\theta\right) + \frac{2\sqrt{3}}{3}i_{be} \cdot \sin\theta$$

$$i_q = -(i_{a1} + i_{ae})\sin\theta + \frac{\sqrt{3}}{3}(i_{a1} + i_{ae}) \cdot \cos\theta + \frac{2\sqrt{3}}{3}(i_{b1} + i_{be}) \cdot \cos\theta$$
$$= i_{a1}\left(\frac{\sqrt{3}}{3}\cos\theta - \sin\theta\right) + \frac{2\sqrt{3}}{3}i_{b1} \cdot \cos\theta + i_{ae}\left(\frac{\sqrt{3}}{3}\cos\theta - \sin\theta\right) + \frac{2\sqrt{3}}{3}i_{be} \cdot \cos\theta \tag{8}$$

Notice how both i_d and i_q have a term that relates to the true phase current and a term that relates to the measurement error ($i_{dq}=i_{dq1}+i_{dqe}$). For this analysis the error terms i_{de} and i_{qe} are of most interest:

$$i_{de} = i_{ae}\left(\cos\theta + \frac{\sqrt{3}}{3}\sin\theta\right) + \frac{2\sqrt{3}}{3}i_{be}\cdot\sin\theta = \frac{2\sqrt{3}}{3}\left(i_{ae}\cdot\sin\left(\theta + \frac{\pi}{3}\right) + i_{be}\cdot\sin\theta\right)$$

$$i_{qe} = i_{ae}\left(\frac{\sqrt{3}}{3}\cos\theta - \sin\theta\right) + \frac{2\sqrt{3}}{3}i_{be}\cdot\cos\theta = \frac{2\sqrt{3}}{3}\left(i_{ae}\cdot\cos\left(\theta + \frac{\pi}{3}\right) + i_{be}\cdot\cos\theta\right)$$

(9)

2.2. Three Phase Measurement

Now consider a 3-phase drive that measures all 3 phase currents. Following the same procedure as with 2 channels the stationary and rotating quantities are derived:

$$i_\alpha = \frac{2}{3}\left(i_{a1} + i_{ae} - \frac{i_{b1} + i_{be} + i_{c1} + i_{ce}}{2}\right) = \frac{2}{3}\left(i_{a1} - \frac{i_{b1}}{2} - \frac{i_{c1}}{2} + i_{ae} - \frac{i_{be}}{2} - \frac{i_{ce}}{2}\right)$$

$$i_\beta = \frac{2}{3}\left(\frac{\sqrt{3}}{2}(i_{b1} + i_{be}) - \frac{\sqrt{3}}{2}(i_{c1} + i_{ce})\right) = \frac{\sqrt{3}}{3}(i_{b1} - i_{c1} + i_{be} - i_{ce})$$

(10)

And in rotating frame:

$$i_d = \frac{2}{3}\left(\left(i_{a1} - \frac{i_{b1}}{2} - \frac{i_{c1}}{2}\right)\cos\theta + \left(i_{ae} - \frac{i_{be}}{2} - \frac{i_{ce}}{2}\right)\cos\theta\right) + \frac{\sqrt{3}}{3}\left((i_{b1}-i_{c1})\sin\theta + (i_{be} - i_{ce})\sin\theta\right)$$

$$= \frac{2}{3}\left(i_{a1} - \frac{i_{b1}}{2} - \frac{i_{c1}}{2}\right)\cos\theta + \frac{\sqrt{3}}{3}(i_{b1}-i_{c1})\sin\theta + \frac{2}{3}\left(i_{ae} - \frac{i_{be}}{2} - \frac{i_{ce}}{2}\right)\cos\theta + \frac{\sqrt{3}}{3}(i_{be} - i_{ce})\sin\theta$$

(11)

$$i_q = -\frac{2}{3}\left(\left(i_{a1} - \frac{i_{b1}}{2} - \frac{i_{c1}}{2}\right)\sin\theta + \left(i_{ae} - \frac{i_{be}}{2} - \frac{i_{ce}}{2}\right)\sin\theta\right) + \frac{\sqrt{3}}{3}\left((i_{b1} - i_{c1})\cos\theta + (i_{be} - i_{ce})\cos\theta\right)$$

$$= \frac{\sqrt{3}}{3}(i_{b1} - i_{c1})\cos\theta - \frac{2}{3}\left(i_{a1} - \frac{i_{b1}}{2} - \frac{i_{c1}}{2}\right)\sin\theta + \frac{\sqrt{3}}{3}(i_{be} - i_{ce})\cos\theta - \frac{2}{3}\left(i_{ae} - \frac{i_{be}}{2} - \frac{i_{ce}}{2}\right)\sin\theta$$

Again the equations have a term that relates to the true phase current (i_{dq1}) and a term that relates to the measurement error (i_{dqe}). The error terms i_{de} and i_{qe} are:

$$i_{de} = \frac{2}{3}\left(i_{ae} - \frac{i_{be}}{2} - \frac{i_{ce}}{2}\right)\cos\theta + \frac{\sqrt{3}}{3}(i_{be} - i_{ce})\sin\theta$$

$$i_{qe} = \frac{\sqrt{3}}{3}(i_{be} - i_{ce})\cos\theta - \frac{2}{3}\left(i_{ae} - \frac{i_{be}}{2} - \frac{i_{ce}}{2}\right)\sin\theta$$

(12)

3. Incorrect Sampling Instant

When a 3-phase motor is fed by a switching voltage-source inverter, the phase current can be seen as two components; a fundamental, and a switching, component, see Figure 2 (A).

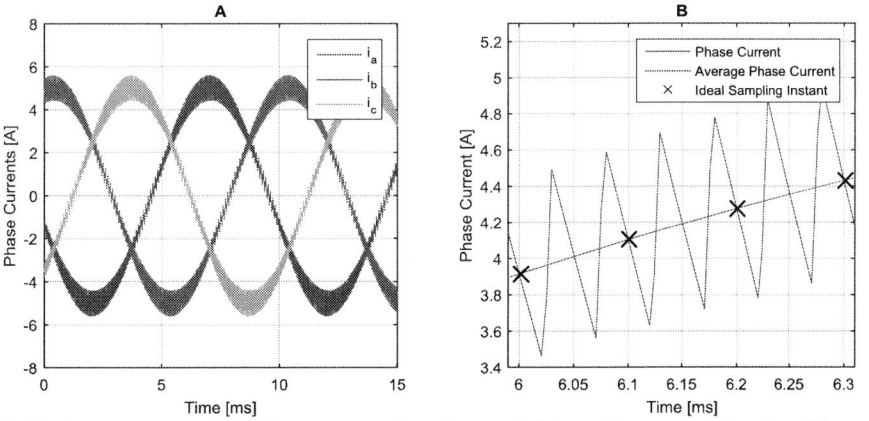

Figure 2: (A) phase currents of a 3 phase motor driven by a switching voltage source inverter. (B) zoom of a phase current illustrating how the current ripple is attenuated through sampling.

For control purposes, the switching component has to be eliminated or it will affect the performance of the current control loop. A common technique to extract the average component is to sample the currents synchronized to the PWM cycle. At the beginning and at the center of a PWM cycle the currents assume their average value and if sampling is tightly synchronized to these instances the switching component is effectively suppressed, as shown

© VDE VERLAG GMBH · Berlin · Offenbach

in Figure 2 (B). However, if the current is sampled with a timing error, aliasing will occur and as a consequence performance of the current loop drops. This section discusses reasons for timing errors, what the effects on the current loop are and finally how to make the system robust to sample timing errors.

3.1. Sample Timing Errors In a Motor Drive

With the fundamental component of a phase current typically in the range of tens of Hz and the bandwidth of the current loop in the range of a few kHz, it seems counter intuitive that a small timing error can affect control performance. However, with only the phase inductance to limit di/dt, even a small timing error will result in significant current distortion. For example, 250V across an inductor of 5mH for 1us, will change the current by 50mA. In addition, assume the system uses a 12-bit ADC with a full-scale of 10A, then the lower 4.3 bits if the ADC are 'lost' because of the timing error. As will be shown, losing bits is the best case scenario. The aliasing can also cause torque ripple as well as gain error in the feedback system.

The most common reasons for an incorrect sampling instant are:

- Insufficient link between the PWM and the ADC making it impossible to sample at the right time.
- A lack of sufficient (2 or 3 depending on the number of phases being measured) independent simultaneous sample and hold circuits.
- Propagation delay on gate drive signals which will bring the motor voltage out phase with the PWM timer.

Generally, anything that can impact di/dt determines how severe an incorrect sampling instant is. Of course the size of the timing error is important but system parameters such as motor speed, load, motor impedance and DC-bus voltage also have a direct influence on the error.

3.2. Effect of Sampling Error on System Performance

Using the equations derived in section 2.1 and 2.2, the effect of sampling error can be determined. With 2-phase current measurement, assume i_a is sampled at the ideal instant (i_{ae}=0) and i_b is sampled with a delay which result in i_{be}≠0. In this case the error terms defined by equation (9) are:

$$i_{de} = \frac{2\sqrt{3}}{3} i_{be} \cdot \sin\theta$$
$$i_{qe} = \frac{2\sqrt{3}}{3} i_{be} \cdot \cos\theta \tag{13}$$

With 3-phase current measurement, assume i_a and i_c are sampled at the ideal instant (i_{ae}=i_{ce}=0) and i_b is sampled with a delay (i_{be}≠0). In this case the error terms defined by equation (12) are:

$$i_{de} = \frac{1}{3} i_{be} (\sqrt{3}\sin\theta - \cos\theta) = \frac{2}{3} i_{be} \sin\left(\theta - \frac{\pi}{6}\right)$$
$$i_{qe} = \frac{1}{3} i_{be} (\sqrt{3}\cos\theta + \sin\theta) = \frac{2}{3} i_{be} \cos\left(\theta + \frac{\pi}{3}\right) \tag{14}$$

From equation (13) and (14) some interesting conclusions can be made. First of all, the Clarke/Park transformation 'gains' the measurement error differently:

$$\frac{\hat{i}_{de,2\,phase}}{\hat{i}_{de,3\,phase}} = \frac{\hat{i}_{qe,2\,phase}}{\hat{i}_{qe,3\,phase}} = \frac{\frac{2\sqrt{3}}{3} i_{be}}{\frac{2}{3} i_{be}} = \sqrt{3} \tag{15}$$

So if the feedback system has a delay on one of the current measurements, the impact on a drive with 2 channels will be 1.73 times higher than if the system had 3 channels.

Using equations (13) and (14), it is also possible to identify the impact of measurement delay on the motor torque. For this analysis it is assumed the phase current is sampled while applying a zero voltage to the motor terminals (V000 or V111) and during this period, the only voltage driving di/dt is the BEMF. With a sinusoidal BEMF, di/dt will also follow a sinusoidal function – that is di/dt=0 at BEMF zero cross and di/dt at maximum when the BEMF peaks. Now, if a phase current is sampled with a fixed delay with respect to the ideal sampling instant the error is sinusoidal:

$$i_{xe} = \hat{i}_e \cdot \sin(\theta - \varphi) \tag{16}$$

© VDE VERLAG GMBH · Berlin · Offenbach

Where x=a,b,c and φ is the phase angle with respect to the dq-reference frame. Using i_{de} from equation (13) as an example:

$$i_{de} = \frac{2\sqrt{3}}{3}\,\hat{\imath_e} \cdot \sin(\theta - \varphi) \cdot \sin\theta = \frac{\sqrt{3}}{3}\,\hat{\imath_e} \cdot (\cos(-\varphi) - \cos(2\theta + \varphi)) \tag{17}$$

The term $\cos(-\varphi)$ is a offset while $\cos(2\theta-\varphi)$ is an AC-component oscillating at twice the fundamental frequency. With these components in the dq-currents, the motor torque will have similar components. Another thing to note is that with 3 current measurements, and the chosen orientation of the dq-frame $\varphi=-\pi$ which means the offset term is zero. That is, no gain error with 3 channels. The difference between a 2- and 3 sensor system is illustrated in Figure 3.

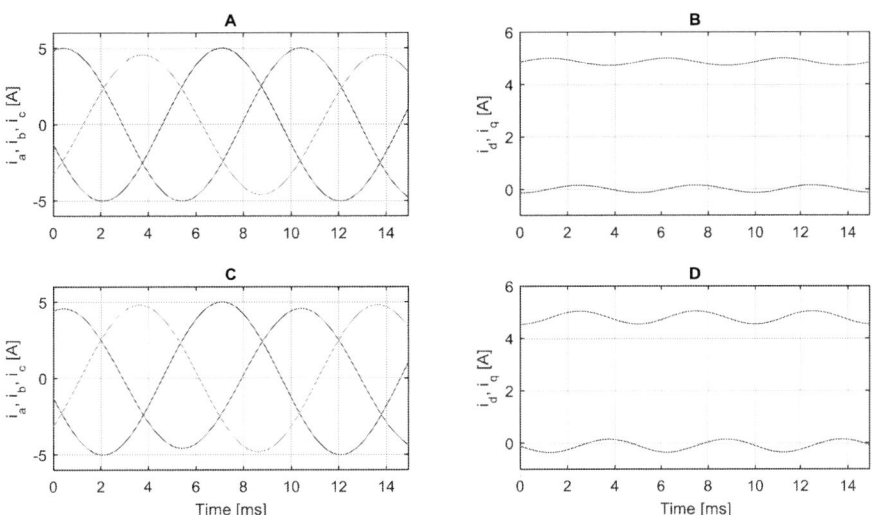

Figure 3: Effect of incorrect sampling instant. i_a,i_b,i_c and i_d,i_q with 2 current sensors A+B and 3 current sensors C+D.

In the 3 sensor case, from Figure 3 A+B, notice how the delay on i_b measurement results in a current (torque ripple) at 2 times the fundamental frequency. Also notice how the DC components of i_d and i_q are not affected.

In the 2 sensor case, from Figure 3 C+D, notice how the delay on i_b measurement results in an AC component which is 1.73 times higher than with 3 sensors. In addition, the DC components of both i_d and i_q are also affected.

3.3. Minimizing the Effect Sample Timing Error

As performance requirements of the control loop increases it becomes desirable to minimize the effect of sample timing errors, especially bearing in mind the trend towards increased ADC resolution. A few years ago 10-12 bit ADCs were common but now resolution of 16-bits is becoming the norm. These additional bits should be utilized, otherwise the value of a high performing ADC will be compromised with the lower bits lost due to delays in the system.

The most effective way to minimize the effect of sampling timing error is to get as close as possible to the ideal sampling instant for all phases. This may result in choosing a controller which is optimized for digital control switching power converters. In addition, optimizing propagation delay/skewing in the gate drive circuit will have a positive impact.

If timing error minimization still does not meet requirements, a significant performance improvement can be achieved by using 3 current sensors and an ADC with 3 independent sample and hold circuits.

4. Offset Error

The derived equations can also describe how the system reacts to an offset on the measured currents. First looking at the 2 sensor case and using i_{de} from equation (9) as an example, the error component can be expressed as:

$$i_{de} = \frac{2\sqrt{3}}{3}\left(i_{a,offset} \cdot \sin\left(\theta + \frac{\pi}{3}\right) + i_{b,offset} \cdot \sin\theta\right) \tag{18}$$

With $i_{a,offset}$ and $i_{b,offset}$ the offset of the a and b channels respectively. As can be seen, the offset will cause an AC component in the currents (and torque) at the fundamental frequency of the

© VDE VERLAG GMBH · Berlin · Offenbach

motor. If the system has offset calibration at startup, any remaining offset will be due to drift. In this case assuming the sensors drift in the same manner, the approximation $i_{a,\text{offset}} = i_{b,\text{offset}} = i_{\text{offset}}$ can be made.

$$i_{de} = \frac{2\sqrt{3}}{3} i_{offset} \left(\sin\left(\theta + \frac{\pi}{3}\right) + \sin\theta \right) = 2 \cdot i_{offset} \sin\left(\theta + \frac{\pi}{6}\right) \tag{19}$$

This means the amplitude of the error component is twice the amplitude of the phase offset. A similar result can be found for the q-axis component of the error current.

Performing the same exercise with 3 current sensors it is found that i_{de} from equation (12) is:

$$i_{de} = \frac{2}{3}\left(i_{a,offset} - \frac{i_{b,offset}}{2} - \frac{i_{c,offset}}{2} \right) \cos\theta + \frac{\sqrt{3}}{3}\left(i_{b,offset} - i_{c,offset} \right) \sin\theta \tag{20}$$

And following the reasoning that initial offset is calibrated out and all sensors drift equally, $i_{a,\text{offset}} = i_{b,\text{offset}} = i_{c,\text{offset}} = i_{\text{offset}}$:

$$i_{de} = \frac{2}{3}\left(i_{offset} - \frac{i_{offset}}{2} - \frac{i_{offset}}{2} \right) \cos\theta + \frac{\sqrt{3}}{3}\left(i_{offset} - i_{offset} \right) \sin\theta = 0 \tag{21}$$

Again, the benefit of having 3 sensors is clear – the offset on the current sensors will have no effect of torque ripple. Even if the sensors do not drift in exactly the same way they are likely to show the same trend. Therefore, a 3 sensor setup will always have significantly lower torque ripple in a system with uncalibrated offset errors.

4.1. Minimizing the Effect of Offset Error

Offset on current feedback is one of the dominating sources of torque ripple in a motor drive and it is desirable to minimize it as much as possible. Generally speaking there are two types of offset errors on current feedback. Firstly, there is the static offset which is present at any point in time and at any temperature. Secondly, there is the offset drift which is a function of parameters such as temperature and time. A common technique to minimize the effect of static offset is to do offset calibration which can be done either at the time of manufacture or every time when the motor current is 0 (typically when the motor is stopped). If this is approach is used, static offset is usually not a concern.

Offset drift is more complicated to handle. As this is a slow drift that usually happens while the motor is operating, it is difficult to do online calibration and stopping the motor is generally not an option. Some online calibration techniques based on observers have been suggested, see for example [4], but the observers rely on models of the motor's electrical and mechanical system. For the online estimation to be effective, exact knowledge of the motor parameters is required and that is usually not the case.

As was discussed in section 4, the most effective mitigation of offset drift is to use 3 current measurements. Assuming the channels use the same type of components the drift of the channels are likely to be similar. If this is the case, the offsets cancel out and will not result in a torque ripple. Even if the channels do not drift at the same rate, as long as they drift in the same direction, the 3 channel approach will have a canceling effect on the offset.

With 2 current measurements, the torque ripple is present even if the channels drift at the same rate. In other words, a 2 sensor system is very sensitive to offset drift. In this case the only way avoid torque ripple is to make sure the drift is kept low which may add cost and complexity to the feedback system. For a given set of performance requirements a 3 channel feedback system may prove to be the cost effective solution.

5. Gain Error

When the system has a gain error on current feedback the error signal, i_{xe} is proportional to the actual phase current i_{x1} (x=a, b, c.):

$$i_{xe} = k_x \cdot i_{x1} \sin(\theta - \varphi) \tag{22}$$

This is a sinusoidal error at the fundamental frequency. As can be seen, the nature of an error due to gain is similar to an error due to incorrect sample timing, see equation (16). Therefore the same conclusions can be drawn:

- If the same gain error is present on all channels, there will be no torque ripple; only a gain error. This applies to 2- as well as 3 channel systems.
- If the gain error varies from one channel to the other, it will result in a torque ripple component at twice the fundamental frequency.

- 2 channel current measurement is 1.73 times more sensitive to gain error than 3 channel current measurement.

6. Experimental Validation

The effect of offset error and gain error on the measured current and output torque are validated in the experimental setup illustrated in Figure 4.

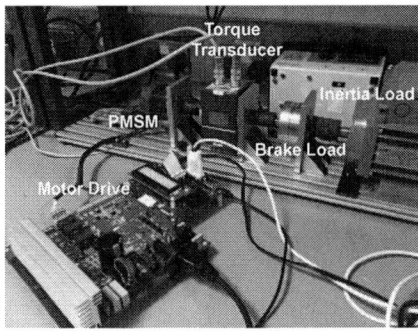

Motor Drive Board	AC Input, 350Vdc, 3-phase closed loop field-oriented controller drive platform from Analog Devices
PM motor	M-2311S-LN-02D Teknic, 4-pole, 0.42Nm, 6000rpm, 3-phase PM synchronous motor
Torque transducer	RWT421-DA, Sensor Technology, +/-2Nm, 0.25% accuracy
Brake Load	Magnetic particle brake, 1.7Nm max
Inertia Load	Unconnected 1kW ABB induction motor

Figure 4: Test Rig Setup

The current feedback circuit in the drive board utilizes Hall-effect transducers in the 3 motor phases. 2-phase or 3-phase current measurement can be selected in the software. Offset calibration is performed when the motor is not running, so under normal operation (without allowing time for drift effects), offset and gain error are quite small. In order to illustrate the effects of these errors, which will typically be present due to temperature drift – in spite of calibration routines – artificial offset and gain errors are introduced in the control software after the calibration routine. The 'measured' quantities as seen by the control algorithm will differ from the actual quantities, which will contain the effects of the errors as discussed in previous sections. Figure 5 illustrates this for a set speed reference of 520rpm – hence a motor electrical frequency of 35Hz.

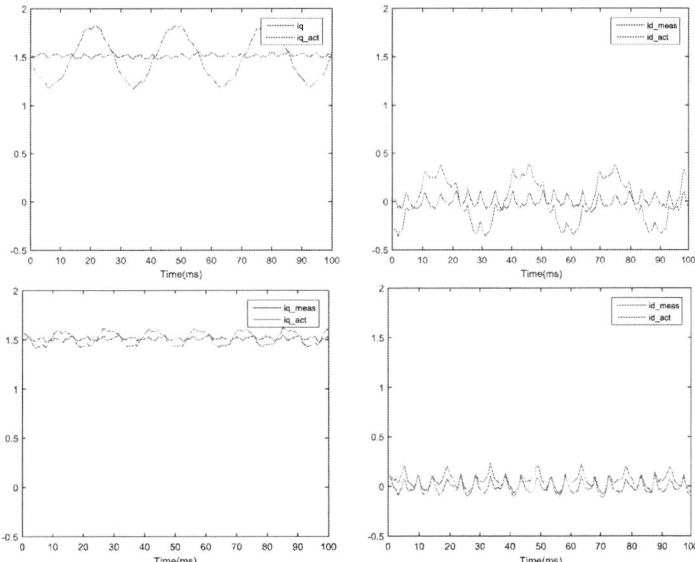

Figure 5: Actual (red) and 'measured' (blue) (top to bottom); iq and id with 1% offset error; iq and id with asymmetric gain error (1.05/0.95)

It is clear that while the drive is controlling the d-axis and q-axis currents to relatively constant values in order to maintain the set speed, the actual currents contain significant harmonic components, especially in the case of offset error. These harmonic components directly impact the output torque ripple. This is illustrated in Figure 6. It must be noted that there is a significant mechanical torque pulsation due to a slight shaft misalignment in the test rig. This is present

© VDE VERLAG GMBH · Berlin · Offenbach

at the mechanical frequency and some low order harmonics. However, the variation in harmonic content related to the offset and gain error sources is still clearly seen. For the offset error, the harmonic component at the electrical frequency (35Hz) increases in proportion with the offset error percentage, as illustrated, while the harmonic content at twice the electrical frequency increases with the gain error asymmetry, as predicted by the theory.

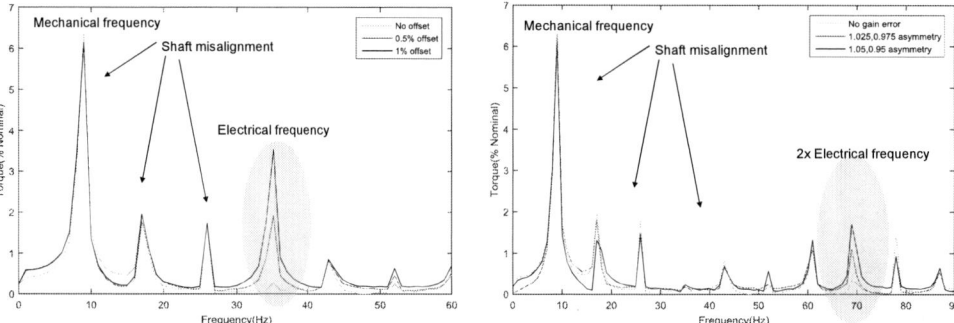

Figure 6: Measured torque ripple as % of nominal with 2-phase current measurement and (left) increasing offset error (right) increasing gain error

Moreover the impact of 3-phase measurement is clearly seen in Figure 7, where the offset error induced torque ripple is completely eliminated and the gain error-induced torque ripple is reduced by a factor of 1.73, again bearing out the theoretical calculations

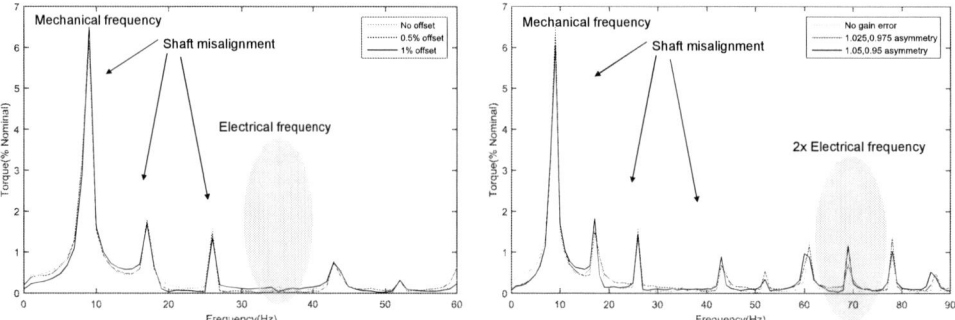

Figure 7: Measured torque ripple as % of nominal with 3-phase current measurement and (left) increasing offset error (right) increasing gain error

7.   Summary

Through analysis and measurement, this paper has shown how non-ideal effects in the current feedback system affect system performance. It has been show how a system with 3 current measurements is significantly more robust towards measurement error then a system with 2 current measurements.

8.   References

[1] Weizhe Qian, S. K. Panda and Jian-Xin Xu, "Torque ripple minimization in PM synchronous motors using iterative learning control," in *IEEE Transactions on Power Electronics*, vol. 19, no. 2, pp. 272-279, Mar. 2004.

[2] Dae-Woong Chung, Seung-Ki Sul and Dong-Choon Lee, "Analysis and compensation of current measurement error in vector controlled AC motor drives," *Industry Applications Conference, 1996. Thirty-First IAS Annual Meeting, IAS '96., Conference Record of the 1996 IEEE*, San Diego, CA, 1996, pp. 388-393 vol.1.

[3] S. Kaitwanidvilai, W. Khan-Ngern and M. Panarut, "The impact of deadtime effect on unwanted harmonics conducted emission of PWM inverters," *Environmental Electromagnetics, 2000. CEEM 2000. Proceedings. Asia-Pacific Conference on*, Shanghai, 2000, pp. 232-237.

[4] Y. Uenaka, M. Sazawa and K. Ohishi, "Fine self-tuning method of both current sensor offset and electrical parameter variations for SPM motor," *IECON 2010 - 36th Annual Conference on IEEE Industrial Electronics Society*, Glendale, AZ, 2010, pp. 841-846.

© VDE VERLAG GMBH · Berlin · Offenbach

Gate driver as part of the inverter safety concept: optimizing the inverter's design

Laurent Beaurenaut
Infineon Technologies, Am Campeon, 1-12 85579 Neubiberg, Germany
laurent.beaurenaut@infineon.com

Peter Sinn
Robert Bosch GmbH, Robert-Bosch-Str. 2, 71701 Schwieberdingen, Germany
Peter.Sinn@de.bosch.com

Abstract

This paper describes the implementation of a powertrain inverter where IGBT gate drivers are part of the inverter's safety concept. It is shown how digitalization of the gate driver leads to a cost effective implementation of safety relevant functions, such as active short circuit and overcurrent detection. It also allows in-application testability of critical functions, which results in a significant improvement of the system diagnostic coverage. Advantages in terms of system costs and compactness of this approach are presented.

Introduction

Electric Vehicles (EV) and Hybrid Electric Vehicles (HEV) are now being produced and commercialized at large scale. This implies stringent requirements in terms of performance, cost and safety for the drivetrain [1]. Relying on standard solutions which have been used in the industrial world for years is not sufficient to meet those challenges. On the contrary innovative inverter solutions shall be developed, that fulfills the highest safety integrity levels as per ISO26262, while still being cost and energy efficient. In order to achieve those three targets at the same time, the partitioning of functions between the key components of the inverter shall be optimized. An effective way to achieve this is to increase the level of functional integration of the gate driver, and integrate it in the safety concept of the inverter.

Digital isolated gate driver

Typical automotive inverter architectures rely on a dedicated gate driver stage between the main logic on the Low Voltage domain (LV) and the power module located in the High Voltage (HV) battery domain. In regular 6-pack systems, the gate driver stage consists of gate driver ICs, post amplifiers, isolated power supplies and additional protection functions (fig X). In order to separate electrically both domains from one another, the gate driver IC (GDIC) embeds a galvanic isolation. Technologies relying on optical, capacitive and inductive principles are today state of the art [2].
The key function of the gate driver stage is to transfer the PWM coming from the LV logic to the IGBT gate. The GDIC shall transfer the PWM signal across the isolation barrier. In order to optimize system performance (e.g. dead time), the propagation time and the jitter generated by the GDIC shall be minimized. The GDIC shall also provide accurate gate voltage levels, since both high and low levels affect directly the losses of the inverter.

© VDE VERLAG GMBH · Berlin · Offenbach

Moreover, faulty voltages may even lead to destructive conditions (such as linear mode conditions).

Some failure modes of the IGBT, such as overcurrent and short circuit conditions, require a very fast reaction of the power electronics. Therefore, the GDIC shall be able to recognize operation outside the regular conditions, and act autonomously to protect the power module. Since a malfunction of the GDIC may lead to the destruction of the power module, it is all the more important to verify during runtime the correct operation of the device.

A cost effective way to increase the overall diagnostic coverage of the complete system is to integrate digital control into the GDIC [3]. A communication link to the LV main logic is used to configure the device at power up, to enable advanced diagnostic and to provide software controlled functions. Figure 1 shows the implementation of a digital isolated gate driver. It supports a SPI communication interface. This communication interface does not control directly the switching behavior of the IGBT. But it is a middle speed (a few Mbps) parallel channel to the regular PWM command.

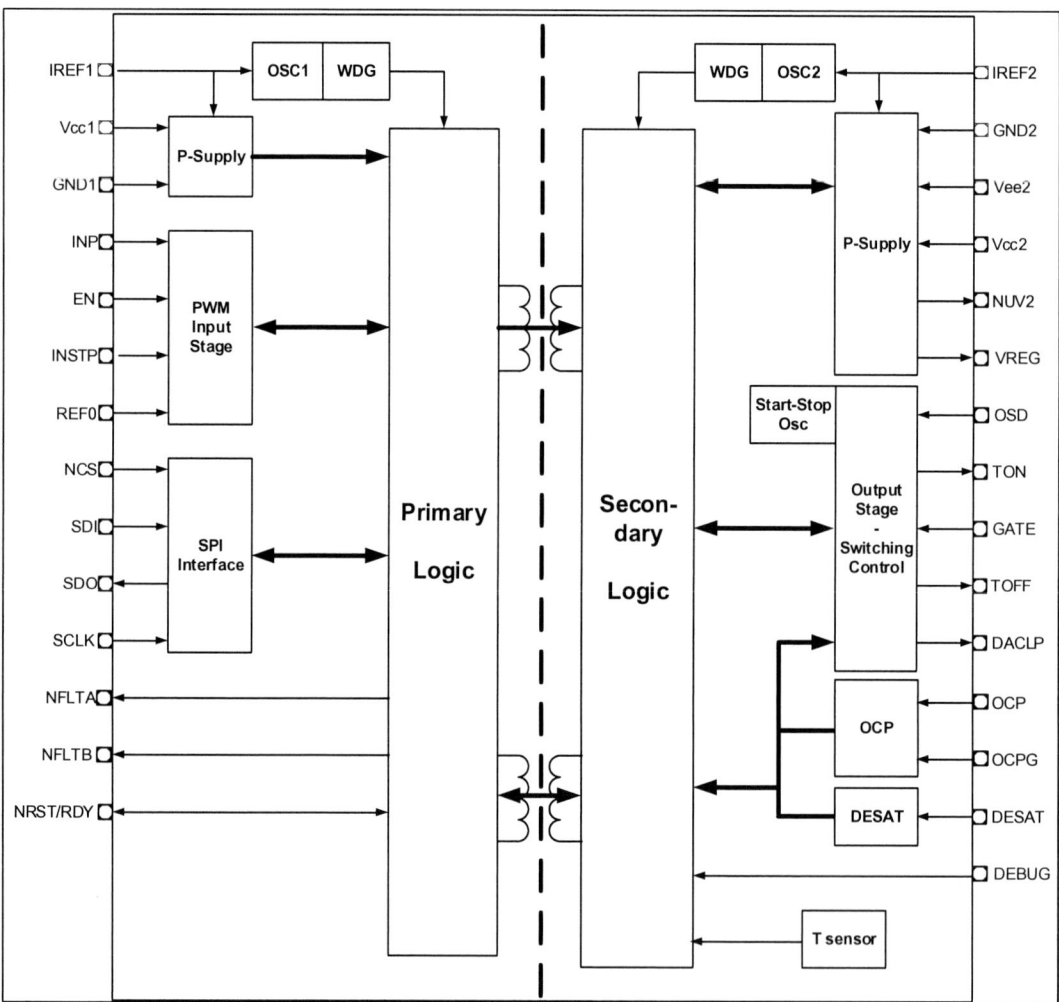

Fig 1: Digital gate driver with SPI interface

The device implements two protection mechanisms against overcurrent and short circuit conditions. The DESAT function aims at protecting the IGBT in case of short circuit [4]. The voltage drop Vce over the IGBT is monitored while the device issues a PWM turn-on signal command (the DESAT function is not active while the output stage is in OFF state). The voltage at pin DESAT (decoupled from high voltage by an external diode) is compared to an internal reference voltage (Figure 2).

When the IGBT goes in full desaturation, the measured voltage is higher than the internal threshold, which initiates an Emergency Turn-Off sequence. The DESAT function can therefore be safety relevant for applications where the destruction of the IGBT leads to a violation of the safety goal. Consequently, the correctness of the DESAT monitor shall be guaranteed during the application lifetime. It is possible to reduce the probability of undetected latent failures by testing regularly the proper reaction of the system to a DESAT event. This can be achieved by injecting regularly, for example at every power on, some "dummy" failures and observing the resulting system reaction. The SPI interface enables a cost effective implementation of such a function.

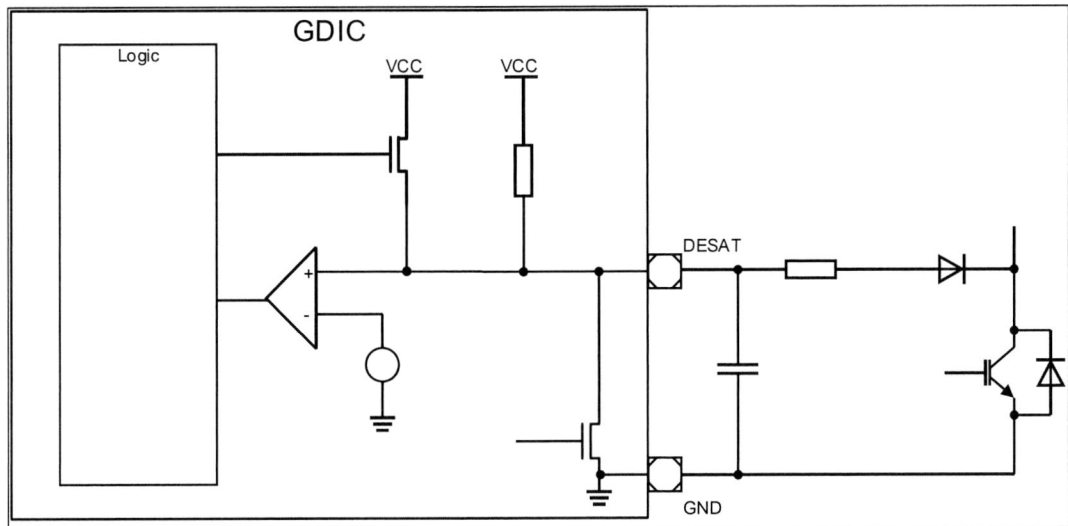

Fig 2: DESAT Implementation

The device supports a special operating mode that can be enabled by the LV logic with a specific sequence of SPI command. This verification mode is intrusive, in the sense that the device is not operating as it would in normal conditions. For this purpose, the device relies on a SPI controlled state machine which ensures that intrusive functions can not be activated during normal operation of the device.

When in verification mode, at the reception of a specific trigger, the GDIC generates a false DESAT failure. There are different ways to generate the dummy failure, depending on which part of the signal chain shall be subject to diagnostic. One straightforward possibility is to use an internal pull up transistor at the input of the comparator in order to exceed the threshold level. If the complete path down to the IGBT collector is to be verified, one can use another mode of operation, where no action is executed by the GDIC output stage at the reception of a PWM turn-on command. The IGBT stays therefore in OFF state. However, the DESAT logic still works normally. It means that after the blanking time has elapsed, the voltage on pin DESAT will exceed the threshold level, leading to a DESAT error. The correct reaction of the device (e.g. emergency turn-off) as well as the reaction time (e.g. notification signals) can be monitored by the LV logic.

The integrated Over Current Protection (OCP) function aims at protecting the IGBT in case of overcurrent conditions that do not lead to full desaturation. The OCP function can be used alone or simultaneously to DESAT, providing a redundant failure detection mechanism. The voltage drop over a sense resistor located on the auxiliary emitter path of the IGBT is monitored while the device issues a PWM turn-on command. The voltage at pin OCP is compared internally to two internal reference thresholds (Figure 3). In case the measured voltage is too high, an Emergency Turn-Off sequence is initiated by the GDIC.

Similarly to DESAT, several failure injection mechanisms can be implemented to trigger dummy failures, in order to ensure the correct operation of the OCP function during application life time. Additionally, it is necessary to verify that current is really flowing through the sense resistance. This is done by implementing an additional comparator, whose "monitoring" threshold voltage lies below the "turn off" sequence. During normal application condition, the current flowing through the IGBT is below the turn-off limit but still high enough to be detected by the monitoring comparator. This sets a sticky flag inside the logic that can be checked regularly by the LV logic during operation. The combination of failure injection and monitoring comparator allows thus a very high level of diagnostic coverage of the OCP function.

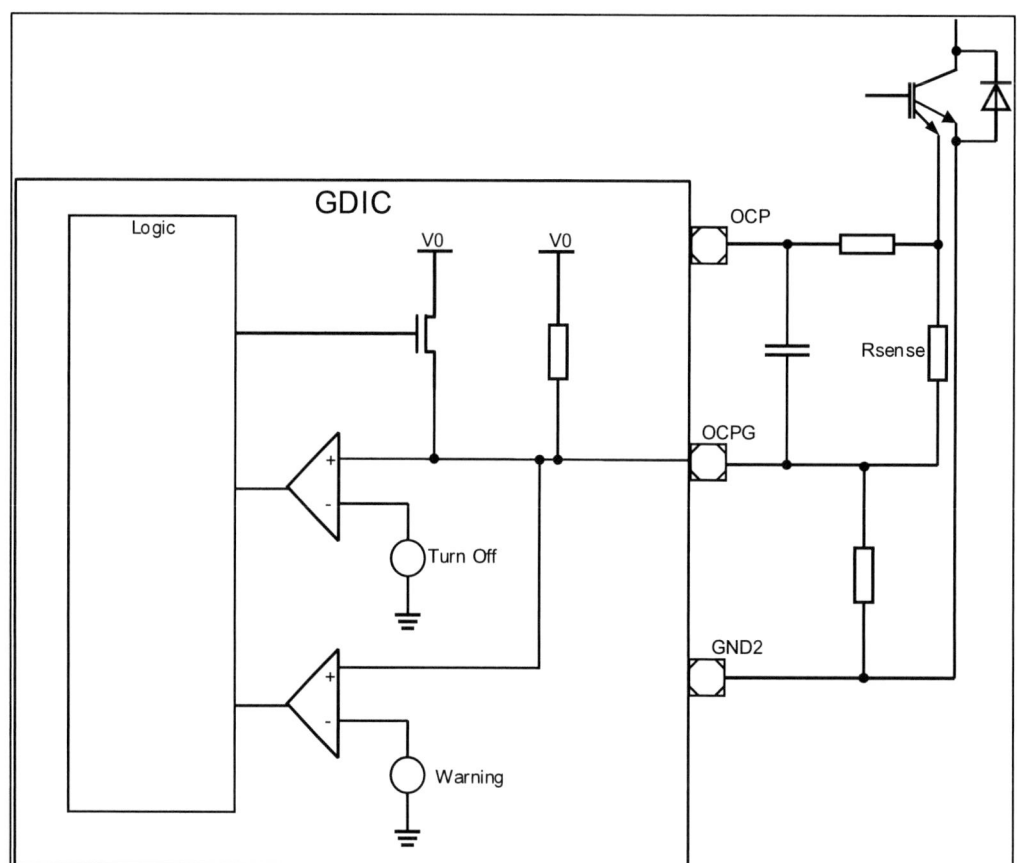

Fig. 3: OCP Implementation

Automotive (H)EV inverter system

Fig. 4 shows the block diagram of the inverter subsystem:

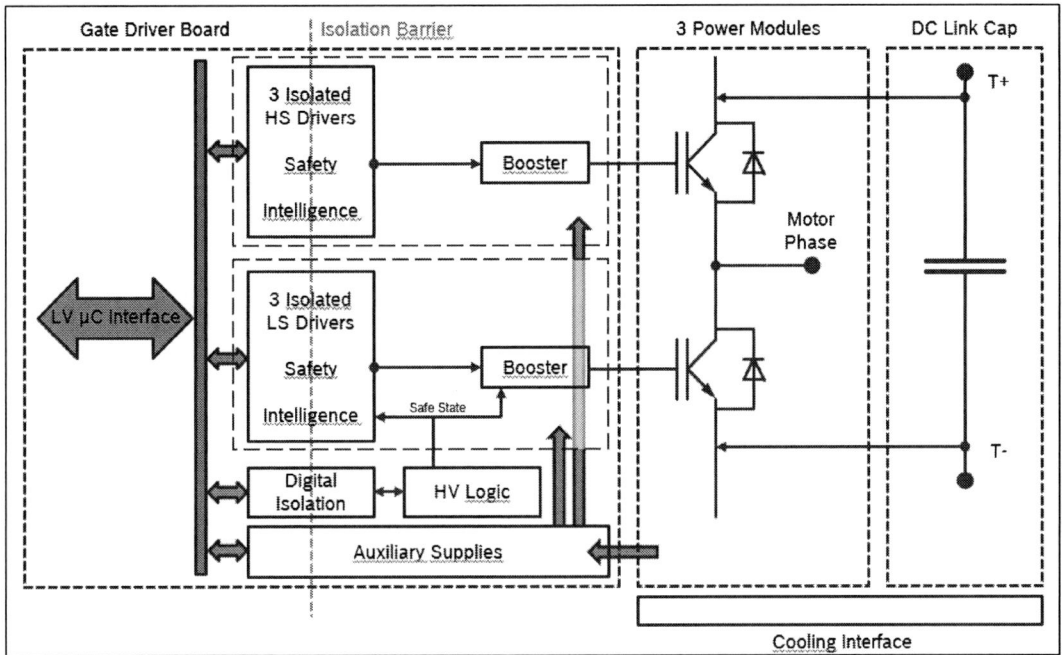

Fig. 4: Inverter subsystem

Safety relevant topics could be addressed by using the implemented protection and diagnosis features of the driver chipset. As one example, the field of overcurrent / short circuit detection will be discussed.

First of all, the well-known DESAT mechanism is used to detect overcurrent situations, which is at the same time a redundant detection path for undervoltage of the driver supply, IGBT defects or ruptures of the connections. In parallel, IGBTs with a Sense Emitter structure are used to achieve a very fast reaction on overcurrent situations, with possible response times below 1 μsec.

Practical realization of the circuits has to consider additional effects such as current sharing between paralleled chips, tolerance of the current sense ratio, dynamic effects including layout parasitics and last but not least noise filtering, in order to achieve high reliability of the short circuit protection.

Both circuits are shown in Fig. 2 and 3; Fig.5 shows the functional overlap of Current Sense and DESAT mechanism, which leads to a high reliability to withstand short circuits.

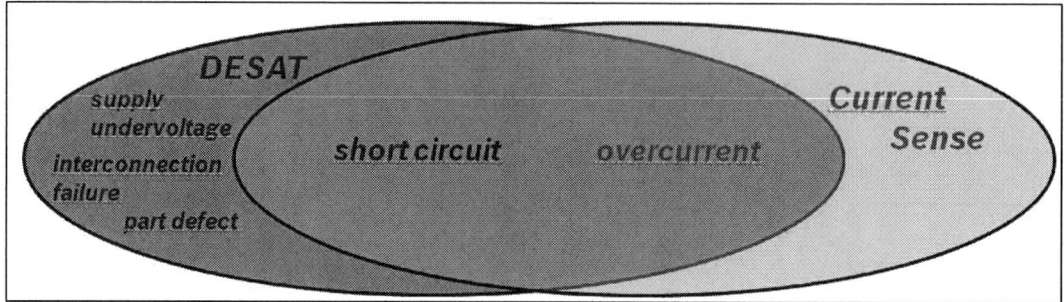

Fig. 5: functional overlap of Current Sense and DESAT principle

Conclusion

This paper shows how an innovative solution has been developed in order to meet the design targets of future automotive powertrain inverters. With the integration of advanced functionality inside the gate driver, the performance of the inverter can be increased while reducing at the same time system costs. An optimized system partitioning is only possible thanks to a close collaboration between component and system suppliers. This implies a tight partnership across the traditional boundaries of the value chain. The development of collaborative models is therefore a decisive contribution to drive e-mobility on the road of success.

References

[1] P. Leteinturier, M. Münzer: Driving e-Power toward automotive mass market acceptance, SIA APE, 2011

[2] L. Beaurenaut, P. Leteinturier: "Coreless Transformer Driver IC for HV Motor Applications" ECPE, November 2010

[3] L. Beaurenaut, K. Scheibert, New System Partitioning for Safe Automotive Inverters, PCIM Europe 2013.

[4] T. Reiter; L. Beaurenaut: Detection and Turn-off of Short Circuits in Automotive IGBT Inverters, EEHE, 2013

PCIM Europe 2016, 10 – 12 May 2016, Nuremberg, Germany

Estimation of ripple and inductance roll off when using powdered iron core inductors

Gautham Ram Chandra Mouli[1], Jos Schijffelen[2], Pavol Bauer[1], Miro Zeman[1]
[1]Department of Electrical Sustainable Energy, Delft University of Technology, Netherlands
(G.R.Chandramouli@tudelft.nl, P.Bauer@tudelft.nl, M.Zeman@tudelft.nl)
[2]Power Research Electronics BV, Netherlands (j.schijffelen@pr-electronics.nl)

Abstract

Magnetic cores like powdered iron cores have a variable permeability, which is dependent on the magnetic motive force of the inductor windings. When inductors are designed using such materials, the inductance varies as a function of the inductor current. This causes a non-linear current to flow through the inductor. In this paper, mathematical derivation of the inductance-current relationship is derived for such inductors. The model is applied for a KoolMµ® powdered iron core and the results are verified experimentally using an interleaved boost converter.

1. Introduction

Inductors are widely used in DC/DC converters as an energy storage element and as a filter. The inductance L can be related to the number of turns of the copper wire N, core material and the dimensions of the core by

$$L = \left(\frac{\mu_0 \mu_r A_c}{l_e}\right) N^2 = A_L N^2 \qquad where \quad A_L = \left(\frac{\mu_0 \mu_r A_c}{l_e}\right) \tag{1}$$

where $\mu_0 = 4\pi x 10^{-7}$, μ_r - relative permeability of the material, A_c - core area, l_e - magnetic path length, A_L is permeance of the material.

When a DC voltage V_L is applied across the inductor, the current through the inductor i_L linearly increases/decreases based on the sign of the voltage where $i_{L(0)}$ is the inductor current at time $t=0$, Δi_L is the current ripple in time Δt:

$$V_L = L\frac{di_L}{dt} \qquad\qquad \Delta i_L = i_L(t) - i_{L(0)} = \frac{V_L}{L}\Delta t \tag{2}$$

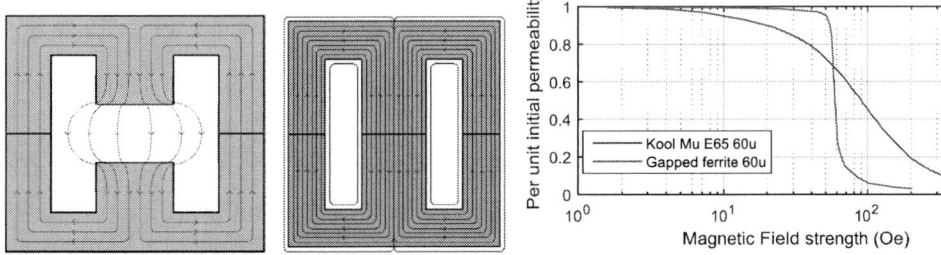

Fig. 1. Path of flux lines through a gapped ferrite core (left) and a powdered iron core (middle). Variation of permeability of core as a function of magnetic field strength for a KoolMµ powdered iron core and gapped ferrite core, both of which have an initial permeability of $\mu_r = 60$ (right) [1]

© VDE VERLAG GMBH · Berlin · Offenbach

1383

PCIM Europe 2016, 10 – 12 May 2016, Nuremberg, Germany

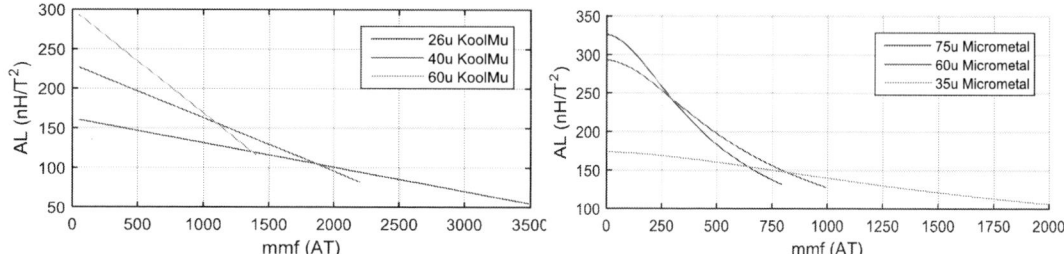

Fig. 2. Permeance A_L of three core of different permeability as a function of the ampere-turns for KoolMµ E65 (left) and Micrometal E-255 (right). The permeance and inductance of the core reduces linearly with increase in current through the inductor.

When inductors are designed using ferrite cores, the core exhibits a practically constant permeability and permeance in the operating region. This is because it has a fixed air gap and the reluctance of the core is primarily characterized by this air gap, as seen in Fig. 1. This means that the inductance does not vary with the magnetic field strength generated by the inductor coils in the operating region. When the core is close to saturation, then there is rapid change in inductance from its initial value to zero as shown in Fig. 1 [1], when μ_r reduces from its initial value to zero.

1.1. Inductors with variable permeability

Powdered iron cores come under the category of distributed air gap cores. They have small air gaps distributed evenly throughout the cores. There a number of powder iron cores that are commercially available – for example Kool Mµ®, MPP, High Flux, XFlux®, AmoFlux® cores from magnetics or powder cores from Micrometals. These cores differ from ferrite cores in a number of ways.

1. The permeability of the core is dependent only on the core material. Powdered iron core thus have a 'fixed' distributed air gap. This is unlike ferrites where the air gap and number of turns can be varied for the same inductance to give the least inductor losses.

2. The powered iron cores have a higher saturation flux density B_{sat} which can be more than twice as that of ferrite cores [2]. 60µ KoolMµ, Epcos N87 ferrite cores have a B_{sat} of 1000mT and 490mT respectively [3]. This means that fewer parallel core sets would be required to build high current inductor in high power density converters [4]–[6].

3. With increase in magnetic field strength, the small pieces of powdered iron gradually saturate one after the other starting with the smallest piece of iron. The process is called soft saturation [7]. This results in the permeability of the core to slowly reduce with increase in inductor current [8], as seen in Fig. 1 and Fig. 2.

4. There is no fringing flux in the air gap, unlike ferrite cores as seen in Fig. 1. This eliminates additional copper losses in the winding at high frequencies.

5. Powdered iron cores have much higher core losses when compared to ferrites, by a factor of ten to fifty times depending on the manufacturer and operating currents [2].

Due to the soft saturation, the permeance of the core depends on the magnetic field strength. With increase in the ampere-turns, the permeance reduces gradually. While this dependency is largely non-linear, the variation can be linearized in the operating region of the inductor, as shown in Fig. 2 for a KoolMµ E65 core of permeability 26µ, 40µ, 60µ and a Micrometal core of permeability 35µ, 60µ and 75µ. The inductance variation can be expressed by (3), where A_{L0} is the permeance at zero ampere-turns and M is the slope of the permeance (or permeability) variation as a function of ampere-turns. As a result, the inductance L linearly varies with inductor current i_L given by (4) where L_0 is the inductance at zero current and K is the slope of inductance reduction with current. K and M can hence be related by a factor of N^3:

© VDE VERLAG GMBH · Berlin · Offenbach

$$A_L = A_{L0} - M(Ni_L) \tag{3}$$

$$L = L_0 - Ki_L \tag{4}$$

$$K = N^3 M \tag{5}$$

It is common that manufacturers provide information regarding the variation of permeance with ampere-turns as shown in Fig. 2. Since the inductance continuously varies with current, it has two important effects.

1. The slope of the current varies with time and this causes non-linear currents through the inductor, unlike what is found in ferrites.

2. Secondly, the reduction in inductance necessitates the oversizing of inductance so that there is sufficient inductance L_{min} at the maximum inductor current [8].

2. Estimation of ripple and inductance roll-off of variable permeability cores based on operating conditions

For DC/DC converters especially boost, buck, buck-boost and flyback converters, two modes of operation are possible – continuous conduction mode (CCM) and discontinuous conduction mode (DCM). In both modes, there is a current ripple ΔI_L through the inductor and in case of CCM, there is a continuous average DC current through the inductor $I_{L(avg)}$. Fig. 3 shows the inductor current waveforms for a boost converter where D is the duty cycle [9].

In CCM,
$$I_{L(max)} = I_{L(avg)} + \Delta I_L/2 \tag{6}$$
$$I_{L(min)} = I_{L(avg)} - \Delta I_L/2 \tag{7}$$

In DCM,
$$I_{L(min)} = 0 \qquad I_{L(max)} = \Delta I_L \tag{8}$$

The current ripple Δi_L is dependent on the inductance as shown in (2). However, when using powdered iron cores, the inductance is itself is dependent on the current through it. This interdependency makes it difficult to directly calculate the either the inductance or the ripple, as in (2). This also means that the determination of the inductor cores losses will be inaccurate as the core losses depend on the ripple estimation [2] and the corresponding peak-peak variation in flux density ΔB, as given by the Steinmetz equation

$$P_{core} = A f_{sw}^a B_{pk}^b V_e \tag{9}$$

$$B_{pk} = \frac{\Delta B}{2} = \mu_0 \mu_r \frac{\Delta H}{2} = \frac{\mu_0 \mu_r}{2} \left(\frac{N \Delta i_L}{l_e} \right) \tag{10}$$

where V_e is volume of core, A, a, b are the Steinmetz parameters, f_{sw} is the switching frequency and $B_{pk} = \Delta B/2$. (Even though the Steinmetz equation is applicable for sinusoidal inductor currents, in the above situation it is being approximated for DC application). In order to manage

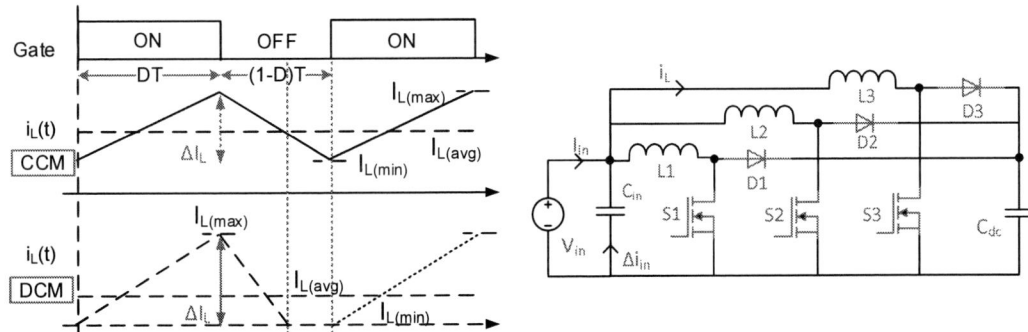

Fig. 3. Inductor current with ripple for a boost converter operating in CCM and DCM modes (left). Topology of interleaved boost converter used for experimental verification (right)

the interdependency between inductance and inductor current when using cores with variable permeability and to determine both the parameters, four approaches can be made:

1. Considering no inductance variation

This case is similar to the case of ferrites. Without considering inductance reduction due to ampere-turn, the inductance will be the estimated by setting $K=0$ in (4). The inductance will be the highest at $L=L_0$ and the estimated ripple will be the lowest, as shown in (11) and (12). This method will lead to under-sizing of the inductor with respect to the actual design requirements and underestimation of the inductor ripple and core losses.

$$L = L_0, \qquad K = 0 \tag{11}$$

$$\Delta I_L = \frac{V_L}{L_0}\Delta t \tag{12}$$

2. Using peak current $I_{L(max)}$ to determine the inductance

In this method, the peak inductor current is used to determine the operational inductance. Using (12), the inductor ripple is estimated assuming no inductance variation. Then using equations (6),(7),(8) and based on the mode of operation, the maximum inductor current $I_{L(max)}$ can be determined. The operational inductance and actual ripple can be estimated as:

$$L_{min} = L_0 - KI_{L(max)} \tag{13}$$

$$L_{max} = L_0 - KI_{L(min)} \tag{14}$$

$$\Delta i_L = \frac{V_L}{L_{min}}\Delta t = \frac{V_L}{L_0 - KI_{L(max)}}\Delta t \tag{15}$$

Estimating the ripple based on $I_{L(max)}$ will give the lowest possible value of inductance and overestimation of the ripple and the core losses. In practice, the measured ripple will be lower than that estimated by (15) but higher than that estimated by (12). This is because, as the current increases from $I_{L(min)}$ to $I_{L(max)}$, the inductance will reduce from L_{max} to L_{min}. Estimating the ripple based on $I_{L(max)}$ will hence lead to over-sizing the required size of passive filters and overestimating the core losses.

3. Using middle current $I_{L(mid)}$ to determine the inductance

Based on the above argument, a simple way to consider the inductance variation is to use the inductor middle current, $I_{L(mid)}$. Using (12), the inductor ripple is estimated assuming no inductance variation. Then using equations (6),(7),(8) and based on the mode of operation, the middle inductor current $I_{L(mid)}$ can be determined by (16):

$$I_{L(mid)} = \left(I_{L(max)} + I_{L(min)}\right)/2$$

For CCM, $\qquad I_{L(mid)} = I_{L(avg)} \tag{16}$

For DCM, $\qquad I_{L(mid)} = \Delta I_L/2$

$$L = L_0 - KI_{L(mid)} \tag{17}$$

$$\Delta i_L = \frac{V_L}{L_{mid}}\Delta t = \frac{V_L}{L_0 - KI_{L(mid)}}\Delta t \tag{18}$$

The operational value of inductance averaged over a time period and the corresponding inductor ripple can be estimated by (17) and (18) respectively. The accuracy of this method is largely dependent on the mode of operation. In CCM with a small ripple Δi_L in relation to the average current, the approximation can be made that $\Delta i_L/I_{L(avg)} \approx 0$. In such a situation, the inductance is largely determined by the DC bias due to the $I_{L(avg)}$. The inductance and ripple can be estimated based on (17) and (18) with high accuracy. However in DCM or BCM, this approximation will never hold true as $\Delta i_L/I_{L(mid)}$ will differ by a factor of two. In such a situation we need to mathematically solve the ripple, inductance dependence as shown in next section.

© VDE VERLAG GMBH · Berlin · Offenbach

4. Differential equation for determination of non-linear inductor current

To get accurate estimation of inductance and ripple when using variable permeability cores, it is essential to mathematically derive the inductance-current dependence. A time dependent variation of inductance can be written based on (4) as

$$L(t) = L_0 - Ki_L(t) \tag{19}$$

$$V_L = L(t)\frac{di_L(t)}{dt} = \left(L_0 - Ki_L(t)\right)\frac{di_L(t)}{dt} \tag{20}$$

The inductor current $i_L(t)$ as a function of time can be expressed as a first order non-linear ordinary differential equation shown above. The solution to this differential equation is

$$V_L t = \left(L_0 i_L(t) - \frac{Ki_L(t)^2}{2}\right) + C_1 \tag{21}$$

Using the initial condition that at $t=0$, $i_L = i_{L(0)}$

$$C_1 = -\left(L_0 i_{L(0)} - \frac{Ki_{L(0)}^2}{2}\right) \tag{22}$$

Using (22) in (21), $$\frac{Ki_L(t)^2}{2} - L_0 i_L(t) + \left(V_L t + L_0 i_{L(0)} - \frac{Ki_{L(0)}^2}{2}\right) = 0 \tag{23}$$

$$i_L(t) = \frac{L_0}{K} - \sqrt{\frac{L_0^2}{K^2} - \frac{2}{K}\left(V_L t + L_0 i_{L(0)} - \frac{Ki_{L(0)}^2}{2}\right)} \tag{24}$$

(23) is a quadratic equation in $i_{L(t)}$ and it has two roots. When V_L is positive, current through the inductor increases. So of the two solutions, the negative solution is correct and is shown in (24). The above equation can hence be used to determine the non-linear current through an inductor with variable permeability. The equation is applicable not only to powdered iron cores but to all cores that exhibit a linear variation in permeability with DC bias.

3. Ripple and inductance estimation applied to a boost converter

Using a boost converter as an example, the derived mathematical model is applied to both CCM and DCM mode of operation. For DCM, $i_{L(0)} = 0$ as seen in Fig. 3 and (24) can be written as (25) where $t=DT$ is the ON time of the switch when inductor current increases. Based on this, the ripple $\Delta i_{L(DCM)}$ and the average inductance $L_{avg(DCM)}$ over a time Δt in DCM can be expressed as in equation (26) and (27) respectively:

$$i_L(t) = \frac{L_0}{K} - \sqrt{\frac{L_0^2}{K^2} - \frac{2V_L t}{K}} \tag{25}$$

$$\Delta i_{L(DCM)} = i_{L(max)} = \frac{L_0}{K} - \sqrt{\frac{L_0^2}{K^2} - \frac{2V_L(DT)}{K}} \tag{26}$$

$$L_{avg(DCM)} = \frac{V_L \Delta t}{\Delta i_L} = \frac{V_L \Delta t}{\left(\frac{L_0}{K} - \sqrt{\frac{L_0^2}{K^2} - \frac{2V_L \Delta t}{K}}\right)} \tag{27}$$

For CCM, with $i_{L(0)} = i_{L(min)}$ the switch is ON till $t=DT$ as seen in Fig. 3. The ripple in CCM $\Delta i_{L(CCM)}$ and average inductance $L_{avg(CCM)}$ over a time Δt can be expressed as:

$$i_L(t) = \frac{L_0}{K} - \sqrt{\frac{L_0^2}{K^2} - \frac{2}{K}\left(V_L t + L_0 i_{L(min)} - \frac{Ki_{L(min)}^2}{2}\right)} \tag{28}$$

© VDE VERLAG GMBH · Berlin · Offenbach

$$\Delta i_{L(CCM)} = i_{L(\max)} - i_{L(\min)} = \left\{ \frac{L_0}{K} - \sqrt{\frac{L_0^2}{K^2} - \frac{2}{K}\left(V_L(DT) + L_0 i_{L(\min)} - \frac{K i_{L(\min)}^2}{2}\right)} \right\} - i_{L(\min)} \qquad (29)$$

$$L_{avg(CCM)} = \frac{V_L \Delta t}{\left(\dfrac{L_0}{K} - \sqrt{\dfrac{L_0^2}{K^2} - \dfrac{2}{K}\left(V_L(\Delta t) + L_0 i_{L(\min)} - \dfrac{K i_{L(\min)}^2}{2}\right)} - i_{L(\min)} \right)} \qquad (30)$$

From a practical design perspective, since $L_{avg(CCM)}$ and $L_{avg(DCM)}$ will be lower than L_0, it is important to increase the number of turns of the inductor so as to compensate for the loss of inductance and increase of ripple magnitude.

3.1. Simulation of four models using KoolMµ inductor in a boost converter

The above four methods to determine the inductor ripple and inductance are applied to an E65 KoolMµ powdered iron core inductor. Using a bobbin of $N=42$ turns, three inductors are built with KoolMµ core of permeability 60µ, 40µ and 26µ. Table 1 shows the core permeability, permeance A_{L0}, permeance variation slope M and calculated inductance at zero current L_0. With $V_L = \pm 700V$ and $\Delta t = 15$ µs, the inductor current and inductance as estimated by the four methods using MATLAB are shown in Fig. 4 for 60µ core and $i_{L(0)}=0$ and 30A respectively. The following observations can be made:

- The first two methods assume a constant inductance as a function of time and do not accurately estimate the inductor current. At $t=15\mu s$, the current estimated by the first and second methods show a difference of more than 10A, as seen in Fig. 4.

- The third method based on the middle current is very good in approximately estimating the inductor current even though it assumes a fixed average inductance. At $t=15\mu s$, the estimated current deviates from that shown by the fourth method by about 1A.

- The fourth method based on the partial differential equation shows a varying inductance as a function of time and estimates a non-linear current. Method 1 and 2 have error of up to 20% compared to 4. Experimental verification presented in the next section proves that this method is most accurate.

4. Experimental verification using KoolMµ core in boost converter

To verify the proposed model for estimating the ripple, a 10kW three leg interleaved boost converter with powdered iron core inductors and MOSFETs is used [6], as shown in Fig. 3 and Fig. 5. It has a switching frequency of $f_{sw}=47kHz$ and an input voltage range of 350V-700V.

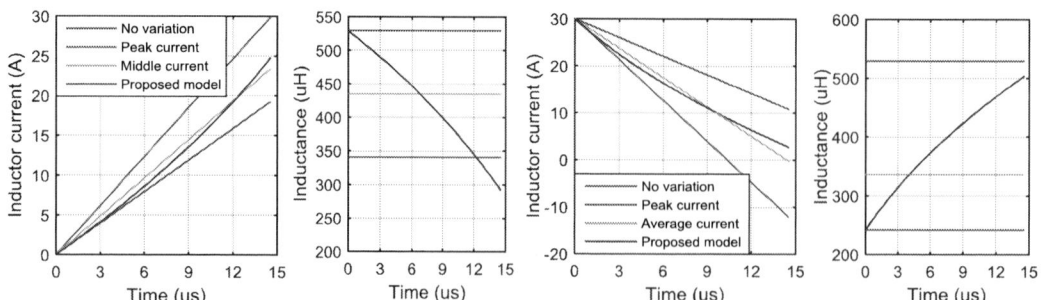

Fig. 4. Inductor current and inductance estimated by four methods for L_0=529.2µH using 60µ KoolMµ core with (left) V_L=700V, Δt=15 µs, i_{L0}=0 and (right) V_L= -700V, Δt=15µs, i_{L0}=30A;

μr	A_{L0} (nH)	M (nH/A)	N	K (nH/A)	L_0 (µH)	$L_{@10A}$ (µH)	$L_{@10A}$ (%)
60µ	300	181/1400	42	9.58	529	433.21	81.89
40µ	230	143/2200	42	4.82	405	356.84	88.11
26µ	162	106/3500	42	2.24	285	262.56	92.13

Table 1 – KoolMµ core inductors with their corresponding permeance and inductance

Fig. 5. Practical setup of interleaved boost converter with three KoolMµ 26µ E65 cores

It is operated at a fixed output voltage of V_{out}=750V. The maximum current through the inductors L1, L2, L3 is $I_{L(avg)}$=10A and it occurs when the input is P_{in}=10kW, V_{in}=350V, I_{in}=30A and the input current I_{in} is shared between the three legs. Table 1 shows the actual and percentage inductance $L_{@10A}$ at 10A. It can be seen that the operational inductance is reduced by 8% to 18% depending on the core permeability.

Experimental measurements of the inductor ripple from the boost converter using the 26µ core with L_0=284µH are shown in Table 2 and Fig. 6. The measurements are compared with estimation of inductor ripple from the four proposed methods in the table. Error of up to 5% is obtained if method 1 and 2 are used for ripple estimation. Method 3 and 4 are close to experimental measurements with less than 0.05% error, showing a ten times reduction in error. The estimates from Method 3 and 4 are very close in value in this case, that it can be concluded that method 3 is an excellent choice for simplified calculations. In situations where high level of accuracy is required in ripple and current estimation, method 4 can be implemented.

V_L(V)	$I_{L(avg)}$ (A)	Duty (%)	Mode	V_{out} (V)	Inductor ripple (A)				
					Method 1	Method 2	Method 3	Method 4	Meas.
350	2.67	30.5	DCM	750	7.94	8.47	8.20	8.21	8.16
400	4.67	38.5	DCM	750	11.46	12.6	12.00	12.03	12.0
500	3.67	22.6	DCM	750	8.41	9.00	8.70	8.71	8.75
500	1.83	17.8	DCM	750	6.62	6.99	6.80	6.80	6.81
600	3.07	15.2	DCM	750	6.79	7.17	6.97	6.98	6.97

Table 2 – Estimated and experimentally measured value of inductor ripple using 26µ KoolMµ

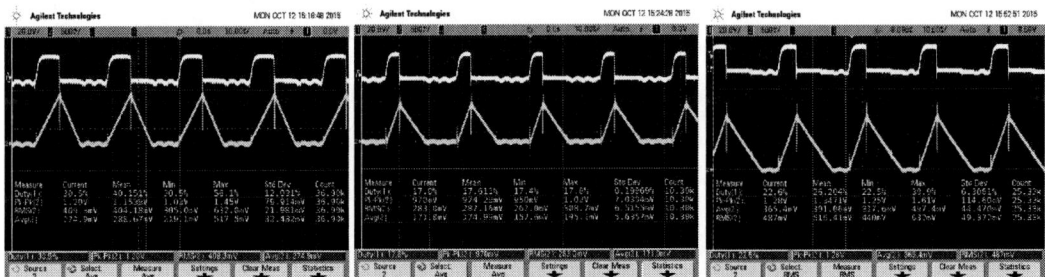

Fig. 6. Gate voltage of the MOSFET and current waveforms of the KoolMµ inductor measured using current probe - V_{in}=350V, $I_{L(avg)}$=8A (left); V_{in} =500V, I_{in} =5.5A (middle); V_{in} =500V, I_{in} =11A (right);

© VDE VERLAG GMBH · Berlin · Offenbach

5. Conclusion

Powdered iron core inductors are excellent choice for use in high power density converters due to their high saturation flux density. The core exhibits a gradual saturation and reduction of inductance with increasing ampere-turns, unlike ferrites that abruptly reduce to zero inductance near the saturation region. As a result of soft saturation, the inductance varies as function of inductor current resulting in non-linear currents. This paper provides a mathematical derivation of this non-linear behavior for both continuous and discontinuous mode of converter operation. The non-linear model is compared with three other simplified approaches, of which the middle-current method gives the closest results. Experimental verification using KoolMµ powered iron core in a boost converter have proven the accuracy of the proposed model.

Acknowledgements

The authors would like to thank the guidance and support of PhD student V. Prasanth from Delft University of Technology; Power Research Electronics B.V, Breda especially to M.Kardolus and M.v.d. Heuvel; and ABB EV Charging Infrastructure, Rijswijk. The research was sponsored by TKI switch2smart grids grant Netherlands.

References

[1] MAGNETICS, "Technical Bulletin MAGNETICS KOOL Mµ® E-CORES," 2005.
[2] M. S. Rylko, K. J. Hartnett, J. G. Hayes, and M. G. Egan, "Magnetic Material Selection for High Power High Frequency Inductors in DC-DC Converters," in *IEEE Applied Power Electronics Conference and Exposition*, 2009, pp. 2043–2049.
[3] "Ferrites and accessories - SIFERRIT material N87," Epcos, pp. 1–7, 2006.
[4] G. R. Chandra Mouli, P. Bauer, M. Zeman, G. R. C. Mouli, P. Bauer, and M. Zeman, "Comparison of system architecture and converter topology for a solar powered electric vehicle charging station," in *2015 9th International Conference on Power Electronics and ECCE Asia (ICPE-ECCE Asia)*, 2015, pp. 1908–1915.
[5] G. R. Chandra Mouli, P. Bauer, and M. Zeman, "System design for a solar powered electric vehicle charging station for workplaces," *Applied Energy*, vol. 168, pp. 434–443, Apr. 2016.
[6] G. Ram, Chandra Mouli, Jos, Schijffelen, Pavol, Bauer, and M. Zeman, "Design and Comparison of a 10kW Interleaved Boost Converter for PV Application Using Si and SiC Devices," *IEEE J. Emerg. Sel. Top. Power Electron. under Rev.*, 2016.
[7] Bong-Gi You, Jong-Soo Kim, Byoung-kuk Lee, Gwang-Bo Choi, Dong-Wook Yoo, B.-G. You, J.-S. Kim, B. Lee, G.-B. Choi, and D.-W. Yoo, "Optimization of powder core inductors of buck-boost converters for Hybrid Electric Vehicles," in *2009 IEEE Vehicle Power and Propulsion Conference*, 2009, pp. 730–735.
[8] J. D. Pollock, W. Lundquist, and C. R. Sullivan, "Predicting inductance roll-off with dc excitations," in *IEEE Energy Conversion Congress and Exposition*, 2011, pp. 2139–2145.
[9] N. Mohan and T. Undeland, "Power electronics: converters, applications, and design," 2007.

PCIM Europe 2016, 10 – 12 May 2016, Nuremberg, Germany

Optimized DC Link for Next Generation Power Modules

Michael A. Brubaker and Terry A. Hosking
SBE, Inc. 81 Parker Road, Barre, Vermont 05641 USA

Tomas Reiter
Infineon Technologies AG, Am Campeon 1-12, 85579 Neubiberg, Germany

Laura D. Marlino and Madhu S. Chinthavali
Oak Ridge National Laboratory, National Transportation Research Center
2360 Cherahala Blvd, Knoxville, Tennessee 37932 USA

Abstract

The market leaders in IGBT technology are now introducing next generation "six-pack" modules to enable increased power density and reduced cost for automotive traction drive applications. However, the potential gains offered by these modules can only be harvested using an optimized DC link with integrated capacitor/bus topology. Two integrated capacitor/bus solutions have been designed to support the new Infineon HybridPACK™ Drive [1] module with the lowest possible µF/kW ratio and minimized equivalent series inductance. Simulation and design results are presented along with third party testing data for a complete inverter.

1. Introduction

Next generation power modules for electric vehicle applications are targeting increased power density and efficiency to reduce drive train cost and weight. In order to fully exploit the capabilities of such modules, an optimized DC link comprised of high performance film capacitors integrated with a suitable bus structure is essential. This approach attacks the two critical factors that limit inverter power density, which are the ripple current per micro-Farad rating of the capacitor and the overshoot voltage across the IGBT switches during turn-off.

The power density of an inverter is traditionally limited by the DC link capacitance value required to safely source the ripple current demanded by the IGBT's during switching. *Using an annular form factor capacitor winding with low losses and low thermal resistance enables a new paradigm where capacitance is defined by the inverter control limit rather than Amperes per micro-Farad.* Note that this approach has recently been demonstrated for wind power where a reduction of DC link capacitance for a smaller µF/kW ratio facilitated fitting a 1MW inverter into a 500kW frame [2].

The DC input voltage for the inverter is constrained by voltage overshoot which results from energy stored in the parasitic inductance seen by the IGBT's during turn-off. This issue is best mitigated by reduction of equivalent series inductance (ESL) combined with intelligent gate control to safely increase the DC voltage closer to the IGBT limit. Annular form factor film capacitors can be "surface mounted" to the bus structure such that one bus conductor becomes a terminal of the capacitor. This approach eliminates redundant copper, thus reducing cost and weight while simultaneously reducing the ESL of the DC link capacitor. The use of optimized connections from the bus structure to IGBT module is essential to get the lowest possible ESL and hence overshoot [3].

The Infineon HybridPACK™ Drive (HP Drive FS820R08A6P2xx) [1] represents a new state-of-the-art IGBT module with new automotive EDT2 IGBT chip generation and a footprint

© VDE VERLAG GMBH · Berlin · Offenbach

reduction of 30% relative to the last benchmark power module Infineon HybridPACK™ 2 with IGBT3 chip generation. Two optimized DC link capacitor/bus topologies have been designed specifically for the HP Drive to match the size reduction. A "horizontal" layout places the DC link capacitor in-line with the IGBT module. A "vertical" layout locates the capacitor/bus underneath the IGBT cooling plate to create a compact geometry. This paper discusses the design of both configurations in detail and inductance measurements are provided in addition to rating calculations based on practical cooling assumptions for a realistic drive cycle. Third party test results are presented for a full inverter using the "horizontal" DC link topology under steady state conditions.

2. DC Link Requirements

The Infineon HP Drive requires a DC link capacitor capable of the following:

$$450V < V_{dc} < 550V \qquad 75Arms < I_{ripple} < 125Arms \text{ continuous} \qquad 50°C < T_{coolant} < 85°C$$

A capacitance of 500µF was specified with a life requirement of 10,000 hours subject to a typical automotive drive cycle. In order to match the module foot print, two annular form factor windings having a height of 32mm and outer diameter of 76mm were selected. Metallized polypropylene film was used for the dielectric to provide the best balance of cost and performance.

3. Horizontal Layout Design

The horizontal configuration DC link is presented in Figure 1a. Two 250µF annular film capacitor windings are directly integrated to a laminated bus structure and share a common "crown" terminal. Note that locating the DC input terminals to the cap/bus on the opposite side of the IGBT connections provides the best utilization of the capacitor and minimizes "current hogging". This configuration provides the shortest possible connection length from capacitor to IGBT inputs combined with "through-hole" connections to achieve the lowest possible ESL. The thermal profile for a single winding subject to 50Arms ripple with an 85°C boundary defined on the capacitor case is presented in Figure 2a.

4. Vertical Layout Design

The vertical configuration DC link is presented in Figure 1b. The capacitor windings and "crown" terminal are identical to that utilized for the horizontal configuration, but the bus interface to the module is modified. While the optimal location of the DC inputs to the cap/bus relative to IGBT inputs is maintained, the ESL will be a few nH higher for this configuration since the path length is increased. A thermal finite element analysis was performed on a single winding for this case using the same methodology as per the horizontal layout. Note that in this case, the 85°C boundary condition was applied to the bus to represent the effect of the IGBT cooling plate. The thermal profile for this case is presented in Figure 2b and the hotspot temperature rise is lower than the horizontal case for the same ripple current. While the ESL of this design is a bit higher than the horizontal layout, the current rating is significantly increased since bus and IGBT losses do not flow through the capacitor.

5. Inductance Measurements

The horizontal capacitor/bus test kit prototype was tested using the SBE ring-out method described previously [4]. This method indicates that the inductance at the IGBT inputs is 8.4nH. Independent measurements performed by Infineon on the optimized capacitor/bus similarly indicate a value of about 8nH. From a practical operating standpoint, Vce must be limited to less than the Vces -40°C specification of the IGBT/diode chipset. While active collector gate clamping is implemented to limit Vce to <710V for the HP Drive, repetitive clamping circuit activation must be avoided to prevent damaging the TVS diodes, which can lead to uncontrolled short circuiting of the IGBT and module failure as a consequence. Higher stray inductances provide a larger overvoltage at turn-off which thus limits the maximum useable working voltage. Alternatively, higher Rg (gate drive resistance) can be adopted with the penalty of lower performance. The best utilization of working voltage can thus be achieved by minimizing the stray inductance.

Comparison of IGBT turn-off measurements between the optimized SBE 8nH capacitor/bus and a conventional 15nH capacitor were performed using the circuit in Figure 3a with the results presented in Figure 3b. The turn-off overvoltage was measured with and without the clamping circuit in order to determine the point of clamping circuit activation. At 500V working voltage, the SBE 8nH capacitor solution showed extremely low overvoltage and the clamping (overvoltage protection) was not active before 1000A. This indicates a usable transient current range up to 1000A at 500Vdc and the full switching speed of the power module can be utilized for working voltages of 500V to 550V. In contrast, the 15nH conventional capacitor testing results showed activation of the clamping circuit at above 500A at 500V. In order to safely operate at 500V with this capacitor, the switching speed has to be slowed down to avoid the risk of damage to the clamping circuit or the IGBT module.

These measurements clearly show that a low inductance capacitor/bus DC link is mandatory for applications that require high working voltages combined with full switching speed at high phase currents to achieve the lowest possible inverter power losses.

6. Inverter Testing Setup

An HP Drive module was supplied by Infineon along with the cooling plate, gate driver and micro-controller to support the inverter testing. The IGBT module was combined with an SBE 700A186 horizontal test kit DC link as shown in Figure 4. Note that an aluminum adapter block has been added to extend from the cooling plate to the capacitor case such that a thermal reference is defined. The capacitor was instrumented with thermocouples at the mid-plane of each winding located inside the core along with mirror image thermocouples on each winding end face. The core measurement is typically very close to the capacitor hotspot and the end face measurements provide a useful indication of the thermal gradient across the windings.

Additional thermocouple measurements were made at the following locations on the inverter:

1) IGBT module output tab for center terminal
2) Positive input to IGBT module on center tab pair
3) DC input tab on the capacitor/bus assembly
4) Aluminum cooling plate under capacitor case
5) Coolant inlet
6) Coolant outlet

Finally, the output of the RTD located on the IGBT module was recorded as a frequency signal that was later correlated to temperature using the relationship provided by Infineon. The complete test setup is presented in Figure 5. A static RL load was connected in a floating Wye across the IGBT module outputs and a Yokogawa WT1800™ power analyzer was utilized to measure the efficiency of the inverter.

For the purpose of this paper the focus will be on a run to thermal equilibrium performed at 22°C ambient temperature with a coolant temperature of 22°C and a flow rate of 6.4 lpm.

The power analyzer output for this test is shown in Figure 6 and the inverter power was 35kW continuous. Note that the inverter was over modulating a bit as the system was adjusted to achieve a phase current of 75Arms without tripping the power supply.

7. Inverter Testing Results

With an inverter system efficiency level of 99% at the 35kW, 400Vdc, 8kHz light load condition, the new automotive EDT2 IGBT chip generation of the HP Drive strongly outperforms last IGBT generations. Further calorimetric based measurements at Infineon validated the recorded efficiency level of the Oak Ridge National Laboratory National Transportation Research Center. The statement, that the new EDT2 IGBT was developed for having an extended EV driving range, was thus clearly proven by this experiment. While the available test power is well below the capabilities of both the Infineon HP Drive and the SBE integrated cap/bus DC link, some very useful information was unfolded from the results. Consider the thermocouple data for the capacitor/bus assembly as compared to the simulation results using the 2D model described earlier with capacitor and capacitor plus bus losses as presented in Figure 7. There are two discrepancies between the measured data and the original design simulation; namely the total hotspot temperature and the thermal time constant. The former issue indicates that some small fraction of the IGBT chip and terminal losses are coupled to the bus via the module power tabs and flow through the capacitor. The latter issue demonstrates that there must be an additional thermal path from the bus to the ambient temperature.

The hotspot temperature can be deconstructed into a simple thermal resistance model as shown below:

$$\Delta T = P_{cap} \times R_{Tcap} + (P_{bus} + P_1) \times R_{Tbus}$$

Note that P_1 denotes the fraction of chip and terminal losses flowing to the IGBT power tabs that influence the capacitor/bus temperature. The simulation can be fitted to the test results by matching the hotspot temperature and thermal time constant. A parallel thermal resistance of 2.45°C/W from the bus to 22°C ambient provides a good match on the time constant. This is attributed to the DC input cables which provide strong coupling to ambient. Combined with an additional 1.96W of power supplied to each winding to account for P_1, the simulation aligns nicely with the data as shown in Figure 8. The relative contributions of the various mechanisms can be summarized in the same equation format as shown on a per winding basis:

$$4.87°C = 0.27W \times 1.95°C/W + (0.34W + 1.96W) \times 1.89°C/W$$

Note that this corresponds to the red curve of Figure 8 after approaching equilibrium at 6000 seconds. The HP Drive design has significantly reduced the amount of heat added to the cap/bus from the power tabs as compared to conventional modules. This is achieved by the ultrasonic welded power tabs to the ceramics substrate.

Now consider scaling up to a more typical power level and coolant temperature for these products. Doubling the power to 70kW with the voltage fixed at 400Vdc requires that the capacitor ripple current and bus current increase by a factor of two such that the losses for both cases increase by a factor of four. The IGBT chip and terminal loss contribution will increase by a factor of 2.5 from the 35kW base case per Infineon calculations. Scaling to 85°C coolant will increase the electrical resistance of the bus conductors, capacitor electrodes, capacitor end spray, and capacitor terminals by the appropriate temperature coefficients. Infineon analysis indicates that the IGBT chip and terminal loss contribution will increase by approximately 6% going from 22°C to 85°C coolant. On a per winding basis, the capacitor losses increase from 0.27W to 1.42W, the bus losses increase from 0.34W to 1.79W, and the IGBT chip and terminal contribution increases from 1.96W to 5.19W. The simple model predicts a temperature rise of 16°C, which matches up nicely with the separately generated SBE capacitor simulation tool output as presented in Figure 9. This result can be readily utilized to generate a rating curve for the horizontal test kit on the basis of the capacitor and bus losses as presented in Figure 10.

Based on life testing results presented previously [5], a life of 10,000 hours can be achieved with a hotspot temperature of 100°C operating at 400Vdc. As such, the capacitor can thus provide 10,000 hour life operating at 84°C coolant and 70kW continuous power. The same scaling assumptions can be applied to further increase the power to 100kW at 400Vdc. Under this condition, assuming continuous operation, the 100°C hotspot limit for 10,000 hour life can be maintained with 70°C coolant. The HP Drive is intended for 50-150kW (peak) electric vehicle applications [1]. These scaling results indicate that the SBE horizontal test kit is fully capable of supporting high performance vehicle applications at the top end of the HP Drive power range. Given the long thermal time constant of the capacitor, peak power of up to 150kW can be readily managed for short durations as part of a realistic drive cycle. It is also important to recognize that the vertical capacitor/bus test kit offers even more capability since the bus bar is tied directly to the IGBT cooling plate. As such, the bus losses and any stray power from the IGBT chip and terminal losses will not flow through the capacitor. This translates to lower capacitor hotspot temperature rise, which will enable the vertical cap/bus kit to support 100kW continuous operation at 85°C coolant or higher.

8. Conclusion

Horizontal and vertical configuration DC link capacitor/bus test kits have been designed and optimized for the Infineon HybridPACK™ Drive power module [1]. Independent measurements for the horizontal kit have confirmed an ESL of 8nH at the module inputs which is about 50% lower than can be achieved with conventional technology. This low inductance minimizes voltage overshoot and enables higher working voltages and faster switching speeds to achieve the best efficiency for the HP Drive. Third party testing of a complete inverter at the Oak Ridge National Laboratory National Transportation Research Center has provided critical thermal data for characterizing the performance of the horizontal test kit. Scaling these results to practical operating conditions indicates that this capacitor/bus can support 100kW continuous operation at 70°C coolant and 400Vdc with a life of greater than 10,000 hours. The SBE 700A186 (500V and 500µF) therefore supports full utilization of the Infineon HP Drive for high performance EV applications. Similar performance can be achieved at even higher coolant temperatures using the vertical configuration test kit which allows even more compact topologies with only a minor penalty in ESL.

Acknowledgements

The authors wish to thank Mr. Larry Seiber and Mr. Steve Campbell at ORNL for their efforts in completing the inverter testing.

References

[1] Infineon, "Compact Power Modules", *Electric & Hybrid Vehicle Technology International*, July 2014.
[2] Michael Brubaker, Terry Hosking, W.-Toke Franke, Edward Sawyer, Dayana El Hage; "Integrated DC link capacitor/bus enables a 20% increase in inverter efficiency", *Proc. Power Conversion and Intelligent Motion*, PCIM 2014, Nuremberg, Germany, May 2014.

[3] Edward Sawyer, Michael Brubaker, and Terry Hosking, "Understanding the Contribution of Switch Input Connection Geometry to Overall DC Link Inductance, *Proc. 17th European Conference on Power Electronics and Applications*, Geneva, Switzerland, September 2015.

[4] E. D. Sawyer, "Low Inductance - Low Temp Rise DC Bus Capacitor Properties Enabling the Optimization of High Power Inverters", *Proceedings of PCIM 2010*, Nuremberg, Germany, May 2010

[5] M. A. Brubaker, T. A. Hosking, and T. F. Von Kampen, "Life Testing of High-Value Annular Form Factor DC Link Capacitors for Applications with 105°C Coolant", *Proceedings of PCIM 2011*, Nuremburg, Germany, May 17-19, 2011

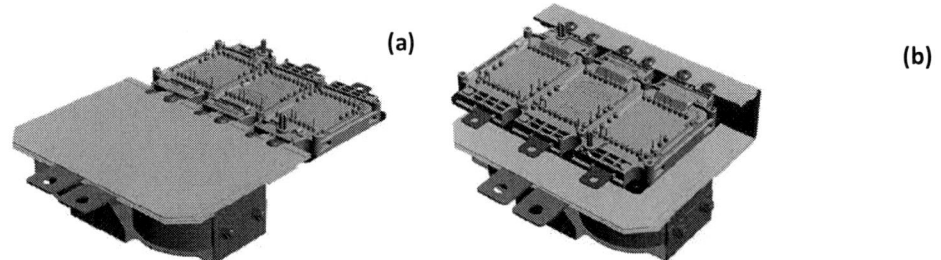

Fig. 1. Integrated capacitor/bus DC link for (a) horizontal and (b) vertical configurations

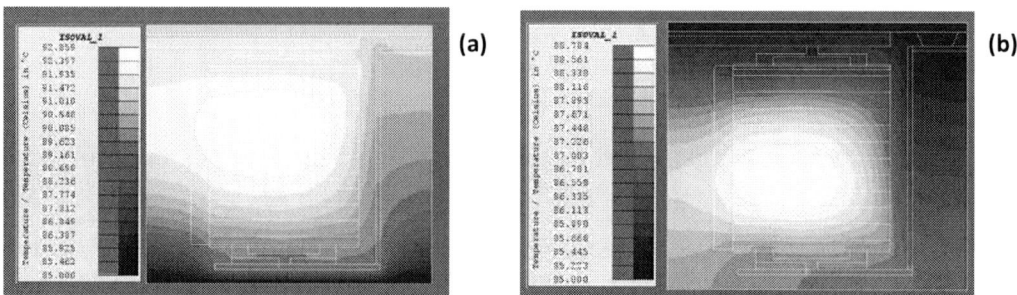

Fig. 2. Thermal profiles for single capacitor winding with (a) horizontal and (b) vertical configurations

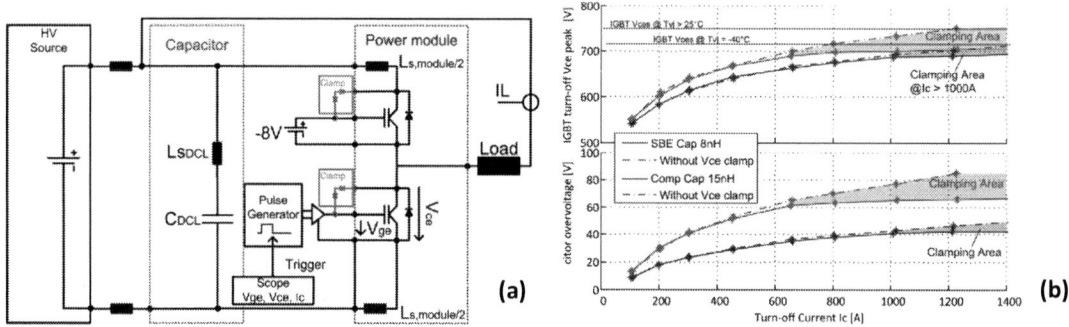

Fig. 3. Overshoot testing DC link capacitors at Infineon (a) test circuit, (b) turn-off overvoltage at 500Vdc working voltage.

© VDE VERLAG GMBH · Berlin · Offenbach

PCIM Europe 2016, 10 – 12 May 2016, Nuremberg, Germany

Fig. 4. Illustration of HybridPACK™ Drive and complete test kit HybridKIT Drive supplied by Infineon for inverter testing

Fig. 5. Complete inverter test setup used at the National Transportation Research Center at Oak Ridge National Laboratory

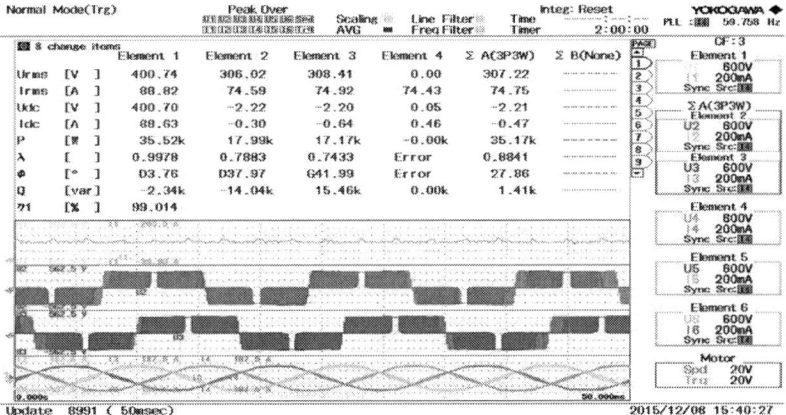

Fig. 6. Yokogawa WT1800™ readout for test run to equilibrium at 50ºC

© VDE VERLAG GMBH · Berlin · Offenbach

PCIM Europe 2016, 10 – 12 May 2016, Nuremberg, Germany

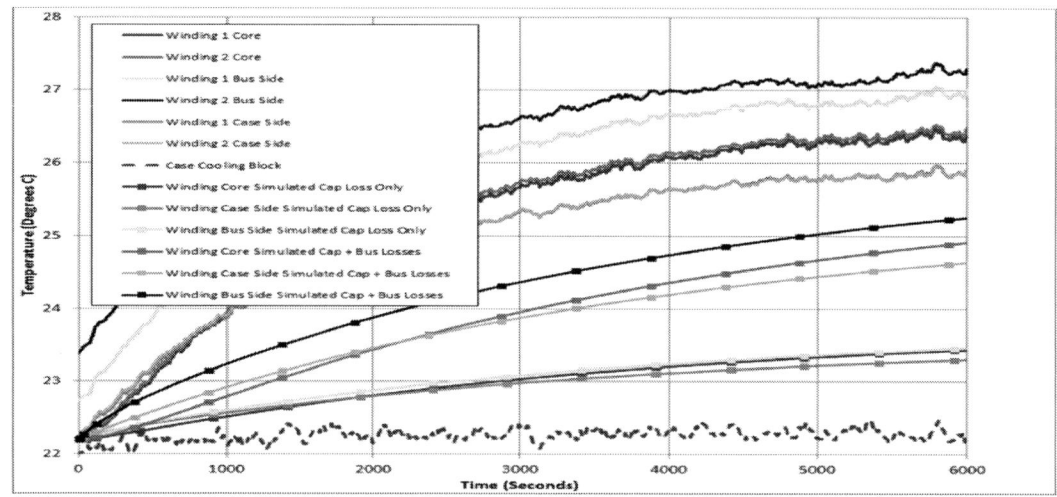

Fig. 7. Capacitor/bus thermocouple data as compared to original design simulation: Top six curves represent measured data, bottom curves show simulation with capacitor losses only, middle curves show simulation with capacitor and bus losses.

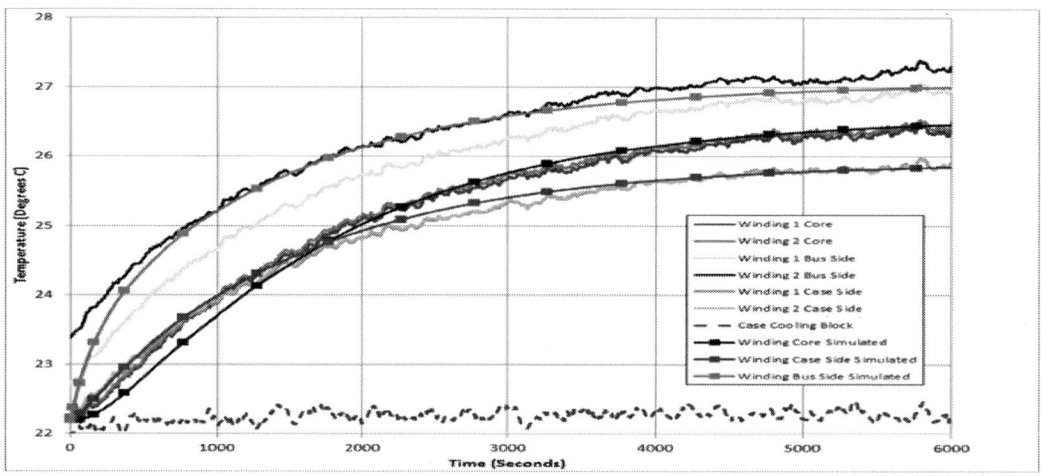

Fig. 8. Improved simulation accounting for additional thermal path and addition of IGBT losses to fit test data

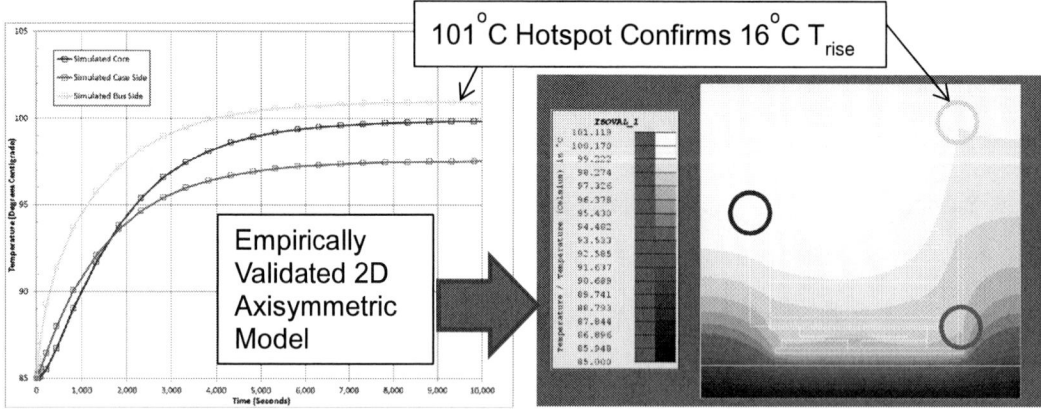

Fig. 9. Capacitor temperatures after scaling inverter test results from 35kW to 70kW

© VDE VERLAG GMBH · Berlin · Offenbach

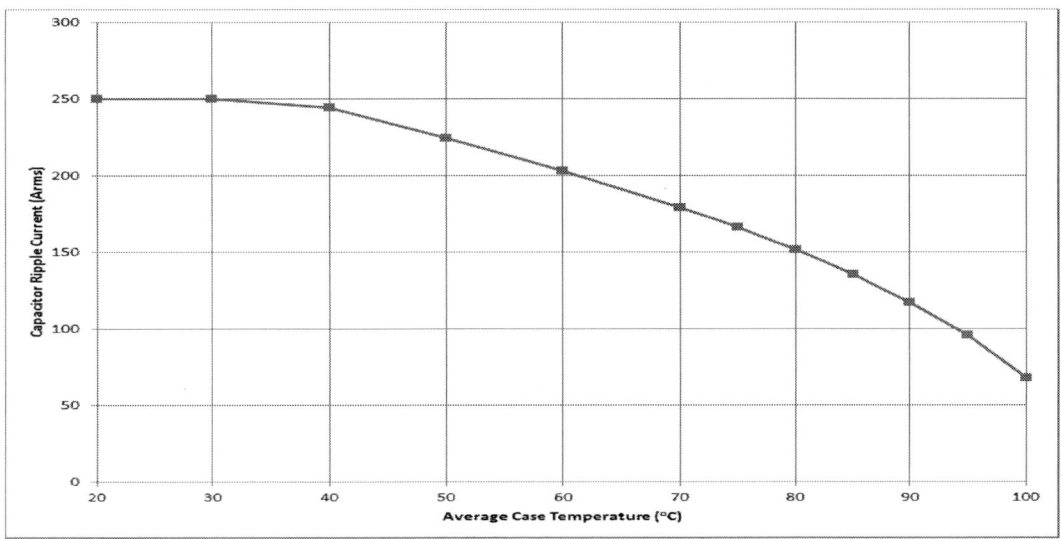

Figure. 10. Continuous current rating curve for the SBE 700A186 horizontal test kit based on inverter test data with capacitor and bus losses only

In-circuit-characterization of ceramic capacitor with anti-ferroelectric material for voltage source inverters

Jürgen Kropp, Department of Mechatronics, Center of Energy Technology
Universitätsstr. 30, 95447 Bayreuth, Germany, juergen.kropp@uni-bayreuth.de
Mark-M. Bakran, Department of Mechatronics, Center of Energy Technology
Universitätsstr. 30, 95447 Bayreuth, Germany, bakran@uni-bayreuth.de

Topic Number: 17.1 Capacitors

Abstract

This paper describes the in-circuit-characterization of ceramic capacitors with anti-ferroelectric material for voltage source inverters. This characterization is based on measurements in a mode identical to the real use in inverters. It will be shown how this method compares to the sinusoidal values based on the datasheet.

1. Introduction

For the DC-Link of voltage source inverters, there's a new capacitor technology available based on PLZT ceramic material (lead-lanthanum-zirconate-titanate) with anti-ferroelectric properties [1]. The advantages of a capacitor made from this material when using it as a DC-Link capacitor in a voltage source inverter are low losses, an increasing capacitance by increasing voltage, low series inductance and a high capacitance density [1]. For specifying the capacitor, normally the capacitor is measured by using sinusoidal signals and varying the frequency. These results are shown in the datasheet of the respective manufacturer. In case of linear capacitor, it is possible to use these results to dimension the inverter. In case of the new non-linear capacitor made from ceramic material it is necessary to perform further measurements to specify the capacitor before designing an inverter DC-Link. In this work, the non-linear capacitor CeraLink^{TM} is examined with the aim to use it in a voltage source inverter. Therefore, the capacitor is measured in-system with pulsed currents. For this characterization, a CeraLink^{TM} 20µF/500V capacitor module is used.

2. Benefit of in-circuit-characterization of a non-linear ceramic capacitor

For the examination of the characteristic of the CeraLink^{TM} capacitor in-circuit, a specific measurement circuit is used to generate the pulsed currents. The measurement circuit is shown in Fig 1. Therefore, the capacitor is supplied by a power supply with adjustable voltage. The capacitor is coupled to a classic H-bridge with a high inductance L_2 between the legs of the bridge which is smoothing the pulsing currents to very small ripples. By switching the IGBTs S1-S4 independent, different switching modes can be formed. The capacitor can be measured with short and long current pulses. Additionally, positive and negative currents to charge and discharge the capacitor can be generated. Also the height of the current pulse is adjustable. The signals for the IGBT-drivers are generated by a FGPA-Controller.
The datasheet of the CeraLink^{TM} specifies three capacitance values: initial capacitance (U_{DC} = 0V), effective capacitance (small signal at nominal voltage) and nominal capacitance (large signal at nominal voltage) [2].

In order to specify the capacitor in-circuit, the differential capacitance is estimated, which is equal to the effective capacitance. Therefore, the current through the capacitor is measured using a Rogowski-coil. The ripple voltage of the capacitor is measured using an oscilloscope. Fig 2 shows an example of the measured waveform of capacitor voltage u_C and current i_C at the ambient temperature $T_C = 25°C$ and at nominal voltage $U_{DC} = 400V$.

The dynamic (during pulse) and static (median) differential capacitance is calculated for each pulse. The measurements are executed for different currents, different pulse width and at different temperatures. The following chapter gives an overview of the differences between the two measuring methods with sinusoidal waveforms and pulse currents.

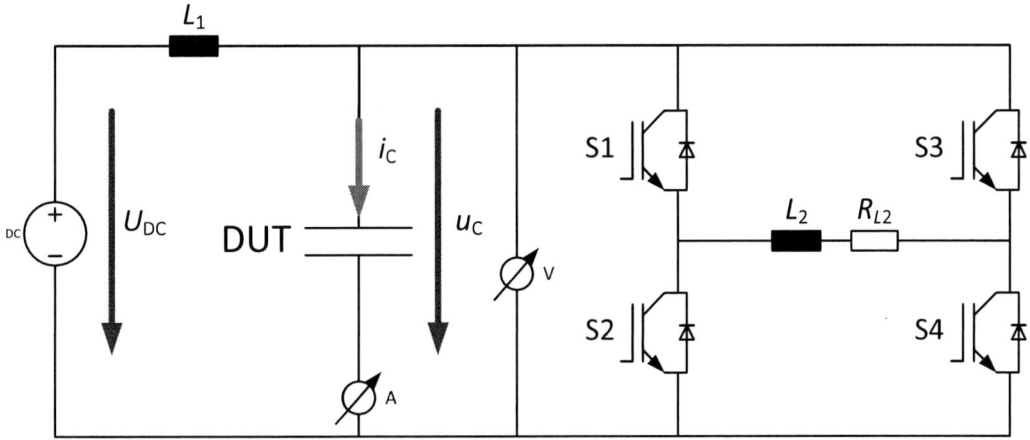

Fig 1: Test circuit CeraLink with H-bridge and power supply

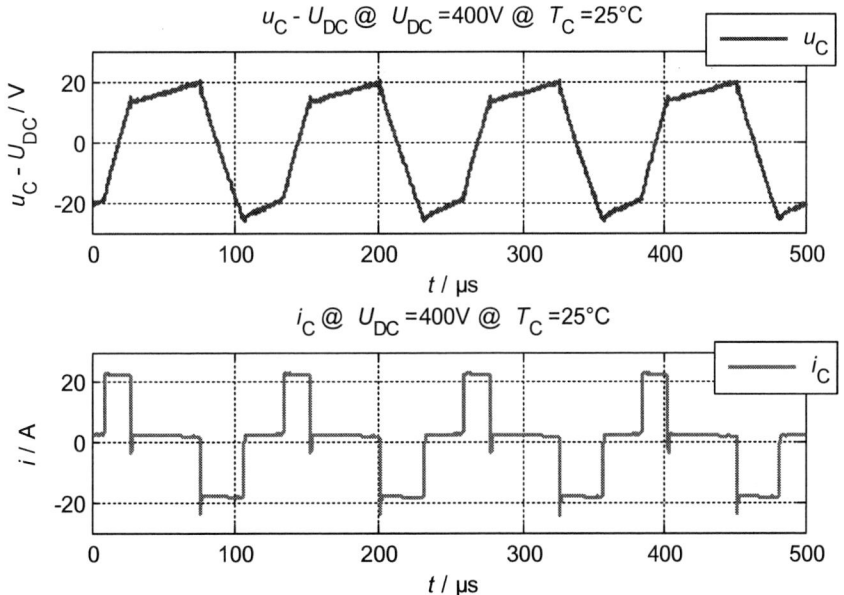

Fig 2: Capacitor Ripple voltage and current @ $U_{DC} = 400V$, $T_C = 25°C$, $I_C = 20A$

PCIM Europe 2016, 10 – 12 May 2016, Nuremberg, Germany

3. Measurement results of characterization a non-linear ceramic capacitor

With these in-circuit measurements, the characterization of the capacitor can be done.
In view of the application in a voltage source inverter, with a varying DC-Link CeraLink[TM] voltage ripple, it is important to determinate $C(U)$ correctly. In this chapter, first the difference in definition of capacitance between datasheet and in-circuit-measurement is described. Subsequently the measurement results of in-circuit capacitance are presented and compared with the values of the datasheet.

3.1. Definition of capacitance

First, it will be examined, how the capacitance value in-circuit can be calculated. As mentioned in chapter 2, the datasheet of the capacitor has three capacitance values for the capacitor. The initial capacitance is not important for the use in-circuit, because of the voltage level of U_{DC} = 0V. The effective capacitance is determined by applying a sinusoidal signal with low ripple voltage of 1V at different voltage levels and temperatures [2, p. 12]. Thereby, with measuring the RMS voltage, RMS current and phase angle between current and voltage, the effective capacitance and losses (ESR) are determined. For the CeraLink[TM] 20µF/500V this effective capacitance is 12µF at U_{DC} = 400V [2, p. 4]. In contrast to this effective capacitance, the nominal capacitance is determined by a voltage signal with slow variation [2, p. 6]. By integration of the measured current signal to the electric charge the characteristic of $Q(U)$ can be calculated. As an effect of hysteresis of the ceramic-dielectric the curve of charge $Q(U)$ and discharge $Q(U)$ are minimal different. The gradient of the median curve $Q(U)$ corresponds to the nominal capacitance [2, p. 12]. For the CeraLink[TM] 20µF/500V, nominal capacitance values higher than 20µF can be reached [2, p. 4].

For in-circuit characterization of the capacitor, it is first necessary to examine how this capacitance value is calculated. The capacitance is given by the quotient of differential charge with the corresponding voltage difference. Therefore, the capacitance could be determined in two ways:

1) Static capacitance "vertex points of voltage signal"
2) Dynamic capacitance of the pulses

This two methods are executed after recording the time signals of capacitor voltage $u_C(t)$ and capacitor current $i_C(t)$. Analog to the determination of nominal capacitance, the current signal $i_C(t)$ is integrated to electric charge $q_C(t)$.

Static capacitance "vertex points of voltage signal"
The method of static capacitance - "vertex points of voltage signal" - is calculated by the difference of charge and voltage between beginning and end of one pulse (cp. Eq. 3.1).

$$c_{diff} = \frac{dQ}{dU} = \frac{\int i_C(t)\, dt}{dU} = \frac{\int_{t0}^{t1} i_C(t)\, dt}{U_1 - U_0}$$

Eq. 3.1

Therefore, the time of start and end of the current pulse is determined. The resulting time of start and end of the current pulse are similar to the vertex points of the voltage signal $u_C(t)$. For every start and end time of the current pulse, a separate linear curve-fitting of the voltage signal $u_C(t)$ is executed. Therefore the line of best fit is calculated separately for the right- and left-side measured values. The intersection of the two resulting lines supplies first the start or

© VDE VERLAG GMBH · Berlin · Offenbach

end of integration of current signal $i_C(t)$ to electric charge $q_C(t)$. Second, this intersection is the voltage at start or end of the current pulse. From the vertex points of voltage signal and the corresponding charge difference during this time of the current pulse, the differential capacitance is determined. In Fig 3 is an example of a voltage signal including the linear best fit for the left (red line) and right-side (black line) measured values for a positive and negative pulse. Additionally the resulting "vertex points of voltage signal" of each fit is marked with a green asterisk (*). Also the voltage difference $U_1 - U_0$ is shown.

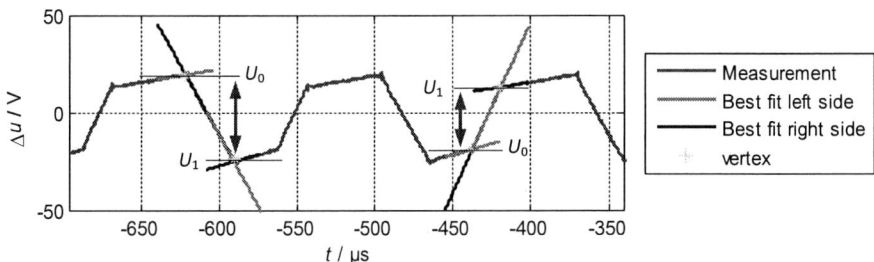

Fig 3: Explanation method of best linear fit for "vertex points of voltage signal" @ U_{DC} = 400V, T_C = 25°C

The advantage of this non-linear ceramic capacitor in a DC-Link compared with conventional capacitors is the increasing capacitance with increasing ripple voltage [3, p. 30]. For dimensioning a DC-Link, with this effect, the capacitance value can be reduced significantly compared with a conventional capacitor. At nominal voltage, a rise of 66% of capacitance in dependence of rising ripple voltage from ΔU = 1V to ΔU = 20V is expected [3, p. 30].

With in-circuit-characterization it is examined, how much of this rising capacitance can be used in a DC-Link application. Therefore, the second method for determining the capacitance value - the dynamic capacitance - is used. The aim of the dynamic method is to examine whether this value shows higher capacitance values during the pulse than the static method.

Dynamic capacitance of the pulses

For determination of the dynamic capacitance time periods of Δt = 5µs during the current pulses are considered. Avoiding spikes in the beginning and end of the current pulse, the first and last 10µs of the current pulse are not considered. Based on the time of start and end of the current signal $i_C(t)$, the electric charge is calculated. In this case no sliding average filter is applied on the signals. With a time period of Δt = 5µs the differential capacitance is calculated (cp. Eq. 3.2):

$$c(t) = \frac{q_C(t + \Delta t) - q_C(t)}{u_C(t + \Delta t) - u_C(t)} \ \ with \ \Delta t = 5\mu s \qquad \text{Eq. 3.2}$$

Comparison of static and dynamic capacitance

Based on the measured signals, the results of the capacitance of these two methods are compared. The method "vertex points of voltage signal" produces stable and repeatable results because of the integral analysis of the time-signals of voltage and current signal. The highest capacitance values are produced by the dynamic capacitance. The difference of the static capacitance value and the dynamic capacitance value is maximum 5% (cp. Fig 4). Including this result, the time-signal of dynamic capacitance shows no significant higher values than the static method. This result of no significant higher values of dynamic capacitance than static capacitance is consistent to the measured voltage signal $u_C(t)$. There, the voltage rises or decreases almost linear during the horizontal current pulses (cp. Fig 2).

© VDE VERLAG GMBH · Berlin · Offenbach

For dimensioning a DC-Link with this non-linear capacitor the static capacitance is the resulting capacitance value. In a DC-Link, the capacitor is dimensioned by the voltage ripple and the current at a specific time difference. So this corresponds to the method of determining the static capacitance value. Due to its importance for dimensioning a DC-Link this static method "vertex points of voltage signal" is used for further analysis of CeraLink™ capacitor in-circuit.

In Fig 4 the difference between static and dynamic capacitance is shown in dependence of voltage level U_{DC}. For this comparison the capacitance values are in relation to the static capacitance defined at voltage level U_{DC} = 400V and T_C = 25°C. This value is comparable to the effective capacitance of sinusoidal signals C_{eff} = 12µF [2, p. 4] from the datasheet. The following figure shows the capacitance at ΔU = 12V for the three methods.

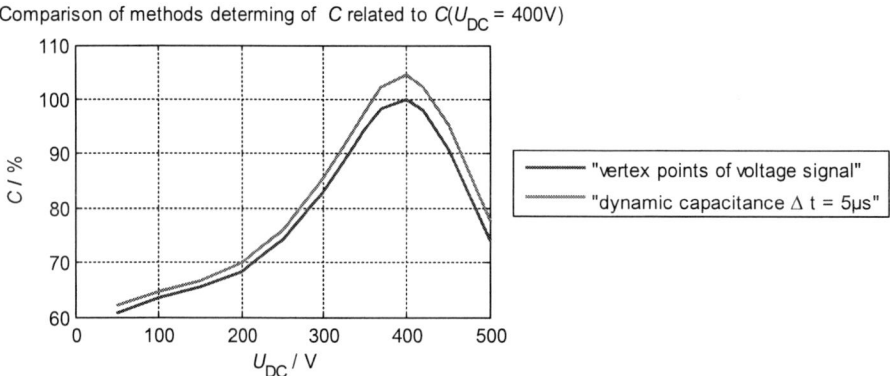

Fig 4: Comparison of methods for destining capacitance in-circuit at T_C = 25°C

Comparing these methods for determining the capacitance with sinusoidal signals, the destining of capacitances and losses can be done with one method. For pulsed signals, this method could not be applied, so the method "vertex points of voltage signal" is only used for destining the capacitance value.

3.2. Characteristics of capacitance of a non-linear ceramic capacitor

The datasheet of the capacitor shows characteristics for small signal in dependence of voltage and temperature and the difference between small and large signal behaviour. In order to test large signal behaviour of the capacitor a low voltage variation is applied [2]. To test the small signal behaviour a sinusoidal signal is applied with a frequency f = 1 kHz.
Relating to the datasheet of the CeraLink™ capacitor at small signals, the capacitor has an increasing capacitance with increasing voltage until nominal voltage. With further increase in voltage, the capacitance decreases steeper than the increase below nominal voltage. The difference of capacitance between small and large signals also shows an increasing capacitance with increasing ripple voltage [3].
For using the capacitor in an inverter the dependence of capacitance is measured for small and large ΔU. So the behaviour of large ΔU is measured in-circuit with long pulses and high currents. The behaviour of small ΔU is measured in-circuit with small pulses and low currents. In order to avoid a change of temperature of the capacitor during measurement, these signals must be applied for only short time.
The result of these measurements for large ΔU is shown in Fig 5. Fig 5a shows the capacitance in dependence of ΔU and U_{DC} related to the capacitance at nominal voltage U_{DC}

= 400V with ΔU = 5V. In Fig 5b each curve relates to the measured capacitance at ΔU = 5V for each U_{DC}. With increasing the ripple voltage from ΔU = 5V to ΔU = 40V at nominal voltage, the capacitance increases 14%. The maximum of capacitance is at the voltage level of U_{DC} = 400V. At voltage levels unequal to nominal voltage, the increase of capacitance in dependence of ripple voltage is less steep as at nominal voltage.

The CeraLink^{TM} 20µF/500V is measured by varying ΔU = 1V to ΔU = 60V at the voltage levels from U_{DC} = 250V to U_{DC} = 500V with steps of 50V. This examination in dependence of temperature is done between T = -40°C and T +95°C.

For voltages with ΔU = 1V the characteristic of capacitance for in-circuit-characterization is equal to sinusoidal capacitance C_{eff} in the datasheet. For current pulses causing ΔU rise to 60V, the capacitance in dependence of ΔU can be fitted by linear curve fitting. For different temperatures and voltage levels U_{DC}, the rise of capacitance in dependence of ΔU is not equal. The capacitance in-circuit for ΔU = 20V at U_{DC} = 400V at T_C = 25°C rises 10% between ΔU = 1V and ΔU = 20V.

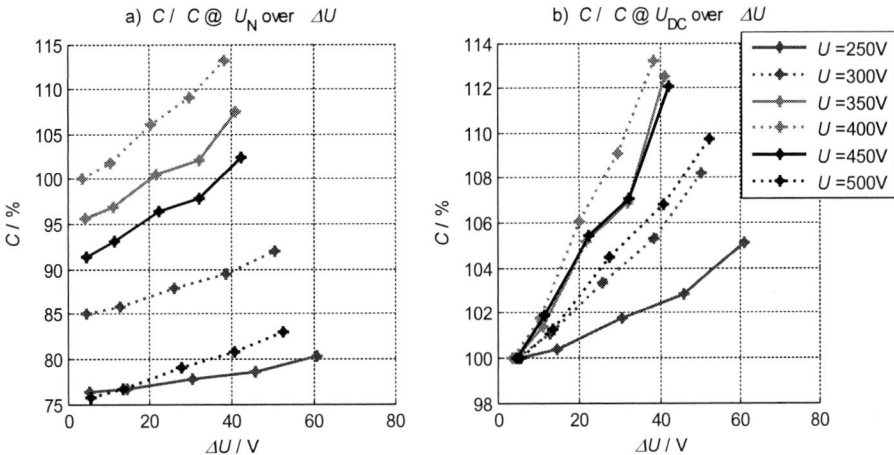

Fig 5: Capacitance C over rippel voltage @ f = 8 kHz, ΔU = 5V to 60V, T_C = 22°C, U_{DC} = 250V to 500V

In paper [4, p. 4], the different temperature characteristics for small and large signal are shown at voltage level U_{DC} = 400V. For small signal, the maximum capacitance is reached at around T_C = 55 °C. For large signal, the maximum capacitance is reached at T_C = 0 °C.

Regarding the capacitance in dependence of temperature for in-circuit-characterization (cp. Fig 6), there is temperature behaviour for all examined voltage levels like the small signal characteristic of CeraLink^{TM}. There, at voltage level U_{DC} = 400V, the maximum capacitance is reached at around T_C = 75°C. In the datasheet for sinusoidal characterization it is shown, that maximum small signal capacitance in dependence of temperatures occurs with rising voltage at lower temperatures (cp. [2, p. 5], $C(T_C)$).

Analog to this behaviour of small signal capacitance for sinusoidal characterization in the datasheet is the temperature behaviour of the in-circuit-capacitance. The highest capacitance in-circuit in dependence of voltage and temperature is achieved at voltage level U_{DC} = 350V and temperature T_C = 75°C. This capacitance is higher than the capacitance at nominal voltage U_{DC} = 400V and so there is an intersection (cp. Fig 6) of the curves $C(T_C, U_{DC} = 400V)$ and $C(T_C, U_{DC} = 350V)$ at T_C = 60°C.

The result of this in-circuit-characterization is that the dependence of capacitance from voltage and temperature must be considered for dimensioning an inverter with the non-linear capacitor as DC-Link capacitor. Regarding the capacitance value at the limits of the operational area of the DC-Link, the capacitance value is reduced relatively to the capacitance at nominal voltage and nominal temperature. The rise of capacitance in dependence of ΔU is measured in-circuit. This effect could compensate this derating of capacitance in dependence of voltage and temperature for several operational points but not for the limits of a DC-Link with a wide temperature and voltage range.

As a result of this derating of capacitance for the complete operational area for a voltage range between U_{DC} = 300V and U_{DC} = 500V only 75% of capacitance value at nominal voltage remains with the maximum voltage as the limiting value (cp. Fig 5). For a wide temperature range from T_C > -40°C and T_C < +95°C the lowest capacitance is determined by the minimum temperature with 60% of the capacitance value at T_C = 25°C (cp. Fig 6). With a ΔU = 40V the capacitance rises 15% compared to capacitance at ΔU = 5V.

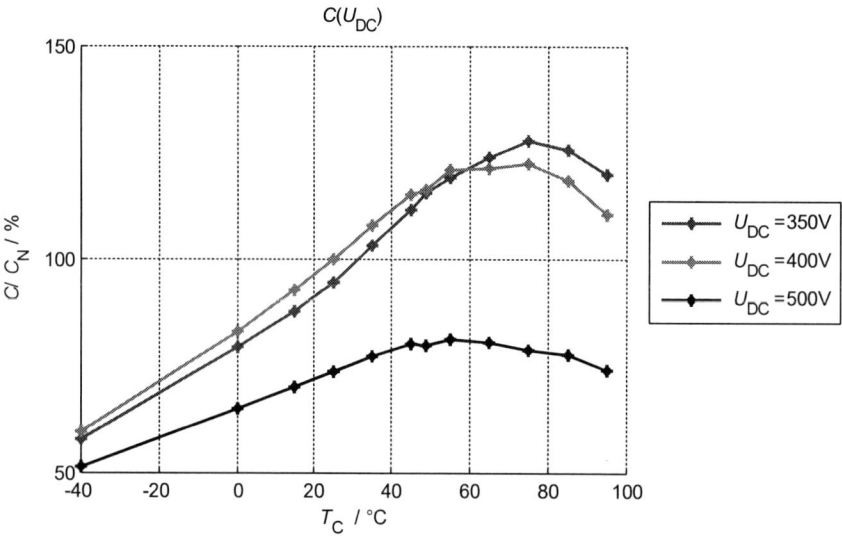

Fig 6: Capacitance C over temperature T_C for different voltage levels U_{DC}

4. Conclusions

Different methods to characterize the capacitance of a non-linear ceramic capacitor were investigated. It was shown that in a real pulsed operation identical to an inverter the capacitance can be determined best with the method "vertex points of voltage signal". It was shown that in contrast to the datasheet values no significant dynamic increase can be observed.

It was also shown, that the pulsed application benefits from the characteristic of a capacitance increase with rising voltage ripple. This can amount to up to 15% more capacitance.

The main adverse effect was formed to the temperature dependence. When having to design an inverter for a temperature range from -40°C to +95°C, one has to accept a derating of the capacitor to just 60% of the nominal value.

5. References

[1] Engel, Koini, Konrad und Schossmann, „A new high current - high voltage ceramic power capacitors," TDK-IPC Corporation, Deutschlandsberg, Austria, 2012.

[2] T. Epcos, „Datasheet - CeraLinkTM capacitor for fast-switching semiconductors 20µF, 500V," Deutschlandsberg, Austria, 2014.

[3] D. Connet, „Passive Embedding for higher Performance and Reliability," *Bodo's Power Systems,* Bde. %1 von %28-15, Nr. August, pp. 30-32, 31 07 2015.

[4] K. S. P. Konrad, „New demands on DC link power capacitors," SIA, Deutschlandsberg, AUSTRIA, 2013.

Operability of Metallized Polypropylene Capacitors under High Pressure

Magnar Hernes, SINTEF Energy Research, Sem Saelands vei 11, Trondheim, Norway, Magnar.Hernes@sintef.no

Ole Christian Spro, SINTEF Energy Research, Sem Saelands vei 11, Trondheim, Norway, OleChristian.Spro@sintef.no

Volker Geitner, ELECTRONICON Kondensatoren GmbH, Keplerstrasse 2, Gera, Germany, V.Geitner@electronicon.com

Abstract

In a research project sponsored by 10 industry partners and the Research Council of Norway, theoretical and experimental work has been done for providing fundamental knowledge to support realization of reliable Pressure Tolerant Power Electronic (PTPE) components and circuits for operation at sea depths down to 3000m. The ambition is to locate the complete power circuit including switching devices like IGBTs, IGBT drivers, capacitors, magnetic components, and various sensors in a pressurized liquid environment, while the main converter control electronics is assumed to be encapsulated in one bar environment.

The present paper reports the results from experiments for evaluating pressure tolerance for metallized polypropylene (PP) DC capacitors. Challenges regarding exposure to a pressurized liquid environment and the experimental methods for assessing these challenges are discussed. The interim conclusions from the experiments are presented. The main conclusion from live experiments with the metallized film capacitors in pressure vessels up to 300bar is that pressure tolerant operation is feasible; however, the important property of self-healing is negatively affected.

1. Introduction

Feasible solutions for PTPE have gained increasing interest in connection with applications such as subsea power grids for offshore wind farms and for subsea oil and gas exploitation.

Until now designs offered by the industry aimed for subsea converter operation have been concepts where the power circuits are completely assembled in steel vessels at atmospheric pressure. As the sea depth and the converter power rating increase, the pressure vessels, usually made of steel, gradually become heavy and unwieldy devices due to need for increasing the wall thickness. Consequently, also the heat conduction from the power electronics components to seawater becomes problematic. Therefore the oil companies are looking for more feasible solutions for subsea power electronics. Enabling PTPE allows the environment for the power circuit to have the same pressure as the ambient sea water. Compared to the required steel vessels for housing MW converters in an environment of atmospheric pressure, pressure compensated solutions open up for new advantages and possibilities such as reduced volume and weight, avoiding pressure barriers for electric penetrators, and improving cooling systems for the power system. These advantages are increasing by increasing sea depth.

In one accomplished [1-2] and one ongoing research project, significant steps have been taken towards the realization of PTPE. In particular, demonstration of reliable operation of live pressurized converter modules up to 6.5kV DC-link voltage has been done. Encouraged by these results, converter manufacturers and oil companies have decided to take a further step towards realization of PTPE components for real applications. However, it is also recognized that in order to achieve PTPE products with sufficient long-term reliability, quite some remaining problems need to be solved. Some of these problems are related to reliable operation of DC capacitors when applied as DC-link components for voltage source converters in a 300 bar liquid environment.

2. Addressing the possible pressure tolerant packaging challenges

2.1. The basic pressure tolerant converter concept

As illustrated in Fig. 1, the research project is assuming that the complete power circuit of the voltage source converter including IGBT modules and DC-link capacitors is located in a dielectric liquid environment with pressure up to 300 bar. This corresponds to a converter location at 3000m sea depth. Furthermore, the pressurized environment is also assumed to include electronic circuitry that need to be located in the vicinity of power components, such as IGBT drivers, auxiliary power supply and monitoring electronics. Circuitry for the converter central control is not included in the present research. A converter control unit is assumed to constitute a small volume compared to the power circuit.

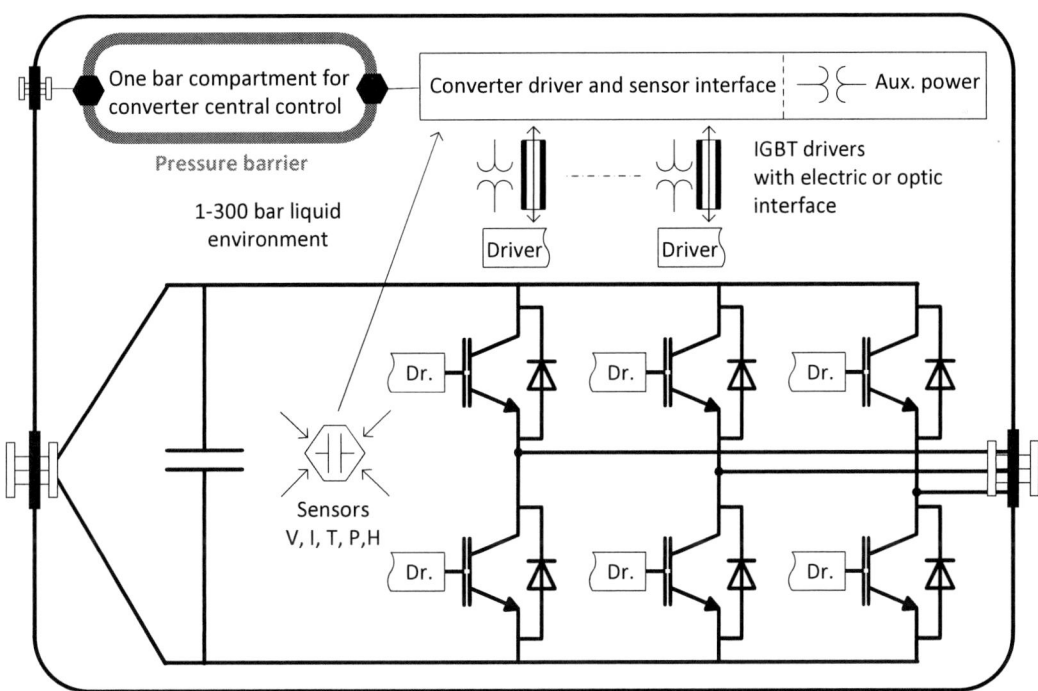

Fig. 1. Converter system with power circuit containing IGBTs, capacitors, drivers, sensors, auxiliary power in pressurized environment. Small one bar unit for the converter central control.

2.2. Assessing the challenges

A roadmap for the research was prepared for addressing the most important challenges assuming the need for a long-term reliable operation of pressurized components, especially highlighting the assumed most critical components such as IGBT modules, IGBT driver circuits and DC-link capacitors.

A plan was set up for the experimental works aiming to reveal the possible problems, and furthermore to propose modification of component materials and packaging, depending on the outcome of the experiments. For this planning work, component and material manufacturers have contributed with valuable inputs. They have also contributed with assessment of findings during the experimental work

The main concerns as regards to reliable operability of PTPE were discussed and are summarized as follows:

© VDE VERLAG GMBH · Berlin · Offenbach

1. Mechanical stress to encapsulations caused by the high pressure environment, and also possibly due to vacuuming processes in connection with liquid filling.
2. Possible change of functionality and characteristics of the power semiconductors IGBTs/ diodes due to pressure, such as changes in switching performance due to impact on driver electronics and/or directly on IGBT parameters.
3. Impact on self-healing performances and interconnection between film- and termination-layer of PP-film capacitors.
4. Possible material compatibility issues due to change in insulating material conditions for the components, such as for the silicone gel normally used for protecting IGBT/ diode chips, when facing various external insulation liquids.
5. Impact from humidity on power semiconductor insulation, giving the requirements to the surrounding filling liquid, or to barriers between the most critical locations and external less critical filling liquid.

The fulfilment of item 1 above is managed by pressure compensation; i.e. by totally relieving the stress from the encapsulations. Then, since the environmental pressure is entering the total interior of the components, special test programs have been executed for revealing possible issues related to item 2 and 3 above. Furthermore special test programs have been executed for investigating the packaging challenges related to item 4 and 5. The goals for the packaging activities have been to find a material or material combinations that can provide the following characteristics:

– Provide 100% filling of solids or liquids
– Electrical insulation properties as good as or better than existing/replacement materials
– The required long term chemical compatibility between the new materials and with the existing component interface materials.
– Maintain the required sealing properties (particles, ions, humidity) assuming liquid environments with various purity conditions.

3. Assessing capacitor technologies for high pressure environment

For converter DC-link operation, metallized polypropylene (PP) film capacitors are about to replace electrolytic capacitors due to recent significant technology improvement for these components and superior reliability compared to electrolytic components. Furthermore, electrolytic capacitors were regarded to be less attainable than film capacitors regarding adaptation to pressure tolerance, and were therefore not considered for the test program.

Separate film/foil capacitors were assessed in an early phase of the project. Compared to PP-film capacitors this technology has a significant less capacitor/volume ration. However, the main reason for not including separate film/ foil capacitors in the test program was lack of contact with manufacturers that could support the research with test objects and consultant work.

A converter also need low voltage buffer capacitors for IGBT drivers, auxiliary power supply etc. Such capacitors are represented by various technologies and packaging. In the project, ceramic capacitors have been included in live experiments of converter modules where such capacitors have served as DC buffer capacitors for the IGBT driver electronics and the power supplies mentioned above. For the ceramic capacitor test objects, the results from live operation up 300 bar pressure have been exclusively positive.

The PP-film capacitor technology has a significant market share of applications such as DC-link capacitors and output filter capacitors for voltage source converters. Therefore the main focus for the pressure investigations has been on this technology. For assessing the possible problems related to reliable operation at high liquid pressure, valuable discussion was made with the capacitor manufacturer Electronicon. Based on these discussions, the following challenges related to a successful high pressure operability was put on the agenda for the experimental work:

PCIM Europe 2016, 10 – 12 May 2016, Nuremberg, Germany

- Would the interconnection between the sprayed polycrystalline zinc termination layer (schoopage) and the metal deposit on the PP-film of the capacitor film rolls have sufficient mechanical strength to sustain a vacuumization and subsequent pressurization, as well as the possible influences by the immersion in a dielectric liquid environment
ref. Fig. 2.
- Would the self-healing properties for the PP-films be maintained under high pressure? The self-healing property of the metallized capacitor film could be affected by pressure and also by unfamiliar liquid environment. Fig. 3 explains a successful and a non-successful self-healing process of an insulation breakdown spot within the capacitor film.
- Could the liquid environment give negative impacts to the capacitor interior, such as degradation of the very thin metallization or reduction of cooling properties?
- Last but not least, could the electrical performance, such as capacitance or loss factor be affected by pressure or possible intrusion of liquid?

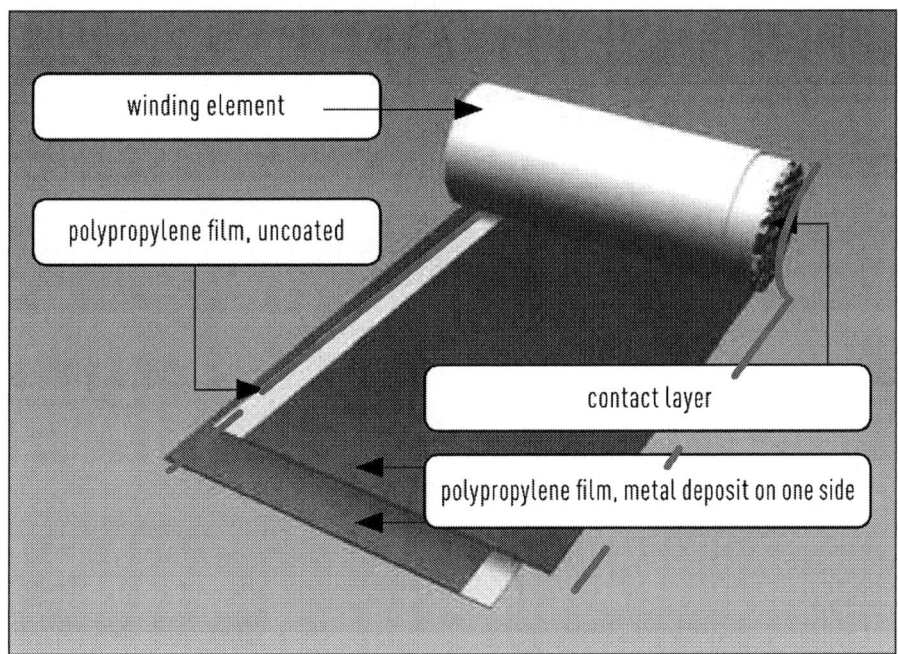

Fig. 2. Inner structure with assumed interior weak points of metallized film capacitor

Self Healing Dielectric Breakdown   **Non-healing Breakdown (Scenario I)**

Discharge at weak spot

Energy of discharge
vaporises and removes
metal coating around
breakdown spot

Breakdown spot insulated,
capacitor continues operation
unchanged

Incomplete insulation,
leakage current rising slowly

step-by-step increase of
temperature around the
breakdown spot

disintegration of PP-Film
Snowball-effect of non-
healing breakdowns
release of gas/pressure rise
inside the capacitor case

Fig. 3. Illustration of successful and non-successful capacitor self-healing process

© VDE VERLAG GMBH · Berlin · Offenbach

4. Experimental methods

4.1. Requirement specification for the high pressure operability

The requirement specification for pressure tolerant operability of PP-film capacitors was that no electrical derating should be required for the components under investigation. Furthermore deviations in electrical parameters, such as capacitance and loss factor, should not exceed the datasheet specifications valid for general industry applications

Pre-and post-characterization of the capacitor test objects was planned to be performed with reference to prevailing standards such as IEC standard 61071:2007: "Capacitors for power electronics" when applicable. Articles in the standard tests found to be irrelevant, such as those related to standard encapsulation etc., were omitted. Additional items, such as possible pressure exposure to the film winding and possible exposure due to liquid were added to the inspection procedure. The test procedures included normal operation up to rated electrical stress levels, as well as electrical overstress situations according to the IEC standard.

Dielectric liquid candidates have been thoroughly examined as part of a separate work package focusing on insulating materials. Based on these investigations the selected candidates for representing the environmental liquid for the component and circuit experiments were natural and synthetic esters, fluorinert and perfluorpolyether. Nevertheless, it must be mentioned that for all experiments involving PP-film capacitor test objects reported in this paper, the liquid environment has been the synthetic ester Midel®7131. One reason for this priority was the good experience from previous research, such as compatibility with other electric component materials and also dielectric properties. A second reason was the good manageability as regards environment and cleanliness.

Based on presumed challenges discussed in chapter 3 and the requirement specification for validating reliability in the high pressure environment, the experimental work has been accomplished along the following main axis, attempting to disclose the assumed possible problems:

- Early warning pressure testing up to target pressure of 300 bar with high pressure slew rates; mainly for exposing weaknesses of encapsulations, internal voids and other internal infirmities.
- Possible effects from liquid on material ageing processes; especially focus on effects on capacitor film metallization.
- Long-term live operation under controlled environment in pressure vessels.

4.2. Experiments with metallized PP-films

A representative selection of PP-film rolls with DC metallization was provided by Electronicon. The film rolls are taken from the production line, meaning that the implementation of the metallization is as for the real capacitors. The PP-films have served as test objects for two separate and parallel running experiments.

1. Liquid compatibility experiments where films have been submerged in the synthetic ester Midel®7131 for several months, while periodical observing possible changes in the film resistivity. The possible effect of voltage polarization has been checked by applying voltage between film metallization and ground.
2. Self-healing experiments where PP-film test objects have been arranged in a special assembly, allowing for visual observation of the films while DC-voltage up to breakdown voltage for the film is applied. Also these experiments have been run in the Midel®7131.

4.4. Experiments with live operation of capacitor elements in pressure vessel

The main efforts have been on live operation of test objects operating in circuits enabling electric stress conditions as in a real converter DC-link. A simplified circuit diagram for the test circuit, allowing for separate control of applied AC current (e.g. for emulating the converter switching ripple current) and applied DC voltage is presented in Fig 4.

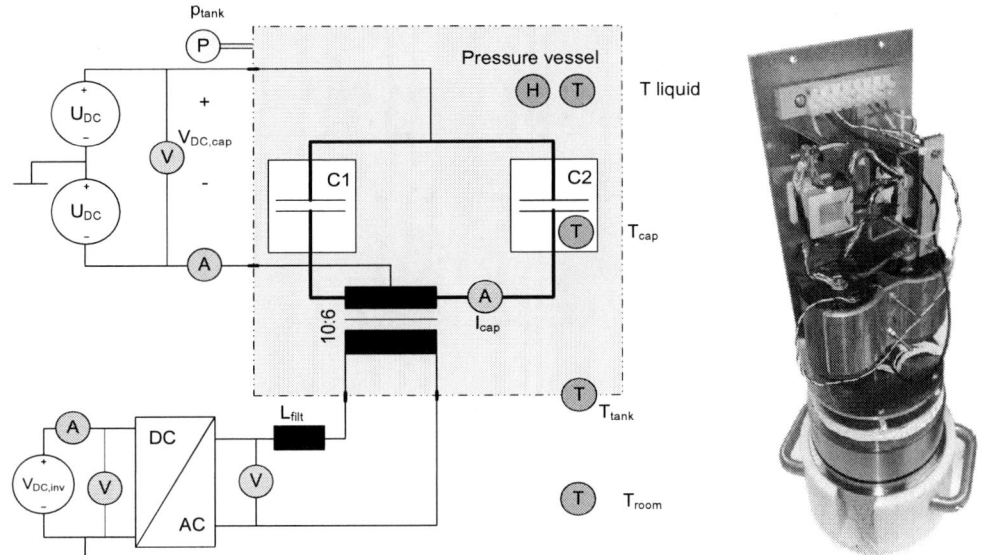

Fig.4. Test circuit for capacitor live experiments and assembly prepared for pressure vessel.

Fig. 4 illustrates a controllable DC voltage source for feeding two test objects C1 and C2 in parallel. This source is characterized by a relatively high internal impedance for limiting the fault current inside the pressure vessel. Furthermore, the voltage control range is well covering the test ranges specified in the IEC standard. The current source is formed by a low impedance controllable voltage source feeding a resonant circuit composed of the transformer short circuit reactance and the two test objects C1 and C2 in series.

5. Findings and interim conclusions

5.1. The investigations of possible erosion of PP-films in liquid environment

A special test assembly was prepared for investigating possible incompatibility between metallized PP-films and the synthetic ester liquid, ref. chapter 4.2. The PP-film metallization contains zinc, which is on the list of potential incompatibility concerns by the manufacturer of Midel®7131. The main motivation for the compatibility experiment was therefore to examine to which extent such possible incompatibility could involve problems for the capacitor metallization in a long-term perspective.

Two different test setups have been prepared; one with film in air environment, and one with film soaked in the synthetic ester liquid. For the one in liquid a DC voltage of 200V was applied between the film metallization and ground in the test box. By February 2016 the test setup with film in air has been running for 15 months, while the setup in liquid has been running for 13 months. The current status for the films is presented in Table 1.The expected increase of film resistance over time for a normal industrial component is also included in the table. As seen the resistance of the film in air has an estimated increase of 3.3 %/year, while the one in liquid with 12.3 %/ year. It should be noticed that there were no special measures for the controlling of the humidity during the film erosion tests. Some sample tests of the liquid during the test indicated relative humidity around 30 %.

© VDE VERLAG GMBH · Berlin · Offenbach

For a normal industrial component, assembled in a dry and sealed case, the estimated increase is 2%/year as an average over a period of 10 years. For an industrial component an increase of internal resistance of 30...50 % would be interpreted as end of life due to increased ohmic losses, depending from the application. Other additional aging processes, like capacitance losses caused by locally oxidation of the metal deposit, driven by the electrical field, may limit the lifetime to a much higher degree. Extrapolating the recorded values for the film test objects indicates that end of life values would be reached after 9...15 years for the film in air, while 2.5....4 years for the film in the synthetic ester liquid.

Table1. Increase of resistance for film in air (after 15 months), film in liquid (after 13 months) and reference, estimated increase per year, and when reaching end of life.

Test condition	Sample avg. resistance increase		End-of–life
	After 13/15 months	Est. increase p.a.	(increase to 30...50%)
Film in air	R15/R0 = 1.048	3.3%	9...15 years
Film in liquid	R13/R0 = 1.184	12.3%	2.5...4 years
Reference (dry, sealed)		<2% in average	> 15...25 years

Investigation of possible negative effects on the PP-film has also been part of the program for live operation of capacitor elements under high pressure and applying the same synthetic ester liquid, ref chapter 5.3. Post-examinations of these test objects show that liquid has penetrated into the capacitor windings. However, no negative effects are found on the film in the liquid affected areas that are indicating deterioration of the metallization. Additionally, it should be noticed that the live experiments in pressure vessel were accomplished under accurate humidity control, varying between 3...30%. Also several other experiments in the project indicate that humidity needs to be kept under control, preferably below 10%.

5.2. Self-healing properties in liquid and pressurized environment

Two parallel experiments have been performed for investigating whether the self-healing property of PP-film would be affected under high pressure, ref chapter 3. One was the live experiments, ref. chapter 5.3. The other was experiment with capacitor PP-films arranged in special test assemblies, ref. chapter 4.2. This test assembly was made up of two transparent 15mm polycarbonate plates where two adjacent film test objects are placed in between. Spring clamps are giving a pretension pressure to the films of about the same magnitude as the winding pressure for a real capacitor. The cross section area for the adjacent films is 80mm x 90mm, giving some capacitance for supporting a self-healing fault current. The complete test assemblies is soaked in the test liquid and then vacuumed.

Successful experiments have been performed in the synthetic ester liquid Midel®7131 in one bar environment, with 4µm and 9.8µm metallized PP films, by applying DC voltage across the films of 300-400 V/µm for both films. According to literature this is the voltage range where self-healing events should be expected and be successful. Visible traces after film punctures were displayed. The bigger scars for the thick film compared to the thin film supports the impression that significant more energy is involved during self-healing of thick film compared to thin film. The self-healing events were also demonstrated for both films by visual glimpses and for the thick film also by electrical current pulses in the external feeding line. Also from the visual glimpses it was obvious that more energy was involved in the punctures of the thick film compared to the thin film.

The conclusion from the film experiments in one bar liquid was that proper self-healing performance was retained. Continuing experiments in pressure vessels was planned. However, these were replaced by the more thorough experiments, to be discussed in chapter 5.3, that already indicate that the self-healing properties may be reduced under high pressure.

5.3. Live operation of capacitor elements under high pressure

The most comprehensive studies of operability in high pressure liquid environment were those applying the test equipment described in Fig 4. These test objects are semi-products of 210µF/ 1100VDC components from the factory, i.e. naked elements with copper strips connected to screw terminal taken out of the production line before filling with polyurethane.

A test procedure was prepared, including plan for the live operation in pressure vessel, and for characterization before and after operation in pressure vessel. The pre- and post-characterization mainly followed the IEC standard 61071, including voltage test between terminals and capacitance and tan delta measurements. The operation in pressure vessel included sequences emulating real converter DC-link operation by applying rated DC voltage and rated AC current (9kHz). These sequences were periodically interrupted by stopping the current circulation and characterizing the capacitance, tan delta and leakage current. During these interrupts, the DC voltage was also elevated to 1.5 times rated voltage for 10 sec.

The live experiments include several runs at 300 bar pressure at rated DC voltage and AC current for the test objects. While under pressure, all electrical characteristics were close to unaffected by pressure. Three test objects have failed to short circuit after 2-4 weeks of operation at 300 bar. Post investigations of these test objects, well assisted by the manufacturer, have given reason to believe that the root cause for these failures is unsuccessful self-healing. The hypothesis is that one abnormal event leads to a cascade of similar events followed by thermal runaway. The PP-film was found to be melted together in the middle of the film towards the center of the roll. Furthermore, some of the investigated film events have been classified as close-to-unsuccessful, meaning that the self-healing stretches over several film layers.

In an attempt to significantly reduce the risk for film punctures, a test run was started with changed operation conditions i.e. reduced DC-voltage. Compared to the test runs where early failure was experienced, this revised test has been running without failure for more than 200 days at 300 bar environment (and is presently still running). This gives support to the hypothesis that the environment is negatively affecting the self-healing mechanism.

5.4. Interim conclusions

The experimental work with metallized PP-film capacitor test objects have substantiated that the electrical characteristics, such as capacitance and tan delta are well maintained in high pressure liquid environment up to 300 bar. Furthermore the mechanical durability such as the film interconnection to the termination layers is maintained. Even though traces from the surrounding liquid are found inside the film roll, there are no indications of deterioration of the film metallization like erosion. The major concern is that the important self-healing mechanism of the metalized PP-film is negatively affected by the high pressure. Continuing experiments applying reduced DC-voltage has been in operation for a significant longer period compared to those resulting in failure. This is taken as an indication that high pressure operability of PP-film capacitors could be feasible provided significant derating of operating voltage. Continuing work is recommended for providing a clearer picture of the pressure effect on the self-healing mechanism, and thereby giving a basis for improved designs enabling the sufficient high pressure operability.

6. References

[1] Hernes M., Pittini R., "Enabling pressure tolerant power electronic converters for subsea applications", European Conference on Power Electronics and Applications, 8-10 September, Barcelona, Spain.

[2] Pittini R., Hernes M., "Pressure Tolerant Power Electronics for Deep and Ultra-Deep Water",Society of Petroleum Engineers Projects, Facilities and Construction Journal, ISSN:1942-2431.

Application of High-voltage 750V Aluminum Electrolytic Capacitor in Inverter

Kezhuang Yu, Mingkai Peng, Mianwei QiU, zeasset@zeasset.com
Shenzhen Zeasset Electronic Technology Co., Ltd.
Building 1, Anle Industrial Park, Hangcheng Road, Gushu, Xixiang, Baoan District, Shenzhen, Guangdong Province, China.

Abstract

In the three-phase 380V inverter, 2 pieces 400VDC capacitors in series are usually used. Zeasset succeeds in developing one single aluminum electrolytic capacitor with voltage endurance up to 750VDC to replace them. The replacement enables the reduction of numbers of parallels and series, the simplication of circuit design for users and higher reliability of system.

1. Introduction

Nowadays AC-DC-AC topology are widely used in converter, UPS etc. Figure 1 shows the structure of AC-DC-AC converter, it makes up of three parts:

Part 1: AC-DC rectifier.
Part 2: Smoothing DC current、 energy storage part.
Part 3: DC-AC inverter.
its working principle is: First, through the AC - DC rectifier power convert AC to DC, then through the DC - AC inverter converts DC voltage to AC with frequency and voltage are adjustable. The DC Bus connect the three parts together

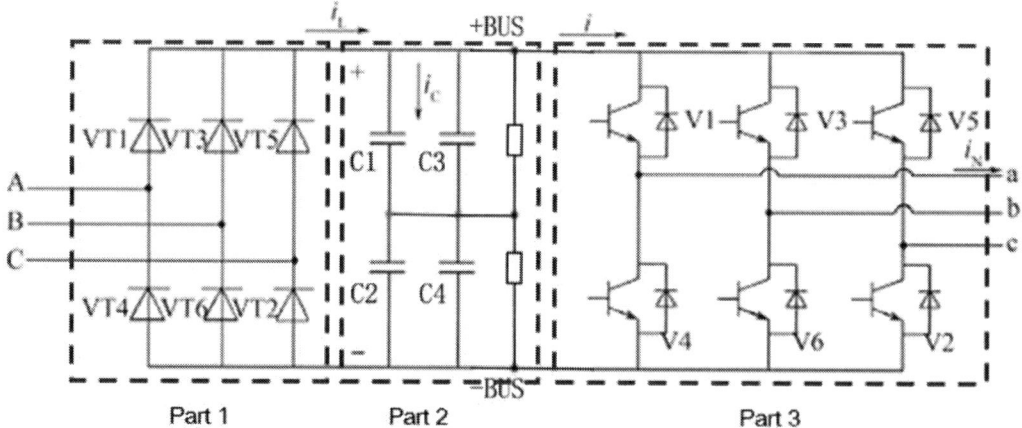

Figure 1:Diagram of circuit structure for AC-DC-AC converter

© VDE VERLAG GMBH · Berlin · Offenbach

2. DC-link Capacitor

2.1 The function of DC-link Capacitor

As the intermediate links between AC - DC and DC - AC energy storage and conversion, DC-link capacitors play a key role:
(1) Stable DC Bus voltage can restrain voltage pulsation and the harmonic current back to the grid.
(2) Compensation the power requirement between DC - AC inverter and AC-DC rectifier input;
(3) Provide i peak power to protect inverter from power grid transient impact;
(4) absorption peak energy of motor brake.

2.2 Requirements of AC - DC - AC conversion circuit for DC-link capacitor:

(1) High voltage, large capacity.
(2) High ripple current.
(3).Long life;
(4) High reliability

In order to achieve the huge capacitance volume and high voltage requirement for conversion circuit, we use capacitors seriesing and paralleling design requirements

The 380 VAC or 400 VAC three-phase inverter, after AC - DC rectify, the DC Bus voltage may be more than 600 VDC. Most designers use two 400V electrolytic capacitors in series, then paralleled them in group. At the same time, resistors with the same pressure are paralleled on both ends of seriesing capacitors, to ensure that the same voltage among the seriesing capacitors.

3. Zeasset LF (Snap in) and TA (Screw) style high voltage DC-link s capacitors

Zeasset developed 750V aluminum electrolytic capacitor, including LF (Snap in) and TA (Screw), used for DC-link smoothing and energy storage, to replace many low voltage in series. In theory, one 750V capacitor can replace 4 pieces 400V capacitors (two in parallel and two in series, Figure 2).

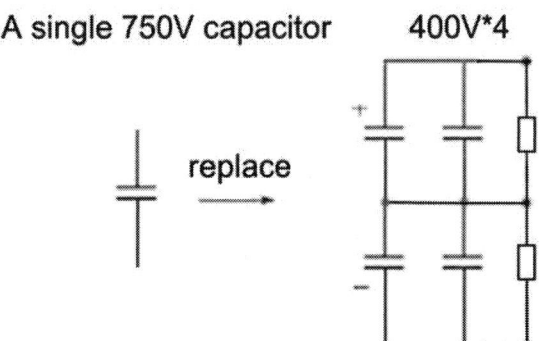

Figure 2: A 750V capacitor replace 4 pieces

3. Advantages of one high voltage capacitor replace four low capacitors in series or in parallel:

(1) Reduce the numbers of capacitors in series or parallel circuit, effectively to reduce the system size and improve power density.
(2) Simplify the design, save materials, reduce cost and improve the reliability of the system.

3.2 Main parameters of Zeasset 750V aluminum electrolytic capacitor

Item	TA (Screw)	LF (Snap-in)
Rated Working Voltage dc	750V	750V
Surge Voltage	800V	800V
Operating Temperature Range	-25~85°C	-25~85°C
Load Life	2000h	2000h
Capacitance	820~3300µF	220~680µF
Diameter	64~89mm	22~50mm
Dissipation Factor	0.20	0.20

Technical Difficulties of 750VDC Capacitor:
•Raw materials
For anode foil:
The common requirement of voltage endurance≥990Vf
Specific capacity≥0.22µF/cm2

Formula of Capacitance C of aluminum electrolytic: $C = 8.855 \times 10_{-12}\varepsilon\, S/d(F)$

ε: related permittivity of media

s: effective area of electrode
d: thickness of dielectric, that is distance between two electrodes

•Etching Technology of Anode Foil
•Forming Technology of Anode Foil

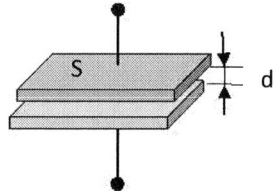

3.3 Characteristics of Zeasset 750V aluminum electrolytic capacitor products

3.3.1 Structure features:

Anti-breakdown:
During the production with high-voltage anode foil, the beginning point of rolling core of capacitors and the top edge are usually the points of failure. It is advisable to use six-layer structure. Breakdown voltage of electrolytic paper is higher than 1500V (Figure 3).

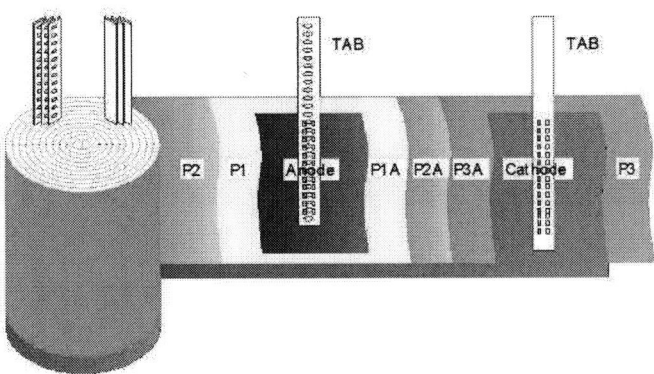

Figure 3: 6-ply paper structure

Prevention of arc discharge:
The arc discharge easily happens when dust gathers between two terminals because of high voltage. Thus, special improvements of terminals should be made to forbid the arc discharge.

3.3.2 Good performance of low temperature operation

Special electrolyte make the capacitor with good low temperature performance.
In order to improve the spark voltage of electrolyte, normally increase the proportion of esters, the proportion of ester content will increase the viscosity of the electrolyte, thus influence on the performance of the electrolyte low temperature. After many tests, we adopt the branched long-chain dibasic ammonium carboxylate and polymer ammonium carboxylate as solute, ethylene glycol, diethylene glycol, polyvinyl alcohol as a mixed solvent. Appropriately adding phosphate, nitrate, spark voltage improve agent. New electrolyte has the high spark voltage, high electrical conductivity and viscosity is moderate, meet the required performance of the ultra-high voltage aluminum electrolytic capacitors.

3.3.3 DC and Pulse mixed aging to improve the storage performance

For ultra-high voltage aluminum electrolytic capacitors, if processed with the normal aging process of aluminum electrolytic capacitors, the capacitor leakage current will recovery fast. Zeasset used DC and Pulse mixed aging, in guarantee of better oxide film density and also restrain the leakage current to recovery fast.

Ways of Aging	Segmented Aging at Room Temperature	High-temperature Aging	Characteristics	Advantages
DC Aging	●	●	Aging slowly, low efficiency	Good compactness of lattice of oxide film, long load life
Pulse Aging	●	●	Aging quickly, high efficiency	Leakage current recovers slowly

4. Conclusion

The DC Bus voltage is 530V after being rectified in the three-phase 380V inverter, which is realized by two 400V aluminum electrolytic capacitors in series. However, the application of 750VDC capacitor can simplify the circuit design, reduce the numbers of components and improve the power density.

References

1. Shenzhen Zeasset Electronic Technology Co., Ltd., "Aluminum Electrolytic Capacitors Application Note".
2. Zhaoan Wang, "Power Electronics", No.5 Version, China, June, 2000.
3. Carl H.Hamann, Andrew Hamnett, Wolf Vielstich, "Electrochemistry", No.2 Version, Germany, 2010.
4. Paul M.S. Monk, "Fundamentals of Electroanalytical Chemistry", UK, 2012.

PCIM Europe 2016, 10 – 12 May 2016, Nuremberg, Germany

Analytic Loss Calculation for E-Core Inductors including the End Windings

Johannes, Heseding, Leibniz University Hannover - Institute for Drive Systems and Power Electronics, Germany, johannes.heseding@ial.uni-hannover.de
Axel, Mertens, Leibniz University Hannover - Institute for Drive Systems and Power Electronics, Germany, mertens @ial.uni-hannover.de

Abstract

This paper presents an analytical approach to calculate the losses of an E-core inductor with litz wire windings for DC/DC-applications with special emphasis on the end winding losses. The central assumption for calculating the end winding losses is the modeling of both end windings as an air coil. The magnetic field of this air coil is calculated by assuming that a single strand can be seen as an infinitely thin conductor. The calculated results are compared to a combined, planar and axisymmetric 2D-FEM model and to measurements. The results show that the losses in the end windings are significant and amount to 15 to 30 percent of the total losses.

1. Introduction

In hybrid electric vehicles, such as the Toyota Prius, bidirectional DC/DC-converters are used to boost the battery voltage to the high voltage bus of the traction inverter in order to reduce motor and inverter currents as well as battery costs and complexity. Main part of these DC/DC converters, besides the IGBT devices, is the inductor in which the energy is stored during a switching cycle. The knowledge of the total losses of an inductor combined with a thermal model can be used to optimize the inductor volume by modifying the allowable current density which affects the minimum possible inductor volume as described in [1] and [2]. For an efficient design process, the speed of calculating the iterative designs is essential, therefore analytical models are preferred compared to 2D or even 3D-FEM calculations. In [2]-[8], several methods are described to calculate the core- and winding losses inside an inductor (Figure 1(a)) analytically. The losses in the end windings of an electric machine, with a similar configuration as in Figure 1b, are estimated in [9] for solid conductors with 3D-FEM method.
This paper focuses on the loss calculation for E-core inductors with rectangular litz wire windings and airgaps in the inner and outer legs of the core. The losses are separately calculated for the DC- and AC- current components carried out by Fourier analysis of the current shape. A comparison to FEM simulations and measurements is performed. The geometry and the used coordinate systems of these inductors are shown in Figure 1.

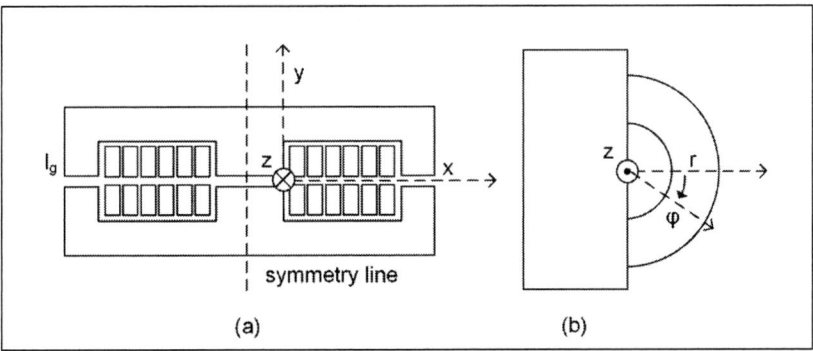

Fig. 1. Geometry and coordinate systems: cartesian coordinates for the inner windings (a), cylindric coordinates for the end windings (b)

© VDE VERLAG GMBH · Berlin · Offenbach

2. Calculation of the core and winding losses

For continuous current operation of a DC/DC converter operating in boost mode, with a given duty cycle calculated by (1), the current shape of the inductor with the inductance L is triangular as it is shown in Figure 2(a) for a switching period T. The input voltage is U_{in}, the output voltage of the converter is U_{out}.

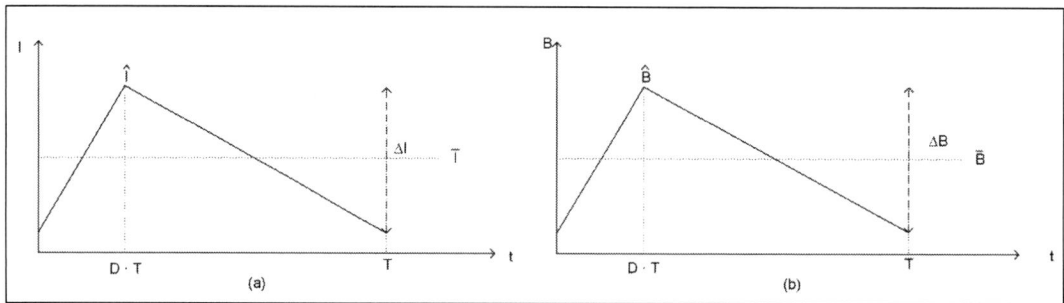

Fig. 2. Current shape for continuous operation with duty cycle D (a) and shape of flux density (b)

$$D = 1 - \frac{U_{in}}{U_{out}} \quad (1)$$

For the following loss calculations, the DC- and AC- current components are calculated for a given output power P by Fourier analysis. $\bar{I}$ is the mean current, ΔI the peak to peak current and $\hat{I}_k$ are the current components at the multiples k of the switching frequency f_s.

$$\bar{I} = \frac{P}{U_{in}} \quad (2)$$

$$\Delta I = \frac{U_{in}D}{L f_s} \quad (3)$$

The inductance L, which is assumed to be linear in this paper, can be obtained by FEM simulation or by solving the magnetic reluctance network described in [8].

$$|\hat{I}_k| = |C_k| \cdot \Delta I \quad (4)$$

C_k is the complex Fourier coefficient of the order k and is calculated by (5) - (7).

$$C_k = A_k + jB_k \quad (5)$$

$$A_k = \frac{2}{T} \left[\int_0^{D \cdot T} \frac{t}{D \cdot T} \cos(2\pi \cdot k \cdot t) \, dt + \int_{D \cdot T}^{T} \left(\frac{-t}{T - D \cdot T} + \frac{T}{T - D \cdot T} \right) \cos(2\pi \cdot k \cdot t) \, dt \right] \quad (6)$$

$$B_k = \frac{2}{T} \left[\int_0^{D \cdot T} \frac{t}{D \cdot T} \sin(2\pi \cdot k \cdot t) \, dt + \int_{D \cdot T}^{T} \left(\frac{-t}{T - D \cdot T} + \frac{T}{T - D \cdot T} \right) \sin(2\pi \cdot k \cdot t) \, dt \right] \quad (7)$$

2.1. Calculation of the core losses

Based on [4], the core losses can be calculated with the Steinmetz equation assuming a current with an equivalent frequency that causes the same core losses as the real current shape. With the triangular shape of flux density in the inductor (see Figure 2(b)), it follows for the equivalent frequency

$$f_{eq} = \frac{2}{\pi^2} \left(\left(\frac{\hat{B} - \bar{B}}{\Delta B} \right)^2 \frac{1}{DT} + \left(\frac{-\Delta B}{\Delta B} \right)^2 \frac{1}{T - DT} \right) \quad (7)$$

With the Steinmetz parameters k, α and β and a temperature coefficient $C = ct_1\tau^2 - ct_2\tau + ct$ with the temperature τ and the core volume V_{core}, the temperature dependent mean core losses are calculated by (8).

$$\overline{P_{v,core}} = \frac{k}{T} \cdot \left(f_{eq}\right)^{\alpha(f_{eq})-1} \cdot \left(\frac{\Delta B}{2\,T}\right)^{\beta(f_{eq})} \cdot C \cdot V_{core} \quad (8)$$

2.2. Calculation of the winding losses

The DC- resistance of the winding is calculated by (9). Litz wire winding is assumed with N turns each consisting of N_s twisted strands with a strand diameter of d_s.

$$R_{s,DC} = \frac{4 \cdot l_{w,ges} \cdot N}{d_s^2 \cdot \pi \cdot \sigma_{cu} \cdot N_s} \quad (9)$$

where $l_{w,ges}$ is the mean value of the individual turn lengths.

Using the orthogonality of the Fourier transformation, the skin and proximity losses can be calculated separately for every Fourier component k for an arbitrary current shape. With the frequency dependent skin depth δ and assuming a homogenous current sharing between the strands, the skin effect losses for a litz wire are given by (10) [7].

$$P_{Skin,k} = R_{s,DC} \cdot \sum_{k=1}^{\infty} \frac{d_s}{8\,\delta} \cdot \frac{(Ber_0\xi - Bei_0\xi)Bei_1\xi - (Ber_0\xi + Bei_0\xi)Ber_1\xi}{(Ber_1\xi)^2 + (Bei_1\xi)^2} \cdot \hat{I}_k^{\,2} \quad (10)$$

with $\xi = \frac{d_s}{\sqrt{2}\delta}$ and the real and imaginary part of the Bessel function $J_n\left(\xi e^{j\frac{3\pi}{4}}\right) = Ber_n + jBei_n$.
The proximity losses of a litz wire can be divided into the inner proximity losses caused by the magnetic field of the single strands inside the wire and the outer proximity losses caused by the external magnetic field of the other windings and the airgap leakage field. In [10] the inner proximity losses for a round litz wire are calculated with the diameter d_{litz} of a round litz wire. To approximate the inner proximity losses in a rectangular wire, the factor $2\pi^2 d_{litz}^{\,2}$ in (11) has to be replaced by $8\pi A_{litz}$ (replacing the round litz wire surface with a rectangular surface). FEM simulations show that this approximation is suitable for common rectangular litz wire geometries.

$$P_{Pi,v} = \frac{R_{s,DC}N_s}{2\pi^2 d_{litz}^{\,2}} \cdot \sum_{k=1}^{\infty} \frac{\xi \pi^2 d_s^2}{2\sqrt{2}} \cdot \frac{(Ber_2\xi - Bei_2\xi)Ber_1\xi - (Ber_2\xi + Bei_2\xi)Bei_1\xi}{(Ber_1\xi)^2 + (Bei_1\xi)^2} \cdot \hat{I}_k^{\,2} \quad (11)$$

To calculate the external proximity effect losses, the external magnetic field H_e consisting of the airgap leakage fields in the inner and outer legs of the core and the windings has to be estimated for the inner core region and the end windings. The method to calculate the external magnetic field $\widehat{H}_e$ for the core region is described in detail in [7]. The external proximity losses are given by

$$P_{Pe,v} = R_{s,DC}N_s^{\,2} \cdot \sum_{v=1}^{\infty} \frac{\xi \pi^2 d_s^2}{2\sqrt{2}} \cdot \frac{(Ber_2\xi - Bei_2\xi)Ber_1\xi - (Ber_2\xi + Bei_2\xi)Bei_1\xi}{(Ber_1\xi)^2 + (Bei_1\xi)^2} \cdot \widehat{H}_e^{\,2} \quad (12)$$

2.3. Calculation of the external magnetic field for the end winding region

To calculate the losses of the end windings, formulas (9)-(11) can be directly used to estimate skin, inner proximity and DC losses of the end windings. To calculate the external magnetic field in the end winding region in order to estimate the external proximity losses in the end windings with (12), the following assumptions are made:

a)   Both end windings together can be seen as a round air coil.

b)   The influence of the leakage field from the air gaps is negligible in the region of the end windings.

c)   A single strand can be seen as infinitely thin compared to the wire diameter

PCIM Europe 2016, 10 – 12 May 2016, Nuremberg, Germany

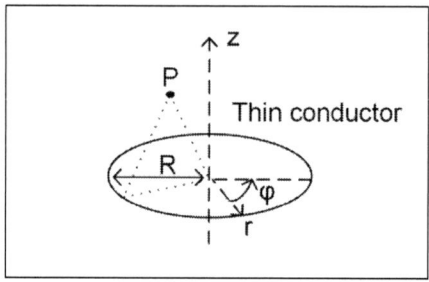

Fig. 3. Infinitely thin conductor in r-φ plane

Due to symmetry (see Figure 3), the magnetic field in the point P has only a z- and r-component. Using Biot Savart Law, the field components in point P resulting from the current I_s in the thin conductor can be calculated with (13) and (14). The derivation of (13) and (14) is given in [11].

$$H_r(P) = \frac{I_s \cdot k}{4\pi \cdot \sqrt{Rr}} \left(\frac{z}{r} \left(-K(k) + \frac{R^2 + r^2 + z^2}{(R-r)^2 + z^2} \cdot E(k) \right) \right) \qquad (13)$$

$$H_z(P) = \frac{I_s \cdot k}{4\pi \cdot \sqrt{Rr}} \left(K(k) + \frac{R^2 - r^2 - z^2}{(R-r)^2 + z^2} \cdot E(k) \right) \qquad (14)$$

with the factor $k = \sqrt{\frac{4Rr}{(R+r)^2 + z^2}}$ and K(k), E(k) being the complete elliptic integrals of first and second kind [12].

This field distribution is calculated only once for R equal to the outer radius of the end winding. For $r = R$ and $z = 0$ the calculation is not possible because the denominator of (13) and (14) becomes zero. For this point H_r is set to zero and a linear transition from $H_z(r = R - \frac{d_s}{2}, z = 0)$ to $H_z(r = R + \frac{d_s}{2}, z = 0)$ is assumed. The fields caused by other strands are approximated by spatial shifting of the results to the position of the respective strand, assuming equal current sharing between the strands. The resulting field of all strands is calculated by superposition.

An error is generated because the center of the wound strand is also shifted. FEM simulations have shown that this error is negligible. To show that these assumptions are valid the external magnetic field of an air coil with one turn in the z=0 plane is calculated. The winding is modeled as a rectangular litz wire (width 2 mm, height 4 mm) consisting of 200 strands with a strand diameter of 0,2 mm. The outer radius of the air coil is 10 mm and the strand current I_s is 0,5 A. Figure 4 shows the comparison of the analytical and FEM calculation results of the magnetic field amplitude in the z=0 plane (resolution in r direction: 0,2 mm). It can be seen, that the linear transition matches the FEM results in the winding region and the error due to the shifted center is small.

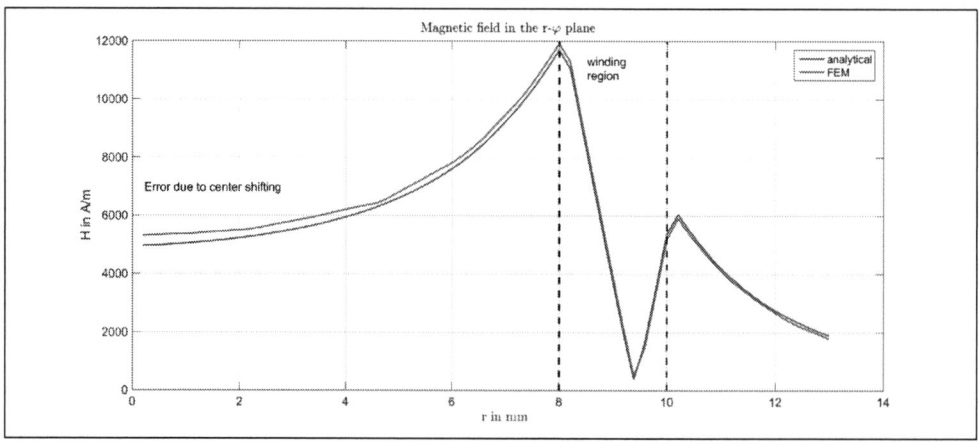

Fig. 4. FEM and analytical solution for the field magnitude in the z=0 for an air coil with one turn

© VDE VERLAG GMBH · Berlin · Offenbach

3. Comparison to FEM simulation and measurements

To validate the analytical calculations, the AC-resistance of a prototype coil shown in Figure 6 is calculated analytically as described above. The used coil has 10 windings. One turn is realized by paralleling three litz wires (width 2 mm, height 4.1 mm) in z-direction (see Figure 1(b) and 5) with 160 strands each and a strand diameter of 0,2 mm. The core consists of 2 E64/10/51 cores with 3C92 material. With the FEM software FEMM the inductor is modeled as a planar problem for the inner windings and as an axisymmetric problem for the end winding region.

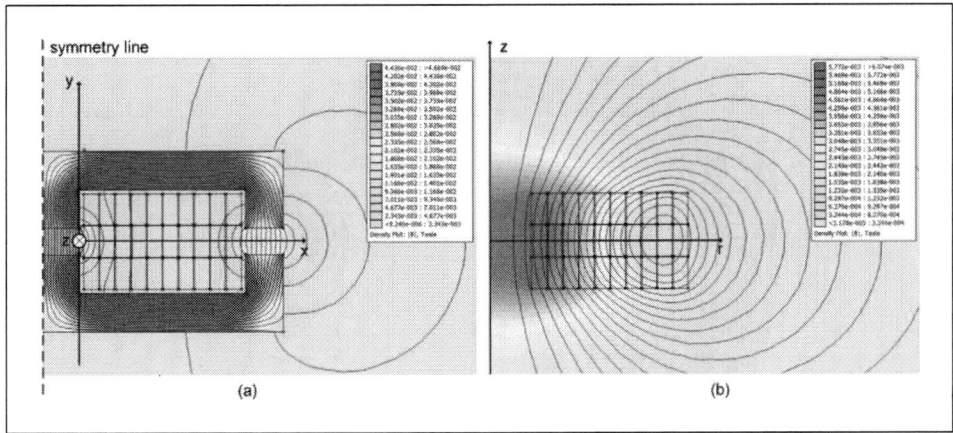

Fig. 5. FEM simulation for the examined coil for 5 A peak current with a frequency of 40 kHz for the inner winding (a) and end winding (b).

Measurements were done with an AGILENT E4980A precision LCR meter. The measured inductance is 33,85 µH. The resulting inductance of the planar and axisymmetric FEM model is 34,89 µH, the analytic calculated inductance is 35,1 µH. The error of the calculated inductance is below 5 %.

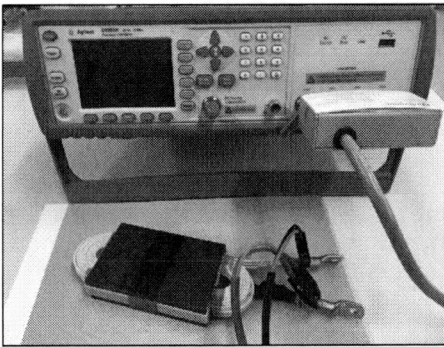

Fig. 6. Test setup

Figure 7 shows good accordance of the results for a frequency range from 10 to 200 kHz. The end winding DC- resistance of the used coil is 40 % of the total resistance, for high frequency the losses inside the coil become more dominant due to the air gap effects. The core losses are not included in the measurements due to the low measurement current of the LCR meter. Therefore the core losses calculated by (8) are excluded from the analytic results and the FEM calculation is done with ideal core material.

© VDE VERLAG GMBH · Berlin · Offenbach

PCIM Europe 2016, 10 – 12 May 2016, Nuremberg, Germany

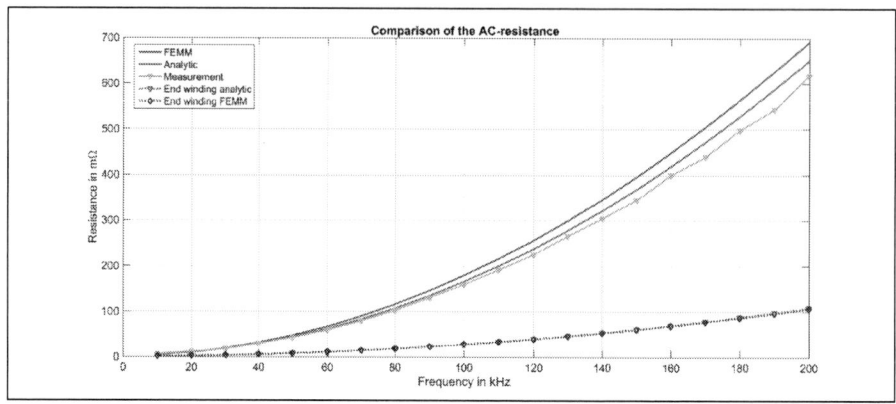

Fig. 7. Comparison of analytic and FEM calculation and measurement

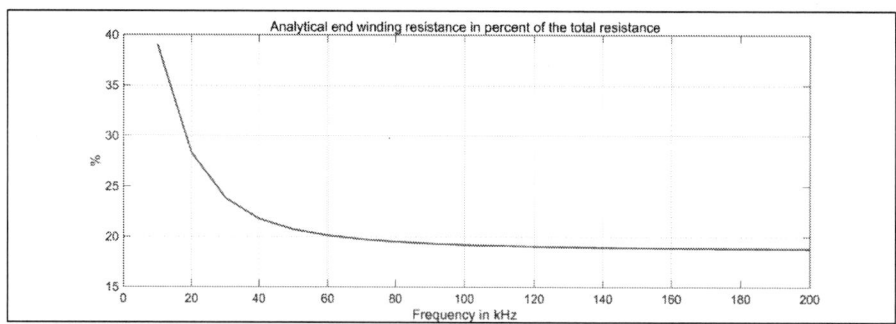

Fig. 8. End Winding Resistance in percent of the total resistance

Figure 8 shows the percentage of the end winding resistance to the total resistance for the analytic values. It can be seen that the increase of the resistance in the end windings due to proximity and skin effect is not negligible for this type of inductor.

4. Conclusion

In this paper a fast analytical approach to calculate the end winding losses in an E-core coil is given. Based on a Fourier analysis of the current shape for a DC/DC converter in continuous current operation, the DC- and AC- losses of the inductor windings are calculated analytically. The results are compared to a combined planar and axisymmetric FEM model and to measurements. The calculated AC resistance shows good accordance to FEM simulations and measurements. The results show that the losses in the end windings should not be neglected for this type of inductor, as they cause 15 to 30 percent (depending on frequency) of the total losses.

© VDE VERLAG GMBH · Berlin · Offenbach

5. References

[1]   R. Chen: Volumetric optimal design of passive integrated power electronics module (IPEM) for distributed power system (DPS) front-end DC/DC converter, IEEE Transactions on Industry Applications, Vol. 41, 2005, S. 9-17

[2]   R. Burkart: Optimal Inductor Design for 3-Phase Voltage-Source PWM Converters Considering Different Magnetic Materials and a Wide Switching Frequency Range, Proc. International Power Electronics Conference (IPEC), 2014

[3]   J. Ferreira: Improved Analytical Modeling of Conductive Losses in Magnetic Components, IEEE Transactions on Power Electronics, Vol. 9, 1994, S.127 -131

[4]   M. Albach, T. Durbaum, und A. Brockmeyer: Calculating Core Losses in Transformers for Arbitrary Magnetizing Currents – A Comparison of Different Approaches, Proc. Power Electronics Specialists Conf. (PESC), Vol. 2, 1996, S. 1463–1468.

[5]   M. Albach: Two-dimensional calculation of winding losses in transformators, Power Electronics Specialists Conference. (PESC) Vol. 3, 2000, S. 1639-1644.

[6]   C. Sullivan: Computationally Efficient Winding Loss Calculation with Multiple Windings, Arbitrary Waveforms, and Two-Dimensional or Three-Dimensional Field Geometry, IEEE Transactions on Power Electronics, Vol. 16, 2001, S.142 -150

[7]   J. Mühlentaler: Modeling and Multi-Objective Optimization of Inductive Power Components, Dissertation, ETH Zürich, 2012

[8]   S. Waffler, Hochkompakter bidirektionaler DC-DC-Wandler für Hybridfahrzeuge, Dissertation, ETH Zürich, 2013

[9]   R. Wrobel: Investigation of end-winding proximity losses in electromagnetic devices, International Conference on Electrial Machines (ICEM), 2010, S. 1-6

[10]   J. Biela, Optimierung des elektromagnetisch integrierten Serien-Parallel- Resonanzkonverters mit eingeprägtem Ausgangsstrom, Dissertation, ETH Zürich, 2005.

[11]   R. A. Schill, General Relation for the Vector Magnetic Field of a Circular Current Loop: A Closer Look, IEEE Transactions on Magnetics, Vol. 39 , 2003, S. 961-967

[12]   G. Merziger, et.al: Formeln und Hilfen zur Höheren Matahematik, Binomi Verlag, Springe S. 116

PCIM Europe 2016, 10 – 12 May 2016, Nuremberg, Germany

The Applicability of Nanocrystalline
Stacked and Block Cores for Power Electronics

Cezary Swieboda, Marian Soinski, Marcin Kwiecien, Magnetic Research Center, Magneto Ltd., Czestochowa, Odlewnikow 43, Poland, cezary.swieboda@magneto.pl

Wojciech Pluta, Czestochowa University of Technology, Czestochowa, Armii Krajowej 17, Poland

Jacek Leszczynski, Department of Hydrogen Energy, AGH University of Science and Technology, Krakow, Poland, jaclesz@gmail.com

Abstract

Core construction methods and thermal treatments allow to control and achieve predefined properties of nanocrystalline magnetic cores under consideration. Different concepts of nanocrystalline magnetic stacked and block cores are discussed in the paper. Obtained results are compared to different magnetic cores used nowadays.

1. Introduction

Nanocrystalline soft magnetic materials are more and more common not only in power electronic but also in power energy conversion applications. Fast development of power electronic and related industrial sectors affects on magnetic cores manufactures to improve their products and technologies and to develop new solutions. To make it possible the research in many areas like e.g. material engineering, thermodynamics, magnetics and electrotechnics must be conduct. Magnetic cores construction has been changing over the years being dependent on application requirements. The most widely used types of magnetic cores are: toroidal cores, toroidal cut cores, oval cores, oval cut cores, stacked and block cores with various configurations. It should be noted that current size limitation for nanocrystalline toroidal cores is around 3 kg per core [1] and for nanocrystalline block cores it is around 3.7 kg per core [2]. Existing solutions of large (bigger than 10 kg) nanocrystalline magnetic cores are made on special demands to specific applications. This fact has a negative influence on productions costs. The next disadvantage is having no feedback information about repeatability of achieved magnetic properties. The demand for large nanocrystalline magnetic cores is determined by rapid development of power electronics and associated with increase of magnetizing current frequency of power electronic devices. Magnetic cores made of nanocrystalline materials are characterized by very good features. The low losses in nanocrystalline magnetic stacked cores were confirmed by the data presented in [3]. This and no limitation per mass in proposed nanocrystalline magnetic stacked and block cores technology gives argument for further nanocrystalline core development. The aim of the paper is to discuss directories to be overcome in exothermal processes during thermomagnetic treatment of Fe-based nanocrystalline stacked and block cores.

2. Concept of nanocrystalline stacked cores (NMSC) and nanocrystalline block cores (NMBC)

Nanocrystalline magnetic stacked (NMSC) and block (NMBC) cores are dedicated mainly to power electronics where large cores masses are required and where frequencies

© VDE VERLAG GMBH · Berlin · Offenbach

of magnetizing current are in the range of 1 kHz – 30 kHz. Different construction methods of NMSC and NMBC are presented in Fig. 1.

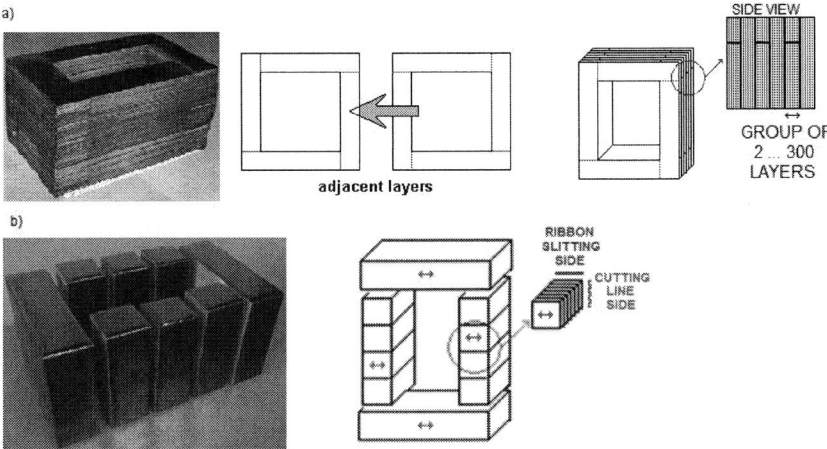

Fig. 1. Proposed nanocrystalline cores construction:
a) NMSC cores, b) NMBC cores.

Proposed technology of stacking NMSC cores and assembling NMBC cores allows to arrange the nanocrystalline rectangular cuts in line or in opposite to the casting direction of the ribbon. Cutting line (Fig. 2a) produces the nanocrystalline rectangular shapes which are formed into desired magnetic core (NMSC or NMBC) by the stacking robot system (Fig. 2b). Three-axis stacking robot system operates from outer sides with four (or more) nozzles, stacking the ready core layer by layer in the robot centre. Subsequent layers are taken from two side magazines. Each magazine has different lamination configuration in the case of NMSC cores and identical in the case of NMBC cores. Operation sequence of robot arm is programmed by the operator. Robot arm allows to grab nanocrystalline layers of the core from magazines on demanded way.

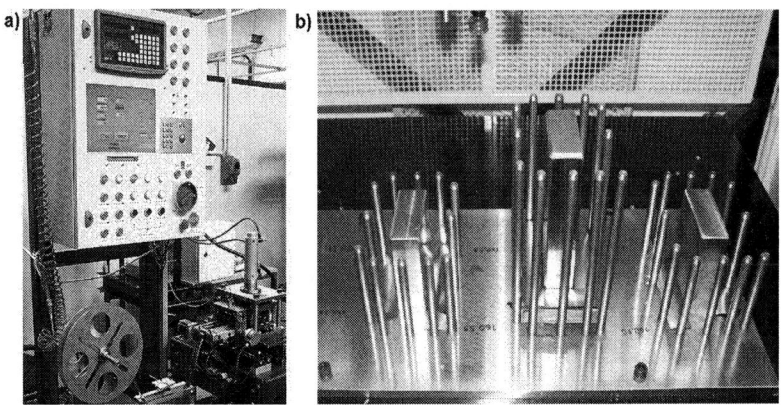

Fig. 2. NMSC and NMBC production devices:
a) cutting line for nanocrystalline tapes,
b) stacking robot system.

When the magnetic core is designed to work in higher frequencies, i.e. f ≥ 8 kHz the interlaminar insulation between every nanocrystalline layer is strongly recommended. Placing that insulation decrease the power loss level of the magnetic core. Furthermore, interlaminar insulation reduces the negative phenomenon of eddy currents which is more noticeable with the increase of magnetizing current frequency [8] and [9]. Another method of improving the uniformity of magnetic field distribution in the core is thermomagnetic T+H treatment. Proper applying of magnetic field on the core being annealed induces in it easy magnetization directions and improves magnetic properties of the core [10]. Specific devices and applications determine the use of proper magnetic core construction. For voltage transformers the most often used are the cores with close magnetic circuit (like NMSC). But when the magnetic chokes are concerned and the calibration of inductance Ls is required the block cores (like NMBC) are applied. The operating point of the magnetic core is evaluated based on magnetic properties of the core (Bs, µ, P, Ls). Ambient conditions like: temperature, presence of acids, oils and dust, vibrations, EMC compatibility are also taken into account.

3. Magnetic properties of NMSC and NMBC cores

Magnetic properties of NMSC and NMBC cores can be controlled by two ways: mechanical forming of the core or by the thermomagnetic treatment. By different mechanical construction of the cores the distributed air gap can be placed into the magnetic circuit in NMSC cores (Fig. 3).

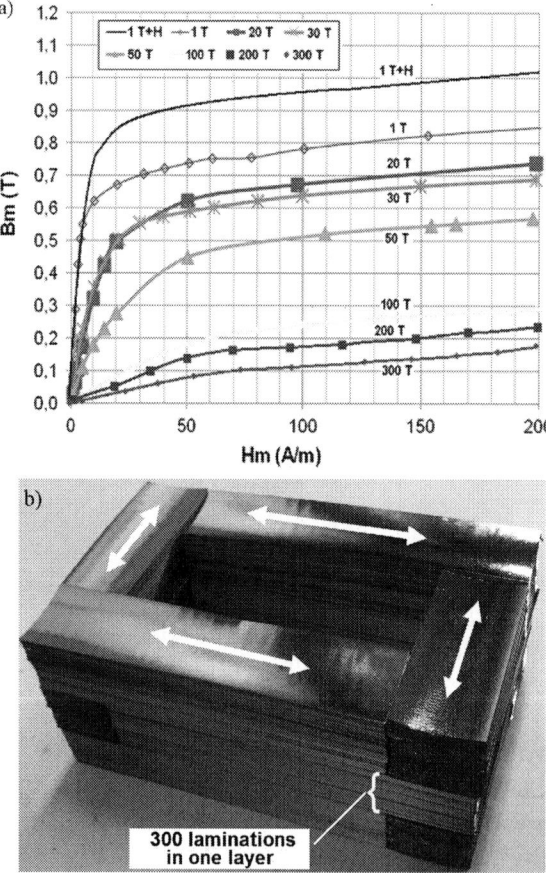

Fig. 3. NMSC cores: a) B-H curves at 50Hz with different numbers of layers: 1, 20, 30, 50, 100, 200 and 300; T+H thermomagnetic treatment, T thermal treatment, b) 6.5 kg NMSC with 300 laminations per one layer; where ↔ means direction of ribbon axis [11].

© VDE VERLAG GMBH · Berlin · Offenbach

PCIM Europe 2016, 10 – 12 May 2016, Nuremberg, Germany

NMSC core stacked layer by layer (NMSC 1) has different arrangement of the layers in every next layer. When there is a need to use the magnetic core with more linear magnetization curve, it is possible to change the number of nanocrystalline layers in identical configuration. Thus, stacking the NMSC core can be realized by groups of 2 ... 300 and more layers. On the other hand, proper annealing of the nanocrystalline material with additional magnetic field presence gives additional way to control the magnetic properties of NMSC and NMBC cores. Some basic properties of NMSC and NMBC cores are presented in the Fig. 4.

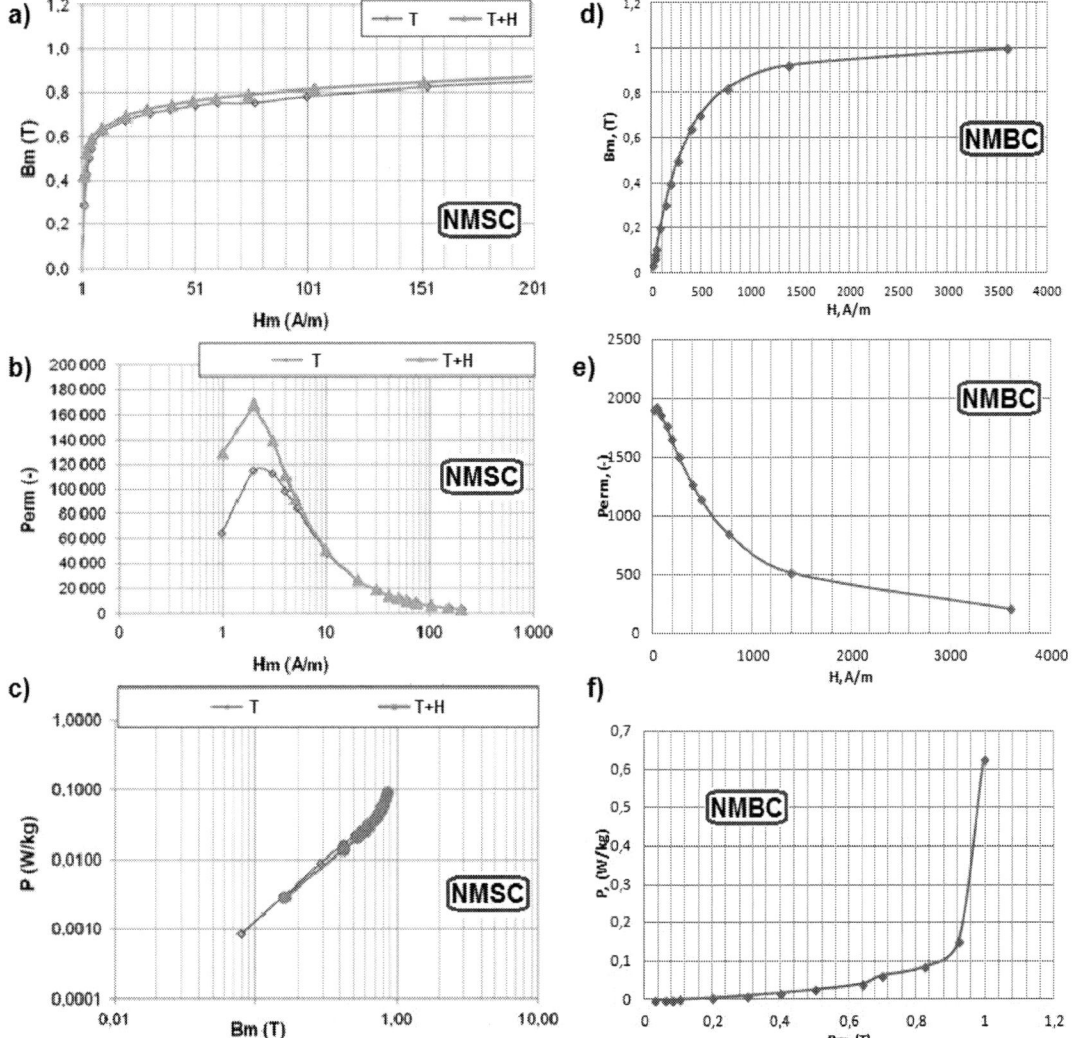

Fig. 4.Magnetic properties of example NMSC and NMBC cores, measured at f = 50 Hz:
a), b), c) 6.5 kg NMSC cores after thermal T and thermomagnetic T+H treatment
for B_m, μ and P respectively,
d), e), f) 5 kg NMBC core after thermal T treatment for B_m, μ and P respectively.

Further improvement of magnetic properties may be achieved by annealing the nanocrystalline material in the presence of magnetic field T+H, as can be seen in the Fig. 4a, 4b and 4c. Grinding and polishing the inner blocks layers of NMBC core decrease

© VDE VERLAG GMBH · Berlin · Offenbach

the air gap size and leads to magnetic properties improve. Only adjacent active layers of inner blocks need to be grinded and polished. However, such processes are long-term processes and are required only in reasonable instances. Significant advantage of nanocrystalline materials over other types of soft magnetic materials is more noticeable with increase of magnetizing current frequency. Power loss levels at f = 10 kHz of magnetic cores made of different materials are presented in Fig. 5.

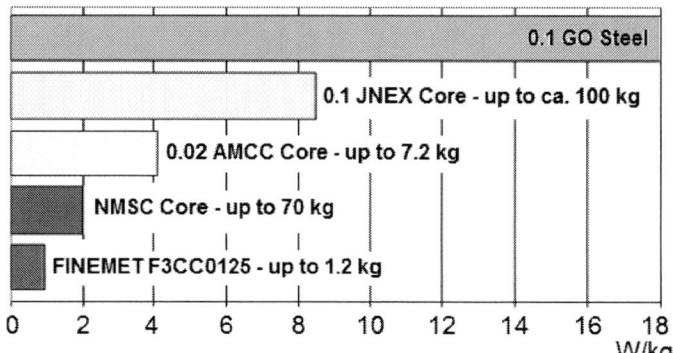

Fig. 5. Power losses of existing magnetic cores, measured at f = 10 kHz and B = 0.1 T (based on catalogue data: JFE Steel Corp. [4], Metglas [5], Hitachi [6] and Magneto Ltd. [7]).

4. SUMMARY

Proposed stacked NMSC and block NMBC cores manufacturing technology is a complex process and real validation of obtained results is possible on final product. Therefore, the very strict control of every realized stage in every detail is required. Presented in the paper results characterize basic magnetic properties of nanocrystalline stacked and block cores. The final properties of the cores can be programmed and controlled by different ways of stacking and by the directional magnetic field presence during annealing process. Proposed production technology of stacked NMSC and block NMBC cores allows to select optimal solution dedicated to specific application. Magnetic cores made of nanocrystalline material are proper solution for high efficiency applications, green transport and for the equipment such as: high frequencies chokes, high frequencies exciter, chokes in power generators, inverters (like in wind/solar power stations), power UPS systems, etc.

5. ACKNOWLEDGMENT

The paper was realized under the project: "Industrial research a new type of magnetic cores made of amorphous and nanocrystalline strips, thin magnetic sheets and composite materials operating in higher frequencies". Project grant supported by The National Centre for Research and Development under European Regional Development Fund in the frame of European Smart Growth Funds. Agreement no. POIR.01.01.01-00-0306/15-00.

REFERENCES

[1] Catalog data: Anhui Astromagnet Co.,Ltd, China, "Amorphous Division". [Available online at www.astromagnet.cn].

[2] Catalog data: Gaotune Technologies Co.,Ltd, China, „Amorphous and nanocrystalline Cores". [Available online at www.gaotune.com].

[3]   M. Soinski, J. Leszczynski, C. Swieboda, M. Kwiecien, Pinkosz P., „Measurement of magnetic features of nanocrystalline stacked and block cores", Poster on 12th International Workshop on 1&2 DM Measurement and Testing Conference in Vienna 3-6 September 2012.

[4]   Catalog data: JFE Steel Corporation, JFE Super Core (electrical steel sheets for high-frequency application). [available online at www.jfe-steel.co.jp].

[5]   Catalog data: Hitachi Powerlite® Inductor Cores, Metglas. [available online at www.metglas.com].

[6]   Hitachi Metals, Tokyo, Japan. Metglas® AMCC Series Cut Core, Power Electronics Components (Catalog). [available online at www.hitachimetals.co.jp].

[7]   Catalogue data: Nanocrystalline magnetic cores. [available online at www.magneto.pl].

[8]   Yu Zhang, Jinliang Liu, „Inter Lamina Insulation Characteristics of Iron-based Nanocrystalline Ribbons under Short Voltage Pulses", IEEE Transaction on Dielectrics and Electrical Insulation, Vol. 20, No. 5, October 2013, pp. 1755-1763.

[9]   Szynowski J., Kolano R., Kolano-Burian A., Polak M., „Reduction of Power Losses in the Tape-Wound FeNiCuNbSiB Nanocrystalline Cores Using Interlaminar Insulation", IEEE Transaction on Magnetics, Vol. 50, Issue 4, April 2014.

[10]   Liu M., Wang Z., Zhang H., "Influence of Magnetic Field Annealing on Soft Magnetic Properties for Nanocrystalline $Fe_{39.4-x}Co_{40}Nb_{2.6}Si_9B_9Cu_x$ Alloys", IEEE Transactions on Magnetics, Volume:50 , Issue: 10, 06.2014.

[11]   Soinski M., Leszczynski J., Swieboda C., Kwiecien., „Nanocrystalline Block Cores for High-Frequency Chokes", IEEE Transactions on Magnetics, vol. 50, November 2014, no. 11, part 1 of 2, article no. 2801904.

The Benefit of Formed or Compacted Litz-Wire Coils

Tobias Appel, STS Spezial-Transformatoren Stockach GmbH & Co. KG, Germany,
tobias.appel@sts-trafo.de
Hans Rossmanith, Friedrich-Alexander-Universität Erlangen-Nürnberg, Germany,
hans.rossmanith@fau

Abstract

This paper shows the analyses, the drawback and benefit of compaction of a litz-wire coil. The experiment of a temperature-rise test shows the good effect of increasing the thermal conductivity. Further measurements with LCR meter shows a drawback in an increasing radio frequency resistance. The thermal conductivity improvement increases the performance of the coil with several components in the current spectrum. But a exact analyse shows also the better behaviour at pure radio frequency current.

Introduction

Modern power electronics tends towards higher switching frequencies, as this allows greater power densities in the passive components. This trend or desire is driven further by the introduction of wide band gap semiconductors, e.g. silicon carbide MOSFET. The state of the art in the power range of a few kW allows switching frequencies of 100 kHz and higher without problems. Passive components such as inductors encounter their limits at even higher frequencies. E.g. using a GaN HEMT, this is impressively shown in [1]. In this example, the inductive component is extremely stressed in contrast to the semiconductor. There are different approaches to achieve high power densities in passive components. One way is to design the passive component to work as effectively as required. This ensures a safe handling of the generated heat. An other way is to make the thermal resistance as small as possible [4]. This technique guarantees that the heat is well dissipated, as it is the case in planar transformers. Here, the compaction of litz wires and coils comes into play.

A coil of round filaments with a orthocyclic winding has a theoretical filling ratio of $\pi/(2\sqrt{3}) = 0.907$. It could be calculated by comprehension of the area of a equilateral triangle as a elementary cell and the area of the $60\,\mathrm{deg}$ sectors of the shape of the filament. This theoretical value could only be improved by using polygonal filaments. Definitely this is not the filling ratio of copper in the coil. The real filling ratio of copper is smaller due to the needed insulation. The smaller the filaments the higher the amount of insulation.

Measurement and Approach Improving the Filling Ratio, the Thermal and Electrical Resistance

The approach of improving the filling ratio is a mechanical compacting process. This could be done with the litz wire as a semi-finished product or with a coil. Off course there is also the possibility to do twice a compaction. The compaction may be realized in various ways. One is to compress the winding in axial direction, as shown in figure 1. By this method, the filling ratio can be adjusted. High pressures lead to an almost optimal filling ratio.

The issue of measuring the filling ratio is quite not clearly. If the focus is on a ratio of the cross-section area of pure copper and needed space for the winding is the filling ratio smaller than with an additional consideration of the insulation. Both filling ratios are calculated and shown in table 1.

Figure 1   : Inductor with non-compacted (left) and compacted Coil (right)

The measurement of the electrical resistance could be easily done with a automatic LCR meter. The coils where measured with different frequencies from DC to 350 kHz.

The evaluation of the temperature-rise test provides the result of the thermal resistance. The conditions of the temperature-rise test are a constant cooling with an air flow of 3 m/s and a constant DC current of 12 A. After a time span of 2 hours is the hot spot temperature of the tested coils stably. The thermal resistance from the hot spot to the ambient could be easily calculated by respect to the temperature coefficient of copper.

Results of Improving the Filling Ratio, the Thermal and Electrical Resistance

A result which is visible to the naked eye in figure 1 is the improve of the filling ratio. In this experiment two identical coils of round litz-wire were used. The coil on the left-hand side in figure 1 is in original condition an on the right is shown a afterwards axial compacted coil. The measured and calculated filling ratios are shown in table 1.
The coil is made with a litz-wire of 245 strains with a diameter of 0.1 mm. The insulation of the strains is according to DIN EN 60317-0-1 with a thickness of $5 \mu m$. Between the layers is a small sheet of insulation. The filling ratio of the litz-wire with round shape and with rectangular shape is also given in table 1.
There are also the practical measured filling ratios of coils with filaments. They can be used to predict the filling ratio of the litz-wire coil.
Non-profiled litz wires usually have filling ratios of 0.5 (copper)or 0.75 (total), as shown in tabel 1. In windings produced by a litz wire, the filling ratio decreases even more, as more dead space accumulates in form of the gussets between the round litz wires. This can be calculated by multiplying the filling ratios of the coil and of the litz wire. The general disadvantage of low filling ratios with litz wires is improved by profiling it. Profiled litz wires are much more compact and have filling ratios of above 0.64 (copper).
In a coil made of rectangular profiled litz wire, there is hardly any additional dead space and the filling ratio does not deteriorate further. Windings with less air space may also be achieved by subsequent compacting of the entire coil made with non profiled litz wire. In this case the filling ratio can not be predicted but it can be adjusted by the preasure.

© VDE VERLAG GMBH · Berlin · Offenbach

Table 1 : Calculation of the filling ratio

	A/A_{Cu}	$A/A_{Cu+Lacquer+Isolation}$
Non profiled litz-wire	0.525	0.659
Profiled litz-wire	0.640	0.802
Coil with solid round filament		0.756
Coil with solid square filament		0.828
Non compacted coil	0.378	0.511
Coil with profiled litz-wire	0.498	0.674
Compacted coil	0.472	0.638

Table 2 : Experimental results

	R_{th} / KW^{-1}	$R_{ac}(100kHz) / \Omega$
Non compacted	6.7	1.56
Compacted	4.3	1.86

Increasing the filling ratio decreases the thermal resistance of the winding, due to the reduction of air space and gaps between good thermal conducting material. This is reflected in the results of a temperature-rise test shown in table 2. In non-compacted windings, the thermal resistance is much higher.

Theory of the RF Resistance Depending on the Filling Ratio

The radio frequency or alternating current resistance of a litz wire is a increase of the direct current resistance by a combination of the skin and proximity effect. The proximity effect it self could be divided into an external and internal proximity effect.

The filling ratio of the coils affects the internal proximity losses of litz wires, as can be seen in equations 1-3 [2]. Also, the entire winding receives worse losses by a higher filling ratio because of the effect to the external proximity effect. This is shown in [3]. In addition, more wires ar presses closer to the air gap and are exposed to a larger external magnetic field by axial compression of the coil. In the experiment, this results in a higher RF resistance as shown in table 2. The increase in RF resistance by compacion reaches here 19.2%, despite identical structure of both coils prior to compaction.

$$J_0 = \frac{k_L \hat{I}}{A_{Cu}} = \frac{k_L 4 \hat{I}}{N_s \pi d_s^2} \tag{1}$$

$$H_i = \frac{\pi r^2 J_0}{2 \pi r} \tag{2}$$

$$P_{prox} = G_R(f) \hat{H}^2 \tag{3}$$

© VDE VERLAG GMBH · Berlin · Offenbach

Linking of AC Resistance and Thermal Resistance

The radio frequency resistance is additionally to the filling ratio influenced by the frequency. This could be seen in equation 3. The factor $G_R(f)$ could be calculated with the skin depth and some modified Bessel functions. If the frequency rises also the resistance rises e.g. with a power function. In this study the resistance of the coils where measured.

Compacting the coil rises not only the RF resistance factor but also the thermal conductivity. The thermal conductivity or the thermal resistance is a constant factor like in table 2 described. This means that on one side more ohmic losses arise, but on the other side more heat can be dissipated. A calculation (eq. 4 can be done to qualify the upcoming heat. In this calculation the RF resistance for example at switching frequency is multiplied with the squared current to calculate the power. The power can be multiplied with the thermal resistance to calculate the temperature rise.

In switched mode power supplies, inductors are used as filters for the switching frequency. In this case, the current spectrum has a DC or a 50 Hz and some high components. These spectral components could be summed up.

Out of this equation could be extracted a figure of merit. It is displayed in equation 5.

$$\Delta T = R_{th} \sum R_{ac}(f)I(f)^2 \tag{4}$$
$$FOM = R_{th}R_{ac} \tag{5}$$

The gain of performance depends strongly on the actual value of the FOMs. For a evaluation of the two build coils the diagram in figure 2 is made. There could be seen the run of the curves of both FOMs and the difference of Δ between them. The Δ-curve shows clearly the range of the advantage of the compacted coil. Despite of an increase of resistance, an inductor operating at high frequency would remain cooler.

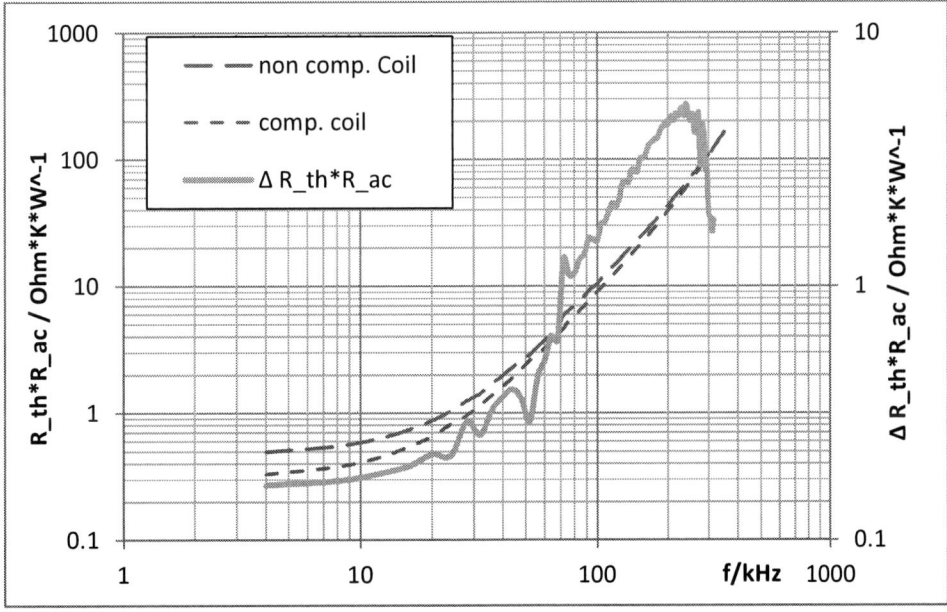

Figure 2   : Diagram of the FOM (eq. 5)of both build coils and its difference

Summary

Compacting coils produces more powerful inductors. First of all, the efficiency of the inductor decreases because the radio frequency resistance increases. But then, also the thermal conductivity of the winding increases, up to a factor of two. This increase in thermal conductivity more than compensates for the increase of radio frequency resistance, leading to a cooler device. Further studies involve the thermal effects on the ohmic resistance of the winding and the core losses. This effects could turn out that compaction could not only lead to more powerful inductors but also to more efficient ones. Depending on the temperature of the device and of the core are more or less iron losses.

Appendix

$A:\ cross-section\ area$

$A_{Cu}:\ cross-section\ area\ of\ copper$

$A_{...+Lacquer+Isolation}:\ cross-section\ area\ of\ insulation$

$R_{th}:\ thermalresitace$

$R_{ac}:\ electrical\ radio\ frequency\ resitance$

$H_i:\ magnetic\ field\ inside\ the\ litz\ wire$

$J_0:\ current\ density\ of\ the\ litz\ wire$

$r:\ radial\ distance\ to\ litz\ wire\ center$

$P_{P,L,int}:\ internal\ proximity\ losses\ in\ litz\ wires$

$N_S:\ number\ of\ strains\ in\ litz\ wire$

$R_DC,E:\ DC-Resistance\ of\ the\ litz\ wire$

$G_R,E:\ AC\ resistance\ factor$

$G_R(f) = \frac{\zeta\pi^2 d^2}{2\sqrt{2}} \frac{ber_2(\zeta)ber_0'(\zeta)+bei_2(\zeta)bei_0'(\zeta)}{ber_0^2(\zeta)bei_0^2(\zeta)}$

$bei, ber:\ modified\ Bessel\ functions$

$\zeta = \frac{d}{\sqrt{2}\delta}$

$\delta = \sqrt{\frac{1}{\pi\kappa\mu f}}$

$\hat{I}:\ current\ amplitude\ in\ litz\ wire$

$d_a:\ diameter\ of\ litz\ wire$

$d_s:\ diameter\ of\ strain$

References

[1] Wienhausen, Arne Hendrik and Kranzer, Dirk, "' 1 MHz Resonant DC/DC-Converter Using 600 V Gallium Nitride (GaN) Power Transistors '", Materials Science Forum 2013 Vol. 740 p.1123–1127

[2] Biela, J, "'Wirbelstromverluste in Wicklungen induktiver Bauelemente'", page 49, 2011

[3] Stadler, Alexander and Huber, Rainer and Stolzke, Tobias and Gulden, Christof, "'Analytical calculation of copper losses in litz-wire windings of gapped inductors'", Magnetics, IEEE Transactions on, volume 50 number 2, page 81–84, 2014

[4] Stadler, Alexander and Stolzke, Tobias and Gulden, Christof, "'Design and simulation of high power filter inductors with minimized thermal resistance'", PCIM Europe 2014; International Exhibition and Conference for Power Electronics, Intelligent Motion, Renewable Energy and Energy Management; Proceedings of, pages 1–7, 2014

PCIM Europe 2016, 10 – 12 May 2016, Nuremberg, Germany

Development of a 100kW, 20 kHz Nanocrystalline Core Transformer for DC / DC Converter Applications

Kapila Warnakulasuriya, Carroll & Meynell Transformers Ltd, UK,kapila@carroll-meynell.com
Farhad Nabhani, Teesside University, United Kingdom, F.Nabhani@tees.ac.uk
Vahid Askari, Teesside University, United Kingdom, v.askari@tees.ac.uk

Abstract

Achievement of high power levels in a single module inverter/converter application always has number of advantages. However, high frequency magnetic components most often become the limiting factor in the achievement of such high power levels in a single module. In this paper the achievement of 100kW in 20kHz transformer is discussed. This development is an extension of a series of developments carried out with high frequency, high power transformers of power and frequency levels such as 40kW, 20kHz and 50kW, 50kHz. The advantages and disadvantages of Ferrite and Nanocrystalline core materials at this power levels are discussed and the choice of Nanocrystalline core is explained. Further naval winding technique that mitigates high frequency conductor losses is presented in this paper with a theoretical explanation on its advantages.

1. Introduction

In order to achieve a high degree of miniaturization, designers move into significantly high operating (switching) frequencies and high power levels in the development of modern high power converters and inverters. There is also a clear trend that the requirement of achieving high power levels in a single module will increase alongside the operating frequencies used in these applications.

Applications like traction, ship and basically any mobile platform with a converter on board require light weight and compact converters to exploit the space available on board more effectively. They often require galvanic isolation for safety and other reasons. Therefore high power high frequency transformers which offer galvanic isolation and a small volume are of increased importance [1]. In addition to the galvanic isolation magnetic components including transformers and inductors perform functions of harmonics filtering, energy storage and parameter matching for power stages as well as control circuitries in a power converter. They often determine the converter size [2], [3]. It has been a long held view that with a continuous increase in operating and/or switching frequency a continuous decrease in physical size of magnetics would follow. However the heat removal surface of the magnetic components decreases as a result of the higher density design; on the other hand, core and winding loss densities increase correspondingly. Therefore attention needs to be paid to magnetic material selection and associated core loss calculations, especially for high frequency high density magnetics and power converter design [2].

In the initial stages of this research authors presented the advantages and disadvantages of Ferrite and Nanocrystalline materials as a core material for high frequency high power transformers. Further mitigation techniques of conductor loses at high power high frequencies with novel winding techniques were presented based on developments such as 40kW,20kHz and 50kW , 50kHz transformers.

Going forward with the findings of these studies an approach was made to achieve 100kW single module transformer that operates at 20kHz, which is used in a DC /DC converter application.

© VDE VERLAG GMBH · Berlin · Offenbach

In the selection of suitable core material for this task several Ferrite grades and Nanocrystalline materials were evaluated with modern approaches of estimating core losses in each type under non sinusoidal excitations of high frequencies. The advantages of Nanocrystalline at 20kHz applications reduce to some extent at 50kHz. The challenges of mitigating high frequency conductor losses at these high current levels are explained. A novel winding technique that mitigates the high frequency losses and constructional challenges of this technique are explained.

Two design options are presented with a Nanocrystalline core for natural convection cooling applications and water plate mounting applications. Housings with specially engineered exterior are used for the natural convection cooling applications and one with specially engineered interior is used in water cooled mounting application.

2. Transformer Design

2.1. Transformer Construction

The transformer is designed to handle a power of 100kW at a square wave excitation of 20kHz. The transformer was designed with two parallel operating secondary windings each handling a power of 50kW.

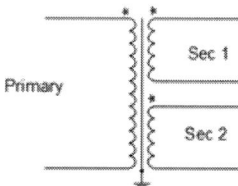

Fig. 1. Schematic of the 100kHz, 20kW transformer

2.1.1. Transformer core selection and turns calculation

The advantage of having a high saturation flux density of Nanocrystalline material was considered. Based on the material properties of Ferrite and Nanocrystalline materials optimal operating flux densities for each material were decided based on the operating frequency of 20 kHz. The Faraday's law with appropriate constants for rectangular wave excitation is used to decide on the primary turns.

2.1.2. Transformer core loss estimation

An improved calculation method of core losses for non-sinusoidal waveforms based on recent publications [7], [8] and [2] was carried out. Modified and generalised versions of Steinmetz equations and a simplified graphical method discussed in these publications were used in loss estimations.

Simplified equation for core loss estimation using Steinmetz parameters of the material, based on [7].

$$P_v = k_i (\Delta B)^{\beta - \alpha} / T \sum_j \left| V_j / (N A_c) \right|^{\alpha} (\Delta t_j)$$

Where,

$$k_i = k / \{2^{\beta + 1} \pi^{\alpha - 1} [0.2761 + 1.7061 / (\alpha + 1.354)]\}$$

N is the number of turns, A_c is the core cross-sectional area, ΔB is the peak to peak flux of the loop under consideration.

2.1.3. Conductor selection and winding loss mitigation.

The option of Litz wires and foil conductors were considered. The application under consideration handles a power of 100kW. The suitability of Litz wires including rectangular cross-section braided Litz wires were considered. However the initial calculations carried out including poor filling factor of Litz wires showed that the foil windings are more suitable for this application. The increased conductor losses at high frequencies as explained in previous publications of the authors were considered and the significance such increased losses at the high power level of interest were considered.

Two approaches to the windings that minimize the high frequency effect on conductor losses were considered. These approaches to transformer windings make the influence of proximity effect has on conductor losses frequency independent. A mathematical explanation on this concept was presented by authors in their previous publications. Sandwich primary and secondary windings id one of these approaches. The second approach is a specially prepared foil arrangement as described in the diagram below. This second approach also provides the flexibility of adjusting the leakage inductance between primary and secondary windings as required by the rest of the parameters in the converter circuit. The mathematical explanation of the controllability of leakage inductance and the mitigation of conductor losses are beyond the scope of this paper and will be presented in a future publication.

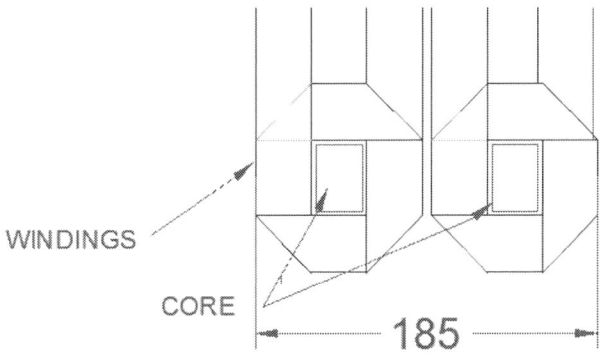

Fig. 2. Novel winding technique that mitigates the conductor losses

3. Effective heat removal of high density magnetics

3.1. Construction for the natural convection cooling

Even though a high level of minimization of losses achieved with the appropriate operating flux levels and winding techniques in this 100kW transformer the high level of miniaturization archived results in considerably high power density. This makes it challenging to maintain the temperature rise under convection conditions.

The surface area of the product or the enclosure is not sufficient for the effective removal of heat under natural convection.

PCIM Europe 2016, 10 – 12 May 2016, Nuremberg, Germany

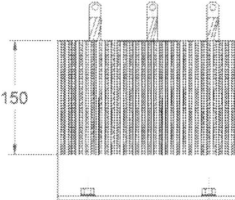 

Fig. 3. Housing construction for the natural convection cooling

Therefor an enclosure made out of heatsinks as shown in the fig.3 considered as a method of heat removal.

3.2. Construction for the water cooled plate mounting

Alternative approach considered for the effective heat removal is the use of water cooling arrangement of the end product for the cooling of magnetic components. Under this the high frequency transformer will be mounted on water cooled plate as shown in the diagram below.

Fig.4. Transformer mounted in water cooled plate

In order to effectively remove the heat generated in the inner parts of the core heat pipe and heat sheet options were considered. However the fact that the heat has to be transferred towards the bottom direction and the short operation life of heat sheets such as 16 years in continuous operation made the use of them non suitable for the application. Therefor an enclosure with specially engineered interior was used.

4. Prototype development and testing

After taking the above discussed factors into consideration a 100kW,20kHz transformer was constructed. A winding arrangement as discussed in 2.1.3 was also constructed. However, it was realised that time required for the new winding arrangement was approximately 30% higher than the sandwiched winding arrangement. The new winding arrangement provides the advantage of controlling the leakage inductance between the primary and the secondary

© VDE VERLAG GMBH · Berlin · Offenbach

winding. Also when duly constructed it provides a slightly better packing factor and improved performance by minimizing proximity effect. However due to the extra production time needed it was identified more research is necessary in production technique point of view for the winding arrangement to be commercially attractive.

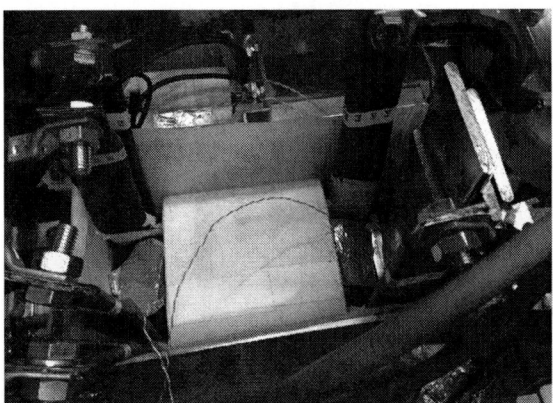

Fig. 5. 100kW, 20Hz HF transformer

A prototype with the secondly mentioned sandwich winding was also contracted as shown in the above diagram. This was tested on a 100kW DC/DC converter. As the initial step the transformer was tested was tested inside a standard housing without any interior modifications. A power of approximately 100kW was archived. Under these conditions as it can be seen in the Fig. 6 below transformer temperature reached of 80 degrees in less than 20minitues time.

Temperature °C

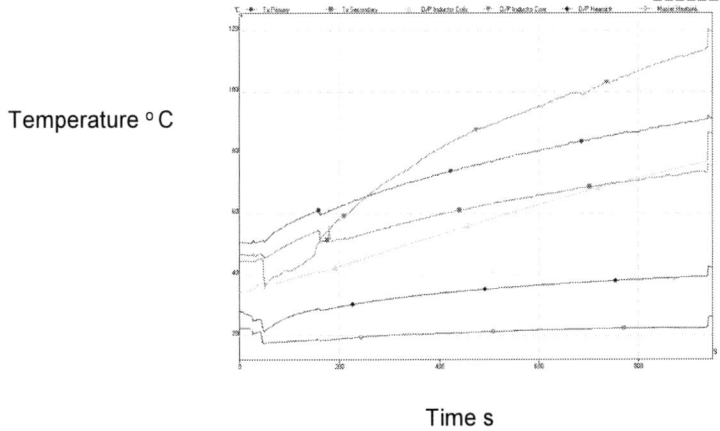

Time s

Fig.6. Temperature rise with a standard housing

Based on these results it was clear that in order to achieve the power of 100kW in the given volume a modified interior housing has to be used. After a further investigation on the thermal model of the arrangement the interior construction of the transformer housing was modified to effective transfer heat from the hotspots of the transformer towards the water cooled plate. After these improvements to the housing construction it could be observed that more than hour of operation at full power of 100kW was necessary for the transformer temperature to reach 80 Degrees. Further it could be seen that the temperature rise gradient was reducing significantly at this time.

© VDE VERLAG GMBH · Berlin · Offenbach

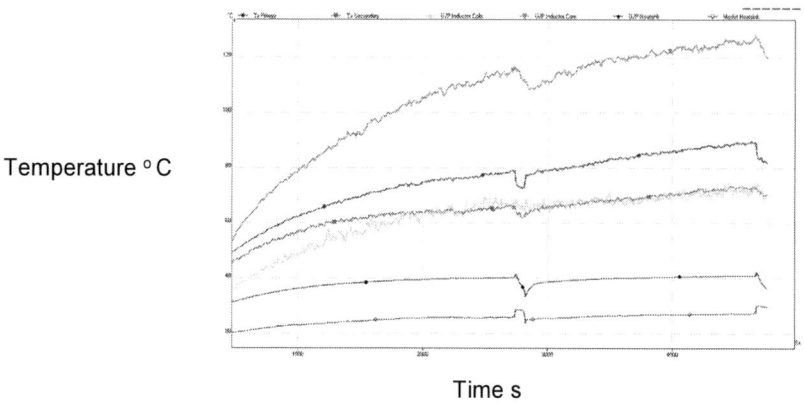

Temperature °C

Time s

Fig.7. Temperature rise with modified interior housing

A father improvement of the situation and achievement of thermal stability below 90 °C could be observed when the modified interior housing was potted.

5. Discussion and Conclusions

Rapid developments in the semiconductor technology and the power electronics industry enable the development of power supplies and power converters of significantly higher power levels. Considerably low switching losses of developments like SiC MOSFETs make it possible to operate at 2 to 5 times higher switching frequencies compared to their conventional counter parts. As the frequency goes high the magnetic components in such system become more and more compact. The magnetics components are usually the bulkiest components in a power electronics system and in most of the situations also the most expensive component, therefor it is of high interest of many designers to develop systems that operate at high frequencies.

The achievement of a high power levels in a single module at high frequencies is next challenge faced by the developers of systems for high power applications. Most often the maximum power achievable with a single module is determined by the magnetic components. As a result it could be seen that constraints in the development of magnetic components make a significant influence to the miniaturization of the power electronics system. Further it is reasonably accurate to state the magnetics components have a great impact on determining the power level and the operating frequencies of power electronics applications

The research discussed in the paper was carried out with the intention of achieving compact, high power, high frequency transformers that are capable of handling power levels in excess of 100kW.

These studies and a serious of the previous developments carried out by the authors have identified the advantages of using the materials such as Nanocrystalline for high power applications and for operating frequency levels in the range of 20kHz to 30kHz. With the appropriate thermal management approaches and designing at optimal operating flux levels Nanocrystalline core gives a considerable volume and weight advantage over the ferrites.

At high power levels the winding design becomes very important for the mitigation of high frequency losses. Two winding constructions are discussed in this paper. The novel method proposed with folded foils showed very good performance in mitigating the high frequency conductor losses. Further it gives the possibility of controlling the leakage inductance between the primary and secondary windings by adjusting the level of overlap. The disadvantage observed in this method was the time consumed for the proper insulation

© VDE VERLAG GMBH · Berlin · Offenbach

process. Through several insulating techniques were considered the amount of time savings archived was not sufficient to adopt this method in mass production. Therefor further research is necessary to improve the production process for the successful implementation of this technique.

The second method proposed with sandwich winding could be seen as a good compromise on the performance and the amount of labour consumption.

The tests carried out with a standard enclosure showed that the amount of heat transferred to the cooled plate was not sufficient to maintain the host spot temperature below the desirable level. However the desirable conditions could be archived with an enclosure with correctly engineered interior.

It could be could be concluded that with the appropriate material selection, winding techniques and with the thermal management techniques the construction of a 100kW high frequency transformer could be archived. The product showed very stable performance and presented a very compact construction giving a volume to power density of over 18W/cm^2 or weight to density of over 6.5kW/kg. This was a considerable reduction in the size and volume against the conventional ferrite construction considered.

Further tests studies carried out showed the possibility of achieving even higher powers such as 150kW or even 200kW in a single high frequency transformer operating at a frequency range of 20kHz to 30kHz with a compact construction is possible in a similar compact structure. In parallel with the industry demand authors intend to continue the research towards these power levels as part of future work.

References

[1] M.Pavlovsky, S.W.H. de Haan, J.A.Ferreira."Partial Interleaving: A method to Reduce High frequency Lossessand to Tune the Leakage Inductance in High Current , High Frequency Transformer Foil Windings" Power Electronics Specialists Conference, 2005. PESC '05. IEEE 36th

[2] Wei Shen, Fei (Fred) Wang."Loss Charaterization and Calculationof Nanocrystalline Cores for High- Frequency Magnetics Applications" IEEE Transactions on Powerelectronics.Vol 23.No.1, January 2008.

[3] A.W. Lotfi and M.A. Wilkowski, "Issues and Advances in High Frequency Magnetics for Switching Power Supplies, " Proc.IEEE,vol.89,no 6,pp.833-845, Jun 2001.

[4] Jeniffier D. Pollock, Tarek Abdallah and Charles R. Sullian. "Easy to use CAD tools for Litz wire winding optimization", Applied Power Electronics Conference and Exposition, 2003. APEC '03. Eighteenth Annual IEEE, 9-13 Feb. 2003 Page(s): 1157 - 1163 vol.2

[5] Linden W. Pierce. "Transformer design and appliocation consuiderations for non sinusodial load currents", Industry Applications, IEEE Transactions on (Volume:32 , Issue: 3) Page(s): 633 – 645

[6] Lloyd H. Dixon. "Eddy Current Losses in Transformer Windings and Circuit Wiring", Citations 1. Electronics Engineer's Handbook, McGraw-Hill- Fink – 1975, 2. Winding Eddy Current Losses in Switch Mode Power Transformers Due to Rectangular Wave Currents – P S Venkatramen – 1984

[7] K. Venkatachalam, C.R. Sullivan, T.Abdallah and H.Tacca "Accurate Prediction of Ferrire Core Loss with Nonsinusoidal Waveforms Using Only Steinmetz Parameters", IEEE Workshop on Computer in Power Electronics, June 2002,pp.

[8] Edward Herbert, Canton CT. User-Friendly Data for Magnetic Core Loss Calculations, November ,2008

[9] Zoya Popovic, Branko D. Popovic. "Introductory Electromagnatics", The Skin Effecit 382-392

[10] Irma Villar. "Multiphysical Characterization of Medium Frequency Power Electronic Transformers", PhD thesis EPFL-Lausanne-Switzerland 2010.

[11]A R Abdul Razak, STaib, I Daut. "Design and Development of High Frequency Highpower Transformerfor Renewable Energy Application", International Conference on Robotics, Vision, Information and Signal Processin. ROVISP2005.

[12] W. Shen, "Design of High density transformers for high frequency high power converters", PhD thesis. 2006 July. Blacksburg, Virginia, Polytechnic institute and State University

[13] Muhammed, Adil Hussein. "High frequency transformer design and modeling using finite element technique", PhD thesis, 2000, Faculty of Engineering. Newcastle University, UK.

[14] Livo Susnjic,Zijad Haznadar and Zvonimir. "3D Finite Element determination of stray losses in power transformers", Elsevier, Electric Power System Research 78(2008) 1814-1818.

[15] A.Stadler, R.Huber, T. Stolzke and C.Gulden. "Analytical Calculation of Copper Lossess in Litz-wire Windings of Gapped inductors",STS Spezial Transformers Stockach GmbH & Co. KG.

[16] Alex Van Den Bossche, Vencislav Cekov Valchev and Georgi Bogomilov Georgiev, "Measurement and Loss Model of Ferrites with Non – sinusoidal Waveforms",35th Annual IEEE Power Electronics Specialists Conference. Aachen , Germany, 2004.

[17] Tommy Kjellqvist,Staffan Norrga,Stefan Ostlund. "Design Considerations for a Medium Frequency Transformer in a Line Side Power Conversion System" 35th Annual IEEE Power Electronics Specialists Conference. Aachen , Germany, 2004.

[18] Anne Berit Mogstad,Marta Molinas,Paal Keim Olsen,Robert Nilsen. "A Power Conversion System for Offshore Wind Parks".2008 IEEE 978-1-4244-1766-7/08.

[19] K.W.E.Cheng. "Computation of theca resistance of Multistranded Conductor Inductors with Multilayers for High Frequency Switching Converters", IEEE Transactions on Magnetics, Vol. 36, No, July 2000.

[20] Ruifang Liu, Chris Chunting Mi,David Wenzhong Gao. "Modeling of Iron Losses of Electrical Machines and Transformers Fed by PWM Inverters". IEEE 2007,1-4244-1298-6/07.

[21] C.R.Sullivan. "Optimal Choice for Number of Strands in a Litz Wire Transformer Winding" IEEE Transactions on Power Electronics, vol. 14, no.2, pp.283-291.

[22] A.F.Picanco, C de Salles, M.L.B.Martinez, P.C. Rosa,H.R.P.M. de Oliveira. "Development of Economical Analysis and Technical Solutions for Efficient Distribution Transformers" Federal University of Itajuba.

[23] G.Ortiz, J.Biela, J.W.Kolar. "Optimized Design of Medium Frequency Transformers with High isolation Requirements" Power Electronics Systems Laboratory, ETH Zurich.

[24] J.A.Ferreira, "Analitical computation of AC resistance of round and rectangular litz wire windings" IEE Proceedings B Electric Power Applications, vol. 139, no.1, pp.21-25,1992

[25] William Gerard Hurley, Eugene Gath, John G. Breslin. "Optimizing the AC Resistance of Multilayer Transformer Windings with Arbitrary Current Waveforms" IEEE Transactions on Power Electronics, vol. 15, no.2, March 2000.

Simulation of a 3-phase common- and differential mode inductor on a four-limb core

Michael Owzareck, BLOCK Transformatoren-Elektronik GmbH, Verden, Germany
Nejila Parspour, IEW University of Stuttgart, Stuttgart, Germany

Abstract

An optimized method to simulate the common- mode and differential mode inductance and the core loss of a four limb inductor is presented here.

Ordinary sinusoidal filters, which have a common- and differential mode filter effects, consist of two inductors (Figure 1). In the case of a four-limb core, the two inductors can be accommodated on a common core. The fourth limb is required to lead the common mode magnetic flux. With only three limbs and a phase balance of the common-mode currents, the magnetic flux would be zero at the magnetic neutral point [1].

Due the non-linearity of the Hysteresis it is not possible to calculate the core losses by using the superposition method. A transient model, which is based on the principal of loss separation, is used, to calculate the core losses of the four-limb inductor. The core losses of electrical steel are measured with a special measurement system, which can simulate typical waveforms for inductors in power electronic applications. Finally, measurement and simulation results are compared.

Introduction

At the development of filter inductors the thermal management must be taken into account. This depends on customer required specifications. Consequently, the inductance and the ohmic losses must be simulated. The winding losses can easily be simulated with the FEM method also for non-sinusoidal magnetizing currents due to their linear superposition and orthogonal loss components, though, it is quite difficult to simulate core losses for non-sinusoidal magnetization currents.

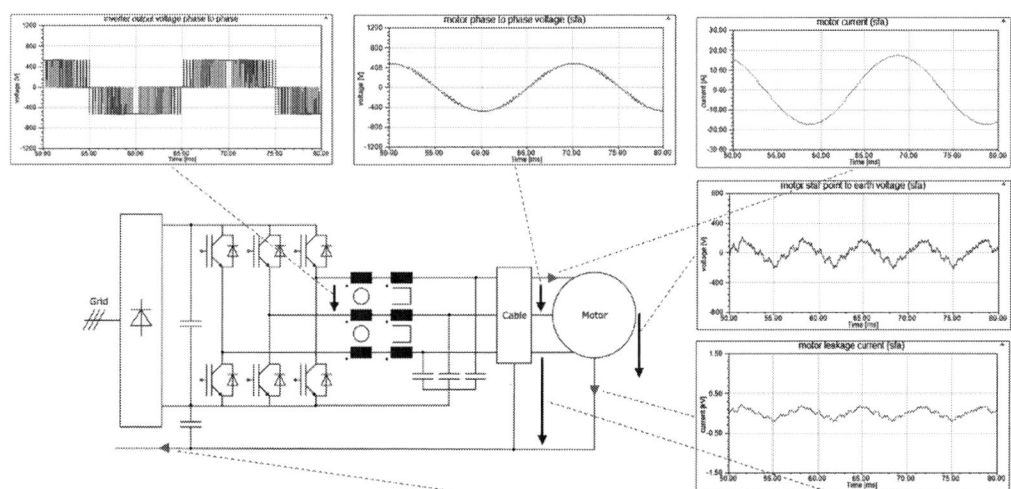

Figure 1: Typical drive system with combined CM and DM inductor

PCIM Europe 2016, 10 – 12 May 2016, Nuremberg, Germany

Modeling of a four-limb core inductor

Each magnetic simulation is based on material properties of the soft magnetic core material. Manufacturer data of the soft magnetic core material are not sufficient to simulate the real application of magnetic components. To optimize the inductance simulation of the four limb inductor, the specific properties of grain orientated steel are measured with the measurement system of [3, 4]. In this way the anisotropic specific material properties and the influence of the waveform of the magnetization current can be considered (Figure 2).

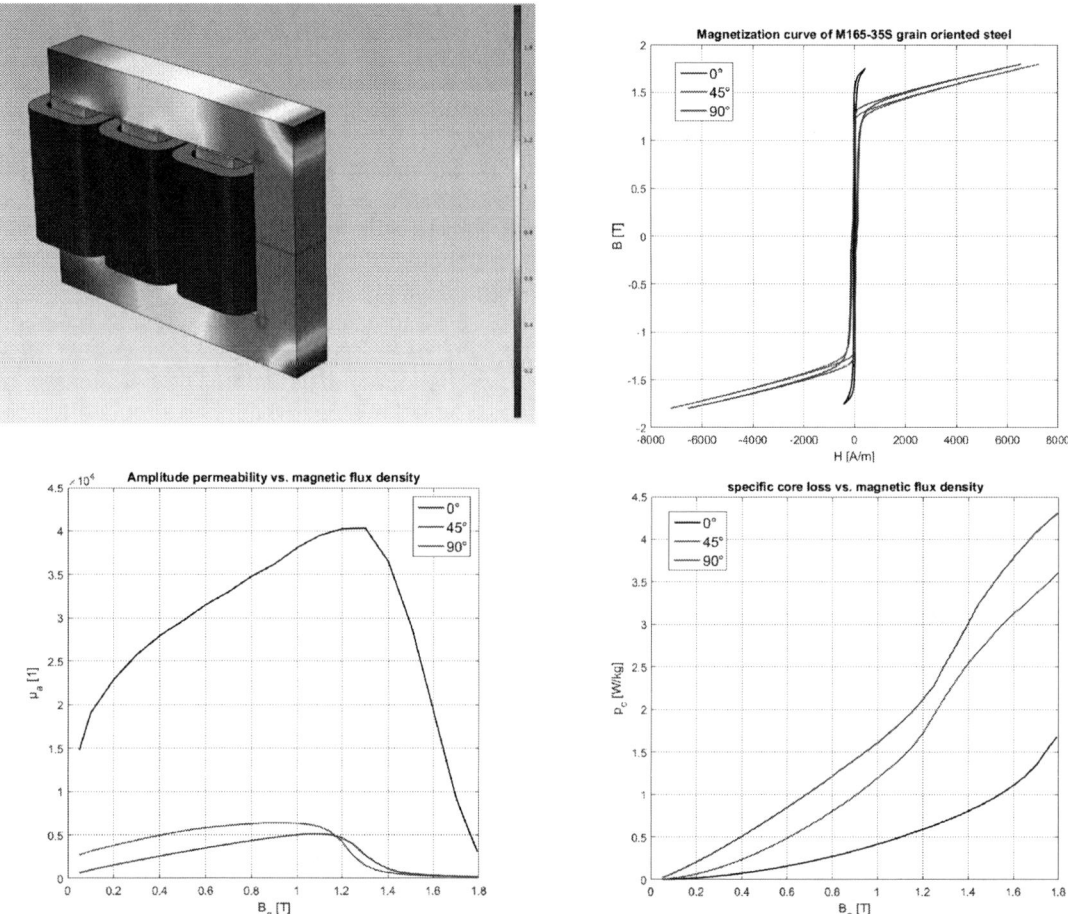

Figure 2: 3D Model of the four limb core differential and common – mode inductor and the specific parameters of the grain oriented electric steel core (For example 50 Hz)

The relative permeability must be described for grain oriented electrical steel by a tensor:

$$\vec{B} = \begin{bmatrix} \mu_x & 0 & 0 \\ 0 & \mu_y & 0 \\ 0 & 0 & \mu_z \end{bmatrix} \vec{H} \tag{1}$$

© VDE VERLAG GMBH · Berlin · Offenbach

In Figure 2 (bottom left) the relative amplitude permeability is shown as a function of magnetic field strength H and as a function of the rolling direction for angles of 0 °, 45 ° and 90 °. (For example 50 Hz).

$$\mu_{a,0°} = \frac{1}{\mu_0} \cdot \frac{\hat{B}_{0°}}{\hat{H}_{0°}} \tag{2}$$

$$\mu_{45°} = \frac{1}{\mu_0} \cdot \frac{\hat{B}_{45°}}{\hat{H}_{45°}} \tag{3}$$

$$\mu_{90°} = \frac{1}{\mu_0} \cdot \frac{\hat{B}_{90°}}{\hat{H}_{90°}} \tag{4}$$

In order to obtain an optimized magnetic simulation of common mode and differential mode inductance as a function of the magnetizing current, these specific material properties must be available. With the soft magnetic measurement system described in [3,4] the normal magnetization curve can be measured under real power electronic magnetization currents. This optimizes the simulation result additionally in relation to the real application of the filter inductor.

FEM program have inability to simulate core losses, only the classic eddy current loss can directly be calculated with FEM Software, because the domain structure of the ferromagnetic material is neglected [3]. The hysteresis loss and the anomalous eddy current loss depend on the domain structure and size and can't be calculated directly. In most cases manufacturer data, which are based on measurements, are used to interpolate the core losses with the FEM method. The difference between simulation and measurement is consequently large, because the normative measurement of specific parameters is based only sinusoidal flux density B.

For the simulation of the specific core loss p_c , the approach from [3] is used:

$$p_C = w_h (f(\hat{B})) \cdot \frac{2}{\pi^2 \cdot \Delta B^2} \int_0^T \left(\frac{dB}{dt}\right)^2 \cdot dt + k_e \cdot \frac{1}{2 \cdot \pi^2} \cdot \frac{1}{T} \int_0^T \left(\frac{dB}{dt}\right)^2 \cdot dt + \frac{k_a}{8,76} \cdot \frac{1}{T} \int_0^T \left|\frac{dB}{dt}\right|^{1,5} \cdot dt \tag{5}$$

with:

$$f_h = \frac{2}{\pi^2 \cdot \Delta B^2} \int_0^T \left(\frac{dB}{dt}\right)^2 \cdot dt \tag{6}$$

For the consideration of inhomogeneous flux density B of specific core loss p_c are integrated over the core volume V and is multiplied with specific material density ρ_m:

$$P_C = \rho_m \cdot \iiint_V p_C \cdot dV \tag{7}$$

To account the influence of the rolling direction for the core loss (Figure 1, bottom right), the loss parameters k_e, k_a and the hysteresis energy w_h must be generated out of measurement data for the different rolling directions. To simulate the core loss of the 4-limb inductor from Figure 1, the losses from equation (7) must be calculated and added for each of the corresponding partial volume V_c of the magnetic field orientation:

$$P_C = \rho_m \cdot \left(\iiint_{V_{C,0°}} p_{C,0°} \cdot dV_{C,0°} + \iiint_{V_{C,45°}} p_{C,45°} \cdot dV_{C,45°} + \iiint_{V_{C,45°}} p_{C,45} \cdot dV_{C,45°} \right) \tag{8}$$

Measurement Setup

For the measurement of common mode and differential mode inductance, the measurement setup in figure 3 has been used.

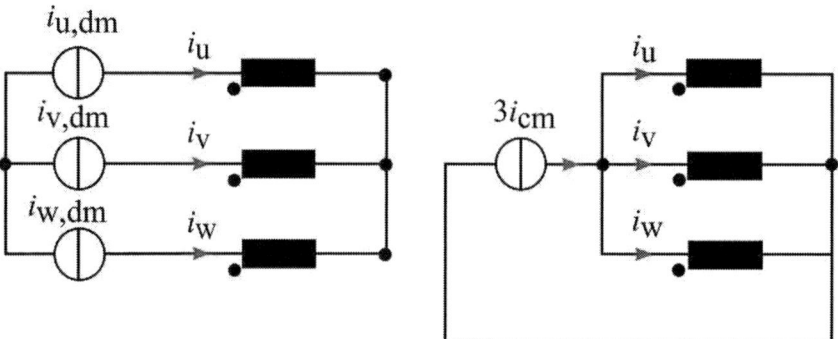

Figure 3: Measurement setup for differential- and common mode inductance
(left side: differential mode setup   right side: common mode setup)

The differential inductance L_{diff} is measured during a switch-on by a rectangular voltage pulses u_{meas}. The change of the magnetizing current di_{meas}/dt is measured and the measured voltage drop across inductor is corrected by the voltage drop across the winding resistance R_w:

$$L_{diff} = \frac{u_{meas}(t) - i_{meas}(t) \cdot R_W}{\dfrac{di_{meas}}{dt}} \tag{9}$$

© VDE VERLAG GMBH · Berlin · Offenbach

Results

Here the results between the measurement of the inductor prototype and the simulation are shown and compared. Figure 4 shows a 2D cut of the 3D model of the 4-limb inductor for the time instant $\omega t = \pi/4$.

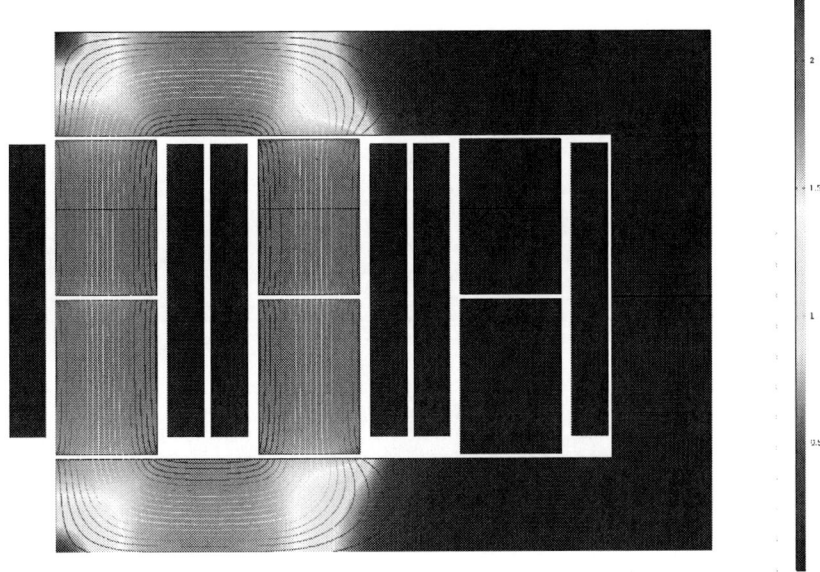

Figure 4: 2D cut in the 3D model for time instant $\omega t = \pi/4$ for differential mode simulation. The magnetic flux density and the contour magnetic vector potential are shown.

For this time instant the differential mode inductance L_{dm} can be simulated for Phase u and v. A comparison between the simulation result and measurement results of the choke prototype are shown in figure 5 and 6.

Figure 7 shows a 2D cut of the 3D model for the simulation of the common mode inductance L_{cm} of the four limb inductor. All Harmonics of the magnetization current of the inductor in a inverter system, which have a multiplying factor of three have the same phase. The common mode current flows in all three limbs of the inductor in figure 7 with the same phase. A comparison between simulation and measurement of the common mode inductance is shown in figure 8.

The core loss were measured with an auxiliary windings, thus to measure the induced main field voltage. The core loss are shown as a function of the maximum value of the magnetic flux density in figure 9.

© VDE VERLAG GMBH · Berlin · Offenbach

PCIM Europe 2016, 10 – 12 May 2016, Nuremberg, Germany

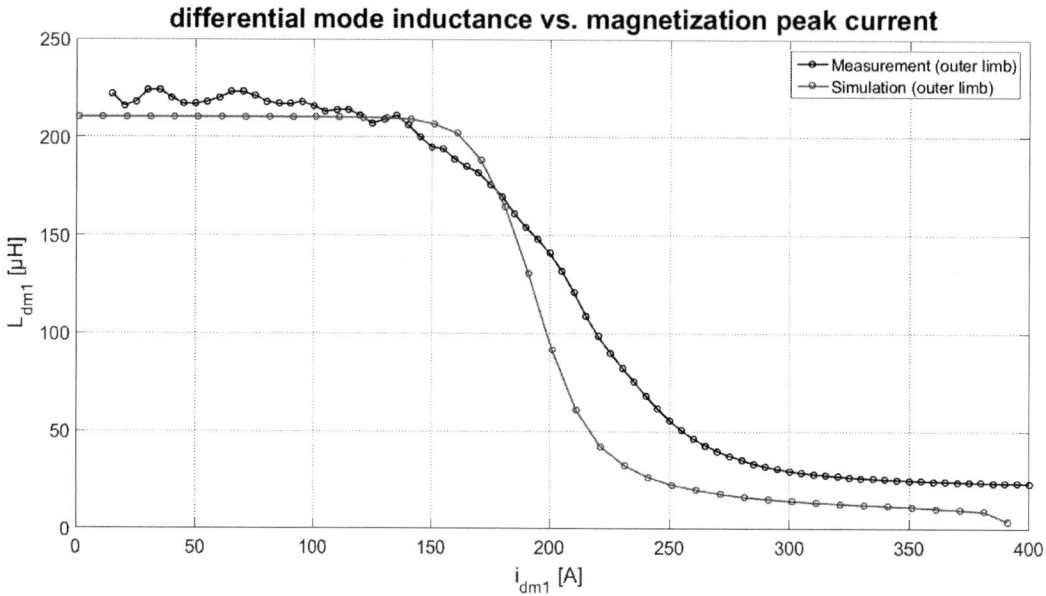

Figure 5: Differential mode inductance vs. peak value of the magnetization current for the outer left limb

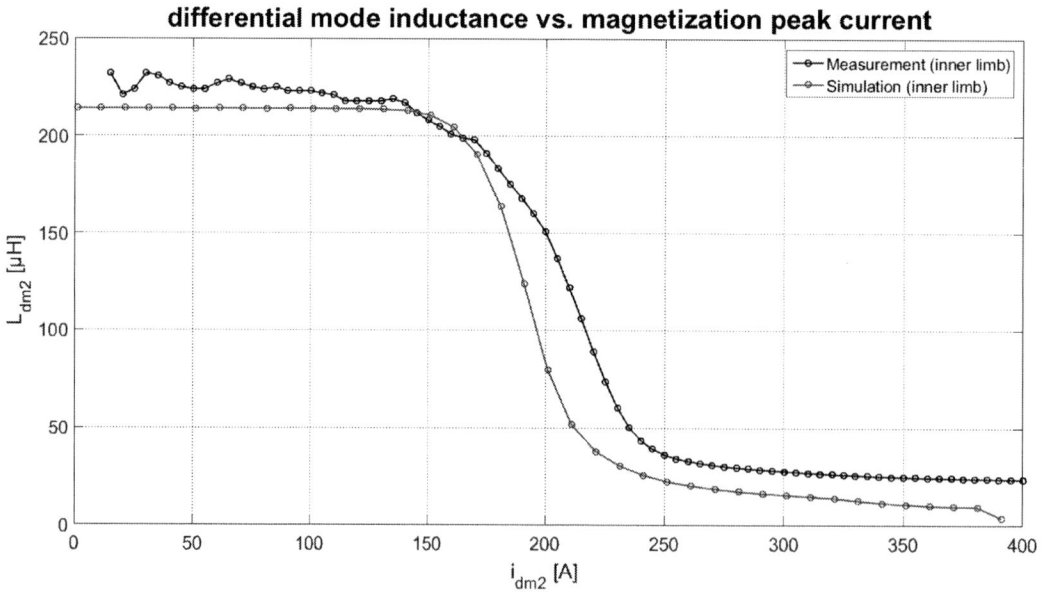

Figure 6: Differential mode inductance vs. peak value of the magnetization current for the middle limb

© VDE VERLAG GMBH · Berlin · Offenbach

PCIM Europe 2016, 10 – 12 May 2016, Nuremberg, Germany

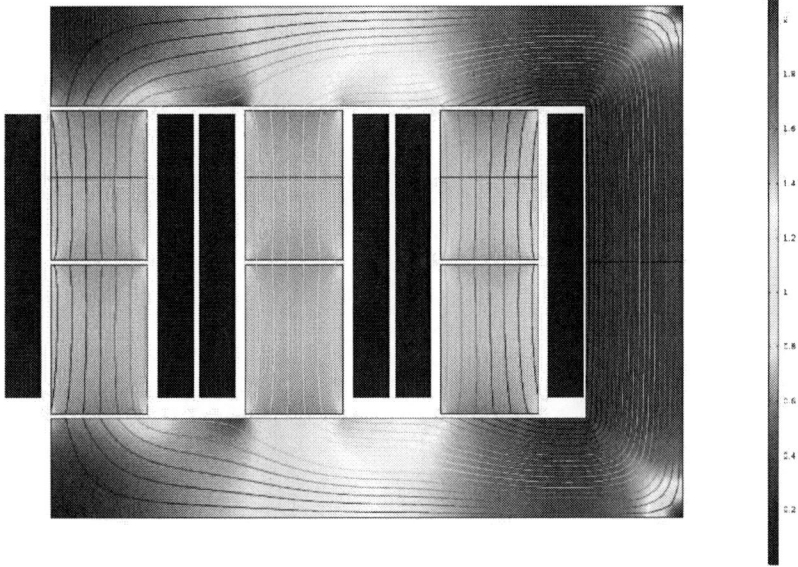

Figure 7: 2D cut in the 3D model for common mode simulation.
The magnetic flux density and the contour magnetic vector potential are shown.

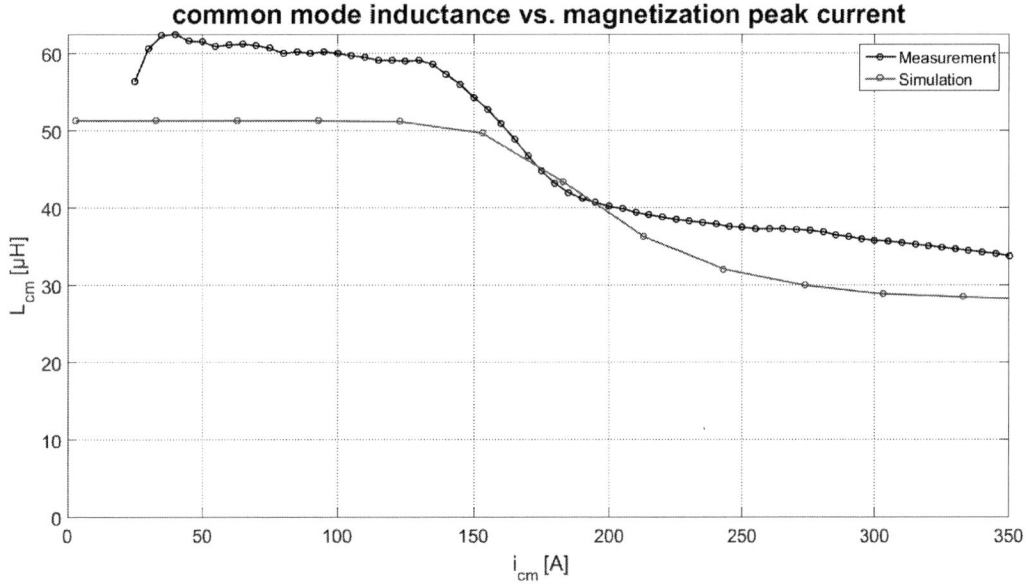

Figure 8: Common mode inductance vs. peak value of the magnetization current

© VDE VERLAG GMBH · Berlin · Offenbach

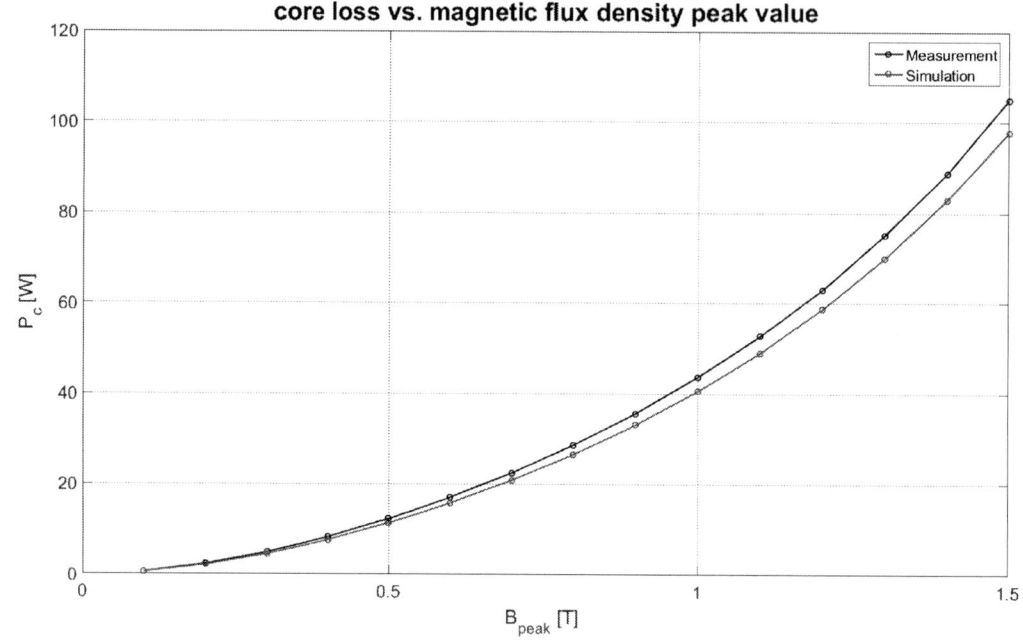

Figure 9: Comparison between measurement and simulation of the core loss

Conclusion

When the specific material properties of the soft magnetic material in a more detail form are available, a realistic simulation of a four-limb inductor can be done. For this purpose, the magnetic anisotropy of the relative permeability must be considered. This improves the calculation of the inductance as a function of the magnetizing current.
An additional measurement of the material properties with actual magnetization currents improves the simulation results compared with the results in the actual application.
The loss coefficients of the core loss model for each of the rolling directions must be available to simulate the core losses of the inductor. Thus, the core loss can be determined for each partial volume. The comparison between simulation results and the measurements on the prototype show good consistency.

References

[1]   Kampen, D.: *Considering mutual impacts of Differential Mode and Common Mode emissions in motor filter design for PWM inverters ,* 13th European Conference on Power Electronics and Applications EPE '09, 2009

[2]   Chen. H.: *Finite-element modeling of saturation effect exited by differential-mode current in a common mode choke,* IEEE Transactions of Power Electronics, Volume:24 , Issue: 3

[3]   Owzareck M.: *Calculation method for core losses of electrical steel inductors in power electronic applications,* PCIM 2015, 19 – 21 May 2015, Nürnberg, Germany, VDE Verlag GmbH

[4]   Owzareck M.: *Measurement method for normal magnetization curve of soft magnetic composites with high magnetization currents,* International Journal of Applied Electromagnetics and Mechanics – Vol. 48 Nos 2,3 2015

© VDE VERLAG GMBH · Berlin · Offenbach

Design procedure for pot-core integrated magnetic component

Martin Foster, Department of Electronic and Electrical Engineering, University of Sheffield, Mappin Street, Sheffield, United Kingdom, m.p.foster@sheffield.ac.uk
Andrew Fairweather & Grant Ashley, Vxl Power Ltd, Station Rd, North Hykeham, Lincoln, LN6 3QY, United Kingdom, andrew.fairweather@vxipower.com, grant.ashley@vxipower.com

Abstract

The side-by-side (primary-secondary) wound integrated magnetic component is often used to combine series inductance and transformer behaviour in LLC resonant converters and dual active bridge converters. This paper presents an easy to follow design procedure for this component, translating the component's electrical specification into a physical design. A combination of reluctance modelling, the application of Ampere's law and the introduction of an alignment factor, obtained through finite element analysis, is used to derive a model for the integrated magnetic component. The model is encapsulated within a design approach accounting for frequency, maximum operating flux density, etc., to ensure the final component is not oversized and meets the design specification.

1. Introduction

Design aesthetics, volume constraints, efficiency improvements, cost and the drive towards tighter system integration pose significant challenges for the power supply engineer, who often finds these drivers are in direct conflict with thermal and packaging constraints. In order to meet these ever increasing demands, the main inductor and transformer of the power converter are often combined into a single integrated magnetic (IM) component in an effort to save volume and cost. The transition to IM has led to the proliferation of modern power supply circuits, such as the LLC resonant converter and the dual active bridge, where the presence of transformer series and magnetizing inductances are readily absorbed into the circuit and provide useful functions, such as current provision for soft-switching. Many integrated magnetic components have been proposed, eg. using a shared primary winding with separate cores to realise both inductance and transformer [1]; placing various windings on the different legs of an E-core structure to construct inductors and transformers [2]; physically separating the primary and secondary windings, side-by-side on the core's centre leg to introduce leakage inductance between the two windings [3]. The side-by-side winding format is particularly attractive as the spacing between the windings can be used to meet the necessary creepage and clearance requirements, and provides the particular focus for this paper.

Traditionally, magnetic equivalent circuit techniques have been employed to model IM components. For certain IM structures magnetic equivalent circuits work very well where the core structure leads to logical partitioning in the equivalent circuit [1,2]. However, for the side-by-side IM structure equivalent circuit modelling leads to inaccurate estimation of the leakage due to the lack of granularity in representing the magnetic field strength distribution. Although a distributed mmf-reluctance model can be employed, where the winding mmf source and coil winding area reluctance contributions are evenly divided over the region of interest [4], the technique is somewhat unwieldy and the developed equivalent circuit model is cumbersome. Therefore, it does not readily lend itself to application in design. A more successful alternative is to equate the energy stored in the IM leakage inductances to energy stored in the winding volume by applying Ampere's law [5]. Although this technique was applied to the side-by-side IM in [3], the application of the resulting model is obfuscated by the various transformer equivalent circuit models described in the manuscript and the numerous definitions for the transformer coupling factor. This paper derives a completely new model based on a combination of reluctance modelling, application of Ampere's law and an alignment factor obtained through finite element modelling. The resulting model is easy to apply and is encapsulated within a design procedure.

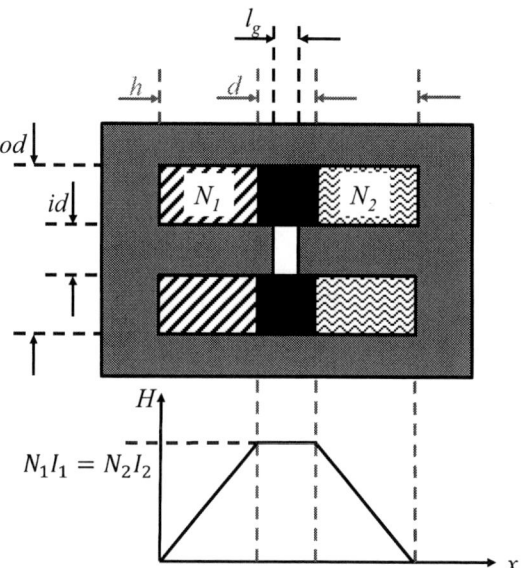

Fig. 1. Integrated magnetic component, geometry and magnetic field intensity

2. Transformer equivalent circuit

The transformer equivalent circuit in Fig. 2 captures the dominant behavior of a transformer using just four components: a primary-side series inductor (L_{lk1}), to model the leakage inductance present on the primary side of the transformer, a shunt inductor (L_m), to model the magnetizing inductance, an ideal transformer with a turn ratio (n_e) and a secondary-side series inductor (L_{lk2}).

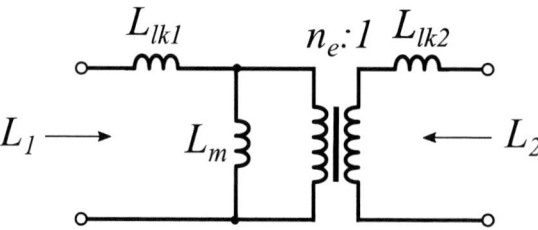

Fig. 2. Transformer equivalent circuit

2.1. Measuring transformer equivalent circuit values

Inductors L_1 and L_2 represent the transformer's primary winding inductance (measured using terminals A & B) and secondary winding inductance (measured using terminals C & D), respectively. The inductance values of both windings can be measured with the other winding connected as an open-circuit (oc) or short-circuit (sc). Using these types of measurements allows the parameters for the equivalent circuit to be found as follows:

© VDE VERLAG GMBH · Berlin · Offenbach

PCIM Europe 2016, 10 – 12 May 2016, Nuremberg, Germany

- $L_{lk1}+L_m=L_{1(oc)}$: L_1 is measured with the secondary winding open circuit
- $L_{lk1}+L_m||n_e{}^2L_{lk2}=L_{1(sc)}$: L_1 is measured with the secondary winding short circuit
- Actual turn ratio $n=N_1/N_2 \approx \sqrt{(L_{1(oc)}/L_{2(oc)})}$ where N_1 and N_2 are the primary and secondary winding turns, respectively
- M: The mutual inductance coupling the primary winding to the secondary winding can be found using $M=\sqrt{((L_{1(oc)}-L_{1(sc)})L_{2(oc)})}$
- k: Coupling coefficient $k= M/\sqrt{(L_{1(oc)} L_{2(oc)})}$
- $L_m=k^2L_{1(oc)}$
- $L_{lk1}=(1-k^2)L_{1(oc)}$
- n_e: equivalent turn ratio $n_e \approx \sqrt{(L_m/L_{2(oc)})}$ – assuming L_{lk2} is negligible

2. Pot-core integrated magnetic component

The objective of this section is to derive a mathematical model to predict the leakage inductance of the IM. Fig. 2 shows a cross section through the centre of the pot-core integrated magnetic component revealing a primary winding of N_1 turns and a secondary winding of N_2 turns. An airgap of length l_g is used to control inductance. It is assumed the primary and secondary windings are wound on a bobbin with an outer diameter od, inner diameter id, height h and with a separator of length d located between the two windings. The plot below the IM cross section shows how the magnetic field intensity H varies across the height the bobbin moving in the x direction.

The inductance for primary winding and the secondary winding with the secondary winding open circuit is,

$$L_{1(oc)} = A_L N_1{}^2 (AF) \tag{1}$$

where A_L is the core specific inductance. AF is an alignment factor that accounts for the effects of coil winding height and its location within the winding window and is described in section 3.
The secondary winding inductance is similarly defined as

$$L_{2(oc)} = A_L N_2{}^2 (AF) \tag{2}$$

The energy stored in the leakage inductance due to the peak primary current $I_{1(pk)}$ is given by

$$W_{Llk1} = 0.5L_{lk1}I_{1(pk)}{}^2 \tag{3}$$

If the permeability of the core is very much higher than that of the coil windings, then it can be assumed that this energy is stored in the winding and so it can be determined from the volume integral of the BH product,

$$W_{Llk1} = \frac{1}{2}\int BH dv = \frac{\mu_0}{2}\int H^2 dv = \frac{\mu_0}{2}\int\int\int H(r,\theta,x)^2 r dr d\theta dx \tag{4}$$

Since the winding can be considered to be a tube, (4) is evaluated over the volume of winding tube. The magnetic field intensity at a specific height (x) in the tube is assumed to be approximately constant across the winding width, $w=0.5(od-id)$, such that,

$$H = \frac{x}{a}\frac{N_1 I_{1(pk)}}{w} \tag{5}$$

for $0<x<a$ where $a=(h-d)/2$.

The contribution from N_1 can be found by using (5) to evaluate (4) as follows,

$$W_{Llk1(N1)} = \frac{\mu_0}{2}\int_{r1}^{r2}\int_0^{2\pi}\int_0^a \left(\frac{x}{a}\frac{N_1 I_{1(pk)}}{w}\right)^2 r dr d\theta dx$$
$$= \frac{\mu_0 \pi}{6}(N_1 I_{1(pk)})^2 \frac{od^2-id^2}{(od-id)^2} \tag{6}$$

Applying this technique to the mmf associated with the secondary winding and winding separator, summing the contributions and using (3) reveals the primary side leakage

© VDE VERLAG GMBH · Berlin · Offenbach

PCIM Europe 2016, 10 – 12 May 2016, Nuremberg, Germany

inductance,

$$L_{lk1} = \frac{\mu_0 \pi (od+id) N_1^2}{6(od-id)} (h + 2d) \tag{7}$$

Using (7) and the equivalent circuit definitions, the magnetising inductance is defined as,

$$L_m = N_1^2 \left[A_L(AF) - \frac{\mu_0 \pi (od+id)}{6(od-id)} (h + 2d) \right] \tag{8}$$

3. Align factor (AF)

In most transformers and inductors the windings are wound over the entire height of the bobbin, and so the assumption is that the flux is uniformly distributed. With integrated magnetic components this is not necessarily the case, and so the standard expressions that are used to calculate inductance are not sufficiently accurate. The flux density plots depicted in Fig. 3 were obtained from a detailed axisymmetric finite element analysis and show the difference between a fully distributed winding consisting of 100 turns on a PC 26/16 core with l_g=1mm (Fig. 3a with L=1.46mH), and a winding covering only 19% of the bobbin height (Fig. 3b with L=1.76mH). As can be seen, with a partial height winding the flux appears to more easily leak out of the air gap and so this has the effect of increasing the inductance (by 20% in this case) and, therefore, needs to be accounted for during the design process. This section presents an alignment factor (AF) that accounts for the effects of coil winding height and its location within the winding window.

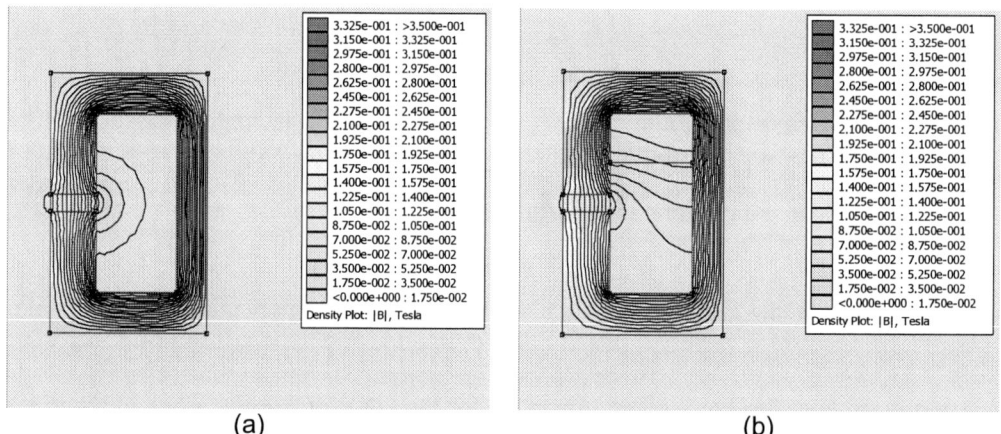

(a)   (b)

Fig. 3 Effect of winding height utilization on flux distribution, a) 100% & b) 19%

Over 1000 data points were generated using finite element analysis simulations (FEMM [6]) of pot cores inductors wound with 100 turns but featuring different airgaps and height values. The pot core dimensions (h, od, id, Ae) were taken from Magnetics Inc. Ferrites Catalogue [7]. For each core type, a specific airgap was inserted and the initial AL value measured using 100% bobbin height. Then the winding utilization was swept and the resulting AL determined . A curve-fit was performed to the data and revealed the alignment factor to be,

$$AF = 1 - 10^{-6} \sqrt{A_e} \ln\left[\frac{(h-d)^2}{4h^2}\right]/A_L \tag{9}$$

Fig. 4 shows the variation in the alignment factor value and also the inductance value for the parameters considered in the FEMM simulation sweep. As can also be seen, the effect of winding geometry can be quite significant. Fig. 4b shows the alignment factor equation predicting the inductance value with good accuracy.

© VDE VERLAG GMBH · Berlin · Offenbach

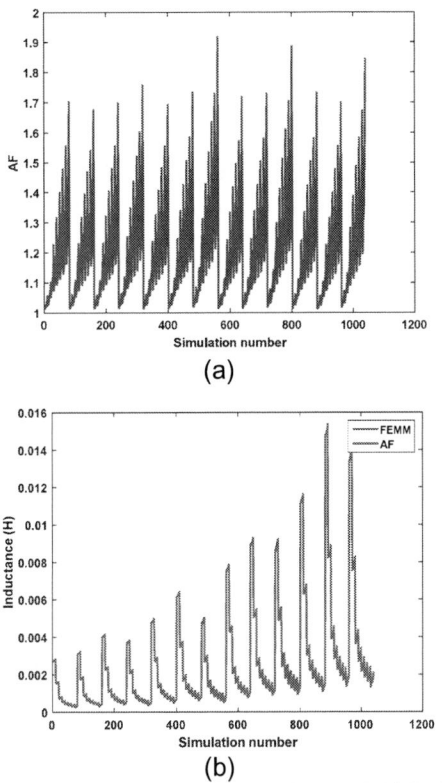

(a)

(b)

Fig. 4 Effect of winding geometry on a) alignment factor & b) inductance

4. Design procedure

The design procedure presented here allows the engineer to choose the minimum core size required to satisfy the design, whilst accounting for flux density and thermal constraints. The transformer primary voltage imposes a volt-second limitation, the high-frequency operation leads to core losses and the primary/secondary windings must be accommodated with the core winding area (W_a), and they also experience copper losses. These concerns are factored into the design process to ensure the transformer operates satisfactorily.

One needs to ensure the primary and secondary coil windings adequately fit within the winding window area W_a along with the winding separator. This leads to the following inequality where K_u ($K_u<1$) is the winding factor accounting for the actual area taken up by the winding and insulation, and J is the current density (typically 4-5 A/mm^2),

$$W_a \geq dw + \frac{1}{K_u}\left(\frac{N_1 I_{1(rms)}}{J} + \frac{N_2 I_{2(rms)}}{J}\right) \approx dw + 2\frac{N_1 I_{1(rms)}}{K_u J} \qquad (10)$$

Substituting for the inductor definition $N A_e B = LI$ (where A_e is the effective core area) gives,

$$W_a \geq dw + 2\frac{L_1 I_{1(rms)} I_{1(pk)}}{K_u A_e B J} \qquad (11)$$

Many ferrite material manufacturers provide detailed design information, including relationships for the limits imposed on maximum flux density level for specific operating frequency and temperature rise. Magnetics Inc., for example, provide a graphical relationship for their P type ferrite material [7] when operating with a 25°C temperature rise. A curve fit to this data (shown in Fig. 5) gives the following relationship between the *optimal* flux density (B_{opt}) and operating frequency (f) over the range 30kHz-1MHz,

$$B_{opt} = 0.0688z^2 - 0.4366z + 0.7054 \tag{12}$$

where $z = \log(f)$ with f in kHz.

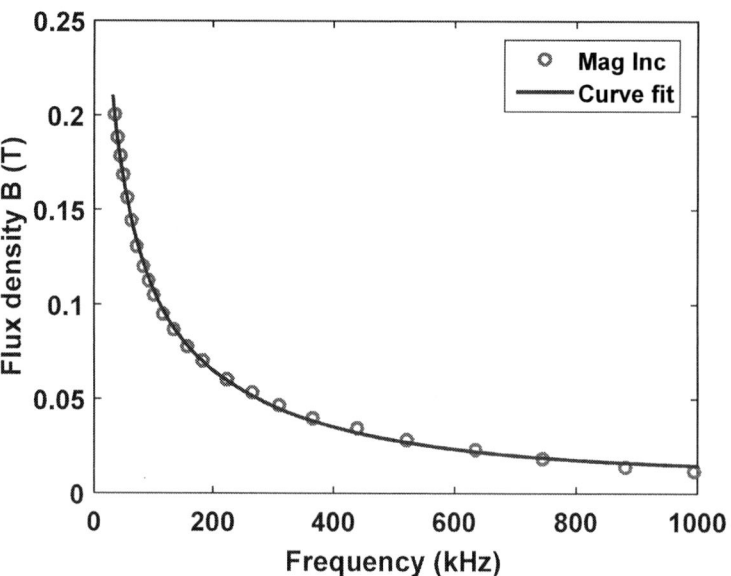

Figure 5: Curve fit to Magnetics Inc. P type ferrite material recommended operating flux density

Rearranging (11) and using (12) gives the minimum area product (AP) required to satisfy the design specification. Note, the inequality in (13) is factored to give the geometry specific relationship on the left hand side and the electrical specification on the right hand side.

$$AP = (W_a - dw)A_e \geq 2\frac{L_1 I_{1(rms)} I_{1(pk)}}{K_u B_{opt} J} \tag{13}$$

The core inductance factor (A_L) required to ensure the core does not saturate can be found by equating the primary winding volt-second product (λ_1) to the flux density swing experienced by the winding.

$$\lambda_1 = \int_0^{T/2} |V_1(t)| dt \tag{14}$$

During a positive half-cycle of the primary voltage, the flux density swings between its positive and negative peaks such that the peak flux density (B_{pk}) can be defined as,

$$B_{pk} = B_{opt} = \frac{\lambda_1}{2N_1 A_e} \tag{15}$$

Thus, rearranging (1) and using (15) provides the required core specific inductance value,

$$A_L = \frac{L_{1oc}}{\left(\frac{\lambda_1}{2B_{opt} A_e}\right)^2 (AF)} \tag{16}$$

The relationship between the effective turn ratio and actual turn ratio is given by,

$$n_e = \frac{N_1}{N_2}\sqrt{\frac{A}{A+1}} \tag{17}$$

where $A = L_m / L_{lk1}$

The above constraints are incorporated into a design procedure as follows,

1. Specify all electrical parameters for the IM and the intended winding separator distance d
2. Determine the optimum flux density (B_{opt}) at which to run the core using (12)
3. Choose the smallest core which meets the AP value in (13). This will then fix the core geometry (h, w=0.5(od-id), A_e)
4. Determine the required A_L value to achieve B_{opt} using (16)
5. Transformer winding turns should be chosen to best meet the desired turn ratio and required inductances
   a. Determine the primary turns N_1 using (1) and then use (17) to determine secondary turns N_2
   b. Or determine the secondary turns N_2 using (2) and then use (17) to determine the primary turns N_1
6. Choose appropriate wire sizes, accounting for high frequency effects as necessary, and confirm the windings fit within the winding window

Should the windings not fit within the winding window it may be necessary to:
1. Choose a larger core
2. Reduce the separation distance d
3. Operate at different flux density
4. Modify the desired inductances and/or turn ratio

5. Design example and validation

To validate the pot core integrated magnetic component design methodology a 25W, 45-55V input to 5V output LLC resonant converter was designed using the LLCDesigner software [8,9]. The specification for the integrated magnetic component is L_{lk1}=5.1μH, L_m=23.1μH, n_e=5.5 with a minimum operating frequency of f=200kHz and a peak primary current of $I_{1(pk)}$=2A. The winding separator distance was chosen as d=5mm. From (12) the optimum flux density was determined as B_{opt}~65mT. Using J=4A/mm and K_u=0.5 the minimum core area product was determined to be,

$$AP = (W_a - dw)A_e \geq 1.23 \times 10^{-9}\text{m}^4 \tag{18}$$

With the required separator distance, a 26/16 pot core with an $AP = 2.9 \times 10^{-9}\text{m}^4$ was chosen. With an inductance ratio A=4.55, the actual turn ratio was determined to be,

$$n = \frac{N_1}{N_2} \approx 6 \tag{19}$$

The actual number of turns on the primary winding was determined as N_1=12, giving N_2=2. Fig. 6 shows the flux density plot taken from FEMM which was used to measure the primary inductance L_1=28.35μH and mutual inductance M=3.69μH. These values were then used to determine the leakage inductance L_{lk1}=6.22μH and magnetizing inductance L_m=22.13μH, which are very close to the design. A prototype IM was built to the same design specification. Experimental measurements taken using an LCR bridge revealed the leakage inductance L_{lk1}=5.86μH, magnetizing inductance L_m=20.76μH and the effective turn ratio n_e=5.3, values which are close to the design specification.

PCIM Europe 2016, 10 – 12 May 2016, Nuremberg, Germany

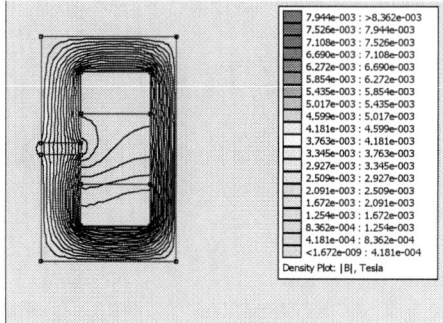

Fig. 6   Flux distribution in the design example

6. Conclusion

A model for a pot core integrated magnetic component has been developed which allows the leakage inductance, magnetizing inductance and effective turn ratio to be predicted. We have introduced an alignment factor (*AF*) which modifies the inductance equation to account for the additional leakage inductance caused by the non-uniformly distributed windings. A design procedure has been presented which allows the engineer to determine the minimum required core size, operating flux density, A_L factor and primary & secondary turns. A design example based on an LLC resonant converter has been given and results obtained from finite element analysis show good agreement.

Acknowledgment

This work was in-part supported by the RAEng under the Industrial Secondment Scheme. Dr Foster would like to thank Vxl Power Ltd for providing the secondment opportunity.

References

[1]   A. Kats, G. Ivensky & S. Ben-Yaakov, 'Application of integrated magnetics in resonant converters', IEEE Applied Power Electronics Conference and Exposition (APEC), pp. 925 – 930, vol.2, 1997.

[2]   B. Yang, R. Chen & F. C. Lee, 'Integrated magnetic for LLC resonant converter', IEEE Applied Power Electronics Conference and Exposition (APEC), pp. 346-351, vol. 1, 2002.

[3]   S. De Simone, C. Adragna, C. & C. Spini, 'Design guideline for magnetic integration in LLC resonant converters', IEEE International Symposium on Power Electronics, Electrical Drives, Automation and Motion (SPEEDAM 2008), pp. 950 – 957, 2008.

[4]   S. A. El-Hamamsy & E. I. Chang, 'Magnetics modeling for computer-aided design of power electronics circuits', IEEE Power Electronics Specialists Conference (PESC), pp. 635 – 645, vol. 2, 1989.

[5]   J. van Rensburg, P. A. & H. C. Ferreira, 'The role of magnetizing and leakage inductance in transformer coupling circuitry', International Symposium. Power-line Communications, pp. 244-249, 2004.

[6]   D. C. Meeker, Finite Element Method Magnetics (FEMM), http://www.femm.info

[7]   Magnetics Inc. Ferrite Cores 2013 Catalog.

[8]   LLCDesigner, http://www.mpfoster.staff.shef.ac.uk/

[9]   M. P. Foster, D. A. Stone & C. M. Bingham, An automated design methodology for LLC resonant converters using a genetic algorithm, Power Electronics, Intelligent Motion, Renewable Energy and Energy Management PCIM 2009 Europe, 2009.

© VDE VERLAG GMBH · Berlin · Offenbach

Investigation of Core Losses under Different Conditions Applying the Cross Power Method

Boris Hudoffsky, PMK Mess- und Kommunikationstechnik GmbH, Germany, info@pmk.de
Chihiro Okinori, IWATSU Test Instruments Corp., Japan, info-tme@iwatsu.co.jp
Jürgen Trüller, HF Instruments GmbH, Germany, info@hf-instruments.com

Abstract

Different materials for power inductor cores are investigated concerning their specific core-losses under different conditions. The characteristic magnetizing curve, the BH-loop, of soft magnetic cores could change dramatically, if the application consists of considerable mean valued currents. Core losses are related to the area in the BH-loop and are expected to be low for high efficiency circuits. In this paper five different types of materials as ring cores are analyzed either without or with a DC-biased excitation current. BH-loops as results show influence as well as immunity of the application of real conditions to the measurements. The measurements are taken on a BH-analyzer system with the capability of the cross power method by varying the DC-bias operated at four different frequencies.

1. Introduction

Understanding the magnetic behavior and calculating losses for magnetic components is a difficult task for power electronic engineers, since there are so many factors of influence and a lack of relevant data. In most design cases these circumstances end in more confusion about this issue than in finding a proper solution. There are many suggestions to measure the magnetic losses [1–3] or to use simulations [4, 5].

One of the most challenging jobs is to get a measurement setup with a high excitation level as well as high precision evaluation, see [6]. This means that high energy is applied to the samples while testing, instead of having low excitation like in spectrum analyzer for electrical circuits or equivalent.

To be able to get reliable measurement data, the magnetic device could be tested in different circuits which will lead to one parameter or another [7]. The use of a dedicated analyzer system offers the availability of any parameter. IWATSU Test Instruments Corp. is manufacturer for systems like this.

1.1. Inductive components in high efficient power electronics

Core losses in power electronics are known to be depended on frequency. The higher the frequency gets the higher the losses will be. The trend to increase the switching frequency of semiconductors to reduce the switching losses leads to a reduction in size and inductance L of the inductive components. Not only their impedance $Z_L \propto f \cdot L$ is increasing with frequency f, it is also their specific core loss P_C:

$$P_C = K_H \cdot f + K_E \cdot f^2 + P_R \tag{1}$$

with K_H as the hysteresis and K_E as the eddy current loss coefficient and P_R the resistive losses. For the reason of the influence of frequency f and the fact that the measurement of low loss

PCIM Europe 2016, 10 – 12 May 2016, Nuremberg, Germany

components is very depended on the accuracy and phase angle stability of the captured values, a measurement setup with flexible exciting frequency and high precision in data acquisition is preferable to be taken for the task of analyzing the AC properties soft magnetic materials, compare [8].

1.2. Cores Under Test (CUT)

Analyzing the material properties has not necessarily to be done in the shape of the final device. There are more options and variables to adjust, if the sample to be investigated is made as an uniform shape. In the case of magnetic components this would be a ring core like in figure 1. The ring core shaped samples do have a constant cross section and an easy to de-

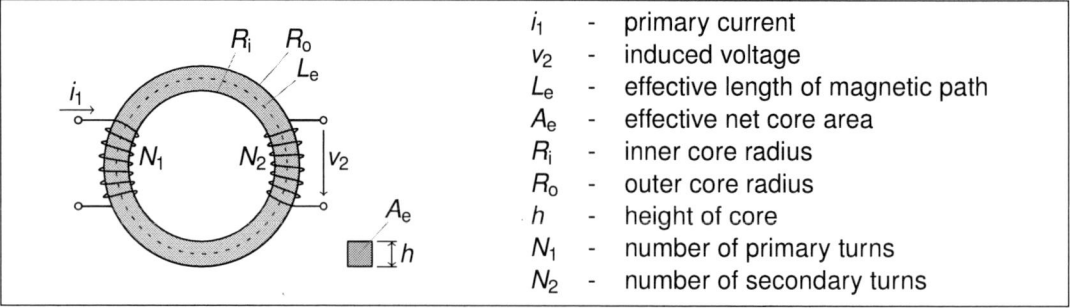

i_1	-	primary current
v_2	-	induced voltage
L_e	-	effective length of magnetic path
A_e	-	effective net core area
R_i	-	inner core radius
R_o	-	outer core radius
h	-	height of core
N_1	-	number of primary turns
N_2	-	number of secondary turns

Fig. 1: CUT (Core Under Test) with two windings for B-H analysis

termine magnetic path length. Besides the listed parameters volume and weight are important as well. The core under test should have one or two windings with an independent number of turns, each. In the case of a two winding configuration, N_1 is the number of turns of the excitation/primary winding and N_2 is the number of turns of the sensing/secondary winding.

2. Principle of B-H analysis

There are different methods to determine the soft magnetic properties and most of them refer to the two winding setup shown in figure 2. With a sensitive capture of the two values of primary current i_1 and secondary voltage v_2 all properties could be calculated. Basically the magnetic field strength H and the flux density B are of great interest as well as the core losses P_C:

$$H(t) = \frac{N_1 \cdot i_1(t)}{L_e} \tag{2}$$

$$B(t) = \frac{1}{N_2 \cdot A_e} \cdot \int v_2(t)dt \tag{3}$$

$$P_C = \frac{N_1}{N_2} \cdot \frac{1}{T} \cdot \int_0^T i_1(t) \cdot v_2(t)dt \tag{4}$$

with $T = 1/f$ as period.

2.1. Multiplying methods

Among some others there are the so called multiplying methods, which are based on the identical voltage-current multiplying principle [9]. These methods are sensitive to errors in phase,

© VDE VERLAG GMBH · Berlin · Offenbach

that means high accuracy is only possible if the phase-shift error is reduced. These methods capture the voltages representing field strength H and flux density B and process them either analogue, digital or mixed way in the time or frequency domain.

2.2. Cross-power method

The cross-power method is one of the multiplying methods and very suitable for error correction in the frequency domain. It

1. converts the time domain waveforms into a frequency domain spectrum and

2. executes integral calculation with no phase difference and

3. could compensate an amplitude or phase error of any detection unit in the frequency domain and

4. calculates the cross-power spectrum from these data and

5. adds up the real parts of the cross-power spectrum at each frequency to get the total power loss and

6. could convert all time dependent data back into time domain.

3. Test setup

In figure 3 the test setup is shown consisting of the main unit (B-H Analyzer) with the digital signal processor (DSP) and the analogue to digital converters (ADC). Also the analogue signal generator to drive the linear power amplifier is in the main unit. To synchronize all time dependent devices there is a clock driving these components. Directly hooked up to the main unit is the measurement POD, which takes the core under test (CUT) and where the current and voltage measurements take place. For high precision and a stable phase angle, there are some transfer functions implemented: G_H and G_B. For AC-excitation there is a linear power amplifier and for additional DC offset there is a DC current source. This arrangement of the two different electrical energy sources is completed with some filters to avoid disturbance.
The current capture by shunt measurement has its own frequency behavior $Z_S(f) = R_S + j \cdot 2\pi f \cdot L_S$ with L_S being the parasitic shunt inductance. The voltage capture has a frequency dependency, too.
This frequency dependent behavior is compensated by knowing the transfer functions of the capturing units $G_H(f)$ and $G_B(f)$ in the frequency domain. With $\omega = 2\pi \cdot f$ formula (2) and (3)

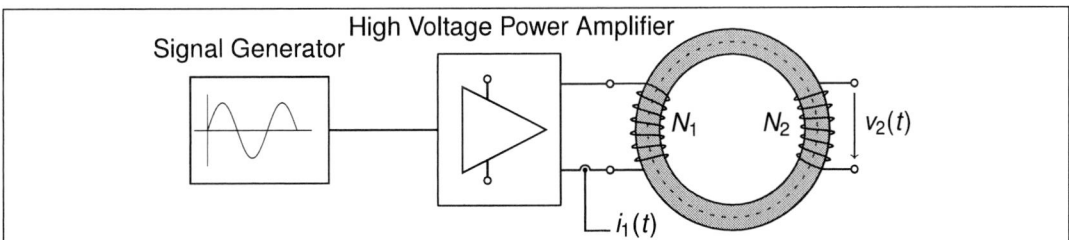

Fig. 2: Principle of B-H analysis

PCIM Europe 2016, 10 – 12 May 2016, Nuremberg, Germany

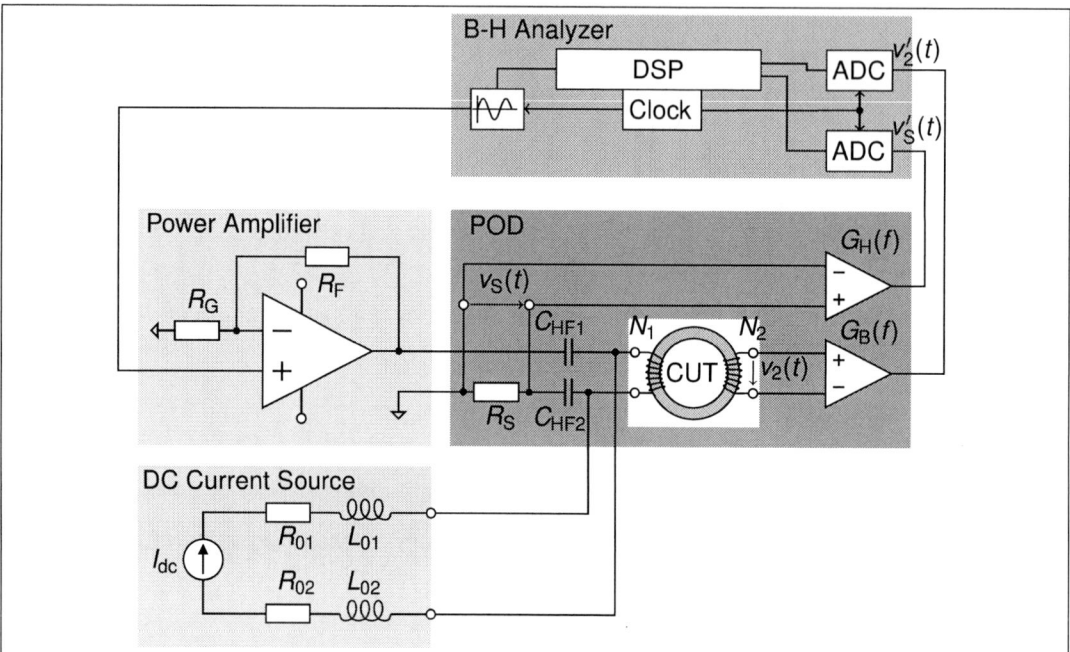

Fig. 3: Test setup of B-H analyzer with DC bias option

could be evaluated following the sketch of figure 3:

$$V'_S(\omega) = \mathrm{FFT}\left\{v'_S(t)\right\} \tag{5}$$

$$V'_2(\omega) = \mathrm{FFT}\left\{v'_2(t)\right\} \tag{6}$$

$$I_1(\omega) = \frac{1}{Z_S(\omega)} \cdot \frac{V'_S(\omega)}{G_H(\omega)} \tag{7}$$

$$V_2(\omega) = \frac{V'_2(\omega)}{G_B(\omega)} \tag{8}$$

$$H(\omega) = \frac{N_1}{L_e} \cdot I_1(\omega) \tag{9}$$

$$B(\omega) = \frac{1}{\mathrm{j}\omega} \cdot \frac{1}{A_e N_2} \cdot V_2(\omega) \tag{10}$$

Furthermore the cross-power spectrum $C(\omega)$ is deduced from these data:

$$C(\omega) = I_1(\omega)^* \cdot V_2(\omega) = C_R(\omega) + \mathrm{j}C_I(\omega) \tag{11}$$

The core loss P_C is obtained by adding the real parts C_R of the cross-power spectrum at each frequency.

$$P_C = \frac{N_1}{N_2} \cdot \sum_\omega C_R(\omega) \tag{12}$$

And all time dependent values are available applying the inverse Fourier Transform towards the frequency domain values i_1, v_2, H and B.

© VDE VERLAG GMBH · Berlin · Offenbach

4. Test results

Five different ring core samples have been tested on the setup described in the section 3 above. The results in form of B-H loops as well amplitude permeability μ_a and core losses P_C versus the level of premagnetization H_{DC} are presented in figures 4 and 5. The different samples, see table 1, have been excited with an AC signal resulting in a swing of $\Delta B = 400\,mT$ each. The excitation frequency has been chosen to 10 kHz, 20 kHz, 50 kHz and 100 kHz with a sinusoidal waveform. Additionally a DC bias offset has been applied in the range of $0 \leq H_{DC} \leq 140\,A/m$.

Core	Material	Initial permeability
0077206A7	Iron powder core	125
T80-70B	Iron Powder core	125
B64290L0632X830	Ferrite N30	4 300
TX362315	Ferrite 3E12	12 000
W376-04	Nanocrystalline	> 10 000

Tab. 1: Properties of sample cores

The low permeability iron powder cores do have a narrow hysteresis and it is neither changing much by increase of frequency nor application of a DC bias. These cores show immunity to the applied premagnetization. The plots of the B-H loops as well as the plots of the core losses do not show an influence of a DC bias field. At higher frequency excitation the core losses increase in the case of material T80-70B. In both cases of iron powder cores the amplitude permeability of the lowest excitation frequency raises at low values of premagnetization. This is caused by the shifting of the hysteresis loop and the measurement of the amplitude permeability.

In the case of the ferrite cores big hysteresis could be seen when excited without or with DC bias field. Here the applied DC bias is enough to drive the core into saturation, which causes a rapid decrease of permeability.

For the tape wound nanocrystalline core saturation is reached after a premagnetization field of approximately $H_{DC} = 40\,A/m$. For this core losses increase by excitation frequency and premagnetization level until saturation is reached.

5. Conclusion

Errors in magnetic property analysis at high-frequency are caused of

- impedance frequency characteristic of current detection resistance,
- magnetic field strength detection circuit's transfer function,
- flux density detection circuit's transfer function

A time domain measurement method such as digitizing method will not be suitable due to its difficulty of compensation for errors. This means it is not suitable for high frequency/low loss materials analysis. A frequency domain measurement method is the most suitable method for high frequency/low loss materials measurement. This method has been proposed in the IEC 62044-3 as the "cross-power method" and is implemented in the B-H analyzer systems of IWATSU Test Instruments Corporation.

PCIM Europe 2016, 10 – 12 May 2016, Nuremberg, Germany

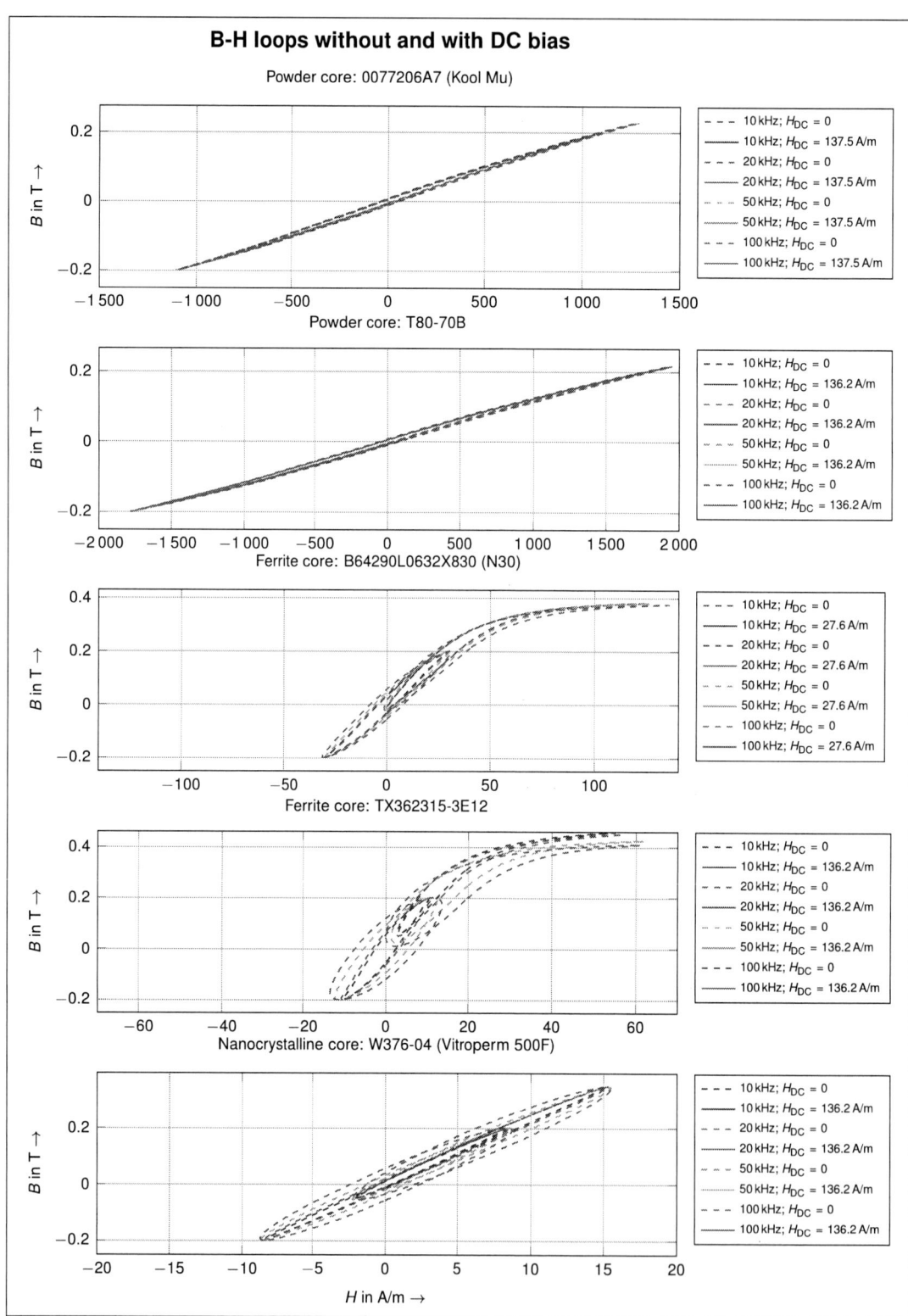

Fig. 4: Results: B-H loops under different conditions

© VDE VERLAG GMBH · Berlin · Offenbach

PCIM Europe 2016, 10 – 12 May 2016, Nuremberg, Germany

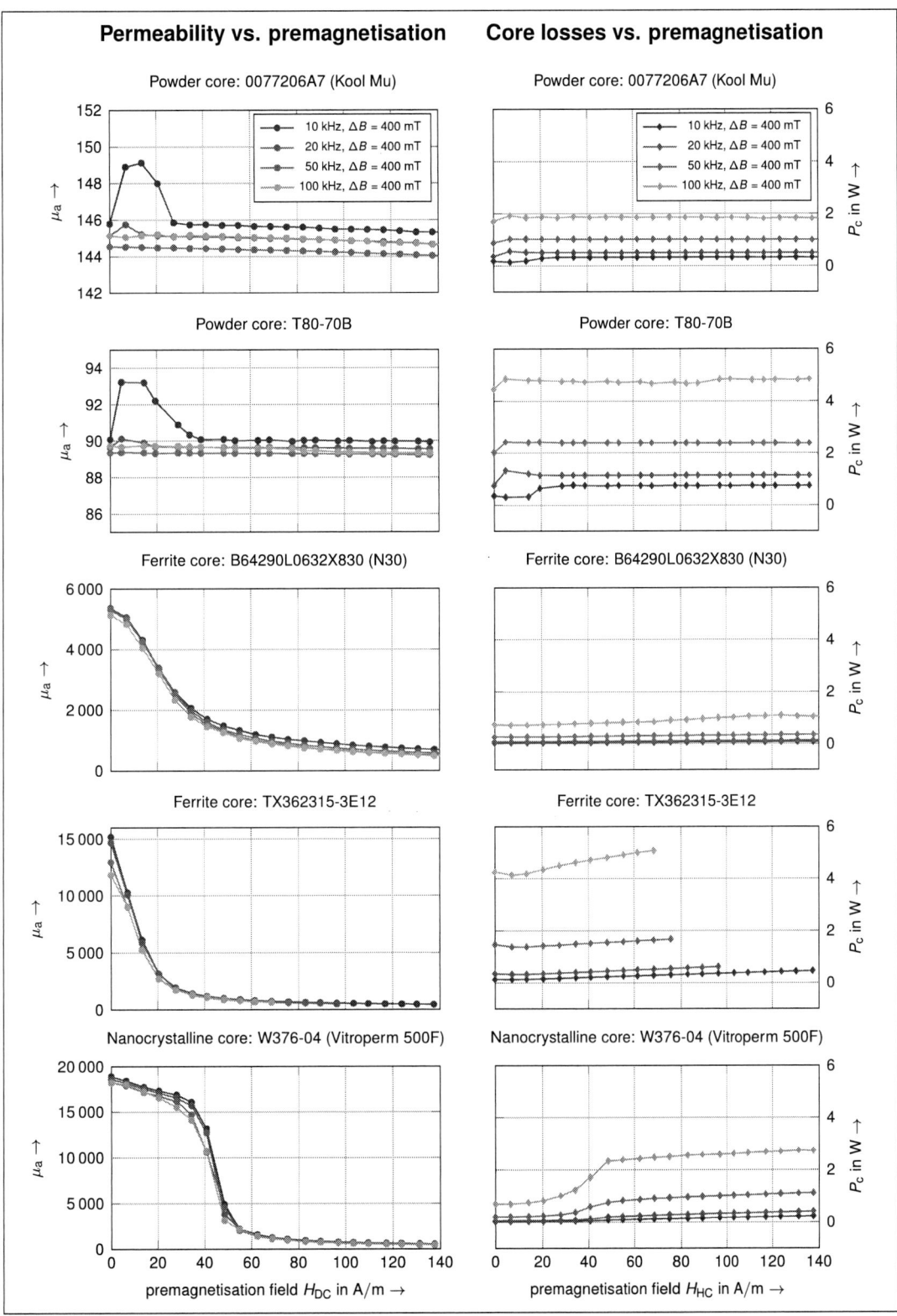

Fig. 5: Results: Amplitude permeability and core losses versus premagnetisation

© VDE VERLAG GMBH · Berlin · Offenbach

6. References

[1] V. Joseph Thottuvelil, Thomas G. Wilson, and Harry A. Owen, Jr. High-frequency measurement techniques for magnetic cores. *IEEE Transactions on Power Electronics*, 5(1):41–53, 1990.

[2] Bruce Carsten. Why the magnetics designer should measure core loss; with a survey of loss measurement techniques and a low cost, high accuracy alternative. In *High Frequency Power Conversion*, pages 103–119, 1995.

[3] Mingkai Mu, Qiang Li, David Joel Gilham, Fred C. Lee, and Khai D. T. Ngo. New core loss measurement method for high-frequency magnetic materials. *IEEE Transactions on Power Electronics*, 29(8):4374–4381, 2014.

[4] Jürgen Reinert, Ansgar Brockmeyer, and Rik W. De Doncker. Calculation of losses in ferro- and ferrimagnetic materials based on the modified steinmetz equation. *IEEE Transactions on Industry Applications*, 37(4):1055–1061, 2001.

[5] Jonas Mühlethaler, Jürgen Biela, Johann Walter Kolar, and Andreas Ecklebe. Improved core-loss calculation for magnetic components employed in power electronic systems. *IEEE Transactions on Power Electronics*, 27(2):964–973, 2012.

[6] C. A. Baguley, B. Carsten, and U. K. Madawala. The effect of dc bias conditions on ferrite core losses. *IEEE Transactions on Magnetics*, 44(2):246–252, 2008.

[7] C. A. Baguley, B. Carsten, and U. K. Madawala. An investigation into the impact of dc bias conditions on ferrite core losses. In *IECON 2007 - 33rd Annual Conference of the IEEE Industrial Electronics Society*, pages 1408–1413, 2007.

[8] A. Brockmeyer. Experimental evaluation of the influence of dc-premagnetization on the properties of power electronic ferrites. In *Applied Power Electronics Conference. APEC '96*, pages 454–460, 1996.

[9] International Electrotechnical Commission (IEC). Cores made of soft magnetic materials; measuring methods; part 3: Magnetic properties at high excitation level, 2000.

PCIM Europe 2016, 10 – 12 May 2016, Nuremberg, Germany

A Finite Element Simulation of Nanocrystalline Tape Wound Cores

Dr. Christian Scharwitz, Dr. Holger Schwenk, Dr. Johannes Beichler, Werner Loges
VACUUMSCHMELZE GmbH & Co. KG, Germany
christian.scharwitz@vacuumschmelze.com

Abstract

In this paper a method is presented for comparing the data of numerical simulations for magnetic nanocrystalline tape wound cores with experimental results. Usually, the Finite Element Method (FEM) is applied to calculate magnetic cores with nonconductive bulk models. However, nanocrystalline material is conductive and these models do not describe the electrical behavior. The electrical eddy current behavior of genuine nanocrystalline tape is determined by its thickness of only 18-25 µm. Such thin structures cannot easily implemented in FEM calculations of cores. Normally the tape dimension in core simulations is in the range of millimeters. Thus, numerical results cannot be directly compared with measurements. The method for the comparison of FEM simulations to experimental results, which is presented here, relies on an elaborated scaling of calculated data. This improves the interpretation and understanding of the experimental data.

1. Introduction

With growing computational power in the last decades numerical simulation techniques got an increasing impact on driving the development in several technological areas.

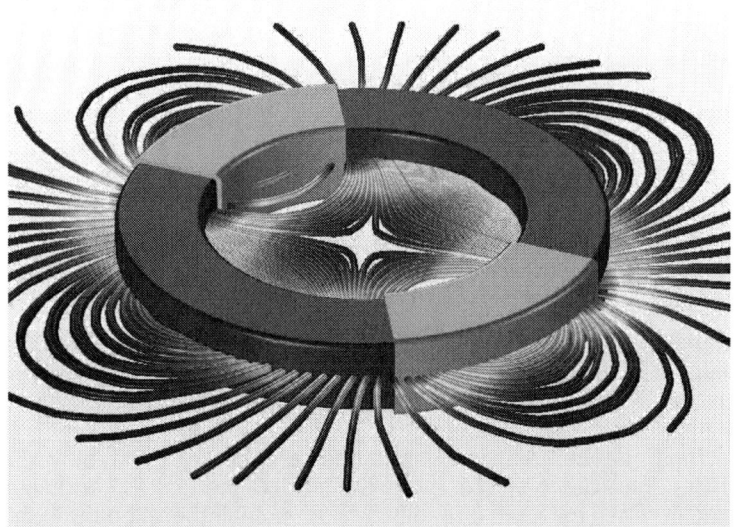

Fig. 1.   Typical pattern of an H-field configuration.

In this work the method of Finite Elements (FEM) is used to deepen the insights in phenomena related to nanocrystalline tape wound cores. Such cores (e.g. of VITROPERM®) offer superior properties for several inductive applications (e.g. common mode chokes in a rather small design or medium frequency transformers featuring high energy efficiency [1]). Typically, the FEM is used to calculate magnetic cores with nonconductive bulk models and usual results are magnetic field configurations, as shown in Figure 1. However, since the nanocrystalline tape is conductive, eddy currents and the cut-off frequency play an important role for the description of its frequency behavior. A nonconductive bulk model cannot be used to

© VDE VERLAG GMBH · Berlin · Offenbach

sufficiently describe the behavior of nanocrystalline tape wound cores. A suitable characterization should include the calculation of electric fields and eddy currents. The major challenge for the investigation of eddy current characteristics is the thin thickness of the genuine nanocrystalline tape of only 18-25 µm. The need to build a mesh for the FEM impedes the implementation of such thin tapes in the simulation geometry of magnetic cores. Depending on the applied computational power, the tape dimension in calculations is normally in the range of millimeters. Thus, numerical results cannot easily compared to measurements. Here, a method is presented to compare the results of FEM calculations to experimental data of nanocrystalline tape wound cores, which relies on an elaborated scaling of the calculated data. This improves the understanding and interpretation of experimental data and helps to further shape the advantages of nanocrystalline materials for future applications.

2. General Considerations

2.1. Model Implementation

To set up the model and to perform the numerical calculations, a commercial software [2] is used. Figure 2 shows an example of a typical FEM model featuring a tape wound core with an outer diameter d_o = 25 mm, an inner diameter of d_i = 16 mm and a height of h = 10 mm.

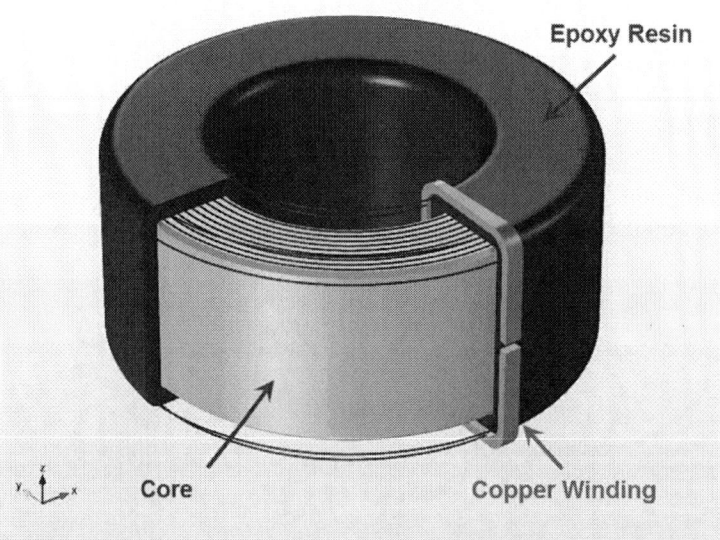

Fig. 2. Geometry to model a nanocrystalline tape wound core.

The tape structure of the core is indicated by black lines on the upside surface of the core, with ten layers implemented. For various implemented core dimensions, models with three to ten ribbon layers are calculated. Then, the thinnest tape thickness in a model is in the order of half a millimeter. The expected wavelengths at the given frequencies f_{sim} are long compared to the geometrical settings, hence, Maxwell equations are set up and solved in quasi-static approximation. The quantities of interest, e.g. the inductance L or the eddy current losses P_e, may be evaluated from the calculated results by numerical integration through

$$W_{mag} = \int_V \underline{H} \cdot \underline{B} \, dV = \int_V \frac{1}{\mu} \cdot \underline{B}^2 \, dV = \frac{1}{2} L I^2, \tag{1}$$

$$P_e = \int_V \underline{J} \cdot \underline{E} \, dV = \int_V \sigma \cdot \underline{E}^2 \, dV = R I^2. \tag{2}$$

The boundary conditions request a source current I_{Source} in the copper winding and electrical insulation on the appropriate interfaces. Typical values for the material parameters can be found in corresponding literature, for example [3] or [4].

2.2. Model Verification and Model Suitability

The FEM model is applied to simulate various characteristic curves. In order to test the implementation, the results are compared to calculations with the classical eddy current theory. In figure 3 the eddy current losses $P_{e,sim}$ are depicted, figure 4 shows the numerically calculated real and imaginary parts of the permeability. The graphs are plotted versus the frequency f_{sim}.

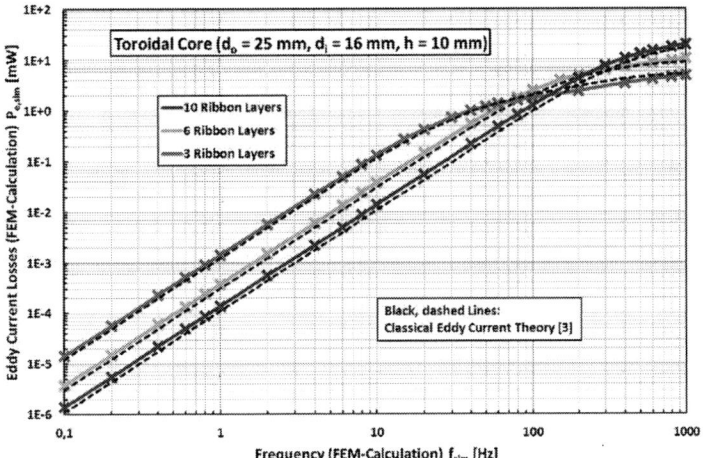

Fig. 3. Calculated eddy current losses versus frequency.

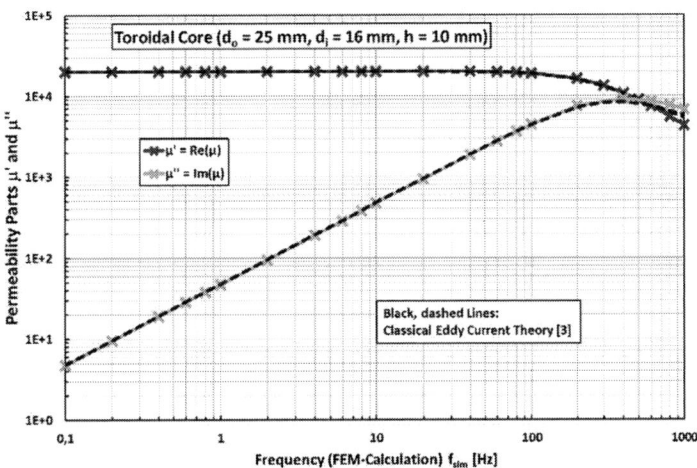

Fig. 4. Calculated permeability curves versus frequency.

The eddy current cut-off can be clearly identified by changes of the curves gradient [3]. Furthermore, curves, which are obtained analytically with the classical eddy current theory [3], are added as black, dashed lines. The numerical simulations match the classical calculations well, the principal usability of the model is proven.

However, since the model set-up features tape thicknesses in the range of millimeters, f_{sim} reaches up to 1 kHz and the eddy current losses $P_{e,sim}$ are located in a range of up to merely 10 mW. For measurements of genuine cores with much thinner tapes higher values for the frequency f_{real} and the eddy current losses $P_{e,real}$ can be expected. Usually, the frequency range of f_{real} starts around 1 kHz and measured eddy current losses $P_{e,real}$ typically range from around 10 mW to more than 100 W, depending on the applied excitation.

In order to compare numerical results to experimental data, a method must be prepared to deal with this different range of values. Thus, an adaption of f_{sim} to f_{real} and $P_{e,sim}$ to $P_{e,real}$ is needed. Furthermore, measurements must be performed to get experimental result for this comparison.

2.3. Experimental Investigations on Genuine Cores

Each of the calculated characteristic curves, which are shown in figure 3 and 4, is also suitable for an experimental determination. The eddy current losses, which are shown in figure 3, describe the large-signal behavior of magnetic cores. They cannot be directly measured. The iron losses P_{Fe} are determined for genuine cores with a measurement set-up according to [3]. Then, the standard method of separation [3] is applied to extract the eddy current losses. The complex parts of the permeability, which are depicted in figure 4, describe the small-signal behavior of magnetic cores. On genuine cores the measurements of the permeability curves are carried out with an impedance analyzer and, at small frequencies, the standard separation method is also applied to the imaginary parts of the permeability. Experimental results can be provided, which are appropriate for a comparison to numerical calculations.

However, the model set-up in figure 2 features complete, perfect insulation between the tape layers. Complete insulation may be implemented in genuine cores with some more or less minor effort, but such an implementation leads to certain disadvantages for other core parameters (e.g. mechanical stress or a reduced filling factor). Thus, in most cases considerations on a balanced core characteristic lead to the decision to apply less elaborated insulation methods. The amount of insulation clearance in genuine cores is formed and controlled by a statistical equal distribution of partially thinner covered spots. Here, a parameter D_E will indicate this distribution of spots, thus, the insulation coverage. Complete insulation with full coverage should be characterized with $D_E = 0$, most genuine cores feature $D_E > 0$.

To use the FEM for an investigation of such genuine cores, the model set-up must be adapted to an insulation with clearances.

2.4. Model Adaption

To achieve usability of the FEM model for a sufficient description of experimental results two challenges must be dealt with. One is the different insulation coverage in the model and in genuine cores. The other is the difference in the ranges of values for the frequencies and the eddy current losses.

The difference in the insulation coverage can be overcome by an introduction of clearances in the surfaces, which represent the insulation in the model set-up. Such clearances are introduced in the model as a pattern of quadratic spots on two levels, as shown in figure 5.

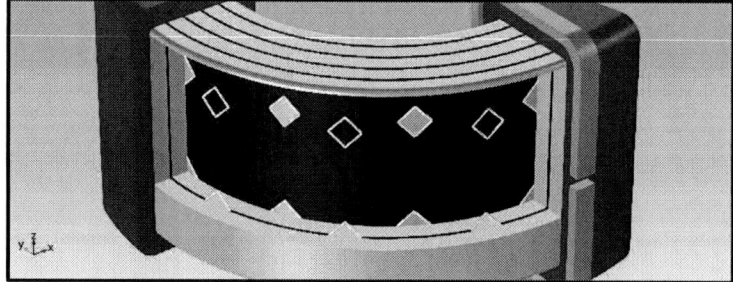

Fig. 5. Quadratic clearances in the electrical insulation.

The parameter to quantify this pattern is the insulation coverage in the model. This is the ratio F of the spots surface to the complete insulation surface. A perfect insulation coverage features F = 0 and calculations with it fix the limiting values for the characteristics of tape wound cores.

The differences in the range of values can be overcome by a scaling of the frequency f_{sim}, which can be performed by normalizing. To execute the normalizing, let $f_{n,sim}$ be a frequency to normalize f_{sim} and $f_{n,real}$ a frequency to normalize f_{real}. Then, the normalized frequency f_n is once $f_n = f_{sim}/f_{n,sim}$ for the simulation and once $f_n = f_{real}/f_{n,real}$ for the experimental data. Assembling the two equations and resolving for f_{real} yields

$$f_{real} = \frac{f_{n,real}}{f_{n,sim}} \cdot f_{sim}. \tag{3}$$

Likewise, the equation $R = 2\pi f \cdot \mu_0 \cdot N^2 \cdot \mu'' \cdot A_{Fe}/l_{Fe}$, which describes the series equivalent resistance of an inductance [3], may be used to provide a similar formula for $P_{e,real}$. Here, f is the frequency, N is the number of turns of the winding, A_{Fe} is the core cross sectional area and l_{Fe} is the mean iron path length. Inserting the equation of the series equivalent resistance in equation (2) and performing some mathematical operations, which are correspondent to the ones carried out for the derivation of equation (3), yields

$$P_{e,real} = \frac{f_{n,real} \cdot \eta_{real}}{f_{n,sim}} \cdot P_{e,sim}. \qquad (4)$$

η_{real} is the filling factor of the genuine core in the experiment, the respective cut-off frequencies are used as $f_{n,sim}$ and $f_{n,real}$.

3. Results

With the achievements gained by the model adaption, the FEM can be used for the description and interpretation of experimental data. This will be demonstrated with two examples.

3.1. Large-Signal and Small-Signal Behavior

Calculations for large-signal data and small-signal data are shown in figure 3 and 4. A transfer of data between large-signal analysis and small-signal analysis is not simple at any rate. While comparing such data, one should take care to avoid misinterpretation, as shown here.

In figure 6 numerically calculated eddy current losses are plotted together with experimental results for a toroidal core with d_o = 40 mm. One of the numerical results is shown for an insulation featuring clearances with F = S and one is plotted for an insulation with F = L. L and S are real numbers, which are determined and fixed to fit the experiment. L is more than a factor of ten greater than S. The calculation is adapted to the measurement with the equations (3) and (4). Classical theory is applied for a tape of 18 µm thickness to calculate the eddy current cut-off frequency, which is used for $f_{n,real}$. The frequency, which is used for $f_{n,sim}$, is determined from a simulation with perfect insulation.

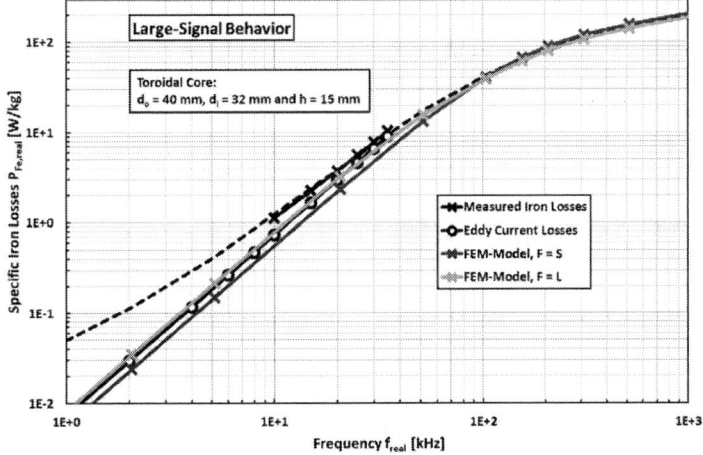

Fig. 6. Calculated and measured eddy current losses.

Below the eddy current cut-off, the curve for F = L fits the experimental results. In this frequency range the curve for F = S is too low.

For the same toroidal core with d_o = 40 mm measured permeability curves are plotted together with numerically calculated curves in figure 7. The numerical results are evaluated from the same simulation as the curves in figure 6. Again one calculated result is shown for an insulation with F = S and one for an insulation with F = L. The same frequencies are used for $f_{n,sim}$ and $f_{n,real}$ to adapt the calculation to the experiment with equation (3).

Both calculated curves for the real parts of the permeability fit the experimental results well below the eddy cut-off frequency. For the imaginary parts, the curve for F = S fits the experimental results well, while the curve for F = L does not fit in this frequency range.

© VDE VERLAG GMBH · Berlin · Offenbach

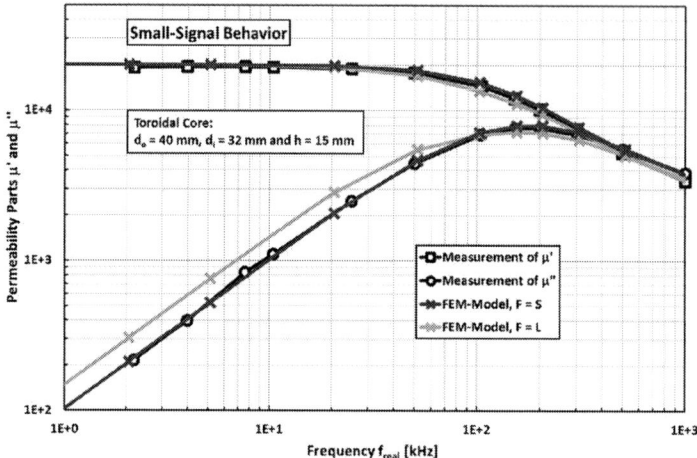

Fig. 7. Calculated and measured permeability curves.

Since the data in both figures apply to the same genuine core, it appears, that for small-signal analysis the amount of insulation coverage is larger than for large-signal analysis. The insulation clearance is composed of partially thinner covered spots and for small signals the thinner insulation cover persists. However, for larger signals, when higher electric fields drop across the tapes, discharges may break through at some of this points.

Thus, in order to tune the working point to the desired amount of insulation coverage the targeted application characteristic has to be considered thoroughly.

3.2. Tuning the Insulation Coverage

For most genuine cores the demand of a balanced performance results in an insulation coverage with clearances and an adjusted working point. Then, to tune the working point to an optimum, an estimation of the amount of insulation clearances in correlation to the expected performance would be extremely useful. This will be investigated here.

Experimentally determined eddy current losses for a toroidal core with $d_o = 100$ mm are plotted together with numerically calculated results in figure 8.

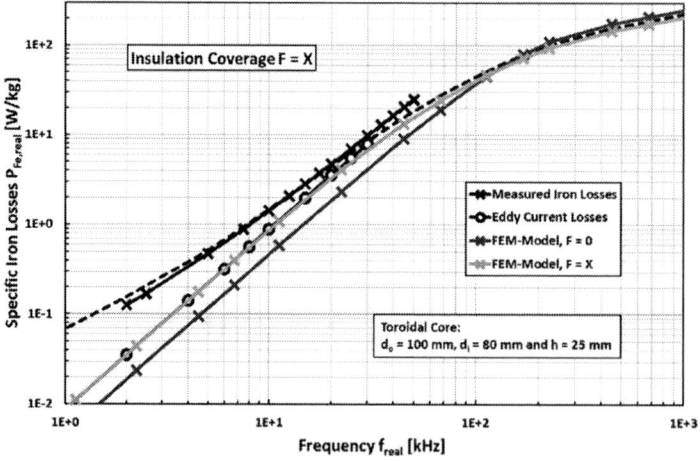

Fig. 8. Measured and calculated eddy current losses.

One of the numerical results is shown for a complete insulation with F = 0 and one is depicted for an insulation featuring clearances with F = X. X is a real numbers, which is fixed to fit the experiment. The adaption of the calculation to the measurement is performed with the equations (3) and (4). The frequency $f_{n,real}$ is calculated with the classical theory for a tape of 18 μm thickness, the frequency $f_{n,sim}$ is determined from the simulation with perfect insulation.

Below the eddy current cut-off, the curve for F = 0 is too low, while the curve for F = X fits the experimental results. For this core the eddy current losses are a factor of around 1.75 higher than the calculated losses for a core with complete insulation coverage.

To estimate and tune the insulation performance, the increase of the simulated $P_{e,sim}$ is compared to the reference F = 0 in figure 9. The model of the toroidal core with d_o = 100 mm is used for the calculation and the edge length of the quadratic clearance areas in the insulation surfaces (see figure 5) is varied. The insulation coverage changes and the percentage rise of the eddy current loss is plotted over F. The point F = X is indicate with a vertical, dashed line.

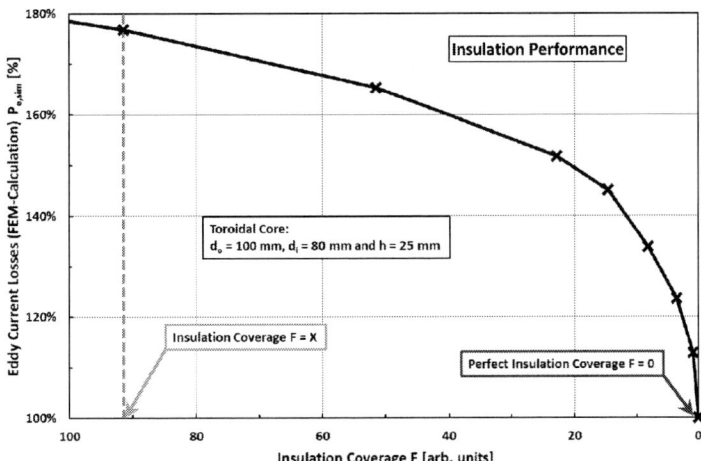

Fig. 9.   Calculated eddy current losses versus insulation coverage.

In the depicted range of F the eddy current loss shows a non-linear behavior. With an increasing insulation coverage a decrease of the eddy current loss can be seen, while the gradient of the decrease is continuously growing.

For given genuine cores such curves may be used to estimate the expected eddy current losses. Then, for example, while adjusting the working point in the design process of a nanocrystalline tape wound core, the number of genuine samples may be reduced to an inevitable minimum.

4. Summary and Outlook

In this paper FEM calculations of nanocrystalline tape wound cores are presented. The numerical results are tested by analytical solutions and, moreover, compared to experimental measurements.

Two drawbacks for the comparison with the experiment are identified and solved. One is the insulation design in a genuine core, which is adjusted in order to feature a balanced core characteristic. It is overcome by the introduction of a spot pattern on the insulation surfaces in the model, to adapt the set-up. The other is the different range of values for the frequencies and the eddy current losses in simulation and experiment. This drawback results from limited computational power leading to different geometrical tape thicknesses in genuine cores and in model set-ups. It is resolved by the preparation of an elaborated scaling technique for the calculated data.

After solving this challenges, the numerical simulation data fit the experimental results well. FEM calculations are made suitable to describe experiments and enhance the interpretation of data. This is demonstrated on two examples.

With the enhancements, which are presented in this paper, the application of FEM calculations to nanocrystalline tape wound cores will lead to various improvements. For example, the product development can receive a fundamental strengthening, especially when challenges occur in a design process, which cannot be handled with established approaches.

Moreover, the potential for a suitable description of experimental data will push basic developments, especially when the data interpretation is difficult or even impossible. One example may be the treatment of cut cores. The introduction of an air gap in the magnetic circuit leads to various advantages [3]. As an example, a calculation of the core from figure 1 is presented in figure 10, with an air gap introduced.

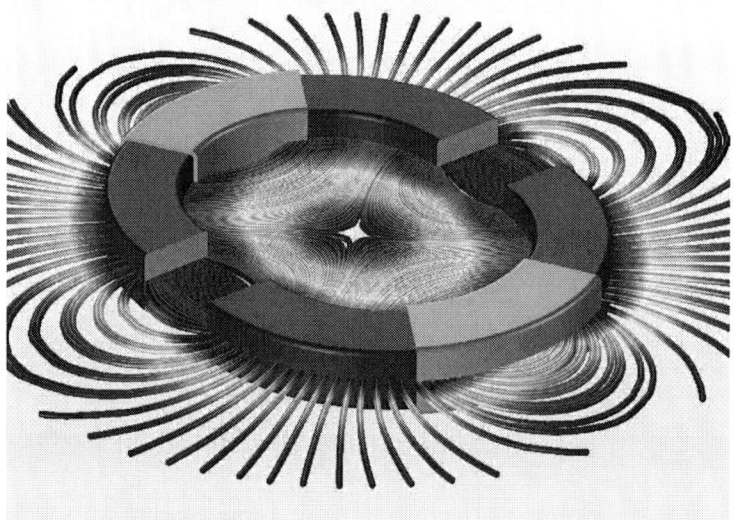

Fig. 10. Typical pattern of an H-field configuration for a gapped cut core.

A change in the magnetic field configuration can be observed and the field strength increases. Thus, an increased amount of magnetic energy is present in the magnetic circuit and enhances potential applications.

The FEM is an appropriate and helpful tool for the investigation of nanocrystalline tape wound cores and will be applied for further development and optimization.

5. References

[1]   H. Schwenk, J. Beichler, W. Loges, C. Scharwitz, "Actual and Future Developments of Nanocrystalline Magnetic Materials for Common Mode Chokes and Transformers", PCIM Europe 2015, Conference Proceedings

[2]   https://www.comsol.de (2016-03-01)

[3]   R. Hilzinger, W. Rodewald, "Magnetic Materials", Publics Publishing, ISBN 978-3-89578-352-4

[4]   H. Stöcker, "Taschenbuch der Physik", Verlag Harri Deutsch, ISBN 3-8171-1720-5

PCIM Europe 2016, 10 – 12 May 2016, Nuremberg, Germany

An Insightful Evaluation of a 650V High-Voltage GaN Technology in Cascode and Stand-Alone Transistors

Jaume, Roig, ON Semiconductor, Belgium, jaume.roig@onsemi.com

Manuel, Fernandez, Universidad de Oviedo, Spain, Manuel.Fernandez@imb-cnm.csic.es

German, Gomez, ON Semiconductor, Belgium, german.gomez@onsemi.com

Frederick, Declercq, ON Semiconductor, Belgium, frederick.declercq@onsemi.com

Diego, G. Lamar, Universidad de Oviedo, Spain, gonzalezdiego@uniovi.es

Filip, Bauwens, ON Semiconductor, Belgium, filip.bauwens@onsemi.com

Abstract

A switching performance comparison between 650V power transistors is carried out in this work by assembling identical GaN DHEMTs as Stand-Alone (GaN-SA) and Cascode (GaN-CS) in TO-220 and SMD packages. To our knowledge, this is the first time that this comparison is reported using identical DHEMTs. Our investigations are focused on hard-switching measurements also including best-in-class Super-Junction MOSFETs. It is experimentally proven that, after proper GaN-CS optimization, waveform ringing and power loss are significantly reduced with respect to a previous GaN technology. Additional physics-based simulations help to elucidate power loss mechanisms, identify key parameters and define suitable operation conditions.

1. Introduction

In the arena of High-Voltage (HV) power transistors for High-Frequency (HF) converters (f_{sw} > 60 kHz, 500 V > BV_{DSS} > 900 V) Si Super-Junction (SJ) technologies are still predominant at low/medium power range (< 1 kW). In spite of this, the first HV GaN transistors have been slowly introduced into the market since 2012 [1] and a co-existence of HV GaN and Si technologies is reckoned in the following years. Among the large diversity of HV GaN technologies, cascode (GaN-CS) and stand-alone (GaN-SA) DHEMTs are two of the most common solutions for normally-off and normally-on transistors, respectively. This work investigates the dynamic performance of these two different approaches in DHEMTs fabricated with our own ON Semiconductor GaN-on-Silicon technology and assembled in Surface Mounted Device (SMD) and TO-220 as indicated in Fig. 1a. Although not explicitly mentioned, electrical data in prior benchmarking exercises between GaN-CS and GaN-SA suggest that DHEMT is not equal [2,3]. The comparative analysis reported in this work follows a methodology that combines experiment and simulation. A boost converter topology has been chosen due to its simplicity to analyze device transient behavior. This topology has been implemented in a dedicated board with high controllability, versatility and sensing capabilities. As a part of this methodology, finite-element structures for SJ and GaN ultimate technologies are calibrated with high precision (see parasitic capacitances in Fig. 1b) and eventually inserted in mixed-mode (MM) simulation using Sentaurus™ [4]. In MM simulations, physics and SPICE equations are consistently solved to investigate the interaction device-circuit and to give insight into the mechanisms originating energy loss during commutation. This is an additional value with respect to other works, where power loss is empirically or behaviorally assessed [5,6,7].

© VDE VERLAG GMBH · Berlin · Offenbach

PCIM Europe 2016, 10 – 12 May 2016, Nuremberg, Germany

2. Experimental Comparison of Waveforms and Power Losses

2.1. Description of test setup and device selection

A set of three dedicated test boards has been developed for versatile comparison between normally-off and normally-on transistors in TO-220 and SMD. Each one of the boards features a boost converter (300 W, $V_{IN}:V_{OUT}$ 100:400 V, Continuous Conduction Mode, f_{sw} = 100 - 800 kHz) with reduced main loop inductance (L_{LOOP} < 30 nH) and accurate current sensing (B_w = 2 GHz). A SSDN-10 from T&M senses I_D whereas P6139A passive and P6251 differential voltage probes from Tektronix measure V_{DS} and V_{GS} in the DUT. Keeping most of the design identical, driver and socket are the main variants between these boards. One of the boards, shown in Fig. 2a, includes a 4-contact slot with Kelvin contact, where small PCBs adapted to SMD footprints are plugged-in (see an small PCB example in Fig. 2c). The other two boards include a TO-220-compatible socket but differ in the type of driver, which is optimized for normally-on or normally-off transistors. For normally-off, the driving range is fixed between 0 V and 12 V whereas for normally-on low voltage is fixed to -15 V and high voltage can be tuned to 0 V, +2 V and +5 V.

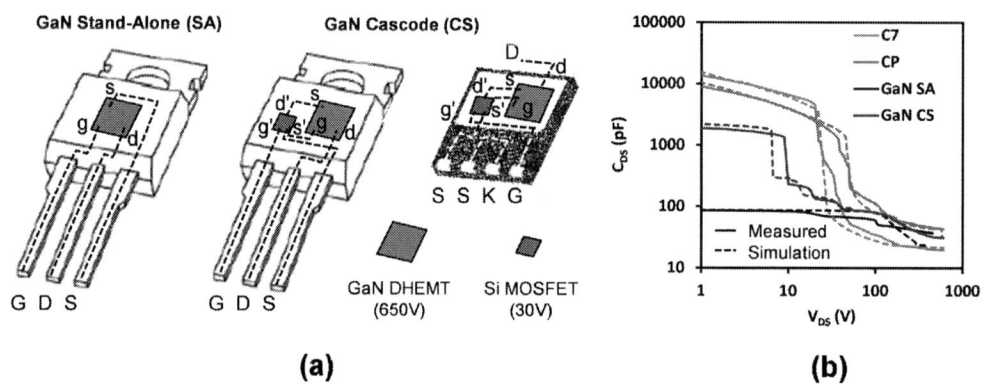

Fig. 1. (a) Schematic description of GaN DHEMTs in Stand-Alone (GaN-SA) and Cascode (GaN-CS) in TO-220 and SMD (8x8 PQFN). (b) Measured (solid) and simulated (dashed) C_{DS} vs. V_{DS} for CP, C7, GaN-SA, and GaN-CS.

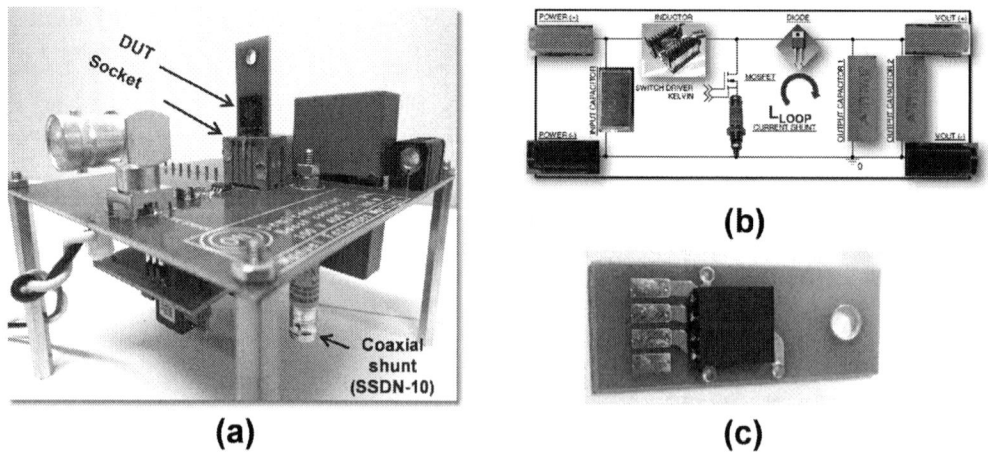

Fig. 2. (a) Dedicated test board with slot and small PCB to test SMD transistors. (b) Circuit scheme of the 300W boost converter. (c) Detail of small PCB board with soldered SMD transistor.

© VDE VERLAG GMBH · Berlin · Offenbach

	BV_{DSS} @25C (V)	R_{ON} @25C (mΩ)	V_{TH} @25C (V)	Q_{GS} (nC)	Q_{GD} (nC)	Q_G (nC)	E_{OSS} @400V (uJ)	R_G (Ω)	Package Type	$(R_{ON}{}^*Q_{GD})^{1/2}$ $(mΩ^*nC)^{1/2}$	$(R_{ON}{}^*Q_{SW})^{1/2}$ $(mΩ^*nC)^{1/2}$	$(R_{ON}{}^*E_{OSS@400V})$ $(Ω^*uJ)$
IPP60R165CP	600	150	3	8.3	13	40	6.6	1.9	TO-220	44.2	79.8	1.0
IPP65R190C7	650	170	3.5	6	6.5	22	3.1	1.1	TO-220	33.2	63.6	0.5
GaN-SA Gen2	650	90	-6	1.9	4	9.6	3.5	0.8	TO-220	19.0	30.3	0.3
GaN-CS Gen2	650	100	6	3.5	3.3	18.2	3.8	1	TO-220	18.2	43.9	0.4

Table 1: List of TO-220 devices under test with their main electrical characteristics and FoMs.

	BV_{DSS} @25C (V)	R_{ON} @25C (mΩ)	V_{TH} @25C (V)	Q_{GS} (nC)	Q_{GD} (nC)	Q_G (nC)	E_{OSS} @400V (uJ)	R_G (Ω)	Package Type	$(R_{ON}{}^*Q_{GD})^{1/2}$ $(mΩ^*nC)^{1/2}$	$(R_{ON}{}^*Q_{SW})^{1/2}$ $(mΩ^*nC)^{1/2}$	$(R_{ON}{}^*E_{OSS@400V})$ $(Ω^*uJ)$
IPL60R199CP	600	180	3	8	11	32	6.1	2	8x8 SMD	44.5	78.7	1.1
IPL65R130C7	650	115	3.5	8	11	35	4.2	1	8x8 SMD	35.6	65.6	0.5
GaN-CS Gen1	600	155	2	2.2	2.3	11.8	5	1	8x8 SMD	18.9	43.9	0.8
GaN-CS Gen2	650	100	6	3.5	3.3	18.2	3.8	1	8x8 SMD	18.2	43.9	0.4

Table 2: List of SMD devices under test with their main electrical characteristics and FoMs.

The list of selected devices for benchmarking is summarized in Tables 1 and 2 for TO-220 and SMD, respectively. Best-in-class Silicon technologies, represented by CP and C7, are compared to ON Semiconductor's first two generations of GaN transistors, called Gen1 and Gen2. Aside to use a new GaN DHEMT process, Gen2 further optimizes the cascode configuration with regard to Gen1. Detailed information on device ruggedness improvements and Q_{GD} reduction are published somewhere else [8,9]. Based on datasheets and own characterization, the main electrical characteristics in Tables 1 and 2 are used to calculate some of the most widely-used Figures-of-Merit (FoMs) for hard-switching. At a glance, CP appears as the less performant technology for any FoM whereas C7 exhibits a very competitive $R_{ON}{}^*E_{OSS@400V}$ in comparison to GaN transistors.

2.2. Experimental results

Measured waveforms for V_{GS}, V_{DS} and I_D are plotted in Figs. 3 and 4 for TO-220 and SMD transistors, respectively. All these waveforms are extracted under identical operating conditions (f_{sw} = 100 kHz P_{OUT} = 300 W) and external gate resistance (R_{G_EXT} = 6.8 Ω). In general, turn-off decay of V_{GS} and Miller plateau times are consistent with the listed capacitances, being both GaN-CS and GaN-SA faster in switching speed. During turn-on, large di/dt combined with high source inductance (L_S) produce severe noise in TO-220 GaN-CS due to the resonance effect in the CS internal loop. At this point, larger V_{TH} in Gen2 (6 V i.o. 2 V) avoids hard accidental turn-off that, for the same test conditions, produces exaggerated power loss and eventual destruction in Gen1. Minimization of L_S and gate bouncing is achieved by using SMD with Kelvin contact. This is translated into a mitigation of turn-on oscillations and suppression of turn-off V_{GS} bump. Dissimilarities in waveforms have a clear impact on FET power loss (P_{FET}) and, subsequently, on the total system power loss (P_{LOSS}). Measured P_{LOSS} is displayed in Figs. 5a and 5b for TO-220 and SMD transistors, respectively. P_{LOSS} variations between boards could be attributed to differences in routing and homemade power inductor; however a higher increase of temperature is also expected when dealing with SMD package. Subsequently, higher contribution of conduction loss (P_C) is also expected in SMD devices.

© VDE VERLAG GMBH · Berlin · Offenbach

PCIM Europe 2016, 10 – 12 May 2016, Nuremberg, Germany

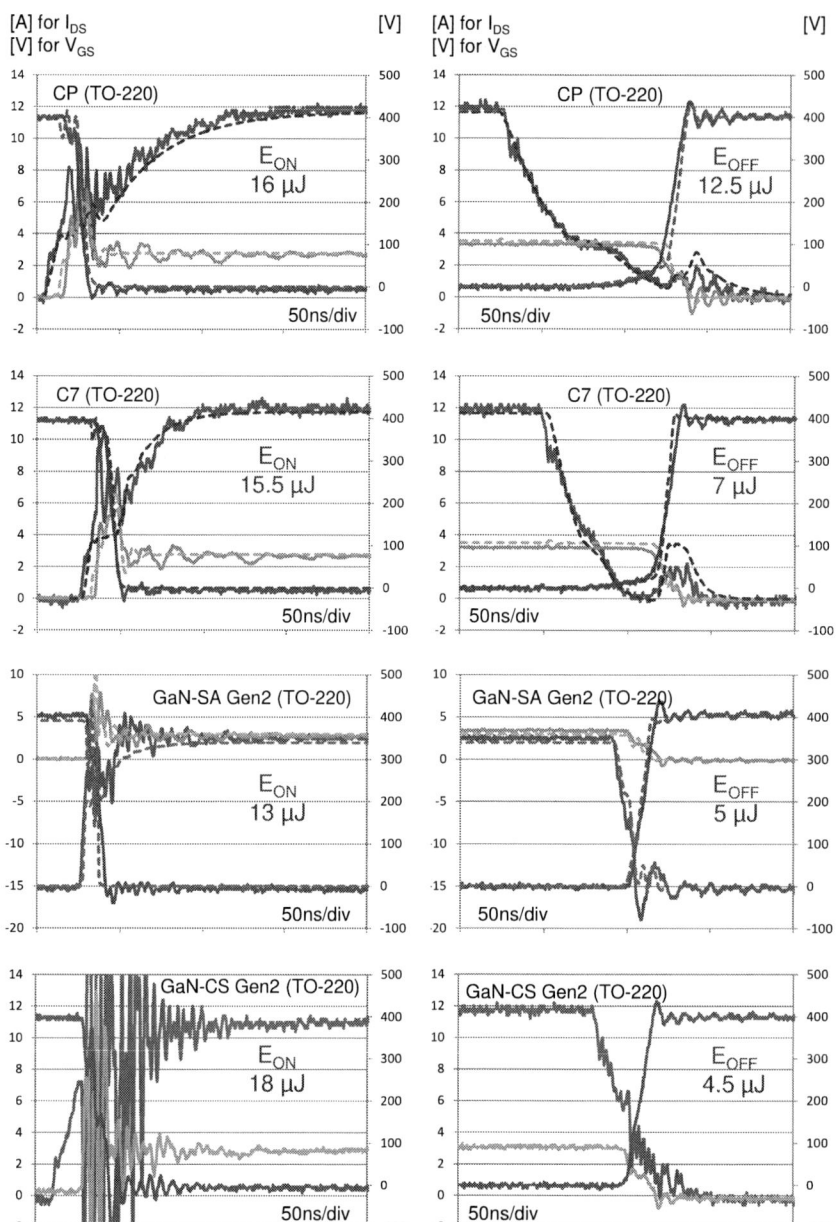

Fig. 3: Measured (solid) and simulated (dashed) curves for TO-220 devices in Table 1 at 300 W, $R_{G\_EXT} = 6.8\ \Omega$ and $V_{DR} = 0$ V to +12 V (red-V_{DS} at right y-axis, blue-V_{GS} and green-I_D at left y-axis). V_{DR} = -15 V to +2 V for GaN-SA. E_{ON} and E_{OFF} are calculated by integrating instantaneous power (I_D*V_{DS}).

In order to filter possible differences between boards and self-heating, switching energy (E_{SW} = E_{ON} + E_{OFF}) is calculated by integrating instantaneous power loss (I_D*V_{DS}) along the turn-on and turn-off switching times, in Figs. 3 and 4. It is worth mentioning that other energy loss contributions due to driving and ringing loss have been neglected. A comparison between measured E_{SW} is also plotted in Fig. 6, showing a clear trend to reduce E_{SW} in SMD devices. At 100 kHz most of P_{LOSS} variations within SMD and TO-220 groups can be explained by E_{SW} variations, except GaN-SA that has been measured in a different board with an exclusive driver. Differently, P_{LOSS} differences at 500 kHz can only be explained by the influence of self-heating that leads to a more accentuated increase of conduction losses in Silicon technologies.

© VDE VERLAG GMBH · Berlin · Offenbach

PCIM Europe 2016, 10 – 12 May 2016, Nuremberg, Germany

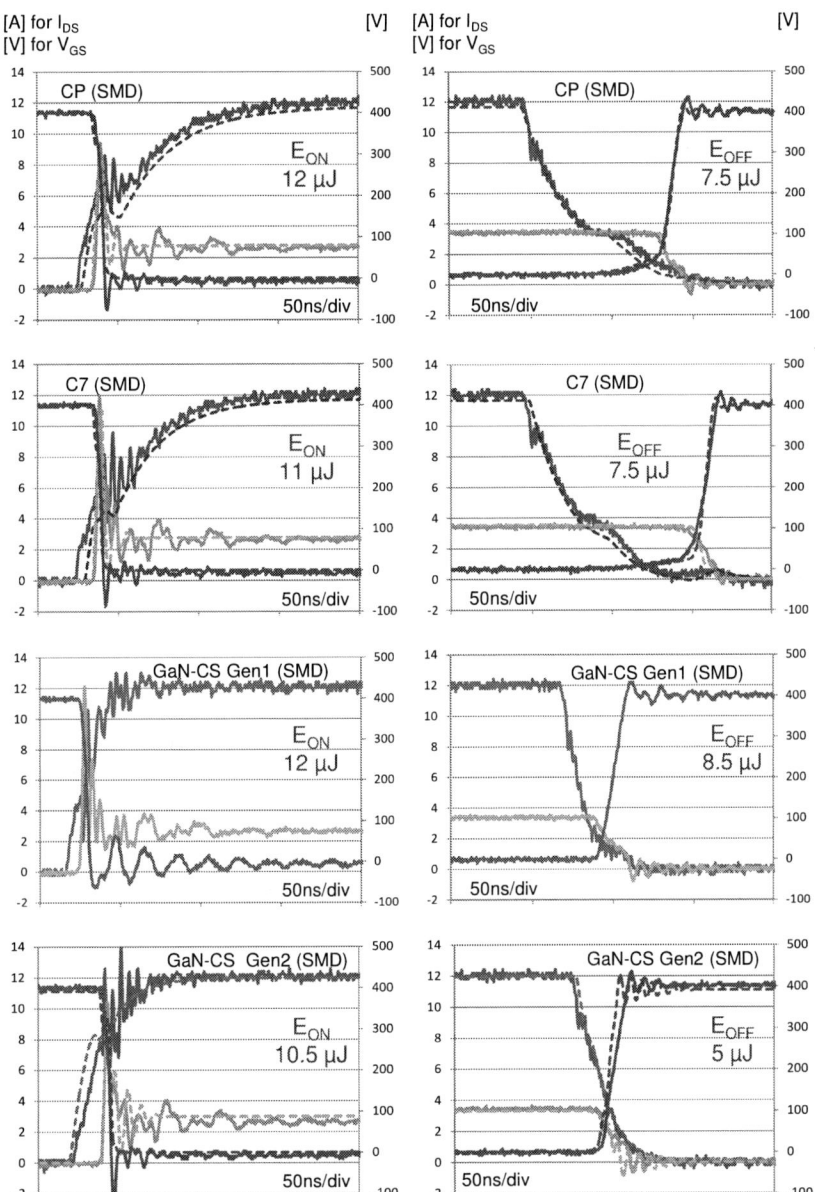

Fig. 4: Measured (solid) and simulated (dashed) curves for SMD devices in Tables 2 at 300 W, R_{G_EXT} = 6.8 Ω and V_{DR} = 0 V to +12 V (red-V_{DS} at right y-axis, blue-V_{GS} and green-I_D at left y-axis). E_{ON} and E_{OFF} are calculated by integrating instantaneous power (I_D*V_{DS}).

The ratio between $R_{ON@150C}$ and $R_{ON@25C}$ is compared in Fig. 5c for different technologies under test with clear disadvantage for Silicon devices. Nonetheless, not all existing GaN technologies exhibit low $R_{ON@150C}$ / $R_{ON@25C}$ ratio [10]. In all cases, the power loss contribution of the power switch (P_{FET}) accounts for 30 % to 50 % of total P_{LOSS}. For the specific case of GaN-SA, an E_{SW} variation of -1.5 µJ and +4.5 µJ is obtained when setting high voltage level of the driver to +5 V and 0 V, respectively. More concretely, this variation occurs in E_{ON} and has its origin in a shorter crossing time between I_D and V_{DS}. All FoMs in Tables 1 and 2 agree that CP shows the worst performance but only $R_{ON}*E_{OSS@400V}$ correctly predicts the performance ranking of the transistors at f_{sw} = 100kHz. This is in agreement with the higher weight of E_{OSS} over E_{SW} at light and medium loads. At f_{sw} = 500 kHz all FoMs become predictive if $R_{ON@150C}$ is used.

© VDE VERLAG GMBH · Berlin · Offenbach

PCIM Europe 2016, 10 – 12 May 2016, Nuremberg, Germany

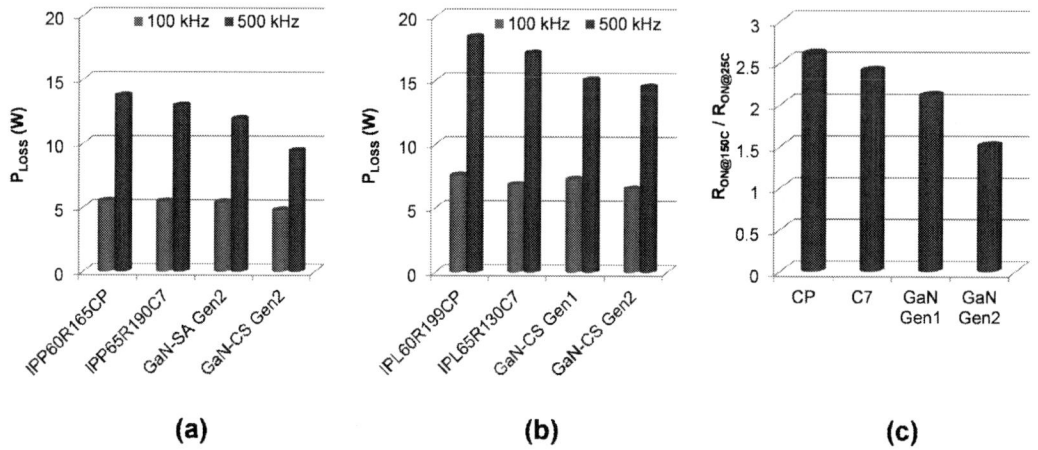

(a)   **(b)**   **(c)**

Fig. 5: Measured P_{LOSS} at an f_{sw} of 100 and 500 kHz for (a) TO-220 and (b) SMD transistors (R_{G_EXT} = 6.8 Ω, V_{DR} = 0 V to +12 V). V_{DR} = -15 V to +2 V for GaN-SA. (c) Ratio between $R_{ON@150C}$ and $R_{ON@25C}$ for different technologies under test.

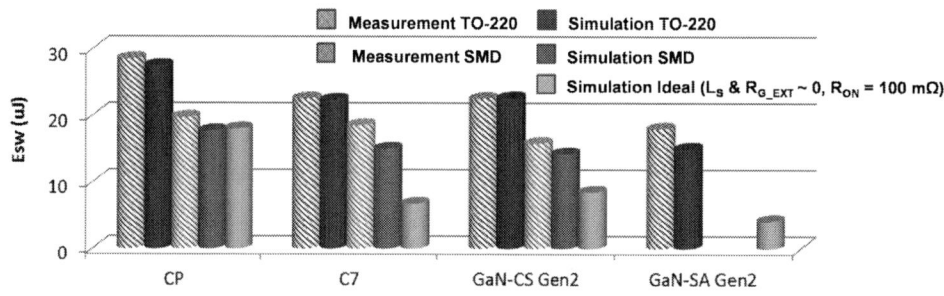

Fig. 6: Measured and simulated E_{sw} for transistors in Tables 1 and 2. Striped and solid columns correspond to TO-220 and SMD transistors, respectively. Green columns correspond to ideal simulation with L_S and R_{G_EXT} ~ 0 (all cases scaled to R_{ON} = 100 mΩ).

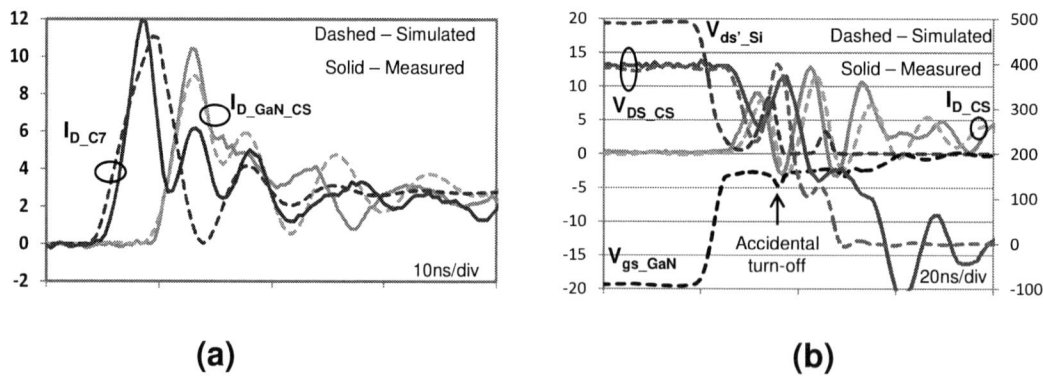

(a)   **(b)**

Fig. 7: (a) Zoom of measured and simulated I_D overshoot during turn-on of C7 and GaN-CS Gen2 (SMD package). (b) Measured and simulated ringing for GaN-CS Gen2 in TO-220 during turn-on. Accidental turn-off is caused by V_{gs_GaN} bouncing.

© VDE VERLAG GMBH · Berlin · Offenbach

3. Elucidation of Power Loss Mechanisms by Mixed-Mode Simulations

3.1. Calibration and validation of MM simulations

After meticulous calibration of device structures and parasitic inductances, the high precision of MM simulations to reproduce measured waveforms is shown by the dashed lines in Figs. 3, 4 and 7. This comparison is not possible for GaN-CS Gen1 due to the unavailability of its device structures. Besides typical commutation, MM also emulates non-conventional switching effects like the prominent turn-on ringing for GaN-CS in TO-220, as shown in Fig. 7b. This effect has been observed to be caused by a repetitive accidental turn-off and it fully disappears by using the SMD package with Kelvin contact. It is inferred from Fig. 7b that, during the turn-on process, V_{gs_GaN} is pulled-down with a subsequent V_{DS_CS} overshoot. This effect has been sometimes observed to occur several times, thus jeopardizing device reliability and degrading system efficiency. The high accuracy of the waveforms is translated to the corresponding E_{SW} extracted from simulation. Small errors below 20 % are demonstrated in Fig. 6, where larger discrepancies are correlated to cases with high di/dt. As a matter of fact, large di/dt produces ΔV across L between sensing voltage point end real electrode and, subsequently, producing an error on measured waveforms. Interestingly, MM simulation can provide correction to switching losses calculated by measured waveforms in conditions where measurement is beyond the confidence limits.

3.2. Extended benchmarking analysis by MM simulation

Once MM simulations are calibrated and validated, the benchmarking analysis between transistors is extended to a wide range of operating conditions. As a matter of example, an ideal scenario without L_S and external gate resistance (R_{G_EXT}) is defined to explore the lower limits for E_{SW}. In a specific study, GaN-CS Gen2, GaN-SA Gen2 and C7 are scaled to R_{ON} ~ 100 mΩ and compared in the previous 300 W boost converter. The resulting E_{SW} values are plotted with green columns in Fig. 6, showing that GaN-SA Gen2 has the largest margin of improvement. From the Silicon side, CP does not benefit from the new conditions whereas C7 takes significant advantage with E_{SW} even lower than GaN-CS Gen2. In a second study, the evolution of E_{SW} with R_{G_TOTAL} ($R_G + R_{G\_EXT} + R_{DR}$) has been investigated. As already observed for SJ transistors in [11,12], GaN-CS and GaN-SA exhibit an E_{SW} non-linear dependence with R_{G_TOTAL}. At small R_{G_TOTAL}, E_{SW} evolution with R_{G_TOTAL} is mainly governed by dV_{DS}/dt stagnation during turn-off. Such stagnation takes place when the time to charge C_{OSS} is larger than the time to charge C_{ISS}. At large R_{G_TOTAL} and I_{LOAD}, E_{SW} raises faster for C7 due to a more predominant role of the Miller plateau in the turn-off energy loss.

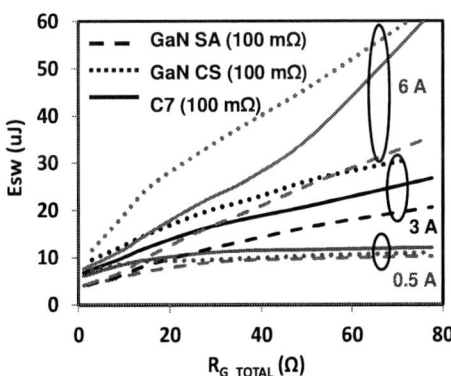

Fig. 8: Simulated E_{sw} vs. R_{G_TOTAL} for GaN-CS Gen2, GaN SA Gen2 and C7 scaled to 100mΩ (I_{LOAD} = 0.5, 3 and 6A)

© VDE VERLAG GMBH · Berlin · Offenbach

4. Conclusions

A benchmarking study between ultimate HV GaN and Silicon technologies is presented in this paper for hard-switching. At high frequencies ON Semiconductor's second GaN generation proves to be clearly superior to Silicon in terms of system power loss. Moreover, the immunity to accidental turn-off in GaN cascode configuration has been improved with respect to the first GaN generation. Interestingly, savings on system power loss are not uniquely attributed to an ultra-fast switching speed but also to a significantly smaller increase of on-state resistance with temperature (halving the Silicon one).

5. Acknowledgements

The work has been performed in the project E2COGaN, co-funded by grants from Belgium, The Netherlands, Germany, France, Italy, Austria, Slovakia, The United Kingdom and the ECSEL Joint Undertaking.

References

[1] P. Parikh, Y.F. Wu, L.K. Shen, "Commercialization of High 600V GaN-on-Silicon Power Devices," In Materials Science Forum, vol. 778, pp. 1174-1179. 2014.

[2] T. Hirose et al., "Dynamic performances of GaN-HEMT on Si in cascade configuration," In Applied Power Electronics Conference and Exposition (APEC), 2014 Twenty-Ninth Annual IEEE, pp. 174-181. IEEE, 2014.

[3] W. Saito et al. "Switching controllability of high voltage GaN-HEMTs and the cascode connection," In Power Semiconductor Devices and ICs (ISPSD), 2012 24th International Symposium on, pp. 229-232. IEEE, 2012.

[4] Sentaurus TCAD Tools Suite. Synopsys 2010.

[5] R. Mitova et al., "Investigations of 600-V GaN HEMT and GaN diode for power converter applications," Power Electronics, IEEE Transactions on 29, no. 5 (2014): 2441-2452.

[6] Z. Liu, X. Huang, F.C. Lee, Q. Li, "Simulation model development and verification for high voltage GaN HEMT in cascode structure," In Energy Conversion Congress and Exposition (ECCE), 2013 IEEE, pp. 3579-3586. IEEE, 2013.

[7] Z. Xu, W. Zhang, F. Xu, F. Wang, L.M. Tolbert, B.J. Blalock, "Investigation of 600 V GaN HEMTs for high efficiency and high temperature applications," In Applied Power Electronics Conference and Exposition (APEC), 2014 Twenty-Ninth Annual IEEE, pp. 131-136. IEEE, 2014.

[8] J. Roig et al., "Unified theory of reverse blocking dynamics in high-voltage cascode devices," In Applied Power Electronics Conference and Exposition (APEC), 2015 IEEE, pp. 1256-1261. IEEE, 2015.

[9] M. Tack, P. Moens, C. Liu, F. Bauwens, "An industrial 650 GaN DHEMT Cascode Technology," ECS Transactions 64, no. 7 (2014): 171-183.

[10] E. A. Jones et al., "Characterization of an enhancement-mode 650-V GaN HFET," In Energy Conversion Congress and Exposition (ECCE), 2015 IEEE, pp. 400-407. IEEE, 2015.

[11] M. Treu et al., "The role of silicon, silicon carbide and gallium nitride in power electronics," In Electron Devices Meeting (IEDM), 2012 IEEE International, pp. 7-1. IEEE, 2012.

[12] Application Note 3994. STMicroelectronics.

Static Characterization of Discrete State-of-the-Art SiC Power Transistors

- real-life properties of SiC devices unveiled -

Michael Meisser, Horst Demattio, Thomas Blank
Institute for Data Processing and Electronics (IPE), Karlsruhe Institute of Technology (KIT)
Karlsruhe, Germany, michael.meisser@kit.edu

Abstract:

Silicon carbide transistors in TO247 packages are wide-spread despite of their limitations regarding parasitic inductance and power dissipation if an electrically insulated heat-sink attach is required. This paper publishes a comprehensive study on the static performance of commercially available SiC discrete transistors. The devices were characterized using a thermally controlled heat sink emulator and precision electrical measurement equipment. The device-specific differences regarding the R_{DSon} stability with regard to current and temperature in forward as well as in reverse operation were investigated.

1. Introduction

For six years, silicon carbide (SiC) discrete power transistors are available in Europe. Today, engineers can choose from a broad variety of SiC devices packaged in the common TO247 package which was originally developed for silicon devices. Being an industrial standard, it offers a huge flexibility regarding topologies and structural setup. On the other hand, the package shows noteworthy drawbacks. Already exhibiting a parasitic inductance of about 10 nH per lead, the standard gate-drain-source connection scheme impedes a low-inductive connection of the gate-driver. Source Kelvin-contacts can reduce the level and impact of parasitic lead inductance [1].

Since the TO247 has no intrinsic electrical insulation to its back side, an external insulation layer is often required adding significant thermal resistance to the thermal path. The ISOPLUS package contains an integrated DCB substrate providing electrical insulation while achieving low thermal resistance.

Datasheets provide valuable information about the maximum ratings of devices and their typical properties under certain general conditions. However, some test environments used in order to obtain datasheet data sets are unrealistic compared to real-world applications.

The classic method for measuring static properties is by pulsed measurements using curve tracers [2-5]. Although, the provided R_{th} together with the R_{DSon} data can be used in order to calculate the junction temperature and the resulting R_{DSon} for a given current and heat sink temperature, the increase of the R_{DSon} due to the heating of the die is not covered. Additionally, the temperature-dependency of the materials forming the total thermal path is not included.

Curve tracers may constantly power the device under test (DUT) while the DUT is connected to a heat sink [6]. In this case, the heat sink temperature is not kept constant but sets depending on the heat flow amplitude, its thermal resistance to ambient R_{thH-A} and ambient temperature.

The thermal path of chips mounted on a DCB substrate was investigated by [7]. Here, the heat-sink temperature is not measured directly underneath the die attach but below a bulk material of known thermal resistance R_{th}.

In contrast to the previously mentioned publications, this work focusses on measurements on SiC TO247 packages under close to real-world conditions. It provides data from continuous static measurements characterizing commercially available SiC transistors at defined heat sink temperatures.

2. Static Characterization Rig for Discrete Devices

For the static measurements the device under test (DUT) is both electrically and thermally connected to a characterization rig. The system is an enhanced version of the setup used in earlier publications [8, 9] and works in continuous mode in contrast to pulsed approaches [10]. The block diagram of the measurement system is depicted in Figure 1A. Figure 1B shows the electro-mechanical setup.

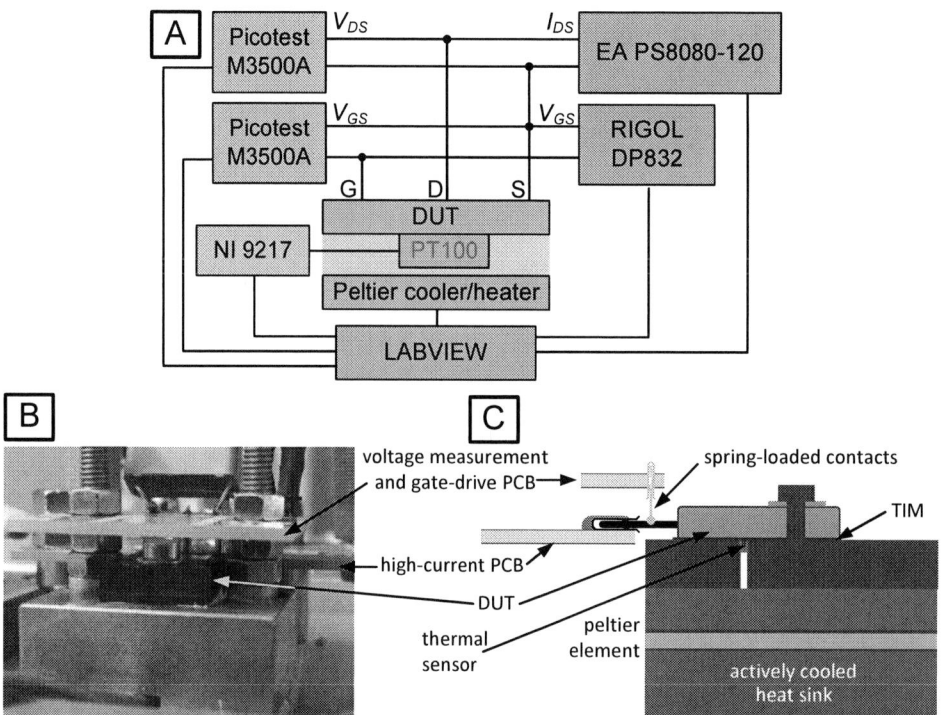

Figure 1: A: Block diagram of the characterization rig including the used equipment. B & C: Physical setup of the chuck, which is electrically and thermally connected to the DUT. The pins of the DUT are Kelvin-connected to the electric power supply and measurement devices. The DUT's backside is pressed onto a temperature-controlled copper surface representing a heat sink. The DUT's case temperature is monitored by a 1.2 x 1.6 mm PT 100 sensor located directly underneath the packaged transistor back plate.

The electric power is transferred to the DUT by a high-current PCB whose receptacles connect to the leads of the TO247 package. An additional PCB hosts spring-loaded contacts that provide Kelvin connections for the measurement of drain-to-source voltage V_{DS} and gate voltage V_G. Both voltages are measured by Picotest 3500A precision meters while the DUT is supplied by an EA PS8080-120 power supply accessing the drain and a DP832 power supply connected to the gate. The electrical power transferred to the DUT is limited to the value of maximum possible power dissipation of the DUT at maximum heat sink temperature. It is selected depending on the type of thermal interface material chosen in order to prevent overheating of the DUT.

The thermal connection is provided by a copper chuck being able to provide and remove thermal energy to the DUT. When the DUT is mounted on the chuck, a PT100 sensor touches the back side of the DUT underneath the die attach area. The sensor required a 1.5 mm hole which provides the exact case temperature of the DUT and only slightly disturbs the heat flow down to the chuck. Due to that, the results presented here should be read as conservative data overestimating the $R_{th,J-C}$. The next section of this paper reports about thermal simulations performed in order to estimate this effect.

The heat sink temperature T_{HS} monitored by the PT100 is used as input parameter for a Labview-based control scheme which accesses Peltier elements and fans in order to keep the chosen T_{HS} constant. The Labview software adjusts the temperature and the electrical input in order to follow a predefined measurement plan. At each reading point, all parameters are measured after T_{HS} stayed within a 0.5 K envelope for 10 s. This time range ensures a thermally settled system. Compared to earlier work [8, 9], the setup used here offers higher precision and higher heat-transfer capability.

3. Thermal Simulation

In order to determine the influence of the used measurement technique on the thermal path, the setup depicted in Figure 1 was modelled and simulated with COMSOL. Wolfspeed's C2M0080120 MOSFET had been taken as the reference device for this simulation. It is assumed that the heat resulting from power dissipation is mainly created close to the top side of the die. According to the grinding pattern shown in Figure 3, the solder layer between die and 2 mm copper back plate has a thickness of 60 µm.

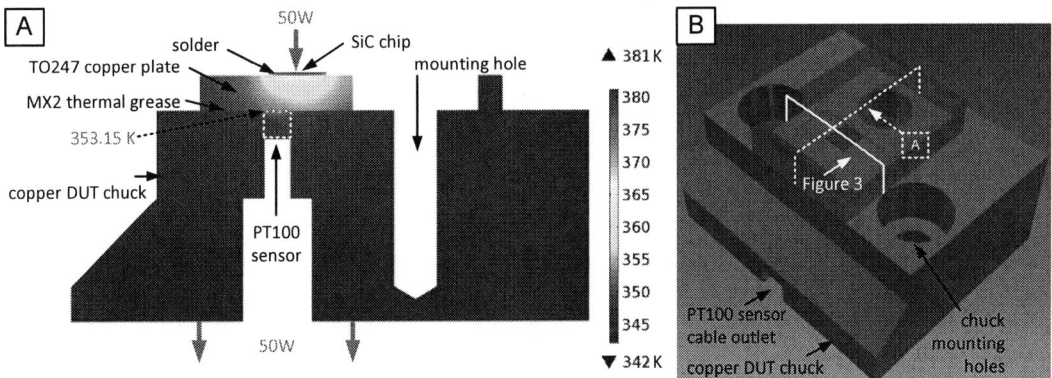

Figure 2: Modelling and thermal simulation of C2M0080120 MOSFET in TO247 package mounted onto the DUT chuck. A: Slice view of the thermal simulation. B: Simulation setup. The setup is modelled using datasheet information and measurements made cross-cutting the package. The CPM2-1200-0080B die is connected via a 60 µm solder layer to the TO247 back plate. The white solid arrow indicates where a grinding pattern according to Figure 3 was taken in order to measure the solder thickness of the die attach. The PT100 sensor is modelled as material with the 0.8 W/mK heat conductivity of the Fischer WLK5 thermal compound used to attach the sensor within the chuck hole.

The DUT chuck is implemented as copper block having the same dimensions as the original while the PT100 sensor is modeled as volume with the 0.8 W/mK heat conductivity given by the used Fischer WLK5 thermal compound. In the simulation 50 W of heat are fed into the top side of the die and drained from the bottom side of the DUT chuck. The sensor's top surface is set to have a fixed temperature of 80 °C. The TO247 package is thermally connected via a 30 µm layer of 5.6 W/mK MX-2 thermal grease.

© VDE VERLAG GMBH · Berlin · Offenbach

Evaluating the die's top side peak temperature, the junction to heat sink thermal resistance of the package is calculated to 0.63 K/W which is very close to the 0.6 W/K given in the MOSFET's datasheet. This result validates the characterization setup.

If the package is attached with a Kerafol® 70/50 1.4 W/mK electrically insulating foil, the thermal resistance increases to R_{thJ-C}=1.62 K/W.

In order to analyze the impact of the 1.5 mm sensor hole in the chuck on the R_{th} estimation, the cylindrical volume, which is in reality filled by the sensor and the WLK5 thermal compound, is defined to be solid copper. The simulated temperature difference was only 100 mK (Kerafol® pad) to 180 mK (MX-2) lower than in case of the cavity filled by WLK5 compound (0.8 W/mK). Hence, the influence of the sensor on thermal path is negligible.

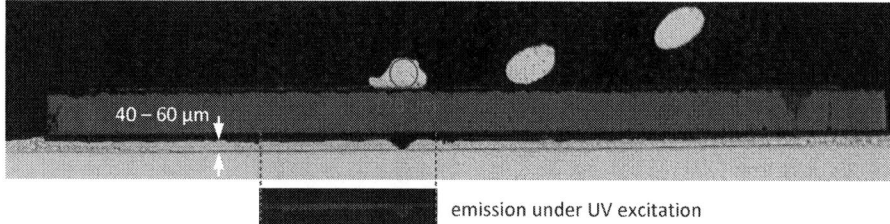

Figure 3: Grinding pattern of a failed C2M0080120 MOSFET in TO247 package. MOSFET die is attached with an 40 - 60 µm solder layer. The in total six bond wires connecting the source are likely to have a diameter of 150 µm. Between the die and the solder layer, an layer of probably organic nature formed. This is indicated by the blue fluorescense excited by UV light radiation.

4. Measurement Results

The characterization campaign included ten devices, eight MOSFETs and two normally-on JFETs as listed with their key properties in Figure 4A. In Figure 4B results for static forward operation at V_{GS} = 12, 16 and 20 V of a SCH2080KE MOSFET are presented. The device is operated at T_{HS} = 60, 80 and 100 °C and power dissipation is limited to below 70 W. In order to get information about the real-world performance of the device, it is mounted to the DUT chuck in two different ways: first, using high-performance MX-2 thermal grease and, second, using an electrically insulating Kerafol® thermal pad with a thermal conductivity of 1.4 W/mK. Although the device can be operated with V_{GS} = 12 V, the R_{DSon} drops by 56 % if V_{GS} is increased to 20 V.

Attached with MX-2 grease and operated at T_{HS} = 100 °C, V_{GS} = 20 V, the R_{DSon} increases only by 21 % if current rises from 1 A to 20 A. In contrast, if the device is mounted using the Kerafol® thermal pad, the current is limited to 17 A and the R_{DSon} increased by 54 %. It should be noted that mounting the TO247 package with a thermal pad is the common method to connect multiple devices to a single heat sink. As shown here, considerable derating is recommended in order to keep the losses low and the device in safe operating mode.

Figure 5 presents the measurement results of all devices listed in Figure 4B at 80 °C heat-sink temperature and 18 V gate voltage in forward and reverse operation mounted with MX-2 thermal grease. In forward conduction mode, the high-current devices APT40SM120B and APT40SM120B show a very flat characteristic in the investigated current range.

Although having a comparatively high R_{DSon} of 260 mΩ, the SCT20N120 also shows a very constant R_{DSon} while the JFET device IJW120R100 shows a significant 60 % increase of R_{DSon} within the investigated current range. The flat curve of the SCT20N120's R_{DSon} in forward operation may be due to the partly significant negative temperature coefficient [11] which was also obtained when operating the device with a Kerafol® insulating pad.

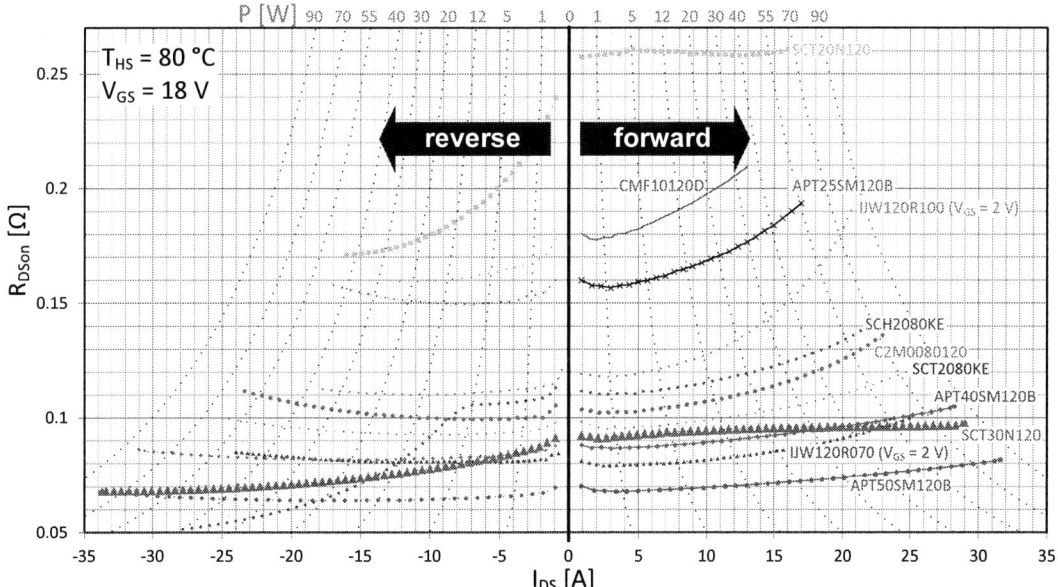

A part #	I_D @ T_C = 25 °C	$R_{DSon,typ}$	R_{thJC}	V_{th}
APT25SM120B	25 A	140 mΩ	0.55 K/W	2.5 V
APT40SM120B	41 A	80 mΩ	0.55 K/W	3 V
APT50SM120B	47 A	50 mΩ	0.55 K/W	2.4 V
CMF10120D	24 A	160 mΩ	0.66 K/W	3.1 V
C2M0080120	36 A	80 mΩ	0.6 K/W	3 V
SCH2080KE	40 A	80 mΩ	0.44 K/W	2.8 V
SCT2080KE	40 A	80 mΩ	0.44 K/W	2.8 V
SCT20N120	20 A	215 mΩ	1 K/W	3.5 V
SCT30N120	45 A	80 mΩ	0.65 K/W	3.5 V
IJW120R100	26 A	80 mΩ	0.78 K/W	-14.5 V
IJW120R070	35 A	55 mΩ	0.63 K/W	-14.5 V

Figure 4: A: List of the investigated SiC transistors with corresponding datasheet specifications. B: Exemplary measurement results showing the impact of the thermal interface material connecting the DUT to the heat sink. Presented is on-state resistance R_{DSon} over drain-source current I_{DS}. Additionally, the power dissipated by the DUT is indicated. MX-2 thermal grease provides significantly better R_{DSon} stability at high currents than the electrically insulating 250 μm 1.4 W/mK Kerafol® sheet. The diagram also shows the impact of different gate voltages and heat sink temperatures. The SCH2080KE shows a clear positive temperature coefficient for both variations of V_G and T_{HS}.

Figure 5: Static characterization results of the 11 tested discrete devices mounted with MX-2 thermal grease in forward and reverse operation. Presented is on-state resistance R_{DSon} over drain-source current I_{DS}. Additionally, the power dissipated by the DUT is indicated. Both Wolfspeed devices show significant increase of their R_{DSon} with rising current. The reverse characteristics are partly dominated by the step-in of the anti-parallel diode (especially SCH2080KE, contains SiC diode).

In this case, the R_{DSon} dropped by up to 50 % when increasing the current from 1 to 10 A at T_{HS} = 60 °C and V_{GS} = 20 V.

In reverse operation, all devices show partly significantly lower R_{DSon} compared to forward operation since the anti-parallel body diode shares a fraction of the total current. However, for the SCT20N120 and the SCH2080KE, the diode characteristic clearly dominates the reverse behavior. In case of the SCH2080KE this is due to the implemented SiC diode die. The diode's low forward voltage drop of only 1.3 V starts to take over current at a level of 7 A. At -20 A, this reduces power loss to 24 W compared to 54 W in forward conduction mode at 20 A – a significant loss reduction. Obviously, this advantage of the implemented SiC diode has the drawback of a slightly higher R_{DSon} increase with rising forward current compared to the transistor-only device SCT2080KE. This may be caused by lower available cross-sectional area for heat transfer since the integrated SiC diode die consumes additional space onto the TO247 copper back plate.

The high R_{DSon} of the SCT20N120 causes visible current sharing already at -1 A reverse current. At -15A, the power loss is reduced to 38 W compared to 58 W in forward conduction mode.

For all other devices, the results underline that in order to reduce power loss in reverse operation, active rectifying, i.e. turning on devices in reverse conduction mode, should be considered. If dead times are minimized in switched operation, active rectifying could additionally reduce switching loss caused by the reverse recovery of the body diodes.

It should be noted that due to their JFET structure, the IJW120R100 as well as the IJW120R070 draw significant gate current if operated in reverse conduction mode. This temperature-dependent gate current depends on I_{DS} and starts to rise significantly *for* $|-I_{DS}|$>5 A. At full negative drain-source current, both devices draw about 1.5 A from the gate driver – a fact which needs to be considered if this operation mode is used in the application.

In Figure 5, two devices, namely the ST Microelectronics parts SCT20N120 and SCT30N120 show a very flat behavior with almost no increase of R_{DSon} with rising current. This phenomenon is investigated in more detail for the higher rated device, the 45 A SCT30N120 specified for up to 200 °C operating temperature. Figure 6 presents the results of a gate drive voltage variation for two mounting methods: thermal grease (A) and thermal pad (B).

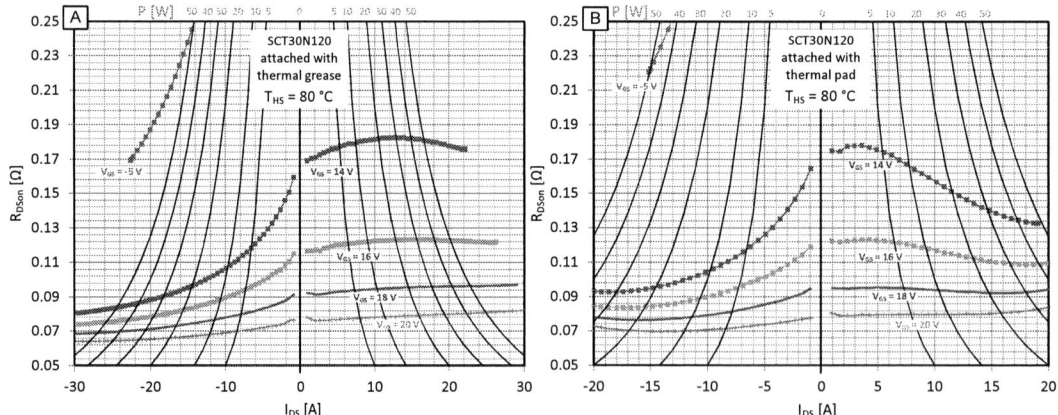

Figure 6: Comparison of the R_{DSon} resulting from different mounting methods for a SCT30N120 device with partly negative temperature coefficient. Presented is on-state resistance R_{DSon} over drain-source current I_{DS}. Additionally, the power dissipated by the DUT is indicated. A: DUT mounted with MX-2 5.6 W/mK thermal grease. B: DUT mounted with Kerafol® 70/50 1.4 W/mK pad.

© VDE VERLAG GMBH · Berlin · Offenbach

The results reveal that in case of the pad attach, the device even shows a negative temperature coefficient at low gate voltage and moderate current. For V_{GS}=14 V, the effect starts at lower current in case of the pad attach compared to the grease attach. This leads to the interesting point that at medium gate voltage, the device mounted with thermal pad may even show much lower losses than that attached with thermal grease. At gate voltage equal or higher than 18 V, the losses are the same for both mounting methods. These properties render this device to be especially suitable to highly efficient, high ambient temperature systems.

5. Conclusion and Outlook

SiC transistors in TO247 package were characterized regarding their static properties. The results disclose the significant differences between the datasheet 25 °C R_{DSon} and device properties when operated attached to a heat sink with a fixed temperature of 80 °C. Especially when mounted with an insulating pad, SiC discrete transistors may require considerable current derating in order to prevent thermal runaway. An anti-parallel SiC Schottky diode included in the transistor package as in the case of Rohm's SCH2080KE is significantly reducing reverse conduction losses. On-state in reverse direction is reducing loss in all tested SIC devices. Devices with dedicated flat R_{DSon} behavior may show significant negative temperature coefficient in certain areas of their characteristic if operated with low gate voltage level. This may lead to problems if those devices are connected in parallel.

6. References

[1] V. Crisafulli, "*A new package with Kelvin Source connection for increasing power density in power electronics design*", presented at the EPE, Geneva, Switzerland, 2015.

[2] C. DiMarino, C. Zheng, D. Boroyevich, *et al.*, "*Characterization and comparison of 1.2 kV SiC power semiconductor devices*", in *15th European Conference on Power Electronics and Applications (EPE)*, pp. 1-10, 2013.

[3] X. Fang, "*Characterization and Modeling of SiC Power MOSFETs*", MSc, Graduate School of The Ohio State University, 2012.

[4] N. Phankong, T. Yanagi, and T. Hikihara, "*Evaluation of Inherent Elements in a SiC Power MOSFET by its Equivalent Circuit*", presented at the EPE 2011, Birmingham, UK, 2011.

[5] T. Funaki, A. S. Kashyap, H. A. Mantooth, *et al.*, "*Characterization of SiC JFET for Temperature Dependent Device Modeling*", in *37th IEEE Power Electronics Specialists Conference (PESC)*, pp. 1-6, 2006.

[6] W. Zhou, X. Zhong, and K. Sheng, "*High Temperature Stability and the Performance Degradation of SiC MOSFETs*", *IEEE Transactions on Power Electronics,* vol. 29, pp. 2329-2337, 2014.

[7] R. OUAIDA, C. BUTTAY, R. e. RIVA, *et al.*, "*Thermal stability of SiC JFETs in conduction mode*", presented at the EPE-ECCE Europe, Lille, France, 2013.

[8] K. Haehre, M. Meisser, F. Denk, *et al.*, "*Characterization and comparision of commercially vailable Silicon Carbide (SiC) power switches*", presented at the PEMD 2012, Bristol, GB, 2012.

[9] M. Meisser, "*Resonant Behaviour of Pulse Generators for the Efficient Drive of Optical Radiation Sources Based on Dielectric Barrier Discharges* ", PhD, Light Technology Institute, Department of Electrical Engineering and Information Technology, Karlsruhe Institute of Technology, 2013.

[10] J. M. Ortiz-Rodriguez, A. R. Hefner, D. Berning, *et al.*, "*Computer-Controlled Characterization of High-Voltage, High-Frequency SiC Devices*", in *IEEE Workshop on Computers in Power Electronics (COMPEL)*, pp. 300-305, 2006.

[11] S. Honggang, C. Zheng, F. Wang, *et al.*, "*Investigation of 1.2 kV SiC MOSFET for high frequency high power applications*", in *APEC*, pp. 1572-1577, 2010.

© VDE VERLAG GMBH · Berlin · Offenbach

Analytical Losses Model for SiC semiconductors dedicated to optimization operations

Gnimdu Dadanema[1], François Costa[4], Yvan Avenas[2], Jean-Luc Schanen[2], Christian Vollaire[3]

[1]Laboratoire SATIE, 41 Avenue du Général Wilson, 94235 Cachan France,

[2]G2ELab, 21 avenue des Martyrs, 38031 Grenoble France

[3]Laboratoire AMPERE, 36 avenue Guy de Collongue, 69134 Ecully France,

[4]Université Paris-Est SATIE-CNRS Cachan, France
Email :gnimdu.dadanema@satie.ens-cachan.fr

Abstract

A detailed analytical losses model prediction dedicated to wide band gap components (SiC, GaN) but also valid for silicon devices is presented. The methodology proposed in this paper is based on switching waveform analysis and do not need any measurement data. Furthermore, this model can easily be used in an optimization process due to its low complexity level. Prior to the losses model development we quickly present a wide band-gap modeling tool that can be used to create a compact model of power MOSFETs and power diodes. The methodology has been demonstrated for Cree Inc. power MOSFET C2M0080120D and Schottky diode C4D20120D.

1. Introduction

Since the 2000s, traditional Silicon based switches for Switch-Mode Power Supply limitations arose. Indeed, limitations in peak voltage, switching speed are related to the material and solutions have to be investigated to go beyond these limitations. Materials such as silicon carbide and Gallium nitride are investigated for this purpose since the past two decades as they seems to be good candidates to replace silicon[1]. One of the most important step in a converter design process is the good estimation of losses. The more accurate the losses estimation will be the less it will cost to design the converter. There are a lot of losses estimation models for semiconductor devices and all of them have advantages and inconvenients. The finite element method for losses estimation is very accurate but on the other hand it's time consuming and so not adapted for optimization[2] . The piecewise linear model or classical analytical model mostly used by designers because of its simplicity is less accurate than the previous finite element model but is very useful for fast losses calculation. This later model is based on the assumption that the parasitic inductance effect is neglected and we only take into account fixed capacitance. Actual devices are designed to have an important switching speed and when working with high voltage and high current, the parasitic element impact can no longer be neglected. Our final purpose is to optimize a converter under EMI and thermal constraint. We chose to use an optimization method based on Sequential quadratic programming (SQP) because of its capability to handle an important number of optimization parameter and its high speed of execution. An important constraint imposed by such algorithm based on the gradients calculation, is that the model they use must be derivable. As a consequence our losses estimation model must be built with analytical derivable expressions. As regards the EMI constraint it is important to be able to accurately reproduce the switching waveform that are observed in a converter. Therefore, the second important characteristic of our model arise. The third characteristic concern the thermal management and it is important that our model be thermosensitive. Regarding these three important characteristics that should have our model, it appears that the best choice is to elaborate an analytical model that will use thermosensitive parameters to build equations that represent the switching waveforms.

In this paper we present an analytical model for losses estimation that take into account parasitic effect and that is not much more complicated than the piecewise linear model. Compared to the piecewise linear model, our model can take into account the ringing effect on losses estimation. Even if it has been shown that the ringing effect [3] is not preponderant in losses estimation, being able to represent ringing effect is very important for EMI analysis.

For validation purpose it's important to compare at a first level, simulation results to those produced by the analytical model. Prior to simulation, we should have accurate compact models of our devices that can be use in a simulation tool. For our simulation task we use a simple model parameter extraction tool to generate a compact thermosensitive model of a SiC power MOSFET or a Schottky diode coded in VHDL-AMS. This modeling tool can use measurement data or datasheet curves to generate component model with parameters related to semiconductors physics. This is a great advantage for designers because they can quickly obtain an accurate model of their component without any measurement. In [5] the methodology used for IGBT model was adapted to establish a silicon carbide MOSFET model. The tool used in these two previous references is a proprietary tool and so not accessible to everyone. In [6] a datasheet based parameter extraction procedure for SiC MOSFET model is presented but equations used in the paper were not clearly defined. The tool presented in this paper is based on simple equations described in [5] and which are related to the semiconductor physics.

The results of the analytical losses model presented in the paper are compared to simulations results and measurement data.

2. The wide band-gap device modeling tool

The wide band-gap device modeling tool presented in this paper is based on datasheet driven model parameter extraction. We present in this paper the case of a SiC Schottky diode and a SiC MOSFET. Analyzing semiconductor physics equations that describe devices behavior we can elaborate a parameters extraction process for Schottky diode and MOSFET.

2.1. Physics-based Schottky diode model

Electrical model of SiC Schottky diode is illustrated by Fig.1

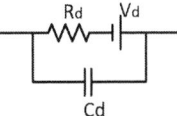

Fig. 1. Electrical model of the Schottky diode

We are not presenting herein all the details of the metal-semiconductor junction physics but it is necessary to recall some basics to make it easy to figure out the model parameter extraction. The electrical model presented on Fig.1 shows the principal elements of the Schottky diode that should be modeled. When a reverse voltage is applied to the Schottky diode, a parasitic capacitance appear and it expression is given by

$$C(V) = \sqrt{\frac{\varepsilon e N_d}{2(V_d - V)}} \qquad (1)$$

The internal resistance of a Schottky diode is related to its drift region, and it can be estimated by:

$$R_d = \frac{L_d}{q\mu_{sic}N_dA} \tag{2}$$

N_d, L_d, A and μ_{sic} are respectively, the donor concentration in N-region, the drift zone length, the diode effective surface and carrier mobility. This last parameter is the one that governs the temperature dependence of the diode resistance. The mobility is given by:

$$\mu_T = \mu_{300}\left(\frac{T}{300}\right)^x \tag{3}$$

Where μ_{300} is the carrier mobility at 300K.

Equations (1), (2), (3), implies that N_d, L_d , A n and x are the model parameters, we have to extract .

2.2. Schottky diode model parameters extraction procedure

The procedure we present in this paper does not require any measurement. We only need the forward characteristics and the capacitance vs reverse voltage. These curves are available on most of SiC Schottky diode datasheets. Fig.2.a shows the example of Cree C4D20120D datasheet curves. The first step is to estimate the resistance by evaluating the slope P_1 of the forward characteristic at room temperature. The second step is to draw and to estimate the slope P_2 of the curve ($1/C^2$) versus reverse voltage. With these first two step we can obtain relations:

$$P_1 = \frac{1}{R_d} = \frac{q\mu_{sic}N_dA}{L_d}, \qquad P_2 = \frac{2}{qN_d\varepsilon A^2} \tag{4}$$

Considering that the Schottky diode drift region is designed to support a blocking voltage of V_{Bmax} we can write:

$$L_d = \sqrt{\frac{2\varepsilon_{sic}V_{Bmax}}{qN_d}} \tag{5}$$

Rearranging equations (4) and(5), we obtain:

$$N_d = \frac{P_1\varepsilon_{sic}\left(V_{Bmax}P_2\right)^{\frac{1}{2}}}{q\mu_{sic}}, \qquad A = \sqrt{\frac{2}{qP_2N_d\varepsilon_{sic}}} \tag{6}$$

The x parameter of the mobility equation is deduced by evaluating the mobility for a different temperature and resolving:

$$x = \frac{\log\left(\dfrac{\mu(T)}{\mu_{300}}\right)}{\log\left(\dfrac{T}{300}\right)} \tag{7}$$

All the parameters related to the Schottky diode structure are identified. If we assume that the thermoelectric emission dominates the diffusion phenomenon, the current versus voltage model of the Schottky diode is:

$$i_{diode} = I_s\left[e^{\frac{qV}{KT}} - 1\right] \tag{8}$$

The saturation current I_S can be obtained by estimating the value at V=0 of the semi logarithmic plot of log (I) versus V.

$$\log\left(i_{diode}\right) = \log\left(I_s\right) + \frac{qV}{kT} \tag{9}$$

Fig.2.a and c show some comparison between datasheet and a model obtain with this procedure.

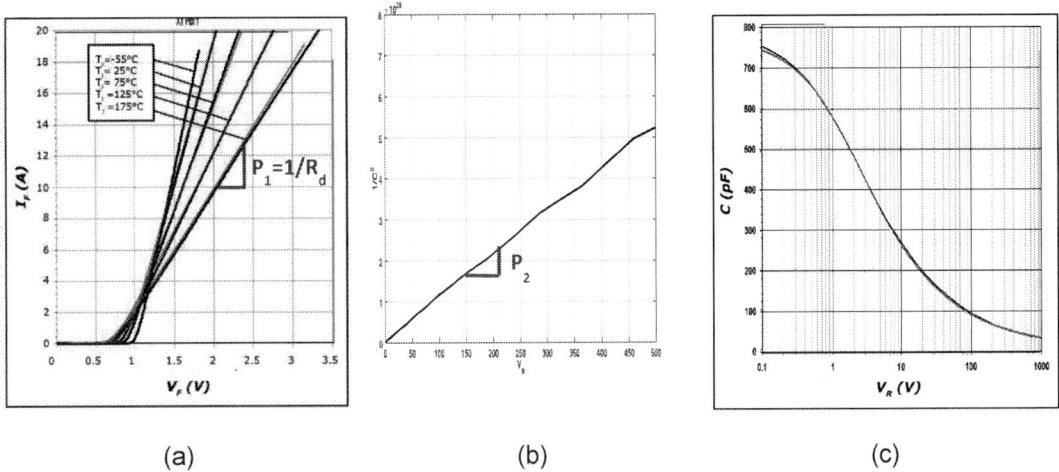

(a)   (b)   (c)

Fig.2. (a) resistance estimation on diode forward characteristic and comparison between datasheet curves and model result (red curves), (b) curve $1/C^2$ vs reverse voltage, (c) comparison of capacitance characteristic

2.3. Physics based power MOSFETs model

Several MOSFET models are available with different complexity level. The one described in [7] offers the advantage of being very accurate and thermosensitive. The inconvenient of that model is that its parameters identification needs measurement data and a proprietary tool. Therefore, we will rewrite the model equation and define a procedure for parameters extraction based entirely on datasheet. We recall the current expression for different working regions of the MOSFET. In linear region:

$$I_{mos} = \frac{K_p K_f \left[V_{ds}\left(V_{gs} - V_{TH}\right) - \dfrac{V_{ds}^{y} P_{vf}^{y-1} \left(V_{gs} - V_{TH}\right)^{2-y}}{y} \right]}{1 + \theta\left(V_{gs} - V_{TH}\right)} \tag{10}$$

In the saturation region current is given by:

$$I_{mos} = \frac{K_p \left(V_{gs} - V_{TH}\right)^2}{2\left(1 + \theta\left(V_{gs} - V_{TH}\right)\right)} \tag{11}$$

We will not redefine all the parameters in this paper but they can be found in [5]. The parameters, to be extracted are V_{TH}, K_p, K_f, θ.

V_{TH} and θ are obtained by analyzing the transfer characteristic. K_p and K_f are obtained on output characteristic. The detail procedure can be found in [6]. Fig. 4 show some comparison between datasheet output characteristic and the obtained model results.

PCIM Europe 2016, 10 – 12 May 2016, Nuremberg, Germany

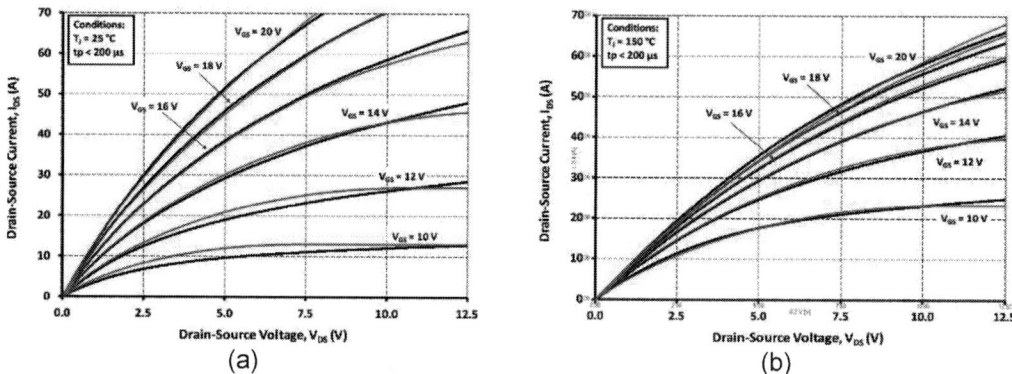

(a)   (b)

Fig.4.(a) output characteristics comparison T= 25°C, (b) output characteristics comparison T=150°C

3. The analytical losses model

We consider here for our study, the commonly used clamped inductive load circuit (Fig.5.a). The important parasitic elements to take into account are the power loop equivalent resistor and the inductance that will impact the current overshoot at the turn-on. The inductance L_s, can slow down the turn-on because the rising drain current create a voltage drop that slow down the gate to source voltage rising. The inductive load can be assimilated to a current source when solving the circuit equations.

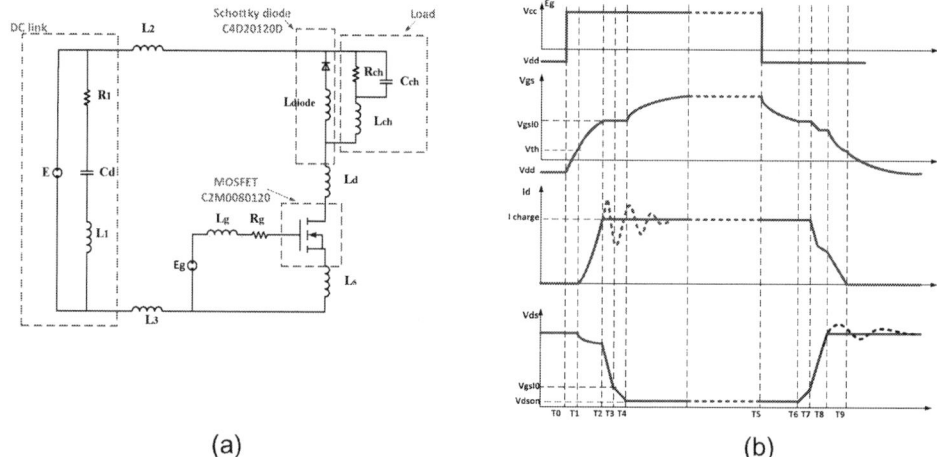

(a)   (b)

Fig.5.(a) test circuit schematic used for analysis, (b) typical switching waveform representation

In this paper the switching period is decomposed in ten steps. For each step the procedure will be the same. We first identify the equivalent circuit associated to the switching step, we resolve the circuit equation to obtain current and voltage values, we integrate the product of current and voltage and the losses P_{losses} are obtain by multiplying the energy by the switching frequency.

$$P_{losses} = \sum_{i=0}^{n-1} \int_{t_i}^{t_{i+1}} V_{ds_i}(t) i_{d_i}(t)\, dt \tag{12}$$

Where, n is the number of steps.

The MOSFET electrical model depends on the switching step, so we can have different values of MOSFET parasitic capacitance. In fact, the nonlinear behavior of MOSFET capacitance can be modeled in different way. [6] Propose to model them by power function but in our case it will add complexity to our equations and make them hard to solve analytically. Thus we will use a simpler methodology. Considering the representation of a

© VDE VERLAG GMBH · Berlin · Offenbach

1498

nonlinear capacitance as shown on fig.6, we will define capacitance with index 1 as the capacitance value to be used when MOSFET is working in saturation region. Similarly, if the MOSFET is working in linear region we will define capacitance with an index 2.

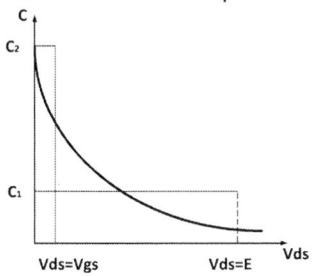

Fig.6. capacitance discrete value selection

The expressions of the electrical quantities in the different working steps are summarized in table 1.

Table 1 : expressions of switching quantities during turn-on/off

Equivalent circuit	Steps	$V_{ds}(t)$, $i_d(t)$
	t_0-t_1	$V_{ds} = E$ $i_d = 0$
	t_9-t_{10}	$V_{ds} = \left(V_{dspic} - E\right)e^{\gamma t}\cos\left(\omega_0 t - \phi\right) + E$ $i_d = C_{oss1}\dfrac{dV_{ds}}{dt}$
	t_1-t_2	$V_{ds} = E - \left(L_s + L_d\right)\dfrac{di_{d_1}}{dt}$ $i_d = g_m\left(V_{cc} - V_{th}\right)\left(1 + \dfrac{r_2}{r_1 - r_2}e^{r_1(t-t_1)} + \dfrac{r_1}{r_2 - r_1}e^{r_2(t-t_1)}\right)$
	t_8-t_9	$V_{ds} = E - \left(L_s + L_d\right)\dfrac{di_{d_1}}{dt}$ $i_d = g_m\left(V_{cc} - V_{th}\right)\left(1 + \dfrac{r_2}{r_1 - r_2}e^{r_1(t-t_1)} + \dfrac{r_1}{r_2 - r_1}e^{r_2(t-t_1)}\right)$
	t_2-t_3	$V_{ds} = -\dfrac{g_m\left(V_{cc} - v_{th}\right) - I_0}{g_m R_g C_{gd_1} + C_{oss_1} + C_d}\left(t - t_2\right) + E_1$
	t_7-t_8	$i_{d_2} = I_0 e^{\gamma(t-t_2)}\left(\gamma\cos\left(\omega_0\left(t-t_2\right) - \phi\right) - \omega_0\sin\left(\omega\left(t-t_2\right) - \phi\right)\right)$
	t_3-t_4	$V_{ds} = -\dfrac{g_m\left(V_{cc} - v_{th}\right) - I_0}{g R_g C_{gd2} + C_{oss2} + C_d}t + V_{gsI_0}$ $i_{d_2} = I_0 e^{\gamma t}\left(\gamma\cos\left(\omega_0 t - \phi\right) - \omega_0\sin\left(\omega t - \phi\right)\right)$
	t_5-t_6	$V_{ds} = -\dfrac{g_m\left(V_{cc} - v_{th}\right) - I_0}{g R_g C_{gd2} + C_{oss2} + C_d}t + V_{gsI_0}$
	t_6-t_7	$i_{d_2} = I_0 e^{\gamma t}\left(\gamma\cos\left(\omega_0 t - \phi\right) - \omega_0\sin\left(\omega t - \phi\right)\right)$

$$r_1 = \frac{-B - \sqrt{B^2 - 4A}}{2A} \quad r_2 = \frac{-B + \sqrt{B^2 - 4A}}{2A} \quad A = R_g C_{gd1} g_m\left(L_s + L_d\right) \quad B = \left(g_m L_s + R_g\left(C_{gs} + C_{gd1}\right)\right) \quad \gamma = \frac{R_d}{2L_d} \quad \omega_0 = \sqrt{\frac{1}{LC} - \gamma^2}$$

4. Experimental validation of the analytical losses model and discussions

First of all, regarding EMI considerations we have to validate the waveform reproduced by the analytical losses model. Fig7 shows comparisons between waveforms obtained in a circuit simulator (Simplorer), those obtained by measurement and the analytical model results.

Secondly, the estimated losses of our analytical model are compared to experimental measure and losses estimated with circuit simulation. The losses measurement method is a calorimetric method similar to the one presented by [8]. The graph Fig.8 shows the comparison results obtained. The waveforms comparison shows that the model cannot reproduce perfect ringing waveform for all bias points. Despite this, we notice a relatively good accuracy in dv/dt and di/dt representation. The difference between measurement and simulation waveform are mainly due to the simplicity of the circuit simulated compared to the real circuit. The common mode impedances are not represented and so the simulation do not take into account the common mode effect that can be observe on measured waveform. We notice a difference between measured losses and estimated losses because the losses calculation in the analytical model is based on the drain to source current but the effective energy dissipated in the MOSFET is due to the channel current.

4.1. Switching waveforms comparison and analysis

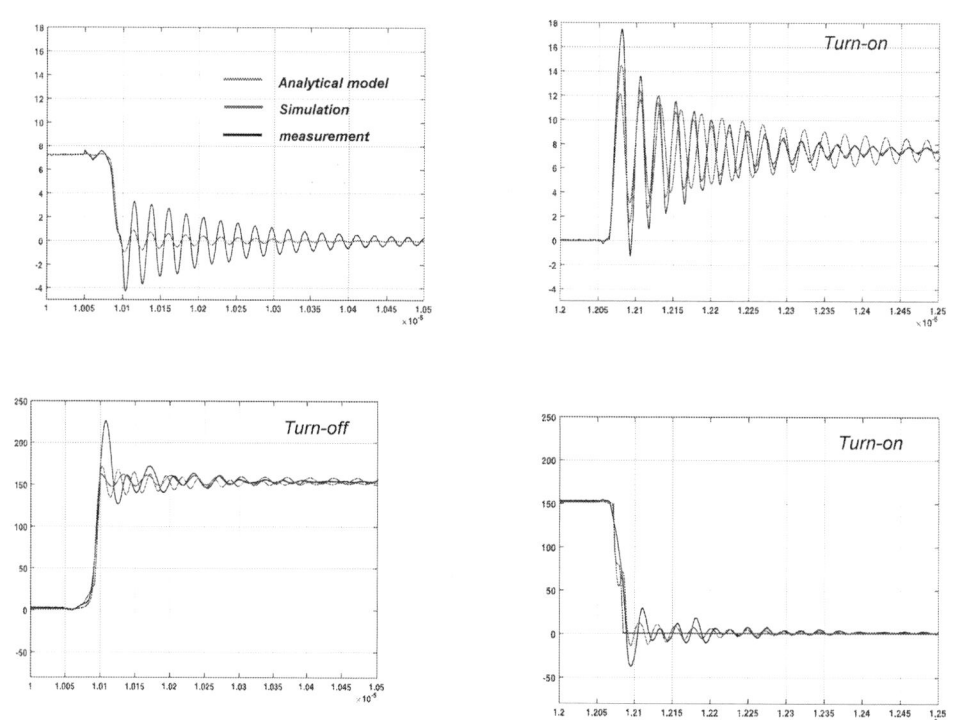

Fig7a. Switching waveform with conditions, V=150V - I_d =7.5A and R_g =12Ω

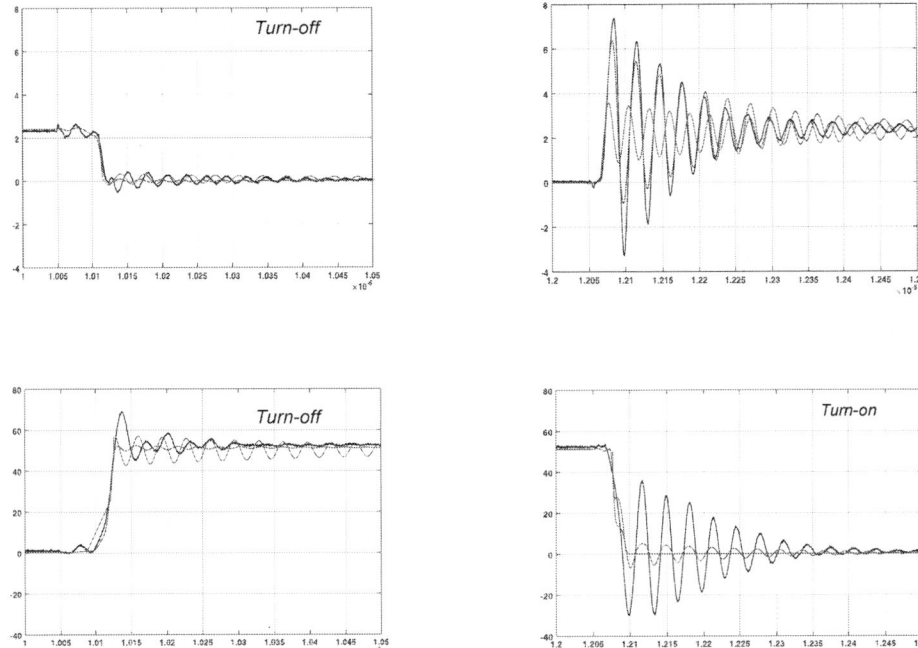

Fig7b. Switching waveform with conditions, V=150V - I_d =7.5A and R_g =12Ω

4.2. Losses comparison and analysis

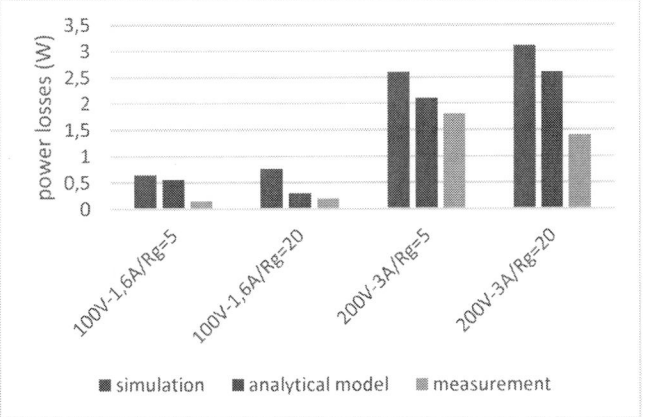

Fig.8. comparison between analytical losses model prediction, circuit simulation estimation and measurement

5. Conclusion

In this paper we investigate an analytical model for losses estimation ad EMI study. The simulation results tend to validate the approach proposed in this paper. The most important advantage of such a model is that it offers a compact model for losses estimation that can easily be used by designer to accurately estimate losses in a silicon carbide based converter. In addition, the model can easily be implemented in an optimization process. The model can also be used to produce current and voltage waveform that are helpful in EMI study of converters.

References

[1] X. Gus, Q. Shui, C. W. Mylesb, and M. A. Gundersens, "COMPARISON OF Si, GaAs, S i c AND GaN FET-TYPE SWITCHES FOR PULSED POWER APPLICATIONS *," pp. 0–3.

[2] M. Rodŕguez, A. Rodríguez, P. F. Miaja, D. G. Lamar, and J. S. Zúniga, "An insight into the switching process of power MOSFETs: An improved analytical losses model," *IEEE Trans. Power Electron.*, vol. 25, no. 6, pp. 1626–1640, 2010.

[3] Z. Zhang, B. Guo, F. Wang, L. M. Tolbert, B. J. Blalock, Z. Liang, and P. Ning, "Impact of ringing on switching losses of wide band-gap devices in a phase-leg configuration," *2014 IEEE Appl. Power Electron. Conf. Expo. - APEC 2014*, pp. 2542–2549, 2014.

[4] A. R. Hefner and S. Member, "Modeling Buffer Layer IGBT ' s for Circuit Simulation," vol. I, no. 2, 1995.

[5] T. R. Mcnutt, A. R. Hefner, H. A. Mantooth, S. Member, D. Berning, and S. Ryu, "Silicon Carbide Power MOSFET Model and Parameter Extraction Sequence," vol. 22, no. 2, pp. 353–363, 2007.

[6] M. Mudholkar, S. Ahmed, S. Member, M. N. Ericson, S. S. Frank, C. L. Britton, and H. A. Mantooth, "MOSFET Model," vol. 29, no. 5, pp. 2220–2228, 2014.

[7] B. J. Baliga, *Fundamentals of power Semiconductor Devices.* 2006.

[8] C. Chen et al., "Power Loss Estimation in SiC Power BJTs," PCIM Europe 2014; International Exhibition and Conference for Power Electronics, Intelligent Motion, Renewable Energy and Energy Management; Proceedings of, Nuremberg, Germany, 2014, pp. 1-8.

PCIM Europe 2016, 10 – 12 May 2016, Nuremberg, Germany

List of Authors of PCIM Europe 2016

Author	Institution	Pages
A		
Abbatelli, Luigi	STMicroelectronics, I	1536
Abdel-Rahman, Sam	Infineon Technologies Americas, USA	1615
Abe, Yasushi	Fuji Electric, J	188
Abl, Reiner	BMW, D	534
Abouchi, Nacer	Institut des Nanotechnologies de Lyon, F	1707
Abronzini, Umberto	University of Cassino and South Lazio, I	379 1217
Adachi, Shinichiro	Fuji Electric, J	1956
Adelmund, Melanie	University of Bremen, D	1232
Adler, Johannes	University of Bremen, D	1324
Adragna, Claudio	STMicroelectronics, I	612
Adrien, Mercier	ENS Cachan – SATIE, F	1647
Aggen, Christian	Danfoss Silicon Power, D	698
Agrawal, Binod	CREE India Private Limited, IN	2114
Ahmad, Jawad	Politecnico di Torino, I	2058
Aizpuru, Iosu	Mondragon University, ES	763 2074
Alberti, Luigi	University of Padova, I	534
Albertin, Thomas	Atherm, F	1132
Ali, Esmail A.	University of Science and Technology Yemen, YE	1046
Allen, Scott T.	Wolfspeed, USA	34
Almai, Farshid	University of Paderborn, D	232
Altstadt, Jochen	Kortec, D	1979
Alvarez, Rodrigo	Siemens, D	588
Alves-Rodrigues, Luis Gabriel	Commissariat à l'énergie atomique et aux énergies alternatives, F	128
Ambacher, Oliver	Fraunhofer-Institute IAF, D	319
Ammann, Ulrich	Esslingen University of Applied Sciences, D	780
An, Bao Ngoc	Karlsruhe Institute of Technology, D	1979
Andenna, Maxi	ABB Switzerland, CH	417
Antivachis, Michael	ETH Zürich, CH	1622
Appel, Tobias	STS, D	1434
Aragues-Penalba, Monica	CITCEA-UPC, ES	2211

PCIM Europe 2016, 10 – 12 May 2016, Nuremberg, Germany

Author	Institution	Pages
Aranzabal, Itxaso	University of the Basque Country, ES	1970
Araujo Vasconcelos, Samuel	University of Kassel, D	573
Arnold, Martin	ABB Switzerland, CH	917
Arpino, Matilde	University of Cassino and South Lazio, I	379 1217
Asada, Shinsuke	Mitsubishi Electric Corporation, J	326
Asfaux, Pascal	Airbus Operation, F	1814
Ashley, Grant	Vxl Power, GB	1455
Askari, Vahid	Teesside University, GB	1439 1685
Attaianese, Ciro	University of Cassino and South Lazio, I	379 1217
Augoustidis, Alex	GrafTech International, USA	1126
Austermann, Johann	University of Applied Sciences Ostwesfalen-Lippe, D	1639
Avenas, Yvan	G2ELAB, F	1494
Azcondo, Francisco	University of Cantabria, ES	1194 1631
Azuma, Katsunori	Hitachi Power Semiconductor, J	348
B		
Baghadadi, Mehdi	University of Cambridge, GB	1845
Baghaie Yazdi, Mehrdad	Fairchild Semiconductor, D	813 2231
Baier, Thomas	Friedrich-Alexander-University Erlangen, D	264
Bakran, Mark-M.	University of Bayreuth, D	272 355 653 683 720 1055 1400 1503 2181
Balke, Christian	Maccon, D	1353
Banerjee, Sujit	Monolith Semiconductor, USA	984
Barater, Davide	Università degli Studi di Parma, I	1073 1584
Barrios, Ernesto L.	Public University of Navarre, ES	1209
Barruel, Franck	Commissariat à l'énergie atomique et aux énergies alternatives, F	128
Barwig, Markus	Friedrich-Alexander-University of Erlangen, D	1693 1701
Baschnagel, Andreas	ABB Switzerland, CH	945

PCIM Europe 2016, 10 – 12 May 2016, Nuremberg, Germany

Author	Institution	Pages
Basler, Thomas	Infineon Technologies, D	180
Basler, Vanessa	Technical University of Munich, D	543
Bäßler, Marco	Danfoss Silicon Power, D	698
Batard, Christophe	University of Nantes, F	712
Bauer, Pavol	Delft University of Technology, NL	1201 1383
Baumann, Michael	SUMIDA Components & Modules, D	1728 1764
Bauwens, Filip	On Semiconductor, BE	1479
Bayer, Martin	ABB Switzerland, CH	432
Bazzano, Gaetano	STMicroelectronics, I	804
Beaucarne, Guy	Dow Corning Europe, B	992
Beaurenaut, Laurent	Infineon Technologies, D	1377 2027
Becker, Karl-Friedrich	Fraunhofer Institute IZM, D	1040
Beichler, Johannes	Vacuumschmelze, D	1471
Beier-Möbius, Menia	Chemnitz University of Technology, D	172 588
Bellingen, Jörg	Technical University of Vienna, AT	2219
Bellini, Marco	ABB Switzerland, CH	911
Bendani, Larbi	Valeo, F	477
Benkendorff, Berthold	Christian-Albrechts-University, D	1942
Berger, Hubert	FH Joanneum, AT	296 2151
Bergogne, Dominique	CEA Leti, F	312
Beringer, Sebastian	Leibniz University Hannover, D	107
Bernardinis, Gabriele	Analog Devices, USA	195
Bernd, Martin	Karlsruhe Institute of Technology, D	1979
Bertelshofer, Teresa	University of Bayreuth, D	272 653 1503
Beuermann, Max	Siemens, D	1888
Beyer, Harald	ABB Switzerland, CH	945
Bhalla, Anup	United Silicon Carbide, USA	1511 1549
Bier, Anthony	Commissariat à l'énergie atomique et aux énergies alternatives, F	128
Billet, Cyrille	Calyos, BE	1139
Bingham, Chris	University of Lincoln, GB	1769
Birkel, Andre	University of Bayreuth, D	2181

PCIM Europe 2016, 10 – 12 May 2016, Nuremberg, Germany

Author	Institution	Pages
Blaabjerg, Frede	Aalborg University, DK	831 1837
Blank, Mathias	Technical University of Vienna, AT	1339
Blank, Thomas	Karlsruhe Institute of Technology, D	1487 1979
Blum, Manuel	Siemens, D	1912 1919
Böcker, Joachim	University of Paderborn, D	232 780 1662 1992
Bödeker, Christian	University of Bremen, D	1232
Böh, Magnus	Technical University of Cologne, D	669
Bolognani, Silverio	University of Padova, I	534
Bolte, Sven	University of Paderborn, D	232 1992
Booth, James	Dynex Semiconductor, GB	704
Borcherding, Holger	University of Applied Sciences Ostwesfalen-Lippe, D	1639
Borecki, Jacek	University of Bremen, D	248
Borghetti, Giovanni	Nidec ASI, I	2175
Boroyevich, Dushan	CPES/Virginia Tech, USA	1286
Bortis, Dominik	ETH Zürich, CH	1622
Bosch, Swen	Heinrich Steinhart, HTW Aalen, D	2129
Böttigheimer, Mike	University of Stuttgart, D	410
Bouguet, Christophe	University of Nantes, F	712
Bourdon, Jeremy	Airbus Operation, F	1814
Boureghda, Monnir	Alpha Assembly Solutions, USA	1027
Bourner, David	Vicor Corporation, USA	1253 1822
Braband, Matthias	ISW – University of Stuttgart, D	256
Bräckle, Dennis	Karlsruhe Institute of Technology, D	788
Bramanpalli, Ranjith	Würth Elektronik eiSos, D	99 1757
Branas, Christian	University of Cantabria, ES	1631
Braun, Michael	Karlsruhe Institute of Technology, D	393 788
Brinkfeldt, Klas	Swerea IVF AB, SE	1063
Brix, Jonathan	Fraunhofer Institute IPA, D	386
Brubaker, Michael	SBE, USA	1391
Budaker, Bernhard	Fraunhofer Institute IPA, D	386

PCIM Europe 2016, 10 – 12 May 2016, Nuremberg, Germany

Author	Institution	Pages
Buetow, Sven	Semikron Elektronik, D	728
Burani, Nicola	Fraunhofer Institute IISB, D	453
Burger, Michael	SEW Eurodrive, D	144
Burgos, Rolando	CPES/Virginia Tech, USA	1286
Burns, Chris	AB Mikroelektronik, AT	969
Buschhorn, Stefan	Infineon Technologies, D	850
Buschkühle, Marc	Infineon Technologies, D	1800
Buttay, Cyril	INSA de Lyon, F	1792
C		
Cabezuelo, David	University of the Basque Country, ES	1970
Camus, Stephane	Aperam Alloys Amilly, F	84
Canales, Jose Mari	Mondragon University, ES	763 2074
Cao, Zhiyu	AEG Power Solutions, D	1662 1677
Carastro, Fabio	GE Research Center, D	645
Carcouet, Sébastien	University Grenoble Alpes CEA LITEN, F	2098
Carvalho, Adriano	Nomad Tech, PT	953
Casady, Jeffrey	Wolfspeed, USA	34
Catalisano, Giuseppe	STMicroelectronics, I	1536
Catellani, Stéphane	Commissariat à l'énergie atomique et aux énergies alternatives, F	128
Cavallaro, Daniela	STMicroelectronics, I	804
Cellier, Remy	Institut des Nanotechnologies de Lyon, F	1707
Cerezo, Jorge	Infineon Technologies Americas, USA	819
Chamaret, André-Philippe	SNCF, F	1103
Chandra Mouli, Gautham Ram	Delft University of Technology, NL	1383
Chatroux, Daniel	University Grenoble Alpes CEA LITEN, F	2098
Chaudhuri, Toufann	ABB Sécheron, CH	1103
Chelghoum, Reda	Valeo, F	477
Chen, Jingxuan	University of Toronto, CA	895
Chen, Zhiyang	ON Semiconductor, USA	1777
Cherix, Nicolas	EPFL – Ecole Polytechnique Fédérale de Lausanne, CH	1949
Cheung, Chun	Intersil Corporation, USA	550
Chinthavali, Madhu S.	Oak Ridge National Laboratory, USA	1391
Choi, Hanna	KCC Corporation, ROK	961
Choo, Byoungho	Infineon Technologies Power Semitech, ROK	863

PCIM Europe 2016, 10 – 12 May 2016, Nuremberg, Germany

Author	Institution	Pages
Choudhary, Vijay	Texas Instruments, USA	2019
Christe, Alexandre	EPFL – Ecole Polytechnique Fédérale de Lausanne, CH	461
Chun, Ho Tae	Hyundai Motors Company, ROK	1262
Chung, Daewoong	Infineon Technologies Power Semitech, ROK	858 863
Ciocia, Alessandro	Politecnico di Torino, I	2058
Cioffi, Philip	GE Research Center, USA	645 1853
Comola, Marco	STMicroelectronics, I	804
Concari, Carlo	Università degli Studi di Parma, I	1584
Conrad, Detlef	Synopsys, D	813
Consentino, Giuseppe	STMicroelectronics, I	804
Corvasce, Chiara	ABB Switzerland, CH	417 432
Costa, Francois	ENS Cachan – SATIE, F	288 1494
Coulinge, Emilien	EPFL – Ecole Polytechnique Fédérale de Lausanne, CH	461
Cousineau, Marc	University of Toulouse, F	371
Crisafulli, Vittorio	ON Semiconductor, D	1178 1865
Curea, Octavian	ESTIA, F	2066
D		
Dadanema, Gnimdu	ENS Cachan – SATIE, F	1494
Dai, Jian	GE Research Center, USA	1853
Dai, Xiaoping	Dynex Semiconductor, GB	704 845 938
Dalessandro, Luca	Schaffner, CH	148
Daly, Michael	Analog Devices, USA	195
Datta, Rajib	GE Research Center, USA	645 1853
Dauchy, Julien	University Grenoble Alpes CEA LITEN, F	2098
Davidson, Colin	Alstom Grid, GB	2197
Davidson, Jonathan	University of Sheffield, GB	1095
de Alegría, Iñigo Martinez	University of the Basque Country, ES	1785 1970
De Bernardinis, Alexandre	French Institute of Science and Technology for Transport, Development and Networks, F	2159
De Michielis, Luca	ABB Switzerland, CH	417

PCIM Europe 2016, 10 – 12 May 2016, Nuremberg, Germany

Author	Institution	Pages
De Monchy, Michiel	Alpha, RUS	957
de Rooij, Michael	Efficient Power Conversion Corporation, USA	304
De Sousa, Luis	Valeo, F	477
de Stoppelaar, Diederik	LightingEurope, BE	604
Deboy, Gerald	Infineon Technologies, AT	1622
Declercq, Frederick	On Semiconductor, BE	1479
Dede, Ercan	Toyota Research Institute North America, USA	72
Delsuc, Vincent	Dow Corning, BE	1017
Demattio, Horst	Karlsruhe Institute of Technology, D	1487 1979
Denk, Fabian	Karlsruhe Institute of Technology, D	1721
Denk, Marco	University of Bayreuth, D	1055
Deviny, Ian	Dynex Semconductor, GB	845
Di Leo, Paolo	Politecnico di Torino, I	2058
Di Monaco, Mauro	University of Cassino and South Lazio, I	379 1217
Diaz Reigosa, Paula	Aalborg University, DK	831
Dieckerhoff, Sibylle	Technical University of Berlin, D	1186
Dietrich, Lothar	Fraunhofer-Institute IZM, D	1155
Dincan, Catalin Gabriel	Aalborg University, DK	2050
Dinis, Corina Maria	Politehnica University Timisoara, RO	2106
Dinulovic, Dragan	Würth Elektronik eiSos, D	107
Djallel, Kerdoun	GLEC Constanine 1, DZ	1785
Dobusch, Julian	Friedrich-Alexander-University of Erlangen, D	1857
Dodge, Jonathan	United Silicon Carbide, USA	484 1549
Domb, Moshe	Infineon, USA	1555
Domes, Daniel	Infineon Technologies, D	53
Doppelbauer, Martin	Karlsruhe Institute of Technology, D	393
Dragan, Mihai	Fraunhofer Institute IPA, D	386
Draghici, Mihai	Infineon Technologies, AT	180
Draugedalen, Eirik	Valeo, NO	113
Dudin, Andrey	Technical University of Ilmenau, D	1896
Dujic, Drazen	EPFL – Ecole Polytechnique Fédérale de Lausanne, CH	461
Dupont, Vincent	Calyos, BE	1139
Dürbaum, Thomas	Friedrich-Alexander-University of Erlangen, D	492 1693 1701 1857

PCIM Europe 2016, 10 – 12 May 2016, Nuremberg, Germany

Author	Institution	Pages
Durham, Jeffrey	Alpha Assembly Solutions, USA	1027
Dworakowski, Piotr	SuperGrid Institute, F	1592
E		
Eberlin, Michael	Fraunhofer Institute ISE, D	1733
Ebli, Michael	Universtiy of Reutlingen, D	210
Eckel, Hans-Günter	University of Rostock, D	581 924 1888
Edelmoser, Karl	Technical University of Vienna, AT	1741 2137
Efika, Ikenna	Alstom Grid, GB	2197
Ehrhardt, Christian	Fraunhofer-Institute IZM, D	1155
Ellinger, Thomas	Technical University of Ilmenau, D	1896
Enami, Hiroji	Dow Corning Toray, J	1017
Endres, Stefan	Fraunhofer Institute IISB, D	526
Engel, Günter	CeraCap, AT	2151
Engelhard, Christoph	HELLA KGaA Hueck, D	669
Epp, Nikolai	AEG Power Solutions, D	1677
Ertl, Hans	Technical University of Vienna, AT	1339 1669 1846
Evans, Kim	Dynex Semiconductor, GB	704
F		
Fahlbusch, Sebastian	Helmut Schmidt University Hamburg, D	1528
Fahlenkamp, Marc	Infineon Technologies, D	620
Fahnert, Holger	AEG Power Solutions, D	1662
Fairweather, Andrew	Vxl Power, GB	1455 2090
Farkas, Gabor	Mentor Graphics,HU	596
Fazio, Gianluca	On Semiconductor Italy, I	1178
Felgemacher, Christian	University of Kassel, D	573
Fernandez, Manuel	Universty of Oviedo, ES	1479
Fernández-Serantes, Luis Alfonso	University of Applied Sciences Joanneum, AT	296
Ferrazza, Francesco	STMicroelectronics, I	612
Fersterra, Fabian	Fraunhofer Institute IISB, D	469
Filho, Faete	Eaton Corporation, USA	1518

PCIM Europe 2016, 10 – 12 May 2016, Nuremberg, Germany

Author	Institution	Pages
Fink, Karsten	Power Integrations, D	48
Fischer, Aaron	Technical University of Ilmenau, D	1896
Fischer, Fabian	ABB Switzerland, CH	945
Fischer, Hermann	Fairchild Semiconductor, D	813
Fortuna, Stefania	STMicroelectronics, I	804
Foster, Martin	University of Sheffield, GB	1095 1455 1769 2090
Franceschini, Giovanni	Università degli Studi di Parma, I	1073
Franco, Augusto	Nomad Tech, PT	953
Frangieh, Tony	GE Research Center, USA	1853
Frank, Wolfgang	Infineon Technologies, D	1615
Franke, Toke	Danfoss Silicon Power, D	1942
Frankeser, Sophia	Technical University of Chemnitz, D	1063
Freitas, Nuno	Nomad Tech, PT	953
Frick, Florian	ISW – University of Stuttgart, D	256
Fröhleke, Norbert	University of Paderborn, D	232 1662 1992
Fuchs, Friedrich W.	Christian-Albrechts-University, D	1942
Fuhrmann, Jan	University of Rostock, D	581 1888
Fujimoto, Mitsunao	ALPS Green Devices, J	91
Funk, Dustin	Technical University of Ilmenau, D	1749
G		
Gaber, Roland	Fraunhofer Institute IWES, D	401
Gafford, James	Mississippi State University, USA	1294
Galai Dol, Lilia	Efficacity, F	2159
Galea, Michael	University of Nottingham, GB	1073
Galek, Marek	Siemens, D	453 1912 1919
Gambino, Giusy	STMicroelectronics, I	1987
Gant, Levi	The University of Alabama, USA	984
Garnier, Laurent	University Grenoble Alpes CEA LITEN, F	2098
Gautier, Cyrille	ENS Cachan – SATIE, F	288
Geissmann, Silvan	ABB Switzerland, CH	417 432

PCIM Europe 2016, 10 – 12 May 2016, Nuremberg, Germany

Author	Institution	Pages
Gekeler, Manfred W.	HTWG Konstanz University of Applied Sciences, D	136
Gerada, Chris	University of Nottingham, GB	1073
Gerlach, Rolf	Infineon Technologies, D	180
Ghossein, Layal	SuperGrid Institute, F	1592
Gierschner, Sidney	University of Rostock, D	1888
Ginot, Nicolas	University of Nantes, F	712
Gleißner, Michael	University of Bayreuth, D	683
Gleitner, Volker	Electronicon Kondensatoren, D	1408
Glose, Daniel	Technical University of Munich, D	1270
Glück, Tobias	Technical University of Vienna, AT	1339
Godbold, Gerald W.	Hyperion Technology, USA	1294
Goikoetxea, Ander	Mondragon University, ES	763 2074
Gomez Suarez, Carlos	Aalborg University, DK	831
Gomez, German	On Semiconductor, BE	1479
Gony, Bashar	Aperam Alloys Amilly, F	84
Gonzalez, Diego	University of Oviedo, ES	1479
González, Manuela	University of Oviedo, ES	1631
Götz, Stefan	Duke University, USA	1904
Grady, Matt	United Silicon Carbide, USA	1549
Grasel, Bernhard	Dewesoft, AT	2204
Greca, Gustavo	Alpha Assembly Solutions, USA	1027
Green, James	University of Sheffield, GB	1095
Grégoire, Luc-André	University of Toulouse, F	371
Grider, Dave	Wolfspeed, USA	34
Grigans, Linards	Riga Technical University, LV	1880
Grimaud, Louis	Safran Group, F	1286
Gritti, Giovanni	STMicroelectronics, I	612
Grohmann, Rolf	HTWK Leipzig, D	2145
Guerre, Vincent	Local Energy Alternative & Fair, F	2066
Guillaume, Herault	ENS Cachan – SATIE, F	1647
Gumaan, Mohammed	Mansoura University, AE	1046
Gustafsson, Emilia	ABB Switzerland, CH	432
H		
Haaf, Peter	Fairchild Semiconductor, D	2231
Hacala, Amélie	ESTIA, F	2066
Hackl, Christoph	Technical University Munich, D	1926

PCIM Europe 2016, 10 – 12 May 2016, Nuremberg, Germany

Author	Institution	Pages
Hähre, Karsten	Karlsruhe Institute of Technology, D	1721
Hain, Stefan	University of Bayreuth, D	720
Hammes, David	University of Rostock, D	1888
Han, SeungHyun	Hyundai Motors Company, ROK	1262
Han, Yunchao	Fraunhofer Institute IISB, D	469
Harel, Jean Claude	Renesas Electronics America, USA	1027
Harfman Todorovic, Maja	GE Research Center, USA	645 1853
Harper, Jonathan	ON Semiconductor, D	877
Hartmann, Michael	Schneider Electric Power Drives, AT	1669
Hartmann, Samuel	ABB Switzerland, CH	945
Hase, Nobuhiro	Rohm Semiconductor, USA	42
Hatae, Shinji	Mitsubishi Electric Corporation, J	677
Haug, Martin	Würth Elektronik eiSos, D	107
Hausberger, Thomas	Technical University of Vienna, AT	1339
Hayakawa, Seiichi	Hitachi Power Semiconductor, J	348
He, Hongtao	Mitsubishi Electric & Electronics, CN	871
He, Weikun	Mentor Graphics, UK	1027
Heer, Daniel	Infineon Technologies, D	53 1800
Heering, Wolfgang	Karlsruhe Institute of Technology, D	1721
Heinemann, Lothar	AEG Power Solutions, D	1677
Heinzel, Thomas	Fuji Electric Europe, D	824 1956
Heldwein, Marcelo	Federal Institute of Santa Catarina – IFSC, BR	621
Helling, Florian	Universität der Bundeswehr Munich, D	1904
Helsper, Martin	Siemens, D	272
Heredero-Peris, Daniel	Universitat Politécnica de Catalunya, ES	2035
Hernandez Gutiérrez, Diego	CH	1178
Hernes, Magnar	SINTEF Energy Research, NO	1408
Herold, Christian	Technical University of Chemnitz, D	588
Herzer, Reinhard	Semikron Elektronik, D	728
Heseding, Johannes	Leibniz University Hannover, D	1421
Hewitt, David	University of Sheffield, GB	1095
Hill, Julia	Aperam Alloys Amilly, F	84
Hiller, Marc	Karlsruhe Institute of Technology, D	788
Himmelstoss, Felix	Technikum Vienna, AT	1741 2137

PCIM Europe 2016, 10 – 12 May 2016, Nuremberg, Germany

Author	Institution	Pages
Hinken, Reiner	Danfoss Silicon Power, D	1934
Hirao, Akira	Fuji Electric, J	1001
Hirayama, Tomohisa	Hitachi Power Semiconductor, J	348
Hochberg, Martin	Karlsruhe Institute of Technology, D	1305 1578
Hoffmann, Klaus F.	Helmut Schmidt University Hamburg, D	202 1300 1528
Hofmann, Daniel	Fuji Electric Europe, D	438
Hofmann, Norbert	University of Applied Sciences Nordwestschweiz, CH	917
Hofmann, Viktor	University of Bayreuth, D	355
Höltgen, Markus	Technical University of Cologne, D	796
Hölzl, Wolfgang	Technical University of Munich, D	543
Homann, Michael	Technical University of Braunschweig, D	216
Honsberg, Marco	Mitsubishi Electric Europe, D	889
Horff, Roman	University of Bayreuth, D	272 653 1503
Hori, Motohiro	Fuji Electric, J	188
Horiuchi, Keisuke	Hitachi, J	1110
Hosking, Terry	SBE, USA	1391
Hossein Khani, Milad Mohammad	Fraunhofer Institute ISE, D	1733
Houzouji, Hiroshi	Hitachi, J	78
Hruska, Miroslav	Skoda Electric, CZ	1808
Hu, Bo	Mitsubishi Electric & Electronic, CN	932
Huang, Jianwei	Dynex Semconductor, GB	845
Huang, Yi	Intersil Corporation, USA	550
Hudoffsky, Boris	PMK Mess- und Kommunikationstechnik, D	1463
Hull, Brett	Wolfspeed, USA	34
Hüning, Felix	University of Applied Sciences Aachen, D	1963
Hunziker, Christoph	University of Applied Sciences Northwestern Switzerland, CH	2167
Hussein, Khalid	Mitsubishi Electric Europe, D	677
I		
Iagar, Angela	Politehnica University Timisoara, RO	2106
Iannuzzo, Francesco	Aalborg University, DK	831
Ichikawa, Hiroaki	Fuji Electric, J	824

PCIM Europe 2016, 10 – 12 May 2016, Nuremberg, Germany

Author	Institution	Pages
Ikawa, Osamu	Fuji Electric, J	438 824
Ikeda, Osamu	Hitachi, J	78
Ikeda, Yoshinari	Fuji Electric, J	188
Immovilli, Fabio	Raw Power, I	1073
Ino, Kazuhide	Rohm, J	42
Inokuchi, Seiichiro	Mitsubishi Electric Corporation, J	677
Inoue, Daisuke	Fuji Electric, J	1956
Inoue, Tomoki	Toshiba, J	566
Iraola, Unai	Mondragon University, ES	763 2074
Irifune, Hiroyuki	Kaga Toshiba Electronics Corporation, J	839
Ishibashi, Hidetoshi	Mitsubishi Electric Corporation, J	342
Ito, Yoichi	Sanken Electric, J	2083
Izuka, Arata	Mitsubishi Electric Corporation, J	677
J		
Jacob, Mathew	Texas Instruments, USA	2019
Jagau, Martin	Technologienetzwerk Allgäu, D	1171
Jang, Hyosang	Infineon Technologies Power Semitech, ROK	863
Jang, JiWoong	Hyundai Motors, ROK	1600
Jang, KiYoung	Hyundai Motors, ROK	1600
Jara, Martin	West Bohemian University, CZ	1808
Jaschke, Reinhard	Helmut Schmidt University Hamburg, D	1300
Jeong, JeeHye	Hyundai Motors Company, ROK	1262
Jeong, KangHo	Hyundai Motors, ROK	1600
Jerinic, Vladan	Danfoss Silicon Power, D	1934
Jeyaprakash, Arun	Technical University of Munich, D	1904
Jiu, Jinting	Osaka University, J	1021
Jones, Steve	Dynex Semiconductor, GB	938
Jonke, Peter	AIT-Austrian Institute of Technology, AT	1846
Joo, JeongHong	Hyundai Motors Company, ROK	1262
Jorge-Ques, Fernando	Universitat Politécnica de Catalunya, ES	2035
Joshi, Shailesh	Toyota Research Institute North America, USA	72
Josso, Stieven	Henkel Electronics, BE	1035
Joubert, Charles	Ampere Laboratory, F	1286 1792
Jun, ChangHan	Hyundai Motors Company, ROK	1262

PCIM Europe 2016, 10 – 12 May 2016, Nuremberg, Germany

Author	Institution	Pages
Jun, Kisoo	KCC Corporation, ROK	961
Jung, Jaehoon	KCC Corporation, ROK	961
Jung, JinHwan	Hyundai Motors Company, ROK	1262 1600
Jung, Marco	Fraunhofer Institute IWES, D	401
Jungwirth, Herbert	SUMIDA Components & Modules, D	1764
K		
Kähr, Christian	University of Applied Sciences Nordwestschweiz, CH	917
Kaiser, Julian	Fraunhofer Institute IISB, D	469
Kaji, Yusuke	Mitsubishi Electric Corporation, J	326
Kakiki, Hideaki	Fuji Electric, J	188
Kako, Naotsugu	Toshiba Corporation Semiconductor, J	839
Kamal, Mustafa	Mansoura University, AE	1046
Kaminski, Nando	University of Bremen, D	1232 1324
Kammerer, Felix	Karlsruhe Institute of Technology, D	788
Kampen, Dennis	BLOCK Transformatoren-Elektronik, D	144
Kanai, Naoyuki	Fuji Electric, J	188
Kaneko, Satoshi	Fuji Electric, J	188
Kapaun, Florian	Universität der Bundeswehr Munich, D	1570
Kapels, Holger	Fraunhofer Institute ISIT, D	202
Kasper, Matthias	ETH Zürich, CH	1622
Kato, Koji	Sanken Electric, J	2083
Kaulfersch, Eberhard	Berliner Nanotest & Design, D	1063
Kawamoto, Noriaki	Rohm, J	42
Kawase, Daisuke	Hitachi Power Semiconductor, J	348 1110
Kawashima, Tetsuya	Fuji Electric, J	895
Ke, Maolong	Dynex Semconductor, GB	845
Kennel, Ralph	Technical University of Munich, D	120 518 629 1248 1270 1926
Keuck, Lukas	University of Paderborn, D	232
Keyse, Richard	Dynex Semiconductor, GB	903
Khani, Toktam	Technologienetzwerk Allgäu, D	2122
Khaselev, Oscar	Alpha Assembly Solutions, USA	1027

PCIM Europe 2016, 10 – 12 May 2016, Nuremberg, Germany

Author	Institution	Pages
Khatri, Danish	Infineon Technologies North America, USA	882
Killeen, Peter	Wolfspeed, USA	34
Kim, Hyunwoo	KCC Corporation, ROK	961
Kim, Taehyun	Infineon Technologies Power Semitech, ROK	858
Kimura, Takashi	Hitachi Automotive Systems, J	331
Kimura, Yoshitaka	Mitsubishi Electric Corporation, J	342
Kinzer, Dan	Navitas Semiconductor, USA	31
Kirchhof, Jörg	Fraunhofer Institute IWES, D	401
Kitamura, Shuichi	Mitsubishi Electric Corporation, J	425
Kizu, Naoyuki	Rohm, J	42
Kjaer, Philip C.	Aalborg University, DK	2050
Klarenbach, Christoph	Beckhoff Automation, D	796
Klauke, Sebastian	Infineon, D	581
Klein, Axel	Technical University of Braunschweig, D	216
Kling, Rainer	Karlsruhe Institute of Technology, D	1721
Klötzer, Sebastian	Helmut Schmidt University Hamburg, D	1528
Kobayashi, Hideto	Fuji Electric, J	1956
Kobayashi, Yasuyuki	Fuji Electric, J	438
Koch, Nelson	EPFL – Ecole Polytechnique Fédérale de Lausanne, CH	1949
Koga, Shunsuke	Osaka University, J	1021
Köhler, Stefan	Technical University Nuremberg Georg Simon Ohm, D	526
Köhnlechner, Benjamin	Technical University of Ilmenau, D	1147
Koike, Yoshihiko	Hitachi, J	78
Koini, Markus	Epcos, D	164
Kolar, Johann Walter	ETH Zürich Power Electronic Systems Laboratory, CH	32 1622
Kolb, Johannes	Schaeffler Technologies, D	1979
Kondo, Satoshi	Mitsubishi Electric Corporation, J	326
Königsmann, Gunter	Semikron Elektronik, D	728
Konishi, Yuichiro	Hitachi, J	1110
Konishide, Masaoki	Yasuhiko Kohno, Hitachi, J	280
Konno, Akitoyo	Hitachi, J	78
Kopta, Arnost	ABB Switzerland, CH	417 432
Körner, Julian	Karlsruhe Institute of Technology, D	1721
Kotani, Raita	Toshiba, J	566
Kouge, Takuma	Fuji Electric, J	1956
Krecek, Tomas	ON semiconductor, CZ	1089

PCIM Europe 2016, 10 – 12 May 2016, Nuremberg, Germany

Author	Institution	Pages
Krenn, Markus	FH Joanneum, AT	2151
Krishna Vytla, Rajeev	Infineon Technologies North America, USA	882
Kroics, Kaspars	Riga Technical University, LV	1880
Kropp, Jürgen	University of Bayreuth, D	1400
Kruschel, Wolfram	Infineon, D	735
Kubera, Sascha	Siemens, D	588
Kübrich, Daniel	Friedrich-Alexander-University of Erlangen, D	1693
Kugi, Andreas	Technical University of Vienna, AT	1339
Kühl, Sascha	Technical University of Munich, D	518
Kusano, Dai	Japan Fine Ceramics, J	64
Kwiecien, Marcin	Magneto, PL	1428
L		
Labrousse, Denis	ENS Cachan – SATIE, F	1647
Lacombe, Bertrand	Safran Group, F	1792
Ladoux, Philippe	University of Toulouse, F	371 2189
Laeuffer, Jacques	Dtalents, F	1541
Lagier, Thomas	SuperGrid Institute, F	2189
Lahl, Peter	Infineon, D	735
Lamo, Paula	University of Cantabria, ES	1194
Lampke, Thomas	Technical University of Chemnitz, D	976
Landsmann, Peter	Technical University of Munich, D	518
Lang, Klaus-Dieter	Fraunhofer Institute IZM, D Technical University of Berlin, D	1040 1155
Langhals, David	Technical University of Cologne, D	796
Larousse, Sebastien	Institut des Nanotechnologies de Lyon, F	1707
Larrañaga, Jon Andreu	University of the Basque Country, ES	1785
Larson, Kent	Dow Corning Corporation, USA	992
Lassmann, Matthias	Infineon, D	735
Laud, Satyavrat	Renesas Electronics America, USA	1027
Laue, Jürgen	Danfoss Silicon Power, D	698
Lautner, Jennifer	Friedrich-Alexander-University of Erlangen, D	1278
Leach, John	Castle, GB	1769
Leary, Alex	Carnegie Mellon University, USA	1518
Lechler, Armin	ISW – University of Stuttgart, D	256
Lee, Brian	AIT, IE	558
Lee, JaeWon	Hyundai Motors Company, ROK	1262

PCIM Europe 2016, 10 – 12 May 2016, Nuremberg, Germany

Author	Institution	Pages
Lee, JeongYun	Hyundai Motors Company, ROK	1262
Lee, Junbae	Infineon Technologies Power Semitech, ROK	858 863
Lee, KiJong	Hyundai Motors, ROK	1600
Lee, Minsub	Infineon Technologies Power Semitech, ROK	858 863
Lefebvre, Stéphane	ENS Cachan – SATIE, F	1647
LeHenaff, Francois	Alpha Assembly Solutions, D	1027
Leibfried, Thomas	Karlsruhe Institute of Technology, D	363
Lemke, Michael	AEG Power Solutions, D	1677
Lemmon, Andrew	The University of Alabama, USA	984
Lempidis, Georgios	Fraunhofer Institute IWES, D	401
Lenz, Kevin	Danfoss Silicon Power, D	1934
Leon-Ramirez, Christian Alejandro	University Simon Bolivar, VE	2211
Leszczynski, Jacek	AGH University of Science and Technology, PL	1428
Lexow, Daniel	University of Rostock, D	924
Leyrer, Benjamin	Karlsruhe Institute of Technology, D	1979
Li, Daohui	Dynex Semiconductor, GB	938
Li, Xueqing	United Silicon Carbide, USA	1511
Lifton, Anna	Alpha Assembly Solutions, USA	957 1027
Lindemann, Andreas	Otto-von-Guericke-University, D	1081 1829
Lindseth, Roar	Valeo, NO	113
List, Hans	FH Joanneum, AT	2151
Liu, Cheng	Chenyang Technologies, D	1248
Liu, Guoyou	Dynex Semconductor, GB	845
Liu, Hui	Valeo, NO	113
Liu, Ji-Gou	Chenyang Technologies, D	1248
Liu, Tao	Lenze SE, D	2007
Liu, Wenduo	Infineon Technologies Americas, USA	1655
Liu, Xiaoshan	ENS Cachan – SATIE, F	288
Lledó-Ponsati, Tomàs	TeknoCEA, ES	2035
Löchel, Jonas	Technical University of Ingolstadt, D	742
Loges, Werner	Vacuumschmelze, D	1471
Loh, Poh Chiang	Aalborg University, DK	1837
Lohner, Andreas	Technical University of Cologne, D	669

PCIM Europe 2016, 10 – 12 May 2016, Nuremberg, Germany

Author	Institution	Pages
Longo, Giuseppe	STMicroelectronics, I	804 1987
López, Felipe	University of Cantabria, ES	1194
López, Jesús	Public University of Navarre, ES	2043
Luo, Haihui	Dynex Semconductor, GB	845
Lura, Shinichi	Mitsubishi Electric Corporation, J	425
Lusiewicz, Anna	University of Stuttgart, D	410
Lutz, Josef	Chemnitz University of Technology, D	172 588
M		
Ma, Xiankui	Mitsubishi Electric & Electronics, CN	871
Machuca, Enrique	Technical University of Ingolstadt, D	742
Mackay, Laurens	Delft University of Technology, NL	1201
Mademlis, Georgios	Aristotle University of Thessaloniki, GR	445
Madiwale, Subodh	Analog Devices, USA	195
Madzharov, Nikolay	Technical University of Gabrovo, BG	1999
Makoschitz, Markus	Technical University of Vienna, AT	1669
Malipaard, Dirk	Fraunhofer Institute IISB, D	453
Mandrusiak, Gary	GE Research Center, USA	645 1853
Mangal, Navneet	CREE India Private Limited, IN	2114
Manier, Charles-Alix	Fraunhofer-Institute IZM, D	1155
Mansour, Madaci	University of the Basque Country, ES	1785
Mantzanas, Panagiotis	Friedrich-Alexander-University of Erlangen, D	1693
Marklein, René	Fraunhofer Institute IWES, D	401
Marlino, Laura D.	Oak Ridge National Laboratory, USA	1391
Marquardt, Rainer	Universität der Bundeswehr Munich, D	1570
Marquart, Janosch	University of Applied Sciences NTB Buchs, CH	501 509
Marroyo, Luis	Public University of Navarre, ES	1209
Martin, Christian	Ampere Laboratory, F	1286 1792
Martin, Jérémy	Commissariat à l'énergie atomique et aux énergies alternatives, F	128
Marxgut, Christoph	Helbling Technik, D	661
März, Andreas	University of Bayreuth, D	272 653 1503

PCIM Europe 2016, 10 – 12 May 2016, Nuremberg, Germany

Author	Institution	Pages
März, Martin	Fraunhofer Institute IISB, D	469 1315
Masana, Francesc	Barcelona Semiconductors, ES	1009
Masayoshi, Nakazawa	Fuji Electric, J	188
Matallana, Asier	University of the Basque Country, ES	1970
Mathieu, Olivier	Rogers Germany, D	1027
Matlok, Stefan	Fraunhofer Institute IISB, D	637
Matocha, Kevin	Monolith Semiconductor, USA	984
Matsushita, Akira	Hitachi Automotive Systems, J	331
Matthias, Sven	ABB Switzerland, CH	417 432
Mau, Matthias	Danfoss Silicon Power, D	698
Mauder, Anton	Infineon Technologies, D	1324
Maurer, Wilhelm	Infineon Technologies, D	1155
Mazzola, Michael	Mississippi State University, USA	1294
McHenry, Michael E.	Carnegie Mellon University, USA	1518
McPherson, Brice	Wolfspeed, USA	34
Meany, Tom	Analog Devices ERDC, IE	1361
Meisser, Michael	Karlsruhe Institute of Technology, D	1487 1979
Meradji, Moudrik	Harbin Institute of Technology, CN	2014
Merlo, Christophe	ESTIA, F	2066
Merten, Jens	National Solar Energy Institute, F	33
Mertens, Axel	Leibniz University Hannover, D	1421
Mesbahi, Tedjani	Ecole Centrale de Lille, F	2014
Mesemanolis, Athanasios	ABB Switzerland, CH	432
Middelstaedt, Lars	Otto-von-Guericke-University, D	1829
Mii, Kenji	Toshiba Corporation Semiconductor, J	839
Mikulla, Michael	Fraunhofer-Institute IAF, D	319
Millington, Alan	Dynex Semiconductor, GB	903
Misra, Sanjay	Henkel, USA	60
Mitic, Gerhard	Siemens, D	2226
Mitova, Radoslava	Technical University of Berlin, D	1155
Miura, Mineo	Rohm, J	42
Miyahara, Satoshi	Mitsubishi Electric, D	996
Miyata, Hiroshi	Fuji Electric, J	1956
Miyazaki, Takaaki	Hitachi, J	78
Mizushima, Takao	ALPS Green Devices, J	91

PCIM Europe 2016, 10 – 12 May 2016, Nuremberg, Germany

Author	Institution	Pages
Mochizuki, Eiji	Fuji Electric, J	188 1001
Moia, Joabel	Federal Institute of Santa Catarina – IFSC, BR	621
Momose, Fumihiko	Fuji Electric, J	1001
Montesinos-Miracle, Daniel	Universitat Politécnica de Catalunya, ES	2035
Morel, Florent	Ampère, F	1592
Morel, Hervé	Ampère, F	1592
Mori, Mutsuhiro	Hitachi, J	78 1110
Morita, Toshiaki	Hitachi, J	78
Motto, Eric R.	Powerex, USA	889
Müller, Christian	Infineon Technologies, D	850
Müller, Florian	Vossloh-Schwabe Lighting Solutions, D	605
Müller, Georg	Karlsruhe Institute of Technology, D	1305 1578
Mumby-Croft, Paul	Dynex Semiconductor, GB	704
Münster, Patrick	University of Rostock, D	924
Müter, Ulf	Helmut Schmidt University Hamburg, D	1528
N		
Nabhani, Farhad	Teesside University, GB	1439 1685
Naeberle, Norbert	Schaffner, CH	148
Nagahara, Teruaki	Mitsubischi Electric Corporation, J	889
Nagao, Shijo	Osaka University, J	1021
Nagashima, Kazuhito	Hitachi Power Semiconductor, J	348
Nagaune, Fumio	Fuji Electric, J	1956
Naitoh, Yutaka	ALPS Green Devices, J	91
Nakagawa, Ryosuke	Mitsubishi Electric Corporation, Japan	1311
Nakamura, Keiichi	Mitsubishi Electric Corporation, J	425
Nakamura, Masato	Hitachi, J	78
Nakanishi, Masaharu	Rohm Semiconductor, D	42
Nakano, Hiroshi	Hitachi, J	78
Nakatsu, Kinya	Hitachi, J	331
Nasadoski, Jeffrey	GE Research Center, USA	645
Nashiki, Masato	Toshiba Corporation Semiconductor, J	839
Nate, Satoru	Rohm, J	42
Neumaier, Klaus	Fairchild Semiconductor, D	2231
Ng, Chiu	Infineon Technologies Americas, USA	819

PCIM Europe 2016, 10 – 12 May 2016, Nuremberg, Germany

Author	Institution	Pages
Ng, Wai Tung	University of Toronto, CA	895
Nicolle, Thomas	Calyos, BE	1139
Niedermayr, Philipp	Alpitronic, I	534
Nigsch, Simon	University of Applied Sciences NTB Buchs, CH	501 509
Nishimura, Yoshitaka	Fuji Electric, J	1001
Nishio, Haruhiko	Fuji Electric, J	895
Nishiura, Akira	Fuji Electric, J	1956
Nitta, Tetsuya	Toshiba, J	566
Nöding, Christian	University of Kassel, D	573
Nogawa, Hiroyuki	Fuji Electric, J	1001
Nojima, Geraldo	Eaton Corporation, USA	1518
Norambuena, Margarita	Technical University of Berlin, D	1186
O		
O'Sullivan, Dara	Analog Devices, IE	1369
Oeder, Christian	Friedrich-Alexander-University of Erlangen, D	1701 1857
Ohara, Ryoichi	Toshiba, J	566
Ohodnicki, Paul	DOE-National Energy Technology Laboratory, USA	1518
Ohta, Hiroshi	Kaga Toshiba Electronics Corporation, J	839
Okinori, Chihiro	IWATSU Test Instruments, J	1463
Olejniczak, Kraig	Wolfspeed, USA	34
Oñederra, Oier	University of the Basque Country, ES	1970
Onno Krah, Jens	Technical University of Cologne, D	796
Onozawa, Yuichi	Fuji Electric, J	824
Oppermann, Hermann	Fraunhofer-Institute IZM, D	1155
Orlik, Bernd	University of Bremen, D	248
Ota, Kenji	Mitsubishi Electric Corporation, J	425
Otsubo, Yoshitaka	Mitsubishi Electric Corporation, J	342 996
Ott, Leopold	Fraunhofer Institute IISB, D	469
Otto, Alexander	Fraunhofer-Institute ENAS, D	1063
Owzareck, Michael	BLOCK Transformatoren-Elektronik, D	1447
P		
Pacas, Mario	University of Siegen, D	224 240 1873

PCIM Europe 2016, 10 – 12 May 2016, Nuremberg, Germany

Author	Institution	Pages
Packwood, Matthew	Dynex Semiconductor, GB	704 938
Pai, Ajay Poonjal	Infineon Technologies, D	1315
Pala, Vipindas	Wolfspeed, USA	34
Palmer, Patrick	University of Cambridge, GB	1845
Palmour, John W.	Wolfspeed, USA	34
Pan, Xinxing	AIT, IE	558
Papadopoulos, Charalampos	ABB Switzerland, CH	432
Parry, John	Mentor Graphics, UK	1027
Parspour, Nejila	University of Stuttgart, D	410 1447
Passmore, Brandon	Wolfspeed, USA	34
Patt, Michael	Technologienetzwerk Allgäu, D	1171 2122
Paulus, Dirk	Technical University of Munich, D	518
Paulwitz, Christian	Epcos, D	156
Pawellek, Alexander	Friedrich-Alexander-University Erlangen, D	492
Pellicone, Devin	Advanced Cooling Technologies, USA	1118
Peng, Mingkai	Shenzhen Zeasset Electronic Technology, CN	1416
Penzel, Michael	Technical University of Chemnitz, D	976
Pereira França, Alex	CPqD, BR	750
Perez, Angel Luis	University of the Basque Country, ES	1785
Perrin, Remi	INSA de Lyon, F	1286 1792
Persson, Eric	Infineon Technologies Americas, USA	1655
Peters, Dethard	Infineon Technologies, D	53
Petkov, Valeri	Technical University of Gabrovo, BG	1999
Petzoldt, Jürgen	Technical University of Ilmenau, D	1896 2145
Pezet, Francois	Nidec ASI, I	2175
Pfost, Martin	University of Innsbruck, AT	210 757
Phung, Van Trang	University of Siegen, D	240
Piepenbreier, Bernhard	Friedrich-Alexander-University of Erlangen, D	264 1278
Pierfederici, Serge	University de Lorraine, F	772
Pietkiewicz, Andrzej	Schaffner, CH	148
Pietrini, Giorgio	Università degli Studi di Parma, I	1073 1584

PCIM Europe 2016, 10 – 12 May 2016, Nuremberg, Germany

Author	Institution	Pages
Pigazo, Alberto	University of Cantabria, ES	1194
Pignataro, Gaetano	STMicroelectronics, I	804
Plumpton, Ashley	Dynex Semiconductor, GB	903
Pluta, Wojciech	Czestochowa University of Technology, PL	1428
Popa, Gabriel Nicolae	Politehnica University Timisoara, RO	2106
Postiglione, Gianluca	Nidec ASI, I	2175
Pottier, Frederic	Aperam Alloys Amilly, F	84
Proulx, Joe	Mentor Graphics, UK	1027
Puff, Markus	Epcos, AT	164
Q		
Qi, Fang	Dynex Semiconductor, GB	938
Qian, Peng	Technical University of Munich, D	1270
Qiu, Mianwei	Shenzhen Zeasset Electronic Technology, CN	1416
Quay, Rüdiger	Fraunhofer-Institute IAF, D	319
Quentin, Nicolas	SAGEM, F	1286 1792
R		
Rädel, Uwe	Technical University of Ilmenau, D	2145
Rahimo, Munaf	ABB Switzerland, CH	417 432 917
Raithel, Stefan	Vossloh-Schwabe Lighting Solutions, D	605
Ramirez Figueroa, Fernando David	University of Siegen, D	224
Ramirez-Elizondo, Laura	Delft University of Technology, NL	1201
Raso, Antonio	Nidec ASI, I	2175
Rathbone, Kevin	Robotae, GB	1845
Razik, Hubert	University Claude Bernard Lyon, F Laboratoire Ampere, F	312 1707
Reddig, Manfred	University of Applied Sciences Augsburg, D	120
Reger, Martin	Rogers Germany, D	1027
Rehm, Markus	IBR Ingenieurbüro Rehm, D	1332
Reimann, Tobias	Technical University of Ilmenau, D	1147 1749
Reiner, Richard	Fraunhofer-Institute IAF, D	319
Reinhold, Andreas	HTWK Leipzig, D	2145
Reiter, Tomas	Infinoen Technologies, D	1315 1391

PCIM Europe 2016, 10 – 12 May 2016, Nuremberg, Germany

Author	Institution	Pages
Remaci, Ahmed	ESTIA, F	2066
Rencz, Marta	BME, HU	596
Revol, Bertrand	ENS Cachan – SATIE, F	288
Richmond, Jim	Wolfspeed, USA	34
Richter, Dennis	Otto-von-Guericke-University, D	1829
Richter, Jan	Karlsruhe Institute of Technology, D	393
Rocha dos Santos, Sender	CPqD, BR	750
Rodriguez, Jose	University Andres Bello, CL	1186
Roesner, Robert	GE Research Center, D	645
Roig, Jaume	On Semiconductor, BE	1479
Rossberg, Matthias	Semikron Elektronik, D	728
Rossmanith, Hans	Friedrich-Alexander-University of Erlangen, D	1434
Rout, Colin	Dynex Semiconductor, GB	903
Rowden, Brian	GE Research Center, USA	645 1853
Ruccius, Benjamin	Fraunhofer Institute IISB, D	453
Rufer, Alfred	EPFL – Ecole Polytechnique Fédérale de Lausanne, CH	445 1713 1949
Rupp, Roland	Infineon Technologies, D	180
Ryu, Sei-Hyung	Wolfspeed, USA	34
Rzepka, Sven	Fraunhofer-Institute ENAS, D	1063
S		
Sack, Martin	Karlsruhe Institute of Technology, D	1305 1578
Sadarnac, Daniel	CentraleSupelec, F	477
Sahli, Nizar	Helmut Schmidt University Hamburg, D	1528
Saito, Katsuaki	Hitachi Europe, GB	348 1110
Saito, Ryuichi	Hitachi Automotive Systems, J	331
Saito, Shoji	Mitsubishi Electric Corporation, J	677
Saitou, Takashi	Fuji Electric, J	1001
Sakai, Shinji	Mitsubishi Electric Corporation, J	336
Sakiyama, Yoko	Toshiba, J	566
Sakurai, Naoki	Yasuhiko Kohno, Hitachi, J	280
Salerno, Paul	Alpha Assembly Solutions, USA	1027
Sanchis, Pablo	Public University of Navarre, ES	1209
Sander, Rene	Karlsruhe Institute of Technology, D	363

PCIM Europe 2016, 10 – 12 May 2016, Nuremberg, Germany

Author	Institution	Pages
Sano, Kenya	Toshiba, J	566
Sarkany, Zoltan	Mentor Graphics, UK	1027
Sárkány, Zoltán	Budapest University of Technology and Economics, HU	596 1155
Sasaki, Masahiro	Fuji Electric, J	895
Sawada, Mutsumi	Fuji Electric, J	824
Schanen, Jean-Luc	G2ELAB, F	1494
Schäning, Björn	Helmut Schmidt University Hamburg, D	1528
Scharwitz, Christian	Vacuumschmelze, D	1471
Schefler, Stefan	Epcos, D	164
Schenk, Kurt	University of Applied Sciences NTB Buchs, CH	501 509
Scherbaum, Markus	University of Applied Sciences Augsburg, D	120
Scheuermann, Uwe	Semikron Elektronik, D	691
Schiele, Jürgen	AEG Power Solutions, D	1662
Schijffelen, Jos	Power Research Electronics, NL	1383
Schilling, Marco	Technical University of Ilmenau, D	1147
Schlenk, Manfred	Infineon Technologies, D	120
Schliewe, Jörn	Epcos, D	156
Schmeller, Markus	SUMIDA Components & Modules, D	1728 1764
Schmidhuber, Michael	SUMIDA Components & Modules, D	1728 1764
Schmies, Stefan	Infineon, D	735
Schmitt, Alexander	Karlsruhe Institute of Technology, D	393
Schmitt-Landsiedel, Doris	University Rosenheim, D	1562
Schnarrenberger, Mathias	Karlsruhe Institute of Technology, D	788
Schnell, Raffael	ABB Switzerland, CH	417 432 945
Schreitmüller, Stefan	HTWG Konstanz University of Applied Sciences, D	136
Schubert, Andreas	Technical University of Chemnitz, D	976
Schuetz, Tobias	GE Research Center, D	645
Schulte-Overbeck, Christian	AEG Power Solutions, D	1677
Schulz, Matthias	Fraunhofer Institute IISB, D	469
Schulz, Nicola	University of Applied Sciences Northwestern Switzerland, CH	2167
Schumacher, Walter	Technical University of Braunschweig, D	216
Schupbach, Marcelo	Cree, USA	2114

PCIM Europe 2016, 10 – 12 May 2016, Nuremberg, Germany

Author	Institution	Pages
Schwalbe, Ulf	ISLE Steuerungstechnik und Leistungselektronik, D	1147 1749
Schweiger, Hans-Georg	Technical University of Ingolstadt, D	742
Schwenk, Holger	Vacuumschmelze, D	1471
Scibilia, Roberto	Texas Instruments, D	1224
Scrimizzi, Filippo	STMicroelectronics, I	804 1987
Seibel, Axel	Fraunhofer Institute IWES, D	401
Seldrum, Thomas	Dow Corning, BE	1017
Seleme Jr., Seleme Isaac	Federal University of Minas Gerais – UFMG, BR	371
Seliger, Bernd	Fraunhofer Institute IISB, D	637
Seliger, Norbert	University Rosenheim, D	1562
Sequeira, Luis	Nomad Tech, PT	953
Shalaby, Rizk	Mansoura University, AE	1046
Shang, Ming	Mitsubishi Electric & Electronics, CN	871
She, Xu	GE Research Center, USA	1853
Sheehan, Cathal	Bourns Electronics, IE	1224
Shekhar, Aditya	Delft University of Technology, NL	1201
Shelton, Ed	Silicon Contact, GB	1845
Shin, SangChul	Hyundai Motors, ROK	1600
Shorten, Andrew	University of Toronto, CA	895
Shousha, Mahmoud	Würth Elektronik eiSos, D	107
Sieweke, Nico	Beckhoff Automation, D	796
Simco, David	Wolfspeed, USA	34
Singer, Arthur	Universität der Bundeswehr Munich, D	1904
Sinn, Peter	Robert Bosch, D	1377
Sirmelis, Ugis	Riga Technical University, LV	1880
Slatter, Rolf	Sensitec, D	1240
Slawinski, Maximilian	Infineon Technologies, D	1800
Soinski, Marian	Magneto, PL	1428
Sokolovs, Alvis	Riga Technical University, LV	1880
Solanki, Jitendra	University of Paderborn, D	1662
Soldati, Alessandro	Università degli Studi di Parma, I	1073 1584
Song, Gaosheng	Mitsubishi Electric & Electronic, CN	871 932
Sorensen, Jens	Analog Devices, USA	1369
Sorrentino, Elmer	University Simon Bolivar, VE	2211

PCIM Europe 2016, 10 – 12 May 2016, Nuremberg, Germany

Author	Institution	Pages
Sorsdahl, Torbjorn	Valeo, NO	113
Spence, Michael	Dynex Semiconductor, GB	903
Spertino, Filippo	Politecnico di Torino, I	2058
Spro, Ole Christian	SINTEF Energy Research, NO	1408
Starks, Ann	ON Semiconductor, USA	1777
Steffen, Jonas	Fraunhofer Institute IWES, D	401
Stegmeier, Stefan	Siemens, D	2226
Steinbring, Manuel	University of Siegen, D	1873
Steinke, Gina	EPFL – Ecole Polytechnique Fédérale de Lausanne, CH	445 1713
Stevanovic, Ljubisa	GE Research Center, USA	645 1853
Stiasny, Thomas	ABB Switzerland, CH	917
Stiegler, Karlheinz	ABB Switzerland, CH	911
Stöckl, Johannes	AIT-Austrian Institute of Technology, AT	1846
Stocksreiter, Wolfgang	FH Joanneum, AT	296 2151
Stone, David	University of Sheffield, GB	1095 1769 2090
Storasta, Liutauras	ABB Switzerland, CH	417
Strauss, Bastian	Otto-von-Guericke-University, D	1081
Streb, Fabian	Infineon Technologies, D	976
Ströbel-Maier, Henning	Danfoss Silicon Power, D	698
Strzalkowski, Bernhard	Analog Devices, D	195
Stubenrauch, Franz	University Rosenheim, D	1562
Stuckmann, Christoph	Maccon, D	1345
Stuckmann, Tim	University of Applied Sciences Ostwesfalen-Lippe, D	1639
Subramanian, Prasanth	GrafTech International, USA	1126
Suganuma, Katsuaki	Osaka University, J	1021
Sumper, Andreas	CITCEA-UPC, ES	2211
Sun, Brian	Infineon Technologies North America, USA	882
Sun, Hui	Chenyang Technologies, D	1248
Suriyah-Jaya, Michael	Karlsruhe Institute of Technology, D	363
Surma, Alexey	Proton-Electrotex JSC, RUS	957
Swieboda, Cezary	Magneto, PL	1428
Szczesny, Paul	GE Research Center, USA	1853

PCIM Europe 2016, 10 – 12 May 2016, Nuremberg, Germany

Author	Institution	Pages
T		
Tadikonda, Ramakrishna	Infineon Technologies Americas, USA	819
Takahashi, Misaki	Fuji Electric, J	438
Takahashi, Takuya	Mitsubishi Electric Corporation, J	342
Takahashi, Yoshikazu	Fuji Electric, J	1001
Takamiya, Yoshikazu	Fuji Electric, J	1956
Takorabet, Noureddine	University de Lorraine, F	772
Tamai, Yuuta	Fuji Electric, J	1001
Tamenori, Akira	Fuji Electric, J	438
Tan, Xiaoya	Schaffner, CH	148
Tanabe, Gen	Japan Fine Ceramics, J	64
Tanaka, Nobuhiko	Mitsubishi Electric Corporation, J	425
Tanioka, Toshikazu	Mitsubishi Electric Corporation, J	336
Tao, Fengfeng	GE Research Center, USA	645
Tchouangue, Georges	Toshiba Electronics Europe, D	566 839
Teixeira Pinto, Rodrigo	CITCEA-UPC, ES	2211
Thal, Eckhard	Mitsubishi Electric Europe, D	48 425
Thesseling, Matthias	Lenze SE, D	2007
Thomas, Tina	Technical University of Berlin, D	1040
Titushkin, Dmitry	Proton-Electrotex JSC, RUS	957
Tiwari, Amit	Newcastle University, GB	2197
Tokuyama, Takeshi	Hitachi Automotive Systems, J	331
Tomasso, Giuseppe	University of Cassino and South Lazio, I	379 1217
Tóssoli de Sousa, Thais	CPqD, BR	750
Trainer, David	Alstom Grid, GB	2197
Treier, Christian	ABB Switzerland, CH	945
Trintis, Ionut	Aalborg University, DK	831
Trüller, Jürgen	HF Instruments, D	1463
Tsang, Chi Wa	University of Lincoln, GB	1769
U		
Uchida, Yoshiyuki	Japan Fine Ceramics, J	64
Ura, Hideyuki	Toshiba Corporation Semiconductor, J	839
Urtasun, Andoni	Public University of Navarre, ES	1209

PCIM Europe 2016, 10 – 12 May 2016, Nuremberg, Germany

Author	Institution	Pages
V		
Van Brunt, Edward	Wolfspeed, USA	34
Vanlathem, Eric	Dow Corning Europe, BE	992 1017
Velasco, David	Public University of Navarre, ES	2043
Vemulapati, Umamaheswara Reddy	ABB Switzerland, CH	917
Verl, Alexander	ISW – University of Stuttgart, D	256
Vershinin, Konstantin	Alstom Grid, GB	2197
Victory, James	Fairchild Semiconductor, D	813
Viera, Juan C.	University of Oviedo, ES	1631
Vijay, Karthik	Indium Corporation, GB	704
Villbusch, Tim	Infineon Technologies, D	1800
Vobecky, Jan	ABB Switzerland, CH	911 917
Vogelsberger, Markus	Bombardier Transportation Austria, AT	1339 2219
Voigt, Gunter	HTWG Konstanz University of Applied Sciences, D	136
Volay, Philippe	Centralp, F	1707
Volke, Andreas	Power Integrations, D	48
Vollaire, Christian	Laboratoire Ampere, F	1494
von Essen, Tobias	Berliner Nanotest & Design, D	2226
Vu, Trong Tue	Eisergy, IE	1163
W		
Wachutka, Gerhard	Technical University of Munich, D	543
Wagner, Bernhard	Technical University Nuremberg Georg Simon Ohm, D	526
Wallscheid, Oliver	University of Paderborn, D	780
Waltereit, Patrick	Fraunhofer-Institute IAF, D	319
Wanderoild, Yohan	CEA Leti, F	312
Wang, Chi-Ming	Toyota Motor Engineering & Manufacturing, USA	1608
Wang, Gang-Yao	Wolfspeed, USA	34
Wang, Gaolin	Harbin Institute of Technology, CN	2014
Wang, Xiongfei	Aalborg University, DK	1837
Wang, Yangang	Dynex Semiconductor, GB	938
Wang, Yazhe	Mitsubishi Electric Corporation, J	336
Warnakulasuriya, Kapila	Carroll & Meynell Transformers, GB	1685 1439

PCIM Europe 2016, 10 – 12 May 2016, Nuremberg, Germany

Author	Institution	Pages
Watabe, Kiyoto	Mitsubishi Electric Corporation, J	336
Wattenberg, Martin	Universtiy of Reutlingen, D	210 757
Weber, Marc	Karlsruhe Institute of Technology, D	1979
Weber, Stefan	Epcos, D	156 164
Wegelin, Viktor	Karlsruhe Institute of Technology, D	1979
Weis, Gerald	FH Joanneum, AT	296 2151
Weiss, Beatrix	Fraunhofer-Institute IAF, D	319
Weiß, Helmut	Technical University of Leoben, AT	2204
Wendt, Hans-Joachim	Lenze Drives, D	2007
Wendt, Michael	Infineon Technologies, D	1615
Werner, Quentin	Daimler, D	772
Wespel, Matthias	Fraunhofer-Institute IAF, D	319
Weyant, Jens	Advanced Cooling Technologies, USA	1118
Weyh, Thomas	Universität der Bundeswehr Munich, D	1904
Wiedemann, Simon	Maccon, D	1353
Wiesner, Eugen	Mitsubishi Electric Europe, D	48 425
Wintrich, Arendt	Semikron, D	1829
Wohlstreicher, Manfred	SUMIDA Components & Modules, D	1728
Wolbank, Thomas	Technical University of Vienna, AT	2219
Wood, John	Silicon Contact, GB	1845
Wright, Nick	Newcastle University, GB	2197
Wu, Xiaomin	Fairchild Semiconductor, D	2231
Wunder, Bernd	Fraunhofer Institute IISB, D	469
Wunderle, Bernhard	Technical University of Chemnitz, D	1155
Würfel, Alexander	Universtiy of Bremen, D	1324
Wurz, Marc C.	Leibniz University Hannover, D	107
Wüthrich, Martin	Schaffner, CH	148
X		
Xu, Dianguo	Harbin Institute of Technology, CN	2014
Xu, Yanan	Chenyang Technologies, D	1248
Y		
Yamada, Junji	Mitsubishi Electric Corporation, J	996
Yamaguchi, Masakazu	Toshiba, J	566

PCIM Europe 2016, 10 – 12 May 2016, Nuremberg, Germany

Author	Institution	Pages
Yamashita, Hiroaki	Toshiba Corporation Semiconductor, J	839
Yang, Gang	Valeo, F	113
Yao, Wenli	Northwestern Polytechnical University, CN	1837
Yasuda, Yuusuke	Hitachi, J	78
Yin, Hang	Technical University of Berlin, D	1186
Yoo, Inpil	Infineon Technologies, D	2027
Yoshida, Hiroshi	Mitsubishi Electric Corporation, J	326 342
Yoshida, Souichi	Fuji Electric, J	438 1956
Yoshiwatari, Shinichi	Fuji Electric, J	824
Young, George	Eisergy, IE	1163
Yu, Kezhuang	Shenzhen Zeasset Electronic Technology, CN	1416
Yu, Zhe	Fraunhofer Institute ISIT, D	202
Yuki, Hata	Mitsubishi Electric Corporation, J	677
Z		
Zacharias, Peter	University of Kassel, D	573
Zeidler, Henning	Technical University of Chemnitz, D	976
Zeltner, Stefan	Fraunhofer Institute IISB, D	637
Zeman, Miro	Delft University of Technology, NL	1383
Zeng, Guang	Technical University of Chemnitz, D	588
Zhang, Hao	Osaka University, J	1021
Zhang, Shirley	United Silicon Carbide, USA	1511
Zhang, Xiaobin	Northwestern Polytechnical University, CN	1837
Zhang, Yuancheng	Mitsubishi Electric & Electronics, CN	871
Zhang, Yuanzhe	Efficient Power Conversion Corporation, USA	304
Zhang, Zhenbin	Technical University of Munich, D	629 1926
Zhou, Haihua	Infineon Technologies Americas, USA	1655
Zhou, Lu	Dow Corning, CN	992
Zhou, Wei	Dynex Semiconductor, GB	938
Zhu, Ke	United Silicon Carbide, USA	1549
Zimmer, Marco	University of Stuttgart, D	410
Zippelius, Bernd	Infineon Technologies, D	180
Zöller, Clemens	Technical University of Vienna, AT	1339 2219
Zschieschang, Olaf	Fairchild Semiconductor, D	1063

Mesago PCIM GmbH
Rotebuehlstrasse 83-85
70178 Stuttgart Germany

ISBN 978-1-5108-2530-7

International Exhibition & Conference for Power Electronics, Intelligent Motion, Renewable Energy and Energy Management (PCIM Europe 2016)

Nuremberg, Germany
10 - 12 May 2016

Volume 3 of 3

International Exhibition & Conference for Power Electronics, Intelligent Motion, Renewable Energy and Energy Management (PCIM Europe 2016)

Nuremberg, Germany
10 - 12 May 2016

Volume 3 of 3

ISBN: 978-1-5108-2530-7

Printed from e-media with permission by:

Curran Associates, Inc.
57 Morehouse Lane
Red Hook, NY 12571

Some format issues inherent in the e-media version may also appear in this print version.

Copyright© (2016) by Mesago PCIM GmbH
All rights reserved.

Printed by Curran Associates, Inc. (2016)

For permission requests, please contact Mesago PCIM GmbH
at the address below.

Mesago PCIM GmbH
Rotebuehlstrasse 83-85
70178 Stuttgart Germany

Phone: 49 711 619 460
Fax: 49 711 619 4690

info@mesago.com

Additional copies of this publication are available from:

Curran Associates, Inc.
57 Morehouse Lane
Red Hook, NY 12571 USA
Phone: 845-758-0400
Fax: 845-758-2633
Email: curran@proceedings.com
Web: www.proceedings.com

PCIM Europe 2016, 10 – 12 May 2016, Nuremberg, Germany

Table of Content PCIM Europe 2016

Keynotes

Keynote: Welcome to the Post-Silicon World: Wide Band Gap Powers Ahead...................................... 31
Dan Kinzer, Navitas Semiconductor, USA

Keynote: Smart Transformers – Concepts-Challenges-Applications...................................... 32
Johann Walter Kolar, ETH Zürich Power Electronic Systems Laboratory, CH

Keynote: Trends of Solar Systems and their Integration in Electricity Networks................. 33
Jens Merten, National Solar Energy Institute, F

SiC Devices

Ultra-low (1.25 mΩ) On-Resistance 900V SiC 62 mm Half-Bridge Power Modules Using New 10 mΩ SiC MOSFETs.................................. 34
Jeffrey Casady, Vipindas Pala, Edward Van Brunt, Brett Hull, Sei-Hyung Ryu, Gang-Yao Wang, Jim Richmond, Scott T. Allen, Dave Grider, John W. Palmour, Peter Killeen, Brice McPherson, Kraig Olejniczak, Brandon Passmore, David Simco, Wolfspeed, USA

Evolution of SiC Products for Industrial Application 42
Naoyuki Kizu, Satoru Nate, Mineo Miura, Noriaki Kawamoto, Kazuhide Ino, Rohm, J; Masaharu Nakanishi, Rohm Semiconductor, D; Nobuhiro Hase, Rohm Semiconductor, USA

Advanced Protection for Large Current Full SiC-Modules.................................. 48
Eugen Wiesner, Eckhard Thal, Mitsubishi Electric Europe, D; Andreas Volke, Karsten Fink, Power Integrations, D

Switching Performance of a 1200 V SiC-Trench-MOSFET in a Low-Power Module.............................. 53
Daniel Heer, Daniel Domes, Dethard Peters, Infineon Technologies, D

Module Materials

Beyond Thermal Grease, Enhancing Thermal Performance and Reliability 60
Sanjay Misra, Henkel, USA

High Thermal Conductivity Silicon Nitride substrate for Power Semiconductor Applications 64
Dai Kusano, Gen Tanabe, Yoshiyuki Uchida, Japan Fine Ceramics, J

Thermal Management of Future WBG Devices using Two-Phase Cooling 72
Shailesh Joshi, Ercan Dede, Toyota Research Institute North America, USA

Highly Reliable and Lead-Free High Power IGBT Modules Using Novel Copper Sintering Die Attachment 78
Akitoyo Konno, Takaaki Miyazaki,Yuusuke Yasuda, Osamu Ikeda, Hiroshi Nakano, Toshiaki Morita, Hiroshi Houzouji, Mutsuhiro Mori, Masato Nakamura, Yoshihiko Koike, Hitachi, J

PCIM Europe 2016, 10 – 12 May 2016, Nuremberg, Germany

Magnetics & Inductors

A New Generation of Nanocrystalline Magnetic Cores with Very Low Magnetic Losses 84
Bashar Gony, Stephane Camus, Julia Hill, Frederic Pottier, Aperam Alloys Amilly, F

**The Fe-based Glassy Alloy Powder Core Inductor for the Boost Converter by GaN HEMT and
SiC SBD 1 MHz Operation** 91
Mitsunao Fujimoto, Yutaka Naitoh, Takao Mizushima, ALPS Green Devices, J

Accurate Calculation of AC Losses of Inductors in Power Electronic Applications 99
Ranjith Bramanpalli, Würth Elektronik Eisos, D

Thin-Film Based Microtransformer Suitable for High Switching Frequency Power Applications 107
Dragan Dinulovic, Mahmoud Shousha, Martin Haug, Würth Elektronik eiSos, D; Sebastian Beringer,
Marc C. Wurz, Leibniz University Hannover, D

DC/AC and AC/DC Converters

**Design of High Efficiency High Power Density 10.5 kw Three Phase On-Board-Charger for
Electric/Hybrid Vehicles** 113
Gang Yang, Valeo, F; Eirik Draugedalen, Torbjorn Sorsdahl, Hui Liu, Roar Lindseth, Valeo Powertrain
Energy Conversion, NO

**A Bridgeless, Quasi-Resonant ZVS-Switching, Buck-Boost Power Factor Correction Stage
(PFC)** 120
Markus Scherbaum, Manfred Reddig, University of Applied Sciences Augsburg, D; Ralph Kennel,
Technical University of Munich, D; Manfred Schlenk, Infineon Technologies, D

**A high efficiency 5.3 kW Current Source Inverter (CSI) Prototype using 1.2 kV Silicon Carbide
(SiC) Bi-Directional Voltage Switches in hard Switching Mode** 128
Jérémy Martin, Anthony Bier, Stéphane Catellani, Luis Gabriel Alves-Rodrigues, Franck Barruel,
Commissariat à l'énergie atomique et aux énergies alternatives, F

**Comparison of the EMC and Efficiency Characteristics of Hard and Soft Switching Three-Level
Inverters** 136
Manfred W. Gekeler, Stefan Schreitmüller, Gunter Voigt, HTWG Konstanz University of Applied
Sciences, D

Special Session "Passive Components"

Drive System Loss Reduction by Allpole Sine Filters 144
Dennis Kampen, BLOCK Transformatoren-Elektronik, D; Michael Burger, SEW Eurodrive, D

Harmonic Filtering in Variable Speed Drives 148
Luca Dalessandro, Xiaoya Tan, Andrzej Pietkiewicz, Martin Wüthrich, Norbert Naeberle, Schaffner, CH

A just Comparison of Ferrite and Nanocristalline Common Mode Chokes 156
Jörn Schliewe, Christian Paulwitz, Stefan Weber, Epcos, D

**Modelling an Anti-Ferroelectric Ceramic Capacitor for Time- and Frequency-Domain Simulations
of Power Systems** 164
Stefan Schefler, Markus Koini, Stefan Weber, Epcos, D; Markus Puff, Epcos, AT

PCIM Europe 2016, 10 – 12 May 2016, Nuremberg, Germany

SiC Reliability

Breakdown of Gate Oxide of 1.2 kV SiC-MOSFETs Under High Temperature and High Gate Voltage 172
Menia Beier-Möbius, Josef Lutz, Chemnitz University of Technology, D

Avalanche Robustness of SiC MPS Diodes 180
Thomas Basler, Roland Rupp, Rolf Gerlach, Bernd Zippelius, Infineon Technologies, D; Mihai Draghici, Infineon Technologies, AT

Compact, Low Loss and High Reliable 3.3 kV Hybrid Module 188
Satoshi Kaneko, Naoyuki Kanai, Motohiro Hori, Nakazawa Masayoshi, Hideaki Kakiki, Yasushi Abe, Yoshinari Ikeda, Eiji Mochizuki, Fuji Electric, J

DC/DC Converters I

Isolated Synchronous Forward Controller with Integrated Feedback Loop and Adjustable Dead Time for High Efficiency DC/C Converter 195
Bernhard Strzalkowski, Analog Devices, D; Subodh Madiwale, Gabriele Bernardinis, Michael Daly, Analog Devices, USA

Extreme High Efficiency Non-Inverting Buck-Boost Converter for Energy Storage Systems 202
Zhe Yu, Holger Kapels, Fraunhofer Institute ISIT, D; Klaus F. Hoffmann, Helmut Schmidt University, D

A High-Efficiency Bidirectional GaN-HEMT DC/DC Converter 210
Michael Ebli, Martin Wattenberg, Universtiy of Reutlingen, D; Martin Pfost, University of Innsbruck, AT

Control Converters

Direct Delta Sigma Signal Processing for Control of Power Electronics 216
Michael Homann, Axel Klein, Walter Schumacher, Technical University of Braunschweig, D

Finite Control Set Model Based Predictive Control of a PMSM with Variable Switching Frequency and Torque Ripple Optimization 224
Fernando David Ramirez Figueroa, Mario Pacas, University of Siegen, D

Frequency- and Mode-Adaptive Control of DC-DC Converter for Efficency Improvement 232
Lukas Keuck, Farshid Almai, Sven Bolte, Norbert Fröhleke, Joachim Böcker, University of Paderborn, D

Control Techniques in Intelligent Motion Systems I

Load Torque Estimation in Repetitive Mechanical Systems by Using Fourier Interpolation 240
Van Trang Phung, Mario Pacas, University of Siegen, D

Fast Current Waveform Calculation Algorithm for a Six Phase Switched Reluctance Machine 248
Jacek Borecki, Bernd Orlik, University of Bremen, D

Process Requirements-Based Adaptive PWM for Improved Efficiency of Machine Tool Feed-Drives 256
Matthias Braband, Florian Frick, Armin Lechler, Alexander Verl, ISW – University of Stuttgart, D

PCIM Europe 2016, 10 – 12 May 2016, Nuremberg, Germany

DC/AC Converters

Comparison of Bidirectional T-Source Inverter and Quasi-Z-Source Inverter for Extra Low Voltage Application 264
Thomas Baier, Bernhard Piepenbreier, Friedrich-Alexander-University Erlangen, D

Benchmarking of SiC JFET and SiC MOSFET Modules for the Application in Medium Power Traction Converters 272
Andreas März, Roman Horff, Teresa Bertelshofer, Mark-M. Bakran, University of Bayreuth, D;
Martin Helsper, Siemens, D

New Bus-bar Topology to Suppress the Current Imbalance of Parallel-connected IGBT Modules for High Power Railway 280
Naoki Sakurai, Masaoki Konishide, Yasuhiko Kohno, Hitachi, J

GaN Converters

EMI Investigation in a GaN HEMT Power Module 288
Xiaoshan Liu, Francois Costa, Bertrand Revol, Cyrille Gautier, ENS Cachan- SATIE, F

Ultra-High Frequent Switching with GaN- HEMTs using the Coss-Capacitances as non-dissipative Snubbers 296
Hubert Berger, Luis Alfonso Fernández-Serantes, Wolfgang Stocksreiter, Gerald Weis, University of Applied Sciences Joanneum, AT

eGaN® FET based 6.78 MHz Differential-Mode ZVS Class D AirFuel™ Class 4 Wireless Power Amplifier 304
Michael de Rooij, Yuanzhe Zhang, Efficient Power Conversion Corporation, USA

High Frequency, High Temperature designed DC/DC Coreless Converter for GaN Gate Drivers 312
Yohan Wanderoild, Dominique Bergogne, CEA Leti, F; Hubert Razik, University Claude Bernard Lyon, F

Monolithic GaN-on-Si Half-Bridge Circuit with Integrated Freewheeling Diodes 319
Richard Reiner, Patrick Waltereit, Beatrix Weiss, Matthias Wespel, Michael Mikulla, Rüdiger Quay,
Oliver Ambacher, Fraunhofer-Institute IAF, D

Module Design

Resin Encapsulation Combined with Insulated Metal Baseplate for Improving Power Module Reliability 326
Shinsuke Asada, Satoshi Kondo, Yusuke Kaji, Hiroshi Yoshida, Mitsubishi Electric Corporation, J

An Experimental Study on the Thermal Performance of Double-Side Direct-Cooling Power Module Structure 331
Akira Matsushita, Ryuichi Saito, Takeshi Tokuyama, Takashi Kimura, Hitachi Automotive Systems, J;
Kinya Nakatsu, Hitachi, J

New Transfer Mold DIPIPM™ Utilizing Silicon Carbide (SiC) MOSFET 336
Yazhe Wang, Kiyoto Watabe, Shinji Sakai, Toshikazu Tanioka, Mitsubishi Electric Corporation, J

A 1700 V-IGBT module and IPM with new insulated metal baseplate (IMB) featuring enhanced Isolation Properties and Thermal Conductivity 342
Takuya Takahashi, Yoshitaka Kimura, Hiroshi Yoshida, Hidetoshi Ishibashi, Yoshitaka Otsubo,
Mitsubishi Electric Corporation, J

PCIM Europe 2016, 10 – 12 May 2016, Nuremberg, Germany

NHPD² (Next High Power Density Dual) with Next Generation Chip Suitable for Low Internal Inductance Package 348
Daisuke Kawase, Kazuhito Nagashima, Tomohisa Hirayama, Katsunori Azuma, Seiichi Hayakawa, Hitachi Power Semiconductor, J; Katsuaki Saito, Hitachi Europe, GB

Power Electronics in Transmission Systems in Smart Grids

An Optimized Hybrid-MMC for HVDC 355
Viktor Hofmann, Mark-M. Bakran, University of Bayreuth, D

Selective HVDC Transmission Line Breaking for Bus Bar Applications under Reduced Expenses 363
Rene Sander, Michael Suriyah-Jaya, Thomas Leibfried, Karlsruhe Institute of Technology, D

Decentralized Controller for Modular Multilevel Converter 371
Seleme Isaac Seleme Jr., Federal University of Minas Gerais – UFMG, BR; Luc-André Grégoire, Marc Cousineau, Philippe Ladoux, University of Toulouse, F

Power Converters for PV Systems with Energy Storage: Optimal Power Flow Control for EV's Charging Infrastructures 379
Mauro Di Monaco, Umberto Abronzini, Ciro Attaianese, Matilde Arpino, Giuseppe Tomasso, University of Cassino and South Lazio, I

Special Session "E-Mobility"

Modular and Comfortable Electromobility 386
Mihai Dragan, Bernhard Budaker, Jonathan Brix, Fraunhofer Institute IPA, D

Power Hardware-in-the-Loop Emulation of Permanent Magnet Synchronous Machines with Nonlinear Magnetics – Concept & Verification 393
Alexander Schmitt, Jan Richter, Michael Braun, Martin Doppelbauer, Karlsruhe Institute of Technology, D

Multimode Charging of Electric Vehicles – A combined Concept with Multiple Use of Components and Strategies for Decreasing Power Losses, Weight and Volume 401
Marco Jung, René Marklein, Georgios Lempidis, Jonas Steffen, Axel Seibel, Jörg Kirchhof, Roland Gaber, Fraunhofer Institute IWES, D

Design of Contactless Energy Transfer System for an Electric Vehicle 410
Mike Böttigheimer, Nejila Parspour, Marco Zimmer, Anna Lusiewicz, University of Stuttgart, D

High Power Semiconductor

3300 V HiPak2 modules with Enhanced Trench (TSPT+) IGBTs and Field Charge Extraction Diodes rated up to 1800 A 417
Chiara Corvasce, Maxi Andenna, Liutauras Storasta, Sven Matthias, Arnost Kopta, Munaf Rahimo, Luca De Michielis, Silvan Geissmann, Raffael Schnell, ABB Switzerland, CH

Durable Design of the New HVIGBT Module 425
Nobuhiko Tanaka, Kenji Ota, Shuichi Kitamura, Shinichi Lura, Keiichi Nakamura, Mitsubishi Electric Corporation, J; Eugen Wiesner, Eckhard Thal, Mitsubishi Electric Europe, D

PCIM Europe 2016, 10 – 12 May 2016, Nuremberg, Germany

The 62Pak IGBT Module Range Employing the 3rd Generation 1700 V SPT++ Chip Set for 175 °C Operation.. 432
Sven Matthias, Chiara Corvasce, Athanasios Mesemanolis, Emilia Gustafsson, Charalampos Papadopoulos, Arnost Kopta, Silvan Geissmann, Martin Bayer, Raffael Schnell, Munaf Rahimo, ABB Switzerland, CH

Extended Power Rating of 1200 V IGBT Module with 7G-RC-IGBT Chip Technologies 438
Misaki Takahashi, Souichi Yoshida, Akira Tamenori, Yasuyuki Kobayashi, Osamu Ikawa, Fuji Electric, J; Daniel Hofmann, Fuji Electric Europe, D

Multi Level Converters

DC-DC Converter based on the Asymmetric Multistage Stacked Boost Architecture with Feed-Forward Control for Photovoltaic Plants 445
Georgios Mademlis, Aristotle University of Thessaloniki, GR; Gina Steinke, Alfred Rufer, EPFL – Ecole polytechnique fédérale de Lausanne, CH

A Novel Submodule Concept for Modular Multilevel Converters 453
Benjamin Ruccius, Nicola Burani, Dirk Malipaard, Fraunhofer Institute IISB, D; Marek Galek, Siemens, D

Electro-Thermal Design of a Modular Multilevel Converter Prototype 461
Emilien Coulinge, Alexandre Christe, Drazen Dujic, EPFL – Ecole Polytechnique Fédérale de Lausanne, CH

DC/DC Converters II

Non-isolated Three-Port DC/DC-Converter for ±380VDC Microgrids 469
Yunchao Han, Julian Kaiser, Leopold Ott, Matthias Schulz, Fabian Fersterra, Bernd Wunder, Martin März, Fraunhofer Institute IISB, D

Wide Voltage Input Range Insulated Current Fed Buck Flyback-Forward for HV/LV Power Conversion in Electric/Hybrid Vehicle 477
Reda Chelghoum, Luis De Sousa, Larbi Bendani, Valeo, F; Daniel Sadarnac, CentraleSupelec, F

SiC JFET Cascode Enables Higher Voltage Operation in a Phase Shift Full Bridge DC-DC Converter 484
Jonathan Dodge, United Silicon Carbide, USA

Lamp Ballasts Lighting Systems

Detailed Comparison of One Stage Topologies for LED Lighting Applications 492
Alexander Pawellek, Thomas Dürbaum, Friedrich-Alexander-University Erlangen, D

Low Cost High Density AC-DC Converter for LED Lighting Applications 501
Simon Nigsch, Janosch Marquardt, Kurt Schenk, University of Applied Sciences NTB Buchs, CH

Design Optimization for a High Power-Density, Wide Output, High Frequency LLC Resonant Converter for Lighting Applications 509
Janosch Marquart, Simon Nigsch, Kurt Schenk, University of Applied Sciences NTB Buchs, CH

Sensorless Motor Control

Computationally efficient Anisotropy-Identification based on a Square-Shaped Injection Pattern 518
Peter Landsmann, Dirk Paulus, Sascha Kühl, Ralph Kennel, Technical University of Munich, D

PCIM Europe 2016, 10 – 12 May 2016, Nuremberg, Germany

Estimation of the Excitation Current and the Rotor Resistance of an Externally Excited Synchronous Machine with an Inductively Supplied Excitation Coil 526
Stefan Köhler, Bernhard Wagner, Technical University Nuremberg Georg Simon Ohm, D;
Stefan Endres, Fraunhofer Institute IISB, D

High Speed Sensorless Control of a Synchronous Motor with Kalman Filter 534
Philipp Niedermayr, Alpitronic, I; Silverio Bolognani, Luigi Alberti, University of Padova, I;
Reiner Abl, BMW, D

Software Tools and Applications

Physical Modeling and High-Fidelity Simulation of the Transient Behavior of Multiply-Contacted Power Busbars 543
Vanessa Basler, Wolfgang Hölzl, Gerhard Wachutka, Technical University of Munich, D

Small-signal Output Impedance Modeling of Intersil's R4TMTechnology 550
Yi Huang, Chun Cheung, Intersil Corporation, USA

An Approach of Reinforcement Learning Based Lighting Control for Demand Response 558
Xinxing Pan, Brian Lee, AIT, IE

Cosmic Ray & Ruggedness

Cosmic Ray Failure Mechanism and Critical Factors for 3.3 kV Hybrid SiC Modules 566
Tetsuya Nitta, Yoko Sakiyama, Raita Kotani, Tomoki Inoue, Ryoichi Ohara, Kenya Sano,
Masakazu Yamaguchi, Toshiba, J; Georges Tchouangue, Toshiba Electronics Europe, D

Benefits of Increased Cosmic Radiation Robustness of SiC Semiconductors in large Power-Converters 573
Christian Felgemacher, Samuel Araujo Vasconcelos, Christian Nöding, Peter Zacharias,
University of Kassel, D

Passive IGBT Turn-off during Short-circuit Type V 581
Jan Fuhrmann, Hans-Günter Eckel, University of Rostock, D; Sebastian Klauke, Infineon, D

High-Current Power Cycling Test-Bench for Short Load Pulse Duration and First Results 588
Guang Zeng, Christian Herold, Menia Beier-Möbius, Josef Lutz, Technical University of Chemnitz, D;
Sascha Kubera, Rodrigo Alvarez, Siemens, D

Issues in Testing Advanced Power Semiconductor Devices 596
Gabor Farkas, Mentor Graphics, HU; Zoltan Sarkany, Marta Rencz, BME, HU

Special Session "Smart Lighting"

Smart Lighting – Requirements for Modern Lighting Systems and Expected Trends 604
Diederik de Stoppelaar, LightingEurope, BE

Special requirements on a SMPS for LED-Lighting Purposes by an Example of an Individual High-End LED-Driver Solution 605
Florian Müller, Stefan Raithel, Vossloh-Schwabe Lighting Solutions, D

The seven challanges of LED lighting 612
Claudio Adragna, Francesco Ferrazza, Giovanni Gritti, STMicroelectronics, I

PCIM Europe 2016, 10 – 12 May 2016, Nuremberg, Germany

Highly Flexible Single Stage Flyback Quasi-Resonant Digital Controller for Advanced LED Applications 620
Marc Fahlenkamp, Infineon Technologies, D

New and Renewable Energy Systems

Carrier-Based Modulation Technique to Reduce Low Frequency Ripple at the Partial Dc-Link Voltages of a Three-Level/-Phase/-Wire NPC Converter Applied to Future Dc Bipolar Active Distribution Networks 621
Joabel Moia, Marcelo Heldwein, Federal Institute of Santa Catarina – IFSC, BR

FPGA Based Direct Model Predictive Power and Current Control of 3L NPC Active Front Ends 629
Zhenbin Zhang, Ralph Kennel, Technical University of Munich, D

Scalable Insulated DC/DC Converters for Safe and Efficient Coupling of Fuel Cells, Electrolyzers and DC Grids 637
Bernd Seliger, Stefan Matlok, Stefan Zeltner, Fraunhofer Institute IISB, D

SiC MW PV Inverter 645
Maja Harfman Todorovic, Ljubisa Stevanovic, Gary Mandrusiak, Brian Rowden, Fengfeng Tao, Philip Cioffi, Jeffrey Nasadoski, Rajib Datta, GE Research Center, USA; Fabio Carastro, Tobias Schuetz, Robert Roesner, GE Research Center, D

Power Electronics in Automotive

A Performance Comparison of a 650 V Si IGBT and SiC MOSFET Inverter under Automotive Conditions 653
Teresa Bertelshofer, Roman Horff, Andreas März, Mark-M. Bakran, University of Bayreuth, D

A Generic Topology for Electrical Energy Storage Systems 661
Christoph Marxgut, Helbling Technik, Dli

Pulse Width- and Frequency Modulated DC/DC Converter for Hybrid- and Electrical Vehicles 669
Magnus Böh, Andreas Lohner, Technical University of Cologne, D; Christoph Engelhard, HELLA KGaA Hueck

New High Power Density Modules for EV/HEV Applications 677
Seiichiro Inokuchi, Shoji Saito, Arata Izuka, Hata Yuki, Shinji Hatae, Mitsubishi Electric Corporation, J; Khalid Hussein, Mitsubishi Electric Europe, D

Module Technology

Fault-Tolerant B6-B4 Inverter Reconfiguration with Fuses and Ideal Short-On Failure IGBT Modules 683
Michael Gleißner, Mark-M. Bakran, University of Bayreuth, D

Statistical Evaluation of Current Imbalance in Parallel Devices 691
Uwe Scheuermann, Semikron Elektronik, D

Batch Purity in Semiconductor Power Modules 698
Christian Aggen, Henning Ströbel-Maier, Matthias Mau, Jürgen Laue, Marco Bäßler, Danfoss Silicon Power, D

PCIM Europe 2016, 10 – 12 May 2016, Nuremberg, Germany

Novel Technique to Reduce Substrate Tilt & Improve Bondline Control between AlN Substrate & AlSiC Baseplate in IGBT Modules 704
James Booth, Paul Mumby-Croft, Matthew Packwood, Kim Evans, Andy Dai, Dynex Semiconductor, GB;
Karthik Vijay, Indium Corporation, GB

Drive Strategies in Power Converters

Communicating Gate Driver for SiC MOSFET 712
Christophe Bouguet, Nicolas Ginot, Christophe Batard, University of Nantes, F

New Ultra Fast Short Circuit Detection Method Without Using the Desaturation Process of the Power Semiconductor 720
Stefan Hain, Mark-M. Bakran, University of Bayreuth, D

High Power, High Frequency Gate Driver for SiC-MOSFET Modules 728
Gunter Königsmann, Reinhard Herzer, Sven Buetow, Matthias Rossberg, Semikron Elektronik, D

Integrating a real-time Tvj calculation into an IPM 735
Stefan Schmies, Peter Lahl, Wolfram Kruschel, Matthias Lassmann, Infineon, D

Energy Storage

12 V Lithium Ion Starter Batteries 742
Hans-Georg Schweiger, Enrique Machuca, Jonas Löchel, Technical University of Ingolstadt, D

Electric Vehicles Batteries Modeling Analysis Based on a Multiple Layered Perceptron Identification Approach 750
Sender Rocha dos Santos, Thais Tóssoli de Sousa, Alex Pereira França, CPqD, BR

An Efficient Implementation of a Reconfigurable Battery Stack with Optimum Cell Usage 757
Martin Wattenberg, Reutlingen University, D; Martin Pfost, University of Innsbruck, AT

Comparative Study and Evaluation of Passive Balancing Against Single Switch Active Balancing Systems for Energy Storage Systems 763
Iosu Aizpuru, Unai Iraola, Jose Mari Canales, Ander Goikoetxea, Mondragon University, ES

Control Techniques in Intelligent Motion Systems II

Voltage Levels Comparison and System Optimization for Electric Drives in Hybrid Vehicles 772
Quentin Werner, Daimler, D; Serge Pierfederici, Noureddine Takorabet, University de Lorraine, F

Real-Time Capable Model Predictive Control of Permanent Magnet Synchronous Motors Using Particle Swarm Optimisation 780
Oliver Wallscheid, Joachim Böcker, University of Paderborn, D; Ulrich Ammann, Esslingen University of Applied Sciences, D

A Modular Multilevel Matrix Converter for High Speed Drive Applications 788
Dennis Bräckle, Felix Kammerer, Mathias Schnarrenberger, Marc Hiller, Michael Braun, Karlsruhe Institute of Technology, D

Smart Supercapacitor based DC-link Extension for Drives offers UPS Capability and acts as an Energy Efficient Line Regeneration Replacement 796
Jens Onno Krah, Markus Höltgen, David Langhals, Technical University of Cologne, D;
Nico Sieweke, Christoph Klarenbach, Beckhoff Automation, D

PCIM Europe 2016, 10 – 12 May 2016, Nuremberg, Germany

MOSFET, IGBTs, Freewheeling Diodes

New LV Wide SOA Power MOSFET Technology for Linear Mode Operation 804
Filippo Scrimizzi, Gaetano Bazzano, Daniela Cavallaro, Marco Comola, Giuseppe Consentino,
Stefania Fortuna, Giuseppe Longo, Gaetano Pignataro, STMicroelectronics, I

Field Stop Trench IGBT Process Parameter Calibration for Advanced Predictive Prototyping 813
Mehrdad Baghaie Yazdi, Hermann Fischer, James Victory, Fairchild Semiconductor, D;
Detlef Conrad, Synopsys, D

Best-in-class 1200 V IGBT for High Frequency Applications 819
Ramakrishna Tadikonda, Jorge Cerezo, Chiu Ng, Infineon Technologies Americas, USA

**Extra Electro-Thermal Performance of 1700V IGBT with the Latest 7th Generation Chipset/
Package Technologies** 824
Thomas Heinzel, Fuji Electric Europe, D; Mutsumi Sawada, Shinichi Yoshiwatari, Hiroaki Ichikawa,
Yuichi Onozawa, Osamu Ikawa, Fuji Electric, J

**Parameter Extraction for PSpice Models by Means of an Automated Optimization Tool –
An IGBT model Study Case** 831
Carlos Gomez Suarez, Francesco Iannuzzo, Paula Diaz Reigosa, Ionut Trintis, Frede Blaabjerg,
Aalborg University, DK

**800 V Super Junction MOSFET (HV-DTMOS IV) with Better Trade-Off Between Switching Loss
and dVDS/dt** 839
Hiroyuki Irifune, Hiroshi Ohta, Kaga Toshiba Electronics Corporation, J; Hiroaki Yamashita, Hideyuki Ura,
Kenji Mii, Masato Nashiki, Naotsugu Kako,Toshiba Corporation Semiconductor, J; Georges Tchouangue,
Toshiba Electronics Europe, D

Highly Robust 1700 V Diodes Fabricated on 8"Line Using Optimized Proton Implanted Buffer 845
Maolong Ke, Haihui Luo, Ian Deviny, Xiaoping Dai, Jianwei Huang, Guoyou Liu, Dynex Semconductor, GB

Loss and Softness Optimized IGBT-Diode System for Fast-Switching Applications 850
Christian Müller, Stefan Buschhorn, Infineon Technologies, D

Intelligent Power Modules

Protection Features of Intelligent Power Module against Transient State 858
Taehyun Kim, Minsub Lee, Junbae Lee, Daewoong Chung, Infineon Technologies Power Semitech, ROK

**New High Level Integrated Intelligent Power Module with Three Phase Inverter and Power Factor
Correction Topologies Optimized for Home Appliance** 863
Hyosang Jang, Byoungho Choo, Junbae Lee, Minsub Lee, Daewoong Chung, Infineon Technologies
Power Semitech, ROK

New DIPIPM+TM Series Module with All-in-one Integrated 871
Yuancheng Zhang, Xiankui Ma, Hongtao He, Gaosheng Song, Ming Shang, Mitsubishi Electric &
Electronics, CN

Improvement of System Level Power Density of 15 A / 600 V Intelligent Power Modules 877
Jonathan Harper, ON Semiconductor, D

Optimization of FREDFET-based µIPMTM for very Low Power Motor Drive Applications 882
Rajeev Krishna Vytla, Danish Khatri, Brian Sun, Infineon Technologies North America, USA

A novel Transfer Molding Intelligent Converter Inverter Brake IGBT Module (DIPIPM+) with Integrated Level Shifting Control ICs 889
Marco Honsberg, Mitsubishi Electric Europe, D; Teruaki Nagahara, Mitsubischi Electric Corporation, J; Eric R. Motto, Powerex, USA

An Automatic IGBT Collector Current Sensing Technique via the Gate Node 895
Jingxuan Chen, Andrew Shorten, Wai Tung Ng, University of Toronto, CA; Masahiro Sasaki, Tetsuya Kawashima, Haruhiko Nishio, Fuji Electric, J

High Voltage Devices

Design and Characterisation of Optimised Protective Thyristors for VSC Systems 903
Michael Spence, Ashley Plumpton, Colin Rout, Alan Millington, Richard Keyse, Dynex Semiconductor, GB

Cathode Emitter vs. Carrier Lifetime Engineering of Thyristors for Industrial Applications 911
Jan Vobecky, Marco Bellini, Karlheinz Stiegler, ABB Switzerland, CH

Experimental Results of a Large Area (91 mm) 4.5 kV "Bi-Mode Gate Commutated Thyristor" (BGCT) 917
Thomas Stiasny, Umamaheswara Reddy Vemulapati, Martin Arnold, Munaf Rahimo, Jan Vobecky, ABB Switzerland, CH; Christian Kähr, Norbert Hofmann, University of Applied Sciences Nordwestschweiz, CH

Effect of Self Turn-On during Turn-On of HV-IGBTs 924
Patrick Münster, Daniel Lexow, Hans-Günter Eckel, University of Rostock, D

An Innovative 6500 V HVIGBT with High Robustness 932
Bo Hu, Gaosheng Song, Mitsubishi Electric & Electronic, CN

New High Power 3.3 kV / 1500 A IGBT Module Packaging 938
Daohui Li, Wei Zhou, Fang Qi, Matthew Packwood, Yangang Wang, Steve Jones, Xiaoping Dai, Dynex Semiconductor, GB

The LinPak High Power Density Design and its Switching Behaviour at 1.7 kV and 3.3 kV 945
Samuel Hartmann, Fabian Fischer, Andreas Baschnagel, Harald Beyer, Raffael Schnell, Christian Treier, ABB Switzerland, CH

Power Converter GTO to IGBT Upgrade – a New Life for Traction Converters 953
Luis Sequeira, Augusto Franco, Adriano Carvalho, Nuno Freitas, Nomad Tech, PT

Packaging Technologies and Materials

Aspects of Reliability Improvement for Large Area Power Semiconductor Devices through Sintering 957
Dmitry Titushkin, Alexey Surma Proton-Electrotex JSC, RUS; Michiel De Monchy, Anna Lifton, Alpha, RUS

Analysis of Interface Structure and Composition of Cu/Al$_2$O$_3$ for the High Stability of DBC (Direct Bonded Copper) 961
Hyunwoo Kim, Jaehoon Jung, Hanna Choi, Kisoo Jun, KCC Corporation, ROK

Power Stack – Advantages and Reliability of an Aluminum Based Stacked Power Module 969
Chris Burns, AB Mikroelektronik, AT

Evaluation of Metal-Matrix composites Baseplates with anisotropic thermal Conductivity Inserts 976
Fabian Streb, Infineon Technologies, D; Henning Zeidler, Michael Penzel, Andreas Schubert, Thomas Lampke, Technical University of Chemnitz, D

PCIM Europe 2016, 10 – 12 May 2016, Nuremberg, Germany

Analysis of Packaging Impedance on Performance of SiC MOSFETs 984
Andrew Lemmon, Levi Gant, The University of Alabama, USA; Sujit Banerjee, Kevin Matocha, Monolith
Semiconductor, USA

Low-Stress Silicone Encapsulant for Reliable Power Conversion Devices 992
Guy Beaucarne, Eric Vanlathem, Dow Corning Europe, B; Lu Zhou, Dow Corning, CN; Kent Larson,
Dow Corning Corporation, USA

Pumping out Failure Free Package Structure 996
Junji Yamada, Yoshitaka Otsubo, Mitsubishi Electric Corporation, J; Satoshi Miyahara, Mitsubishi
Electric, D

High power IGBT Module with New AIN Insulated Substrate 1001
Hiroyuki Nogawa, Akira Hirao, Yoshitaka Nishimura, Takashi Saitou, Yuuta Tamai, Fumihiko Momose,
Eiji Mochizuki, Yoshikazu Takahashi, Fuji Electric, J

Nanosilver Paste for Low Pressure Die Attach: A Turn Key Process 1009
Francesc Masana, Barcelona Semiconductors, ES

New Silicone Gel Enabling High Temperature Stability for next Generation of Power Modules 1017
Thomas Seldrum, Eric Vanlathem, Vincent Delsuc, Dow Corning, BE; Hiroji Enami, Dow Corning Toray, J

**A New Ag Paste Composed by Nano and Micro-Ag Particles prepared Simultaneously and
Application as Die-attachment Materials** 1021
Katsuaki Suganuma, Jinting Jiu, Hao Zhang, Shunsuke Koga, Shijo Nagao, Osaka University, J

Packaging and Reliability

Reliability of Double Side Silver Sintered Devices with various Substrate Metallization 1027
Francois LeHenaff, Alpha Assembly Solutions, D; Gustavo Greca, Paul Salerno, Oscar Khaselev,
Monnir Boureghda, Jeffrey Durham, Anna Lifton, Alpha Assembly Solutions, USA; Olivier Mathieu,
Martin Reger, Rogers Germany, D; Zoltan Sarkany, Weikun He, Joe Proulx, John Parry, Mentor Graphics,
UK; Jean Claude Harel, Satyavrat Laud, Renesas Electronics America, USA

New Interconnect Materials: For Future High Reliable Power Module Assembly 1035
Stieven Josso, Henkel Electronics, BE

**Encapsulation of Smart Power Electronic Devices – Thermal Degradation and Dielectric
Behavior** 1040
Tina Thomas, Technical University of Berlin, D; Karl-Friedrich Becker, Klaus-Dieter Lang, Fraunhofer
Institute IZM, D

**Improvement of the Mechanical Properties of Sn-Ag-Sb Lead-Free Solders: Effects of Sb
Addition and Rapidly Solidified** 1046
Mohammed Gumaan, Rizk Shalaby, Mustafa Kamal, Mansoura University, AE; Esmail A. Ali, University
of Science and Technology Yemen, YE

Health-Monitoring of IGBT Power Modules using repetitive Half-sinusoidal Power Losses 1055
Marco Denk, Mark-M. Bakran, University of Bayreuth, D

Reliability Investigation on SiC BJT Power Modules 1063
Alexander Otto, Sven Rzepka, Fraunhofer-Institute ENAS, D; Eberhard Kaulfersch, Berliner Nanotest &
Design, D; Sophia Frankeser, Technical University of Chemnitz, D; Klas Brinkfeldt, Swerea IVF AB, SE;
Olaf Zschieschang, Fairchild Semiconductor, D

PCIM Europe 2016, 10 – 12 May 2016, Nuremberg, Germany

Investigation of the Influence of Ageing Processes on Thermal Characteristics of an IGBT Power Module by Means of Transient Thermal Analysis 1072
Tobias von Essen, Berliner Nanotest & Design, D; Stefan Stegmeier, Gerhard Mitic, Siemens, D

Test Setup for Multistress Characterization of Insulation Degradation Mechanisms in Electric Drives 1073
Davide Barater, Alessandro Soldati, Giorgio Pietrini, Giovanni Franceschini, Università degli Studi di Parma, I; Chris Gerada, Michael Galea, University of Nottingham, GB; Fabio Immovilli, Raw Power, I

Integration of a Measurement Circuit to determine Junction Temperatures of IGBTs in a Three-phase Converter 1081
Bastian Strauss, Andreas Lindemann, Otto-von-Guericke-University, D

A Recuperation Topology for Power Device Testing 1089
Tomas Krecek, ON semiconductor, CZ

Electrolytic Capacitor Age Estimation using PRBS-Based Techniques 1095
David Hewitt, James Green, Jonathan Davidson, Martin Foster, David Stone, University of Sheffield, GB

On-line Monitoring for Diagnosis on Traction Transformer for Rolling Stock Application 1103
André-Philippe Chamaret, SNCF, F; Toufann Chaudhuri, ABB Sécheron, CH

Cooling

Direct-Water-Cooled Next High Power Density Dual (nHPD2) Considering Inverter Layout 1110
Keisuke Horiuchi, Yuichiro Konishi, Mutsuhiro Mori, Daisuke Kawase, Hitachi, J; Katsuaki Saito, Hitachi Europe, GB

Heat Pipes used as Heat Flux Transformers and for Remote Heat Rejection 1118
Devin Pellicone, Jens Weyant, Advanced Cooling Technologies, USA

New Class of Graphite TIMs provide Performance and Reliability 1126
Prasanth Subramanian, Alex Augoustidis, GrafTech International, USA

Heat Pipe System Development for Railway Application working with speed Motion Convection 1132
Thomas Albertin, Atherm, F

High Performances Passive Two-Phase Loops for Power Electronics Cooling 1139
Vincent Dupont, Cyrille Billet, Thomas Nicolle, Calyos, BE

Thermal Modelling and Management for increasing the Power Density in High Current Power Electronic Systems 1147
Marco Schilling, Benjamin Köhnlechner, Ulf Schwalbe, Tobias Reimann, Technical University of Ilmenau, D

Packaging and Characterization of Silicon SiC-based Power Inverter Module with Double Sided Cooling 1155
Charles-Alix Manier, Hermann Oppermann, Lothar Dietrich, Christian Ehrhardt, Fraunhofer-Institute IZM, D; Zoltán Sárkány, Budapest University of Technology and Economics, HU; Bernhard Wunderle, Technical University of Chemnitz, D; Wilhelm Maurer, Infineon Technologies, D; Radoslava Mitova, Klaus-Dieter Lang, Technical University of Berlin, D

PCIM Europe 2016, 10 – 12 May 2016, Nuremberg, Germany

Sensors, Control and Protection

Digital Adaptive Control Approach to Dynamic Response Improvement for Compact PFC Rectifiers 1163
Trong Tue Vu, George Young, Eisergy, IE

Nonlinear Output Characteristic of DAB Converter caused by ZVS Transition 1171
Martin Jagau, Michael Patt, Technologienetzwerk Allgäu, D

Efficiency Maximization for Half-Bridge LC Converter through Automatic Dead Time Tuning 1178
Vittorio Crisafulli, ON Semiconductor, D; Gianluca Fazio, On Semiconductor Italy, I;
Diego Hernandez Gutiérrez, CH

Improved Finite Control Set Model Predictive Control with Fixed Switching Frequency for Three Phase NPC Converter 1186
Margarita Norambuena, Hang Yin, Sibylle Dieckerhoff, Technical University of Berlin, D;
Jose Rodriguez, University Andres Bello, CL

Current Sensorless Totem-pole Bridgeless Power Factor Corrector 1194
Felipe López, Francisco Azcondo, Paula Lamo, Alberto Pigazo, University of Cantabria, ES

State Space Model for n-Parallel Connected DC-DC Converters with Predictive Current Control Strategy 1201
Aditya Shekhar, Pavol Bauer, Laurens Mackay, Laura Ramirez-Elizondo, Delft University of Technology, NL

Parameter-Independent Battery Voltage Control Based on Virtual Capacitor Emulation 1209
Andoni Urtasun, Ernesto L. Barrios, Pablo Sanchis, Luis Marroyo, Public University of Navarre, ES

FPGA Digital Control for VSI Nonlinearity Effect Compensation 1217
Mauro Di Monaco, Umberto Abronzini, Ciro Attaianese, Matilde Arpino, Giuseppe Tomasso, University of Cassino and South Lazio, I

Offline Non Isolated Converter Protection 1224
Cathal Sheehan, Bourns Electronics, IE; Roberto Scibilia, Texas Instruments, D

Optimisation of Shunt Resistors for Fast Transients 1232
Melanie Adelmund, Christian Bödeker, Nando Kaminski, University of Bremen, D

High Bandwidth Current Sensors as an Enabler for Advanced Control Techniques 1240
Rolf Slatter, Sensitec, D

Rotational Speed Measurement Based on Avago ADNS-9800 Laser Mouse Sensor 1248
Cheng Liu, Yanan Xu, Ji-Gou Liu, Hui Sun, Chenyang Technologies, D; Ralph Kennel, Technical University of Munich, D

Low EMI high efficiency converters

Practical EMI Control in a Power Component Design Space 1253
David Bourner, Vicor Corporation, USA

Converter Switching Noise Reduction for Enhancing EMC Performance in HEV and EV 1262
Ho Tae Chun, SeungHyun Han, ChangHan Jun, JeongYun Lee, JaeWon Lee, JeeHye Jeong,
JeongHong Joo, JinHwan Jung, Hyundai Motors Company, ROK

Efficiency and Vibration Observations of a Symmetrical Six-Phase Drive applying Interleaved Space Vector Modulation 1270
Daniel Glose, Peng Qian, Ralph Kennel, Technical University of Munich, D

PCIM Europe 2016, 10 – 12 May 2016, Nuremberg, Germany

High Efficiency Three-Phase-Inverter with 650 V GaN HEMTs 1278
Jennifer Lautner, Bernhard Piepenbreier, Friedrich-Alexander-University of Erlangen, D

A Large Input Voltage Range 1 MHz Full Converter with 95 % Peak Efficiency for Aircraft Applications 1286
Nicolas Quentin, SAGEM, F; Remi Perrin, INSA de Lyon, F; Christian Martin, Charles Joubert, Ampere Laboratory, F; Louis Grimaud, Safran Group, F; Rolando Burgos, Dushan Boroyevich, CPES/Virginia Tech, USA

Integrating Depletion-Mode SiC VJFETs into Production Motor Drives 1294
Michael Mazzola, James Gafford, Mississippi State University, USA; Gerald W. Godbold, Hyperion Technology, USA

Higher Light Efficacy in LED-Lamps by Lower LED-Current 1300
Reinhard Jaschke, Klaus F. Hoffmann, Helmut Schmidt University Hamburg, D

Synchronized Switching and Active Clamping of IGBT Switches in a Simple Marx Generator 1305
Martin Sack, Martin Hochberg, Georg Müller, Karlsruhe Institute of Technology, D

High Efficient and Lightweight Auxiliary Power supply with new SiC Power Device 1311
Ryosuke Nakagawa, Mitsubishi Electric Corporation, Japan

A New Behavioral Model for Accurate Loss Calculations in Power Semiconductors 1315
Ajay Poonjal Pai, Tomas Reiter, Infineon Technologies, D; Martin März, Fraunhofer Institute IISB, D

High Speed Electronic Over Current Breaker for DC-Grids without Additional Sensing 1324
Alexander Würfel, Johannes Adler, Nando Kaminski, University of Bremen, D; Anton Mauder, Infineon Technologies, D

Wireless Power Transmission with High Efficiency for Extensive Applications 1332
Markus Rehm, IBR Ingenieurbüro Rehm, D

Motors and Motor Drives

Prevention of Traction Drives Stator Insulation Faults Based on Overvoltage Reduction Utilizing Active Edge Shaping 1339
Clemens Zöller, Thomas Hausberger, Mathias Blank, Tobias Glück, Hans Ertl, Andreas Kugi, Technical University Vienna, AT; Markus Vogelsberger, Bombardier Transportation Austria, AT

Noise & Vibration Levels of modern Electric Motors 1345
Christoph Stuckmann, Maccon, D

Development Platform and Techniques for the Rapid Implementation of High Performance Drives 1353
Christian Balke, Simon Wiedemann, Maccon, D

Functional Safety for Integrated Circuits used in Variable Speed Drives 1361
Tom Meany, Analog Devices ERDC, IE

A Sytem Approach To Understanding The Impact of Non-ideal Effects In A Motor Drive Current Loop 1369
Jens Sorensen, Analog Devices, USA; Dara O'Sullivan, Analog Devices, IE

Gate Driver as Part of the Inverter Safety Concept: Optimizing Inverter's Design 1377
Laurent Beaurenaut, Infineon Technologies, D; Peter Sinn, Robert Bosch, D

PCIM Europe 2016, 10 – 12 May 2016, Nuremberg, Germany

Passive Components

Estimation of Ripple and Inductance Roll off when using Powdered Iron Core Inductors 1383
Gautham Ram Chandra Mouli, Pavol Bauer, Miro Zeman, Delft University of Technology, NL;
Jos Schijffelen, Power Research Electronics, NL

Optimized DC Link for Next Generation Power Modules 1391
Michael Brubaker, Terry Hosking, SBE, USA; Tomas Reiter, Infinoen Technologies, D;
Laura D. Marlino, Madhu S. Chinthavali, Oak Ridge National Laboratory, USA

**In-Circuit-Characterization of Ceramic Capacitor with Anti-Ferroelectric Material for Voltage
Source Inverters** 1400
Jürgen Kropp, Mark-M. Bakran, University of Bayreuth, D

Operability of Metallized Polypropylene Capacitors under High Pressure 1408
Magnar Hernes, Ole Christian Spro, SINTEF Energy Research, NO; Volker Gleitner, Electronicon
Kondensatoren, D

Application of High-Voltage 750 V Aluminum Electrolytic Capacitor in Inverter 1416
Kezhuang Yu, Mingkai Peng, Mianwei Qiu, Shenzhen Zeasset Electronic Technology, CN

Analytic Loss Calculation for E-Core Inductors including the End Windings 1421
Johannes Heseding, Axel Mertens, Leibniz University Hannover, D

The Applicability of Nanocrystalline Stacked and Block Cores for Power Electronics 1428
Cezary Swieboda, Marian Soinski, Marcin Kwiecien, Magneto, PL; Wojciech Pluta, Czestochowa
University of Technology, PL; Jacek Leszczynski, AGH University of Science and Technology, PL

The Benefit of Formed or Compacted Litz-Wire Coils 1434
Tobias Appel, STS, D; Hans Rossmanith, Friedrich-Alexander-University of Erlangen, D

**Development of a 100 kW, 20 kHz Nanocrystalline Core Transformer for DC / DC Converter
Applications** 1439
Kapila Warnakulasuriya, Carroll & Meynell Transformers, GB; Farhad Nabhani, Vahid Askari,
Teesside University, GB

Simulation of a 3-Phase Common- and Differential Mode Inductor on a Four-Limb Core 1447
Michael Owzareck, BLOCK Transformatoren-Elektronik, D; Nejila Parspour, University of Stuttgart, D

Design Procedure for Pot-Core Integrated Magnetic Component 1455
Martin Foster, University of Sheffield, GB; Andrew Fairweather, Grant Ashley, VxI Power, GB

Investigation of Core Losses under Different Conditions Applying the Cross Power Method 1463
Boris Hudoffsky, PMK Mess- und Kommunikationstechnik, D; Chihiro Okinori, IWATSU Test Instruments, J;
Jürgen Trüller, HF Instruments, D

A Finite Element Simulation of Nanocrystalline Tape Wound Cores 1471
Christian Scharwitz, Holger Schwenk, Johannes Beichler, Werner Loges, Vacuumschmelze, D

SiC and GaN

**An Insightful Evaluation of a 650 V High-Voltage GaN Technology in Cascode and Stand-Alone
Transistors** 1479
Jaume Roig, German Gomez, Frederick Declercq, Filip Bauwens, On Semiconductor, BE;
Manuel Fernandez, Diego Gonzalez, University of Oviedo, ES

PCIM Europe 2016, 10 – 12 May 2016, Nuremberg, Germany

Static Characterization of Discrete State-of-the-Art SiC Power Transistors 1487
Michael Meisser, Horst Demattio, Thomas Blank, Karlsruhe Institute of Technology, D

Analytical Losses Model for SiC Semiconductors dedicated to Optimization Operations 1494
Gnimdu Dadanema, Francois Costa, ENS Cachan - SATIE, F; Jean-Luc Schanen, Yvan Avenas,
G2ELAB, F; Christian Vollaire, Laboratoire Ampere,F

**Current Measurement and Gate-Resistance Mismatch in Paralleled Phases of High Power SiC
MOSFET Modules** 1503
Roman Horff, Teresa Bertelshofer, Andreas März, Mark-M. Bakran, University of Bayreuth, D

Gate Drive Strategies of SiC Cascodes 1511
Anup Bhalla, Xueqing Li, Shirley Zhang, United Silicon Carbide, USA

State-of-the-art of HF Soft Magnetics and HV/UHV Silicon Carbide Semiconductors 1518
Geraldo Nojima, Faete Filho, Eaton Corporation, USA; Paul Ohodnicki, DOE-National Energy
Technology Laboratory, USA; Alex Leary, Michael E. McHenry, Carnegie Mellon University, USA

**Comparison of Unipolar Silicon Carbide Power Transistors Used in High Switching Frequency
Inverter Topologies** 1528
Sebastian Fahlbusch, Nizar Sahli, Sebastian Klötzer, Ulf Müter, Björn Schäning, Klaus F. Hoffmann,
Helmut Schmidt University Hamburg, D

ST SiC MOSFETs in 1 MHZ DC-DC Converter 1536
Luigi Abbatelli, Giuseppe Catalisano, STMicroelectronics, I

Towards a One Nano-Henry Power Module for SiC and GaN 1541
Jacques Laeuffer, Dtalents, F

Scalable SiC Cascode Power Blocks 1549
Jonathan Dodge, Matt Grady, Ke Zhu, Anup Bhalla, United Silicon Carbide, USA

**High Power Density, High Efficiency 380v to 52v LLC Converter Utiliziing Emode GaN
Switches** 1555
Moshe Domb, Infineon, USA

Gate Drive Units

**A Low Impedance Drive Circuit to Suppress the Spurious Turn On in High Speed Wide
Band-Gap Semiconductor Halfbridges** 1562
Franz Stubenrauch, Norbert Seliger, Doris Schmitt-Landsiedel, University Rosenheim, D

Isolated Gate Driver for High Current/ High Speed FET-Converters 1570
Florian Kapaun, Rainer Marquardt, Universität der Bundeswehr Munich, D

Simple Gate-boosting Circuit for Reduced Switching Losses in Single IGBT Devices 1578
Martin Hochberg, Martin Sack, Georg Müller, Karlsruhe Institute of Technology, D

**Stability and Performance Analysis of a Voltage Controlled Resistor Circuit for Wide Band-gap
Device Gate Drivers** 1584
Alessandro Soldati, Giorgio Pietrini, Davide Barater, Carlo Concari, Università degli Studi di Parma, I

State of the Art of Gate-Drive Power Supplies for Medium and High Voltage Applications 1592
Layal Ghossein, Piotr Dworakowski, SuperGrid Institute, F; Hervé Morel, Florent Morel, Ampère, F

PCIM Europe 2016, 10 – 12 May 2016, Nuremberg, Germany

The Optimized Gate Driver Design Techniques for IGBT Properties and Downsizing in Eco-Friendly Vehicle 1600

KangHo Jeong, SangChul Shin, KiYoung Jang, JinHwan Jung, KiJong Lee, JiWoong Jang, Hyundai Motors, ROK

A Revisit to Resonant Gate Driver and a New Driver to Improve EMI vs. loss Tradeoff for SiC MOSFET 1608

Chi-Ming Wang, Toyota Motor Engineering & Manufacturing, USA

Application and Design Considerations of CoolMOS™ CFD2 and EiceDRIVER™ IC in Motor Drive Application 1615

Wolfgang Frank, Michael Wendt, Infineon Technologies, D; Sam Abdel-Rahman, Infineon Technologies Americas, USA

AC-DC Converters and Power Supplies

4D-Interleaving of Isolated ISOP Multi-Cell Converter Systems for Single Phase AC/DC Conversion 1622

Matthias Kasper, Michael Antivachis, Dominik Bortis, Johann Walter Kolar, ETH Zürich, CH; Gerald Deboy, Infineon Technologies, AT

Battery Charger Based on a Triple-LCp Resonant Converter 1631

Christian Branas, Francisco Azcondo, University of Cantabria, ES; Juan C. Viera, Manuela González, University of Oviedo, ES

High Efficient Flyback Converter with SiC-MOSFET 1639

Johann Austermann, Tim Stuckmann, Holger Borcherding, University of Applied Sciences Ostwesfalen-Lippe, D

PCB Integration of a Magnetic Component dedicated to a Power Factor Corrector Converter 1647

Herault Guillaume, Mercier Adrien, Stéphane Lefebvre, Denis Labrousse, ENS Cachan-SATIE, F

Evaluation of TCM and CrCM modulation for Totem Pole PFC 1655

Haihua Zhou, Wenduo Liu, Eric Persson, Infineon Technologies Americas, USA

System Concept and Model-Based Optimization of High-Current Variable-Voltage Chopper-Rectifiers 1662

Zhiyu Cao, Holger Fahnert, Jürgen Schiele, AEG Power Solutions, D; Jitendra Solanki, Norbert Fröhleke, Joachim Böcker, University of Paderborn, D

Evaluation of a Unidirectional Three-Phase Rectifier Based on the Third Harmonic Injection Concept in Comparison to a VIENNA Rectifier 1669

Markus Makoschitz, Hans Ertl, Technical University of Vienna, AT; Michael Hartmann, Schneider Electric Power Drives, AT

SiC Improves Switching Losses, Power Density and Volume in UPS 1677

Nikolai Epp, Christian Schulte-Overbeck, Zhiyu Cao, Michael Lemke, Lothar Heinemann, AEG Power Solutions, D

Optimization of 12 Pulse and 18 Pulse Rectifier Systems by the Selection of Optimum Parameters for Magnetics 1685

Kapila Warnakulasuriya, Carroll & Meynell Transformers, GB; Farhad Nabhani, Vahid Askari, Teesside University, GB

PCIM Europe 2016, 10 – 12 May 2016, Nuremberg, Germany

DC-DC Converters I

Analysis of the Flyback Converter Utilizing a Transformer with Stepped Air-Gap 1693
Panagiotis Mantzanas, Daniel Kübrich, Markus Barwig, Thomas Dürbaum, Friedrich-Alexander-University of Erlangen, D

Novel Method for the Estimation of Switching Losses in Resonant Converters 1701
Christian Oeder, Markus Barwig, Thomas Dürbaum, Friedrich-Alexander-University of Erlangen, D

Active Dead-Time Optimization for wide Range Flyback Active-Clamp Converter 1707
Sebastien Larousse, Nacer Abouchi, Remy Cellier, Institut des nanotechnologies de Lyon, F;
Hubert Razik, Laboratoire Ampere, F; Philippe Volay, Centralp, F

**Energetic Macroscopic Representation (EMR) and Control Scheme for the Asymmetric 4-Stage
MSBA** 1713
Gina Steinke, Alfred Rufer, EPFL - Ecole Polytechnique Fédérale Dde Lausanne, CH

**Adjustable 20 kW Full-SiC Electronic Load with Energy Recovery for Medium-frequency
Inverter** 1721
Fabian Denk, Karsten Haehre, Julian Koerner, Rainer Kling, Wolfgang Heering, Karlsruhe Institute
of Technology, D

A New High Frequency Transformer for UPS 1728
Michael Schmidhuber, Manfred Wohlstreicher, Michael Baumann, Markus Schmeller, SUMIDA
Components & Modules, D

DC/DC-Converter for Modular Coupling of 48 V Battery Packs to a High Voltage DC Bus 1733
Michael Eberlin, Milad Mohammad Hossein Khani, Fraunhofer Institute ISE, D

High Dynamic Current Source for LED Light and Data Transmission Applications 1741
Karl Edelmoser, Technical University of Vienna, AT; Felix Himmelstoss, Technikum Vienna, AT

**Design Methods for LLC Converter considering Buck and Boost Mode with Limited Frequency
Range for Wide Input Voltage Range** 1749
Dustin Funk, Tobias Reimann, Technical University of Ilmenau; Ulf Schwalbe, ISLE Steuerungstechnik
und Leistungselektronik, D

The Behavior of Electro-Magnetic Radiation of Storage Inductor in DC-DC Converters 1757
Ranjith Bramanpalli, Würth Elektronik Eisos, D

DC-DC Converters II

A New High Frequency Ferrite Material for Gan Applications 1764
Herbert Jungwirth, Michael Schmidhuber, Michael Baumann, Markus Schmeller, SUMIDA Components
& Modules, D

Multi-Stage LLC Resonant Converters designed for Wide Output Voltage Ranges 1769
Chi Wa Tsang, Chris Bingham, University of Lincoln, GB; Martin Foster, Dave Stone, University of
Sheffield, GB; John Leach, Castle, GB

Application Advantages and Disadvantages of Modern Fast Switching MOSFETs in VRM 1777
Zhiyang Chen, Ann Starks, ON Semiconductor, USA

**Medium to Low Voltage DC/DC Resonant Converter with SiC SCRs and Nanocrystalyne
Magnetic Core Transformer** 1785
Iñigo Martinez de Alegria, Angel Luis Perez, Madaci Mansour, Jon Andreu Larrañaga, University of
the Basque Country, ES; Kerdoun Djallel, GLEC Constanine 1, DZ

PCIM Europe 2016, 10 – 12 May 2016, Nuremberg, Germany

GaN Active-Clamp Flyback Converter with Resonant Operation Over a Wide Input Voltage Range 1792
Nicolas Quentin, SAGEM, F; Remi Perrin, Cyril Buttay, INSA de Lyon, F; Christian Martin, Charles Joubert, Ampere Laboratory, F; Bertrand Lacombe, Safran Group, F

Demonstration of superior SiC MOSFET Module performance within a Buck-Boost Conversion System 1800
Maximilian Slawinski, Tim Villbusch, Daniel Heer, Marc Buschkühle, Infineon Technologies, D

High Efficiency and High Power Density Boost / Buck Converter with SiC JFET Modules for Advanced Auxiliary Power Supply in Trolleybuses 1808
Miroslav Hruska, Skoda Electric, CZ; Martin Jara, West Bohemian University, CZ

Development of a 12 kW isolated and bidirectional DC-DC Converter dedicated to the More Electrical Aircraft: The Buck Boost Converter Unit (BBCU) 1814
Pascal Asfaux, Jeremy Bourdon, Airbus Operation, F

Reverse Mode Application of Sine Amplitude Converters 1822
David Bourner, Vicor Corporation, USA

DC-AC Converters

Influence of the Configuration of the Load Cable on Switching Characteristics of IGBTs 1829
Lars Middelstaedt, Dennis Richter, Andreas Lindemann, Otto-von-Guericke-University, D; Arendt Wintrich, Semikron, D

Improved Power Decoupling Scheme for Single-Phase Grid-Connected Differential Inverter with Realistic Mismatch in Storage Capacitances 1837
Wenli Yao, Xiaobin Zhang, Northwestern Polytechnical University, CN; Xiongfei Wang, Poh Chiang Loh, Frede Blaabjerg, Aalborg University, DK

Technical Approach: Interleaved, Folding, Interpolating Dual-Path Adiabatic Autotransformer Based Power Converter 1845
John Wood, Ed Shelton, Silicon Contact, GB; Kevin Rathbone, Robotae, GB; Mehdi Baghadadi, Patrick Palmer, University of Cambridge, GB

Design and Performance Evaluation of a Three Phase AC Power Source with Virtual Impedance for Validation of Grid Connected Components 1846
Peter Jonke, Johannes Stöckl, Hans Ertl, AIT-Austrian Institute of Technology, AT

Design and Testing of a Modular SiC based Power Block 1853
Maja Harfman Todorovic, Rajib Datta, Ljubisa Stevanovic, Xu She, Philip Cioffi, Gary Mandrusiak, Brian Rowden, Paul Szczesny, Jian Dai, Tony Frangieh, GE Research Center, USA

A Novel Method to simulate the Control-to-output Transfer Function of Resonant Converters 1857
Julian Dobusch, Christian Oeder, Thomas Dürbaum, Friedrich-Alexander-University of Erlangen, D

A Study of the Thermal and Parasitic Optimization of a Large Current Density Highly Parallelized Three-Phase Reference Board for Motor Drive Applications 1864
Mehrdad Baghaie Yazdi, Xiaomin Wu, Peter Haaf, Klaus Neumaier, Fairchild Semiconductor, D

PCIM Europe 2016, 10 – 12 May 2016, Nuremberg, Germany

AC-AC and Multilevel Converters

Trends in Residential and Industrial Induction Cooking: Topologies and Power Devices for High Efficiency 1865
Vittorio Crisafulli, ON Semiconductor, D

Direct Power Control for a Grid Connection of a Three Phase Z-Source Inverter 1873
Manuel Steinbring, Mario Pacas, University of Siegen, D

Interleaved Series Input Parallel Output forward Converter with Simplified Voltage Balancing Control 1880
Kaspars Kroics, Alvis Sokolovs, Linards Grigans, Ugis Sirmelis, Riga Technical University, LV

Fault-Tolerant Behaviour of the Three Level Advanced-Active-Neutral-Point-Clamped Converter ... 1888
Sidney Gierschner, David Hammes, Jan Fuhrmann, Hans-Günter Eckel, University of Rostock, D;
Max Beuermann, Siemens, D

Cell Voltage Balancing Controller for the Modular Multilevel Converter Arm using Symmetrical Transformation 1896
Andrey Dudin, Aaron Fischer, Thomas Ellinger, Jürgen Petzold, Technical University of Ilmenau, D

Isolated low-power multi-output DC-DC Converters with Heterogeneous Loads for an Efficient Supply of Modular Power Electronics Systems 1904
Arthur Singer, Thomas Weyh, Florian Helling, Universität der Bundeswehr Munich, D; Arun Jeyaprakash,
Technical University of Munich, D; Stefan Götz, Duke University, USA

A wire based communication interface for Medium and High-Voltage Converters 1912
Marek Galek, Manuel Blum, Siemens, D

An Auxillary Power Supply with integrated Communication Capability for Medium and Highvoltage Applications 1919
Manuel Blum, Marek Galek, Siemens, D

FPGA Based Direct Model Predictive Current Control of PMSM Drives with 3L-NPC Power Converter 1926
Zhenbin Zhang, Christoph Hackl, Ralph Kennel, Technical University Munich, D

IGBT Power Module in Three-Level Neutral Point Clamped Type 2 (NPC2, T-NPC, Mixed Voltage) Topology in Short Circuit Modes 1934
Vladan Jerinic, Kevin Lenz, Reiner Hinken, Danfoss Silicon Power, D

Efficiency Verification Power Circulation Method of a High Power Low Voltage NPC Converter for Wind Turbines 1942
Berthold Benkendorff, Friedrich W. Fuchs, Christian-Albrechts-University, D; Toke Franke, Danfoss
Silicon Power, D

Control of the Actively Balanced Capacitive Voltage Divider for a Five-Level NPC Inverter-Estimation of the Intermediary Levels Currents 1949
Alfred Rufer, Nelson Koch, Nicolas Cherix, EPFL – Ecole Polytechnique Fédérale de Lausanne, CH

Automotive Applications

Automotive Power Module Technologies for High Speed Switching 1956
Shinichiro Adachi, Takuma Kouge, Souichi Yoshida, Hiroshi Miyata, Daisuke Inoue, Yoshikazu Takamiya,
Hideto Kobayashi, Akira Nishiura, Fumio Nagaune, Fuji Electric, J; Thomas Heinzel, Fuji Electric Europe, D

PCIM Europe 2016, 10 – 12 May 2016, Nuremberg, Germany

Power Semiconductors for the Automotive 48 V Board Net 1963
Felix Hüning, University of Applied Sciences Aachen, D

Status and Advances in Electric Vehicle's Power Modules Packaging Technologies 1970
Itxaso Aranzabal, Asier Matallana, Oier Oñederra, Iñigo Martinez de Alegría, David Cabezuelo,
University of the Basque Country, ES

A Highly Integrated Full SiC Six Pack Power Module for Automotive Applications 1979
Bao Ngoc An, Viktor Wegelin, Martin Bernd, Benjamin Leyrer, Michael Meisser, Horst Demattio,
Thomas Blank, Marc Weber, Karlsruhe Institute of Technology, D; Johannes Kolb, Schaeffler Technologies, D;
Jochen Altstadt, Kortec, D

**Automotive-grade P-channel Power MOSFETs for Static, Dynamic and Repetitive Reverse
Polarity Protection** 1987
Filippo Scrimizzi, Giuseppe Longo, Giusy Gambino, STMicroelectronics, I

Isolated On-Board DC-DC Converter for Power Distribution Systems in Electric Vehicles 1992
Sven Bolte, Joachim Böcker, Norbert Fröhleke, University of Paderborn, D

Innovative Solution of Static and Dynamic Contactless Charging Station for Electrical Vehicles 1999
Nikolay Madzharov, Valeri Petkov, Technical University of Gabrovo, BG

Combining an External Rotor Motor with Vernier Concept for Drives in Intralogistics 2007
Matthias Thesseling, Tao Liu, Lenze SE, D; Hans-Joachim Wendt, Lenze Drives, D

**Dynamic Modeling and Optimal control for Series-parallel Drivetrain based on Lithium-ion
battery** 2014
Tedjani Mesbahi, Ecole Centrale de Lille, F; Moudrik Meradji, Gaolin Wang, Dianguo Xu, Harbin
Institute of Technology, CN

**Smart Diode and 4-Switch Buck-Boost Provide Ultra High Efficiency, Compact Solution for
12-V Automotive Battery Rail** 2019
Vijay Choudhary, Mathew Jacob, Texas Instruments, USA

On-Chip Temperature Measurement: A new Approach for Optimizing Automotive Inverters 2027
Laurent Beaurenaut, Inpil Yoo, Infineon Technologies, D

Renewable Energy Systems

Resonant load Emulator for Distributed Energy Resources to test Anti-islanding Algorithms 2035
Daniel Heredero-Peris, Fernando Jorge-Ques, Daniel Montesinos-Miracle, Universitat Politécnica de
Catalunya, ES; Tomàs Lledó-Ponsati, TeknoCEA, ES

Low Voltage Ride Through (LVRT) Capability of an Enhanced DFIG System 2043
David Velasco, Jesús López, Public University of Navarre, ES

Control and Modulation for Loss Minimization for Dc/Dc Converter in Wind Farm 2050
Catalin Gabriel Dincan, Philip C. Kjaer, Aalborg University, DK

**A Variable Step Size Perturb and Observe Method Based MPPT for Partially Shaded
Photovoltaic Arrays** 2058
Jawad Ahmad, Filippo Spertino, Paolo Di Leo, Alessandro Ciocia, Politecnico di Torino, I

**Renewable Electricity Conversion and Storage: Focus on Power to Gas process, EMR
Modelling and Simulation** 2066
Ahmed Remaci, Octavian Curea, Christophe Merlo, Amélie Hacala, ESTIA, F; Vincent Guerre, Local
Energy Alternative & Fair, F

PCIM Europe 2016, 10 – 12 May 2016, Nuremberg, Germany

Balancing Current and Efficiency Modelling of Single Switch Active Balancing Systems for Energy Storage Systems 2074
Iosu Aizpuru, Unai Iraola, Jose Mari Canales, Ander Goikoetxea, Mondragon University, ES

A Control Strategy for Multiple Energy Storage Devices for Power Leveling of Renewable Energy Systems 2083
Koji Kato, Yoichi Ito, Sanken Electric, J

Examining Contrasting Excitation Modes within Battery Characterisation using Maximum Length Sequences 2090
Andrew Fairweather, VxI Power, GB; David Stone, Martin Foster, University of Sheffield, GB

Comparison Between Standard and Innovative Solutions to exchange Energy between High Energy Storage Systems 2098
Laurent Garnier, Daniel Chatroux, Sébastien Carcouet, Julien Dauchy, University Grenoble Alpes CEA LITEN, F

EMI, Harmonics, Filters

Simulation and Experimental Analysis of Non-Linear Loads from Residential and Educational Buildings 2106
Gabriel Nicolae Popa, Angela Iagar, Corina Maria Dinis, Politehnica University Timisoara, RO

A Digital Predictive Constant Frequency Controller for High Frequency 3-Phase Silicon Carbide PFC Rectifier 2114
Marcelo Schupbach, Cree, USA; Binod Agrawal, Navneet Mangal, CREE India Private Limited, IN

DC-link Harmonic Content in Double Two-Level Inverter for Permanent Magnet Synchronous Motor Drive Systems – Comparison and Analysis 2122
Toktam Khani, Michael Patt, Technologienetzwerk Allgäu, D

Hybrid Filter With an Optimized Switching Method of the Compensation Capacitors and Predictive Active Filter Control 2129
Swen Bosch, Heinrich Steinhart, HTW Aalen, D

Active Mains Filters with Combined Feed-Forward and Feed-Back Control 2137
Felix Himmelstoss, Technikum Vienna, AT; Karl Edelmoser, Technical University of Vienna, AT

Influence of the Zero Sequence Voltage on the Design of a Series Active Filter 2145
Andreas Reinhold, Rolf Grohmann, HTWK Leipzig, D; Uwe Rädel, Jürgen Petzoldt, Technical University of Ilmenau, D

Electromagnetic Emissions in High Density and Fast GaN Switched Half Bridges with Resonance Filter Structures 2151
Wolfgang Stocksreiter, Hans List, Hubert Berger, Gerald Weis, Markus Krenn, FH Joanneum, AT; Günter Engel, CeraCap, AT

Energy Transmission and Grid

AC or DC Grid for Railway Stations? 2159
Lilia Galai Dol, Efficacity, F; Alexandre De Bernardinis, French Institute of Science and Technology for Transport, Development and Networks, F

Solid-State Transformer Modeling for Analyzing its Application in Distribution Grids 2167
Christoph Hunziker, Nicola Schulz, University of Applied Sciences Northwestern Switzerland, CH

PCIM Europe 2016, 10 – 12 May 2016, Nuremberg, Germany

Hybrid Reactive Power Compensation System for Grid Code Compliance in Renewable Energy Power Plants 2175
Gianluca Postiglione, Antonio Raso, Giovanni Borghetti, Francois Pezet, Nidec ASI, I

A new shunt connected HVDC Tap Based on a Highly Efficient Resonant Cascade Converter 2181
Andre Birkel, Mark-M. Bakran, University of Bayreuth, D

Analysis of Voltage and Current Unbalance in a Multi-Converter Topology for a DC-Based Offshore Wind Farm 2189
Thomas Lagier, SuperGrid Institute, F; Philippe Ladoux, University of Toulouse, F

Experimental Demonstration of a Solid-StateDamping Resistor for HVDC Applications 2197
Konstantin Vershinin, Ikenna Efika, David Trainer, Colin Davidson, Alstom Grid, GB; Nick Wright, Amit Tiwari, Newcastle University, GB

High Precision Loss Measurement at HVDC Converter 2204
Helmut Weiß, Technical University of Leoben, AT; Bernhard Grasel, Dewesoft, AT

A Fast Methodology for Solving Power Flows in Hybrid AC/DC Networks: The European North Sea Supergrid Case Study 2211
Rodrigo Teixeira Pinto, Monica Aragues-Penalba, Andreas Sumper, CITCEA-UPC, ES; Christian Alejandro Leon-Ramirez, Elmer Sorrentino, University Simon Bolivar, VE

Panel Discussion "The smart future of power electronics"

Using Smart Converter to obtain Traction-Machine Insulation Health State Information 2219
Markus Vogelsberger, Bombardier Transportation Austria, AT; Clemens Zöller, Jörg Bellingen, Thomas Wolbank, Technical University of Vienna, AT

Manuscripts which were handed in late

Investigation of the Influence of Ageing Processes on Thermal Characteristics of an IGBT Power Module by Means of Transient Thermal Analysis 2226
Tobias von Essen, Berliner Nanotest & Design, D; Stefan Stegmeier, Gerhard Mitic, Siemens, D

A Study of the Thermal and Parasitic Optimization of a Large Current Density Highly Parallelized Three-Phase Reference Board for Motor Drive Applications 2231
Mehrdad Baghaie Yazdi, Xiaomin Wu, Peter Haaf, Klaus Neumaier, Fairchild Semiconductor, D

PCIM Europe 2016, 10 – 12 May 2016, Nuremberg, Germany

Current Measurement and Gate-Resistance Mismatch in Paralleled Phases of High Power SiC MOSFET Modules

Roman Horff, Teresa Bertelshofer, Andreas März, Mark-M. Bakran
roman.horff@uni-bayreuth.de

University of Bayreuth, Department of Mechatronics, Center of Energy Technology (ZET),
Universitätsstr. 30, 95447 Bayreuth, Germany

Abstract

This paper deals with the problems of current measurement in paralleled phases. A solution based on low inductive shunts is proposed. The influence of gate-resistance mismatch on the switching behaviour and the resulting switching losses are analysed for high power SiC MOSFET modules.

1. Introduction

In order to reach higher power capability of SiC MOSFET inverter systems and to achieve at once a certain scalability in converter design processes, it is desirable to connect the phases of a three-phase SiC MOSFET module in parallel. Therefore the current distribution and the resulting distribution of power losses due to component tolerances have to be analysed. This paper deals with current measurement and gate-resistor unbalance of high power modules, in contrast to [1-4], where parallelisation of discrete SiC MOSFET devices and single chips is investigated.

2. Current measurement – problems and solutions

To characterise the parallel connection, a double pulse test is performed using high power three-phase SiC MOSFET modules with a blocking voltage of 1200 V and a nominal current of 200 A (see fig. 1). The voltage can easily be measured with a voltage probe, but current measurement with a high bandwidth is posing a challenge.

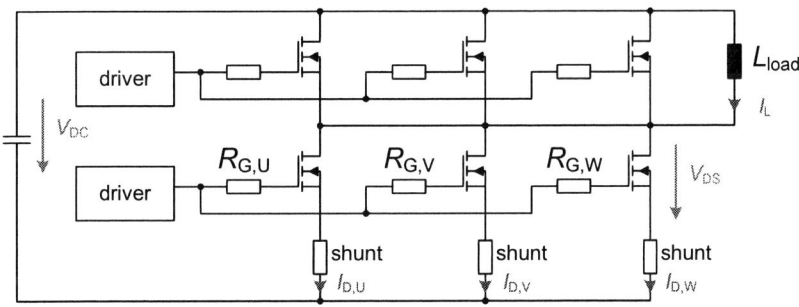

Fig. 1   Test setup of the double pulse test

Closed Rogowski coils were regarded to measure the current in the paralleled phases. To eliminate the influence of nearby phases, a large number of windings is required to achieve a homogenous current linkage. This leads to a low bandwidth of the Rogowski coils.
So high bandwidth current sensors based on the Rogowski principle with only three windings were tested. Because the current sensors are not ring-shaped, different varying coupling factors occur depending on the sensor position and orientation. The voltage of the Rogowski sensor signal consists of the voltage that is proportional to the $\frac{di}{dt}$ of the phase current to be measured and the voltage induced by the nearby phases:

© VDE VERLAG GMBH · Berlin · Offenbach

$$\begin{pmatrix} v_{\text{Rogowski,U}} \\ v_{\text{Rogowski,V}} \\ v_{\text{Rogowski,W}} \end{pmatrix} = \begin{pmatrix} L_{\text{k,U}} & -M_{\text{VU}} & -M_{\text{WU}} \\ -M_{\text{UV}} & L_{\text{k,V}} & -M_{\text{WV}} \\ -M_{\text{UW}} & -M_{\text{VW}} & L_{\text{k,W}} \end{pmatrix} \cdot \frac{d}{dt} \begin{pmatrix} i_{\text{D,U}} \\ i_{\text{D,V}} \\ i_{\text{D,W}} \end{pmatrix}$$

Identifying the coupling factors is laborious and has to be repeated for every series of measurement. Furthermore the calculation of the DC current distribution is inaccurate using Rogowski coils, because the rising current during the first pulse of the double pulse test measured by the Rogowski coil has a very low slew rate determined by the inductive load, so the output voltage of the Rogowski sensor is too low to calculate an exact DC current distribution.

For this reason shunts are taken into account for current measuring. Low inductive coaxial shunts have a low inductance between the points of measurement, but mounting a coaxial shunt leads to a high additional inductance in the power circuit because of its disadvantageous pins. Alternatively SMD shunts are used. The parasitic inductance of the shunts cannot be neglected. In addition to the parasitic inductance of the shunts there is an inductive coupling of the phase current to the measuring signal (fig. 2).

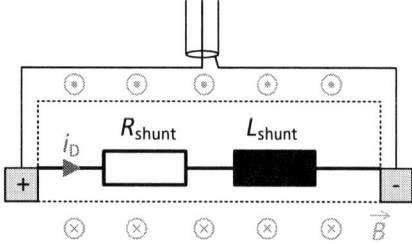

Fig. 2   Equivalent circuit of the SMD shunt with parasitic inductance and the magnetic induction caused by the load current

Because of the high $\frac{di}{dt}$ during the switching transient of the SiC MOSFETs, the measured voltage has to be corrected by a PT1 element:

$$v_{\text{shunt}}(t) = R_{\text{shunt}} \cdot i_{\text{D}}(t) + (L_{\text{shunt}} + L_{\text{coupling}}) \cdot \frac{di_{\text{D}}(t)}{dt}$$

$$\Rightarrow\ I_{\text{D}}(s) = \frac{1}{s \cdot R_{\text{shunt}} \cdot (L_{\text{shunt}} + L_{\text{coupling}}) + 1} \cdot V_{\text{shunt}}(s)$$

The sum of parasitic shunt inductance and coupling inductance of the measurement loop has to be determined after each modification of the measurement system.

The comparison of the current measured by a high bandwidth Rogowski coil and the corrected signal of the shunt voltage in single phase operation shows, that the measurement method using SMD shunts and correcting the measurement signal by a PT1 element is suitable enough to calculate the switching losses (fig. 3).

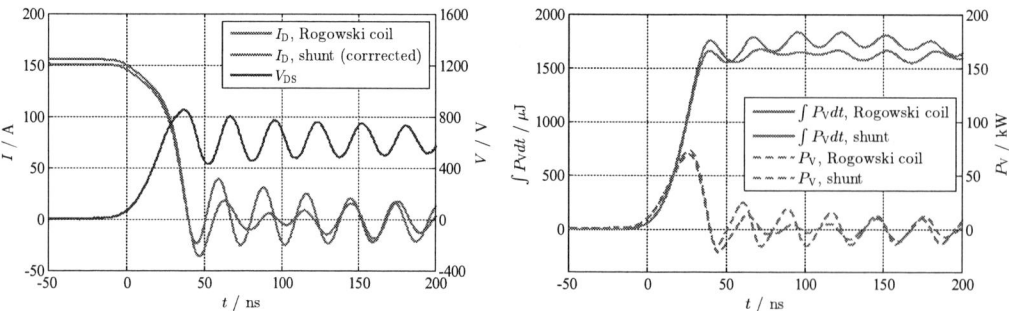

Fig. 3   Comparison of the corrected shunt signal and the measurement with a high bandwidth Rogowski coil for single phase operation

© VDE VERLAG GMBH · Berlin · Offenbach

PCIM Europe 2016, 10 – 12 May 2016, Nuremberg, Germany

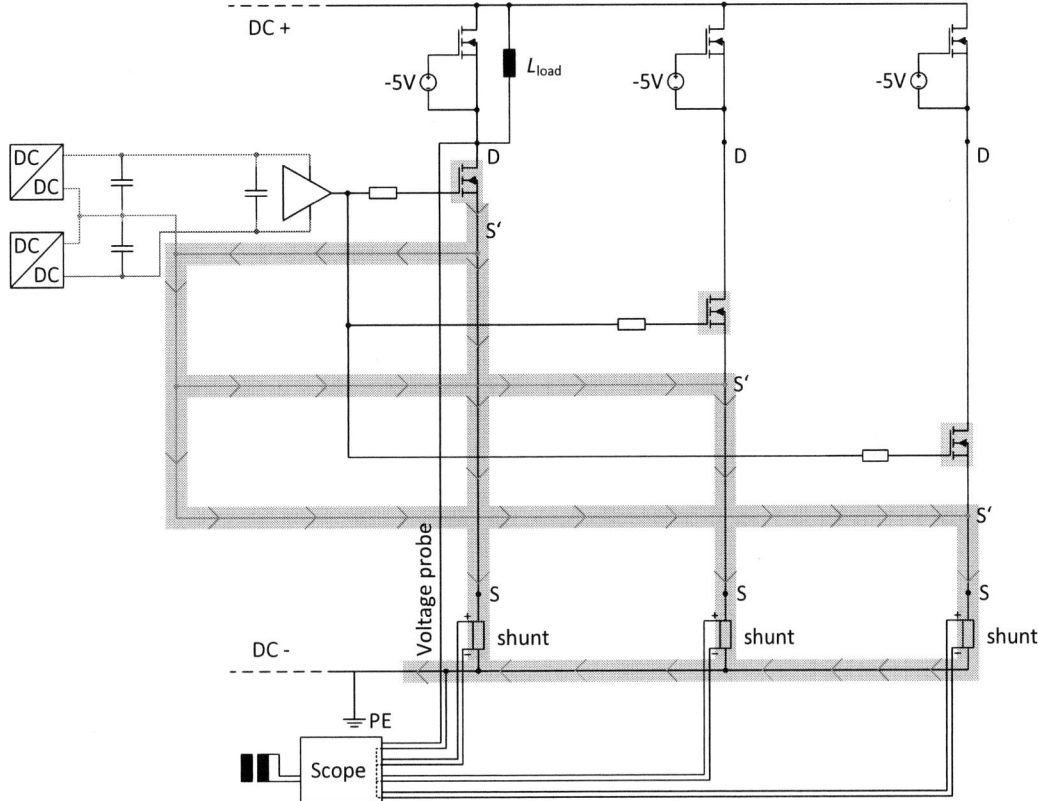

Fig. 4   Circulating current mismatch via auxiliary source connection and gate-drive circuit

Using three shunts in a parallel arrangement of three phases and connecting the gate-drivers via auxiliary source contact, a circulating current might occur, so the measured current at the shunt differs from the MOSFET's drain current. Fig. 4 shows a schematic of the test bench with the highlighted path of the circulating current mismatch via auxiliary source connection and gate-drive circuit.

Because the driving voltage of the circulating current is the difference between the shunt voltages, the circulating current mismatch is only noticeable for high DC current mismatch, e.g. switching only one phase as shown in fig. 4.

In parallel operation the circulation current is very low since the DC current mismatch is low. The difference in shunt voltages due to transient current mismatch is negligible because of the inductance of the driver circuit and the high switching speed of the SiC MOSFETs.

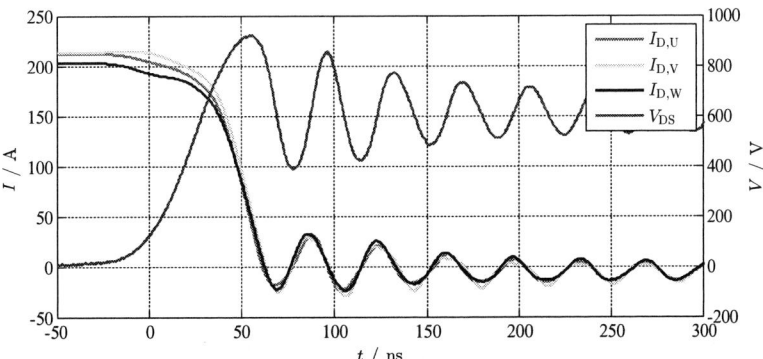

Fig. 5   Turn-off process at $I_{D,tot} = 600$ A and $V_{DC} = 600$ V for symmetrical gate-resistance

© VDE VERLAG GMBH · Berlin · Offenbach

3. Switching behaviour in parallel operation

Fig. 5 shows a turn-off switching operation at $V_{DC} = 600\,V$ with nominal drain current $I_{D,tot} = I_{D,U} + I_{D,V} + I_{D,W} = 600\,A$ for symmetrical external gate-resistances. There is a small static and transient current mismatch that leads to a deviation of about 2.7% of the total turn-off losses from the average turn-off switching losses.

In order to quantify the influence of current mismatch in parallel operation the switching losses are calculated for different operating points. Fig. 6 shows the turn-off losses of the three phases in parallel operation in comparison to the turn-off losses in single phase operation. The parallel connection of multiple phases leads to higher turn-off losses. In contrast to that the turn-on losses in parallel operation are lower (fig. 7).

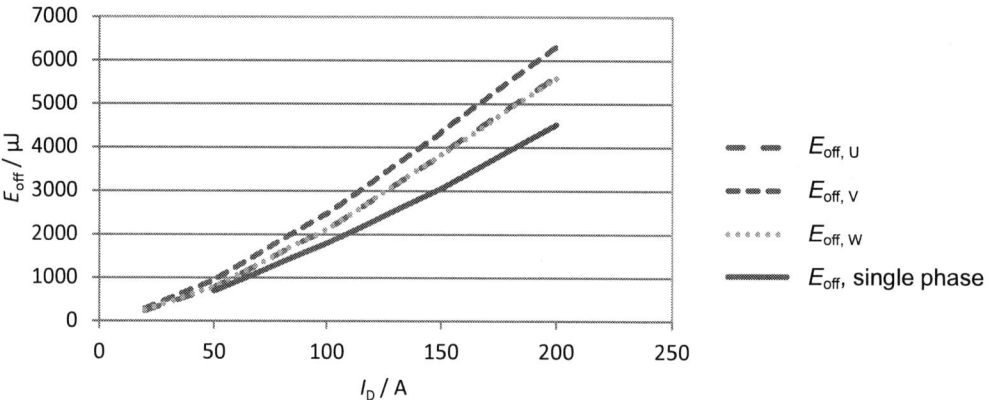

Fig. 6 Comparison of the turn-off losses in single and parallel operation with symmetrical gate-resistance ($V_{DC} = 600\,V$, $T_J = 150°C$)

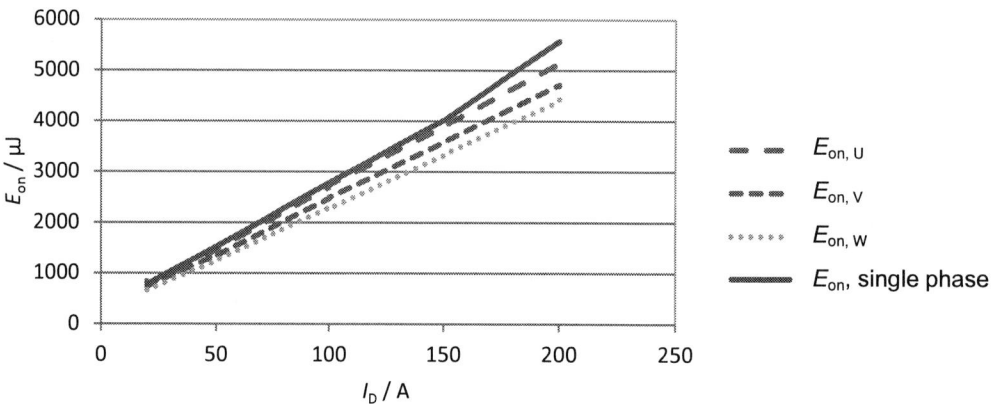

Fig. 7 Comparison of the turn-on losses in single and parallel operation with symmetrical gate-resistance ($V_{DC} = 600\,V$, $T_J = 150°C$)

A comparison of the total switching losses in single phase operation and in parallel operation of three phases is shown in fig. 8. It can be seen that the average switching losses in the three phases correspond to the total switching losses in single phase operation, so there are no additional losses due to parallel operation of multiple phases.

The reason therefor is that there is 3-times the $\frac{di}{dt}$ in the common part of the DC link in three-phase parallel operation in comparison to single phase operation. As a result the turn-off losses are higher due to the higher turn-off overvoltage, and the turn-on losses are lower because of the major inductive voltage drop during turn-on transient.

PCIM Europe 2016, 10 – 12 May 2016, Nuremberg, Germany

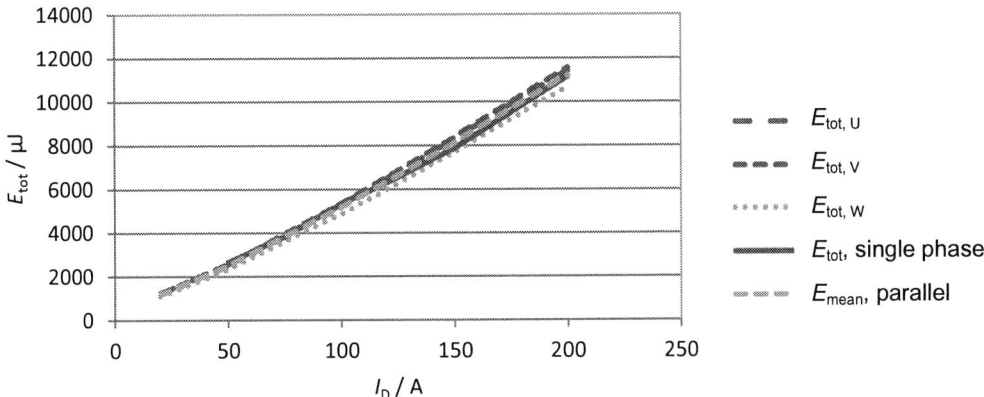

Fig. 8   Total losses of the of the single phase in comparison to the switching losses of three phases in parallel with symmetrical gate-resistance ($V_{DC} = 600\,V$, $T_J = 150°C$)

4. Variation of external gate-resistance

In order to evaluate the influence of the tolerances of internal and external gate-resistors, the value of the external gate-resistor was varied to provoke an asymmetric switching.

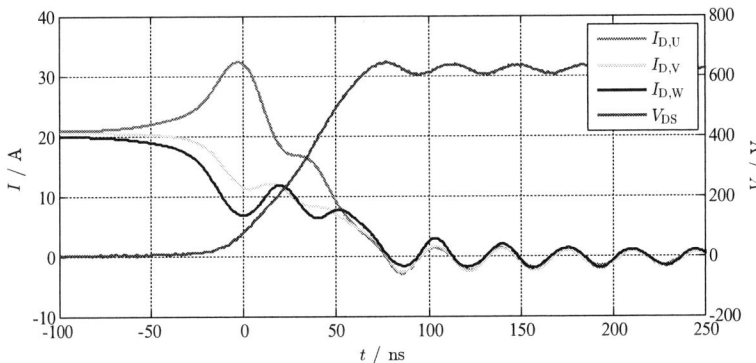

Fig. 9   Turn-off   process   at $I_{D,tot} = 60\,A$ and $V_{DC} = 600\,V$ for   asymmetrical   gate-resistance ($\Delta R_{G,U} = 1\,\Omega$)

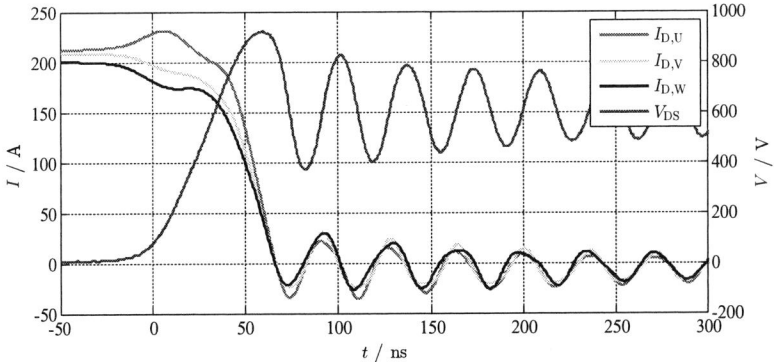

Fig. 10 Turn-off   process   at $I_{D,tot} = 600\,A$ and $V_{DC} = 600\,V$ for   asymmetrical   gate-resistance ($\Delta R_{G,U} = 1\,\Omega$)

For the switching turn-off processes in fig. 9 and 10 the semiconductor switch of the phase U was slowed down by an increase of the external gate-resistance of 1 Ω. At a total drain

© VDE VERLAG GMBH · Berlin · Offenbach

current of $I_{D,tot} = 60\,A$ there is a significant current mismatch due to the unsymmetric switching behaviour (fig. 9). The decelerated phase U has to switch off more than 1.5-times the DC phase current. The turn-off process at a total drain current of $I_{D,tot} = 600\,A$ is shown in fig. 10. The maximum current peak due to the slowed down turn-off of phase U in this operating point is just 10%. To qualify the influence of a gate-resistance unbalance the resulting switching losses have to be analysed.

Fig. 11 shows the turn-off losses of the three phases while one of the gate-resistances is increased by 1 Ω in each case. It is obvious that the decelerated phase has the highest turn-off losses. The not decelerated phases have the same turn-off losses, regardless of which of the other two phases is slowed down. So the deviation due to the variance of the gate-resistance is considered as superposition to the inherent existing loss balance that can be observed switching without variation of external gate-resistance. Analysing the turn-on losses while accelerating one of the phases the same effect is found.

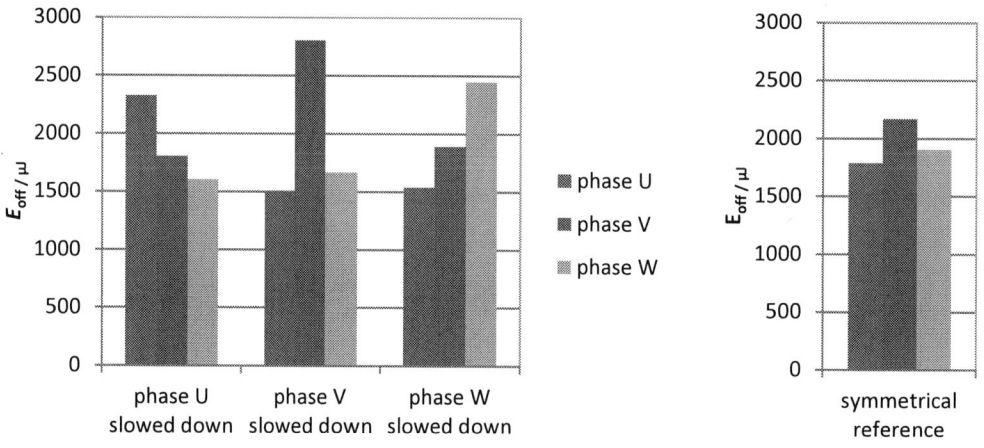

Fig. 11 Turn-off loss distribution for the three phases for different gate-resistor increase ($\Delta R_G = 1\,\Omega$)

The absolute aberration of the unbalanced phase after eliminating the inherent existing loss unbalance is shown in fig. 12 for turn-off (one phase slowed down) and in fig. 13 for turn-on (one phase accelerated) respectively. It can be stated, that both turn-on and turn-off losses maldistribution is proportional to the difference of the gate-resistance ΔR_G.

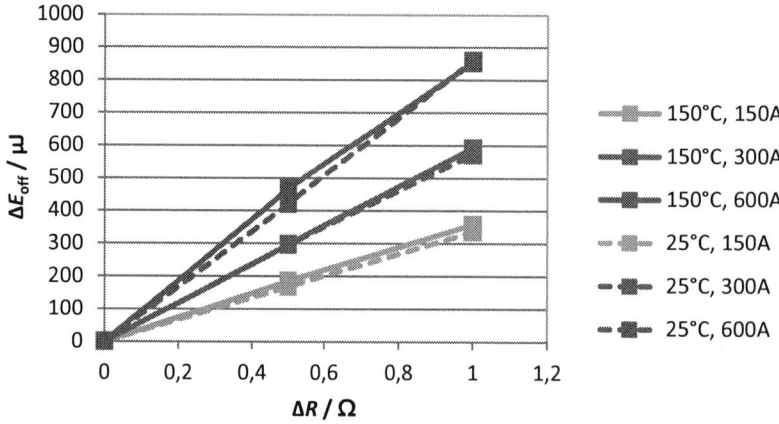

Fig. 12 Absolute aberration of turn-off losses due to increase of R_G ($V_{DC} = 600\,V$)

© VDE VERLAG GMBH · Berlin · Offenbach

PCIM Europe 2016, 10 – 12 May 2016, Nuremberg, Germany

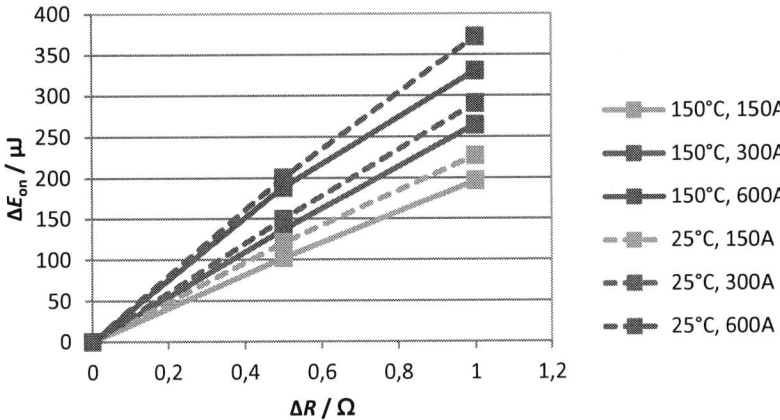

Fig. 13 Absolute aberration of turn-on losses due to decrease of R_G ($V_{DC} = 600$ V)

In fig. 14 and 15 the relative turn-off and turn-on losses are shown. The unbalance of the gate-resistance is normalised by the temperature-sensitive value of the internal damping resistor network located on the DCB. In this plots the resulting unbalance of the maximum internal gate-resistor tolerance can be identified: For a maximum tolerance of 10% the unbalance in turn-off losses will be about 5% assuming that all of the internal damping resistors of the decelerated phase showing the maximum aberration. Regarding the minimum limit of the gate-resistor, the accelerated turn-on process has to be considered as critical operation (fig. 15). The maximum maldistribution in this case is less than 2% of the turn-on losses.

This leads to the conclusion that the tolerance of the gate-resistors is not the critical point in converter design, even if a maximum disadvantageous arrangement of the gate-resistors is assumed.

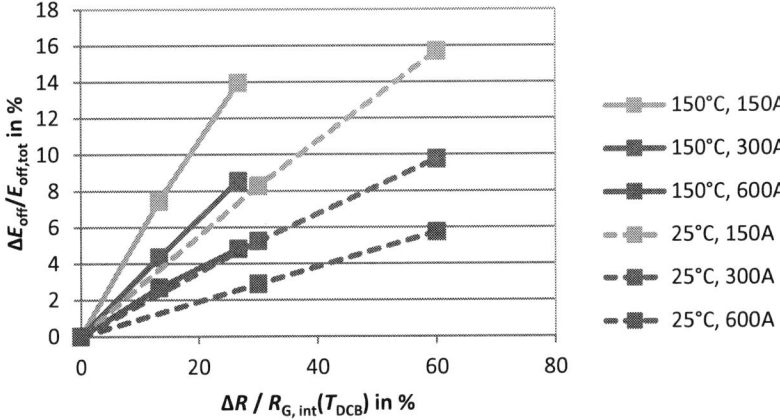

Fig. 14 Relative aberration of turn-off losses due to increase of R_G ($V_{DC} = 600$ V)

© VDE VERLAG GMBH · Berlin · Offenbach

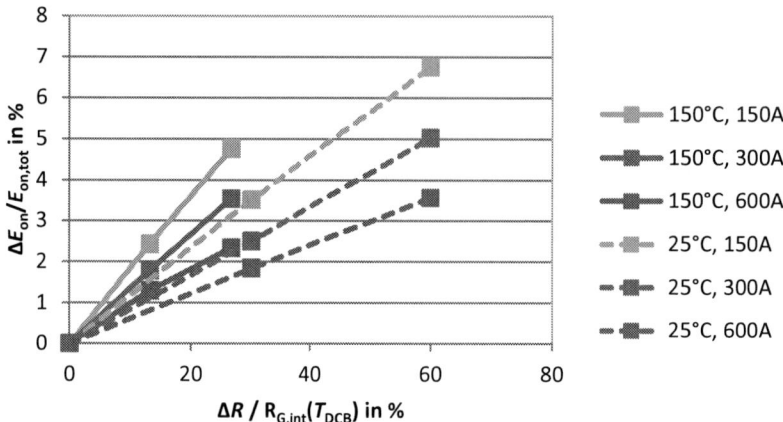

Fig. 15 Absolute aberration of turn-on losses due to decrease of R_G ($V_{DC} = 600$ V)

5. Conclusion

Problems of current measurement in parallel connection of high power SiC MOSFETs were discussed. Using SMD shunts and correcting the parasitic inductance and the inductive coupling of the drain current was found to be the most feasible kind of current measurement to analyse the switching behaviour and quantify the switching losses in parallel operation.
The influence of the variation of the gate-resistance on current and loss distribution was presented and analysed. It was shown that the deviation of switching losses is proportional to the variance of the gate-resistor. Regarding a disadvantageous arrangement of gate-resistors with large aberrations of the nominal value it can be stated that gate-resistor tolerance is not critical in parallelisation of multiple phases of a SiC MOSFET power module, since deviation of switching losses due to gate-resistor unbalance is lower than 5%.

6. References

[1] Gangyao Wang; Mookken, J.; Rice, J.; Schupbach, M., "Dynamic and static behavior of packaged silicon carbide MOSFETs in paralleled applications", APEC, 2014
[2] Sadik, D.-P.; Colmenares, J.; Peftitsis, D.; Jang-Kwon Lim; Rabkowski, J.; Nee, H.-P., "Experimental investigations of static and transient current sharing of parallel-connected silicon carbide MOSFETs", EPE, 2013
[3] Helong Li; Munk-Nielsen, S.; Cam Pham; Beczkowski, S., "Circuit mismatch influence on performance of paralleling silicon carbide MOSFETs", EPE, 2014
[4] Li, H.; Munk-Nielsen, S.; Wang, X.; Maheshwari, R.; Beczkowski, S.; Uhrenfeldt, C.; Franke, W., "Influences of Device and Circuit Mismatches on Paralleling Silicon Carbide MOSFETs", IEEE Transactions on Power Electronics, 2016

Gate Drive Strategies of SiC Cascodes

Xueqing Li, United Silicon Carbide, Inc., USA, lli@unitedsic.com
Hao Zhang, United Silicon Carbide, Inc., USA, szhang@unitedsic.com
Anup Bhalla, United Silicon Carbide, Inc., USA, abhalla@unitedsic.com

Abstract

The USCi SiC cascode is a composite power switch formed by series-connecting a high-voltage normally-on SiC JFET and a low-voltage Si MOSFET. The SiC cascode has more complex switching processes than a standalone MOSFET or JFET. The low voltage Si MOSFET may be driven into avalanche breakdown during turn-off process and the resonant tank formed by the parasitic inductances of the bond wires and capacitances of MOSFET and JFET may cause large oscillations. All of these issues must be carefully considered in the design to ensure reliable and stable operation of the SiC cascode. These issues can be mitigated or even eliminated by using proper gate drive approach. This work will discuss the impact of the different gate drive strategies on the SiC cascode switching performance.

1. Introduction

As a composite switch, the SiC cascode inherits excellent properties from the Si MOSFET, such as reliable MOS gate and simple gate drive, and from the SiC JFET, such as high voltage, high speed and high temperature capabilities. These properties provide the SiC cascode with significant advantages over SiC MOSFET and Si Superjunction MOSFET. Compared to SiC MOSFET, the SiC cascode needs only a conventional IGBT/MOSFET gate driver with a power supply of +12V to drive the gate, while SiC MOSFET usually requires a gate voltage of 18-20V for full turn-on. The SiC cascode usually has a much smaller gate charge than SiC MOSFET and Si superjunction MOSFET because the gate charge of the SiC cascode is mainly determined by the low-voltage Si MOSFET which typically has a blocking voltage of 20-30V and has a much smaller gate charge than high-voltage SiC MOSFET and Si Superjunction MOSFET for the same current rating. For example, for 1,200V-45mΩ class devices, the SiC cascode (UJC1206K) has a total gate charge of 45nC at V_{GS} = 0V to +12V and V_{DS} = 800V, while the SiC MOSFET (C2M0040120D) has a total gate charge of 105nC at V_{GS} = 0V to +20V and V_{DS} = 800V [1]. For 650V-45mΩ class devices at V_{GS} = 0V to +12V and V_{DS} = 400V, the SiC cascode (UJC06506K) has a total gate charge of 45nC, while the Si superjunction MOSFET (IPB65R045C7) has a total gate charge of 110nC [2]. This means the SiC cascode has a much low figure-of-merit of on-resistance times gate charge (R_{DS}(on)*Q_G) translating into lower driving losses. The threshold voltage of the SiC cascode is also determined by the Si MOSFET. High threshold voltage of >3.5V at 150°C can be easily realized without compromising Rdson to achieve high noise immunity. By way of comparison, SiC MOSFET typically has a threshold voltage of 2.0V or less at 150°C junction temperature, explaining why a turn-off V_{GS} of -5V is recommended by the manufacturers. In addition, the JFET drain terminal is electrically and physically isolated from the Si MOSFET gate terminal, which makes the Miller capacitance C_{GD} negligible and substantially reduces the chance of C_{GD}*dV/dt induced false turn-on when a rapid voltage transition occurs on the drain side of the switch during off-state.

Another advantage of the SiC cascode is that it has an excellent Schottky-like built-in antiparallel diode with a knee voltage of only about 0.7V. The SiC MOSFET has a built-in pn-junction body diode with a knee voltage of higher than 2.0V, presenting a significant conduction energy loss and a significant switching energy loss due to its large reverse recovery charge Q_{rr} especially at high temperatures. Si Superjunction MOSFET has a built-in pn-junction body diode with a knee voltage of about 0.7V, but the body diode has a very high reverse recovery charge and a very low diode commutation speed (di/dt) of typically less than 70A/us for 650V CoolMOS C7 and 900A/us for 650V CoolMOS CFD2. Poor body diode performance makes Si Superjunction MOSFET not desirable for use in the next generation high-speed power conversion applications requiring body-diode current flow in the power switches.

These superior features make SiC cascode an ideal power component for high voltage and high power density power conversion applications. At present, 1.2kV and 650V co-packaged SiC cascodes have been made commercially available [3]. Significant efforts have been invested to investigate the performance and reliability of the co-packaged SiC cascodes, and the results show that these SiC cascodes have very fast switching speed and high avalanche and short-circuit ruggedness [4, 5], and are easy to use and parallel [6].

As a composite device, the SiC cascode contains more parasitic capacitances and inductances as illustrated in Fig.1, and has more complex switching processes than single power switches like MOSFETs and IGBTs. Large voltage/current spikes or even sustained oscillations may occur under certain conditions and should be carefully considered to ensure a reliable operation of the cascode. The gate drive approach has significant effects on the switching processes of the SiC cascodes. The simplest gate drive approach is the standard gate drive approach for the devices packaged in a 3-lead package such as TO-247-3L. The standard gate drive approach uses a gate resistor to control the device switching speed and works well for the co-packaged SiC cascodes [4]. One disadvantage of the standard gate drive approach is that a large gate resistor is usually needed to control the voltage and current rings during the turn-off process of the cascode. More complex gate drive approaches can be used to improve the switching process control. This work investigates the effects of three gate drive approaches on the switching processes of the SiC cascode: the standard gate drive approach for the cascode with a 3-lead package such as TO-247-3L, the kelvin source gate drive approach for the cascode with a 4-lead package such as TO-247-4L, and a dual-gate drive approach for the cascode with the JFET gate terminal accessible.

2. SiC Cascode Gate Drive Approaches

Fig.1 shows the equivalent circuit of the SiC cascode with the standard gate drive approach including all parasitic inductances and capacitances. L_{SD} is the inductance of the bond wire between the JFET source and the MOSFET drain. L_{SD} can be eliminated by placing MOSFET die directly on the top source pad of the JFET die forming a stack cascode [3]. Stack SiC cascode is still under development, so this work focuses on the side-by-side cascode [7] only. L_{S1} is the inductance of the source bond wire of the MOSFET. L_{S2} is the inductance of the package source lead. L_{SD}, L_{S1} and L_{S2} are all in the power loop and may generate large voltage spikes during switching process due to the rapid current change in the power loop. These spikes have significantly affected the switching performance of the cascode. During turn-on, the current in the cascode flowing from the drain to the source increases rapidly and a voltage is developed across each of the inductances L_{SD}, L_{S1}, and L_{S2}. The voltages across L_{S1} and L_{S2} de-bias the MOSFET gate and the voltage across L_{SD} de-biases the JFET gate, these effects decrease the turn-on speed and increase the turn-on energy loss. In addition, L_{S1} and L_{S2} also lie in the gate resonant loop (Loop 2 in Fig.1), large oscillations/spikes caused by L_{S1} and L_{S2} during switching transients can feedback to the gate drive or control circuits and cause malfunction of the gate drive circuit. EMI is also a concern because the SiC cascode is a very fast switch. The effects of L_{S1} and L_{S2} on the gate drive and control circuits can be eliminated by using a kelvin source connection for the gate driver to move L_{S1} and L_{S2} to the outside of the gate resonant loop, as shown in Fig.2. However, the package must have 4 leads now. The lead SS is the source sensing lead for the gate drive only. Since L_{S1} and L_{S2} are not in the gate drive loop 2, much cleaner gate waveforms can be obtained and the turn-on energy loss can also be reduced.

It should be pointed that, since L_{SD}, L_{S1} and L_{S2} are all in the power loop, they tend to limit the di/dt rate during turn-on process and decrease current ringing's. But during the turn-off process, the di/dt rate is not limited by these parasitic inductances and could be very high due to the high transconductance of the cascode. This high di/dt rate may excite large oscillations in the resonant circuit of the loop 1 inside the cascode as shown in Fig.1 and Fig.2. R_{J_G} is the JFET internal gate resistance and L_{J_G} is the inductance of the JFET gate bond wire. R_{J_G} is the damping element in the loop 1. The oscillations in the loop 1 can be effectively reduced by increasing R_{J_G}, but this comes at the cost of higher switching loss.

© VDE VERLAG GMBH · Berlin · Offenbach

PCIM Europe 2016, 10 – 12 May 2016, Nuremberg, Germany

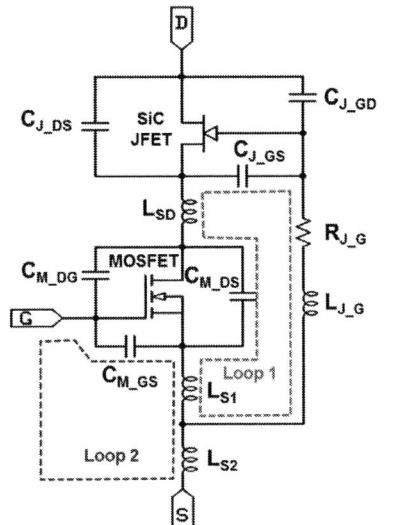

Fig.1. Equivalent circuit of the SiC cascode in standard TO-247-3L package including all of the parasitic inductances and capacitances.

Fig.2. Equivalent circuit of the SiC cascode in TO-247-4L package with a kelvin source lead for gate driving circuit.

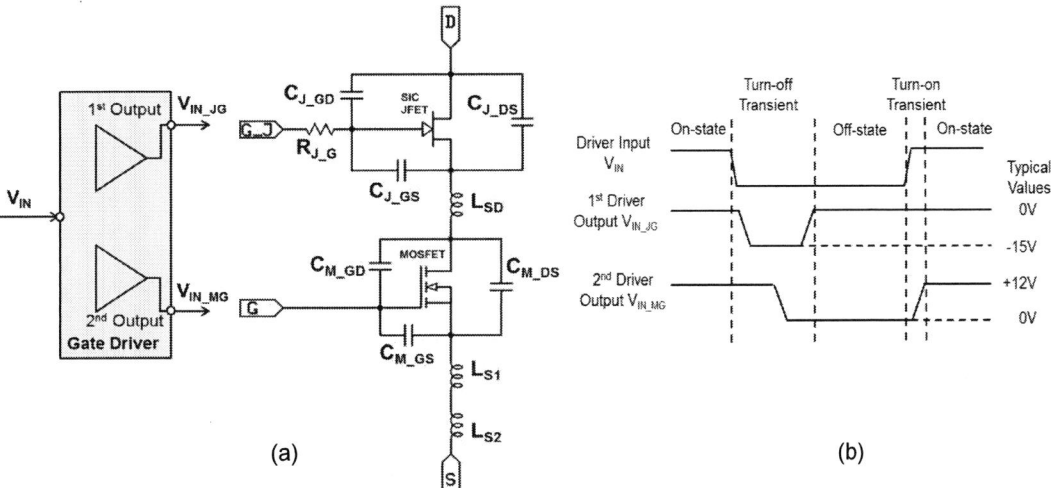

(a)

(b)

Fig.3. Gate driver and the equivalent circuit (a) and the control logic (b) of the SiC cascode with dual-gate drive control.

A dual-gate drive approach has been proposed in [8] to provide a better control on the turn-off process of the cascode, as shown in Fig.3. In this approach, the gate driver provides two outputs. The first output V_{IN_JG} is connected to the gate terminal of the SiC JFET, and the second output V_{IN_MG} is connected to the gate terminal of the MOSFET. In such a way, the MOSFET and JFET are switched separately. Fig.3b shows the control logic of this approach. In on-state, the gate driver produces a high level signal on both outputs V_{IN_JG} and V_{IN_MG}, typically $V_{IN\_JG} = 0V$ and $V_{IN\_MG} = +12V$ to $+15V$, so that both the JFET and the MOSFET are kept in on-state. In off-state, the gate driver produces a low level signal on the output V_{IN_MG}, and still a high level signal on V_{IN_JG}, typically $V_{IN\_JG} = 0V$ and $V_{IN\_MG} = -5V$ to $0V$. Under this condition, the device is kept in off-state according to the operation principle of the standard cascode, and more importantly, the device is still able to conduct a reverse current in the same way as the standard cascode. The excellent built-in antiparallel diode of the cascode is preserved in this approach. This is the most important difference between this approach and the other dual-gate drive approaches [9, 10].

To turn-off the cascode, the gate driver first pulls down the first output V_{IN_JG} to a low voltage level, typically -15V, to actively turn-off the JFET. Once the JFET is turned off, the whole

© VDE VERLAG GMBH · Berlin · Offenbach

1513

cascode will enter off-state. During turn-off transient, the MOSFET is kept in on-state and, therefore, will not experience any voltage spike. This means the turn-off process of the cascode is controlled solely by the JFET, which can be very well controlled through the JFET gate resistor. After the turn-off process of the cascode is complete, the gate driver pulls down the second output V_{IN_MG} to the low level, typically -5V to 0V, to actively switch off the MOSFET, and then pulls up V_{IN_JG} to the high level state so that the normally-on JFET or the cascode is allowed to conduct a reverse current.

To turn-on the cascode, the gate driver pulls up the second output V_{IN_MG} to the high level to actively turn on the MOSFET while the V_{IN_JG} stays at the high level throughout the entire turn-on process, meaning the cascode is turned on in the same way as the standard cascode. The proposed dual-gate drive approach separates the controls of the turn-on and turn-off processes, eliminates the possibility of the MOSFET being driven into avalanche breakdown, and breaks the resonant loop 1 in Fig.1 and Fig.2 and makes the turn-off process more controllable. Detailed descriptions of this approach are presented in [8]. In this approach, the JFET gate terminal must be made accessible, so the package suitable for this dual-gate drive approach requires 4 leads.

3.   Experimental Results

In this work, the experiments have been designed and performed to investigate the effects of the standard gate drive approach, the kelvin source gate drive approach, and the dual-gate drive approach on the switching performance of the SiC cascodes.

3.1.   Standard and Kelvin Source Gate Drive Approaches

The popular package for the discrete high voltage power switches is TO-247. The 3-lead TO-247-3L package is used for the standard gate drive, and the 4-lead TO-247-4L package is generally used for the kelvin source gate drive. In order to compare the effects of the standard and kelvin source gate drive approaches, a special PCB is designed and manufactured to simulate a TO-247-4L package. The PCB has a soft gold finish suitable for die attach of SiC JFET and aluminum wire bonding. Different gate drive approaches for the SiC cascode can be constructed with this PCB as illustrated in Fig.4.

A SiC cascode is built on the PCB board with a kelvin source lead (SS) using a 1.2kV-50mΩ SiC JFET die and a custom designed 20V-4.5mΩ Si MOSFET die. The JFET die and MOSFET die are sealed in silicone gel to avoid arc at high voltage. This device can be driven with the standard gate drive approach if the kelvin source lead (SS) is not used or with the kelvin source gate drive approach if the kelvin source lead (SS) is used. In this way, the comparisons of the standard gate drive approach and the kelvin source gate drive approach can be made on exactly the same device.

Fig.5 compares the measured turn-on switching waveforms of the same SiC cascode using the kelvin source gate drive and the standard source gate drive under 800V inductive load condition. The commercial 1.2kV-10A SiC JBS diode (UJ2D1210T) is used as the freewheeling diode. The turn-on gate resistor R_{G_ON} is 1.1Ω. The gate driver output voltage is from 0V to +12V. It is clearly seen that both gate drive approaches produce clean drain voltage (V_{DS}) and drain current (I_D) waveforms, but the kelvin source gate drive gives much smaller oscillations on the gate voltage waveform (V_{GS}) at both low current and high current load conditions. For the standard gate drive, the large oscillations on the gate terminal are caused by the source inductances L_{S1} and L_{S2}. The de-basing effect of the source inductances L_{S1} and L_{S2} increases the turn-on energy loss. Under 800V-30A load condition, the turn-on energy loss is 800uJ for the standard gate drive approach, and 700uJ for the kelvin source gate drive approach, more than 12% reduction in the turn-on energy loss when the kelvin source gate drive is used.

Fig.6 compares the measured turn-off switching waveforms of the same SiC cascode using the kelvin source gate drive and the standard source gate drive under 800V inductive load condition. The turn-off gate resistor R_{G_OFF} is 47Ω. The kelvin source gate drive again presents much smaller oscillations on the gate voltage waveform V_{GS} at both low current and high current load conditions. It is also seen that the kelvin source gate drive approach does

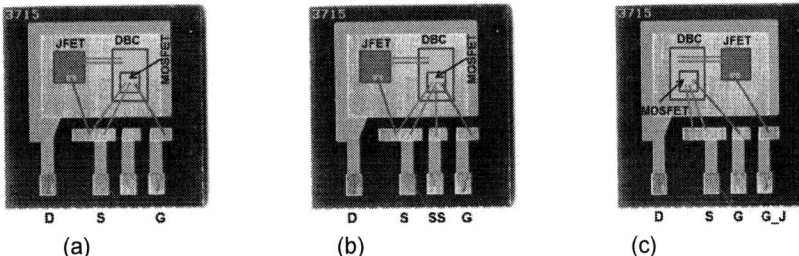

(a)   (b)   (c)

Fig.4. Schematic drawing of SiC cascodes built on the PCBs to simulate a TO-247-4L package with (a) standard gate drive configuration, (b) kelvin source gate drive configuration, and (c) dual-gate drive configuration.

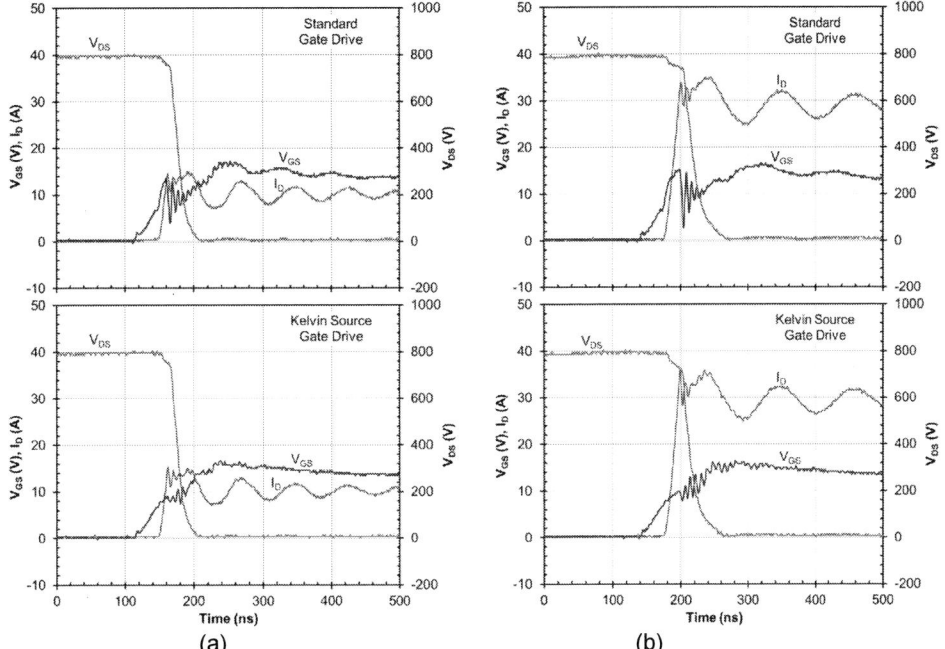

(a)   (b)

Fig.5. Measured turn-on waveforms of the 1.2kVSiC cascode with standard gate drive and with kelvin source gate drive under (a) 800V-10A and (b) 800V-30A inductive load conditions. Freewheeling diode: UJ2D1210T and R_{G_ON} = 1.1Ω.

not seem to have much effect on the waveforms of the drain voltage (V_{DS}) and drain current (I_D). These two gate drive approaches have almost the same the turn-off energy loss. For example, under 800V-30A load condition, the turn-off energy loss is 476uJ for the standard gate drive approach, and 472uJ for the kelvin source gate drive approach.

3.2. Dual-Gate Drive Approach

A discrete cascode formed with a 1.2kV-50mΩ SiC JFET in TO-247 and a 25V-5mΩ Si MOSFET in DPAK is used to investigate the operation principle of the proposed dual-gate drive approach. A gate driver is developed to generate two output signals as shown in Fig.3b for driving the dual-gate cascode. Fig.7a shows the schematic of the driving circuit used in the tests. R_{G_M} is the external gate resistor of the MOSFET; R_{G_J} is gate resistor of the JFET; a ferrite bead is connected to the JFET gate to limit the possible oscillations on the JFET gate; R2 is used to connect the JFET gate to the ground GND so that the cascode operates in normally-off mode during startup and gate driver failure, and D1 is used to block the current flowing through R2 during the turn-off transient when the negative pulse appears on the JFET gate. Fig.7b presents the measured waveforms under 600V-22A inductive load condition with R_{G_J} = 0Ω and R_{G_M} = 2.3Ω, clearly showing how the proposed dual-gate drive approach works. It is seen that, during turn-off transient from t=1100ns to t=1200ns, the

© VDE VERLAG GMBH · Berlin · Offenbach

PCIM Europe 2016, 10 – 12 May 2016, Nuremberg, Germany

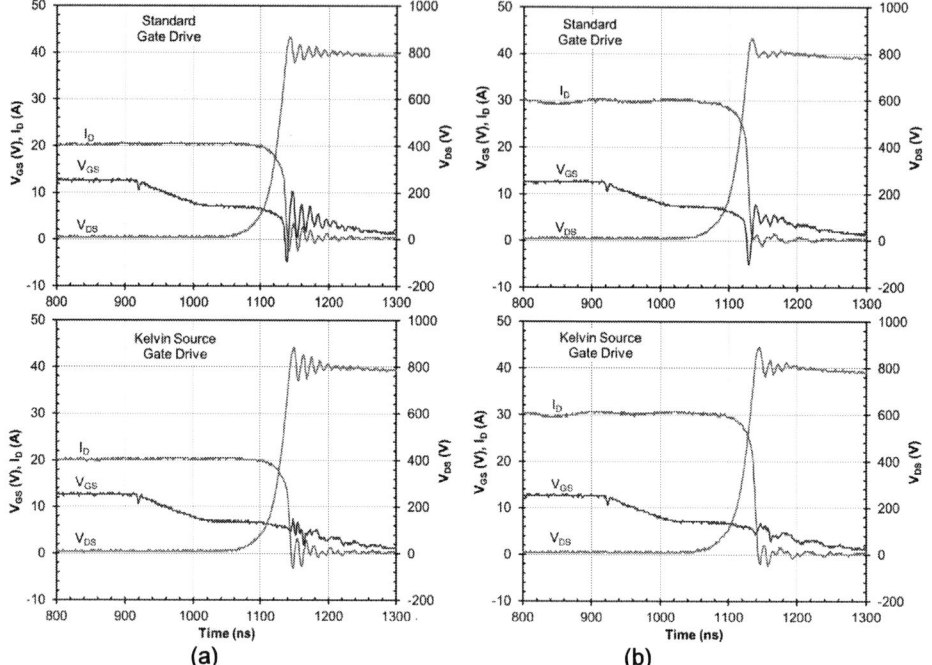

(a)   (b)

Fig.6. Measured turn-off waveforms of the 1.2kV SiC cascode with standard gate drive and with kelvin source gate drive under (a) 800V-20A and (b) 800V-30A inductive load conditions. Freewheeling diode: UJ2D1210T and R_{G_OFF} = 47Ω.

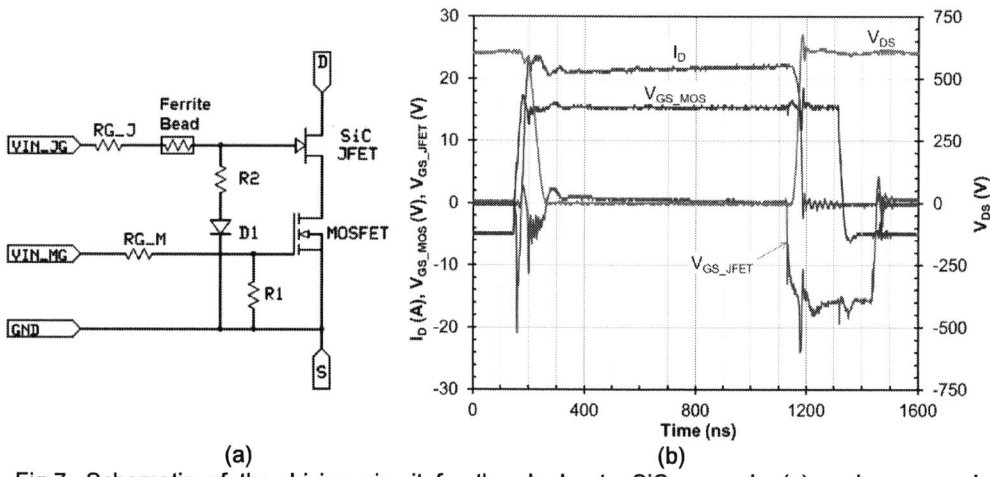

(a)   (b)

Fig.7. Schematic of the driving circuit for the dual-gate SiC cascode (a) and measured switching waveforms of the 1.2kV SiC cascode under 600V-22A inductive load condition. R_{G_J} = 0Ω and R_{G_M} = 2.3Ω.

MOSFET is in on-state because its gate is maintained at +15V. At t=1300ns, the cascode already enters the off-state and the drain current I_D is zero. At this moment, the MOSFET is switched off, which will not cause oscillations or avalanche breakdown since the MOSFET is turned off at zero current. After the MOSFET enters the off-state, the JFET gate voltage is pulled up to 0V at about t=1440ns to allow the JFET to conduct a reverse current.

The standard gate drive approach uses the MOSFET gate resistor R_{G_M} to control the turn-off speed and the dual-gate drive approach uses the JFET gate resistor R_{G_J} to control the turn-off speed. Fig.8a shows the measured turn-off di/dt and dv/dt at different JFET gate resistors (R_{G_J}) of the 1.2kV-50mΩ SiC discrete cascode using the dual-gate drive approach. Fig.8b shows the measured turn-off di/dt and dv/dt at different MOSFET gate resistors (R_{G_M}) of a 1.2kV-50mΩ SiC co-packaged cascode in TO-247-3L package using the standard gate drive approach. For the dual-gate drive approach as depicted in Fig.8a, when the R_{G_J} is increased

© VDE VERLAG GMBH · Berlin · Offenbach

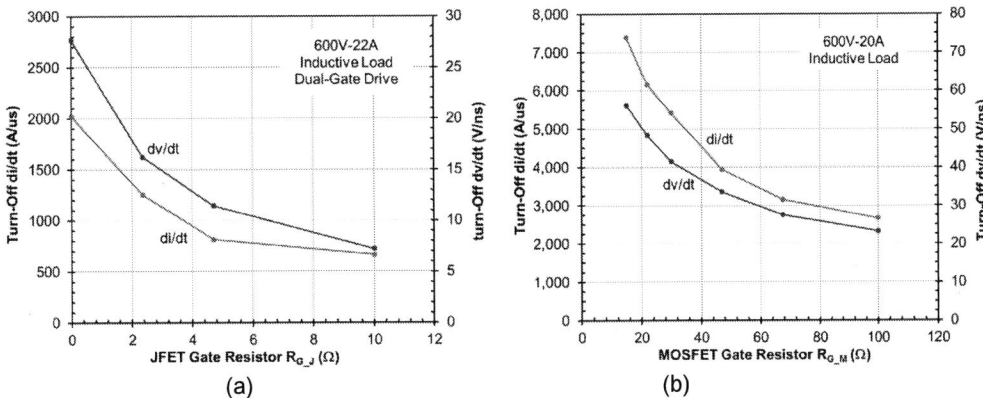

(a)   (b)

Fig.8. Measured turn-off di/dt and dv/dt rates of (a) the 1.2kV-50mΩ SiC discrete cascode with the dual-gate drive approach, and (b) the 1.2kV-50mΩ SiC side-by-side cascode in TO-247-3L package with the standard gate drive.

from 0 Ω to 5Ω, the turn-off dv/dt rate is decreased substantially from about 28V/ns to 11V/ns and the turn-off di/dt rate is decreased substantially from 2,000A/us to 800A/us. Clearly, the dual-gate drive technique allows the cascode to be operated over a wider range of di/dt and dv/dt control with obvious EMI benefits.

4. Conclusions

In this work, three gate drive approaches for driving the SiC cascodes have been discussed including the standard gate drive approach, the kelvin source gate drive approach and the dual-gate drive approach. Experiments have been designed and performed to investigate the effects of these three gate drive approaches on the switching performance of the SiC cascode. The experimental results show both the standard gate drive approach and the kelvin source gate drive approach work well for the SiC co-packaged cascode, but the kelvin source gate drive approach gives smaller oscillations on the gate of the low-voltage Si MOSFET. The dual-gate drive approach can provide a wider range of turn-off di/dt and dv/dt control for the SiC cascode than the standard gate drive approach.

References

[1]   Datasheets of UJC1206K and C2M0040120D.

[2]   Datasheets of UJC06506K and IPB65R045C7.

[3]   http://unitedsic.com/cascodes/

[4]   Anup Bhalla, Xueqing Li, and John Bendel, "Switching Behavior of USCi's SiC Cascodes" Magazine of Bodo's Power Systems, June 2015, pp.22-26.

[5]   Anup Bhalla, John Bendel, and Xueqing Li, "Robustness of SiC JFETs and Cascodes," Magazine of Bodo's Power Systems, May 2015, pp.48-50.

[6]   Matt O'Grady, Ke Zhu, Xueqing Li, and John Bendel, "Paralleling SiC Cascodes for High Performance, High Power Systems," Magazine of Bodo's Power Systems, September, 2015, pp.22-26.

[7]   Xueqing Li, Anup Bhalla, Peter Alexandrov, John Hostetler, Leonid Fursin, "Investigation of SiC Stack and Discrete Cascodes," PCIM Europe 2014, 20 – 22 May 2014, Nuremberg, Germany, pp.448-455.

[8]   United States Patent 9,083,343.

[9]   D. Domes, X. Zhang: CASCODE LIGHT – normally-on JFET standalone performance in a normally-off Cascode circuit. Proc. PCIM, Nuernberg, 2010.

[10]   US patent application: US2013/0335134.

PCIM Europe 2016, 10 – 12 May 2016, Nuremberg, Germany

State-of-the-art of HF Soft Magnetics and HV/UHV Silicon Carbide Semiconductors

Paul Ohodnicki, National Energy Technology Laboratory, USA,
paul.ohodnicki@NETL.DOE.GOV
Alex Leary, Carnegie Mellon University, USA, aleary@andrew.cmu.edu
Michael E. McHenry, Carnegie Mellon University, USA, mm7g@andrew.cmu.edu
Faete Filho, Eaton Corporation, USA, faetefilho@eaton.com
Geraldo Nojima, Eaton Corporation, USA, geraldonojima@eaton.com
Allen Hefner, National Institute of Standards and Technology, allen.hefner@nist.gov

Abstract

The post silicon era has begun with the introduction of wide bandgap semiconductors. Among wide bandgap semiconductors being researched, Silicon Carbide (SiC) has been shown so far as the best candidate to high power and high voltage applications. SiC have a superior performance compared to its counterpart (Si) and this brings new challenges to vital ancillary components such as transformers, inductors, drivers as well as addressing issues such as leakage, clearance, EMI and material challenges. A particular challenge for the magnetics components are increasing demands being placed upon higher operational frequency and higher operational temperature conditions.

This paper briefly discusses the current status of high voltage SiC devices and soft magnetic materials and cores for high power and high frequency (1-50kHz) applications comprised of amorphous and nanocrystaline nanocomposite alloys.

1. Introduction

The advantages of HV SiC over state-of-the-art HV Si semiconductors are well known.

Simpler topologies for MV converters with more compact packages that can reach output fundamental frequencies of 1kHz are some of the features promised by HV SiC. These features will be demonstrated with the DOE-AMONGEM project.

Like any game changing upcoming technology, there are three main barriers for HV SiC widespread acceptance: technology readiness, cost and availability of supporting passives and peripherals.

2. Breaking the barriers for wide penetration of SiC semiconductor technologies: Technology Readiness

Compare the previous example with current status of SiC technology. Research has been ongoing for a number of years [28]. In 2002, 600-1700V SiC became widely available, then the first generation SiC 10kV MOSFET was first developed in the 2007-2008 timeframe, and an improved third generation with lower specific resistance in 2015 (Figure 1). More recent advances have been demonstrated in 10kV SiC MOSFET with reliable body diode, as shown in Figure 2, and enhanced short circuit capability, shown in Figure 3, where a short circuit at 5kV is safely turned-off for a time greater than 13usec.

Additionally, recent reports show that in the past few months there have been solid solutions for previously considered impossible to solve HV SiC wafer and die problems, namely basal plane dislocation and short circuit capability. Continuous improvements in wafer manufacturing, processing and inspection have enabled cost effective SiC MOSFET

© VDE VERLAG GMBH · Berlin · Offenbach

body diode to be used reliably. This eliminates the need of SiC antiparallel diode and series Si Schottky diode dies which enables the SiC HV MOSFET modules to attain unprecedented current capability such as a 140mm x 190mm 10kV 500A SiC MOSFET module.

Figure 1 - 10kV SiC MOSFET Next

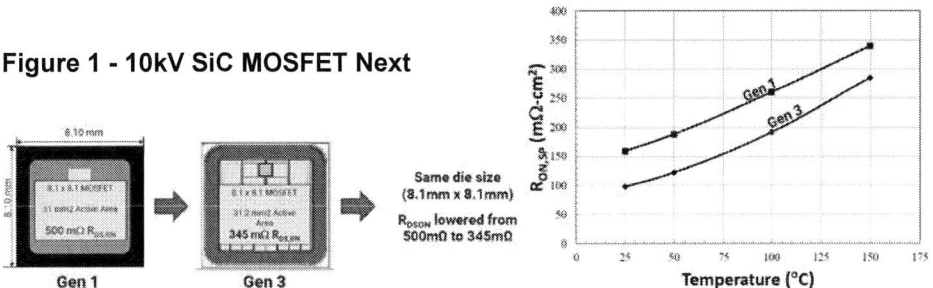

Figure 2 - Stability of Body Diode After Stressing at 10 A

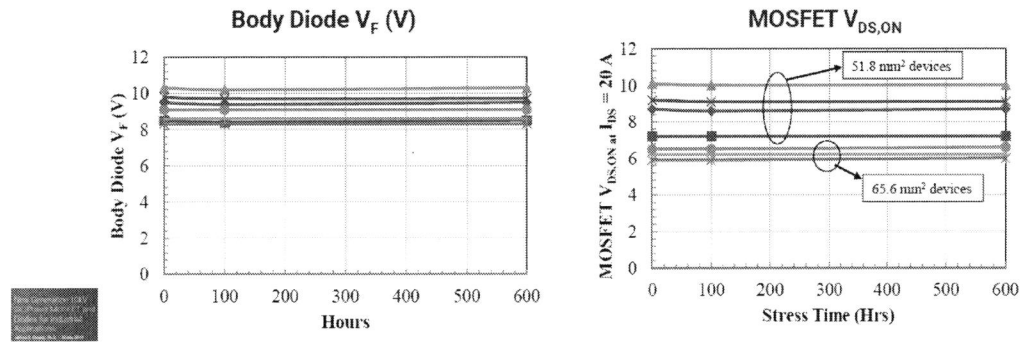

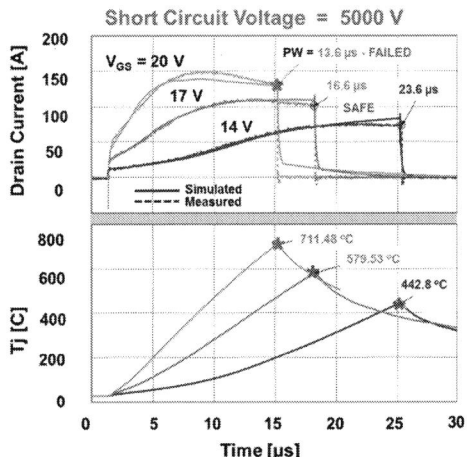

Fig. 3 – Recent SiC MOSFET development with short circuit capability.

3. Breaking the barriers for wide penetration of SiC semiconductor technologies: Cost

The second barrier is the cost conundrum, the technology is not adopted because the price is too high but the price is too high because the technology is not widespread enough to increase production volume.

SiC Die manufacturers claim that Swanson's law, which shows that the cost of solar PV modules drops 20% when the cumulative shipped volume doubles, is applicable to SiC dies.

Initiatives such as government funded programs help to startup this demand process

3.1 Key applications for penetration of HV SiC semiconductors

In the realm of high fundamental frequency at medium voltage level for megawatts class motors running at 15000RPM to 20000RPM, the required fundamental frequency of >500Hz becomes the minimum target frequency. Two level and three level topologies cannot be implemented to reach such high fundamental frequencies and thus bulky topologies and/or mechanical gearbox using low/medium voltage IGBTs have to be used to achieve them.

Wide bandgap HV and UHV SiC power semiconductors such as MOSFETs and IGBTs are emerging as the solution for the above issue. For example, a high voltage 10kV SiC MOSFET can be used in a two-level inverter to achieve 5kVrms output voltage and hard switching at 10kHz while a 15kV SiC MOSFET or IGBT would yield 6.9kVrms output voltage at 5kHz switching frequency.

Such high fundamental frequency applications combined with industry expertise and government funding allow addressing current challenges on SiC technology as cost, and technology maturity. Key applications such as gas compression and gas separation used in the oil and gas sector [24], subsea gas compression, onshore and offshore topside compression and air separation, all of which require 3MW to 20MW high speed, high voltage compact and efficient motors and drives can greatly benefit from HV SiC technology.

Another area of interest is the MV and HV transmission and distribution by providing efficient devices with a better voltage utilization ratio such as static circuit breakers [25] and subsea DC-DC converters [26-27].

Increasing power density in these devices drives innovation in many material classes including magnetics, semiconductors, dielectrics, and packaging as described in technology roadmaps [1, 2]. Figure 4 shows the application area intersections for various soft magnetic materials with semiconductors used in diodes and active switches. In [1], the application frequency of MW scale 6 kV Si-based switches is given as <200 Hz and this theoretical limit is further described in [3]. Wide bandgap semiconductors extend this frequency range into the kHz region with the 2014 announcement of a CREE 15 kV SiC MOSFET switching 60 kW at 10 kHz [4]. These application boundaries change rapidly and with improved wide bandgap materials processing (SiC, GaN, diamond), single level switching networks will enable efficient MW scale power converters with direct connection to medium voltage distribution grids. For inductive materials, silicon steels are widely used in high power applications but typically limited to frequencies below 1 kHz, due to relatively low resistivity resulting in excessive eddy current losses. Ferrites are used at higher frequencies but are limited to lower power density due to reduced saturation induction. Metal amorphous nanocomposite soft magnetic materials with high induction, high resistivity, tunable permeability, and low losses allow for high power density devices that fully realize the potential value offered by SiC-based wide bandgap semiconductors in high power converter applications [5]. These converters will provide for integrated control in grid systems that incorporate distributed power sources/storage and decrease vulnerability of centralized components [6, 7]. Amorphous and nanocomposite materials have also been identified for their potential in rotor or stators in motor applications operating at high rotational speeds [55, 56]. Describing the vast scale of prototype topologies is beyond the scope of this work, but

PCIM Europe 2016, 10 – 12 May 2016, Nuremberg, Germany

[8, 9] provide system topology reviews and [10] describes performance trends in power electronics.

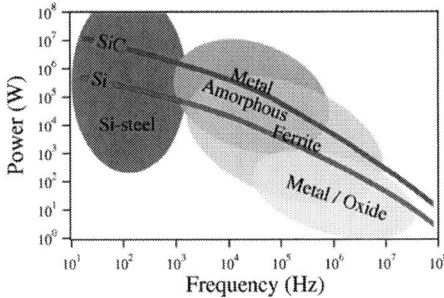

Fig. 4 –Application areas for soft magnetic materials compared to Si and Si-C based semiconductor components.

4. Breaking the barriers for wide penetration of SiC semiconductor technologies: Enhanced Passives and Peripherals - Soft Magnetic Materials in High Power and High Power Density Converters

Barriers to widespread integration of high power, high power density power converters include device complexity, development cost, and product consistency/quality. Components used in these applications require tuned impedances to withstand the thermal, electric, and mechanical requirements. There is a number of interconnected considerations for transformer design, and this complexity is apparent in solid state transformers (SSTs) used in many power converters that require galvanic isolation [11,12]. The power density for low frequency Si-steel transformers is around 0.2 kVA/kg with high power amorphous designs typically below 1 kVA/kg. Nanocomposite transformer prototypes operating at around 10-100kHz are approaching the 20 kVA/kg as can be seen in Fig. 5. This figure shows that nanocomposite prototypes are approximately 5X larger in power density than the Si-steel design [13] that operate at 1 kHz and 100X larger than low frequency Si-steel applications. Power density values vary depending on the inclusion of a cooling system and are shown here in kW/kg to eliminate the variability in box volume measurements. This scaling considers only the core material, which can occupy a significant portion of the weight/volume of a power converter. Despite significantly higher mass density, nanocomposites show higher kW/kg power density compared to ferrites in prototypes >10kW.

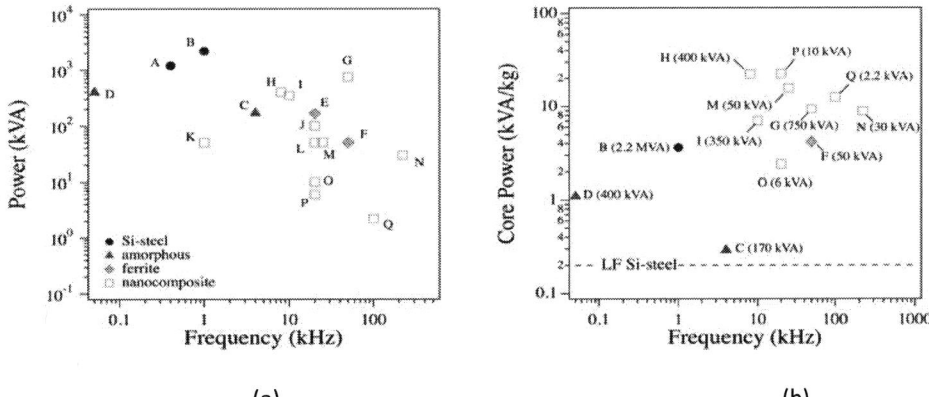

(a)   (b)

Fig. 5 – (a) Power converter prototypes with galvanic isolation and (b) core power of the transformer core showing power density limits.

© VDE VERLAG GMBH · Berlin · Offenbach

Table 1 – A sample of power converters with galvanic isolation using different magnetic materials

Label	Reference	Material	$V_1:V_2$ (kV)	kVA	kHz	kg
A	[32]	Si-steel	1.8 : 1.8	1200	0.4	-
B	[31]	Si-steel	5 : 5	2200	1	607
C	[33]	Metglass 2605SA1	15 : 1.8	170	4	145
D	[34]	FeSi amorphous	15 : 0.4	400	0.05	364
E	-	N87 ferrite	0.3 : 0.3	160	20	-
F	[35]	PC40 ferrite	0.2 : 2	50	50	12
G	[36]	nanocomposite	0.7 : 13.8	750	50	80
H	[37]	Nanocomposite	2.8 : 2.8	400	8	18
I	[38]	Vitroperm 500F	3 : 3	350	10	50?
J	[39]	Finemet	0.75 : 0.75	100	20	-
K	[40]	Gammamet	3 : 3	50	1	-
L	[41]	Nanocomposite	0.38 : 0.38	50	20	-
M	[30]	Vitroperm 500F	0.75 : 0.6	50	25	3.2
N	[29]	FT-3M	0.3 : 10	30	220	3.4
O	[42]	Finemet	0.2 : 0.06	6	20	2.5
P	[43]	FT-3M	0.35 : 0.35	10	20	0.45
Q	[44]	FT-3M	0.024 : 0.36	2.2	100	12.5

5. Recent Developments in Nanocomposite Alloys and Processing

Achieving high power density at high power requires materials with low losses at high frequency and/or high temperature stability. The emergence of wide-bandgap based switching devices that enable high power density power converters results in an increasing demand for magnetic components that operate at high frequency and high temperature and a variety of soft magnetic nanocomposite compositions have been developed to meet these application needs [14]. Alloy composition engineering plays a critical role, but properties that are demonstrated by the final fabricated magnetic core are the ones that must be optimized.

In transformers, high permeability cores with high magnetizing inductance require low magnetizing current and produce good voltage regulation. Magnetizing inductance also affects the soft switching envelope [15] and the sensitivity of the transformer to flux imbalances created by asymmetric switching [16]. These flux imbalances can be monitored and corrected using a control algorithm to prevent high losses resulting from core saturation [17]. In contrast, low permeability may be needed in interphase transformers that accommodate ripple currents between parallel rectifier outputs. Low permeability can also be required in the case of high power inductors which are required to serve as a filter or an energy storage element, particularly for operation at high DC bias levels.

To modify core permeability, an air gap is often added with high permeability core material to increase reluctance and optimize the core permeability for a particular application. This standard practice increases losses due to (1) mechanical deformation and stresses of the core, (2) local flux concentrations and fringing fields, and (3) the creation of flux paths with a component transverse to the ribbon plane so that eddy currents circulate in the ribbon plane [18,19]. Alternatively through alloy composition engineering and advanced materials processing, nanocomposite cores with induced anisotropy by field [20,21] and stress annealing [22,23] can allow for tunable permeability ranging from $10-10^5$ so that magnetizing inductance can be controlled in tape wound cores without the need for air gaps.

Nanocomposites are complex multi-component alloys formed by a rapid solidification process to obtain an amorphous precursor followed by partial devitrification through controlled annealing treatments [5,14]. While a broad range of alloy compositions have been investigated, the majority of technologically relevant alloys can be described in terms of the following general composition:

$$(Fe_{1-a}Co_a)_b Nb_c Si_d B_x Cu_y, \quad \text{Eq. 1}$$

For optimized alloy compositions, ferromagnetic transition metal elements such as (Fe,Co) are present at the highest possible level to maximize the saturation induction of the alloy while still retaining a nanocrystalline microstructure for realization of suitable soft magnetic properties. Relatively large transition metal elements such as Nb contribute to glass formability of the alloy upon rapid quenching and also play a critical role in the metastability of the optimized nanocrystalline microstructure. Metalloids such as Si and B further enhance the glass formability and Si, in particular, plays an important role in tuning the magnetostriction and resistivity of the amorphous and crystalline phases in the final nanocomposite microstructure.

Several nanocomposite alloy archetypes are based upon the stoichiometric formula described above and dictated by the relative content of Fe and Co in Equation 1 including: (1) Fe-based alloys, (2) FeCo-based alloys, and (3) Co-based alloys. These alloys have properties that can be distinguished based upon characteristic figures of merit including saturation induction, magnetostriction, responses to anisotropic processing for permeability engineering (strain and field annealing), high temperature stability, mechanical ductility, and raw material costs. Emerging classes of alloys are also exploring modifications to the glass former compositions in order to increase saturation induction by maximizing the ferromagnetic transition metal elements. Table 2 illustrates the relative advantages and disadvantages of each archetype alloy class with the understanding that property optimization within each class is accomplished through adjustment of alloy chemistry and processing. Subsequent sections briefly review the advantages and disadvantages of each alloy class and discuss applications for which they will provide the greatest benefits as advanced core materials. We conclude with a brief discussion of emerging research opportunities for new alloy chemistry and alloy processing development in this class of materials.

Table 2 – Advantages and disadvantages of common archetype nanocomposite soft magnetic alloys

	Bs	λ_s	Field Annealing Response	Strain Annealing Response	Initial μ Range	High Temperature Stability	Mechanical Ductility	Raw Material Costs
Fe-Based	1.1-1.3 T	<|10ppm|	Low	Moderate	10^4-10^6	Low	Brittle	Low
FeCo-Based	1.3-1.5 T	>10ppm	Moderate	Moderate	10^2-10^4	Moderate	Brittle	Moderate
Co-Based	0.8-1 T	<|10ppm|	High	High	10^1-10^4	High	Ductile	High
High Bs (glass former modified)	1.3-1.8 T	Depends on FeCo-Ratio and glass former composition, tends to have lower temperature stability than standard glass former compositions						

5.1 Fe-Based Nanocomposite Alloys

A combination of low magnetostriction (and hence low losses), high saturation induction, and relatively low raw material costs make Fe-Based alloys the primary class of commercially available nanocomposite alloys [45]. Optimized alloys exhibit high initial permeabilities approaching $\sim 10^6$ with saturation inductions ranging from ~ 1.1-$1.3T$ and core losses improved as compared to commercially available amorphous alloys at switching frequencies in the 10's of kHz range. Fe-based alloys are therefore most attractive for medium frequency transformer applications in which high magnetizing inductance combined with low core losses and low cost are the primary considerations. Extremely brittle mechanical properties and weak responses to field annealing makes the manufacturing of tuned permeability cores through field and strain annealing processing a significant challenge [46-47]. As a result, core impregnation and gapping remains the primary technique for reducing permeability of Fe-based alloy nanocomposite cores in applications where reduced magnetizing inductance and/or higher saturation fields are required.

5.2 FeCo-Based Alloys

FeCo-based alloys show a large saturation induction relative to Fe-based alloys ranging from ~ 1.3-$1.5T$ and enhanced Curie temperatures of the constituent amorphous and nanocrystalline phases but tend to exhibit significantly increased raw material costs (Co is much more expensive than Fe) and increased magnetostriction making theme sensitive to core fabrication including impregnation and gapping [5,14]. Optimized FeCo-based alloys exhibit initial permeabilities spanning $\sim 10^2$-10^4 and can be adjusted significantly in this range through field annealing [20]. However they are not typically suitable for strain annealing as the mechanical properties are comparable with Fe-based alloys showing extreme brittleness. FeCo-based alloys show significant technical advantages for inductors and transformers which are to be optimized for high temperatures and / or the highest possible power densities and offer an attractive alternative to Fe-based alloys when these performance metrics outweighs additional raw material costs and the potential for increased losses [48].

5.3 Co-Based Alloys

Co-based alloys exhibit reduced saturation inductions relative to Fe- and FeCo-based alloys ranging from ~ 0.8-$1T$, but also significantly improved high temperature stability of the constituent amorphous phase relative to Fe-based alloys [21,49]. Co-based alloys can also be engineered to exhibit near-zero magnetostriction in the crystalline and amorphous phases and enhanced responses to anisotropic field and strain annealing allowing for realization of an unprecedented range of permeabilities from ~ 10-10^4[22-23,50-51]. Dramatically improved mechanical ductility of these alloys enable advanced core fabrication and processing allowing for the realization of tuned permeability cores at scales that are relevant for high power inductors and transformers [52]. Co-based alloys are therefore particularly attractive for applications in which relatively low permeability and magnetizing inductance is required.

5.4 Alloys with Modified Glass Former Compositions

In addition to variation of the Fe and Co-ratio, several new Fe-based and FeCo-based alloy classes have emerged in recent years for which the glass former compositions have been modified significantly from that of standard commercially available Fe-based alloys as presented in Eq. 1. In all cases, a primary motivation has been to achieve higher saturation inductions than obtainable in standard compositions by maximizing the content of the ferromagnetic transition metal elements (Fe,Co) and reducing or substituting for the most commonly employed glass formers Nb, B, and Si [53,54]. In such alloys, higher saturation induction can typically be achieved. In many cases, this increase in saturation induction has been reported to come at the expense of alloy temperature stability and core losses at high frequencies due to non-optimized microstructures [53].

5.5 Future Developments in Nanocomposite Alloys and Processing

In recent years, the importance of combining alloy chemistry optimization with advanced processing approaches such as field annealing, strain annealing, and rapid annealing has become apparent. Moving forward, advanced processing of nanocomposite alloys will play an increasingly important role in optimizing core permeability and losses for enabling a broad range of SiC-based power conversion applications. Several key areas of future research and development opportunities in this area include:

1) Large-scale rapid solidification and processing to achiever thinner, high-quality ribbons of all compositions of interest for reduced eddy current losses.
2) Engineered alloys with optimized magnetic properties and ductile mechanical properties that allow for improved manufacturability of cores.
3) Alloy compositions with reduced raw material costs.
4) Alternative and advanced anisotropic processing approaches.

6. Summary

The recent advances in both SiC semiconductors and soft magnetics components have been shown. Every two or three years a new development on those technologies are reported, making it better, more reliable and closer to the threshold where it can become widely adopted by industry. Funding opportunities plays a key role in the initial development as they have to initially break the initial technological challenges, become more reliable, compact and cost effective.

There are today key applications such as high speed medium voltage megawatt class motors, among others, that can directly benefit from those technologies and provide the justification to application of this technology as has been shown. A few funding opportunities exists today and in the coming years some of the barriers that still exist for these technologies are expected to be broken.

7. References

[1] R. H. Wolk, "Proceedings of the High Megawatt Power Converter Technology R&D Roadmap Workshop," *National Institute of Standards and Technology*, Gaithersburg, MD, Tech. Rep., 2008.
[2] "SunShot Vision Study," *U.S. Department of Energy*, Tech. Rep. February 2012.
[3] A. Nakagawa, Y. Kawaguchi, and K. Nakamura, "Silicon limit electrical characteristics of power devices and ICs," *9th International Seminar on Power Semiconductors* (ISPS 2008), vol. 84, pp. 25–32, 2008.
[4] J. W. Palmour, L. Cheng, V. Pala, E. V. Brunt, D. J. Lichtenwalner, G. Y. Wang, J. Richmond, M. O'Loughlin, S. Ryu, S. T. Allen, A. A. Burk, and C. Scozzie, "Silicon carbide power MOSFETs: Breakthrough performance from 900V up to 15 kV," *Proceedings of the International Symposium on Power Semiconductor Devices and ICs*, pp. 79–82, 2014.
[5] Michael E. McHenry, Matthew A. Willard, and David E. Laughlin, "Amorphous and nanocrystalline materials for applications as soft magnets," *Progress in Materials Science*, 44(4):291–433, 1999.
[6] P. W. Paformak, "Physical Security of the U.S. Power Grid: High-Voltage Transformer Substations," *Congressional Research Service*, Tech. Rep. Jun, 2014.
[7] R. Smith, "Transformers Expose Limits in Securing Power Grid," *The Wall Street Journal*, March 4, 2014.

© VDE VERLAG GMBH · Berlin · Offenbach

PCIM Europe 2016, 10 – 12 May 2016, Nuremberg, Germany

[8] F. Blaabjerg, Z. Chen, and S. B. Kjaer, "Power Electronics as Efficient interface in Dispersed Power Generation Systems in Dispersed Power Generation Systems," *IEEE Transactions on Power Electronics*, vol. 19, no. 5, pp. 1184–1194, 2004.

[9] P. Jain, M. Pahlevaninezhad, S. Pan, and J. Drobnik, "A review of high-frequency power distribution systems: For space, telecommunication, and computer applications," *IEEE Transactions on Power Electronics*, vol. 29, no. 8, pp. 3852–3863, 2014.

[10] J. W. Kolar, J. Biela, S. Waffler, T. Friedli, and U. Badstuebner, "Performance Trends and Limitations of Power Electronic Systems," in Invited Plenary Paper at the *6th International Conference on Integrated Power Electron-ICs Systems* (CIPS), Nuremberg, 2010, pp. 16–18.

[11] X.She, A.Q.Huang, and R.Burgos,"Review of Solid-State Transformer Technologies and Their Application in Power Distribution Systems," *IEEE Journal of Emerging and Selected Topics in Power Electronics*, vol. 1, no. 3, pp. 186–198, Sep 2013.

[12] B. Zhao, Q. Song, W. Liu, and Y. Sun, "Overview of dual-active-bridge isolated bidirectional DC-DC converter for high-frequency-link power-conversion system," *IEEE Transactions on Power Electronics*, vol. 29, no. 8, pp. 4091–4106, 2014.

[13] N. Soltau, H. Stagge, R. W. De Doncker, and O. Apeldoorn, "Development and demonstration of a medium-voltage high-power DC-DC converter for DC distribution systems," 2014 *IEEE 5th International Symposium on Power Electronics for Distributed Generation Systems* (PEDG), no. 978, pp. 1–8, 2014.

[14] Matthew A Willard and Maria Daniil, "Nanocrystalline Soft Magnetic Alloys: Two Decades of Progress," vol. 21. Elsevier B.V., 2013.

[15] M. H. Kheraluwala, R. W. Gascoigne, D. M. Divan, and E. D. Baumann, "Performance characterization of a high-power dual active bridge DC-to-DC converter," *IEEE Transactions on Industry Applications*, vol. 28, no. 6, pp. 1294–1301, 1992.

[16] V. Va¨isa¨nen, T. Riipinen, and P. Silventoinen, "Effects of switching asymmetry on an isolated full-bridge boost converter," IEEE Transactions on Power Electronics, vol. 25, no. 8, pp. 2033–2044, 2010.

[17] G. Ortiz, L. Fassler, J. W. Kolar, and O. Apeldoorn, "Application of the magnetic ear for flux balancing of a 160kW/20kHz DC-DC converter transformer," Conference Proceedings - *IEEE Applied Power Electronics Conference and Exposition* - APEC, pp. 2118–2124, 2013.

[18] M. S. Rylko, J. G. Hayes, and M. G. Egan, "Experimental investigation of high-flux density magnetic materials for high-current inductors in hybrid-electric vehicle DC-DC converters," *IEEE Vehicle Power and Propulsion Conference*, sept 2010, pp. 1–7.

[19] S.Odawara,N.Denis,S.Yamamoto,K.Sawatari,K.Fujisaki,Y.Shindo,N.Yoshikawa,andT.Konishi, "Impact of Material on the Iron Losses of a Reactor With Air Gap," *IEEE Transactions on Magnetics*, vol. 9464, no. c, pp. 1–1, 2015.

[20] F. Johnson, C. Y. Um, M. E. McHenry, and H. Garmestani, "The influence of composition and field annealing on magnetic properties of FeCo-based amorphous and nanocrystalline alloys," J. Magn. Magn. Mater., vol. 297, no. 2, pp. 93–98, Feb. 2006.

[21] P. R. Ohodnicki, D. E. Laughlin, M. E. McHenry, V. Keylin, and J. Huth, "Temperature stability of field induced anisotropy in soft ferromagnetic Fe,Co-based amorphous and nanocomposite ribbons," *Journal of Applied Physics*, vol. 105, no. 7, p. 07A322, 2009.

[22] S. J. Kernion, P. R. Ohodnicki, J. Grossmann, A. Leary, S. Shen, V. Keylin, J. F. Huth, J. Horwath, M. S. Lucas, and M. E. McHenry, "Giant induced magnetic anisotropy In strain annealed Co-based nanocomposite alloys," *Applied Physics Letters*, vol. 101, no. 10, p. 102408, 2012.

[23] A. M. Leary, V. Keylin, P. R. Ohodnicki, and M. E. McHenry, "Stress induced anisotropy in CoFeMn soft magnetic nanocomposites," *Journal of Applied Physics*, vol. 117, no. 17, p. 17A338, 2015.

[24] Funding Opportunity Announcement, DE-FOA-0001208, "Next Generation Electric Machines: Megawatt Class Motors," https://eere-exchange.energy.gov

[25] Broad Agency Announcement, ONRBAA13-016, "High Power Solid State Circuit Protection for Power Distribution and Energy Storage," http://www.onr.navy.mil

[26] G. Nojima, G. Braga, L. O. Barros, S. T. S. Lima, A. M. Oliveira, L. S. Rezende, "A Methodology for the Development of s Subsea Electrical Power Transmission and Distribution System," Offshore Technology Conference, Houston TX, May, 2013.

[27] Department of Energy, DE-FOA-0001467, "Next Generation Electric Machines: Enabling Technologies," http://eere-exchange.energy.gov

[28] D. Grider, M. Das A. Agarwal and J. Palmour, "10 kV/120A SiC DMOSFET half H-bridge power modules for 1 MVA solid state power substation," *IEEE Elec. Ship Tech. Symp, 2010, pp. 131-134.*

[29] W. Shen, F. Wang, D. Boroyevich, and W. Tipton, "High-Density Nanocrystalline Core Transformer for High-Power High-Frequency Resonant Converter," IEEE Transactions on Industry Applications, vol. 44, no. 1, pp. 213–222, 2008.

[30] M. Pavlovsky, S. W. H. de Haan, and J. A. Ferreira, "Reaching High Power Density in Multikilowatt DCDC Converters With Galvanic Isolation," IEEE Transactions on Power Electronics, vol. 24, no. 3, pp. 603–612, Mar 2009.

© VDE VERLAG GMBH · Berlin · Offenbach

PCIM Europe 2016, 10 – 12 May 2016, Nuremberg, Germany

[31] N. Soltau, H. Stagge, R. W. De Doncker, and O. Apeldoorn, "Development and demonstration of a medium-voltage high-power DC-DC converter for DC distribution systems," 2014 IEEE 5th International Symposium on Power Electronics for Distributed Generation Systems (PEDG), no. 978, pp. 1–8, 2014.

[32] N. Hugo, P. Stefanutti, M. Pellerin, and A. Akdag, "Power electronics traction transformer," in 2007 European Conference on Power Electronics and Applications, EPE, 2007.

[33] T. Kjellqvist, S. Norrga, S. O ̈stlund, and K. Ilves, "Thermal evaluation of a medium frequency transformer in a line side conversion system," Power Electronics and Applications, 2009. {EPE} '09. 13th European Conference on, pp. 1–10, 2009.

[34] R. Kolano, A. Kolano-Burian, M. Polak, and J. Szynowski, "Application of Rapidly Quenched Soft Magnetic Materials in Energy-Saving Electric Equipment," IEEE Transactions on Magnetics, vol. 50, no. 4, pp. 1–4, Apr 2014.

[35] M. H. Kheraluwala, R. W. Gascoigne, D. M. Divan, and E. D. Baumann, "Performance characterization of a high-power dual active bridge DC-to-DC converter," IEEE Transactions on Industry Applications, vol. 28, no. 6, pp. 1294–1301, 1992.

[36] E. Limpaecher, G. Deffley, and F. Hoffmann, "750kW AC-link power converter for renewable generation and energy storage applications," in IEEE 2011 EnergyTech, 2011, pp. 1–6.

[37] M. Steiner and H. Reinold, "Medium frequency topology in railway applications," 2007 European Conference on Power Electronics and Applications, EPE, 2007.

[38] L. Heinemann, "An actively cooled high power, high frequency transformer with high insulation capability," in Seventeenth Annual IEEE Applied Power Electronics Conference and Exposition, 2002, pp. 352–357.

[39] H. Akagi, T. Yamagishi, N. M. L. Tan, S.-i. Kinouchi, Y. Miyazaki, and M. Koyama, "Power-Loss Breakdown of a 750-V, 100-kW, 20-kHz Bidirectional Isolated DC-DC Converter Using SiC-MOSFET/SBD Dual Modules," IEEE Transactions on Industry Applications, vol. 51, no. 1, pp. 420–428, 2013.

[40] D. Vinnikov, J. Laugis, and I. Galkin, "Middle-frequency isolation transformer design issues for the high-voltage DC/DC converter," 2008 IEEE Power Electronics Specialists Conference, pp. 1930–1936, Jun 2008.

[41] A. Prasai, H. Chen, R. Moghe, Z. Wolanski, K. Chintakrinda, A. Zhou, J. C. Llambes, and D. Divan, "Dyna-C: Experimental results for a 50 kVA 3-phase to 3-phase solid state transformer," 2014 IEEE Applied Power Electronics Conference and Exposition - APEC 2014, pp. 2271–2277, Mar 2014.

[42] N. M. L. Tan, T. Abe, and H. Akagi, "Topology and application of bidirectional isolated dc-dc converters," 8th International Conference on Power Electronics - ECCE Asia, pp. 1039–1046, May 2011.

[43] S. Inoue and H. Akagi, "A Bidirectional Isolated DC DC Converter as a Core Circuit of the Next-Generation," IEEE Transactions on Power Eletronics, vol. 22, no. 2, pp. 535–542, 2007.

[44] Y. Wang, S. W. H. de Haan, J. a. Ferreira, and S. W. H. D. Haan, "Design of low-profile nanocrystalline transformer in high-current phase-shifted DC-DC converter," in Energy Conversion Congress and Exposition (ECCE). IEEE, Sep 2010, pp. 3001–3008.

[45] Y. Yoshizawa, S. Oguma, and K. Yamauchi, "New Fe-based soft magnetic alloys composed of ultrafine grain structure", Journal of Applied Physics 64, 6044 (1988).

[46] Y. Yoshizawa and K. Yamauchi, "Induced Magnetic Anisotropy and Thickness Dependence of Magnetic Properties in Nanocrystalline Alloy Finemet", IEEE Translation Journal on Magnetics in Japan 5 (11) 1070 (1990).

[47] G. Herzer, V. Budinsky, and C. Polak, "Magnetic properties of nanocrystalline FeCuNbSiB with huge creep induced anisotropy", Journal of Physics: Conference Series 266 (1), 012010 (2011).

[48] J. Long, M. E. McHenry, D. P. Urciuoli, V. Keylin, J. Huth, and T. E. Salem, "Nanocrystalline material development for high-power inductors", Journal of Applied Physics 103, 07E705 (2008).

[49] P. R. Ohodnicki, Y. L. Qin, M. E. McHenry, D. E. Laughlin, and V. Keylin, "Transmission electron microscopy study of large field induced anisotropy (Co1-xFex)89Zr7B4 nanocomposite ribbons with dilute Fe-contents", Journal of Magnetism and Magnetic Materials 322 (3), 315-321 (2010).

[50] P. R. Ohodnicki, J. Long, D. E. Laughlin, M. E. McHenry, V. Keylin, and J. Huth, "Composition dependence of field induced anisotropy in ferromagnetic (Co,Fe)89Zr7B4 and (Co,Fe)88Zr7B4Cu1 amorphous and nanocrystalline ribbons", Journal of Applied Physics 104 (11), 113909 (2008).

[51] S. Fujii, Y. Yoshizawa, D. H. Ping, M. Ohnuma, and K. Hono, Materials Science and Engineering A 375-377, 207 (2004).

[52] M. Daniil, P. R. Ohodnicki, M. E. McHenry, and M. A. Willard, "Shear band formation and fracture behavior of nanocrystalline (Co,Fe)-based alloys", Philosophical Magazine 90 (12), 1547-1565 (2010).

[53] S. Kernion, K. J. Miller, S. Shen, V. Keylin, J. Huth, and M. E. McHenry, "High Induction, Low Loss FeCo-Based Nanocomposite Alloys with Reduced Metalloid Content", IEEE Transactions on Magnetics 47 (10), 3452-3455 (2011).

[54] P. Sharma, X. Zhang, Y. Zhang, and A. Makino, "Competition driven nanocrystallization in high Bs and low coreloss Fe-Si-B-P-Cu soft magnetic alloys", Scripta Materialia 95, 3-6 (2015).

[55] Silveyra, P. Xu, V. Keylin, V. DeGeorge, A. Leary, "Amorphous and Nanocomposite Materials for Energy-efficient Electric Motors." Accepted for the Journal of Electronic Materials, 2015.

[56] J. M. Silveyra, A. M. Leary, V. DeGeorge, S. Simizu, M. E. McHenry, "High Speed Electric Motors Based on High Performance Novel Soft Magnets." Journal of Applied Physics, 17A319-21, 2014, http://dx.doi.org/10.1063/1.4864247.

© VDE VERLAG GMBH · Berlin · Offenbach

PCIM Europe 2016, 10 – 12 May 2016, Nuremberg, Germany

Comparison of Unipolar Silicon Carbide Power Transistors Used in High Switching Frequency Inverter Topologies

S. Fahlbusch, N. Sahli, S. Klötzer, U. Müter, B. Schäning, K. F. Hoffmann
Helmut Schmidt University, Faculty of Electrical Engineering, Department of Power Electronics
Email: Sebastian.Fahlbusch@hsu-hh.de

Abstract

Unipolar silicon carbide (SiC) power transistors are a promising alternative in power electronic applications at dc-link voltages above 600 V, in which IGBTs are predominantly used. SiC enables a significant increase of switching frequencies in such applications. The objective of this paper is to analyse potentials of commercially available unipolar SiC power transistors and their suitability for high frequency operation. For this purpose, SiC-MOSFETs and SiC-JFETs are investigated in a hard switching inverter half-bridge topology.

1. Introduction

In comparison to silicon, SiC offers a combination of higher blocking capability, higher maximum switching frequency due to smaller parasitic capacitances and lower on-resistance as well as the potential for high temperature operations. These properties enable unipolar devices with their fast switching performances to be applied in topologies with dc-link voltages in which silicon IGBTs are typically used. Therefore, SiC has become a promising alternative to conventional silicon solutions and allows significant increase of switching frequency for several power electronic applications.

However, conventionally used packaging technologies prevent that SiC's thermal advantages are fully exploited [1]. Thus, minimization of switching energy losses becomes the key element when utilizing silicon carbide power transistors in high frequency applications. The potential of SiC-MOSFETs for high frequency operation in a half-bridge topology has already been analysed [2]. SiC-JFETs with comparable on-resistance and blocking voltage are commercially available. This investigation focuses on the high switching frequency potential of SiC-JFETs and particularly their third quadrant switching performance in comparison to SiC-MOSFETs.

2. Power devices

In the present analysis, four TO-247 packaged normally-on SiC-JFETs (JFET1 to 4) with blocking capability of 1200 V are analysed and compared to a SiC-MOSFET (MOS1), which has shown the best performance in a previous analysis [2]. Fundamental parameters of all investigated devices are summarized in table 1.The maximum junction temperature $T_{j,max}$ of all JFETs is 175 °C while MOS1 is specified up to 150 °C. JFET1 and 2 are devices with lateral channel structures and intrinsic body diodes whereas JFET3 and 4 have vertical channels (VJFET) without body diodes [3]. Simplified equivalent circuits and structures of the used JFETs are depicted in figure 1. Unlike MOSFETs with an isolated gate electrode, JFETs show bipolar pn-junction like characteristics at the gate. These are represented as gate-drain and gate-source diodes. If the gate-source voltage v_{gs} exceeds the diode's forward voltage, a significant current will flow into the gate and increase driver losses. At sufficient negative v_{gs} the gate-source-junction enters reverse breakdown with a corresponding current flow across the junction. Both effects limit the operational gate-source voltage range. The specified static gate-source voltage limits of all JFETs can be exceeded temporarily in order of nanoseconds. In contrast to lateral JFETs, vertical JFETs (VJFETs) include no body diodes. It can be shown that these devices also possess reverse conduction capability due to forced channel turn-on which can be used during dead time [4]. In forward direction, SiC-JFETs are conducting if the applied controlling voltage drop v_{gs} exceeds the threshold voltage v_{th} at positive v_{ds}.

© VDE VERLAG GMBH · Berlin · Offenbach

1528

PCIM Europe 2016, 10 – 12 May 2016, Nuremberg, Germany

Table 1: Comparison of the fundamental parameters of analysed SiC-JFETs and MOSFET

Device	$R_{\text{ds(on)}}$ mΩ	I_{d} A	V_{gs} V	V_{th} V	Q_{g} nC	Q_{gs} nC	Q_{gd} nC	$R_{\text{th,jc}}$ K/W
JFET1	100	26	-19.5/+2	-15.7	72	16	32	0.78
JFET2	70	35	-19.5/+2	-15.7	92	21	51	0.63
JFET3	55	38	-20/+3	-10	107	10	74	0.65
JFET4	95	21	-20/+3	-10	62	6	42	1.10
MOS1	98	36	-10/+25	1.8	49.2	10.8	18	0.65

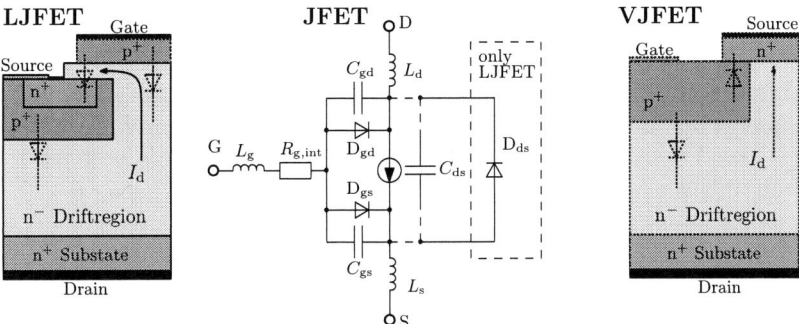

Figure 1: Simplified lateral and vertical SiC-JFET structures and corresponding equivalent circuits

Due to the symmetrical device structure, a similar dependency can be described with v_{gd} as controlling voltage for reverse conduction [5]. Analogous to the previous case, reverse conduction occurs if v_{gd} exceeds the threshold voltage v_{th} in combination with a positive value of v_{sd}. This can occur in bridge configurations when a load inductance forces current in reverse direction through an off-state VJFET ($v_{\text{gs}} < v_{\text{th}}$). The inductive load increases v_{sd} until the sum of v_{sd} and the negative applied v_{gs} exceeds $v_{\text{gd}} = v_{\text{th}}$. The channel is forced to open which leads to reverse conduction. The channel behaves like a diode with a threshold voltage depending on the applied negative bias across v_{gs}. This reverse conduction mode can be used as a freewheeling path which allows diode-less inverter operation [4].

3. Preliminary considerations

The SiC transistors are analysed in a commonly used half-bridge inverter with R-L-load and sinusoidal pulse width modulation (SPWM) at a dc-link voltage of 600 V. The sinusoidal output current is $I_{\text{L,rms}} = 10$ A. Therefore, varying currents of approximately -14 A to 14 A will flow through the devices and stress the transistors in forward and reverse direction. Additionally, turn-on and turn-off switching transients occur in first as well as third quadrant operation. Thus, the bridge configuration allows a detailed analysis of the devices because it includes a wide range of stresses. The objective of this analysis is to investigate the suitability of SiC-JFETs for high switching frequency operation. It can be assumed that at high switching frequencies f_{s} the switching losses P_{s} will significantly exceed the conduction losses P_{c} and dominate the total power losses P_{t} (see equation (1)). In order to achieve high switching frequencies, a good thermal conductivity is required to avoid exceedance of maximum junction temperatures of the devices. The total thermal resistance depends on the thermal resistance from junction to case, the cooling system and the thermal interface materials. In the experimental setup, the same cooling units and thermal interface materials are used for all tested devices.

$$P_{\text{t}} = P_{\text{c}} + P_{\text{s}} = P_{\text{c}} + f_{\text{s}} \cdot E_{\text{s}} = \frac{T_{\text{j}} - T_{\text{a}}}{R_{\text{th,total}}} \Rightarrow f_{\text{s,max}} = \frac{1}{E_{\text{s}}} \cdot \left(\frac{T_{\text{j,max}} - T_{\text{A,max}}}{R_{\text{th,total}}} - P_{\text{c}} \right) \qquad (1)$$

© VDE VERLAG GMBH · Berlin · Offenbach

All devices are housed in TO-247 packages. Thus, neither the maximum junction temperature $T_{\mathrm{j,max}}$ nor the total thermal resistance $R_{\mathrm{th,jc}}$ can be significantly influenced. At constant ambient temperature T_{a} the only remaining factor to increase the switching frequency is to decrease the switching energy losses E_{s}. Therefore, an optimized gate impedance to achieve low switching energy loss E_{s} with data sheet compliant v_{gs} in first quadrant is required. This leads to very fast switching transients with high $\mathrm{d}v_{\mathrm{ds}}/\mathrm{d}t$ and $\mathrm{d}i/\mathrm{d}t$ slew rates. The risk of parasitic effects like miller-induced turn-on or induced voltages caused by common-source inductance will significantly grow with steeper slew rates. An optimization of the switching energy losses E_{s} may lead to unsafe switching performance which has to be countered.

During turn-off, a high positive $\mathrm{d}v_{\mathrm{ds}}/\mathrm{d}t$ slew rate leads to a quick charging of the device's capacitances (see figure 2a). The charging current through C_{gd} will divide into one part which flows out of the gate back into the driver. The second part flows through C_{gs} and charges the capacitor. This results in unwanted rise of v_{gs}. If an insufficient amount of charge is bypassed through the gate driver, the gate-source capacitor C_{gs} may be charged beyond v_{th} which leads to unwanted turn-on of the device. This may result in short circuit and potentially the destruction of devices, especially in bridge topologies. During turn-on (negative $\mathrm{d}v_{\mathrm{ds}}/\mathrm{d}t$), the same effect leads to a shift to negative v_{gs} which tends to turn-off the device again. This scenario is considered to be less critical. The susceptibility of power devices for Miller induced turn-on is indicated by the Miller charge ratio $Q_{\mathrm{gd}}/Q_{\mathrm{gs}}$ which is shown in table 2 for all analysed devices. A $Q_{\mathrm{gd}}/Q_{\mathrm{gs}}$-ratio of less than one will guarantee theoretical $\mathrm{d}v_{\mathrm{ds}}/\mathrm{d}t$ immunity. As represented in table 2, all devices possess a miller charge ratio above one (especially JFET3 and JFET4). Therefore, none of the analysed SiC transistors possess $\mathrm{d}v_{\mathrm{ds}}/\mathrm{d}t$ immunity.

Table 2: Comparison of the Miller charge ratios and gate off to threshold voltage margins

Device	Miller charge Q_{gd}/ nC	Gate charge Q_{gs}/ nC	Miller charge ratio $Q_{\mathrm{gd}}/Q_{\mathrm{gs}}$	Max Off-margin $\lvert v_{\mathrm{th}} - v_{\mathrm{gs,min}} \rvert$ / V
JFET1	32	16	2	3.8
JFET2	51	21	2.43	3.8
JFET3	74	10	7.4	10
JFET4	42	6	7	10
MOS1	18	10.8	1.67	11.2

In order to counter this effect, sufficient amount of Q_{gd} has to be bypassed through the gate driver path. The bypass capability of the driver path depends on the effective gate impedance consisting of the internal gate resistance $R_{\mathrm{g,int}}$, external gate resistance $R_{\mathrm{g,ext}}$, driver sink resistance R_{sink} as well as the internal and external parasitic gate inductance L_{g}. Assuming that the internal gate resistance, driver sink resistance and gate inductance cannot be significantly influenced, the effective gate impedance can only be adjusted by decreasing the external $R_{\mathrm{g,ext}}$ for turn-off resulting in lower damping and higher risk of oscillations. Another method is to provide an additional bypass path. This can be realized by adding an external bypass capacitor $C_{\mathrm{gs,ext}}$ between the gate and source terminals of the device. Additionally, a suitable Zener diode can be utilized to protect against reverse breakdown. Both measures will reduce slew rates and hence increase switching energy and driver losses. The effects of these countermeasures strongly depend on the internal gate resistance and inductances.

Another parasitic effect is caused by high $\mathrm{d}i/\mathrm{d}t$ slew rates in combination with the common-source inductance. If a rising current flows in reverse direction through an off-state device (e.g. third quadrant turn-on), a positive step voltage across the common-source inductance L_{s} will be induced. The result is a voltage drop in opposite, negative direction at v_{gs} (see figure 2b). In combination with insufficient damping in the gate path, the gate-source voltage resonates. If the oscillation at v_{gs} exceeds the threshold voltage, unintended turn-on and shoot-through may

© VDE VERLAG GMBH · Berlin · Offenbach

occur. In order to counter this effect, the common-source inductance should to be reduced. This approach is not applicable because all analysed devices are housed in TO-247 packages which restrict significant influence on L_s. Another method is to sufficiently increase the damping of the gate impedance during turn-off. This can be realized by increasing the external gate resistance $R_\mathrm{g,ext}$. However, increasing $R_\mathrm{g,ext}$ will result in negative impact on the $\mathrm{d}v_\mathrm{ds}/\mathrm{d}t$ immunity. Alternatively, $\mathrm{d}i/\mathrm{d}t$-immunity can be increased by adding an additional freewheeling path. A SiC-Schottky diode in parallel to the device can bypass a significant amount of the reverse flowing current which reduces the internal $\mathrm{d}i/\mathrm{d}t$ resulting in lower feedback to v_gs. The additional Schottky diode will increase the total output capacitance and thus will reduce slew rates of the switching transients resulting in higher total switching energy losses.

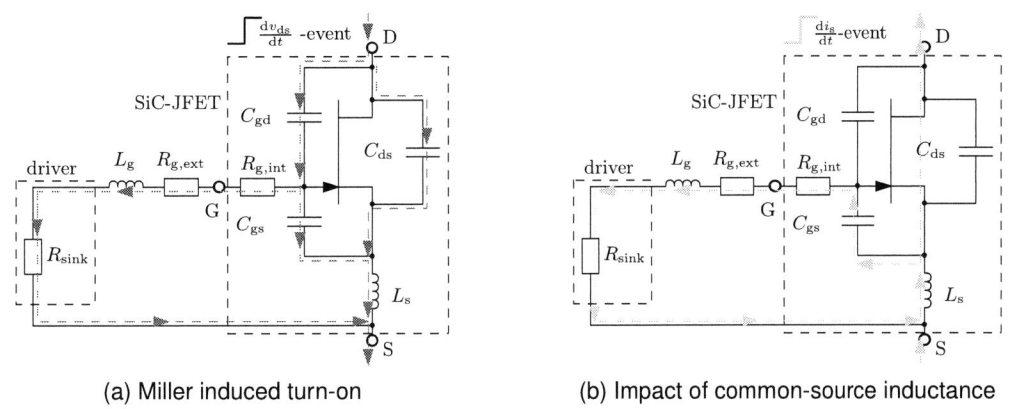

(a) Miller induced turn-on (b) Impact of common-source inductance

Figure 2: Parasitic effects during switching transients

A general measure to enhance immunity against parasitic turn-on is to maximize the device's gaps between applied negative gate-off bias $v_\mathrm{gs,min}$ and the threshold voltage v_th (off-margin). For Si devices it is a simple method because of their high $|v_\mathrm{gs}\text{-}v_\mathrm{th}|$ margin. This option is hardly applicable for SiC devices. As summarized in table 1 and depicted in figure 3, the JFET's negative gate-source dc biases are limited by reverse breakdown. In combination with their threshold voltages close to the lower gate-source dc limit (especially for JFET1 and 2), the maximum off-margins are smaller in comparison to MOS1 (see table 2). Hence, significant gate-off bias adjustments are limited. Their smaller off-margins and larger miller charge ratios indicate that all JFETs show higher risk of parasitic turn-on than MOS1.

Another approach is to utilize SiC-JFETs in a cascode configuration to build a virtual normally-off device [6]. In a cascode configuration, JFETs are indirectly controlled by driving the cascode's low voltage MOSFET. Their higher operational $v_\mathrm{gs,mos}$ range simplifies adjustments to increase safe operation. However, a cascode configuration requires a specific and more complex PCB layout resulting in higher parasitic inductances. In order to avoid unwanted oscillations higher damping is required resulting in lower switching speeds and higher switching energy losses.

The presented methods are used to guarantee safe operation within data sheet ratings without unwanted turn-on during the intended off-state. First, an external resistor is applied to provide enough damping at the gate to avoid under- or overshoot caused by charging or discharging the gate. Second, the impact on v_gs after forced turn-on or turn-off caused by active switching of the opposite transistor is analysed. Depending on the occurring parasitic effects, additional Zener diodes, external capacitors and external freewheeling diodes are added in order to reduce gate feedback and to achieve low switching energy losses. All limitations for JFETs which have to be taken into account are depicted in figure 3.

© VDE VERLAG GMBH · Berlin · Offenbach

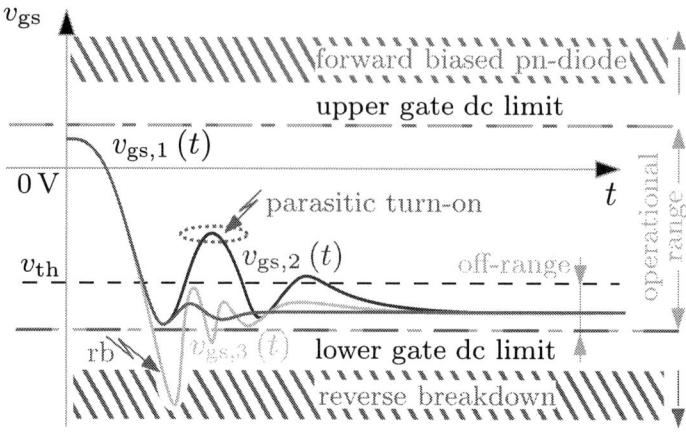

Figure 3: Exemplarily shown v_{gs} turn-off processes and gate limitations of the analysed SiC-JFETs

4. Experimental setups and results

A half-bridge is used in a clamped inductive switching configuration to analyse the device's switching performance in the first and third quadrant. The maximum achievable switching frequencies are determined in the same half-bridge inverter with R-L-load. This test bench operates with a sinusoidal pulse width modulation (SPWM), a dc-link voltage of 600 V, a load current of 10 A_{rms} and water cooling.

4.1. Results of measured first and third quadrant switching performances

The best individual switching performance with highest efficiency is summarized in table 3 and illustrated in figure 4a. JFET1 and 2 have lower switching energy losses than JFET3 and 4 but higher than MOS1. The v_{gs} impact on the opposite SiC-transistor caused by fast transients (forced turn-on/-off) is shown in figure 4b. Although the devices are in off-state at $v_{gs,low} \approx -19\,\text{V}$, significant parasitic v_{gs} spikes ($v_{gs,low}$ in figure 4b) temporarily exceed the individual threshold voltage v_{th} of each JFET. Additionally, the devices' v_{gs} reach reverse breakdown (see also parasitic v_{gs} impact in table 3). MOS1 is also affected but the impact on v_{gs} is still within data sheet ratings and considered to be uncritical. Due to the parasitic effects caused by common-source inductance and insufficient bypass of Miller charge, unsafe operation of the JFETs are probable. Therefore, the JFET's optimized gate parameters for low switching energy loss E_s obtained from first quadrant optimization are not applicable in the half-bridge test bench. In order to avoid short circuit during SPWM tests, the investigated JFETs require additional countermeasures to increase $\mathrm{d}v_{ds}/\mathrm{d}t$ and $\mathrm{d}i/\mathrm{d}t$ resistance resulting in lower switching speeds, efficiency and higher dead times in comparison to MOS1.

Table 3: Switching energy losses for 1st, 3rd quadrant optimization and cascode configuration

| Device | Optimized for 1st quadrant | | | Optimized for 3rd quadrant | | | Cascode |
| | E_S / µJ | parasitic v_{gs} impact | | E_S / µJ | parasitic v_{gs} impact | | E_S / µJ |
		$v_{gs,min/max}$ / V			$v_{gs,min/max}$ / V		
JFET1	159	-38	-2.8	544	-21	-16	686
JFET2	192	-33	-12	556	-19.5	-15	723
JFET3	183	-35	3.8	390	-19.5	-11	482
JFET4	225	-33	5	602	-18.5	-10.5	655
MOS1	132	-7	0.5	132	-7	0.5	n.a.

The switching performances and energies after 3rd quadrant optimization are shown and summarized in figure 5 and table 3. Additional bypass capacitors, freewheeling diodes and partially Zener diodes are utilized to reduce impact on v_{gs} (see figure 5b). Due to the parallel freewheeling diodes, commutation starts within the device but after a short time the source current i_s is redirected to the freewheeling diode. The source current changes from the freewheeling diode path to the device's channel in reverse direction again after the device is turned-on ($v_{gs} > v_{th}$).

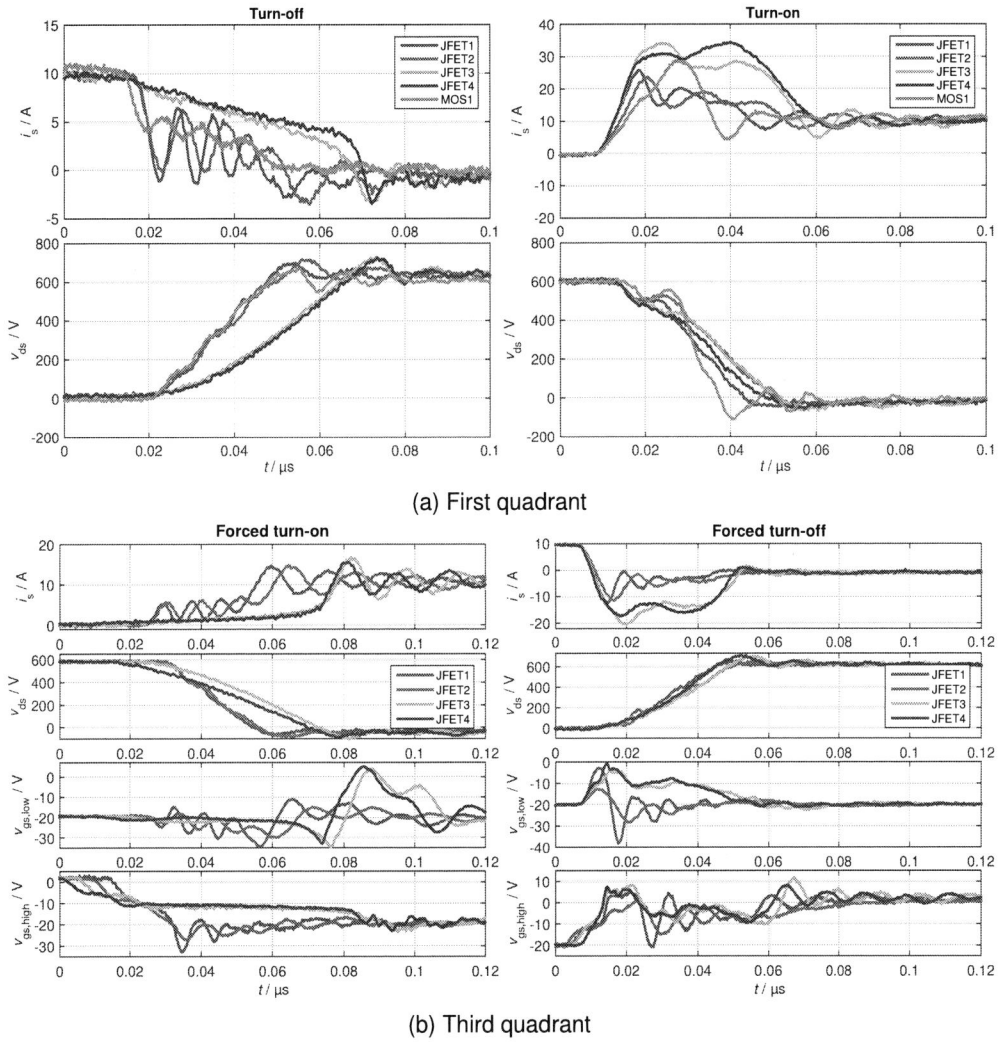

(a) First quadrant

(b) Third quadrant

Figure 4: First and third quadrant switching performance optimized for 1st quadrant operation

The measures to reduce parasitic impact on v_{gs} lead to lower $\mathrm{d}v_{ds}/\mathrm{d}t$ and $\mathrm{d}i/\mathrm{d}t$ transients and higher switching energy losses E_s. Additionally, higher dead times from 300 ns up to 500 ns have to be used, while a 50 ns dead time for MOS1 is still appropriate. Instead of driving JFETs in half-bridge configurations directly, another approach is to apply the devices in a cascode configuration. As shown in table 3, the achieved cascode switching energy losses from first quadrant optimization cannot compete with the results of the directly driven JFET after third quadrant optimization. Thus, this approach is not applicable to achieve highest possible switching frequencies.

PCIM Europe 2016, 10 – 12 May 2016, Nuremberg, Germany

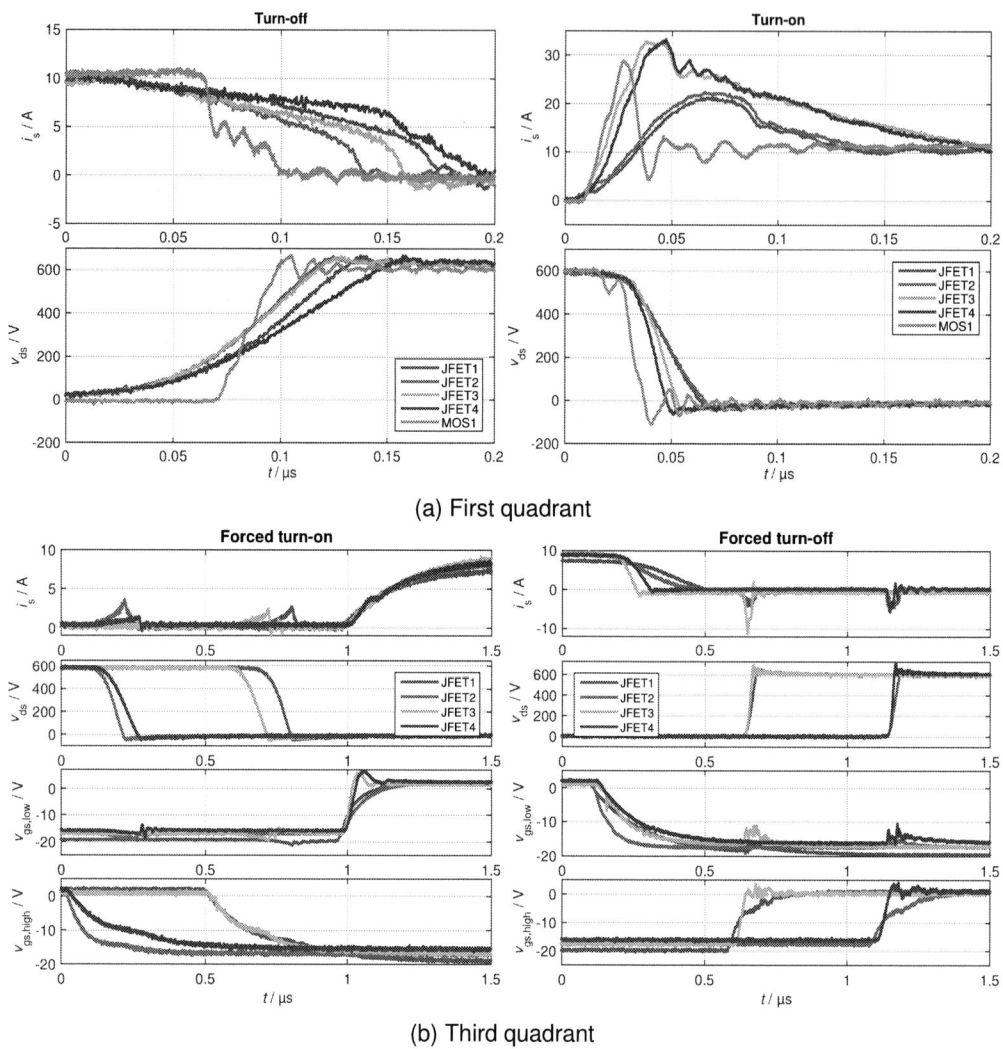

(a) First quadrant

(b) Third quadrant

Figure 5: First and third quadrant switching performance after 3rd quadrant optimization

4.2. Results of the maximum switching frequency determination

In order to achieve the individual maximum switching frequency, the JFET's and MOSFET's parameters with lowest E_s and suppressed impact on v_{gs} are chosen (third quadrant optimization). While MOS1 can operate safely with uncritical feedback on v_{gs} at its thermal limit at stable SPWM switching frequency of 1.6 MHz (see figure 6), the JFETs have to be operated at much lower maximum switching frequencies. Due to their high required dead times, the SPWM-based duty-cycle variations are massively restricted at higher frequency operation. At increasing switching frequencies, the modulated duty-cycles lead to decreasing turn-on/-off times comparable to the length of the dead times. At their maximum frequency limit, the JFETs are not able to charge/discharge their gate fast enough which leads to skipped turn-on/-off pulses. The achievable maximum switching frequencies of the four tested SiC-JFETs vary from 300 kHz (JFET1,4), 400 kHz (JFET2) to 550 kHz (JFET3) depending on the necessary measures to achieve safe operation. The long dead times prevent that the JFETs can be operated up to their thermal limit in this scenario. These results show that JFETs have high switching frequency potentials (see low switching energy losses E_s in table 3) but cannot be fully exploited in half-bridge configurations.

© VDE VERLAG GMBH · Berlin · Offenbach

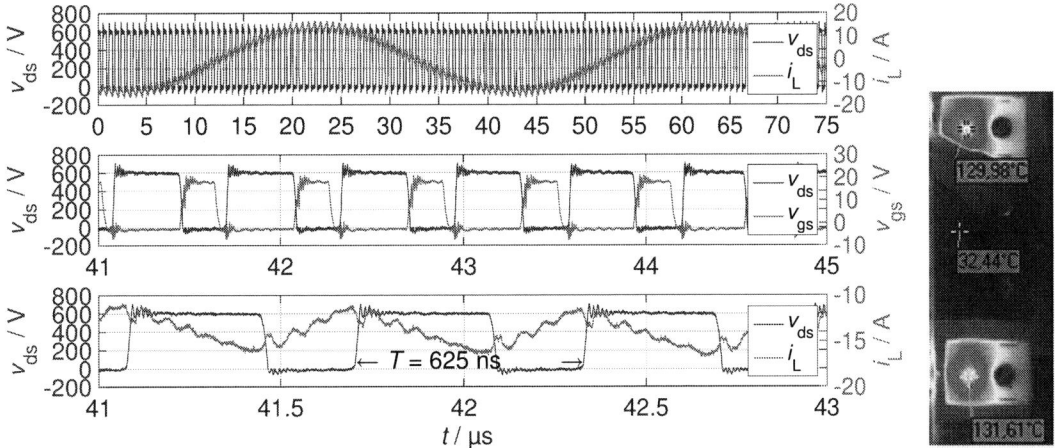

Figure 6: Stable 1.6 MHz SPWM operation of the tested SiC-MOSFET and case temperatures

5. Summary

In this analysis high switching frequency potentials of TO-247 packaged SiC-JFETs and a MOSFET are investigated at a dc-link voltage of 600 V and a load current of 10 A_{rms}. Directly driven JFETs can achieve low switching energy losses comparable to SiC-MOSFETs. When applied in a half-bridge, JFETs are much more affected by parasitic effects caused by common-source inductance and miller-induced turn-on than MOSFETs. In order to guarantee safe operation, countermeasures are required which drastically decrease the achievable switching frequencies of the tested JFET's between 300 kHz to 550 kHz. In contrast, the SiC-MOSFET can be operated safely and stable at 1.6 MHz switching frequency. It is only limited by conduction, switching and thermal performance. TO-247 packaged JFETs, however, cannot exploit their full potential in bridge configurations due to measures necessary to avoid parasitic effects.

6. References

[1] J. Lutz. Packaging and Reliability of Power Modules. In *Integrated Power Systems (CIPS), 2014 8th International Conference on*, pages 1–8, Feb 2014.

[2] S. Fahlbusch, U. Müter, and K.F. Hoffmann. Analysis of SiC-MOSFETs utilised in Hard Switching Inverter Topologies with Switching Frequencies up to 1 MHz. In *PCIM Europe 2015; International Exhibition and Conference for Power Electronics, Intelligent Motion, Renewable Energy and Energy Management; Proceedings of*, pages 1–8, May 2015.

[3] R. Siemieniec and U. Kirchner. The 1200V Direct-Driven SiC JFET power switch. In *Power Electronics and Applications (EPE 2011), Proceedings of the 2011-14th European Conference on*, pages 1–10, Aug 2011.

[4] R. Ouaida et al. SiC Vertical JFET Pure Diode-Less Inverter Leg. In *Applied Power Electronics Conference and Exposition (APEC), 2013 Twenty-Eighth Annual IEEE*, pages 512–517, March 2013.

[5] G. Kampitsis, P. Stefas, N. Chrysogelos, S. Papathanassiou, and S. Manias. Assessment of the Reverse Operational Characteristics of SiC JFETs in a Diode-Less Inverter. In *Industrial Electronics Society, IECON 2013 - 39th Annual Conference of the IEEE*, pages 477–482, Nov 2013.

[6] B.J. Baliga. *Fundamentals of Power Semiconductor Devices*. Springer Publishing Company, Incorporated, 2010.

© VDE VERLAG GMBH · Berlin · Offenbach

PCIM Europe 2016, 10 – 12 May 2016, Nuremberg, Germany

ST SiC MOSFETs in 1MHZ DC-DC converter

Luigi Abbatelli, STMicroelectronics, Stradale Primosole, 50, 95121, Catania, Italy
luigi.abbatelli@st.com

Giuseppe Catalisano, STMicroelectronics, Stradale Primosole, 50, 95121, Catania, Italy,
giuseppe.catalisano@st.com

Abstract

Wide-bandgap semiconductors are semiconductor materials that permit devices to operate at much higher voltages, frequencies and temperature than conventional semiconductor materials. Hence WBG devices allow more power conversion solution to be built which are cheaper and more energy efficient. Today power electronics experts know that 1200V SiC MOSFET can replace 1200V silicon IGBTs allowing frequency range enlargement up to dozens of kHz. What will happen in the power conversion field if new SiC MOSFET technologies, featuring extremely low $R_{DS(on)}xQ_g$ Figure-of-Merit, will be released into the market? In the present work the SiC MOSFET potential to afford very-high-frequency power conversion (up to 1MHz) will be investigated by looking at the results of the latest ST SiC devices. The analysis will be carried out also by comparing ST SiC MOSFET different trade-off solutions able to sustain very high switching frequency values.

1. Introduction

The aim of this work consists in deepening the limits of ST SiC MOSFET devices, SCT50N120 and SCT20N120, in very high frequency operation and exploring at the same time the influence of other parameters like the temperature. These characteristics, both high frequency and high temperature, for example are needed for Down-Hole-Oil-Drilling applications where the power converters work at high frequency (in order to optimize the size since one of the main requirements is the power converter to be located as close as possible to the down hole application) and at high temperature due to the higher depth (the average depth of exploratory and development oil wells has reached 5.964km in 2008 and in the same conditions may exceed 180°C).
There are trade-offs to designing high-frequency switching converters. Some of the advantages are smaller size, faster transient response, and smaller voltage over-/undershoots. The main penalties paid for these are reduced efficiency and increased heat
In order to explore the behavior of ST devices at above limit conditions (high frequency and high temperature) ST has performed bench measurements at several values of switching frequency up to 1MHz. According with the current level a single SiC MOSFET or the parallel of two SiC MOSFETs have been tested, so including in the very high frequency investigation the impact of current unbalance, parasitic tolerances, etc..
The test circuit used for the preliminary test is a CCM hard-switched BOOST converter whose output voltage is up to 900V. In order to have a good level of current balance even when the DUTs work in parallel, devices with a small delta on threshold voltage $V_{GS(th)}$ at 100°C have been chosen.
The test conditions are: V_{IN}=400V, V_{OUT}=800V, R_{GON}=R_{GOFF}=2.2 Ohm, V_{GS}=+20V/-2V, I_{L-AVG}= [0-10A], P_{IN} up to 4kW. The block diagram of the test circuit is reported in Fig. 1.

© VDE VERLAG GMBH · Berlin · Offenbach

PCIM Europe 2016, 10 – 12 May 2016, Nuremberg, Germany

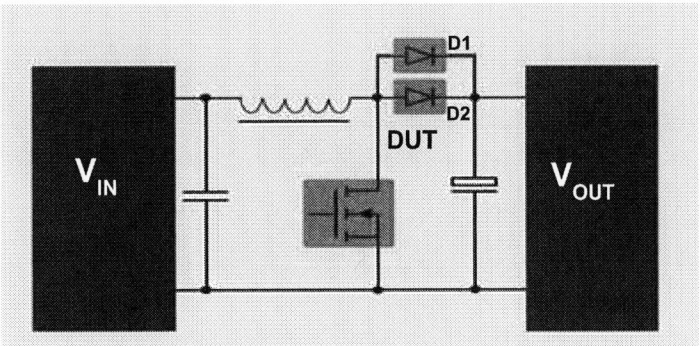

Figure 1: 4kW BOOST converter block diagram

2. A possible trade-off between $R_{DS(on)}$ and gate charge?

Developing a switching power converter for 1MHz operation requires special care in managing parasitic elements, second order effects (like the energy dissipated due to the gate driving) and the risk of unbalance during switching times. Moreover SiC MOSFET is sensitive to Miller turn-on, hence a negative gate bias voltage has been applied between gate and source during off-time. Moreover, before a wide input-voltage DC/DC converter for high-frequency applications is selected, the manufacturer's datasheet should be checked for important specifications such the MOSFET resistance, and the MOSFET switching loss.

The test results performed in a hard-switching BOOST converter confirm the gate charge is the most relevant parameter when the devices are switched up to 1MHz. The ST SiC MOSFET first generation planar structure allow to target very low gate charge values thus enabling ultra high speed switching.

In some cases a trade-off between the MOSFET size and the capability of switching at very high speed must be found. In fact it could happen that a lower static on-voltage drop is not convenient because it leads to higher switching losses which represent the largest portion of the total ones. However the results of the test performed in the 4kW Boost converter showed that ST SiC MOSFET switching losses are almost independent of the chip-size and the on resistance as well. Hence it is possible to achieve the lowest conduction losses without any significant impact on the switching performance. This is mainly a consequence of the excellent SiC figure-of-merit $R_{DS(on)}$xArea that allows to achieve very low $R_{DS(on)}$ with very small die-size, nevertheless it is also due to the optimized layout of ST 1st generation SiC MOSFETS that improves the switching speed also at high current levels. This behavior represents a big advantage for the application designer who is facilitated in dimensioning the SiC solution for very high frequency converters since there's no need to find a trade-off between conduction and switching losses. In fact the losses decrease monotonically with the chip-size, thus the only trade-off to be found is between cost and performance once the target switching frequency has been identified.

The following table report the main electrical parameters of two different $R_{DS(on)}$ level part-numbers belonging to ST SiC MOSFET first generation.

© VDE VERLAG GMBH · Berlin · Offenbach

1537

PCIM Europe 2016, 10 – 12 May 2016, Nuremberg, Germany

1200V SiC MOSFET			
	SCT20N120	SCT30N120	SCT50N120
Current	20 A	45 A	65 A
$R_{DS(on)typ}$	< 169 mΩ	< 80 mΩ	< 55 mΩ
$Q_{g(typ)}$	< 45nC	< 105nC	< 130nC

Figure 2: 1200V 1st Generation ST SiC MOSFE family

In spite of different gate charge values (SCT20N120 has almost 3 times smaller Q_g), the use of an adequate gate driver featuring high current capability allows to maintain the switching losses almost independent of $R_{DS(on)}$. Hence the SCT50N120 offering the lowest on-resistance performs better than SCT20N120 in all the frequency range, as shown in Fig. 3.

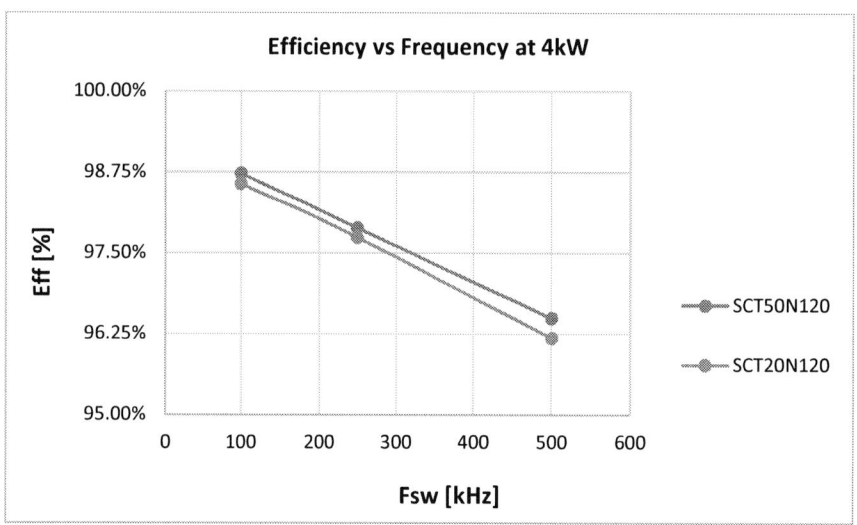

Figure 3: ST SiC MOSFETs efficiency comparison at 4kW, V_{in}=400V, V_{out}=800V,

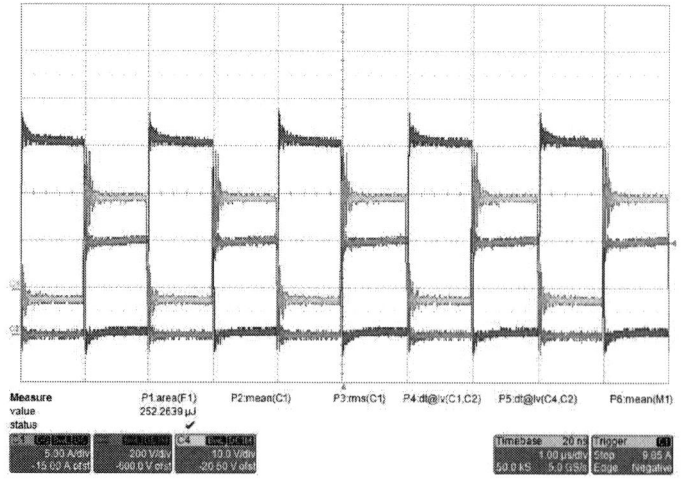

Figure 4: electrical waveforms at 4kW, V_{in}=400V, V_{out}=800V, 500kHz

© VDE VERLAG GMBH · Berlin · Offenbach

3. Test results in CCM BOOST converter up to 1MHz

At very high frequency the conduction losses are much lower than the switching losses. Thus, even if the SCT50N120 switching losses are comparable to those of SCT20N120, the considerations about cost make preferable the latter one since SCT20N120 surely represents a cost-effective solution to afford the 1MHz switching operation.

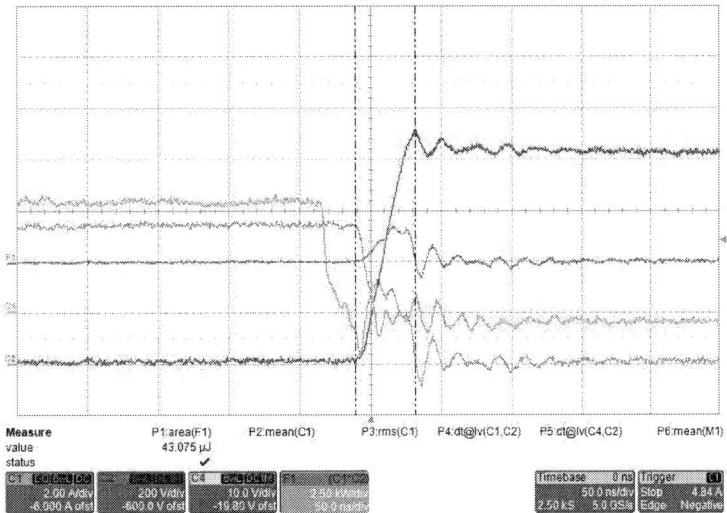

Figure 5: turn-off waveforms at 2kW, V_{in}=400V, V_{out}=800V, 1MHz

The turn-off switching energy must be as low as possible in order to reduce the case temperature and enlarge the frequency range operation. In fact the power losses estimation show that above roughly 250kHz switching losses predominate over the conduction. The weight of switching losses is highlighted in Fig. 5.

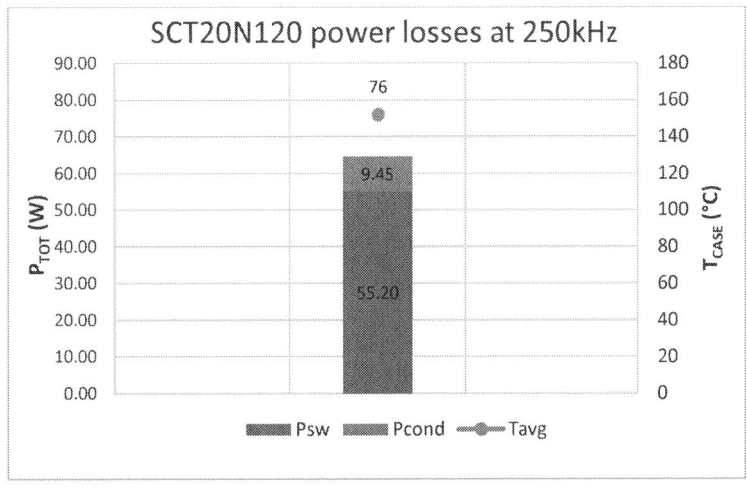

Figure 6: SCT20N120 total losses split at 4kW, V_{in}=400V, V_{out}=800V, 1MHz

4. 1MHz converter based on 1200V SiC MOSFET: achievable efficiency

The target to reduce the switching losses by changing the profile of the electrical waveforms is achievable in few years thanks to the continuous improvement of the SiC MOSFET technology, ST targets to develop trench gate SiC MOSFET within the end of the second decade of 21st century thus achieving unimaginable FOM $R_{DS(on)}$xArea and enabling ultra-fast switching with minimal gate current requirements. The main challenge will consist in managing high power at very high frequency with a very small semiconductor area. The minimization of the parasitic capacitances, of the gate charge and definitively of the total switching energy will be of paramount importance. The second generation of ST SiC MOSFETs represent a step forward in this direction as shown in the next picture.

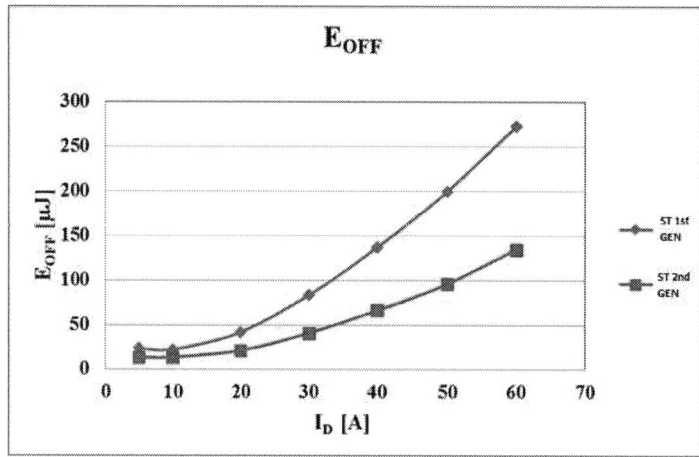

Figure 7: E_{off} losses vs. I_D (V_{DS}=400V, 2.2Ω,20V/-2V)

5. Conclusions and next steps

Finally ST SiC MOSFET shows the potentiality to work at very high frequency thanks to its inherent exceptional switching performance. Moreover the switching losses are almost independent of the die area, thus removing the need to tailor different trade-off for each application/frequency. Basically the losses monotonically decrease with the on-resistance and, once the right $R_{DS(on)}$ has been chosen also considering cost constraints, they monotonically increase with the switching frequency.

The continuous improvement of all the electrical parameters in the ST SiC second generation will make possible to further explore the high frequency operation mode so making closer and closer the possibility to work in switching mode in the several MHz range.

References

[1] "Direct Comparison of Silicon and Silicon Carbide Power Transistors in High-Frequency Hard-Switched Applications" , John S. Glaser, Jeffrey J. Nasadoski, Peter A. Losee, Avinash S. Kashyap, Kevin S. Matocha, Jerome L

[2] "Direct comparison among different technologies in Silicon Carbide" , Bettina Rubino, Michele Macauda, Massimo Nania, Simone Buonomo, PCIM 2012.

[3] "Application Considerations for Silicon Carbide MOSFETs", Bob Callanan, Cree.

Towards a One Nano-Henry Power Module for SiC and GaN

Jacques LAEUFFER, Dtalents, France, jacques.laeuffer@gmail.com

Abstract

Decreased commutation times, especially introducing SiC and GaN, require much reduced inductances inside power modules. This paper proposes an optimal geometry of strip line layout at every step of construction of the power module. For each commutation loop, return conductors are made of copper foils mounted face to face in front of chips boundings. Optimization includes reduction of length, increase of width of conductors, and decrease of insulation layer thickness. For a half bridge module including six paralleled MOS, inductance is estimated around only one Nano-Henry.

Introduction

Strong efforts are done from decades to reduce stray inductances inside fast commutation loops, especially for high current converters. Reducing conductor length is already done.

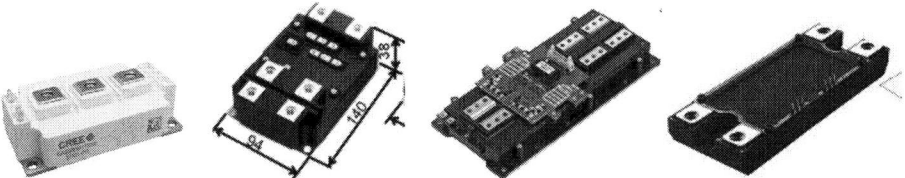

Further improvements require an optimized strip line structure for all the module construction, even including the transistors and diodes chips interconnection by i.e. boundings and DCB.

1. The Wiring Inductance Issue

Figure 1 shows a hard-switching commutation cell includes a transistor, a free-wheeling diode and a DC link capacitor. These three components are series connected. As every loop, this one makes an inductance, and fast switching of the cell makes an overvoltage on the turned-off semiconductor, leading to a voltage derating of the semiconductors, in order not to exceed their maximum voltage rating.

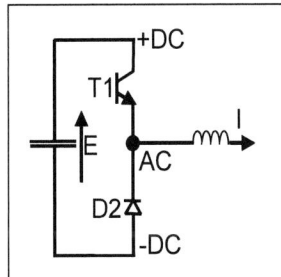

Figure 1: Commutation cell

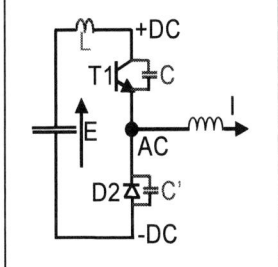

Figure 2: Cell with L-C

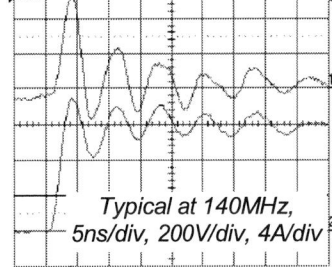

Typical at 140MHz, 5ns/div, 200V/div, 4A/div

Figure 3: Ringing to be avoided

Figure 2 and 3 show loop inductance (L) is able to initiate a very high frequency ringing with stray output capacitances (C) of semiconductors, when commutation time is shorter than the $(LC)^{1/2}$ time constant, leading to further overvoltage, and noisy electromagnetic interferences.

These two issues are worst as commutation times are reduced, for instance introducing faster IGBTs and MOS, and especially SiC and GaN devices. Higher current densities increase also the problem, as the numerator of the (di/dt) ratio is increased, for a same package size. Figure 4 shows silicon diodes turn-off recovery may make huge losses, but on the other hand these losses provide a strong damping of the above mentioned (LC) oscillation.

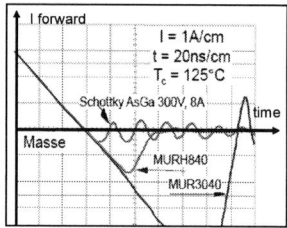

Figure 4: Diodes recovery

Introduction of wide bandgap diodes suppresses this damping, making even more necessary to reduce the (L) wiring inductance, in such a way $(L.C)^{1/2}$ time constant be smaller than the reduced commutation time, in order not to excite this potential ringing, as shown on Figure 5.

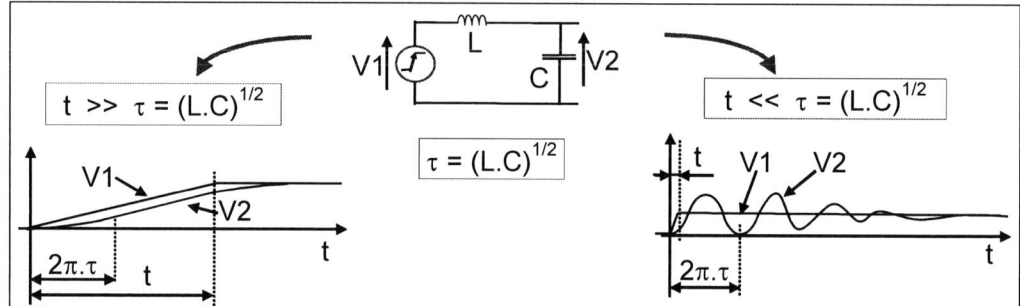

Figure 5: Making or not a ringing

2. What makes the Inductance Value?

Inductance corresponds to a stored magnetic energy ($\frac{1}{2}.L.I^2$). This energy equals ($\iiint H.B.dv$), where (H) is magnetic field and (B=µo.H) is magnetic induction density, linked to the former current (I) and to the loop geometry. Product (H.B) is integrated on the space.

Values of inductance are calculated by transmission lines analysis. As well known, inductance is proportional to the length of the line, leading to requirement of short wiring. Now, inductance is decreased with the distance between way and return conductors, leading to the requirement of close together conductors. Last but not least, inductance is also decreased when the face to face width of way and return conductors is increased. This leads to the optimal geometry of "strip line" shown on Figure 6.

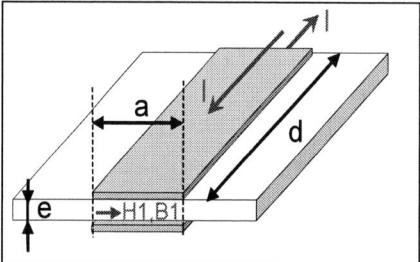

Figure 6: Strip-line design

© VDE VERLAG GMBH · Berlin · Offenbach

A strip line is thus made of two planar and parallel conductors, way and return, with flat rectangular cross section. Face to face width of conductors is (a). Conductors are spaced by an insulation layer of thickness (e). Strip line structure is achieved as soon as width (a) is large compared to thickness (e). Now, inductance depends on magnetic fields. Figure 7 shows magnetic field distribution around conductors, with same amplitudes above and under. (Figure shows less field lines above conductors only to clear space for labels.)

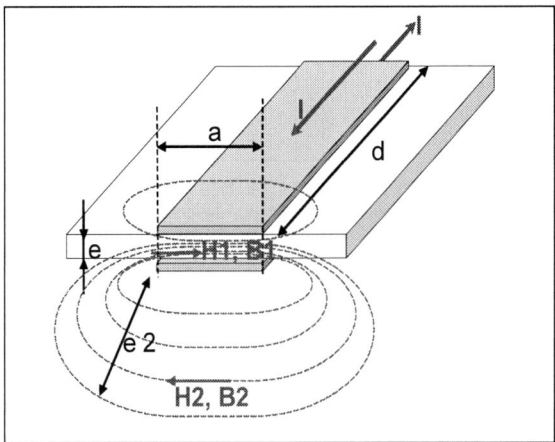

Figure 7: Strip-line field distribution

Flux (φ) conservation means: $\qquad \varphi = \iint_{S1} B1.dS = \iint_{S2} B2.dS$

With   (B1) induction density between conductors, and (B2) induction density outside
     S1 = e . d   and   S2 = (two half spaces outside strip line) . d

Thus   S1 << S2   thus   B1 >> B2

Now   B1 = μo.H1   and   B2 = μo.H2   thus   **H1 >> H2**

Thus, neglecting (H2), magnetic field circulation around a conductor is: $\int$ H.dx = **H1.a = I**

A/ As an impedance, inductance may be calculated according to:

Voltage across is: V = dφ/dt = (e.d) . dB1/dt = (μo.e.d) . dH1/dt = (μo.e.d / a) . dI/dt

Now   V = L . dI/dt   thus   **L = μo.e.d / a**

B/ As an energy coefficient, inductance may also be calculated according to:

Energy stored is: Enm = $\iiint \frac{1}{2}$. H.B . dv = $\iiint \frac{1}{2}$. H.B . dS.dl

Now   H1 >> H2   B1 >> B2   S1 << S2

There are two reason (H1 and B1) to have a large global energy between the conductors, against only one reason (S2) to have a large global energy outside. Thus main part of energy is stored between the conductors, and:

Enm = $\frac{1}{2}$. H1.B1 . e.a.d = $\frac{1}{2}$.e.a.d . μo.H1^2 = $\frac{1}{2}$.(e.a.d.μo / a^2) . I^2

Now   Enm = $\frac{1}{2}$.L.I^2 thus   **L = μo.e.d / a**

As expected, reduced (L) means reduced (d) and (e), and increased (a). Additionally, the much reduced magnetic energy is even concentrated between the two conductive plates, making external space free of EMI, a very important feature to increase frequencies.

3. Implementation inside a Half-Bridge Module

[1], [2] and [3] use this strip line arrangement for power module outputs. [4] and [5] propose to extend the concept in the whole module, using new materials for connection. This paper proposes to apply the strip line arrangement from dies to output terminals, analyzing currents and fields distribution, evaluating inductances values according to defined geometries, using conventional or new wiring technologies.

Conventional boundings connections are made of parallel aluminum round wires, flowed by parallel currents, making a geometry – from a magnetic point of view – close from a conductive foil. Main idea is to provide a return way for current by a real conductive foil, face to face parallel and close to boundings.

Figure 8 shows the electronic schematic of a half-bridge. Note free-wheeling diode for transistor T1 is diode D2, and free-wheeling diode for T2 is D1. Thus loop inductances to be minimized are (E – T1 – D2) on one hand, and (E – T2 – D1) on the other hand.

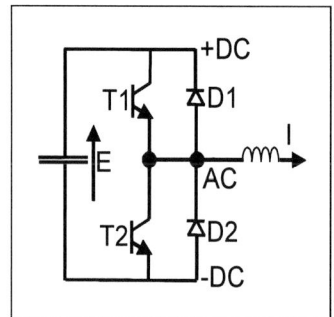

Figure 8: Electronic schematic

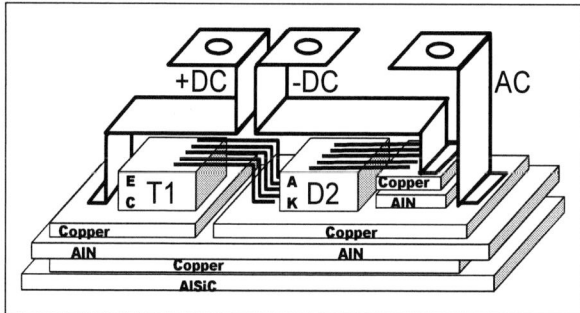

Figure 9: Layout including T1 and D2

Figure 9 shows a perspective view of a possible layout including T1 and D2, where thicknesses of layers and chips are expanded to write labels. Paralleled boundings cover the whole width of chips, and copper foils are as large as chips. Minimizing distance between boundings and copper foils, strip line structure is achieved everywhere inside the module.

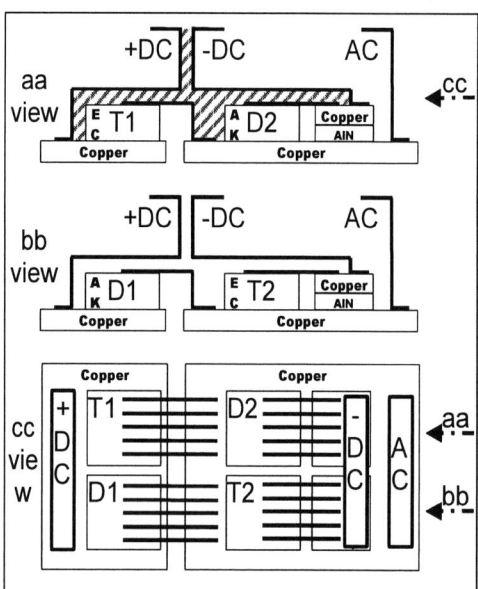

Figure 10: Half-bridge module layout

Figure 10 shows a top view (cc) of the half bridge module, including T1, D2, T2, D1, with two cross sections (aa) and (bb) showing the two commutation loops.

Figure 11 shows a typical electronic circuit inside and outside the module. During commutation, current (Ia) in the load remains constant, due to large load inductance. When T1 is on, current (Ia) encircles both red and blue hatched areas, making a magnetic flux in these two areas. When D2 is on, current (Ia) encircles only blue hatched area.

Figure 12 shows same electronic circuit, with geometrical implementation inside the module. When current (Ia) is reported from T1 to D2, magnetic flux must be removed only from the red hatched area. This paper focuses only on the part of this red flux included in the module.

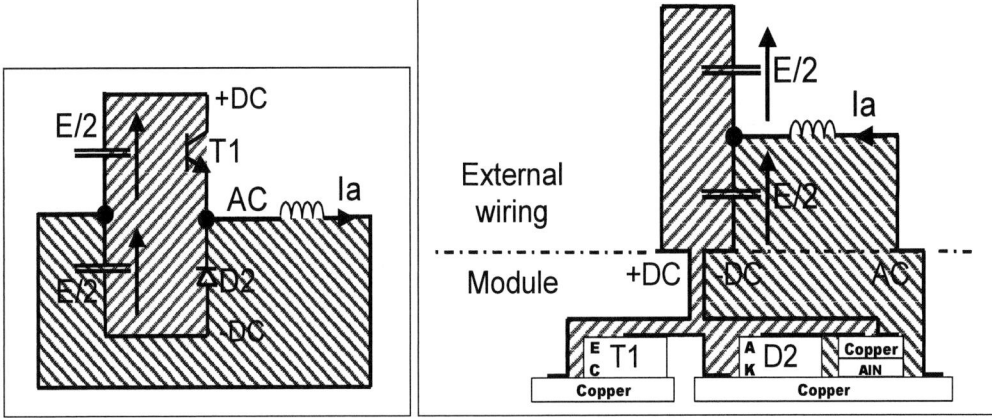

Figure 11: Typical electronic circuit   Figure 12: With module implementation

4. Infinitesimal Propagation inside Insulating Layer

For a more comprehensive understanding of the phenomena, we consider things now from an infinitesimal point of view, a local point of view of what happens in a very small space, during time, showing continuity of magnetic energy movement, that is power.

Figure 13 shows idealized waveforms during transistor T1 turn off and diode D2 turn on. During the (di/dt) process, overvoltage ($\Delta V'$) is applied on transistor, including inductive process inside and outside module. During this process, magnetic energy is removed i.e. from the strip line part encircled on figure 14.

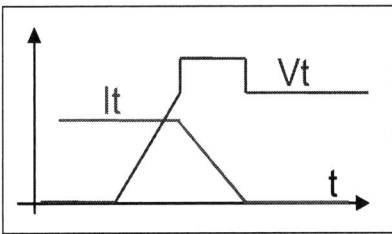

Figure 13: Waveforms

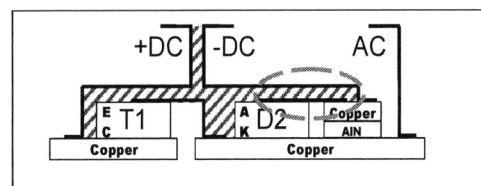

Figure 14: Strip line part

Figure 15 shows this strip line part enhanced. As already discussed: $\int H.dx = \mathbf{H.a = I}$

Magnetic energy to be removed is located between the two conductors, with a volume density ($\frac{1}{2}.\mu o.H^2$). As (H) is uniform in the strip line section, energy stored is:

$Enm = \iiint (\frac{1}{2}.\mu o.H^2).dv$ $\mathbf{Enm = (\frac{1}{2}.\mu o.H^2).e.a.d}$

$(Enm = (\frac{1}{2}.\mu o.I^2/a^2).e.a.d = (\frac{1}{2}.\mu o.e.d/a).I^2 = \frac{1}{2}.L.I^2$ as expected.)

© VDE VERLAG GMBH · Berlin · Offenbach

Electric field (E) at the opening (S) of the strip line is according to:

E.e = ∫ E.dl = dφ/dt   with (φ) magnetic flux encircled: φ = ∬$_{S2}$ B.dS = μo.H.e.d

Thus   E = (1/e).(μo.e.d).(dH/dt) = μo.d.(dH/dt)   **E = μo.d.(dH/dt)**

Now, power crossing the opening (S) of the strip line section is:

P = ∬$_S$ Py.dS   with (Py) the power surface density and (S = e.a)

According to Poynting theorem   Py = E∧H = E.H = μo.d.(dH/dt).H

Now, as (Py) is uniform on (S):   P = μo.d.(dH/dt).H.e.a

Thus   P = μo.e.a.d.H.(dH/dt)   P = e.a.d . d(½.μo.H²)/dt

P = d(e.a.d.½.μo.H²)/dt   **P = d(Enm)/dt**

This verifies that all the magnetic energy (Enm) - stored in the strip line - gets out from it, as power (P), through surface (S) on the left.

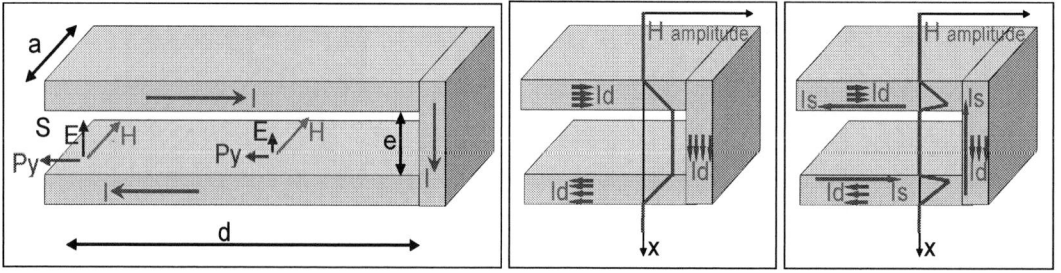

Figure 15: Strip line section   Fig. 16: Before turn off   Fig. 17: After turn off

5. Infinitesimal Propagation inside Copper and Aluminum

Figure 16 shows (H) field amplitude as a function of vertical direction (x), inside both insulations and conductors, before transistor current fall.

Copper foil and aluminum boundings have a thickness of about (1 mm), a similar order of magnitude than insulating layer. Thus, a significant part of magnetic field (H) - so of magnetic energy - is inside copper and aluminum.

If current fall occurs within (100 ns), equivalent frequency is   f = 1 / (2.5 x 100ns) = 4MHz

Skin thickness in copper at (4MHz) is   δ = [2.ρ / (2.π.f.μo)]$^{1/2}$

With   ρ = 1.85.10^{-8} copper resistivity   f = 4MHz   thus   **δ = 34 μm**

Similar very small skin thickness occurs also in aluminum, compared to metal and insulation thicknesses. Skin thicknesses becomes even smaller with faster commutation.

This means, during transistor turn off time, current is globally cancelled in the conductor, but the deep current (Id) cannot be cancelled so fast. As magnetic energy is removed through insulating layer between conductors, only magnetic energy located on the inner skin of conductors may be removed during transistor current fall.

Figure 17 shows distribution of (H) field just at the end of transistor current fall. This distribution means a reverse surface current (Is), with huge density, equilibrates the deep current (Id). Both currents will cancel each other's in the following times.

This means also magnetic energy inside conductors is very few changed during transistor current fall, thus do not participate significantly to the process of power flow during this time, so do not participate to overvoltage, and thus must not be included in the "H.F. inductance".

© VDE VERLAG GMBH · Berlin · Offenbach

6. Inductance Optimization for a High Power Module

Figure 18 recalls the half-bridge configuration, and Figure 19 shows (3) paralleled transistors and diodes dies, each of them i.e. 15 mm large. Transistors alternate with diodes to optimize thermal dissipation. Thus global strip line for (T1 – D2) loop includes a width of about (a = 45 mm). Length of horizontal part of strip line is about (45 mm) also, and length of vertical part about (15 mm). Thickness of strip line should be reduced down to about (e = 1 mm).

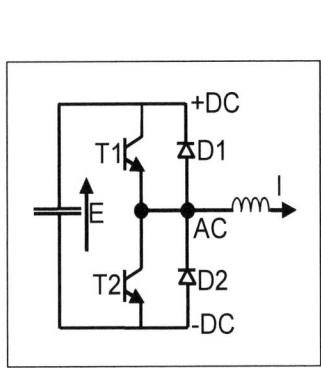

Figure 18: IGBT Half-bridge

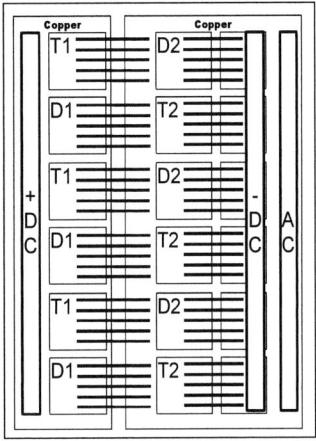

Figure 19: 2nH module

Conventional boundings are bend, thus return copper foil should be pre-bended accordingly before mounting, as shown on Figure 20. Of course, thicknesses of layers and chips are still expanded on figure, to write labels.

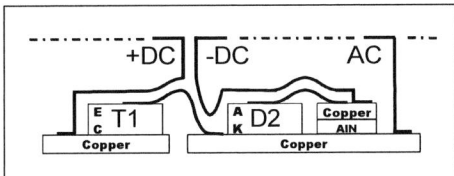

Figure 20: Bend strip line

A foil of insulation material could insure voltage distance. Both insulation foil and copper foil could include holes, to ease gel impregnation, voltage strength being thus achieved by gel.

According to ($L = \mu o.d.e/a$), this would lead to an inductance about:

$$L \approx (4.\pi.10^{-7}).(45mm + 15mm).(1mm)/(45mm) = 1.68 \text{ nH}$$

Figure 21 shows parallel round wires make magnetic lines a bit more expanded than real plan foil, increasing slightly inductance. This should lead to about two Nano-Henry.

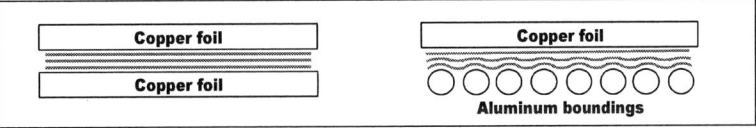

Figure 21: Parallel boundings compared to plan foil

7. SiC and GaN Applications

Furthermore, SiC and GaN MOS are reverse conducting devices. In this case, T1 and D1 are a same chip M1, and T2 and D2 are a same chip M2. Figure 22 recalls the MOS half bridge configuration, and Figure 23 shows six paralleled MOS, thus width (a) of commutation loop is multiplied by 2, thus inductance is divided by 2, and could become only one Nano-Henry. Note this applies also to smaller chips, as both directions (d) and (a) are reduced together.

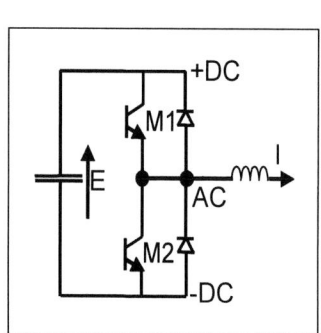

Figure 22: MOS Half-bridge

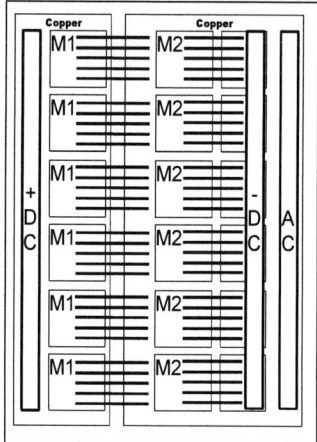

Figure 23: 1nH module

Conclusion

Strip line structure – applied for all the module internal layout – leads to a much reduced inductance. Parallel aluminum round boundings make a geometry – from a magnetic point of view – close from a conductive foil. Main idea is to provide a return way for current by a real conductive foil, face to face parallel and close to boundings.

Currently, typical overvoltage due to module inductance alone maybe about:

$\Delta V = L.di/dt = (20nH).(3\times200A)/100ns = 120V$.

In the future, it could be typically:

$\Delta V = L.di/dt = (2nH).(3\times400A)/20ns = 120V$,

With much decreased inductance, increased current density, reduced commutation time, and similar induced voltage (ΔV). Note this (ΔV) does not include any busbars and DC capacitor inductance.

Additionally, this much reduced magnetic energy is even concentrated between the two conductive plates, making external space free of EMI, a very important feature to increase frequencies.

Reference

[1] KAWASE D. et al., "High voltage module with low internal inductance for next chip generation…", Hitachi Power Semiconductor Device, Ltd, Japan, PCIM Europe 2015.

[2] BORGHOFF G., "Implementation of low inductive strip line concept for symmetric switching in a new high power module", Infineon Technologies AG, Germany, PCIM Europe 2013.

[3] SCHNELL R. et al., "Linpack, a new low inductive phase-leg IGBT module with easy paralleling for high power density converter design", ABB Ltd, Switzerland, PCIM Europe 2015.

[4] BECKEDAHL P. et al., "Breakthrough into the third dimension - Sintered multi-layer flex for ultra-low inductance power modules", Integrated Power Systems (CIPS), 2014 8th International Conference.

[5] SCHEUERMANN U., "Packaging and Reliability of Power Modules – Principles, Achievements and Future Challenges", Semikron Elektronic GmbH & Co, KG, Germany, PCIM Europe 2015.

PCIM Europe 2016, 10 – 12 May 2016, Nuremberg, Germany

Scalable SiC Cascode Power Blocks

Jonathan Dodge, P.E., United Silicon Carbide, Inc., USA, jdodge@unitedsic.com

Matt O'Grady, United Silicon Carbide, Inc., USA, mogrady@unitedsic.com

Ke Zhu, United Silicon Carbide, Inc., USA, kzhu@unitedsic.com

Anup Bhalla, United Silicon Carbide, Inc., USA, abhalla@unitedsic.com

Abstract

A design methodology that leverages the performance and reliability of SiC cascodes in TO-247 packages to cost effectively implement high performance, high power converters is described and demonstrated. By paralleling complete converter legs, which include gate drive circuit, DC bus capacitance, and cascodes in TO-247 packages, a converter design that features high speed switching with minimal ringing and well balanced power dissipation can be easily achieved. The design principle is introduced and test results from a hard switched half bridge converter are presented.

1.   Introduction

SiC cascodes have demonstrated exceptional device level performance and reliability [1]. Individual cascodes compatible with standard gate drivers are available in TO-247 packages. These factors make them suitable as drop-in replacements for silicon switches in many applications. For high power applications, where it is necessary to parallel multiple switches, cascodes exhibit the same conduction behavior as silicon MOSFETs; the positive $R_{DS(on)}$ temperature dependence tends to balance conduction losses between devices and helps prevent thermal runaway (Figure 1). Furthermore, cascode delay times and switching losses are highly independent of temperature as shown in Figure 1 where there is a negligible difference in switching characteristics between two parallel cascodes despite an intentional junction temperature difference of 40 °C. These temperature characteristics enable cascodes to be easily paralleled without excessive derating as is required with some other switch technologies.

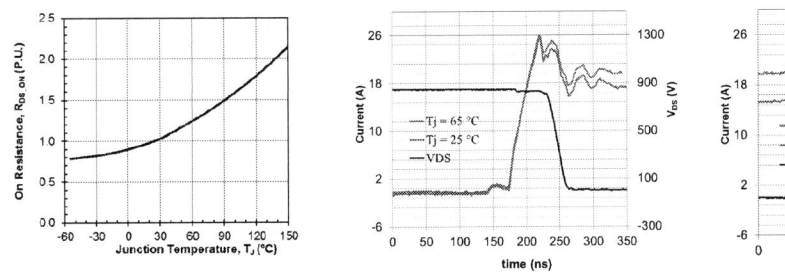
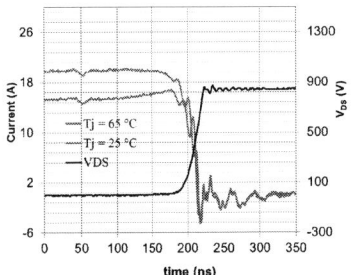

Figure 1. SiC Cascode temperature characteristics. Left) normalized on-resistance versus temperature; Center & Right) turn-on and turn-off of two parallel cascodes with 40 °C junction temperature difference

The main challenge when integrating SiC switches into high power converter designs comes from the increase in switching speed compared to silicon switches. The significantly higher switching speeds make SiC based circuits more susceptible to ringing from stray inductance in both the power and gate loops. When paralleling cascodes, careful attention to both the

© VDE VERLAG GMBH · Berlin · Offenbach

total parasitic inductance and the matching of inductance between parallel paths is needed to achieve maximum performance while minimizing stress on the switching devices and other circuit components. In a conventional module based design, a single capacitor at the module DC terminals and a single gate driver external to the module are used as shown in Figure 2. In this configuration there are parasitic inductances in both the power and gate drive loops that are common to the devices of all legs of the circuit in the module. With high speed switching of SiC this will cause voltage overshoot and ringing.

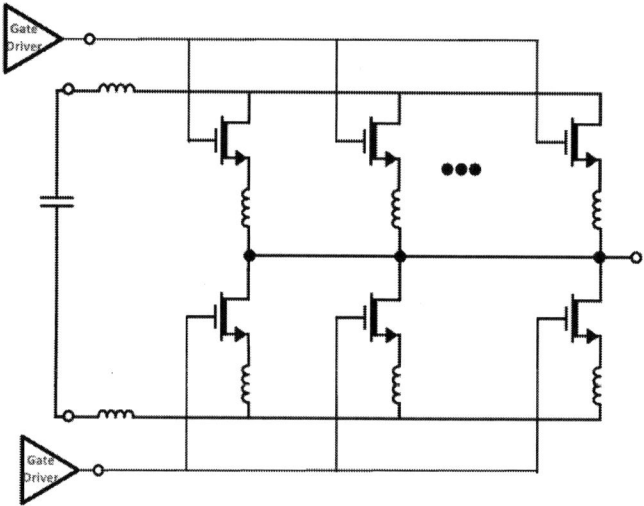

Figure 2. Conventional module based design, including lumped parasitic inductances

This paper presents an alternative design approach that enables implementation of low and matched stray inductance when using paralleled switches. By using TO-247 based switches and standard PCB technologies the methodology can be used to cost effectively implement high power converters. This allows system level performance advantages enabled by SiC devices to be realized without the nonrecurring engineering costs and development time required for custom module development and qualification. The design technique also enables the SiC switches to be operated with minimal voltage derating due to extremely low V_{DS} overshoot during switching.

2. Design Methodology

The guiding principle used to both ease the converter layout design and achieve high performance is to optimize the design of a single leg of the converter, including gate drive circuit, bus capacitance, switches, and bus bar connection; and then array multiple converter legs in parallel so that parasitic system inductances seen by each remains low and balanced. The bus bar itself consists of closely spaced DC bus bars and film capacitors distributed within the converter legs. This paralleled leg approach enables optimizing the parasitic inductance on a single leg, and then matched parasitic inductance and distributed bus capacitance are automatically achieved when converter legs are paralleled. The result is a circuit of the type in Figure 3.

With this design the high frequency switching is largely confined within each converter leg as the impedance of the local bus capacitor is lower than the external DC bus connection. As each converter leg effectively operates independently at high frequencies, the switching behavior does not degrade as additional converter legs are added. From the system point of view, this means that as converter legs are added the di/dt for the power block scales directly with the number of converter legs but the overshoot remains constant. This allows designs to be easily scaled to high power levels without excessive switch derating. This is in contrast to the conventional module design wherein increased module di/dt appears across a common stray inductance leading to increased V_{DS} overshoot.

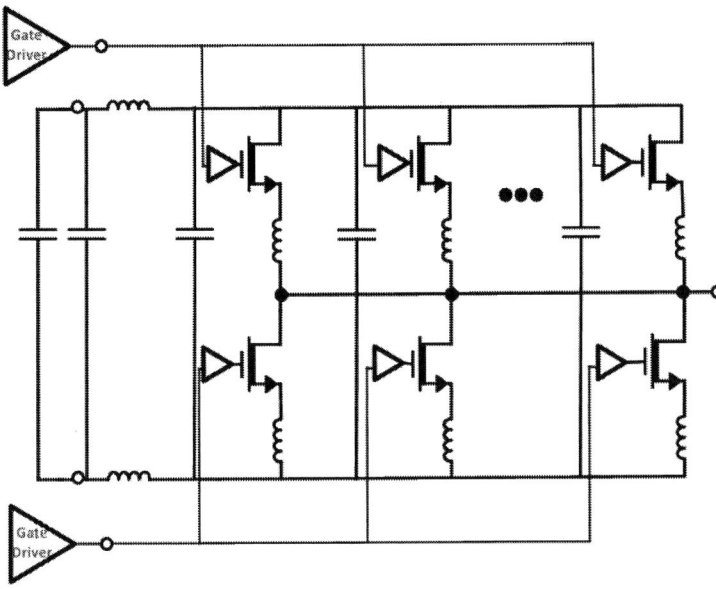

Figure 3. Power block layout with parallel converter legs

The detailed design methodology is as follows:

1) Place the switches and surface mount ceramic bus capacitor(s) in the PCB layout so as to minimize the power loop area. This is accomplished by placing them as close as possible while maintaining creepage distances required in the application. The TO-247 leads may be formed to allow the package body to be attached parallel or perpendicular to the PCB so that the layout inductance can be minimized while providing sufficient space for heatsinking each device. Use of a common heatsink for parallel switches helps leverage the device $R_{DS(on)}$ positive temperature coefficient to counteract any potential on-state current imbalance. The ability to make electrically conductive attachment from the TO-247 to the heatsink can be used to improve the thermal performance over conventional module based designs which contain an insulating layer.

2) Provide connection for a PCB bus bar. This allows the DC bus to be made from low cost PCB laminate while allowing additional metal film DC bus capacitors to be easily distributed along the bus. The close spacing of the DC bus supply and return lines across the PCB thickness (as shown in top of center panel in Figure 4) minimizes the inductance of the DC bus and incorporation of additional capacitors provides a low impedance DC bus connection at high frequencies (bottom of center panel in Figure 4). As the PCB need not contain any fine pitched devices or traces very thick copper can be used to provide low impedance at low frequencies. Additional external copper bus bar can be easily incorporated if needed to further reduce bus bar copper losses.

3) Place a gate drive buffer circuit directly adjacent to each TO-247 on one side of the PCB. On the reverse side of the PCB place copper connected to the source voltage under the gate drive buffer circuit to provide a source kelvin connection. The use of a gate drive buffer including bypass capacitors to the source kelvin connection limits the high frequency gate loop to a compact area which reduces ringing and increases

© VDE VERLAG GMBH · Berlin · Offenbach

electromagnetic compatibility. It also decouples the device switching performance from transmission line effects which can occur when transmitting the PWM input signal across relatively long signal lines.

4) Determine the number of circuit legs needed to meet the current and power requirements of the application. Select a pitch for the circuit legs based on the desired power density and the heat spreading needed to meet the thermal requirements. Array the single converter leg in the layout.

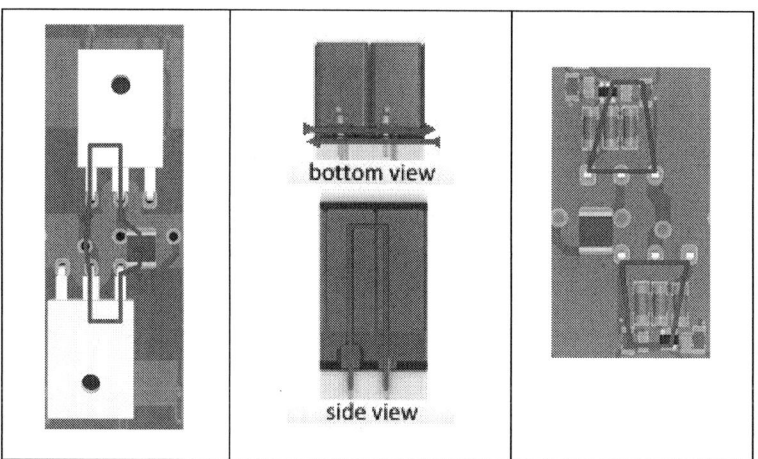

Figure 4. Left) Power loop layout on bottom of PCB; Center) bus bar layout; Right) gate drive loop layout on top of PCB

3. Prototype Construction

Figure 5 shows a half bridge implementation designed and constructed as described above. The switches in TO-247 packages are bent under the PCB to provide heatsinking. The ability to attach the switches to the heatsink without an intervening insulating layer and the ability to accomplish heat spreading through spacing of the legs is used to provide low thermal impedance. The switch node bus bar passes between the heatsinks. The DC bus is inserted perpendicular to the PCB holding the switches. Metal film capacitors are mounted along its length and terminals are provided for DC bus voltage monitoring.

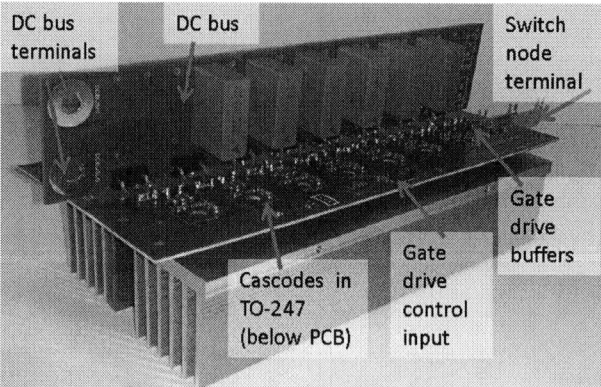

Figure 5. Prototype Half Bridge Power Block

While the half bridge shown here consists of 8 parallel legs of UJC1210K devices, the approach is extremely flexible and the choice of device type and number of devices can be tailored to meet the performance and environmental requirements for an application in a way that optimizes the converter semiconductor cost while incurring very low nonrecurring engineering costs. Also, the entire assembly consists of off-the-shelf components and PCBs fabricated with widely available PCB fabrication technology. The power blocks are easily assembled with automated surface mount assembly followed by through hole device assembly to provide low manufacturing costs.

4.   Demonstration and Experimental Results

A prototype power stage using four parallel UJC1206K based half bridge legs was tested. First, the switching behavior was tested in a double pulse test configuration using passive probes directly at the device terminals. V_{GS} and V_{DS} waveforms for turn-on and turn-off with 80 A load current are shown in Figure 6. The device current is not shown as introduction of an accurate current sensor would increase the loop inductance. The V_{DS} waveform (light blue, 200 V/div) exhibits negligible (< 20 V or 2.5%) overshoot and ringing despite the 40 kV/µs transition rate. The resulting overshoot is substantially less than that seen in both conventional modules and those specifically designed for low inductance [2].

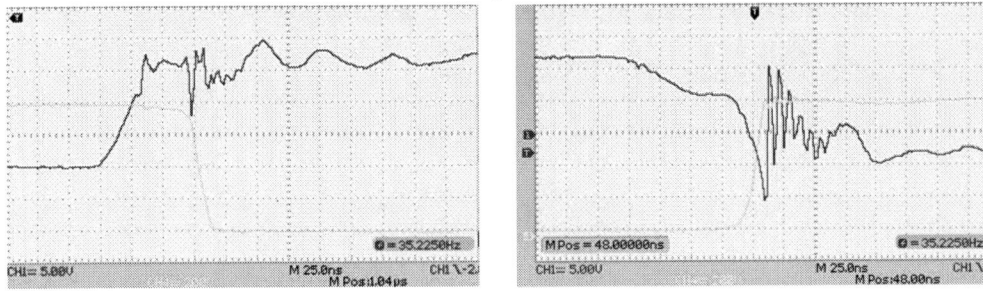

Figure 6. Turn-on (left) and turn-off (right) switching waveforms at 80 A and 800 V for 4 parallel legs at 25 ns/div. Light blue V_{DS}(200 V/div), dark blue V_{GS}(5V/div)

While some gate ringing occurs, it is mostly due to the fast di/dt across the TO-247 source lead inductance which causes V = L di/dt induced ringing that is capacitively coupled to the gate. This ringing does not represent the potential across the internal cascode gate to source junction and it does not stress the gate. With proper gate drive design, including compact gate loop area, the potential for false turn-on or EMC issues due to gate ringing can be mitigated.

A 20 kW, 20 kHz synchronous buck converter was constructed to demonstrate the half bridge under continuous operation. For this test, isolated differential probes were used. The long probe leads add inductance to the measurement loop resulting in increased ringing on the measured signals compared to the double pulse test results above. Despite the additional overshoot introduced by the measurement, the switch node waveform (light blue) still exhibits minimal overshoot and ringing while maintaining the high dv/dt demonstrated in pulse testing.

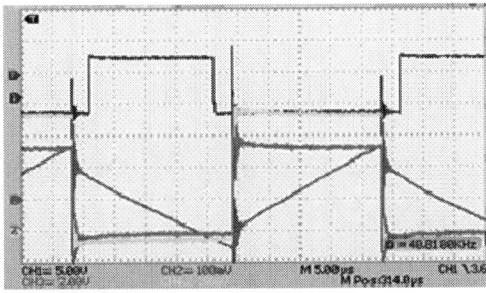

Figure 7. 10 kW buck converter continuous operation (red = inductor current (5 A/div), dark blue = low side V_{GS} (10V/div) , magenta = high side V_{GS} control signal, light blue = V_{DS} (200 V/div)

© VDE VERLAG GMBH · Berlin · Offenbach

5. Summary

A cost effective, high performance design approach for high power converters based on SiC cascodes has been introduced. A half bridge converter was demonstrated to show the combination of fast switching with low voltage overshoot (2.5%) that can be achieved in practical applications. Together with the low conduction loss and standard gate drive compatibility of SiC cascodes the parallel leg design approach enables efficient high power converters to be easily implemented using off-the-shelf TO-247 SiC cascodes. Power converter designs using this approach can be easily scaled to different power levels at low cost by increasing the number of parallel stages. While a half bridge based power block was demonstrated, the approach can be extended to other circuit topologies.

6. References

[1] Li, Xueqing, Bhalla, Anup, Peter Alexandrov, John Hoestetler, Leonid Fursin, "Investigation of SiC Stack and Discrete Cascodes", PCIM Europe 2014

[2] Hatsukawa, Satoshi, Tsuno, Takashi, Toyoshima, Shigenori, Hirakata, Noriyuki "Low inductance full SiC power module", PCIM Europe 2014

PCIM Europe 2016, 10 – 12 May 2016, Nuremberg, Germany

High power density, high efficiency 380v to 52v LLC converter utilizing E-Mode GaN switches.

Moshe Domb .
Senior Application Manager .
Infineon Technologies .
101 N Sepulveda Blvd , El Segundo , CA 90245
moshe.domb@infineon.com

Abstract : A high efficiency(98.3%) , high power density, LLC 380v to 52v dc/dc converter utilizing 600V 0.07ohm **E-M**ode GaN switches will be presented. The **E-M**ode GAN enables high frequency switching frequency, while achieving both **140w/ inch^3 power density &** 98.3% efficiency. The circuit design will be discussed , emphasizing the fast switching property of the E mode GAN which renders both high power density and high efficiency. The LLC circuit utilizing **E-M**ode GaN in the primary will be discussed as well.

1. Introduction

The E-Mode GaN devices are the next generation for high efficiency , high power density enablers devices. A high power, 3000W, telecom dc/dc converter was chosen to evaluate the GaN switching properties. The LLC was used as the best topology for achieving higher efficiency among soft switching topologies. Due to the high switching capability of E-Mode GaN devices there is no need for two LLC stages,90 degree apart, as will be shown in the following.

2. Schematic of LLC 380V to 52V dc/dc , advantages of FB in pri & sec.

Fig. 1 below depicts the 3000w LLC 380V to 52V converter . Among the many variations of the LLC topology the Full Bridge(FB) for **both** primary and secondary, was chosen. This version yields the highest efficiency for high output voltage supplies.

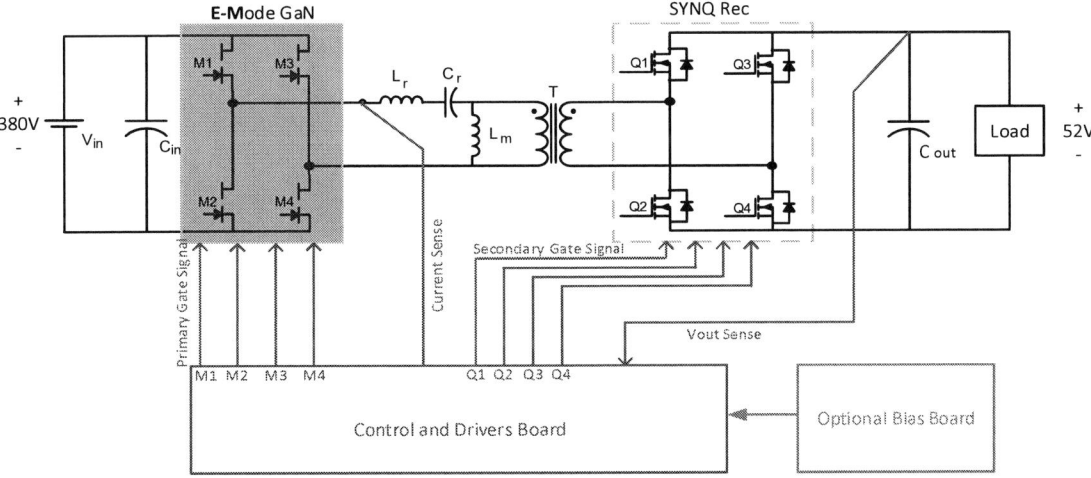

Fig. 1.   Simplified schematic of 3000W LLC 380V to 52v converter, utilizing E-Mode GaN devices

© VDE VERLAG GMBH · Berlin · Offenbach

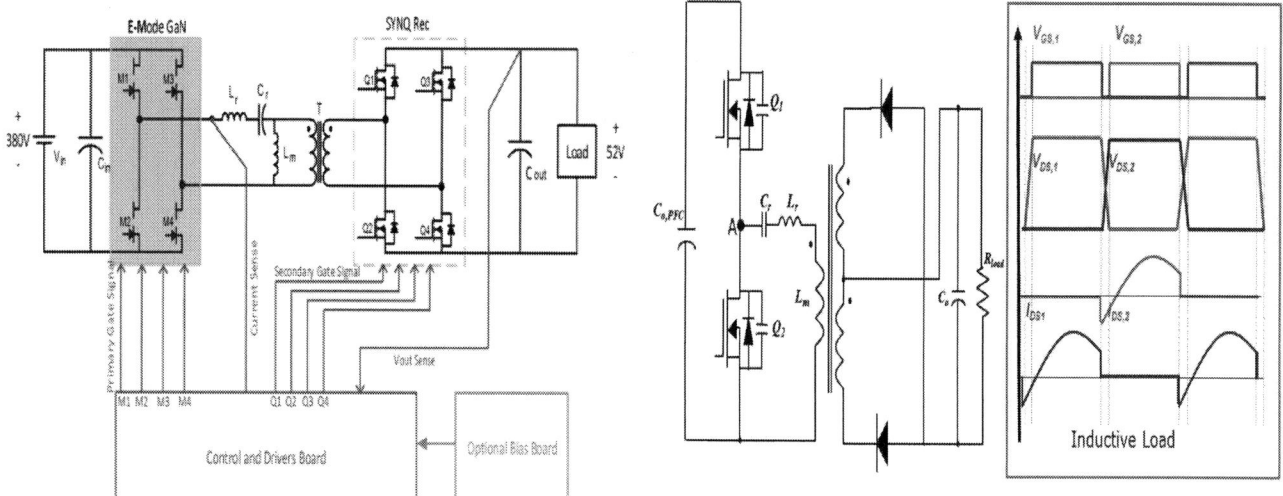

Fig. 2a. LLC with FB in primary and secondary   Fig. 2b. LLC with HB, diodes shown in sec for simplicity

Lets look at : Fig 2a,(repeat of Fig.1), and Fig. 2b,(Half Bridge LLC, with diodes for simplicity in the secondary).   The advantages for LLC with **FB in both primary & sec are**:

1. Reverse voltage across Secondary sync rectifiers is only the output voltage with almost no spikes, as they are clamped to Vout , so a voltage rating is 80v was chosen.

This is not the case for the center tape LLC , where the

voltage rating of the sync rectifier fets is: 2*Vout +Vspike =150V .

2. Primary LLC current is half compared to the center tape LLC , as the turns ratio is 8 to 1 in the FB , compared to 4 to 1 in the HB.
3. In the HB center tape LLC , the transformer is not fully utilized, as only one secondary winding(out of two), is conducting current. Also asymmetry between secondary winding causes mismatch between two half cycles(loss of efficiency).

3. Benefits of multiple Transformers , primary in series and secondary's in parallel

Fig. 3. shows the actual transformers arrangements in the 3000W design. Two transformers , T1 &T2, are used .

The primaries are in series and secondary's are in parallel. Each secondary carries half of the output current. This usage of the multiple transformers has many advantages,

1. Current sharing is automatically happening in the secondary , since primary current are equal.
2. Each one of the secondary's has a local set of 100uf ceramics(Cx &Cy in Fig. 3). No switching current flows from one half to the second. Low output PARD and high current capability are assured. No need for two LLC stages , 90 degrees apart, for low PARD.

This is a modular arrangement, so for instance, 4500W output power, can be implemented with three transformers , primaries in series, secondary's in parallel.

© VDE VERLAG GMBH · Berlin · Offenbach

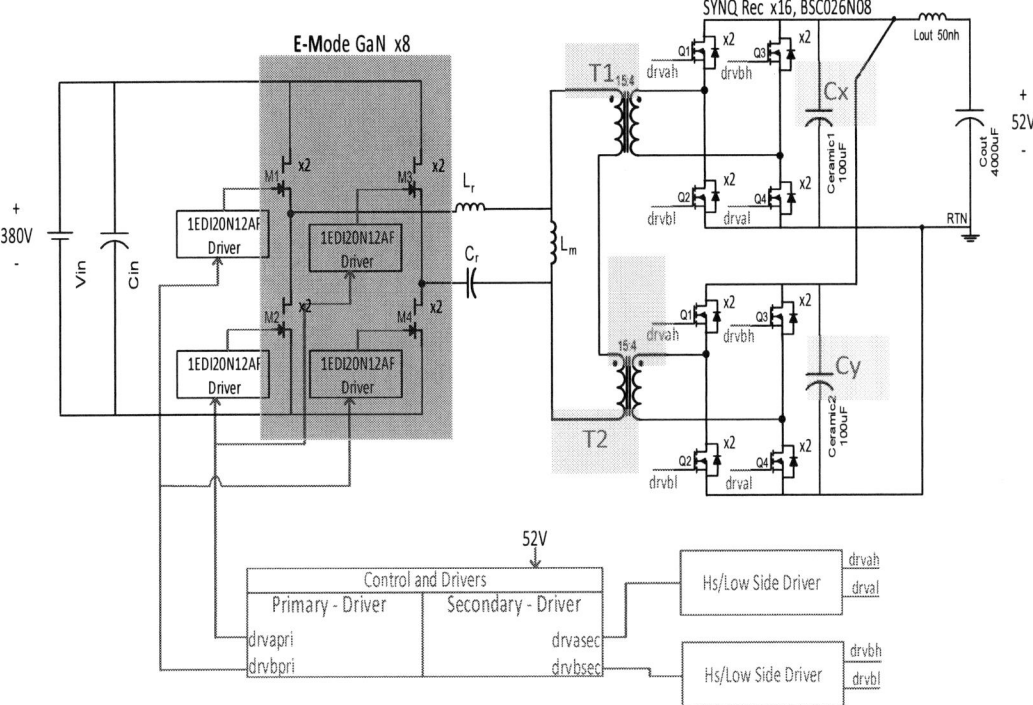

Fig. 3. 3000W LLC with Multiple Transformers , primary in series and secondary's in parallel

4. Efficiency for E-Mode GaN and CoolMos

The 3000W LLC 380V to 52V Eval Brd ,(using E-Mode GaN 600V 0.07ohm), data was taken for Vin = 385V ,Vout = 52V, Iout full load = 58amp. The efficiency data are presented in Table 1.

Iout	350 Vdc	360 Vdc	370 Vdc	385 Vdc	390 Vdc	400 Vdc
	Eff	Eff	Eff	Eff	Eff	Eff
7 amps	94.54%	95.24%	95.56%	96.06%	94.84%	88.50%
15 amps	97.17%	97.38%	97.55%	97.78%	97.0%	94.75%
31 amps	97.78%	97.87%	98.00%	98.30%	97.82%	96.97%
45 amps	97.62%	97.60%	97.80%	98.20%	97.82%	97.31%
58 amps	96.88%	97.10%	97.42%	97.70%	97.70%	96.87%

Table 1. Efficiency vs load and input line

The 3000W LLC was changed to use CoolMos IPB65R110CFD (0.11ohm 650V) .The efficiency curves for both E-Mode GaN and CoolMos are shown in Fig. 4.

© VDE VERLAG GMBH · Berlin · Offenbach

PCIM Europe 2016, 10 – 12 May 2016, Nuremberg, Germany

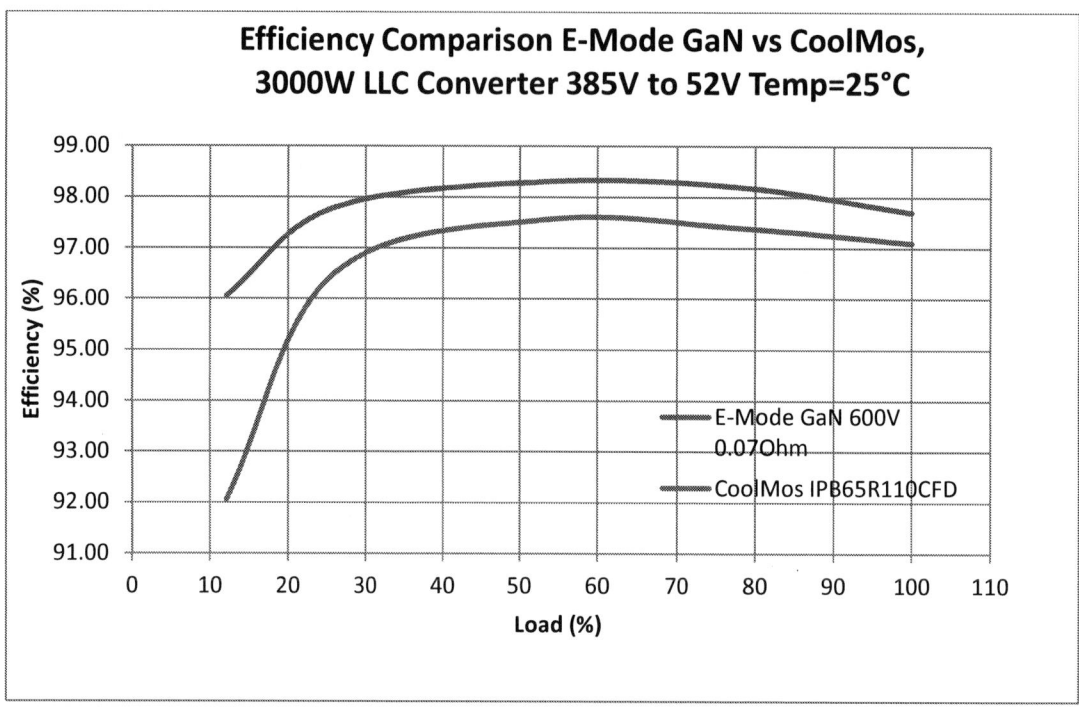

Fig. 4. Efficiency vs load , 3000W LLC, 385V to 52v , Vin =385V

It is worth mentioning that for E-Mode GaN at light load, the efficiency is still >96%, peaking to 98.3% at 50% load. Even at 100% the efficiency is close to 98% !!!. This is well above Titanium standard efficiency. This is **NOT** the case for CoolMos where, for 10% & 20% load ,Titanium standard efficiency can NOT be achieved.

5. Switching waveforms of E-Mode GaN device

Fig. 5. depicts the Vgs and Vds of 600V 0.07ohm E-Mode GaN device. Those are waveforms for top & bottom devices in the primary LLC FB. Only 130ns is allocated for "dead" time when both switches are off. The Vgs switches from -6V (off state) , to 3v(on state) in an extremely fast time,10ns .

On the other hand, looking at the same waveforms for CoolMos ,(600V ,0.11 ohm IPB65R011CFD mosfet), in Fig. 6. , we need to have 350ns "dead time " , and turn on/off of Vgs are 50-60 ns. The switching is much slower,(mostly due to non-linear very large Coss when Vds is lower than 40v). This entails the larger turn off and turn on loss and a lower efficiency as seen in Fig. 4.

Fig. 7. depicts the Ilr(primary resonance inductor current), Vds sec, Vds primary and Vgs primary of a switching cycles at half load(~1500W). Even at a switching frequency of 350khz, the waveforms look very "clean".

© VDE VERLAG GMBH · Berlin · Offenbach

PCIM Europe 2016, 10 – 12 May 2016, Nuremberg, Germany

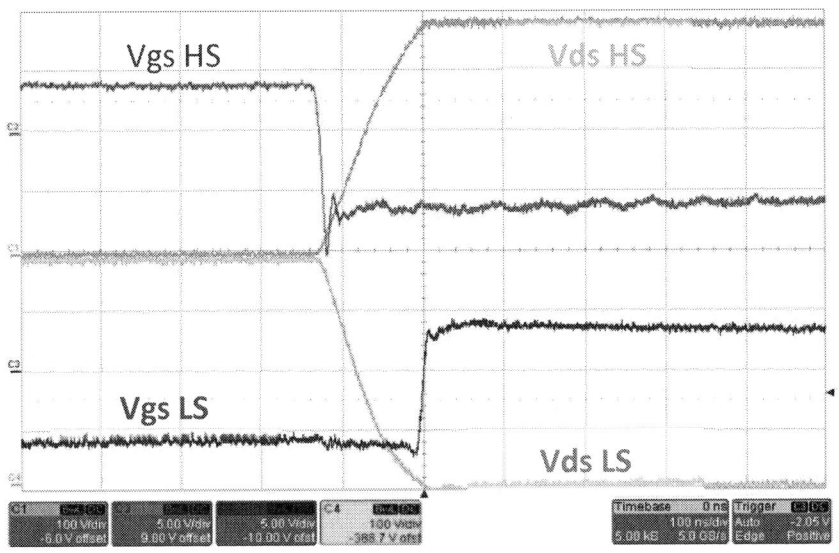

Fig. 5. Switching Waveforms at 50% load ,31amp & 52v, for E-Mode GaN

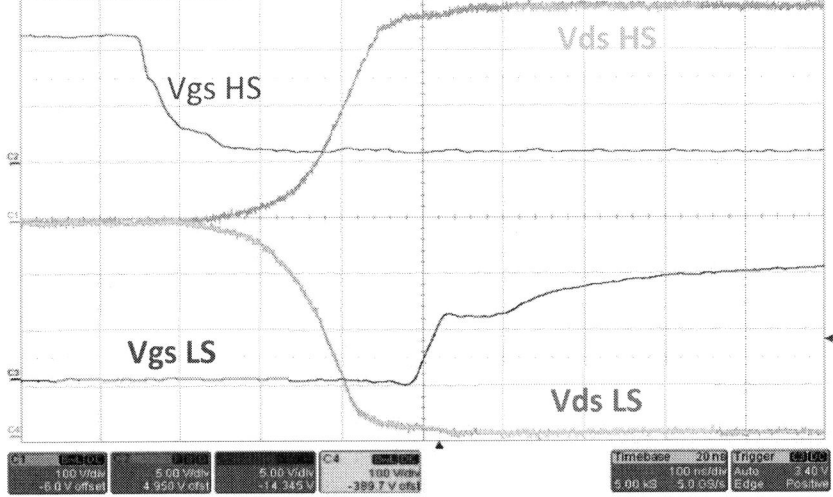

Fig. 6. Switching Waveforms at 50% load ,31amp & 52v, for CoolMos IPB65R110CFD

PCIM Europe 2016, 10 – 12 May 2016, Nuremberg, Germany

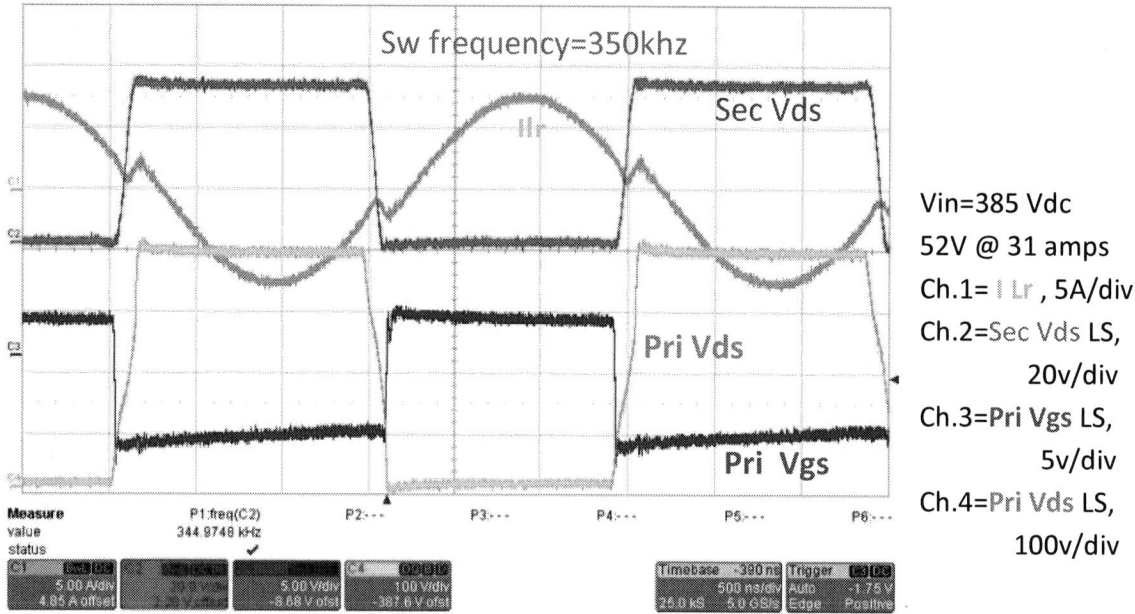

Fig. 7. 3000w LLC Switching waveforms at 52v & 31amp

6. 3000W LLC photo & Power density

The 3000W LLC is shown in Fig. 8 , with it's dimensions .The power density is 140W/inch^3, which is about 2-3 times better compared to CoolMos mosfets designs. The fan is not included the dimensioning.

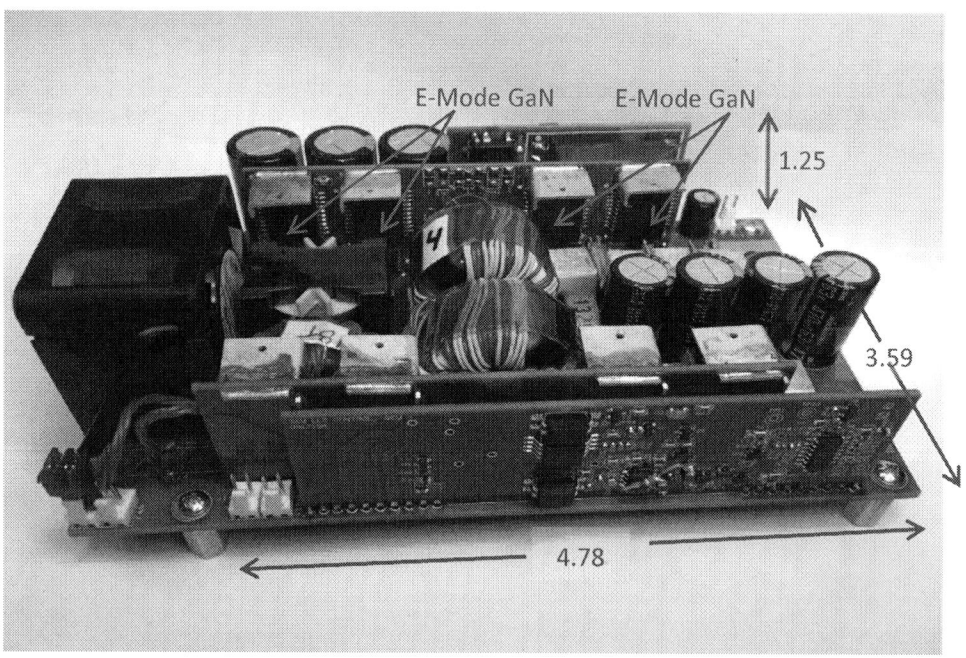

Fig. 8. 3000W LLC converter and dimensions of converter

© VDE VERLAG GMBH · Berlin · Offenbach

7. Conclusions

The E-Mode GaN is an enabler technology that allows pushing switching frequency and power density, while maintaining the same efficiency achieved a lower power density.

This was manifested in the 3000W LLC 380V to 52v dc/dc converter, where a world class efficiency of 98.3% was achieved with a power density of 140W/inch^3. This power density is 2-3 times better compared to Silicon CoolMos fets.

A Low Impedance Drive Circuit to Suppress the Spurious Turn-On in High Speed Wide Band-Gap Semiconductor Halfbridges

Franz, Stubenrauch, Hochschule Rosenheim, Germany, franz.stubenrauch@fh-rosenheim.de
Norbert, Seliger, Hochschule Rosenheim, Germany
Doris, Schmitt-Landsiedel, TUM, Germany

Abstract

A gate drive circuit for gallium nitride (GaN) enhancement mode (e-mode) transistors is presented, which avoids parasitic turn-on of the power devices in the halfbridge configuration. New e-mode GaN devices turn on at very low threshold voltages between 1V and 2V. This makes the transistors highly sensitive to spurious turn-on and thus reduces the required safety margin of the gate drive signals. To avoid this parasitic turn-on, a very low gate loop impedance is required. This prevents the halfbridge against bridge shorts during the switching events and guarantees stable gate drive control with increased switching efficiency. The new gate drive circuit is developed in a SPICE simulation environment and verified in a prototype setup by a double pulse test. The simulation matches very well with the experimental result and demonstrates the suppression of parasitic semiconductor turn-on with the proposed gate drive. Furthermore the dissipated switching energy is reduced, compared to a standard gate drive circuit. High DCDC converter efficiency of 98.67% at 1kW output power is achieved by using the driving circuit for a buck converter prototype with 200kHz switching frequency.

1. Introduction

Due to the development of wide bandgap power semiconductor device technologies like GaN, faster switching speed can be achieved. This leads to smaller switching loss and shorter dead times in the halfbridge configuration, enabling power converters to work at higher switching frequencies. As a further benefit high control bandwidth and high power density is achieved. The maximum performance can be reached by very short switching times, which are limited only by the physical behavior of the power semiconductor. In practical applications the switching speed is additionally limited due to parasitic inductances formed by the electric connections inside the semiconductor package [1] and the Miller capacitance [2] [3]. In this paper the main effects and sources of parasitic semiconductor turn-on of high voltage e-mode GaN semiconductors in halfbridge configuration are analyzed. The analysis shows that the intrinsic Miller capacitances combined with high voltage switching transients are the main cause for the spurious triggering. This problem is only solvable on semiconductor device level or by gate drive optimization, which is done here. The developed low impedance gate drive circuit effectively prevents parasitic turn-on of the halfbridge semiconductors. This is demonstrated by a SPICE simulation and an inverse double pulse prototype setup. As a benefit the switching losses can be reduced. Therefore high efficient power conversion can be achieved, which is validated in a buck converter prototype.

© VDE VERLAG GMBH · Berlin · Offenbach

2. Impact of the Gate Drive Circuit on the Spurious Triggering Pulse

To investigate and verify the parasitic semiconductor turn-on a clamped inductive load switching circuit, which is shown in Fig. 1 is analyzed. With the inverse double pulse circuit a verification of the parasitic turn-on can be performed at the rated values of the power semiconductor devices. The advantage of an inverse double pulse is high bandwidth measurement of all relevant voltages with respect to ground potential. The prototype circuit is shown in Fig. 2. Drain-Source-Voltage overshoots will be suppressed by a low inductive coplanar switching cell layout, consisting of the two semiconductor switches on the bottom side of the PCB and the DC-link ceramic capacitors on the top side of the PCB. A DC voltage V_{DC} of 400V is supplied as specified for the final application. The High-Side-Switch is turned on for $T2_{On1} = 370ns$ to get approximately an inductor current increase of 10A for an inductor value of $L = 15\mu H$. After that the High-Side-Switch is turned off and the inductor current is free-wheeling in the time interval $T2_{Off}$ through the reverse conduction capability or by active turn-on of the Low-Side-Switch T1. In the time interval $T2_{On2}$ the semiconductors can be switched on and off at rated current and voltage values. To prevent the setup from bridge shorts caused by conduction of both switches at the same time, a small dead time of 15ns is chosen between the opposite gate signals. A possible spurious triggering may arise due to the rising edges of the gate signal

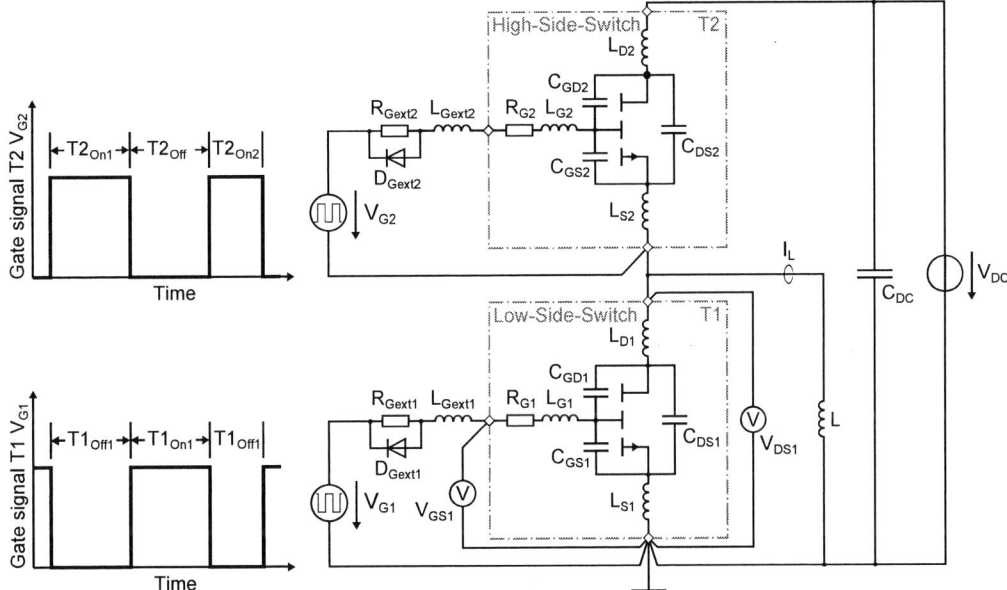

Fig. 1: Schematic of the inverse double pulse circuit with the gate drive control signals.

V_{G2} of the High-Side-Switch T2 and can be observed at the gate signal V_{G1} of the Low-Side-Switch. Due to the small dead time, the Low-Side-Switch is already in the Off state, when this transition occurs. Therefore the gate signal V_{G1} is actively pulled down to ground potential by the gate drive circuit. Ideally the gate source capacitor C_{GS1} is connected in parallel with the low pull down resistance of the driving circuit and fixes the gate potential to ground. This mechanism eliminates the capacitive voltage divider between the Miller capacitor C_{GD1} and the gate capacitor C_{GS1}, which is shorted by the gate drive circuit. Due to the nonzero overall driver impedance, the power semiconductor package and the connections from the driver to the power semiconductor, the circuit operation can be quite different from the ideal behavior.

© VDE VERLAG GMBH · Berlin · Offenbach

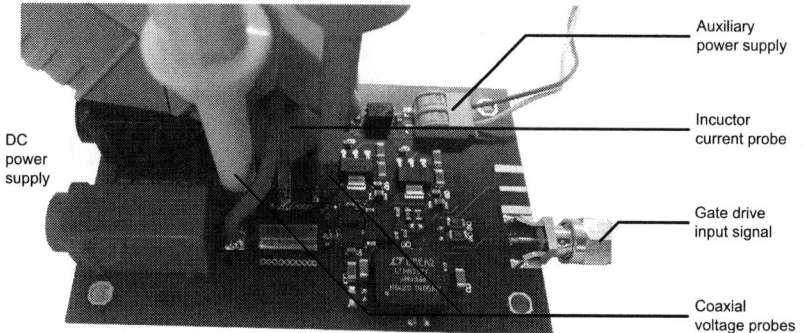

Fig. 2: Setup of the inverse double pulse test.

2.1. Standard Gate Drive Circuit

The verification of a classical gate drive circuit is done by a low resistive driver with 1Ω pull down resistance. To avoid an impact of the common source inductance of the power semiconductor a low impedance package of [4] is chosen for the power semiconductors. Additionally the gate loop connections are separated and placed perpendicular to the power loop. This minimizes the magnetic coupling between these circuit parts. To limit the turn-on speed of the power semiconductors, external gate resistors of $R_{Gext} = 22\Omega$ are used, which protects the circuit against instability caused by the parasitic turn-on. High speed turn-off is achieved by adding a Schottky-Diode in parallel to the external gate resistance depicted in Fig. 1. The resulting SPICE circuit simulation and the measured waveforms are shown in Fig. 3. As a result of the

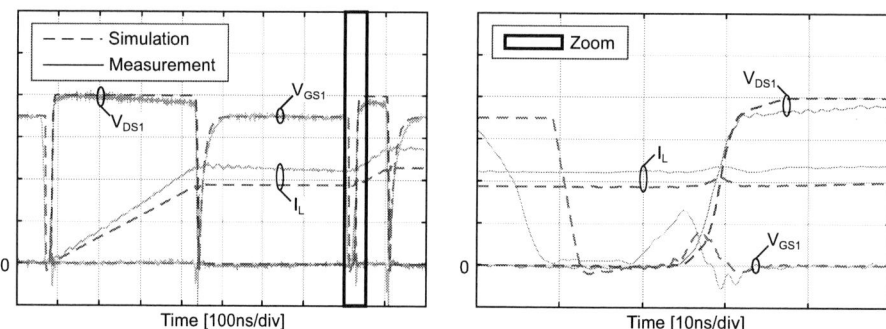

Fig. 3: Standard gate drive circuit waveform at an input voltage $V_{DC} = 400V$ of the inverse double pulse test. The y-axis scalings are $V_{GS1} = 2V/div$, $V_{DS1} = 100V/div$ and $I_L = 5A/div$.

nonzero pull down impedance of the gate loop and drive circuit a gate voltage peak induced by the Miller capacitance C_{GD1} arises at the Low-Side-Switch when the High-Side-Switch is turned on, which causes a drain-source-voltage transient V_{DS1} depicted in Fig. 3 (right). For a short time interval of less than 10ns the Low-Side-Switch turned on again, because the threshold voltage of 1.6V is exceeded. This parasitic turn-on causes a bridge short and adds additional switching losses in both switches. To avoid this mechanism, the turn-on speed of the power semiconductors must be reduced by high values of gate resistors in the order of 20 to 40Ω to guarantee stable operation. To reduce the induced voltage of the Miller capacitance a very low impedance current path ($R < 1\Omega, L < 5nH$) to the gate input is necessary for the Off intervals of both switches.

© VDE VERLAG GMBH · Berlin · Offenbach

2.2. Optimized Low Impedance Gate Drive Circuit

Minimizing the pull down impedance in the time intervals where the switches are in the Off state improves the gate drive circuit for GaN semiconductors. A possible solution is to place an additional low resistive, and low inductive small signal N-Channel MOSFET as near as possible to the gate terminals of the power semiconductor, which is shown in Fig. 4. This circuit reduces the gate pull down impedance ($R < 0.5\Omega, L < 1nH$). To control the additional MOSFET a second driver output is needed with a complementary signal to the primary output. A comparison of both driver circuits is done with the same external resistance of $R_{Gext} = 22\Omega$ and a paralleled Schottky-Diode. But the Schottky-Diode is no longer required, because the N-Channel MOSFET already dominates the very short turn-off time. The resulting waveforms of the new drive circuit in the inverse double pulse test are shown in Fig. 5.

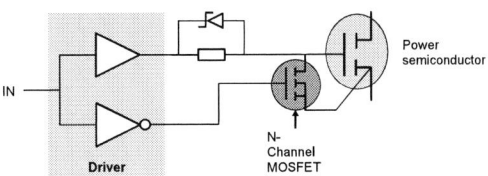

Fig. 4: Schematic of the proposed low impedance gate drive circuit.

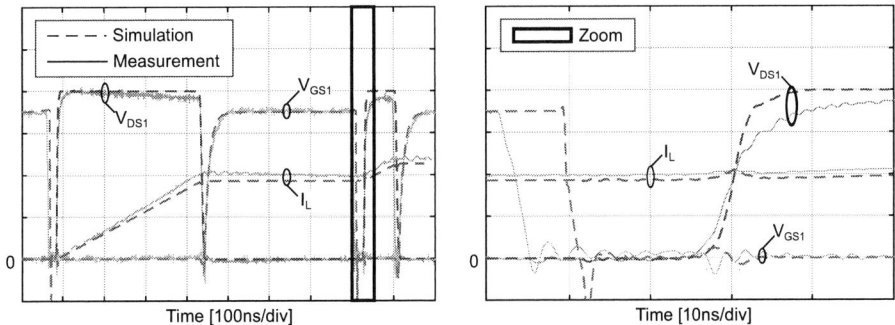

Fig. 5: Low impedance gate drive circuit waveform at an input voltage $V_{DC} = 400V$ of the inverse double pulse test. The y-axis scalings are $V_{GS1} = 2V/div$, $V_{DS1} = 100V/div$ and $I_L = 5A/div$.

Comparing to Fig. 3 it can be seen in Fig. 5 (right) that the minimized pull down impedance of the gate loop and drive circuit prevents the gate against a voltage peak induced by the Miller capacitance C_{GD1} during the switching events.

3. Switching loss comparison

A more complex gate drive circuit like in Fig. 4 is only justifiable, if the resulting circuit benefits are noticeable. The most relevant property is the impact of the driver circuit on the switching loss of the power semiconductor. Faster switching results in higher circuit efficiency, representing a key characteristic in power electronics. The switching energy dissipated during the switching event is therefore analyzed for both gate driving circuits.

Before the High-Side-Switch turns on the drain-source-voltage V_{DS2} equals the blocking voltage of 400V and the load current I_L equals 10A. At High-Side turn-on the load current commutates from the Low-Side to the High-Side-Switch. After the commutation the drain-source-

PCIM Europe 2016, 10 – 12 May 2016, Nuremberg, Germany

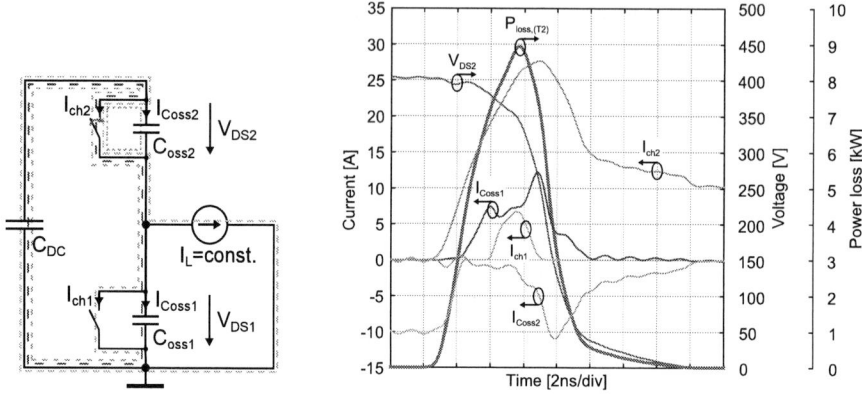

Fig. 6: Modeled turn-on waveform of the High-Side-Switch.

voltage V_{DS2} decreases. Thus the High-Side-Switch channel must carry additionally the displacement current I_{Coss1} of the Low-Side-Switch, which flows through the intrinsic output capacitance C_{oss1} (see Fig. 6 left). Concurrently the High-Side-Switch has to discharge its own intrinsic output capacitance C_{oss2}. If a parasitic turn-on occurs at the Low-Side-Switch a further current will be added to the High-Side-Switch channel I_{ch2}. The turn-on power loss of the High-Side-Switch is the sum of all currents multiplied by the switch drain-source-voltage. A simplified circuit schematic of the turn-on process, the modeled current transients and the calculated power loss are depicted in Fig. 6 left and right side. The dominant part of turn-on power loss $P_{loss,(T2)}$ is generated by the displacement current I_{Coss1} due to the presence of simultaneous high voltage and current operation condition. Based on this modeling the turn-on energies for both gate drive circuits are simulated with parametric variation of gate resistance and gate loop inductance respectively. The dissipated power semiconductor switching energy for both gate drive circuits is shown in Fig. 7. To consider the parasitic turn-on, the dissipated energy of both

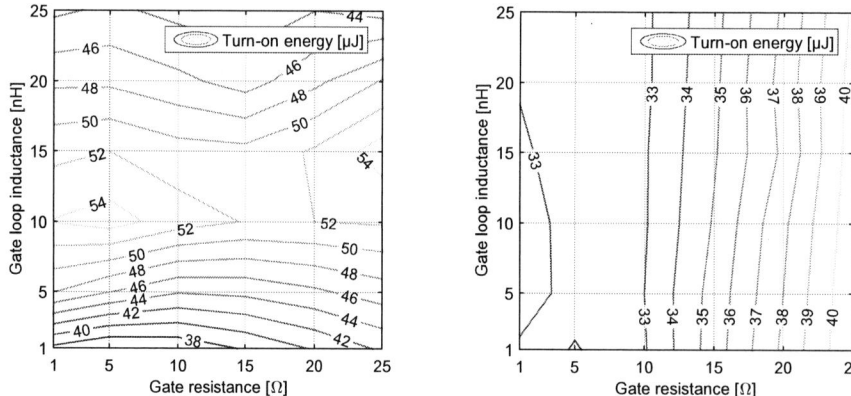

Fig. 7: Simulated power semiconductor turn-on energy using the standard driver (left) and the low-impedance-driver (right) for different values of external gate resistances and loop inductances.

switches is summed up to the total turn-on energy. A comparison of both gate drive circuits shows that the low impedance driver achieves less switching energy in the power semiconductors for a broad variation of parameters. This is attributed to the effective suppression of spurious triggering for gate resistor values above 5Ω. A variation of the gate loop inductance

© VDE VERLAG GMBH · Berlin · Offenbach

PCIM Europe 2016, 10 – 12 May 2016, Nuremberg, Germany

has no effect on this behavior, because the loop inductance between the power semiconductor and the N-Channel MOSFET is fixed to values below $1nH$. This must be realized by a proper layout to achieve stable and efficient operation. Note that low gate loop impedance yields to less turn-on energy for the standard driver as well. But the absolute value of dissipated energy as shown in Fig. 7, resulting from spurious triggering for small values of gate resistance is higher in that case.

4. Driver influence in buck converter operation

The modeling results are verified on an experimental buck converter prototype, operating at a switching frequency of 200kHz. Since we have realized a low inductance layout ($L \approx 5nH$) for the standard driver, based on multilayer coplanar traces, the additional losses of 1.6W due to spurious triggering are well below the main loss contributors. Therefore only the standard driver is used for efficiency comparison between the model and the prototype. The buck converter under test consists of two active semiconductor switches ($T1, T2$), an input and output capacitor (C_{in}, C_{out}) and a power output inductor L_{out}. The schematic is depicted in Fig. 8. GaN e-

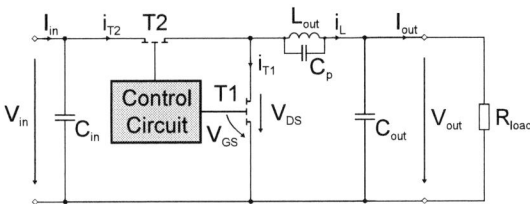

Fig. 8: Buck converter schematic.

mode HEMT's [4] are used for the active switches. Both capacitors are splitted up in ceramic and film capacitors, which form together $10\mu F$ at the input and output. The power output inductor is built by two stacked iron powder toroid cores wound with 38 turns of solid copper wire (Ø1.4mm), which leads to an inductance of $40\mu H$. Furthermore the input voltage V_{in} is fixed to 400V at a constant duty cycle of 0.5, producing approximately an output voltage V_{out} of 200V. The resulting LTSpice [5] simulated and measured buck converter waveforms and efficiency are shown in Fig.9. The measurement with a 500MHz oscilloscope shows a very

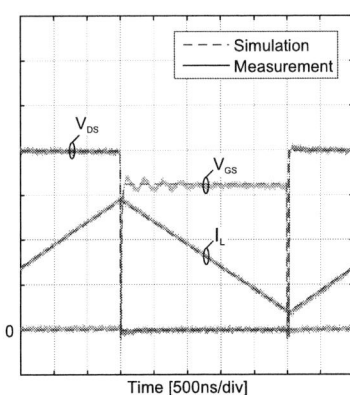
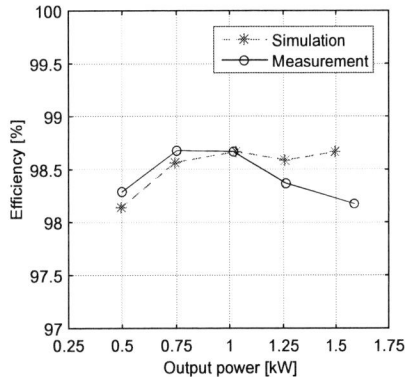

Fig. 9: Buck converter waveforms at an output power of 1.6kW, and efficiency for different output power levels. The y-axis scalings of the waveforms are $V_{GS} = 2V/div$, $V_{DS} = 100V/div$ and $I_L = 5A/div$.

© VDE VERLAG GMBH · Berlin · Offenbach

1567

good matching between the simulated and measured waveforms in one switching period of $5\mu s$. To minimize the measurement error for the efficiency calculation two high precision current shunt resistors (Burster model 1282) are used in combination with four high precision digital multimeter's (Keysight 34401A). The comparison of the measured and simulated converter efficiency of Fig. 9 (right) shows a good agreement for the output power range from 0.5 to 1kW. The deviation for high output power levels results from thermal heating in the real setup, which is not considered in the LTSpice simulation. An overall efficiency above 98% is achieved over the complete output power range. The measured peak efficiency is 98.67% at an output power of 1kW. Although the efficiency is nearly constant over the whole output power range, the loss distribution shown in Fig. 10 (left) changes dramatically. The main reason of loss

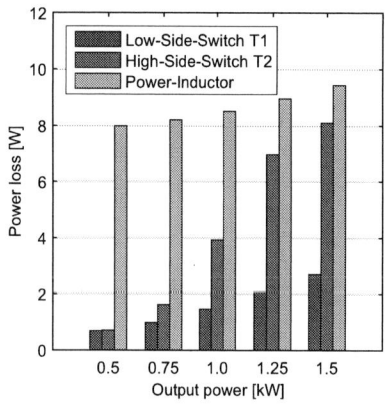 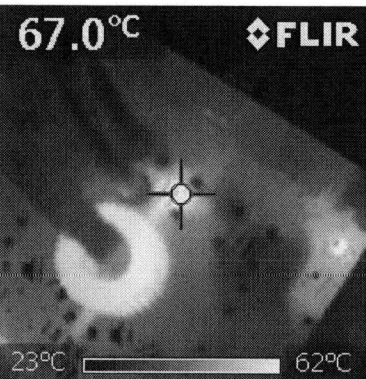

Fig. 10: Simulated loss distribution over the full output power range and measured infrared temperature at 1.6kW output power. The High-Side-Switch temperature is approximately at $67°C$.

distribution variation is attributed to the different modes of buck converter operation. While for output power levels of 0.5 and 0.75kW the circuit operates in triangular current mode (TCM) it operates in continuous conduction mode (CCM) from 1 to 1.5kW. Triangular current mode means, that the inductor current gets positive and negative in every switching period. The inductor current charges and discharges the intrinsic semiconductor capacitances during the short dead time period, which protects the switches against cross conduction. Therefore both halfbridge semiconductors switch at zero voltage condition, resulting in nearly zero switching loss. For short dead times T_d the inductor can be approximated as an ideal constant current source. The minimum necessary current to achieve zero voltage switching (ZVS) is given by Eq. 1, and will be achieved for a calculated output power up to 674W.

$$I_{min,res} = 2 \cdot C_{o(tr)}(V_{in}) \cdot \frac{V_{in}}{T_d} = 2 \cdot 108pF \cdot \frac{400V}{30ns} = 2.9A \tag{1}$$

$C_{o(tr)}$ represents the so called time related capacitance, which shows the same charging time behavior than a voltage independent constant capacitor. The value can be calculated by Eq. 2.

$$C_{o(tr)}(V_{in}) = \frac{1}{V_{in}} \int_0^{V_{in}} C_{oss}(V_{DS}) dV_{DS} \tag{2}$$

The ZVS method is often used in power factor correction (PFC) rectifiers [6] and high frequency buck converters [7] to increase the light load efficiency. Due to an increased output power the resulting DC current content of the output inductor increases. Above 1kW the negative inductor current vanishes and the High-Side-Switch T2 moves from ZVS to hard switching. Therefore the total losses of the High-Side-Switch T2 strongly increase by the amount of switching loss.

© VDE VERLAG GMBH · Berlin · Offenbach

Compared to that, the Low-Side-Switch T1 is acting with zero voltage switching over the total operating range. The power loss of T1 is only attributed to conduction loss, which has the same value for T2 in the special case of a duty cycle of 0.5 as was used here. Splitting up the output inductor losses in DC copper loss, AC copper loss and hysteresis loss explain the light loss increase with increasing output power. The first two mechanisms are the dominating loss parts and they are almost constant over the entire output power range. Only the DC copper loss changes due to a higher DC inductor current at increased output power levels. The temperature increase of the High-Side-Switch Fig. 10 (right) for an output power of 1.6kW measured by infrared tomography is compared with the computed one based on loss distribution and thermal resistance of the assembly. Good agreement is found confirming the loss model.

5. Summary

The paper presents the challenges in controlling the gate of high voltage GaN semiconductors in a hard switching halfbridge environment and shows the parasitic turn-on problematics for state of the art MOSFET drivers. A low impedance gate drive circuit is developed, which effectively suppresses the parasitic turn-on of the power GaN semiconductors. The required low gate loop impedance is achieved by an additional small signal N-Channel MOSFET, which is located as close as possible to the power semiconductor gate and thereby minimizes the inductive part of the gate connection. Thus the circuit prevents parasitic turn-on of the halfbridge in the short time period of the switching event. A detailed analysis on the switching losses of two different gate drive circuits illustrates the application range and limits of the proposed low impedance driver.

6. References

[1] F. Merienne, J. Roudet, and J.L. Schanen. Switching disturbance due to source inductance for a power MOSFET: analysis and solutions. In *Proc. of the 27th Annual IEEE Power Electronics Specialists Conference (PESC)*, volume 2, pages 1743–1747, Jun 1996.

[2] J. Wang and H.S.-H. Chung. Impact of parasitic elements on the spurious triggering pulse in synchronous buck converter. *IEEE Transactions on Power Electronics*, 29(12):6672–6685, Dec 2014.

[3] Z. Zhang, W. Zhang, F. Wang, L.M. Tolbert, and B.J. Blalock. Analysis of the switching speed limitation of wide band-gap devices in a phase-leg configuration. In *Procs. of the IEEE Energy Conversion Congress and Exposition (ECCE)*, pages 3950–3955, Sept 2012.

[4] GS66506T 650V enhancement mode GaN transistor preliminary datasheet. www.gansystems.com, Sept 2015.

[5] M. Engelhardt. LTSpice/SwitcherCAD IV. *Linear Technology Corporation*, 2011.

[6] C. Marxgut, F. Krismer, D. Bortis, and J.W. Kolar. Ultraflat interleaved triangular current mode (TCM) single-phase PFC rectifier. *IEEE Transactions on Power Electronics*, 29(2):873–882, Feb 2014.

[7] J. Wang, F. Zhang, J. Xie, S. Zhang, and S. Liu. Analysis and design of high efficiency quasi-resonant buck converter. In *Procs. of the IEEE International Power Electronics and Application Conference and Exposition (PEAC)*, pages 1486–1489, Nov 2014.

© VDE VERLAG GMBH · Berlin · Offenbach

Isolated gate driver for high current/ high speed FET-Converters

Florian Kapaun, Universität der Bundeswehr München, Germany, Florian.Kapaun@unibw.de
Rainer Marquardt, Universität der Bundeswehr München, Germany, Rainer.Marquardt@unibw.de

Abstract

Expanding the power of FET-based converters from several ten to more than 100kW presents an interesting development trend. Major advantages, aimed at, are ultra low loss, increased switching frequencies and a higher level of integration. For semiconductors in the voltage class up to 650V, PCB-based converters with "all components on board" become feasible.
Minimized board space and high reliability are enforcing the application of integrated circuits for gate drive and auxiliary power supply. True galvanic isolation of these components is desirable, too, in order to enable clean and fast switching of the power devices. A gate drive concept, suitable for fast switching of many paralleled chips, using IC-level [1] transformers is presented. It makes use of a new power boost circuit in order to overcome the limited power range of the integrated (IC-level) transformers.

1. State of the Art

Common, integrated circuits for driving halfbridges are using level shifters and bootstrap circuits for the high side transistor. For fast switching of high currents, however, true galvanic isolation for both high side and low side transistors is strictly advisable, in order to avoid disturbing voltage differences between the source connections and common ground. In addition, the peak charging currents of bootstrap circuits are causing severe distortion of the gate voltages [2].
Furthermore, discrete power supplies with separate transformers are bulky and tend to have high parasitic capacitances across the isolation barrier. Therefore, integrated circuits with IC-level transformers are desirable. Unfortunately, their maximum power is strictly limited.

2. Gate drive circuit

According to these issues, the new gate drive concept uses IC-level transformers with full galvanic isolation and very low parasitic capacitance for all driver channels. In order to boost the output power of these ICs, a special boost circuit is used, which doesn't disturb the isolation barrier.

The whole circuit can be divided into a central part (left side in Fig. 1) and multiple local turn-off circuits (LTO), which are connected by a low-inductive bus-bar. As only the high side driver of the IC is truly galvanically isolated, both transistors have their own driver IC. Solely the high side driver channel is used.
The individual parts of the circuit and their mode of operation are discussed in the following sections.

© VDE VERLAG GMBH · Berlin · Offenbach

PCIM Europe 2016, 10 – 12 May 2016, Nuremberg, Germany

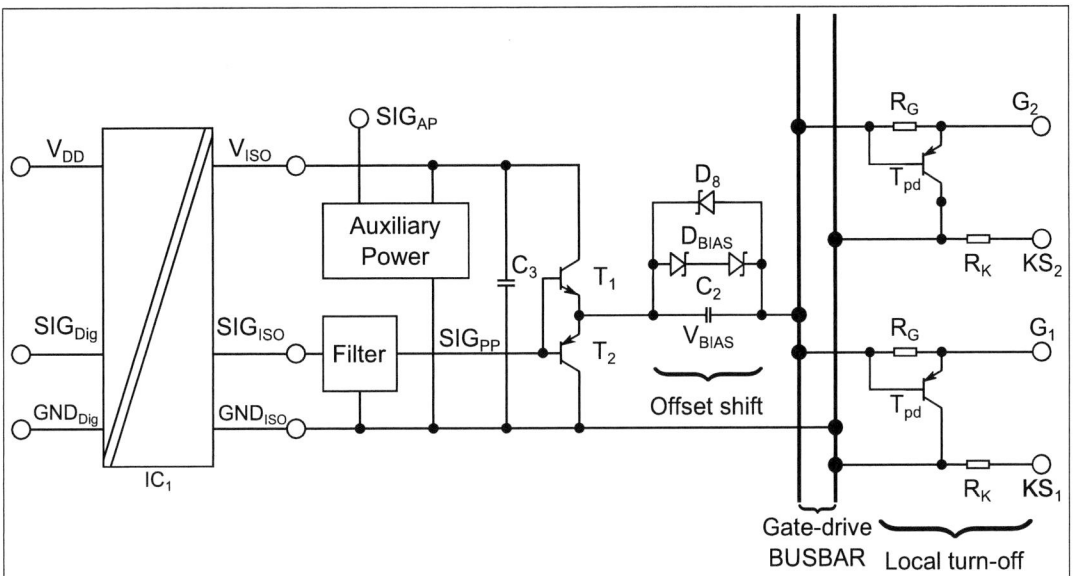

Fig. 1: Drive circuit with galvanic isolated signals and power supply

Integrated Driver Core

As explained, true galvanic isolation power is very desirable. Only few integrated circuits - based on IC-chip scale "coreless" transformers - are available. Analog devices type [3] - which includes a galvanically isolated DC-DC converter with a maximum output power of approximately $P = 0,25W$ - has been chosen. The isolation barrier is characterized by a high isolation voltage ($V_{RMS} = 3,7kV$) and an extremely low parasitic capacitance ($C_k \approx 2pF$). This superior isolation leads to high flexibility when combining multiple converters and helps to mitigate EMV-disturbances.

Slew Rate Filter

In highspeed switching applications, definite measures to control or limit dv/dt are very desirable. Generally, mainly the choice of gate resistor values, is the known measure to adjust switching speed. In the present application this measure has turned out to be insufficient.

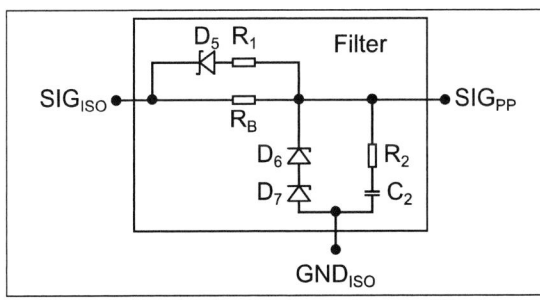

Fig. 2: Slew Rate Filter

Therefore, a "Slew rate filter" has been introduced in the gate driver in order to enable additional degrees of freedom. This enables to combine the desired dv/dt with very low gate resistor values. Low gate resistor values are advantageous for minimized delay times and safe suppression of parasitic turn-on. Fig. 2 shows the electrical set up of the filter. Principle components are the resistors R_B, R_2 and the capacitor C_2. These three parts form a lowpass filter to limit the voltage slew rate (1) at the input of the push-pull (T_1, T_2) stage.

© VDE VERLAG GMBH · Berlin · Offenbach

1571

$$V_{B_T}(t) = V_{driver} \left[1 + e^{-\frac{t}{\tau}} \left(\frac{R_2}{R_B + R_2} - 1 \right) \right] \tag{1}$$

$$\tau = (R_B + R_2) \cdot C_2 \tag{2}$$

Gate-source voltage V_{GS} of the controlled FET has to reach a specific threshold voltage before conduction starts. In order to minimize dead time, an initial step of SIG_{PP} is generated with the help of R_2. To avoid time delay at turn-off, R_B is bypassed by R_1 and D_5.

Apart from the slew rate filter, the zener diodes D_6 and D_7 are introduced. With the help of D_6 and D_7 the gate voltage can be stabilized at a sufficient value. This helps to stabilize the switching delay times. Without this measure, the output voltage of the driver IC is output-power-dependant and fluctuates between 12V and 15,5V.

Auxiliary Power

As mentioned above, the IC-level transformers limit the power to approximately 250 mW. In converter structures with high switching frequencies f_{sw} and many (N) paralleled semiconductors the necessary driver power P_D may be considerably higher. It can be estimated by (3):

$$P_{sw} = 0,5 \cdot N \cdot f_{sw} \cdot V_{iso} \cdot \sum Q_g \tag{3}$$

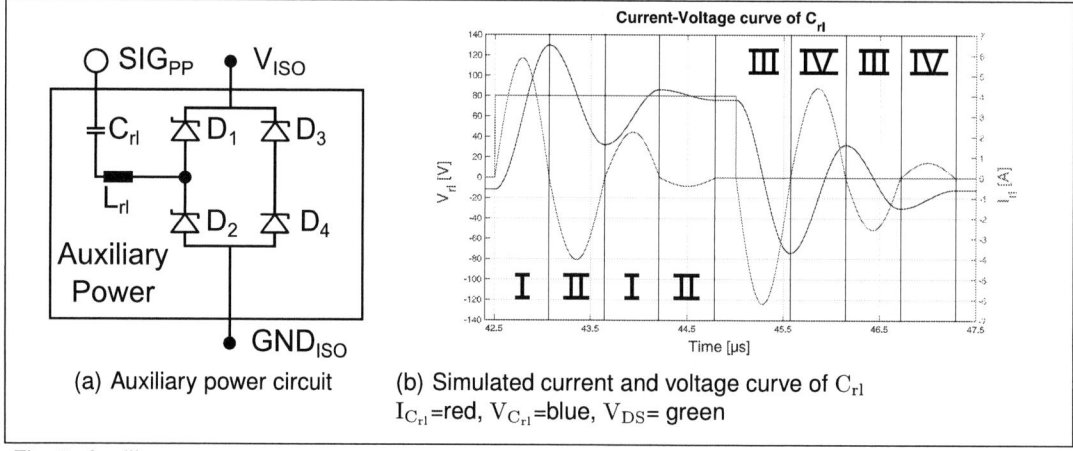

(a) Auxiliary power circuit

(b) Simulated current and voltage curve of C_{rl}
$I_{C_{rl}}$=red, $V_{C_{rl}}$=blue, V_{DS}= green

Fig. 3: Auxiliary power

Therefore, the driver circuit needs to be powered, additionally. Supplying with external DC-DC converters leads to additional board space and an increased coupling capacity C_C. For that reason a concept, where the energy is "taken" from the DC-link is chosen. Thus, the isolation barrier is not disturbed at all.

The circuit shown in fig. 3(a) delivers energy without influencing quality of the isolation barrier. The resonant circuit, consisting of L_{rl} and C_{rl}, is activated by each step change of V_{DS}. During positive half-wave of I_{rl}, D_1 conducts and energy is supplied from the resonant circuit into V_{iso}. During negative half-wave D_2 conducts. Thus, a nearly "lossless" free-wheel path for the resonant circuit is generated. Fig. 3(b) shows the typical voltage and current curve of the reloading capacitor C_{rl}. It can be seen that nearly the complete energy of the capacitor can be supplied within 1,5 (n_{sw}) oscillations of capacitor voltage $V_{C_{rl}}$. The necessary minimum pulse width T_{min} at a given frequency can be estimated with equation (4).

$$T_{min} \geq 2 \cdot \pi \cdot \sqrt{LC} \cdot f_{sw} \cdot n_{sw} \tag{4}$$

The inductor L_{rl} limits the peak current $\hat{I}_{rl}$ and can be chosen in a very wide range. Resonant frequency of the reload circuit and the minimum acceptable pulse width are affected, by choosing the inductance:

$$\hat{I}_{rl} = V_{DC} \cdot \sqrt{\frac{C_{rl}}{L_{rl}}} \tag{5}$$

The amount of supplied energy is independent of L_{rl} and only influenced by C_{rl}. The capacitor voltage and currents can be described with the help of the two equations (6) and (8).

$$V_c(t) = \Delta V \left(1 - \cos(\omega t)\right) + V_c(0) \cdot \cos(\omega t) \tag{6}$$

$$\Delta V = V_{DC} - V_{iso} \tag{7}$$

$$i_c(t) = \sqrt{\frac{C}{L}} \cdot K_{\mathrm{x}} \cdot sin(\omega t) \tag{8}$$

Due to the diodes D_1 and D_2, $V_c(t)$ must be divided into four sectors (I-IV, Fig. 3(b)) for analysis. The whole supplied energy is:

$$E_{rl} = V_{iso} \int i_c(t) dt \tag{9}$$

The relevant sections for energy extraction are I and IV.

$$E_{rl} = V_{iso} \int_0^{\pi} \left((i_{c_I}(t) + i_{c_{IV}}(t)\right) dt \tag{10}$$

Solving the integral delivers:

$$\int i_{c_x}(t) dt = 2 \cdot C_{rl} \cdot K_x \tag{11}$$

The required capacity C_{rl} to compensate the driver losses E_{sw} can be computed from equation (12). K_I and K_{IV} are given by equations (13) and (14):

$$C_{rl} = \frac{E_{sw}}{2 \cdot U_{iso} \cdot (K_I + K_{I_V})} \tag{12}$$

$$K_I = V_{DC} - V_{iso} - V_{c_I}(0)) \tag{13}$$

$$K_{IV} = V_{iso} + V_{c_{IV}}(0) \tag{14}$$

It presents an essential advantage, that the supplied power is proportional to the switching frequency, as needed by the gate drive. For correct operation, however, a minimum value $V_{DC_{min}}$ (15) is required:

$$V_{DC_{min}} \geq \frac{E_{sw} - E_{iso}}{2 \cdot C \cdot V_{iso}} - K_{II} + V_{iso} + U_{c_I}(0) \tag{15}$$

© VDE VERLAG GMBH · Berlin · Offenbach

PCIM Europe 2016, 10 – 12 May 2016, Nuremberg, Germany

Gate drive busbars

Paralleling multiple FETs requires very exact timing to avoid unbalanced switching stress.

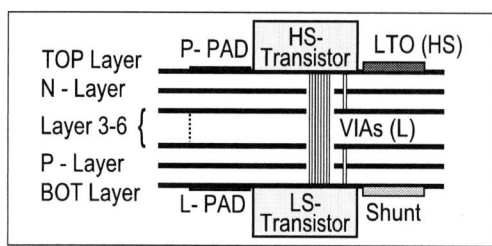

Fig. 4: Buildup-sequence of the gate drive bus-bar

Therefore the gate signals have to reach each FET simultaneously. The inevitable differences can be minimized if there is a very low-inductive busbar between the local driver and all connected switches.

To achieve that, all FETs are tied to the local drive circuit by a broad busbar with nearly zero inductance. The gate signals are large-area routed on nearby layers (layer 3-6, Fig. 4). Shielding by P- and N-layers (positive and negative DC-link) is accomplished, too.

Local turn off- (LTO) and offset shift- circuit

In addition to the zero-delay signal distribution, a fast and safe turn off of each FET is necessary. In general, negative gate voltages from separate negative gate voltage supply, are used to meet this need. An alternative method is the use of local turn-off circuits (Fig. 5(a)) which have to

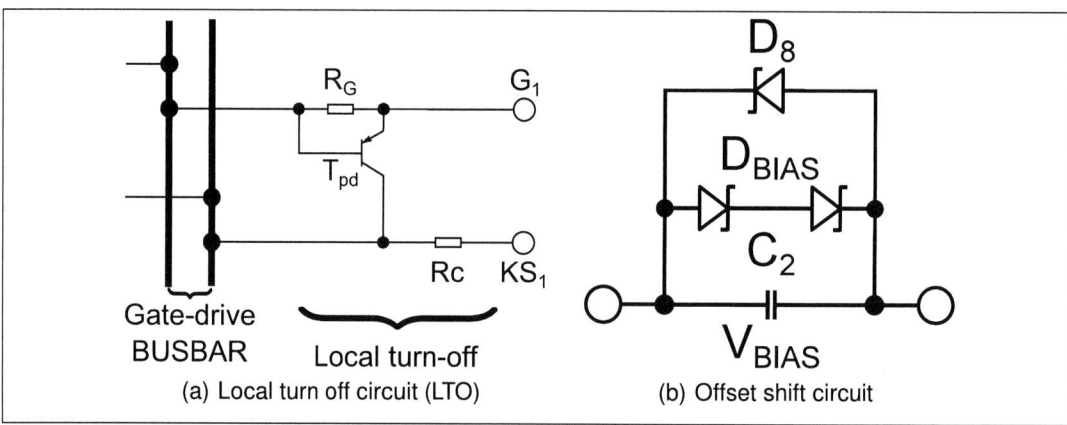

(a) Local turn off circuit (LTO)   (b) Offset shift circuit

Fig. 5: Local turn off- and offset shift- circuit

be placed to each FET (Fig. 6). They consist of a pull down transistor T_{pd} and a decoupling resistor R_c. T_{pd} directly "shorts" the gate-source terminals. R_c decouples the different source signals from the common driver-ground-potential.

To compensate the voltage drop V_{BE} (during turn-off) across the base-emitter junction of T_2 and T_{pd} an additional offset shift circuit is strictly advisable. During on-state of the driver circuit, a positive voltage drop V_{BIAS} occurs above D_{BIAS}. The paralleled capacitor C_2 charges to this voltage level.

During turn-off, the voltage drop V_{BE} ($\Rightarrow$ positive voltage remain between gate and source terminal V_{GS} of the FET) is compensated by V_{BIAS}.

© VDE VERLAG GMBH · Berlin · Offenbach

$$V_{GS} = V_{EB_{T_2}} + V_{EB_{T_{pd}}} - V_{BIAS} \tag{16}$$

In order to avoid a collector-base junction breakdown of T_{pd}, V_{BIAS} must be limited to (17).

$$V_{BIAS} \leq \left(V_{EB_{T_2}} + V_{EB_{T_{pd}}} + 0.4V\right) \tag{17}$$

3. Experimental set-up

For the following measurements, a high frequency converter based on switching cells [4] was realized. It consists of four half-bridges, the associated drive circuits and further electronics

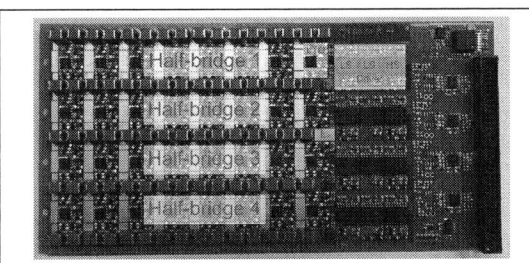

Fig. 6: Realized drive circuit on a multilayer PCB

for voltage and current measurement. Each halfbridge consists of eight paralleled switching cells.

On the right side of the multilayer PCB (Fig. 6) the drive circuit for each half-bridge can be seen. According to the asymmetrical design (one high side (HS) FET and two low side (LS) FETs) of each switching cell, two low side driver ICs are paralleled to manage the increased driving power.

4. Measurement results for auxiliary power boost circuit

An essential part of the drive circuit is the auxiliary power supply. Fig. 7(a) shows the typical voltage and current curves of C_{rl} for a single reload operation which takes place once per period. It can be seen that nearly the whole energy balancing is terminated after 1,5 oscillations

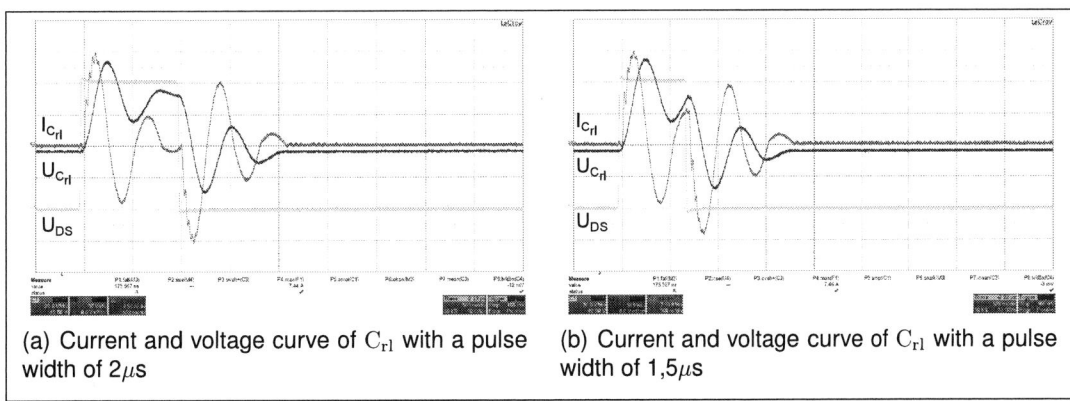

(a) Current and voltage curve of C_{rl} with a pulse width of $2\mu s$

(b) Current and voltage curve of C_{rl} with a pulse width of $1,5\mu s$

Fig. 7: Measurement results for auxiliary power ($U_{C_{rl}}$= blue (50V/div), $I_{C_{rl}}$= red (2,5A/div), U_{DS}= green (20V/div))

of $V_{C_{rl}}$. The capacitor voltage $V_{C_{rl}}$ remains at the level of V_{DC}. The second part is initiated by the following falling edge of $V_{DS}(HS)$. Here, the balanced state is reached after the same time, too. $V_{C_{rl}}$ becomes nearly zero at the end of the switching period.

The efficiency of the auxiliary power circuit is very high ($> 90\%$). Only small losses occur in the choke L_{rl} and the rectifying diodes D_1 and D_2.

5. Measurement results for gate slew rate filter

The influence of the filter circuit can be seen in figures 8(a) and 8(b).

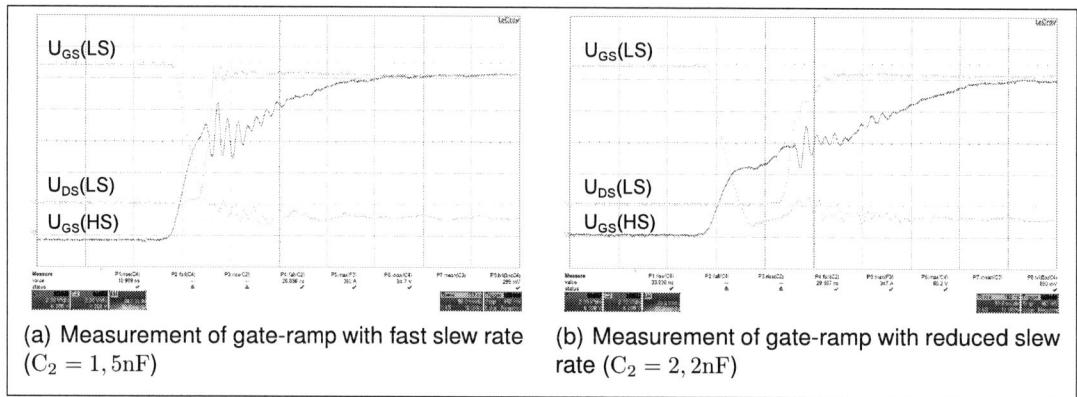

(a) Measurement of gate-ramp with fast slew rate ($C_2 = 1, 5nF$)

(b) Measurement of gate-ramp with reduced slew rate ($C_2 = 2, 2nF$)

Fig. 8: Impact of filter circuit on gate signal ($U_{GS}(LS)$ = blue (2V/div), $U_{GS}(HS)$ = red (2V/div), $U_{DS}(HS)$ = green (20V/div), t = 50ns/div)

Both measurements have been done at equal conditions with a load current of ≈ 350 A and a DC voltage of 80 V. In Fig. 8(a) the switching behaviour for a very fast gate slew rate is displayed. The rise time ($10\% - 90\%$) of drain-source voltage V_{DS} is very short ($\approx 11ns$). Some overshoot (V_{DS}) and moderate oscillations of the gate voltages can be seen.

In contrast to that, the second picture (Fig. 8(a)) shows the same measurement with a slightly reduced gate slew rate (Increased value of C_2, same gate resistors). The rise time of V_{DS} increases considerably ($\approx 30ns$). Parasitic gate oscillations and dv/dt of V_{DS} are further reduced, as expected.

For most applications the adjusted switching behaviour maybe suitable in a range between case (a) and (b).

6. Conclusion

This paper explains the realization of an essentially integrated gate drive circuit for high current/ high speed switching applications.

True galvanic isolation is realized for signals and power supply. Halfbridges that are driven with this circuit can easily be connected in parallel (current increase) or in series (for multilevel systems)[5]. Due to the boosted auxiliary power supply, even converters with high gate drive power requirements can be handled.

The variable slew rate filter circuit allows to actively adjust the dv/dt in the power circuit and enables stable and clean switching of the FETs.

7. References

[1] B. Chen. icoupler products with isopower technology: Signal and power transfer across isolation barrier using microtransformers. 2006.

[2] L. Balogh. Design and application guide for high speed mosfet gate drive circuits. Technical report, Texas Instruments, 2001.

[3] Analog Devices. Isolated half-bridge gate driver with integrated isolated high-side supply. Technical report, Analog Devices, 2012.

[4] M. Schulz, F. Kapaun, and R. Marquardt. Scalable high frequency converters for drives based on switching cells. *CIPS*, 2014.

[5] M. Schulz, L. Lambertz, and R. Marquardt. Dimensioning of modular high frequency converter for drives. *ECCE Asia*, 2013.

Simple Gate-boosting Circuit for Reduced Switching Losses in Single IGBT Devices

Martin Hochberg, Martin Sack, Georg Mueller
Karlsruhe Institute of Technology, Institute for Pulsed Power and Microwave Technology
Hermann-von-Helmholtz-Platz 1
76344 Eggenstein-Leopoldshafen, Germany
Martin.Hochberg@kit.edu

Abstract

Rise and fall time of current and voltage across an IGBT determine the switching losses under hard switching conditions. Due to parasitic inductances in the gate drive circuit and the device leads, gate charging is delayed and hence switching speed is limited. The simple gate-boosting circuit under investigation enables fast charging of the gate capacitance by applying 80 V external gate drive voltage while not exceeding the specified maximum voltage across the internal gate capacitance. Under hard switching conditions at a collector current of 80 % of the maximum rated pulse current, reduction of turn-on speed and turn-on losses in the order of 90 % could be achieved.

1. Introduction

Under hard switching conditions, current and voltage occur across the switch at the same time, causing switching losses. Switching losses, as product of voltage and current, can be decreased by decreasing the switching time. One method is to employ inherently fast switching elements such as MOSFETs. However, their conduction losses are comparably high. One alternative are IGBTs exhibiting lower conduction losses. The switching speed of IGBTs (and as well of MOSFETs) might be increased by the application of increased voltage to the gate terminal (so called gate boosting), [1]. While other researchers ([1], [2]) employ complex gate drive circuits, the proposed gate drive circuit in this work has been designed without additional active elements to operate at a driver voltage of up to 80 V for single IGBT devices.

2. Gate drive circuitry

One major challenge in fast switching of IGBTs is the fast charging of the gate capacitance. The parasitic inductance of the device leads and gate drive circuit hinders the charging process and reduces the switching speed [3]. For faster charging, a higher voltage at the gate terminal is necessary. However, for operation within the datasheet limits, the maximum voltage across the internal gate capacitance must not be exceeded. As a result, the upper limit for the charge delivered to the gate is the value given in the datasheet. The gate drive circuit under investigation as shown in fig. 1 is basically a damped capacitive divider with damping resistors R_D, dividing the driver voltage V_{GD} between the coupling capacitor C_C and the internal gate capacitance C_G of the IGBT. In the circuit diagram, the IGBT is considered to be an ideal component, whereas its internal components such as gate capacitance C_G, parasitic lead inductance L_P and internal gate resistor R_P are displayed in gray. To compensate charge loss for longer pulses, additional

PCIM Europe 2016, 10 – 12 May 2016, Nuremberg, Germany

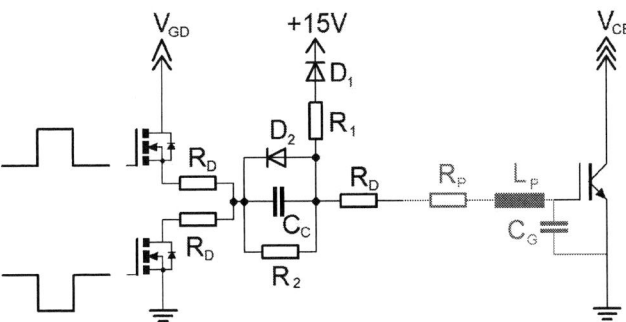

Fig. 1: The gate-boosting circuit under investigation, using drive voltages V_{GD} of up to 80 V. The IGBT's internal/parasitic components are shown in gray.

charge is supplied via resistor R_2 (several kΩ). Diode D_1 and resistor R_1 ensure that the gate voltage settles to approximately 15 V in steady on-state. To prevent gate charging during off-state, diode D_2 offers a low resistance path to ground when the low-side driver is switched on.

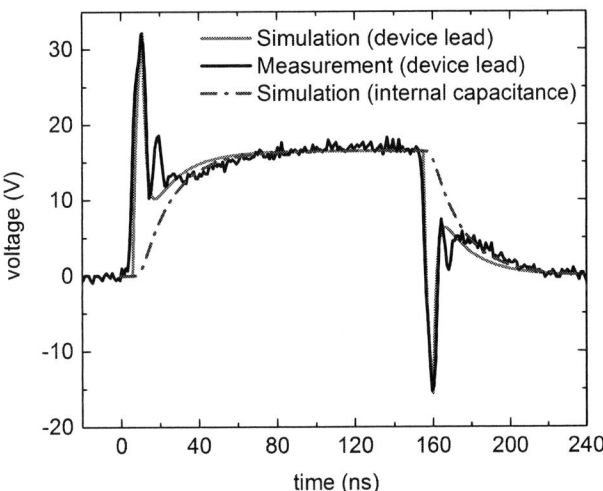

Fig. 2: SPICE simulation results together with measurements on the device leads. Parasitic inductances cause a voltage overshoot which does not affect the internal gate capacitance.

The black trace in figure 2 shows the measured gate signal at the device leads using a gate drive voltage of 80 V. The voltage overshoot of around 30 V as seen in the measurement is caused by the parasitic inductance of the circuit. For reliable operation of the device, however, the voltage across the internal gate capacitance must be limited to the value given in the datasheet(<20 V, [4]) at all times to prevent damage to the device. To estimate the voltage across the internal gate capacitance, a SPICE simulation based on ideal components as shown in fig. 1 has been performed. The value of capacitance C_G was derived from measurements to

© VDE VERLAG GMBH · Berlin · Offenbach

be 18 nF. Values for R_P and L_P have been iteratively adjusted such, that the simulated voltage across the device leads (red solid line in fig. 2) matched the measurement(black solid line in fig. 2). A good agreement was achieved for 8 nH and 0.8 Ω. Under this assumption, the voltage across the internal gate capacitance (blue dashed line in fig. 2) can be derived from the simulation. As can be seen, it stays well within the specified limits. Therefore, the proposed gate drive circuit is considered to be suitable for operation even at high driving voltage.

3. Experimental results

The following results were obtained for a standard IGBT in a TO-247 housing. To ensure hard switching conditions, the total inductance of the test circuit was kept below 60 nH. As the IGBT showed a pronounced increase of rise time for high currents, the measurements were conducted at 25 %, 50 % and 80 % of the maximum rated pulse current in single pulse operation (the device under test: 320 A, [4]). The different currents were achieved by adjusting the load resistance, while the collector-emitter voltage was kept constant at 1 kV. Figure 3 shows collector-emitter current as well as collector-emitter voltage across the IGBT during switching. As both curves coincide, the inductance of the pulse circuit can be neglected.

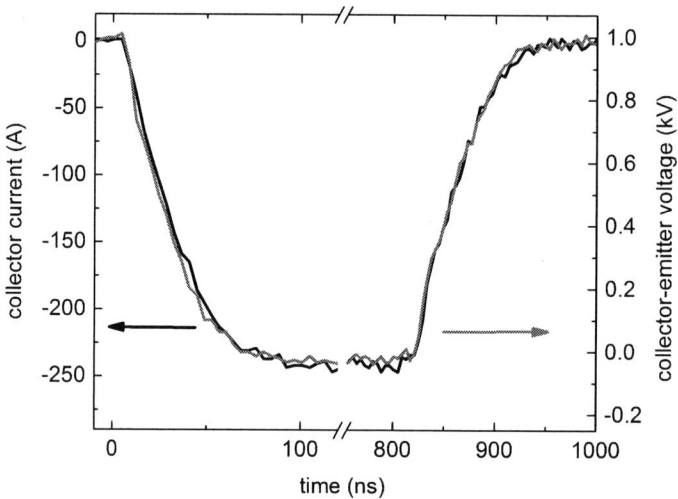

Fig. 3: Collector current and collector-emitter voltage across the IGBT during the pulse. The low inductance of the pulse circuit ensures hard switching conditions.

The driver voltage has been increased up to 80 V. As reference, the coupling capacitor C_C was shorted for applying a driving voltage of 15 V, representing the standard gate drive. In figure 4, the collector current for different driver voltages is shown. From the graph it becomes evident that the switching time significantly decreases for higher driver voltage. Figure 5(a) shows the rise time dependence on the driver voltage for different collector currents. As can be seen, the rise time is strongly increased for higher currents. However, with high driver voltage it is possible to speed up switching by approximately the same order of magnitude for high and low currents. For 80 % maximum pulsed current (256 A) a reduction from 400 ns to 50 ns

PCIM Europe 2016, 10 – 12 May 2016, Nuremberg, Germany

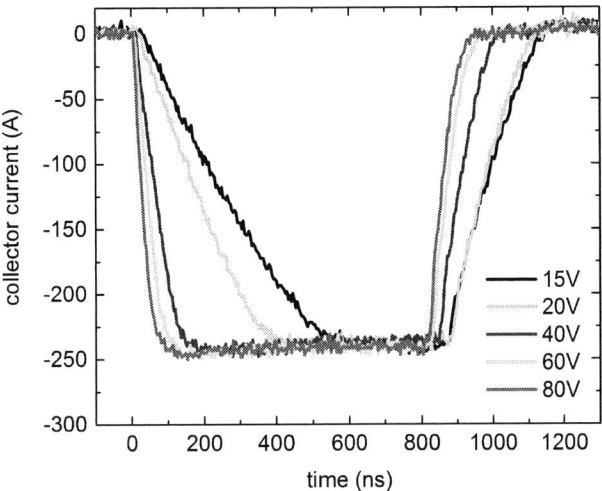

Fig. 4: Collector current for different boosting voltages. A strong increase in switching speed is visible.

was achieved, which is comparable to the reduction from 125 ns to 18 ns for 25 % maximum pulsed current (80 A). The decrease in fall times, however, as displayed in fig. 5(b) is not as pronounced as for the rise times. In addition, the turn-off times are less susceptible to the magnitude of the gate drive voltage for low currents.

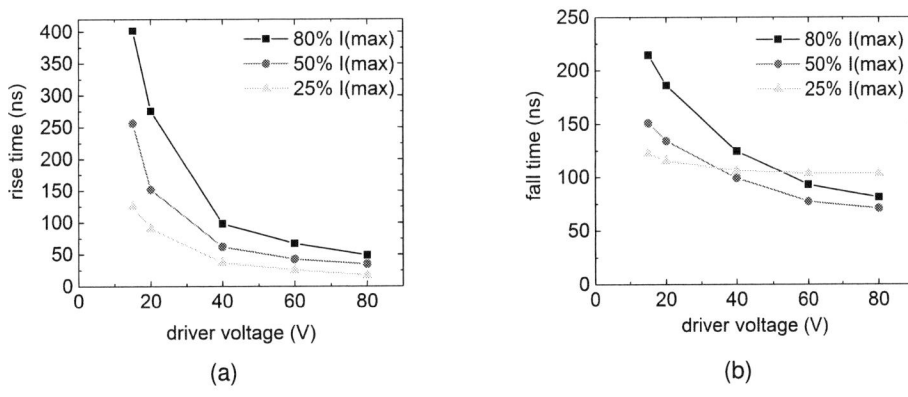

Fig. 5:
(a): Collector current rise time and
(b): fall time for different gate drive voltages at 25 %, 50 % and 80 % of the maximum rated pulse current. The rise time can be influenced to a greater extend by high gate drive voltages.

The resulting switching losses at 80 % maximum rated pulse current are shown in fig. 6(a) for different driver voltages. This data was obtained using a low inductance shunt resistor (as presented in [5]) and a fast high voltage probe. Both measurements were synchronized within 2 ns. Figure 6(b) displays the loss energy normalized to the losses using the standard gate drive. As could be expected from the huge decrease in switching time, the turn-on losses are

© VDE VERLAG GMBH · Berlin · Offenbach

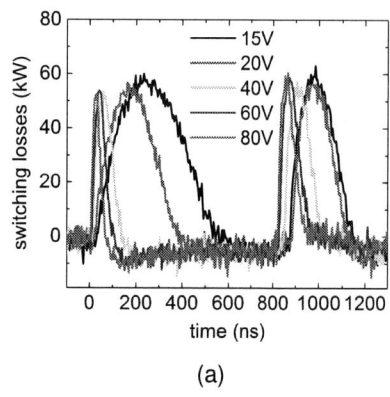

(a)

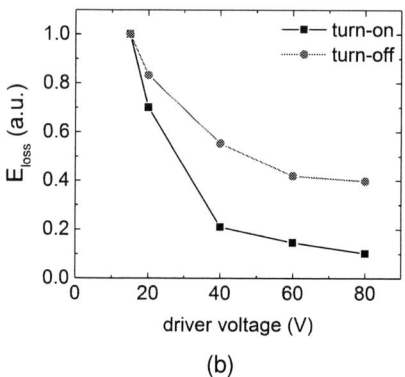

(b)

Fig. 6:
(a): Switching losses at 80 % maximum rated pulse current (256 A) for different driver voltages.
(b): Dependence of normalized switching losses on the driver voltage during turn-on and turn-off.

decreased by 90 % for a driver voltage of 80 V. In agreement with the slower fall time, the turn-off losses can be influenced to a smaller extend as compared to the turn-on losses. However, the reduction of turn-off losses of about 50 % is possible for 80 V driver voltage.

4.   Summary and Conclusion

To overcome the limitations of IGBTs under hard switching conditions with respect to switching speed and switching losses, a simple gate-boosting circuit has been investigated. It allows fast gate charging by the application of up to 80 V driver voltage. Single trench IGBT devices have been operated at up to 80 % maximum rated pulse current in a low-inductance pulse circuit in single pulse mode. SPICE simulations revealed, that even at 80 V driver voltage, the specifications for the internal gate capacitance were not exceeded. For 80 % maximum rated pulse current it was possible to achieve a reduction of current rise time from 400 ns to 50 ns and accordingly a reduction of turn-on losses in the order of 90 %. It was found that the turn-off losses are not as strongly dependent on the driver voltage. Hence, a decrease of turn-off losses by 50 % was achievable for high currents only. Further work will be devoted to in-depth analysis and optimization of the resulting gate signal with respect to switching speed and reliable operation. Another focus will be the applicability of the proposed circuit to different switching elements.

5. References

[1] M. N. Nguyen, R. L. Cassel, J. E. deLamare, and G. C. Pappas. Gate drive for high speed, high power igbts. *PPPS-2001: Pulsed Power Plasma Science 2001, Vols I and II, Digest of Technical Papers*, 2001.

[2] Andreas Volke and Michael Hornkamp. *IGBT modules: Technologies, driver and application*. Infineon Technologies AG, Munich, 2011.

[3] R. Bayerer and D. Domes. Parasitic inductance in gate drive circuits. *PCIM Europe Conference Proceedings*, pages 473–477, 2012.

[4] ON Semiconductor. Datasheet for IGBT NGTB40N120IHLWG, http://www.onsemi.com/pub/collateral/ngtb40n120ihlw-d.pdf, Accessed: Nov. 15, 2015.

[5] M. Sack and G. Müller. *EMC aspects in the layout of a fast semiconductor switch*. Presented at: Electromagnetic Compatibility in Automotive Engineering, GMM Specialist Conference 2014; Dusseldorf; Germany; 11 March 2014 through 13 March 2014.

Stability and Performance Analysis of a Voltage Controlled Resistor Circuit for Wide Band-gap Device Gate Drivers

Alessandro Soldati, Università degli Studi di Parma, Italy, alessandro.soldati@studenti.unipr.it
Giorgio Pietrini, Università degli Studi di Parma, Italy, giorgio.pietrini@studenti.unipr.it
Davide Barater, Università degli Studi di Parma, Italy, davide.barater@unipr.it
Carlo Concari, Università degli Studi di Parma, Italy, carlo.concari@unipr.it

Abstract

Wide band-gap devices are making inroads in the power converters scenario, and specific circuits to drive these components are actively under development. The purpose of this paper is to analyze, from the stability and dynamic performance point of view, a Voltage Controlled Power Resistor (VCPR), that can be used to control the gate resistance of the device driver with values over a continuous range. Parametric analysis, SPICE simulations and experimental outcomes are presented, in order to determine circuit characteristics. Results show that the proposed topology is stable under a wide range of electric parameters, and suggest that the circuit bandwidth can be tuned in order to benefit from the VCPR in a wide band-gap device gate driver.

1. Introduction

Emerging wide band-gap devices (WBD) require dedicated gate drivers in order to fully exploit their improved capabilities with respect to standard silicon power switches. These capabilities include, but are not limited to, active thermal control, faster transitions, reduced waveform distortion, EMI control and lower losses [1].

This paper addresses stability and preliminary design issues of a circuit devoted to finely control the current driving a WBD control terminal. The circuit is presented in Fig. 1 and can be regarded as a Voltage Controlled Power Resistor (VCPR): it can set the device gate to a particular voltage, controlling continuously the equivalent resistance of the generator during the turn on/turn off transient. The resistance value is determined by a low-swing voltage signal, while the output can sink/source (with the complete, complementary design) up to 2 A, with a voltage swing of 30 V.

The circuit, originally presented in [2], derives from the basic op-amp operated current generator (highlighted in green), where the voltage control signal is substituted with a voltage feedback coming from the output node, transferred by the voltage divider in the orange box. The JFET transistor (in the blue rectangle) is added to provide the design with voltage control over resistor ratio, hence over the equivalent resistance seen at output terminal G (main power switch control terminal).

This circuit, allowing precise control of the resistor in series with the gate-source capacitance, can be part of an active gate driving system, enabling an advanced control of WBD performance. This circuit can also be regarded as the continuous values improvement of traditional circuits including diode- or time-switched resistors for gate drivers [3].

© VDE VERLAG GMBH · Berlin · Offenbach

PCIM Europe 2016, 10 – 12 May 2016, Nuremberg, Germany

In order to address the stability and performance study of the circuit, a small-signal model was studied, first. This is reported in Fig. 2, and is based on simple linearization of component characteristics. It must be noted that, while the output MOSFET M_2 and the OA operate in saturation and linear high-gain region, respectively, the JFET J_1 operates in the linear region, with a reduced drain-source voltage and an appreciable output resistance (incorporated in R_1).

2. Stability analysis

In order to carry out a more detailed symbolic analysis of the circuit and to handle feedback loops properly, a graphical method [4, 5] based on Signal-Flow Graphs (SFG) and Driving-Point Impedances (DPI) was applied. Unlike the matrix based approaches, best suited to computer processing, this technique makes it possible to manually find the transfer functions of the dynamic system. SFG/DPI provides designers with insights into the circuit behavior and useful information during circuit sizing and performance optimization activity. In addition, using this technique, even the most hidden loops in the circuit are highlighted, thus preventing errors and omissions.

Looking at the Fig. 3, which shows the SFG of the small signal circuit in Fig. 2, many feedback loops can be identified, therefore an in-depth stability analysis is mandatory. The feedback loops are reported in the denominator terms section of Tab. 1.

2.1. Transfer function

The most important transfer function in a circuit of this type is the voltage controlled equivalent resistance $R_{eq} = v_{out}/i_{in}$. The small-signal parameters of J_1 change with its DC gate voltage, as is well known, so R_{eq} is a function of JFET V_g though this parameter is not explicitly present in the analysis proposed.

R_{eq} is obtained by means of the Mason's rule applied to the SFG in Fig. 3. The analytic expression returned by the SFG/DPI method are quite complex (6th-order), therefore symbolic computation and simplification has to be performed with a mathematical environment like MATLAB in order to make feasible the pole-zero stability analysis.

It is worth noticing that single-pole approximation is used to take into account the op-amp's dynamic behaviour, so $A_v(s)$ is a transfer function in the complex frequency domain. Moreover, R_g incorporates not only a gate resistor in the circuit but also the output resistance of the op-amp.

$$R_{eq} = \frac{T_{1n} \cdot (1 + T_{1d} + T_{6d} + T_{7d} + T_{1d}T_{6d} + T_{1d}T_{7d})}{1 + \sum_{i=1}^{11} T_{id} + T_{1d} \cdot (T_{2d} + T_{3d} + T_{4d} + T_{6d} + T_{7d} + T_{9d} + T_{10d} + T_{11d}) + T_{8d} \cdot (T_{6d} + T_{7d})} \tag{1}$$

2.2. Analytic model simulation

In order to assess the stability of the overall system, pole-zero diagrams were plotted (Fig. 4) sweeping the small-signal resistance r_{ds1} and transconductance g_{m1} of the JFET in a way similar to varying its gate voltage. Fig. 4, shows that all poles have a negative real part as required for stability (some poles and zeroes at high frequencies were omitted from the diagram

© VDE VERLAG GMBH · Berlin · Offenbach

PCIM Europe 2016, 10 – 12 May 2016, Nuremberg, Germany

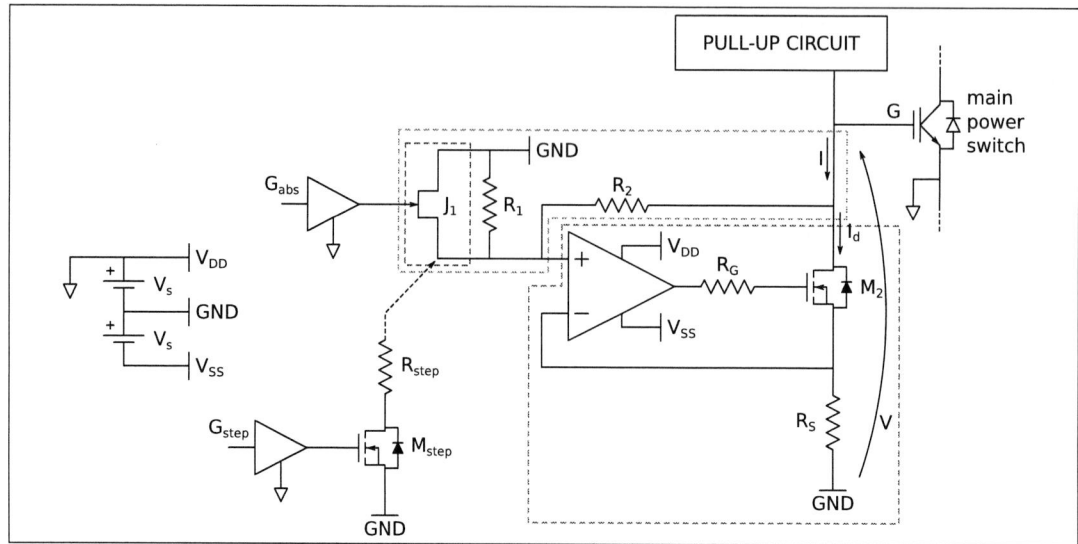

Fig. 1: Schematic representation of the analyzed Voltage Controlled Power Resistor circuit.

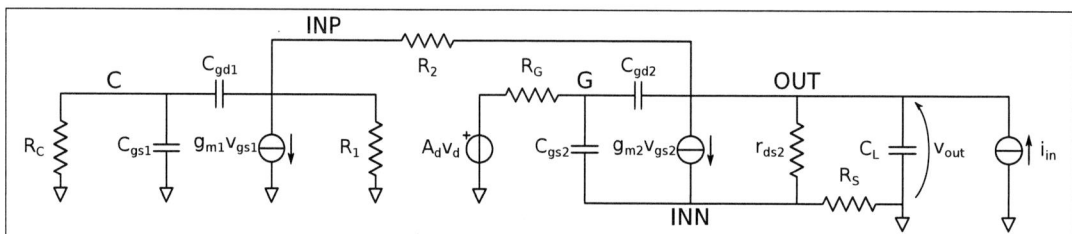

Fig. 2: Small signal model of the analyzed circuit.

Base terms	
$d_1 = R_c^{-1} + s\left(C_{gs1} + C_{gd1}\right)$	$d_2 = R_1^{-1} + R_2^{-1} + r_{ds1}^{-1} + s\,C_{gd1}$
$d_3 = R_2^{-1} + s\left(C_l + C_{gd2}\right) + r_{ds2}^{-1}$	$d_4 = R_g^{-1} + s\left(C_{gs2} + C_{gd2}\right)$
$d_5 = R_s^{-1} + r_{ds2}^{-1} + s\,C_{gs2} + g_{m2}$	
Denominator terms	
$T_{1d} = -\dfrac{s\,C_{gd1}\cdot\left(s\,C_{gd1}-g_{m1}\right)}{d_1 d_2}$	$T_{2d} = -\dfrac{s\,C_{gd2}\cdot\left(s\,C_{gd2}-g_{m2}\right)}{d_3 d_4}$
$T_{3d} = -\dfrac{s\,C_{gd2}\cdot\left(s\,C_{gs2}+g_{m2}\right)\left(r_{ds2}^{-1}+g_{m2}\right)}{d_3 d_4 d_5}$	$T_{4d} = -\dfrac{r_{ds2}^{-1}\left(r_{ds2}^{-1}+g_{m2}\right)}{d_3 d_5}$
$T_{5d} = -\dfrac{R_2^{-1}\cdot A_v\cdot R_g^{-1}\cdot\left(s\,C_{gs2}-g_{m2}\right)\left(r_{ds2}^{-1}+g_{m2}\right)}{d_2 d_3 d_4 d_5}$	$T_{6d} = -\dfrac{R_g^{-1}\cdot A_v\cdot\left(s\,C_{gs2}+g_{m2}\right)}{d_4 d_5}$
$T_{7d} = -\dfrac{s\,C_{gs2}\cdot\left(s\,C_{gs2}+g_{m2}\right)}{d_4 d_5}$	$T_{8d} = -\dfrac{R_2^{-2}}{d_2 d_3}$
$T_{9d} = -\dfrac{R_2^{-1}\cdot A_v\cdot R_g^{-1}\cdot\left(s\,C_{gd2}-g_{m2}\right)}{d_2 d_3 d_4}$	$T_{10d} = \dfrac{r_{ds2}^{-1}\cdot A_v\cdot R_g^{-1}\cdot\left(s\,C_{gd2}-g_{m2}\right)}{d_3 d_4 d_5}$
$T_{11d} = -\dfrac{r_{ds2}^{-1}\cdot s\,C_{gs2}\cdot\left(s\,C_{gd2}-g_{m2}\right)}{d_3 d_4 d_5}$	
Numerator terms	
$T_{1n} = \dfrac{1}{d_3}$	

Tab. 1: Sub-expressions of the transfer function.

for convenience). The range of r_{ds1} values (from $2.3\,\Omega$ to $23\,\text{k}\Omega$) applied to the transfer function (1) of the analytical model is suitable for producing the typical dv/dt slopes of modern power devices. Of course, the g_{m1} was varied in accordance with r_{ds1}.

© VDE VERLAG GMBH · Berlin · Offenbach

PCIM Europe 2016, 10 – 12 May 2016, Nuremberg, Germany

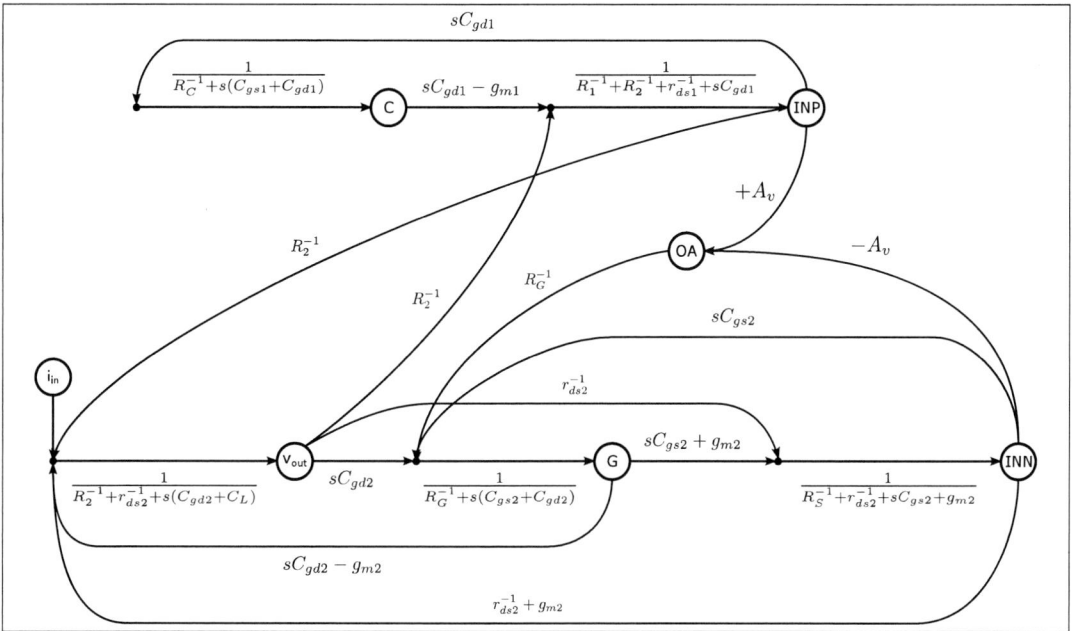

Fig. 3: Signal Flow Graph of the equivalent circuit.

Although the circuit involves many feedback loops, it exhibits stable behaviour over a wide range of values for the parameters, and the experimental tests confirmed this conclusion.

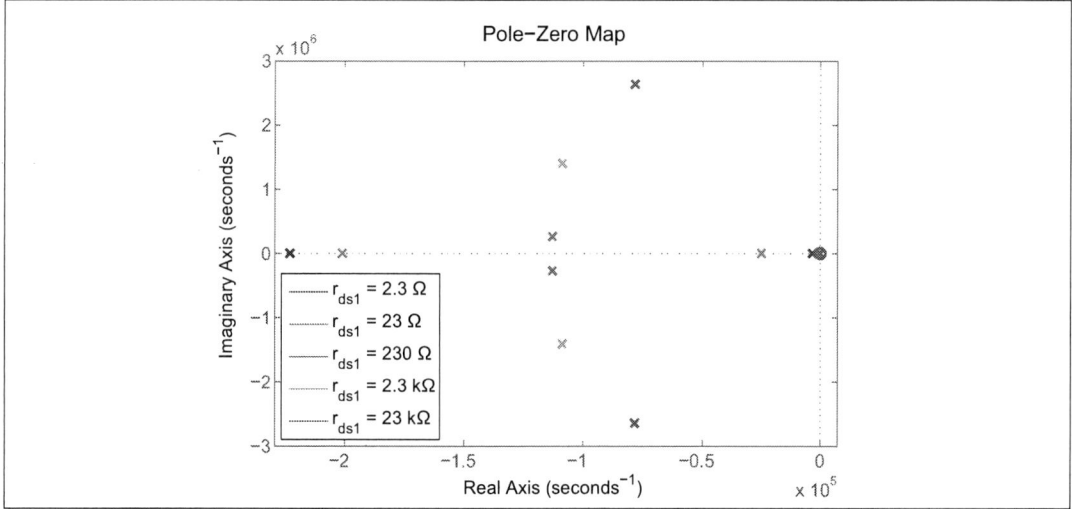

Fig. 4: Pole-zero diagram of transfer function.

3. Electric simulation and experimental results

Preliminary electric (SPICE) simulations, pertaining the full symmetric circuit, presented in [2], were used to verify the model depicted in Sec. 2.

The circuit described in this paper is partly different from the symmetric one, in order to in-

© VDE VERLAG GMBH · Berlin · Offenbach

1587

vestigate stability in detail. With respect to Fig. 1, the high-side circuit is not the symmetric counterpart of the pull-down, but a simple pMOS. Moreover, JFET J_1 is replaced by the series of an n-channel MOSFET and a resistor, in order to generate abrupt steps. Resistor R_1 is replaced with a trimmer, to manually control the actuated resistance.

Due to a shortage of specific commercial devices, others with similar characteristics as those of [2], but generally lower bandwidth, were employed.

3.1. Tests setup

A new set of electrical simulations with models of the exerted components were carried out, to compare more fairly with experimental evidence. The components employed are: *OPA4350* operational amplifier, *IRFRC20* nMOS (both for M_2 and M_{step}), *VP2410L* pMOS as active pull-up. Passive components are $R_1 = 50\,\text{k}\Omega$ trimmer, $R_2 = 47\,\text{k}\Omega$, $R_G = 100\,\Omega$, $R_S = 1\,\Omega$, $R_{step} = 1\,\Omega$, $C_L = 1\,\text{nF}$ (to mimic the input capacitance of the main power switch).

OPA4350, despite having $35\,\text{MHz}$ bandwidth and $22\,\text{V}/\mu\text{s}$ slew-rate, imposed some limitations on voltage range, being single supply (thus, $V_{SS} \equiv GND$) and withstanding only $7\,\text{V}$. Using such component, no negative output voltage, with respect to the main power switch source terminal, could be supplied.

Switch commands and stimulus signals were delivered to the circuit using a *TTi TG101A* function generator, while retrieving voltages and currents with a *LeCroy HDO6034* oscilloscope. Analysis in the frequency domain was carried out off-line.

3.2. Test results

The first test to be performed was the generation of large signal commanded steps; associated results are shown in Fig. 5, where generation of steps with four different fall times is presented. Despite some minor ringing immediately before and after the voltage transition, the shape of the transient response is merely a pure exponential one: being the load capacitive, it means that the circuit is behaving like a commanded resistor.

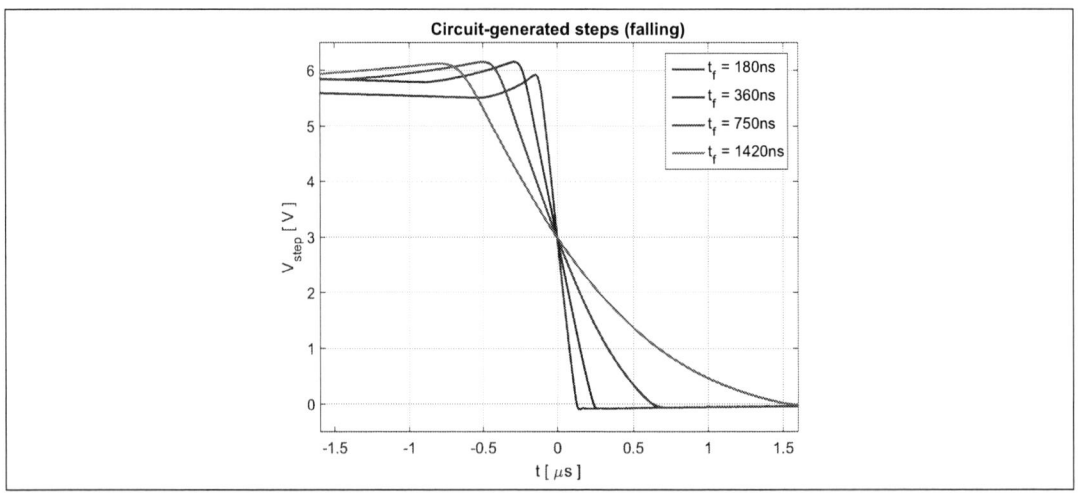

Fig. 5: Experimental measurement of the circuit generated steps, with command resistance as parameter, generating different falling times.

It is important to observe that during the non-driving phase (generating high rail voltage, for the pull-down circuit analyzed) the operational amplifier should not be driven open loop, in order to maintain a fast response. What is being done is to force the generation of a high, yet finite, resistance. In this way, the amplifier operates always with low differential voltage and fast switching response.

The best falling time recorded is $180\,\mathrm{ns}$, which is nearly ten times slower than the speed needed for the effective use of the circuit with wide band-gap devices. It is believed that this performance can be hugely improved by an accurate selection of amplifier and MOSFET: the op-amp output saturates in this extreme condition, limiting the waveform steepness. Higher MOSFET conductivity can obviously be obtained with larger overdrive: greater MOS gain, lower threshold voltage and wider op-amp supply are several means by which the reduction of fall time can be achieved.

Other experiments involved small signal response and bandwidth evaluation. These measurements were done setting the feedback trimmer to a particular value, and forcing a small amplitude ($500\,\mathrm{mV}$-wide) falling step. Results over time are plotted in Fig. 6(a), where an equivalent resistance of around $R_{eq} = 220\,\Omega$ is being actuated. This value holds only during the steady-state phase: close to the triggering event, the limited bandwidth translates to a different value of effective resistance.

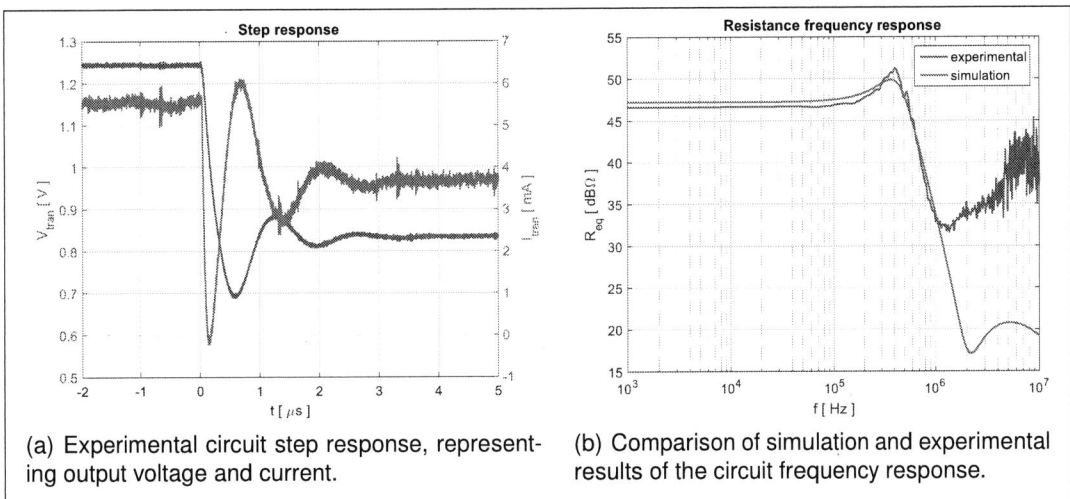

(a) Experimental circuit step response, representing output voltage and current.

(b) Comparison of simulation and experimental results of the circuit frequency response.

Fig. 6: Transient and frequency responses of the presented circuit.

Fig. 6(b) depicts the experimental frequency response of the VCPR (in blue) and Bode diagram obtained from SPICE simulation (in red). These results are obtained as follows: voltage and current values from Fig. 6(a) are each directly transformed using FFT, gathering the Discrete Fourier Transform of the impulse (i.e. step derivative) response divided by frequency. The ratio of these transforms is then computed, making the divide-by-frequency term, common to both, disappear. The frequency response obtained in this way is equivalent to directly dividing the transform of voltage by that of current *impulse* responses. The value of $47\,\mathrm{dB}\Omega$ at low frequency is indeed correspondent to $R_{eq} = 220\,\Omega$ in DC.

The Bode diagram (in red) is directly obtained from small signal AC analysis in SPICE simulator. Fig. 6(b) shows a very good agreement between real world and simulation. A significant difference is noticeable at high frequencies ($f > 1\,\mathrm{MHz}$); this is believed to be ascribable to inaccurate modeling of high frequency capacitive and inductive parasitic components in SPICE.

The bandwidth, in this case, is defined by the authors as the difference from the DC value, regardless of the Bode diagram drifting upwards or downwards; the usual $3\,\mathrm{dB}$ threshold is adopted. This definition allows to spot errors in equivalent resistance even in presence of resonant-like behavior. Table 2 shows bandwidth measurements for different actuated R_{eq} values.

Equivalent DC resistance R_{eq} [Ω]	10	30	100	300
Bandwidth BW [kHz]	37	95	200	176

Tab. 2: Experimental measurement of the circuit-generated resistance bandwidth.

A strong dependence of the bandwidth on the equivalent DC resistance is noticeable, with slower response in correspondence with lower resistance values, hence when device is supposed to switch faster. A non-monotonic trend is also visible, and it could be referable to the appearance of resonant behavior in some cases.

4. Conclusion

The analysis presented in this paper aims to support designers in making well-founded decisions during the circuit sizing phase. The goal is to attain a bandwidth as wide as possible, combined with good stability, so to let the gate driver circuit be compatible with wide band-gap power devices.

Accordance between electric simulation and experimental results prove modeling is correct and that the circuit is stable under a wide range of operating conditions. The primary limit connected to stability is the resonant behavior occurring when generating some particular equivalent resistance values: this behavior needs to be kept under control, in order to avoid undesired reactions from the controlled main power switch. Both time and frequency analysis show the effectiveness of the proposed architecture in implementing a Controlled Power Resistor.

The main concerns regard the achieved bandwidth and fastest generated fall time: the latter is too slow, while the former is narrow. These limitations can be overcome acting on the operational amplifier and output MOSFET component choice: high bandwidth and high slew rate are a must for the first requirement, low threshold and large gain are desirable for the second.

In the future, effects of JFET limitations on circuit speed must be investigated, as well as main switch driving parameters, such as supply magnitude and sign: many WBG devices require wide and sometimes negative voltage rails.

5. Acknowledgments

This work is carried out under the funding of the EU Clean Sky Joint Undertaking, in the framework of the Clean Sky call SP1-JTI-CS-2013-03 (topic: SGO-02-088) with the proposal no. 641496: ALEA - Accelerated Life tests for Electrical drives in Aircraft.

6. References

[1] G. Wang, F. Wang, G. Magai, Y. Lei, A. Huang, and M. Das, "Performance comparison of 1200V 100A SiC MOSFET and 1200V 100A silicon IGBT," in *Energy Conversion Congress and Exposition (ECCE), 2013 IEEE*, pp. 3230–3234, Sept 2013.

[2] A. Soldati, D. Barater, C. Concari, M. Galea, and C. Gerada, "A voltage controlled power resistor circuit for active gate driving of wide-bandgap power devices," in *Industrial Electronics Society, IECON 2015 - 41st Annual Conference of the IEEE*, pp. 2445–2450, Nov. 2015.

[3] K. Onda, A. Konno, and J. Sakano, "New concept high-voltage IGBT gate driver with self-adjusting active gate control function for SiC-SBD hybrid module," in *Power Semiconductor Devices and ICs (ISPSD), 2013 25th International Symposium on*, pp. 343–346, May 2013.

[4] A. Ochoa, "On stability in linear feedback circuits," in *Circuits and Systems, 1997. Proceedings of the 40th Midwest Symposium on*, vol. 1, pp. 51–55 vol.1, Aug 1997.

[5] A. Ochoa, "A systematic approach to the analysis of general and feedback circuits and systems using signal flow graphs and driving-point impedance," *Circuits and Systems II: Analog and Digital Signal Processing, IEEE Transactions on*, vol. 45, pp. 187–195, Feb 1998.

State of the Art of Gate-Drive Power Supplies for Medium and High Voltage Applications

Layal, GHOSSEIN, SuperGrid Institute – Univ Lyon, INSA Lyon, Ampère UMR CNRS 5005, France, Layal.Ghossein@supergrid-institute.com

Florent, MOREL, Univ Lyon, École Centrale de Lyon, Ampère UMR CNRS 5005, France, Florent.Morel@ec-lyon.fr

Hervé, MOREL, Univ Lyon, INSA Lyon, Ampère UMR CNRS 5005, France, Herve.Morel@insa-lyon.fr

Piotr, DWORAKOWSKI, SuperGrid Institute, France, Piotr.Dworakowski@supergrid-institute.com

Abstract

Gate-drive power supplies in medium voltage and high voltage direct current (MVDC / HVDC) applications require medium to high voltage insulation. In MVDC and HVDC power converters there are a number of different power supply techniques that can be used for powering the gate-drives of floating power switches, which are commonly found in such converters. This paper presents the state of the art of gate-drive power supply techniques. Contactless energy transfer methods and self-powering techniques used for powering the gate-drives of power switches are presented. The power range obtained and isolation range for the different solutions are then presented. Finally, the most promising techniques for MV and HV applications are highlighted.

1. Introduction

The main component of any power converter is a power module, containing power semiconductor devices. The power module is driven by a gate-drive board for which the main role is the conditioning of the control signal, but usually it also realises some protection functions. In general, the gate-drive requires a low-power supply providing somewhere in the range of 2-100 of watts at a voltage of 15-25 volts. For low voltage applications (up to a few kilovolts), the gate-drive supply can be realised using isolation transformers. For example, a solution was presented in [1] to supply up to 6 gate-drives with an isolation voltage of 15 kV. However, for medium voltage (tens of kilovolts) and high voltage (hundreds of kilovolts) applications, more advanced techniques are required. This article considers a study case of a high voltage DC-DC converter based on front-to-front connected Alternated Arm Converters (AAC) (cf. Fig. 1), which is an evolution of the Modular Multilevel Converter (MMC) [2]; hereafter, this topology will be named AAC-MMC. In such a converter, the control signals can be transmitted to the gate-drive using fibre optics, but the gate-drive supply technique needs to be more deeply studied. This paper deals with the power supply of classic gate-drive boards. It is for this reason that the gate-drive specific circuits like the one reported in [3] or light triggered thyristors are not considered.

PCIM Europe 2016, 10 – 12 May 2016, Nuremberg, Germany

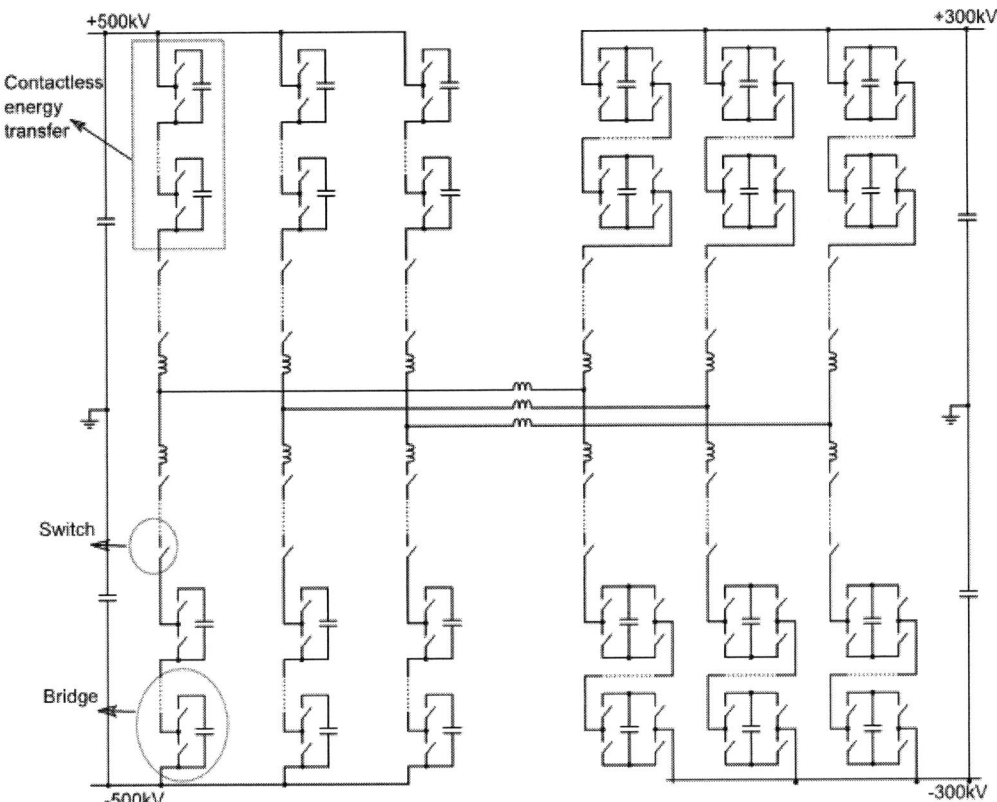

Fig. 1. Location of need of gate-drive power supply in an Alternated Arm Converter (AAC) topology [2]

In this article, different possible techniques for gate-drive supply are classified (cf. Fig. 2), after which some of them are analysed in detail. The first class of gate-drive supply (contactless energy transfer, paragraph 2) considers that the energy source is outside of the converter and referenced to ground; as a consequence, medium/high voltage insulation is required. The second class (energy harvesting, paragraph 3) uses the energy of the power converter circuit. In this case the insulation to ground is not applicable, as the gate-drive supply is referenced to the same floating potential as the power module. The last paragraph gives the conclusion about what appears to be the most promising gate-drive supply state of the art techniques for medium and high voltage applications.

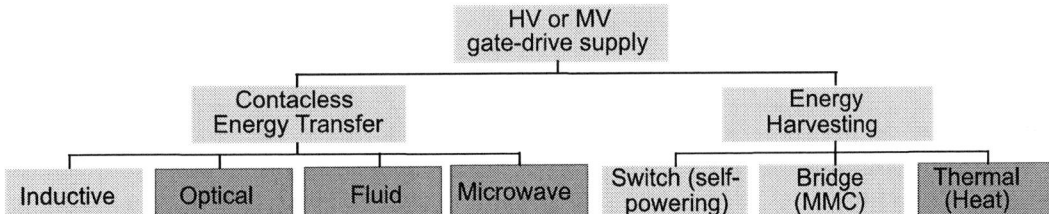

Fig. 2. Classification of MV and HV gate-drive supply possible techniques, with those that are detailed in this article highlighted in green

© VDE VERLAG GMBH · Berlin · Offenbach

2. Contactless energy transfer

Many research efforts have been carried out on developing wireless power transfer (WPT) using various physical phenomena. To the best knowledge of the authors, none of these phenomenon (e.g. optical, fluid and microwave), excluding induction, have been investigated for powering gate-drives.

Power can be transferred over optic fibers but the efficiency and the maximum transferred power are too low. A compressed-air-powered electric generator can be found in [4] but the life time is too low for industrial applications. For microwaves, efficiency and maximum power are low and the converter environment with many metallic parts is not favorable.

Inductive power transfer (IPT) is a WPT technology that uses coils with large air gaps (up to several cm). So, it provides a voltage insulation in the range of tens of kV. As it can be used to transfer several tens of watts with a good efficiency, it has been found to be a promising solution for gate-drive power supply in MV applications.

2.1. Supply single gate drive

An ICPT (Inductive Coupled Power Transfer) circuit was shown in [5] based on a DC-DC converter circuit using a coreless transformer with insulation up to 35 kV. The adopted circuit (cf. Fig. 3) achieved 100 W at the output with more than 80% efficiency.

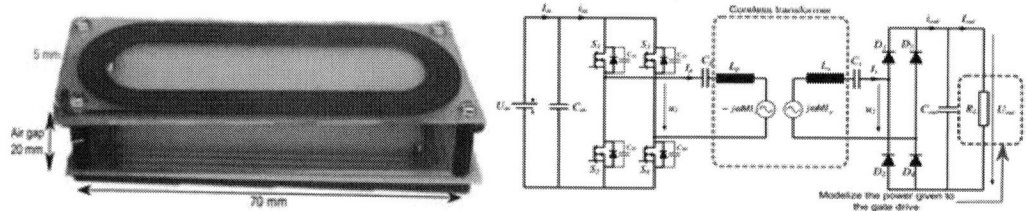

Fig. 3. Coreless transformer and the DC-DC converter used to replace the ICPT [5]

A recent study [6], using a schematic close to what has been shown in Fig. 3, also used an IPT solution to supply the auxiliary power of an MV converter. The transformer employed was made out of two concentric cylinders. The system achieved a 30 W output power with 90% efficiency.

2.2. Supply of multiple gate-drives

While the studies listed in the previous section used only one power supply for each gate-drive, other studies showed the possibility of using a single power supply for several gate-drives providing an interesting solution for multilevel converters where many gate-drives at various floating potentials are needed and for converters where there are several switches connected in series.

Also, a galvanic isolation solution using printed circuit boards has been presented in [7]. The latterly mentioned circuit used one transmitting board and six receiving boards. The galvanic isolation was assured using the required air gap between the boards (cf. Fig. 4, left image). This solution was used for an MV inverter, which had an input and output voltage of 6.6 kV. The total power received by the receiving boards was around 16 W.

The solutions that have been described in [8, 9] and [10] used a "loop wire" (cf. Fig. 4, right image) where a resonant power supply created a medium frequency current (tens of kHz) in the loop wire. The energy was distributed and rectified on each gate-drive that was magnetically coupled with the loop wire. It was considered to transfer the control signal and the energy on the same loop wire [8, 9]. Another recent article [11] proposed to use this method to supply the gate-drives in an MMC operating at MV (up to 9 kV input voltage and 35 kV output voltage, at a rated power of 250 kVA). The loop wire passed through a number of ring-core-based transformers to provide an insulation to ground level in the range of 20 kV.

© VDE VERLAG GMBH · Berlin · Offenbach

PCIM Europe 2016, 10 – 12 May 2016, Nuremberg, Germany

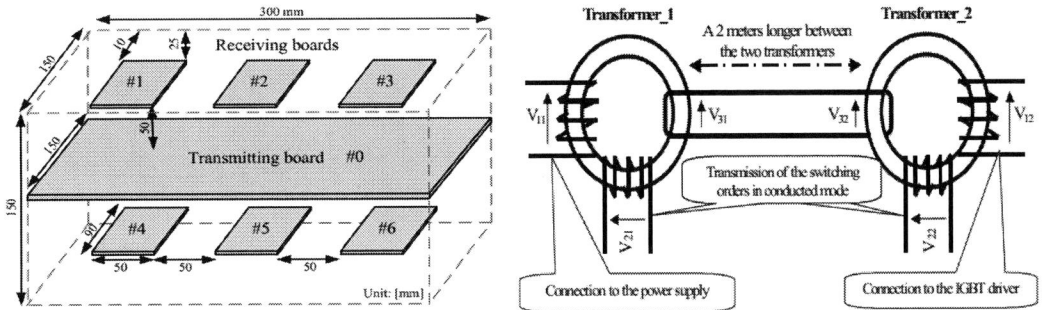

Fig. 4. Schematic view of the galvanic isolation solution using printed circuit boards for gate-drive power supply (left) [7], DGIT (Double Galvanic Insulated Transformer) with three windings (right) [9]

3. Energy harvesting

An alternative solution to contactless energy transfer is the energy harvesting from the power circuit.

An energy harvesting solution has been studied in [12] showing the possibility of using a thermoelectric module (TEM) attached to the power switch. In this study, for the best case, a small amount of power (0.5W) was harvested and the author claims that it could be used to supply the gate-drive. Also the TEM allows the reduction of the switch temperature for 4°C to 7°C. But the main drawback is the necessity to wait for the switch to be sufficiently heated in order for the TEM to work. Also the power that can be harvested depends on the switch temperature, then on the average current in the switch.

Depending on the converter topology a local energy storage, as in the capacitor in the AAC-MMC submodule (bridge in Fig. 1), may be available. Alternatively the voltage across the power switch (switch in Fig. 1) can be used to supply its own gate-drive (self-powering). For self-powering, a circuit is connected in parallel to the power switch to harvest the energy used to supply the gate-drive of this switch. Consequently, no external power source is required. Generally the energy is gathered during the switch turn-off transition, but it can also be made during turn-on transition [13], on state or even during off state [14]. The solutions applicable for self-powering may be considered for bridge capacitor but not inversely. The major difference is the impact of the circuit on the power switch behavior.

3.1. Integrable self-powering topology

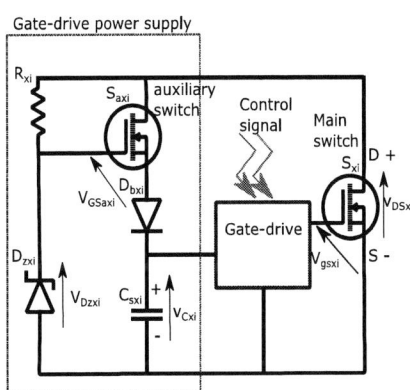

Fig. 5. Self-powering gate-drive (SPGD) circuit [27]

The most studied topology for low voltage applications is the MOSFET/MOSFET topology [15] (cf. Fig. 5).

During the turn-off of the main switch a current charging C_{sxi} flows through the auxiliary MOSFET, then some energy is stored in capacitor C_{sxi}.

The voltage between gate-source of S_{axi} is given by $V_{GSaxi} = V_{Dzxi} - V_{Csxi}$. When this voltage becomes lower than the threshold voltage, S_{axi} is switched off. The Zener diode D_{zxi} is polarized by the resistor R_{xi}. As a result, the voltage across the capacitor is regulated by the forward voltage of D_{zxi}. The diode D_{bxi} prevents the discharging of C_{sxi}, via the internal diode of the auxiliary MOSFET, when the main power switch S_{xi} is on.

The same topology was adapted to drive an IGBT in [16]. Similar topologies were based on JFET/MOSFET [15] for auxiliary/main switch and others used JFET/JFET topology [17].

© VDE VERLAG GMBH · Berlin · Offenbach

1595

These solutions have the following main advantages:

- All the components could be integrated according to the technological process of the main switch. Some studies showed solutions of integrating the self-powering components in the same package [18, 19, 20] where the insulation is given by the semiconductor package.
- It contains a small amount of components.

The important drawback of these solutions is the very low efficiency. For instance, with a DC power source V_{DS}=400 V and a regulated voltage level V_{Csxi}=20 V, the gate-drive supply efficiency reaches 5 % [18]. However, [21] claims that with replacing in Fig. 5 the auxiliary MOSFET and the polarization branch by a Darlington structure, the efficiency was much higher, making the system suitable for MV and HV applications. In [22], a performance analysis of this circuit and an optimized circuit was done in order to study the feasibility of this circuit for higher voltage applications. If moderated turn-off transient and current levels are acceptable, then the performance of the design is satisfying. In this test a 600V CoolMOS is selected as the main MOSFET. The study showed that the efficiency of the system would be improved if the design was integrated in the package of the main MOSFET since the equivalent leakage inductance of the circuit could be minimized.

3.2. Self-powered resonant gate-drive

The topology presented in [23] is for a resonant SPGD that operates for a wide range of frequency (from hundreds of Hz to hundreds of kHz). The gate-drive gets energy from the main switch. The power stage is formed of the complementary P-channel and N-channel power MOSFETs, resonant gate inductor, and self-powered unit (SPU) (Fig. 6 right image). This SPU is formed of a HV capacitor C_1, a LV storage capacitor C_2, a clamp capacitor C_3 which is used to clamp the gate voltage, auxiliary resistors R_1 and R_2 that are used to make a path for the current to charge C_2 and an inductor L.

The main drawback of this topology is that the line capacitor C_1 (Fig. 6, left image) has to be at the same voltage rating as the main switch.

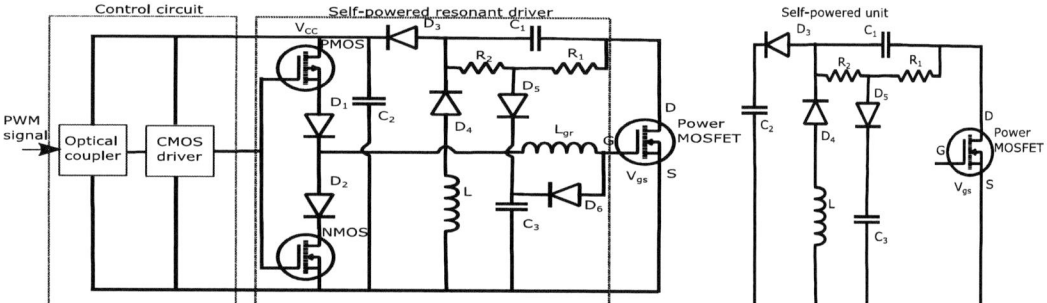

Fig. 6. High frequency resonant SPGD topology (Complete circuit, left) and its self-powered unit (Part of the circuit, right) [23]

3.3. Snubber capacitor self-powered gate-drive

In [13], another self-powered topology was proposed for thyristors. The proposed circuit drives power from energy stored in the snubber capacitor at turn-on and from the voltage across the thyristor at turn-off to supply its gate-drive.

Also in [14], a self-powered circuit for a GCT gate-drive was presented. The supply uses the energy at the snubber circuit of the GCT switch and then provides the gate-drive a regulated DC voltage.

3.4. Bridge capacitor supplied gate-drive

An AAC-MMC submodule is formed of one (cf. Fig. 7, left image) or two half-bridges. Each submodule includes a capacitor providing a voltage source that can be used to supply the gate-drive. Typically, the voltage at the DC capacitor can reach few kV which could be challenging for the designing of a suitable step-down voltage ratio converter. Indeed, the dedicated DC-DC converter should be able to withstand some kV at the input and provide some V at the output to supply the gate-drive. The use of the input DC capacitor as a floating power supply is not a recent idea, and amelioration of respective auxiliary power supply (APS) topologies or new topologies are continuously proposed. This method is used to solve the problem of supplying the gate-drive in a bridge (cf. Fig. 1).

An APS for submodules was shown in [24]. The topology used is a tapped-inductor buck converter (TI-buck converter) (cf. Fig. 7 right). It is a non-isolated step-down DC-DC converter with an input voltage of 3 kV and a 100 W power rating. S_1 is a power switch and it is formed of series of switches in order to withstand the high voltage.

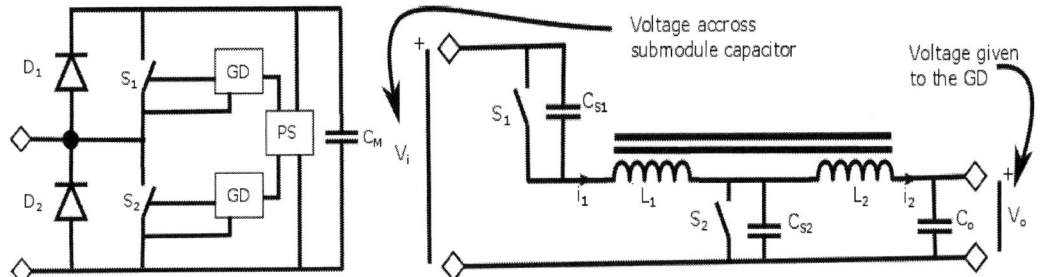

Fig. 7. Half-bridge submodule showing auxiliary power supply (PS) and gate-drive (GD) (left). Tapped inductor buck converter topology (right). [24]

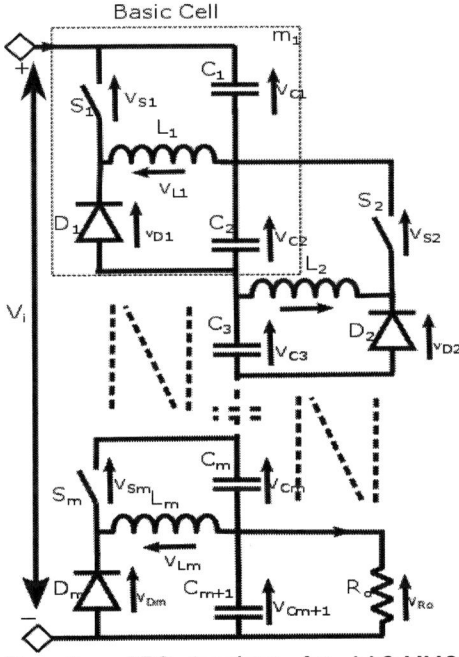

Fig. 8. APS topology for AAC-MMC submodule gate-drive formed of m number of cells. [25]

In [25], a non-isolated and non-regulated multi-cell converter was proposed (Fig. 8). In this converter, the switches are ON at the same time, in the first phase, and then are switched off, in the second phase. When the switches are on, the voltage source provides energy to the load R_0 (modelling the gate-drive) and the inductances are charged and the capacitors that were previously charged are discharging. When the switches are off, the diodes are on, and the energy stored in the inductors is transferred to the load. Then, the voltage source charges the capacitors. This topology can be used in applications where ultrahigh step-down converter is needed. It is then well adapted to the application considered in this section. The complete converter has a high step-down voltage ratio multi-cell stage followed by a regulated and isolated stage to supply an AAC-MMC submodule gate-drive. Such a converter was realized with seven cells and with a nominal input voltage range of 2.8 kV, almost 200 W was harvested at the output with a nominal voltage of 350 V. The second stage must include a DC-DC converter in order to have the low voltage needed at the gate-drive (order of 20V).

4. Conclusion

This paper presents the state of the art of gate-drive power supply techniques for MV and HV applications. The methods shown in this paper are able to provide the power required to supply the gate-drive (some tens of W). For MV applications, IPT is a promising technology but the insulation voltage in the range of hundreds of kV would lead to air gaps so large that the efficiency would be seriously decreased or the coil size would be dramatically increased. Other contactless energy transfer techniques based on microwaves or laser transmission would allow the system to operate at higher voltages but the power density of such a supply is quite low [26] and, to the best knowledge of the authors, until now no paper has been published on a gate-drive powered in such a way.

The application case of the AAC-MMC submodule is quite specific since the capacitor in each submodule provide a voltage source with the same potential reference as the power switches. Self-powering methods have the possibility to harvest energy during the turn-off transition (in general) of the power switch or from snubbers to supply the gate-drive. But as silicon carbide (SiC) components are expected to be used in HV converters and as these components will have reduced commutation durations, the energy recovered during turn-off and the energy stored in snubbers could be too low. So, innovative self-supply topologies are a potential candidate for future high voltage applications involving power switches in series.

References

[1] J. Afsharian, N. Zaragari and B. Wu, "A special high-frequency soft-switched high-voltage isolated DC/DC power supply for multiple GCT gate drivers," in *Energy Conversion Congress and Exposition (ECCE), 2010 IEEE*, 2010.

[2] T. Lu th, M. Merlin, T. Green, C. Barker, F. Hassan, R. Critchley, R. Crookes and K. Dyke, "Performance of a DC/AC/DC VSC system to interconnect HVDC systems," in *AC and DC Power Transmission (ACDC 2012), 10th IET International Conference on*, 2012.

[3] H. Fujita, "A Resonant Gate-Drive Circuit With Optically Isolated Control Signal and Power Supply for Fast-Switching and High-Voltage Power Semiconductor Devices," *IEEE Transactions on Power Electronics*, vol. 28, no. 11, pp. 5423-5430, 2013.

[4] Metal work pneumatic, "Pneumo-power," [Online]. Available: http://www.metalwork.it/eng/pneumo-wireless/index.html.

[5] R. Steiner, P. Steimer, F. Krismer and J. Kolar, "Contactless energy transmission for an isolated 100W gate driver supply of a medium voltage converter," in *Industrial Electronics, 2009. IECON '09. 35th Annual Conference of IEEE*, 2009.

[6] B. Wunsch, J. Bradshaw, I. Stevanovic, F. Canales, W. Van-der-Merwe and D. Cottet, "Inductive power transfer for auxiliary power of medium voltage converters," in *Applied Power Electronics Conference and Exposition (APEC), 2015 IEEE*, 2015.

[7] K. Kusaka, M. Kato, K. Orikawa, J.-I. Itoh, I. Hasegawa, K. Morita and T. Kondo, "Galvanic isolation system for multiple gate drivers with inductive power transfer --- Drive of three-phase inverter," in *Energy Conversion Congress and Exposition (ECCE), 2015 IEEE*, 2015.

[8] S. Brehaut and F. Costa, "Gate driving of high power IGBT by wireless transmission," in *Power Electronics and Motion Control Conference, 2006. IPEMC 2006. CES/IEEE 5th International*, 2006.

[9] S. Brehaut and F. Costa, "Gate driving of high power IGBT through a Double Galvanic Insulation Transformer," in *IEEE Industrial Electronics, IECON 2006 - 32nd Annual Conference on*, 2006.

[10] H. Wen, W. Xiao and Z. Lu, "Current-Fed High-Frequency AC Distributed Power System

for Medium--High-Voltage Gate Driving Applications," *IEEE Transactions on Industrial Electronics,* vol. 60, no. 9, pp. 3736-3751, 2013.

[11] D. Peftitsis, M. Antivachis and J. Biela, "Auxiliary power supply for medium-voltage modular multilevel converters," in *Power Electronics and Applications (EPE'15 ECCE-Europe), 2015 17th European Conference on,* 2015.

[12] Y. Tian, D. Vasic and S. Lefebvre, "Application of thermoelectricity to IGBT for temperature regulation and energy harvesting," in *Industrial Electronics (ISIE), 2012 IEEE International Symposium on,* 2012.

[13] D. M. Raonic, "SCR self-supplied gate driver for medium-voltage application with capacitor as storage element," *Industry Applications, IEEE Transactions on,* vol. 36, no. 1, pp. 212-216, 2000.

[14] W. Hu, B. Wu, N. Zargari and Z. Cheng, "A Novel Self-Powered Supply for GCT Gate Drivers," *IEEE Transactions on Power Electronics,* vol. 24, no. 4, pp. 1093-1099, 2009.

[15] R. Mitova, "Intégration de l'alimentation de la commande rapprochée d'un interrupteur de puissance à potentiel flottant," *Institut National Polytechnique de Grenoble-INPG,* 2005.

[16] T. V. Nguyen, P. Jeannin, E. Vagnon, D. Frey and J.-C. Crebier, "Series Connection of IGBTs With Self-Powering Technique and 3-D Topology," *IEEE Transactions on Industry Applications,* vol. 47, no. 4, pp. 1844-1852, 2011.

[17] D. Peftitsis, "On gate drivers and applications of normally-on SiC JFETs," 2013.

[18] J.-C. Crébier and N. Rouger, "Loss free gate driver unipolar power supply for high side power transistors," *Power Electronics, IEEE Transactions on,* vol. 23, no. 3, pp. 1565-1573, 2008.

[19] N. Rouger, J.-C. Crebier, R. Mitova, L. Aubard and C. Schaeffer, "Fully integrated driver power supply for insulated gate transistors," in *Power Semiconductor Devices and IC's, 2006. ISPSD 2006. IEEE International Symposium on,* 2006.

[20] N. Rouger and J.-C. Crebier, "Toward Generic Fully Integrated Gate Driver Power Supplies," *IEEE Transactions on Power Electronics,* vol. 23, no. 4, pp. 2106-2114, 2008.

[21] N. Rouger, J.-C. Crebier, H. T. Manh and C. Schaeffer, "Toward integrated gate driver supplies: Practical and analytical studies of high-voltage capabilities," in *Power Electronics Specialists Conference, 2008. PESC 2008. IEEE,* 2008.

[22] S. Busquets-Monge, D. Boroyevich, R. Burgos and Z. Chen, "Performance analysis and design optimization of a self-powered gate-driver supply circuit," in *Industrial Electronics (ISIE), 2010 IEEE International Symposium on,* 2010.

[23] H. Wang and F. Wang, "A self-powered resonant gate driver for high power MOSFET modules," in *Applied Power Electronics Conference and Exposition, 2006. APEC '06. Twenty-First Annual IEEE,* 2006.

[24] T. Modeer, S. Norrga and H.-P. Nee, "High-voltage tapped-inductor buck converter auxiliary power supply for cascaded converter submodules," in *Energy Conversion Congress and Exposition (ECCE), 2012 IEEE,* 2012.

[25] G. Tibola, J. Duarte and A. Blinov, "Multi-cell DC-DC converter with high step-down voltage ratio," in *Energy Conversion Congress and Exposition (ECCE), 2015 IEEE,* 2015.

[26] S. Sohr, R. Rieske, K. Nieweglowski and K. J. Wolter, "Laser Power Converters for optical power supply," in *Electronics Technology (ISSE), 2011 34th International Spring Seminar on,* 2011.

[27] S. Busquets-Monge, J. Rocabert, J.-C. Crebier and J. Peracaula, "Diode-clamped multilevel converters with integrable gate-driver power-supply circuits," in *Power Electronics and Applications, 2009. EPE '09. 13th European Conference on,* 2009.

PCIM Europe 2016, 10 – 12 May 2016, Nuremberg, Germany

The Optimized Gate Driver Design Techniques for IGBT Properties and Downsizing in Eco-Friendly Vehicle

KangHo Jeong, Hyundai Motor Company, Republic of Korea, kangho.j@hyundai.com
KiJong Lee, Hyundai Motor Company, Republic of Korea, kjlee1@hyundai.com
JiWoong Jang, Hyundai Motor Company, Republic of Korea, jangjw@hyundai.com
SangChul Shin, Hyundai Motor Company, Republic of Korea, tiang@hyundai.com
KiYoung Jang, Hyundai Motor Company, Republic of Korea, kyjang@hyundai.com
JinHwhan Jung, Hyundai Motor Company, Republic of Korea, jhjung10@hyundai.com

Abstract

The three-phase PWM inverter, which is used for eco-friendly vehicles, requires following fea tures such as low power conversion loss, high fault protection reliability and downsizing. The gate drive unit to operate IGBT (Insulated Gate Bipolar Transistor) module of the inverter is the important component which is developed to operate these properties. In this paper, optimized gate drive unit design techniques are presented. First, the over-current protection technique with temperature compensation is proposed to reduce inverter switching loss for increasing the efficiency of the PWM inverter. After then, the separate protection methods to enhance the protection reliability of IGBT module are described in any abnormal conditions. At last, the power supply circuit design of gate drive unit contributing to downsize inverter is presented.

1. Introduction

The main components of eco-friendly vehicle electric driven train (hybrid or electrical system) can be divided into motor, high-voltage battery and inverter. The inverter is a conversion system that can supply DC power from the high-voltage battery to the AC motor and reverse the energy flow to recharge high-voltage battery. IGBT module is mostly used as the main switching device in the inverter. The drive units of inverter are configured to motor controller unit and gate drive unit for turning on and off a gate of IGBT. The gate driver IC, which amplifies the control signal by boosting the voltage and current level in the gate drive unit, is also necessarily required. Fundamentally, on/off operation of the gate driver IC is same as to charge and discharge the input and reverse transfer capacitance of the IGBT. The switching loss is determined by a gate-resistor connected between the driver IC and IGBT. Reducing

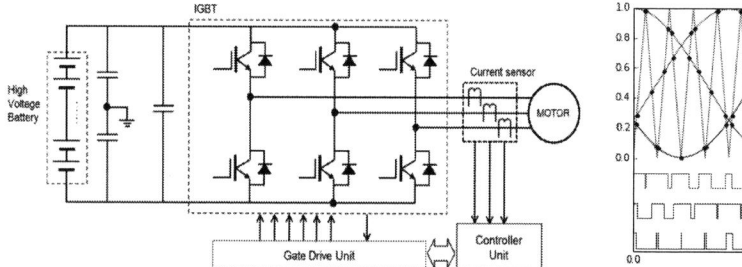
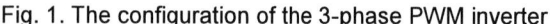
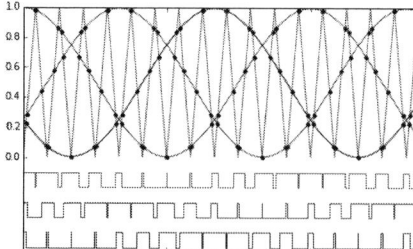

Fig. 1. The configuration of the 3-phase PWM inverter

© VDE VERLAG GMBH · Berlin · Offenbach

the switching loss is an important role of the gate drive unit, which contributes to reduce inverter conversion loss and increase fuel efficiency of the eco-friendly vehicle. In addition, the gate drive unit must have functions to check the status of the IGBT module for the various protection modes, such as module temperature monitoring for preventing fault under high-temperature and over-current detection function to prevent overload, etc. Consequently, gate drive unit design is a major factor to improve car efficiency, minimize the inverter size and protect drive system of eco-friendly vehicles.

2. The optimized over-current detection techniques

2.1. Design considerations

Over current sensing

The gate drive unit of the inverter must have functions that prevent breakout or burnout of the IGBT module from over-current. The gate drive unit is used to detect an over-current state by providing feedback to the sensing circuit which can safely turn-off the IGBT before damaged. In the past, de-saturation detection method, which monitors the voltage between the collector and the emitter of the IGBT at the over-current state, has been used in various applications. But as the current density of the IGBT becomes higher, the on-chip detection method is preferred to detect over-current of the IGBT due to requirement the short SOA (Safety Of Area) and to avoid power stress in serious desaturation state. This method is implemented by separation of a defined number of IGBT cells (emitter sense-IGBT) from power emitter metallization. The current flowing through the emitter sense-IGBT is a fraction of the main-IGBT, it measures the amount of collector current. Detection of the over-current is configured that emitter sense-IGBT current (Is) flows the low ohmic resistor mounted on the gate drive unit, and a system detects over-current state by setting threshold voltage. Fig.2 shows a typical structure. The voltage across the low ohmic resistor is compared with a set reference voltage of comparator to report an over current fault output signal. With appropriate design of the low ohmic resistor and set reference voltage it can effective protect to IGBT and robust short circuit protection that avoids severe fault conditions. Furthermore, it is important techniques for improving reliability of the drive-system of eco-friendly vehicles.

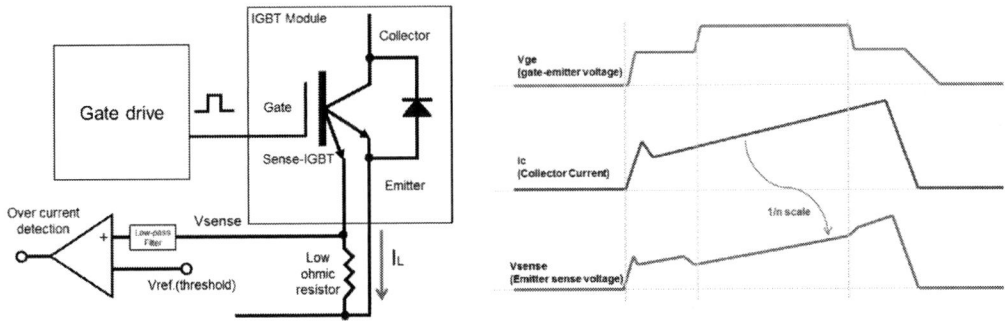

Fig. 2. Typical structure of on-chip emitter sense current and waveform

Temperature dependency of on emitter sense-IGBT

The current sensing voltage (Vsense) has large dependency from temperature. This characteristic is due to the difference in output characteristics of the main IGBT and the emitter sense-IGBT. The main reason is the difference of current density between main IGBT and sense IGBT. While main IGBT cells are surrounded by other cells, sense IGBT cells are isolated so carrier density of the sense IGBT is lower than the main IGBT. It induces imbalance of the detection level in the low and high temperature characteristics. In figure 3, output characteristics of main-IGBT and emitter sense-IGBT is shown. As temperature rises, emitter sense-IGBT conducts more proportion current than actual conducted collector current, which makes negative correlation between the junction temperature and over-current detection level (Fig. 4)

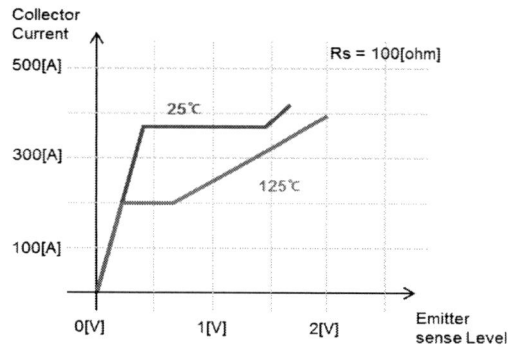

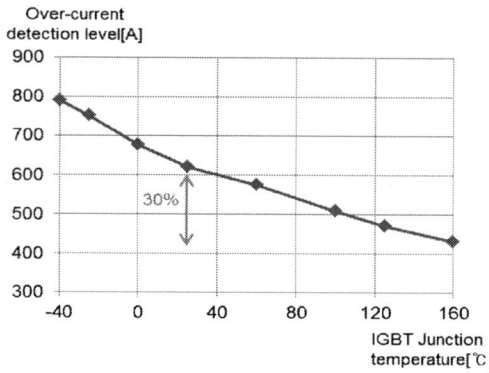

Fig. 3. Output characteristics of main-IGBT and emitter sense-IGBT

Fig. 4. OC detection level by temperature

IGBT junction temperature sensing

To protect IGBT chip and freewheeling diode from the over-heat, the gate drive unit has function to monitor junction temperature of main chip from the temperature sensor mounted to the IGBT module. The temperature sensors, NTC (Negative Temperature Coefficient) resistor or diode is mainly used. The NTC is located on the substrate of IGBT module. There are disadvantages that some inaccuracy because it is not sensing value from directly in the IGBT-chip or freewheeling diode. In addition, the resolution of the sensing value is decrease at a high temperature due to the NTC resistance characteristic. The NTC not suitable for fast over-temperature protection because it is designed for detection of a long time overload conditions. On the other hand, diode sensor can integrated to the IGBT chip so it is possible to accurately and fast measure the junction temperature of IGBT chip and freewheeling diode. In addition, diode sensor is more accurate at high-temperature than NTC-resistor because diode has the linear forward voltage drop over entire temperature range. Fig.6 shows the comparing of these two sensors.

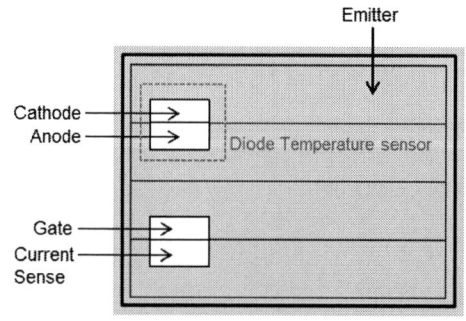

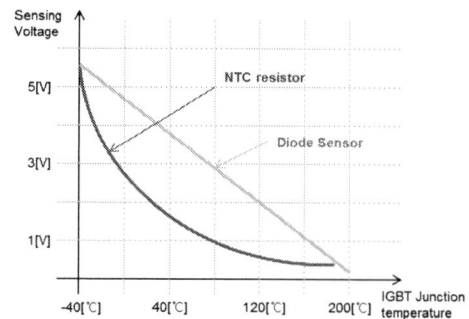

Fig. 5. Diode temperature sensor on IGBT chip

Fig. 6. Ouput characteristics of NTC and Diode

© VDE VERLAG GMBH · Berlin · Offenbach

Gate resistor design for minimizing switching loss

It is important to minimize the switching loss of the IGBT in order to increase the efficiency of the inverter because conduction loss of the IGBT is fixed by material but switching loss is possible to optimize by design of gate drive unit. Gate on and off resistors determine the switching loss of the IGBT. As the gate off resistance is small, the switching off loss is reduced but which induce to increase switching peak voltage (Vce, Collector-Emitter voltage)(Fig.7). For this reason, at the same time to satisfy the maximum Vce of the IGBT module, it is necessary to design that can minimize switching loss.

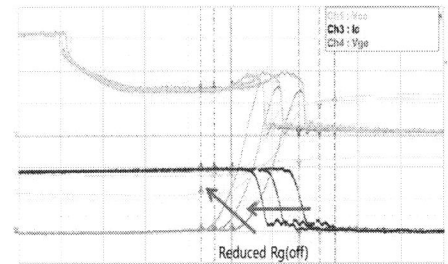

Fig. 7. Switching off loss by gate off resistors

2.2. Over-current detection with temperature compensation for reducing switching loss

As described above, as junction temperature goes high, the over-current detection level goes down for a fixed threshold level of gate drive unit. This feature affects negative impact on optimized gate drive switching operation. The (1) line of Fig.9 shows the measuring switching collector current when the overcurrent is detected by the gate drive unit over the vehicle operating temperature region. The design values of the low ohmic resistor for over-current protection must be selected within a range not affected the normal operating maximum current of the IGBT at high temperature because it is possible to erroneous detection at the normal operating switching current. As can be seen from the (1) line of Fig.9, if the maximum peak drive current of PWM inverter is 400A, switching current when overcurrent is detected at a normal temperature (25degree) is about 600A from the low ohmic resistor set in consideration of the worst case of high-temperature. Therefore the increasing of protection level is required, which leads to increase switching collector current. In order to meet this, no choice but to take the design of large value for gate off resistance to satisfy the maximum collector-emitter breakdown voltage (Vce). This design induces a decrease in the overall efficiency by increasing the off loss of the PWM inverter. It is not optimized switching operating design of gate drive unit.

In order to improve this problem, a gate drive unit used a technique to compensate for over-current level is set by a circuit as a function of temperature. Using the characteristics having the opposite tendency of the diode temperature sensor and on chip current sensor depends on the temperature, we developed the circuit that compensate for the two sensing value. Fig. 8 shows circuit configuration. The operation principle is that voltage compensation of the over current sensing voltage by the sensing voltage of the diode temperature sensor. Configuration of the circuit can be realized by PNP or NPN transistor, diode and resistor. It can realize uniform overcurrent detection level over the entire temperature range. Fiq.9 shows the effectiveness of the temperature compensation. Without compensation, the deviation of over current trip level is about 30% ((1) line of Fig. 9) and gate off resistance should be designed in

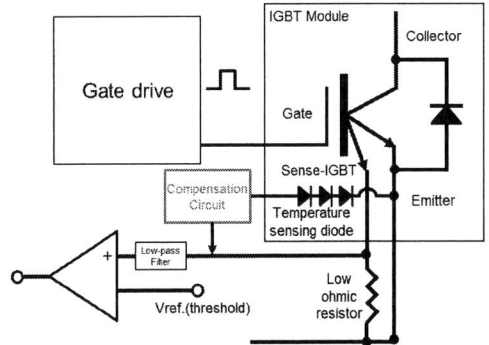

Fig. 8. Over-current detection with temperature compensation

consideration of maximum switching current at the lowest temperature (800[A]). Also low ohmic resistor should be designed not to effect normal switching current region at the high temperature. However applying the temperature compensation, the deviation was reduced to under 5% ((2) line of Fig. 9). Over-current level in the entire junction temperature has a constant value (about 600[A]). Gate off resistance can be designed based on 600[A] switching current. Another advantage is that it is possible to reduce the over current detection level because coming out same properties at high temperature. As a result, the over-current detection level can be optimized reduced.

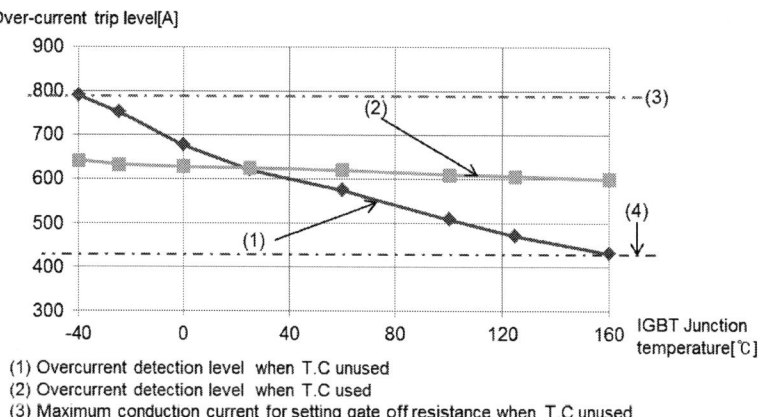

(1) Overcurrent detection level when T.C unused
(2) Overcurrent detection level when T.C used
(3) Maximum conduction current for setting gate off resistance when T.C unused
(4) Maximum conduction current for setting gate off resistance when T.C used
※ T.C : Temperature Compensation

Fig. 9. The effect of the temperature compensation (Over-current trip level over temperature)

Fig. 10 shows comparing of switching off loss without temperature compensation (left side) and with temperature compensation (right side). The gate off resistance via the temperature compensation is reduced by 30%. As a result, switching off loss could be reduced by 38%. The overall efficiency of inverter has increased from the previous application.

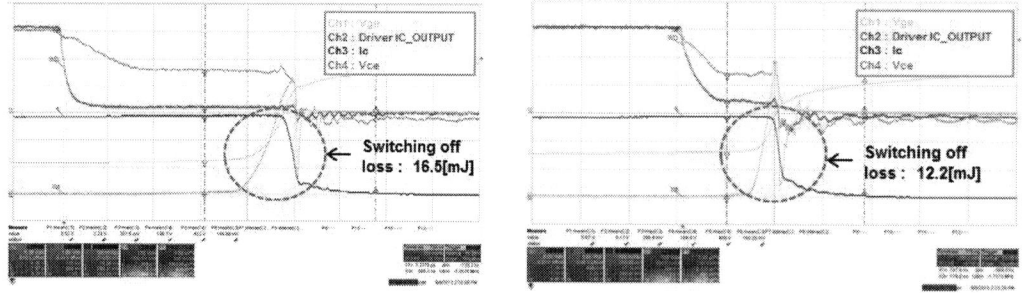

Fig. 10. Switching off loss without temperature compensation (left side) and with temperature compensation (right side)

2.3. Separate detections of over current for enhancing fault reliability

Over-current fault conditions of the IGBT are largely divided by a load over current (OC) and short circuit current (SC) from arm-short. The OC fault is an over current state that occurs in a state in which the connection of inductor (motor) is maintained so IGBT collector current rising speed (di / dt) is not fast in a fault condition. Also IGBT is in saturation state when gate drive unit operates protection such as soft shutdown so gate turn-off action takes a lot of time due to miller period where gate voltage keeps constant.

© VDE VERLAG GMBH · Berlin · Offenbach

On the other hand, the SC fault is an over current state that occurs in a state in which failure in internal of IGBT module or 3-phase output lines is shorted or when the same gate-on signal is applied to top and bottom side of half bridge. Recent IGBT, the current density of chip is increased due to reduction size, has reduced the short-circuit capability. Consequently the SC fault state demands very fast fault recognition and shutdown action since the di / dt is very fast. Also IGBT is active state when a fault condition occurs because most of the failure situations that occur on the state of turn on initially.

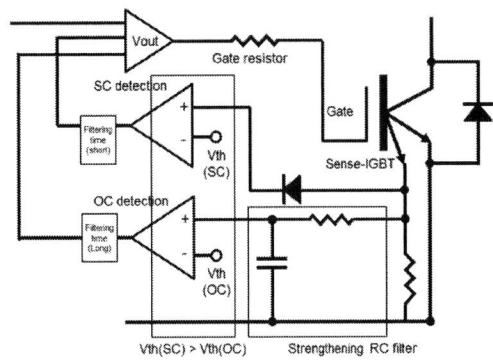

Fig. 11. Configuration of the separate detections of over current

Fig. 11 shows the optimized detection configuration of the separate detections over current. OC protection has strong external RC low pass filter to protect from a lot of noise induced by fast power switching operation and compact inverter package in order to prevent stopping driving vehicle due to false detection. Also it has purpose to minimize wrong diagnosis from the blip of the emitter current sense voltage generated at initial turn on. On the other hand SC protection has no external RC low pass filter because SC state requires fast turn off time. However it minimizes false detection due to noise by rising overcurrent detection threshold. Fig.12 shows the gate drive unit operation of the arm-short state. By separate detections of over current, it is possible to implement a more robust overcurrent detection function to vehicle noise and it prevents burnout of the IGBT module by fast protecting operation. It enhances the reliability of the fault protection of PWM inverter, it is possible to minimize the inverter parts replacement.

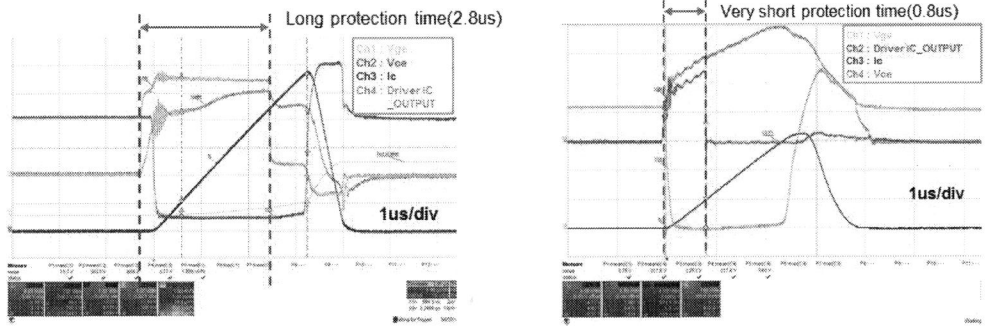

Fig. 12. Arm-short protection operation of normal gate drive unit (left side) and proposed gate drive unit (right side)

3. Power supply circuit design of gate drive unit for downsizing

3.1. Design considerations

The gate drive unit included to isolated power supply design because electrical potential between the motor controller unit (microcontroller) and switching device to drive high power is not same. Generally flyback DC-DC converter including transformer is many used. The gate drive unit design considering insulation is important because the coexistence of high-voltage and low-voltage. This should ensure an insulation distance that meets the international standard (IEC-60664-1). Optimal design of the power supply enables to effectively use the PCB area and it is an important way of miniaturization.

3.2. Distributed power supply structure

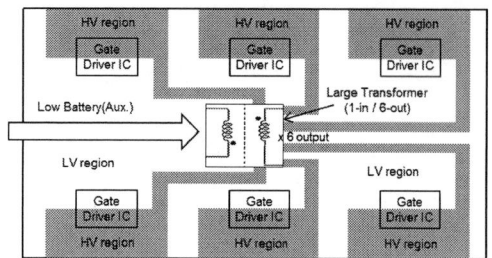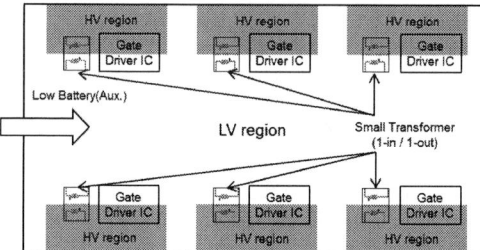

Fig. 13. PCB layout configuration of centralized (left side) and distributed power structure(right side)

In order to satisfy to the downsized inverter and power module, distributed power supply structure is applied to our gate drive board unlike conventional gate drive board using centralized power structure. Fig.13 shows the power configuration comparison between centralized and distributed power structure. In case of centralized configuration, six output power traces from the transformer are routed across over the entire PCB area because it is necessary to apply power to the IGBT driving stage which is arranged in the entire PCB. Since the separation distance must be kept for isolation safety, it causes an increase unusable PCB area and overall gate drive board size. On the other hand, distributed power supply structure is more flexibility in designing the circuit layout. Low voltage region can be distinguished and isolated easily from the high voltage region. Overall PCB routing becomes simple and straight forward which enhance the signal integrity and avoid unfavorable noise disturbance to the signal line. In order to implemented distributed power supply structure, we use the specialized gate driver IC integrated DC-DC flyback controller (ACPL-32JT, Avago tech). This has big effect on the optimized component placement and reducing the amount of passive component of the drive board. Fig.14 shows gate drive board design of our car project. The one board is consisted of six drive stage for driving one motor. The figure on the left is the gate drive board using centralized power structure. Because of the high-voltage power traces out from one transformer, area cannot be utilized is increased and low-voltage area is reduced. As a result, it is difficult to downsize the drive board. On the other hand, the figure on the right is the gate drive board using distributed power supply structure. Compared to centralized power supply structure, it has achieved reduction of approximately OO% of the board size but the total Low voltage area is increased. As a result, it has contributed to the size reduction of the PWM inverter which adapted new eco-friendly car model.

As well as, another advantage of having distributed power supply is cost saving as overall material cost can be minimized. It brings the effect of reduction cost by PCB miniaturization

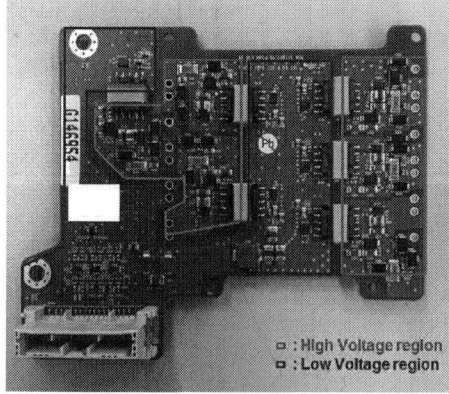

Fig. 14. Our gate drive units of centralized (left side) and distributed power structure(right side)

and reduction of the PCB layer. This can be a big benefit in eco-friendly car market having purpose to reduction of material costs. We have achieved a cost reduction about OO% than before previous car model.

A large transformer (one input – six output) using centralized power structure typically emits a lot of more EMI noise than six small (one input – one output) transformers. The measurement result fig.13 shows the comparing of EMI emission of each structure. Over the entire test frequency domain, the centralized structure shows a low amount of noise radiation than the distributed structure. This would contribute to reduction radiation noise of the PWM inverter and enhance EMI performance.

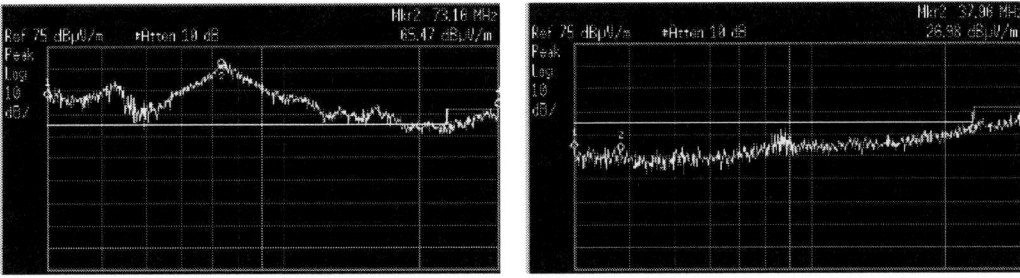

Fig. 15. EMI emissions of centralized (left side) and distributed power structure(right side)

4. Conclusion

Optimization of the over-current level by temperature compensation has functions to implement the design of the lower inverter loss. In actual vehicle application case, there is an advantage of reducing the switching off loss up to 30%. And optimized protection in accordance with the failure situation is to reduce wrong detection from noise and it served to effectively protect the IGBT module for fast protection in adverse condition. It also reduces replacement costs due to the IGBT failure, which eventually leads to cost reduction in eco-friendly vehicles. Finally, application of the distributed power supply structure is a lot of contribution to the downsizing of our inverter

5. Reference

[1] Daniel Domes, Ulrich Schwarzer. "IGBT-Module integrated Current and Temperature Sense Features based on Sigma-Delta Converter", PCIM 2009

[2] M.Kudoh, Y.Hoshi(*), S.Momota, T.Fujihi1.a and K.Sakurai "Current Sensing IGBT for Future Intelligent Power Module", IEEE 1996

[3] Eric R. Motto, John F. Donlon, Powerex Incorporated, "IGBT Module with User Accessible On-Chip Current and Temperature Sensors"

[4] Hiroaki Ichikawa, Takeshi Ichimura, Shin Soyano, "IGBT Modules for Hybrid Vehicle Motor Driving", FUJI ELECTRIC review

[5] Roy Tan, Choo Mei Zhen, "Distributed DCDC Power Supply Enable Greater Design Simplicity, Robustness, EMI Performance and Cost Saving", AVAGO Tech.

A Revisit to Resonant Gate Driver and a Hybrid Driver to Improve EMI vs. loss Tradeoff for SiC MOSFET

Chi-Ming Wang, Toyota Motor Engineering & Manufacturing - North America, United States, chi-ming.wang@toyota.com

Abstract

In the past two decades, numerous active gate drivers were designed or invented to achieve 'defined switching behavior' of Si IGBT [1]. The tradeoff between switching loss and electromagnetic interference (EMI) has always been a challenge. With SiC MOSFET having switching speed (drain voltage V_{DS} or drain current I_D transient) that could be tenfold ore more compared with Si IGBT, it is not feasible to use closed-loop active gate driver to control EMI generated from SiC MOSFET. In this paper, the concept of using high gate impedance and open-loop driver to control EMI is first discussed. Subsequently, a hybrid driver is invented to further reduce switching loss in addition to EMI. Simulation results confirm that tradeoff between switching loss and EMI can be improved for SiC MOSFET.

1. Introduction

With the advent of SiC MOSFET and more advanced cooling technology, higher switching speed and switching frequency are becoming the trend [2]. Compared with Si IGBT, SiC MOSFET can be switched at much higher speed without incurring more loss due to its bandgap and unipolar nature [3,4]. However, the well-known tradeoff between switching loss and EMI still limits SiC MOSFET switching speed in any application sensitive to EMI noise. An example from simulation shows this tradeoff in Fig. 1. SiC MOSFET produces only a third of switching loss compared to Si IGBT, but it generates about 20dB additional EMI noise across the frequency range of 1 – 100 megahertz.

In the literature, many different techniques were proposed to reduce EMI and the summary of these techniques can be found in [5,6]. Among these methods, modification of the slopes of drain voltage (dV_{DS}/dt) and drain current (dI_D/dt) provides benefits such as no additional filter is required, radiation noise reduction, and weight reduction. Several active gate drivers implementing closed-loop feedbacks to control dV_{DS}/dt and dI_D/dt were proposed to achieve loss or EMI reduction for Si IGBT [1,7-9]. However, for SiC MOSFET, the target switching speed is much higher than Si counterparts, making closed-loop control infeasible due to A/D sampling speed and controller/gate IC delay limitations. For papers that use active gate driver to reduce EMI generated from SiC MOSFET, switching speed is much lower than optimal values [10]. Clearly, a new driver has to be designed to reach maximum potential of utilizing SiC MOSFET power applications.

The goal of this paper is to provide a more practical gate driver solution to improve switching loss vs. EMI tradeoff for SiC MOSFET. The limitations of conventional voltage driver are discussed in Chapter 2. The concept of using high gate impedance open loop driver to reduce EMI is explained in Chapter 3. Chapter 4 shows the proposed new hybrid driver structure and simulation results under specific operating conditions. In comparison to conventional voltage driver, the proposed hybrid driver reduces EMI and switching loss simultaneously.

© VDE VERLAG GMBH · Berlin · Offenbach

PCIM Europe 2016, 10 – 12 May 2016, Nuremberg, Germany

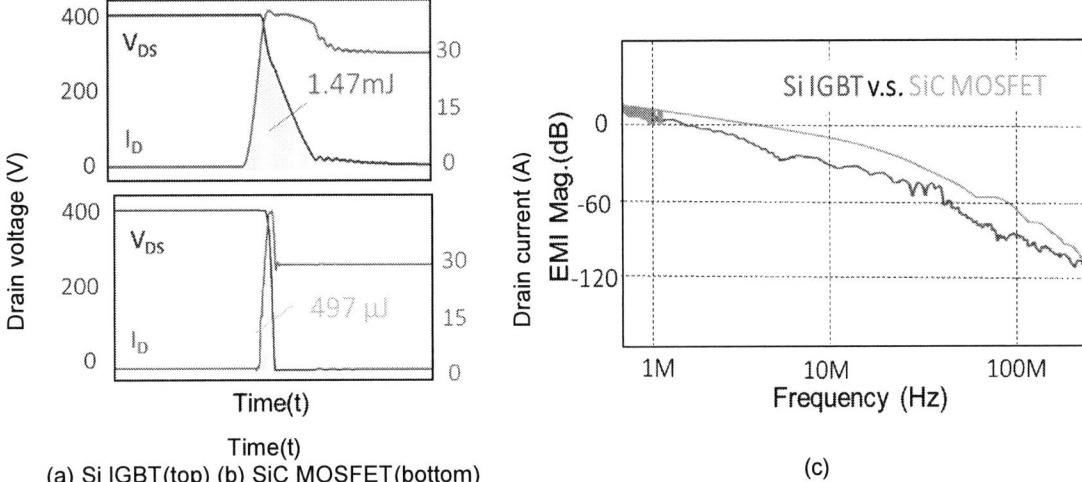

(a) Si IGBT(top) (b) SiC MOSFET(bottom)   (c)

Fig. 1. Tradeoff between switching loss vs. EMI. (a) Si IGBT switching loss (b) SiC MOSFET switching loss. (c) EMI comparison. From this figure, Si IGBT has higher switching loss/lower EMI and vice versa.

2. Limitations of conventional voltage driver

Due to its simplicity and MOSFET gate characteristics, conventional voltage driver [11] are widely used to drive Si IGBT and SiC MOSFET. Driver ICs, a gate resistor, and a unipolar or bipolar power source are the only required components to constitute a voltage driver. However, several limitations of it exist and are reported in [12]. One of the big limitations is that during device switching transient, gate charging and discharging currents are determined by impedance of gate driving loop Z_g and internal nonlinear parameters of the switching device. The gate current during switching transient determines device switching waveforms dV_{DS}/dt and dI_D/dt. According to [1], dV_{DS}/dt affects EMI noise generated from the device; therefore, to control EMI generation internally from the device, gate charging and discharging current have to be controlled externally.

Based on conventional voltage driver structure, it has no control over gate current other than using different gate resistor R_g values to change driving loop impedance. For conventional driver for Si devices, gate resistor is chosen to be large enough to damp ringing and overshoot in the gate and power loop. Without loss of generality, in this configuration, driving loop impedance is dominated by gate resistor R_g, and gate current $i_g(t)$ is almost constant during drain voltage swing interval except initial V_{DS} drop period. This is demonstrated using simulation in Fig. 2(a) and selecting $R_g = 15\Omega$. Whereas, for SiC devices, high speed driving is more desired with low gate resistor to fully benefit from its advantages. In this case, the gate loop impedance is dominated by the MOSFET reverse capacitance C_{rss}. As shown in Fig. 2(b) with $R_g = 1.5\Omega$, the gate current $i_g(t)$ increases rapidly during drain voltage V_{DS} swing as reverse capacitance C_{rss} increases rapidly following $i_g(t) = C_{rss} \times (dV_{DS}/dt)$. It behaves very differently compared with (a). If small gate resistance is used, conventional drivers cannot control gate current shape, and in turn cannot control EMI generation and is not suitable for SiC MOSFET operation.

To eliminate the effect of C_{rss} (since it is an internal parameter that users have no control) and design a feasible gate driver, a high gate loop impedance together with no closed loop feedback should be present during voltage swing dV_{DS}/dt transient. This viewpoint makes open loop current source driver a proper choice that offers EMI reduction capability.

© VDE VERLAG GMBH · Berlin · Offenbach

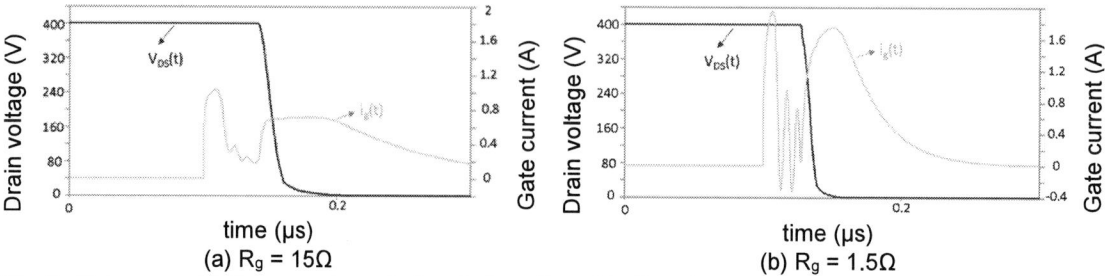

(a) $R_g = 15\Omega$ (b) $R_g = 1.5\Omega$

Fig. 2. Gate current $i_g(t)$ comparison during drain voltage $V_{DS}(t)$ swing using (a) large gate resistor vs. (b) small gate resistor

3. A revisit to resonant current gate driver

In the literature, two types of open loop current source gate drivers have been reported, namely the 'constant current gate driver' and the 'resonant current gate driver,' Two configurations of them were described in [12] and [13] respectively. Compared with constant gate driver, the structure of resonant current driver is simpler and does not require four additional switches. This resonant driver was selected for simulation to demonstrate EMI reduction capability. Fig. 3 shows the driver configurations for comparison. EMI noise results using drain voltage V_{DS} spectrum are shown in Fig. 4. The supply voltage V_{dri2}, resonant inductor L_r, and gate resistor R_{g2} were selected and tuned to make sure rail to rail gate voltage applied to SiC MOSFET are the same in both cases. Therefore, supply voltages and gate resistors are different for these drivers (herein use subscript 1,2 to distinguish one from another). Switching frequency was selected to be around tens of kHz. SiC MOSFET TO-247 package Spice model from Cree was calibrated, modified, and used for simulation.

The results show that from 10M-100MHz, resonant current driver has about 5-10dB EMI reduction with about 7% additional switching loss compared with voltage driver. In addition to EMI reduction, this driver preserves driver loss reduction capability as well. To bring switching loss back to about the same as conventional driver or further reduce it, switching loss mechanism was investigated. One problem of this configuration is that the initial sinusoidal current rises slowly relative to conventional driver, causing additional drain current and drain voltage overlap (switching loss). To solve this problem, a new 'hybrid driver' structure is proposed in this paper to realize reduction of both switching loss and EMI to improve tradeoff between them together.

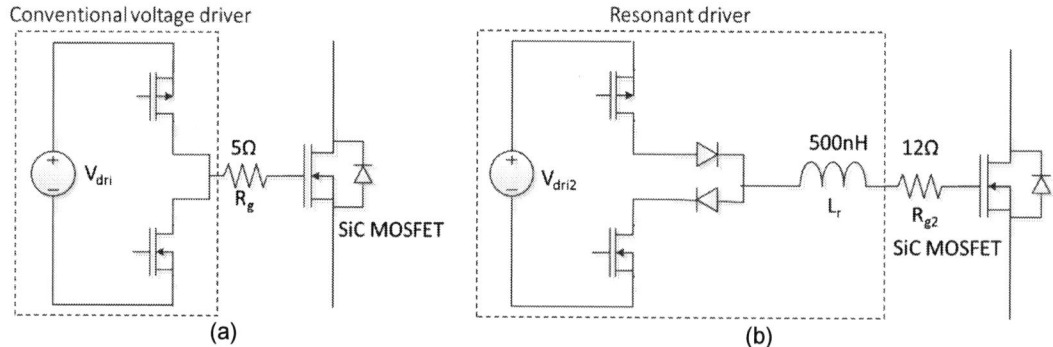

Fig. 3. Simplified circuit diagram used in simulation for switching loss vs. EMI comparison between (a) conventional voltage driver and (b) resonant current gate driver

PCIM Europe 2016, 10 – 12 May 2016, Nuremberg, Germany

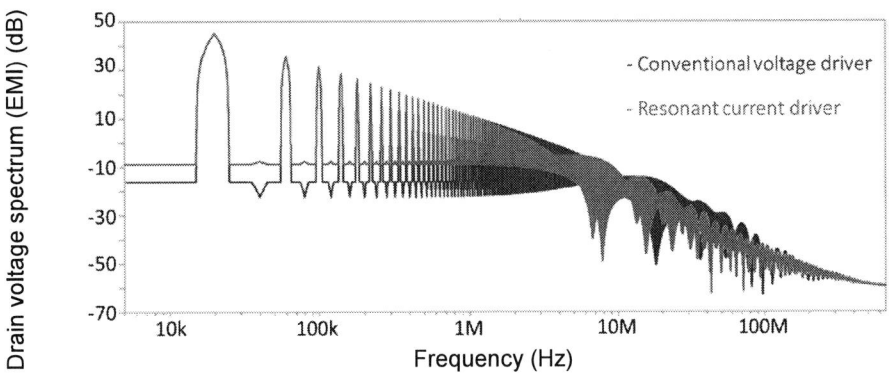

Fig. 4. EMI noise (drain voltage spectrum) comparison between conventional voltage driver (blue) and resonant current driver (red)

4. Hybrid Driver

To further reduce switching loss of resonant current driver, a 'hybrid gate driver' is proposed in this section. Fig. 5 shows the structure of this hybrid driver. The configuration is based on resonant current driver from [13] with some modifications. The resonant inductor L_r in standard resonant driver is replaced with an inductor L_r coupled to a DC biasing coil that shares same magnetic core. This coupled inductor design is similar to what was used in [14]. Before turning on or turning off SiC MOSFET, the DC biasing current $I_{bias}(t)$ is provided and it saturates the coupled magnetic core that makes resonant inductor L_r equivalent to a short circuit. Therefore, during the start of each switching transient, the hybrid driver behaves like a conventional voltage driver shown in Fig. 6 (a). Right before drain voltage V_{DS} swing, this DC biasing current is removed and the hybrid driver becomes a resonant current driver that has high gate loop impedance shown in Fig. 6 (b). Simulation waveforms of gate current and gate voltage of this driver is shown in Fig. 7. It can be seen that the gate current has a high initial peak current like a conventional voltage driver and a sinusoidal current subsequently that resembles resonant current driver. Saturation of the core is selected to be 0.5A and $I_{bias}(t)$ is 1A in simulation.

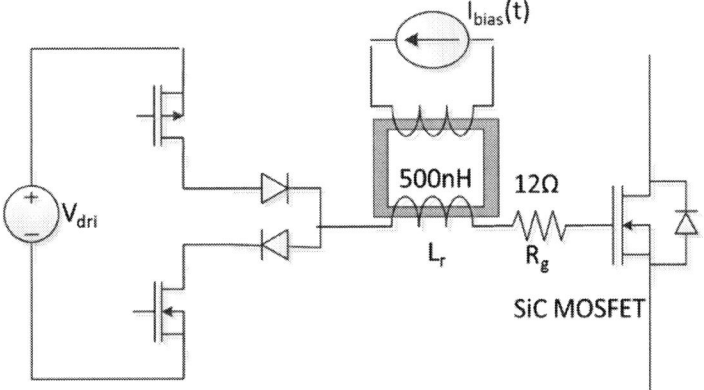

Fig. 5. The structure of proposed hybrid driver

© VDE VERLAG GMBH · Berlin · Offenbach

PCIM Europe 2016, 10 – 12 May 2016, Nuremberg, Germany

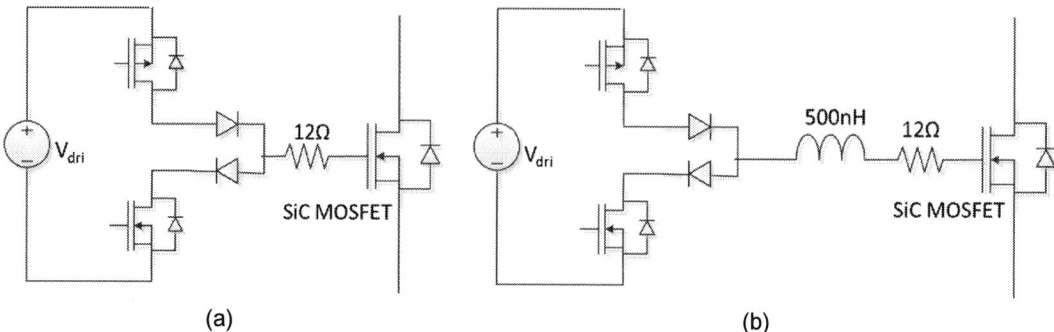

Fig. 6. Hybrid driver equivalent circuit diagram when (a) L_r is saturated by DC biasing current and (b) DC saturation current is removed

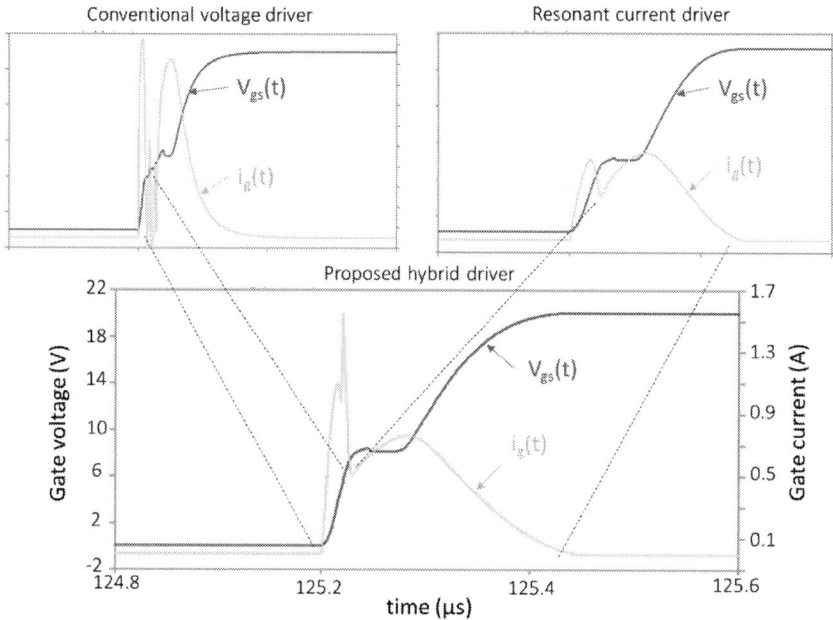

Fig. 7. Gate current $i_g(t)$ of hybrid driver shows that it combines the gate current feature of conventional voltage driver and resonant current driver

By controlling the resonant inductor between zero biasing and saturation, proposed hybrid driver reduces turn-on/off delay compared with resonant driver and decreases switching loss by about 25% compared with conventional voltage driver in simulation. To make a fair comparison, the value of supply voltage V_{dri}, L_r and R_g are tuned to have the same steady-state rail to rail gate voltage $V_{gs}(t)$ applied to the switching device (SiC MOSFET). At the same time, with high impedance controlled sinusoidal gate current, EMI level (drain voltage spectrum) is kept the same as resonant current driver which is 5-10dB lower than voltage driver. Fig. 8 shows the overall comparison among all three drivers. As can be seen from Fig. 8, proposed hybrid driver improves both EMI and switching loss performances without implementing any feedback loop by combining the benefits of both open loop voltage driver and current driver. Since operating conditions will change driver performances, the emphasis of this paper is the new hybrid structure rather than percentage of EMI and switching loss reduction.

This driver also has another advantage since resonant inductance value can be changed by DC biasing level, making it applicable to wider range of operating conditions.

© VDE VERLAG GMBH · Berlin · Offenbach

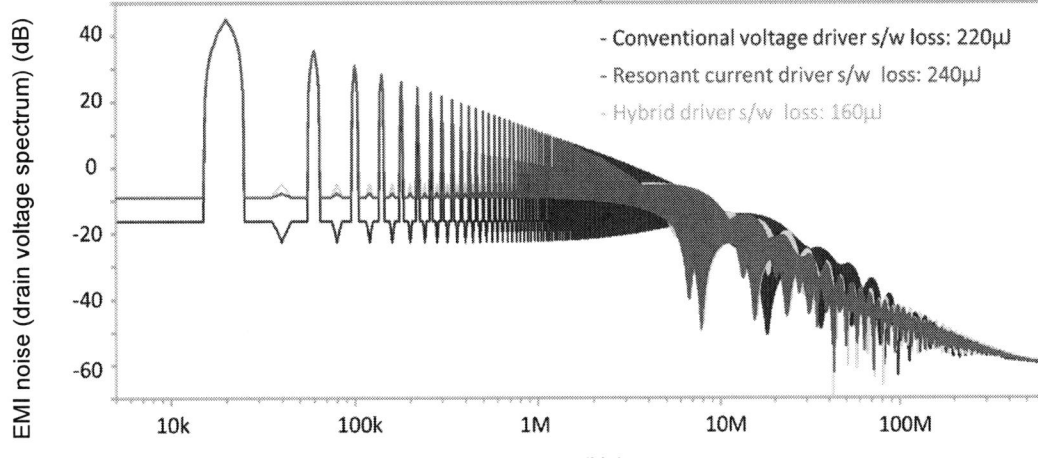

Fig. 8. Loss vs. EMI comparison among all three drivers

5. Conclusion

In this paper, the limitations of conventional voltage driver and closed-loop active gate driver were first discussed. After that, another benefit of resonant gate driver to reduce EMI noise in addition to driver loss saving was simulated. It was shown that by providing high driving loop impedance, resonant gate current can be used to decrease EMI generation. Finally, the main contribution of this paper is that a new and simple hybrid driver structure was proposed to further improve the tradeoff between switching loss and EMI. By combining the benefit of open loop voltage and current driver, this new driver is more suitable for SiC MOSFET applications compared with other driver structures.

References

[1] Lobsiger, Y. and Kolar, J.W., 2015. Closed-Loop d d and d d IGBT Gate Driver. *Power Electronics, IEEE Transactions on, 30*(6), pp.3402-3417.

[2] Oswald, N., Anthony, P., McNeill, N. and Stark, B.H., 2014. An experimental investigation of the tradeoff between switching losses and EMI generation with hard-switched all-Si, Si-SiC, and all-SiC device combinations. *Power Electronics, IEEE Transactions on, 29*(5), pp.2393-2407.

[3] Glaser, J.S., Nasadoski, J.J., Losee, P.A., Kashyap, A.S., Matocha, K.S., Garrett, J.L. and Stevanovic, L.D., 2011, March. Direct comparison of silicon and silicon carbide power transistors in high-frequency hard-switched applications. In *Applied Power Electronics Conference and Exposition (APEC), 2011 Twenty-Sixth Annual IEEE* (pp. 1049-1056). IEEE.

[4] Chen, Z., Boroyevich, D. and Li, J., 2013, March. Behavioral comparison of Si and SiC power MOSFETs for high-frequency applications. In *Applied Power Electronics Conference and Exposition (APEC), 2013 Twenty-Eighth Annual IEEE* (pp. 2453-2460). IEEE.

[5] Mainali, K. and Oruganti, R., 2010. Conducted EMI mitigation techniques for switch-mode power converters: A survey. *Power Electronics, IEEE Transactions on, 25*(9), pp.2344-2356.

[6] Yazdani, M.R., Farzanehfard, H. and Faiz, J., 2011. Classification and Comparison of EMI Mitigation Techniques in Switching Power Converters-A review. *Journal of Power Electronics, 11*(5), pp.767-777.

[7] Yang, X., Yuan, Y., Zhang, X. and Palmer, P.R., 2015. Shaping High-Power IGBT Switching Transitions by Active Voltage Control for Reduced EMI Generation. *Industry Applications, IEEE Transactions on, 51*(2), pp.1669-1677.

© VDE VERLAG GMBH · Berlin · Offenbach

[8] Chen, L. and Peng, F.Z., 2009, February. Closed-loop gate drive for high power IGBTs. In *Applied Power Electronics Conference and Exposition, 2009. APEC 2009. Twenty-Fourth Annual IEEE* (pp. 1331-1337). IEEE.

[9] Fink, K. and Bernet, S., 2013. Advanced Gate Drive Unit With Closed-Loop Control. *Power Electronics, IEEE Transactions on*, *28*(5), pp.2587-2595.

[10] Riazmontazer, H., Rahnamaee, A., Mojab, A., Mehrnami, S., Mazumder, S.K. and Zefran, M., 2015, March. Closed-loop control of switching transition of SiC MOSFETs. In *Applied Power Electronics Conference and Exposition (APEC), 2015 IEEE* (pp. 782-788). IEEE.

[11] CPWR-AN10, Rev-C, SiC MOSFET Isolated Gate Driver, Cree application note

[12] Eberle, Wilson, Zhiliang Zhang, Yan-Fei Liu, and AndParesh C. Sen. "A current source gate driver achieving switching loss savings and gate energy recovery at 1-MHz." *Power Electronics, IEEE Transactions on* 23, no. 2 (2008): 678-691.

[13] De Vries, Ian D. "A resonant power MOSFET/IGBT gate driver." In *Applied Power Electronics Conference and Exposition, 2002. APEC 2002. Seventeenth Annual IEEE*, vol. 1, pp. 179-185. IEEE, 2002

[14] Ahsanuzzaman, S.M., McRae, T., Peretz, M.M. and Prodić, A., 2012, February. Low-volume buck converter with adaptive inductor core biasing. In*Applied Power Electronics Conference and Exposition (APEC), 2012 Twenty-Seventh Annual IEEE* (pp. 335-339). IEEE.

PCIM Europe 2016, 10 – 12 May 2016, Nuremberg, Germany

Application and Design Considerations for CoolMOS™ CFD2 and EiceDRIVER™ IC in Motor Drive Application

Wolfgang Frank, Infineon Technologies Germany, Am Campeon 1-12, 85579 Neubiberg, Wolfgang.Frank@infineon.com

Michael Wendt, Infineon Technologies Germany, Am Campeon 1-12, 85579 Neubiberg, Michael.Wendt@infineon.com

Sam Abdel-Rahman, Infineon Technologies Americas, 419 Davis Drive, Morrisville-NC 27560, USA, Sam.Abdel-Rahman@infineon.com

Abstract

While IGBT switches have been conventionally applied in motor drives such as refrigeration compressor drives, their threshold voltage ($V_{CE,SAT}$) is a hurdle towards higher efficiency, especially in the mid and partial load region. In contrast, MOSFETs such as CoolMOS™ CFD can solve this problem since they have on-state resistance instead of a constant voltage drop. However, the high *di/dt* during reverse recovery of the body diode requires detailed analysis of the commutation cell inside the inverter's half bridges. This paper discusses the switching behavior of a MOSFET-based inverter, properties of Super Junction MOSFETs, proposes a gate driver circuit design to properly control CoolMOS™ CFD2 and recommends driver layout guidelines for reliable drive systems.

1. MOSFET switching in motor drive application

A BLDC motor driver is composed of three half bridges; each bridge is switching and delives power for one third of the motor's cycle, as shown in Figure 1.

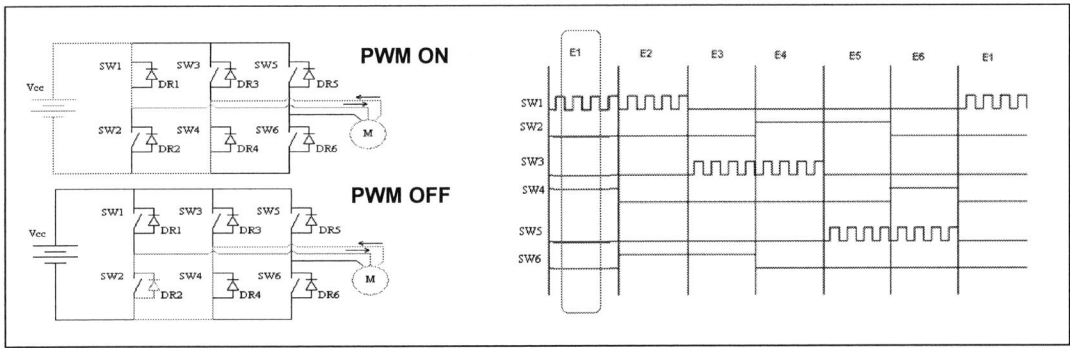

Figure 1: BLDC motor driver circuit and switching modes

A half-bridge cell as depicted in Figure 2 is enough to represent and study the switching behaviour of all FETs. S1 and S2 are switching in a buck configuration, S4 is turned on (not switching) to provide the return current path. S1 is the active switch. The inductor current ramps up when S1 is turned on. S2 is the rectifier switch, its body diode carries the inductor current when S1 is off and the the inductor current ramps down. This mode of operation takes place for one third of the motor's cycle. Then the same operation moves to the next half-bridge. It seems a simple operation, in reality there are several practical switching considerations.

© VDE VERLAG GMBH · Berlin · Offenbach

These are related to the FET's parameters, package, layout parasitics and gate driver design, as shown in Figure 3.

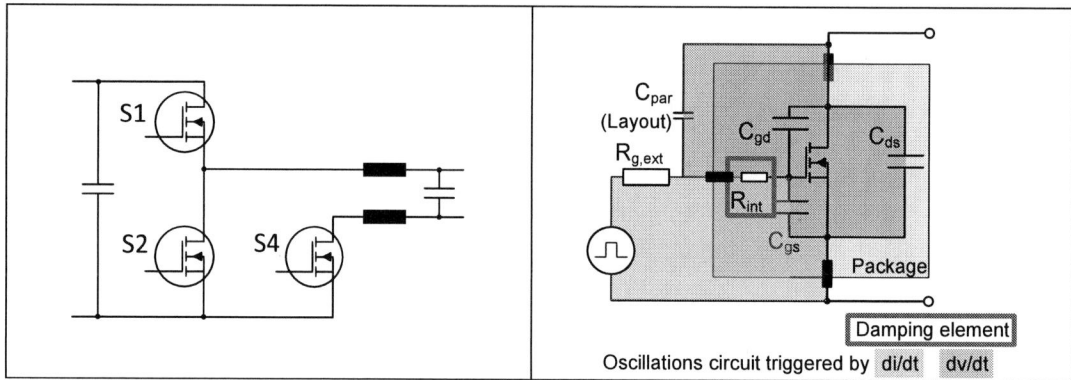

Figure 2: Switching cell

Figure 3: Parasitic elements of the MOSFET, package, layout and driver

The main concerns during fast switching transitions are high *dv/dt* and high *di/dt*, that can cause shoot-through or gate oscillation:

- In shoot through, S1 is turned on, the resulting *dv/dt* across S2 couples to its gate and a voltage spike appears. If the coupling spike is high enough to reach the FET's gate threshold voltage, then both FETs in the bridge will be on for a short period of time, causing failures.
- Gate oscillation can be triggered by two events:
  1. At turn-on, high *di/dt* causes a voltage drop across the source inductance of the FET's package and layout; this voltage forms a negative feedback to the driving voltage, causing the FET's gate to resonate (
  2. Figure **4**).
  3. At turn-off, high dv_{ds}/dt can couple to the gate trough the drain-gate capacitance, causing oscillations.

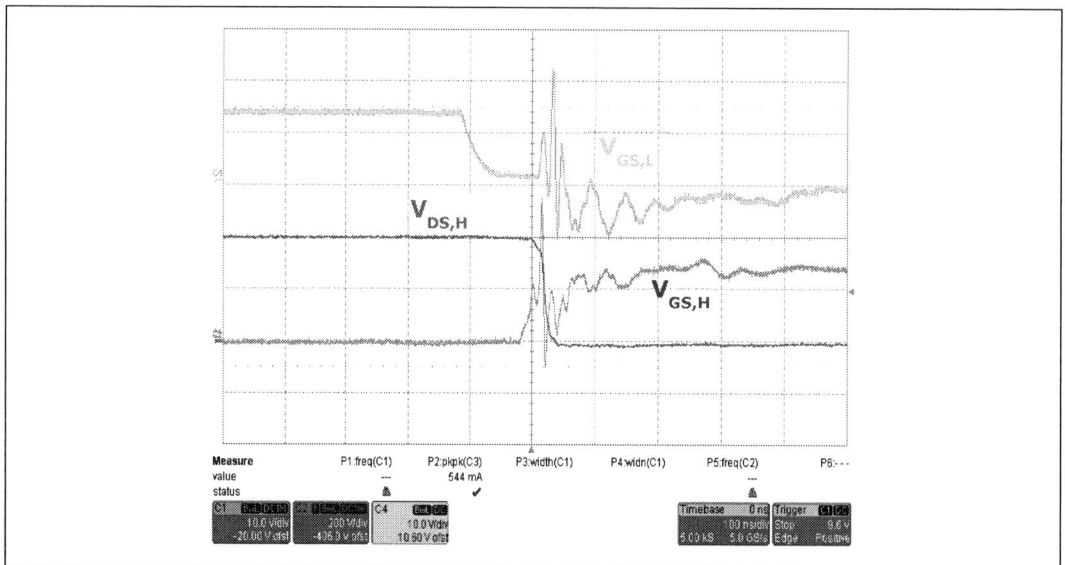

Figure 4: Strong oscillations during turn-on of a MOSFET

1.1. *dv/dt* origin and factors affecting its magnitude

Generally, *dv/dt* takes place during the charging period of the reverse capacitance C_{rss}, represented by the charge Q_{gd}. Hence, the C_{rss} value and the charging current level are two factors that affect *dv/dt*. Higher gate resistance (R_g) value means charging C_{rss} with lower current, which extends the miller plateau time and reduces *dv/dt*.

Reverse recovery behavior of S2's body diode is another factor that increases *dv/dt*. The current change rate dI_{rec}/dt generates a voltage across the parasitic source inductance and this creates a positive feedback to the driving voltage, which increases the driving current. This causes a faster charging of C_{rss} and higher *dv/dt*. Diodes with snappy recovery behavior lead to higher *dv/dt*. Moreover, the snappyness increases the voltage overshoot on S2, caused by the loop inductance.

During turn-off, the non-linear capacitance C_{oss} in Super-Junction FETs (CoolMOS[TM] CFD2), and the low C_{oss} values at high V_{ds} lead to increased *dv/dt*. This provides a low switching loss and fast voltage transition but also requires careful layout- and gate driver design techniques considering the higher *dv/dt*.

1.2. Negative impacts of high *dv/dt*

MOSFETs are closed loop controlled devices. As high *dv/dt* appears on S2's drain, it will inject a current to the gate through the miller capacitance C_{rss} and any other external capacitances. The magnitude of such feedback depends on the FET's structure itself, the layout and the gate drive circuit. This drain-gate coupling can either trigger the gate of S2 to turn on or cause gate resonance.

The FET's sensitivity to high dv/dt coupling is a function of its forward gain, particularly the transconductance g_{sm} ($\Delta I_d/\Delta V_{gs}$). Newer FET technologies tend to achieve higher transconductance to accommodate fast switching applications. This requires carefully designing the drive, layout and choice of package to avoid drain-gate coupling problems. The main factors affecting the feedback and gate coupling are:

- FETs reverse capacitance to input capacitance ratio (C_{rss}/C_{iss}): It is desired to have this ratio as small as possible, since both capacitances form a voltage divider.
- The gate drive loop and its impedance determine how solid the gate is connected to the source during high frequency perturbations like the high *dv/dt* event. The impedance seen from the gate side to the driver side is in parallel with the FET's C_{iss}. If such impedance is high then coupling is worse since most of the current charging C_{rss} will be charging C_{iss} and will cause gate voltage spikes. Gate loop impedance not only depends on the driver output stage impedance and external components values such as gate resistors, but also is related to layout and package inductances. Therefore it's essential to place the driver close to the FET to get the lowest possible gate loop impedance.

2. CoolMOS™ CFD2 benefits in motor drive applications

Motor drive typically is a hard switching application with hard commutation stressing the body diode. Therefore, it requires a robust body diode with low reverse recovery charge Q_{rr} and soft recovery. CoolMOS[TM] CFD2 is a 650V technology with integrated fast body diode. As aforementioned, fast diode recovery and softness are important factors affecting the *dv/dt* and limit the voltage overshoot during hard commutation.

Figure 5 compares the body diode recovery current and softness of CoolMOS™ CFD2 (IPW65R080CFD) to the older CFD technology (SPW47N60CFD). It illustrates the lower Q_{rr} and softer recovery of the CFD2 CoolMOS™.

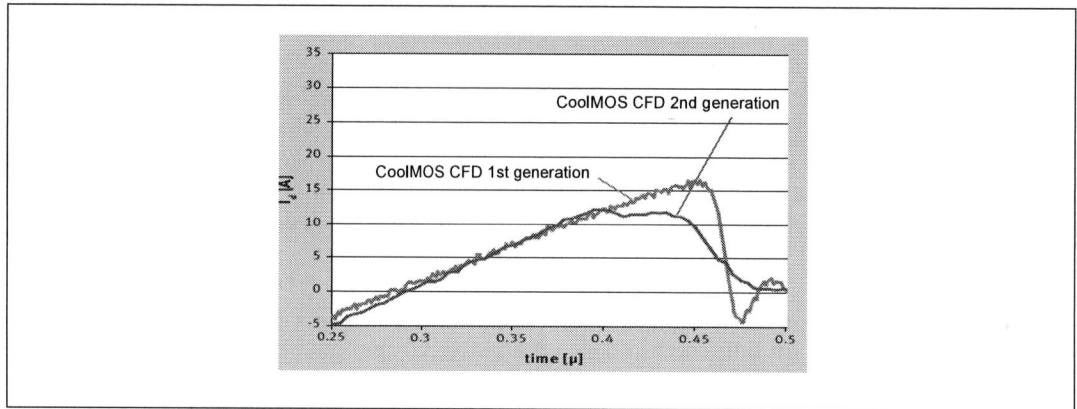

Figure 5: Body diode recovery current waveform of CoolMOS™ CFD2 and first generation

2.1. Proposed gate circuit design for CoolMOS™ CFD2

Figure 6 shows a proposed design for the switching cell with CoolMOS™ IPD65R420CFD. Mainly a capacitor (C_{ds}=0.47nF) is added to each bridge's switch node to limit and linearize the *dv/dt*, which is the most reliable approach to prevent the shoot-through and resonance problems discussed earlier in this paper. In motor drive application, the typical switching frequency is low (<10 kHz), therefore the added C_{ds} capacitor only has a minor impact on switching losses. Other driving parameters are chosen considering:

- R_{on}=1000 Ω → slower turn on, longer plateau, reduced *dv/dt*.

- R_{off}=0 Ω → Lower impedance to GND when turned off, lower voltage coupling spike.

- C_{gs}=0.47nF → reduced C_{rss}/C_{iss} ratio, reduced drain-gate (Miller) coupling gain.

- C_{ds}=0.47nF → controlled / linearized *dv/dt* at turn on, this has the benefit of removing gate oscillation and reduce EMI.

The circuit is operated by a 2EDL05N06PF EiceDRIVER™ IC. Based on a SOI technology, this IC provides excellent robustness against negative transient voltages [3]. The excellent properties of the integrated bootstrap diode supports the requirements of high power density and cost-performance ratio.

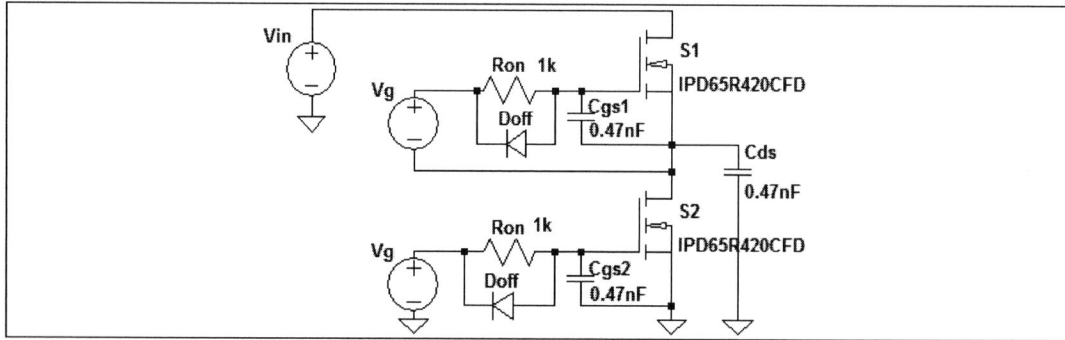

Figure 6: Proposed gate circuit design for CoolMOS™ CFD2

2.2. Layout recommendations

In general, there are a few layout guidelines that are recommended for reducing noise and resonance in the gate drive loop:

- Gate driver as close as possible to the gate.
- Minimum external capacitance gate to drain.
- Slow down *dv/dt* by properly choosing gate resistor R_g.
- Separate power ground from gate driver ground.
- R_g as close as possible to the gate pin.
- Use thick trace between gate driver and gate.

Figure 7 depicts a layout with minimized stray inductance due to the short distance between high side source terminal and low side drain terminal. The low side transistors on the bottom layer are shifted to the left side with respect to the high side transistors on the top layer. This leads to a thermal decoupling of both transistors. Furthermore, the low side transistors move even closer to the respective gate resistors. The shift also allows that the drain terminals of the low side transistors move directly underneath the source terminals, so that an appropriate number of vias provides a close connection to the high side source terminals. Hence, the loop inductance is minimized. A double sided assembly can be avoided when placing the low side transistors appropriately onto the top layer. This of course will lead to higher area consumption.

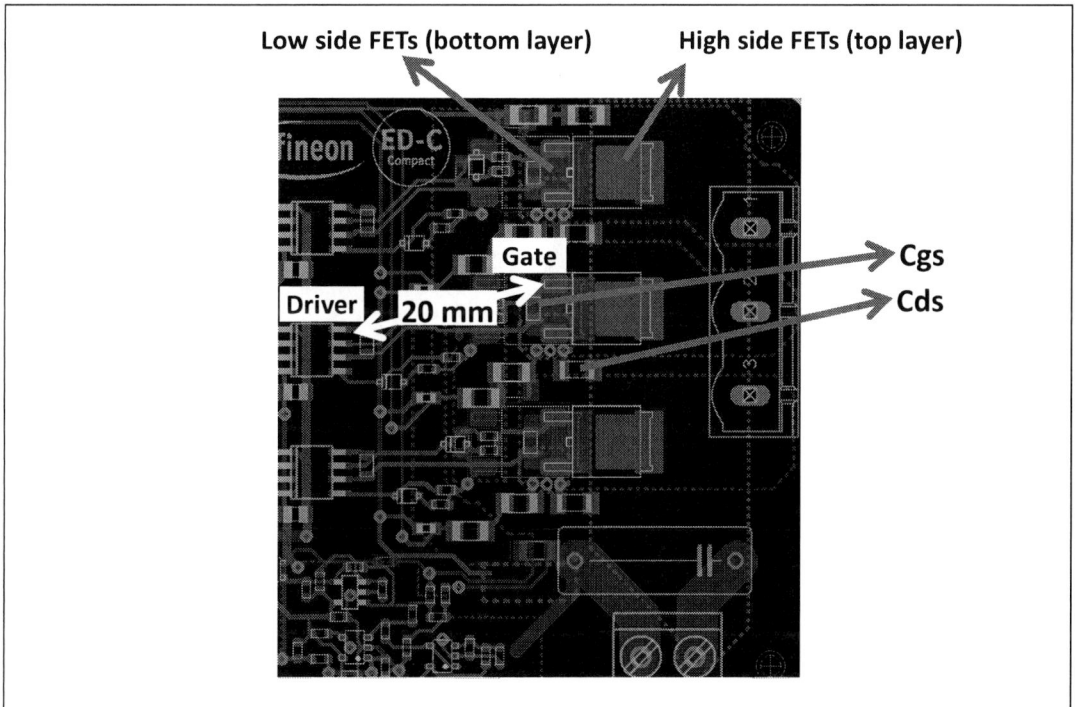

Figure 7: Example of a layout with minimized stray inductances by means of double side assembly

The physical proximity of the gate resistors to the gate terminals in combination with the reduced stray inductances leads to an improved performance and excellent switching behavior of the CoolMOS™ transistors. A turn-on waveform of the proposed driving circuit design is given in Figure 8. It shows a clean gate signal free of oscillations and the drain-source voltage slowly ramping down to 0V during the miller plateau region, too.

The same behavior can be expected for the other two switching bridges given that the layout and driving circuit is the same.

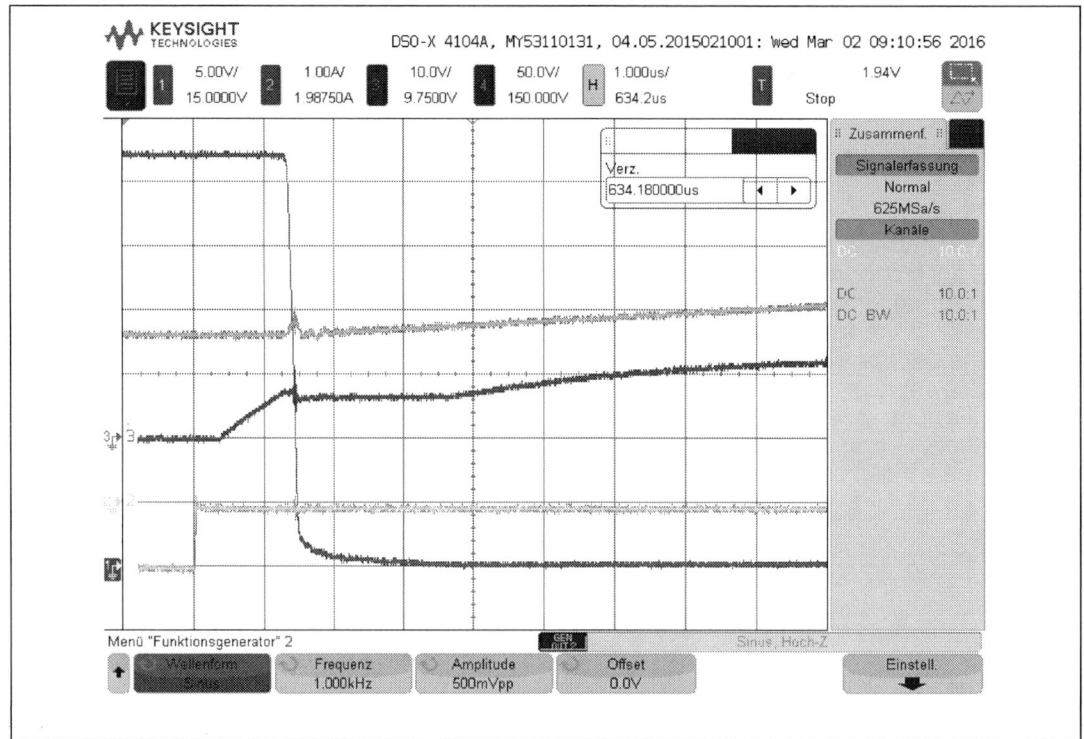

Figure 8: Turn-on waveforms at DC-link voltage V_{DC} = 320 V and load current I_L = 2.5 A.
V_{DS} (red, 50 V/div), I_L (green, 1 A/div), V_{GS} (blue, 10 V/div), PWM (yellow, 5 V/div), time scale 1 µs/div]

3. Conclusion

CoolMOS™ CFD2 offers efficiency benefits in motor drive applications. The proposed gate drive circuit design ensures an oscillation-free switching, leading to the conclusion that CoolMOS™ CFD can be efficiently and reliably operated in drives systems. Additionally, the SOI technology which is used for the 2EDL family provides a high robustness with respect to *dv/dt* in motor drive applications and an excellent controllability of CoolMOS™ CFD.

4. References

[1] R. Mente, F. Di Domenico, M.A. Kutschak, A. Steiner: CoolMOS™ CFD2 first 650 V rated super junction mosfet with fast body diode suitable for resonant topologies, Application Note, Infineon Technologies, February 2011.

© VDE VERLAG GMBH · Berlin · Offenbach

[2] W. Choi, D. Son, M. Hallenberger S. Young: Driving and Layout Requirements for Fast Switching MOSFETs, Fairchild Semiconductor Power Seminar 2010-2011, Fairchild, 2011.

[3] J. Song, W. Frank: Robustness of level shifter gate driver ICs concerning negative gate voltages; Proceedings of PCIM 2015, Nuremberg, Germany, 2015.

PCIM Europe 2016, 10 – 12 May 2016, Nuremberg, Germany

4D-Interleaving of Isolated ISOP Multi-Cell Converter Systems for Single Phase AC/DC Conversion

Matthias Kasper, Power Electronic Systems Lab., ETH Zurich, CH, kasper@lem.ee.ethz.ch
Michael Antivachis, Power Electronic Systems Lab., ETH Zurich, CH
Dominik Bortis, Power Electronic Systems Lab., ETH Zurich, CH
Johann W. Kolar, Power Electronic Systems Lab., ETH Zurich, CH
Gerald Deboy, Infineon Technologies Austria AG, Villach, Austria

Abstract

The multi-cell converter approach allows to break the performance barriers of conventional systems by leveraging the advantages of using multiple interleaved low voltage and/or low current converter cells. In this digest, a fourth dimension of interleaving is proposed which considers the time dependent degrees of freedom in the control of the entire multi-cell converter system. This new control concept is based on the possibility to decouple the operation of the series connected input stages from the parallel connected output stages by using the energy storage capability of the DC-link capacitors. This digest shows how the 4D-interleaving concept improves the system performance such as the efficiency which will be demonstrated with measurement results on a hardware prototype system in the final paper.

1. Introduction

Driven by the rising demand for highly efficient telecom and server power supplies resulting from the global trend of cloud computing, the industry research efforts for new converter topologies with improved performance concerning efficiency and power density have been considerably increased. A very promising approach described in (1) (cf. **Fig. 1(b)** and **Fig. 2(a)**) towards a hyper-efficient and super-compact telecom rectifier ($\eta = 98\%$, $\rho = 2.2\text{kW/dm}^3$, $V_{\text{in,RMS}} = 230\,\text{V}$, $V_{\text{out}} = 48\,\text{V}$, $P_{\text{rat,tot}} = 3.3\,\text{kW}$) is based on a multi-cell converter concept with ISOP arrangement of the converter cells. Each converter cell contains an AC/DC rectifier input stage which is a full-bridge operated with Totem-Pole modulation and an isolated DC/DC converter output stage consisting of a phase-shifted full bridge converter. This multi-cell ISOP configuration allows to share the input voltage among the converter cells and thus enables the use of low-voltage semiconductors with superior Figure-of-Merits throughout the converter cells. In addition, the cells can be operated in an interleaved fashion in order to increase the effective switching frequency of the entire system up to a multiple of a single converter cell's switching frequency. These advantages of the ISOP converter approach provide significant benefits in terms of reduced conduction and switching losses and smaller volumes of inductive components and heat sinks (2).

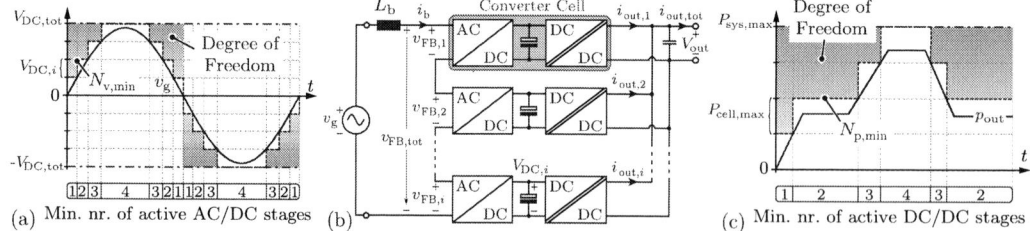

Fig. 1: 4D interleaving concepts (**(a)** and **(c)**) for an ISOP multi-cell telecom power supply module **(b)**. The available degrees of freedom in the operation of the system arise from the fact, that for each operating point of input voltage and output power only a certain minimum number of active AC/DC and DC/DC converter stages - $N_{\text{v,min}}$ and $N_{\text{p,min}}$ - is required. Any number of stages between the minimum and the maximum available number of cells can be operated at the **(a)** AC/DC and **(c)** DC/DC stages. Furthermore, the number of active cells can be different for both stages, as the DC-link capacitors employed in the cells decouple the power flow between the input and the output stages.

© VDE VERLAG GMBH · Berlin · Offenbach

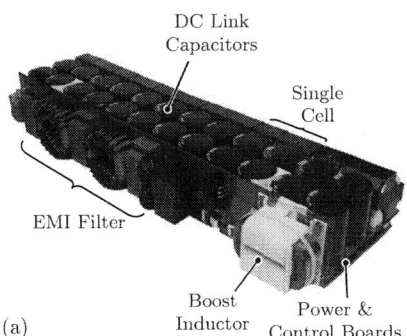 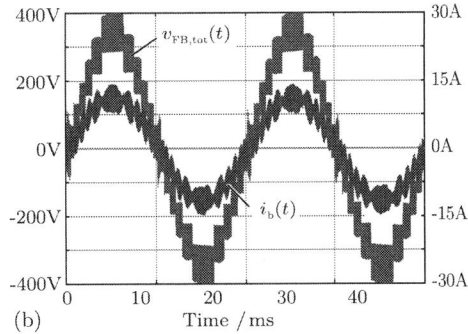

(a)   (b)

Fig. 2: (a) Hardware demonstrator of a multi-cell telecom power supply module (3) with $N_{\text{cells}} = 6$ cells and (b) measured waveform of the converter input voltage $v_{\text{FB,tot}}$ and the current i_{b} in the input side boost inductor for the operation with $N_{\text{PWM}} = 6$ PWM cells at a power level of $P_{\text{out}} = 1.5\,\text{kW}$.

The ISOP configuration itself can be considered as a spatial, i.e. 3-dimensional interleaving of multiple converter cells, where the cells are operated with a fixed operation mode and/or constant time-invariant interleaving. In this paper, a 4th dimension (4D) of interleaving the converter cells is introduced, where the operation mode (PWM vs. fundamental frequency modulation) and/or the number of active cells is varied over time depending on the input voltage and the load power level. In **Sec. 2** the basic degrees of freedom in the operation of the multi-cell converter of **Fig. 1(b)** which provide the possibility for 4D-interleaving are introduced. Furthermore, different 4D-interleaving concepts are proposed for the series connected AC/DC input stages and the parallel connected DC/DC output stages of the multi-cell system. In **Sec. 3** the best suited 4D-interleaving concepts are identified for both stages by means of a comprehensive Pareto-optimization and analytical derivations, respectively. Furthermore, the performance improvements are quantified in comparison to a conventional system control and a balancing scheme for the DC-link voltages is introduced.

2. 4D-Interleaving of ISOP Converters

For conventional 3D-interleaving multi-cell converter systems with ISOP configuration are usually operated with a common duty cycle for all AC/DC rectifier stages and all DC/DC converter stages, since this ensures an equal voltage sharing at the series connected inputs and an equal current sharing at the parallel connected outputs of the converter cells (4; 5).

However, this modulation scheme, does not take into account that the time-varying input voltage and load power in addition to the presence of large DC-link capacitors provide additional degrees of freedom for the operation of the system. Depending on the input voltage and output power, the number of active rectifier stages and output stages can vary over time as long as the following three conditions are fulfilled:

- *Input voltage*: At any given time the AC/DC rectifier stages have to provide a voltage $v_{\text{FB,tot}}(t)$ which is defined by the grid voltage, i.e. $v_{\text{g}}(t) \approx \overline{v}_{\text{FB,tot}}(t)$ over a switching period, where $\overline{v}_{\text{FB,tot}}(t) = \sum_{i=1}^{N_{\text{cells}}} \overline{v}_{\text{FB},i}(t) = \sum_{i=1}^{N_{\text{cells}}} V_{\text{DC},i} \cdot m_i(t)$ with m_i denoting the modulation index of a cell. This allows to derive a minimum number of active rectifier stages $N_{\text{v,min}}$ that are required for the operation of the system: $N_{\text{v,min}} = \left\lceil \frac{v_{\text{g}}(t)}{V_{\text{DC},i}} \right\rceil$ (cf. **Fig. 1(a)**).

- *Output current*: The total output current $i_{\text{out,tot}}(t)$ which is provided by the parallel connected DC/DC converters has to be equal to the load current, i.e. $i_{\text{load}}(t) = i_{\text{out,tot}}(t) = \sum_{i=1}^{N_{\text{cells}}} i_{\text{out},i}(t)$. Since each converter cell has a maximum power and thus also a maximum current rating $I_{\text{DC/DC,max}}$, a minimum number $N_{\text{p,min}}$ of converter cells required for the power transfer can be derived as $N_{\text{p,min}} = \left\lceil \frac{i_{\text{load}}(t)}{I_{\text{DC/DC,max}}} \right\rceil$ (cf. **Fig. 1(c)**).

- *DC link capacitors*: The DC-link capacitance is typically sized for a hold-up time requirement.

© VDE VERLAG GMBH · Berlin · Offenbach

PCIM Europe 2016, 10 – 12 May 2016, Nuremberg, Germany

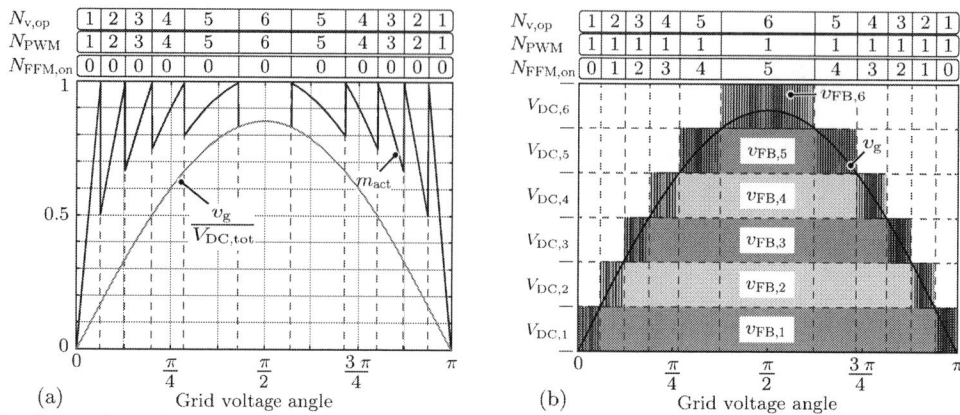

(a)   Grid voltage angle   (b)   Grid voltage angle

Fig. 3: Examples of the 4D-interleaving concepts for the series connected AC/DC rectifier stages: (a) operation with a variable number of PWM modules where all active rectifier stages ($N_{v,op}$) are operated with PWM with the same modulation index m_{act}. The minimum number of active PWM stages is set to $N_{PWM,min} = 1$; (b) operation with a fixed number of PWM stages (i.e. $N_{PWM} = 1$) where the remaining cells are operated with fundamental frequency and are either turned on or off ($N_{FFM,on} = N_{v,op} - N_{PWM}$).

Therefore, especially at low-power levels, this capacitance allows to have a power flow of the AC/DC rectifier stages that is different from the power flow of the DC/DC converters. The difference in the power flow is then stored in or supplied by the DC-link capacitors. This means, that the number of active AC/DC rectifier stages $N_{v,op}$ at a given time can be different from the number of active DC/DC converter stages $N_{p,op}$ as long as the net power flow of the DC-link capacitors of each cell is zero over a certain time interval, e.g. some milliseconds.

Based on the above mentioned degrees of freedom, a selection of modulation schemes making use of the 4th dimension of interleaving is presented in the following paragraphs.

2.1. AC/DC Rectifier Input Stages

The total sinusoidal cell input voltage $v_{FB,tot}$ can be distributed among the AC/DC rectifier stages in different ways. Apart from the number of active modules ($N_{v,op} \in [N_{v,min}, N_{cells}]$) the mode of operation of each active module can also be chosen. Namely, an active module can be either operated with PWM modulation or with fundamental frequency modulation (ON-OFF). Based on these degrees of freedom, two essentially different modulations schemes can be identified:

- *Variable number of PWM cells*: In this modulation scheme only the minimum number $N_{v,op} = N_{v,min}$ of required cells is operated with PWM while the remaining cells are turned off, as shown in **Fig. 3(a)**. This means, that the number of active cells, $N_{v,op}$, changes with the time-varying value of the grid voltage. All active cells are operated with a fixed frequency PWM, i.e. $N_{PWM} = N_{v,op}$, with the same modulation index and a phase shift between the cells which depends on the number of active cells, $\phi = 360°/N_{PWM}$. This leads to a varying effective switching frequency due to the varying number of active cells. In order to limit the difference between the maximum and minimum effective switching frequency, a lower limit for the number of active PWM cells, $N_{PWM,min}$, can be defined.

- *Fixed number of PWM cells*: Instead of changing the number of PWM cells during a grid cycle, this modulation scheme operates with a defined number of PWM cells at all times ($N_{PWM} = $ const), as depicted in **Fig. 3(b)** for the case of a single PWM cell ($N_{PWM} = 1$). The remaining cells are operated with fundamental frequency which means they are either turned on or off ($N_{FFM,on} = N_{v,op} - N_{PWM}$). As a result, in this operation mode only a fixed number of cells (N_{PWM}) exhibits switching losses, whereas the other cells only have conduction losses. A disadvantage, however, is the reduced effective switching frequency which requires an increase of the boost inductance value in order to limit the peak-to-peak current ripple of the

© VDE VERLAG GMBH · Berlin · Offenbach

PCIM Europe 2016, 10 – 12 May 2016, Nuremberg, Germany

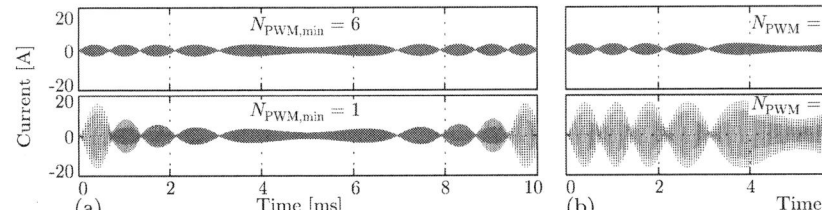

Fig. 4: Comparison of simulated boost inductor current ripple waveforms (i.e. current i_b without the $50\,\text{Hz}$ low frequency component) for different 4D-interleaving schemes: (a) variable number of PWM modules ($N_{\text{PWM,min}}$ denotes the minimum number of active PWM modules); (b) fixed numbers N_{PWM} of PWM modules where the remaining cells are operated as fundamental frequency modules. (System with 6 cells and a switching frequency of the PWM cells of $f_{\text{sw,PWM}} = 20\,\text{kHz}$, a boost inductance of $L_b = 25\,\mu\text{H}$, and a DC-link voltage of each cell of $V_{\text{DC,cell}} = 66\,\text{V}$.)

input current.

For all of the above cases, the instantaneous cell power values are unequal and the individual DC-link capacitor voltages inherently fluctuate. Thus, it is required to properly permute the operation of the cells, e.g. by cyclically changing the selection of PWM cells. Especially, in order to achieve an accurate cancellation of harmonics by interleaving multiple PWM cells, the DC-link voltages of these PWM cells should be kept as close as possible to the same value.

2.2. DC/DC Converter Output Stages

Similarly to the voltage distribution among the input side series connected AC/DC converter stages the output current can be distributed in different ways among the parallel connected DC/DC converter output stages:

- *Equal current sharing*: The number of active DC/DC stages varies depending on the output current. All cells are operated with the same current reference value and therefore equally share the output current. The current reference of a single cell can be calculated by dividing the total output current by the number of active cells. As a drawback, however, the current reference of each cell will exhibit a step change every time the number of active cells changes, which requires a highly dynamic current controller in each cell.

- *Unequal current sharing*: In this modulation scheme the output current is unequally divided among the active cells. This might be achieved e.g. if an additional cell is only activated when the previously actived cell has reached its maximum current value. This has the advantage that the current reference values of all cells are continuously changed over time without any step changes (under the assumption that there are no load steps). As a drawback, the stress on the DC-link capacitors is heavily unbalanced as some cells operate at their maximum power levels while others are in stand-by mode.

A minimum number of DC/DC converter stages can be defined to reduce the total current ripple at the output by interleaving the active DC/DC converter stages. This can be applied to both concepts, since the current ripple at the output of an individual cell is independent from the current level of the cell (i.e. the average value of the output current). Thus, even the interleaving of two cells with unequal current sharing reduces the output current ripple.

3. Simulation and Optimization Results

In this chapter the different 4D-interleaving modulation schemes are comparatively evaluated in order to identify the best possible modulation schemes for the AC/DC and the DC/DC converter stages. In the following, the 4D-interleaving modulation schemes are applied to an ISOP multi-cell telecom power supply module with $N_{\text{cells}} = 6$ converter cells where each cell has a rated power

© VDE VERLAG GMBH · Berlin · Offenbach

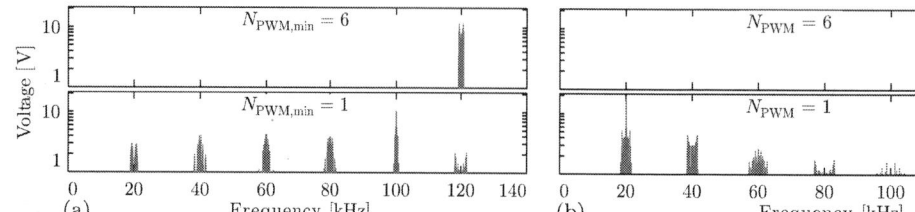

Fig. 5: Comparison of the harmonic spectrum of the input voltage $v_{\text{FB,tot}}$ of the series connection of AC/DC converter cells (without the $50\,\text{Hz}$ low frequency component) for different 4D-interleaving schemes: (a) variable number of PWM modules ($N_{\text{PWM,min}}$ denotes the minimum number of active PWM modules); (b) fixed numbers of PWM modules N_{PWM} where the remaining cells are operated as fundamental frequency modules. (System with 6 cells and a switching frequency of the PWM cells of $f_{\text{sw,PWM}} = 20\,\text{kHz}$ and a DC-link voltage of each cell of $V_{\text{DC,cell}} = 66\,\text{V}$.)

level of $P_{\text{rat}} = 550\,\text{W}$ and a DC-link capacitance of $C_{\text{DC}} = 8.8\,\text{mF}$ with a nominal DC-link voltage of $V_{\text{DC,cell}} = 66\,\text{V}$.

3.1. AC/DC Converter Stages

The modulation scheme of the AC/DC stages takes direct influence on the harmonic spectrum of the input current waveform, the RMS value of the input current, and also on the total switching losses. The effect of the selected modulation scheme on the input current waveform is shown for both modulation schemes (variable and fixed numbers of PWM cells) in **Fig. 4** for two selected scenarios. The modulation with a variable number of PWM modules and a value of $N_{\text{PWM,min}} = 6$ is equal to the operation with a fixed number of $N_{\text{PWM}} = 6$ cells which is the standard 3D interleaving operation as measured in **Fig. 2(b)**.

The harmonic spectrum of the generated converter input voltage is shown for both of the aforementioned modulation schemes in **Fig. 5**. It can be seen that the modulation with the variable number of PWM modules (**Fig. 5(a)**) leads to a more even distribution of the switching frequency harmonics over the frequency range, due to its varying effective switching frequency.

As a remark, for the generation of the plots in **Fig. 4** and **Fig. 5** the DC/DC stages were operated with conventional (3D) modulation where all stages are activated and equally share the total output power.

Since the choice of the modulation scheme affects several system parameters, a comprehensive optimization of the entire input stage comprising the EMI filter, the boost inductor and the MOSFET chip area of the AC/DC full bridges has to be performed to take into account the dependencies between the different elements and the modulation schemes. This optimization allows to identify the Pareto-limit of the trade-off between efficiency and power-density of the entire rectification stage by considering all available degrees of freedom given for the design:

- *EMI filter*: Regardless of the selected modulation scheme, the system has to comply with the CISPR Class B directive which specifies limits for the noise emissions in the frequency range of $f_{\text{CISPR}} = 150\,\text{kHz} - 30\,\text{MHz}$. This requires the calculation of the harmonic spectrum for each modulation scheme and the identification of the Pareto-optimal filter designs which can be found by considering different degrees of freedom like the number of filter stages and the choice of different values for the filter elements.

- *Boost Inductor*: The value of the boost inductance is another optimization parameter since a small value leads to a large input current ripple which increases the RMS value and also the losses of the inductor and conduction losses of the MOSFETs. However, a large current ripple also results in lower switching losses of the MOSFETs as the hard-switching instants occur at lower current levels. In addition, the design of the boost inductor represents a Pareto-optimization problem by itself as for a given set of electrical parameters different inductor designs can be found due to the possibility of employing e.g. different types of core material, core shapes, winding types (e.g. litz or foil windings), number of turns and air gap lengths.

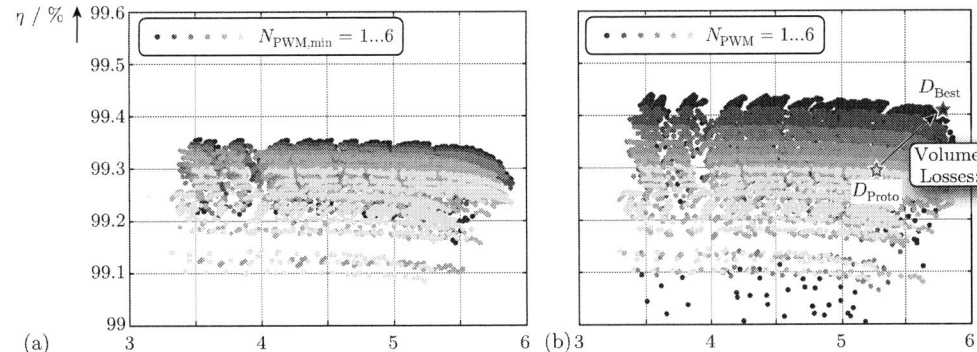

Fig. 6: η-ρ efficiency vs. power-density Pareto-optimization results of the entire rectification stage (incl. EMI filter, boost inductor, AC/DC full bridges, and DC-link capacitors): (a) Performance trade-off with a variable number of PWM cells for different minimum numbers $N_{PWM,min}$ of PWM cells, and (b) performance space for a fixed number N_{PWM} of PWM cells where the remaining cells are operated with fundamental frequency modulation. Compared to a conventional approach employed in the prototype system (D_{Proto}, $N_{PWM} = 6$) a selected design on the Pareto-front (D_{Best}, $N_{PWM} = 1$) shows 10% lower total volume and 17% lower losses.

- *Switching Frequency*: The switching frequency directly impacts the switching losses but also the shape of the input current for a given boost inductance. This also has an effect on the inductor design and the EMI filter design. Regarding the MOSFETs, a high switching frequency leads to larger switching losses but to a lower current ripple and thus to a lower current RMS value which decreases the conduction losses.

- *MOSFET Chip Areas*: The chip area selection of the MOSFETs allows to trade-off conduction losses which decrease with a larger chip area and (turn-on) switching losses which increase with a larger chip area.

The results of the Pareto-optimization of the entire rectification stage, including the EMI filter, boost inductor, full-bridge MOSFETs of the AC/DC stages, DC-link capacitors (selected for a hold-up time of $t_{holdup} = 10\,\text{ms}$), resistive PCB losses, and constant control losses, for operation at rated power ($P_{rat} = 3.3\,\text{kW}$) are plotted in **Fig. 6(a)** and **6(b)** for the modulation scheme with a variable number of PWM modules and the modulation scheme with a fixed number N_{PWM} of PWM modules, respectively. It can be concluded that the highest performance can be achieved by operating only one cell with PWM and the remaining cells with fundamental frequency modulation. Compared to the design of the hardware demonstrator (D_{Proto})(3) with a continuous operation of $N_{PWM} = 6$ cells, the 4D-interleaving concept allows to simultaneously improve the efficiency and the power density of the rectification stage; the losses are reduced by -17% and the volume by -10%, as shown for design D_{Best} in **Fig. 6(b)**. More details about both designs can be found in **Tab. 1**.

3.2. DC/DC Converter Stages

For the parallel connected DC/DC converter stages the most efficient 4D interleaving modulation scheme can be analytically derived by assuming a generic loss function of each DC/DC converter

Tab. 1: Comparison of the parameters of the AC/DC converter stages of the prototype design D_{Proto} and the Pareto-optimal 4D-interleaving design D_{Best} (cf. Fig. 6).

Variable	D_{Proto}	D_{Best}
N_{PWM}	6	1
$f_{sw,PWM}$	$20\,\text{kHz}$	$24\,\text{kHz}$
L_b	$25\,\mu\text{H}$	$90\,\mu\text{H}$
EMI Filter Stages	3	3
Parallel MOSFETs	2	5

PCIM Europe 2016, 10 – 12 May 2016, Nuremberg, Germany

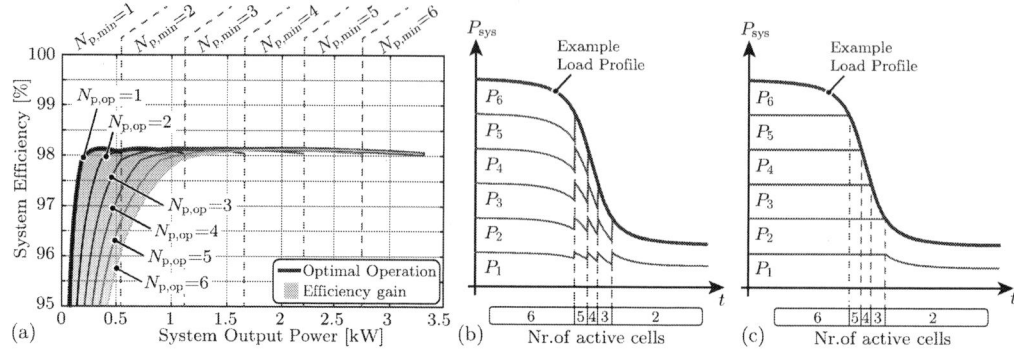

Fig. 7: 4D-interleaving of the parallel connected DC/DC converter stages of an ISOP system with $N_{\text{cells}} = 6$ converter cells which leads to a higher part-load efficiency (cf. (a)) by always operating only a subset of converter cells. (b) Operation mode with equal power reference values for all active cells which results in discontinuous changes of the power levels of the individual active cells. (c) Alternative operation with unequal power levels of the converter cells but smooth changes of the power levels of the individual cells.

depending on the output power as

$$p_{\text{loss},i} = k_0 + k_1 \cdot p_{\text{out},i} + k_2 \cdot p_{\text{out},i}^2 \tag{1}$$

where k_0 models the constant losses (e.g. auxiliary power), $k_1 \cdot p_{\text{out},i}$ the linearly dependent losses (e.g. a diode voltage drop) and $k_2 \cdot p_{\text{out},i}^2$ the quadratically dependent losses (e.g. resistive losses). The total losses of the DC-DC converter stage comprising N DC/DC converters can thus be written as

$$p_{\text{losses,tot}} = N \cdot k_0 + k_1 \cdot \sum_{i=1}^{N} p_{\text{out},i} + k_2 \cdot \sum_{i=1}^{N} p_{\text{out},i}^2 \ . \tag{2}$$

By applying the Lagrangian method to the minimization problem of (2) under the constraint of $p_{\text{out,tot}}(t) = \sum_{i=1}^{N_{\text{cells}}} p_{\text{out},i}(t)$ following solution can be found for the general case

$$p_{\text{out,1,opt}} = p_{\text{out,2,opt}} = ... = p_{\text{out,N,opt}} = \frac{p_{\text{out,tot}}}{N} \tag{3}$$

under which the total losses are minimized. By inserting the solution of (3) in (2) the total losses can be calculated as

$$p_{\text{losses,tot}} = N \cdot k_0 + k_1 \cdot p_{\text{out,tot}} + k_2 \cdot \frac{p_{\text{out,tot}}^2}{N} \ . \tag{4}$$

The remaining optimization parameter is the optimum number of active DC-DC converters $N = N_{\text{opt}} \in [N_{\text{p,min}}, N_{\text{cells}}]$ used for the power transfer for maximum efficiency. By differentiating (4) with respect to N and setting the derivative equal to zero, the optimal number of active cells can be found as

$$N_{\text{opt}} = \sqrt{\frac{k_2}{k_0}} \cdot p_{\text{out,tot}} \ . \tag{5}$$

Since the number N_{opt} is typically a rational number, the two nearest integer values have to be considered and the integer value which still satisfies $N_{\text{opt,int}} \in [N_{\text{p,min}}, N_{\text{cells}}]$ and leads to lower losses according to (4) has to be considered. If none of the nearest integer values satisfies the condition of $N_{\text{opt,int}} \in [N_{\text{p,min}}, N_{\text{cells}}]$ (typically for systems with larger constant losses than quadratic losses, i.e. $k_0 \gg k_2$), the value of $N_{\text{opt,int}} = N_{\text{p,min}}$ has to be chosen.

As a conclusion, this means, that the most efficient 4D-interleaving modulation scheme for the parallel connected DC/DC stages is obtained by equally sharing the total power and/or the total output current among the active DC/DC cells. This allows to extend the level of highest efficiency also to very low power levels as shown in **Fig. 7(a)** for the multi-cell telecom rectifier of **Fig. 2(a)**.

© VDE VERLAG GMBH · Berlin · Offenbach

PCIM Europe 2016, 10 – 12 May 2016, Nuremberg, Germany

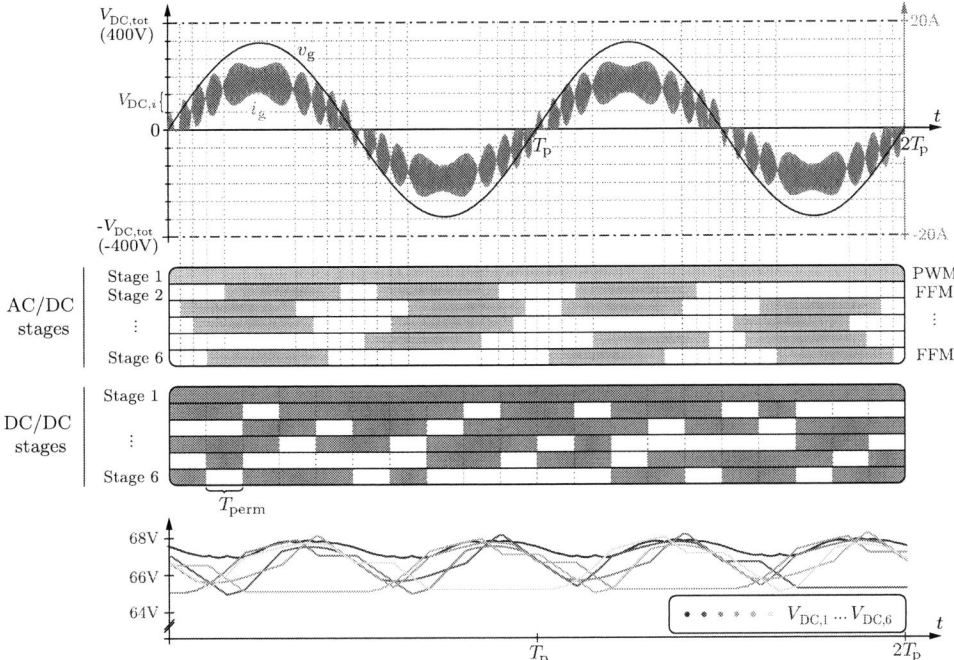

Fig. 8: Simulation results of the 4D-interleaving modulation of the ISOP multi-cell telecom rectifier system with $N_{cells} = 6$ at the operating point of $P_{out} = 1.5\,kW$. At the input side only one AC/DC stage (Stage 1) is modulated with PWM while the remaining five stages are operated with fundamental frequency modulation (FFM). At the selected power level of $P_{out} = 1.5\,kW$ the most efficient operation of the DC/DC stages is achieved with $N_{p,op} = 4$ active DC/DC stages. The balancing algorithm selects the active AC/DC and DC/DC stages based on the DC-link voltages of the cells. The permutation time interval for the DC/DC stages is chosen as $T_{perm} = 2\,ms$.

Compared to the standard modulation where all six DC/DC stages are operated at all power levels, significant efficiency gains can be achieved at power levels below 30% of the rated power $P_{rat,tot}$.
In order to avoid step changes of the power reference values of the cells every time the number of cells changes (cf. **Fig. 7(b)**) it is preferable to allow unequal reference values during transient load changes (cf. **Fig. 7(c)**).

3.3. DC-Link Voltage Balancing

The 4D-interleaving modulation schemes lead to an unequal stress of the DC-link capacitors since, on the one hand, at low input voltages and output power only a fraction of the AC/DC stages and DC/DC stages is active at a given time and, on the other hand, the number $N_{v,op}$ of active AC/DC input stages can be different from the number $N_{p,op}$ of active DC/DC output stages. Thus, in order to balance the DC-link voltages during the operation with 4D-interleaving a proper permutation algorithm has to be employed which activates/deactivates the AC/DC and DC/DC stages of the cells in such way that a minimal voltage ripple on the DC-link capacitors is obtained.
One possible permutation algorithm for the AC/DC stages is described below for a system with $N_{cells} = 6$ converter cells and a fixed number of $N_{PWM} = 1$ PWM AC/DC stages and $N_{FFM} = 5$ AC/DC stages with fundamental frequency modulation (FFM). In contrast to the balancing scheme proposed in (6), the AC/DC stage which is selected for PWM operation is always the uppermost cell in the stack of converter cells, i.e the cell which is connected to the input side boost inductor L_b, since this minimizes the common-mode currents in the system caused by the switching operation of the PWM cell. The proposed permutation algorithm works in such way, that an additional FFM AC/DC stage is activated every time the modulation index of the PWM cell reaches its upper limit and is deactivated when the lower limit of the modulation index of the PWM stage is

© VDE VERLAG GMBH · Berlin · Offenbach

1629

reached, similar to the concept shown in **Fig. 3(b)**. The decision about which FFM AC/DC stage to activate/deactivate is based on the deviation of the DC-link voltages of the cells from the set-point voltage ($V_{DC,set} = 400\,V/N_{cells}$) in such a way that the AC/DC stage of the cell with the lowest DC-link voltage is always the next one to be activated and the AC/DC stage of the cell with the highest voltage is the next one to be deactivated.

Apart from the AC/DC stages of the cells the DC/DC stages can also be utilized for DC-link voltage balancing by means of permutation. For a system which operates at a constant power level, however, there is no natural event when DC/DC stages have to be activated/deactivated like there is for the AC/DC stages, due to the time-varying input voltage. Thus, a constant time interval of e.g. $T_{perm} = 2\,ms$ is chosen after which the selection of active stages is re-evaluated based on the DC-link voltage values of the cells. This means, that the DC/DC stages of the $N_{p,op}$ cells with the highest DC-link voltages are activated.

Consequently, the balancing of the DC-link voltage is achieved by permutation of the active AC/DC and DC/DC stages which is shown in **Fig. 8** for the operation at a constant output power level of $P_{out,tot} = 1.5\,kW$ where the most efficient operation is achieved with $N_{p,op} = 4$ active DC/DC stages. As can be seen, the ISOP system can be operated in a stable condition with 4D-interleaving at the AC/DC and DC/DC stages while the DC-link voltages are effectively balanced with only small deviations of around a maximum of $\Delta V_{DC} \approx 3\,V$.

4. Conclusions

A new dimension of interleaving the operation of the converter cells of an ISOP multi-cell telecom power supply module is presented. The proposed concept is based on a time-varying activation/deactivation of individual AC/DC input and DC/DC output stages of different cells and utilizes the decoupling of the input and output sides of the cells provided by the energy storage capability of the DC-link capacitors. Different 4D-interleaving operation schemes are discussed for the series connected AC/DC stages and the parallel connected DC/DC stages and evaluated by means of a comprehensive η-ρ Pareto optimization and analytical calculations. Compared to the standard 3D-interleaving the losses of the entire rectifier stage can be reduced by 17% and the volume can be decreased by 10% and a very flat efficiency vs. output power characteristic of the parallel connected DC/DC converter stages can be achieved.

Considering the increasing importance of higher part-load efficiency of telecom power supplies, the multi-cell converter approach in combination with the proposed 4D-interleaving concept therefore provides an interesting solution for future implementations.

5. References

[1] M. Kasper, D. Bortis, J. W. Kolar, and G. Deboy, "Hyper-Efficient (98%) and Super-Compact (3.3kW/dm³) Isolated AC/DC Telecom Power Supply Module based on Multi-Cell Converter Approach," in *Proc. of the IEEE Energy Conversion Congress and Exposition (ECCE USA)*, pp. 150–157, 2014.

[2] M. Kasper, D. Bortis, and J. Kolar, "Scaling and Balancing of Multi-Cell Converters," in *Proc. of the International Power Electronics Conference (IPEC ECCE Asia)*, 2014.

[3] M. Kasper, C.-W. Chen, D. Bortis, J. Kolar, and G. Deboy, "Hardware Verification of a Hyper-Efficient (98%) and Super-Compact (2.2kW/dm³) Isolated AC/DC Telecom Power Supply Module Based on Multi-Cell Converter Approach," in *Proc. of the IEEE Applied Power Electronics Conference and Exposition (APEC)*, pp. 65–71, 2015.

[4] R. Giri, V. Choudhary, R. Ayyanar, and N. Mohan, "Common-Duty-Ratio Control of Input-Series Connected Modular DC-DC Converters with Active Input Voltage and Load-Current Sharing," *IEEE Trans. Ind. Appl.*, vol. 42, no. 4, pp. 1101–1111, 2006.

[5] W. van der Merwe and T. Mouton, "Natural Balancing of the Two-Cell Back-to-Back Multilevel Converter with Specific Application to the Solid-State Transformer Concept," in *Proc. of the 4th IEEE Conf. on Industrial Electronics and Applications (ICIEA)*, pp. 2955–2960, 2009.

[6] H. Iman-Eini, J.-L. Schanen, S. Farhangi, and J. Roudet, "A Modular Strategy for Control and Voltage Balancing of Cascaded H-Bridge Rectifiers," *IEEE Trans. Power Electron.*, vol. 23, no. 5, pp. 2428–2442, 2008.

© VDE VERLAG GMBH · Berlin · Offenbach

Battery Charger Based on a Triple-LC$_p$ Resonant Converter

Christian Brañas, University of Cantabria, Spain, branasc@unican.es

Juan C. Viera, University of Oviedo, Spain, viera@uniovi.es

Francisco J. Azcondo, University of Cantabria, Spain, azcondof@unican.es

Manuela González, University of Oviedo, Spain, mgonzalez@uniovi.es

Abstract

In this paper, the analysis and design of a multiphase resonant converter suitable for high-current battery charger applications is presented. The inverter stage of the converter is obtained from the parallel connection of three class D LC_p resonant inverters, increasing the output current capability of the circuit. The regulation of the charging current is implemented at constant frequency by adjusting the phase displacement angles of the drive signals of the inverter sections. The battery charger is designed to provide a maximum charging current, I_o=15 A to a high-performance Absorbent Glass Mat (AGM) battery; widely used in micro-hybrid vehicles.

1. Introduction

Micro-hybrid vehicles include a system that automatically shuts down and restarts the internal combustion engine in order to reduce the emissions. This system, known as stop-start, requires high-performance batteries, which accept repetitive high power peaks of discharge [1]. In addition, when the vehicle is shut down, all auxiliary systems are supplied by the battery. Currently, the lead acid battery technology is being upgraded and redesigned to supply the power required for micro-hybrid cars at lower cost than others advanced technologies [2]. The high-performance AGM battery is an example of this development. These batteries present lower internal resistance for additional power, deeper discharges and quicker recharging, higher corrosion resistance for longer service life, less sulfation, and reduced self-discharge when stored on the shelf [3].

High output current capability is a desirable feature of any battery charger to fully charge the battery as fast as possible. From the design point of view, the battery charger is a circuit with a relative low output voltage in contrast to its high output current. In order to achieve the desired charging current capability, this paper presents the design of a battery charger based on a multiphase resonant converter [4]. In this converter, the resonant inverter section is composed by paralleled class D sections. The output current is distributed among the paralleled stages of the inverter. Overall efficiency, size and weight are improved with this approach due to the reduction of the conduction losses in the inverter stage. The output stage consists of a center-tap rectifier with Schottky diodes which allows reducing both, switching and conduction losses.

1.1. Charging Method

The main features of the battery Varta Silver Dynamic AGM, specially designed for vehicles with advanced Start-Stop functionality with regenerative braking are: 12 V nominal voltage, 105 Ah nominal capacity at C/20 measured at 25^{0}C and low internal resistance. Conventional constant current - constant voltage (CC-CV) based charging method is shown

© VDE VERLAG GMBH · Berlin · Offenbach

in Fig. 1, using three experimental profiles with a constant current stage set at 10 A, 15 A and 20 A respectively.

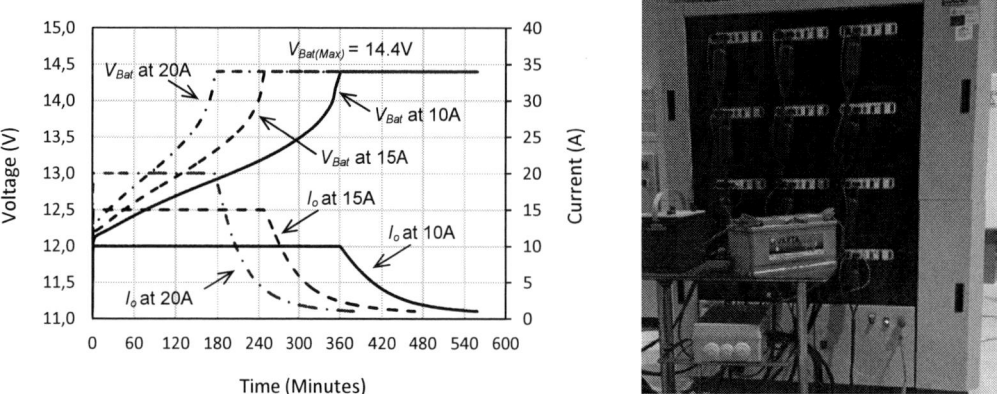

Fig. 1. Charging profiles at 10 A, 15 A and 20 A for an AGM battery. Picture on the right shows the Battery Laboratory where tests were carried out and the battery under test (Varta Silver Dynamic AGM).

Voltage and current profiles were obtained from the battery using the Test Equipment SBT 10050. From Fig. 1, it is observed that during the constant-current stage the battery accepts the highest current level. The experimental study of the battery under test determined that the maximum battery voltage must be limited at $V_{Bat(Max)}$=14.4 V in order to avoid electrolyte loss. In this way, during the constant-voltage stage, the voltage is set to the maximum, $V_{Bat(Max)}$=14.4 V, so that the battery current decreases as the charging process continues. Finally, it should be taken into account that in order to avoid an overcharge situation, gas production and consequently the battery degradation, limitation of the current ripple is mandatory.

2. Triple LC$_p$ Resonant Converter

The circuit of the proposed battery charger is shown in Fig.2. The battery is modeled in steady state by its internal impedance, r_{Bat}, plus an ideal voltage source, V_{Bat}.

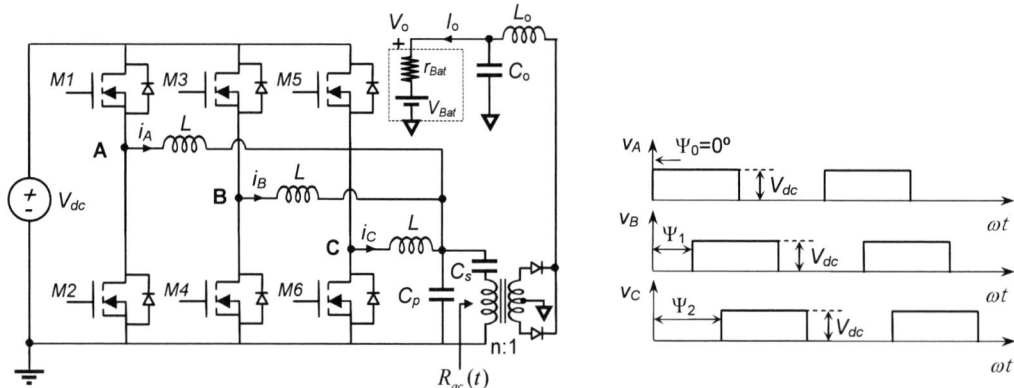

Fig.2. Battery charger based on a multiphase LC_p resonant converter and midpoint voltages of each inverter section.

The AC side is a multiphase resonant inverter which consists of three paralleled LC_p class D sections [2]. The DC side is a center tap rectifier where the transformer provides insulation and rises the output current. The control parameters of the proposed converter are the angles of phase displacement [5], Ψ_1 and Ψ_2, as depicted in Fig. 2. Making use of the fundamental approximation [5-6], the input voltages v_A, v_B and v_C are represented by the phasors given in (1-3).

$$\mathbf{V_a} = \frac{2V_{dc}}{\pi} \qquad (1) \qquad \mathbf{V_b} = \frac{2V_{dc}}{\pi} \cdot e^{-j(\Psi_1)} \qquad (2) \qquad \mathbf{V_c} = \frac{2V_{dc}}{\pi} \cdot e^{-j(\Psi_2)} \qquad (3)$$

The rectifier stage imposes an equivalent resistive load, R_{ac}, for the resonant inverter stage. Since the output filter removes the high frequency ripple, the rectifier stage is modeled in steady state using the averaged variables [6]. With the first harmonic of a square waveform, the relationship between the AC and DC side currents is given by (4). On the other hand, the output voltage, V_o, is obtained as the mean value of the full wave rectified voltage (5).

$$I_o = \frac{n\pi}{4} \cdot \hat{I}_{ac} \qquad (4) \qquad\qquad V_o = \frac{2}{n\pi} \cdot \hat{V}_{ac} = r_{Bat} \cdot I_o + V_{Bat} \qquad (5)$$

From (4) and (5), the battery model is represented in the AC side by (6). From (4) and (6), the rectifier stage is reflected to the AC side as the equivalent resistance given in (7).

$$\hat{V}_{ac} = \frac{n^2\pi^2}{8} \cdot r_{Bat} \cdot \hat{I}_{ac} + \frac{n\pi}{2} \cdot V_{Bat} \qquad (6) \qquad R_{ac} = \frac{\pi^2}{8} n^2 R_o = \frac{\pi^2}{8} n^2 \left(r_{Bat} + \frac{V_{Bat}}{I_o} \right) \qquad (7)$$

Finally, by using (1-7), the simplified circuit of the resonant converter is depicted in Fig. 3.

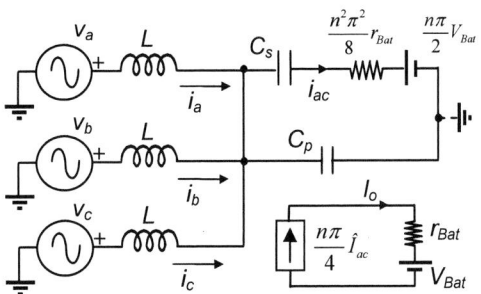

Fig.3. Simplified circuit by using the fundamental approximation for circuit analysis purposes.

The analysis of the proposed resonant converter is carried out by using the simplified circuit shown in Fig.3. The main features of the resonant converter are defined by parameters shown in Table I.

TABLE I.
PARAMETERS OF THE MULTIPHASE LC_p INVERTER

Parallel Resonant Frequency	Parallel Characteristic Impedance	Parallel Quality Factor
$\omega_p = \dfrac{1}{\sqrt{LC_p/3}}$	$Z_p = \omega_p L = \dfrac{3}{\omega_p C_p}$	$Q_p = \dfrac{3R_{ac}}{Z_p}$

The proposed LC_p resonant converter is designed to behave as a current source in order to limit the charging current by fixing $\omega = \omega_p$, where ω_p is the parallel resonant frequency, given in table I. The analysis of the resonant converter is carried out considering the angles Ψ_1 and Ψ_2 as control parameters. The inductors' current phasors are given in (8-10),

$$\mathbf{I_a} = \frac{2V_{dc}}{\pi Z_p} \times \left\{ \frac{Q_p}{3} \left[1 + \cos\Psi_1 + \cos\Psi_2\right] - j\left[1 + \frac{Q_p}{3}\left(\sin\Psi_1 + \sin\Psi_2\right)\right] \right\} \qquad (8)$$

$$\mathbf{I_b} = \frac{2V_{dc}}{\pi Z_p} \times \left\{ \frac{Q_p}{3} \left[1 + \cos\Psi_1 + \cos\Psi_2\right] - \sin\Psi_1 - j\left[\cos\Psi_1 + \frac{Q_p}{3}\left(\sin\Psi_1 + \sin\Psi_2\right)\right] \right\} \qquad (9)$$

$$\mathbf{I_c} = \frac{2V_{dc}}{\pi Z_p} \times \left\{ \frac{Q_p}{3} \left[1 + \cos\Psi_1 + \cos\Psi_2\right] - \sin\Psi_2 - j\left[\cos\Psi_2 + \frac{Q_p}{3}\left(\sin\Psi_1 + \sin\Psi_2\right)\right] \right\} \qquad (10)$$

The output current phasor seen from the AC side, $\mathbf{I_{ac}}$, is calculated by (11).

$$\mathbf{I_{ac}} = \frac{2V_{dc}}{\pi Z_p} \left\{ \sin\Psi_1 + \sin\Psi_2 + j\left[1 + \cos\Psi_1 + \cos\Psi_2\right] \right\} \qquad (11)$$

The charging current, I_o, is obtained from (11) and (4).

$$I_o = \frac{nV_{dc}}{2Z_p} \cdot \sqrt{\left(\sin\Psi_1 + \sin\Psi_2\right)^2 + \left(1 + \cos\Psi_1 + \cos\Psi_2\right)^2} \qquad (12)$$

Working with (12), the maximum charging current is achieved at $\Psi_0 = \Psi_1 = \Psi_2 = 0°$ and is given by (13),

$$I_o = \frac{3nV_{dc}}{2Z_p} \qquad (13)$$

2.1. Pattern for adjusting the control angles

During the constant-voltage stage, the charging current must be adjusted to avoid that the battery voltage exceeds the maximum value, $V_{Bat(Max)}$. The current is modulated through the control angles Ψ_1 and Ψ_2, while keeping constant the switching frequency. Different patterns for adjusting Ψ_1, Ψ_2 are possible. In this paper, the selected control pattern is the simplest one by shifting only one phase so that, $\Psi_0 = \Psi_1 = 0°$, $\Psi_2 = \Psi$. In this case, only two independent control signals are required, which means a simplification of the control circuit. Thus, the charging current is calculated according to (14).

$$I_o = \frac{nV_{dc}}{2Z_p} \cdot \sqrt{5 + 4\cos\Psi} \qquad (14)$$

From (14), the minimum charging current corresponds to $\Psi = 180°$, being one third of the maximum value given in (13). The reduction of the charging current, while keeping constant the battery voltage, leads to a significant increment of the equivalent resistance, R_o. Consequently, the reflected impedance in the AC side, R_{ac}, and the quality factor, Q_p, also increases. Assuming $V_{Bat(Max)}$ constant and working with (14), (7) and (5), the variation of the quality factor is given in (15),

$$Q_p = \frac{\pi^2 nV_{Bat(Max)}}{V_{dc}\sqrt{5 + 4\cos\Psi}} \qquad (15)$$

The normalized amplitude of the charging current, I_o, is depicted in Fig. 4 as a function of the control angle, Ψ, taking into account the pattern $\Psi_0 = \Psi_1 = 0°$, $\Psi_2 = \Psi$. The variation of the quality factor due to the current adjustment is also represented in Fig. 4.

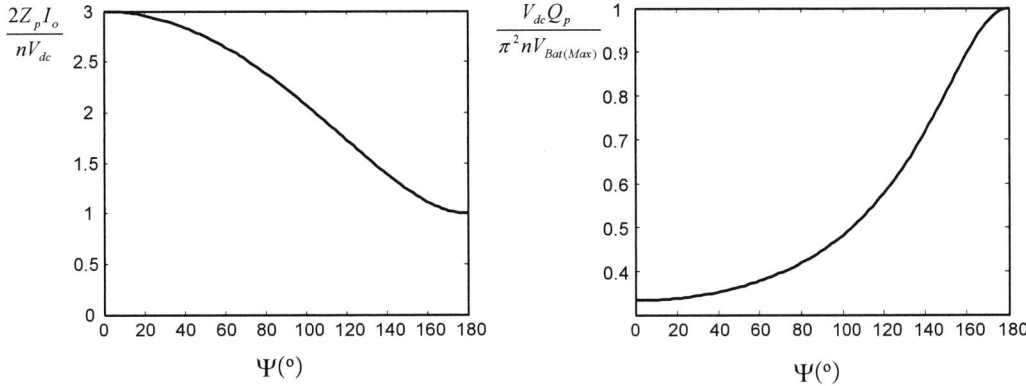

Fig. 4. Left: normalized amplitude of the charging current as a function of the control angle, Ψ. Right: quality factor variation during charging at constant voltage

3. Switching mode and complex power

In order to minimize the switching loss the converter is designed to achieve zero-voltage switch (ZVS). The ZVS mode requires a phase delay [5-7] of the resonant current, referred to the input voltage. In order to determine the power factor angle, ϕ_a, ϕ_b, and ϕ_c of each transistors leg and study the power distribution, the complex power for each generator v_a, v_b and v_c is calculated considering the control pattern $\Psi_0 = \Psi_1 = 0^\circ$, $\Psi_2 = \Psi$ in (16-17).

$$\mathbf{S_a} = \mathbf{S_b} = \frac{2V_{dc}^2}{\pi^2 Z_p} \times \left[\frac{Q_p}{3}(2 + \cos\Psi) + j\left(1 + \frac{Q_p}{3}\sin\Psi\right) \right] \tag{16}$$

$$\mathbf{S_c} = \frac{2V_{dc}^2}{\pi^2 Z_p} \times \left[\frac{Q_p}{3}(1 + 2\cos\Psi) + j\left(1 - \frac{2Q_p}{3}\sin\Psi\right) \right] \tag{17}$$

The total power handled by the converter is calculated as $\mathbf{S} = \mathbf{S_a} + \mathbf{S_b} + \mathbf{S_c}$ in (18),

$$\mathbf{S} = \frac{2V_{dc}^2}{\pi^2 Z_p} \times \left[\frac{Q_p}{3}(5 + 4\cos\Psi) + 3j \right]. \tag{18}$$

From (16-18) the power factor angle, ϕ_i is obtained as $\phi_i = \text{angle}(\mathbf{S_i})$ and is strongly dependent on Q_p. For maximum output current, i.e. $\Psi = 0^\circ$, the power factor angles are,

$$\phi_a = \phi_b = \phi_c = \phi = \arctan\left(\frac{1}{Q_p}\right) = \arctan\left(\frac{3V_{dc}}{\pi^2 n V_{Bat(Max)}}\right) \tag{19}$$

The value of the quality factor defines the ratio active power to reactive power handled by each inverter section. In this way, a high value of Q_p reduces the reactive energy in the resonant converter and improves the efficiency. From (15) and (19), it is observed that by increasing the transformer turns ratio, n, the reactive energy in the resonant circuit can be reduced. However, a minimum amount of reactive energy must be accepted for assuring the ZVS mode of all transistors. The minimum value of power factor angle ϕ_{zvs} [7], given in (20), depends on the dead time, t_d, of the driver transistors and the switching frequency.

$$\phi_{zvs} = \frac{t_d}{T_s} \cdot 360^\circ \tag{20}$$

© VDE VERLAG GMBH · Berlin · Offenbach

The power factor angles, ϕ_a, ϕ_b, and ϕ_c are plotted in Fig. 5 as a function of the control angle, Ψ, and the quality factor, Q_p.

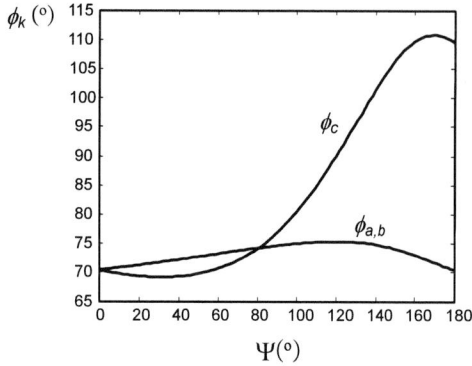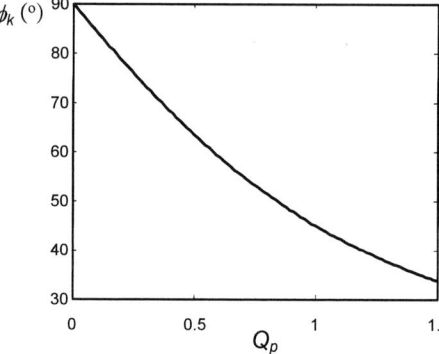

Fig. 5. Left: power factor angles ϕ_a, ϕ_b, and ϕ_c as a function of the control angle, Ψ. Right: Initial value of the power factor angles as a function of the parallel quality factor Q_p

4. Design of the resonant converter

This section contains a step-by-step summary of the design procedure for a multi-phase controlled LC_p converter.

1) The electrical data of the battery are: V_{Bat}=12 V nominal voltage, 105 Ah nominal capacity at C/20, $V_{Bat(Max)}$=14.4 V and the maximum charging current is set at I_o=15 A. The converter supply voltage is V_{dc}=400 V, which is the output voltage of a previous front-end power factor correction stage.

2) The switching frequency is set at ω_p=2π(125kHz). Considering a typical value of transistor driver dead time t_d=700ns and the switching frequency ω_p=2π(125kHz), the minimum value of power factor angle is ϕ_{zvs}=(700ns/8µs)*360°=31.4°

3) The quality factor is chosen to be close to Q_p=0.5, which yields an initial value of the power angles ϕ_a=ϕ_b=ϕ_c=63°, well above of the minimum value ϕ_{zvs}. According to (15), Q_p=0.47 is achieved for a transformer turn ratio n=4.

4) For n=4, the characteristic impedance is obtained from (13): Z_p=160Ω.

5) The reactive components are: L=Z_p/ω_p=203µH and C_p=3/$\omega_p Z_p$=23.8nF. The capacitor C_s is chosen high enough for blocking the DC current through the transformer, C_s=1µF.

6) The output filter $L_o C_o$ is configured with L_o=150µH, which is enough to remove the high frequency current ripple, and C_o=100µF.

5. Experimental Results

The practical implementation was carried out using the low cost transistor IRF840LC. The resonant tank elements are L=200µH and C_p=22nF. The resonant inductors were implemented using the core RM12 of 3C96 material. The drive signals were generated by a digital circuit which is able to set up to 15 possible values of the control angle Ψ. The experimental values of the charging current, battery voltage, quality factor and efficiency as a function of the control angle, Ψ, are shown in Fig. 6. The experimental charging current is in agreement with the theoretical one, shown in Fig. 4. The maximum value was I_o=13.87 A, close to the theoretical maximum of 15 A. The minimum value of the experimental charging current was I_o=4.8 A, slightly below of the theoretical minimum of 5 A.

© VDE VERLAG GMBH · Berlin · Offenbach

PCIM Europe 2016, 10 – 12 May 2016, Nuremberg, Germany

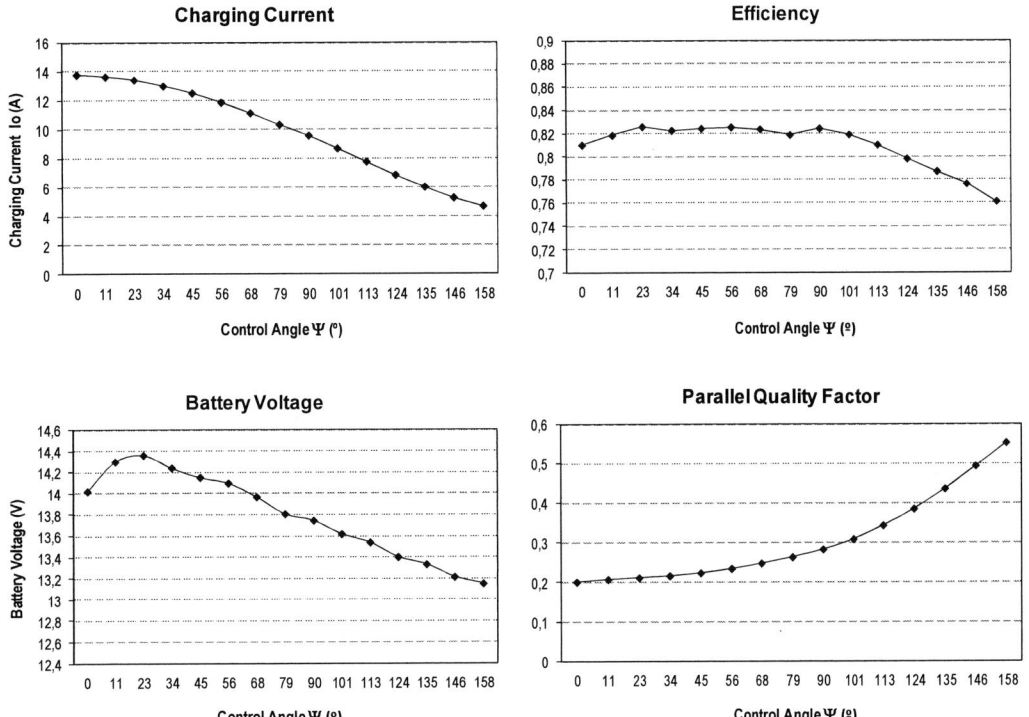

Fig. 6. Experimental charging current, efficiency, battery voltage and quality factor as a function of the control angle, Ψ.

The efficiency of the charger, including both stages, has little variation for the whole range of the charging current. The maximum value of efficiency was approximately 0.83. The battery voltage is kept below the maximum value of $V_{Bat(Max)}$=14.4 V by adjusting the charging current. The variation of the quality factor of the resonant circuit is also in agreement with the theoretic calculation shown in Fig. 4. Experimental waveforms, charging at maximum and minimum current are shown Fig. 7, where it can be observed that both voltage and current ripples are almost negligible.

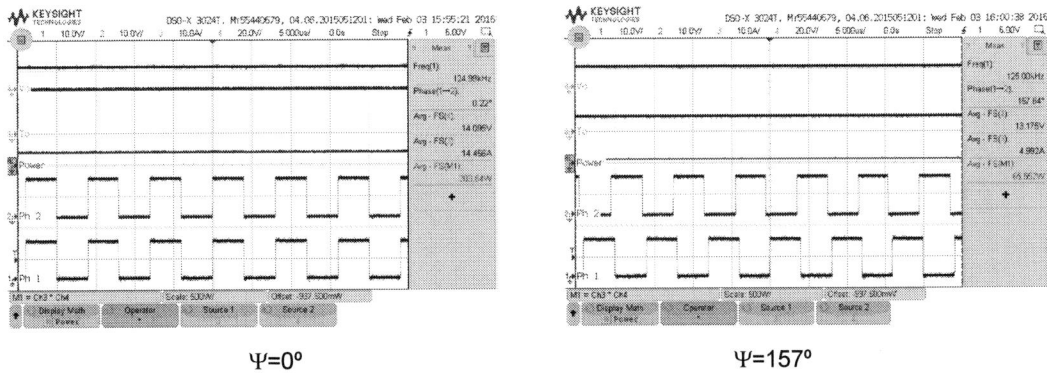

Ψ=0º   Ψ=157º

Fig. 7. Experimental waveforms charging at maximum (Ψ=0º) and minimum (Ψ=157º) current rate. From bottom to top: Drive signals, battery power, charging current and battery voltage.

© VDE VERLAG GMBH · Berlin · Offenbach

6. Conclusions

The analysis and design of a multiphase resonant converter for battery charger applications has been presented. Since the load current is shared among three equal sections, the circuit presents high output current capability using low cost power MOSFETs and small magnetic components. The control is carried out at constant frequency, by adjusting the phase displacement of the drive signals. The maximum value of experimental efficiency of the proposed charger was η=83%. Further efficiency increments are possible by using synchronous rectifiers instead of diodes although it should be considered that the cost and complexity of the circuit increases. The increment of the quality factor during the charging process is a good feature of the resonant converter leading to a little variation of the overall efficiency of the circuit. With an output current capability from 15 A to 5 A, the converter meets the specifications of the battery charging profile to extend its service life and preserve the state of health.

Acknowledgements

This work is sponsored by the Spanish Ministry of Economy and Competitiveness, the Principality of Asturias Government and the European Union (ERFD), under Research Grants FC-15-GRUPIN14-073 and DPI2013-46541-R and the Spanish Ministry of Science and the EU through the project TEC2014-52316-R: 'Estimation and Optimal Control for Energy Conversion with Digital Devices' ECOTRENDD.

7. References

[1] Pillot Christophe, "Micro hybrid, HEV, P-HEV and EV market 2012-2025. Impact on the battery business" EVS27 International Battery, Hybrid and Fuel Cell Electric Vehicle Symposium, Barcelona, Spain, November 17 - 20, 2013.

[2] Joern Albers, Eberhard Meissner and Sepehr Shirazi "Lead-acid batteries in micro-hybrid vehicles" Journal of Power Sources 196, pages 3993–4002. 2011.

[3] Eckhard Kardena, Paul Shinnb, Paul Bostockc, James Cunninghamc, Evan Schoultzd, Daniel Koka, "Requirements for future automotive batteries – a snapshot" Journal of Power Sources 144, pages 505–512. 2005.

[4] C. Branas, F.J. Azcondo, R. Casanueva, "A Generalize Study of Multiphase Parallel Resonant Inverters for High-Power Applications", Circuits and Systems I: Regular Papers, IEEE transactions on, Vol. 55, No7, pp2128-2138, Aug. 2008.

[5] D. Czarkowski, M.K. Kazimierczuk, "Phase-Controlled Series-Parallel Resonant Converter," IEEE Trans. on Power Electronics, Vol.8, No.3, July 1993. pp. 309-319.

[6] R. Erickson, D. Maksimovic, "Fundamentals of Power Electronics" Second Edition, Springer Ed. 2001.

[7] Lopez, V.M.; Navarro-Crespin, Alejandro; Schnell, Ryan W.; Branas, C.; Azcondo, Francisco Javier; Zane, Regan A. "Current Phase Surveillance in Resonant Converters for Electric Discharge Applications to Assure Operation in Zero-Voltage-Switching Mode", Power Electronics, IEEE Transactions on, Volume: 27 , 2012 , pp. 2925 – 2935

High Efficient Flyback Converter with SiC-MOSFET

Johann Austermann, Ostwestfalen-Lippe UAS, Germany, johann.austermann@hs-owl.de
Tim Stuckmann, Ostwestfalen-Lippe UAS, Germany, tim.stuckmann@hs-owl.de
Holger Borcherding, Ostwestfalen-Lippe UAS, Germany, holger.borcherding@hs-owl.de

Abstract

This paper introduces a compact and cost-efficient AC/DC-converter with 200 W nominal power to generate 48 V DC from the three phase 400 V line-voltage. The converter essentially consists of a line-side uncontrolled rectifier and a downstream flyback converter. The paper shows how a good efficiency can be achieved using a silicon carbide MOSFET in combination with a good transformer design and active rectification. After a short description of the circuit, the operation of the converter will be explained with the help of measured current and voltage curves.

1. Introduction

Low-voltage drive systems are commonly used for adjustment and positioning functions. In the industrial sector where 400 V AC supply voltage can be seen as standard, very often converters from 400 V AC to 48 V DC are needed to supply these low-voltage drives. The converters typically have a line-side rectifier and a downstream galvanic isolated DC/DC-converter. Flyback converters are commonly used for galvanic isolated DC/DC-conversion with relatively low power [1]. Due to the low number of components the flyback topology is ideally suited for cost- and space- sensitive applications. The most expensive component usually is the flyback transformer, whose size depends on the selected switching frequency. New wide-bandgap semiconductors based on silicon carbide allow high switching frequencies without the disadvantage of high switching losses [2]. Unfortunately, the efficiency of the flyback converter decreases with increasing switching frequency due to the non-ideal coupling of the primary and secondary side of the transformer. In simple flyback topologies the energy stored in the leakage inductance of the primary side is wasted in the snubber circuit. Hence, special attention has to be paid to the design of the storage transformer to achieve an acceptable efficiency of the converter.

2. Circuit description

The topology of the converter shown in Fig. 1 essentially consists of a flyback converter that is fed with the DC-link voltage generated by a line-side uncontrolled rectifier. Hence the DC-link voltage depends on the line-voltage and is typically between 500 and 600 V DC for standard European grids. An EMC-filter and a inrush current limiter (not shown) are integrated between the rectifier and the DC-link capacitors in order to comply with the EMC regulations and to protect the rectifier diodes against inrush current.

The flyback converter is regulated by a standard current mode PWM controller [3] which allows 100 kHz switching frequency. Due to the high switching frequency a small storage transformer can be used. The transformer design will be explained in chapter 3. The control loop is closed via an optocoupler and a regulator circuit on the secondary side of the transformer. The PWM controller works in current control discontinous conduction mode (DCM). The design of those control loops is shown in [4] and [5].

For the transformer design a turn-on time of $t_{on} = t_{off} = 0,5 \cdot T_s$ at 200 W load is assumed. The turns ratio can be determined with this assumtion:

$$\frac{U_1 \cdot t_{on}}{N_p} = -\frac{U_2 \cdot t_{off}}{N_s} \text{ with } t_{on} = t_{off} \Rightarrow -\frac{U_1}{U_2} = \frac{N_p}{N_s} \tag{1}$$

In practical applications, the turns ratio must be selected slightly smaller in order to ensure working in discontinous current mode:

$$-\frac{N_p}{N_s} < \frac{U_1}{U_2} \tag{2}$$

The voltage U_{DS} of the MOSFET Q_1 can be calculated by summing up the input- and the reflected output voltage:

$$U_{DS} = U_1 + \frac{N_p}{N_s} \cdot U_2 \tag{3}$$

With the assumtions above, the voltage U_{DS} over the mosfet will be twice the input voltage during the off-phase. Considering the ratings in Tab.1 and a safety margin for transient overshoots, a MOSFET with a blocking cabability over 1200 V is required.

Tab. 1: Rating of the flyback converter

Symbol	Description	Value
U_1	input voltage flyback	565 V
U_2	output voltage	48 V
P_N	nominal power	200 W
I_{out}	nominal output current	4,18 A

The SiC-MOSFET SCT2H12NY (ROHM) is well suited for flyback converters with high input voltage, because it delivers 1700 V blocking voltage and low switching losses. As shown in Fig.1, an additional snubber circuit, build up with TVS diodes, is used in order to clamp the MOSFETs voltage U_{DS} to its maximum blocking voltage. The TO-268-2L SMD-package allows easy mounting and cooling by the surface of the circuit board.

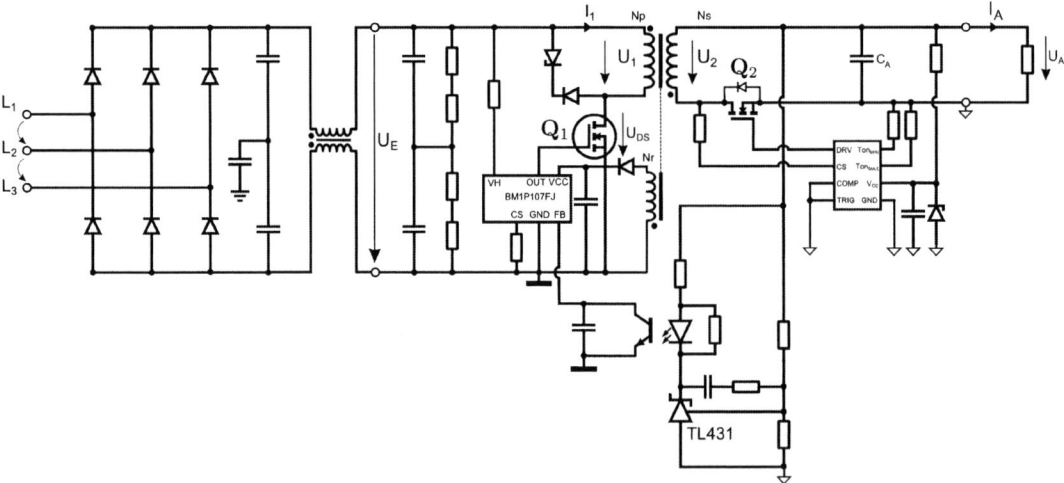

Fig. 1: Topology of the AC/DC-Converter

The efficiency of a flyback converter can be improved significantly by using a synchronous rectifier on the secondary side of the transformer. During the conduction phase, current flows through the MOSFET Q_2 instead of through a rectifier diode. This results in a significantly lower forward voltage drop and thus a better efficiency. In this approach the NCP4303A IC [6] is used to control the MOSFET Q_2. The NCP4303A detects the voltage U_{DS2} over Q_2 and turns it on when U_{DS2} becomes negative.

3. Design of the Storage Transformer

Special attention has to be paid to the transformer design because it determines the efficency, the volume and isolation of the converter. A good nesting of the primary and secondary windings generates low leakage inductance and thus good efficiency of the converter. The arrangement of the windings is shown in Fig.2a [1] S.601.

Due to the high du/dt, which can be above $50\,kV/\mu s$ in applications with SiC-MOSFETs, partial discharges can occure that damage the isolation of the winding (see Fig.2b). Partial discharges can be prevented by inserting isolation foil between the individual layers of the winding.

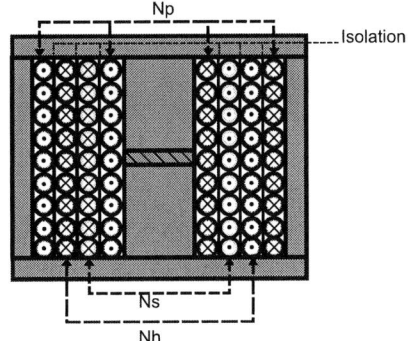

(a) Arrangement of the transformers winding

(b) Breakdown of the isolation

Fig. 2: Transformer design

The copper losses can be significantly reduced by using litz wire for the winding. For the determination of the litz wire it is helpful to describe $i_1(t)$ and $i_2(t)$ (see Fig.7) as a fourier series:

$$i_1(t) = \frac{\hat{i}_1}{4} + \sum_{k=1}^{\infty} \left[\frac{\hat{i}_1}{(k\cdot\pi)^2} \cdot \left((-1)^k - 1\right) \cdot \cos(k\cdot\omega_s t) + \left(-\frac{\hat{i}_1}{k\cdot\pi}\right) \cdot (-1)^k \cdot \sin(k\cdot\omega_s t) \right] \quad (4)$$

$$i_2(t) = \frac{\hat{i}_2}{4} + \sum_{k=1}^{\infty} \left[\frac{\hat{i}_2}{(k\cdot\pi)^2} \cdot \left((-1)^{k+1} - (-1)^k\right) \cos(k\cdot\omega_s t) + \frac{\hat{i}_2}{k\cdot\pi} \cdot (-1)^k \sin(k\cdot\omega_s t) \right] \quad (5)$$

Fig. 3 shows the resistance of round copper conductors as a function of the frequency. This data base on sectionally defined formulars shown in [7] pp.B14ff.. The litz wire can be selected using Fig.3 and the fourier coefficients of $i_1(t)$ and $i_2(t)$. For the converter described in this paper, 20 x 0,1 mm litz wire was chosen for the primary and 120 x 0,1 mm for the secondary winding.

© VDE VERLAG GMBH · Berlin · Offenbach

Tab. 2: Fourier Coefficients

Order	f_s	% of $\hat{\imath}_1$	% of $\hat{\imath}_2$
0	0 kHz	25,0 %	25,0 %
1	100 kHz	37,7 %	37,7 %
2	200 kHz	15,9 %	16,7 %
3	300 kHz	10,8 %	10,9 %
4	400 kHz	8,0 %	8,1 %
5	500 kHz	6,4 %	6,4 %

Tab. 3: Design data of the transformer

Symbol	Description	Value
$\hat{\imath}_1$	primary peak current	1,5 A
$\hat{\imath}_2$	secondary peak current	23,3 A
N_1	turns primary	120
N_2	turns secondary	8
N_2	turns auxillary	2
L_1	primary inductance	1,69 mH
δ	air gap for ETD34 core	1 mm

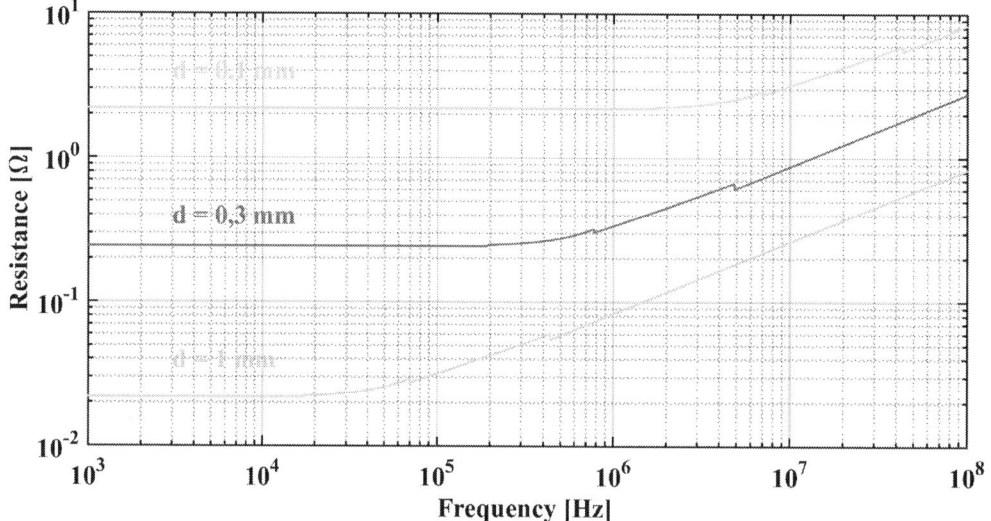

Fig. 3: Skineffect

4. Secondary side active rectifier

A large proportion of the losses in flyback converters typically arises at the rectification on the secondary side. As mentioned in the previous chapter "circuit description", these losses can be reduced significantly by using an active rectification. Therefore extremely low-resistance MOS-FETs are used as synchronous rectifiers. For this purpose it is a necessary ability of MOSFETs to conduct current bi-directionally. The used circuit structure for synchronous rectification on the output side of the flyback is shown in Fig. 4. As it can be seen the MOSFET is oriented such that its body diode has the same orientation as that of the rectifier diode commonly used in flyback topologies. Therefore, the body diode is conductive during blocking phase. The necessary blocking voltage can be calculated using the reflected input voltage:

$$U_2 = U_1 \cdot \frac{N_s}{N_p} + U_A = 565\,V \cdot \frac{8}{120} + 48\,V \approx 86\,V \tag{6}$$

With consideration of a safety margin, a 100 V-type MOSFET was chosen.
As driver circuit serves a regulator IC specially developed for synchronous rectification applications [6]. The supply voltage is clamped to 27V by a zener diode. The resistors R_{TMIN} and

R_{TMAX} are used to adjust the minimum time that must elapse after the corresponding switching operation before the MOSFET can be switched again. This function helps to prevent switching operations caused by interferences. By means of the CS pin the voltage drop across the drain-source path of the MOSFET is monitored. When the voltage drop becomes negative the MOSFET is driven. Thus the body diode of the MOSFET only conducts current until the IC detects a negative voltage drop at the CS pin. During switch-on the voltage drop across the drain-source path is only the forward voltage of the body-diode, consequently the switch-on losses are very low. The MOSFET remains conductive until a polarity change is detected at CS pin. The polarity of V_{DS} changes when the flyback converter passes into the conductive phase. When the regulator IC detects the polarity change the MOSFET is turned off. For the turn-off losses there are nearly the same conditions as for the turn-on losses, hence the turn-off losses are also very low. The resistor R_{CS} can be used to adjust the turn-on and turn-off threshold. It is important to set the thresholds correctly because they also determine how effective the synchronous rectification is. The voltage waveforms shown in Fig.5 serve for a better understanding of the individual functions of the synchronous rectification controller.

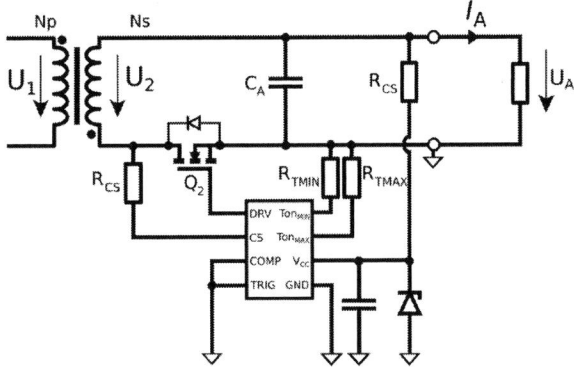

Fig. 4: Active rectifier circuit

The synchronous rectification firstly helps to improve efficiency but secondly the losses in the rectifier diode would be very high and therefore a very large heatsink would be needed. Thus there are advantages in efficiency and size of the flyback converter by using synchronous rectification. The influence of the $R_{DS(on)}$ of Q_2 on the efficiency is exemplary shown in Fig. 8.

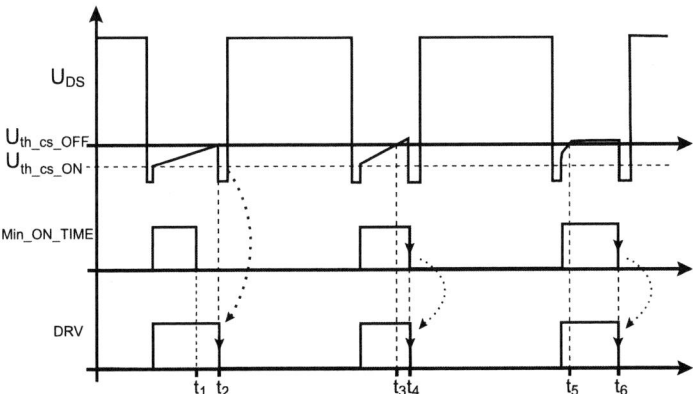

Fig. 5: Determination of the control signal from U_{DS}

5. Practical Setup and Measurement Results

Fig.6 shows a practical setup of the converter with 200 W nominal power. The SMD components (line-side rectifier, switching MOSFET, synchronous rectifier) are mounted on the bottom side in order to reduce the total space of the converter and the THT components (capacitors, EMC-components, storage transformer) are located on the top side of the circuit board. The storage transformer is build up using an epcos ETD34 core and highfrequency litz wire. Fig.7 shows the measured voltage and current waveforms of the switching MOSFET Q_1 on the primary side and of the rectifier MOSFET Q_2 on the secondary side.

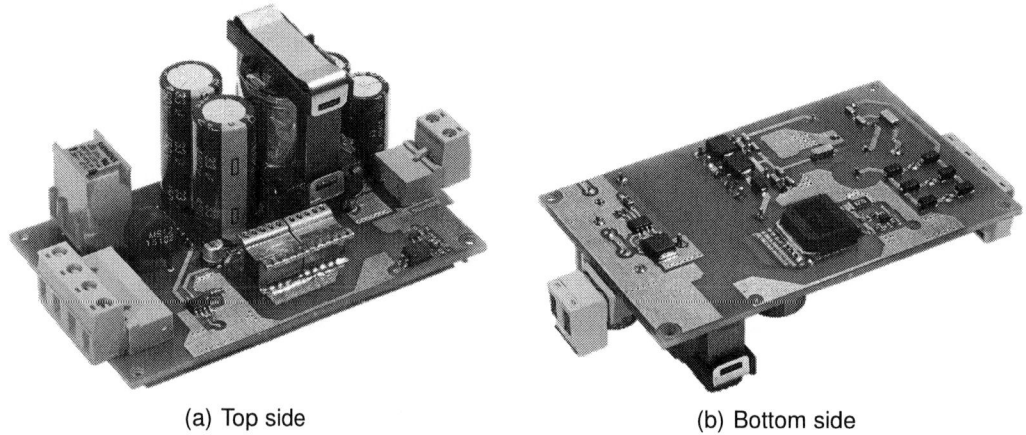

(a) Top side   (b) Bottom side

Fig. 6: Practical setup of the converter

The blocking voltage u_{DS1} over the MOSFET during off-phase is in the range of 1200 V. Transient overvoltages which occure immediately after switching on the MOSFET are even close to the maximum blocking voltage. The current of the primary winding was measured using a clamp ammeter (tektronix TCP312). This method can not be used for the secondary current i_{D2} because the additional conductor loop would affect the synchronous rectifier. Hence the current waveform i_{D2} was measured with a special current probe (Aim I-prober 520), that allows to measure the current by placing its tip onto the conductor. Due to this technique the waveform of the current could be detected, but not the exact value.

As a result of the fast switching of the SiC-techology only two little SMD-heatsinks are necessary in order to ensure adequate cooling of the MOSFET Q_1. The overall efficiency of the AC/DC-converter as a function of the

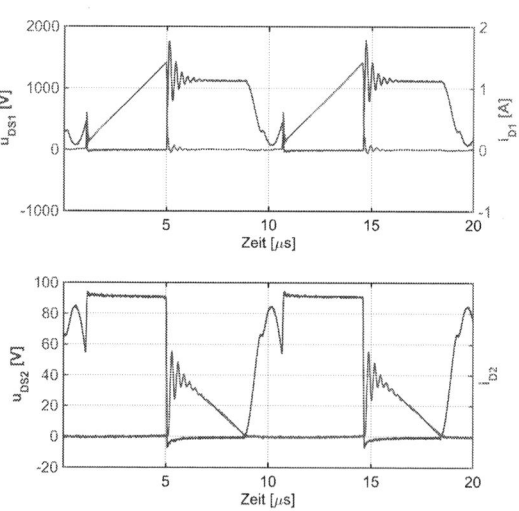

Fig. 7: Voltage and Current on the Primary and Secondary Side of the Transformer

load is shown in Fig.8. Fig.8 displays the overall efficiency using an uncontrolled rectifier (red), a synchronous rectifier whose MOSFET has 24 mΩ R_{DSon} (green) and a synchronous rectifier whose MOSFET has 14 mΩ R_{DSon} (blue). It can be seen clearly that the efficiency significantly increases with lower voltage drop in the rectifier on the secondary side. Thanks to the SiC-MOSFET in combination with a good transformer design and a synchronous rectifier an efficiency of over 90 % at full load was achieved.

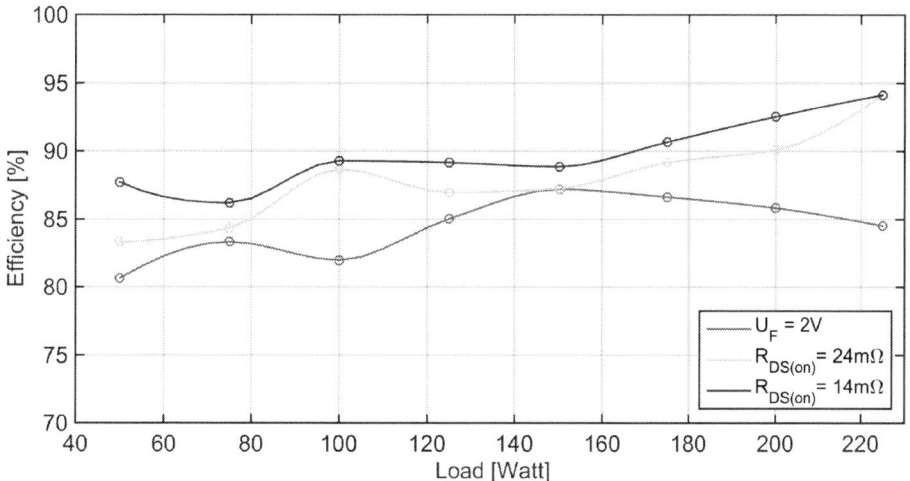

Fig. 8: Efficiency of the AC/DC-Converter

The thermal hotspots of the circuit board can be seen in Fig. 10. Here it is particulary noticeable that the inrush current limiter is the hottest device. Optimizations on this part of the circuit would cause an additional increase of the efficiency. Further hotspots are the SiC-MOSFET, the rectifier MOSFET and the snubber circuit. Thanks to the SiC-technologie and the active rectifier no bulky heatsink is required.

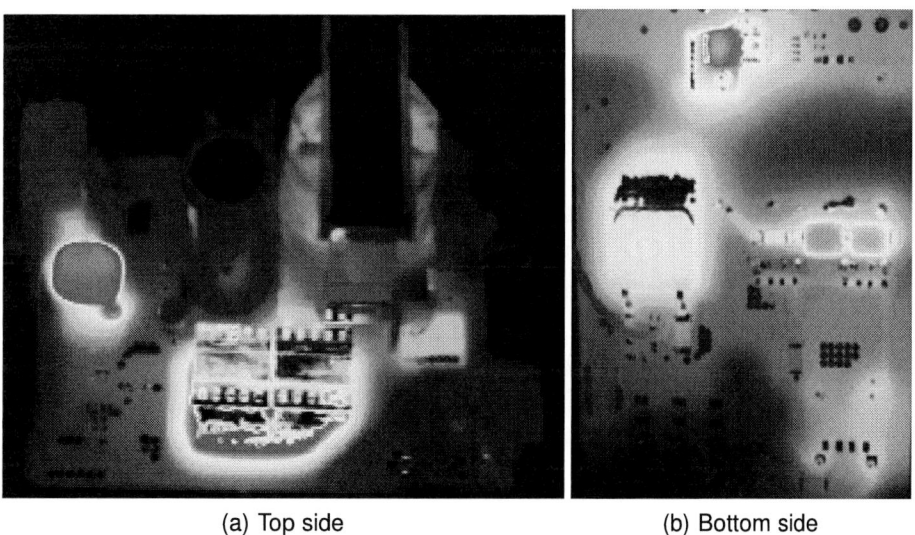

(a) Top side   (b) Bottom side

Fig. 10: Thermal images

6. Conclusion

- The presented circuit converts line-voltage (3~ 400 V AC) in 48 V DC with 200 W rated power.

- The converter is build up using a line-side uncontrolled rectifier and a flyback converter with synchronous rectifier.

- Due to a high switching frequency (100 kHz) and 1700 V blocking capability of the SiC-MOSFET a small transformer design was possible.

- The efficiency can be increased significantly by the use of an active rectifier.

- Thanks to the SIC-MOSFET and the synchronous rectifier an efficiency over 90 % at 200 W load was achieved.

7. References

[1] Ralf Kories and Heinz Schmidt-Walter. *Taschenbuch der Elektrotechnik: Grundlagen und Elektronik*. Deutsch, Frankfurt am Main, 2010.

[2] J. Glaser and et ali. *Direct Comparison of Silicon and Silicon Carbide Power Transistors in High-Frequency Hard-Switched Applications*. IEEE, 2011.

[3] ROHM. Datasheet pwm control ic bm1p107fj.

[4] Christophe P. Basso. *Switch-mode power supplies: SPICE simulations and practical designs*. McGraw-Hill, New York, NY, 2. ed. edition, 2014.

[5] Christophe P. Basso. The tl431 in switch-mode power supplies loops.

[6] ON Semiconductor. Datenblatt ncp4303a: Secondary side synchronous rectification driver for high efficiency smps topologies.

[7] Hans Heinrich Meinke and Klaus Lange. *Taschenbuch der Hochfrequenztechnik*. Springer, Berlin, 1986.

PCIM Europe 2016, 10 – 12 May 2016, Nuremberg, Germany

PCB integration of a magnetic component dedicated to a power factor corrector converter

Guillaume Hérault[1], Denis Labrousse[1], Adrien Mercier[1], Stéphane Lefebvre[1,2]
[1] SATIE, CNRS, ENS Cachan, Cnam, Paris-Saclay University, 61 av. Président Wilson, F-94230 Cachan, France, [2] TPU Tomsk Polytechnic University

Abstract

This paper focuses on the PCB integration of the magnetic component of a high switching frequency PFC converter (from 2 MHz to 8 MHz) constituted by a single-phase diode rectifier and a boost converter. In a first prototype the rectifier and the active components of the boost converter are reported on the top layer of the PCB. The maximum ripple current following in the inductance is 5 A (using self-oscillating current control) for a power of about 400 W. The paper describes the sizing of the inductance, the choice of the magnetic material and the technological process of integration. The implementation of these components into the PCB and the very high switching frequency allowed by the performances of the magnetic material and the GaN semi-conductor components make possible the design of a very high power density converter (about 10 kW/l).

1.   Description and analysis of the structure

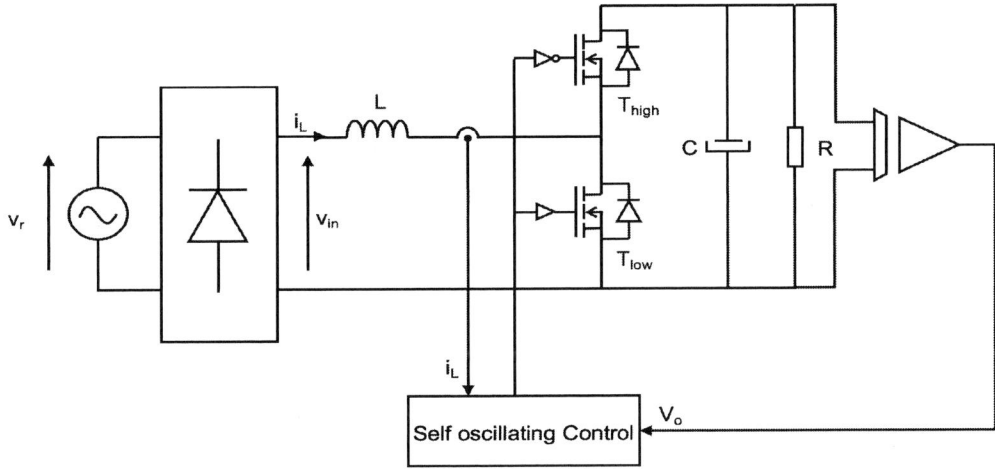

Fig. 1. Converter schematic

The PFC converter is designed for an output power of about 400 W and 500 V output voltage. The RMS input voltage V_{in} is 230V, therefore, the RMS input current I_{eff} is 1.7 A, and the maximum current I_{max} in the inductance is 4.9 A ($I_{max} = I_{eff}.2.\sqrt{2} = 4.9A$). Figure 2a, shows the idealized shape of the current in the inductance (i_L). The maximum value of the current tracks a sinusoidal current reference and the switching frequency depends on the instantaneous value of the reference.

With this control strategy, an instantaneous duty cycle α_i could be defined like this:

$$\alpha_i = 1 - \frac{V_{max}.|\sin(\omega t_i)|}{V_o} \qquad \text{(eq. 1)}$$

© VDE VERLAG GMBH · Berlin · Offenbach

1647

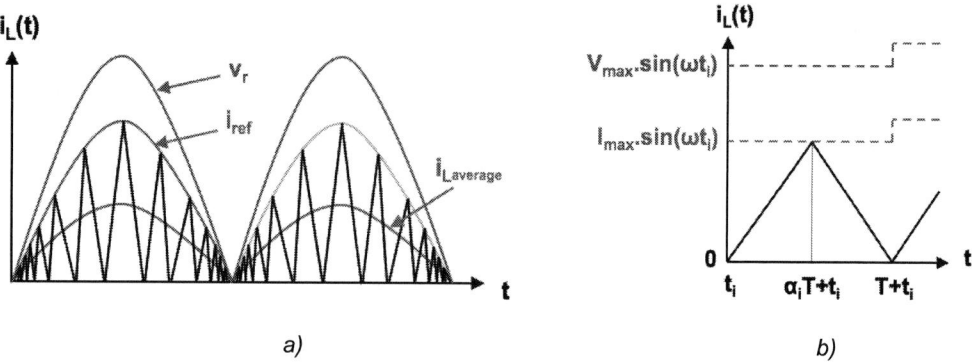

Fig. 2. Current in the inductance with a self-oscillating control a) over the main grid period, b) over a switching period.

The reference current $i_{ref}(t)$ has a sinusoidal rectified shape and is in phase with the input voltage. Therefore, input voltage and reference current are given by the two following relationships:

$$v_{in}(t) = V_{max}.|\sin(\omega t)| \qquad \text{(eq. 2)}$$
$$i_{ref}(t) = I_{max}.|\sin(\omega t)| \qquad \text{(eq. 3)}$$

With a very high switching frequency, the current reference $i_{ref}(t)$ and input voltage $v_{in}(t)$ can be considered as constant over a switching period, (figure 2b). In this conditions, the evolution of the inductance current i_L is given by the following equations:

On $[t_i , t_i + \alpha_i T]$,
$$i_L(t) = \frac{V_{max}|\sin(\omega t_i)|}{L}(t - t_i) \qquad \text{(eq. 4)}$$

On $[t_i + \alpha_i T, t_i + T]$,
$$i_L(t) = \frac{V_{max}|\sin(\omega t_i)| - V_o}{L}\left(t - (t_i + \alpha_i T)\right) + I_{max}|\sin(\omega t_i)| \qquad \text{(eq. 5)}$$

Considering equation 4, with t= $t_i + \alpha_i T$, $\alpha_i T$ is defined by equation 6. It shows that the conduction time $\alpha_i T$ is constant over the time and depends on the maximum reference current, the maximum input voltage and the inductance value.

$$\alpha_i T = \frac{I_{max}.L}{V_{max}} \qquad \text{(eq. 6)}$$

On time t = $t_i + T$ and using equation 5, the off-state time ($T-\alpha_i T$) is expressed in equation 7. It shows that the time ($T-\alpha_i T$) is time dependent. In these conditions, using equations 6 and 7, equation 8 gives the evolution of the switching period versus time.

$$T - \alpha_i T = \frac{L.I_{max}.|\sin(\omega t_i)|}{V_o - V_{max}|\sin(\omega t_i)|} \qquad \text{(eq. 7)}$$

$$T = \frac{I_{max}.L}{V_{max}} + \frac{L.I_{max}.|\sin(\omega t_i)|}{V_o - V_{max}|\sin(\omega t_i)|} \qquad \text{(eq. 8)}$$

With α_i the time dependent duty cycle on each switching period, the switching period T is given by equation 9. The maximum switching frequency is obtained for a theoretical duty cycle α_i of 100%.

$$T = \frac{I_{max}.L}{\alpha_i.V_{max}} \qquad \text{(eq. 9)}$$

© VDE VERLAG GMBH · Berlin · Offenbach

The maximum switching frequency will not exceed 8 MHz in order to limit the switching losses in the power devices. This frequency limitation allows to calculate the inductance equal to about 8 µH.

The minimum duty cycle α_i is determine by the following relationship $\alpha_{i_{min}} = 1 - \frac{V_{max}}{V_o}$ with $V_{max} = 325$ V and $V_o = 500$ V, which give a minimum duty cycle α_i of 35%. In this case the minimum switching frequency is 2.8 MHz.

2.   Choice of the magnetic material

The choice of the magnetic material depends on the switching frequency and the maximum magnetic induction. According to previous studies [1, 2], when the switching frequency is upper than 50 MHz, the magnetic core is ineffective. Indeed, the volume of the magnetic circuit, the manufacturing difficulty and core losses are not justified compared to performances of a coreless inductance.

Due to the switching frequency limitation of the power semiconductor devices, a maximum switching frequency of about 10 MHz may be considered for several hundred Watts. Therefore, a magnetic circuit can be justified provided to find a magnetic core with low losses and an easy inductance manufacturing [4]. The magnetic core must be not thicker than 1 mm (1.5 mm for entire PCB) in order to be integrated in the PCB (Printed Circuit Board). It must also withstand lamination pressure of the PCB and must have a sufficient magnetic permeability.

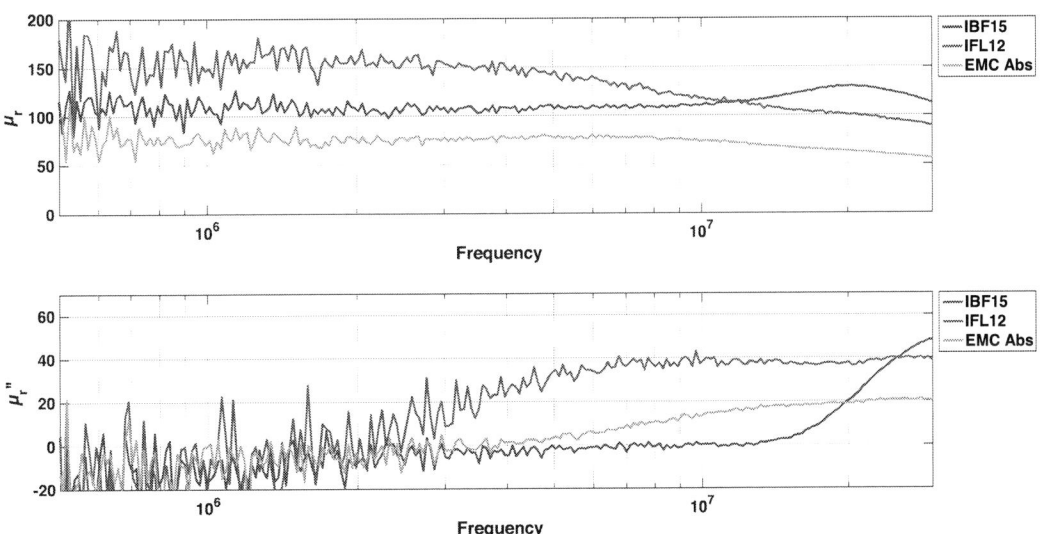

Fig. 3. Measured relative permeability μ_r' and μ_r'' vs frequency for different magnetic materials

Several magnetic materials that meet these criteria are available. We compared the behavior of three thin materials from two different manufacturers (EMC Absorber from 3M and IFL12 and IBF15 from TDK). The EMC Absorber material is a polymer resin with iron flake filler, the IFL12 material is a magnetic powder with polymeric resin and the IBF15 material is a super-thin sintered ferrite platen.

Figure 3 compares the measured relative permeability of these three materials. The imaginary part μ_r'' is representative of the losses in the material. Figure 3 shows that IBF15 material does not have significant losses up to 13 MHz, while the two others present a significant increase of μ_r'' value respectively from 2 MHz or 3 MHz for IFL12 Material and

EMC Absorber. As the switching frequency of the converter is varying between 2 MHz to 8 MHz, IBF15 material is most suitable for losses considerations. Moreover, the permeability of the IBF15 material ($\mu_r' = 110$) is relatively high for a thin material allowing to minimize the core size.

The saturation induction has been measured using samples of the different materials in a Vibrating Sample Magnetometer (VSM). In Table 2 we can see that the EMC Absorber material has a relatively high saturation induction compared to the others. However, the magnetic losses begin to be significant from 2 MHz as shown before. This material would have been interesting for switching frequencies below 1 MHz. Note that the selected IBF15 material has a saturation induction similar to the various ferrite cores (about 300mT).

Tab. 2. Saturation induction of the different magnetic materials

Magnetic material	Saturation induction	μ_r	F_{max} (before μ'' increase)
EMC Absorber material (polymer resin with iron flake filler)	Bsat = 561mT	70	2 MHz
IFL12 material (Magnetic powder + polymeric resin)	Bsat = 261mT	150	4 MHz
IBF15 material (thin sintered ferrite platen)	Bsat = 349mT	110	13 MHz

3. Sizing of the magnetic core

After choosing the magnetic material, the magnetic circuit dimensions, the number of turns and the winding section were calculated to meet specification of the structure and to minimize the volume of the component. The theoretical value of the inductance is about 8 µH for a maximum input current ripple of about 5 A. In order to minimize the leakage magnetic field, a toroid magnetic core shape with minimum space between wires (Figure 4a) was chosen [3]. Considering the inductance value, a full thickness of less than 1 mm, and minimizing the volume of the inductance, the dimensions of the core were calculated to obtain a theoretical value of 8 µH. The geometrical characteristics of the inductance are: 42 mm for outer diameter, 18 mm for inner diameter, a thickness of 1080 µm, and a number of turns of 21.

The design of the core has been implemented on a PCB design tool and simulated using Maxwell software. The computed value of the inductance is 6.8 µH with a relative material permeability of 110. Figure 4a shows the design of the inductance and Figure 4b the simulation of the magnetic field in the core. Figure 4b shows that the induction into the magnetic material is less than 300mT for excitation current of 5 A, avoiding the saturation of the material.

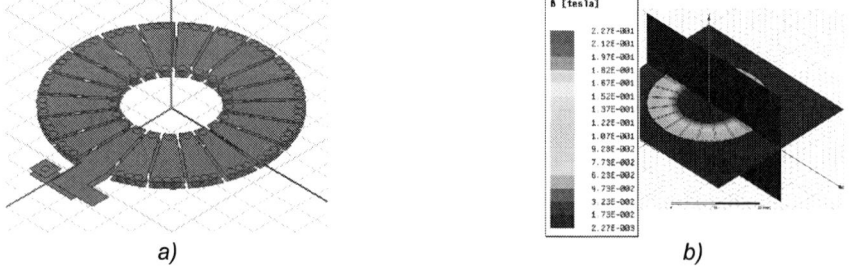

a)   b)

Fig. 4. a) Inductance design on Maxwell software, b) simulation of the magnetic induction in the magnetic material.

4. Implementation of the inductance prototype

In order to integrate the magnetic component into the PCB, we have previously etched copper tracks (top and bottom layers) on two different thin PCB (thickness = 0.2mm). After lamination of the magnetic core under layers of prepreg, electrical connections between top and bottom layers were realized in a first step using LPKF ProConduct® paste to validate the design. Figure 6 presents a schematic view of the process.

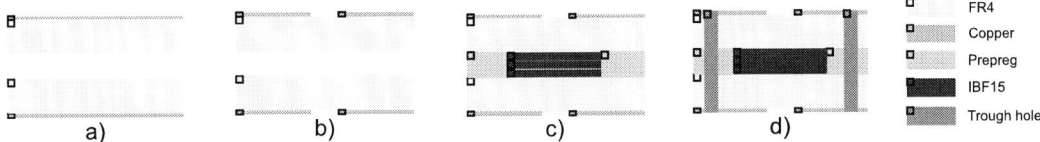

a)   b)   c)   d)

FR4
Copper
Prepreg
IBF15
Trough hole

Fig. 6. Inductance realization process, a) thin PCB, b) etched copper tracks, c) lamination of the PCB, d) connections through metallized holes

The inductance prototype is shown on Fig 7. Measurements of self inductance and winding resistance have been realized using an impedance bridge (HP4194A). These measurements were realized before and after the laminated process. Figure 8 shows that the inductance is roughly constant until 10 MHz.

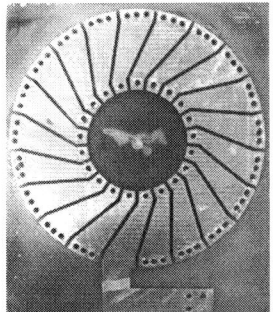

Fig. 7. Photography of the inductance prototype

However, a significant difference exists between pre-laminating and post-laminating inductance values. Indeed, before laminating the inductance, a value of 7,6 µH was measured for the self inductance. This value is only 4.2 µH after the lamination process. This difference could be explaining by the modification of the relative permeability of the material during lamination process of the PCB. Moreover, the lamination could modify the thickness of the magnetic material and thus the reluctance.

In order to confirm these assumptions, the characteristics of the IBF15 material was measured after lamination. Indeed, the PCB lamination process requires a pressure of 200 N.cm^{-2} and a temperature of 180°C during 60 minutes. Figure 9 compares IBF15's relative permeability before and after lamination. It shows that the lamination seems to affect the properties of the material.

Indeed, the relative permeability modified from 110 to 95. A microscope view shows that IBF15 material consists of fragments of ferrite in a polymer (Fig. 10a). After lamination (Fig. 10b) the material is degraded, fragments are separated creating an additional distributed air gap.

© VDE VERLAG GMBH · Berlin · Offenbach

PCIM Europe 2016, 10 – 12 May 2016, Nuremberg, Germany

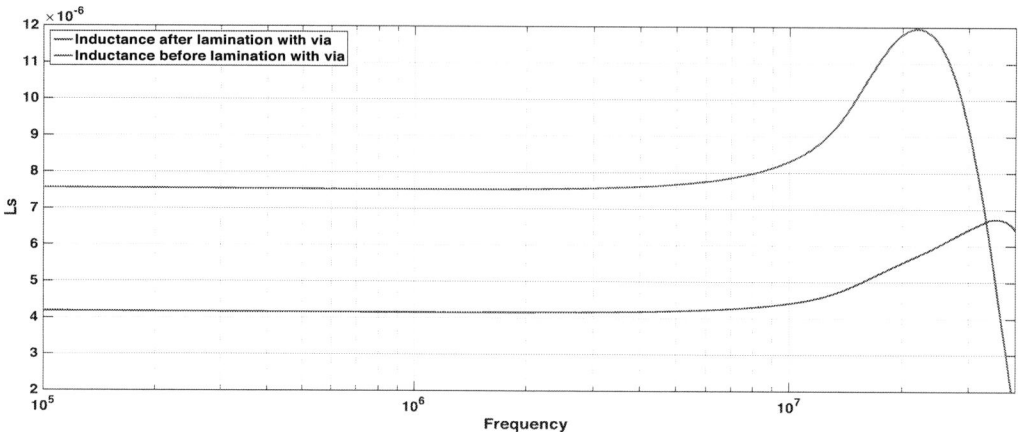

Fig. 8. Comparison between the value of the inductance before and after lamination

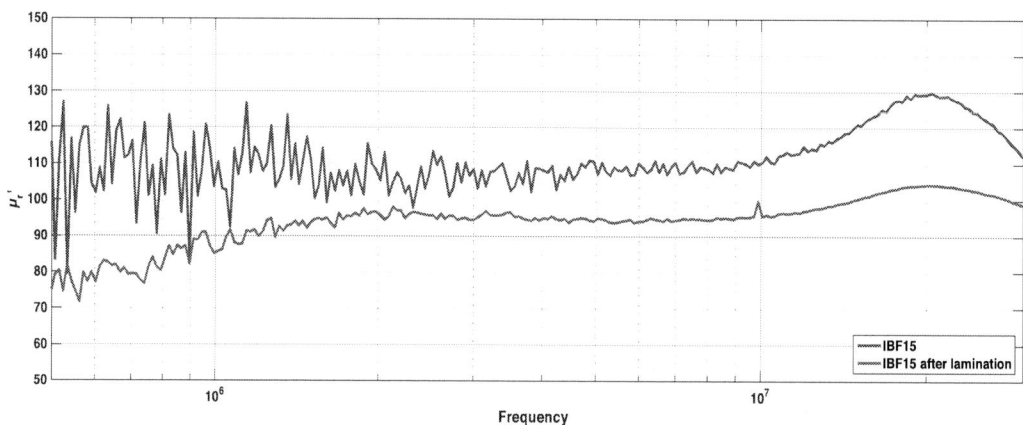

Fig. 9. Comparison of relative permeability of IBF15 material before and after lamination process

Moreover, additional studies are needed to verify the stability in time of the pressed material and to precisely analyze the effects of pressure and temperature during the lamination process on the material performances.

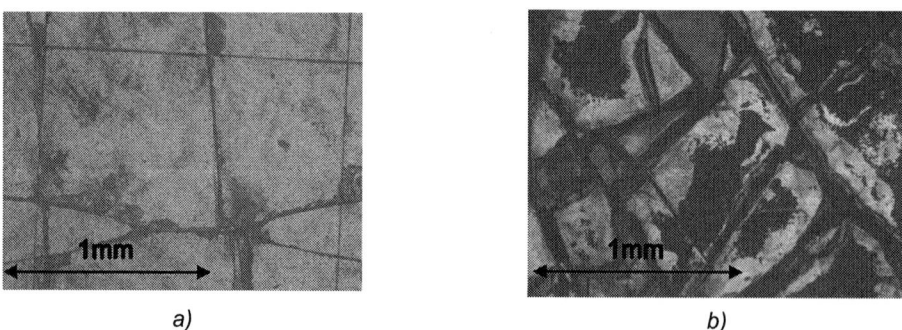

Fig. 10. Structure of the IBF15 material a) before lamination, b) after lamination

© VDE VERLAG GMBH · Berlin · Offenbach

5. Test of the boost converter

For this first implementation and for electrical characterizations simplicity raisons, three different PCB have been developed in order to realize the full converter. The first PCB integrates the diode bridge rectifier. The inductance is realized in a second PCB and finally a third PCB integrates the power devices (GaN transistors) and their drivers. Photography of the test bench is presented on figure 11a.

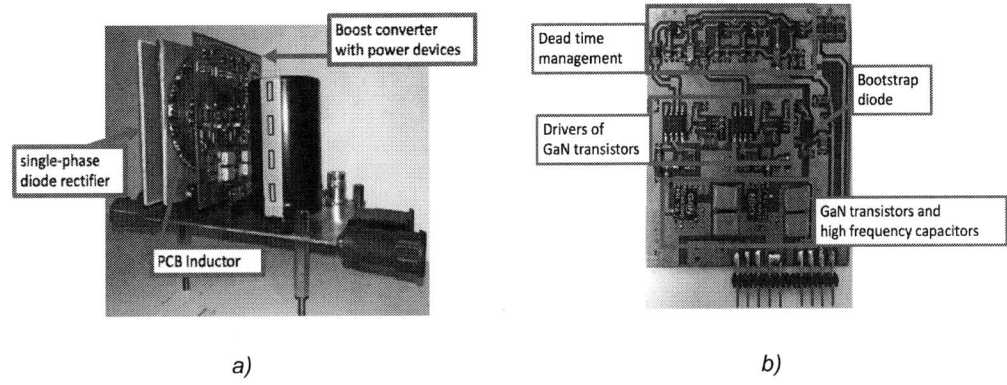

a)   b)

Fig. 11. a) Photography of the test bench of the full PFC prototype, b) photography of the boost converter with power devices and their drivers

In this first realization, GaN power dies have been soldered on a PCB with copper tracks of 70 microns thick without additional cooling system (Figure 11b). The poor thermal performances of this first realization limits the power of the converter to about 200 W.

The converter is controlled in open loop. In order to check the operation of the converter, the boost converter was supplied by a constant voltage source V_{IN} and the switching frequency was fixed in a first step to 1 MHz with a duty cycle of 50%.

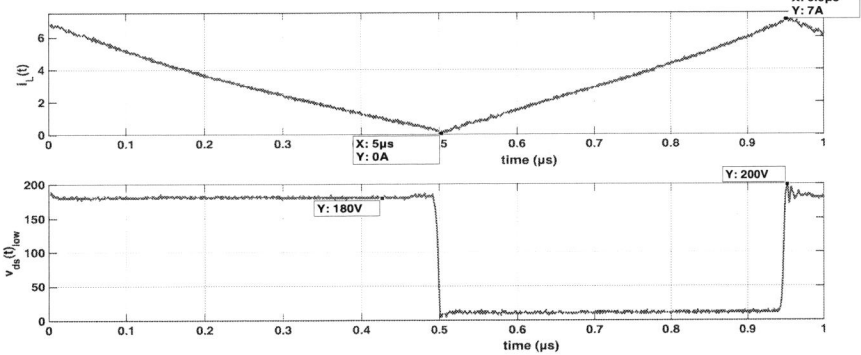

Fig. 12. Experimental waveforms of the current in the inductance and voltage V_{ds} of the low side transistor for the boost converter in a self-oscillating control mode at a switching frequency of 1MHz and a duty cycle of 50% (V_{IN} = 100V, V_O = 180V)

Figure 12 shows the current in the inductance for an input voltage of 100 V, the maximum ripple current value is fixed to 7 A. These measurements allow determining the value of the inductance : L=6.4 µH. Figure 12 shows also Drain to Source voltage V_{ds} on the low side GaN transistor, with an overvoltage at turn off limited to only about 20 V.

© VDE VERLAG GMBH · Berlin · Offenbach

Figure 13 shows turn-on and turn-off times of the low side transistor, allowing the estimation of the dV/dt generated by the transistor switching at turn-off and turn-on.

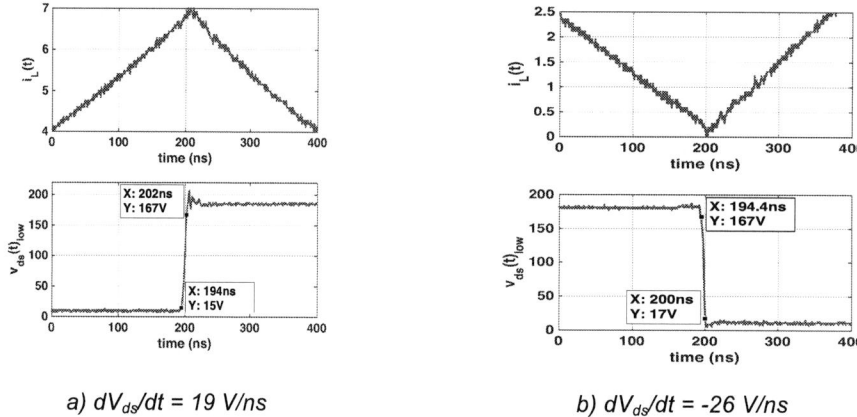

a) $dV_{ds}/dt = 19$ V/ns

b) $dV_{ds}/dt = -26$ V/ns

Fig. 13. Turn on and turn off time of the low side transistor

The poor performances of the die thermal management do not allow to increase the power over 200 W. Nevertheless, the operating principle of the structure was validated with reduced power and showing high speeds of switching and low over-voltage at turn-off.

6.   Conclusion

This paper deals with the development of an AC/DC converter working at very high switching frequency (8 MHz). It will be supplied by a main voltage of 230V RMS and will be able to deliver an output power of about 400W for a DC output voltage of 500V. The paper focuses on the inductance's ferrite core integrated into a PCB. IBF15 material has been chosen due to its low losses and its high relative permeability until 10 MHz. The sizing of the inductance was chosen in order to minimize the volume of the component. Results show that the laminating process of the PCB is responsible for structural change in the material that affects its performances decreasing the permeability. Several additional characterizations have to be made in order to precisely evaluate the effects of the process temperature and pressure on the material degradation. A prototype has been nevertheless realized, but insufficient cooling of the power dies in this first realization did not allow to achieve the desired output power

References

[1] R. Meere, N. Wang, T. O'Donnell, S. Kulkarni, S. Roy, and S. C. O'Mathuna, "Magnetic-core and air-core inductors on silicon: A performance comparison up to 100 MHz," *IEEE Trans. Magn.*, vol. 47, no. 10, pp. 4429–4432, Oct. 2011.

[2] S. Kelly, C. Collins, M. Duffy, F. M. F. Rhen, S. Roy, "Core Materials for High Frequency VRM Inductors," *IEEE*, 2007.

[3] M. Nigam and C. R. Sullivan, "Multi-layer folded high-frequency toroidal inductor windings," in *Proc. IEEE Appl. Power Electron. Conf.*, Feb. 2008, pp. 682–688.

[4] Mingkai Mu, "High Frequency Magnetic Core Loss Study," *Dissertation submitted to the faculty of the Virginia Polytechnic Institute and State University in partial fulfillment of the requirements for the degree of Doctor of Philosophy in Electrical Engineering,* Feb. 2013.

[5] Yang Zhou, Xiaoming Kou, Paul Parsons, Asif Warsi Muhammad, Hao Zhu1, Stoyan Stoyanov1, and John Q. Xiao, "Magnetic Substrates for Power Supply on a Chip", *Department of Physics & Astronomy University of Delaware Newark, DE 19716.*

© VDE VERLAG GMBH · Berlin · Offenbach

Evaluation of TCM and CrCM modulation for Totem Pole PFC

Haihua Zhou, Wenduo Liu, and Eric Persson

Infineon Technologies Americas Corp, 101 N Sepulveda Blvd, El Segundo, CA, 90245, USA, haihua.zhou@infineon.com, wenduo.liu@infineon.com and eric.persson@infineon.com

Abstract

Recently totem pole FPC becomes popular due to the minimum component number and less common mode EMI noise. With the availability of GaN switches, high frequency switching is applied in totem pole PFC to achieve high power density. Triangular Current Mode (TCM) and Critical Current Mode (CrCM) modulation minimize switching loss and are good for high frequency. In this paper, two modulations are evaluated. Simulation results show CrCM is simple for implementation but does not achieve Zero Voltage Switching (ZVS) in full range with wide frequency variation at different load conditions; while TCM requires complex calculation for on/off timings to ensure ZVS at full range with narrow frequency variation. CrCM has higher THD than TCM due to discontinuous input current. Power switches loss is measured in simulation by Simetrix. Device with lower Q_{oss} is preferred.

1. Background

To increase power conversion efficiency at low line input for universal line applications, totem pole PFC as Fig. 1 are proposed [1]. Here Q_1/Q_2 are fast switches and Q_3/Q_4 are line frequency switches. Compared with conventional PFC, conducted devices reduce from three to two at any time. Unlike bridgeless converter, its Common Mode (CM) noise is minimized because the output is clamped to the input by line frequency switches during each half line cycle [1]. At any half cycle, totem pole PFC behaves the same as boost PFC circuit. Modulations for conventional boost PFC design can be used. Continuous Current Mode (CCM) is good for higher power application but high frequency operation is limited by switch's body diode performance. To operate higher frequency, Discontinuous Current Mode (DCM) and Critical Current Mode (CrCM) are two approaches. At both DCM and CrCM, synchronous switch turns off at zero current which minimizes the reverse recovery effect from body diode. DCM does not fully utilize the full switching period and only good in power level less than 100W. Turn on of control FET in CrCM is ZVS or valley switching. Triangular Current Modulation (TCM) in [4] introduces additional negative current and ensures ZVS turn on for all switches at full range. Extra peak current is required to compensate the negative current and additional conduction loss occurs. This paper is to understand pros and cons for CrCM and TCM modulation.

PCIM Europe 2016, 10 – 12 May 2016, Nuremberg, Germany

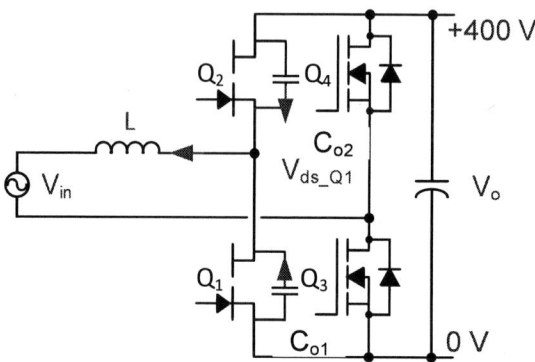

Fig. 1 Topology of totem pole PFC

2. Basics of CrCM and TCM

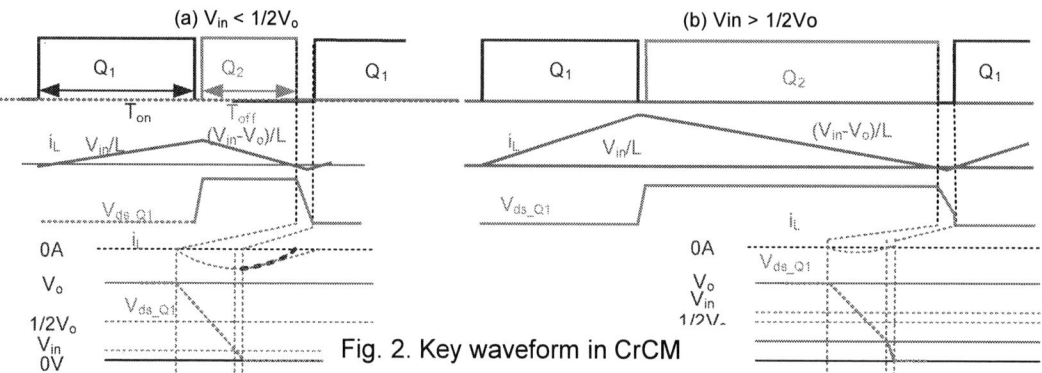

Fig. 2. Key waveform in CrCM

2.1 Basics for CrCM modulation

For totem pole PFC, during positive half cycle, Q_1 is the control FET and Q_2 is the synchronous FET. Q_1 and Q_2 switch their roles during negative half cycle.

In CrCM, as Fig. 2, when Q_1 is on, inductor current increases linearly with the slope (V_{in}/L). Here V_{in} is instantaneous input voltage and L is inductor value. On time T_{on} controls PFC and is fixed for each line cycle. During on time of Q_2, inductor current decreases linearly with the slope ($V_o - V_{in})/L$. This sequence terminates when inductor current reaches zero. Off time T_{off} is designed to achieve voltage regulation [3]. T_{on} and T_{off} are described as below

$$T_{on} = 2L \frac{P_o}{V_{ac}^2} \quad [1] \quad and \quad T_{off}(t) = \frac{V_{in}(t)}{(V_o - V_{in}(t))} \cdot T_{on} \quad [2]$$

where $V_{in}(t) = \sqrt{2}V_{ac} \sin(2\pi f_{line} \cdot t)$; V_{ac} is the RMS value of input voltage and f_{line} is line frequency ; V_o is output voltage. Inductor value can be calculated as

© VDE VERLAG GMBH · Berlin · Offenbach

$$L = \frac{(V_o - \sqrt{2}V_{ac}) \cdot V_{ac}^2 \cdot \eta}{2 \cdot f_s \cdot V_o \cdot P_o} \qquad [3]$$

P_o is output power and η is power converter efficiency and f_s is desired minimum switching frequency at specified input voltage.

The frequency for CrCM is then can be calculated as

$$f_s = \frac{V_{ac}^2 \cdot \eta}{2 \cdot L \cdot P_o} \cdot (1 - \frac{\sqrt{2}V_{ac}\sin(2\pi f_{line} \cdot t)}{V_{out}}) \qquad [4]$$

f_s consist of two parts: (1) $\frac{V_{ac}^2 \cdot \eta}{2 \cdot L \cdot P_o}$ shows frequency varies with test conditions (V_{ac}, P_o) and

(2) $(1 - \frac{\sqrt{2}V_{ac}\sin(2\pi f_{line} \cdot t)}{V_{out}})$ varies with sinusoidal input voltage.

2.2 Zero Voltage Switching conditions in CrCM

After inductor current reaches zero, Q_2 turns off. Before Q_1 turns on, the output capacitors of Q_1 and Q_2, C_{o1} & C_{o2} are resonant with inductor L as shown in Fig. 1. When $V_{in} < \frac{1}{2} V_o$ (as Fig. 2.(a)), V_{co1} drops from V_o to zero and Q1 achieves ZVS within dead time. The minimum dead time to achieve ZVS varies with input voltage. Real time dead time updating requires additional calculation. T_{on} and T_{off} are required to be adjusted for transferring same amount of the power. The other way is to select a fixed maximum dead time as $\frac{1}{2 \cdot 2\pi\sqrt{L \cdot 2C_{oss}}}$. Here C_{oss} is switch's body capacitor value.

It would result in complex calculation if dead time needs to be updated for each switching cycle and thus T_{on} and T_{off} needs to be adjusted in time as well. Practically, a fixed dead time $\frac{1}{2 \cdot 2\pi\sqrt{L \cdot 2C_{oss}}}$ is applied for each line cycle. Here C_{oss} is output capacitor of the switch.

3. TCM modulation

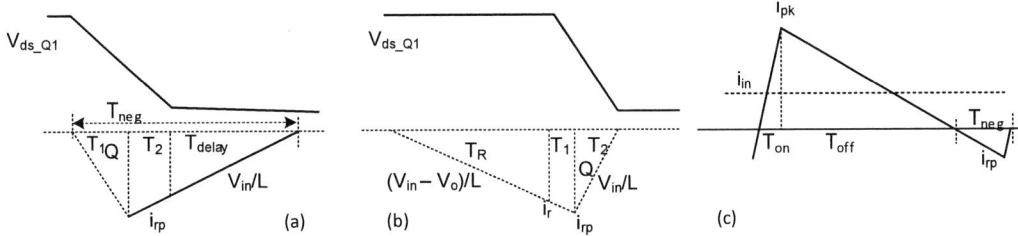

Fig. 3. Key waveform in TCM

An issue of CrCM operation is that when input voltage is larger than half of output voltage, the switch can't achieve ZVS, instead it turns on with valley voltage, where the resonant current reaches zero and V_{ds} reaches minimum value. To overcome the limitation, TCM [4] is introduced to ensure ZVS at all the switches in the full range. T_{on}, T_{off} and negative current time T_{neg} are all programmed real time. When ($V_{in} < \frac{1}{2} V_o$) as Fig. 3(a), dead time is set to equal to negative current time T_{neg}. It is mainly determined by the device output charge Q_{oss} and input voltage. To achieve ZVS when ($V_{in} > \frac{1}{2} V_o$) as Fig. 3(b), after i_L reaches zero, Q_2 does not turn off until reaching desired reverse current i_r as described below,

$$I_r = \sqrt{2 \cdot Q \frac{(2\sqrt{2} V_{ac} \sin(2\pi f_{line} \cdot t) - V_o)}{L}} \qquad V_{in} > \frac{1}{2} V_o \quad [5]$$

T_{on} and T_{off} then are calculated to provide desired average output current.

4. Performance evaluation

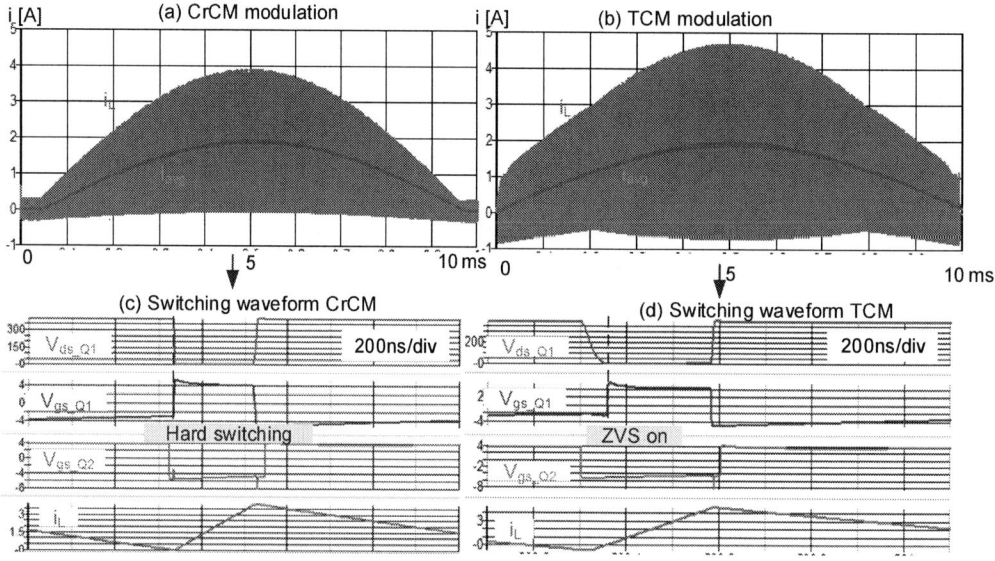

Fig. 4. Typical waveform for CrCM and TCM

Table 1 Specifications

L [uH]	V_o [V]	f_{line} [Hz]	V_{ac} [V]	Q_{oss} [nC]	P_o [W]	Dead time at CrCM[ns]
15	385	50	85 to 265	15	300	DT1 = 20ns DT2 = 100ns

In comparison of these two modulations, the same inductor is used for both CrCM and TCM modulations. Simulation circuits are implemented in Simetrix. Table. 1 lists test specifications. Fig. 4 (a)&(b) show inductor current and its average waveform in half line cycle with CrCM and TCM respectively. For CrCM modulation as Fig. 4(a), average current i_{avg} is zero when input voltage around zero area while i_{avg} is continuous in TCM as Fig. 4(b). Discontinuous current causes higher THD in CrCM than TCM. In [5][6], extend on time is used in CrCM at voltage cross zero range to improve THD. TCM utilize resonant to achieve ZVS. Higher negative current and peak current are observed in TCM (Fig. 4(b)) than CrCM (Fig. (a)).

© VDE VERLAG GMBH · Berlin · Offenbach

Fig. 4(c)&(d) are key switching waveforms at peak input voltage for CrCM and TCM. As Fig. 4(c), Q_2 turns off when input current equals zero. With fixed dead time, Q_1 turns on at hard switching. With Fig. 4(d), Q_2 does not turn off immediately when input current equals zero. Instead it turns off at minimum reverse current point i_r and V_{ds} drops to zero right before Q_1 gate is enabled. ZVS is fully achieved at Fig. 4(d).

Design parameters are generated using Mathcad for CrCM from [3] and TCM from [4] respectively. For CrCM, as described in Equation. (4), its frequency varies with test condition (V_{ac}, P_o) and with sinusoidal input voltage. Fig. 5(a) is the frequency vs V_{ac}. At t=0s, frequency at V_{ac} =265V is ~8 times higher than the case V_{ac} = 85V. Fig. 5(d) shows the frequency vs the output power. Same at t = 0s, frequency at 20% of full load is ~8 times higher the case at full load P_{load}.

Fig.5 (b) shows T_{on} and T_{off} for two modulations vs time. At CrCM, T_{on} is constant and T_{off} increases with input voltage. For TCM, T_{on} initially is high when input voltage equals zero and reduces to relative constant value when V_{in} approaches its peak. T_{off} has the same trend as that in CrCM but slightly larger to compensate the negative current. T_{on} and T_{off} variation verify frequency variation at Fig. 5(a). Fig. 5(c) shows the peak current when different Q_{oss} device is applied. Small Q_{oss} results in smaller negative current and is critical to power stage performance.

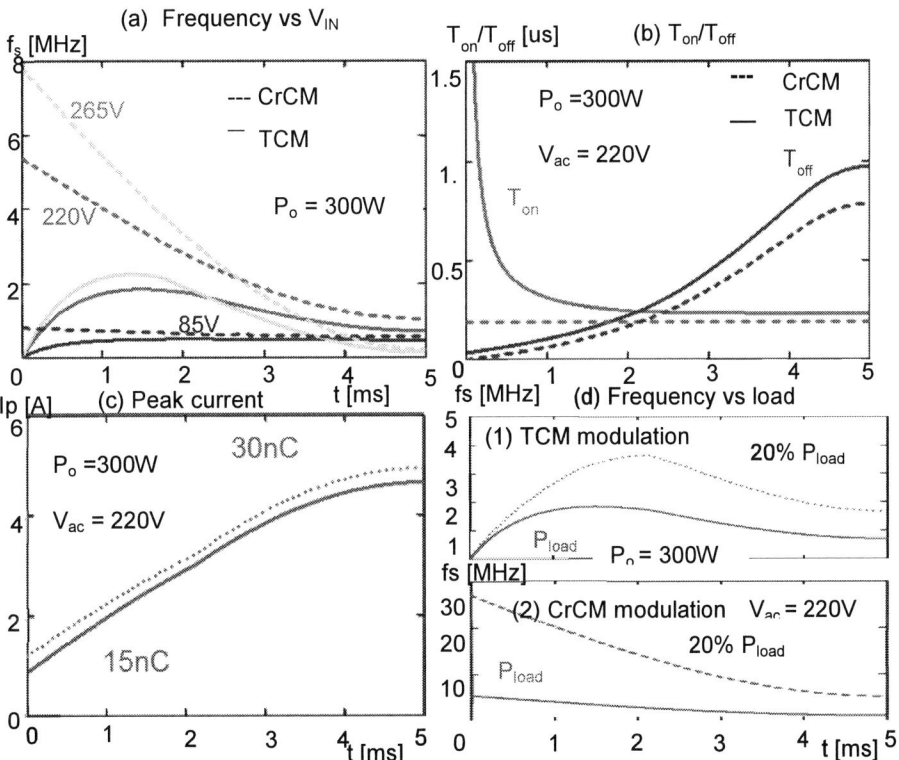

Fig. 5. Design parameter comparison at TCM and CrCM modulation

PCIM Europe 2016, 10 – 12 May 2016, Nuremberg, Germany

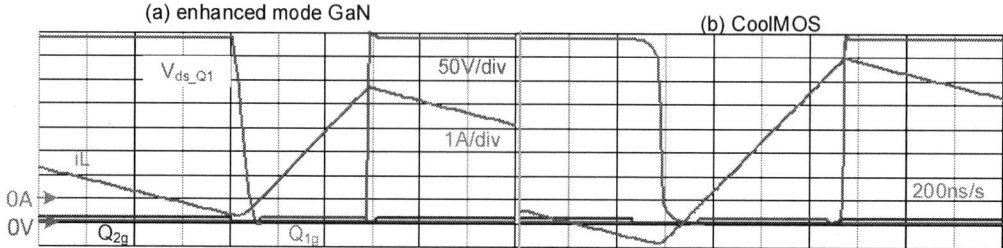

Fig. 6. Switching comparison for enhancement mode GaN and CoolMOS

Fig. 6 shows the switching waveform with enhancement mode GaN and CoolMOS when input voltage is at its peak in TCM modulation. Their Q_{oss} are 15nC and 50nC respectively. With higher Q_{oss}, higher negative current and peak current is observed. For enhancement mode GaN, V_{ds} drops immediately after the gating signal Q_{2g} is off. This Q_{2g} is signal from controller directly. For CoolMOS case, V_{ds} does not drop immediately due to its strong nonlinear capacitor. This slightly changes the inductor current shape at resonant period to triangular than sinusoidal.

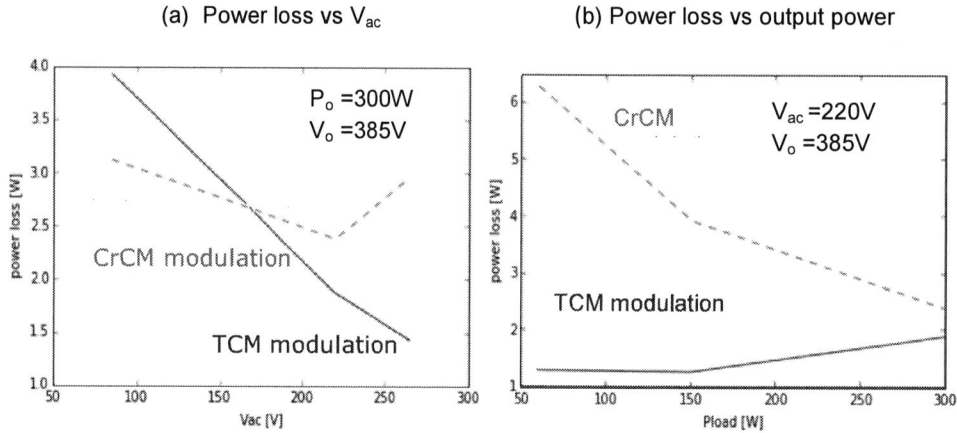

Fig. 7. Switching loss CrCM vs TCM (exclude inductor)

Fig. 7 documents the switching loss from totem pole PFC using CrCM and TCM modulation at different input voltage and power loading condition.

Fig. 7 (a) shows CrCM has smaller power loss than TCM at 85Vac. For 300W output, with low line voltage, conduction loss dominates. Conduction loss in CrCM is smaller than TCM modulation. When input voltage increases to 260Vac, switching loss contributes higher in total loss. CrCM has higher total loss than TCM. Because TCM achieve ZVS turn-on for all conditions, switching loss is negligible. Its total loss equals to conduction

© VDE VERLAG GMBH · Berlin · Offenbach

loss. When V_{ac} increases, current become smaller results total power loss decreases. This can be observed by blue TCM modulation curve in Fig. 7(a).

Fig. 7 (b) shows power loss vs output power. As shown in Fig. 5(d), CrCM has extremely high switching frequency when load decreases. Significantly switching loss contributes the total loss at light loading condition. CrCM frequency reduces when load increases thus loss decreases when load power increases. At the other hand, TCM achieves ZVS turn-on for the full power range. Its total loss is proportional conduction loss. Total losses increase when output power increase as blue curve.

Summary

Totem Pole PFC presents good efficiency with fewer components than competing topologies. TCM and CrCM are two modulation strategies good for high frequency operation. CrCM is simple in implementation. Its design parameter has large variations at different test condition. Power loss increases significantly at light loading. TCM has less device loss than CrCM when output power is 300W at MHz application. It has narrower frequency variation range than CrCM modulation. Q_{oss} is critical in TCM. Lower Q_{oss} enables less device loss and less reverse conducting time.

References

[1] Huber, L.; Yungtaek Jang; Jovanovic, M.M., "Performance Evaluation of Bridgeless PFC Boost Rectifiers," in Power Electronics, IEEE Transactions on , vol.23, no.3, pp.1381-1390, May 2008

[2] Firmansyah, Eka, Satoshi Tomioka, Seiya Abe, Masahito Shoyama, and Tamotsu Ninomiya. "An interleaved totem-pole power factor correction converter." Research reports on information science and electrical engineering of Kyushu University 15, no. 1 (2010).

[3] Onsemi, "Power Factor Correction Stages Operating in Critical Conduction Mode", in Nov. 2014, application note AND8123/D

[4] Marxgut, Christoph Berndt, "Ultra-flat isolated single-phase AC-DC converter systems" , Diss. ETH Zürich, Nr. 20946, 2013.

[5]J. W. Kim, S. M. Choi and K. T. Kim, "Variable On-time Control of the Critical Conduction Mode Boost Power Factor Correction Converter to Improve Zero-crossing Distortion," *Power Electronics and Drives Systems, 2005. PEDS 2005. International Conference on*, 2005, pp. 1542-1546.

[6] Z. Liu, X. Huang, M. Mu, Y. Yang, F. C. Lee and Q. Li, "Design and evaluation of GaN-based dual-phase interleaved MHz critical mode PFC converter," *Energy Conversion Congress and Exposition (ECCE), 2014 IEEE*, Pittsburgh, PA, 2014, pp. 611-616.

PCIM Europe 2016, 10 – 12 May 2016, Nuremberg, Germany

System Concept and Model-Based Optimization of High-Current Variable-Voltage Chopper-Rectifiers

Zhiyu Cao[1], Holger Fahnert[1], Jürgen Schiele[1], Jitendra Solanki[2], Norbert Fröhleke[2], Joachim Böcker[2]

1 AEG Power Solutions GmbH, Warstein-Belecke, Germany,
  zhiyu.cao@aegps.com, holger.fahnert@aegps.com, juergen.schiele@aegps.com
2 University of Paderborn, Paderborn, Germany
  ejitendra@gmail.com, froehleke@lea.upb.de, boecker@lea.upb.de

Abstract

Recent studies show that in the power range from 100 kW to 10 MW the multi-phase chopper technology is the best solution for high-current variable-voltage rectifiers. A modular system concept for this kind of rectifier systems is introduced in this paper to enable engineering and manufacturing cost minimization, while reserving the design freedom for efficiency optimization. For the task of efficiency optimization (for various customized specification) a computer-aided optimization tool is developed. This tool is verified by measurements and applied in customized rectifier design and optimization.

1. Introduction

High-current variable voltage (HCVV) rectifiers are required in many industrial processes, especially in metal and chemical industries. Typical voltage requirements for these processes are of the order of few hundred volts at several kilo amperes current. A recent state of the art review summarized the most applied technologies of HCVV rectifiers versus power (Figure 1) [1]. Thanks to the high efficiency, good power factor, low input current harmonics and high reliability the multi-pulse diode or thyristor rectifier with mechanical on-load tap-changing (OLTC) transformer is the dominating technology for power ratings above 10 MW (up to few GW) till now. For applications with low power rating up to few hundreds kW, the switched-mode power supply (SMPS) operating at high switching frequency is the most favorite technology (Figure 1).

For the power rating from 100 kW to tens of MW the multi-pulse thyristor rectifier is the dominating technology for HCVV applications. The output power is controlled via changing the firing angle. Essential drawbacks of this technology are high reactive power and input current harmonics. In order to meet the technical conditions for connection (TAB) to medium-voltage (MV) [2] or to low-voltage (LV) network [3], a harmonic filter is to be implemented. A review of the state-of-the-art harmonic filters shows that a hybrid filter combining a passive and an active filter can achieve the best trade-off between cost and performance ratio, i.e. high and controllable power compensation, low system loss and minimum installation power of the active part [4].

Thanks to the rapid development of insulated-gate bipolar transistor (IGBT) technology, more and more HCVV rectifiers in this power rating are implemented with multi-phase IGBT chopper technology. A quantitative comparison of thyristor rectifier with hybrid filter and chopper rectifier

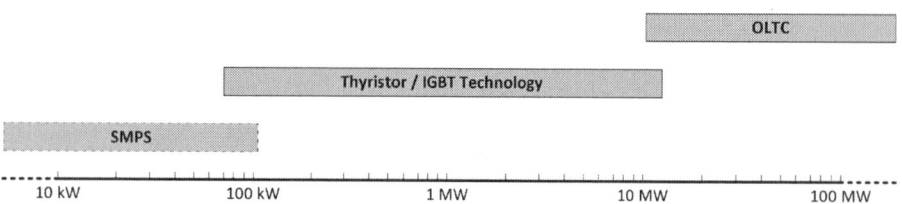

Figure 1: State-of-the-art HCVV technologies vs. power range

© VDE VERLAG GMBH · Berlin · Offenbach

for a 1 MW system is given in [5] and [6]. Comparison results show that the chopper rectifier has higher efficiency, smaller footprint, lower cost and easy connection to MV grid.

Based on the results of the state-of-the-art review discussed above, the chopper-based HCVV rectifier is chosen to be focused at AEG Power Solutions GmbH, addressing a power range of 100 kW to 10 MW and load voltage range up to few hundreds volts, i.e. low-voltage application. In the remaining contribution, the system concept and the model-based design and optimization of customized HCVV rectifier based on chopper solution are discussed in more detail.

2. System Concept and Technical Specification

After fixing the topology, a number of pilot projects were performed. Experiences summarized from these projects are:

1. to meet various specifications, design customizations are always required.
2. customized design means high efforts in engineering, construction, production, qualification and documentation that lead to longer delivery time and higher product cost.

In order to deal with this challenge a standardized modular system concept is applied, i.e. the converter is standardized and the customized specification is fulfilled by means of optimized design of transformer and operation scheme. As shown in Figure 2, a compact solution for the power rating of up to 200 kW is provided; while a modular solution can reach a power rating up to 10 MW by means of parallel operation. A brief technical specification is listed in Table 1, and a single-line diagram (SLD) of the modular chopper rectifier is illustrated in Figure 3. Both for the modular system and for the compact system the power circuit excl. power transformer

Table 1: Brief technical specification

	Modular system	Compact system
Grid connection	Low-voltage / medium-voltage	Low-voltage
Transformer integrated	No	Yes
Dimension (W x H x D)	1200 x 2200 x 600 mm	1200 x 2200 x 800 mm
Upper limit of DC current / voltage / power	1800 A / 550 V / 450 kW per module	1000 A / 550 V / 200 kW
Cooling	Water cooling	Forced air (FA)
Parallel operation	Yes, up to 20 units	No

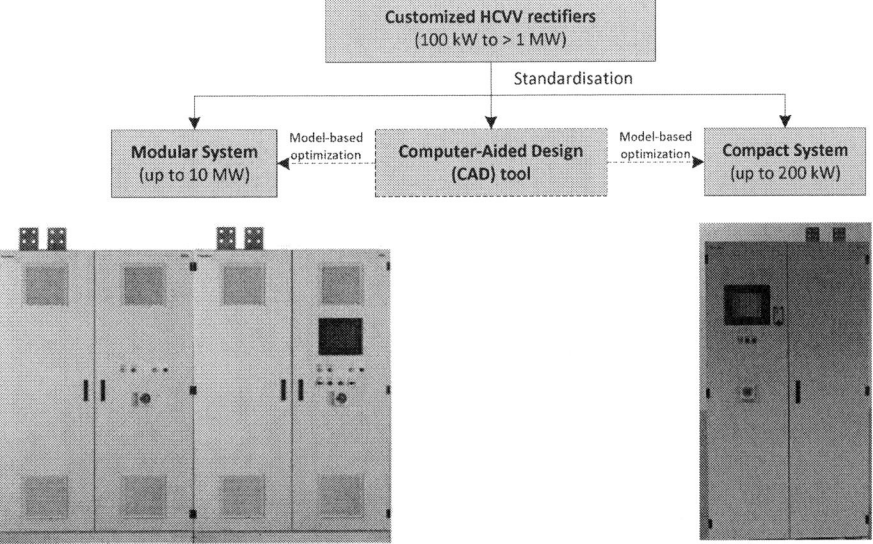

Figure 2: System concept for HCVV chopper rectifiers

© VDE VERLAG GMBH · Berlin · Offenbach

PCIM Europe 2016, 10 – 12 May 2016, Nuremberg, Germany

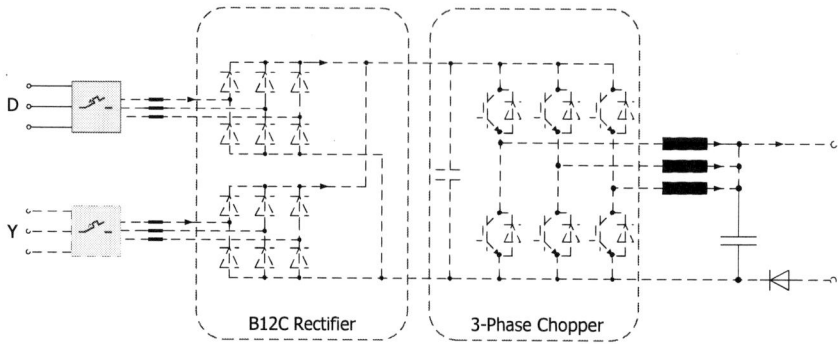

Figure 3: Single line diagram (SIL) of a modular chopper rectifier

is standardized. Customized specifications are fulfilled by means of optimal design of the transformer voltage ratio and optimized operation scheme, which is supported by a computer-aided design (CAD) tool.

3. Modeling of Chopper Rectifier

As shown in Figure 3, a chopper module consists of two 3-phase AC chokes, a 12-pulse rectifier (B12 rectifier), a 3-phase interleaved chopper, three DC chokes, an output capacitor bank and a diode at the output to avoid back flow of current from the load (e.g. electrolyzer stack). Steady-state model and loss calculation are performed for all components.

3-phase IGBT chopper: Each phase consists of four paralleled IGBT modules (1200 V, 300 A). The modeling procedure is as follows [7], [8]:

- Steady-state model for continuous and discontinuous current modes (CCM and DCM)

- Calculation of root mean square (RMS) current I and average current $\bar{\imath}$ of IGBT and freewheeling diode (FWD). Conduction losses of IGBT and FWD are calculated by

$$P = U_0(T_j)\bar{\imath} + R_{on}(T_j)I, \tag{1}$$

  where T_j denotes the junction temperature of the chips, U_0 and R_{on} denote the on-state equivalent forward voltage and resistance, respectively [7].

- Calculation of switching-on and switching-off currents of semiconductors. Switching losses of IGBT and reverse recovery loss of FWD are deduced by means of interpolation approach [7]. Basic data for the interpolation are obtained from semiconductor data sheet or by means of double-pulse measurement [8].

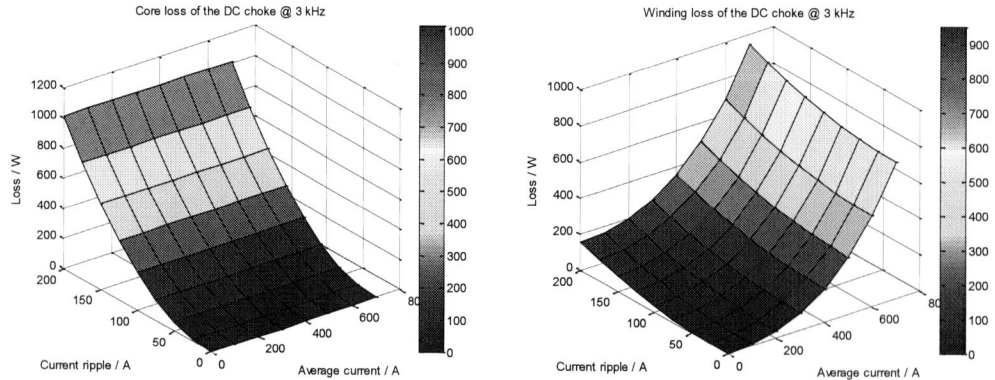

Figure 4: Total DC choke loss versus average current and current ripple

© VDE VERLAG GMBH · Berlin · Offenbach

1664

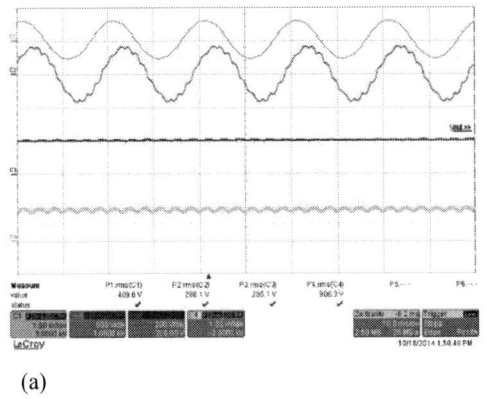 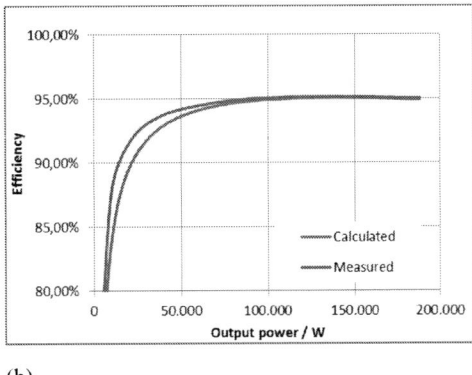

(a)   (b)

Figure 5: (a) waveforms of a 188 kW compact chopper rectifier: Ch.1 (yellow): input line-to-line voltage (1 kV/div.), Ch.2 (red): input phase current (500 A/div.), Ch.3 (blue): output voltage (200 V/div) and Ch.4 (green): output current (1 kA/div), time scale: 10 ms/div., (b) verification of rectifier efficiency

B12 rectifier: The steady-state RMS and mean currents of the diode are calculated as described above [9]. The diode conduction loss is obtained by (1).

DC choke: The DC choke is designed and optimized for the rated current of 600 A and the rated frequency of approx. 3.2 kHz. The core loss is calculated by means of Bertotti formula, which consists of three parts: hysteresis, eddy current and additional losses. Correction factors obtained by measurements and experience are applied for improving the calculation accuracy. The winding loss is obtained by calculation and finite-element-method (FEM) simulation. The proximity and skin effects for medium frequency operation are considered. The additional losses depending on the geometry of the air gap are also considered in said simulation, which contributes a significant part of the total winding loss. The final calculation/simulation results for the operating frequency of 3 kHz are illustrated in Figure 4.

AC choke and transformer: As the components are operated with grid frequency (50/60 Hz), they are modeled by means of standard method, i.e. R_{Fe} for core loss and R_{Cu} for winding loss, which are provided by the manufacturer.

CAD tool and verification: All analytical models deduced above are integrated in a CAD tool and verified by a 188 kW, 900 A rectifier (compact system incl. transformer), which shows a good agreement between calculation and measurement (Figure 5). Measurement results show also very high AC and DC power quality in a wide operation range without any additional filter, which cannot be achieved by Thyristor technique discussed in Chapter 1:

- High power factor (PF): PF > 0.96 from 50% to 100% of full load
- Low current total harmonic distortion (THDi): THDi = 7% at full load and THDi < 15% at 50% of full load
- Very low DC current ripple: < 1%

4. Model-Based Converter Design and Optimization

As discussed in Chapter 2 the chopper rectifier is standardized. The customized specification is to be fulfilled by means of optimized design of transformer and operation scheme. In this chapter the both optimization measurements are discussed based on a case study to optimize a 250 V / 1400 A three-phase chopper rectifier. With help of the CAD tool the design and optimization procedure is performed in three steps.

4.1. Lower and Upper Limit of DC-Link Voltage

The minimum required DC-link voltage is limited by the output voltage and maximum allowed duty ratio D_{max} of the chopper stage. In this example the minimum required DC-link voltage is 265.5 V if D_{max} is selected 0.95, where the voltage drops on the IGBT and on the Oring diode is considered.

In design of chopper-rectifiers many engineers select this minimum DC-link voltage as the optimum one to achieve a maximum duty ratio of IGBT, like in dimensioning of Thyristor-based rectifiers. In this case, as shown in Section 4.2 later, it is the worst design.

The maximum allowed DC-link voltage is majorly limited by the DC choke, because higher DC-link voltage results in increased current ripple and loss on the choke. Figure 6 shows the relationship between DC choke loss and DC-link voltage , where $P_{Choke,rat}$ denotes the choke loss at rated point of operation (250 V / 1400 A) and $P_{Choke,max}$ denotes the maximum choke loss, i.e. the points of operation with maximum current and maximum current ripple on it ($D = 0.5$). As the DC choke is designed for maximum losses of 800 W, in this example the upper limit of the DC-link voltage is 569.6 V.

4.2. Optimization of Full Load Efficiency

As shown in Figure 7 high DC-link voltage leads to lower input current, which results in lower losses in the AC choke and B12 rectifier. However, a higher DC-link voltage results in enhanced switching losses of IGBTs and increased losses on the DC choke. With help of the CAD tool discussed in Chapter 3 the optimum DC-link voltage can be fund. For this case the total loss can be minimized selecting a DC-link voltage of 378.6 V. Compared to a non-optimized design 783 W loss can be reduced in the worst case (9424 W vs. 8641 W total losses, more than 8% loss reduction).

For a ±10% grid voltage variation [10] the highest stress and the thermal requirement of all semiconductors are also computed by the CAD tool:

- Highest stress of B12 rectifier and AC choke at 90% of grid voltage (worst case)

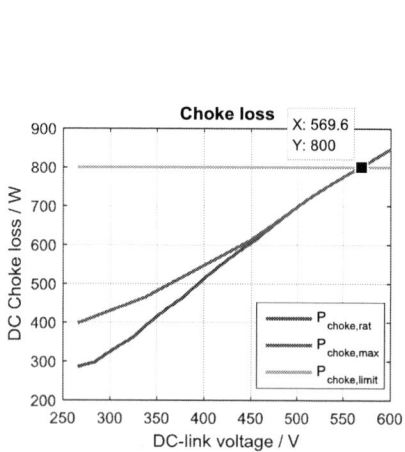

Figure 6: DC choke loss versus DC-link voltage

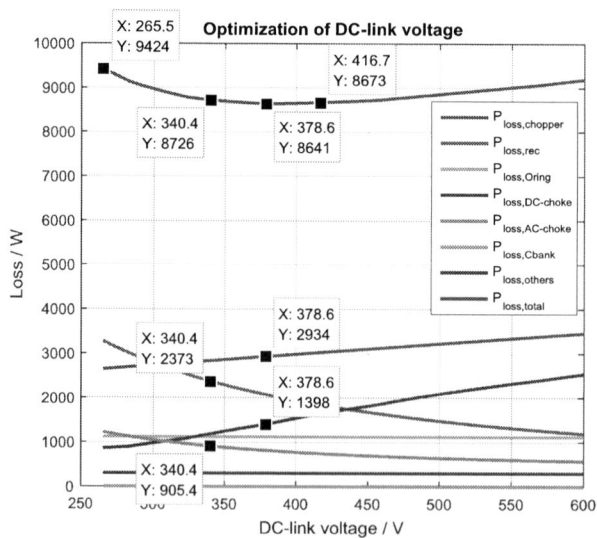

Figure 7: Model-based optimization of the DC-link voltage for the point of operation 250 V, 1400 A

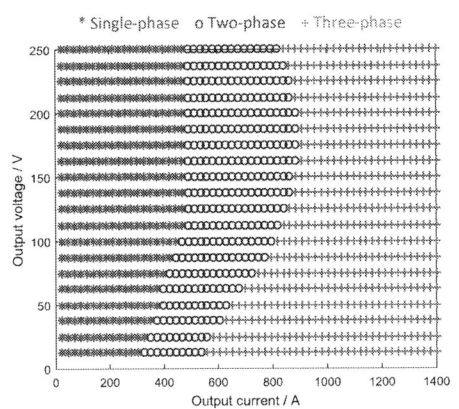

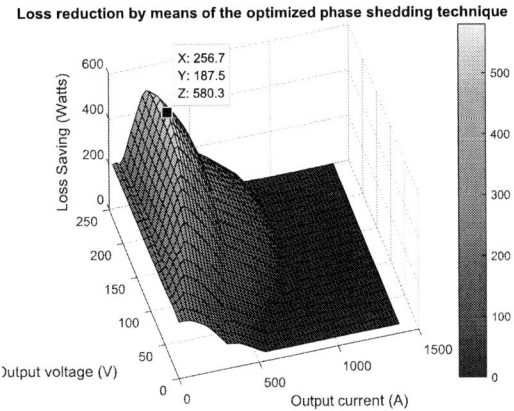

Figure 8: Optimum number of active chopper phases

Figure 9: Loss reduction by means of phase shedding optimization

- Highest stress of IGBT chopper and DC choke at 110% of grid voltage
- Max. allowed heat sink temp. (incl. 10 °C reserve): 112 °C (IGBT stack), 101 °C (B12 stack)

4.3. Optimization of Partial Load Efficiency

As the single design freedom of the power circuit (DC-link voltage) is fixed to minimize total loss at rated point of operation, an optimization of partial load efficiency can be only obtained by software means. Phase shedding is one of popular techniques for this task. With help of the CAD tool the optimum phase numbers is calculated and given in Figure 8. Compared to the three-phase operation a loss reduction up to 580 W is achieved (Figure 9).

5. Conclusions and Future Work

State-of-the-art review and comparison show that in the power range from 100 kW to 10 MW the multi-phase chopper technology is the best solution for high-current variable-voltage rectifier. By standardizing and modularizing the power circuit of the chopper rectifier excl. the transformer a reduction of engineering and manufacturing cost and an optimization of rectifier efficiency can be achieved at the same time.

In order to optimize the converter efficiency a computer-aided tool is developed and verified by measurement. A case study shows that more than 8% loss can be reduced at the ratted point of operation by performing a model-based optimization of the DC-link voltage (transformer voltage ratio), and in partial load condition a further loss reduction up to 580 W can be achieved by means of optimized phase shedding technique. The proposed system concept and the CAD tool are successfully applied in several customized DC power supply projects.

In the current version of the CAD tool the loss calculation of the IGBT chopper is based on the output and switching characteristics provided in the datasheet. The missing data (i.e. switching losses in very light load condition, switching losses at different DC-link voltages and different junction temperatures, etc.) are obtained by interpolation. The deviation between real operation and test condition defined in the datasheet is compensated by correction factors. In order to achieve a higher calculation accuracy, characterization of parallel operated IGBTs under real operation condition is planned in the near further.

© VDE VERLAG GMBH · Berlin · Offenbach

Acknowledgment

Thanks belong to colleagues Dueppe Gregor, Samir Elgharib, Kai Becker and Olaf Linke who contributed to high-quality electrical, mechanical and software designs in the discussed high-current variable-voltage chopper-rectifiers.

References

[1]   J. Solanki, N. Fröhleke, J. Böcker, A. Averberg and P. Wallmeier, "High-current variable voltage rectifiers: state of the art topologies" *IET Power Electronics*, vol. 8 no. 6, pp. 1068-1080, June 2015.

[2]   *Technical conditions for connection to the medium-voltage network,* German Association of Energy and Water Industries, 2008.

[3]   *Technical conditions for connection to the low-voltage network,* German Association of Energy and Water Industries, 2007.

[4]   J. Solanki, N. Fröhleke, J. Böcker, A. Averberg and P. Wallmeier, "Analysis, design and control of 1MW, high power and high current rectifier system," in proc. of *IEEE Energy Conversion Congress and Exposition (ECCE), Raleigh, USA,* 2012.

[5]   J. Solanki, "Comparison of thyristor-rectifier with hybrid filter and chopper-rectifier for high-power, high-current application," in conf. proc. of *PCIM Europe,* 2013.

[6]   J. Solanki, *High power factor high-current variable-voltage rectifiers,* Dissertation, University of Paderborn, Germany, 2015.

[7]   Z. Cao, *Model-Based Development of DC-DC-Converters with Wide Operation Range and High Dynamics,* Dissertation, University of Paderborn, Shaker Verlag Aachen, Germany, 2014.

[8]   U. Nicolai, A. Wintrich, *Determining switching losses of SEMIKRON IGBT modules,* Rev. 00, Application Note AN1403, www.semikron.com, 2014.

[9]   A. Steimel, *Grundlagen der Leistungselektronik (Fundamentals of Power Electronics),* Lecture Notes, Ruhr-University Bochum, 2001.

[10]   *Merkmale der Spannung in öffentlichen Elektrizitätsversorgungsnetzen,* European Standards DIN EN 50160:2011-02, 2011.

© VDE VERLAG GMBH · Berlin · Offenbach

PCIM Europe 2016, 10 – 12 May 2016, Nuremberg, Germany

Evaluation of a Unidirectional Three-Phase Rectifier based on the Third Harmonic Injection Concept in Comparison to a VIENNA Rectifier

M. Makoschitz, M. Hartmann*, H. Ertl

Vienna University of Technology, Institute of Energy Systems and Electrical Drives,
Power Electronics Section, Gusshausstrasse 27-29, A-1040 Wien, Vienna, Austria
*) Schneider Electric Power Drives GmbH, Vienna, Austria
Email: markus.makoschitz@tuwien.ac.at

Abstract

One of the most attractive rectification circuit for three-phase AC-to-DC conversion is the very well known three-phase diode (B6) bridge rectifier. Simplicity of circuit and design, high efficiency as well as robustness and cost-efficiency are major beneficial outcomes of this type of rectifier.

For applications which require high input current quality (low THD and a high power factor) active three-phase rectifiers (e.g. VIENNA rectifier) have to be used which in general are dedicated systems fully replacing the passive rectifier.

For specific applications which do not require a controlled output voltage (e.g., AC drives), however, a concept seems to be attractive, which opens the opportunity to optionally extend an existing B6 rectifier to a low harmonic input stage. In order to emphasize additional benefits and drawbacks of the optional third harmonic injection circuit (employing two half-bridge branches) compared to an active unidirectional rectifier commonly used in industry (e.g. the VIENNA rectifier), this work is engaged in an opposed comparison of different performance indices as system efficiency, switching/conduction losses, rated inductor power etc.

1. Introduction

The passive three-phase diode bridge rectifier (B6) is an attractive solution if a highly efficient, robust and simple low-cost rectification circuit is required which is operable at a fixed output voltage level (dependent on mains input voltage levels). Major field of interests for such a rectifier are e.g. AC drives, switch-mode power supplies, charging stations for electric vehicles etc. The B6 (as indicated in **Fig. 1**) mainly consists of a three-phase diode bridge (D_1-D_6), a smoothing inductor (L_{DC} - mainly located at the DC-side) and an electrolytic capacitor output stage (C_o). Due to the rather few component count and the purely passive implementation of the circuit, it is indeed a very cost effective mains input stage. The electrical characteristics of the topology are however limited by a power factor of $0.9...0.95$ and a rather high harmonic input current distortion ($\mathrm{THD_i}$) of up to $48\,\%$ (nominal load operation).

If a low harmonic circuit on the other hand is required (e.g. by regulation) without fully redesigning the

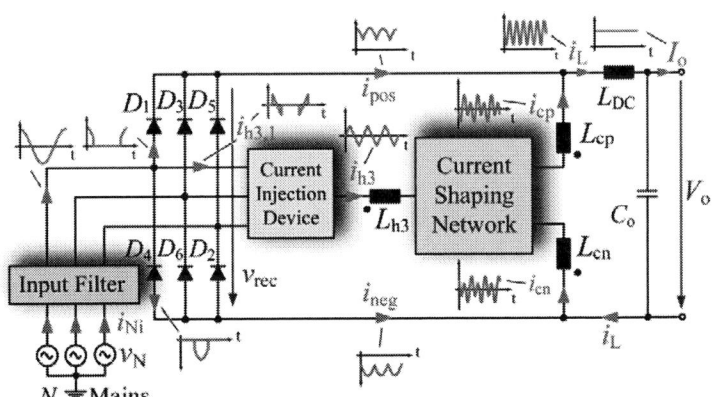

Fig. 1: Basic concept of the third harmonic injection principle. The optional active upgrade consists of a "current shaping network" (CSN) and a current injection device (CID) which are responsible for obtaining properly shaped sinusoidal mains currents i_{Ni}.

passive bridge an optional active current shaping concept may be of interest which is formally known as "third harmonic injection" (THI) principle (discussed in [1]). The idea is to separately design an active circuit which is fitting the electrical characteristics of the already existing passive three-phase rectifier and acts as optional upgrade for the B6 bridge. The equipped converter system offers an adequate solution to provide unity power factor ($\lambda = 1$) and elim-

© VDE VERLAG GMBH · Berlin · Offenbach

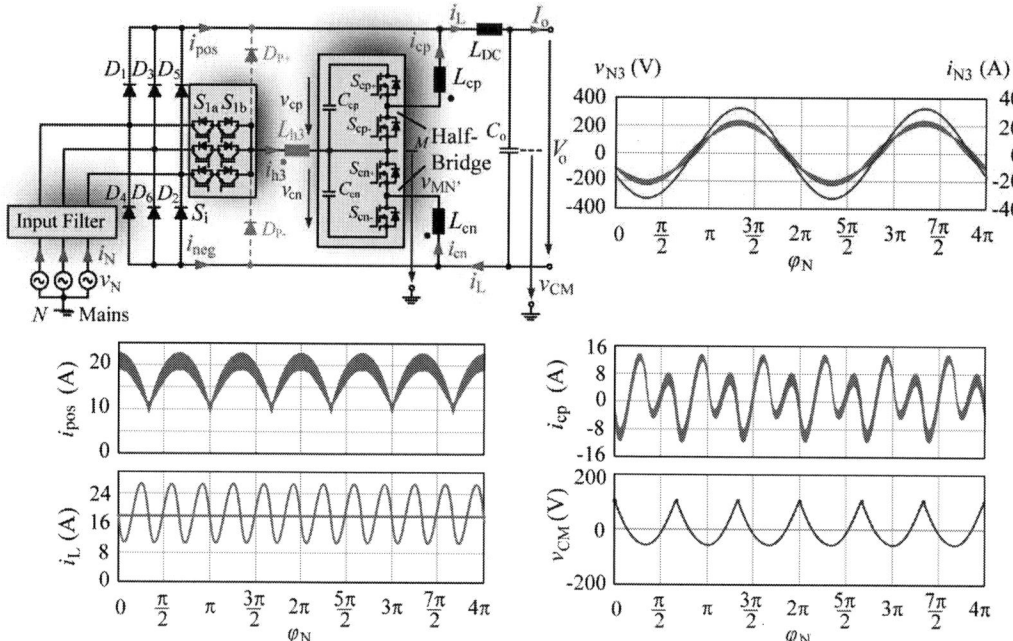

Fig. 2: Third harmonic injection concept employing 2 half-bridges branches as proposed in [2,3] and appropriate simulated most relevant voltage and current signals v_{N3}, i_{N3}, i_{pos}, i_{cp}, i_L, v_{CM} and I_o for $10\,\text{kW}$ rated output power.

inates low harmonic currents at the AC-side input ($\text{THD}_i < 5\,\%$) for a wide power range.

The optional active power electronics network, is going to be connected to the passive bridge at five points of the rectifier, which are typically accessible. The upgrade generates currents i_{cp} and i_{cn} which are going to be injected into the positive and negative bus bar of the passive bridge (current shaping network). These currents are required in order to compensate the unfavorable $300\,\text{Hz}$ harmonic current of the LC output filter current i_L and furthermore form $150\,\text{Hz}$ sinusoidal wave shapes. Purely sinusoidal input currents are finally achieved by filling up the naturally (due to passive rectification) evoked zero current gaps by a third harmonic current i_{h3} which is going to be injected (current injection device) into the appropriate mains phase.

Several active circuits which are able to fulfill these requirements have been proposed and presented already (cf., [2]).

2. Pertained Competitive Rectifiers

One feasible implementation of the third harmonic injection rectifier is depicted in **Fig. 2**. The black box which was labeled as "current shaping network" (CSN) has e.g. been replaced by a series connection of two half-bridge legs. Each half-bridge can either consist of 2 SiC - MOSFETs (or IGBTs). The DC - link of the cell is defined by $600\,\text{V}$ for proper operation ($M = 3V_{\text{Npk}}/V_c = 0.8125$) of the circuit, which consequently leads to $900\,\text{V}/1200\,\text{V}$ SiC devices. The current injection device (3 bidirectional switches) which forms the interconnection between AC-side input and DC-side current shaping network can be implemented by **(i)** back-to-back connected IGBTs/SiC-MOSFETs (reverse conducting IGBTs applicable and only 1 DC/DC converter required for 1 bidirectional switch gate drive), **(ii)** reverse blocking IGBTs, **(iii)** one active switch (e.g. IGBT/MOSFET) and 4 diodes or **(iv)** 2 diodes and 2 IGBTs/MOSFETs.

In-between CSN and CID an optional third inductor L_{h3} is considered in order to further improve the input current quality of system (as discussed in [3]). A switching frequency of $72\,\text{kHz}$ finally leads to inductance values for L_{cp}, L_{cn} and L_{h3} of $190\,\mu\text{H}$ (for 10kW nominal output power).

Alternatively to the proposed rectifier concept, one of the most attractive PFC circuits for unidirectional power flow, is the **_VIENNA rectifier_** (e.g. [4-6]). There are different topologies available. The VIENNA rectifier depicted in **Fig. 3** has a very low number of active and passive switches and is, therefore,

PCIM Europe 2016, 10 – 12 May 2016, Nuremberg, Germany

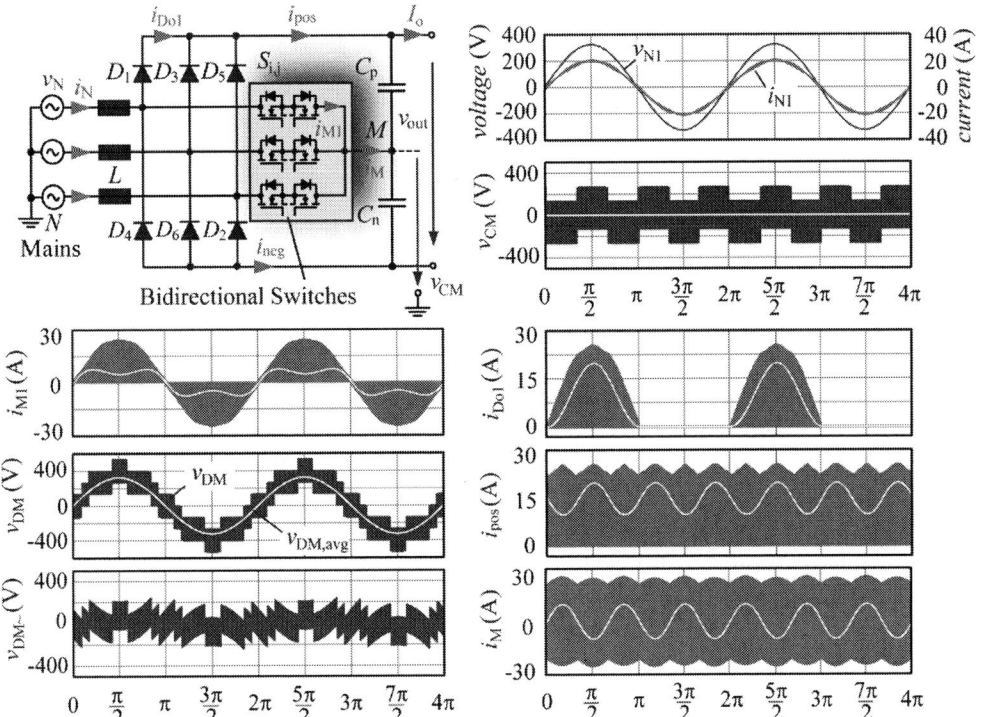

Fig. 3: VIENNA rectifier with 6 600 V MOSFETs which form the interconnection between AC-side and output capacitors midpoint M and diode bridge (typically implemented fast recovery diodes).

going to serve as competitive partner for this investigation. It has to be noted that a comparison of these two active rectifiers should NOT identify a superior concept (which is indeed not possible due to completely different attributes of both rectifiers). The investigation should merely reveal benefits and drawbacks of the THI circuit by comparing the upgradeable rectifier to a today well established industrial standard PFC concept.

3. Benchmarking Specifications

A reliable comparison between the two active systems shall allow an advanced benchmarking of the THI rectifier. Therefore, different performance indices (which are going to be used for this comparison) are introduced, listed and shortly described in the following.

A. Diode/MOSFET VA-Ratings: The VA-ratings of a switching component is defined by multiplication of maximum blocking voltage and peak current of the appropriate semiconductor. This factor hence gives a prediction of the relative maximum stress of diodes (μ_D^{-1}) and transistors (μ_S^{-1}) in a power electronics circuit and can hence be determined by

$$\mu_S^{-1} = \frac{\sum_n v_{S,\max,n} i_{S,\max,n}}{P_o} \qquad \text{and} \qquad \mu_D^{-1} = \frac{\sum_n v_{D,\max,n} i_{D,\max,n}}{P_o}.$$

B. Rated Inductor Power/Capacitor Current Stress: Two major performance aspects in order to classify inductors and capacitors in an electrical way, are to introduce a rated inductor power index ρ_L and a normalized capacitive current stress equivalent ρ_C of all implemented inductors and capacitor stages within the converter system.

$$\rho_L = \frac{\sum_n I_{L,n} \Delta I_{L,pkpk,n} L_{c,n} f_s}{P_o}, \qquad \rho_C = \frac{\sum_n I_{C,rms,n}}{I_o}.$$

C. Inverse Power Density of Inductors: Besides electrical characteristics of inductors, also a volume dependent design might be of interest for the final implementation of the total rectifier system. Therefore, the power density of the implemented inductors is selected as promising benchmark indexing.

© VDE VERLAG GMBH · Berlin · Offenbach

$$\varepsilon_{\mathrm{L}}^{-1} = \frac{\sum_{\mathrm{n}} V_{\mathrm{L,n}}}{P_{\mathrm{o}}}$$

D. Switching Losses of Semiconductors: Furthermore, the switching losses of all impaired components are compared to each other with respect to the total output power P_{o}. τ_{p} can be qualitatively determined by

$$\tau_{\mathrm{p}} = \frac{\sum_{\mathrm{n}} I_{\mathrm{S,avg,n}} V_{\mathrm{S,n}}}{P_{\mathrm{o}}}$$

E. Conduction Losses of Diodes and Switches: Conduction losses of switches and diodes can be opposed by evaluation of averaged and rms current values related to the output current (cf., [7]).

$$\tau_{\mathrm{c}} = \frac{\sum_{\mathrm{n}} I_{\mathrm{S,rms,n}}}{I_{\mathrm{o}}}, \qquad \delta_{\mathrm{c}} = \frac{\sum_{\mathrm{n}} I_{\mathrm{D,avg/rms,n}}}{I_{\mathrm{o}}}.$$

F. System Efficiency: Finally the performance of the total system is going to be determined by comparing total losses of both active systems.

$$1 - \eta = \frac{P_{\mathrm{in}} - P_{\mathrm{o}}}{P_{\mathrm{in}}}.$$

4. Comparative Values - THI

In this section, the previously described performance indices are now going to be derived analytically. The VA-ratings of the implemented semiconductor is the first parameter of interest in this evaluation. The THI rectifier diodes ($D_{1\text{-}6}$) are stressed with a maximum blocking voltage which is dependent according to the peak line-to-line voltage situation ($v_{\mathrm{D,max}} = \sqrt{2}\,V_{\mathrm{LL}}$) and a maximum current which is equivalent to I_{Npk}. The diodes of one bidirectional switch, however, obtain their maximum while the two remaining bidirectional switches initiate the current commutation of i_{h3}. This commutation however takes place before the applied line-to-line voltage attains its maximum value. The maximum blocking voltage of the diodes (and thus also transistors) of bidirectional switches is hence defined by $3V_{\mathrm{Npk}}/2$, and the maximum current is defined according to i_{h3} (whose peak value is defined by $I_{\mathrm{Npk}}/2$).

The switches of each half-bridge are stressed with the peak current of injection currents i_{cp}, i_{cn} and the DC voltage of C_{cp}, C_{cn}, respectively. The VA ratings of switches and diodes hence lead to

$$\mu_{\mathrm{S}}^{-1} = \frac{9\hat{V}_{\mathrm{N}}\hat{I}_{\mathrm{N}} + 16V_{\mathrm{c}}\hat{I}_{\mathrm{cp}}}{4P_{\mathrm{o}}} \qquad \mu_{\mathrm{D}}^{-1} = \frac{6\hat{V}_{\mathrm{N}}\hat{I}_{\mathrm{N}}(8\sqrt{3}+3)}{8P_{\mathrm{o}}}.$$

In order to evaluate the current stress of capacitors, 3 different capacitor banks have to be evaluated (C_{cp}, C_{cn} and C_{o}). C_{o} is characterized by $I_{\mathrm{L,ac,pk}}/2$. Current stress of half-bridge capacitors C_{cp}, C_{cn}, can only be determined numerically and result in $3.6\,\mathrm{A_{rms}}$ for an output power of $10\,\mathrm{kW}$. The normalized capacitive current stress ρ_{C} of all implemented capacitors of the THI rectifier then results in

$$\rho_{\mathrm{C}} = \frac{\hat{I}_{\mathrm{L,ac}} + 2\sqrt{2}I_{\mathrm{ccp}}}{\sqrt{2}I_{\mathrm{o}}}, \quad \text{with} \quad \hat{I}_{\mathrm{L,ac}} = \frac{\sqrt{3}\hat{V}_{\mathrm{N}}}{\omega_{\mathrm{N}}L_{\mathrm{DC}}}\left(\sin\left(\arccos\left(\frac{3}{\pi}\right)\right) - \frac{3}{\pi}\arccos\left(\frac{3}{\pi}\right)\right).$$

The rated inductor power (of the appropriate implemented chokes L_{cp}, L_{cn}, L_{h3} and L_{DC}) is dependent on current rms values I_{cp}, I_{cn}, I_{h3}, I_{L}, and the appropriate maximum peak-to-peak current ripples ΔI_{cp}, ΔI_{cn}, ΔI_{h3}, ΔI_{L}, at the operated frequencies f_{s}, $6f_{\mathrm{N}}$. The required appropriate parameters are defined by

$$I_{\mathrm{cp/n}} = \hat{I}_{\mathrm{N}}\sqrt{\frac{2\pi^3 - 24\pi + 9\sqrt{3}}{24\pi} + \left(\frac{\hat{I}_{\mathrm{L,ac}}}{\sqrt{2}\hat{I}_{\mathrm{N}}}\right)^2}, \qquad I_{\mathrm{h3}} = \frac{\hat{I}_{\mathrm{N}}}{2}\sqrt{\frac{2\pi - 3\sqrt{3}}{\pi}}, \qquad I_{\mathrm{L}} = \frac{\hat{I}_{\mathrm{N}}}{2}\sqrt{\frac{2\pi^2}{3} + 2\left(\frac{\hat{I}_{\mathrm{L,ac}}}{\hat{I}_{\mathrm{N}}}\right)^2},$$

$$\Delta I_{\mathrm{cp/n}} = \frac{2MV_{\mathrm{c}}(1-M)}{3f_{\mathrm{s}}L_{\mathrm{cp/n}}}, \qquad \Delta I_{\mathrm{h3}} = \frac{V_{\mathrm{c}}M}{3\sqrt{3}f_{\mathrm{s}}L_{\mathrm{cp/n,h3}}}, \qquad \Delta I_{\mathrm{L}} = 2\hat{I}_{\mathrm{L,ac}}.$$

The rated inductor power hence calculates to

$$\rho_{\mathrm{L}} = \frac{2I_{\mathrm{cp/n}}\Delta I_{\mathrm{cp/n}}L_{\mathrm{cp/,n}}f_{\mathrm{s}} + I_{\mathrm{h3}}\Delta I_{\mathrm{h3}}L_{\mathrm{h3}}f_{\mathrm{s}} + 6I_{\mathrm{L}}\Delta I_{\mathrm{L}}L_{\mathrm{DC}}f_{\mathrm{N}}}{P_{\mathrm{o}}}.$$

In order to identify conduction losses of switches (rms) and diodes (avg), the current stress of corresponding components is required which can be finally determined by

$$I_{\mathrm{D1\text{-}6,avg}} = \frac{\sqrt{3}\hat{I}_{\mathrm{N}}}{2\pi}, \qquad I_{\mathrm{Diab,avg}} = \frac{\hat{I}_{\mathrm{N}}}{2\pi}\left(2 - \sqrt{3}\right), \qquad I_{\mathrm{Siab,rms}} = \hat{I}_{\mathrm{N}}\sqrt{\frac{2\pi - 3\sqrt{3}}{24\pi}}.$$

The current stress of both half-bridge semiconductors can again only be calculated numerically and eventually result in $3.6\,\text{A}_{\text{rms}}$ and $5\,\text{A}_{\text{rms}}$ for the positive (S_+) and negative switch (S_-) of one half-bridge, respectively ($10\,\text{kW}$ nominal load operation). If all derived characteristics are considered, conduction losses of both, diodes and transistors thus result in

$$\delta_{\text{c}} = \frac{6\,\hat{I}_{\text{N}}}{\pi\,I_{\text{o}}}, \qquad \tau_{\text{c}} = \frac{3\hat{I}_{\text{N}}\sqrt{2\pi - 3\sqrt{3}} + 2\sqrt{6\pi}\,(I_{S_+} + I_{S_-})}{\sqrt{6\pi}I_{\text{o}}}.$$

The switching losses of the system can be "qualitatively" determined by utilizing the averaged pulsed current of the each switch (assuming linear dependency of the switching losses on the switched current) and applied drain source voltage (during the off state of the transistor). It has to be noted, that the bidirectional switches of the THI rectifier are only turned on and off twice during one mains period. Switching losses of these device are hence negligibly small. The only devices which are operated with switching frequency f_{s} are the half-bridge semiconductors (S_+, S_-). The averaged values of these components can be determined numerically and lead to $0\,\text{A}_{\text{avg}}$ and $1.6\,\text{A}_{\text{avg}}$, respectively. τ_{p} therefore can be assessed and yields

$$\tau_{\text{p}} = \frac{2I_{S_-}V_{\text{c}}}{P_{\text{o}}}.$$

The coils of the injection cell are designed such to allow a maximum ripple of $20\,\%\,I_{\text{Npk}}$ which results in an inductance value of $\sim 190\,\mu\text{H}$ for the required power level of 10kW. The implemented inductor (stacked core assembly) leads to a core volume of $0.074\,\text{dm}^3$ for each inductor (stacked core: 2xT184-14, windings: 59). The necessary passive three-phase rectifier choke is a $300\,\text{Hz}$ inductor and hence considerably larger than the injection coils (as it is typically designed for a maximum input current THD_{i} value of $48\,\%$ for B6 standalone operation). The volume of the implemented inductor results in $0.57\,\text{dm}^3$ (not optimized). The reciprocal power density of all inductors of the THI rectifier is defined by

$$\varepsilon_{\text{L}}^{-1} = \frac{3V_{\text{L}} + V_{\text{LDC}}}{P_{\text{o}}}$$

The efficiency of the total system was calculated to be $\sim 98\,\%$ and $97.8\,\%$ has been measured ($10\,\text{kW}/72\,\text{kHz}$ prototype) for a nominal load of $10\,\text{kW}$. The total losses of the system can therefore be assessed and result in $1 - \eta = 2.2\,\%$.

5. Comparative Values - VIENNA

The maximum blocking voltage of the VIENNA rectifier diodes (D_{1-6}) is dependent on the required (and controllable) output voltage V_{o}. The maximum diode current is (as for the THI rectifier topology) the peak mains input current I_{Npk}. Blocking voltages levels of each diode of the bidirectional switches only have to withstand $V_{\text{o}}/2$. Similar assumptions as discussed for diodes of the bidirectional switches also apply for appropriate MOSFETs. The VA-ratings of implemented switches therefore results in

$$\mu_{\text{D}}^{-1} = \frac{9V_{\text{o}}\hat{I}_{\text{N}}}{P_{\text{o}}} \qquad \text{and} \qquad \mu_{\text{S}}^{-1} = \frac{3V_{\text{o}}\hat{I}_{\text{N}}}{P_{\text{o}}}.$$

The capacitor current stress of the VIENNA rectifier output capacitors (both connected to the midpoint M) can be evaluated in an analytical form. The normalized capacitive current stress ρ_{C} of the required output capacitor stage of the rectifier is therefore defined by

$$\rho_{\text{C}} = 2\frac{\hat{I}_{\text{N}}}{I_{\text{o}}}\sqrt{\frac{5\sqrt{3}M}{4\pi} - \frac{9M^2}{16}} \quad \text{with} \quad M = \frac{2\hat{V}_{\text{N}}}{V_{\text{o}}}.$$

In order to characterize the rated inductor power of the VIENNA rectifier, it is assumed that the maximum current ripple does appear at $\varphi_{\text{N}} = 0°$, which is valid for a modulation index $M > 0.85$ ($V_{\text{o}} < 760\,\text{V}$). For smaller values of M, $\Delta I_{\text{L,max}}$ is located at $\varphi_{\text{N}} = 30°$ which is neglected for the comparison at hand. The inductor rms current is defined by the input current rms and results in $I_{\text{L,rms}} = I_{\text{N,rms}}$. The rated inductor power then leads to a ρ_{L} of

$$\rho_{\text{L}} = \frac{3}{4\sqrt{2}}\frac{\hat{I}_{\text{N}}V_{\text{o}}}{P_{\text{o}}}\left(\frac{8}{3}M - M^2 - \frac{4}{3}\right).$$

The boost inductors of an 800V VIENNA, are implemented as Schott 193 type coils. The total volume of all 3 chokes results in $0.285\,\text{dm}^3$, which leads to an reciprocal power density of $0.029\,\text{dm}^3/\text{kW}$.

© VDE VERLAG GMBH · Berlin · Offenbach

In a next step the conduction losses of transistors and diodes are going to be calculated. The transistor conduction losses are represented by the rms value of its drawn current for one mains period. The diodes conduction losses are, however, characterized by its averaged current values. The evaluated values of transistors and diodes conduction losses eventually result in

$$\tau_c = 6\frac{\hat{I}_N}{I_o}\sqrt{\frac{1}{4} - \frac{2M}{3\pi}} \qquad \text{and} \qquad \delta_c = \frac{6}{\pi}\frac{\hat{I}_N}{I_o}.$$

The switching losses can be again "qualitatively" identified by the averaged current of the appropriate semiconductor. The switching losses of S_{1-6} hence result in

$$\tau_p = 3\frac{\hat{I}_N V_o}{P_o}\left(\frac{1}{\pi} - \frac{M}{4}\right).$$

The efficiency of the VIENNA rectifier for $800\,V$ output voltage was documented in [7] and has been evaluated to be 97.3%.

6. Comparison of Results

Results of the performance evaluation for both rectifiers (implemented THI prototype cf., **Fig. 4(a)**) are going to be summarized in two different "radar diagrams" (VIENNA rectifier with $650\,V$ and $800\,V$ regulated output voltage V_o) as depicted in **Fig. 4(b)** and e.g. used in [8]. The output voltage levels of the VIENNA rectifier are chosen such to provide comparison values for a minimum achievable output voltage ($650\,V$- if midpoint voltage control is not considered) and one commonly used output voltage level ($800\,V$) of the VIENNA rectifier. It is however important to bear in mind that the THI rectifier is characterized by a fixed output voltage according to the mains voltage situation ($537\,V$). The variation of the output voltage of the VIENNA rectifier should therefore only illustrate the modification of the selected parameters of the VIENNA in comparison to the THI rectifier. Predefined specifications of the rectifier systems which are used for benchmarking are given in **TABLE I**. Results of the calculated performance indices are listed in **TABLE II**. The performance parameters are chosen such, that a smaller value emphasizes a good system behaviour. One conspicuous feature of the THI rectifier is the small values of conduction and switching losses of diodes and transistors. This is mainly evoked due to the fact that the additional active converter stage only has to transfer a small amount of output power ($\sim 6\%$) and process approximately 17% of reactive power. Additionally, it has to be considered that, although rated inductor power of both circuits seem to be of similar value, the implemented inductor volume of the THI rectifier is definitely higher than that of a VIENNA rectifier. This underlies the fact, that the THI rectifier system requires an additional 4th inductor (due to passive rectification). This coil is however electrically characterized by a $300\,Hz$ current ripple and its size majorly defined according to the specified THD_i for B6 standalone operation. Hence, no shrinking due to increased switching frequency is possible for the volume and design of L_{DC}.

Furthermore, as can be read from **TABLE II**, both rectifiers obviously show strong- and weak-points. Drawbacks of the THI rectifier are for example

- the high inductor volume, which is majorly determined due to the $300\,Hz$ choke (which belongs to the original passive rectifier),
- the relatively large component count (4 additional switching devices) compared to the VIENNA rectifier (however it has to be noted that 6 switches S_{ij} of the VIENNA rectifier are stressed with switching frequency, but merely 4 switches $S_{cp/n\pm}$ of the THI rectifier),
- higher complexity of the total system
- no regulation of output voltage V_o available (fixed according to mains voltage situation).

On the other hand, however, there are numerous aspects which militates in favor of a THI rectifier system implementation, as e.g.

- + robustness (still operable in B6 standalone mode even if injection cell has to be turned off e.g. due to malfunction) which is primarily attributed to its
- + optional implementation,
- + no high frequency common mode voltage v_{CM} (see **Fig. 2**) at the output of the total system

PCIM Europe 2016, 10 – 12 May 2016, Nuremberg, Germany

TABLE I: Specifications of the two benchmarked three-phase rectifier systems.

Mains voltage:	$V_{LL} = 400\ \text{V}_{rms}$
Mains frequency:	$f_N = 50\ \text{Hz}$
Active Rectifier:	VIENNA
THI Rectifier (Shaping circuit):	2 Half-Bridge Branches
THI Rectifier (Injection circuit):	3 Bidirectional Switches
Switching frequency:	$f_s = 72\ \text{kHz.}$
THI Cell DC-link voltage:	$V_{cp} = V_{cn} = 600\ \text{V}$
THI output voltage (uncontrolled):	$V_o = 537\ \text{V}$
VIENNA output voltage (controlled):	$V_o = 650\ \text{V}...800\ \text{V}$
Output power:	$P_o = 10\ \text{kW}$

TABLE II: Results of calculated performance indices of a VIENNA and THI rectifier system.

	VIENNA (650V)	VIENNA (800V)	THI
μ_D^{-1}	11.94	14.7	8.43
μ_S^{-1}	3.98	4.90	4.38
τ_c	1.53	2.71	1.73
τ_p	0.27	0.56	0.2
δ_c	2.53	3.12	2.09
ρ_c	0.94	1.41	0.68
ρ_L	0.24	0.15	0.15
ε_L (dm^3/kW)	0.029	0.016	0.057
1-η (%)	-	2.7	2.2

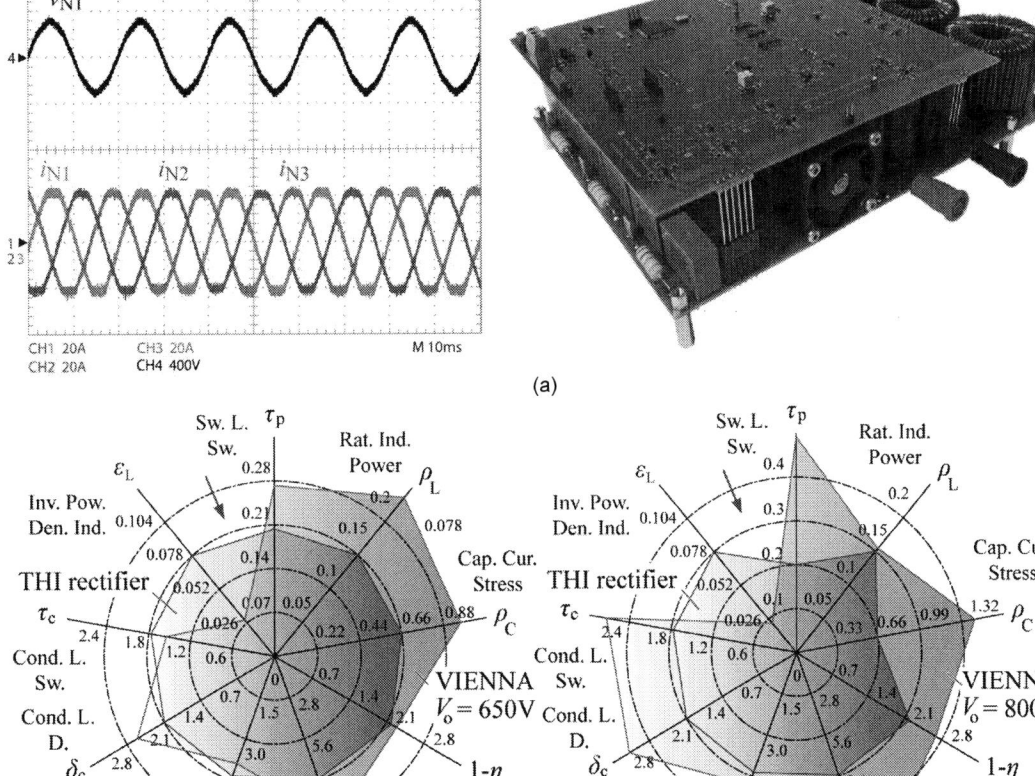

Fig. 4: (a) Measurement results and constructed laboratory prototype of a 72 kHz/10 kW THI rectifier (cf., [9]) at 10 kW nominal load. The system input is characterized by a power factor of 0.999 and a THD$_i$ between 2-3 % (b) Radar diagram (smaller values characterize a better system behaviour - indicated by blue arrow) consisting of main performance indices comparing a 10 kW VIENNA rectifier with 650V...800V output voltage and a THI rectifier.

© VDE VERLAG GMBH · Berlin · Offenbach

+   only semiconductors $S_{cp\pm}$ and $S_{cn\pm}$ are stressed with switching frequency.

+   the active upgrade only has to process some fraction of output power

The VIENNA rectifier topology is already widely used in industry and therefore a detailed discussion about pros and cons of this valuable and attractive low harmonic rectifier is not required in this work.

7. Conclusion

Focus of this paper is a comparative evaluation of a selected unidirectional PFC rectifier (VIENNA rectifier) and one specific realization of a rectifier system based on the third harmonic injection principle (2 half-bridge branches). It has once again to be mentioned that, this comparison should merely reveal advantages and drawbacks of the THI circuit by comparing electrical parameters of the upgradeable THI rectifier to an industrial standard PFC concept (VIENNA). It is hence important to notice and/or keep in mind that both topologies originally provide different rectifier attributes (e.g. controllable/no controllable output voltage etc.). Besides benefits and drawbacks also several performance indices are chosen, in order to allow a more reliable characterization and benchmarking of the THI system. The higher circuit complexity of the THI rectifier circuit is compensated by a high efficiency, high robustness and the fact that the THI rectifier shows no high-frequency CM voltage at the DC-output. The circuit does not offer a controlled output voltage, however, the topology reuses the main elements of a passive three-phase rectifier circuit with DC-side located smoothing inductor which allows the extension of an existing passive rectifier circuit to a rectifier circuit with low harmonic input.

Acknowledgement

The authors are very much indebted to the Austrian Research Promotion Agency (FFG) which generously supports the work of the Vienna University of Technology Power Electronics Section (Institute of Energy Systems and Electrical Drives).

References

[1] J. W. Kolar, T. Friedli, "*The Essence of Three-Phase Rectifier Systems*", 33rd Int. IEEE Telecommunications Energy Conference (INTELEC 2011), Amsterdam, Netherlands, Oct. 9-13, pp. 1-27, 2011.

[2] M. Makoschitz, M. Hartmann, H Ertl, "*Topology Survey of DC-side Enhanced Passive Rectifier Circuits for Low-Harmonic Input Currents and Improved Power Factor*", Proceedings of the Conference for Power Electronics, Intelligent Motion, Power Quality (PCIM), Nuernberg, Germany, May 19-21 2015.

[3] M. Makoschitz, M. Hartmann, H. Ertl, R. Fehringer, "*A Passive Three-Phase Rectifier Enhanced by a DC-side High Switching Frequency Add-On SiC-Converter Stage for Unity Power Factor Applications*," in Proceedings of the Conference for Power Electronics, Intelligent Motion, Power Quality (ECCE/EPE), Geneva, Switzerland, September 8-10 2015.

[4] J. W. Kolar, F. Zach, "*A Novel Three-Phase Three-Switch Three-Level PWM Rectifier*," in Proceedings of the Conference for Power Electronics, Intelligent Motion, Power Quality (PCIM), Nuernberg, Germany, June 28-30 1994, pp. 125-138.

[5] J. W. Kolar, H. Ertl, F. Zach, "*Design and experimental investigation of a three-phase high power density high efficiency unity power factor PWM (VIENNA) rectifier employing a novel integrated power semiconductor module*," in Proceedings of the 11th Annual Applied Power Electronics Conference and Exposition (APEC), pp. 514-523, 1996.

[6] M. Hartmann, S. D. Round, H Ertl, J. W. Kolar, "*Digital Current Controller for a 1 MHz, 10kW Three-Phase VIENNA Rectifier*," IEEE Transactions on Power Electronics, pp. 2496 - 2508, 2009.

[7] T. Friedli, M. Hartmann, J. W. Kolar, "*The Essence of Three-Phase PFC Rectifier Systems - Part II*", IEEE Transactions on Power Electronics, Vol. 29, No. 2, February 2014.

[8] T. Soeiro, J. W. Kolar, "*Comparative Evaluation of Bidirectional Buck-Type PFC Converter Systems for Interfacing Residential DC Distribution Systems to the Smart Grid*," Proceedings of the 38th Annual Conference of the IEEE Industrial Electronics Society (IECON 2012), Montreal, Canada, October 25-28, 2012.

[9] M. Makoschitz, M. Hartmann, H Ertl, "*Hardware Implementation and Characterization of a SiC-Based Hybrid Three-Phase Rectifier Employing Third Harmonic Injection*," accepted paper at the 31th Annual Applied Power Electronics Conference and Exposition (APEC), Long Beach, USA, March, 2016.

© VDE VERLAG GMBH · Berlin · Offenbach

SiC improves switching losses, power density and volume in UPS

Nikolai Epp, Christian Schulte-Overbeck, Zhiyu Cao, Michael Lemke, Lothar Heinemann

AEG Power Solutions GmbH – Emil-Siepmann-Str.32, Germany,

nikolaiepp@gmail.com, christian.schulte-overbeck@aegps.com, zhiyu.cao@aegps.com,

michael.lemke@aegps.com, lothar.heinemann@aegps.com

Abstract

The state-of-the-art transformerless uninterruptible power supplies (UPS) for the power rating from 50 kVA to hundreds of kVA are with silicon IGBT technology. In order to achieve high efficiency and good power quality, 3-level topology is state-of-the-art. Thanks to the significantly reduced switching losses of silicon carbide (SiC) semiconductors and the availability of high-current SiC metal-oxide-semiconductor field-effect transistor (MOSFET) module, a SiC-based UPS solution is more and more attractive for this power range. In this contribution a comparative study of Si-based and SiC-based UPS is presented first. For characterizing the new component, a test setup is developed. The test results, open issues and future works are discussed.

1. Introduction

Today the implicitness of the high demand on security of power supply for all areas of energy applications force the engineering to do research in uninterruptible power supply (UPS). Loss of data and cost intensive downtimes can be avoided by an UPS. According to application field the UPS system can be divided into two categories: industry and commercial UPS. Because of the rough field environment and the industry safety requirements, low voltage battery and transformer between inverter and output are required. Conversely, due to the round-the-clock "on-line" character high efficiency, high power quality and high power density are the major requirements for the commercial UPS. Hence, high-voltage battery, high-voltage DC-link and transformerless configuration are usually applied. In order to achieve a good tradeoff among all three requirements above, silicon (Si) insulated-gate bipolar transistor (IGBT) based three-level topology is usually selected, which represents the state-of-the-art. In the front end side, neutral point clamped (NPC) three-level rectifier or Vienna rectifier is applied, while in the output side the NPC three-level inverter is applied. Thanks to the cost reduction of commercial silicon carbide (SiC) metal-oxide field-effect transistors (MOSFET), SiC-based photovoltaic (PV) inverter, power factor correction (PFC) rectifier and UPS for the power rating up to 10 kW are presented in the last years [1] [2] [3]. For the power rating above tens of kW a hybrid solution was provided by many semiconductor manufacturers, i.e. semiconductor module consists of Si IGBT and SiC Schottky diode [4] [5] [6]. Recent technical progress in packaging technology and cost reduction of SiC material production make it possible to apply full SiC components in a higher power range. Many semiconductor manufacturers are starting to provide full SiC MOSFET module with a voltage rating of 1.2 kV/1.7 kV and a current rating up to a few hundreds Amperes, either as engineering samples or as series product [7] [8] [9].

© VDE VERLAG GMBH · Berlin · Offenbach

In this contribution a feasibility study of full SiC-based UPS for the power rating above 50 kVA is provided. In the following chapters a simulative study of the Si IGBT based three-level UPS with the power rating of 250 kVA is performed first, which provides a reference for the later design and comparison. In Chapter 3 a full SiC-based UPS is dimensioned and compared with Si IGBT solution. For power circuit design and properly operation of the new semiconductor module, a first test setup for characterization of SiC MOSFET module is developed. Measurement results and some open issues are discussed in chapter 4. In chapter 5 the intermediary results and future works are given.

2. Simulative Study of a Si-IGBT-based UPS

In order to perform a simulative study of the state-of-the-art high-power commercial UPS, one of the "best in class" commercial UPS is selected as reference (s. Figure 1). A draft technical specification of the selected UPS is given in Table 1. As shown in Figure 2 three-level NPC topology is applied both in the front end side and in the inverter side. Each leg consists of three parallel NPC IGBT modules (650 V, 300 A, Trench IGBT 4).

By performing an iterative design process a good tradeoff among efficiency (up to 96 % in online operation), switching frequency (10 kHz), power quality (THDi ≤ 3 %, PF > 0.99) and volume (H x W x D = 1915 x 1000 x 960 mm) are achieved.

Table 1 Technical specification of the Si- based reference UPS

Specification	Rated value
Power	250 kVA
f_s	10 kHz
U_{in} / U_{out} (RMS)	400 V
I_{in} / I_{out} (RMS)	360 A
Efficiency (on-line mode)	96 %
Input THDi	≤ 3 %
Input PF	>0.99
Output THD (linear load)	< 3 %

Figure 1 State-of-the-art higher-power commercial UPS system

Target of the simulation is to study the steady-state characters of the AFE rectifier and the inverter to provide a reference for the comparison between Si IGBT solution and SiC MOSFET solution. For this objective MATLAB/Simulink and the system-level power electronic simulation tool SimpowerSystems is applied. Ideal switcher model is selected to calculate the semiconductor losses. The voltage drops and the conduction losses on IGBTs and diodes are calculated based on the output characteristics given in the datasheet. The switching losses of IGBTs and the reverse recovery losses of diodes are deduced by means of interpolation of the switching energies specified in the datasheet. The junction temperatures of all semiconductors, i.e. IGBTs, diodes and SiC MOSFETs, are set by 125 °C constant in the simulation/calculation. In the system-level simulation the chokes in the AFE side and in the inverter side are modeled as ideal components. The choke losses are estimated separately by means of analytical methods. The steady-state simulation results at the rated point of operation are listed in Table 2. The simulated/calculated efficiency at the rated point of operation has a good agreement to the measurement (96.3 % simulated/calculated vs. 96 % measured). The small deviation is majorly caused by the switching losses calculation, because the switching energies specified in the datasheet are usually measured on a very low-inductive double-pulse test bed (PCB design) and extremely low gate resistance is applied. In practical application the leakage

© VDE VERLAG GMBH · Berlin · Offenbach

inductance of IGBT stacks is much higher. Hence, larger gate resistance is required to reduce the switching speed and the voltage overshoot, which results enhanced switching losses. The other reason of the deviation is that the auxiliary power supply is not considered in the simulation. However, in order to perform a fair comparison, in simulations the datasheet specified values are applied. Simulation results show also a good agreement to measurement in the input THDi, output THD, current and power factor as specified in the datasheet. Therefore these values can be used in the ongoing process to reduce the choke dimensions with a higher frequency and get the same switching losses as before.

In the latter comparison with SiC technique the values of the power losses, the choke ripple current and the THDi are important.

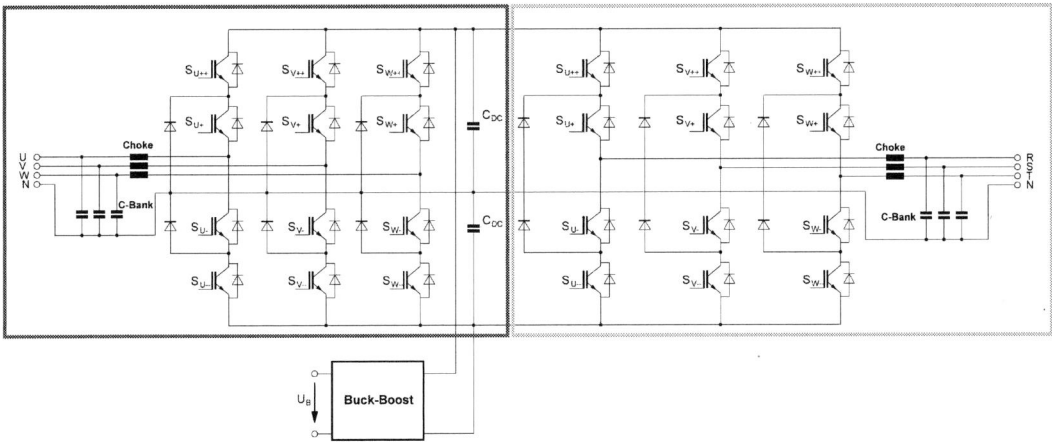

Figure 2 Equivalent diagram of the reference UPS

Table 2 Simulation and Calculation results

	AFE rectifier	Inverter	Total
THDi on choke	3.17 %	7.92 %	N.A.
Choke ripple current	62.81 A	139.4 A	N.A.
Semiconductor losses	3586 W	3570 W	7156 W
Choke losses	930 W	1190 W	2120 W
Efficiency	98.2 %	98.1 %	96.3 %
THD	THDi = 3.17 %	THDu = 0.52 %	N.A.

Further simulation and calculation show because of the high switching losses of Si IGBT and diode an increasing of switching frequency (i.e. volume reduction in passive components) is not possible (s. Figure 3). Thanks to the ultra-low switching losses and good on-state characters, SiC MOSFET provides the possibility to achieve a higher power density.

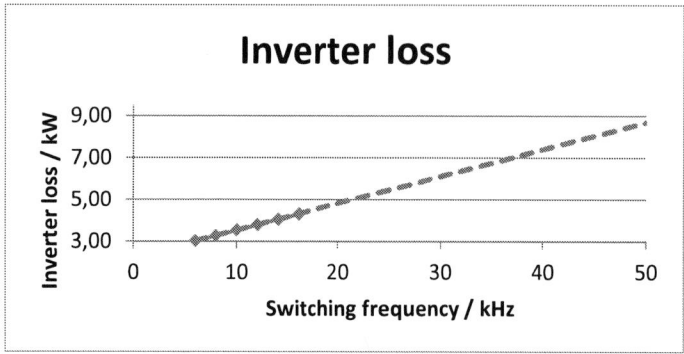

Figure 3 Inverter loss vs. switching frequency

3. Simulative Analysis of a SiC-based UPS

Because of the very fast switching the directly parallel operation of SiC MOSFET modules is very difficult. Hence a modular system concept is applied. The rated power for the simulated model amount 50 kW with an input and output voltage of 400 V (phase to phase). Higher power rating is achieved by parallel operation of UPS modules. A simplified circuit diagram of the proposed UPS module is illustrated in Figure 4. Thanks to the very low switching losses of SiC components, two-level topology and 1.2 kV / 300 A half bridge SiC MOSFET modules are selected.

Interactive simulations show, that identical normalized semiconductor losses as the Si IGBT solution is generated, if the SiC MOSFET UPS is operating at 50 kHz switching frequency. In order to keep identical (normalized) current ripple on the choke, the required choke inductance is calculated by [10]:

$$L_{in/out} = \frac{U_1}{4 \cdot \Delta i_{L\,max} \cdot f_{sw}} = \frac{750\,V}{4 \cdot 28\,A \cdot 50\,kHz} = 134\,\mu H \tag{1}$$

The capacity is reduced linear to $C_{out\,2L\,SiC} = 83\,\mu F$ and $C_{in\,2L\,SiC} = 50\,\mu F$.

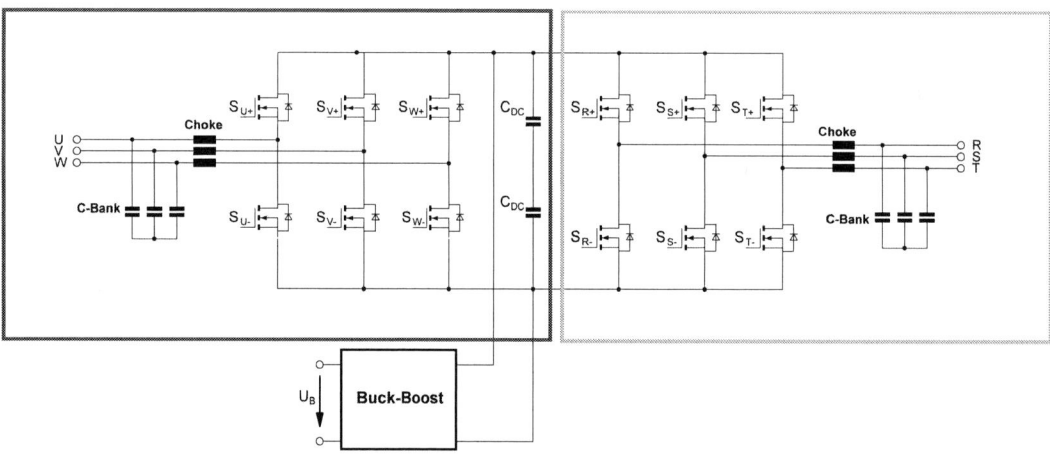

Figure 4 Equivalent circuit diagram of the proposed SiC-based UPS

The comparison results between Si-IGBT-based and SiC-MOSFET-based UPS are summarized in Figure 5 and Figure 6. Generating the same semiconductor losses, the SiC-MOSFET-based UPS can switch 5 times faster than the three-level Si IGBT solution. And the character value $L \cdot I^2$ of the choke can be reduced to 40 %, which means a volume reduction of magnetic components up to approx. 60 %.

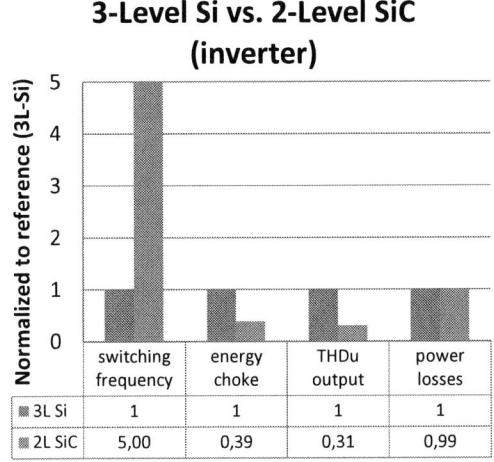

Figure 5 Comparison inverter normalized

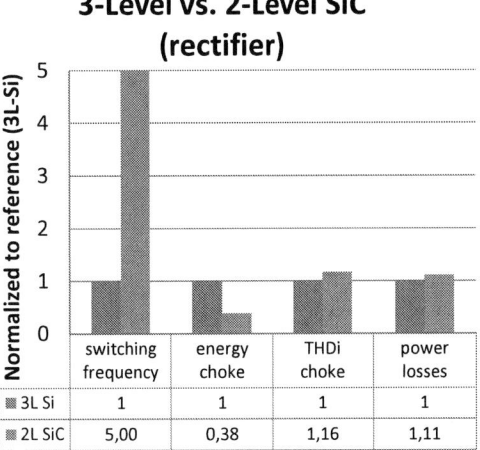

Figure 6 Comparison rectifier normalized

4. Hardware Test Bed for SiC MOSFET Module

For power circuit design and properly operation of the new semiconductor module, a first test setup is developed. Because of the very high switching frequency of the SiC component, low-inductive PCB design is applied to avoid high voltage overshoot (s. Figure 7). The major objective of the test setup is to characterize the SiC MOSFET module by means of double-pulse test method. Electrical parameters like gate resistance, snubber capacitance, etc. can be designed and optimized with the help of this test bed. High-frequency passive probes (100 MHz -3dB bandwidth) are applied for the gate-source and the drain-source voltage measurement. In order to measure the drain current, high-frequency Rogowski coil with a -3dB bandwidth up to 30 MHz is selected. Delay caused by the Rogowski coil is compensated in the evaluation of measurement data [11] [12] . An inductive load with the value of 70 µH is applied.

Table 3 Switching losses of Si IGBT module

	Datasheet	Calculated acc. IEC 60747-9
E_{on}	19 mJ	21.14 mJ
	(600 V 450 A)	(600 V 441 A)
E_{off}	26 mJ	28.18 mJ
	(600 V 450 A)	(600 V 437 A)

Figure 7 Low-inductive PCBA Test setup

In order to check the test setup a standard Si IGBT module (FF450R12KT4) is measured by means of double-pulse test method. The measurement result and the zoomed turn-on transient are given in Figure 8 and Figure 9, respectively. Measurement result evaluation according to standard IEC 60747-9 is given in Table 3, which shows a good agreement to the datasheet specification.

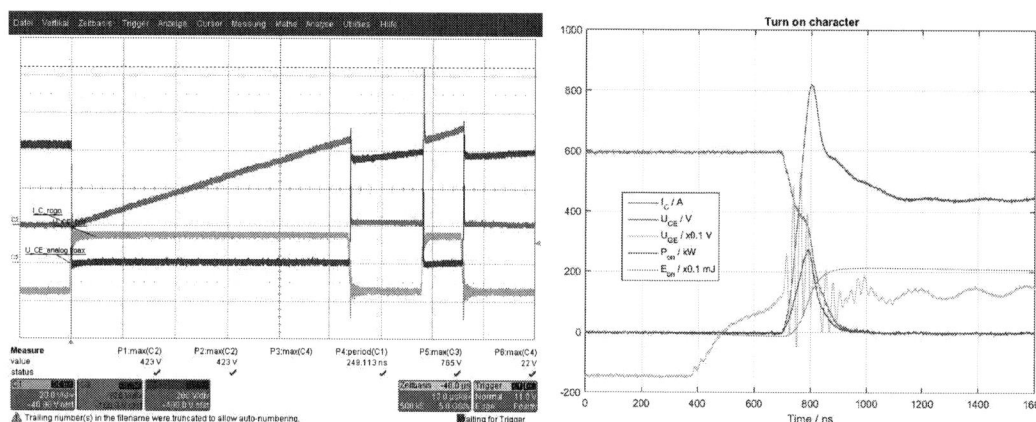

Figure 8 Scope Print operating point 600 V 450 A R_G 1 Ω 22 °C

Figure 9 Turn on transient of the Si IGBT

The double-pulse test for SiC MOSFET module is performed with the same test setup. The measurement result for transient are given in Figure 10 and Figure 11. Thanks to the negligible reverse recovery of the SiC diode, the current spike during the turn-on transient is significantly reduced. The oscillation during the switching transient is caused by the resonant circuit built by the parasitic inductance and capacitance. The calculated switching energy according to the standard IEC 60747-8 is given in Table 5, which show a relative large deviation (especially the turn-on losses) to the values provide in the datasheet. Root reasons of this deviation are still to be investigated. Table 4 shows also that different standards results different calculation results of the switching energy of the SiC MOSFET module. Due to the very high-frequency oscillation of current and voltage, a new standard for the fast-switching is desired.

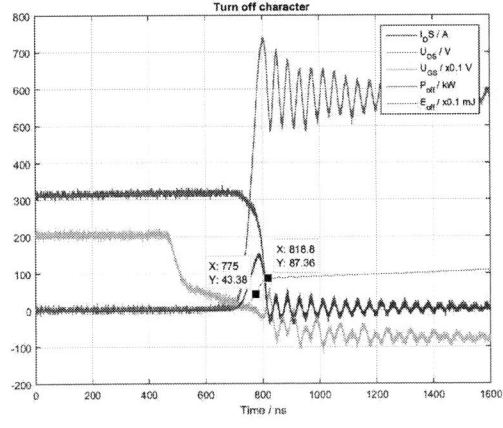
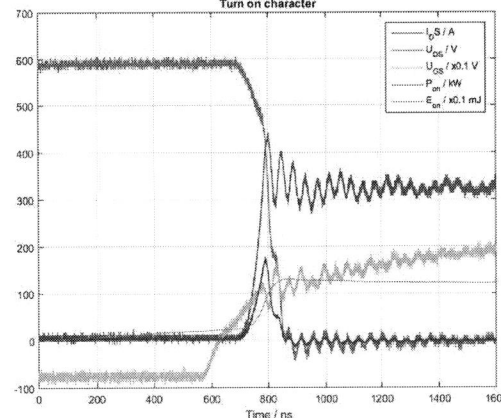

Figure 10 Turn off transient of the SiC MOSFET module (600 V 300 A)

Figure 11 Turn on transient of the SiC MOSFET module (600 V 300 A)

Table 4 Switching losses of SiC MOSFET module

	Datasheet	Calculation acc. (IEC 60747-8)	Calculation incl. oscillation
E_{on}	6.5 mJ (600 V 300 A)	13.19 mJ (600 V 324 A)	13.19 mJ
E_{off}	5.95 mJ (600 V 300 A)	4.55 mJ (600 V 313 A)	8.74 mJ

5. Conclusion and Future Works

In this contribution a feasibility study of a SiC MOSFET module for high-power UPS application is performed. Simulation results show that SiC MOSFET technique enables 5-times higher switching frequency without resulting higher semiconductor losses compared to the state-of-the-art Si IGBT technique. Hence, a volume reduction of magnetic components up to 60% can be achieved.

For characterizing and properly operation of the SiC MOSFET module, a hardware test bed is developed and a measurement setup is built. The test setup is first verified by a standard Si IGBT module, and then applied for characterizing the SiC MOSFET module. Measurement results show a relatively big deviation of switching losses between the measurement and the datasheet specifications. Root reason of this deviation is being investigated and final converter design and optimization will be performed in the future works.

As both the IGBT standard IEC 60747-9 and the MOSFET standard IEC 60747-8 are not suitable for high-current SiC MOSFET module, new standard is required for this kind of high-speed semiconductor module.

Acknowledgment

The research leading to these results has received funding from the program "IKT 2020 - Forschung für Innovation" of the Federal Ministry of Education and Research (German: Bundesministerium für Bildung und Forschung), abbreviated BMBF, under grant agreement number 16ES0205K.

Literature

[1] D. J. Liu, K. L. Wong and P. Kierstead, *Increase Efficiency and Lower System Cost with 100KHz, 10kW Sili- con Carbide (SiC) Interleaved Boost Circuit Design,* Cree, 2013.

[2] D. K. B. B. C. Wilhelm, *Development of a Highly Compact and Efficient Solar Inverter with,* Nuremberg: CIPS, 2010.

[3] M. S. S. W. B. W. a. J. K. J. Biela, *SiC vs. Si Evaluation of Potentials for Performance Improvement of Power Electronics Converter Systems by SiC Power Semiconductors,* Zurich: ETH Zurich, 2010.

[4] Semikron, "technical datasheet, SKM200GB12T4SiC," [Online]. Available: https://www.semikron.com/. [Accessed 08 03 2016].

[5] Infineon, "technical datasheet, FF600R12IS4F," [Online]. Available: http://www.infineon.com/. [Accessed 08 03 2016].

[6] Mitsubishi, "technical datasheet, CMH1200DC-34S," [Online]. Available: http://www.mitsubishielectric.com/. [Accessed 08 03 2016].

[7] Cree, "technical datasheet, CAS300M17BM2," [Online]. Available: http://www.cree.com/. [Accessed 03 03 2016].

[8] Rohm, "Technical Datasheet, BSM300D12P2E001," [Online]. Available: http://rohmfs.rohm.com/. [Accessed 09 03 2016].

[9] Semikron, "technical datasheet, SKM500MB120SC," [Online]. Available: https://www.semikron.com/. [Accessed 09 03 2016].

[10] P. D.-I. J. Böcker, *Leistungselektronik,* Paderborn, 2014.

[11] Athena Energy, "Application Note AE-010," [Online]. Available: http://www.athenaenergycorp.com/. [Accessed 10 03 2016].

[12] Power Electronic Measurements Ltd, "Technical notes - 001," [Online]. Available: http://www.pemuk.com/.

Optimization of 12 and 18 pulse rectifier systems by the selection of optimum parameters for magnetics

Kapila Warnakulasuriya, Carroll & Meynell Transformers Ltd, UK,kapila@carroll-meynell.com
Farhad Nabhani, Teesside University, United Kingdom, F.Nabhani@tees.ac.uk
Vahid Askari, Teesside University, United Kingdom, v.askari@tees.ac.uk

Abstract

In this paper an approach is made to arrive at optimum designs for 12 and 18pluse rectifier systems by defining the optimum parameters and configurations of magnetics. The approach is based on the definition of the parameters for magnetics considering the system impact on actual magnetics designs and the performance of the system with the set parameters of magnetic components. The several possible situations of unbalanced supply are also considered and optimum magnetic parameters to minimize the effect of the supply imbalance are discussed. Based on a series of simulations which were verified by practical testing a quantitative explanation of the size of the magnetics against the quality of the output and robustness to supply imbalances is presented. This explanation gives a measure on the degree of compromise that can be made on magnetics and the quality of the output. Situations where the magnetic designs are made larger or the requirement of using special magnetic materials that make the magnetic components more expensive are also discussed.

1. Introduction

As technology grows, the study of power systems has shifted its direction to power electronics to produce the most efficient energy conversion [1]. Large harmonics, poor power factor and high total harmonic distortion (THD) in the utility interface are common problems when nonlinear loads such as adjustable speed drives. Power supplies, induction heating systems, UPS systems and aircraft converter systems are connected to the electric utility [2]. In most power electronics applications, diode rectifiers are commonly used in the front end of power converter as an interface with the electrical utility. The nonlinear operation of the diode bridge rectifiers causes highly distorted input current. The non-sinusoidal shape of the input current drawn by the rectifiers causes a number of problems in the sensitive electronic equipment [3]. A number of methods have been proposed for harmonics reduction in utility line currents. A conventional 12-pulse diode bridge rectifier results in 5th & 7th harmonics cancellation in utility line current. Many multi-pulse converters have been introduced to achieve clean power such as 12-pulse 18-pulse & 24-pulse systems. These multi-pulse converters are formed by combination of 6-pulse converters & isolation transformer [4].

Multi-pulse rectifier systems contain a front end poly phase transformer. When it is required to achieve galvanic isolation this has to be made as an isolation transformer, otherwise a poly phase auto transformer can be used which gives significant advantage on the size of the transformer. This point is described in the system with actual design details.

Two or three of three phase rectifier bridges depending 12 pulse or 18 pulse system are connected via an inter-phase balancing reactor. Improper selection of inductance value for the interface transformer can result in unnecessarily larger unit or a unit that is subjected to higher order harmonics. Therefor considering the actual magnetic design in deciding the values of inter phase reactors give an additional dimension for the optimization of multi pulse rectifier systems.

PCIM Europe 2016, 10 – 12 May 2016, Nuremberg, Germany

A series of simulations were carried out with different inter phase transformer reactance values and respective magnetic designs were carried out based on the inductance value and the waveform that these magnetic components are subjected at those particular values of inductance.

Simulation results were verified with a practical prototype built and the conclusions are presented in this paper. Further the effect of an unbalanced three phase input is evaluated for the different interphase reactance values considered above and the results were verified with the prototype built up.

2. Magnetics selection

2.1. Poly phase transformer selection.

2.1.1. Isolation transformer vs Auto transformer

It is required to use an isolation transformer if the system requires the galvanic isolation. However depending on the input output voltage ratio auto-transformer can give a significant weight and size advantage.

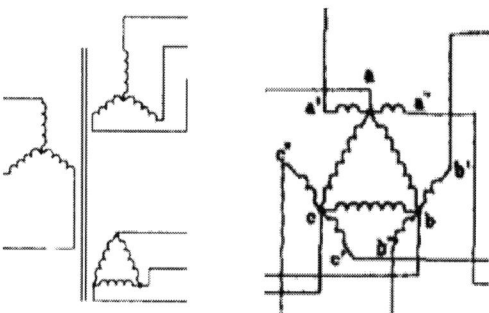

Fig. 1. Isolation and Auto poly phase transformer

2.1.2. Delta vs Star transformer

Another consideration in designing a 12pulse system is the section between the delta and star confirmation.

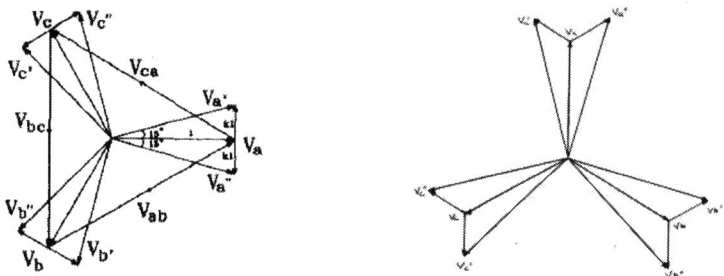

Fig. 2. Vector diagram for Delta and Star Auto poly phase transformers

The selection of an Auto Delta transformer can result in over 35% smaller transformer compared to the selection of an Auto Star transformer. Below table makes a comparison of

© VDE VERLAG GMBH · Berlin · Offenbach

the sizes of an auto Delta 12 pulse transformer and an auto star 12pulse transformer based on actual designs carried out.

Transformer weight(kg) / power	10kVA	50kVA	500kVA
Auto Delta Transformer	33.5	105	615
Auto Star Transformer	45	163	990
Percentage saving in weight in Delta Configuration	25.5%	35.5%	37.8%

Fig. 3. Weight Comparison of Delta and Star Transformers in 12 Pulse Rectifications

It could be seen from the above table that the weight advantage of the star configuration become significant as the power increases.

2.2.1. Interphase transformer for 12 pulse system

In order to active the high power requirements the parallel operation the two six pulse rectifier converter systems that form the 12 pulses is required. In this parallel operation of two systems which have waveforms with a phase displacement of 30^0 proper consideration has to be made to address any voltage variations. These voltage variations can result in one of the consisting 6 pulse systems getting overloaded causing them to the thermal runway situations and other failures. This situation can be effectively addressed by the use of interphase rectifiers which ae also called as interphase transformers.

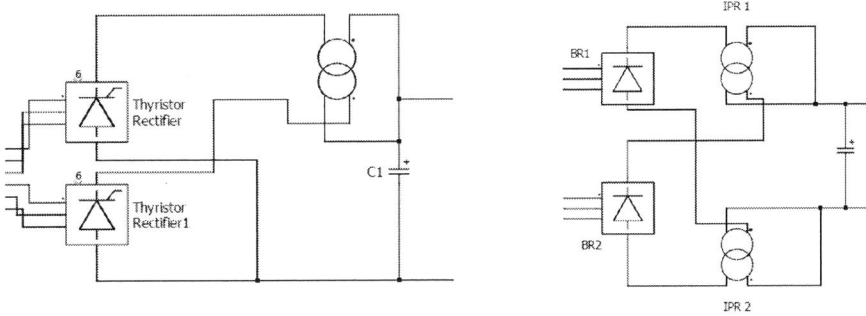

Fig. 4. Parallel connection of two 6 pulse converters with the use of one and two interphase reactors

As shown in the above Fig.3 this can be archived with a use of one or two interphase reactor arrangements.

2.2.2. Interphase transformer for 18 pulse system

18 pulse systems are also used in many applications such as heating, air conditioning etc. These systems are achieved by the parallel or series connection of three 6 pulse converter systems. In the series configuration three, six-pulse systems having a third of the output DC

voltage are connected in series. The series connection does not have the issues associated with current sharing and a requirement of having an interspace reactor. However this has the disadvantage of the requirement of having high current capability rectifier systems. 18pulse systems with parallel connected 6 pulse rectifier systems require appropriate interphase reactors and these con be connected in a number of configurations

3. Simulations

During the study a number of simulations were carried out for different conditions of the supply voltages. Balanced and unbalanced supply voltages are considered and the performances of the system under these conditions are evaluated for different interface reactor values. The changes in the interphase current waveform under different interphase reactor vales were studied and impact of these current on the actual magnetic design of such interphase reactors were investigated.

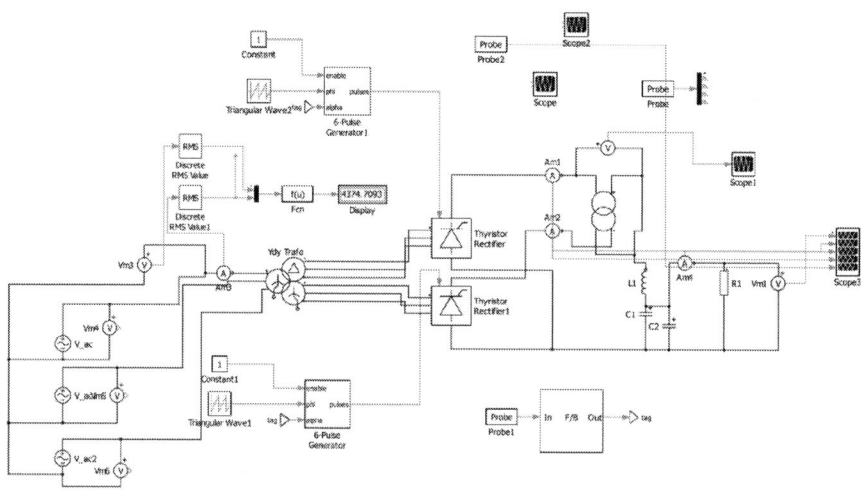

Fig. 5. Simulation arrangement

3.1 Simulations of a balanced three-phase input

Simulations were carried with the inter phase reactor values of 50µH, 200µH and 800µH with a balanced three phase input. The harmonic content of the inter phase reactor currents were studied

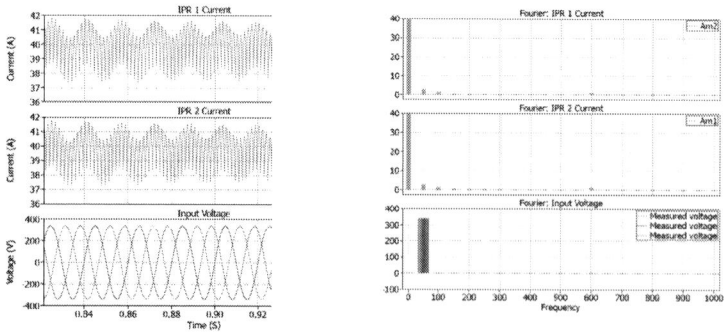

Fig. 6. Simulation results with a 50µH interphase reactor

PCIM Europe 2016, 10 – 12 May 2016, Nuremberg, Germany

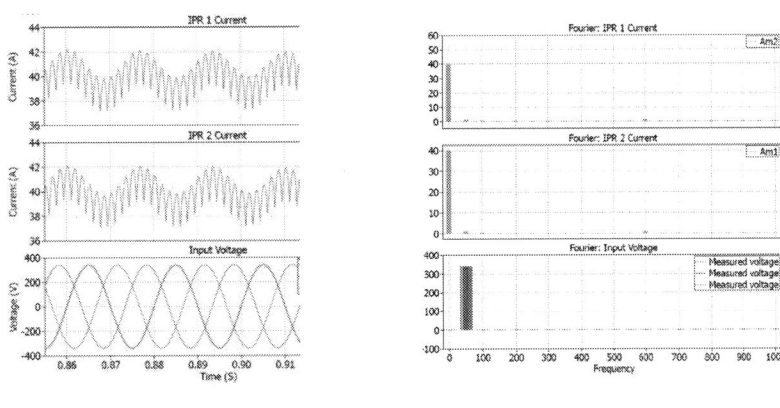

Fig. 7. Simulation results with a 200µH interphase reactor

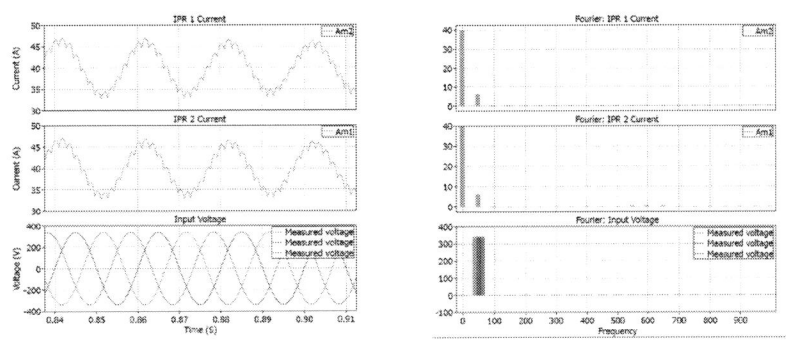

Fig. 8. Simulation results with an 800µH interphase reactor

3.2 Simulations of an unbalanced three-phase input

The performance of the system was evaluated with the interphase reactor values of 50µH, 200µH, 600µH and 800µH when the three phase input is unbalanced. The harmonic content of the inter phase reactor currents were studied.

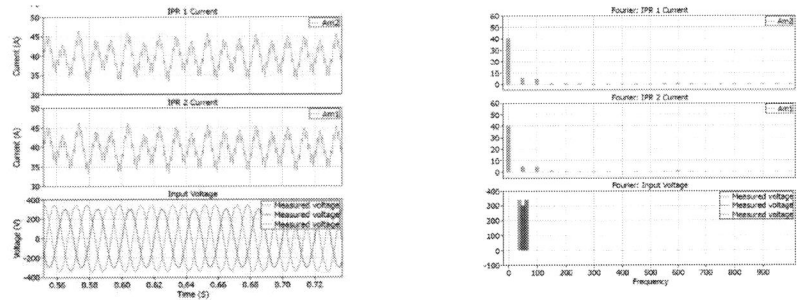

Fig. 9. Simulation results with a 50µH interphase reactor with unbalanced input

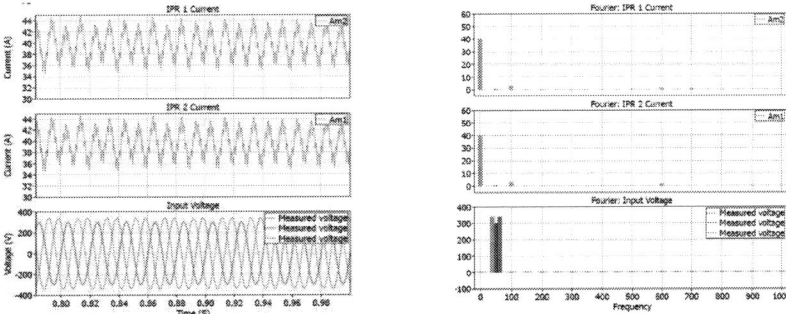

Fig. 10. Simulation results with a 200µH interphase reactor with unbalanced input

© VDE VERLAG GMBH · Berlin · Offenbach

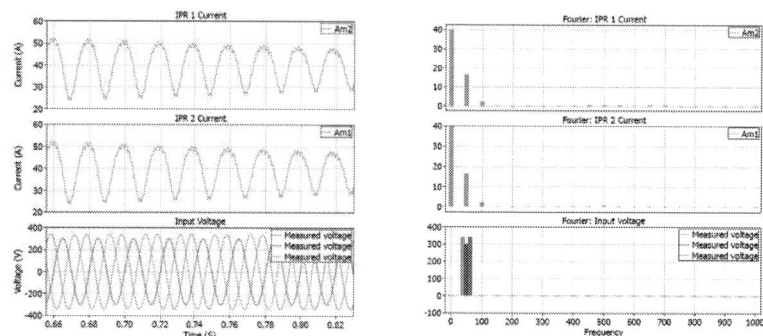

Fig. 11. Simulation results with a 600µH interphase reactor with unbalanced input

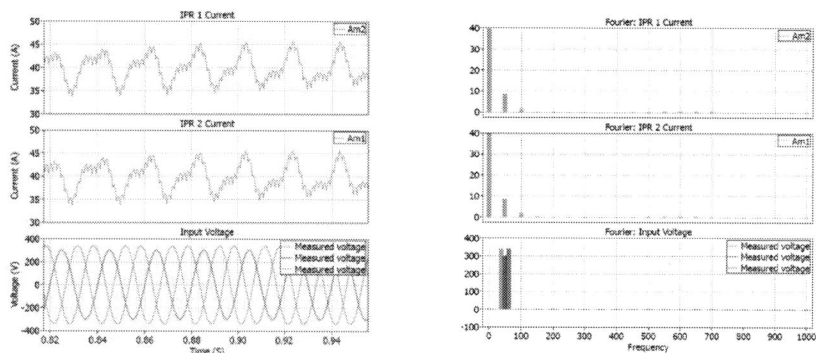

Fig. 12. Simulation results with an 800µH interphase reactor with unbalanced input

4. Prototype development and testing

After the detailed analysis of the situation with the use of simulation tools two prototype interphase reactors were built. These were constructed such that the inductance of each inter phase reactor can be adjusted to the values of 50µH, 200µH, 600µH and 800µH by adjusting the air gap thickness in the magnetic path. These were developed to be suitable for a 2kW, 12pulse rectifier system. In designing the interphase reactors appropriate selection of magnetic materials was made based on their exciting waveforms obtained through simulations. As it could be seen from the waveforms and their Fourier spectrums the presence of the higher order harmonic currents content was not very high. Therefor the requirement of using high performance magnetic core materials was not identified.

However attention was given to the increase in the physical size and the weight of the interphase reactors with the increased values of their inductance in contrary to contribution given by the high inductance for the reduction of harmonic content.

The 12pulse rectifier system was tested with these interphase reactors for each of the above mentioned inductance values with a balanced three phase input voltage. Same tests were carried out for the system with a three phase input supply of unbalanced conditions. The phase imbalance conditions were achieved with the use of a three phase variable transformer. Obtained results were recorded and compared with the simulated results.

© VDE VERLAG GMBH · Berlin · Offenbach

Fig.13. Prototype and testing arrangement

5. Discussion and Conclusions

Magnetic components play a vital role in any power electronic system. Usually they become the bulkiest components in a system and in most occasions they become a major contributor for the cost of the system. This study was carried out with the intention identifying the optimum parameters for the magnetic components used in the multi-pulse rectifier systems such as 12pulse and 18pulse systems. In the case of designing multi-pulse rectifier systems it is a common practice in the industry to finalise the parameters for magnetic components based on the desirable output conditions and with a rough understanding of the physical size based on equivalent power rating.

In this study an approach is made to add an additional dimension for the multi-pulse system optimization. That is by taking the actual magnetic designs into consideration in defining the parameters for the vital magnetic components such as interphase reactors, rather than deciding them only based on the equivalent power approach. A series of simulations were carried out and the current waveforms through the inter phase reactor was analysed. The test results obtained on a 2kW, 12 pulse rectifier system showed the simulation results were very much in line the actual values obtained for the current waveforms.

The considered 2kW, 12pulse system showed reasonable performance in the simulation as well as in the actual testing even with a 50µH inductance inter phase reactor. Actual magnetic design showed a weight less than a 1kg in such an inter phase reactor. However, when the unbalanced supply situation was considered it could be seen that 100Hz and 600Hz components of the current wave form increase. Though an actual magnetic design of a considerably small size can be active even with this, increased harmonic content especially in high power application, it is advisable to increase the inductance and reduce the harmonic content which enables a reduction in high frequency losses in magnetic components.

The increase the inductance up to about 200µH resulted in a further reduction in the harmonic content of the current and this situation could be seen both in the simulation and the actual measurements. This results in an actual component size of about little over 2kg for the system under consideration. Further increase of the interphase reactor values up to 600µH and 800µH resulted in a considerable drop in the 600Hz and higher harmonic content. However increase of actual size of the interphase reactor makes it less attractive to increase the inductance values to this level.

Based on the series of testing and simulations carried out it could be concluded that an optimum value of inductance can be achieved for a system and it is advisable to do it by taking the actual designs of magnetic components. A smaller value that may still give the required total harmonic distortion can result in a higher high frequency content in the inter phase rector current making it more loss making and on the other hand a larger value will result in a much bigger component size. For the considered example of 2kW,12 pulse rectifier it could be seen that the best compromise is archived little over 200μH level considering both system performance and actual magnetic designs.

The authors have identified the requirement of continuing the study for several other configurations of 12 and 18 pulse system and for several power levels and arrive at table of optimum interphase reactor values that gives the best system performance and optimum magnetic components designs and it is expected to carry out this as part of future research work.

References

[1]Hernadi, A.; Taufik; Anwari, M., "Modeling and Simulation of 6-Pulse and 12-Pulse Rectifiers under Balanced and Unbalanced Conditions with Impacts to Input Current Harmonics," in Modeling & Simulation, 2008. AICMS 08. Second Asia International Conference on , vol., no., pp.1034-1038, 13-15 May 2008

[2]Sewan Choi; von Jouanne, A.R.; Enjeti, P.N.; Pitel, I.J., "Polyphase transformer arrangements with reduced kVA capacities for harmonic current reduction in rectifier type utility interface," in Power Electronics Specialists Conference, 1995. PESC '95

[3]Anandpara, M.; Panchal, T.; Patel, V., "An active interphase transformer for 12-pulse rectifier system to get the performance like 24-pulse rectifier system," in Power Systems Conference (NPSC), 2014 Eighteenth National , vol., no., pp.1-6, 18-20 Dec. 2014

[4]Lee, B.S.; Hahn, J.; Enjeti, P.N.; Pitel, I.J., "A robust three-phase active power-factor-correction and harmonic reduction scheme for high power," in Industrial Electronics, IEEE Transactions on , vol.46, no.3, pp.483-494, Jun 1999

[5] Karnath, G.R.; Benson, D.; Wood, R., "A novel autotransformer based 18-pulse rectifier circuit," in Applied Power Electronics Conference and Exposition, 2002. APEC 2002. Seventeenth Annual IEEE , vol.2, no., pp.795-801 vol.2, 2002

[6] Salmon, J.; Roberge, I., "Performance assessment of 18-pulse 3-phase rectifiers using harmonic reducing auto-transformers," in Electrical and Computer Engineering, 2005. Canadian Conference on , vol., no., pp.637-640, 1-4 May 2005

[7]Yii-Shen Tzeng; Nanming Chen; Ruay-Nan Wu, "Modes of operation in parallel-connected 12-pulse uncontrolled bridge rectifiers without an interphase transformer," in Industrial Electronics, IEEE Transactions on , vol.44, no.3, pp.344-355, Jun 1997

[8] Miyairi, Shota; Iida, S.; Nakata, Kiyoshi; Masukawa, S., "New Method for Reducing Harmonics Involved in Input and Output of Rectifier with Interphase Transformer," in Industry Applications, IEEE Transactions on , vol.IA-22, no.5, pp.790-797, Sept. 1986

[9]Kim, S.; Enjeti, P.; Rendusara, D.; Pitel, I.J., "A new method to improve THD and reduce harmonics generated by a three phase diode rectifier type utility interface," in Industry Applications Society Annual Meeting, 1994., Conference Record of the 1994 IEEE

[10]Yasuyuki Nishida, "A 12-pulse diode rectifier using 3-phase bridge 6-pulse diode rectifier with 2 additional diodes and an auto-transformer," in Power Electronics and Drive Systems, 1999. PEDS '99. Proceedings of the IEEE 1999 International Conference on , vol.1, no., pp.75-79 vol.1, 1999

[11]Basic, D.; Ramsden, V.S.; Muttik, P.K., "Harmonic filtering of high-power 12-pulse rectifier loads with a selective hybrid filter system," in Industrial Electronics, IEEE Transactions on , vol.48, no.6, pp.1118-1127, Dec 2001

Analysis of the Flyback Converter Utilizing a Transformer with Stepped Air-Gap

Panagiotis Mantzanas, panagiotis.mantzanas@fau.de
Daniel Kübrich, daniel.kuebrich@fau.de
Markus Barwig, markus.barwig@fau.de
Thomas Dürbaum, thomas.duerbaum@fau.de

Chair of Electromagnetic Fields, University of Erlangen-Nuremberg, Germany

Abstract

Saturable inductors are employed in various power electronics applications. For instance, ref. [1] deals with the use of a stepped air-gap inductor in a Buck converter leading to increased efficiency. This paper demonstrates the analytical calculation of the flyback converter with a stepped air-gap transformer, which enables a first estimation of converter losses. Thus, the influence of a stepped air-gap transformer on converter efficiency can be examined. Measurements have confirmed that this innovative topology leads to a total loss reduction of up to 6.4%. Consequently, the efficiency can be increased noticeably in comparison to a conventional flyback converter.

1. Introduction

Traditionally, flyback converters employ a linear transformer. In combination with a wide input voltage range, however, the use of a linear transformer constitutes a drawback as a high saturation current is required. This results either in a low magnetizing inductance value or in a large number of turns, which in turn implies a low efficiency. Moreover, the calculations included in this paper show that a stepped air-gap transformer with identical size and number of turns possesses a higher magnetizing inductance in the unsaturated state than the conventional transformer. Consequently, the stepped air-gap transformer could be used to increase the efficiency of the flyback converter in a certain voltage and load range. This paper deals with the analytical calculation of the flyback converter with a stepped air-gap transformer.

2. Stepped Air-Gap Transformer

A non-linear inductor can be characterized by its effective inductance [2]:

$$u_L(t) = L_\Delta\big(i_L(t)\big)\frac{di_L(t)}{dt}. \tag{1}$$

Measurements have demonstrated that the effective inductance of a stepped air-gap inductor (Fig. 1), can be approximated by the piecewise constant function depicted in Fig. 2 [3]. If the current through the inductance i_L is smaller than i_{1s}, no saturation occurs leading to a great value for the inductance. For currents greater than i_{1s} the cylindrical part with the radius of r_1 and the height of $l_{g2} - l_{g1}$ saturates causing an abrupt decline in inductance. If the current i_L exceeds i_{2s}, the center leg with the radius r begins to saturate. Thus, the inductance becomes extremely small. The center leg of the core saturates if

$$B_{center\_leg}(i = i_{2s}) = B_{sat}. \tag{2}$$

© VDE VERLAG GMBH · Berlin · Offenbach

PCIM Europe 2016, 10 – 12 May 2016, Nuremberg, Germany

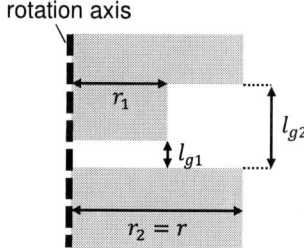

Fig. 1. Center leg with a stepped air-gap

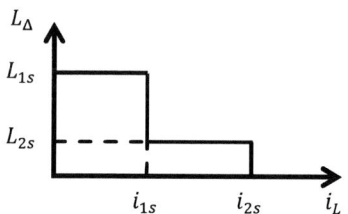

Fig. 2. Effective inductance

Using the approximation from Fig. 2 as well as (2), it can be shown that the inductance values and the saturation currents of the stepped air-gap transformer are linked by the expression

$$B_{sat} = \frac{L_{1s}i_{1s} + L_{2s}(i_{2s} - i_{1s})}{N_p A_{min}}, \qquad (3)$$

where N_p represents the number of turns on the primary side, $A_{min} \neq \pi r_1^2$ represents the minimum cross-sectional area of the core and B_{sat} the saturation flux density of the core material. A_{min} and B_{sat} can be found in the datasheet of the core type and core material, respectively.

3. Analysis of the Flyback Converter

Fig. 3 shows the basic schematic of the flyback converter. In order to enable an analytical calculation, some simplifications have to be introduced. In the first step, any parasitic resistances as well as parasitic capacitances and inductances are neglected. Furthermore, the leakage inductance of the transformer is not taken into account. Additionally, the input and the output voltage are assumed to be constant. Moreover, the boundary conduction mode (BCM) with valley switching and BCM with valley skipping [4], [5] is approximated by classical BCM and by discontinuous conduction mode (DCM), respectively.

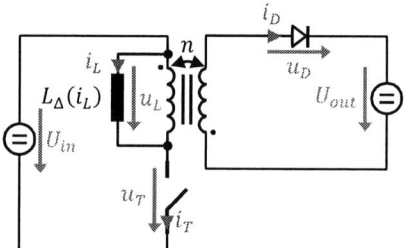

Fig. 3. Basic schematic of the flyback converter

Based on those simplifications, the voltage u_L can be derived and the current waveforms can be calculated by solving (1). In contrast to the conventional flyback converter, the magnetizing inductance of the flyback converter with a stepped air-gap transformer is current-dependent. Thus, (1) represents a nonlinear equation. However, using the approximation depicted in Fig. 2, an analytical calculation of i_L is possible by splitting the calculation in several subintervals in which the magnetizing inductance is constant. Depending on the current i_{max}, two different solutions for the current i_L are obtained, which are illustrated in Fig. 4 and Fig. 5.

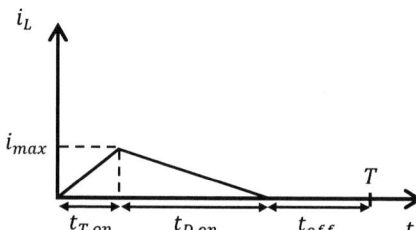

Fig. 4. Magnetizing current i_L for $i_{max} \leq i_{1s}$: No saturation occurs

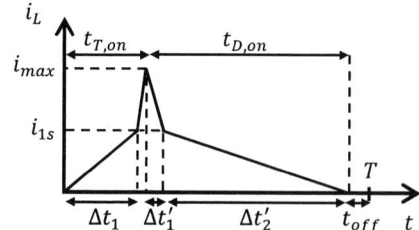

Fig. 5. Magnetizing Current i_L for $i_{max} > i_{1s}$: Saturation occurs for $i_L > i_{1s}$

© VDE VERLAG GMBH · Berlin · Offenbach

Similar to the conventional flyback converter, it can be shown that the on-time of the MOSFET and the diode are linked by

$$t_{D,on} = \frac{U_{in}}{nU_{out}} t_{T,on},$$

(4)

where n represents the transmission ratio of the transformer. Since the efficiency of a flyback converter decreases with increasing leakage inductance [6], a highly coupled transformer is desired. In this case, the transmission ratio n is nearly equal to the turns ratio:

$$n \approx \frac{N_p}{N_s}.$$

(5)

N_p and N_s represent the number of turns on the primary and secondary side, respectively.
The flyback converter can operate in discontinuous conduction mode (DCM), boundary conduction mode (BCM) or continuous conduction mode (CCM). This paper focuses on the analysis in DCM and BCM. The next two subsections describe how the current waveforms as well as their rms values can be calculated in DCM and BCM.

3.1. Discontinuous conduction mode (DCM)

In DCM the switching frequency $f_{S,DCM}$ and the period time T_{DCM} are fixed and do not depend on the operating point. In this paper, DCM is used as an approximation for BCM with valley skipping [4], [5]. The fixed switching frequency $f_{S,DCM}$ varies from controller to controller and is specified in the datasheet [5].
The maximum current i_{max} through the magnetizing inductance can be calculated to

$$i_{max} = \begin{cases} \sqrt{\dfrac{2T_{DCM}P_{out}}{L_{1s}}}, & P_{out} \leq \dfrac{L_{1s}i_{1s}^2}{2T_{DCM}} \\[3ex] \sqrt{\dfrac{2T_{DCM}P_{out} - L_{1s}i_{1s}^2}{L_{2s}} + i_{1s}^2}, & P_{out} > \dfrac{L_{1s}i_{1s}^2}{2T_{DCM}}, \end{cases}$$

(6)

where P_{out} represents the output power. The inductances L_{1s} and L_{2s} as well as the saturation current i_{1s} are defined in Fig. 2. The on-time of the MOSFET can be determined to

$$t_{T,on} = \begin{cases} \dfrac{i_{max}L_{1s}}{U_{in}}, & i_{max} \leq i_{1s} \\[3ex] \dfrac{(i_{max} - i_{1s})L_{2s} + i_{1s}L_{1s}}{U_{in}}, & i_{max} > i_{1s}. \end{cases}$$

(7)

Once the on-time of the MOSFET $t_{T,on}$ is known, (4) can be used to calculate the on-time of the diode $t_{D,on}$. The rms values of the currents through the MOSFET and the diode are given by

$$i_{T,rms} = \begin{cases} i_{max}\sqrt{\dfrac{t_{T,on}}{3T_{DCM}}}, & i_{max} \leq i_{1s} \\[3ex] \sqrt{\dfrac{1}{3T_{DCM}}\left[i_{1s}^2\Delta t_1 + (i_{1s}^2 + i_{1s}i_{max} + i_{max}^2)(t_{T,on} - \Delta t_1)\right]}, & i_{max} > i_{1s} \end{cases}$$

(8)

$$i_{D,rms} = \begin{cases} ni_{max}\sqrt{\dfrac{t_{D,on}}{3T_{DCM}}}, & i_{max} \leq i_{1s} \\[4mm] n\sqrt{\dfrac{1}{3T_{DCM}}[(i_{1s}^2 + i_{1s}i_{max} + i_{max}^2)\Delta t_1' + i_{1s}^2\Delta t_2']}, & i_{max} > i_{1s} \end{cases} \tag{9}$$

with

$$\Delta t_1 = \frac{i_{1s}L_{1s}}{U_{in}} \tag{10}$$

$$\Delta t_1' = \frac{(i_{max} - i_{1s})L_{2s}}{nU_{out}} \tag{11}$$

$$\Delta t_2' = \frac{i_{1s}L_{1s}}{nU_{out}}. \tag{12}$$

Operation in DCM is only possible if the condition

$$t_{T,on} + t_{D,on} \leq T_{DCM} \tag{13}$$

is fulfilled (Fig. 4 or Fig. 5). Thus, the maximum MOSFET on-time in DCM can be calculated by using (4) and (13) to

$$t_{T,on,DCM,max} = \frac{nU_{out}}{U_{in} + nU_{out}}T_{DCM}. \tag{14}$$

If the MOSFET on-time $t_{T,on}$ exceeds the maximum on-time $t_{T,on,DCM,max}$, the converter operates in boundary conduction mode (BCM). BCM is used in this paper as an approximation for BCM with valley switching (or quasi-resonant operating mode according to [5]).

3.2. Boundary conduction mode (BCM)

The equations given in the previous subsection are also valid in BCM. By replacing the period time T_{DCM} with T_{BCM}, those equations can be used to calculate the waveforms in BCM. Nevertheless, in BCM the period time T_{BCM} is not fixed but depends on the operating point. The period time in BCM can be obtained from

$$T_{BCM} = \begin{cases} \dfrac{2P_{out}L_{1s}(U_{in} + nU_{out})^2}{(nU_{out}U_{in})^2}, & P_{out} \leq \dfrac{L_{1s}i_{1s}^2nU_{out}}{2(U_{in} + nU_{out})\Delta t_1} \\[5mm] \dfrac{-K_2 \pm \sqrt{K_2^2 - 4K_1K_3}}{2K_1}, & P_{out} > \dfrac{L_{1s}i_{1s}^2nU_{out}}{2(U_{in} + nU_{out})\Delta t_1} \end{cases} \tag{15}$$

with

$$K_1 = \frac{(nU_{in}U_{out})^2}{2L_{2s}(U_{in} + nU_{out})^2} \tag{16}$$

$$K_2 = \frac{i_{1s}U_{in}nU_{out}}{(U_{in} + nU_{out})} - \frac{\Delta t_1 nU_{out}U_{in}^2}{L_{2s}(U_{in} + nU_{out})} - P_{out} \tag{17}$$

$$K_3 = \frac{L_{1s}i_{1s}^2}{2} + \frac{(U_{in}\Delta t_1)^2}{2L_{2s}} - i_{1s}U_{in}\Delta t_1. \tag{18}$$

It should be noted that only one solution of the second line of (15) fulfills the condition

$$\left(1 + \frac{U_{in}}{nU_{out}}\right)\Delta t_1 < T_{BCM} < \left(1 + \frac{U_{in}}{nU_{out}}\right)\left(\Delta t_1 + \frac{(i_{2s} - i_{1s})L_{2s}}{U_{in}}\right) \tag{19}$$

and is therefore the right solution for the period time.

For a given operating point (U_{in}, U_{out} and P_{out}) the current waveforms as well as the rms values of the flyback converter with stepped air gap can be calculated in DCM and in BCM by using the equations given in this section. These results can be used to calculate the losses and estimate the converter efficiency.

4. Evaluation

The calculations presented in this paper show that a stepped air-gap transformer with identical size and number of turns possesses a higher magnetizing inductance in the unsaturated state than a conventional transformer. Accordingly, the advantage of higher inductance can be used in a certain voltage and load range. Fig. 6 and Fig. 7 depict a comparison between the flyback converter with stepped air-gap transformer and the conventional flyback converter. For both transformers the following data have been used: $N_p = 44$, $N_s = 7$, $A_{min} = 55.4\,\mathrm{mm}^2$ and $B_{sat} = 385$ mT. The frequency in DCM has be chosen to $f_{S,DCM} = 125$ kHz [5]. The indices **s** and **c** indicate whether a stepped air-gap or a conventional air-gap is employed. The output voltage is $U_{out} = 20\,\mathrm{V}$. The conventional transformer is designed in such a way that no saturation occurs for input voltages higher than the minimum input voltage $U_{in,min} = 60\,\mathrm{V}$ at full load ($P_{out,max} = 65\,\mathrm{W}$). Consequently, the conventional transformer does not saturate over the full voltage and load range. The stepped air-gap transformer has been designed (for each λ_2-value) in such a way that the first step saturates ($i_{max} > i_{1s}$) when the input voltage is smaller than the nominal input voltage $U_{in,nom} = \sqrt{2} \cdot 230\,\mathrm{V}$ at full load. The second step saturates ($i_{max} > i_{2s}$) when the input voltage decreases under $U_{in,min} = 60\,\mathrm{V}$ at full load. The converter specification has been chosen on the basis of a commercial notebook power supply. The case of minimum input voltage $U_{in,min}$ occurs at mains dips operation.

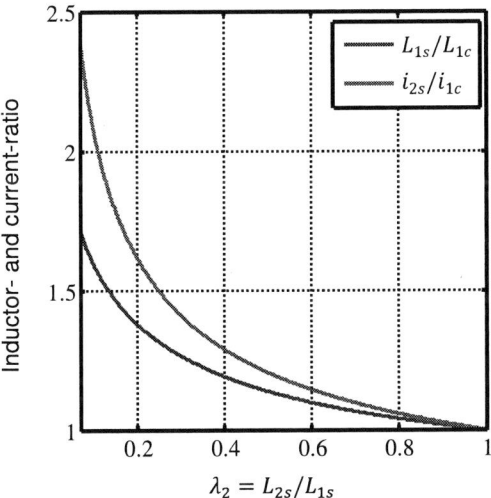

Fig. 6. Comparison of the stepped air-gap flyback converter with the conventional

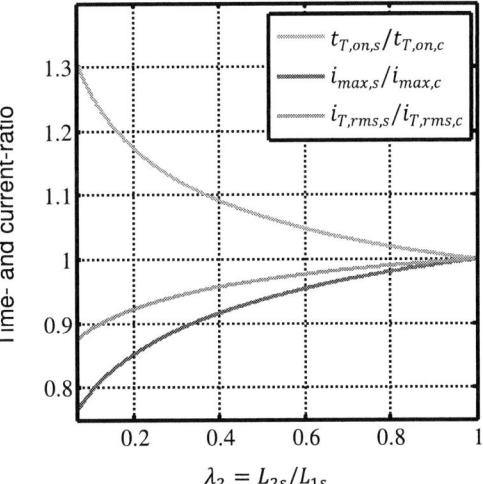

Fig. 7. Comparison of the stepped air-gap flyback converter with the conventional for $U_{in} = U_{in,nom}$ and $P_{out} \leq P_{out,max}$ (DCM, $f_S = 125$ kHz)

As mentioned above, Fig. 6 illustrates that the magnetizing inductance L_{1s} of the stepped-air gap transformer in the unsaturated state is higher than the inductance L_{1c} of the conventional transformer. Furthermore, the second saturation current i_{2s} of the stepped air-gap transformer is higher than the saturation current i_{1c} of the conventional transformer. This is due to the fact that at minimum input voltage and full load the first step of the stepped air-gap transformer saturates. This results in a lower magnetizing inductance than the conventional transformer and therefore in a higher current. Fig. 7 compares the MOSFET on-time, the maximum current through the magnetizing inductance as well as the rms value of the MOSFET current for operating points $U_{in} = U_{in,nom}$ and $P_{out} \leq P_{out,max}$. Calculations show that the converter operates in DCM at those operating points. Furthermore, at these operating points no saturation occurs in the stepped air-gap transformer. As a result, the waveform of the current i_L in case of a stepped air-gap is similar to that one of the conventional flyback converter (Fig. 4). However, the stepped air-gap transformer possesses a higher magnetizing inductance in the unsaturated state than the conventional transformer. A higher value for the magnetizing inductance results in higher values for $t_{T,on}$ and lower values for i_{max}, $i_{T,rms}$ and $i_{D,rms}$. Lower values for i_{max}, $i_{T,rms}$ and $i_{D,rms}$ imply lower winding losses in the transformer, lower MOSFET and diode conduction losses as well as lower losses in the clamping circuit [6]. Furthermore, higher values for $t_{T,on}$ result in lower values for t_{off} (duration of the third interval in Fig. 4). Consequently, the high frequency oscillation occurring in this subinterval causes less losses in the transformer [4]. Moreover, the frequency of this oscillation is lower due to the higher value of the magnetizing inductance. Unfortunately, the specific core losses in the transformer rise with increasing values for $t_{T,on}$ [7]. Nevertheless, measurements have demonstrated that the employment of a stepped air-gap transformer noticeably reduces the losses of the flyback converter. Apparently, the reduction of the losses mentioned above outweighs the increase of the specific core losses in this investigated design.

According to Fig. 7, most converter losses decrease with a decreasing value for λ_2. Hence, a low value for λ_2 is desired for high converter efficiency. However, a small value for λ_2 leads to a low value for the inductance L_{2s}. Therefore, very small values for λ_2 result in extremely high currents when the first step of the stepped air-gap transformer saturates ($i_{max} > i_{1s}$). These high currents cause increased losses which could lead to thermal instability of the converter. For further investigations $\lambda_2 = 0.3$ is chosen. Table 1 lists the resulting inductances and saturation currents for both the

L_{1c}	293.2 µH
i_{1c}	3.2 A
L_{1s}	371.3 µH
L_{2s}	111.4 µH
i_{1s}	1.67 A
i_{2s}	4.52 A

Table 1: Calculated inductances and saturation currents

conventional and the stepped air-gap transformer. Fig. 8 to Fig.12 show a comparison between the flyback converter with stepped air-gap transformer and the conventional flyback converter as a function of the output power and for different input voltages. While solid lines represent the flyback converter with stepped air-gap transformer, dashed lines indicate the conventional flyback converter. It should be noted that for $U_{in} = \sqrt{2} \cdot 230$ V, the switching frequency of both converters is 125 kHz over the full load range. Thus, for reasons of clarity, those two curves have been omitted in Fig. 8. Furthermore, it should be noted that the dashed red and the green curve in Fig.10 overlay exactly. The black solid line in Fig. 9 and Fig.10 represents the saturation current i_{1s} of the stepped air-gap transformer.

As expected, Fig.10 illustrates that at $U_{in} = \sqrt{2} \cdot 230$ V, the stepped air-gap transformer does not saturate over the full load range. The higher magnetizing inductance of the stepped air-gap transformer leads to lower maximum and rms currents and therefore to a loss reduction. At $U_{in} = \sqrt{2} \cdot 110$ V this effect can be observed only up to approximately 60 W. Due to saturation, i_{max} of the flyback

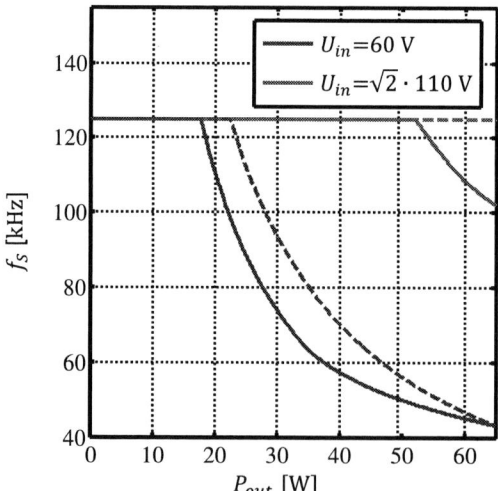

Fig. 8. Switching frequency (solid: stepped air-gap, dashed: conventional, DCM: $f_S = 125$ kHz, BCM: $f_S < 125$ kHz)

PCIM Europe 2016, 10 – 12 May 2016, Nuremberg, Germany

converter with stepped air-gap transformer becomes higher than i_{max} of the conventional flyback converter for loads higher than 60 W. At $U_{in} = 60$ V an efficiency improvement can only be achieved up to approximately 33 W, as shown in Fig. 9 and Fig.11. At higher loads the maximum and rms currents becomes very high resulting in a low efficiency. Nevertheless, low voltages such as $U_{in} = 60$ V occur only in case of mains dips which lasts only for a very short time. Consequently, a high efficiency at mains dips is not necessary.

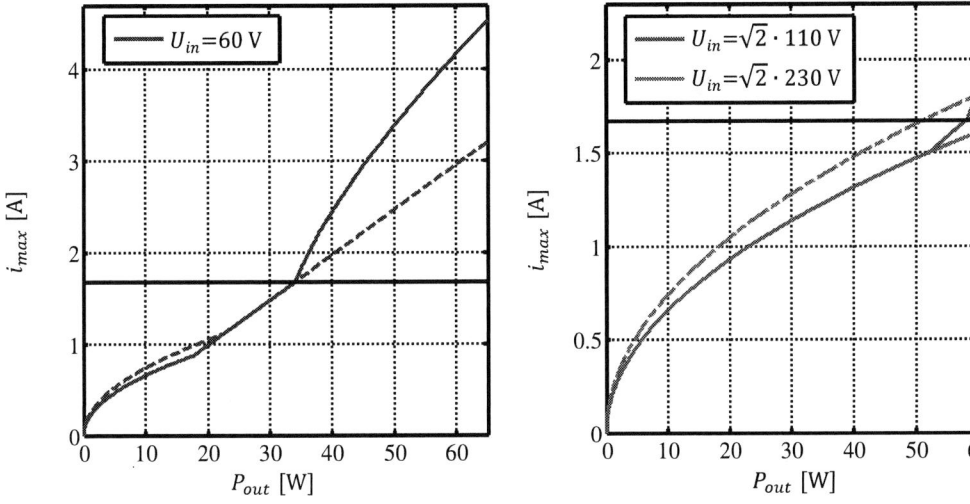

Fig. 9.   Maximum current (solid: stepped air-gap, dashed: conventional)

Fig.10.   Maximum current (solid: stepped air-gap, dashed: conventional)

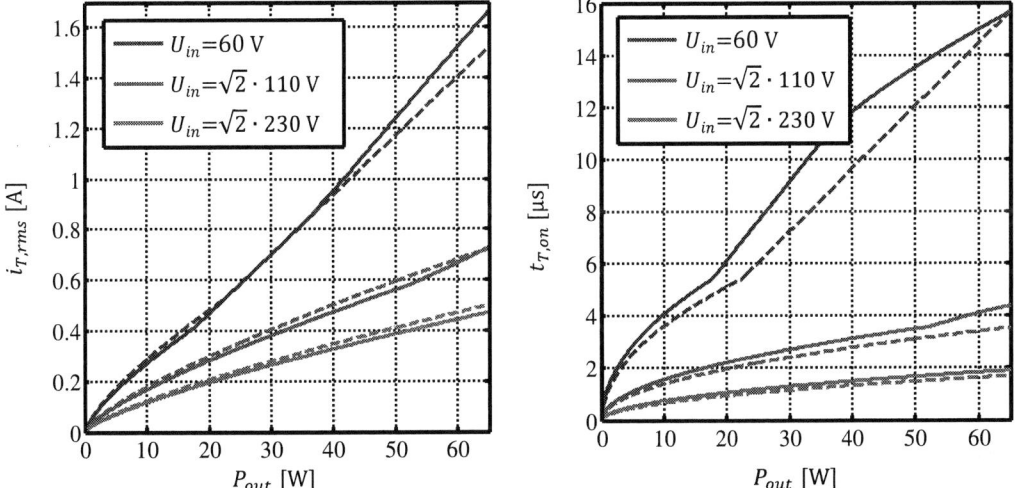

Fig.11.   Rms value of the MOSFET current (solid: stepped air-gap, dashed: conventional)

Fig.12.   MOSFET on-time (solid: stepped air-gap, dashed: conventional)

5. Measurement results

To demonstrate the improvement of the converter efficiency, a prototype has been developed for the specification mentioned above. This prototype can either be equipped with a standard RM-core or a RM-core with stepped air-gap. Thus, the layout as well as the transformer

© VDE VERLAG GMBH · Berlin · Offenbach

winding is identical for both flyback converters and a fair comparison is guaranteed. Fig.13 depicts the used RM-core with stepped air-gap.

Table 2 lists the measured relative total loss reduction at different operating points that can be achieved by replacing the conventional transformer with a stepped air-gap transformer. A negative value for the relative loss reduction indicates that the total losses are higher in case of the flyback converter with stepped air-gap. The measurement results show that the stepped air-gap transformer improves the efficiency at $U_{in} = \sqrt{2} \cdot 230$ V over the entire load range. At $U_{in} = \sqrt{2} \cdot 110$ V, the efficiency can be improved only up to a certain load.

Fig.13. Stepped air-gap

U_{in} [V] \ P_{out} [W]	16.25	32.5	48.75	65
$\sqrt{2} \cdot 230$	6.4%	5.03%	5.52%	5.55%
$\sqrt{2} \cdot 110$	6.43%	5.95%	5.22%	−0.79%

Table 2: Relative loss reduction

Conclusion

This paper presents an analytical description of the flyback converter with a stepped air-gap transformer. Calculations as well as measurements demonstrate that a higher magnetizing inductance in the unsaturated state can be obtained by employing a stepped air-gap transformer instead of a conventional transformer. As the converter operates in BCM with valley skipping at high input voltages, the increased magnetizing inductance causes a reduction of almost all losses of the flyback converter. Measurement results verify a total loss reduction of up to 6.4%. Thus, the converter efficiency is significantly improved.

References

[1]   J. Sun, M. Xu, Y. Ren, and F. Lee, "Light-load efficiency improvement for buck voltage regulators," *Power Electronics, IEEE Transactions on*, vol. 24, no. 3, pp. 742–751, March 2009.

[2]   J. Stahl, A. Fetzer, A. Pawellek, and M. Albach, "Measurement set-up to characterize the current dependent inductivity of nonlinear inductances," in *Control and Modeling for Power Electronics (COMPEL), 2012 IEEE 13th Workshop on*, June 2012, pp. 1–6.

[3]   W. Wolfle, W. Hurley, and S. Arnold, "Power factor correction for ac-dc converters with cost effective inductive filtering," in *Power Electronics Specialists Conference, 2000. PESC 00. 2000 IEEE 31st Annual*, vol. 1, 2000, pp. 332–337 vol.1.

[4]   M. Doebroenti, M. Schmid, and T. Duerbaum, "Matlab based fast and accurate simulation of a power factor corrector switch-mode power supply in boundary conduction mode with valley skipping," in *Control and Modeling for Power Electronics (COMPEL), 2010 IEEE 12th Workshop on*, June 2010, pp. 1–8.

[5]   NXP, "Tea1552 hv start-up flyback controller datasheet," June 2012.

[6]   M. Schmid, "Untersuchung von Netzteilen mit limitierter Bauhoehe und hoher Effizienzanforderung," Ph.D. dissertation, Friedrich-Alexander-Universitaet Erlangen-Nuernberg, 2012.

[7]   T. Duerbaum and M. Albach, "Core losses in transformers with an arbitrary shape of the magnetizing current," in *EUROPEAN CONFERENCE ON POWER ELECTRONICS AND APPLICATIONS*, vol. 1. PROCEEDINGS PUBLISHED BY VARIOUS PUBLISHERS, 1995, pp. 1–171.

Novel Method for the Estimation of Switching Losses in Resonant Converters

C. Oeder, University Erlangen-Nuremberg, Germany, christian.oeder@fau.de

M. Barwig, University Erlangen-Nuremberg, Germany, markus.barwig@fau.de

T. Duerbaum, University Erlangen-Nuremberg, Germany, thomas.duerbaum@fau.de

Abstract

Resonant converters exhibit the ability to achieve nearly lossless switching by using zero current or zero voltage switching (ZCS/ZVS) for the input bridge. Since most switched-mode power supplies (SMPS) use MOSFETs as semiconductor devices today, a ZVS operation is typically preferred. However, with higher switching frequencies reaching of ZVS becomes more difficult. Thus, a (partial) loss of ZVS might be the consequence, making the need for a reasonable estimation of the occurring losses simply inevitable. Due to the half- or full-bridge configuration, this estimation becomes much more complex compared to a traditional hard-switched boost or buck converter. This paper gives a detailed insight into the switching transients of resonant converters and proposes a proper test circuit. A novel method is presented to estimate the generated switching and driving losses during the turn-on event of the switches in case of incomplete ZVS.

1. Introduction

Resonant converters exhibit several benefits compared to hard-switched converters [1]. Especially their ability to achieve nearly lossless switching makes this converter class very attractive for high switching frequency applications [2]. Based on the principle of zero current and zero voltage switching (ZCS/ZVS) the circuit designer is able to get rid of switching losses during the turn-on or turn-off event of the input switches. As a consequence, most publications simply refer to this ability without considering switching losses at all.

However, the ongoing pursuit of higher switching frequencies complicates this situation and makes the achievement of ZVS more and more difficult in resonant converters [3]. Even if ZVS is only slightly lost under these circumstances, the generated losses might strongly heat up the device and reduce the converter's efficiency. In order to cope with this upcoming problem, this paper gives a detailed insight into the switching transients of resonant converters. Moreover, a novel method is presented, allowing the estimation of switching and driving losses, which is only based on the MOSFET's datasheet parameters. No additional information regarding its internal device characteristic is necessary.

2. Proposed test circuit

For the classic investigation of switching losses in SMPS the clamped inductive switching circuit [4] is usually used, which is illustrated in Fig. 1. Although this test circuit is well-suited for hard-switched converters, it cannot be directly applied to resonant converters. Thus, the authors extend the test circuit according to Fig. 2 for a half-bridge configuration.

In order to ensure a symmetric operation of resonant converters, parasitic components should be avoided as far as possible. In this case, the switching transients of the top and bottom MOSFETs can be assumed to be completely identical. Hence, it is sufficient to investigate only the switching behavior of the bottom one M_b. In this context, it is important to ensure that the top MOSFET M_t stays in its off-state (symbolized by the shortened gate-source contact of M_t in Fig. 2) for all investigated transients. While U_{in} represents the input voltage of the converter, the resonant current I_{res} flows to the half-bridge node. Since the switching transients are usually short compared to the converter's switching period, a significant current change in the resonant tank during both transitions is prevented. Therefore, this current is assumed to be constant. The same assumption is typically made for the simple clamped inductive switching circuit [4].

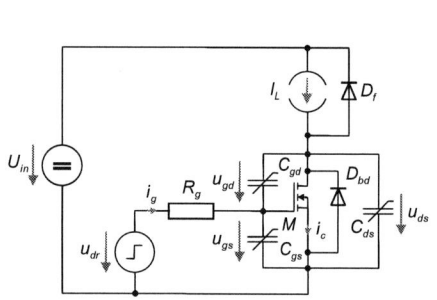

Fig. 1.   Clamped inductive switching circuit

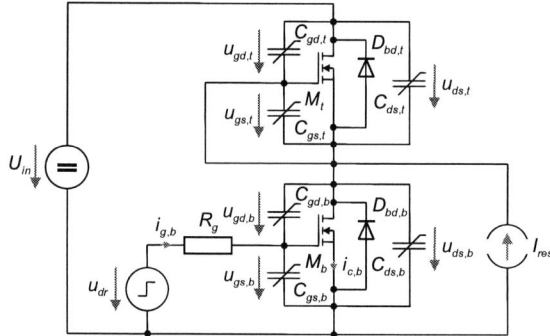

Fig. 2.   Proposed test circuit for the investigation of switching losses in resonant converters

3. Estimation of switching losses

For the optimization of SMPS it is of vital importance for the circuit designer to identify the power losses in (almost) every component. While conduction losses due to ohmic components are relatively simple to determine, an accurate prediction of switching losses is rather complicated for semiconductor devices due to their nonlinear device characteristics [4,5]. In case of MOSFETs, these losses can be separated into turn-off losses, turn-on losses as well as driving losses.

3.1.   Turn-off losses

When using MOSFETs in the input bridge, a ZVS operation mode is usually preferred. This operation allows a nearly lossless turn-on, but consequences losses during the turn-off event [6]. Fig. 3 shows a turn-off simulation for the device IPP60R280P6 (level 3 model) from [7] with U_{in} = 400 V, I_{res} = 1 A, $U_{dr,on}$ = 12 V, $U_{dr,off}$ = 0 V and different external gate resistors R_g.

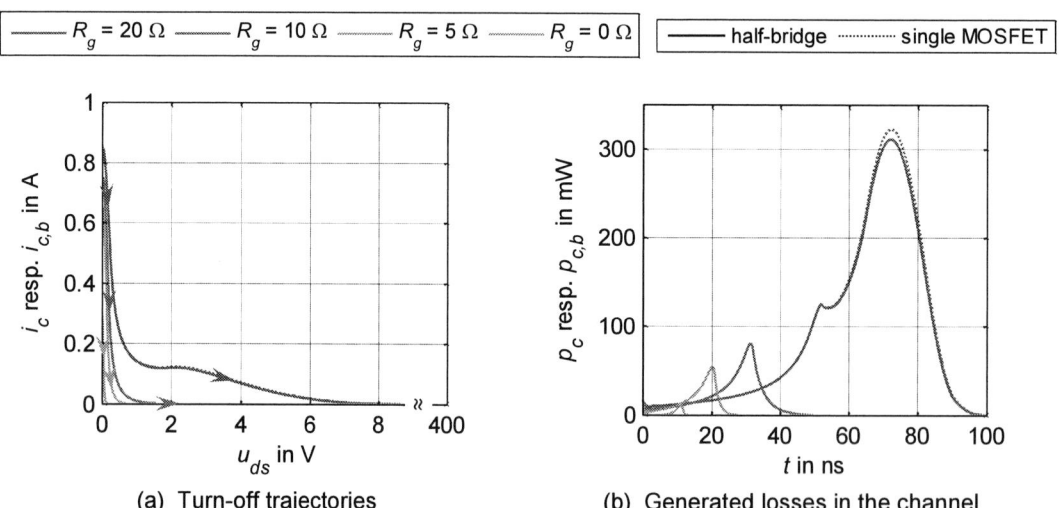

(a) Turn-off trajectories

(b) Generated losses in the channel

Fig. 3.   Simulated waveforms in LTSpice for the turn-off event of IPP60R280P6 with U_{in} = 400 V, I_{res} = 1 A, $U_{dr,on}$ = 12 V, $U_{dr,off}$ = 0 V and different external gate resistors

© VDE VERLAG GMBH · Berlin · Offenbach

With smaller values for the external gate resistor the channel current i_c in Fig. 3 (a) falls faster and promotes a "rapid channel turn-off" [4]. By integrating the given loss curve in Fig. 3 (b), an energy value of 750 pJ can be identified in case of $R_g = 5.5\,\Omega$, which fits to the specified output resistance (during turn-off event) of the commercial resonant driver IC NCP1397 from OnSemi [8]. Even when operating at its maximum switching frequency with $f_{s,max} = 500$ kHz, the turn-off losses in the input bridge are less than 10 mW for this configuration. Thus, there is no need to further model turn-off losses of resonant converters under these circumstances. It is to be noted that the losses during this switching transition (solid lines) are almost identical to a "single MOSET" (dotted lines) within the clamped inductive switching circuit. This is due to the fact that the value of the top MOSFET's output capacitance is almost negligible for high drain-source voltages and hence, according to [4] also its influence on the switching transient of M_b.

3.2. Turn-on losses

As long as ZVS is reached, the turn-on losses are almost zero and negligible. If ZVS is lost, the MOSFET is turned on at drain-source voltages U_v higher than 0 V and significant losses might be generated. Fig. 4 shows the simulated waveforms in LTSpice during the turn-on transient for an operating point within the inductive region of a resonant converter (needed for ZVS operation) for several turn-on voltages with $U_{in} = 400$ V, $I_{res} = -50$ mA and $R_g = 13\,\Omega$ (output resistance during turn-on event for NCP1397 [8]). This time significant differences can be recognized between the half-bridge configuration in Fig. 2 and the clamped inductive switching circuit in Fig. 1 ("single MOSFET") proving the importance of the modified test circuit. In case of a complete loss of ZVS with $U_v = 400$ V the dissipated energy is 26.6 µJ per MOSFET in the half-bridge and more than eight times higher than for a "single MOSFET" with 3.1 µJ. This huge difference might be astonishing at first sight, but the following estimation approach will explain this effect in detail. Additional information regarding this significant difference can also be found in [9].

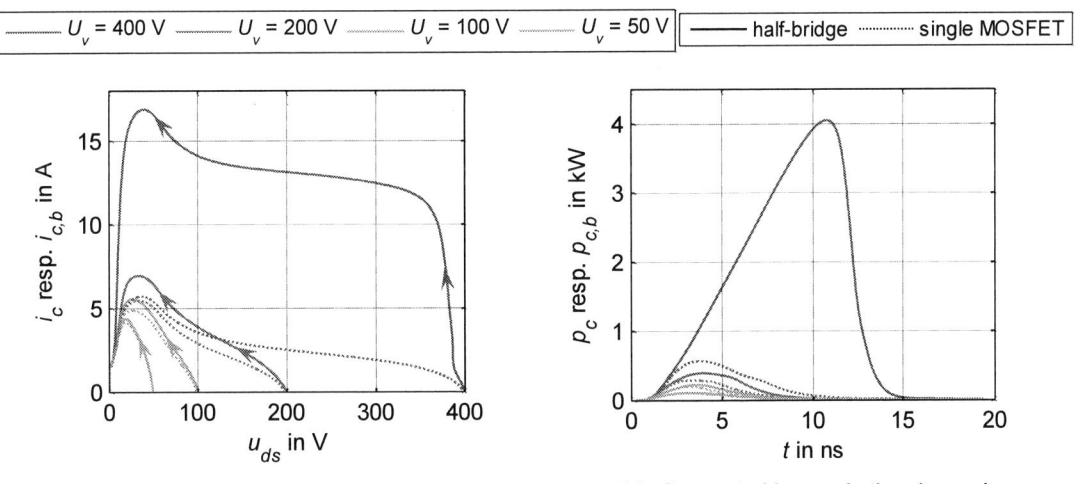

(a) Turn-on trajectories   (b) Generated losses in the channel

Fig. 4. Simulated waveforms in LTSpice for the turn-on event of IPP60R280P6 with $U_{in} = 400$ V, $I_{res} = -50$ mA, $U_{dr,on} = 12$ V, $U_{dr,off} = 0$ V, $R_g = 13\,\Omega$ and different turn-on voltages U_v

As already utilized in [10], the turn-on losses of a MOSFET device are strongly influenced by the stored amount of charge and energy in its output capacitance C_{oss}. Thus, in order to model the turn-on losses generated by a half-bride configuration, we have to determine the total charge (see Fig. 5 (a)) out of the MOSFET's datasheet with

$$Q_{oss,b}\left(U_{ds,b}\right) = \int_{0\,V}^{U_{ds,b}} C_{oss,b}\left(\upsilon_{ds,b}\right)\mathrm{d}\upsilon_{ds,b} \ \text{ resp. } \ Q_{oss,t}\left(U_{ds,t}\right) = \int_{0\,V}^{U_{ds,t}} C_{oss,t}\left(\upsilon_{ds,t}\right)\mathrm{d}\upsilon_{ds,t} \,, \tag{1}$$

© VDE VERLAG GMBH · Berlin · Offenbach

which is needed to charge the output capacitance of a MOSFET from 0 V to U_{ds}. Turning on the bottom MOSFET M_b at $u_{ds,b} = U_v$ directly leads to the dissipation of the stored energy

$$E_1(U_v) = Q_{oss,b}(U_v) \cdot U_v - \int_{0\,V}^{U_v} Q_{oss,b}(v_{ds,b}) \mathrm{d}v_{ds,b} \,. \tag{2}$$

In order to explain the significant deviation for the dissipated energy in Fig. 4 (b), a second loss mechanism has to be considered, which is based on the additional top MOSFET in the half-bridge. By studying the turn-on transient of M_b in detail it becomes quite obvious that the output capacitance of M_t has to be charged at the same time as the output capacitance of M_b is discharged, since their device voltages are linked according to

$$U_{in} = u_{ds,b} + u_{ds,t} \,. \tag{3}$$

Due to the (hard) turn-on of M_b at $U_v > 0$ V, the "missing charge" $Q_{\Delta oss,t^*}$ for the top MOSFET cannot be provided by the resonant current I_{res}, but must be taken from the voltage source U_{in}. This charge is illustrated in Fig. 5 (b) and depends on the stored charge Q_{oss,t^*} in M_t with

$$Q_{\Delta oss,t^*}(U_{ds,b}) = Q_{oss,t}(U_{in}) - Q_{oss,t^*}(U_{ds,b}) = Q_{oss,t}(U_{in}) - Q_{oss,t}(U_{in} - U_{ds,b}) \,. \tag{4}$$

Using the charge theory in (2) and the expression in (4) for the "missing charge" of the top MOSFET, the second portion of the dissipated turn-on energy E_2 (see Fig. 5 (b)) is given by

$$E_2(U_v) = Q_{\Delta oss,t^*}(U_v) \cdot U_v - \int_{0\,V}^{U_v} Q_{\Delta oss,t^*}(v_{ds,b}) \mathrm{d}v_{ds,b} \,, \tag{5}$$

which dominates especially for high values of U_v.

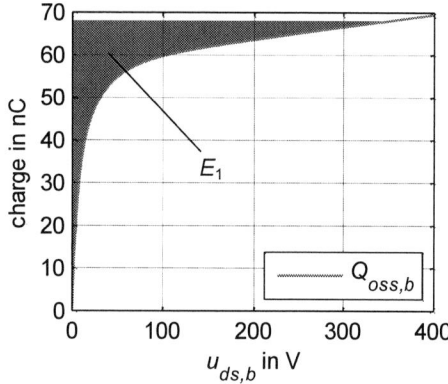 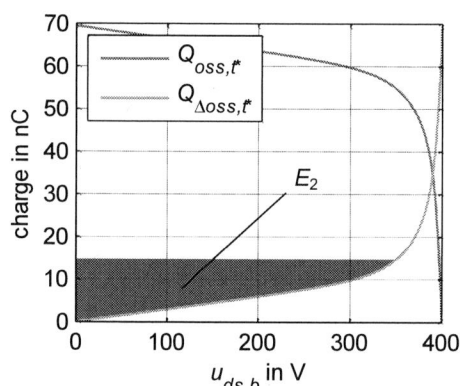

(a) Stored charge in the bottom MOSFET M_b (b) Stored charge in the top MOSFET M_t

Fig. 5. Stored charge in IPP60R280P6 for a half-bridge configuration with $U_{in} = 400$ V (blue area symbolizes the dissipated energies for an exemplified turn-on voltage of $U_v = 350$ V)

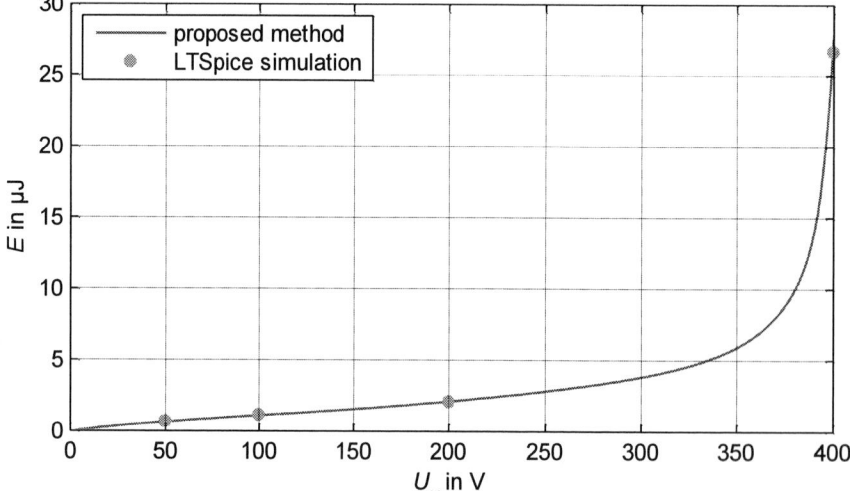

Fig. 6. Dissipated energy for a half-bridge configuration with IPP60R280P6 and $U_{in} = 400$ V (additional LTSpice parameters $I_{res} = -50$ mA, $U_{dr,on} = 12$ V, $U_{dr,off} = 0$ V and $R_g = 13\ \Omega$)

Fig. 6 compares the predicted energy dissipation $E = E_1 + E_2$ of the proposed method to the simulation results from Fig. 4 (b).

It turns out that a very high accuracy can be achieved only by using the MOSFET's datasheet parameters. As a consequence and as already stated by [10], a capacitance modelling approach is well suited for the estimation of switching losses in resonant converters. Considering a half-bridge configuration with two (identical) MOSFETs, the total turn-on losses can be finally approximated by

$$P_{turn,on} = 2 \cdot f_s \cdot E(U_v) \tag{6}$$

taking the switching frequency f_s of the input bridge into account.

3.3. Driving losses

In addition to the turn-on losses, especially the driving losses are of vital importance when the resonant converter approaches no-load. Many datasheets provide the total gate charge Q_g, which is needed to charge the input capacitance of the MOSFET from 0 V to a certain gate-source voltage U_{gs}. The total energy provided by the driving stage is given by

$$E_{dr} = U_{gs} \cdot Q_g(U_{gs}). \tag{7}$$

While this amount of energy is almost fixed in a hard-switched converter, this is no longer the case for a resonant converter. This is due to the fact that the total gate charge Q_g depends not only on the gate-source voltage U_{gs}, but also on the turn-on voltage U_v of the MOSFET device, which varies in case off a (partial) loss of ZVS. For the P6 series from Infineon [7] for example, deviations up to 10 % - 20 % might occur, if this effect is left unconsidered. Unfortunately, only limited information can be found in datasheets regarding this influence.

Due to this lack of information, the authors propose a relative simple method to overcome this problem and to model the driving losses more accurately. Although this method is based on simplified assumptions, the predicted charge values depend again only on datasheet parameters and are quite close to the simulated ones (see Fig. 7), which are extracted again from LTSpice simulations with the highly accurate level 3 model of IPP60R280P6 [7]. With respect to $U_v = 0$ V, the total amount of gate charge in Fig. 7 rises by 7 % for $U_v = 5$ V and already by 10 % for $U_v = 10$ V revealing the importance of this effect with increasing switching frequencies.

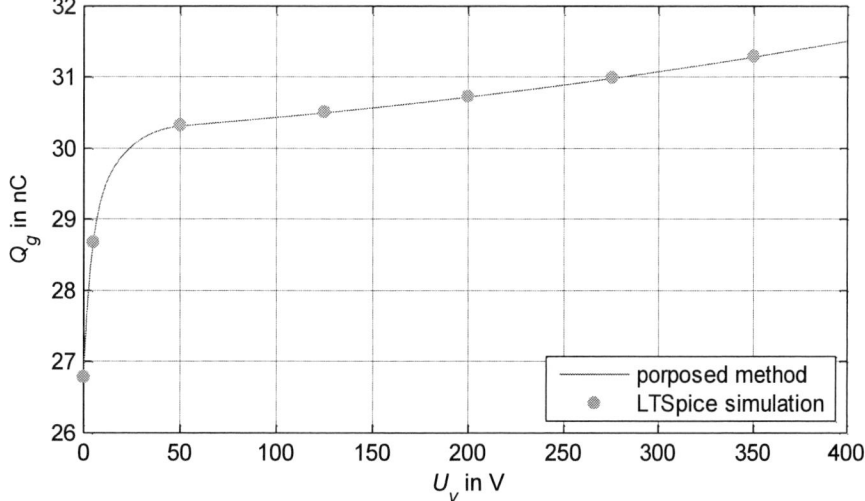

Fig. 7. Influence of the turn-on voltage U_v on the total gate charge Q_g for a half-bridge configuration with IPP60R280P6 and $U_{in} = 400$ V (additional LTSpice parameters $I_{res} = -50$ mA, $U_{dr,on} = 12$ V, $U_{dr,off} = 0$ V and $R_g = 13$ Ω)

The proposed method assumes that the voltage dependency of the total gate charge Q_g is purely based on the voltage dependency of the charge Q_{gd}, which is necessary to charge the

MOSFET's gate-drain capacitance C_{gd} from 0 V to U_v. This charge can be easily calculated extracting the datasheet information with

$$Q_{gd}(U_v) = \int_{0\,V}^{U_v} C_{gd}(v_{ds})\mathrm{d}v_{ds} \,.$$

(8)

Since the datasheet usually specifies the value of the total gate charge $Q_{g,DS}$ (at least) for one fixed value of the drain-source voltage $U_{ds,DS}$, the expression

$$Q_{gd}(U_v) = Q_{g,DS}\Big|_{U_{ds}=U_{ds,DS}} + \int_{U_{ds,DS}}^{U_v} C_{gd}(v_{ds})\mathrm{d}v_{ds}$$

(9)

reveals the simplified voltage dependency of Q_g over the entire voltage range $0 < U_v < U_{in}$. Although the datasheet information about the $C_{gd}(u_{ds})$-shape is only valid for $U_{gs} = 0$ V in fact, the accuracy of the corresponding result in Fig. 7 is quite astonishing. Obviously,

$$P_{dr} = 2 \cdot f_s \cdot U_{gs} \cdot \left[Q_{g,DS}\Big|_{U_{ds}=U_{ds,DS}} + \int_{U_{ds,DS}}^{U_v} C_{gd}(v_{ds})\mathrm{d}v_{ds} \right]$$

(10)

is a very simple and useful expression to estimate the occurring driving losses for a (resonant) half-bridge configuration, even in case of incomplete ZVS transitions.

4. Conclusion

This paper presents a powerful method to estimate the switching and driving losses in resonant converters. While an accurate prediction of the driving losses becomes important when the converter approaches no-load, switching losses are generated in case of an incomplete ZVS transition. The simulation results prove that a very high accuracy can be reached only by taking the MOSFET's output capacitance curve into account, which is given in almost every datasheet. No additional information regarding internal device characteristics is necessary allowing a reasonable and fast comparison of different MOSFET devices.

5. References

[1] A. Pawellek, "Low-Profile Power Adapter Based on a Resonant LCC Converter", IEEE International Symposium on Industrial Electronics 2010, ISIE 2010, p. 3894 – 3899

[2] H. de Groot, E. Janssen, R. Pagano, K. Schetters, "Design of a 1 MHz LLC Resonant Converter based on a DSP-driven SOI Half-Bridge Power MOS Module", IEEE Power Electronics Specialists Conference 2006, PESC 2006, p. 1 – 14

[3] C. Oeder, T. Duerbaum, "A Novel Method to Predict ZVS Behavior of LLC Converters", International Exhibition and Conference for Power Electronics, Intelligent Motion, Renewable Energy and Energy Management 2014, PCIM 2014, p. 1507 – 1514

[4] D. Kuebrich, T. Duerbaum, A. Bucher, "Investigation of Turn-Off Behaviour under the Assumption of Linear Capacitances", International Conference of Power Electronics Intelligent Motion Power Quality 2006, PCIM 2006, p. 239 – 244

[5] Y. Xiong, S. Sun, H. Jia, P. Shea, Z. Shen, "New Physical Insights on Power MOSFET Switching Losses", IEEE Transactions on Power Electronics 2009, p. 525 – 531

[6] R. Steigerwald, "A Comparison of Half-Bridge Resonant Converter Topologies", IEEE Transactions on Power Electronics 1988, vol. 3, p. 174 – 182

[7] CoolMOS™ simulation models, refer to http://www.infineon.com/

[8] Resonant Control IC from ON Semiconductor, refer to http:// www.onsemi.com/pub_link/Collateral/NCP1397-D.PDF

[9] R. Elferich, "General ZVS Half Bridge Model Regarding Nonlinear Capacitances and Application to LLC Design", Energy Conversion Congress and Exposition 2012, ECCE 2012, p. 4404 – 4410

[10] S. Hamamsy, R. Fisher, "Inclusion of Nonlinear Output Capacitor Behavior in Zero-Voltage Switched Circuit Design", IEEE Power Electronics Specialists Conference 1997, PESC 1997, p. 1424 – 1430

Active Dead-time optimization for wide range Flyback active-clamp converter

Sébastien Larousse, INL, Bâtiment Curien, 43 bd. Du 11 novembre 1918, 69100 Villeurbanne France, slarousse@ieee.org

Hubert Razik, Laboratoire Ampère, 43 bd. du 11 novembre 1918, 69100 Villeurbanne, France, hubert.razik@univ-lyon1.fr

Remy Cellier, INL, Bâtiment Curien, 43 bd. Du 11 novembre 1918, 69100 Villeurbanne France, remy.cellier@cpe.fr

Nacer Abouchi, INL, Bâtiment Curien, 43 bd. Du 11 novembre 1918, 69100 Villeurbanne France, nacer.abouchi@cpe.fr

Philippe Volay, Centralp, 21 rue Marcel Pagnol, 69200 Vénissieux, France, pvolay@centralp.fr

Abstract

The Flyback Active-clamp topology can work on a wide input voltage range. But the parasitic capacitor of the main MOSFET used in the soft switching is affected by the input voltage variation and then reduces the efficiency of the converter. In this article, the strategy used to maintain an optimal valley switching by adapting PWM dead-times is described. We have tested our strategy on a 20 V to 120 V input / 30 W converter operating at 200 kHz We have obtained a 2% efficiency increase above 60 V compared to a fixed dead-time control.

1. Introduction

For embedded application, DC/DC converters have to sustain a wide input voltage range, in order to compensate power bus voltage variations and local standards. In transportation applications, this input voltage swing can reach a factor 5 between minimal and maximal input voltage, which varies from 24 V to 110 V. For such a large voltage range, a Flyback converter is the most common solution, with input voltage range capacities beyond a factor 8. The main limitation of this strategy is the hard switching of the MOSFET, which reduces the possibility of high frequency switching. The use of active-clamp strategy to reduce the commutation loss in the converter allows us a significant gain in switching frequency while maintaining a wide input voltage range [1]. This topology uses the energy stored in an inductor to discharge the parasitic capacitor of the main MOSFET. Still, in low input current conditions, this energy does not permit a zero voltage switching (ZVS) [2], but only a resonance which can be used for valley switching. As this resonance involves the Drain-Source Capacitor (C_{DS}) of a MOSFET, this resonance frequency will change with the variations of C_{DS}.

In this article the behavior of the drain voltage resonance of the MOSFET in Flyback active-clamp over a wide voltage range is analyzed, and an optimization strategy to overcome the resonance frequency fluctuation and maintain a minimal commutation voltage of the main switch is proposed. This strategy has been implemented on a digital controller and tested on a Flyback converter with a 20-120 V input voltage and a 12 V/30 W output.

2. Converter and resonance behavior

2.1. Converter behavior

The Flyback active clamp converter uses a clamp capacitor in parallel with the primary side of the transformer to impose a reverse voltage on it, and thus inverting the current direction in the input of the transformer (fig.1). In this topology, the leakage inductance is used to discharge the parasitic capacitor between the drain and the source of the bottom MOSFET (C_{DS}), so as to allow a ZVS of the main MOSFET (S_1).

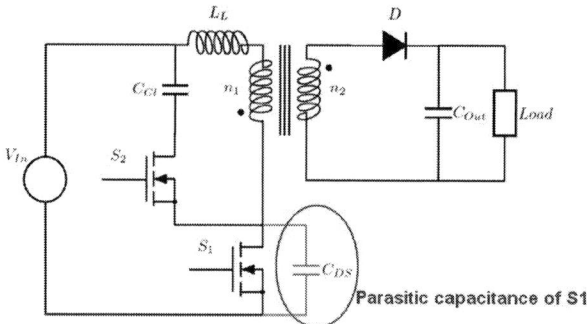

Fig. 1. Flyback Active-Clamp circuit

Here, the energy stored in the inductor at the moment of the commutation is:

$$E_L = \frac{1}{2}L_L I_L^2,$$

whereas the energy in the drain capacitor is:

$$E_{Cds} = \frac{1}{2}C_{DS}V_{DS}^2.$$

So, as E_L must be higher than E_{Cds} to unload C_{DS} [2], the ZVS conditions gets more difficult to reach with low values of I_l and high values of the drain voltage V_{DS}, which occur at high input voltages Thus, the soft switching conditions cannot be reached for the whole working range of a converter working from 20 V to 120 V input voltage with a reasonable value of L_L. Indeed, ZVS over such a wide range would only be possible with an L_L value larger than the magnetizing inductance of the transformer, meaning that an additional series inductance would be necessary. Such an added component would be counter-productive in an industrial application, adding cost, volume and conduction loss to the converter. If the ZVS conditions are not reached, the resonance between L_L and C_{DS} will not make V_{DS} reach zero, but V_{DS} will have a sinusoidal form (fig.2), which frequency is :

$$f_r = \frac{1}{2\pi\sqrt{L_L C_{DS}}}$$

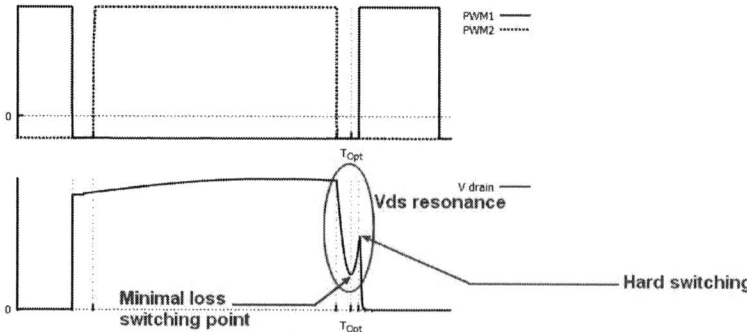

Fig. 2. Main waveforms

This resonance can be used to reduce the switching loss in the converter, even if the loss reduction is lower than in ZVS. This commutation strategy, known on conventional Flyback as Quasi-resonant mode [3], can ensure significant power loss reduction compared to hard-switching. As the commutation loss due to C_{DS} is equal to the energy stored in the capacitor at the instant of the commutation, the energy loss is proportional to the square of V_{Drain}. The value V_{Drain} at the instant of the commutation has to be minimal to ensure the best efficiency for the converter when the ZVS is impossible, and for this purpose, the knowledge of f_r is vital to know the dead time between PWM1 and PWM2 (fig.2) controlling respectively S1 and S2 (fig.1).

2.2. Resonance frequency variation

Within narrow voltage range utilization, the dead time length necessary for a valley-switching would be constant as the period of the resonance depends on the supposed constant elements L_L and C_{DS}. But, in our wide input voltage range application, V_{DS} varies by a factor 7. With such a wide voltage variation, the capacitor C_{DS}, cannot be considered as constant anymore in a power MOSFET but it decreases with input voltage [4]. This resonant capacitor value variation is causing a reduction of the resonance period at high input voltage. This resonance period can have a strongly reduced value at high voltage, down to -60% from 60 V_{IN} to 120 V_{IN}, as we can see on the curve V_{min} on fig. 4, describing the evolution of the minimal value of V_{ds}. A fixed commutation delay can then cause a commutation voltage increase compared to minimal voltage value, and thus an increase of the commutation loss. To control these resonance frequency variations, the voltage dependent value of C_{DS} has been linearized using the charge equivalent value of the capacitor [5], which gave us the function $C_{DS}(V_{DS})$, fig.3. This function has been used to simulate the behavior of C_{DS} during the resonance and to estimate the optimal dead-time value for a minimal switching voltage, using a MOSFET that fits with our application, Infineon IPW60R041C6 [6].

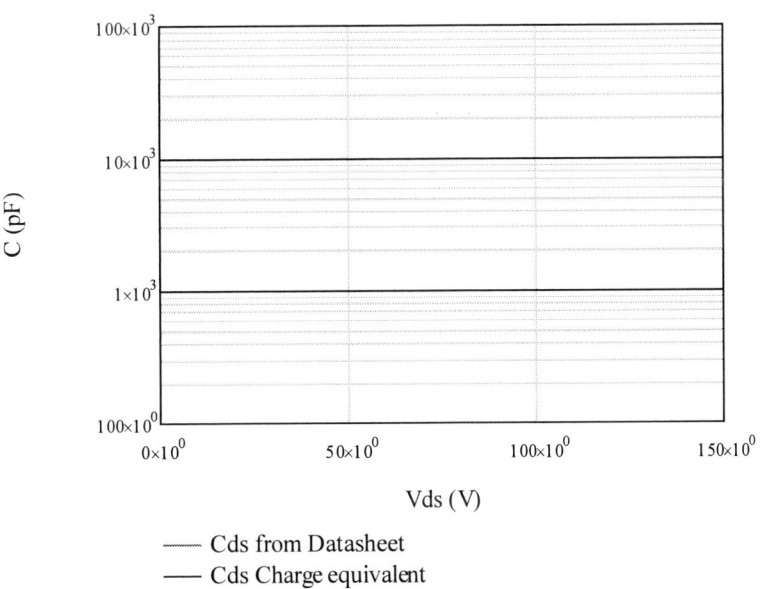

Fig. 3. Linearized function of $C_{DS}(V_{DS})$ compared to C_{DS} value from the datasheet.

PCIM Europe 2016, 10 – 12 May 2016, Nuremberg, Germany

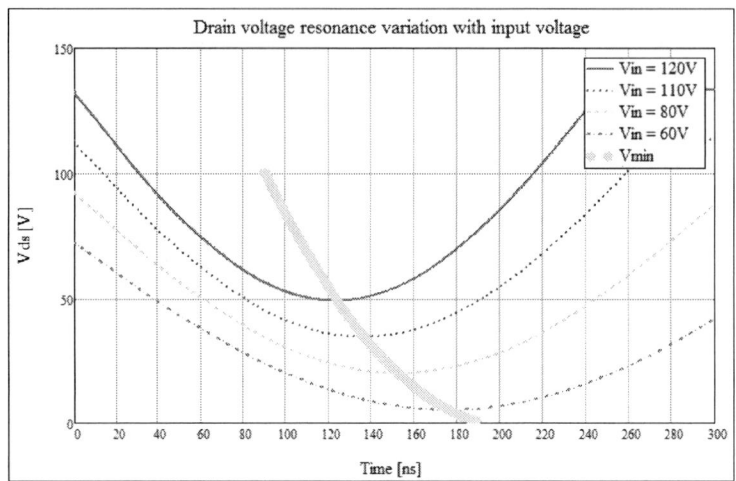

Fig. 4. Simulated drain voltage resonance for different input voltage values at 30W.

Our simulations of the behavior of the resonance between C_{CL} and L_L(fig. 4), have confirmed the frequency variation of the resonance and the impossibility to keep a commutation at the minimal drain voltage at the same dead-time on the whole input voltage range. To avoid a commutation after the drain voltage dip, and thus a hard switching of the main MOSFET, the delay between the beginning of the resonance and the commutation of the MOSFET, corresponding to the dead time of the PWM, has to be adapted so as to make the commutation of the transistor at the lowest drain voltage, with the lowest commutation loss.

3. Dead time adaptation strategy

Once the variation of C_{DS} with the voltage is known using MOSFET datasheet information and C_{DS} linearization [3], the optimal dead time, meaning the lowest commutation voltage, can be determined by the measurement of the converter's input voltage. This measurement can easily be performed by a microcontroller which can adjust the PWM dead-time according to predetermined values stored in a look up table. This adaptation has to be faster than the input voltage variation to maintain the best dead-time, but slower than the output voltage regulation to avoid causing instability of the regulation. Thus, the dead-time regulation loop is a low frequency process which requires few resources and could be implemented on a microcontroller designed for SMPS regulation with no extra cost. With this control strategy, the commutation voltage can stay close to the lowest value of V_{DS}, reducing the commutation loss on the MOSFET, increasing the overall efficiency and reducing the electromagnetic perturbation.

4. Measurements and results

The dead time adaptation strategy has been tested on a 30 W / 12 V output converter, working on a 20 V to 120 V input voltage range at a switching frequency of 200 kHz. The variation of the resonance caused by the input voltage has been analyzed to determine optimal dead-time for each input voltage. Then we have applied our dead time adaptation strategy and compared it to a classical fixed dead time control with a dead time calibrated to ensure the widest ZVS range possible (200ns dead-time length). The regulation of the converter is ensured by a microcontroller which also controls the dead time by applying the calculated optimal dead-time value adapted to the drain-source voltage value measured before the beginning of the resonance. The converter efficiency measurement on our prototype shows an efficiency improvement higher than 2% (fig.5) compared to a fixed dead time control once working in valley switching mode, while maintaining same performance at low voltage, when the converter is working in ZVS mode.

© VDE VERLAG GMBH · Berlin · Offenbach

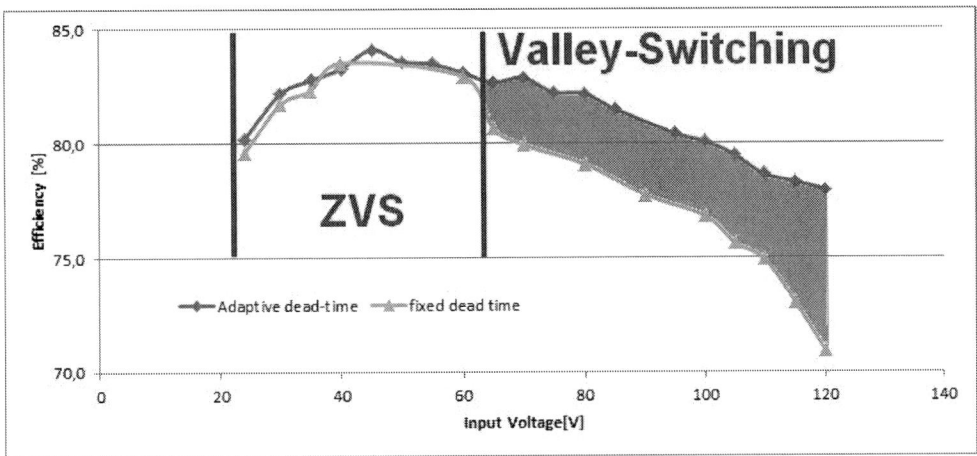

Fig. 5. Efficiency comparison at 30W-12V output

This efficiency improvement is due to lower commutation loss on the MOSFET occurring at the minimal drain voltage (fig.6), contrary to the fixed dead-time commutation mode which cannot optimize commutation delay at low and high voltage at the same time (fig.7).

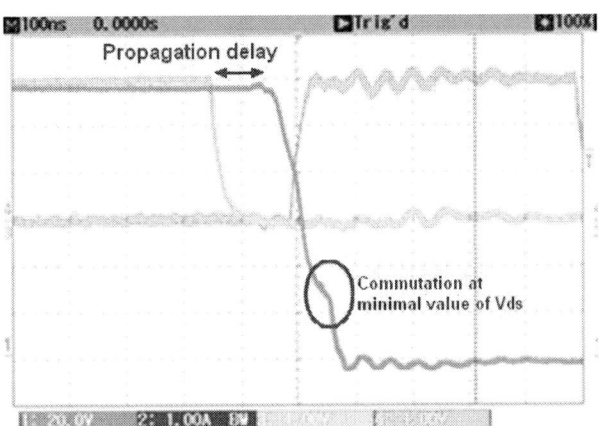

Fig. 6. PWM signals 1 & 2(Green & Blue) and V_{DS} (Orange) at V_{IN}=100V, Optimal dead-time

The efficiency difference increases with the input voltage elevation, as the difference between the optimal dead-time and the fixed dead time value grows. At the highest input voltage, 120V, the performance difference is up to 7%, showing a dangerous heating of the MOSFET S1 on fixed dead-time mode, while the performance of the converter in variable dead-time mode stay reasonable with 5% efficiency difference between best efficiency point and maximal input voltage point. The increase of the switching voltage in the converter in the case of a high switching voltage also impacts transmitted noise caused by the commutations, which is increased in the case of the non-optimal switching (fig. 7) in which the commutation noise is transmitted up to the PWM signals.

PCIM Europe 2016, 10 – 12 May 2016, Nuremberg, Germany

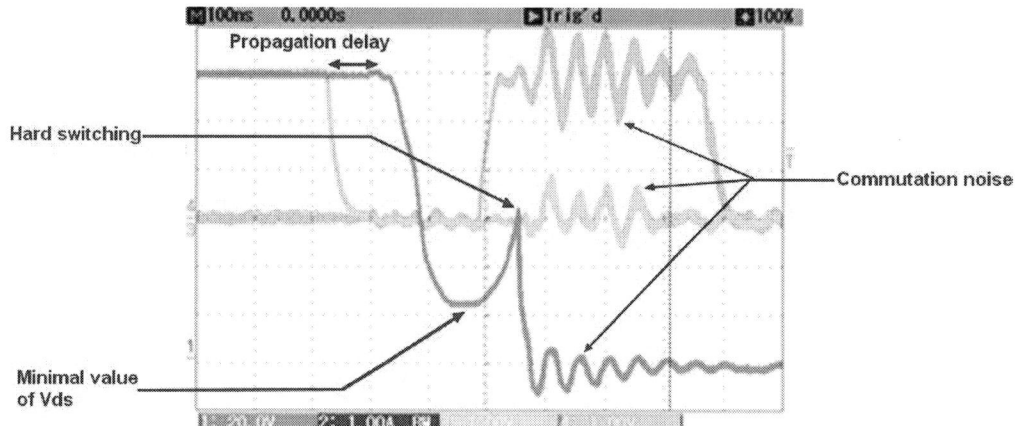

Fig. 7. PWM signals 1 & 2(Green & Blue) and V_{DS} (Orange) at V_{IN}=100V, fixed dead-time

5. Conclusion

The Flyback converter offers wide input voltage range capabilities, useful in embedded devices such as in railway applications. The efficiency of this converter can be improved with active clamp, enabling zero voltage switching. But this topology cannot maintain ZVS on a very wide input voltage range. In this article, we have applied active clamp to a 20-120 V input voltage range converter, using both ZVS and valley switching commutation strategies. We have calculated and observed the resonance variations caused by the voltage dependence of the intrinsic drain-source capacitor of the MOSFET, used as resonant capacitor. To maintain minimal commutation loss of the MOSFET and optimal conversion efficiency in valley switching mode, we have defined a simple dead time control strategy which was tested on our 30W-12V_{OUT} prototype. This active dead-time control increased the efficiency of the converter by 2 to 7% on high input voltages compared to a fixed dead time control strategy, allowing a wide input voltage range from 20 to 120V.

6. References

[1] P. Alou, O. Garcia, J. Cobos, J. Uceda and M. Rascon, "Flyback with active clamp: a suitable topology for low power and very wide input voltage range applications," in *Applied Power Electronics Conference and Exposition, 2002. APEC 2002. Seventeenth Annual IEEE*, 2002.

[2] B.-R. Lin, H.-K. Chiang, K.-C. Chen and D. Wang, "Analysis, design and implementation of an active clamp flyback converter," in *Power Electronics and Drives Systems, 2005. PEDS 2005. International Conference on*, 2005.

[3] R. Stracquadaini, "Mixed Mode control (Fixed off Time amp; Quasi Resonant) for flyback converter," in *IECON 2010 - 36th Annual Conference on IEEE Industrial Electronics Society*, 2010.

[4] L. Aubard, G. Verneau, J. C. Crebier, C. Schaeffer and Y. Avenas, "Power MOSFET switching waveforms: an empirical model based on a physical analysis of charge locations," in *Power Electronics Specialists Conference, 2002. pesc 02. 2002 IEEE 33rd Annual*, 2002.

[5] D. Costinett, D. Maksimovic and R. Zane, "Circuit-Oriented Treatment of Nonlinear Capacitances in Switched-Mode Power Supplies," *Power Electronics, IEEE Transactions on,* vol. 30, no. 2, pp. 985-995, Feb 2015.

[6] Infineon, *CoolMOS C6 Data Sheet,* 2010.

© VDE VERLAG GMBH · Berlin · Offenbach

PCIM Europe 2016, 10 – 12 May 2016, Nuremberg, Germany

Energetic Macroscopic Representation (EMR) and control scheme for the asymmetric 4-stage MSBA

Gina K. Steinke, Ecole Polytechnique Fédérale de Lausanne, EPFL, CH1015 Lausanne, Switzerland, gina.steinke@epfl.ch
Alfred Rufer, Ecole Polytechnique Fédérale de Lausanne, EPFL, CH1015 Lausanne, Switzerland, alfred.rufer@epfl.ch

Abstract

The Multistage Stacked Boost Architecture is a step-up converter topology intended for renewable energy generation systems or distribution networks. This paper concentrates on the Energetic Macroscopic Representation of this topology and the simulation of this representation as well as showing a functional control and a working prototype.

1. Introduction

The Multistage Stacked Boost Architecture (MSBA) is a non-isolated step-up converter topology intended for renewable energy generation systems or distribution networks. In previous papers the functionality and intended applications for this topology have been discussed ([1-3]); therefore this paper will concentrate on the Energetic Macroscopic Representation of the topology as well as a functional control and a working prototype. The Energetic Macroscopic Representation (EMR) is a graphical as well as functional description of an energetic system based on the physical causality. The EMR highlights the energetic properties of the subsystems of a whole system in order to deduce its control scheme in a systematic way [4-5].

2. Building the EMR

Since the structure of the asymmetric MSBA with four stages is quite complex, the approach is to represent one stage and then by using the superposition theorem adding the other stages one by one, starting with the top stage (Figure 1).

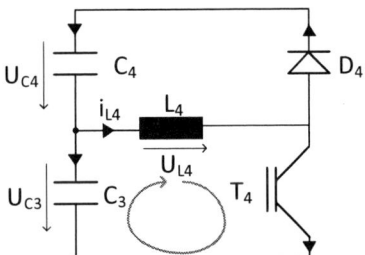

Figure 1: Elementary MSBA stage

Considering the picture above the equation of the lower loop can be represented as:

$$u_{L4} = -d_4 \cdot (u_{C4} + u_{C3}) + u_{C3}$$

© VDE VERLAG GMBH · Berlin · Offenbach

Then the voltage applied to the inductor (Figure 2) can be separated into two different terms:

$$u_{L4} = -d_4 \cdot u_{C4} - d_4 \cdot u_{C3} + u_{C3}$$

$$= -d_4 \cdot u_{C4} + (1 - d_4) \cdot u_{C3}$$

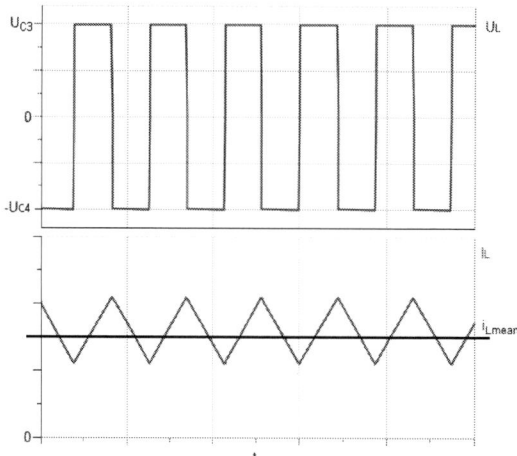

Figure 2: waveform of current and voltage in the inductor of the MSBA stage

This can now be split into

$$u_{Lmean}^- = -d_4 \cdot u_{C4}$$

where u_{Lmean}^- is the contribution of the voltage mean value coming from the upper capacitor C_4, which decreases the inductor current and

$$u_{Lmean}^+ = (1 - d_4) \cdot u_{C3}$$

which is the contribution of the voltage mean value coming from the lower capacitor C_3, that increases the inductor current. According to the aforementioned formulas the system can also be represented as shown in Figure 3.

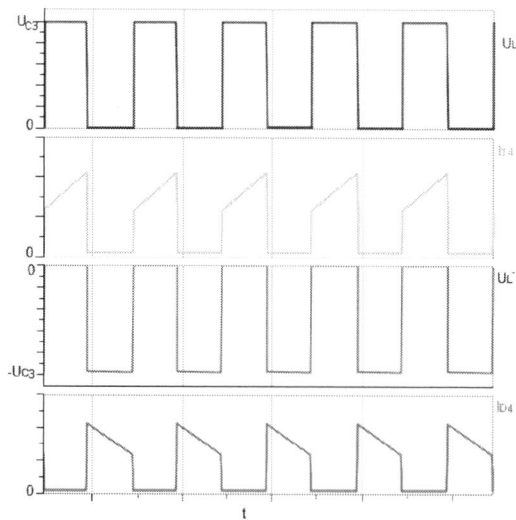

Figure 3: Waveform representation of the MSBA structure

So the mean values of the currents can further be defined as:

$$i_{Lmean}{}^{D4} = i_{D4} = (1-d_4) \cdot i_{L4}$$

$$i_{Lmean}{}^{T4} = i_{T4} = d_4 \cdot i_{L4}$$

This leads to the following formulas for the accumulation elements:

$$u_{C4} = \frac{1}{C_4} \int d_4 \cdot i_{L4} dt = \frac{1}{C_4} \int i_{T4} dt$$

$$i_{L4} = \frac{1}{L_4} \int \left(-d_4 \cdot u_{C4} + (1-d_4) \cdot u_3 \right) dt = \frac{1}{L_4} \int \left(u_{D4} - u_{T4} \right) dt$$

$$u_{C3} = \frac{1}{C_3} \int -i_{D4} dt$$

It is important to keep in mind that the diodes and switches act complementary to each other and for the whole representation the value d has two alternative interpretations. In the case used here, d has a value between 0 and 1 and the inductor current and capacitor voltage are average values. The other case would be using harmonic modeling where d has to be 0 or 1 and the current and voltage values are instantaneous values since d is the trigger order for the switches. Using these formulas results in the EMR of the elementary stage that is shown in Figure 4.

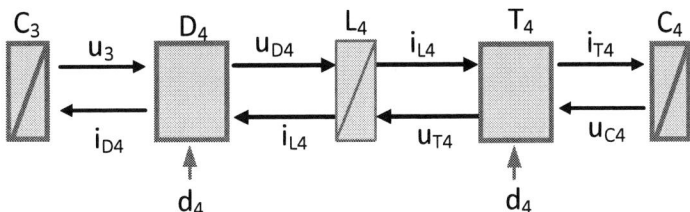

Figure 4: EMR of the elementary MSBA stage

The blocks between the accumulation elements are conversion elements, representing the currents of the switches and the parts of the inductor current coming from the input and the capacitor using the following formulas:

$$u_{T4} = d_4 \cdot u_{C4}$$

$$u_{D4} = (1-d_4) \cdot u_{C3}$$

With this basic principle the EMR was expanded to a 4-stage MSBA converter, which is shown in Figure 5.

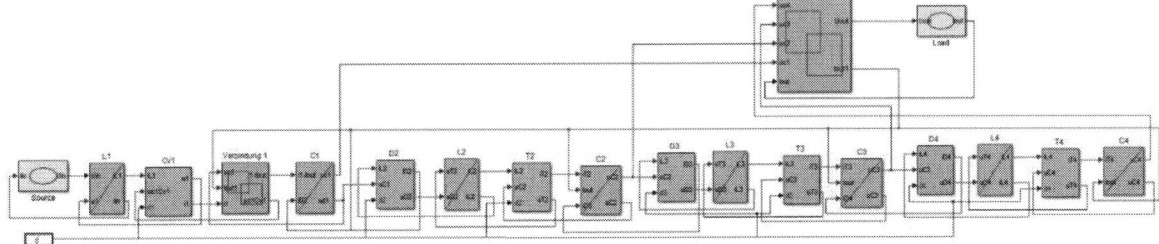

Figure 5: EMR of the 4-stage MSBA converter

© VDE VERLAG GMBH · Berlin · Offenbach

3. Control

A dedicated control structure using the principle of Feed-Forward to control the MSBA converter and avoid resonant oscillations was implemented with the EMR and simulated. The implementation for the elementary MSBA stage can be seen in Figure 6.

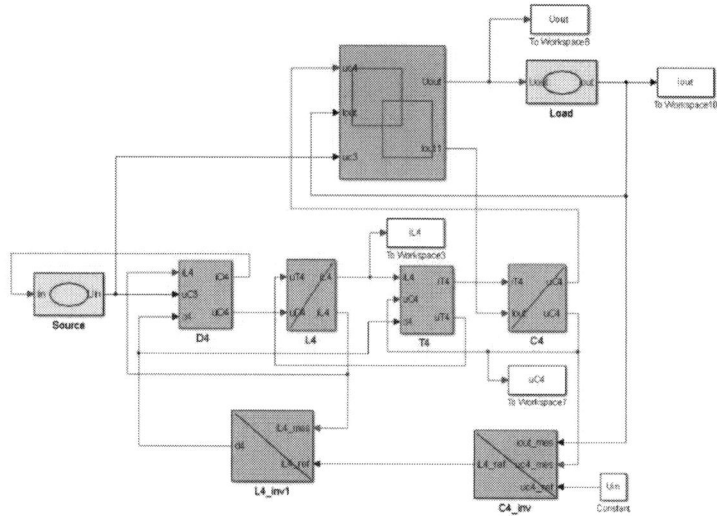

Figure 6: Example of the control structure for the elementary MSBA stage in MATLAB Simulink

The control structure used for the single MSBA stage and the boost stage are represented in Figure 7. The current in the inductor is controlled with a simple proportional controller and the converter-part itself is commanded through a PWM modulator. Due to the presence of a pure integral component for the system to be controlled, the current controller doesn't need any integral component.

The superimposed voltage balancing controller is also a single proportional controller. Even if the output load represents a perturbation for the voltage control loop, the need of an integral component in the controller itself can be avoided through feedforward of the perturbation current that is measured. The feedforward signal must of course represent an anticipation quantity for the inductor current. This signal will be calculated from the measured output current.

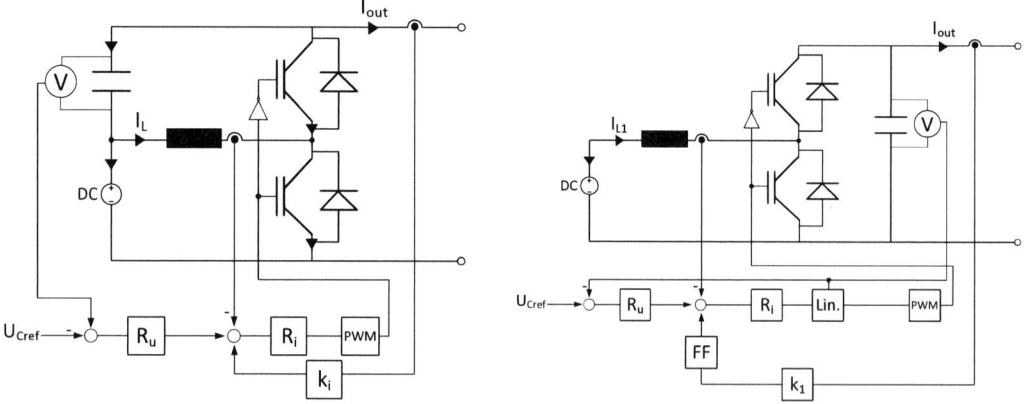

Figure 7: Control circuits of the single MSBA stage (left) and the boost stage (right)

The complete control structure for the 4-stage MSBA is shown in Figure 8. In order to properly protect the current solicitation of the power semiconductor devices and the passive components, it became apparent that the control structure must present the possibility to limit the current during transient operation. Therefore, between the voltage balancing controller with its feedforward signal and the current controller, a limitation element has been inserted in all stages except the boost stage (Figure 7 right) since it is not needed there.

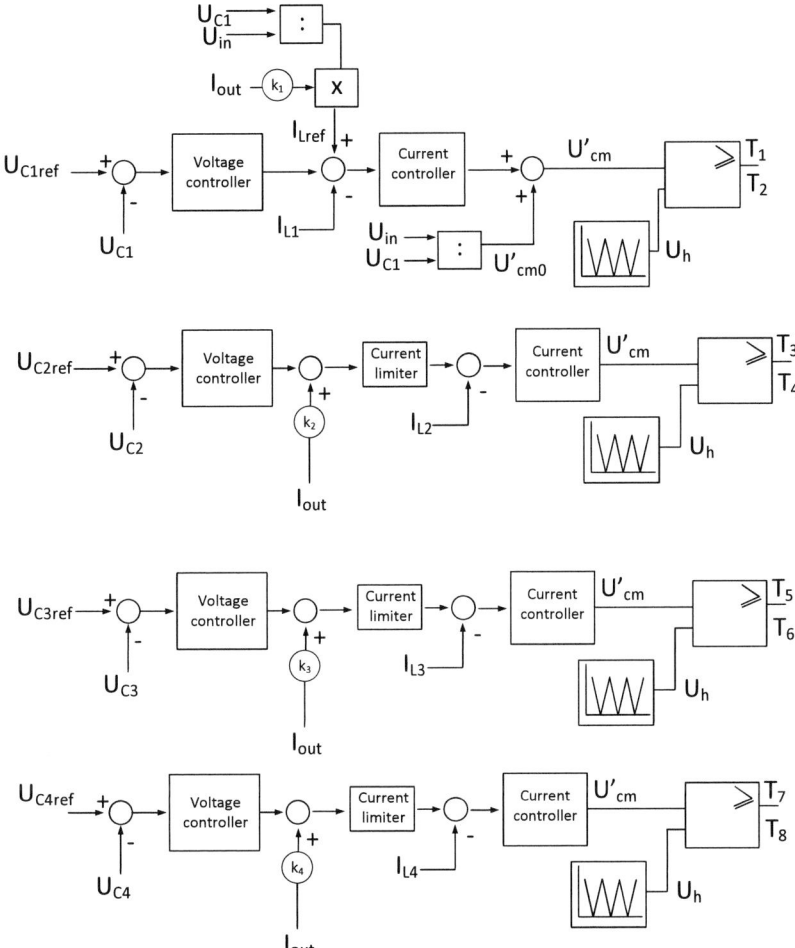

Figure 8: Control for the 4-stage MSBA

4. Simulation and prototyping of the controlled system

For the simulation of the controlled system the load is set to 40Ω, the input voltage to 100V and for the switches a switching frequency of 20kHz is chosen. Figure 9 shows the capacitor voltages and the inductor currents of the EMR simulation of the 4-stage MSBA converter with implemented control.

PCIM Europe 2016, 10 – 12 May 2016, Nuremberg, Germany

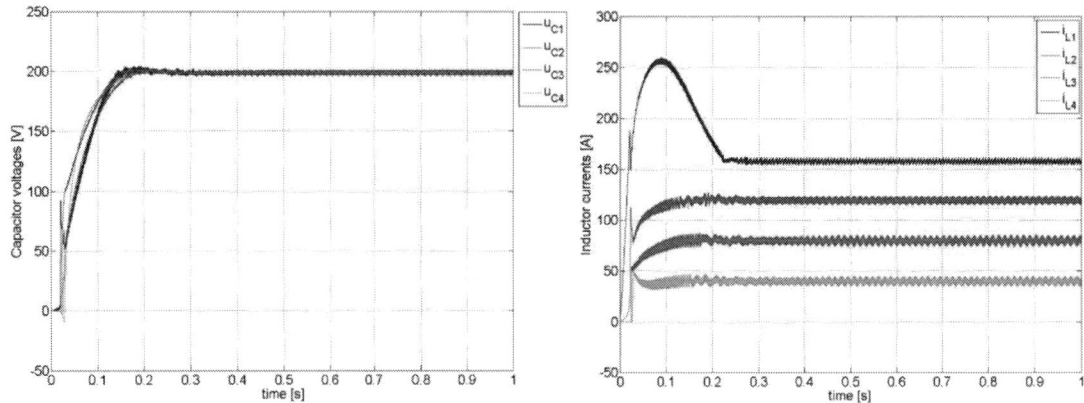

Figure 9: Capacitor voltages (left) and inductor currents (right) of the 4-stage MSBA

Figure 10 shows the output voltage and the output current of the 4-stage MSBA converter.

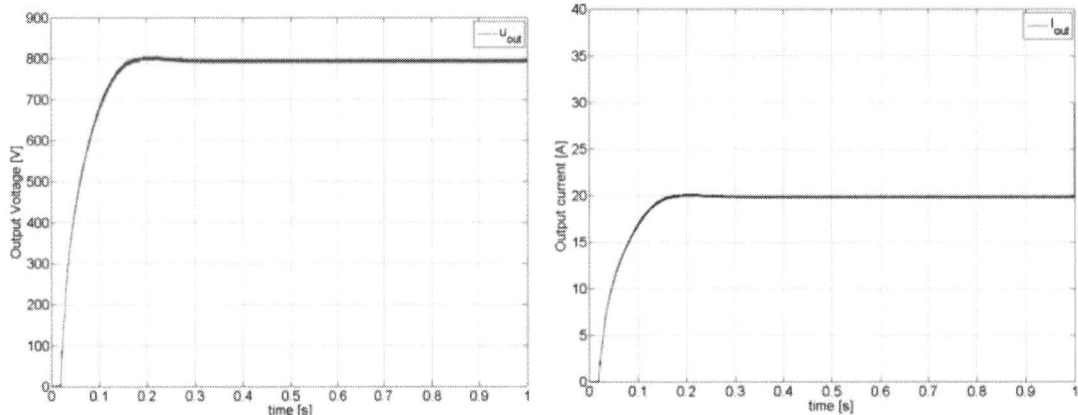

Figure 10: Output voltage (left) and output current (right) of the 4-stage MSBA

To further verify the functionality of the control a laboratory prototype using standardized building blocks [6] has been build (Figure 11).

Figure 11: 4-stage MSBA prototype

This prototype is using optical receivers and LEM transducers as a compatible interface concept for the numeric controller. The converters is built to work with an input voltage of 100V using a step up ratio of 1:8. More details about the prototype can be found in Table 1.

© VDE VERLAG GMBH · Berlin · Offenbach

PCIM Europe 2016, 10 – 12 May 2016, Nuremberg, Germany

Table 1: Prototype ratings

	Converter		IGBT
V_{in}	100V	V_{max}	600V
V_{out}	800V	I_C	72
P_{out}	1.5kW	E_{on}	4.4mJ
C_{1-4}	1.5mF	E_{off}	3.45mJ
L_{1-4}	1.7mH	V_{CE0}	1.7V
f_s	20kHz	R_{Gate}	1Ω

Experimental results during steady-state operation are shown in Figure 12 (left).

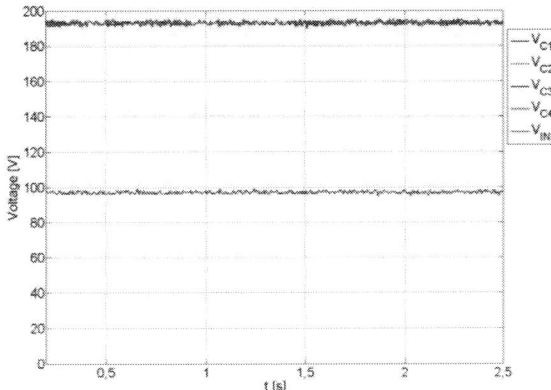

Figure 12: Steady-state measurement

It is apparent that the voltage levels of all capacitors are successfully kept at same level. In this case the voltage correspondents to the set reference value of the doubled input voltage.

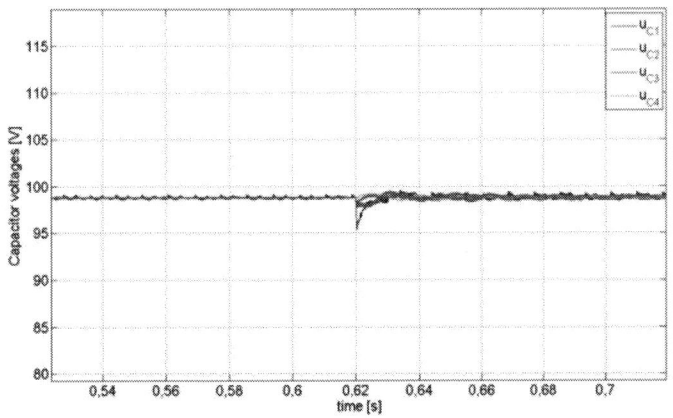

Figure 13: Load step measurement

Experiments testing the prototype in case of perturbations for example at the output because of a load change also showed satisfying results and a good response from the control and measurements can be seen in Figure 13.

© VDE VERLAG GMBH · Berlin · Offenbach

5. Conclusion

The paper showed the Energetic Macroscopic Representation (EMR) for the 4-stage MSBA converter. The converter representation was build up from top to bottom starting with an elementary MSBA stage. The different formulas for the components inside a stage that are used in the EMR were shown as well as their implementation with MATLAB/Simulink. The same was done for the boost stage and finally for a complete 4-stage MSBA converter with a source and a load. Open loop simulations were carried out to ensure the correct implementation. Finally a control based on the feed-forward of the output current was implemented for all stages and the blocks used to represent this were explained. Last simulations were done to verify that the EMR and control structure of the boost stage, the MSBA stage and the 4-stage MSBA converter function correctly during normal operations and with perturbations at the output. Finally a prototype was built to further verify these results. The measurements verified that the EMR concept and the implemented control work reliably.

References:

[1] G. K. Steinke, A. Rufer, "Use of a DC-DC step up converter in photovoltaic plants for increased electrical energy production and better utilization of covered surface area", PCIM 2015, Nuremberg Germany, 19-21 May 2015

[2] Rufer, P. Barrade, G. Steinke, Voltage Step-Up Converter based on Multistage Stacked Boost Architecture (MSBA), IPEC 2014: International Power Electronics Conference, Hiroshima, Japan, 18-20 May 2014

[3] G. K. Steinke, A. Rufer "Comparison of Multistage Stacked Boost Architecture topologies in regards of duty cycle, stage number and efficiency relationship", IECON 2015

[4] http://www.emrwebsite.org

[5] J. P. Hautier, P. J. Barre, "The causal ordering graph - A tool for modelling and control law synthesis", Studies in Informatics and Control Journal, vol. 13, no. 4, December 2004, pp. 265-283.

[6] http://imperix.ch/products/power/modules/peb6035-phase-leg

PCIM Europe 2016, 10 – 12 May 2016, Nuremberg, Germany

Adjustable 20 kW full-SiC electronic load with energy recovery for medium-frequency inverter

Fabian Denk, Karsten Hähre, Julian Körner, Rainer Kling, Wolfgang Heering,
Karlsruhe Institute of Technology (KIT) – Light Technology Institute (LTI), Engesserstr. 13,
76131 Karlsruhe, Germany, fabian.denk@kit.edu

Abstract

An adjustable electronic load with energy recovery is required for the development and the characterization of high power medium-frequency (MF) inverters. Therefore, in this work we present a DC-DC-converter operating as an electronic load that can handle up to 20 kW of input-power at frequencies over 2.5 MHz. The input resistance can be varied between 5 and 40 Ω. We implemented energy recovery in order to minimize the energy consumption during long term testing of new MF-inverters. The supplied AC-power is fed back into the DC-link of the MF-inverter. To achieve efficiencies over 97 % at high power densities, exclusively SiC devices are utilized.

1. Introduction

Wide band gap semiconductor devices, e.g., SiC MOSFETs, are ideal for high power, high voltage and medium-frequency applications like inductive heating, plasma generation or edge-zone hardening, as well as resonant-mode DC-DC-converters. Examples of medium-frequency inverters for plasma generation are given in [1] and [2], and are in the scope of actual research. However, during the investigations, the delivered output power of the medium-frequency inverters is converted to heat without further purpose in an ohmic load. However, an adjustable load can simplify the development and the characterization of medium-frequency inverters. Therefore, we developed an adjustable electronic load with energy recovery that can handle input powers up to 20 kW AC at frequencies of more than 2.5 MHz. Its resistance can be adjusted between 5 and 40 Ω. Furthermore, the implemented energy recovery reduces the power consumption of our laboratory equipment and cooling efforts drastically compared to a conventional load. Figure 1 illustrates the energy flow of the measurement setup. In simple words, the AC output power of the medium-frequency inverter is rectified and fed back into its DC link.

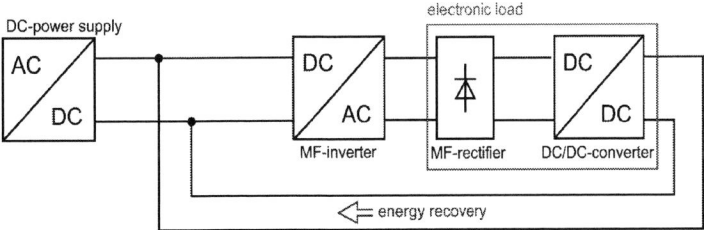

Figure 1: Block diagram of the electronic load to illustrate the energy flow of the measurement setup.

© VDE VERLAG GMBH · Berlin · Offenbach

2. Method and Materials

The schematic diagram of the electronic load is shown in Figure 2. It can be divided into three parts: Firstly, the MF-rectifier ($D_1 - D_4$) secondly, the full-bridge converter ($S_1 - S_4$) with a 1:1 transformer (T_1) for galvanic isolation and the corresponding low frequency rectifier ($D_5 - D_8$) and thirdly, a boost converter (S_5 and D_9).

Between the MF-rectifier and the full-bridge converter, a CLC filter, C_1, L_{1a}, L_{1b} and C_2, is implemented to separate the medium- from the low-frequency part. The choke L_{1a} and L_{1b} of this filter is implemented as a common mode choke to compensate common-mode noise.

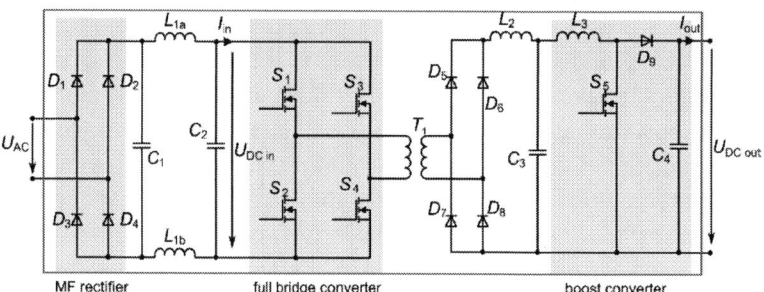

Figure 2: Schematic diagram of the electronic load's power stage, consisting of the MF-rectifier (D_1 - D_4), the full bridge converter (S_1 - S_4), with transformer (T_1) and rectifier (D_5 - D_8), and the boost converter (S_5, D_9).

The desired input resistance can be adjusted by varying the duty cycle α of the boost converter and the phase shift angle β of the full bridge converter. Hence, the rectified input voltage $U_{\text{DC\_in}}$ can be calculated using (1) and the input current using (2), as given in [4, 5], respectively.

$$U_{\text{DC in}} = \frac{U_{\text{DC out}} \cdot (1 - \alpha)}{\sqrt{\beta/\pi}} \quad (1) ; \qquad I_{\text{in}} = \frac{1}{2L_3} \cdot U_{\text{DC out}} \cdot \alpha \cdot (1 - \alpha) \cdot T_{\text{boost}} . \quad (2)$$

Consequently, the output voltage $U_{\text{DC\_out}}$ is a fixed value determined by the DC power supply. To prove the design concept and to dimension the cooling system, a PLECS-simulation (Plexim GmbH, Switzerland) is performed. To gather system efficiency from PLECS, a detailed model of every utilized semiconductor has to be developed. Most of the models are based on the values given by the datasheets of the devices. The first part of the simulation aims at finding a good ratio between the switching frequency and the needed value of inductances as well as the efficiency. Thereby, we decide to choose 30 kHz for the full-bridge and 100 kHz for the boost-converter. The semiconductor losses as a function of the output power are

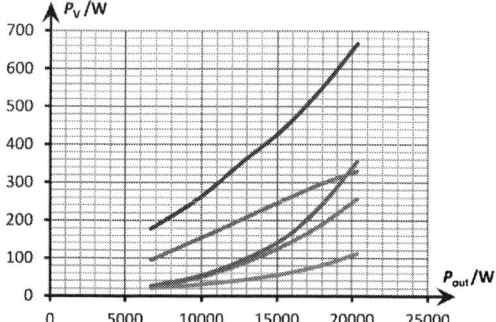

Figure 3: Comparison of the simulated semiconductor losses as a function of the output power for $f_{\text{full-bridge}} = 30$ kHz and $f_{\text{boost}} = 100$ kHz. There, the red curve represents the switching and conduction losses of the full-bridge, the orange curve the losses of the low frequency rectifier, the blue curve the losses of the booster

depicted in Figure 3. The several heatsink and the overall cooling system can be dimensioned with respect to these curves.

To estimate the dependence of the output power on the parameters α and β, the output power, as well as the efficiency, for the complete electronic load is plotted in Figure 4 (a) and (b) as a function of α and β.

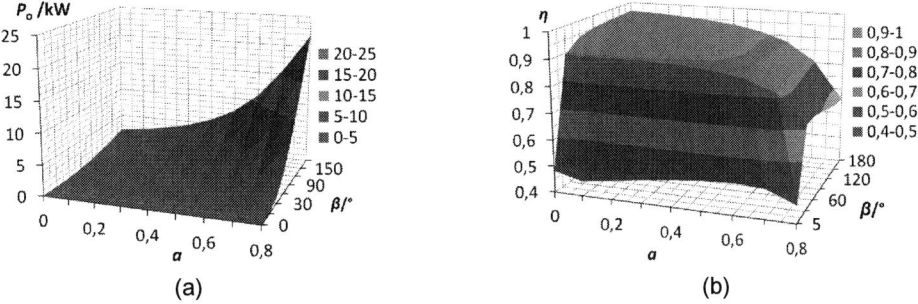

(a) (b)

Figure 4: 3D-plot of the simulated output power P_0 and the corresponding efficiency η as a function of the duty cycle, α, of the boost converter and the phase shift angle, β, of the full bridge converter with a DC input voltage of $U_{DC\,in}$ = 350 V are presented in (a) and (b), respectively.

To guarantee safe operation of the MF-inverter, an algorithm is implemented that enables constant resistance control at the input of the electronic load. The parameters of the transfer function can be extracted from the PLECS simulation and implemented into the control unit by following [5]. The control unit is implemented on a microcontroller-board (TWR K60F120, Freescale Semiconductor - NXP, Netherlands). A block diagram of the control unit is shown in Figure 5.

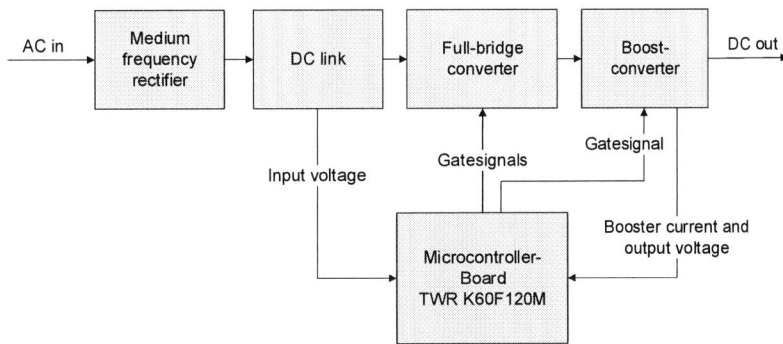

Figure 5: Block diagram of the control unit. The gate signal are generated on a microcontroller board based on the in- and output-voltage as well as the booster current.

3. Experimental Setup

The different parts of the electronic load are built as several printed circuit boards, which are connected via copper bus bars or cables. Thus, the parts can be tested separately and replaced easily. Figures 5 – 7 show photographs of the main parts of the electronic load. For a compact and efficient cooling, all power semiconductors are placed on liquid-cooled heat sink. The MF-rectifier ($D_1 - D_4$) in Figure 6 is built with four 1200 V 20 A SiC Schottky diodes (C4D20120A, Cree Inc., USA) in TO-220-2 package. The relatively small TO-220-2 package exhibits less parasitic capacitances compared to the TO-247-3 package, making it more suitable for the MF-frequency. For the full-bridge converter (S_1 - S_4) in Figure 8, two 1200 V SiC MOSFET half-bridge modules (CAS100H12AM1, Cree) are utilized. The low frequency bridge rectifier (D_5 - D_8) is built with four 1200 V 40 A SiC Schottky diodes (C4D40120D, Cree) in TO-247-3 package. These diodes are suitable to be placed on the same water-cooler as the parts of the boost converter, which are also in TO-247-3 package, as shown in Figure 7. The boost converter (S_5, D_9) is built with two 1200 V 40 mΩ SiC MOSFETs (C2M0040120D, Cree) in parallel and two 1200 V 40 A SiC Schottky diode (C4D40120D, Cree).

The transformer (T_1) for the galvanic isolation is wound with high frequency litz wire on an Epcos PM 114/93 N27 core. Hence, a magnetizing inductance of L_m = 3.5 mH and a leakage inductance of L_σ = 1.1 µH could be achieved. For the common mode choke (L_{1a} and L_{1b}) of the CLC filter an Epcos N30 toroid with enameled copper wire winding was used. The DC choke (L_2) is built with an Epcos PM 74/59 N87 core with an enameled copper wire winding and the booster choke (L_3) with high frequency litz wire on a Micrometals T520-2 core.

  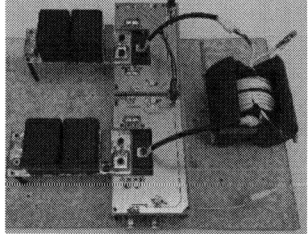

Figure 6: Photograph of the liquid cooled MF-rectifier.

Figure 7: Photograph of the liquid cooled boost converter with the low frequency rectifier.

Figure 8: Photograph of the liquid cooled full-bridge converter including the 1:1 transformer.

The measurement setup consists of a high-bandwidth oscilloscope (RTO 1044, 4 GHz, 20 GSa/s, Rohde & Schwarz, Germany) to measure the gate-source U_{GS} and the drain-source voltages U_{DS} of the transistors, as well as the voltages across the rectifier diodes and the currents. For measuring U_{GS}, 10:1 (RT-ZP10, Rohde & Schwarz), and U_{DS}, 1000:1 (RT- ZH11, Rohde & Schwarz), passive voltage probes are used, respectively. The currents are measured using current probes (RT-ZC20, Rohde & Schwarz) and current transformers (6600 wideband current monitor, Pearson Electronics, USA). For the characterization, the electronic load is tested, as a normal DC-DC-converter with a DC source (EA-PS 91500-30 3U, EA Elektro-Automatik GmbH & Co. KG, Germany) and DC load. The DC load is an array of halogen lamps with a nonlinear resistance. To determine the efficiency, the in- and output powers are measured using a three phase precision wattmeter (LMG 310, ZES ZIMMER Electronic Systems GmbH, Germany).

4. Results and Discussion

The characterization is performed in several steps: First, only the full bridge and afterwards the full bridge including the boost converter is tested. The measured waveforms of the full-bridge converter including the transformer and the low frequency rectifier are presented for a phase shift angle of β = 83° and β = 180° in Figure 9 (a) and (b), respectively. There, one can see that a reduction of the phase shift angle β results in a reduction of the effective value of the secondary voltage and current, which is equivalent to a reduction of the output power. The efficiency of this setup in the β = 180° working point has reached more than 97.5%.

© VDE VERLAG GMBH · Berlin · Offenbach

PCIM Europe 2016, 10 – 12 May 2016, Nuremberg, Germany

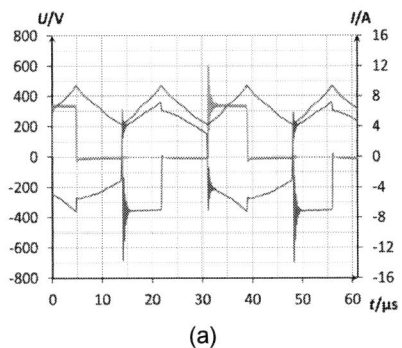 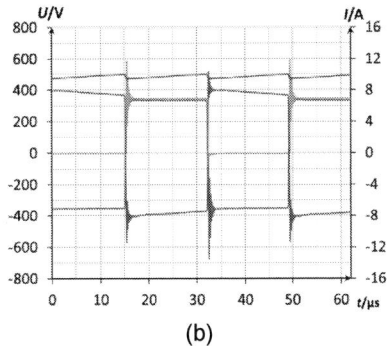

(a)   (b)

Figure 9: Measured waveforms of the full-bridge converter at $U_{DC\,in}$ = 350 V for β = 83° (a) and for β = 180° (b). There, the purple curve represents the current through the DC choke L_2, the blue curve the voltage and the red one the current through the secondary winding of the transformer and green curve the voltage across the low frequency rectifier diode D_7.

The next step is to run the full-bridge converter inclusive the booster. The waveforms of these measurement are shown in Figure 10 (a) for α = 14% and in (b) for α = 44%. It can be seen that generally the circuit is working, but during the investigation two problems occurred. Firstly, the DC-choke saturates and secondly, the transformer winding gets too hot. The problem with the saturation of L_2 can be solved by adding an increased airgap to the choke core, but the transformer has to be redesigned.

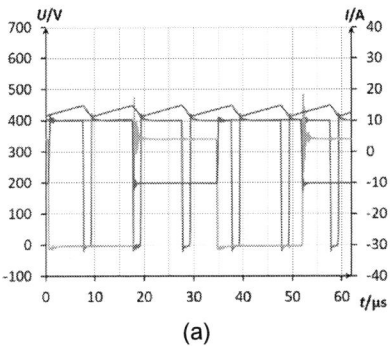 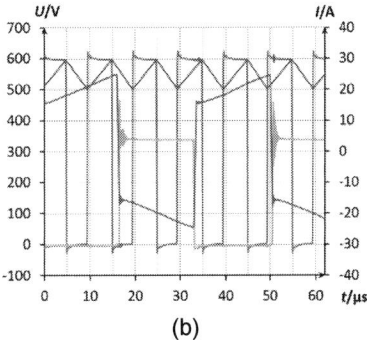

(a)   (b)

Figure 10: Measured waveforms of the full-bridge converter and the booster at $U_{DC\,in}$ = 350 V for β = 180° and α = 14% in (a) and for α = 44% in (b). The blue curve represents the voltage across the MOSFET S_5 of the booster, the green one the voltage across the low frequency rectifier diode D_7, the red one the current through the secondary winding of the transformer and the purple curve the current throw the booster coke L_3.

The initial version of the transformer was built with layer windings, but during the first measurements the litz wire got almost 100°C hot at an output power of only P_{out} = 8.2 kW, this can be seen in the thermal image in Figure 11. For that reason an improved version with a disc winding was produced. Due to the minimized parasitic couple capacity between the windings, the ringing of the secondary current could be reduced, which can be seen in Figure 12 (a) and (b). However, this reduction only decreases the losses in the low frequency rectifier and not in the transformer. Thus, the maximum output power could not be exceeded with the new transformer winding.

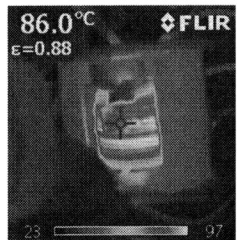

Figure 11: Thermal image of the transformer at P_{out} = 8.2 kW.

© VDE VERLAG GMBH · Berlin · Offenbach

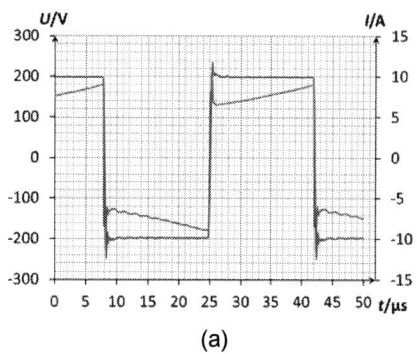 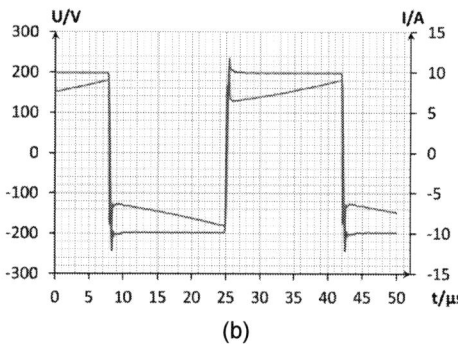

(a)   (b)

Figure 12: Comparison of the measured secondary current and voltage waveforms of the transformer with layer winding in (a) and disc winding in (b) at $U_{DC\,in}$ = 200 V and β = 180°.

The results of the efficiency measurements can be seen in Figure 13. There, the efficiency as a function of the output power with the corresponding values of the duty cycle α of the boost converter and the phase shift angle β of the full bridge converter are plotted. However, the desired output power of 20 kW could not be achieved due to the inadequate transformer. Nevertheless, the electronic load is working up to an output power of 8.2 kW with a maximum efficiency of 97.4%.

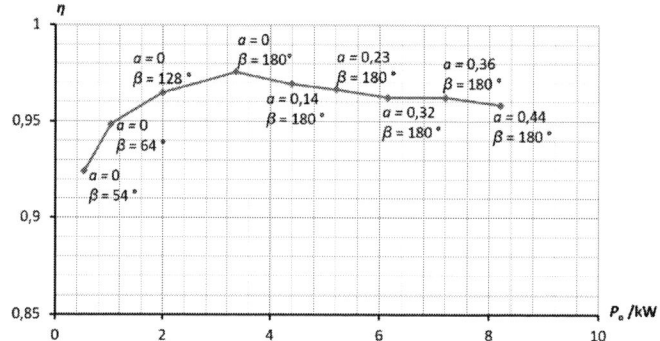

Figure 13: Measured efficiency as a function of the output power for different values of the duty cycle α of the boost converter and the phase shift angle β of the full bridge converter at $U_{DC\,in}$ = 350 V.

5. Conclusion

This paper presents the development of an adjustable electronic load with energy recovery. The resistance can be adjusted between 5 and 40 Ω. To achieve high efficiency, only silicon carbide semiconductors are utilized. To prove the design concept and to dimension the cooling system, we have performed a detailed PLECS-simulation. The parameters for the control unit ware also extracted from this simulation and afterword the controller is implemented in a microcontroller. The different parts of the electronic load are built as several board, thus they can be tested separately. For the whole system an efficiency of 97.4% could be measured.

6. Acknowledgment

The German Federal Ministry of Education and Research (BMBF) supported this work in the context of the public funded project 16ES0204 "MMPSiC".

7. References

[1] F. Denk, K. Haehre, W. Heering, and R. Kling, "Minibloc SiC-MOSFET in a Resonant Half-Bridge Inverter Operating in the MHz-Range," in *Power Conversion and Intelligent Motion Conference (PCIM Europe 2015)*, 2015.

[2] K. Haehre, T. Lueth, F. Denk, R. Kling, and W. Heering, "Normally-on SiC-JFET Cascode under ZVS conditions," in *Power Conversion and Intelligent Motion Conference (PCIM Europe 2015)*, 2015.

[3] A. Pressman, K. Billings, and T. Morey, *Switching Power Supply Design, 3rd Ed.* 2007.

[4] M. H. Rashid, *Power Electronics Handbook*, Second Ed. 2007.

[5] M. K. Kazimierczuk, *Pulse-width Modulated DC-DC Power Converters.* 2008.

A new high frequency transformer for UPS

Michael Schmidhuber, Manfred Wohlstreicher, Michael Baumann, Markus Schmeller

SUMIDA Components & Modules GmbH, Dr. Hans-Vogt-Platz 1, 94130 Obernzell

mschmidhuber@eu.sumida.com

Abstract

Within the scope of a governmental funded project (FKZ: 16ES0208) new technologies for stationary storages or non-centralized energy sources shall be developed. Amongst other focal points the electronic DC-DC power converter should work in the range from 50kVA – 250kVA. A suitable method to reach the goals of high energy density, compactness, high efficiency and moderate costs is the significant increment of the switching frequency. SiC based devices in a hard switching circuit topology are enablers to drive the high frequency transformer which is purposed for galvanic isolation. The reduction of switching losses in the ferrite based transformer core material together with high current conductor winding techniques lead to a significant improvement of the system effectiveness. The paper describes an interim development status of this novel passive transformer unit.

1. Synopsis

Around 650.000 new Uninterruptible Power Supplies (UPS) in a power range > 10kW are installed. The average life time of those systems is estimated to 10 years. The enhancement of the system effectiveness by only 2 percent results in an annual reduction of 10TWh energy consumption. In parallel more than 10 megatons of CO_2 can be avoided. The effective savings could even be higher because older types of UPS have a lower efficiency (>85%). This project aims to reduce costs by a lower dedicated material effort. Reducing the volume also leads to a more attractive power converter system and thus to a sustainable energy footprint. The most bulky element is the transformer which galvanically isolates the battery source to the output circuitry. Thus the biggest potential to save resources is to minimize its size by increasing the switching frequency according to the basic transformer law:

$$U \sim B \cdot f \cdot N \cdot A$$

Saturation flux density (B) within the core material and frequency (f) build the magnetic product which has to be high, when the geometric product consisting of the winding number (N) and the area (A) shall be small. The challenge of realizing a high power DC-DC converter is driven by lowering core and winding losses.

This paper describes which effects are to be considered on a transformer design when low skin and proximity effects, high temperatures, big core masses, high flux densities and best couple factors or very low leakage inductance are required, by means of an interim status of the overall project.

2. Transformer Design

2.1. Requirements and first assessment

For the design of magnetic components an optimal compromise between loss and weight/volume is important. The first and most challenging part when developing a high efficient power transformer is the minimization of the dissipation power in the core as well as in the windings. The second demand resulting from the hard switching B6-topology is minimizing the leakage inductance to avoid the reduction of the output voltage at the transformer inductivities. The converter has a nominal output power of 100 kW at 500 V input and 750 V nominal output voltage at a frequency of 50 kHz. The first sample had neither housing nor any casting compound to acquire thermal resistances and temperature hotspots by direct measurement.

For the development of scientific solutions for the inductor design, a software tool was created. It summarizes the existing methods for the theoretical calculation of power inductors and supplements it additionally to the "Method of McLyman" (see Figure 1). Sizing the converter, the selection of core shapes, enameled wires and material parameters are available. The core geometry coefficient can then be selected in accordance with a volume-optimized design. The result is a converter with an EE ferrite core with a cross-sectional area of 16 cm², a flat wire with a cross section of 50 mm², calculated losses of 271 W and a total weight of about 12 kg.

Frequency [kHz]	40	50	60	40	50	60
Power [kW]		50			100	
Min. input voltage [V]			500			
Nom. output voltage [V]			750			
Current density [A/mm³]	3	3	3	3	3	3
Primary winding number	19	15	13	13	10	10
Input current [A]	100,5	100,5	100,5	201,0	201,0	201,0
Primary winding cross section [mm²]	34,0	32,6	32,2	67,0	66,6	67,0
Primary winding losses [W]	27,6	22,8	18,2	58,0	37,1	32,2
Secondary winding number	30	24	21	24	18	16
Secondary winding cross section [mm²]	22,5	21,6	21,3	44,4	44,1	44,4
Secondary winding losses [W]	28,9	22,3	19,5	61,5	36,8	34,1
Total winding losses [W]	56,5	44,1	37,7	119,5	73,9	66,3
Calculated core losses [W]	57,8	74,6	82,8	147,3	197,4	237,0
Total losses [W]	114,3	118,9	120,5	266,8	271,3	303,3
Core weight [g]	1184	2976	2550	3064	7872	7296
Copper weight [g]	2940	2936	1771	6092	4286	1381
Total weight [g]	9124	5112	4321	14130	12138	10672
Dimensions LxBxH[mm]	240x120x160	230x120x160	220x120x150	120x120x180	310x120x180	280x120x180
Volume [l]	4,61	4,42	3,96	0,91	6,7	6,05

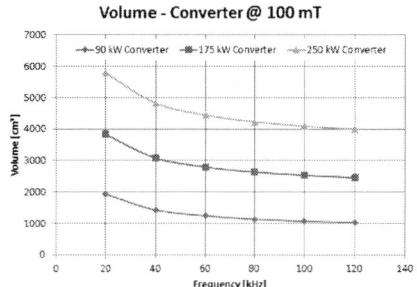

Figure 1: Mathematical simulation of volume, losses and electrical transformer parameters for various power levels

2.2. Ferrite core selection

For preselection and evaluation of different core materials basic investigations have been carried out on different Mn-Zn ferrite materials. The goal was to estimate the losses depending on the magnetic mass in large-volume cores which will be additionally a matter of the thermal resistance and its cooling.

For this purpose a series of measurements were performed on a R38 core with different cross-sectional areas A at variable frequencies f and a fixed saturation flux density B. The sum of the hysteresis and eddy current losses increases thereby linearly with the frequency and the magnetic cross section (see Figure 2). These results provide essential input on mathematical core loss calculation models and are also the basis for developing low loss optimized ferrite materials for large volume transformers. Sumida ferrite core material Fi335 has been selected.

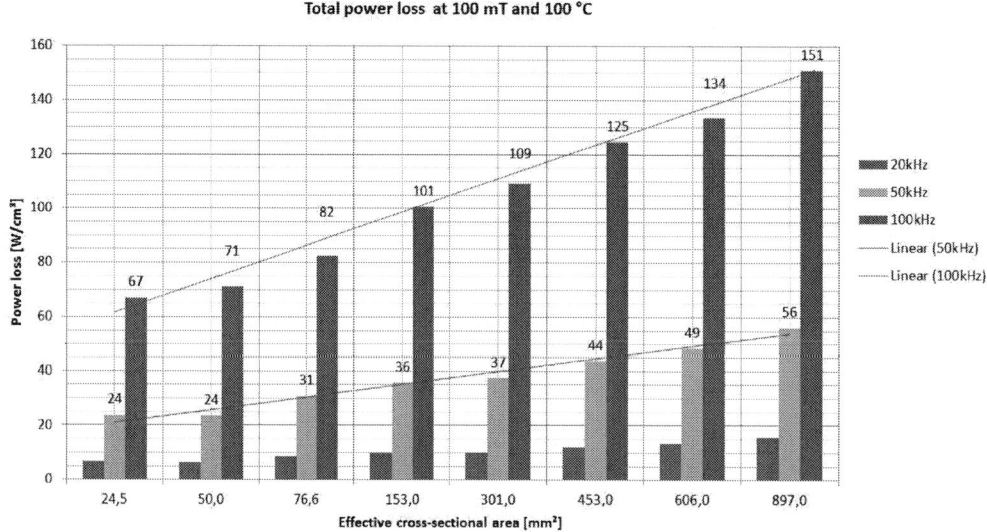

Figure 2: Total power loss of a ferrite material by different cross-sectional areas and frequencies

2.3.   Wire and winding selection

Due to the high operating frequency the eddy current effects (skin and the proximity) are increasing. To minimize these eddy current losses litz wire is the first choice, following the equation: $\delta = (\pi * f * \mu * \kappa)^{-0.5}$ for the penetration depth.

Nevertheless the very small value of leakage inductance of less than 1.4 µH has not been fulfilled in the first design. Simulations showed that the coupling factor was too less due to the winding structure. As a result the current pulse rate of rise could not be achieved.

In this case rectangular wire with a special wire wrap technique delivered a value of 0.6 µH which was acceptable. The skin influence has been partially compensated by the proximity effect, which leads to a current flux density in the inner part of each wire cross section of more than 5 A/mm² (see Figure 3).

Figure 3: Power loss of flat wire in nested winding

2.4. Simulation and results

Various designs have been modelled and simulated within the magnetic analysis software JMAG Designer 6. The lowest losses in the core material are calculated and also simulated to 91 mW per cm³, whereas the core volume adds up to 699 cm³. The winding losses contribute with 292 W to the overall system. The coupling capacitance is calculated to 20 pF and the resonance frequency has been measured at 581 kHz. An equivalent circuit diagram of the transformer is shown in Figure 4. The first draft of the DC-DC converter is shown in Figure 5.

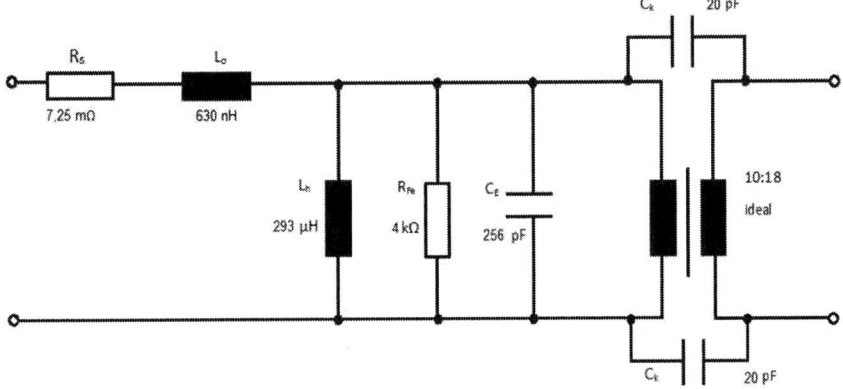

Figure 4: Equivalent circuit diagram of interims design of the transformer

Figure 5: CAD model and photo of the interims design of the transformer

3. Summary and outlook

The leakage inductance could be reduced in the development of the first converter design from 2.7 µH to 0.6 µH by a special wire wrap technique and placement. At a test rig the converter has been successfully operated at 50 kHz with a new ferrite material in terms of its transmission behavior up to the maximum possible surface temperature without casting compound/housing. In the next steps cooling technology and housing design have to be verified to integrate the converter into the overall power converter system. In accordance with the project partner also the terminal geometry has to be discussed to reduce the E-field in the bus bars towards the SiC switches.

References

[1] C. W. T. McLyman: Transformer and Inductor Design Handbook; Kg Magnetics; 2004; ISBN 0-8247-5393-3

[2] O. Kilgenstein: Schaltnetzteile in der Praxis; Vogel-Fachbuchverlag; 1992; ISBN 3-8023-1436-0

[3] T. Bülo: Methode zur Evaluation leistungselektronischer Schaltungstopologien für die Anwendung in dezentralen Netzeinspeisern kleiner Leistungen; KDEE; 2011; ISBN 978-3-86219-094-2

[4] R. W. Erickson, Dragan Maksimovi: Fundamentals of Power Electronics; 2004; ISBN 0-7923-7270-0

Acknowledgements

The research and development presented here was supported by the German Ministry of Education and Research (BMBF Grant No. 16ES0208).

PCIM Europe 2016, 10 – 12 May 2016, Nuremberg, Germany

DC/DC-converter for modular coupling of 48 V battery packs to a high voltage DC bus

Michael Eberlin, Milad Khani,
Fraunhofer Institute for Solar Energy Systems ISE, Germany,
michael.eberlin@ise.fraunhofer.de, milad.khani@ise.fraunhofer.de

Abstract

A DC/DC dual active bridge converter for connecting modular low voltage battery packs to a high voltage DC bus is presented. It enables easy scalability of the power output. The converter topology allows the use of zero voltage switching which enables a reduction of the switching losses. But in some operation points this feature is lost. To maximize the efficiency of the converter a variable frequency modulation scheme was developed. The experimental results show an improvement of the efficiency curve.

1. Introduction

In conventional battery packs, higher voltages are achieved by hard-wired series connection of the single cells, meaning that the weakest cell limits the performance of all cells. Moreover, one defective cell makes the whole battery pack unusable. Within the "'Cell-Booster'" project (BMBF reference number: 16ES0140), a modular concept combined with intelligent power electronics is being investigated. It optimizes the energy and cost efficiency of the battery storage systems that are common today and also tries to extend their operating duration and lifetime. The intended innovation addresses system technology, with a focus on a novel embedded power electronics, the so-called Cell-Booster. These power electronics are responsible for direct impedance matching and DC/DC conversion, which decouples the 48 V battery packs and forms a well-defined module. A nominal battery voltage of 48 V was chosen to meet the safety extra-low voltage (SELV) limits which reduces safety precautions for handling, transport and manufacturing. Battery pack and DC/DC converter form one module with output voltages between 680 V and 710 V. The number of modules in the storage system is variable so the capacity of the system can be changed easily. Thus it is possible to adapt the storage volume to the application which leads to highly flexible storage systems.

The modular construction not only improves the system efficiency but also makes it easy to implement a hybrid battery. In this way, storage systems combining lead-acid and lithium-ion batteries can be constructed directly and cycled individually, depending on their individual properties. Maintenance of the system is simple, as individual modules can be replaced even during operation. This ensures a long system lifetime, as a module with a defective cell can easily be replaced by a battery pack of the most recent generation with completely different properties.

2. DC/DC-converter

Each module consists of a lithium-ion battery pack and the DC/DC-converter. The converter represents the link between the 48 V battery pack and the high voltage three phase inverter input. Due to the high voltage gain a DC/DC-converter with a high frequency transformer has been chosen. A bidirectional power flow is needed to charge and discharge the battery pack.

© VDE VERLAG GMBH · Berlin · Offenbach

Due to these requirements the Dual Active Bridge (DAB) has been picked as the switching topology for the converter. This topology has been introduced in [1]. The equivalent circuit diagram can be seen in fig. 1. The topology consists of eight active switches (S_1 to S_8), two DC-link capacitors, one inductor and the transformer. The switches transform the DC voltage V_{HV} and V_{LV} into the symmetrical square wave voltages v_{ac1} and v_{ac2}.

For symmetrical loading of the transformer a voltage-fed full-bridge was used on the high- and low-voltage side. This allows the use of the full magnetisation range of the transformer core from the positive to the negative limit (push-pull principle) and therefore significantly reduces the transfomers volume and weight.

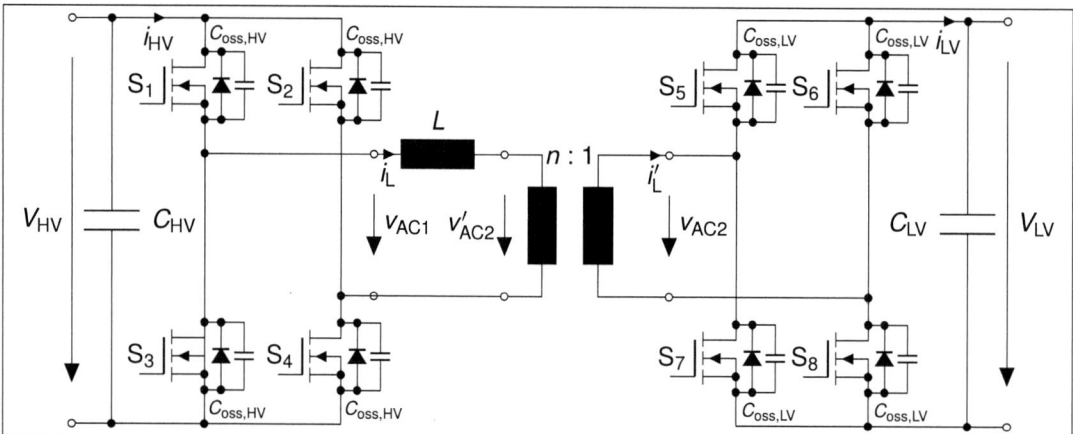

Fig. 1: Equivalent circuit diagram of the DAB with parasitic MOSFET output capacitors

The inductor L decouples the primary and secondary voltage sources and integrates the modulated voltage blocks to a continuous current. The transformer design integrates the inductor L into the leakage inductance to ensure a small module size and a low number of components.

The circuit diagram of the DAB can be simplified by referring all voltages and currents to the high voltage side. The simplified equivalent circuit diagram (Fig. 2a) is made up of only two square wave voltage sources and the inductor L. The simplified equivalent circuit diagram will be used for all further studies.

2.1. Phase shift modulation

The voltages v_{ac1} and v'_{ac2} need to be modulated to enable a power transfer between the two terminals. The most common modulation scheme for the DAB is the phase shift modulation (PSM). In this modulation scheme the primary and secondary full bridges are driven by a constant duty cycle of 50%. The power flow is controlled by introducing a phase shift between the two square wave voltages v_{ac1} and v'_{ac2}. The sign of the phase shift determines the power flow direction. To avoid transformer saturation, the symmetric modulation of the transformer current must be ensured at all times. The typical voltages and current of the PSM can be seen in fig. 2b. In this case d represents the phase shift between the two voltages and is defined between −0.5 and 0.5 and is equivalent to a phase shift between −90° and 90°.

PCIM Europe 2016, 10 – 12 May 2016, Nuremberg, Germany

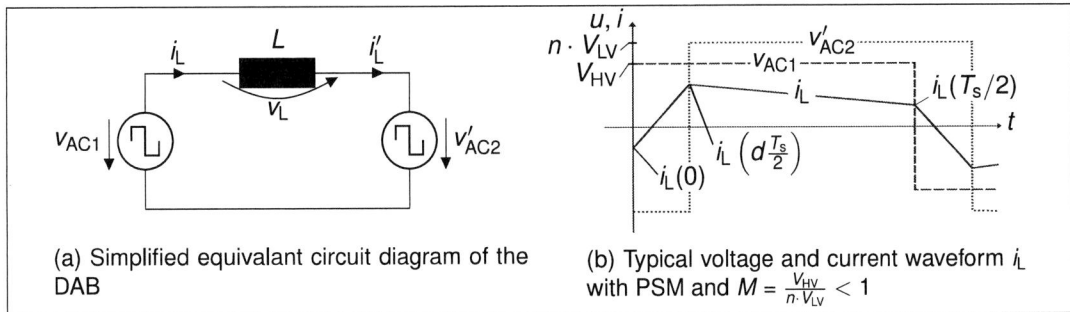

(a) Simplified equivalant circuit diagram of the DAB

(b) Typical voltage and current waveform i_L with PSM and $M = \frac{V_{HV}}{n \cdot V_{LV}} < 1$

Fig. 2: Analytical model of the DAB with PSM

With the current i_L from fig. 2b it is possible to describe the transferred power with the following equation:

$$P = \frac{n \cdot V_{LV} \cdot V_{HV} \cdot d \cdot (1 - |d|)}{2 \cdot f_s \cdot L} \quad (1)$$

n represents the transformer ratio and f_s the switching frequency. The maximum power flow occurs at a phase shift d of 0.5 or −0.5.

The voltages on the DAB determine the shape of the current waveform. Reactive currents will occur at voltage gains besides $M = \frac{V_{HV}}{n \cdot V_{LV}} = 1$. This will distort the current waveform and reduce the overall efficiency. Reactive currents will also be generated at higher phase shifts. Therefore it is suggestive to design the converter in a way, making the operation near the maximum of 0.5 unnecessary.

2.2. Zero voltage switching

The DAB enables the use of zero voltage switching (ZVS), which significantly reduces the switching losses. From now on MOSFETs will be considered for the active power switches.

ZVS means that the current at the turn-on instant flows through the antiparallel body diode of the MOSFET. This reduces the switch-on voltage to the forward voltage of the antiparallel diode and eliminates almost any overlap between voltage and current. This minimizes the turn-on losses and allows a higher overall efficiency of the system. For ZVS the current must commutate during the dead time onto the antiparallel diode of the next ongoing MOSFET. The commutation is driven by the inductor L. This is only possible if the current of the MOSFET is negative before turning-on and the parasitic output capacitor C_{oss} of the MOSFET has already been discharged to the voltage of the antiparallel body diode. Not only the sign of the current is important, but also the amount of stored energy in the inductor L. The inductor current must be able to completely discharge the parasitic output capcacitors during the dead time. Therefore a minimal current $I_{ZVS,x}$ can be defined in which the energy stored in the inductor L will be sufficient to achieve ZVS.

The current $I_{ZVS,x}$ depends on the inductor L and the parasitic output capacitors of the MOSFETs. It can be determined to [2]:

$$I_{ZVS,x} = 2 \cdot \sqrt{\frac{n \cdot V_{LV} \cdot V_{HV} \cdot C_{oss,x}}{L}} \quad (2)$$

© VDE VERLAG GMBH · Berlin · Offenbach

1735

with $C_{oss,x}$ representing either the parasitic output capacitor of the high or low voltage side which are both referred to the high voltage side.

Now it is possible to describe the ZVS conditions for each full bridge depending on the current through the inductor at the switching instant. The following conditions are derived for a power flow from the high to the low voltage side:

$$i_L(0) < -I_{ZVS,HV} \quad \text{(HV)} \qquad i_L\left(d \cdot \frac{T_s}{2}\right) > I_{ZVS,LV} \quad \text{(LV)} \tag{3}$$

ZVS cannot be achieved in every operating point by using phase shift modulation. The ZVS conditions depend on the inductor current profile. Especially at low power, these conditions are not fulfilled, which leads to losses and a lower efficiency. As already mentioned, the applied voltages influence the shape of the current and therefore the current at the switching instants. For instance at operating points with $M \neq 1$, only one of the both ZVS conditions can be fulfilled at light loads.

To describe this effect it is necessary to determine the current at the switching instants. With fig. 2b it is possible to derive the following equations for the currents:

$$i_L(0) = i_{L,0} = \frac{n \cdot V_{LV} - V_{HV} - 2 \cdot d \cdot n \cdot V_{LV}}{4 \cdot f_s \cdot L} \tag{4}$$

and

$$i_L\left(d \cdot \frac{T_s}{2}\right) = \frac{n \cdot V_{LV} - V_{HV} + 2 \cdot d \cdot V_{HV}}{4 \cdot f_s \cdot L} \tag{5}$$

Now it is possible to visualize the current at the switching instant when the phase shift d and therefore the power P is changed.

As already mentioned the ZVS conditions cannot be fulfilled at light loads with $M \neq 1$. In fig. 3 the voltage and current waveforms can be seen for three different loads with an operating point of $M > 1$. The voltage v_{ac2} is being phase shifted in respect to v_{ac1}. This represents a power flow from the high to the low voltage side. At nominal power (blue curve in fig. 3) both ZVS conditions are fulfilled. The critical switching instants for this operating point are marked with arrows. When the power flow is now being decreased, the current at the switching instants will get smaller. So, less energy will be stored in the inductor for the commutation. As a consequence, the ZVS conditions for the low voltage side cannot be fulfilled anymore and ZVS will be lost there. The current (green arrow) even changes its sign by further reducing the power.

With the ZVS conditions from eq. (3) it is possible to determine the ZVS area in dependence of the voltage gain M and the transferred power P (fig. 5). The parameters of the DAB used in fig. 5 are listed in table 1. The conditions and the ZVS area are only valid for a power flow from the high to the low voltage side. For a power flow in the other direction the conditions have to be swapped.

Fig. 5 shows the loss of ZVS at low power levels because at least one of the ZVS conditions cannot be satisfied. Depending on the voltage gain M either the high voltage side or low voltage side bridge loose ZVS. Even at $M = 1$ ZVS cannot be achieved at very small loads. In these operating points the current at the switching instants is very small due to extremely low reactive current. So the parasitic output capacitors of the MOSFETs cannot be discharged durig the dead time.

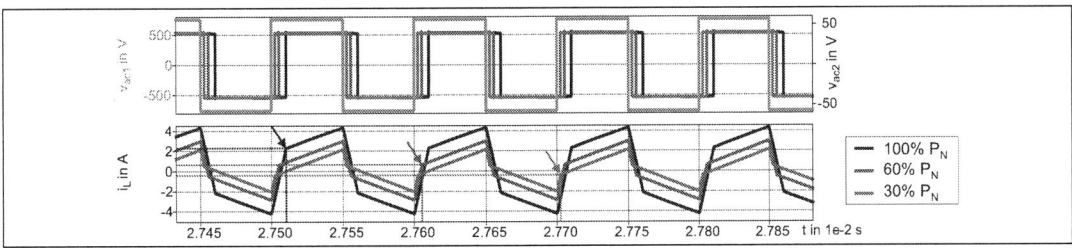

Fig. 3: Voltages and current with PSM for different loads at $V_{HV} = 760\,V$ and $V_{LV} = 40\,V$ ($M > 1$)

It must be mentioned, that for simplification the ZVS area was calculated for a constant parasitic output capacitor of the MOSFETs. Of course this assumption is not valid for reality, as there will be variations due to manufacturing. Also the applied voltage will have a strong impact on the amount of charge being stored in the parasitic output capacitance.

2.3. Power loss distribution

An analytical loss model has been designed to describe the losses in the DAB. The model primarily estimates the losses produced in the MOSFETs. Three operating points at different battery voltages have been compared. These points are marked in fig. 5 as red dots. The power in all three operating points is equal ($P = 500\,W$). Also the voltage on the high voltage side is constant at $V_{HV} = 710\,V$. The result of the examination can be seen in fig. 4. Also the corresponding current waveforms can be seen beside each loss bar chart.

At a battery voltage of 40 V the ZVS requirements for the LV side from eq. (3) cannot be fulfilled and ZVS is lost. This can also be seen in the current waveform, as the current at the switching instant of the low voltage side bridge is negative. But according to the ZVS conditions this current has to be positive to achieve ZVS. Therefore turn-on losses exist on the low voltage side. The high voltage side operates under ZVS because the condition is fulfilled. In this case the turn-off losses also play a major role in regards to the total losses. As the highest current occurs at the switching instant of the high voltage side bridge, the highest turn-off losses occur there, too. If ZVS would not be achieved in this operating point even higher turn-on losses would occur and the overall efficiency would be worse.

At a battery voltage of 47 V both bridges operate under ZVS. The currents at the switching instants are relatively small because of the absence of reactive current. So, the turn-off losses are smaller. Also, smaller conduction losses hint to lower reactive currents in the system. But there is enough energy stored in the inductor for the commutation.

At 55 V the high voltage side bridge doesn't operate under ZVS. Both turn-on and turn-off losses can be seen. The low voltage side bridge now has to switch at the highest current which results in big turn-off losses.

The self consumption of the converter is constant throughout all operating points. The self consumption consists of the supply of the microcontroller, the digital and analogue circuits, the gate drivers and the fan.

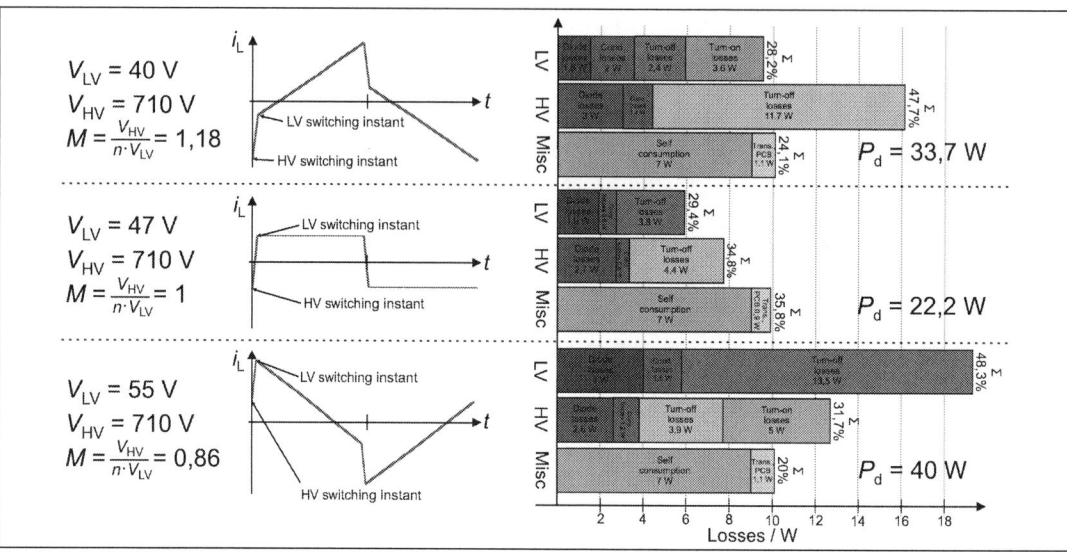

Fig. 4: Calculated loss distribution with $P = 500\,\text{W}$, $V_{HV} = 710\,\text{V}$, various battery voltages and the corresponding current waveforms

2.4. Variable frequency modulation

The ZVS conditions depend on the switching frequency of the circuit. Therefore it is possible to influence the ZVS area by varying the switching frequency. The ZVS area can be extended by raising the switching frequency. This effect can be seen in fig. 5. A disadvantage of higher switching frequency are more switching losses. It is therefore advisable not to operate at a higher switching frequency all the time. It also has to be considered that a higher switching frequency will extend the area in which none of the two bridges will operate with ZVS. A lower switching frequency will result in a smaller ZVS area but will also result in smaller switching losses. Due to these reasons the switching frequency should be changed in a way satisfying the ZVS conditions on the point.

If the ZVS conditions from eq. (3) and eq. (1) are combined it is possible to calculate the minimal switching frequency which is required to achieve ZVS. At very small loads this will result in very high frequencies. It is therefore necessary to define an upper limit which is dependent on the maximum switching frequency of the chosen switches. At higher power levels the frequency can be lowered to achieve lower switching losses. But a lower limit also has to be defined because too low frequencies will drive the transformer into saturation.

3. Implementation

The proposed modulation scheme was implemented on a prototype which can be seen in fig. 6. The dimensions of the prototype are $270\,\text{mm} \times 210\,\text{mm}$. The parameters of the prototype are listed in table 1. The MOSFETs on the low voltage side use silicon technology whereas the high voltage side uses silicon carbide MOSFETs due to the high voltage. The microcontroller TMS320F28069 from Texas Instruments is being used for the PWM generation and the control algorithm.

The calculation of the optimal switching frequency depends on the battery voltage and the

PCIM Europe 2016, 10 – 12 May 2016, Nuremberg, Germany

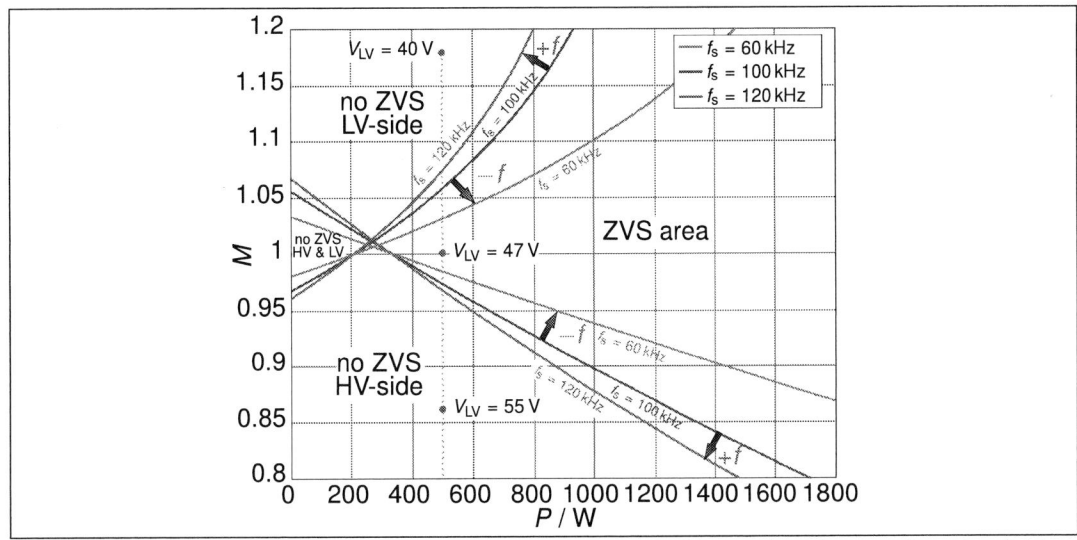

Fig. 5: ZVS area of the DAB in dependece of the voltage gain M, transferred power P and switching frequency f_s (V_{HV} = 710 V)

transferred power which consists of a nonlinear function. Calculating these values during runtime would take up too many resources. A lookup table with the precalculated values has been implemented into the microcontroller. During the runtime it is only necessary to use linear interpolation to calculate the actual value between the data points.

Fig. 6: Bidirectional DC/DC-converter of the Cell-Booster

Parameter	Value
Microcontroller	Texas Instruments TMS320F28069
HV-MOSFETs	4x Cree C2M0280120D $C_{s,HV}$ = 23 pF
LV-MOSFETs	4x Infineon IPB020N08 $C_{s,LV}$ = 1950 pF
Winding ratio n	15
Inductance L	200 μH
Switching frequency f_s	100 kHz
Voltage V_{HV}	710 V
Voltage V_{LV}	40 V – 55 V
Nominal power P_N	1800 W

Tab. 1: Protoype parameters

4. Results and conclusion

First of all the efficiency of the converter with the PSM had been measured. The efficiency also considers the losses produced from the microcontroller, fans and other circuitry. Then the efficiency with the variable frequency PSM had been analysed. All measurements were done for a battery voltage of 40 V and 47 V and a constant 710 V on the high voltage side. The results can be seen in fig. 7 and fig. 8.
As expected, the measurement at M = 1 shows no improvement in the low to mid load region. For these operating points the ZVS area is not enlarged by increasing the frequency (see fig. 5). But at higher loads the reduction of the switching frequency leads to lower switching losses. For

© VDE VERLAG GMBH · Berlin · Offenbach

PCIM Europe 2016, 10 – 12 May 2016, Nuremberg, Germany

this reason the efficiency can be improved in this region.

At M = 1.18 the efficiency could be improved in the low to mid load range by increasing the frequency. This is caused by achieving ZVS for lower loads as the ZVS area is enlarged. Summing up, the efficiency for almost every operating point could be improved by varying the switching frequency. On the one hand the ZVS area could be enlarged by higher switching frequency for low and mid power range. On the other side switching losses had been reduced by a reduction of the switching frequency for high power range.

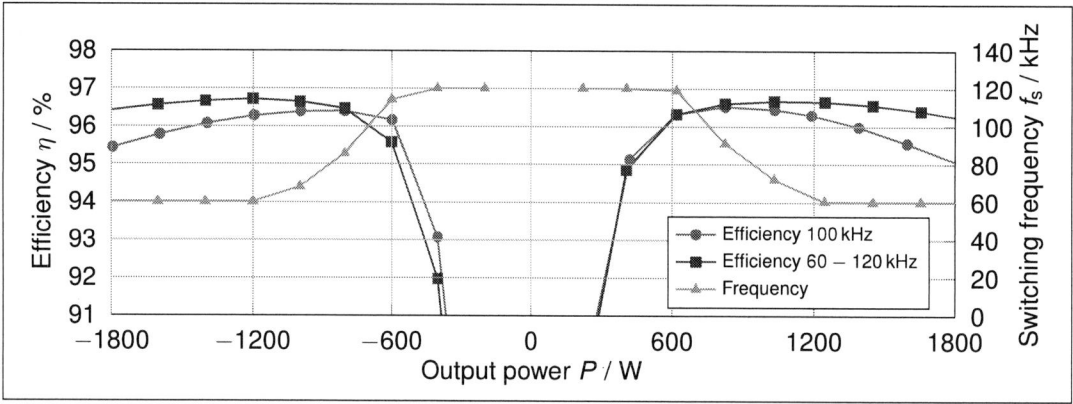

Fig. 7: Efficiency of the DAB with PSM and variable frequency PSM at V_{LV} = 47 V and V_{HV} = 710 V (M = 1)

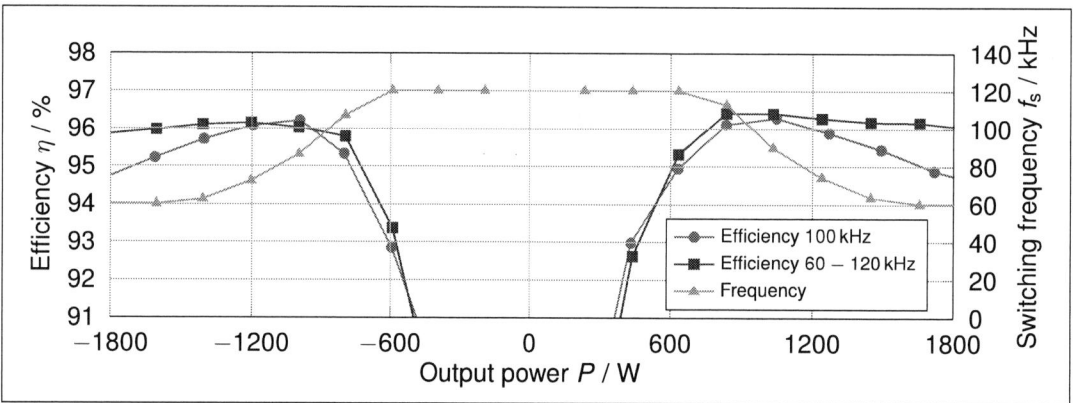

Fig. 8: Efficiency of the DAB with PSM and variable frequency PSM at V_{LV} = 40 V and V_{HV} = 710 V (M = 1.18)

5. References

[1] R. W. de Doncker, D. M. Divan, and M. H. Kheraluwala. A three-phase soft-switched high power density dc/dc converter for high power applications. In *Industry Applications Society Annual Meeting, 1988., Conference Record of the 1988 IEEE*, pages 796–805, 1988.

[2] R. T. Naayagi, A. J. Forsyth, and R. Shuttleworth. Performance analysis of dab dc-dc converter under zero voltage switching. In *Electrical Energy Systems (ICEES), 2011 1st International Conference on*, pages 56–61, 2011.

© VDE VERLAG GMBH · Berlin · Offenbach

High Dynamic Current Source for LED Light and Data Transmission Applications

K. H. Edelmoser, Institute of Electrical Drives and Machines, Technical University Vienna, Gusshausstr. 27-29, A-1040 Wien, AUSTRIA, karl.edelmoser@tuwien.ac.at

F. A. Himmelstoss, University of Applied Science Technikum Wien, Hoechstaedtplatz 6, A-1200 Wien, AUSTRIA, himmelstoss@technikum-wien.at

Abstract

In the field of high power LED and laser drive application there is a growing need for fast switching current sources with stable and reliable output levels as well as reproducible switching shapes in the range of several amps. An additional application is the hidden data transmission possibility with RGB-LEDs in combination with conventional lightning applications. The proposed approach discussed in this paper uses a combined structure of a DC-to-DC converter with some auxiliary switches to fulfill the given requirements: A wide input voltage range, and minimized output current ripple, which reduces the problems with EMC. The necessary switching frequencies of a suitable converter are derived.

1.   Introduction

Based on the well-known buck- and boost converter topologies, a solution with two switching legs has been chosen for a high dynamic laser pulser (Fig. 1). The advantage of this rather simple approach is the possibility to set up the full range of the load current in the output inductor before activating the load. So, especially at low output voltage levels, the current drop in case of load switching can be minimized. This operating principle establishes the facilities of rapid current switching in the output leg. (It should be marked that the parallel usage of switching stages given in Fig. 1 can be used to achieve a multi-level current signal in the load.) Especially when rather low (up to 10V) input voltage levels are used, a very high (in the range of several MHz) switching frequency in the buck section can be obtained to achieve a very smooth output current.

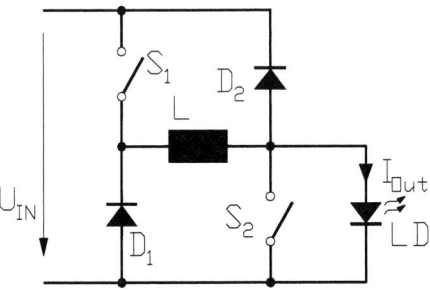

Fig. 1. Switching mode energy source for LED and laser supply

The starting point of our investigations was a DC-to-DC converter for voltage to current level adaptation according to Fig. 4, with the main DC-to-DC converter given in Fig. 1 as LED and laser diode pulse current power supply. The converter is extended with the possibility of individual power control for each light source in the output leg, enhanced by additional switches.

2. The Basic Converter Analysis

2.1. The Basic Operation Principle

For a detailed converter analysis a simulation model was derived and the system behavior was estimated. Fig. 2 shows the results.

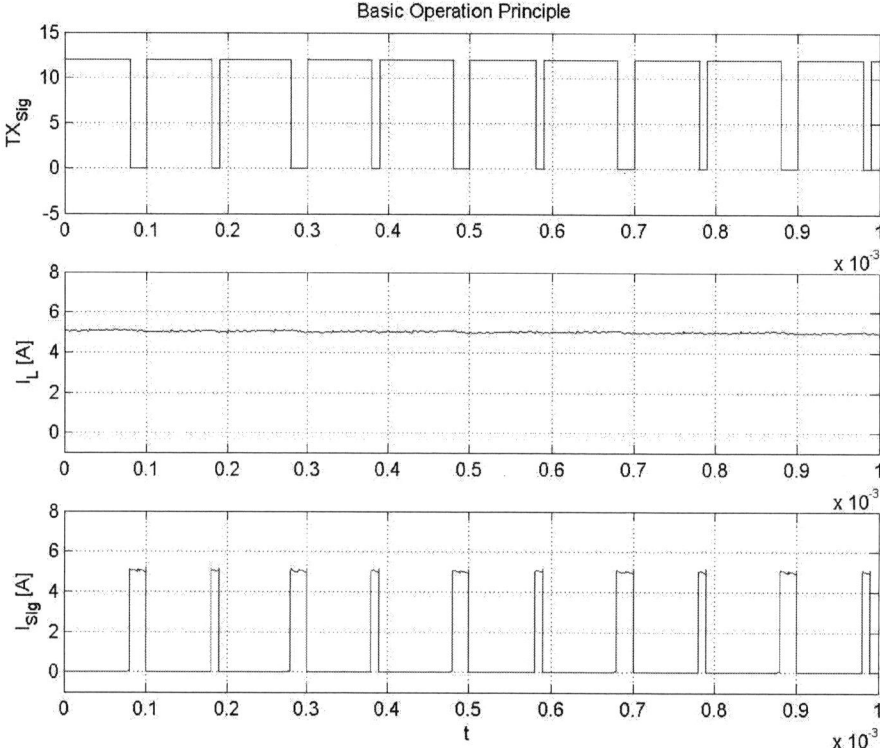

Fig. 2. Switching behavior of the current pulse source. From top to bottom: transmission data, inductor current, and load current

Figure 3 shows the waveforms more in detail. One can see the control signals and the load current as well as the control response.

To get a precise model of the converter, we include the loss resistors of the devices. R_L is the series resistor of the coil L, R_{LED} is the equivalent resistor of the LEDs, V_{LED} is the knee-voltage of the LEDs, R_S the on-resistor of the active switch, and R_D the differential resistor of the diode which has the knee voltage V_D. Depending on the switching state four modes can be selected:

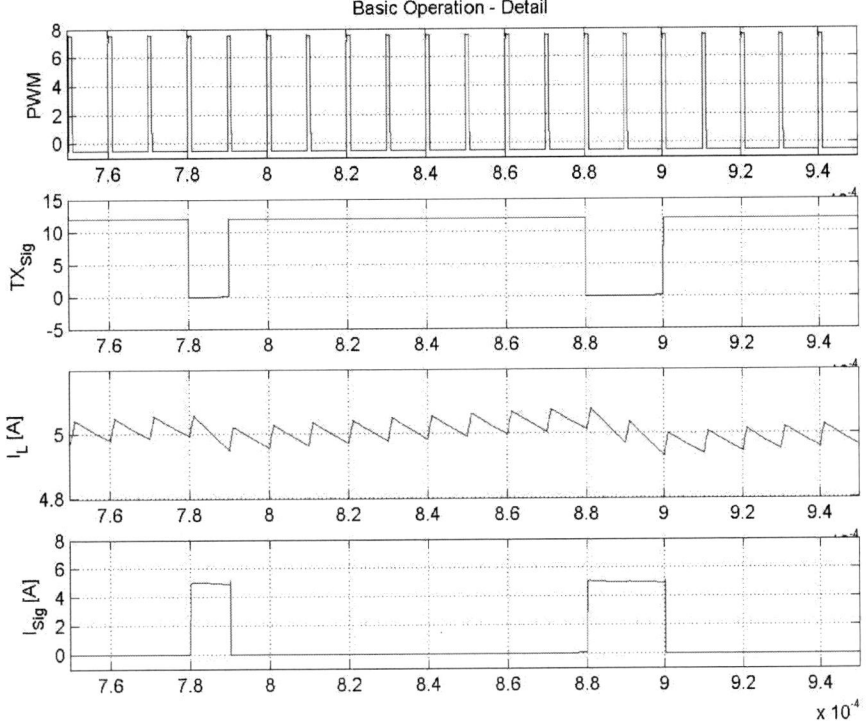

Fig. 3. Switching behavior of the current pulse source. From top to bottom: PWM signal of the current source, transmission data, inductor current, and load current

Stage 1 - charging phase energizing: both active switches S_1 and S_2 are turned on and the current through the inductor L increases according to

$$\frac{di_L}{dt} = \frac{U_{IN} - (R_L + 2R_S)i_L}{L} \; .$$

Stage 2 – charging phase free-wheeling: the active switch S_1 is turned off, the free-wheeling diode D_1 turns on and the current through the inductor decreases according to

$$\frac{di_L}{dt} = -\frac{(R_L + R_S + R_D)i_L + V_D}{L} \; .$$

Stage 3 – driving phase charging: the active switch S_1 is on, S_2 is off, the current increases according to

$$\frac{di_L}{dt} = \frac{U_{IN} - U_{LED} - (R_L + R_{LED} + R_S)i_L}{L} \; .$$

Stage 4 – driving phase free-wheeling: both active switches S_1 and S_2 are off, the free-wheeling diode is on; the current decreases according to

$$\frac{di_L}{dt} = \frac{-V_D - U_{LED} - (R_D + R_L + R_{LED})i_L}{L} .$$

The converter can be controlled by a bang-bang controller. With a mean value of the inductor current I_L and a current ripple of ΔI, one can calculate the on- and off-time of the switches. For the charging mode we get the on-time of S_1 and S_2 to

$$T_{on} = L \frac{\Delta I}{U_{IN} - (R_L + 2R_S)I_L} .$$

When S_2 is still on and S_1 is turned off (this starts a free-wheeling stage) one gets

$$T_{off} = L \frac{\Delta I}{(R_L + R_S + R_D)I_L + V_D} .$$

For the driving phase with S_1 on and S_2 off we get

$$T_{on} = L \frac{\Delta I}{U_{IN} - V_{LED} - (R_L + R_{LED} + R_S)I_L} ,$$

And with S_1 off, S_2 off, and D on

$$T_{off} = L \frac{\Delta I}{V_D + V_{LED} + (R_D + R_L + R_{LED})I_L} .$$

The switching frequency in the charging mode and the driving mode can be calculated with the appropriate on- and off-times to

.

$$f = \frac{1}{T_{on} + T_{off}} .$$

Based on the laser driver described above, a LED-lightning system for data transmission with multiple diodes has been derived.

2.2. Multi-LED Pulse Current Source

Several applications show the necessity of modulating several LED-sources e.g. RGB or RGBW arrangements. An approach is to shunt the inactive devices while the common current is controlled by the driving source. This solution can be used for LED-dimming and data transmission in parallel. So e.g. the brightness control can be realized by a simple PWM algorithm, while the data are packed in the sequence of driving the LEDs. Figure 4 shows the basic operation principle.

The switches S_b, S_g, and S_r, which are connected in parallel to the appropriate LED, guide the current when the diodes are turned off. If the input voltage U_{IN} is high enough, no special high- side driver is necessary.

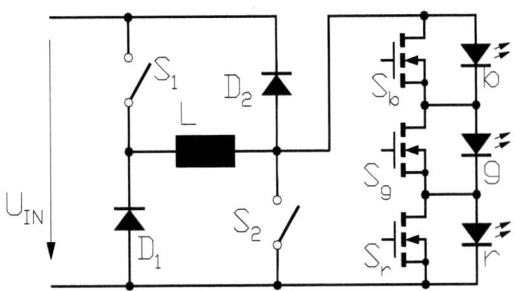

Fig. 4. Series topology of the LED or laser supply arrangement

A sample of the control is shown in Fig. 5. One can see the transmission data packed into the LED-lightning.

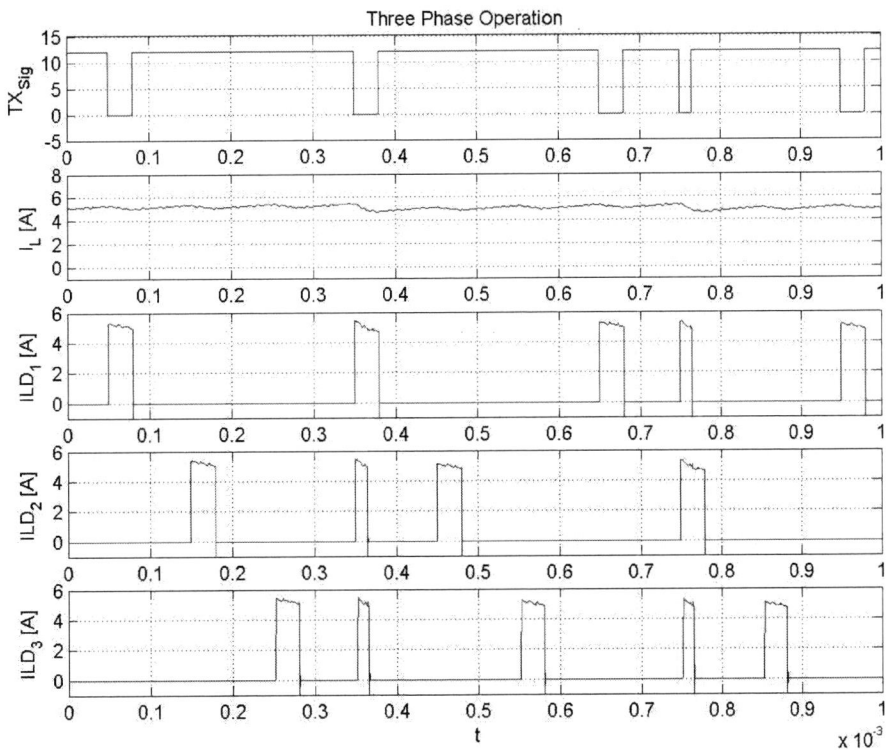

Fig. 5. Multi-phase LED control, operational principle. From top to bottom: transmission signal, inductor current, current of LED b, LED g and LED r

The solution shown above requires a high dynamic control of the load current. Thus a PWM system operating in the range of 1MHz makes sense. In addition the input voltage in relationship with the inductor has to maintain the current drop of the load. The main problem in the practical realization is the acquisition of the current.

3. Tapped Inductor Converter

3.1. Operation Principle

To achieve a better control performance (to minimize the load voltage stepping), a parallel connection of the LED was chosen as given in Fig. 6. Here the load voltage variation is reduced leading to a better performance and a reduced current drop in the driving phase. The used tapped inductors form a transformer which doubles the load current in case of equal windings. It offers an additional point of optimization and can be used to improve the duty cycle of the converter.

Figure 6 shows an improved arrangement for a parallel control scheme. The structure is well suited for a wide range of voltage level adoption between input and output side.

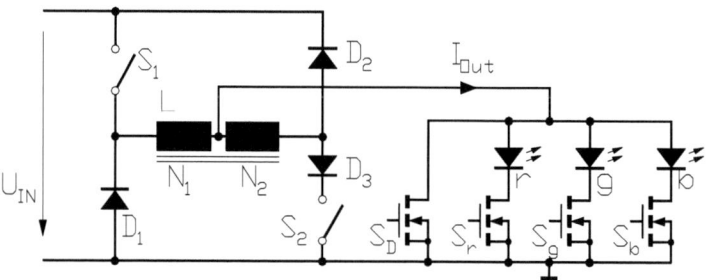

Fig. 6. Parallel arrangement with interleaved control and improved supply

3.2. Tapped Inductor Converter - Analysis

To estimate the system behavior, the topology shown in Fig.6 was simulated on circuit level. Figure 7 depicts the operation principle of the current transformer.

The diode D_2 enables to demagnetize the coils, when no load is connected. There are also different possibilities to magnetize the coils. We can magnetize it be closing S_1 and S_2, or by closing S_1 and S_D, or by closing S_1 and one or more of the colored LEDs. We can also demagnetize over D_1 and one or more of the LEDs connected at the output. The two coils should be coupled very well to avoid an overvoltage due to the stray inductance. The best way is to use bifilar windings.

When S_1 and S_2 are turned on, the input voltage is across both windings and the current increases. When S_1 is turned off, D_1 turns on and there is a free-wheeling stage, the current is now nearly constant (it decreases according to the losses). When both active switches are turned off and one of the load switches is turned on, the current in the left coil N_1 has to jump to produce a steady flux in the magnetic core. When the number of windings of the two coupled coils is equal, which is especially useful when bifilar windings are used, the current through the winding N_1 doubles.

One can see the voltage overshoot at the right switching leg resulting from the stray inductor of the current transformer. A peak-voltage of less than -35V can be reached and therefore measures (e.g. snubber circuits across D_3 and S_2) have to be taken to limit the voltage overshoot.

The great advantage of this topology is the perfect combination of well-known designs and the optimal application and integration into the converter system. This leads to a rather simple and cheap design, which is well suited for rugged field applications.

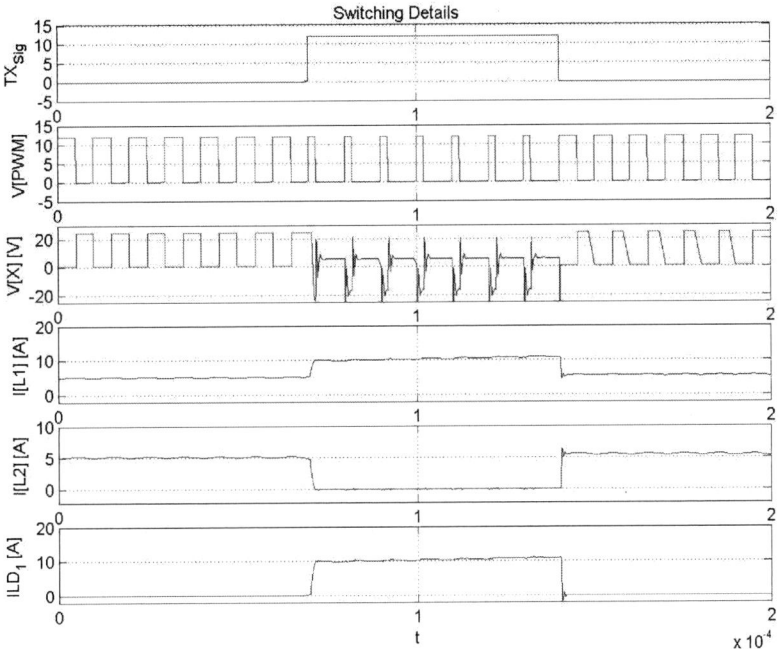

Fig. 7. Multi-phase LED control, operational principle. From top to bottom: transmission signal, PWM control, right leg voltage, inductor current L_1 and L_2, current of the load

3.3. Optimization: Energy Recuperation

As described above the stray energy of the transformer has to be maintained. An innovative solution is to recuperate this energy into the source by an active controlled converter. Figure 8 shows this approach and Fig. 9 the results. There are many possibilities to control this converter circuit.

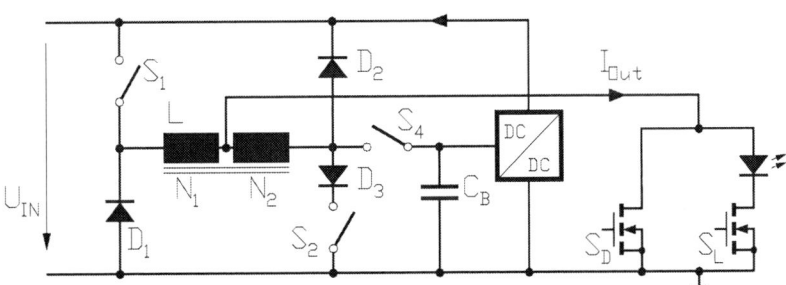

Fig. 8. Parallel arrangement with active energy recuperation

The voltage overshoots are stored in the buffer capacitor C_B. The voltage across the capacitor is controlled by an additional DC/DC converter to a constant level (-20V in our case) and fed back into the source. Figure 9 shows the resulting voltage shapes. As one can see in comparison to Fig. 7, the ringing is minimized leading to reduced voltage stress and better control performance

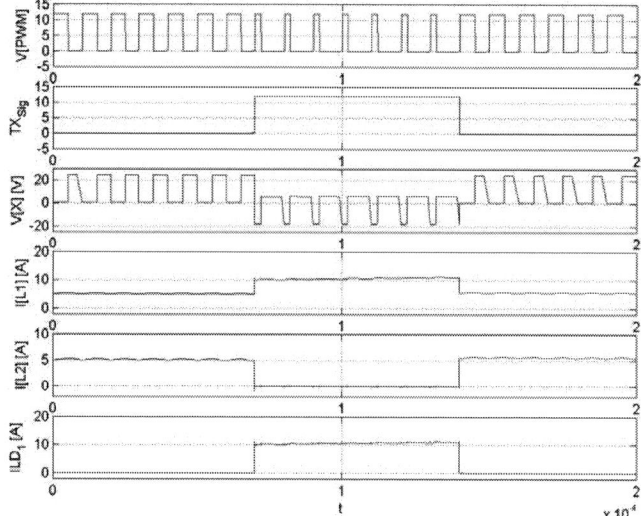

Fig. 9. Multi-phase LED control, operation details. From top to bottom: transmission signal, PWM control, right leg voltage, inductor current L_1 and L_2, and current of the load

4. Conclusion

The proposed approach discussed in this paper uses a combined structure of two converters to fulfill the given requirements: High dynamic switching of the load current with good control performance, a wide input voltage range and minimized output current ripple to minimize EMC-problems. It can be built up with low voltage power MOSFETs as well as with GaAs-FETs to reach the goal of high load current dynamics. A tapped inductor can be used to optimize the duty-cycle of the switching stage. It is well suited for combined lightning and data transmission applications. The whole design can be built up in surface mounting technology leading to a compact and efficient system design. To reduce the losses the free-wheeling diode of the basic converter could be shunted. This is not necessary for D_2, because it is working only in case of an error or in the case of fast demagnetizing of the coils. Common used power supply for LED-lightning could easily be modified to supply the ambient with digital information in parallel to nearly independent light dimming.

5. Literature

[1] N. Mohan, T. Undeland and W. Robbins, Power Electronics, Converters, Applications and Design, 3rd edition., New York: W. P. John Wiley & Sons, 2003.

[2] F. Zach: Leistungselektronik. 4th edition, Wien: Springer, 2010.

[3] Y. Rozanov, S. Ryvkin, E. Chaplygin, P. Voronin, "Power Electronics Basics," CRC Press, 2016.

[4] P. Narra, D. S. Zinger, "An Effective LED Dimming Approach," IAS Annual Meeting, October 2-6, 2005,Hong Kong, pp. 1671-1676.

[5] C. Yandong, M. Zhiqiang, L. Jinglin, J. Heping: "High-Precision Infrared Pulse Laser Ranging for Active Vehicle Anti-collision Application," International Conference on Electric Information and Control Engineering (ICEICE), 15-17 April 2011, Wuhan, ISBN: 978-1-4244-8036-4 pp. 1404 – 1407.

[6] F.A. Himmelstoss, , and , K.H. Edelmoser, "Converters for pulsing UV-diodes," 14th International Conference on Optimization of Electrical and Electronics Equipment, OPTIM'14, Brasov, Romania, 2014, May 22-23, pp. 566-569.

PCIM Europe 2016, 10 – 12 May 2016, Nuremberg, Germany

Design Method for LLC Resonant Converter Considering Buck and Boost Mode with Limited Frequency Range for Wide Input Voltage Range

Dustin Funk, University of Technology Ilmenau, Germany, dustin.funk@4nuts.de
Ulf Schwalbe, ISLE GmbH Ilmenau, Germany, u.schwalbe@isle.de
Tobias Reimann, University of Technology Ilmenau, Germany, tobias.reimann@tu-ilmenau.de

Abstract

LLC resonant converters are well known for their ability of resonant switching and reduced switching losses over a wide load range. Different specifications make it difficult to implement a design. Input and output voltage range, frequency width, size and stress of passive components are just a few of them. This contribution deals with a design mythology considering the below and above resonance mode of LLC resonant converter by a limited frequency range as well as a maximum resonant tank voltage. The paper describes two ways for designing the resonant tank parameters. Every method is based on a theoretical analysis followed by a simulative investigation. A worst-case analysis is also included to cover the impact of component tolerances and the major voltage drops. Furthermore advantages and drawbacks of both methods are pointed out. The result is verified on a prototype.

1. Introduction

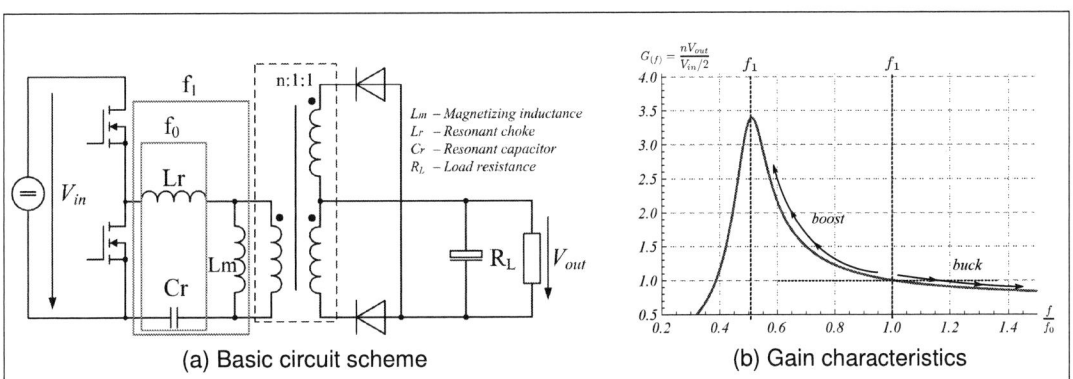

(a) Basic circuit scheme   (b) Gain characteristics

Fig. 1: LLC resonant converter basic circuit with typical transfer behaviour

Due to increasing power density design engineers want to reduce passive component size by increasing the system switching frequency. Thus, switching losses in the semiconductors increase. Resonant converters get more importance in comparison with hard switching topologies because of their soft switching ability. Moreover, a high efficiency over a wide load range can be achieved by using the LLC resonant topology (Fig. 1a). Furthermore, a wide input voltage range is possible through the buck and boost ability of the resonant tank (Fig. 1b) which is often used for an improved hold-up time[1] or a varying input voltage [1] [2] [3]. Existing methods [1] [2] only works below resonance or have a free frequency range. An optimized buck mode

---

[1] The amount of time that a power supply unit can maintain output within the specified voltage range after line dropout.

© VDE VERLAG GMBH · Berlin · Offenbach

modulation strategy is presented in [4]. This paper describes two methods for a LLC resonant converter design with a variable input voltage and limited frequency range. In addition, the above resonance mode is included. The design is investigated and optimized by simulation. Furthermore, component tolerances are included. Based on the presented methodology a 20W LLC resonant converter with 24-55V input voltage range and 12V output is designed to verify the theoretical analysis. The frequency range is limited from 40 to 145kHz. The prototype is to be used in science to investigate the different operating modes. Thus a system voltage of 60V should not exceed due to safety reasons.

2. Theoretical analysis

This chapter shortly describes the fundamental basics of the LLC topology followed by two design methods. The contribution deals only with specifics. A general design is explained in [1] and [2]. As already mentioned LLC converters have the ability to buck or boost its input voltage. An amplification occurs between the parallel resonant frequency f_1 (Eq. 2) and the serial resonant frequency f_0 (Eq. 1). Attenuation occurs above f_0 (cf. Fig. 1b).

$$f_0 = \frac{1}{2\pi\sqrt{C_r L_r}} \quad (1) \qquad f_1 = \frac{1}{2\pi\sqrt{C_r(L_r + L_m)}} \quad (2) \qquad \frac{f_0}{f_1} = \sqrt{1 + L_n} \quad (3) \qquad L_n = \frac{L_m}{L_r} \quad (4)$$

The ratio of both resonant frequencies is shown in Eq. 3 where L_n is the inductor ratio (Eq. 4). The LLC stage transfer behavior (Eq. 8) strongly depends on each resonant tank component.

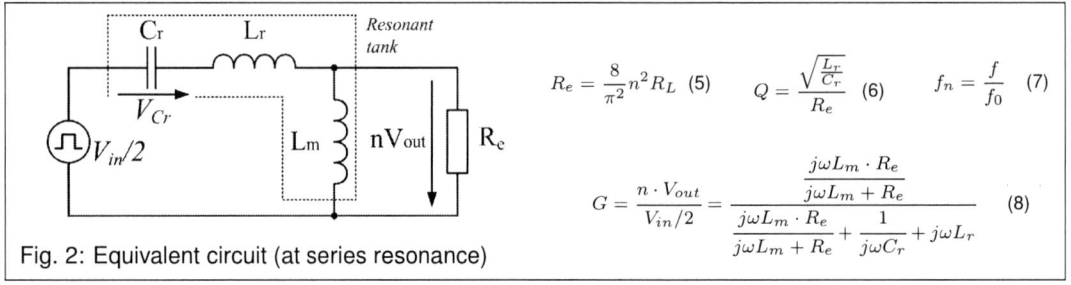

$$R_e = \frac{8}{\pi^2}n^2 R_L \quad (5) \qquad Q = \frac{\sqrt{\frac{L_r}{C_r}}}{R_e} \quad (6) \qquad f_n = \frac{f}{f_0} \quad (7)$$

$$G = \frac{n \cdot V_{out}}{V_{in}/2} = \frac{\dfrac{j\omega L_m \cdot R_e}{j\omega L_m + R_e}}{\dfrac{j\omega L_m \cdot R_e}{j\omega L_m + R_e} + \dfrac{1}{j\omega C_r} + j\omega L_r} \quad (8)$$

Fig. 2: Equivalent circuit (at series resonance)

A simplified consideration of the equivalent circuit (Fig. 2) using a first harmonic approximation (FHA) combined with Eq. 4 to 7 leads to the normalized transmission behavior (Eq. 9) where R_e is the equivalent load resistance of the secondary side at a sinusoidal current, n is the turns ratio of the transformer and Q the quantity. f_n is the normalized frequency.

$$G(f_n) = \left| \frac{L_n f_n^2}{L_n f_n^2 + (f_n^2 - 1)(1 + j f_n L_n Q)} \right| \quad (9)$$

When the switching frequency moves away from the resonant frequency f_0 the current through the rectifiers is no longer sinusoidal. Thus Eq. 9 becomes an error but finally the maximum gain at below resonance is higher. Nevertheless the equation can be used for design. The final results need to be verified by simulation or hardware test. As shown in Fig. 3a the load has a big impact on the total gain G_{max} - the higher the load the greater the attenuation of the resonance tank. Thus the worst gain at below resonance occurs at full load and the worst attenuation at above resonance occurs at light load which is important for designing the buck and boost ability. Another parameter of Eq. 9 is L_n. The impact is shown in Fig. 3b. It will be seen that a small L_n furthers the gain below f_0 and the attenuation above f_0. Furthermore, the resonant frequencies f_0 and f_1 are closer together which decrease the used frequency range.

© VDE VERLAG GMBH · Berlin · Offenbach

PCIM Europe 2016, 10 – 12 May 2016, Nuremberg, Germany

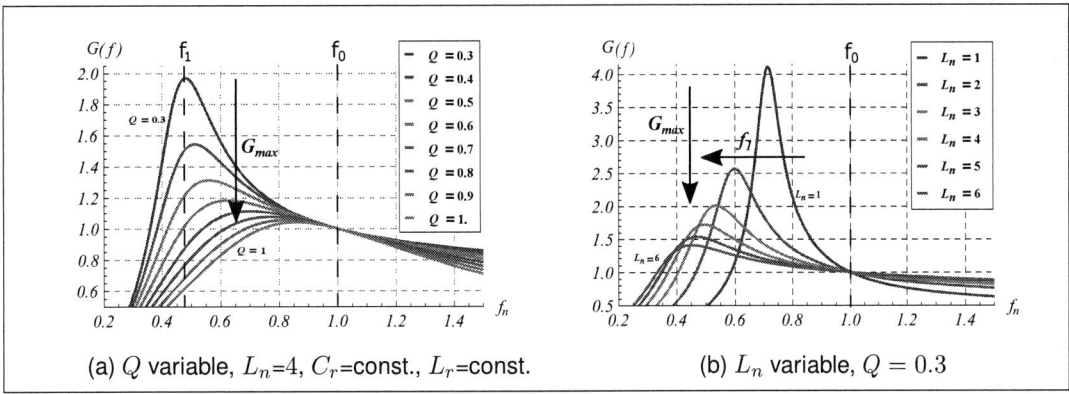

(a) Q variable, $L_n=4$, C_r=const., L_r=const.

(b) L_n variable, $Q = 0.3$

Fig. 3: Impact of Q and L_n on the gain characteristics

2.1. Design Methodology

The presented design methods differ in the used frequency range. Where the first method uses the whole frequency range, the second method keeps it as small as possible (Fig. 4). For both methods it is important that the maximum gain of the resonant tank is only as high as necessary. Anything beyond this leads to lower L_n values (cf. Fig. 3b) and thus to a higher turn off current of the semiconductors in resonance mode, therefore, the switching losses increase. For a better determination of the following equations at light load condition a minimum load should be used (e.g. R_{Lmax}=1$k\Omega$). Subsequently a design example for each method will given. Advantages and drawbacks will figured out in the simulation section.

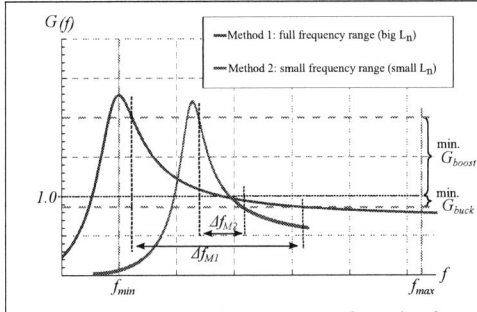

Fig. 4: Principle of the presented methods

Input voltage	24..55V
Output voltage	12V
Output power	0..20W
Frequency width	40-145kHz
$\hat{V}_{C_{r\,max}}$	60V
Transformer turns ratio	2

Tab. 1: Specification for the design example and the investigated prototype

Method 1: By using the whole frequency range the inductor ratio L_n is as big as possible. Thus, the resonances f_0 and f_1 are distanced furthest apart (Eq. 3). As already mentioned the maximum gain should not be greater than necessary at the maximum load condition (R_{Lmin}). This results in three independent equations (Eq. 10-12) by which the resonant parameters L_r, L_m and C_r are uniquely determined. The gain at f_{min} is big as required (Eq. 10) and the first derivative is zero (Eq. 11). The third equation (Eq. 12) indicates that the required attenuation for the buck capability must be achieved at the upper frequency limit and minimum load (R_{max}).

$$\text{i: } G\left(L_r, L_m, C_r\right)\Big|_{f_{min}, R_{Lmin}} = G_{boost} \qquad (10)$$

$$\text{ii: } G'\left(L_r, L_m, C_r\right)\Big|_{f_{min}, R_{Lmin}} = 0 \qquad (11)$$

$$\text{iii: } G\left(L_r, L_m, C_r\right)\Big|_{f_{max}, R_{Lmax}} = G_{buck} \qquad (12)$$

© VDE VERLAG GMBH · Berlin · Offenbach

Solving the system of equations give the first resonant trio value. For this purpose the absolute transfer behavior is necessary (Eq. 13) which is followed by substituting Eq. 4 to 7 in Eq. 9.

$$G_{(f,L_r,L_m,C_r)} = 4\pi^2 \sqrt{\frac{C_r^2 f^4 L_m^2}{1 + 16 C_r^2 f^4 L_r^2 \pi^4 \left(1 + \frac{2L_m}{L_r} + \frac{L_m^2 \left(1 - \frac{L_r \pi^4}{512 C_r R_L^2}\right)}{L_r^2}\right) - 4 C_r f^2 L_r \pi^2 (2 + \frac{2L_m}{L_r} - \frac{L_m^2 \pi^4}{1024 C_r L_r R_L^2}) - \frac{C_r^2 f^6 L_m^2 L_r^2 \pi^{10}}{16 R_L^2}}} \tag{13}$$

Assuming the specification of Tab. 1 the required boost and buck goals are $G_{boost}=2$ and $G_{buck}=0.873$. The load resistance at full load is $R_{Lmin}=7.2\Omega$. The maximum load resistance is assumed to be $R_{Lmax}=1k\Omega$. Solving the above mentioned system of equations e.g. with *Mathematica* yields $L_r=12.1\mu H$, $L_m=52.4\mu H$ and $C_r=282nF$.

Method 2: The second method based on a small inductor ratio L_n. At first the smallest L_n and the associated Q must be determined. These values may be limited e.g. by a maximum voltage $\hat{V}_{cr}$ or AC-stress of the resonant capacitor (Eq. 14).

$$\hat{V}_{Cr} = \frac{4}{\pi} n V_0 \left| \frac{1}{L_n f_n^2} + j \frac{Q}{f_n} \right| = \frac{4}{\pi} n V_0 \frac{1}{L_n f_n^2} \sqrt{1 + L_n^2 Q^2 f_n^2} \tag{14}$$

However L_n and Q describe the transmission behavior only normalized in relation to f_n and thus there is no unique solution for the resonant tank parameters (cf. Eq. 9). Therefore, the second part of method two consists to calculate the frequency followed by the resonant tank parameters. Fig. 5a shows two ways for determining the absolute frequency. The resonant frequency f_0 of M2(i) is uniquely determined by reaching the maximum gain at f_{min} and $R_{L_{min}}$ whereas M2(ii) must reach the buck goal at f_{max} and $R_{L_{max}}$. Any other solution between M2(i) and M2(ii) is possible. Since the values of L_n and Q are equal the turn-off current through the semiconductors is also equal which results in higher switching losses at greater frequencies. Therefore, only M2(i) is investigated further in this contribution.

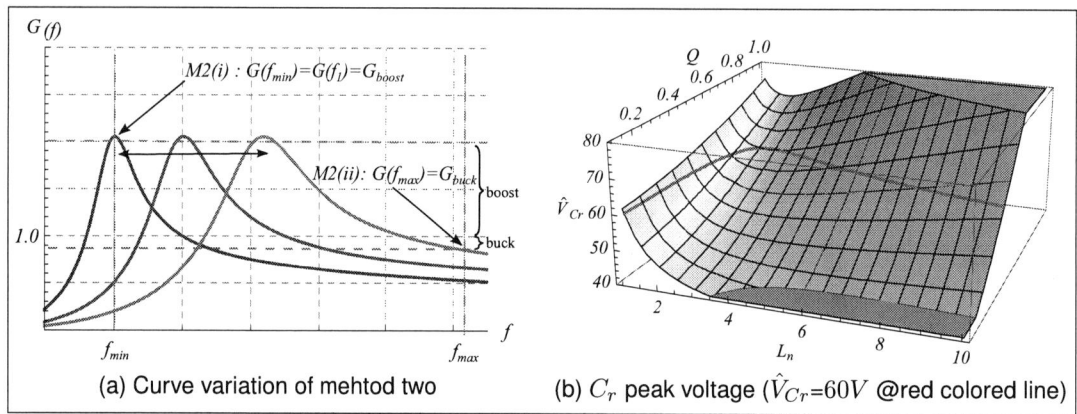

Fig. 5: Design Method 2 (small inductor ratio L_n)

As already mentioned L_n can be limited by the resonant capacitor voltage $\hat{V}_{cr}$. The maximum voltage occurs at the minimum switching frequency. Due to L_n the ratio of both resonant frequencies is defined (Eq. 3). Thus $\hat{V}_{cr}$ can be simplified to Eq. 15.

$$\hat{V}_{Cr}\Big|_{f_n=f_1/f_0} = \frac{96(1+L_n)}{L_n \pi} \sqrt{1 + \frac{L_n^2 Q^2}{1+L_n}} \tag{15}$$

© VDE VERLAG GMBH · Berlin · Offenbach

In this design example the resonant capacitor peak voltage should not exceeds 60V (Tab. 1). Fig. 5b shows $\hat{V}_{cr}$ at different Q-L_n values. The 60V limit is red colored. To have a small margin $L_n = 1.5$ and $Q = 0.55$ will be chosen. For the resonant tank parameters the serial resonant frequency f_0 must be calculated (Eq 16).

$$f_{0_{M2(i)}} = \frac{f_1}{f_n} = \frac{f_1}{\sqrt{\frac{1}{1+L_m}}} = \frac{40kHz}{\sqrt{\frac{1}{1+1.5}}} = 63kHz \qquad (16)$$

Now the resonant tank parameters can be determined to L_r=32μH, L_m=48μH and C_r=220nF.

3. Simulation

Both results of the above presented methods are verified by simulation. For this propose *SPICE* is used. Fig. 6 shows the simulation results of the already calculated resonant tank parameters.

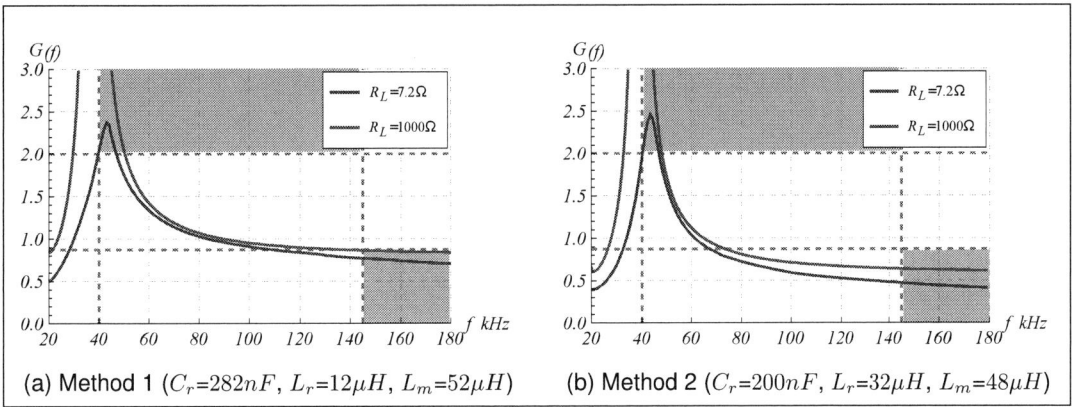

(a) Method 1 (C_r=282nF, L_r=12μH, L_m=52μH)   (b) Method 2 (C_r=200nF, L_r=32μH, L_m=48μH)

Fig. 6: *SPICE* simulation results at full load and no load condition (target area gray shaped)

Tab. 2 gives more insight into the results. It will be seen that both methods are within the frequency range. Method one has a small margin to the frequency limits which can be improved by decreasing the magnetizing inductance L_m or the resonant capacitor C_r which decrease L_n or Q. Both favor the maximum gain as well as the damping and give more margin at the upper frequency limit (cf. Fig. 3). The second method exceeds the capacitor voltage at below resonance a bit due to the FHA. In this case Q or L_n need to be reduced e.g. by increasing C_r a little (cf. Eq. 15).

V_{out}=12V	Method 1: C_r=282nF,L_r=12μH,L_m=52μH					Method 2: C_r=200nF,L_r=32μH,L_m=48μH				
Load :	$R_L = 7.2\Omega$				R_L =1kΩ	$R_L = 7.2\Omega$				R_L =1kΩ
V_{in}	f	$I_{sw_{off}}$	$\hat{V}_{cr}$	Γ^a	f	f	$I_{sw_{off}}$	$\hat{V}_{cr}$	Γ^a	f
24V	47,2 kHz	0,95A	45,1V	44,8	50,8kHz	47,7kHz	1,43A	61,1V	68,2	49kHz
36V	60,6 kHz	1,53A	38,5V	92,7	63,4kHz	54,4kHz	1,97A	55,8V	107,2	56kHz
48V	85 kHz	1,36A	36,5V	115,6	92kHz	62,8kHz	1,98A	54V	124,3	63,3kHz
55V	111 kHz	1,8A	36,2V	199,8	145kHz	68,8kHz	2,15A	53,2V	147,9	75,2kHz
Δf	97,8kHz					27,5kHz				

^{a} $\Gamma = f[kHz] \cdot I_{sw_{off}}$

Tab. 2: *SPICE* simulation results of method one and two at characteristic operating points

Tab. 2 also points out to the advantages and drawbacks. Where method one has a overall smaller turn off current the Γ value is greater at above resonance. Γ as a result of the frequency multiplied by the turn off current indicates a factor for the expected turn off losses of the primary semiconductors when they are assumed to be linear. Method one also has a smaller peak voltage across the resonant capacitor. This also results in less AC-stress whereas method two works at lower frequencies. The smaller frequency range of method two leads to a bigger margin to the frequency limits and allows better optimizations of the passive components. In case of using snubber networks on the primary or secondary side method two would be preferable because lower frequencies and thus less snubber losses. A summary is shown in the following table.

Component	Method 1	Method 2
Losses	• lower losses at below resonance (boost mode) and resonance	• lower losses at above resonance
Resonant capacitor	• overall lower peak voltage • high frequencies in buck mode	• lower frequencies and a smaller frequency range
Resonant inductance	• smaller size due to smaller inductance values	• a high resonant inductance leads to higher system voltage drops
Passive components	• can operate partially more efficient	• can be optimized for a smaller frequency range
Frequency behavior	• flat gain decrease in buck mode can cause in leaving the frequency range at no load	• the specified frequency range is not as vulnerable to component tolerances

Tab. 3: Advantages and drawbacks of the presented methods based on the simulation results of Tab. 2

The parameters should now be optimized considering component tolerances and major voltage drops. Method one is investigated further because of a lower resonant capacitor peak voltage. Moreover, less switching losses at below resonance and resonance mode are expected.

3.1. Simulative optimization

The simulative optimization include the component tolerances to ensure the functionality in the worst case scenario. The production-related tolerances will continue accepted as $C_r=\pm 5\%$, $L_r=\pm 10\%$ and $L_m=\pm 5\%$. To find the worst case tolerance scenario every constellation was simulated. The worst result occurs at $C_r=-5\%$, $L_r=-10\%$ and $L_m=+5\%$. The reason is that a higher L_n due to the tolerances has a greater effect on the transmission behavior than Q. As consequence the gain and attenuation deteriorates. The next step is to include the major voltage drops. In this contribution the coupling factor k of the transformer and the forward voltage V_{fw} of the secondary side rectifiers are considered. Tab. 4 shows the worst case constellation for full load and light load.

Parameter	full load	light load
V_{in}	24V	55V
k	0.95	1
V_{fw}	0.5V	0V
R_L	7.2Ω	1kΩ
Tol. $C_r/L_r/L_m$	−5% / −10% / +5%	

Tab. 4: Simulation parameters for worst case scenario

Results	full load		light load
$C_r/L_r/L_m$	G_{max}	$f_{|G=2}$	$f_{|G=0.873}$
(a): 282/12/52	2,27	46kHz	>180kHz
(b): 330/13/49	2,44	45kHz	137kHz
(c): 220/16/45	2,29	53kHz	133kHz

Tab. 5: Results of the worst case analysis with different resonant tank parameters (respected Tab. 4)

The worst case simulations of the precalculated parameters show that under consideration of the voltage drops and component tolerances the specification are not met (Tab. 5(a)). Tab. 5(b) and (c) shows different optimization possibilities in which (b) is based on an increasing Q and (c) on a larger L_n. Both improve the transmission behavior and the specification is met.

4. Experimental Implementation

To verify the proposed design methodology a prototype is built (Fig. 7a). In addition a synchronous rectification was built-in to investigate effects of different resonance modes and enhance efficiency. Applying the specification of Tab. 1 and the mentioned methods above the final design parameters are $L_r=17\mu H$, $L_m=42\mu H$ and $C_r=222nF$. Fig. 7b reveals the gain

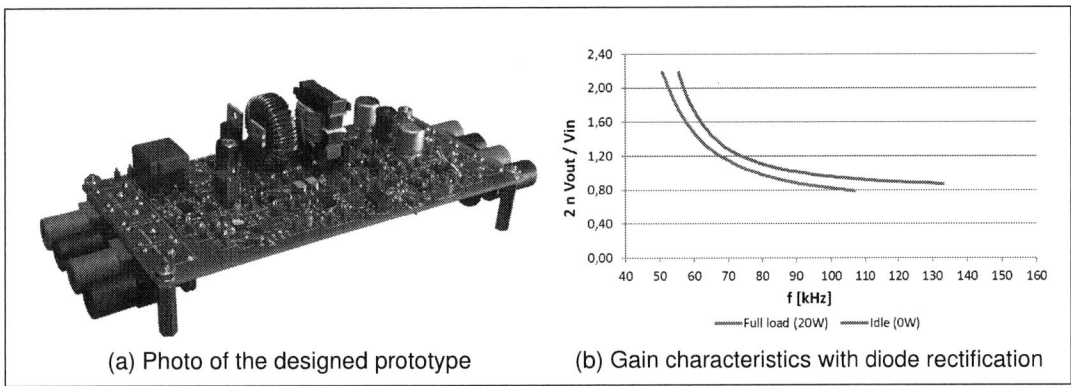

| (a) Photo of the designed prototype | (b) Gain characteristics with diode rectification |

Fig. 7: LLC resonant converter prototype ($C_r=222nF$, $L_r=17\mu H$, $L_m=42\mu H$)

characteristics. In Fig. 8a the efficiency of both rectification methods is shown. As expected the efficiency of the diode rectification increases from 24V (below resonance) to 48V (quasi resonance) and decreases until 55V (above resonance). Due the current gap at below resonance on the secondary rectifiers the voltage begins to oscillate due the parasites of the semiconductors and the transformer. To ensure the synchronous rectification functionality a snubber is installed. This seems to be the reason why the efficiency begins to decrease at 42V with synchronous rectification due to the increasing frequency and thus higher snubber losses.

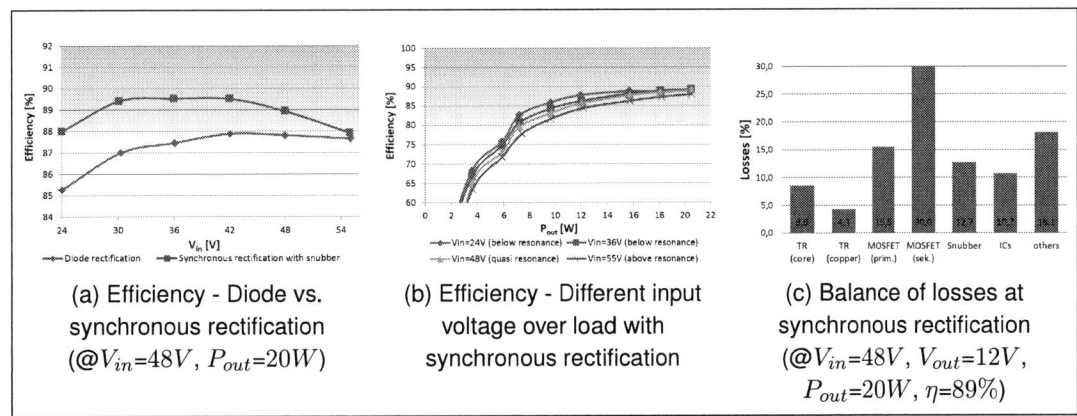

| (a) Efficiency - Diode vs. synchronous rectification (@V_{in}=48V, P_{out}=20W) | (b) Efficiency - Different input voltage over load with synchronous rectification | (c) Balance of losses at synchronous rectification (@V_{in}=48V, V_{out}=12V, P_{out}=20W, η=89%) |

Fig. 8: LLC resonant converter prototype ($C_r=222nF$, $L_r=17\mu H$, $L_m=42\mu H$, schottky: $STPS3L60U$, synchronous rectifier: $BSC093N04LSG$, $C_{snubber}$=10nF, $R_{snubber}$=30Ω)

© VDE VERLAG GMBH · Berlin · Offenbach

Fig. 8a shows the efficiency of the resonant converter with synchronous rectification at different resonance modes over the output power. It will be seen that a high efficiency over a wide load range achieves up to 89,4%. The snubber effect is also visible which is the reason why the efficiency at below resonance is partially greater then in resonance mode. At 6W output power the synchronous rectification is turned off due to light load function (cf. Fig. 8b). The major losses are shown in Fig. 8c where the main parts are conduction losses of the secondary side rectifiers.

5. Conclusion

This contribution presents two methods for designing a LLC resonant Converter. The methods are based on a mathematical relationship for the resonant components followed by a simulative optimization including component tolerances. The methods respect above and below resonance and the resonant tank voltage. They differ in the used frequency range. Advantages and drawbacks were pointed out. The result is verified on a 20W LLC Converter which achieves a high efficiency over a wide load range in different resonance modes. The prototype can be used for further investigations and achieves an efficiency up to 89,4%. The proposed methods can be used for further designs.

6. References

[1] Bing. Lu et al. Optimal design methodology for llc resonant converter. In *Twenty-First Annual IEEE Applied Power Electronics Conference and Exposition, 2006. APEC '06*, pages 533–538, March 19, 2006.

[2] Yu Fang et al. Design of high power density llc resonant converter with extra wide input range. In *PEC 07 - Twenty-Second Annual IEEE Applied Power Electronics Conference and Exposition*, pages 976–981.

[3] H. Figge et al. Overcurrent protection for the llc resonant converter with improved hold-up time. In *2011 IEEE Applied Power Electronics Conference and Exposition - APEC 2011*, pages 13–20.

[4] Christian P. Dick et al. Optimized buck - mode modulation strategy and control of a llc - type resonant converter in a solar application. In *2014 16th European Conference on Power Electronics and Applications (EPE'14-ECCE Europe)*, pages 1–9.

[5] Ulf Schwalbe. *Vergleichende Untersuchungen dreistufiger Schaltnetzteiltopologien im Ausgangsleistungsbereich bis 3 kW.* dissertation, University of Technology Ilmenau, 2009.

The Behavior of Electro-Magnetic Radiation of Storage Inductor in DC-DC Converters

Ranjith Bramanpalli
Würth Elektronik eiSos GmbH, Max-Eyth Str.1, 74638, Waldenburg, Germany
ranjith.bramanpalli@we-online.com

Abstract

This paper focuses on the behavior of the Electro-Magnetic (EM) radiation of a storage inductor, in DC-DC converters, which is dependent on several parameters such as - ripple current, switching frequency, rise & fall time of a switch, the core material, it´s permeability and the style of coil winding.

1. Introduction

DC-DC converters are widely used in power management applications where the inductor is one of the key component. But the usual focus is on its performance like Rdc, Rac, and core loss. The other key factor vastly ignored is the material it is made of, windings and the electro-magnetic radiation performance.

Due to the availability of advanced technologies in molding, the core of an inductor is now made in various particle size, material composition, forms and shapes. This is applicable for coil windings too. Based on the mechanical and electrical property requirements, appropriate core is then placed around the coil to confine and store the energy.

Switch Mode Power Supplies, especially DC-DC converters are moving towards higher frequencies to reduce the system size and fast turn on and turn off of switching devices to reduce switching loses. As switching frequency increases, DC/DC converters also employ faster rise and fall times of the switching device to keep switching losses low. But this creates steep switch-node transitions, accompanied by switch-node ringing and spikes. The resulting switch-node ringing will also be present in the Inductor voltage as shown in Fig 1.

As the shielding material of the Inductor is designed to attenuate fundamental frequencies, the effectiveness of shielding for ringing frequencies is overlooked which is in the range of 100 MHZ to 200MHZ.

To better shield the ringing frequencies, near field characteristics of the inductor, its core material and external shielding solutions are discussed.

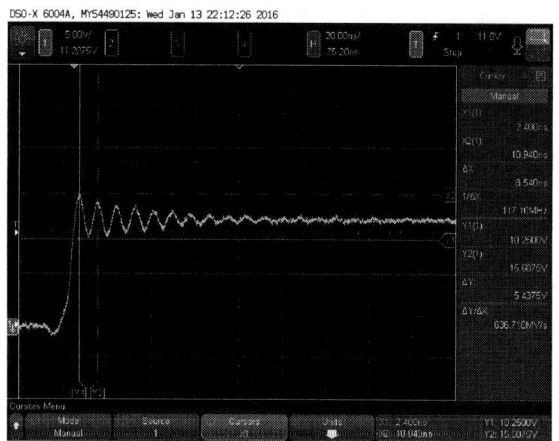 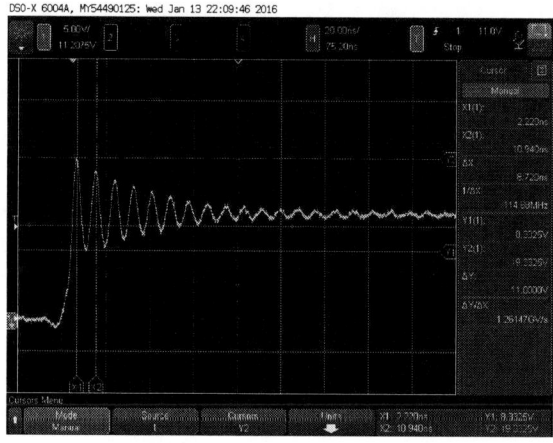

Fig 1: Switch-node ringing at various magnitudes

2. EM radiation (near fields) in DC-DC Converter

In a DC-DC converter the voltage across a storage inductor is AC in nature. This AC voltage across the coil which acts like a dipole antenna produces E-field and the current produces H-field. H & E-fields are perpendicular to each other. These fields are categorized into near-field and far-fields depending on the distance from the source. Since inductor windings acts as loop antenna, H-field is dominant than E-field for obvious reasons.

One has to keep in mind that the fields near the coil are spherical or curved and they propagate away from the antenna perpendicular to each other. They also regenerate each other along the way depending upon the victim. How well the fields are confined to inductor are totally dependent on the material of the core, winding structure and the orientation of the inductor.

The core is made of various materials like Iron powder, MnZn, NiZn, Iron alloys, etc., and the electro-magnetic properties like permeability and Magnetic flux density are then dependent on the material mixture and their characteristics are shown in Fig 2. Every material has its own advantage & disadvantage. For example, if the inductor is made up of iron powder & iron alloy core, it exhibits characteristics such as soft saturation and ensures stable operation with varying system conditions & can be made in to smaller parts with ease, but the permeability is low and losses are relatively on higher side due to the iron content. MnZn & NiZn core exhibits characteristics otherwise.

The H-field radiation (30MHz to 1GHz) of a storage inductor of various cores is measured in DC-DC (Buck) converter are shown in figures from 1 to 4. The DC-DC converter is switched at 1MHz frequency. The fields are measured at a distance of 1mm from the DUT.

The iron powder, iron alloy MnZn and NiZn cores have the dimensions of 10x10x4, 3x3x2, 18x18x9 and 10x10x5 mm respectively. As a rule of thumb iron powder & iron alloy cores offer good shielding at lower frequencies, MnZn cores are usually good for mid-range and NiZn cores are good for higher frequencies. The frequency at which peak of the radiation occurs is equal to the ringing frequency at the drain during the turn off of the switch. This is when Inductor acts as a source to the load in buck converter. The effectiveness of H-field shielding does not depend on core size but only on the material but on the other hand E-field radiation is dependent on size of the coil too.

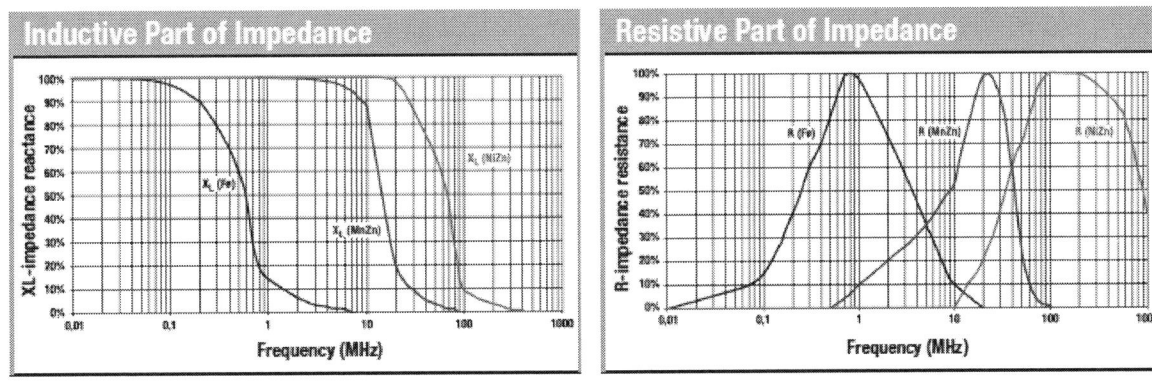

Fig 2: Real & imaginary Impedance curves for various materials

3. Shielding:

Shielding of an inductor is an option to prevent excess radiation due to high frequencies and fast switching. Shielding can be done by choosing the inductor of appropriate material or by adding external magnetic or metal shielding to the inductor. Metal & Magnetic shielding solutions can be done optimally based on the application.

Magnetic shielding materials are sheets or plates made of NiZn, MnZn or Iron to place above the source of radiation and the brief characteristics are shown in Fig 2.

Metal shielding materials are made of Copper, Aluminum, metal alloys and composite mixture. The metal shielding is usually an enclosed over the source to reflect/absorb the noise. The shielding characteristic of metals are shown in Fig 3. Thickness and material can be chosen based on shielding effectiveness, frequency of attenuation and cost.

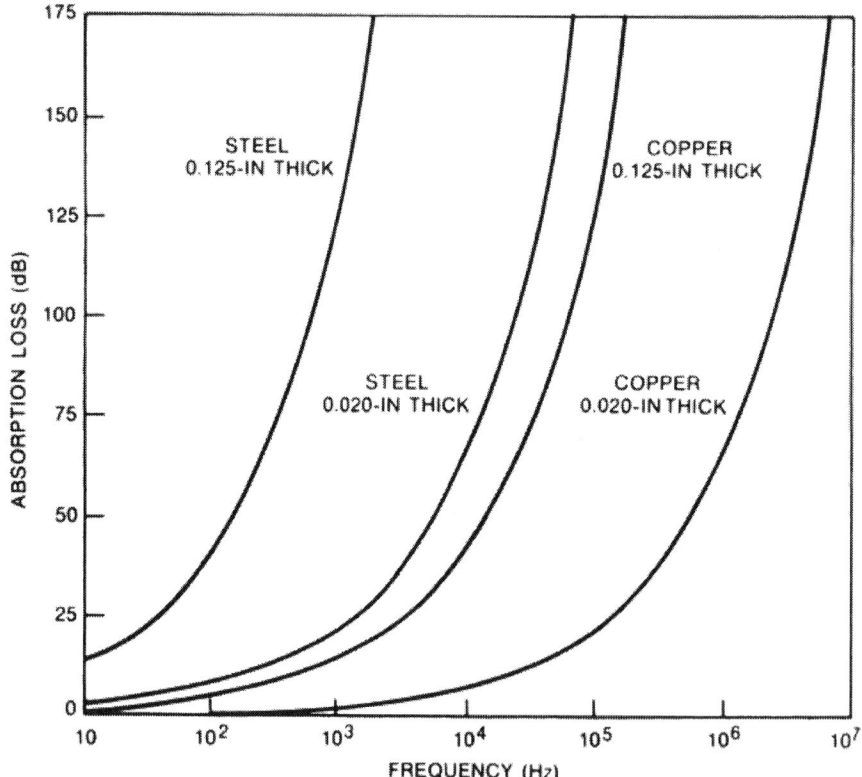

Fig 3: Absorption characteristics of metals

4. Experimental Results

Inductors made of higher shielding materials can be used to mitigate some radiation but for example to utilize the advantages of distributed gapped iron alloy & iron powder cores there are several approaches to suppress the radiation at high switching frequencies. One approach is placing a flexible shielding material contains MnZn or NiZn or ferrite plates on the surface of the inductor depending on the frequency. The other approach is to enclose the inductor with metal, so that the fields can reflect and contain within the enclosure.

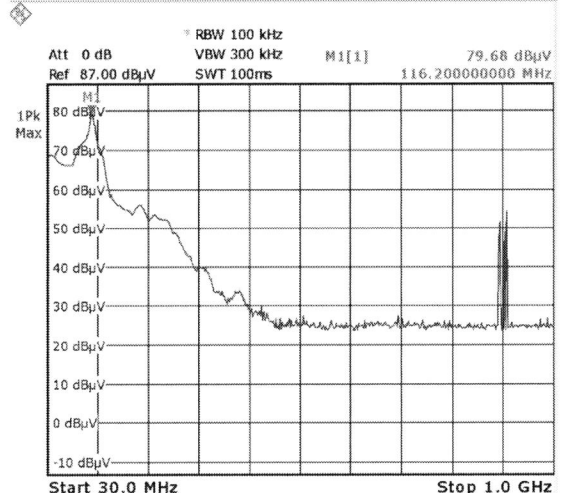

Fig 1: H-field of Iron Powder core @ Fsw - 1MHz

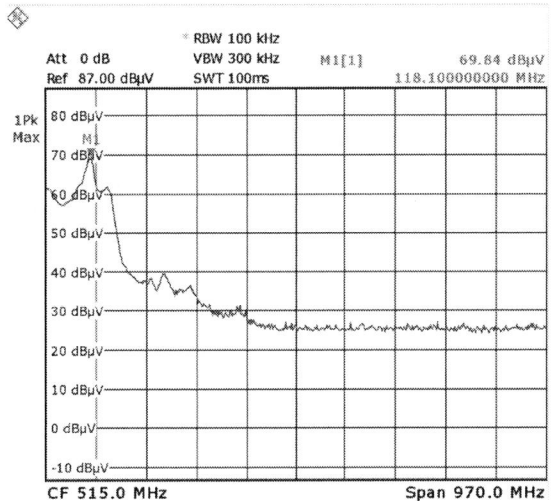

Fig 2: H-field of MnZn core @ Fsw – 1MHz

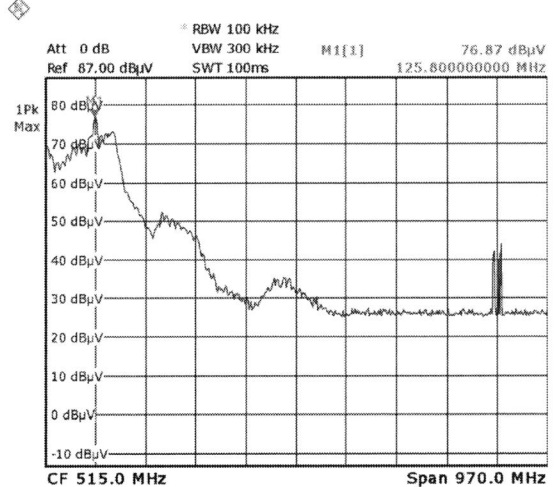

Fig 3: H-field of NiZn core @ Fsw – 1MHz

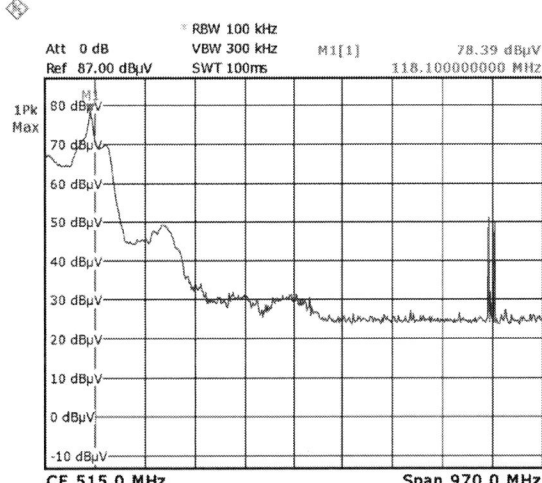

Fig 4: H-field of Iron alloy core @ Fsw – 1MHz

© VDE VERLAG GMBH · Berlin · Offenbach

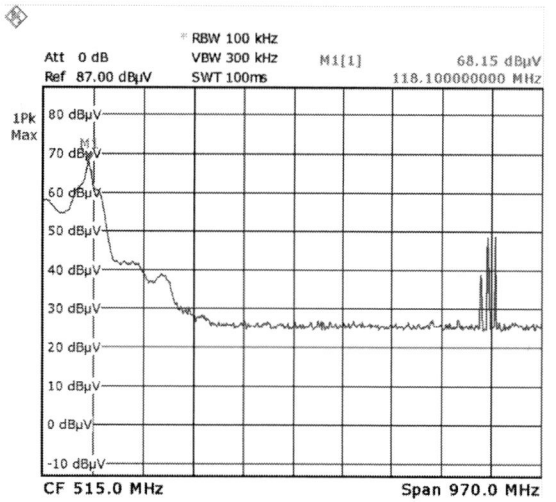

Fig 5: H-field of iron powder core with Al enclosure

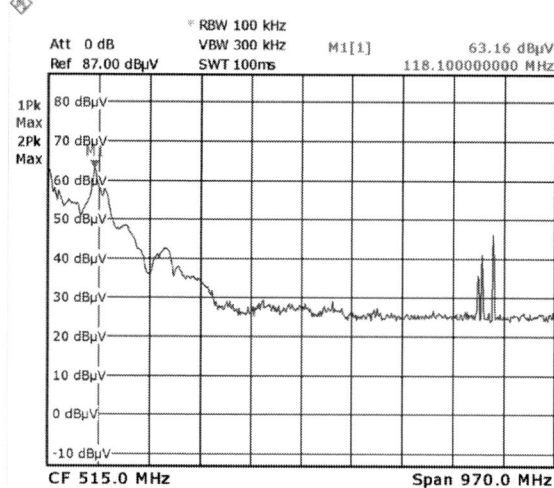

Fig 6: E-field of iron powder core @ 1MHz Fsw

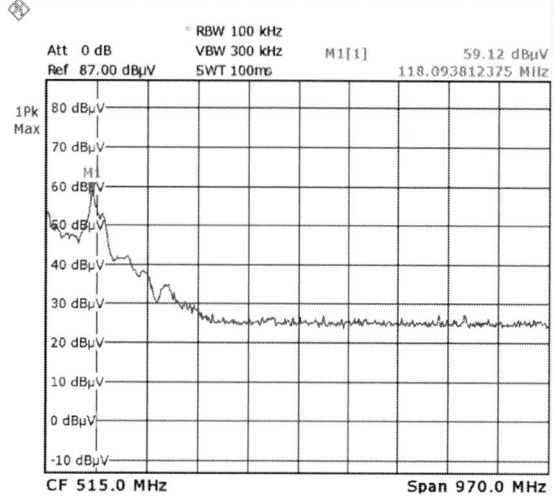

Fig 7: E-field of iron powder core with Al enclosure @ 1MHz Fsw

Figures 5 & 7 shows EM radiation (near fields) improvement of 10 dBµv for iron powder cores, enclosed with an Aluminum shielding Cabinet. Fig: 1&6 are H & E-fields respectively of the iron powder core without Al enclosure. By employing shielding materials, one can also observe considerable attenuation but they must be used according to the attenuation frequencies.

5. Summary & Future work:

With the arrival of new technology in Mosfets, the switching frequency & rise times are only going to increase further, hence behavior of EM radiation is going to be very vital for Electro-Magnetic Compatibility.

EM radiation is a very vast topic as even a small variation in any one of parameters influence source antenna, the near field response also varies. To characterize the near fields for various types of inductors needs many experiments & observations. In future the focus will be on the system performance in far field with respect to the performance of inductor in near fields.

References:

1. *Electromagnetic Compatibility Engineering by Henry W. Ott*
2. *Selecting and Using Ferrite Beads for Ringing Control in Switching Converters BY CHRISTOPHER RICHARDSON & RANJITH BRAMANPALLI*

A new high frequency ferrite material for GaN applications

Herbert Jungwirth, Michael Schmidhuber, Michael Baumann, Markus Schmeller
SUMIDA Components & Modules GmbH, Dr. Hans-Vogt-Platz 1, 94130 Obernzell
hjungwirth@eu.sumida.com

Abstract

Efficiency is not exclusively related to the electrical efficiency of power electronics, but can also be increased by weight reduction in the case of mobile systems. Due to the savings in weight and the related savings of fuel the overall system efficiency of an aircraft and the especially harmful ejection of exhaust gases at high altitude can be reduced. This allows direct economic savings and environmental damages are reduced.

With this converter, fundamental insights into the operation of power converters with higher operating voltage, current and a frequency beyond 1 MHz can be achieved. These parameters will be challenging in terms of operational reliability as well as the EMI of the operating network in terms of conducted and radiated interference.

The rapid progress in GaN and SiC power semiconductors will lead to a further miniaturization of power electronic assemblies and subsystems. Inductive components have a significant impact here. The drastically increased frequency requires improved ferrite materials with lowest losses and transformer designs with unique construction technologies.

Within the governmental funded project "GaN-resonant" (FKZ: 16ES0075) a new DC-DC resonant converter should operate up to 3 kVA @ f >>1 MHz. The paper describes an interim development status of this novel passive resonant transformer unit and the new improved ferrite material.

1. Introduction and background

The development of switching power supplies usually aims to the highest possible electrical conversion efficiency. A review of the cumulative energy balance of those systems includes also the transportation energy of the device itself. Higher electrical losses can sometimes be compensated by savings in weight in mobile systems and the related savings in fuel, for example considering a whole airplane.

Passive components like capacitors and magnetic components determine mostly the overall size of a power system. To achieve a miniaturization of the magnetic devices and thus the DC-DC converter the following options are to be considered:

Lowering the electrical losses result in an increased energy density due to the lower wire resistance. Magnetic losses can be reduced by the introduction of a new core material. In the application described here standard Mn-Zn ferrites will not be applicable because of dominating eddy currents and the resulting heat dissipation. Simulations show that temperatures over 1000°C would appear. As a result the commonly known map (see Figure 1) of choosing magnetic materials must be extended towards higher frequencies and lower losses at comparable high inductions.

© VDE VERLAG GMBH · Berlin · Offenbach

PCIM Europe 2016, 10 – 12 May 2016, Nuremberg, Germany

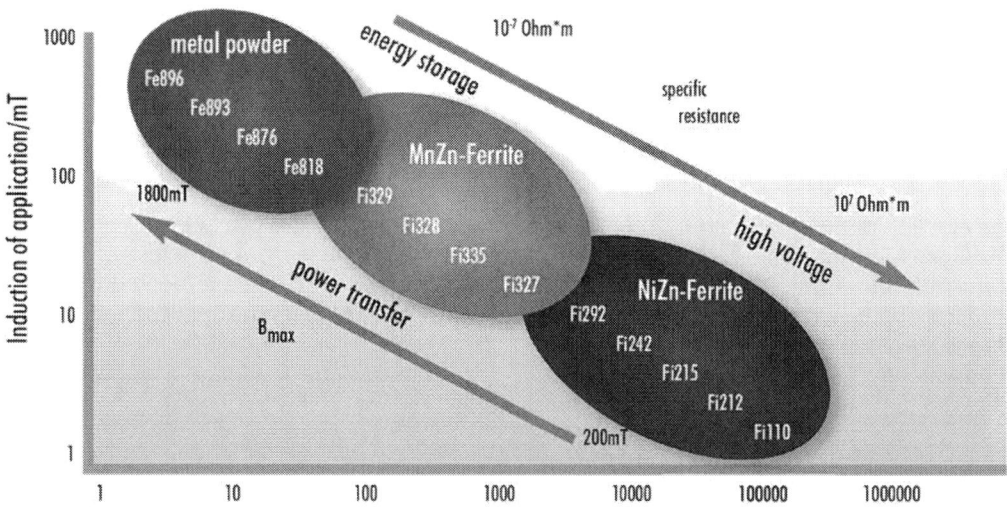

Figure 1: Map of standard SUMIDA ferrites for power transfer applications

2. Requirements and first design

The usual switching frequency of standard resonant DC-DC converters – at present up to 350 kHz – has to be raised significantly to more than one megahertz. Such operating frequencies are well known in DC-DC converters, but with much lower voltage ratings. A significant improvement in the magnetic properties of the core material and innovative winding structures lead to a first prototype, shown in Figure 2. The specification required a transformation ratio of 8:1:1 at an input voltage of 280 V. The primary inductance was defined to about 8 µH whereas the leakage inductance should be around 2 µH, resulting from simulation results of the resonant circuitry.

Figure 2: First prototype of the resonant 2 MHz GaN transformer

© VDE VERLAG GMBH · Berlin · Offenbach

3. Ferrite material development

A new ferrite material enabled a miniaturized transformer design at same magnetic gain. Due to the reduction of size and volume a much higher thermal resistance and also a higher self-heating was expected. During project run-time the optimization of the core material is aligned in such a way as to minimize the specific material losses with respect to the flux density Bmax and the switching frequency.

The specific losses have been determined at the beginning of the project to 375 mW/cm³ at 100°C and 25 mT at 1.5 MHz. For this new material mixtures are to explore, leading to optimized manufacturing technology, pressing and sintering processes to new atomic pattern structures with significantly improved properties. It also was observed that the induced eddy current losses are much higher than the hysteresis losses.

Consequently, it was necessary to improve the grain structure. Figure 3 shows the core losses of two ferrites with different particle sizes depending on the frequency.

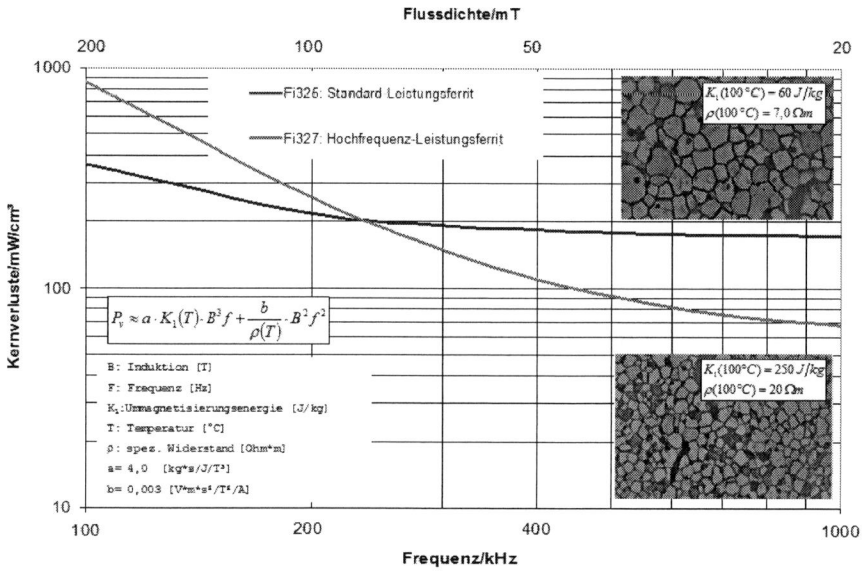

Figure 3: Core loss of different materials over the frequency

The new developed core material within this governmental funded research project is able to reduce the losses to at least 100 mW/cm³ at 100°C and 25 mT at 1.5 MHz. This result shows drastically improvements in ferrite material technology. More than 70% of the magnetic losses compared to standard high frequency ferrite materials could be reduced. The next Figure 4 shows the progress of loss reduction during the project life time.

PCIM Europe 2016, 10 – 12 May 2016, Nuremberg, Germany

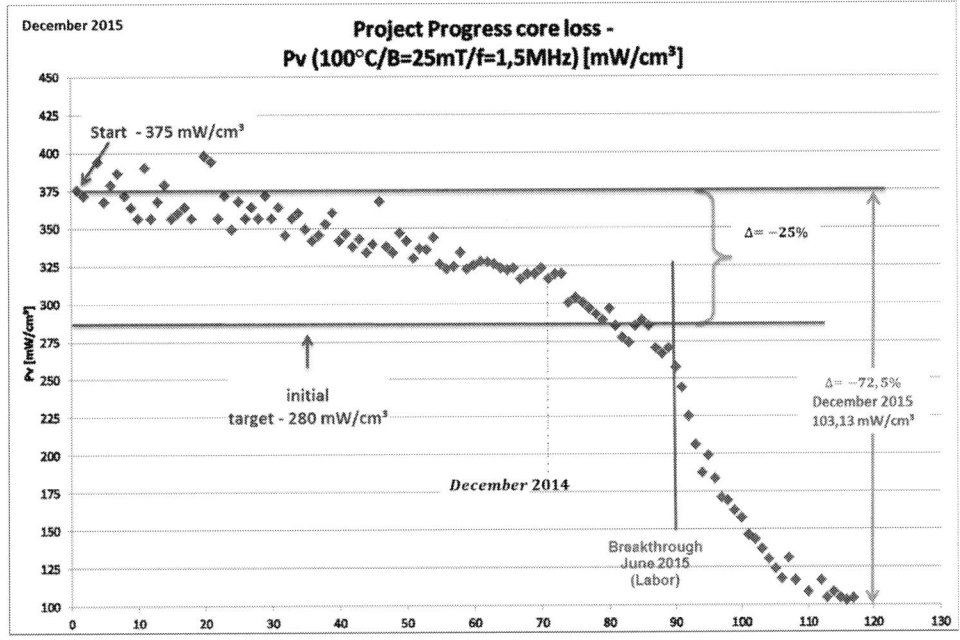

Figure 4: Core loss progress over project time

4. Results

The progress in ferrite material development can be presented visibly by the so-called Performance Factor (see Figure 5). This factor is the Flux Density – Frequency product (B x f, in mT x kHz), for a fixed core loss Pv = 300 mW/cm³ and also fixed temperature T = 100°C, plotted as a function of frequency. It is normally used to find the optimum material for the target switching frequency.

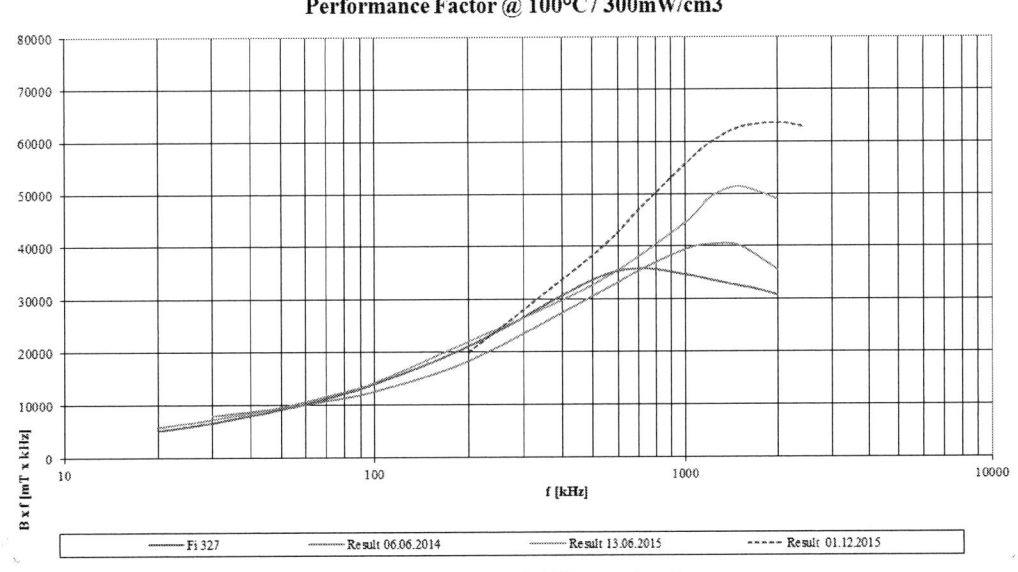

Figure 5: Performance Factor of different development steps

© VDE VERLAG GMBH · Berlin · Offenbach

1767

5. Summary and outlook

At a test rig the converter has been successfully tested with the new ferrite material with more than 2 MHz switching frequency. The driving circuitry actually operates at a lower output power of about 0.8 kW to prove the concept. In the next steps the temperature rise and efficiency of the resonant transformer has to be verified at the maximum transferable power of up to 3 kW at more than 1.5 MHz.

References

[1] W. Hirschmann/A. Hauenstein: Schaltnetzteile; Siemens Aktiengesellschaft; 1990; ISBN 3-8009-1550-2
[2] O. Kilgenstein: Schaltnetzteile in der Praxis; Vogel-Fachbuchverlag; 1992; ISBN 3-8023-1436-0
[3] D. Schöder: Leistungselektronische Schaltungen – Funktion, Auslegung und Anwendung; Springer Vieweg; 2012; ISBN 978-3-642-30103-2
[4] F. Zach: Leistungselektronik – Ein Handbuch; Springer Wien New York; 2010; ISBN 978-3-211-89213-8

Acknowledgements

The research and development presented here was supported by the German Ministry of Education and Research (BMBF Grant No. 16ES0075).

Multi-stage LLC resonant converters designed for wide output voltage ranges

Chi-Wa, Tsang, University of Lincoln, UK, ctsang@lincoln.ac.uk
C. Bingham, University of Lincoln, UK, cbingham@lincoln.ac.uk
M.P. Foster, University of Sheffield, UK, m.p.foster@sheffield.ac.uk
D.A. Stone, University of Sheffield, UK, d.a.stone@sheffield.ac.uk
J. Leach, Castlet Ltd, john.leach@castletltd.com

Abstract

The paper describes a novel multi-stage LLC resonant converter topology for facilitating wide output voltage ranges. This is achieved by combining the gain range of a capacitor-diode clamped LLC resonant converter with that of a traditional LLC resonant converter. A prototype converter is designed and commissioned to illustrate the design procedure and demonstrate resulting operational characteristics. Experimental results are used to show operational characteristics of the proposed converter.

1. Introduction

A challenge for many application sectors is the availability of controllable power converters that facilitate the production of wide output voltage ranges—the increasing use of high-power LED-based systems obtained through the connection of multiple low-power LEDs (in series or parallel configurations), provide a candidate example [1]. This is particularly acute when the full benefits of maintaining high efficiency across the voltage range is required [2].

Resonant converters have become increasingly popular candidates due to their soft-switching characteristics that reduce switching losses [3] and, as a result, improve the overall converter efficiency. While series and parallel resonant converters support the soft-switching characteristics needed for high efficiency, multi-resonant converters [3] have more favorable operating characteristics (e.g. narrow operating frequency). The LLC resonant converter, in particular, utilizes the parasitic elements found in a transformer to allow high power densities to be achieved. Nevertheless, one of the main impediments to their widespread adoption is that at their nominal operating point (the independent load point) an excessive current can flow if the load is not controlled, potentially leading to damage of both the converter and the load. A capacitor-diode clamp configuration has been considered in [5] to change the resonant converter characteristics by switching the resonant tank components, thereby facilitating a reduced voltage, and hence current, when subject to overloading conditions. The main advantage of the method in [5] in comparison to methods reported in [6][7] is that the output voltage is reduced autonomously without the need of any controller action.

This paper utilizes the unique gain reduction characteristic of the capacitor-diode clamped LLC resonant converter to allow a controllable wide output-voltage range to be achieved. A second LLC resonant converter is then connected in series with the first stage to further shape the current to produce the constant current characteristic, required in LED applications for instance.

2. Multi-stage LLC resonant converter

The schematic of the proposed converter is shown in Fig. 1, consists of two resonant converters connected in series. While both stages are of a LLC resonant converter type, the first stage has an additional capacitor-diode clamp.

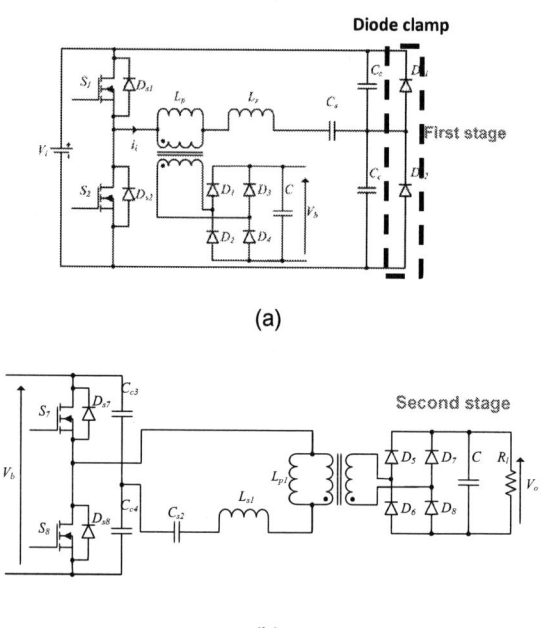

(a)

(b)

Fig. 1 Schematic diagram of the proposed converter. (a) First stage (b) Second stage

2.1. Converter's configuration

Three main functional parts are typically found in LLC resonant converters, viz. 1) a full- or half-bridge DC chopper which converts the DC input into a pulsed AC waveform, 2) a resonant tank which function is to block all but the fundamental component of the pulsed AC waveform to the output. The two inductors in the converter can be convenient provided by the components of a transformer, which also provides galvanic isolation, and 3) a half- or full-bridge rectifier with a bulk capacitor which rectifies then smooth's the waveform from the resonant tank to form a DC output.

A half-bridge DC chopper is chosen for the both converter stages in this instance, and is formed by MOSFET S_1 and S_2 for the first stage and by MOSFET S_7 and S_8 for the second stage. Resonant inductors, L_p and L_s of the first stage, L_{p1} and L_{s1} of the second stage, are the magnetizing and leakage inductances of the isolation transformers T_1 and T_2 , respectively. C_{c1}, C_{c2}, C_s and C_{c3}, C_{c4}, C_{s2} are the resonant capacitors of the first and second stage, respectively. The three resonant capacitors of each stage can be combined to form a single resonant capacitor using (1):

$$C_r = C_s + 2C_c \tag{1}$$

Diodes D_1, D_2, D_3 and D_4 form the full-bridge rectifier for the first stage. Similarly, diodes D_5, D_6, D_7 and D_8 form the full-bridge rectifier for the second stage. The internal body diodes D_{s1}, D_{s2}, D_{s7}, D_{s8} are also shown in the figure, they are the critical components for zero voltage switching (ZVS) to take place.

Two additional diodes (D_{c1}, D_{c2}) across the resonant capacitors C_{c1}, C_{c2} can be found in the first stage. These become active whenever the voltage across C_{c1}, C_{c2} rises above the input voltage or below zero volt, at which point the current is bypassed through the diodes, reducing the power transfer to the output, and hence reducing the effective gain without the need for any alteration in switching frequency.

3. Equivalent circuit model

With only the fundamental component of the input current passing through to the output, the equivalent circuit model of a non-capacitor-diode clamped (i.e. second stage) LLC resonant converter can be found using fundamental harmonic analysis (FHA) [3]. The magnitude of the input current (I_i) and output voltage (V_o) can be calculated using (2) and (3), respectively (n.b. ideal diodes are assumed and components parasitic resistances are neglected).

$$I_i = \frac{2V_i}{\pi Z_1} \tag{2}$$

$$V_o = \frac{\pi I_i (R_{eq}||sL_p)}{4n} \tag{3}$$

where $Z_1 = R_{eq}||sL_p + sL_s + 1/sC_s + 1/2sC_c$, is the input impedance of the resonant circuit and the load. $s = j\omega_s$, is the complex frequency and $R_{eq} = 8n^2 R_l/\pi^2$ is the equivalent resistance presented by the rectifier, output filter and load reflected through the transformer; n is the transformer primary to secondary turns ratio.

With the capacitor C_c excited by a sinusoidal input current, its voltage at any given instant is found from (4). Substituting $\omega_s t$ for θ (where $\omega_s = 2\pi f_s$ is the angular switching frequency), the capacitor voltage v_c at any θ is then given by (5):

$$v_c(t) = \frac{1}{2C_c} \int I_i \sin(\omega_s t)\, dt$$

$$= -\frac{I_i}{2\omega_s C_c} cos(\omega_s t) + V_n \tag{4}$$

$$v_c(\theta) = -\frac{I_i}{2\omega_s C_c} cos(\theta) + V_n \tag{5}$$

where V_n is the initial condition for a given conduction state starting at $\theta = n$.

Using (2) and (3), the output voltage for a given set of resonant tank components, load condition and operating frequency can be calculated. It is of benefit to be able to study the converter's voltage gain characteristics independently from the resonant tank component selection and input/output voltages. This can be achieved using the nominalized gain M_g which is obtained by substituting (2) into (3), and rearranging in terms of $(2nV_o)/V_i$, and then normalising against the three parameter: inductor ratio $A = L_p/L_s$, loaded quality factor $Q = \sqrt{L_s/C_r}/R_{eq}$ and normalised switching frequency $f_n = f_s/f_o$. The result is given in (6):

$$M_g = \frac{2nV_o}{V_i} = \frac{Af_n^2}{Af_n^2 + f_n^2 - 1 + j(f_n^3 QA - f_n QA)} \tag{6}$$

Characteristics of the capacitor-diode clamped LLC resonant converter can also be found using FHA after the equivalent impedance of the clamped resonant capacitor is obtained. This involves the following three step procedure:

PCIM Europe 2016, 10 – 12 May 2016, Nuremberg, Germany

Step 1: obtain the piecewise equation describing the capacitor voltage under clamping conditions, shown in Fig. 2. The four equations covering each period are given in (7):

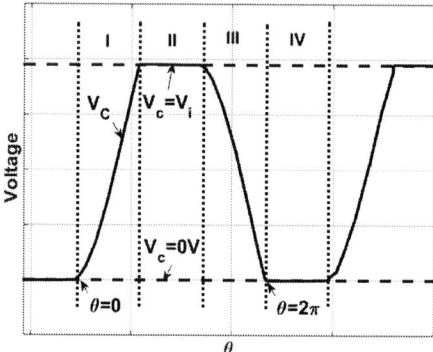

Fig. 2 Resonant capacitor voltage waveform under diode clamp conditions.

$$v_C(\theta) = \begin{cases} \frac{I_i}{2\omega_s C_c}(1 - cos(\theta)) & 0 < \theta \le \delta \\ V_i & \delta < \theta \le \pi \\ V_i - \frac{I_i}{2\omega_s C_c}(1 + cos(\theta)) & \pi < \theta \le \pi + \delta \\ 0 & \pi + \delta < \theta \le 2\pi \end{cases} \tag{7}$$

The diode-clamp non-conduction angle, δ, is found by substituting $v_C(\delta) = V_i$ into (7) and rearranging in term of δ:

$$\delta = cos^{-1}\left(1 - \frac{2\omega_s C_c V_i}{I_i}\right) \tag{8}$$

Step 2: obtain the fundamental component of V_c using Fourier series, as in (9):

$$f(t) = a_0 + \sum_{n=1}^{\infty}\left(a_n cos(n\omega_s t) + b_n sin(n\omega_s t)\right) \tag{9}$$

$$a_0 = \frac{1}{2\pi}\int_{-T/2}^{T/2} f(t)dt$$

$$a_n = \frac{1}{\pi}\int_{-T/2}^{T/2} f(t)cos(n\omega_s t)dt$$

$$b_n = \frac{1}{\pi}\int_{-T/2}^{T/2} f(t)sin(n\omega_s t)dt$$

By substituting (5) into (7) the fundamental component of v_c is given by:

$$v_C(\theta) = \left[\frac{2}{\pi}V_i \cos(\delta) + \frac{I_i}{2\pi\omega_s C_c}\left(1 + \cos(\delta)(\cos(\delta) - 2)\right)\right]sin(\theta)$$

$$+ \left[-\frac{2}{\pi}V_i \sin(\delta) - \frac{I_i}{2\pi\omega_s C_c}\left(\delta + \sin(\delta)(\cos(\delta) - 2)\right)\right]cos(\theta) \tag{10}$$

© VDE VERLAG GMBH · Berlin · Offenbach

Step 3: obtain the equivalent impedance of the diode-capacitor combination using the transform $cos(\theta) = jsin(\theta)$ and then dividing by resonant current, as follows:

$$Z_{\text{C}} = \left[\frac{2V_{\text{i}}}{\pi I_{\text{i}}}\cos(\delta) + \frac{1}{2\pi\omega_{\text{s}}C_{\text{c}}}\left(1 + \cos(\delta)\left(\cos(\delta) - 2\right)\right)\right]$$

$$+j\left[-\frac{2V_{\text{i}}}{\pi I_{\text{i}}}\sin(\delta) - \frac{1}{2\pi\omega_{\text{s}}C_{\text{c}}}\left(\delta + \sin(\delta)(\cos(\delta) - 2)\right)\right] \tag{11}$$

With the equivalent impedance Z_{C} of the clamped capacitor identified, where the effect of the diode-clamp is accommodated, the input impedance of the capacitor-diode clamped LLC resonant converter is given by $Z_2 = R_{\text{eq}}||sL_{\text{p}} + sL_{\text{s}} + 1/sC_{\text{s}} + Z_{\text{c}}$. The magnitude of the resonant tank current under clamping is again found by FHA (12). The output voltage can be found by substituting (12) into (3).

$$I_{\text{i}} = \frac{2V_{\text{i}}}{\pi Z_2} \tag{12}$$

$$M_{\text{g(clmp)}} = \frac{V_{\text{o}}}{V_{\text{i}}}2n = \frac{R_{\text{eq}}||sL_{\text{p}}}{R_{\text{eq}}||sL_{\text{p}}+sL_{\text{s}}+\frac{1}{sC_{\text{s}}}+Z_{\text{c}}} \tag{13}$$

Since (12) cannot be solved analytically since I_{i} is unknown, and δ and Z_{c} depend on I_{i}, an iterative procedure is employed, as follows: Firstly, estimate the resonant tank current I_{i} using (1), assuming the diode-clamp is inactive. Using this estimated value, the non-conduction angle, δ, and capacitor-diode clamp equivalent impedance, Z_{c}, are estimated using (8) and (11). Next, instead of using (1) during the next iteration, using (12) for I_{i}, then (8) and (11) for the δ and Z_{c} until convergence ensues.

Similar to the second stage, for the purpose of studying the voltage gain characteristic, the first stage converter gain, $M_{\text{g(clmp)}}$, is obtained through substituting (12) into (3) and rearrange in terms of $(2nV_{\text{o}})/V_{\text{i}}$ as in (6), after which substituting $s = j\omega$, $Z_{\text{c}} = R + jX$ (where R and X are the real and imaginary part of Z_{c} in (11) respectively), and introducing the $j\omega C_{\text{r}}$ term into both the numerator and denominator as in (14):

$$M_{\text{g(clmp)}} = \frac{j^2\omega^2 L_{\text{p}}C_{\text{r}}}{j^2\omega^2 L_{\text{p}}C_{\text{r}}+\frac{j^3\omega^3 L_{\text{s}}L_{\text{p}}C_{\text{r}}}{R_{\text{eq}}}+j^2\omega^2 L_{\text{s}}C_{\text{r}}+\frac{j\omega L_{\text{p}}C_{\text{r}}}{C_{\text{s}}R_{\text{eq}}}+\frac{C_{\text{r}}}{C_{\text{s}}}+\frac{j^2\omega^2 RL_{\text{p}}C_{\text{r}}}{R_{\text{eq}}}+Rj\omega C_{\text{r}}+\frac{j^3\omega^2 C_{\text{r}}XL_{\text{p}}}{R_{\text{eq}}}+j^2X\omega C_{\text{r}}} \tag{14}$$

The final step involves substituting the following normalising factors $L_{\text{p}} = AL_{\text{s}}$, $L_{\text{s}}C_{\text{r}} = 1/{\omega_0}^2$, $L_{\text{s}}/R_{\text{eq}} = Q/\omega_0$, $f_{\text{n}} = \omega/\omega_0$ and $B = C_{\text{r}}/C_{\text{s}}$, to obtain:

$$M_{\text{g(clmp)}} = \frac{{f_{\text{n}}}^2 A}{{f_{\text{n}}}^2 A+{f_{\text{n}}}^2-B+k_{\text{r}}+j({f_{\text{n}}}^3 AQ-f_{\text{n}}AQB+k_{\text{i}})} \tag{15}$$

where $k_{\text{r}} = \omega C_{\text{r}}(f_{\text{n}}AQR + X)$ and $k_{\text{i}} = \omega C_{\text{r}}(-R + f_{\text{n}}AQX)$, are terms accounting for the change in the effective impedance of the C_{c} caused by the diode-clamp when it is active, and it is assumed that the values for R and X have converged.

PCIM Europe 2016, 10 – 12 May 2016, Nuremberg, Germany

4. Operation of the proposed converter

As part of the first stage (see Fig. 1 (a)), the capacitor-diode clamp limits the voltage across the resonant capacitor to reduce the overall voltage gain of the stage. Using (15), the gain of the clamped LLC resonant converter under different loading conditions at different operating frequencies can be found, as shown in the example Fig. 3 (a). Results show that as the current demand increases, the gain around the load independent point (LIP) reduces. The V-I characteristic at the operating point is given by the dashed-line in Fig. 3 (b).

The second stage (see Fig. 1 (b)) is included to shape the current of the first stage to that desired at the output. This stage is then operating at or above the resonant frequency (LIP) to achieve zero voltage switching (ZVS) and to reduce the required voltage gain range, since the gain reduces in a manner inversely proportional to the load—the desired constant current characteristic can therefore be obtained.

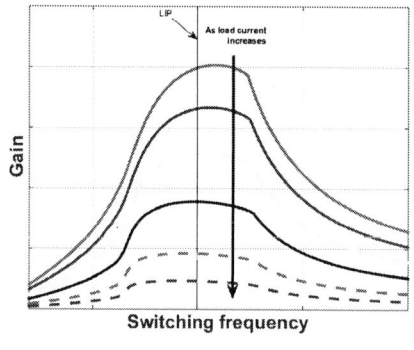

(a)

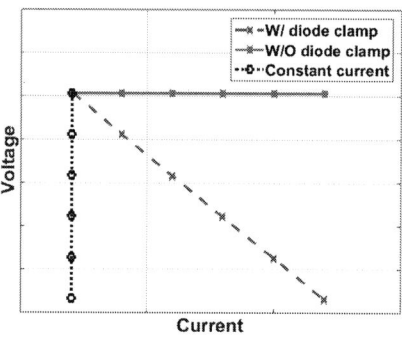

(b)

Fig. 3 Gain and V-I characteristics of the converter. (a) Gain characteristic of the first stage (b) V-I characteristic

5. Design example

A 20 W (12 V, 1.5 A) prototype is commissioned and built to demonstrate the operation of the proposed converter. Using (1), and with a chosen resonant frequency of 100 kHz and transformer turns ratio of 3:1, the resulting resonant tank components for the first stage are: L_p = 197 μH, L_s = 39 μH, C_{r1} = 32.5 nF. The second stage is also operating at the same switching frequency and its tank components are, by design, L_{p1} = 36 μH, L_{s1} = 7 μH, C_{r2} = 179 nF. The commissioned first and second stages of the converter are shown in Fig. 4 (a) and (b) respectively. Experimental measurements from the proposed converter are

© VDE VERLAG GMBH · Berlin · Offenbach

shown in Fig. 4 (c). The solid and dashed lines are the V-I characteristics of the first stage, and both stages, respectively; cf. Fig. 3 (b). The results show that the desired constant current characteristic can indeed be obtained by changing the switching frequency, hence the gain of the second stage. This makes it suitable for use in LEDs where the output current drawn is constant irrespectively of their different impedance.

(a)

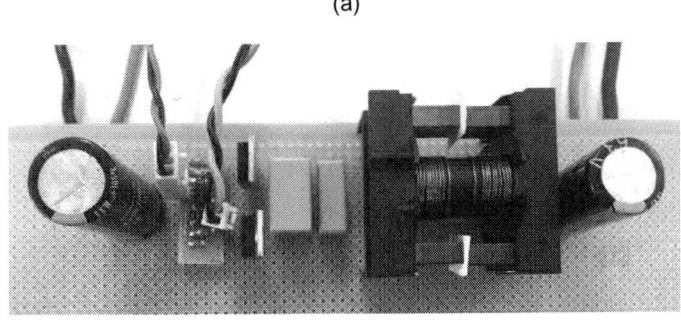

(b)

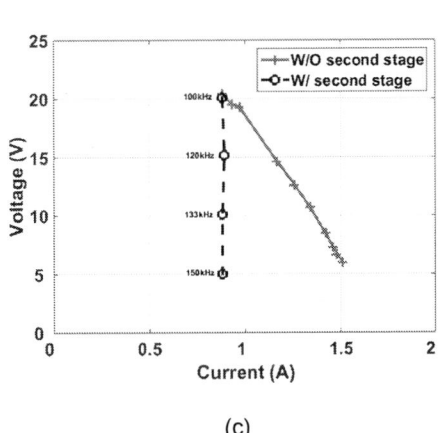

(c)

Fig. 4 The V-I characteristics of the proposed converter. (a) first stage (b) second stage (c) P-I curve

6. Conclusion

This paper proposes connecting two LLC resonant converters in series to achieve a wide output voltage range. The first stage utilizes the unique characteristics of a diode-clamp to reduce the voltage gain found in a traditional LLC resonant converter, and hence the overload current. The second stage further shapes the current to obtain the desired constant current characteristic during normal operation. A design example is given and prototype built to show the feasibility of the proposed multi-stage converter, the results show that a constant current characteristic can be obtained, allowing the same current to be drawn by LEDs with different impedance.

7. Literature

[1] Daek, J.-I., Kim, J.-K., Lee, J.-B., Youn, H.-S., Moon, G.-W., 'Integrated Asymmetrical Half-Bridge Zeta (AHBZ) Converter for DC/DC Stage of LED Driver with Wide Output Voltage Range and Low Output Current', IEEE Trans. Ind. Electron., 2015, 1 Vol. PP.

[2] Musavi, F., Craciun, M., Gautam, D.S., Eberle, W., Dunford, W.G., 'An LLC Resonant DC-DC Converter for Wide Output Voltage Range Battery Charging Applications' IEEE Trans. Power Electronics., 2013, 5437-5445, Vol.28.

[3] Steigerwald, R.L., 'A Comparison of Half-bridge Resonant Converter Topologies', IEEE Trans. Power Electron., 1988, pp. 174-182, Vol. 2.

[4] Batarseh, I., 'Resonant converter topologies with three and four energy storage elements' IEEE Trans. Power Electron., 1994, pp. 66-73, Vol. 9.

[5] Tsang, C.W., Foster, M.P., Stone, D.A., Gladwin, D.T., 'Analysis and Design of LLC Resonant Converters with Capacitor-Diode Clamp Current Limiting', IEEE Trans. Power Electronics., 2015, 1345-1355 Vol.30.

[6] Gould, G.R. Bingham, C.M., Foster, M.P., Stone, D.A., 'CLL resonant converters with output short-circuit protection strategy', IEE Electric power applications, pp. 1296-1306, Vol, 152

[7] Xie, X., Zhang, J., Zhao, C., Zhao, Z., Qian, Z., 'Analysis and optimization of LLC resonant converter with a novel over-current proection for LLC converter', IEEE Trans. Power Electronics., 2015, 435-443 Vol.30.

Application Advantages and Disadvantages of Modern Fast Switching MOSFETs in VRM

Zhiyang Chen, ON Semiconductor, USA, zhiyang.chen@onsemi.com
Ann Starks, ON Semiconductor, USA, ann.starks@onsemi.com

Abstract

Advancements in integrated MOSFET silicon and packaging technologies have increased switching speed and improved over-all system performance for VRM applications. However, fast-switching can also lead to voltage overstress and an increased dv/dt and di/dt turn-on susceptibility that previously was not as prominent a concern. VRM application advantages and disadvantages associated with fast switching are analyzed, and methods to minimize dv/dt-induced and di/dt-induced turn-on are provided. Finally, an optimized PCB layout that takes full advantage of fast switching MOSFETs is presented.

1. Introduction

The synchronous buck converter is the dominant topology for dc:dc power conversion in VRM applications. Computers, servers, tablets, game consoles, and smart phones are but a few examples of this application space. Today's market seeks to increase power delivery while simultaneously increasing overall system efficiency. With the latest technology advances of MOSFETs, it is possible to achieve higher power density capability and more efficient power conversion than ever before. Switching speed is one of the key parameters affecting the efficiency of the power converter.

In order to achieve faster switching, both the silicon and packaging technology must be optimized. From a silicon standpoint, this means attaining a lower on resistance while minimizing the switching losses of the die. Parameters that directly affect switching losses include capacitances and gate charge of the device. At light load the switching losses can be significant. By optimizing the silicon technology, capacitances can be minimized, creating a fast switching MOSFET. For the packaging, this means minimizing two inductances: loop inductance L_{LOOP}, and high side MOSFET source inductance L_{SH}, as shown in Figure 1. The switching speed, dv/dt, and di/dt of the power switching loop are limited by whichever inductance component is slower. The PCB layout can be optimized to minimize the coupling of the main switching loop and the gate drive loop, minimizing the common source inductance and improving the switching speed of the high side MOSFET [1][2].

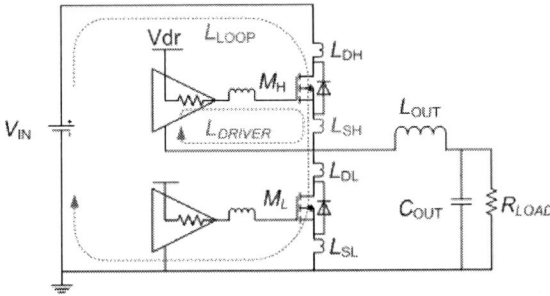

Figure 1: VRM Circuit with Parasitic Inductance

The combined silicon and packaging advancements has resulted in the modern fast switching MOSFET, providing an integrated solution with the high side and low side MOSFETS contained in a single package (Figure 2). Integrating the MOSFETs into the PowerPhase package provides an over-all reduction in L_{LOOP} of over 50%, and reduces L_{SH} from approximately 1.5nH to nearly 0nH. This results in a dramatic increase in MOSFET switching speed for both the high side and low side MOSFETs[3][4].

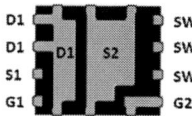

Figure 2: PowerPhase Modern Fast Switching MOSFET Pair for VRM

Table 1 lists four key inductance components that are reduced by using the integrated PowerPhase package. In order to highlight the characteristics of the modern fast switching MOSFETs, two pairs of power MOSFETs are evaluated in this paper. Device A is a fast-switching discrete pair in the SO8-FL package, and Device B is an integrated fast-switching pair in the PowerPhase package.

Table 1: Comparison of VRM Parasitic Inductances

Inductance Type	Device A	Device B
Switching loop inductance, L_{LOOP}	2 nH ~ 3 nH	1 nH ~ 2 nH
Common source inductance, L_{SH}	~1 nH	~ 0 nH
High side gate inductance, L_{GH}	10 nH ~ 20 nH	10 nH ~ 15 nH
Low side gate inductance, L_{GL}	15 nH ~ 25 nH	10 nH ~ 20 nH

The two pairs were selected to have as similar equivalent resistance, R_{EQ}, as possible. Device B silicon and packaging have been optimized to minimize switching and conduction losses, and is representative of the modern fast-switching MOSFET pair for VRM. Table 2 shows the typical $R_{DS(on)}$ of the MOSFET pairs, as well as an equivalent resistance, R_{EQ}, as defined in Equation 1, where D is the duty cycle of the converter:

$$R_{EQ}=D*R_{DS(on)HS}+(1-D)*R_{DS(on)LS} \hspace{2cm} EQ\ (1)$$

Table 2: Summary of Device Parameters for Device A and Device B

No.	Package (HS / LS)	$R_{DS(on)}$ at 4.5V HS/LS (mΩ)	R_{EQ} at 4.5V (mΩ)	HS Capacitances (pF)			LS Capacitances (pF)		
				Ciss	Coss	Crss	Ciss	Coss	Crss
Device A	SO8-FL/SO8-FL	5.0/1.2	1.58	1988	1224	71	10144	5073	148
Device B	PowerPhase	7.2/1.4	1.98	592	262	13	3667	1439	68

2. Advantages of Fast Switching MOSFETs

2.1 Switching Speed Improvement

According to the model presented in [1], MOSFET switching time is calculated as:

$$t_{SW}=\frac{(C_{GD}+C_{GS})*\sqrt{\frac{2*I_{LOAD}}{K_P}+\frac{L_S*I_{LOAD}}{R_G}}}{I_G} \hspace{2cm} EQ\ (2)$$

PCIM Europe 2016, 10 – 12 May 2016, Nuremberg, Germany

As can be seen in Equation 2, total switching time can be reduced by optimizing device capacitances, gate resistance, and source inductance. Moving from the discrete implementation of Device A to the integrated fast-switching Device B, the total switching time is reduced significantly, as shown in Figure 3. Device B has optimized both silicon and packaging technology to optimize capacitance, gate resistance, and source inductance of both the high side and low side MOSFET. As a result, Device B total switching time is 30% – 40% that of Device A.

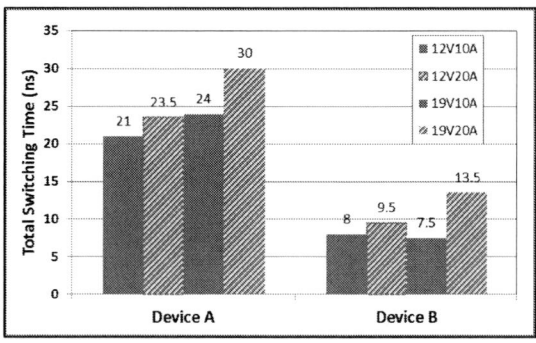

Figure 3: Switching Speed Reduction of Integrated Fast-Switching MOSFET

2.2 Efficiency improvement

Figure 4 illustrates the efficiency improvement that results from the reduction in switching losses, achieved by minimizing silicon and packaging parasitic losses. Device B achieves at least 2% higher peak efficiency at light load (10A) due to the optimization of C_{GD}, C_{GS}, R_G, L_S of each device. At higher load currents Device B efficiency slope rolls off faster than Device A because of the thermal response of the integrated device. Even so, the integrated fast-switching Device B is a competitive solution for modern VRM applications, providing additional savings in total board space. With the latest optimization of silicon technology, switching loss plays a more significant role in system performance, for both light load and heavy load. It becomes a balancing act between system efficiency and switching waveform quality.

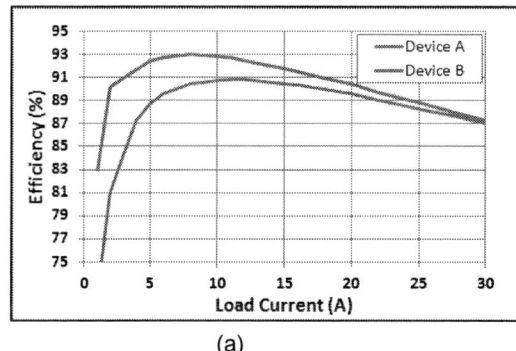 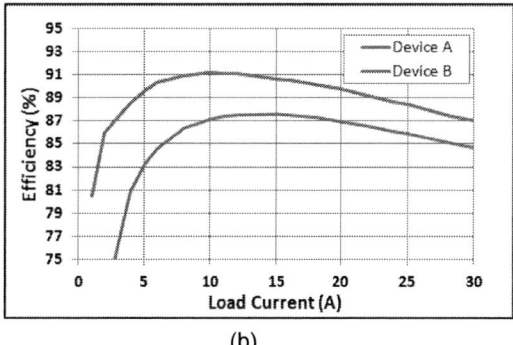

(a)   (b)

Figure 4: Efficiency Improvement of Integrated Fast Switching MOSFET (a)12V Input Voltage; (b) 19V Input Voltage

© VDE VERLAG GMBH · Berlin · Offenbach

3. Disadvantages of Fast Switching MOSFETs

3.1. Voltage Stress

Improving the switching speed of modern fast-switching MOSFETs increases the turn-on and turn-off transition of the MOSFET gate signal, inducing a higher voltage overshoot on the switch node that can cause a significant increase in the drain-to-source voltage stress experience by the low side MOSFET. This voltage overstress is observed at high side MOSFET turn-on/off, as seen in Figure 5. The modern fast-switching integrated MOSFET pair, Device B, experiences an overshoot that is 60% - 90% above the final switch node voltage. The additional reduction in parasitic source inductance of Device B makes the MOSFETs more susceptible to voltage overstress, resulting in the need to add more voltage margin to the system solution, or the need to slow down the MOSFET to reduce the voltage overshoot.

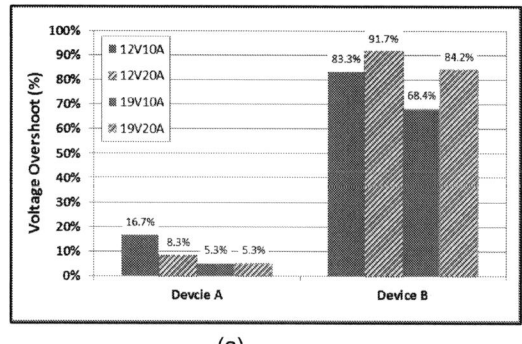
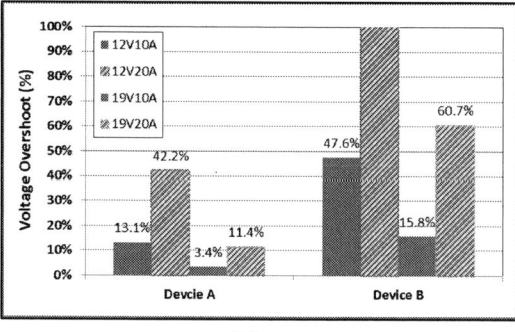

(a)   (b)

Figure 5: Voltage Stress on Low Side MOSFET at: (a) High Side MOSFET Turn-On; (b) High Side MOSFET Turn-Off

3.2. MOSFET Susceptibility to dv/dt and di/dt Turn-On

Gate bounce is defined as the induced voltage transient seen on the gate of the MOSFET as a result of a dv/dt or di/dt event. Gate bounce on the low side MOSFET is a primary concern for designers of modern VRM systems today. If the induced gate voltage is high enough for a long enough transient, the device can begin to turn on for a brief period of time. This unexpected turn-on can impact both the system performance and reliability. Modern fast-switching MOSFETs are most susceptible to this unexpected turn-on because of the significant reduction in device turn on time.

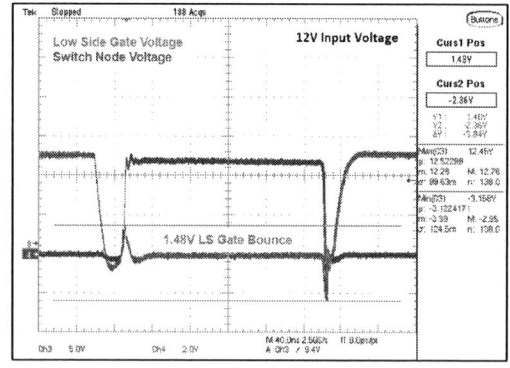
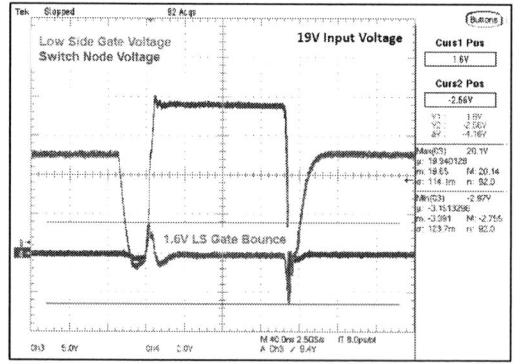

Figure 6: Measured Gate Bounce for Fast Switching MOSFET at 12V Input and 19V Input

© VDE VERLAG GMBH · Berlin · Offenbach

Gate bounce can be caused by: faster high side MOSFET turn-on, higher input voltage of the converter, low C_{OSS} of the low side MOSFET, and small loop inductance, L_{LOOP}. Figure 6 shows an example of measured low side gate bounce for a two-phase VRM application.

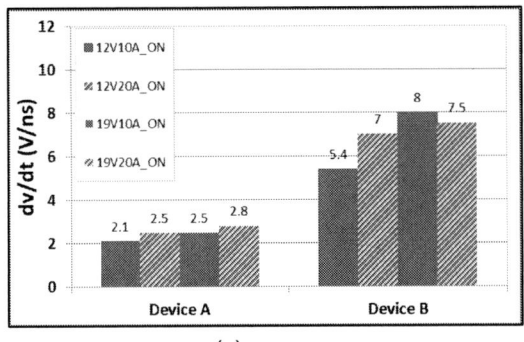 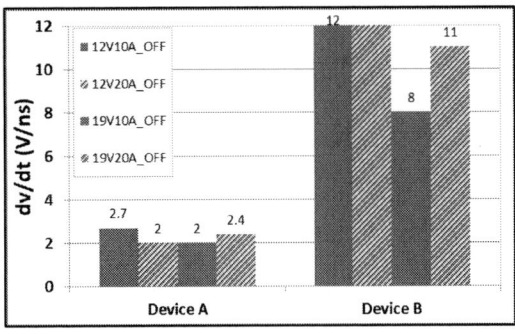

(a)   (b)

Figure 7: MOSFET dv/dt at High Side MOSFET: (a) Turn-On; (b) Turn-Off

The significant reduction in MOSFET turn-on and turn-off time for modern fast-switching MOSFETs creates excessive dv/dt and di/dt, which can turn-on the low side MOSFET by inducing a gate-to-source voltage comparable to the threshold voltage of the device [5]. Dv/dt and di/dt susceptibility was not as significant an issue with discrete MOSFET implementations due to slower switching characteristics. As seen in both Figure 7 and Figure 8, dv/dt and di/dt of the fast-switching MOSFET Device B are significant.

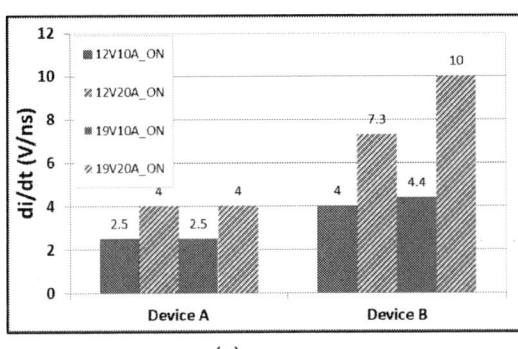 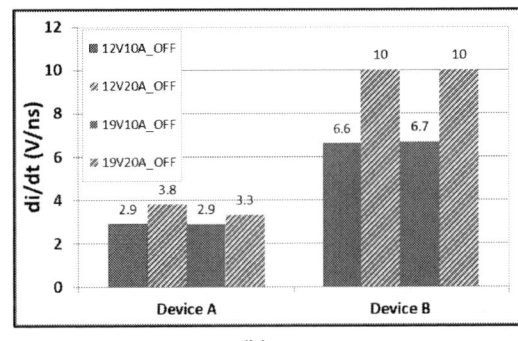

(a)   (b)

Figure 8: MOSFET di/dt at High Side MOSFET: (a) Turn-On; (b) Turn-Off

Device B fast-switching FET pair experiences significantly higher dv/dt and di/dt on the gate of the device. The faster transition (slope) of HS turn-on induces a ringing experienced on the gate of the LS device that can cause turn-on. Turn-on susceptibility becomes even more significant with lower threshold of modern fast-switching LS FETs.

4. Recommendations for dv/dt and di/dt Immunization

Until recently, the focus has been on the optimization of the high side MOSFET R_G and C_{RSS} to optimize the high side turn-on. If the high side MOSFET R_G is too low, the HS can induce device turn-on that negatively impacts system performance and reliability. However, equally important is the R_G of the low-side MOSFET. Traditionally, low side MOSFT optimization has focused on reducing R_G, balancing the ratio of Q_{GD}/Q_{GS}, and minimizing C_{OSS}. With modern fast-switching MOSFETs, the low-side MOSFET R_G has been reduced to less than one Ohm. The combination of optimized capacitances and significantly low R_G of the LS MOSFET has led to a new set of application considerations that were not as prominent before, due to the slower switching capability of devices.

Dv/dt occurs during the high-side MOSFET turn-on as well as during the high-side MOSFET turn-off, and is limited by I_{LOAD} / C_{OSS}. Turn-on dv/dt occurs during the high side MOSFET turn-on, when the switch node is pulled up to the input voltage. During this time, the C_{OSS} of the low side MOSFET is charged to the input voltage. Turn-off dv/dt occurs when the high side MOSFET is turning off, when the C_{OSS} of the high side MOSFET is charged from 0V to the input voltage. Device B experiences higher dv/dt than Device A due to a significantly reduced C_{OSS}.

To minimize sensitivity to high dv/dt events, capacitance can be added between the gate and source to modify the gate-to-source voltage divider to a lower value. This can be accomplished via an external capacitor in the circuit or via the internal silicon design. Figure 9(a) illustrates the dv/dt sensitivity improvement for the fast-switching integrated MOSFET pair, Device B. Adding C_{GS} to the low side MOSFET reduces the sensitivity to dv/dt events, minimizing unexpected low side turn-on, improving the converter performance.

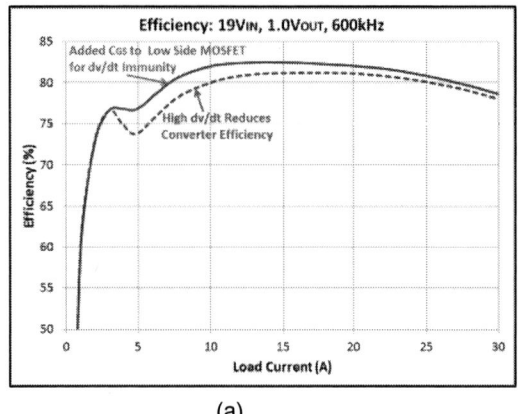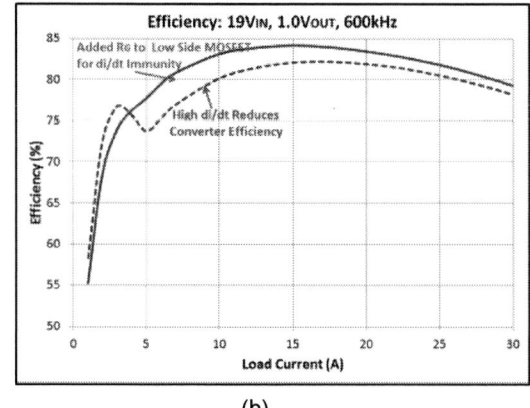

(a)   (b)

Figure 9: Practical Application of: (a) dv/dt Immunization; (b) di/dt Immunization

Di/dt occurs during the charge / discharge intervals for the low side MOSFET C_{OSS}, and is limited by V_{IN} / L_{LOOP}. As the high side MOSFET turns on, the C_{OSS} of the low side MOSFET charges, and the current through the low side MOSFET reduces as the current through the high side MOSFET increases. Device B experiences a higher di/dt than Device A due to a significantly reduced C_{ISS} and L_{SH}. When combined with low R_G, undesirable low side MOSFET turn-on can occur.

Sensitivity to di/dt can be reduced by adding resistance to the gate of the low side MOSFET. This can be accomplished by modifying the silicon, or through the use of an external resistor. Figure 9(b) illustrates the di/dt sensitivity improvement for the fast-switching integrated MOSFET pair, Device B. Adding R_G to the low side MOSFET reduces the sensitivity to di/dt events, minimizing unexpected low side turn-on, improving the converter performance.

5. Recommended PCB Layout

For VRM applications with fast switching integrated MOSFETs, care must be taken when laying out the printed circuit board (PCB). To maximize the performance of the integrated MOSFETs as well as to balance voltage stress, dv/dt and di/dt sensitivity, the PCB layout must be optimized to:

© VDE VERLAG GMBH · Berlin · Offenbach

- Minimize switching loop inductance – creates faster current transit between the high side and low side MOSFET, increasing the MOSFET switching speed
- Minimize source inductance – reduces di/dt sensitivity of the low side MOSFET
  Minimize gate inductance – reduces sensitivity to dv/dt-induced turn-on
- Optimize gate resistance – reduces sensitivity to di/dt-induced turn-on
- Isolate the drive loop from the main power loop – speeds up the high side MOSFET and minimizes the distance between the MOSFET and the driver

To minimize the switching loop inductance, common source inductance and gate inductances, the PCB layout must be optimized, as shown in Figure 10. The grounding of the driver is also a key for the PCB layout. To minimize the switching loop inductance, it is recommended to route the return path of low side MOSFET source to the ground of the input capacitor through the first internal layer. To minimize the common source inductance, the ground of the MOSFET driver must be connected to the source of the low side MOSFET directly through a single-point connection to eliminate PCB trace inductance coupling. To minimize gate inductances, it is recommended to minimize the trace distance, and the MOSFET driver should be as close to the MSOFET gate as possible. As can be seen, the input signals are neatly separated from the output signals, and the drive loop is decoupled from the main current-carrying loop. This layout provides cleaner gate signals for the MOSFETs and helps to optimize the converter performance.

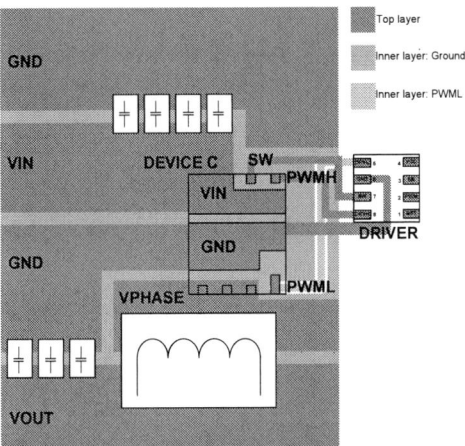

Figure 10: Recommended PCB Layout for PowerPhase Fast-Switching MOSFET

6. Conclusion

Modern fast-switching MOSFETs in VRM applications provide high efficiency solutions. With the proper PCB layout and circuit modifications, optimal performance can be ensured while minimizing voltage stress and sensitivity to dv/dt-induced and di/dt-induced turn-on.

7. References

[1] An inductive-switching loss model accounting for source inductance and switching loop inductance, Applied Power Electronics Conference and Exposition (APEC), 16-20 March 2014, pp 497–504

[2] An insight into the switching process of power MOSFETs: An improved analytical losses model, IEEE Transactions on Power Electronics, Vol 25, No 6, June 2010,

pp 1626-1640

[3] New thermally enhanced packages for power MOSFETs in battery pack applications, Electronic Components and Technology Conference, 2002., Page(s):1762 - 1764 ISSN :0569-5503

[4] Novel power MOSFET packaging technology doubles power density in synchronous buck converters for next generation microprocessors, APEC 2002. Page(s):106 - 111 vol.1, ISBN:0-7803-7404-5

[5] AN-7019, Limiting Cross-Conduction Current in Synchronous Buck Converter Designs Alan Elbanhawy, Fairchild Semiconductor

Medium to low voltage DC/DC resonant converter with SiC SCRs and nanocrystalyne magnetic core transformer.

Iñigo Martinez de Alegria, Electronic Technology Department, University of the Basque Country (UPV/EHU), inigo.martinezdealegria@ehu.eus
Madaci Mansour, Department of electrical engineering, University Constantine 1, Algeria
Angel Luis Perez, Electronic Technology Department, University of the Basque Country (UPV/EHU)
Kerdoun Djallel, Department of electrical engineering, University Constantine 1, Algeria
Jon Andreu Electronic Technology Department, University of the Basque Country (UPV/EHU), jon.andreu@ehu.eus

Abstract

Offshore wind farms can benefit from direct connection of their wind turbines to medium voltage DC distribution lines. The power converter cost can be low if the connection is unidirectional and only diodes are used in the high voltage side. However the wind turbines require electric power to operate their control electronics and pitch and yaw systems. This paper presents a power converter to feed wind turbines from a medium voltage bipolar DC distribution line using high voltage SiC SCR-s operating in resonant mode at high frequencies. The maximum switching frequency is only limited by the circuit commutated turn off time, tq, of the SiC SCR-s. With the introduction of SiC semiconductors devices, Medium Voltage High Frequency Nanocrystalyne Core Transformers are a future promising solution for the different power network levels (generation, transportation and distribution) and smart grid applications. Compact size, high reliability and more efficiency are promising features that could be provided by such devices. Analytical and simulation results are explained and validations with experimental results of a low power prototype are presented.

1. Introduction

Connection of offshore wind turbines and other marine energy generators to DC distribution lines within the wind farm can reduce the cost of energy. The use of high frequency transformers can also reduce the volume and weight inside the nacelles, which is a critical parameter of marine systems [1, 2, 3, 4]. Unidirectional power converters with only diodes in the high voltage side can be very cost effective in these systems, but offshore wind turbines also need power to be supplied when there is no wind in order to feed their control and yaw and pitch systems. Because the power bidirectionality is asymmetrical (power generated is in the megawatt range, whereas only tens of kW are needed to feed the turbine when it is not in generation mode) costs can be reduced using a lower power converter to feed the wind turbine. Although study of DC/DC converter topologies is very extended in the scientific community, studies of DC/DC converter in the kV range are limited. There are special applications where the use of high level dc voltages (thousands of volts) is necessary. Typical examples are laser-based systems, medical and industrial x-rays and telecommunications equipment with travelling wave tube (TWT), utilized in communication satellites. In this area low to high voltage unidirectional converters prevail, with diode rectification in the high voltage output stage, whereas scientific production in high voltage (kV range) to low voltage DC/DC converters is scarce [5, 6, 7]. This paper presents a high to low voltage DC/DC converter fed from a bipolar DC line with zero

© VDE VERLAG GMBH · Berlin · Offenbach

ground current based on SiC SCR-s, high frequency nanocrystaline transformer and resonant operation. Although the idea is based on the classical paper by Schwarz [9], bipolar operation, elimination of the input capacitors and their voltage swing and the availability of new materials (SiC and nanocrystaline cores) make this old idea attractive for the offshore power market and other niche applications requiring medium voltage to low voltage DC/DC conversion in the smart grids of the future.

2. Power converter topology

Figure 1 shows the power circuit topology for the bipolar DC resonant transformer. The main basic idea for a DC resonant transformer consists in the use of a resonant charging pulse to feed power to the output capacitor. The resonance is obtained using the series connection of the transformer leakage inductance and the output capacitor.

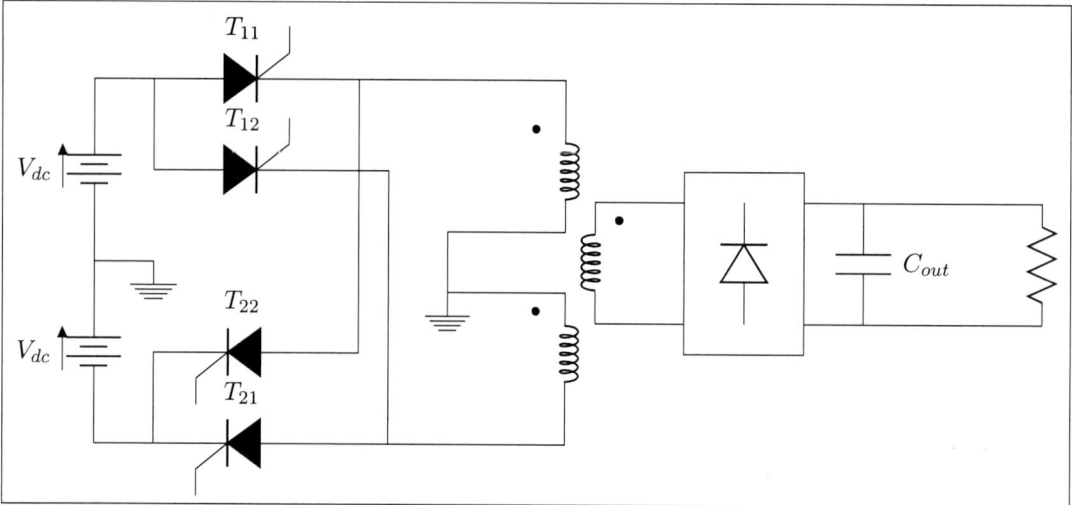

Fig. 1: Power converter topology

High voltage can be achieved by conventional series connections of thyristors, but with 6.5 kV SiC thyristors only a small number of thyristors in series is necessary to achieve medium voltage capability. The voltage level at which offshore power generators operate at is one of the major factors in the design of the power converters. The offshore wind industry has adopted 33 kV 50Hz AC as a preferred collection voltage but manufacturers of offshore generators are using voltages between 3.3 kV and 11 kV. Higher voltages place a large burden on the design because extra insulation and separation of components has to be considered. In the short and medium-term, marine power generating devices are expected to operate at between 3.3 kV and 11 kV [10]. State of the art high voltage semiconductors in the market are 6.5 kV IGBTs and SCRs with current ratings too high and low switching frequencies for this type of applications, or high voltage MOSFETS with too low current rating as for example IXTH03N400 from IXYS rated 4 kV and 0.3 A. The use of SiC SCR-s with 10 kV, 40 A and low turn off times seems the best alternative in the near future for medium voltage power converters in the 10kW-100kW range (see fig. 2).

© VDE VERLAG GMBH · Berlin · Offenbach

Fig. 2: 6.5 kV 40 A SiC thyristor (Courtesy of GeneSic and SandiaNational Laboratories.

2.1. Resonant operation

Thyristors T_{11} and T_{21} are fired simultaneously. Each one of them generates a resonant circuit using the leakage inductance of the transformer, L_{leak}, and the output capacitor, C_{out}, formed series LC tank, as shown in the equivalent circuit during SCR turn on (figure 3(a)). The load discharge current is discarded in this section for simplification. Figure 3(b) shows the theoretical waveforms of the SCR charging current pulse and the output capacitor voltage during this period. It must be born in mind that the capacitor current is the sum of the current delivered by both transistors. Because the currents are equal, the ground current is zero.

The resonant pulse naturally turns the SCR-s off and additional forced commutation circuit is not required. The resonant charging current value depends on the voltage difference between the dc bus and the capacitor voltage. This voltage difference is mainly determined by the capacitor discharge through the load during SCR turn off period. The frequency of operation is limited by the circuit commutated turn off time, tq, of the SCR-s. In the case of 6.5 kV SiC SCRs tq can be as low as 4.7 us [?][10] and fast switching operation can be achieved.

For simplicity the transformer turns ratio is considered 1:1. If the initial capacitor voltage, at $t = t_{c0}$, is

$V_c(t_{c0}) = V_{dc} - \Delta V_1$, then

$$i_{T_{11}} = \frac{(V_{dc} - V_1(t_0))}{L_{leak} w_0} \sin w_0 t$$

The current is a sinusoidal positive pulse (see figure 3(b)) with a maximum value of

$$I_{max} = \sqrt{\frac{C_{out}}{L_{leak}}} \times (V_{dc} - V_c(t_{c0}))$$

where

$$w_0 = \frac{1}{\sqrt{L_{leak} C_{out}}}$$

at the end of the charging phase, at $t = t_{cf}$, the capacitor voltage will be

© VDE VERLAG GMBH · Berlin · Offenbach

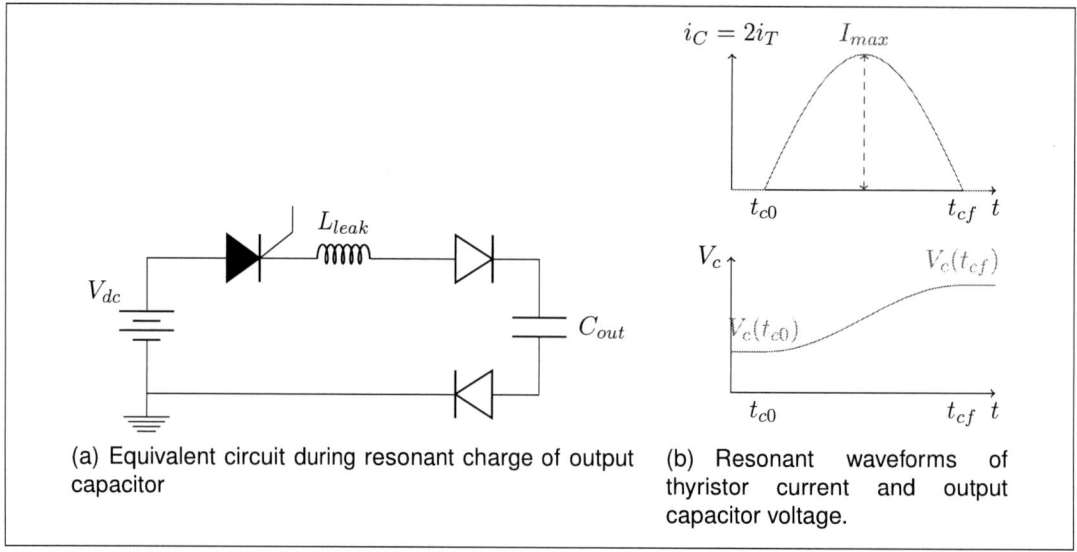

(a) Equivalent circuit during resonant charge of output capacitor

(b) Resonant waveforms of thyristor current and output capacitor voltage.

Fig. 3: Equivalent circuit and waveforms

$V_c(t_{cf}) = V_{dc} + \Delta V_1$.

The pulse duration is

$t_{cf} - t_{c0} = \pi \sqrt{L leak C_{out}}$

The energy delivered to the capacitor during the charging phase is

$E_{charge} = C_{out} \times (V_{dc}^2 - V_1^2(t_{c0}))$

As the resonant current goes to zero both thyristor open and the magnetizing current in the transformer is discharged through the output rectifier to the output capacitor.

Until any one of the transformers is fired again the load is supplied by the output capacitor, and the capacitor is dicharged as a RC exponetial circuit.

Thyristors T_{11}-T_{21} and T_{12}-T_{22} are fired sequentially, achieving a zero ground current operation from a bipolar DC line.

At thyristor turn off, the rectifier diodes transfer the transformer magnetizing current to the output capacitor. During this interval the thyristor is subject to the input dc bus voltage plus the output voltage reflected to the primary.

3.   Validation of the proposed converter

In order to validate the theoretical design, a low voltage low power prototype has been simulated, built and tested before proceeding with the construction of a high voltage high power converter. The prototype is built using scratch material for the lab and the transformer parameters, thyristor ratings and driver circuit are not optimized for the application. The driver circuit could not fire the thyristor gates with frequencies higher than 800 Hz but the proof of concept could be carried away at very low power levels. The following sections show the validity of the proposal and the match between simulation and experimental results.

The simulation and experimental prototype use the following parameters:

- Input voltage: $\pm 15 V_{dc}$ bipolar power supply
- Transformer turns ratio: $50 : 33$
- Transformer leakage inductance (measured in primary): $460 \mu H$
- Transformer magnetizing inductance(measured in primary): $76 mH$
- Output capacitor: $10 \mu F$
- Load: 100Ω

3.1. Simulation results

Simulations of the system were carried away using PSIM software. Figure 4(a) shows the capacitor charge and discharge ripple and the resonant current pulses in the capacitor, as expected from the analytical study.

Figure 4(b) shows in detail the current and voltage in the thyristor with zero current switching at turn on and turn off. The turn off current is not exactly zero because of the hold current of the thyristor.

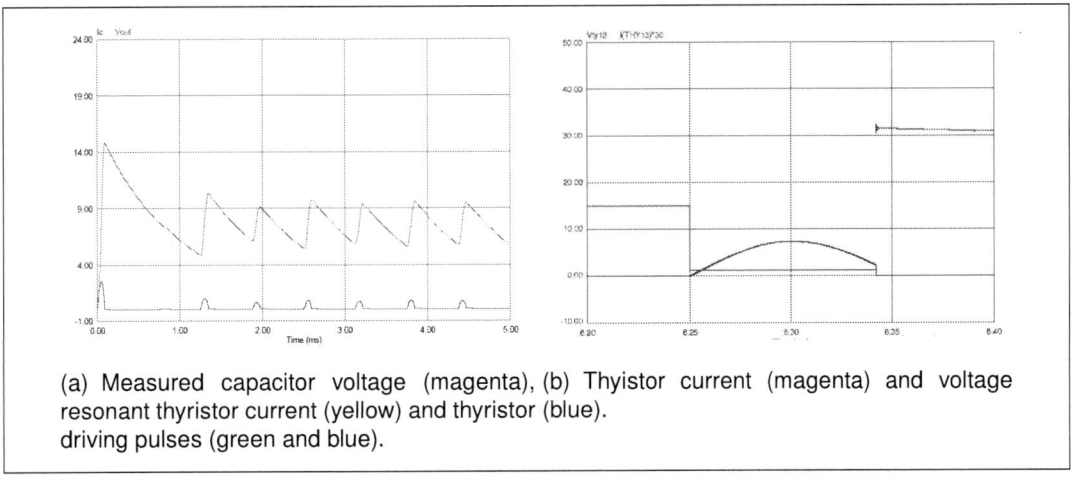

(a) Measured capacitor voltage (magenta), (b) Thyistor current (magenta) and voltage resonant thyristor current (yellow) and thyristor (blue).
driving pulses (green and blue).

Fig. 4: Simulation results.

3.2. Experimental results

After simulations the system was built and tested, resulting in behaviour very close to simulation results. Figure 5(a) shows the capacitor charge and discharge ripple and the resonant current pulses in the capacitor as expected from simulation. The behaviour of the output voltage is very similar to the output of a conventional acdc rectifier stage in power supplies.

Figure 5(b) shows in detail the current and voltage in the thyristor. It is clear from the figure that at turn on and turn off the thysristors present a zero current switching behaviour. At turn off there is a minimum residual current due to the hold current of the thyristors.

© VDE VERLAG GMBH · Berlin · Offenbach

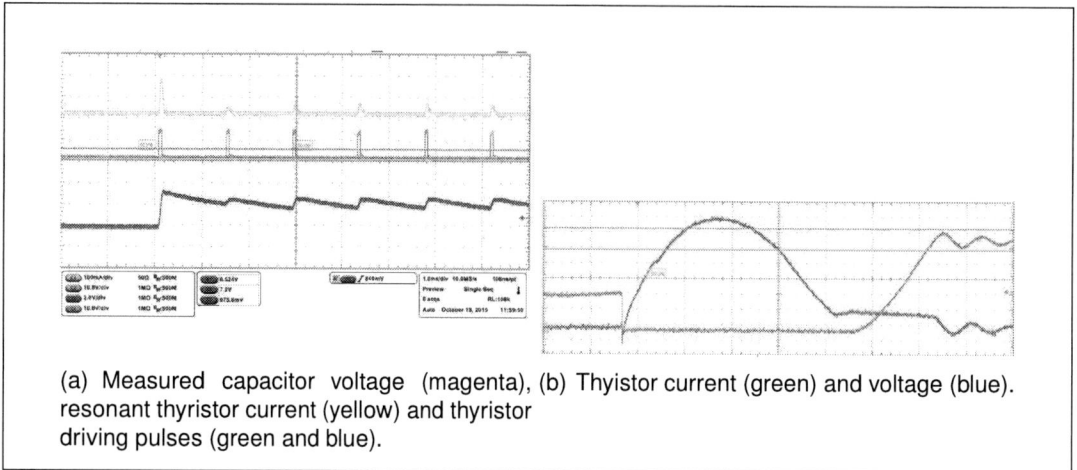

(a) Measured capacitor voltage (magenta), resonant thyristor current (yellow) and thyristor driving pulses (green and blue).

(b) Thyistor current (green) and voltage (blue).

Fig. 5: Experimental results.

4. Conclusions

The bipolar resonant DC transformer can be an attractive solution for the introduction of high frequency transformers in DC distribution systems. The uses of SiC SCRs can be a suitable choice for Medium Voltage to Low voltage DC/DC power conversion in the 10kW-100kW range to power up multimegawat marine generators from medium voltage DC lines due to their high frequency capability, high voltage rating and robustness. Simulation and experimental results show the resonant operation of the converter, with very low switching power loss at turn on and turn off of the semiconductors.

5. Acknowledgments

This work has been supported by the Government of the Basque Country within the research program ETORTEK as the project FUTUREGRIDS-2020 (IE14-389).

6. References

7. References

[1] I. Martinez de Alegria *Study on full Direct Current Offshore Wind Farm*, PhD. Thesis, Universidad del Pais Vasco 2012.

[2] J.W. Kolar, G.I. Ortiz, *Solid State Transformers: Key components of Future Traction and Smart Grid Systems*, Proceedings of the International Power Electronics Conference IPEC, Hiroshima, Japan, 2014.

[3] M. Stieneker, J. Riedel, N. Soltau, H. Stagge, R.W. De Doncker *Design of series-connected dual-active bridges for integration of wind park cluster into MVDC grids*, Power Electronics and Applications (EPE'14-ECCE Europe), 2014 16th European Conference on, pp.1,10, 26-28 Aug. 2014.

[4] D. Ricchiuto, K. Schönleber, S. Ratés, L. Trilla, J.L. Dominguez, O. Gomis-Bellmunt, *Overview Of High-Power Medium-Frequency DC/DC Converter Topologies for Wind Turbines Interfaced to a MVDC Collection Grid*, Proceedings of the EWEC 2015.

[5] R. Singh *Ultra High Voltage SiC Bipolar Devices for Reduced Power Electronics Complexity*, ARPA Power Electronics in Photovoltaic Systems Workshop, 2011.

[6] S. Lu, M.A. El-Sharkawi, H. Kirkham, B.M. Howe, *NEPTUNE Power System: Startup Power Supply for 10 kV to 400 V Dc-Dc Converter*, Applied Power Electronics Conference and Exposition, 2006.

[7] N.N. Lopatkin, G.S. Zinoviev, H. Weiss, *High-Voltage Bi-Directional DC-DC-Converter for Advanced Electric Locomotives* EPE 2009.

[8] G. Ortiz, D. Bortis, J. Biela and J.W. Kolar, *Optimal Design of a 3.5 kV/11 kW DC-DC Converter for Charging Capacitor Banks of Power Modulators*, Pulsed Power Conference, 2009.

[9] F.C. Schwarz, *A method of resonant pulse modulation for power converters* IEEE Transactions on Industrial Electronics, vol. IECI 17, n° 3, may, 1970.

[10] A. Mason, R. Driver *Marine Energy Electrical Architecture Report 3: Optimum Electrical Array Architectures* Ctapult offshore renewable energy, September 2015.

GaN Active-Clamp Flyback Converter with Resonant Operation Over a Wide Input Voltage Range

Nicolas Quentin, University of Lyon/Ampere/Safran Group, France, nicolas.quentin@sagem.com
Remi Perrin, INSA Lyon/Ampere, France, remi.perrin@insa-lyon.fr
Christian Martin,University of Lyon/Ampere, France, christian.martin@univ-lyon1.fr
Charles Joubert, University of Lyon/Ampere, France, charles.joubert@univ-lyon1.fr
Bertrand Lacombe, Safran Group, France, bertrand.lacombe@sagem.com
Cyril Buttay, INSA Lyon/Ampere, France, cyril.buttay@insa-lyon.fr

Abstract

This paper presents an active-clamp Flyback converter with a resonant operation which can realize soft-switching on all power elements with the minimum of additional components compared to the classical hard-switched flyback topology. This converter is a good candidate for wide input voltage and load range application. The operational principle and design procedure are presented and verified experimentally with a $50\ W$ prototype converter at $1\ MHz$ switching frequency, 18 to $80\ V$ input voltage range and $15\ V$ output voltage. In this switching frequency range, GaN transistors and planar transformer are used in order to improve the efficiency.

1. Introduction

In the design of high-power density DC/DC converter, high-switching frequency is proposed for the reduction of the passive components size at the penalty of excess switching losses in power devices. In this case, soft-switching is a reasonable technique to limit these power losses. For a wide input voltage range operation, specific topology including components with low losses is required to achieve efficiency higher than $85\ \%$.

One of the most popular isolated soft-switching converter is the LLC resonant converter [1, 2] which can achieve Zero Voltage Switching (ZVS) at the primary side and Zero Current Switching (ZCS) at the secondary side. This converter suffers from a difficulty with respect to modeling. At low input voltage, low magnetizing inductance is needed to achieve ZVS operation. In this condition, the magnetizing current has a significant value compared to the resonant current. This has an important impact on the conduction and magnetic losses. LLC resonant converter is therefore seldomly considered at low-power and low voltage.

In this paper, a variant of the Flyback converter with the addition of an Active-Clamp circuit is proposed. For the high frequency application, a specific design is detailed in compliance with the use of GaN transistors. This specific design allows the output rectifier to be turned-off softly by using ZCS condition over the whole load range, like in a resonant converter. Besides, there are less current stresses on the resonant capacitor and primary switches compared to the LLC converter.

2. Description of the Resonant Active-Clamp Flyback Converter

Fig. 1(a) shows the simplified schematic of the proposed converter. Based on a Flyback Active-Clamp structure [3], the active-clamp circuit is composed of an auxiliary switch $S2$ and a clamp capacitor C_r. In the proposed operating principle, this capacitor acts also as a resonant capacitor by using the leakage inductance of the transformer, L_r. Each transistor, SX, is modeled as the drain to source capacitance C_{SX}, the anti-parallel diode D_{SX} for the reverse conduction and an ideal switch T_{SX}. Regarding the simplicity of the architecture, this topology has a few number of components which is an important criteria for low power application.

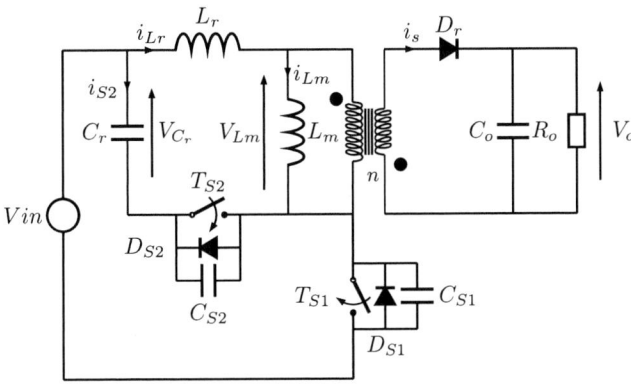

(a) Schematic of Flyback Active-Clamp topology

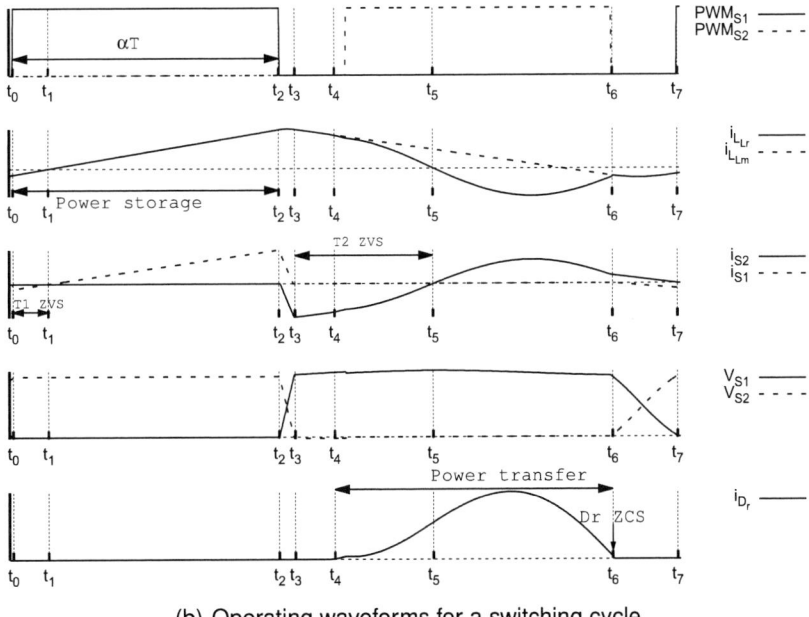

(b) Operating waveforms for a switching cycle

Fig. 1: Architecture description

Fig. 1(b) shows currents and voltages waveforms in the proposed topology. The waveforms are slightly different from a classical Flyback converter. The steady-state operation of the converter includes 7 modes in one switching period, T. The detailed operation as well as the state equations are described as follow.

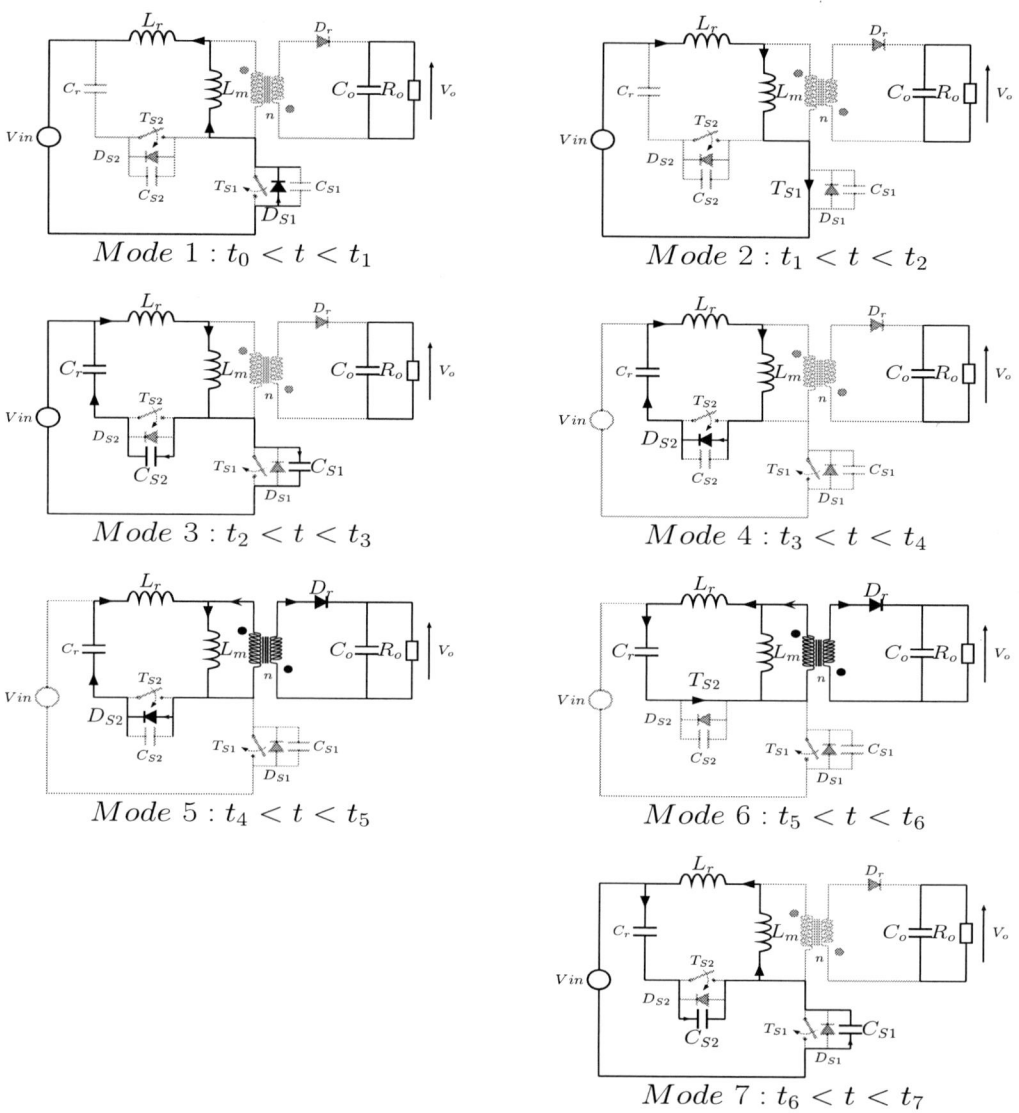

Fig. 2: Active-Clamp Flyback topology operating states.

Mode 1 and Mode 2: Inductance charging state $t_0 < t < t_2$: At t_0, the rectifier is blocked and the magnetizing current i_{L_m} is negative and will flow through the body diode of T_{S1}, resulting in a positive voltage across the magnetizing inductance, L_m. Thus, the magnetizing current i_{L_m} linearly increases with the following equation

$$i_{L_m}(t) = \frac{V_{in}}{L_m}(t - t_0) + i_{L_m}(t_0) \tag{1}$$

The ZVS condition is realized due to the body diode conduction and the gate signal of T_{S1}

should be applied during mode 1. For the beginning of mode 2 at t_1, the magnetizing current reaches zero and T_{S1} is softly turned on. During this mode, the magnetizing current, i_{Lm}, still increases with the same slope as in mode 1. This mode ends at t_2 ($t_2 - t_0 = \alpha T$) when T_{S1} is turned off.

Mode 3: Dead time, $t_2 < t < t_3$**:** At t_2, T_{S1} is turned off. The magnetizing current, i_{Lm}, which is equal to the resonant current, i_{Lr}, acts as a constant current source because of an important L_m. Similarly C_r is large enough and the voltage, V_{Cr}, acts as a constant voltage source.

When T_{S1} is turned off, the drain source parasitic capacitance of T_{S1} is charged to $-V_{Cr} + V_{in} \approx \frac{V_o}{n} + V_{in}$ with $n = \frac{n_2}{n_1}$ and the drain-source parasitic capacitance of T_{S2} is discharged ($0\ V$). During the first part of the dead-time, capacitances C_{S1} and C_{S2} will be respectively charged and discharged. This mode ends when the capacitances are fully charged or discharged.

Mode 4: Resonant phase, $t_3 < t < t_4$**:** At t_3, C_{S2} is fully discharged and both T_{S1} and T_{S2} are in off-state. Therefore, the inductor current, i_{Lr}, is flowing through the body diode D_{S2} thus $V_{Lm} \approx V_{Cr}$. During this mode, the voltage across the magnetizing inductance doesn't allow the transfer of power to the secondary side ($V_{Cr} > -\frac{V_o}{n}$) and the rectifier diode is blocked. L_m is thus free to participate to the resonance. It will form a resonant circuit where L_m is in series with L_r and both resonate with C_r. The voltage across the magnetizing inductance decreases and this mode ends when the rectifier is turned on when $V_{Lm} \approx V_{Cr} \approx -V_o/n$.

Mode 5 and Mode 6: Power transfer, $t_4 < t < t_6$**:** At t_4, V_{Lm} reaches $-V_o/n$. The primary transformer voltage is clamped to $-\frac{V_o}{n}$. Therefore i_{Lm} decreases linearly because of the output voltage reflected to the primary side. As a consequence, L_m no longer participates to the resonance. During this mode, the circuit operates like a LLC resonant converter with a resonance between L_r and C_r. This new resonance will force the secondary side rectifier, D_r, to conduct the output current, i_s with a sinusoidal shape. During mode 5, the ZVS condition is realized due to T_{S2} body diode conduction and the gate signal of T_{S2} should be applied during this time. This mode finishes when $i_{Cr} = i_{S2}$ reaches zero.

At t_5, the current i_{Cr} reaches zero. T_{S2} is softly turned on. Mode 6 ends with the ZCS commutation of D_r, which happens when the magnetizing current, i_{Lm}, is equal to the resonant current, i_{Lr}.

Mode 7: Dead time, $t_6 < t < t_7$**:** At t_6, T_{S2} is turned off. This mode is equivalent to mode 3 with different initials conditions in the current levels.

3. Soft-switching operation: ZVS at the primary and ZCS at the secondary

Contrarily to the classical Active-Clamp Flyback converter [4, 5], the soft-switching at the primary side (discharge of the drain-source capacitors of the power transistors) is exclusively achieved with the energy stored in the magnetizing inductance during the dead time [t_6-t_7] for $S1$ and during [t_2-t_3] for S2. Regarding the energy stored into the leakage inductance, it is fully transferred to the secondary side during [t_4-t_6]. This behavior is similar to the LLC resonant converter.

The ZCS condition for D_r is satisfied when the secondary side current, i_s, reaches zero before the end of the switching period, T. From the primary side, this condition occurs when the resonant current, i_{Lr}, reaches the magnetizing current, i_{Lm} at t_6. This condition occurs when

PCIM Europe 2016, 10 – 12 May 2016, Nuremberg, Germany

the transfer time, $(1 - \alpha)T$, is higher than $\frac{3}{4}$ of the resonant period imposed by the resonant elements (L_r, C_r). At the end, the ZCS condition on D_r is

$$f_{sw} = \frac{4}{3} \frac{1 - \alpha_{max}}{2\pi \sqrt{L_r C_r}} \tag{2}$$

With f_{sw} the switching frequency.

4. Steady-State Analysis: A transfer function suitable for a wide input voltage range

this topology applies the operating modes of active-clamp and resonant topologies at different times. First ($[t_0; t_2]$), the energy is stored in the magnetizing inductance, as is the case in the active-clamp topology. Next ($[t_4; t_6]$), the power flows through the secondary with a sinusoidal waveform.

From a structural viewpoint, the active-clamp mode and resonant mode correspond respectively to the configurations when T_{S1} is on and T_{S2} is on. The converter large signal modeling

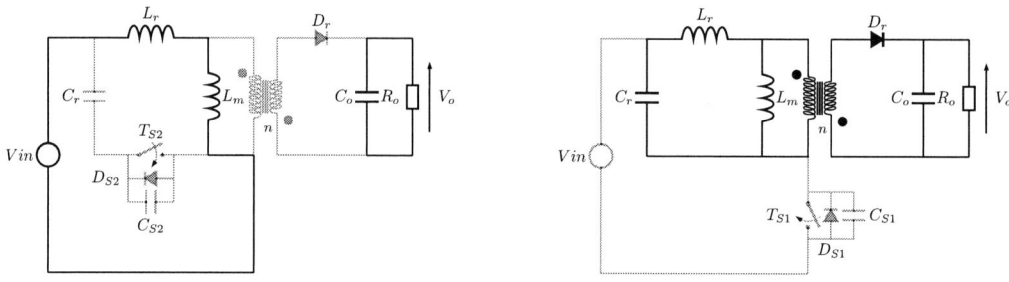

Energy storage - active-clamp mode   Energy transfer - resonant mode

Fig. 3: Operating modes

can be derived by applying the volt-seconds balance principle across the magnetizing inductance with the Fundamental Harmonic Approximation [6] commonly used to describe the DC characteristic of LLC resonant converters yields:

$$V_{in}\alpha \frac{L_m}{L_m + L_r} = \frac{V_o}{n}(1 - \alpha)\frac{Z_{C_r} + Z_{L_r} + \frac{R_{ch}}{n^2}//Z_{L_m}}{\frac{R_{ch}}{n^2}//Z_{L_m}}$$

$$with \ Z_{L_m} = jL_m 2\pi \frac{f_{sw}}{1 - \alpha}$$

$$Z_{L_r} = jL_r 2\pi \frac{f_{sw}}{1 - \alpha} \tag{3}$$

$$Z_{C_r} = \frac{1}{jC_r 2\pi \frac{f_{sw}}{1-\alpha}}$$

Assuming that $L_m >> L_r$ thus $\frac{L_m}{L_m + L_r} \approx 1$ this leads to the following transfer function :

© VDE VERLAG GMBH · Berlin · Offenbach

$$\frac{V_o}{V_{in}} = \frac{n\alpha}{1-\alpha} \frac{\frac{R_{ch}}{n^2}//Z_{L_m}}{Z_{C_r} + Z_{L_r} + \frac{R_{ch}}{n^2}//Z_{L_m}} \tag{4}$$

Equation 4 is the product of 2 transfer functions. On one hand the active-clamp Flyback transfer function: $\frac{n\alpha}{1-\alpha}$ and on the other hand the transfer function of a LLC resonant topology $\frac{\frac{R_{ch}}{n^2}//Z_{L_m}}{Z_{C_r}+Z_{L_r}+\frac{R_{ch}}{n^2}//Z_{L_m}}$. According to the transfer function and Fig. 4, this converter has 2 control variables which are the duty cycle and the switching frequency. Regarding the wide input voltage requirement, it is better to use the duty cycle to improve the dynamic of the system and to keep constant the frequency. Besides, the LLC transfer function has the benefit to add a DC gain offset (higher than 1 to operate in ZCS) with the choice of the resonant frequency. With this offset the transformer ratio (number of turns) could be reduced.

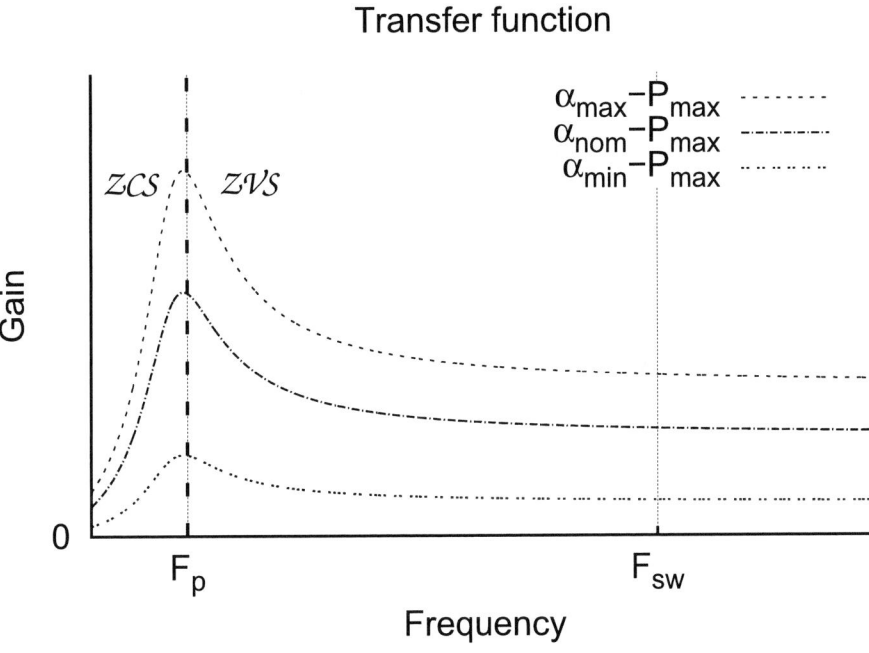

Fig. 4: DC Behavior with duty cycle variations

5. Experimental results

The power stage circuit of the proposed converter is shown in Fig. 5. Fig. 6(a) depicts the operating waveforms of the main voltages and currents from the primary side at $1\ MHz$-$50\ W$. This converter includes advanced technologies dedicated to high frequency and high efficiency operation with the use of GaN transistors from EPC and $EQ20 - 3F45$ planar transformer. However, using half-bridge GaN transistor in a wide input voltage application has the drawback to complicate the driver architecture. Because to respect the $200\ V$ requirement the driver part now includes 2 low side drivers, one of whom isolated with a bootstrap system (diode

PCIM Europe 2016, 10 – 12 May 2016, Nuremberg, Germany

Fig. 5: Prototype with GaN: EPC2010C, Driver: UCC27611, Cr: 200 V MLCC X7R 1206

and capacitor) for the power and a digital isolator for its PWM signal. The board dimensions

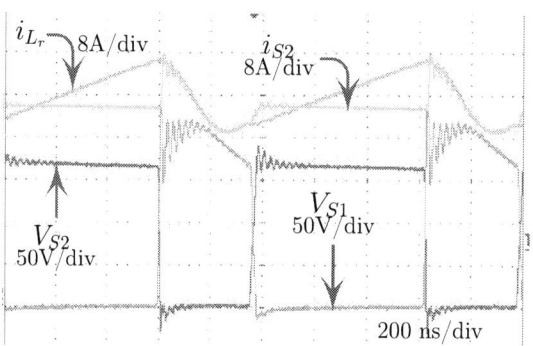

(a) Experimental waveforms at Vin= 28 V, P=50 W and f= 1 MHz with Vout= 15 V

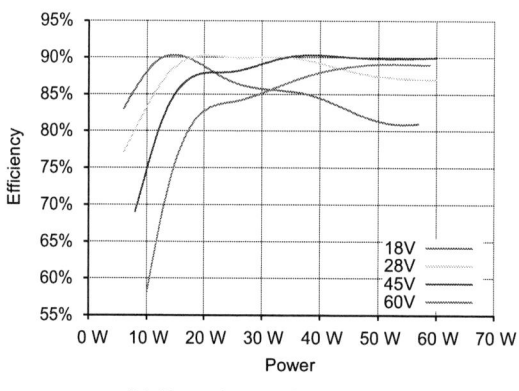

(b) Experimental efficiency

Fig. 6: Experimental results from the proposed prototype

are 85x85mm including power-cell, feedback system and PWM controller. These experimental waveforms globally match with the theoretical analysis. It is demonstrated in Fig. 6(b) that for a heavy load (28V-50W), the efficiency reaches $89 - 90\%$ due to the soft-switching on all the power elements. The converter can achieve, for this load range and at this frequency, a higher efficiency than other classical converters such as the hard-switched flyback topology.

6. Conclusion

The particularity of this paper is the design of the converter with the wide input voltage issues which is solved at different levels:

1- The topology is suitable for wide input voltage application with a step-down and step-up transfer function. With the additional benefits of having a low number of components and a simple control (duty cycle at fixed frequency) which is better for low power application.

2- The specific design with the resonant operating mode allows to reduce the stress on the components especially on the resonant capacitor and also provides soft switching on the rectifier. Another possible benefit is the output voltage stability against the step-up load if the converter operates at the resonant frequency as it is the case in a resonant LLC converter.

3- The technologies suitable for high frequency operation with the use of planar transformer and GaN transistors.

At the end, the power stage is both efficient and compact because it can achieve ZVS on the primary and ZCS on the secondary, with a good use of the transformer where the energy stored into the leakage inductance is transfered. All these features in only one topology which has a few number of components and stay simple to control.

7. References

[1] Xue Zhang, Wei You, Wei Yao, Shen Chen, and Zhengyu Lu. An improved design method of llc resonant converter. In *Industrial Electronics (ISIE), 2012 IEEE International Symposium on*, pages 166–170, May 2012.

[2] Hangseok Choi. Analysis and design of llc resonant converter with integrated transformer. In *Applied Power Electronics Conference, APEC 2007 - Twenty Second Annual IEEE*, pages 1630–1635, Feb 2007.

[3] Bor-Ren lin, Huann-Keng Chiang, Kao-Cheng Chen, and David Wang. Analysis, design and implementation of an active clamp flyback converter. In *Power Electronics and Drives Systems, 2005. PEDS 2005. International Conference on*, volume 1, pages 424–429, 2005.

[4] T. LaBella, B. York, C. Hutchens, and Jih-Sheng Lai. Dead time optimization through loss analysis of an active-clamp flyback converter utilizing gan devices. In *Energy Conversion Congress and Exposition (ECCE), 2012 IEEE*, pages 3882–3889, Sept 2012.

[5] R. Watson, F. C. Lee, and G.-C. Hua. Utilization of an active-clamp circuit to achieve soft switching in flyback converters. In *Power Electronics Specialists Conference, PESC '94 Record., 25th Annual IEEE*, pages 909–916 vol.2, Jun 1994.

[6] Robert W. Erickson and Dragan Maksimovic. *Fundamentals of Power Electronics*. Springer, 2ed edition, 2001.

Demonstration of superior SiC MOSFET Module performance within a Buck-Boost Conversion System

Maximilian Slawinski, Infineon Technologies AG, Germany, maxilimilian.slawinski@infineon.com
Tim Villbusch, Infineon Technologies AG, Germany, tim.villbusch@infineon.com
Daniel Heer, Infineon Technologies AG, Germany, daniel.heer@infineon.com
Marc Buschkühle, Infineon Technologies AG, Germany, marc.buschkuehle@infineon.com

Abstract

This paper shows the performance of Infineon's new SiC MOSFET power module operating in a buck-boost conversion system. The fast switching characteristics of the module will be illustrated with the help of double pulse measurements which show dv/dt levels above 50kV/µs. A conversion efficiency of 99% has been measured at a switching frequency of 100 kHz. In addition a comparison to a Si IGBT based system has been made to evaluate the potential performance gain of using the Infineon SiC MOSFET module. Finally the performance increases achieved by the use of synchronous rectification, which has been tested up to a switching frequency of 500 kHz, will be delineated.

1. Introduction

As silicon (Si) based power semiconductors are reaching their technical limits in terms of switching performance, reliability and power density, power switches based on compound semiconductor materials are becoming the focus of a lot of interest for several applications like photovoltaics (PV), uninterruptable power supplies (UPS) and motor drives. Silicon carbide (SiC) is widely regarded as the most promising semiconductor material for power devices in the voltage range above 600V.
While SiC diodes are widely used in many applications[1], existing SiC MOSFETs have not yet achieved a similar level of acceptance in the power electronics market.
The main reasons for this lower acceptance level are the higher cost, unproven long term reliability [1] and additional design effort required to replace IGBTs with SiC MOSFETS [2].

This paper demonstrates the performance of a bidirectional buck-boost converter based on a new SiC MOSFET power module manufactured by Infineon. After characterization of the basic switching behavior using double pulse testing, functional tests have been performed to investigate the DC converter's performance under different operating conditions. A Si IGBT based converter is used as a reference system to compare the performance of equivalent Si and SiC based converters. The goal of this work is to evaluate the potential performance gains possible using a SiC MOSFET. A further target is the demonstration of the reduced

---

[1] E.g. Solar power inverters.

implementation effort for the shown power module in terms of driver circuit design and driving.

1.1. SiC MOSFET based DC/DC converter

The tested SiC module consists of a 50 A 1200 V half-bridge topology with each switch built with two 25 A chips in hard parallel connection. As integrated body diodes are implemented in the SiC MOSFET chips, no additional freewheeling diodes are necessary. An Easy1B [3] power module serves as package.

The passive components and PCB of the buck-boost converter have been designed with respect to the targeted switching frequency of 100 kHz. Figure 1 draws the circuit and implemented components of the converter.

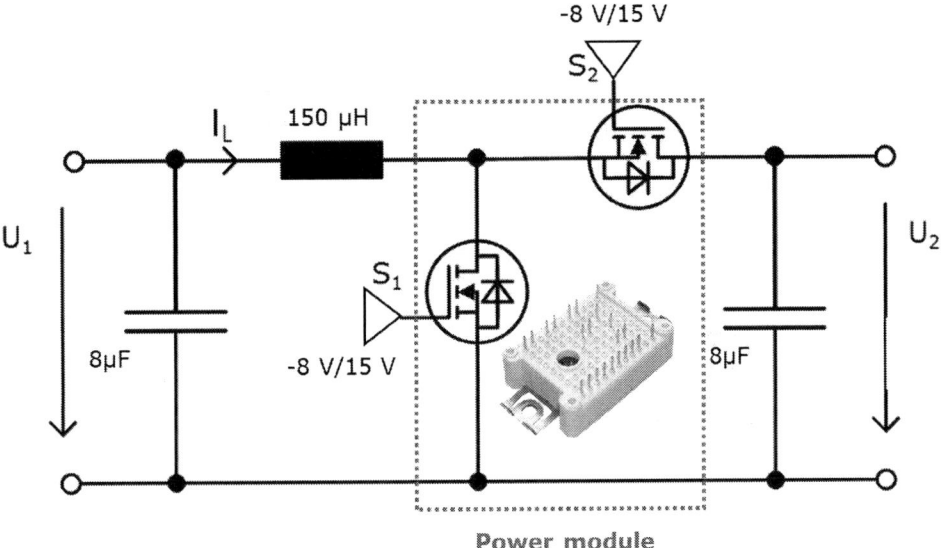

Fig. 1. Circuit of the investigated bidirectional buck-boost converter including the values of the implemented passive components and the driver voltages.

Film capacitances from Epcos [4] serve as input and output capacitors. A 150 μH inductor from Sumida [5], based on ferrite core technology, has been used as the DC choke. The two switches of the SiC MOSFET module are driven at -8 V/15 V using standard EiceDRIVER™ Compact [3] devices. The PCB consists of four 35 μm copper layers using FR4 as matrix material.

All the test data shown are with the system operating in boost or step up mode i.e. with the current flow as shown in figure 1.

2.   Double pulse characterization

The following section describes double pulse measurements of the converter testing the switching behavior of S_1. In this configuration a resistive load is connected across U_2 and S_2 is held off with $V_{GS}=-8$ V. The external gate turn-on and turn-off resistance values are 1 Ohm.

© VDE VERLAG GMBH · Berlin · Offenbach

2.1. Switching behavior

Figure 2 presents the turn-on and turn-off behavior of S_1 with U_1 at 600 V, I_d at 5 A, and a junction temperature of 25 °C i.e. room temperature RT.

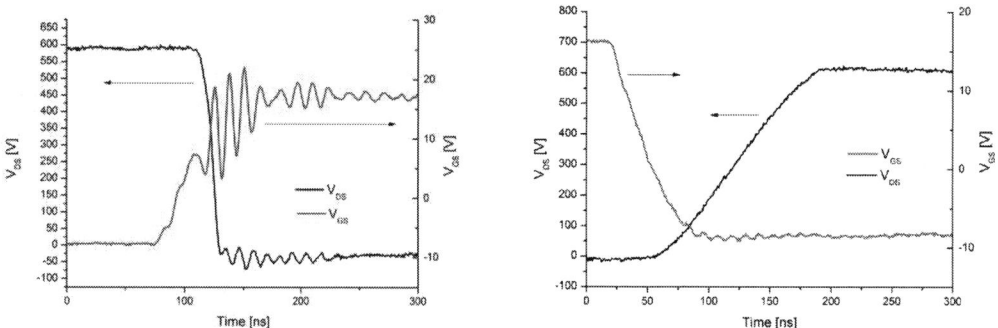

Fig. 2. (Left) Turn-on and (Right) turn-off behavior of S_1 at 5 A.

The gate-source waveform at turn-on shows some oscillatory behavior. These oscillations are caused by module internal inductance between the chip and the source auxiliary terminal where the measurements have been taken [6]. The switches themselves are turning-on properly which can be seen through the very low oscillations of V_{DS} during turn on. The dv/dt shown is at a high value of 40 kV/µs.

The turn-off at 5 A is much softer than the turn-on, as it can be seen in the right side of figure 2. Almost no oscillations of V_{GS} or V_{DS} are occurring. The dv/dt is at 5 kV/µs which is relatively low for a SiC switch.

As the turn-on switching is already very fast, the slow turn-off at 5 A does not represent the turn-off performance, which is possible with the presented SiC MOSFET. That is the reason why a further double pulse measurement has been performed at 30 A, which represents 60 % of the modules current rating. Figure 3 shows the waveforms of these measurements.

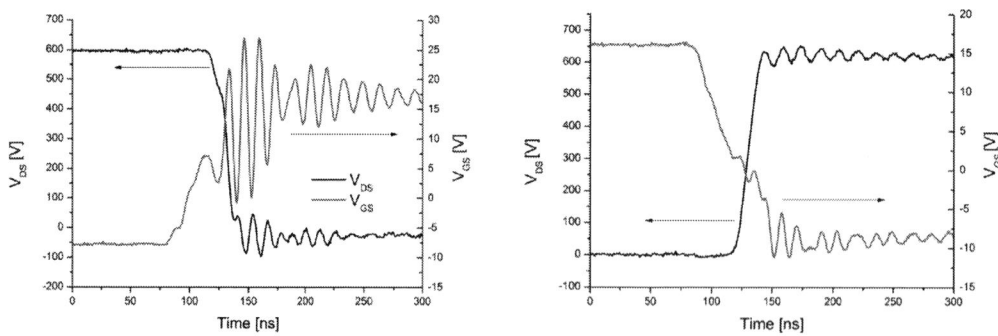

Fig. 3. (Left) Turn-on and (Right) turn-off behavior of S_1 at 30 A.

Again, strong oscillations are measured at the gate-source voltage during turn-on.

Nevertheless V_{DS} shows only a low level of oscillations indicating proper operation of the switches. The dv/dt values are 55 kV/µs during this turn-on and 34 kVµs during turn-off showing the very strong influence of current and dv/dt during turn-off for this device.

Table 1 gives measurement values taken at different current levels.

Current [A]	dv_{on}/dt [kV/µs]	$\Delta V_{DS, on}$ [V]	dv_{off}/dt [kV/µs]	$\Delta V_{DS, off}$ [V]
5	40	-71	5	+24
10	52	-79	11	+33
20	50	-89	22	+40
30	55	-97	34	+52

Table 1. Over-voltage levels and dv/dt values at different currents.

As expected, the overvoltage increases with higher currents caused by higher di/dt levels during turn-on and turn-off. At turn-on currents higher than 10 A, dv/dt reaches a constant value of about 50 kV/µs. In the case of turn-off the dv/dt shows a large increase with the higher values of the switched current.

2.2 Investigation of parasitic high side turn-on

As the switching behavior of the investigated MOSFET is very fast, which can be seen by the high dv/dt values, it is important to investigate the possibility of parasitic turn-on of the upper switch. Figure 4 presents the gate-source voltage of the upper MOSFET during the turn-on of the lower switch.

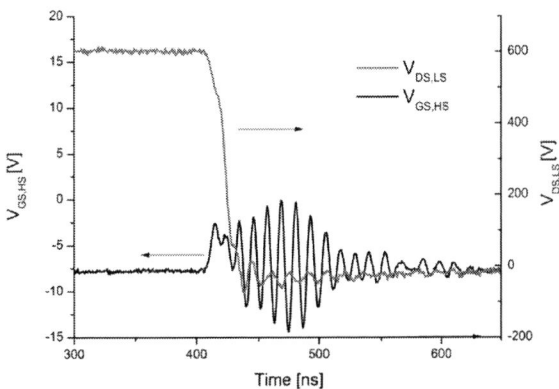

Fig. 4. Behavior of upper gate-source during turn-on of lower switch. I_d 30 A, U_1 at 600 V, RT and R_g 1 Ohm.

Under the conditions shown the dv/dt is >50 kV/µs. The high dv/dt causes voltage oscillations at the upper gate-source voltage. Nevertheless the measured gate-source voltage of the upper switch stays below 0 V and no parasitic turn-on is occurring as the threshold voltage is above 3 V. This indicates that at these high dv/dt levels -8 V is a good choice for the gate driver negative voltage.

© VDE VERLAG GMBH · Berlin · Offenbach

As a conclusion of the double pulse investigations the fast switching of the investigated SiC MOSFET has been proven within the buck-boost converter. No parasitic turn-on of the upper switch could be detected even at 55 kV/µs. The measured gate-source oscillations are a consequence of the source inductance within the driver loop internal to the module and can be reduced with a negative feedback of the source potential towards the gate [7] connection. The switching performance of the MOSFET itself is not affected by this phenomenon.

3. Conversion Performance

For the measurement of the boost conversion efficiency a 100 Ohm resistor has been connected to the output. The input and output voltages and currents have been measured to calculate input and output power. The DC/DC converter is driven in conventional boost operation using S_1 as active switch and the integrated body diode of S_2 for freewheeling ($V_{GS,S2}$=-8 V). The switching frequency is 100 kHz using a duty cycle of 25 %. Figure 5 shows the slope of the efficiency at different input power levels.

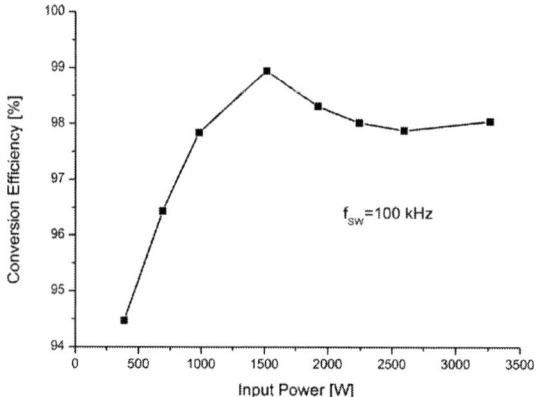

Fig. 5.   Boost conversion efficiency as a function of input power.

A very high conversion efficiency of about 99 % at a rated input power of 1500 W can be measured. For a higher power rating up to 3500 W, the determined conversion efficiency stays about 98 %. No efficiency derating can be observed up to 3500 W. The slight efficiency increase between 2500 W and 3500 W occurs from the faster turn-off switching at higher current rating causing lower turn-off losses.

The slope of the power-efficiency curve may change for different duty cycles but the underlying high conversion efficiency of the converter at 100 kHz has been shown. These results give an impression about the potential of the SiC MOSFET module to operate at both high switching frequencies and conversion efficiency at the same time.

4. Benchmark performance comparison with Si IGBT based converter

The following section will compare the SiC MOSFET based unit with one using a Si IGBT. The focus of this study is the comparison of system size and efficiency.

4.1 Si IGBT reference characteristics

The Si IGBT based buck-boost converter uses the same circuit and passive component technologies as the SiC MOSFET based converter. As the switching losses of the selected IGBT4 and the antiparallel EC4 diode module are higher than the SiC MOSFET module losses, the targeted switching frequency of the Si IGBT solution is 16 kHz.

Consequently the size of the inductor as well as the size of the capacitors have to be increased to achieve the same voltage ripple as the SiC MOSFET solution (<1 %). Table 2 and table 3 compare the component dimensioning, the geometrical dimensions and the weight of both converters.

	Si IGBT based system	SiC MOSFET based system
f_{SW}	16 kHz	100 kHz
L	1000 µH	150 µH
C_{input}	42 µF	8 µF
C_{output}	40 µF	8 µF

Table 2. Component dimensioning of the Si IGBT and SiC MOSFET based DC/DC converter.

		SI IGBT Reference	SiC MOSFET Demonstrator
L	Size (H x W x L)	125 mm x 95 mm x 70 mm	70 mm x 55 mm x 40 mm
	Weight	3.20 kg	0.50 kg
Total	Size (H x W x L)	139 mm x 300 mm x 185 mm	84 mm x 206 mm x 134 mm
	Weight	3.77 kg	0.93 kg

Table 3. Weight and size of inductors and the total converters.

As the dimensioning of the passive components for the Si IGBT based converter is larger, the size as well as the weight is much higher compared to the SiC MOSFET based system. The main increase of weight and size is related to the larger inductor required for the Si IGBT reference.

Both systems have the same full power rating which results in the passive components for the SiC MOSFET solution having a four times higher power density (in W/kg).

A picture of both systems is presented in figure 6 to give an idea of the physical sizes of both designs.

Fig. 6. (Left) Si IGBT and (right) SiC MOSFET based buck-boost converter.

4.2 Comparison of conversion performance

The test conditions for the comparison of conversion efficiency are similar to the conditions in section 3. Again, a 100 Ohm resistor serves as load. Both systems are driven as boost converters.

The conversion efficiency has been measured over a range of switching frequencies at an input power rating of 2000 W. The SiC MOSFET measurements have been taken under two different operating conditions. The first operating mode uses conventional switching as described in section 3. The second operating mode is to use S_2 in synchronous rectification mode to reduce the conduction losses of the body diode during the freewheeling portion of the cycle [8]. Figure 7 draws the efficiency results of these tests in a semi-logarithmic scale.

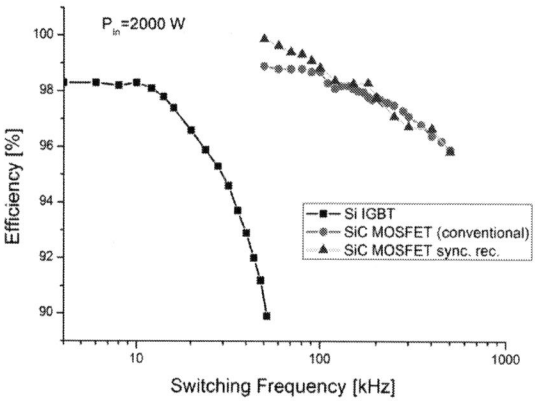

Fig. 7. Plot of the conversion efficiency the Si IGBT and SiC MOSFET based systems at different switching frequencies.

The measurement demonstrates the impressive performance jump from the Si IGBT to the SiC MOSFET. The Si IGBT based reference system only reaches an efficiency of 90 % at 50 kHz while the SiC MOSFET system achieves efficiencies beyond 99 % under the same conditions. The SiC MOSFET converter has been tested up to 500 kHz where the efficiency is still higher than 95 %.

The conversion efficiency of the SiC MOSFET demonstrator in the frequency regime up to 250 kHz can be increased by the use of synchronous rectification. This is due to the lower conduction losses as shown in Figure 8

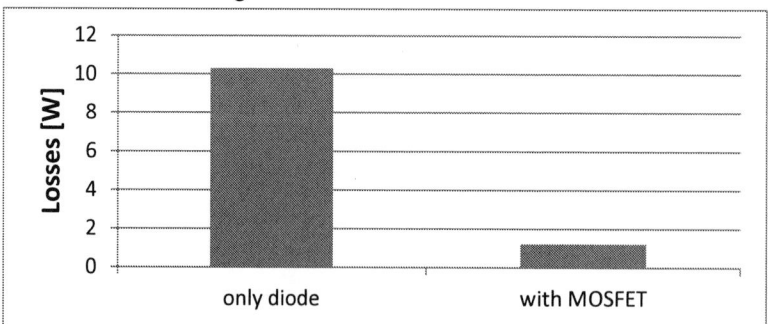

Fig. 8. Losses of the SiC demonstrator in bypass mode (S_1 turned off) with and without use of S_2.

The presented results are for the case of 300 V input voltage and 6 A current. If both MOSFETs are turned-off, only the upper diode is conducting, leading to a loss of 10 W caused by the voltage drop at the body diode. If S_2 is turned-on, the channel of the MOSFET conducts the freewheeling current, which leads to a reduction of the voltage drop and the losses in the given example reduce to 1 W.

5. Summary and Conclusion

This paper gives an overview of the performance of Infineon's new SiC MOSFET power module operating in a DC/DC converter. The device demonstrates fast switching behavior of the module in the application without parasitic turn-on. The unit has achieved an efficiency of up to 99 % at a switching frequency of 100 kHz. The comparison with a Si IGBT based system clearly shows the potential advantages in terms of efficiency, size and weight.

In addition the gate driver of the SiC MOSFET was realized with a standard IGBT driver IC (EiceDRIVER™ Compact) at conventional drive voltages of -8 V/15 V. Even with this standard drive components; performance has been measured for operation up to 500 kHz.

Infineon is setting a milestone in power electronics with the introduction of its new generation of SiC MOSFET 1200 V modules. The outstanding performance and straight forward implementation are a big step towards improving the power density and performance in a wide range of power electronic devices like solar inverters, drive converters or power supplies.

The device and module design is targeting towards applications using switching frequencies up to 200 kHz, although the chip technology is able to be used at even higher frequencies provided the application will benefit from such an operation mode and auxiliary elements are available to run the full system at those values.

It has to be mentioned that we have investigated an early prototype SiC MOSFET module. The final product will be optimized in layout and performance with respect to effects like the source current feedback and other.

6. Acknowledgements

The authors like to thank the people of Infineon in Villach, Erlangen, Warstein and North America for the great collaboration resulting in the results of this work.

7. References

[1] Lelis, Aivars J., et al. *Electron Devices, IEEE Transactions on* 55.8 (2008): 1835-1840.

[2] Fabre, Joseph, Philippe Ladoux, and Michel Piton. *Power Electronics, IEEE Transactions on* 30.8 (2015): 4079-4090.

[3] Infineon Technologies AG, Neubiberg Germany

[4] EPCOS AG, Munich Germany

[5] Sumida AG, Obernzell Germany

[6] Yin, Shan, et al *Power Electronics and Drive Systems (PEDS), 2015 IEEE 11th International Conference on.* IEEE, 2015.

[7] Grbovic, Petar J. *Power Electronics, IEEE Transactions on* 23.2 (2008): 643-652.

[8] Ji, H. K., and H. J. Kim. *Power Electronics Specialists Conference,* PESC'94 Record., 25th Annual IEEE. IEEE, 1994.

© VDE VERLAG GMBH · Berlin · Offenbach

PCIM Europe 2016, 10 – 12 May 2016, Nuremberg, Germany

High Efficiency and High Power Density Boost / Buck Converter with SiC JFET Modules for Advanced Auxiliary Power Supplies in Trolleybuses

Miroslav, Hruska, SKODA ELECTRIC a. s., Czech Republic, miroslav.hruska@skoda.cz
Martin, Jara, University of West Bohemia in Pilsen, Czech Republic, jara@rice.zcu.cz

Abstract

In the field of auxiliary power supplies for the light traction vehicles such as trolleybuses, trams and eventually hybrid busses the customers' demands increasingly focuses on low volume and mass of the converters. The high efficiency, power density and low audible noise are not of lower importance while EMC regulations become tougher. One way to cope with such requirements in the auxiliary drives design is to utilise new topologies and/or state-of-the-art power devices which would allow operating the converters at higher frequencies and temperatures with comparable or even lower losses. Typical case is the proliferation of the SiC devices in various areas of power electronics. This paper describes the design points and achieved results of the bi-directional Boost / Buck DC converter based on SiC JFET modules which was intended as an input voltage stabilizer within the auxiliary power supplies structure.

1. Introduction

Auxiliary power supplies (APS) are an important integral part of the electrical equipment of light traction vehicles such as trolleybuses, trams etc. Their main purpose is usually to generate the on board power grid 3x400 V_{AC} and to charge the on - board battery. Galvanic isolation and the capability to operate under significantly fluctuating catenary voltage are common requirements in this case as well.

An example of an APS structure is shown on Fig.1a. Both branches, the 3 – phase inverter and the battery charger, contain its own voltage stabilizer followed by an isolating converter. Due to cost and complexity issues these converters can be designed as a buck and a simple half-bridge hard switching converter. Major disadvantages of such power supplies are their low efficiency and low frequency operation resulting in higher mass and possible high audible noise because of low switching frequency 1700 V device in the buck regulators.

One of the goals of the new generation electrical equipment for trolleybuses development in SKODA ELECTRIC a. s. was to develop an optimum APS solution with following constraints:

1. Input voltage within full catenary range 400 V_{DC} -1000 V_{DC}.
2. Galvanic isolation in both branches.
3. Nominal power of on board power grid 3x400 V_{AC} / 50 Hz to be 10kW with a short-term overload up to 12.5 kW.
4. Battery charger power to be 10 kW.
5. Capability to reverse the power flow from the on-board battery to the DC-link capacitors (pre-charging at start).
6. Forced air cooling.

© VDE VERLAG GMBH · Berlin · Offenbach

7. Compared to the SKODA ELECTRIC a. s. previous generation APS lower volume, mass and higher efficiency.

The need of reliable performance under these conditions especially with respect to the mass and volume constraint led us to the proposal of the converter structure shown on Fig.1b.

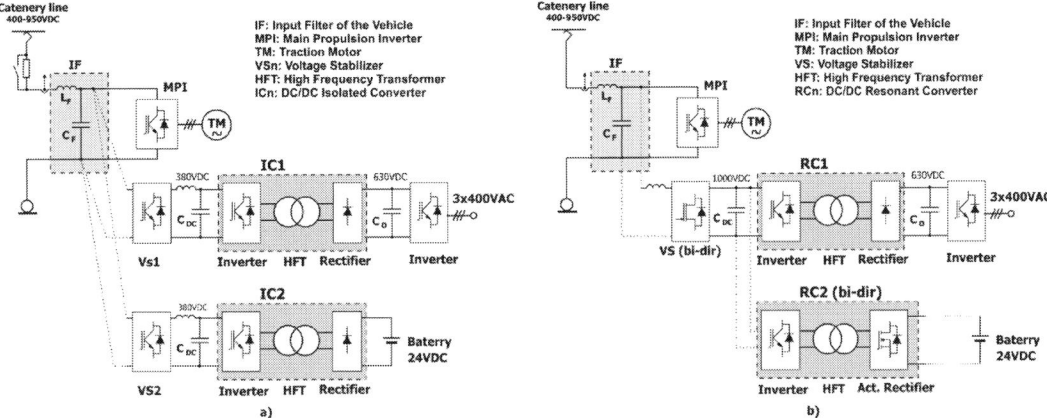

Fig.1. An example of the previous APS structure (a), newly developed structure (b)

This new concept is based on several basic ideas. The reverse power flow is achieved by using a bi-directional input voltage stabilizer (VS) and a bi-directional battery charger. Single VS improves the APS overall volume. Isolating function is provided by HF transformers in two DC/DC resonant converters (RC1 and RC2). The high output voltage of VS requires 1700 V devices to be used in RCn inverters. On the other hand the device currents are lowered and they operate under soft switching conditions. That, together with active rectification of the charger converter contributes to the overall efficiency [1].

The separation of stabilizing and isolating converters provides optimum operating conditions for the isolating DC/DC resonant converters although the number of components and circuit complexity rise. But thanks to the fact that they can work at the switching frequency of 20 kHz allowed good optimization of the transformers' dimensions [2].This paper further focuses on the development of the VS converter.

2. Development of the Boost / Buck DC-DC converter

A common half-bridge bi-directional DC-DC converter topology fitted well the reverse power flow need. With respect to the maximum input voltage the 1700 V devices seemed to be a choice. But their power losses under hard switching boost and buck operations would not allow reaching the design goals. Their switching frequency would be less than 10 kHz - well within the audible noise and also the DC choke parameters would be inappropriate in the term of mass and volume reduction.

Due to such difficulties the serial connection of two bi-directional power stages was used as the basic converter topology (Fig. 5). Consequently, based on power loss analysis (as of 2012), the Infineon's FF45R12W1J1_B11 power module was selected [3]. These modules contain two switches in half-bridge configuration in a proprietary Easy1B package. Each switch consists of a 1200 V / 45 A SiC normally-on JFET transistor and the appropriate Si P-MOSFET transistor in series with separated gates which allows maximally efficient switching. Despite the low values of the switching energies the modules had to be paralleled in each power section because of the targeted power. These devices are capable of reverse channel conduction typical for the FET devices so that the complementary control of the transistor pair with minimized dead-times greatly enhances the efficiency [4].

Optimal and reliable control of the power switches is provided by the newly developed drivers – two channel driver unit S4201C1. Specific Infineon's 1EDI30J12CP integrated drivers provide proper gating of both SiC JFET and P-MOSFET transistors in Cascode Light

© VDE VERLAG GMBH · Berlin · Offenbach

configuration (Direct Drive JFET Topology) [5][6]. These units provide galvanic isolation of all the control and error signals as well as various protective functions. That involves overcurrent / short circuit, interlock, over temperature and UVLO protection. The driver is assembled with the power supply module S6303C1.

There are two main purposes of the reverse power flow feature. The first purpose is to make possible pre-charging of the propulsion DC-link capacitors at start of vehicle and also supplying the 3 – phase inverter DC link capacitor with low yet sufficient power when the vehicle is found on catenary isolated sections. The second purpose is to provide the possibility of emergency drive of a trolleybus in case of temporary catenary supply blackout. This solution with the reverse power flow feature has a big benefit that the usual pre-charge circuit (bulky pre-charge resistor and contactor) could be omitted.

For the regular direction (power flow from catenary to the battery charger and 3 – phase inverter) when the converter operates in boost mode the desired these parameters are:

Input voltage V_{IN} = 400 V_{DC} – 1000 V_{DC}
Output voltage V_{OUT}= 950 V_{DC} – 1050 V_{DC}
Power (nominal) P_{NOM} = 25 kW
Power (maximum) P_{MAX} = 30 kW / 5 min
Switching frequency f_{SW} = min. 30 kHz

For the reverse mode (power flow from the on – board battery) when converter operates in buck mode the desired these parameters are:

Input voltage V_{IN} = 900 V_{DC} – 950 V_{DC}
Output voltage V_{OUT} = 400 V_{DC} – 750 V_{DC}
Power (nominal) P_{NOM} = 5 kW
Power (maximum) P_{MAX} = 10 kW / 3 min
Switching frequency f_{SW} = min. 30 kHz

2.1 Experimental results

A prototype of the converter was developed and built in cooperation of SKODA ELECTRIC a. s. and the University of West Bohemia (RICE) in the year 2013. Consequently the prototype was submitted to a number of tests and measurements where the power parameters, safety and reliability were verified. That involved the performance check of the paralleled FF45R12W1J1_B11 modules, proper function of the driver units and, last but not least, the EMC characteristics and their compliance with the relevant regulations.

The examples of the operating waveforms are shown on Fig. 2 and Fig. 3. The first one presents drain-source voltages of the concurrently turned on switches in modules V11 and V21 (see Fig. 5), DC choke current and input voltage in case of minimal catenary voltage level and close to nominal output power. The second picture shows the same waveforms in case of maximum catenary voltage.

Converter efficiency in the boost mode involving the DC choke losses is documented on Fig. 4. The efficiency peaks at 98.5% under the load of 22kW and at the highest permissible catenary voltage of 950 V_{DC}. While catenary voltage is at the lowest permissible level (400 V_{DC}) the efficiency is over 97.5% within in the load range above 5 kW till the nominal load of 25 kW.

PCIM Europe 2016, 10 – 12 May 2016, Nuremberg, Germany

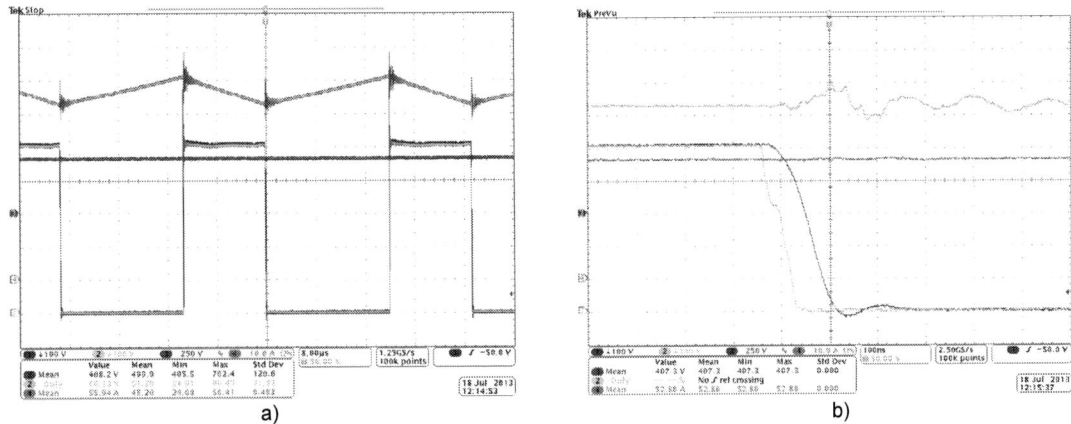

Fig. 2. Boost mode (a) at V_{IN} = 400 V_{DC}, P_{OUT} = 22 kW, , f_{SW} = 30 kHz; detail (b)
Ch1 – $V_{DS(V21A)}$ 100 V/div; Ch2 - $V_{DS(V11B)}$ 100 V/div;
Ch3 – V_{IN} 250 V/div; Ch4 – I_{IN} 10 A/div

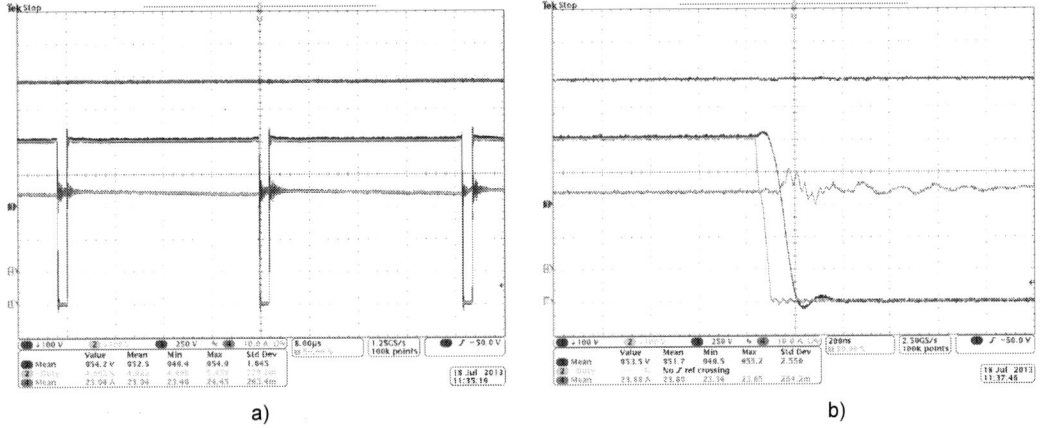

Fig. 3. Boost mode (a) at V_{IN} = 950 V_{DC}, P_{OUT} = 22 kW, f_{SW} = 30 kHz; detail (b)
Ch1 – $V_{DS(V21A)}$ 100 V/div; Ch2 - $V_{DS(V11B)}$ 100 V/div;
Ch3 – V_{IN} 250 V/div; Ch4 – I_{IN} 10 A/div

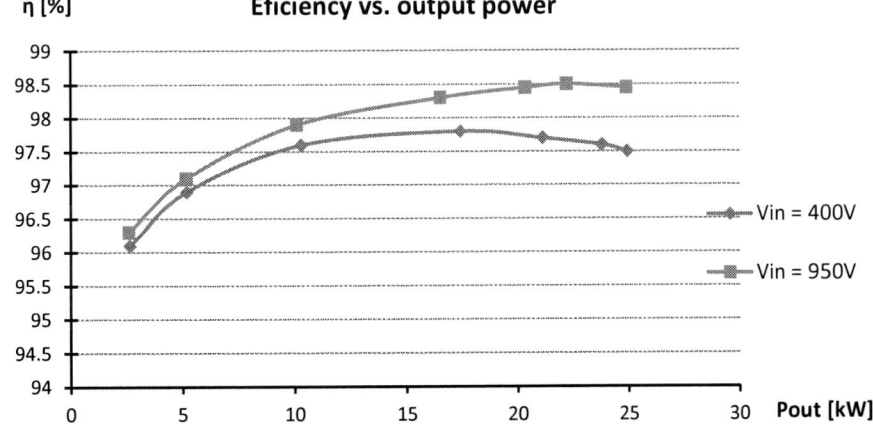

Fig. 4. Efficiency of the boost / buck converter in boost mode including DC choke losses

© VDE VERLAG GMBH · Berlin · Offenbach

The converter development was finalized by a successful type test and later it was introduced into serial production by the end of 2014 under the type name 2QC2 (see Fig. 6 and Fig. 7). Currently it has been assembled into the new generation roof unit (type name SJ10) as a catenary voltage stabilizer for 3 – phase inverter and battery charger according to the development goals (see Fig. 8). The roof unit is a core part of the electrical equipment of the 12 m and 18 m long trolleybuses 26Tr, 27Tr, 30Tr and 31Tr produced by SKODA ELECTRIC a. s.

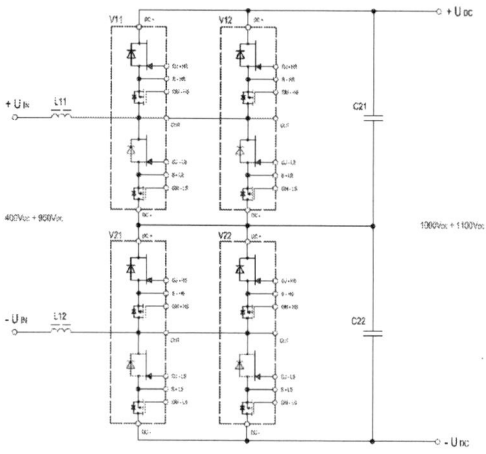

Fig. 5. Topology of the boost / buck DC converter

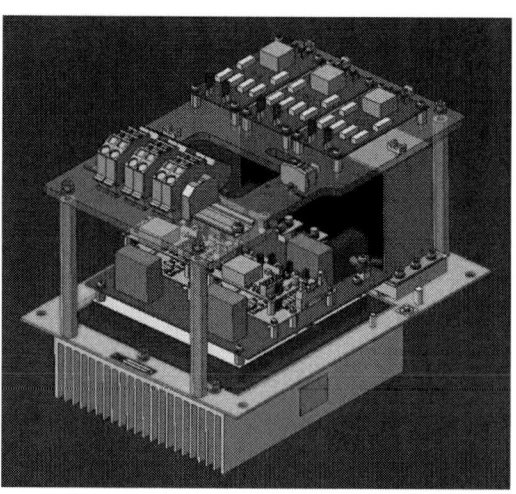

Fig. 6. 3D Model of the boost / buck DC converter with SiC JFET modules

Fig. 7. Power block of boost / buck DC Converter

Fig. 8. Power block 2QC2 (left) and 3-ph. inverter (right) in the roof unit SJ10.1

3. Conclusion

An advanced input voltage stabilizer for the trolleybus' auxiliary power supply was successfully developed and introduced into production. Conceptually the converter is a part of the SKODA ELECTRIC a. s. new generation of electrical equipment for light traction vehicles and fits the newly developed roof unit SJ10.

The converter parameters fully meet the demanding requirements imposed on modern auxiliary power supply converters for the light traction vehicles. It is characterised by high efficiency, sufficient overload capability, and low volume. These features were achieved while the switching frequency was set far above the audible range. The bi-directional capability of the converter is another distinguished feature as it enhances the auxiliary power supply system functionality. These outstanding properties, especially in comparison with the previous generation of voltage stabilizers with IGBTs, were achieved thanks to the use of modern SiC JFET power devices.

Further enhancement can be seen in the power loss and thermal management optimization of the double isolated DC choke under poor cooling conditions, effective control strategy (interleaving) and additional power density increase by construction means.

4. Acknowledgement

A special thanks to Dr. Peter Friedrichs from Infineon for technical support in SiC problems and Mr. Vladimir Zizek from Infineon for obtaining of the SiC JFET module samples.

References

[1] O. Deblecker, A. Moretti, and F. Vallee, "Comparative Study of Soft-Switched Isolated DC-DC Converters for Auxiliary Railway Supply," *IEEE Trans. Power Electron.,* vol 23, Issue 5, pp. 2218 – 2229, Sept. 2008.

[2] CWmT, McLyman. "Transformer and inductor design handbook". Electrical and computer engineering. Marcel Dekker, Inc.; 2004.

[3] see datasheet of FF45R12W1J1_B11 at www.infineon.com

[4] H. Zhang and L. M. Tolbert, "Efficiency of SiC JFET-Based Inverters," *Industrial Electronics and Applications*, 2009. ICIEA 2009. 4th IEEE Conference on, Xi'an, 2009, pp. 2056-2059.

[5] Domes, D. and Zhang, Xi : CASCODE LIGHT – normally-on JFET stand alone performance in a normally-off Cascode circuit. *Proc. PCIM*, Nuernberg 2010

[6] G. Kasebacher : „Application Note AN2013-17 EiceDRIVER Enhanced 1EDI30J12Cx", Infineon, 2013

PCIM Europe 2016, 10 – 12 May 2016, Nuremberg, Germany

Development of a 12kW isolated and bidirectional DC-DC Converter dedicated to the More Electrical Aircraft: The Buck Boost Converter Unit (BBCU)

Pascal Asfaux(pascal.asfaux@airbus.com) - Jérémy Bourdon(jeremy.bourdon@airbus.com) – Airbus Operation SAS, Avionics and Simulation Products – Toulouse, France

Abstract - This paper introduces the development steps of a specific converter dedicated to the More Electrical Aircraft (MEA). This is a 12kW DC-DC converter called BBCU (Buck Boost Converter Unit), isolated, bidirectional and designed in the context of an increased embedded electrical power in aircraft. In recent years, two prototypes have been created to answer the challenges of the BBCU function, but also to reach the maturity and reliability needed for aerospace implementation. Retained technologies and solutions are presented in order to answer to the constraints related to integration, power density and future aircraft performance.

1. Power Electronics in the More Electrical Aircraft (MEA)

a. The More Electrical Aircraft (MEA) Context and Challenges

The context of this communication is the More Electrical Aircraft (MEA), already presented in numerous publications [1]. Because of the increasing demand of electrical applications in the current and future aircraft, a coherent energy transition is needed by partial or full abandonment of the pneumatic and hydraulic sources in favor of the electrical one [2]. Some challenges of this transition relate to Power Electronics, with the creation of new electrical networks and static converters to support and supply them. The challenges on these new systems are in our case their increasing integration and power density and their decreasing losses, weight and time of development, as shown in figure 1. Our application, the BBCU (Buck Boost Converter Unit), is a perfect example of a power converter that can be integrated in the MEA, but also an example of design and embedded technologies optimization.

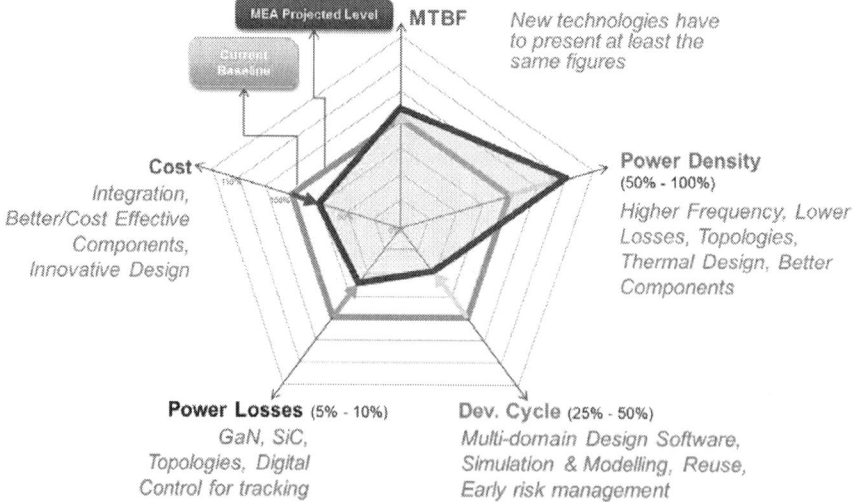

Figure 1: Power Electronics Challenges in the MEA

© VDE VERLAG GMBH · Berlin · Offenbach

b. One specific function of the MEA: the BBCU and its specification

In a few words, the BBCU is a bidirectional and isolated DC-DC converter dedicated to the energy conversion between an embedded network +/- 270 VDC – currently studied for an implementation in the MEA - and a 28 VDC network, which is the classical distribution basis. The BBCU offers transfer power up to 12 kW, which makes its design quite tricky to size. The global principle is exposed in Figure 2. In addition, numerous operation modes answer to the mission profiles of the aircraft, including a wide range of input and output voltage that imply major difficulties for the design and the control scheme. The Airbus specification gives more details on the BBCU background and specificities.

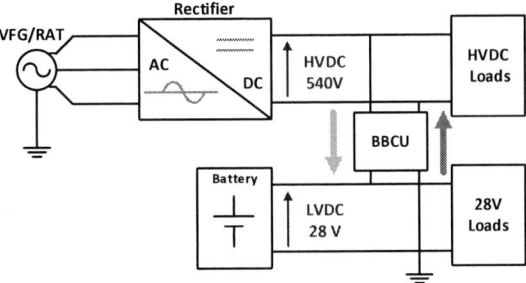

Figure 2: BBCU Principle Schematic

2. BBCU - Version 1

a. Presentation of the first prototype: context and maturity

This chapter presents the complete design of the first BBCU converter. This prototype has been developed by Airbus in a research activity to demonstrate its feasibility in the electrical architecture of the MEA. The work started in 2012 with the aim to optimize individually the different technologies of the design. We have already presented in detail a part of this work in the paper [1]. This first prototype was achieved in 2015. It was compliant to all the norms and answered to the different challenging aspects (electrical, EMC, mechanical, thermal etc.). This prototype, shown in
Figure 4 with its different operation stages, is currently under test in an aircraft bench. This means that the concept is totally demonstrated with a high maturity (Technology Readiness Level (TRL) 5), proving the possibility of implementation in future Airbus programs. The key information of this converter are a weight of 25kg for 12kW, and a power density of 0,5 kW/kg with an efficiency of 85% (Figure 3).

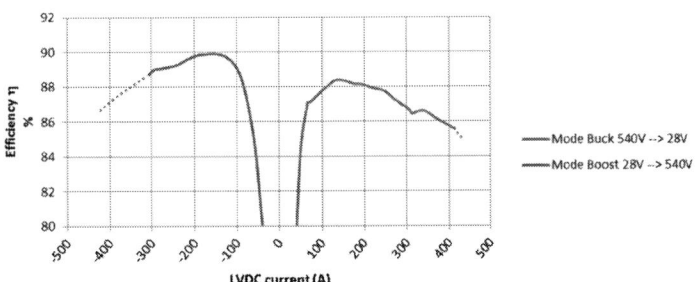

Figure 3: BBCU v1.0 Efficiency Measurement

© VDE VERLAG GMBH · Berlin · Offenbach

PCIM Europe 2016, 10 – 12 May 2016, Nuremberg, Germany

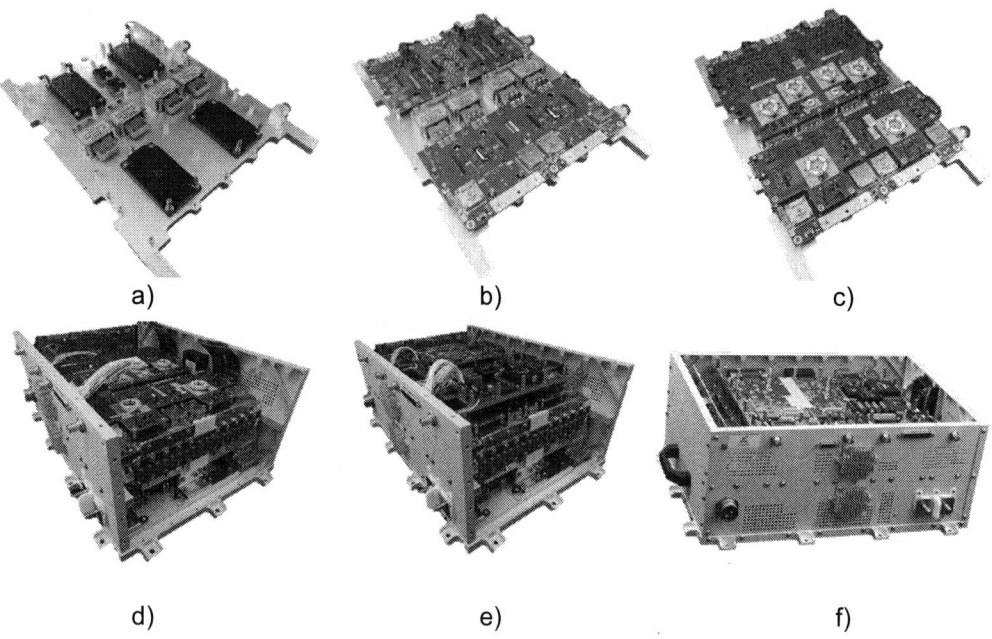

a)   b)   c)

d)   e)   f)

Figure 4: Different Operation Stages of the Buck Boost Converter Unit
a) Cold Plate, Power Modules, Transformers - b) Power Boards - c) Driver Boards d) Packaging, HVDC LU, LVDC Busbar, Damping e) Control board f) Complete BBCU

b. Topological choices

The topology is composed of four parallel 3kW cells which are designed using the current sharing concept, i.e. each brick receives the same current. Each brick is composed with the following pattern: the HV side contains a two-stages conversion with a double serial-buck converter. It is connected to a full bridge converter (current-fed full-bridge type) in order to better adapt the operation point to the DC-AC conversion imposed by the presence of a transformer for the galvanic isolation. On the LV side, we find a double push-pull topology. A filtering stage is added for each power side in compliance with the required input/output impedance and aeronautical norms ABD100.1.8, DO160G and Airbus HVDC directive. A principle schematic is given in Figure 5.

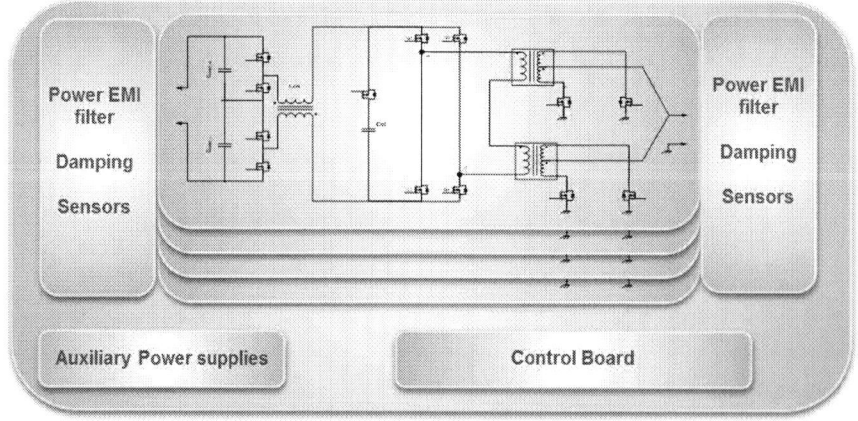

Figure 5: Electrical Topology of the BBCU

© VDE VERLAG GMBH · Berlin · Offenbach

c. Power Control

The control is realized thanks to a classical phase-shift method. Because of the different operation missions during a flight, the BBCU has to comply with constrained rules and operations imposed by the aircraft bench in order to test all main systems, realize their certification and control their industrial feasibility. It induces different tests: Under voltage /Overvoltage, Power cut, Load impact, Soft start, Battery interface, Inrush current, etc. Complex control voltage and current loop has been set up according to 4 axes: HVDC regulation, Boost power way, Buck power way and battery charge (Figure 6).

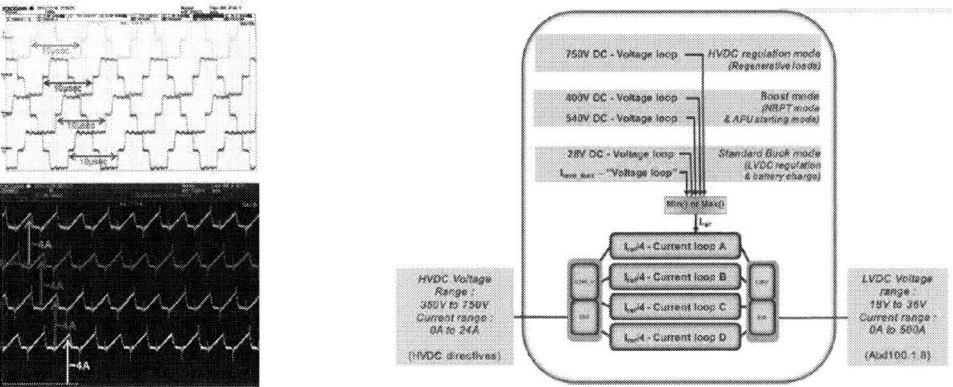

Figure 6: Measurements (transformer voltage & inductor current) and strategy of BBCU control

d. Work on Integrated Technologies

The BBCU Prototype was the opportunity to evaluate existing or emerging solutions and to assess the capability of PE suppliers to offer their products for an industrial production. The first technological example relates to the power modules and allows significant integration improvements with a proven reliable solution. Its contents is up to 42 integrated dies. We used the last generation of available high speed Mosfet semiconductor achieving a 100kHz switching with anti-parallel SiC diodes. The module packaging is realized thanks to AlSiC Baseplate (142g Lightweight), which is compliant with the Aeronautic Standard lifetime and mission profile, insuring a high reliability. In addition, this technology is mature because it reaches TRL6 and is already used in automotive applications to achieve semiconductor integration.

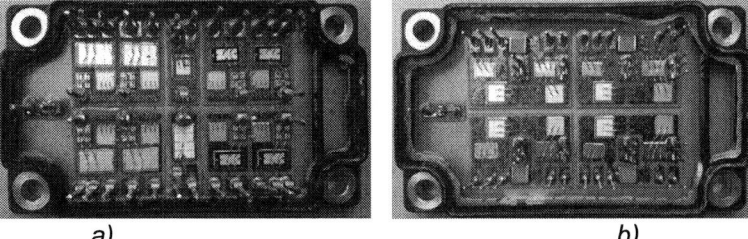

a)   b)

Figure 7: 3kW Power modules (62x108mm) - a) HVDC side– b) LVDC side

The second example relates to the cooling. Liquid cold plates have been investigating to play the role of the key element to manage power losses. Two independent loops of folded fin technology have been integrated to comply with the cooling loss and ensure good performance (Figure 8). The flow is realized through a standard liquid, the Propylene Glycol Water (PGW), while standard thermal performance is reached with low pressure drop, low thermal resistance (Rth) and low thermal gradient. The final integration has a light structure ~ 3.5kg (heat Sink & Structural part). Moreover, various shapes are achievable at a reasonable cost. Tests have been achieved to assess the robustness of the cold plate and connectors assembly at 10 Bar. This technology is today mature enough, easily replicable and already used on one specific application in A380. Thus, liquid cold plate technology is now available and mastered.

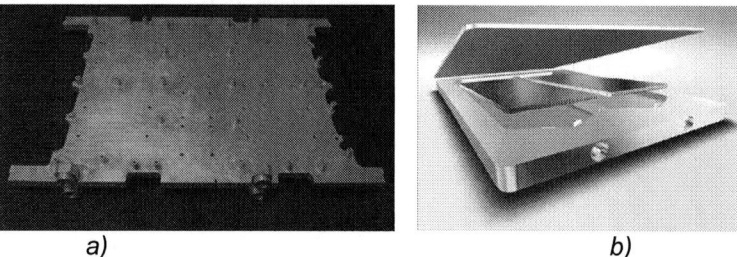

a)   b)

Figure 8: a) Water Cold Plate – b) Example of two loops of folded fin technology

The third example is about planar transformer, which is a key component for efficient power transfer and galvanic isolation (Figure 9-a). First of all, low primary leakage inductances were measured (500nH). Then, considering the planar form factor, better thermal characteristics such as a lower thermal resistance compared to conventional wire wound transformer (up to 50%) have been achieved compared to conventional wire wound transformer. Very low profile EE43/10/28 ferrite are used for a 210g weight. In addition, higher repeatability and higher control of parasitic elements (capacitance & leakage inductance) than conventional wire wound transformer are possible because the industrial process contains a minimum of handcraft operation. The maturity is certified for planar technology, with current application for automotive equipment. As a conclusion, with the emergency of DC/DC topologies in the mass market (automotive and renewable energy), planar transformers become an unavoidable technology for future development in power electronics.

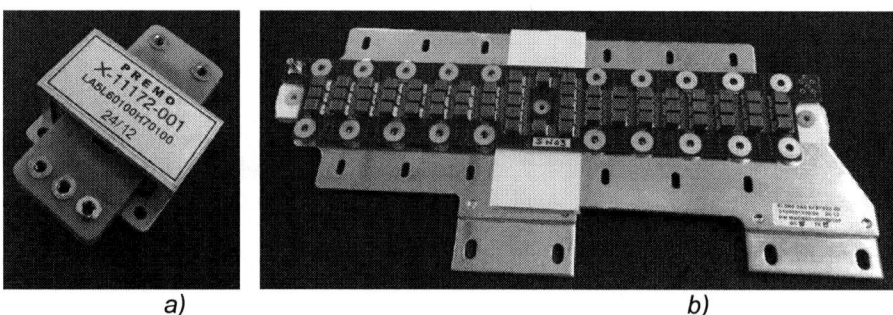

a)   b)

Figure 9: a) Planar Transformer – b) Stacked Capacitors integrated in copper bus bar

The last example relates to the BBCU LVDC side, which implies high current and low voltage range. Stacked SMD ceramic capacitors have been used with less parasitic elements and more compactness than film ones (<1nH versus 10th nH). In addition, they were integrated in the copper busbar technology (Figure 9-b) because it is more suitable for high current (500A) versus traditional PCB. The maturity has been proven for the laminated busbar technology used in A/C equipment and for Stack ceramic used in the automotive equipment.

© VDE VERLAG GMBH · Berlin · Offenbach

3. BBCU - Version 2

a. Presentation of the first prototype and its maturity

In parallel of the first BBCU feasibility demonstration, a research project gathering Airbus and academics named ETHAER has been launched in 2013 [3]. The aim was to improve the efficiency and power density of the BBCU converter by different ways: 1) Global Optimization of the converter, to establish the best choice of topology and get the best technological trade-offs and sizing - 2) Integration of semiconductors and passive components in order to improve the losses, thermal management and electrical parasitic effects – 3) Creation of a new prototype considering both previous work packages and current new technologies such as the wide band gap switches for instance. This whole project leads to a new converter prototype, expected early 2016. Currently, we already have an elementary brick of 3kW realizing the same operation as the final converter and proving the interest of this new development (Figure 10). The key information of this 3kW converter are a power density of 1 kW/kg and an efficiency up to 95%.

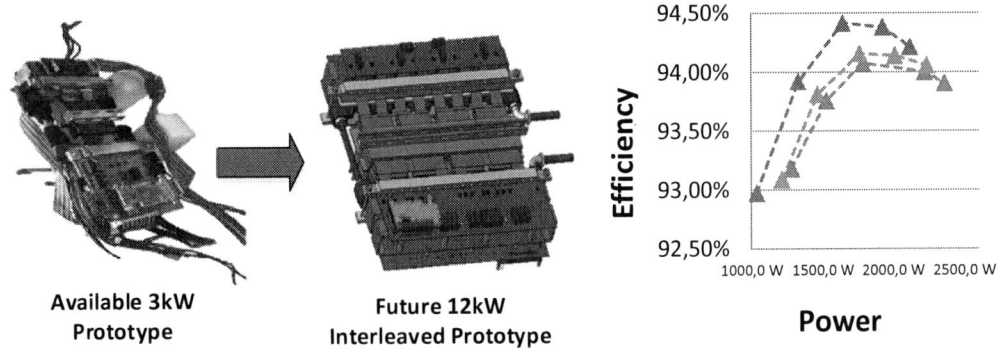

Available 3kW Prototype　　　　**Future 12kW Interleaved Prototype**　　　　**Power**

Figure 10: Picture of a 3kW BBCU v2.0 and an example of efficiency measurements in Buck Mode

a. Topological choices

In order to define the best topology (among 8 candidates) and technologies to realize the BB CU conversion, calculation and prototypes have been realized. The different parts of the equi pment on local and global optimization have been assessed. An example of the multilevel int erleaving trade-off is presented in Figure 11. Similar work has been applied on the filtering par t for differential and common modes.

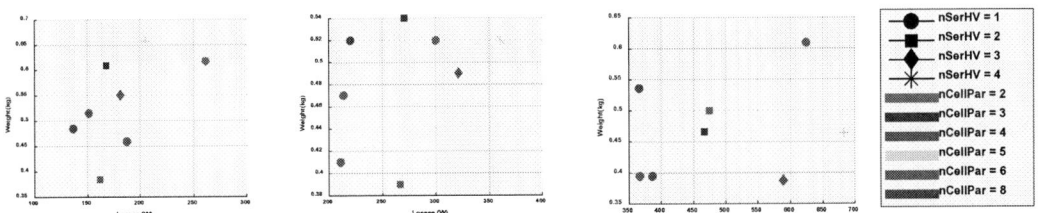

Figure 11: Optimization of the current doubler topology for different multi-level interleaving

Calculation and experiences led to choose the current doubler topology (Figure 12), which is s et in multi-level parallel configuration with 8 cells, and uses a "neutral common" configuration to realize the final converter. In addition, CALC (active clamp circuit) have been added to imp rove the switching operation of semiconductors and so to reach a better efficiency. The relati ve and complete work of this design and sizing is explained in [4].

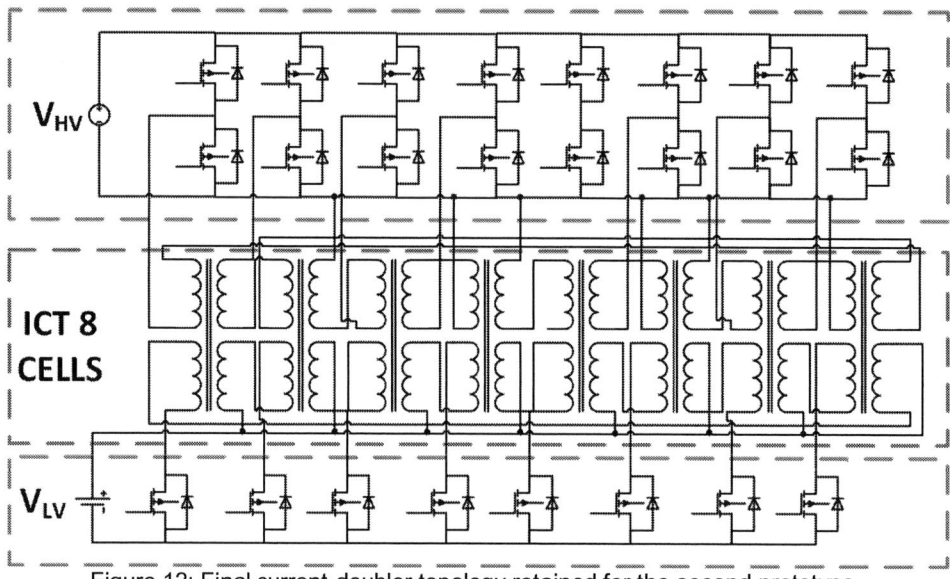

Figure 12: Final current-doubler topology retained for the second prototype

a. Work on Integrated Technologies

Technologies of the first prototype were reported on the second one. In addition, a deeper study mainly focused on the semiconductor, and package and driver integrations was realized. As an example, we investigated several thermal management techniques for GaN transistors with a Wafer-Level Packaging (WLP), mount on Direct-Bonded Copper (DBC) ceramic substrates to detail the manufacturing process and get thermal simulations and experimental results.

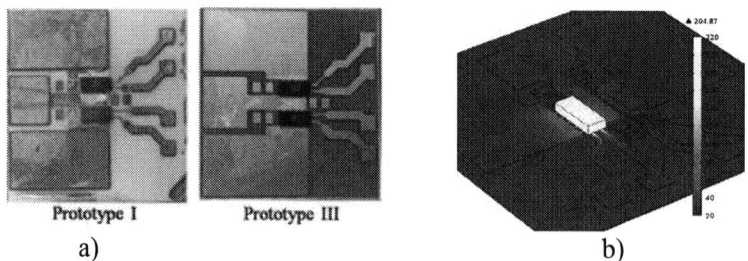

a)   b)

Figure 13: a) Photographs of the two integrated switching cell prototypes - b) Temperature distribution by thermal simulation.

The current work is focused on embedding several dies at once (diodes and transistors) in order to form a complete switching cell (including DC capacitors and drivers). More details can be found in two publications [5][6], where demonstrators are presented and that show satisfying results and characteristics on the packaged die.

4. Conclusion and Expectations for the future

As a result, we can provide the key values of the different BBCU prototypes on the weight and power density by weight and volume. In addition, we can present other research and commercial prototypes, sold for the automation industry in order to make comparisons with our prototypes [7] [8]. To conclude, the initial objective on the BBCU equipment was to reach a power density of 1kW/kg (Table 1), in order to potentially embed it in the future airbus MEA. This objective was achieved on the second prototype (estimation from 3kW available prototype), with impressive advancements due to improved topological choices and additional switching circuits, orientated by preliminary studies and global sizing. A parallel integration study of drivers and semiconductors has been conducted, which let think that the present results can be improved with these technologies.

	BBCU v1.0	BBCU v2.0 (estimation from 3kW available prototype)
Weight (kg)	24,5	12
Ext. power Density (kW/kg)	0,49	1
Total volume (L)	43	13
Ext. power Density (kW/L)	0,27	0.92
Efficiency	85%	95 %

Table 1: Synthesis of both BBCU prototypes

References

1   Bourdon Jérémy, Asfaux Pascal, Morentin Etayo Alvaro "Review of power electronics opportunities to integrate in the more electrical aircraft" ESARS. Aachen, 2015.

2   Roboam Xavier "New trends and Challenges of Electrical Networks embedded in "more electrical aircraft"".

3   ETHAER, ANR. "Advanced Power Electronics for Aeronautical Applications". 2013. www.agence-nationale-recherche.fr/en.

4   Brunello Julien „Conception de convertisseurs de puissance DC-DC isolés pour l'avion plus électrique" 2015 PhD Thesis

5   Y. Chenjiang, C. Buttay, E. Labouré, V. Bley, and C. Combettes, "Highly integrated power electronic converters using active devices embedded in printed-circuit board," presented at the 4th Micro/Nano-Electronics, packaging and assembling, design and manufacturing forum MiNaPAD 2015, Grenoble, France, 2015.

6   Y. Chenjiang, E. Laboure, and C. Buttay, "Thermal management of lateral GaN power devices," in Integrated Power Packaging (IWIPP), 2015 IEEE International Workshop on, 2015, pp. 40-43.

7   Brandelero Julio Cezar « Conception et réalisation d'un convertisseur multicellulaire DC/DC isolé pour application aéronautique» 2015 PhD Thesis.

8   Krismer Florian "Modeling and Optimization of Bidirectional Dual Active Bridge DC–DC Converter Topologies" 2010.

Reverse Mode Application of Sine Amplitude Converters

David Bourner : Vicor Corp, 25 Frontage Road Andover MA 01844 USA   dbourner@vicr.com

Abstract

The power electronics industry is seeing the re-emergence of DC high voltage distribution in place of AC systems of power transmission within advanced machines and installations [1]. A modular DC-DC converter is available in a number of different package and power formats that can form a bridge from low voltage systems to ones that run at levels in the 400 - 1000 V range. Experiments and feasibility studies [1,4] point the way to new product variants of both the BCM (bus converter module) and VTM (voltage transformation module) engines – both of which manifest mature SAC (sine amplitude converter) technology. Efforts are now focused on qualifying parts for bidirectional power flow use. Three operational topology variants are described. Early experimental objectives and outcomes are outlined in the article, along with pertinent techniques that can be used to overcome new challenges implied in this novel deployment of SAC parts. AExamples of applications drawing on the technique are outlined in the final part of this article.

1. Descriptions of Application Spaces

1.1 Motivation

The BCM has seen gradual and significant improvements in efficiency over the past decade [1]. Concurrent to deploying the BCM in new packages is a drive toward a novel way of using the SAC engine, manifest also as the VTM. The SAC –is a resonant, ratio-metric, constant power, isolated, near-ideal DC-DC transformer topology. It can draw power into its secondary port, boosting applied secondary voltage by parameter K as high as 32 in a single step. It transfers most of the energy received in the secondary port, directly to the primary port with very little power loss. The inverse of the transformation ratio, K, a field of integer numbers, is also being expanded as new products are offered. The levels of the HV link to be supported in future products are also set to follow industry trends.

1.2 Classifying New SAC Operational Modes

This article deals in three new topology types associated with the use of SAC engines – besides forward mode, the conventional operational paradigm: These are reverse, mirror and bi-directional mode. It should be pointed out that all the SAC-based solutions may involve any number of parts arrayed in parallel for the purposes of increasing power throughput. So, any mention of a module in what follows may also imply an array of paralleled, identical devices. Figure 1 shows all the topologies and their governing equations.

In *reverse* mode, the source is applied to the SAC secondary power port. Following start-up, the SAC delivers a voltage boosted according to the K factor associated with the particular module. This it presents to a load connected across its primary power port.

A pair of SAC engines is used in *mirror* mode. A source is processed with reverse-mode boost in the first SAC unit, and the other to which it is connected via the primary port interface, works in forward mode, bucking the HV rail to a level that is required in the customer's end equipment.

In *bi-directional* mode, a single SAC is deployed with the intent that once excited, the module is to deliver power in either the forward or reverse direction, depending on the way in which the SAC is being actively driven at a given instant in time.

2. Experiments and Outcomes

2.1 Establishing Reverse Mode Operation

Preliminary results were gathered from two test setups devised for steady state and transient condition assessments. A steady-state test rig, designed for a long-term testing on the bench is pictured in figure 2. Prior to start-up, the BCM's secondary power port was back-biased - a steady voltage was applied to the secondary side of the BCM - something it tolerates without any adverse consequences. The BCM was then energized at the primary side with a unidirectional, low-current high-voltage source sufficient to bias the primary referenced controller internal to the BCM. The sources used in the experiments have large capacitor banks which can tolerate some reverse current flow prior to running a load on the primary side of the SAC. With the controller on the primary side of the part's isolation barrier, there is a need to hold off the primary load whilst starting up the BCM, and to then prevent high levels of primary current in-rush into the load, which is to be engaged only after BCM startup has been completed.

In a real application, where the secondary back bias source connected to the BCM is likely the only available power source, an auxiliary power stage would be needed to boost from the secondary side of the BCM, delivering charge at much higher potential to the primary port of the BCM, without compromising its isolation barrier.

A primary inrush test setup and result are shown in figures 3 and 4 respectively.

2.2 Review of the Application Space

On the basis of the outcomes developed in the first experimental phase, BCM components were successfully deployed in an automotive energy harvesting and adaptive suspension system proof-of-concept, with the bidirectional BCM implementation based on learnings gleaned from the first experiments.

Since this first demonstration, customers have shared demanding applications for power transmission in tethered under-water and airborne vehicles.

These use a high voltage link, made with narrow gauge wire to pass power to the AUV or UAV at the other end of a long tether, using the mirror topology mode running from source to load. Typical power levels are in the 1-2 kW range. An airborne applications example is shown in figure 5 in a Vicor Whiteboard, a web-based tool maintained for the benefit of customers on the Vicor website. The proposed alternative being described here extends the well-tried Power Component approach which offers modularity of design, low weight, and high efficiency and power density metrics.

3. Summary

The use of BCMs in reverse, mirror and bidirectional topologies has been described. Some preliminary results have been shown for an experimental test rig used to verify consistent operation of standard product operated in reverse mode. A couple of application spaces have been described, one in the automotive regenerative braking / active suspension arena and another for tethered UAVs to show the potential way forward with other, as yet untried, practical uses of the new SAC topologies for power processing mentioned in this report.

PCIM Europe 2016, 10 – 12 May 2016, Nuremberg, Germany

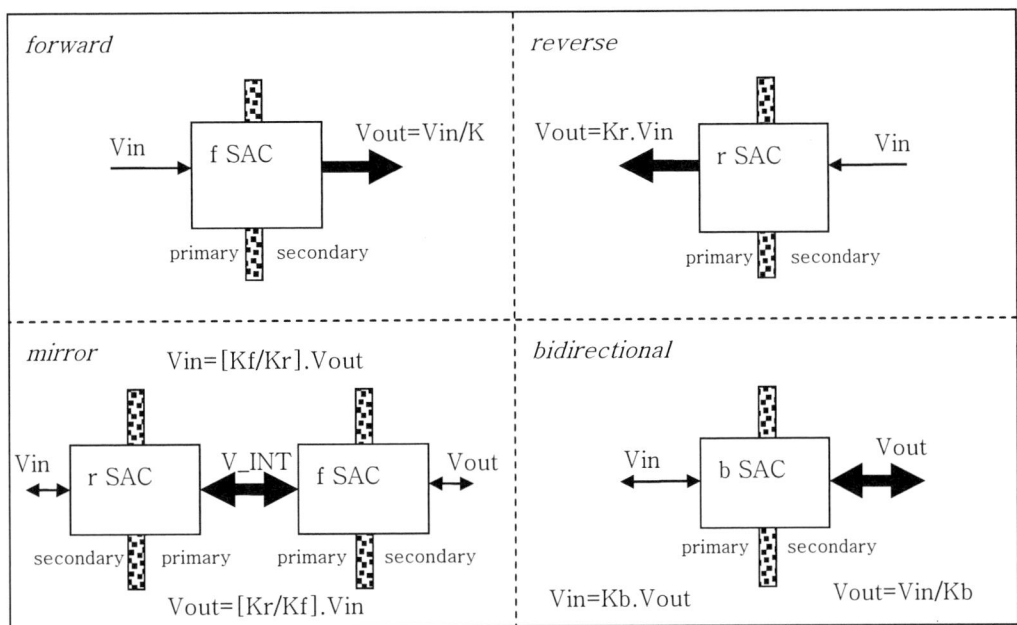

Figure 1 – SAC application topology options. The SACs operate as 'f' (forward), 'r' (reverse) or 'b' (bidirectional) power converters which have isolation barriers set between their primary and secondary power ports. The grey bar represents the SAC inbuilt galvanic isolation barrier between the input and output power ports.

Figure 2 – Reverse BCM in steady-state operation on the bench following BCM start-up

© VDE VERLAG GMBH · Berlin · Offenbach

Current State of Long-Term S/S Test Rig				
K for the MBCM270F450(F/T)M270A00 = 1/6				
V_LV (V)	57.7	LV Rail voltage setting (38–55V in Forward mode)		
V_HV (V)	338.61	HV rail voltage level in reverse mode		
R_HV (ohms)	333	Load set across MBCM's HV bus		
K'	1/5.86	Calculated Voltage Transformation Ratio		
I_HV (A)	1.06	338.61/333 =	1.017	Calculated HV load current
P_HV (W)	V_HV*I_HV	338.61*1.06=	344.31	Power in Load
I_LV		1.06 * 6 =	6.36	Calculated LV source current. Set Current limit to > 7A

Table 1 – Test settings, calculations corresponding to the setup shown in figure 2

LV Rail Voltage (V DC)	HV Rail Voltages (V DC)		LV Rail Currents (A DC)		No-Load
	Reverse-mode startup threshold	Active reverse mode level, K factor	Active reverse mode level	Post HV rail pre-charge completed, load connected	Power from LV Rail (W)
11.00	306	304.97 27.72	0.408	0.455	5.01
11.75	306	325.73 27.72	0.438	0.489	5.75
12.00	306	332.66 27.72	0.446	0.497	5.96
12.50	306	346.48 27.72	0.459	0.519	6.49
13.00	306	360.28 27.71	0.485	0.536	6.97
13.50	306	374.05 27.71	0.497	0.540	7.29
14.00	306	387.87 27.71	0.519	0.561	7.85

Table 2 – Test results of DC startup action for a standard BCM

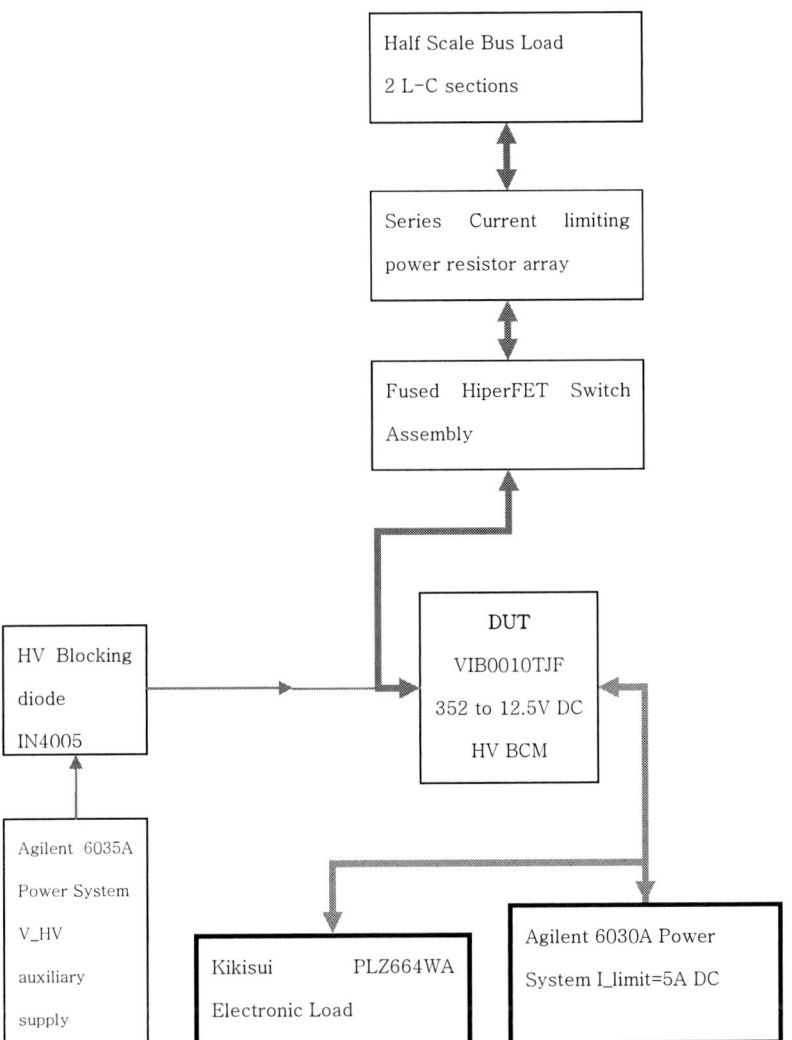

Figure 3 – Reverse BCM Transient test setup

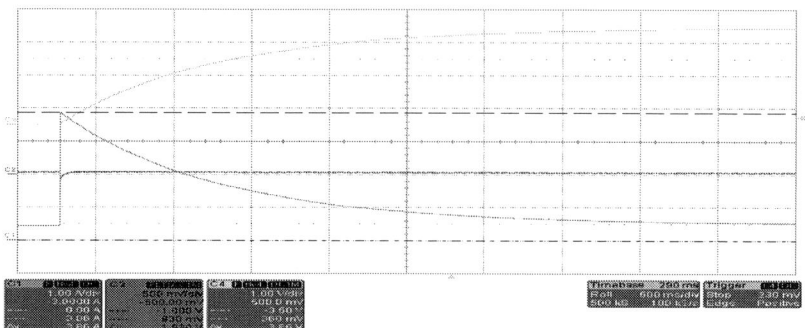

Figure 4 – Transient test outcome. Green trace: boosted output with x100 attenuation, red trace: LV bus voltage; olive green trace: controlled secondary side inrush current.

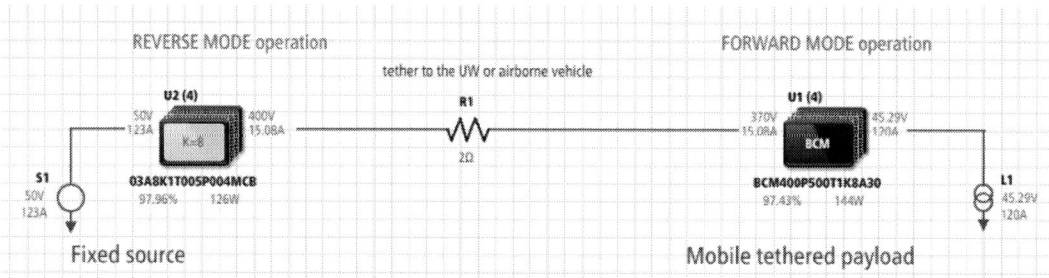

Figure 5 – Airborne power transmission example uses HV link in a tether, a system that is realized with reverse and forward-mode operated BCMs. This is an example of a SAC mirror topology, albeit one that is operated with a unidirectional power flow.

References

[1] M Salato "The Sine Amplitude Converter™ Topology Provides Superior Efficiency and Power Density in Intermediate Bus Architecture Applications, Vicor Corporation June 2011.

[2] Vicor Company Private Document "VTM II: Theory of Operation"

[3] P Yeaman VIC Application Note AN:016 "Using BCM Bus Converters in High Power Arrays"

[4] A Patel "An Isolated Step-Up DC-DC Converter Using Series Connected Sine Amplitude Converters " APEC2015, ISBN 978-1-4799-6735-3/15

[5] D Bourner "Back-Driven Vicor Bus Converter Modules – Startup Characteristics" Vicor Internal report Sep 2012

PCIM Europe 2016, 10 – 12 May 2016, Nuremberg, Germany

Influence of the Configuration of the Load Cable on Switching Characteristics of IGBTs

Lars Middelstaedt [1], Dennis Richter[1], Andreas Lindemann [1], Arendt Wintrich[2]

[1] Otto-von-Guericke-University Magdeburg, Universitätsplatz 2, 39106 Magdeburg, Germany
[2] Semikron, Sigmundstrasse 200, 90431 Nuremberg, Germany
email: lars.middelstaedt@ovgu.de

Abstract

While the negative influence of long cables between an inverter and an AC motor on the motor has been investigated in literature the effect on the switching energy of the IGBTs is often not considered. This leads to a discrepancy between the expected losses derived from data sheet measurements with pure inductive load and the real application.

This work focuses on evaluating the switching energy of IGBTs in a three-phase inverter depending on the geometry of the cable. The effects of different lengths of a shielded 4 core cable (up to 50 m) are investigated and compared with non-shielded cables. Critical paths of parasitic capacitive leakage currents are defined and evaluated by measurement. Additionally, the influence of different grounding concepts are discussed.

The measurement results show, that especially the large capacitance between shield and phase leads to an increased turn-on current and therefore turn-on energy, while the turn-off energy is reduced. However, the overall switching energy increases with the length. Additionally, two significant effects within the switched current and voltage occur depending on cable length and current amplitude. A simulation model is able to investigate and explain these effects.

1. Introduction

A widely used power electronic system in industrial applications consists of a power electronic inverter, filters, a load cable of many meters and a motor. Unwanted motor transients as overvoltage, bearing currents and ringing can be linked to the fast switching transients in conjunction with the long load cable and are usually optimized by filters between the inverter and motor and different inverter control strategies [1–6]. However, the influence of the periphery of the power electronic circuit on the switching characteristics of the IGBTs is not discussed in detail. The fundamentals of switching characteristics of IGBTs are well understood and are described in literature [7–9], where the main parasitic elements of the power circuit are taken into account but not the periphery, which has a significant influence on the switching speed and therefore on the switching energy as well.

This paper focuses on the IGBT switching characteristics of a three phase DC-AC inverter with a load cable with variable length powering an AC motor. The IGBT switching energy depending on different cable configurations is compared. The most critical capacitive coupling paths are derived and cable dependent switching effects are discussed.

© VDE VERLAG GMBH · Berlin · Offenbach

2. Equipment under Test

2.1. Power Circuit

The investigations were carried out by setting up a DC-source, 3-phase DC-AC inverter, load cable and an AC motor. The 1200 V-IGBT module has a nominal current of 8 A, which corresponds to an inverter power rating of up to 5 kW. For better controllability of the point of operation and therefore for a better reproducibility of the switching characteristics, a double pulse operation mode was applied. Accordingly, the low side IGBT of the first branch T2 is switched, while the other five IGBTs are constantly in either on (T3, T5) or off (T1, T4, T6) state, as shown in Fig. 1. Hence, phase V and W are constantly on the same high potential,

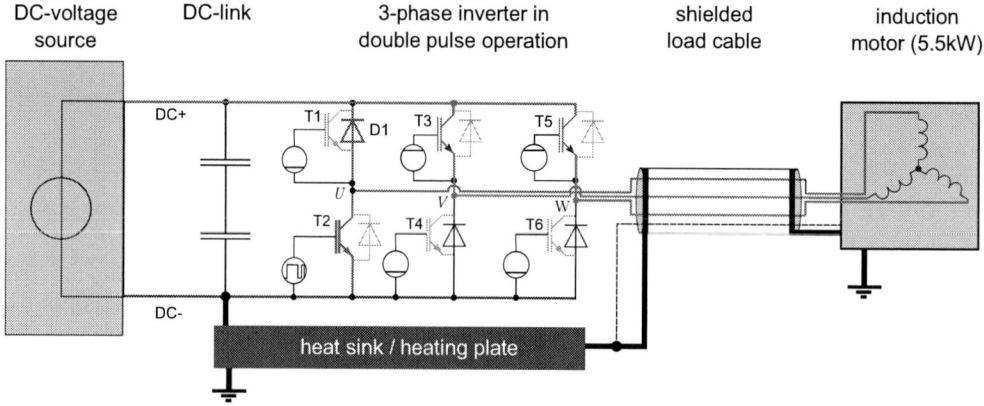

Fig. 1: Circuit diagram of the 3-phase inverter in double pulse operation mode

while the potential of phase U is changing between maximum and minimum.

In most of the three phase power electronic applications, a three phase inverter with ground potential close to the middle of the DC-link is used. In a first test it was confirmed that a circuit with grounded midpoint and ground at DC- have the identical results with regards to the switching characteristic at the IGBT and the capacitive currents. Because of the easier handling an inverter with ground at DC- is investigated in this work.

To investigate the influence of different cable lengths on the switching energy of the IGBT a four wire, shielded, 4 mm² TOPFLEX®-EMV-2YSLCY-J cable of different lengths of 2 m, 5 m, 10 m, 23 m and 50 m was used (see Fig. 2). Additionally, a non-shielded four wire cable (wires in close proximity) and four single wires of the same material with a length of 10 m were applied and the switching results compared to evaluate the difference in coupling capacitances between shield and wires.

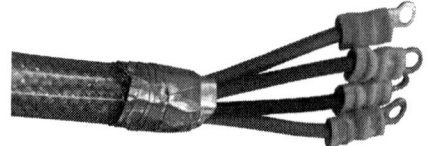

Fig. 2: Shielded four wire cable applied to experimental setup

2.2. Measurement Setup

To evaluate the switching behavior and the capacitive current paths within the equipment under test, the voltages of T2 between gate and emitter U_{ge} and between collector and emitter U_{ce} were measured. Additionally, following currents were measured:

- $I_{DC(-)}$ at DC-
- $I_{DC(+)}$ at DC+
- I_{gnd} from heatsink to ground
- $I_{shield,inv}$ at the shield at the inverter
- I_U in phase U and I_{VW} in phases V, W at the inverter
- $I_{shield,m}$ at the shield at the motor
- I_m, ground current from the motor

The measurement points are displayed in Fig. 3.

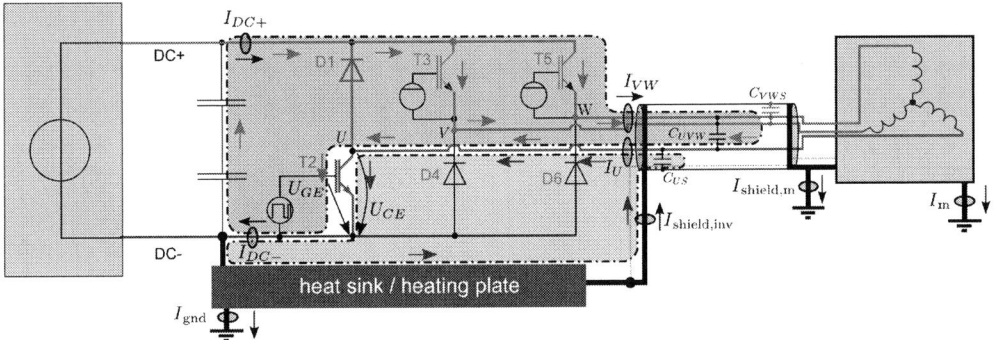

Fig. 3: Measurement setup for different currents and voltages, including possible current paths during turn-on

3. Results

Most of the investigations are carried out by analyzing measurement results. In the following parasitic current paths, cable-dependent switching effects and switching energy will be compared. A simple simulation model has been developed to explain the major observations.

3.1. Simulation Model

The explanations and findings in this section are derived or verified by a basic simulation model with a lumped element cable model, a simple diode model and an IGBT model based on [8]. The parameters for the cable were derived by different open circuit and short circuit impedance measurements. Using nodal analysis the parameters for the different intra cable capacitances, wire inductances and resistances per unit length were extracted and implemented in n cable segments. One segment is shown in Fig. 4. Figure 5 shows a block diagram of the simulation

PCIM Europe 2016, 10 – 12 May 2016, Nuremberg, Germany

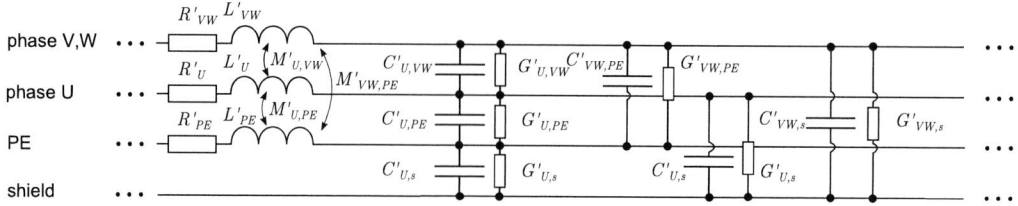

Fig. 4: One segment of the cable model under investigation

model. The model is able to reproduce most of the measured effects and the values of the parasitic capacitances correlate with the data sheet of the cable.

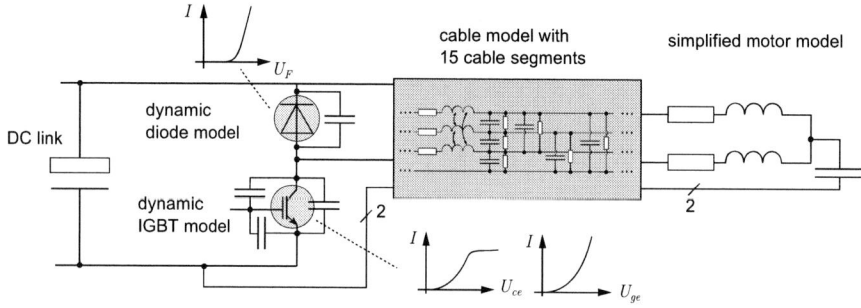

Fig. 5: Block diagram of the simulation model

3.2. Analysis of the Capacitive Current Paths

In Fig. 6(a) and Fig. 6(b) the measurement results for different currents measured during turn-on and turn-off are displayed. I_{diff} is the difference of $I_{DC-} - (I_{DC+} + I_{\text{shield,inv}} + I_{gnd})$ and is approximately zero, showing, that the proposed current paths in Fig. 3 are plausible. $I_{\text{shield},m}$ and I_m are negligible for the equipment under test (EUT).

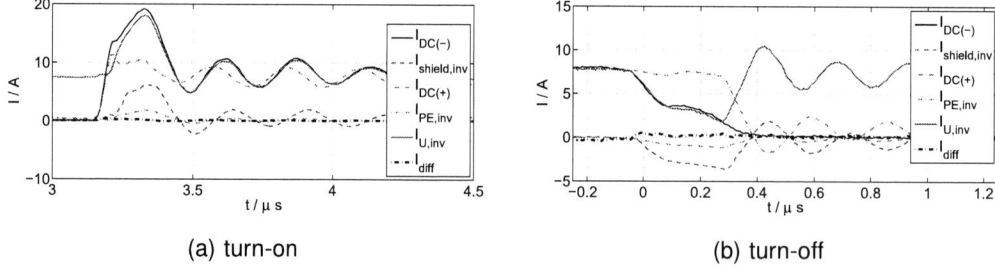

| (a) turn-on | (b) turn-off |

Fig. 6: Measurement results of different currents

During the **turn-on** process the capacitances between phase U and shield $C_{U,s}$ are discharged. This discharging current loop between phase U, IGBT T2, heat sink and shield corresponds to the blue current loop in Fig. 3 and explains the considerable current flow within the shield in Fig. 6(a). The current $I_{DC(-)}$ is influenced only slightly. A small peak due to the reverse recovery current of the diode is observed. The subsequent oscillation propagates within phase U and

© VDE VERLAG GMBH · Berlin · Offenbach

therefore DC(-) as well as phases V, W and therefore DC(+), via the DC-link capacitors and IGBT T2. Thus, the diode has to be treated as open circuit. This is plausible since the diode is in blocking state at this point. The difference in the oscillation amplitudes is found within the shield current. Hence, the oscillation currents can be described by

$$I_{\text{shield,inv}} = I_{\text{DC(-)}} - I_{\text{DC(+)}} = I_U - I_{V,W} \tag{1}$$

The charging of the capacitances between phase U and phases V, W $C_{U,VW}$ results in a slightly increased $I_{\text{DC(+)}}$ (see Fig. 6(a) and Fig. 3 red loop). Since the capacitance between the phases is smaller than $C_{U,s}$ this capacitive current is smaller as well. Measurements have confirmed a smaller capacitance by the factor of 0.1. The deviation to the datasheet ratio of 0.5 results from the different setup, where PE and shield are short circuited as well as phase V and phase W.

During **turn-off** the capacitive currents change directions, since $C_{U,s}$ is now charged and $C_{U,VW}$ is discharged. Hence, a decrease in $I_{\text{DC(-)}}$ and a negative shield current is observed in Fig. 6(b). The oscillation has changed as well. Now the diode is conducting and the IGBT equals an open connection which is why the oscillation propagates between phase U and phase V, W via the diode. The difference between the oscillating currents is found in $I_{\text{shield,inv}} = -I_{\text{DC(+)}}$, resulting in the oscillation currents described by the equation

$$I_{\text{DC(-)}} = I_U - I_{V,W} \tag{2}$$

3.3. Comparing Switching Behavior as a Function of Cable Length

As can be seen from Fig. 7(a) and 7(b) the switching characteristics strongly depend on the length of the cable. With increasing length, the parasitic cable capacitances between the wires and the shield increase and hence the charging and discharging currents of those capacitances during the switching process increase as well. These currents add up to the reverse recovery current of the diode during turn-on (see Fig. 7(a)). At 50 m this current keeps the IGBT desaturated for several 100 ns. Thus, the amplitude of the plateau depends on the transfer characteristic of the IGBT and cannot exceed a certain level. Since the necessary over all charge Q of the cable capacitances is dictated by the cable, the duration of the current plateau correlates with the charging time of the cable capacitance. The additional capacitive current increases the turn-on losses considerably.

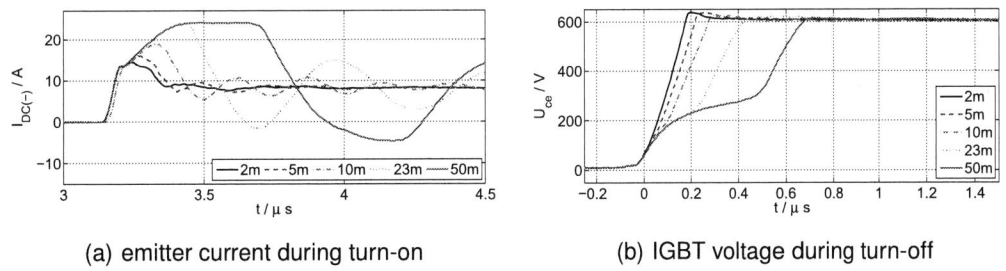

(a) emitter current during turn-on (b) IGBT voltage during turn-off

Fig. 7: Measurement results for different lengths of cable

In Fig. 7(b) a plateau within the rise of u_{ce} is observed. This effect as well correlates with the length of the cable. With increasing length, the length of the plateau increases as well, which is due to the propagation delay and reflection at the end of the not matched cable termination.

The propagation speed v_p of an electromagnetic wave for a cable without loss is defined in [10] by

$$v_p = \frac{1}{\sqrt{L'C'}},\qquad(3)$$

where L' and C' are the inductance and capacitance per unit length of the cable. Their values were determined by carrying out open circuit and short circuit impedance measurements between phase U and short circuited phases V and W. Accordingly, the derived values for L' and C' are the equivalent over all inductance and the over all capacitance per unit length of the cable, that are effective at this port. From (3) the propagation delay t_p is calculated using the cable length l_c

$$t_p = \frac{l_c}{v_p}\qquad(4)$$

which corresponds to the inverse of the first resonant frequency of the cable under investigation. The delay time for a 50 m cable with $C' = 104\,\text{pF/m}$ and $L' = 289{,}6\,\text{nH/m}$ was determined to be 0,27 μs. A reflected wave therefore travels for 100 m, resulting in $t_p = 0{,}54\,\mu\text{s}$. This time corresponds directly with the length of the plateau measured in Fig. 7(b). Simulation results confirm this result. Using the simulation model displayed in Fig. 5 the influence of different parameters on the plateau was investigated. The capacitance values of the semiconductors at the input of the cable determine the shape and the rise time of the voltage. With large capacitances the rise time increases. Additionally, they act like a buffer capacitance at the inverter output. Hence, the plateau becomes less obvious and the voltage characteristic resembles more a linear slope. In case of small semiconductor capacitances the plateau changes to a stronger ringing. Next to the capacitances of the semiconductors, the load current has a major influence as well. With a smaller current the capacitances are charged and discharged more slowly and thus voltage rise time increases. Furthermore, the amplitude of the plateau decreases with the current and multiple plateaus might be observed, as shown in Fig. 8. Again, a time constant of 0,54 μs is observed.

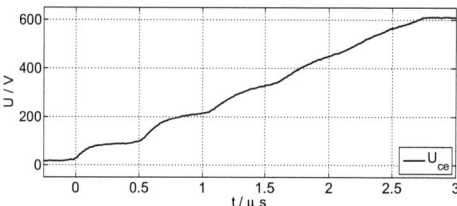

Fig. 8: Measurement result of multiple plateaus during turn-off of the IGBT within the collector emitter voltage at a load current of 2 A and a cable length of 50 m

3.4. Comparing Switching Energy as a Function of Cable Length

In table 1 an overview over measured turn-on energy E_{on}, turn-off energy E_{off} and the overall switching energy E_{all} as a function of the cable length is given at 25 °C. Very similar dependencies are observed for the switching characteristic at 125 °C. While E_{on} increases with the length of the cable due to an increase in $I_{\text{shield,inv}}$, E_{off} decreases since the absolute value of $-I_{\text{shield,inv}}$ increases as well during turn-off opposing $I_{\text{DC-}}$. Therefore, I_{DC-} is reduced. However, the overall energy E_{all} increases considerably with the length of the cable. An increase of switching energy of more than 30 % is observed when comparing the cable length of 2 m and 50 m. Additionally, an increase in stress on motor and inverter due to current oscillations with high amplitudes has to be pointed out. Hence, it is vital to use a cable, that is as short as possible or an output inductor in order to increase the efficiency and reduce the strain of the whole system.

Tab. 1: Switching Energy depending on the length of the cable using a shielded four core cable, V_{cc}= 600 V, I_c=8 A, T_j=25 °C

Cable length	E_{on} / mJ	E_{off} / mJ	E_{all} / mJ	ΔE_{all} / %
2 m	0.82	0.38	1.20	0.0
5 m	0.94	0.37	1.31	9.1
10 m	1.06	0.34	1.40	16.6
23 m	1.26	0.28	1.54	28.3
50 m	1.38	0.22	1.60	33.3

3.5. Cable Design

Most of all the shield of the cable and the corresponding capacitances to the wires are the main reason for the observed phenomena. In some cases, it might be a useful alternative to use an unshielded cable with twisted or solitary wires. The experimental setup of a 10 m cable was changed by removing the shield, resulting in an unshielded four core cable with the four wires in very close proximity. Afterwards, the four wires were separated so that four single wires were applied. Figure 9(a) and Fig. 9(b) show the corresponding measurement results.

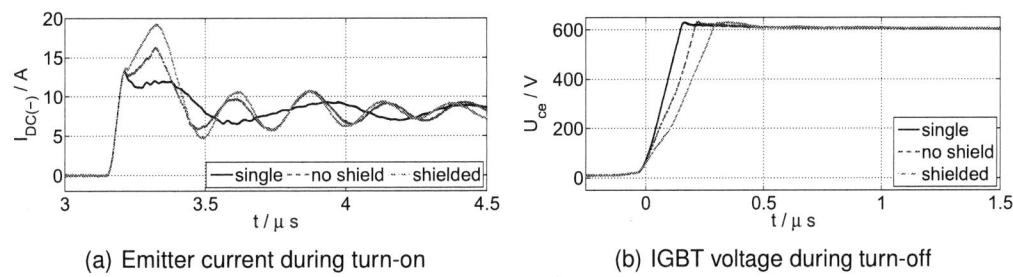

(a) Emitter current during turn-on   (b) IGBT voltage during turn-off

Fig. 9: Measurement results for different cable designs

The turn-on current and therefore the turn-on energy is reduced considerably when the capacitive coupling is minimized for single wires. Additionally, the plateau within the voltage during the turn-off is reduced and a faster turn-off process is possible. The oscillation frequency is the same for the four core cable with no shield and the shielded cable, which proves that the oscillation is evoked by the intra wire capacitances and inductances as stated above. Even though that the switching energy is improved by non shielded cables, a shielded cable is often required due to electromagnetic interference issues.

4. Conclusion

In this paper the influence of the length of the load cable on the switching energy of the inverter IGBTs is discussed. It is shown, that especially for long cables the turn-on losses increase, while the turn-off losses decrease. Since the turn-on losses dominate the overall losses increase with the length of a shielded cable. The switching energy is increased by 33 % when using a 50 m cable compared to a 2 m cable. Mainly, this is due to the phase-shield capacitance within the cable, that is charged or discharged during the switching process. Additional effects due to the capacitances and the corresponding capacitive currents are observed. One or multiple u_{ce} voltage plateaus during the turn-off are evoked depending on the length of the cable and the nominal current. The parasitic currents add to or even dominate the reverse recovery current, increasing the turn-on current through the IGBT considerably. For a length of 50 m the additional capacitive current of the cable leads to an over all current amplitude, that results in a current plateau due to the desaturation of the IGBT in accordance to the transfer characteristic. Hence, the time for the turn-on process increases. When using an unshielded cable or even single wires with some space in between them, these unwanted effects can be minimized. However, when using unshielded cables electromagnetic compatibility issues need to be taken into account.

References

[1] J.M. Bentley and P.J. Link. "Evaluation of motor power cables for PWM AC drives." In: *Pulp and Paper Industry Technical Conference, 1996., Conference Record of 1996 Annual.* June 1996, pp. 55–69. DOI: 10.1109/PAPCON.1996.535983.

[2] R.J. Kerkman, D. Leggate, and G.L. Skibinski. "Interaction of drive modulation and cable parameters on AC motor transients." In: *Industry Applications, IEEE Transactions on* 33.3 (May 1997), pp. 722–731. ISSN: 0093-9994. DOI: 10.1109/28.585863.

[3] A. von Jouanne et al. "Filtering techniques to minimize the effect of long motor leads on PWM inverter-fed AC motor drive systems." In: *Industry Applications, IEEE Transactions on* 32.4 (July 1996), pp. 919–926. ISSN: 0093-9994. DOI: 10.1109/28.511650.

[4] A. von Jouanne and P. Enjeti. "Design considerations for an inverter output filter to mitigate the effects of long motor leads in ASD applications." In: *Applied Power Electronics Conference and Exposition, 1996. APEC '96. Conference Proceedings 1996., Eleventh Annual.* Vol. 2. Mar. 1996, 579–585 vol.2. DOI: 10.1109/APEC.1996.500499.

[5] H. Akagi and S. Tamura. "A Passive EMI Filter for Eliminating Both Bearing Current and Ground Leakage Current From an Inverter-Driven Motor." In: *Power Electronics, IEEE Transactions on* 21.5 (Sept. 2006), pp. 1459–1469. ISSN: 0885-8993.

[6] N.O. Cetin and A.M. Hava. "Interaction between the filter and PWM units in the sine filter configuration utilizing three-phase AC motor drives employing PWM inverters." In: *Energy Conversion Congress and Exposition (ECCE), 2010 IEEE.* Sept. 2010, pp. 2592–2599. DOI: 10.1109/ECCE.2010.5617995.

[7] Josef Lutz et al. *Semiconductor Power Devices. Physics, Characteristics, Reliability.* Berlin Heidelberg: Springer-Verlag, 2011, ISBN: 978-3-642-11124-2. DOI: 10.1007/978-3-642-11125-9.

[8] Arendt Wintrich. "Verhaltensmodellierung von Leistungshalbleitern für den rechnergestützten Entwurf leistungselektronischer Schaltungen." PhD thesis. TU Chemnitz, 1997.

[9] Arendt Wintrich et al. *Applikationshandbuch Leistungselektronik.* ISLE Verlag, Ilmenau, 2010. ISBN: 978-3-938843-56-7.

[10] Frieder Strauss. *Grundkurs Hochfrequenztechnik. Eine Einführung.* Vieweg+Teubner Verlag, Wiesbaden, 2012. ISBN: 978-3-8348-1242-1.

PCIM Europe 2016, 10 – 12 May 2016, Nuremberg, Germany

Improved Power Decoupling Scheme for Single-Phase Grid-Connected Differential Inverter with Realistic Mismatch in Storage Capacitances

Wenli, Yao, Northwestern Polytechnical University, China, ywl0158@mail.nwpu.edu.cn

Xiongfei, Wang, Aalborg University, Denmark, xwa@et.aau.dk, pcl@et.aau.dk

Poh Chiang, Loh, Aalborg University, Denmark, pcl@et.aau.dk

Xiaobin, Zhang, Northwestern Polytechnical University, China, dgl907@126.com,

Frede, Blaabjerg, Aalborg University, Denmark, fbl@et.aau.dk

Abstract

The single-phase differential inverter has been introduced in grid connected application with a differential mode for power transfer and a common mode for actively decoupling the second-order power oscillation. This capability is referred to as power decoupling, which when implemented properly, may prolong the lifespan of the dc source. With applying this topology, the existing works have however done with ideally matched storage capacitance, which is not realistic since the capacitance may suffer parameter drifts duo to time and thermal effect. It is therefore the intention of this paper to analysis the performances degradation with capacitance mismatch, shown that both AC and DC have significant distorted. As a result, a simple improved scheme is then proposed for raising performance of the differential inverter, following the identified mitigation criterion. Simulation and experimental results provided have verified the computation and control scheme developed.

1. Introduction

Single-phase inverters have been widely used with the Photovoltaic (PV) systems and other small distributed power generators [1]. However, the inherent second-order ripple power may challenging the system efficiency and performance. Traditionally, a bulky electrolytic capacitor is used in the DC-link to buffer the ripple power, but its short life time may bring in reliability issues [2]. To avoid the use of an electrolytic capacitor, active power decoupling techniques have thus been developed, which is generally realized with an additional energy storage element and switches to absorb the pulsating power flowing into the dc link [2]-[7]. Yet, the additional component not only increases the total cost of the inverter, but also complicates the system control algorithm.

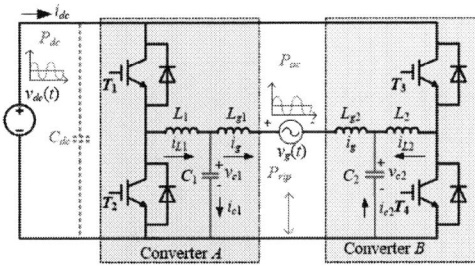

Fig. 1. Schematic of the buck differential inverter for power decoupling.

To further reduce the cost, a single-phase differential buck inverter is recently introduced in [4] for realizing the power decoupling without using additional switches and storage elements. The differential inverter is basically composed of two dc-dc buck converters, which are operating in two modes: 1) the Differential Mode (DM) for active power transfer; and 2) the Common Mode (CM), which is for compensating the second order ripple power. The work in [4] is limited to islanded UPS systems, and in [6], the differential buck inverter was used in the grid-connected applications with an LCL-filter, as shown in Fig. 1. The existing power decoupling and grid current controllers designed in [6-7] show a good performance under the assumption that filter capacitance are perfect matched.

© VDE VERLAG GMBH · Berlin · Offenbach

1837

However, this is not realistic since the capacitance may has thermal drifts and aging affects [8], even the film capacitor also has a maximum ±20% capacitance tolerance (depend on the capacitance tolerance code) [9-10]. It has been shown in [5] that ±10% Capacitance variation will significant increasing the ripple reduction factor. But unfortunately, these existing methods still assume that similar variation for both capacitances. It is therefore appropriate to conclude that control and performance of a differential inverter under storage capacitance mismatch have neither been fully studied nor quantified. Thus, in this paper, a theoretical analysis is firstly developed to investigate the effects of capacitance mismatch, shown that the capacitance create serious harmonic distortion in both AC and DC terminal, which are proportional to the amount of capacitance drift. These deteriorations may degrade the source, and breach grid standards like the IEEE 1547 (second to tenth even harmonics < 1%) [11]. As a result, some mitigation criteria are identified for resolving the complications distortion, before realizing them as an improved control scheme. The improved scheme can generate the required control references autonomously without knowing the amount of mismatch between the two storage capacitances. It has also been pointed out that the identified grid complications can only be resolved by feeding back the grid current, and not the inverter current. Feeding back the grid current may however require additional active damping, which is also discussed. Experimental results captured for verifying the control effectiveness and theoretical quantification are then presented to show the effectiveness of theoretical analysis and the proposed control approach.

2. System description and operation principle

2.1. Second-order ripple power in dc-link

The inverter shown in Fig. 1 is composed of two buck converters, which can be denoted as the converter x (x = 1 or 2) formed by the switch $T_{x.1}$, $T_{x.2}$, inductor L_x and L_{gx}, capacitor C_x, The midpoint of the capacitors C_x is connected back to the negative dc-rail, providing a circuit path for compensating second-order ripple power. Assuming the grid voltage v_g, capacitor voltage v_{c1} and v_{c2}, and grid current $i_g(t)$ are denoted as

$$v_g = V_g \sin\omega t\,, i_g = I_g \sin(\omega t + \alpha), v_{c1} = V_d + 0.5V_m \sin(\omega t + \delta), v_{c2} = V_d - 0.5V_m \sin(\omega t + \delta) \quad (1)$$

where V_m and I_m represents the magnitude of v_g and i_g respectively, α represents their phase difference, ω is the grid angular frequency, δ is the phase difference between the capacitor and grid voltages. Ignoring losses, this power, together with powers flowing through the two capacitors, is then equal to the power supplied by the dc source. Further assuming that the dc source voltage V_{dc} is constant and $C_1 = C_2 = C$, input current i_{dc} to the inverter can successively be expressed as [5]:

$$i_{dc} = \frac{I_g V_g}{2V_{dc}}\cos\alpha + \frac{V_g}{2V_{dc}}\sqrt{I_g^2 + \frac{1}{4}\omega^2 C^2 V_g^2 + I_g V_g \omega C \sin\alpha} \cdot \cos(2\omega t - \theta), \quad \tan\theta = -\frac{\left(I_g \sin\alpha + \frac{1}{2}\omega C V_g\right)}{I_g \cos\alpha} \quad (2)$$

The second term of (2) is obviously an unwanted input ripple current oscillating at 2ω.

2.2. Power decoupling

Power decoupling, in case of the differential buck inverter shown in Fig. 1, is to control C_1 and C_2 to produce the required ripple power to the ac grid, rather than drawing it from the dc source. To do that, voltages across C_1 and C_2 must have a CM second harmonic term, in addition to those given in (1). Such addition has previously been referred to as waveform control in [5], as.

$$v_{c1} = V_d + kV_g \sin\omega t + B\sin(2\omega t + \varphi), \quad v_{c2} = V_d + (k-1)V_g \sin\omega t + B\sin(2\omega t + \varphi) \quad (3)$$

where B and φ are amplitude and arbitrary phase of the common second harmonic component, and k is a constant commonly set to 0.5. Assuming the inverter to be lossless, input power to each elementary converter can be set equal to its output power，leads to the total input current can be

$$i_{dc} = \frac{I_g V_g}{2V_{dc}}\cos\alpha + \frac{\left[C_1 k + C_2(k-1)\right]}{V_{dc}}\left[\omega V_d V_g \cos\omega t - \frac{1}{2}\omega V_g B\sin(\omega t + \varphi)\right] + \frac{2\omega(C_1 + C_2)V_d B}{V_{dc}}\cos(2\omega t + \varphi)$$

$$+ \frac{\omega\left[C_1 k^2 + C_2(k-1)^2\right]V_g^2}{2V_{dc}}\sin 2\omega t - \frac{I_g V_g}{2V_{dc}}\cos(2\omega t + \alpha) + \frac{3\omega\left[C_1 k + C_2(k-1)\right]V_g B}{2V_{dc}}\sin(3\omega t + \varphi) + \frac{\omega(C_1 + C_2)B^2}{V_{dc}}\sin(4\omega t + 2\varphi)$$

$$(4)$$

Simplification can next be performed by setting $C_1 = C_2 = C$, which in the literature, is a popular assumption imposed. This assumption is however not realistic as time progresses, even though it may be possible at the initial design stage. Not considering the variation first, the simplified total input current is given in (5) [5], after substituting $C_1 = C_2 = C$ to (4).

$$i_{dc} = \frac{I_g V_g}{2V_{dc}}\cos\alpha + \frac{4\omega C V_d B}{V_{dc}}\cos(2\omega t + \varphi) + \frac{V_g}{2V_{dc}}\sqrt{I_g^2 + \frac{1}{4}\omega^2 C^2 V_g^2 + I_g V_g \omega C \sin\alpha} \cdot \cos(2\omega t - \theta) + \frac{2\omega C B^2}{V_{dc}}\sin(4\omega t + 2\varphi) \quad (5)$$

In (5), the sum of the second and third terms however corresponds to a ripple power oscillating at 2ω. Its prominence can be adjusted by tuning magnitude B and phase φ of the added CM voltage across C_1 and C_2, and can, in fact, be nullified by solving the equality provided in (6). Corresponding expressions obtained for B and φ are given in (7).

$$4\omega C V_d B\cos(2\omega t + \varphi) + \frac{V_g}{2}\sqrt{I_g^2 + \frac{1}{4}\omega^2 C^2 V_g^2 + I_g V_g \omega C \sin\alpha} \cdot \cos(2\omega t - \theta) = 0 \quad (6)$$

$$B = \frac{V_g}{8\omega C V_d}\sqrt{I_g^2 + \frac{1}{4}\omega^2 C^2 V_g^2 + I_g V_g \omega C \sin\alpha}, \quad \tan\varphi = -\tan\theta = \frac{I_g \sin\alpha + \frac{1}{2}\omega C V_g}{I_g \cos\alpha} \quad (7)$$

The intended power decoupling has hence been realized with the second-order power generation transferred from the dc source to C_1 and C_2. The tradeoff is an unintended generation of fourth-order ripple power by the dc source, which in (5), is the fourth term. This fourth-order ripple is usually much smaller, and it can similarly be eliminated by adding another CM voltage at 4ω to (3).

2.3. Control structure of the system

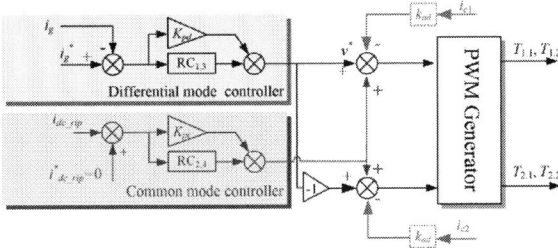

Fig. 2. Overall control structure of the system [6].

To date, many control schemes have been developed for performing power decoupling with most focusing on the computations capacitor voltage reference given in (3) and (7). These schemes are subsequently mentioned as condition and parameter dependent. An alternative approach is thus proposed in [5], whose tracking accuracy has been improved in [4] and [7] by using better tracking controller. This is followed by [6], where the power decoupling scheme has been modified for grid connection. The control objective now is grid current regulation, which means earlier determination of capacitor voltage references cannot be applied directly.

As an example, the scheme from [6] is shown in Fig. 2, where two paralleled control loops can distinctively be identified. The first is the DM loop with Resonant Controllers (RCs) tuned to ω and 3ω for regulating the grid current with unity power factor and zero third harmonic. The latter may flow if there are any residual second-order ripples not fully compensated at the inverter input terminals. The second is the CM loop for removing second and fourth harmonics from the dc input current without using (3) and (7) for computing the capacitor voltage references. It is therefore for power decoupling by first extracting ripple component i_{dc_rip} from the input dc current, before regulating it to zero by two RCs tuned to 2ω and 4ω.

3. The effects of capacitance mismatch

Despite that Fig. 2 is thus to avoid such sensitivity towards parameter variation, this schemes and many existing schemes [4] and [7] for differential inverters have assumed $C_1 = C_2 = C$, which will gradually drift as time progresses. In the more realistic case of $C_1 \neq C_2$, some characteristic harmonics will generated at the dc side of the differential inverter, and they can clearly be seen with (8)

$$i_{dc} = \frac{I_g V_g}{2V_{dc}}\cos\alpha + \frac{C(m-1)}{2V_{dc}}\left[\omega V_d V_g \cos\omega t - \frac{1}{2}\omega V_g B \sin(\omega t + \varphi)\right] + \frac{2\omega C(m+1)V_d B}{V_{dc}}\cos(2\omega t + \varphi)$$

$$+ \frac{\omega C(m+1)V_g^2}{8V_{dc}}\sin 2\omega t - \frac{I_g V_g}{2V_{dc}}\cos(2\omega t + \alpha) + \frac{3\omega C(m-1)V_g B}{4V_{dc}}\sin(3\omega t + \varphi) + \frac{\omega C(m+1)B^2}{V_{dc}}\sin(4\omega t + 2\varphi) \tag{8}$$

With the control shown in Fig. 2 (or those in [4] and [7]) next applied, the second-order ripple in (8) will gradually reduce to zero in the steady state. When it happens, the second-order capacitor voltage magnitude and phase can be determined by setting the second harmonic component in (8) to zero. The resulting expressions obtained are provided as follows.

$$B = \frac{V_g}{4\omega C(m+1)V_d}\sqrt{I_g^2 + \frac{1}{16}\omega^2 C^2(m+1)^2 V_g^2 + \frac{1}{2}I_g V_g \omega C(m+1)\sin\alpha} \;,\; \tan\varphi = \frac{I_g \sin\alpha + \frac{1}{4}\omega C(m+1)V_g}{I_g \cos\alpha} \tag{9}$$

Substituting (9) to (8) and reorganizing, magnitudes of the first and third harmonics at the dc side of the inverter can be determined as follows, where $A_{\omega,dc}$ and $A_{3\omega,dc}$ are their respective magnitude notations.

$$A_{\omega,dc} = \frac{1}{2}C(m-1)\omega V_g \sqrt{V_d^2 + \frac{B^2}{4}} \;,\; A_{3\omega,dc} = \frac{3(m-1)V_g^2}{16(m+1)V_d}\sqrt{I_g^2 + \frac{1}{16}\omega^2 C^2(m+1)^2 V_g^2 + \frac{1}{2}I_g V_g \omega C(m+1)\sin\alpha} \tag{10}$$

Substituting values from Table I, (9) and (10) are plotted in Fig. 3 in terms of m. As expected, when m = 1 or $C_1 = C_2$, magnitudes of the first and third harmonics at the dc side are zero. In contrast, when m decreases to 0.8 or increases to 1.2 (±20% variation of C_1), magnitudes of these components increase to their respective maximums of around 0.3 A and 0.12 A, which if not removed, will prevent the dc current from being ripple-free, even with power decoupling enabled.

The distorted dc current will next cause second harmonic to flow to the grid, which unfortunately, cannot be eliminated by the control scheme shown in Fig. 2. For illustration, Fig. 4 is referred to, where an average model of the differential buck inverter is shown. The model uses v_{inv1} and v_{inv2} for notating the immediate terminal voltages of the inverter, and v_{c1} and v_{c2} for notating voltages across the two ac capacitors. If control scheme in Fig. 2 is now applied, only its CM loop will introduce second harmonics to v_{inv1} and v_{inv2}. These harmonics, notated as $v_{inv1,2\omega}$ and $v_{inv2,2\omega}$, are always equal since they are from the same CM controller in Fig. 2. If the LCL filters, including C_1 and C_2, are next assumed similar, second harmonic currents drawn by them, and voltage drops across L_1 and L_2 will be equal. Second harmonic voltages $v_{c1,2\omega}$ and $v_{c2,2\omega}$ across C_1 and C_2 will hence also be equal, like demanded in (3). These harmonics cancel when the capacitor voltages subtract to give the DM capacitor voltage.

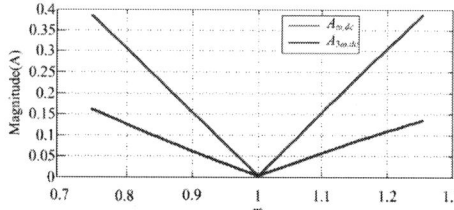

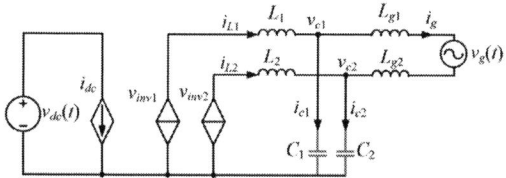

Fig. 3. Variations of $A_{\omega,dc}$ and $A_{3\omega,dc}$ with capacitance mismatch factor m.

Fig. 4. Average model of differential buck inverter.

The same study can be repeated with non-matching LCL parameters like $C_1 \neq C_2$, $L_1 = L_2 = L$ and $L_{g1} = L_{g2} = L_g$ considered as an example. If the same control scheme in Fig. 2 is applied, the same second harmonic voltages ($v_{inv1,2\omega} = v_{inv2,2\omega}$) will again be inserted by the CM loop, which despite being equal, will supply different second harmonic currents to the two non-matching LCL filters. Second harmonic voltages $v_{c1,2\omega}$ and $v_{c2,2\omega}$ across the two capacitors will then differ by $\Delta v_{c,2\omega}$, as

$$\Delta v_{c,2\omega} = v_{c1,2\omega} - v_{c2,2\omega} = \frac{v_{inv1,2\omega}}{1+(2\omega)^2 LC_1} - \frac{v_{inv2,2\omega}}{1+(2\omega)^2 LC_2} \tag{11}$$

Assuming the grid to be ideal with only fundamental voltage across it, the second grid harmonic current $i_{g,2\omega}$ can eventually be derived as:

$$i_{g,2\omega} = \frac{\Delta v_{c,2\omega}}{4\omega L_g} = v_{inv1,2\omega}\frac{\omega LC - \omega L \cdot mC}{L_g \cdot \left[1+(2\omega)^2 L \cdot mC\right]\left[1+(2\omega)^2 LC\right]} \tag{12}$$

This second harmonic current must be mitigated since most grid standards would require the second to tenth even harmonics to be capped. For IEEE 1547, the limit has been set to 1% [11], which unfortunately, may not always be met by the control scheme shown in Fig. 2. To illustrate, (12) and I_g = 5 A are used to plot Fig. 5, from which the ratio of $|i_{g,2\omega}| / I_g$ has been noted to easily exceed 1%, so long as $m \neq 1$. The control scheme in Fig. 2 is therefore not satisfactory, so long as a slight mismatch between C_1 and C_2 (or L_1 and L_2) exists. The main reason is attributed to its ($v_{inv1,2\omega} = v_{inv2,2\omega}$) condition imposed by its CM loop, which with unequal filter parameters, will lead to $v_{c1,2\omega} \neq v_{c2,2\omega}$. This is against (3), which requires $v_{c1,2\omega} = v_{c2,2\omega}$, and hence $v_{inv1,2\omega} \neq v_{inv2,2\omega}$ under parameter mismatch.

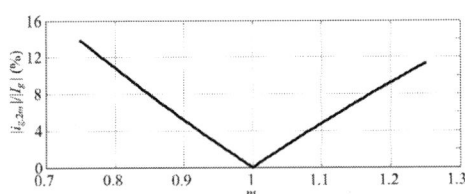

Fig. 5. Variation of $|i_{g,2\omega}| / I_g$ with capacitance mismatch factor m.

TABLE I. PARAMETERS OF DIFFERENTIAL INVERTER

Nominal power	P_n	800 VA
Switching frequency	f_{sw}	10 kHz
Line frequency	f_n	50 Hz
Line voltage	$V_m \approx V_g$	325 V
DC-link voltage	V_{dc}	500 V
Inverter-side inductors	L_1 and L_2	1.5 mH
Grid-side inductors	L_{g1} and L_{g2}	0.6 mH
Filter capacitors	C_1 and C_2	$C_1 = mC$ $C_2 = C = 60$ µF

4. Mitigation the side effect of capacitance mismatch

In this work, we particularly focus on the effects of capacitance mismatch, and identified the performance of both dc current and ac grid injection current will be degraded. Hence, apart from power decoupling, there are two issues arise when the capacitance mismatched, one is the ω and 3ω current in the dc-link; the other is the 2ω current injected into grid.

4.1. ω and 3ω current in dc-link

Like introduce a double-line frequency components voltage on both capacitors to mitigate the 2ω ripple current in dc-link, an addition fundamental voltage $nV_m\sin(\omega t+\delta)$ is add on both capacitor. Where n is the ratio of the amplitude of the additional fundamental voltage against the amplitude of the differential voltage of capacitor V_m. By doing so, the emerged ω and 3ω ripple current will be compensated, and the capacitor voltage become

$$\begin{cases} v_{c1} = V_d + 0.5V_m\sin(\omega t + \delta) + B\sin(2\omega t + \varphi) + nV_m\sin(\omega t + \delta) \\ v_{c2} = V_d - 0.5V_m\sin(\omega t + \delta) + B\sin(2\omega t + \varphi) + nV_m\sin(\omega t + \delta) \end{cases} \tag{13}$$

Refer back to (3), the relationship between k and n is $k=0.5+n$. Since the eq. (4) give out the general form of dc current, the zero ω and 3ω current will be achieved in case of $C_1k - C_2(k-1)=0$. Assuming the capacitor C_1 has a parameter drifting, and its capacitance become $C_1 =mC$, result in the ratio of addition common mode fundamental voltage n will be $0.5*(1-m)/(1+m)$. By modifying the capacitor voltage as (13), the ω and 3ω current in dc-link will be eliminated simultaneously.

4.2. 2ω current in grid current

As indicated in eq.(12), in case of $v_{a,2\omega} = v_{b,2\omega}$, the 2ω harmonic current appear in grid current when mismatch of two capacitors. The 2ω harmonic current can be eliminated by precisely control capacitor voltage to (11). Hence, the 2ω component of inverter output voltage $v_{a,2\omega}$ and $v_{b,2\omega}$ become

$$v_{a,2\omega} = B\sin(2\omega t + \varphi) - 4\omega^2 LC_1 B\sin(2\omega t + \varphi), \quad v_{b,2\omega} = B\sin(2\omega t + \varphi) - 4\omega^2 LC_2 B\sin(2\omega t + \varphi) \tag{14}$$

And the 2ω component of invert side inductor current $i_{L1,2\omega}$ and $i_{L2,2\omega}$ will be

$$i_{L1,2\omega} = 2\omega C_1 B\sin(2\omega t + \varphi), \quad i_{L2,2\omega} = 2\omega C_2 B\sin(2\omega t + \varphi) \tag{15}$$

Since $C_1 \neq C_2$, a 2ω voltage will appear in differential mode inverter output voltage, and it can be calculated as

$$\Delta V_{ab,2\omega} = v_{a,2\omega} - v_{a2,2\omega} = 4\omega^2 LC(m-1)B\cos(2\omega + \varphi) \tag{16}$$

The common mode and differential mode inverter side inductor current at frequency 2ω are given as

$$i_{Ldiff,2\omega} = \omega(C_1 - C_2)B\sin(2\omega t + \varphi), \quad i_{Lcom,2\omega} = \omega(C_1 + C_2)B\sin(2\omega t + \varphi) \tag{17}$$

Thus, to achieve no second order distortion in grid current, $v_{c1,2\omega} = v_{c2,2\omega}$ must be guaranteed, leading to a seconder order voltage as shown in (16) have to introduced in differential mode output of inverter. But the equation (16) clearly is highly related to the mismatch factor m, which is an unknown parameter, make it is hard to realize control $v_{c1,2\omega} = v_{c2,2\omega}$ precisely. Since the differential mode control loop shown in Fig. 2 is a current control loop, which is design to make the feedback current signal to follow its reference i^*. Thus, if differential mode inverter side current i_{Ldiff} is employed as feedback variable, $i_{Ldiff, 2\omega}$ should be calculated and then plus with reference i^* to get the inductor current reference. However, $i_{Ldiff, 2\omega}$ still depended on the unknown mismatch factor m. Another way is using i_g as control variable, because the current reference i^* is a pure sinusoidal at fundamental frequency, hence if the differential mode controller can achieve zero tracking, the 2ω harmonic current in i_g can be reduce to zero.

5. Controller design

As presented before, three mitigation criterions are emphasized, these criterions collectively state that the inverter immediate terminal voltages v_{inv1} and v_{inv2} must have both DM and CM harmonics at ω, 2ω, 3ω and 4ω, in order to achieve satisfactory performance when the capacitances drift. It is also only when v_{inv1} and v_{inv2} have DM and CM harmonics at those frequencies, before the desired capacitor voltages in (13) can be arrived at autonomously. Unfortunately, this requirement cannot be met by the control scheme in Fig. 2, which generates only CM harmonics at 2ω and 4ω, and DM harmonics at ω and 3ω for v_{inv1} and v_{inv2}.

An improved scheme for the single-phase differential inverter is thus necessary, and is presented in Fig. 6 with both DM and CM loops. It can generate additional DM harmonics at 2ω and 4ω, and CM harmonics at ω and 3ω for v_{inv1} and v_{inv2}, before using them as references for pulse-width modulation (PWM). In terms of controllers, it means additional DM RCs at 2ω and 4ω, and additional CM RCs at ω and 3ω for producing the necessary v_{inv1} and v_{inv2}. These are the differences noted between Fig. 2 and Fig. 6, where the full expressions for the DM and CM controllers in Fig. 6 are provided as follows [12].

$$G_{PR\_d}(s) = k_{pd} + \sum_{h=1,2,3,4} \frac{k_{id}s}{s^2 + (h\omega)^2} \quad , \quad G_{PR\_c}(s) = k_{pc} + \sum_{h=1,2,3,4} \frac{k_{ic}s}{s^2 + (h\omega)^2} \tag{18}$$

where k_{pd}, k_{id}, k_{pc} and k_{ic} are controller gains, respectively. Additional details about the control loops are provided as follows.

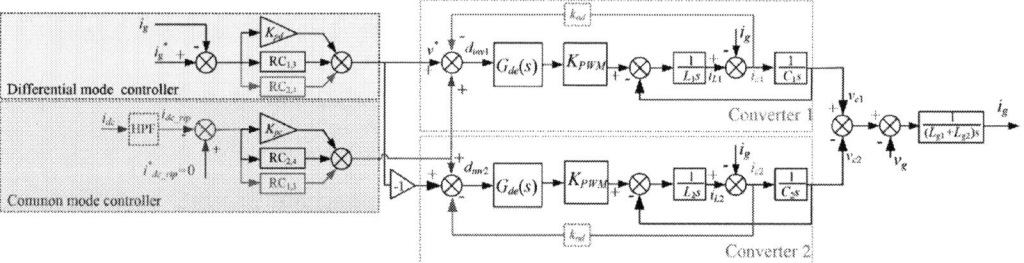

Fig. 6. Proposed control scheme with more comprehensive DM and CM signal generation.

5.1. DM grid current loop

The DM loop in Fig. 6 is for regulating the grid current, which is regulated by a RC tuned to ω. References for the other harmonics regulated by RCs tuned to 2ω to 4ω are simply all zero, rather than computed from (15) if the capacitor voltages are regulated. This is, no doubt, an advantage of grid current regulation, which depending on the ratio of LCL resonance frequency f_{res} to sampling frequency f_s, may require additional active damping. The critical frequency ratio for comparison is 1/6, if the typical time delay of $1.5T_s$ ($T_s = 1 / f_s$) is assumed [13]. In other words, if $f_{res}/f_s < 1/6$, active damping is mandatory, which will most likely, be the case for differential inverter, whose capacitances must be large for storing the ripple power. The larger capacitances then give rise to a lower f_{res}, which according to values provided in Table I, is equal to 839 Hz ($< f_s/6$).

Explicit active damping has therefore been shown in Fig. 6 by feeding back the capacitor currents through proportional gains k_{ad} for emulating virtual resistances across the two capacitances. This method is presently the simplest and most widely adopted. It will hence not be described further. Instead, it should be emphasized that the second option of feeding back the DM inverter current i_L (= $i_{L1} - i_{L2}$ in Fig. 1), rather than the grid current i_g, should not be encouraged when $C_1 \neq C_2$. The reason is that the DM inverter voltage v_{inv} (= $v_{inv1} - v_{inv2}$) and DM capacitor voltage v_c (= $v_{c1} - v_{c2}$) carry different harmonics, which will hence not be cancelled. The un-cancelled harmonics appear across inductances L_1 and L_2 in Fig. 1, which then cause i_L to carry significant harmonics. These harmonics depend on the extent of capacitance mismatch, and are hence not readily determinable, if the capacitances are not measured frequently. On the other hand, if i_L is controlled to track a fundamental sinusoidal reference, harmonics will appear in i_g, making it tough to meet grid standards.

5.2. CM dc current loop

Without power decoupling and with $C_1 \neq C_2$, Table I emphasizes that the inverter dc input current will carry first and third harmonics, in addition to the usual second and fourth harmonics. These harmonics must all be nullified, before a ripple-free dc input current can be achieved. The CM loop is therefore realized in Fig. 6 by measuring the dc input current, extracting its harmonics, and feeding them through RCs tuned to ω to 4ω. No explicit references have been indicated in the figure because they are all zero in case of nullifying harmonics. The CM loop is therefore parameter independent like the DM loop, which also does not rely of explicit reference computation.

6. Experiment results

Experimental testing was performed using a 2.2 kW Danfoss inverter with LCL filter parameters provided in Table I. The inverter is also tied to the 230-V, 50-Hz grid through an isolation transformer, whose total leakage inductance of 1.2 mH formed the grid-side inductance ($L_{g1} + L_{g2}$) of the LCL filters. The inverter-side inductances L_1 and L_2 are, on the other hand, chosen based on a peak-to-peak ripple ΔI_{Lx} of 40% of its rated current. The inductances computed are then $L_1 = L_2 = 1.4$ mH, which because of availability in the laboratory, is chosen as 1.5 mH instead. The inverter control is next implemented with a dSPACE DS1007 system complemented by externally built voltage and current sensors. The capacitances of the LCL filters have also been intentionally set to $C_1 = 1.2C_2$ for testing a 20% drift of C_1. Parameter m is therefore 1.2.

Fig.7 shows the experiment result by using the power decoupling scheme in Fig. 2 in case of $C_1 = 1.2C_2$. It can be seen that the capacitor voltages v_{c1} and v_{c2} have been intentionally distorted by second harmonic in accordance to (3). The differential capacitor voltage v_c has however remained undistorted, which certainly, is anticipated. The spectrum in Fig. 7(b) has also confirmed the

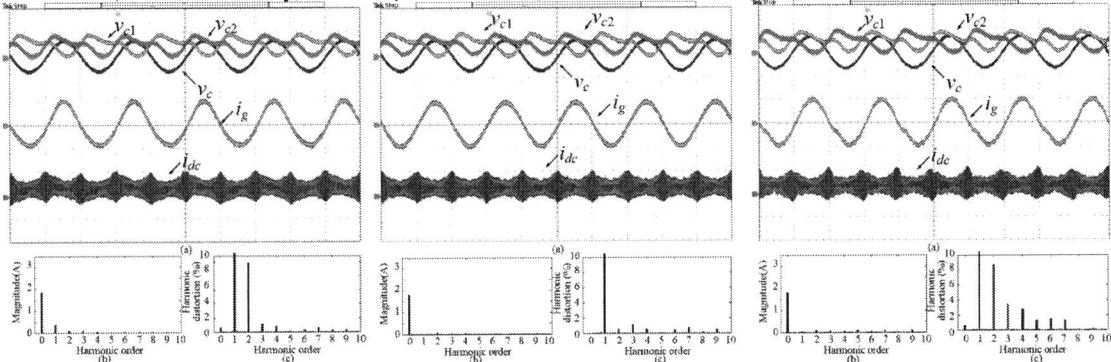

Fig. 7. Experiment steady-state results with the power decoupling control under the condition (C_1=1.2C; C_2=C), dc input current (i_{dc}: 5 A/div), voltage of capacitor C_1 and C_2, and their difference v_c (v_{c1}, v_{c2}, v_c: 500 V/div) and grid current (i_g: 5 A/div): (a) current and voltage of AC and DC side; (b) Spectrum of dc current i_{dc}; (c) Spectrum of grid current i_g

Fig. 8. Experiment steady-state results with the power decoupling and ac&dc side harmonic compensation control under the condition (C_1=1.2C; C_2=C) , dc input current (i_{dc}: 5 A/div), voltage of capacitor C_1 and C_2, and their difference v_c (v_{c1}, v_{c2}, v_c: 500 V/div) and grid current (i_g: 5 A/div) ; Grid current feedback control. (a) current and voltage of AC and DC side; (b) Spectrum of dc current i_{dc}; (c) Spectrum of grid current i_g

Fig. 9. Experiment steady-state results with the power decoupling and ac&dc side harmonic compensation control under the condition (C_1=1.2C; C_2=C) , dc input current (i_{dc}: 5 A/div), voltage of capacitor C_1 and C_2, and their difference v_c (v_{c1}, v_{c2}, v_c: 500 V/div) and grid current (i_g: 5 A/div); Inverter current feedback control. (a) current and voltage of AC and DC side; (b) Spectrum of dc current i_{dc}; (c) Spectrum of grid current i_g

© VDE VERLAG GMBH · Berlin · Offenbach

reduction of second harmonic in the dc input current, but not its first harmonic introduced by mismatch. Moreover, the grid current in Fig. 7(c) is found to contain sizable second harmonic, because of reasons provided in Section III. The existing scheme in Fig. 2 is therefore not satisfactory since it can hardly keep all even grid harmonics below 1%.

For better dc and ac conditioning, the proposed scheme in Fig. 6 with grid current feedback is thus recommended. Its results obtained are shown in Fig. 8, where as anticipated, all harmonics in the dc current have almost been nullified in Fig. 8(b). The second harmonic in the grid current has also been suppressed to 0.49% within the limit of 1% demanded by IEEE 1547. This performance is however possible with grid current feedback only. To strengthen this claim, the experiment is repeated with the inverter DM current fed back for tracking a sinusoidal reference. No harmonic reference has been added since it depends on the amount of capacitance mismatch, and is hence not practically realizable. The results obtained are shown in Fig. 9(a) to (c), where the spectrum in Fig. 9(c) shows that the grid current is significantly distorted, even though the dc input current in Fig. 9(b) is almost ripple-free. To some extent, this is expected since the grid current is not directly controlled. Inverter current feedback is therefore not generic for differential inverter control, since it does not work properly when the capacitances drift.

7. Conclusion

In this paper, the effects of capacitance mismatch experienced by single-phase buck differential inverter are explored. The complications found are additional first and third harmonics in the dc input current, as a result, the second and fourth harmonic currents are detected in the grid current, which if not mitigated, may not meet grid standards. It is therefore necessary to eliminate harmonic content at the dc input, which existing control schemes can only satisfy partially. Thus, an improved scheme with grid current feedback is therefore proposed, and tested experimentally to produce ripple-free dc current and fundamental ac grid current only. The same performance cannot be achieved by feeding back the inverter DM current, which carries harmonics that are hard to be determined. The proposed scheme may additionally resolve problems related to a distorted grid, but even if the grid is ideal, the proposed scheme must be implemented to address realistic mismatches that can occur in the differential inverter. The proposed scheme is hence more generic than existing schemes which function only with perfectly matching capacitances.

Reference

[1] B. N. Singh, A. Chandra, K. Al-Haddad, A. Pandey, and D. P. Kothari, "A review of single-phase improved power quality AC-DC converters," *IEEE Trans. Ind. Electron.*, vol. 50, no. 5, pp. 962-981, Oct. 2003

[2] M. Su, X. Long, Y. Sun, and J. Yang, "An active power decoupling method for single-phase AC/DC converters," *IEEE Trans. Ind. Informat.*, vol. 10, no. 1, pp. 461-468, Jan. 2014.

[3] H. Li, K. Zhang, H. Zhao, S. Fan, and J. Xiong, "Active power decoupling for high-power single-phase PWM rectifiers," *IEEE Trans. Power Electron.*, vol. 28, no. 3, pp. 1308-1319, Mar. 2013.

[4] I. Serban, "Power Decoupling Method for Single-Phase H-bridge Inverters with no Additional Power Electronics," *IEEE Trans. Ind. Electron.*, to be published.

[5] G-R. Zhu, S-C. Tan, Y. Chen, and C. K. Tse, "Mitigation of low-frequency current ripple in fuel-cell inverter systems through waveform control," *IEEE Trans. Power Electron.*, vol. 28, no. 2, pp. 779-792, Feb. 2013.

[6] W. Yao, X. Wang , X. Zhang, Y. Tang, P.C. Loh, and F. Blaabjerg, "A Unified Active Damping Control for Single-Phase Differential Mode Buck Inverter with *LCL*-Filter," *in Proc. IEEE PEDG 2015 Conf.*, pp. 7–14, Jun. 22-25, 2015.

[7] Y. Tang, W. Yao, H. Wang, P. C. Loh, and F. Blaabjerg, "Transformerless Photovoltaic Inverters with Leakage Current and Pulsating Power Elimination," *in Proc. IEEE ECCE-Asia 2015 Conf.*, Jun. 1-5, 2015.

[8] Z. Li, H. Li, F. Lin, Y. Chen, D. Liu, B. Wang, H. Li, and Q. Zhang., "Lifetime investigation and prediction of metallized polypropylene film capacitors," *Microelectron.*, vol. 53, no. 12, pp. 1962–1967, Jun. 2013. "

[9] *"Film Capacitors: Metallized Polyester Film Capacitors (MKT),"* EPCOS film capacitor datasheet.

[10] *"Film Capacitors: Metallized Polypropylene Film Capacitors (MKP),"* EPCOS film capacitor datasheet.

[11] *IEEE Standard for Interconnecting Distributed Resources with Electric Power Systems*, IEEE Standard 1547.2, 2008.

[12] R. Teodorescu, F. Blaabjerg, M. Liserre, and P. C. Loh, "Proportional resonant controllers and filters for grid connected voltage source converters," *Proc. Inst. Elect. Eng.-Elect. Power Appl.*, vol. 153, no. 5, pp. 750-762, Sep. 2006.

[13] S. Parker, B. P. McGrath, and D. G. Holmes, "Region of active damping control for *LCL* filters," *IEEE Trans. Ind. Appl.*, vol. 50, no. 1, pp. 424-432, Jan./Feb. 2014.

© VDE VERLAG GMBH · Berlin · Offenbach

For information on the presentation

Technical Approach: Interleaved, Folding, Interpolating Dual-Path Adiabatic Autotransformer Based Power Converter

please contact the speaker directly:

Dr. John Wood

Silicon Contact
Engineering
9 JJ Thomson Avenue
CB3 0FA Cambridge
United Kingdom

E-Mail: john.wood@siliconcontact.com

Design and Performance Evaluation of a Three Phase AC Power Source with Virtual Impedance for Validation of Grid Connected Components

Peter Jonke[1], Johannes Stöckl[1], Hans Ertl[2]

[1] Austrian Institute of Technology, Electric Energy Systems,
Giefinggasse 2, A-1210 Wien, Vienna, Austria
[2] Vienna Universtiy of Technology, Institute of Energy Systems an Electrical Drives,
Gusshausstrasse 27-29, A-1040 Wien, Vienna, Austria

Abstract

The paper reports design and performance evaluation of a three-phase AC power source which is mainly used for the development, validation and testing of grid connected converter system and components. The main objective of the designed AC power source is the controllability of the output (grid) voltage amplitude and frequency, harmonic components as well as of the grid impedance characteristic. The small signal bandwidth of the proposed system is specified up to 2 kHz. To fulfill this requirement at very low output noise voltage levels, the target switching frequency has to be in the region of 200 kHz. For achieving acceptable efficiency 1200V silicon carbide (SiC) MOSFETs are used. The paper presents a comparison of estimated and measured semiconductor losses for discrete as well as module packaging and discusses the implementation of the gate driver stage, of the DC link and of the EMI output filter.

1 Introduction

Increasing penetration of smart grid components for distributed generation requires advanced testing procedures. Hence high sophisticated test equipment is needed and especially high performance AC power sources (AC simulators) for power grid emulation. The adjustment of different grid parameters such as voltage amplitude, nominal frequency, harmonics are mandatory for testing purposes. Furthermore, emulation of a power grid also includes specifically defined grid impedance characteristic, which is typically implemented using additional power choke coils and resistors at the AC simulator output. As an alternative, this work shows the design of an AC simulator with integrated impedance emulation, which is based on output current measurement and proper control of the output voltage. Due to efficiency reasons, switched mode amplifiers are preferred compared to linear amplifiers. As basic analyses show [1] switching and sampling frequencies in the region of up to 200 kHz are required to achieve the intended dynamic behavior (e.g., impedance emulation up to 2 kHz) at low output voltage noise levels. Several approaches for achieving such high effective switching frequencies are known. With standard Si semiconductors (e.g. IGBT), interleaved topologies could be used to achieve such high switching frequencies. The disadvantage of such an approach however is the high amount of semiconductors and gate drivers and necessary balancing measures. This paper therefore presents a different approach based on silicon carbide (SiC) MOSFETs, which enable fast switching at comparatively low switching losses.

The paper is organized as follows: In Section 2 specifications of the AC power source and used topology are presented. The design concept as well as analyses of the switching and conduction losses is described in Section 3. Measurement results of the power stage prototype and a comparison of module and discrete packaging are presented in Section 4. Finally, discussion of results and impacts to the further development are discussed in Section 5.

2 Specification and Topology

As described before, an advanced testing of smart grid components requires also considering grid impedance characteristics, usually implemented using power choke coils and power resistors. In contrast to such a bulky, expensive and non-flexible solution, the proposed power source emulates the required output impedance characteristic exclusively by its control avoiding any additional power components. The basic concept for this is shown in Figure 1, where the voltage drop and voltage phase shift of the (virtual) output impedance (R_V, L_V) is calculated based on the actual output current i "correcting" the voltage control of the amplifier.

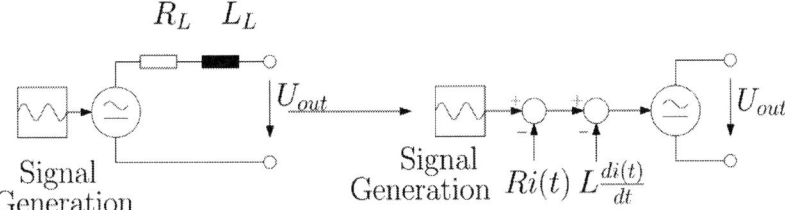

Figure 1 Equivalent circuit of a) conventional grid emulators b) the concept of the AC power source with virtual impedance [1]

Figure 2 shows the schematic of the implemented hardware. The voltage source converter (VSC) is based on the three phase two –level SiC module fed by a DC-Link (film capacitor array) with midpoint connection and nominal apparent power of 8.8 kVA [1]. The systems output filter is implemented using three non-coupled inductors and in total six filtering capacitors. The advantage of this topology is that each phase can be controlled independently by a control loop without any cross coupling effects. The damping of the filter resonance is implemented by digital control based on capacitor current feedback (active damping). A dedicated common-mode filter at the output side is not used (the three filter inductors in this configuration cover differential- as well as common-mode voltage components).

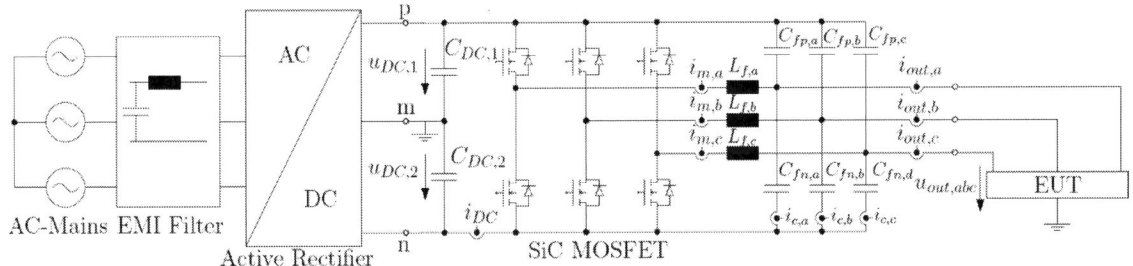

Figure 2 Schematic of SiC MOSFET based Three Phase AC Power Source with Virtual Impedance

Table 1 AC Power Source Specification Summary

Parameter	Value
Phases	3P+N
Operation Mode	4-Quadrant
Nominal output power	8.8 kVA
Nominal output voltage	230 V (L-N)
Nominal DC-link voltage	900 V
Nominal Frequency	50 Hz
Small-signal Bandwidth	2 kHz
Switching frequency	200 kHz
Sample time	5 µs
Emulated Resistance	0 Ω -1 Ω
Emulated Inductance	0 mH – 5 mH

The analysis performed in [1] shows that a signal bandwidth of the power amplifier of about 2 kHz is needed that the desired impedance emulation can be achieved. Due to this an output filter of ≈20 kHz cut-off frequency is required such that the filter does not essentially affect the impedance emulation. On the other hand to achieve a sufficient switching frequency ripple a switching frequency up to 200 kHz is required, to meet specified voltage ripple as well as current ripple requirements. A summary of the specification is given in Table 1, where also the emulated resistance and emulated inductance range is defined.

3   Construction Concept

3.1   Power Semiconductor Evaluation

SiC MOSFETs are providing low switching loses, due to fast switching behavior as well as low conduction loses at high blocking voltage rates. For demonstration a bridge leg based on the semiconductor module CCS050M12CM2 (standard six pack package) and alternatively using discrete MOSFETs C2M0080120D (TO247 package) from CREE were tested with a double pulse setup to characterize switching behavior. Table 2 shows a comparison of the evaluated SiC MOSFETS.

Table 2 Comparison of CCS050M12CM2 and C2M0080120D

Parameter	CCS050M12CM2	C2M0080120D
Voltage Rating	1200 V	1200 V
Current Rating	59 A	36 A
Power Dissipation	312 W	192 W
On State Resistance	25 mΩ	80 mΩ
Turn-On Switching Energy (800 V; 20 A)	900µJ	265 µJ
Turn-Off Switching Energy (800 V; 20 A)	300 µJ	135 µJ

The measurement results of both devices are described in 4 Experimental Results. However, due to the fact that the construction as well as cooling of an (isolated) module-based design is advantageous, the module based concept is preferred.

3.2   Gate Drive

To implement fast switching (as a condition for achieving a switching frequency of 200 kHz) low impedance gate driver circuits according to [2], [3] are used.
Figure 3 shows the designed driver circuit, which allows fast switching during minimizing switching losses E_{on} and E_{off}.
The Silicon Labs Si8233 gate drive IC including two totally isolated drivers for a bridge leg allows fast switching at 4 A output peak driver currents and a maximum propagation delay of 60 ns being sufficient for this application. The effective gate resistor during turn-on was set to $R_{g,2} = 10\ \Omega$ and the dominating gate resistor during turn-off was set to $R_{g,2} = 5\ \Omega$ to consider

© VDE VERLAG GMBH · Berlin · Offenbach

the different driver voltage level (≈20 V for turn-on, ≈ -5 V for turn-off, provided by zener diode ZD.

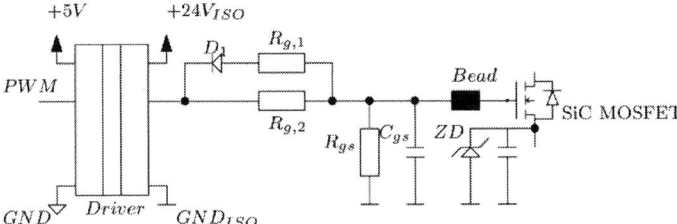

Figure 3 Schematic of designed gate drive circuit

An isolated voltage supply for each drive circuit is provided by a small 24 V DC/DC converter. The capacitor C_{gs} and the resistor R_{gs} will reduce ringing on the gate pins according to [3].

3.3 MOSFET Losses

The estimated switching losses for the relevant switching parameters DC-Link voltage level V_{DC}=900 V, gate resistor R_g=10 Ω, drain current I_D=13 A (peak) are calculated in the following: According to [4] and the data sheet specifications of the SiC module for V_{DC}=800 V, R_g=20 Ω the estimated turn-on switching energy is scaled to E_{on}=0,376 mJ. Similarly, the estimated turn-off switching energy is scaled to E_{off}=0,0623 mJ. With this, the resulting switching losses are calculated as

$$P_{SW} = (E_{on}+E_{off}) \cdot f_{SW} = 0.44 \text{ mJ} \cdot 200 \text{ kHz} = 87,79 \text{ W}$$

The conduction losses which are calculated per bridge leg by

$$P_{cond} = R_{DS,on} \cdot I_{rms}^2 = 25 m\Omega \cdot (9,02A)^2 = 2,03W$$

are almost negligible related to the switching losses. In total semiconductor module losses of about 282 W will appear for the three-phase system (i.e., ≈3,13% losses).
The same procedure is applied for the discrete SiC MOSFET where the turn-on switching energy is scaled to E_{on}=0,201 mJ and turn-off switching energy is scaled to E_{off} = 0,0753 mJ. Table 3 shows a comparison of estimated losses.

Table 3 Comparison of estimated semiconductor losses

Parameter	CCS050M12CM2	C2M0080120D
P_{SW}	87,79 W	55,26 W
P_{cond}	2,03 W	6,51 W
Overall Losses	282 W (3,13%)	204 W (2,33%)

For the thermal management of the laboratory prototype a cooling aggregate with axial fan and air flow chamber with thermal resistance of R_{th}=0.04 K/W was selected. For advanced operations, however, a liquid cooling with lower thermal resistance will be required.

3.4 Output Filter

Table 4 Output Filter Parameters

The output filter of the AC power source is implemented as three individual (decoupled) 2nd-order LC single-stage filters. This gives the advantage of an independent control of each

Parameter	Value
Filter capacitor	265 nF
Filter inductor	335 µH

phase and that an additional common-mode filter is not required (in fact common-mode as well as differential-mode voltage components are covered by filter inductor and will influence its dimensioning).

© VDE VERLAG GMBH · Berlin · Offenbach

Furthermore the design of filter inductors was also evaluated, with different types of core shapes and magnetic materials such as E-core, I-Core and Toroid –Core as well as different core materials regarding size, construction and loses.

For all different inductor types a FEM simulation was carried out using FEMM simulation environment in [6]. The result of E-core and I-Core show a significant amount of fringe flux around the gap of the core. The fringe flux would cause eddy current induction in the inductor winding, which would result in higher losses. Therefore the knowledge of the appearing fringe flux is mandatory to designing the inductor and its coil former. The toroidal powder core with distributed air gap provides the best trade-off between fringe flux, size and construction afford.

3.5 Experimental Prototype

The designed 8.8kVA laboratory prototype system (Figure 4) consists of two parts, the power board (DC-Link, SiC module, AC output filter capacitors) and the control board (dual core controller card, signal conditioning). Exclusively the output filter inductor is not integrated on PCB and has to be connected externally. For the power board a planar arrangement based on 4-layer PCB is used (. Since precise current measurement is essential for the control loops of the AC Power Source, LEM LAX 100-NP current transducers were used showing a bandwidth of 300 kHz.

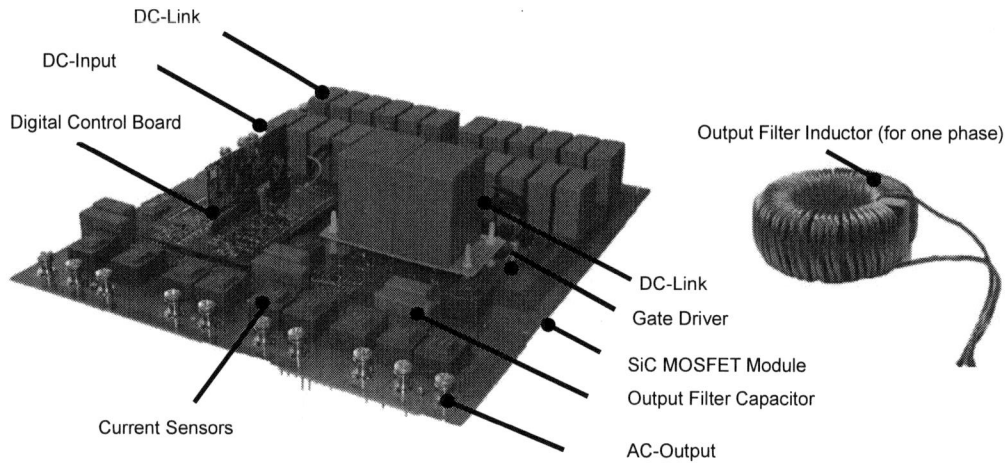

Figure 4 Design of 8.8 kVA three phase AC Power Source with SiC MOSFETs and Single Phase Output Filter Inductor

4 Experimental Results

The following measurements demonstrate the switching behavior of the SiC power module CCS050M12CM2 and the discrete SiC MOSFET C2M0080120D. A two pulse test setup with 335 µH inductor in parallel to the high side switch was used. Figure 5 shows the turn-on and the turn-off switching waveform of SiC power module, where turn-on and turn-off time of 50 ns was achieved. Figure 6 shows the turn-on and the turn-off switching waveform of SiC MOSFET, where faster turn-on and turn-off times about 40 ns were achieved. It can be observed that the SiC module shows significant voltage ringing, obviously caused by the stray inductances of the module case (originally used for IGBT in drive applications).

PCIM Europe 2016, 10 – 12 May 2016, Nuremberg, Germany

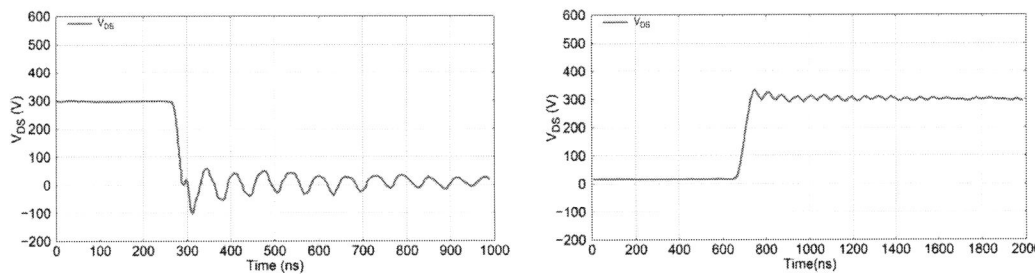

Figure 5 a) SiC Module CCS050M12CM2 turn-on b) turn-off switching waveforms (I_D=10 A)

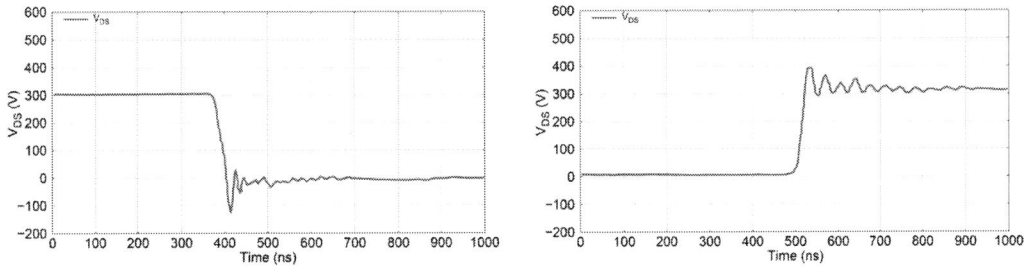

Figure 6 SiC MOSFET C2M0080120D turn-on b) turn-off switching waveforms (I_D=10 A)

For the measurement of the semiconductor loses a double pulse test fixture was used with V_{DC}=900 V and I_D=10 A. Due to the voltage ringing of the SiC module the measured losses are significantly increased compared with estimated losses.

Table 5 Comparison of measured semiconductor losses

Parameter	CCS050M12CM2	C2M0080120D
E_{on}	566 µJ	301 µJ
E_{off}	92,3 µJ	70,4 µJ
P_{SW}	131,6 6W	74,28 W

Figure 7a gives measurements of a 50 Hz output current and voltage signals at resistive load (switching frequency 200 kHz). Figure 7b shows the according inductor current indication a ripple of $I_{pp,ripple}$ = 850 mA.

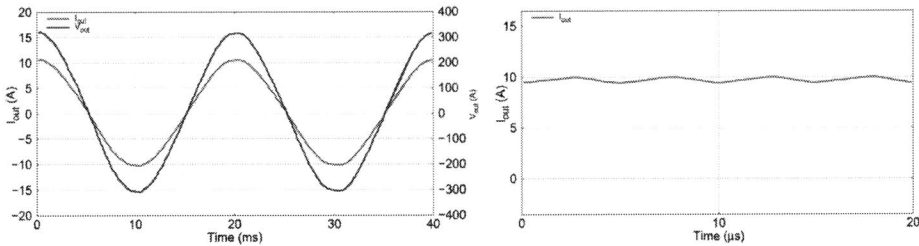

Figure 7 Experimental results of 8.8 kVA/200 kHz AC- Power source using SiC Module CCS050M12CM2 two-level topology. a) Single phase 50 Hz output voltage and output current for testing filter inductors (10A peak). b) Output current ripple and gate signals. The measured ripple is below 10% of peak current.

© VDE VERLAG GMBH · Berlin · Offenbach

5 Conclusion and Outlook

This paper shows the design an construction of a 8 kVA three phase AC power source, which is capable for emulation of a virtual output impedance. The presented prototype operates at a switching frequency of 200 kHz and has a small signal bandwidth of 2 kHz. A comparison of estimated and measured switching losses for the SiC Module CCS050M12CM2 and discrete SiC MOSFET C2M0080120D has shown, that the module packaged semiconductors has a significant voltage ringing. Due to, the switching losses are quite higher compared to the discrete SiC MOSFET and therefore further investigations with discrete packaged SIC MOSFET will be made. Also an implementation of controls for the virtual impedance is subject to ongoing research and will be described in future work.

Acknowledgement

This work was part of an independent research project: Development of a Three Phase AC Power Source with Virtual Output Impedance for Tests of Grid Connected Components supported by Austrian Ministry for Transport Innovation and Technology.

References

[1] P. Jonke, J. Stöckl and H. Ertl, "Concept of a Three Phase AC Power Source with Virtual Output Impedance for Tests of Grid," in *EDST*, Vienna, 2015.

[2] T. Friedli, S. Round, D. Hassler and J. Kolar, "Design and Performance of a 200 kHz All-SiC JFET Current Source Converter," in *IEEE Industry Applications Society Meeting*, 2008.

[3] C. P. Aplications, "CREE MOSFET Evaluation CRD-8FF127P-1," in *Application Note*, 2014.

[4] D. Graovac, M. Pürschel and A. Kiep, "MOSFET Power Losses Calculation Using the Datasheet Parameters," in *Application Note*, 2006.

[5] J. Liu, K. L. Wong, S. Allen and J. Mookken, "Performance Evaluations of Hard-Switching Interleaved DC/DC Boost Converter with New Generation Silicon Carbide MOSFETs," 2013.

[6] M. Fuchs, Filterdesign for a 3-phase multilevel topology AC-Simulator, Wien: FH Technikum, 2015.

[7] CREE Power Applications, "KIT8020-CRD-8FF1217P-1 CREE MOSFET Evaluation Kit Manual," CREE Power Applications, CREE USA, 2014.

[8] CREE Power Applications Inc., "Design Considerations for Designing with Cree SiC Modules Part 1.," CREE Power Applications Inc., Cree, Inc. 4600 Silicon Drive Durhame, 2015.

PCIM Europe 2016, 10 – 12 May 2016, Nuremberg, Germany

Design and Testing of a Modular SiC based Power Block

Maja Harfman Todorovic, GE Research Center, USA, harfmanm@ge.com
Rajib Datta, GE Research Center, USA, Rajib.Datta@ge.com
Ljubisa Stevanovic, GE Research Center, USA, stevanov@ge.com
Xu She, GE Research Center, USA, Xu.She@ge.com
Philip Cioffi, GE Research Center, USA, cioffi@ge.com
Gary Mandrusiak, GE Research Center, USA, mandrusi@ge.com
Brian Rowden, GE Research Center, USA, rowden@ge.com
Paul Szczesny, GE Research Center, USA, szczesny@ge.com
Jian Dai, GE Research Center, USA, daij@ge.com
Tony Frangieh, GE Research Center, USA, frangieh@ge.com

Abstract

The paper describes a modular, power electronics building block based on GE's industry-leading SiC MOSFET technology and vertically integrated design approach to achieve maximum performance in power applications ranging from solar, wind and HEV/EV to Industrial and Traction drive applications. This plug-and-play unit enables customers to launch high-performance products more quickly, and at lower cost versus conventional design approaches. The Power block is comprised of low-inductance SiC modules with high performance gate drives, optimized bus bar with DC capacitors, air or liquid-cooled thermal management, sensors feeding low-level controls and protection, a standardized control interface, and connectivity to GE's Predix engine for data analytics. It uses a simple, two-level topology capable of switching up to 40kHz with a DC link voltage up to 1500V.

1. Introduction

SiC as a semiconductor device opens up a new design space by moving to higher frequency, higher voltage and higher temperature at high power. In today's Si-based LV (i.e. below 1000Vac) systems, IGBTs constitute roughly 10-15% of the converter cost in a MW-scale inverter; the balance of plant including thermal management, controls, filters, breakers, bus-bars constitute the bulk of the cost. Even though as a device SiC is expected to be more expensive than Si, it can have an appreciable impact on the system cost by reducing and simplifying the thermal management system, reducing filter requirements by operating at higher frequency, increasing power capability by operating at higher junction temperature. The impact of all these together leads to higher power density and reduces per unit cost of power conversion. In addition, higher efficiency, compared to IGBT-based solutions, can be easily achieved, particularly at part load.

In order to extract the full performance entitlement of SiC, it is important to have an integrated design approach from the chip to the system. Various interdependent design

© VDE VERLAG GMBH · Berlin · Offenbach

parameters need to be optimized simultaneously and iteratively to enable a SiC based high performance system at lower cost. These include low inductance commutation loops, body diode recovery, gate drive optimization, common mode noise reduction etc. GE's power block attempts to address these challenges with an integrated solution that can be interfaced easily with the rest of the converter system without substantial NRE (non-recurring engineering) cost. The schematic of the power block and the functional block diagram are shown in Fig. 1(a) and 1 (b) respectively. It uses a simple, two-level topology capable of switching up to 40kHz with a DC link voltage up to 1500V. It can be viewed as a 3-port network: (i) Electrical port processing power between DC and AC terminals comprising high performance low-inductance SiC modules with optimized gate drives, bus bar with DC capacitors, (ii) Thermal port comprising heat sink and air or liquid-cooled thermal management, (iii) Control and Communication port comprising voltage and current sensing, switch and bridge-level protection, integrated PWM and optional current regulator, a standardized control interface, and connectivity to GE's Predix engine for data analytics.

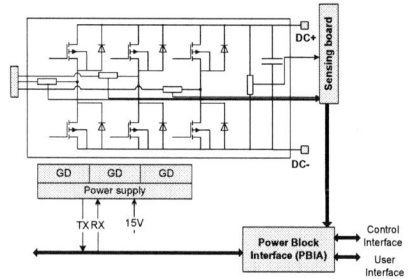

Fig. 1 (a) Schematic of the power block

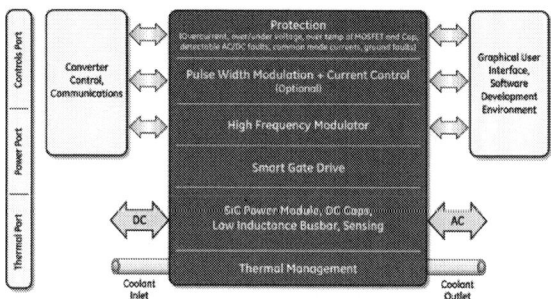

Fig. 1(b) Functional block diagram of the power block

a. Low-inductance modules with Intelligent Gate Drive

At the core of GE's SiC Power block are 1700V, 450A SiC MOSFET modules that are designed in the industry standard housing with patent-pending, low-inductance, laminated internal bus bar. The DC terminal inductance is 4.5nH. The intelligent gate drive, optimized for fast switching, low EMI/EMC and short-circuit protection, enables optimal switching performance of the module. Multiple modules can be switched synchronously with less than 1 ns jitter, resulting in almost perfect dynamic sharing of currents and voltages. This enables easy scaling for higher power system applications. Fig. 2 shows a 3-module configuration of the power block (PB3) with air-cooling. Double pulse tests conducted on PB3 shows almost perfect sharing of currents and voltages between the modules in dynamic and steady state, as in Fig. 3.

Fig. 2 Power block configuration with 3 modules (PB3)

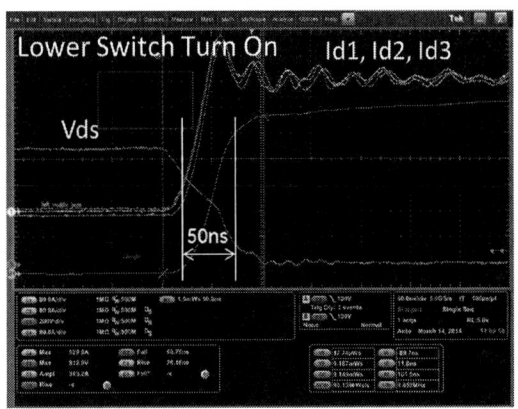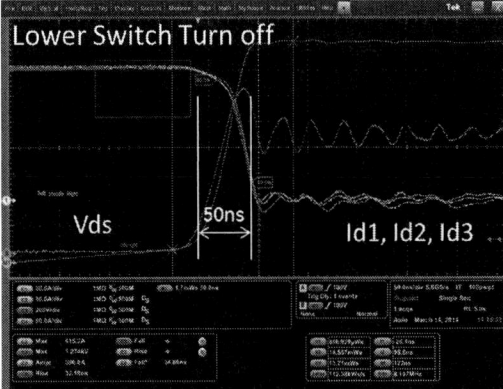

Fig. 3 Double-pulse test results with PB3

It should be noted that the SiC MOSFET offers symmetric reverse conduction with positive gate bias in the 3rd quadrant. This allows for synchronous rectification and eliminates the need for separate anti-parallel diodes within the module. The body diode of the MOSFET only comes into operation during the dead time between the switching of the upper and lower devices. Extensive reliability testing has been conducted to prove the robustness of the body diode for low duty cycle operation.

b. Controls and Protection

GE's SiC Power block employs a local digital controller, plus sensing for current, voltage and temperature, which enables fast protection of the power block. The controller is capable of generating high-frequency pulse width modulation (PWM) up to 40kHz locally, while also allowing quick and seamless interfacing to legacy system-level controllers.

c. SiC User Interface

The well-designed USB interface allows the user to configure and monitor the Power block during operation, and perform event-triggered analysis using capture buffers. It provides the capability to perform common device tests as well as perform remote firmware updates. The SiC user interface can share data with PLECS simulation tools for data analysis, and provides a gateway to GE's cloud-based Predix engine for remote analytics, prognostics and servicing.

2. Power Capability and Efficiency

GE's SiC Power block is designed for both air-cooled and liquid-cooled configurations in PB3 (three modules per phase) and PB6 (six modules per phase) arrangements. Designed for maximum power density in this class of power conversion, the power block can extract twice as much power at 5kHz and four times as much power at 10kHz, compared to state-of-the-art IGBT technology as shown in Fig. 4. This drastic loss reduction yields an unprecedented efficiency of 99.4% at 10kHz under rated conditions; a MW-scale solar inverter powered by GE's SiC power block has been demonstrated to achieve 99% CEC efficiency.

© VDE VERLAG GMBH · Berlin · Offenbach

PCIM Europe 2016, 10 – 12 May 2016, Nuremberg, Germany

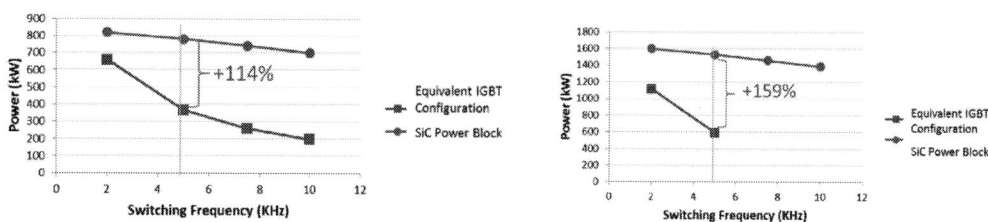

Fig. 4 Comparison of power capability between SiC power block and equivalent IGBT solution

3. Summary

SiC provides a new cost and performance curve by simplification of topologies, higher system performance and higher power density. However, there are challenges in design associated with fast switching transients such as commutation loops, body diode recovery, gate drive optimization, common mode noise reduction etc. Designed for maximum power density and highest efficiency in this class of power conversion, the SiC power block provides an integrated solution that addresses these challenges and enables quick adoption in industrial and renewable application.

© VDE VERLAG GMBH · Berlin · Offenbach

A novel method to simulate the control-to-output transfer function of resonant converters

Julian, Dobusch, julian.dobusch@fau.de
Christian, Oeder, christian.oeder@fau.de
Thomas, Dürbaum, thomas.duerbaum@fau.de
Chair of Electromagnetic Fields, Cauerstraße 7 91058 Erlangen, Germany

Abstract

Nowadays, resonant converters represent a viable option for highly efficient power conversion. Typically, the control of resonant converters uses frequency variation. However, the determination of an accurate control to output transfer function, needed to improve the dynamic behavior, appears to be a challenging task for most designers. Therefore, this paper describes a fast and yet precise method to obtain the desired transfer function by means of simulation (in an early design phase) or by means of measurements (on existing prototypes). This paper applies the method based on a time-discrete model in case of the well-known resonant LLC-converter. In order to demonstrate on one hand the accuracy of the modeling approach and on the other hand the applicability of the same method during measurement of transfer functions, a hardware version of the converter is built. The practical setup shows that by using digital generated gate signals and a modern oscilloscope engineers can obtain the transfer function at their workbench.

Introduction

Figure 1 depicts the resonant LLC-converter - a promising converter topology for highly efficient power conversion due to the possibility of Zero Voltage Switching. Compared to other resonant converters the LLC exhibits advantages in case of wide ranges of input voltage and load [1]. Engineers rely on the easy to use **F**irst **H**armonic **A**pproximation [2] to calculate the static input to output transfer-function for different load situations. Unfortunately, the FHA doesn't help designing a controller including the dynamics of the resonant tank. For this purpose, the engineer has to know the control-to-output transfer function.

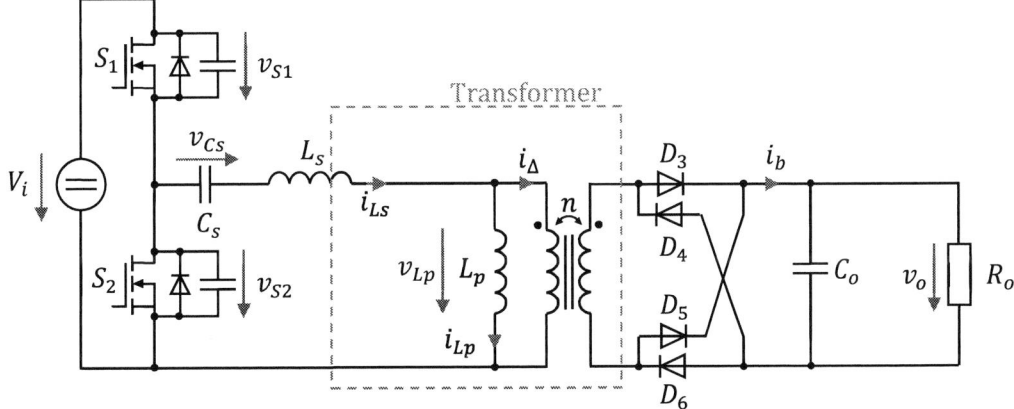

Figure 1: Circuit of the LLC-converter.

In [3] an optimized digital control is designed based on the measured transfer function obtained by using the correlation of the output voltage excited by perturbing the duty cycle with white noise. This paper uses a similar multi frequency excitation, both for the simulation model and for the practical measurement on a real converter.

Simulation Model

The simulation bases on a discrete-time model [1], [4]. This approach combines a quite high accuracy with acceptable simulation speed.

A resonant converter is in general a time-variant, nonlinear system. For minimizing the simulation effort it is common practice to linearize the converter in its operation point. A common example is the exponential diode behavior that can be approximated through a piecewise linear model consisting of a voltage source, a resistor and an ideal diode [5]. The remaining time-variant behavior of ideal switches and diodes can be simulated using the discrete-time model. Therefore the converter of Figure 1 is split up into six different networks like in [6] (though different numeration). Figure 2 shows one of these linear time invariant networks including parasitic resistances.

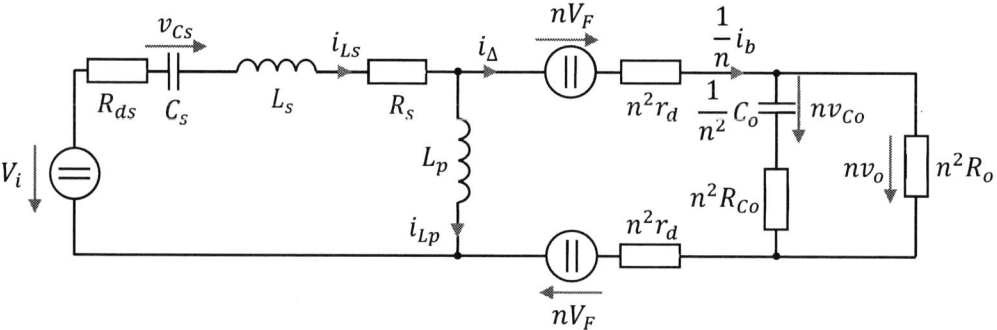

Figure 2: Network A of the discrete-time model.

Not included in the calculation are the parasitic capacitances of the MOSFETs. Therefore the networks have a constant input voltage either equivalent to V_i (Network A to C) or equal to 0 V (Network D to E). The output rectifier has three states, it can be positive (Network A and D), negative (Network C and E - compared to Figure 2 the part right to i_Δ is inverted) or non-conducting (Network B and D - compared to Figure 2 the left and the right part are separated). Each of these LTI networks can be described through a state-space representation

$$\dot{x}(t) = Ax(t) + Bu(t) \text{ with } x = \begin{bmatrix} v_{Cs} \\ v_{Co} \\ i_{Ls} \\ i_{Lp} \end{bmatrix} \text{ and } u = \begin{bmatrix} V_i \\ V_F \end{bmatrix} \tag{1}$$

using network analysis.

The simulation evaluates the LTI networks depending on the state of the switches and diodes starting at the time t_0 with the initial values $x(t_0)$. Depending on the state variables, transition times from one network to another can be determined. The values of the state variables at the time t_{end} can be calculated by

$$x(t_{end}) = e^{A(t_{end}-t_0)}x(t_0) + \int_{t_0}^{t_{end}} e^{A(t_{end}-\tau)}Bu(\tau)\mathrm{d}\tau, \tag{2}$$

delivering the initial condition for the next network.

The switching moments of the MOSFETs are defined by the switching frequency. The switching of the rectifier has to be determined by monitoring the current i_b which cannot be negative or the voltage across the diodes in forward direction which cannot be positive. As a result Figure 3 shows a typical simulated output voltage during the startup.

The steady state of the converter is found when the difference in the state vector x after one period is negligible small. Figure 4 depicts one switching period of the output voltage under steady state.

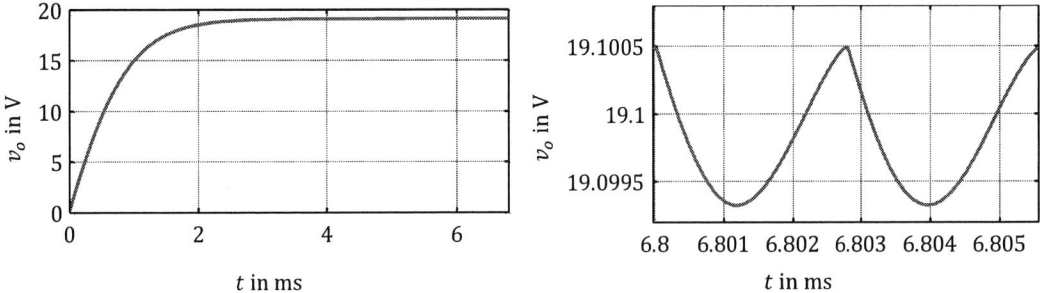

Figure 3:   Simulated startup.

Figure 4:   One period at the end of startup.

As described in [1] and [6] when the converter has reached steady state, it works in different modes depending on the frequency. The specific sequence of linear networks characterizes such a mode. While letters identify these linear networks as described, a group of letters defines the resulting mode, e.g. CADF. Figure 5 shows as an example the typical relation between switching frequency and the average output voltage including in different colors the corresponding modes.

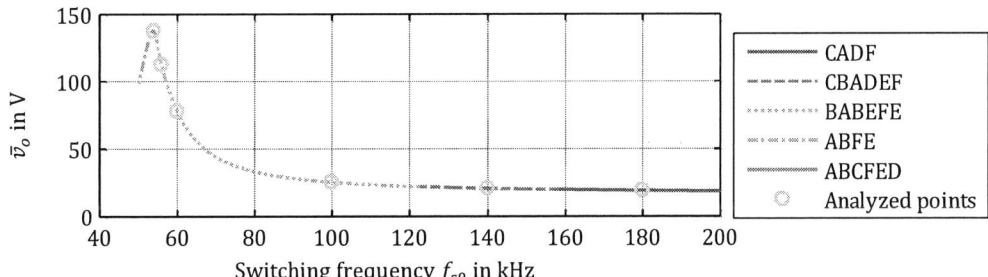

Figure 5:   Averaged output voltage depending on the frequency. The figure shows furthermore the different modes with different colors.

Simulating the Transfer Function

Resonant converters use the switching frequency to control the output voltage. However, looking at implementations available control ICs apply the switching period as the control variable. Thus Eq. (3) defines the control-to-output transfer function $G_T(s)$:

$$G_T(s) = \frac{U_o(s)}{T_s(s)} \tag{3}$$

Following the idea of [3] the method described in this paper applies a multi-frequency perturbation of the control variable T_s in order to obtain the transfer function at several frequencies within one simulation run. Instead of noise [3] this paper perturbs the switching period by means of a rectangular (thus a deterministic) signal. Figure 6 shows the resulting

© VDE VERLAG GMBH · Berlin · Offenbach

simulation model. First the operation point of $T_{s0} = 1/f_{s0}$ is calculated using the discrete-time modelling described earlier. Then a perturbation is added to T_s in form of a periodical increase and decrease. The correspondingly perturbed system itself has, similar to the not perturbed system, to reach a steady state – the signals have to be periodically with respect to the perturbation period T_p [7].

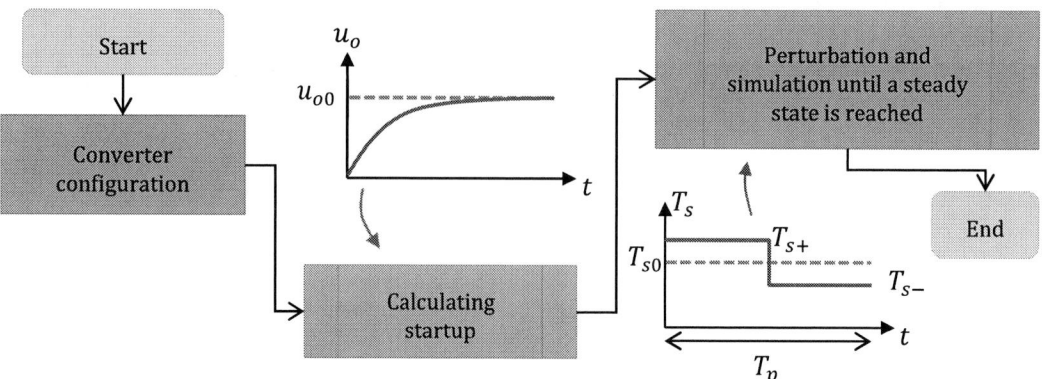

Figure 6:   Flowchart that shows the simulation of the perturbed converter.

The last step is to perform a Laplace-transformation on these steady state waveforms of $u_o(t)$ and $T_s(t)$ to obtain $U_o(s)$ and $T_s(s)$ and thus $G_T(s)$.

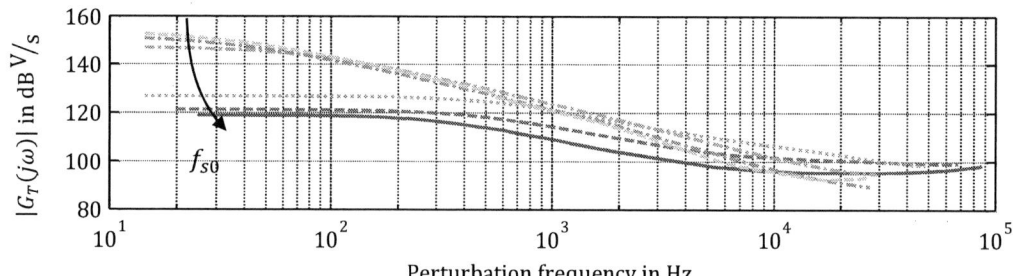

Figure 7:   Simulated gain of the control-to-output transfer function. The operation points for the different graphs are marked in Figure 5.

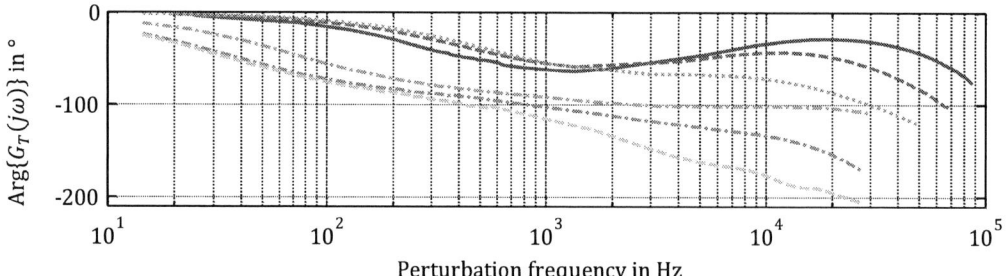

Figure 8:   Simulated phase of the control-to-output transfer function. The lines have the same style as in Figure 7.

For the operation points marked with circles in Figure 5, Figure 7 and Figure 8 show the resulting transfer functions. The Figures reveal that an approach based on FHA, which is not able to distinguish between the different modes, cannot correctly predict the different shapes of the transfer functions.

The problem of using a square wave as perturbation is the decreasing excitation amplitude at higher frequencies. Therefore, it is advisable to construct the complete transfer function through the use of different perturbation periods especially in case of measurement. E.g. the periods in Figure 11 (comparison with measurements) are scaled by a factor of ten.

Measuring the Transfer Function

Figure 9: Inverting half-bridge.

Figure 10: Resonant tank with capacitive output filter.

While offering the possibility to investigate the influence of design parameters upfront to hardware design, simulation require certain simplifying assumptions. It is good practice to verify the results. For this purpose the converter shown in Figure 9 and Figure 10 was built.

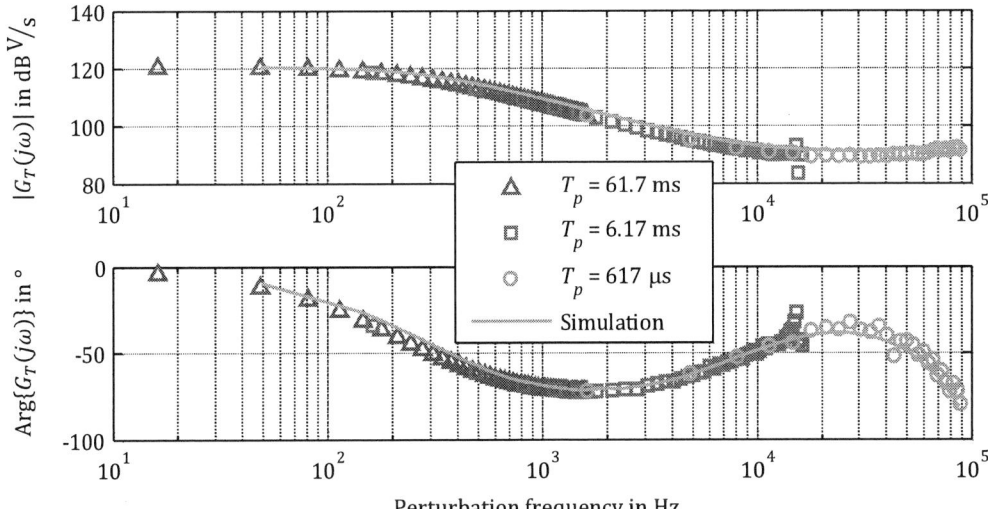

Figure 11: Comparison of the simulated and the measured transfer function for a converter working at $f_{s0} = 180$ kHz (mode CADF). The measured results are shown as points for each frequency while the simulated results are plotted as a line.

A DSP from TI (TMS320F28335) generates the control signals for the half-bridge. Using a digital controller offers the possibility to manipulate the gate signals in the required fashion. Especially such a controller allows implementing a perturbation identical to the simulation and therefore a different approach to the measurement than the one executed in [7]. An oscilloscope measures the response of this perturbation on the output voltage. With the information of perturbation and the measured voltage a transfer function can be calculated the same way as with the data of the simulation. Figure 11 shows the results of the converter with the parameters of Table 1. Three perturbations with different T_p are used because the information at higher frequencies vanishes in noise. For comparison the simulation is added as a solid line. Figure 12 shows another operation point.

The results match very well in both cases.

Conclusions

The method described in this paper offers a fast method to simulate and measure the control-to-output transfer function. Both, simulation and measurement use the same rectangular perturbation of the switching period. Experiences have shown that only three different perturbation periods spaced by a factor of ten are sufficient to obtain good transfer functions over a wide frequency range. Comparison of simulation and measurement confirms the validity of the depicted approach.

V_i	410	V
C_s	47,6	nF
L_s	45,8	µH
L_p	153,2	µH
n	7,37	
C_o	1,73	mF
R_o	6	Ω
R_{ds}	900	mΩ
R_{Ls}	0	Ω
R_{Cs}	70,9	mΩ
R_{prim}	100	mΩ
R_{sec}	2	mΩ
r_d	27,2	mΩ
R_{Co}	17,1	mΩ

Table 1: Converter parameters. R_{prim} and R_{sec} are the primary and the secondary side resistance of the transformer.

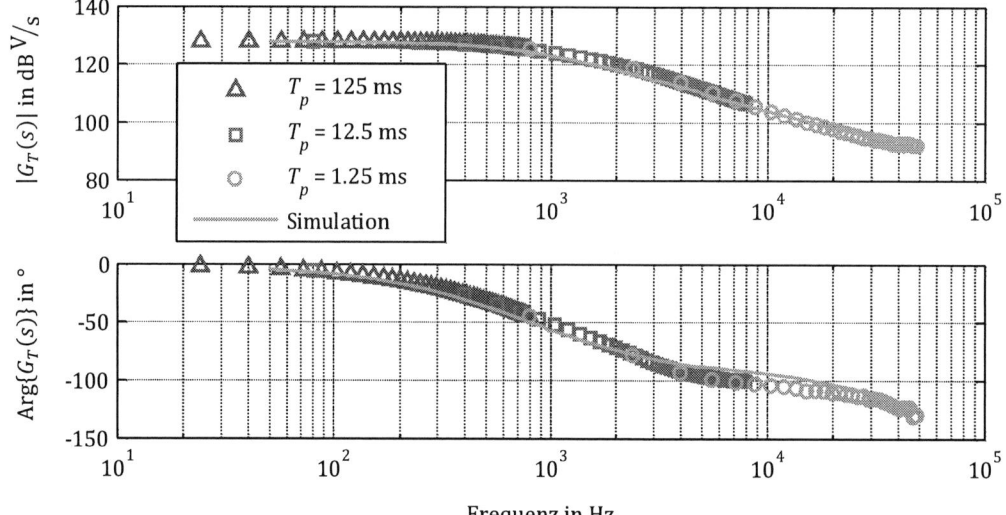

Figure 12: Comparison of the simulated and the measured transfer function at $f_{s0} = 100\ \text{kHz}$ (mode BABEFE).

References

[1] C. Oeder, "Analysis and design of a low-profile llc converter," in *Industrial Electronics (ISIE), 2010 IEEE International Symposium on*, July 2010, pp. 3859–3864.

[2] R. Steigerwald, "A comparison of half-bridge resonant converter topologies," in *Applied Power Electronics Conference and Exposition, 1987 IEEE*, March 1987, pp. 135–144.

[3] B. Miao, R. Zane, and D. Maksimovic, "System identification of power converters with digital control through cross-correlation methods," *Power Electronics, IEEE Transactions on*, vol. 20, no. 5, pp. 1093–1099, Sept 2005.

[4] A. Brown, "Sampled-data modeling of switching regulators," in *Power Electronics Specialists Conference, 1981 IEEE*, June 1981, pp. 349–369.

[5] L. Iannelli, F. Vasca, and G. Angelone, "Computation of steady-state oscillations in power converters through complementarity," *Circuits and Systems I: Regular Papers, IEEE Transactions on*, vol. 58, no. 6, pp. 1421–1432, Jun. 2011.

[6] J. F. Lazar and R. Martinelli, "Steady-state analysis of the LLC series resonant converter," in *Applied Power Electronics Conference and Exposition, 2001. APEC 2001. Sixteenth Annual IEEE*, vol. 2, 2001, pp. 728–735 vol.2.

[7] J. Stahl, H. Steuer, and T. Duerbaum, "Discrete modeling of resonant converters - steady state and small signal description," in *Energy Conversion Congress and Exposition (ECCE), 2012 IEEE*, pp. 1578–1584.

© VDE VERLAG GMBH · Berlin · Offenbach

PCIM Europe 2016, 10 – 12 May 2016, Nuremberg, Germany

For information on the presentation

A Study of the Thermal and Parasitic Optimization of a Large Current Density Highly Parallelized Three-Phase Reference Board for Motor Drive Applications

Mr. Mehrdad Baghaie Yazdi

Fairchild Semiconductor
TDC - Modeling / Power Solution Center
Einsteinring 28
85609 Aschheim

Phone: +49 (89) 998876 171
E-Mail: mehrdad.baghaie@fairchildsemi.com

Please note that the following manuscript
has been handed in late.

→ please go to the following page 2231

© VDE VERLAG GMBH · Berlin · Offenbach

Trends in Residential and Industrial Induction Cooking: Topologies and Power Devices for High Efficiency

Vittorio, Crisafulli, ON Semiconductor Germany GmbH, Munich, Germany,
vittorio.crisafulli@onsemi.com

Abstract

Energy consumption and safety are essential considerations in our daily activities. Power electronics subsystems in mass market and consumer accessible products such as home appliances have to be designed with special attention to these issues. In recent years, popularity of induction cooktops has led to development activities that address the safety and efficiency of these appliances. In this paper, an overview of the trends in residential and industrial induction cooking will be presented. Furthermore, topologies and power devices for high efficiency will be discussed. Test results will be presented and influence of IGBT parameters will be also analyzed.

1. Introduction

In many countries, today, energy demand and increase of safety are considered the most important political/economical/demographic issues. For instance, in Ecuador, the government has initiated a national campaign with the aim of replacing 3 million existing gas stoves with induction stoves. The main objective of this project is to move from consumption of liquefied gas to electricity and thus increase the safety by avoiding the gas consumed in homes. Many countries are moving in the same direction as Ecuador. Induction cooking offers high efficiency compared to the standard technology (gas or ceramic) and high reliability/safety (heat is generated only if the pot is over the coil). Induction cooking has a very long history. In fact, typing cooking into the search engine of EPO (European Patent Office) one can discover that the first patent related to induction heating (IH) was published in 1906: GB190612333 "Improvements in or relating to Apparatus for the Electrical Production of Heat for Cooking and other purposes" by Berry Arthur Francis. Notwithstanding this history, the induction heating technology has become popular just recently. Induction cooking is perhaps one of the most revolutionary advancements in cooking technology in the last century. Although the technology is already popular in Europe and Asia, it is still largely unknown in the U.S. However, it seems falling prices and ever-growing consumer awareness might finally help the technology in gaining share against the traditional technologies. Induction is fundamentally unique in that it uses electromagnetic energy to directly heat pots and pans. Induction technology is employed not only in residential application (stove up to 4 kW), but also for the industrial applications covering professional stoves for restaurants (> 5 kW for single stove [1]). In comparison, gas and electric cooktops heat indirectly, using either a burner or a heating element to heat the cookware from underneath. Induction cooktops do not use heating elements or burners; they employ a series of magnets that excite the iron atoms in a pan to generate heat [2]. Some international organizations claim that it is far more efficient to heat the cookware directly than indirectly.

PCIM Europe 2016, 10 – 12 May 2016, Nuremberg, Germany

Induction Cooking Block Diagram

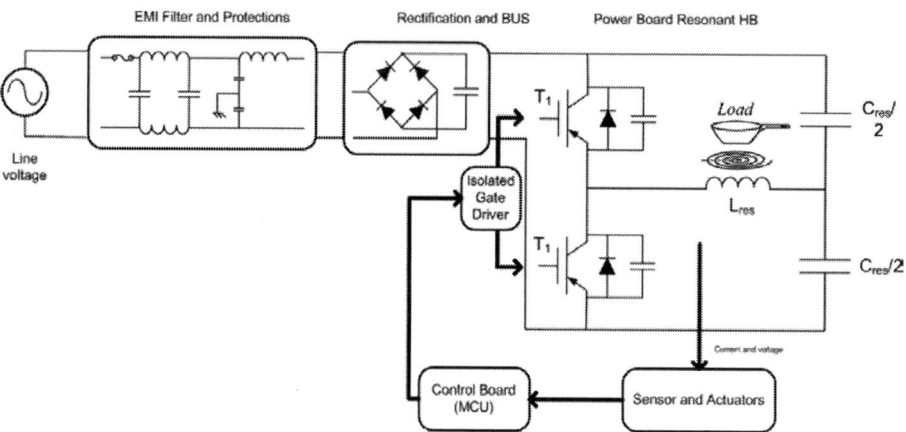

Fig. 1: Block Diagram of a Residential Induction Cooktop

Induction is able to deliver roughly 80 to 90 percent of its electromagnetic energy to the food i n the pan. Compare that to gas, which converts a mere 38 percent of its energy, and electric, which can only manage roughly 70 percent. That means induction cooktops not only heat up much faster, but their temperature controls are also far more precise. One of the exciting fea tures of the high-end products from many induction cooktop manufactures is the temperature control, which will settle the pot content temperature to the chosen level, thus enabling new cooking techniques such as simmering melting and sous vide, which require stable and preci se under boiling temperatures. To achieve this temperature control, two different paths could be taken: external sensors which have additional cost and communication issues and comple x algorithms, or using the burners' safety sensors and the power given to the pot, to estimate the content temperature with a certain precision. The trend of all the Induction manufacturer s is to reduce the price in order to supply the market with a very reliable (safe), efficient and c heap product. Each company is trying to get rid of components, surrogating them through a b etter control technique (like sensorless or some exotic modulation in order to comply with the standard or increase regulation etc.) [3] [4]. The core of this technology (Fig. 1) is the transis tors, and today the best choice amongst all the switches is the Insulated Gate Bipolar Transis tor, also called IGBT. Since the IGBT forms the core of the application (along with the Control Unit), its choice requires careful consideration.

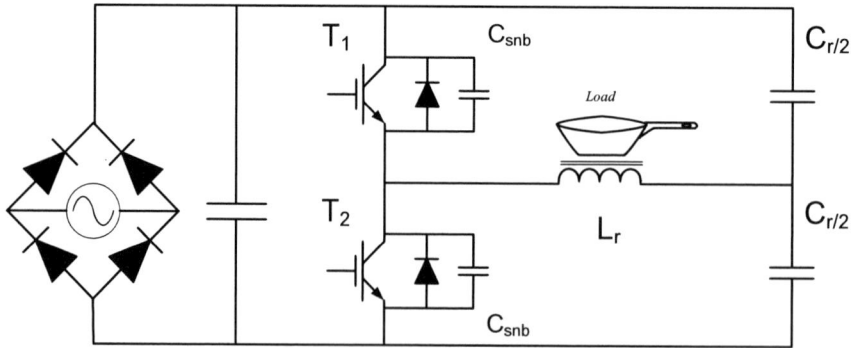

Fig. 2: Resonant Half Bridge.

© VDE VERLAG GMBH · Berlin · Offenbach

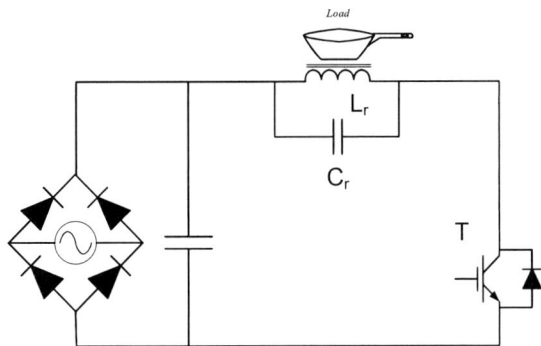

Fig. 3: Quasi Resonant Single Switch or Single ended.

2.   Topologies considerations

In the last two decades, many investigations have been carried out in this field for identifying appropriate topologies and devices. The aim of the investigations was to improve safety, imp rove efficiency and reduce cost. Fig. 2 and Fig. 3 show the most popular topologies: resonant half-bridge (more popular in the European market), and quasi resonant-single ended (more popular in the Asian market). The main requirements of an induction cooking converter are a s follows:

- High frequency switching
- Wide load range

High Frequency is needed because the energy transfer is done through Electromagnetic indu ction and using high frequency it allows having smaller passive component in the circuit. Wid e load range is needed because there is any limitation for the user in term of pan/pot.

The resonant topologies fit these requirements very well. The main benefit of using resonant converter is the soft switching, which allows to reduce losses and increase frequency. Furthe r, these converters are able to work with wide load range without efficiency degradation. Des pite the soft switching technique applied to the induction cooking inverter, some barriers still have to be overcome. In fact, most of these converters can operate only below a certain freq uency. This is due to the losses generated by the switches during their normal operations.

2.1.   Resonant Half Bridge

The Resonant half-bridge inverter is the most employed topology in induction cookers due to its simplicity, its cost-effectiveness, and the electrical requirements of its components. The eq uivalent load is a resonant tank, which consists of the inductive element the coil and capacitiv e element the resonant capacitor and the resistive one, the pan. Induction-coil-and-pan coupl ing it can be modeled as a series connection of an inductor and a resistor, based on the anal ogy of a transformer, and it is defined by the values of L_r and R_{load}. These values vary with th e switching frequency applied to the switches, pan material, temperature, and inductor–pan c oupling. The resonant half bridge belongs to the resonant converter family. It is similar to a st andard half bridge, where the two half bridge capacitors are set in accordance with the coil fo r resonating a certain frequency (the so called resonant frequency). The power stage is depic ted in Fig. 2 and it is composed by a two switches with antiparallel diode, two capacitors and a coil. Fig. 4 shows the output power versus switching frequency. Fig. 5 shows operational w aveforms.

© VDE VERLAG GMBH · Berlin · Offenbach

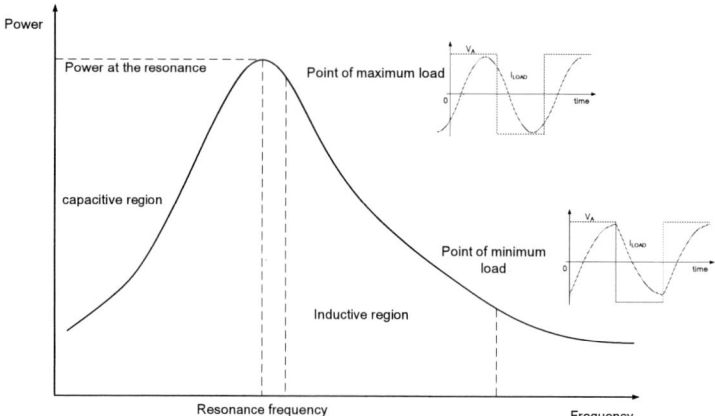

Fig. 4: Output power versus switching frequency for maximum load and minimum load.

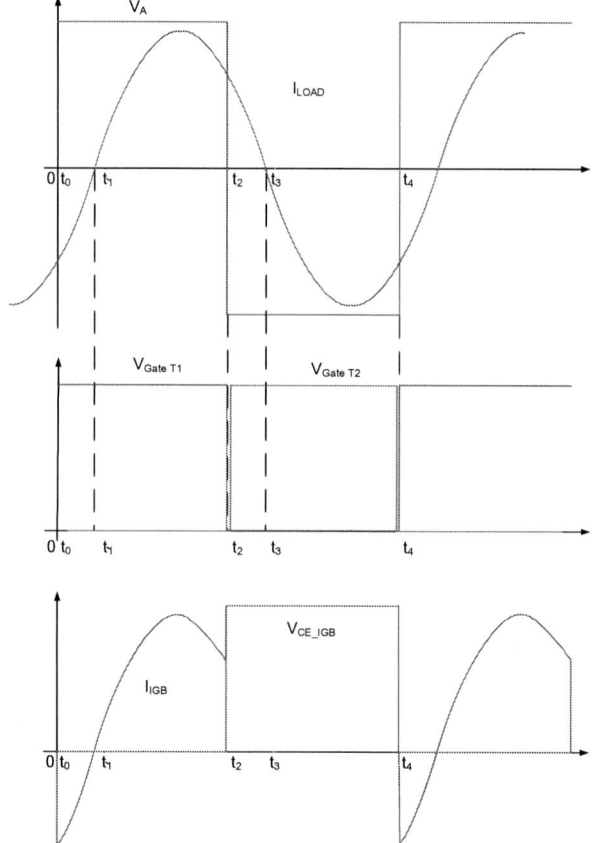

Fig. 5: Resonant Half Bridge waveforms.

2.2. Quasi Resonant

Single ended or Quasi resonant converters (shown in Fig. 3) [13][14][15] are widely used as AC power supplies like induction heating cooktop or microwave inverter applications for supp lying the magnetron. Such converters are quiet attractive for the domestic appliances becaus e it requires only one switch, usually an IGBT, and only one resonant capacitor. QR converte rs might be considered as a good compromise between cost and energy conversion efficienc y. One drawback of this family of converter is the not wide regulation range, which is commo nly defined through the ratio between the maximum power settable, and the minimum power settable. While in one side the maximum power is limited by the maximum voltage capability of the power device to withstand without fail, the minimum power deliverable is limited by the hard switching region at the turn on, which at a certain level may cause the destruction of the device and the inverter due to thermal causes. To achieve ZCS/ZVS operation at the turn-on in a QR converter several conditions have to be met, and not all of them are under the desig n control when the QR is for induction heating applications. As a consequence, QR converter cannot always be operated in soft-switching conditions. This not only nullifies the aforementi oned benefits, but it may also lead an increasing into the EMI, possibly exceeding the maxim um level set by standard.

3.   Device consideration

In the past decades many new device technologies have been explored in induction cooking applications, like SiC MOSFET, SIC JFET[5] [6] and GaN HEMT, but the technology advanta ge of using these was not enough to justify the cost increase for the overall solution, compar ed to the classic IGBT. Recent advances in IGBT Trench Field Stop technology have solidifie d the case for the IGBTs further [7] [8] [9]. The field stop technology not only brings better per formance in conduction and switching losses but also a lower manufacturing cost, which can be translate into a cost reduction for the end customer.

The IGBT characteristic are influencing the performance in application. That's why the right d evice has to be chosen in a wise way. This kind of application has a huge constrain in term of cost. First the device has to comply with a certain price range, which implies to use devices with embedded diode (or the so called shorted anode). These devices offer an ok performan ce in term of diode characteristic for soft switching topology working in inductive region (wher e the contribution of Q_{rr} doesn't affect the losses). Another important consideration during the IGBT selection is how to select the tradeoff between switching and conduction losses. Looki ng at the device listed in table Table 1 it is very difficult to decide the right one for the applicat ion.

Table 1: Comparison among different Devices (normalized to Device B)

Part	VCEsat@150C	VF@150C	Hard Switching Eoff	Capacitive Eoff 22 nF
Device A	1.3	0.91	0.98	0.83
Device B	1	1%	1	1
Device C	1.11	1.27	1.12	1.04
Device D	0.97	1.32	1.70	1.18

For Example Device A seems more optimized for fast application compare the other. Its lowe r Eoff is also notable in capacitive switching (Switching mechanism in the Resonant Half brid ge). On the contrary the Device D seems more suitable for lower switching frequency. But wh en we consider the capacitive switching the Eoff losses the different is not as big as the class ical hard switching.

© VDE VERLAG GMBH · Berlin · Offenbach

4. Experimental Results

A 3.7 kW Resonant half bridge prototype (Fig. 6) has been built and all the devices have been tested in order to see how different devices perform in the applications. The new Field Stop 2.5 IGBT technology [10] [12] was the most promising candidate to match the application requirement in the different working conditions. This new technology may be selected to replace previous generation IGBTs, while operating at superior switching frequency. This new technology is a good compromise in term of conduction losses and switching characteristic.

This new technology has been tested in a 3.7 kW Resonant Half Bridge prototype (Test Specifications in Table 3). In Table 1 are shown the different static and dynamic parameters of the two latest ON Semiconductor IGBT technologies: FS2 and FS2.5. It is evident from Fig. 7 and Fig. 8 what is the advantage to choose the right IGBT in terms of parameters. Even though the new technology suffers slightly in terms of diode performance and turn-off performance, it provides an overall improvement in the end application by reducing the losses by >4 W.

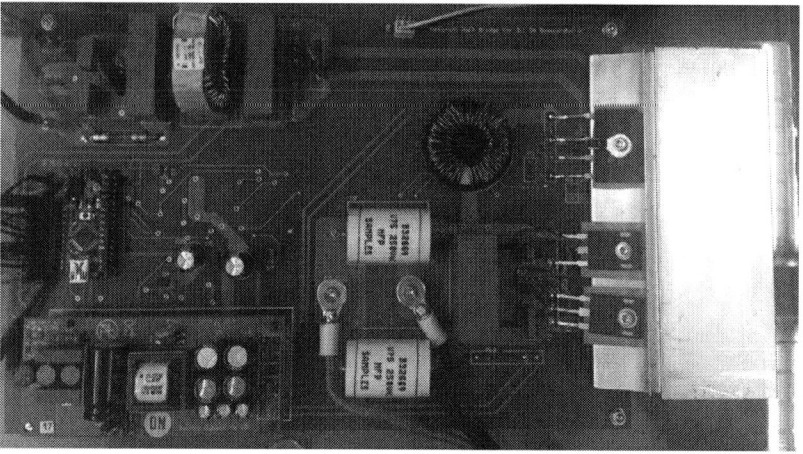

Fig. 6: Resonant Half Bridge Prototype.

Table 2: Comparison between FSII and FSII.5

Part	VCEsat@ 25C	VF@25C	VCEsat@ 150C	VF@150 C	Eoff@25C (mJ)	Eoff 22 nF
FS2	1.77	1.2	2.21	1.17	0.43	0.25
FS2.5	1.57	1.47	1.91	1.63	0.48	0.311

Table 3: Test conditions

Resonant Cap	2*750	nF
Coil	2.9507E-05	H
	1.5	Ω
Snubber Capacitors	22	nF
POT	~Normative	

PCIM Europe 2016, 10 – 12 May 2016, Nuremberg, Germany

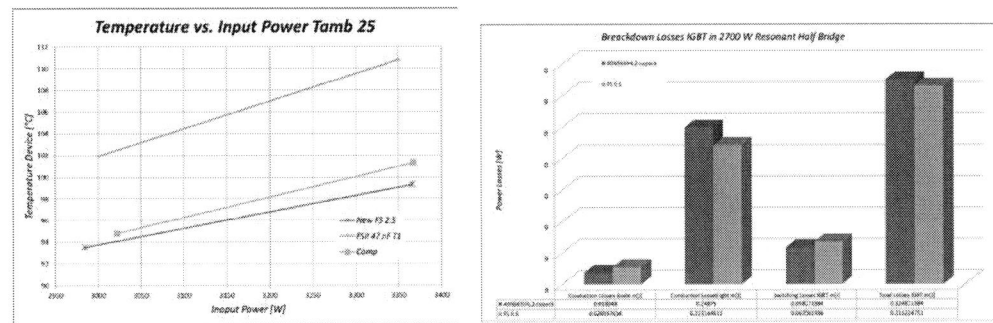

Fig. 7: Case temperature (Left side) and power losses (Right side of the FSII IGBT and the new FS2.5 and a competitor.

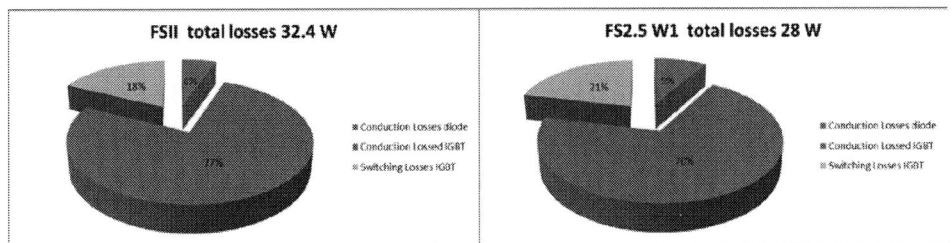

Fig. 8 Losses Distribution for the FSII IGBT and the new FS2.5.

5.   Conclusion

The paper presents the latest trend in induction cooking technologies: topologies/control and power devices. The mix of both allows for an increase of efficiency and safety compared to th e standard technologies. Further is to give guidelines in selecting the right topology, based o n the requested power level and also the right selection of the IGBT parameters. In order to make the right choice, model based design is needed. It is important to build a model able to predict the behavior of the device in the application.

References

[1] Nagai, S.; Hiraki, E.; Arai, Y.; Nakaoka, M., "New phase-shifted soft-switching PWM series resonant inverter topologies and their practical evaluations," in Power Electronics and Drive Systems, 1997. Proceedings., 1997 International Conference on , vol.1, no., pp.318-322 vol.1, 26-29 May 1997

[2] AND9166/D Induction Cooking Everything You Need to Know On semiconductor 2014.

[3] V. Crisafulli, C. V. Pastore, "New control method to increase power regulation in a AC/AC quasi resonant converter for high efficiency induction cooker," Power Electronics for Distributed Generation Systems (PEDG), 2012 3rd IEEE International Symposium on , vol., no., pp.628,635, 25-28 June 2012

[4] Hirokawa, T.; Okamoto, M.; Hiraki, E.; Tanaka, T.; Nakaoka, M., "A novel type time-sharing high-frequency resonant soft-switching inverter for all metal IH cooking appliances," in IECON 2011 - 37th Annual Conference on IEEE Industrial Electronics Society , vol., no., pp.2526-2532, 7-10 Nov. 2011

[5] Gaudo, P.M.; Bernal, C.; Otin, A.; Burdio, J.-M., "Silicon carbide JFET resonant inverter for induction heating home appliances," IECON 2011 - 37th Annual Conference on IEEE Industrial Electronics Society , vol., no., pp.2551,2556, 7-10 Nov. 2011

© VDE VERLAG GMBH · Berlin · Offenbach

[6] Avellaned, J.; Bernal, C.; Otin, A.; Molina, P.; Burdio, J.M., "Half-bridge resonant inverter with SiC cascode applied to domestic induction heating," Applied Power Electronics Conference and Exposition (APEC), 2013 Twenty-Eighth Annual IEEE , vol., no., pp.122,127, 17-21 March 2013

[7] A. Salih, "IGBT for high performance induction heating applications," IECON 2012 - 38th Annual Conference on IEEE Industrial Electronics Society, vol., no., pp.3274,3280, 25-28 Oct. 2012.

[8] F. Blaabjerg, U. Jaeger, S. Munk-Nielsen, J.K. Pedersen, "Comparison of NPT and PT IGBT-devices for hard switching applications," Industry Applications Society Annual Meeting, 1994., Conference Record of the 1994 IEEE , vol., no., pp.1174,1181 vol.2, 2-6 Oct 1994

[9] Millán, I.; Puyal, D.; Burdio, J.-M.; Lucia, O.; Palacios, D., "IGBT selection method for the design of resonant inverters for domestic induction heating," Power Electronics and Applications, 2009. EPE '09. 13th European Conference on , vol., no., pp.1,7, 8-10 Sept. 2009.

[10] Crisafulli, Vittorio, "New IHR Field Stop II IGBT technology, the best efficiency for high frequency Induction Cooking Applications," PCIM Europe 2014; International Exhibition and Conference for Power Electronics, Intelligent Motion, Renewable Energy and Energy Management; Proceedings of , vol., no., pp.1,8, 20-22 May 2014

[11] V. Crisafulli, A. Gallivanoni, C. V. Pastore, "Model based design tool for EMC reduction using spread spectrum techniques in induction heating platform," Optimization of Electrical and Electronic Equipment (OPTIM), 2012 13th International Conference on , vol., no., pp.845,852, 24-26 May 2012

[12] Crisafulli, Vittorio; Antretter, Marco, "Design Considerations to Increase Power Density in induction cooking applications using the new Field stop II technology IGBTs," in PCIM Europe 2015; International Exhibition and Conference for Power Electronics, Intelligent Motion, Renewable Energy and Energy Management; Proceedings of , vol., no., pp.1-8, 19-20 May 2015

[13] Quasi-Resonant Dual Mode Soft Switching PWM and PDM High-Frequency Inverter with IH Load Resonant Tank Ahmed, N.A. ; Eid, A. ; Hyun Woo Lee ; Nakaoka, M. ; Miura, Y. ; Ahmed, T. ; Hiraki, E. ; Electr. Energy Saving Res. Center, Kyungnam Univ., Masan This paper appears in: Power Electronics Specialists Conference, 2005. PESC '05. IEEE 36th Issue Date : 16-16 June 2005 On page(s): 2830 - 2835

[14] Hirota, I.; Omori, H.; Chandra, K.A.; Nakaoka, M.; , "Practical evaluations of single-ended load-resonant inverter using application-specific IGBT and driver IC for induction-heating appliance ," Power Electronics and Drive Systems, 1995., Proceedings of 1995 International Conference on , vol., no., pp.531-537 vol.1, 21-24 Feb 1995

[15] Hirota, I.; Omori, H.; Nakaoka, M.; , "Performance evaluations of single-ended quasi-load resonant inverter incorporating advanced-2nd generation IGBT for soft switching ," Industrial Electronics, Control, Instrumentation, and Automation, 1992. Power Electronics and Motion Control., Proceedings of the 1992 International Conference on , vol., no., pp.223-228 vol.1, 9-13 Nov 1992

Direct Power Control for a Grid Connection of a Three Phase Z-Source Inverter

Dipl.-Ing. Manuel Steinbring, Universität Siegen, Hölderlinstraße 3, 57068 Siegen, Germany
manuel.steinbring@uni-siegen.de

Univ.-Prof. Dr.-Ing. Mario Pacas, Universität Siegen, Hölderlinstraße 3, 57068 Siegen, Germany
pacas @uni-siegen.de

1. Abstract

The Z-Source inverter (ZSI) has proven to be a robust and easy to control inverter. As a drawback it has increased requirements towards the PWM modulator to insert the shoot through states that are necessary for the boost operation. To overcome that restriction a novel direct power control (DPC) scheme that does not require a PWM has been developed and successfully validated by experiment. It works similarly to the well-known direct torque control (DTC) with the difference that the output voltage of the inverter does not directly control the flux and the torque of an electrical machine but the active and reactive power fed into the grid. In addition, the DPC control had to be extended to also manage the shoot through states i.e. the boosting of the ZSI.

2. Introduction

In many research works, the ZSI has shown to be a good and robust power electronic converter. Its modified DC link enables the unique feature of the Z-source inverter that combines the properties of a buck and of a boost converter in one inverter stage. Thus, a short circuit in one inverter leg is no longer a catastrophic failure; moreover, it is necessary for enabling the boost operation of the Z-source inverter. Since the ZSI delivers a constant voltage even if fed from sources of variable voltage and it has increased immunity to short circuits in the legs of the inverter, it can be considered for distributed electric power generation in the lower power range e.g. in remote areas. Therefore, the single-phase ZSI was investigated in previous works and presented as low cost, easy to control and robust solution [1] - [4]. On the other hand, it has a reduced efficiency if compared with standard two stage concepts.

If the amount of harvested energy is increased or if a grid connection is considered, a single-phase concept is no longer suitable, instead a three-phase topology should be chosen. For the control of the three phase ZSI many concepts have been presented in the literature [5] - [7]. To cope with the requirements of remote and rugged areas, a simple, low-cost solution should be the aim. Nevertheless, the PWM features of standard microcontrollers are limited and not suitable for this task. In particular, the PWM – units implemented in microcontrollers are usually not able to generate the pulses as they are required in the well-known control strategies of ZSIs. On the other hand an additional processing unit (e.g. CPLD) would increase the cost and the complexity of the system that becomes less attractive and not suitable for harvesting energy in remote and rural areas.

In the following, a DPC control structure for a three phase ZSI without any PWM-unit is proposed that can be used to feed the generated power into a grid.

3. Z-Source inverter

Fig. 1 shows the structure of the three phase ZSI as it is used in the present work. Instead of a regenerative source of variable voltage a variable transformer connected to a rectifier is used to provide the input voltage and to operation the inverter in defined conditions for the investigation. The output is connected to the grid. Between the input rectifier and the output inverter the special DC - link with Z-topology is shown. The explanation of the principle of operation of the ZSI can be found in previous papers or in the standard literature [1][5]. Compared to a conventional voltage source inverter, the ZSI exhibits an additional switching state (shoot through) that has to be included in the generation of the switching patterns. The short circuiting of the DC link, which is necessary for the boost operation, has the same effect on the inverter side like a zero voltage vector of a standard VSI. Therefore, any zero voltage vector can be replaced by a shot through vector without influencing the modulation scheme of the output voltage.

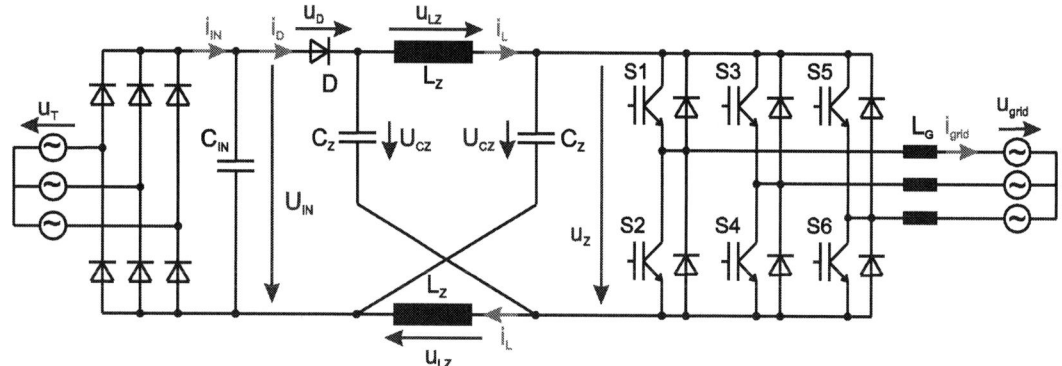

Fig. 1: Proposed setup of a three phase ZSI for grid connection

4. Direct Power Control for ZSI

The principal scheme of DPC is similar the well-known DTC which is widely used in motor drives. In comparison to DTC, where the stator flux and torque of the motor are controlled, the DPC is designed to control the active and reactive power fed into the grid. At every sampling time, the instantaneous active and reactive power are calculated and the algorithm chooses the right switching state, to manipulate the system into the right direction [8].

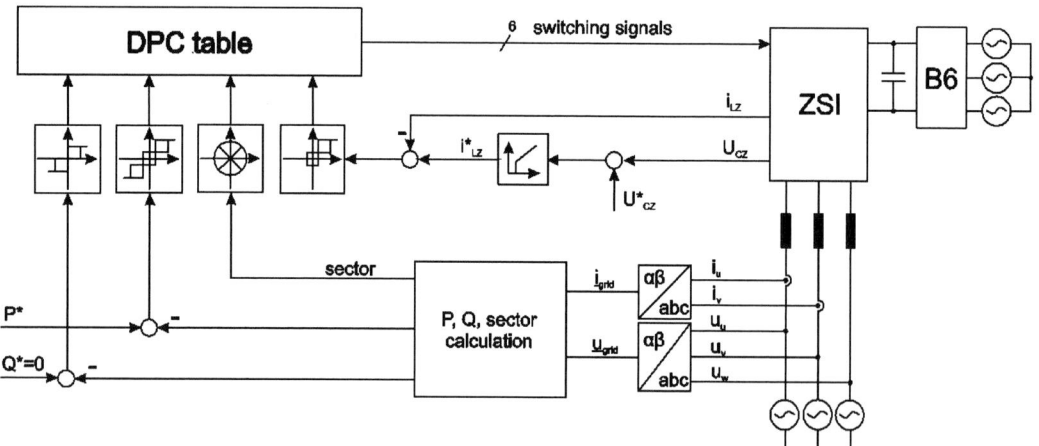

Fig. 2: Proposed PDC control structure for a three phase ZSI

In the proposed setup, as depicted in Fig. 2, the grid voltage and the currents injected into the grid are measured. After transforming both voltage and current into the α-β-coordinate frame, the space phasors $\underline{u}_{grid}$ and $\underline{i}_{grid}$ are obtained and the instantaneous active and reactive power are calculated as:

$$P = u_{grid,\alpha} \cdot i_{grind,\alpha} + u_{grid,\beta} \cdot i_{grid,\beta} \tag{1.1}$$

$$Q = u_{grid,\beta} \cdot i_{grind,\alpha} + u_{grid,\alpha} \cdot i_{grid,\beta} \tag{1.2}$$

The set point for the control of the active power P* is obtained from a superimposed control according to the available power of the source. The set point for Q* is set to zero in order to have a unity power factor. For the control of the active power a three level hysteresis control is used, while the reactive power is controlled by a four level hysteresis controller.

For the control of the output voltage of the ZSI in boost operation a cascaded control structure with an underlying inner current control loop is used. For that purpose the measurement of the voltage of the ZSI capacitors U_{CZ} and of i_{LZ}, the current through one of the main inductors are required. The reference value for the capacitor U_{CZ}^* has to be set according to the control requirements. The current control is performed by a three level hysteresis controller with a reference value obtained from the outer control loop i.e. from the PI-voltage controller. The manipulated variable of the current controller determines how the shoot through states are inserted into the pulse pattern of the inverter.

If the output of the current controller is zero (mode 0), it indicates that the level of U_{CZ} is sufficient and no boosting is required. In this case, the switching of the ZSI is performed like in the case of a standard VSI and no special provisions are required.

If the voltage $U_{CZ} < U_{CZ}^*$ it has to be boosted and the control demands a higher current I_{LZ}. As a result, the current controller will switch to the operation mode 1. In this mode of operation the zero vectors are replaced by the shoot through vector and the capacitor voltage U_{CZ} is boosted. Like DTC, DPC is a non-deterministic control and it is not possible to predict, which switching pattern is chosen next. In case of a high load of the inverter or of a low DC link voltage, it might happen that only very few zero states are available to be replaced by shoot through states. If that happens, the whole control algorithm can become unstable and cannot properly function. This situation is worsened because the amount of available zero states decreases under high load or at low DC link conditions. This case would correspond to the case of higher modulation index in PWM.

To overcome this problem, a third stage (mode 2) at the output of the current controller is implemented that inserts the shoot through states in a deterministic way. From previous papers it is known that the theoretical maximum boost (infinity) occurs at a shoot through duty cycle of 0.5 corresponding to a DPC operation in which every second switching state is a shoot through. Based on previous experiences, a shoot through duty cycle of 0.333 < 0.5 shows good results and was chosen for the implementation. In every third DPC cycle a shoot through state is introduced. Between two shoot through states the normal DPC operation (mode 0) is applied. In this way, the shoot through that is necessary for the boost operation of the ZSI can be achieved even under heavy load or at low DC-link voltage.

Fig. 3 shows the single-phase equivalent circuit of an inverter connected to the grid across an inductor. Fig. 4 shows the resulting phasor diagram. As mentioned the angle φ is controlled to zero to achieve unity power factor and to completely inject the available energy into the grid. The phase and the amplitude of the current $\underline{i}_{grid}$ and can be controlled by manipulating $\underline{u}_{Lgrid}$ by applying the proper voltage space phasor $\underline{u}_{Inv}$ delivered by the inverter. If the voltage $\underline{u}_{Lgrid}$ has a component in direction of $\underline{i}_{grid}$, the amplitude of $\underline{i}_{grid}$ will change. If

$\underline{u}_{L\,grid}$ has a component orthogonal to $\underline{i}_{grid}$, the phase of $\underline{i}_{grid}$ can be influenced. The angle φ has an impact on Q while the amplitude of the current $\left|\underline{i}_{grid}\right|$ determines P. During operation $\underline{i}_{grid}$ is rotating with line frequency, yet the available six active space phasors as depicted in Fig. 6 do not move. At every control cycle the states of the three hysteresis controllers are evaluated and the right switching state is chosen from a table in a way that $\underline{i}_{grid}$ follows the desired trajectory.

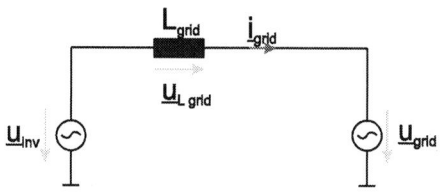

Fig. 3: Simplified equivalent circuit

Fig. 4: corresponding phasor diagram

Table 1 lists the available switching states and gives the effect of each of the voltage spaces phasors on P and Q. "+" indicates an increase of the value of the corresponding variable, "-" indicates a decrease and finally a "0" indicates no change.

space phasor	$\underline{u}_1$	$\underline{u}_2$	$\underline{u}_3$	$\underline{u}_4$	$\underline{u}_5$	$\underline{u}_6$	$\underline{u}_7$	$\underline{u}_8$	$\underline{u}_9$
$\left|\underline{i}_{Grid}\right| \approx P$	+	+	-	-	-	-	-	-	-
$\varphi \approx Q$	-	+	+	+	-	-	0	0	0
boosting	no	no	no	no	no	no	no	no	yes

Table 1: Available inverter states and their effect on the active power P, on the reactive power Q and on the boosting of the ZSI for sector 0

The effect of the vectors on $\underline{i}_{grid}$ depends on the sector in which $\underline{i}_{grid}$ is located. The influence of the six active vectors changes circular. If $\underline{i}_{grid}$ moves from sector 0 to sector 1, $\underline{u}_2$ has the effect $\underline{u}_1$ had in sector 0. The plane is divided into six sector as it can be seen in Fig. 5. The knowledge of the sector in which $\underline{i}_{grid}$ is, is also necessary for the evaluation of the DPC algorithm

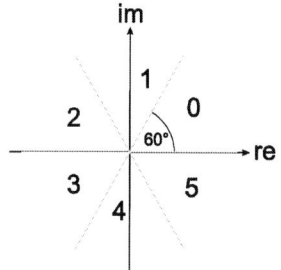

Fig. 5: Arrangement of the sectors used for the DPC

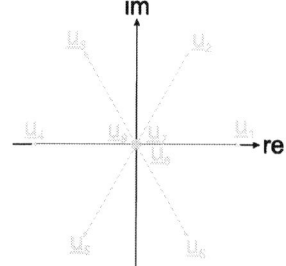

Fig. 6: Available voltage space phasors utilized in the DPC

© VDE VERLAG GMBH · Berlin · Offenbach

5.   Dimensioning of the main inductance L_Z

The dimensioning of the main inductance L_Z has in the case of DPC a higher importance than in PWM based control schemes and has to be examined with some detail. As previously mentioned, the inductance current I_{LZ} is increasing during the shoot through. In PWM based control schemes the duration of the shoot though can be finely adjusted and as a consequence the current increase can be limited. In the proposed control method, the states of the inverter will be applied for at least the duration of one DPC cycle. During that time I_{LZ} will constantly increase. For a voltage U_{CZ} = 560V, as it is required to feed a grid with a line to line voltage of 400V with a DPC sampling frequency of 40 kHz i.e. a sampling time of 25μs and for an inductance L_Z = 1.2mH the current will increase by

$$\Delta i_{LZ} = \frac{\Delta t_{DPC} \cdot U_{LZ}}{L_Z} = \frac{25\mu s \cdot 560V}{1,2mH} = 11,7A \tag{1.3}$$

during one DPC cycle. The design of the inductance has to take this high current ripple into account. If a high fluctuation of the current is not acceptable, a solution would be to increase the DPC sampling frequency or the inductance or both. For the proposed setup, the frequency was therefore increased to 66 kHz. That reduces the current ripple but still gives enough time for the microcontroller to calculate the required data. The real switching frequency of each MOSFET is not necessarily increased by this measure. The real switching frequency mainly depends on the values of the inductances and on the choice of the hysteresis levels.

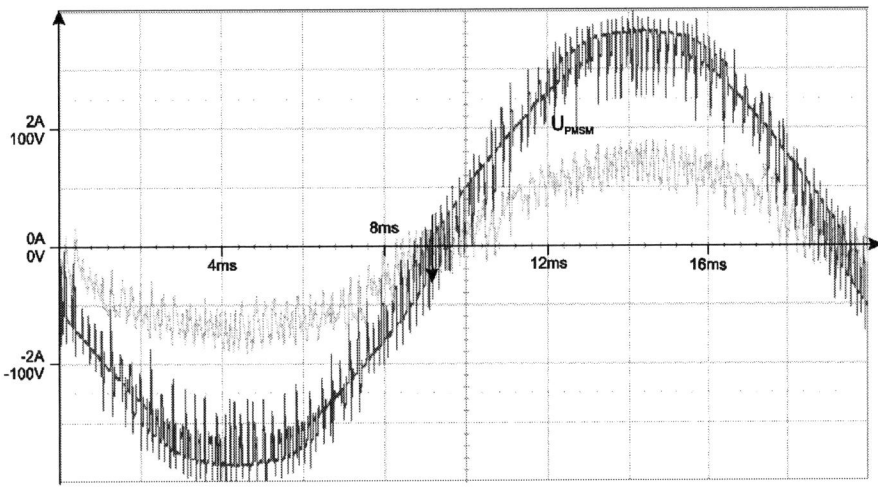

Fig. 7: Grid voltage u_{grid} (blue, phase voltage) an injected current i_{grid} (green).

6.   Experimental Results

For the experimental validation of the proposed control scheme the inverter was connected to grid with a voltage of 230V RMS line to line. In order to be able to inject a current into the grid the DC link voltage U_{CZ} was set to 350V. The main inductance was L_Z = 1.2mH, the gird coupling inductance was L_{GRID} = 18mH. Fig. 7 shows the grid voltage $\underline{u}_{grid}$ (phase voltage) as well as $\underline{i}_{grid}$, the current injected into the grid. The current exhibits a ripple characteristic for the DPC and the hysteresis control keeps it within a tolerance band around the reference value. The RMS value of $\underline{i}_{grid}$ is 1A and approximately 400W are fed into the grid. The power control algorithm keeps the power factor λ= 1 so voltage and current are in phase.

Fig. 8 gives a general view of the functioning of the DPC. The current I_{Lz} is shown by the green trace. Every time the voltage U_{cz} (yellow trace) drops below the threshold shoot through states are applied. This is clearly visible as the current I_{Lz} increases. The shoot through is activated until the level of U_{cz} is sufficient again. This procedure repeats periodically to keep U_{cz} at level.

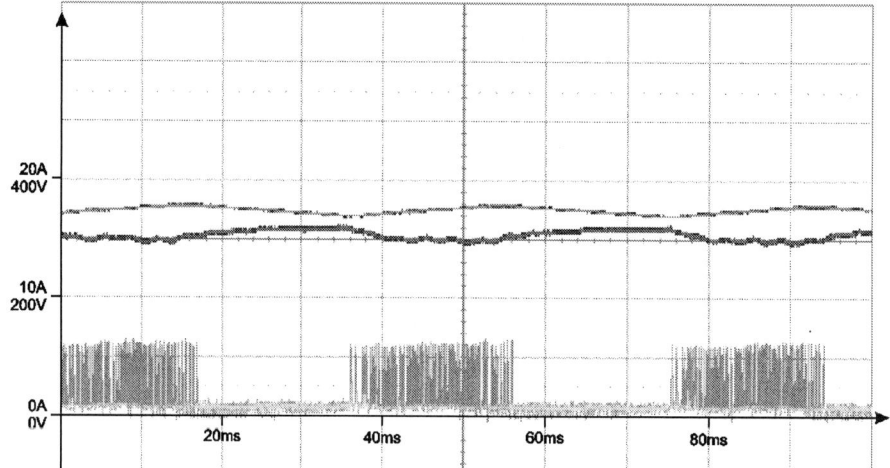

Fig. 8: View of the boosting behavior of the DPC; U_{IN} (red), U_{cz} (yellow), I_{Lz} (green)

Fig. 9 shows the same context but with a zoom into the boosting (same color coding as above). It can be seen that the current increases for the duration of one DPC cycle, after that it falls back to zero. Within the 15μs it increases by more than 5A. This can be calculated according to (1.3) but the line to line voltage of the grid is only 230V so U_{cz} of only 350V is sufficient resulting in a smaller current peak. The voltage U_{cz} is increased above the level of U_{IN} (red trace), indicating a boosting. Discontinuous mode as it is described in [1] is not a problem, because it is automatically corrected by the control.

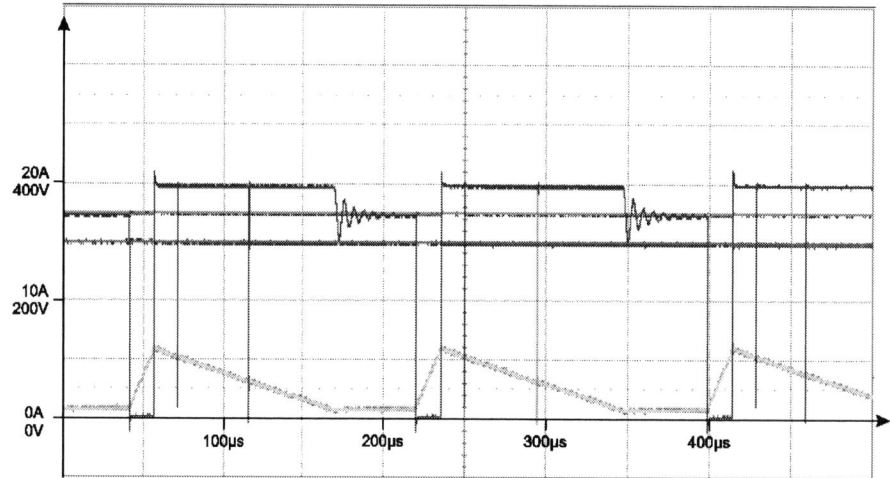

Fig. 9: Detailed view of the boosting behavior of the DPC; U_{IN} (red), U_{cz} (yellow), U_z (blue), I_{Lz} (green)

7. Conclusion

This paper presents a novel PDC control structure that is based on the idea of DTC. The proposed control method extends the well-known DPC scheme and includes the boosting operation of the ZSI. The appropriate switching states of the inverter are selected from a table in order to control the instantaneous values of P and Q in a way that only active power is fed into the grid. The advantage of this method is that it does not require a PWM modulator that can cause problems in other control schemes. One main advantage of the proposed scheme is the low computational effort: in fact it could be implemented on a simple dsPIC µController platform with a sampling time of 15µs. For validating the proposed control method by experiment active power was injected into the grid by successfully including the boosting of the ZSI.

8. Reference

[1] Mario Pacas, Manuel Steinbring, *Modified Control Structure for Single Phase Z-Source-Inverter and Efficiency Analysis,* PCIM 2012, May 2012

[2] Mario Pacas, Manuel Steinbring, Mohammed Allnajar, *Emulation of a Micro-Hydro-Turbine for Standalone Power Plants with Z-Source Inverter,* IECON 2012, October 2012

[3] Mario Pacas, Manuel Steinbring, *Resonant circuit for the reduction of the power pulsation in the DC-link of a single phase ZSI, IECON 2014,* Dallas, TX 2014

[4] Mario Pacas, Manuel Steinbring, *Increasing the Efficiency of a Single Phase Z-Source Inverter by utilizing SiC – MOSFETS,* PCIM 2014, May 2014

[5] Fang Zheng Peng, *Z-Source-Inverter,* IEEE Transactions on industry applications, Vol. 39, No. 2, Mar./Apr. 2003

[6] Fang Zheng Peng, Miaosen Shen, Zhaoming Qian, *Maximum Boost Control of the Z-Source Inverter,* IEEE transactions on power electronics, Vol. 20, No. 4, July 2005

[7] Miaosen Shen, Jin Wang, Alan Joseph, Fang Zheng Peng, Leon M. Tolbert, Donald J. Adams, *Constant Boost Control of the Z-Source Inverter to Minimize Current Ripple and Voltage Stress,* IEEE transactions on industry applications, Vol. 42, No. 3, May/June 2006

[8] M. Malinowski, M. P. Kazmierkowski, S. Hamen, F. Blaabjerg, G. Marques, *Virtual Flux Based Direct Power Control of Three-phase PWM Rectifiers,* Industry Applications Conference, 2000, Rome

[9] Mariusz Malinowski, Sensorless Control Strategies for Three - Phase PWM Rectifiers, Ph.D. Thesis, Warsaw University of Technology, 2001, Warsaw

Interleaved series input parallel output forward converter with simplified voltage balancing control

Kaspars Kroics, Riga Technical University Cesis Branch, Latvia, kaselt@inbox.lv

Alvis Sokolovs, Riga Technical University Cesis Branch, Latvia, alvis.sokolovs@rtu.lv

Ugis Sirmelis, Riga Technical University Cesis Branch, Latvia, ugis.sirmelis@rtu.lv

Linards Grigans, Riga Technical University Cesis Branch, Latvia, linards.grigans@gmail.com

Abstract

The interleaved multilevel DC-DC converters have advantages of low voltage stress of the switches and diodes and reduction of filter size. Particularly series input parallel output (ISOP) configuration is well suited for high output voltage and large output current application. While multilevel topology offers many new features, it also necessitates a balance control of the input capacitors. The paper discusses design of the ISOP DC-DC converter with independent capacitor voltage balancing. The paper describes the operating principles of the balancing circuit, analyzes the fundamental relationships, introduces with control principles and experimental results based on a 3 kW prototype with 600 V input voltage and 30 V 100A output .

Introduction

The input-series output-parallel (ISOP) configuration consists of two modular DC-DC converters connected in series at the input and in parallel at the output, enabling the use of high switching frequency metal oxide semiconductor field effect transistors (MOSFETs) with low voltage ratings, which leads to a high power density and a high conversion efficiency. As output current ripple frequency is twice of the switching frequency, size and costs of the output filter can be reduced. MOSFETs with high voltage ratings have high on state resistance that leads to high conduction losses so by using series connection of the input transistors efficiency of the converter can also be improved.

In literature [1]–[6] different transformer demagnetizing schemes of the forward converter were discussed. The using of a third demagnetizing winding or RCD network leads to high voltage on the MOSFET. The two-switch forward converter have the lowest voltage stress on MOSFET [7] therefore this topology was selected for particular converter as high side driver for multilevel topology anyway is necessary. There are a lot of other resetting circuits but usually those are complex.

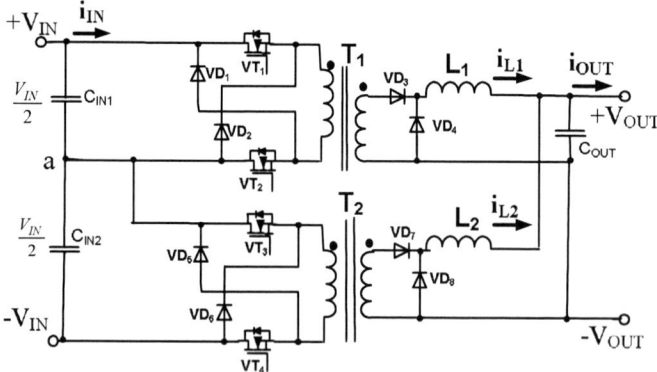

Fig. 1. Interleaved ISOP forward DC-DC converter

Figure 1 illustrates circuit configuration of the two level forward interleaved ISOP converter. When two converters are connected in series, a small mismatch in parameters always exists between them. This leads to unequal individual output currents of the converters. As a result, charge balance of capacitors is caused by the unbalanced output currents. Therefore one of the biggest challenge in such converter design is to ensure equal voltage sharing at input capacitors of the both input legs. Even output current sharing do not ensure equal input voltages [8]. If the input voltage is not balanced, the input power of the modules will be unbalanced, the input voltage of one of the module can exceed voltage rating of the capacitor or the MOSFET which will cause system failure. To keep input voltage sharing balance circuit or control method is necessary.

In [9], [10] combination of two transformers in one magnetic core is used to balance the input voltages, in [11] equal duty cycle is utilized to balance the input voltages: the module with higher input voltage supplies more current. For interleaved converter common duty cycle in the transient process can not be realized. In papers [12]–[16] input voltage feedback loops are applied to balance the input voltage. This complicates the design of the control system, isolated high voltage sensing circuit is required. This article will highlight the balancing method, that allows equal input voltage sharing in a simple way without complex control system.

Theoretical analysis of voltage misbalance

Figure 2 shows simple capacitive voltage divider that is used to obtain two voltage levels. If the system is balanced the voltage of the capacitor is equal to $V_{IN}/2$ and current i_N is zero.

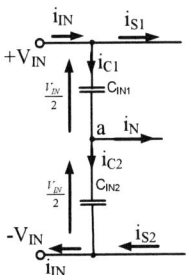

Fig. 2. Simplified circuit of the voltage divider used in theoretical analysis

To understand the capacitor voltage behaviour, it is necessary to study the capacitor current while the capacitor voltage and current are directly related. Analyzing the current through the capacitors of the circuit shown in figure 2, the following equations can be obtained:

$$i_{C1} = i_{IN} - i_{S1} ,$$ (1)

$$i_{C2} = i_{IN} - i_{S2} ,$$ (2)

$$i_N = i_{C1} - i_{C2} ,$$ (3)

$$i_{C1} = C_{IN1} \frac{dv_{C1}}{dt} .$$ (4)

By substitution (4) into (3) , it is obtained:

$$i_N = C_1 \cdot \frac{dv_{C1}}{dt} - C_2 \frac{dv_{C2}}{dt} .$$ (5)

As input voltage is the sum of the capacitor voltages, it follows:

$$\frac{dV_{IN}}{dt} = \frac{dv_{C1}}{dt} + \frac{dv_{C2}}{dt} \tag{6}$$

The input voltage is considered constant, free of oscillations, so its derivative is equal to zero, then equation (6) can be simplified:

$$\frac{dv_{C1}}{dt} = -\frac{dv_{C2}}{dt} \tag{7}$$

Considering $C_1 = C_2$ and replacing (7) in (6), it is obtained the relationship between i_N and v_{C1}, as follows:

$$i_N = 2C_1 \frac{dv_{C1}}{dt}. \tag{8}$$

By rearranging of (8), we obtain:

$$\frac{dv_{C1}}{dt} = \frac{i_N}{2C_1} \tag{9}$$

From the other side, current i_N can be expressed through the load current as follows:

$$i_N = i_{S2} - i_{S1} \tag{10}$$

From these equations can be concluded that difference in currents through transistors (i_{S1}, i_{S2}) causes misbalance of the capacitors voltage. The difference of currents of the both legs i_{S1}, i_{S2} can be called by unequal inductance of the inductors, transformers, unlike parameters of semiconductors or different duty cycle, specially in the transient process.

Misbalance in the voltage causes larger current in the leg with higher voltage so slow process of stabilization will take place but mostly it is not enough to hold voltage of the capacitor in the desired level. Therefore it is necessary some solution to prevent situation when voltage of the capacitor reaches dangerous value and damages semiconductors or passive elements. The solution must compensate current i_N thus preventing misbalance of the capacitor voltages. The scheme must be sufficiently simple with as small as possible power losses.

The traditional voltage balancing methods

The simplest method to balance voltages of the capacitors is using of a voltage divider which consists of two resistors R_B, such circuit is shown in figure 3. For practical application value of this resistors must be selected such that the expected misbalance current i_N is small in comparison to the current through resistors by equal voltages on the capacitors divider to achieve voltage deviation ΔU as close as possible to zero. Even more the resistors R_B must be selected taking into account worst-case misbalance current i_N. If the actual value of i_N is smaller, losses in balancing resistors still remain permanently high.

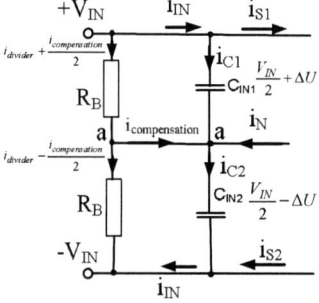

Fig. 3. Passive voltage balancing circuit

As difference of the leakage currents of the both capacitances are small ($i_{C1}\sim=i_{C2}$) in comparison to misbalance current i_N, it follows:

$$i_N = i_{compensation} \cdot \tag{11}$$

The voltage deviation of the capacitor can be expressed as follows:

$$\Delta V = \frac{1}{2} \cdot i_{compensation} \cdot R_B = \frac{1}{2} \cdot i_N \cdot R_B \cdot \tag{12}$$

The normalized deviation by the nominal voltage $V_{IN}/2$ of the capacitors is equal to:

$$\Delta v = \frac{\Delta V}{\frac{1}{2} V_{IN}} = \frac{\frac{1}{2} i_N R_B}{i_{divider} \cdot R_B} = \frac{1}{2} \frac{i_N}{i_{divider}} \tag{13}$$

If the deviation of the voltage of the capacitor Δv shall be less than 20 percents to do not breakdown transistors and capacitors, a value of the R_B has to be selected to ensure $i_{divider} > 2,5\ i_N$. Then the losses of the balancing resistors can be calculated as:

$$P_B = 2 \cdot i_{divider}^2 \cdot R_B + \frac{R_B}{2} \cdot i_N^2 = 2 \cdot R_B \cdot i_{divider}^2 (1 + \Delta u^2) \tag{14}$$

A misbalance current i_N=100 mA at V_{IN}=600 V DC results in balancing losses of P_B=156 W! If the converter is used often it leads to very large loses of the electrical energy. This suggests the development of an advanced balancing method.

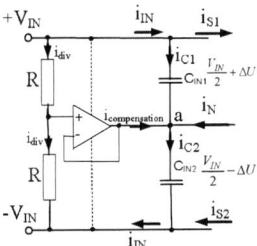

Fig. 4. Active voltage balancing

The active balancing method shown in figure 4 is proposed in [17] allows to avoid permanently losses in the resistors, the losses of the active balancing can be calculated:

$$P_{balancing} = \frac{1}{2} \cdot V_{IN} \cdot i_N \tag{15}$$

But still losses remain high as input voltage is high, additionally this balancing circuit requires a high-voltage operational amplifier which is an expensive.

Principle of the proposed balancing circuit

To balance voltages of the input voltage capacitive divider in the pulse transformer is placed additional winding as shown in figure 5. This winding is connected to the opposite input capacitor. Figure 5 explains operational principle of the proposed balancing circuit. If the voltage of the balancing winding (equal to the voltage of the opposite capacitor) is higher than of the capacitor, current during the on time of the corresponding transistor flows from balancing winding to the capacitor and partly compensates current i_N thus gradually equalizing the voltage levels of the both capacitors. The main task of the balancing is to prevent dangerous values of the voltage, additionally there is realized circuit that switch of the converter if the voltage of one of the capacitors becomes critically high.

© VDE VERLAG GMBH · Berlin · Offenbach

PCIM Europe 2016, 10 – 12 May 2016, Nuremberg, Germany

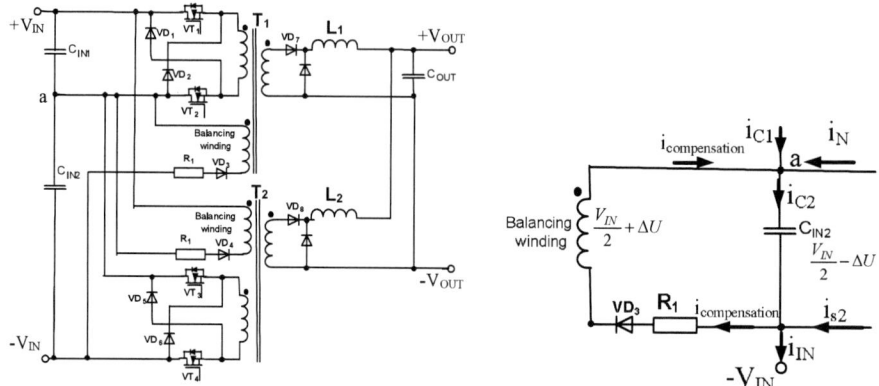

Fig. 5. Interleaved input series output parallel dc-dc converter with balancing circuit and operational principle of the balancing circuit

Figure 6 shows structure of the proposed transformer which consists of round ferrite core and three windings. Number of turns of the secondary winding and balancing winding are equal to 27. For the balancing winding wire with cross sectional area equal to 0,2 mm^2 is used. For the secondary winding litz wire with cross sectional area equal to 7 mm^2 is used. The balancing resistor in series with balancing winding must be selected according to desired rapidity of voltage balancing process. Must be considered that resistor with low resistance leads to higher current therefore cross sectional area of the balancing winding must be increased. In the particular case balancing resistor has value around 800 Ω.

Fig. 6. The transformer with integrated balancing winding

The power losses in the resistor can be expressed as follows:

$$P_{balancing} < 2 \cdot \Delta U \cdot i_N . \tag{16}$$

The power losses in this circuit is low therefore compact balancing resistor can be used with power of less than watt, additional balancing windings can be created by using wire with small cross-sectional area. Figure 7 shows the oscillogram with voltage balancing process. The voltage equalization circuit is not provided to hold equal voltages in all conditions. The main task of the balancing winding is to ensure that voltage does not reach dangerous value. In particular case there is voltage shifting margin of safety equal to 70 volts as maximum allowed voltage of the input capacitor is equal to 350 V. And other task is equate voltages when misbalance current is over. In the particular case the main voltage

© VDE VERLAG GMBH · Berlin · Offenbach

misbalance cause is unequal current in both output inductors as the control loop regulates only summary output current.

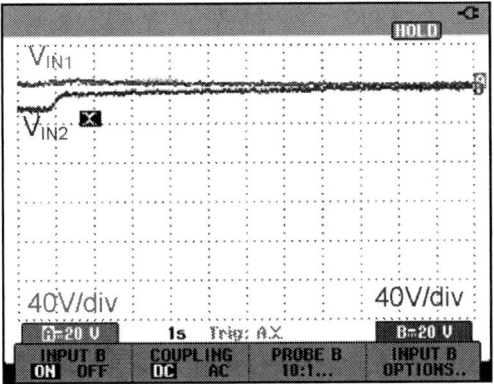

Fig. 7. The process of voltage balancing

In the oscillogram shown in the figure during first moment the converter works with full load therefore voltage balancing circuit can not fully compensate misbalance current but the voltage stays in the safety margin. After reducing of the load balancing circuit equalize both voltages. The oscilogram is measured in the short circuit and no load conditions in which the voltage equalizing process is the worst as duty cycle in these modes is small. But even in the following conditions balancing circuit works satisfactorily.

The hardware of the ISOP converter

The particular converter is designed for input voltage 600 V DC and output voltage from 0-30 volts, output current 100 A. The control of the converter is realized in two ways - by using analogue switch mode PWM circuit or by using DSP microcontroller. Both of the control methods was tested and both are suitable for such application. Fig. 8 shows hardware of the converter

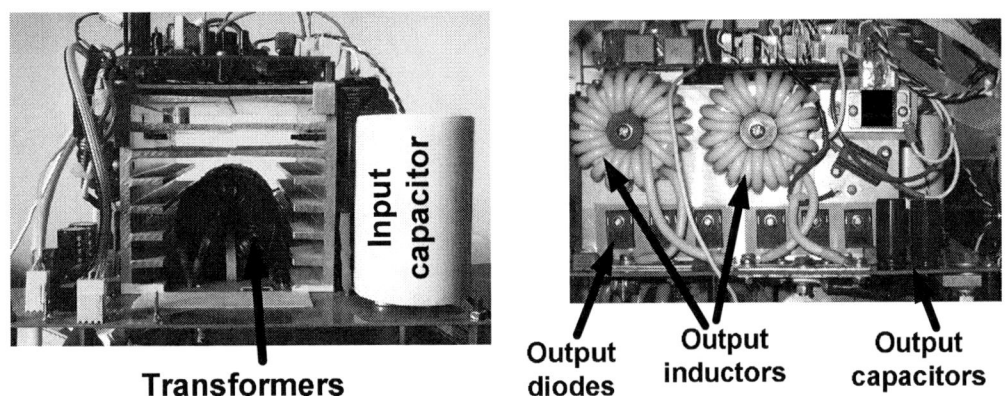

Fig. 8. The hardware realization of the ISOP DC-DC converter

For transformer design it is used N87 material two ring ferrite cores to increase effective cross sectional area of the core. For inductor it is used metal powder cores. Output diodes are connected 3 in parallel to minimize power losses. Input capacitors have high value as converter also was tested from the 3 phase rectified AC power. The transformer based drivers were designed to drive MOSFET transistors. The efficiency of the converter is approximately 90 percent but it is possible efficiency improvement by using semiconductors with better parameters and by implementing soft-switching circuits.

PCIM Europe 2016, 10 – 12 May 2016, Nuremberg, Germany

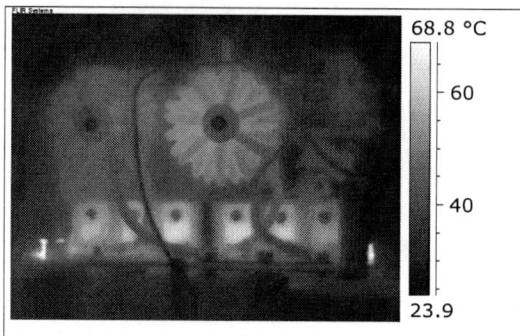

Fig. 9. The thermal image of the converter working with nominal load

Figure 9 shows thermal image of the converter wit the hottest spots. To cool down the semiconductors and magnetic elements the converter is equipped with radiator and electric air exhauster.

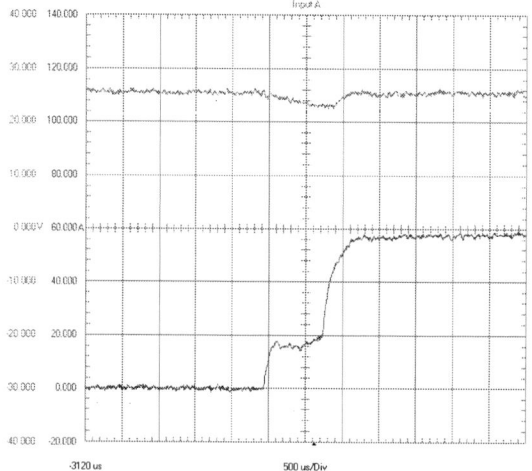

Fig. 8. The transient process of load changing: from no-load condition to 400 W and to 1500W
(red line - output voltage; blue line - output current)

As input voltage balancing works independently from the control system, the control system of the ISOP converter is simpler. As input voltage is controlled only the output voltage and total output current must be controlled. The control of the converter can be realized by conventional PWM controller or digital controller. Figure 8 shows transient process of the converter. The process is stable, if faster transient response is required the control loop can be easily optimized.

Conclusion

An ISOP topology enables the use of MOSFETs with lower voltage rating and lower $R_{DS\,ON}$. The voltage balancing of the capacitors of the multilevel converter can be big challenge. In the paper proposed method to solve this problem without implementation of the digital control. The circuit is simple and cheap, it allows prevent rising of the capacitor voltage to the dangerous voltage with negligible power losses. The proposed circuit is useful for the isolated multilevel converters. The balancing method is validated experimentally of the 3 kW ISOP forward converter. An important advantage of the proposed input balancing circuit is that sensing of input voltages of the converter is no required.

© VDE VERLAG GMBH · Berlin · Offenbach

References

[1]   E. H. Wittenbreder, V. D. Baggerly, and H. C. Martin, "A duty cycle extension technique for single ended forward converters," in Applied Power Electronics Conference and Exposition, 1992. APEC '92. Conference Proceedings 1992., pp. 51–57.

[2]   N. Murakami and M. Yamasaki, "Analysis of a resonant reset condition for a single-ended forward converter," in , 19th Annual IEEE Power Electronics Specialists Conference, 1988. PESC '88 Record, 1988, pp. 1018–1023.

[3]   A. K. S. Bhat and F. D. Tan, "A unified approach to characterization of PWM and quasi-PWM switching converters: topological constraints, classification and synthesis," in 20th Annual IEEE Power Electronics Specialists Conference, 1989. PESC '89 Record, 1989, pp. 760–767.

[4]   Q. M. Li and F. C. Lee, "Design consideration of the active-clamp forward converter with current mode control during large-signal transient," IEEE Transactions on Power Electronics, vol. 18, no. 4, 2003, pp. 958–965.

[5]   K. Harada and H. Sakamoto, "Switched snubber for high frequency switching," in , 21st Annual IEEE Power Electronics Specialists Conference, 1990. PESC '90 Record, 1990, pp. 181–188.

[6]   F. D. Tan, "The forward converter: from the classic to the contemporary," in Seventeenth Annual IEEE Applied Power Electronics Conference and Exposition, 2002. APEC, 2002, vol. 2, pp. 857–863.

[7]   T. Jin, K. Zhang, K. Zhang, and K. Smedley, "A New Interleaved Series Input Parallel Output (ISIPO) Forward Converter With Inherent Demagnetizing Features," IEEE Transactions on Power Electronics, vol. 23, no. 2, Mar. 2008, pp. 888–895.

[8]   X. Ruan, L. Cheng, and T. Zhang, "Control Strategy for Input-Series Output-Paralleled Converter," in 37th IEEE Power Electronics Specialists Conference, PESC '06, 2006, pp. 1–8.

[9]   S. Yang, Y. Fang, X. Qiu, and C. Gong, "Voltage Sharing Control for Interleaving Series-Parallel Dual Two-Transistor Forward Converter," in 33rd Annual Conference of the IEEE Industrial Electronics Society, 2007. IECON 2007, 2007, pp. 1896–1900.

[10]   T. Jalakas and J. Zakis, "Experimental verification of light electric vehicle charger multiport topology," in 2015 9th International Conference on Compatibility and Power Electronics (CPE), 2015, pp. 415–418.

[11]   R. Giri, V. Choudhary, R. Ayyanar, and N. Mohan, "Common-duty-ratio control of input-series connected modular DC-DC converters with active input voltage and load-current sharing," IEEE Transactions on Industry Applications, vol. 42, no. 4, 2006, pp. 1101–1111.

[12]   D. V. Ghodke and K. Muralikrishnan, "ZVZCS, dual, two-transistor forward DC-DC converter with peak voltage of Vin/2, high input and high power application," in Power Electronics Specialists Conference, 2002, vol. 4, pp. 1853–1858.

[13]   W. Chen, G. Wang, X. Ruan, W. Jiang, and W. Gu, "Wireless Input-Voltage-Sharing Control Strategy for Input-Series Output-Parallel (ISOP) System Based on Positive Output-Voltage Gradient Method," IEEE Transactions on Industrial Electronics, vol. 61, no. 11, 2014, pp. 6022–6030.

[14]   G. Xu, D. Sha, and X. Liao, "Decentralized Inverse-Droop Control for Input-Series Output-Parallel DC DC Converters," IEEE Transactions on Power Electronics, vol. 30, no. 9, 2015, pp. 4621–4625.

[15]   R. Giri, R. Ayyanar, and E. Ledezma, "Input-series and output-series connected modular DC-DC converters with active input voltage and output voltage sharing," in Nineteenth Annual IEEE Applied Power Electronics Conference and Exposition, 2004, pp. 1751–1756.

[16]   A. Bhinge, N. Mohan, R. Giri, and R. Ayyanar, "Series-parallel connection of DC-DC converter modules with active sharing of input voltage and load current," in Seventeenth Annual IEEE Applied Power Electronics Conference and Exposition, APEC 2002, 2002, vol. 2, pp. 648–653.

[17]   H. Ertl, T. Wiesinger, and J. W. Kolar, "Active voltage balancing of DC-link electrolytic capacitors," IET Power Electronics, vol. 1, no. 4, 2008, pp. 488–496.

Fault-Tolerant Behaviour of the Three-Level Advanced-Active-Neutral-Point-Clamped Converter

Sidney Gierschner, University of Rostock, GERMANY, sidney.gierschner@uni-rostock.de
David Hammes, University of Rostock, GERMANY, david.hammes@uni-rostock.de
Jan Fuhrmann, University of Rostock, GERMANY, jan.fuhrmann@uni-rostock.de
Max Beuermann, Siemens AG, GERMANY, max.beuermann@siemens.com
Hans-Günter Eckel, University of Rostock, GERMANY, hans-guenter.eckel@uni-rostock.de

Abstract

Multilevel converters in medium-voltage applications reduce the cable cross-section, the voltage step size and the necessary filter effort in comparison to the conventional two-level converter. While at low-voltage level, several converters can be connected in parallel at low cost, at medium-voltage level this redundancy is missing. In case of a failure, the whole system cannot continue to operate. This paper presents a three-level converter topology as solution for medium voltage. A three-level converter based on half-bridges is divided into two partial converters, which are high-inductively coupled. A failure is limited to one half of the converter and half of the DC-link voltage remains. Redundant operation at reduced power is available. The protective inductance is determined, based on measurements on a 6.5 kV-IGBT .

1. Introduction

For high-power applications at low voltage, several converters in parallel are realized at low costs. Due to splitting the total power to several systems, the surge currents in the machine in case of a failure in the converter are reduced. At the same time redundancy is provided, which enables operating at lower power when losing one system. But high power at low voltage leads to high currents and large cable cross-sections as well as high cable losses. Therefore, a change to medium voltage will be an appropriate way. Converters at medium-voltage level have smaller cable cross-sections and lower cable losses. But at higher DC-link voltages the common two-level converter needs a serial connection of power semiconductors to obtain sufficient blocking capability. Therefore multilevel converters, which enable a higher DC-link voltage at the same blocking capability of the IGBTs, are widespread for medium-voltage applications. But by the complex topology of multilevel converters, a parallelisation of several systems is expensive and therefore avoided. When feeding a machine, no parallel systems and so no splitting of the total power will lead to high surge currents in the machine as well as high impact moments for the whole drive train in case of a failure in the converter. This may lead to consequential damages. Due to no redundancy, the whole system cannot continue to operate.

A new topology based on the Advanced-Active-Neutral-Point-Clamped converter (AANPC converter) presented in [1] is investigated within this paper. In case of a short-circuit failure, the new topology brings down the risk of consequential failures of the remaining semiconductors as well as the risk of damaging the whole drive train, due to a reduction of the surge currents. In addition, the new topology enables operating at reduced power after a failure by providing still half of the DC-link voltage.

2. The Advanced-Active-Neutral-Point-Clamped Converter

The three-level Active-Neutral-Point-Clamped (ANPC) converter presented in [2] uses active switches instead of clamping diodes for the connection to the midpoint of the DC-link. Thus, several current paths to the midpoint of the DC-link are available, independent of the current direction. The more complex commutation circuits of the three-level converter, compared to the two-level converter, involve several power semiconductors and conceivably a DC-link capacitor. Due to higher parasitic inductances the switching losses are higher [3].

An ANPC converter based on half-bridge modules is presented in [1]. As shown in Fig. 1(a) the switches T_{11}/D_{11} and T_{10}/D_{10} respectively T_{20}/D_{20} and T_{21}/D_{21} form half-bridges with a low-inductive connection to the DC-link capacitor as used in two-level converters. The load terminals of the half-bridges, which are high-inductive, are connected to the output via T_{12}/D_{12} respectively T_{22}/D_{22}. Considering the three-phase topology, the so-called AANPC converter consists of two independent two-level converters. An output half-bridge for each phase connects both.

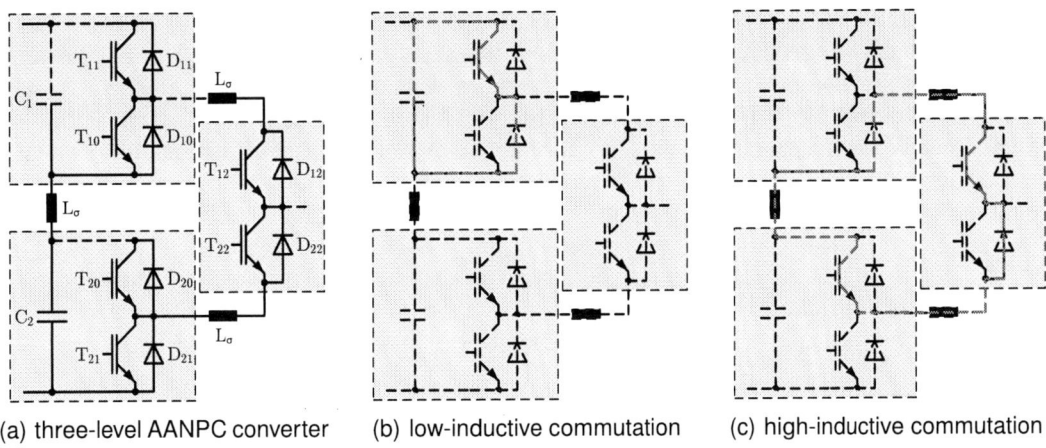

(a) three-level AANPC converter   (b) low-inductive commutation   (c) high-inductive commutation

Fig. 1: topology of one phase of the AANPC converter (a), commutation during normal operation (b), (c)

For the conventional ANPC converter, there are several modulation schemes [4]. In contrast, not all available switching states are allowed for the AANPC converter. Due to the high-inductive connection of upper and lower half-bridge to the output stage, switching states that lead to a commutation circuit involving the high inductance and a DC-link capacitor are forbidden to avoid overvoltages across the power semiconductors. Only two types of commutation circuits are allowed during normal operation. As depicted in Fig. 1(b), within the upper respectively the lower half-bridge, there is a low-inductive commutation circuit that involves half of the DC-link voltage. A transition from the upper to the lower half-bridge and vice versa is only allowed by the use of the high-inductive commutation circuit without any DC-link voltage, as illustrated in Fig. 1(c). A modulation scheme based on space-vector modulation, which guarantees save operation by avoiding forbidden transitions, is presented in [5].

© VDE VERLAG GMBH · Berlin · Offenbach

3. Short-Circuit Failure in Three-Level Converter

During normal operation the switching losses should be low. Therefore the bus bar of the converter has to be designed as low-inductive as possible. But this is a drawback in case of a short-circuit failure of a semiconductor. In most cases, the failure of a semiconductor will lead to additional failures of semiconductors of the same phase, which are involved into the short-circuit path, due to the high occurring di/dt and the high overvoltage.

The occuring short circuits can be subdivided in dependence on current direction and switching state of the involved semiconductors. The short-circuits type I (SC I) and II (SC II) concern the IGBT. A detailed description can be found in [6]. An occuring short circuit during diode conduction mode is named short-circuit type III (SC III), while the anti-parallel IGBT is switched ON [7], respectively short-circuit type IV (SC IV), while the anti-parallel IGBT is switched OFF [8]. A new short-circuit type V (SC V) with a passive turn-OFF of the IGBT, which can occur in the ANPC converter, is presented first in [9].

There are only a few publications regarding the short-circuit behaviour of three-level converters. The short-circuit failure-modes in the NPC converter are investigated in [10] for low-voltage IGBTs. But only short-circuit types that affect the IGBT are concerned. The SC IV, most important for the diode is neglected. In [11] there are possibilities shown to continue operation after a short circuit. Both presuppose no subsequent failures caused by a short circuit, but this is very likely in actual applications. Another important fact about short circuits in three-level converters is that different short-circuit types can occur at the same time, due to the higher number of power semiconductors and the more complex commutation circuits.

As said before, due to the low-inductive bus bar, a primary failure of a semiconductor will lead to subsequent failures of other semiconductors. The appearance of such secondary and tertiary failures caused by a short circuit in the ANPC converter is shown in Fig. 2.

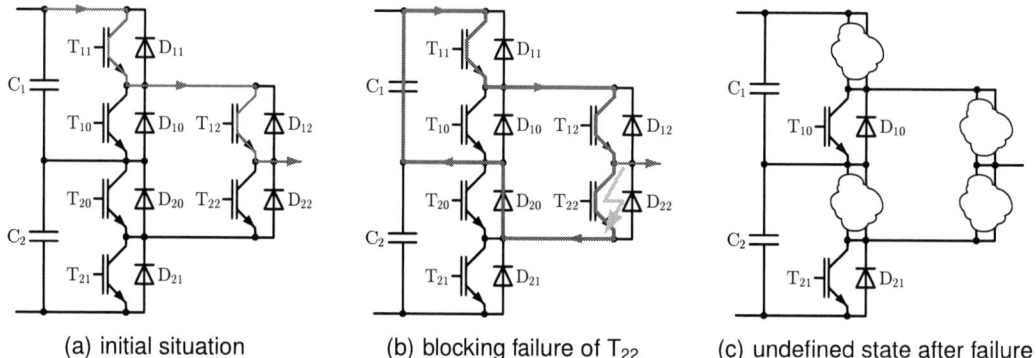

(a) initial situation   (b) blocking failure of T_{22} (c) undefined state after failure

Fig. 2: subsequent failures of semiconductors in ANPC converter

Before the failure, the IGBTs T_{11} and T_{12} are conducting the current [Fig. 2(a)]. The initiating event is a blocking failure of IGBT T_{22}, which leads to a short circuit inside the converter. The upper half of the DC-link drives a short-circuit current through T_{11}, T_{12}, T_{22} and D_{20} [Fig. 2(b)]. Since the parasitic inductance of the bus bar is small to prevent overvoltages during switching, both IGBT T_{11} and T_{12} are in a low-inductive SC II, which is the most critical situation for them. The current rise is determined by the parasitic inductance of the short circuit. The short-circuit current will rise until the IGBTs are desaturated. In most cases, the high resulting power loss will lead to the destruction of T_{11} and T_{12}. Discharging of the upper half of the DC-link causes a surge current of several hundred kA, which will destroy diode D_{20}. The primary failure of IGBT T_{22} causes secondary and tertiary failures of three additional semiconductors and therefore four switches are damaged respectively in an undefined state after the failure [Fig. 2(c)].

4. Fault-Tolerant Behaviour of the AANPC Converter

For the same blocking failure of T_{22}, the AANPC converter has a higher inductance L_σ in the short-circuit path. The current rise is lower and so the stress of the IGBTs T_{11} and T_{12} is decreased. This brings down the risk of destruction. The primary blocking failure of T_{22} causes a short-circuit current through T_{11}, T_{12}, T_{22} and D_{20} [Fig. 3(a)]. IGBT T_{11} is turned OFF and the current commutates to diode D_{10}. The DC-link capacitor C_1 is no longer part of the short circuit. The remaining current in the free-wheeling path is only driven by the inductance [Fig. 3(b)].

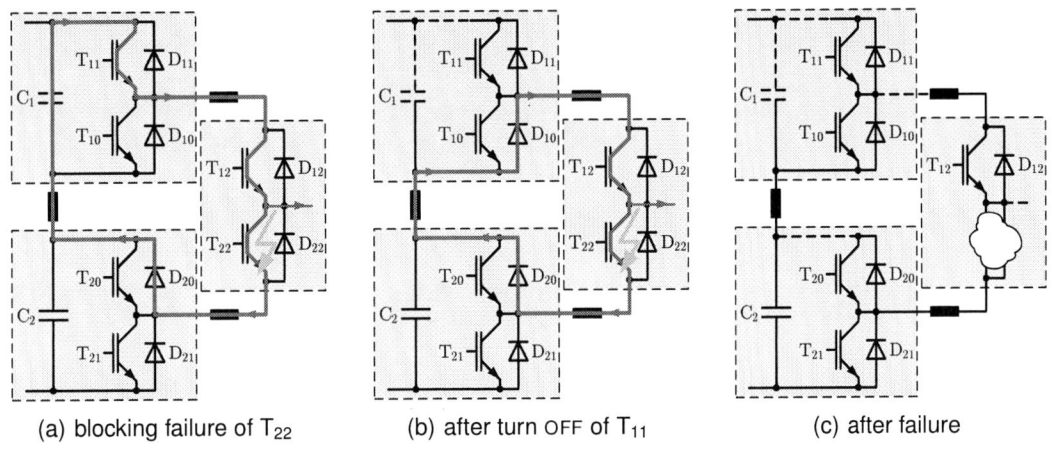

(a) blocking failure of T_{22} (b) after turn OFF of T_{11} (c) after failure

Fig. 3: behaviour of AANPC converter during short circuit

Because of the higher inductance involved into short circuit, subsequent failures of semiconductors, caused by the primary blocking failure of T_{22}, are avoided. Since the AANPC converter consists of half-bridges, only one module has to be changed instead of the whole phase. The required inductance has to be chosen in the right way. Due to the considered topology, the AANPC converter can continue to operate at half of the DC-link voltage, as a two-level converter, after switching OFF the short circuit.

4.1. Determination of required Inductance

On the one hand, the required inductance L_σ has to be as high as needed to limit current rise, to guarantee to switch OFF the short circuit quite safely. But, on the other hand, it has to be as low as possible to do not affect switching actions in a negative way. To determine the inductance L_σ, the most critical event has to be considered, which is SC IV for the diode.

© VDE VERLAG GMBH · Berlin · Offenbach

According to the data sheet, the free-wheeling diode of the tested IGBT module is sized to withstand a maximum power dissipation of 3 MW at a junction temperature of 125 °C. This value is a direct indication of the level of the electric field strength. Exceeding the maximum power loss leads to possible failures of the diode during switching, due to exceeding critical field strength. For determination of L_σ, the diode must not exceed maximum power loss during short circuit. SC IV for diode D_{21} caused by a blocking failure of T_{11} is shown in Fig. 4. As initial situation T_{20} is switched OFF and the current commutates to D_{21}. During the interlock time T_{21} is in OFF-state [Fig. 4(a)]. Due to a blocking failure of T_{11}, the voltage drop across T_{12} will rise to full DC-link voltage without any measures to limit it. Therefore, an active gate-clamping limits the voltage drop across T_{12} and a short-circuit current can flow through T_{11}, T_{12}, T_{22} and D_{21}. Diode D_{21} suffers SC IV [Fig. 4(b)]. Then the current commutates to D_{20} and the capacitor C_2 is no longer involved [Fig. 4(c)]. As last step, the IGBTs T_{12} and T_{22} are switched OFF.

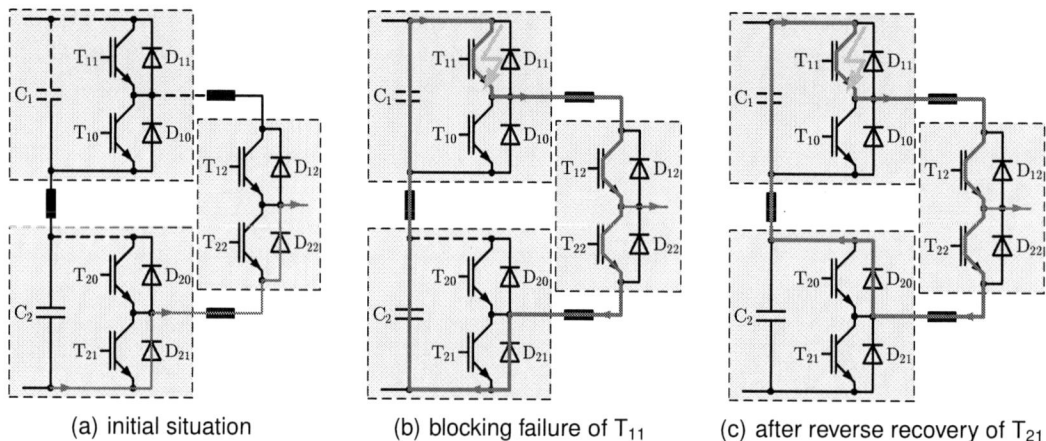

(a) initial situation   (b) blocking failure of T_{11} (c) after reverse recovery of T_{21}

Fig. 4: short circuit type IV of D_{21} for determination of required inductance

Considering a converter equipped with 6.5 kV-IGBTs, a typical DC-link voltage is 7.2 kV. Determination of the inductance L_σ has to be done regarding the worst case, which is at maximum allowed DC-link voltage. Assuming a maximum DC-link voltage of 8.2 kV and a limitation of the voltage drop across T_{12} to 6 kV by the active clamping, the maximum voltage drop across D_{21} is 2.2 kV. This worst case value is used for determination of the inductance L_σ within the test bench shown in Fig. 5. First, the load inductance L_{load} is charged by switching ON T_1 [Fig. 5(a)]. After turn OFF of T_1, the current is in free-wheeling mode through D_2 [Fig. 5(b)]. Next, T_1 is switched ON again much faster than in normal operation. Therefore, no normal commutation, but a SC IV for diode D_2 occurs [Fig. 5(c)].

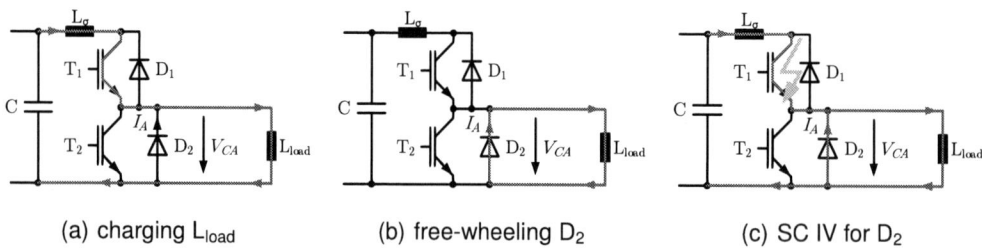

(a) charging L_{load} (b) free-wheeling D_2 (c) SC IV for D_2

Fig. 5: test bench for determination of required inductance L_σ

Inductance L_σ is varied to fulfil requirements for maximum short-circuit power loss of the diode under worst case conditions of $V_{CA} = 2.2\,\text{kV}$ and $I_C = 1.5\,\text{kA}$. Measurements are done at a junction temperature of $T_j = 25\,°\text{C}$ as well as $T_j = 125\,°\text{C}$.

Fig. 6 shows the measurement of a SC IV to determine the inductance L_σ. When the short circuit occurs at $0.7\,\mu\text{s}$, the current flowing through the diode decreases. As the current crosses zero, the diode is able to take voltage. During reverse recovery, the current becomes negative and the charge is extracted from the diode. The maximum peak value of the reverse-recovery current is about $1\,\text{kA}$, when V_{CA} becomes the DC-link voltage. Afterwards, the positive di/dt leads to an overvoltage of $1.7\,\text{kV}$, due to the parasitic inductance. The short-circuit power loss does not exceed the limit of $3\,\text{MW}$. During tail current, the remaining charge is extracted. The gradient of the voltage across the diode is more than twice the value, which is reached during normal switching conditions. Also the overvoltage is much higher. Therefore, the stress of the semiconductor during short circuit is significantly higher.

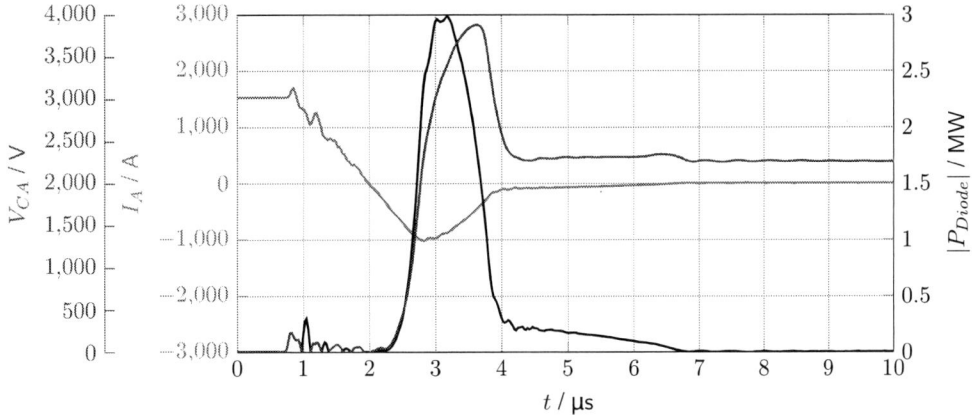

Fig. 6: measurement of SC IV, 6.5 kV / 750 A IGBT-module, $I_C = 1.5\,\text{kA}$, $V_{CA} = 2.2\,\text{kV}$, $T_j = 25\,°\text{C}$

According to the measurement the required inductance L_σ for the AANPC converer is determined to about $1.7\,\mu\text{H}$. A lower value will lead to an exceed of power loss during SC IV at the conditions mentioned above. Measurement at $125\,°\text{C}$ shows a slightly increase of the power dissipation up to $10\,\%$.

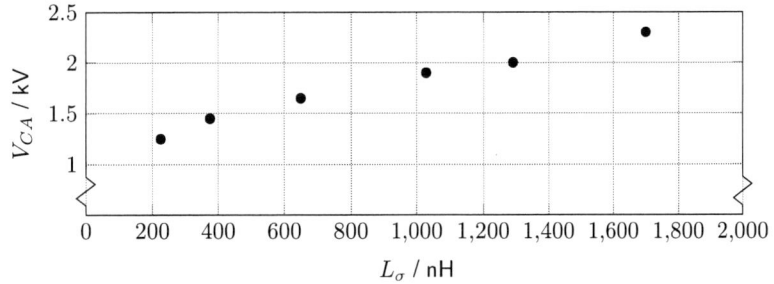

Fig. 7: dependence of V_{CA} and L_σ at constant short-circuit power of $3\,\text{MW}$

An almost linear dependence of the short-circuit inductance L_σ and the allowable short-circuit voltage at constant power of $3\,\text{MW}$ is shown Fig. 7.

© VDE VERLAG GMBH · Berlin · Offenbach

4.2. Redundant Operation at reduced Power

By dividing the converter into several partial converters, there is the opportunity of redundant operation at reduced power after the failure of a semiconductor. As shown above, the improved behaviour, in case of a short-circuit, avoids subsequent failures of semiconductors. The blocking failure, described above, only affects the upper half of the converter. By including a switch between the upper and the lower half of the converter, both can be disconnected. This switch can be either electrical or mechanical.

As illustrated in Fig. 8, the switch is opened after the failure and the upper half is disconnected. When IGBT T_{22} is switched ON, as well as IGBT T_{12} is switched OFF, permanently, the lower half of the converter can operate as a two-level converter with half of the DC-link voltage. Therefore, the whole system can still operate at reduced power until maintenance and downtime is reduced.

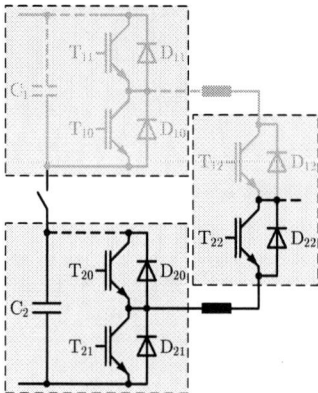

Fig. 8: redundant operation at reduced power after failure

5. Conclusion

A new three-level ANPC converter topology based on half-bridge modules is investigated. While a short-circuit failure will cause subsequent failures for conventional ANPC converter, the presented topology provides a high-inductive short-circuit path that limits the di/dt in case of a failure. Secondary and tertiary failures of semiconductors are avoided.

Determination of required inductance L_σ is done for most critical short-circuit failure of the diode at worst case conditions. Measurements show that an inductance L_σ of about 1.7 µH is needed to fulfil requirements regarding maximum power loss of the diode.

By introducing a switch between upper and lower half of the converter, both can be disconnected after a failure. Still half of the DC-link voltage is applied and an operation as two-level converter, at reduced power, is still possible. Therefore, downtime of the whole system is reduced.

© VDE VERLAG GMBH · Berlin · Offenbach

6. Acknowledgement

This project is a cooperation between the University of Rostock, the Wind to Energy GmbH and the Siemens AG, supported by the Federal Ministry for Economic Affairs and Energy on the basis of a decision by the German Bundestag.

References

[1] J. Fuhrmann and H.-G. Eckel: "Advanced Active Neutral Point Three Level Inverter with Standard Half-Bridge Modules." In *International Exhibition and Conference for Power Electronics, Intelligent Motion, Renewable Energy and Energy Management (PCIM Europe)*. 2014.

[2] T. Brückner and S. Bernet: "Loss balancing in three-level voltage source inverters applying active NPC switches." In *Power Electronics Specialist Conference (PESC)*. 2001.

[3] A. Bryant, K. Vadlapati, J. Starkey, A. Goldney, S. Kandilidis, and D. Hinchley: "Current distribution in high power laminated busbars." In *14th European Conference on Power Electronics and Applications (EPE)*. 2011.

[4] D. Floricau, E. Floricau, and M. Dumitrescu: "Natural doubling of the apparent switching frequency using three-level ANPC converter." In *International School on Nonsinusoidal Currents and Compensation (ISNCC)*. June 2008, pp. 1–6.

[5] J. Fuhrmann and H.-G. Eckel: "Implementation of a Space Vetor Modulation for an advanced ANPC Three Level Inverter." In *IEEE International Conference on Industrial Technology (ICIT)*. 2015.

[6] H.-G. Eckel and L. Sack: "Experimental investigation on the behaviour of IGBT at short-circuit during the on-state." In *20th International Conference on Industrial Electronics, Control and Instrumentation (IECON)*. 1994.

[7] J. Lutz, R. Dobler, J. Mari, and M. Menzel: "Short circuit III in high power IGBTs." In *13th European Conference on Power Electronics and Applications (EPE)*. 2013.

[8] S. Pierstorf and H.-G. Eckel: "Short-circuit behavior of diodes in voltages source inverters." In *International Exhibition and Conference for Power Electronics, Intelligent Motion, Renewable Energy and Energy Management (PCIM Europe)*. 2012.

[9] J. Fuhrmann and H.-G. Eckel: "Passive IGBT Turn-off during Short-circuit Type V." In *International Exhibition and Conference for Power Electronics, Intelligent Motion, Renewable Energy and Energy Management (PCIM Europe)*. 2016.

[10] M. Sprenger, R. Alvarez, M. Tannhäuser, and S. Bernet: "Experimental investigation of short-circuit failures in a three level neutral-point-clamped voltage-source converter phase-leg with IGBTs." In *IEEE Energy Conversion Congress and Exposition (ECCE)*. Sept. 2013, pp. 4067–4075.

[11] J. Li, A. Huang, Z. Liang, and S. Bhattacharya: "Analysis and Design of Active NPC (ANPC) Inverters for Fault-Tolerant Operation of High-Power Electrical Drives." In *IEEE Transactions on Power Electronics* 27.2 (Feb. 2012), pp. 519–533.

PCIM Europe 2016, 10 – 12 May 2016, Nuremberg, Germany

Cell voltage balancing controller for the modular multilevel converter arm using symmetrical transformation

Andrey Dudin, Aaron Fischer, Thomas Ellinger, Jürgen Petzoldt
Technische Universität Ilmenau, Germany
andrey.dudin@tu-ilmenau.de, aaron.fischer@tu-ilmenau.de,
thomas.ellinger@tu-ilmenau.de, juergen.petzoldt@tu-ilmenau.de

Abstract

This work describes a method for balancing of the cell capacitor voltages in a modular multilevel converter arm with phase-shift modulation. The voltage equalization control structure is developed using symmetrical transformation approach under consideration of the cell insertion index limitations. Proposed method is compared to sorting equalization algorithm in terms of converter control dynamics, equalization behavior and losses.

1. Phase-shift modulated modular multilevel converter

To control a maximum current frequency f_{max} in a pulse-width modulated (PWM) converter the switching frequency f_{sw} is set n times f_{max}, where n is often set 10 for a satisfactory current control. In the Modular Multilevel Converter (MMC, Fig. 1(a)) with m cells per arm (Fig. 1 (b)) the effective cell switching frequency is not always nf_{max} since it depends on the cell capacitor voltage equalization method. For the sorting algorithms a carrier switching frequency of f_{sw} results in the total arm switching frequency $f_{sw.arm} = \psi f_{sw}$, the factor ψ ($1 < \psi < m$) is estimated per experiment or simulation [1], [2], [3]. Phase-shift modulation (PSM) can be applied in converters with $m < n$, it offers a good current ripple and a well-predictable voltage spectrum. Disadvantageous are the arm voltage distortions when one or more cell insertion indices are in limitation as well as a slower cell capacitor voltage equalization dynamics.

To equalize the cell voltages in a PWM-MMC arm individual cell controllers are used [4], [5]. Fig. 1 (c) shows a control structure with individual cell controllers G_{Rz} having the arm current controller G_{Ri} in parallel. Controllers G_{Riarm} and G_{Ruarm} are a simplified representation of the MMC current and energy controllers known from the state of the art [4], [5].

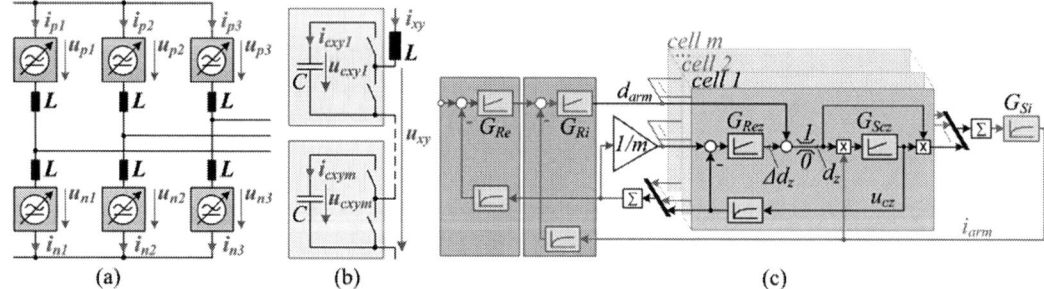

Fig. 1: (a) MMC circuit. (b) MMC arm. (c) Control structure with individual cell voltage controllers.

For a stable operation controller G_{Rez} should be significantly slower than G_{Ri} which leads to a worse equalization compared to the sorting algorithms. Additionally, if the sum of the arm insertion index d_{xy} ($x \in \{p,n\}$, $y \in \{1,2,3\}$) and the cell voltage controller output Δd_z crosses

© VDE VERLAG GMBH · Berlin · Offenbach

the physical boundaries $[0, 1]$ of a cell insertion index d_z (e.g. at transients), the arm voltage does not correspond to its reference value leading to arm current distortions.

2. Transformation-based control

2.1. Decoupled control in an idealized average model

To decouple the cell voltage equalizer from the arm current control, a transformation of the cell values into one zero sequence and $(m - 1)$ symmetrical components is applied:

$$[x]_M = [x_0 \quad x_{s1} \quad \cdots \quad x_{s(m-1)}]^T = [T]^{-1}[x]_N \tag{1}$$

$$[x]_N = [x_1 \quad \cdots \quad x_m]^T = [T][x]_M \tag{2}$$

$$[T]^{-1} = \frac{1}{m}\begin{bmatrix} 1 & 1 & 1 & \cdots & 1 \\ m-1 & -1 & -1 & \cdots & -1 \\ -1 & m-1 & \ddots & & \vdots \\ \vdots & \ddots & \ddots & -1 & -1 \\ -1 & \cdots & -1 & m-1 & -1 \end{bmatrix}, [T] = \begin{bmatrix} 1 & 1 & 0 & \cdots & 0 \\ 1 & 0 & 1 & \ddots & \vdots \\ \vdots & \vdots & \ddots & \ddots & 0 \\ 1 & 0 & \cdots & 0 & 1 \\ 1 & -1 & \cdots & -1 & -1 \end{bmatrix} \tag{3}$$

with m-size real cell values vector $[x]_N$, its symmetrical component representation $[x]_M$ and an $m \times m$ transformation matrix $[T]$. Assumed the cell capacitances and their voltages are equal ($u_{cxy1} = u_{cxy2} = \cdots = u_{cxym} = u_c$) and the arm current actual value is i_{xy}, the cell capacitor current vectors $[i_{cxy}]_N$, $[i_{cxy}]_M$ and the arm voltage depend on the cell insertion index vectors $[d_{xy}]_N$, $[d_{xy}]_M$:

$$[i_{cxy}]_N = i_{xy}[d_{xy}]_N \tag{4}$$

$$[i_{cxy}]_M = i_{xy}[T][d_{xy}]_M \tag{5}$$

$$u_{xy} = \sum_{z=1}^{m} u_{xyz}d_{xyz} = u_{cxy0}md_{xy0} \tag{6}$$

with u_{cxy0}, d_{xy0} - zero sequence components of the capacitor voltage and insertion indices vector respectively. Fig. 2(b) shows an equivalent circuit representation of one zero- and $(m - 1)$ symmetrical equivalent circuits obtained by the transformation of the averaged arm model in Fig. 2(a) using Eqn. (1) for the case of equal capacitor voltages.

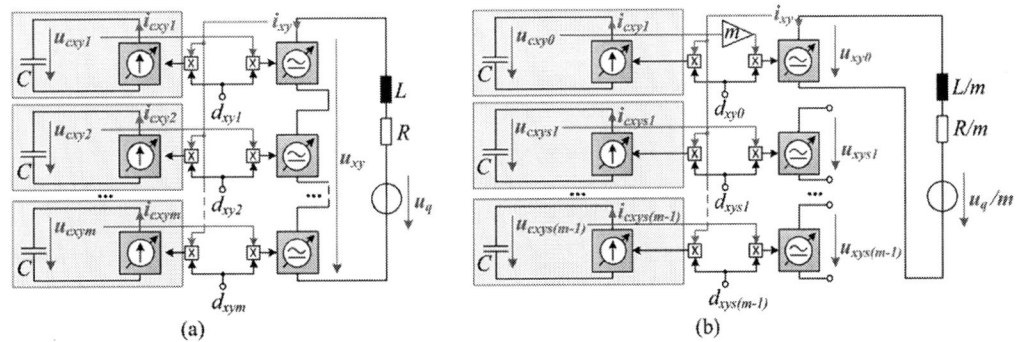

Fig. 2: Averaged MMC arm model. (a) Natural-values model. (b) Transformed-values model.

The arm current is controlled by the zero-sequence component d_{xy0}, the zero-sequence capacitor voltage u_{cxy0} represents the average cell voltage of the arm. Symmetrical components d_{xys1} to $d_{xys(m-1)}$ do not affect either arm current or the average arm energy. The real capacitor voltages are equalized if their symmetrical components u_{cxys1} to $u_{cxys(m-1)}$ are all zero.

2.2. Capacitor voltage inequalities

Eqn. (6) is only applicable under assumption of equal cell capacitor voltages. To avoid distortions caused by capacitor voltage inequalities, transformed cell powers are considered:

$$\begin{bmatrix} p_{xy0} \\ p_{xys1} \\ \vdots \\ p_{xy(m-1)} \end{bmatrix} = [T]^{-1} \begin{bmatrix} u_{cxy1} d_{xy1} i_{xy} \\ \vdots \\ u_{cxym} d_{xym} i_{xy} \end{bmatrix} = \frac{i_{xy}}{m} \begin{bmatrix} u_{cxy1} d_{xy1} + \cdots + u_{cxym} d_{xym} \\ (m-1) u_{cxy1} d_{xy1} - \displaystyle\sum_{\substack{z \in \{1,2,\ldots,m\}, \\ z \neq 1}} u_{cxyz} d_{xyz} \\ \vdots \\ (m-1) u_{cxy(m-1)} d_{cxy(m-1)} - \displaystyle\sum_{\substack{z \in \{1,2,\ldots,m\}, \\ z \neq (m-1)}} u_{cxyz} d_{xyz} \end{bmatrix} \tag{7}$$

Using Eqn. (7), cell insertion indices necessary to set the transformed powers references can be obtained:

$$[d_{xy1} \quad \cdots \quad d_{xym}]^T$$

$$= \frac{1}{i_{xy}} \begin{bmatrix} \dfrac{p_{xy0}}{u_{cxy1}} & \cdots & \dfrac{p_{xy0}}{u_{cxy(m-1)}} & \dfrac{p_{xy0}}{u_{cxym}} \end{bmatrix}^T + \frac{1}{i_{xy}} \begin{bmatrix} \dfrac{p_{xys1}}{u_{cxy1}} & \cdots & \dfrac{p_{xys(m-1)}}{u_{cxy(m-1)}} & -\dfrac{\sum_{j=1}^{m-1} p_{xysj}}{u_{cxym}} \end{bmatrix}^T \tag{8}$$

After applying the power balance condition $m p_{cxy0} = u_{xy}^* i_{xy}$ where u_{xy}^* is the reference arm voltage, Eqn. (8) is modified to

$$[d_{xy1} \quad \cdots \quad d_{xym}]^T$$

$$= \begin{bmatrix} \dfrac{u_{xy}^*}{m u_{cxy1}} & \cdots & \dfrac{u_{xy}^*}{m u_{cxy(m-1)}} & \dfrac{u_{xy}^*}{m u_{cxym}} \end{bmatrix}^T + \frac{1}{i_{xy}} \begin{bmatrix} \dfrac{p_{xys1}}{u_{cxy1}} & \cdots & \dfrac{p_{xys(m-1)}}{u_{cxy(m-1)}} & -\dfrac{\sum_{j=1}^{m-1} p_{xysj}}{u_{cxym}} \end{bmatrix}^T \tag{9}$$

2.3. Insertion index boundaries

The symmetrical transformation described before is only valid as long as each of the cell insertion indices is within the interval of [0,1]. If one of the real insertion indices is limited without adjusting the corresponding symmetrical components, both zero-sequence and symmetrical components are distorted. According to Eqn. (9) each of the symmetric powers $p_{sj}, j \in \{1, \ldots, (m-1)\}$ affects insertion indices d_{xyj} and d_{xym}. To find symmetrical component boundaries, Eqn. (9) is resolved for p_{xysj}:

$$p_{xysj} = d_{xyj} i_{xy} u_{cxyj} - p_{xy0} \tag{10}$$

$$\sum_{j=1}^{m-1} p_{xysj} = p_{xy0} - d_{xym} u_{cxym} i_{xy} \tag{11}$$

Inserting $d_{xyjmax} = 1$ and $d_{xyjmin} = 0$ in Eqn. (10), maximum and minimum symmetric component boundaries $p_{xysjmax}, p_{xysjmin}$ with respect to the insertion indices d_{xyj} are found:

$$p_{xysjmaxdj} = \begin{cases} i_{xy} u_{cxyj} - p_{xy0} \ \forall \ i_{xy} \geq 0 \ (d_{xyj} = 1) \\ -p_{xy0} \ \forall \ i_{xy} < 0 \ (d_{xyj} = 0) \end{cases} \tag{12}$$

$$p_{xysjmindj} = \begin{cases} -p_{xy0} \ \forall \ i_{xy} \geq 0 \ (d_{xyj} = 0) \\ i_{xy} u_{cxyj} - p_{xy0} \ \forall \ i_{xy} < 0 \ (d_{xyj} = 1) \end{cases} \tag{13}$$

To find boundaries with respect to d_{xym} Eqn. (11) is adjusted assuming that all of the symmetrical powers are equal ($p_{xys1} = \cdots = p_{xys(m-1)} = p_{xys}$):

$$p_{xys} = \frac{p_{xy0} - d_{xym}u_{cxym}i_{xy}}{(m-1)} \tag{14}$$

Now, symmetrical power boundaries with respect to d_{xym} are

$$p_{xysjmaxdm} = \begin{cases} \dfrac{p_{xy0}}{(m-1)} & \forall\ i_{xy} \geq 0\ (d_{xyj} = 0) \\[2ex] \dfrac{p_{xy0} - u_{cxym}i_{xy}}{(m-1)} & \forall\ i_{xy} < 0\ (d_{xyj} = 1) \end{cases} \tag{15}$$

$$p_{xysjmindm} = \begin{cases} \dfrac{p_{xy0} - u_{cxym}i_{xy}}{(m-1)} & \forall\ i_{xy} \geq 0\ (d_{xyj} = 1) \\[2ex] \dfrac{p_{xy0}}{(m-1)} & \forall\ i_{xy} < 0\ (d_{xyj} = 0) \end{cases} \tag{16}$$

Depending on the operating point specified by the actual values of u_{xy}^*, i_{xy} each of the pairs $p_{xysjmaxdj}, p_{xysjmaxdm}$ and $p_{xysjmindj}, p_{xysjmindm}$ has one active boundary:

$$max(p_{xysjmindj}, p_{xysjmindm}) \leq p_{xysj} \leq min(p_{xysjmaxdj}, p_{xysjmaxdm}) \tag{17}$$

Fig. 3 shows the maximal and minimal values from Eqns. (12)-(13) and (15)-(16) visualized by 1-dimensional vectors for a converter arm with three cells.

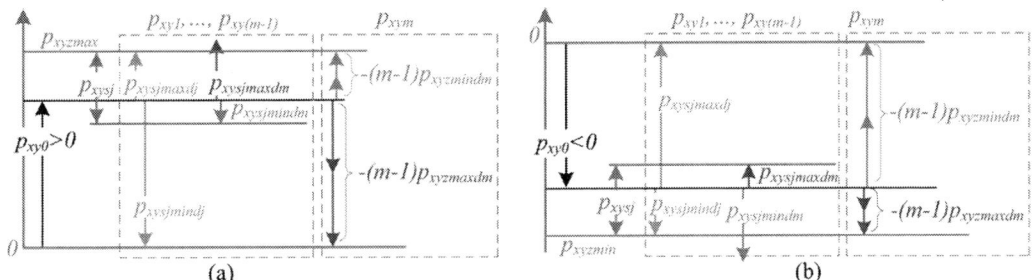

Fig. 3: Visualization of the symmetrical power boundaries in an operating point for an arm with three cells. (a) Positive p_{xy0} (b) Negative p_{xy0}.

2.4. Control structure

Control structure with the symmetrical component controller (SCC) is shown in Fig. 4.

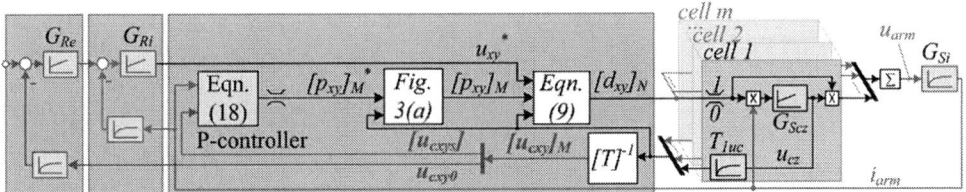

Fig. 4: (a) Control structure using symmetrical transformation. (b) P-controller with dynamic gain.

Symmetrical components of the transformed cell voltage vector are directed to the proportional cell voltage equalization controller with arm current-dependent gain K_{Peqxy}:

$$K_{Peqxy} = \frac{u_c^*}{2T_{1uc}|i_{xy}|}\, sign(i_{xy}) \tag{18}$$

where T_{1uc} is the time constant of the voltage measurement. It is obvious that the arm current value should be limited before inserting it in Eqn. (14) to avoid infinite K_{Peqxy}. The zero-sequence component u_{cxy0} is forwarded to the arm energy controller. The symmetrical power references at the controller output are limited according to Eqn. (17) and are then used to calculate cell insertion indices with Eqn. (9), thereafter the resulting indices are transformed into natural coordinates and sent to the cells.

3. Simulation results

3.1. Insertion indices sampling

The SCC was tested on a model of a low-voltage 25 kW MMC with MOSFET switches and 4 cells per arm. The switching frequency f_{sw} was set 5 kHz.

When applying the complete sorting modulation algorithm all of the cell carrier signals have the same phase and the sampling is performed with intervals of $1/f_{sw}$ (Fig. 5, left). For PSM two methods were tested. The first option (later SCC1) is to sample an insertion index when the respective carrier signal is at zero, resulting an 'overlapped' sampling (Fig. 5, middle). This kind of sampling assures that each of the cells switches no more than twice per switching period. The second option (SCC2) is to sample all of the insertion indices when any of the cell carriers is at zero (Fig. 5, right). As will be shown in next sections, this method results a slightly higher amount of switching operations but also has a more robust large-signal behavior.

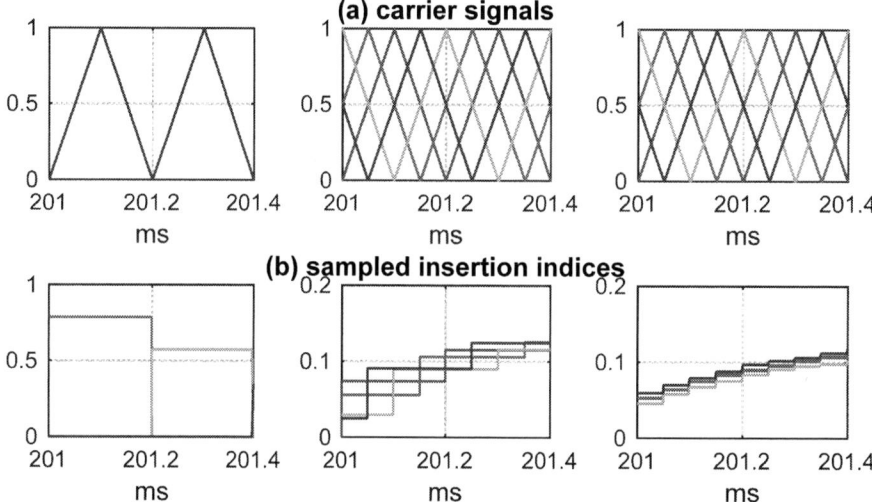

Fig. 5: (a) Carrier signals and (b) sampled insertion indices of the cells in the arm $p1$. Left: complete sorting, middle: SCC1, right: SCC2.

3.2. Current and voltage control

Fig. 6 shows a comparison of the three modulation methods described in Section 3.1. The well-known full sorting algorithm shows robust current control and voltage equalization behavior. Both SCC1 and SCC2 show good capacitor voltage equalization (Fig. 6 (d)) as well. A disadvantage of SCC1 is seen at the arm currents that have distortions at their zero-axis crossings, caused by the 'overlapped' sampling of the results of Eqn. (9). In SCC2 this problem is solved. Through the faster and synchronized sampling SCC2 is closer to the averaged model described in Section 2 that can also be seen on the insertion index limitations performed with no current disturbances.

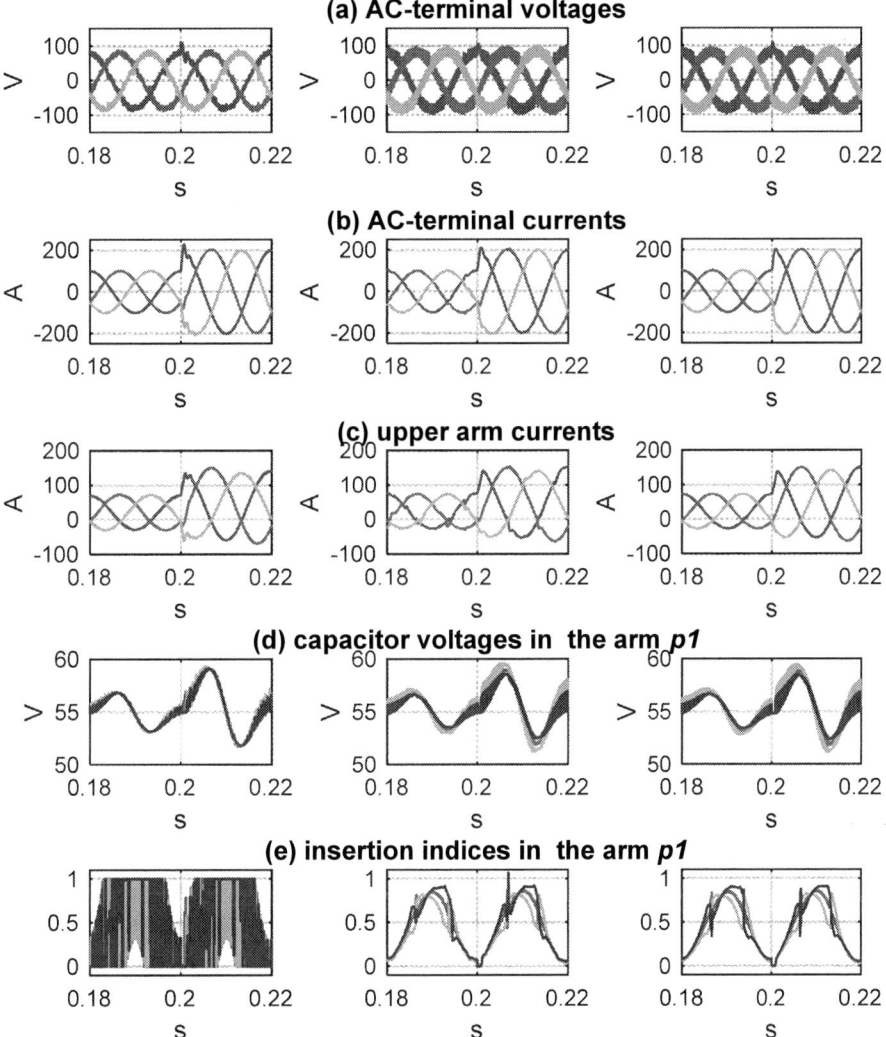

Fig. 6: A comparison of complete sorting modulator (left), SCC1 (middle) and SCC2 (right). (a) AC-terminal voltages. (b) AC-terminal currents. (c) Upper arm currents. (d) Capacitor voltages in the arm $p1$. (e) Insertion indices in the arm $p1$.

3.3. Power losses

The loss distribution within a converter arm was analyzed since it is affected by the modulation method. Cell loss components wire estimated for the complete sorting algorithm [1], SCC1 and SCC2. Fig. 7 (a) and (b) shows cell capacitances and their respective equivalent series resistances (ESR) for the cells of the arm $p1$, depicted for cells 1 to 4 in each plot from the right to the left respectively. The ESR-values are indirectly proportional to the capacitance values as it is usual for real capacitors. Switch parameters were assumed equal for all cells.

For the complete sorting algorithm the capacitor and conduction losses are obviously directly proportional to the capacities. SCC1 shows an indirect proportionality of the capacitor losses and direct proportion of conduction losses with respect to the cell capacitances. For SCC2 no connection between the capacities, while conduction losses are directly proportional to the capacitances. Switching loss variations are negligible in considered design.

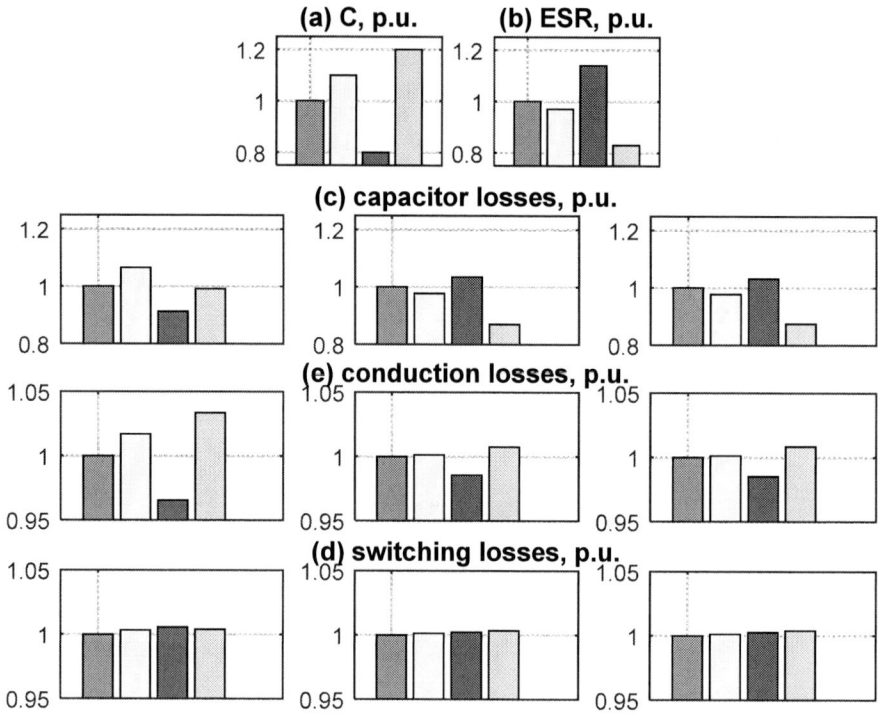

Fig. 7: Switching losses in the four cells of the considered MMC for different balancing methods

Table 1 shows the average arm switching frequency $f_{sw.av}$, and average cell power loss components over an AC-side frequency period for the three methods. PSM-based SCC methods have 23 per cent more total capacitor losses. The total conduction losses are equal for the three methods.

	$f_{sw.av}, kHz$	$P_{Carm}, p.u.$	$P_{condarm}, p.u.$	$P_{swarm}, p.u.$
complete sorting	8,725	1	1	1
SCC1	19,975	2,17	1	0,99
SCC2	20,125	2,16	1	0,99

Table 1: Comparison of equalization methods on an MMC arm with 4 cells

The switching losses are higher for the SCC methods because of the higher switching frequency. As shown in Section 3.1, SCC2 has a higher sampling frequency than sorting and switching losses can be reduced by factor $1/m$ for the same sampling frequency as sorting. Then the switching losses would be lower than those of the sorting algorithm. However, due to the carrier phase shifts a minimal inductance is required for SCC2, otherwise it behaves similar to SCC1.

Summary

This work describes a method of symmetrical components for the cell capacitor voltage equalization in an MMC-arm with phase-shift modulation. The method offers a comprehensible description of the power flows in the arm and can be taken a basis for the solution of the insertion index limitation problem. The limitation of the method is the minimal inductance or switching frequency necessary for a robust operation caused by the phase-shifted carriers.

References

[1] S. Rohner, S. Bernet, M. Hiller und R. Sommer, „Modulation, Losses, and Semiconductor Requirements of Modular Multilevel Converters," in *Industrial Electronics, IEEE Transactions on (Volume:57 , Issue: 8)*, Sept. 2009.

[2] J. Kolb, „Optimale Betriebsführung des Modularen Multilevel-Umrichters als Antriebsumrichter für Drehstrommaschinen," KIT Scientific Publishing, 2014.

[3] A. Hassanpoor, L. Angquist, S. Norrga, K. Ilves und H.-P. Nee, „Tolerance Band Modulation Methods for Modular Multilevel Converters," *IEEE Transactions on Power Electronics,* Bd. 30, Nr. 1, p. 311–326, 2015.

[4] M. Hagiwara und H. Akagi, „PWM Control and Experiment of Modular Multilevel Converters," in *Proc. PESC 2008. IEEE*.

[5] X. Li, Q. Song, J. Li und W. Liu, „Capacitor voltage balancing control based on CPS-PWM of Modular Multilevel Converter," p. 4029–4034.

[6] Q. Tu und Z. Xu, „Impact of Sampling Frequency on Harmonic Distortion for Modular Multilevel Converter," in *Power Delivery, IEEE Transactions on (Volume:26 , Issue: 1)* , Jan. 2011 .

[7] Y. Li, E. A. Jones und F. F. Wang, „The Impact of Voltage-Balancing Control on Switching Frequency of the Modular Multilevel Converter," *IEEE Transactions on Power Electronics,* Bd. 31, Nr. 4, p. 2829–2839, 2016.

Isolated low-power multi-output DC-DC converters with heterogeneous loads for an efficient supply of modular power electronics systems

Arthur Singer, Universität der Bundeswehr München, Germany, arthur.singer@unibw.de
Arun Jeyaprakash, Technische Universität München, Germany, arun.jeyaprakash@tum.de
Stefan Goetz, Duke University, Durham, North Carolina, USA, stefan.goetz@duke.edu
Florian Helling, Universität der Bundeswehr München, Germany, florian.helling@unibw.de
Thomas Weyh, Universität der Bundeswehr München, Germany, thomas.weyh@unibw.de

Abstract

Appropriate isolated supply solutions for recent power converters, such as distributed smart-grid units or modular multilevel converters, have to provide sufficient power reserve, but often suffer from low efficiency at partial load, causing unnecessary large power loss in the converter control unit. Particularly in modular multilevel converters, the power supply does not only affect the overall efficiency in partial-load conditions, but with its input voltage range importantly determines the module capacitor size and thus a major cost factor. This contribution will present a low-power isolated multi-output DC-DC converter with wide input voltage range (4.5 V to 30 V) for the application in power electronics supply. The presented solution includes a design routine for planar transformers, eliminating the need for custom-made wound transformers and cutting down costs with an all-SMD approach. Experimental results show an almost flat efficiency curve above 80% from 0.3 W to 3 W for an input voltage from 10 V to 20 V. The prototype implements 6 heterogeneous outputs rated from 5 V, 100 mA to 15 V, 60 mA and four times 15 V, 20 mA.

1. Design Goals and Application

Though typically neglected, the power supply of controllers and gate drivers in power electronics components plays a major role for the choice of the topology, the overall system efficiency, and the cost. Particularly, the recent trend for modularization of modern power electronics, such as modular multilevel converters (M2C) [1] and modular smart-grid appliances, and the high number of switches that have to be operated galvanically isolated at different voltage levels suggest focusing on efficient isolated supply. Especially as MOSFETs, instead of IGBTs, are used in M2C, the power demand for switching is significantly reduced.

Most-recently proposed modular multilevel circuits provide increased flexibility at the price of up to five different voltage levels per module [2–7]. Although the increasing modularity increases the number of supplies per system, off-the-shelf solutions are often inappropriate for the required power levels, expensive, and inefficient at partial load.

In addition to the galvanically isolated generation of multiple output voltages such as –3 V, 5 V, and 18 V, key aims for the presented DC-DC converter for power supply are a wide power range, the fluctuations of recent adaptive switching control, and a wide input voltage range. The limits of the input voltage immediately determine the capacitor voltage lift, i.e., the upper and lower voltages the module capacitor may reach during one power line cycle of the converter input and/or output terminals, and thus the capacitor size, a major factor of module cost and size. For otherwise constant load, a by 50% reduced lower input voltage, for instance, can require to quadruple the module capacitance.

1.1 Requirements of the Power Supply

We require the power supply to provide multiple, galvanically isolated voltage levels, a high partial load efficiency, a limited part count and ease of production. The galvanic isolation can either be necessary for security reasons or because the electrical potential of different parts of the system can dynamically change, as it might happen in modular applications. Multiple output voltages are needed, when supplying different digital circuits and ICs with different supply needs; higher positive voltage levels for faster gate switching or lower gate resistances, negative voltage levels for active gate blocking, etc. High partial load efficiency is essential in systems with increased operation time at partial load range. There are fixed parts as the supply of ICs, holding currents, etc., and variable parts which only occur on certain events, e.g. the loading and unloading the gate charge of power MOSFETs which is proportional on the switching frequency. A low part count is beneficial for prototyping and production.

1.2 Topology selection and DC-DC Converter Design

In this article, a modular multilevel converter acts as a reference application for the discussed DC-DC converter. The requirements of these modular-multilevel-converter modules are shown in Table 1. Commercially available isolated DC-DC converters with the lowest power (at 15 V input voltage) have power ratings of 1 W, which is notably above the needed range, limiting the achievable efficiency with such commercial solutions. To achieve simplicity and to reduce size, cost and energy waste in control circuits a topology with low active switches count is favorable. Especially in high power applications transformer utilization has a heavy impact on size and cost. For lower power, other factors like semiconductors begin to dominate, so that a single ended topology, where only some of the windings operate at a given time is thinkable. Multiple-output topologies gain over-proportionately from secondary sides with a low number of components, limiting cost, loss, and size. Regarding the aspects above we decided to use a Flyback topology (see Figure 1), as it utilizes only one switch, without feedback from the secondary side, as opto-isolators waste output power and increase the part count or extra transformer windings increase its size.

Table 1: Requirements for the 6 isolated power supplies

Output	Voltage (V)	Load max (W)	Load min (W)
1	5	0.5	0.08
2	15	0.9	0.18
3	15	0.3	0.03
4	15	0.3	0.03
5	15	0.3	0.03
6	15	0.3	0.03

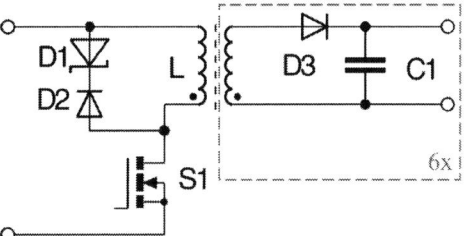

Figure 1: Basic Flyback topology with diode snubber and six secondary outputs

2. Planar Transformer Design

To provide the required galvanic isolation between the input and the various outputs, the DC-DC converter depends on a transformer. Key design criteria for the transformer in the application of the supply of power electronics control are size and losses, as core losses due to hysteresis and eddy currents, winding DC losses due to ohmic resistance, and winding AC losses due to skin and proximity effects [8]. Reflecting the implicit nonlinear relationship of the key parameters, we designed a fast-converging eight-step process, shown in Figure 2. The main relationships [9] that govern switching mode power supply (SMPS) transformer design are

$$L = N^2 * A_L \tag{1}$$

$$\Delta B = \frac{L_{pri} * I_{ripple,pri}}{N_{pri} * A_e} \tag{2}$$

with the inductance of the coil (H) L, number of turns N, inductance index A_L which is a datasheet value of ferrite cores denoting inductance per squared turns, and with the change in flux density (T) ΔB, transformer primary inductance (µH) L_{pri}, transformer primary ripple current (A) $I_{ripple,pri}$, transformer primary number of turns N_{pri} and the cross sectional area of core (mm^2) A_e.

Core size, number of turns, air gap and core material are all related in a single equation (2). Hence arriving at a compact and efficient design becomes an iterative process. The following procedure gives one possible solution to achieve an efficient and compact transformer.

2.1. Design Procedure

To design a DC-DC converter, one has to determine the inductor size appropriate for the power requirements and voltage and current ripple constraints. Power requirements also define the conductor size due to losses and temperature constraints. The operating frequency influences the choice of core material and shape. Summarized, it leads to the following list of steps:

1. Determine the inductance values that are required for the DC-DC converter.
2. Determine the turns ratio required for the DC-DC converter.
3. Calculate the conductor size from temperature and skin effect limits. [10]
4. Select a core material that fits the converter operation frequency. (Several choices may be available. It will be narrowed down in following steps)
5. Calculate the total cross sectional area of all the conductors. This area in addition with factor of safety of 30% forms the minimum window area of the core.
6. Select a core shape from the list of cores available using manufacturer guides.
7. Set the number of turns of each winding to the minimum whole number value.

Transformer design is a critical stage in high-efficiency converter design. After obtaining the general parameters for the transformer, an actual, application-wise optimal transformer has to be designed in detail. To facilitate this process, we propose an iterative, 8 step process as depicted in the flowchart (see Figure 2).

1. Calculate required A_L using equation (1).
2. From the list of available cores available from manufacturers, select the smallest one with appropriate A_L.
3. Check if the window area of the core is large enough to accommodate the windings.
4. If the window area is not enough, use a bigger core with larger window area.
5. Use equation (2) to calculate the maximum operational flux density (B_{max}) that could be reached. Using the maximum-current of the switch in the converter could be one safe method.
6. Check if this B_{max} is smaller than B_{sat} (saturation flux density) of the core material.
7. If B_{max} is not lesser than B_{sat}, increase the number of turns of each winding (Maintaining the turns ratio).
8. Calculate the minimum window area required to accommodate the conductors.

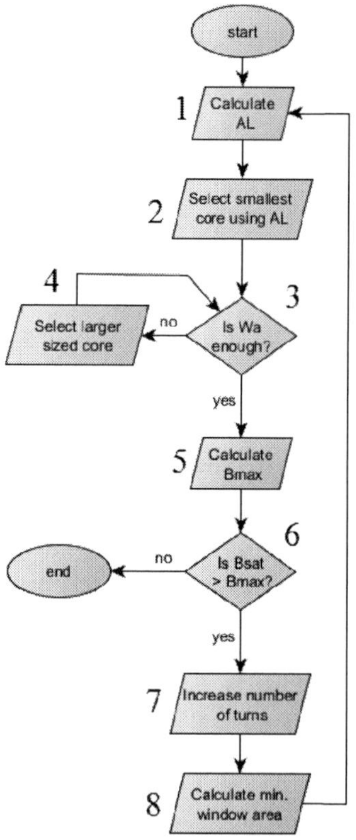

Figure 2: 8 step transformer design process

The first iteration yields a transformer design with high efficiency but large size. Repeating the process using stages 7 and 8 (while stage 6 is satisfied) will reduce the transformer's size. When an acceptable size is reached, the iterations can be stopped. An optimal transformer design has been found.

2.2 Losses

One has to minimize the losses for a high-efficiency operation. Whereas in high-power DC-DC converters, multiphase designs have become standard for achieving high partial-load efficiency [12, 13], the constraints on size, cost, and complexity for the application in question asks for a circuit with a low component count. For reasons of clarity, we will separate the losses in transformer losses and losses in the electronics, consisting of the losses in the switch, the driver and the snubber.

Transformer losses

A SMPS transformer-core, because of its high frequency operation, is generally made of hard magnetic materials like ferrites whereas the low frequency power transformers mostly use soft magnetic materials like silicon steel [11]. Ferrites have very high ohmic resistance and the area enclosed under the hysteresis loop of their B-H magnetization curve is significantly lower than that of silicon steel. As a result, even at very high frequency operation, the hysteresis and eddy current losses are low. Low hysteresis loss is due to less B-H loop area and low eddy current loss is due to very high resistivity of the core material. As the core losses depend on the material selected, the conductor losses depend on one's design.

Besides the conduction losses due to DC-resistance especially the high frequency losses are of interest in SMPS transformers. One is the "skin effect", which is a consequence of Lenz's law. It describes, that a high-frequency AC current's induction field causes a current of opposite direction. This leads the current density to decrease exponentially from the surface of the conductor, effectively reducing its cross-sectional area and thus increasing its AC resistance. The usual approach to reduce the skin effect is to use more, but finer, wires in parallel, e.g. high-frequency litz wires, which is not very practicable for planar transformers, since copper thickness and trace width are defined by technology.

A second, but very similar effect is the "proximity effect", wherein the effective cross-sectional area is also reduced. In this case the induction field is caused by a nearby conductor. One can calculate the effects using Dowell's formula [14] from 1966 or more precise approaches based on FEM modelling [15]. Interleaving transformer windings can reduce the proximity loss significantly, when primary and secondary currents are in phase [16] – a practice which should be used with planar transformers.

Electronics losses

Losses in the electronics consist of conduction losses, switching losses, driving losses and losses in the snubber circuit. Conduction losses are low due to low current whereas the latter three are frequency dependent and thus dominant. Therefore the switching frequency is a main parameter to be controlled. For high efficiency operation, an adaptive switching frequency control is beneficial, whose principle of operation is shown in Figure 3 a) [17]. In high load events, the converter is operating in continuous current mode (CCM) with higher on-times of the switch. If the power demand decreases, the frequency rises as the switch's on-time is reduced. Two restrict switching losses, the switching frequency is cropped at f_{max} – in our case 400 kHz – between P_{disc} and P_{cont}. and the device is operating in discontinuous current mode (DCM) [18]. As the load decreases, also the switching frequency falls. For even lower loads (or standby) below P_{burst}, a burst mode control scheme even further reduces the losses.

In our application, the switching frequency was at 100 kHz in the power electronics standby mode and went up to 250 kHz at full load. For an input voltage of 10 V the device enters CCM (see Figure 3 b): The switching frequency decreases while the on-time increases. The same behavior is implemented for higher input voltages at higher loads.

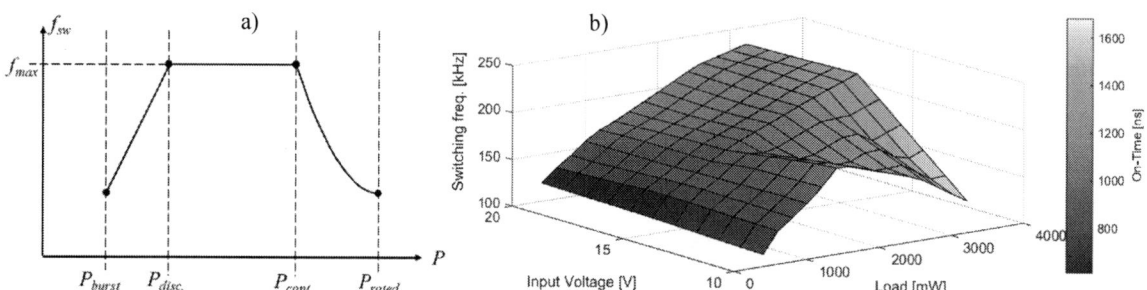

Figure 3: a) Principle of adaptive switching rate dependent on load (derived from [13]) and b) measured switching frequency and on-time of switch over input voltage and load (right)

3. Experimental Results

3.1 Planar Transformer

Following our 8-step process we designed a planar transformer on a standard 4-layer FR4 (with a dielectric strength of 140 V/µm as given by the manufacturer) PCB with 35 µm of copper and the parameters shown in Table 2. For the primary side this yields a cross section of 0.133 µm², for

Table 2: Transformer properties

Number of turns (P:S1:S2:S3:S4:S5:S6)	6:1:3:3:3:3:3
Core shape and size	E22/6/16R
Material	3F4 (Ferroxcube)
Primary inductance	65 µH
Primary trace width	15 mils (35 µm Cu)
Secondary 1 & 2 trace width	10 mils (35 µm Cu)
Secondary 3-6 trace width	6 mils (35 µm Cu)

secondary 1-2 0.053 µm² and for secondary 3-6 0.027 µm², respectively. Interleaving has been used to reduce the effective AC resistance. Impedance measurements have been made from 50 Hz up to 500 kHz (see Figure 4). One can easily see, that the primary's impedance is much higher than the other's due to the higher number of turns. When short circuiting the secondary outputs, the primary's impedance drops to 1.46 Ω at 100 kHz (corresponding to a leakage inductance of 2.32 µH) compared to 88.5 Ω in the open circuits case. An air gap is used to get different values of AL in cores. Increasing the air gap reduces the AL of the core at the cost of fringe flux and worse EMI. In our design an air gap was not necessary, so we attached the two parts of the ferrite core directly on each other.

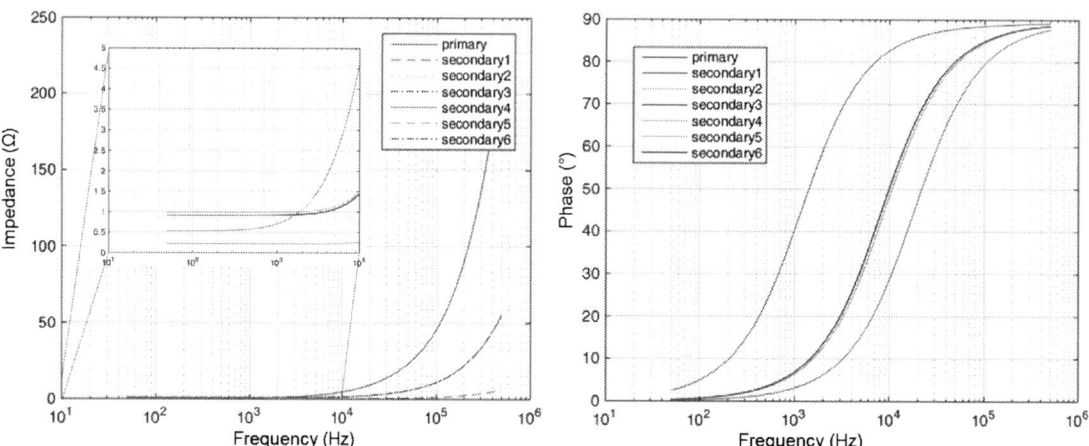

Figure 4: Inductance and phase over frequency of the primary and secondary windings of the designed planar transformer

Following the same design procedure, we also produced an optimized wound transformer over several iterations. Both transformers were used in the same flyback converter test setup. The setup with hand-wound transformer had a slightly higher efficiency at a total load of 3.5 W

(83% vs. 80%) whereas the leakage inductances between primary and secondary windings were about the same for both transformers. Regarding the footprint, the wound transformer was smaller (208 mm² vs. 606 mm²), but higher in profile (10.5 mm² vs. 6 mm²). Although the wound transformer provides better results, the planar transformer was chosen for the final design due to its lower profile and better ease of production [19]. Manufacturing custom transformers, especially in limited numbers, is expensive. As the planar transformers are integrated in the PCB of the converter, only the cores have to be attached.

3.2 Operational Efficiency

A prototype with the transformer specified above and an integrated controller (LT8301, Linear Technologies, Milpitas, CA, USA) was built and tested together with the power electronics module it supplies (see Figure 5). For the transformer core operated between 100 and 250 kHz, we used a hard-magnetic ferrite (E22/6/16/R-3F4, Ferroxcube, Taiwan, permeability $\mu_e \sim 770$, inductance index $A_L \sim 2.9$ µH), which resulted in low core losses in our measurements at high flux usage. With a package volume of 4.2 cm³, a PCB footprint of 16 cm², and a weight of 21 g, it is comparable with the most typical commercial alternative (TMA1515S and TMA1505S, Traco Power Co., Baar, Switzerland) with 6.9 cm³, 7.1 cm², and 15.3 g, respectively.

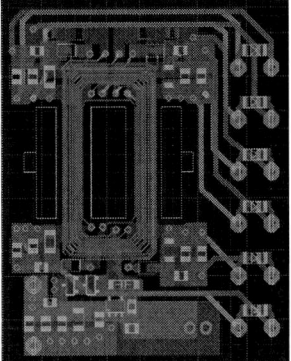

Figure 5: Layout of proposed DC-DC-converter (left) and populated prototype stacked on power electronics PCB as replacement for the 6 Traco DC-DC-converters (right)

However, the overall efficiency was superior. It exceeds 80% over the entire power range from 0.3 W to 1.7 W measured at an input voltage of 15 V, whereas the overall efficiency of the off-the-shelf solution with six isolated DC-DC converters starts below 30% for low loads and peaks at 60% at nominal power (see Figure 6). The minimum power was defined by the standby state in our power electronics. Then the MOSFETs where activated and swept up to a switching frequency of 40 kHz leading to a power consumption of 1.7 W.

The higher efficiency is also obvious when comparing the infrared images (see Figure7) taken after 15 minutes of standby operation. On the left, the peak temperature is 53.3 °C. One can easily see, that the heat source are the six Traco DC-DC converters. On the right, the Traco converters have been replaced by our prototype as shown on Figure 5. Here, the maximum temperature is only 30.4 °C and located on the MOSFET drivers – thus the load itself. The bright spots on the image are reflections by unpopulated copper pads.

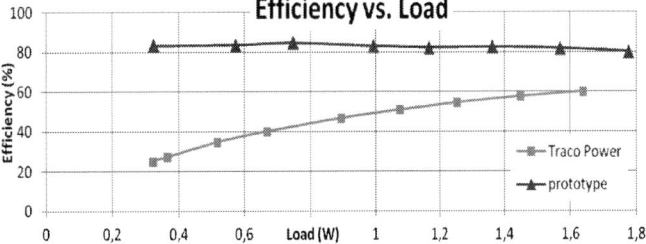

Figure 6: Efficiency vs. Load of Traco Converters compared to our prototype

The efficiency of the Traco variant is so low, because each of the supplies is loaded way below the rated power (1 W) of a single Traco converter. As it needs a certain output current to regulate its output voltage, it itself creates this current internally, which – seen from the application side – is a loss. Additionally it doesn't feature an adaptive switching rate and operates constantly at 100 kHz. Six individual DC-DC converters are used which means there are six transformers and at least six switches in the current path. Even though they are of smaller built than the ones used in our prototype, the losses add up and are dissipated at every cycle of operation.

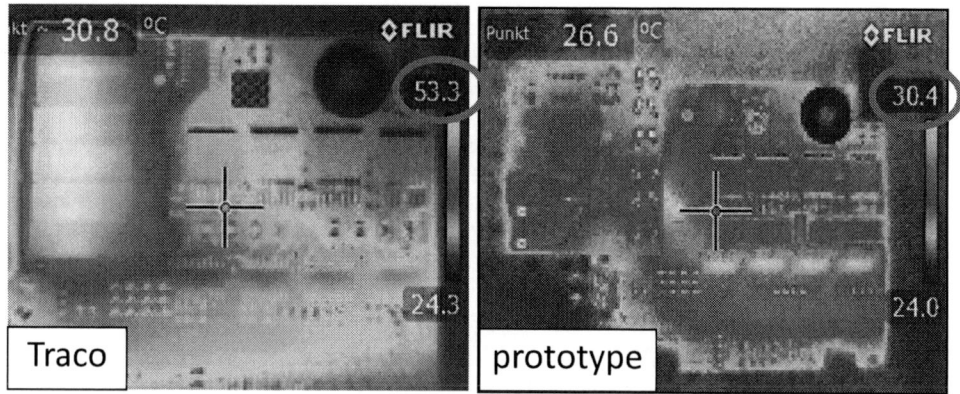

Figure 7:Infrared image of heat dissipation in the power module in standby. One can easily see that the Traco converters (left) reach a temperature of 53.5 °C, illustration the energy waste at low loads. The proposed DC-DC-converter prototype (right) exhibits no augmented temperature. The yellow dots on the left are reflections from the unpopulated footprints. The highest temperature of 30.4 °C is located in the MOSFET drivers on the right.

4. Summary

This paper presents an iterative 8 step SMPS transformer design process, which was used to design a planar transformer with multiple-outputs for a low power application. Furthermore we provide guidelines, what to consider when building a low power, isolated DC-DC converter. These considerations led to a prototype which far exceeds the off-the-shelf solution with six dedicated, isolated DC-DC converters (1515S and 1505S, Traco Power, each 1 W rated power) used so far. Our solution displays an easily adaptable design with almost flat efficiency at over 80% over the tested range from 0.3 W to 3 W at an input voltage range from 10 V to 20 V, a small voltage ripple of less than <1% and cost-effective all-SMD approach with planar transformer.

5. Acknowledgement

The authors gratefully acknowledge financial support from the Deutsche Bundesstiftung Umwelt (DBU, German Federal Foundation for the Environment).

6. References

[1]   M. Glinka and R. Marquardt, "A new AC/AC-multilevel converter family applied to a single-phase converter," in *Fifth International Conference on Power Electronics and Drive Systems*, pp. 16–23.

[2]   S. M. Goetz, A. V. Peterchev, and T. Weyh, "Modular Multilevel Converter With Series and Parallel Module Connectivity: Topology and Control," *IEEE Trans. Power Electron*, vol. 30, no. 1, pp. 203–215, 2015.

[3]   S. M. Goetz, M. Pfaeffl, J. Huber, M. Singer, R. Marquardt, and T. Weyh, "Circuit topology and control principle for a first magnetic stimulator with fully controllable waveform," (eng), *Conference proceedings : ... Annual International Conference of the*

IEEE Engineering in Medicine and Biology Society. IEEE Engineering in Medicine and Biology Society. Annual Conference, vol. 2012, pp. 4700–4703, 2012.

[4]   S. Debnath, J. Qin, B. Bahrani, M. Saeedifard, and P. Barbosa, "Operation, Control, and Applications of the Modular Multilevel Converter: A Review," *IEEE Trans. Power Electron,* vol. 30, no. 1, pp. 37–53, 2015.

[5]   V. Dargahi, A. K. Sadigh, M. Abarzadeh, S. Eskandari, and K. A. Corzine, "A New Family of Modular Multilevel Converter Based on Modified Flying-Capacitor Multicell Converters," *IEEE Trans. Power Electron,* vol. 30, no. 1, pp. 138–147, 2015.

[6]   F. Helling, S. Gotz, and T. Weyh, "A battery modular multilevel management system (BM3) for electric vehicles and stationary energy storage systems," in *2014 16th European Conference on Power Electronics and Applications (EPE'14-ECCE Europe),* pp. 1–10.

[7]   F. Helling, A. Singer, S. Gotz, and T. Weyh, "Optimization of Hydrogen Gas Turbine Power Supply by means of a novel Modular Multilevel Parallel Converter (M2PC)," in *2014 5th International Renewable Energy Congress (IREC),* pp. 1–5.

[8]   M. Sippola and R. E. Sepponen, "Accurate prediction of high-frequency power-transformer losses and temperature rise," *IEEE Trans. Power Electron,* vol. 17, no. 5, pp. 835–847, 2002.

[9]   Siyang Zhao, Junming Zhang, and Yang Shi, "A low cost low power Flyback converter with a simple transformer," in *2012 7th International Power Electronics and Motion Control Conference (IPEMC 2012),* pp. 1336–1342.

[10] A. A. Ahmed Abdelrahman, E. E. Omer Elfaki, and H. A. ElnazirAdam, "Design of high frequency transformer for switch mode power supply," in *2015 International Conference on Computing, Control, Networking, Electronics and Embedded Systems Engineering (ICCNEEE),* pp. 129–135.

[11] M. T. Quirke, J. J. Barrett, and M. Hayes, "Planar magnetic component technology-a review," *IEEE Trans. Comp, Hybrids, Manufact. Technol,* vol. 15, no. 5, pp. 884–892, 1992.

[12] H. Mao, L. Yao, C. Wang, and I. Batarseh, "Analysis of Inductor Current Sharing in Nonisolated and Isolated Multiphase dc–dc Converters," *IEEE Trans. Ind. Electron,* vol. 54, no. 6, pp. 3379–3388, 2007.

[13] P. Zumel, O. Garcia, J. A. Cobos, and J. Uceda, "Tight magnetic coupling in multiphase interleaved converters based on simple transformers," in *APEC 2005. Twentieth Annual IEEE Applied Power Electronics Conference and Exposition,* pp. 385–391.

[14] P. L. Dowell, "Effects of eddy currents in transformer windings," *Proc. Inst. Electr. Eng. UK,* vol. 113, no. 8, p. 1387, 1966.

[15] F. Robert, P. Mathys, and J.-P. Schauwers, "A closed-form formula for 2-D ohmic losses calculation in SMPS transformer foils," *IEEE Trans. Power Electron,* vol. 16, no. 3, pp. 437–444, 2001.

[16] G. Sen, Z. Ouyang, O. C. Thomsen, M. A. E. Andersen, and L. Moller, "A high efficient integrated planar transformer for primary-parallel isolated boost converters," in *2010 IEEE Energy Conversion Congress and Exposition (ECCE),* pp. 4605–4610.

[17] G.-C. Huang, T.-J. Liang, and Kai-Hui Chen, "Losses analysis and low standby losses quasi-resonant flyback converter design," in *2012 IEEE International Symposium on Circuits and Systems - ISCAS 2012,* pp. 217–220.

[18] Jian Sun, D. M. Mitchell, M. F. Greuel, P. T. Krein, and R. M. Bass, "Averaged modeling of PWM converters operating in discontinuous conduction mode," *IEEE Trans. Power Electron,* vol. 16, no. 4, pp. 482–492, 2001.

[19] A. I. Maswood and Lim Keng Song, "Design aspects of planar and conventional SMPS transformer: a cost benefit analysis," *IEEE Trans. Ind. Electron,* vol. 50, no. 3, pp. 571–577, 2003.

A wire based communication interface for Medium and High-Voltage converters

Marek Galek, Siemens AG, Corporate Technology, Munich Germany,
marek.galek@siemens.com

Manuel Blum, Siemens AG, Corporate Technology, Munich Germany,
manuel.blum@siemens.com

Abstract

This work focuses on the design and the implementation details of a wire based network communication interface for modular power electronic devices. For this purpose the isolation requirements needed in such an area are presented in detail. In consequence, two implementations of the hardware layer are shown and discussed regarding their characteristics in the environment of power electronics.

In order to prove the concept of distributed isolation barriers and to investigate its capability, one specific network interface was selected and implemented. The circuitry has been implemented in an existing low voltage M2C where the data rates and the Isolation distribution have been measured. Finally, this paper demonstrates the possibility to reach even the higher voltage range with low voltage communication devices. For this purpose, both the potential and the limits of this new communication approach are shown.

1.   Introduction

Due to the rapid development prices in signal electronics even extremely complex converter topologies are moving into the focus of today's developments. As an example, the modular multi-level topology (MMC / M2C) and the Series-Connected-H-bridge topology (Siemens Perfect Harmony) may be mentioned here in particular.

Since even more complex topologies result in even more complex wirings, new structures in the system architecture have been recently presented [2][5][6]. Due to the high isolation requirements of the communication interface, the current wire-based technologies can only be used up to about 1000 Volts. It is clear that these limitations are mainly caused by the limited dielectric strength and the break down voltage of the isolation barriers. For this reason, the isolation by use of plastic optical fiber represents the preferred choice in the higher voltage range. In order to increase the voltage range of wire based communication this paper presents possible solutions with distributed isolation barriers. This enables the usage of a common copper based communication interface, even in the higher voltage ranges.

1.1.   State of the art

The structure of today's inverter systems is strongly influenced by the setup of the common two-level topology. The power switches, which can be seen as converter actuators, are usual directly controlled by the central processor. Therefore, this communication structure represents a standard star topology which can be particularly easily isolated. In the higher power range the isolation is realized by the use of fiber optic communication. In a more complex topology such as the M2C, this wiring scheme leads to a drastic increase of the wiring complexity. In order to reduce the wiring effort, new control and wiring methods have been presented recently [2] [3] [4] [6]. All these approaches are based on a communication link by means of copper cables.

Due to the high isolation requirements this copper wire based communication is usually limited to the low-voltage applications. In order to work around this limitation it should be noted that a hybrid communication structure was already presented in [5]. As the isolation in this work is based on an optical communication, this solution has still to deal with problems like aging and reliability issues due to pollution in the optical interfaces.

1.2. Isolation shortcomings with common communication interfaces

In order to decrease the expenses on wiring, M2C's can be controlled with a bus-network (see Fig. 1). Each module is connected to a common bus potential whereby the communication interface has to be isolated by a digital isolator in relation to main inverter control [2] [3]. Under worst case conditions the communication interface has to block the total DC-Link voltage. Therefore the isolation barriers must be designed to withstand this voltage. For higher voltages that leads to comparatively big and cost intensive isolators. That means that common bus potential represents the main problem of these approaches, since the maximum operating voltage is limited to the isolation capability of the communication transformers.

In order to overcome this limitation it is desirable to create an (distributed) isolation barrier that scales in the same voltage range as the modules do. Regarding the modular multi level converter this would be the voltage rating of the modules.

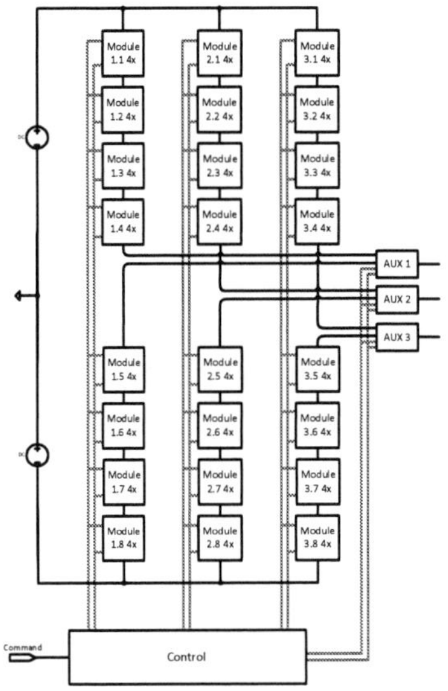

Fig. 1 M2C with network based control

1.3. Proposed solution

The presented solution allows to overcome the voltage limitation of the known circuitry by eliminating the common bus potential, which enables the use of a „distributed"- isolation barrier. As the voltage reference of the subdivided isolation can be freely chosen, it will be set to a fixed potential of the sub-module. Thus, the communication interface of the modules must only be able to block the output voltage of the module. This allows to usage of well known bus topologies even in a higher voltage range. One possible implementation of a communication network for a three phase converter with four sub-modules per phase is shown in Fig. 2. The "low voltage" isolation barrier between the sub-modules is marked in blue. In order to keep the structure as simple as possible, the connection between the control and the modules is also realized by use of the low voltage interface.

A single high voltage isolation barrier for the user interface is sufficient to ensure a safe operation of the converter (marked in red **Fig. 3**).

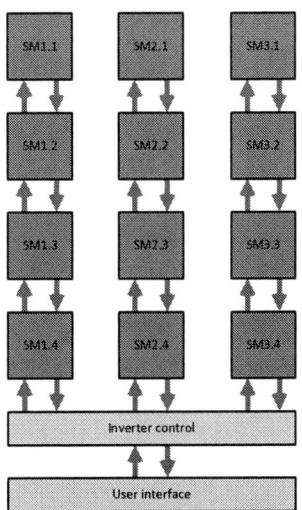

Fig. 2 Distributed isolation concept

© VDE VERLAG GMBH · Berlin · Offenbach

PCIM Europe 2016, 10 – 12 May 2016, Nuremberg, Germany

1.4.   Possible implementations

This subsection presents two possible implementations of the communication interface with a distributed isolation barrier. The suggested structure is based on a half duplex communication which easily can be extended to full duplex communication as well.

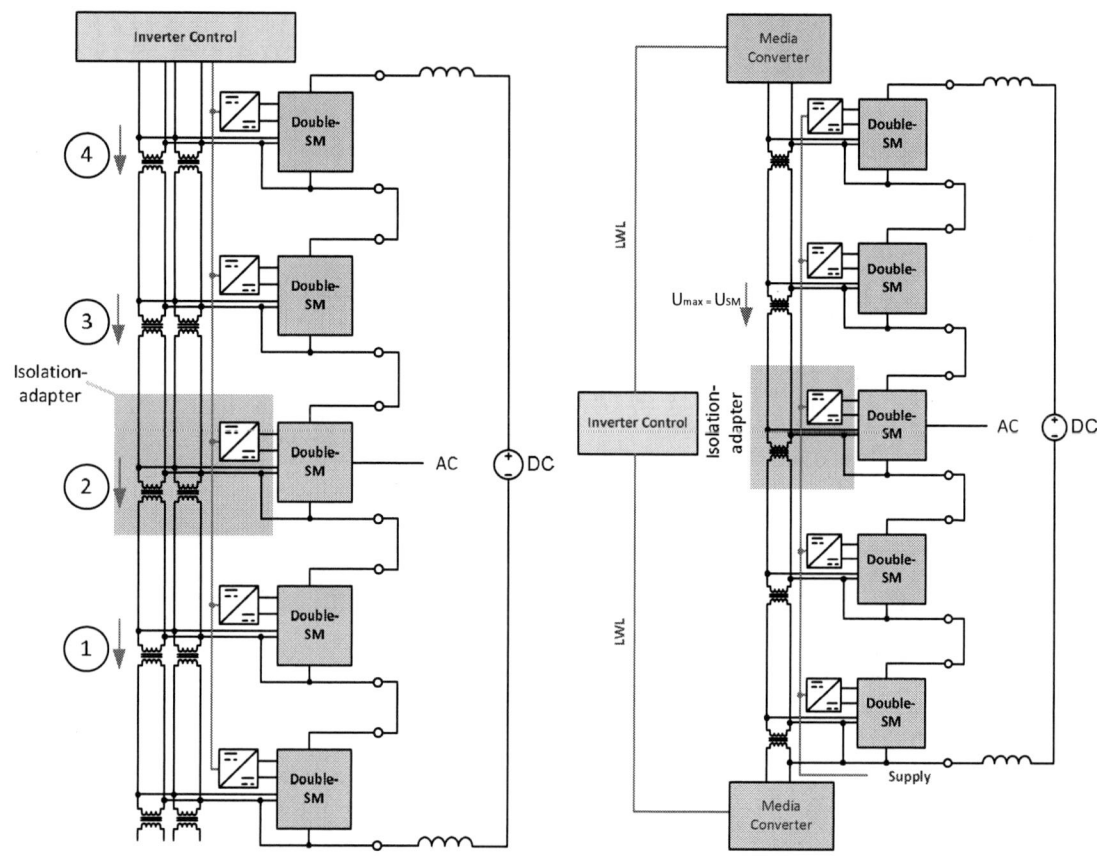

Fig. 3 Fully transformer isolated communication

Fig. 4 Communication with an hybrid isolation approach

The proposed wiring has been applied to an existing low voltage M2C converter which uses a network-based control already presented in [2] and shown in Fig. 1. For this reason the presented interconnection of the network was implemented using an isolation adapter (see green shaded area Fig. 3 and Fig. 4). In this particular case the signal isolation has been realized by signal transformers.

As already mentioned, the potential of the isolated subsections can be chosen almost freely. By referencing the corresponding isolated part-section to the negative potential of the module, the maximum voltage drop across the isolation barrier (see red arrows in Fig. 3 and Fig. 4) is limited to the maximal output voltage of the sub-module.

As discussed in [2] redundant communication paths are essential in this application field. In the first implementation (see Fig. 3), the redundancy is achieved by doubling the existing communication path. Due to the low isolation capability of the individual sections, the reference potential of the control cannot be freely selected and has to be located in the voltage range of the corresponding sub-module. That's why this approach completely avoids the usage of highly

© VDE VERLAG GMBH · Berlin · Offenbach

isolated communication paths. In return this approach leads to a higher wiring effort and limits the choice of the reference potential of the control within certain limits. These drawbacks can be avoided by the structure depicted in Fig. 4. The alternative wiring scheme achieves the redundancy requirement by a bilateral feed-in of the control signals in the communication bus. Since the two ends of the bus typically do not share the same voltage level this approach requires at least one highly isolating communication path in order to achieve double-sided feed-in of the control signal. In order to avoid voltage level limitations of control both control signal feeding points have to be realized in a highly isolating manner. These isolation requirements are normally met by fiber communication.

In order to connect the optical fiber to the wired communication, a media converter is required. Therewith this structure reduces the required network connections and reduces the required isolation barriers to a minimum. By contrast this approach requires only two communication sections with a high isolation capability.

Due to the eliminated fiber optic cell communication (typically used in high end HV applications) both structures represent an extremely robust and cost effective solution up to a working voltage of 1 kV per isolation barrier. Therefore this concept is very interesting for cell based structures, as the communication scales in the same way as the working voltage of the converter does.

2.   Results and Measurements

The proposed concept, as mentioned before was implemented and tested with an experimental low voltage M2C converter. In this particular case the proof of concept was achieved by use of the circuitry shown in Fig. 3.

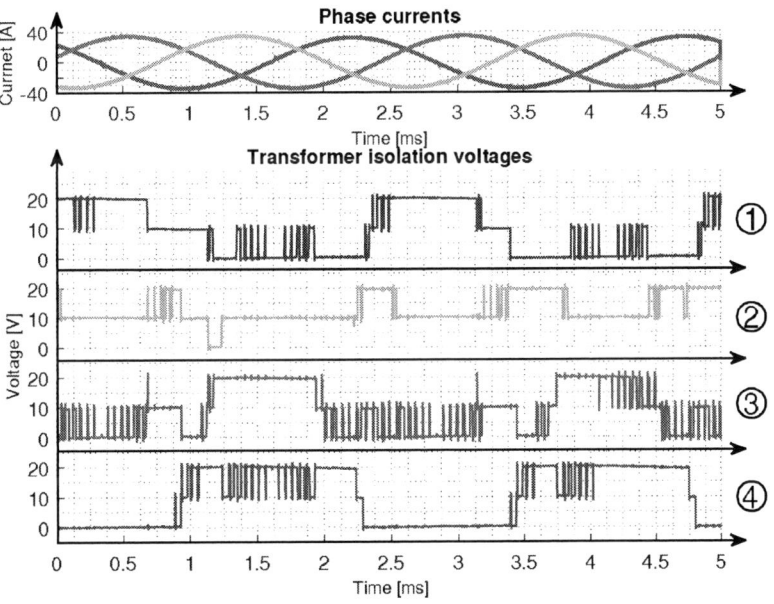

Fig. 5 Voltage distribution along the communication bus

The experimental low voltage M2C is based on five cells per arm which are organized in double sub-modules, five isolation adapters have been used for this test. In order to ensure an

PCIM Europe 2016, 10 – 12 May 2016, Nuremberg, Germany

balanced voltage distribution the voltages across the isolating transformers has been measured. The measurement points are marked with numbers from 1 to 4 in Fig. 3. The corresponding voltage traces are shown in Fig. 5.

In this setup the reference voltage of the transformer is set to the negative potential of the modules. That's why the maximal voltage applied to the transformer isolation should not exceed the nominal voltage of a double module. As shown in Fig. 5 the voltages applied to the segmented communication path perform very well and the measured maximum voltage is limited to the value of the included cells. In this experimental setup the DC-link voltage was limited to a voltage of 50V.

In a next step the additional degradation of the data signal was analyzed. Due to limited coupling of each transformer and each additional bus loads the isolated section lead to a continuous amplitude reduction of the control signals along the bus. In order to evaluate the amount of this damping the data signal amplitude was measured at the beginning end of the communication bus.

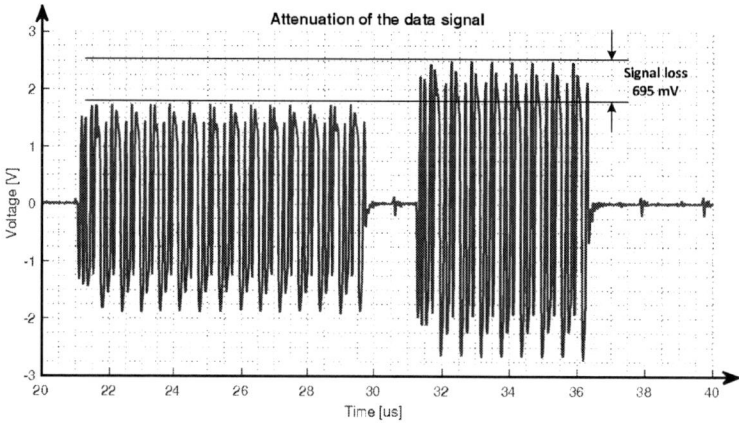

Fig. 6 Signal damping across the communication bus

The result can be seen in Fig. 6. The first frame represents the data packet from the central control unit and is subject to maximal damping of the communication path. The second frame corresponds to the response frame form the last double sub-module and can be considered as an undamped signal. The signal loss in this setup was found to be 700mV along the entire chain.

$$a_{Gain\,margin}(dB) = 20 \cdot log\left(\frac{U_{max}}{U_{min}}\right) = 28,0dB \qquad \text{Eq. 1}$$

$$a_{signal\,client}(dB) = \frac{20 \cdot log\left(U_{Sig\,Master}\right) - 20 \cdot log\left(U_{Sig\,end}\right)}{n_{Client}} = 0,569dB \quad n_{client} = 5 \qquad \text{Eq. 2}$$

$$n_{Client\,max} = \frac{a_{Gain\,margin}}{a_{Signal\,client}} = 49,1 \qquad \text{Eq. 3}$$

Considering a minimal signal magnitude of 100mV for a stable receiver output, the gain margin can be calculated according to Eq. 1. To calculate the maximal number of receivers which can

© VDE VERLAG GMBH · Berlin · Offenbach

be connected in this manner, the damping per section has to be calculated according to Eq. 2. Since the measurement in Fig. 6 already represents the attenuation over 5 interfaces, the calculated damping level can be treated as a sufficient realistic and averaged value. By use of this value the maximal number of clients can be calculated according to Eq. 3.

It should be mentioned that this number can be significantly increased by use of a higher communication voltage level. This means in this setup a shift from 3.3V to 5V would lead to a maximal number of 57 clients.

3.　Conclusion

This paper presents a new (copper) wire based communication approach which allows to overcome the voltage limitations of the communication interface used in networked power electronic converters. The proposed structure allows to reduce the dielectric strength on the isolating components and offers therefore the potential to reduce the costs and to increase the reliability of the converter-internal bus system. This is especially important as the communication system represents the backbone of this kind of converter systems.

Two different wiring schemes have been presented enabling the use of wire based communication even in the higher voltage ranges without losing the redundancy aspect. Measurements have been taken on a working 150kVA modular multilevel converter (see Fig. 7) in order to prove the concept and to investigate the voltage distribution and the damping along the communication chain.

Fig. 7 Low voltage M2C prototype

4.　References

[1] Galek, Marek: "MOSFET-based Modular Multilevel Converters: Design Challenges and Implementation Details" ECPE Workshop Advanced Multicell Multilevel Power Converters 2014.

[2] Galek, Marek, Manuel Blum, and Hamza Mlayeh. "A fault tolerant communication interface for modular and distributed power electronics." PCIM Europe 2015; International Exhibition and Conference for Power Electronics; Proceedings of. VDE, 2015.

[3] Toh, C. L., and L. E. Norum. "Implementation of high speed control network with fail-safe control and communication cable redundancy in modular multilevel converter." Power Electronics and Applications (EPE), 2013 15th European Conference on. IEEE, 2013.

[4] Broadmeadow, Mark AH, and Geoffrey R. Walker. "A LIN inspired optical bus for signal isolation in multilevel or modular power electronic converters." Power Electronics and Drive Systems (PEDS), 2015 IEEE 11th International Conference on. IEEE, 2015.

[5] Cottet, Didier, et al. "Integration Technologies for a Fully Modular and Hot-Swappable MV Multi-Level Concept Converter." PCIM Europe 2015; International Exhibition and Conference for Power Electronics; Proceedings of. VDE, 2015.

© VDE VERLAG GMBH · Berlin · Offenbach

[6] Dirk Malipaard "Innovative, cost effective control bus solutions for multi-level converters" http://www.iisb.fraunhofer.de/en/research_areas/energy_electronics/multi_level_inverter/research_development.html 2016.

PCIM Europe 2016, 10 – 12 May 2016, Nuremberg, Germany

An auxillary power supply with integrated communication capability for medium and highvoltage applications

Manuel Blum, Siemens AG, Corporate Technology, Munich Germany,
manuel.blum@siemens.com

Marek Galek, Siemens AG, Corporate Technology, Munich Germany,
marek.galek@siemens.com

Abstract

This work focuses on the design and implementation of a isolating power supply with integrated communication capability for medium and highvoltage applications. As a first step the isolation requirements and problems of conventional interfaces are shown. The basic idea to overcome these problems is to combine the power supply and communication capability into one isolating transformer. The proposed solution combines communication and power. In order to create such a combined device, the proper methods for both, communication and power transfer, have to be identified and adapted.

Finally a working prototype is built and basic operation of both power- and data-transfer is shown.

1. Introduction

Due to the decreasing prices in electronics, complex converter topologies are moving into the focus of today's developments. As an example, the modular multi-level topology (MMC / M2C) [1] and the Series-Connected-H-bridge topology (Siemens Perfect Harmony) may be mentioned here in particular.

Since more complex topologies result in even more complex wiring schemes, new system architecture structures have been presented recently [2][5][6]. Due to the isolation requirements of the communication interface, the current wire-based technologies can only be used up to 1000 Volts. It is clear that these limitations are mainly set by the limited dielectric stress and the break down voltage of the isolation barriers. Due to this fact an optical communication and a power supply fed by the local DC-Link voltage, represent the preferred choice in the high voltage inverters (e.g. HVDC).

2. Problems of conventional interfaces

The conventional method to supply local gate-driver and measuring devices would be a wide range DC/DC converter, fed by the local DC-Link voltage of the switching-cell.

Although this method has no problems regarding isolation-barriers, it is not the ideal choice. The main drawbacks are: low efficiency, expensive, complex and place-consuming hardware.

An externally fed power supply would avoid these disadvantages.

Moreover it would be especially desirable to stay in control over the submodule during failure and hot plugging events (meaning very low or no DC-Voltage).

The typical isolation-barrier of the communication would be via a fiber optic link. While this offers an excellent isolation, it has some major disadvantages. These are the complex point-to-point

© VDE VERLAG GMBH · Berlin · Offenbach

wiring, high costs of fiber optic communication and degradation problems. In the low voltage range however it is possible to use a wired bus or ring-bus communication topology [1]. This offers a simple and cost-effective solution.

From a communication point of view the ideal topology would therefore be a wired bus-based communication.

3. Proposed solution

This section presents a solution for the problems mentioned in 2). The proposed solution combines a bus-based communication network with an external power supply.

Since an externally fed power supply has to include the isolation-barrier between the floating switching-cell-potential and the common primary potential, it is not possible to use a conventional transformer.

The necessary isolation barrier for a highvoltage application requires a transformer with an increased creepage- and clearance-distance between primary and secondary winding. This can be achieved either by using specially coated windings and core material or by using a large air gap transformer.

In this work the isolation is achieved with a coreless, 15mm air-gap transformer.

Since that reduces the achievable coupling coefficient, a resonant topology (Figure 1) is necessary in order to transfer the required amount of power.

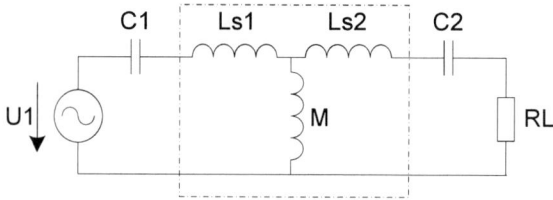

Figure 1: Resonant Topology

By the use of external resonance-capacitors (C1 and C2) it is possible to compensate the large leakage inductances (Ls1 and Ls2) at a specific frequency. With the appropriate tuning of these resonant structures it is possible to overcome the bad coupling within a small frequency band and transfer power to the load.

By the use of this resonant approach, it is even possible to transfer power over an air-gap that is large enough to isolate several kV [3].

Since it is possible to transfer power inside a specific frequency-band, the switching frequency of U1 is modulated according to the data-stream, in order to combine power and information-transfer into one device. This is comparable to the well known frequency shift keying (FSK) of mobile-communication signals [4].

4. Power-Supply

In order ensure a high isolation capability of the power supply, an air gap transformer prototype was built (see Figure 2).

Figure 2: Built transformer   RX- &TX-Board (different scale)

The coupling coefficient was found to be around 29% with the following transformer parameters:

$$Ls1 = 238uH \qquad Ls2 = 4.3uH \qquad M = 26.7uH$$

It is obvious that the leakage-inductances (especially Ls1) would dominate the overall impedance if it would remain uncompensated. As a result power-transfer would not be possible. As introduced in 3) the resonance-capacitors C1 and C2 are used to overcome these stray-effects.

$$XLs1 = j\omega Ls1 \ \& \ XC1 = -j\frac{1}{\omega C1} \qquad \text{with} \ \omega = 2\pi f$$

To eliminate these effects at the working frequency, the capacitor C1 has to be tuned to Ls1 and C2 to Ls2.

At the chosen resonance frequency f_{res} the stray-component $XLs1$ should be compensated by $XC1$:

$$XLs1 = XC1 \qquad => \qquad C1 = \frac{1}{\omega^2 Ls1} \qquad \& \qquad C2 = \frac{1}{\omega^2 Ls2}$$

For the prototype at a working-frequency of around f=100 kHz the following two capacitors have therefore been selected:

$$C1 = 10.6nF \qquad C2 = 590nF$$

The resulting frequency-variable transferred power can be seen in Figure 3.

© VDE VERLAG GMBH · Berlin · Offenbach

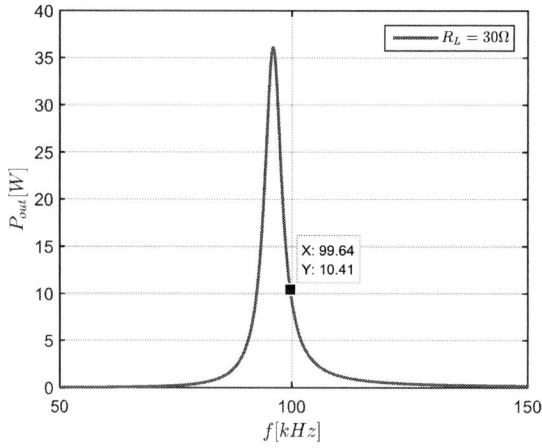

Figure 3: Transferred power over frequency (24Vac input Voltage)

The proposed compensation technique can only be an approximation since the system resonance-frequency also depends on the coupling inductance M and the actual load R_L of the circuit.

The actual operating points should be chosen above the actual resonance frequency to enable soft switching and further enhance system-efficiency.

5. Communication Interface

Since the communication interface is supposed to use the same hardware as the power-supply, the modulation-scheme has to be selected carefully.

As mentioned in 3) an FSK like modulation was selected to use the frequency-specific transformer-behavior.

The basic principle of FSK modulation maps the information to different frequencies (f0 and f1). In case of the proposed solution this means, that the power supply has to work in two operating points.

The two frequencies have to be far enough apart to detect the differences on the secondary side. On the other hand they have to be close enough together to stabilize the transferred power.

In this work the secondary-side data is decoded with a 50 MHz counter. That counter is used to evaluate the frequency of the transformator-output-voltage. The base-frequency of f1=1/t1≈100 kHz (see 4)) would therefore result in a counter-value of t1_inc=502 increments.

To have a reasonable large margin for error, the difference of t1_inc to t0_inc shall be at least 10 increments. Taking this margin into account the second frequency should therefore be at approximately 97 kHz.

Logic-Value	Power Supply Frequency fx	Secondary-Side Counter Value tx_inc	Approx Transferred Power
"1"	99.6 kHz	502 increments	10 W
"0"	97.6 kHz	512 increments	12 W

Table 1: FSK-Coding

PCIM Europe 2016, 10 – 12 May 2016, Nuremberg, Germany

Without any further modulation the transferred power would be heavily dependent on the transferred data (see table above).

Therefore the data stream has to be encoded with a DC-free algorithm. This algorithm has to ensure an even distribution of logic "ones" and "zeros" over a predefined period of time. One possible algorithm would be Manchester coding [7].

In combination with the proposed FSK this results in a constant power transfer of 11 W.

6. Measurements

To prove the proposed concept, a working prototype was built. It consists of the wide airgap transformer, one TX and one RX-Board (see Figure 2).

The TX-board includes an FPGA for data-stream encoding (Manchester) and one halfbridge DCAC converter including the primary side resonance capacitors.

The RX-board includes an FPGA for signal decoding, a comparator for signal conditioning, compensation network and a rectifying circuit.

The input voltage was set to 48Vdc. With the working-frequencies chosen in 5) and a ~35Ohms dummy-load the secondary-side DC voltage was approx 15V.

The resulting power output was found to be around 6.5W. With 8W input power this leads to an efficiency of ~80%.

The bitstream was Manchester-Coded as described in Table 2.

In order to guarantee a correct frequency detection, the system needs to settle to the new frequency. This requires a specific time with a constant frequency. In this implementation the frequency was held constant for 16 repetitions.

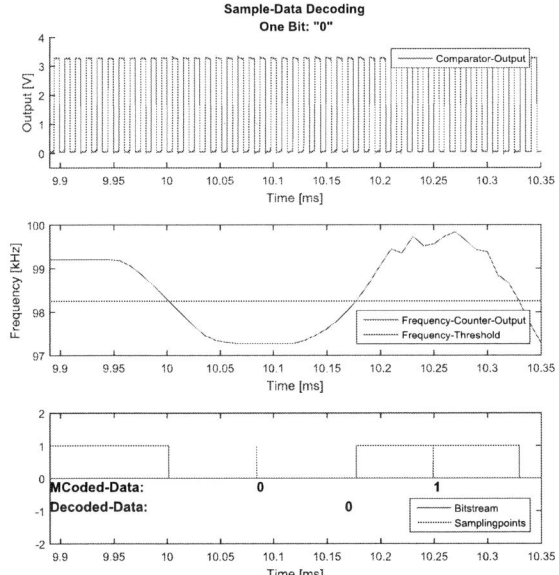

Figure 4: Data-Decoding: Single Bit

Input-Bit-Value	Manchester Coded
"1"	"10"
"0"	"01"

Table 2: Manchester datacoding

Figure 4 shows the data-decoding steps for an example bit ("0"). To decode the bitstream the secondary side input AC-voltage was first compared to Vdc/2 using a comparator (signal see Figure 4 top). This digital signal is then used to determine the current input-frequency with a FPGA. In order to extract the Manchester-coded bits the sensed-frequency is compared to a

© VDE VERLAG GMBH · Berlin · Offenbach

threshold (Figure 4 middle and bottom) and then sampled. In this implementation the encoded sequence "01" represents a logic "0".

To test the described techniques several test-patterns were transferred and analyzed. Figure 5 shows the results for two example streams.

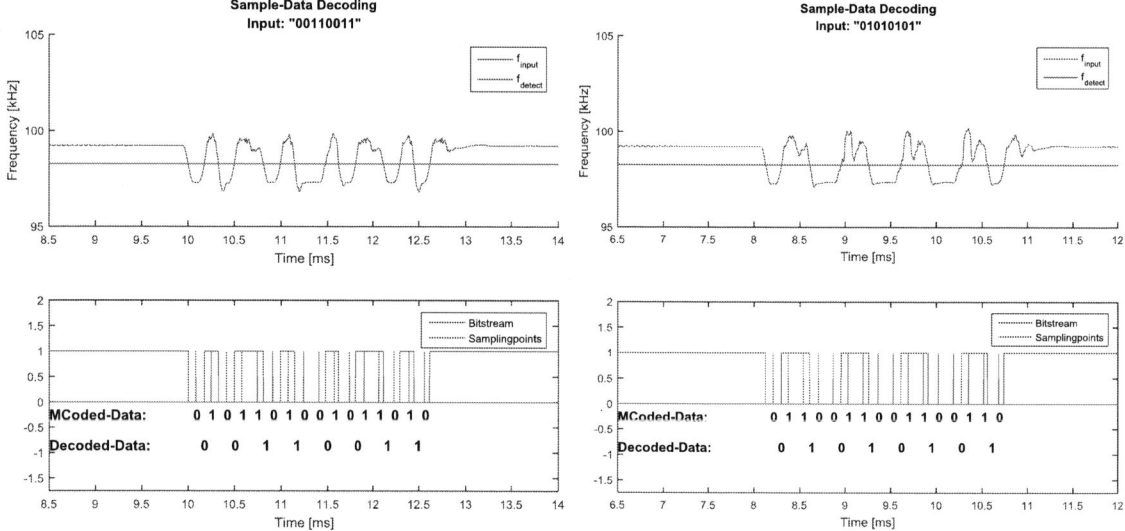

Figure 5: Data-Decoding: test patterns

In order to simplify datastream synchronization in this demonstration, a relatively large preamble was chosen (and is not fully shown here).
After the preamble a datastream of 16bits can be observed. These 16bits represent 8bits of payload and are decoded according to Table 2. After decoding the test-pattern is fully recovered and can be further processed.

The datarate of this prove of concept demonstrator is 8bit/3ms = 2,5kbit/s.

7. Conclusion

This paper presents a novel power-supply concept. It combines power and data-transfer over the same transformer. This enables many new application fields. One example is the use in a highvoltage modular converter like the M2C. This enables the use of a cost effective network-based communication even in converter-systems with high-isolation demands.
Therefore the proposed structure can be seen as a key-component for a wire-based communication in highvoltage converters.
In order to use this kind of communication interface for control purposes, the data-rate should be improved. This can be achieved using a higher switching frequency in the MHz range, using GaN technology for example. Furthermore the coding scheme could be changed to a 16b20b pattern, which would reduce the amount of overhead needed to guarantee a continuous powertransfer significantly.

References

[1] Galek, Marek: "MOSFET-based Modular Multilevel Converters: Design Challenges and Implementation Details" ECPE Workshop Advanced Multicell Multilevel Power Converters 2014.

[2] Galek, Marek, Manuel Blum, and Hamza Mlayeh. "A fault tolerant communication interface for modular and distributed power electronics." PCIM Europe 2015; International Exhibition and Conference for Power Electronics, Intelligent Motion, Renewable Energy and Energy Management; Proceedings of. VDE, 2015.

[3] Komma Thomas, Poebl Monika "Characterization Of Large-Air-Gap Transformer Systems By Two-Port-Theory" PCIM Europe 2013; International Exhibition and Conference for Power Electronics, Intelligent Motion, Renewable Energy and Energy Management; Proceedings of. VDE, 2013.

[4] Federau, Joachim. "Operationsverstaerker." Vieweg, Wiesbaden (1998).

[5] Cottet, Didier, et al. "Integration Technologies for a Fully Modular and Hot-Swappable MV Multi-Level Concept Converter." PCIM Europe 2015; International Exhibition and Conference for Power Electronics, Intelligent Motion, Renewable Energy and Energy Management; Proceedings of. VDE, 2015.

[6] Dirk Malipaard "Innovative, cost effective control bus solutions for multi-level converters" http://www.iisb.fraunhofer.de/en/research_areas/energy_electronics/multi_level_inverter/research_development.html 2016.

[7] Tanenbaum, Andrew S. "Prof. David J. Wetherall: Computernetzwerke."

PCIM Europe 2016, 10 – 12 May 2016, Nuremberg, Germany

FPGA Based Direct Model Predictive Current Control of PMSM Drives with 3L-NPC Power Converters

Zhenbin, Zhang, Technische Universität München, Germany, james.cheung@tum.de
Christoph, Hackl, Technische Universität München, Germany, christoph.hackl@tum.de
Ralph, Kennel, Technische Universität München, Germany, ralph.kennel@tum.de

Abstract

Direct control techniques are interesting alternatives, in particular, for drive systems with multi-level and/or multi-phase power converters, for which the modulator design becomes rather complex. This work presents a *direct model predictive current control* (DMPCC) method for permanent-magnet synchronous machine (PMSM) drives fed by three level neutral point clamped (3L-NPC) power converters. The proposed DMPCC method takes the currents, neutral point voltage balancing and switching frequency regulations as control targets and is implemented on a FPGA based real-time controller. Its control performances are compared with the conventional direct torque control method with switching table (DTC-TB). Experimental results confirm that the proposed DMPCC scheme may outperform the conventional DTC-TB technique in steady-state with reduced tuning efforts.

1. Introduction

Three-level neutral-point (diode) clamped (3L-NPC) power converters are widely used in medium voltage drives. This topology allows for more than two voltage levels and the required amount of components is drastically less than for five-level topologies. Permanent-magnet synchronous machines (PMSM) show nice features such as, high efficiency, compact size and simple control, hence PMSMS are a viable choice for high performance drives or energy generation.

Fig. 1 depicts the topology of a PMSM drive system with 3L-NPC power converter. Considering whether a modulator is required or not, control schemes to deal with such systems are divided into **two** groups (see [1, 2]): **(a)** Modulator based (linear) control schemes, such as: field-oriented control (FOC); direct torque control with modulator, and deadbeat like model predictive control (DBC); **(b)** (Nonlinear) direct control methods, such as: direct torque control with look-up table (DTC-TB) and direct model predictive control (DMPC) with cost-function. Control methods with modulator (which is usually based on the principle of "timed-average-approximation" [3]) is straight-forward for two-level power converter based systems. However, for multi-level power converters (e.g., 3L-NPC power converters), in particular when considering neutral point voltage balancing or common-mode voltage minimization requirements, modulator design becomes (rather) complex. Moreover, a power conversion/drive system driven by a multi-level power converter is, in essence, a highly *nonlinear* and *switched* system. From the concept point of view, by approximating such a system as a linear plant and using a modulator to emulate the continuous commands will constrain the potential dynamics of the system.

On the contrary, direct control methods — such as DMPC — take the switched nature of the power converter into account and combine modulation and switching sequence decision process into one single step, hence no "averaging" approximation of the modulation method is required and a modulator is obsolete. DMPC can easily include multiple nonlinear constraints

© VDE VERLAG GMBH · Berlin · Offenbach

into a customer designed cost function. Its implementation has appealing features, such as: straight-forward concept, flexible design and fast control dynamics, etc. Therefore, this technique has been subject to extensive research in the last decade(s). However, high computational load is regarded *the* short-coming in comparison with conventional techniques (e.g., switching table based DTC). Real-time implementation using FPGA-based systems provides a (feasible) choice to tackle the heavy computational loads.

This work presents a (nonlinear) direct model predictive current control (DMPCC) method based on the DMPC concept for a PMSM drive system fed by 3L-NPC power converter. No extra weightings are required for the targeting set control. The achieved control performance is compared to conventional DTC-TB techniques **by experiments**. The presented method is **implemented on a fully FPGA-based real-time system**.

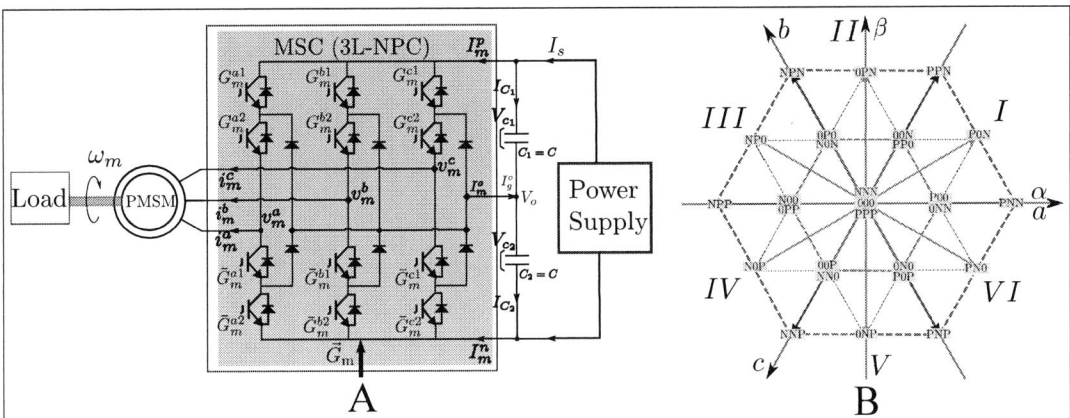

Fig. 1: Simplified electrical circuit of a 3L-NPC PMSM(G) drive system

2. Physical system description and system modeling

Fig. 1 illustrates the considered physical system. This section revisits the detailed modeling of the system in discrete time. A short description for all the symbols used in the figures and the models are as follows: $\vec{v}_m^{abc} = (v_m^a, v_m^b, v_m^c)^\top$ [V]^3 is the output voltage vector of the power converter, $\vec{i}_m^{abc} = (i_m^a, i_m^b, i_m^c)^\top$ [A]^3 are the PMSM stator current vectors, ω_e, ω_m [rad/s] are the rotor electrical and mechanical rotational speed ($\omega_e = N_p \omega_m$, N_p is the number of pole pairs), ϕ_e [rad] is the electrical angle (position) of the rotor flux, $\vec{\psi}_s^{abc} = (\psi_s^a, \psi_s^b, \psi_s^c)^\top$ [Vs]^3, ψ_{pm} [Vs], R_s [Ω] and L_s [Vs/A] are the stator and permanent-magnet flux linkages, stator resistance and inductance, respectively, V_d, V_{c1}, V_{c2} [V] are the DC-link, upper and lower capacitor voltages, respectively; $V_o := V_{c1} - V_{c2}$, [V] is the neutral point voltage, $I_s, I_m^{p,n}$ are the source, positive and negative currents of the DC-link (see Fig. 1), Θ_m, [kgm^2] and B, [Nms/rad] are the total moment of inertia and viscous friction coefficient[1], respectively, $T_{t[k]}$, [Nm] is the load side torque. All quantities in abc frame can be transformed to $\alpha\beta$ reference frame by invoking the Clarke transformation given by (with Clarke transformation matrix $\mathbf{T}_C$)

$$\vec{x}^{\alpha\beta} = \underbrace{\sqrt{\tfrac{2}{3}} \begin{bmatrix} 1 & -\tfrac{1}{2} & -\tfrac{1}{2} \\ 0 & \tfrac{\sqrt{3}}{2} & -\tfrac{\sqrt{3}}{2} \end{bmatrix}}_{=:\boldsymbol{T}_c \ (\text{Clarke transformation})} \vec{x}^{abc}. \tag{1}$$

---

[1] Note that viscous friction is mostly described by a constant.

2.1. PMSM modeling in $\alpha\beta$ frame

Assuming that the rotor flux position θ_e will not change within a very small control interval $T_s \ll 1\,\text{s}$, i.e., $\theta_{e[k+1]} \approx \theta_{e[k]}$, applying the Euler-forward method yields the following discrete-time model of the PMSM in $\alpha\beta$ reference frame [4]:

$$
\left.
\begin{aligned}
\vec{i}^{\,\alpha\beta}_{m[k+1]} &= \left(1 - \tfrac{T_s R_s}{L_s}\right)\vec{i}^{\,\alpha\beta}_{m[k]} + \tfrac{T_s}{L_s}\left(\vec{v}^{\,\alpha\beta}_{m[k]} - \underbrace{\begin{pmatrix} -\psi_{pm}\omega_{e[k]}\sin\left(\theta_{e[k]}\right) \\ \psi_{pm}\omega_{e[k]}\cos(\theta_{e[k]}) \end{pmatrix}}_{=:\vec{e}^{\,\alpha\beta}_{m[k]}}\right). \\
\vec{\psi}^{\,\alpha\beta}_{s[k+1]} &= \vec{\psi}^{\,\alpha\beta}_{s[k]} + \left(\vec{v}^{\,\alpha\beta}_{m[k]} - R_s\vec{i}^{\,\alpha\beta}_{m[k]}\right)\cdot T_s, \\
\omega_{e[k+1]} &= \omega_{e[k]} + \tfrac{T_s}{\Theta_m N_p}\left(\underbrace{T_{t[k]} - N_p\left(\psi^\alpha_{s[k]}i^\alpha_{m[k]} - \psi^\beta_{s[k]}i^\beta_{m[k]}\right)}_{=:T_{e[k]}} - B\cdot\omega_{m[k]}\right)
\end{aligned}
\right\}.
\tag{2}
$$

2.2. 3L NPC power converter

Considering the 3L-NPC converter depicted in Fig. 1, for $x \in \{a, b, c\}$ and $i \in \{1, 2\}$, the gate signals of the upper IGBTs are introduced as G^{xi}_m. Considering the non-shoot-through operation principle of such a topology, the gate signals of the lower IGBTs shall be complementary (negated) to the upper ones (see Fig. 1), i.e, given the upper switching state is G^{xi}_m, the lower shall be $\bar{G}^{xi}_m$. Therefore, neglecting all the uncertain switching combination[2], we can define the switching state vector by

$$
u^x_m := \begin{cases} 1\,(P) & \text{if}: G^{x1}_m = 1 \wedge G^{x2}_m = 1 \\ 0\,(0) & \text{if}: G^{x1}_m = 0 \wedge G^{x2}_m = 1 \\ -1\,(N) & \text{if}: G^{x1}_m = 0 \wedge G^{x2}_m = 0 \end{cases}
\tag{3}
$$

for phase x. Then the switching state vector has the following form $\vec{u}^{\,abc}_m = [u^a_m, u^b_m, u^c_m]^\top$ and $\in \mathcal{U}_{27} := \{NNN, NN0, \ldots, PP0, PPP\}$ of 27 admissible switching states. Hence, for DC-link voltages V_{c1} and V_{c2} (see Fig. 1), the phase voltages (for only balanced/symmetric situation) of the converter can be described by [2]:

$$
\vec{v}^{\,abc}_m = \begin{bmatrix} v^a_m \\ v^b_m \\ v^c_m \end{bmatrix} = \frac{1}{2}\cdot\frac{(V_{c1}+V_{c2})}{3}\underbrace{\begin{bmatrix} 2 & -1 & -1 \\ -1 & 2 & -1 \\ -1 & -1 & 2 \end{bmatrix}}_{:=T_{SW}}\vec{u}^{\,abc}_m + \frac{1}{2}\cdot\frac{(V_{c1}-V_{c2})}{3}\underbrace{\begin{bmatrix} 2 & -1 & -1 \\ -1 & 2 & -1 \\ -1 & -1 & 2 \end{bmatrix}}_{:=T_{SW}}|\vec{u}^{\,abc}_m|
$$

$$
\tag{4}
$$

2.3. DC-link and neutral point voltage dynamics

For a 3L-NPC power converter, the DC-link modeling includes both the DC-link charging/discharging equation and also the DC-link capacitor voltage difference equation. They are introduced separately in the following.

---

[2]Uncertain switching combination means that the combination which lead to uncertain output voltages (e.g., $G^{x1}_m = 1 \wedge G^{x2}_m = 0$).

DC-link (charging/discharging) equation:

The DC-link voltage equation, when considering the current flow of the converter (see Fig. 1), can be modeled in the discrete time as follows

$$V_{\mathrm{d}[k+1]} = V_{\mathrm{d}[k]} + \frac{T_s}{C}\left(I_{\mathrm{s}[k]} - I_{\mathrm{m}[k]}\right), \tag{5}$$

where $I_{\mathrm{m}[k]} = \vec{i}_{\mathrm{m}[k]}^{\mathrm{abc}\top} \cdot \vec{u}_{\mathrm{m}[k]}^{\mathrm{abc}}$ and I_{s} are DC-link current components of the source and load sides, respectively.

DC-link capacitor voltage difference equation:

To achieve voltage balancing, the difference voltage $V_o(t) := V_{c_1}(t) - V_{c_2}(t)$ should be zero (for all time) and it can be controlled through the power converter. As can be observed in Fig. 1, V_o depends on the charging state of the two DC-link capacitors C_1 and C_2 (where $C := C_1 = C_2$) and will only change when currents I_{m}^o (which is the neutral point current, see Fig. 1) is drawn from it (see Fig. 1), i.e., when $\vec{u}_{\mathrm{m}}$ contains "zero" elements. For a given phase current vector $\vec{i}_{\mathrm{m}}^{\mathrm{abc}} := [i_{\mathrm{m}}^{\mathrm{a}}, i_{\mathrm{m}}^{\mathrm{b}}, i_{\mathrm{m}}^{\mathrm{c}}]^\top$, the positive and negative currents of

$$I_{\mathrm{m}}^{\mathrm{p}} = \frac{1}{2}(|\vec{u}_{\mathrm{m}}^{\mathrm{abc}}| + \vec{u}_{\mathrm{m}}^{\mathrm{abc}})\,\vec{i}_{\mathrm{m}}^{\mathrm{abc}}, \quad I_{\mathrm{m}}^{\mathrm{n}} = \frac{1}{2}(|\vec{u}_{\mathrm{m}}^{\mathrm{abc}}| - \vec{u}_{\mathrm{m}}^{\mathrm{abc}})\,\vec{i}_{\mathrm{m}}^{\mathrm{abc}} \tag{6}$$

can be computed. Therefore, the dynamics of V_o are given by

$$\frac{\mathrm{d}V_o}{\mathrm{d}t} = \frac{\mathrm{d}V_{c1}}{\mathrm{d}t} - \frac{\mathrm{d}V_{c2}}{\mathrm{d}t} = \frac{1}{C}\left\{(I_{\mathrm{m}}^{\mathrm{p}} + I_{\mathrm{m}}^{\mathrm{n}})\right\} = \frac{1}{C}\,|\vec{u}_{\mathrm{m}}^{\mathrm{abc}}|\,\vec{i}_{\mathrm{m}}^{\mathrm{abc}\top} \tag{7}$$

Applying the forward Euler approximation yields the voltage difference equation for the DC-link voltage difference as follows

$$V_{\mathrm{o}[k+1]} = V_{\mathrm{o}[k]} + \frac{T_s}{C}\,|\vec{i}_{\mathrm{m}[k]}^{\mathrm{abc}\top} \cdot \vec{u}_{\mathrm{m}[k]}^{\mathrm{abc}}|. \tag{8}$$

3. Control Algorithms

3.1. DMPC and the proposed direct model predictive current control

Classical DMPC schemes evaluate a given cost function of the following form

$$J_{\mathrm{DMPC}}(\vec{u}_{i,j}) = \underbrace{\sum_{i=1}^{m} \gamma_{\mathrm{TS}_i} |TS_{\mathrm{i}[k+1]}^* - TS_{\mathrm{i}[k+1]}^p(\vec{u}_i)|}_{=:J_{\mathrm{TS}}} + \underbrace{\sum_{j=1}^{n} \gamma_{\mathrm{CS}_j} |CS_{\mathrm{j}[k+1]}^* - CS_{\mathrm{j}[k+1]}(\vec{u}_j)|}_{=:J_{\mathrm{CS}}}, \tag{9}$$

which describes the aimed at control objectives consisting in general of two parts: J_{TS_i} and J_{CS_j} (with corresponding weighting factors γ_{TS_i} and γ_{CS_j}). J_{TS_i} and J_{CS_j} represent sub-costs for the *Target Set* TS_i (such as: reference tracking of current, torque, or power with reference TS_i^*) *and* the *Constraint Set* CS_j (such as: current/torque, or power constraints with reference CS_j^*), respectively. $\mathrm{M}_{\mathrm{TS}_i}$ *and* $\mathrm{M}_{\mathrm{CS}_j}$ are abbreviations for the *prediction model for the Target Set* TS_i and the *prediction model for the Constraint Set* CS_j, respectively. For the considered

© VDE VERLAG GMBH · Berlin · Offenbach

3L-NPC converter, the control set is $\vec{u}_{\mathrm{m}} \in \mathcal{U}_{27}$. After evaluating and minimizing cost-function (9) for $\vec{u}_{\mathrm{m}} \in \mathcal{U}_{27}$, the optimal gate vector $\vec{G}$ will be applied to the converter.

The proposed direct model predictive current control takes the stator current tracking as the **target set** (i.e., $J_{\mathrm{TS_m}}$ in (10)) and the DC-link voltage balancing requirement (J_{V_o}) and switching frequency regulation (J_{sfm}) as the **constraint set** (i.e., $J_{\mathrm{CS_m}}$ in (10)). DMPCC is designed in the $\alpha\beta$ frame to eliminate the Park transformation. Therefore, the cost function is defined by

$$J_{\mathrm{DMPC}}^{\mathrm{m}}(\vec{u}_{\mathrm{m}}) = \underbrace{\left(i_{\mathrm{m}}^{\alpha*} - i_{\mathrm{m}[k+1]}^{\alpha}(\vec{u}_{\mathrm{m}})\right)^2 + \left(i_{\mathrm{m}}^{\beta*} - i_{\mathrm{m}[k+1]}^{\beta}(\vec{u}_{\mathrm{m}})\right)^2}_{=:J_{\mathrm{TS_m}}} + \underbrace{\overbrace{\gamma_{V_\mathrm{o}}\left(V_{\mathrm{o}[k+1]}(\vec{u}_{\mathrm{m}})\right)^2}^{J_{V_\mathrm{o}}} + \overbrace{\gamma_{\mathrm{sf}}\Delta_{\vec{u}_{\mathrm{m}}}}^{J_{\mathrm{sfm}}}}_{=:J_{\mathrm{CS_m}}}, \quad (10)$$

where $\gamma_{V_\mathrm{o}}[1], \gamma_{\mathrm{sf}}[1]$ are the weighting factors, $\Delta_{\vec{u}_{\mathrm{m}}}$ is responsible for the switching frequency regulation, and is defined as

$$\Delta_{\vec{u}_{\mathrm{m}}} = |u_{\mathrm{m}[k+1]}^{\mathrm{a}} - u_{\mathrm{m}[k+1]}^{\mathrm{a}}| + |u_{\mathrm{m}[k+1]}^{\mathrm{b}} - u_{\mathrm{m}[k]}^{\mathrm{b}}| + |u_{\mathrm{m}[k+1]}^{\mathrm{c}} - u_{\mathrm{m}[k]}^{\mathrm{c}}|. \quad (11)$$

In Eq. (10) the predicted values of the currents $\vec{i}_{\mathrm{m}[k+1]}^{\alpha\beta}(\vec{u}_{\mathrm{m}})$ and the neutral point voltage are obtained by evaluating the prediction models (2) and (8), respectively, with $\vec{u}_{\mathrm{m}} \in \mathcal{U}_{27}$. After evaluating and minimizing cost-function (10), the obtained optimal gate vectors $\vec{G}_{\mathrm{m}}$ will be applied to the converter. The overview of the proposed control method is depicted in Fig. 2-a.

3.2. Direct torque control with switching table (DTC-TB)

The first direct torque control (DTC) scheme was proposed in the 1970s by Depenbrock [5], Takahashi and Noguchi [6] for controlling a two-level converter-fed induction machine. Then the same concept was extended to PMSM in 1995. The extended DTC for 3L-NPC power converters was firstly reported in [7]. The switching table is designed off-line and is indexed by the position of the flux vector, the outputs of torque, neutral point voltage difference and flux hysteresis controllers. It directly outputs the optimal switching vector. In this paper, the basic idea of DTC schemes are not discussed. For more details please refer to [7] and the reference therein. For the implementation, the standard switching table for 3L-NPC power converters (as are introduced in [8,9]) were adopted. The overall control structure is illustrated in Fig. 2-b.

4. Experimental verification

To verify and compare the control performance of DMPCC and DTC-TB, both schemes are implemented on a NI-FPGA based real-time system and are tested on a self-constructed 3L-NPC power converter system with DC-link capacitors $C := C_1 = C_2 = 1100e^{-6}[\mathrm{F}]$ (see Fig. 4). The relevant state estimations, cost function calculation/minimization procedures are implemented as subroutines and optimized with the Single-Cycle-Timed-Loop method [10] using the available NI-FPGA technique. For a fair comparison, both schemes utilize the same outer loop controllers and a cycle time of $T_{\mathrm{s}} = 50e^{-6}[\mathrm{s}]$. The upper and lower bandwidths of the DTC-TB hysteresis controllers are set to 5% and 2.5% of the relevant nominal values; to achieve DC-link balancing, γ_{V_o} is set to 0.5, while $\gamma_{\mathrm{sf}} = 0.03285$ to achieve a switching frequency (around 4kHz during steady-state) similar to that of the DTC-TB method.

Fig. 5 illustrates the overall control performance using the proposed DMPCC method. The testing scenarios are: a reference speed which changes from 600 rpm to -800 rpm and -300

© VDE VERLAG GMBH · Berlin · Offenbach

PCIM Europe 2016, 10 – 12 May 2016, Nuremberg, Germany

rpm and then back to 600 rpm, with a slope of 1860 rpm/min, while a rated torque is applied to the machine. As can be seen, a good steady-state and dynamic control performance is achieved. The capacitor voltage differences are also quite small during the whole experiment.

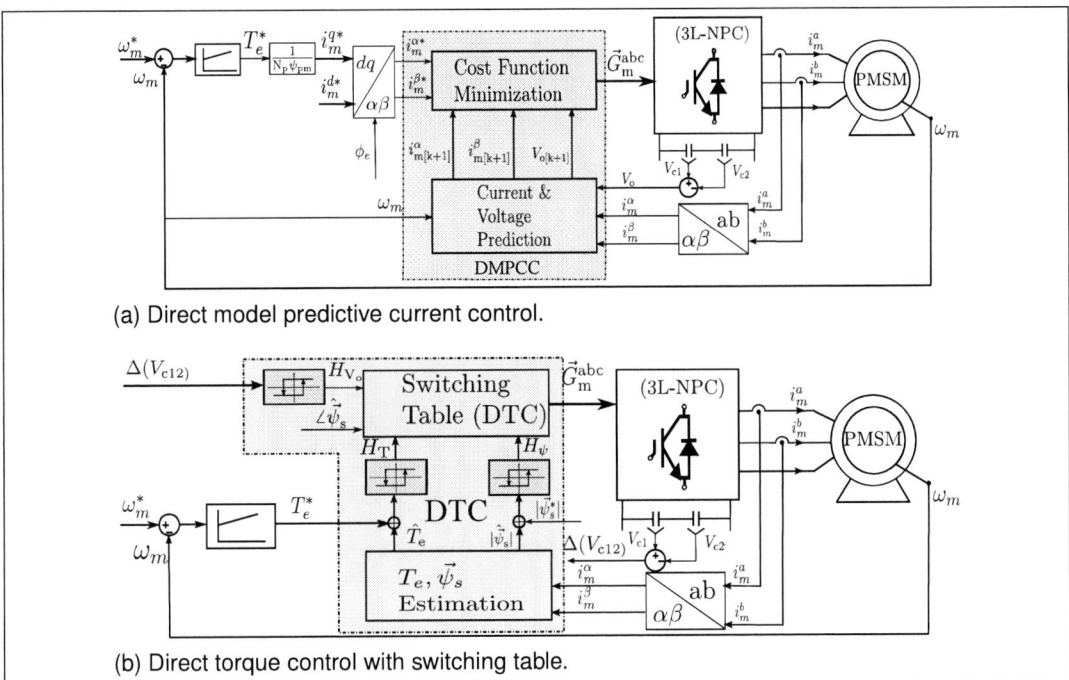

(a) Direct model predictive current control.

(b) Direct torque control with switching table.

Fig. 2: Control structures of the proposed direct model predictive current control scheme and the conventional direct torque control method.

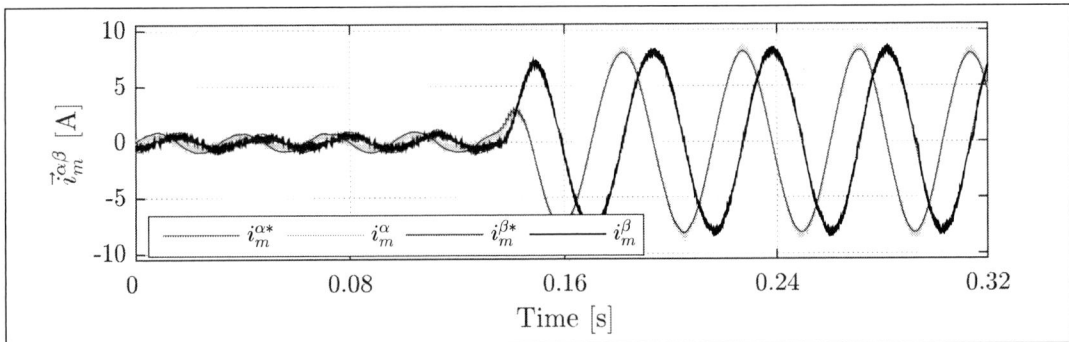

Fig. 3: Current tracking performances of the proposed direct model predictive current control.

Fig. 3 illustrates the current tracking performance of the proposed DMPCC. Good racking performance, smooth waveforms with small currents ripples are obtained for both 10% and 100% of the rated torque. Fig. 6 shows the performance comparison between the proposed and the conventional DTC-TB methods during steady- state[3]. As can be clearly seen, the proposed method shows much smaller ripples in the stator currents than the DTC-TB method. The reason is their differing switching patterns: At the same operating point, DMPCC "selects" much

---

[3]Similar dynamic/transient phase performances will be achieved by both methods during to their similar direct switching vector selection principles.

© VDE VERLAG GMBH · Berlin · Offenbach

smoother switching patterns and almost no full voltage pattern (i.e., the "P" position of the phase switches) is selected; while DTC-TB synthesizes much more "noisy" switching patterns with lead to higher instantaneous voltage magnitudes (see the last sub-figure in Fig. 6), although the magnitude of the fundamental is the same as for the DMPCC method (i.e., both the filtered command voltages from DMPCC and DTC-TB have a similar magnitude).

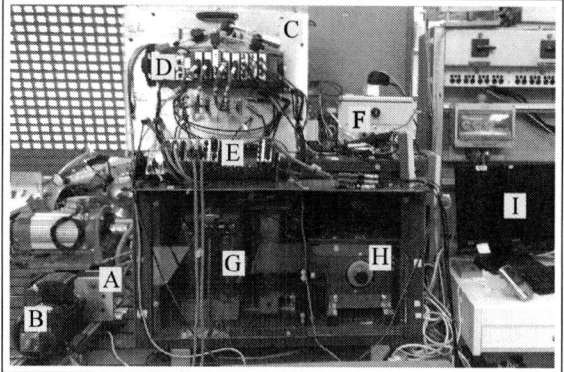

Fig. 4: Test-bench for experimental verification. A: PMSM, B: AC-Drive, C: self-constructed 3L-NPC power converter, D: NI-CRIO FPGA based real-time controller, F: protection devices, H: Variac and I: PC for software development and data logging.

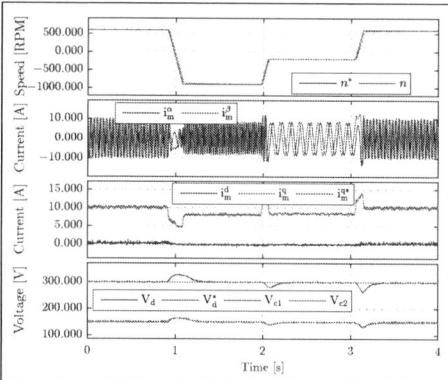

Fig. 5: Experimental results: Overall performances of the proposed DMPCC control scheme. From top to bottom: Speed, currents in $\alpha\beta$ and dq frame and DC-link voltages, respectively.

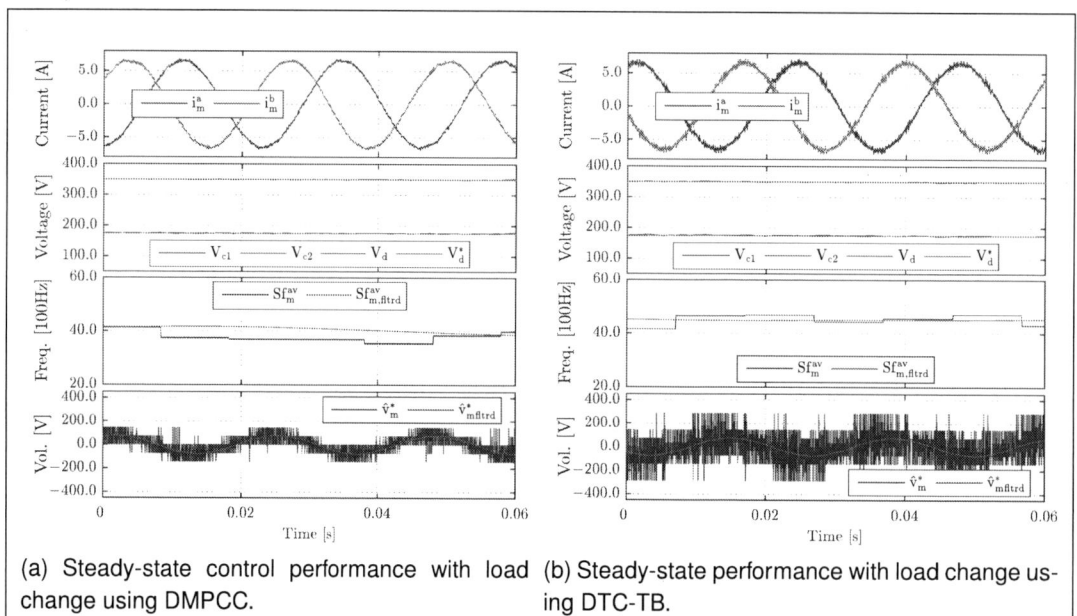

(a) Steady-state control performance with load change using DMPCC.

(b) Steady-state performance with load change using DTC-TB.

Fig. 6: Experimental results: Steady-state performance comparison of the proposed DMPCC control scheme and DTC-TB method, where $\mathrm{Sf}_m^{av}, \hat{v}_m^*$ are the average switching frequency (updated every 10ms) and command voltage (in α-axis); while $\mathrm{Sf}_{m,fltrd}^{av}, \hat{v}_{mfltrd}^*$ are their filtered values with cut-off frequencies of 5Hz and 300Hz, respectively.

© VDE VERLAG GMBH · Berlin · Offenbach

5. Conclusion

This work has presented and experimentally validated a direct model predictive current control (DMPCC) method for a PMSM drive system fed by a three-level neutral-point clamped (3L-NPC) power converter. The proposed solution is implemented in the $\alpha\beta$ frame. No specially tuned weighting factors are required for the targeting set, which significantly simplifies the tuning process. The proposed method belongs to the (nonlinear) direct control class. Its steady-state control performance was compared with its well-known counter-part, i.e., the switching table based direct torque control solution (during similar experimental conditions). The comparison is illustrated by experimental results. Moreover, both schemes have been implemented on a fully FPGA-based real-time system and validated at a self-constructed test-bench. The results confirm that better steady-steady control performances are achieved using DMPCC.

Acknowledgment: This work is supported by DFG founding (No.: KE817/32-1). Zhenbin Zhang is the corresponding author and would like to express his gratefulness to Mr. Marc Back-meyer from National Instruments and Dr. Martin Schulz from Infineon for their help during his test-bench construction.

References

[1] M. Kazmierkowski, L. Franquelo, J. Rodriguez, M. Perez, and J. Leon, "High-performance motor drives," *Industrial Electronics Magazine, IEEE*, vol. 5, no. 3, pp. 6–26, Sept 2011.

[2] Z. Zhang and R. Kennel, "Direct Model Predictive Control of Three-Level NPC Back-to-Back Power Converter PMSG Wind Turbine Systems Under Unbalanced Grid," in *Predictive Control of Electrical Drives and Power Electronics (PRECEDE 2015), Valparaiso, Chile.*, 2015.

[3] S. Kouro, M. A. Perez, J. Rodriguez, A. M. Llor, and H. A. Young, "Model predictive control: Mpc's role in the evolution of power electronics," *IEEE Industrial Electronics Magazine*, vol. 9, no. 4, pp. 8–21, Dec 2015.

[4] Z. Zhang, C. Hackl, F. Wang, Z. Chen, and R. Kennel, "Encoderless model predictive control of back-to-back converter direct-drive permanent-magnet synchronous generator wind turbine systems," in *Power Electronics and Applications (EPE), 2013 15th European Conference on*, Sept 2013, pp. 1–10.

[5] M. Depenbrock, "Direct self-control (dsc) of inverter-fed induction machine," *Power Electronics, IEEE Transactions on*, vol. 3, no. 4, pp. 420–429, Oct 1988.

[6] I. Takahashi and T. Noguchi, "A new quick-response and high-efficiency control strategy of an induction motor," *Industry Applications, IEEE Transactions on*, vol. IA-22, no. 5, pp. 820–827, Sept 1986.

[7] J. Steinke, "Control strategy for a three phase ac traction drive with three-level gto pwm inverter," in *Power Electronics Specialists Conference, 1988. PESC '88 Record., 19th Annual IEEE*, April 1988, pp. 431–438 vol.1.

[8] H. Alloui, A. Berkani, and H. Rezine, "A three level npc inverter with neutral point voltage balancing for induction motors direct torque control," in *Electrical Machines (ICEM), 2010 XIX International Conference on*, Sept 2010, pp. 1–6.

[9] I. Messaif, E. Berkouk, and N. Saadia, "Performances of dtc system fed by a three-level npc vsi," in *Power Engineering, Energy and Electrical Drives (POWERENG), 2013 Fourth International Conference on*, May 2013, pp. 1471–1476.

[10] Z. Zhang and R. Kennel, "Fully fpga based direct model predictive power control for grid-tied afes with improved performance," in *Industrial Electronics Society, IECON 2015 - 41th Annual Conference of the IEEE*, Nov 2015.

PCIM Europe 2016, 10 – 12 May 2016, Nuremberg, Germany

IGBT Power Module in Three-Level Neutral Point Clamped Type 2 (NPC2, T-NPC, Mixed Voltage) Topology in Short Circuit Modes

Vladan, Jerinic, Danfoss Silicon Power, Germany, vladan.jerinic@danfoss.com
Kevin, Lenz, Danfoss Silicon Power, Germany, kevin.lenz@danfoss.com
Reiner, Hinken, Danfoss Silicon Power, Germany, reiner.hinken@danfoss.com

Abstract

In NPC2 topology a very high current and a transient current peak (I_{tp}) can occur in a special NPC2 short circuit mode (SC_{NPC2}) [1]. A detailed description of that effect can help to protect IGBT power modules in this topology against malfunction and can help during a post mortem analysis to understand the root cause. With a full understanding of that effect the lifetime of the inverter can be increased and the time and costs for qualification reduced. In this paper we discuss how typical application parameters like temperature and gate driver can influence that effect.

1. Introduction

Three level topologies are implemented in an increasing number of applications. Especially Solar, UPS and Active Filter [2] use the benefit of low switching losses for a high efficient inverter or use higher switching frequencies to reduce the filter size. Motor drive and Wind turbine applications further benefit of lower dv/dt and bearing currents that can extend the lifetime of the motor or generator [3]. Better shape of sinusoidal current, smaller common mode voltage and less dv/dt can help saving money in filter design for all applications. In literature [2,4], two types of neutral point clamped (NPC) three-level topologies are described.

2. IGBT Power Module Concept P3L

The Danfoss P3L IGBT power module package is known as a standard in multi-level power application. A full neutral point clamped Type 2 (NPC2) topology with low stray inductance commutation paths offers the three level benefits for high power applications. The same P3L package is also able to accommodate Type 1 (NPC1) topologies [5].

The internal module topology was optimized to create an improved NPC2 topology. To minimize inductive loop in the switches with the highest switching rate the topology is split into a high side bridge (T1, D1, T2, D2) and a low side bridge (T4, D4, T3, D3). To address the reverse recovery current of the diodes D2 and D3, protection diodes D5 and D6 are required. They have to handle only the reverse recovery current of D2, respectively D3, and can therefore be very small compared to the other semiconductors, e.g. 90A instead of 900A.

Device under test (DUT):

DP900N1200TU104204, with I_c=900A, T1/T4 1200V chip and T2/T3 650V chip

© VDE VERLAG GMBH · Berlin · Offenbach

DP700N1700TU104202 with Ic=700A, T1/T4 1700V chip and T2/T3 1200V chip

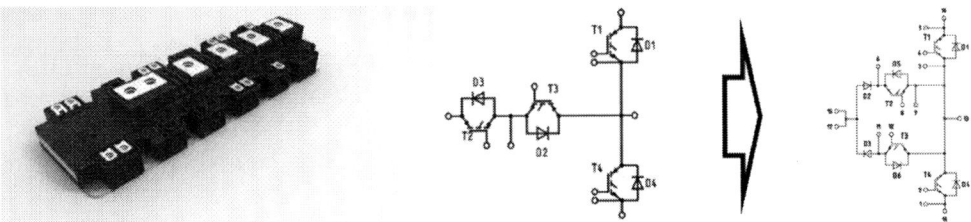

Fig. 1. (left) Danfoss P3L module, (right) for higher power density optimized NPC2 topology

3. Investigated short circuit effect

In most cases the short circuit test of an IGBT module will be conducted as follows: IGBT T1 is turned on until desaturation. After turning off the IGBT the current commutates into the freewheeling diode D4. T2 has to commutate to D4, T3 and T4 to D1, making that always one IGBT is switching.

In regards to a typical three level PWM pattern a test with two turned on devices was completed. In this mode T2 and T1 were turned on at the same time (Fig. 2). This simulates a typical pulse pattern of a three level topology [7]. The IGBT T2 was switched on for 1µs and not in short circuit mode before T1 turned on.

The current through T1 rises until it reaches the desaturation level. At this moment T2 turns on too. Now both T1 and T2 conduct current and the load current rise further. At the moment when T2 desaturates the diode D4 starts to conduct current, in the same time I_{T2} decreases. This effect was described in [1].

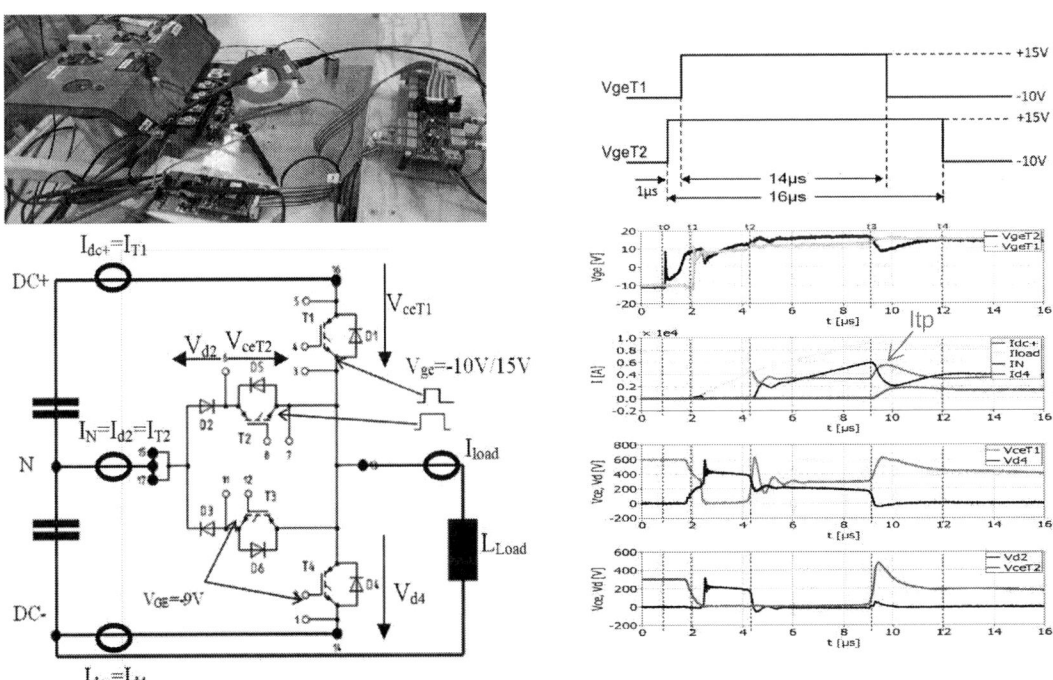

Fig. 2. Test settings summary, pane 1: V_{geT1} (green line), V_{geT2} (black line), pane 2: I_{dc+} (lila line), I_{tp} (transient peak current), I_{load} (yellow line), I_N (black line), I_{dc-} (red line), pane 3: V_{ceT1} (blue line), V_{d4} (black line), pane 4: V_{ceT2} (red line), V_{d2} (black line); R_{gonT2}=5R, R_{goffT2}=15R, R_{gonT1}=0R56, R_{goffT1}=3R3, V_{DC}=600V, $T_{junction}$=25°C, no gate clamping

t0:   Transistor T2 gate switch on.

t1:   Transistor T1 gate switch on

t2:   T1 in desaturation, as expected the current I_{T2} doesn't rise further on. But the load current is rising further until t3. V_{ceT1} is $V_{dc}/2=300V$ in that period. V_{ceT2} rise until $V_{dc}/2$. Now T2 can take over current. The current doesn't commutate from T1 in T2 – both IGBTs conducting right now and the load current (I_{load}) is the sum of I_{T1} and I_{T2}.

t3:   T2 desaturates. I_{T2} and I_{load} don't rise further on. The current in T2 falls and commutate in T1 and D4.

t4: I_{T1} and I_{T2} are in desaturation with each $V_{dc}/2$. The diode D4 conducts further on.

The behavior of the system after the time t4 depends on the turn off sequence of the IGBTs. The IGBT which turns off first commutates the current in the other IGBT and the diode D4. When both IGBTs are turned off the diode will take over the current. Further measurements on other NPC2 IGBT power modules have shown that the described effect appears in those IGBT power modules too. So it's not a bug or a feature only of the so called P3L module.

Further measurements show that this happens also in a standard two level IGBT halfbridge. In a halfbridge the current can commutate into the freewheeling diode at the moment where the IGBT desaturates. In a three level topology this commutation current can commutate additionally into the opened IGBT (in this case T1). Therefore current is flowing in T1, T2 and D4 at the same time. This explains the current-dip in Fig.3 between t3 and t4.

4. Investigation

For a better understanding of the described effect the influence of different parameters are investigated during the short circuit event and the results are discussed in this paper. The turn off behavior is not discussed here.

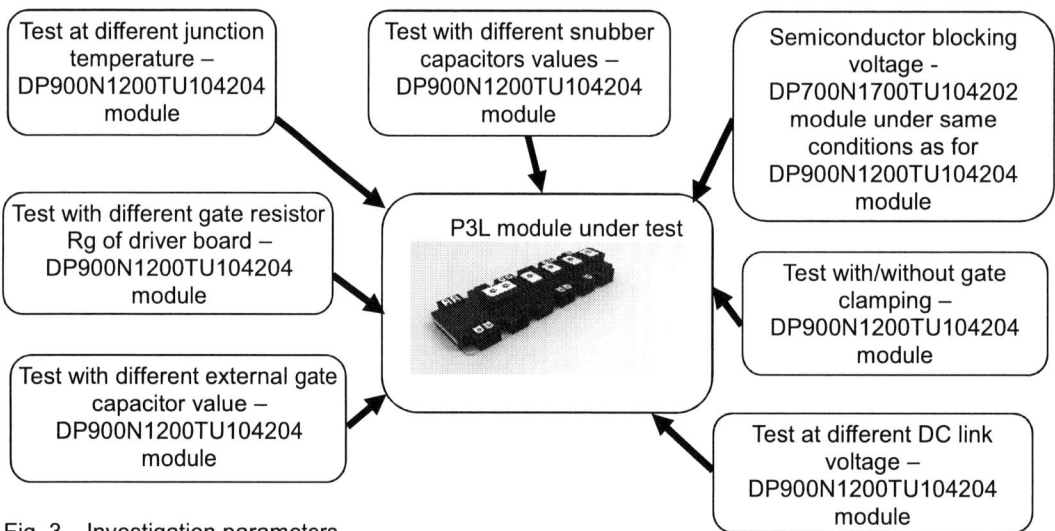

Fig. 3.   Investigation parameters

4.1.   Investigation of gate driver parameter impact on IGBT behavior during short circuit

4.1.1. Turn off gate resistor (R$_{goff}$)

In order to influence turn off event the gate resistor can be changed [11].

During the investigation it has been observed that the possibility to influence the short circuit behavior by varying the R_{goff} value is not significant compare to the turn on event. [10].

4.1.2. Turn on gate resistor (R_{gon})

Different Turn on gate resistor T1 value (R_{gonT1})

In order to influence turn on event the gate resistor can be changed [11].

This test step was conducted with three different turn on gate resistor values and as it can be seen on the Fig.4, by increasing R_{gon} the turn on switching event is smoother with switching delay at turning on and off.

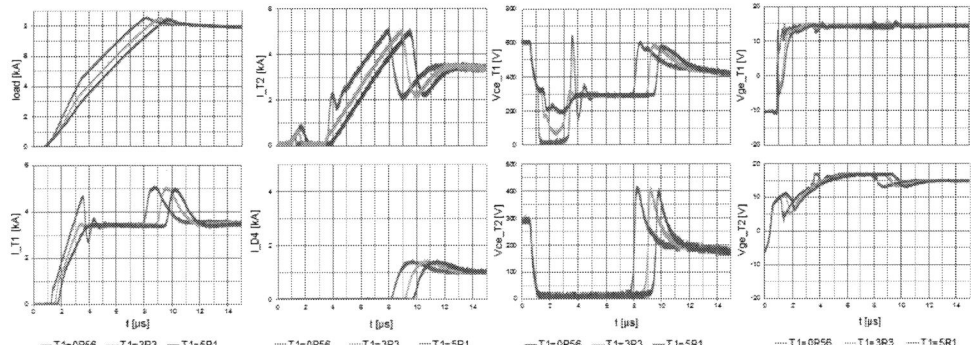

Fig. 4. Turn on resistor R_{gonT1}

→ Load current di/dt less and I_{T2} delayed
→ I_{tp} (transient peak current) and I_{d4} similar value and shape, but it comes earlier

Different Turn on gate resistor T2 value (R_{gonT2})

Following test step was conducted in similarity to the previous one and in order to give us full picture of impact of the different gate resistor during the short circuit event.

As it can be seen from the attached Fig. 5, by increasing resistor value, we'll lower the V_{ceT2} voltage spike and lower the current through the T2 and diode D4.

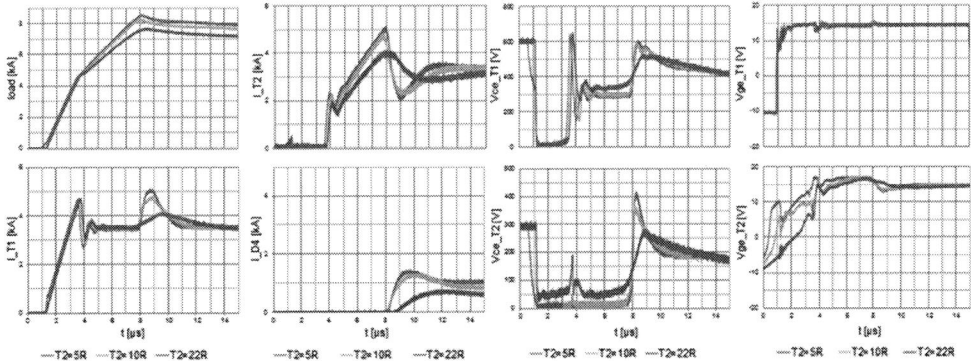

Fig. 5. Turn on resistor R_{gonT2}

→ Current I_{load} di/dt less from the moment when T2 turn on. di/dt of T2 is less

→ I_{tp} and I_{d4} have a lower value.

→ Voltage spike of T1 and T2 is less in the moment when T2 desaturates (t=8 μs)

4.1.3. Gate clamping

To reduce V_{ce} peak during turn off event in overload conditions like short circuit, so called clamping can be used. There are different solutions described [9,10]. In our experiments we used only solution with a Zener-diode, called the Basic clamping.

Module behavior during the short circuit where 300V Zener diode is being connected between T2 collector and gate was compared against same circuit without any clamping, shown in Fig. 6.

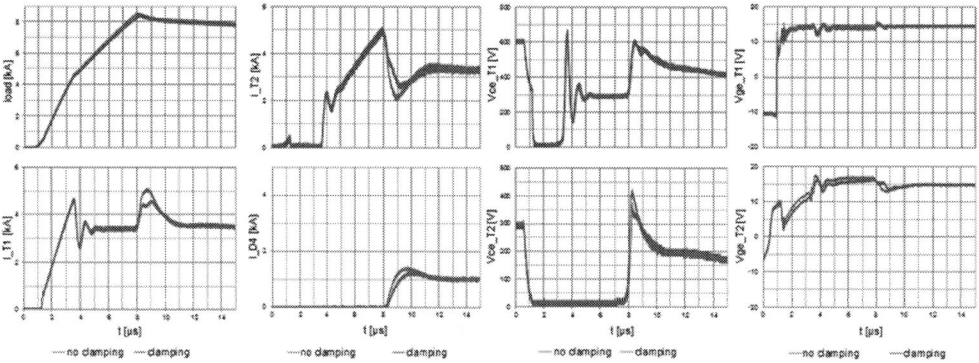

Fig. 6. Gate clamping

→ V_{ceT2} peak is cut (376V instead of 424V) (t=8μs)

→ V_{geT2} is turned on for longer time (t=8-9μs)

→ T2 conducts more current (t=8-9μs)

→ I_{tr} and I_{d4} are less

→ It is expected that with a more optimized clamping circuit the described influence will be bigger

4.1.4. External gate-emitter capacitor

A common way to achieve a soft turn-on and increase dV_{ce}/dt an additional capacitance is connected between gate and emitter. [10].

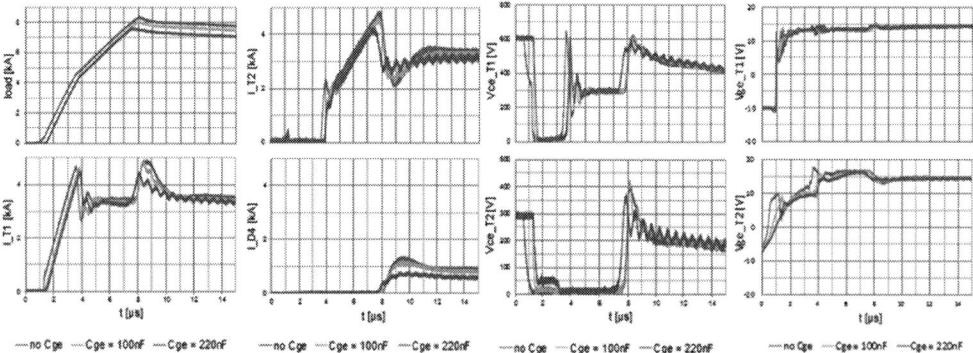

Fig. 7. External gate-emitter capacitor

By increasing the value of external gate-emitter capacitor, we achieved lower voltage overshoot values and lower peak currents, as per Fig.7 [10].

→ Less I_{sc} regarding higher gate capacitor value and delayed
→ di/dt values similar
→ I_{D4} and I_{tp} reduced and both start earlier

4.2. Different snubber capacitor

In order to reduce V_{ce} during the IGBT turn off very often in application external film capacitors are introduced, known in literature as snubber capacitors [10,13].

It has been selected three different test setups, first without snubber capacitor and other two with different values of snubber capacitors helping to explore the influence. Results are shown in Fig. 8 below, where we can see that by increasing snubber capacitor we will lower collector-emitter overshoot, but we'll have more oscillation.

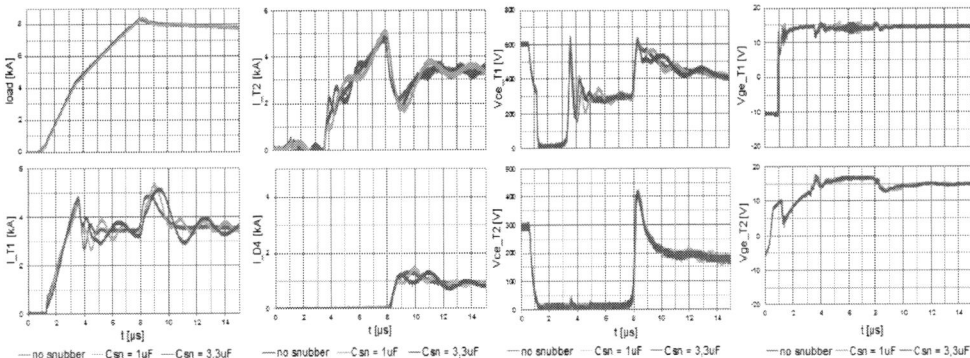

Fig. 8. Snubber capacitor

→ Current I_{load} insignificant change, I_{tp} and I_{d4} of T1 similar value, same time but more ringing can occur.
→ In completed experiments the snubber doesn't bring a change in the described effect but it brings ringing on I_{T1}, I_{T2}, I_{D4} and V_{ceT1}

4.3. Different DC link voltage

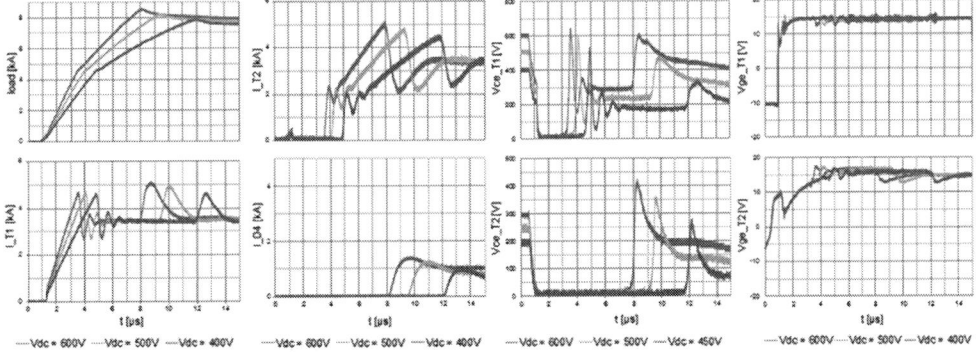

Fig. 9. DC link voltage

© VDE VERLAG GMBH · Berlin · Offenbach

By lowering DC link voltage collector-emitter overshoot is lowered and increased switching delay, as per Fig. 9.

→ di/dt of T1, T2 and I_{load} is less regarding the lower voltage (di/dt = V_{dc}/L)

→ regarding the lower di/dt the desaturation moment of each IGBT is later on, so V_{ceT1} and V_{ceT2} rise further, I_{tr} and I_{d4} are delayed

→ regarding the lower maximum I_{load} the I_{tp} and I_{d4} are less

4.4. Different junction temperature

The impact when junction temperature is being increased in a range from 25C° to 150C° can be seen in Fig.10 where current spike is lower at high temperatures, but at the same time voltage overshoot is higher, while switching delay is smaller.

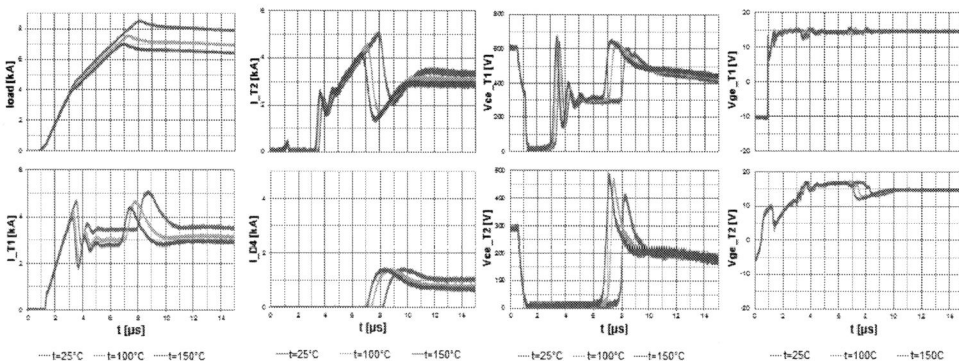

Fig. 10. Different junction temperature

→ less I_{sc} regarding high temp [13]

→ desaturation and I_{tp} start earlier

4.5. Semiconductor blocking voltage

To investigate whether the described effect is unique we take a module from same P3L family with other voltage ratings. By selecting DP700N1700TU104202 where is the halfbridge semiconductors 1700V instead of 1200V and the bidirectional switches are 1200V semiconductors instead of 650V conducting the short circuit test we've shown that this effect is not only related to the 1200V IGBT family, see Fig.11.

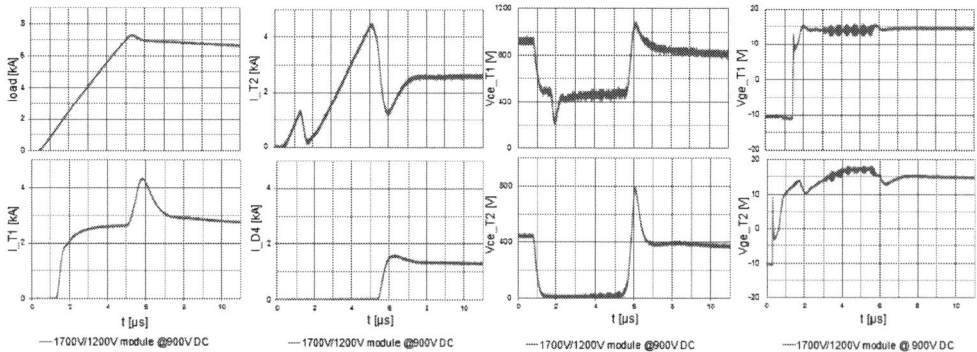

Fig. 11. DP700N1700TU104204 module

© VDE VERLAG GMBH · Berlin · Offenbach

1940

→ less I_{scT1} and I_{scT2} regarding higher blocking voltage of IGBTs [13]

→ I_{sc} is sum of short circuit currents of transistors T1 and T2, like we had with DP900N1200TU104204 module

→ I_{tp} also occurs

5. Conclusion

During the testing we have investigated impact of different parameters on two DUT´s (DP700N1700TU104202 and DP900N1200TU104204) behavior during the short circuit SC_{NPC2}. It has been determined that the effect does not depend on module selection. Overall short circuit current (I_{sc}), which is a sum of I_{scT1} and I_{scT2}, can´t be avoided and it is a factor which we have to keep in mind during the application design.

There are influences on gate drive or setup which can change the short circuit SC_{NPC2} behavior. Also, we have found out that there are several parameters which have no or less impact.

This test should be done in every inverter, because when application and operational point is known, by varying with the enlisted above parameters we are able to evaluate the destructive influences and select the optimum for the safe and reliable application work.

In order to get a better picture and investigate deeper SC_{NPC2} behavior during the short circuit our next step will be evaluating turn off sequence with a focus on overvoltages.

Literature

[1]   K.Lenz, V.Jerinic, R. Hinken, Investigation of short circuit in a IGBT power module with Three-Level Neutral Point Clamped Type 2 topology, APEC 2016

[2]   T. B. Soeiro, M. Schweizer, J. Linner, P. Ranstad, W. Kolar, Comparison of 2- and 3-level Active Filters with Enhanced Bridge-Leg Loss Distribution

[3]   Yaskawa Product Application Note, Motor Bearing Current Phenomen and 3-Level Inverter Technology

[4]   A. Nagae, I. Takahashi, H. Akagi, A new neutral-point-clamped PWM inverter, IEEE 1981

[5]   K. Lenz, J. Rudzki, F. Osterwald, U. Pandey, M. Poech, New IGBT Power Module concept in NPC Topology with Extended Reliability, PCIM 2015

[6]   J. Lutz, H. Schlangenotto, U. Scheuermann, R. De Doncker, Semiconductor Power Devices, ISBN 978-3-642-11125-9

[7]   I. Staudt, Semikron Application Note AN-11001

[8]   V. Bolloju, J. Yang, Influence of Short Circuit conditions on IGBT Short circuit current in motor drives, Whitepaper

[9]   O. Garcia, J. Thalheim, N. Meili, Safe Driving of Multi-Level Converters Using Sophisticated Gate Driver Technology, PCIM 2013

[10]   A. Volke, M. Hornkamp : IGBT Modules – Technologies, Driver and Application, Infineon Technologies AG, ISBN 978-3-00040134-3

[11]   M. Hermwille, Gate Resistor – Principles and Applications, Semikron Application Note AN-7003

[12]   J. Schumann, S. Pierstorf, H.-G. Eckel. Influence of the Gate Drive on the Short-Circuit Type II and Type III Behaviour of HV-IGBT, PCIM Nuremberg 2010

[13]   V. Bolloju, J. Yang, Influence of Short Circuit conditions on IGBT Short circuit current in motor drives, Whitepaper

Efficiency Verification Power Circulation Method of a High Power Low Voltage NPC Converter for Wind Turbines

Berthold Benkendorff[1], Toke Franke[2], Friedrich W. Fuchs[1]

[1] University of Kiel, Institute for Power Electronics, Kaiserstrasse 2, 24143 Kiel, Germany
Email: bb@tf.uni-kiel.de, fwf@tf.uni-kiel.de
[2] Danfoss Silicon Power GmbH, Husumer Str. 251, 24941 Flensburg, Germany
Email: toke.franke@danfoss.com

Abstract

Development and research for power converters is concentrated on increasing power density, higher modularity, lower costs as well as higher reliability and efficiency. The use of multilevel converters, like Neutral Point Clamped (NPC) Converters, in high power applications in the range of 1.5 MW up to 6 MW can be favourable because of the filter size reduction and therefore reduction of weight and total volume of the system. This paper evaluates a high accurate practical efficiency verification in a circulation power testrun for a three level NPC converter with an apparent power of approximately 1 MVA.

1 Introduction

Today, high performance wind turbines are built as variable speed systems with compact converters. Two generator concepts for wind energy applications are common: the Doubly-Fed-Induction-Generator with gearbox and the full-scale converter with gearless synchronous machine. [1, 2] So far, converters in wind turbines are positioned in the nacelles of the wind turbine and the optimizing of the power converters for wind turbines is a continuous task.

In this configuration, the most frequently used converter type of the wind turbine from 1.5 MW up to 6 MW is a pulse-width modulated voltage source converter with a two level output voltage. Typically, the primary output voltage level is 690 V and the DC-link voltage is 1100 V. This leads to a high RMS current and a high cross section of the cables, which have high length between the nacelle and the basement of the turbine and thus high costs. Therefore, a high converter output voltage is favourable. Three level converters are typically used for higher output power (> 6 MW) and built as medium voltage versions. When using the three level neutral point clamped topology for low-voltage wind turbine application, it is possible to use semiconductors with a considerably lower blocking voltage. This is possible because the semiconductors have only to withstand half of the DC-link voltage. One limit is 1000 V AC, as for higher values expensive and bulky medium-voltage equipment is necessary. At the same time, it is possible to remain inside the low-voltage application operating area with minimised safety requirements for the construction of the converter, compared to a system using 3.3 kV semiconductors.

For evaluating the losses and the efficiency in this power range different methods are well know. These methods are generally divided into two sections, like electrical and thermal methods. In this paper an approach, corresponding to the electrical method, is taken into account for evaluating at the same time two different inverters in one single operation.

The paper is structured in the following way. The first section shows the investigated converter and the investigated operation points. The second section presents the verification method and the according formulas. The third section shows the circulation method and the evaluated measured results of the circulated power. Finally, a conclusion is given.

2 Investigated NPC Converter and analyzed operation points

New improvements of three level IGBT-modules are a decisive factor in developing a compact high power converter in the low-voltage neutral point clamped (NPC) topology. The internal electric schematic of the converter is shown in figure 1. The parameters for the three level NPC Power-Stack is presented in table 1.

Table 1: Parameter for generator-/grid-side connected inverter and investigated operation points

Name	Value
Primary side voltage	950 V
Rated current	600 A
DC-link voltage	1500 V
Nominal IGBT switching frequency	4.5 kHz
Rotor frequency	variable
Semiconductor type	IGBT
IGBT voltage class	1200 V
Internal topology	Three level NPC
Investigated operation point	current $i_{rms} = 100\,A...600\,A$
	$cos\varphi \approx 1.0$ with $m = 0.25...1$ for grid-operation
	$cos\varphi \approx -1.0$ with $m = 0.5...1$ for generator-operation

3 Verification Method and according Formulas

For verifying the converter at a high power level it is most of the time a big challenge. Lots of attempts have been made to minimize the consumption of energy while testing an AC/DC converter. If small power will be tested a very good method is to test the converter setup with a resistive load. With higher power ($>100\,kW$), the consumption of energy will be very high and burning the energy is typically an inconvenient method. The basic approach with higher power is, to connect two converters in back-to-back operation. [3, 4] The aim is to verify the converter efficiency and perform a loss measurement in a very accurate way. Two converters are supplied by one single source, the first converter operates as a generator and the second one as a receptor with the purpose to circulate a high power through both systems. For testing two directions of energy flow through the converters, they have to be reversible. Different methods are possible to supply this configuration. Two methods are mainly used, like direct fed from the AC mains without using an inductor or like fed from the DC supply. [5] The version with DC-Source was chosen in the following investigation. Both AC-sides of the two converters are connected via an inductor. The connected inductor limits the current variation between the two converters. Likewise both DC-links are connected, via short bus bar configuration. The DC-supply has mainly to power the losses of both converters. For having a very high accuracy a new approach is taken into account. Usually two equal converters will be tested with this method. If two different converters are used, one converter is typically very well known. [6]

The three level NPC converter [7] can be tested and verified. This converter was set as a receptor. A second equal three-level inverter was not available at that time. That is why an equivalent two level inverter with similar power is taken as a generator. With this approach it is possible to characterize two unknown converters at the same time, which have not been verified before. Therefore different voltage

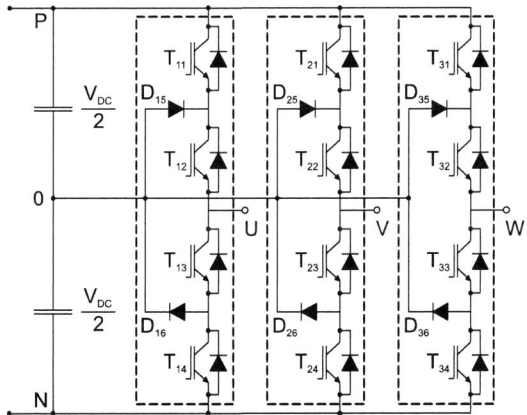

Figure 1: Schematic diagram of neutral point clamped (NPC) converter

© VDE VERLAG GMBH · Berlin · Offenbach

PCIM Europe 2016, 10 – 12 May 2016, Nuremberg, Germany

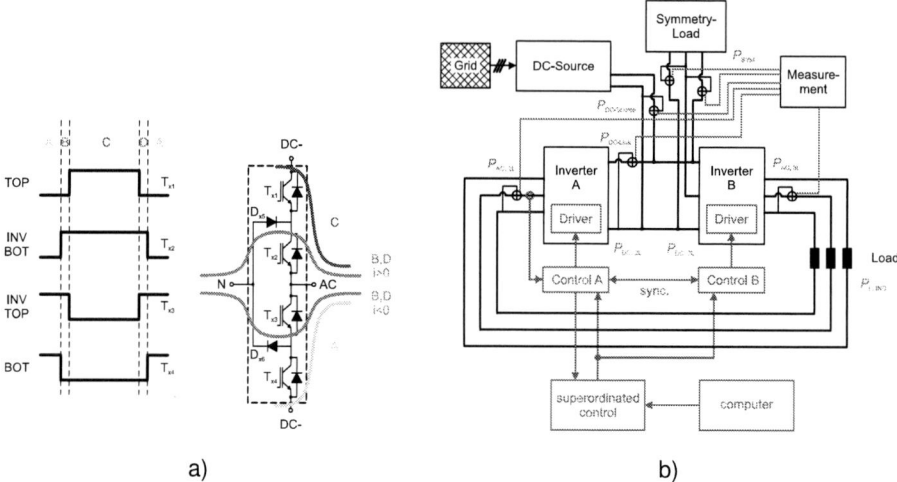

Figure 2: Laboratory Setup for circulation test run b) driver signals versus time of three-level modulation for three-level converter a) detailed scheme of laboratory setup with 'Inverter A' as the generator and 'Inverter B' as the receptor

and current measurements are important to do, as shown in 2.

Measuring the efficiency higher than 98% are always difficult because the losses are very small compared to the total power. Small errors of the measurement typically have a high influence on the calculated efficiency. Therefore a different approach is taken. Instead of measuring the input and output power of the power stack and calculating the losses by P_{in}-P_{out} the total power losses of the test setup are measured and distributed to the different components of the test system. The total losses are equal to the power that need to be feed into the DC-link and is therefore easy and precise measurable. The following components are identified causing power losses:

• Three level inverter (receptor)

• Two level inverter (generator)

• AC inductors

• DC resistor chain to achieve an equal voltage distribution between $DC+$ to $DC0$ and $DC0$ to $DC-$

The DC resistor chain (symmetry-load) is only required due to the fact that the inverter control is not jet able to achieve a symmetrical DC-link, as shown in figure 2a. With this circuit the neutral-point potential of the three level converter is kept balanced. Later this circuit will be obsolete and therefore its static losses are measured and subtracted from the total power losses. Since both inverters share the same AC inductor its losses are equally distributed to both power stacks. For calculating the distribution of the remaining power losses the power is measured on the AC and DC side of each inverter.

Since the sum of the measured losses is not equal to the total power losses a correction factor CF, given in equation 6 is derived that normalizes the sum of the single losses in regards to the total losses. This reduces the overall error significantly since it could be assumed that due to the same measurement equipment the error for both inverters is similar. This leads to the following equations for the power losses of the power stacks:

$$P_{\text{AC,2L}} - P_{\text{AC,3L}} = P_{\text{L,IND}} \tag{1}$$
$$P_{\text{L,total}} = P_{\text{DC-Link}} - P_{\text{SYM}} \tag{2}$$
$$P_{\text{L,INV2L\&3L}} = P_{\text{L,total}} - P_{\text{L,IND}} \tag{3}$$

© VDE VERLAG GMBH · Berlin · Offenbach

PCIM Europe 2016, 10 – 12 May 2016, Nuremberg, Germany

$$P_{\mathrm{L,3L}} = P_{\mathrm{L,INV2L\&3L}} \cdot \frac{P_{\mathrm{AC,3L}}}{P_{\mathrm{DC,3L}}} \cdot CF \tag{4}$$

$$P_{\mathrm{L,2L}} = P_{\mathrm{L,INV2L\&3L}} \cdot \frac{P_{\mathrm{AC,2L}}}{P_{\mathrm{DC,2L}}} \cdot CF \tag{5}$$

With

$$CF = \frac{1}{\frac{P_{\mathrm{AC,3L}}}{P_{\mathrm{DC,3L}}} + \frac{P_{\mathrm{AC,2L}}}{P_{\mathrm{DC,2L}}}} \tag{6}$$

$P_{\mathrm{AC,3L}}$ and $P_{\mathrm{AC,2L}}$ are the AC power of each inverter and $P_{\mathrm{DC,3L}}$ and $P_{\mathrm{DC,2L}}$ are the DC power measurements of the inverters. $P_{\mathrm{DC-Link}}$ is the circulating DC-power in the DC-link. With P_{SYM} the losses for the DC resistor chain (symmetry-load) is named. With $P_{\mathrm{L,IND}}$ the losses of the AC inductor are named. $P_{\mathrm{DC-Source}}$ is the applied power out of the DC-Source.

There $P_{\mathrm{L,total}}$ are the total losses of the laboratory setup. In this case, the total losses are the losses of the three-level NPC power-stack combined with the losses of the two-level inverter, $P_{\mathrm{L,INV2L\&3L}}$.

With these measured values and the correction via the correction factor CF the losses of the three level NPC converter and the two level converter can be analyzed at the same time. So, with $P_{\mathrm{L,3L}}$ and $P_{\mathrm{L,2L}}$ the losses of each system are calculated very precisely.

4   Circulation method and the evaluated measured results of the circulated power

Figure 2 shows the laboratory setup. 'Inverter A' is used as the generator, in this test setup the two level inverter is used for it. As the receptor ('Inverter B') the three level NPC inverter is used. Both inverters are controlled via a computer which controls the superordinate control. Each inverter itself has an individual control unit, which have to be synchronized. The converter control unit drives the driver boards of the inverter. The DC-link is supplied via a DC-Source, with an approximate maximum power of 70 kW.

For the exact verification of the converter losses and the efficiency, for this setup six current and six voltage sensors are needed for the measurement. For the evaluation of the losses the AC-phase line-to line voltage and the AC-current on each output side of the inverters are measured. Additionally the DC-Link voltage and the DC-link current is measured. To identify the supply power, the DC-Source voltage and current is measured. To be very accurate the losses of the symmetry-load is evaluated by the current and voltage sensing to and from the symmetry circuit.

Table 2 shows the used equipment for the laboratory setup. For the analysis of the losses and the efficiency a very precise power analyzer is applied. To supervise the output voltage and current a scope recorder is installed. As a load and a decoupling of both sides three parallel three-phase inductors, with 0.53 mH for a single inductor, are used. Figure 3 shows the experimental results of voltage and current the three level NPC inverter. The measured line-to-line voltage is approx. 770 V at a phase current of 600 A.

Table 2: Used equipment for laboratory setup

Name	Value
Power Analyser	Yokogawa WT 1800
Voltage Sensors	Integrated in Yokogawa WT 1800
Current Sensors	MCTS Transducer with LEM IT 700-S Ultrastab
Scope Recorder	DL 850 with Analog Voltage Input Module HS10M12
Voltage Sensors	TESTEC TT-SI 9010 with ratio 1:1000
Current Sensors	PEM Rogowski CWT 30 mini, 1 mV/A
DC-Source	approx. 70 kW, $U_{\mathrm{DC}} = 100\,\mathrm{V}...1500\,\mathrm{V}$
Symmetry-Load	Linare operating IGBT connected to load-resistors
Load-Inductor	three phase, 690 V, 450 A, 0.53 mH (single inductor)
	3 times parallel connected per phase

© VDE VERLAG GMBH · Berlin · Offenbach

PCIM Europe 2016, 10 – 12 May 2016, Nuremberg, Germany

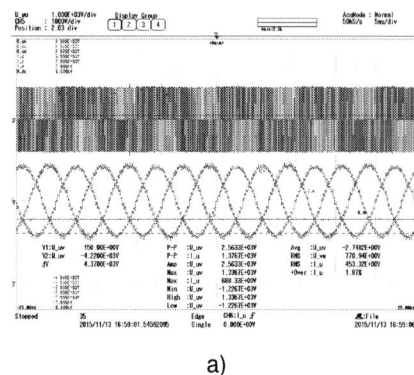 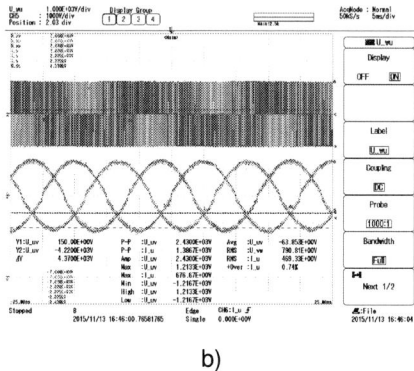

a)   b)

Figure 3: Experimental results of the NPC measured AC-line to line phase voltage and AC-phase current ($f_{sw} = 4500\,\text{Hz}$, $i_{rms} = 600\,\text{A}$, $T_a = 25°$, $U_{DC} = 1200\,\text{V}$ for $m = 1$) a) generator-side operation $f_N = 100\,\text{Hz}$ with $cos\varphi \approx -1.0$ b) grid-side operation $f_N = 50\,\text{Hz}$ with $cos\varphi \approx 1.0$

Experimental results of the total losses of the NPC inverter measured with the approach in circulation power test is shown in figure 4. The DC-link voltage of the circulation test is limited to a maximum of 1200 V, caused by the maximum limit of the DC-link voltage of the two level inverter. Pointed out are the losses for seven different phase currents from 100 A to the full rated current of 600 A for a DC-link voltage of 1200 V each for grid-side and generator-side operation. The losses for the grid-side operation are for 600 A a bit below 10 kW for the NPC inverter. For the generator-side operation the losses are higher, caused by the higher base frequency of 100 Hz and the major commutation over the diodes.

The experimental efficiency evaluation for the generator-side operation of the NPC converter is shown in figure 5. With the higher losses at generator-side operation the efficiency of the overall NPC inverter at 600 V, 600 A is approx. 97,6 % and at 1200 V, 600 A is approx. 98,1 %. These efficiencies are lower than for the grid-side operation caused by the changed base frequency and the major commutation over the diodes of the module.

The experimental efficiency evaluation for the grid-side operation of the NPC converter is shown in figure 6.

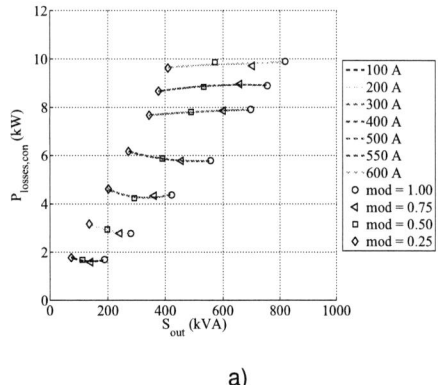 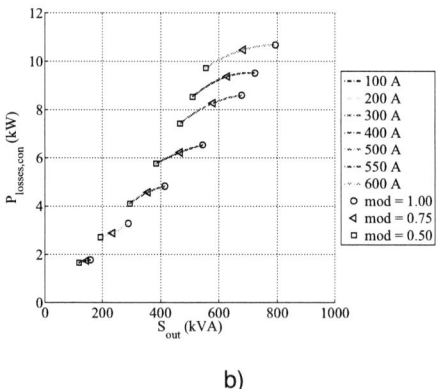

a)   b)

Figure 4: Experimental results of the total losses of the NPC measured with the approach in circulation power test at ($f_{sw} = 4500\,\text{Hz}$, $f_N = 50\,\text{Hz}$, $T_a = 25°$ for $i_{rms} = 100\,\text{A} ... 600\,\text{A}$) with (a) in grid-side operation with $m = 0.25 ... 1$, $cos\varphi \approx 1.0$ and with (b) in generator-side operation with $m = 0.5 ... 1$, $cos\varphi \approx -1.0$ (a), (b) $U_{DC} = 1200\,\text{V}$

© VDE VERLAG GMBH · Berlin · Offenbach

PCIM Europe 2016, 10 – 12 May 2016, Nuremberg, Germany

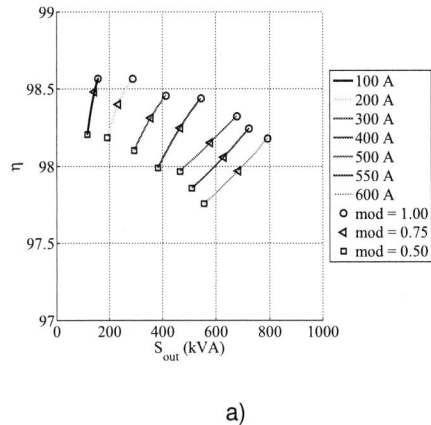
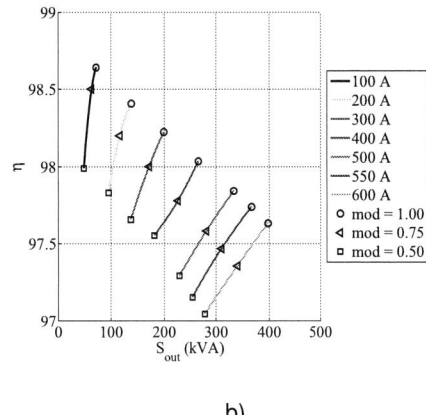

a)   b)

Figure 5: Efficiencies of the NPC measured with the approach in circulation power test (f_{sw} = 4500 Hz, f_{N} = 100 Hz, T_{a} = 25 ° for i_{rms} = 100 A...600 A and m = 0.5 ...1 with $cos\varphi \approx -1.0$) (a) U_{DC} = 1200 V (b) U_{DC} = 600 V

At grid-side operation the total losses of the inverter is lower over the whole investigated range, which causes a slightly better efficiency. The efficiency for grid-side of the overall NPC inverter at 600 V, 600 A is approx. 97,8 % and at 1200 V, 600 A is approx. 98,3 %.

The experimental efficiency evaluation for the generator-side and the grid-side operation of the NPC inverter show good results for the evaluation of the NPC inverter. For the circulation power test the emulated three level modulation, as shown in figure 2 is used.

Caused by this emulated three level modulation generated out of a two level modulation the power losses are different to the losses of a classic three level modulation. For a classic three level modulation the switching frequency of the inner IGBTs (T_{x2}) and (T_{x3}) is typically switched with the base frequency. With this adapted modulation the switching frequency of the inner IGBTs is the same as for the outer IGBTs, which is equal to the switching frequency of the inverter, which causes clearly higher losses and a generally less good efficiency of the NPC inverter.

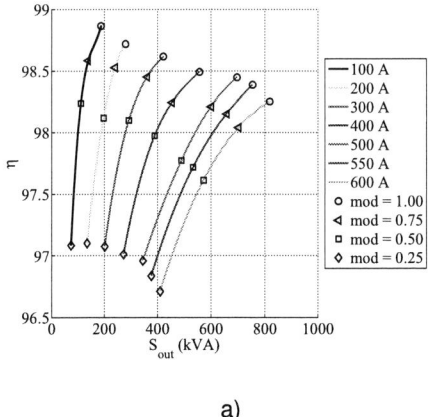
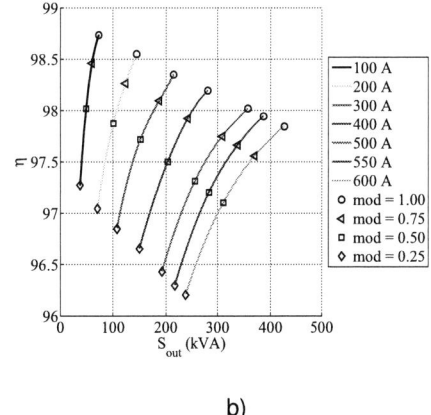

a)   b)

Figure 6: Efficiencies of the NPC measured with the approach in circulation power test (f_{sw} = 4500 Hz, f_{N} = 50 Hz, T_{a} = 25 ° for i_{rms} = 100 A...600 A and m = 0.25 ...1 with $cos\varphi \approx 1.0$) (a) U_{DC} = 1200 V (b) U_{DC} = 600 V

© VDE VERLAG GMBH · Berlin · Offenbach

5 Conclusion

The approach of the circulation method for power loss measuring shows that a very precise evaluation of total losses and the efficiency of two different not yet evaluated inverters is at the same time possible with reduced measurement equipment. The method can be easily applied. As shown for the results of the NPC inverter with an approximate output power of 1 MVA, this measurement method is suitable for higher power and lower evaluation of different inverters.

6 Acknowledgement

The authors would like to thank the Fraunhofer-Gesellschaft and the state of Schleswig-Holstein, which mainly founded this project. This project is run as a multi-partner project in the KLSH Innovation Cluster Schleswig-Holstein. Further partners are Danfoss Silicon Power GmbH, Reese & Thies GmbH, FT-Cap GmbH, Applied Science University of Kiel, Applied Science University of Westkueste, Senvion SE, Vishay Siliconix Itzehoe GmbH.

References

[1] F. Blaabjerg, M. Liserre, and K. Ma. Power electronics converters for wind turbine systems. *Industry Applications, IEEE Transactions on*, 48(2):708–719, 2012.

[2] Zhe Chen, J.M. Guerrero, and F. Blaabjerg. A Review of the State of the Art of Power Electronics for Wind Turbines. *Power Electronics, IEEE Transactions*, (8), 2009.

[3] Shyh-Jier Huang and Fu-Sheng Pai. Design and operation of burn-in test system for three-phase uninterruptible power supplies. *Industrial Electronics, IEEE Transactions on*, 49(1):256–263, Feb 2002.

[4] O.S. Senturk, L. Helle, S. Munk-Nielsen, P. Rodriguez, and R. Teodorescu. Converter structure-based power loss and static thermal modeling of the press-pack igbt-based three-level anpc and hb vscs applied to multi-mw wind turbines. In *Energy Conversion Congress and Exposition (ECCE), 2010 IEEE*, pages 2778–2785, Sept 2010.

[5] JS Siva Prasad and G Narayanan. Apparatus and method for heat-run test on high-power pwm converters with low energy expenditure. *Sadhana*, 38(3):359–375, 2013.

[6] F. Forest, J.-J. Huselstein, S. Faucher, M. Elghazouani, P. Ladoux, T.A. Meynard, F. Richardeau, and C. Turpin. Use of opposition method in the test of high-power electronic converters. *Industrial Electronics, IEEE Transactions on*, 53(2):530–541, April 2006.

[7] Berthold Benkendorff, Friedrich W. Fuchs, Detlef Friedrich, Joern Hinz, Max Poech, Klaus Kohlmann, Hagen Reese, Heinz-Hermann Letas, Christoph Weber, Roland Eisele, Zeno Mueller, Michael Berger, Jacek Rudzki, Frank Osterwald, and Tobias Mono. Bottom up research and development for a low-voltage three level npc converter. In *Proceedings of PCIM Europe 2015; International Exhibition and Conference for Power Electronics, Intelligent Motion, Renewable Energy and Energy Management*, pages 1–8, May 2015.

PCIM Europe 2016, 10 – 12 May 2016, Nuremberg, Germany

Control of the Actively Balanced Capacitive Voltage Divider for a Five-Level NPC Inverter - Estimation of the Intermediary Levels Currents

A. Rufer, N. Koch, N. Cherix, EPFL, Ecole Polytechnique Fédérale de Lausanne, Switzerland,
alfred.rufer@epfl.ch

Abstract

This article addresses the control of a Multistage Stacked Boost Architecture (MSBA) feeding a five-level NPC inverter in a grid-tied application. A suitable control strategy was previously presented, in which estimated/reconstructed intermediary currents are used as feed-forward signals for the voltage controllers that balance the intermediate busses. While the reconstruction of the intermediary currents from the measured output current is easy to implement when the instantaneous switching signals of the NPC inverters are known, the real-time implementation on a digital controller must cope with the fact that these switching signals are generally not available at the base interrupt rate, but are usually generated in a peripheral unit, typically an FPGA-based modulator. The present paper describes an estimation/reconstruction method that overcomes this limitation and allows an appropriate estimation of the intermediary currents at the base sampling rate. The quality of the estimated signals is presented and analysed through comparisons of the measured and the estimated signals.

1 Introduction

A 5 Level inverter for PV applications has been presented in [1], using a so called MSBA (Multistage Stacked Buck/Boost Architecture) as the input circuit. This MSBA stage works as an actively balanced capacitive voltage divider/multiplier, stabilizing the PV panel voltage through a conventional boost stage. Then, it multiplies the pre-stabilized voltage by four between the top positive and bottom negative rails. Since two positive (+1 and +2) and two negative (-1 and -2) rails can be used as voltage sources, a 5 Level NPC inverter is chosen

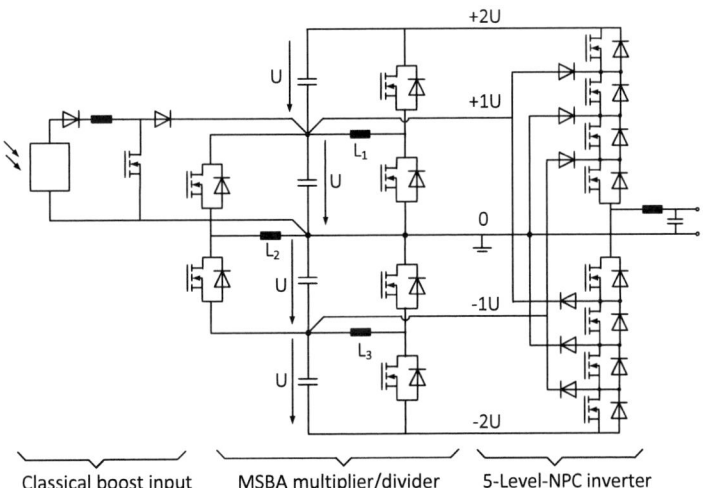

Fig. 1 The 5-Level Inverter with MSBA

© VDE VERLAG GMBH · Berlin · Offenbach

for the AC output circuit. In this configuration, all DC levels are used for powering the 5 level NPC-inverter (Fig. 1). The proposed topology has first the advantage to produce an output voltage with a better resolution and reduced output filter, but it can also use low voltage devices for an application where the output voltage magnitude is above their blocking capability. Figure 1 shows the complete topology with its different conversion stages. The input stage is a conventional boost converter used for the MPPT (Maximum Power Point Tracking) function of the PV generator and simultaneously for the stabilization of the input voltage of the MSBA stage. Then, the stabilized input voltage is elevated to a +2U and -2U DC voltage system.

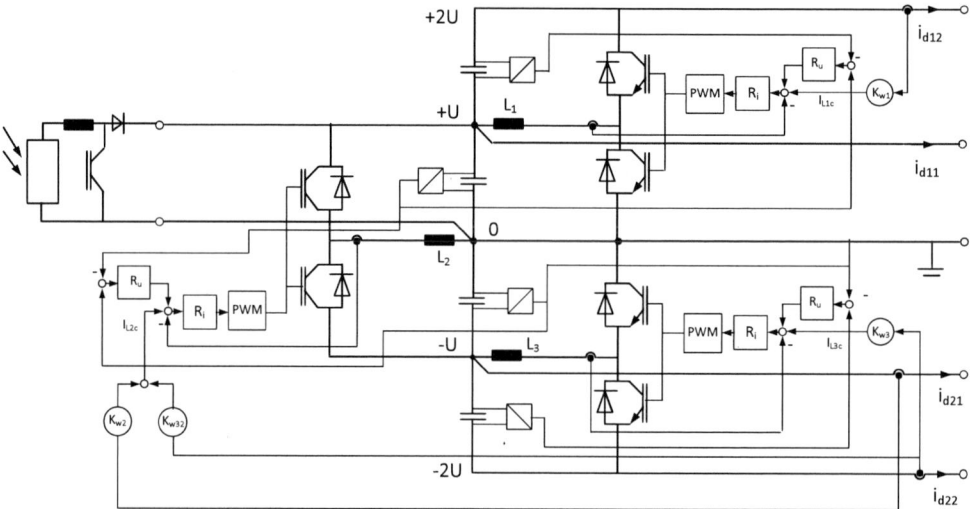

Fig. 2   Control structure for the MSBA stage with local feed-forward signals

The principle of the Multistage Stacked Boost Architecture has been proposed in [2] as a DC-DC step-up converter dedicated to extra-large photovoltaic plants. Originally, the principle of the capacitive voltage divider had been developed for the active balancing of series-connected supercapacitors [3]. Other application examples of the actively controlled voltage divider are described in [4] and [5], where the converters are used as asymmetric or symmetric voltage step-down conversion circuits. In these references, control strategies are described using cascaded current and voltage balance controls, completed with a feed-forward strategy for the current references (Fig. 2). The main benefits of the use of feed-forward current signals is that the superimposed voltage balance controllers can be designed as simple P-controllers, due to the fact that the static errors are suppressed. The second benefit is that the balance controllers can achieve robust and high dynamic performances. As a consequence, the direct use of the intermediate voltage levels for the feeding of a multilevel NPC output stage is possible.

2 Estimation versus reconstruction of the levels currents
Figure 3 show the intermediate level current reconstruction method using the switching signals of the 5 level NPC inverter and its measured output current.

Figure 4 shows the generated waveforms corresponding to instantaneous values of the reconstructed signals. This method is based on the knowledge of the output signals of the modulator. However, these signals are generally unavailable the level of the interrupt-based digital control, because they are generated in peripheral timing circuit (Fig.5). These modulation circuits typically receive only duty cycle commands and generate the firing signals independently.

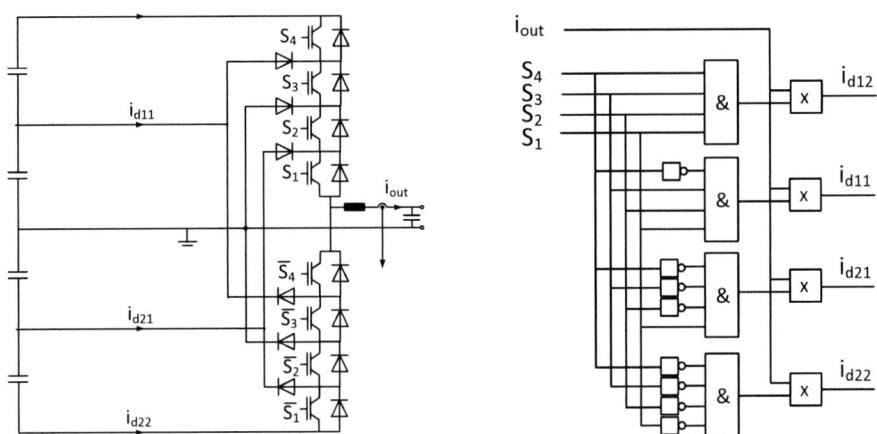

Figure 3: Reconstruction of the intermediary-level currents from the inverter output current with the help of the switching signals

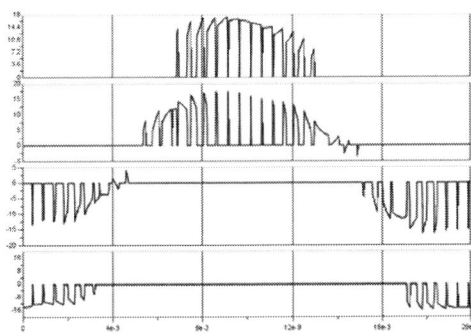

Figure 4: Intermediate-level currents reconstructed with switching signals

This section describes an estimation method for the intermediate level current signals that are needed for the calculation of the feedforward signals used for the stabilization of the DC voltages of the MSBA. The feedforward signals result from the multiplication of the corresponding factors (K_{wx}) by the DC levels currents themselves (I_{dxy}), (Fig.2).

The estimation is based on the modulation indexes that are available at the level of the sampled control. Figure 5 shows the structural diagram of the complete system including the

MSBA control (only one channel is represented (Sy) and the 5-Level inverter control (5_L PWM). In this diagram, all PWM functions are implemented in peripheral timers.

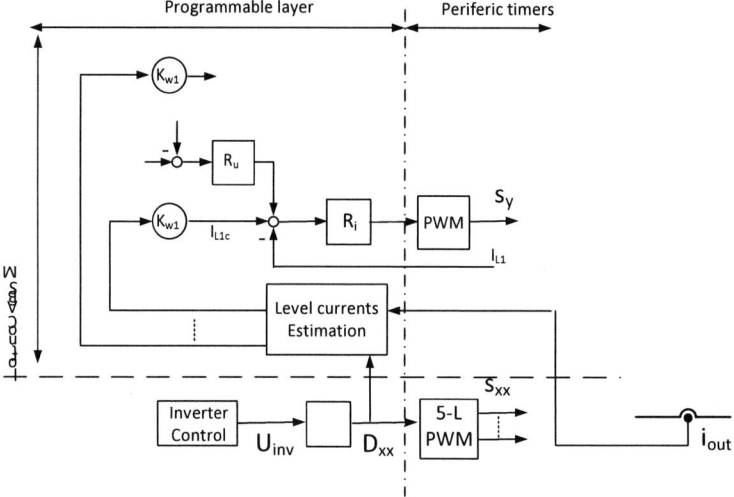

Figure 5: Estimation of the intermediary-level currents from the inverter current with the duty-cycles D of the inverter control

According to the signal definition given in Fig. 3, the intermediate level currents are calculated with the algorithm given through relations (1) to (4).

$$i_{d11} = (Duty_B - Duty_A) \cdot I_{out} \tag{1}$$

$$i_{d12} = (Duty_A) \cdot I_{out} \tag{2}$$

$$i_{d21} = (Duty_D - Duty_C) \cdot I_{out} \tag{3}$$

$$i_{d22} = (1 - Duty_D) \cdot I_{out} \tag{4}$$

The computation of the intermediary level currents is based on the duty cycles $Duty_A$, $Duty_B$, $Duty_C$, $Duty_D$, that correspond to the conduction states and that are coded via the switching signals combinations A, B, C, and D defined through table 1.

States	Voltage level	S1	S2	S3	S4
A	+2U	1	1	1	1
B	+U	0	1	1	1
	0	0	0	1	1
C	-U	0	0	0	1
D	-2U	0	0	0	0

Table 1: Conduction states and switching signals

The duty cycles are subsequently calculated from the reference voltage Uref, with the expressions (5) to (8). Uref is the sinusoidal reference signal for the generated output voltage of the inverter

$$Duty_A = (Uref - 0.5) * 2 \qquad (5)$$
$$Duty_B = (Uref) * 2 \qquad (6)$$
$$Duty_C = (Uref + 0.5) * 2 \qquad (7)$$
$$Duty_D = (Uref + 1) * 2 \qquad (8)$$

when the peripheral PWM modulation stage (FPGA) uses a triangular carrier defined between 0 and 1. The corresponding theoretical modulation curves are represented in Fig. 6. In this representation, the PD modulation method uses carrier signals that are vertically shifted and are of reduced amplitude from 0 to 0.5, and from 0.5 to 1, also in the negative direction. The reference voltage has then to be shifted and corrected in its amplitude as indicated in rel. (5) to (8).

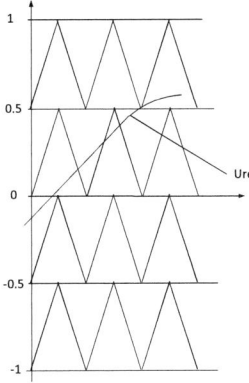

Figure 6 Modulation curves of the 5-Level NPC inverter.

3 Experimental results

The intermediate-level current estimation algorithms have been implemented in a dedicated control platform where DSP and FPGA-based are available [6], [7]. The whole experimental set-up is shown in Fig. 7. (1) represents the control platform with its fiber-optic outputs and dedicated measurement inputs. (2) correspond to the 5-level NPC inverter, (3) is the load of the inverter.

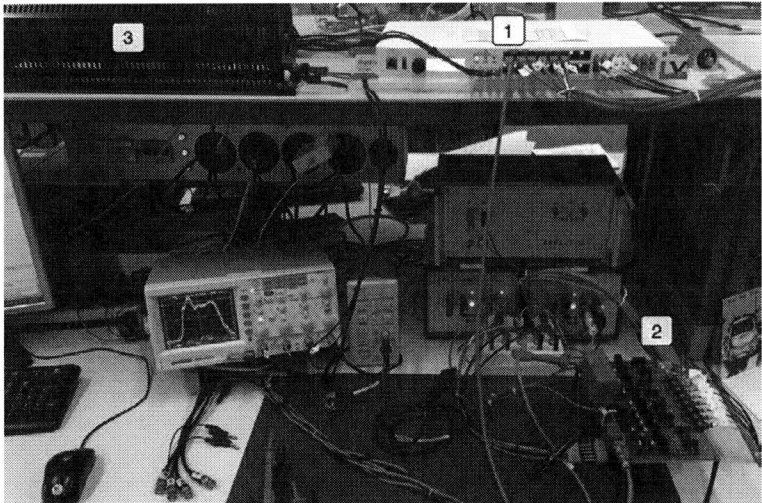

Figure 7 Experimental set-up

Figure 8 presents the results obtained with the experimental converter for all the intermediary-level currents. Fig. 8a) shows the real-time current signals of the intermediate levels. Fig. 8b) shows the estimated averaged signals of the same quantities that are further used for the calculation of the feed-forward signals for the corresponding voltage controllers. Fig. 9 gives the details and illustrates the quality and accuracy of the estimated quantities.

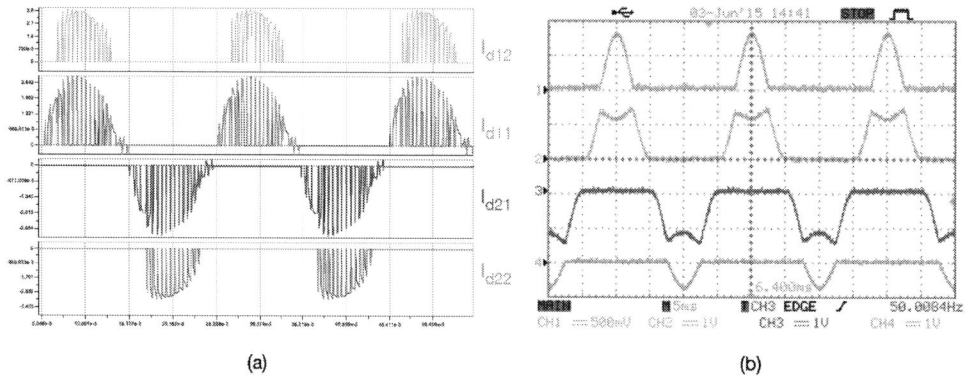

(a)   (b)

Figure 8: Estimated currents of the intermediary levels

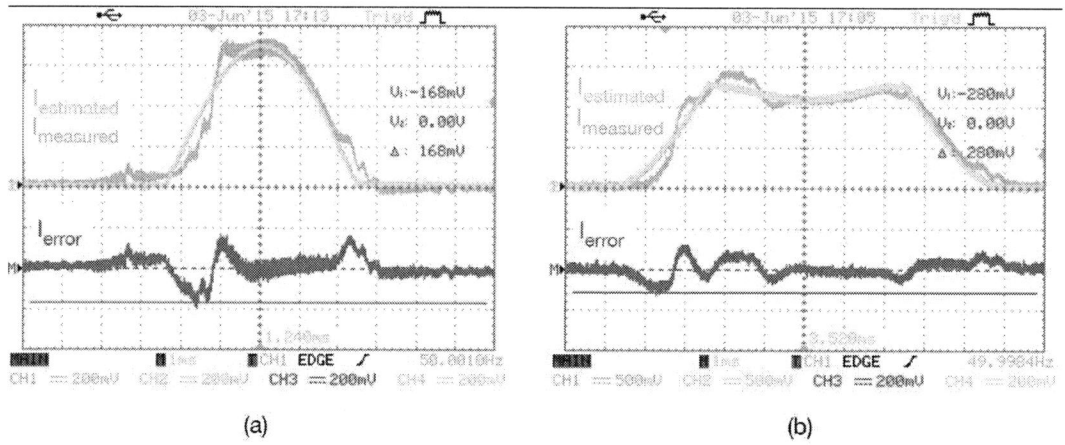

(a)   (b)

Figure 9: Comparison of estimated / reconstructed value of Id12 (a) (yellow) with the actual measurement made with a current probe (blue). The estimation error is shown in (b), (red).

Conclusion

The paper has presented an estimation method for the calculation of the feedforward quantities needed for the control of an MSBA converter feeding a 5 level NPC inverter. The intermediate-level current estimation is based on a dedicated duty-cycle calculation for the conduction states, and has been successfully implemented on a dedicated control platform using a DSP and FPGA-based modulation circuit.

References

[1] A. Rufer, A Five-Level NPC Photovoltaic Inverter with an Actively Balanced Capacitive Voltage Divider, PCIM Conference , Nueremberg, 2015

[2] Rufer A., Barrade P., Steinke G., *"Voltage step-up converter based on Multistage Stacked Boost Architecture (MSBA),"* IPEC 2014, The 2014 International Power Electronics Conference, Hiroshima, Japan, 18-21 May 2014.

[3] Barrade P., Pittet S., Rufer A., *"Energy storage system using a series connection of supercapacitors with an active device for equalizing the voltages,"* IPEC 2000, International Power Electronics conference, Tokyo, Japan, 3-7- April 2000.

[4] Kutkut N.H., *"A modular non-dissipative current diverter for EV battery charge equalization,"* APEC '98, Applied Power Electronics Conference and Exposition, 1998.

[5] Barrade, P., Rufer, A., Non-Isolated DC-DC Converters for High Power Appli-cations - Control of the Capacitive Voltage Divider, PCIM 2014 : International Exhibition and Conference for Power Electronics, Intelligent Motion, Renewable Energy and Energy Management, Nuremberg, Germany, 20-22 May 2014

[6] http://imperix.ch/products/control/boombox/intro

[7] Cherix, N., Delalay, S.; Barrade, P., *"Fail-safe Modular Control Platform for Power Electronic Applications in R&D Environments,"* EPE 2013 : 15th European Conference on Power Electronics and Applications, Lille, France, 3-5 September 2013.

Automotive power module technologies for high speed switching

Shinichiro Adachi[*1] , Souichi Yoshida[*1], Hiroshi Miyata[*1], Takuma Kouge[*1], Daisuke Inoue[*1], Yoshikazu Takamiya[*1], Fumio Nagaune[*1], Hideto Kobayashi[*1], Thomas Heinzel[*2], Akira Nishiura[*1]

[*1]Fuji Electric Co., Ltd, 4-18-1, Tsukama, Matsumoto, Nagano, Japan

[*2]Fuji Electric Europe GmbH., Goethering 58. 63067 Offenbach, Germany

Email : adachi-shinichiro@fujielectric.com

Abstract

IGBT module for EV(Electric Vehicle) and HEV(Hybrid Electric Vehicle) are required high power density. To increase power density of IGBT modules, downsizing of power module and reduction of power loss are necessary. We have developed RC-IGBT(Reverse Conducting IGBT) by using latest thin wafer technology to meet high power density of IGBT module. RC-IGBT which is IGBT and FWD fabricated on single die can significantly downsize IGBT module. Thin RC-IGBT technology can decrease steady-state loss and switching loss reduction is also important for power loss reduction in inverter operation. In this paper, the design of thin RC-IGBT technology and package structure for high speed switching are presented.

1. Introduction

Downsizing is important for automotive IGBT module form the point of view of keeping enough cabin space and its loading flexibility of vehicles. In addition, requirement for output power range is up to 120kW in automotive inverter application. Therefore, high power density of automotive IGBT module is required. To increase power module of IGBT module, higher thermal conductivity, higher operation temperature and reducing power loss are necessary. Especially, reducing power loss is important from the viewpoint of electrical efficiency of inverter system which affects the cruising distance in EV/HEV. We have developed the RC-IGBT by thin wafer technology to meet the demand [1],[2]. RC-IGBT which is IGBT and FWD fabricated on single die can significantly downsize IGBT module. One of the methods of loss reduction is to reduce the thickness of Si chip. Because of steady-state loss are proportional to the thickness of Si chip. But it is important to reduce power loss not only reducing steady-state loss but also reducing switching loss. Although switching loss reduction is realized by high speed switching, basically switching speed is limited by at turn-off or reverse recovery operation of IGBT module. The spike voltage is caused by stray inductance of package and superimposed spike voltage which is generated in two phase switching during turn-off and reverse recovery switching. This presents newly developed RC-IGBT and package design for high speed switching and its effect of reducing spike voltage and power loss.

© VDE VERLAG GMBH · Berlin · Offenbach

2. Design of RC-IGBT

2-1. Feature of RC-IGBT

RC-IGBT for automotive application has been developed based on the field stop (FS) IGBT, which has alternating IGBT and FWD regions arranged in stripes. Figure 1 shows a schematic structure of the RC-IGBT. Newly developed cutting-edge thin wafer processing technology has enabled to make wafers thinner and reduced power loss. In addition, optimized surface structures, such as trench pitch, channel density and contact has improved the RC-IGBT performance. Figure 2 shows the output characteristics of conventional 6th generation IGBT/FWD and newly developed RC-IGBT with the same active area. Due to thin wafer technology and optimize surface structure, Vcesat and Vf have decreased dramatically compared with conventional IGBT and FWD.

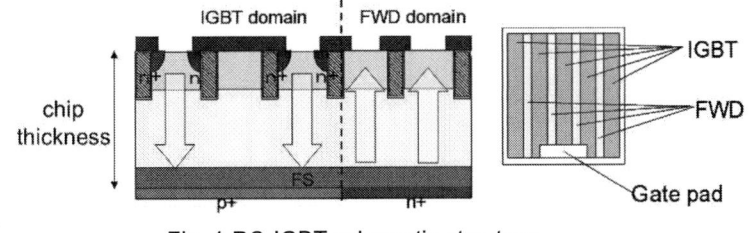

Fig. 1 RC-IGBT schematic structure

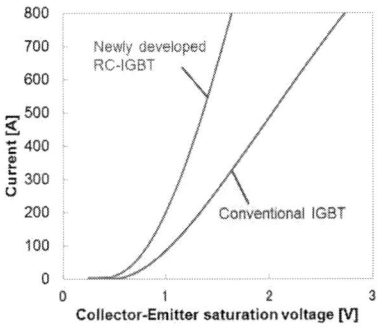
(a) Collector-Emitter voltage – Collector current

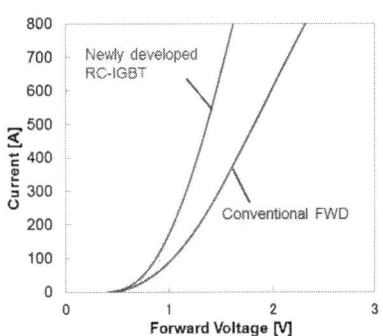
(b) Forward voltage- Forward current

Fig. 2 The output characteristics of conventional 6th gen IGBT/FWD and RC-IGBT.

	Conventional IGBT and FWD	Newly developed RC-IGBT
A general Substrate layout of half-bridge circuit		
Ratio of substrate size	1	0.75
Ratio of length between P and N terminal	1	0.78

Fig. 3   A general substrate layout of half-bridge circuit

RC-IGBT technology has enabled to downsize the package size, because IGBT and FWD are fabricated on single die. Newly developed RC-IGBT can achieve an output power equivalent to that in the size of 70 % of the total chip size of the conventional chips [2]. Figure 3 shows a general substrate layout of half-bridge circuit. In the case of the RC-IGBT, it is possible to reduce substrate area and current path from P to N line to each of 75 % and 78 % compared to IGBT module consists of the conventional IGBT and FWD. The stray inductance of IGBT module is dependent on current path from P to N line, width of current path and distance between P and N line. Since IGBT module consisted of IGBT and FWD, there is a limit to reduce current path. Therefore parallel chip connection to make current path wide and laminates bus bar to make distance between P and N line short are usually applied to decrease the stray inductance as a package design [3],[4],[5],[6]. But these approaches tend to package size increases. RC-IGBT can significantly reduce the stray inductance by shortening the current path. This is useful in automotive application which required the downsizing of the package.

2-2. Field stop layer design for thinner RC-IGBT chip

When switching speed in turn-on increases in order to reduce turn-on switching loss, reverse recovery current increases. As a result, since dir/dt becomes higher, recovery spike voltage becomes higher and oscillation is generated. Especially, dir/dt is likely to be bigger in thin chip. Therefore it is important to optimize field-stop (FS) layer for reducing dir/dt. Figure 4 shows the reverse recovery wave form and simulation result of hole carrier concentration at maximum reverse recovery spike voltage. A deep FS layer helps to prevent extending of depletion layer. For this reason, the hole carriers are stored in the cathode-side at the time of reverse recovery, which is lead to the improvement of recovery characteristic and reduction of recovery spike voltage.

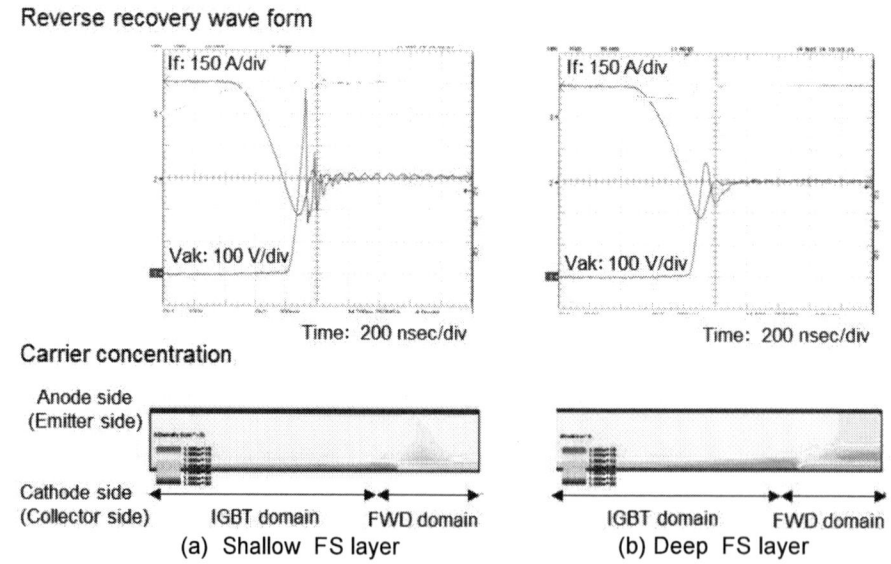

Fig. 4 (a) Shallow FS layer,(b) Deep FS layer. Top side is reverse recovery wave form of RC-IGBT. Bottom side is simulation result of the hole carrier concentration at maximum reverse recovery spike voltage. White line shows a depletion layer.

3. Packaging design for lower spike voltage

As well known, low stray inductance of package leads to reduce the turn-off and recovery spike voltage. New package (M653) [2] is to half the stray inductance compared to conventional package (M652) [7] by applying the RC-IGBT and optimization of internal layout. But it is important to reduce not only stray inductance but also superimposed spike voltage in inverter operation. In the three-phase inverters, the spike voltage is generated in the P-N terminal of power module with the connection of smoothing capacitor and power module at turn-off operation. When the turn-off operation is occurred in U-phase and another phase such as V-phase, the spike voltage generated between P and N terminal is superimposed. Figure. 5 shows the comparison of spike voltage between P and N terminal in the two cases of P-N, that is conventional package(M652) and three pairs P-N connection structure with new package(M653). Film capacitor is usually used for DC-link in the automotive inverters. It is obvious that spike voltage between P and N terminal is dramatically reduced in new package case in spite of 1.5 times higher of turn-off switching speed (-di/dt). The spike voltage is superimposed easily, because P and N lines for each phase are common. On the other hand, the spike voltage between P and N terminal is very small, because P and N terminal for each phase are independent in new package.

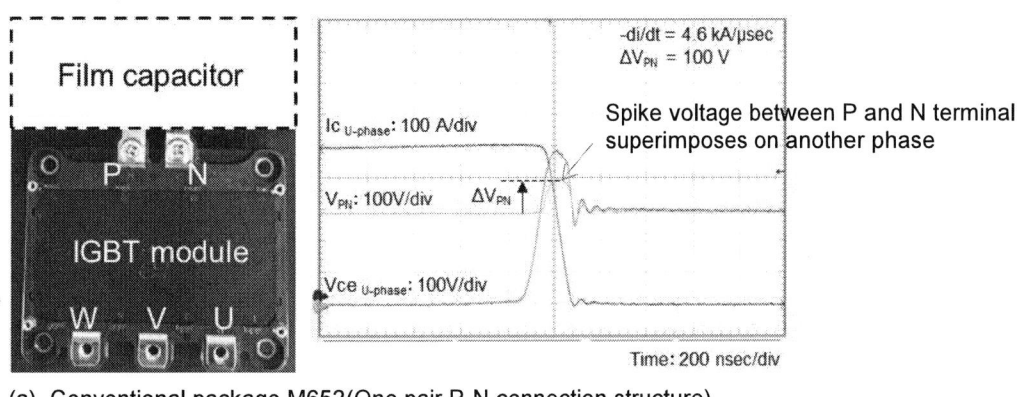

(a) Conventional package M652(One pair P-N connection structure)

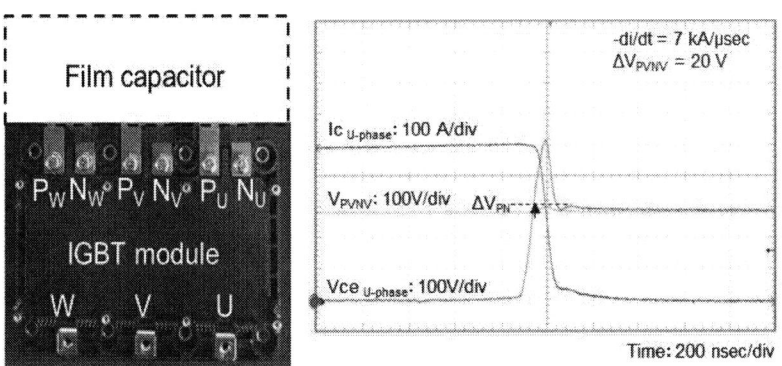

(b) New package M653(Three pairs P-N connection structure)

Fig. 5 Comparison of spike voltage between P and N terminal

For evaluation of superimposed spike voltage, we measured spike voltage in two phase switching. Evaluation circuit of superimposed spike voltage in two phase switching is shown in Figure 7. It is difficult to measure each phase current for restriction of package structure in conventional package. So current was measured total current of two phases. Figure 8 shows turn off waveform, in left side waveforms U-phase is only switched and in right side waveforms U and V phase are switched at the same time. Spike voltage of two phase switching increases by 54 V compared with one phase switching in conventional package. On the other hands, the spike voltage in new package is almost the same between one phase switching and two phase switching. Furthermore, spike voltage in new package is lower than in conventional package even though turn-off switching speed (-di/dt) is 1.5 times higher. This results show that new IGBT module package enable to increase switching speed over 1.5 times than the conventional package under the condition of the same battery voltage and the device breakdown voltage. The superimposed spike voltage is also generated in reverse recovery operation. Consequently, new package can also increase turn-on switching speed.

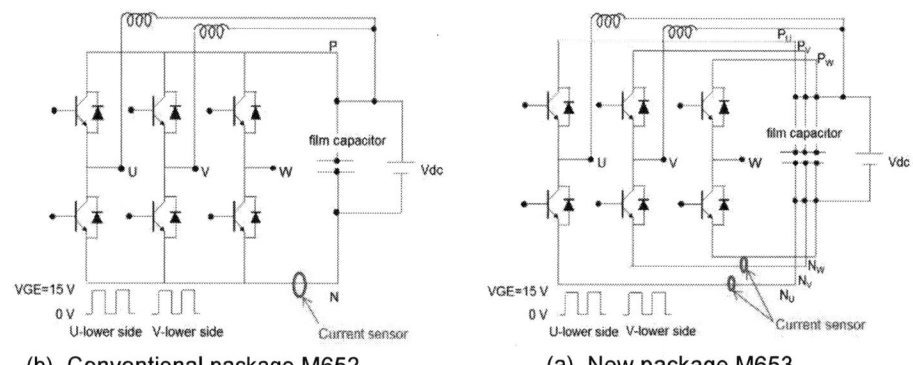

(b) Conventional package M652   (a) New package M653

Fig. 7 Evaluation circuit of superimposed spike voltage in two phase switching

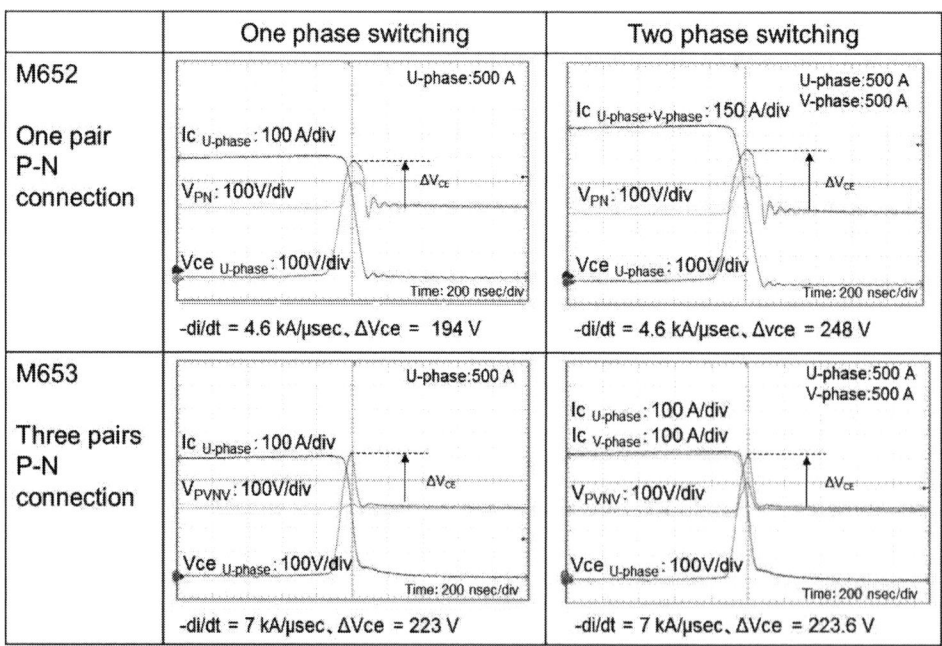

Fig. 8 Superimposed spike voltage in two phase switching

4. Effect of new module combining the RC-IGBT and new package

Figure 9 shows comparison of power loss conventional IGBT module and new module combining the RC-IGBT and new package during inverter operation under the following conditions: Vcc = 400 V, output current = 400 Arms and switching frequency fc = 10 kHz. Turn-on di/dt and turn-off –di/dt were set so that spike voltage included superimposed spike voltage in both are the same voltage. Die size of the RC-IGBT is in the size of 70 % of the total chip size of the conventional chips. As shown in Figure 9, switching loss decreases by 30 % because of high speed switching.

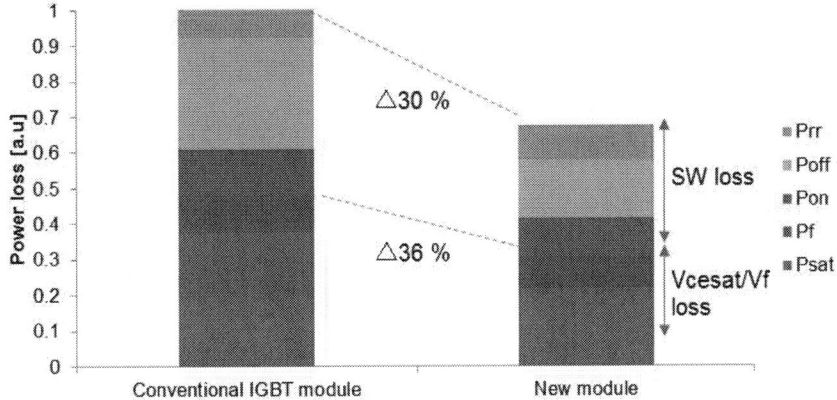

(a) Power loss in traction mode.
Vcc = 400 V, Iout = 400 Arms, fsw =10kHz, fout=300 Hz, m=0.86, cos Φ =+0.68

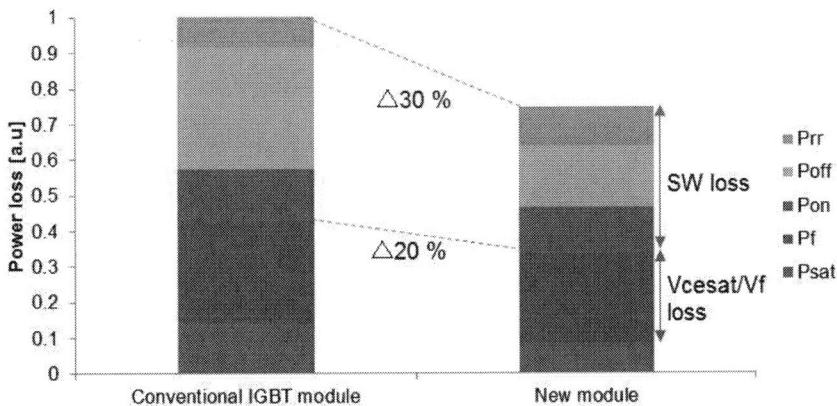

(b) Power loss in generation mode.
Vcc = 400 V, Iout = 400 Arms, fsw =10kHz, fout=300 Hz, m=0.86, cos Φ =-0.68

Fig. 9 Power loss calculation result of conventional IGBT module installed conventional chips and new module combining the RC-IGBT and new package. Die size of the RC-IGBT is in the size of 70 % of the total chip size of the conventional chips.

5. Conclusion

In this paper design of thin RC-IGBT and package for high speed switching were described. In order to improve softness of reverse recovery characteristics, deep FS layer was applied in newly developed RC-IGBT. This design is enable us to increase turn-on switching speed. And stray inductance of new package was decreased to 50 % compared with conventional package by applying the RC-IGBT and optimization of internal layout. Furthermore superimposed spike voltage was reduced in new package which has a P-N connection for each of the three-phase. For these technologies new module was reduced switching loss by 30 % compared with conventional module. M653 [2] which installed these technologies has great possibilities for high power density EV/HEV inverter system.

6. Reference

[1] S. Noguchi, S. Adachi, S. Yoshida, "RC-IGBT for Mild Hybrid Electric Vehicles", Fuji Electric review Vol. 60-No.4, 2014, pp. 224-227.

[2] K. Higuchi, et al,"New standard 800A/750V IGBT module technology for Automotive application", PCIM Europe 2015, pp.1137-1144.

[3] C. Muller, S. Buschhom, " Power-module optimizations for fast switching a comprehensive study", PCIM Europe 2015, pp. 434-441

[4] D. Kawase, M. Inaba, K.Horiuchi, K. Saito,"High voltage module with low internal inducance for next chip generation- next High Power Density Dual-", PCIM Europe 2015, pp.217-223.

[5] G. Borghoff,"Implementation of low inductive strip line concept for symmetric switching in a new high power module", PCIM Europe 2013, pp.185-191

[6] R. Bayerer, D. Domes," Power circuit design for clean switching", CIPS2010.

[7] S. Adachi, et al,"High thermal conductivity technology to realize high power density IGBT modules for electric and hybrid vehicles", PCIM Eurpe 2012, pp. 1378-1384.

Power semiconductors for the automotive 48V board net

Felix Hüning, University of Applied Science Aachen, Eupener Strasse 70, 52066 Aachen, Germany, Huening@fh-aachen.de

Abstract

To cope with increasing electrical power demand in modern cars a 48 V board net is introduced by German car makers. This voltage level requires power semiconductors with higher voltage rating compared to the 12 V board net. The power demand for different applications is analyzed and the requirements for the power semiconductors like PowerMOSFET are extracted. Using a standard D2PAK package (TO-263 or TO-263-7) current technologies are analyzed in terms of power and current capability. The comparison of these data yields a maximum power for the 48 V board net using existing technologies and proposals for new devices to increase the power range.

1. Introduction

Electrification of cars is one of the major trends of the automotive industry to reduce fuel consumption and emissions and to realize new applications. The electrification is not only related to the power train like for hybrid and electrical vehicles (HEV/EV) but also affects auxiliary equipment like electric power steering (EPS), water pump or HVAC (heating, ventilation and air conditioning) in conventional cars with an internal combustion engine. The power requirements of these components might be high and it is difficult or even impossible to realize the desired functionality with the standard 12 V board net. As a consequence a higher voltage board net is needed to realize applications with higher power demand.

In case of pure electric vehicles the power demand for the traction motor is in general high (>50 kW) and requires a high voltage board net (HV, e.g. 400 V) in addition to the 12 V board net. As this high voltage level is well above 60 V high efforts are necessary to ensure maximum safety, like protection against contact or an insulation monitor [1]. In addition only people trained for HV are allowed to work with this voltage level.

Due to these high efforts for the HV board net an additional mid voltage board net is introduced. Already 12 years ago a middle voltage board net of 42 V was about to be introduced [2]. That time the disadvantages and the technical challenges like electric arcs were too high to bring this board net into series production. Now a 48 V is entering the market, in particular for mirco and mild hybrid cars. Main target in the beginning is to optimize the mild and micro hybrid technologies by the 48 V board net, e.g. an advanced start-stop function or regenerative breaking [5]. Second target is to electrify new applications that are currently pure mechanically driven due to their high power demand, like the climate compressor or the engine fan. Third target is to improve existing systems that were already electrified, like electric power steering (EPS). These improvements could cover simplification of the electric systems as well as simplified assembly technologies and reduced system cost.

2. 48 V board net

The voltage of the new 48 V board net is well below the 60 V DC level, reducing the efforts for this new additional electrical system drastically. No special protection against contact, no insulation monitor is needed [3]. In addition it offers the possibility to realize applications with higher power demand that were hard to realize with the 12 V board net – or even impossible.

Today German OEMs agreed on a standard for the 48 V electrical system. Requirements and testing procedures are summarized in VDA320 [4]. Besides the nominal 48 V level overvoltage and undervoltage pulses are defined, for DC as well as for transient pulses. Maximum DC voltage is specified as 60 V, but with respect to the power semiconductors the transient overvoltage pulse E48-02 is more important, refer to Fig. 1. This overvoltage pulse simulates the switching off of loads and can reach up to 70 V for 40 ms. This voltage defines the minimum breakdown voltage of the power semiconductors that can be used in the 48 V electrical system.

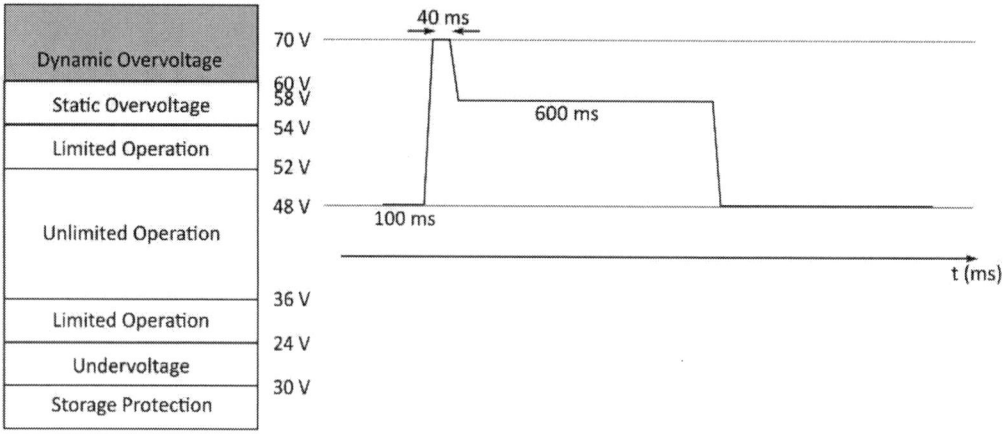

Fig. 1   DC voltage levels (left) and transient pulse E48-02 according to VDA320 [4]

3. Power semimconductors

As shown before the minimum breakdown voltage of the power semiconductors has to be greater than 70V. For this voltage level PowerMOSFETs with a voltage rating of 75V, 80V or 100V are used as power semiconductors to add some margin on top of the maximum specified voltage. PowerMOSFET provide low conduction losses with an easy gate drive capability and fast switching times.

In general there are two main properties of the PowerMOSFET that limit the maximum power that can be switched: the current capability and the power dissipation within the PowerMOSFET:

The current capability is on the one hand determined by the internal structure of the device, in particular the bonding wires. The number of wires and their diameter is limited due to package constraints, e.g. 4 bond wires with 500 µm diameter in a TO-263-7 package.

On the other hand the maximum power dissipation P_{max} within a PowerMOSFET is determined by internal as well as external properties and is limited by the maximum temperature rise ΔT:

$$P_{max} = \frac{\Delta T}{R_{th}} = R_{DS}(on) \cdot I^2$$

One major property of PowerMOSFET is the On-State resistance $R_{DS}(on)$. The thermal resistance Rth is composed of the thermal resistance of the device itself and the thermal resistance of the mounting conditions, e.g. on FR-4 PCB. For given Rth and $R_{DS}(on)$ the current I determines the maximum power.

© VDE VERLAG GMBH · Berlin · Offenbach

For optimal cooling conditions (reflected in the data sheet value Rthj-c) the current capability of low voltage PowerMOSFET (30 V/40 V breakdown voltage) is determined by the internal structure for ambient temperatures below about 120 °C. For higher temperatures the power dissipation is limiting due to small values of ΔT. But if real mounting conditions are considered yielding significantly higher values for Rth also the maximum power dissipation can be the limiting factor.

Fig. 2   Internal structure of a PowerMOSFET with one thin gate wire (left) and four thick source wires (by courtesy of Renesas Electronics)

R_{DS}(on) is strongly correlated to the size of the silicon die and the breakdown voltage of the device. Fig. 3 exemplifies the R_{DS}(on) vs BVDSS dependence for automotive PowerMOSFET in a TO-263-7 package using Infineon's Optimos T2 technology. This technology is used for the analysis as it provides the best performance as of today. Based on their thermal resistance from junction to case Rthj-c of about 0.5 K/W all devices but the 60 V device have a similar die size. The Rthj-c of the 60 V device is 0.6 K/W and its die size is somewhat higher. It is clearly visible in Fig. 3 that the R_{DS}(on) increases with increasing breakdown voltage BVDSS. Compared to PowerMOSFETs with a breakdown voltage of 40 V (which are used in the 12 V board net) the R_{DS}(on) for 80 V PowerMOSFET is roughly a factor of 2.5 higher.

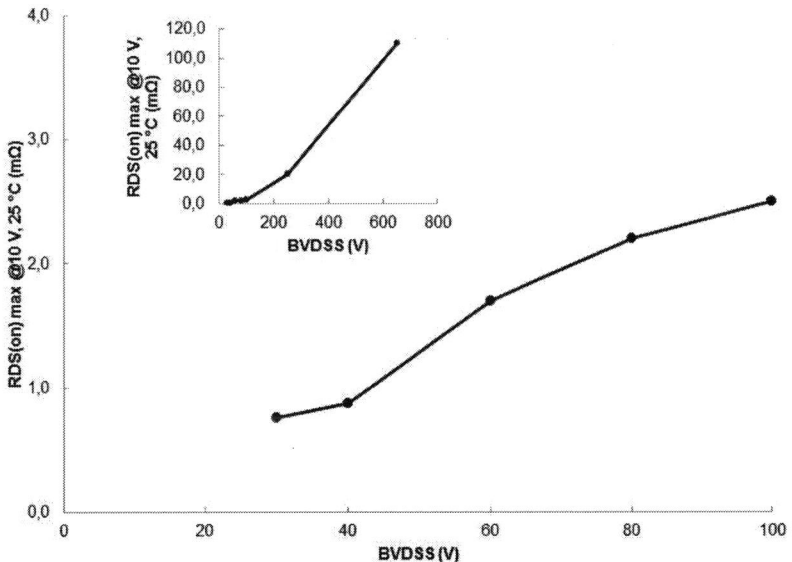

Fig. 3   On-State resistance RDS(on) vs breakdown voltage BVDSS for PowerMOSFETs in TO-263-7 package using Infineon's Optimos T2 technology [6]

For a comparison of devices with different breakdown voltages table 1 lists some important parameters for PowerMOSFETs in TO-263-7 package using Infineon' Optimos T2 technology. At an ambient temperature T_A of 100 °C the current limit for the D2PAK package itself is as high as 240A for the 40 V deivce , 180 A and 171 A for the 80 V and 100 V device respectively. But this limit is valid for an optimum Rth value of 0,5 K/W only. Depending on mounting of the device a thermal resistance of about 10 K/W is realistic yielding a maximum power dissipation of 7.5 W. This maximum power dissipation results in a much lower current capability of 70 A, 43 A and 38 A respectively. Therefore the current at T_A = 100 °C is limited by the power dissipation, not by the package.

From load point of a maximum power in the 48 V board net can be reached that is a factor of four higher than in the 12 V board net. But the reduced current capability of the higher voltage PowerMOSFET has a direct impact on the maximum power that can be switched by the devices. For the 40 V device in the 12 V board net a current of 70 A can be switched by a single device with Rth = 10 K/W resulting in a load power of about 850 W. For the 80 V PowerMOSFET the maximum current is reduced by a factor of 0.62 and hence the maximum power that can be switched with the same assembly like in the 12 V board net is just a factor of 2.5 higher. Therefore the maximum load power that can be switched in the 48 V board net by a single 80 V PowerMOSFET is about 2 kW. For the 100 V PowerMOSFET the current is reduced by a factor of 0.54 resulting in a maximum load power of about 1.8 kW. Of course higher power can be switched using several devices in parallel.

Parameter / Device	IPB240N04S4-R9	IPLU300N04S4-R8	IPB180N08S4-02	IPB180N10S4-02
Package	TO-263-7	H-PSOF-8-1	TO-263-7	TO-263-7
Rthj-c [K/W]	0.5	0.35	0.54	0.5
BVDSS [V]	40	40	80	100
Current limit by package [A] @ 100°C	240	300	180	171
Max. RDS(on) [mOhm] @ 175°C	1.5	1.3	4	5.2
Maximum power [W] @ 100°C (R_{th} =10K/W)	7.5	7.5	7.5	7.5
Maximum current [A] @100°C	70	76	43	38
Typ. gate charge Q_G [nC]	220	221	128	156

Tab. 1   Parameters of PowerMOSFETs using Infineon's Optimos T2 technology [6]

There are two options to increase the load power that can be switched by a single device. First option is to develop new silicon technologies to reduce the On-State resistance R_{DS}(on). This R_{DS}(on) reduction directly influences the power loss within the device and hence enables higher currents that can be switched. But a new technology also has an impact on other properties of the PowerMOSFET like the switching behavior or the Avalanche capability. In general the switching behavior and the Avalanche capability will be worse for new technologies due to the silicon physics. Therefore for new silicon technologies a careful trade-off has to be done.

A second option is to use new and more sophisticated package which have to fulfil some requirements for automotive applications:

- Surface Mount Technology (SMD)
- Large die sizes within the package to reduce R_{DS}(on)
- Low impedance thermal path to provide a low Rth
- Automotive reliability and AEC-Q101 qualification, in particular for excessive temperature cycles (TC) in the application
- Capability of Automatic Optical Instection (AOI)

The current analysis is based on the standard SMD TO-263-7 package. This package is widely used in automotive high power applications. But the ratio of package size to maximum die size is rather bad. The package size is 150 mm² whereas the maximum die size is about 25 mm².

New packages that provide a better die size to package size ration are of SOF type (Small Outline Flat Lead). Examples for this package type are HSON-8 (Renesas Electronics) and SO-8FL (OnSemi) as replacement for standard TO-252 packages and H-PSOF-8-1 by Infineon Technologies as TO-263-7 replacement. As required these packages are SMD packages. For comparison of H-PSOF-8-1 package with a TO-263-7 package parameters of a 40 V PowerMOSFET in an H-PSOF-8-1 package are also listed in Tab. 1. Both devices use the same 40 V Optimos T2 technology and according to the total gate charge Q_G the die size is the same for both devices. Therefore the differences in the parameters of the two devices are due to the package. H-PSOF-8-1 shows some important improvements compared to TO-263-7:

- Thermal impedance reduced by 30 %
- Higher current capability, both for package and power limit
- R_{DS}(on) reduced by 0.2 mΩ
- Better die to package ratio, package size reduced by 25 %

Currently no 100 V PowerMOSFET in an H-PSOF-8-1 are available. But high power applications in the 48 V board net can benefit from this package: the mounting area can be reduced, the R_{DS}(on) is slightly lower (about 5 %) and the package provides a better thermal path.

4. Applications for the 48 V board net

Table 2 lists some applications that might be supplied by the 48V board net. Some of these applications can be realized also within the 12V board net, but as soon as the power exceeds about 800 W (like for rack type EPS (Electric Power Steering) systems) the realization within the 12 V board net requires a very high effort and bare die assembly or dedicated power modules. Every power demand higher than about 1kW will be rather impossible to realize with a 12V electrical system using SMD packages without parallelization of PowerMOSFET. To enable higher power applications within the 12 V board net more sophisticated assembly methods are used. For example for the rack type EPS with a peak power of 1.4 kW bare die assembly is used. Here bare dies are mounted onto a special substrate like DBC (Direct Bonded Copper) to achieve an excellent thermal path and lower resistance by using better bonding connections. But this assembly is rather sophisticated and expensive.

80 V PowerMOSFET in the 48 V board net can switch up to 2.5 times the power compared to 12 V board net. Using the maximum current capability of 43 A for the 80 V device applications with a power demand up to about 2.5 kW can now be realized. Therefore high power applications like the rack type EPS could be realized using packaged PowerMOSFET on standard PCB instead of bare die assembly on DBC. Here standard TO-263-7 or new H-PSOF-8-1 could be used. This change simplifies the assembly and reduces the cost significantly.

For even higher power demands like climate compressor or e-charger an optimized thermal mounting with lower Rth values or lower On-State resistance is needed. To achieve these targets the H-PSOF-8-1 package is advantageous. It simplifies the optimization of the thermal path to about 5 K/W and provides, even for same die size, slightly lower R_{DS}(on). In addition a parallelization of two PowerMOSFET can be used to drive these high power applications, even without the need for bare die assembly.

Application	Power [W]	Current [A] @12 V	Current [A] @48 V
Electric power window	250 (peak)	20	5
Column type EPS	600 (peak)	50	12,5
Engine fan	1000 (cont.)	80	20
Rack type EPS (bare die assembly for 12 V board net)	1400 (peak)	120	30
Climate compressor	4000 (peak)	320	80
e-charger	4000 (peak)	320	80

Tab. 2   Power and current demand for selected automotive applications

5. Conclusion

High power applications can benefit from the 48V board net using 80 V PowerMOSFETs in D2PAK as the maximum power that can be switched can be increased by a factor of 2.5 roughly. By this improvement more applications like rack type EPS can be realized using standard assembly with packaged devices on PCB. This results in a significant reduction of assembly complexity, supply chain and cost.

To benefit from the 48 V board net standard TO-263-7 packages can be used, but new packages like H-PSOF-8-1 even improves the benefits given by the 48 V board net. Here in particular packages like the H-PSOF-8-1 combine superior thermal and electrical properties with other automotive requirements like AEC-Q101 qualification and AOI capability.

As the current is limited by the power dissipated within the device new silicon technologies that reduce On-State Resistance could increase current capability and hence enable even higher power applications. But a trade-off has to be done between the $R_{DS}(on)$ improvement and worsening of other parameters like the switching behavior or the Avalanche capability.

6. References

[1] H. Potdevin, Isolationsüberwachung in Hochvolt-Bordnetzen von Elektro- und Hybridfahrzeugen, ATZelektronik 06/2009, Jahrgang 4

[2] A. Graf, Halbleiter im 42V-Bordnetz, VDE, 5. Internationaler ETG-Kongress 2001, Nürnberg, 23.24.Oktober 2001

[3] R. Friedrich, Das 48V Bordnetz. Pflicht oder Kür? BMW Group, 31.01.2013

[4] VDA Empfehlung 320, Elektrische und elektronische Komponenten im Kraftfahrzeug 48V-Bordnetz, 08/2014

[5] T. Dörsam, S. Kehl, A. Klinkig, A. Radon, O. Sirch, The new voltage level 48V for vehicle power supply, ATZelektronik 01/2012

[6] Data sheets by Infineon Technologies: IPB240N04S4-R9 data sheet Rev. 1.1, 2014-04-07; IPB180N10S4-02 Datenblatt Rev. 1.0, 2013-01-30, IPLU300N04S4-R8 data sheet Rev. 1.0, 2014-08-12, IPB180N08S4-02 data sheet Rev. 1.0, 2014-06-20

Status and advances in Electric Vehicle's power modules packaging technologies

Itxaso Aranzabal, University of the Basque Country (UPV/EHU), Spain, itxaso.aranzabal@ehu.eus
Asier Matallana, University of the Basque Country (UPV/EHU), Spain, asier.matallana@ehu.eus
Oier Oederra, University of the Basque Country (UPV/EHU), Spain, oier.oinederra@ehu.eus
Inigo Martinez de Alegria, University of the Basque Country (UPV/EHU), inigo.martinezdealegria@ehu.eus
David, Cabezuelo, University of the Basque Country (UPV/EHU), Spain, dcabezueloromero@gmail.com

Abstract

The technology used in the design of the high power density Electric Vehicle (EV) power inverters packaging (power modules electrical, thermal and thermo-mechanical properties) directly affect the power inverter performance, cost, reliability, efficiency and power density. In this paper the technical trends and advances in EV power module packaging technologies will be reviewed and actual manufactured automotive power modules will be evaluated. Advantages and disadvantages of such modules will be highlighted.

1. Introduction

High power density Electric Vehicle (EV) power inverter modules contain multiple power semiconductor switches (IGBTs) and diodes.The module packaging provides electrical interconnections, thermal management, and mechanical support to these semiconductors dies.

In order to develop a power module for EV applications (with excessive ambient temperatures, humidity, vibration and dirt in the EV engine compartment among others) new solutions for module integration and packaging technology are needed. Conflict requirements as maximum power density, efficiency and reliability at low cost, can however only be achieved if the right choice of components is made, innovative solutions and technology developed, and thermal and electrical properties optimized. In this paper the technical trends and advances in EV power module packaging technologies will be reviewed. The aim of all new technologies is to improved technical parameters as thermal impedance, maximum operating temperature, parasitic inductance, thermal conductivity, lifetime, etc. Besides actual manufactured automotive power modules are evaluated and advantages and disadvantages of such modules are discussed in term of these technical parameters.

2. Packaging technologies

Fig. 1 shows a general cross-sectional view of packaging components stack. Denpending of the package configuration desing, some of the parts of the package stack could be eliminated.

The most important parts of the package stack to consider in the development of new packaging designs are: die or semiconductor device, die-attach, DBC, substrate-attach , interconnection technology and the cooling method.

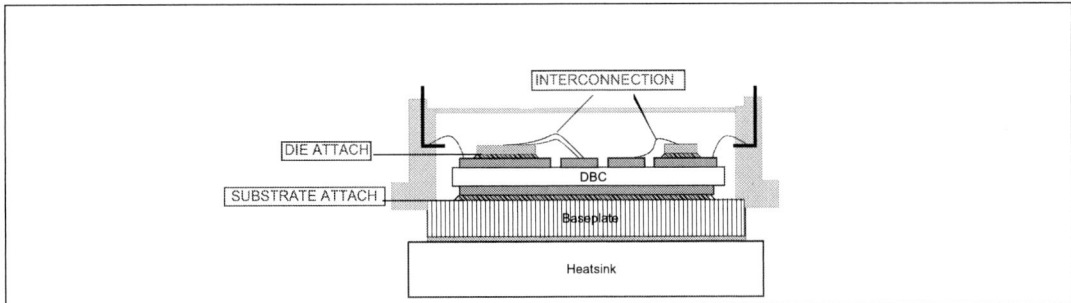

Fig. 1: Cross-sectional view of packaging components stack.

The figure 3 shows a brief comparison of some of actual manufactured automotive power modules packaging technologies.

2.1. Die

There are a lot of different developments to improve dies performance, as advanced architectures on silicon as Trench Field Stop. However, silicon has some physic limits which are difficult to solve. Manufactures are researching into other semiconductor materials, such as SiC and GaN, to obtain more efficiency in the minor size of die.

2.2. Die-attach

Sintering technology is a well established technology today and has started to replace soldering of chips to DBC substrates already in mass production [1, 2]. Thanks to its unprecedented reliability and thermal behavior this joint technology makes power modules better suited for EV application.

Sintered layers exhibit better thermal, electrical and mechanical properties than solder layers [3]. In Table 1 the advantage of Ag sinter layers are shown.

	Ag sinter layer	Ag Solder layer	Factor
Melting point (C)	961	221	4
Elec.conduc.(MS/m)	41	7.8	5
Ther.conduc.(W/mK)	250	70	4
Density (g/cm_3)	8.5	8.4	1
CTE (μm/mK)	19	28	1
Tensil strength (Mpa)	55	30	2

Tab. 1: Comparison of important properties of solder layer and sintered silver layer.

Semicron SKIM 63/93 is one of the EV power module which does not contain any soldering interconnect. The chips are sintered to a DBC substrate, and the power and auxiliary contacts are pressed to this substrate. The module does not have a base plate and the substrate is in direct contact with the heat sink [4].

© VDE VERLAG GMBH · Berlin · Offenbach

	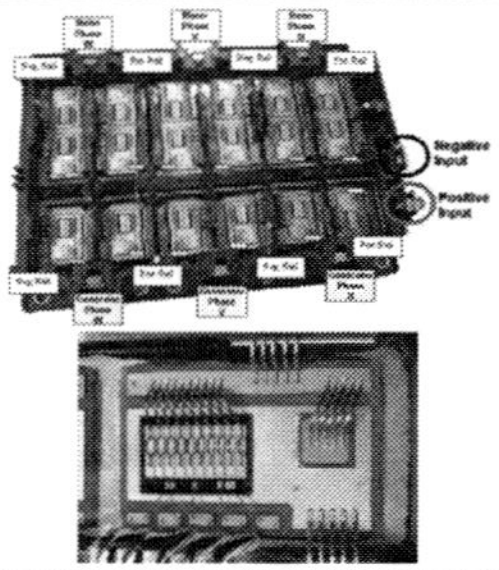 Toyota Prius 2004	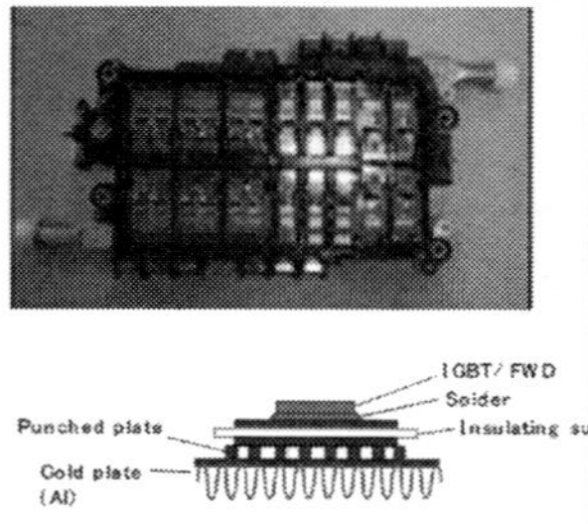 Toyota Prius III 2010	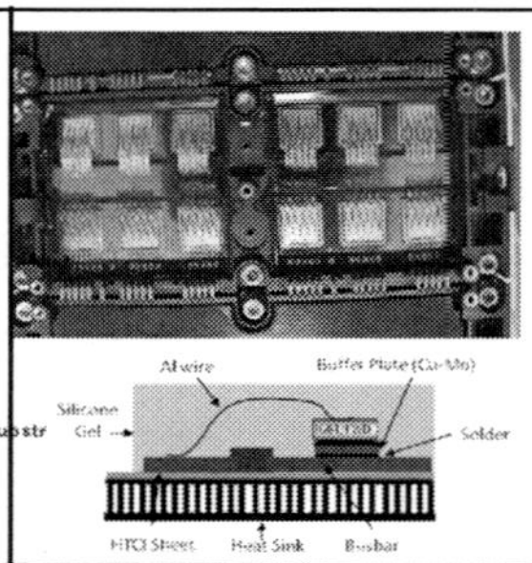 Nissan LEAF	 HybridPack 1 >30kW	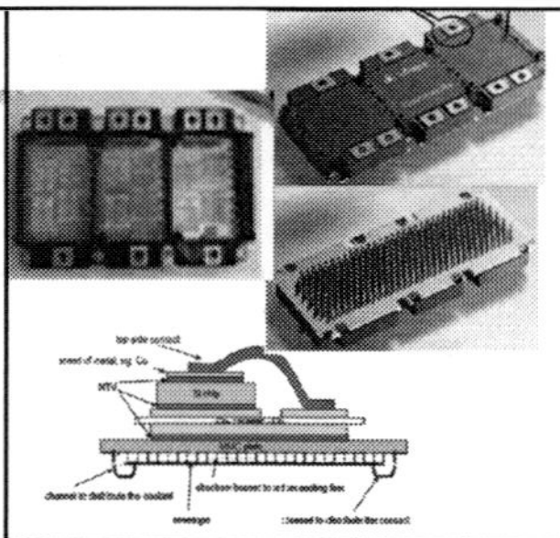 HybridPack 2 >80kW
IGBT-die	Die	Die	Die	Die	Die
	Wire bonding (Al wires)	Ribbon wiring (Al)	Wire bonding (Al wires)	Wire bonding (Al wires)	Al Wire bonds /copper bonded terminals
Die-attach	Solder	Solder	Solder Lead (Pb)-free	Solder	Solder
DBC substrate	DBA Al	DBA Al	Placa de Cu-Mo	DBC Cu	DBC Cu
	AlN	AlN	Solder Lead (Pb)-free	DBC Al2O3	DBC Al2O3
	DBA Al	DBA Al	Cu bar	DBC Cu	DBC Cu
	Solder	Directly bonded by brazing	No base plate	Solder	Solder
	Baseplate	No base plate (3)		Copper base plate	Coopper base plate with pin-fins
TIM	ZnO Thermal Paste	No TIM	No TIM. New HTCI sheet	TIM	No TIM
Heat sink	Al cold plate	Direct bond cooler. Cold plate	Cold plate	Al cold plate	No Cold plate.
Cooling method	Indirect cooling	Direct Cooling	Microchannel cold plate.	Indirect Cooling	Direct cooling. Pin fins base plate.
Advantages		* No base plate. * No TIM layer. * Al Ribbond bonds. * Integrated cooler structure. * Achieves 30% improvement in thermal performance.	* No base plate. * No TIM layer. * HTCI layer: (Heat radiating and insulating sheet) * Cu-Mo layer Compensate CTE difference between the bus bar and die. *Solder Lead (Pb)-free power module.		* No capa TIM. * No Cold plate. * Direct cooled base plate. Integrated cooler.
Disadvantages	* Standar packaging	*Buffer layer between DBA and cold plate worsen thermal conductivity .	*Large electrical parasitic parameters.	* Standar packaging	* Difficulty in pin fin manufacture. * Difficult integration of cooler. * Large electrical parasitic parameters.

(1) DLB_ Direct Lead Bonded. Cu leads are soldered directly on top of all switches dies (direct lead). These interconnection componets reduce the package parasitic resistance. Such bonding by direct Cu soldering requires a so-called solderable front metal (SFM) of die top electrodes.

(2) SKIN technology: comprises the sintering of power chips to a substrate, a top side sintering of the power chips to a flexible circuit board and the sintering of the substrate to a pin-fin heat sink .

(3) A buffer plate with punched holes, which was inserted to release the stresses between the cooler and DBA caused by the CTE mismatch.

Fig. 2: Comparison of some of actual manufactured automotive power modules packaging technologies.

	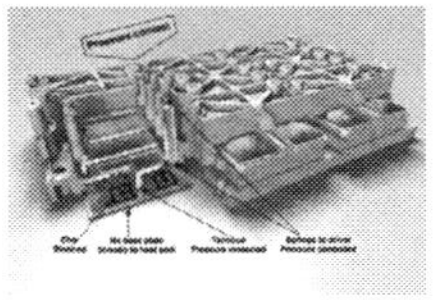 Semicron SKIM 63/93	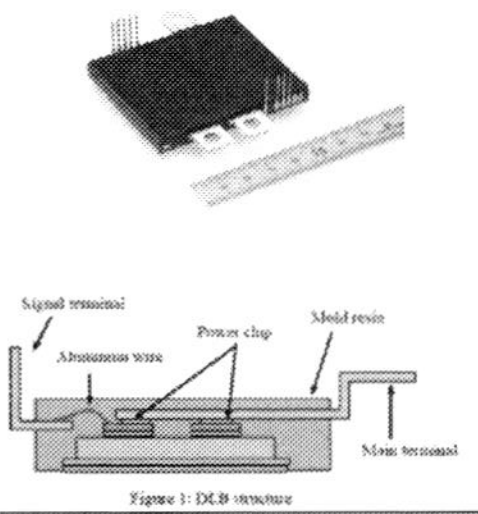 Módulo J1-Serie . Mitsibishi Electric	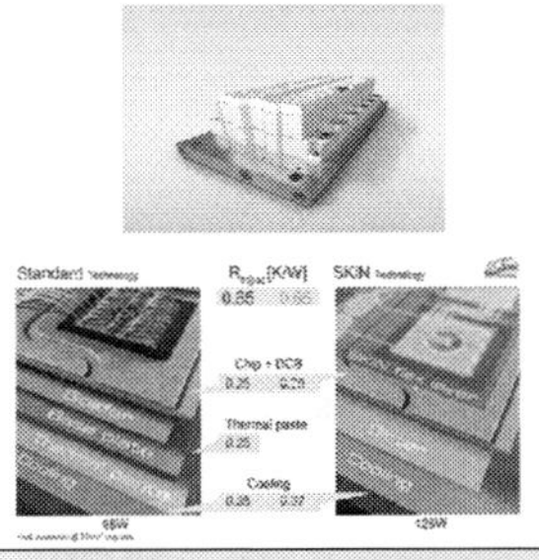 Semicron Skiip 4	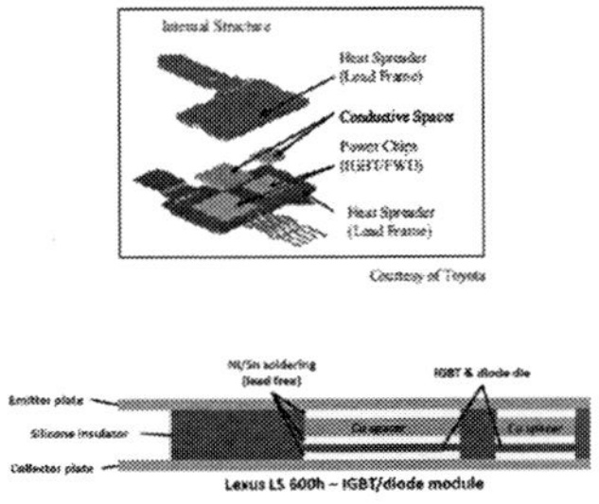 Toyota Lexus 600h
IGBT-die	Die	Die (SFM) (1)	Die	Die (SFM) (1)
Die-attach	Al Wire bonding	Cu Direct Lead Bonding. DLB	No wire bonding (SKIN Technology) (2)	Planar Cu plates. Gate by wire bonding
	Sintering (Skinnter Technology)	Solder Lead (Pb)-free power module	Sintering (Ag sintering)	Solder Lead (Pb)-free power module
DBC substrate	DBC Cu	Cu	DBC Cu	Cu
	DBC Al2O3/AlN	T-PM	DBC Al2O3/AlN	
	DBC Cu	Cu	DBC Cu	Cu
	Pressure contact	Solder Lead (Pb)-free power module	Sintering technology	Solder Lead (Pb)-free power module
	No base plate		No base plate	
TIM	Pressure contact technology	TIM	No TIM	Doble TIM
Heat sink	Cold plate.	Cold plate	Cold plate	Cold plate
Cooling method	Indirect Cooling	Doble side cooling.	Indirect cooling	Double side cooling
Advantages	* No base plate. * Solder free Ag sintered die-attach (Skinnter Technology). * Pressure points close to the chips provides low thermal resistance.	*DLB : Direct PlanarLead Bonded * Double side cooling. * No DBC. A thick Cu/TCIL/Cu DB structure replace de DBC.	* No wire bonding. SKIN technology (2) * No base plate. * Ag sintered die-attach. (Skinnter Technology). * NoTIM	* No base plate. * Double side planar interconnection. (3) * Double side cooling.
Disadvantages	* Large electrial parasitic parameters. * Difficult integration of cooler.	* Module level assembly needed. * Doble TIM layer. * Poor thermal propieties of TCIL.		* Ceramic slice insulation and double TIM layer. * Complex inverter (electrical and thermal). assembly.

(1) DLB_Direct Lead Bonded. Cu leads are soldered directly on top of all switches dies (direct lead). These interconnection componenets reduce the package parasitic resistance. Such bonding by direct Cu soldering requires a so-called solderable front metal (SFM) of die top electrodes.

(2) SKIN technology: comprises the sintering of power chips to a substrate, a top side sintering of the power chips to a flexible circuit board and the sintering of the substrate to a pin-fin heat sink .

(3) A buffer plate with punched holes, which was inserted to release the stresses between the cooler and DBA caused by the CTE mismatch.

Fig. 3: Comparison of some of actual manufactured automotive power modules packaging technologies.

2.3. Interconnection technologies

Interconnectors add extra power losses by parasitic electric inductance, resistance, and capacitance. To reduce these parasitic parameters and improve the reliability, new interconnection techniques have been developed: Ribbon bonding, Direct Lead bonding (DLB) and Copper bonding. Changing the wire interconnection configuration to a planar or symmetric package will bring enormously comprehensive benefits.

In the Toyota Prius Hybrid III 2010 Toyota employed Al ribbons to replace Al wires which helps improve the reliability and electric parasitic parameters of die interconnections.

Figures 4(a) and 4(b) show Toyota Prius Hybrid 2004 and Toyota Prius Hybrid III 2010 power modules. Furthermore, figures 4(c) and 4(d) show in detail wire bonding and ribbon bonding interconection technologies.

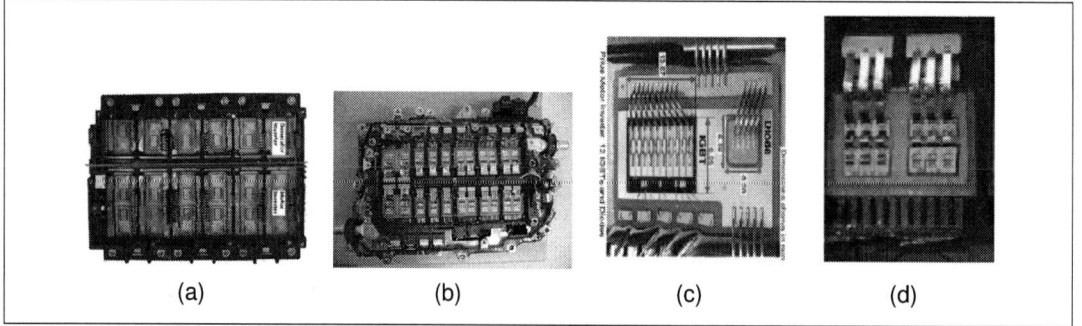

Fig. 4: Power modules interconnecton technologies (a) *Wire bonds*, Toyota Prius Hybrid 2004. (b)*Ribbon bonding*, Toyota Prius Hybrid 2010. (c) *Wire bonding* detail. (d) *Ribbon bonding* detail.

Most recently, Copper(Cu) wire bonding has been introduced as a new alternative contact technology. This technology reveals several significant advantages as higher thermal conductivity, higher electrical conductivity and lower cost, over Aluminium bond wires [5, 6]. Infineon, employed Copper wire bonds in the "Infineon .XT" technology [7].

Table 2 shows a general comparison of the relevant material properties of both Copper and Aluminium.

	Aluminium	Copper	Copper/Aluminium
Elec. Resistivity	2,7 $\mu\Omega \cdot cm$	1,7 $\mu\Omega \cdot cm$	-40%
Elec.conduc.	220 $W/m \cdot K$	400 $W/m \cdot K$	$+5\%$
CTE	25 ppm	16,5 ppm	-35%

Tab. 2: Comparison of important properties of Copper and Aluminium

SEMIKRON also has developed the "SKIN Technology" [8]. In this architecture wire bonds are replaced by a flexible board which is sintered onto the chip surface. The SKiN flex layers take over the function of the bond wires. They allow an increase of about 25% surge current in the power module due to the sintered layer on the chip tops. Compared to conventional power modules the additional performance allows an approximate doubling of the current density. Excellent thermal and electrical properties of the sintered layers increase the module lifetime up to tenfold [3]. Figure 5 shows the comparison of standard connection technology and SKIN technology.

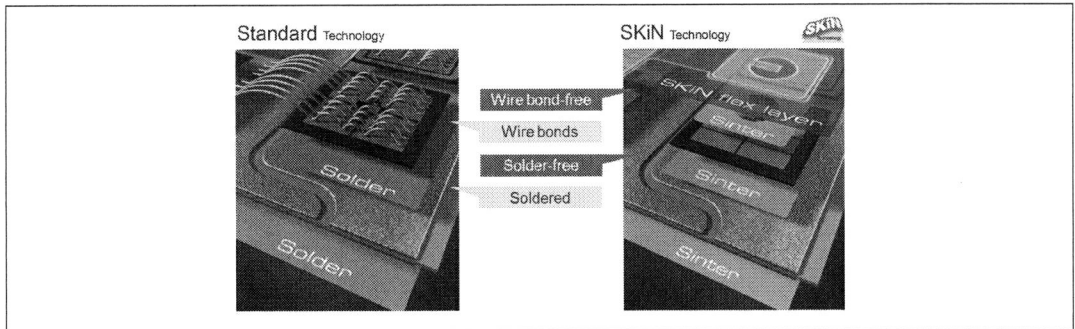

Fig. 5: Comparison of standard connection technology and SKIN technology. Picture is from [4]

Other example is the CooliR [9] platform from International Rectifier. The solderable front metal (SFM) allows soldered or sintered die attachment on both sides of the die. This enables wire bondless packaging techniques replacing it by DLB technology. Mitsubishi also developed a Cu lead bonded TPM automotive module [10].

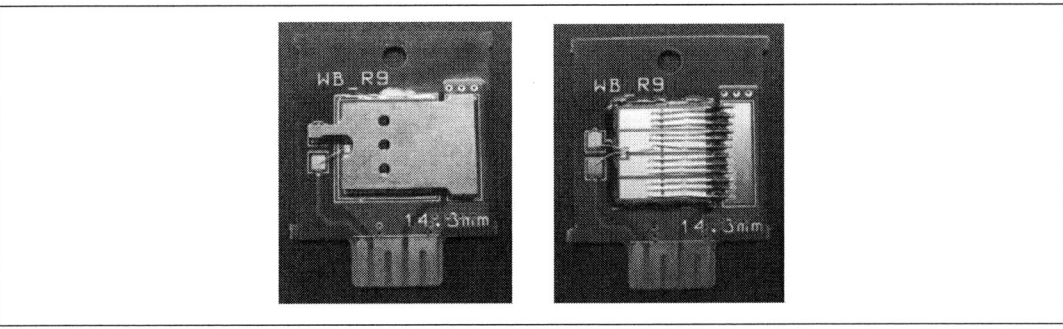

Fig. 6: International Rectifier packaged and wire bond packaged power IGBTs.

2.4. Power substrate DBC

The DBC is the medium on which are located the semiconductors (IGBTs and diodes). Its aim is to provide a mechanical construction, electrical insulation and evacuate adequately heat from semiconductors. The DBC (Direct Bond Copper) or DBA is composed of three layers, the first and last one metal (Cu or Al) and the intermediate ceramic (electrically insulating).

The trend is to make the ceramic layer (electrically insulating) thinner and the metal layer (Cu or Al) thicker in order to improve the heat transport. However, the CTE mismatch between the ceramic layer and the metal layer is an issue. Therefore, some new power substrates schemes have been made.

Based on EV application requirements, in the TPM module developed by Mitsubishi Electric a thick Cu / thin TCIL /thin Cu structure replaces the DBA. The TCIL (Thermal Conductive Insulation layer) is made of especial insulation resin and has a good thermal conduction capability [10]. The power module in Nissan LEAF has the same configuration. The IGBT is soldered to a thick Co-Mo spacer for CTE matching. Curamic also introduces a new hybrid substrate (where the cooling channels are integrated directly into a DBC substrate) comprising of cooper, ceramic and aluminum [11].

3. Cooling methods

Conventional power modules cooling options based on natural air cooled and forced air cooled heat sinks are not able to meet the existing demand for cooling the power modules used in traction applications such as EV. Very high current density, excessive ambient temperatures, humidity, vibration and dirt in the engine compartment among others can lead to premature rupture of the power module if the cooling system is not designed properly.

The following factors influence the reliability of the cooling options [7]:

- The contact area to the coolant.

- Turbulence in the water flow.

- The volumetric flow rate as a function of the pressure drop.

- The heat storage capability of the coolant.

- The coolant temperature.

- Heat conduction and spreading in the heat sink.

Liquid cooling solutions can be divided into two groups: indirect and direct liquid cooling.

Indirect cooling means that the power module is assembled on a closed cooler, e.g. a cold plate. When dealing with cold plates, it is necessary to apply a layer of TIM between the power module and the cold plate, which significantly reduces the cooling system performance [12].

In direct liquid cooling systems the coolant is in direct contact with the surface to be cooled and eliminates the layer of TIM. The cooling efficiency is improved by increasing the surface area. Various designs can be distinguished:

- Pin fins base plates. Liquid flow through pin fins formed directly on base plates [13].

- Spray cooling. This method used the principle of spraying the liquid coolant onto the surface either as droplets or jet. Cooling may be applied from one side or both sides [14].

- Jet impingement cooling [15]. Danfoss Silicon Power has developed a system called "Shower power" which is based on this technique. Here, a plastic insert with many parallel holes in the heat sink opening creates turbulent and vertical flow ensures good and even cooling [16].

- Microchannel coolers built into cold plate or integrated with DBC substrate or into specially customized package design [17].

- Two-phase cooling. Oak Ridge National Laboratory (ORNL) has developed a system based on Two-phase cooling technology with automotive air conditioning R-134a coolant [18].

- Double side cooling. This enables a further reduction of the thermal resistance. It doubles the effective area of the heat dissipation. Denso in cooperation with Universities of Cambridge, Oxford and Nottingham [19], Fraunhofer Institute IZM [20] and Semikron (SKIN Technology) among others, have developed a module based on this technology.

© VDE VERLAG GMBH · Berlin · Offenbach

4. Conclusions

The combination of a reliable die attach technique, a reliable top contact technology, as well as a good thermal, thermo-mechanical and electrical design, associated with the successful combination of different materials are crucial factors to integrate successfully power electronic in EV propulsion system. Trends are to replace soft solder contacts by sintering and diffusion soldering, to replace Al wire bonds by Direct Lead bonding (DLB) and Copper bonding, and to employed new liquid cooling methods (microchannel, thermoelectric cooling, two phase cooling, double side cooling) for improving the thermal performance of EV high-power modules.

5. Acknowledgment

This work has been carried out inside the Research and Education Unit UFI11/16 of the UPV/EHU and supported by the Department of Education, Universities and Research of the Basque Government within the fund for research groups of the Basque university system IT394-10 and by the University of the Basque Country. This work has been supported by the Ministerio de Economa y Competitividad of Spain within the project DPI2014-53685-C2-2-R and FEDER funds.

6. References

[1] T. Stockmeier. From packaging to un-packaging - trends in power semiconductor modules. In *Proc. of International Symposium Power Semiconductor Devices and Integrated Circuits (ISPSD)*, pages 12–19, 2008.

[2] K. Guth, N. Oeschler, L. Boewer, R. Speckels, G. Strotmann, N. Heuck, S. Krasel, and A. Ciliox. New assembly and interconnect technologies for power modules. In *Proc. of International Conference on Integrated Power Electronics Systems (CIPS)*, pages 1–5, 2012.

[3] C. Gobl and J. Faltenbacher. Low temperature sinter technology die attachment for power electronic applications. In *Proc. of International Conference on Integrated Power Electronics Systems (CIPS)*, pages 1–5, 2010.

[4] A. Wintrich, U. Nicolai, W. Tursky, and T. Reimann. *Application Manual Power Semiconductor*. Semikron International GmbH, 2011.

[5] D. Siepe, R. Bayere, and R. Roth. The future of wire bonding is? wire bonding! In *Proc. of International Conference on Integrated Power Electronics Systems (CIPS)*, 2010.

[6] R. Ott, M. Bable, R. Tschirbs, and D Sierpe. New superior assembly technologies for modules with highest power densities. In *Proc. of International Conference on Power Electronics Systems and Applications (PESA)*, pages 528–531, 2010.

[7] R. Tschirbs, G. Borghoff, T. Nubel, W. Rusche, and G. Strotmann. Ultrasonic metal welding as contact technology for state-of-the-art power modules. In *Proc. of International Exhibition and Conference for Power Electronics, Intelligent Motion, Renewable Energy and Energy Management (PCIM)*, 2008.

© VDE VERLAG GMBH · Berlin · Offenbach

[8] T. Stockmeier, P. Beckedahl, C. Gobl, and T. Malzer. Skin: Double side sintering technology for new packages. In *Proc. of International Symposium Power Semiconductor Devices and Integrated Circuits (ISPSD)*, pages 324–327, 2011.

[9] J. Marcinkowski. Dual-sided cooling of power semiconductor modules. In *Proc. of International Exhibition and Conference for Power Electronics, Intelligent Motion, Renewable Energy and Energy Management (PCIM)*, pages 1–7, 2014.

[10] Hui Han and Gaosheng Song. Consideration on igbt module lifetime for electrical vehicle (ev) applications. In *Proc. of International Exhibition and Conference for Power Electronics, Intelligent Motion, Renewable Energy and Energy Management (PCIM)*, pages 1–7, 2014.

[11] Xinhe Tang, Andreas Meyer, Karsten Schmidt, Ulrich Voeller, and Manfred Goetz. Hybrid substrate - a future material for power semiconductor modules. In *Proc. of International Exhibition and Conference for Power Electronics, Intelligent Motion, Renewable Energy and Energy Management (PCIM)*, pages 1–5, 2014.

[12] S.S. Kang. Advanced cooling for power electronics. In *Proc. of International Conference on Integrated Power Electronics Systems (CIPS)*, pages 1–8, 2012.

[13] Zhihong Liang and Lei Li. Hybridpack2 - advanced cooling concept and package technology for hybrid electric vehicles. In *Proc. of Vehicle Power and Propulsion Conference (VPPC)*, pages 1–5, 2008.

[14] H. Bostanci, D. Van Ee, B.A. Saarloos, D.P. Rini, and L.C. Chow. Thermal management of power inverter modules at high fluxes via two-phase spray cooling. *IEEE Transactions on Components, Packaging and Manufacturing Technology,*, 2:1480–1485, 2012.

[15] K. Gould, S.Q. Cai, C. Neft, and A. Bhunia. Liquid jet impingement cooling of a silicon carbide power conversion module for vehicle applications. *IEEE Transactions on Power Electronics*, 30:2975–2984, 2015.

[16] K. Olesen, F. Osterwald, M. Tonnes, R. Drabek, and R. Eisele. Direct liquid cooling of power modules in converters for the wind industry. In *Proc. of International Exhibition and Conference for Power Electronics, Intelligent Motion, Renewable Energy and Energy Management (PCIM)*, pages 742–747, 2010.

[17] J. Schulz-Harder. Efficient cooling of power electronics. In *Proc. of International Conference on Power Electronics Systems and Applications (PESA)*, pages 1–4, 2009.

[18] J.B. Campbell, L.M. Tolbert, C.W. Ayers, B. Ozpineci, and K.T. Lowe. Two-phase cooling method using the R134a refrigerant to cool power electronic devices. *Industry Applications, IEEE Transactions on*, 43(3):648–656, 2007.

[19] C. Buttay, J. Rashid, C.M. Johnson, P. Ireland, F. Udrea, G. Amaratunga, and R.K. Malhan. High performance cooling system for automotive inverters. In *Proc. of European Conference on Power Electronics and Applications (EPE)*, pages 1–9, 2007.

[20] Martin Schneider-Ramelow, Thomas Baumann, and Eckart Hoene. Design and assembly of power semiconductors with double-sided water cooling. In *Proc. of International Conference on Integrated Power Electronics Systems (CIPS)*, pages 1–7, 2008.

PCIM Europe 2016, 10 – 12 May 2016, Nuremberg, Germany

A highly integrated full SiC six-pack power module
for automotive applications

Bao Ngoc An, Viktor Wegelin, Martin Bernd, Benjamin Leyrer, Michael Meisser, Horst Demattio, Thomas Blank, Marc Weber, Institute for Data Processing and Electronics (IPE), Karlsruhe Institute of Technology (KIT), Karlsruhe, Germany, bao.an@kit.edu

Johannes Kolb, SHARE am KIT – Schaeffler Technologies AG & Co. KG, Karlsruhe, Germany, johannes.kolb@schaeffler.com

Jochen Altstadt, kortec Industrieelektronik GmbH & Co. KG, j.altstadt@kortec.de

Abstract

A thick-film-based, full SiC, low inductance six-pack power module with integrated gate-drive circuits and DC-link capacitors is presented. Thermal simulations are conducted with COMSOL Multiphysics in order to determine the temperature distribution within the power module as well as its thermal resistance. For thermal characterisation the surface temperature distribution of the power module is scanned by infrared photography and the thermal resistance from junction to the thermal interface layer is determined. The measured values differ from the simulated results by 12 % to 20 % due to the deviating thermal properties of the materials and the measurement error of the utilised infrared camera.

1. Introduction

For automotive applications the need for power modules with high power density is growing, due to restricted design space in hybrid (HEVs) or battery electric vehicles (BEVs). Present HEVs mostly have two separate cooling circuits, one for the combustion engine, and another for the electric motor and its power electronic system. Merging of both coolant loops could increase the system power density and lower the overall system cost. However, the high coolant temperature of the engine of up to 105 °C requires power devices, which can operate at junction temperature higher than 125°C, in order to meet the thermal management requirements. Therefore, this application impedes the use of standard silicon (Si) power devices [1]. Wide bandgap silicon-carbide (SiC) power semiconductors can endure a junction temperature greater than 220 °C and replace Si devices in high temperature applications [2]. SiC has 10x higher breakdown voltage than Si due to the larger bandgap. In comparison to Si, its thermal conductivity is 3x better and, therefore, enhances the heat dissipation [3–5].
SiC MOSFETs offer higher switching frequency and lower conduction loss compared to Si-IGBTs. However, the performance of SiC can fully be exploited if the parasitic inductances are minimised by integrating DC-link capacitors and the gate-drive circuits into the power module. Conventional power module structures are built up using direct copper bonding (DCB) substrate. The presented power module is based on a copper-thick film substrate, which is used in order to create fine-line conductor traces for applying electrical contact to the gate-drive ICs [6].

2. Module Design

The presented six-pack power module consists of three parallel connected, thick-film based substrates, each carrying a full SiC half bridge module. Figure 2 shows one half bridge, which includes two integrated gate-drive circuits and two DC-link capacitors.

© VDE VERLAG GMBH · Berlin · Offenbach

PCIM Europe 2016, 10 – 12 May 2016, Nuremberg, Germany

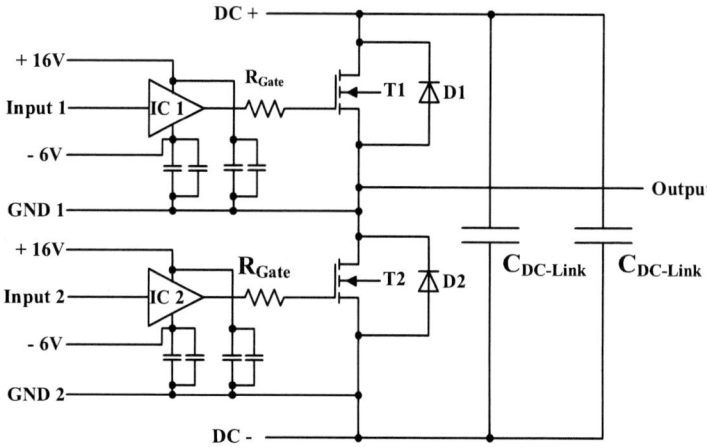

Figure 1: Circuit the half bridge power module with integrated gate-drive ICs.

The integrated DC-link capacitors minimise the commutation inductance whereas the integrated gate-driver shortens the gate path length and, therefore, reduces parasitic inductances [7]. Each gate-drive circuit consists of one IXYS IXDD609D2 gate-drive IC and four decoupling capacitors at the voltage supply. Each gate driver requires a four-wire connection to the power supply and control system with an external galvanic insulation. Both the high side and low side switches are equipped with one Cree SiC MOSFET (CPM300900065B) and one Cree SiC diode (CPW41200S020B). An external SiC diode is required in order to minimise losses during the reverse recovery phase. Due to the high forward voltage (4.8 V) of the integrated antiparallel body diode of the SiC MOSFET, the external diode carries the main part of the current and thus determines the recovery behaviour.

3. Manufacturing Process and Quality Inspection

Figure 2 shows the thick-film layer composition of the power module in detail. The full SiC half bridge module is built on a 380 µm ceramic substrate, which is 31 x 24 mm² in size. It is embedded between two copper thick-film layers. The top-side copper layer is structured whereas the bottom-side layer covers the whole area of the ceramic. The gate-drive circuit islands on the top-side are built with multi-layer copper thick-film technology.

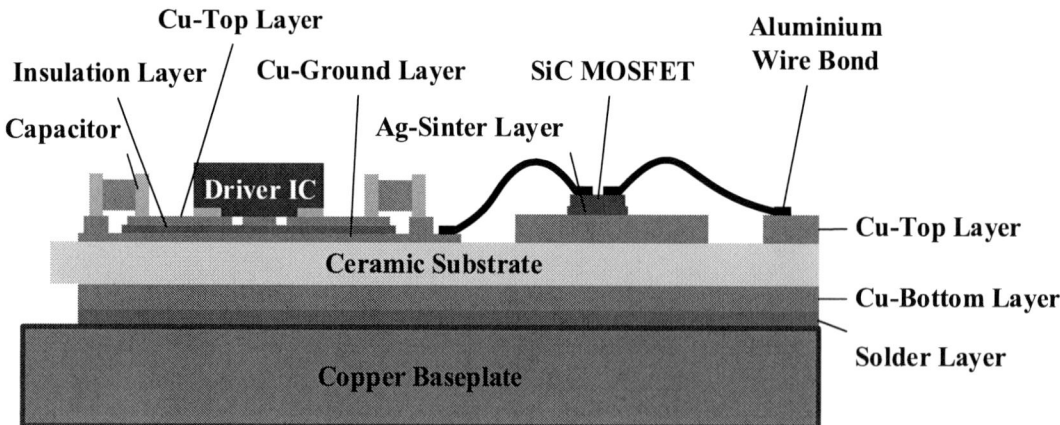

Figure 2: Cross section of one half bridge power module.

© VDE VERLAG GMBH · Berlin · Offenbach

Initially, a 40 μm thick copper seed layer, which adheres to the ceramic and provides the base for the following layers, is screen printed on both sides of the Al_2O_3 ceramic substrate. The top-side copper layer is filled up to 250 μm whereas the bottom-side layer is built up to 200 μm, in order to minimise thermo-mechanical stress and warpage during the production process. After the firing process in nitrogen atmosphere at 925 °C three insulation layers of thick-film dielectric with a total thickness of 100 μm are deposited on the ground layer of both gate-drive circuits. Next, a 40μm thick structured copper top layer of the gate-drive circuit is printed on the insulation layer and the copper areas of the power circuit underneath the power semiconductors are built up to 300 μm thickness, which provides for better heat spread underneath the dies and lower ohmic losses. An 80μm thick silver sinter layer is stencil-printed onto the thick-film substrate and dried for 3 min at 100 °C. The SiC MOSFETs and SiC diodes are attached onto the sinter layer and are sintered at 14.7 bar and 260 °C for 30 minutes. Shear tests reveal an excellent average shear value of 52.2 MPa in comparison to comparable Ag-sintering experiments, which achieve a shear value of 47 MPa [8].

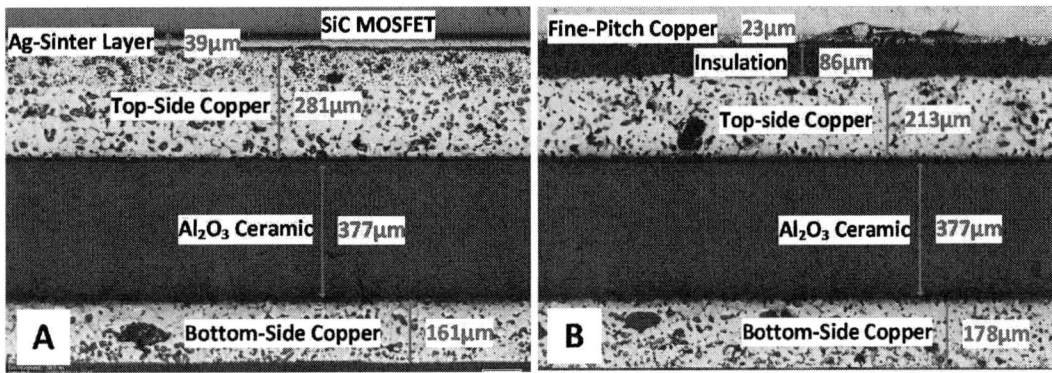

Figure 3: Cross-sectional micrographs of one half bridge module at the location of the power circuit through a SiC MOSFET switch (A) and at the location of the gate-driver island (B).

Power Circuit	
Top-side Copper	31.4%
Bottom-side Copper	29.3%
Gate-driver Circuit	
Insulation	31%
Top-side Copper	27.8%
Bottom-side Copper	32.9%

Table 1: Porosity of the fired copper and insulation layers.

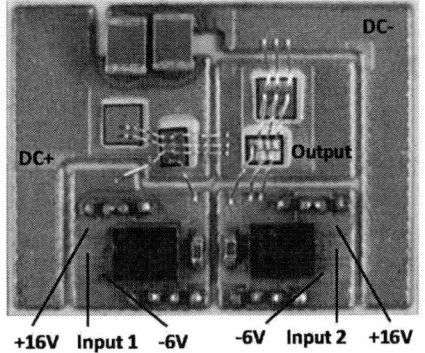

Figure 4: Half bridge power module with integrated driver ICs and DC-link capacitors.

Figure 3 shows the effective thickness of all copper and insulation layers after the sinter process. Table 1 reveals that the copper and insulation layers have in average a porosity of about 30%. Compared to standard substrate type e.g. DCB or AMB, the thick-film substrate has a higher elasticity and, therefore, an increased ruggedness against temperature cycling. However, this porous structure deteriorates the thermal and electrical properties of the thick-film layers. Figure 4 shows the driver-ICs and the capacitors, which are then soldered onto the substrate by a vapour phase soldering process at 230 °C peak temperature. The connections between the power switches and the gate drivers as well as the interconnections between the high-side switch and low-side switch are established by 200 µm thick Heraeus aluminium wire bonds for automotive and power applications. The pull test reveals wire-break at the average breaking load of 251.2 cN, which exceeds the manufacturer's guarantied breaking load of 150 cN to 230 cN. Next, low voiding KOKI S3X48-M500 solder paste (Sn96.5 Ag3.0 Cu0.5) with a thickness of 270 µm is stencil printed onto a copper baseplate. Three half bridge power modules are then mounted onto the baseplate to a six-pack module via vacuum vapour phase soldering at 230 °C. The X-ray inspection in Figure 5 proves a low voiding rate of 0.2 % in the solder layer between the power modules and the baseplate. This voiding rate is excellent in comparison to other baseplate vacuum soldering experiments, which could achieve only a voiding rate of less than 2 % [9].

Figure 5: X-ray image of the six-pack power module.   Figure 6: Six pack module with Infineon EconoPACK™ case.

Figure 6 shows a standard Infineon Econopack™ case with attached copper pins which is assembled with the thick-film substrates. The DC-link connections between the half bridge modules as well as the connection from power modules to the pins are established by 300 µm aluminium wire bonds. Finally, the six-pack power module is sealed with silicone gel in order to provide electrical insulation of the components inside the six-pack power module.

4. COMSOL Multiphysics Simulation

A thermal analysis with COMSOL Multiphysics is performed with the objective to determine the thermal resistance from the MOSFET die to the thermal interface material layer of one half bridge power module as a function of the dissipated power. The simulation model consists of one half bridge module soldered onto the baseplate and attached to a temperature-controlled platform, which is defined as a heat sink with a constant temperature. The thermal interface material is modelled in COMSOL as a thin thermally resistive layer located at the boundary between the baseplate and the temperature-controlled platform. The MOSFET of the low-side or high-side switch is defined as the heat source. It is assumed that the driver IC's, decoupling capacitors and DC-link capacitors do not significantly influence the thermal behaviour of the power module. Therefore, these components are not included in the simulation model. This modelling considers heat conduction from the surface of the MOSFET towards the heat sink while heat radiation and convection of the power switch are neglected due to their minor contribution to the total heat transfer. The silver sinter layer and the thick-film copper layer have a lower thermal conductivity compared to pure solid silver or copper due to their porous structure. For this simulation the thermal conductivity of the silver sinter layer is set to 150 W/mK in accordance with the manufacturer's data. Different properties of the thick-film copper were analysed in the previous work [10]. The density and specific heat capacity are determined by Archimedes' principle and

Differential Scanning Calorimetry. Whereas, the thermal diffusivity, which is needed for the calculation of the thermal conductivity, is contactless measured using photothermal beam deflection method. The thermal conductivity of the copper thick-film supposes to be 147.38 W/mK equal to 38.7 % of the thermal conductivity of solid copper. The thermal conductivity of the WLK5 thermal interface material is specified to be 5.6 W/mK in the datasheet and its thickness is approximated to be 100 µm.

Figure 7: Cross-sectional temperature distribution of a half bridge module at 80 °C heat sink temperature and 25 W total power dissipation.

The thermal simulations are performed for different assumed heat sink temperatures from 80 °C to 110 °C in 10 °C steps and power dissipation range from 5 W to 25 W in 1W steps. Figure 7 displays one exemplary results of the thermal analysis under COMSOL Multiphysics in order to approximate the junction temperature of the power MOSFET.

5. Thermal Characterisation of the six-pack power module

The objectives of the characterisation of the six-pack module are to measure the thermal resistance from the junction of the SiC MOSFET to the thermal interface Layer as a function of the dissipated power or heat sink temperature.

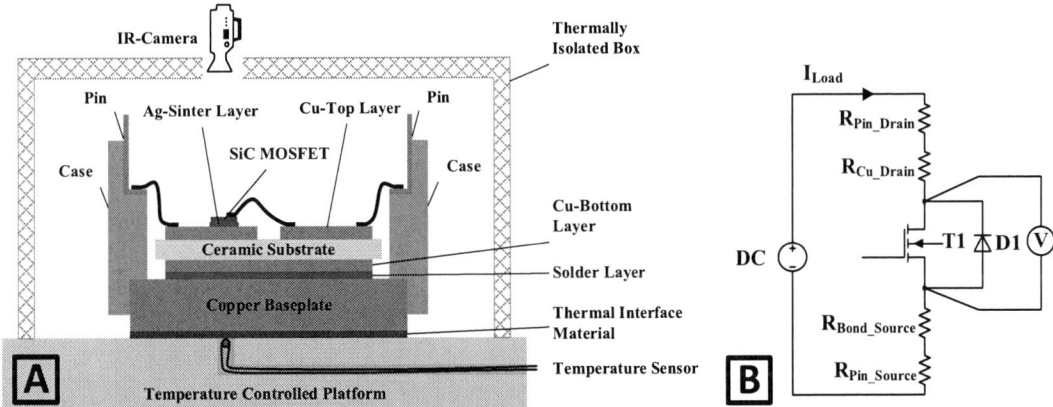

Figure 8: The experimental setup for the characterisation of the six-pack power module is drafted in (A). Subfigure (B) shows the connection of the MOSFET to the measurement system via four-wire technique.

Figure 8 shows the six-pack power module mounted on a temperature-controlled platform equipped with Peltier elements. A PT100 sensor is placed underneath the thermal interface layer and is positioned vertically underneath the respective MOSFET die in order to record the heat sink temperature at the top side of the temperature-controlled platform. This temperature is precisely controlled and is kept constant. The whole power module is enclosed by a thermally isolated box, which helps to create an isolated system by minimising heat transfer from the power module to the ambiance. The surface temperatures of the MOSFETs are assumed to be approximately equal the junction temperatures, due to the proximity of the junction to the chip surface. Therefore, the maximum junction temperature can be measured by infrared photography. For the measurement, the drain and the source connections of the high-side switch are connected to a DC power supply, which operates as a current source. First, the measurement is performed turning the MOSFET on with a defined gate voltage level. Then, the temperature-controlled platform is set to a defined temperature and the voltage drop across the power switch is measured using four-wire technique in order to adjust the input of the electric power. The surface of the MOSFET is scanned by the infrared camera to record the surface temperature distribution. The measurements are performed for heat sink temperatures varied from 80 °C to 110 °C in 10 °C steps and for varied drain current values from 5 A to 20 A. The same measurements are conducted for the low-side switch. Finally, the measured surface temperature and the thermal resistance of both MOSFET switches are compared to the calculated results of the thermal simulation with the objective to validate the thermal simulation results.

6. Results

Figure 9 shows that the thermal resistance of the high-side and low-side switch rises steadily with increasing heat sink temperature. The increase becomes slower at higher heat sink temperature and the reason for that could be the slower decrease of the thermal conductivity of the utilized materials. The measurement shows that the high-side switch is thermally better connected to the heat sink compared to the low-side switch. This tiny discrepancy in thermal resistance is related to the production process of the power module.

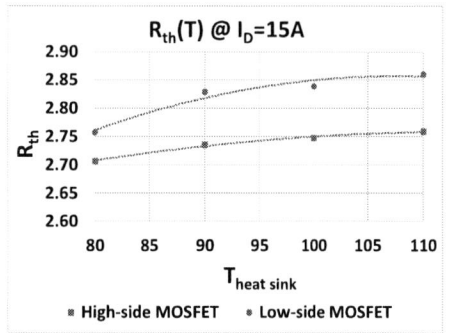

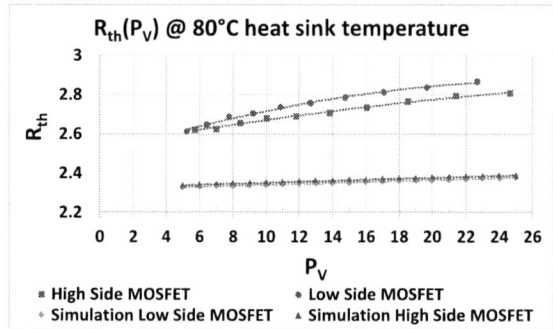

Figure 9: Thermal resistance of the high-side and low-side switch as a function of the heat sink temperature.

Figure 10: Thermal resistance of the high-side and low-side switch as a function of the dissipated power.

Figure 10 reveals that the measured thermal resistances are approximately 12 % to 20 % higher than the simulated values. Both the measured and simulated thermal resistances of the high-side and low-side switch rise with increasing dissipated power. This increase corresponds to the increase in thermal resistance of the materials with rising temperature. However, in comparison to the simulation the increase of the measured thermal resistance is higher due to the reason that in the simulation all material parameters and the dependency of the material parameters on temperature are entirely approximated. The thermal conductivity of the silver sinter and thick-film copper layer are dependent on their porosity, which could vary during the printing, drying and sintering process. The temperature dependency of the thermal conductivity of the Al_2O_3 ceramic substrate is only linearized. Furthermore, the thickness of the thermal interface material layer is approximated and this layer is ideally assumed to be void-free. Moreover, the measurement error of the infrared photography lies between 1K and 2K and needs to be considered. The simulation shows that both the high-side and low-side switch have a similar thermal resistance. This is consistent with the design of the half bridge module, which offers nearly the same copper area for both MOSFETs for heat spread. In contrast, the measurement indicates that the high-side switch is thermally better connected to the heat sink.

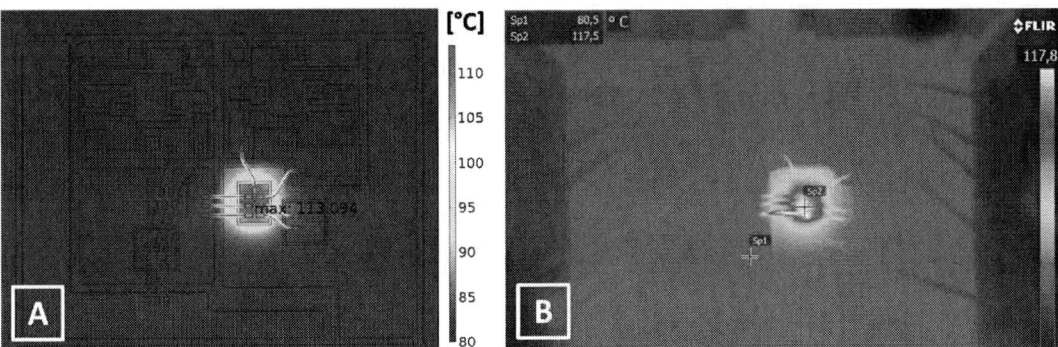

Figure 11: The simulated surface temperature (A) and the infrared image (B) of the half bridge power module during the on-state of the high side MOSFET at 80 °C heat sink temperature.

The comparison of the simulated and the measured surface temperature in Figure 11 shows the peak temperature on the power switch and the temperature profile declines to the heat sink temperature at a greater distance to the SiC MOSFET die. The simulation predicts a maximum junction temperature of 113.1 °C at the centre of the die, whereas the infrared image reveals a maximum temperature of 117.5 °C closed to the wire bonding area. Both the simulation as well as the infrared image display a heat spread in the area surrounding the heat source due to the lower thermal conductivity of the Al_2O_3 ceramic substrate. However, the infrared image shows a larger heat spread area which indicates that the thermal conductivity of the Al_2O_3 ceramic is basically lower than the supposed value in the thermal simulation model. Figure 3 reveals the porous structure of the Al_2O_3 substrate, which reduces the thermal conductivity of the ceramic layer. Additionally, the thermal interface layer can include voids, which causes a strong increase in thermal resistance and, therefore, promotes more heat spread.

7. Conclusion

A thick-film-based, full SiC six-pack power module with integrated gate-drive circuit is designed and built up. The structure of the fired copper thick-film and insulation layers are investigated and the results show a porosity of about 30 %. This reduces the thermal conductivity of the copper thick-film by 61.3 % compared to the thermal conductivity of solid copper. However, this porous structure has a lower young's modulus and, therefore, provides a higher flexibility, which is needed to reduce mechanical stresses between the copper thick-film layer and the Al_2O_3 ceramic substrate caused by the CTE mismatch. Furthermore, the copper thick-film technology provides fine-pitch and fine line capability allowing the integration of gate-drive circuits. A thermal simulation of the presented six-pack module is conducted under COMSOL Multiphysics and a thermal characterisation of the power module is performed. The measurement yields a 12 % to 20 %

higher thermal resistance compared to the results of the simulation. This difference is caused by the limited accuracy of the estimated material parameters and the approximation of the temperature dependency of the thermal conductivity of these materials as well as the measurement error of the infrared photography. In summary, the measurement values could validate the thermal simulation results.

8. Outlook

In future works, the thick-film layers of the gate-drive circuits are printed on a DCB substrate, in order to combine both advantages of the higher thermal conductivity of a conventional DCB substrate and the fine-pitch capability of the thick-film technology [6]. A further approach to reduce the thermal resistance of the copper layer could be to place copper lead frames in the chip position areas directly on the wet copper thick-film layer and sintered them in the firing process. These lead frames enhance the heat spread underneath the power switches and, therefore, optimise the thermal management of the six-pack power module.

References

[1] M. Chinthavali, L. M. Tolbert, H. Zhang, J. H. Han, F. Barlow, and B. Ozpineci, "High power SiC modules for HEVs and PHEVs," in *2010 International Power Electronics Conference (IPEC - Sapporo)*, pp. 1842–1848.

[2] Hee Yeoul Yoo, Byung Hoon Moon, Jae Sung Kwak, Cheol Woo Kwak, Ji Young Lee, and Thomas J Borghard, "NOVEL DIE ATTACH ADHESIVE FOR THIN QUAD FLAT PACKAGE," http://www.henkelna.com/us/content_data/113838_tqfpadh.pdf.

[3] K. Gould, S. Q. Cai, C. Neft, and A. Bhunia, "Liquid Jet Impingement Cooling of a Silicon Carbide Power Conversion Module for Vehicle Applications," *IEEE Trans. Power Electron,* vol. 30, no. 6, pp. 2975–2984, 2015.

[4] D. R. M. Woo, H. H. Yuan, J. A. J. Li, H. S. Ling, L. J. Bum, and Z. Songbai, "High power SiC inverter module packaging solutions for junction temperature over 220°C," in *2014 IEEE 16th Electronics Packaging Technology Conference (EPTC)*, pp. 31–35.

[5] J. Lutz, *Halbleiter-Leistungsbauelemente: Physik, Eigenschaften, Zuverlässigkeit,* 2nd ed. Berlin: Springer, 2012.

[6] Michael Meisser, "Highly integrated power modules based on copper thick-film-on-DCB for high frequency operation of SiC semiconductors," in *EPE '15 ECCE Europe*

[7] M. Meisser, M. Schmenger, and T. Blank, "Parasitics in Power Electronic Modules: How parasitic inductance influences switching and how it can be minimized," in *PCIM Europe 2015; International Exhibition and Conference for Power Electronics, Intelligent Motion, Renewable Energy and Energy Management; Proceedings of*, 2015, pp. 1–8.

[8] T. Blank, B. Leyrer, T. Maurer, M. Meisser, M. Bruns, and M. Weber, "Copper thick-film substrates for power electronic applications," in *Electronics System-Integration Technology Conference (ESTC), 2014*, 2014, pp. 1–6.

[9] N. Doug DeVoto, *Physics of Failure of Electrical Interconnects.* Available: http://energy.gov/sites/prod/files/2014/03/f10/ape036_devoto_2012_p.pdf.

[10] Bao Ngoc An, Martin Bernd, Benjamin Leyrer, Thomas Blank, Marc Weber, "Full SiC power module with substrate integrated liquid cooling for battery electric vehicles," in *Integrated Power Electronics Systems CIPS 2016.*

PCIM Europe 2016, 10 – 12 May 2016, Nuremberg, Germany

Automotive-grade P-channel Power MOSFETs for Static, Dynamic and Repetitive Reverse Polarity Protection

Filippo Scrimizzi, Giuseppe Longo, Giusy Gambino. STMicroelectronics, Stradale Primosole 50, Catania, Italy, filippo.scrimizzi@st.com, giuseppe-mos.longo@st.com, giusy.gambino@st.com

Abstract

The latest P-channel trench technology qualified for automotive applications (according to AEC Q101) from STMicroelectronics is able to provide an outstanding performance in terms of:
- low Q_{rr} which allows a quick current flowing interruption in case of reverse polarity events
- high ruggedness versus repetitive pulse sequence, according to **ISO 7637** Pulse 1 condition [1, 2].

Introduction

To avoid that reverse voltages can be applied to automotive systems thus causing fatal errors on the electronics of a car and even the ECUs burning because of the permanent short circuit of the battery (body-drain diodes of the power stage FETs permanently ON due to forward bias), different solutions can be adopted.

As well known in literature, for high power applications the use of a diode is not feasible for reverse battery protection, since power losses are very high.

The N-channel MOSFET solution offers the highest efficiency with the drawback of additional circuit requirements like a charge pump circuit and EMI filter.

A very simple solution still with an excellent efficiency would be the P-channel approach since requires nearly no additional circuit effort compared to the diode and only a slightly worse efficiency in comparison to the N-channel MOSFET [3, 4].

By using the Automotive-grade P-channel Power MOSFETs from STMicroelectronics, three different reverse polarity conditions have been evaluated with experimental data: static, dynamic, and repetitive reverse polarities.

Reverse Polarity Condition Testing

The P-channel Power MOSFET is used with the function to protect the TCUs (Transmission Control Units) when a reverse polarity event occurs. In general the device is connected in series between the battery and TCU, as shown in Fig.1.

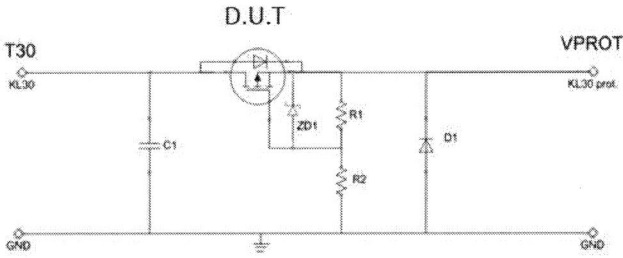

Figure 1: Circuit schematic with a P-Channel power MOSFET used for reverse polarity protection.

© VDE VERLAG GMBH · Berlin · Offenbach

Under normal operation the P-channel device is turned-on and offers a low impedance path thanks to its low R_{DSon}. When the reverse polarity occurs, the device has to be switched off in order to stop the short circuit and avoid the TCU damage.

The P-channel Power MOSFETs used are the following (Tab.1):
• STD28P3LLH6AG
• STD15P6F6AG

	STD28P3LLH6AG	STD15P6F6AG
Package	DPAK (TO-252)	DPAK (TO-252)
BV$_{DSS}$ @ - 250 mA	- 30V	- 60V
R$_{DSon}$	@ V$_{GS}$ = - 4.5V, I$_D$ = - 6A max 50 mΩ @ V$_{GS}$ = - 10V, I$_D$ = - 6A max 30 mΩ	@ V$_{GS}$ = -10V, I$_D$ = - 5A max 160 mΩ
V$_{GS(th)}$ @ - 250 mA	- 1V ÷ - 2.5V	- 2V ÷ - 4V

Table 1: STD28P3LLH6AG and STD15P6F6AG main parameters.

The three different reverse polarity conditions have been tested at the following conditions, respectively:
a) static reverse polarity – a disconnected TCU is connected to reverse polarity for 1 minute
   • T30 = 0 V (not connected device)
   • T30 = -17 V @ 1 min
b) dynamic reverse polarity – a TCU in active operation is connected to reverse polarity for 1 minute
   • T30 = 13.5 V (TCU active)
   • T30 = -17 V @ 1 min
   c) repetitive reverse polarity – **ISO 7637** Pulse 1 conditions are applied to the T30 line as shown in Fig.2

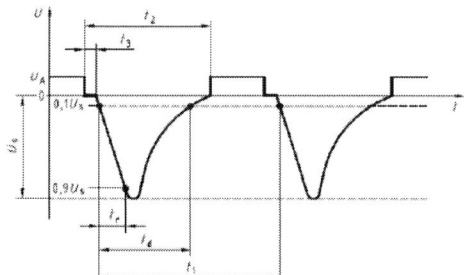

U_A = 13.5 V ± 0.5 V
U_S = -100 V
t_d = 2 ms
t_r = 1 µs
t_1 = 5 s
t_2 = 200 ms
t_3 < 100 µs
number of pulses: 5000

Figure 2: ISO 7637 Pulse 1 conditions applied to test the repetitive reverse polarity protection.

Measured Data for Reverse Polarity Conditions

In order to perform both the static and dynamic test, a specific circuit has been implemented in lab able to produce a positive and negative voltage supply. This voltage supply is applied to the terminals of the testing schematic shown in Fig.1.

The voltage applied for the dynamic test is shown in Fig.3, where the voltage is initially +13.5V and after is reversed at -17.5V.

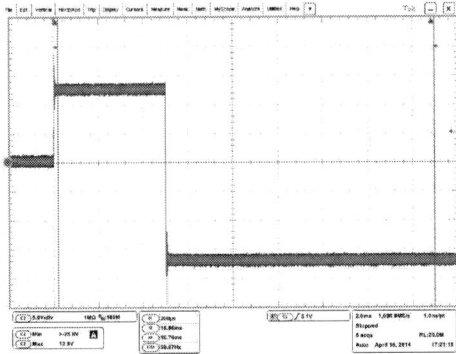

Figure 3: The voltage signal applied for the dynamic test.

The voltage applied for the static test is different because the testing circuit is initially submitted to 0V and after to -17.5V.
In both cases, the reverse polarity voltage has to be sustained by D.U.T. for 60s time.
By using the P-channel power MOSFET STD15P6F6AG, the measured waveforms for both the static and dynamic test are shown in Fig.4 (with a zoom over the red area).

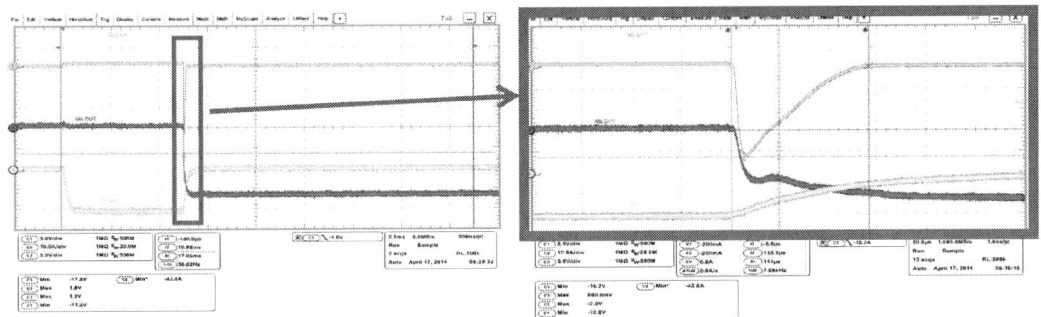

Figure 4: The measured waveforms for both static and dynamic test by using automotive STD15P6F6AG P-channel power MOSFET.

The above measured curves are relevant to I_{DS}, V_{DS}, and V_{GS} of STD15P6F6AG.
When the voltage is reversed, STD15P6F6AG switches off in about 140µs with a -43.5A current peak. After that, the negative voltage is sustained for all the remaining time.
If the P-channel power MOSFET STD28P3LLH6AG is used, the measured waveforms for both the static and dynamic test are shown in Fig. 5 (with a zoom on the switching).

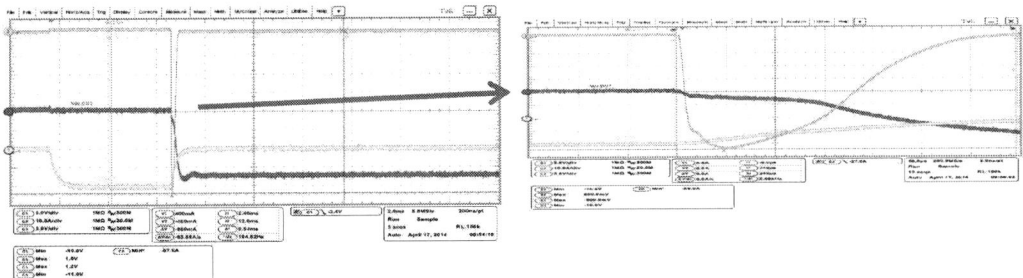

Figure 5: The measured waveforms for both static and dynamic test by using automotive STD28P3LLH6AG P-channel power MOSFET.

© VDE VERLAG GMBH · Berlin · Offenbach

The above measured curves are relevant to I_{DS}, V_{DS}, and V_{GS} of STD28P3LLH6AG.
When the voltage is reversed, STD28P3LLH6AG switches off in about 346us with a -87A current peak. After that, the negative voltage is sustained for all the remaining time.

Repetitive reverse polarity – ISO 7637 Pulse 1 test

In order to check whether the device is able to sustain this kind of stress, a 5000 pulse sequence has been submitted and the main electrical parameter values have been measured before and after the stress sequence.
The provided pulse has to match the below reported conditions (Tab. 2):

Pulse type	Number	U_S (V)	t_d (µs)	t_r (µs)	Generator R (Ω)		Note
					12V	42V	
Pulse 1	5000	-100	2000	1	4	10	t_2 = 200 ms

Table 2: ISO 7637 Pulse 1 conditions.

with the relevant shape (Fig. 6).

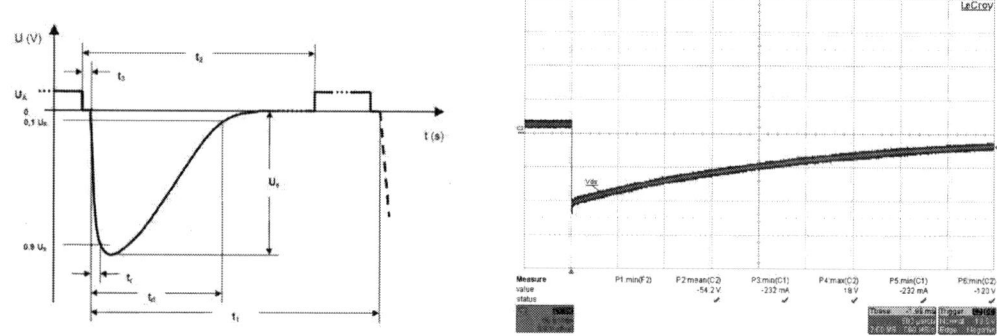

Figure 6: ISO 7637 Pulse 1 voltage and timing details (spec requirement on the left, actual waveform on the right).

The following picture (Fig. 7) shows the measured waveforms relevant to the single pulse event applied to the automotive P-channel MOSFET.
A constant 2.5A current flows through the body-drain diode of the power MOSFET under the forward bias condition.
Once the negative voltage pulse is applied, the device clamps the voltage to the avalanche value which is set by the body-drain diode. This clamping voltage is about 48V when the current peak reaches about 7A.

PCIM Europe 2016, 10 – 12 May 2016, Nuremberg, Germany

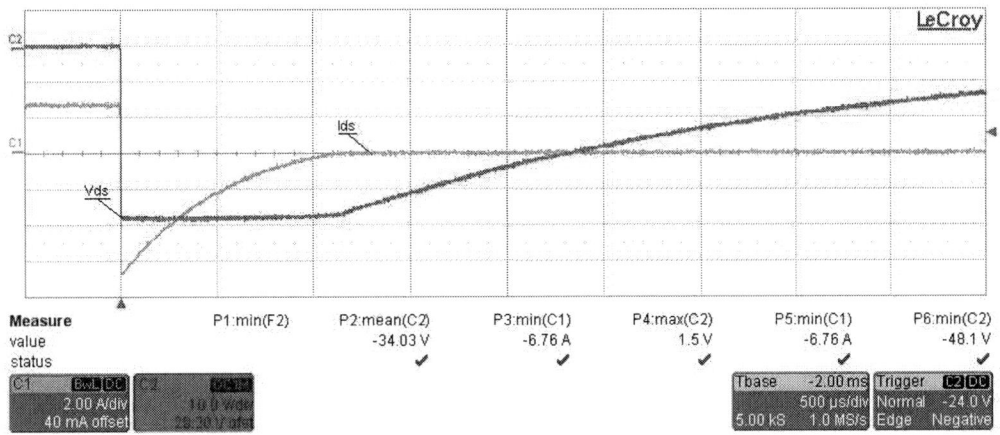

Figure 7: Clamping voltage and reverse current peak of the reverse polarity MOSFET.

This pulse type has been applied 5000 times to the device diode with a time gap of 5s.
In order to verify the device ruggedness, the main electrical parameters have been measured before and after the pulse sequence and no significant variations have been observed (Tab. 3).

	Pre-Test	Post-Test
V_{th} @ 250μA [V]	1.68	1.69
BV_{dss} @ 250μA [V]	41.1	41.35
V_{sd} @ 25mA [mV]	592	600

Table 3: Measured power MOSFET parameters before and after the pulse sequence.

Then the tested automotive P-channel power MOSFET can sustain a 5000 pulse sequence at -100V, according to the **ISO Pulse Test 1** (spec. ISO_7637_2 - 2004 ed.) without any drift on the main electrical values.

Conclusions
it has been verified that the two Automotive-grade P-channel power MOSFETs STD28P3LLH6AG and STD15P6F6AG from STMicroelectronics can easily withstand the negative voltages in static, dynamic and repetitive reverse polarity.
Then the new STMicroelectronics trench P-channel devices are good candidates to be used as reverse polarity protection switches for automotive applications.

References
[1] F. Frisina, "Dispositivi di Potenza a semiconduttore", Edizione DEL FARO, Prima Edizione, Giugno 2013
[2] B. Jayant Baliga, "Fundamentals of Power Semiconductor Devices", Springer Science, 2008
[3] N. Mohan, T. M. Undeland, W. P. Robbins, "Power Electronics Converters, Applications and Design", 2nd edition J. Wiley & Sons NY, 1995
[4] B. Murari, F. Berrotti, G. A. Vignola " Smart Power ICs: Technologies and Applications", 2nd Edition

© VDE VERLAG GMBH · Berlin · Offenbach

PCIM Europe 2016, 10 – 12 May 2016, Nuremberg, Germany

Isolated On-Board DC-DC Converter for Power Distribution Systems in Electric Vehicles

Sven Bolte, Norbert Fröhleke, Joachim Böcker,
Paderborn University, Power Electronics and Electrical Drives, Paderborn, Germany,
bolte@lea.upb.de

Abstract

In this contribution, a 2 kW isolated on-board DC-DC converter for power distribution systems in electric vehicles is proposed. It consists of two power stages which combine a hard-switching buck converter with silicon carbide MOSFETs and a soft-switching resonant converter. A model of the DC-DC converter including conduction and switching losses of semiconductors, copper and core losses of magnetic components is developed and utilized to identify the best design by numeric optimization of number of turns, turns ratio and length of air gap. In order to obtain a high accuracy of loss calculations, the loss model was calibrated by means of calorimetric measurements. Simulation results show an efficiency of about 98.2 % at full load. At light load, the efficiency can be increased by 0.8 % with Phase Shedding.

1. Introduction

Traction batteries with a maximum voltage of about 400 V are typically used as energy storage in electric vehicles (EV). Some appliances in the EV can be adapted to this voltage. This leads to smaller cable cross sections due to the decreased current compared to the traditional 14 V electrical system. However, there is still a demand for a 14 V electrical system for appliances like lighting, electronic control units (ECU) and smaller electrical drives [1]. In this paper, a 2 kW isolated DC-DC converter for an on-board electrical system is proposed (s. Fig. 1).

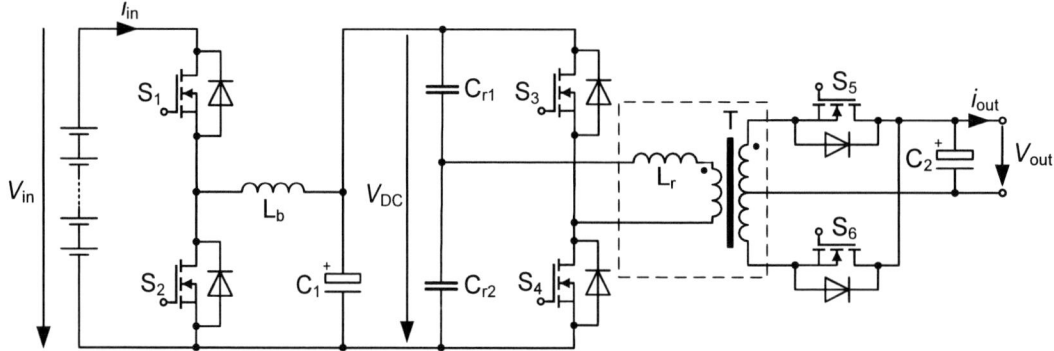

Fig. 1. Schematic of one leg of the DC-DC converter with two power stages

Because of various advantages over other battery technologies like high specific energy and power, lithium-ion batteries are commonly used in EVs. The battery voltage changes in relation to the state of charge. Thus, the converter has to facilitate a wide input voltage range.

© VDE VERLAG GMBH · Berlin · Offenbach

2. Converter Concept

The converter consists of two power stages: A hard-switching buck converter for output voltage regulation and a series resonant converter with constant voltage ratio for galvanic isolation. The buck converter is characterized by low demands on control efforts and a high dynamic performance. Two legs with P_{max} = 1 kW each are connected in parallel for power scaling.

A hard-switching half-bridge with silicon MOSFETs would suffer from high switching losses due to the reverse recovery behavior of the body diodes [2]. Consequently, silicon carbide (SiC) MOSFETs with comparatively low reverse recovery charge are used for the switches S_1, S_2. Hence, the switching frequency can be increased which allows to reduce the volume of the buck inductor L_b. Since the output voltage is controlled by the buck converter, the voltage ratio of the DC-DC stage can be constant. A series resonant converter operating with fixed frequency and duty cycle of 50 % is suitable for this application. It achieves zero-current switching (ZCS) at turn-on and turn-off regardless of input and output variation and load condition [3]. The operation at fixed frequency allows to optimize the resonant tank (L_r, C_{r1}, C_{r2}) where the resonant inductance L_r is represented by the leakage inductance of the transformer T. Thus, the inductance value does not change by saturation of the core material and the power density is increased because no additional inductor is needed. Due to the relatively high output current, a center-tapped secondary winding with two MOSFETs as synchronous rectifiers (S_5, S_6) are deployed which helps to reduce the conduction loss.

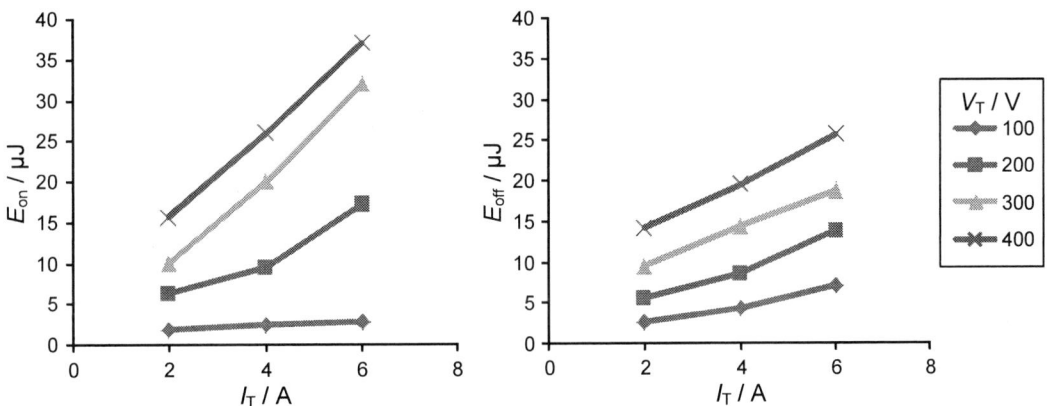

Fig. 2.   Measured switching losses of MOSFETs S_1, S_2 for all relevant voltages and currents

3. Loss Modeling

3.1 Semiconductor Losses

The switching losses of the transistors depend on many parameters like voltage, current, temperature and commutation inductance. As not all operation points are covered in the datasheet, the switching losses of the switches S_1, S_2 were measured with the double-pulse test for all relevant operation points [4]. The switch-on and switch-off losses of the MOSFETs S_1, S_2 are depicted in Fig. 2.

To compute the switching losses it is necessary to determine the voltage and the current at the semiconductors for every switching event which is applied with a numerical calculation script. It turned out that the measurement data can be fitted with reasonable residual errors to a 2-dimensional polynomial of the follow form:

$$E_{sw}(V_T, I_T) = (a_2 V_T^2 + a_1 V_T + a_0)(b_2 I_T^2 + b_1 I_T + b_o) \qquad (1)$$

The conduction losses of diode and transistors are derived from the RMS and mean current values

$$P_{cd,D} = V_f \bar{\imath} + r_d I^2, \quad P_{cd,T} = R_{DSon} I^2 \qquad (2)$$

with the parameters threshold voltage V_f and differential resistance r_d of diodes and channel on-resistance R_{DSon} of the MOSFETs (s. Tab. 1.).

Buck Converter MOSFET S_1, S_2	R_{DSonb} = 65 mΩ
Resonant Converter MOSFET S_3, S_4	R_{DSonr} = 18 mΩ
Synchronous MOSFET S_5, S_6	R_{DSonsr} = 0.5 mΩ
Threshold Voltage Body Diode Synchronous MOSFET S_5, S_6	V_{fsr} = 0.9 V

Tab. 1. On-resistance and threshold voltage of semiconductors

3.2 Magnetic Component Losses

The very low loss angle in the magnetic components causes high demands on the resolution of the measurement equipment since the amplitudes of the measured voltage and current at the inductor stands mostly for reactive power. Hence, a small inaccuracy of the phase measurement results in a huge error of the derived power loss.

Consequently, a calorimetric measurement method [5] is applied to determine the power losses in the boost inductor L_b.

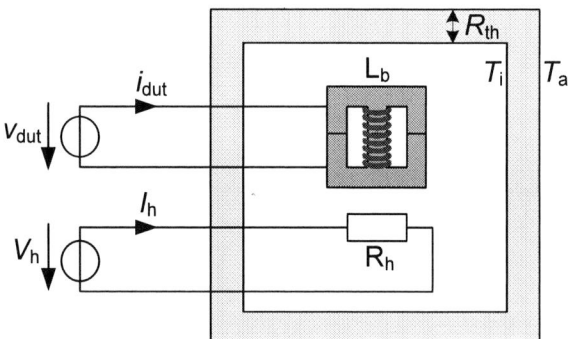

Fig. 3. Calorimetric measurement of inductor losses

The power losses of the inductor which consist of core loss and copper loss are dissipated as heat:

$$P_{th} = P_{fe} + P_{cu} \qquad (3)$$

The inductor is put in a heat-isolated box (s. Fig. 3) with a relatively large thermal resistance R_{th}. The heat flow P_{th} through the walls can be estimated by the temperature difference $\Delta T = T_i - T_a$, where T_a is the ambient temperature and T_i the temperature inside the box:

$$P_{th} = \frac{T_i - T_a}{R_{th}} \tag{4}$$

The inductor is supplied with a triangular-shaped current with a certain amplitude and frequency. After approximately four hours, the temperature difference stays constant and the dissipated power can be calculated. In order to calibrate the calorimeter, an additional heating resistor R_h is mounted inside the box. This allows eliminating errors caused by the thermal resistance, which is not constant due to changes in air humidity and the absolute ambient temperature. Directly after the current through the inductor is switched off, the heating resistor is fed with a DC current in a way that the temperature difference stays constant. The dissipated power of the resistor gives thus the value for the power losses of the inductor.

The core loss P_{fe} of the inductor still has to be separated from the copper loss P_{cu} which is computed with the ohmic resistance of the copper winding and the RMS value of the current through the inductor.

An exemplary inductor was constructed using an ETD29/16/10 core with ferrite material N87. The losses were measured at three different operation points (two values for frequency and peak-to-peak flux density) to extract the Steinmetz parameters of Eq. (5):

$\alpha = 1.68$	$\beta = 2.13$	$k = 5.26$ kW/m³

Tab. 2. Measured Steinmetz parameters of ferrite core inductor L_b

The waveform of the flux density can be separated in piecewise linear intervals. These intervals represent minor loops in the magnetization curve of the core material. The core loss P_{fe} is then calculated utilizing the improved generalized Steinmetz Equation [6],

$$P_{Vfe} = \frac{2}{T_m} \int_0^{T_m/2} k_i \left| \frac{dB}{dt} \right|^\alpha (\Delta B)^{\beta - \alpha} dt \tag{5}$$

where P_{Vfe} is the power density, which still has to be multiplied with the volume V_{fe} of the core and ΔB is the peak-to-peak flux density. Finally, the copper loss P_{cu} is calculated by scaling the resistance of the winding of the exemplary inductor with the new number of turns:

$$R'_{cu}(N') = \frac{N'}{N} R_{cu} \tag{6}$$

Due to the proximity effect, this interpolation is only true in a small range around the original value N.

4. Simulation Results and Optimization

4.1 Switching Frequency and Buck Inductor

In this section, the numeric optimization of the efficiency is described. Parameters for the optimization are the switching frequency f_{swb} of the buck converter the number of turns N and length of air gap l_d of the buck inductor L_b. The simulation parameters are given in Tab. 3. A sweep through switching frequencies in the range of $f_{swb} = 25 \ldots 75$ kHz with 1 kHz step size showed that the maximum efficiency for the selected components appears at $f_{swb} = 50$ kHz.

Input voltage	$V_{in} = 400$ V
DC-link voltage	$V_{DC} = 250$ V
Output voltage	$V_{out} = 14$ V
Output power	$P_{out} = 2$ kW
Switching frequency	$f_{swb} = 25 \ldots 75$ kHz
Number of turns L_b	$N = 50 \ldots 150$
Length of air gap L_b	$l_d = 2.00 \ldots 3.00$ mm

Tab. 3. Simulation Parameters

The flux density in the inductor L_b is limited to $B_{max} = 300$ mT. Under this restriction, an inductor with number of turns $N = 110$ and length of air gap $l_d = 2.90$ mm delivers the lowest losses. In this configuration the inductance is $L_b = 400$ µH. The efficiency η as function of output power P_{out} for the complete range of input voltage V_{in} is depicted in Fig. 4. Hereby, the two legs are operated in parallel also for light load.

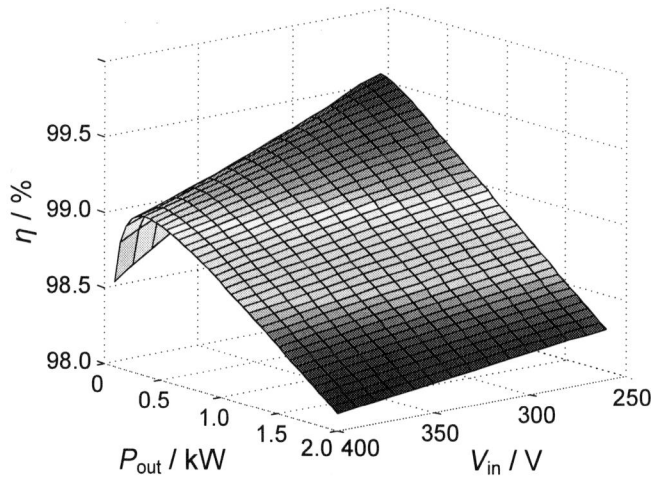

Fig. 4.   Efficiency η versus output power P_{out} and input voltage V_{in}

The converter still reaches efficiencies over 98 % at full load $P_{out} = 2$ kW with a maximum efficiency of about 98.2 % at the lowest input voltage. At light load operation $P_{out} \leq 500$ W even more than 99.0 % can be reached.

© VDE VERLAG GMBH · Berlin · Offenbach

4.2 Power Balancing and Phase Shedding

In order to enlarge the output power range, two legs are connected in parallel. The buck converters are operated with 180° phase shift of the switching signals. With the desired input voltage range and DC-link voltage, the duty-cycle of the buck converters does not fall below 50 %. Otherwise the input current becomes discontinuous which is not desirable in regard to the input filter.

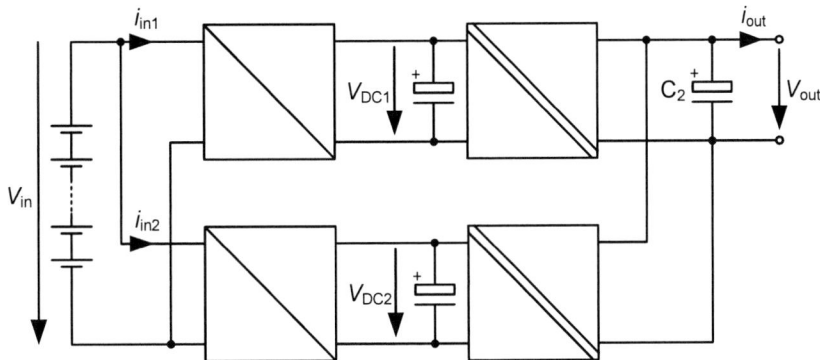

Fig. 5.   Two converter legs in parallel operation

Since the resonant converters are operated with fixed frequency and duty-cycle, device tolerances would lead to unbalanced power distribution. However, that can be compensated by different DC-link voltages V_{DC1} and V_{DC2}. Therefore, the DC-link capacitors cannot be connected in parallel (cf. Fig. 5). The power balancing has then to be applied with the buck converters.

At light load, the efficiency suffers from load-independent losses like charging losses of the power semiconductors. Consequently, only one of the two legs should be active until the intercept point, where parallel operation yields a higher efficiency, is reached.

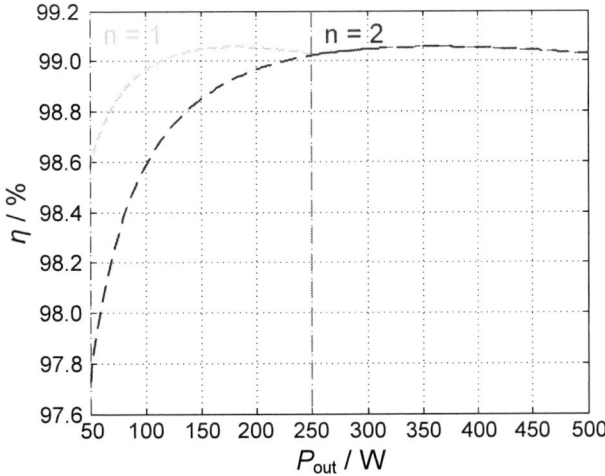

Fig. 6.   Efficiency η at light load P_{out} with Phase Shedding

Fig. 6. shows that the efficiency can be increased by 0.8 % for $n = 1$ active leg at an output power P_{out} = 50 W. If the output power exceeds P_{out} = 250 W parallel operation of the $n = 2$ legs provides higher efficiency.

5. Conclusion

By means of a systematic optimization of an isolated 2 kW DC-DC converter for power distribution systems in electric vehicles, a maximum efficiency of about 98.2 % at full load was achieved. At light load, the efficiency can even be increased by 0.8 % through implementation of Phase Shedding. These results recommend the two-staged converter as promising candidate for future power distribution systems.

6. Acknowledgement

This research and development project is funded by the German Federal Ministry of Education and Research (BMBF) within the Leading-Edge Cluster "Intelligent Technical Systems OstWestfalenLippe" (it's OWL) and managed by the Project Management Agency Karlsruhe (PTKA). The authors are responsible for the contents of this publication.

7. References

[1] W. Schmidt, "DCDC Converter for Hybrid Vehicle Applications," International Conference on Integrated Power Systems (CIPS), 2008.

[2] T. Funaki, M. Matsushita, M. Sasagawa, T. Kimoto, T. Hikihara, "A Study on SiC Devices in Synchronous Rectification of DC-DC Converter," Applied Power Electronics Conference (APEC), 2007.

[3] R. Steigerwald, "A comparison of half-bridge resonant converter topologies," Applied Power Electronics Conference (APEC), 1987.

[4] S. Bolte, J. Baurichter, C. Henkenius, N. Fröhleke, J. Böcker, H. Figge, "Verlustmodellierung und Effizienzoptimierung einer hart schaltenden, netzfreundlichen Pulsgleichrichterstufe (PFC)," Internationaler ETG Kongress, Berlin, 2013.

[5] E. Ritchie, J. Pedersen, F. Blaabjerg, P. Hansen, "Calorimetric measuring systems," IEEE Industry Applications Magazine, 2004.

[6] J. Mühlethaler, J. Kolar, A. Ecklebe, "Loss modeling of inductive components employed in power electronic systems," Power Electronics and ECCE Asia (ICPE ECCE), 2011.

Innovative solution of static and dynamic contactless charging station for electrical vehicles

assoc prof. Nikolay Madzharov,
Technical University – Gabrovo, Bulgaria, madjarov@tugab.bg
mag. eng. Valeri Petkov
Technical University – Gabrovo, Bulgaria, valeri_2202@abv.bg

Abstract

Inductive Power Transfer (IPT) for advanced power distribution system offering significant benefits in modern automation systems. Important research has been done over last two decades in the development of efficient electric vehicles (EV) charging methods [15]. One of the main problems for the mass bring EVs is into use is how far can drive on a single charge and the battery charging duration. This paper deals with the design, development and real tests of 35 kW inductive charging station for EVs. Attention is paid to develop a uniform IPT module for high-power transferring which has to implement two modes of charge- static when EV is stationary and dynamic when the EV is moving. Dynamic charging increasing the travel range of an EV and decrease price and weight of the battery. Additionally, several major aspects were discussed - high peak power transfer capability with high efficiency, realization of battery charging via primary converter control algorithm and last but not least - implementation of the system in real urban environment.

1.   Introduction

The interest in the use of IPT technologies has increased in recent years. Their major areas of applications are: Consumer Electronics, Biomedical, EVs, Industrial, Solar Energy Harvesting, Aerospace [1,2]. A preliminary review of publications in this field showed that 20% are based on contactless charging of EV [14,15]. There are several research groups in various universities [1] and projects [9] which have very good research and real results in this field. Also, some of the major automotive companies GM, Tesla Motors, Nissan, Toyota, etc. have shown interest in wireless charging technology and have announced that in near future will roll out IPT chargers for their EVs [1,5,6,14,15].

The problems about development of IPT technology for EVs can be divided into two main groups - infrastructure integration (charging stations) and mobile integration-EV. Important aspect is the electrical design of IPT module that has strong connection with dimensions of transmitting, receiving coils and the air gap between them. While the distance between the coils is defined by the specific application of the IPT module then their geometrical dimensions are subject of optimization and have great influence on the configuration of the equivalent IPT circuit and the value of the equivalent inductances and matching elements.

This paper aims to present the design and tests of fast inductive charging applications of EVs system concerning both static and on-route charging. Special attention is paid for the IPT module, which has to transfer 35 kW through app. 10cm air gap and should be applicable for static and dynamic charging of EV.

2.   IPT modeling and optimization

The functional chain of static and dynamic contactless charging station for EVs has been carefully scrutinized in order to ensure an optimal, safe and sustainable solution: battery charging, EV performance and safety, EV range, communication between the EV and the charging station, connection of charging stations to the grid (EMC), grid management and energy supply, intelligent coordinated systems. Such an approach for both static and on-route charging infrastructure can greatly help cities and local authorities to deal with the investment problem linked with the integration of heavy cost infrastructures. The investment and the construction work can be distributed along the years and avoid huge one-shot investments. The modular approach allows implementing easier stationary and/or on-route

© VDE VERLAG GMBH · Berlin · Offenbach

charging solutions and also decreases the overall cost of the infrastructure, because it relies on mass-production of a single element. The general concepts are presented in figure 1.

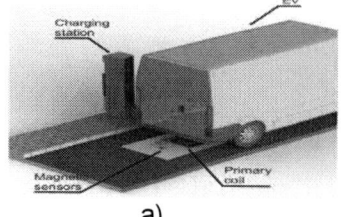 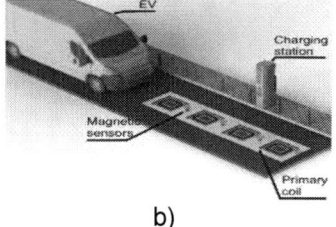

a) b)

Fig. 1. Charging infrastructure – a) static; b) dynamic.

The charging station provides a high frequency voltage to the transmitting coil, which is necessary to transfer sufficiently high power to the vehicle. The high frequency current of the transmitting coil causes a magnetic flux to pass through the surface of the receiving charging coil, installed in the vehicle. In this way is induced an AC voltage, which is rectified in order to charge the vehicle's battery. The sensors gave information for the right positioning of EV and "START" charging process.

The key point in the developed power electronics has electrical and mechanical design of IPT module. Taking into account the main idea - to develop fast inductive charging solutions for EVs, concerning both static and dynamic and after deep analyses of state of the art up-to-date information in this area [1,5,6] the coils with rectangular shape were chosen. Dynamic charging infrastructure incorporates the use of many "segments", with a design similar to that used in static inductive charging (see figure 1 b).

The IPT module is conducted in harmony to "Qi" standard [15] and its draft is sown in figure 2. The main structural components of primary and secondary winding are: HF windings–flat spiral coils, made from LITZ wire; ferrite core, created form ferrite bars and aluminum shield. This construction has 3 main degrees of freedom (X-Y-Z) though a yawing angle can have a certain influence, which depends upon the supporting mechanical system design. For a fixed distance Z a number of degrees of freedom drops to 2. It is important that only few (one or two) degrees of freedom are typically essential for the energy transfer. Other degrees, such as tilting and rolling angles, can play a role of "perturbing" factors with small effect on coupling coefficient of the receiver and transmitter and, respectively, on the power transfer.

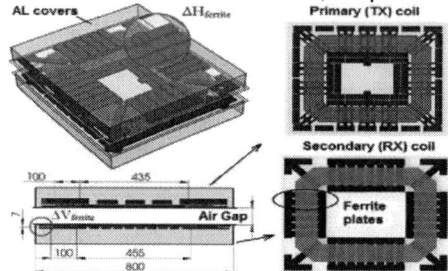

Fig. 2. IPT coils - construction

Several parameters, which are artistic summarized in figure 3, potentially impact the IPT module efficiency and EM field strength. They can be conditionally separated on electricals and mechanicals and their calculation is widely discussed in [3,4,15]. These parameters are analyzed to determine the extent of their impact on system efficiency and EM field strength. The optimization process between them is too complex and depends on a number of restrictive conditions.

The first and the most important is the ratio of magnetic coupling *k* between transmitting and receiving coils. The physical essence is related with the distribution of the magnetic flux between the coils, i.e. what part of the generated magnetic flux by the primary coil induces the voltage in the secondary coil. The generated flux could be calculated by:

$$\psi = \sum_N \Phi_N = N.\Phi = N.\mu_0.\mu_r.H.S \quad , \quad \text{where} \tag{1}$$

Φ_N - magnetic flux for the number of turns, covering the transmitting area *[Wb]*; *N* - number of turns for the area that crosses the magnetic flux ; $\mu_0 = 4.\pi.10^{-7}$; μ_r - relative magnetic permeability; *H* - field intensity, *[A/m]*; *S* - area, which is crossed by the magnetic flux *[m²]*

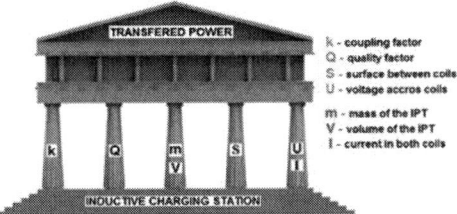

Fig. 3. IPT module – important parameters

Mutual inductance between the receiving and transmitting coils is represented by the expression:

$$M_{TX-RX} = M_{RX-TX} = M = \oint_{S_{RX}} \frac{\Phi_{TX-RX}(I_{TX})}{I_{TX}} dS_{RX} = k\sqrt{L_{TX}.L_{RX}} \tag{2}$$

S_{RX}-area of the receiver side, which is crossed by the magnetic flux [m²]; Φ_{TX-RX}-mutual magnetic flux [Wb]; M_{TX-RX}-mutual inductance [µH]; I_{TX}-current through the transmitting coil (Tx), [A].
From the expressions (1) and (2) can be seen that one of the important factors for *k* and *M* is the area S_{RX}. The developed IPT module with closely spaced rectangular coils, suggests several extreme restrictions. While the inductances can acquire values in a wide range using a different number of turns or supplementary ferrite cores, the coupling coefficient depends mainly on the dimensions of the two coils. One of the main problems associated with its practical determination is the complexity of the magnetic circuit, in this type of electromagnetic devices. At rectangular coils the magnetic coupling coefficient is equal to:

$$k(Z) \approx \left(2n/(4n^2 + 12z^2)\right)^{3/2} , n = (a+b)/2 \tag{3}$$

a, b - external sizes of the transmitting and receiving coils [m]; z - the vertical distance between the two coils, [m].
The inductance of the coil with a rectangular configuration [3,4,14,15], is calculated by the expression:

$$L \approx N^2.\mu_0.\mu_r.T_1 / \pi \qquad\qquad T_1 = t_1 + t_2 + t_3 \tag{4}$$

T_1 is represented as the sum of three main parts:

$$t_1 = -2(a+b) + 2\sqrt{a^2 + b^2} \tag{5}$$

$$t_2 = -a.\ln\left((a+\sqrt{a^2+b^2})/b^2\right) - b.\ln\left((b+\sqrt{a^2+b^2})/a^2\right) \tag{6}$$

$$t_3 = a.\ln(2a/r_w) + b.\ln(2b/r_w) \tag{7}$$

With the expressions (1) to (7) can be calculate all of the electrical parameters of the equivalent "Y" or "Δ" circuit and then proceed to IPT module matching with HF generator from the charging station. Another key point, directly related to the efficiency of the IPT module is coil quality factor Q_{ind}:

$$Q_{ind} = 2.\pi.f.L / R = X_L / R \tag{8}$$

Its physical definition is the ratio of the stored energy of the resonant circuit and the energy consumed by the load each cycle multiplied by 2, i.e. the higher of the quality factor, the greater of the energy that can be transmitted. When the input voltage or input current of the circuit is constant, the reactive power in the circuit is very large if the reactance of the circuit is large, thereby the output power and efficiency of the system reduced. The reactance of the

© VDE VERLAG GMBH · Berlin · Offenbach

circuit would be zero when the circuit is in resonance state. Only, if the transmitting circuit and the reception circuit achieve resonance at the same time, the efficiency of the wireless transmission system is maximum. The ratio of coil loses $-P_{LOSS}$ toward the output power P_{OUT} is equal to:

$$\lambda = P_{LOSS} / P_{OUT} = 2 \cdot \left(1 + \sqrt{1+(kQ_{ind})^2}\right) / (kQ_{ind})^2 \tag{9}$$

From expression (9) can be see that to achieve better efficiency it is necessary $\lambda \ll 1$. The value of losses increases dramatically at $k<0.1$ and $Q_{ind} < 10$. The implemented analysis proved that for the reliable operation of an IPT module it is necessary $k> 0.2$ and Qind>20 ÷ 50, in other words $k.Q_{ind} \gg 10 \div 30$. From this it follows that equation (9) could be simplified using the above mentioned values:

$$\lambda_{IPT-EV} = P_{LOSS} / P_{OUT} \approx 2 \cdot \left(1 + \sqrt{\beta_{IPT}^2}\right) / \beta_{IPT}^2 = 2(1+\beta_{IPT}) / \beta_{IPT}^2 \approx 2 / \beta_{IPT} \quad , \text{ where} \tag{10}$$

$k.Q_{ind} = \beta_{IPT} \qquad \beta_{IPT} \gg 2$.

If the coupling coefficient k has low value (k<0.2) it is possible by optimizing Q_{ind} (increasing the inductance and/or the cross section of wires) to keep the ratio $k.Q_{ind} \gg 10$. Otherwise the IPT module will have bad economic indicators. Therefore, the main goal at the designing of IPT module is achieving optimal coupling coefficient and quality factor. This condition strongly depends of the dimensions of the IPT coils (a in X and b in Y directions) and distance between them Z. If the air gap $z < (a+b)/4$, the transmission efficiency will become greater than 80%, if z/((a+b)/2)<0.25, the transmission efficiency could be close to 90-92 %.

3. Concept of static and dynamic charging infrastructure

As was mention above the main aims of the paper is to present a cost-effective modular infrastructure bought for static and dynamic chagrin. Along with the development of the IPT module, important point is the power circuit of the charging station. The proposed dynamic inductive charging infrastructure contains four successively placed primary coils. The problem is how to implement cost efficient supply of the transmitting coils in dynamic charging. Three methods were investigated. In the first method, each primary coil is supplied by a separate HF inverter and this common module is built in the roadway. HF inverter is feed only by grid AC voltage and control signals, i.e. missing charging station installed near the roadway. In the second method one charging station supplied with HF energy several transmitting coils. The power electronics topology consists of a common dual IGBT module and four dual IGBT modules, each one connected to one of the four primary coils. The common IGBT module is continuously enabled given that the charging station is in operation. Concerning the other four modules, they are enabled successively such that only one module operates each time forming a full bridge inverter with the common module. In the third method power electronics circuit contains full bridge HF inverter with energy dosing (ED) and four electronic switches – figure 4. Each el. switch has only one transistor and one series capacitor [7,8,13]. This capacitor is a part of the primary matching circuit of the coli. Every capacitor from the electronic switches are charged up to U_{MAX} by reverse diode of the transistor when transistor is switched off. Later, when the transistor is switched on the corresponding capacitor will be add to the AC circuit by the transistor and its reverse diode.

The economic cost of the second and third method is considerably decreased, therefore they are used in real tests. Especially, in this paper is presented the test results from the third version – figure 4. From the possible operating modes the most suitable for application is this one, when the working frequency is smaller than the resonance frequency of the alternating circuit. The energy stored in the capacitor Cd in each half-period is given by the following expression:

$$W = 2.C_d.E^2 \tag{11}$$

The power P, transmitted by the power supply, through IPT to the EV, is equal to

$$P = 4E^2 f C_K = E I_{in} = U_{out} I_{out} = const \tag{12}$$

The first conclusion that can be drawn from the expressions for W and P is that, when the working frequency, the input voltage E and the dosing capacitors Cd values are unchanged, the power transmitted to the EV is constant and does not depend on the battery parameters. Supporting constant power means that the output voltage is matched with the battery voltage in order to supply the specified power.

The second special feature of the power supply is obtained by replacing the expression $I_{out} = U_{out} / Z_L$ in (12). After some transformations a correlation giving the connection between the output and the input voltage is obtained.

$$U_{out} = 2.E\sqrt{C_K.Z_T /(2.T)} \tag{13}$$

The supporting and the regulation of the output voltage or current are realized by feedback, which changes the working frequency or/and capacitor C_d value. The analytic dependence of the regulation law can be obtained as the expression for the output power (4) is differentiated in relation to the time t.

$$U_{out} di_{out} + I_{out} du_{out} = 4.C_d.E^2 df \tag{14}$$

Using the expressions $du_{out} = di_{out}.(Z_L //1/\omega.C_F)$ and $I_{out} = U_{out} / Z_L$ is obtained

$$du_{out} df = 2E\sqrt{C_K.Z_K /(2.f)}.1/(1 + \omega.C_F.Z_T / 2) \tag{15}$$

This expression is the operating function of the control system and gives the law, by which the changing frequency, should be modified the load parameters in order to maintain constant output voltage.

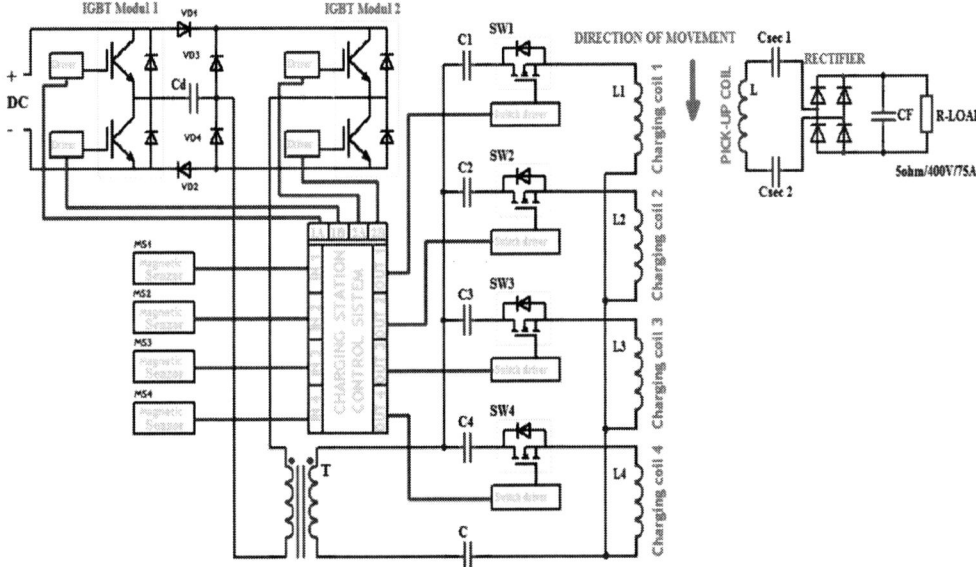

Fig. 4. Dynamic charging infrastructure

In order to ensure the activation of one primary coil at a time, magnetic sensors are placed before each primary coil which is activated as the EV passes over them – fig.1. Summarizing all investigation results [9], magnet sensors type MGT 201 production of company IFM have been chosen as reliable solution. Their position determines maximum specified misalignment between primary and secondary coils. Permanent magnet that creates a magnetic field for switching the sensor is situated in the rear part of the secondary coil. In order to concentrate the magnetic field in the switching area of the sensor, the magnetic circuit was used. Its length determines the maximum horizontal displacement between the two coils.

4. Test results

The developed static and dynamic inductive charging stations were tested. In dynamic infrastructure four primary coils were built in a road and one for static – figure 5. For this reason, each primary coil is integrated in a protective polymer concrete box. The thicknesses of the cover is 4cm in order to satisfy specified distance between primary and secondary coils and also need to have very good mechanical strength to cover requirements of road standard for class B125 (12,5 tone loading). The cover was reinforced without using metal.

a)

b)

Fig. 5. Real tests – a) static and b) dynamic

The main parameters of IPT charging station, including EV's side, is summarized in table 1.

Table1. Main input parameters for the developed contactless charging system – electrical and mechanical.

Charging Station Converter		Secondary side - EV	
Parameter	Value	Parameter	Value
Nominal output power	35kW	Max. dimensions of HF rectifier	500mm x 350mm x 180mm
Peak output power	45kW / 1min	Output parameters regulations	Via primary side converter
Nominal HF current	60A÷90A HF AC	Nominal output voltage	310÷375V DC
Nominal input voltage	3x400V AC/50Hz	Nominal output current	60÷ 90A DC
Efficiency, [%]	≥80%	Weight secondary + HF rectifier	28kg
Tx and Rx coils		**Battery type: LiFePo4 [9]**	
Parameter	Value	Parameter	Value
Max. dimensions of Tx coil	800 x 700 x 90mm	Nominal energy	18,7kWh (67.32Mj)
Max. dimensions of Rx coil	800 x 700 x 60mm	Nominal capacity	56.7Ah
Vertical air gap span	dz=75÷100mm	Nominal capacity	56.7Ah
Horizontal misalignment	dx=dy= ±150mm	Nominal charging current	60÷ 90A DC
Type of positioning	Autonomous system	Max. output current	300A for 10sec

The tests of dynamic charging (fig.5 b) was performed at a vehicle speed of 15 – 20 km/h and the obtained results are presented in Table 2. Although the charging infrastructure is capable of providing 30kW, due to EV battery operational constraints (see table 1), an average power of 23.73kW was achieved during the tests.

This result is particularly important when looking at the potential deployment of this technology, with the ultimate goal of realizing dynamic charging lanes, with speeds higher than what has been tested. When such a technology will be implemented the EVs could undergo a radical breakthrough, since charge-while-driving could reduce the amount of energy to store in the battery pack by design, implying consistent weight and cost reduction, since this component is one of the main drivers for both parameters. At the same time, the driving range could increase dramatically with respect to the 150-200 km that nowadays the majority of EV models in the market propose.

Table 2. Test results of dynamic charging station.

Primary winding number	1	2	3	4
Charging station Input power, W	26179	26719	26179	26179
Battery DC voltage, V	349	352	349	349
Charging DC current, A	68	69	68	68
Charging Battery DC power, W	23732	24288	23732	23732
Efficiency, %	90,65	90,9	90,65	90,65

The same tests were performed at static charging. The purpose of these tests is to demonstrate the maximum output power of the developed charging station at a 80mm distance between the coils. Because of the limit of the maximum capacity of the battery as a load in this tests were used active resistive load with the same impedance as battery has - 5.2 Ohm. The main measured parameters are summarized in Table 3.

Table 3. Test results of static charging station with active load

Primary winding parameters	Static charging Zero misalignment	Static charging Misalignment ± 15cm
Charging station Input power, W	34428	34400
Output DC voltage, V	392,3	372,6
Output DC current, A	80,32	76,35
Output DC power, W	31509	28448
Efficiency, %	91,49	82,7

This test validate and prove, that developed static charging station, without misalignment between coils, can deliver maximum output charging power 31.5kW at 34.4kW input power, that define 91.49% efficiency. In this operating mode with 15 cm horizontal misalignment between primary and secondary coils, at the same input power, the output power is 28.45 kW. Nevertheless, that in charging mode with 15 cm misalignment the coupling factor between coils is strongly worsen, the static charging station operate with 82.7% efficiency, which value can evaluate as a good and prove the right design of IPT module.

During design of primary and secondary coils an important tests are taken for shielding of coils again electromagnetic field exposure. These design measures have respected the proper Directive of EC and its latest version from 2013. It is closely based on the guideless published by ICNIRP-in the case of power frequencies (ICNIRP 2010). In accordance with all these documents a new limit values for electromagnetic field has been defined-"27µT for general public and 100µT for occupational exposure" [11].

Concerning the electromagnetic safety, the magnetic flux density around the coils was measured at power value 24.5kW. In table 4 there are five measurements of electromagnetic field intensity values, which are well below the limits set by the ICNIRP guidelines.

Table 4. EMC measurements

Distance from coil center [cm]	60	70	85	90	110
B-field [µT]	13,8	5,4	3	2,2	2,1

The EMC tests validate the present design of coils shielding again electromagnetic field exposure. The magnetic flux density is further reduced when considering a greater distance from the center of the coil.

Conclusion

By developed charging station the transfer of energy in static and dynamic mode was tested. The results of the tests can be summarized as follows:

- demonstration of contactless transfer of electricity at 7.5 - 10 cm air gap;
- sustainability and duration of operation in static mode and achieving the output power of 31.5kW at 91,49 % efficiency;
- sufficient charging time during in on route charging and obtaining output power of 24.29 kW at 90.9 % efficiency;
- sensorless algorithm for switching On and Off of transmitter windings during charge in motion;
- safe electromagnetic radiation and the lack of heating of the ferromagnetic materials close to the magnetic field;
- ability to peak overload of the system (up to +40%), which can be used under dynamic charging.
- selection of reinforcing non-magnetic material with adequate strength, but with the ability to work in a high magnetic field strength;
- design and manufacture of reinforced concrete housing of transmitting windings having the necessary strength, taking account the real weather urban conditions.

References

[1] Swagath Chopra, „Contactless Power Transfer for Electric Vehicle" Charging Application",Delft University of technology, July 2011

[2] Throngnumchai K., Kai T. and Minagawa Y., „A Study on Reciever Circuit Topology of a Cordless Battery Charger for Electric Vehicles"–Nissan Research Center, Nissan Motor Company.

[3] Danila E., Livinţ G., Lucache D.D.,"Dynamic modelling of supercapacitor using artificial neural network technique",EPE2014-Proceedings of the 2014 International Conference and Exposition on Electrical and Power Engineering,Publisher EEE,2014,DOI:10.1109/ICEPE.2014.6969988, p.642-645.

[4] Danila E., Sticea D., Livint G., Lucache D.D., "Hybrid backup power source behaviour in a microgrid", EPE 2014 - Proceedings of the 2014 International Conference and Exposition on Electrical and Power Engineering, Publisher IEEE, 2014, DOI: 10.1109/ICEPE.2014.6969987, Pages 637-641.

[5] Karfopoulos, E.L. Voumvoulakis, E.M. Hatziargyriou, N.," Energy Management System for fast inductive charging network: The FastInCharge project", Medpower, 2014.

[6] Karakitsios, I. Karfopoulos, E.L.; Hatziargyriou, N., "Static and dynamic fast inductive charging: The FastInCharge project concept", Medpower 2014.

[7] Kraev G., N. Hinov, D. Arnaudov, N. Ranguelov and N. Gradinarov, „Multiphase DC-DC Converter with Improved Characteristics for Charging Supercapacitors and Capacitors with Large Capacitance", Annual Journal of Electronics, V6,B1,TU of Sofia, Faculty of EET, ISSN 1314-0078, pp.128-131, 2012.

[8] Bankov,N.,Al.Vuchev,G.Terziyski. Operating modes of a series-parallel resonant DC/DC converter. – Annual Journal of Electronics, Sofia, 2009, Volume 3, Number 2, ISSN 1313-1842, pp.129-132.

[9] "Innovative fast inductive chargingsolution for electric vehicle" - Smart infrastructures and innovative services for electric vehicles in the urban grid and road environment, part of 7th Framework Program of EU, www.fastincharge.eu.

[10] Nikolay D. Madzharov, RaychoT.Ilarionov, Anton T.Tonchev, "System for Dynamic Inductive Power Transfer", Indian Journal of applied research, Vol. 4, Issue 7, 2014.

[11] International Commission on non-ionizing radiation protection, "ICNIRP Guidelines for Limiting Exposure to Time-Varying Electric, Magnetic and Electromagnetic Fields (1-100kHz)", HEALTH PHYSICS 99(6), 2010, http://www.icnirp.de/documents/LFgdl.pdf.

[12] Nikolay D. Madzharov – Anton T. Tonchev, "Inductive high power transfer technologies for electric vehicles", Journal of ELECTRICAL ENGINEERING, Vol. 65, No. 2, pp. 125–128, 2014.

[13] Madzharov N.D, Ilarionov R.T,Battery Charging station for electromobiles with inverters with energy dosing, PCIM'11,Power Conversion, Nurnberg, Germany, 2011.

[14] Covic G., Elliott J., Raabe S., Boys J.T., "Multiphase Pickups for Large Lateral Tolerance Contactless Power-Transfer Systems", Transactions on industrial electronics,Vol.57, No.5, 2010

[15] Covic G., Boys J.T., "Modern Trends in Inductive Power Transfer for Transportation Applications", IEEE Selected Topics in Power Electronics, Vol.1, No,1, March 2013.

PCIM Europe 2016, 10 – 12 May 2016, Nuremberg, Germany

Combining an External Rotor Motor with Vernier Concept for Drives in Intralogistics

Matthias Theßeling, Lenze SE, Germany, matthias.thesseling@lenze.com
Tao, Liu, Lenze SE, Germany, Tao.Liu.de@lenze.com
Hans-Joachim, Wendt, Lenze Drives GmbH, Germany, hans-joachim.wendt@lenze.com
. . .

Abstract

In this paper the application conveying is analyzed in detail to derive the requirements in the intralogistics industry. Technically high torque density and high overload capability are the most decisive requirements, which were met by the external rotor concept in combination with a vernier motor. Furthermore the external rotor concept has advantages regarding mechanical design. All advantages are explained in detail and the proposed motor is compared to conventional machine designs with the FE-method. The FEM-simulations are verified by measument results.

1. Introduction

When developing drives for special applications, it is necessary to analyze the application in order to derive the requirements. Conveying in the context of the intralogistics industry can be characterized by the load profile of the individual transporting system, for example a roller conveyor (cp. figure 1(a)). When dimensioning the drive the acceleration of the maximum load to nominal speed in a given duration has to be considered. When the acceleration is finished and the conveyor runs at constant speed, the drive has only to provide the power needed to overcome the friction of the bearings in the rollers. A special use case in the context of conveying can be found, when considering an accumulating roller conveyor (cp. figure 1(b)). This system is used to buffer goods during the transporting process. Drives in this application never operate in nominal point, accelerating is the usual operating point. This leads to the disadvantage, that a drive in a roller conveyor works either in partial load operation point (usual conveying) or it accelerates (accumulating roller conveyor) during most of the time in its life cycle. In partial load operation the efficiency is significantly lower than in nominal point [1]. So for the named application a low-power drive, working near to its nominal power with high overload capability is the better solution compared to larger drives. Furthermore the usage of low-power drives leads to the advantage of accurate dimensioning, because scaling of the drive power is easier. The idea of the drive proposed in this paper is to develop a low-power drive with a high overload capability. The acceleration from zero to nominal speed is independent of the load condition and ranges typically from 0.2 s to 0.5 s. This leads to comparable low dynamic requirements for the speed controller. The drive has to be capable of being integrated into a roll to ensure an easy mounting process. In order to reach the requirements of the intralogistics market, the drive has to be very reliable. Of course this goal has to be reached by a good ratio of torque and resources, leading to acceptable pricing for the drive.

© VDE VERLAG GMBH · Berlin · Offenbach

PCIM Europe 2016, 10 – 12 May 2016, Nuremberg, Germany

(a) normal conveying   (b) accumulating roller conveyor[2]

Fig. 1: Use cases for roller conveyors: Horizontal conveing (left) and buffing goods in an accumulating roller conveyor (right). The accumulating roller conveyor transports the good to the next slot, when this is empty.

2. Basic concept

2.1. External Rotor Motor

The drive concept described in this paper is based on an external rotor motor. External rotor motors have several advantages, which make them suitable for applications in the field of the intralogistics industry. One major advantage is the simple and robust mechanical design of external rotor motors: When combining an external rotor motor with a gearless drive structure it is possible to use the rotor directly to transfer the mechanical energy to the rollers of the conveyor. This decreases the number of mechanical parts and therefore increases the reliability. The permanent magnets, which are used for excitation, will not be stressed by centrifugal forces, because they are pressed to the surface of the rotor by the radial force. Considering the inner stator design, it can be seen, that the winding end turns can be arranged radially across the stator, which leads to a compact coil ending and therefore to increased efficiency. Due to the special mechanical design, less volume is necessary for external rotor motors than for internal rotor designs [3], assuming the same air gap for both motors (cp. figure 2). Comparing the inner and outer rotor design, it is obvious, that cooling of the permanent magnets is better, when using an outer rotor design. This leads to higher overload capability, because the peak current can be higher without destroying the magnets. The disadvantage of this kind of motor design is, that the thermal resistance from the stator to the environment is higher than for internal rotor motors leading to less nominal power. This disadvantage is acceptable, because the nominal power is not the decisive factor in the described application in the intralogistics industry. The better ratio of overload capability to resource efficiency is more important.

2.2. Vernier motor

To achieve high torque at low speed mechanical gears are the most common solution. Magnetical gears have been proposed, but not used in commercial drive systems so far due to the poor ratio between volume and torque [4]. The concept of a vernier motor combines the technology of synchronous motors with magnetical gears [5]. This strategy enables the proposed motor to be used as direct drive. To realize a vernier motor, it is necessary to transform the outer pole

© VDE VERLAG GMBH · Berlin · Offenbach

PCIM Europe 2016, 10 – 12 May 2016, Nuremberg, Germany

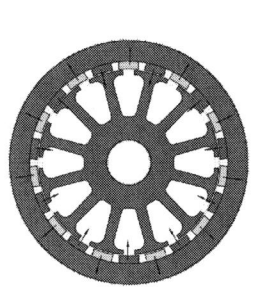
(a) external motor

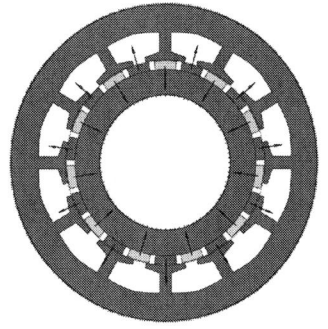
(b) internal motor

Fig. 2: Comparison between external and internal rotor motors. In both designs the air gap diameter, the rotor jaw, stator jaw and the groove area are identical.

pairs to the inner pole pairs by magnetical modulators (cp. equation 1) [6]:

$$Z_s = p \pm Z_r \qquad (1)$$

with:

Z_s = number of magnetical modulators(here 12)

p = number of stator pole pairs (here 1)

Z_r = number of rotor pole pairs (here 11)

The novel idea of the concept proposed in this paper is, that the stator teeth are directly used modulators (cp. figure 3(a)). This leads to a very simple stator geometry and therefore to an easy and fast production process. The mentioned gearing effect can be observed, when analyzing the magnetic field for different rotor angle positions (cp. figure 3(a) and 3(c)). It can be seen, that a rotation of the rotor leads to eleven times faster rotation of the inner stator field. As expected in [7], the direction of the rotor rotation is in opposite direction to the stator rotation.

Comparing a conventional PM-synchronous motor with a vernier motor two major effects are observable in view to the torque/volume ratio:

- The Rotor jaw can be very slim due to the high number of rotor pole pairs (cp. figure 5).

- The induced voltage and therefore the torque are higher for the identical winding number (cp. figure 4). The higher induced voltage can be explained by the high derivation of the fast changing stator field. In this case the torque produced by a vernier motor is three times higher than for comparable conventional motors.

3. Measurement Results

To verify the concept proposed in this paper a first prototype (cp. figure 6) has been realized with a nominal power of 42W. The rotor is made of non-laminated steel, which is no problem regarding to eddy currents during the synchronous motor technology. The active part of the machine is 16mm, which is very short compared to the overall motor length.

The test bench for the external rotor motor has to be able to couple the external rotor with an internal rotor motor used as load machine. This has been realized by clamping the stator only

© VDE VERLAG GMBH · Berlin · Offenbach

PCIM Europe 2016, 10 – 12 May 2016, Nuremberg, Germany

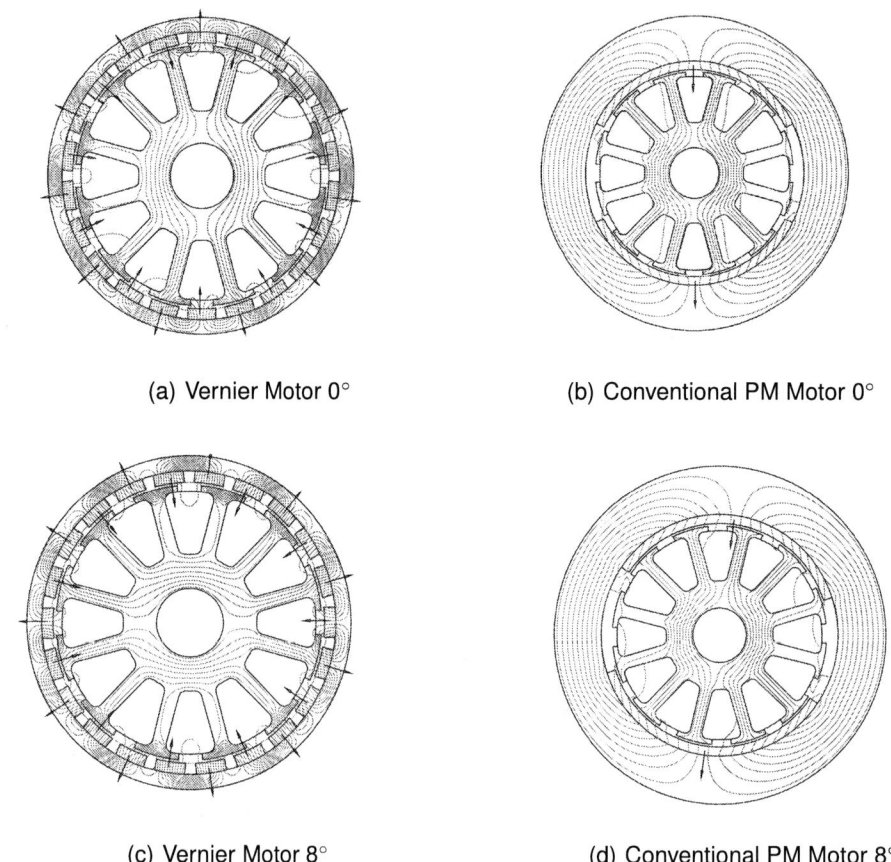

(a) Vernier Motor 0° (b) Conventional PM Motor 0°

(c) Vernier Motor 8° (d) Conventional PM Motor 8°

Fig. 3: Field calculation of the proposed vernier motor for different rotor angles ((a) and (c)). The inner stator field is rotating in the opposite direction with eleven times higher angular frequency than the rotor. For illustration purpose the same field calculation has been performed for a conventional PM Motor showing no magnetical gearing effect ((b) and (d))

at one side and use the other side as output shaft (cp. figure 7(a)). This construction enables also the usage of standard measurement equipment for internal rotors, in particular encoder and torque measurement devices (cp. figure 7(b)).

In a first step the motor was driven by the load machine and the induced voltage was measured. The measurement result shows a good accordance to the FEM-simulation with a discrepancy of 10% (cp. figure 8(a)). The vernier concept and the magnetical gear effect can be observed by the frequency. To verify the overall drive system including all measurement equipment a stepwise load change has been performed (cp. figure 8(b)).

© VDE VERLAG GMBH · Berlin · Offenbach

PCIM Europe 2016, 10 – 12 May 2016, Nuremberg, Germany

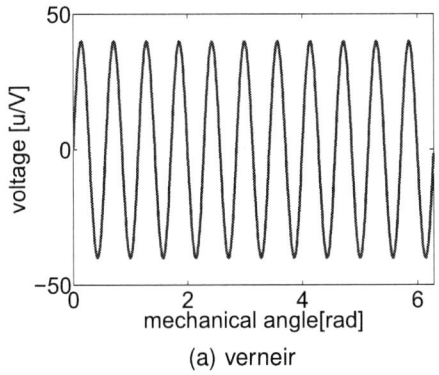

(a) verneir

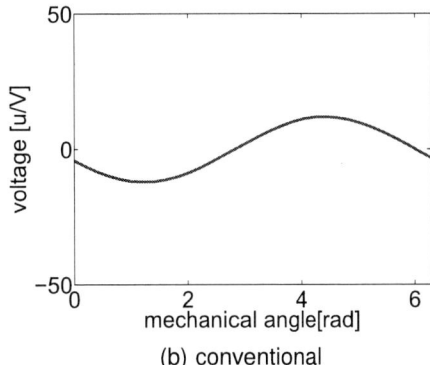

(b) conventional

Fig. 4: Comparison of the induced voltage of vernier motors (a) and conventional PM motors (b) based on FEM-simulation. The mechanical speed is identical in both simulations (750 rpm), so that the magnetical gearing effect can be observed due to the high induced voltage frequency of the vernier motor. The voltage peak is approximately three times higher within the vernier concept leading to three times higher torque.

Fig. 5: Comparison between vernier motor (left) and conventional PM motor (middle and right). In conventional motors the rotor jaw has to be thicker (right), otherwise the rotor would be saturated in an unacceptable way (middle). In all geometry designs the magnet volume is identical.

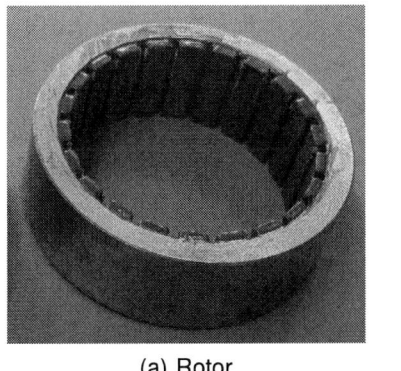

(a) Rotor

(b) Stator

Fig. 6: First prototype of the novel motor concept. The rotor (left) is made of steel and the stator is made of laminated steel. The active part is only 16mm. The overall motor diameter is 50mm which is the most common dimension for rolls in roller conveyors.

© VDE VERLAG GMBH · Berlin · Offenbach

PCIM Europe 2016, 10 – 12 May 2016, Nuremberg, Germany

Load machine   Torque   Encoder   Motor
                measurment   under test

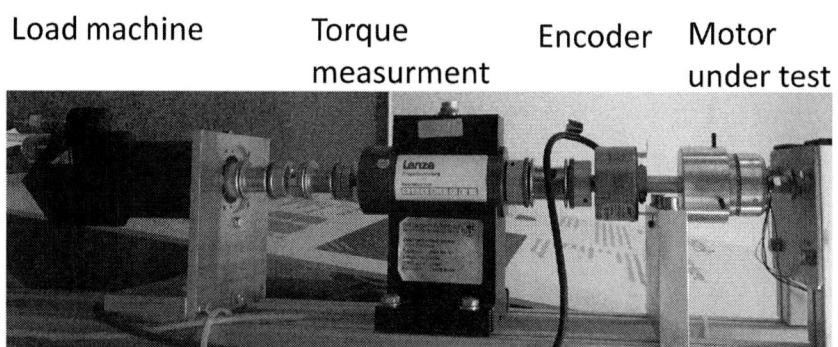

(a) Mechanical
coupling of the
external rotor

(b) Test bench

Fig. 7: For mechanical coupling of the external rotor motor an adapter is necessary for using internal rotor motor test equipment (a). Within the test bench it is possible to couple a load machine and measure torque and the rotor angle (b).

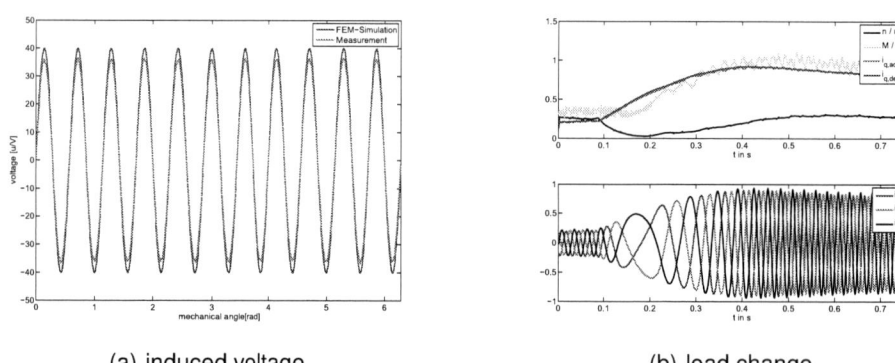

(a) induced voltage

(b) load change

Fig. 8: Comparison between induced voltage from FEM-simulation and measurement. The discrepancy can be explained by the comparable large coil ending dimension. Measurement result of a stepwise load change for zero to nominal torque.

© VDE VERLAG GMBH · Berlin · Offenbach

4. Conclusion

The motor concept proposed in this paper achieves all requirements regarding to overall costs, overload capability and functionality demanded by the intralogistics industry. In particular the demanded high overload capability could be reached by combining the vernier concept with an external rotor motor. This concept has been proven by FEM-simulation. First measurement results show good accordance between simulation and reality. In further steps the magnet geometry will be manipulated in a way that the motor is observable so that an encoderless control also for zero and low speed is possible. This research and development project is funded by the German Federal Ministry of Education and Research (BMBF) within the Framework Concept Research for Tomorrows Production and managed by the Project Management Agency Karlsruhe (PTKA). The author is responsible for the contents of this publication.

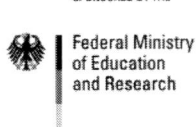

SPONSORED BY THE

Federal Ministry
of Education
and Research

5. References

[1] Hohnsbein T., Klaus U., and Kiel E. Energieeffiziente antriebe in der foerdertechnik. *Proceeding SPS/IPC/Drives Nuermberg,*, 2010.

[2] Transnorm System GmbH. Internal report.

[3] Heinzelmann G., Liebhard G., , and Rosskamp H. Energy efficient drive train for a high-performance battery chain saw. *Electric Drives Production Conference (EDPC)*, pages 140–145, 2011.

[4] Atallah K. and Howe D. A novel high-performance magnetic gear. *IEEE TRANSACTIONS ON MAGNETICS*, 2001.

[5] Ronghai Q. and Dawei L. Jin W. Relationship between magnetic gears and vernier machines. *Electrical Machines and Systems (ICEMS), Beijing*, 2011.

[6] Toba A. and T.A. Lipo. Generic torque-maximizing design methodology of surface permanent-magnet vernier machine. *Industry Applications, IEEE Transactions on (Volume:36 , Issue: 6)*, 2002.

[7] Byungtaek K. and Lipo T. Operation and design principles of a pm vernier motor. *Industry Applications, IEEE Transactions on (Volume:50 , Issue: 6)*, 2014.

PCIM Europe 2016, 10 – 12 May 2016, Nuremberg, Germany

Dynamic Modeling and Optimal Control for Series-parallel Drivetrain Based on Lithium-ion Battery

Tedjani Mesbahi *, Moudrik Meradji**, Gaolin Wang**, Dianguo Xu**,Nassim Rizoug*
(*)S2ET-Ecole Supérieure des Techniques Aéronautiques et de Construction Automobile
Rue Georges Charpak BP76121, 53061 LAVAL Cedex 9 – France
(**)Harbin Institute of Technology, Harbin, China

Abstract

A new modeling approach of the series-parallel drivetrain architecture focusing on optimal control and using Lithium-ion battery is presented. First, the drivetrain structure components are modeled in detail using basic Matlab/Simulink blocks. Then the optimal control of the engine, battery and the interior permanent magnet synchronous machine (IPMSM) is established in the wide speed region. The vehicle dynamic response and the efficiency of the control are verified by simulation.

1. Drivetrain Structure

The series-parallel drivetrain structure represented in fig.1 can be modeled by taking into account the following effects:

- Driver: It will determine the reference torque requested from the motorization by comparing the vehicle speed to the desired one;
- Transmission: This block is representing the link between motorization and road;
- Vehicle dynamics: Through the consideration of three resistant forces applied to the vehicle (rolling resistance, gravitational force and aerodynamic drag force), the resistant torque applied to the transmission is determined.
- Motorization: Representing the different vehicle components and their interactions (Fig.2).

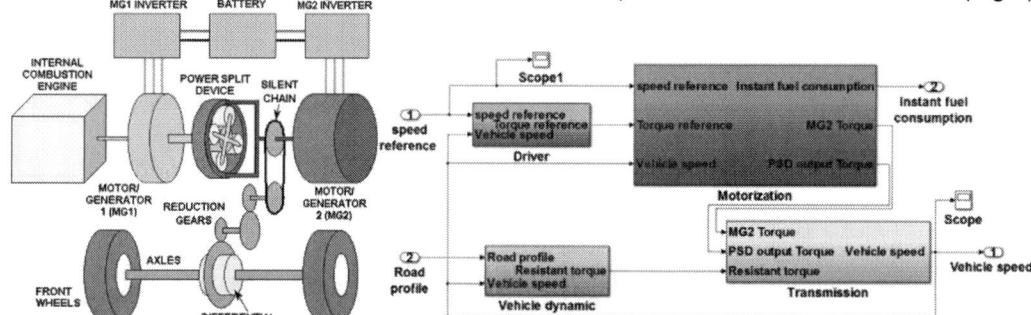

Fig 1:   a) Series-Parallel drive train structure;   b) Simulink model of the vehicle

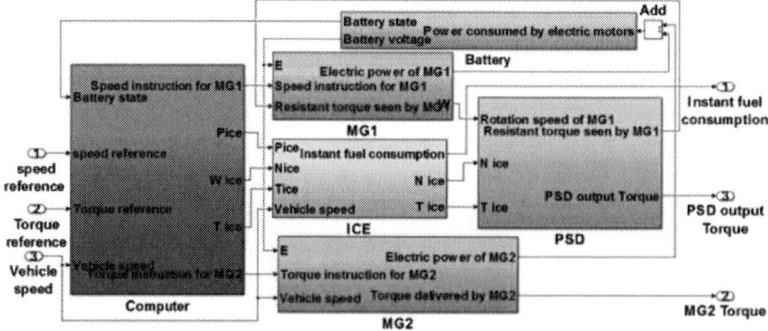

Fig. 2. Motorization Simulink block

In order to understand the adopted optimal control process, Table I. is resuming the most important motorization characteristics [1].

© VDE VERLAG GMBH · Berlin · Offenbach

TABLE I. MOTORIZATION COMPONENTS AND THEIR CHARACHTERISTICS

Component	Characteristics
Battery	Discharge Power ($P_{Bat,dis}$)= 22000W
	Maximum charging power ($P_{Bat,ch}$)= 13680W
	Operating range of the battery depending on state of charge (SOC) : 40-80%
Internal Combustion Engine (ICE)	Rotation speed (ω_{ICE})
	Torque developed by the engine (T_{ICE})
	Optimal Operation Power ($P_{ICE,opt}$)= 25918 W and speed
	Optimal Torque ($T_{ICE,opt}$)=90n.m
	optimal engine speed ($\omega_{ICE,opt}$)= $\frac{P_{ICE,opt}}{T_{ICE,opt}}$=2750rpm
Electric machines : MG1 and MG2	Rotation speed of MG1 (ω_{MG1}) and MG2 (ω_{MG2})
	Resistant torque of MG1 (T_{MG1}) and developed torque by MG2 (T_{MG2})
Power Split Device (PSD)	Epicyclical train ratio (λ)= -2.6
	Rotation speed relation: $\omega_{MG1} = (1-\lambda)\omega_{ICE} + \lambda\,\omega_{MG2}$
	Torque at the output of PSD: $T_{PSD} = \frac{\lambda}{(\lambda-1)}T_{ICE}=-\lambda\,T_{MG1}$

2. Lithium-ion Battery

For traction applications, there are different battery technologies possible: Nickel-Cadmium (Ni-Cd), Nickel-Metal Hydride (Ni-MH), and Lithium-Ion (Li-Ion) [4, 6]. The characteristics of this last technology are high voltage, light mass, low self-discharge and prolonged lifetime [5].

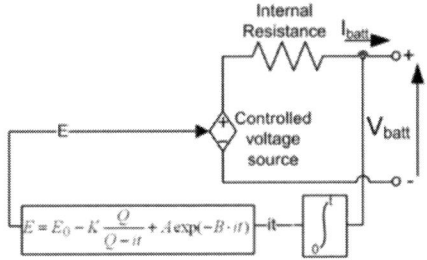

Fig. 3. Non-linear battery model

E=no-load voltage (V)
E_0= battery constant voltage (V)
K= polarisation voltage (V)
Q = battery capacity (Ah)
$\int i\,dt$ = actual battery charge (Ah)
A=exponential zone amplitude (V)
B=exponential zone time constant inverse (Ah)$^{-1}$
V_{bat}= battery voltage (V)
R_{int}=internal resistance (Ω)
i=battery current (A)
The following system has been realized on Matlab Simulink

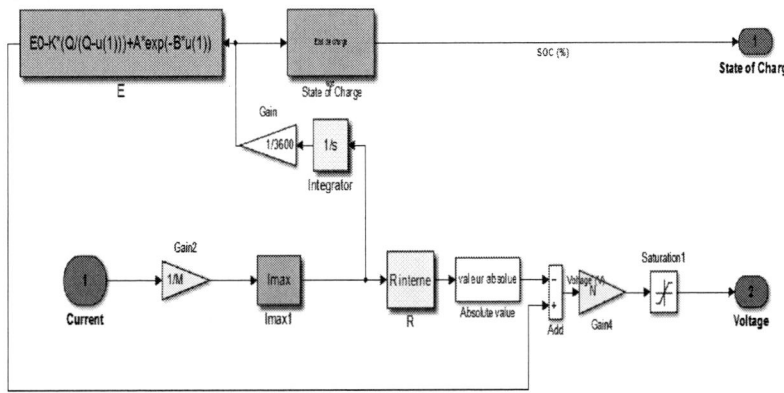

Fig. 4. Li-ion battery Simulink model

© VDE VERLAG GMBH · Berlin · Offenbach

3. Engine Optimal Control

According to the battery SOC and the requested power, the appropriate engine operational state is determined. Table II. is summarizing these states. For instance, the engine is in its optimal operation when the requested power is in the range$[P_{ICE,opt} - P_{Bat,ch}, P_{ICE,opt} + P_{Bat,dis}]$. The excess or lack of engine power is then eliminated, respectively, by charging or discharging the battery. The approach for controlling the engine is simulated using a Stateflow program (Fig. 5) with two levels tests: on the battery SOC and the vehicle power requested.

TABLE II.   ENGINE OPERATIONAL STATE

	REQUESTED POWER (W)			
	$P_{ICE,opt} - P_{Bat,ch}$ 12238W	$P_{Bat,dis}$ 22000W	$P_{ICE,opt}$ 25918W	$P_{ICE,opt} + P_{Bat,dis}$ 47918W
60 (SOC %)	STATE3	STATE 2 "OPTIMAL OPERATION"		STATE4 "DEGRADED OPERATION HIGH POWER DEMAND"
		CHARGING BATTERY		
	STATE3 "DEGRADED LOW POWER DEMAND"	STATE 2 "OPTIMAL OPERATION" $T_{ICE,opt}, \omega_{ICE,opt}$		STATE4
80		CHARGING BATTERY	DISCHARGE	
	STATE1 "FULL ELECTRIC" ENGINE STOPPED: $T_{ICE} = \omega_{ICE} = 0$	STATE 2 "OPTIMAL OPERATION"		STATE4
	DISHARGING BATTERY			

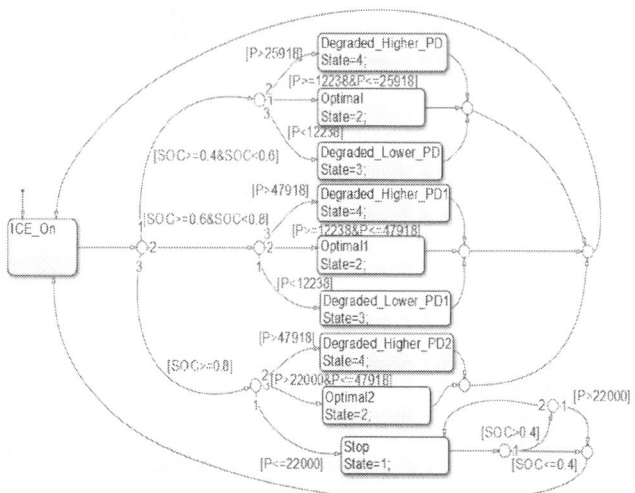

Fig. 5. Stateflow programming for engine control

4. Electric Machines Wide Speed Control

The electric machines are interior permanent magnet machines (IPMSM). Thus, the wide speed control is considered using Maximum Torque Per Ampere (MTPA) for base speed and flux-weakening above [2, 3]. MG1 is speed controlled while MG2 is controlled by torque. The references are directly obtained using the PSD relations present in Table I. The different control process equations are summarized into the following table:

TABLE IV.   MOTORIZATION COPMONENTS AND THEIR CHARACHTERISTICS

Control name	Equations
Maximum torque per Ampere (MTPA)	$$i_d^s = \frac{\psi_m - \sqrt{\psi_m^2 + 4(L_q - L_d)^2 i_q^2}}{2(L_q - L_d)}$$

Constant power (CP)	$$i_d^s = -\frac{\psi_m}{L_d} + \frac{1}{L_d}\sqrt{\frac{V_{max}^2}{\omega_e^2} - (L_q i_q^s)^2}$$
Maximum power per voltage (MPPV)	$$i_d^s = -\frac{\psi_m}{L_d} + \Delta i_d^s \text{ With} \quad \Delta i_d^s = \frac{L_q\psi_m - \sqrt{(L_q\psi_m)^2 + 8(L_q-L_d)^2\left(\frac{V_{max}}{\omega_e}\right)^2}}{4L_d(L_q-L_d)}$$

where,

v_d^s and v_q^s : d-axis and q-axis stator voltages;

i_d^s and i_q^s : d-axis and q-axis stator currents;

r_s : stator resistance;

L_d and L_q : d-axis and q-axis stator inductances;

ω_e: electrical angular velocity;

ψ_m: permanent magnet flux linkage;

p : derivative operator.

5. Simulation Results

In order to verify the performance of the proposed model and its control, two tests have been proposed with the following vehicle usage profile:

- Fixed speed reference of 90km/h on a flat road: Results are present in Fig. 3.a;
- Speed reference: 80km/h then a deceleration to 40km/h followed by an acceleration to 140km/h. A 5% slope is passed from 80 to 150s. The model response for these conditions is represented in Fig. 3.b.

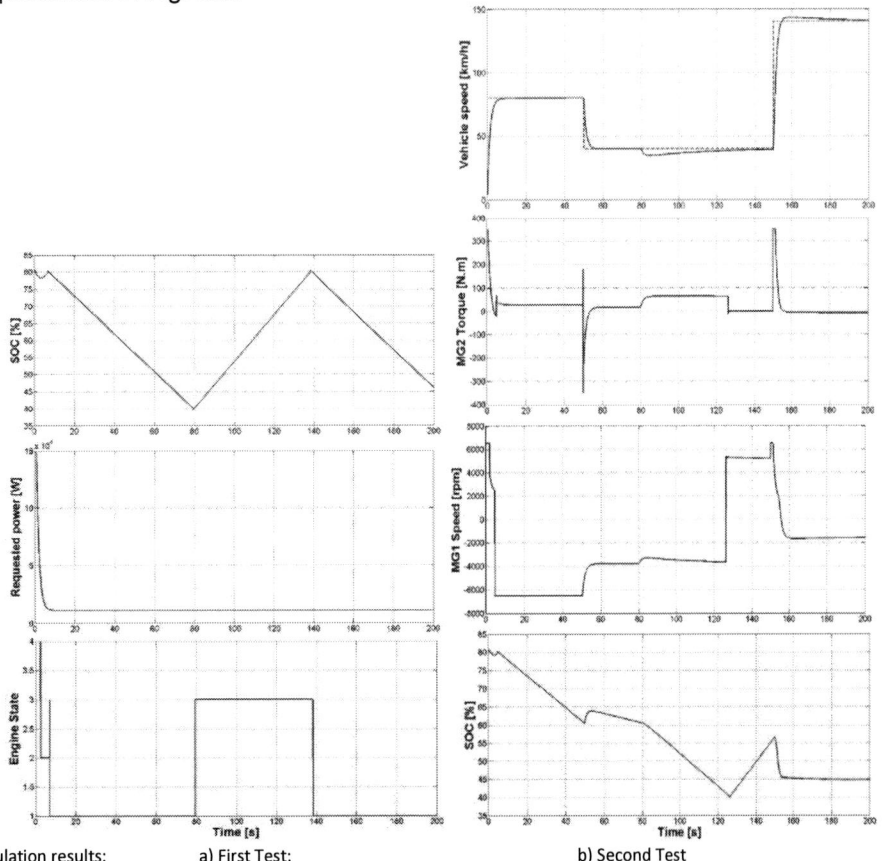

Fig. 3. Simulation results: a) First Test;   b) Second Test

© VDE VERLAG GMBH · Berlin · Offenbach

It is observed that the battery SOC is in the safe operating range (40-80%). Furthermore, the engine state is in accordance with the expected results such as the State transitions (4 to 2) and (2 to 3) observed when the requested power is crossing the steps 47918W and 12238W, respectively. After that, the engine control will keep changing vehicle mode from full electric to charging battery mode.

In the second test, it is observed that the model speed response is coherent with the reference and has good stability. Torque characteristic of MG2 is pointing out its assistance to the engine especially during speed change. Furthermore, it is possible to observe the change of mode in 126s when MG2 became a generator and MG1 a motor to charge the battery. Thus, MG1 and MG2 optimal speed and torque control is verified in the wide speed range.

6. Summary

The series-parallel drivetrain architecture has been modeled and the proposed optimal control process of both engine and electric machines has been verified by simulation in the wide range of operating points. The modeling and simulation results show a good performance of this control and confirm their correctness. Future work is currently undergoing to improve the proposed control strategy and carry out its experimental validation on the real hybrid vehicle or HIL test bench.

References

[1] Ghayebloo, Abbas, and Ahmad Radan. "Superiority of Dual Mechanical Ports machine based structure for Series-Parallel Hybrid Electric Vehicle Applications." (2013);

[2] XU, Z., DATTA, R., YIN, G. X., et al. Optimal stator-current trajectory control of an IPM synchronous machine for hybrid electric vehicles. In: Power Electronics and ECCE Asia (ICPE & ECCE), 2011 IEEE 8th International Conference on. IEEE, 2011. p. 1958-1963;

[3] K. Hoang, J. Wang, M. Cyriacks, A. Melkonyan, and K. Kriegel, "Feedforward torque control of interior permanent magnet brushless AC drive fortractionapplications," inProc. Int. Electric Mach. Drives Conf.,2013, pp. 152–159.

[4] Mesbahi, T., Rizoug, N., Bartholomeus, P. and Le Moigne, P., 2013, October. Li-ion battery emulator for electric vehicle applications. In Vehicle Power and Propulsion Conference (VPPC), 2013 IEEE (pp. 1-8). IEEE.

[5] Yiu, K., 2011, June. Battery technologies for electric vehicles and other green industrial projects. In Power Electronics Systems and Applications (PESA), 2011 4th International Conference on (pp. 1-2). IEEE.

[6] Mesbahi.T, Khenfri.F, Rizoug.N, Chaaban.K, Bartholomeüs.P, and Le Moigne.P, "Dynamical modeling of Li-ion batteries for electric vehicle applications based on hybrid Particle Swarm–Nelder–Mead (PSO–NM) optimization algorithm," Electr. Power Syst. Res., vol. 131, pp. 195–204, Feb. 2016.

PCIM Europe 2016, 10 – 12 May 2016, Nuremberg, Germany

Smart Diode and 4-Switch Buck-Boost Provide Ultra High Efficiency, Compact Solution for 12-V Automotive Battery Rail.

Vijay, Choudhary, Texas Instruments, USA, Vijay.Choudhary@ti.com

Mathew, Jacob, Texas Instruments, USA, Mathew.Jacob@ti.com

Abstract

DC-dc converters operating off the 12-V auto rail have to deal with reverse battery condition as well as wide input voltage range. The traditional solutions include PFETs for reverse battery and a buck-boost solution utilizing two stages or multiple windings. Both of these result in higher losses and bulky solution size. This paper presents and approach consisting of smart diode and 4-switch single-inductor buck-boost converter. The smart diode provides an efficient diode or PFET replacement while the single inductor buck-boost provides ultra-high efficiency buck-boost conversion stage in a small solution size.

1. Introduction

The number of applications operating from automotive 12-V battery rail is increasing with each new generation of vehicles. The increased emphasis on high efficiency and better utilization of space places stringent demands on the dc-dc conversion stage. There are two sources of loss in power conversion. The first is in the reverse battery protection that is needed to protect against accidental reversal of positive and negative polarity connections, for example, during a jump start. The second part is the dc-dc converter itself which maintains a regulated output voltage as the 12-V rail sees wide variations from somewhere around 4.5V on the low end (cranking) to 42V on the high end (load dump).

Traditionally the reverse battery protection is done either using a diode for low power applications or a PFET for medium to high power levels. Diodes are limited to lower current levels because of high forward voltage drop. P-channel MOSFETs are preferred to n-channel

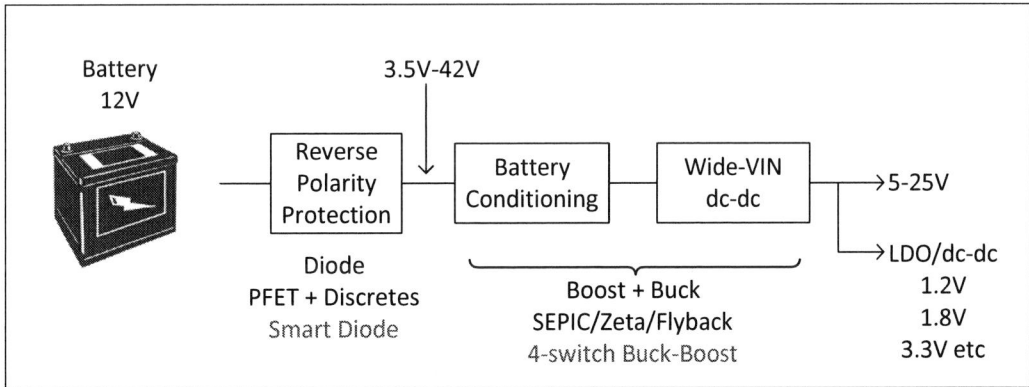

Fig. 1. Off-battery dc-dc conversion stage

MOSFETs because they are simple to drive and do not require a higher voltage than the input supply to turn them on. PFETs typically have higher conduction losses than the n-channel MOSFETs which is undesirable. The smart diode presented in this paper (section 2) utilizes an n-channel MOSFET with gate drive integrated on chip that provides the performance of an n-channel MOSFET and the simplicity of a diode.

© VDE VERLAG GMBH · Berlin · Offenbach

To be able to deal with wide input voltage (V_{IN}) range of the 12-V auto rail the dc-dc conversion stage often needs to be able to both buck and boost. Practical implementations of buck-boost converters have often involved a two stage solution (boost + buck) or higher order converters (SEPIC, Zeta) with multiple windings that are difficult to design, harder to compensate, and have lower efficiency. This paper presents a 4-switch buck-boost converter (section 3) that avoids the double conversion inherent in cascaded approach, and also avoids the higher winding and rms losses in multi-winding topologies.

A complete dc-dc conversion stage consisting of single inductor 4-switch buck-boost with a smart diode protection stage was built to demonstrate the efficiency and size advantages of these new building blocks (section 4).

2. Smart Diode

Smart diode (LM74610-Q1) provides protection for the system against reverse polarity protection and blocks reverse current flowing back. It mimics a diode's behavior without the associated power loss [9].

A p-channel MOSFET has been conventionally used for high current applications since it does not need a charge pump. However the R_{DSON} of the p-channel MOSFET gets much higher at low input voltages (Cold crank or start stop voltage dips) and it does not prevent reverse current from flowing back into the input. It also needs additional circuitry and signals to turn it off to reduce quiescent current (I_Q). Anytime the input dips relative to the output of the PMOS solution, reverse current flows back into the input. This results in the output capacitor getting robbed out of charge in situations like voltage interruptions, cold crank or start stop.

A typical application circuit for automotive front end systems is shown in Fig 2. LM74610-Q1 (smart diode controller) along with an n-channel MOSFET and the charge pump capacitor makes up the smart diode solution.

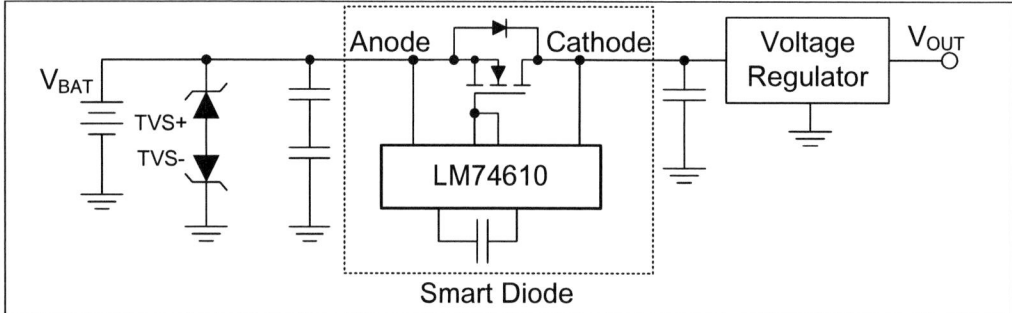

Fig. 2. Automotive front end system with smart diode building block

Because the smart diode (LM76410-Q1) does not need any control signals, it mimics a two-terminal device and is not ground-referenced. "Hook it up like a diode and it acts like a MOSFET" could be a marketing slogan.

The key advantage of not being ground-referenced is that the smart diode solution consumes zero quiescent current. When applying reverse voltage, the body diode of the MOSFET is not turned on, so it does not turn on the smart diode controller. When applying a normal polarity voltage, the body diode conducts, and the internal charge-pump circuitry starts up with the diode voltage and generates voltage for the MOSFET to turn on. Periodically (at 98% duty cycle), the FET turns off to replenish the charge pump. A protected circuit would see a 0.6V drop at periodic intervals at that 98% duty cycle. With a 2.2µF cap used as the charge-pump capacitor, the MOSFET turns off for approximately 25ms once every 2.5 seconds. Fig 3 shows the block diagram of the LM74610-Q1.

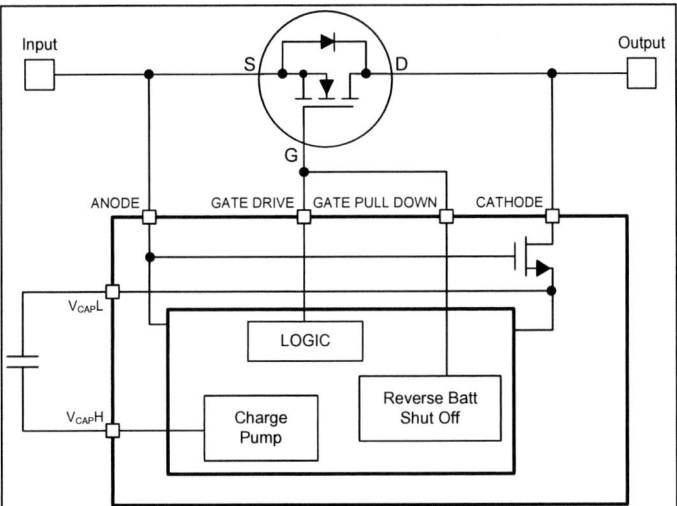

Fig 3.   Smart diode block diagram

One of the inherent properties of a diode is that it blocks reverse voltage and does not allow reverse current to flow. The smart diode controller mimics this behavior and has very fast turnoff during reverse currents (typically 2µs, see fig 4). This is an important feature to pass automotive testing as per ISO 7637. The specifications call for electronic modules to be subjected to negative voltage pulses dynamically while operating at 12V.

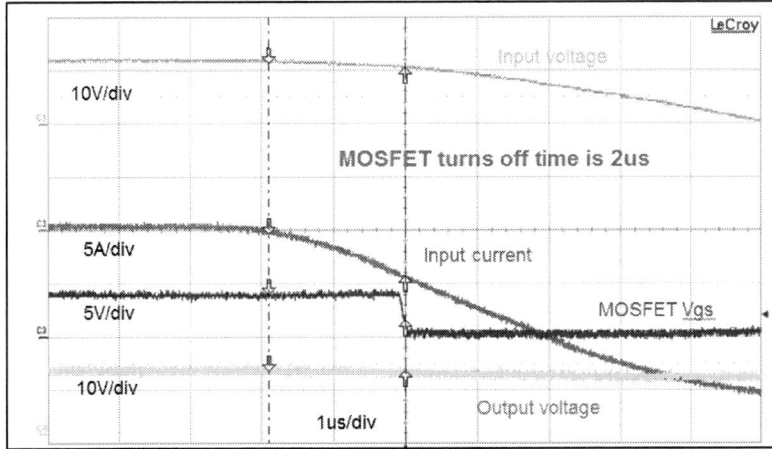

Fig 4.   Reponse time of the smart diode.

A slow response to the reverse voltage can cause the output to go negative or discharge significantly during the pulse. The impact of negative-going output is typically a dead board. The impact of discharging the caps is to interrupt the electronics or use bigger bulk capacitors. Lab tests have also verified that the smart diode controller is much faster than a PMOS based scheme. Fig 4 shows the fast acting response of smart diode to reverse polarity enabling it to meet an ISO 7637 pulse 1 with a small 4.7uF output capacitor as shown in fig 5.

© VDE VERLAG GMBH · Berlin · Offenbach

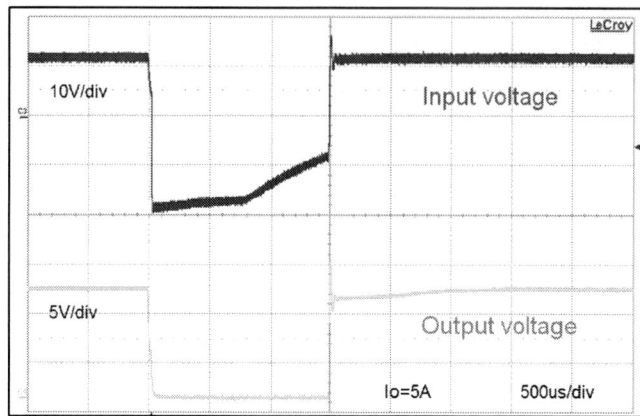

Fig 5. Smart diode output voltage response to ISO 7637 pulse 1 input

3. 4-Switch Buck-Boost Converter

Numerus specifications including ISO-7637 and many OEM specific standards [1-6] provide guidelines and test conditions to simulate the supply variations in the 12-V battery bus. The dc-dc converters designed to operate from a 12-V battery rail must take into account these supply transients. Typically auto 12-V rail can vary from 4.5V or lower (cold-crank) to 42V (load-dump). A wide-VIN buck-boost power conversion stage is often needed to survive the transients at the high end and at the same time provide power at the low end of input voltage.

A comparative study of buck-boost converters is presented in reference [8]. A 4-switch buck-boost converter provides compact and highly efficient buck-boost conversion by avoiding double conversion or multi-winding transformers or inductors.

3.1. 4-switch buck-boost operation

Fig 6 shows a simplified implementation of a 4-switch buck-boost. It consists of a buck switching leg, a boost switching leg, and an inductor connecting the two switch nodes. When input voltage sufficiently higher than the output voltage, the buck leg switches in manner identical to a buck converter. The boost high side switch (QH2) is in pass through mode. This mode of operation is similar to a buck converter (Fig 7-8).

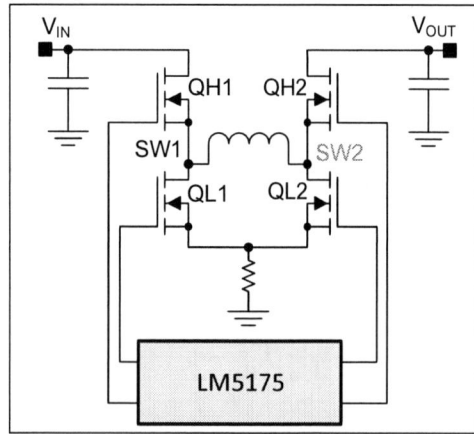

Fig. 6. Buck-Boost power stage with 4-switch buck-boost controller

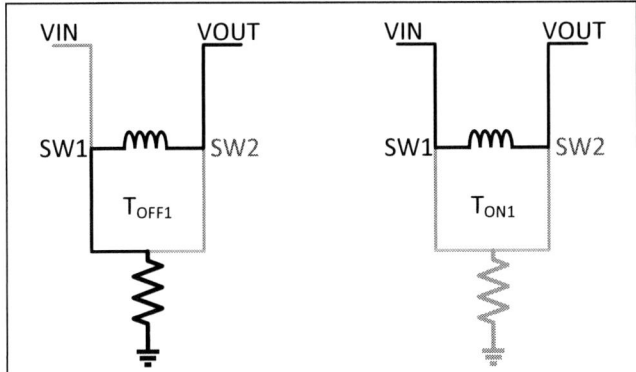

Fig. 7. Buck switching cycle sub-intervals.

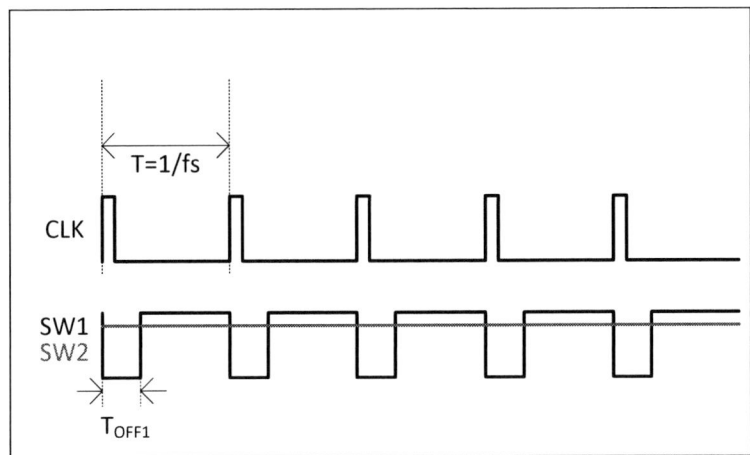

Fig. 8. Buck switching cycle switch node waveforms

When the input voltage is sufficiently lower than the output voltage, the boost leg (QL2, QH2) switches and the buck high side switch (QH1) is in pass through mode. This mode of operation is similar to a boost converter (Fig 9-10).

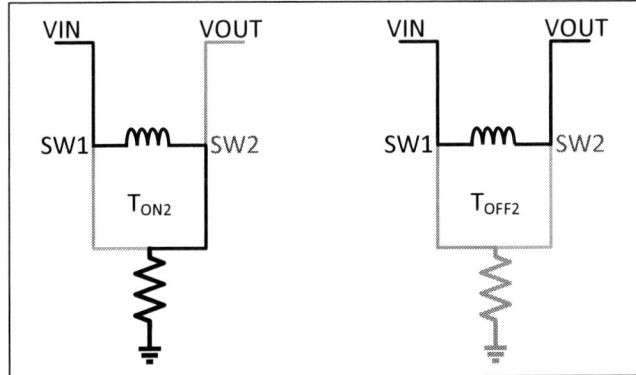

Fig 9. Boost switching cycle sub-intervals.

PCIM Europe 2016, 10 – 12 May 2016, Nuremberg, Germany

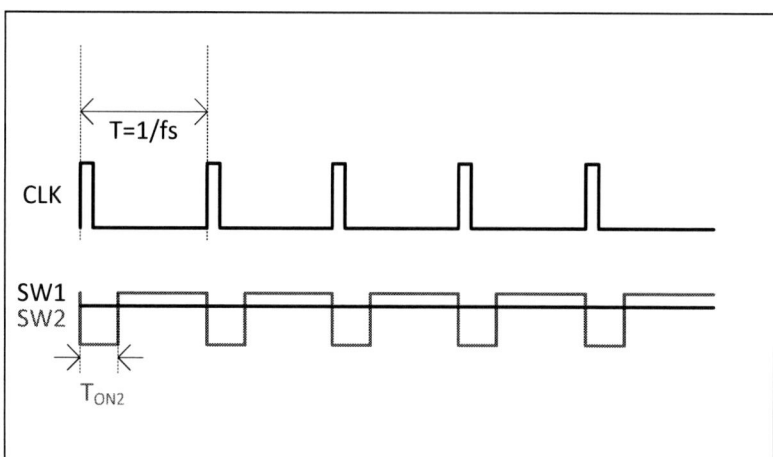

Fig. 10. Boost switching cycle switch node waveforms

When V_{IN} is close to V_{OUT}, either buck or boost mode alone cannot regulate the output voltage in an effective way because of minimum pulse width requirements on the switch nodes. For $V_{IN} \sim V_{OUT}$ condition, the 4-switch buck-boost alternates between buck and boost pulses, effectively operating in a 4-switch mode. This allows smooth regulation of output voltage even at $V_{IN} = V_{OUT}$ condition.

3.2. Higher Efficiency

The 4-switch buck-boost converter has significantly higher efficiency compared to a two stage or a multi-winging approach. The reasons are discussed in detail in [7] and can be attributed to the following:
1. Single switching event per cycle (vs cascade approach)
2. Utilization of lower voltage switches for given V_{IN} and V_{OUT} (vs. SEPIC or Flyback)
3. Lower circulating currents in boost mode (vs. all SEPIC or Flyback).

The combination of fewer switching event, lower switch voltages, and lower circulating currents result in 4-switch buck-boost converters maintain high efficiency even at higher power levels.

4. Results for Smart Diode and 4-Switch Buck-Boost Solution

4.1. Smart Diode and 4-Switch Buck-Boost

The proposed high efficiency system with 4-switch buck-boost converter and a smart diode front end for reverse polarity connection is shown in Fig 11. The smart diode protects the downstream converter circuit from reverse polarity voltages while allowing the use of an efficient N-channel MOSFET. The 4-switch switch buck-boost provides a compact, highly efficient dc-dc stage with regulated output rail from widely varying automotive 12-V battery rail.

4.2. Efficiency

As discussed in section 3.2, the 4-switch buck-boost topology provides higher efficiency than alternate approaches. The efficiency of the 4-switch buck-boost converter is shown in Fig 12. For a nominal V_{IN} of 12V, the converter has a peak efficiency of greater than 98%.

© VDE VERLAG GMBH · Berlin · Offenbach

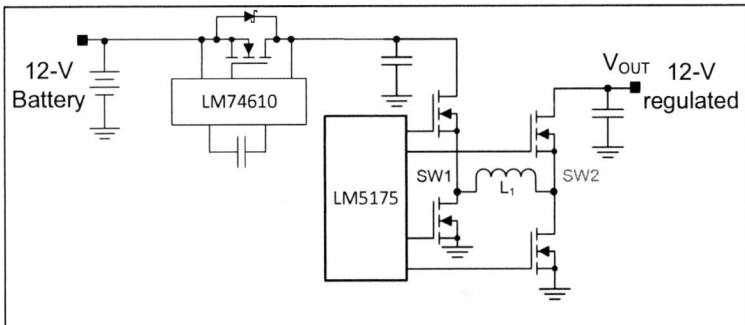

Fig 11. An ultra-high efficiency compact power stage for automotive 12-V rail consisting of a smart diode and a 4-switch buck-boost

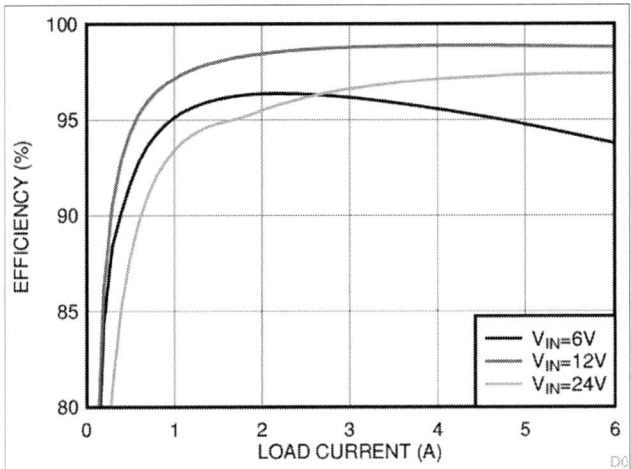

Fig 12. Efficiency of the 4-switch buck-boost converter (V_{OUT} = 12 V, f_{sw} = 300 kHz)

4.3.   Reverse Battery Protection

Fig 13 shows the operation of the system with smart diode in reverse battery connection. Even as the input voltage (V_{IN}) changes from 12 V to -12 V, the smart diode prevents the downstream converter from the negative potential. The input and output voltage of the buck-boost converter is limited to a minimum of 0V.

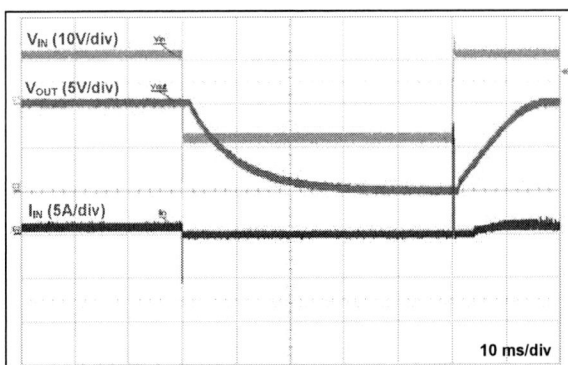

Fig 13. Reverse battery operation of smart diode + buck-boost system.

4.4 Cold-Crank Performance

Fig 14 shows the cold-crank response on the combined system consisting of smart diode and 4-switch buck-boost.

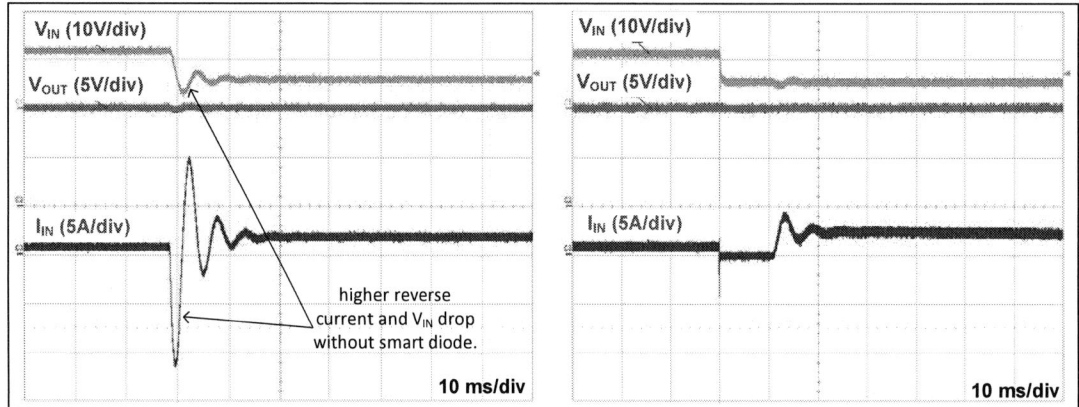

Fig 14. Output voltage regulation and input reverse current of 4-switch buck-boost without (left) and with (right) smart diode as the input voltage falls from 12 V to 5 V in a simulated cold-crank condition.

5. Conclusion

The paper presented a front end power stage building block operating from automotive 12-V rail consisting of smart diode and 4-switch buck-boost. The smart diode combines the simplicity of a diode for reverse polarity protection with the efficiency of an N-channel MOSFET. The 4-switch buck-boost provides regulated output voltage from 12-V battery rail in the presence of wide voltage variations from cold-crank to load dump while achieving higher efficiency and smaller solution size than existing buck-boost solutions.

References:

[1] ISO 7637-2: 2011 Electrical disturbances from conduction and coupling – Part 2: Electrical transient conduction along supply lines only

[2] ISO 16750-2: 2012: Road vehicles – Environmental conditions and testing for electrical and electronic equipment – Part 2: Electrical loads

[3] BMW Group Standard GS 95024-2-1 and GS 95024-2-2: Electrical and electronic components in motor vehicles – Electrical requirements and tests

[4] DaimlerChrysler Corporation Performance Standard PF-10541: Electrical specifications for electrical and electronic modules and motors – 2004 E/E architecture

[5] GM General Specification All Vehicle GMW3100: General Specification for Electrical/Electronic Components and Subsystems; Electromagnetic Compatibility

[6] Volkswagen AG Group Standard VW 80101: Electrical and Electronic Assemblies in Motor Vehicles: General Test Conditions

[7] Four-switch buck-boost controller delivers high power and efficiency (https://e2e.ti.com/blogs_/b/powerhouse/)

[8] V. Choudhary, "Selecting the right buck-boost converter for wide-VIN rails," Electronic Products and Technology, June 2015

(http://www.ept.ca/features/selecting-the-right-buck-boost-converter-for-wide-vin-rails/)

[9] Reverse-polarity protection comparison: diode vs. PFET vs. a smart diode solution.
http://e2e.ti.com/blogs_/b/behind_the_wheel/archive/2015/12/21/shoot-out-between-diode-and-pfet-smart-diode

On-chip temperature measurement: a new approach for optimizing automotive inverter.

Laurent Beaurenaut, Principal Engineer, laurent.beaurenaut@infineon.com
Inpil Yoo, Staff Engineer, inpil.yoo@infineon.com
Infineon Technologies, Am Campeon, 1-12 85579 Neubiberg, Germany

Abstract

This paper presents the potential of integrating an On Chip Temperature Sensor on an IGBT compared to the classical use of an NTC. A possible implementation of the sensor, using a temperature diode, is described. Simulation and measurement results of the temperature diode sensor are discussed. Finally a new concept for signal acquisition is presented. It relies on a digital gate driver integrating an ADC on the high voltage side. System accuracy is discussed based on experimental results.

Introduction

One crucial aspect when dimensioning an inverter for automotive traction applications is to ensure the system can sustain the worst case peak power conditions. For those particular cases, the design shall ensure that the maximum junction temperature of the power switches (in most cases today an IGBT) is not exceeded. The key tuning parameter, for a given cooling concept, is the chip area per switch inside the power module, since larger area means a better spreading of the heat and finally a lower junction temperature. On the other side, it also means higher system costs, not only induced by the higher chip costs but also due to the impact on the size of the inverter. That is why it may be important to have a realtime accurate measurement of the junction temperature of the switch. Indeed, it allows to minimize the design margin to the maximal temperature conditions, and therefore to meet the best trade-off between system performance and cost.

Limitations of classical NTC solution

State of the art inverters make use of NTC to measure the temperature of the NTC. It is generally located in the power module a few centimeters away from the power switches [1]. In practice, it is loosely coupled thermally to the IGBT. This is especially true for power module thermally optimized, where it is wished to have most of the heat flowing vertically to the coolant (Fig.1).
The drawback of such an NTC concept is not only that it inaccurately correlates with the temperature of the IGBT in steady state conditions. The thermal time constant of the NTC is also significant, in the range of a few seconds typically. It means that the response time of the sensor to a power pulse lags behind the actual response of the IGBT. In typical automotive traction mission profiles, where current conditions are changing fast due to acceleration and deceleration, the NTC can offer a good level of protection of the system against failures, but can mot monitor the exact junction temperature.

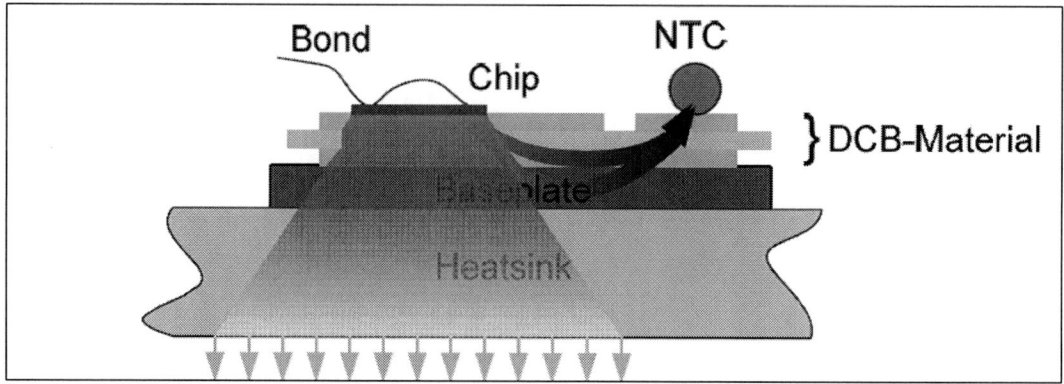

Fig. 1: Flow of thermal energy inside a power electronic module

Implementation of an On-Die Temperature Sense

In order to obtain accurate temperature of the IGBT with less time constant, the temperature sensor can be integrated on the IGBT chip directly. A possible approach is to use a poly diode as On Die Temperature Sensor (ODTS). The diode is electrically isolated from the IGBT. Fig. 2 shows the I-V characteristic of the diode for different temperatures.

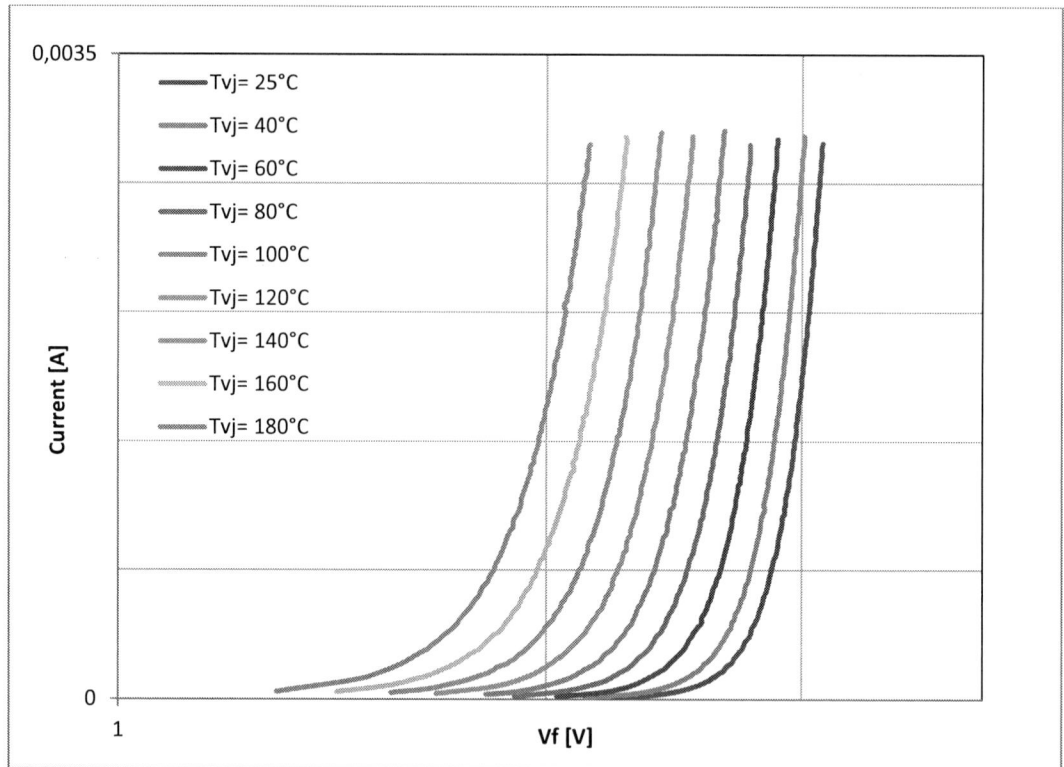

Fig. 2: I-V characteristic of the ODTS

The advantage of the poly diode is that, as a first approximation, it has a linear characteristic over temperature. Fig. 3 shows the measured response for a fixed current.

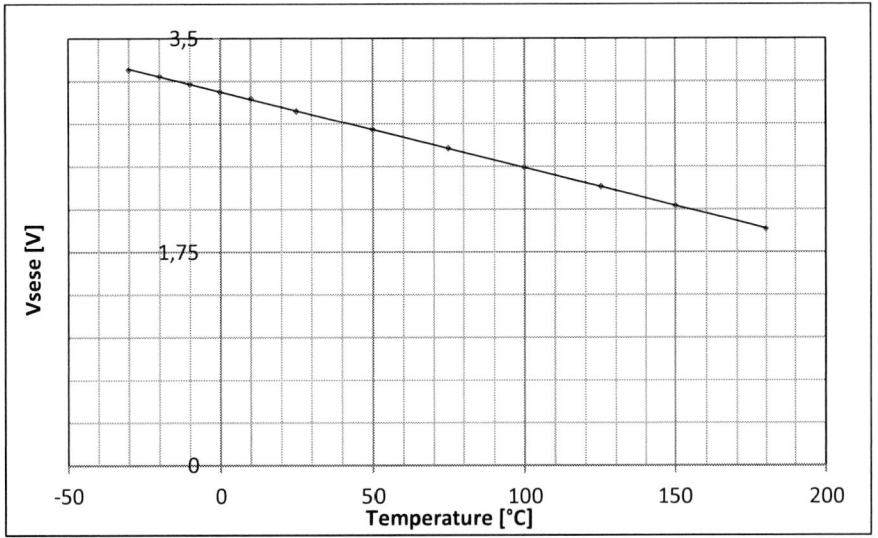

Fig. 3: Temperature response of the ODTS

For an optimal use of the ODTS, it is important to decide what should be measured by temperature sense. The limiting factor at application level is to ensure that the average temperature across the complete die does not exceed the maximum rated temperature. Therefore, the ODTS position shall be selected carefully, so that it correlates with the average junction temperature of the die.   The following criteria shall therefore be taken into account:

- average junction temperature can be sensed
- the correlation between sensed temperature and average junction temperature
- the correlation should be consistent in every operational condition
- the correlation should be independent on external influence (cooling temperature)

Fig. 4a and 4b show examples of "bad" (left) and "good" (right) positions of the ODTS.

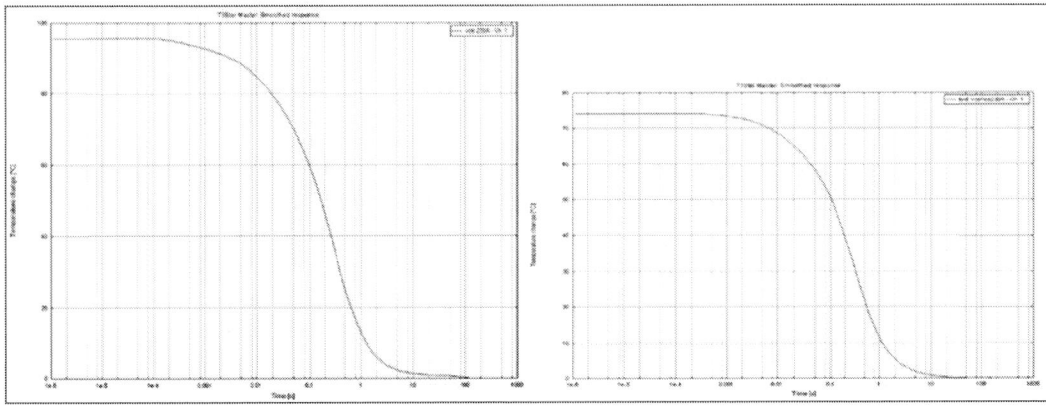

Fig. 4a: Impact of the ODTS position ("bad" position).

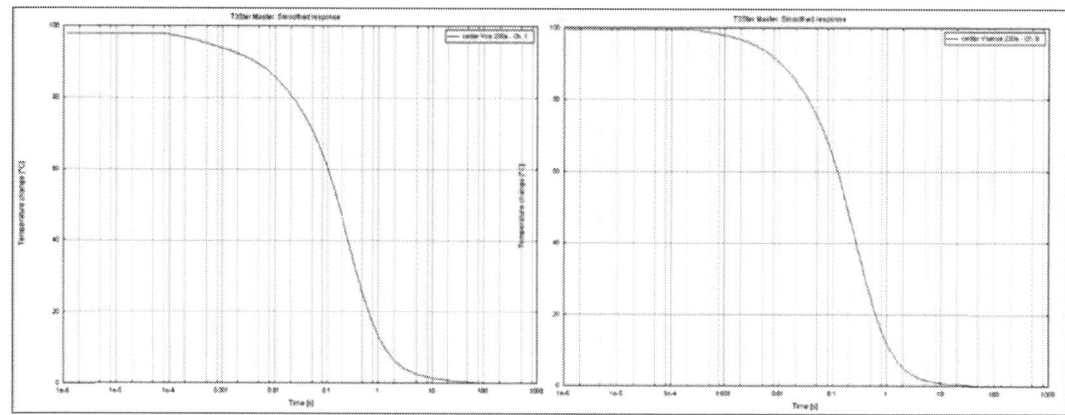

Fig. 4b: Impact of the ODTS position: "good" position.

Since the diode is directly on the IGBT the time response of the sensor is small compared to the power module thermal time constant. A transient thermal analysis (Fig. 5) shows less than 30ms of time delay.

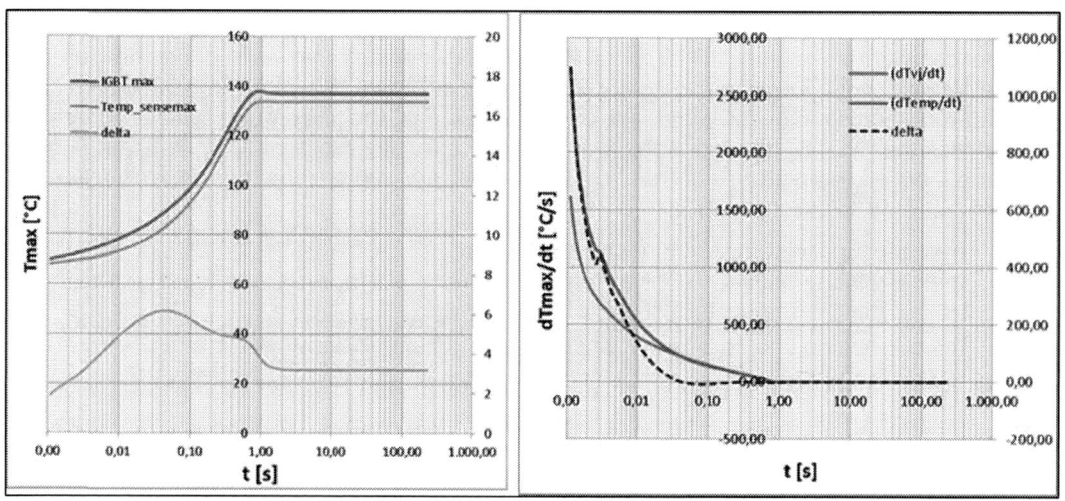

Fig. 5: Simulated Time Response of the ODTS

Experimental results (Fig. 6) show even a smaller time delay.

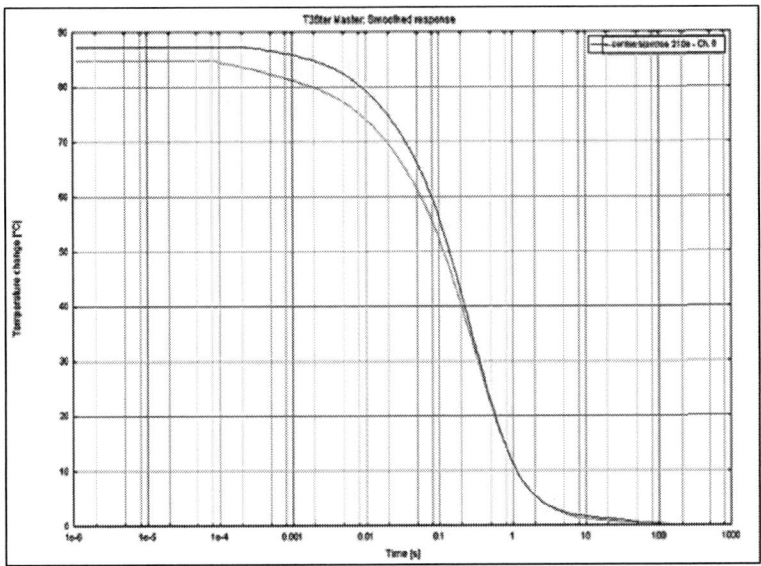

Fig. 6: Measured Time Response of the ODTS.

Optimized Signal Acquisition

In standard inverter architectures, there are typically to ways process the ODTS signal. One is to use a comparator function that is used to turn off autonomously the IGBT automatically in case of an overtemperature condition. The main drawback of this approach is that it shows no flexibility of usage. For example, it may be advantageous to define several thresholds levels (warning level, protection level) dependent on the operating mode of the inverter. Specific operating conditions such as for example active short circuit may lead to a higher operating temperature of the IGBT compared to normal operation.

Since failure modes associated with overtemperature conditions are quite "slow"- in the range of 100ms or more- another approach is to use an analog isolator for the ODTS signal, and read it by using one of the ADC channels of the LV microcontroller. This allows for more flexibility, since the application software can decide of the most appropriate reaction of the system to the sensed value. However, this approach increases significantly the Bill of Material of the inverter, since discrete analog isolator are expensive components.

A way to reduce system costs and increase the number of functions at inverter level is to increase the level of integration of the gate driver. This can be done by placing part of the digitalizing part of the functions of the isolated gate drivers (IGDR) [2].

The core functions of an IGDR is to drive the gate of the IGBT in a reliable and accurate way, provide galvanic isolation between low and high voltage domains, and protect the IGBT against typical failure modes such as overvoltage, overcurrent , etc. Further functional integration can be made to "reuse" the integrated isolation, in order to transfer more information across the isolation barrier.

Fig. 7 shows as an example of such an integration. The 1EDI2010AS EiceSense is a 2nd Generation IGDR based on Infineon's Coreless transformer technology [3]. One of its main characteristic is to integrate an 8-bit Analog Digital Converter on its secondary side (high voltage). An integrated current source forces a DC current of 1 mA in the ODTS. When it receives a trigger, the ADC acquires the voltage of the ODTS and stores it in one register.

© VDE VERLAG GMBH · Berlin · Offenbach

PCIM Europe 2016, 10 – 12 May 2016, Nuremberg, Germany

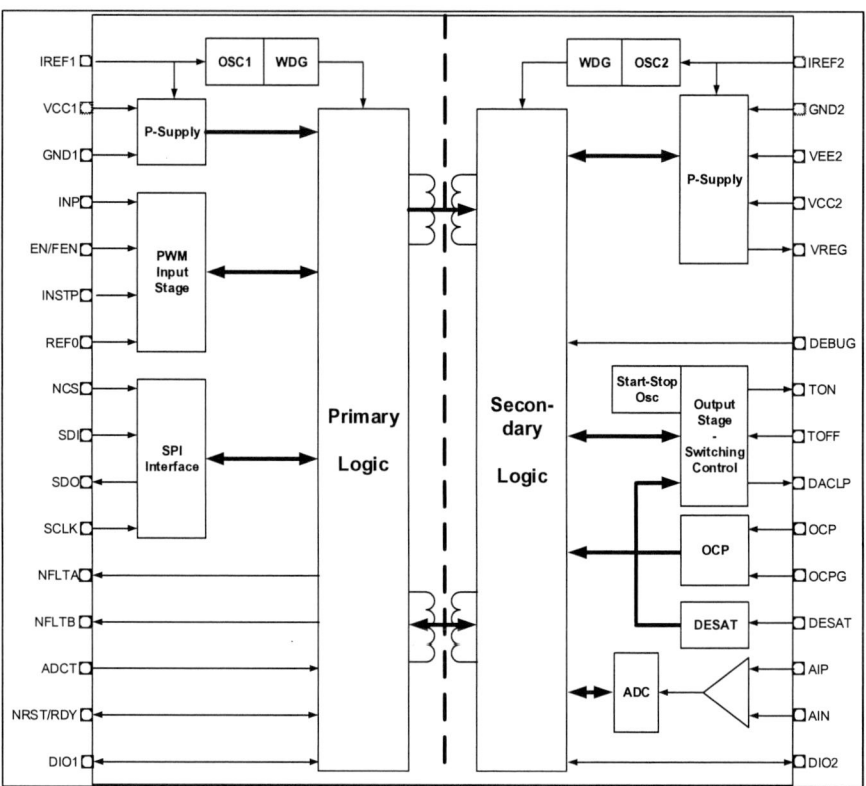

Fig 7: Digital gate driver with integrated ADC

The result of the conversion can then be read by the main microcontroller on the LV side via the SPI interface. Gain and offset compensation of the DAC input stage can also be configured via the SPI link, in order to optimize tailor the ADC response to the sensor's characteristic (Fig. 8).

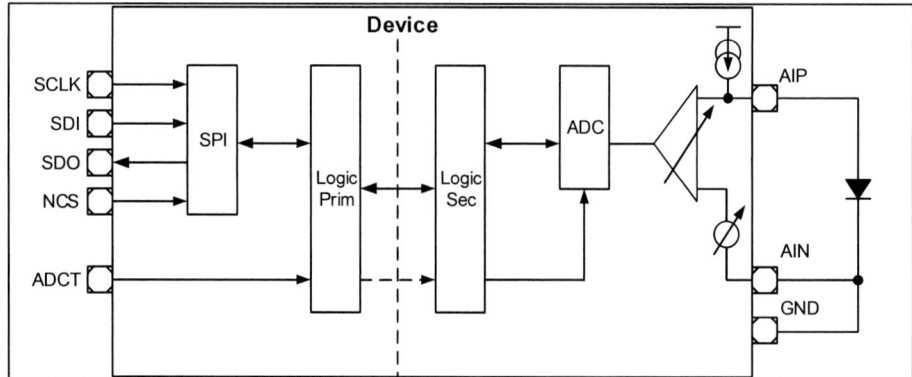

Fig 8: Typical application example using the integrated ADC

Fig. 9 shows experimental results of the ADC characteristic versus voltage, and Fig. 10 shows experimental results of the total system response versus temperature.

© VDE VERLAG GMBH · Berlin · Offenbach

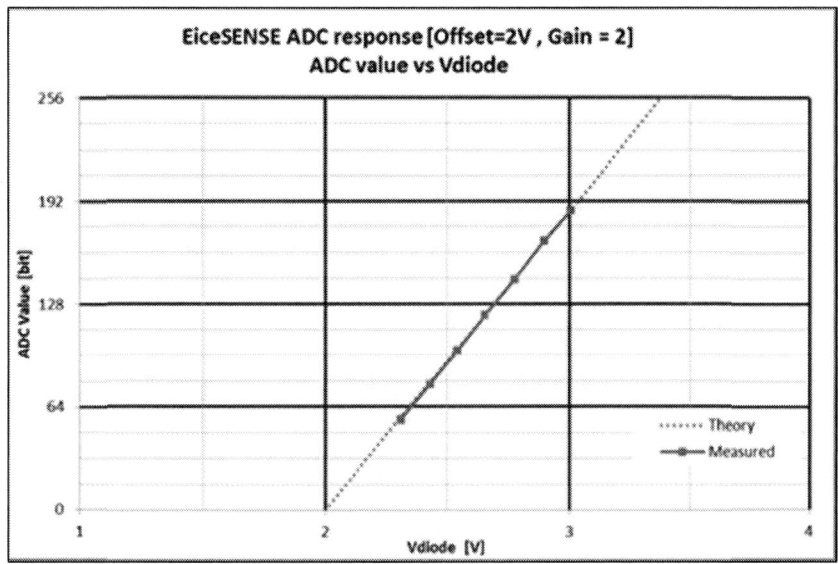

Fig. 9: Experimental results of the ADC characteristic

With a one point calibration, the total accuracy, including process variation is expected to reach +/-5 °C, with a signal resolution higher than 1 bit / °C. This new concept combines the advantages of the conventional solution and still provides new ones:

- Autonomous notification (boundary checker)
- Flexibility of operation at minimal BOM
- Support all types of sensor characteristics.
- Support calibration strategy.
- Adaptive switching

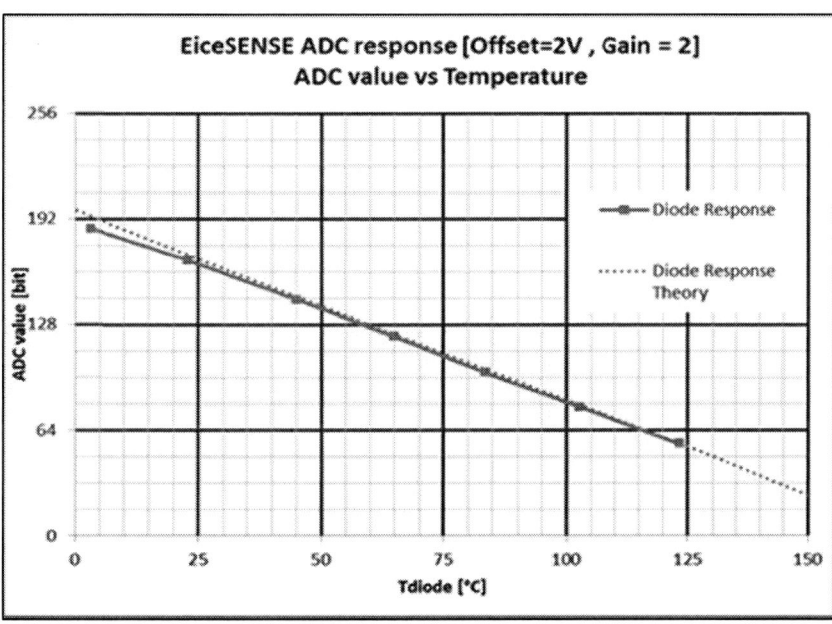

Fig. 10: Experimental results of the system response versus temperature

Conclusion

Adding an ODTS provides an accurate estimation of the junction temperature of the power switch, which allows a more efficient dimensioning of the inverter, by reducing margin at system level and allowing for more complex derating strategies. Integrating of an ADC in the isolated gate driver is a cost effective implementation of the signal acquisition circuit. It also opens the path to future control strategies, for example adaptive switching of the IGBT.

References

[1] Application note, AN 2009-10 Using the NTC inside a power electronic module, Infineon Technologies website.

[2] L. Beaurenaut, K. Scheibert, New System Partitioning for Safe Automotive Inverters, PCIM Europe 2013.

[3] L. Beaurenaut, P.Leteinturier: "Coreless Transformer Driver IC for HV Motor Applications" ECPE, November 2010

Acknowledgements

The authors would like to thank Daniel Lee, Infineon Technologies, for his contribution to this paper.

PCIM Europe 2016, 10 – 12 May 2016, Nuremberg, Germany

Resonant load emulator for distributed energy resources to test anti-islanding algorithms

Daniel Heredero-Peris, Fernando Jorge-Ques, Daniel Montesinos-Miracle, Centre d'Innovació Tecnològica en Convertidors Estàtics i Accionaments (CITCEA-UPC), Departament Enginyeria Elèctrica, Universitat Politècnica de Catalunya. ETS d'Enginyeria Industrial de Barcelona Av. Diagonal, 647, Pl. 2. 08028 Barcelona, Spain
Tomàs Lledó-Ponsati, TeknoCEA, Tecnologies de control de l'electricitat i automatització S.L., Barcelona, Spain, tomas.lledo@teknocea.cat

Abstract

A resonant load emulator for distributed energy resource inverters to test anti-islanding algorithms is presented. This emulator allows to change the operation conditions that can be applied during development, test and certification of an inverter under resonant loads. It is based on a three-phase three-wire inverter that exchanges the required currents imitating the behavior of real resonant loads.

1.   Introduction

During the last decades, power electronic emulators have been investigated as a way of imitate the behaviour of different electrical systems. The aim is to obtain a flexible device to test novel algorithms or certificate a third device without the use of real systems. Some examples are load emulators, grid emulators, Distributed Energy Resource (DER) emulators or even storage system emulators [1, 2, 3, 4]. Thus, it is possible to avoid premature degradations, use less space, change the parameters to operate in different frameworks in a few steps or replicate tests with reduced time.

The increasing use of renewable resources such photovoltaic panels (PV) or wind turbines derives in the need of improved Current Controlled-Voltage Source Inverters (CC-VSI), also named Grid Supply Inverters (GSI), to integrate DER into the grid. Nowadays, an inverter has to accomplish with several standards to be certified. Integration regulations such as VDE4105 [5], IEEE1547 [6] or IEC61727 [7] require that when the grid goes from a grid-connected to a grid-disconnected situation, the inverter should detect this occurrence in a predefined time. The main purpose of this action is to avoid the creation of electrical islands, known as islanding situation. In this sense, it is limited to continue supplying energy to the local loads preventing non adequate voltage regulation at the Point of Common Coupling (PCC) or dangerous situations for maintenance tasks. The PCC is defined as the common node where coexists the grid, the loads and the inverter(s). The cited regulations propose the use of the called Resonant Load (RL) to test if the GSI are able to detect the mains loss thanks to any implemented anti-islanding algorithms [8].

This paper presents an AC load emulator that is able to emulate the required resonant loads to test anti-islanding algorithms on GSI, i.e., a resonant load emulator (RLE).

© VDE VERLAG GMBH · Berlin · Offenbach

2. Anti-islanding considerations

The resonant load emulator should be developed according to standards specifications. The detection time of the mains loss, a proper understanding of the resonant load and the anti-islanding methods are the key elements.

2.1. The clearing time

Specific detection times of the mains loss can be found in VDE4105 [5], IEEE1547 [6] or IEC61727 [7]. This time is also known as clearing time. The clearing time is dependent of the type of load connected at the PCC distinguishing between conventional and resonant loads. The clearing time is also function of a voltage threshold percentage over the considered rated voltage and a deviation on the rated frequency. Voltage displacements implies clearing times from 50 ms to 2 s and frequency displacements from 160 ms to 200 ms. The voltage and frequency tolerable displacement constitutes a Non-Detection-Zone (NDZ). When resonant loads are considered, detection times are lax, going from 2 to 5 s depending on the standard.

2.2. The resonant load

A resonant load can be defined as a load that sets null power flux between the grid and the PCC. The resonant load is described by an equivalent resistance R that consumes all the active power P delivered by the inverter, maintaining the voltage level, and a parallel LC load that resonates at grid frequency. This LC circuit maintains the grid frequency when grid-disconnected consuming null reactive power Q during all operation. When a resonant load is locally connected to the inverter and the utility is disconnected, neither the voltage nor the frequency will change significantly producing a non-grid non-detectable situation outside the NDZ.

2.3. The applied anti-islanding method

An anti-islanding method can be defined as a monitoring process of the relevant electrical magnitudes of the PCC, i.e., voltage, frequency or impedance, used to detect the mains loss. It is possible to find in the literature three categories for anti-islanding detection methods:

- Passive methods. Methods based on variable observation to take decisions.
- Active methods. Methods based on perturb and observe to take decisions.
    - Positive feedback strategy.
    - Impedance measurement strategy.
- Based on communications. Related with intentional grid disconnections.

The passive methods result ineffective under resonant loads presence. For this reason an active method should be selected to be applied to the inverter. This active method has to be adequate for a comparison between the ideal and the emulated behaviour of the resonant load. For this reason, in this study, it is proposed to use the active anti-islanding method named Slip-Mode-Shift (SMS) [8]. SMS is an active-antislanding positive feed-back method that tries to

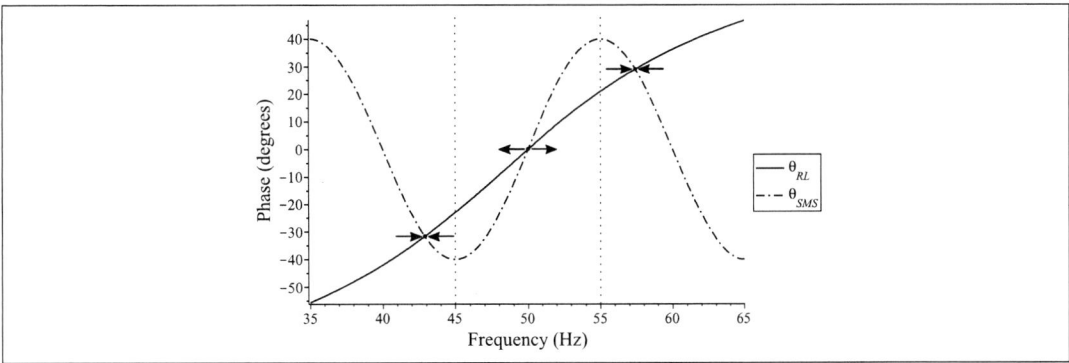

Fig. 1: SMS perturbation phase with $f_m = 55$ Hz and $\theta_m = 40°$

destabilise the inverter current by changing the controlled frequency by means of the phase. It is based on the phase characteristic of the resonant load as

$$\theta_{RL} = \arctan\left(-q\left(\frac{f}{f_r} - \frac{f_r}{f}\right)\right) \tag{1}$$

where θ_{RL} is the phase of the current absorbed by the resonant load in grid-disconnected mode, f is the controlled frequency and f_r and q are the resonant frequency and quality factor of the local resonant load. It is usual in RLC loads to define the quality factor as

$$q = R_{RL}\sqrt{\frac{C_{RL}}{L_{RL}}} \tag{2}$$

In order to make the frequency unstable, a perturbation on the current injected phase is added

$$\theta_{SMS} = \theta_m \sin\left(\frac{\pi}{2}\frac{f - f_r}{f_m - f_r}\right) \tag{3}$$

where θ_{SMS} is the phase of the referenced current and θ_m and f_m are the SMS parameters that represent the maximum phase displacement and the particular frequency. As it can be seen in Figure 1, representing Equation 1 (solid line) and Equation3 (dashed line), when the inverter is grid-connected, the utility holds the frequency closed to its fundamental value (50 Hz). When grid-disconnected, the resonance frequency results an unstable equilibrium point ($\leftarrow \cdot \rightarrow$) and the frequency is forced to move out from the NDZ to a new stable equilibrium point ($\rightarrow \cdot \leftarrow$), detecting the islanding status.

3. Resonant load emulator design

Real R, L and C passive components can be used for a resonant load implementation, but the amount of space, lack of flexibility and energy consumed makes more convenient to use regenerative loads. The use of real R, L and C components makes also impractical the re-configuration of the resonant load under different PQ set-points of the inverter under test. This section presents the relevant aspects for the proposed resonant load emulator.

PCIM Europe 2016, 10 – 12 May 2016, Nuremberg, Germany

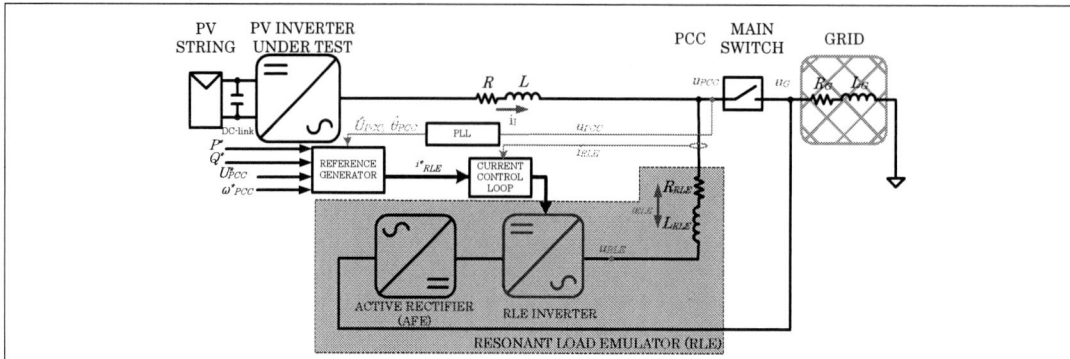

Fig. 2: Connection diagram of the resonant load emulator and inverter during anti-islanding test

3.1. Resonant load emulator considerations

The proposed equipment is based on a three-phase three-wire GSI that emulates the consumption of a resonant load by synthesizing the appropriate currents (i_{RLE}^*), as can be seen in the gray region of Figure 2. An active rectifier (or Active Front End (AFE)) would be the responsible to feed-back to the grid the energy drawn from the inverter under test, making the system more efficient than using real passive components. Using this configuration, only the losses of the two converters are consumed.

The power consumption of a RLC resonant load connected at the PCC can be defined as

$$P_{RL} = U_{PCC}^2 \frac{1}{R_{RL}} \tag{4a}$$

$$Q_{RL} = U_{PCC}^2 \left(\frac{1}{\omega_{PCC} L_{RL} - \omega_{PCC} C_{RL}} \right) \tag{4b}$$

where P_{RL} and Q_{RL} are the active and reactive power at PCC of the resonant load, respectively. R_{RL}, L_{RL} and C_{LR} are the resistance, the inductance and the capacitance of the resonant load. The angular frequency at PCC is defined by ω_{PCC}.

The anti-islanding algorithm of the inverter under test is tested according to its rated power (P^*) and, depending on the standard, the quality factor for the test (q^*) has to be set between 1 and 2. Also, the rated rms voltage and angular frequency of the PCC (U_{PCC}^*, ω_{PCC}^*) are required test inputs. Using Equations 2 and 4 the theoretical R_{RL}, L_{RL} and C_{LR} input values for the proposed test are computed offline. A Phase-Locked Loop (PLL) algorithm is proposed to be used in order to obtain the current rms voltage, the current angular frequency and the voltage phase at the PCC ($\hat{U}_{PCC}$, $\hat{\omega}_{PCC}$, $\hat{\theta}_{PCC}$) from sensing the instantaneous voltages u_{PCC}, as shown in Figure 3(a). Then, the P_{RL} and Q_{RL} power values can be calculated using the $\hat{U}_{PCC}$, $\hat{\omega}_{PCC}$, R_{RL}, L_{RL} and C_{LR} pre-computed values applying Equation 4. Finally, the current reference for the RLE (i_{RLE}^*) is calculated from P_{RL}, Q_{RL}, $\hat{U}_{PCC}$ and the voltage phase $\hat{\theta}_{PCC}$ at the PCC. Figure 3(b) depicts the exposed procedure.

In this paper, a synchronous reference frame has been assumed for the current control loop of the RLE, as can be seen in Figure 3(c). Thus, conventional PI controllers are used for the direct and quadrature ($I_{RLE_{dq}}$) current control, obtained from i_{RLE}, $\hat{\theta}_{PCC}$ and using the Park transform. It has been considered a totally decoupled system and feed-forward to be more robust against grid voltage disturbances.

© VDE VERLAG GMBH · Berlin · Offenbach

2038

PCIM Europe 2016, 10 – 12 May 2016, Nuremberg, Germany

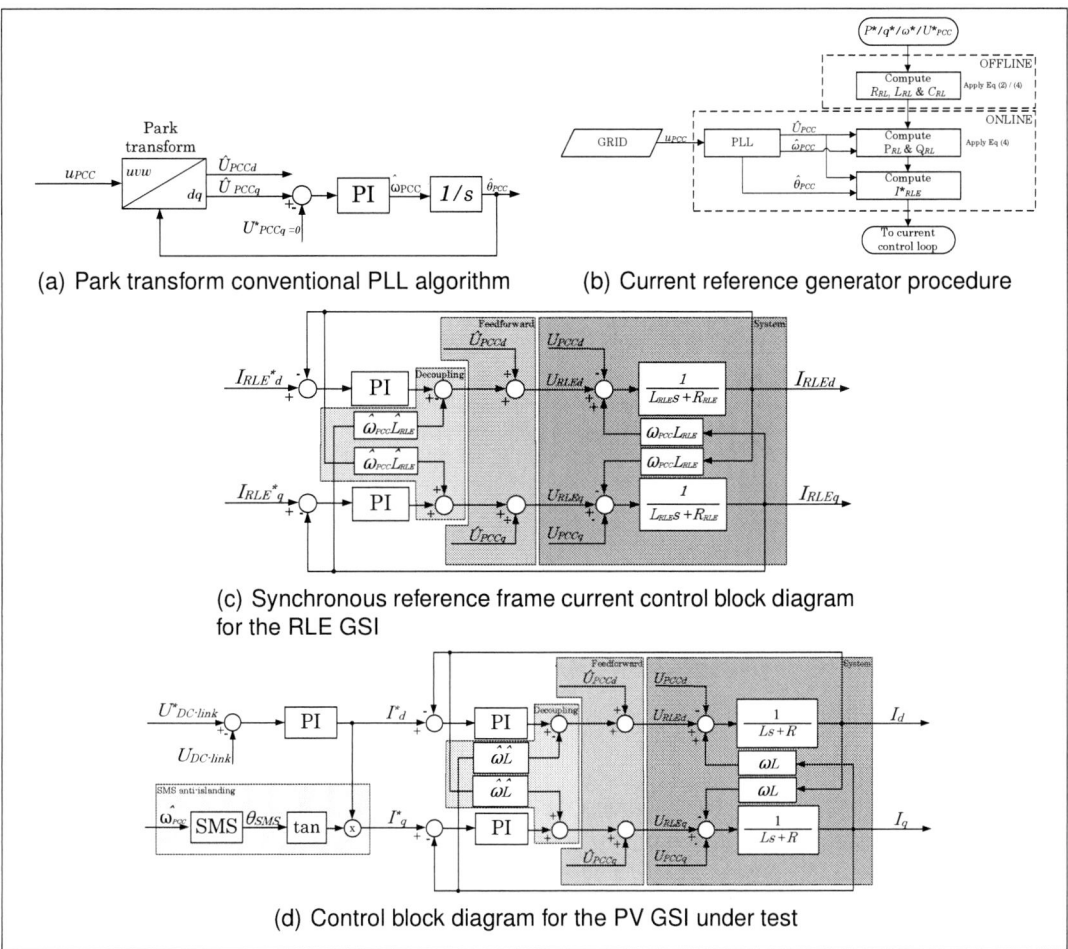

(a) Park transform conventional PLL algorithm

(b) Current reference generator procedure

(c) Synchronous reference frame current control block diagram
for the RLE GSI

(d) Control block diagram for the PV GSI under test

Fig. 3: Control schemes. $\hat{x}$ variables are estimations or obtained by calculation

3.2. Tested inverter considerations

The PV GSI under test also considers a synchronous reference frame for the inner control loop.
An outer control loop manages the DC-link voltage by means of a PI controller. The control
action of this DC voltage controller results into the direct current reference (I_d^*). Figure 3(d)
represents the control scheme in block diagrams for considered inverter under test. The SMS
anti-islanding is incorporated in the control scheme considering that the phase of the current
can be computed as

$$\theta_I = \arctan\left(\frac{I_q}{I_d}\right) \tag{5}$$

Then, considering Equation 5 and substituting θ_I by θ_{SMS}, the quadrature current reference
(I_q^*) can be computed as

$$I_q^* = \tan(\theta_{SMS})I_d^* \tag{6}$$

© VDE VERLAG GMBH · Berlin · Offenbach

2039

Parameter	Value	Units	Parameter	Value	Units
R	0.20	Ω	DC-link voltage loop k_i	5.00	
L	3.00	mH	SMS f_m frequency	50.16	Hz
DC-link capacitor	1.02	mF	SMS θ_m phase	0.0087	rad
AC current loop k_p	13.8		PLL k_p	0.50	
AC current loop k_i	880		PLL k_i	5.00	
DC-link voltage loop k_p	0.10				

Tab. 1: PV GSI parameters

Parameter	Value	Units
R_{RLE}	0.25	Ω
L_{RLE}	1.00	mH
DC-link capacitor	1.02	mF

Tab. 2: RLE GSI parameters

4. Results

The proposed set-up is shown in Figure 2. The voltage U_G is set to 100 V phase-to-phase and the grid angular frequency ω_G to 314.15 rad/s. Then, $U^*_{PCC} = U_G/\sqrt{3}$ and $\omega^*_{PCC} = \omega_G$.

A 2.3 kW three-phase three-wire PV GSI is proposed. Table 1 summarizes the PV GSI hardware and software parameters.

The resonant load is set to have a quality factor q^* of 1 and an active power P^* of 2.26 kW, matching with the maximum active power of the proposed PV GSI. To validate the operation, a 3 kW three-phase three-wire RLE GSI is proposed. The hardware parameters can be seen in Table 2. For simplicity, the AFE proposed in Figure 2 is substituted by a DC power source of 650 VDC. Different controller constants for the RLE current loop and for the PLL algorithm have been considered. The different controller gains implies different settling times and different bandwidths either for the current loop or the PLL. The considered gains or bandwidths are shown in the respective legends of Figure 4. Figure 4 shows the frequency response at the PCC, comparing a real passive RLC circuit and the proposed emulator. When the RLC circuit is considered the frequency is obtained from the PV GSI's PLL. The islanding situation is intentionally set at time t equal to 0.5 s.

In Figure 4(a) can be seen that the variation of the current loop controller gains of the RLE practically does not affect to the RLE performance. In this case, when the RLE is considered, the frequency is obtained from the RLE's PLL assuming $k_{p_{PLL\ RLE}} = 2.44$ and $k_{i_{PLL\ RLE}} = 244.8$. On the other hand, as can be seen in Figure 4(b), the determination of wider PLL bandwidths improve the time response being closer to the expected theoretical resonant load response.

With the aim of checking the operation of the emulator in a more real framework, a white noise is introduced into voltage and current measures. Under these assumptions, different simulations are conducted at different powers inputs (P^* set to 300 W, 600 W and 2.65 kW) maintaining the operation power of the PV GSI at 2.26 kW. All simulations are done considering a RLE's PLL bandwidth of 22.5 Hz. The islanding situation is set at time t equal to 1.5 s. Figures 5(a) to 5(f) show the obtained PV GSI frequency response with noise overlapped and filtered. It can be observed that the RLE emulates satisfactorily the theoretical resonant load.

PCIM Europe 2016, 10 – 12 May 2016, Nuremberg, Germany

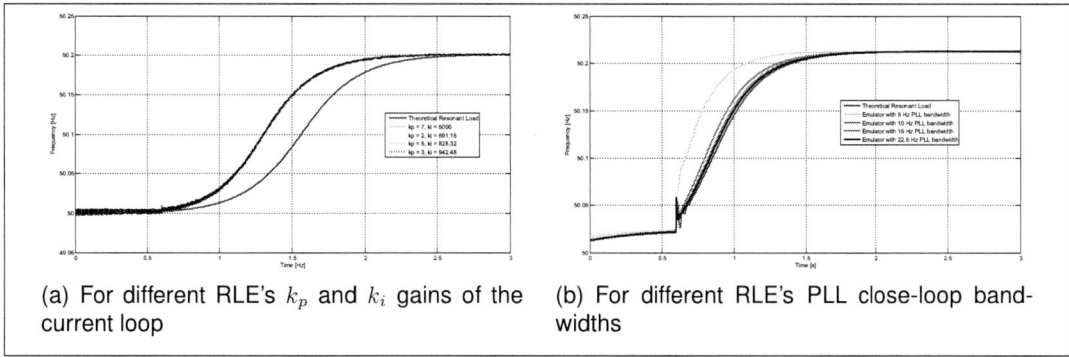

(a) For different RLE's k_p and k_i gains of the current loop

(b) For different RLE's PLL close-loop bandwidths

Fig. 4: Comparison of the performance on the frequency response of the SMS anti-islanding algorithm when connected to the RLE or to the theoretical resonant load. Theoretical RLC resonant: R_{RL} = 4.42 Ω, L_{RL} = 14.1 mH and C_{RL} = 720 μF (Star connected)

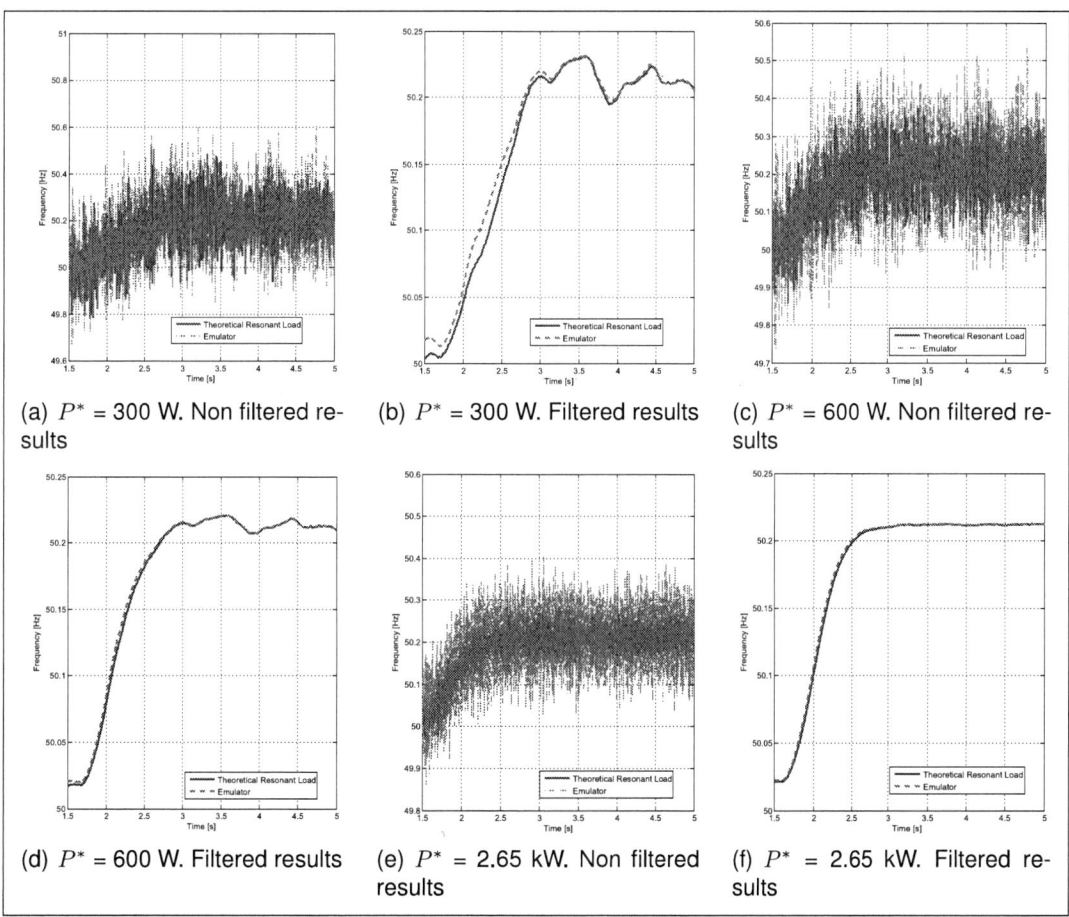

(a) P^* = 300 W. Non filtered results

(b) P^* = 300 W. Filtered results

(c) P^* = 600 W. Non filtered results

(d) P^* = 600 W. Filtered results

(e) P^* = 2.65 kW. Non filtered results

(f) P^* = 2.65 kW. Filtered results

Fig. 5: Simulation of the frequency response at the PCC for different theoretical RLC resonant loads and the corresponding RLE behavior. Different P^* are considered when the PV GSI is operated at 2.26 kW

© VDE VERLAG GMBH · Berlin · Offenbach

5. Conclusions

A resonant load emulator has been proposed based on the characterization of the suggested resonant load in the regulations. This characterization is translated into current references for a current controlled voltage source inverter based on a widely extended three-phase three-wire inverter. The resonant load emulator is evaluated considering a three-phase three-wire photovoltaic inverter operated at the maximum point. Also, the slip-mode-shift anti-islanding algorithm is considered as a mechanism of comparison between the resonant theoretical and emulated behavior.

The effect of the resonant load emulator controller gains and the PLL bandwidth has been studied. It has been concluded that the most relevant influence of the resonant load emulator concerns to PLL tuning parameters.

Different active power references for the resonant load emulator inverter have been simulated. The obtained results supports that it is possible to obtain a satisfactory behavior to emulate resonant loads for distributed energy resource inverters and test anti-islanding algorithms.

6. References

[1] Y. Srinivasa Rao and M. Chandorkar. Electrical load emulator for unbalanced loads and with power regeneration. In *Industrial Electronics (ISIE), 2012 IEEE International Symposium on*, pages 320–327, May 2012.

[2] A. Collet, J. Crebier, and A. Chureau. Multi-cell battery emulator for advanced battery management system benchmarking. In *Industrial Electronics (ISIE), 2011 IEEE International Symposium on*, pages 1093–1099, June 2011.

[3] D. Heredero-Peris, M. Capo-Lliteras, C. Miguel-Espinar, T. Lledo-Ponsati, and D. Montesinos-Miracle. Development and implementation of a dynamic PV emulator with HMI interface for high power inverters. In *Power Electronics and Applications (EPE'14-ECCE Europe), 2014 16th European Conference on*, pages 1–10, Aug 2014.

[4] R.V. Gokhale, S.M. Mahajan, B.W. Abegaz, and R.P.M. Craven. Development of a real time wind turbine emulator based on RTDS using advanced perturbation methods. In *Environment and Electrical Engineering (EEEIC), 2015 IEEE 15th International Conference on*, pages 1713–1718, June 2015.

[5] VDE-AR-N 4105 2011-08 Power generation systems connected to the low-voltage distribution network - Technical minimum requirements for the connection to and parallel operation with low-voltage distribution networks, 2011.

[6] IEEE Standard 1547 Standard for Interconnecting Distributed Resources with Electric Power Systems, 2003.

[7] IEC 61727. Photovoltaic (PV) systems - Characteristics of the utility interface, 2004.

[8] Michael Bower, Ward Ropp. Evaluation of islanding detection methods for utility-interactive inverters in photovoltaic systems. *Sandia Report*, 2002.

Low Voltage Ride Through (LVRT) Capability of an enhanced DFIG System

David Velasco, Jesús López.
Department of Electrical and Electronic Engineering – Institute of Smart Cities
Public University of Navarre (UPNa), Pamplona, Spain.
Email: david.velasco@unavarra.es;

Abstract

Deep sags commonly induce in DFIG-based systems high transient currents and torque spikes, putting the wind turbine under stress. An enhanced DFIG system derived by performing several modifications on the conventional one is proved to present several benefits regarding LVRT. Special effort has been employed in decreasing the mechanical stress on the shaft by reducing torque spikes produced during grid faults. For this purpose, different software and hardware-based measures involving the Grid-Side Converter (GSC) have been applied on the enhanced system, achieving torque reductions up to 81% with regard to the conventional DFIG system.

1. Introduction

As widely known, voltage dips induce high rotor currents in the DFIG which can damage the MSC semiconductors and raise the DC voltage level to unacceptable limits [1] [2]. Moreover, the high currents obtained cause an electromagnetic torque transient peak that can reach the value of several times the rated mechanical shaft torque, favoring the stress on the whole system [1]. In this context, an enhanced DFIG system, namely xFIG, is presented in this work. This topology is proved to withstand properly LVRT and reduce significantly the torque spikes caused by grid faults on the mechanical shaft.

2. xFIG system

At the enhanced system proposed the DFIG rotor is connected to a back-to-back converter, as happens in the conventional DFIG system. However, the Grid-Side Converter, which is typically connected to the electrical grid, is connected in the enhanced system to a Permanent Magnet Machine (PMG) coupled to the mechanical shaft [3]. A scheme of the conventional DFIG topology is shown in Fig. 1, while the aforementioned enhanced system is shown in Fig. 2.

Since the GSC is not connected to the grid, a substantial reduction in harmonic injection is achieved with the enhanced system in comparison to a conventional DFIG application, and the corresponding grid harmonic filter can be suppressed. In addition, a medium-voltage power evacuation at the Point of Common Coupling (PCC) terminal is feasible in this topology, avoiding the material and space costs of the commonly-used power transformer, increasing this way the efficiency of the system.

PCIM Europe 2016, 10 – 12 May 2016, Nuremberg, Germany

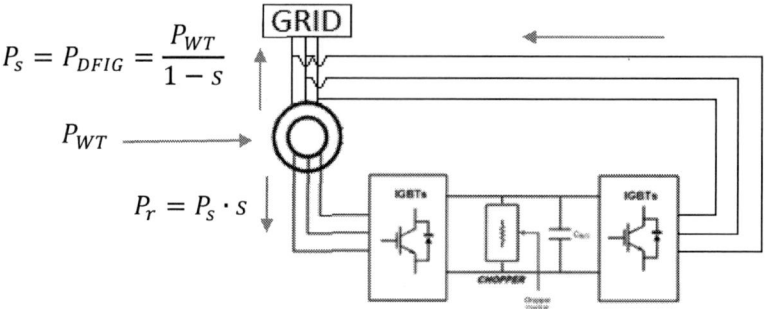

$$P_s = P_{DFIG} = \frac{P_{WT}}{1-s}$$

$$P_{WT} \longrightarrow$$

$$P_r = P_s \cdot s$$

Figure 1: Power flow at DFIG topology

Concerning the LVRT, the enhanced system generates voltage at all times between PMG terminals. In this manner, when a grid fault occurs, this feature helps the GSC at maintaining the DC Bus voltage control capability. Another remarkable advantage to highlight is the controlled torque generation by means of the PMG during a transient event, which can be of great value in order to reduce torque spikes on the mechanical shaft. This last feature is reported in detail throughout this paper.

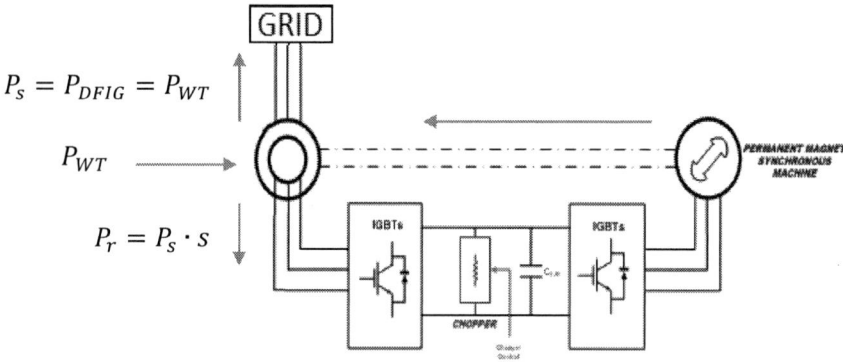

$$P_s = P_{DFIG} = P_{WT}$$

$$P_{WT} \longrightarrow$$

$$P_r = P_s \cdot s$$

Figure 2: Power flow at xFIG topology

The fact of requiring a PMG may also be considered as a drawback. However, it is worthy to notice that the sizing of this machine represents only a small fraction of the power of the main generator: the one that flows through the rotor. In addition, the generator of the enhanced system must be sized for full wind turbine power, and not for a fraction of it, as happens in the conventional DFIG system. As can be observed in Figs. 1-2, while in the DFIG system just a fraction of the total mechanical power flows through the stator, in the xFIG system the total amount of this power must flow through it. This result is due to the fact that the power flowing through the rotor is absorbed by the PMG, which in hyper-synchronism works as motor, feeding back this power through the mechanical shaft. Therefore, the torque of the main generator is the addition of the wind torque and the PMG torque, and the generator must be sized therefore for full wind turbine power.

© VDE VERLAG GMBH · Berlin · Offenbach

2.1 LVRT Capability of the proposed topology

A currently-installed 2.5 MW commercial DFIG wind turbine has been taken as a reference for this study. In this regard, an equivalent xFIG system has been designed and implemented in MATLAB, with the criteria of maintaining the same mechanical power as the one of the conventional DFIG application.

For the purpose of proving the LVRT capability, a zero-voltage three-phase fault has been considered. A suitable chopper has been sized and integrated in the xFIG topology in order to prevent the DC bus voltage level to reach dangerous limits and consequent overvoltages on the semiconductor devices. As shown in Fig. 3a, the DC bus voltage is kept under the conservative value of 1.1 pu, which is commonly affordable for any manufacturer. The rotor currents are shown in Fig 3b and have been referred to the rated current value of the converter. As can be seen, the currents are at all times under the limit value of three times the rated current of the converter, and therefore no damage is produced to the MSC during the transient event [4] .

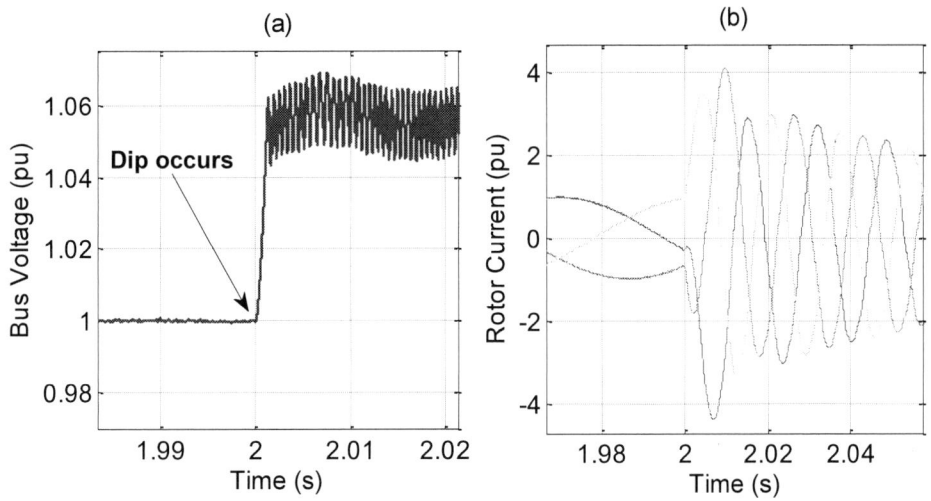

Figure 3: Bus Voltage (a) and Rotor Currents (b) evolution after the simulated grid fault in the xFIG system.

2.2 Torque reduction on the mechanical shaft

As mentioned before, the high currents obtained after a voltage sag provoke an electromagnetic torque transient peak that can reach the value of several times the mechanical shaft rated torque, putting the wind turbine under stress [1] [5]. Therefore, the shaft must be designed taking into account these transient states [6]. In the particular case of the DFIG system considered, the torque spike reaches the value of 277% the rated value. One of the objectives of this work is to analyze whether, during a voltage sag, a reverse torque to the one made by the DFIG can be performed by means of the PMG, in order to reduce the total mechanical shaft torque i.e. the sum of the main generator torque and the PMG torque. In this case, a less demanding design on the drive train could be considered, decreasing this way the associated costs.

As soon as the voltage sag is detected in the xFIG application, a very high torque reference in the opposite sense to the one that takes place in the main generator is set to the current-loop of the PMG. Despite this generator is sized for a small fraction of the wind turbine power, it is affordable to force high currents through it without damaging the machine, taking into account the short dynamics of the torque transient produced by the grid fault (a few ms). However, when the high torque reference is set to the PMG, the current loop of the GSC demands a voltage not feasible by the converter, reaching saturation, and not obtaining therefore the desired torque.

In order to increase the torque reduction in the xFIG system, different software and hardware-based measures regarding the system control and the GSC are taken.

2.2.1 Software-based measures

The first measure analyzed regarding the system control is the effect of changing the *dq* voltage distribution applied by the GSC immediately after the voltage sag. By this moment, the GSC is under heavy saturation. This situation is illustrated in Fig. 4. The voltage demanded by the GSC is plotted in green, while the circle represents the maximum voltage achievable with the converter.

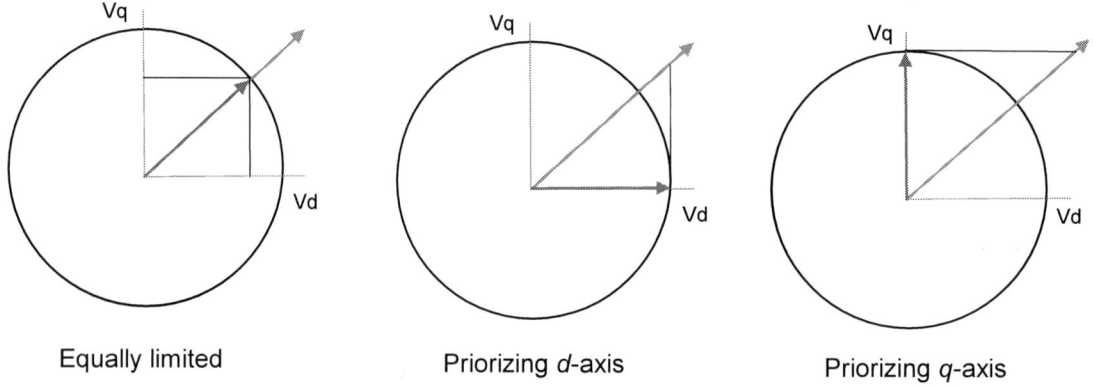

Equally limited   Priorizing *d*-axis   Priorizing *q*-axis

Figure 4: GSC Voltage limiting strategies studied for torque reduction

The voltage demanded by the current loop after the voltage sag must be reduced in order not to exceed the maximum voltage available. To this purpose, the *dq* components of the voltage applied by the converter can be limited equally (left-side figure, red arrow). However, another strategies can be considered. In this study, a strategy consisting on priorizing one of the two axes has been analyzed. Fig. 4 represents two extreme situations: priorizing completely the *d*-axis voltage (middle figure, red arrow) and priorizing completely the *q*-axis (right-side figure, red arrow).

In order to get the *dq* voltage distribution which optimizes the torque reduction, a battery of simulations is performed, where different *dq* voltage distributions are set on the GSC. For ease of understanding, these simulations are reported in this paper by Pareto Limit diagrams, a way of representing optimal points where there is not one but two objective variables involved. In this case, it turns out interesting to obtain high torque reduction (*y*-axis) with minimum current (*x*-axis), in order not to need to oversize the GSC.

The torque reductions obtained are shown in Fig. 5. The most interesting cases are therefore the ones that lie on the colored curve, which represents the maximum torque achievable for a specific current.

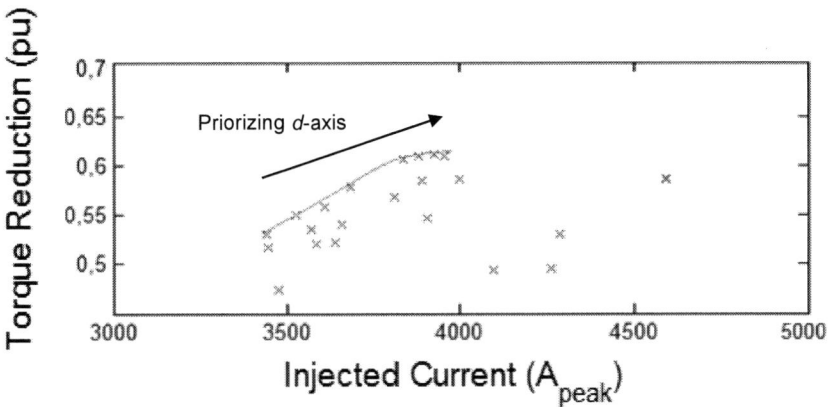

Figure 5: Torque reduction obtained through control variations

As can be observed, the major reductions are obtained when priorizing the *d*-axis (indicated by the black arrow). Injecting demagnetizing current (negative *d*-axis current) helps on relieving the converter from saturation, and therefore, a greater *q*-axis current can be achieved through the PMG during the transient event of a grid fault. This circulating current allows the PMG to perform a high torque during the first most-demanding moments of the voltage dip.

2.2.2 Hardware-based measures

A hardware-based measure regarding the converter is analyzed: having a higher DC voltage available is proved to allow the PMG current-loop injecting a higher current immediately after the voltage sag, and therefore bigger reductions can be achieved.

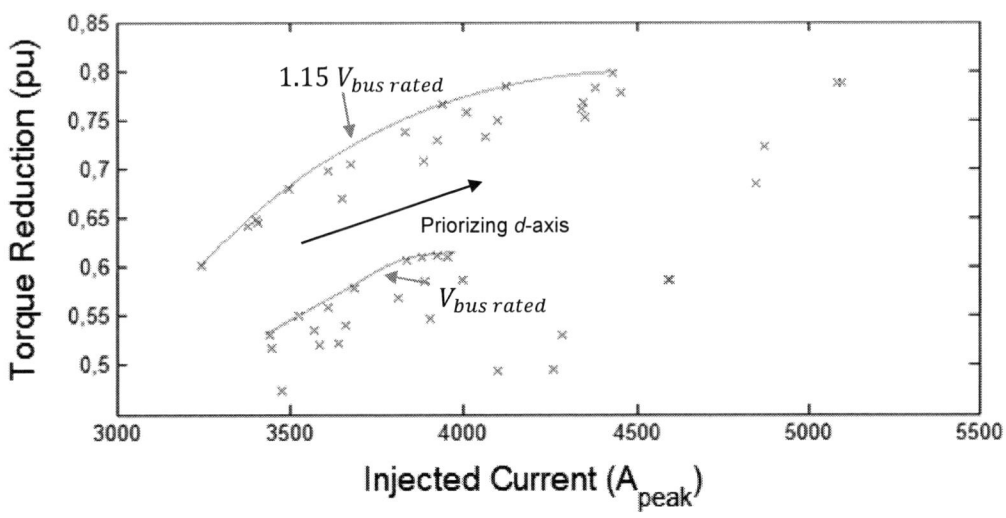

Figure 6: Torque reductions obtained through control variations for $V_{bus\,rated}$ and for $1.15\,V_{bus\,rated}$

Considering the aforementioned measures, a battery of simulations is performed. The torque reductions obtained are shown in Fig. 6 for rated DC bus voltage and $1.15\ V_{bus\ rated}$. As can be observed, the major reductions are obtained again when priorizing the d-axis (indicated by the black line) and considering $1.15\ V_{bus\ rated}$ on the back-to-back converter.

It is important to notice that not only the torque spikes are reduced but also the associated mechanical energy, which is not absorbed by the shaft thanks to the PMG performance. This energy can be the cause of multiple problems i.e. mechanical vibrations and resonances.

Summary

This paper begins with the introduction of the xFIG topology, an enhanced DFIG application that presents several benefits respect to the conventional one. Among the outstanding benefits can be highlighted the grid-harmonics reduction, the suppression of the grid-harmonics filter, and the torque reductions during grid faults. The LVRT capability of this application is then proved.

Furthermore, for the purpose of reducing the mechanical stress on the shaft, special effort has been employed in decreasing torque spikes. To this purpose, the Permanent Magnet Generator coupled to the mechanical shaft of the enhanced system is used. Different measures involving the system control and GSC bus voltage have been analyzed. The simulation proves that the major torque reductions are obtained when the *d-axis* is priorized and the GSC DC voltage level raised, obtaining reductions up to 81% with regard to the conventional DFIG system.

Acknowledgements

This work was supported in part by the Spanish Ministry of Economy and Competitiveness under grant DPI2013-42853-R.

The authors gratefully acknowledge INGETEAM POWER TECHNOLOGY for their financial and permanent support.

References

[1] P. S. Flannery and G. Venkataramanan, "A Fault Tolerant Doubly Fed Induction Generator Wind Turbine Using a Parallel Grid Side Rectifier and Series Grid Side Converter," *IEEE Trans. Power Electron.*, vol. 23, no. 3, pp. 1126–1135, May 2008.

[2] G. Abad, J. Lopez, M. A. Rodriguez, L. Marroyo, and G. Iwanski, *Doubly Fed Induction Machine.* 2011.

[3] P. Es and J. Perez, "Patent Application Publication US 2007/0216164 A1," no. 19, 2007.

[4] O. Abdel-Baqi and A. Nasiri, "Series Voltage Compensation for DFIG Wind Turbine Low-Voltage Ride-Through Solution," *IEEE Trans. Energy Convers.*, vol. 26, no. 1, pp. 272–280, Mar. 2011.

[5] Xiangwu Yan, G. Venkataramanan, P. S. Flannery, and Yang Wang, "Evaluation the effect of voltage sags due to grid balance and unbalance faults on DFIG wind turbines," in *2009 International Conference on Sustainable Power Generation and Supply*, 2009, pp. 1–10.

[6] H. Karmaker, M. Ho, D. Kulkarni, and E. Chen, "Design studies for a 10 MW direct drive superconducting wind generator," in *IECON 2014 - 40th Annual Conference of the IEEE Industrial Electronics Society*, 2014, pp. 497–501.

PCIM Europe 2016, 10 – 12 May 2016, Nuremberg, Germany

Control and modulation for loss minimization for dc/dc converter in wind farm

Catalin Dincan, Department of Energy Technology, Aalborg University, cgd@et.aau.dk
Philip Kjær, Department of Energy Technology, Aalborg University, pck@et.aau.dk

Topic Number:10.1; Topic Name: New and Renewable Energy Systems; Wind Farms.
Preferred Presentation Form: Poster Presentation

Abstract

For a DC wind turbine, a single phase series-resonant converter for unidirectional power is studied. This paper aims to identify and compare impact on electrical losses and component ratings from the choice of three candidate control strategies. The evaluation is purely based on circuit simulations and offline post-processing of losses. The initial findings indicate that lower losses are obtained in discontinuous current mode (DCM) conditions, with drawbacks on transformer design.

Series Resonant Converter

HVDC offshore wind farms with MVDC collection grid (Fig.1) promise an increase in efficiency and a reduced bill of materials [1]. One of the key components is the DC/DC converter used in the wind turbine. The proposed DC/DC topology in this paper is a unidirectional series resonant converter (SRC), composed of: inverter, resonant tank, monolithic transformer and medium voltage (MV) rectifier built with series connected diodes. The concept is illustrated in Fig.2 and design specifications are in Table 1.

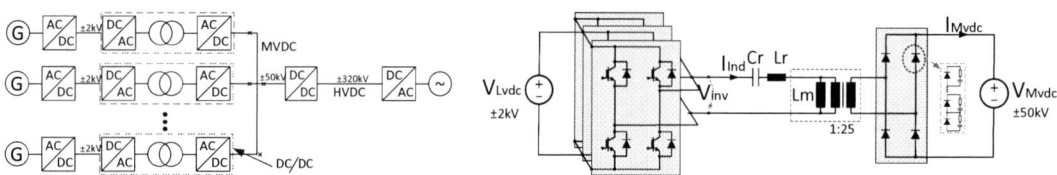

Fig. 1 DC Wind farm diagram   Fig. 2 Converter circuit

Despite the SRC's limitations in control range, it offers an attractive trade-off between component cost (ratings) and electrical losses. Principal advantages are ZCS at turn-off for LV and MV devices, transformer sinusoidal currents and reduced content of harmonics. Bidirectional single phase topology was reported in [2] for railway applications and in [3] a three phase topology was introduced.

Nominal Power	P_n	10 MVA	DC/AC Device	3x4 x IGBT (6500V x 750A)[4]
Input DC Voltage	V_{LVDC}	±2 kV	AC/DC Device	4x40 x Diode (6500V x 750A)[4]
Output DC Voltage	V_{MVDC}	±50 kV	Isolation level	150 kV
Resonant Capacitor	C_R	78 uF	$I_{Short\_circuit\_MVDC}$	50 x I_{MVDC}
Resonant Inductor	L_R	250 uH	E_{cap}	2500 J
Magnetizing Inductor	L_m	20 mH	E_{ind}	2500 J
Table 1 Converter Nominal Specifications				

© VDE VERLAG GMBH · Berlin · Offenbach

Problem formulation

The goal of this paper is to apply a methodology which evaluates the impact of different control strategies and modes of operation on losses, stress and component ratings. The evaluation is based purely on circuit simulations.

Methodology

The overall system (control and model of the converter) are performed by using a circuit simulator [5]. The SRC is modelled to estimate the current flow through the devices. It uses ideal switches; stray inductance, capacitance and dead time are not included. Analytical expressions are used to build the semiconductors loss model, while transformer losses are calculated in offline post processor. Finally a comparison of efficiencies and component ratings for different control methods is achieved and ranking of most stress components is performed. These results will impact the resonant converter specifications and design drivers.

Control methods

Three different control methods are discussed: frequency control [6], phase shift [7] and the dual control [8]. Other methods are discussed in [9]. By frequency control of input voltage, the resonant tank impedance is changed. The phase shift method is controlling applied voltage to the resonant tank by changing the duty cycle of the square wave, while having constant excitation frequency. The third method is achieved by combining the two previous methods, being able to control output voltage and the switching current. In [8] phase shift control is implemented on a single phase, bidirectional topology. In [2], the half cycle discontinuous mode series resonant converter (HC-DCM-SRC) is analyzed for traction application.. On the other hand, in [3] a three phase topology was introduced, claiming that frequency control in resonant and super resonant conditions offer highest efficiency.

Modes of operation

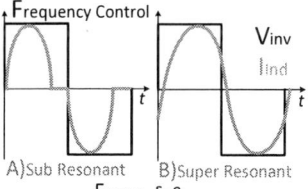
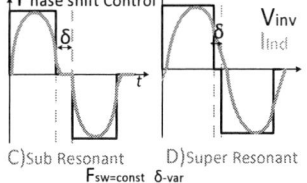
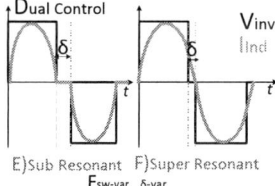

Fig. 3 Modes of operation

For each control strategy, two modes of operation are investigated: sub-resonant and super-resonant (Fig. 3). For sub-resonant, only DCM is considered. In sub-resonant mode, tank current leads applied inverter voltage (Fig. 3 *A,C,E*), while in super resonance it lags (Fig.3 *B,D,F*). In sub-resonant mode, ZCS is possible at turn off, while in super-resonant mode, ZVS is present at turn on. Phase shift control allows passives design at constant frequency, while in frequency control, they have to be designed for lowest switching frequency. Dual control claims to reduce losses compared to phase shift in the entire operational range. No former publication comparing SRC losses with these control strategies and two modes of operation has been identified. The goal of this paper is to fill the gap.

© VDE VERLAG GMBH · Berlin · Offenbach

Converter controller

Controller structure is shown in Fig. 4. As the turbine's generator rectifier maximizes the wind energy extraction, the dc/dc converter controls the dc-link voltage. Changes in captured wind power disturb the dc-link voltage, in turn requiring changes to the dc/dc converter power transfer. In the case of sub-resonant frequency control, output power is dependent on the amount of energy transferred to the MVDC link. For a given frequency and duty cycle, output power is a function of number of energy pulses transferred to the output. DC link voltage control is achieved by transferring a certain integral number of pulses to the output. In this case two controllers are used. P_{ref} comes from the dc/dc converter's own dc-link voltage controller and when divided by the measured V_{MVDC}, provides the current reference I_{MVDC}. This current is the output state variable, which is measured and averaged. The feedforward controller equation is determined by the mode of operation, while the PI feedback controller complements the feed-forward controller by correcting its inaccuracies. For frequency control, the output control signal for modulator block is frequency, while for phase shift, it is duty cycle. In case of dual control, both frequency and duty cycle are control variables.

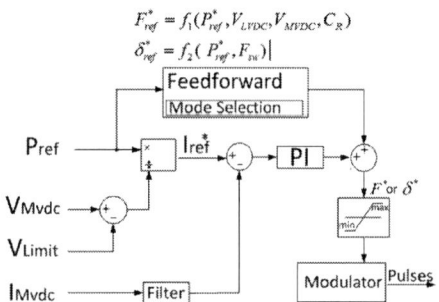

Fig. 4 Controller structure

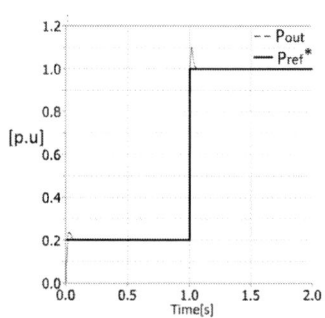

Fig. 5 Output power step response

Fig. 5 shows the output power step response for frequency control, operating in sub-resonant mode. Further on, in Fig.6 and Fig.7, the function of output power to frequency and duty cycle is presented. Two phenomena related to frequency control are noteworthy: as the converter is designed to operate in discontinuous mode in sub-resonance, output power can easily be controlled by frequency in a linear mode. On the other hand, in sub resonance mode, the frequency range is from 1160 to 1200Hz for operation between 1 pu and 0.2 pu output power, and it's not a linear characteristic. Fig.7 indicates that output power can indeed be controlled at constant frequency, by changing the duty cycle of the applied inverter voltage, but in a very limited range. Considering that a change in duty cycle from 0 to 0.225% changes output power from 1 to 0.2pu, the sensitivity of the modulator's impact on output power is a challenge.

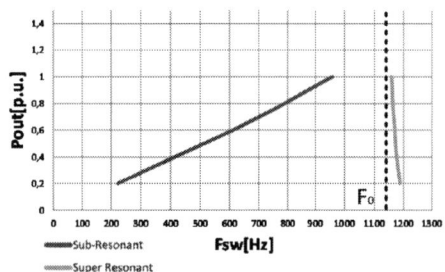

Fig. 6 Pout = f(Fsw)

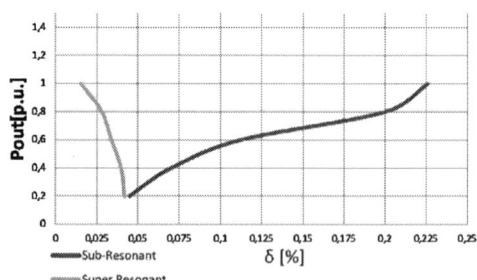

Fig. 7 Pout = f(δ)

Transformer design

The transformer is expected to be the key component in determining the circuit losses. As no standard component is available for the application in mind, dimensions for a preliminary design were used as input to a simplified calculation of transformer losses. Design methodology is similar to [10],[11] and [12]. Only one transformer is used in this converter, with a single primary and a single secondary winding. As no validation has yet been performed in laboratory and to reduce loss model errors, identical transformer designs are proposed for all three modes of operation analyzed. In this manner the same active material is employed. Final transformer specifications are listed in Table 2.

A standard C-core structure, based on amorphous material is pre-selected, with the maximum available size of [13]. In Fig. 8 the winding arrangement is shown. Each leg has a LV and HV winding. Copper foils are used for both. The LV winding is built with a layered construction of Np turns, while HV winding has a structure consisting of 37 layers and 25 turns.

The insulation level is set to 150kVAC. NOMEX paper is proposed for windings insulation and mineral oil as main insulation material and coolant. Voltage level and insulator dielectric strength determines minimum distance between primary and secondary $Dins$. According to [11], the insulation level is calculated with:

$$D_{ins} = \frac{V_{ins}}{\lambda \cdot E_{ins}} \tag{1}$$

where E_{ins} is the dielectric strength of the isolation material (in this case oil and paper) and λ is used as a safe margin parameter. In this paper, dielectric losses are neglected, but it needs to be mentioned that increasing the operating frequency and the insulation requirements, the losses can not be neglected in the actual prototype.

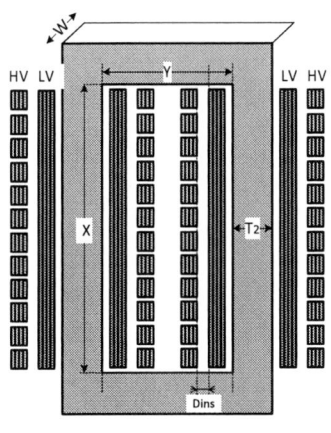

Fig. 8 Transformer drawing

Np (Primary turns)	37 (37 layers x 1 Turn),foil
Ns (Secondary turns)	925 (37 layers x 25 Turns),foil
X (Window height)	1 m
Y (Window width	0.2240 m
T2 (core build)	0.160 m
W(ribbon width)	0.213 m
Core material	Amorphous [13]
Weight of core material	820 kg
Weight of windings	750 kg
Dins	0.0250 m
Bsat	1.63 T
Isolation level	150 kV
Oil dielectric strength	10 kV/mm

Table 2 Transformer specifications

Loss Model

Semiconductors and transformer loss models are presented in Fig. 9 and Fig. 10.

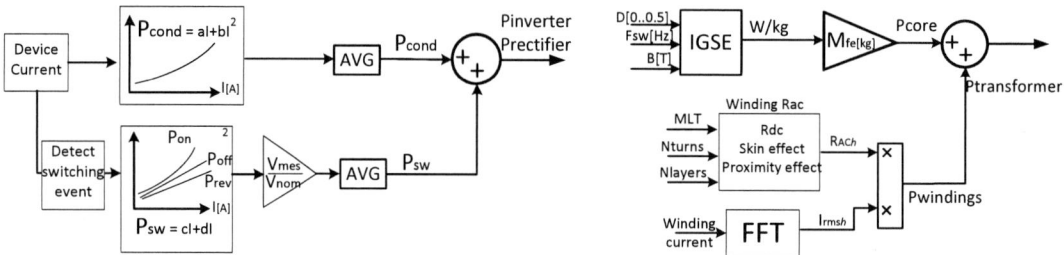

Fig. 9 IGBT and Diode loss model   Fig. 10 Transformer loss model

Conduction losses

Methodology proposed in [14],[15] is used for conduction loss modeling for both IGBTs and diodes. The current is multiplied with the according voltage directly from the data sheet for the highest acceptable temperature, e.g. $T=125°C$ to extract conduction power loss. Afterwards, the curve is approximated with 2nd order polynomial fitting curves and is described in (2), which uses the current through the ideal switch as input and outputs conduction loss of the device during the simulation. The output is averaged for 1 switching cycle.

Switching losses

Switching losses are determined in similar way like in [14],[15]. Current dependent E_{ON}, E_{OFF} and E_{REC} are given in the device datasheet and are considered for a maximum junction temperature of $T=125°C$. This dependency is approximated with a second order polynomial fitting curve described in   (3), and multiplied with voltage factor V_{mes}/V_{nom}, where V_{nom} is datasheet parameter and V_{mes}, actual applied voltage. Whenever a switching event occurs, losses are calculated and then averaged for 1 switching cycle.

$$P_{Cond} = a \cdot I + b \cdot I^2 \qquad (3) \qquad\qquad P_{Sw} = c \cdot I + d \cdot I^2 \qquad (4)$$

Core losses

Different methods have been compared in [10],[11],[12],[16] for core losses. In the present loss model, the *Improved Generalized Steinmetz Equation* (IGSE*)* described in [11] was used with K_i, α and β determined from [16].

$$P_{Core} = K_i \cdot 2^{\alpha+\beta} \cdot f^{\alpha} \cdot \hat{B}^{\beta} \cdot D^{1-\alpha}$$

Winding losses

Foil winding losses are calculates, as according to [10],[11]. The expression from (4) is explained in [10]. The overall losses are calculated by summing the effect of every current harmonic. Skin effect losses are frequency dependent, while proximity losses are influenced by the number of layers. D is foil thickness, δ is skin depth and m is number of layers.

$$P_{Winding} = R_{DC} \cdot \frac{D}{\delta} \left[\frac{\sinh\left(\frac{D}{\delta}\right) + \sin\left(\frac{D}{\delta}\right)}{\cosh\left(\frac{D}{\delta}\right) - \cos\left(\frac{D}{\delta}\right)} + \frac{2 \cdot (m^2-1)}{3} \frac{\sinh\left(\frac{D}{\delta}\right) - \sin\left(\frac{D}{\delta}\right)}{\cosh\left(\frac{D}{\delta}\right) + \cos\left(\frac{D}{\delta}\right)} \right] \cdot I_{rms}^2 \qquad (5)$$

Results

Comparison of losses for the SRC, operated with three different control strategies (frequency, phase shift and dual control) and two modes of operation (sub-resonant and super resonant) are illustrated in Fig. 11 and Fig. 12, for 1pu and 0.5pu output power. First remark in both figures is that transformer core and rectifier losses are similar in all cases, while inverter and winding losses are smaller with frequency control strategy and sub resonant mode.

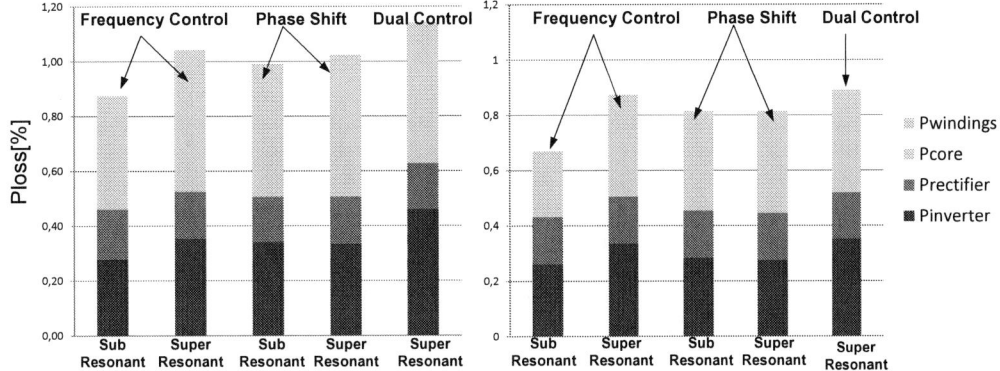

Fig. 11 Losses at 1pu output power   Fig. 12 Losses at 0.5pu output power

Resonant tank stress was evaluated at 1pu output power. Peak inductor current and capacitor voltage are presented in Fig. 13. Highest stress level is present in frequency control-sub resonant, regardless of output power, while for the rest is load dependent. Peak tank energy is show in Fig. 14.

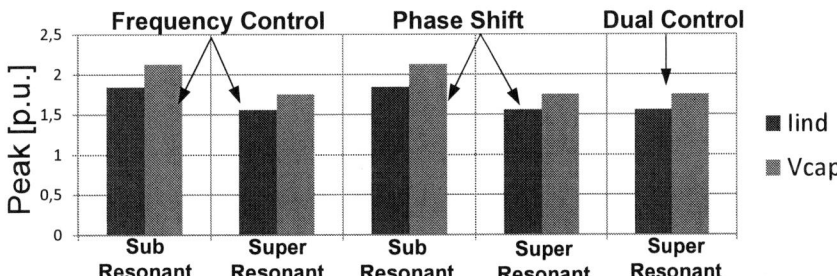

Fig. 13 Inductor current and capacitor voltage pu

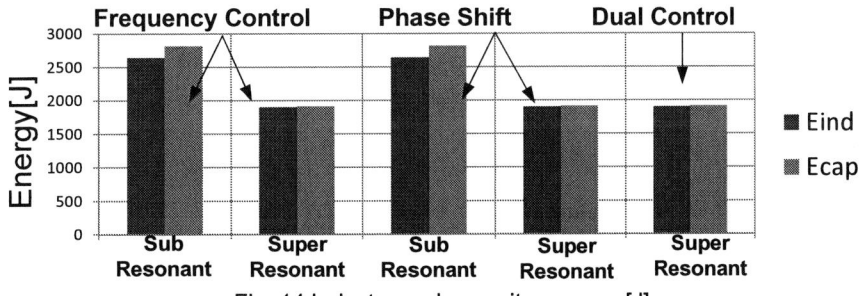

Fig. 14 Inductor and capacitor energy [J]

Discussions

The results in Figures 11 and 12 are now discussed. Firstly, the following were neglected in the loss analysis: (i) Rectifier snubber and resonant tank losses; (ii) ZCS at turn-off has been assumed ideal (truly zero), yet [2] has shown the turn-off losses are not negligible. Planned future characterization of the chosen semiconductors will allow extraction of soft-switching parameters r and k_s for better loss modelling; (iii) the dual control strategy running in sub-resonance mode is not added to the comparison, as the strategy limits operation above 0.5pu output power. For the SRC, windings losses will be predominant, as they are linear with the squared rms current; therefore transformer characterization is necessary to decrease uncertainties on the windings loss model.

Lowest total losses are found in frequency control sub-resonant mode (0.85%), while the highest total losses occur with dual control in super-resonant mode (1.14%), both at 1pu output power. The difference in simulated losses between all modes is relatively small and is subjective to errors in the loss model, the highest uncertainty being the winding loss model. According to [17], windings loss model is overestimated with ≈15% error. Losses in frequency control sub resonant mode appear smaller, due to small turn on current, while at turn-off they are zero. In super-resonant mode, for all control strategies, ZVS appears at turn on, but turn off losses are considerably higher and winding losses are increased due to higher ac resistance. At elevated frequency, skin and proximity effects increase the winding ac resistance. A strong limitation of frequency control-sub resonant mode is the influence on transformer design, which needs to be designed for the lowest operating frequency. [18] States that no voltage is applied to the transformer during the discontinuous subintervals below 0.5p.u operation. This is true, if the ratio input voltage/output reflected voltage is 2. In current design, the ratio is 1 and it is able to operate in the range from 500Hz and above, as seen in Fig. 16.

Phase shift control can allow designing the transformer for a constant frequency, but the narrow range of duty cycle to output power function will bring sensitivity issues and very low variations in the duty cycle could trigger high power fluctuations. The authors have not yet studied to which degree this sensitivity can be managed in practice. On the other hand, phase shift control does offer an extension to the operational range of frequency control in sub resonant and super resonant mode. Below 0.5pu, phase shift control could be applied, as seen in Fig. 15. Regarding resonant tank stress, in both modes, the tank should be designed for peak values of voltage and current. The peak value is constant in sub-resonant mode (2pu for capacitor voltage and 1.8p. for inductor current) and load dependent in super-resonant mode (1.7pu for capacitor voltage and 1.5pu for inductor current). Peak values will determine selection of semiconductors and number of parallel inverters.

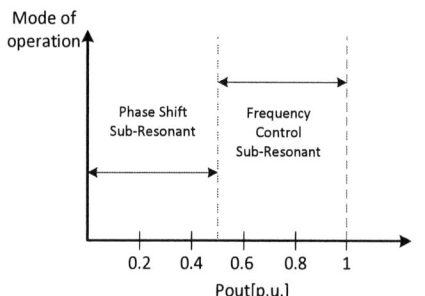

Fig. 15 Range increase in sub-resonant

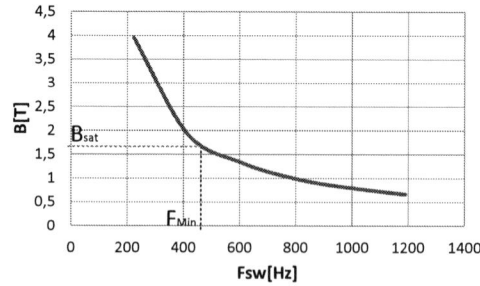

Fig. 16 $B_{pk} = f(Fsw)$

Conclusions

A single phase, monolithic and unidirectional series resonant converter has been studied for applications in DC wind turbines. A methodology of evaluating the impact of three different control strategies (frequency, phase shift, dual control) on losses and stress has been applied. Semiconductors losses and tank stress were investigated with circuit simulator, while transformer losses calculated in off-line post processor. Frequency control operated in sub-resonance mode has the potential of minimum losses (0.85% in our study, some 10%-25% lower than the compared control strategies) while paying the price of designing the transformer for the lowest operating frequency, which will yield a bulkier design. Tank stress is around 20% higher in frequency control sub-resonant mode, compared to the other strategies and modes. Phase shift and dual control strategies have a narrow control range and could generate high power fluctuations for this particular design. The difference of losses between all modes is small and subjective to loss model error. The highest uncertainty is the windings loss model, which are predominant.

References

[1] C. Meyer, Key Components for Future Offshore DC Grids. 2007.

[2] G. Ortiz, H. Uemura, D. Bortis, S. Member, J. W. Kolar, and O. Apeldoorn, "Modeling of Soft-Switching Losses of IGBTs in DC / DC Converters,"

[3] J. Jacobs, A. Averberg, and R. De Doncker, "A novel three phase series resonant converter for high power Applications," 2004.

[4] "www.infineon.com."

[5] "www.plexim.com." .

[6] R.L.Steigerwald, "A comparison of Hald Bridge Resonant Converter Topologies,"

[7] M. Nakaoka, S. Nagai, Y. J. Kim, Y. Ogino, and Y. Murakami, "The state-of-the-art phase-shifted ZVS-PWM series and parallel resonant DC-DC power converters using internal parasitic circuit components and new digital control,"

[8] P. Ranstad, H. H.-P. H. Nee, and J. Linner, "A novel control strategy applied to the series loaded resonant converter," 2005 Eur. Conf. Power Electron. Appl., pp. 1–10, 2005.

[9] R. Oruganti and F. C. Lee, "Resonant power processors. II - Methods of control," vol. I, no. 6, pp. 1461–1471, 1984.

[10] I. Villar, A. Garcia-Bediaga, U. Viscarret, I. Etxeberria-Otadui, and A. Rufer, "Proposal and validation of medium-frequency power transformer design methodology," [11] G. Ortiz, J. Biela, and J. W. Kolar, "Optimized design of medium frequency transformers with high isolation requirements,"

[12] M. a Bahmani and T. Thiringer, "Design Methodology and Optimization of a Medium Frequency Transformer for High Power DC-DC Applications,"

[13] "http://www.hitachi-hqt.cn/en/core/index.html."

[14] J. W. Kolar, "A General Scheme for Calculating Switching- and Conduction-Losses of Power Semiconductors in Numerical Circuit Simulations of Power Electronic Systems."

[15] K. Lee, Y. Suh, and Y. Kang, "Loss Analysis and Comparison of High Power Semiconductor Devices in 5MW PMSG MV Wind Turbine Systems,"

[16] R. U. Lenke and R. U. Lenke, E . ON Energy Research Center A CONTRIBUTION TO THE DESIGN OF ISOLATED DC-DC CONVERTERS FOR UTILITY APPLICATIONS A Contribution to the Design of Isolated DC-DC Converters for Utility Applications. .

[17] I. Villar, U. Viscarret, I. Etxeberria-Otadui, and A. Rufer, "Global loss evaluation methods for nonsinusoidally fed medium-frequency power transformers,"

[18] R. W. Erickson, Fundamentals of Power Electronics, Second edition. .

A Variable Step Size Perturb and Observe Method Based MPPT for Partially Shaded Photovoltaic Arrays

Jawad Ahmad, F. Spertino, P. Di Leo, A. Ciocia
Energy Department, Politecnico di Torino,
Corso duca degli Abruzi 24, Torino, Italy

Abstract

Due to the effect of partial shading, multiple peaks appear in the Power vs. Voltage (*P-V*) characteristics of Photovoltaic (PV) array. If a conventional Maximum Power Point Tracking (MPPT) algorithm is applied to a partially shaded PV array it may result in the tracking of a Local Peak (LP) instead of Global Peak (GP). It is therefore important to apply special MPPT technique for tracking the GP under partial shading conditions. Many methods reported in the literature perform blind scanning of the *P-V* curve in order to find the GP. This type of scanning takes a long time during which the array is not operated at the MPP which results in a considerable amount of power loss. In this paper an MPPT algorithm is proposed which quickly tracks the GP among various LPs. Simulation results show that the proposed algorithm tracks the GP in a much shorter time than other methods reported in the literature. A modified form of the conventional Perturb and Observe method is employed to accelerate the tracking process.

1. Introduction

Due to environmental pollution and depletion of fossil fuels, interest in renewable energy has increased in recent years. Among all the renewable sources of energy, solar Photovoltaic (PV) energy has seen a rapid growth due to the free and abundant availability of sunlight. PV cells directly convert sunlight into electrical energy. One of the shortcomings of the PV systems is their low conversion efficiency and nonlinear Power vs. Voltage (*P-V*) characteristics. For optimal utilization of the solar arrays Maximum Power Point Trackers (MPPT) are used. Some of the well-known conventional MPPT techniques are Incremental Conductance, Perturb and Observe (P&O), and Ripple Correlation. These techniques have been surveyed in [1].

For getting a required level of power various PV modules are connected together in series, parallel, and series-parallel combination to form PV arrays. If a module in a PV array receives less irradiance, it behaves like a load and dissipates some of the power generated by the array. The dissipation of power inside shaded module causes local overheating and hot spot problems. If power dissipation inside the shaded module exceeds a certain limit, irreversible damage to the PV cells may occur. To overcome this problem, a bypass diode is used across a certain number of cells inside a PV module. This method allows the generated current to flow in correct direction, even if some of the modules do not receive full irradiance.

The use of bypass diodes, however, results in another problem in electrical characteristic of a PV array. Because of the action of these diodes, multiple peaks appear in the *P-V* curve during non-uniform irradiance. Partial shading is one of the common problems in the building integrated PV systems due to the shadows cast by surrounding buildings, trees, and utility poles. When an MPPT algorithm tracks a Local Peak (LP) instead of a Global Peak (GP) the resulting power loss can be as high as 70% [2]. Out of 1000 building integrated PV systems installed in Germany, 41% suffered from partial shading [3].

For dealing with partial shading conditions, various techniques have been developed which are surveyed in [4] and [5]. The techniques proposed in the literature can be broadly classified as Hardware based and Software based. Some of the well-known hardware techniques are: Submodule Integrated MPPT Controllers, Parallel Connected MPPTs, Multi-level Converters, and Power Electronic Equalizers etc. [6]. Software based techniques include: Stepped-up Chaos Optimization, Differential Evolution Technique, Artificial Neural Network based, Ant Colony Optimization, Fuzzy Logic control based, Fibonacci Search Algorithm, and Particle Swarm Optimization based algorithms. The disadvantages of using hardware MPPTs are decreased reliability and efficiency, increased system complexity, and implementation cost. Software techniques usually have complex algorithms and may need powerful microcontrollers which results in increased system cost [7].

PV arrays are periodically scanned for the presence of partial shading and finding the GP. During the process of tracking the GP, the array is not operated at the Maximum Power Point (MPP) which inevitably results in wastage of power. The time required for performing the scanning is of critical importance. According to [8] such a scanning can take as long as 8.6s. Techniques proposed in [8] and [9] avoid complete scanning of the P-V curve and the time required for tracking GP is reduced to 1.1s and 1.2s respectively. The MPPT method presented in [10] takes about 5s. The algorithm proposed in this paper needs much shorter time for tracking a GP during partial shading. The proposed algorithm utilizes modified version of P&O algorithm to speed-up the tracking process. The algorithm can identify that whether the array is under uniform irradiance or partial shade and also determines the number and location of peaks present and thus avoids blind scanning of the P-V curve.

This paper is further organized as follows:

In section 2 we discuss the characteristics of a PV array under partial shading conditions. Section 3 deals with modified P&O algorithm. In section 4 the proposed algorithm is described in detail. Section 5 gives the simulation results and section 6 concludes this paper.

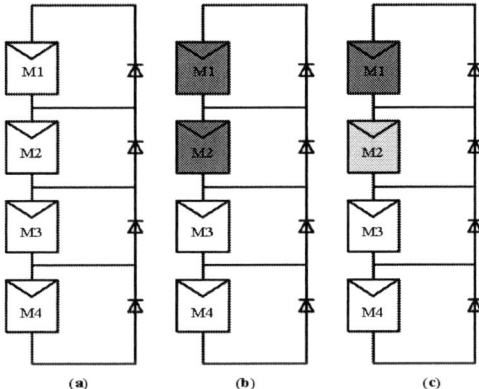

Fig. 1 PV array under different shading conditions. (a) All the modules receiving 1000W/m². (b) Partially shaded array with M1 and M2 receiving 400W/m². (c) M1 receiving 400W/m², M2 700W/m², M3 and M4 receiving 1000W/m²

2. Characteristics of Photovoltaic Arrays under Partial Shading Conditions

For tracking of GP under partial shading conditions, the P-V curve of an array is periodically scanned. To avoid unnecessary scanning of the curve, it is important to devise a mechanism for determining the presence of partial shading condition. Here we briefly discuss some of the characteristics of PV array under non uniform irradiance. Let us consider Fig. 1(a) in which all the modules inside the array receive 1000W/m2 irradiance. The characteristic curves of this array are shown in Fig. 2. It can be noticed that the P-V curve has a single peak namely P1 and the I-V has a single step. When the array is operated at any point, each of the module works as a source and has identical voltage across it which is an evidence for uniform irradiance. However, this situation changes when partial shading occurs.

Consider the case of Fig. 1 (b) in which two of the modules M1 and M2 in the array are receiving 400W/m² irradiance and M3 and M4 receive 1000W/m². Fig. 3(a) and 3(b) show the P-V and I-V curve of the array respectively; whereas Fig. 3(c) shows the voltage of each of the modules as a function of array voltage. Referring to Fig. 3 if the array is operated at P1, all of the series connected modules M1, M2, M3 and M4 supply power to the load. In this case the current of the array is controlled by modules M1 and M2 which receive lesser irradiance. As shown in Fig. 3(c) the voltage across the modules M1 and M2 (i.e. VM1 and VM2 respectively) is identical, whereas that across M3 and M4 (VM3, and VM4) is the same. However, the voltage across the modules receiving different irradiance is different. From the difference of voltages across the modules the proposed algorithm determines the number of peaks and their location on the *P-V* curve which is explained in detail in section 4.

Looking at Fig. 3(b), if we move along the curve from right to left we see that at about 62V, the current of the array begins to rise and modules M1 and M2 cannot supply the current. In these conditions when the array is operated below 62V the modules are bypassed and the voltages VM1 and VM2 become equal to the negative of the diode voltage. However M3 and M4 continue to supply power to the load. We can also observe from Fig. 3 that the number of peaks present on the *P-V* curve depend upon the number of different irradiance levels that are falling on the array. In practical PV systems the situation similar to Fig. 3 is usually encountered. However in some extreme cases more than two peaks may appear. This situation is depicted in Fig. 1(c) with the corresponding characteristic curves shown in Fig. 4.

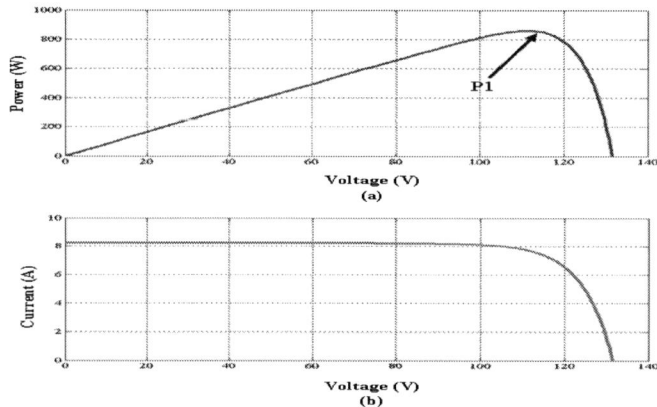

Fig. 2 Characteristic curves of the array in Fig. 1 (a). (a) P-V curve, (b) I-V curve

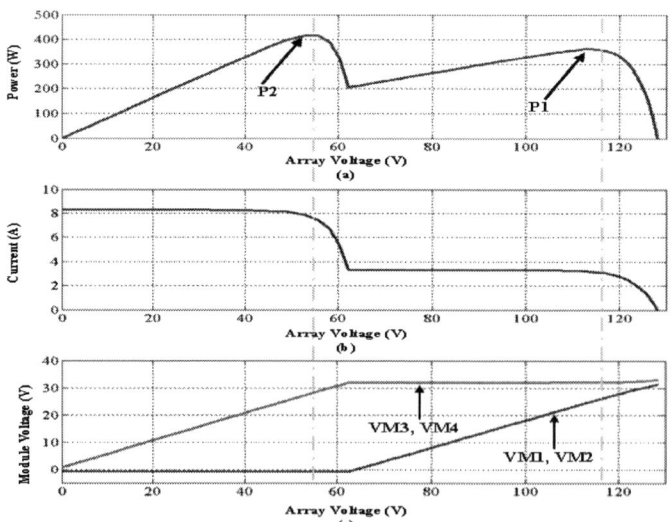

Fig. 3 Characteristic curves of the array corresponding to Fig. 1 (b). (a) P-V curve, (b) I-V curve, (c) Modules voltages as function of array voltage

PCIM Europe 2016, 10 – 12 May 2016, Nuremberg, Germany

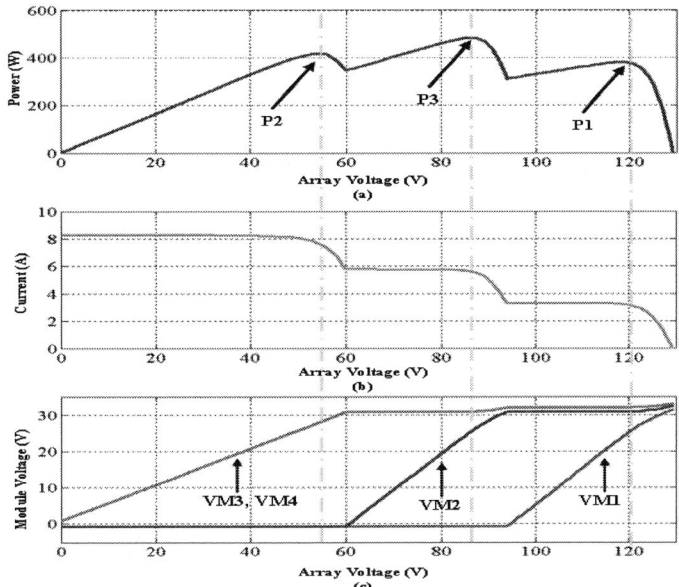

Fig. 4 Characteristic curves of the array corresponding to Fig. 1 (c). (a) P-V curve, (b) I-V curve, (c) Modules voltages as function of array voltage

3. Modified P&O Algorithm

Conventional P&O algorithm is a well-known MPPT method which is commonly employed for PV arrays operating under uniform irradiance. The basic principle of the method is that the operating voltage of PV array is perturbed in a certain direction. If there is an increase in the output power as a result of the perturbation, it means that the operating point of the array has moved towards the MPP. In this way, next perturbation is applied in the same direction. If, however, the power drawn from array is reduced, it means that the operating point has moved away from the MPP and next perturbation is applied in the opposite direction. This process of changing the operating point of array and observing the output power is repeated until the MPP is tracked.

The MPPT algorithm proposed in this paper uses a modified form on P&O method to accelerate the GP tracking. In the modified P&O algorithm, an MPP is tracked in two modes i.e., the Voltage search and the MPP search mode. The modified P&O algorithm is explained with the help of Fig. 5, which shows the MPP tracking process for uniform irradiance. The voltage search mode is employed for getting close to a certain reference voltage. The reference voltage gives approximate location of a certain peak (LP or GP in case non uniform irradiance). The MPP search mode performs fine tuning and it is employed for taking the operating point of the PV array to the nearest peak. The size of the duty ratio perturbation step (ΔD) is kept large during the voltage search mode. The value of ΔD monotonically decreases as the operating point gets closer to a reference voltage. After getting sufficiently close to the MPP at t1 (in Fig. 5) the MPP search mode begins in order to converge to the peak. The MPP search mode is also based on the principle of monotonically decreasing P&O method [11-12]. Each time the operating point of the PV system passes through the MPP the size of ΔD decreases. This mode lasts till the system converges to the actual peak at time t2. At this stage if the tracked peak turns out to be GP (which is the case in Fig. 5) and there is no other peak to be tracked, a constant step size P&O (conventional P&O) method is employed to begin the steady state operation around the GP. On the other hand, if there is another peak that is to be tracked then the proposed algorithm goes through the process of Voltage search mode and the MPP search mode to track the remaining peak/peaks as explained above. This process goes on till all the peaks are tracked one by one. The system then operates around the GP in the steady state.

© VDE VERLAG GMBH · Berlin · Offenbach

PCIM Europe 2016, 10 – 12 May 2016, Nuremberg, Germany

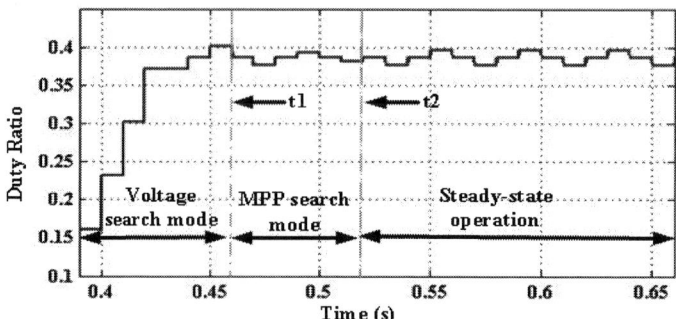

Fig. 5 Variation of Duty ratio in variable step size P&O method

4. Proposed Algorithm

The operation of the proposed algorithm is explained with the help of the flow chart shown in Fig. 6. At the start of the algorithm the Reference Voltage (V_{REF}) is set at about $0.80*V_{OC}$, where V_{OC} is the open circuit voltage of the array. This is the approximate location of the peak closest to the V_{OC} of the array. This peak is tracked first by applying modified P&O algorithm discussed in section 3. When this peak is tracked, the voltage across each of the module in the array is measured. If there is uniform irradiance all of the modules in the array will have almost equal voltage. In this case, the conventional P&O algorithm with constant ΔD is invoked for operation of the system around the MPP in the steady-state. The GP tracking is only performed if there is an indication of the presence of partial shading. In this way unnecessary tracking of GP is avoided.

During the steady state operation, in case of the timer overflow or if the absolute difference in the output power of the array and the MPP power exceeds a certain value 'ε' the GP track subroutine is called by the main program. As shown in Fig. 6, the peak that is at $0.80*V_{OC}$ is tracked first. If the array is found to be under partial shade, the value of 'b' is calculated. The variable b represents the number of modules which receive lesser irradiance than the module(s) that receive the highest irradiance. Referring to Fig. 3(b), the voltage V2 at which P2 occurs is then determined by the following relation [8]:

$$V2 = \left(1 - \frac{b}{N}\right) * 0.85 * VOC \tag{1}$$

Where N is the total number of the series connected modules in the array.

In some extreme cases, there may be more than two peaks present on the characteristic curve of the array. This is shown in Fig. 1 (c) and the corresponding *P-V* curve is shown in Fig 2 (c). The location of P1 and P2 are determined in the manner discussed in the previous paragraph. To determine the presence and location of any other peak between P1 and P2, the algorithm takes the subset of the total number of the PV modules which receive lesser irradiance than the module(s) that receive the maximum irradiance. This subset contains b No. of modules. From the subset a number c is determined. This number corresponds to the number of modules receiving lesser irradiance than the maximum irradiance in the subset with b No. of modules. Thus, with reference to Fig. 2(c), the voltage V3 at which P3 occurs is determined using the relation:

$$V3 = \left(1 - \frac{c}{N}\right) * 0.85 * VOC \tag{2}$$

© VDE VERLAG GMBH · Berlin · Offenbach

PCIM Europe 2016, 10 – 12 May 2016, Nuremberg, Germany

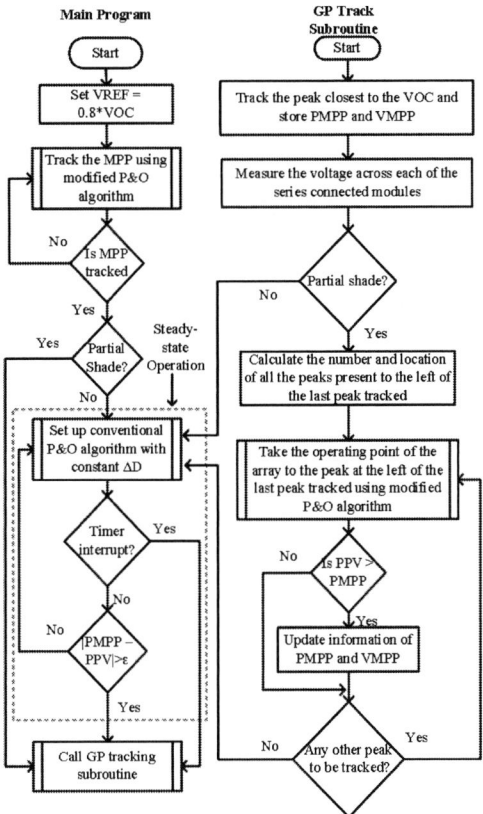

Fig. 6 Flowchart of the proposed algorithm

In this way we can determine the number and location of any number of peaks present due to partial shading. However, in this paper we will restrict our analysis to the presence of two and three peaks for simplicity.

In order to understand the operation of the proposed algorithm with example, consider the case of Fig. 1 (b) with the $P\text{-}V$ curve shown in Fig. 3 (a). When this situation occurs, the main program calls for the GP track subroutine and P1 is detected using modified P&O method. The values of the power corresponding to that peak P_{MPP}, the duty ratio of the DC-DC converter D_{MPP}, and voltage V_{MPP} are recorded. The algorithm measures the voltage of each of the series connected modules and finds (VM1 = VM2 = 24.85V, and VM3 = VM4 = 32.01V) which indicates the presence of two irradiance levels. As we can see from the voltage of the modules, M3 and M4 receive the maximum irradiance. We then take the subset which contains the modules that receive lesser irradiance than M3 and M4. This subset is composed of M1 and M2 which means that b = 2. As M1 and M2 receive the same irradiance, from which we obtain c = 0. The voltage and power at P1 are already known. With V_{OC} = 132V the location of P2 is found to be at 56V using (1). The algorithm then proceeds to track this peak using modified P&O algorithm discussed in section 3. After tracking P2 the power corresponding to the peak is compared to the value at P1 for finding the GP. The algorithm then starts operation at the GP in the steady state.

Let us consider the case of Fig. 1 (c) with the characteristic curve shown in Fig. 4. In the manner discussed previously, the MPPT first tracks P1. The voltages of the modules are found to be (VM1 = 21.79V, VM2 = 30.64V, VM3 = 31.98V, VM4 = 31.98V). This indicates three different voltage levels which mean that there are three peaks present on the P-V curve. Similar to the previous case we obtain b = 2, as M3 and M4 have the highest irradiance. This means that one of the peaks (P2 in Fig. 4(c)) is at V2 = 56V which is obtained using (1). From the subset containing M1 and M2 we see that M2 receives the highest irradiance than M1 from which c is found to be 1 which indicates the presence of P3 and its location is at 84V using (2). Having already known the values of power at P1, the

© VDE VERLAG GMBH · Berlin · Offenbach

algorithm operates the system respectively at P3 and P2. Out of these peaks the GP is determined followed by the steady-state operation.

5. Simulation Results

The proposed MPPT algorithm was simulated using MATLAB/SIMULINK. The PV array used in simulations has four series connected modules. The rated V_{OC} of the array is 132V and the rated short circuit current is 8.26A. The sampling period is 10ms. DC-DC buck converter with switching frequency of 100 kHz is used. The simulation results are shown in Fig. 7 . Suppose that the system is operating at its MPP in the steady state with each of the four modules receiving 1000W/m^2 irradiance. At t1 a step change in irradiance occurs and two of the modules now receive 400W/m^2. This is similar to the situation described in Fig. 1(b) with the corresponding P-V curve shown in Fig. 3 (a). The resulting decrease in the array output power is sensed during the next sampling interval at about 0.28s. The GP tracking process starts and the algorithm tracks both the peaks P1 and P2. By comparing the powers at the two peaks, P2 turns out to be the GP and the PV system is operated at this peak in the steady state. The whole process takes about 0.23s. This time duration is far less than 1.1s and 1.2s, and 5s taken by the algorithms in [8], [9] and [10] respectively. Fig. 8 shows the GP tracking process when three peaks are present. At t = 0.255s a partial shading occurs which is indicated by a sudden decrease in power from the array. The tracking process starts in the next sampling period and lasts till t2 at 0.58s. In this case, P3 turns out to be the GP and the steady state operation begins at 0.59s. It means that in the extreme case of three peaks the process takes only 0.32s.

6. Conclusion

Multiple peaks appear in the P-V characteristics curve of solar arrays under partial shading conditions. Tracking of GP in the presence of multiple peaks can be challenging. In this paper an MPPT algorithm is proposed which quickly tracks the GP of a PV array under non uniform irradiance conditions. The performance of the proposed algorithm has been confirmed through simulations. The simulation results indicate that the proposed algorithm merely takes 0.23s which is far shorter time than that of the methods reported in [8], [9], and [10].

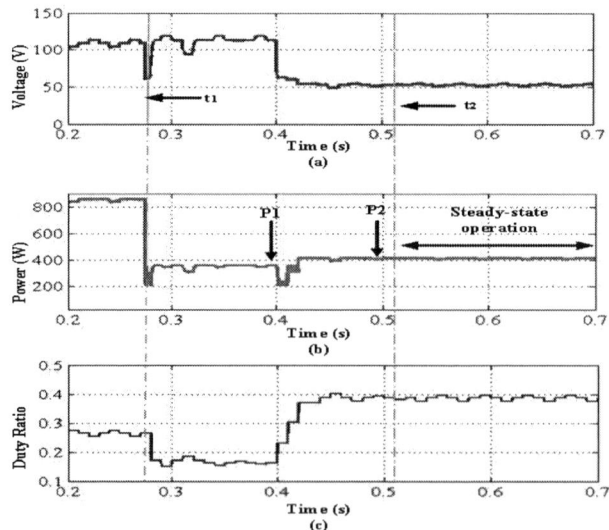

Fig. 7 GP tracking for the case of Fig. 1(b). (a) Voltage variation during tracking, (b) Power variation, (c) Duty ratio variation

PCIM Europe 2016, 10 – 12 May 2016, Nuremberg, Germany

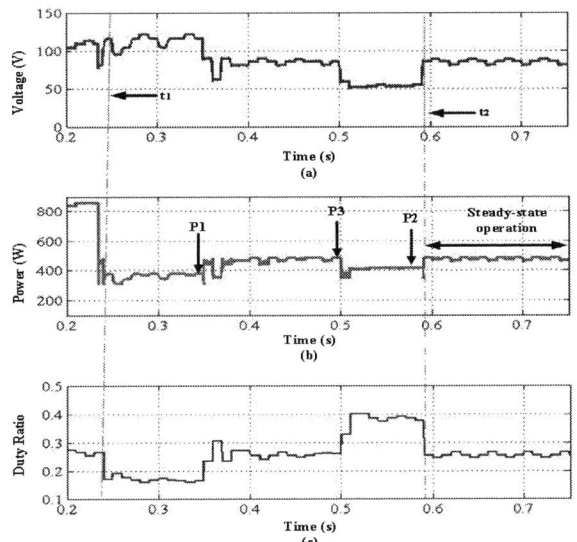

Fig. 8 GP tracking for the case of Fig. 1(c). (a) Voltage variation during tracking, (b) Power variation, (c) Duty ratio variation

References

1. T. Esram, P.L. Chapman, "Comparison of photovoltaic array maximum power point tracking techniques," IEEE Transaction on Energy Conversion 22 (2), 2007, PP. 439–449.
2. G. Petrone, G. Spagnuolo, R. Teodorescu, M. Vitelli, "Reliability issues in photovoltaic power processing systems", IEEE Transactions on Industrial Electronics, Vol. 55, No. 7, pp. 2569-2580, July 2008.
3. E. Koutroulis, "A new technique for tracking the global maximum power point of PV arrays under partial shading conditions," IEEE Journal of Photovoltaics, Volume 2, Issue 2, April 2012, pp. 184-190.
4. K. Ishaque, Z. Salam, "A review of maximum power point tracking techniques of PV system for uniform insolation and partial shading condition," Renewable and Sustainable Energy Reviews 19(2013), pp. 475-488.
5. Ali Bidram, Ali Davoudi, Robert S. Balog, "Control and circuit techniques to mitigate partial shading effects in photovoltaic arrays", IEEE Journal of Photovoltaics, Vol. 2, No. 4, October 2012.
6. Mutlu Boztepe, Francese Guinjoan, Guillermo Velasco-Quesada, Santiago Silvestre, Aissa Chouder, and Engin Karatepe, "Global MPPT scheme for photovoltaic string inverter based on restricted voltage window search alogorithm," IEEE Transactions on Industrial Electronics, vol. 61, Issue 7, pp. 3302-3312, 2014.
7. Ali Murtaza, Marcello Chiaberge, Filippo Spertino, Diego Boero, Mirko De Giuseppe, "A maximum power point tracking technique based on bypass diode mechanism for PV arrays under partial shading," Energy and Buildings, vol. 73, pp. 13-25, 2014.
8. Kai Chen, Shulin Tian, Yuhua Cheng, Libing Bai, "An improved MPPT controller for photovoltaic system under partial shading conditions," IEEE Transactions on Sustainable Energy, Vol. 5, No. 3, July 2014, pp. 978-985.
9. H. Patel, V. Agarwal, "Maximum power point tracking scheme for PV systems operating under partially shaded conditions," IEEE Transactions on Industrial Electronics, vol. 55, Issue 4, pp. 1689-1698, 2008.
10. K. M. Tsang, W. L. Chan, "Maximum power point tracking for PV systems under partial shading conditions using current sweeping," Energy Conversion and Management 93 (2015), pp. 249-258.
11. H. Kim, S. Kim, C. K. Kwon, Y. J. Min, C. Kim, and S. W. Kim, "An energy efficient fast maximum power point tracking circuit in an 800 μW photovoltaic energy harvester," *IEEE Trans. Power Electron.* vol. 28, pp. 2927-2935, Jun, 2013.
12. Jawad Ahmad, F. Spertino, A. Ciocia, P. Di Leo, "A maximum power point tracker for module integrated PV systems under rapidly changing irradiance conditions," Accepted for publication in Proc. IEEE ICSGCE Oct, 2015.

© VDE VERLAG GMBH · Berlin · Offenbach

Renewable Electricity conversion and storage: Focus on Power to Gas process, EMR modelling and simulation

Ahmed, REMACI, IMS laboratory CNRS UMR 5218351, 33405 Talence, France, ESTIA Institute of Technology, 64210 Bidart, France, ahmed.remaci@u-bordeaux.fr

Cristophe, MERLO, IMS laboratory CNRS UMR 5218351, 33405 Talence, France, ESTIA Institute of Technology, 64210 Bidart, France, c.merlo@estia.fr

Octavian, CUREA, ESTIA Institute of Technology, 64210 Bidart, France, o.curea@estia.fr

Amelie, HACALA, ESTIA Institute of Technology, 64210 Bidart, France, o.curea@estia.fr

Vincent, GUERRE, Local Energy Alternative & Fair, 33295 Blanquefort, France, vincent.f.guerre@gmail.com

Abstract

Nowadays, the world attends an energy transition, which is motivated by the massive use of fossil energies that is at the origin of greenhouse gas emissions, in particular carbon monoxide/dioxide. Therefore, this energy transition promotes the large-scale use of renewable energy sources. The intermittence and the fluctuation of these sources make their integration difficult; this is why the incorporation of energy storage can provide additional beneficial features and aid in its further growth. This paper gives an overview about the electrical energy conversion and storage. A particular attention is paid to Power to Gas.

To perform an optimized sizing of power to gas system, there must be a modeled, which approximates the behavior of the real physical system. To do this multi-physics modeling is required. The macroscopic energy performance will be used in modeling each system component. EMR is a graphical modeling tool that facilitate understanding of the model of the energy chain. EMR modeling approach is a reliable option for real-time energy management of energy system macroscopically.

1. Introduction

All around the world the energy production and consumption model admits its limits. The intensive use of fossil energies has led to increasing greenhouse gas emissions, which enhanced the development and the use of Renewable Energy Sources (RES) and waste recovery at the expense of oil and natural gas. In 2012 [1], oil remains the most solicited source with an average rate of 40.7%. Electricity is the second-largest energy source with an average rate of 18.1%. The renewable electricity production reached 4699.2 Twh per year, exceeding 20% of the total electricity production but fossil fuels remain the core of global production. Even if the RES seem the most adapted solution to overcome the pollution problems, they also have limits. More than their availability problems, the most critical point is the time shift between the power generation and the consumption needs (peak PV production at midday while the peak demand is mainly in the evening) [2], [3]. Therefore, the step of electrical energy storage is compulsory.

The majority of primary energies can be stored easily, however it is very difficult to store electrical energy in large quantities. In this case, the energy storage consists in converting the electrical energy supplied by different sources or power grids into other forms of intermediary energy. The stored energy can be directly used or transformed again into electrical energy to support the power grid when other sources are unable to respond to high demands, when the cost of generation is high or in case of failure of other means of production [4], [5]. The storage duration is limited due to losses associated with Energy Storage systems (ESs).

© VDE VERLAG GMBH · Berlin · Offenbach

2. Power to gas process

Against global warming caused by the massive greenhouse emissions, the Power to gas (PtG) appears as a pragmatic solution, which allows valorizing the CO_2.
PtG can be a one-step or two-steps process:

One step: electrical power $\xrightarrow{\text{Water electrolysis}}$ Hydrogen (H2)

Two steps: electrical power $\xrightarrow{\text{Water electrolysis}}$ Hydrogen (H2) $\xrightarrow{\text{Methanation}}$ Methane (CH4)

The one-step process consists in converting electrical power (electricity that can be generated by renewable energy sources) into H_2 by Water Electrolysis (WEL). The two-steps process consists in converting the H_2 produced by the one-step process into CH_4 by methanation process; it consists in combining the H_2 with recovered CO_2. As illustrated by Fig.1, several applications are possible. These produced gases can be used directly or converted into electricity via fuel cells or gas turbines. The efficiency of PtG is 60%-70% [6]. Concerning the discharge phase (conversion from gases to electricity), the losses are higher compared to other electrical energy storage systems. However, the discharge time and storage capacity are the most interesting. Furthermore, it is possible to enhance the performance by recovering the waste heat, which may offers other opportunities of application. Furthermore, the establishment of a methodology for the energy management can also improve the system performance. Several techniques exist for energy management in distributed smart grids such as multi-agents systems or fuzzy logic for example. Therefore, the dynamic model of the system is compulsory because it allows taking into account the transient behavior. Hence, the simulation results will be closest to the real system behavior. In this work, EMR (Energetic Macroscopic Representation) has been used as modelling tool.

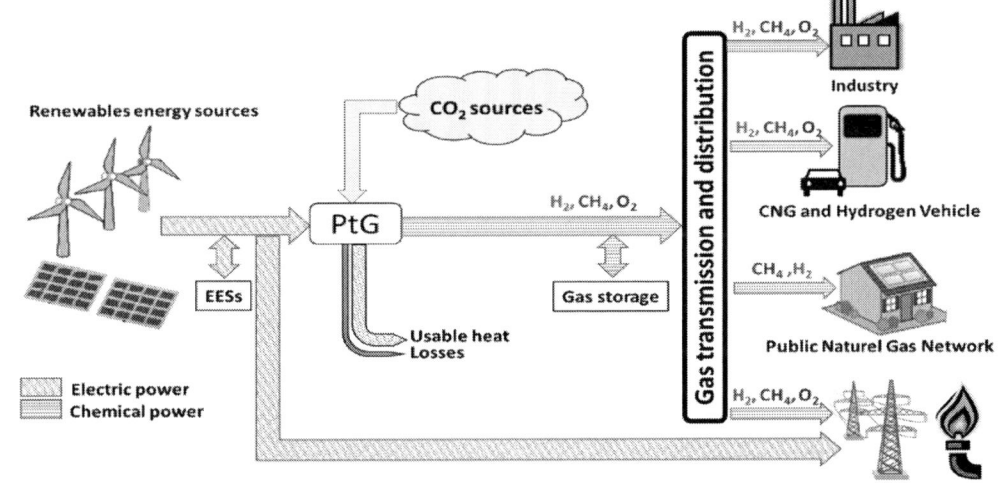

Fig. 1. Power to Gas process and its applications

3. The energetic macroscopic representation EMR

The energetic macroscopic representation is a synthetic representation for complex and multi-physical systems, based on the principle of action and reaction. Considering its control predilection, the EMR admits only the physical causality "integral causality". It consists of three types of symbols (the source elements, the Multi/Mono physical conversion elements, the accumulation elements, Multi/Mono physical domain coupling device) Fig.2 [7], [8].

It therefore allows following the different steps of conversion undergone by the power supplied to a system. The variables of action / reaction linking each transition between two elements shows the state of the transmitted energetic flow. To simulate the EMR model, we used the EMR library created under Matlab/Simulink. This library is available on the website of EMR [7], [9], [10].

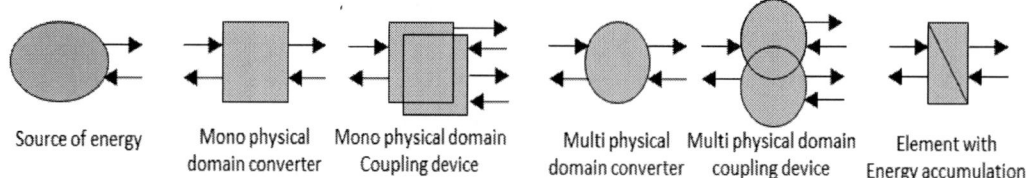

Fig. 2. EMR basic pictograms

In this article, the system to be modeled is a stationary system, which consists in supplying hydrogen to a methanation reactor. This hydrogen is produced by an electrolyzer powered by photovoltaic source. All the generated electricity is converted into hydrogen. Fig.3 is a synoptic of the power to gas system.

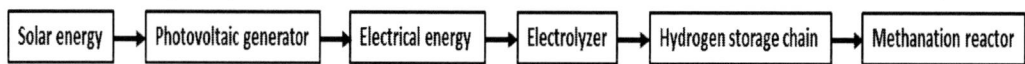

Fig. 3. Synoptic scheme of the system

3.1 Photovoltaic generator

The electricity produced is the result of coupling the thermal and radiometric domains. A dynamic modelling requires the inclusion of these two quantities. In this EMR model, thermal phenomena are involved. As for the radiometric domain, just the intensity of illuminance is represented. For more details of the radiometric modelling, please refer to [7].

The photovoltaic generators is a nonlinear source. For this work, a moderate complexity model is used. The relations describing the photovoltaic system are [7], [11]:

$$I_{PV} = I_L - I_D = I_L - I_{sat}\left[\exp\left(\frac{V_{PV} + IR_S}{nV_T}\right) - 1\right] \tag{1}$$

with $V_T = \frac{kT_{PV}}{q}$ \hfill (2)

V_{PV} is the output voltage and R_s is the series resistance. n is the ideality factor, k the Boltzmann constant and q is the charge of the electron. I_L is the irradiance current, it can be calculated as:

$$I_L = \left(\frac{E}{E_{ref}}\right)\left[I_{L,ref} + \mu_{I,sc}\left(T_{PV} - T_{PV,ref}\right)\right] \tag{3}$$

E is the solar irradiance and E_{ref} is the solar irradiance at the reference temperature $T_{PV,ref}$ (Temperature in the standard operating conditions). $\mu_{I,sc}$ is the variation coefficient of the short circuit current which depends on the temperature.
All the constants in the above equations could be determined using the photovoltaic panel datasheet and data from the characteristics curves. A portion of the solar radiation received by the module is converted into electricity, while the rest is dissipated in the module environment

as heat [7], [12], [13], [14]. The dynamic of thermal activity within the module can assess the time course of the module temperature. To do this, all input and output heat flows of the module are considered such as:

$$C_{PV}\frac{dT_{PV}}{dt} = \sum \dot{Q}_{in} - \sum \dot{Q}_{out} \qquad (4)$$

C_{PV} is the thermal capacity of the module, $\dot{Q}_{in}$ and $\dot{Q}_{out}$ are respectively the input and output heat flows. The main source of input thermal energy is the one from the solar irradiance. It may be expressed as follows:

$$\dot{Q}_{in} = S_{PV}E(1 - \eta_{PV}) \qquad (5)$$

S_{PV} is the module surface and η_{PV} is the module efficiency. We can consider the exchange of heat flow between the module and the environment as a convective flow, such as [15]:

$$\dot{Q}_{env/PV} = \frac{1}{R_{env/PV}}(T_{PV} - T_{amb}) \qquad (6)$$

$R_{env/PV}$ is the thermal resistance and T_{amb} is the ambient temperature. The convective heat loss is calculated as [12]–[20]:

$$R_{env/PV} = \frac{1}{hS_{PV}} \qquad (7)$$

Considering the thermal inertia, the module can be seen as a heat accumulation element. In order to calculate the thermal capacity, the module is considered as three layers of material. A flat sheet of PV cells laminated within a Polyester / Tedlar trilaminate, behind a glass face. The heat capacity of the module is the sum of the capacities of the three layers of material [12]–[25]:

$$C_{PV} = \sum_{PV} S_{PV}.d_{PV}.\rho_{PV}.c_{PV} \qquad (8)$$

$d_{PV}, \rho_{PV}, c_{PV}$ are respectively the dimension, the density and the heat capacity of each photovoltaic module.
The photovoltaic generator is not connected directly to the DC bus because of the fluctuation of the produced electricity. Therefore, the addition of an CL filter is necessary. This filter is represented by the following equations:

$$I_{PV} - I_l = C\frac{dV_{PV}}{dt} \qquad (9)$$

$$V_{PV} - U_l = L\frac{dI_l}{dt} + ri_l \qquad (10)$$

An MPPT (maximum power point tracking) algorithm controls the DC-DC converter at the output of photovoltaic generator.

3.2 Hydrogen production unit

A PEM (Proton exchange membrane) electrolyzer provides the production of hydrogen, the electrolysis of water decompose it into H2 and O2 molecule. This method of producing hydrogen is considered as the cleanest method with 98% of gas purity at output. To ensure electrochemical reaction, a DC voltage is applied at both electrodes. The electrolyzer uses a part of this electrical energy to overcome potential of the Gibbs free energy. During this operation, heat is generated due to various losses. In this article, the electrolyzer will be represented by a simple EMR model. For a complete study of the PEM electrolyzer please refer to [10]. The equations used are [26]–[34] :

$$V_{EL} = E + V(I_{EL}) \tag{11}$$

V_{EL} is the DC voltage applied at both electrodes, E is the voltage that is assigned to the free energy of Gibbs. $V(I_{EL})$ Represents the sum of the overvoltage, which are related to the current I_{EL} that crosses the electrolyser. Regarding the thermal model, a total energy balance is established and which is expressed by the equation:

$$C_{EL} \frac{dT_{EL}}{dt} = \dot{Q}_{th} - \dot{Q}_{EL/env} - \dot{Q}_{EL/water} \tag{12}$$

$\dot{Q}_{th}$ is the thermal power generated by the reaction, $\dot{Q}_{EL/env}$ and $\dot{Q}_{EL/water}$ are respectively the exchanged flow with the external environment and exchanged flux with the feed water.

3.3 Hydrogen storage system

The produced hydrogen is intended to supply methanation reactor. For the system stability, an intermediate storage step is necessary. For this power to gas, the storage by compression is chosen because it is currently the most simple, the most used and the most effective for storing hydrogen until 200 bars.

The moto-compressor is constituted by a compression stage driven by an electric motor (permanent magnet motor). The EMR model of moto- compressor that comprises two parts, one part for the electric motor and a part for the compression stage, which is quasi-static. The equation, which links between these two parts is [9]:

$$P_{Mec} = \frac{P_{mec}}{\Omega_{shaft}} \, \dot{m}_{in} \, C_p \, T_{in} \left(\left(\frac{P_{out}}{P_{in}} \right)^{\frac{k-1}{k}} - 1 \right) \tag{13}$$

P_{mec} is the mechanical power of the moto-compressor, Ω_{shaft} is the drive speed and C_p is the Thermal capacity of the hydrogen.

In order to maintain the pressure stable within the storage system, the use of a pressure regulator is required. The regulator ensures the control of the pressure based on a reference pressure. Regarding the tank, it is characterized by the following equation [9]:

$$\dot{m}_{BT}(t_0 + \Delta t) = \int_0^{t_0 + \Delta t} \dot{m}_{BT}(\tau) \, d(\tau) + \dot{m}_{BT}(t_0) \tag{14}$$

Fig.4 is the EMR model of the whole system. In this paper, the methanation reactor is considered as a load.

PCIM Europe 2016, 10 – 12 May 2016, Nuremberg, Germany

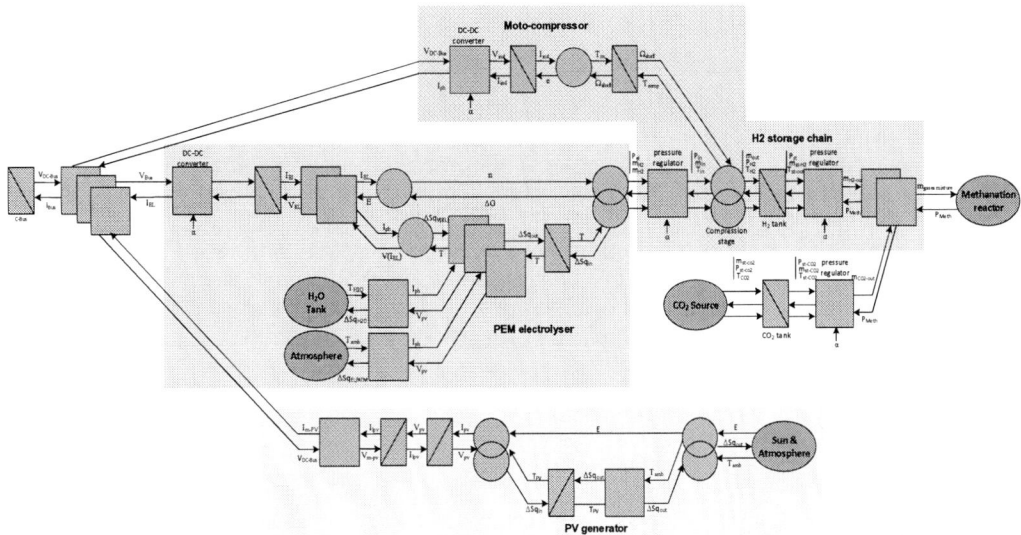

Fig. 4. EMR of the whole system

4. Simulation results of EMR model

Fig.5 illustrates the simulation results of a PV module made by BP Solar (BP485). It also shows the influence of irradiance and temperature on the electrical parameters of the module. The module power curve shows that the use of a control algorithm is required to obtain the optimal point.

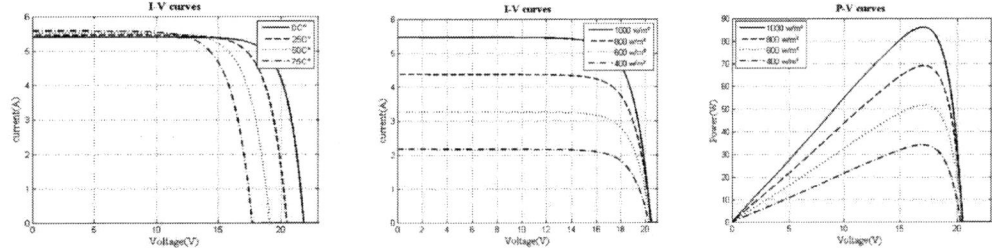

Fig. 5. Photovoltaic module

Fig. 6 illustrates the dynamic behavior of the PEM electrolyzer which is composed by 10 stacks. The electrolyzer is powered by a current of 10 A during 4000 seconds. For the same current and different values of temperature, the stacks voltage decreases when the temperature increases.

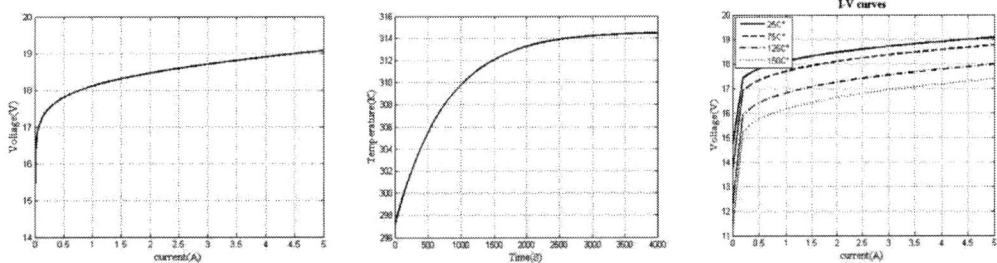

Fig. 6. Electrical curves and temperature behavior of PEM electrolyzer

© VDE VERLAG GMBH · Berlin · Offenbach

2071

5. Conclusion

This paper gives a short overview about the electrical energy conversion and storage, and then focus on power to gas (PtG) process. This process has been analyzed and modelled using the Energetic Macroscopic Representation. The EMR modelling approach is useful and provides in this work an adaptable model. Regarding the photovoltaic model, the interaction between temperature and electrical parameters allowed to obtain simulation results similar to the real photovoltaic module behavior. This modeling approach will allows us to develop a strategy for the energy management of the whole system. This strategy will be based on the technique of multi-agent systems and will be presented in a future paper.

References

[1] International Energy Agency, "Key World Energy" Paris: Chirat, 2014.

[2] A. Etxeberria, I. Vechiu, H. Camblong, J. Vinassa, "Hybrid energy storage systems for renewable energy sources integration in microgrids: A review," IPEC, 2010 Conf. Proc., pp. 532–537, 2010.

[3] Muljadi, E.; Bialasiewicz, J.T., "Hybrid power system with a controlled energy storage," Industrial Electronics Society, 2003. IECON '03. The 29th Annual Conference of the IEEE , vol.2, no., pp.1296,1301 Vol.2, 2-6 Nov. 2003.

[4] H. Chen, T. N. Cong, W. Yang, C. Tan, Y. Li, and Y. Ding, "Progress in electrical energy storage system: A critical review," Prog. Nat. Sci., vol. 19, no. 3, pp. 291–312, Mar. 2009.

[5] H. Ibrahim, R. Beguenane, and A. Merabet, "Technical and Financial Benefits of Electrical Energy Storage" pp. 86–91, 2012.

[6] M. Götz, R. Reimert, D. Buchholz, and S. Bajohr, "Storage of volatile renewable energy in the gas grid applying 3-phase methanation" in International Gas Research Conference Proceedings, MARCH 2011, vol. 2, pp. 1283–1297.

[7] K.S. Agbli, M-C. Péra, D. Hissel, I. Doumbia. EMR modelling for the forecasting of the actual energy delivered by a photovoltaic system. ELECTRIMACS2011, Cergy-Pontoise, France ,6-8th June, 2011.

[8] L. Boulon, D. Hissel, A. Bouscayrol, and M. C. Péra, "From modeling to control of a PEM fuel cell using energetic macroscopic representation" IEEE Trans. Ind. Electron., vol. 57, no. 6, pp. 1882–1891, 2010.

[9] K. S. Agbli, D. Hissel, M.-C. Péra, and I. Doumbia, "EMR modelling of a hydrogen-based electrical energy storage" Eur. Phys. J. Appl. Phys., vol. 54, no. 2, p. 23404, 2011.

[10] K. S. Agbli, M. C. Péra, D. Hissel, O. Rallires, C. Turpin, and I. Doumbia, "Multiphysics simulation of a PEM electrolyser: Energetic Macroscopic Representation approach" Int. J. Hydrogen Energy, vol. 36, pp. 1382–1398, 2011.

[11] M. Al-refai, "Matlab / Simulink Simulation of Solar Energy Storage System," vol. 8, no. 2, pp. 304–309, 2014.

[12] A. D. Jones, C. P. Underwood, "A thermal model for photovoltaic systems," Sol. Energy, vol. 70, no. 4, pp. 349–359, 2001.

[13] S. Armstrong and W. G. Hurley, "A thermal model for photovoltaic panels under varying atmospheric conditions," Appl. Therm. Eng., vol. 30, no. 11–12, pp. 1488–1495, 2010.

[14] S, C. W. Krauter, "Solar Electric Power Generation: Photovoltaic energy system". Springer-Verlag Berlin Heidelberg, 2006.

[15] G. Notton, C. Cristofari, M. Mattei, and P. Poggi, "Modelling of a double-glass photovoltaic module using finite differences," Appl. Therm. Eng., vol. 25, no. 17–18, pp. 2854–2877, 2005.

[16] J. A. Palyvos, "A survey of wind convection coefficient correlations for building envelope energy systems' modeling," Appl. Therm. Eng., vol. 28, no. 8–9, pp. 801–808, 2008.

[17]   M. Mattei, G. Notton, C. Cristofari, M. Muselli, and P. Poggi, "Calculation of the polycrystalline PV module temperature using a simple method of energy balance," Renew. Energy, vol. 31, no. 4, pp. 553–567, 2006.

[18]   E. Skoplaki and J. A. Palyvos, "On the temperature dependence of photovoltaic module electrical performance: A review of efficiency/power correlations," Sol. Energy, vol. 83, no. 5, pp. 614–624, 2009.

[19]   J. A. Duffie, W. A. Beckman, and W. M. Worek, Solar Engineering of Thermal Processes, 4nd ed., vol. 116. 2003.

[20]   P. Henshall, P. Eames, F. Arya, T. Hyde, R. Moss, and S. Shire, "Constant temperature induced stresses in evacuated enclosures for high performance flat plate solar thermal collectors," Sol. Energy, vol. 127, pp. 250–261, 2016.

[21]   A. D. L. Vollaro, G. Galli, and A. Vallati, "CFD analysis of convective heat transfer coefficient on external surfaces of buildings," Sustain., vol. 7, no. 7, pp. 9088–9099, 2015.

[22]   E. Sartori, "Convection coefficient equations for forced air flow over flat surfaces," Sol. Energy, vol. 80, no. 9, pp. 1063–1071, 2006.

[23]   G. M. Tina and R. Abate, "Experimental verification of thermal behaviour of photovoltaic modules," MELECON 2008 - 14th IEEE Mediterr. Electrotech. Conf., pp. 579–584, 2008.

[24]   H. Matsukawa and K. Kurokawa, "Temperature Fluctuation Analysis of Photovoltaic Modules at Short Time Interval," Photovolt. Spec. Conf. 2005. Conf. Rec. Thirty-first IEEE, pp. 1816–1819, 2005.

[25]   V. B. Sharma, Wind induced heat losses from outer cover of solar collectors,"
Renewable Energy, vol 10, issue 4, April 1997, pp 613-616.

[26]   D. An, Q. Li, X. Wang, H. Yang, and L. Guo, "Characterization on hydrogen production performance of a newly isolated Clostridium beijerinckii YA001 using xylose," Int. J. Hydrogen Energy, vol. 39, issue. 35, pp. 19928–19936, 2014.

[27]   L. An, T. S. Zhao, Z. H. Chai, P. Tan, and L. Zeng, "Mathematical modeling of an anion-exchange membrane water electrolyzer for hydrogen production," Int. J. Hydrogen Energy, vol. 39, pp. 19869–19876, 2014.

[28]   O. Atlam and M. Kolhe, "Equivalent electrical model for a proton exchange membrane (PEM) electrolyser," Energy Convers. Manag., vol. 52, no. 8–9, pp. 2952–2957, 2011.

[29]   A. Awasthi, K. Scott, and S. Basu, "Dynamic modeling and simulation of a proton exchange membrane electrolyzer for hydrogen production," Int. J. Hydrogen Energy, vol. 36, no. 22, pp. 14779–14786, 2011.

[30]   F. Barbir, "PEM electrolysis for production of hydrogen from renewable energy sources," Sol. Energy, vol. 78, pp. 661–669, 2005.

[31]   O. Bendaíkha and S. Larbi, "Hydrogen production system analysis using direct photo-electrolysis process in Algeria," Renewable Energy Research and Applications (ICRERA), 2013 International Conference on, Madrid, 2013, pp. 1123-1128.

[32]   I. Bolvashenkov and H. G. Herzog, "Highly effective hydrogen production process based on Tolman-Stewart effect: Feasibility of its implementation," Ecological Vehicles and Renewable Energies (EVER), 2013 8th International Conference and Exhibition on, Monte Carlo, 2013, pp. 1-4.

[33]   E. Cetin, A. Yilanci, Y. Oner, M. Colak, I. Kasikci, and H. K. Ozturk, "Electrical analysis of a hybrid photovoltaic-hydrogen/fuel cell energy system in Denizli, Turkey," Energy Build., vol. 41, pp. 975–981, 2009.

[34]   S. A. Chattanathan, S. Adhikari, M. McVey, and O. Fasina, "Hydrogen production from biogas reforming and the effect of H_2S on CH_4 conversion," Int. J. Hydrogen Energy, vol. 39, no. 35, pp. 19905–19911, 2014.

PCIM Europe 2016, 10 – 12 May 2016, Nuremberg, Germany

Balancing Current and Efficiency Modelling of Single Switch Active Balancing Systems for Energy Storage Systems

Iosu, Aizpuru, Mondragon University, Loramendi 4 20500 Arrasate-Mondragón, Spain, iaizpuru@mondragon.edu
Unai, Iraola, Mondragon University, Loramendi 4 20500 Arrasate-Mondragón, Spain, uiraola@mondragon.edu
Jose Mari, Canales, Mondragon University, Loramendi 4 20500 Arrasate-Mondragón, Spain, jmcanales@mondragon.edu
Ander, Goikoetxea, Mondragon University, Loramendi 4 20500 Arrasate-Mondragón, Spain, agoikoetxeaa@mondragon.edu

Abstract

High Power battery packs are a series/parallel connection of single energy storage cells. The series connection of cells affects in the whole battery pack energy and power due to the unbalancing behavior of each different cell. To improve the whole battery pack energy a balancing system is connected. Passive balancing systems are wasteful and do not increase the Battery pack energy during discharge. Active balancing systems are efficient but most topologies add big complexity to the BMS work. Single switch active balancing systems are simple and reliable. During this work the main balancing current modelling and efficiency equations are presented for the most important single switch active balancing systems. These expressions permit to evaluate the balancing power of the system, and to know how much energy is balanced during the balancing system process between the Weak cell and the Strong cells. Experimental validation of each balancing system is implemented to compare balancing power, efficiency and behavior.

1. Introduction

Driven by nowadays increase of power and energy demand several continental, national and regional energy strategy plans have been developed with the main issue of decreasing primary energy sources and increase insertion of renewable energies, decrease CO_2 emissions and increase energy efficiency [1], [2].

Energy storage systems *ESS* are the main technology to help increasing the impact of renewable energy, as wind energy and photovoltaic generation [3]–[10], or the insertion of the hybrid and electric car [11]–[16].

Inside *ESS* systems battery based energy systems are one of the most promising technologies. High power battery packs are a series/parallel connection of single battery cells or small battery modules [17]–[22].

Series connection of battery cells have an unbalancing effect in the battery pack, due to differences between cells [23]–[26], inducing a total battery energy and performance reduction [27], [28].

Passive balancing is the most used balancing technique, but it is wasteful and the balancing procedure is only useful during charge process [29]–[31].

Active balancing systems are more efficient but are complex, with high number of switching devices, which increase the price and decreases the competitiveness of these devices [32]–[39]. To decrease the complexity of active balancing systems single switch balancing topologies are a great option [40], [41]. A single switch is used to balance all the cells, with the possibility of controlling the active switch with an open loop strategy.

In the present work, a generalized description of main single switch topologies for active balancing systems will be given. The main goal of the present work is to model the current of the active balancing systems shunted from each single cell. It is the main parameter in the design and sizing of the balancing system. The efficiency of the balancing system is also modelled, to know the power loss respected to the ideal current model.

© VDE VERLAG GMBH · Berlin · Offenbach

As the first step in section II, main single switch balancing topologies and modelling equations will be presented. These equations are the main part for the balancing system sizing.

In section III, power losses of the single switch balancing circuits are modelled. The power losses of balancing circuits are of great impact and interest in balancing systems because they manage low balancing power and the losses have big influence on the balancing currents. With the power losses the balancing currents are recalculated.

During section IV, experimental results of real prototypes will be presented and compared to modelling results. Special interest will be presented regarding power losses and efficiency of the prototypes.

2. Ideal modelling of single switch balancing topologies

Opel loop switching control (no feedback control) is possible in single switch balancing systems guaranteeing the Discontinuous Conduction Mode DCM during the full working range. During the first step the most important issues to guarantee the DCM will be presented. After the main issues of current balancing will be presented to finish with the ideal balancing current models.

The critical duty cycle ratio D_{cri} is the maximum working duty cycle D to guarantee the DCM. D_{cri} is defined when the value of the zero current period D_2 is zero.

$$D_{cri} = 1 - D_1, \quad \text{for } D_2 = 0 \tag{1}$$

The only requirement to guarantee the DCM mode is to keep D behind or equal to D_{cri}.

$$D \le D_{cri}, \quad \text{for DCM} \tag{2}$$

2.1. Ideal current modelling of single switch balancing topologies

Single switch balancing system topologies perform voltage based balancing. The balancing system equals all the voltages of the series connected cells by the natural topology behavior.

When the voltage is equal in all n series connected $V_{B,i}$ cells, no balancing current $I_{B,i}$ is injected nor extracted from the cells.

$$V_{B,1} = V_{B,2} = ... = V_{B,n-1} = V_{B,n}, \qquad I_{B,i} = 0 \tag{3}$$

When a voltage unbalance occurs, the cell with the lowest voltage $V_{B,W}$, for now on *Weak* cell, receives energy from the higher voltage cells $V_{B,S}$, henceforth *Strong* cells.

$$V_{B,W} < V_{B,i}, \qquad I_{B,W} > 0 \text{ receive energy}$$
$$V_{B,S} > V_{B,W}, \qquad I_{B,S} < 0 \text{ give energy} \tag{4}$$

During the ideal current modelling, the balancing current $I_{B,i}$ of each single series connected cell will be defined under a voltage unbalanced condition, with one *Weak* cell and *(n-1) Strong* cells.

The maximum D_{cri} will be defined respect to the *Weak* cell voltage $V_{B,W}$ and the *Strong* cell voltages $V_{B,S}$.

The battery pack voltage V_{PACK} is defined as the sum of each single cell voltage $V_{B,i}$, or the sum of the *Weak* cell voltage $V_{B,W}$ with the Strong cell voltages $V_{B,S}$.

$$V_{PACK} = \sum_{i=1}^{n} V_{B,i} = V_{B,W} + \sum_{i=1}^{n-1} V_{B,Si} \tag{5}$$

During a complete switching period the generalized equations that determine the mean AV and the RMS of the variables are expressed by

$$X_{AV_T} = \sum_{i=1}^{n} X_{AV_i} D_i = X_{AV_1} D_1 + X_{AV_2} D_2 + ... + X_{AV_n} D_n$$

$$X_{RMS_T} = \sqrt{\sum_{i=1}^{n} X_{RMS_i}^2 D_i} = \sqrt{X_{RMS_1}^2 D_1 + X_{RMS_2}^2 D_2 + ... + X_{RMS_n}^2 D_n} \tag{6}$$

© VDE VERLAG GMBH · Berlin · Offenbach

The topologies under study are 3 flyback based topologies and 3 buck-boost. The main parameters of these topologies regarding number of components and balancing behavior are presented in TABLE I.

TABLE I

CHARACTERISTICS OF THE MAIN SINGLE SWITCH ACTIVE BALANCING SYSTEMS: NUMBER OF COMPONENTS AND BALANCING BEHAVIOR

Topology	S	D	C	L	T	Selective
FLY_1PnS	1	n	–	–	1	Yes
FLY_nPnSP	1	n	–	–	n	Yes
FLY_nPnSS	1	n	–	–	n	No
SEPIC	1	n	n	n+1	–	Yes
ZETA	1	n	n	n+1	–	Yes
Isolated_CUK	1	n	n+1	n+1	1	Yes

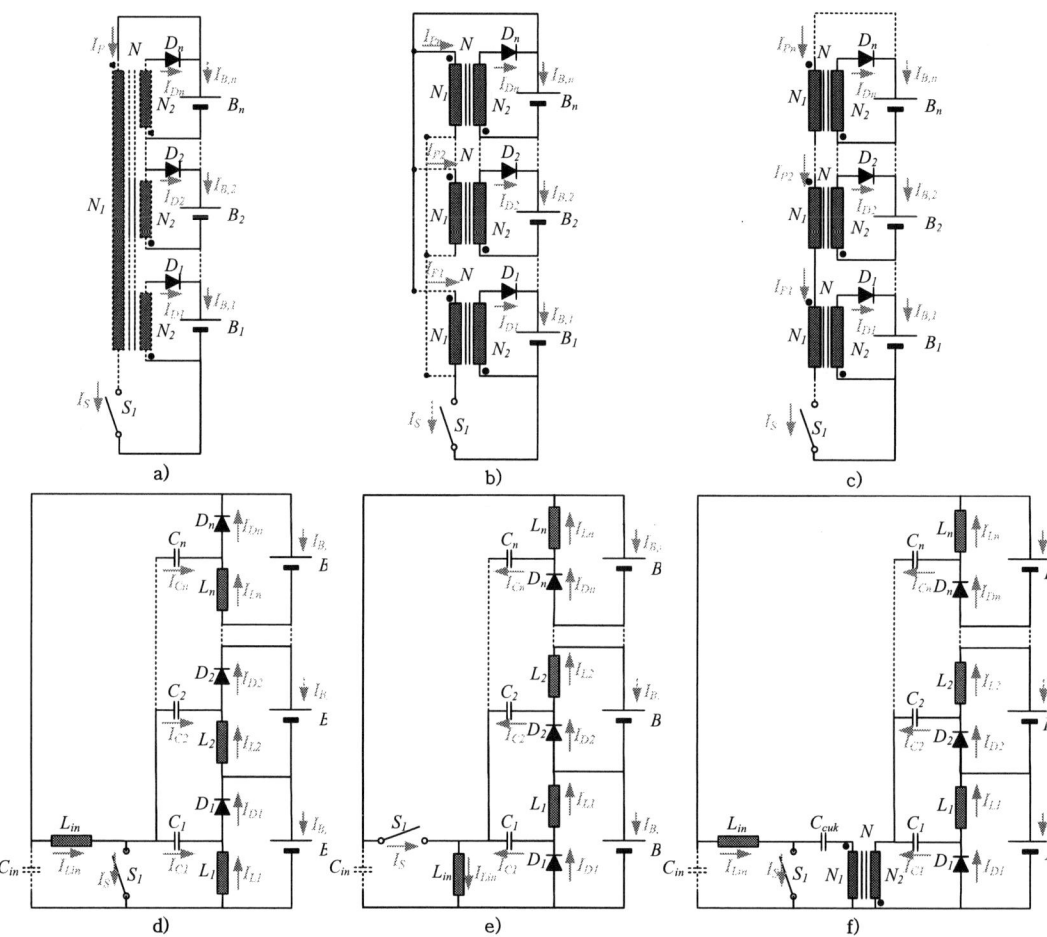

Fig. 1. Main single switch active balancing topologies schematics. a) Flyback with 1 primary and *n* secondaries with single core *FLY_1PnS* b) Flyback with *n* primaries and *n* secondaries with the primaries connected in parallel *FLY_nPnSP*. c) Flyback with *n* primaries and *n* secondaries with primaries connected in series *FLY_nPnSS*. d) The multistacked Sepic topology *SEPIC*. e) The multistacked Zeta topology *ZETA*. f) The multistacked Cuk topology in the isolated configuration *Isolated_CUK*.

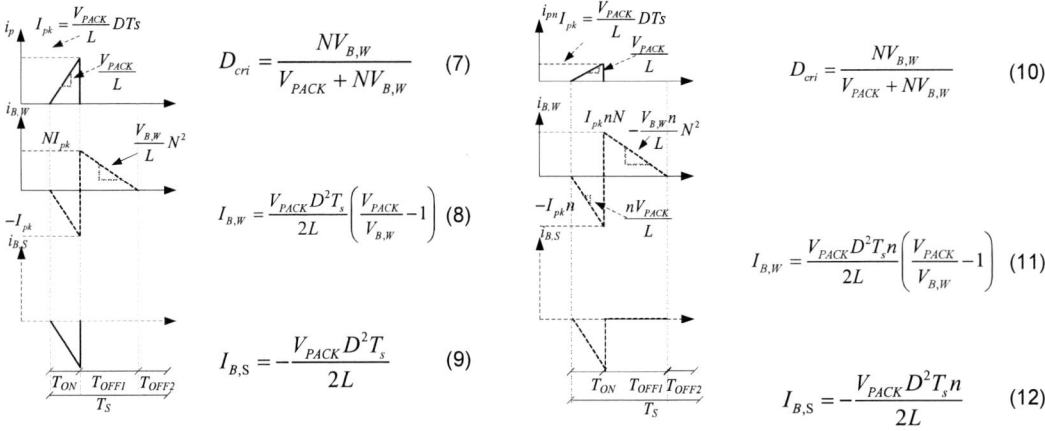

$$D_{cri} = \frac{NV_{B,W}}{V_{PACK} + NV_{B,W}} \quad (7)$$

$$I_{B,W} = \frac{V_{PACK}D^2T_s}{2L}\left(\frac{V_{PACK}}{V_{B,W}} - 1\right) \quad (8)$$

$$I_{B,S} = -\frac{V_{PACK}D^2T_s}{2L} \quad (9)$$

$$D_{cri} = \frac{NV_{B,W}}{V_{PACK} + NV_{B,W}} \quad (10)$$

$$I_{B,W} = \frac{V_{PACK}D^2T_s n}{2L}\left(\frac{V_{PACK}}{V_{B,W}} - 1\right) \quad (11)$$

$$I_{B,S} = -\frac{V_{PACK}D^2T_s n}{2L} \quad (12)$$

Fig. 2 *FLY_1PnS* main balancing current waveforms and equations.

Fig. 3. *FLY_nPnSP* main balancing current waveforms and equations.

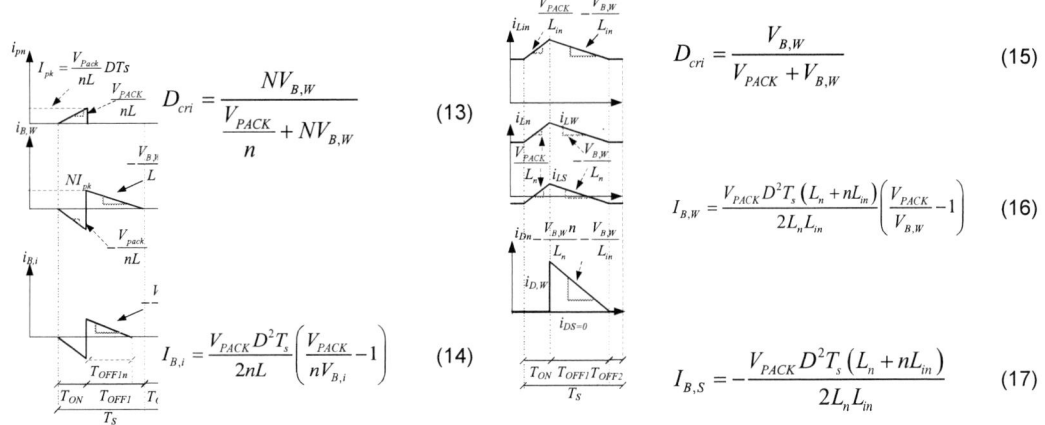

$$D_{cri} = \frac{NV_{B,W}}{\dfrac{V_{PACK}}{n} + NV_{B,W}} \quad (13)$$

$$I_{B,i} = \frac{V_{PACK}D^2T_s}{2nL}\left(\frac{V_{PACK}}{nV_{B,i}} - 1\right) \quad (14)$$

$$D_{cri} = \frac{V_{B,W}}{V_{PACK} + V_{B,W}} \quad (15)$$

$$I_{B,W} = \frac{V_{PACK}D^2T_s\left(L_n + nL_{in}\right)}{2L_nL_{in}}\left(\frac{V_{PACK}}{V_{B,W}} - 1\right) \quad (16)$$

$$I_{B,S} = -\frac{V_{PACK}D^2T_s\left(L_n + nL_{in}\right)}{2L_nL_{in}} \quad (17)$$

Fig. 4. *FLY_nPnSS* main balancing current waveforms and equations

Fig. 5. *SEPIC* main balancing current waveforms and equations.

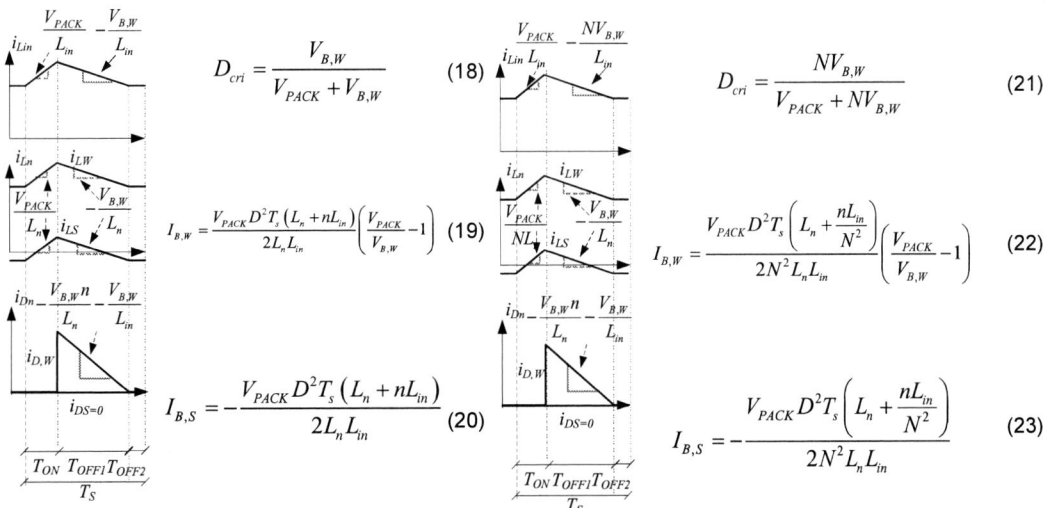

$$D_{cri} = \frac{V_{B,W}}{V_{PACK} + V_{B,W}} \quad (18)$$

$$I_{B,W} = \frac{V_{PACK}D^2T_s\left(L_n + nL_{in}\right)}{2L_nL_{in}}\left(\frac{V_{PACK}}{V_{B,W}} - 1\right) \quad (19)$$

$$I_{B,S} = -\frac{V_{PACK}D^2T_s\left(L_n + nL_{in}\right)}{2L_nL_{in}} \quad (20)$$

$$D_{cri} = \frac{NV_{B,W}}{V_{PACK} + NV_{B,W}} \quad (21)$$

$$I_{B,W} = \frac{V_{PACK}D^2T_s\left(L_n + \dfrac{nL_{in}}{N^2}\right)}{2N^2L_nL_{in}}\left(\frac{V_{PACK}}{V_{B,W}} - 1\right) \quad (22)$$

$$I_{B,S} = -\frac{V_{PACK}D^2T_s\left(L_n + \dfrac{nL_{in}}{N^2}\right)}{2N^2L_nL_{in}} \quad (23)$$

Fig. 6. *ZETA* main balancing current waveforms and equations.

Fig. 7 *Isolated_CUK* main balancing current waveforms and equations.

2.1.1 FLY_1PnS

The *FLY_1PnS* is a flyback converter of 1 primary and *n* secondaries wounded in the same magnetic core. The main advantage is that a single core is used to balance all the cells. The main disadvantages are that is not modular and deviations occur due to differences between windings. Good transformer design, by coaxial windings [42] or the use of commercial multiwinding transformers is necessary to avoid deviation between windings.

2.1.2 FLY_nPnSP

The *FLY_nPnSP* is a flyback based design with *n* magnetic cores connected in parallel in the primary. The secondary of the *n* magnetics core is connected to each series connected cell. The main advantages are the modularity of the design and the high current capability, although it has high power losses due to high current necessity to magnetize all transformers.

2.1.3 FLY_nPnSS

The flyback converter *FLY_nPnSS* is a series connection of *n* magnetic cores. The transformers are serialized and all magnetized by the same current from the battery pack. The main advantage is the modularity, but the most important drawback is the non-selective behavior.

The magnetizing inductors *L* of each transformer are charged during T_{ON}. During T_{OFF1}, the energy introduced to each transformer, is inserted back to each cell with a variable T_{OFF1} time depending on each cell Voltage $V_{B,i}$.

2.1.4 SEPIC

The multistacked based Sepic converter *SEPIC* is a Sepic converter topology with *n* outputs stacked in series. The current ripple of the current is always bigger in the bottom connected cell. The balancing current is modeled by analyzing the inductor current I_{Lin} and diode current I_{Di}

The main advantages are the transformer free solution and an easy modularity. The disadvantage is the different instantaneous current through the different series connected cells.

2.1.5 ZETA

The *ZETA* is also a multistacked based buck-boost topology. It has the same advantages and disadvantages of the *SEPIC* converter. The instantaneous currents are also different depending on the cell position of the series connected stack. The balancing current ripple is bigger in the top connected cell. Another disadvantage is also the high side switch of the topology.

2.1.6 Isolated_CUK

The *Isolated_CUK* topology is a multistacked topology based on the Cuk converter in its isolated configuration to reverse the polarity.

The biggest advantage of the *Isolated_CUK* is the modularity and the degree of freedom given by the transformer ratio *N*, to determine D_{cri}. The disadvantages are the requirement of extra components as the transformer and the split capacitor C_{cuk}.

3   Power loss modelling

Active balancing systems have higher efficiency than passive systems; they transfer charge between cells instead of burning the energy in a power resistor. However the efficiency of these systems is not 100% and a piece of the energy transfer of the ideal modelling is wasted in power losses.

Due to this power loss the energy transfer is reduced, and has to be taken into account. Due to DCM operation of the single switch balancing systems the efficiency modelling is harder than in CCM because

$$I_{AV} \approx I_{RMS} \rightarrow \text{CCM}$$
$$I_{AV} \neq I_{RMS} \rightarrow \text{DCM}$$

(24)

And each component RMS current should be calculated to correctly determine the power losses and efficiency. The efficiency and power losses are dependent on the working point and the parameters of the converter.

PCIM Europe 2016, 10 – 12 May 2016, Nuremberg, Germany

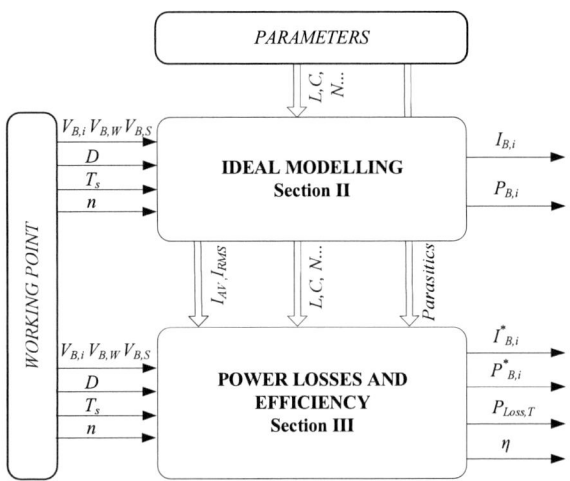

Fig. 8 Current balancing modelling strategy, with an ideal power loss modelling and an adjustment based on efficiency and power losses calculation.

The power losses $P_{Loss,T}$ are mainly due to the high balancing current redirected to the weak cell $I^*_{B,W}$. To recalculate the real balancing currents $I^*_{B,i}$ the power $P_{Loss,T}$ is only subtracted to the balancing power of the Weak cell, so

$$P_{Loss,T} \xrightarrow{affect} I^*_{B,W} \qquad I^*_{B,W} = \frac{V_{B,W}I_{B,W} - P_{Loss,T}}{V_{B,W}} \qquad I^*_{B,S} = \frac{V_{B,S}I_{B,S}}{V_{B,S}} = I_{B,S} \tag{25}$$

For the non-selective case of the FLY_nPnSS the power losses are distributed to all the cells. The lowest voltage cell has bigger power losses that the $Strong$ cells. The power losses $P_{Loss,T}$ are distributed inversely to the voltage of the cells as

$$P_{Loss,T} \xrightarrow{FLY\_nPnSS} I^*_{B,i} \ redistributed \ to \ all \ cells \qquad I^*_{B,i} = \frac{V_{B,i}I_{B,i} - \dfrac{P_{Loss,T}\sum\limits_{\substack{j=1 \\ j\neq i}}^{n}V_{B,j}}{V_{PACK}}}{V_{B,i}} \tag{26}$$

For a balanced situation the equations are different because the ideal balancing power $P_{B,i}=0$, because there is not balancing current in a balanced situation (3).

4. Experimental results

To evaluate the ideal modelling and the power loss and efficiency models presented in Section 2 and Section 3 the 6 balancing systems under study have been designed to evaluate modelling equations.

TABLE II

DESIGN AND IDEAL PROTOTYPE PARAMETERS FOR A WORKING POINT OF ONE WEAK CELL $V_{B,W}$=2 V AND 3 STRONG CELLS $V_{B,S}$=3.65 V FOR 4S1P BATTERY PACK n=4 SWITCHING PERIOD T_s=10 μS

Topology	Design					Prototype				
	$I_{B,W}$[mA]	D [ ]	L_n [μH]	L_{in} [μH]	N [ ]	$I_{B,W}$[mA]	D [ ]	L_n [μH]	L_{in} [μH]	N [ ]
FLY_1P4S	400	0.134	16	–	1	500	0.13	12	–	1
FLY_4P4SP	400	0.134	63	–	1	600	0.13	40	–	1
FLY_4P4SS	100	0.382	15	–	1	144	0.38	10	–	1
SEPIC	400	0.134	66	427	–	365	0.13	68	470	–
ZETA	400	0.134	66	427	–	365	0.13	68	470	–
Isolated_CUK	400	0.134	66	427	1	365	0.13	68	470	1

4.1 Model and experimental comparison

Experimental results and model equations are going to be compared under an unbalancing situation. Main parameters and results are resumed in Table III.

© VDE VERLAG GMBH · Berlin · Offenbach

2079

TABLE III
IDEAL, MODELLED AND REAL DATA COMPARISON. TOTAL BALANCING POWER $P_{B,T}$, TOTAL POWER LOSSES $P_{Loss,T}$, EFFICIENCY η AND BALANCING CURRENT OF THE *WEAK* CELL $I_{B,W}$. WORKING POINT UNBALANCED $V_{B,W}$=2 V AND $V_{B,S}$=3.65 V.

	$P_{B,T}$ [W]			$P_{Loss,T}$ [W]			η [%]			$I_{B,W}$ [A]		
	Ideal	Model	Real	Ideal	Model	Real	Ideal	Model	Real	Ideal	Model	Real
FLY_1P4S	1.997	1.672	1.708	0	0.325	0.328	100	83.71	83.88	0.499	0.337	0.345
FLY_4P4SP	2.198	1.809	1.991	0	0.388	0.535	100	82.33	78.82	0.599	0.305	0.364
FLY_4P4SS	0.578	0.324	0.366	0	0.254	0.29	100	56.05	55.78	0.145	0.037	0.02
SEPIC	1.461	1.175	1.163	0	0.286	0.195	100	80.43	85.64	0.365	0.222	0.242
ZETA	1.461	1.175	1.256	0	0.286	0.245	100	80.43	83.81	0.365	0.222	0.254
Isolated_CUK	1.461	1.174	1.508	0	0.286	0.332	100	80.4	81.96	0.365	0.222	0.294

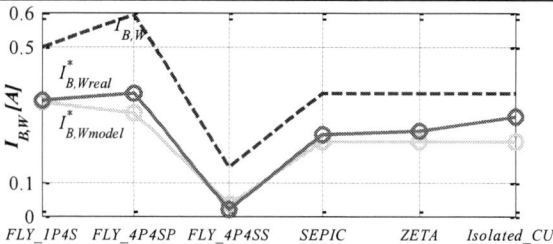

Fig. 9. Ideal *Weak* balancing current $I_{B,W}$ in dashed blue. Modelled real *Weak* balancing current $I^*_{B,W}$ filled circle green. Real experimental *Weak* balancing current $I^*_{B,W}$ filled red circle.

5. Conclusions

Single switch active balancing systems present high simplicity due to a unique active device is used. They also have the natural behavior to balance series connected cells with an open loop control strategy. For that issue, DCM mode must be guaranteed during all the working points.

During Section II the ideal balancing behavior and equations are presented for the 6 most representative single switch balancing systems. The ideal modelling is focused on the energy transfer between the *Strong* cells and the *Weak* cell of a series connected energy storage system. The Ideal modelling equations give the main parameters for the design of a single switch active balancing system.

In Section 3 the power losses $P_{Loss,T}$ of the converters are taken into account to measure the power reduction of the converter. This approach gives a more accurate model of the balancing currents $I^*_{B,i}$ and the balancing power $P^*_{B,T}$.

For the experimental results presented in Section 4 the 6 different balancing systems have been designed and prototyped. During unbalancing situation it has been seen that the *Weak* cell balancing current $I^*_{B,W}$ is always bigger that the *Strong* cells balancing current $I^*_{B,S}$. During balanced situation it has been demonstrated that the balancing power of the system only contributes energy to compensate the power losses of the system.

The balancing system efficiency η for a big unbalance case is kept above 75% for selective topologies.

The non-selective *FLY_4P4SS* is the lowest efficiency system, and is not selective, decreasing the balancing speed of the system. This topology is the worst located for future balancing system designs.

Between the selective topologies, the efficiencies and the behavior are similar for a 400 mA $I_{B,W}$ case.

The transformer based topologies are good candidates thanks to the degree of freedom of the turn ratio when D_{cri} is low due to high number of series connected cells.

References

[1] E. Comission, "Energy 2020 - A strategy for competitive, sustainable and secure energy," *Brussels*, 2010. .

[2] EVE, "Basque country energy strategy 2020 (E2020)," p. 236, 2012.

[3] J. M. Carrasco, L. G. Franquelo, J. T. Bialasiewicz, E. Galvan, R. C. P. Guisado, M. A. M. Prats, J. I. Leon, and N. Moreno-Alfonso, "Power-Electronic Systems for the Grid Integration of Renewable Energy Sources: A Survey," *Industrial Electronics, IEEE Transactions on*, vol. 53, no. 4. pp. 1002–1016, 2006.

[4] B. Hartmann and A. Dan, "Methodologies for Storage Size Determination for the Integration of

Wind Power," *Sustainable Energy, IEEE Transactions on*, vol. 5, no. 1. pp. 182–189, 2014.

[5] N. Mendis, K. M. Muttaqi, and S. Perera, "Management of Battery-Supercapacitor Hybrid Energy Storage and Synchronous Condenser for Isolated Operation of PMSG Based Variable-Speed Wind Turbine Generating Systems," *Smart Grid, IEEE Transactions on*, vol. 5, no. 2. pp. 944–953, 2014.

[6] K. Strunz, E. Abbasi, and D. N. Huu, "DC Microgrid for Wind and Solar Power Integration," *Emerging and Selected Topics in Power Electronics, IEEE Journal of*, vol. 2, no. 1. pp. 115–126, 2014.

[7] C. X. Wu, C. Y. Chung, F. S. Wen, and D. Y. Du, "Reliability/Cost Evaluation With PEV and Wind Generation System," *Sustainable Energy, IEEE Transactions on*, vol. 5, no. 1. pp. 273–281, 2014.

[8] F. Diaz-Gonzalez, F. D. Bianchi, A. Sumper, and O. Gomis-Bellmunt, "Control of a Flywheel Energy Storage System for Power Smoothing in Wind Power Plants," *Energy Conversion, IEEE Transactions on*, vol. 29, no. 1. pp. 204–214, 2014.

[9] S. Teleke, M. E. Baran, A. Q. Huang, S. Bhattacharya, and L. Anderson, "Control Strategies for Battery Energy Storage for Wind Farm Dispatching," *Energy Conversion, IEEE Transactions on*, vol. 24, no. 3. pp. 725–732, 2009.

[10] G. Xu, L. Xu, D. J. Morrow, and D. Chen, "Coordinated DC Voltage Control of Wind Turbine With Embedded Energy Storage System," *Energy Conversion, IEEE Transactions on*, vol. 27, no. 4. pp. 1036–1045, 2012.

[11] A. F. Burke, "Batteries and Ultracapacitors for Electric, Hybrid, and Fuel Cell Vehicles," *Proceedings of the IEEE*, vol. 95, no. 4. pp. 806–820, 2007.

[12] A. Khaligh and Z. Li, "Battery, Ultracapacitor, Fuel Cell, and Hybrid Energy Storage Systems for Electric, Hybrid Electric, Fuel Cell, and Plug-In Hybrid Electric Vehicles: State of the Art," *Vehicular Technology, IEEE Transactions on*, vol. 59, no. 6. pp. 2806–2814, 2010.

[13] O. Briat, J.-M. Vinassa, W. Lajnef, S. Azzopardi, and E. Woirgard, "Principle, design and experimental validation of a flywheel-battery hybrid source for heavy-duty electric vehicles," *Electric Power Applications, IET*, vol. 1, no. 5. pp. 665–674, 2007.

[14] E. Schaltz, A. Khaligh, and P. O. Rasmussen, "Influence of Battery/Ultracapacitor Energy-Storage Sizing on Battery Lifetime in a Fuel Cell Hybrid Electric Vehicle," *Vehicular Technology, IEEE Transactions on*, vol. 58, no. 8. pp. 3882–3891, 2009.

[15] E. Tara, S. Shahidinejad, S. Filizadeh, and E. Bibeau, "Battery Storage Sizing in a Retrofitted Plug-in Hybrid Electric Vehicle," *Vehicular Technology, IEEE Transactions on*, vol. 59, no. 6. pp. 2786–2794, 2010.

[16] H. Yoo, S.-K. Sul, Y. Park, and J. Jeong, "System Integration and Power-Flow Management for a Series Hybrid Electric Vehicle Using Supercapacitors and Batteries," *Industry Applications, IEEE Transactions on*, vol. 44, no. 1. pp. 108–114, 2008.

[17] C.-S. Moo, K. S. Ng, and Y.-C. Hsieh, "Parallel Operation of Battery Power Modules," *Energy Conversion, IEEE Transactions on*, vol. 23, no. 2. pp. 701–707, 2008.

[18] C.-S. Moo, K.-S. Ng, and J.-S. Hu, "Operation of battery power modules with series output," *Industrial Technology, 2009. ICIT 2009. IEEE International Conference on*. pp. 1–6, 2009.

[19] T. Kim, W. Qiao, and L. Qu, "A series-connected self-reconfigurable multicell battery capable of safe and effective charging/discharging and balancing operations," *Applied Power Electronics Conference and Exposition (APEC), 2012 Twenty-Seventh Annual IEEE*. pp. 2259–2264, 2012.

[20] F. Jin and K. G. Shin, "Pack Sizing and Reconfiguration for Management of Large-Scale Batteries," *Cyber-Physical Systems (ICCPS), 2012 IEEE/ACM Third International Conference on*. pp. 138–147, 2012.

[21] T. Kim, W. Qiao, and L. Qu, "Power Electronics-Enabled Self-X Multicell Batteries: A Design toward Smart Batteries," *Power Electronics, IEEE Transactions on*, vol. PP, no. 99. p. 1, 2012.

[22] H. Qian, J. Zhang, J.-S. Lai, and W. Yu, "A High-Efficiency Grid-Tie Battery Energy Storage System," *Power Electronics, IEEE Transactions on*, vol. 26, no. 3. pp. 886–896, 2011.

[23] J. R. Belt, C. D. Ho, T. J. Miller, M. A. Habib, and T. Q. Duong, "The effect of temperature on capacity and power in cycled lithium ion batteries," *J. Power Sources*, vol. 142, no. 1–2, pp. 354–360, Mar. 2005.

[24] S. Santhanagopalan and R. E. White, "Quantifying Cell-to-Cell Variations in Lithium Ion

Batteries," *Int. J. Electrochem.*, vol. 2012, pp. 1–10, 2012.

[25] M. Uno and K. Tanaka, "Influence of High-Frequency Charge-Discharge Cycling Induced by Cell Voltage Equalizers on the Life Performance of Lithium-Ion Cells," *Vehicular Technology, IEEE Transactions on*, vol. 60, no. 4. pp. 1505–1515, 2011.

[26] I. Aizpuru, U. Iraola, J. M. Canales, E. Unamuno, and I. Gil, "Battery pack tests to detect unbalancing effects in series connected Li-ion cells," *Clean Electrical Power (ICCEP), 2013 International Conference on*. pp. 99–106, 2013.

[27] C. Martinez, "Cell Balancing Maximizes The Capacity Of Multi-Cell Li-Ion Battery Packs," *Intersil. Inc.*

[28] Y. Barsukov, "Battery cell balancing: what to balance and how," *Texas Instruments*, 2005.

[29] W. C. Lee, D. Drury, and P. Mellor, "Comparison of passive cell balancing and active cell balancing for automotive batteries," *Vehicle Power and Propulsion Conference (VPPC), 2011 IEEE*. pp. 1–7, 2011.

[30] B. Lindemark, "Individual cell voltage equalizers (ICE) for reliable battery performance," *Telecommunications Energy Conference, 1991. INTELEC '91., 13th International*. pp. 196–201, 1991.

[31] S. Moore, "A review of cell equalization methods for lithium ion and lithium polymer battery systems," 2001.

[32] C.-H. Kim, M. Kim, J.-H. Kim, and G.-W. Moon, "Modularized charge equalizer with intelligent switch block for lithium-ion batteries in an HEV," *Telecommunications Energy Conference, 2009. INTELEC 2009. 31st International*. pp. 1–6, 2009.

[33] C.-H. Kim, M.-Y. Kim, Y.-D. Kim, and G.-W. Moon, "A modularized charge equalizer using battery monitoring IC for series connected Li-Ion battery strings in an electric vehicle," *Power Electronics and ECCE Asia (ICPE & ECCE), 2011 IEEE 8th International Conference on*. pp. 304–309, 2011.

[34] C.-H. Kim, M.-Y. Kim, H.-S. Park, and G.-W. Moon, "A Modularized Two-Stage Charge Equalizer With Cell Selection Switches for Series-Connected Lithium-Ion Battery String in an HEV," *Power Electronics, IEEE Transactions on*, vol. 27, no. 8. pp. 3764–3774, 2012.

[35] C. Kim, H. Park, G. Moon, and A. C. Description, "A Modularized Two-Stage Charge Equalization Converter for Series Connected Lithium-Ion Battery Strings in an HEV," pp. 992–997, 2008.

[36] H.-S. Park, C.-E. Kim, C.-H. Kim, G.-W. Moon, and J.-H. Lee, "A Modularized Charge Equalizer for an HEV Lithium-Ion Battery String," *Industrial Electronics, IEEE Transactions on*, vol. 56, no. 5. pp. 1464–1476, 2009.

[37] H.-S. Kim, K.-B. Park, S.-H. Park, G.-W. Moon, and M.-J. Youn, "A new two-switch flyback battery equalizer with low voltage stress on the switches," *Energy Conversion Congress and Exposition, 2009. ECCE 2009. IEEE*. pp. 511–516, 2009.

[38] H.-S. Park, C.-H. Kim, K.-B. Park, G.-W. Moon, and J.-H. Lee, "Design of a Charge Equalizer Based on Battery Modularization," *Vehicular Technology, IEEE Transactions on*, vol. 58, no. 7. pp. 3216–3223, 2009.

[39] J. Ewanchuk, D. Yague, and J. Salmon, "A modular balancing bridge for series connected Li-ion batteries," *Energy Conversion Congress and Exposition (ECCE), 2011 IEEE*. pp. 2908–2915, 2011.

[40] M. Uno and K. Tanaka, "Single-Switch Cell Voltage Equalizer Using Multistacked Buck-Boost Converters Operating in Discontinuous Conduction Mode for Series-Connected Energy Storage Cells," *Vehicular Technology, IEEE Transactions on*, vol. 60, no. 8. pp. 3635–3645, 2011.

[41] M. Uno and K. Tanaka, "Single-Switch Multi-Output Charger Using Voltage Multiplier for Series-Connected Lithium-Ion Battery/Supercapacitor Equalization," *Industrial Electronics, IEEE Transactions on*, vol. PP, no. 99. p. 1, 2012.

[42] N. H. Kutkut, H. L. N. Wiegman, D. M. Divan, and D. W. Novotny, "Design considerations for charge equalization of an electric vehicle battery system," *IEEE Trans. Ind. Appl.*, vol. 35, no. 1, pp. 28–35, 1999.

A Control Strategy for Multiple Energy Storage Devices for Power Leveling of Renewable Energy systems

Koji Kato, Keisuke Nakano, Toshihiro Shimao, Youichi Ito, Sanken Electric Co., Ltd., Japan, k.kato@sanken-ele.co.jp
Hitoshi Haga, Nagaoka University of Technology, Japan
Kenji Arimatsu, Katsuhiro Matsuda, Tohoku Electric Power Co., Inc., Japan

Abstract

The paper proposes an energy storage system (ESS) with multiple energy storage devices such as a LiB and an EDLC for reducing the cost. The ESS is integrated into a power grid to compensate for the power fluctuation from renewable energy systems. The ESS which combines two types of energy devices is effective for low-cost and small size of the storage system. The proposed control method divides the amplitude of the compensation power compensation command into the LiB and the EDLC. Experimental results demonstrate that the proposed control method effectively compensated for the power fluctuation from the renewable energy system.

1. Introduction

The renewable energy such as a photovoltaic cell and a wind turbine is introduced into the power grid by reconstructions of the earthquake disaster 2011 in Japan. There is a ministry plan that the renewable energy will be introduced to 50 % of the consumed power until 2050 in Japan. Renewable energy systems, especially a wind turbine and a photovoltaic cell, generated a power fluctuation due to the meteorological conditions. Therefore problem of renewable energy into power grid is frequency and voltage fluctuation of power grid. To avoid frequency fluctuation of power grid, the energy storage system (ESS) is connected to the power grid [1]-[8]. It is expected that ESSs are installed in a remote mountain area to compensate the weak power grid. Therefore ESS requires the environment resistance and long lifetime.

There are a lot of energy storage devices such as lead acid battery (Lab), lithium ion battery (Lib), electric double layer capacitor (EDLC), Flywheel (FW) and etc. The energy storage devices such as FWs or EDLCs can be used to provide fast frequency regulation, and power leveling. However, these energy storage devices are bulky and expensive. The LaB is lifetime-limiting by factor of the storage system. The energy-storage system which combines two types of energy devices is effective for low-cost and small size of the storage system. In [9], the compensation power is divided at frequency. High frequency component of the compensation power is supplied by the EDLC, and low frequency component of the compensation power is supplied by the LaB. The system is effective for small size of the storage-system.

In [10]-[11], the authors propose a high efficiency and cost minimization method for grid power fluctuation compensation system. The proposed system is consists of the LiB and the EDLC for the energy-storage device. The proposed control method divides the amplitude of the compensation power command. However, control parameter of proposed method isn't carried out. Hence, this paper presents optimization of control parameter for the proposed systems. Experimental results and simulation results demonstrate that the proposed control method effectively compensated for the power fluctuation.

2. Energy Storage System

2.1. Energy Storage devices

Figure 1 shows comparison with power storage devices for power density and energy density. Vertical axis of Figure 1 is the storage energy [Wh]. Horizontal axis is the use range of each storage device. The chemical batteries such as Lib are used for the energy storage of long time in the lower right area of Figure 1. EDLC and FW are used for the energy storage of the short time in upper left area of the Figure 1. There are a lot of case that FW is put in operational service in Japan; the largest FWES in the world, which is used for the power source of the nuclear fusion reactor JT60 [12][13]. However, there is

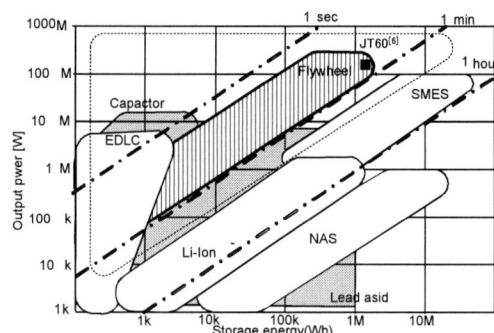

Figure 1. Comparison with power storage devices for power density and energy density.

no case that FW is industrialized as the generic product. On the other hand, FWES is industrialized as UPS in U.S. and Europe. In addition, they have over 10 years of experience [14][15].

Table 1 shows the comparison with the power storage devices for the cycle life and the storage capacity. FW and EDLC, which has long cycle life, are suitable to compensate the short cycle frequency fluctuation. However, EDLC is weak against heat and cold. It is difficult to use the EDLC in the cold area and warm area. On the other hands, FW doesn't degrade with temperature variation and charge-discharge cycle.

Table 1. Cycle life and the storage capacity.

	LaB	LiB	FW	EDLC
Energy capacity [Wh/kg]	50	100	1	1
Cycle life [times]	1500	3000	300000	100000

2.2. Proposed Energy Storage System Configuration

Figure 2 shows the system configuration of proposed power fluctuation compensation system with multiple energy storages (PFC‐MuES). The renewable energy system such as PV connects to the three-phase power-grid. The power compensation system is connected to the power grid to compensate the power fluctuation from the PV system. The PFC‐MuES consists of AC/DC converters, bi-directional DC/DC converters, LiBs and EDLCs.

The power flow of the energy storage devices are controlled by the bidirectional DC/DC converters. The AC/DC converter controls the DC link voltage constantly.

© VDE VERLAG GMBH · Berlin · Offenbach

PCIM Europe 2016, 10 – 12 May 2016, Nuremberg, Germany

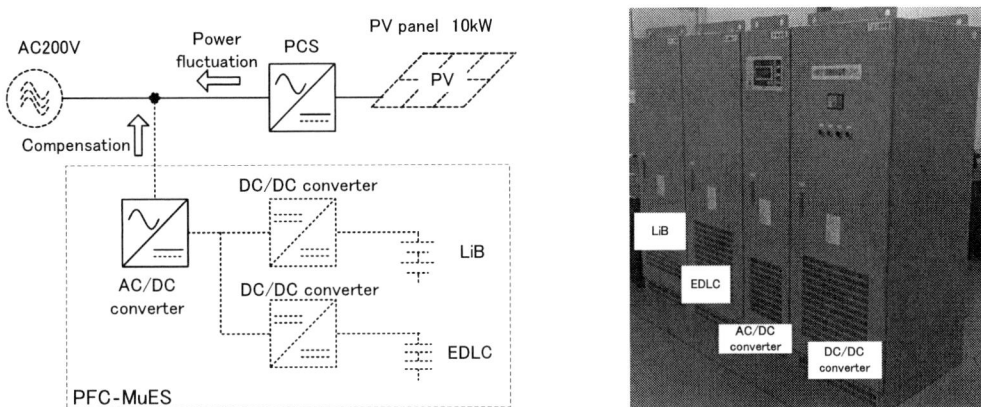

Figure 2. Proposed Energy Storage System Configuration

3. Control Strategy

3.1. Conventional methods

Figure 3 shows the conventional control method (frequency division allotment method: FDAM). The power compensation command $P_{comp}{}^*$ is calculated by rate limiter. The compensation power command $P_{comp}{}^*$ is divided at frequency. For example, the $P_{comp}{}^*$ is divided by using low-pass filter (LPF). The low frequency component of $P_{comp}{}^*$ is supplies to the energy storage device 1 as $P_{ES1}{}^*$, and high frequency component of $P_{comp}{}^*$ is supplied to the energy storage device 2 as $P_{ES2}{}^*$.

The EDLC is used for the energy storage device2 to contain the high frequency component. The EDLC, which has long cycle life, are suitable to compensate the short cycle frequency fluctuation. On the other hands, the energy storage device 1 requires the high energy capacity. The energy storage device 1 is used at the LiB to compensate the low frequency component.

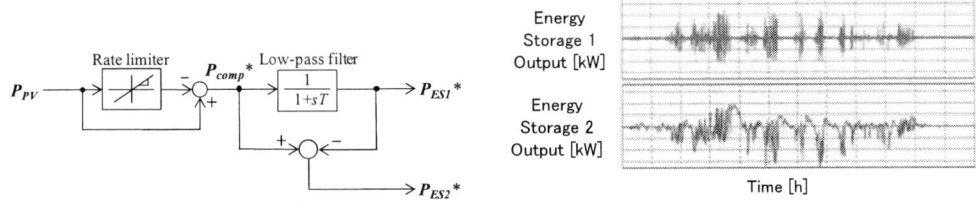

(a) Control block diagram   (b) Example of power command

Figure 3. Control method for the conventional method.

3.2. Proposed methods

Figure 4 shows the proposed control method (amplitude division allotment method: ADAM). In the proposed control method, the compensation power command $P_{comp}{}^*$ is calculated by rate limiter same as conventional method. Then, the compensation power command $P_{comp}{}^*$ is divided by amplitude limiter. For example, the command which is lower than the limiter level

© VDE VERLAG GMBH · Berlin · Offenbach

is supplied to the energy-storage device 1 as $P_{ES1}{}^*$. The command which is higher than the limiter level is supplies to the energy storage device 2 as $P_{ES2}{}^*$. In the proposed control method, the power converter for the energy storage device 1 needs electricity capacity of limiter level. The converter for the energy storage device 2 needs electricity capacity of (1-limiter level). Hence, the proposed control method can reduce the power converter rating power.

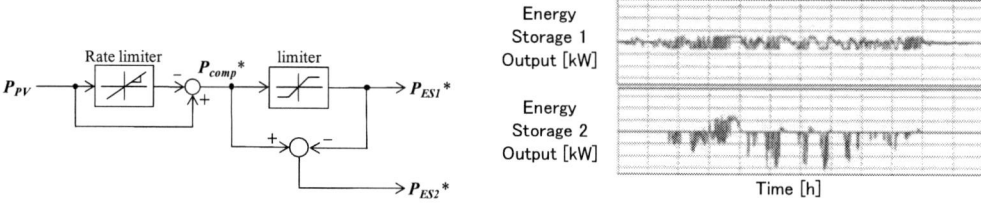

    (a) Control block diagram   (b) Example of power command

Figure 4. Control method for the proposed method.

4. Experimental Results

4.1. System specifications

Table 2 shows the specification of the power leveling system. The system rating power is 10kVA to compensate the power fluctuation from the PV system of 10kW. In order to interface energy storage device to AC/DC converter, the DC/DC converter is used. The DC/DC converter control the power flow of the energy storage device by power commands of $P_{ES1}{}^*$ and $P_{ES2}{}^*$.

Table 2 System specifications

AC/DC converter	Rating power	10kVA
	Rating voltage	200V
DC/DC converter	Rating power	10kW
	High side rating voltage	300V - 400V
	Low side rating voltage	150V - 250V
Energy storage device	LiB	4kWh
	EDLC	100kJ

4.2. Power compensation characteristics

Figure 5 shows the experimental result of the power leveling. The experimental result of conventional control method shown in fig.5(a). In the figures, the output power of the PV system, which is shown in green line, includes the power fluctuation. The red line shows the compensated power of power grid. The power fluctuation components are compensated by the EDLC and the LiB. The EDLC absorb the high frequency power fluctuation and the LiB absorbs the low frequency power fluctuation. The FRR(Fluctuation Reduction Rate) of 46.9% are obtained by the conventional method.

© VDE VERLAG GMBH · Berlin · Offenbach

This paper uses FRR in order to evaluate the compensation ability of the system. The calculating formula of FRR is expressed in the following expression.

$$FRR = \frac{(S_{PV} - S_o)}{S_{PV}} \cdot 100 [\%] \qquad (1)$$

where S_{PV} is the total of the spectrum in the evaluation section of the output power from PV system, S_O is the total of the spectrum in the evaluation section of the grid power. The grid power fluctuation is controlled effectively so that FRR value is large.

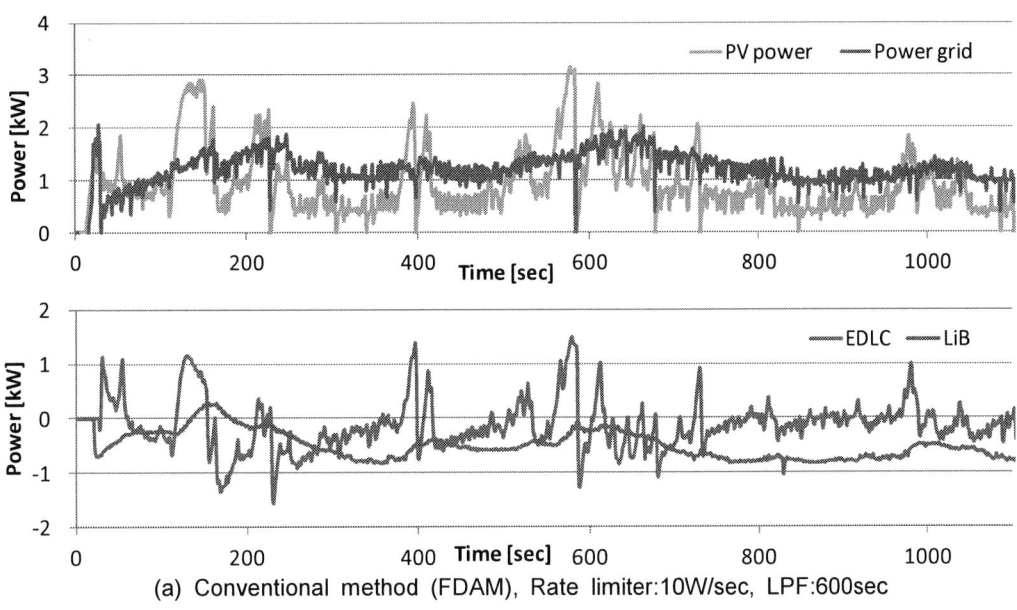

(a) Conventional method (FDAM), Rate limiter:10W/sec, LPF:600sec

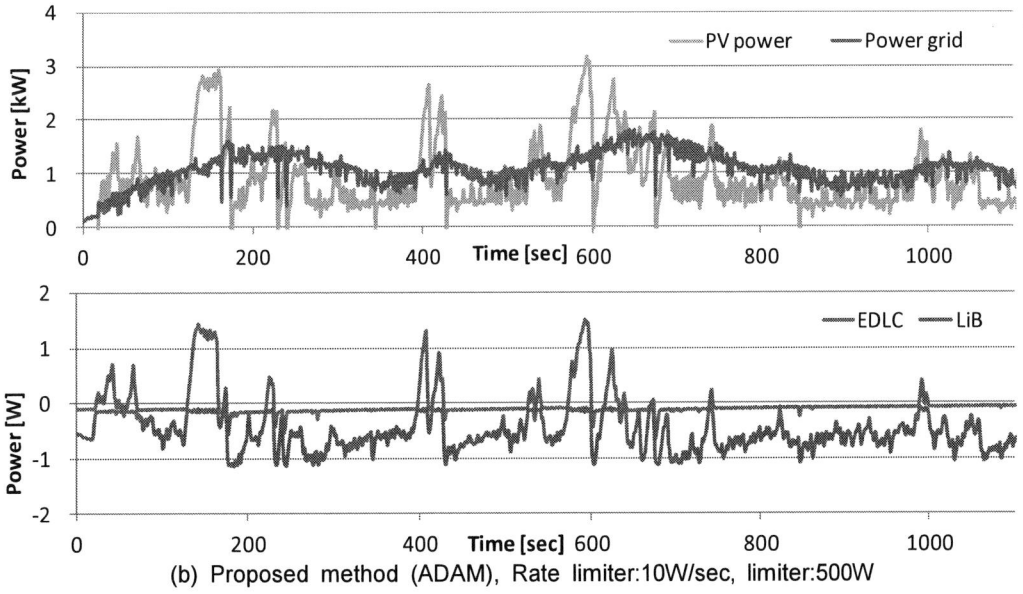

(b) Proposed method (ADAM), Rate limiter:10W/sec, limiter:500W

Figure 5. Experimental results of proposed system

Figure 5(b) shows the experimental results of proposed method. The LiB compensates power fluctuation of PV system which is less than the limiter value. The EDLC compensates power fluctuation which is greater than the limiter value. The proposed method compensates for power fluctuation from the PV system. The FRR of 55.0% are obtained by proposed method. The FRR of proposed method is 55%. The FRR of conventional method is 47%. The FRR of ADAM is higher 8% than that of FDAM.

The rate of change of the grid power which caused by PV is controlled to constant or less. This paper changes the threshold and the time constant variously. However, the grid power is controlled approximately in the same level.

4.3.   Rate limiter design method

Figure 6 shows the comparison between rate limiter and FRR. In this figure, the x-axis shows the power fluctuation level. For example, right side of the figure means sunny and rainy with low power fluctuation. Left side of the figure means cloudy with high power fluctuation. The y-axis shows the FRR by rate limiter parameter changing.

It is difficult to design the rate limiter parameter. Because the rate limiter parameter is depend on the weather condition. For example, the rate limiter parameter has to set the 90-100%/sec to achieve the FRR of 50% in cloudy with high power fluctuation. On the other hands there is no power fluctuation in sunny and rainy. The proposed system doesn't work. The rate limiter parameter is designed by weather forecast, insolation condition, and etc.

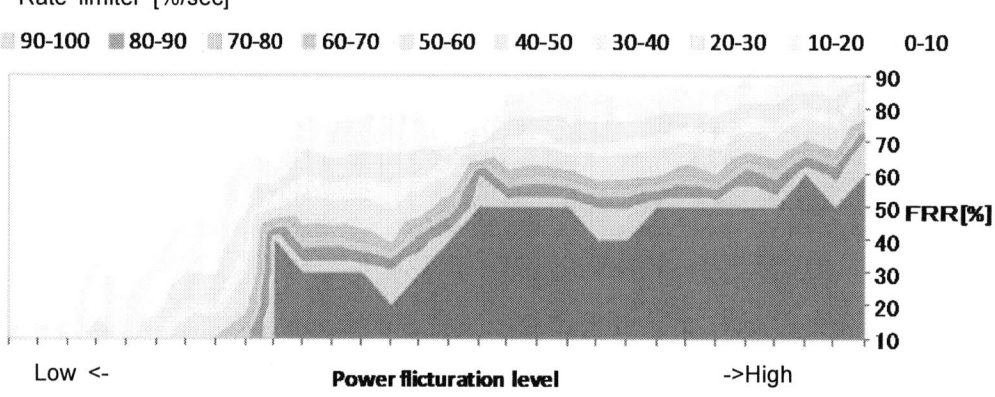

Figure 6 Comparison between rate limiter and FRR.

5.   Conclusion

The paper proposes the ESS with multiple energy storage devices such as the LiB and the EDLC for reducing the cost. The ESS is integrated into a power grid to compensate for the power fluctuation from renewable energy systems. The ESS is connected to the power grid to compensate the power fluctuation from the PV system. The proposed system consists of AC/DC converters, bi-directional DC/DC converters, LiBs and EDLCs. The proposed control method divides the amplitude of the compensation power compensation command into the

LiB and the EDLC. Experimental results demonstrate that the proposed control method effectively compensated for the power fluctuation from the renewable energy system.

6. Reference

[1] O. Wasynczuk, D.T. Man, and J.P. Sullivan: "Dynamic Behavior of a Class of Wind Turbine Generators During Random Fluctuations" IEEE Transactions on Power Apparatus Sysr., Vol. PAS-100, No. 6, pp2837-2854, 1981

[2] J. Baba, et al:" Combined power supply method for micro grid by use of several type of distributed power generation systems" Proc. of EPE 2005, CD-ROM, Dresden, Germany, 2005

[3] T.Tanabe, S.Suzuki, Y.Ueda, T.Ito, S.Numata, E.Shimoda, T.Funabashi, and R.Yokoyama: "Control Performance Verification of Power System Stabilizer with an EDLC in Islanded Microgrid" IEEJ Trans. PE, Vol.129, No.1, pp.139-147, 209(in Japanese).

[4] M. J. E. Alam, K. M. Muttaqi, and D. Sutanto, "Mitigation of Rooftop Solar PV Impacts and Evening Peak Support by Managing Available Capacity of Distributed Energy Storage Systems," *IEEE Trans. On Power Systems*, vol.28, no.4, pp.3874-3884, 2013.

[5] A. Canova, L. Giaccone, F. Spertino, and M. Tartaglia, "Electrical impact of photovoltaic plant in distributed network," *IEEE Trans. Ind. Appl.*, vol. 45, pp. 341–347, 2009.

[6] R. A. Walling, R. Saint, R. C. Dugan, J. Burke, and L. A. Kojovic, "Summary of distributed resources impact on power delivery systems," *IEEE Trans. Power Del.*, vol. 23, pp. 1636–1644, 2008.

[7] M. Takagi, Y. Iwafune, K. Yamaji, H. Yamamoto, K. Okano, R. Hiwatari, and T. Ikeya, "Economic Value of PV Energy Storage Using Batteries of Battery-Switch Stations," *IEEE Trans. On Sustainable Energy*, vol.4, no.1, pp.164-173, 2013.

[8] S.Tamura, "Economic Analysis of Hybrid Battery Energy Storage System Applied to Frequency Control in Power System," *The transactions of the Institute of Electrical Engineers of Japan. B*, vol.135, No.1, pp.2-8 (2015).

[9] S. Suzuki, S. Maeshima, K. Morino, E. Shimoda, H. Sugihara, and T. Esaki, "Site Test in Micro-Grid into which a large amount of PV power generation system are introduced," *The 2008 Annual Conference of Power and Energy Society, IEEJ*, no. 166, pp. 11-12, 2008.

[10] H. Haga, T. Shimao, S. Kondo, K. Kato, Y. Itoh, K. Arimatsu, K. Matsuda, "High-efficiency and Cost-minimization Method of Energy Storage System with Multi Storage Devices for Grid Connection", *Proceedings of the 2014 International Power Electronics Conference* (IPEC-Hiroshima 2014), 19P5-3, 2014.

[11] H. Haga, T. Shimao, S. Kondo, K. Kato, Y. Itoh, K. Arimatsu, K. Matsuda, "Loss Analysis of Energy-Storage System with Multiple Storage Devices for Grid Connection", *Proceedings of the 37th International Telecommunications Energy Conference* (Intelec-Osaka 2015), TS11-1, pp270-275, 2015.

[12] H. Kameno, A. Kubo and R. Takahata, "Basic Design of 1kWh Class Compact Flywheel Energy Storage System", Koyo Engssineering Journal No.163, March 2003

[13] T.Matukawa, M.Kanke, R.Shimada, et al:" A 215MVA Flywheel Moter-generator with 4GJ Discharge Energy for JT-60 Troidal Field Coil Power Supply System " IEEE Trans. on Energy Conv. EC-2,262, 1987.

[14] http://www.cat.com/power-generation/ups

[15] http://www.activepower.com/CompanyInformation/

PCIM Europe 2016, 10 – 12 May 2016, Nuremberg, Germany

Examining Contrasting Excitation Modes within Battery Characterisation using Maximum Length Sequences

Andrew Fairweather, Vxl Power Ltd, Station Road, North Hykeham, Lincoln, LN6 3QY, United Kingdom, andrew.fairweather@vxipower.com

David Stone, Electrical Machines and Drives Research Group, Department of Electronic and Electrical Engineering, The University of Sheffield, Mappin Street, Sheffield, S1 4DT, United Kingdom, d.a.stone@sheffield.ac.uk

Martin Foster, Electrical Machines and Drives Research Group, Department of Electronic and Electrical Engineering, The University of Sheffield, Mappin Street, Sheffield, S1 4DT, United Kingdom, m.p.foster@sheffield.ac.uk

Abstract

This paper extends on previous work involving the use of maximum length sequences as tools for parameter estimation within electrochemical batteries[1-4] and comprises a study of the modes of application of a perturbation signal to batteries in the form of discharge, charge and bipolar charge/discharge arrangements driven by a frequency rich signal.

Examinations are made of the contemporary techniques, and by the use of experiments over a range of States-of-Charge, the advantages of each method are presented.

Using a series of experiments, impedance responses are produced using the tests batteries under perturbation from each of the test modes. Analysis of these results under comparative test conditions allows observations to be made regarding the application of each mode, and proposals are made for the complementary inclusion of the test regimes in a hybrid SoC/SoH prediction system.

1. Introduction

Electrochemical cells remain a key enabling technology for the progression of the technologies which are at the forefront of renewable energy and electric vehicles. Measurement of State-of-Function (SoF), State-of-Health (SoH) and State-of-Charge (SoC) of the battery or cell, and in turn, the electrochemical "fuel gauge" are notoriously difficult to realise. SoC reporting methods employing measurement of terminal voltage [5] can be effective if the load is constant, but typically require implementation of an algorithm to allow for cell degradation, whilst existing methods involving Coulomb counting[6] have been successful in consumer electronics, they are often subject to periodic recalibration to maintain accuracy.

Electrochemical cells and batteries employ chemical reactions in order to affect charge storage and delivery of current and as such exhibit different characteristics during charge and discharge [7]. Additionally, a battery at higher SoC accepts additional charge less readily and experiences a characteristic increase in impedance [7]. It follows therefore that applying perturbation signals in different modes could facilitate improved state identification over a range of charge states.

The motivation for this work was therefore focused on investigating then benefits and applicability of a battery perturbation test technique using differing methods (Charge, Discharge and Bipolar) to apply the test signal.

The investigation was concerned with establishing how the methods examined could be applied practically as individual testing schemes, or as a combined three-mode test signal which could be used to identify specific battery states.

2. Pseudo Random Binary Sequences as a perturbation signal

Pseudo Random Binary Sequences (PRBS) offer a digitally generated signal which on inspection appears random in nature, but is actually periodic, and therefore has properties which are extremely useful in several application areas. There are a class of PRBSs termed Maximum Length Sequence (MLS) that exhibit properties similar to white-noise and the sequences have extensively been used to establish audio frequency response[8], and system frequency response analysis generally[9].

© VDE VERLAG GMBH · Berlin · Offenbach

A PRBS generator can be seen in figure 1, using shift registers with modulo 2 (XNOR) feedback at predetermined "tap" positions, with the number of shift registers defining the bit order, n[10].

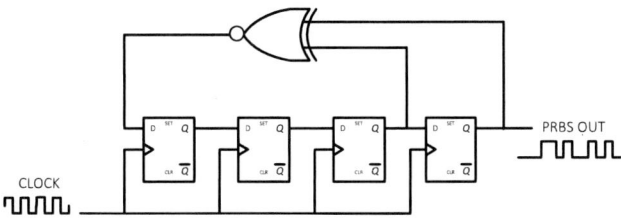

Fig. 1.   4-bit PRBS generator constructed from shift registers with determined "tap" positions and XNOR feedback

2.1.   Effects of test current amplitude and voltage thresholds

In the course of previous investigations[1], the selection of test current amplitude was considered in order to achieve an appropriate level of excitation to the test battery. Levels close to the rated capacity $(c_r)/20$ rate were found to be a good compromise between being intrusive to battery state and providing a reasonable level of voltage information for analysis[11]. Additionally, as the experimental set up was based on a constant current charger, the upper charge voltage limit for this system was very relevant to the battery under test. Charge voltages for the test battery from manufacturer's data[12] were taken into consideration during the application of the PRBS charge profile (during both charge mode and bipolar tests), as it was important that the battery should not be overcharged. It was decided also that by using the manufacturer's recommended voltage levels, temperature compensated, observations at near to 100% state of charge could lead to distinct SoC indicators [3].

3. PRBS test investigation

The overall test system block diagram is shown in Fig. 2 with the PRBS test system photograph in Fig. 3.

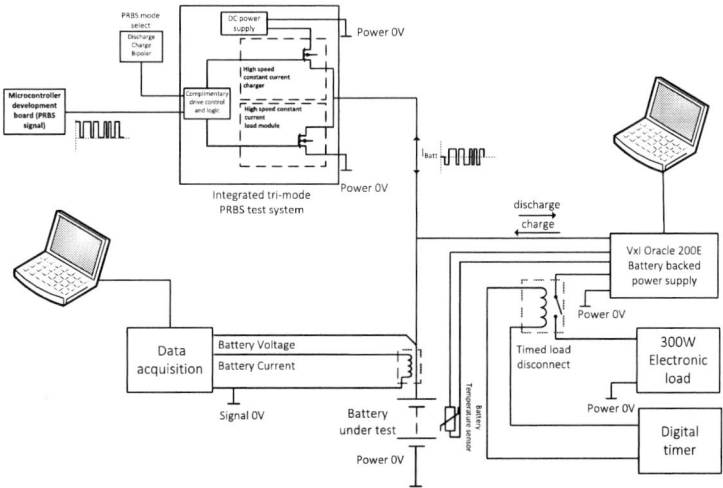

Fig 2. PRBS test system block diagram

© VDE VERLAG GMBH · Berlin · Offenbach

PCIM Europe 2016, 10 – 12 May 2016, Nuremberg, Germany

Fig. 3. PRBS test system photograph

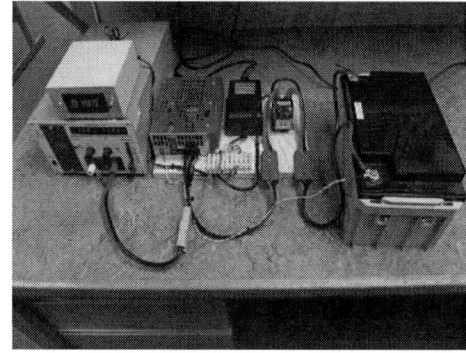

Fig. 4. Controlled charge/discharge system photograph

3.1. Test procedure

A controlled charge and discharge system (Fig. 4) was used to remove a pre-determined amount of energy from the test battery, separately to the PRBS system used in applying the test perturbation. Discharge and charge power stages are driven by an amplitude offset drive circuit in order to generate a PRBS perturbation signal centred around zero current. The hardware was configured in order that switching between the 3 modes of test (charge, discharge and bipolar) could be carried out easily.

The battery used during the tests was a 65Ah 12V Valve Regulated Lead Acid (VRLA) (Yuasa NP65-12i) type, which was conditioned with a number of charge and discharge cycles before being charged to 100% SoC using the temperature compensated Lead-Acid charger within the test controlled charge/discharge apparatus (figure 4). The bipolar PRBS was developed with prior work in mind and the test level used was +/- 4A, using an extended bandwidth hybrid PRBS sequence. The upper voltage limit for the charge pulse was fixed before the test, using previously established limits for such excitation[3]. PRBS tests (discharge, charge and bipolar) were then carried out on the battery at 100% SoC before it being discharged at 5 amps for 2 hours to remove around 15% of the rated capacity in preparation for the next test at 85% SoC. The third stage was to discharge the battery at the 20 hour discharge rate to the manufacturers specified End-of-Discharge (EoD) Voltage before carrying out the final PRBS test at close to 0% SoC.

3.2. Battery model development

In previous work carried out by the authors in battery characterisation the Randles' model[13] was used, and this was developed during further investigations [3, 4, 14]. These investigations, and examination of the work by Salameh et al[15] led to development of the model shown in Fig. 5 which was used in the analysis.

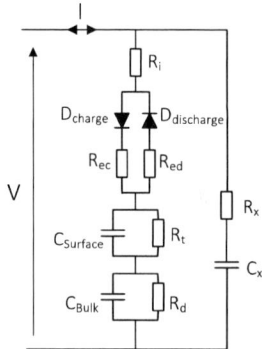

Fig. 5. Developed model

D_{charge} and $D_{discharge}$ are ideal diodes with no forward volt drop, in series with R_{ec} and R_{ed} respectively which represent the electrolyte resistance[15] which in conjunction with R_i (ohmic resistance) represent the significant series resistance of the battery.

© VDE VERLAG GMBH · Berlin · Offenbach

PCIM Europe 2016, 10 – 12 May 2016, Nuremberg, Germany

R_x and C_x were used in development of models in the wider part of this research [14] and have been shown to improve the curve fitting of the impedance response to simulation. These parameters represent a parallel branch element of $C_{Surface}$ and R_t which have found to be applicable to this model development.

3.3. Test results

Example current and voltage data are seen in figures 6 and 7 at 85% SoC one of the dynamic charge (PRBS) tests.

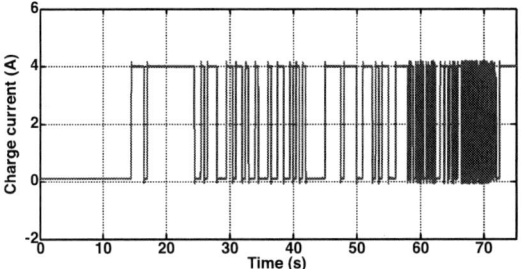

Fig. 6. Current waveform, 85% SoC, charge test

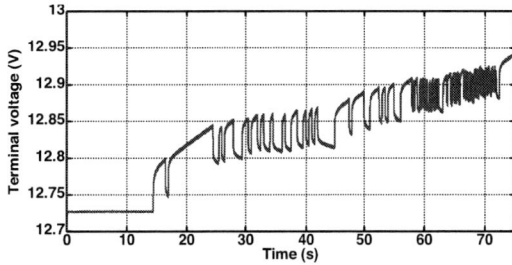

Fig. 7. Voltage response, 85% SoC charge test

Example current and voltage waveforms are seen in figures 8 and 9 for the bipolar test at 85% SoC.

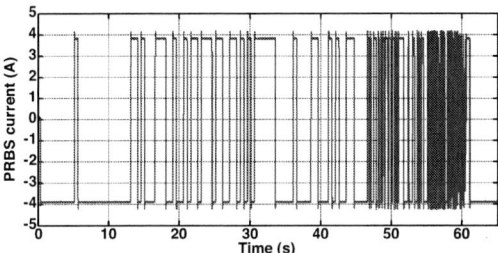

Fig. 8. Bipolar PRBS test current waveform, 85% SoC.

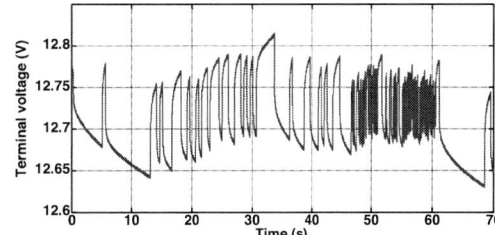

Fig. 9. Bipolar PRBS test voltage response, 85% SoC

The impedance information obtained from the tests are shown in Figs. 10 to 18. Transfer function analysis of the adopted model (Fig. 5) was employed to obtain a curve fit for each of the results.

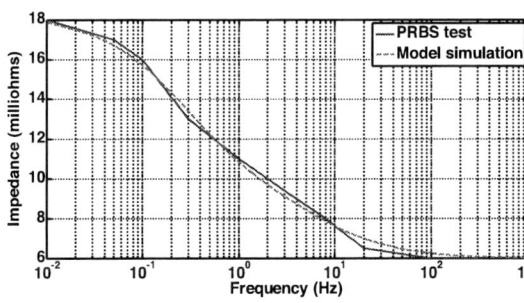

Fig.10. 100% SoC, discharge mode PRBS

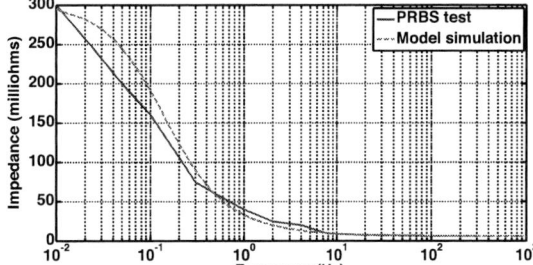

Fig. 11. 100% SoC, charge mode PRBS

© VDE VERLAG GMBH · Berlin · Offenbach

PCIM Europe 2016, 10 – 12 May 2016, Nuremberg, Germany

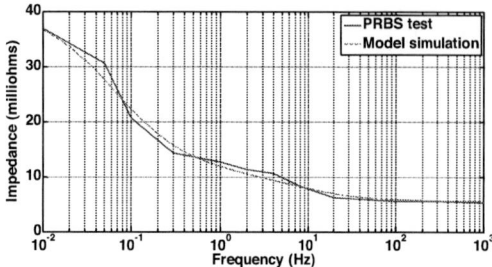

Fig. 12. 100% SoC, Biplolar mode PRBS

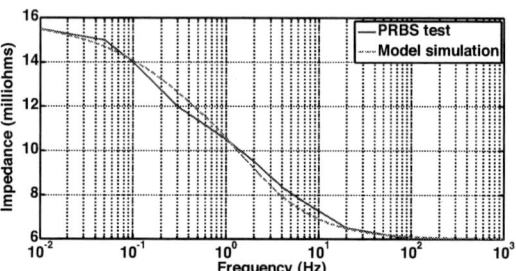

Fig. 13. 85% SoC, discharge mode PRBS

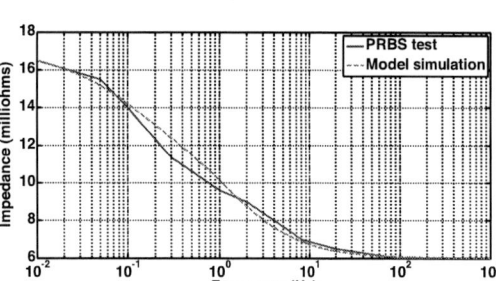

Fig.14. 85% SoC, charge mode PRBS

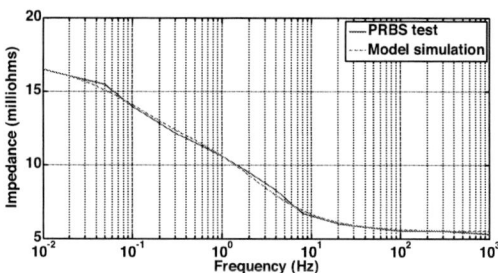

Fig.15. 85% SoC , Bipolar mode PRBS

Figures 10 to 12 illustrate the comparative results at 100% SoC. As mentioned, testing at this state of charge can lead to an indeterminate result, as the battery may not be at a steady state terminal voltage. This has been addressed previously in battery pulse testing by applying a preload to the battery[16], and in PRBS discharge tests by disregarding initial data sets until a pseudo steady-state voltage envelope is observed. However, the PRBS charge technique does not have this facility as the test mode inherently charges the battery. This led to an elevation of terminal voltage during the 100% SoC test which resulted in "clipping" in the PRBS charge current. Importantly, this is observed in figure 11 as the high magnitude of low frequency impedance, whilst the high frequency impedance approaches the expected level. This phenomenon clearly shows detection of end of charge, whilst showing healthy impedance results for the higher frequency part of the response.

Figures 13 and 14 show the test results at 85% SoC for the discharge and charge test modes. Both results show similar results but the differences in the charge and discharge processes are apparent in elements of the curve fitting.

Figure 15 shows the impedance plot for the bipolar test at 85% SoC. This is consistent with the discharge and charge PRBS methods, within the quasi-linear area of operation of the battery. The low frequency impedance of the battery is therefore more representative of typical performance, and the plot generally would be used to indicate battery SoH.

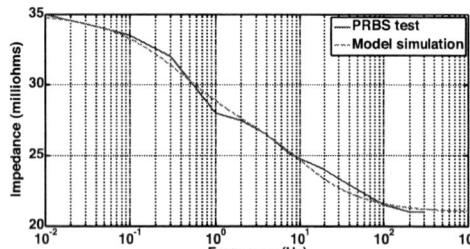

Fig. 16. 0% SoC, discharge mode PRBS

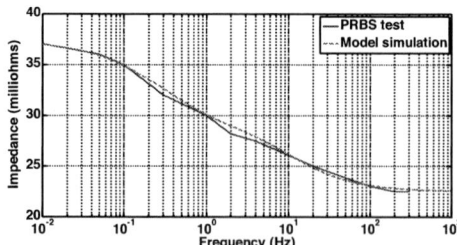

Fig. 17. 0% SoC, charge mode PRBS

Figures 16 and 17 show the discharge and charge test results for 0% SoC. The battery shows an elevation in impedance across the test frequency range which shows both test modes

© VDE VERLAG GMBH · Berlin · Offenbach

exhibit similar results but with some differences for the two test methods over the frequency range, with the discharge mode showing a generally lower impedance.

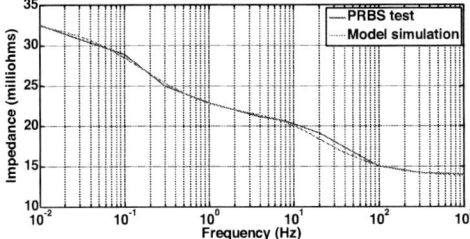

Figure 18. 0% SoC, Bipolar PRBS

Figure 18 shows the bipolar impedance plot for the test at 0% SoC. The battery in this state shows an elevation in impedance across the test frequency range. This gives a distinct indication of a discharged battery and can be compared to the 100% SoC results (Figure 13) which show similarly the increased LF impedance, but preserve the healthy HF result. The test impedance results are presented together in Figs. 19 and 20.

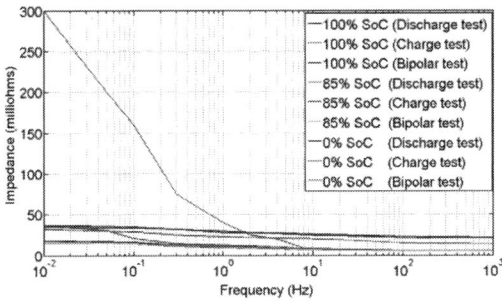

Fig.19. Comparative impedance results

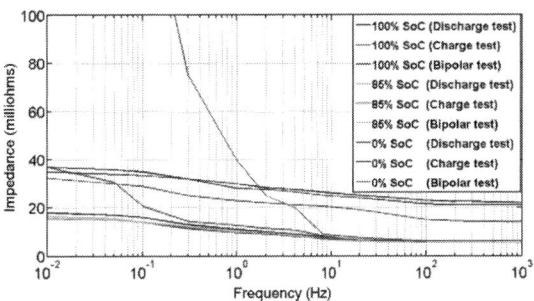

Fig. 20. Expanded scale to show detail

In spite of this observed increase in low frequency impedance, the higher frequency impedance for the battery appears healthy, allowing both SoC and SoH to be reported, provided the upper voltage threshold for the PRBS charge is carefully selected.

3.4. Battery model parameters

Table 1 shows the parameters obtained for the curve fitting using the bipolar tests, with the corresponding discharge and charge test parameters.

R_i, R_{ec} and R_{ed} are combined parameters for the bipolar tests as no distinction can be made between the individual components, and are therefore compared to R_i+R_{ec} for the charge mode tests and R_i+R_{ed} for the discharge mode tests.

At 85% SoC the three modes of test show similarly impedance and parameter results, but it is at the extremes of charge states 0% and 100% where the individual methods show the significant differences.

Across the test methods, the elements of surface capacitance ($C_{Surface}$ and C_x) remain indicators of the ability of the battery to deliver energy, and further profiling of this capacitance against actual discharge tests may reveal direct correlations to bulk capacity.

The results in the table show differences the charge and discharge technique mainly related to the elements of $C_{Surface}$. This demonstrates somewhat the different reactions involved in the charge and discharge processes [7] and overall observations on the validity of the charge technique are satisfied in that clear results are observed for the various charge states as compared to the discharge technique with trends that are recognisable for both methods.

© VDE VERLAG GMBH · Berlin · Offenbach

Table 1. Obtained model parameters

BATTERY STATE AND TEST MODE	R_I,R_{EC},R_{ED} (mΩ)	R_I+R_{EC} (mΩ)	R_I+R_{ED} (mΩ)	R_T (mΩ)	$C_{SURFACE}$ (F)	C_X (F)	R_X (mΩ)
100% (BIPOLAR)	5.5	-	-	35.3	3.9	60	5.8
100% SOC (DISCHARGE)	-	-	6	12	6	34	4
100% SOC (CHARGE)	-	6	-	300	4	2	9
85% (BIPOLAR)	5.5	-	-	11.3	12	60	10
85% SOC (DISCHARGE)	-	-	6	9.5	16	35	9
85% SOC (CHARGE)	-	6	-	10.75	20	60	9
0% (BIPOLAR)	13.7	-	-	19.2	1	31	12.5
0% SOC (DISCHARGE)	-	-	21	13.8	2.5	16	9
0% SOC (CHARGE)	-	22.5	-	14.6	2.5	22	9

4. Conclusion

This work explores the respective benefits of frequency rich signals applied as charge, discharge and bipolar modes to obtain state indicators and equivalent circuit parameters for Lead-Acid batteries.

Applying a perturbation to the test battery as a discharge is the most straightforward method of applying this signal, and produces useful results. Long tests are precluded due to the overall effect on the test battery terminal voltage, but this is a trade off against a straightforward test.

The effect of applying a bipolar PRBS perturbation signal with an average value of zero facilitates longer tests at lower frequency and/or increased bit length. Additionally, the bipolar system was applicable at 100% SoC, and showed repeatable results at this state, with a characteristic, elevated low frequency impedance at the fully charged state.

Within the charge mode PRBS tests, it was discovered that by control of the charge voltage headroom high states of charge were clearly identified, and whilst reporting this SoC, the charge based PRBS was also suitable for reporting SoH, indicating via high frequency impedance the current delivering capability of the battery.

The charge PRBS could be used to measure impedance over the full range of charge and therefore finds application as a state evaluation system that can be incorporated with a battery charger with minimal additional hardware, predominantly requiring an embedded processor to carry out the analysis.

The investigations described within this paper form part of a body of ongoing research by the author and co authors and will lead to further publications in this field.

5. References

1. Fairweather, A.J., M.P. Foster, and D.A. Stone, *Battery parameter identification with Pseudo Random Binary Sequence excitation (PRBS).* Journal of Power Sources, 2011. **196**(22): p. 9398-9406.
2. Fairweather, A.J., M.P. Foster, and D.A. Stone, *Modelling of VRLA batteries over operational temperature range using Pseudo Random Binary Sequences.* Journal of Power Sources, 2012. **207**(0): p. 56-59.
3. Fairweather, A.J., M.P. Foster, and D.A. Stone, *Application of Maximum Length Sequences to Battery Charge Programming for Parameter Estimation in Lead-Acid Batteries,* in *PCIM Europe 2013.* 2013: Nuremberg, Germany.
4. Fairweather, A.J., M.P. Foster, and D.A. Stone, *Bipolar Mode Pseudo Random Binary Sequence Excitation for Parameter Estimation in Lead-Acid Batteries,* in *PCIM Asia.* 2013: Shanghai, China.
5. Coleman, M., et al., *State-of-Charge Determination From EMF Voltage Estimation: Using Impedance, Terminal Voltage, and Current for Lead-Acid and Lithium-Ion Batteries.* Industrial Electronics, IEEE Transactions on, 2007. **54**(5): p. 2550-2557.
6. Nguyen, K.S., et al., *Enhanced coulomb counting method for estimating state-of-charge and state-of-health of lithium-ion batteries.* Applied Energy, 2009. **86**(9): p. 1506-1511.
7. Linden, D. and T.B. Reddy, *Handbook of batteries.* 4th ed. McGraw-Hill handbooks. 2010, New York: McGraw-Hill. 1 v. (various pagings).
8. Jamieson, D.G. and T. Schneider, *Electroacoustic evaluation of assistive hearing devices.* Engineering in Medicine and Biology Magazine, IEEE, 1994. **13**(2): p. 249-254.
9. Vermeulen, H.J., J.M. Strauss, and V. Shikoana. *On-line estimation of synchronous generator parameters using PRBS perturbations.* in *Power Engineering Society General Meeting, 2003, IEEE.* 2003.
10. Davies, W.D.T., *System identification for self-adaptive control.* 1970, London, New York,: Wiley-Interscience. xiv, 380 p.
11. Fairweather, A.J., D.A. Stone, and M.P. Foster, *Evaluation of UltraBattery™ performance in comparison with a battery-supercapacitor parallel network.* Journal of Power Sources, 2013. **226**(0): p. 191-201.
12. Yuasa Battery Europe. *Yuasa NP Valve Regulated Lead Acid Battery Manual.* NP VRLA Application Manual [Application Manual] 1999 1/12/99; 1:[1, 2, 5, 6, 7, 8, 9, 12, 22, 24, 27, 29]. Available from: http://www.yuasa-battery.co.uk/industrial/downloads.html.
13. Fairweather, A.J., M.P. Foster, and D.A. Stone, *VRLA battery parameter identification using pseudo random binary sequences (PRBS),* in *IET Conference Publications.* 2010. p. TU244.
14. Fairweather, A.J., *State-of-Health (SoH) and State-of-Charge (SoC) Determination in Electrochemical Batteries and Cells Using Designed Perturbation Signals* in *Electronic and Electrical Engineering.* 2015, University of Sheffield: Sheffield, UK.
15. Salameh, Z.M., M.A. Casacca, and W.A. Lynch, *A mathematical model for lead-acid batteries.* Energy Conversion, IEEE Transactions on, 1992. **7**(1): p. 93-98.
16. Coleman, M., W.G. Hurley, and L. Chin Kwan, *An Improved Battery Characterization Method Using a Two-Pulse Load Test.* Energy Conversion, IEEE Transactions on, 2008. **23**(2): p. 708-713.

PCIM Europe 2016, 10 – 12 May 2016, Nuremberg, Germany

Comparison between standard and innovative solutions to exchange energy between high energy storage systems

Laurent GARNIER, Daniel CHATROUX, Sébastien CARCOUET, Julien DAUCHY

CEA, LITEN, DEHT F-38054 Grenoble, France

laurent.garnier@cea.fr

Abstract

Electrification of always more powerful systems is usually correlated to higher needs in reliability, service continuity and energy exchanges between sources. In the field of energy storage systems, these needs are often addressed by parallelization of batteries, which are automatically disconnected in case of fault. The service continuity is thus simply ensured. If the disconnection of two batteries in parallel is an instantaneous process well controlled, the connection is a longer process more complicated, which requires adapted power electronics solutions. Based on an existing case of application and thanks to simulations, we propose in this paper to compare different solutions to exchange energy between two lithium ion battery systems.

1. Presentation of the application

The "Easily diStributed Personal Rapid Transit" (ESPRIT) H2020 project aims to develop a purpose-built, light weight, L category electric vehicle that can be stacked together to gain space. Thanks to pioneering coupling systems, up to 8 ESPRIT vehicles can be nested together in a road train, seven being towed, for an efficient redistribution of fleets and a smartly-balanced and cost efficient transport system. During redistribution or parking, energy can be exchanged between vehicles to maximize and secure the state of charge of the first vehicles which will be taken first by a user.

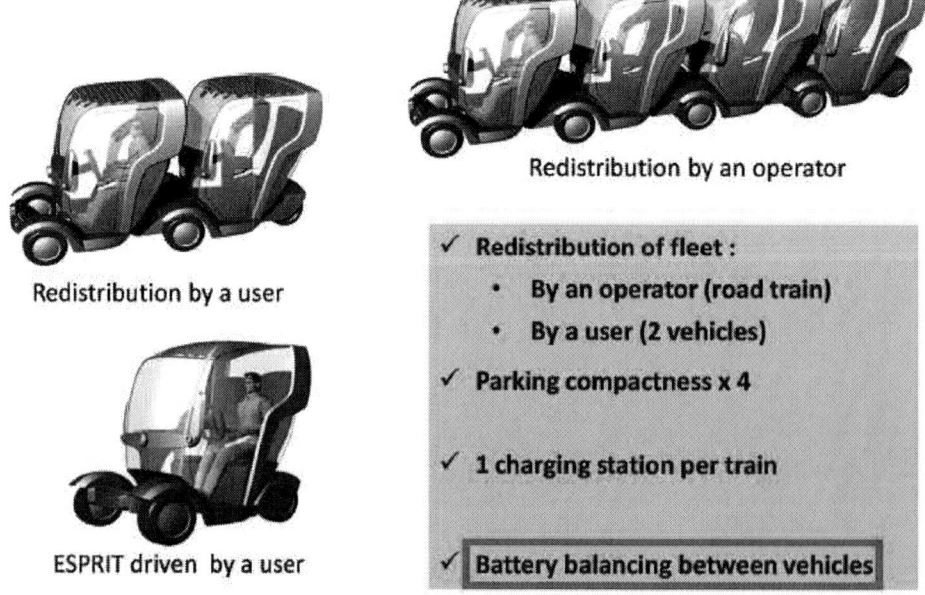

Figure 1 - Aims of the ESPRIT project

© VDE VERLAG GMBH · Berlin · Offenbach

2. Problem presentation

Inside each electric vehicle presented just above, we have a battery system and a power exchange solution to balance energy between vehicles. The objective is to maximize and secure as far as possible the state of charge of the first vehicle of the road train because it will be the first one to leave the charging station. We can thus optimize the availability of a charged vehicle for the user.

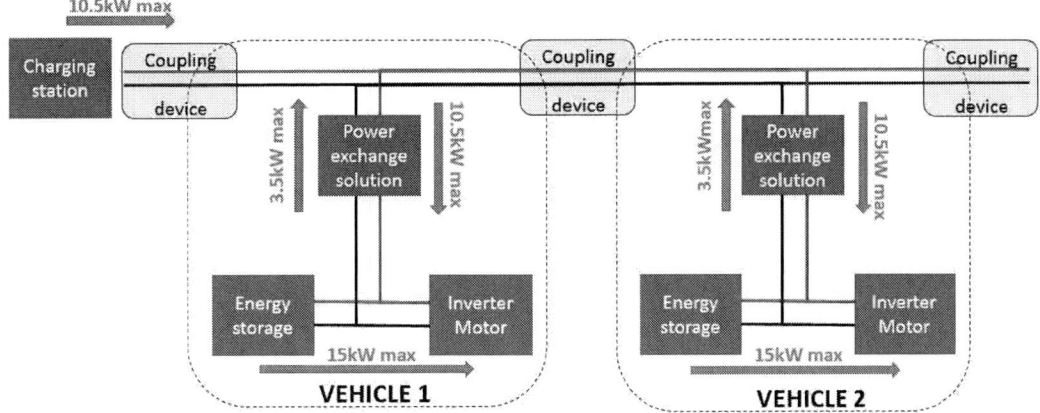

Figure 2 - main electrical architecture

The main specifications of the energy storage system and power exchange converter are:
- Li-ion chemistry: LiFePO4
- Unity cells format: 2.3Ah 26650, power cell
- Battery system composed of 12 modules 8S5P in series (320V, 3.5kWh)
- Maximum charge current : 3C ➔ 34,5A
- Maximum exchange current : 1C ➔ 11,5A

The main question we will address in this paper is:

Regarding the main criteria, balancing duration, losses and efficiency, volume and weight, cost and simplicity, behavior in case of short-circuit, what is the best power electric solution to exchange energy between LiFePO4 Li-ion batteries?

3. Battery model

To perform these comparisons a battery model, which includes the parameters capacity, internal resistance and voltage versus state of charge (SOC) is used:

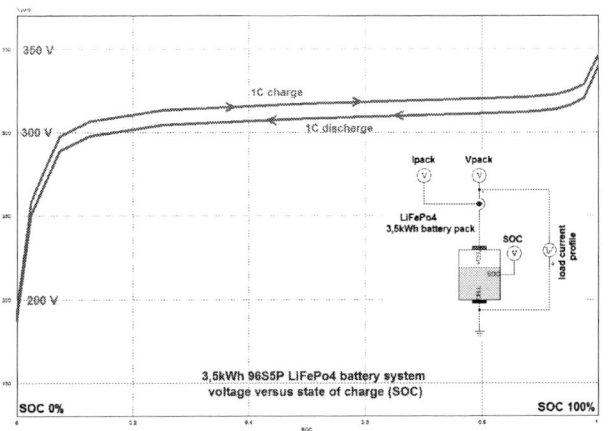

Figure 3 – Charge discharge battery cycle

4. Comparisons of five different power conversion solutions

In this part, after having detailed what happened in case of connection between two batteries in parallel without any adapted power electronics system, we will compare fifth solutions. The main objectives of these different solutions are firstly to compensate a difference of voltage between two sources and secondly to control the current flowing between these two sources.

To perform the comparisons, we use the following case of study:

- Battery 1 with an initial state of charge (SOC) of 5%
- Battery 2 with an initial state of charge (SOC) of 95%
- The balancing sequence is ended when the difference of SOC is less than 10%

4.1 Behavior of the system without any control of energy

Each battery system is protected against short-circuits and overloads with dedicated fuses. For example, we can use for our application a fuse OHEV040 (40A) from Littlefuse, specified with a resistance of about 2mΩ and a melting energy of 1495 A²s. The length of the cable between the two battery systems is estimated at 2 m, with thus a resistance of about also 2mΩ. The internal resistance of each battery is 192 mΩ. With all these assumptions the equivalent circuit is the following:

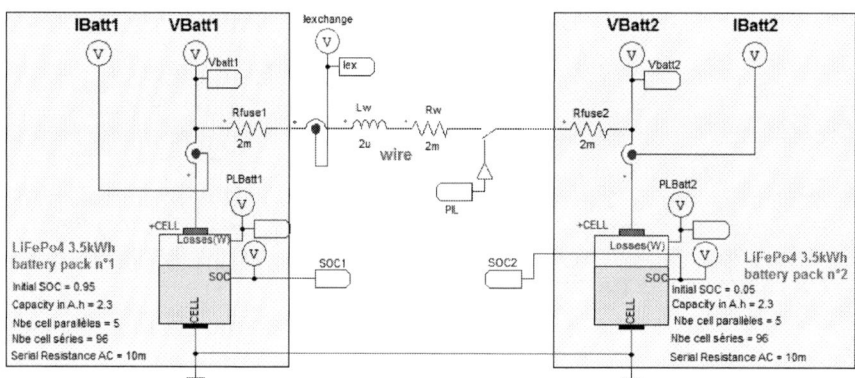

Figure 4 – reference simulations with two batteries in parallel

With initial states of charge of 95% and 5%, the starting exchange current is higher than 100A.

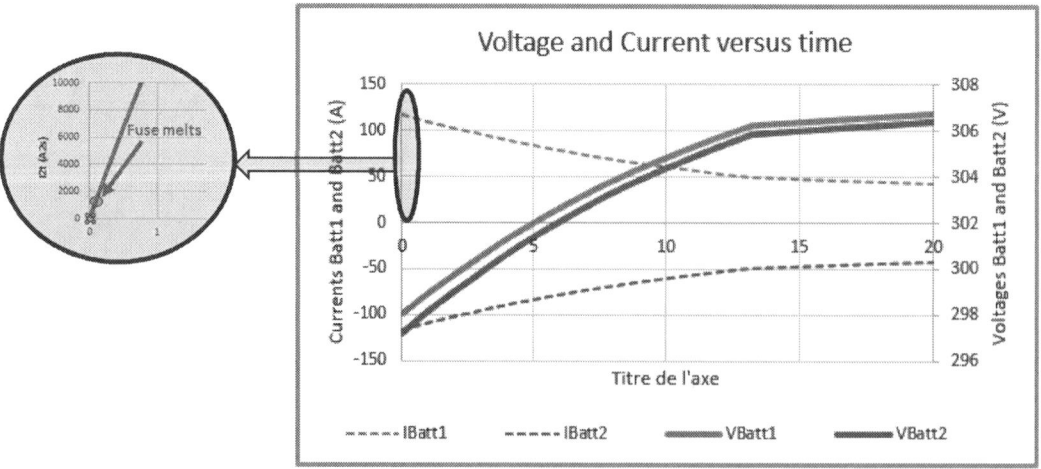

Figure 5 - waveforms without limitation device

© VDE VERLAG GMBH · Berlin · Offenbach

With these simple simulations we can easily understand that the current is here only limited by the internal resistances of the two battery systems, which are much higher than the wiring resistances. Without any external control of the current, the fuse melts very quickly in less than 1s. To bear such currents, we could oversize the fuses to allow the exchange of currents. But the high current levels (up to 10C) generate thermal losses inside battery systems and has an impact on battery life time.

In this reference simulation we can calculate balancing time and energy losses:

- Balancing time : **2600 s**
- Energy exchanged : **5147kJ (1429Wh)**
- Losses inside the two battery systems : **160 kJ**
- Losses inside cabling : **1794J**

4.2 First control solution: full power bidirectional DC/DC boost converters

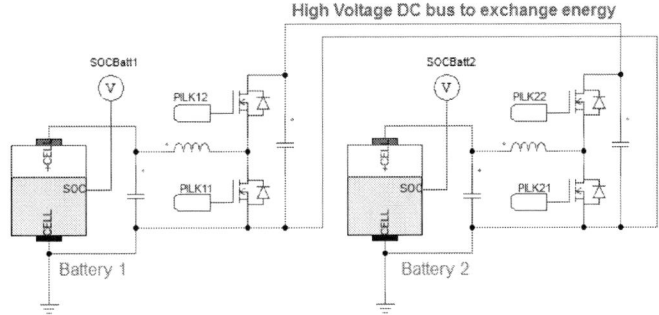

Figure 6 - solution 1

The first solution proposed to exchange energy is a solution with two bidirectional converters which can be used in boost (direct) or buck (reverse) modes. For instance, a solution to exchange energy between Battery 1 and Battery 2, if the voltage level of Battery 1 is higher than voltage level of Battery 2, is to close K12, to open K11 and to use K21, K22 in buck configuration mode.

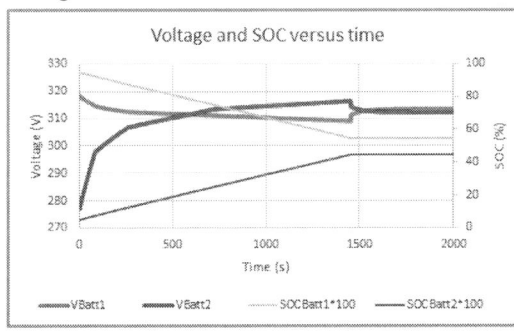
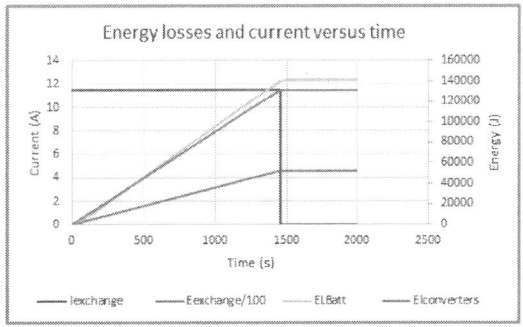

Figure 7 - waveforms solution 1

Unfortunately, the two converters can't be used to protect the system against short circuits on the high voltage bus. Because of the diodes D12 or D22 the currents flowing through the converter can't be interrupted. With an estimated efficiency of 97.5% (2% of losses in the converter and 0.5% in the switch K12), we can calculate balancing time and energy losses:

- Balancing time : **1455 s**
- Energy exchanged : **5200 kJ (1447 Wh)**
- Losses inside two batteries systems : **141 kJ**
- Losses inside converters : **131 kJ**

© VDE VERLAG GMBH · Berlin · Offenbach

4.3 Second control solution: full power bidirectional DC/DC buck converters

In the configuration with two bidirectional buck converters, each converter could be either used either in buck mode (direct) or in boost mode (reverse) . A solution to exchange energy between Battery 1 and Battery 2, if the voltage level of Batt1 is higher than voltage level of Batt2, is to close K22, open K21 and to use the switch K11, K12 in buck configuration mode.

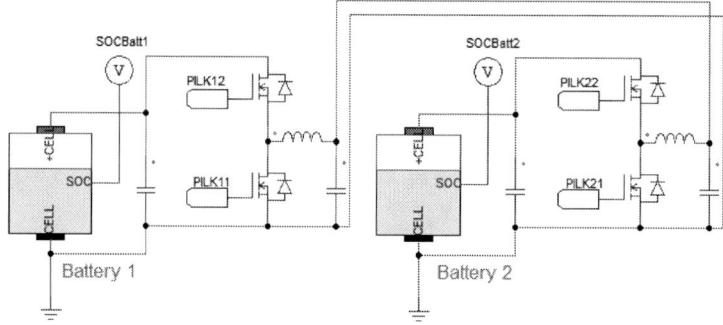

Figure 8 – solution 2

In contrast of the previous solution, the two converters can be used to protect the system against short circuits on the high voltage DC bus. With an estimated efficiency of 97,5% (2% of losses in the converter and 0,5% in the switch K12), we can calculate balancing time and energy losses, which are the same as in the previous case.

4.4 Third control solution: "low power" isolated bidirectional converters

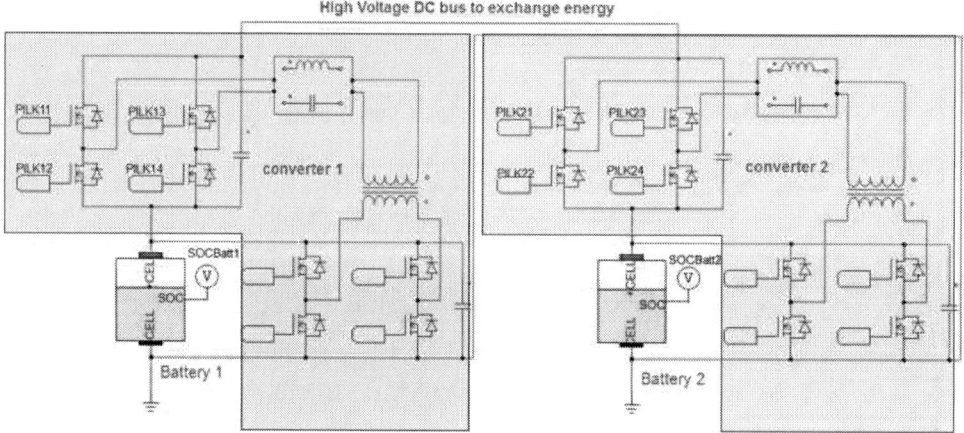

Figure 9 - solution 3

In this configuration the idea is to design a small power converter, sized only for the voltage difference between the battery packs. The power is much lower. A bidirectional isolated power converter is used. It allows a voltage conversion on each battery pack. With a correct regulation, the current exchanged between Battery 1 and Battery 2 can be easily controlled. The drawback is that this system, without any other modification, can't be protected against short circuit.

PCIM Europe 2016, 10 – 12 May 2016, Nuremberg, Germany

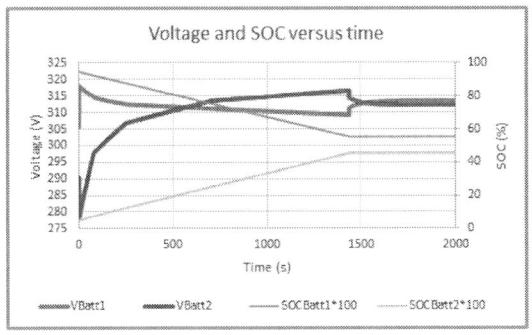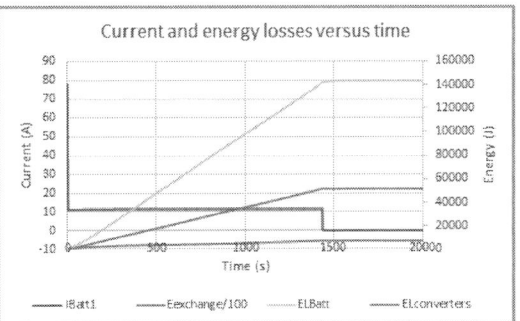

Figure 10 - waveforms solution 3

The losses inside the two converters are very low, in comparison of the energy exchanged. The main losses are located into the two batteries. At the beginning of the exchange, only the converter 2 is used to create a voltage difference and then limit the current. During this phase the switches K11, K12, K13, K14 stay closed and are used in synchronous mode to minimize the losses. At the end of this phase, when the voltage difference is too low to obtain the right level of current, the converter 1 is used to create a higher voltage difference. It allows to continue to push the current in Battery 2.

The peak power transmitted by converters is about 500W, whereas the mean power is less than 100W (see figure 11)

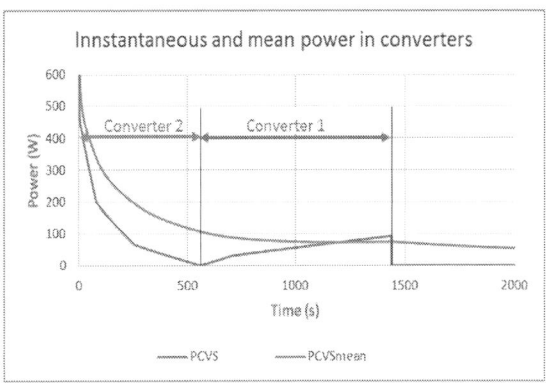

Figure 11 - power in converters

With an estimated efficiency of about 95% (5% of losses in the converters and 10mΩ to take into account of the closed switches in synchronous modes), we can calculate balancing time and energy losses:

- Balancing time : **1455 s**
- Energy exchanged : **5150 kJ (1430 Wh)**
- Losses inside two batteries systems : **143 kJ**
- Losses inside converters : **7,2 kJ**

We can remark that the losses inside converters are very low in comparison with the previous solutions. This can be explained by the optimized power of the solution and by the very low resistance of power MOSFET in synchronous mode.

4.5 Fourth control solution: "low power" regulated modules

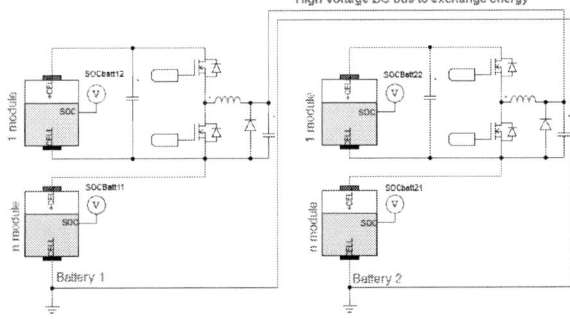

Figure 12 - solution 4

© VDE VERLAG GMBH · Berlin · Offenbach

2103

In this configuration each battery is divided into two parts. A first part with n modules in series (n=11 in our case) and a second one with only one module associated with one converter. Like in solution 3, converters are sized to compensate only the differences of voltage between batteries and not for the all voltages. But the main problem of this solution is the need of a specific module with also a specific software algorithm for SOC calculations and balancing. This solution will not be more detailed in this paper.

4.6 Fifth control solution: linear regulators

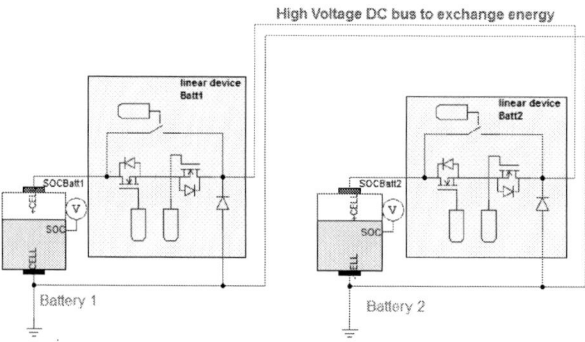

Figure 13 - solution 5

In this configuration, we use the flat curve (voltage versus SOC) of Lithium-ion Iron Phosphate batteries and we try to evaluate if there is an interest to use simple bidirectional transistors in linear mode to limit the current with a relay in parallel.

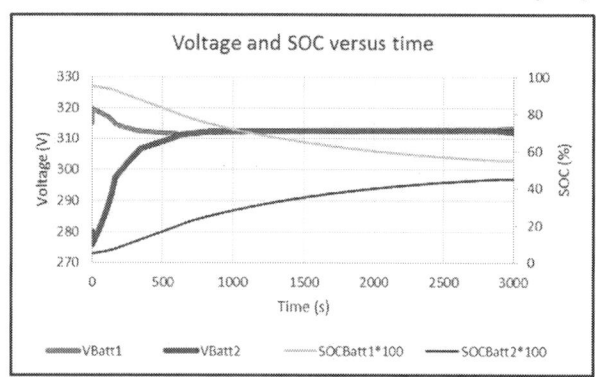
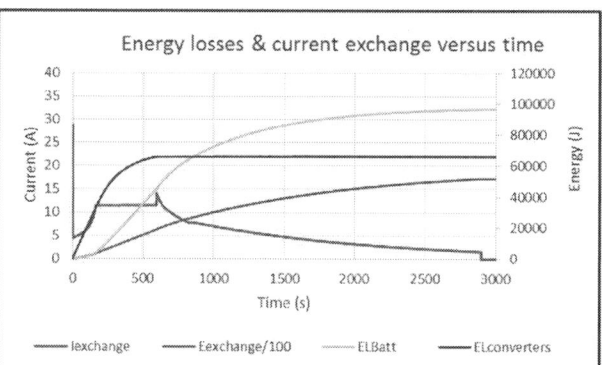

Figure 14 - Simulation of current exchange with solution 5

In this last configuration we can calculate balancing time and energy losses:

- Balancing time : **2900 s**
- Energy exchanged: **5177kJ (1438Wh)**
- Losses inside two battery systems : **96 kJ**
- Losses inside converters : **66kJ**

The main drawback of this solution is the balancing time which is higher than with converters solutions. That can be explained, by the impossibility to create an artificial voltage difference between battery 1 and battery 2, to accelerate the energy exchange. However this solution has many advantages. It seems to be the simplest one, it presents a high efficiency and it allows to limit the current in case of short circuit. The sizing of limitation device must be precisely studied, especially regarding transient thermal considerations.

4.7 Results of the comparisons

	S0-Reference without current limitations	S1-High power boost converters	S2-High power buck converters	S3-Low power isolated bidirectional converters	S4-Low power regulated module	S5-Linear regulators solution
Max/mean power of balancing device (W)	No limitation	3974 / 3532	3974 / 3532	500 / 74	NE(*)	10000 / 1785
Losses in battery systems (kJ)	160	141	141	143	NE(*)	96
Losses in balancing devices (kJ)	No device	131	131	7,3	NE(*)	66
Energy exchange (kJ)	5147	5200	5200	5150	NE(*)	5177
Mean estimated efficiency (%)	NE	97,5	97,5	99,8 %	NE(*)	98.7
Balancing duration (s)	2600	1455	1455	1438	NE(*)	2900
Short circuit limitation	No	No	Yes	No	No	Yes
Complexity	NE(*)	++	++	+++	+++	+
Volume, weight, cost	NE(*)	+++	+++	+	++	+
Specific module	No	No	No	No	yes	No

Figure 15 - comparison table between solutions

*NE=non estimated

5. Impact of battery technology and usage on the preferred solution

The battery technology and the usage have an impact on the preferred solution because the sizing of the balancing device is directly linked with the following criteria:
- Specification of balancing time (usage)
- Usage of the battery (micro-cycles or cycles) which impact the voltage variation
- Battery technology which impacts directly the curve voltage in function of SOC and thus the maximum power of the balancing device

6. Conclusion

In this paper, we compared different solutions to exchange energy between high voltage iron phosphate Li-ion (LiFePO4) battery packs. Two of them seem very interesting for our application (see table figure 15). The solution 3 "**Low power isolated bidirectional converter**" presents a very high efficiency and is certainly the most compact one. The only important drawback is its impossibility to protect the system against short circuits by limiting the current. The solution 5 "**Linear regulator**" is certainly the simplest one, and presents an efficiency higher than standard solutions with buck or boost converters for iron phosphate Lithium-ion batteries. It allows to protect the system against short circuits. However the transient thermal aspects must be studied carefully in order to obtain a performant solution.

7. References

[1] Jiunchun Jiang, "Topology of a bidirectional converter for energy interaction between electric vehicles and the grid", Energies 2014, 7, 4858-4894

PCIM Europe 2016, 10 – 12 May 2016, Nuremberg, Germany

Simulation and Experimental Analysis of Non-Linear Loads from Residential and Educational Buildings

Gabriel Nicolae Popa, Politehnica University Timișoara, Romania, gabriel.popa@fih.upt.ro
Angela Iagăr, Politehnica University Timișoara, Romania, angela.iagar@fih.upt.ro
Corina Maria Diniș, Politehnica University Timișoara, Romania, corina.dinis@fih.upt.ro

Abstract

The paper analyses the power quality of non-linear loads from residential and educational buildings. Analysis of voltage and current harmonics for the representative consumers (fluorescent lamps, PC, LCD monitor), operating in an isolated mode, was performed by modelling and simulation using PSCAD-EMTDC program. Validation of simulations was done by laboratory measurements, using power quality analyser CA 8334B. To assess the cumulative effect (harmonics, unbalance, power factor) produced by the analysed consumers, measurements in the power substation were performed, using power quality analyser. The measurements demonstrate the power factor improving, when using low power capacitive loads, balanced on phases.

1. Introduction

Power quality is becoming more and more important, owing the continuous increase of electric users susceptible to the electromagnetic disturbances. Among the most important issues of power quality are voltage instability, harmonic distortion and reactive power burden [1, 2]. Widespread use of non-linear and time-varying single-phase loads increasingly affects the operation of distribution networks in residential, commercial, and industrial areas. The current distortion of the single-phase loads depends on design of power supplies, the voltage level and current-voltage characteristics of the loads. Harmonics can interfere with control, communication, or protection equipment, causing additional losses and decreasing the equipment lifetime. Non-linear loads can cause current and voltage fluctuations in the point of common coupling (PCC, i.e. electrical station). Variations in current magnitude lead to overloads or additional losses [1-4].
The switching mode power supplies (SMPSs) with large electrolytic capacitors are the main design of the power supplies used for single-phase loads such as computers, monitors, laptops, electronic ballasts for fluorescent lamps, etc. The currents of SMPSs are strongly deformed. In the three-phase electrical network the connection of single-phase non-linear loads creates unbalanced loads, with increasing current into neutral conductor. The unbalanced loads cause voltage drops on the electrical network [5-8].
Harmonic distortion decreases the power factor. The power factor (PF) improvement reduces the network losses and increases the electric capabilities of the cable, and, also, maintains the desire voltage level [4-6].
For power energy providers the power factor improvement reduces network losses and increases the electrical capacity for productive, and, also, it can help to maintain the voltage to desire level. The consumers with low power factor are penalize, applying tariff clauses. For a constant active power, if the power factor decrease, the required apparent power, and, also, the electrical system losses increase.

2. Case study

The electric measurements were made in PCC at power substation of residential and educational buildings. The residential and educational campus has six buildings and the surface is 12800 m^2 of floor spaces. Fig.1.a shows the electrical diagram for typical residential and educational buildings. The circuit has two main power transformers T$_1$ (400 kVA, 6/0.4 kV, D/y 11) and T$_2$ (250 kVA, 6/0.4 kV, D/y 11). Also, the circuit has two main low

PCIM Europe 2016, 10 – 12 May 2016, Nuremberg, Germany

voltage (LV) branches that supply the electrical loads through two main breakers Q_1 (for power transformer T_1) and Q_2 (for power transformer T_2). A transversal circuit breaker Q_3 can connect the two main branches. At this branches are connected some linear and a lot of non-linear electrical loads. In the most situations, the electrical loads are connected to Q_1 and Q_3 circuit breaker from power transformer T_1. Initially, the electrical non-linear loads were distributed non-equally across three-phases. To evaluate the electric parameters was used a three-phase power quality analyser CA 8334B connected at PCC, after circuit breaker Q_1. The voltage were measured with four voltage probes (three phases L_1, L_2, L_3 and neutral 0) and currents were measured with AmpFlex probes (3000 A) [9].

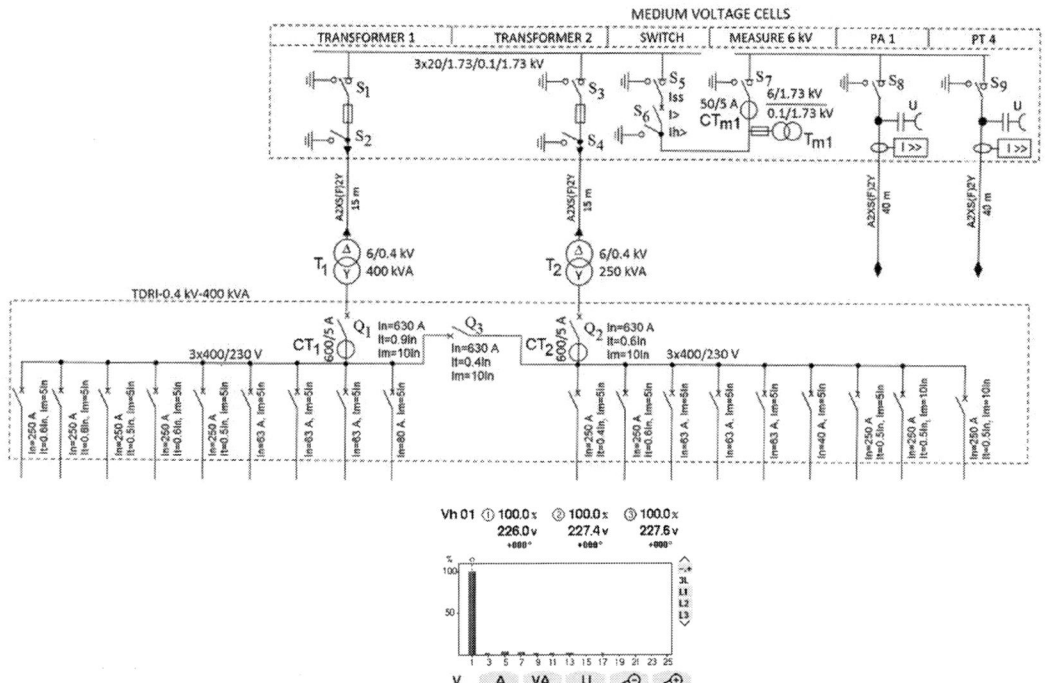

Fig.1. Electrical circuit diagram for residential and educational buildings, and harmonic spectra of three-phase voltages (L_1-0, L_2-0, L_3-0) measured on LV Line

Fig.1.b shows the harmonic spectra (up to 25 orders) of three-phase voltages measured on LV Line. The most important harmonics are 5th and 7th.
The disturbances can be classified in fast and slow switching disturbances. Some impulses may be imposed on the current shape during the switching. Fig.2 presents the fast voltage transients from electrical power substation.

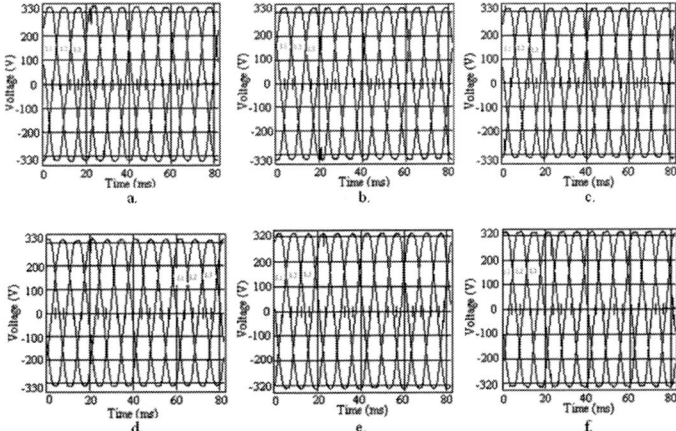

Fig.2. Fast voltage transients from electrical power substation

2107

© VDE VERLAG GMBH · Berlin · Offenbach

The voltage and current unbalance in three-phase power systems occurs due unbalanced loads, or persistence of defects (i.e. two-phase operation of three-phase loads). In addition to household single-phase electrical loads, which are very numerous, but with small powers, the main loads with major powers can cause unbalances in power system. Although all power systems shall be provided with balancing equipment, these missing or malfunctioning due to changes in parameters and operations for these electrical loads.

3. Modelling and simulations of non-linear loads

Non-linear loads from residential and educational buildings generate current harmonics that can causes malfunctioning of the sensitive loads connected at the PCC, decreasing their lifetime, and reducing the efficiency of the electrical installation. Simulation results must be verified by laboratory experiments to validate the accuracy of modelling. If the models are accurate, computer simulation further allow the study of harmonic currents injected at PCC by several non-linear loads that simultaneously operating. This section presents the modelling and simulations of most representative non-linear loads from analysed residential and educational buildings using PSCAD-EMTDC program [10].

3.1. Personal Computer (PC)

Personal computers have negative impacts on power quality due to the using of SMPS. SMPS has a large capacitor which maintains approximately constant voltage for the DC bus in the power supply.

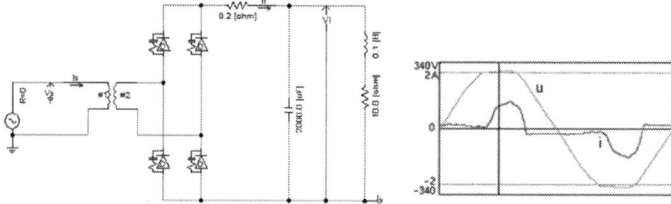

Fig.3. PC model used in PSCAD-EMTDC simulation; real voltage (u) and current (i) through PC

Fig. 3 presents the PC electrical model. SMPS contains a diode bridge, a DC storage capacitor (C=2000μF), a resistance (R=0.2Ω) and a series radio frequency interference (RFI) choke (represented by inductance L=0.1H) [11].

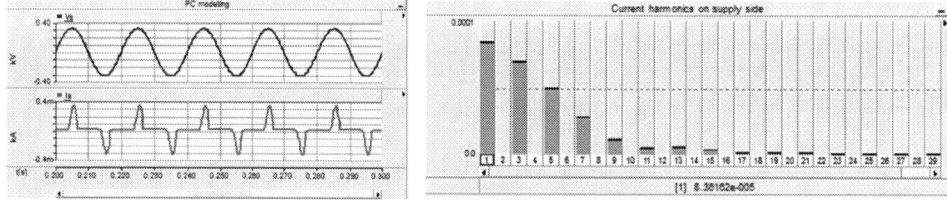

Fig.4. Simulated voltage, current and current harmonic spectrum for PC

Simulation results are presented in Fig. 4. Waveform of supply voltage is sinusoidal, but waveform of PC current is highly non-linear (total harmonic distortion for current is THD_I=109.345%). Because the capacitor of SMPS is charged only during the peak of the voltage waveform, large current pulse appears in the current draw by PC, at the peak of the voltage waveform [12]. Harmonic spectrum of the current, obtained by FFT analysis, reveals large amounts of third and higher order harmonics.

3.2. Compact fluorescent lamps

Compact fluorescent lamps (CFLs) are increasingly being used because of their low-energy consumption and long lifetime. High quality CFLs contain active filtering circuits; these lamps have high power factor and low THD_I, but are expensive. CFLs with passive filtering circuit have average current harmonic spectra, and those with valley-fill circuit have good current harmonic spectra. Most manufacturers still produce low cost and low quality CFLs, with simple ballast circuit, which have low power factor and high harmonic distortion. CFLs with THD_I greater than 100% can be found on the market because the EMC standards [13] are not particularly strict due to their low consumption [14-16].

Figs. 5 and 6 present the first electrical model of CFL used in PSCAD-EMTDC [11], and the simulation results. Electronic ballast, that feeding the fluorescent tube by high frequency current, is modelled by a PWM half-bridge inverter, feed from a full-wave rectifier and smoothing capacitors, in this case. CFLs draw about 75% less current than incandescent lamps, but their integrated ballasts dramatically distort the waveform of the current. Harmonic spectrum of the current (THD_I=147.431%) contains all odd harmonics, with extremely high levels. THD_I can result in a considerable increase of harmonic voltage level in distribution network, with a negative impact on power quality.

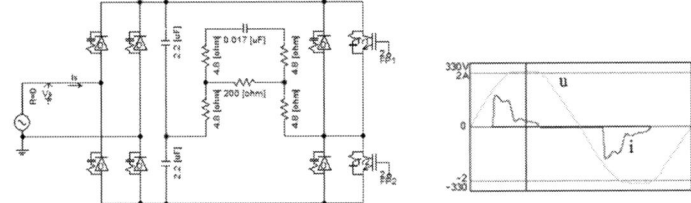

Fig.5. CFL first electrical model used in PSCAD-EMTDC simulation; real voltage (u) and current (i) through CFL

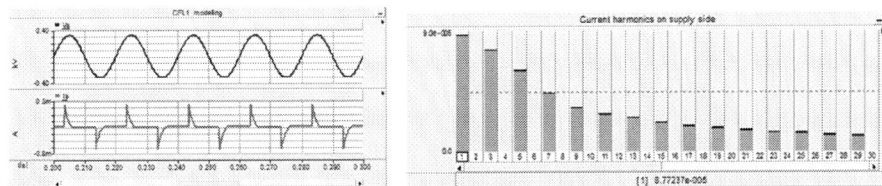

Fig.6. Simulated voltage, current and current harmonic spectrum for the first model of CFL

The second electrical model of CFL consists of a diode rectifier and a capacitive output filter (Fig.7).

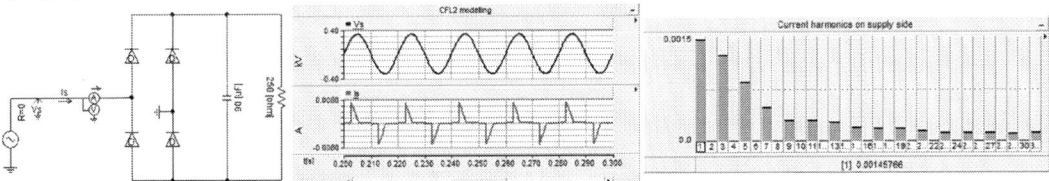

Fig.7. CFL second electrical model used in PSCAD-EMTDC and simulation results

Harmonic spectrum of the current (THD_I=116.57%) for the second electrical model of CFL (Fig. 7) contains all odd harmonics, but with lower levels than in previous case. THD_I of CFL current is less than in the first simulation model, but exceeds 100%.

3.3. LCD Monitor

The model of LCD monitor used in simulation consists of a network transformer, a diode rectifier and a capacitive filter (Fig.8).

PCIM Europe 2016, 10 – 12 May 2016, Nuremberg, Germany

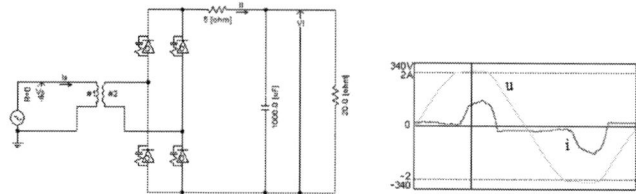

Fig.8. LCD monitor electrical model used in PSCAD-EMTDC; real voltage (u) and current (i) through LCD monitor

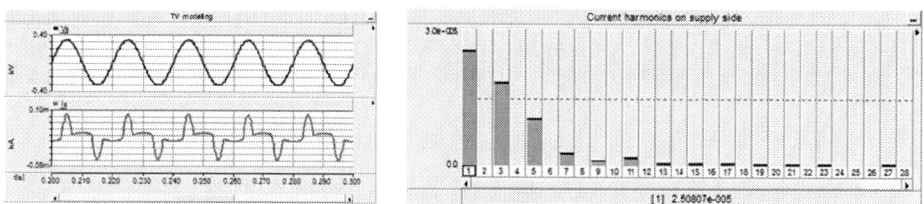

Fig.9. Simulated voltage, current and current harmonic spectrum for LCD monitor

LCD monitor current (Fig. 9) has a lot of odd harmonics. THD_I in the current waveform for LCD monitor has a high value, equal to 84.63%.

4. Experimental measurements of current through the neutral conductor

The goal of experiments is to study the neutral current. Measurements were made for three identical non-linear loads with star connection. Each lamp was connected between a phase and neutral conductor (L_1-0, L_2-0, and L_3-0). In the first case, were used three identical magnetic ballast fluorescent lamps (MBFLs), in classical version, with rated power of 40 W. In the second case, were used three identical CFLs, with rated power of 18 W. Using power quality analyser CA 8334B [9], were measured the voltage between L_1 and neutral conductor (0), and the current through neutral conductor with MN 93 A (5 A) probe. To make experimental setup, in series with each fluorescent lamp was connected a breaker in order to disconnect/connect each lamp.

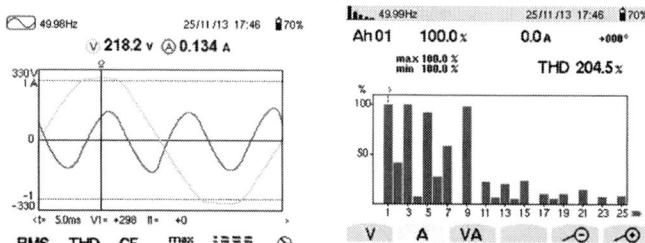

Fig.10. The phase voltage, the neutral current (upside down), and current harmonic spectrum when three identical MBFLs were connected in star connection (between L_1-0, L_2-0, L_3-0)

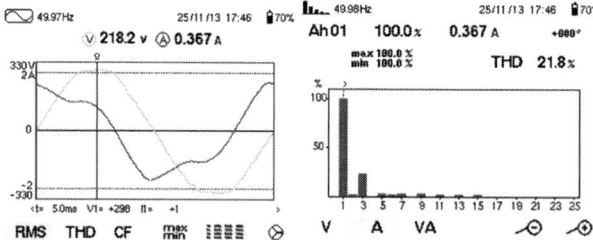

Fig.11. The phase voltage, the neutral current (upside down), and current harmonic spectrum when two identical MBFLs were connected (L_1-0, L_2-0; L_3-0 without MBFL)

2110   © VDE VERLAG GMBH · Berlin · Offenbach

PCIM Europe 2016, 10 – 12 May 2016, Nuremberg, Germany

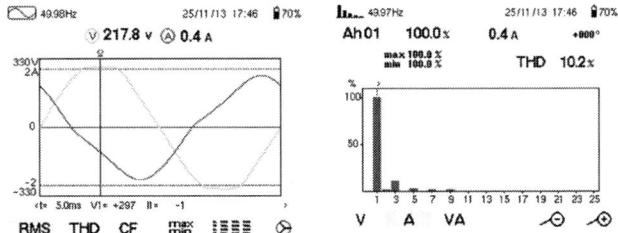

Fig.12. The phase voltage, the neutral current (upside down), and current harmonic spectrum when only one MBFL were connected (L_1-0; L_2-0 and L_3-0 without MBFLs)

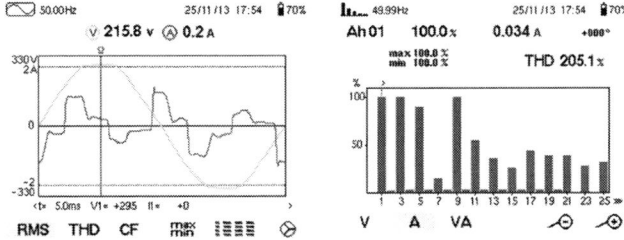

Fig.13. The phase voltage, the neutral current and current harmonic spectrum when three identical CFLs were connected in star connection (between L_1-0, L_2-0, L_3-0)

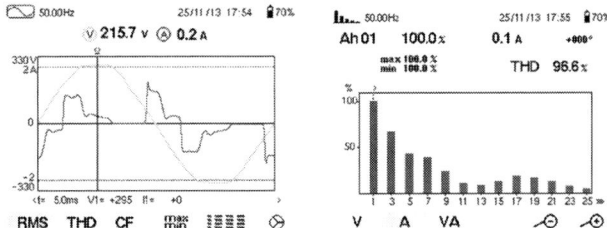

Fig.14. The phase voltage, the neutral current and current harmonic spectrum when two identical CFLs were connected (L_1-0, L_2-0; L_3-0 without CFL)

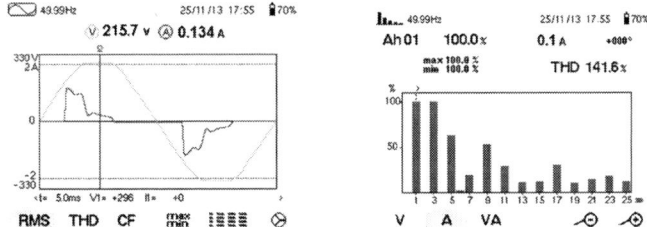

Fig.15. The phase voltage, the neutral current (upside down) and current harmonic spectrum when only one CFL were connected (L_1-0; L_2-0 and L_3-0 without CFLs)

At star connection, current through the fluorescent lamps (MBFLs or CFLS) is strongly deformed and there are multiple current harmonics: 3rd order and multiply 3rd (9th, 15th, 21st). From Figs.10-15 it is found that the neutral current is less deformed and has higher values for MBFLs type compared with CFLs type. Currents harmonics (3rd order and multiply 3rd) of MBFLs and CFLs are zero sequence, and they add algebraically through neutral conductor.
Unbalanced electric loads on three phases produce currents through neutral conductor that are comparable with each electric load. For more non-linear loads the effect is more pregnant (i.e. CFL compare with MBFL).
So, for non-linear loads, even if they are balanced on the three phases will circulate an important neutral current, that cannot be eliminated or reduced. Usually, the current through the neutral conductor is zero if electrical loads are the same and linear.

© VDE VERLAG GMBH · Berlin · Offenbach

PCIM Europe 2016, 10 – 12 May 2016, Nuremberg, Germany

5. Measurements in the substation

To assess the cumulative effect produced by the analysed consumers, the monitoring equipment, with power quality analyser CA 8334B, were connected in PCC at power substation of residential and educational buildings, after circuit breaker Q_1 (Fig.1.a). The measurements were focused on powers and power factor (Figs.16-19) in two days: first day is a work week day and the second day is a week-end day.

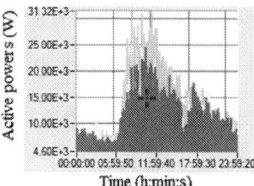 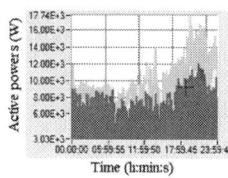

Fig.16. Active powers in the first and second days, measured in PCC at electrical station

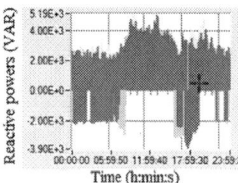 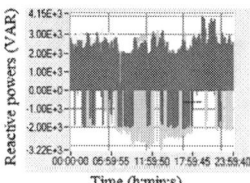

Fig.17. Reactive powers in the first and second days, measured in PCC at electrical station

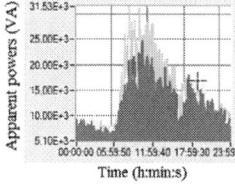 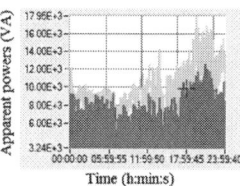

Fig.18. Apparent powers in the first and second days, measured in PCC at electrical station

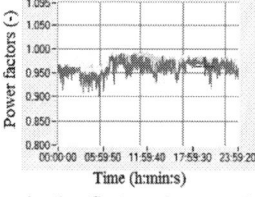 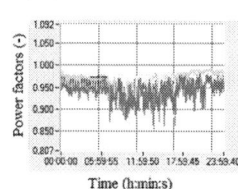

Fig.19. Power factors in the first and second days, measured in PCC at electrical station

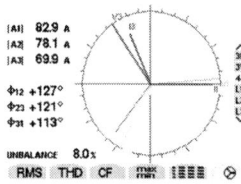 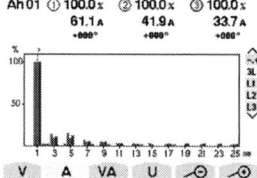

Fig.20. Fresnel diagrams and currents spectra measured in PCC at electrical station

The electrical measurements were made along 24 h. From Figs.16-18 there are differences between amplitude and shape between the first day and the second day of monitoring. Active power is larger during the work day comparable with week-end day. The reactive powers are inductive and capacitive type, depending on electrical loads that are connected to the network. The apparent power in the work day is almost twice larger than in the week-end day.

The PCC does not have connected capacitor banks or a power factor regulator with capacitors banks.

From Fig.19 it can be seen that power factor is bigger than 0.95, especially during the work day (between 8.00 h and 20.00 h). For the same day, in the rest of the interval of time the

2112 © VDE VERLAG GMBH · Berlin · Offenbach

power factor is bigger than 0.92 (that is the neutral power factor for low voltage distribution network). In the week-end day, when are a few electrical loads, the power factor is bigger than 0.9 (for all three phases).

In the almost all situations, the current unbalance is lower than 10% (Fig.20). The current spectra contain 5th, 3rd, 7th and 9th harmonics.

6. Conclusions

The unbalance of non-linear loads creates additional losses, supplementary charged into neutral conductor, affect the electrical energy measurements, and occurs bigger voltage drops. Some single-phase loads have inductive character (single-phase induction motors, MBFLs, CFLs), and other have capacitive character (LED lamps, loads with SMPS). Using uniform arrangement of the electric loads on the phases, can be improved power factor without using capacitor banks into PCC.

References

1. A.D. Aquilla et al., "New Power-Quality Assessment Criteria for Supply Under Unbalanced and Nonsinusoidal Conditions", IEEE Trans. on Power Delivery, Vol.19, no.3, July, 2004, pp.1284-1290.
2. M.M.A. Aziz et al., "Practical Considerations Regarding Power Factor for Nonlinear Loads", IEEE Trans. on Power Delivery, Vol.19, no.1, January, 2004, pp..337-341.
3. A. Elnady, Y.F. Liu, "A Practical Solution for the Current and Voltage Fluctuation in Power Systems", IEEE Trans. on Power Delivery, Vol.27, no.3, July, 2003, pp.1339-1349.
4. A.E. Emanuel, "Apparent Power Definitions for Three-Phase Systems," IEEE Trans. on Power Delivery, Vol. 14, no. 3, July 1999, pp. 767–772.
5. A. von Jouanne, B.B. Banerjee, "Assessment of Voltage Unbalance", IEEE Trans. on Power Delivery, Vol.16, no.4, October, 2001, pp.782-790.
6. M.A.S. Masoum, P.S. Moses, A.S. Masoum, "Derating of Asymmetric Three-Phase Transformers Serving Unbalanced Nonlinear Loads", IEEE Trans. on Power Delivery, Vol.23, no.4, October, 2008, pp.2033-2041.
7. P.J. Moore, I. Portugués, "The Influence of Personal Computer Processing Modes on Line Current Harmonics", IEEE Trans. on Power Delivery, Vol.18, no.4, October, 2003, pp.1363-1368.
8. J.A. Pomilio, S.M. Deckmann, "Characterization and Compensation of Harmonics and Reactive Power of Residential and Commercial Loads", IEEE Trans. on Power Delivery, Vol.22, no.2, April, 2007, pp.1049-1055.
9. ***, "Three Phase Power Quality Analyzer CA 8334B", User's Guide, Chauvin-Arnoux, France, 2007.
10. ***, "EMTDC. Transient Analysis for PSCAD Power System Simulation. User's Guide", Manitoba HVDC Research Center, Manitoba, Canada, 2005.
11. C. Venkatesh, D. Srikanth Kumar, D.V.S.S. Siva Sarma, M. Sydulu, "Modelling of Nonlinear Loads and Estimation of Harmonics in Industrial Distribution System", Fifteenth National Power Systems Conference (NPSC), IIT Bombay, December 2008, pp. 592-597.
12. J. Meyer, P. Schegner, K. Heidenreich, "Harmonic Summation Effects of Modern Lamp Technologies and Small Electronic Household Equipment", 21st International Conference on Electricity Distribution, Frankfurt, 6-9 June 2011, pp.1-4.
13. IEC 61000-3-2, Electromagnetic compatibility (EMC) - Part 3-2: Limits - Limits for harmonic current emissions (equipment input current ≤ 16 A per phase).
14. J. Molina, L.Sainz, "Model of Electronic Ballast Compact Fluorescent Lamps", IEEE Trans. on Power Delivery, Vol. 29, Issue 3, October 2013, pp. 1363 – 1371.
15. J. Slezingr,, J.Drapela, R. Langella, A. Testa, "Model Of Electronic Ballast Compact Fluorescent Lamps", IEEE 15th International Conference on Harmonics and Quality of Power (ICHQP), 17-20 June 2012, Hong Kong, pp. 835 – 841.
16. H. Farooq, C. Zhou, M. E. Farrag, "Analyzing the Harmonic Distortion in a Distribution System Caused by the Nonlinear Residential Loads", International Journal of Smart Grid and Clean Energy, Vol. 2, No. 1, January 2013, pp. 46-51.

PCIM Europe 2016, 10 – 12 May 2016, Nuremberg, Germany

A Digital Predictive Constant Frequency Controller For High Frequency 3-Phase Silicon Carbide PFC Rectifier

Marcelo Schupbach, Marcelo.schupbach@wolfspeed.com, Wolfspeed
Binod Agrawal, Binod.agrawal@wolfspeed.com, Wolfspeed
Navneet Mangal, navneet.mangal@wolfspeed.com, Wolfspeed,

Abstract

Commerciality available SiC devices allow for a 2-5× increase in switching frequencies leading to a substantial reduction of the value of control inductor required between the grid and the converter in grid-connected applications such as 3-phase PFC rectifiers. While this is beneficial from a cost and power density perspective, it may also lead to control challenges due to the reduction of the impedance between the grid and the power converter. These challenges are aggravated in the case of polluted grids (i.e., harmonics). These control challenges limit how much the mentioned grid inductors can be reduced on a practical power converter. As such, improved control schemes are needed to work effectively under low grid-side impedance conditions. This paper presents a digital implementation of Constant Frequency Predictive (CFP) control based on system parameters for 3-phase, six-switch PFC rectifier which leads to a robust control similar to the robustness afforded by analog hysteresis controllers. The new control Scheme is simulated in PLECS and experimentally verified on a 7.5kW, 3-phase PFC rectifier using a 2-level six-switch SiC Module.

1. Introduction

A 2-level six-switch type active front end rectifiers (AFE) as shown in Fig. 1 (left) and Vienna rectifier like the one in Fig. 1 (right) are two of the most popular topologies at present for grid-connected 3-phase rectifiers. For unidirectional power flow, the Vienna rectifier topology delivers a good solution due to its 3-level nature, which allows for a higher effective switching frequency in grid-side inductors as well as the reduction of voltage stress across the active devices (enabling the use of 650V technology in most cases). For bidirectional power flow, the 2-level six-switch AFE topology is generally a good choice and is used widely across the industry. Furthermore, when combined with SiC MOSFETs, the 2-level six-switch AFE topology delivers very high efficiency, potentially higher than the Vienna, making this topology a good candidate for high-efficiency unidirectional applications, as well as bidirectional applications. The high efficiency of the 2-level six-switch AFE topology with SiC MOSFETs stem from the fact that SiC devices deliver 2-5× lower switching losses. Additionally, the conduction losses of a 2-level topology (such as a six-switch AFE) are inherently lower than the conduction losses of a 3-level topology (such as Vienna). Lastly, the use of SiC MOSFETs, instead of IGBTs, allow the reduction of conduction losses within the anti-parallel diodes due to the possibility of using synchronous rectification and only using the anti-parallel diodes during dead time (100-500 ns) delivering a very low-loss converter.

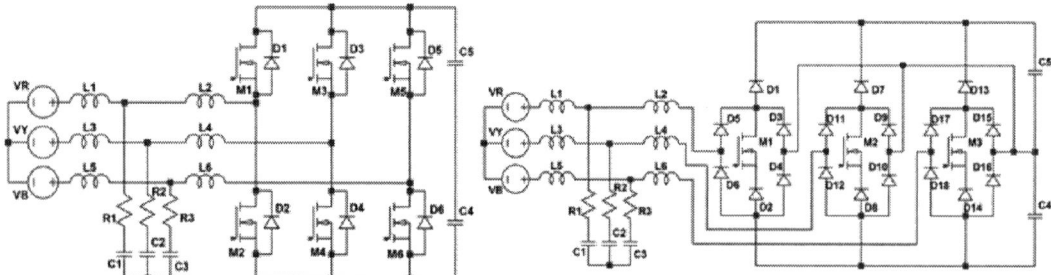

Fig. 1. 2-level six-switch AFE (left) and 3-level 3-switch Vienna rectifier (right).

The grid-side impedance of the grid-tied converter must be carefully designed as it has significant influence on the overall performance (i.e., efficiency, power density, ruggedness, etc.) and cost of the converter. The main functions of this filter are two-fold: one, to provide current attenuation to meet Total Harmonic Distortion (THD) requirements and two, to

© VDE VERLAG GMBH · Berlin · Offenbach

provide 'sufficient' impedance between the grid and the power converter to allow for stable and rugged operation). Several options are possible including: L, L-C or L-C-L. However, L-C-L filter topology is one of the most widely used due to smaller size and consequent cost reduction. A detailed design procedure has been provided in [1-2]. The values (and hence size and cost) of these filtering components is inversely proportional to switching frequency. However as the switching frequency increases beyond the typical 20 kHz, a linear reduction in filter size is not realized due to control issues. This problem may become more severe under unbalance and/or distorted grid conditions since these conditions could lead to computational errors of control variables.

Various control methods have been reported in literature to shape the desired sinusoidal current in the inductors in order to meet THD requirements. The d-q controllers in synchronously rotating frame (SRF) provide good flexibility in control of active and reactive power separately and adapt very well to DSP implementation. However, it is not very robust under distorted grid conditions. Decoupled Double Synchronous Frame (DDSRF) is good for unbalanced grid conditions but requires extensive computation and a very fast control loop leading to significant controller requirements, in particular at high switching frequencies [3-4]. Analog hysteresis controllers have a robust and fast dynamic response but under variable frequency operation makes the output and EMI filters difficult to design [5]. Lastly, predictive controllers based on system parameters allow for fast response while minimizing hardware requirements [6-8]. However, they require knowledge of system parameters within a tolerance band. Error on system parameter values could lead to improper operation. Lastly, A new control theory known as currents' physical components can which can successfully decouple real power, reactive power, unbalanced contents, harmonic content, scattered content, etc. However, this approach required heavy computations and the practical implementation of such control approach is not well known in industry yet.

To address the issues that standard control approaches experience under very low grid impedance conditions, this paper proposes a new control scheme for grid-connected converters based on improvements to known control approaches. The proposed approach utilizes a DDSRF Phase Lock Loop (PLL) to provide robust synchronization with the grid even under non-ideal conditions. The high-level power flow control problem (i.e., maintain DC-link voltage is addressed and low input current THD) is addressed by controlling the DC-link voltage via adjustable, positive, balanced, in-phase set of sinusoidal currents pull from the grid, even if grid voltage is distorted. The reference to these set of balanced current is internally generated in the a-b-c reference frame to minimize computation time. A Proportional Integral (PI) controller is used to adjust the amplitude of these references current. The low-level control problem (i.e., how to control the semiconductor switches to make the references actually happen) is implemented using an average current control strategy to minimize current sensing speed requirements. Lastly, a predict component and a proportional controller are added to the low-level control equations to increase control accuracy and minimize error. Note that while no attempt has been made to analyze the coupling effect or stability margins of such system, simulation and hardware testing have been carried out showing good performance and stability under nominal conditions, distorted grid voltage conditions, and with large model parameter uncertainties.

The remaining of the paper is organized as follow. Section 2 introduces and describes in details the proposed control approach. Section 3 discusses control-level simulation (PLECS) results for the following: 1) steady-state, 2) step-load change 3) distorted grid condition and 4) up to 100% error on system parameters. Section 4 presents results collected while testing a 7.5 kW 2-level six-switch AFE hardware setup with SiC MOSFET. Lastly, Section 5 provides a summary and concluding remarks.

2. Proposed CFP controller based on system parameters

2.1. Controller Overview

Fig. 2 shows a block-level view of the proposed CFP controller including all key components. The high-level control goal of the CFP controller is to regulate the DC link voltage by injecting/absorbing (an adjustable) balanced, in-phase set of sinusoidal currents into/from the grid, even if the grid voltage is distorted. The proposed control method operates as follows. The 3-phase voltages are measured, low-pass filtered, isolated and then fed to the DDSRF

PLL which generates the required angle θ locked with the grid. The DDSRF PLL only locks with the fundamental positive sequence component of the grid voltage even under distorted grid conditions. The angle θ is used to generate the three sinusoidal current references, I_r^*, I_y^*, I_b^*, in the a-b-c reference frame which are synchronized with their respective grid phase voltages. The a-b-c reference frame implementation requires 30% less computational power than a d-q implementation. The magnitude of this sinusoidal current reference, I^*, is controlled by a PI controller in order to maintain DC-link voltage as shown in Fig. 2.

The three phase currents I_r, I_y, and I_b, are measured and low-pass filtered in order to obtain the average value within the switching frequency range of interest. These currents are then subtracted to the reference current and feed to three independent controllers set on the a-b-c stationary reference frame [9]. A model-predictive average-current-mode control approach is used to instantiate the individual phase currents. Switch turn-on -off equations are used to control the devices using the difference in phase current (averaged within a switching cycle), the phase voltages and the dc-link voltage. The derivation of those control equation is discussed next.

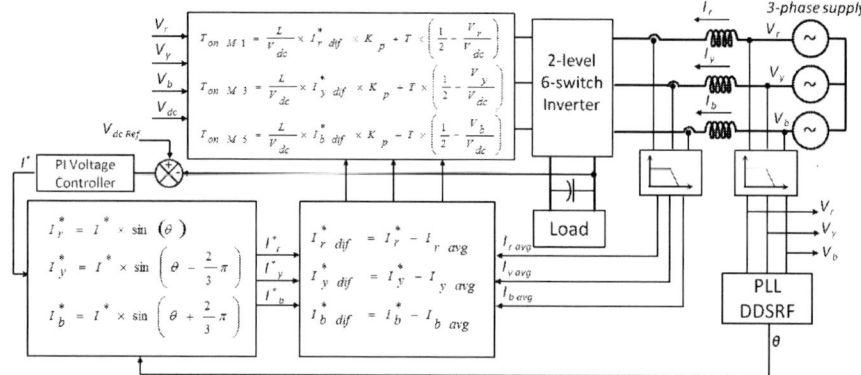

Fig. 2. High-level description of proposed CFP controller based on system parameters

2.2.   Current Controllers and Generation of Turn-on and Turn–off times

The current controller model is shown in Fig. 3 a). This model assumes that the grid neutral is connected to the converter's dc-link midpoint. In actuality, if the converter's midpoint is not tied to the grid, this will generate an error that can be minimized by a proportional controller as explained further on. Turn-on and turn-off equivalent circuits shown in Fig. 3 b) and c) respectively, where L is the converter-side inductor (phase currents are being sensed).

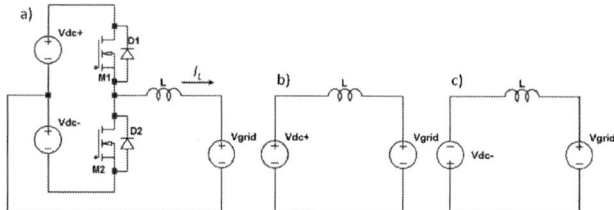

Fig. 3. a) Simplified equivalent circuit for the current controller. b) Equivalent circuit for current controller during on-time and c) during off-time

Assuming phase current is positive as shown in Fig.3 a), change of current during the on time on M1 (and off-time of M2) can be expressed by equation 1. Similarly, change of current during the off time of M1 (and on-time of M2) can be expressed by equation 2.

$$d_{i1} = \frac{1}{L} \times T_{On} \times \left(V_g + \frac{V_{dc}}{2} \right) \tag{1}$$

$$d_{i2} = \frac{1}{L} \times T_{Off} \times \left(V_g - \frac{V_{dc}}{2} \right) \tag{2}$$

Where V_g is the current grid voltage, V_{dc} is the current dc-link voltage. Additionally, because the converter operated under constant switching frequency with complementation PWM,

$$T = T_{On} + T_{Off} \tag{3}$$

Where T is the switching period. Typically equations (1) and (2) are solved for T_{on} and T_{off} by calculating the net change in the current inductor d_{i1} and d_{i2} (instantaneous quantities). This requires, however, the exact value of the inductor current at the beginning of the switching cycles, as well the exact value (or at least a good approximation) of L. The generation of such a high speed current feedback signal (switching speed range) and precise value of L represent a challenge and limits the practical implementation of many predictive control approaches. To eliminate these obstacles, which are particularly difficult when using device capable of very high switching frequencies, the inductor average current (approximately within a switching period) is used instead of the exact inductor instantaneous current as shown in Fig 4 a) and described by equation (4).

$$d_i = d_{i1} + d_{i2} \tag{4}$$

This approximation to the inductor average current (via low-pass filtering of the sensed currents) introduces an error as which is then minimized via a proportional feedback loop discussed later. Using, d_i one can solve for T_{on} using equations (1), (2) and (3). Lastly, using equation (3), T_{off} can also be calculated.

$$T_{On} = \frac{L}{V_{dc}} \times d_i + T \times \left(\frac{1}{2} - \frac{V_{dc}}{V_g} \right) \tag{5}$$

Fig. 4b) shows the inductor reference current, I_{ref}, its net chance, $d_{i\ ref}$, along with the instantaneous trajectories of inductor current during turn-on and –off, the averaged inductor current and its net change. The goal of the control approach is for the average inductor current to follow the reference phase current. At the instant T_1, the averaged inductor current $I_{avg\ T1}$, the reference $I_{ref\ T1}$, the inductor reference at instant T_2, $I_{ref\ T2}$, as well as the net change in inductor reference current $d_{I\ ref}$ ($I_{ref\ T2}$- $I_{ref\ T1}$), are known quantities. The error at T_1 is defined as $I_{ref\ T1}$- $I_{avg\ T1}$. The total change in average current required to reach the target at T_2, can be expressed as d_i,=$d_{I\ ref}$ + error = $d_{I\ ref}$ + $I_{ref\ T1}$- $I_{avg\ T1}$ = $I_{ref\ T2}$- $I_{ref\ T1}$ + $I_{ref\ T1}$- $I_{avg\ T1}$ = $I_{ref\ T2}$- $I_{avg\ T1}$. By substituting this into equation (5) we obtain:

$$T_{On} = \frac{L}{V_{dc}} \times \left(I_{ref\ T2} - I_{avg\ T1} \right) \times K_p + T \times \left(\frac{1}{2} - \frac{V_{dc}}{V_g} \right) \tag{6}$$

where K_p is the constant of the proportional controller. Note that this proportional feedback loop is also able to compensate for changes (or errors) in inductance value estimations as those inductance value errors translate into current errors in the inductors. As such the proportional feedback loop is able to desensitize the controller to inductance estimations error. Note that RC low-pass used on voltage and current measurements introduces a phase lag that can be compensated to minimize control error. Current errors (or differences) are computed used the current references and actual sensed phase currents (after lag compensation) and this information along with the phase voltages and output dc voltage is used to compute the switches turn-on and -off times.

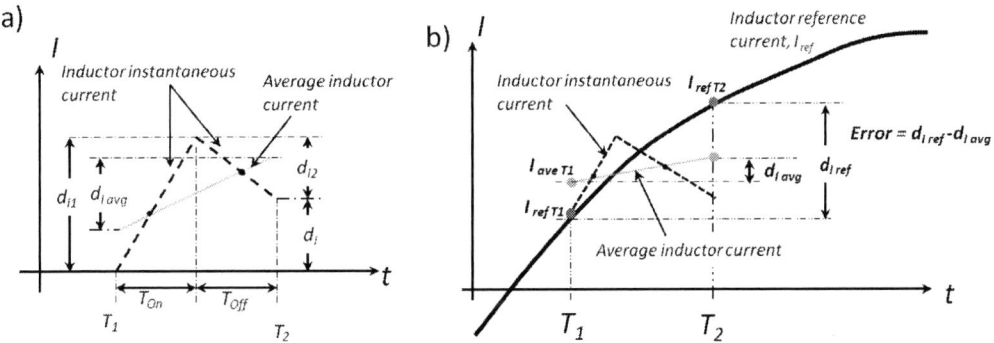

Fig. 4. a) Description of instantaneous inductor current trajectories and net change in inductor current. b) Description of net change in inductor current, actual inductor current reference and error introduced by using the proposed simplifying assumption.

3. Control-Level Simulations

The control scheme introduced in Section 2 has been simulated in PLECS for a 3-phase 2-level AFE converter as shown in Fig. 1 (left). Table 1 summarizes the parameters of 20kW 2-level six-switch AFE used to validate the proposed control method. Fig 5 (left) shows the simulation results of the converter under steady-state operation with a phase current reference of 15 Amps. For this simulation, L1, L3 and L5 were set to 1.5 mH, L2, L4 and L6 were set to 200 µH and switching frequency was set at 20 kHz. Lastly, Kp was set to 0.5. Simulation results show a THD lower than 1.5%. Fig 5 (left) shows the transient response to a current reference chance from 15Amps to 30 Amps.

Input Voltage	Input Freq.	DC-link Voltage	Switching Freq.	C4, C5, C5
415 Vll	50 Hz	680 V	20, 40kHz	4.7 µF
L1, L3, L5	**L2, L4, L6**	**Nominal Power**	**R1, R2, R3**	**RL1, RL3, RL5**
1.5, 1.2, 1.0, 0.8 mH	200, 100 µH	20 kW	1 Ohm	70 mOhm

Table 1.   Parameters of the 20kW 2-level six-switch AFE hardware.

Fig. 6 (left) shows the simulation results of the converter under steady-state operation with a phase current reference of 15 Amps. In this case, L1, L3 and L5 were reduced 0.8 mH; however, the value on the control code was keep at 1.5mH introducing 100% error on inductor value. L2, L4 and L6 were reduced to 100 µH and switching frequency was increased to 40 kHz. Lastly, K_p was set to 0.5. Simulation results indicate proper operation of the proposed CFP control with a THD lower than 1.5%. Lastly, Fig 6 (right) show simulation results under distorted grid conditions (~5% THD). All system parameters are the same as the one used to get the results in Fig. 6 (left). Despite this distortion on grid voltages, currents have a THD value of < 3.0%

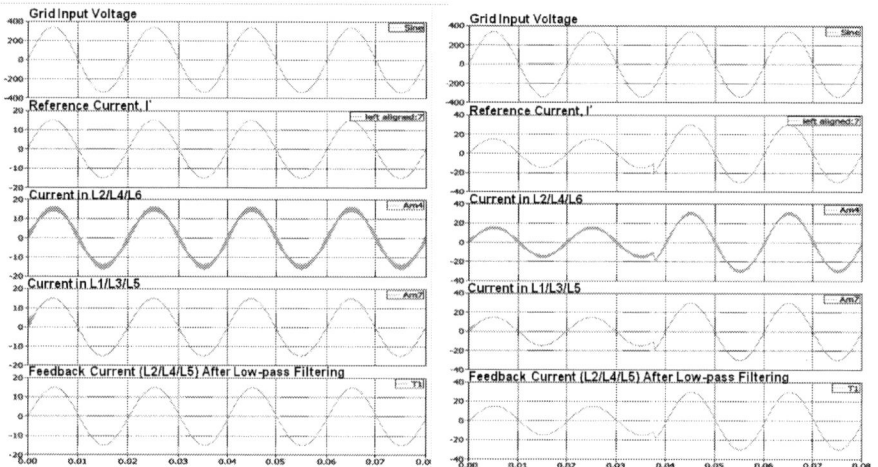

Fig. 5. Simulation for 15 Amps stead-state operation showing less than 1.5% THD (left) and simulation for reference current step change from 15 Amps to 30 Amps (right)

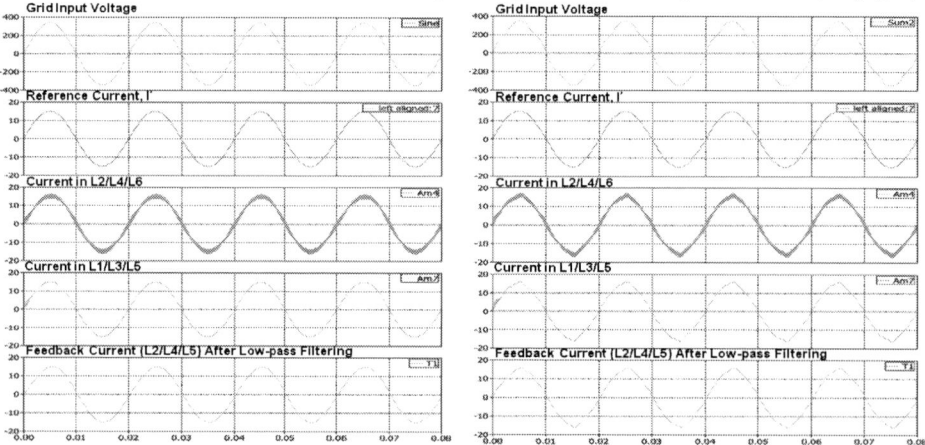

Fig. 6. Simulation for 15 Amps stead-state operation showing normal operation: (left) under 100% error on inductor value (actual=1.5mH, Code=.75mH) and (right) under 6% THD voltage distortion on the grid (5th and 7th harmonics).

© VDE VERLAG GMBH · Berlin · Offenbach

4. Experimental Setup and Results

An experimental AFE system was built in the lab with Cree CCS050M12CM2, 1200V/50A silicon carbide module. The module includes 25mΩ 1200V SiC MOSFETs and 1200V 50A SiC Schottky diodes in anti-parallel per switch position. The SiC devices were driven with a 10 Ohm gate resistor for both turn on and turn off and a V_{GS} voltage of +20V/-5V as recommended by the manufacturer. Complete control for the AFE system was built around TI DSP 320F28335 floating point processor and both controls (traditional d-q and proposed CFP) were built using the same PLL based on DDSRF. Voltage sensing was implemented using standard voltage divider techniques and then isolated before being fed into the DSP. Current sensing was implemented using hall-effect sensors. Fig. 7 shows a picture of the AFE controller and power stage. Bulk dc-link capacitor and L-C-L input filter are not included in this picture were mounted off-board for easy of change during testing and are not included in the figure.

Fig. 7. Photograph of the 20kW AFE hardware fabricated for control validation

Converter side inductors (i.e., L1, L3 and L5) were built with AMCC80 Hitachi core. Note that the same core was used for all tests (i.e., 1.5, 1.0 and 0.8 mH). This is acceptable since the main aim of this paper was to demonstrate the proposed control scheme. However, in an industrial application, the inductor design would be optimized for each value allowing for substantial cost and size reduction of the smaller inductors values. Grid-side inductors (i.e., L2, L4 and L6) were built using ferrite core with values of 100 µH and 200 µH. Initial aim was to test the inverter up to 20 kW. However, due to lab power limitation testing up to 7.5 kW only was carried out. Despite this lab limitation, the control characteristics can be clearly compared and contrasted at the tested power level.

All measurements of THD and efficiency were recorded with YOKOGAWA 1800 power analyzer and waveforms were recorded with YOKOGAWA DLM 2024 oscilloscopes. Experiments were conducted at 20 kHz and 40 kHz switching frequencies and varying values of converter-side inductors. Results have been tabulated in Table 2 along with the THD and efficiency values at the 7.5 kW condition.

Type of controller	L1-C1-L2 (µH-µF-mH)	Sw. Freq. (kHz)	THD (%)	Eff. (%)	Ref Fig.
d-q controller	200-4.7-1.5	20	3.7	99.46	Fig. 8 a)
d-q controller	200-4.7-1.5	40	3.4	99.08	Fig. 8 c)
CFP – controller	200-4.7-1.5	20	2.2	99.43	Fig. 8 b)
CFP – controller	200-4.7-1.5	40	2.4	99.01	Fig. 8 d)
CFP – controller	100-4.7-1.0	40	2.7	99.03	Fig. 9 a)
CFP – controller	100-4.7-0.8	40	2.9	98.99	Fig. 9 b)

Table 2. Summary of measured THD and efficiency results under various input filter values and switching frequencies for a traditional d-q and proposed CFP control.

Note: 1. Inductor resistance does not change here even if the inductor values changed because for lower inductance, some turns were unwound but not cut off. Lower inductance is the grid side inductor.
2. Note 2: Efficiency measurements do not include the controller board power and isolated power supplies for gate drives of the MOSFETs which were supplied separately.

© VDE VERLAG GMBH · Berlin · Offenbach

PCIM Europe 2016, 10 – 12 May 2016, Nuremberg, Germany

Fig. 8 shows measured results collected on the described experimental setup when implementing d-q and CFP control approaches. For these results L1-C1-L2 filter values were set at 200µH-4.7µF-1.5mH and switching frequency was changed from 20 kHz (Fig. 8 a-b) to 40 kHz (Fig. 8 c-d). Fig 8 a) and c) correspond to a d-q control implementation while Fig 8 b) and d) correspond to a CFP control implementation. As summarized in Table 2, CFP control implementation delivers superior THD performance allowing for a reduction of the input filter requirements 200µH-4.7µF-1.5mH to 100µH-4.7µF -0.8mH. Such a reduction in filter requirements was also experimentally validated; the results are summarized at the bottom of Table 2 and collected waveforms are shown in Fig 9 a) and b. Even with a converter-side inductance of 0.8 mH and a grid-side inductance of 100µH, the measured THD is below 3%. All waveforms show that the hardware is able to operate under near-unity power factor for all component values and control methods.

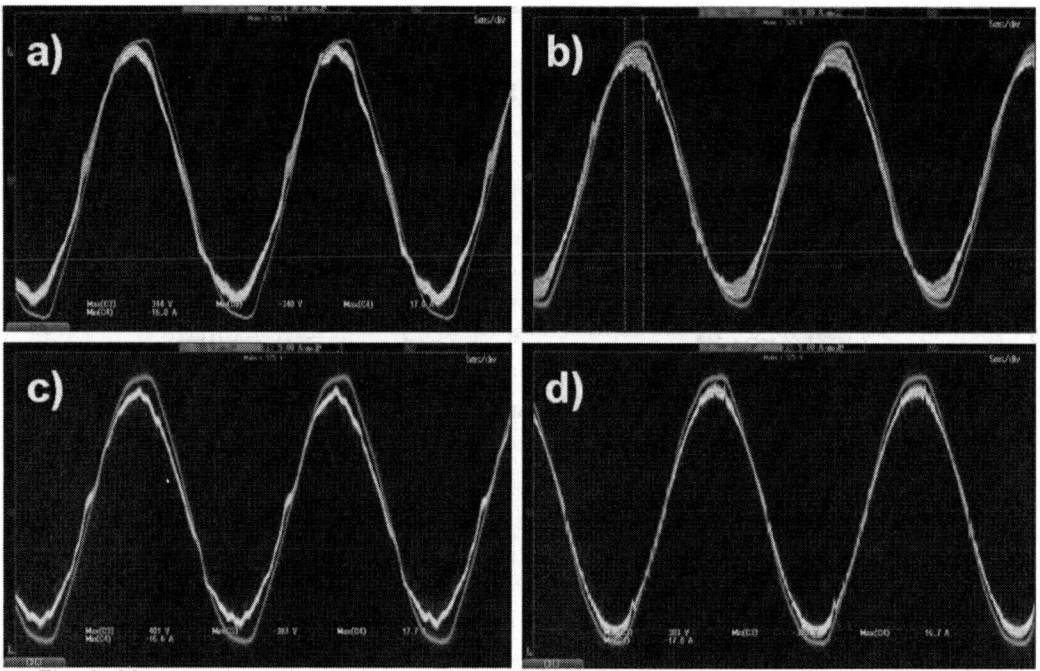

Fig. 8. Measured result 7.5 kW operation of the AFE under d-q control (a) and c)) and proposed CFP control (b) and d)) for 20 kHz (top figures) and 40 kHz (bottom figures). Input filter values was set at 200µH-4.7µF-1.5mH

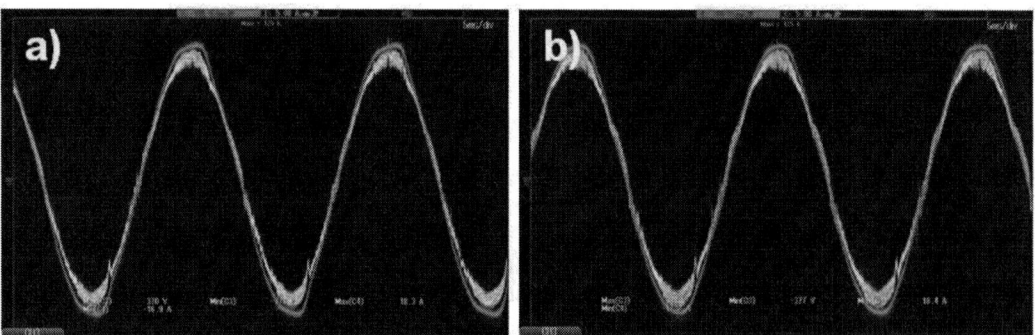

Fig. 9. Measured result 7.5 kW operation of the AFE under proposed CFP control at 40 kHz operating under reduced input filter conditions. a) 100µH-4.7µF-1.0mH and b)100µH-4.7µF-0.8mH

While efficiency was not a focus of the initial prototype, greater than 99% efficiency was measured. Inductor optimization will deliver increased efficiency. Lastly, note that hardware was tested under distorted grid conditions (~2% THD as shown in Fig 8 and 9. Despite this, the proposed CFP control scheme operates well drawing a sinusoidal current with low THD.

© VDE VERLAG GMBH · Berlin · Offenbach

5. Summary

Commercially available SiC devices now allow for an important increase in switching frequency (2-5×) leading to a considerable reduction on filtering components. However, in some applications such as grid-connected converters, the reduction in filtering components is sometimes limited by control stability or other robustness issues hindering the full potential and cost saving capabilities of SiC-based grid-connected converters. Many discussed control methods rely on a minimum impedance between the power converter and the grid in order to operate reliably over a wide range of conditions.

To address this, the paper introduces a new CFP control scheme implemented in the a-b-c stationary reference frame. The proposed CFP control scheme generates internal sinusoidal current references which are synchronized with the grid voltage via a PLL implemented in DDSRF. In this manner the PLL is very robust to grid harmonics and the current references are always pure sinusoids since they are internally generated. Phase currents closely track the reference at every switching cycle. The average phase current is managed by a PI controller used for closed loop control of the converter's DC link voltage. The proposed CFP control scheme is able to pull (or feed) sinusoidal current from (into) the grid even in the present of significant grid voltage distortion.

Simulation and experimental results showed that when compared with standard d-q control the proposed CFP control schemes allows for a substantial reduction of output filtering components (close to 2×) while delivering very good THD and efficiency performance. Inductor core weight was reduced by 40% and cooper weigh by 35% allowing for an estimated 33% cost reduction on inductors.

References

[1] P. Cortés, J. Rodríguez, P. Antoniewicz, M. Kazmierkowski, "Direct Power Control of an AFE Using Predictive Control". IEEE Transactions on Power Electronics, vol. 23, no. 5, pp. 2516-2553, Sep. 2008.

[2] Z. Dawei, X. Lie, amd B.W. Williams, "Model-Based Predictive Direct Power Control of Doubly Fed Induction Generators," IEEE Transactions on Power Electronics, vol.25, no.2, pp.341-351, Feb. 2010.

[3] P. R. e. al, "Decoupled Double Synchronous Reference Frame PLL for Power Converters Control," IEEE Transaction on Power Electronics, vol. 22, no. 2, 2007.

[4] Blaabjerg, F., R. Teodorescu, M. Liserre, and A. Timbus, "Overview of Control and Grid Synchronization for Distributed Power Generation Systems," IEEE Transactions on Industrial Electronics, vol 53, no 5, Oct 2006.

[5] P. R. &. R. T. Marco Liserre, Grid Converters for Photovoltaic and Wind Power Systems, John Wiley & Sons Ltd., 2011.

[6] Teodorescu, R.; Blaabjerg, F.; Liserre, M.; Loh, P.C. "Proportional-resonant controllers and filters for grid-connected voltage-source converters", IEE Proceedings Electric Power Applications,Volume.153, Issue.5, pp.750, 2006, ISSN: 13502352

[7] Yaramasu, V.; Bin Wu "Predictive Control of a Three-Level Boost Converter and an NPC Inverter for High-Power PMSG-Based Medium Voltage Wind Energy Conversion Systems", Power Electronics, IEEE Transactions on, On page(s): 5308 - 5322 Volume: 29, Issue: 10, Oct. 2014

[8] Rodriguez, J.; Kazmierkowski, M.P.; Espinoza, J.R.; Zanchetta, P.; Abu-Rub, H.; Young, H.A.; Rojas, C.A. "State of the Art of Finite Control Set Model Predictive Control in Power Electronics", Industrial Informatics, IEEE Transactions on, On page(s): 1003 - 1016 Volume: 9, Issue: 2, May 2013

[9] Hartmann, M.; Ertl, H.; Kolar, J.W. "Current Control of Three-Phase Rectifier Systems Using Three Independent Current Controllers", Power Electronics, IEEE Transactions on, On page(s): 3988 - 4000 Volume: 28, Issue: 8, Aug. 2013

DC-link Harmonic content in Double two-level inverter for Permanent Magnet Synchronous Motor Drive systems – comparison and analysis

M.Eng. Toktam, Khani, Technologie Netzwerk Allgäu, Germany, Bahareh.khani@tn-allgaeu.de

Prof. Dr.-Ing. Michael, Patt, Technologie Netzwerk Allgäu, Germany, Michael.patt@tn-allgaeu.de

Abstract

Battery driven vehicles with Voltage Source Inverters (VSIs) are widely used in power electronics for many years. Meanwhile the Pulse Width Modulation (PWM) Techniques to improve efficiency, output voltage quality, current harmonic spectrum and etc. is the topic which gain massive attention through these years and many studies are accomplished in this regard. Nevertheless, DC-link capacitor as the main bulky and expensive element which is affected by generated current harmonics is the focus of this paper. Here we investigate this factor for single and double inverter with continuous and discontinuous PWM pattern and compare the capacitor values for each case.

1. Introduction

Since the introduction of first PWM integrated circuit in 1976 [1], its techniques have been developed and improved in so many different ways. Regardless of all their advantages, these fast switching elements, will introduce current harmonics on the DC-link. These harmonics are one of the key factors in DC-link capacitor selection.

As it is also explained in [2], considering the series resistance of the capacitor, the RMS current will generate the power loss which consequently increases the temperature across it and affect the capacitor lifetime as it can be seen from (1) and (2).

$$\Delta T = I_{c,rms}^2 \cdot R_{ESR} \cdot R_{th} \tag{1}$$

$$L = L_B \cdot f(T_M - T_c) \cdot f(V) \tag{2}$$

Where R_{ESR} is the equivalent series resistance of the capacitor, R_{th} thermal resistance, L is the estimated life time, L_B is the base life at elevated maximum core temperature T_M, T_c is the actual core temperature in the application and V is the applied DC Voltage.

During the past decades, many different modulation strategies introduced. In Modern carrier-based PWM techniques, different type of zero sequence is added to the modulator signal with the aim of improving losses, harmonics, voltage gain, common mode issues and etc. A detailed description of all these different modulators can be found in [3]. These different modulation strategies will generate different current harmonic spectrum in DC-link.

As the voltage ripple across the capacitor depends on the current which pass through it, for a known constant voltage ripple, by knowing the current, the capacitance can be estimated using (3). In next sections we consider Sinusoidal-PWM (SPWM), Discontinues PWM1 (DPWM1), and Discontinues PWM2 (DPWM2) and compare the calculated capacitance value for the same voltage ripple.

$$i = C \cdot \frac{dv}{dt} \tag{3}$$

In this article the analysis is done for 18 kW, two-level three phase inverter which is supplied with 58V Battery for drive of permanent Magnet Synchronous Motor (PMSM) as it is illustrated in fig. 1. For the calculations Mathcad software is used. Fast Fourier Transform (FFT) is applied on the capacitor current in order to determine the current spectrum for three different modulation strategies which enable us to calculate the capacitor value for voltage ripple of 2 V_{p-p}.

In section 2 we start with the principle of the pulse width modulation as it is the basis for our analysis. In section 3 we continue with the analysis of DC- link considering SPWM, DPWM1 and DPWM2. It also includes comparison of current spectrum and capacitor value for different strategies. In section 4 we consider the system with double inverter configurations. With an open winding Motor, voltage drop on motor winding would be double of the single method. With this configuration smaller DC-link capacitor value is expected. This section will be followed by a conclusion.

2. Single inverter

2.1. PWM Principle

Three phase voltage source inverter is schematically shown in Fig. 1. For pulsing pattern, different methods can be used; here we consider a carrier based PWM in which a modulator signal - sinusoidal or non-sinusoidal waveform - will compare to a high frequency triangle waveform (carrier) – in our case 7.5 kHz - (fig. 2). The intersection will define the switching instances of semiconductors. Many articles can be found in this regard in literature. In [3] the pros of this method are mentioned as low harmonic distortion waveform characteristics with well-defined harmonic spectrum, fixed switching frequency and implementation simplicity.

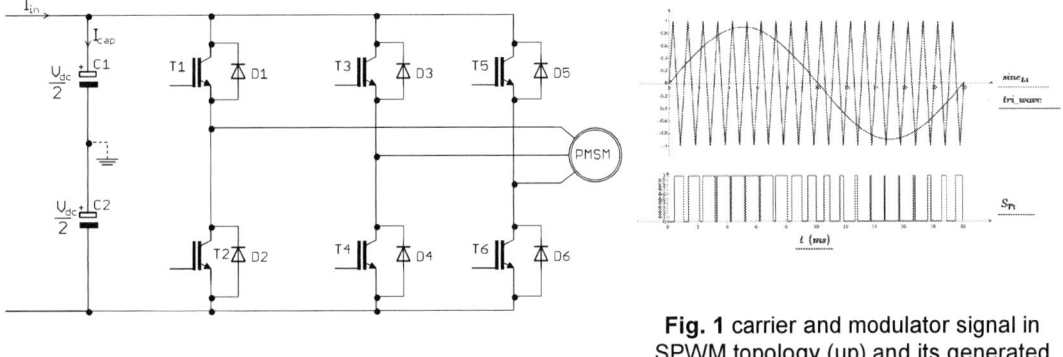

Fig. 1 Three phase VSI for driving PMSM

Fig. 1 carrier and modulator signal in SPWM topology (up) and its generated pulse pattern (down)

2.2. Single inverter with SPWM

Sinusoidal waveform is the earliest modulator pattern which is developed for pulse width modulation. Although it has limited linear range, because of its simplicity it is still widely used in different applications.

For this project, regardless of the load, we considered a completely sinusoidal phase current with 0°, 120° and 240° phase shift for different phases. Switching actions are defined by SPWM. Assumptions are: V_{dc} equal to 58 V, output power of 18 kW and no AC current in the battery. The current through the Capacitor is calculated:

$$i_{DC} = i_{T1} + i_{T2} + i_{T3} \tag{4}$$

$$I_{cap} = i_{DC} - i_{DCav} \tag{5}$$

© VDE VERLAG GMBH · Berlin · Offenbach

Fig. 3 and Fig. 4 respectively depict the capacitor current I_{cap} and its harmonic spectra by implementing the forward Fourier Transform (FFT). As it can be seen we have peak at PWM switching frequency and twice of it. For simplicity the magnitude is normalized to the peak value of phase current. Using (3), having the current waveform through the capacitor, it is calculated for 2 V ripple across its terminals, 6800 µF is needed.

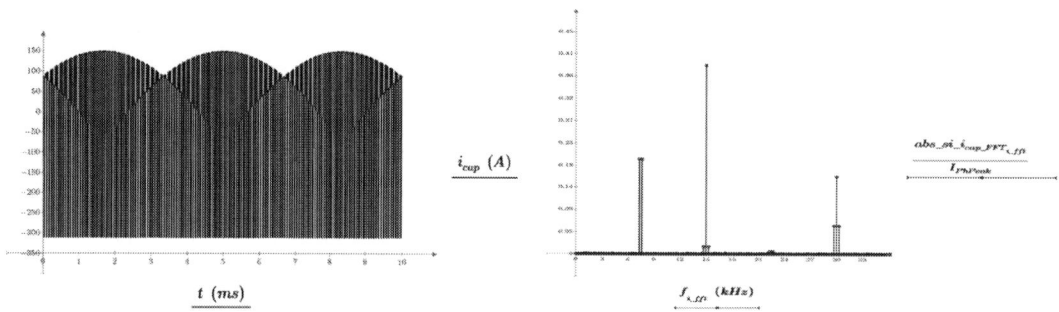

Fig. 3 Capacitor current using SPWM

Fig. 4 normalized capacitor current harmonics

2.3. Single inverter with DPWM

As it is already mentioned, to improve the characteristics of the VSIs zero sequence will be added to the sinusoidal waveform. There are infinite possibilities how to add these zero sequence. Here we explore the effect of discontinues PWM (DPWM). According to definition, in this method the modulation signal will have at least one part in which it is clamped to positive and/or negative DC-bus for maximum of 120°. Using [3] zero sequence for generalized DPWM (GDPWM) calculated and added to the sinusoidal waveform. Defining Ψ as modulator phase angle which starts at intersection of two sine waves at 30°, for Ψ=30° we will have the modulator as it is shown by Fig. 5 (a) which is called DPWM1. For Ψ=60° fig. 5 (b) is generated which is called DPWM2 modulator. The same as SPWM, here also the intersection area with carrier will specify switching actions of semiconductors.

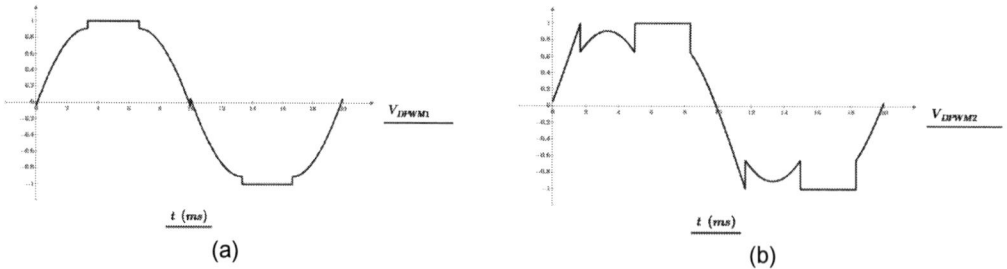

(a)

(b)

Fig. 5 Modulator waveforms for (a) DPWM1, (b) DPWM2

© VDE VERLAG GMBH · Berlin · Offenbach

The resultant DC currents and their FFTs are plotted as bellow;

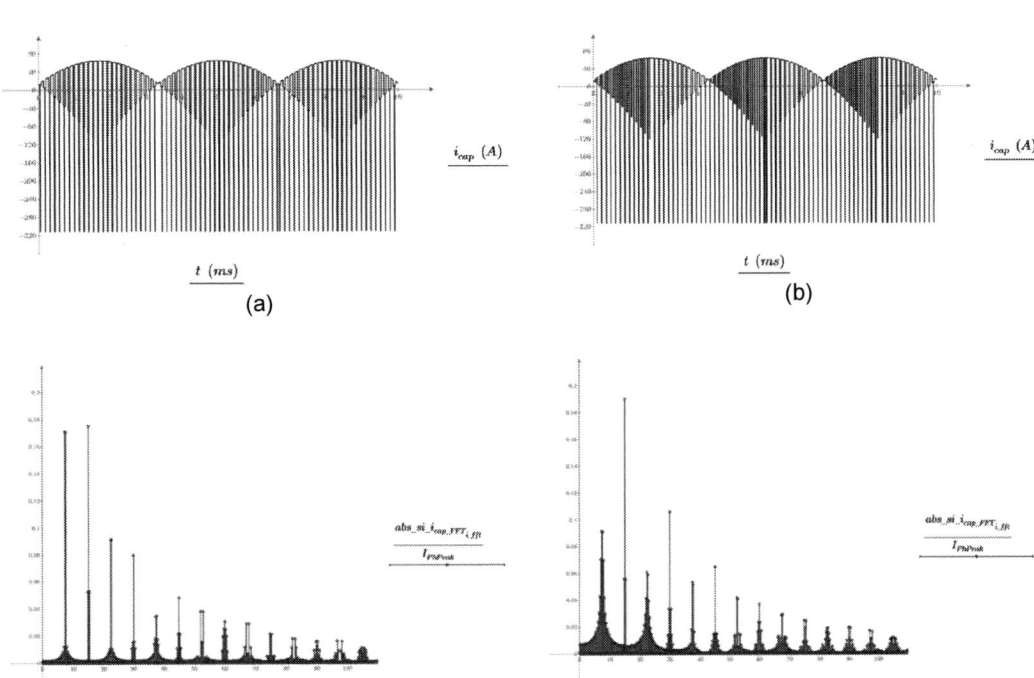

(a)

(b)

(c)

(d)

Fig. 6 Single inverter: (a) Capacitor current for DPWM1, (b) Normalized capacitor current harmonics for DPWM1, (c) Capacitor current for DPWM2, (d) Normalized capacitor current harmonics for DPWM2

In DPWM patterns the phase which carries the largest current will not switch so the switching losses will significantly decrease. As it is stated by [4], DPWM1 is suitable for utility interface applications and AC permanent magnet motor applications while DPWM2 is preferred for application like induction motor drives. They also give the possibility to increase modulation index from 0.9 in SPWM to 1.1 and still benefit from the voltage linear range.

Using the same calculation method for capacitor, the capacitance-value decreases 35.8% in the case of DPWM1 and 11.76% for DPWM2 in comparison to SPWM for having 2 V ripple across its terminals. Fig. 5 shows the voltage drop of capacitor in both mentioned cases.

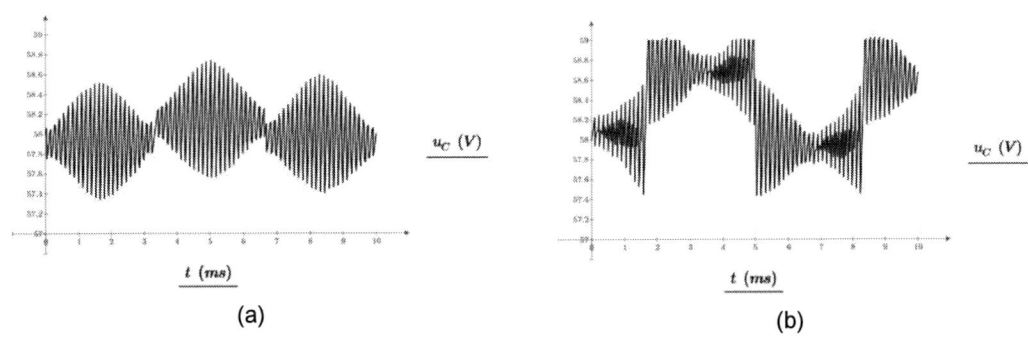

(a)

(b)

Fig. 7 Capacitor Voltage for (a) DPWM1, (b) DPWM2

3. Double inverter

3.1. Double inverter with SPWM

Another configuration which can improve the characteristics of the power device is double two-level inverter with open winding on the motor side. In this way, inverters would be parallel from DC-side and in series at motor windings. The scheme of the dual inverter is depict in fig. 8 with an open winding, voltage drop on motor winding would be double of the single method. This will reduce the current through the system.

The double inverter, lead to cancellation of odd harmonics and as mentioned already, lead to reduction in the RMS current through capacitor and finally smaller DC-link Capacitor size. Also with this topology the partial efficiency can be improved. Its advantages and modulation strategies are discussed in [5]. For double inverter with overall power of 18 kW, the modulation signal for second inverter should have 180° phase shift.

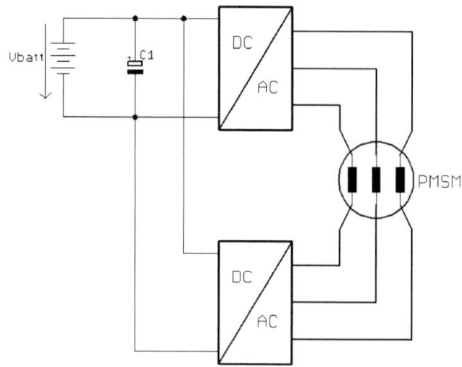

Fig. 8 Double inverter configuration with open winding motor

Same as procedure in section 2.2, DC-link current equals to the sum of all upper switches, without average current.

Fig. 9 (b) shows the current spectra of DC-link Capacitor for SPWM for double inverter. As it can be seen from the graph, for double inverter odd harmonic cancellation happens which have a great effect on Capacitor Value. Also the magnitude is significantly decreases. In the case of double inverter with SPWM, using a capacitor with 3290 μF generate 2 V ripple at input which is half of the single inverter capacitor value. Graph in Fig. 9 (a) illustrate the current through capacitor.

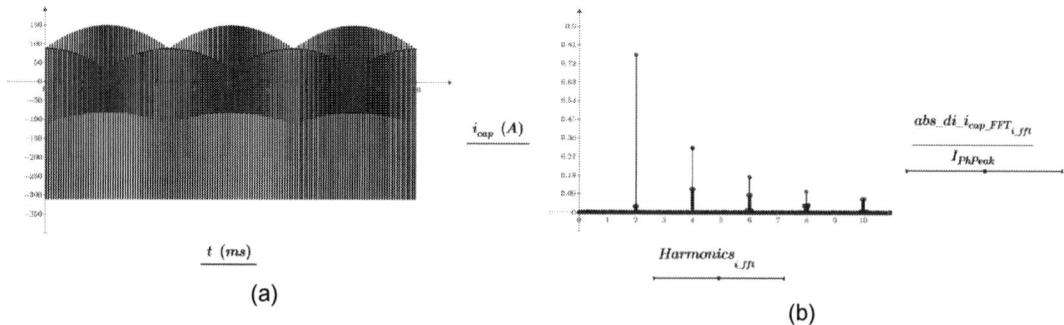

Fig. 9 (a) Capacitor current in Double-inverter with SPWM, (b) normalized capacitor current harmonics for double inverter with SPWM

© VDE VERLAG GMBH · Berlin · Offenbach

3.2. Double inverter with DPWM

Adding the zero system to modulator aims better performance of the whole system. The fact is, by changing the modulator the current spectra will change and consequently it changes the capacitor value. Using DPWM1 and DPWM2 pattern with 180° phase shift for pulsing the second inverter, the capacitor value will even shrink more as there is cancelation in current harmonic spectrum which can be seen in fig. 10 (b) With DPWM1 as modulator and (d) with DPWM2. In both cases for gaining 2 V ripple, respectively 1600 µF and 1770 µF will suffice.

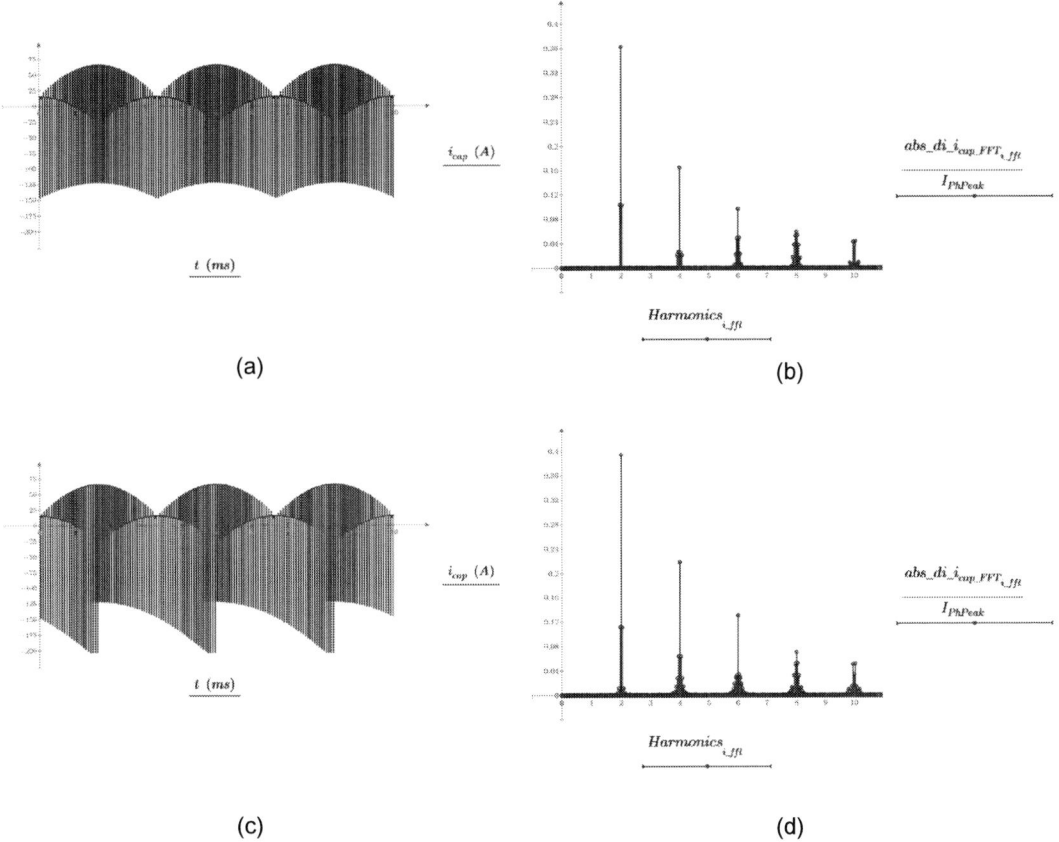

Fig. 10 Double inverter: (a) Capacitor current for DPWM1, (b) Normalized capacitor current harmonics for DPWM1 (c) Capacitor current for DPWM2, (d) Normalized capacitor current harmonics for DPWM2

4. Conclusion

In modern power electronics, saving money, energy and volume is done with the cost of complexity especially in PWM patterns of semiconductors. For a designer there is always a trade of among all these parameters. As it can be seen with capacitor value calculation for this case. Table 1 is an overview of all the values with different PWM.

DC-link capacitor value	SPWM	DPWM1	DPWM2
Single- inverter	6800 µF	4365 µF	6000 µF
Double- inverter	3290 µF	1600 µF	1770 µF

Table 1: Capacitor value for different methods and system configuration

Apart from the advantages of double inverter topology and despite capacitor value reduction, it cannot be considered as final conclusion. Still other parameters needed to be regarded such as number of elements and its overall price and control complexity.

5. References

[1] Dorin O. Neacsu, "Switching Power Converters: Medium and High Power," Boca Raton, Fla.: CRC, 2014. Print.

[2] G. Sam, Jr. Parler, "Deriving life multipliers for electrolytic Capacitors," IEEE Power electronics Society Newsletter, vol. 16, no. 1, Feb. 2004, pp. 11-12

[3] M. Hava, R. J. Kerkman and T. A. Lipo, "Simple analytical and graphical tools for carrier based PWM methods," Power Electronics Specialists Conference, 1997. PESC '97 Record., 28th Annual IEEE, St. Louis, MO, 1997, pp. 1462-1471 vol.2.

[4] A. M. Hava, R. J. Kerkman and T. A. Lipo, "A high-performance generalized discontinuous PWM algorithm," in *IEEE Transactions on Industry Applications*, vol. 34, no. 5, pp. 1059-1071, Sep/Oct 1998.

[5] G. Grandi, A. Lega, C. Rossi, and D. Casadei, "Double Inverter as a Multilevel Converter: Circuit Topology, Modulation strategies and Applications," International Conference on Power Electronics and Intelligent Control for Energy Conservation, Pelincec 2005

PCIM Europe 2016, 10 – 12 May 2016, Nuremberg, Germany

Hybrid filter with an optimized switching method of the compensation capacitors and predictive active filter control

Swen Bosch[1], swen.bosch@htw-aalen.de;

Heinrich Steinhart[1], heinrich.steinhart@htw-aalen.de

[1] Aalen University, Laboratory of Power Electronics and Electrical Drives,

Anton-Huber-Straße 25, D-73430 Aalen, Germany

Abstract

A hybrid filter, consisting of a parallel-connected reactive power compensator (RPC) and a shunt active power filter (APF), is presented.
The RPC is used to compensate inductive reactive power by need-based switching capacitors to the grid. At this, a charging circuit and a switching method, which is causing no inrush currents and no transient currents, are introduced.
The APF is based on a voltage source inverter and is used for the compensation of distortion power. A simple, but very efficient predictive control scheme is presented.
Both filters are built up as three-phase four-wire systems, so that the presented methods are also suitable for single-phase systems.

1. Introduction

In industrial facilities, many loads are not only receiving active power, but also reactive power and distortion power, whereby all transmission devices, such as transformers, are burdened in addition to the active power. Because most of the linear loads, like induction machines, have a resistive-inductive behavior, the reactive power can be compensated by switching capacitors in parallel to the load. Non-linear loads, like diode rectifiers supplying e.g. variable speed drives, are causing harmonics in the grid current, and by this, distortions in the voltage at the point of common coupling (PCC). A voltage source inverter can be used to inject the needed current harmonics in phase opposition to the harmonics caused by the load and thereby compensate the distortion power.

2. System overview and hardware design

A system overview of the hybrid filter, the grid and the load is given in Fig. 1. Both RPC and APF are connected in parallel to the grid, so that both filters can operate independently of each other.

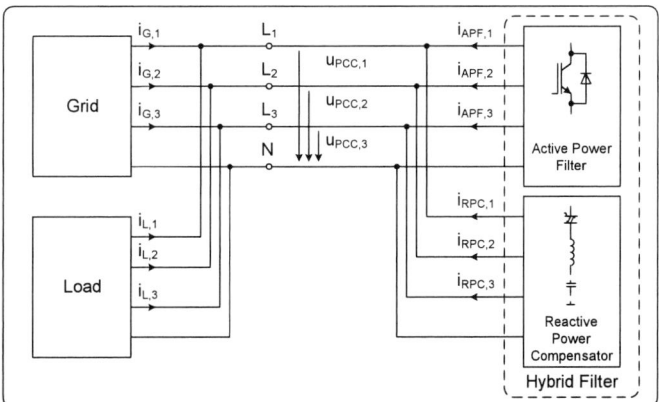

Fig 1. System overview of the hybrid filter, the grid and the load

© VDE VERLAG GMBH · Berlin · Offenbach

2.1. Reactive Power Compensator

The RPC consists of three binary graded compensation stages (index κ, κ=1,2,3) per phase μ (μ=1,2,3) with a ratio of 1:2:4. Thereby it is possible to adjust the capacitive reactive power of the RPC in eight equidistant steps per phase. The hardware design is shown in Fig. 2a.

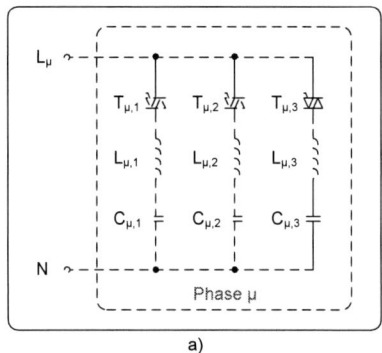
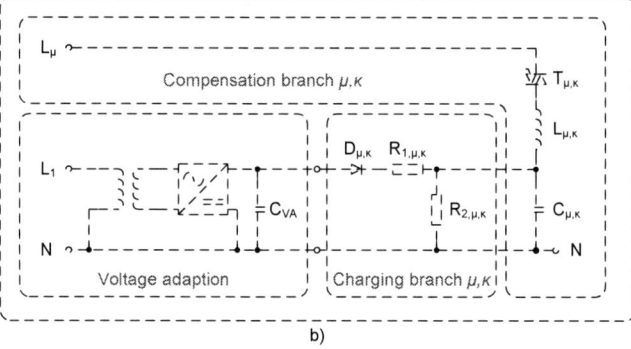

a) b)

Fig. 2. a) Hardware setup of the RPC; b) Charging circuit

As protection against resonance, the capacitors are choked. Thyristors are used as switches. In the following, three different methods of switching capacitors to the grid are summarized:

1) By using mechanical contactors it is not possible, to switch at a defined point in time. Hereby, transient currents as big inrush currents, caused by the difference between the PCC voltage and the capacitor voltage, occur [1].

2) Applying thyristors as switches makes it possible to switch at a defined point in time. By this, the capacitors can be switched to the grid at the time when the PCC voltage and the capacitor voltage are equal. This avoids inrush currents, but due to discharging of the capacitors when they are not connected to the grid, transient currents will occur [1][2, p. 422].

3) Using thyristors and precharging the capacitors to the maximum capacitor voltage in the steady state. By switching the precharged capacitors to the grid at the time of the positive maximum of the PCC voltage, the circuit is immediately in the steady state, because the capacitor current in the steady state, at this time, is also zero. Inrush currents and transient currents can be avoided completely.

Method	Inrush currents	Transient currents
1	Yes	Yes
2	No	Yes
3	No	No

Tab. 1. Occurrence of inrush and transient currents depending on different switching methods

The RMS value of the capacitor voltage in the steady state can be calculated to

$$U_{C,Stat,RMS} = U_{PCC,RMS} \cdot \frac{1}{1 - (\omega_N^2 \cdot L \cdot C)},$$ (1)

where ω_N is the angular frequency of the grid, L the inductance of the choking and C the capacitance of the compensation capacitor. Hence, the capacitor preload voltage is:

$$U_{C,Preload} = \sqrt{2} \cdot U_{C,Stat,RMS}$$ (2)

Because of the choking and the consequential voltage superelevation, the capacitor preload voltage is always higher than the maximum PCC voltage. That is why a separate charging circuit, as shown in Fig. 2b, is needed. The transformation ratio of the transformer has to be chosen in that way, that the rectified voltage at the smoothing capacitor C_{VA} exceeds the required preload voltage. The adjustment of the preload voltage can be done by the voltage divider, taking the voltage drop at the diode into account.

2.2. Active Power Filter

The APF is based on a three-phase, two-level voltage source inverter with an L-filter, as depicted in Fig.3.

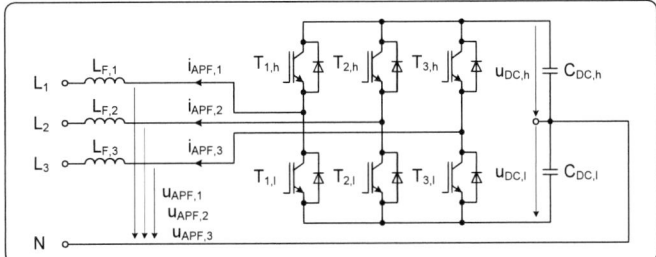

Fig. 3. Hardware setup of the APF

The neutral is connected to the midpoint of the DC link. With this, the filter currents $i_{APF,\mu}$ can be controlled separately. IGBT's are used as switches.

3. Software and Control Design

Because the APF and the RPC are connected in parallel to the grid, they can work completely independent from each other. The control structures of both filters are realized as three single-phase control structures, so they are also suitable for single-phase systems.

3.1. Reactive Power Compensator

Since the control algorithm is based on three single-phase control structures, the reactive power of each phase can be controlled separately. For this, a power calculation and a controller are needed.
The power calculation is determining the active and reactive power of the grid, the load, and the RPC.
Depending on the intended power factor, the controller is generating the switching signals for the thyristors. To avoid inrush currents and transient currents, the compensation capacitors are precharged to their maximum voltage in the steady state and switched to the grid at the time of the positive maximum of the PCC voltage, as described above as method 3.

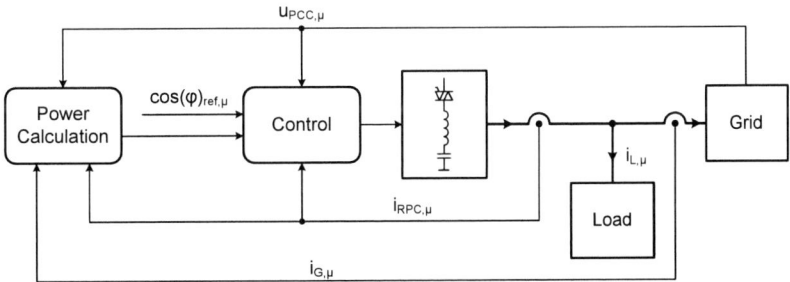

Fig. 4. Phase µ control structure of the RPC

3.2. Active Power Filter

The control structure of the APF is shown in Fig 5. It is based on three single-phase current controllers in the natural frame. A phase-locked loop (PLL) is providing the phase angle and furthermore keeps the number of samples N per fundamental period constant at $N = 400$ by adapting the switching frequency f_{SW}. The three current references are given by the DC link control and a current reference generator. Depending on the distortion power caused by the load, a reference current is calculated for each phase. In combination, the current of the load and the APF have to result in a sinusoidal grid current.

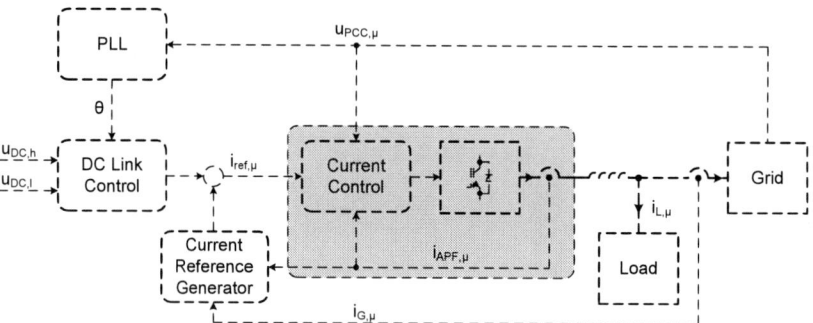

Fig. 5. Control structure of the APF

As it can be seen in Fig. 5 and 6, no transformations in rotating frames with the frequency of each harmonic [3][4] or a resonant controller for each harmonic [5][6] are used, as it is usually done in the control of grid connected voltage source inverters [7].

Especially, disturbances in the PCC voltages related to the commutation of three-phase diode rectifiers supplying a RL load can only be sufficiently damped, if the APF is able to compensate harmonics up to the 49th or even higher orders [2, p. 426]. As it can be seen, using the methods mentioned before results in an enormous computational effort, because controllers for each harmonic are needed.

The applied current control structure of phase µ is shown in Fig 6. Because the PI controllers used are only able to regulate DC values with zero steady state error, an additional structure is needed to inject the required current harmonics with zero steady state error to the grid.

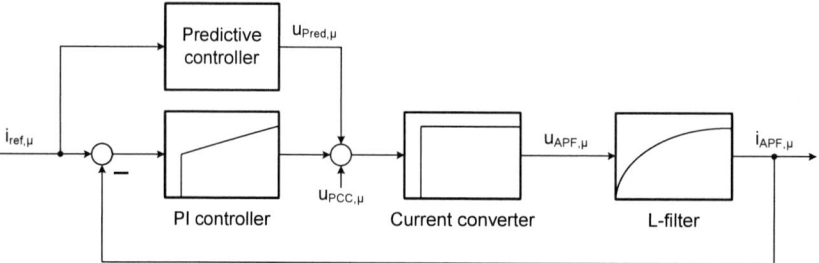

Fig. 6. Phase µ current control structure (compare Fig. 5, grey highlighted part)

The predictive controller is doing this by using an inverse model of the voltage source inverter and the L-filter to compensate the delay caused by the measurement and the control dead time. A similar approach, but in the synchronous reference frame, is done in [8], where the reference current is corrected. The transfer function of the dead time, the L-filter and predictive controller in the discrete z-domain are

$$G_{DT}(z) = \frac{1}{z} \ , \quad G_F(z) = K_F \cdot \frac{z}{z - e^{-T_{SW}/\tau_F}} \tag{3}$$

and

$$G_{Pred}(z) = \frac{u_{Pred}(z)}{i_{ref}(z)} = \frac{1}{G_{DT}(z) \cdot G_F(z)} = \frac{1}{K_F} \cdot \left[z^1 - e^{-T_{SW}/\tau_F} \right] . \tag{4}$$

At this is

$$K_F = \frac{1}{R_F} \ , \quad \tau_F = \frac{L_F}{R_F} \quad \text{and} \quad T_{SW} = \frac{1}{f_{SW}} \ . \tag{5}$$

The transfer function of the predictive controller contains a future value of the input signal $i_{ref,\mu}$, which is not known at the present time, to calculate the output signal $u_{Pred,\mu}$. In the steady state, the reference current is periodical with the time of the fundamental period. This means, that a future value is equal to the correlating past value:

$$i_{ref,(k+1)} = i_{ref,(k+1-N)} \ , \tag{6}$$

where N is the number of samples per fundamental period and k is the number of the actual sample. Thus this past value can be used. In total, the predictive controller consists only in two delays, two multiplications, and one addition. This results in a huge reduction of the computational effort.

4. Simulative and experimental results

The control algorithms for the RPC are implemented on a TI Concerto F28M35H52C1 DSP in fixed-point arithmetic. For the APF, a STM STM32F407VG DSP and floating-point arithmetic are used. The simulations were done with MATLAB/Simulink and the toolbox SimPowerSystems for modelling the hardware. The parameters of the APF and the RPC are listed in Tab. 1.

Active Power Filter		Reactive Power Compensator	
DC Link Voltage	U_{DC} = 750V	Reactive Power	Q = 8.3 kvar
DC Link Capacitance	C_{DC} = 7 mF	Choking rate	p = 7 %
Switching/Control freq.	f_{SW} = 20 kHz	Control frequency	f_C = 10 kHz
Max. inverter current	$i_{AF,\mu,max}$ = 26 A	Switches	Thyristors
Filter Inductance	$L_{f,\mu}$ = 1.4 mH	Preload voltage	$U_{C,Preload}$ = 350 V
Switches	IGBTs		

Tab. 2. System parameters of the APF and the RPC

4.1. Reactive Power Compensator

A comparison of the switching methods for the RPC is shown in Fig 7. Switching of discharged capacitors when the PCC voltage is zero is causing no inrush current, because the capacitor voltage equals the PCC voltage, but transient currents occur (Fig. 7a), which are burdening the APF additionally. The damping of these transient currents is depending on the choking and the grid parameters. By precharging the capacitors and switching them to the grid at the time of the positive maximum of the PCC voltage, these transient currents can be avoided completely, as shown in Fig. 7b.

© VDE VERLAG GMBH · Berlin · Offenbach

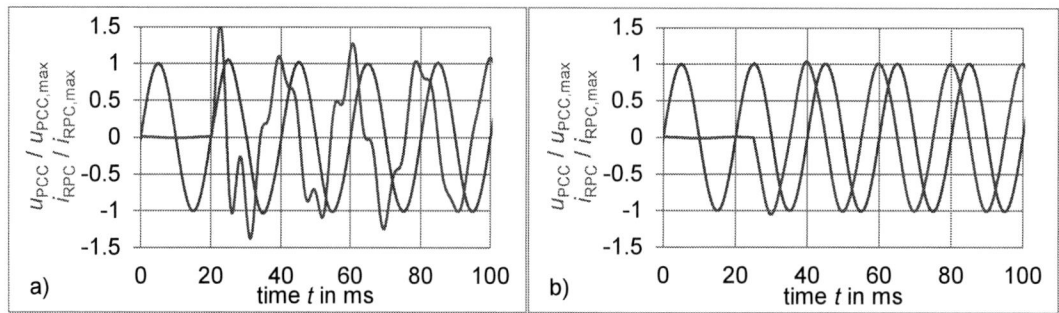

Fig. 7. Simulated compensation current i_{RPC} and PCC voltage u_{PCC} at: a) switching of discharged capacitors when the grid voltage is zero b) switching of precharged capacitors to the grid at the time of the positive maximum of the PCC voltage

For pointing out the differences of the switching methods, the PCC voltage in the simulation was chosen to be completely sinusoidal ($THD_U = 0$ %). In the majority of cases this holds not true with reality and the PCC voltage is distorted. This is causing harmonics in the current of the RPC, which can be seen in the measurement depicted in Fig. 8. Both the current of the RPC in the steady state and while switching are shown.

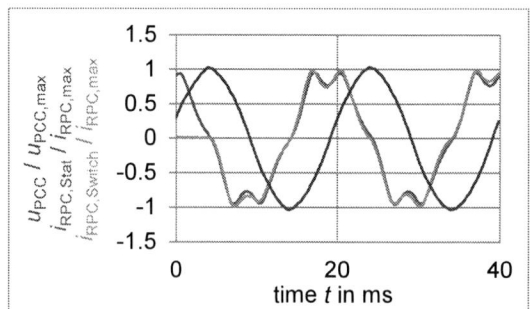

Fig. 8. Measured compensation current i_{RPC} in the steady state ($i_{RPC,Stat}$) and while switching ($i_{RPC,Switch}$) at distorted u_{PCC} ($THD_U = 3.1\%$)

Immediately after switching, the compensation capacitors are in the steady state. Especially for the compensation of fast and often changing loads like welding machines, cranes or elevators this fact is very advantageous, because no additional disturbances are caused by this switching method. Depending on the dynamic of the algorithms used for the power calculation, the RPC is able to compensate the reactive power of a load within the time of one fundamental period.

4.2. Active Power Filter

As a harmonic producing load, a three-phase diode rectifier with a resistive-inductive load was chosen ($R = 50$ Ω, $L = 0.7$ mH). Measurements of different compensation states are depicted in Fig. 9. The grid current without compensation is shown in Fig. 9a. Due to the small inductance, there are rapid changes in the grid currents, which are causing rapid changes in the reference currents of the APF. While the commutation of the current from one phase to the other, the reference currents are changing by about 40 % of $i_{APF,max}$ within two sampling times. Fig. 9b is showing the grid current while compensating with PI control. Due to the restricted dynamic of the control and the control dead time, big glitches appear when the grid current is changing rapidly. By using the predictive controller in addition to the PI control, the control dead time can be compensated and by this the glitches can be significantly reduced (Fig. 9c).

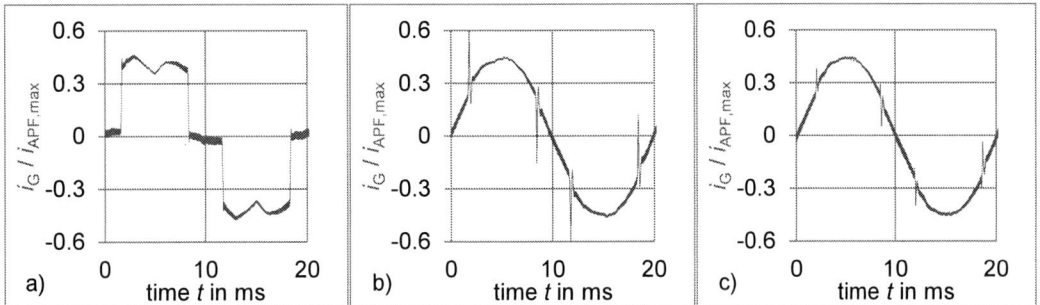

Fig. 9. Measured grid current i_G: a) uncompensated, THD_i = 28.8 %; b) compensated with PI control, THD_i = 8.8 %; c) compensated with PI and predictive control; THD_i = 4.7 %; measurement taken with DEWE-2600, sampling rate 200 kS/s, no filter for THD calculation, filter for depiction: 10 kHz

The glitches cannot be compensated completely, because there are limitations in the changing rate of the APF current caused by the inductance of the L-filter and the limited DC link voltage.
In addition, the L-filter cannot be modelled precisely. On the one hand, changes in the inductance and resistance related to e.g. temperature changes should be taken into account. On the other hand, grid parameters, which are variable and not known exactly, should be considered in the model.

5. Summary

The paper presents an optimized method for switching compensation capacitors to the grid. This is done by precharging the compensation capacitors to their maximum positive voltage in the steady state and switching them to the grid at the time of the positive maximum of the PCC voltage. Thus, the circuit is immediately in the steady state. Inrush and transient currents can be avoided completely. The functionality of this method was proven by simulation and measurement.
Furthermore, an easy to implement control structure for active power filters with low computational effort and good performance is presented. The predictive control is based on an inverse model of the voltage source inverter and the L-filter and is compensating the delay caused by the control dead time.

6. Acknowledgements

This work was funded by the Baden-Württemberg Ministry of Science, Research and the Arts (MWK) as part of the program „Leistungsorientierte Förderung des akademischen Mittelbaus für Forschungsgruppen an Hochschulen für angewandte Wissenschaften in Baden-Württemberg".

7. References

[1]   Alcaide, V.; Goldstrass, P., "Compensation of fast changing loads," in Electrical Power Quality and Utilisation, 2007. EPQU 2007. 9th International Conference on

[2]   Just, W.; Hofmann, W., "Blindstromkompensation in der Betriebspraxis", VDE Verlag, 2003

[3] Mattavelli, P.; Fasolo, S., "A closed-loop selective harmonic compensation for active filters," in Applied Power Electronics Conference and Exposition, 2000. APEC 2000. Fifteenth Annual IEEE , vol. 1, pp. 399-405, 2000

[4] Newman, M.J.; Zmood, D.N.; Holmes, D.G., "Stationary frame harmonic reference generation for active filter systems," in Industry Applications, IEEE Transactions on , vol. 38, no. 6, pp. 1591-1599, Nov/Dec 2002

[5] Xiaoming Yuan; Allmeling, J.; Merk, W.; Stemmler, H., "Stationary frame generalized integrators for current control of active power filters with zero steady state error for current harmonics of concern under unbalanced and distorted operation conditions," in Industry Applications Conference, 2000. Conference Record of the 2000 IEEE , vol. 4, pp.2143-215, Oct 2000

[6] Chuan Xie; Chao He; Hui Yan; Guozhu Chen; Hua Yang, "Digital Generalized Integrators of current control for three-phase Active Power filter with selective harmonic compensation," in Applied Power Electronics Conference and Exposition (APEC), 2012 Twenty-Seventh Annual IEEE , pp.748-753, 5-9 Feb. 2012

[7] Blaabjerg, F.; Teodorescu, R.; Liserre, M.; Timbus, A.V., "Overview of Control and Grid Synchronization for Distributed Power Generation Systems," in Industrial Electronics, IEEE Transactions on , vol. 53, no. 5, pp. 1398-1409, Oct. 2006

[8] Routimo, M.; Salo, M.; Tuusa, H., "A novel simple prediction based current reference generation method for an active power filter," in Power Electronics Specialists Conference, 2004. PESC 04. 2004 IEEE 35th Annual , vol.4, pp.3215-3220, 2004

Active Mains Filters with Combined Feed-Forward and Feed-Back Control

F. A. Himmelstoss, University of Applied Science Technikum Wien, AUSTRIA,
himmelstoss@technikum-wien.at
K. H. Edelmoser, Institute of Electrical Drives and Machines, Technical University Vienna,
AUSTRIA, kedel@pop.tuwien.ac.at

Abstract

An active filter concept to reduce disturbances of the mains is presented and converters for this purpose are described. The input voltage is processed with AC/AC converter structures based on different buck, boost and buck-boost topologies. The system can be used as active filters for sensitive loads, but can also be used to produce disturbances superimposed on a harmonic mains voltage for test systems. The control is done by a combination of a feed-forward and a feed-back control. The whole system is explained, the converter topologies are described, the control laws are deduced, and dimensioning hints are given.

1. Introduction

Disturbances in the mains can in some cases lead to perturbations in the load. Passive filters made of large and heavy inductors and capacitors can reduce the influence of the disturbances on the mains supply. Active filters with semiconductor switches, however, lead to a more compact design and additionally can control the power flow. There exists an extensive literature for active and hybrid filters e.g. [1] till [8].

The figures Fig. 1 a...d show several basic concepts for the power part of the active filter. They are shown for one phase, by duplicating them they can be used for more-phase systems. The mains supply is indicated by U_{11} and the load is connected to the output U_{21}. The switches S_{11} and S_{21} are voltage bidirectional switches realized with active and passive switches. The converter topologies are for example the buck-boost structure with an input filter (Fig. 1.a), a Cuk converter (Fig. 1.b), a buck converter with input filter (Fig. 1.c), and a special boost converter (Fig. 1.d). So we have three types of filters. Type 1 is based on the buck converter and can only reduce the input voltage, type 2 is based on the step-up concept and produces a higher output voltage, and type 3 uses buck-boost structures and enables us to produce an output voltage which is higher or lower than the input voltage.

2. Control Basics

With a harmonic input voltage $u_1(t) = \hat{U}_1 \cdot \cos \omega t$ and with d as the duty cycle of switch S_{11} (S_{21} is controlled in push-pull mode with the inverse duty cycle), the output voltage of the converter for the different types is

$$\text{Type 1: } u_2(t) = \hat{U}_1 \cdot d(t) \cdot \cos \omega t \ ,$$

Type 2: $u_2(t) = \hat{U}_1 \cdot \dfrac{1}{1-d(t)} \cdot \cos\omega t$,

Typ 3: $u_2(t) = \hat{U}_1 \cdot \dfrac{d(t)}{1-d(t)} \cdot \cos\omega t$.

Fig. 2 shows some possibilities to realize the voltage bidirectional switch.

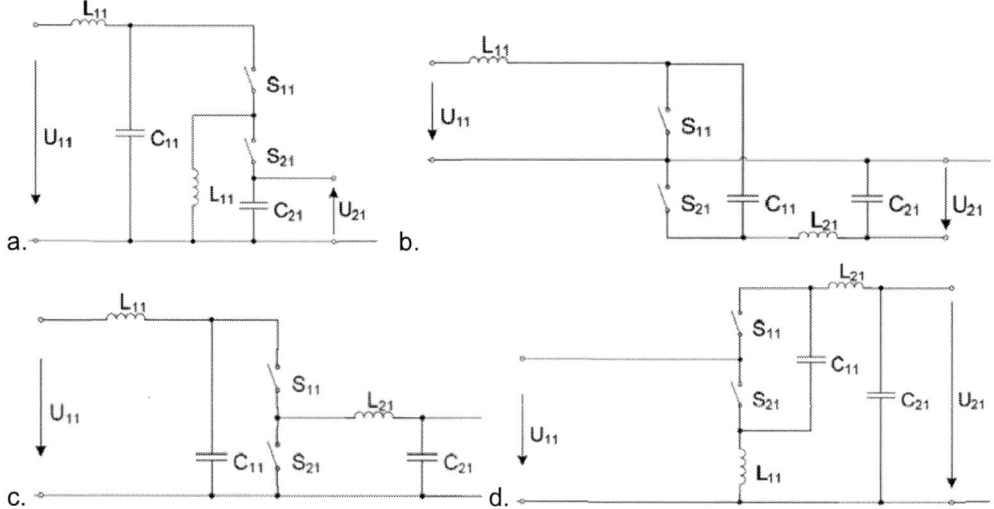

Fig. 1. Circuit diagrams for the power part of the active filter based on the (a) buck-boost, (b) Cuk, (c) buck, and (d) on a special boost converter

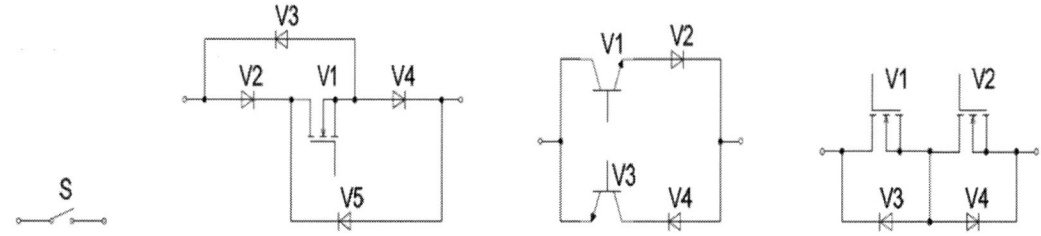

Fig. 2. Several realizations of the voltage bidirectional switch S

It is therefore possible to influence the output voltage with the duty cycle. A constant duty cycle only changes the input voltage according to a fixed value, but by changing the duty cycle the form of the output voltage can be different in relation to the input voltage. The converter can also be used to produce disturbances on a sinusoidal waveform e.g. for a test equipment.

The active filter works as depicted in the block diagram Fig. 3. The input voltage U_1 is reduced by a voltage sensor or by a voltage divider (e.g. with the factor 100) to get an image of the input voltage for the control electronics. With a phased-locked-loop PLL a clean harmonic voltage of the same frequency and phase as the mains supply is produced. The superimposed control system delivers the reference amplitude value. With a multiplier the

reference signal $u_{2,REF}(t) = \hat{U}_2 \cdot \cos\omega t$ is produced.

In the next block the duty ratio is calculated corresponding to the filter type and the input voltage according to

Type 1: $d(t) = \hat{U}_2 \cdot \dfrac{\cos \omega t}{u_1(t)}$,

Type 2: $d(t) = \dfrac{\hat{U}_2 \cdot \cos \omega t - u_1(t)}{\hat{U}_2 \cdot \cos \omega t}$,

Type 3: $d(t) = \dfrac{\hat{U}_2 \cdot \cos \omega t}{\hat{U}_2 \cdot \cos \omega t + u_1(t)}$.

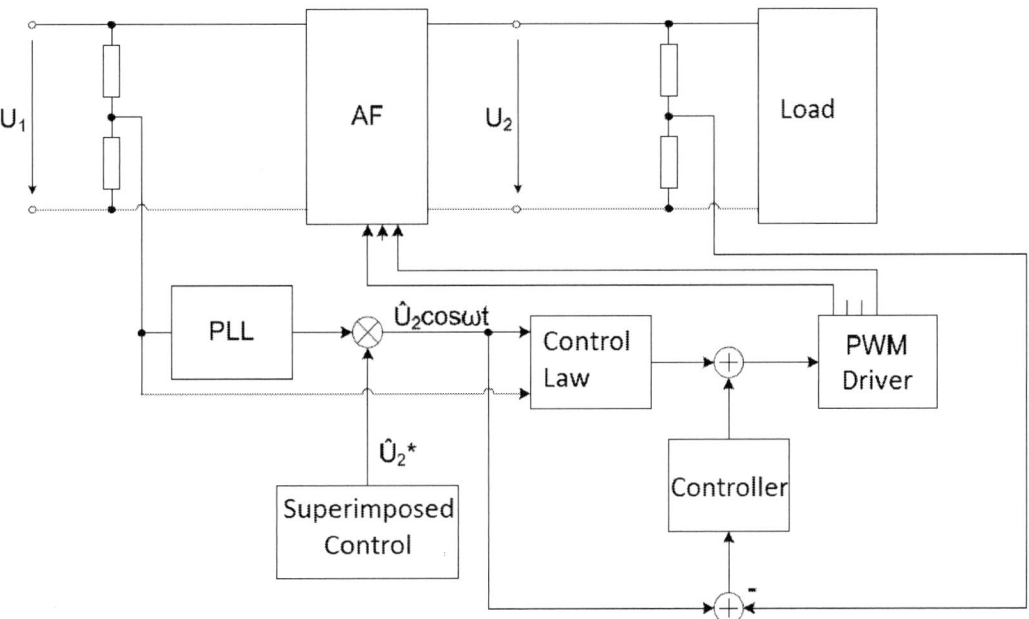

Fig. 3. Control concept of the active filter system

When an additional feed-back control is used to correct the error of the feed-forward control, a summing point is necessary to combine the two signals. The block PWM Driver produces the control signals for the active switches of the power section AF. At the output of the power section a voltage divider is drawn to match the control electronics. This signal is subtracted from the reference value to get the error which is processed by the feed-back controller. This controller only corrects the inaccuracy of the feed-forward control and contributes to the resulting duty cycle only by a small fraction.

3. Modelling

We show the derivation of the model for the converter according Fig. 1.c. This converter has an input filter to avoid disturbances of the mains due to the switching process. The output filter produces a clean output voltage. As the circuit is based on the buck converter, the output voltage has to be lower than the input voltage. To achieve a precise model we have to include the loss resistors of the components. We use R_{L1}, R_{L2} for the resistors of the coils L_1 and L_2, R_{C1} and R_{C2} for the equivalent series resistors of the capacitors C_1 and C_2, R_{S1} and R_{S2} for the on-resistors of the active switches S_1 and S_2. The switching transient of the switches is very fast and can be neglected. First the two switching states are described by state-space descriptions. These two descriptions can be time averaged to get the large signal model. Using the disturbance approach the linear small signal model can be achieved.

3.1. Large signal model of the converter

The converter has two stages. Either S_1 is conducting and S_2 is turned off, or it is the other way around. For stage 1 S_1 is turned on (and S_2 is turned off) we get

$$
\frac{d}{dt}\begin{pmatrix} i_{L1} \\ i_{L2} \\ u_{C1} \\ u_{C2} \end{pmatrix} = \begin{bmatrix} -\dfrac{R_{C1}+R_{L1}}{L_1} & \dfrac{R_{C1}}{L_1} & -\dfrac{1}{L_1} & 0 \\ \dfrac{R_{C1}}{L_2} & -\dfrac{R_{C1}+R_{L2}+R_{S1}+R//R_{C2}}{L_2} & \dfrac{1}{L_2} & -\dfrac{R}{L_2(R+R_{C2})} \\ \dfrac{1}{C_1} & -\dfrac{1}{C_1} & 0 & 0 \\ 0 & \dfrac{R}{C_2(R+R_{C2})} & 0 & -\dfrac{1}{C_2(R+R_{C2})} \end{bmatrix} \begin{pmatrix} i_{L1} \\ i_{L2} \\ u_{C1} \\ u_{C2} \end{pmatrix} + \begin{bmatrix} \dfrac{1}{L_1} \\ 0 \\ 0 \\ 0 \end{bmatrix} (u_1).
$$

For stage 2 S_2 is turned on (and S_1 is turned off) we get

$$
\frac{d}{dt}\begin{pmatrix} i_{L1} \\ i_{L2} \\ u_{C1} \\ u_{C2} \end{pmatrix} = \begin{bmatrix} -\dfrac{R_{C1}+R_{L1}}{L_1} & 0 & -\dfrac{1}{L_1} & 0 \\ 0 & -\dfrac{R_{L2}+R_{S2}+R//R_{C2}}{L_2} & 0 & -\dfrac{R}{L_2(R+R_{C2})} \\ \dfrac{1}{C_1} & 0 & 0 & 0 \\ 0 & \dfrac{R}{C_2(R+R_{C2})} & 0 & -\dfrac{1}{C_2(R+R_{C2})} \end{bmatrix} \begin{pmatrix} i_{L1} \\ i_{L2} \\ u_{C1} \\ u_{C2} \end{pmatrix} + \begin{bmatrix} \dfrac{1}{L_1} \\ 0 \\ 0 \\ 0 \end{bmatrix} (u_1).
$$

Weighting the two systems by their duration, leads to the large signal model of the converter (d is the duty cycle of switch S_1):

$$
\frac{d}{dt}\begin{pmatrix} i_{L1} \\ i_{L2} \\ u_{C1} \\ u_{C2} \end{pmatrix} = \begin{bmatrix} -\dfrac{R_{C1}+R_{L1}}{L_1} & \dfrac{R_{C1}d}{L_1} & -\dfrac{1}{L_1} & 0 \\ \dfrac{R_{C1}d}{L_2} & -\dfrac{R_{C1}d+R_{L2}+R_S+R//R_{C2}}{L_2} & \dfrac{d}{L_2} & -\dfrac{R}{L_2(R+R_{C2})} \\ \dfrac{1}{C_1} & -\dfrac{d}{C_1} & 0 & 0 \\ 0 & \dfrac{R}{C_2(R+R_{C2})} & 0 & -\dfrac{1}{C_2(R+R_{C2})} \end{bmatrix} \begin{pmatrix} i_{L1} \\ i_{L2} \\ u_{C1} \\ u_{C2} \end{pmatrix} + \begin{bmatrix} \dfrac{1}{L_1} \\ 0 \\ 0 \\ 0 \end{bmatrix} (u_1).
$$

3.2. Small signal model of the converter

With the disturbance approach (capital letters and a zero in the index represent the working point values and small letters with a small roof over them represent the disturbances) we get the small signal model of the converter. The variables f (inductor currents, voltage across the capacitors, duty cycle, and input voltage $i_{L1}, i_{L2}, u_{C1}, u_{C2}, d, u_1$) are represented by

$$ f = F_0 + \hat{f} \ . $$

With the small signal model the linear control theory can be used (e.g. transfer functions, Bode plots).

$$\frac{d}{dt}\begin{pmatrix}\hat{i}_{L1}\\\hat{i}_{L2}\\\hat{u}_{C1}\\\hat{u}_{C2}\end{pmatrix}=\begin{bmatrix}-\dfrac{R_{C1}+R_{L1}}{L_1} & \dfrac{R_{C1}D_0}{L_1} & -\dfrac{1}{L_1} & 0\\[2mm]\dfrac{R_{C1}D_0}{L_2} & -\dfrac{R_{C1}D_0+R_{L2}+R_S+R/\!/R_{C2}}{L_2} & \dfrac{D_0}{L_2} & -\dfrac{R}{L_2(R+R_{C2})}\\[2mm]\dfrac{1}{C_1} & -\dfrac{D_0}{C_1} & 0 & 0\\[2mm]0 & \dfrac{R}{C_2(R+R_{C2})} & 0 & -\dfrac{1}{C_2(R+R_{C2})}\end{bmatrix}\begin{pmatrix}\hat{i}_{L1}\\\hat{i}_{L2}\\\hat{u}_{C1}\\\hat{u}_{C2}\end{pmatrix}+\begin{bmatrix}\dfrac{1}{L_1} & \dfrac{R_{C1}I_{L20}}{L_1}\\[2mm]0 & \dfrac{R_{C1}(I_{L10}+I_{L20})+U_{C10}}{L_2}\\[2mm]0 & -\dfrac{I_{L20}}{C_1}\\[2mm]0 & 0\end{bmatrix}\begin{pmatrix}\hat{u}_1\\\hat{d}_1\end{pmatrix}$$

The working point duty cycle D_0 can be calculated by the simple control law

$$D_0=\hat{U}_2\cdot\frac{\cos\omega t}{u_1(t)}.$$

4. Dimensioning

4.1. Converter according to Fig. 1.c

We discuss the dimensioning of the converter according to Fig. 1.c. The input low-pass filter consisting of L_1 and C_1 must be dimensioned in such a way that the switching frequency is blocked from the mains supply. If the resonance frequency is about one decade lower than the switching frequency, the switching frequency is damped by -40 dB and has an amplitude of only one percent. With a chosen frequency f and a chosen inductor value for L_1, the necessary capacitor is

$$C_1=\frac{1}{4\pi^2L_1f^2}.$$

The characteristic impedance of the filter is

$$Z=\sqrt{\frac{L_1}{C_1}}.$$

The transfer function of the output filter can be determined to

$$G(s)=\frac{1}{s^2L_2C_2+\dfrac{L_2}{R}s+1}.$$

With the Bode nominal form

$$G(s)=\frac{1}{\dfrac{s^2}{\omega_0^2}+\dfrac{2d}{\omega_0}s+1}$$

the characteristic angular frequency of the asymptotic approximation and the damping can be calculated.

Fig. 4 shows exemplarily the voltage across C_1 (input filter voltage) and across C_2 (output voltage) with a constant duty cycle. (L_1 = 100 µH, L_2 = 150 µH, C_1 = 10 µF, C_2 = 33 µF)
There are nearly no disturbances even around the zero crossing of the input voltage.

PCIM Europe 2016, 10 – 12 May 2016, Nuremberg, Germany

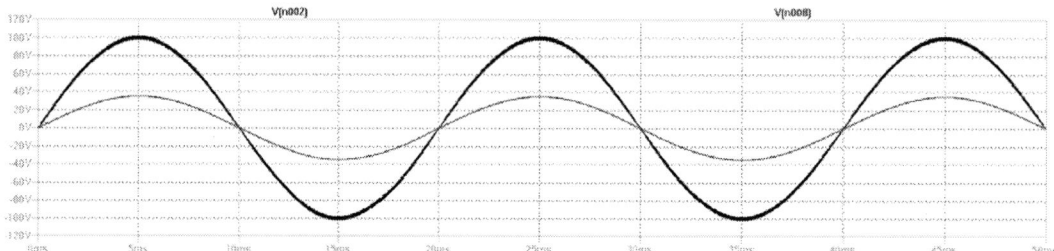

Fig. 4. Converter Fig.1.c: voltage across input filter (black) and output voltage (blue)

4.2. Converter according to Fig.1.a

The converter according to Fig. 1.a. is a buck-boost converter with an input filter. The input filter reduces the backlash to the input source considerably. It can be designed in the same way as for the converter according to Fig. 1.c. The output capacitor has to feed the load during the time when the inductor L_2 is charged. Therefore it must be chosen larger.

Fig. 5 shows exemplarily the voltage across C_1 (input filter voltage) and across C_2 (output voltage) with a constant duty cycle for the converter according to Fig.1.a. ($L_1 = 100$ µH, $L_2 = 150$ µH, $C_1 = 10$ µF, $C_2 = 33$ µF). The plot shows also the inverting function of the converter.

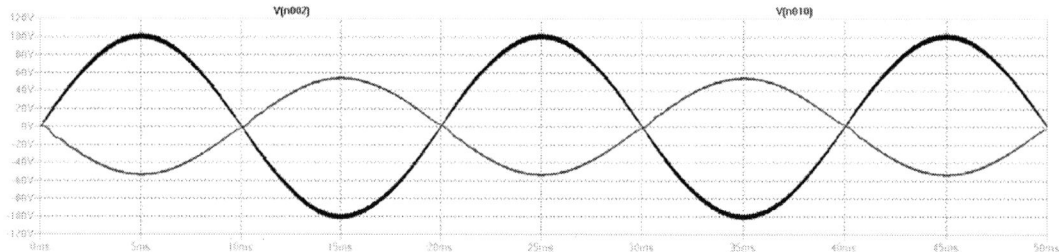

Fig. 5. Converter Fig.1.a: voltage across input filter (black) and output voltage (blue)

4.3. Converter according to Fig.1.b

The converter according to Fig. 1.b. is based on a Cuk converter. The Cuk converter is also an inverting converter and has the advantage of continuous input and output current. The output stage is identical to that of converter 1.c. The input inductor reduces the backlash to the input source. It can be designed in a similar way as for the converter according Fig. 1.c (L_1, C_1).

Fig. 6 shows exemplarily the voltage across C_1 (coupling capacitor) and across C_2 (output voltage) with a constant duty cycle for the converter according to Fig.1.b. ($L_1 = 100$ µH, $L_2 = 150$ µH, $C_1 = 10$ µF, $C_2 = 33$ µF).

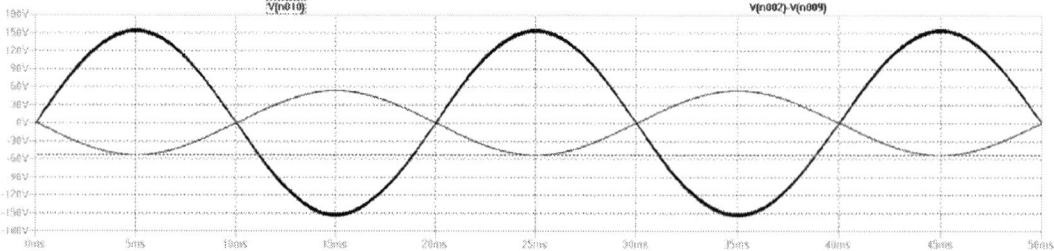

Fig. 6. Converter Fig.1.b: voltage across coupling capacitor (black) and output voltage (blue)

© VDE VERLAG GMBH · Berlin · Offenbach

4.4.   Converter according to Fig.1.d

The converter according to Fig. 1.d is an interesting kind of a boost converter. It has an output filter and the input current is continuous. The input current is always the sum of the two inductor currents independent of which AC-switch is on or off.

Fig. 7 shows exemplarily the voltage across C_1 (capacitor in parallel to the two ac-switches) and across C_2 (output voltage) with a constant duty cycle for the converter according to Fig.1.d. ($L_1 = 100\ \mu H$, $L_2 = 150\ \mu H$, $C_1 = 10\ \mu F$, $C_2 = 33\ \mu F$).

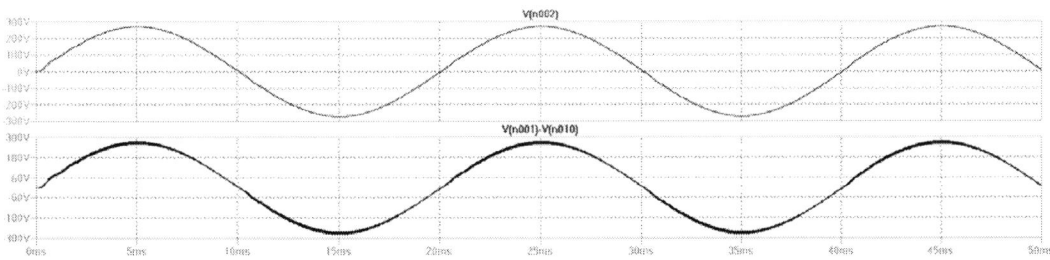

Fig. 7.   Converter Fig.1.d: voltage across capacitor in parallel to the ac-switches (black, lower trace) and output voltage (blue, upper trace)

5. Conclusion

Systems for AC/AC conversion were described. Instead of simple choppers, four kinds of converters are shown. The converter transforms the energy and additionally works as a filter. The systems can be feed-forward controlled by a control law. If a very precise output voltage has to be achieved (accuracy better than 5 %), an additional feed-back controller can be implemented. This controller only has to correct the error due to the not exact control law.

References

[1]   N. Mohan, T. Undeland and W. Robbins: Power Electronics, Converters, Applications and Design, 3nd ed. New York: W. P. John Wiley & Sons, 2003.

[2]   F. Zach: Power Electronics, in German: Leistungselektronik, Wien: Springer, 4th ed., 2010.

[3]   Yuriy Rozanov, Sergey Ryvkin, Evgeny Chaplygin, Pavel Voronin, "Power Electronics Basics," CRC Press, 2016.

[4]   C.-S. Lam, M.-C. Wong, W.-H. Choi, X.-X. Cui, H.-M. Mei, and J.-T. Liu, "Design and Performance of an Adaptive Low-DC-Voltage-Controlled LC-Hybrid Active Power Filter with a Neutral Inductor in Three-Phase Four-Wire System," *IEEE Trans. Indust. Electronics*, vol. 61, no. 6, pp.2635-2647, June 2014.

[5]   P. Dang, T. Ellinger, and J. Petzold, "Dynamic Interaction Analysis of APF Systems," *IEEE Trans. Indust. Electronics*, vol. 61, no. 9, pp. 4467-4473, Sept. 2014.

[6]   F. Himmelstoss, "Aktive Netzfilter," Öst. Patentamt, AT 505460 B1, filed July 2007.

[7]   S. Ben-Yaakov, Y. Hadad, and N. Diamantstein, "A Four Quadrants HF AC Chopper with no Deadtime," *IEEE Applied Power Electronics Conference and Exposition, APEC '06*, pp. 1461-1465, 2006.

[8]   Z. Fedyczak, R. Strzelecki, G Benysek, "Singlephase PWM AC/AC Semiconductor Transformer Topologies and Applications," Conf. Rec. on Power Electronics Specialists Conf. 2002, pp. 1048 – 1053.

© VDE VERLAG GMBH · Berlin · Offenbach

Appendix 1: Small signal models of the converters

Type according Fig. 1.a

$$\frac{d}{dt}\begin{pmatrix}\hat{i}_{L1}\\\hat{i}_{L2}\\\hat{u}_{C1}\\\hat{u}_{C2}\end{pmatrix} = \begin{bmatrix} -\dfrac{R_{C1}+R_{L1}}{L_1} & \dfrac{R_{C1}D_0}{L_1} & -\dfrac{1}{L_1} & 0 \\[2mm] \dfrac{R_{C1}D_0}{L_2} & -\dfrac{R_{C1}D_0+R_{L2}+R_s+R//R_{C2}(1-D_0)}{L_2} & \dfrac{D_0}{L_2} & \dfrac{R(D_0-1)}{L_2(R+R_{C2})} \\[2mm] \dfrac{1}{C_1} & -\dfrac{D_0}{C_1} & 0 & 0 \\[2mm] 0 & \dfrac{R(1-D_0)}{C_2(R+R_{C2})} & 0 & -\dfrac{1}{C_2(R+R_{C2})} \end{bmatrix}\begin{pmatrix}\hat{i}_{L1}\\\hat{i}_{L2}\\\hat{u}_{C1}\\\hat{u}_{C2}\end{pmatrix} + \begin{bmatrix} \dfrac{1}{L_1} & \dfrac{R_{C1}I_{L20}}{L_1} \\[2mm] 0 & \dfrac{R_{C1}(I_{L10}-I_{L20})+R//R_{C2}\cdot I_{L20}+U_{C10}-\dfrac{RU_{C20}}{R+R_{C2}}}{L_2} \\[2mm] 0 & -\dfrac{I_{L20}}{C_1} \\[2mm] 0 & -\dfrac{RI_{L20}}{C_2(R+R_{C2})} \end{bmatrix}\begin{pmatrix}\hat{u}_1\\\hat{d}_1\end{pmatrix}$$

Type according Fig. 1.b

$$\frac{d}{dt}\begin{pmatrix}\hat{i}_{L1}\\\hat{i}_{L2}\\\hat{u}_{C1}\\\hat{u}_{C2}\end{pmatrix} - \begin{bmatrix} -\dfrac{R_{C1}(1-D_0)+R_{L1}+R_S}{L_1} & \dfrac{R_S}{L_1} & \dfrac{D_0-1}{L_1} & 0 \\[2mm] -\dfrac{R_S}{L_2} & -\dfrac{R_{C1}D_0+R_{L2}+R_S+R//R_{C2}}{L_2} & \dfrac{D_0}{L_2} & -\dfrac{R}{L_2(R+R_{C2})} \\[2mm] \dfrac{1-D_0}{C_1} & -\dfrac{D_0}{C_1} & 0 & 0 \\[2mm] 0 & \dfrac{R}{C_2(R+R_{C2})} & 0 & -\dfrac{1}{C_2(R+R_{C2})} \end{bmatrix}\begin{pmatrix}\hat{i}_{L1}\\\hat{i}_{L2}\\\hat{u}_{C1}\\\hat{u}_{C2}\end{pmatrix} + \begin{bmatrix} \dfrac{1}{L_1} & \dfrac{R_{C1}I_{L10}+U_{C10}}{L_1} \\[2mm] 0 & \dfrac{R_{C1}I_{L20}+U_{C10}}{L_2} \\[2mm] 0 & -\dfrac{I_{L10}+I_{L20}}{C_1} \\[2mm] 0 & 0 \end{bmatrix}\begin{pmatrix}\hat{u}_1\\\hat{d}_1\end{pmatrix}$$

Type according Fig. 1.d

$$\frac{d}{dt}\begin{pmatrix}\hat{i}_{L1}\\\hat{i}_{L2}\\\hat{u}_{C1}\\\hat{u}_{C2}\end{pmatrix} = \begin{bmatrix} -\dfrac{R_{C1}(1-D_0)+R_{L1}+R_S}{L_1} & -\dfrac{R_S}{L_1} & \dfrac{D_0-1}{L_1} & 0 \\[2mm] -\dfrac{R_S}{L_2} & -\dfrac{R_{C1}D_0+R_{L2}+R_{S2}+R//R_{C2}}{L_2} & \dfrac{D_0}{L_2} & -\dfrac{R}{L_2(R+R_{C2})} \\[2mm] \dfrac{1-D_0}{C_1} & -\dfrac{D_0}{C_1} & 0 & 0 \\[2mm] 0 & \dfrac{R}{C_2(R+R_{C2})} & 0 & -\dfrac{1}{C_2(R+R_{C2})} \end{bmatrix}\begin{pmatrix}\hat{i}_{L1}\\\hat{i}_{L2}\\\hat{u}_{C1}\\\hat{u}_{C2}\end{pmatrix} + \begin{bmatrix} \dfrac{1}{L_1} & \dfrac{R_{C1}I_{L10}+U_{C10}}{L_1} \\[2mm] 0 & \dfrac{-R_{C1}(I_{L20})+U_{C10}}{L_2} \\[2mm] 0 & -\dfrac{I_{L10}+I_{L20}}{C_1} \\[2mm] 0 & 0 \end{bmatrix}\begin{pmatrix}\hat{u}_1\\\hat{d}_1\end{pmatrix}$$

PCIM Europe 2016, 10 – 12 May 2016, Nuremberg, Germany

Influence of the zero sequence voltage on the design of a series active filter

Andreas Reinhold [1], Uwe Rädel [2], Rolf Grohmann [1], Jürgen Petzoldt [2]

[1] HTWK Leipzig, University of Applied Sciences, Faculty Electrical Engineering and Information Technology, POBOX 301166, D-04251 Leipzig, Germany

[2] Technische Universitaet Ilmenau, Department of Electrical Engineering and Information Technology, POBOX 100565, D-98684 Ilmenau, Germany

Abstract

This paper discusses a series active power filter (APF) for a three-phase diode rectifier. The series APF reduces the AC side harmonic current which is drawn by the rectifier and generates a constant DC current. The design of the APF depends on the waveform of the compensation voltage which is injected into the network. The paper demonstrates that the compensation voltage can be changed significantly through a zero sequence voltage without influencing the functioning of the filter and the rectifier. Therefore, the zero sequence voltage is a degree of freedom in the design of the APF. Simulation results are shown.

The proposed series active power filter

Fig. 1 shows the proposed topology of the transformerless series APF. The series APF consists of three voltage source inverters (VSI) and requires three isolated DC link capacitors [1]. The topology of the filter corresponds to a full bridge sub-module of a modular multilevel converter (MMC) [2, 3].

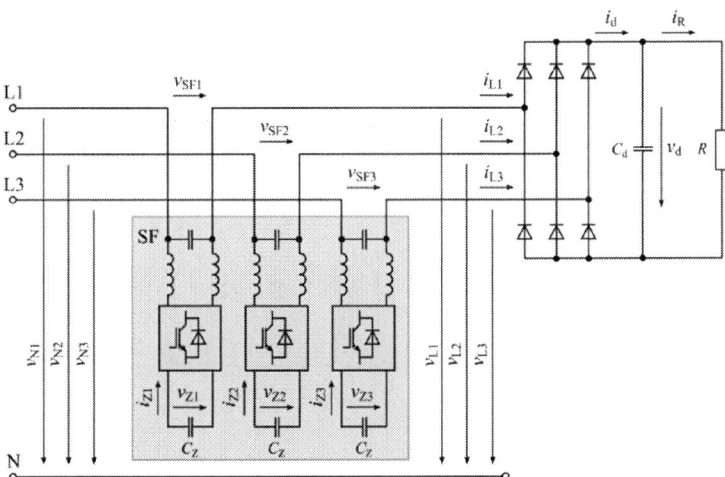

Fig. 1: Basic block diagram of the proposed transformerless series active power filter

The desired current and voltage waveforms

Diode rectifiers draw fundamental reactive current and harmonic current in the ac mains. These currents can be reduced and modified with a series APF [4-6]. The series APF acts as a controlled voltage source to achieve a constant dc-voltage at the rectifier. An inner control loop of the series APF controls its load current i_L to achieve the desired waveform [4].

Fig. 2 shows the investigated reference load current waveform i_L. The current waveform converts into a constant DC current i_d at the rectifier. Under steady-state conditions the

© VDE VERLAG GMBH · Berlin · Offenbach

DC voltage v_d is assumed to be constant (Eq. 2). This is the case when the currents i_d and i_R are equal (Eq. 1).

$$i_d(t) = i_R(t) = I_d = const. \tag{1}$$

$$v_d(t) = V_d + \tilde{v}_d(t) = V_d + \frac{1}{C_d}\int_0^t (i_d(\tau) - i_R(\tau))\,d\tau = V_d = const. \tag{2}$$

By setting the reference current waveform and with the assumptions (1) and (2) the waveform of the load voltage is not clearly defined. Fig. 2 shows the basic voltage waveform $v`_L$ which is investigated within this paper.

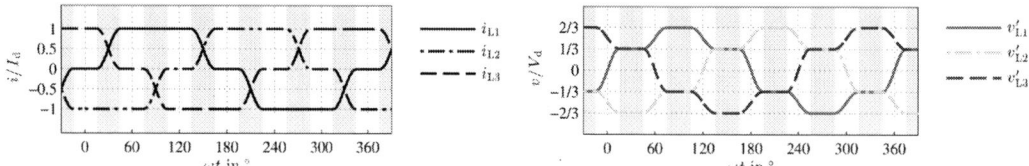

Fig. 2: The reference waveforms: load current i_L (left) and load voltage $v`_L$ without zero sequence voltage (right)

The zero sequence voltage

In modular multilevel converter topologies, it is possible to reduce the voltage ripple of the sub-modules by injecting circulating currents [7, 8]. Here, this solution is not possible because the zero sequence component of load current does not exist (Eq. 3). However, there is the possibility to inject a zero sequence voltage with the series APF.

$$i_{L1}(t) + i_{L2}(t) + i_{L3}(t) = 0 \tag{3}$$

The basic load voltage waveform $v`_L$, which is shown in Fig (2), has no zero sequence component.

$$v`_{L1}(t) + v`_{L2}(t) + v`_{L3}(t) = 0 \tag{4}$$

Notably, the voltage $v`_L$ can be modified by adding a zero sequence voltage v_0 without influencing the performance of the filter and the rectifier.

$$v_{L1}(t) = v`_{L1}(t) + v_0(t) \tag{5a}$$
$$v_{L2}(t) = v`_{L2}(t) + v_0(t) \tag{5b}$$
$$v_{L3}(t) = v`_{L3}(t) + v_0(t) \tag{5c}$$

However, this zero sequence component influenced the waveform of the compensation voltage v_{SF} which is injected by the series APF.

$$v_{SF1}(t) = v_{N1}(t) - v_{L1}(t) = v_{N1}(t) - v`_{L1}(t) - v_0(t) \tag{6a}$$
$$v_{SF2}(t) = v_{N2}(t) - v_{L2}(t) = v_{N2}(t) - v`_{L2}(t) - v_0(t) \tag{6b}$$
$$v_{SF3}(t) = v_{N3}(t) - v_{L3}(t) = v_{N3}(t) - v`_{L3}(t) - v_0(t) \tag{6c}$$

For the following considerations, the voltage v_{SF} should be decomposed into the voltage without a zero sequence component $v\grave{}_{SF}$ and the zero sequence voltage v_0.

$$v_{SF1}(t) = v\grave{}_{SF1}(t) - v_0(t) \tag{7a}$$
$$v_{SF2}(t) = v\grave{}_{SF2}(t) - v_0(t) \tag{7b}$$
$$v_{SF3}(t) = v\grave{}_{SF3}(t) - v_0(t) \tag{7c}$$

The zero sequence voltage is a degree of freedom in the design of the series APF. The next two sections show that with an appropriate waveform of the zero sequence voltage the required capacitor value C_Z or the required voltage V_Z of the DC link circuit can be reduced.

Reducing the required capacitor value

The required value C_Z of the DC link capacitor of the series APF can be reduced by the following approach. The required capacitor value can be calculated as [9]:

$$C_Z = \frac{\Delta w_Z}{(V_{Z,min} + 0.5\Delta v_Z)\Delta v_Z} \tag{8}$$

The equation shows that the capacitor value is proportional to the energy deviation Δw_Z and inversely proportional to the allowed voltage deviation Δv_Z.

The stored energy of the capacitor can be decomposed into the average and alternating component $\overline{w}_Z$ and $\widetilde{w}_Z$. The energy deviation is defined by the difference between the maximum and the minimum value of alternating energy.

$$w_Z(t) = \overline{w}_Z + \widetilde{w}_Z(t) \tag{9}$$

$$\Delta w_Z = \widetilde{w}_{Z,max} - \widetilde{w}_{Z,min} \tag{10}$$

It is assumed that the series APF is lossless and the control of series APF generates no average power (Eq. 12). Based on this condition the alternating energy can be calculated with the instantaneous power:

$$\widetilde{w}_Z(t) = \int_0^t p_Z(\tau)\, d\tau \tag{11}$$

$$P_Z = \frac{1}{T}\int_0^T p_Z(t)\, dt = 0 \tag{12}$$

If the passive filter of the series APF is neglected, the DC side power p_Z can be calculated from the product of the AC side current and voltage i_L and v_{SF}.

$$p_{Z1}(t) = p_{SF1}(t) = v_{SF1}(t)i_{L1}(t) \tag{13a}$$
$$p_{Z2}(t) = p_{SF2}(t) = v_{SF2}(t)i_{L2}(t) \tag{13b}$$
$$p_{Z3}(t) = p_{SF3}(t) = v_{SF2}(t)i_{L3}(t) \tag{13c}$$

With the following approach a zero sequence voltage is calculated which reduces the instantaneous power. The zero sequence voltage influences all three series APFs at the same time. Therefore, the three instantaneous powers must be included in the calculation. The approach uses the method of least squares:

$$f(v_0) = \sum_{k=1}^{3} p_{Zk}^2(t) = p_{Z1}^2(t) + p_{Z2}^2(t) + p_{Z3}^2(t) \tag{14}$$

The zero sequence voltage v_{01} that minimizes the sum can be calculated with the derivation of the function:

$$\frac{\partial f(v_0)}{\partial v_0} = \frac{\partial}{\partial v_0} \sum_{k=1}^{3} (i_{Lk}(t)v_{SFk}(t))^2 = \frac{\partial}{\partial v_0} \sum_{k=1}^{3} i_{Lk}^2 (v`_{SFk}(t) - v_0(t))^2 = 0 \tag{15}$$

$$v_{01}(t) = v_0(t) = \frac{i_{L1}^2(t)v`_{SF1}(t) + i_{L2}^2(t)v`_{SF2}(t) + i_{L3}^2(t)v`_{SF3}(t)}{i_{L1}^2(t) + i_{L2}^2(t) + i_{L3}^2(t)} \tag{16}$$

Fig. 3 shows the zero sequence voltage and Fig. 4 shows the impact on the instantaneous power and thus on the energy deviation. The energy deviation Δw_Z in case II ($v_0 = v_{01}$) is considerably smaller than in case I ($v_0 = 0$) and therefore, the required value C_Z of the DC link capacitor can be reduced.

$$\Delta w_Z(v_0 = v_{01}) < \Delta w_Z(v_0 = 0) \tag{17}$$

Reducing the required capacitor voltage

The approach shown above results in an increase of the required DC link voltage V_Z (Fig. 3). In order to reduce this voltage the following approach can be used.

The DC link voltage V_Z must be at least as large as the maximum absolute value of the voltage v_{SF} which is injected by the series APF.

$$V_Z \geq V_{Z,\min}(v_0) = \max\{|v_{SF}(t)|\} = \max\{|v`_{SF}(t) - v_0(t)|\} \tag{18}$$

With the zero sequence voltage v_{02} this maximum value can be minimized for any instant of time. The equation (21) shows how the zero voltage is calculated. The voltages $v_{\min}$ and $v_{\max}$ are the minimum and the maximum values of the three voltages without zero sequence component $v`_{SF1}, v`_{SF2}$ and $v`_{SF3}$.

$$v_{\min}(t) = \min\{v`_{SF1}(t), v`_{SF2}(t), v`_{SF3}(t)\} \tag{19}$$

$$v_{\max}(t) = \max\{v`_{SF1}, v`_{SF2}, v`_{SF3}\} \tag{20}$$

$$v_{02}(t) = \frac{1}{2}(v_{\min}(t) + v_{\max}(t)) \tag{21}$$

PCIM Europe 2016, 10 – 12 May 2016, Nuremberg, Germany

Fig. 3 shows the zero sequence voltage and the impact on the maximum absolute value of the voltage v_{SF}. The required DC link voltage $V_{\mathrm{Z,min}}$ in case III ($v_0 = v_{02}$) is smaller than in case I and in case II. Because of this reduced value it is possible to install semiconductors which are designed for lower voltage.

$$V_{\mathrm{Z,min}}(v_0 = v_{02}) < V_{\mathrm{Z,min}}(v_0 = 0) \tag{22}$$

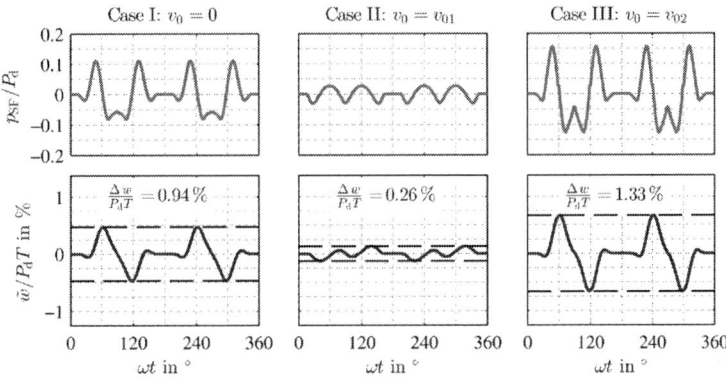

Fig. 3: Voltage waveforms for different zero sequence voltages v_0 and their impact on the required DC link voltage $V_{\mathrm{Z,min}}$ of the series APF

Fig. 4: Instantaneous power p_{SF} and energy deviation Δw_{Z} for different zero sequence voltages v_0

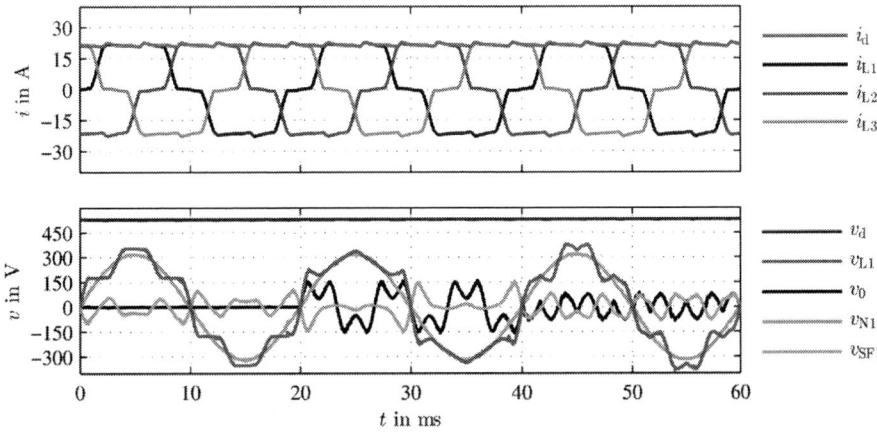

Fig. 5: System response to zero sequence voltage changes

© VDE VERLAG GMBH · Berlin · Offenbach

Simulation results

Matlab/Simulink was used as a simulation tool. Fig. 5 depicts the system response when the zero sequence voltage v_0 was changed. The voltage changed every 20ms. The above cases I ($v_0 = 0$), II ($v_0 = v_{01}$) and III ($v_0 = v_{02}$) are performed successively. The zero sequence voltage had no influence on the performance of the filter. The load current i_L and the load voltage v_L achieved the desired waveform while the load current remained the same even though the voltage was changed. Furthermore, the zero sequence voltage had no influence on the operations of the rectifier. The simulation proved that the voltage v_d and the current i_d at the rectifier can be kept nearly constant for the whole time.

Summary

The paper presents the influence of the zero sequence voltage on the design of a series active power filter. Simulation results show that the zero sequence voltage has no influence on the performance of the filter and therefore, the zero sequence voltage is a degree of freedom. Two different optimized zero sequence voltages are calculated. With the appropriate voltage, either the energy deviation or the voltage of the DC link of the series active power filter can be reduced. Therefore, a smaller DC link capacitor or semiconductors which are designed for lower voltage can be installed.

References

[1] J. G. Pinto, H. Carneiro, B. Exposto, C. Couto, and J. L. Afonso, "Transformerless series active power filter to compensate voltage disturbances," in *Power Electronics and Applications (EPE 2011), Proceedings of the 2011-14th European Conference on*, 2011, pp. 1-6.

[2] R. Marquardt, "Modular multilevel topologies: State of the art and new developments," in *PCIM Europe*, 2011.

[3] M. A. Perez, S. Bernet, J. Rodriguez, S. Kouro, and R. Lizana, "Circuit Topologies, Modeling, Control Schemes, and Applications of Modular Multilevel Converters," *Ieee Transactions on Power Electronics,* vol. 30, pp. 4-17, 2015.

[4] A. Reinhold, U. Raedel, R. Grohmann, and J. Petzoldt, "AC- and DC-Power Quality improvement of diode rectifiers due to parallel-series active filtering," in *PCIM Europe 2015; International Exhibition and Conference for Power Electronics, Intelligent Motion, Renewable Energy and Energy Management; Proceedings of*, 2015, pp. 1-6.

[5] M. Aredes and R. M. Fernandes, "A dual topology of Unified Power Quality Conditioner: The iUPQC," in *Power Electronics and Applications, 2009. EPE '09. 13th European Conference on*, 2009, pp. 1-10.

[6] B. W. Franca and M. Aredes, "Comparisons between the UPQC and its dual topology (iUPQC) in dynamic response and steady-state," in *IECON 2011 - 37th Annual Conference on IEEE Industrial Electronics Society*, 2011, pp. 1232-1237.

[7] R. Picas, J. Pou, S. Ceballos, J. Zaragoza, G. Konstantinou, and V. G. Agelidis, "Optimal injection of harmonics in circulating currents of modular multilevel converters for capacitor voltage ripple minimization," in *ECCE Asia Downunder (ECCE Asia), 2013 IEEE*, 2013, pp. 318-324.

[8] A. Perez-Basante, S. Ceballos, J. Pou, J. Pou, A. G. d. Muro, A. Pujana, and P. Ibanez, "SiC Modular Multilevel Converters: Sub-Module Voltage Ripple Analysis and Efficiency Estimations," in *PCIM Europe 2014; International Exhibition and Conference for Power Electronics, Intelligent Motion, Renewable Energy and Energy Management; Proceedings of*, 2014, pp. 1-8.

[9] J. Kolb, "Optimale Betriebsführung des Modularen Multilevel-Umrichters als Antriebsumrichter für Drehstrommaschinen," KIT Scientific Publishing, 2014.

© VDE VERLAG GMBH · Berlin · Offenbach

Electromagnetic Emissions in High Density and Fast GaN Switched Half Bridges with Resonance Filter Structures

W. Stocksreiter[1], H. List[1], H. Berger[1], G. Weis[1,2], M. Krenn[1], G. Engel[1,3]

Mail to: wolfgang.stocksreiter@fh-joanneum.at
[1]FH JOANNEUM, University of Applied Sciences, Werk-VI-Straße 46, 8605 Kapfenberg, Austria.
[2]AT&S Austria Technologie & Systemtechnik AG, 8700 Leoben Hinterberg, Fabriksgasse 13, Austria.
[3]CeraCap Technology & Innovation Consulting, Kapellenweg 38, 8430 Leibnitz, Austria.

Abstract:

The wide bandgap semiconductor GaN is a promising material for future extended use in high power electronic applications due its fast and low loss switching. For high-density power electronic systems with miniaturized system size, a balancing of the semiconductor switching speed, duty cycles ranges, frequency of switching, current density, ripple and overshoot voltages are necessary.

In order to assess the design rules of GaN switch circuits in connection with the supporting passive components, in particular the DC-link and filter capacitors, simultaneous measurements of the electromagnetic emissions and the waveforms of the voltage and currents in the capacitors and also in the switched node in a 650V GaN half bridge followed by a resonant filter structure were made. Results are obtained for different switching frequency ranges, mainly between 50 kHz and 1000 kHz, different interlock delay and capacitor technologies. The results are interpreted in terms of the correlation between the voltage and current waveforms and the emitted radiation for varying input voltage levels.

It was found that for the half bridge with RC filter – circuit without load, that film capacitors show higher amplitudes of voltage and current transients at high switching frequency, resulting in increased levels of EMI, than the ceramic ones. Also the interlock delay time setting in conjunction with duty cycle parameters and the rise time of the gate pulse showed considerable influence on EMI levels. For a LC combination in a buck converter structure it has been established that no optimized interlock delay time could be found where ripple currents and EMI emissions could be reduced while maintaining a fast rise at the GaN, independent of load. The results from a GaN – capacitor combination test circuit showed that clamping of gate voltage either passively, actively, or by topology is necessary for avoiding unwanted switching actions in fast GaN circuits.

Introduction:

Fast switching of semiconductors when used for high efficient power electronics circuit is accompanied with high change rate of currents. The inherent ripple current flowing through unavoidable inductances generates voltage overshoots that have to limited for reason of danger of breakthrough of all the components involved. Also, the change of accelerating charges is responsible for emission of radiation (EMI), which has to be kept within certain limits by design.

The relation between the ringing frequency of the current through the switches and the frequency spectrum of emission of radiation is discussed for silicon (Si) devices in general terms e.g. in Ref [1]. A fast change of both switch currents and voltages is considered to be the source of EMI. Also, larger reverse recovery currents increase the EMI levels, as is discussed for IGBT devices. The analysis in both the time- and the frequency domains is suggested. The use of passive RC or LC filters is suggested to reduce conducted EMI.

The corresponding recommendations are widely and successfully used today in Si – based power converters, and they are expected to be valid at least partly also for the new wide band gap semiconductors (SiC, GaN). However, the high speed and fast slope switching will result in new challenges.

On one hand, the extremely fast switching enables smaller and less expensive energy storage elements lower switching losses, thus increased efficiency and reduced cooling requirements. In addition, for GaN devices, for cascode arrangements, there are relatively simple drive requirements using Si MOSFETs, and anti-parallel diodes. On the other hand, there are specific difficulties encountered with the high frequencies and slopes. First to mention is the difficulty in the measurement methods, due to the requirement of fast current and voltage oscilloscopes, probes, and interconnects.

The need for adapted design rules is evident. Whereas in low frequency (e.g. 16kHz to 30kHz) Si- based devices, the effect of parasitics of gate drivers and passive components is well explored, not so for the higher frequencies. A resonance between the loop inductance and the transistor capacitance most often results in ringing. The capacitance is fixed, but has a parasitic inductance (ESL, equivalent series inductance), which adds to that of the interconnecting PCB planes and connectors. The higher the switching frequency and rising slopes, the more ringing is expected.

In fast SiC and GaN switches, there is very low reverse recovery current and thus very low effect on EMI from this cause, but on the other hand, any semiconductor device that switches in nano- or even picoseconds range is likely to generate large amounts of EMI due to inverse proportionality of EMI level to the timescale of rise time. There are only few reports on details of EMI found in the recent literature. It was shown e.g. in Ref. [2], that in a low voltage GaN system (step down in a half bridge from 10 V to 1.2 V), the voltage ringing at a frequency of 217 MHz was correlated to a conducted EMI resonance at 220 MHz, at levels that the author considered as problematic. It was concluded, that with appropriate design methods to lower the involved inductances, the ringing amplitude, and thus the EMI, will be lessened.

Further means to reduce ripple are mentioned in Ref.[3], where a two-phase interleaving is recommended to reduce current ripple and EMI noise.

The target of the present paper is to investigate the relation of the EMI levels of 650V – GaNs, switching in simple half bridge topology with different switching frequencies from 50 kHz to 1 MHz, interlock delay settings, and capacitor technologies.

Since the inductance is composed by the sum of inductances of all components in the commutation loop, plus the inductance of the conductors themselves, the amount of emissions depends on the geometry of the commutation loops and also on the parasitic properties of the components. In order to find out the main influencing parameters in the case of GaN as the fast switch device to EMI level, an experiment was designed to gain insight into the cause of EMI emissions with the help of a simple two switch half bridge stage with a resonant filter structure. The performance in the switching circuit is easily related to the parameters of the capacitors, especially when increasing the switching frequency and slopes of the GaN.

650V rated GaN was used from GaN Systems. The control circuit was a fast FPGA – evaluation platform from National Instruments. The circuit was designed such, that easy access to all measurement points was possible and all components were easily exchangeable.

Emphasis was placed on low parasitic inductances and simplicity to keep the experiment as applicable to practical applications as possible. With respect to capacitor technologies emphasis is laid on a comparison between film and ceramic capacitors. For the latter, two types of dielectric were taken, firstly the BaTiO3 –based commercial high voltage MLCC, secondly the PLZT – based MLCC with antiferroelectric phase and copper inner electrodes.

Experimental Setup:

There were several circuits used for study the switching performance and the EMI emissions. The first circuit used is schematically shown in Figure 1, the device hardware is shown in Figure. 2, a further test equipment is shown schematically in Figure.3, the corresponding hardware in Figure 4.

The following equipments and components were used in the measurement setup:

- Current measurement devices: Rogowski sensor and RF near field probes.

PCIM Europe 2016, 10 – 12 May 2016, Nuremberg, Germany

- Capacitors used as filter capacitors and DC-Link: Ceramic MLCC (BaTiO3 –based and antiferroelectric PLZT) or metallized film capacitors e.g. 1 µF.
- Inductor: e.g. Lr = 470 µH.
- Filter - Resistor: R= 0 or 200 Ohm.
- Oscilloscope: Picoscope D3206 with a measurement bandwidth of 200 MHz.
- Thermocouples.
- (Shielded anechoic chamber using standardized equipment e.g. EMI receiver of Rohde & Schwarz (ESU 8). All measurements were made with the antenna vertically and horizontally orientated, within the range 30 MHz to 1 GHz. The testing device was placed 3m in front of the antenna.

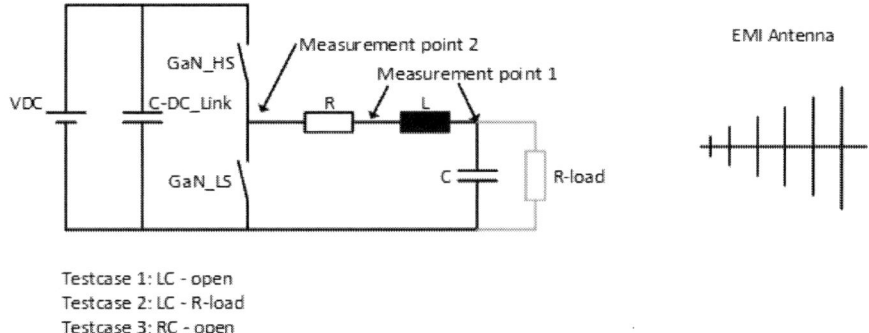

Testcase 1: LC - open
Testcase 2: LC - R-load
Testcase 3: RC - open

Fig. 1: Schematic measurement setup (half bridge with resonant filter), measurement points and EMI measurement.

The DC-Link was supplied with a voltage between 30 and 400 V. The typical settings of the driver were 50kHz, 100kHz, 250kHz and 500kHz as the repetition rate of switching, mostly with a highly asymmetric duty cycle regime D=0,9 to 0,1 high side on to off or rather D=0,5.
A suitable interlock-delay time was set in order to safely avoid overlap of on-times of the two switches and varied in order to observe the effect on the EMI and according to the switching frequency.

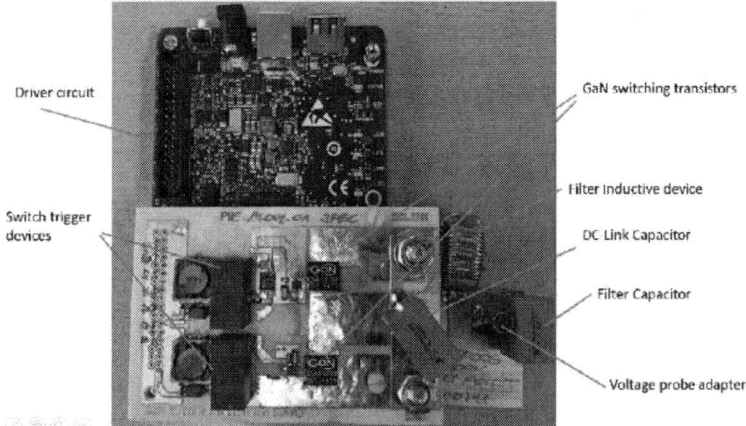

Fig. 2: Test circuit hardware. Note that there is a hidden low inductance DC-link capacitor on the backside.

© VDE VERLAG GMBH · Berlin · Offenbach

PCIM Europe 2016, 10 – 12 May 2016, Nuremberg, Germany

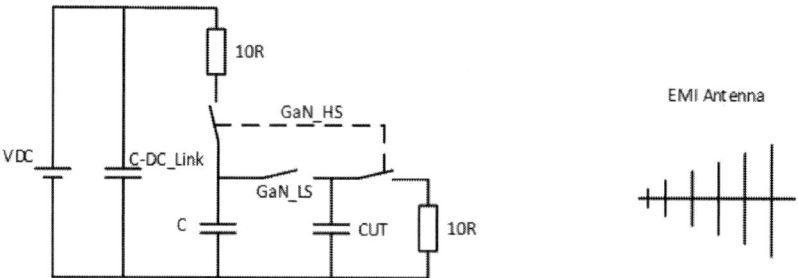

Fig.3: Schematic of the capacitor- GaN combination circuit.

Fig.4: Test circuit hardware for this combination. T1,T2,T3 GaN switches, R1,R2 resistor 10Ohm, C1, CUT 220nF capacitors

Results and Discussion:

Typical measured waveforms for the measurement setup of Fig.1 with LC filter (R=0, test case 1) and with load are shown in the Fig. 5a, and in 5b for RC filter (R=200 Ohm, L=0, test case 2) without load.

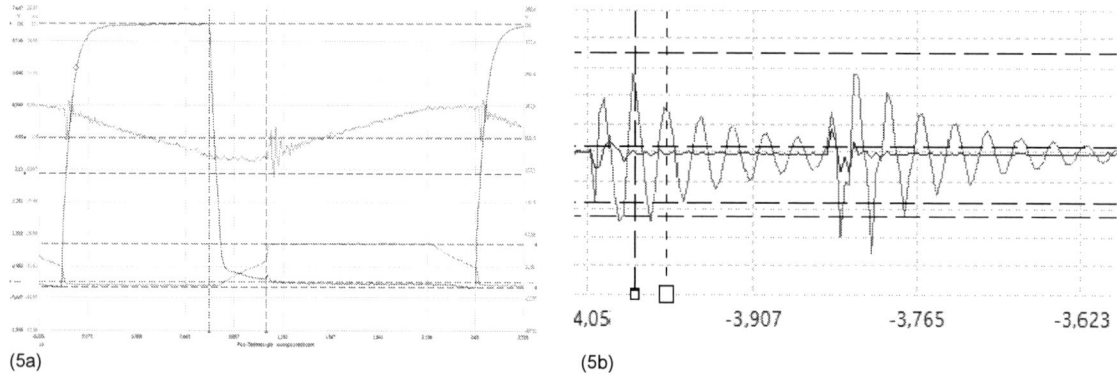

(5a)   (5b)

Fig. 5: (a) Typical waveform for the LC filter circuit: Voltage (blue) and current (red) at the switch node (meas. point 1), DC-Link capacitor is a Ceralink (~1.4uF at the voltage used), filter film cap 1uF, L=470uH, resistive load, IDL ~ 340ns.
(b) Typical ripple measurement for the RC filter circuit, current (red) and voltage (blue) at the filter capacitor (CeraLink), DC-link is a film cap. The current ripple is at approx. 35,5 MHz.

© VDE VERLAG GMBH · Berlin · Offenbach

2154

PCIM Europe 2016, 10 – 12 May 2016, Nuremberg, Germany

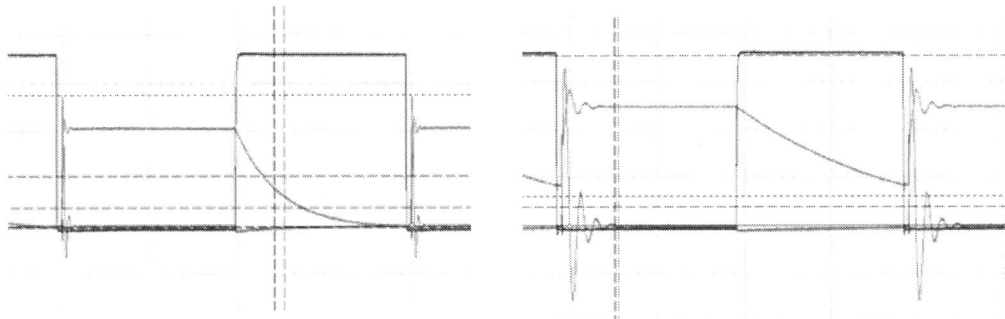

Fig. 6: (a) Typical waveform for the capacitor – GaN test circuit: Voltage (blue) and current (red) at the GaN, capacitor voltage (green). Capacitor is a 220nF CeraLink, load voltage 30 Volt over 10 Ohm, 50 kHz, high currents up to 80A.
(b) Same as (a) for a 1 µF film cap, loading 20V, showing high peak currents and ripple amplitudes.

The ripple frequencies were determined from subsequent maxima or minima. In most cases the ripple amplitudes were sufficiently high for frequency determination.
Emission spectra for the RC circuit are shown in Fig. 7 and have peaking levels at the same frequencies as found as the (initial) ripple frequency. Since the correspondence of pulse frequencies and EMI frequencies remains the same with changing input voltage levels, there is a good matching between the time signals of voltage and current and the EMI emissions.
By systematic variation of parameters, the following known facts were again verified:

- Emission get stronger with increasing amplitudes of voltage and current ripple.
- Emission get stronger with increasing parasitic inductances, especially in connection with the DC-link.
- Emission get stronger with voltage overshoot at the GaNs and the switch node.
- Emission get stronger with decreasing interlock delay time.

A new finding was for the RC filter - circuits that film capacitors showed higher amplitudes of voltage and current transients at high switching frequency, resulting in increased levels of EMI, than the ceramic ones (see Figs. 7a and b).

For varying interlock delay times, the following facts were found for the no load case: For ILD = 90 nsec there is a significant increase of EMI level at the switching pulse frequencies, which decreases for ILD= 150 and 200 nsec.

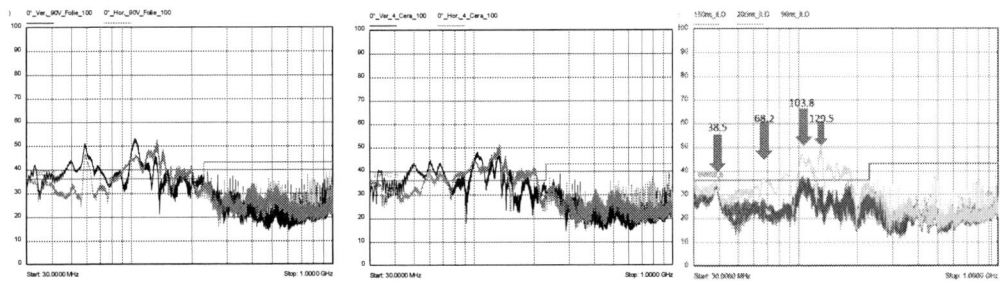

Figure 7a,b: Emission spectra (magenta: horizontal, black: vertical) a: film capacitor, b: ceramic (CeraLink), input voltage 90 Volt, 100KHz, IDL=100 nsec. Duty cycle 90%. Figure 7c: EMI spectra for different interlock delay times

Examples for measurement of the emission spectra for the LC circuit are shown in Fig. 8. In Fig 8a there was an antiferroelectric ceramic capacitor (Ceralink from TDK-EPOS, see e.g. [4] for the technology background) used in the filter, in Fig. 8b there was a film capacitor (TDK-EPCOS). The DC- link capacitor was in film technology in both cases.

© VDE VERLAG GMBH · Berlin · Offenbach

PCIM Europe 2016, 10 – 12 May 2016, Nuremberg, Germany

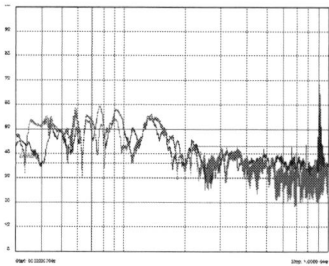 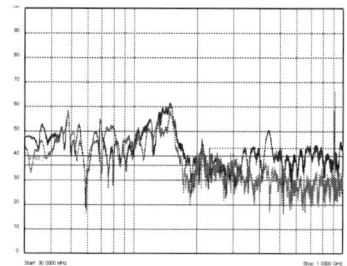

Fig 8a: Fig. 8b:

Fig 8a: Spectra for 100 kHz, film DC-link, ceramic as filter cap. Fig. 8b: Spectra for 100 kHz, film- DC-link and filter capacitor. Load~18 Ohm. Gate resistance 33Ohm.
Red : vertical, blue : horizontal antenna polarisation.

It was found that the emission levels increase with increasing load current and are too high already at low input voltage of ~ 50 Volt (e.g. EN 55011). The variation of the interlock delay time also did not ad to decreased level, even for very long delay times > 350nsec. A decreasing gate resistance (from 56 Ohm down to 15 Ohm, with rise times decreasing down to 8 nsec) resulted in higher parasitic gate currents due to higher dU/dt at the switched node and thus higher EMI level. This effect was present both in turn-on and in turn-off case. An active or a passive clamping at the low side transistor was not yet tried to overcome this effect. The highside transistor is not affected, because there is no as high parasitic currents flowing over gate to source capacity to the switched node in the buck converter constellation.

The fast rise of the voltage at the gate resulting in the « Miller effect » (unintended switching of the low side transistor, see Fig. 9), is present even at the low switching frequencies of 50 kHz. It is concluded that the short rise time only is the cause for high emissions, whereas switching frequency has lower effect, when rise time is high. There was always continuous current mode for the duty cycles used (50/50 to 10/90), under load.

Again we found good correlations between the emission frequencies and the ripple frequencies of current and of voltage. Emissions at 140 MHz however, probably had a source outside the measuring range of the Rogowski coil.The emission results did not show great change for variations of capacitors in the filter or in the DC- link, so we conclude that either active or passive clamping, or the change of the topology to interleaved legs is needed for effective reduction of emissions.

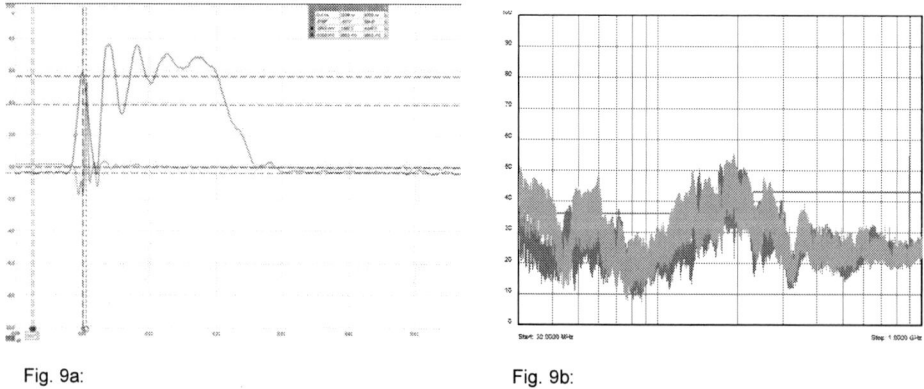

Fig. 9a: Fig. 9b:

Fig. 9a: GaN gate voltage showing parasitic switching in turn –on case low side due to too high ripple voltage (250 kHz, 300 V input voltage, IDL=200 nsec, duty cycle 90%), Fig. 9b : Emissions, green : vertical, red : horizontal. No load. Gate resistance 56Ohm.

© VDE VERLAG GMBH · Berlin · Offenbach

The GaN – capacitor circuit showed following results : The levels of the emission spectra for the GaN – capacitor combination were low for the reason that the load voltage was also low, and despite high current peaks, the emission spectra correspondingly showed low levels. Fig.10 shows a typical result. The high current peaks did not allow to increase the load voltage to higher levels for the moment. Although acc. Fig.6 different EMI levels are expected, for different capacitor technologies, they were not found most probalbly due to the low voltage levels of input voltage.

Due to the high current peaks and the high di/dt, the voltage rise at the capacitor is not always smooth, as is shown in Fig. 10a:

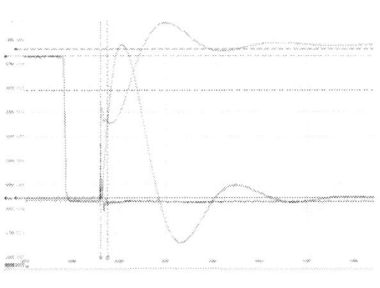

Fig. 10a:

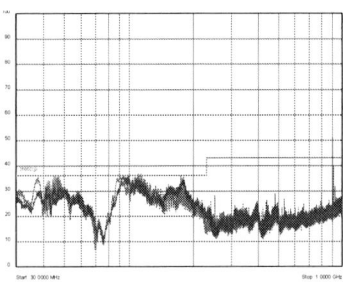

Fig. 10b:

Fig. 10a: Waveforms of gate voltage (blue), capacitor voltage (green) and current (red) in the GaN – capacitor test circuit. In a period of ~50 nsec, there is a different slope in the capacitor than for the rest of the voltage rise. 10b: Corresponding emissions, red: film cap 1µF, blue: Ceralink ~ 220nF, DC link voltage 30 Volt each.

The Fig. 10 suggests, that the "Miller effect" leading to unwanted switching actions is present also in the simple GaN – capacitor combination even at low voltages. The emission peaks between 30-40 MHz are likely to be caused by this effect.

It is evident, that clamping of the gate voltage would help to avoid the unwanted switching also in this very simple case, which is to be preferred over filtering of the emitted frequencies.
In the half – bridge with LC filter resonator and load, the parasitic switching causes high ripple (Fig.11a) and emissions too, at frequencies between ca. 35 MHz until 160 MHz (Fig.11b), caused by switch node current.

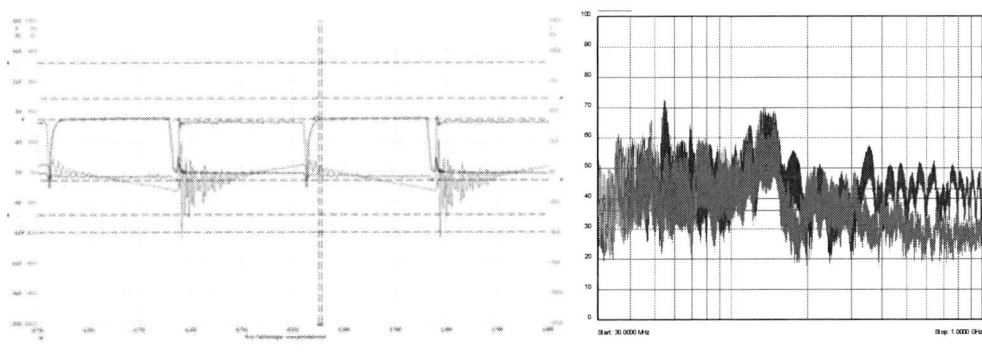

Fig.11a: Fig.11b:

Fig.11a: Ripples due to parasitic switching. Red: Switch node voltage, Blue: gate voltage low side GaN, Green: switch node current. Fig.11b: Emissions (film cap as the filter capacitor) Red: Vertical, Blue: Horizontal. Load ~13 Ohm, 400 kHz, duty cycle 50%, 40 Volt input, gate resistance 33 Ohm.

Conclusions and Outlook:

Simultaneous measurements of the electromagnetic emissions and the waveforms of the voltage and currents in the capacitors and also in the switched node in a 650V GaN half bridge followed by a resonant filter structure were made. Results were obtained for switching frequencies between 50 kHz and 1000 kHz, interlock delay times between ~90 nsec and 350 nsec, and between the capacitor technologies metallized film and ceramic multilayer.

It was found for the RC filter - circuits that film capacitors showed higher amplitudes of voltage and current transients at high switching frequency, resulting in increased levels of EMI, than the ceramic ones. Also the interlock delay time setting in conjunction with duty cycle parameters and the rise time of the gate pulse showed considerable influence on EMI levels.

For LC resonator circuits it was found that no optimized interlock delay time could be set where ripple currents and EMI emissions could be reduced while maintaining a fast rise at the GaN. In the first approach, the reason was identified as unwanted switching actions in fast GaN, due to unstable gate voltage.

From the findings, we firstly suggest to investigate means for stabilizing the gate voltages at the transistors in the half bridge, i.e. active or passive clamping of the gate voltage, or modify the topology by introducing interleaving legs. Secondly we suggest, after solving the parasitic switching issues, to investigate the influence of capacitor technologies on switching performance again.

References:

[1] K. M. Muttaqi and M. E. Haque, « Electromagnetic Interference Generated from Fast Switching Power Electronic Devices », International Journal of Innovations in Energy Systems and Power, Vol. 3, no. 1 (April 2008).

[2] Kenneth Wyatt, "GaN technology and the potential for EMI", EDN website, http://www.edn.com/electronics-blogs/the-emc-blog/4439839/2/GaN-technology-and-the-potential-for-EMI, July, 2015.

[3] Zhengyang Liu, Fred C. Lee, Qiang Li, Yuchen Yang, Xiucheng Huang, "GaN-based MHz Totem-Pole PFC Rectifier", CPES Power Electronics Annual Conference, April 12-14, 2015, Virginia Tech Center for Power Electronics, Blacksburg VA, Tutorial T2.2, Virginia Tech Center for Power Electronics

[4] G.F. Engel, „Material Requirements for Power and High Temperature Multilayer Ceramic Capacitors (MLCC)", Proc. IMAPS/ACerS, 11th Int. CICMT Conference, Dresden, Germany.

AC or DC grid for railway stations?

Lilia GALAI DOL, Efficacity, France, l.galai-dol@efficacity.com
Alexandre DE BERNARDINIS, SATIE TEMA / IFSTTAR, Efficacity, France, alexandre.de-bernardinis@ifsttar.fr

Abstract

The energy consumption of urban railway stations is very large and ever increasing. To reduce the national electrical grid consumption, the first step is optimizing the equipment size and the second step is adding other available electrical sources like the train residual braking energy. Many of local energies productions use Direct Current (DC. The actual internal station grid is in Alternative Current (AC) but the majority of the equipment has an AC/DC converter to be supplied in DC. This paper describes the project led by Efficacity Institute, which addresses the use of additional existing energies and to develop a more adapted station grid to reducing the daily energy consumption peak. One Efficacity energetic concept aims to store the braking energy of the trains with a stationary electrical saving system. This energy is integrated to the power supply of a railway station thanks to a microgrid. First step of this microgrid are studied in this paper to develop the more adapted grid structure.

1. The station electrical network

Nowadays, the network is AC, but it should be in DC in order to be freed of the AC network transient effect and well adapted to electrochemical storage systems. A study led by M. Iordache [1] and S. Nasr [2] for the OSIRIS project demonstrates that a DC network would be an appropriate solution for urban railway stations. A measurement campaign led on one subway station in Paris, shows the two main load contributors are the light and the ventilation (it will be more detailed in the final paper). If we use LED technology for the light and asynchronous motors with frequency control, we could argue that the DC grid is the more appropriate [3], [4]. But some grid actuators still being AC like the local electricity distribution.

The present paper compares these two kinds of grid, and study the storage of train braking energy in a hybrid storage system (composed of batteries and super-capacitors cells) and the judicious restitution at different moments of the day (during peak energy consumption hours) to several kind of station loads.

1.1. Station main equipment electrical modeling

Measurements were made in a Parisian Metro station which allow to know its real consumption: almost 3 MWh. We can see in figure1 the split of this consumption in main station equipment.

PCIM Europe 2016, 10 – 12 May 2016, Nuremberg, Germany

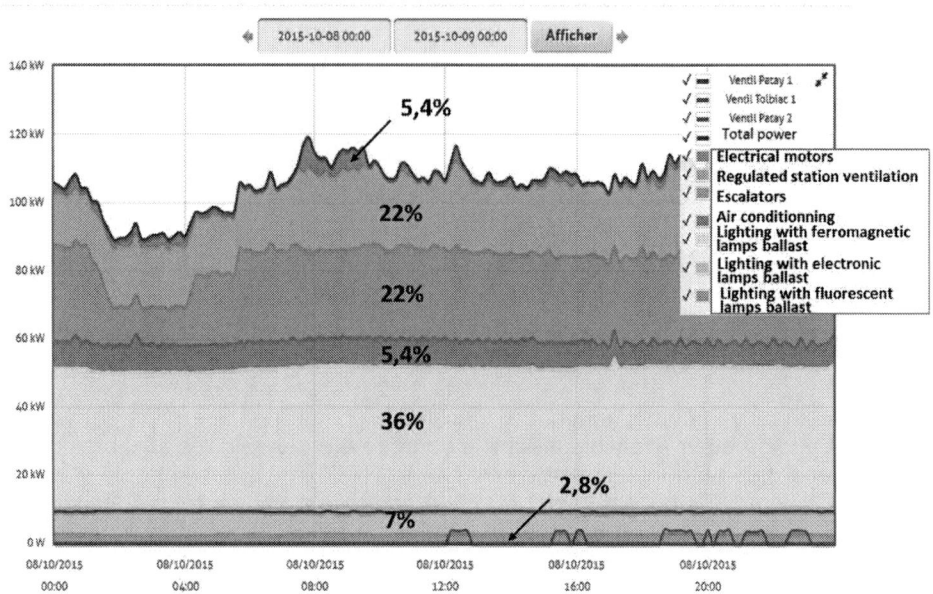

Fig. 1.   A typical RATP station consumption for one day for each main equipment

In figure 1 we can see that most part of electrical consumption is constant during the day. It means we can save more energy by a good management of some equipment like HVAC.

In order to investigate the interaction of these main loads and the braking energy storage system, we developed a study case and an electrical model.

1.2.   Description of the station equipment and proposed solutions

Many investigations were made to reuse the lost braking energy for the railway electrical vehicle. Nevertheless, there are no works on the railway station potential use. In the literature we can find 4 methods to regenerate braking energy (in Figure 2):

- Receptivity: the energy stays in the railway grid for the starting-up of other vehicles.

- Restitution to the grid: with a reversible AC/DC converter the energy can be restituted to the medium voltage grid.

- Electrical storage: thanks to the technology constant evolution [9]

- Heat: The lost energy is transformed to heat dissipated into resistors.

In this study, we focus on the electrical storage solution.

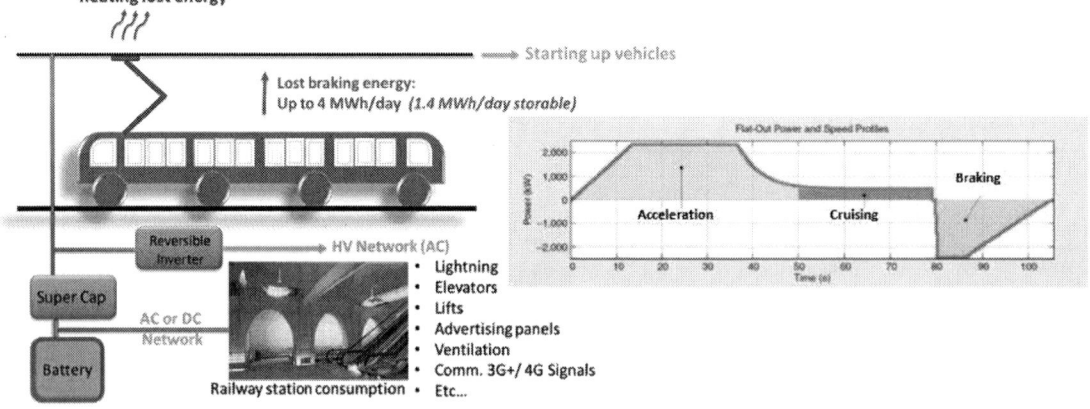

Figure 2: Energy train profile [6] and station grid concept

© VDE VERLAG GMBH · Berlin · Offenbach

PCIM Europe 2016, 10 – 12 May 2016, Nuremberg, Germany

At present in urban railway network, the braking energy is redirected to other rail vehicles thanks to the good line receptivity (capacity of the energy exchange between trains). The better daily receptivity can rise almost 96%. Taking into account the power electronics limitations, the available energy is almost 1.4 MWh.

A storage system with a mix of super capacitors (SC) and batteries allows the use of the residual braking energy at an arbitrary daily time.

At first, the super capacitors are charged by the residual train braking current. At the end of braking phase, the super capacitors are discharged into a battery. Then the energy stored can be immediately used (it depends on the daily energy demand).

2. Station electrical behavior with a storage system

The kind of grid (DC or AC) depends on the equipment. All main equipment works with a DC voltage but the input is dimensioned for an AC voltage. So it would be difficult to change all the already existing equipment but it can be interesting to design it for new stations.

The typical stress linked to the train braking energy is the large amount of energy produced leading to voltage and current line variations.

2.1. Transients, constraints and frequency of train braking energy

At the daily peak time the frequency of braking is 1 per 200s.

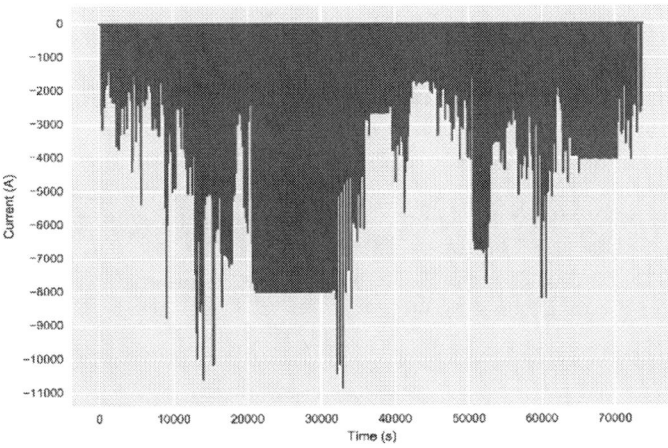

Figure 3: Metro line residual braking energy estimation

Averaged measurements on a metro line give the power profile versus the running phases of the metro vehicle (fig 3).

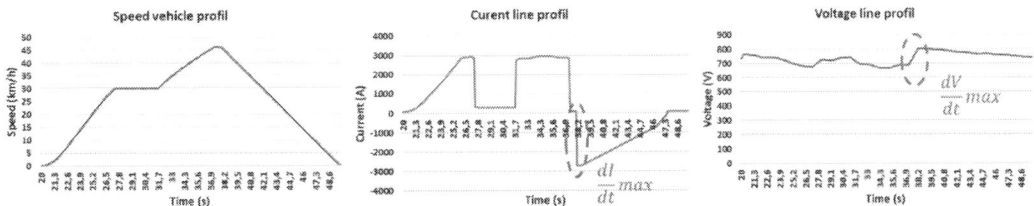

Figure 4: Speed and electrical metro average profiles

© VDE VERLAG GMBH · Berlin · Offenbach

As shown in figure 4, the typical speed profiles is a succession of acceleration, constant speed and braking which is associated a consumption and a generation of current causing a voltage drop and an over-voltage.

In figure 4, two points are highlighted because they have a high and fast amplitude variation:

$$\frac{dV}{dt} max = \frac{100\ (V)}{1\ (s)}$$

We can observe that the dV/dt is not really a problem for the power switches when one braking occurs on the metro line due to its rather low voltage level.

$$\frac{dI}{dt} max = \frac{2800\ (A)}{0.1\ (s)}$$

However the dI/dt is more constraining. The use of supercapacitors is one way to reduce its influence on the electrical circuit. For conventional supercapacitors the charging current is limited to 1kA (depending on the component depth of discharge). The differential current is sent back to the metro supply line.

2.2. DC grid study

For a DC grid the storage system is the same as for an AC one. The difference occurs in the part of the grid where the energy is fed to the loads.

The storage system is composed by a supercapacitors (800V-54F) with a parallel battery (700V-300A/h).

The residual current was represented for the most constraining context with a fast current rising (1000A at 1s) and a 10s braking duration.

The two main equipment in a station are the lighting and electrical motors. Electrical motors are the essential compound of HVAC with a vectoriel speed control.

The HVAC is represented by an asynchronous machine with a flux PI regulation acting on PWM inverter control. This control is based on vectoriel control identically to actual metro HVAC. The active power was simulated on the subway equipment at 10kW for a 40m^3/s air flow, leading to a mechanical Torque of 300Nm.

Loads are connected to the storage system by a buck converter. The simulation was made with the SimPowerSystem Matlab® toolbox.

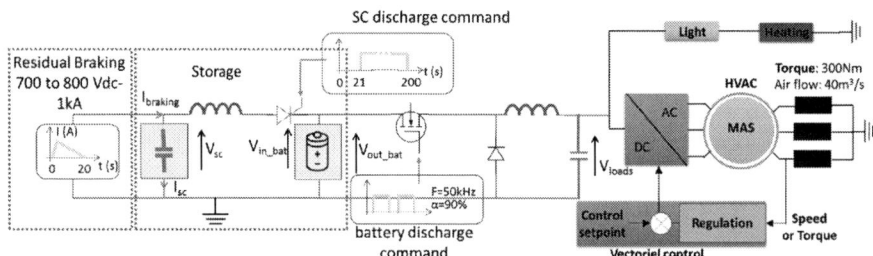

Figure 5: DC grid with mains station loads

Initial conditions were imposed to SC (Vsc=800V), to battery (SOC=90%), light and heating loads (50Ω) and motor speed (100 rad/s).

PCIM Europe 2016, 10 – 12 May 2016, Nuremberg, Germany

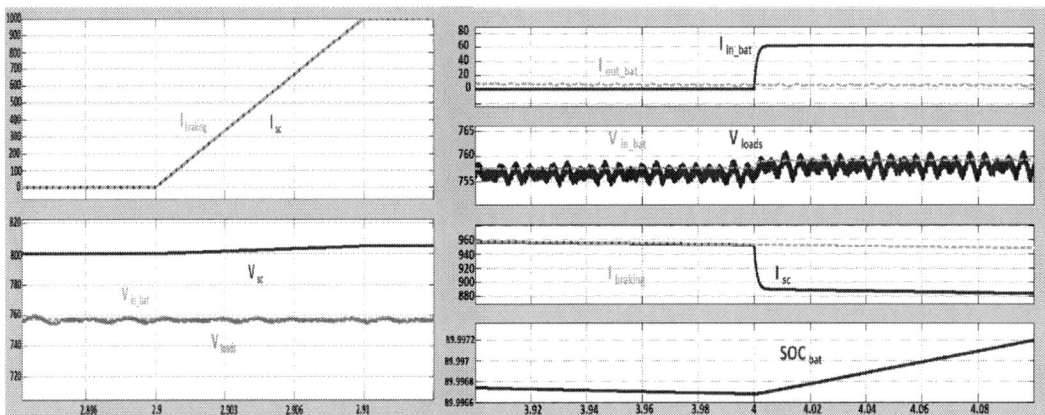

Figure 6: Simulation of the storage and the restitution of the braking energy in a DC grid

At figure 6, the results of the simulation are observed:

• At the left: we can see the residual braking current rising phase. The super capacitor current (I_{sc}) follows the residual braking current ($I_{braking}$). The super capacitor voltage (V_{sc}) increases by 5 V in 100 ms.

• At the right: we can see the residual braking current dropping phase. At 4 s, the thyristor before the battery is switching on. At this time, the input battery current (I_{in_bat}) rising to 60A with a time response about 2 ms. I_{sc} follows $I_{braking}$ tills 4s and drop of 60A. At the same time, the battery voltage (V_{in_bat}) increases insignificantly. There is no influence on load voltage.

So the high dI/dt has no impact on the grid thanks to super capacitors proprieties.

2.3. AC grid study

The modelizations of the storage, the residual braking current and the loads are same at the DC study case. The two only changes are the type of converter at the battery output and the rectifier before the HVAC.

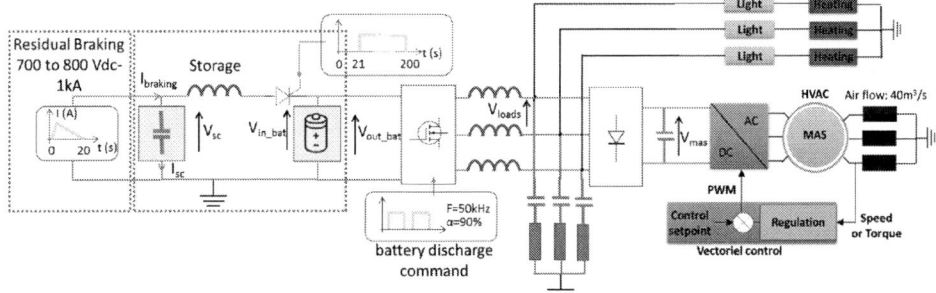

Figure 7: AC grid with mains station loads

The initial conditions are same that the DC study case.

© VDE VERLAG GMBH · Berlin · Offenbach

2163

PCIM Europe 2016, 10 – 12 May 2016, Nuremberg, Germany

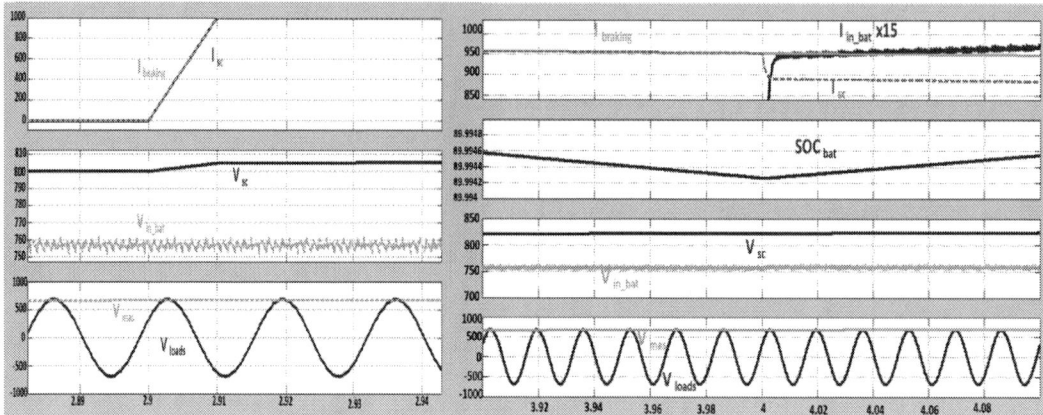

Figure 8: Simulation of the storage and the restitution of the braking energy in an AC grid

At figure 8, the results of the simulation are observed:

•   At the left: we can see the residual braking current rising phase. The super capacitor current (I_{sc}) follows the residual braking current ($I_{braking}$). The super capacitor voltage (V_{sc}) increases by 5 V in 100 ms (like in the DC grid). The voltage at the motor input (V_{mas}) is constant without any perturbation linked to the braking current.

•   At the right: we can see the residual braking current dropping phase. At 4 s, the thyristor before the battery is switched on. At this time, the input battery current (I_{in_bat}) rises to 60 A with a time response about 2 ms. I_{sc} follows $I_{braking}$ even after 4s. At the same time, the battery voltage (V_{in_bat}) increases insignificantly. There is no influence on load voltage.

In AC mode the thyristor placed before the battery sustain same current as in DC mode.

Thanks to the super capacitors, there is no EMI disturbance in the grid

The filters needed are conventional ones. These filters have to reduce the impact of the PWM (Pulsed Width Modulation) HVAC control system.

3. DC and AC network comparison

3.1.   Equipment and efficiency

At first, we can compare the number of electrical equipment needed to supply the same loads.

We can observe this difference in the schematics of the next station generation hereunder (including electrical vehicle charging station).

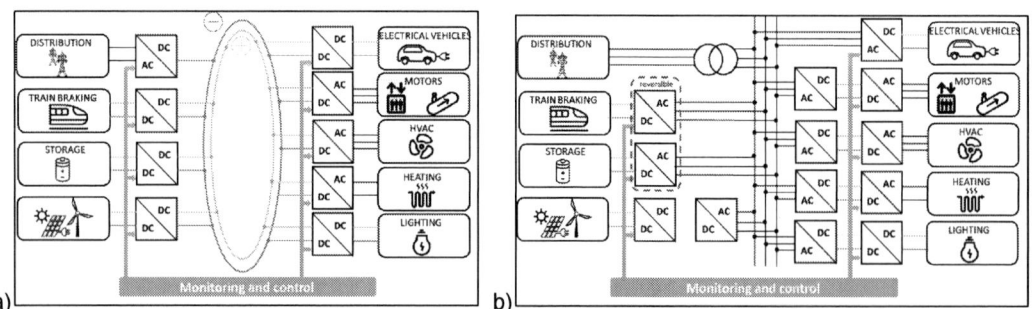

Figure 9: DC (a) and AC (b) microgrid electrical diagrams.

© VDE VERLAG GMBH · Berlin · Offenbach

To control some electrical systems like motors, a power electronics layer has to be added. For example, AC/DC and DC/AC converters are needed in order to act on the supply frequency for the speed regulation of a controlled AC motor.

We can see in figure 9 that these systems can quickly complexify the stations electrical grid.

At the second step, we have studied the energy consumption. We made this comparison with the case studied before, including the train braking energy, an electrical storage, an HVAC, a heating and a lighting systems.

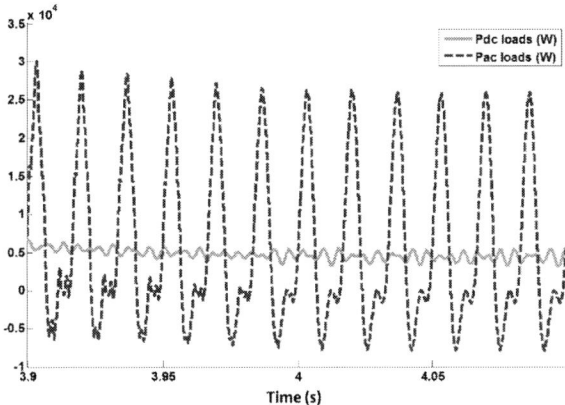

Figure 10: Simulation results of DC and AC microgrid power consumption

The simulation results of the AC and DC power consumption was calculated thanks to loads voltage and loads current represented in cases studies at the steady state.
The calculation of the AC and DC energy for 200ms shows a difference of 20% leading the DC solution as the better one.

This results was recently shown with a grid including renewable energies and energy storage system [8].

3.2.   Optimized control

In order to be completed, this schema has to be coupled to a heat network. Thus, a mixing of thermal and electrical grids will allow to reduce the global energy consumption with the regeneration of ALL wasted energies. The next step will be dedicated to develop an electro thermal micro grid.

This study case is the first step to a multi physical micro grid, which is a path to the micro grid and the urban living of the future as reported by J. Jarass and D. Heinrichs [4]. In the future, the micro grid will integrate many kind of energy resources and need to be managed thanks to an optimal demand response controller.

4. Conclusion

We can see in this paper the behavior of railway station loads and one alternative electric supply: the metro braking energy. The train braking energy shows high dI/dt be copped with the storage system.

The study of its storage and the use of this DC energy is described and compared for two cases of station network: AC and DC.

The AC network need more energy conversion than the DC one which lead to an energy consumption difference about 20% between.

© VDE VERLAG GMBH · Berlin · Offenbach

Today the DC solution is possible only for high power equipment because the DC current still not be safe. Recently some studies of the no-zero crossing of the current [7] are emerging making the DC network safety possible.

5. References

[1] M. Iordache, "Smart grid system definition, system studies and modeling, technologies evaluation", OSIRIS Rail EU project deliverable, 2013.
[2] S. Nasr; "Smart DC Grid integration in railway systems"; JCGE'2014 - SEEDS, June 2014, Saint-Louis, France.
[3] T. Ahmed, A. Mohamed Ahmed, A. Mohamed Osama, "DC microgrids and distribution systems: An overview" Electric Power System Research, Elsevier, 2015.
[4] Estefanía Planas, Jon Andreu. "AC and DC Technology in Microgrids: A Review." Renewable and Sustainable Energy Reviews 43 , 2015, on page(s): 726–49.
[5] R.R. Pecharroman, A. Lopez-Lopez, A. P. Cucala, A. Fernandez-Cardador, " Riding the rails to DC Power Efficiency", IEEE Power and Energy Society, 2014, on page(s): 32 – 38
[6] D. Cornic, « Efficient recovery braking energy though a reversible dc substation », IEEE, Electrical Systems for Aircraft, Railway and Ship Propulsion (ESARS), 2010, on.page(s) 1 – 9
[7] Tironi, E.; Corti, M.; Ubezio, G. "DC networks including multi-port DC/DC converters: Fault analysis", IEEE 15th International Conference on Environment and Electrical Engineering (EEEIC), 2015, On page(s): 1109 – 1114
[8] Sanjay K. Chaudhary; Josep M. Guerrero ; Remus Teodorescu, "Enhancing the Capacity of the AC Distribution System Using DC Interlinks—A Step Toward Future DC Grid", IEEE Transactions on Smart Grid (Volume:6 , Issue: 4), July 2015, on page (s) 1722 – 1729
[9] K. Ogura, Kobe, Japan ; K. Nishimura, T. Matsumura, C. Tonda, E. Yoshiyama, M. Andriani, W. Francis, R. A. Schmitt, A. Visgotis, N. Gianfrancesco, "Test Results of a High Capacity Wayside Energy Storage System Using Ni-MH Batteries for DC Electric Railway at New York City Transit", IEEE Green Technologies Conference (IEEE-Green) 2011, on page(s) 1 - 6

PCIM Europe 2016, 10 – 12 May 2016, Nuremberg, Germany

Solid-State Transformer Modeling for Analyzing its Application in Distribution Grids

Christoph Hunziker, Nicola Schulz
University of Applied Sciences Northwestern Switzerland, Switzerland,
christoph.hunziker@fhnw.ch

Abstract

This work describes the modeling of a solid-state transformer (SST) in DIgSilent Power Factory. As various models and tools are used for SST or converter design in general, this work provides the basis for studying SST applications in realistic grid environments. The model is based on the dynamic average technique and is compatible with three-phase, four-wire technology on low-voltage side. It describes the controlled bidirectional power flow for symmetric and asymmetric load situations, as well as for faults in the external grid. SST functionality and the performance of its control have been analyzed and compared with standard low frequency transformers (LFT).

1. Introduction

In recent years, SSTs have been highlighted as an enabling technology for a modernization of the power distribution system in the context of a forecasted increase of renewable energy in-feed [1, 2, 3]. Compared to conventional low-frequency transformers (LFT), SSTs have several advantages such as: compensation or generation of reactive power, accessible DC-port for direct connection of battery storage or other DC-based loads and generation devices, dynamic voltage control, possibility for emulating inertia, active limitation of fault currents, etc. DIgSilent Power Factory is a state-of-the-art power system analysis software for applications in generation, transmission and distribution of electric power [4]. It is widely used for grid simulations by utility companies and in academia. It incorporates models for standard operating equipment and also allows to define user defined controls. The use of detailed switching models for power-electronic devices leads to very short time-steps, typically below 1 μs. In combination with extended and complex grid models, this leads to long simulation times and limits the potential of grid simulations. Thus, in order to simulate extended grids containing a plurality of SSTs, a simplified, but - from grid perspective - realistic and fully functional SST model is required. In this study, this is achieved by using dynamic average models of the powerelectronic elements. Thus, the switching effects are neglected while preserving the dynamic behavior.

2. SST model and its implementation in DIgSilent

A typical three-stage SST topology consists of a high voltage rectifier, a DC-DC isolation stage that links the high voltage DC-link to the low voltage DC-link, followed by a low-voltage inverter [1]. The SST allows bidirectional power flow from high-voltage to low-voltage side and vice versa. Since the DC-DC conversion in the second stage of the SST ideally does not influence the dynamic behavior seen at the SST terminals, it is neglected here.

© VDE VERLAG GMBH · Berlin · Offenbach

PCIM Europe 2016, 10 – 12 May 2016, Nuremberg, Germany

The SST model in DIgSilent is based on existing standard elements. This includes a three-phase PWM converter component on HV side with floating star point attached to the DC-link (see Fig. 1). Two current sources represent the load/infeed from an attached battery storage and the low-voltage output, respectively. The LV side is modeled as a three-phase voltage source with neutral wire on ground representing a four leg inverter.

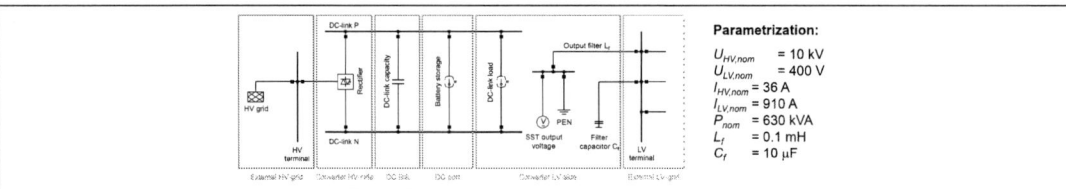

Fig. 1: DIgSilent model of an SST consisting of a high voltage converter stage, high voltage DC-link, DC-port, as well as a controlled voltage source representing the converter on low voltage side.

3. Implementation of control

The SST model in DIgSilent has been implemented as a composite model *(ElmComp)* which combines internal elements (see also Sec. 2 above) with general models *(ElmDsl)* describing transient behavior of user defined elements. The composite model is defined using a composite frame which describes the basic structure of the model (see Fig. 2). This approach has the advantage that individual control algorithms or transfer functions can be exchanged in a flexible way in order to compare their impact on the grid.

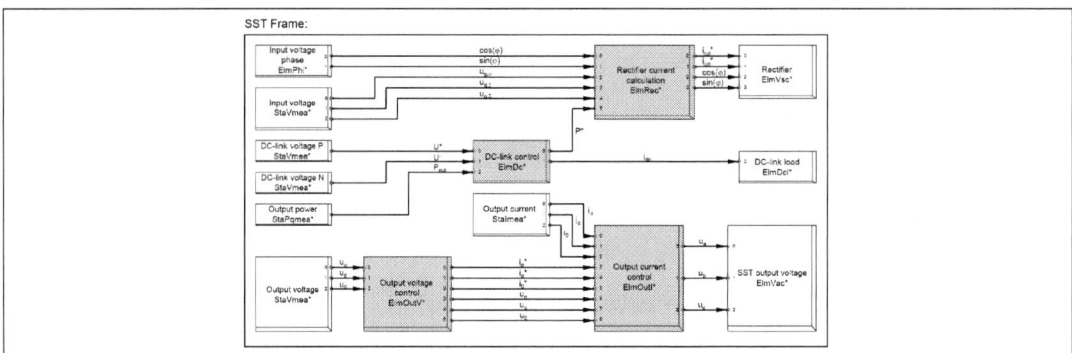

Fig. 2: Composite frame describing the structure of the SST model in DIgSilent. The four blocks implementing control functions for the rectifier stage, the DC-link voltage, as well as the voltage- and current control at the inverter output are shaded.

The implemented control reflects an operation mode where the SST is tied to the grid on HV side. The low-voltage side represents a standalone converter, feeding the external grid. The implemented control for the different converter stages are described in the subsections below.

3.1. DC-link control *(ElmDc)*

The DC-link voltage control is based on the stored energy in the DC-link capacitor as reference quantity [5]. A simple PI controller determines the necessary power reference P^* required for a stable DC-link voltage at the actual consumption/ generation rate of the attached low voltage grid P_{out}.

© VDE VERLAG GMBH · Berlin · Offenbach

3.2. Rectifier current calculation *(ElmRec)*

On HV-side, the converter control determines the reference current to be drawn from the HV-grid, based on the grid voltage $u_{g,\alpha\beta0}$ as well as on the active and reactive power reference P^* and Q^*, respectively. For the simulations in this work, Q^* was parametrized manually, however, it can be determined based on other control goals in the grid as well. The calculated phase currents $i_{g,dq}$ are forwarded to the internal rectifier model *(ElmVsc)* in a dq synchronously rotating reference frame. DIgSilent allows either to directly impress the calculated currents (method chosen for this work) or use an internal PI controller for current regulation.

The grid voltage u_g is assumed sinusoidal, though not necessarily symmetric. The control strategy ensures constant, non-oscillating power flow given by P^*, reactive power flow according to Q^* as well as sinusoidal phase currents. In order to fulfil the requirement of sinusoidal phase currents, the calculation is based on a complex representation of u_g and i_g in $\alpha\beta$-form.

Instantaneous reactive power theory (IRP-theory) can be used to describe instantaneous power as follows [6]:

$$p(t) \;=\; \frac{3}{2}[u_\alpha(t)i_\alpha(t) + u_\beta(t)i_\beta(t)] \qquad (1) \qquad\qquad q(t) \;=\; \frac{3}{2}[-u_\alpha(t)i_\beta(t) + u_\beta(t)i_\alpha(t)] \qquad (2)$$

These equations can be rewritten using rotating phasors with $x(t) = \Re(\bar{x} \cdot \exp(j\omega t)) = \Re((x_r + jx_i) \cdot \exp(j\omega t))$ (x to be replaced by u_α, u_β, i_α, and i_β). The momentary quantities calculated by the simulation were transformed to rotating phasors with $x_r = x(t)$ and $x_i = x(t - T/4)$ where $T = \omega/2\pi$ is obtained from the PLL measurement of the grid voltage u_g.

$$p(t) = \frac{3}{4}\underbrace{(u_{\alpha r}i_{\alpha r} + u_{\alpha i}i_{\alpha i} + u_{\beta r}i_{\beta r} + u_{\beta i}i_{\beta i})}_{P_{dc}}$$
$$+\frac{3}{4}\underbrace{(u_{\alpha r}i_{\alpha r} - u_{\alpha i}i_{\alpha i} + u_{\beta r}i_{\beta r} - u_{\beta i}i_{\beta i})}_{P_{ac1}} \cdot \cos(2\omega t)$$
$$-\frac{3}{4}\underbrace{(u_{\alpha r}i_{\alpha i} + u_{\alpha i}i_{\alpha r} + u_{\beta r}i_{\beta i} + u_{\beta i}i_{\beta r})}_{P_{ac2}} \cdot \sin(2\omega t) \quad (3)$$

$$q(t) = \frac{3}{4}\underbrace{(-u_{\alpha r}i_{\beta r} - u_{\alpha i}i_{\beta i} + u_{\beta r}i_{\alpha r} + u_{\beta i}i_{\alpha i})}_{Q_{dc}}$$
$$+\frac{3}{4}\underbrace{(-u_{\alpha r}i_{\beta r} + u_{\alpha i}i_{\beta i} + u_{\beta r}i_{\alpha r} - u_{\beta i}i_{\alpha i})}_{Q_{ac1}} \cdot \cos(2\omega t)$$
$$-\frac{3}{4}\underbrace{(-u_{\alpha r}i_{\beta i} - u_{\alpha i}i_{\beta r} + u_{\beta r}i_{\alpha i} + u_{\beta i}i_{\alpha r})}_{Q_{ac2}} \cdot \sin(2\omega t) \quad (4)$$

Eqs. 3 and 4 each contain one term for the mean value as well as two terms describing the alternating value of $p(t)$ and $q(t)$. Setting $P_{dc} = P^*$, and $P_{ac1} = P_{ac2} = 0$ fulfills the requirement for a non-oscillating active power flow. The required mean reactive power can be calculated by setting $Q_{dc} = Q^*$. The mentioned four conditions were used to calculate the phasor components of the phase currents given by $i_{\alpha r}$, $i_{\alpha i}$, $i_{\beta r}$, and $i_{\beta i}$.

Non-oscillating reactive power flow can only be reached for symmetric grid voltage u_g. The additional conditions $Q_{ac1} = Q_{ac2} = 0$ lead to an over-determined equation system and cannot be fulfilled in general when requiring sinusoidal phase currents. However, for symmetric grid voltages reactive power is non-oscillating as well. Solving the four conditions yields the components of the phase current phasors. Once the phasor components have been calculated, they can be transformed into dq signals using the measured momentary phase of the grid voltage.

3.3. Control of output inverter

The four leg voltage source inverter is implemented in DIgSilent by replacing the four leg switching network by its average model as shown in Fig. 1. The average model has been derived using a cycle-by-cycle averaging process for the pulsating voltages u_{an}, u_{bn}, and u_{cn}, and i_{dc}

© VDE VERLAG GMBH · Berlin · Offenbach

PCIM Europe 2016, 10 – 12 May 2016, Nuremberg, Germany

[7]. A cascaded control scheme with an outer voltage- and an inner current control loop has been implemented for the output inverter as proposed in [8, 9] as described below.

Output voltage control *(ElmOutV)*

The chosen implementation in DIgSilent using a composite frame to describe the control structure (see also Fig. 2) has the advantage of allowing to flexibly change controllers and compare their performance and effects onto the grid. In this work, a similar control method as proposed in [10, 8] has been implemented. The method defines the reference frequency and voltage internally and controls the necessary currents accordingly (see Figure 3a). In order to have DC reference signals even in the case of asymmetric load in the LV grid, the measured voltage is first transformed in symmetrical components. Each sequence is then controlled individually in a dq reference frame. Fig. 3b shows a block diagram describing the mathematical process of the sequence decomposition. The measured momentary phase voltages are transformed to complex phasors and then decomposed into a positive, negative and zero sequence. The decomposed sequences are then converted into dq-signals.

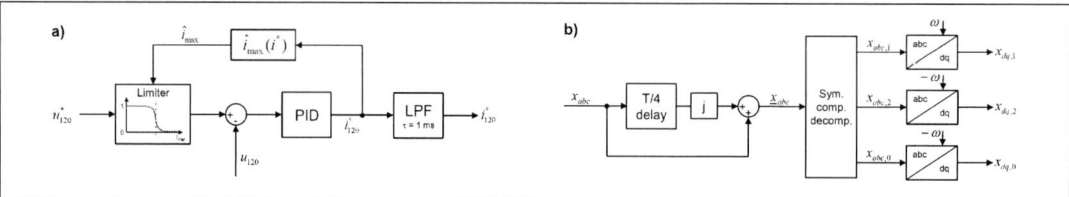

Fig. 3: a) Control scheme for the outer voltage control loop. A limiter has been implemented which reduces reference voltage as soon as the maximum phase current reaches 1.5 pu. b) Block diagram for sequence decomposition.

The voltage control loop includes a sinusoidal current limitation to 1.5 pu in case of grid faults or short circuits. This protection mechanism is an important advantage compared to conventional transformers and has strong implications on grid protection concepts.

The open-loop frequency response of the voltage control loop has been analyzed in order to understand the control performance. The analysis has been constrained to the symmetric case, where the three-phase model can be replaced by a single-phase model. The control loop is depicted in Fig. 4a. The inner current loop must feature a 5 to 10 times higher bandwith and can therefore be neglegted. The single-phase representation of the grid is shown in Fig. 4b. The used control method requires complex phasors to enable the sequence decomposition (see also Fig. 3b). The phasor is constructed by adding the delayed signal as imaginary part to the momentary value (the delay equals a quarter grid period). This introduces non-ideal effects in the frequency response similar to the behavior of a comb filter.

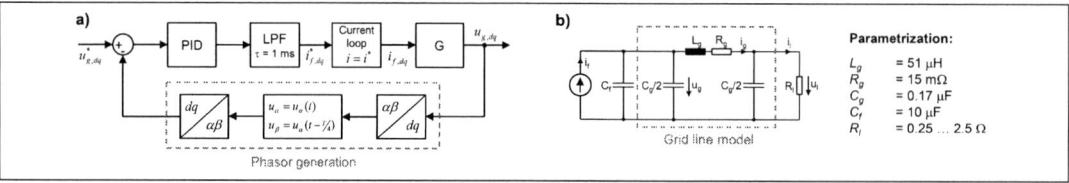

Fig. 4: a) Schematic of the outer voltage control loop valid for the symmetric grid. The influence of the sequence decomposition of control method A is marked with a red box. b) Single-phase grid model representing the symmetric grid G.

© VDE VERLAG GMBH · Berlin · Offenbach

2170

For the sequence decomposition to be calculated, the measured grid voltage is transformed to a complex phasor in $\alpha\beta$-space. The Hilpert transform is the ideal transformation to generate the β-component from a momentary α-signal since it represents an all-pass filter with a phase of $-\pi/2$. In the model described here, a simple implementation has been chosen based on a time delay of a quarter period (see also Fig. 4a) [10]. Strictly, this approach yields correct behavior only at the fundamental grid frequency of 50 Hz. The inaccuracy of this approximation increases for deviating frequency components.

The transformation in the box of Fig. 4a can be written as

$$
\begin{bmatrix} u_{d,out} \\ u_{q,out} \end{bmatrix} = \begin{bmatrix} \sin(\omega_0 t) & -\cos(\omega_0 t) \\ \cos(\omega_0 t) & \sin(\omega_0 t) \end{bmatrix} \cdot \begin{bmatrix} 1 & 0 \\ e^{-sT/4} & 0 \end{bmatrix} \cdot \begin{bmatrix} \sin(\omega_0 t) & \cos(\omega_0 t) \\ -\cos(\omega_0 t) & \sin(\omega_0 t) \end{bmatrix} \cdot \begin{bmatrix} u_{d,in} \\ u_{q,in} \end{bmatrix}
$$

$$
= \begin{bmatrix} \sin^2(\omega_0 t) u_{d,in}(t) + \cos^2(\omega_0 t) u_{d,in}(t - T/4) \\ \sin^2(\omega_0 t) u_{q,in}(t - T/4) + \cos^2(\omega_0 t) u_{q,in}(t) \end{bmatrix}
$$

$$
+ \begin{bmatrix} \sin(\omega_0 t)\cos(\omega_0 t) u_{q,in}(t) - \sin(\omega_0 t)\cos(\omega_0 t) u_{q,in}(t - T/4) \\ \sin(\omega_0 t)\cos(\omega_0 t) u_{d,in}(t) - \sin(\omega_0 t)\cos(\omega_0 t) u_{d,in}(t - T/4) \end{bmatrix}, \tag{1}
$$

where ω_0 is the fundamental grid frequency. The second term in Eq. 1 represents crosscoupling from q to d-channel and vice versa. Setting $u_{q,in}(t) = 0$ and $u_{d,in} = \cos(\omega t)$ yields

$$
u_{d,out}(t) = \sin^2(\omega_0 t)\cos(\omega t) + \cos^2(\omega_0 t)\cos(\omega(t - T/4)) \tag{2}
$$

$$
= \frac{1}{2}(1 + \cos(\omega T/4))\cos(\omega t) + \frac{1}{2}\sin(\omega T/4)\sin(\omega t) \tag{2a}
$$

$$
+ \frac{1}{4}(\cos(\omega T/4) - 1)[\cos((\omega + 2\omega_0)t) + \cos((\omega - 2\omega_0)t)] \tag{2b}
$$

$$
+ \frac{1}{4}(\sin(\omega T/4) - 1)[\sin((\omega + 2\omega_0)t) + \sin((\omega - 2\omega_0)t)]. \tag{2c}
$$

Eq. 2 shows that the approach generates frequency components at $\omega \pm 2\omega_0$ which are detrimental to control. The magnitude of the linear terms in Eq. 2a is given as

$$
|u_{d,out}(t)|_{lin} = \sqrt{\frac{1}{2} + \frac{1}{2}\cos(\omega T/4)} = \cos(\omega T/8). \tag{3}
$$

The open-loop frequency response with the non-ideal influence of phasor generation is shown in Fig. 5a. The linear terms of Eq. 2a have been added to the loop. The magnitude, therefore, incorporates the behavior given in Eq. 3. The gain is always smaller compared to the same loop without phasor generation which limits closed-loop performance. Points marked with crosses have been validated in the DIgSilent model with an open voltage control loop and a sinusoidal reference signal at the corresponding frequency. Different load situations have a large influence on the open-loop response as shown in Fig. 5b. In order to further extend the range of good control performance also during grid fault events, control parameters might have to be adapted based on the load situation.

Output current control *(ElmOutl)*

Current control has been implemented in a $dq0$ stationary frame with a PI controller for each component. The mutual coupling between d and q component due to the filter inductance L_f has been compensated. The control approach used for voltage control based on symmetrical components has not been applied for the inner control loop since a higher bandwidth is required here which leaves more room for the non-ideal behavior described in the Sec. 3.3. Each channel of the $dq0$ frame is controlled with an individual PI controller. The bandwidth has been chosen to be at least five times larger than the one of the outer voltage loop.

PCIM Europe 2016, 10 – 12 May 2016, Nuremberg, Germany

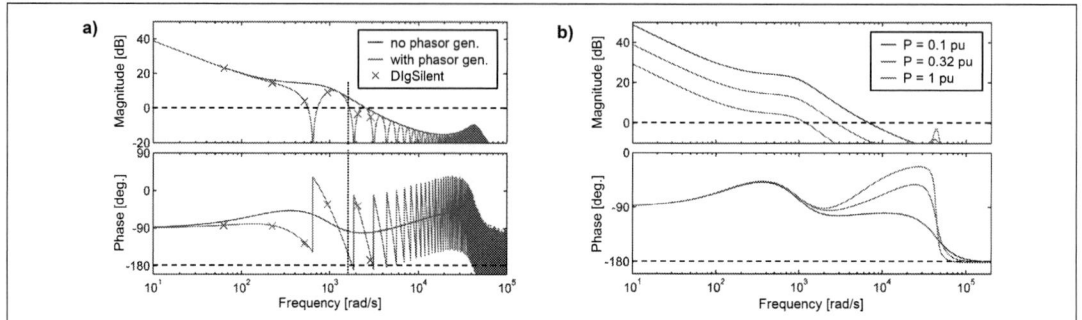

Fig. 5: a) Open loop frequency response of the outer voltage control loop. The non-ideal behavior of the chosen control method leads to a gain similar to a comb filter. b) Envelope of the open-loop controller for three different load situations. The bandwidths result to 196, 445, and 1120 Hz.

4. Simulation results

Simulations have been conducted for a simple low voltage grid model as shown in Fig. 6. For stable and accurate simulation results, a timestep of 0.1 ms was used. The implementation of the SST model in DIgSilent allows to simulate different kinds of events in the grid like e.g. various short circuits or load steps in a very flexible way, which is the strength of a grid simulation tool like DIgSilent. SST operation has been compared with the behavior of a standard low-frequency transformer. Fig. 7 shows phase voltages and currents of the SST's low-voltage side during the event of a phase-to-neutral (L_1-N) short circuit triggered at time $t = 0.1$ s at *bus 1a*. The SST limits the output current to 1.5 pu by reducing the voltage. The system features symmetric phase voltages even in the presence of highly asymmetric currents due to the short circuit. In the case of the LFT, the short circuit current is solely limited by the voltage drop over transformer and lines. Phase voltages and currents are not symmetric in the grid.

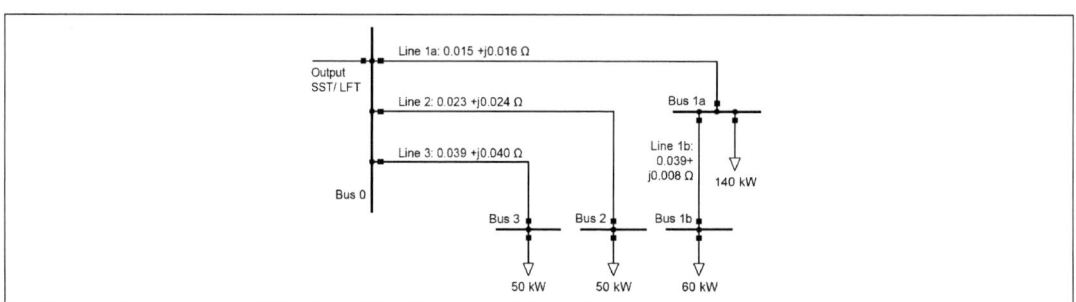

Fig. 6: Schematic of low voltage grid used for simulation in DIgSilent. The parameterization of Line 1a corresponds to the model in Fig. 4b.

In contrast to a conventional, iron-based LFT where short-circuit currents can be e.g. 25 times the nominal current (i.e., $\varepsilon_c = 4$ %), the current is limited to e.g. 1.5 pu. Hence, for grids operated behind a SST different protection concepts (e.g. automatic fuses) are required.

Fig. 8 shows the power flow on LV as well as on HV side for the same short circuit event as described above. Power oscillations with high amplitude are present on both sides of the LFT. In contrast, the SST is capable of eliminating the power oscillations during the short circuit event entirely on HV side.

Fig. 9 shows the simulation results for the event of an asymmetric voltage sag on high voltage

© VDE VERLAG GMBH · Berlin · Offenbach

2172

PCIM Europe 2016, 10 – 12 May 2016, Nuremberg, Germany

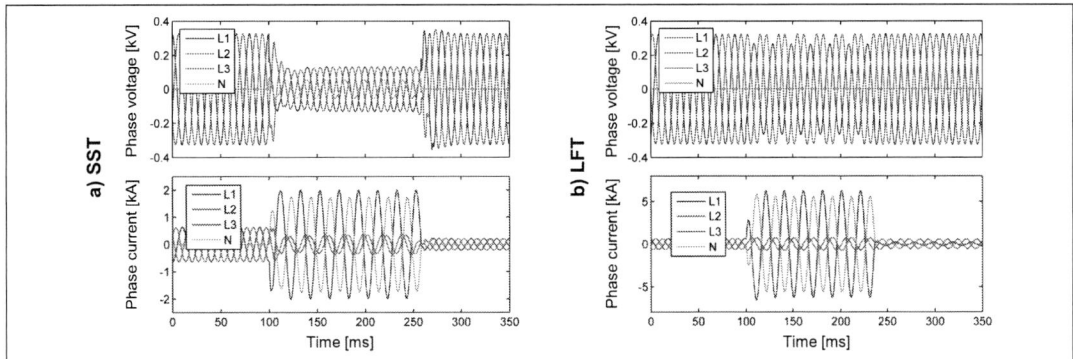

Fig. 7: Phase voltages and currents during a phase-to-neutral short circuit event. The fault is cleared for the network fed by the SST by opening the breaker of the corresponding feeder at *Bus 0* after 0.15 s. The short circuit fed by the LFT is cleared by the corresponding fuse at the same position.

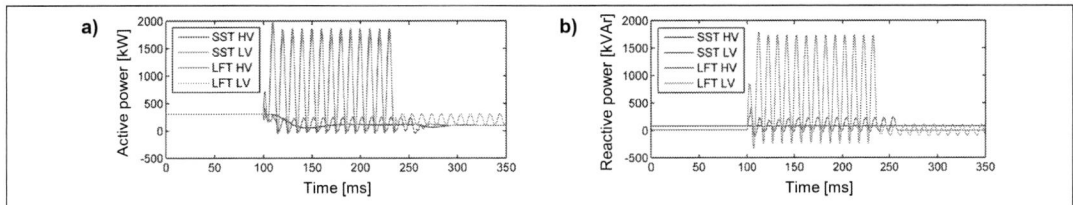

Fig. 8: Comparison of the power flow during a phase-to-neutral short circuit for an SST and LFT, respectively. In the case of the SST a reactive power reference of 80 kVAr has been activated.

side of the transformer during the time interval $0.1 < t < 0.25$ s ($u_{g,L1} = 94.2$ pu, $u_{g,L2} = 89.6$ pu, $u_{g,L3} = 84.4$ pu). The resulting phase currents shown in b) and c) yield the power flows shown in d) and e), respectively.

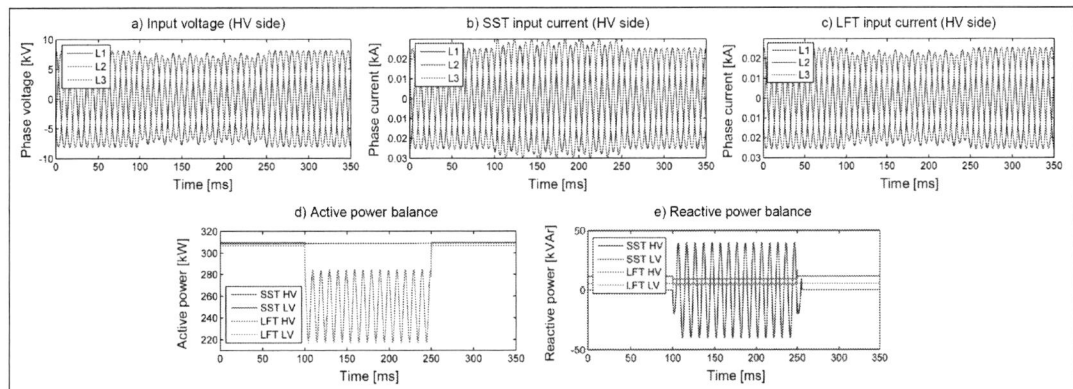

Fig. 9: a) Asymmetric grid voltage. b), c) Resulting currents for SST and LFT, respectively. d), e) Resulting active and reactive power flows on HV and LV side.

5. Conclusion

In conclusion, we have shown a SST model in DIgSilent for dynamic grid simulations covering situations with strongly asymmetric loads and short circuit events. The SST model can be implemented into complex grid models to simulate the behavior of large-scale electric grids

© VDE VERLAG GMBH · Berlin · Offenbach

incorporating a plurality of SSTs, and featuring the relevant SST functionalities.

6. Acknowledgment

This research project is part of the National Research Programme "Energy Turnaround" (NRP 70) of the Swiss National Science Foundation (SNSF). Further information on the National Research Programme can be found at www.nrp70.ch.

7. References

[1] I. Roasto, E. Romero-Cadaval, J. Martins, and R. Smolenski. State of the art of active power electronic transformers for smart grids. In *IECON 2012 - 38th Annual Conference on IEEE Industrial Electronics Society*, pages 5241–5246, Oct 2012.

[2] L. Heinemann and G. Mauthe. The universal power electronics based distribution transformer, an unified approach. In *Power Electronics Specialists Conference, 2001. PESC. 2001 IEEE 32nd Annual*, volume 2, pages 504–509 vol.2, 2001.

[3] A.Q. Huang, M.L. Crow, G.T. Heydt, J.P. Zheng, and S.J. Dale. The future renewable electric energy delivery and management (freedm) system: The energy internet. *Proceedings of the IEEE*, 99(1):133–148, Jan 2011.

[4] *DIgSilent PowerFactory 15 - User manual*. DIgSilent GmbH, 1 edition, May 2013.

[5] Youyuan Jiang, L. Breazeale, R. Ayyanar, and Xiaolin Mao. Simplified solid state transformer modeling for real time digital simulator (rtds). In *Energy Conversion Congress and Exposition (ECCE), 2012 IEEE*, pages 1447–1452, Sept 2012.

[6] J. L. Afonso, M. J. S. Freitas, and J. S. Martins. p-q theory power components calculations. In *Industrial Electronics, 2003. ISIE '03. 2003 IEEE International Symposium on*, volume 1, pages 385–390 vol. 1, June 2003.

[7] R. Zhang, V.H. Prasad, D. Boroyevich, and F.C. Lee. Three-dimensional space vector modulation for four-leg voltage-source converters. *Power Electronics, IEEE Transactions on*, 17(3):314–326, May 2002.

[8] Juan A. Martinez-Velasco, Salvador Alepuz, Francisco Gonzlez-Molina, and Jacinto Martin-Arnedo. Dynamic average modeling of a bidirectional solid state transformer for feasibility studies and real-time implementation. *Electric Power Systems Research*, 117(0):143 – 153, 2014.

[9] V. Ramachandran, A. Kuvar, U. Singh, S. Bhattacharya, and M. Baran. A system level study employing improved solid state transformer average models with renewable energy integration. In *PES General Meeting — Conference Exposition, 2014 IEEE*, pages 1–5, July 2014.

[10] I. Vechiu, O. Curea, and H. Camblong. Transient operation of a four-leg inverter for autonomous applications with unbalanced load. *Power Electronics, IEEE Transactions on*, 25(2):399–407, Feb 2010.

Hybrid reactive power compensation system for Grid Code compliance in renewable energy power plants

Gianluca Postiglione, Nidec ASI S.A., France, Gianluca.Postiglione@nidec-asi.com
Antonio Raso, Nidec ASI S.p.A., Italy, Antonio.Raso@nidec-asi.com
Giovanni Borghetti, Nidec ASI S.p.A., Italy, Giovanni.Borghetti@nidec-asi.com
François Pezet, Nidec ASI S.A., France, Francois.Pezet@nidec-asi.com

Abstract

During recent years, the sizes of the Renewable Energy (RE) plants have grown in a way that their effect to the connecting grid can no longer be neglected. To cope with these problems, the Grid operators are forced to review and to tighten their grid connection rules – also known as Grid Code. Key issues are: steady state and dynamic reactive power capability and voltage control. This paper presents a straightforward solution, based on the combination of passive devices (MSC and MSR) + STATCOM, that allows getting Grid Code compliance. The proposed developed control strategy allows to avoid reactive power steps and to damp voltage and current oscillations initiated by the switching events of MSC.

1. Introduction

Nowadays fundamental changes are affecting the electrical utilities: the energy markets are being deregulated and liberalized; the urbanization is continuing around the world, accompanied by a constantly growing demand of electrical energy. This demand is often faced thanks to the installation of Renewable Energy plants.

The RE plants, in fact, can offer several advantages: they can be locally produced, are safer to produce and maintain, and have minimum environmental impact.

On the other hand, one of major drawback arising from connecting RE plants to the power network is the difficulty to achieve the balance between energy supply and demand.

For consumers of electrical power it is important that the power is available with a stable voltage and frequency. To make sure that this aim is accomplished, grid operators issue requirements for connecting power generating equipment to the grid. Collectively these requirements are known as the grid code, and equipment that meets these requirements is called grid compliant.

A typical requirement by the grid operators is that generators should be able to vary their reactive power output dependent on the grid voltage level. This and other requirements necessitate the deployment of devices that can control reactive power [6].

Passive devices, such as Mechanically Switched Capacitors and Reactors (MSC and MSR respectively), are interesting because of the low cost and the low losses of the equipment; moreover, the dynamics related to the connection and disconnection are compatible with the required response time. On the other hand, it is not possible to guarantee the steady state accuracy; moreover, each connection or disconnection causes severe voltage and current transients on the network due to the step variation of the reactive power.

Connecting a dynamic compensation system (D-STATCOM) dimensioned for the full compensation reactive power would satisfy all requirements, but would need large initial investments; moreover, a solution entirely based on power electronics converter would not guarantee the maximum efficiency of the power plant, because of the high losses of the semiconductor devices [5].

The sum of the advantages and drawbacks of the previous solutions suggests the realization of a Hybrid Reactive Power Compensation System (RPCS), made of a D-STATCOM and a suitable number of MSCs and MSRs. If the required response time is compatible with the

© VDE VERLAG GMBH · Berlin · Offenbach

one of the MSCs and MSRs, it is possible to choose a D-STATCOM dimensioned for ¼ or ⅓ of the total amount of requested reactive power.

The paper will be organized as follows: after an overview of reactive power compensation equipment, the section 3 illustrates the control strategy especially developed to avoid reactive power steps initiated by the switching events of MSC/MSR. Finally a case study will be presented in the section 4.

2. Reactive Power Compensation Equipment for Grid Code Compliance

2.1. MSC & MSR

Mechanically Switched Capacitors and Reactors (MSC and MSR respectively), are interesting because of the low cost, the low losses and the small footprint of the equipment; the dynamics related to the connection and disconnection are compatible with the response time required for voltage supporting in RE plants (1 ÷ 2 seconds). It must be remarked, however, that the reactive current depends linearly on the grid voltage (Fig. 1. a), and consequently the reactive power decreases with the square of the grid voltage: for this reason it is not possible to guarantee the steady state accuracy [1] [2]. From a dynamic point of view, each connection or disconnection causes severe voltage and current transients on the network due to the step variation of the reactive power. It is mandatory to analyze how the network impedance changes when the MSCs and MSRs are connected or disconnected. The anti-resonance frequencies and their amplitudes must be carefully checked, in order to avoid dangerous grid voltage oscillations every time the devices are switched. Moreover, when a MSC is switched off, it is not possible to switch it on again, unless it is discharged: this means a waiting time of about 5 ÷ 10 minutes for a MV MSC.

Some of the drawbacks of the switching transients can be reduced. For MSCs, a suitable series inductance can be used to change the frequency and the amplitude of the anti-resonances; of course, the additional reactor increases the cost and the required space. State-of-the-art fast switches, coupled with a synchronization device, guarantee the switching of the device at the best point of the voltage wave, so that no grid voltage oscillations (for MSCs) or transient DC current (for MSRs) occur; the problem of the unavailability due to the trapped charge is also overcome. These kind of fast switches, however, have no breaking capacity, so a back-up circuit breaker must be installed.

2.2. STATCOM

The STATCOM is the most recent solution [1] [3] for the dynamic reactive power compensation. Based on VSC technology, it allows the injection of reactive current independently of the grid voltage (according to Fig. 1. b), and the fastest dynamic performances. This means that a STATCOM dimensioned for the full compensation reactive power would perfectly satisfy the response time and steady-state accuracy requirements. The control system is designed [4] to avoid reactive power step on the network, in order to improve grid stability. The major drawbacks of this solution are the large investment. Moreover, depending on the topology of converter the losses of the semiconductor devices (operating costs), can affect the efficiency of the whole system.

2.3. Hybrid Reactive Power Compensation System (H – RPCS)

The sum of the advantages and drawbacks of the previous solutions suggests the realization of an Hybrid Reactive Power Compensation System (H – RPCS), made of a STATCOM and a suitable number of MSCs and MSRs. Considering the response time requirement, it is possible to choose a STATCOM dimensioned for ¼ of the total amount of requested reactive power, while the remaining ¾ is obtained with passive devices.

PCIM Europe 2016, 10 – 12 May 2016, Nuremberg, Germany

Nidec ASI has defined a control strategy to coordinate the elements of a Hybrid RPCS (the new "SILCOVAR-D" D-STATCOM, MSCs and MSRs), especially developed to take the best from the two technologies, and employ the dynamic capability of the VSC technology to avoid reactive power step on the grid. The strategy will be detailed described in the following Section.

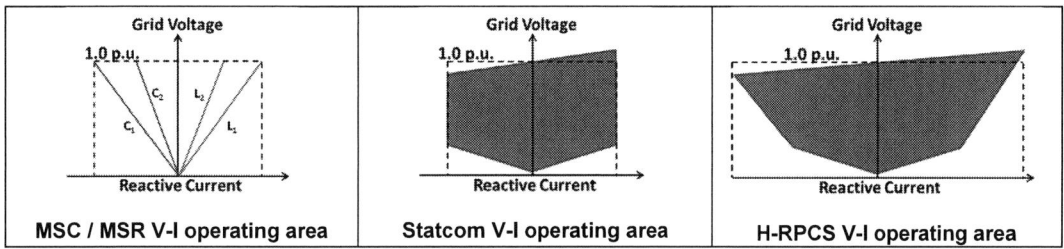

Fig. 1. Comparison on V-I plan of MSC / MSR (a), Statcom (b) and H – RPCS (c)

3. Case Study: H – RPCS for 100 MW wind farm

In order to help 100MW wind park to achieve Grid Code compliance in term of: Steady state reactive power supply, voltage control and dynamic reactive power supply it was decided to add a hybrid reactive compensation system.

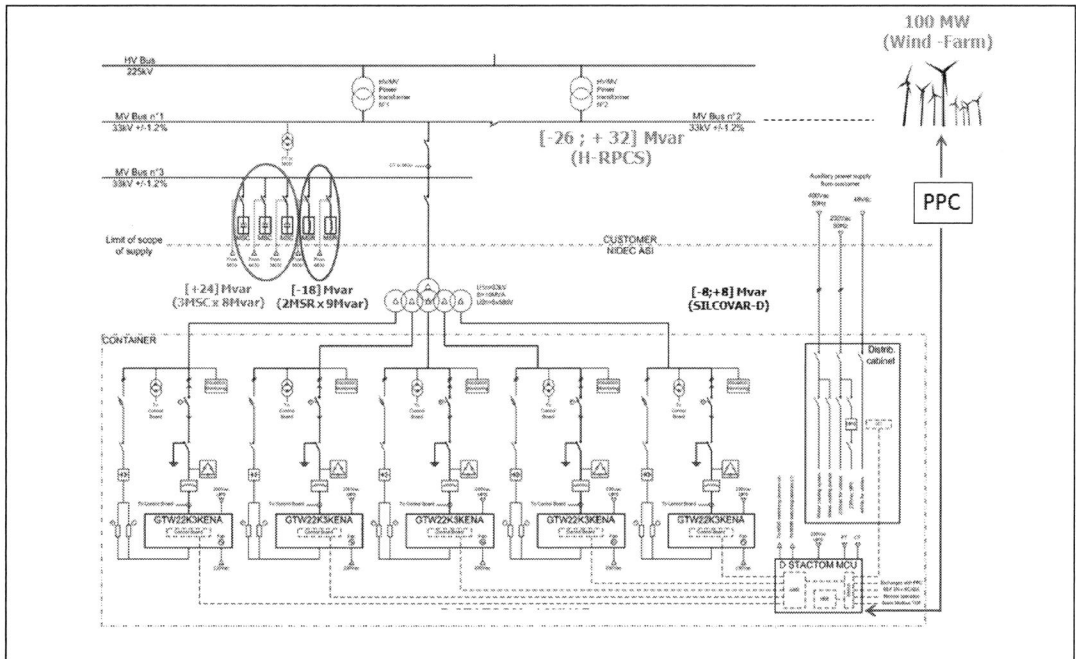

Fig. 2. layout of H – RPCS for 100 MW wind farm

The 100 MW wind park is split in two parts, each one connected to the 225 kV bus through a 90 MVA 225 kV / 33 kV step down transformer. The Hybrid Reactive Power Compensation System is sized [-26 MVAr ;+32 MVAr] at 33 kV and consists of:

- 1 SILCOVAR-D, able to operate in the reactive power range [-8;+8] MVAr. The SILCOVAR-D, consisting of 5 Voltage Source Converters, is connected to the 33 kV Medium Voltage Bus (MV Bus n°3 in the Fig. 2) using a transformer (33 kV / 580V) with 5 secondary windings, one for each VSC. It is designed with an advanced

© VDE VERLAG GMBH · Berlin · Offenbach

2177

redundancy operating mode to guarantee an uninterrupted operation in case of module failure, with N-1 capability.

- 3 MSCs connected to 33kV bus, each one designed for reactive power of +8 MVAr.
- 2 MSRs connected to 33kV bus, each one designed for reactive power of -9 MVAr.

3.1. Control strategy

A remote Power Plant Controller (PPC) defines the total amount of reactive power needed to control the voltage at the Point of Common Coupling. Then, it splits the reference of reactive power into 2 contributions: one is for the wind farm turbines, and one for the H – RPCS.

The following example will explain how the control strategy works, when a step reference of +32 MVAr is received from the PPC. First of all, the reference is always processed by a ramp filter. The chosen slope is equal to 16 MVAr/s, to obtain an overall response time of 2 seconds in correspondence of a 0 – 100% reference step. Initially no MSC or MSR are connected, so the reactive power which can be injected into the grid is provided only by the SILCOVAR-D converter. The SILCOVAR-D follows the filtered reference. When the reference exceeds the threshold for the connection of one MSC (Qt_{hMSC} = 4 MVAr), then the first MSC is connected. Just before the connection, the reactive power injected into the grid Q_{GRID} is:

$Q_{GRID} = Q_{CONV} = Q_{thMSC}$ = 4 MVAr

After the connection of the MSC, to avoid reactive power steps, it must be verified that the reactive power on the grid doesn't change:

$Q_{GRID} = Q_{CONV} + Q_{MSC}$ = 4 MVAr

Since Q_{MSC} is equal to 8 MVAr, then Q_{CONV} must rapidly "jump" from +4 MVAr to –4 MVAr.

Note that with one MSC connected the reactive power that can be managed by the RPCS is in the range [0 ÷ +16] MVAr. After the "jump", the reactive power of the converter rises to follow the reference. When Q_{CONV} is again equal to Q_{thMSC} (Q_{GRID} = 12 MVAr), the second MSC is connected. Again, Q_{CONV} jumps from +4 MVAr to –4 MVAr. Now the reactive power that can be managed by the RPCS is in the range [+8 ÷ +24] MVAR. The third MSC is connected with the same logic. When all MSCs are connected the reactive power that can be managed by the RPCS is in the range [+16 ÷ +32] MVAR. The same strategy is used to disconnect the MSCs when the reference decreases; Fig. 3. shows the reactive power control strategy in case of capacitive reactive power reference.

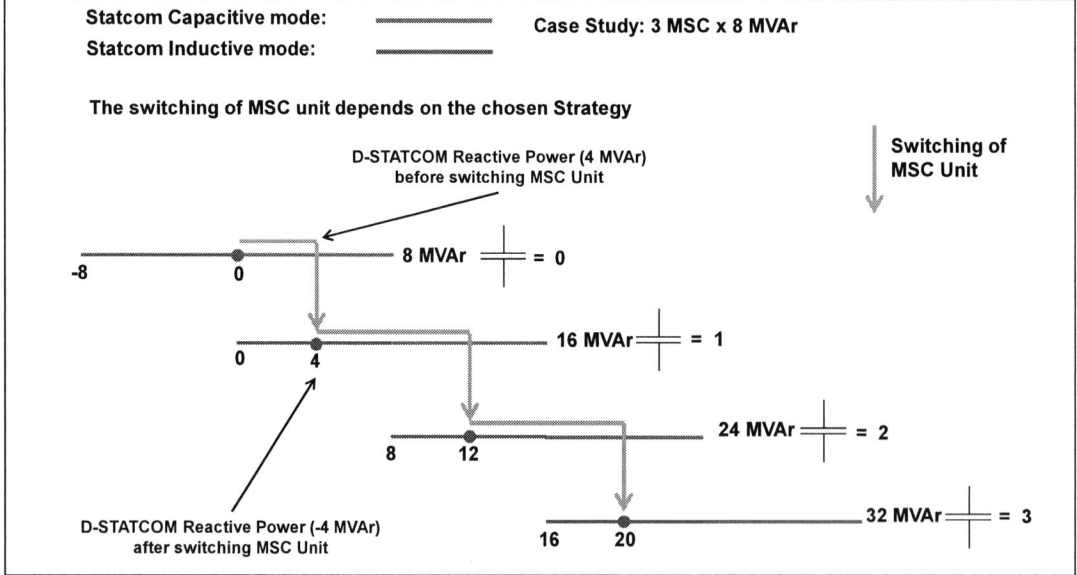

Fig. 3. Reactive power control strategy for a capacitive reactive power reference

The same strategy is used to connect and disconnect the MSRs, with and inductive reactive power reference. The reactive power control strategy in case of inductive reactive power reference is shown in Fig. 4., where Q_{thMSR} = 4.5 MVAr.

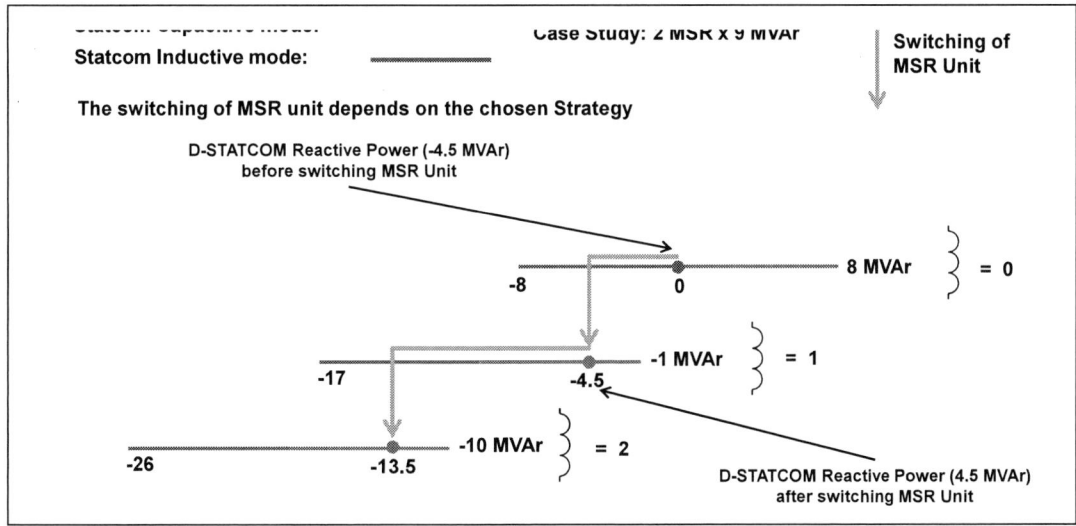

Fig. 4. Reactive power control strategy for a capacitive reactive power reference

Thus, the first task of the Control Strategy is to determine how many MSCs – MSRs must be connected. Moreover, the control logic takes into account how many times the devices have been switched on and off: when one MSC or one MSR is requested, the one with the lower number of operation is connected, in order to guarantee a suitable sharing of the stresses on the shunt devices.

3.2. Performances of the Hybrid Reactive Power Compensation System

The following figure shows the performances of the Hybrid RPCS, for a reactive power ramp reference; the figure shows the measures of reactive power effectively supplied to the grid, the MV bus voltages, and the MSC currents.

The sum of the step reactive power provided by the passive elements (MSC, red line) and the fast dynamic compensation (STATCOM, green line) allows to exchange reactive power with the grid in a linear way (blue line), with high accuracy and with damped grid voltage and current oscillations.

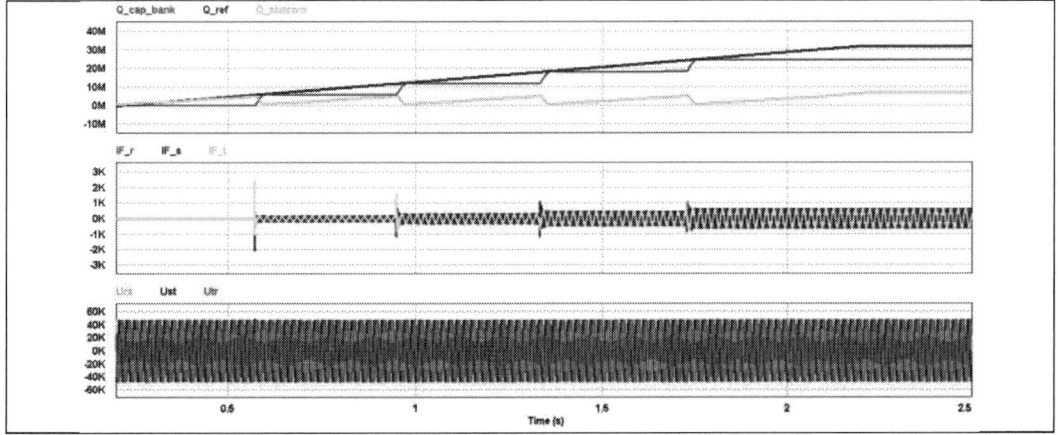

Fig. 5. Performances of Hybrid RPCS – top: Q of MSCs (red), Q reference (blue), Q of SILCOVAR-D (green); center: MSCs currents; bottom: grid voltages

4. Conclusions

The comparison between advantages and drawbacks of the MSC/MSR and STATCOM lead to the use of a Hybrid Reactive Power Compensation System. The control strategy developed by Nidec ASI has been presented in detail, together with the case study of the H – RPCS for a 100 MW wind farm.

Finally, a Hybrid RPCS, designed with the criteria described above, allows achieving the following main advantages:

- Increase of dynamic grid stability: during voltage dips the H-RPCS helps to support the grid voltage.
- Low Power Losses: low losses of semiconductor devices thanks to the small SILCOVAR-D power, high efficiency of the MSCs and MSRs.
- High steady state accuracy: the dynamic requirements of the Grid Codes are met, without notable overshoots or oscillations.
- Fast response time.
- Low Equipment Cost.

5. References

[1] Narain G. Hingorani, Laszlo Gyugyi: Understanding FACTS: Concepts and Technology of Flexible AC Transmission Systems, Wiley-IEEE Press, 1999

[2] T. J. E. Miller: Reactive Power Control in Electric Systems, John Wiley & Sons, 1982.

[3] G. Postiglione, G. Borghetti, G. Torre, P. Bordignon, "Transformerless STATCOM based on multilevel converter for grid voltage restoring", PCIM 2011

[4] G. Postiglione, G. Borghetti, G. Torre, F. Pezet, E. Gatti, O. Laurence, "D-STATCOM: A Dynamic Local Reactive Power Reserve for Correction of System Voltage Problems", PCIM 2014

[5] K. Berringer, J. Marvin, P. Perruchoud: "Semiconductor Power Losses in AC Inverters", 30th Annual Meeting of the IEEE Industry Application Society (IAS), 1995.

[6] D. F. Opila, A. M. Zeynu, I. A. Hiskens, "Wind Farm Reactive Support and Voltage Control". 2010 IREP Symposium- Bulk Power System Dynamics and Control – VIII (IREP), August 1-6, 2010, Buzios, RJ, Brazil

PCIM Europe 2016, 10 – 12 May 2016, Nuremberg, Germany

A new shunt connected HVDC tap based on a highly efficient resonant cascade converter

Andre Birkel, Mark-M. Bakran, University Bayreuth,
Department of Mechatronics, Center of Energy Technology, Germany
andre.birkel@uni-bayreuth.de

Abstract

This paper presents a new concept for HVDC tapping applications based on a resonant cascade DC/DC converter. The scalable DC/DC converter is characterized by soft switching condition for all switches and diodes. Furthermore a continuous current load for all half bridge modules offers a good utilization of the installed semiconductors. The paper starts with a demonstration of the converter's structure and its fundamental functionality. For the validation of the operation principle a simulation of the DC/DC converter is done. This is followed by a loss calculation and some considerations about the converter design for tapping in HVDC systems. Finally, some thoughts on the passive components are given.

1. Introduction

For transmitting massive amounts of electrical power AC power transmission systems capability is limitated and so the most promising choice for a new transmission system are HVDC lines or even grids [1]. Existing HVDC transmission systems are usually designed for long distance transmission. Some of these HVDC transmission systems pass over relatively small communities with no connection to major power transmission systems. It is most desirable to find methods of economically connecting these communities to the HVDC system, to supply them with electric energy. This has been an engineering focus for the last few years [2], [3].

The challenge is to tap a small amount of power from a high rated HVDC system. In order to tap approximately 10 % to 20 % of the HVDC system rated power, a full scale voltage source converter station would be oversized. Therefore, topologies with a reduced demand of power electronic devices are necessary. These HVDC taps can be classified in two main groups: Series connected circuits, which tap the power from the load current, and shunt or parallel connected systems [4]. A shunt connected tap has to withstand the full HVDC link voltage, but it operates with a smaller input current than serial systems. The system presented in this paper is a shunt tapping concept. Fig. 1 shows the basic structure of a shunt tapping concept.

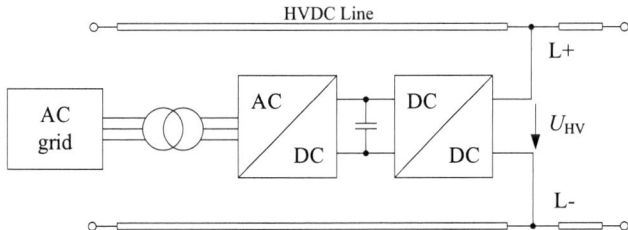

Fig. 1: Shunt tapping in HVDC systems

The main component of this tapping concept is the DC/DC converter which has to withstand the full HVDC link voltage U_{HV}. A standard medium power voltage source converter (VSC)

© VDE VERLAG GMBH · Berlin · Offenbach

PCIM Europe 2016, 10 – 12 May 2016, Nuremberg, Germany

can be used to supply the local AC grid. The main focus of this paper is the description of the DC/DC converter, which is a resonant cascade converter.

2. Structure of the resonant cascade converter

Like the most modern converter topologies for HVDC applications, the following tapping concept is also based on a modular approach with several cells connected to each other. Fig. 2 shows the design of a basic modular cell. Each cell consists of a half bridge with two IGBTs T_{pk1} and T_{pk2} and a filter capacitor C_{pk}. The half bridges themselves are connected in a cascaded structure to each other. A LC-circuit with the resonant inductor L_{prk} and the resonant capacitor C_{prk} connects the epicenters of the half bridges and transports the energy from the high voltage side to the low voltage side.

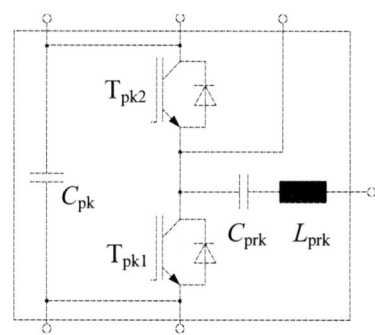

Fig. 2: Cell structure

The DC/DC converter is a bidirectional modification of the non-isolated resonant switched-capacitor step-up converter presented in [5] and [6]. For tapping applications, the diodes of the step up converter need to be replaced by IGBTs. By using this modification, the topology can also operate as a step-down converter. In Fig. 3 the structure of the DC/DC converter is shown. On the low voltage side, a simple rectifier with two switches T_p and T_n is integrated. If only step-down operation is required, simple diodes instead of IGBTs are sufficient to rectify the current. The circuit is controlled by the IGBTs in the half bridges of the cells. As well, T_{pk1} and T_{pk2} are controlled to complementary switch with 50% duty cycle. Several cells in series can be summarized to a module path.

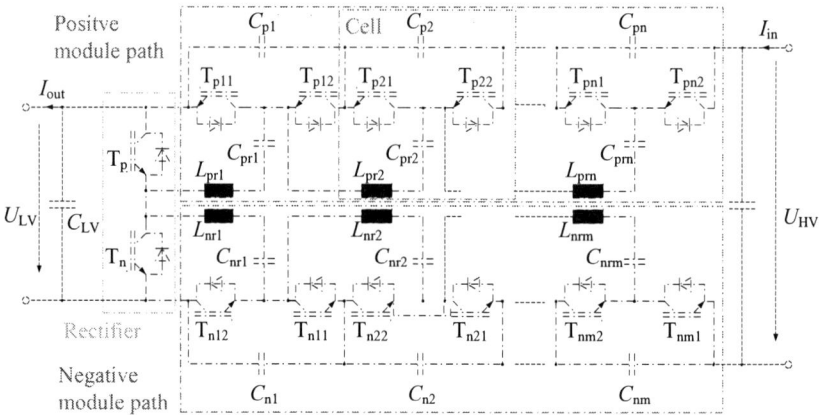

Fig. 3: Structure of the resonant cascade DC/DC converter

3. Operation principle of the DC/DC converter

3.1. Analytical description

The output current strongly depends on the converter configuration. The rectified sine wave can be described by Eq.(1). The currents of the rectifier diodes on the low voltage side are shown

© VDE VERLAG GMBH · Berlin · Offenbach

in Eq.(2) and Eq.(3). In general, the converter works with a fixed voltage transformation ratio of $r = m + n + 1$, where m and n are the numbers of negative and positive modular cell, respectively. There are several different combinations between m and n. When the sum of them is constant, even asymmetrical designs are possible. Furthermore, the converter's output current I_{out} get an asymmetrical character. Besides, for an asymmetrical operation, it is preferred to choose n and m as close as possible. It is important to highlight here that the converter offers a fixed transformation ratio, which cannot be dynamically modified during tapping operation. Only by increasing or decreasing the number of cells in the converter the ratio can be modified. The low voltage can be calculated by Eq.(4). The current in the cells semiconductor, described by Eq.(5), depends only on the input current and has the same magnitude in each cell. So the cells semiconductors have a very low current stress. The resonant frequency is given in Eq.(6).

$$I_{out} = \begin{cases} I_{in} + |I_{in} \cdot n \cdot \pi sin(\omega_r t)| & \text{for } 0 \leq t \leq T_s/2 \\ I_{in} + |I_{in} \cdot m \cdot \pi sin(\omega_r t)| & \text{for } T_s/2 \leq t \leq T_s \end{cases} \quad (1) \qquad U_{LV} = \frac{U_{HV}}{n + m + 1} \qquad (4)$$

$$I_{Tn} = \begin{cases} |I_{in}(n + m) \cdot \pi sin(\omega_r t)| & \text{for } 0 \leq t \leq T_s/2 \\ 0 & \text{for } T_s/2 \leq t \leq T_s \end{cases} \quad (2) \qquad \hat{i}_{T_{pnj}} = \pi \cdot I_{in} \qquad (5)$$

$$I_{Tp} = \begin{cases} 0 & \text{for } 0 \leq t \leq T_s/2 \\ |I_{in}(n + m) \cdot \pi sin(\omega_r t)| & \text{for } T_s/2 \leq t \leq T_s \end{cases} \quad (3) \qquad f_r = \frac{1}{2\pi \cdot \sqrt{L_r \cdot C_r}} \qquad (6)$$

In the resonant LC-circuit the current has a sinusoidal shape. It can be observed from Eq.(7) and Eq.(8) that the current through the LC-circuits is not equal from cell to cell but cumulative with the cells closer to the low voltage side.

$$I_{pLCk} = (n + 1 - k) \cdot \pi \cdot I_{in} \cdot sin(\omega_r t) \qquad (7) \qquad I_{nLCk} = I_{in}(m + 1 - k) \cdot \pi sin(\omega_r t) \qquad (8)$$

As shown in Eq.(1) the output current I_{out} strongly depends on the converters configuration. If a symmetrical converter design with $m = n = 5$ is used, the current has the waveform as shown in Fig. 4. For a full asymmetrical design, for example with $n = 10$ and $m = 0$, the current in the low voltage DC circuit has a very high peak value in the first half-period as shown in Fig. 5.

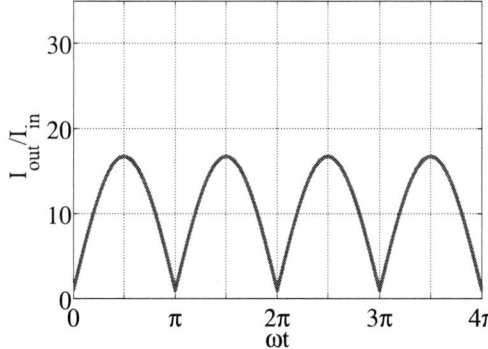

Fig. 4: Output current waveform for a symmetrical cascade converter

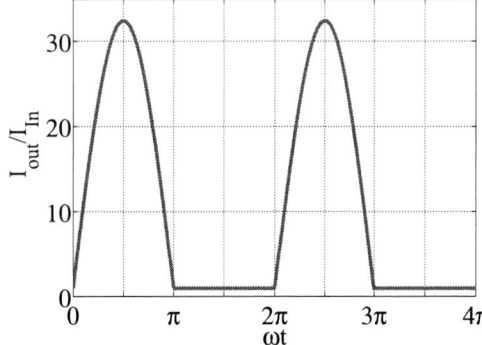

Fig. 5: Output current waveform for a unsymmetrical cascade converter

3.2. Simulation

For the validation of the DC/DC converters functionality, a symmetrical 1 MW 7-level type is simulated. It is composed of 3 positive modular cells and 3 negative modular cells. The resonant frequency is set to 8 kHz, the filter capacitors are $C_p = C_n = 1$ mF, all resonant capacitors are $C_r = 10\ \mu F$ and all resonant inductors are set to $L_r = 40\ \mu H$.

The high voltage level is U_{in} = 14 kV. On the low voltage side and in each cell the voltage is set to 2 kV. In Fig. 7 the simulated waveforms of the converter are shown. On the low voltage side there is a very smooth voltage ripple. In Fig. 8, the simulated waveforms of the converters currents are shown. On the high voltage side, the converter has a smooth DC input current. The currents in the resonant circuit have a sinusoidal shape with a growing magnitude closer to the low voltage side. Furthermore, it can be seen that all switches and diodes operate under zero current condition.

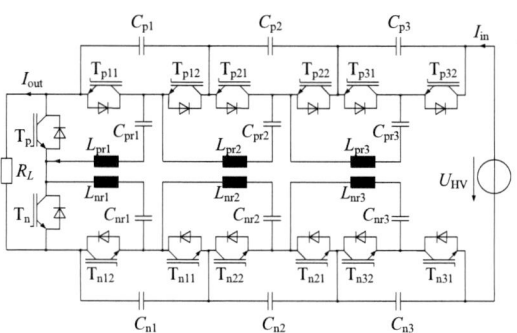

Fig. 6: Simulation model

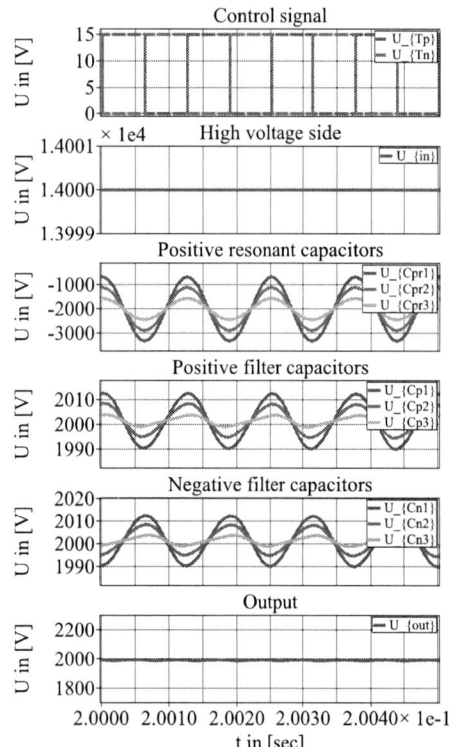

Fig. 7: Simulated voltage waveforms

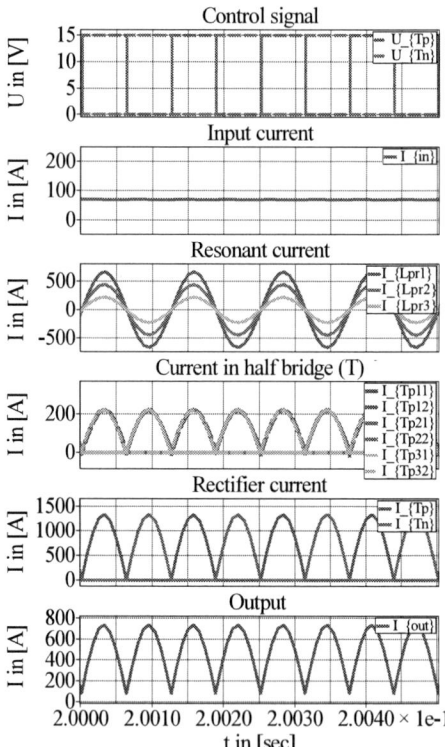

Fig. 8: Simulated current waveforms

4. Performance of the DC/DC converter

To evaluate the performance of the DC/DC converter it is necessary to determine the maximum current at the thermal stress limit of the semiconductors. Due to the fact that tapping applications require only a small amount of power of a full rated HVDC system and the application over the full DC-link voltage, the input current of the tapping converter has to be very low. For this reason, the calculation of the converter losses and the maximum current ratings of the semiconductors a virtual IGBT of a 400 A / 4.5 kV type is used. This is a modification of the datasheet of the 1200 A /4,5 kV IGBT (CM1200HC-90R, Mitsubishi, [7]). A junction temperature of 125°C is assumed, while the baseplate temperature is 80°C. For the maximum current rating the semiconductors in the cells T_{pk1} and T_{pk2} are used. All these switches see a half-wave in each modulation period with the magnitude, shown in Eq.(5). With the used downscaled semiconductors, the maximum DC-input current of the DC/DC converter is I_{in} = 244 A. Each IGBT in each cell can handle this current load for the given current waveform. Furthermore, the voltage at the DC/DC converters output has to be inverted. Therefore a simple inverter is used. For the loss calculation of the inverter the method described in [8] is used. For the full rated HVDC system the DC/DC converter of the shunt tap has to be scaled with a sufficient number of cells. Each cell has a blocking voltage capability of 2,2 kV. This blocking voltage is also the value of the low voltage terminal.

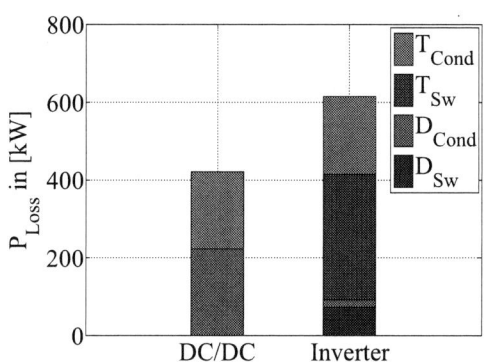

Fig. 9: Losses of the converter

The regarded converter has to be designed for a HVDC reference system, shown in Tab. 1. It has a nominal power of P_{HVDC} = 450 MW and a DC voltage level of U_{DC} = 320 kV. This is the input voltage of the tapping system. The aim is to tap a power of 10-20 % of the nominal HVDC system power.

Nominal power P_{HVDC}	450 MW
Voltage U_{HVDC}	320 kV
DC current I_{HVDC}	1406,25 A

Tab. 1: HVDC reference system

Fig. 9 shows the losses of the regarded tapping system. It can be seen, that the DC/DC converter has no switching losses at all. In the DC/DC converter cells, only the IGBTs produce conduction losses. With the given high voltage value and the cell voltage of 2.2 kV, the number of cells for the DC/DC converter is 145. The diodes have to rectify a very large current on the low voltage terminal, so there are relevant conduction losses too. For the inverter, the switching frequency is set to f_{inv} = 750 Hz. Hence, the switching losses for the IGBTs and diodes are dominant. The total losses of the tapping concept are P_{Loss} = 1036 kW.

Maximum DC-input-current I_{DC} in [A]	Maximum tapping-power $P_{Tap,Max}$ in [MW]	Number of cells No. in [-]	Efficiency of the DC/DC converter $\eta_{DC/DC}$ in [%]	Efficiency of the tapping system η_{Tap} in [%]
244,5	78,24	145	99,46	98,67

Tab. 2: Performance data of the tapping concept for 320 kV

In Tab. 2 the performance values of the tapping concept are summarized. With a maximum DC input current of $I_{DC,max}$ = 244,5 A, the maximum tapping power is $P_{Tap,Max}$ = 78,24 MW. This is 17,3 % of the nominal value of the HVDC reference system. The overall system efficiency is η_{Tap} = 98,67 %. The efficiency of the stand-alone DC/DC converter is $\eta_{DC/DC}$ = 99,46 %.

5. Converter design for HVDC application

The main problem of shunt connected tapping concepts for HVDC systems is the fact, that a rising voltage step-down ratio of a converter leads to a very high current on the low voltage terminal of the tapping system. So it is necessary to qualify the converters components for high currents on low voltage side.

It is important to note that the current in the diodes on the low voltage terminal defined by Eq.(2) and Eq.(3), have to rectify a current, which depends on the total number of cells in the converter. Hence, for a high number of cells, a parallel connection of several diodes is necessary. Fig. 10 show the parallelization of devices in the rectifier circuit of the DC/DC converter. Compared to the current in the cells semiconductors, the current magnitude is very high. For simplicity, only diodes are shown. Without the need of a bidirectional load flow, diodes instead of IGBTs in the rectifier circuit are sufficient. This leads to a special design for the rectifier circuit and the inverter.

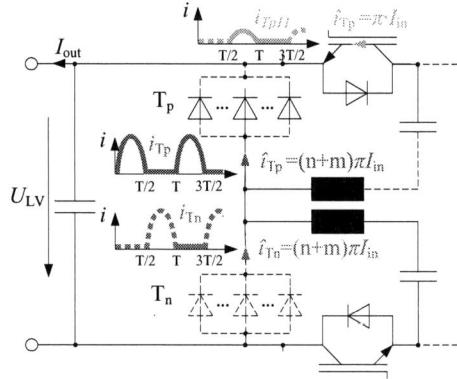

Fig. 10: Parallelization of the rectifier-diodes on the low voltage side

Fig. 11 shows the final converter design for an overall tapping concept which is based on the introduced DC/DC converter. The modular DC/DC converter works as a DC transformer between U_{HV} and U_{LV}. For inverting the voltage U_{LV} in the DC-circuit, a parallelization of several, simple inverters can be used. The final value for the AC voltage can be applied by a transformer on the AC side.

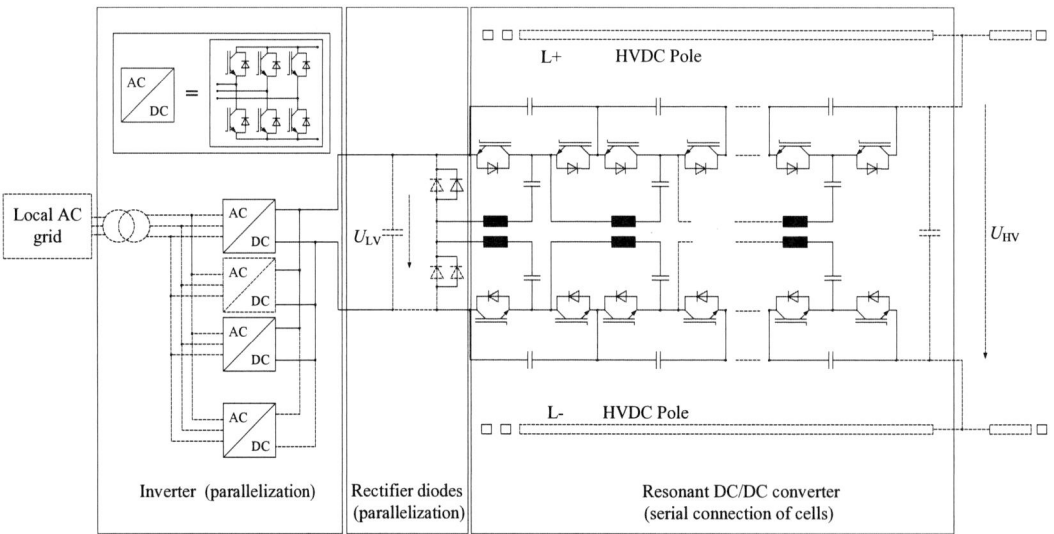

Fig. 11: Converter design for the overall tapping concept for a HVDC transmission systems

© VDE VERLAG GMBH · Berlin · Offenbach

6. Passive components

6.1. Filter capacitors

The filter capacitor C_{pk} of one cell can be calculated by Eq.(9). Every capacitor has a DC-voltage with a sinusoidal voltage ripple, where $k_{u,max}$ is the maximum ripple voltage and $k_{u,min}$ is the minimum ripple voltage. This voltage ripple is set to 10 % and the DC cell voltage is U_{Cell} = 2,2 kV. ΔW_C is the energy hub of one resonant period.

$$C_{pn} = \frac{2 \cdot \Delta W_C}{U_{Cell}^2 \cdot (k_{u,max}^2 - k_{u,min}^2)} \qquad (9)$$

Fig. 12 shows the values of the filter capacitors in the cell of one module path which is next to the low voltage terminal. This cell has the highest current stress. It can be seen that the size of the capacitor grows linear with number of cells in one module path. Furthermore, the size of the capacitors strongly depends on the resonant frequency. Due to the linear development of the capacitor in the cell next to the low voltage terminal, the energy of all capacitors in a module path increases in a square manner. Fig. 13 shows this energy development of one module path for different resonant frequencies depending on the number of cells n. This is a major disadvantage of this topology, because unlimited scalability is not possible.

6.2. LC circuit

The values of the resonant components strongly depend on the number of cells in a module path. For a given resonant frequency, the energetic minimum level of the LC components is given, when the AC-peak voltage $\hat{u}_{AC}$ of the capacitor is the same as the cell voltage U_{Cell}. With this resonant capacitor size, the inductor can be calculated by Eq.(10).

$$L_r = \frac{1}{\omega_r^2} \cdot \frac{1}{C_r} \qquad (10)$$

Fig. 14 shows the amount of energy of a single cell for the LC-circuit at its energetic minimum, the energy in the filter capacitor W_C and the total module energy W_{Module} as the sum of them.

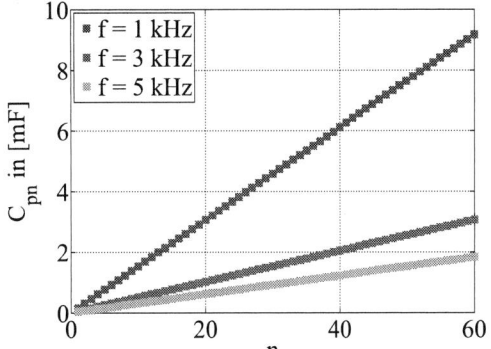

Fig. 12: Size of capacitor in one cell

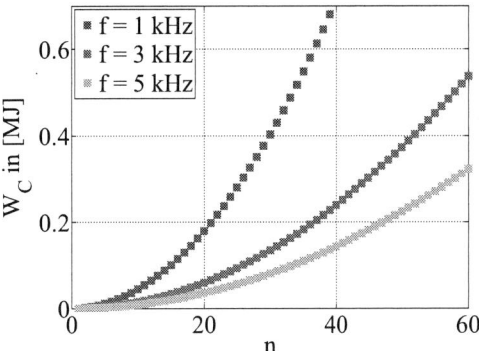

Fig. 13: Energy development of the filter capacitors in one module path

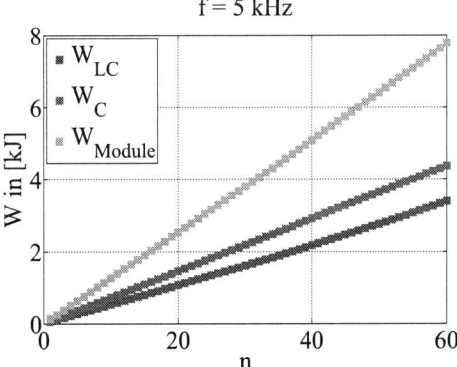

Fig. 14: Energy amounts of a single cell depending on the number of cells

7. Conclusion

A new resonant cascade DC/DC converter for a shunt connected HVDC tap is presented in this paper. The combination of this DC/DC converter with a simple inverter structure allows the supply of small AC grids along a HVDC transmission line. Like most converters for HVDC applications, a modular structure for a better scalability is used. For the validation of the DC/DC converters functionality, a computer simulation of the DC/DC converter is done. The calculation of the converters performance has shown that the concept is able to tap a power of under 20 % of a full rated HVDC system. However, due to the linear development of the sizes of the cell capacitors, an arbitrary high scalability of the converter is not possible. But for HVDC systems with a voltage up to 320 kV, this concept may be a viable solution for tapping applications.

References

[1] Noman Ahmed, Staffan Norrga, H-P Nee, A. Haider, D. van Hertem, Lidong Zhang, and Lennart Harnefors, editors. *HVDC SuperGrids with modular multilevel converters — The power transmission backbone of the future: Systems, Signals and Devices (SSD), 2012 9th International Multi-Conference on*, 2012.

[2] Aghaebrahimi, M. R. and R. W. Menzies. Small power tapping from hvdc transmission systems: a novel approach. *Power Delivery, IEEE Transactions on*, 12(4):1698–1703, 1997.

[3] M. Bahram, M. Baker, J. Bowles, R. Bunch, J. Lemay, W. Long, J. McConnach, R. Menzies, J. Reeve, and M. Szechtman. Integration of small taps into (existing) hvdc links. *Power Delivery, IEEE Transactions on*, 10(3):1699–1706, 1995.

[4] J. Maneiro, S. Tennakoon, and C. Barker, editors. *Scalable shunt connected HVDC tap using the DC transformer concept: Power Electronics and Applications (EPE'14-ECCE Europe), 2014 16th European Conference on*, 2014.

[5] Wu Chen, A. Huang, Chushan Li, and Gangyao Wang, editors. *A high efficiency high power step-up resonant switched-capacitor converter for offshore wind energy systems: Energy Conversion Congress and Exposition (ECCE), 2012 IEEE*, 2012.

[6] W. Chen, Huang, A. Q., C. Li, G. Wang, and W. Gu. Analysis and comparison of medium voltage high power dc/dc converters for offshore wind energy systems. *Power Electronics, IEEE Transactions on*, 28(4):2014–2023, 2013.

[7] MITSUBISHI ELECTRIC. Datasheet hvigbt module cm1200hc-90r.

[8] A. Wintrich, U. Nicolai, W. Tursky, and T. Reimann. *Applikationshandbuch Leistungshalbleiter*. SEMIKRON International GmbH, 2010.

PCIM Europe 2016, 10 – 12 May 2016, Nuremberg, Germany

Analysis of voltage and current unbalance in a multi-converter topology for a DC-based offshore wind farm.

Thomas LAGIER, SuperGrid Institute, Villeurbanne, France
LAPLACE, Université de Toulouse, CNRS, INPT, UPS, Toulouse, France,
thomas.lagier@supergrid-institute.com
Philippe LADOUX, LAPLACE, Université de Toulouse, CNRS, INPT, UPS, Toulouse, France,
philippe.ladoux@laplace.univ-tlse.fr

Abstract

High Voltage Direct Current (HVDC) is increasingly being used for electric power transmission over long distances. In the case of offshore wind farms, HVDC becomes more competitive than HVAC beyond a distance of 80 km. In this paper, a DC-based offshore wind farm is under consideration. In order to step-up the voltage before the transmission line, an isolated DC-DC converter is used. This converter is achieved by associating several elementary isolated DC-DC converters. The key issue with this multi-converter topology is the balancing of currents and voltages between the elementary converters. This paper proposes an analysis of the unbalance by taking into account the disparities on the components used within the elementary converters. After this study, a solution is proposed to achieve a satisfactory balancing of currents and voltages.

1. Introduction

Today, it is well established that a HVDC link can be more competitive than a High Voltage Alternating Current (HVAC) link. In the case of submarine cables, the break-even distance is about 80 km [1]. The study realized in [2] has shown that a full DC offshore wind farm without a platform would offer a lower cost and a higher efficiency than actual solutions. In [3], a full DC offshore wind farm arrangement was proposed. Figure 1 shows the corresponding diagram. Several DC-DC converters are used in order to step up the voltage before the transmission line.

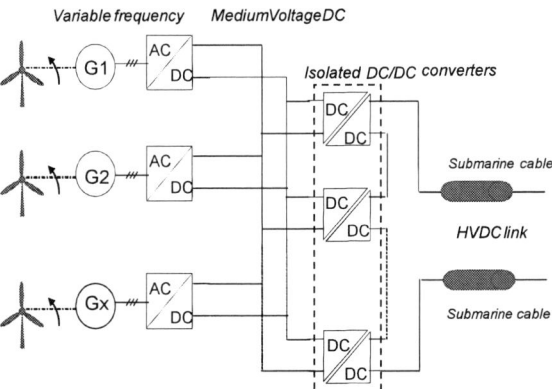

Fig. 1: DC-based offshore wind farm arrangement.

The choice of the DC-DC converter topology is linked to the implementation of the power semiconductors. To achieve a high voltage operation, the series association of switches is the simplest. Nevertheless, compared to multilevel topologies, this solution suffers from higher switching losses and higher harmonic distortion for voltage and current [4]. The Modular

© VDE VERLAG GMBH · Berlin · Offenbach

Multilevel Converter based topologies [5], [6] show higher efficiencies [7] and higher quality voltage waveforms. However, since the power of one DC-DC converter is important, the use of medium frequency is not permitted. Consequently, the footprint of the passive components can be huge. In [3], elementary DC-DC converters were associated in order to form a DC-DC converter with high current and high voltage capabilities. Those elementary DC-DC converters, which operate with lower voltage and power (< 2 MW), can have a medium switching frequency (some kHz) in order to reduce the size of the passive components.

According to the system requirements and the rating of the elementary DC-DC converters several types of association can be implemented in the DC-DC converter. Fig. 2 presents the two types of association which have an interest in our case.

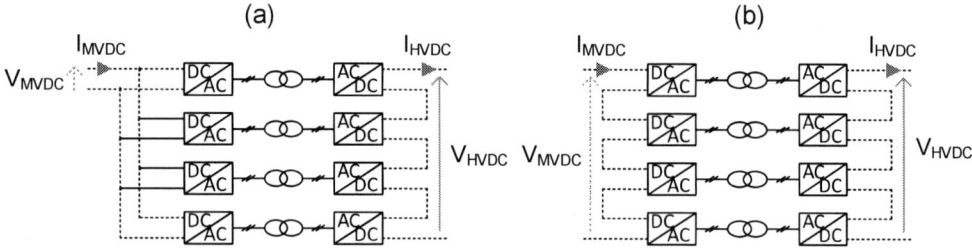

Fig. 2: Types of elementary DC-DC converters associations

The first type corresponds to the Input-Parallel-Output-Series (IPOS) association and is depicted in Fig. 2a. The second corresponds to the Input-Series-Output-Series (ISOS) association and is depicted in Fig. 2b. The main advantages of these solutions are their modularity and the redundancy.

During their operation, because of disparities on the components used in the elementary DC-DC converters, voltage and current unbalance can occur. For the reliability of the elementary DC-DC converters, they should be limited. Consequently, a study, taking into account the disparities of the components, is required. Nevertheless, due to the high number of elementary DC-DC converters, simulation circuits can be complex and time-consuming. Therefore, this paper proposes an analytical method to estimate the unbalance and a method to correct it.

2. DC-DC converter topologies

In [3], three topologies were proposed for the elementary converters. The most efficient was the Series Resonant Converter (SRC). The diagram is depicted in Fig. 3.

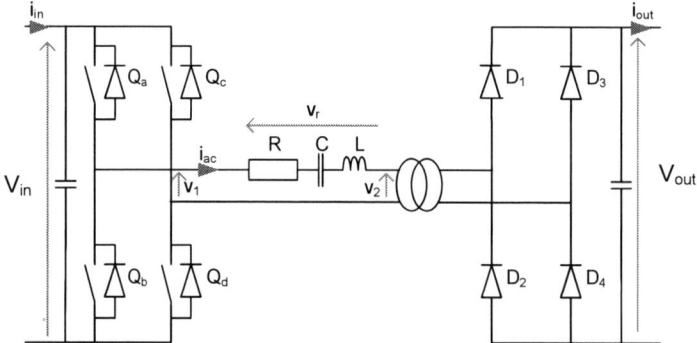

Fig.3: Series Resonant Converter (SRC) topology.

This topology can be controlled by the phase shift of the inverter legs or by a variable switching frequency. For the first solution, closed loop operation is required in order to control the power and to balance the currents and the voltages. To ensure these two requirements, specific control loops have to be implemented. In our study, the wind turbines are considered as constant power sources and we assume that the HVDC line voltage (V_{HVDC}) is fixed by the onshore receiving substation. Consequently, from the point of view of the DC transmission system, the elementary converters behave like classical AC transformers. Thus, they can operate at fixed switching frequency and theoretically with an open loop control.

According to the choice of the resonant frequency, the output voltage and the input current, several conduction modes can be obtained [8]. In our study, two conduction modes are considered, that is the Continuous Conduction Mode (CCM) and the Discontinuous Conduction Mode (DCM). Voltage and current waveforms are shown in Fig.4. The considered specifications are presented in Table 1.

Input voltage - V_{in} (kV)	7~7.2
Output voltage - V_{out} (kV)	7
Output current - I_{out} (A)	215
Transformer's turn ratio - m	1
Switching frequency - f_s (kHz)	10
Leakage inductor - L (µH)	145
Resonant capacitor in CCM - C (µF)	2.3
Resonant capacitor in DCM - C (µF)	1.32
Series resistor - R (mΩ)	120

Table 1: specifications for the elementary DC-DC converters.

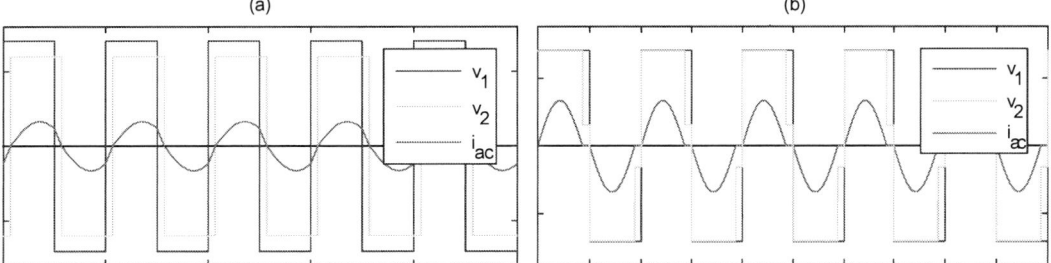

Fig.4: waveforms for the two conduction modes considered.

2.1 Continuous Conduction Mode

In this case, the switching frequency has to be higher than the resonant one so that the current in the AC link flows continuously. By neglecting the harmonic distortion on the current waveform and using a vector diagram, the voltage-current characteristic of the converter can be found [9]. This characteristic is calculated with (1).

$$V_{in}^2 = (V_{out})^2 + \left(\frac{\pi^2 X}{8}\right)^2 I_{out}^2 \tag{1}$$

Where X is the reactance of the resonant circuit at the switching frequency f_s given by (2).

$$X = 2\pi L f_s - \frac{1}{2\pi C f_s} \tag{2}$$

The converter operating in CCM shows an elliptical output characteristic (Fig. 5a).

2.2 Discontinuous Conduction Mode

When the converter operates in DCM, the current is clamped to zero after each half cycle of the AC current. Voltage and current waveforms are shown in Fig. 4b .To ensure this mode, the switching frequency has to be lower than the resonant one. Moreover, there is a condition on the output voltage and the resonant capacitor peak voltage. To perform DCM, the voltage, across the resonant capacitor has to be low enough to avoid the spontaneous turn-on of the diodes at the end of each current half-cycle. By neglecting the threshold voltage of the diodes, this condition can be expressed by (3).

$$V_{out} > V_{in} - v_{cmax} \tag{3}$$

The peak voltage across the resonant capacitor is linked to the AC current by relation (4):

$$v_{cmax} = i_{acmax}\sqrt{\frac{L}{C}} \tag{4}$$

The average output current is given by (5)

$$I_{out} = \frac{2i_{acmax}f_s}{\pi f_0} \tag{5}$$

Integrating (5) and (4) in (3), we obtain the operating condition in DCM:

$$V_{out} > \frac{I_{out}}{4Cf_s} - V_{in} \tag{6}$$

By using the state plan analysis [8], the characteristic of the converter in DCM can be found (R is the series resistor of the resonant circuit):

$$V_{in} = V_{out} + I_{out} \cdot \frac{f_0}{f_s} \cdot \frac{\pi^2}{8} \cdot \frac{R}{1 - \frac{\pi^2}{2}f_0RC} \tag{7}$$

The converter operating in DCM has a linear characteristic (see Fig. 5b).

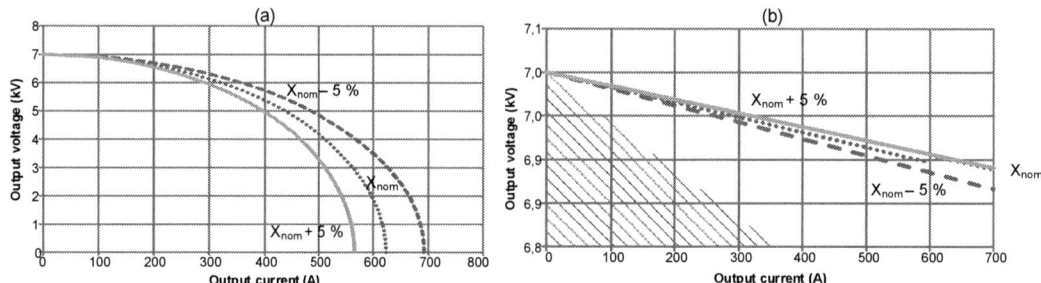

Fig. 5: characteristics of the SRC in CCM (a) and in DCM (b).

2.3 Impacts of the disparities on the characteristics of the elementary converters

In Fig. 5, the converter output characteristics are plotted for the two operating modes. Considering the specifications of Table 1, disparities of ± 5 % have been included in the resonant circuit parameters. In CCM, voltage and current variations are bigger than DCM and will cause a larger unbalance.

3. Input-Parallel-Output-Series association (IPOS)

In IPOS association, input voltage and output current are the same for all the elementary DC-DC converters. Consequently, the following relations can be written.

$$V_{MVDC} = V_{in-i}, i = \{1,2 \dots p\} \tag{8}$$

$$V_{HVDC} = \sum_{i=1}^{p} V_{out-i} \tag{9}$$

$$I_{MVDC} = \sum_{i=1}^{p} I_{in-i} \tag{10}$$

$$I_{HVDC} = I_{out-i}, i = \{1,2 \dots p\} \tag{11}$$

Where p, is the number of elementary DC-DC converters associated in IPOS. Considering, the characteristic of the elementary converters and the previous relations, the global characteristic for p elementary converters in IPOS association can be found. However, the relations (1) and (7) have to be rearranged:

$$I_{in}^2 = I_{HVDC}^2 \left(1 - \frac{I_{HVDC}^2}{V_{MVDC}^2} \left(\frac{\pi^2 X_i}{8} \right)^2 \right) \text{ in CCM} \tag{12}$$

$$I_{in} = I_{HVDC} \left(1 - \frac{I_{HVDC}}{V_{MVDC}} \cdot \frac{f_0}{f_s} \cdot \frac{\pi^2}{8} \cdot \frac{R}{1 - \frac{\pi^2}{2} f_0 RC} \right) \text{ in DCM} \tag{13}$$

By integrating the relations (12) and (13) in (10), the characteristics of p elementary converters associated in IPOS can be found.

$$I_{MVDC} = \sum_{i=1}^{p} \sqrt{I_{HVDC}^2 \left(1 - \frac{I_{HVDC}^2}{V_{MVDC}^2} \left(\frac{\pi^2 X}{8} \right)^2 \right)} \text{ in CCM} \tag{14}$$

$$I_{MVDC} = I_{HVDC} \left(p - \frac{I_{HVDC}}{V_{MVDC}} \sum_{i=1}^{p} \frac{f_{0-i}}{f_{s-i}} \cdot \frac{R_i}{1 - \frac{\pi^2 f_{0-i} R_i C_i}{2}} \right) \text{ in DCM} \tag{15}$$

4. Input-Series-Output-Series association (ISOS)

In ISOS association, input and output currents are the same for all the elementary DC-DC converters. Consequently, the following relations can be written.

$$V_{MVDC} = \sum_{p=1}^{n} V_{in-i} \tag{16} \qquad V_{HVDC} = \sum_{i=1}^{n} V_{out-i} \tag{17}$$

$$I_{MVDC} = I_{in-i}, i = \{1,2 \dots n\} \tag{18} \qquad I_{HVDC} = I_{out-i}, i = \{1,2 \dots n\} \tag{19}$$

Where n, is the number of elementary converters in ISOS association. By arranging the characteristic (1) and (7), the input voltage of the elementary converters can be expressed:

$$V_{in} = \sqrt{\frac{\left(\frac{\pi^2 X}{8} \right)^2 I_{HVDC}^2}{1 - \left(\frac{I_{MVDC}}{I_{HVDC}} \right)^2}} \text{ in CCM} \tag{20}$$

$$V_{in} = \frac{I_{HVDC}}{1 - \frac{I_{MVDC}}{I_{HVDC}}} \cdot \frac{f_0}{f_s} \cdot \frac{\pi^2}{8} \cdot \frac{R}{1 - \frac{\pi^2}{2} f_0 RC} \text{ in DCM} \tag{21}$$

By integrating the relations (20) and (21) in (16), the characteristics of p elementary converters in ISOS association can be found:

$$V_{MVDC} = I_{HVDC} \sum_{i=1}^{n} \sqrt{\frac{\left(\frac{\pi^2 X_i}{8} \right)^2}{1 - \left(\frac{I_{MVDC}}{I_{HVDC}} \right)^2}} \text{ in CCM} \tag{22}$$

$$V_{MVDC} = \frac{\pi^2}{8} \cdot \frac{I_{HVDC}}{1 - \frac{I_{MVDC}}{I_{HVDC}}} \sum_{i=1}^{n} \frac{f_{0-i}}{f_{s-i}} \cdot \frac{R_i}{1 - \frac{\pi^2}{2} f_{0-1} R_i C_i} \text{ in DCM} \tag{23}$$

5. Theoretical analysis of the voltage and current unbalance

The previous expressions allowed the input characteristic ($I_{MVDC} = f(V_{MVDC})$ or $V_{MVDC} = f(I_{MVDC})$) of several elementary converters associated in IPOS or in ISOS to be found. By considering a given power level, and by neglecting the losses in the elementary converters, the output current is known. Consequently, the characteristics of the converters can be plotted and matched with those of the wind turbines. The intersection of the two characteristics corresponds to the operating point and allows estimating input current and input voltage (I_{MVDC} and V_{MVDC}). Knowing the input current, relations (1) and (7) or (20) and (21) can be used in order to calculate the voltages of each elementary converter.

In our case, tolerances of ± 5 % are assumed of the resonant circuit parameters. These values were chosen randomly. Moreover, calculations were checked with time domain simulations for a DC-DC converter with a lower number of elementary converters. Fig. 6 presents the operating points for 20 converters associated in IPOS and in ISOS. The HVDC link voltage was set to 140 kV and the wind turbines power to 30 MW in order to satisfy the rating of the elementary converter (see Table 1).

Around the operating point, the input voltage of the converters rises suddenly. Because of the diode on the secondary side on the elementary converters, the output voltage cannot be higher than the input one. As a result, operation with the output current lower than the input one is not permitted. This condition will introduce a pole in $I_{MVDC} = I_{HVDC}$ in the input characteristics. Mathematically, this will introduce an important voltage rise when the input and output current become closer. Table 2 presents the input voltage and current estimated by the previous method.

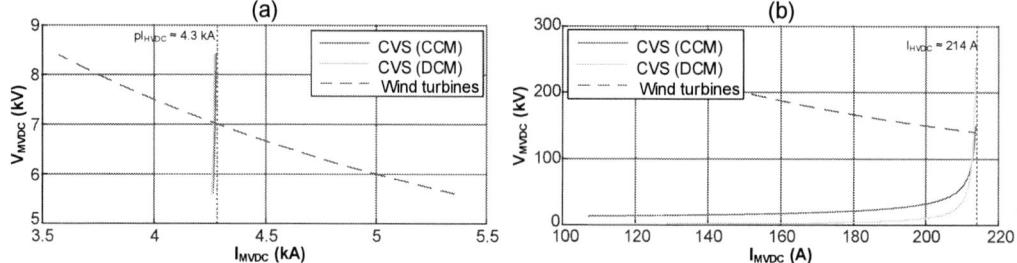

Fig. 6: operating point of the wind turbines and 20 converters in IPOS (a) and in ISOS (b) associations.

	IPOS		ISOS	
	CCM	DCM	CCM	DCM
Input voltage - V_{MVDC} (kV)	7.02	7.03	141.4	140.7
Input current - I_{MVDC} (A)	4275	4264	212.2	213.2

Table 2: operating point for 20 converters.

For the IPOS association, input currents and output voltages of each elementary converter are shown in Fig.7.

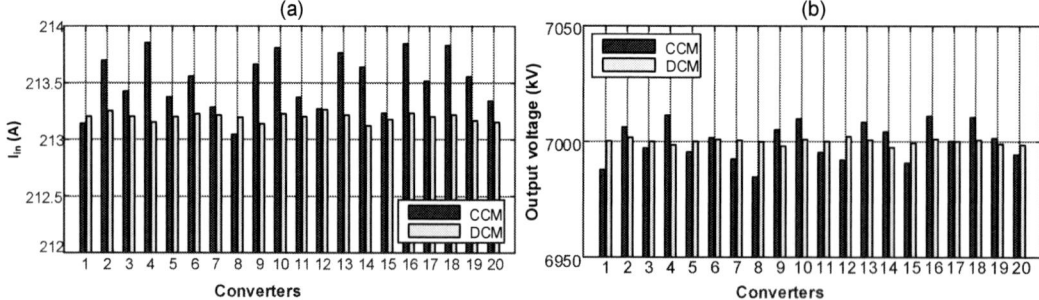

Fig. 7: input current (a) and output voltage (b) of 20 elementary converters in IPOS association.

For the ISOS association, the input voltage (a) and output voltage (b) sharing are shown in Fig. 8.

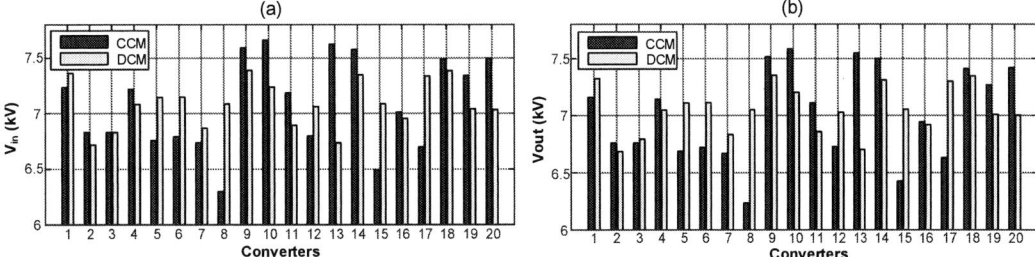

Fig. 8: input voltage (a) and output voltage (b) of 20 elementary converters in ISOS association.

In these figures, it can be seen that the CCM is more sensitive to the disparities in the resonant circuit parameters than the DCM. Indeed, since the CCM shows a higher voltage drop (Fig. 5.a), a small variation on the reactance of the resonant circuit will cause a significant changing in the characteristic, compared to the DCM.

In IPOS association, the voltage and current unbalance is limited and would not require a correction. However, this is not the case in ISOS association, because the voltages can reach 20 % of the nominal voltage for the CCM and 5 % for the DCM.

6. Correction of the voltage and current unbalance

In order to balance the voltage and the current, the switching frequency of each elementary converter is adjusted in order to get the same series impedance. Considering the values of the resonant circuit components, the new switching frequencies are calculated in order to respect the following condition, for each elementary converter:

$$V_{in} - V_{out} = DV \qquad (24)$$

Where DV is the nominal voltage drop obtained with the theoretical resonant circuit parameters and for a given power level. Considering (25), the new switching frequencies are presented in Fig. 9.

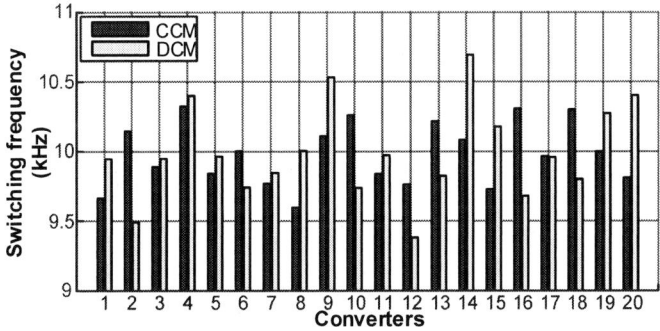

Fig.9: switching frequencies of the elementary converters

The theoretical switching frequency range is [9.6~10.3] kHz in CCM and [9.5~10.7] kHz in DCM. These ranges are acceptable and would not require an over rating of the passive components (transformers, filtering capacitors). With the new switching frequencies, the output voltages of each elementary converter are shown in Fig. 10.

With the switching frequency correction, voltage and current unbalance is canceled. In this study, only variations on the series impedances were considered. Nevertheless, other elements in the converter may cause a larger unbalance (magnetizing inductance, filtering capacitors, on state resistance of the switches, etc.) and a closed loop operation would be necessary. Time domain simulations have been performed for a DC-DC converter with a lower number of elementary converters. They took into consideration variations on the values of all the components. After all, the results were close to the calculations previously performed and the frequency correction showed good results.

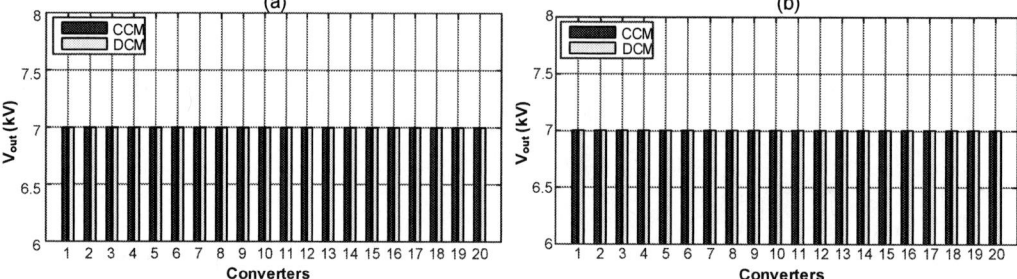

Fig. 10: output voltages of the elementary converters in IPOS (a) and in ISOS (b) associations with the frequency correction

7. Conclusion

Multi converter topologies can be a good solution for high voltage and high power applications. Several elementary DC-DC converters can be associated in IPOS or in ISOS. Due to disparities on the components used within the elementary converters, voltage and current unbalance may occur. Consequently, a dedicated control system is required in order to rebalance voltages or currents. Because of the huge number of components, the simulation of such a topology is complex. This paper has proposed a theoretical analysis based on the characteristics of the series resonant converter in continuous or discontinuous conduction mode. The calculations have shown that the ISOS association causes the largest unbalance. Moreover, the disparities have a more significant impact on the unbalance when elementary converters operate in CCM. To balance currents or voltages, the switching frequency of each elementary converter is adapted considering the value of the series impedance. This method will be validated on a small scale prototype in order to take into account the influence of all the components.

8. References

[1] K. Meah and S. Ula, "Comparative Evaluation of HVDC and HVAC Transmission Systems," in *Power Engineering Society General Meeting, 2007. IEEE*, 2007, pp. 1–5.

[2] J. Pan, S. Bala, M. Callavic, and P. Sandeberg, "DC Connection of Offshore Wind Power Plants without Platform," in *13th Wind Integration Workshop, Berlin, Germany*.

[3] T. Lagier and P. Ladoux, "A comparison of insulated DC-DC converters for HVDC off-shore wind farms," in *International Conference on Clean Electrical Power (ICCEP), Toarmina, Sicily*, 2015, pp. 33–39.

[4] A. M. Massoud, S. J. Finney, and B. W. Williams, "Multilevel converters and series connection of IGBT evaluation for high-power, high-voltage applications," in *Second International Conference on Power Electronics, Machines and Drives (PEMD)*, 2004, vol. 1, pp. 1–5 Vol.1.

[5] J. Ferreira, "The Multilevel Modular DC Converter," *IEEE Transactions on Power Electronics,*, vol. 28, no. 10, pp. 4460–4465, Oct. 2013.

[6] T. Luth, M. M. C. Merlin, T. C. Green, F. Hassan, and C. D. Barker, "High-Frequency Operation of a DC/AC/DC System for HVDC Applications," *IEEE Transactions on Power Electronics*, vol. 29, no. 8, pp. 4107–4115, Aug. 2014.

[7] P. Ladoux, N. Serbia, and E. Carroll, "On the Potential of IGCTs in HVDC," *IEEE Journal of Emerging and Selected Topics in Power Electronics*, vol. 03, no. 99, pp. 1–1, 2015.

[8] Erickson, *Fundamentals of Power Electronics*. Springer US, 2013.

[9] H. Foch, P. Ladoux, and H. Piquet, "Association de convertisseurs assurant une liaison énergétique," *Techniques de l'ingénieur Convertisseurs électriques et applications*, vol. base documentaire : TIB253DUO., no. ref. article : d3178, 2016.

PCIM Europe 2016, 10 – 12 May 2016, Nuremberg, Germany

Experimental Demonstration of a Solid-State Damping Resistor for HVDC applications

Konstantin Vershinin, GE's Grid Solutions Business, UK, konstantin.vershinin@ge.com
Ikenna Efika, GE's Grid Solutions Business, UK
David Trainer, GE's Grid Solutions Business, UK
Colin Davidson, GE's Grid Solutions Business, UK
Nick Wright, University of Newcastle, UK, nick.wright@ncl.ac.uk
Amit Tiwari, University of Newcastle, UK

Abstract

Significant drivers for Silicon Carbide based semiconductor devices focus on the areas of increased power density, voltage and thermal rating. Combination of high breakdown field, good thermal conductivity and high power densities of SiC opens opportunities for innovative device concepts. This work presents the experimental investigations undertaken to characterise the performance of 3.3 kV SiC-based Solid State Damping Resistor installed within a High Voltage Direct Current Line Commutated Converter Valve as a replacement for the passive damping resistor in a valve auxiliary circuit.

1. Introduction

GE's Grid Solutions Business has provided High Voltage Direct Current (HVDC) solutions to support Electrical Power Networks for more than 50 years. A key component is the Line Commutated Converter (LCC) Valve which is the core power converter technology for the traditional, and mature, LCC HVDC market. Market analysis shows LCC technology will remain a strong HVDC market segment and is the preferred choice for bulk power transmission projects even with recent interest in Voltage Source Converter technology. In a typical application each converter valve can comprise 100 or more series-connected power thyristors. In practice, these thyristors are grouped within Valve modules cascaded to make the LCC valve. An example of GE's Grid Solutions Business LCC Valve module is shown in Figure 1 and the string of series-connected thyristors can be easily seen.

A thyristor is a very robust device can handle thousands of amperes of current during on-state conduction. Construction of the thyristor optimised for low conduction loss but leads to vulnerability during initial stages of the turn-on process. A current limiting reactor, usually of saturating type, is installed in the thyristor current path to limit the rate of rise of current below a critical $\delta i/\delta t$ value specified by the device manufacturer. Variability in the thyristor manufacturing process leads to thyristors exhibiting different amounts of reverse recovery charge which causes voltage unbalance when the devices are series-connected. A damping circuit built from a series connection of Resistors (R_D) and Capacitors (C_D) is installed to equalise such off-state voltage variation across the thyristors. The capacitor value C_D is derived from the maximum desired voltage variation across series-connected devices and the resulting electrical circuit for the valve is as shown in Figure 1a. In practice this circuit is split into two branches.

During thyristor turn-on the damping capacitor discharges a current into the thyristor with amplitude only limited by the resistance of the damping branch. Such current can damage the thyristor during the initial stages of turn-on when the thyristor is not homogenously conducting. To prevent this, the damping resistance R_D is specified to limit the maximum allowed in-rush current from the damping capacitor during the initial stages of thyristor turn-on. The resistor experiences this current each time there is a voltage change across thyristor. In practice, eight such pulses occur due to the four commutation overlaps that occur during every fundamental AC cycle with the pulses at thyristor turn-on and turn-off transients

© VDE VERLAG GMBH · Berlin · Offenbach

2197

being the most demanding. This resistor operating in pulsed regime is a bulky specialist component. The authors have proposed earlier to replace this resistor with a solid-state device based on Silicon Carbide material [1]. Simulation studies of LCC converter equipped with new device named Solid State Damping Resistor (SSDR) have demonstrated promise [2].

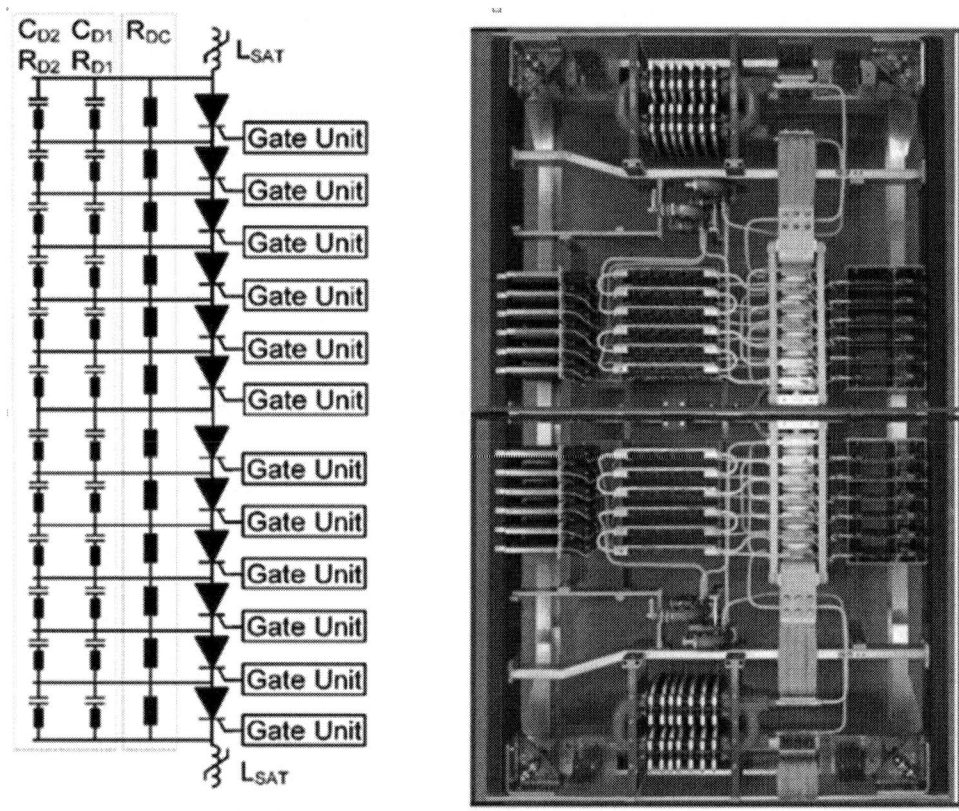

a) Electrical   b) Physical

Figure 1.   Alstom Grid LCC Valve module: electrical (a) and physical (b) circuit.

In this paper initial results of experimental evaluation of 3.3 kV SiC SSDR device are presented. In section 2 production and characterisation of SSDR die and SSDR device is discussed. Section 3 presents HVDC thyristor level test setup configuration. In section 4 experimental results of SSDR performance during thyristor turn-on and turn-off are presented and discussed. Conclusions are drawn in section 5.

2. SSDR Prototype

2.1. SSDR die

The 3.3kV SSDR die utilised is based on a commercial 3.3 kV SiC JFET device but with a modified electrode structure used to connect the JFET's source and drain contact terminals on the chip level to create a two terminal device [4]. These die have been packaged in the reusable pressure contact package for initial electrical characterisation. Figure 2 shows typical IV characteristics for different ambient temperatures from 25 degC up to 175 degC. As expected, the device shows positive temperature coefficient and starts the transition from linear (ohmic) to saturation region at applied voltage of ~20 V. In the linear region devices show equivalent on resistance in the region of 0.5 Ohm for 25 degC which rises to ~1.5 Ohm for 175 degC. Saturation current measured at 35 V across device reduces from 27.5 A at 25 degC to 16 A at 175 degC.

© VDE VERLAG GMBH · Berlin · Offenbach

PCIM Europe 2016, 10 – 12 May 2016, Nuremberg, Germany

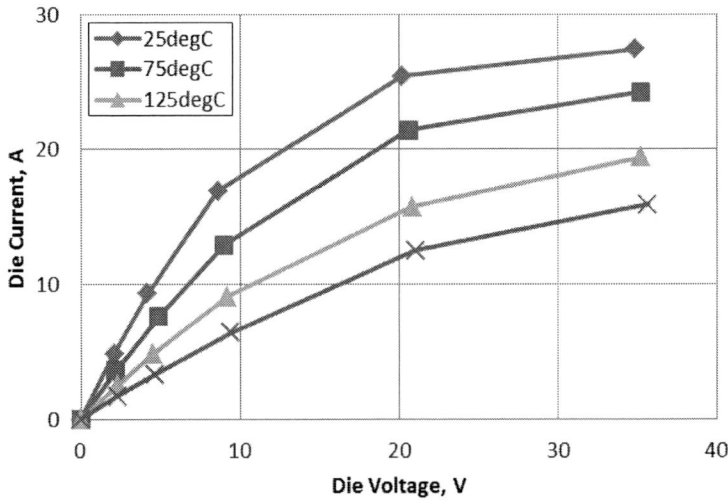

Figure 2. Temperature Dependence of 3.3 kV SSDR die IV Characteristics

TCAD simulations [5] of the device structure have been performed to understand expected change in the IV characteristics with increase in the device rated voltage to 10 kV which would be required to implement SSDR within commercial HVDC thyristor valve. Drift region thickness and doping concentration have been adjusted to achieve the required breakdown voltage while the rest of the structure was kept the same. Simulated IV characteristics for 3.3 kV and 10 kV device are shown in Figure 3.

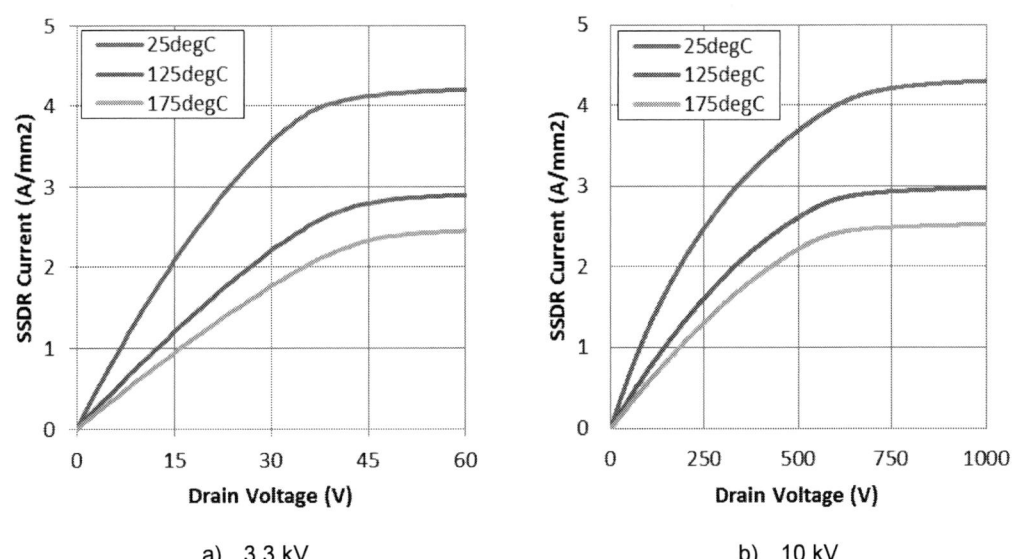

a) 3.3 kV   b) 10 kV

Figure 3. Simulated IV characteristics for 3.3 kV and 10 kV SSDR devices

The main difference which can be observed from the graph in Figure 3 is on set of saturation characteristic occurs much later for a 10 kV component which is due to a much higher resistance of the drift region. In practice, this means that device will behave as a resistor for much higher range of applied voltage. The top structure of the device in the simulations have remained the same for both devices which results in nearly identical saturation current level for devices. This results in the same area requirements for 10 kV device as for 3.3 kV device. Based on this assumption equivalent resistance of 10 kV device can be estimated at ~28 Ohm for 25 degC rising to ~45 Ohm at 175 degC. These values are comparable with the value of passive resistor used within standard thyristor damping circuit.

© VDE VERLAG GMBH · Berlin · Offenbach

PCIM Europe 2016, 10 – 12 May 2016, Nuremberg, Germany

2.2. SSDR device

Following initial IV characterisation 3.3 kV SSDR die has been matched in pairs on the level of saturation current observed at applied voltage of 20 V and packaged in a two die prototype package as shown in Figure 4.

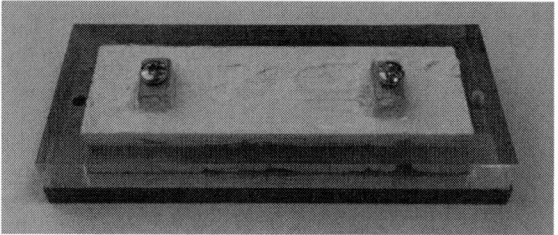

Figure 4.   SSDR 50A 3.3kV prototype assembly

Within the package each SSDR die has been attached to copper base plate which forms the drain of the device. The source of each die has been connected to a separate source pillar as shown in Figure 4. This arrangement allows SSDR tests with two different current levels of 25 A and 50 A. Following assembly the SSDR device has been connected within the thyristor level of the test circuit to assess its performance in application. In this paper single device connection have been used during all tests.

3. Experimental Setup

The main stress is experienced by damping circuit during turn on and turn off of the thyristor valve. A test circuit has been built to replicate these conditions. The circuit has been based on a cascaded arrangement of two LC discharge circuits as shown in Figure 5a. The 'Main Pulse' circuit is used to simulated turn off of the valve. To simulate appropriate turn off condition the circuit should provide required turn off di/dt as well as ensure conduction within valve thyristors is fully established. Due to construction of the HVDC thyristor it takes around 1ms for the conduction to spread across the full device area. Therefore, the 'Main Pulse' circuit should provide a pulse of the sufficient duration. A pulse of ~2.5 ms has been used in the current study. The 'Turn on Pulse' circuit emulates effect of stray capacitances found within LCC converter and provides a much faster turn on transient in the range of 10 us.

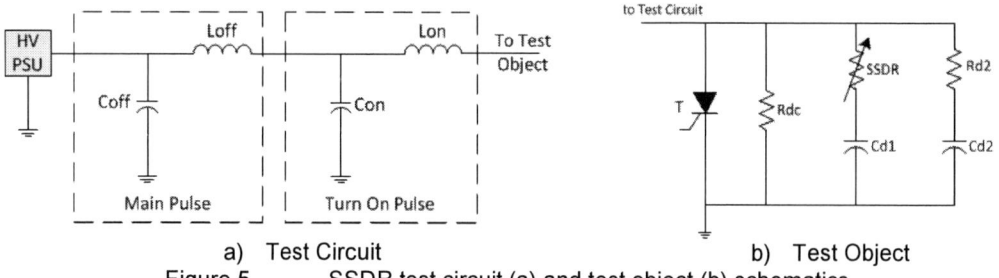

a)   Test Circuit   b)   Test Object
Figure 5.   SSDR test circuit (a) and test object (b) schematics

A test object based on the standard design of LCC valve with one thyristor level has been used within this study. Within the main damping branch the passive resistor has been replaced with a SSDR as shown in Figure 5b. During thyristor turn on and turn off current flows out of the damping circuit. Therefore, SSDR has been connected with its drain tied to the capacitor and source wired to the thyristor anode.

During the test, capacitors within test circuit are charged to initial voltage by HV power supply unit. Once the test voltage is reached the power supply is disconnected and thyristor within

© VDE VERLAG GMBH · Berlin · Offenbach

thyristor level is fired. This generates turn on transient on the thyristor and associated SSDR. Following quasi-sinusoidal current thyristor turns off once current becomes negative. This charges test circuit capacitors to negative voltage and introduces turn off stress within SSDR device. Test voltage has been varied from 100 V to 1500 V with 100 V increments.

4. SSDR Experimental Results

4.1. Performance during turn-on

Turn-on transient represents the highest stress for SSDR. At lower voltage SSDR does not enter saturation and typical RC discharge characteristics are observed. Once the test voltage is high enough to force SSDR into saturation it exhibits typical voltage and current characteristics shown in Figure 6.

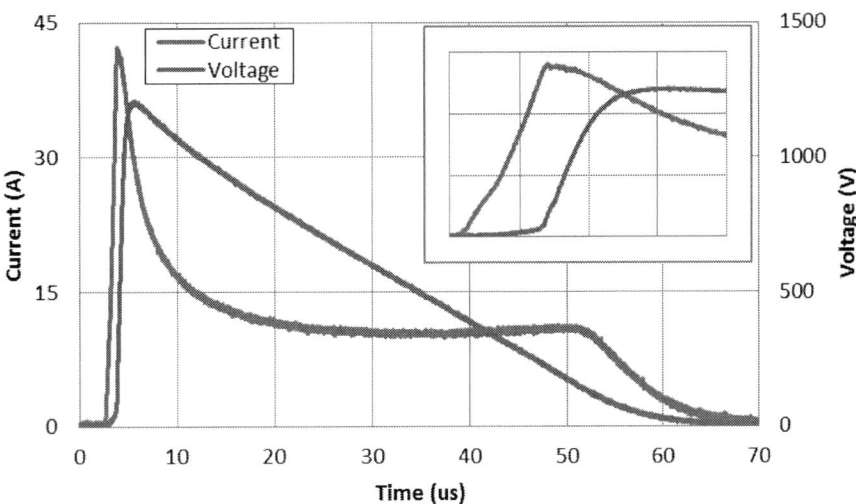

Figure 6. Typical SSDR turn-on current and voltage characteristics. Insert: horizontal resolution 1 us/div; vertical 15 A/div and 500 V/div.

The current waveform is characterised by initial fast rise to the peak value and a slow decay. The peak value is independent of the test voltage as can be seen from Figure 7 which shows a family of SSDR current and voltage waveforms for different test voltages. The decay is attributed to internal device heating. Remarkable is the difference in the saturation level of ~25 A observed during pretest and peak current level of SSDR ~42 A during application test. The cause for this difference is under investigation.

The voltage waveform is characterised by the step rise and nearly linear decay. Profile of the initial voltage rise is determined by the thyristor turn on and associated voltage fall profile. It can be noted that the peak SSDR voltage is lower than test voltage which is due to partial discharge of the damping capacitor prior SSDR entering saturation region. Linear decay profile is indicative of the constant limiting current by SSDR.

© VDE VERLAG GMBH · Berlin · Offenbach

PCIM Europe 2016, 10 – 12 May 2016, Nuremberg, Germany

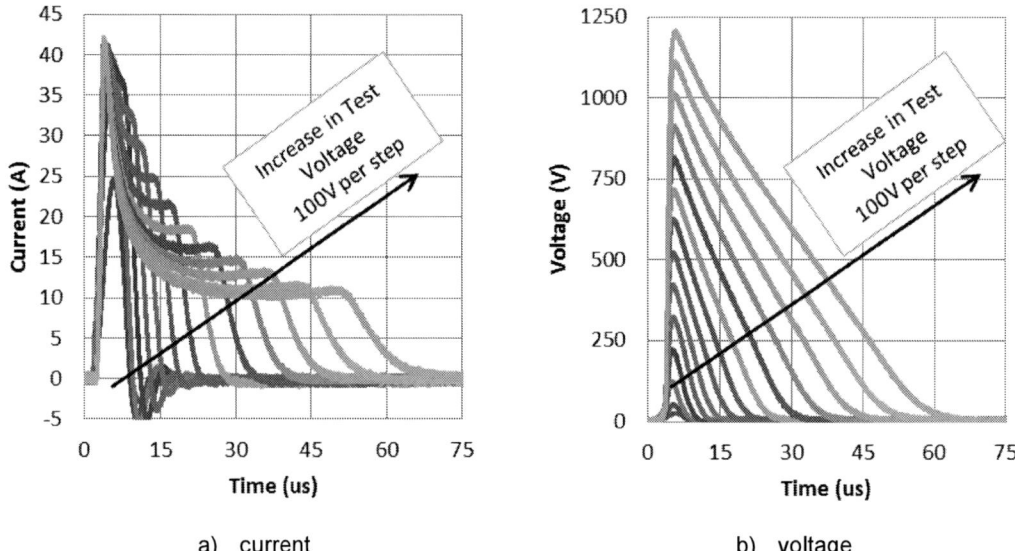

a) current   b) voltage
Figure 7.   SSDR turn-on current and voltage characteristics evolution with test voltage

At test voltage of 1500 V SSDR has successfully withstood turn-on transient for more than 50 us dissipating 0.45 J of energy within the die area of 2.8 mm by 2.8 mm.

4.2.   Performance during turn-off

During turn-off the thyristor recovery voltage is smaller than test voltage during turn-on due to losses observed within test circuit. In the current test setup at 1500 V test voltage recovery voltage of 850 V is expected. Typical current and voltage characteristics of SSDR device during turn-off are shown in Figure 8.

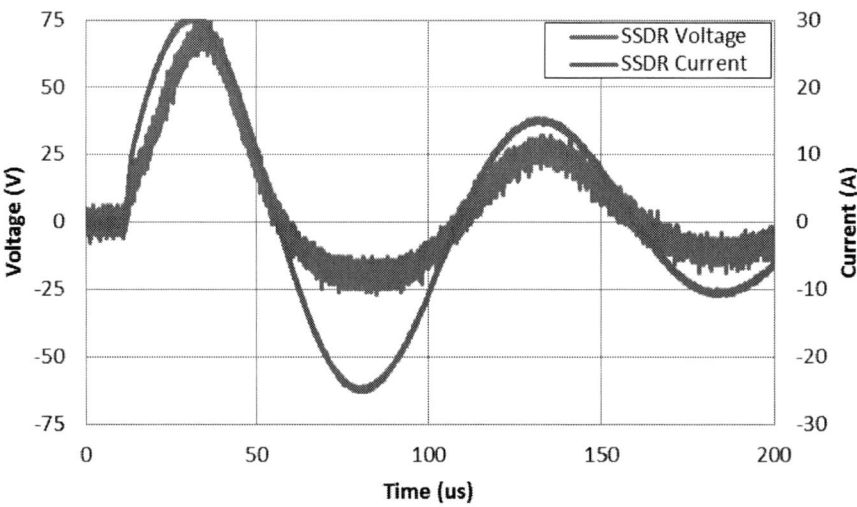

Figure 8.   SSDR turn-off current and voltage characteristics

The voltage change profile is much slower during turn-off in comparison to turn-on. Therefore, SSDR does not enter its saturation region and operates in its linear region during thyristor reverse recovery. In its linear region resistance of 3.3 kV SSDR is quite small. Therefore, at the low voltage SSDR does not provide sufficient damping and oscillations are observed. These oscillations result in increase in thyristor recovery voltage overshoot.

© VDE VERLAG GMBH · Berlin · Offenbach

2202

For 10 kV part the equivalent resistance in the linear region is comparable with passive damping resistor used within damping circuit and it is expected that oscillations will be suppressed and performance will be similar to the standard passive network.

5. Conclusions

In this paper experimental results of the first evaluation of 3.3 kV SiC SSDR device installed within HVDC thyristor valve assembly are presented. 25 A 3.3 kV SSDR devices have been used during the study and tested at voltages of upto 1500 V.

Results show excellent performance during thyristor turn-on with SSDR limiting current level to 42 A for test voltage above 400 V. At the highest test voltage SSDR withstood discharge current for more than 50 us dissipating 0.45 J. Turn-off results show that SSDR operates in its linear region thus reducing effectiveness of the damping circuit due to low value of SSDR equivalent resistance. A more practical 10 kV part would increase the SSDR equivalent resistance in the linear region to the value comparable to passive resistor used in the current configuration of the valve and provide effective damping during turn-off.

Further investigations are under way to investigate thermal aspects of SSDR operation.

6. References

[1]   I. B. Efika, D.R. Trainer, C.C Davidson and N. G. Wright. "Design and evaluation of a SiC asymmetric resistor for HVDC applications". iPower 2013 Conference, University of Warwick. 27-28 November 2013.

[2]   I. B. Efika, D.R. Trainer, C.C Davidson, K, Vershinin, N. G. Wright, A. Tiwari. "Investigation of a Solid-State Damping Resistor for HVDC applications". PCIM 2015

[3]   N. M. Macleod, C.C. Davidson and M.L Woodhouse. "Design and Testing of Thyristor Valves for 800kV HVDC Projects". IEC/CIGRE UHV Symposium Beijing 2007-07-21.

[4]   R. Elpelt, P. Friedrichs , J. Biela, "Fast switching with SiC VJFETs – influence of the device topology", International Conference on Silicon Carbide and Related Materials, 2009.

[5]   TCAD Sentaurus User Manual, 2014.

PCIM Europe 2016, 10 – 12 May 2016, Nuremberg, Germany

High Precision Loss Measurement at HVDC Converter

Helmut Weiss, Montanuniversitaet Leoben, Austria, helmut.weiss@unileoben.ac.at
Bernhard Grasel, Dewesoft GmbH, Austria, bernhard.grasel@dewesoft.com

Abstract

In energy transmission, high power converters are widely used because they offer advanced opportunities for controlling power flow at very high efficiency. Although electrical losses of these converters are rather low yet they add over the lifetime of equipment (e.g. 30 years) and reach the range of the investment cost of a plant. Therefore, every decision on a certain manufacturer to deliver the plant includes a careful evaluation of the calculated losses and a strict comparison to the actual losses which must be measured correctly. Measurement error margins are to be kept small as they are subtracted from the indicated values. While units with two blocks allow a back-to-back operation a special measurement procedure has to be defined for precise loss measurements at single block units. Usually, the power flow direction at converters can be changed easily by circuit capabilities and control. A special selection of power flow forward and backward is used for internal compensation of measurement system errors to a high extent. The loss measurement procedure at a single block yielding very small error margins is described and the hardware used is evaluated.

1. High Power Energy Conversion

1.1. Circuit Configurations

By today, the requirement of power transmission and power flow control in connection with power conversion is fulfilled by various power electronics circuit configurations.

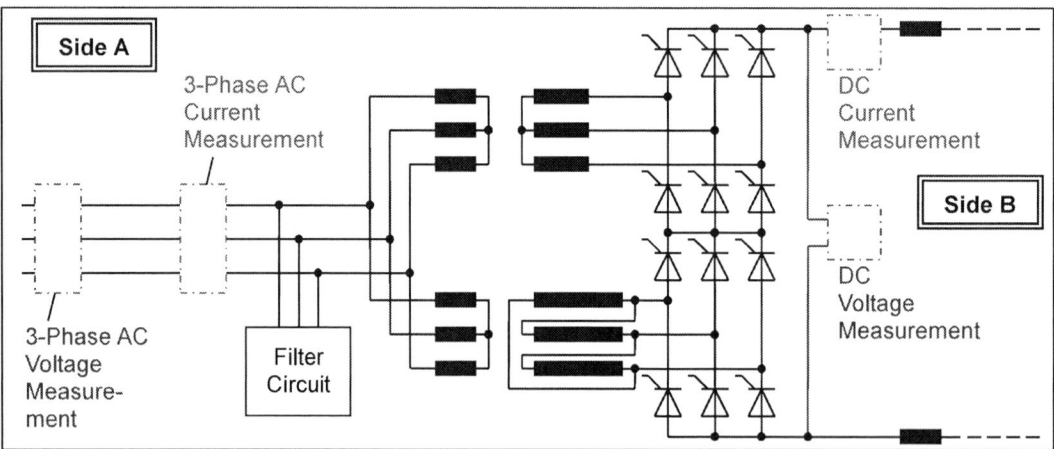

Fig. 1. Standard High Voltage Direct Current (HVDC) base station with thyristor semiconductors

For GigaWatt power, the standard current source DC transmission from point to point is used. This traditional HVDC (high voltage direct current) power conversion system is

© VDE VERLAG GMBH · Berlin · Offenbach

composed of filter circuits, transformers, thyristor rectifier/inverter and a DC choke (Fig. 1). In order to calculate losses, power measurements are inserted at side A and side B.

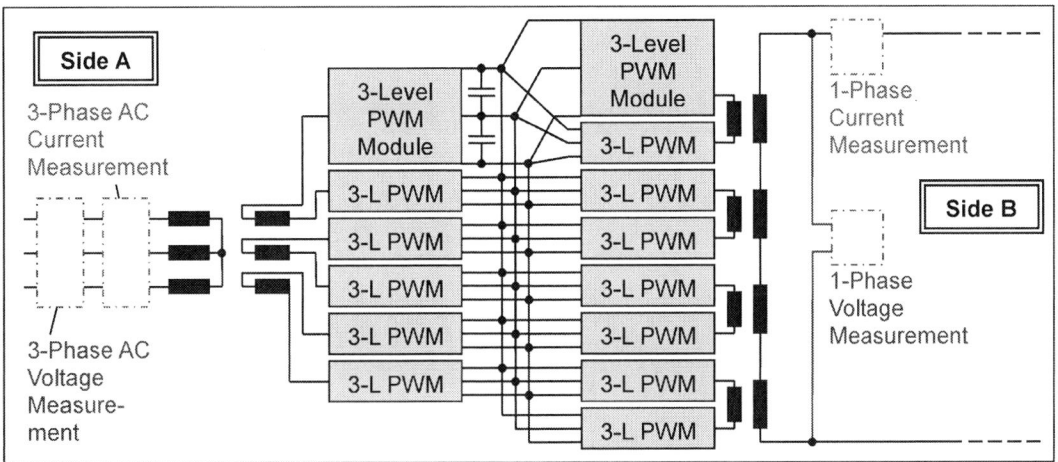

Fig. 2. Three-phase to single-phase three-level converter for railway supply (16,7 Hz or 25 Hz)

New developments in turn-off devices like IGCTs and IGBTs led the way to advanced voltage source circuits. Three-level inverters as building blocks (with concentrated or distributed DC link capacitors) are common with 3-phase (side A) to single-phase (side B) converters for railway supply applications (Fig. 2). "High voltage DC light" or "high voltage DC plus" are designs using PWM modules (Half modules) interconnected in series by output terminals only with free level DC side capacitors (Fig. 3). Such circuits yield a unipolar output voltage.

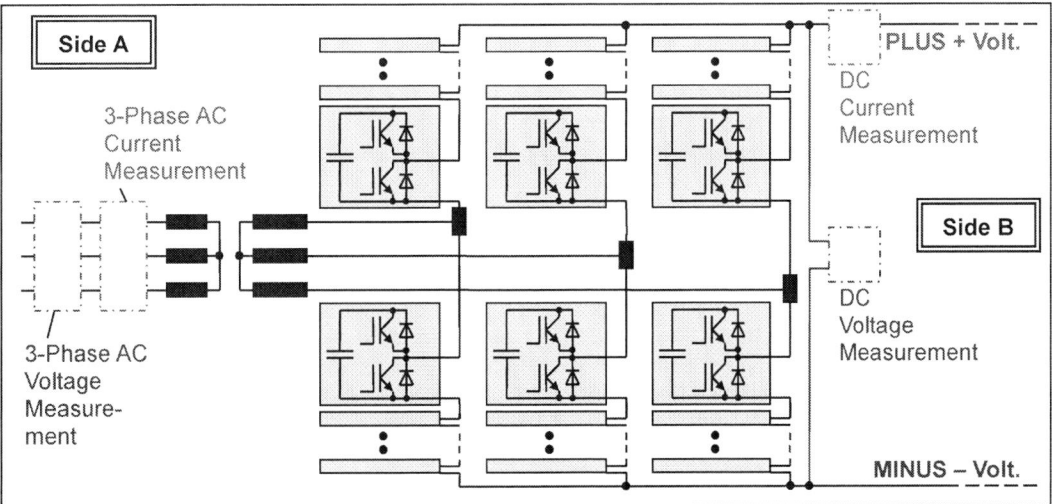

Fig. 3. Unipolar PWM-based high voltage source AC/DC rectifier/inverter

Employing full PWM modules rather than half modules we get an output voltage of both polarities at side B, thus a 3-phase to 1-phase converter which is also used for railway supply systems (inter-ties from 50 Hz or 60 Hz to 16.7 Hz or 25 Hz).

All these circuits have in common that low losses have to be obtained. Measurement of losses requires measurement of 3 voltages and 3 currents at side A and 1 voltage and 1 current at side B, in total 8 signals. Additional converter consumptions are added.

© VDE VERLAG GMBH · Berlin · Offenbach

1.2. Overview on Electrical Losses

Electrical losses in converter components are based on physical dissipative effects or standard consumption of cooling system, control equipment, and up to air-conditioning of converter building. The main loss types are:

- Ohmic losses (in conductors caused by idling or load current, respectively): *Copper Losses*
- Alternating magnetization losses (in ferromagnetic active material due to magnetic hysteresis and due to eddy currents in ferromagnetic active material): *Iron Losses*
- Additional losses (in construction elements (mainly of transformers) due to alternating fields, induced voltages through these fields and dissipative currents in construction elements caused by these voltages, also by eddy current effects in conductors): *Additional Losses*
- Semiconductor On-state losses (in power electronics components due to current flow): *On-state Losses (in power electronics)*
- Semiconductor Switching losses (in power electronics components due to switching between conduction state and off-state and vice versa): *Switching Losses (in power electronics)*
- Dielectric and internal conduction losses (also equivalent serial resistance) in DC-link capacitors of PWM inverters or inductors to achieve current source behavior (applicable if capacitors are present, or inductors, respectively) and/or converter-internal connections: *DC-link Losses*
- Dielectric and conduction losses in filter equipment (applicable if filters at line side are present: filter capacitors, filter inductors, filter damping resistors, damping effects by conductors): *Filter Losses*
- Cooling system consumption (for forced removal of heat from converter power electronics components): *Cooling System Consumption*
- Control system consumption (electrical power required for operation of open loop and closed loop control system): *Control System Consumption*
- Air conditioning consumption (for removing waste heat coming out of system parts in power section and control units with heat entering air in building): *Temperature Conditioning Consumption*

Losses depend to certain amounts on temperature and load, also on voltages (line, DC link) and frequency. Losses can be pre-calculated within certain error margins. Some influences, e.g. actual magnetic material loss factor based on material and iron sheet cutting and placement, are found out only at the end of the manufacturing process, when the transformer is subjected to the idling and short circuit test and heat up process. Appropriate design and careful calculations supplemented by experience allow manufactures to provide accurate values for the losses at fully defined operating conditions (voltage, temperature, active and reactive power). To be on the safe side, manufacturers add a "safety contribution" to calculated loss values when defining the guarantee values for losses they place into offers.

1.3. Loss Evaluation

Eventually at the final acceptance test, the actual losses are measured and evaluated against the guarantee values. This requires a suitable and efficient high precision measurement as well as a correct method in obtaining the "actual loss value". At the time of measurement, a certain voltage and a given outside temperature are present and cannot be changed to the "nominal" value. Also some difference between "nominal" power and actual power has to be taken into consideration. Therefore, a calculation is to be performed in order to obtain the "nominal" value of the losses. Calculations are based on loss models which include e.g. temperature effects and voltage deviations and power differences and estimations about influence of harmonics in line voltage.

PCIM Europe 2016, 10 – 12 May 2016, Nuremberg, Germany

2. Loss Measurement Methodology

2.1. Voltage and Current Transducer Requirements

Electrical power is calculated from voltage and current data. Voltage and current transducers shall have very low error margins in amplitude and phase error, consequently. However, there are physical and especially cost limits for these transducers. Standard measurement transducers (usually passive transducers based on voltage and current transformer principle due to very high voltage level of plant at 110 kV or 220 kV) at class 0.1 (i.e. 0.1% maximum error related to rated value) are a reasonable choice.

Errors can be minimized by use of a calibration curve for each transducer, again a cost raising issue. Calibration can be limited around the power measurement points for the current transducers (at operating voltage level) and around general operating voltage for the voltage transducers.

However, transducers are quite linear in their characteristics by operation principle. We may assume that a certain, small and constant amplitude transfer ratio as well as a certain, small and constant phase difference exist for each transducer. This becomes essential for realizing high precision loss measurements at a single block plant.

2.2. Basic Loss Calculation

By ideal measurement equipment, losses can be simply calculated as described for a 50 Hz / 16,7 Hz inter-tie following Fig. 4.

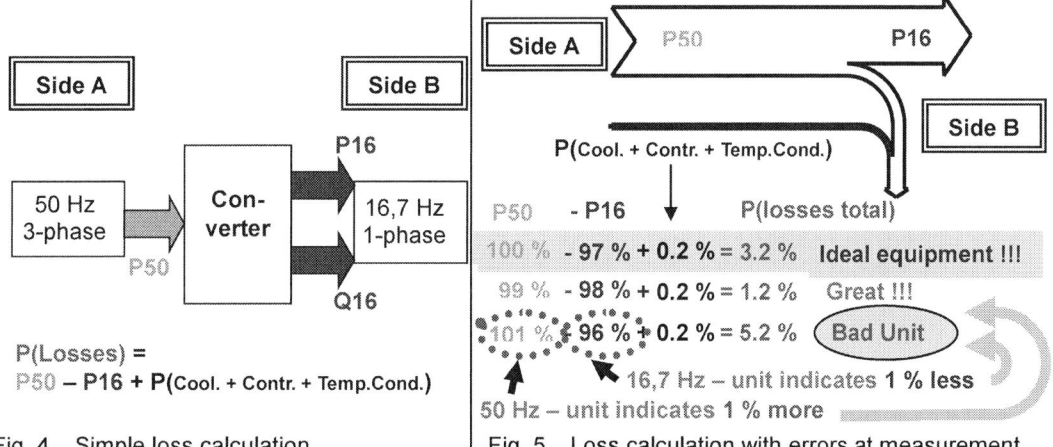

Fig. 4. Simple loss calculation   Fig. 5. Loss calculation with errors at measurement

Provided the measurement equipment works without errors we obtain correct results with the calculation of the losses. Unfortunately, errors in measurement at the A side may go into different direction compared to errors at the B side: Generally, errors will NOT compensate each other. The last example for possible error condition in Fig. 5 explains that with inter-ties of very high efficiency we easily get completely misleading results when using practical equipment with certain error margins and reasonable cost: At certain conditions the plants appears producing excessive losses as 5.2% in our example. Therefore, we need to improve measurement methodology.

2.3. Circular Power Flow at Plants with Two Independent Blocks

Very often, power conversion plants are composed of individual blocks in order to obtain some remaining power conversion capability in case of malfunctions or scheduled maintenance time of one block. This plant configuration allows a circular power flow: The output power of block 1 at side B is re-inserted into block 2 at its side B and re-appears at

© VDE VERLAG GMBH · Berlin · Offenbach

side A of block 2 but under subtraction of losses of both blocks. To compensate the losses, exactly this power loss has to be reinserted at side A of both blocks. This power can be easily measured with high accuracy and yields very good results for efficiency checks. The efficiency check is based on pre-calculated guarantee values for exactly this type of operation. Fig. 6 explains the function.

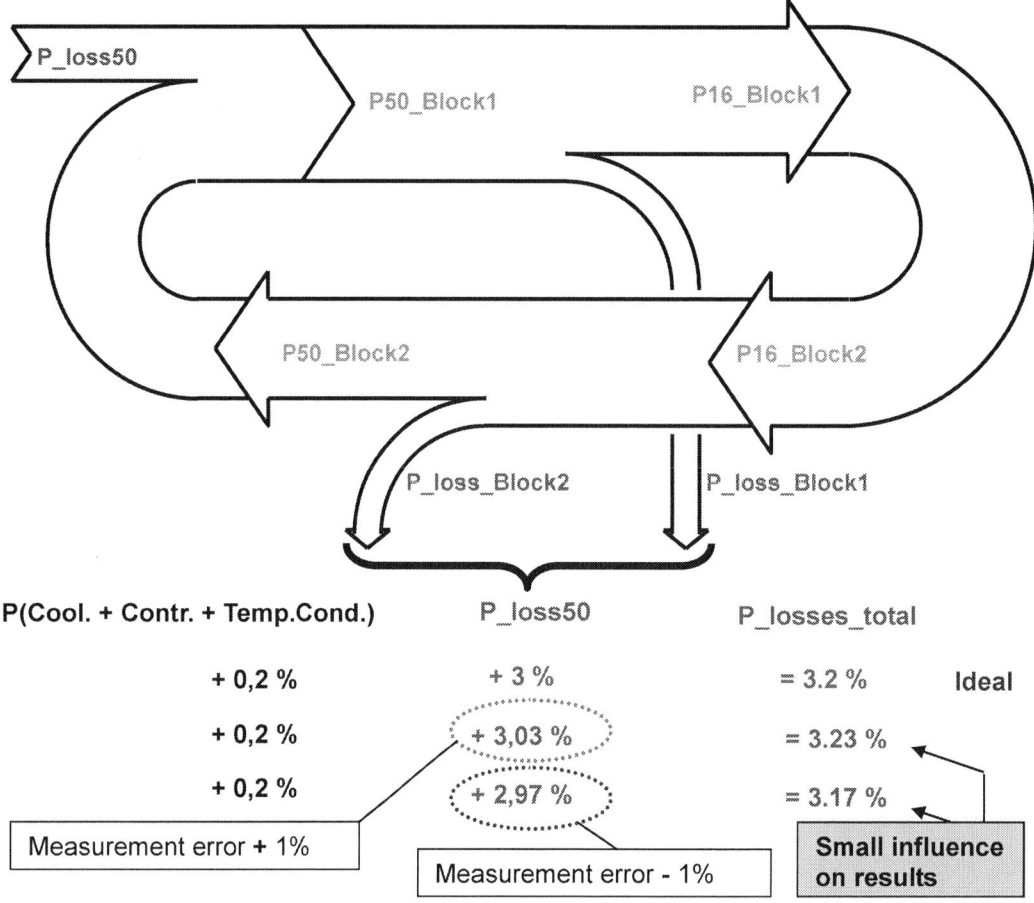

Fig. 6. Loss acquisition in circular power flow with error considerations

2.4. Electrical Loss Measurement at a Single Block

In case that there is only a single block to be evaluated for losses we need a different methodology. A measurement principle according to Fig. 4 becomes necessary. However, we need to avoid the huge measurement uncertainty. Based on the good linearity of voltage and current transducers with reproducible and constant error factors for amplitude and phase and the capability of reversed power flow at such a converter we propose a forward + backward power flow in real measurement at well defined and multiple power ratings. An extensive analysis was undertaken to obtain values for the total errors in calculating losses out of measurements with inherent and unknown but constant errors factors of transducers up to digitizing equipment for power calculation. The result is that even 1% and 1 degree error margins can be accepted while still obtaining high precision results for losses. The method for error compensation simply requires adding certain measurement values. While the individual loss value calculated from difference of A side power to B side power might include a huge error, the sum of a defined group of losses is very accurate despite sensor and amplifier errors.

To use this method the predefined operation points (defined ratings for active and reactive power at A side or B side) are used for loss calculation by manufacturer and within loss measurements. The comparison on fulfilling guarantee losses requirements is finally based only on the sum of losses: sum of losses guaranteed should be less than (or may be equal) to the sum of measurement values of same operating points.

Unfortunately, the actual voltage level and temperature for the time of measurement cannot be predicted. Therefore, guarantee losses are requested for e.g. 2 voltage levels (over and under standard operation voltage level) and e.g. 2 temperatures. Using reasonable models for the dependency of losses on voltage, or temperature, respectively, the guarantee values for the actual measurement environment (mainly voltage, temperature) can be calculated and the final comparison is feasible at low uncertainty.

It is essential to keep the total uncertainty low because the total uncertainty has to be subtracted from the measurement results before the comparison to guarantee values is done because measurement equipment might in just this case display to high losses and this cannot be imposed on the manufacturer.

As a standard, the measurement methodology MUST be defined completely by the customer before any correct offer of a possible manufacturer can be evaluated.

3. Loss Measurement Hardware

3.1. Connections of Power Measurement System to Plant Signals

Generally, access to real signals of the plant is within the control unit area. Usually, this place is rather small as the room is not occupied by operating personnel in general. Only at the acceptance test all free space is occupied. Compact as well as precise and extendable measurement units provide a great advantage over clumsy equipment when it comes to the decisive loss measurement at the final acceptance test (Fig. 7). Interconnections from signals to the power measurement unit are made provisionally inside the control cubicle at a very limited space using standard 4mm contacts (Fig. 8).

Fig. 7. Power measurement and display unit   Fig. 8. Connection plant signals - measurement

3.2. Signal Acquisition and Power Calculation

Today's equipment is digital. Anyway, there are quite different opportunities and capabilities inside the measurement equipment that is available for purchase. The traditional systems are integrated systems: Input are voltage and current values, output is the average power over e.g. 1 second. Output values are transferred via Ethernet (or GPIB or RS232 at older systems) to a host computer handling automatic data storage. There are few opportunities to interfere for the measurement personnel. These standard units can be used very efficiently and include some advanced features of e.g. harmonic analysis of signals. Everything is done by internal software operating on an internal CPU of limited calculation power, and the user cannot influence the internal software.

Exploiting all capabilities of software requires an open system composed of high performance CPUs and high performance software. Then, an application oriented design of screen and display layout, long term data storage, advanced on-line display of harmonics etc. can be performed.

3.3. Additional Display Opportunities

As a very nice example for an "open system" we see the opportunities for providing on-line high speed measurement, power calculation and signal analysis inside a high quality and versatile software package coming with a compact and rugged industry style powerful computer (Fig. 9, Fig. 10).

Fig. 9. Power measurement display

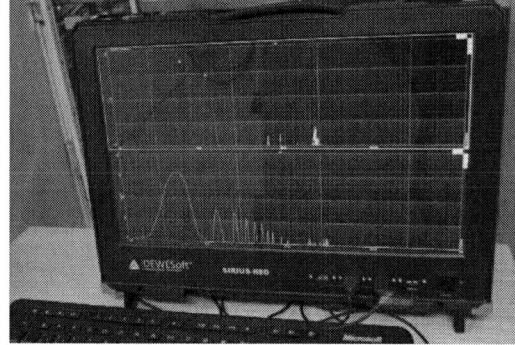

Fig. 10. On-line FFT analysis display

This instrument combines the functionality of a couple of instruments like: Power Analyser, Scope, Spectrum Analyser, Power Quality Analyser, Data logger etc. The possibility to store raw data (voltage and current signals) in the full sampling rate and the powerful math library (basic math operations up to sophisticated data analysis) allows analysis of losses already during measurement. Preconfigured screens can be defined based on the requirements of user or customer, also nicely used for instant documentation purposes without time consuming work after the measurement activities.

3. Summary

Today's opportunities in signal acquisition and real-time computing provide excellent equipment for fulfilling fast final acceptance test measurements like loss computation for comparison to guarantee values while providing accurate on-line results under correct and advanced methodology options. User and application oriented screens are predefined for instant displays.

Compact equipment is necessary to accomplish such a final acceptance test for efficiency (or losses, respectively) at the generally limited space with control systems where all signals are accessible. Much more complicated and requiring more time in final data handling is necessary if different places for signals of input side and output side are used.

Losses shall be computed at very low measurement uncertainty. A problem arises if a single block is to be evaluated. Using extremely cost-raising calibration of sensors to very low error margins which might be really critical for very high voltage transducers can be avoided by a error compensation measurement methodology based on summing up individual loss values of special pre-selected operating points finally yielding the results with a low amount of expenditure concerning cost and time. The deviation between actual operating environment concerning voltages and temperature on one hand and the offer-based environment on the other hand can be handled by appropriate loss models so that correct results at the final acceptance test are possible. Modern software speeds up the decision process nicely as external calculations become obsolete.

© VDE VERLAG GMBH · Berlin · Offenbach

PCIM Europe 2016, 10 – 12 May 2016, Nuremberg, Germany

A fast methodology for solving power flows in hybrid ac/dc networks: The European North Sea Supergrid case study

Rodrigo Teixeira Pinto, CITCEA-UPC, Spain, rodrigo.teixeira@citcea.upc.edu
Christian Alejandro, León-Ramírez, Universidad Simón Bolívar, Venezuela, 01-34040@usb.ve
Mònica Aragüés-Peñalba,CITCEA-UPC, Spain, monica.aragues@citcea.upc.edu
Andreas Sumper, CITCEA-UPC, Spain, sumper@citcea.upc.edu
Elmer Sorrentino, Universidad Simón Bolívar, Venezuela, elmersor@usb.ve

Abstract

This paper presents a sequential algorithm for power-flow studies in hybrid ac-dc networks. The algorithm developed, based on the Newton-Raphson method, can be applied to any power system configuration, without constraining the topology of either the ac or of the dc grids. Combinations of multiple non-synchronized ac systems and multiple dc systems can be solved. The algorithm includes simplified HVDC converter representations, for both VSC and LCC technology, needed to study the steady-state interaction between ac and dc systems. The proposed algorithm for hybrid ac-dc power flow calculations is based on MATLAB and uses the Newton-Raphson power-flow routine from MATPOWER to solve both the ac and the dc power flows. This paper presents the algorithm principles, its flowchart diagram and implementation aspects. Finally, through the developed tool a study on the benefits and impacts when connecting large amounts of offshore wind energy to the European transmission network is studied to demonstrate the applicability of the tool to large scale systems.

Keywords

HVDC, MTDC networks, VSC, ac/dc hybrid power flow

1. Introduction

The increasing need of transmitting electrical bulk power through large distances has lead to the a rise in the number of HVDC projects worldwide [1]. In Europe, the possibility to interconnect asynchronous power systems, to integrate offshore wind energy and the additional control flexibility are the main drivers for building new HVDC lines. In this scenario, HVDC supergrids could be one of the key elements to help power systems face its main future challenge, which is meeting the rising electricity demand in a manner that is sustainable, secure, efficient and competitive [2].

To study the interactions between ac and dc grids, it is clearly necessary to study the combined power flows. Many studies in this direction can be presently found in the literature [2, 3, 4, 5]. From these studies, it results that there are two main approaches for calculating power flows in hybrid ac-dc networks: the Sequential and Unified approaches [2, 5].

This paper presents an algorithm that uses the sequential approach but without the need for iterations between the ac and dc power-flow solutions. This means that the ac and the dc

© VDE VERLAG GMBH · Berlin · Offenbach

power-flow solutions are calculated independently. In theory, the lack of iterations between the ac and dc power flow generates a minor deviation introduced by the inaccurate calculation of the HVDC converter losses. However, in practice, generic HVDC converter loss-calculation formulae also suffer from parametrization and validation inaccuracies, thus the impact of the losses calculation without iteration can be neglected.

The developed algorithm allows the application of the tool to any power system configuration, without constraining the topology of either the ac or of the dc grids. Therefore, combinations of multiple non-synchronized ac systems and multiple dc systems can be solved. The algorithm relies on a simplified ac-dc converter representation to model the steady-state interaction between the ac and dc systems. The converter losses are estimated through a quadratic expression as proposed in [6].

The paper is structured as follows: section 2 discussed the principles behind the proposed power-flow algorithm for hybrid ac-dc grids and presents its flowchart diagram. In Section 3, to demonstrate the validity of the proposed algorithm, a test on the WSCC 3-machine 9-bus system is performed and the obtained results are compared with those obtained using MAT-ACDC and DIgSILENT PowerFactory v15.2. Section 4 presents a power-flow study performed on a possible scenario in Europe: a MTDC grid operating in the North Sea (The North Sea Transnational Grid) in conjunction with the network model of continental Europe connected to UK through HVDC links. A scenario where UK is importing power is analyzed. Finally, the conclusions of this work are discussed and recommendations for future research are given.

2. Power Flow in hybrid ac/dc systems

Power-flow algorithms for hybrid ac-dc systems have been studied previously and classified into two types: sequential and unified. Sequential power-flow routines solve one system first and then, using interface variables, it solves the remaining systems sequentially. On the other hand, unified power-flow routines solve the entire ac and dc systems together. In this case, usually the dc equations are solved as a set of additional constraints to the ac system power-flow equations. The methodology proposed in this paper uses a sequential approach, where the dc system is solver first and the remaining ac systems are solved afterwards.

Using the Newton-Raphson formulation, the power flow in ac and dc grids can be solved. The classical ac power flow formulation has been reviewed in [7, 8, 9]. Instead, a dedicated Newton-Raphson formulation for dc grids with multiple slack nodes is given in [1].

The proposed algorithm for hybrid ac-dc power flow calculations is based on MATLAB and uses the Newton-Raphson power-flow routine from MATPOWER to solve both the ac and the dc power flows. Figure 1 shows the proposed algorithm flowchart with the methodology steps.

First, the systems parameters of the ac and the dc grids are loaded and the user specifies the control mode of all the HVDC converters in the system. Possible converter control modes are: $(P_{ac};Q_{ac})$ and $(V_{dc};Q_{ac})$. Since this is a power-flow tool, controllers looking into the ac system frequency, e.g. $(f_{ac};V_{ac})$, are not included. The control modes $(P_{ac};V_{ac})$ and $(V_{dc};V_{ac})$ are not included because knowledge of the converter reactive power is necessary to estimate the converter losses when using VSC technology and LCC technology will always consume reactive power.

Afterwards, a dc power flow is performed on the HVDC system considering as inputs the demand (or generation) on power controlled converters and the direct voltages on voltage con-

PCIM Europe 2016, 10 – 12 May 2016, Nuremberg, Germany

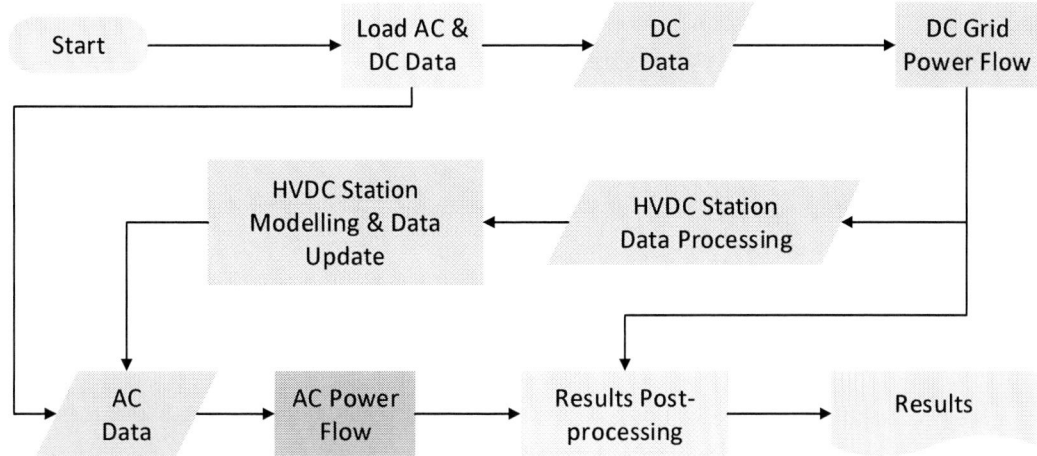

Figure 1: Flowchart of the algorithm

trolled stations. The algorithm then calculates the voltages and the power injections on all the nodes in the dc grid and uses them to estimate the converter losses in VSC-HVDC stations through a quadratic function of the current through the converter [6]:

$$\begin{cases} P_{loss}^{VSC_i} = a_i + b_i \cdot I_{VSC_i} + (c_i + r_i)I_{VSC_i}^2 \\ I_{VSC_i}^* = S_{VSC_i}/V_{VSC_i} \approx \sqrt{(P_{dc\_VSC_i}^2 + Q_{VSC_i}^2)}/V_{VSC_i} \end{cases} \quad (1)$$

where: $P_{loss}^{VSC_i}$, represents losses at converter i; I_{VSC_i}, S_{VSC_i}, $P_{dc\_VSC_i}$ and Q_{VSC_i} represent, respectively, the alternate current, the apparent power, the dc power and the reactive power through VSC converter i; a_i, b_i, c_i, are per unit VSC-HVDC converter losses; and r_i, represents the resistance at converter i.

Table 1: Per unit VSC-HVDC station loss coefficients [6].

converter mode	a	b	c	r
Rectifier	11.033×10^{-3}	3.464×10^{-3}	4.400×10^{-3}	1.000×10^{-3}
Inverter	11.033×10^{-3}	3.464×10^{-3}	6.667×10^{-3}	1.000×10^{-3}

For LCC-HVDC stations, the converter losses are considered to be 0.75% of the converter active power. The converter reactive power is instead calculated as:

$$Q_{ac} = \frac{P_{dc}}{V_{dc}cos(\alpha)} \sqrt{(V_{dc} + R_c I_{dc})^2 - (V_{dc}cos(\alpha))^2} \quad (2)$$

For nominal operating conditions, i.e. P_{dc} = 1 pu and V_{dc} = 1 pu; assuming R_c = 0.1 pu and α = 15 deg., the converter absorbed reactive power will be equal to Q_{ac} = 0.55 pu.

After calculation of the HVDC stations losses, the algorithm then processes the HVDC transmission losses and the ac active and reactive power net injections, which are then used to form the input for the ac power system power-flow study. Finally, during the ac power flow computation, the algorithm considers the HVDC technology and calculates the absorbed reactive

© VDE VERLAG GMBH · Berlin · Offenbach

power from LCC-HVDC stations according to (2) and it applies the specified user control for the reactive power of all the VSC-HVDC stations.

3. Validation of the proposed algorithm

The methodology is first tested on the WSCC 3-machine 9-bus system [10], overlayed with a multi-terminal dc (MTDC) network with 3 terminals. Figure 2 (a) shows the topology of the network. All converters are assumed to use VSC-HVDC technology. The HVDC grid cables resistance is assumed to be 0.0150 Ω/km and the grid rated voltage is set at 400 kV ($\pm$ 200 kV). Figure 2 (b) shows a comparison between the bus voltages and angles in the ac network obtained with the proposed methodology, MATACDC and DIgSILENT Power Factory 15.2. It can be seen from the results that there is a very good agreement between the different methods.

The smallest difference between results given by the proposed algorithm and the other tools in terms of apparent power through the transmission lines is $\pm$ 0.05%, whereas the largest is $\pm$ 5%. However, in terms of computational time performance, this networks has been solved by the proposed algorithm in 0.78 seconds and in 0.94 seconds by MATACDC; whereas DIgSILENT Power Factory® took over 1 second to finish the computation. This gain in speed can be particularly important when multiple power-flow runs are necessary, e.g. when there is need for optimization of the combined ac-dc network through optimal power-flow (OPF) routines. Since optimization routines are iterative themselves, a further advantage in this case is that the VSC-HVDC converter power losses calculation can be made more precise by using the ac power-flow results from the previous optimization time step. Additionally, it is plausible to expect that the gain in computational speed would be even bigger for larger networks.

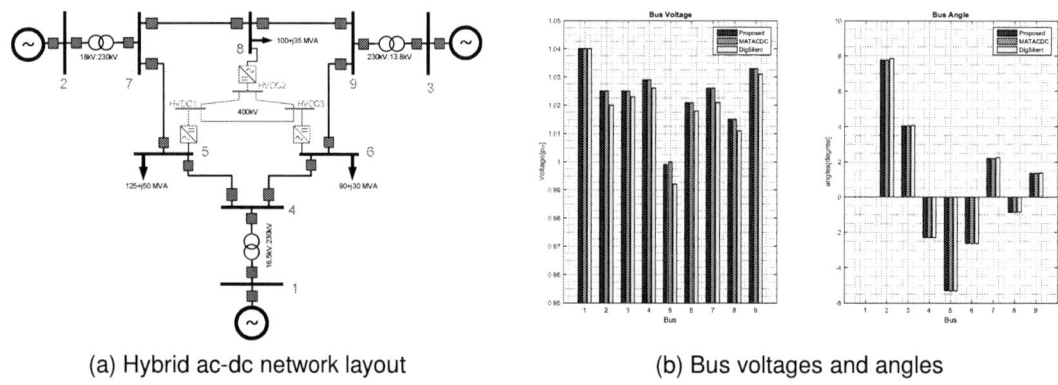

(a) Hybrid ac-dc network layout   (b) Bus voltages and angles

Figure 2: 9-bus WSCC system overlayed with a MTDC network with 3 terminals

4. Large-Scale Simulation

After the validation presented in Section 3, the proposed approach is then applied to a possible scenario in Europe: a 20-bus MTDC grid operating in the North Sea (the so-called North Sea Transnational Grid) in conjunction with the 1494-bus network model (suitable only for simplified ac power-flow calculations) of continental Europe connected to a 29-bus GB grid model through HVDC links.

© VDE VERLAG GMBH · Berlin · Offenbach

The 29-bus GB grid model is a reduced network model found in [11]. The 19-bus MTDC grid operating in the North Sea (The North Sea Transnational Grid) was proposed in [1]. The network model of continental Europe is an ac simplified model which can be found in [12]. An illustrative scheme of the network in a possible scenario where the 29-bus GB grid is importing power through HVDC links from continental Europe and wind-farms is shown in Figure 3. Table 2 gives the main circuit parameters of the considered HVDC links in the studied network.

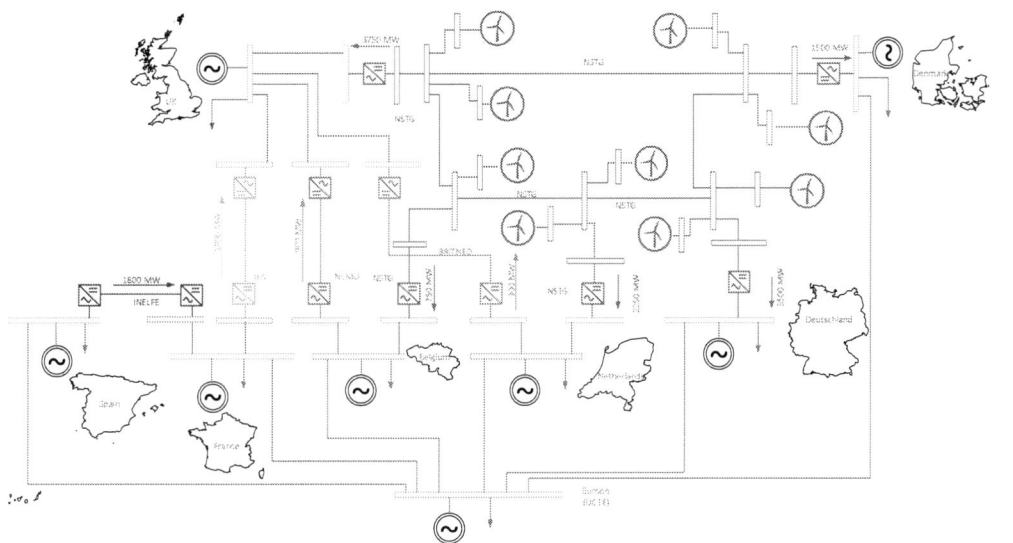

(a) Combined ac-dc network in Europe

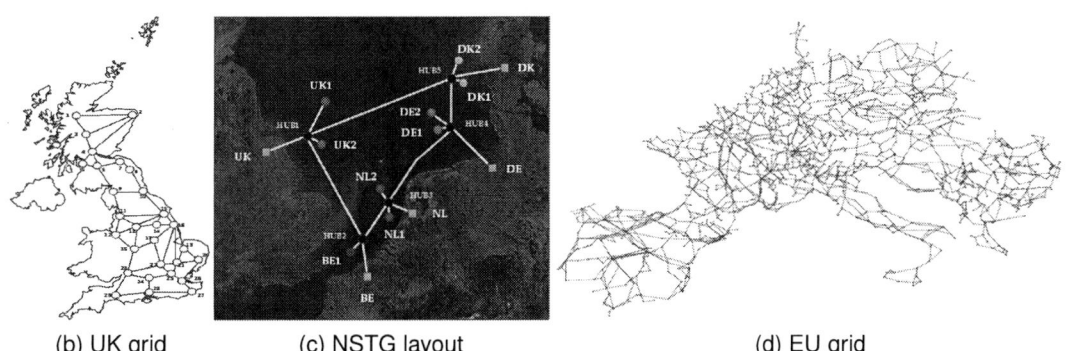

(b) UK grid   (c) NSTG layout   (d) EU grid

Figure 3: Illustrative scheme of the European supergrid showing HVDC connection buses

In the large-scale power-flow calculation, the considered scenario is a high offshore wind generation which connects to the UK 3750 MW of offshore wind from the NSTG (in UK bus 15) and a total of 3600 MW from France (IFA in UK bus 27), Belgium (NEMO in UK bus 26) and Netherlands (BRITNED in UK bus 26). The HVDC Links from Europe to Great Britain are set to transmit power at 90% of their rated capacities and the slack bus for the NSTG is set at the dc node onshore connected to Germany, which controls the MTDC network voltage at 1 pu. The network model has four point-to-point HVDC links representing three different HVDC systems in operation (INELFE, IFA, BRITNED) and one under construction (NEMO). For simplicity, only one case is here presented: the GB grid is importing power through three HVDC links (IFA, NEMO, BRITNED) and extracting power from NSTG. The INELFE HVDC link has been set so

PCIM Europe 2016, 10 – 12 May 2016, Nuremberg, Germany

Table 2: HVDC-Grid Characteristics

HVDC Station	Converter Technology	Transmission Length [km]	Rated Power [MW]	Rated Voltage [kV]	Number of poles	Cable Resistance $[\Omega/km]$
IFA	Thyristor	73	2000	270	4	0.0150
BritNed	Thyristor	250	1000	450	2	0.0150
Nemo	IGBT	140	1000	400	1	0.0150
INELFE	IGBT	65	2000	320	2	0.0150
NSTG	IGBT	2290	11000	300	2	0.0230

that Spain is exporting power to France.

Figure 4 displays the UK bus voltages and angles for the cases with and without the NSTG. With the UK already importing power from continental EU, no significant impact from the additional power-flow from the NSTG can be seen in the British grid, as shown in Figure 4. In total, accounting for the transmission losses and converter losses, 7158 MW are injected into the British grid. It is worth noticing that voltage problems do not arise as values in all buses are kept within ± 3% of the rated values.

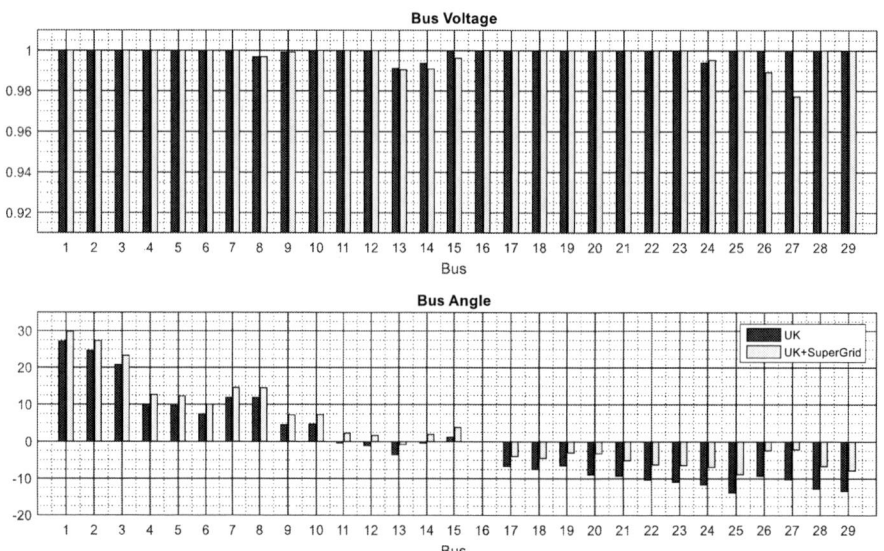

Figure 4: 29-bus GB system state

In the NSTG, since the German onshore node was chosen as the slack node, its direct voltage is controlled at 1 pu. To control the direct voltage inside the MTDC network, the reference node will have to account for the system transmission losses. The total network losses in the NSTG amount to 570 MW and the losses on the individual HVDC lines result always below 3%. However, it is important to note that the addition of the NSTG – and the operating point in which it is controlled – can have an impact on the grids of all European countries. Figure 5 displays the change in the load angles at the frontier buses with and without the NSTG. As it can be expected, Germany is one of the most affected country as it was serving as a slack node for the dc grid. There is also a large variation of the load angle in Denmark due to the large amount

© VDE VERLAG GMBH · Berlin · Offenbach

PCIM Europe 2016, 10 – 12 May 2016, Nuremberg, Germany

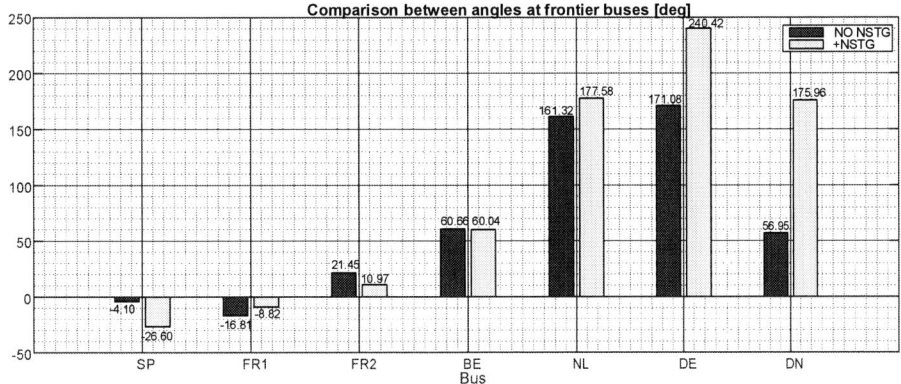

Figure 5: Load angles at the frontier buses with and without the NSTG

Table 3: Resulting transmitted power and voltages on both ends of the studied HVDC links.

HVDC system	Transmitted Power [MW]	Transmission Losses [MW]	Vac UK [pu]	Vac Europe [pu]	Vdc UK [pu]	Vdc Europe [pu]
IFA	1800	24.334	0.977 ∠ -2.137°	1 ∠ 10.97°	1	0.986
BritNed	900	15.622	0.989 ∠ -2.429°	1 ∠ 177.58°	1	0.983
Nemo	900	10.639	0.989 ∠ -2.429°	1 ∠ 60.04°	1	0.988

HVDC system	Transmitted Power [MW]	Transmission Losses [MW]	Vac ES [pu]	Vac FR [pu]	Vdc ES [pu]	Vdc FR [pu]
INELFE	1800	30.880	1 ∠ -26.60°	1 ∠ -08.82°	1	0.983

Table 4: Direct voltages and power in the NSTG network.

NSTG BUS	V_{dc} [kV]	Injected P [MW]	NSTG BUS	V_{dc} [kV]	Injected P [MW]	NSTG BUS	V_{dc} [kV]	Injected P [MW]
1	1.062	2250.00	7	1.059	1500.00	13	1.047	0 (HUB)
2	1.042	1500.00	8	1.047	750.00	14	1.043	0 (HUB)
3	1.043	750.00	9	1.048	750.00	15	0.976	-3750.00
4	1.059	1500.00	10	1.035	0 (HUB)	16	1.029	-750.00
5	1.044	750.00	11	1.039	0 (HUB)	17	1.021	-2250.00
6	1.054	1500.00	12	1.041	0 (HUB)	18	1.000	-2430.39
						19	1.015	-1500.00

of offshore wind coming from the NSTG. Table 3 presents the resulting transmitted power and voltages on both ends of the studied HVDC links, whereas Table 4 shows the resulting direct voltages and powers flowing in the NSTG network.

5. Conclusion

A novel tool for solving power flow in hybrid ac/dc systems has been presented. Power flow results from study cases have been compared with two other tools to show the validity of the

© VDE VERLAG GMBH · Berlin · Offenbach

proposed methodology. Since the proposed algorithm does not iterate between the dc and the ac power-flow routines, a small error is possibly introduced in the converter losses calculations, but this error is very small on an already very small converter loss, in general lower than 1%, and can thus be safely neglected. Therefore, in terms of computational speed, the algorithm is faster than the other analyzed tools. For this reason, this algorithm is better suited for application to large systems such as the presented scenario of the European supergrid and for contingency and optimization studies where many power flow calculations have to be performed. The main contribution of this work is the development of a sequential algorithm which can be easily integrated into current ac power flow programs. As future research the efforts could be on implementing the other remaining control modes and on substituting the converter losses calculation by pre-calculated look-up table for efficiency curves.

6. References

[1] R. Teixeira Pinto. *Multi-Terminal DC Networks: System Integration, Dynamics and Control.* Dissertation, Technische Universiteit Delft, 2014.

[2] F Gonzalez-Longatt, JM Roldan, and CA Charalambous. Solution of ac/dc power flow on a multiterminal hvdc system: Illustrative case supergrid phase i. In *Universities Power Engineering Conference (UPEC), 2012 47th International*, pages 1–7. IEEE, 2012.

[3] J Reeve, G Fahny, and B Stott. Versatile load flow method for multiterminal hvdc systems. *Power Apparatus and Systems, IEEE Transactions on*, 96(3):925–933, 1977.

[4] J Arrillaga and P Bodger. Integration of hvdc links with fast-decoupled load-flow solutions. In *Proceedings of the Institution of Electrical Engineers*, volume 124, pages 463–468. IET, 1977.

[5] Jef Beerten, Stijn Cole, and Ronnie Belmans. A sequential AC/DC power flow algorithm for networks containing multi-terminal VSC HVDC systems. *IEEE PES General Meeting, PES 2010*, pages 1–7, 2010.

[6] Johan Rimez and Ronnie Belmans. A combined ac/dc optimal power flow algorithm for meshed ac and dc networks linked by vsc converters. *International Transactions on Electrical Energy Systems*, 2014.

[7] J. Grainger and William Stevenson. *Power System Analysis*. McGraw-Hill, 1994.

[8] Hadi Saadat. *Power system analysis*. WCB McGraw-Hill, 1999.

[9] Elmer Sorrentino, Angynes Zavala, and Jenny Rodriguez. Web application for load flow problems. *Latin America Transactions, IEEE (Revista IEEE America Latina)*, 12(6):1094–1100, 2014.

[10] Paul M Anderson and Aziz A Fouad. *Power system control and stability*. John Wiley & Sons, 2008.

[11] M. Belivanis and K. R. W. Bell. Representative gb network model. Technical report, Dep.of Electronic and Electrical Engineering, University of Strathclyde, Glasgow, 2011.

[12] Neil Hutcheon and Janusz W. Bialek. Updated and validated power flow model of the main continental European transmission network. *2013 IEEE Grenoble Conference PowerTech, POWERTECH 2013*, pages 1–5, 2013.

Using Smart Converter to Obtain Traction-Machine Insulation Health State Information

M.A. Vogelsberger[2], C. Zoeller[1], J. Bellingen[3] and T. M. Wolbank[1]

[1] TU Wien, Institute of Energy Systems and Electrical Drives, Gusshausstraße 25-29, 1040 Vienna, Austria; Email: thomas.wolbank@tuwien.ac.at

[2] Bombardier Transportation (Austria) GmbH, Hermann Gebauer Straße 5 1220 Vienna, Austria; E-Mail: markus.vogelsberger@rail.bombardier.com

[3] Bombardier Transportation (Switzerland) AG, Brown-Boveri-Strasse 5, 8050 Zurich

Abstract

In traction applications, inverter-fed machines are widely used, where they operate near and even above their rated values leading to high strains on the machine. Especially the winding insulation of the machine suffers from increased stress caused by high inverter switching frequencies and fast voltage rise times (high dv/dt rates) through the continuous development of new semiconductor technologies (silicon carbide SiC or gallium nitride GaN). A smart converter, which is a power inverter with integrated control, monitoring and protection functions, enables a safe and reliable operation of a drive. With an integrated smart online monitoring method by evaluating the current response after voltage step excitation, the detection of incipient insulation degradation is possible and enables that further maintenance tasks can be defined if necessary. With the usage of the integrated current sensors of the inverter and a low ADC sampling rate due to economic reasons, a special signal acquisition strategy is implemented to improve the performance of the method despite the limited hardware and software resources imposed by product constraints.

1. Introduction

In traction applications medium voltage drives operate with voltage source inverters (VSI) at voltage levels of approximately 3 kV covering power ratings of few megawatt (MW). The inverter generally converts the DC voltage to a three-phase AC voltage with adjustable magnitude and frequency. Induction motors are being extensively used in modern traction drives where fast and accurate control of speed is needed using vector control technology. Basically, the design of the induction machine is very simple, with windings in the stator which forms electrical poles, carrying the supply current to induce a magnetic field that penetrates the rotor, wherein the rotor is designed as a squirrel cage with closed bars. Especially in public and cargo transportation systems a reliable and safe operation of the drive is demanded over decades to prevent high costs. With focus on the machine side after bearing faults, stator related faults are the second most common faults, causing an outage of the drive system [1, 2]. A high fraction of these stator faults are based on a failure of the insulation system. The modern voltage source inverters with fast switching transients fed the induction machine on the stator side. Literature clearly states, that the magnitude of the applied voltage and the temperature are most influencing the insulation status life-time [3]. Additionally, modern voltage source inverters with fast switching transients in today's traction drive applications lead to transient overvoltage, stressing the machine's insulation system. Additionally new emerging semiconductor technologies with high switching frequencies and high dv/dt rates increase the stress for the motor winding insulation. This leads to insulation deterioration and a loss of the electrical strength of the system, which finally results in an insulation breakdown. The insulation degradation is usually a slowly proceeding process, usually affecting at first turn-to-turn insulation, followed by a complete phase-phase and

© VDE VERLAG GMBH · Berlin · Offenbach

finally phase-to-ground short circuit. In addition to the aforementioned stress factor several further stresses can be stated, e.g. thermal, mechanical or environmental stress, that all lead to aging of the insulation system. These main causes responsible for a machine breakdown have been analyzed in [3]-[7]. To increase the reliability of the drive system different strategies like fault tolerant design, electrical filters etc. can be implemented. However, filters are bulky, cause space requirements and lead to additional costs.

An integrated smart online monitoring method by evaluating the transient current response after voltage step excitation enables the detection of incipient insulation degradation. With the proposed method the converter is able to analyze the insulation state of the machine using the integrated current sensors, also used for the control of the machine. This is one of the key requirements and targets of the proposed online insulation monitoring technique. The current sensors are of standard industrial type transducers based on the Hall-effect based closed loop measurement principle, with low bandwidth specification of about $f_{3dB}\sim 150$ kHz and the di/dt with 50A/µs. With the opportunity of the observation of the insulation state during the operation or at least at startup or shutdown of the drive without additional signal injection and sensor equipment and without disassembling of the drive unexpected down times can be avoided and maintenance on demand scheduled. The analysis of the transient current response requires sufficient resolution in time during the sampling process. The additional hardware limitation of a low sampling ADC unit (sampling rate max. 1MS/s) requires a sophisticated sampling strategy to improve the performance of the method and reach the necessary resolution which is multiple times greater than 1MS/s.

2. Online Insulation Monitoring with Smart Converter Topology

In Fig. 1 (a) the scheme of the smart converter as part of the main components of a drive system inverter, cabling and machine is depicted. The inverter consists of three half bridges, at which every power switch is controlled by its own gate drive unit (GDU). The built in current sensors are standard industrial current transformers (hall-effect based closed loop) with a bandwidth of 150 kHz and a di/dt of 50 A/µs and are normally used for the control of the machine only. The current signals are sampled with an ADC unit with a maximum sampling rate of 1MS/s. The machine can be represented by the equivalent circuit diagram (Fig.1 b), consisting of resistances, inductances and capacitances. The test machine is a 1.4MW induction machine designed for railway applications equipped with taps for further investigations of a winding insulation change. In this work the degradation of the insulation system is emulated with a capacitor placed parallel to the winding system or parts of it. Due to the impedance mismatch of the cables and machine and high dv/dt-rates of modern semiconductor switching devices, transient overvoltages appear at the machine terminals after inverter switching. These oscillations with decaying amplitude are also visible in the current signal. The occurring frequency components of the transients are defined by the electrical components of the machine-cabling system and are unique for every setup. Thus the frequency response can be considered as the fingerprint of the system. According to [3] and [7] different studies show that the parasitic capacitances of the machine's winding system change after a high number of thermal- or mechanical aging cycles have been applied. Thus, in this work the degradation of the insulation system can be emulated with a capacitor placed parallel to the winding system or parts of it using a special machine equipped with taps (depicted in Fig.1 b with "*Cfault*"). The capacitance value of the complete winding system of the test machine (1.4 MW) measured to ground is given with 63nF. In Fig. 2 (a) – upper subfigure, the time signals of the current response of the machine, measured on phase L1, as a reaction of a voltage step applied with the inverter from lower short circuit to high DC-link voltage is depicted for two machine insulation states. The healthy machine condition (blue trace) and the machine with a capacitor placed parallel to the first coil of phase L1 (red trace) representing the machine condition with insulation degradation are shown. After detection of the accurate switching time instant by the trigger detection unit (Fig.1 a) and the elimination of the current slope to prevent influences through e.g. slotting, the time domain data is transformed into the frequency domain, depicted in Fig. 2 (a) – lower subfigure.

If insulation degradation occurs, also the current step response changes and can be compared to the initial measurements stored in the storage device of the smart converter. Based on the history data of the machine, the system is able to check the insulation condition before each startup of the drive. This is done by alternatively applying a voltage step in every phase and evaluation of the transient current response of the machine and their frequency components. By calculation of an indicator to assess the severity of the insulation recommendations for maintenance actions can be advised.

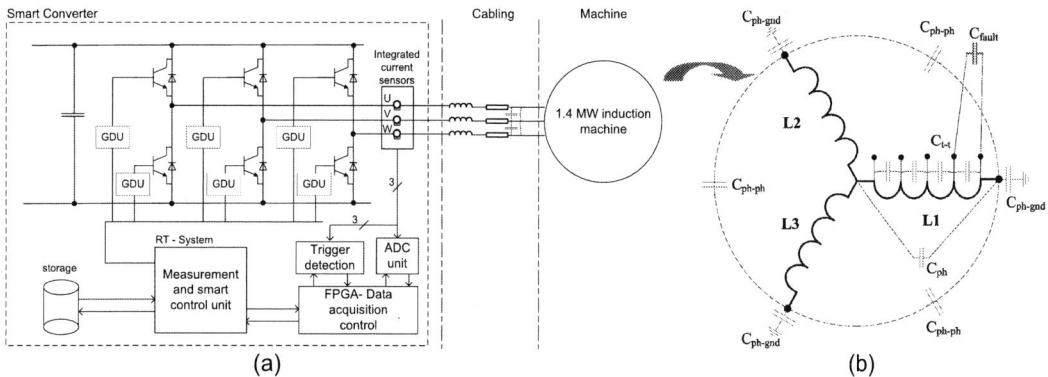

Fig. 1 a) Smart converter topology, b) machine electrical parameter and parasitic components.

The used converter system is a three phase IGBT-inverter. With the measurement and smart control unit as depicted in Fig 1 (a) the inverter starts an insulation test at the startup of the drive, with all three phases connected to low side of the dc potential and then initiate a transition of the corresponding inverter lag by turning off the low side transistor and turn on the high side one of the corresponding phase. The process is repeated several times for every phase to enable a statistical representation of the measurement. Due to the fact that the capacitance change of the winding system is the dominant parameter to indicate the insulation health state the position and the value of the capacitor "C_{fault}" is varied from 3nF to 15nF, cf. Fig. 2 (a) and (b), to emulate a trend in the severity of the insulation degradation. Regarding the nominal voltage levels of the test machine, the presented measurements in this work are made with lower DC link voltage of about 440 V.

In the target application the nominal power of 1.4MW is reached with a DC link voltage of around 2.8 kV, which imply that the current levels in this work are quite small and the currents will be about 6.4 times higher. However with the low DC-link voltage of 440V the sensitivity of the current transducer responses is still high enough to analyze the insulation health state.

© VDE VERLAG GMBH · Berlin · Offenbach

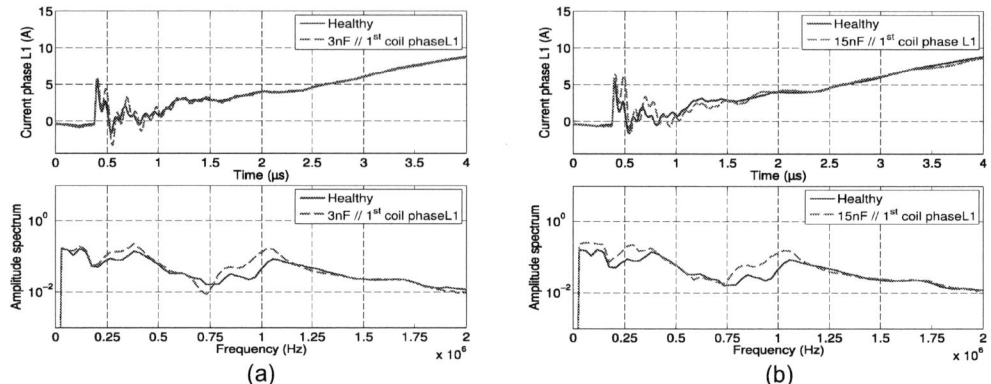

Fig. 2 Transient current response after voltage step excitation by the inverter, with emulated insulation degradation (a) 3nF parallel to phase L1 and (b) 15nF parallel to phase L1.

3. Introduction of an Indicator to Assess Winding Insulation State

First it is important to perform measurements on a healthy machine to determine the initial insulation state. This is done at the first startup of the drive system. This measurement is repeated several times for statistical analysis and serves as a reference which represents the healthy machine state. This measurement is compared to later measurements in operation to assess the machine's actual insulation condition respectively possible deterioration. With the introduction of the Insulation State Indicator (ISI) the assessment of the insulation condition for each phase is possible. In a comparison process each equidistant value of the amplitude spectrum of the investigated current machine state is compared with the reference (healthy machine condition) based on the Root Mean Square Deviation (RMSD), see equation (1). With equation (2) the mean value of the repeated measurements is obtained. The Fourier components Y_{ref} represent the reference obtained at the healthy machine at the initial operation. The further Fourier components Y_{con} represent the condition assessment after several operation of the drive.

$$ISI_{p,k} = RMSD_{p,k}(x_1,x_2) = \sqrt{1/(n_{high} - n_{low}) \sum_{g=n_{low}}^{n_{high}} \left(Y_{ref,p}(f) - Y_{con,p,k}(f)\right)^2} \tag{1}$$

$$ISI_p = \frac{\sum_{k=1}^{m} ISI_{p,k}}{m} \tag{2}$$

The difference of the variable "n_{high}-n_{low}" represents the analysis frequency range defined by sampling frequency and FFT-window length (50-500kHz). The index p defines the investigation phase (L1, L2, L3) and m indices the number of measurement repetitions.
It should be noted that the Insulation State Indicator (ISI) magnitude correlates with the severity of insulation degradation, and is hence suited to act as the final monitoring value.
In order to assign the insulation degradation to a spatial location the linear combination of the previous calculated phase ISI values is introduced with equation (3). In the further course of this work it is denoted with Spatial Insulation State Indicator (SISI). With this indicator changes of the high-frequency behavior due to temperature variation are eliminated as these would lead to zero-sequence components.

$$SISI = ISI_{L1} + ISI_{L2} \cdot e^{j\frac{2\pi}{3}} + ISI_{L3} \cdot e^{j\frac{4\pi}{3}} \tag{3}$$

In Fig. 3 (a) different machine scenarios with the 1.4MW test machine with emulated insulation degradation are analyzed based on the calculated indicators. An increase in the indicator by use of different increasing capacitance values (3nF, 6.8nF and 15nF) can be equated with an increasingly deteriorated insulation, and this is equivalent to a change in the capacitance of the winding. The capacitors are placed separately parallel to phase L1. Thus an increase mainly of the indicator ISI_L1 is expected and observable. In Fig. 3 (b) the spatial indicator for the analyzed machine states are represented. It can be seen that the values points in the direction of phase L1 as the change of the machine's high frequency has been carried out there

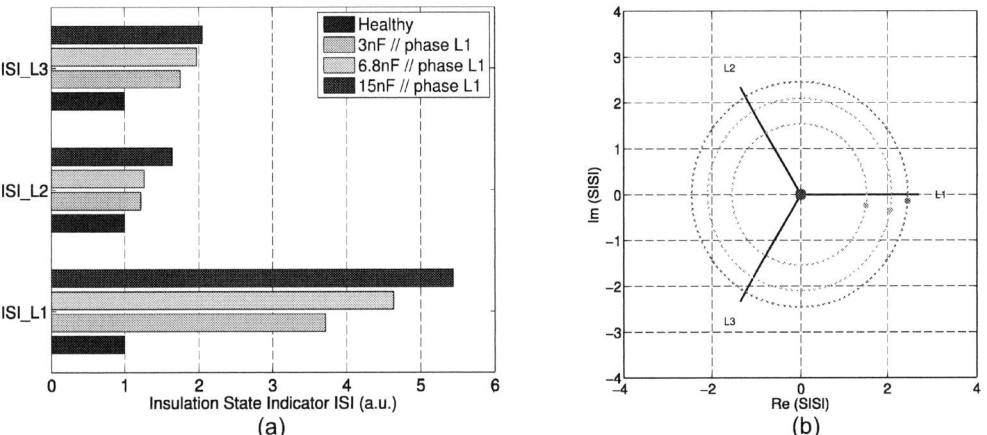

(a) (b)

Figure 3 (a) Insulation State Indicator (ISI) for different investigated machine scenarios. (b) Spatial Insulation State Indicator (SISI).

4. Influences due to Hardware Restrictions

Due to the fact that the interesting frequency components of the step response are typically in the range of some tens kHz to MHz, depending on different factors e.g. machine type, cabling etc., a strategy to enable the monitoring using built-in ADC-unit (1MS/s) is proposed. The time resolution for the test machine in this work (1.4MW induction machine) is aimed 15MS/s. For the extension of the frequency range a smart sampling technique based on equivalent time-sampling, in the following referred to as repetitive sampling, was implemented on an FPGA to enhance the frequency resolution. Instead of gathering all samples for a waveform with one trigger event the system acquires the data with several trigger events over multiple (15) measurements each shifted by 66.6 ns (1/(15 MS/s)), as depicted in Fig. 4 (a). The sub-signals are assembled to the final current response.

Furthermore, time jitter is an unwanted behavior that the system posses which is unavoidable. With the usage of repetitive sampling to increase measurement bandwidth, jitter is an important issue that has to be taken into account. A timing error between the actual switching transition and the trigger of the ADC unit can have different reasons, for instance, jitter of the gate drive units GDU's. It was analyzed that the target system in which the monitoring method has to be implemented has a time jitter of up to 50ns, which seems to be realistic compared to other systems. Fig. 4 (b) shows the shifted current responses occurring after consecutive voltage step excitation by the inverter. Simulations of the influence of up to 50ns jitter are conducted and compared to the results of the measurement system.

PCIM Europe 2016, 10 – 12 May 2016, Nuremberg, Germany

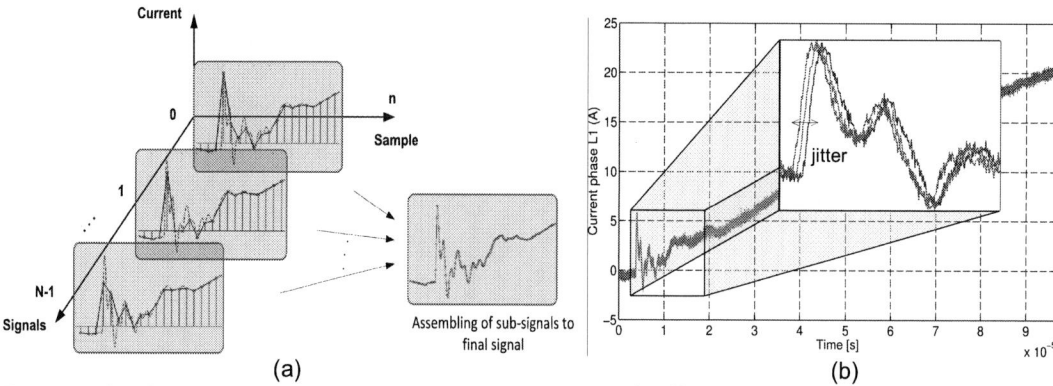

(a)   (b)

Figure 4 (a) Scheme of the repetitive sampling process (b) effect of time jitter on the current response.

4.1. Experimental Results with Sampling approach

In this section the influence of the jitter on the results of the indicators are analyzed. For this purpose one complete measured signal recorded with a very high sampling rate of 120MS/s (sample points every 8.3ns) is taken as the basic signal. A maximum jitter of 50 ns has been added randomly to the equidistant sample points of the basic signal. In Fig 5 (a) the blue traces represent fifteen spectra calculated out of measurements with jitter of this healthy machine state. With jitter added in the simulation a widespread trend is observable above 1.5MHz. The red curve depicts the reference signal where all the step responses are combined into one 120 MHz signal.

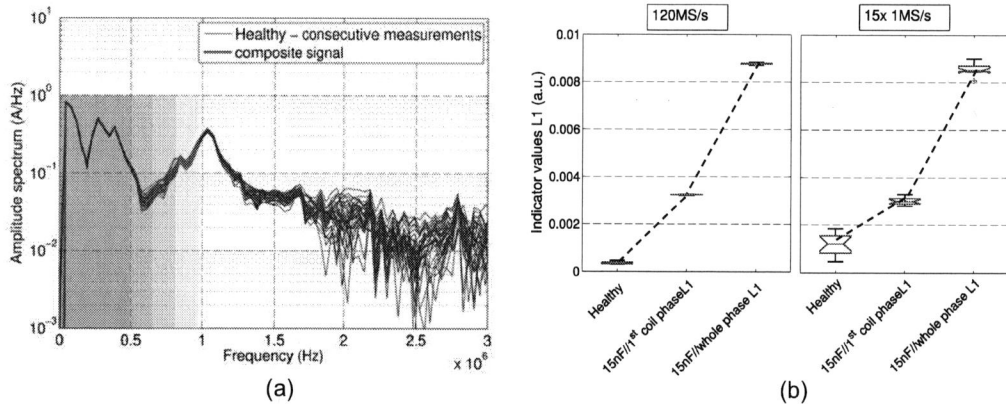

(a)   (b)

Figure 5 (a) Spectra of 15 signals with randomly added jitter (blue traces) and combined signal out of the 15 signals (b) box-plot representation of different machine states with emulated insulation degradation in phase L1 based on signals with hgih sampling rate (120MS/s) and signals with low sampling rate (1MS/s) and repetitive sampling

It is difficult to determine what is acceptable regarding the frequency variation. Therefore, no statements will be made regarding how much time jitter is manageable. The the highest frequency current component of interest focused from the authors for this machine type is located at 0.5 MHz, but can be enhanced to 1MHz.

© VDE VERLAG GMBH · Berlin · Offenbach

Within the interesting frequency range (up to 500 kHz) the deviations are negligible and the results show the applicability of the sampling method for systems with a jitter value up to 50ns.

In Fig. 5 (b) the measurements are statistically analyzed with the box-plot representation. In these two subfigures only the ISI values of phase L1 are analyzed. It shows the distribution of the resulting indicator values from the repeated measurement and shows the difference between the results of the high sampled data and the repetitive sampling results. It can be seen that the results analyzed with a resolution of 120 MS/s show a smaller variance than the data analyzed with repetitive sampling and a final resolution of 15 MS/s. However, all tested fault scenarios are still detectable.

5. Conclusion

A smart converter system with integrated online health state monitoring of the machine's insulation system has been presented. The detection of incipient insulation degradation by evaluating the current response after predefined voltage pulse excitations prevents unexpected outages and enables scheduling of further maintenance tasks. The method was enhanced and tested with a special sampling technique and tests are conducted on a 1.4 MW induction machine with different insulation degradation scenarios. The intention was to determine whether the standard industrial current sensors and standard ADC units (1 MS/s) can be used with the presented smart concept. It can thus be stated and concluded that the approach of the proposed online insulation condition monitoring method is well working and applicable to inverter fed AC machines.

References

[1] IEEE Committee Report; "Report of large motor reliability survey of industrial and commercial installation, Part I," *IEEE Transactions on Industry Applications*, vol.21, no.4, pp.853–864, 1985.

[2] IEEE Committee Report; "Report of large motor reliability survey of industrial and commercial installation, Part II," *IEEE Transactions on Industry Applications*, vol.21, no.4, pp.865–872, 1985.

[3] G. C. Stone, E. E. Boulter, I. Culbert, and H. Dhirani, "Electrical Insuation for Rotating Machines". *IEEE Press*, 2004.

[4] Farahani, M.; Gockenbach, E.; Borsi, H.; Schaefer, K.; Kaufhold, M.; "Behavior of machine insulation systems subjected to accelerated thermal aging test," *IEEE Trans. on Dielectrics and Electrical Insulation*, vol.17, no.5, pp.1364-1372, 2010.

[5] Nussbaumer, P.; Vogelsberger, M.A.; Wolbank, T.M.; "Induction Machine Insulation Health State Monitoring Based on Online Switching Transient Exploitation," *IEEE Trans. on in Industrial Electronics,* vol.62, no.3, pp.1835-1845, 2015.

[6] D Yang, J; Cho, J.; Lee, S.B.; Yoo, J.-Y.; Kim, H.D.; "An Advanced Stator Winding Insulation Quality Assessment Technique for Inverter-Fed Machines," *IEEE Trans. on Industrial Appl.*, vol.44, no.2, pp.555-564, 2008.

[7] Sumislawska, M.; Gyftakis, K.N.; Kavanagh, D.F.; McCulloch, M.; Burnham, K.J.; Howey, D.A.; "The Impcact of Themal Degradation on Electrical Machine Winding Insulation," *10th IEEE International Symposium on Diagnostics for Electric Machines, Power Electronics and Drives (SDEMPED)*, 2015.

[8] Zoeller, C.; Vogelsberger, M.A.; Fasching, R.; Grubelnik, W.; Wolbank, Th.M., "Evaluation and current-response based identification of insulation degradation for high utilized electrical machines in railway application," *10th IEEE International Symposium on Diagnostics for Electrical Machines, Power Electronics and Drives (SDEMPED)*, pp.266-272, 2015.

Investigation of the Influence of Ageing Processes on Thermal Characteristics of an IGBT Power Module by Means of Transient Thermal Analysis

Tobias von Essen, Berliner Nanotest und Design GmbH, Germany, vonessen@nanotest.eu

Stefan Stegmeier, Siemens AG, Germany, stefan.stegmeier@siemens.com

Gerhard Mitic, Siemens AG, Germany, gerhard.mitic@siemens.com

1. Abstract

Power electronics demand high reliability and lifetime to prevent high costs and downtimes due to unplanned maintenance. This paper uses a transient thermal FE simulation to analyze the thermal behavior of an IGBT power converter module. Different failure modes are in the focus of investigation: die attach constriction, wire bond lift-off and solder as well as thermal grease degradation. The simulation is optimized with the help of thermal impedance measurements of a device in power cycling. With the help of FE simulation different failure modes are possible to identify in the structure function. Extended use of these results is presented.

2. Motivation

The application fields of power electronic devices are various and usually all applications require high reliability and long lifetime. But increasing power density and temperature load act contrary to longer MTTFs (mean time to failure). Thermal cycles and the resulting thermo-mechanical loads limit the lifetime, leading to unexpected early failures and require maintenance and the replacement of failed devices. Unplanned maintenance causes high costs and/or downtimes and must be prevented.

The government-funded project EHLMOZ (07/2014 - 12/2016) deals with highly efficient and reliable power converter modules for the energy network of the future. Together with other German partners from both industry, research institutions and universities both more reliable and more performant modules are in development. Investigations on reliability enhancement and failure detection as well as prevention are key part of the work in the consortium.

The mechanisms that are causing devices to fail are well-known. Die attach constriction, solder degradation and cracks, wirebond lift-off and thermal grease pump- and dry-out are some of the most common effects that create thermal

Figure 1: The future energy supply requires efficient power electronic solutions

bottlenecks and eventually result in overheating and in irreversible failure of the device. Taking these mechanisms as a basis predictive analysis is possible e.g. by finite elements simulation.

The aim of this work is to gain understanding of how the thermal behavior of the device under test is changing due to ageing processes. In this paper thermal impedance analysis supported by FE simulations is presented. Transient thermal analyses allow identifying different failure modes which has already been shown in [1] and [2]. The interpretation of structure functions with the help of FE simulations are considered powerful in [3]. Three relevant failure modes are in focus of this paper. Important problems are: (i) detectability of wire bond lift-offs (ii) device failure due to increasing die attach constriction (iii) weighting of concurrent degradation effects. The FE model is optimized with the help of power cycling measurements of a converter module.

3. Method and approach

Target is to find out about the influence of different ageing processes inside a power electronics package on the thermal behavior of the device. It is important to learn about to what extent each single degradation mechanism causes temperature rises inside a device and how sensible common methods are. This poster presents a study about a thermal finite element model of a high power inverter module which different failures are induced into. The method of transient thermal analysis is utilized to detect the failures in the simulated thermal behavior.

3.1. Transient Thermal Analysis

Transient thermal analysis is a field of characterization techniques and methods that allows to extract structural information of systems and sub-systems from transient temperature curves. When applying the method to a real application, key requirement is being able to measure the temperature inside the active layer of the device. For transistors this is usually done by determining the

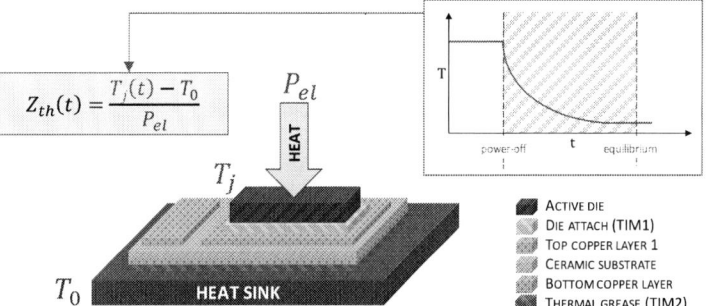

Figure 2: Schematic of a sub-system illustrating the heat path and all necessary parameters as well as the formula for Zth(t) calculation

device-specific temperature sensitivity of the reverse diode's forward voltage. At first, from a measured temperature response $T(t)$ to a discrete change in power dissipation ΔP the thermal impedance $Z_{th}(t)$ is calculated using the equation shown in Figure 2. The thermal impedance curve is starting point for various analyses. The mathematical principles are well-documented in JESD51-14 [4]. The equation in Figure 2 illustrates the simplicity of the approach as the necessary measurement is limited to just a transient temperature and a power dissipation.

3.2. Finite Elements Simulation

Thermal FE modelling allows to reconstruct an application case of a system or sub-system in order to enable transient thermal analysis based on simulation results. First, taking the device geometry and material data as input, the device is implemented as a FE model. As primary boundary condition the power step is set according to the modelled application and as secondary all heat paths need to comply with the application case, including convection and, in some cases, radiation losses as well. The results are interpreted as data directly gathered from real application measurements.

© VDE VERLAG GMBH · Berlin · Offenbach

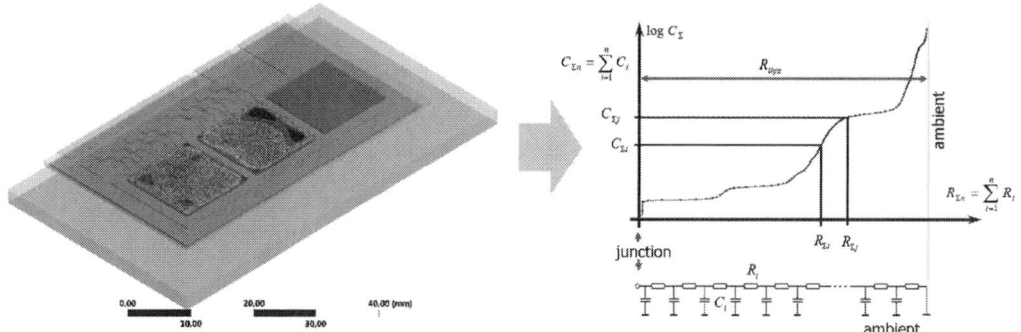

Figure 3: Simulation results are used to perform transient thermal analysis similar to measurements in real application

3.3. Preliminary study

To validate comparability between measurement and simulation results a preliminary study is performed. An off-the-shelf TO220 MOSFET is being both measured transiently and simulated as described above. To model the MOSFET with best possible exactness of geometry a cross-section has been performed at one device to determine the actual thicknesses of the material layers. (see Figure 4). No extended investigations by means of transient thermal analysis are done but both transient temperature step response curved are compared to cover as good as possible. Only a valid simulation model and setup will allow a gain of information.

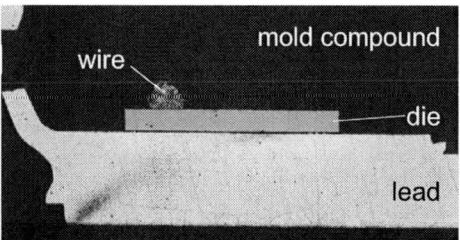

Figure 4: Light microscopy cross-section of TO220 MOSFET

3.4. Main study

Figure 5: power inverter sub-system: four IGBTs and two diodes on ceramic substrate

Subject to the main study is an IGBT converter module with four silicon IGBTs and two silicon bias diodes. The chips are soldered on top on a DCB–substrate (AlN) and interconnected with Al-wirebonds. This ceramic substrate is soldered on top of a copper baseplate which is mounted with thermal grease on a heat sink. In this study only the IGBT junctions dissipate heat and function as temperature sensing elements.

These substrates are part of 4.5 kV converter modules and are applied in a 2 x 3 matrix on a water-cooled heat sink. The device is polymer-encased and filled with silicone. For symmetry reasons and to keep computing times of the model low it is sufficient to model just half a substrate. As evidence for the validity of the reduced model a single substrate is measured for a comparison to the simulation results.

© VDE VERLAG GMBH · Berlin · Offenbach

PCIM Europe 2016, 10 – 12 May 2016, Nuremberg, Germany

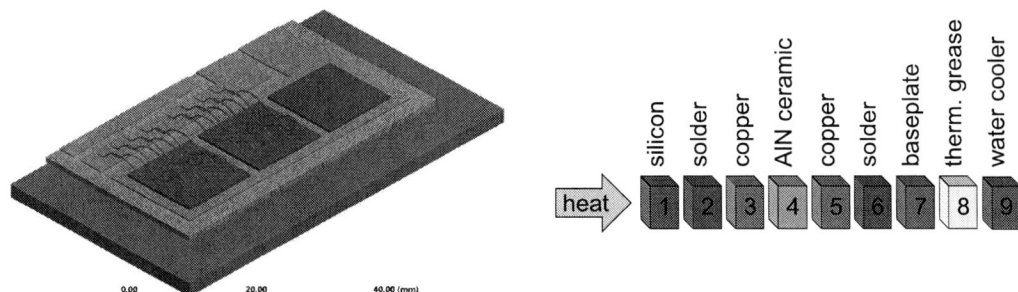

Figure 6: FE-model of the converter module

Figure 7: Different material layers in the thermal path between junction and cooler

The creation of the FE model is part of the paper as there have been important issues to consider. These issues, including mesh density, boundary conditions, symmetry and area-weighted temperature averaging are discussed in detail. The resulting model is shown in Figure 6. The degradation is taken into account with suitable assumptions, too.

As reference data for the FE model a power cycle test has been done to provide thermal impedance data for different stages of ageing. Within about 200 000 cycles ($t_{on} = t_{off} = 5$ s) and transient thermal measurements every 5000 cycles, corresponding stages of degradation have been recorded. The first and undamaged curve has been used as reference for the initial FE model. The given data from the power cycle tests is summarized in Figure 8: (b) shows four different thermal impedance curves from the power cycling series (correlating measurement number indicated with line in (a)) that have been used for correlation between simulation and experiment.

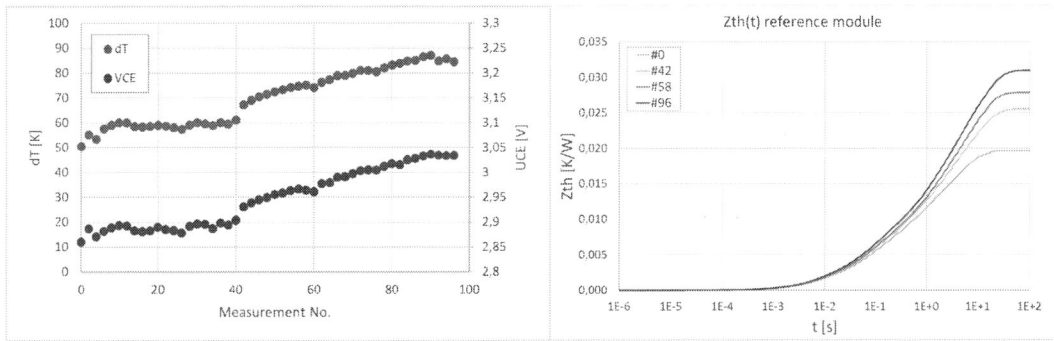

Figure 8: **(a)** Measurement data, power cycling, dT: temperature rise, VCE: collector emitter voltage
(b) Zth(t) curves for four different cycle steps

As first result it is shown that the aluminum bond wires do not contribute to the (measurable) transient thermal behavior of the system at all. Therefore, a wirebond lift-off is – despite the great number of bond wires per chip and the large wirebond diameter – not possible to detect using transient thermal measurement techniques.

Further analyses are performed with respect to different ageing mechanisms at three different layers: first is the die attach constriction, which usually is considered the dominating degradation factor because of its close distance to the heat dissipating junction and the highest temperature variation. Second ageing effect under consideration is degradation of the system solder, interconnecting the substrate with the copper baseplate (layer 6 in Figure 7). The third effect under investigation is degradation of the thermal interface between baseplate and water cooler. Thermal grease is affected by dry-out and pump-out and hence increases its effective thermal resistance over a certain amount of cycles.

© VDE VERLAG GMBH · Berlin · Offenbach

4. Conclusion and outlook

In this paper it is shown that the simulated ageing mechanisms of die attach, system solder and thermal interface layer can be detected in the structure function. Therefore, the thermal impedance curves were translated into structure functions and compared to each other (see Figure 9). The first derivative – the slope – of the structure function represents the correlation between thermal resistance and thermal capacitance throughout the thermal path. Different segments of the structure function are characteristic for different material layers and their ageing progress. As the interpretation of differences between structure functions is not trivial, FE simulations can provide a great tool for the evaluation of structure functions. The three degradation mechanisms have also been attested for the reference device and match the simulation results.

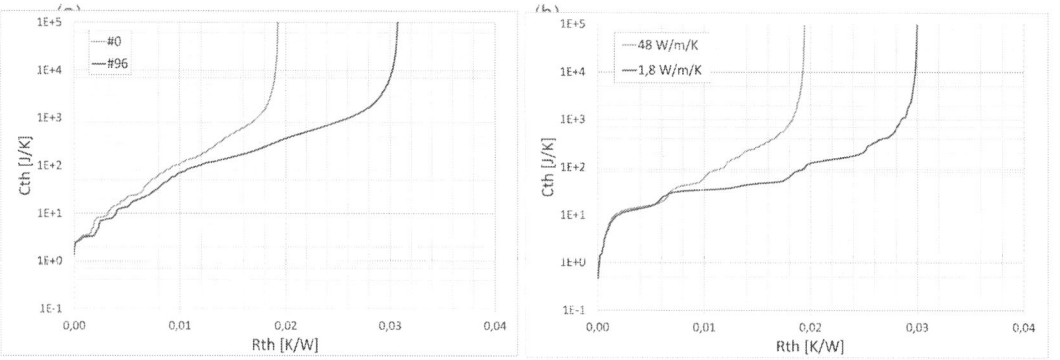

Figure 9: Comparison of structure functions: **(a)** Measurement #0 vs. #96
(b) Simulation initial vs. aged system solder layer

To evaluate the FE-model a single substrate with four IGBTs and two diodes is measured using a measurement system for transient thermal measurements developed in-house. This hardware provides power-off measurements within single digit µs range and therefore allows to measure thermal impedance data with high time resolution. It is shown that the FE model reproduces a single substrate's thermal behavior as well as the reference measurement data, generated with a whole module.

Finally, an outlook for the extended use of such approach is given. The extended use in focus includes geometry optimization, health monitoring and lifetime prediction.

5. References

[1] Hensler, *Lastwechselfestigkeit von Halbleiter-Leistungsmodulen für den Einsatz in Hybridfahrzeugen*, PhD Thesis, Chemnitz: Technische Universität Chemnitz, 2013.
[2] O. Wittler, A. Mazloum Nejadari and B. Michel, *Detection of Degradation in Die-Attach Materials by In-Situ Monitoring of Thermal Properties*, 11th Electronics Packaging Technology Conference, p. 153ff., 2009.
[3] D. Schweitzer, H. Pape and L. Chen, *Transient Thermal Measurement of the Junction-To-Case Thermal Resistance Using Structure Functions: Chances and Limits*, IEEE Semi-Therm Synopsium, no. 24, p. 191ff., 2008.
[4] JESD51-14, *Transient Dual Interface Test Method for the Measurement of the Transient Resistance Junction to Case of Semiconductor Devices with Heat Flow Through a Single Path*, Arlington: JEDEC, 2010.

PCIM Europe 2016, 10 – 12 May 2016, Nuremberg, Germany

A Study of the Thermal and Parasitic optimization of a large current density highly parallelized three-phase reference Board for Motor Drive Applications

Mehrdad, Baghaie Yazdi, Fairchild Semiconductor, Germany,
Mehrdad.Baghaie@fairchildsemi.com
Xiaomin, Wu, Fairchild Semiconductor, Germany, Xiaomin.Wu@fairchildsemi.com
Peter, Haaf, Fairchild Semiconductor, Germany, Peter.Haaf@fairchildsemi.com
Klaus, Neumaier, Fairchild Semiconductor, Germany, Klaus.Neumaier@fairchildsemi.com

Abstract

This paper discusses the challenges and solutions of designing a three phase motor drive board for applications requiring 100 A or higher output current. The approach of using multiple parallel MOSFETs in an ultra-low inductive half bridge package is described. A unique reference design PCB is presented that is constructed to meet both thermal as well as symmetry requirements when it comes to handling the parallelization of multiple MOSFETs. Thermal simulation and electro-thermal measurements are used to verify the designs and to further deepen the know-how on compact high density PCBs.

1. Introduction

Motor drive applications using power semiconductors have seen an increasing growth over the past years. Ranging from power tools, bidirectional charger, robotics to E-scooter there is a great need for compact high power (5 kW and more) designs that enables fast switching and good thermal management. This paper demonstrates a possible solution for the challenges met in designing such PCBs. Based on the FDMD85100 dual N-channel PowerTrench® MOSFET (fig1.) from Fairchild Semiconductor a three-phase 100V / 100 A reference design for high power applications is demonstrated. Thermal simulation using Ansys finite element software is employed to better understand losses and improve thermal design. Given the nature of signal delay and threshold voltage deviation among used devices, the necessity for highly symmetric designs is discussed and demonstrated using methods such as time domain reflectometry (TDR) of the relevant PCB tracks. Parasitic inductance is studied using both measurements and simulation based on physically scalable electro-thermal (PSET) SPICE models.

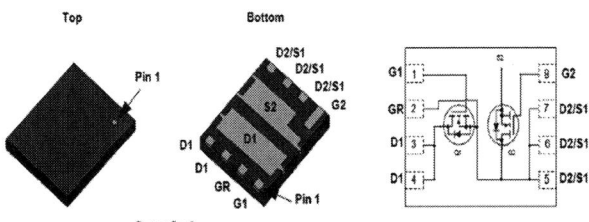

Figure 1 schematic and pin out of the FDMD85100 dual N-channel PowerTrench® MOSFET module

© VDE VERLAG GMBH · Berlin · Offenbach

2. PCB-Design and layout

The FDMD85100 is a unique package that accommodates two Fairchild Semiconductor N-channel PowerTrench® MOSFETs, which are rated for at 100 V/48 A. These are connected in a half-bridge layout and can therefore be directly used in any half-bridge topology. Furthermore the SMD package enables low inductance soldering to any PCB, and given the fact that the two MOSFETs are in one package, the Source-drain parasitic between the two MOSFETs is also dramatically reduces. To take full advantage of this package and the MOSFETs a design had to be created that makes use of the thermal and parasitic advantages of this package. Furthermore given the compact dimensions of this package (5 x 6 mm^2), we aimed at maximizing the integration density of half-bridges while maintaining operational symmetries. The result of these requirements will be presented in the following.

2.1. PCB layout

In figure 2 a typical three-phase topology is illustrated. This simple configuration is at the core of our PCB design; however each phase consists of ten parallel half-bridges, these being the FDMD85100. The 10 parallel FDMD85100s can be operated with a total current up to 480 A if needed, we have , for now, limited the use to 100 A per phase. Each phase has its own gate driver, which simultaneously controls the 10 parallel FDMD85100s in each phase. These operation conditions lead to the definition of our PCB requirements, namely:

1. Low parasitic inductance, in order to reduce oscillations and voltage overshoots in this highly parallelized operation.

2. High level of symmetry, given the number of paralleled devices and the production process given variation in the properties of the MOSFETs, one has to design a PCB that is highly symmetric for a safe operation.

3. Each phase being able to operate at 100 A with the large density of MOSFETs in parallel, requires a design that is thermally efficient and capable of handling such thermal dissipation.

4. A compact design that can be fitted in the vicinity of a motor.

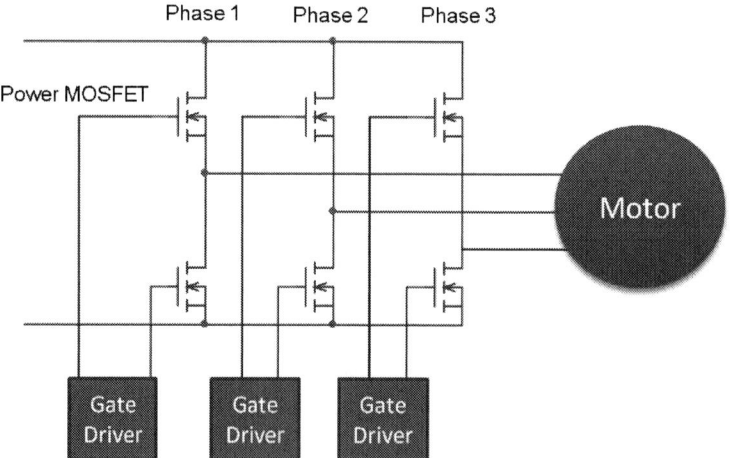

Figure 2 illustration of a three-phase motor drive configuration for use in typical modern drive application

The result of these considerations can be seen in the PCB layout presented in figure 3. The two layer design has a round shape with a diameter of 113 mm. The circular layout was chosen as it provides a large degree of symmetry for all interconnects; furthermore given the typical configuration of an electric motor it becomes a natural choice of shape. Some novel structures have been introduced to further enhance critical connection points in mutli-MOSFET operation. Such as the branch like connection of the FDMD module to the output metal, these branches aid in equally distributing the current from the parallel half-bridges to the output. Further a driver signal bus is introduced, this bundles the signal pathways coming from the gate driver on the top side, then distributes it to the bottom layer and reroutes it to the gate resistors, each devices has a total of 4 separately configurable resistors) on the top layer again. All of this is done in such a manner that the signal pathway to each FDMD85100 module is maintained as equal in length as possible. To increase thermal dissipation the conduction and ground metal are also used as cooling bodies, furthermore openings to these metals are provided as "thermal pad" and can be used to dock additional cooling. In figure 4 the full PCB layout for all the three phases can be seen.

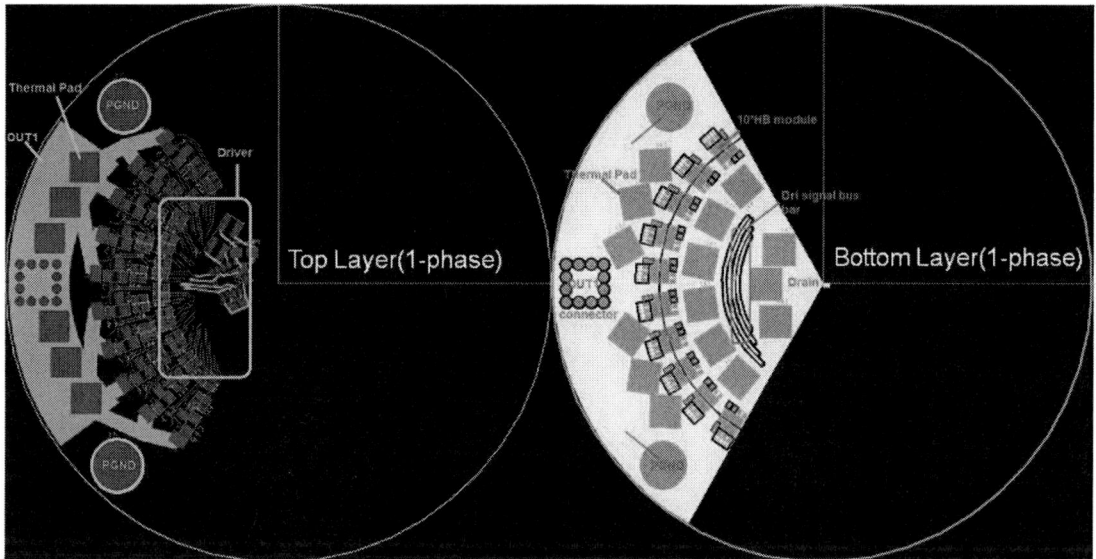

Figure 3 layout for one phase of the PCB. On the left side the top layer is shown with its connection to the gate driver and the gate resistances, on the right side the bottom layer, which hosts the FDMD.

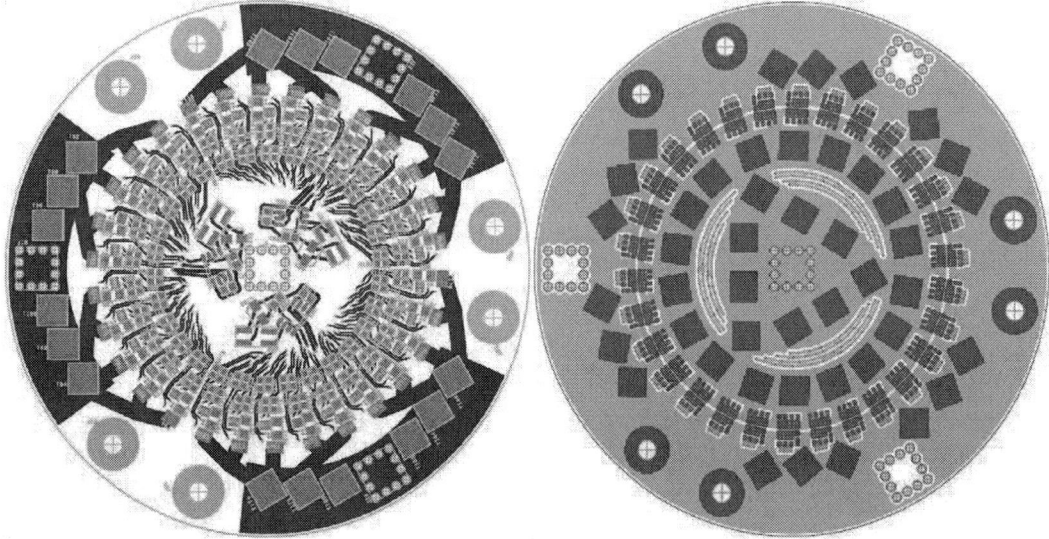

Figure 4 showing the layout for the full 3-Phase PCB, left top side and on the right bottom side.

3. Experimental Setups and Results

In the following simulation and experimental work will be presented, that was used to optimize and verify the PCBs performance. We will begin by verifying the predicted symmetries of the design in respect to crucial control signals, then by means of simulation and measurements various parasitic of the circuit are determines, followed by switching performance tests. To conclude this section, we perform thermal measurements and simulation.

3.1. Verifying Symmetry

In order to study the propagation of signals on the PCB, various methods where applied. The most sensitive being time domain reflectometry (TDR), the details of this method can be found in [1]. Using TDR the gate signal propagation of the various tracks was tested for possible time delays. For parallel MOSFET operation it is crucial that all devices receive the gate driver signal simultaneously, any delay due to PCB track variation could lead to different turn on timing of the parallel transistors and thereby cause damage to the system. The fastest pulse on the gate driver signal path will be the turn on and off rise times (dv/dt, di/dt), taking this as our fastest signal a pulse of the equivalent length was used to conduct the TDR experiment.

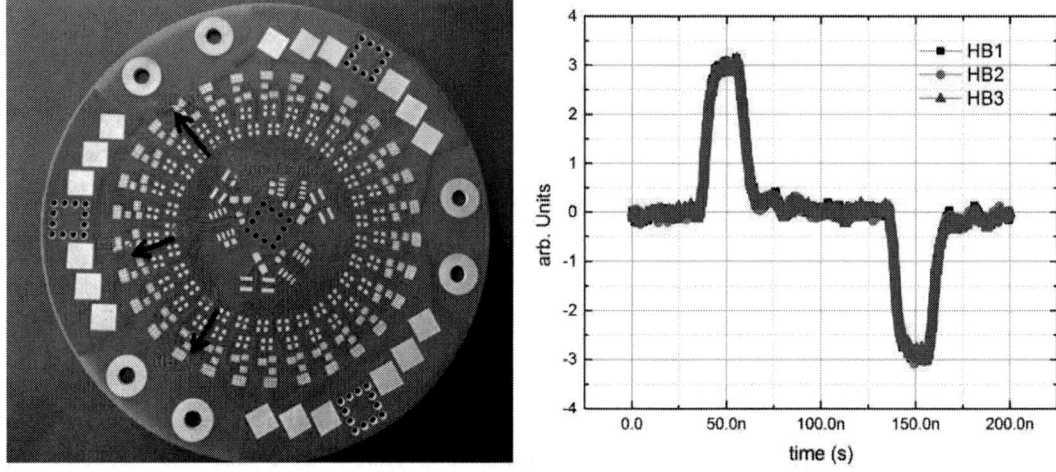

Figure 5 left, PCB gate driver – to gate signal tracks marked has HB1-HB3 (black arrows); right, TDR waveform showing identical signal propagation along the tracks.

In figure 5 the measured signal path ways and the resulting TDR waveform are shown. The measurement shows the full propagation path of the signal from the signal generator through the connecting cables and back into the signal generator. It is evident that along all the measured points the signal is identical and therefore a high degree of symmetry is guaranteed within the timescale of the switching signal coming from the gate driver.

3.2. Electric Simulation and Measurement of Parasitic Inductances

Due to the complexity of the system, we reduced the number of active modules to one FDMD85100. By doing so the parasitic inductance along various paths could be singled out and measured. Later using physically scalable electro-thermal (PSET) SPICE models [2], the measurements and the resulting parasitic where simulated, in order to obtain a circuit model that can be later used to simulate the full more complex system.

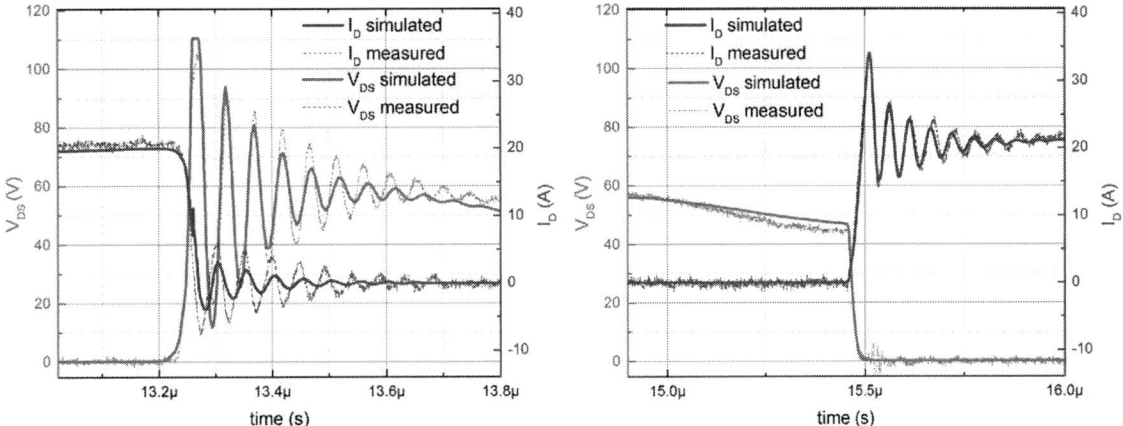

Figure 6 PSAT model simulation of FDMD85100 (model can be downloaded from Fairchild Webpage) vs measurements[1]. The simulations are used to calibrate the circuit parasitics.

Using the both measurement and simulation we have derived a parasitic inductance model, that proves that the total loop inductance seen by one FDMD85100 module is less than 10 nH. As can be seen from figure 6, a remarkable match between simulation and measurements can be seen, indicating the right choice and measurement of circuit parasitic. In figure 7, some of the track parasitics and their values are visualized.

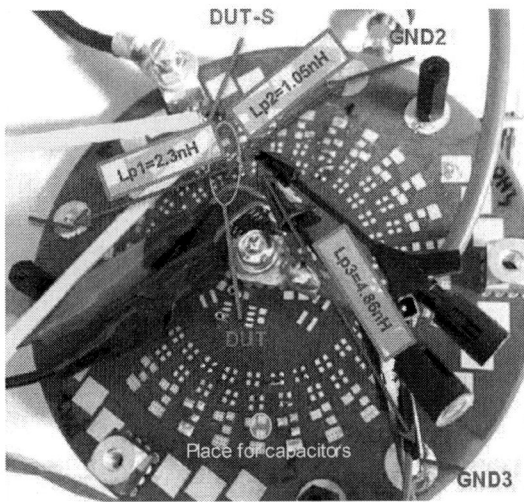

Figure 7 Measured parasitic inductances along various paths.

Currently investigations using a full 3-D model are ongoing to study and understand the full extent of parasitic inductances introduced by the massive parallel operation. However as a first degree approximation the PSET models and a well defined circuit model result in accurate SPICE simulation results that are encouraging for use in studying the more complex three-phase operation, which becomes difficult to measure especially due to the inaccessibility of current measurements once fully mounted with 30 modules.

---

[1] Oscillations will be fully dampened in the final application with input capacities parallel and in very close vicinity of the half bridges (see fig 7 capacitor place holder)

3.3. Investigating Performance

The switching performance of one FDMD85100 on the PCB was tested to determine optimal choice of gate resistance (on and off R_{gs} where set to the same value); the results are summarized in table 1.

R_g	Turn-off		Turn-On	
	E_{off} (µJ)	di/dt (A/ns)	E_{on} (µJ)	di/dt (A/ns)
10	6.99	1.19	2.4	0.83
4.7	4.24	1.21	1.32	0.87
1	3.32	1.18	0.99	0.86

3.4. Thermal Simulation and Verification

Finally the thermal performance of the board was tested and simulated. Again the approach of simplifying the experiment to a well controllable condition was used. The PCB was tested in a still air environment as defined by JDEC (see fig. 8), within a defined volume of air seen in figure 8. Simulation was performed using Ansys Icepak, the model refers to the experimental setup with only a single active package on top side of the PCB (assuming the PCB is flipped, so that back side becomes top side, see fig. 3). The PCB thermal model includes separate layer representation and the thermal vias; each layer represents the local equivalent thermal conductivities computed from E/CAD data. Furthermore the package model of the FDMD85100 is a fully detailed 3-D model. Both measurements and simulation were performed under the steady-state thermal conditions, for power dissipations of 0.22 W, 0.43 W, 0.66 W, 0.93 W and 1.19 W (see fig. 9, left), for one FDMD85100 module, with the assumption of natural convection and radiation.

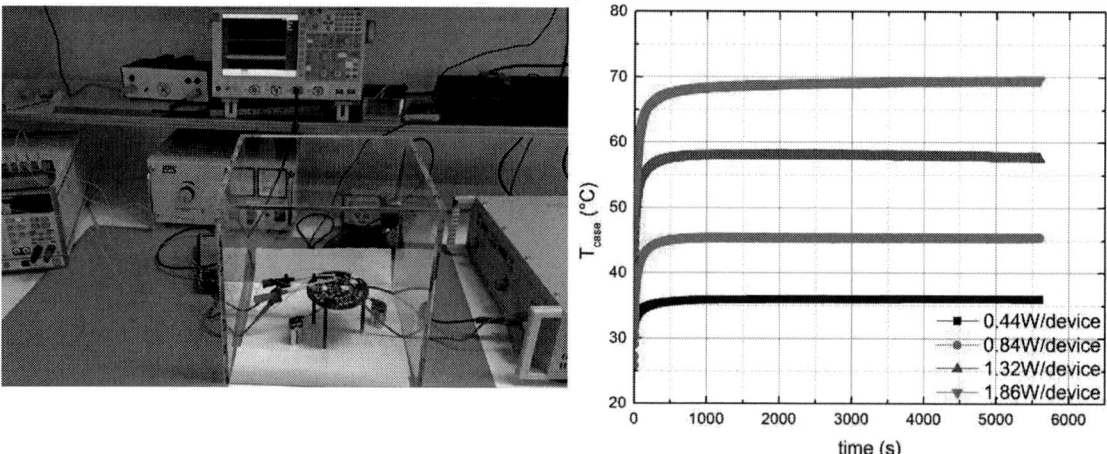

Figure 8 left, thermal measurement setup with PCB enclosed in a JDEC standard volume of air. Right, thermocouple measurement of case temperature for different power dissipation of one FDMD85100.

The simulation relative to measurements, shows an offset of about three degree, although the root cause for this could not be identified, there are multiple factors which are complicated to adjust or control, such as the ambient air temperature, Thermal vias effect and material radiation. However given the consistency between simulation offset and measurements, it is safe to assume that an even better match can be achieved with further improvements of simulation and experimental methods

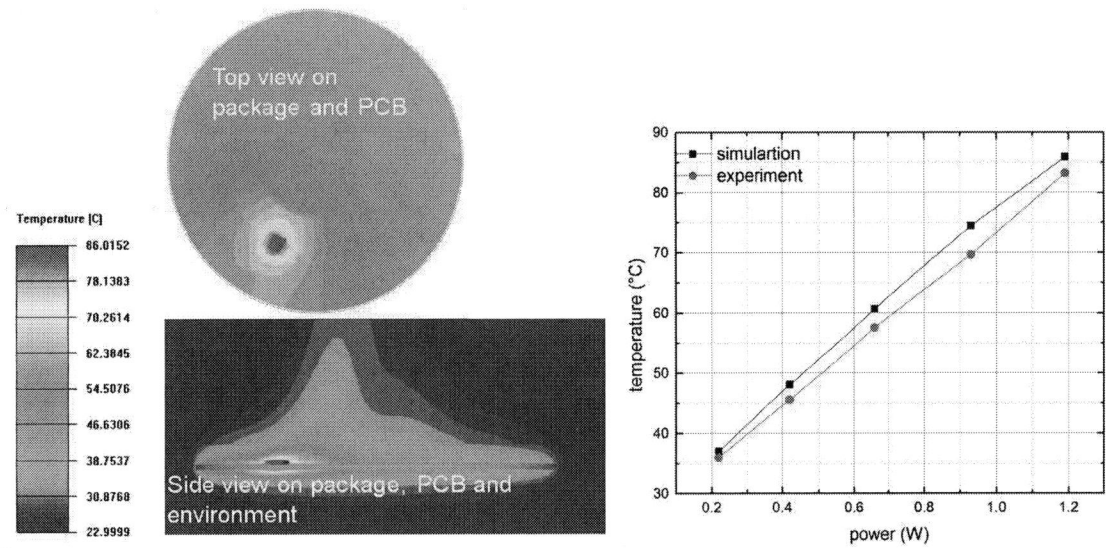

Figure 9 left, simulation results for highest power dissipation, right simulation vs. experimental results.

Furthermore the offset can be accounted for when drawing the other conclusions. In this light, the thermal simulation was conducted for a fully assembled PCB to understand the thermal performance under three-phase operation. Figure 10, shows that due to the design and thermal management features layout the heat is very homogeneously distributed and a high degree of thermal dissipation is already achieved under natural convection. This concludes that a simple and cheap cooling system can already provide sufficient thermal relieve for maximum performance operation.

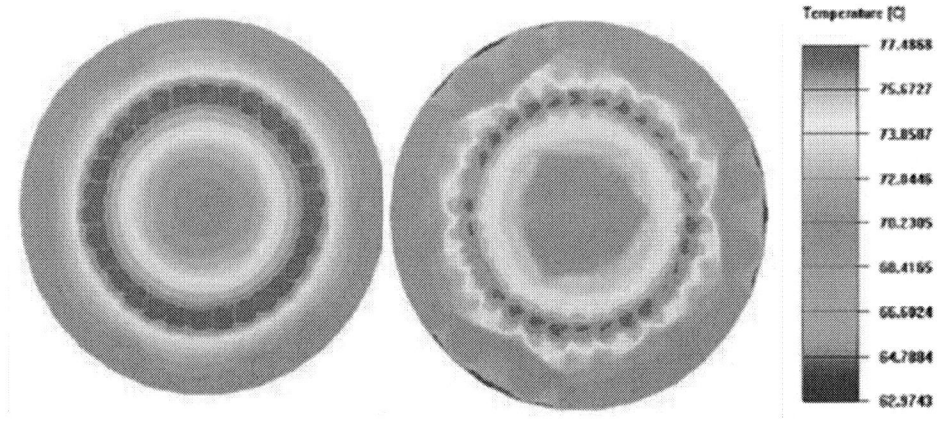

Figure 10 thermal simulation of PCB with 30 FDMD85100 Half-bridge modules all operating.

4. Conclusion

In this paper we have demonstrated a novel approach to designing and testing PCBs for multi-MOSFET parallel operations. We have achieved, by carefully designing and an extensive use of simulation a PCB that offers a high degree of symmetry, is thermally well balanced and has a low power loop inductance hence is perfectly suited for the parallel operation of power MOSFETs. Furthermore the circular design is well suited for the placement in the vicinity of the motor drives, and the intelligent thermal design requires little cooling in order to achieve optimized efficiency and maximized power output. Furthermore the design as such is scalable and can be extended to other SMD packages.

© VDE VERLAG GMBH · Berlin · Offenbach

References

[1] Huibin Zhu, A. R. Hefner and J. S. Lai, "Characterization of power electronics system interconnect parasitics using time domain reflectometry," in *IEEE Transactions on Power Electronics*, vol. 14, no. 4, pp. 622-628, Jul 1999.

[2] J. Victory, D. Son, T. Neyer, K.Lee, E. Zhou, J. Wang, and M. Baghaie Yazdi "A Physically Based Scalable SPICE Model for High-Voltage Super-Junction MOSFETs," *PCIM Europe 2014; International Exhibition and Conference for Power Electronics, Intelligent Motion, Renewable Energy and Energy Management; Proceedings of*, Nuremberg, Germany, 2014, pp. 1-8.

PCIM Europe 2016, 10 – 12 May 2016, Nuremberg, Germany

List of Authors of PCIM Europe 2016

Author	Institution	Pages
A		
Abbatelli, Luigi	STMicroelectronics, I	1536
Abdel-Rahman, Sam	Infineon Technologies Americas, USA	1615
Abe, Yasushi	Fuji Electric, J	188
Abl, Reiner	BMW, D	534
Abouchi, Nacer	Institut des Nanotechnologies de Lyon, F	1707
Abronzini, Umberto	University of Cassino and South Lazio, I	379 1217
Adachi, Shinichiro	Fuji Electric, J	1956
Adelmund, Melanie	University of Bremen, D	1232
Adler, Johannes	University of Bremen, D	1324
Adragna, Claudio	STMicroelectronics, I	612
Adrien, Mercier	ENS Cachan – SATIE, F	1647
Aggen, Christian	Danfoss Silicon Power, D	698
Agrawal, Binod	CREE India Private Limited, IN	2114
Ahmad, Jawad	Politecnico di Torino, I	2058
Aizpuru, Iosu	Mondragon University, ES	763 2074
Alberti, Luigi	University of Padova, I	534
Albertin, Thomas	Atherm, F	1132
Ali, Esmail A.	University of Science and Technology Yemen, YE	1046
Allen, Scott T.	Wolfspeed, USA	34
Almai, Farshid	University of Paderborn, D	232
Altstadt, Jochen	Kortec, D	1979
Alvarez, Rodrigo	Siemens, D	588
Alves-Rodrigues, Luis Gabriel	Commissariat à l'énergie atomique et aux énergies alternatives, F	128
Ambacher, Oliver	Fraunhofer-Institute IAF, D	319
Ammann, Ulrich	Esslingen University of Applied Sciences, D	780
An, Bao Ngoc	Karlsruhe Institute of Technology, D	1979
Andenna, Maxi	ABB Switzerland, CH	417
Antivachis, Michael	ETH Zürich, CH	1622
Appel, Tobias	STS, D	1434
Aragues-Penalba, Monica	CITCEA-UPC, ES	2211

PCIM Europe 2016, 10 – 12 May 2016, Nuremberg, Germany

Author	Institution	Pages
Aranzabal, Itxaso	University of the Basque Country, ES	1970
Araujo Vasconcelos, Samuel	University of Kassel, D	573
Arnold, Martin	ABB Switzerland, CH	917
Arpino, Matilde	University of Cassino and South Lazio, I	379 1217
Asada, Shinsuke	Mitsubishi Electric Corporation, J	326
Asfaux, Pascal	Airbus Operation, F	1814
Ashley, Grant	Vxl Power, GB	1455
Askari, Vahid	Teesside University, GB	1439 1685
Attaianese, Ciro	University of Cassino and South Lazio, I	379 1217
Augoustidis, Alex	GrafTech International, USA	1126
Austermann, Johann	University of Applied Sciences Ostwesfalen-Lippe, D	1639
Avenas, Yvan	G2ELAB, F	1494
Azcondo, Francisco	University of Cantabria, ES	1194 1631
Azuma, Katsunori	Hitachi Power Semiconductor, J	348
B		
Baghadadi, Mehdi	University of Cambridge, GB	1845
Baghaie Yazdi, Mehrdad	Fairchild Semiconductor, D	813 2231
Baier, Thomas	Friedrich-Alexander-University Erlangen, D	264
Bakran, Mark-M.	University of Bayreuth, D	272 355 653 683 720 1055 1400 1503 2181
Balke, Christian	Maccon, D	1353
Banerjee, Sujit	Monolith Semiconductor, USA	984
Barater, Davide	Università degli Studi di Parma, I	1073 1584
Barrios, Ernesto L.	Public University of Navarre, ES	1209
Barruel, Franck	Commissariat à l'énergie atomique et aux énergies alternatives, F	128
Barwig, Markus	Friedrich-Alexander-University of Erlangen, D	1693 1701
Baschnagel, Andreas	ABB Switzerland, CH	945

PCIM Europe 2016, 10 – 12 May 2016, Nuremberg, Germany

Author	Institution	Pages
Basler, Thomas	Infineon Technologies, D	180
Basler, Vanessa	Technical University of Munich, D	543
Bäßler, Marco	Danfoss Silicon Power, D	698
Batard, Christophe	University of Nantes, F	712
Bauer, Pavol	Delft University of Technology, NL	1201 1383
Baumann, Michael	SUMIDA Components & Modules, D	1728 1764
Bauwens, Filip	On Semiconductor, BE	1479
Bayer, Martin	ABB Switzerland, CH	432
Bazzano, Gaetano	STMicroelectronics, I	804
Beaucarne, Guy	Dow Corning Europe, B	992
Beaurenaut, Laurent	Infineon Technologies, D	1377 2027
Becker, Karl-Friedrich	Fraunhofer Institute IZM, D	1040
Beichler, Johannes	Vacuumschmelze, D	1471
Beier-Möbius, Menia	Chemnitz University of Technology, D	172 588
Bellingen, Jörg	Technical University of Vienna, AT	2219
Bellini, Marco	ABB Switzerland, CH	911
Bendani, Larbi	Valeo, F	477
Benkendorff, Berthold	Christian-Albrechts-University, D	1942
Berger, Hubert	FH Joanneum, AT	296 2151
Bergogne, Dominique	CEA Leti, F	312
Beringer, Sebastian	Leibniz University Hannover, D	107
Bernardinis, Gabriele	Analog Devices, USA	195
Bernd, Martin	Karlsruhe Institute of Technology, D	1979
Bertelshofer, Teresa	University of Bayreuth, D	272 653 1503
Beuermann, Max	Siemens, D	1888
Beyer, Harald	ABB Switzerland, CH	945
Bhalla, Anup	United Silicon Carbide, USA	1511 1549
Bier, Anthony	Commissariat à l'énergie atomique et aux énergies alternatives, F	128
Billet, Cyrille	Calyos, BE	1139
Bingham, Chris	University of Lincoln, GB	1769
Birkel, Andre	University of Bayreuth, D	2181

PCIM Europe 2016, 10 – 12 May 2016, Nuremberg, Germany

Author	Institution	Pages
Blaabjerg, Frede	Aalborg University, DK	831 1837
Blank, Mathias	Technical University of Vienna, AT	1339
Blank, Thomas	Karlsruhe Institute of Technology, D	1487 1979
Blum, Manuel	Siemens, D	1912 1919
Böcker, Joachim	University of Paderborn, D	232 780 1662 1992
Bödeker, Christian	University of Bremen, D	1232
Böh, Magnus	Technical University of Cologne, D	669
Bolognani, Silverio	University of Padova, I	534
Bolte, Sven	University of Paderborn, D	232 1992
Booth, James	Dynex Semiconductor, GB	704
Borcherding, Holger	University of Applied Sciences Ostwesfalen-Lippe, D	1639
Borecki, Jacek	University of Bremen, D	248
Borghetti, Giovanni	Nidec ASI, I	2175
Boroyevich, Dushan	CPES/Virginia Tech, USA	1286
Bortis, Dominik	ETH Zürich, CH	1622
Bosch, Swen	Heinrich Steinhart, HTW Aalen, D	2129
Böttigheimer, Mike	University of Stuttgart, D	410
Bouguet, Christophe	University of Nantes, F	712
Bourdon, Jeremy	Airbus Operation, F	1814
Boureghda, Monnir	Alpha Assembly Solutions, USA	1027
Bourner, David	Vicor Corporation, USA	1253 1822
Braband, Matthias	ISW – University of Stuttgart, D	256
Bräckle, Dennis	Karlsruhe Institute of Technology, D	788
Bramanpalli, Ranjith	Würth Elektronik eiSos, D	99 1757
Branas, Christian	University of Cantabria, ES	1631
Braun, Michael	Karlsruhe Institute of Technology, D	393 788
Brinkfeldt, Klas	Swerea IVF AB, SE	1063
Brix, Jonathan	Fraunhofer Institute IPA, D	386
Brubaker, Michael	SBE, USA	1391
Budaker, Bernhard	Fraunhofer Institute IPA, D	386

PCIM Europe 2016, 10 – 12 May 2016, Nuremberg, Germany

Author	Institution	Pages
Buetow, Sven	Semikron Elektronik, D	728
Burani, Nicola	Fraunhofer Institute IISB, D	453
Burger, Michael	SEW Eurodrive, D	144
Burgos, Rolando	CPES/Virginia Tech, USA	1286
Burns, Chris	AB Mikroelektronik, AT	969
Buschhorn, Stefan	Infineon Technologies, D	850
Buschkühle, Marc	Infineon Technologies, D	1800
Buttay, Cyril	INSA de Lyon, F	1792
C		
Cabezuelo, David	University of the Basque Country, ES	1970
Camus, Stephane	Aperam Alloys Amilly, F	84
Canales, Jose Mari	Mondragon University, ES	763 2074
Cao, Zhiyu	AEG Power Solutions, D	1662 1677
Carastro, Fabio	GE Research Center, D	645
Carcouet, Sébastien	University Grenoble Alpes CEA LITEN, F	2098
Carvalho, Adriano	Nomad Tech, PT	953
Casady, Jeffrey	Wolfspeed, USA	34
Catalisano, Giuseppe	STMicroelectronics, I	1536
Catellani, Stéphane	Commissariat à l'énergie atomique et aux énergies alternatives, F	128
Cavallaro, Daniela	STMicroelectronics, I	804
Cellier, Remy	Institut des Nanotechnologies de Lyon, F	1707
Cerezo, Jorge	Infineon Technologies Americas, USA	819
Chamaret, André-Philippe	SNCF, F	1103
Chandra Mouli, Gautham Ram	Delft University of Technology, NL	1383
Chatroux, Daniel	University Grenoble Alpes CEA LITEN, F	2098
Chaudhuri, Toufann	ABB Sécheron, CH	1103
Chelghoum, Reda	Valeo, F	477
Chen, Jingxuan	University of Toronto, CA	895
Chen, Zhiyang	ON Semiconductor, USA	1777
Cherix, Nicolas	EPFL – Ecole Polytechnique Fédérale de Lausanne, CH	1949
Cheung, Chun	Intersil Corporation, USA	550
Chinthavali, Madhu S.	Oak Ridge National Laboratory, USA	1391
Choi, Hanna	KCC Corporation, ROK	961
Choo, Byoungho	Infineon Technologies Power Semitech, ROK	863

PCIM Europe 2016, 10 – 12 May 2016, Nuremberg, Germany

Author	Institution	Pages
Choudhary, Vijay	Texas Instruments, USA	2019
Christe, Alexandre	EPFL – Ecole Polytechnique Fédérale de Lausanne, CH	461
Chun, Ho Tae	Hyundai Motors Company, ROK	1262
Chung, Daewoong	Infineon Technologies Power Semitech, ROK	858 863
Ciocia, Alessandro	Politecnico di Torino, I	2058
Cioffi, Philip	GE Research Center, USA	645 1853
Comola, Marco	STMicroelectronics, I	804
Concari, Carlo	Università degli Studi di Parma, I	1584
Conrad, Detlef	Synopsys, D	813
Consentino, Giuseppe	STMicroelectronics, I	804
Corvasce, Chiara	ABB Switzerland, CH	417 432
Costa, Francois	ENS Cachan – SATIE, F	288 1494
Coulinge, Emilien	EPFL – Ecole Polytechnique Fédérale de Lausanne, CH	461
Cousineau, Marc	University of Toulouse, F	371
Crisafulli, Vittorio	ON Semiconductor, D	1178 1865
Curea, Octavian	ESTIA, F	2066
D		
Dadanema, Gnimdu	ENS Cachan – SATIE, F	1494
Dai, Jian	GE Research Center, USA	1853
Dai, Xiaoping	Dynex Semiconductor, GB	704 845 938
Dalessandro, Luca	Schaffner, CH	148
Daly, Michael	Analog Devices, USA	195
Datta, Rajib	GE Research Center, USA	645 1853
Dauchy, Julien	University Grenoble Alpes CEA LITEN, F	2098
Davidson, Colin	Alstom Grid, GB	2197
Davidson, Jonathan	University of Sheffield, GB	1095
de Alegría, Iñigo Martinez	University of the Basque Country, ES	1785 1970
De Bernardinis, Alexandre	French Institute of Science and Technology for Transport, Development and Networks, F	2159
De Michielis, Luca	ABB Switzerland, CH	417

PCIM Europe 2016, 10 – 12 May 2016, Nuremberg, Germany

Author	Institution	Pages
De Monchy, Michiel	Alpha, RUS	957
de Rooij, Michael	Efficient Power Conversion Corporation, USA	304
De Sousa, Luis	Valeo, F	477
de Stoppelaar, Diederik	LightingEurope, BE	604
Deboy, Gerald	Infineon Technologies, AT	1622
Declercq, Frederick	On Semiconductor, BE	1479
Dede, Ercan	Toyota Research Institute North America, USA	72
Delsuc, Vincent	Dow Corning, BE	1017
Demattio, Horst	Karlsruhe Institute of Technology, D	1487 1979
Denk, Fabian	Karlsruhe Institute of Technology, D	1721
Denk, Marco	University of Bayreuth, D	1055
Deviny, Ian	Dynex Semconductor, GB	845
Di Leo, Paolo	Politecnico di Torino, I	2058
Di Monaco, Mauro	University of Cassino and South Lazio, I	379 1217
Diaz Reigosa, Paula	Aalborg University, DK	831
Dieckerhoff, Sibylle	Technical University of Berlin, D	1186
Dietrich, Lothar	Fraunhofer-Institute IZM, D	1155
Dincan, Catalin Gabriel	Aalborg University, DK	2050
Dinis, Corina Maria	Politehnica University Timisoara, RO	2106
Dinulovic, Dragan	Würth Elektronik eiSos, D	107
Djallel, Kerdoun	GLEC Constanine 1, DZ	1785
Dobusch, Julian	Friedrich-Alexander-University of Erlangen, D	1857
Dodge, Jonathan	United Silicon Carbide, USA	484 1549
Domb, Moshe	Infineon, USA	1555
Domes, Daniel	Infineon Technologies, D	53
Doppelbauer, Martin	Karlsruhe Institute of Technology, D	393
Dragan, Mihai	Fraunhofer Institute IPA, D	386
Draghici, Mihai	Infineon Technologies, AT	180
Draugedalen, Eirik	Valeo, NO	113
Dudin, Andrey	Technical University of Ilmenau, D	1896
Dujic, Drazen	EPFL – Ecole Polytechnique Fédérale de Lausanne, CH	461
Dupont, Vincent	Calyos, BE	1139
Dürbaum, Thomas	Friedrich-Alexander-University of Erlangen, D	492 1693 1701 1857

PCIM Europe 2016, 10 – 12 May 2016, Nuremberg, Germany

Author	Institution	Pages
Durham, Jeffrey	Alpha Assembly Solutions, USA	1027
Dworakowski, Piotr	SuperGrid Institute, F	1592
E		
Eberlin, Michael	Fraunhofer Institute ISE, D	1733
Ebli, Michael	Universtiy of Reutlingen, D	210
Eckel, Hans-Günter	University of Rostock, D	581 924 1888
Edelmoser, Karl	Technical University of Vienna, AT	1741 2137
Efika, Ikenna	Alstom Grid, GB	2197
Ehrhardt, Christian	Fraunhofer-Institute IZM, D	1155
Ellinger, Thomas	Technical University of Ilmenau, D	1896
Enami, Hiroji	Dow Corning Toray, J	1017
Endres, Stefan	Fraunhofer Institute IISB, D	526
Engel, Günter	CeraCap, AT	2151
Engelhard, Christoph	HELLA KGaA Hueck, D	669
Epp, Nikolai	AEG Power Solutions, D	1677
Ertl, Hans	Technical University of Vienna, AT	1339 1669 1846
Evans, Kim	Dynex Semiconductor, GB	704
F		
Fahlbusch, Sebastian	Helmut Schmidt University Hamburg, D	1528
Fahlenkamp, Marc	Infineon Technologies, D	620
Fahnert, Holger	AEG Power Solutions, D	1662
Fairweather, Andrew	Vxl Power, GB	1455 2090
Farkas, Gabor	Mentor Graphics,HU	596
Fazio, Gianluca	On Semiconductor Italy, I	1178
Felgemacher, Christian	University of Kassel, D	573
Fernandez, Manuel	Universtiy of Oviedo, ES	1479
Fernández-Serantes, Luis Al-fonso	University of Applied Sciences Joanneum, AT	296
Ferrazza, Francesco	STMicroelectronics, I	612
Fersterra, Fabian	Fraunhofer Institute IISB, D	469
Filho, Faete	Eaton Corporation, USA	1518

PCIM Europe 2016, 10 – 12 May 2016, Nuremberg, Germany

Author	Institution	Pages
Fink, Karsten	Power Integrations, D	48
Fischer, Aaron	Technical University of Ilmenau, D	1896
Fischer, Fabian	ABB Switzerland, CH	945
Fischer, Hermann	Fairchild Semiconductor, D	813
Fortuna, Stefania	STMicroelectronics, I	804
Foster, Martin	University of Sheffield, GB	1095 1455 1769 2090
Franceschini, Giovanni	Università degli Studi di Parma, I	1073
Franco, Augusto	Nomad Tech, PT	953
Frangieh, Tony	GE Research Center, USA	1853
Frank, Wolfgang	Infineon Technologies, D	1615
Franke, Toke	Danfoss Silicon Power, D	1942
Frankeser, Sophia	Technical University of Chemnitz, D	1063
Freitas, Nuno	Nomad Tech, PT	953
Frick, Florian	ISW – University of Stuttgart, D	256
Fröhleke, Norbert	University of Paderborn, D	232 1662 1992
Fuchs, Friedrich W.	Christian-Albrechts-University, D	1942
Fuhrmann, Jan	University of Rostock, D	581 1888
Fujimoto, Mitsunao	ALPS Green Devices, J	91
Funk, Dustin	Technical University of Ilmenau, D	1749
G		
Gaber, Roland	Fraunhofer Institute IWES, D	401
Gafford, James	Mississippi State University, USA	1294
Galai Dol, Lilia	Efficacity, F	2159
Galea, Michael	University of Nottingham, GB	1073
Galek, Marek	Siemens, D	453 1912 1919
Gambino, Giusy	STMicroelectronics, I	1987
Gant, Levi	The University of Alabama, USA	984
Garnier, Laurent	University Grenoble Alpes CEA LITEN, F	2098
Gautier, Cyrille	ENS Cachan – SATIE, F	288
Geissmann, Silvan	ABB Switzerland, CH	417 432

PCIM Europe 2016, 10 – 12 May 2016, Nuremberg, Germany

Author	Institution	Pages
Gekeler, Manfred W.	HTWG Konstanz University of Applied Sciences, D	136
Gerada, Chris	University of Nottingham, GB	1073
Gerlach, Rolf	Infineon Technologies, D	180
Ghossein, Layal	SuperGrid Institute, F	1592
Gierschner, Sidney	University of Rostock, D	1888
Ginot, Nicolas	University of Nantes, F	712
Gleißner, Michael	University of Bayreuth, D	683
Gleitner, Volker	Electronicon Kondensatoren, D	1408
Glose, Daniel	Technical University of Munich, D	1270
Glück, Tobias	Technical University of Vienna, AT	1339
Godbold, Gerald W.	Hyperion Technology, USA	1294
Goikoetxea, Ander	Mondragon University, ES	763 2074
Gomez Suarez, Carlos	Aalborg University, DK	831
Gomez, German	On Semiconductor, BE	1479
Gony, Bashar	Aperam Alloys Amilly, F	84
Gonzalez, Diego	University of Oviedo, ES	1479
González, Manuela	University of Oviedo, ES	1631
Götz, Stefan	Duke University, USA	1904
Grady, Matt	United Silicon Carbide, USA	1549
Grasel, Bernhard	Dewesoft, AT	2204
Greca, Gustavo	Alpha Assembly Solutions, USA	1027
Green, James	University of Sheffield, GB	1095
Grégoire, Luc-André	University of Toulouse, F	371
Grider, Dave	Wolfspeed, USA	34
Grigans, Linards	Riga Technical University, LV	1880
Grimaud, Louis	Safran Group, F	1286
Gritti, Giovanni	STMicroelectronics, I	612
Grohmann, Rolf	HTWK Leipzig, D	2145
Guerre, Vincent	Local Energy Alternative & Fair, F	2066
Guillaume, Herault	ENS Cachan – SATIE, F	1647
Gumaan, Mohammed	Mansoura University, AE	1046
Gustafsson, Emilia	ABB Switzerland, CH	432
H		
Haaf, Peter	Fairchild Semiconductor, D	2231
Hacala, Amélie	ESTIA, F	2066
Hackl, Christoph	Technical University Munich, D	1926

PCIM Europe 2016, 10 – 12 May 2016, Nuremberg, Germany

Author	Institution	Pages
Hähre, Karsten	Karlsruhe Institute of Technology, D	1721
Hain, Stefan	University of Bayreuth, D	720
Hammes, David	University of Rostock, D	1888
Han, SeungHyun	Hyundai Motors Company, ROK	1262
Han, Yunchao	Fraunhofer Institute IISB, D	469
Harel, Jean Claude	Renesas Electronics America, USA	1027
Harfman Todorovic, Maja	GE Research Center, USA	645 1853
Harper, Jonathan	ON Semiconductor, D	877
Hartmann, Michael	Schneider Electric Power Drives, AT	1669
Hartmann, Samuel	ABB Switzerland, CH	945
Hase, Nobuhiro	Rohm Semiconductor, USA	42
Hatae, Shinji	Mitsubishi Electric Corporation, J	677
Haug, Martin	Würth Elektronik eiSos, D	107
Hausberger, Thomas	Technical University of Vienna, AT	1339
Hayakawa, Seiichi	Hitachi Power Semiconductor, J	348
He, Hongtao	Mitsubishi Electric & Electronics, CN	871
He, Weikun	Mentor Graphics, UK	1027
Heer, Daniel	Infineon Technologies, D	53 1800
Heering, Wolfgang	Karlsruhe Institute of Technology, D	1721
Heinemann, Lothar	AEG Power Solutions, D	1677
Heinzel, Thomas	Fuji Electric Europe, D	824 1956
Heldwein, Marcelo	Federal Institute of Santa Catarina – IFSC, BR	621
Helling, Florian	Universität der Bundeswehr Munich, D	1904
Helsper, Martin	Siemens, D	272
Heredero-Peris, Daniel	Universitat Politécnica de Catalunya, ES	2035
Hernandez Gutiérrez, Diego	CH	1178
Hernes, Magnar	SINTEF Energy Research, NO	1408
Herold, Christian	Technical University of Chemnitz, D	588
Herzer, Reinhard	Semikron Elektronik, D	728
Heseding, Johannes	Leibniz University Hannover, D	1421
Hewitt, David	University of Sheffield, GB	1095
Hill, Julia	Aperam Alloys Amilly, F	84
Hiller, Marc	Karlsruhe Institute of Technology, D	788
Himmelstoss, Felix	Technikum Vienna, AT	1741 2137

PCIM Europe 2016, 10 – 12 May 2016, Nuremberg, Germany

Author	Institution	Pages
Hinken, Reiner	Danfoss Silicon Power, D	1934
Hirao, Akira	Fuji Electric, J	1001
Hirayama, Tomohisa	Hitachi Power Semiconductor, J	348
Hochberg, Martin	Karlsruhe Institute of Technology, D	1305 1578
Hoffmann, Klaus F.	Helmut Schmidt University Hamburg, D	202 1300 1528
Hofmann, Daniel	Fuji Electric Europe, D	438
Hofmann, Norbert	University of Applied Sciences Nordwestschweiz, CH	917
Hofmann, Viktor	University of Bayreuth, D	355
Höltgen, Markus	Technical University of Cologne, D	796
Hölzl, Wolfgang	Technical University of Munich, D	543
Homann, Michael	Technical University of Braunschweig, D	216
Honsberg, Marco	Mitsubishi Electric Europe, D	889
Horff, Roman	University of Bayreuth, D	272 653 1503
Hori, Motohiro	Fuji Electric, J	188
Horiuchi, Keisuke	Hitachi, J	1110
Hosking, Terry	SBE, USA	1391
Hossein Khani, Milad Mohammad	Fraunhofer Institute ISE, D	1733
Houzouji, Hiroshi	Hitachi, J	78
Hruska, Miroslav	Skoda Electric, CZ	1808
Hu, Bo	Mitsubishi Electric & Electronic, CN	932
Huang, Jianwei	Dynex Semconductor, GB	845
Huang, Yi	Intersil Corporation, USA	550
Hudoffsky, Boris	PMK Mess- und Kommunikationstechnik, D	1463
Hull, Brett	Wolfspeed, USA	34
Hüning, Felix	University of Applied Sciences Aachen, D	1963
Hunziker, Christoph	University of Applied Sciences Northwestern Switzerland, CH	2167
Hussein, Khalid	Mitsubishi Electric Europe, D	677
I		
Iagar, Angela	Politehnica University Timisoara, RO	2106
Iannuzzo, Francesco	Aalborg University, DK	831
Ichikawa, Hiroaki	Fuji Electric, J	824

PCIM Europe 2016, 10 – 12 May 2016, Nuremberg, Germany

Author	Institution	Pages
Ikawa, Osamu	Fuji Electric, J	438 824
Ikeda, Osamu	Hitachi, J	78
Ikeda, Yoshinari	Fuji Electric, J	188
Immovilli, Fabio	Raw Power, I	1073
Ino, Kazuhide	Rohm, J	42
Inokuchi, Seiichiro	Mitsubishi Electric Corporation, J	677
Inoue, Daisuke	Fuji Electric, J	1956
Inoue, Tomoki	Toshiba, J	566
Iraola, Unai	Mondragon University, ES	763 2074
Irifune, Hiroyuki	Kaga Toshiba Electronics Corporation, J	839
Ishibashi, Hidetoshi	Mitsubishi Electric Corporation, J	342
Ito, Yoichi	Sanken Electric, J	2083
Izuka, Arata	Mitsubishi Electric Corporation, J	677
J		
Jacob, Mathew	Texas Instruments, USA	2019
Jagau, Martin	Technologienetzwerk Allgäu, D	1171
Jang, Hyosang	Infineon Technologies Power Semitech, ROK	863
Jang, JiWoong	Hyundai Motors, ROK	1600
Jang, KiYoung	Hyundai Motors, ROK	1600
Jara, Martin	West Bohemian University, CZ	1808
Jaschke, Reinhard	Helmut Schmidt University Hamburg, D	1300
Jeong, JeeHye	Hyundai Motors Company, ROK	1262
Jeong, KangHo	Hyundai Motors, ROK	1600
Jerinic, Vladan	Danfoss Silicon Power, D	1934
Jeyaprakash, Arun	Technical University of Munich, D	1904
Jiu, Jinting	Osaka University, J	1021
Jones, Steve	Dynex Semiconductor, GB	938
Jonke, Peter	AIT-Austrian Institute of Technology, AT	1846
Joo, JeongHong	Hyundai Motors Company, ROK	1262
Jorge-Ques, Fernando	Universitat Politécnica de Catalunya, ES	2035
Joshi, Shailesh	Toyota Research Institute North America, USA	72
Josso, Stieven	Henkel Electronics, BE	1035
Joubert, Charles	Ampere Laboratory, F	1286 1792
Jun, ChangHan	Hyundai Motors Company, ROK	1262

PCIM Europe 2016, 10 – 12 May 2016, Nuremberg, Germany

Author	Institution	Pages
Jun, Kisoo	KCC Corporation, ROK	961
Jung, Jaehoon	KCC Corporation, ROK	961
Jung, JinHwan	Hyundai Motors Company, ROK	1262 1600
Jung, Marco	Fraunhofer Institute IWES, D	401
Jungwirth, Herbert	SUMIDA Components & Modules, D	1764
K		
Kähr, Christian	University of Applied Sciences Nordwestschweiz, CH	917
Kaiser, Julian	Fraunhofer Institute IISB, D	469
Kaji, Yusuke	Mitsubishi Electric Corporation, J	326
Kakiki, Hideaki	Fuji Electric, J	188
Kako, Naotsugu	Toshiba Corporation Semiconductor, J	839
Kamal, Mustafa	Mansoura University, AE	1046
Kaminski, Nando	University of Bremen, D	1232 1324
Kammerer, Felix	Karlsruhe Institute of Technology, D	788
Kampen, Dennis	BLOCK Transformatoren-Elektronik, D	144
Kanai, Naoyuki	Fuji Electric, J	188
Kaneko, Satoshi	Fuji Electric, J	188
Kapaun, Florian	Universität der Bundeswehr Munich, D	1570
Kapels, Holger	Fraunhofer Institute ISIT, D	202
Kasper, Matthias	ETH Zürich, CH	1622
Kato, Koji	Sanken Electric, J	2083
Kaulfersch, Eberhard	Berliner Nanotest & Design, D	1063
Kawamoto, Noriaki	Rohm, J	42
Kawase, Daisuke	Hitachi Power Semiconductor, J	348 1110
Kawashima, Tetsuya	Fuji Electric, J	895
Ke, Maolong	Dynex Semconductor, GB	845
Kennel, Ralph	Technical University of Munich, D	120 518 629 1248 1270 1926
Keuck, Lukas	University of Paderborn, D	232
Keyse, Richard	Dynex Semiconductor, GB	903
Khani, Toktam	Technologienetzwerk Allgäu, D	2122
Khaselev, Oscar	Alpha Assembly Solutions, USA	1027

PCIM Europe 2016, 10 – 12 May 2016, Nuremberg, Germany

Author	Institution	Pages
Khatri, Danish	Infineon Technologies North America, USA	882
Killeen, Peter	Wolfspeed, USA	34
Kim, Hyunwoo	KCC Corporation, ROK	961
Kim, Taehyun	Infineon Technologies Power Semitech, ROK	858
Kimura, Takashi	Hitachi Automotive Systems, J	331
Kimura, Yoshitaka	Mitsubishi Electric Corporation, J	342
Kinzer, Dan	Navitas Semiconductor, USA	31
Kirchhof, Jörg	Fraunhofer Institute IWES, D	401
Kitamura, Shuichi	Mitsubishi Electric Corporation, J	425
Kizu, Naoyuki	Rohm, J	42
Kjaer, Philip C.	Aalborg University, DK	2050
Klarenbach, Christoph	Beckhoff Automation, D	796
Klauke, Sebastian	Infineon, D	581
Klein, Axel	Technical University of Braunschweig, D	216
Kling, Rainer	Karlsruhe Institute of Technology, D	1721
Klötzer, Sebastian	Helmut Schmidt University Hamburg, D	1528
Kobayashi, Hideto	Fuji Electric, J	1956
Kobayashi, Yasuyuki	Fuji Electric, J	438
Koch, Nelson	EPFL – Ecole Polytechnique Fédérale de Lausanne, CH	1949
Koga, Shunsuke	Osaka University, J	1021
Köhler, Stefan	Technical University Nuremberg Georg Simon Ohm, D	526
Köhnlechner, Benjamin	Technical University of Ilmenau, D	1147
Koike, Yoshihiko	Hitachi, J	78
Koini, Markus	Epcos, D	164
Kolar, Johann Walter	ETH Zürich Power Electronic Systems Laboratory, CH	32 1622
Kolb, Johannes	Schaeffler Technologies, D	1979
Kondo, Satoshi	Mitsubishi Electric Corporation, J	326
Königsmann, Gunter	Semikron Elektronik, D	728
Konishi, Yuichiro	Hitachi, J	1110
Konishide, Masaoki	Yasuhiko Kohno, Hitachi, J	280
Konno, Akitoyo	Hitachi, J	78
Kopta, Arnost	ABB Switzerland, CH	417 432
Körner, Julian	Karlsruhe Institute of Technology, D	1721
Kotani, Raita	Toshiba, J	566
Kouge, Takuma	Fuji Electric, J	1956
Krecek, Tomas	ON semiconductor, CZ	1089

PCIM Europe 2016, 10 – 12 May 2016, Nuremberg, Germany

Author	Institution	Pages
Krenn, Markus	FH Joanneum, AT	2151
Krishna Vytla, Rajeev	Infineon Technologies North America, USA	882
Kroics, Kaspars	Riga Technical University, LV	1880
Kropp, Jürgen	University of Bayreuth, D	1400
Kruschel, Wolfram	Infineon, D	735
Kubera, Sascha	Siemens, D	588
Kübrich, Daniel	Friedrich-Alexander-University of Erlangen, D	1693
Kugi, Andreas	Technical University of Vienna, AT	1339
Kühl, Sascha	Technical University of Munich, D	518
Kusano, Dai	Japan Fine Ceramics, J	64
Kwiecien, Marcin	Magneto, PL	1428
L		
Labrousse, Denis	ENS Cachan – SATIE, F	1647
Lacombe, Bertrand	Safran Group, F	1792
Ladoux, Philippe	University of Toulouse, F	371 2189
Laeuffer, Jacques	Dtalents, F	1541
Lagier, Thomas	SuperGrid Institute, F	2189
Lahl, Peter	Infineon, D	735
Lamo, Paula	University of Cantabria, ES	1194
Lampke, Thomas	Technical University of Chemnitz, D	976
Landsmann, Peter	Technical University of Munich, D	518
Lang, Klaus-Dieter	Fraunhofer Institute IZM, D Technical University of Berlin, D	1040 1155
Langhals, David	Technical University of Cologne, D	796
Larousse, Sebastien	Institut des Nanotechnologies de Lyon, F	1707
Larrañaga, Jon Andreu	University of the Basque Country, ES	1785
Larson, Kent	Dow Corning Corporation, USA	992
Lassmann, Matthias	Infineon, D	735
Laud, Satyavrat	Renesas Electronics America, USA	1027
Laue, Jürgen	Danfoss Silicon Power, D	698
Lautner, Jennifer	Friedrich-Alexander-University of Erlangen, D	1278
Leach, John	Castle, GB	1769
Leary, Alex	Carnegie Mellon University, USA	1518
Lechler, Armin	ISW – University of Stuttgart, D	256
Lee, Brian	AIT, IE	558
Lee, JaeWon	Hyundai Motors Company, ROK	1262

PCIM Europe 2016, 10 – 12 May 2016, Nuremberg, Germany

Author	Institution	Pages
Lee, JeongYun	Hyundai Motors Company, ROK	1262
Lee, Junbae	Infineon Technologies Power Semitech, ROK	858 863
Lee, KiJong	Hyundai Motors, ROK	1600
Lee, Minsub	Infineon Technologies Power Semitech, ROK	858 863
Lefebvre, Stéphane	ENS Cachan – SATIE, F	1647
LeHenaff, Francois	Alpha Assembly Solutions, D	1027
Leibfried, Thomas	Karlsruhe Institute of Technology, D	363
Lemke, Michael	AEG Power Solutions, D	1677
Lemmon, Andrew	The University of Alabama, USA	984
Lempidis, Georgios	Fraunhofer Institute IWES, D	401
Lenz, Kevin	Danfoss Silicon Power, D	1934
Leon-Ramirez, Christian Alejandro	University Simon Bolivar, VE	2211
Leszczynski, Jacek	AGH University of Science and Technology, PL	1428
Lexow, Daniel	University of Rostock, D	924
Leyrer, Benjamin	Karlsruhe Institute of Technology, D	1979
Li, Daohui	Dynex Semiconductor, GB	938
Li, Xueqing	United Silicon Carbide, USA	1511
Lifton, Anna	Alpha Assembly Solutions, USA	957 1027
Lindemann, Andreas	Otto-von-Guericke-University, D	1081 1829
Lindseth, Roar	Valeo, NO	113
List, Hans	FH Joanneum, AT	2151
Liu, Cheng	Chenyang Technologies, D	1248
Liu, Guoyou	Dynex Semconductor, GB	845
Liu, Hui	Valeo, NO	113
Liu, Ji-Gou	Chenyang Technologies, D	1248
Liu, Tao	Lenze SE, D	2007
Liu, Wenduo	Infineon Technologies Americas, USA	1655
Liu, Xiaoshan	ENS Cachan – SATIE, F	288
Lledó-Ponsati, Tomàs	TeknoCEA, ES	2035
Löchel, Jonas	Technical University of Ingolstadt, D	742
Loges, Werner	Vacuumschmelze, D	1471
Loh, Poh Chiang	Aalborg University, DK	1837
Lohner, Andreas	Technical University of Cologne, D	669

PCIM Europe 2016, 10 – 12 May 2016, Nuremberg, Germany

Author	Institution	Pages
Longo, Giuseppe	STMicroelectronics, I	804 1987
López, Felipe	University of Cantabria, ES	1194
López, Jesús	Public University of Navarre, ES	2043
Luo, Haihui	Dynex Semconductor, GB	845
Lura, Shinichi	Mitsubishi Electric Corporation, J	425
Lusiewicz, Anna	University of Stuttgart, D	410
Lutz, Josef	Chemnitz University of Technology, D	172 588
M		
Ma, Xiankui	Mitsubishi Electric & Electronics, CN	871
Machuca, Enrique	Technical University of Ingolstadt, D	742
Mackay, Laurens	Delft University of Technology, NL	1201
Mademlis, Georgios	Aristotle University of Thessaloniki, GR	445
Madiwale, Subodh	Analog Devices, USA	195
Madzharov, Nikolay	Technical University of Gabrovo, BG	1999
Makoschitz, Markus	Technical University of Vienna, AT	1669
Malipaard, Dirk	Fraunhofer Institute IISB, D	453
Mandrusiak, Gary	GE Research Center, USA	645 1853
Mangal, Navneet	CREE India Private Limited, IN	2114
Manier, Charles-Alix	Fraunhofer-Institute IZM, D	1155
Mansour, Madaci	University of the Basque Country, ES	1785
Mantzanas, Panagiotis	Friedrich-Alexander-University of Erlangen, D	1693
Marklein, René	Fraunhofer Institute IWES, D	401
Marlino, Laura D.	Oak Ridge National Laboratory, USA	1391
Marquardt, Rainer	Universität der Bundeswehr Munich, D	1570
Marquart, Janosch	University of Applied Sciences NTB Buchs, CH	501 509
Marroyo, Luis	Public University of Navarre, ES	1209
Martin, Christian	Ampere Laboratory, F	1286 1792
Martin, Jérémy	Commissariat à l'énergie atomique et aux énergies alternatives, F	128
Marxgut, Christoph	Helbling Technik, D	661
März, Andreas	University of Bayreuth, D	272 653 1503

Author	Institution	Pages
März, Martin	Fraunhofer Institute IISB, D	469 1315
Masana, Francesc	Barcelona Semiconductors, ES	1009
Masayoshi, Nakazawa	Fuji Electric, J	188
Matallana, Asier	University of the Basque Country, ES	1970
Mathieu, Olivier	Rogers Germany, D	1027
Matlok, Stefan	Fraunhofer Institute IISB, D	637
Matocha, Kevin	Monolith Semiconductor, USA	984
Matsushita, Akira	Hitachi Automotive Systems, J	331
Matthias, Sven	ABB Switzerland, CH	417 432
Mau, Matthias	Danfoss Silicon Power, D	698
Mauder, Anton	Infineon Technologies, D	1324
Maurer, Wilhelm	Infineon Technologies, D	1155
Mazzola, Michael	Mississippi State University, USA	1294
McHenry, Michael E.	Carnegie Mellon University, USA	1518
McPherson, Brice	Wolfspeed, USA	34
Meany, Tom	Analog Devices ERDC, IE	1361
Meisser, Michael	Karlsruhe Institute of Technology, D	1487 1979
Meradji, Moudrik	Harbin Institute of Technology, CN	2014
Merlo, Christophe	ESTIA, F	2066
Merten, Jens	National Solar Energy Institute, F	33
Mertens, Axel	Leibniz University Hannover, D	1421
Mesbahi, Tedjani	Ecole Centrale de Lille, F	2014
Mesemanolis, Athanasios	ABB Switzerland, CH	432
Middelstaedt, Lars	Otto-von-Guericke-University, D	1829
Mii, Kenji	Toshiba Corporation Semiconductor, J	839
Mikulla, Michael	Fraunhofer-Institute IAF, D	319
Millington, Alan	Dynex Semiconductor, GB	903
Misra, Sanjay	Henkel, USA	60
Mitic, Gerhard	Siemens, D	2226
Mitova, Radoslava	Technical University of Berlin, D	1155
Miura, Mineo	Rohm, J	42
Miyahara, Satoshi	Mitsubishi Electric, D	996
Miyata, Hiroshi	Fuji Electric, J	1956
Miyazaki, Takaaki	Hitachi, J	78
Mizushima, Takao	ALPS Green Devices, J	91

PCIM Europe 2016, 10 – 12 May 2016, Nuremberg, Germany

Author	Institution	Pages
Mochizuki, Eiji	Fuji Electric, J	188 1001
Moia, Joabel	Federal Institute of Santa Catarina – IFSC, BR	621
Momose, Fumihiko	Fuji Electric, J	1001
Montesinos-Miracle, Daniel	Universitat Politécnica de Catalunya, ES	2035
Morel, Florent	Ampère, F	1592
Morel, Hervé	Ampère, F	1592
Mori, Mutsuhiro	Hitachi, J	78 1110
Morita, Toshiaki	Hitachi, J	78
Motto, Eric R.	Powerex, USA	889
Müller, Christian	Infineon Technologies, D	850
Müller, Florian	Vossloh-Schwabe Lighting Solutions, D	605
Müller, Georg	Karlsruhe Institute of Technology, D	1305 1578
Mumby-Croft, Paul	Dynex Semiconductor, GB	704
Münster, Patrick	University of Rostock, D	924
Müter, Ulf	Helmut Schmidt University Hamburg, D	1528
N		
Nabhani, Farhad	Teesside University, GB	1439 1685
Naeberle, Norbert	Schaffner, CH	148
Nagahara, Teruaki	Mitsubischi Electric Corporation, J	889
Nagao, Shijo	Osaka University, J	1021
Nagashima, Kazuhito	Hitachi Power Semiconductor, J	348
Nagaune, Fumio	Fuji Electric, J	1956
Naitoh, Yutaka	ALPS Green Devices, J	91
Nakagawa, Ryosuke	Mitsubishi Electric Corporation, Japan	1311
Nakamura, Keiichi	Mitsubishi Electric Corporation, J	425
Nakamura, Masato	Hitachi, J	78
Nakanishi, Masaharu	Rohm Semiconductor, D	42
Nakano, Hiroshi	Hitachi, J	78
Nakatsu, Kinya	Hitachi, J	331
Nasadoski, Jeffrey	GE Research Center, USA	645
Nashiki, Masato	Toshiba Corporation Semiconductor, J	839
Nate, Satoru	Rohm, J	42
Neumaier, Klaus	Fairchild Semiconductor, D	2231
Ng, Chiu	Infineon Technologies Americas, USA	819

PCIM Europe 2016, 10 – 12 May 2016, Nuremberg, Germany

Author	Institution	Pages
Ng, Wai Tung	University of Toronto, CA	895
Nicolle, Thomas	Calyos, BE	1139
Niedermayr, Philipp	Alpitronic, I	534
Nigsch, Simon	University of Applied Sciences NTB Buchs, CH	501 509
Nishimura, Yoshitaka	Fuji Electric, J	1001
Nishio, Haruhiko	Fuji Electric, J	895
Nishiura, Akira	Fuji Electric, J	1956
Nitta, Tetsuya	Toshiba, J	566
Nöding, Christian	University of Kassel, D	573
Nogawa, Hiroyuki	Fuji Electric, J	1001
Nojima, Geraldo	Eaton Corporation, USA	1518
Norambuena, Margarita	Technical University of Berlin, D	1186
O		
O'Sullivan, Dara	Analog Devices, IE	1369
Oeder, Christian	Friedrich-Alexander-University of Erlangen, D	1701 1857
Ohara, Ryoichi	Toshiba, J	566
Ohodnicki, Paul	DOE-National Energy Technology Laboratory, USA	1518
Ohta, Hiroshi	Kaga Toshiba Electronics Corporation, J	839
Okinori, Chihiro	IWATSU Test Instruments, J	1463
Olejniczak, Kraig	Wolfspeed, USA	34
Oñederra, Oier	University of the Basque Country, ES	1970
Onno Krah, Jens	Technical University of Cologne, D	796
Onozawa, Yuichi	Fuji Electric, J	824
Oppermann, Hermann	Fraunhofer-Institute IZM, D	1155
Orlik, Bernd	University of Bremen, D	248
Ota, Kenji	Mitsubishi Electric Corporation, J	425
Otsubo, Yoshitaka	Mitsubishi Electric Corporation, J	342 996
Ott, Leopold	Fraunhofer Institute IISB, D	469
Otto, Alexander	Fraunhofer-Institute ENAS, D	1063
Owzareck, Michael	BLOCK Transformatoren-Elektronik, D	1447
P		
Pacas, Mario	University of Siegen, D	224 240 1873

PCIM Europe 2016, 10 – 12 May 2016, Nuremberg, Germany

Author	Institution	Pages
Packwood, Matthew	Dynex Semiconductor, GB	704 938
Pai, Ajay Poonjal	Infineon Technologies, D	1315
Pala, Vipindas	Wolfspeed, USA	34
Palmer, Patrick	University of Cambridge, GB	1845
Palmour, John W.	Wolfspeed, USA	34
Pan, Xinxing	AIT, IE	558
Papadopoulos, Charalampos	ABB Switzerland, CH	432
Parry, John	Mentor Graphics, UK	1027
Parspour, Nejila	University of Stuttgart, D	410 1447
Passmore, Brandon	Wolfspeed, USA	34
Patt, Michael	Technologienetzwerk Allgäu, D	1171 2122
Paulus, Dirk	Technical University of Munich, D	518
Paulwitz, Christian	Epcos, D	156
Pawellek, Alexander	Friedrich-Alexander-University Erlangen, D	492
Pellicone, Devin	Advanced Cooling Technologies, USA	1118
Peng, Mingkai	Shenzhen Zeasset Electronic Technology, CN	1416
Penzel, Michael	Technical University of Chemnitz, D	976
Pereira França, Alex	CPqD, BR	750
Perez, Angel Luis	University of the Basque Country, ES	1785
Perrin, Remi	INSA de Lyon, F	1286 1792
Persson, Eric	Infineon Technologies Americas, USA	1655
Peters, Dethard	Infineon Technologies, D	53
Petkov, Valeri	Technical University of Gabrovo, BG	1999
Petzoldt, Jürgen	Technical University of Ilmenau, D	1896 2145
Pezet, Francois	Nidec ASI, I	2175
Pfost, Martin	University of Innsbruck, AT	210 757
Phung, Van Trang	University of Siegen, D	240
Piepenbreier, Bernhard	Friedrich-Alexander-University of Erlangen, D	264 1278
Pierfederici, Serge	University de Lorraine, F	772
Pietkiewicz, Andrzej	Schaffner, CH	148
Pietrini, Giorgio	Università degli Studi di Parma, I	1073 1584

PCIM Europe 2016, 10 – 12 May 2016, Nuremberg, Germany

Author	Institution	Pages
Pigazo, Alberto	University of Cantabria, ES	1194
Pignataro, Gaetano	STMicroelectronics, I	804
Plumpton, Ashley	Dynex Semiconductor, GB	903
Pluta, Wojciech	Czestochowa University of Technology, PL	1428
Popa, Gabriel Nicolae	Politehnica University Timisoara, RO	2106
Postiglione, Gianluca	Nidec ASI, I	2175
Pottier, Frederic	Aperam Alloys Amilly, F	84
Proulx, Joe	Mentor Graphics, UK	1027
Puff, Markus	Epcos, AT	164
Q		
Qi, Fang	Dynex Semiconductor, GB	938
Qian, Peng	Technical University of Munich, D	1270
Qiu, Mianwei	Shenzhen Zeasset Electronic Technology, CN	1416
Quay, Rüdiger	Fraunhofer-Institute IAF, D	319
Quentin, Nicolas	SAGEM, F	1286 1792
R		
Rädel, Uwe	Technical University of Ilmenau, D	2145
Rahimo, Munaf	ABB Switzerland, CH	417 432 917
Raithel, Stefan	Vossloh-Schwabe Lighting Solutions, D	605
Ramirez Figueroa, Fernando David	University of Siegen, D	224
Ramirez-Elizondo, Laura	Delft University of Technology, NL	1201
Raso, Antonio	Nidec ASI, I	2175
Rathbone, Kevin	Robotae, GB	1845
Razik, Hubert	University Claude Bernard Lyon, F Laboratoire Ampere, F	312 1707
Reddig, Manfred	University of Applied Sciences Augsburg, D	120
Reger, Martin	Rogers Germany, D	1027
Rehm, Markus	IBR Ingenieurbüro Rehm, D	1332
Reimann, Tobias	Technical University of Ilmenau, D	1147 1749
Reiner, Richard	Fraunhofer-Institute IAF, D	319
Reinhold, Andreas	HTWK Leipzig, D	2145
Reiter, Tomas	Infinoen Technologies, D	1315 1391

PCIM Europe 2016, 10 – 12 May 2016, Nuremberg, Germany

Author	Institution	Pages
Remaci, Ahmed	ESTIA, F	2066
Rencz, Marta	BME, HU	596
Revol, Bertrand	ENS Cachan – SATIE, F	288
Richmond, Jim	Wolfspeed, USA	34
Richter, Dennis	Otto-von-Guericke-University, D	1829
Richter, Jan	Karlsruhe Institute of Technology, D	393
Rocha dos Santos, Sender	CPqD, BR	750
Rodriguez, Jose	University Andres Bello, CL	1186
Roesner, Robert	GE Research Center, D	645
Roig, Jaume	On Semiconductor, BE	1479
Rossberg, Matthias	Semikron Elektronik, D	728
Rossmanith, Hans	Friedrich-Alexander-University of Erlangen, D	1434
Rout, Colin	Dynex Semiconductor, GB	903
Rowden, Brian	GE Research Center, USA	645 1853
Ruccius, Benjamin	Fraunhofer Institute IISB, D	453
Rufer, Alfred	EPFL – Ecole Polytechnique Fédérale de Lausanne, CH	445 1713 1949
Rupp, Roland	Infineon Technologies, D	180
Ryu, Sei-Hyung	Wolfspeed, USA	34
Rzepka, Sven	Fraunhofer-Institute ENAS, D	1063
S		
Sack, Martin	Karlsruhe Institute of Technology, D	1305 1578
Sadarnac, Daniel	CentraleSupelec, F	477
Sahli, Nizar	Helmut Schmidt University Hamburg, D	1528
Saito, Katsuaki	Hitachi Europe, GB	348 1110
Saito, Ryuichi	Hitachi Automotive Systems, J	331
Saito, Shoji	Mitsubishi Electric Corporation, J	677
Saitou, Takashi	Fuji Electric, J	1001
Sakai, Shinji	Mitsubishi Electric Corporation, J	336
Sakiyama, Yoko	Toshiba, J	566
Sakurai, Naoki	Yasuhiko Kohno, Hitachi, J	280
Salerno, Paul	Alpha Assembly Solutions, USA	1027
Sanchis, Pablo	Public University of Navarre, ES	1209
Sander, Rene	Karlsruhe Institute of Technology, D	363

PCIM Europe 2016, 10 – 12 May 2016, Nuremberg, Germany

Author	Institution	Pages
Sano, Kenya	Toshiba, J	566
Sarkany, Zoltan	Mentor Graphics, UK	1027
Sárkány, Zoltán	Budapest University of Technology and Economics, HU	596 1155
Sasaki, Masahiro	Fuji Electric, J	895
Sawada, Mutsumi	Fuji Electric, J	824
Schanen, Jean-Luc	G2ELAB, F	1494
Schäning, Björn	Helmut Schmidt University Hamburg, D	1528
Scharwitz, Christian	Vacuumschmelze, D	1471
Schefler, Stefan	Epcos, D	164
Schenk, Kurt	University of Applied Sciences NTB Buchs, CH	501 509
Scherbaum, Markus	University of Applied Sciences Augsburg, D	120
Scheuermann, Uwe	Semikron Elektronik, D	691
Schiele, Jürgen	AEG Power Solutions, D	1662
Schijffelen, Jos	Power Research Electronics, NL	1383
Schilling, Marco	Technical University of Ilmenau, D	1147
Schlenk, Manfred	Infineon Technologies, D	120
Schliewe, Jörn	Epcos, D	156
Schmeller, Markus	SUMIDA Components & Modules, D	1728 1764
Schmidhuber, Michael	SUMIDA Components & Modules, D	1728 1764
Schmies, Stefan	Infineon, D	735
Schmitt, Alexander	Karlsruhe Institute of Technology, D	393
Schmitt-Landsiedel, Doris	University Rosenheim, D	1562
Schnarrenberger, Mathias	Karlsruhe Institute of Technology, D	788
Schnell, Raffael	ABB Switzerland, CH	417 432 945
Schreitmüller, Stefan	HTWG Konstanz University of Applied Sciences, D	136
Schubert, Andreas	Technical University of Chemnitz, D	976
Schuetz, Tobias	GE Research Center, D	645
Schulte-Overbeck, Christian	AEG Power Solutions, D	1677
Schulz, Matthias	Fraunhofer Institute IISB, D	469
Schulz, Nicola	University of Applied Sciences Northwestern Switzerland, CH	2167
Schumacher, Walter	Technical University of Braunschweig, D	216
Schupbach, Marcelo	Cree, USA	2114

PCIM Europe 2016, 10 – 12 May 2016, Nuremberg, Germany

Author	Institution	Pages
Schwalbe, Ulf	ISLE Steuerungstechnik und Leistungselektronik, D	1147 1749
Schweiger, Hans-Georg	Technical University of Ingolstadt, D	742
Schwenk, Holger	Vacuumschmelze, D	1471
Scibilia, Roberto	Texas Instruments, D	1224
Scrimizzi, Filippo	STMicroelectronics, I	804 1987
Seibel, Axel	Fraunhofer Institute IWES, D	401
Seldrum, Thomas	Dow Corning, BE	1017
Seleme Jr., Seleme Isaac	Federal University of Minas Gerais – UFMG, BR	371
Seliger, Bernd	Fraunhofer Institute IISB, D	637
Seliger, Norbert	University Rosenheim, D	1562
Sequeira, Luis	Nomad Tech, PT	953
Shalaby, Rizk	Mansoura University, AE	1046
Shang, Ming	Mitsubishi Electric & Electronics, CN	871
She, Xu	GE Research Center, USA	1853
Sheehan, Cathal	Bourns Electronics, IE	1224
Shekhar, Aditya	Delft University of Technology, NL	1201
Shelton, Ed	Silicon Contact, GB	1845
Shin, SangChul	Hyundai Motors, ROK	1600
Shorten, Andrew	University of Toronto, CA	895
Shousha, Mahmoud	Würth Elektronik eiSos, D	107
Sieweke, Nico	Beckhoff Automation, D	796
Simco, David	Wolfspeed, USA	34
Singer, Arthur	Universität der Bundeswehr Munich, D	1904
Sinn, Peter	Robert Bosch, D	1377
Sirmelis, Ugis	Riga Technical University, LV	1880
Slatter, Rolf	Sensitec, D	1240
Slawinski, Maximilian	Infineon Technologies, D	1800
Soinski, Marian	Magneto, PL	1428
Sokolovs, Alvis	Riga Technical University, LV	1880
Solanki, Jitendra	University of Paderborn, D	1662
Soldati, Alessandro	Università degli Studi di Parma, I	1073 1584
Song, Gaosheng	Mitsubishi Electric & Electronic, CN	871 932
Sorensen, Jens	Analog Devices, USA	1369
Sorrentino, Elmer	University Simon Bolivar, VE	2211

PCIM Europe 2016, 10 – 12 May 2016, Nuremberg, Germany

Author	Institution	Pages
Sorsdahl, Torbjorn	Valeo, NO	113
Spence, Michael	Dynex Semiconductor, GB	903
Spertino, Filippo	Politecnico di Torino, I	2058
Spro, Ole Christian	SINTEF Energy Research, NO	1408
Starks, Ann	ON Semiconductor, USA	1777
Steffen, Jonas	Fraunhofer Institute IWES, D	401
Stegmeier, Stefan	Siemens, D	2226
Steinbring, Manuel	University of Siegen, D	1873
Steinke, Gina	EPFL – Ecole Polytechnique Fédérale de Lausanne, CH	445 1713
Stevanovic, Ljubisa	GE Research Center, USA	645 1853
Stiasny, Thomas	ABB Switzerland, CH	917
Stiegler, Karlheinz	ABB Switzerland, CH	911
Stöckl, Johannes	AIT-Austrian Institute of Technology, AT	1846
Stocksreiter, Wolfgang	FH Joanneum, AT	296 2151
Stone, David	University of Sheffield, GB	1095 1769 2090
Storasta, Liutauras	ABB Switzerland, CH	417
Strauss, Bastian	Otto-von-Guericke-University, D	1081
Streb, Fabian	Infineon Technologies, D	976
Ströbel-Maier, Henning	Danfoss Silicon Power, D	698
Strzalkowski, Bernhard	Analog Devices, D	195
Stubenrauch, Franz	University Rosenheim, D	1562
Stuckmann, Christoph	Maccon, D	1345
Stuckmann, Tim	University of Applied Sciences Ostwesfalen-Lippe, D	1639
Subramanian, Prasanth	GrafTech International, USA	1126
Suganuma, Katsuaki	Osaka University, J	1021
Sumper, Andreas	CITCEA-UPC, ES	2211
Sun, Brian	Infineon Technologies North America, USA	882
Sun, Hui	Chenyang Technologies, D	1248
Suriyah-Jaya, Michael	Karlsruhe Institute of Technology, D	363
Surma, Alexey	Proton-Electrotex JSC, RUS	957
Swieboda, Cezary	Magneto, PL	1428
Szczesny, Paul	GE Research Center, USA	1853

PCIM Europe 2016, 10 – 12 May 2016, Nuremberg, Germany

Author	Institution	Pages
T		
Tadikonda, Ramakrishna	Infineon Technologies Americas, USA	819
Takahashi, Misaki	Fuji Electric, J	438
Takahashi, Takuya	Mitsubishi Electric Corporation, J	342
Takahashi, Yoshikazu	Fuji Electric, J	1001
Takamiya, Yoshikazu	Fuji Electric, J	1956
Takorabet, Noureddine	University de Lorraine, F	772
Tamai, Yuuta	Fuji Electric, J	1001
Tamenori, Akira	Fuji Electric, J	438
Tan, Xiaoya	Schaffner, CH	148
Tanabe, Gen	Japan Fine Ceramics, J	64
Tanaka, Nobuhiko	Mitsubishi Electric Corporation, J	425
Tanioka, Toshikazu	Mitsubishi Electric Corporation, J	336
Tao, Fengfeng	GE Research Center, USA	645
Tchouangue, Georges	Toshiba Electronics Europe, D	566 839
Teixeira Pinto, Rodrigo	CITCEA-UPC, ES	2211
Thal, Eckhard	Mitsubishi Electric Europe, D	48 425
Thesseling, Matthias	Lenze SE, D	2007
Thomas, Tina	Technical University of Berlin, D	1040
Titushkin, Dmitry	Proton-Electrotex JSC, RUS	957
Tiwari, Amit	Newcastle University, GB	2197
Tokuyama, Takeshi	Hitachi Automotive Systems, J	331
Tomasso, Giuseppe	University of Cassino and South Lazio, I	379 1217
Tóssoli de Sousa, Thais	CPqD, BR	750
Trainer, David	Alstom Grid, GB	2197
Treier, Christian	ABB Switzerland, CH	945
Trintis, Ionut	Aalborg University, DK	831
Trüller, Jürgen	HF Instruments, D	1463
Tsang, Chi Wa	University of Lincoln, GB	1769
U		
Uchida, Yoshiyuki	Japan Fine Ceramics, J	64
Ura, Hideyuki	Toshiba Corporation Semiconductor, J	839
Urtasun, Andoni	Public University of Navarre, ES	1209

PCIM Europe 2016, 10 – 12 May 2016, Nuremberg, Germany

Author	Institution	Pages
V		
Van Brunt, Edward	Wolfspeed, USA	34
Vanlathem, Eric	Dow Corning Europe, BE	992 1017
Velasco, David	Public University of Navarre, ES	2043
Vemulapati, Umamaheswara Reddy	ABB Switzerland, CH	917
Verl, Alexander	ISW – University of Stuttgart, D	256
Vershinin, Konstantin	Alstom Grid, GB	2197
Victory, James	Fairchild Semiconductor, D	813
Viera, Juan C.	University of Oviedo, ES	1631
Vijay, Karthik	Indium Corporation, GB	704
Villbusch, Tim	Infineon Technologies, D	1800
Vobecky, Jan	ABB Switzerland, CH	911 917
Vogelsberger, Markus	Bombardier Transportation Austria, AT	1339 2219
Voigt, Gunter	HTWG Konstanz University of Applied Sciences, D	136
Volay, Philippe	Centralp, F	1707
Volke, Andreas	Power Integrations, D	48
Vollaire, Christian	Laboratoire Ampere,F	1494
von Essen, Tobias	Berliner Nanotest & Design, D	2226
Vu, Trong Tue	Eisergy, IE	1163
W		
Wachutka, Gerhard	Technical University of Munich, D	543
Wagner, Bernhard	Technical University Nuremberg Georg Simon Ohm, D	526
Wallscheid, Oliver	University of Paderborn, D	780
Waltereit, Patrick	Fraunhofer-Institute IAF, D	319
Wanderoild, Yohan	CEA Leti, F	312
Wang, Chi-Ming	Toyota Motor Engineering & Manufacturing, USA	1608
Wang, Gang-Yao	Wolfspeed, USA	34
Wang, Gaolin	Harbin Institute of Technology, CN	2014
Wang, Xiongfei	Aalborg University, DK	1837
Wang, Yangang	Dynex Semiconductor, GB	938
Wang, Yazhe	Mitsubishi Electric Corporation, J	336
Warnakulasuriya, Kapila	Carroll & Meynell Transformers, GB	1685 1439

PCIM Europe 2016, 10 – 12 May 2016, Nuremberg, Germany

Author	Institution	Pages
Watabe, Kiyoto	Mitsubishi Electric Corporation, J	336
Wattenberg, Martin	Universtiy of Reutlingen, D	210 757
Weber, Marc	Karlsruhe Institute of Technology, D	1979
Weber, Stefan	Epcos, D	156 164
Wegelin, Viktor	Karlsruhe Institute of Technology, D	1979
Weis, Gerald	FH Joanneum, AT	296 2151
Weiss, Beatrix	Fraunhofer-Institute IAF, D	319
Weiß, Helmut	Technical University of Leoben, AT	2204
Wendt, Hans-Joachim	Lenze Drives, D	2007
Wendt, Michael	Infineon Technologies, D	1615
Werner, Quentin	Daimler, D	772
Wespel, Matthias	Fraunhofer-Institute IAF, D	319
Weyant, Jens	Advanced Cooling Technologies, USA	1118
Weyh, Thomas	Universität der Bundeswehr Munich, D	1904
Wiedemann, Simon	Maccon, D	1353
Wiesner, Eugen	Mitsubishi Electric Europe, D	48 425
Wintrich, Arendt	Semikron, D	1829
Wohlstreicher, Manfred	SUMIDA Components & Modules, D	1728
Wolbank, Thomas	Technical University of Vienna, AT	2219
Wood, John	Silicon Contact, GB	1845
Wright, Nick	Newcastle University, GB	2197
Wu, Xiaomin	Fairchild Semiconductor, D	2231
Wunder, Bernd	Fraunhofer Institute IISB, D	469
Wunderle, Bernhard	Technical University of Chemnitz, D	1155
Würfel, Alexander	University of Bremen, D	1324
Wurz, Marc C.	Leibniz University Hannover, D	107
Wüthrich, Martin	Schaffner, CH	148
X		
Xu, Dianguo	Harbin Institute of Technology, CN	2014
Xu, Yanan	Chenyang Technologies, D	1248
Y		
Yamada, Junji	Mitsubishi Electric Corporation, J	996
Yamaguchi, Masakazu	Toshiba, J	566

Author	Institution	Pages
Yamashita, Hiroaki	Toshiba Corporation Semiconductor, J	839
Yang, Gang	Valeo, F	113
Yao, Wenli	Northwestern Polytechnical University, CN	1837
Yasuda, Yuusuke	Hitachi, J	78
Yin, Hang	Technical University of Berlin, D	1186
Yoo, Inpil	Infineon Technologies, D	2027
Yoshida, Hiroshi	Mitsubishi Electric Corporation, J	326 342
Yoshida, Souichi	Fuji Electric, J	438 1956
Yoshiwatari, Shinichi	Fuji Electric, J	824
Young, George	Eisergy, IE	1163
Yu, Kezhuang	Shenzhen Zeasset Electronic Technology, CN	1416
Yu, Zhe	Fraunhofer Institute ISIT, D	202
Yuki, Hata	Mitsubishi Electric Corporation, J	677
Z		
Zacharias, Peter	University of Kassel, D	573
Zeidler, Henning	Technical University of Chemnitz, D	976
Zeltner, Stefan	Fraunhofer Institute IISB, D	637
Zeman, Miro	Delft University of Technology, NL	1383
Zeng, Guang	Technical University of Chemnitz, D	588
Zhang, Hao	Osaka University, J	1021
Zhang, Shirley	United Silicon Carbide, USA	1511
Zhang, Xiaobin	Northwestern Polytechnical University, CN	1837
Zhang, Yuancheng	Mitsubishi Electric & Electronics, CN	871
Zhang, Yuanzhe	Efficient Power Conversion Corporation, USA	304
Zhang, Zhenbin	Technical University of Munich, D	629 1926
Zhou, Haihua	Infineon Technologies Americas, USA	1655
Zhou, Lu	Dow Corning, CN	992
Zhou, Wei	Dynex Semiconductor, GB	938
Zhu, Ke	United Silicon Carbide, USA	1549
Zimmer, Marco	University of Stuttgart, D	410
Zippelius, Bernd	Infineon Technologies, D	180
Zöller, Clemens	Technical University of Vienna, AT	1339 2219
Zschieschang, Olaf	Fairchild Semiconductor, D	1063

Mesago PCIM GmbH
Rotebuehlstrasse 83-85
70178 Stuttgart Germany

ISBN 978-1-5108-2530-7